DECIMAL & MILLIMETER EQUIVALENTS

INCH	INCH	MM
1/64	.015625	.397
1/32	.03125	.794
3/64	.046875	1.191
1/16	.0625	1.587
5/64	.078125	1.984
3/32	.09375	2.381
7/64	.109375	2.778
1/8	.125	3.175
9/64	.140625	3.572
5/32	.15625	3.969
11/64	.171875	4.366
3/16	.1875	4.762
13/64	.203125	5.159
7/32	.21875	5.556
15/64	.234375	5.953
1/4	.25	6.350
17/64	.265625	6.747
9/32	.28125	7.144
19/64	.296875	7.541
5/16	.3125	7.937
21/64	.328125	8.334
11/32	.34375	8.731

INCH	INCH	MM
23/64	.359375	9.128
3/8	.375	9.525
25/64	.390625	9.922
13/32	.40625	10.319
27/64	.421875	10.716
7/16	.4375	11.113
29/64	.453125	11.509
15/32	.46875	11.906
31/64	.484375	12.303
1/2	.5	12.700
33/64	.515625	13.097
17/32	.53125	13.494
35/64	.546875	13.890
9/16	.5625	14.287
37/64	.578125	14.684
19/32	.59375	15.081
39/64	.609375	15.478
5/8	.625	15.875
41/64	.640625	16.272
21/32	.65625	16.669
43/64	.671875	17.065

INCH	INCH	MM
11/16	.6875	17.462
45/64	.703125	17.859
23/32	.71875	18.265
47/64	.734375	18.653
3/4	.75	19.050
49/64	.765625	19.447
25/32	.78125	19.884
51/64	.796875	20.240
13/16	.8125	20.637
53/64	.828125	21.034
27/32	.84375	21.431
55/64	.859375	21.828
7/8	.875	22.225
57/64	.890625	22.622
29/32	.90625	23.019
59/64	.921875	23.415
15/16	.9375	23.812
61/64	.953125	24.209
31/32	.96875	24.606
63/64	.984375	25.003
1		25.400

Special Service Tools

Throughout this manual references are made and illustrations may depict the use of special tools required to perform certain jobs. These special tools can generally be ordered through the dealers of the make vehicle being serviced. It is also suggested that you check with local automotive supply firms as they also supply tools manufactured by other firms that will assist in the performance of these jobs. The vehicle manufacturers special tools are supplied by:

American Motors & General Motors . . . . Service Tool Division
Kent-Moore Corporation
1501 South Jackson Street
Jackson, Michigan 49203

Chrysler Corp. . . . . Miller Special Tools
A Division of Utica Tool Co.
32615 Park Lane
Garden City, Michigan 48135

Ford Motor Co. . . . . Owatonna Tool Company
Owatonna, Minnesota 55060

Your steps to success.

STEP 1: Register

All you need to get started is a valid email address and the access code below. To register, simply:

1. Go to www.campbellbiology.com
2. Click the appropriate book cover.
   Cover must match the textbook edition being used for your class.
3. Click "**Register**" under "**First-Time User?**"
4. Leave "**No, I Am a New User**" selected.
5. Using a coin, scratch off the silver coating below to reveal your access code.
   Do not use a knife or other sharp object, which can damage the code.
6. Enter your access code in lowercase or uppercase, without the dashes.
7. Follow the on-screen instructions to complete registration.
   During registration, you will establish a personal login name and password to use for logging into the website. You will also be sent a registration confirmation email that contains your login name and password.

Your Access Code is:

Note: If there is no silver foil covering the access code, it may already have been redeemed, and therefore may no longer be valid. In that case, you can purchase access online using a major credit card. To do so, go to www.campbellbiology.com, click the cover of your textbook, click "**Buy Now,**" and follow the on-screen instructions.

STEP 2: Log in

1. Go to www.campbellbiology.com and click the appropriate book cover.
2. Under **"Established User?"** enter the login name and password that you created during registration. *If unsure of this information, refer to your registration confirmation email.*
3. Click **"Log In."**

STEP 3: (Optional) Join a class

Instructors have the option of creating an online class for you to use with this website. If your instructor decides to do this, you'll need to complete the following steps using the Class ID your instructor provides you. By "joining a class," you enable your instructor to view the scored results of your work on the website in his or her online gradebook.

To join a class:

1. Log into the website. For instructions, see "STEP 2: Log in."
2. Click **"Join a Class"** near the top left.
3. Enter your instructor's **"Class ID"** and then click **"Next."**
4. At the Confirm Class page you will see your instructor's name and class information. If this information is correct, click Next.
5. Click **"Enter Class Now"** from the Class Confirmation page.

- *To confirm your enrollment in the class, check for your instructor and class name at the top right of the page. You will be sent a class enrollment confirmation email.*
- *As you complete quizzes on the website from now through the class end date, your results will post to your instructor's gradebook, in addition to appearing in your personal view of the Results Reporter.*

To log into the class later, follow the instructions under "STEP 2: Log in."

Got technical questions?

Visit http://247.aw.com. Email technical support is available 24/7.
Call 1-800-677-6337. Phone support is available 8:00am – 8:00pm Eastern, Monday-Friday and 5:00pm - 12:00am (midnight) Eastern, Sunday.

SITE REQUIREMENTS

For the latest updates on Site Requirements, go to www.campbellbiology.com, choose your text cover, and click Site Reqs.

WINDOWS
366 MHz CPU
Windows 98/2000/XP
64 MB RAM
800 x 600 screen resolution, thousands of colors
Browser: Internet Explorer 5.0/5.5/6.0; Netscape 6.2.3 and 7.0
Plug-Ins: Shockwave Player 8, Flash Player 7
Internet Connection: 56k modem minimum

MACINTOSH
266 MHz PowerPC
OS 9.2/10.2.4/10.3.2
64 MB RAM
800 x 600 screen resolution, thousands of colors
Browsers: Internet Explorer 5.1/5.2; Netscape 6.2.3
Plug-Ins: Shockwave Player 8, Flash Player 7
Internet Connection: 56k modem minimum

Register and log in

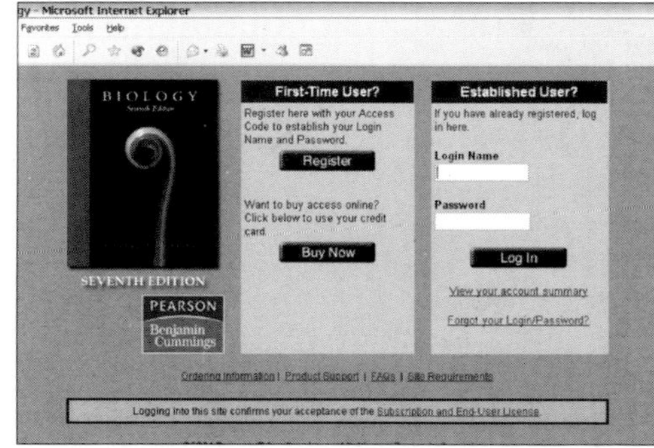

Join a class

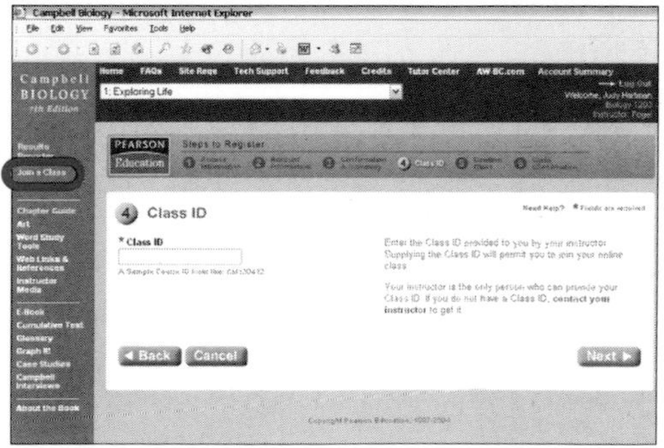

Important: Please read the Subscription and End-User License agreement, accessible from the book website's login page, before using the Campbell Biology website and CD-ROM. By using the website or CD-ROM, you indicate that you have read, understood, and accepted the terms of this agreement.

Features of the Campbell Biology CD-ROM and Website

RESULTS REPORTER AND GRADEBOOK

Track your progress with the new Results Reporter. When you take quizzes on the website, your scores are recorded in the Results Reporter. If your professor provides a class ID and you join the online class, your scores will also be recorded in your professor's Gradebook.*

CHAPTER GUIDE

Review each chapter efficiently and effectively through a focus on key concepts. Activities, Investigations, Videos, graphing exercises, and the E-Book* are linked directly to each numbered key concept in the text. The Chapter Guide also includes each chapter's Quizzes.

ACTIVITIES: Explore over 230 interactive activities, including animations, virtual labs, review exercises, and videos.

INVESTIGATIONS: Experiment, investigate, and analyze 55 case studies that develop scientific thinking skills.

VIDEOS: See biology come to life with 85 new videos.

GRAPH IT!: Learn how to build and interpret graphs with 11 new graphing exercises.

QUIZZES: Assess understanding with four different multiple-choice quizzes for each chapter: a Pre-Test to diagnose current knowledge, an online version of the Self-Quiz* from the book, an Activities Quiz with graphics that tests understanding of the media Activities, and a comprehensive Chapter Quiz. Most quizzes have links to the E-Book,* hints, and feedback for right and wrong answers. All quizzes are graded automatically. On the website, students can see their grades in the Results Reporter and instructors can track students' progress with an online Gradebook.*

ART

View and print art from the textbook, available with and without labels.*

WORD STUDY TOOLS AND GLOSSARY

Flashcards, Word Roots, Key Terms, and a Glossary will all help you improve your biology vocabulary. Includes selected audio pronunciations.

WEB LINKS AND REFERENCES

Extend your knowledge through links to relevant websites with descriptions of the sites.* Access news links on recent developments in biology, an archive of biology articles, and further readings for each chapter.*

E-BOOK

Refer to a convenient online version of the book while you study. Quiz questions are linked to specific concepts in the E-Book.*

CUMULATIVE TEST

Test yourself on multiple chapters at once to best prepare for an exam, choosing the chapters and the number of questions. Each question provides a hint, a reference to the relevant concept in the text, and immediate feedback.*

CAMPBELL INTERVIEWS

Learn how scientists think and why they love what they do. Access in-depth interviews with eminent biologists from past editions of Campbell/Reece *Biology*.

ABOUT THE BOOK

Learn more about Campbell/Reece *Biology*, the best-selling college science textbook in the world.*

*Requires a live internet connection

months[9] = 'Septembe

BIOLOGY

Seventh Edition

Neil A. Campbell
University of California, Riverside

Jane B. Reece
Berkeley, California

CONTRIBUTORS AND ADVISORS

Lisa Urry	Mills College, Oakland, California
Manuel Molles	University of New Mexico, Albuquerque
Carl Zimmer	Science writer, Guilford, Connecticut
Christopher Wills	University of California, San Diego
Peter Minorsky	*Plant Physiology* and Mercy College, Dobbs Ferry, New York
Mary Jane Niles	University of San Francisco, California
Antony Stretton	University of Wisconsin-Madison

PEARSON
Benjamin Cummings

San Francisco Boston New York
Cape Town Hong Kong London Madrid Mexico City
Montreal Munich Paris Singapore Sydney Tokyo Toronto

Editor-in-Chief: *Beth Wilbur*

Senior Supervising Editor: *Deborah Gale*

Supervising Editors: *Pat Burner and Beth N. Winickoff*

Managing Editor: *Erin Gregg*

Art Director: *Russell Chun*

Photo Editor: *Travis Amos*

Marketing Managers: *Josh Frost and Jeff Hester*

Developmental Editors: *John Burner, Alice E. Fugate, Sarah C. G. Jensen, Matt Lee, Suzanne Olivier, Ruth Steyn, and Susan Weisberg*

Developmental Artists: *Hilair Chism, Blakeley Kim, Kenneth Probst, Carla Simmons, and Laura Southworth*

Biology Media Producer: *Christopher Delgado*

Media Project Manager: *Brienn Buchanan*

Project Editor: *Amy C. Austin*

Photo Coordinator: *Donna Kalal*

Permissions Editors: *Sue Ewing and Marcy Lunetta*

Publishing Assistants: *Trinh Bui and Julia Khait*

Illustrations: *Precision Graphics, Russell Chun, Phil Guzy, and Steve McEntee*

Text and Cover Designer: *Mark Ong*

Photo Researchers: *Brian Donnelly, Donna Kalal, Ira Kleinberg, Robin Samper, and Maureen Spuhler*

Copy Editor: *Janet Greenblatt*

Production Management, Art Coordination, and Design Support: *Morgan E. Floyd, Robert R. Hansen, Sherrill Redd, S. Brendan Short, and Kirsten Sims at GTS Companies*

Compositor and Prepress: *GTS Companies*

Manufacturing Manager: *Pam Augspurger*

Cover Printer: *Phoenix Color*

Printer: *Von Hoffmann Press, Inc.*

On the cover: Photograph of bird's nest fern, *Asplenium nidus*: Linda Broadfoot. Special thanks to Dennis High, Center for Photographic Art, Carmel, California, for his advice and assistance with cover research. Credits continue following the appendices.

Library of Congress Cataloging-in-Publication Data

Campbell, Neil A.

Biology / Neil A. Campbell, Jane B. Reece. – 7th ed.

p. cm.

Includes bibliographical references.

ISBN 0-8053-7171-0

1. Biology. I. Reece, Jane B. II. Title.

570—dc21 2001047033

Student Package ISBN 0-8053-7146-X

Student Text Component ISBN 0-8053-7171-0

P-copy Package ISBN 0-8053-7147-8

P-copy Text Component ISBN 0-8053-7166-4

3 4 5 6 7 8 9 10—VHC—08 07 06 05

www.aw-bc.com

San Francisco Boston New York
Cape Town Hong Kong London Madrid Mexico City
Montreal Munich Paris Singapore Sydney Tokyo Toronto

About the Authors

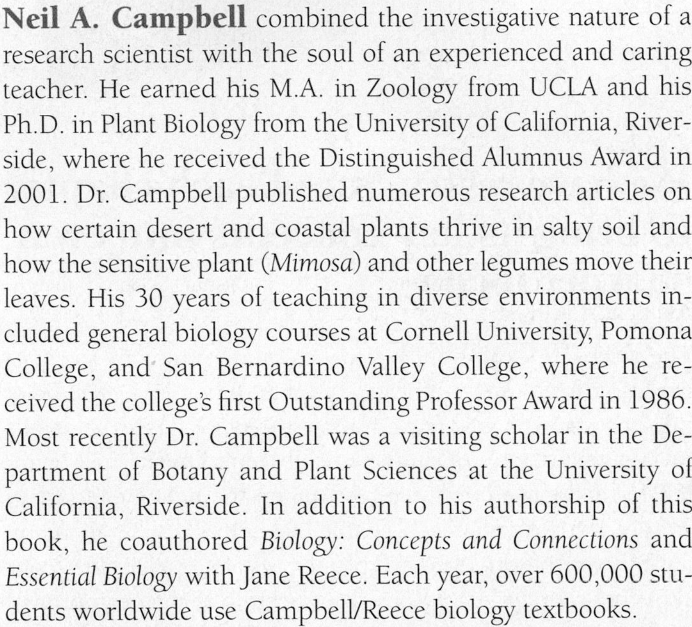

Neil A. Campbell combined the investigative nature of a research scientist with the soul of an experienced and caring teacher. He earned his M.A. in Zoology from UCLA and his Ph.D. in Plant Biology from the University of California, Riverside, where he received the Distinguished Alumnus Award in 2001. Dr. Campbell published numerous research articles on how certain desert and coastal plants thrive in salty soil and how the sensitive plant (*Mimosa*) and other legumes move their leaves. His 30 years of teaching in diverse environments included general biology courses at Cornell University, Pomona College, and San Bernardino Valley College, where he received the college's first Outstanding Professor Award in 1986. Most recently Dr. Campbell was a visiting scholar in the Department of Botany and Plant Sciences at the University of California, Riverside. In addition to his authorship of this book, he coauthored *Biology: Concepts and Connections* and *Essential Biology* with Jane Reece. Each year, over 600,000 students worldwide use Campbell/Reece biology textbooks.

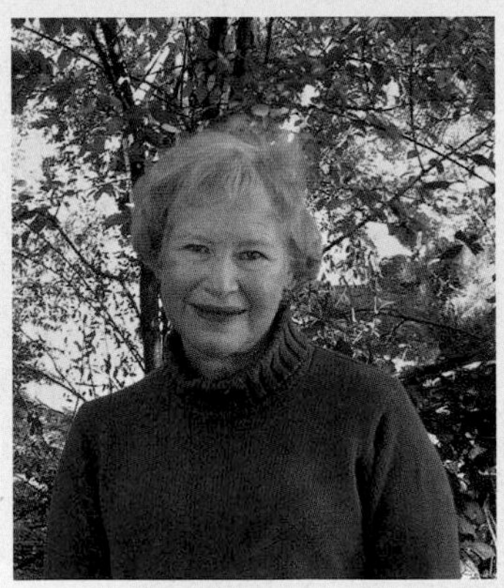

Jane B. Reece has worked in biology publishing since 1978, when she joined the editorial staff of Benjamin Cummings. Her education includes an A.B. in Biology from Harvard University, an M.S. in Microbiology from Rutgers University, and a Ph.D. in Bacteriology from the University of California, Berkeley. At UC Berkeley, and later as a postdoctoral fellow in genetics at Stanford University, her research focused on genetic recombination in bacteria. She taught biology at Middlesex County College (New Jersey) and Queensborough Community College (New York). As an editor at Benjamin Cummings, Dr. Reece played major roles in a number of successful textbooks. In addition to being a coauthor with Neil Campbell on *BIOLOGY, Biology: Concepts and Connections*, and *Essential Biology*, she coauthored *The World of the Cell*, Third Edition, with W. M. Becker and M. F. Poenie.

To Rochelle and Allison, with love

—N.A.C.

To Paul and Daniel, with love

—J.B.R.

NEIL A. CAMPBELL
died October 21, 2004 after finishing work on this revision.
He is deeply mourned by his many friends and colleagues
at Benjamin Cummings and
throughout the biology community.

Preface

Charles Darwin described evolution as a process of "descent with modification." It is a phrase that also fits the continuing evolution of *BIOLOGY*. This Seventh Edition is our most ambitious revision of the book since its origin—a new textbook "species" with several evolutionary adaptations shaped by the changing environment of biology courses and by the astonishing progress of biological research. But these adaptive modifications are still true to the two complementary teaching values at the core of every edition of *BIOLOGY*. First, we are dedicated to crafting each chapter from a framework of key concepts that will help students keep the details in place. Second, we are committed to engaging students in scientific inquiry through a combination of diverse examples of biologists' research and opportunities for students to practice inquiry themselves.

These dual emphases on concept building and scientific inquiry emerged from our decades of classroom experience. It is obviously gratifying that our approach has had such broad appeal to the thousands of instructors and millions of students who have made *BIOLOGY* the most widely used college science textbook. But with this privilege of sharing biology with so many students comes the responsibility to continue improving the book to serve the biology community even better. As we planned this new edition, we visited dozens of campuses to hear what students and their instructors had to say about their biology courses and textbooks. What we learned from those conversations about new directions in biology courses and the changing needs of students informed the many improvements you'll find in this Seventh Edition of *BIOLOGY*.

We have restructured each chapter to bring its key concepts into even sharper focus

The discovery explosion that makes modern biology so exciting also threatens to suffocate students under an avalanche of information. The past few editions of *BIOLOGY* set the details in a context of key concepts, typically ten to twenty per chapter. In this new edition, we have taken the next evolutionary step of restructuring each chapter to help students focus on fewer, even bigger ideas—typically just five or six key concepts per chapter. A new Overview section at the beginning of each chapter sets an even broader context for the key concepts that follow. And at the end of each of the concept sections, a Concept Check with two or three questions enables students to assess whether they understand that concept before going on to the next. Answers to the Concept Check questions are located in Appendix A, as are the answers to the Self-Quizzes from the Chapter Review at the end of each chapter.

In our ongoing interactions with students and instructors, they have responded enthusiastically to our new organization and pedagogy. Compared to other textbooks, including earlier editions of our own, students have found the new chapter structure and design of *BIOLOGY*, Seventh Edition, to be more inviting, more accessible, and much more efficient to use. But in achieving these goals, we have not compromised the depth and scientific accuracy the biology community has come to expect from us.

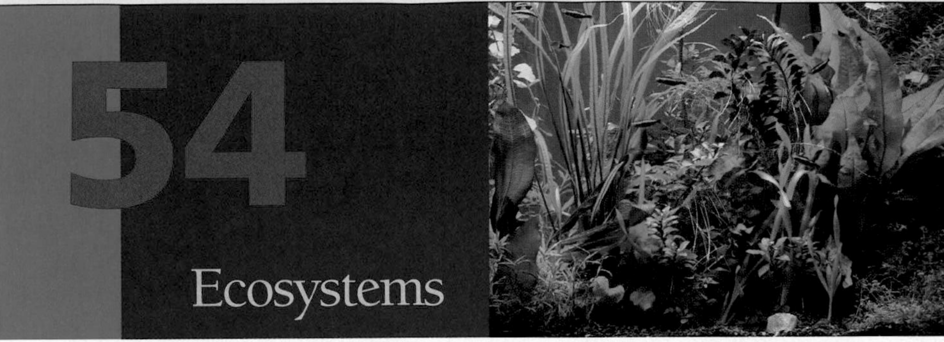

▲ Figure 54.1 **An aquarium, an ecosystem bounded by glass.**

Key Concepts keep the supporting details in context.

Key Concepts

54.1 Ecosystem ecology emphasizes energy flow and chemical cycling
54.2 Physical and chemical factors limit primary production in ecosystems
54.3 Energy transfer between trophic levels is usually less than 20% efficient
54.4 Biological and geochemical processes move nutrients between organic and inorganic parts of the ecosystem
54.5 The human population is disrupting chemical cycles throughout the biosphere

The **Overview** sets the stage for the rest of the chapter.

Figure references in color help students move easily between text and figures.

Overview

Ecosystems, Energy, and Matter

An **ecosystem** consists of all the organisms living in a community as well as all the abiotic factors with which they interact. Ecosystems can range from a microcosm, such as the aquarium in **Figure 54.1**, to a large area such as a lake or forest. As with populations and communities, the boundaries of ecosystems are usually not discrete. Cities and farms are examples of human-dominated ecosystems. Many ecologists regard the entire biosphere as a global ecosystem, a composite of all the local ecosystems on Earth.

Regardless of an ecosystem's size, its dynamics involve two processes that cannot be fully described by population or community processes and phenomena: energy flow and chemical cycling. Energy enters most ecosystems in the form of sunlight. It is then converted to chemical energy by au-

among abiotic and biotic components of the ecosystem. Photosynthetic organisms assimilate these elements in inorganic form from the air, soil, and water and incorporate them into organic molecules, some of which are consumed by animals. The elements are returned in inorganic form to the air, soil, and water by the metabolism of plants and animals and by other organisms, such as bacteria and fungi, that break down organic wastes and dead organisms.

Both energy and matter move through ecosystems via the transfer of substances during photosynthesis and feeding relationships. However, because energy, unlike matter, cannot be recycled, an ecosystem must be powered by a continuous influx of energy from an external source—in most cases, the sun. Energy flows through ecosystems, while matter cycles within them.

Resources critical to human survival and welfare, ranging from the food we eat to the oxygen we breathe, are products of ecosystem processes. In this chapter, we will explore the dynamics of energy flow and chemical cycling in ecosystems and consider some of the impacts of human activities on these processes.

Each numbered **Concept Head** announces the beginning of a new concept section.

Concept 54.1

Ecosystem ecology emphasizes energy flow and chemical cycling

Ecosystem ecologists view ecosystems as transformers of energy and processors of matter. By grouping the species in a community into trophic levels of feeding relationships (see ... transformation of energy in ... ments of chemical elements

▲ Figure 54.3 **Fungi decomposing a dead tree.**

consumers in an ecosystem. In a forest, for example, birds might feed on earthworms that have been feeding on leaf litter and its associated prokaryotes and fungi. But even more important than this channeling of resources from producers to consumers is the role that detritivores play in making vital chemical el... available to producers.

D... the organic material in an e... ents in inor...

Concept Check 54.1

1. Why is the transfer of energy in an ecosystem referred to as energy flow, not energy cycling?
2. How does the second law of thermodynamics explain why an ecosystem's energy supply must be continually replenished?
3. How are detritivores essential to sustaining ecosystems?

For suggested answers, see Appendix A.

Concept Check Questions at the end of each concept section encourage students to assess their mastery of the concept.

Concept 54.2

Physical and chemical factors limit primary production in ecosystems

The amount of light energy converted to chemical energy (organic compounds) by autotrophs during a given time period is an ecosystem's **primary production.** This photosynthetic product is the starting point for studies of ecosystem metabolism and energy flow.

Ecosystem Energy Budgets

Most primary producers use light energy to synthesize energy-rich organic molecules, which can subsequently be broken down to generate ATP (see Chapter 10). Consumers acquire their organic fuels secondhand (or even third- or fourthhand) through food webs such as that in Figure 53.13. Therefore, the extent of photosynthetic production sets the spending limit for the energy budget of the entire ecosystem.

...lobal Energy Budget

... bombarded ...

New "Exploring Figures" provide efficient access to many complex topics

Biology is a visual science. Thus we have always authored *BIOLOGY's* graphics and narrative side by side to coordinate their message. In the Seventh Edition, this text-art integration reaches its next evolutionary level with a new feature called "Exploring Figures." Each of these large figures is a learning unit that brings together a set of related illustrations and the text that describes them. The Exploring Figures enable students to access dozens of complex topics much more efficiently, now that the textual and visual components have merged.

The Exploring Figures represent core chapter content, not to be confused with some textbooks' "boxes," which feature content that is peripheral to the flow of a chapter. Modern biology is challenging enough without diverting students' attention from a chapter's conceptual storyline. Thus, each Exploring Figure is referenced in the main text body where it fits into the development of a concept, just as the text points students to all the other supporting figures at the appropriate places in the narrative.

In **Exploring Figures**, art, photos, and text are fully integrated.

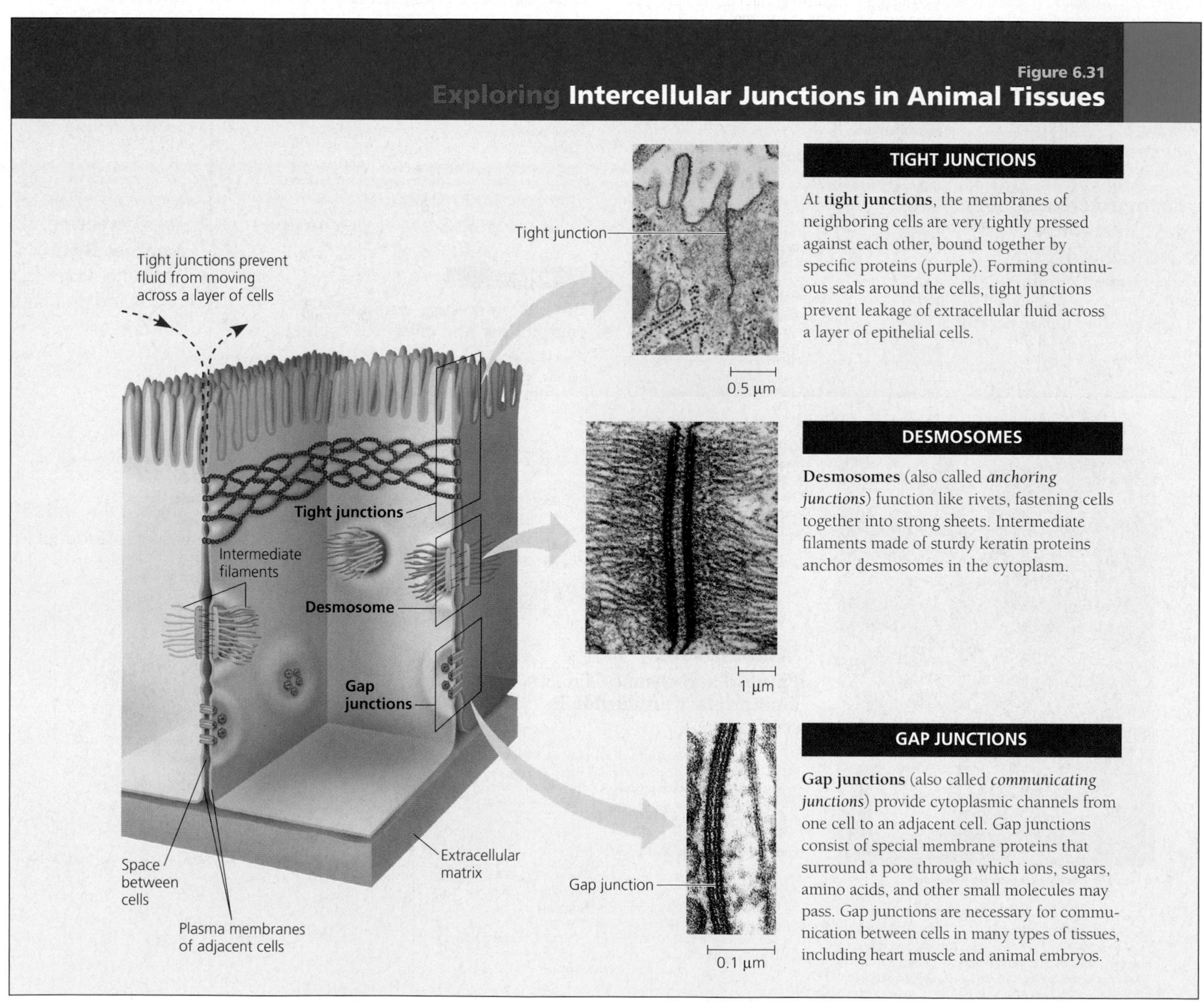

Figure 6.31
Exploring Intercellular Junctions in Animal Tissues
Tight junctions prevent fluid from moving across a layer of cells
Tight junctions
Intermediate filaments
Desmosome
Gap junctions
Space between cells
Plasma membranes of adjacent cells
Extracellular matrix
Tight junction
0.5 µm
1 µm
Gap junction
0.1 µm
TIGHT JUNCTIONS
At **tight junctions**, the membranes of neighboring cells are very tightly pressed against each other, bound together by specific proteins (purple). Forming continuous seals around the cells, tight junctions prevent leakage of extracellular fluid across a layer of epithelial cells.
DESMOSOMES
Desmosomes (also called *anchoring junctions*) function like rivets, fastening cells together into strong sheets. Intermediate filaments made of sturdy keratin proteins anchor desmosomes in the cytoplasm.
GAP JUNCTIONS
Gap junctions (also called *communicating junctions*) provide cytoplasmic channels from one cell to an adjacent cell. Gap junctions consist of special membrane proteins that surround a pore through which ions, sugars, amino acids, and other small molecules may pass. Gap junctions are necessary for communication between cells in many types of tissues, including heart muscle and animal embryos.

Scientific inquiry is more prominent than ever in *BIOLOGY* and its supplements

One objective for many biology instructors is for students to learn to think as scientists. In both the lecture hall and laboratory, colleagues are experimenting with diverse approaches for involving students in scientific inquiry, in which questions about nature focus strategic investigation and analysis of data. New textbook features and new inquiry-based supplements make this edition of *BIOLOGY* more effective than ever as a partner to instructors who emphasize the process of science.

Figure 1.29

Inquiry Does the presence of poisonous coral snakes affect predation rates on their mimics, king snakes?

EXPERIMENT David Pfennig and his colleagues made artificial snakes to test a prediction of the mimicry hypothesis: that king snakes benefit from mimicking the warning coloration of coral snakes *only* in regions where poisonous coral snakes are present. The Xs on the map below are field sites where the researchers placed equal numbers of artificial king snakes (experimental group) and brown artificial snakes (control group). The researchers recovered the artificial snakes after four weeks and tabulated predation data based on teeth and claw marks on the snakes (see Figure 1.28).

RESULTS In field sites where coral snakes were present, predators attacked far fewer artificial king snakes than brown artificial snakes. The warning coloration of the "king snakes" afforded no such protection where coral snakes were absent. In fact, at those field sites, the artificial king snakes were *more* likely to be attacked than the brown artificial snakes, perhaps because the bright pattern is particularly easy to spot against the background.

Key
% of attacks on artificial king snakes
% of attacks on brown artificial snakes
X Field site with artificial snakes

In areas where coral snakes were absent, most attacks were on artificial king snakes.
17%
83%
North Carolina
South Carolina
16%
84%
In areas where coral snakes were present, most attacks were on brown artificial snakes.

CONCLUSION The field experiments support the mimicry hypothesis by not falsifying the key prediction that imitation of coral snakes is only effective where coral snakes are present. The experiments also tested an alternative hypothesis that predators generally avoid all snakes with brightly colored rings, whether or not poisonous snakes with that coloration live in the environment. That alternative hypothesis was falsified by the data showing that the ringed coloration failed to repel predators where coral snakes were absent.

Modeling Inquiry by Example

Scientific inquiry has always been one of *BIOLOGY*'s unifying themes. Each edition has traced the history of many research questions and scientific debates to help students appreciate not just "what we know," but "how we know," and "what we do not yet know." In *BIOLOGY*, Seventh Edition, we have strengthened this theme by making examples of scientific inquiry much more prominent throughout the book.

The increased emphasis on inquiry begins in Chapter 1, where we have thoroughly revised the introduction to the many ways that scientists explore biological questions. Chapter 1 also introduces a new feature called "Inquiry Figures," which showcase outstanding examples of experiments and field studies in a format that is consistent throughout the book. Complementing the Inquiry Figures are the new "Research Method Figures," which walk students through the techniques and tools of modern biology. You can find a list of the Inquiry and Research Method Figures on pages xx-xxi. These new features, like the Exploring Figures, are integral to chapter flow rather than being appended as boxed asides.

New **Inquiry Figures** and **Research Method Figures** help students learn to think like scientists.

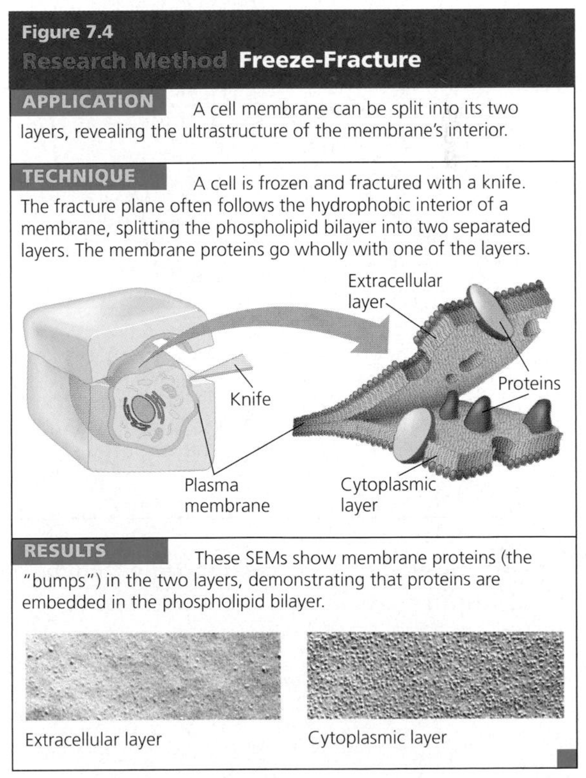

Figure 7.4
Research Method Freeze-Fracture
APPLICATION A cell membrane can be split into its two layers, revealing the ultrastructure of the membrane's interior.
TECHNIQUE A cell is frozen and fractured with a knife. The fracture plane often follows the hydrophobic interior of a membrane, splitting the phospholipid bilayer into two separated layers. The membrane proteins go wholly with one of the layers.
Extracellular layer
Proteins
Knife
Plasma membrane
Cytoplasmic layer
RESULTS These SEMs show membrane proteins (the "bumps") in the two layers, demonstrating that proteins are embedded in the phospholipid bilayer.
Extracellular layer
Cytoplasmic layer

Learning Inquiry by Practice

Modeling scientific inquiry by example has only ephemeral impact unless students have an opportunity to apply what they have learned by asking their own biological questions and conducting their own investigations. On a small scale, *BIOLOGY*, Seventh Edition, encourages students to practice thinking as scientists by responding to "Scientific Inquiry" questions in the Chapter Review at the ends of chapters.

On a much bigger scale, new supplements build on the textbook to provide diverse opportunities for students to practice scientific inquiry. One example is *Biological Inquiry: A Workbook of Investigative Cases*, by Margaret Waterman of Southeast Missouri State University and Ethel Stanley of Beloit College, which is available without cost to students whose instructors request it as a supplement to the textbook. This innovative new workbook offers eight case studies, coordinated with the eight units of chapters in *BIOLOGY*. In each case, a realistic scenario sets up a series of inquiry-based activities. The cases work well either as class-discussion projects or as take-home assignments for students working alone, or better, in small groups.

Another student-centered supplement is *Practicing Biology*, by Jean Heitz, University of Wisconsin, Madison, which is also available without additional cost upon request of instructors using *BIOLOGY*, Seventh Edition. This workbook supports various learning styles with a variety of activities—including modeling, drawing, and concept-mapping—that help students construct an understanding of biological concepts.

Students will find still more opportunities for active learning at *www.campbellbiology.com* and the CD-ROM that is included with each book. And the excellent *Student Study Guide*, by Martha Taylor of Cornell University, continues to be a proven learning tool for students.

The Campbell/Reece Interviews: A Continuing Tradition

Scientific inquiry is a social process catalyzed by communication among people who share a curiosity about nature. One of the many joys of authoring *BIOLOGY* has been the privilege to humanize science by interviewing some of the world's most influential biologists. Eight new interviews that introduce the eight units of the textbook provide students with windows to inquisitive minds that are driving progress in biology and connecting science to society. The interviewees for this edition are listed on page xxiii.

Scientific Inquiry

When bacteria infect an animal, the number of bacteria in the body increases in an exponential fashion (graph A). After infection by a virulent animal virus with a lytic reproductive cycle, there is no evidence of infection for a while. Then, the number of viruses rises suddenly and subsequently increases in a series of steps (graph B). Explain the difference in the growth curves.

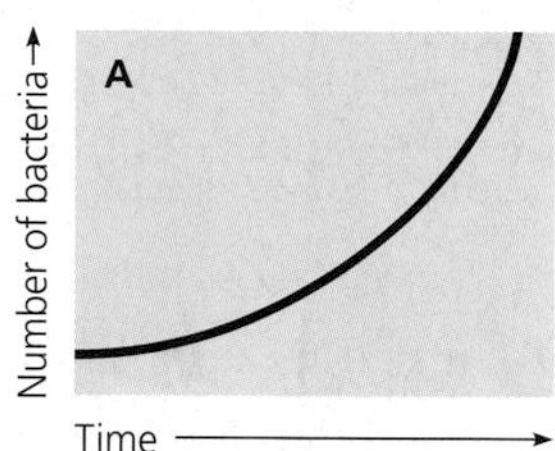

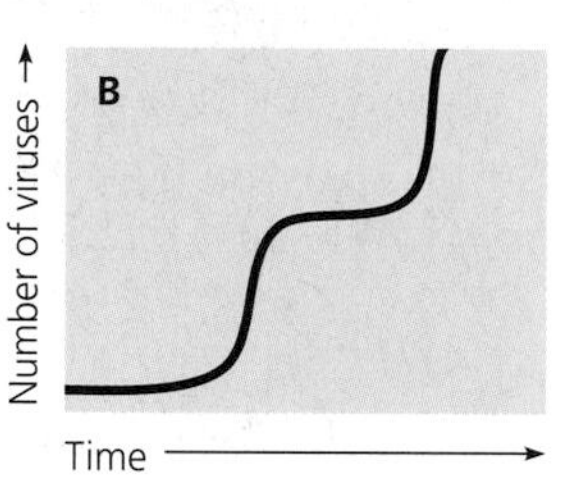

Biological Inquiry: A Workbook of Investigative Cases *Explore West Nile virus in the case "The Donor's Dilemma."*
Investigation *What Causes Infections in AIDS Patients?*
Investigation *Why Do AIDS Rates Differ Across the U.S.?*
Investigation *What Are the Patterns of Antibiotic Resistance?*

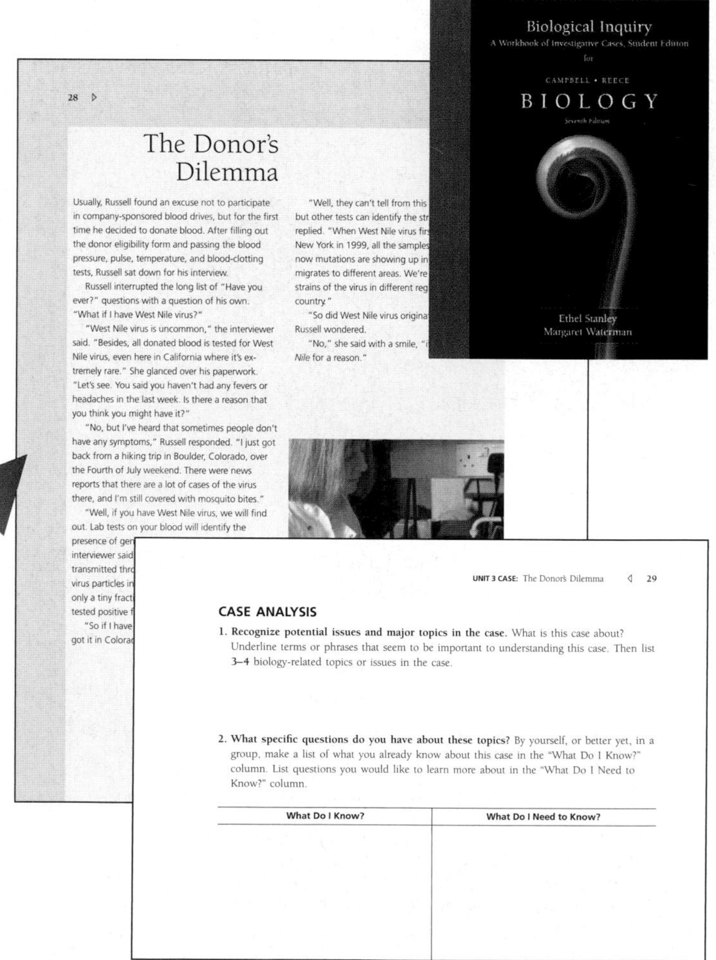

28

The Donor's Dilemma

Usually, Russell found an excuse not to participate in company-sponsored blood drives, but for the first time he decided to donate blood. After filling out the donor eligibility form and passing the blood pressure, pulse, temperature, and blood-clotting tests, Russell sat down for his interview.

Russell interrupted the long list of "Have you ever?" questions with a question of his own. "What if I have West Nile virus?"

"West Nile virus is uncommon," the interviewer said. "Besides, all donated blood is tested for West Nile virus, even here in California where it's extremely rare." She glanced over his paperwork. "Let's see. You said you haven't had any fevers or headaches in the last week. Is there a reason that you think you might have it?"

"No, but I've heard that sometimes people don't have any symptoms," Russell responded. "I just got back from a hiking trip in Boulder, Colorado, over the Fourth of July weekend. There were news reports that there are a lot of cases of the virus there, and I'm still covered with mosquito bites."

"Well, if you have West Nile virus, we will find out. Lab tests on your blood will identify the

Russell wondered.

UNIT 3 CASE: The Donor's Dilemma 29

CASE ANALYSIS

1. **Recognize potential issues and major topics in the case.** What is this case about? Underline terms or phrases that seem to be important to understanding this case. Then list 3–4 biology-related topics or issues in the case.

2. **What specific questions do you have about these topics?** By yourself, or better yet, in a group, make a list of what you already know about this case in the "What Do I Know?" column. List questions you would like to learn more about in the "What Do I Need to Know?" column.

What Do I Know?	What Do I Need to Know?

Inquiry Questions, Media Investigations, and the new **Biological Inquiry Workbook** help students practice scientific inquiry.

Balancing Inquiry with a Conceptual Foundation

Although this new edition of *BIOLOGY* showcases the process of science more prominently than ever, there are two good reasons to avoid overstating the power of inquiry-based content in any biology textbook.

First, those of us who advocate more inquiry in biology courses mainly have student-centered inquiry in mind, not textbook-centered inquiry. As a mostly passive experience, reading about inquiry in a textbook should be merely an entryway to a variety of active experiences that are promoted by inquiry-based supplements, by investigative labs, and by activities that instructors create to support student-centered inquiry.

Second, the most important way a textbook can support student inquiry is by providing context with clear, accurate explanation of the key biological concepts. Just as biologists generally study the scientific literature as background for their own inquiry, students will be much more successful in their personal inquiry if it emerges from a basic understanding of the relevant biology. Thus, *BIOLOGY*, Seventh Edition, is *not* a "reform textbook" of the genre that replaces a careful unfolding of conceptual content with a stream of relatively unconnected research examples, requiring beginning students to put it all together for themselves. We believe that such an unbalanced reaction to the call for inquiry-based reform is likely to leave most students frustrated and ill-equipped to practice active inquiry in their labs, course projects, class discussions, and Socratic lecture environments. In *BIOLOGY*, Seventh Edition, we have carefully integrated the inquiry-based content into the development of each chapter's main ideas so that the research examples reinforce rather than obscure the conceptual framework.

BIOLOGY supports a diversity of courses and serves students throughout their biology education

Even by limiting our scope to a few key concepts per chapter, *BIOLOGY* spans more biological territory than most introductory courses could or should attempt to cover. But given the great diversity of course syllabi, we have opted for a survey broad enough and deep enough to support each instructor's special emphases. Students also seem to appreciate *BIOLOGY*'s breadth and depth; in this era when students sell many of their textbooks back to the bookstore, more than 75% of students who have used *BIOLOGY* have kept it after their introductory course. In fact, we are delighted to receive numerous letters and emails from upper-division students and graduate students, including medical students, expressing their appreciation for the long-term value of *BIOLOGY* as a general resource for their continuing education.

Just as we recognize that few courses will cover all 55 chapters of *BIOLOGY*, we also realize that there is no one "correct" sequence of topics for a general biology course. Though a biology textbook's table of contents must be linear, biology itself is more like a web of related concepts without a fixed starting point or a prescribed path. Diverse courses can navigate this network of concepts starting with molecules and cells, with evolution and the diversity of organisms, or with the big-picture ideas of ecology. We have built *BIOLOGY* to be versatile enough to support various syllabi. The eight units of the book are largely self-contained, and most of the chapters within each unit can be assigned in a different sequence. For example, instructors who integrate plant and animal physiology can merge chapters from Unit Six (Plant Form and Function) and Unit Seven (Animal Form and Function). Instructors who begin their course with ecology and continue with this "top-down" approach can assign Unit Eight (Ecology) right after Chapter 1, which introduces the unifying themes that provide students with a panoramic view of biology no matter what the topic order of the course syllabus.

Evolution and *BIOLOGY*'s other themes connect the concepts and integrate the whole book

The first chapter articulates 11 themes that provide touchstones for students throughout the book and distinguish our approach in *BIOLOGY* from an encyclopedic topical approach. In this Seventh Edition, we have added the theme of "biological systems" to integrate a variety of research initiatives based on high-throughput data collection and readily available computing power. But as in all previous editions, the central theme is evolution, which unifies all of biology by accounting for both the unity and diversity of life. The evolutionary theme is woven into every chapter of *BIOLOGY*. Evolution and the other whole-book themes work with the chapter-level concepts to help students construct a coherent view of life that will serve them long after they have forgotten the details fossilized in any biology textbook.

Neil Campbell and Jane Reece

Acknowledgments

One of the eminent scientists interviewed in this new edition pointed out that much of the fun of doing biology comes from working with a diversity of talented people. The same can be said for making a biology textbook. Fortunately for us, this Seventh Edition of *BIOLOGY* is the product of the talents, dedication, and enthusiasm of a large and varied group of people. The authors wish to express their deepest thanks to the numerous instructors, researchers, students, publishing professionals, and artists who have contributed to this edition.

As authors of both past and present editions of this text, we are mindful of the daunting challenge of keeping up to date in all areas of our rapidly expanding subject. We are particularly grateful to the seven Contributors and Advisers listed on the title page, whose expertise has ensured that the book is current and enlivened with fresh examples. We worked especially closely with developmental biologist Lisa Urry, who had major responsibility for updating content and implementing our new format and features for Units 1–3 and Chapter 47. Her rigorous scholarship and attention to detail in the areas of biological chemistry, cell and molecular biology, genetics, and developmental biology were a great boon. We thank her for her commitment and enthusiasm, relentless hard work, punctuality, and good cheer throughout the process. Equally helpful was ecologist Manuel Molles, who brought his scientific and teaching expertise to the revision of Unit 8, enhancing the structure of the unit and its verbal and visual presentation of ecology; he played a major role rewriting the behavioral ecology chapter, which is essentially new. He also helped provide a more ecological perspective to Chapters 40, 42, and 44, in the unit on animal form and function. Science writer Carl Zimmer contributed many improvements and new perspectives to Unit 5, the diversity unit. Evolutionary biologist Christopher Wills helped us tackle the challenge of improving and updating Unit 4, the evolution unit, and Chapter 26. Plant biologist Peter Minorsky helped bring Unit 6 up to date. And neurobiologist Antony Stretton advised us on the revision of Chapters 48 and 49. As in earlier editions, immunologist Mary Jane Niles organized and implemented the significant revision of Chapter 43.

Thanks also to the instructors who suggested revised or new Concept Check and Chapter Review questions. These include (in alphabetical order) Bruce Byers, Jean Heitz, William Hoese, Tom Owens, Mark Lyford, Randy Phillis (special thanks), Mitch Price, Fred Sack, Richard Showman, and Elspeth Walker. It's not easy to write good questions, and we appreciate the time and effort these dedicated educators contributed to enhancing the effectiveness of our book's questions.

Further helping us improve *BIOLOGY's* scientific accuracy and pedagogy, about 240 biologists and teachers, cited on the list that follows these Acknowledgements, provided detailed reviews of one or more chapters for this edition. Special thanks to Lawrence Brewer, Richard Brusca, Anne Clark, Douglas Eernisse, Mark Kirk, Walter Judd, Mike Levine, Diane Marshall, Nick Money, Tom Owens, Kevin Padian, Daniel Papaj, Mitch Price, Bruce Reid, and Alistair Simpson for their guidance.

Thanks also to the numerous other professors and their students, from all over the world, who offered suggestions by writing directly to the authors. In addition, we appreciate the candid and specific feedback we received from the students and faculty who participated in group discussions held at Skyline College, Mills College, and Indiana University. Last but not least, we thank our coauthors on our nonmajors texts, Eric Simon and Marty Taylor, for providing rigorous feedback on a number of chapters. Of course, we alone bear the responsibility for any errors that remain in the text, but the dedication of our contributors, advisers, reviewers, and correspondents makes us especially confident in the accuracy of this edition.

Many scientists have also helped shape this Seventh Edition by discussing their research fields with us, answering specific questions in their areas of expertise, and, often, sharing their ideas about biology education. Neil Campbell thanks the many University of California, Riverside, colleagues who have influenced this book, including Ring Carde, Richard Cardullo, Mark Chappell, Darleen DeMason, Norman Ellstrand, Anthony Huang, Bradley Hyman, Tracy Kahn, Elizabeth Lord, Carol Lovatt, Eugene Nothnagel, John Oross, Timothy Paine, David Reznick, Rodolfo Ruibal, Clay Sassaman, William Thomson, John Trumble, Rick Redack, Mike Adams, and the late John Moore (whose "Science as a Way of Knowing" essays have had such an important influence on the evolution of *BIOLOGY*). Jane Reece thanks members of the Mills College Biology and Chemistry/Physics Departments, especially Elisabeth Wade, as well as Fred Wilt, John Gerhart and Kris Niyogi from the University of California, Berkeley, for their assistance to contributor Lisa Urry.

Interviews with prominent scientists have been a hallmark of *BIOLOGY* since its inception, and conducting these interviews was again one of the great pleasures of revising the text. To open the eight units of this Seventh Edition, we are proud to include interviews with Lydia Makhubu, Peter Agre, Eric Lander, Kenneth Kaneshiro, Linda Graham, Natasha Raikhel, Erich Jarvis, and Gene Likens.

The value of *BIOLOGY* as a learning tool is greatly enhanced by the supplementary materials that have been created for instructors and students. We recognize that the dedicated authors of these materials are essentially writing mini (and not so mini) books. We much appreciate the hard work and creativity of the following: Margaret Waterman and Ethel Stanley (authors of the new *Biological Inquiry: A Workbook of Investigative Cases*); Jean Heitz (*Practicing Biology,* 2nd edition); Joan Sharp (*Instructor's Guide*); Janet Lanza (*New Designs for Bio-Explorations*); Chris Romero (PowerPoint Lectures); Laura Zanello (*Spanish Glossary*); and Judith Morgan and Eloise Brown Carter (*Investigating Biology Lab Manual,* 5th Edition). We thank Bill Barstow for heading up the test bank team, and we wish to acknowledge the test bank contributors: Jean DeSaix, Michael Dini, Conrad Firling, Peter Follette, Mark Hens, Janice Moore, Tom Owens, Marshall Sundberg, Robert Yost, and Ed Zalisko. Thanks also to Bill Wischusen, who compiled our Active Learning Questions and wrote discussion points. Once again, we thank our long-time colleague Marty Taylor for her excellent and student-focused work on the *Student Study Guide;* she has now completed seven editions of this popular student aid. In addition, we are grateful to the many other people—biology instructors, editors, artists, production experts, and narrators—who are listed in the credits for the impressive electronic media that accompany the book.

BIOLOGY, Seventh Edition, results from an unusually strong synergy between a team of scientists and a team of publishing professionals. An all-new design, the comprehensive revision of the illustration program as well as the text, the addition of major new pedagogical features, and a rich package of supplements, both printed and electronic, combined with a tight schedule to create unprecedented challenges for the publishing team.

The members of the core book team at Benjamin Cummings brought extraordinary talents and extraordinarily hard work to this project. Our leader, Editor-in-Chief Beth Wilbur, is a full colleague in the book's creation and a respected advocate for biology education in general and our book in particular in the academic community. Enthusiastic, creative, endlessly supportive of us and the other members of the team, Beth is a wonderful person and a pleasure to work with. Unflappable under pressure, she navigates difficult situations gracefully—a major asset in a project of this complexity.

The incomparable Deborah Gale, Director of Development, managed the entire project on a day-by-day basis. Deborah coproduced the first and second editions of the book, along with the developmental editing of the second edition, and we have been delighted with her return. Amazingly, Deborah is able to combine a totally professional, no-nonsense management style and a willingness to dig into the nitty-gritty with a sense of humor that kept the rest of us happily slaving away at her direction.

Supervising editors Pat Burner and Beth Winickoff had the awesome responsibility of overseeing in detail the work of the contributors, developmental editors, and developmental artists. Together, Pat and Beth carefully read every single chapter and checked every illustration, doing whatever was necessary to make this edition the most effective biology textbook available—and, we think, exceeding the high standards established in previous editions. We are immensely grateful to Pat, the multitalented and tireless Developmental Manager for Biology who has been our colleague for many years, for her incredible dedication, sound editorial judgment, and extraordinary attention to detail. The exceptionally talented Beth Winickoff, new to this Seventh Edition, was the originator of our new process of book development and production. Beth brought fresh perspectives on process, pedagogy, and editorial approach—in addition to her superb hands-on editing of six chapters. We look forward to working again with Pat and Beth on subsequent editions (after they recover from this one, of course!).

The responsibilities of the developmental editors for this edition were especially challenging. Almost all the chapters were heavily revised, requiring intensive editorial involvement from initial planning through production. We were fortunate to have on our team some of college publishing's top developmental editors. In alphabetical order, they were John Burner (Units 5–7), Alice Fugate (Units 4 and 7), Sarah Jensen (Units 2, 3, and 7), Matt Lee (Units 5 and 7), Suzanne Olivier (Units 1 and 2), Ruth Steyn (Units 3 and 7), and Susan Weisberg (Units 7 and 8). In addition to their other tasks, John Burner, Matt Lee, and Ruth Steyn brought their specific content expertise to bear on their chapters' revisions.

The support of our bright, efficient, and good-natured Publishing Assistants, Trinh Bui and Julia Khait, is much appreciated. What would we all have done without them?

We also want to thank someone who doesn't fit neatly into any of our publishing categories: our colleague, former editor, and friend Robin Heyden. Robin brought her imaginative energy and dedication to biology education to the Seventh Edition in several ways. These include early planning for the development of the media for this edition and the conception and developmental management of the new case study workbook by Margaret Waterman and Ethel Stanley. Robin also organized the first Benjamin Cummings Biology Leadership Conference, which brought us a fresh supply of creative teaching ideas from outstanding biology educators.

Once again the book has benefited greatly from the work of Russell Chun, our Senior Producer, Art and Media. Russell established a vibrant new art style for this edition that met the requirements of the content and exceeded our expectations for pedagogical and aesthetic excellence. Under his direction were the developmental artists, who developed all the new figures and redesigned many of the older ones to make them clearer and more appealing. These skilled and creative illustrators were Hilair Chism (Units 1–3 and 7), Blakeley Kim (Unit 8), Kenneth Probst (Units 4 and 5), and Laura Southworth (Units 3 and 7). Carla Simmons (Units 5 and 6) has contributed her artistic and pedagogical talents to every edition of this textbook. Final rendering of the hundreds of new and revised illustrations was carried out by Russell, Phil Guzy, Steve McEntee, and the artists of Precision Graphics. Meanwhile, Photo Editor Travis Amos led a team of photo researchers in finding hundreds of new photos for this edition. The photo researchers were Brian Donnelly, Donna Kalal, Ira Kleinberg, Robin Samper, and Maureen Spuhler. The efficient Donna Kalal also coordinated the ordering of photos from a multitude of sources. We are indebted to the entire art and photo team and to the book's talented text and cover designer, Mark Ong, for the most beautiful and visually effective edition ever. In addition to creating the stunning design, Mark was involved in laying out every chapter, and his artistic sensibility reinforced all of our goals for this revision.

The book production team had the crucial responsibility of converting the text manuscript and illustrations to pages ready for the printer. Many thanks to Managing Editor Erin Gregg, who was responsible for overseeing the complex design and production process, including the management of both in-house and freelance employees. At GTS Companies (the compositor), we particularly want to thank Rob Hansen, Brendan Short, Morgan Floyd, and Sherrill Redd, who provided expertise and solutions to complicated production challenges with good humor, and designer Kirsten Sims, who helped us improve the appearance and pedagogical utility of the Exploring Figures. Finally, we thank Manufacturing Manager Pam Augspurger, without whose work you would not be holding a physical copy of the book in your hands.

We are pleased to thank the topnotch publishing professionals who worked on the book's printed supplements. Amy Austin, Robin Heyden, Ginnie Simione Jutson, and Joan Keyes developed these supplements, and Vivian McDougal and Jane Brundage were responsible for their production.

In regard to the excellent package of electronic media that accompanies the book, we offer special thanks to Brienn Buchanan, who creatively pulled together all the elements of the student CD-ROM and website, and Christopher Delgado, who produced all of the instructor media resources, as well as the Art Notebook.

Linda Davis, President of Benjamin Cummings Publishing, has shared our commitment to excellence and provided strong support for three editions now, and we are happy to thank her once again. We also want to thank the Addison Wesley/Benjamin Cummings President, Jim Behnke (who was the editor of the first edition of this book), for his support of our new developmental process, and Editorial Director Frank Ruggirello for his vigorous commitment to the book's success.

Both before and after publication, we are fortunate to have experienced Benjamin Cummings marketing professionals on our book team. Senior Marketing Manager Josh Frost and Director of Marketing Stacy Treco provided consistent support and useful input throughout the entire development of this edition. Thanks, also, to Jeff Hester, who has recently joined the marketing team. We much appreciate the work of the talented Lillian Carr and her marketing communications team, who have created stunning brochures, posters, and other materials that have helped get the word out about this new edition. And thanks to Mansour Bethany for developing the ebrochure and other assistance.

The Addison Wesley/Benjamin Cummings field staff, which represents *BIOLOGY* on campus, is our living link to the students and professors who use the text. The field representatives tell us what you like and don't like about the book, and they provide prompt service to biology departments. The field reps are good allies in science education, and we thank them for their professionalism in communicating the features of our book.

Finally, we wish to thank our families and friends for their encouragement and for enduring our continuing obsession with *BIOLOGY*.

Neil Campbell and Jane Reece
October 2004

Reviewers of the Seventh Edition

Thomas Adams, *Michigan State University*
Shylaja Akkaraju, *Bronx Community College of CUNY*
Bonnie Amos, *Angelo State University*
Jeff Appling, *Clemson University*
J. David Archibald, *San Diego State University*
David Armstrong, *University of Colorado at Boulder*
Mary Ashley, *University of Illinois at Chicago*
Karl Aufderheide, *Texas A&M University*
Ellen Baker, *Santa Monica College*
Susan Barman, *Michigan State University*
Andrew Barton, *University of Maine, Farmington*
David Bass, *University of Central Oklahoma*
Bonnie Baxter, *Hobart & William Smith College*
Tim Beagley, *Salt Lake Community College*
Margaret E. Beard, *College of the Holy Cross*
Chris Beck, *Emory University*
Patricia Bedinger, *Colorado State University*
Tania Beliz, *College of San Mateo*
Robert Blanchard, *University of New Hampshire*
Andrew Blaustein, *Oregon State University*
Allan Bornstein, *Southeast Missouri State University*
Lisa Boucher, *University of Nebraska-Omaha*
Robert Bowker, *Glendale Community College (Arizona)*
Barbara Bowman, *Mills College*
Sunny Boyd, *University of Notre Dame*
Lawrence Brewer, *University of Kentucky*
Paul Broady, *University of Canterbury*
Carole Browne, *Wake Forest University*
David Bruck, *San Jose State University*
Rick Brusca, *Arizona-Sonora Desert Museum*
Howard Buhse, *University of Illinois at Chicago*
Arthur Buikema, *Virginia Polytechnic Institute and State University*
Al Burchsted, *College of Staten Island*
Bruce Byers, *University of Massachusetts*
Alison Campbell, *University of Waikato*
Frank Cantelmo, *St John's University*
John Capeheart, *University of Houston-Downtown*
Robert Carroll, *East Carolina University*
David Champlin, *University of Southern Maine*
Giovina Chinchar, *Tougaloo College*
Anne Clark, *Binghamton University*
Greg Clark, *University of Texas, Austin*
Randy Cohen, *California State University, Northridge*
Jim Colbert, *Iowa State University*
Robert Colvin, *Ohio University*
Elizabeth Connor, *University of Massachusetts*
Joanne Conover, *University of Connecticut*
Greg Crowther, *University of Washington*
Karen Curto, *University of Pittsburgh*
Anne Cusic, *University of Alabama at Birmingham*
Larry Davenport, *Samford University*
Teresa DeGolier, *Bethel College*
Roger Del Moral, *University of Washington*
Veronique Delesalle, *Gettysburg College*
Daniel Dervartanian, *University of Georgia*
Jean DeSaix, *University of North Carolina at Chapel Hill*
Michael Dini, *Texas Tech University*
Biao Ding, *Ohio State University*
Stanley Dodson, *University of Wisconsin-Madison*
Mark Drapeau, *University of California, Irvine*
Gary Dudley, *University of Georgia*
Douglas Eernisse, *California State University, Fullerton*
Brad Elder, *University of Oklahoma*
Norman Ellstrand, *University of California, Riverside*
Dennis Emery, *Iowa State University*
John Endler, *University of California, Santa Barbara*
Gerald Esch, *Wake Forest University*
Frederick B. Essig, *University of South Florida*
Mary Eubanks, *Duke University*
Paul Farnsworth, *University of Texas at San Antonio*
Kim Finer, *Kent State University*
Frank Fish, *West Chester University*
Steven Fisher, *University of California, Santa Barbara*
Lloyd Fitzpatrick, *University of North Texas*
William Fixsen, *Harvard University*
James Franzen, *University of Pittsburgh*
Frank Frisch, *Chapman University*
Bernard Frye, *University of Texas at Arlington*
Chandler Fulton, *Brandeis University*
Michael Gaines, *University of Miami*
J. Whitfield Gibbons, *University of Georgia*
J. Phil Gibson, *Agnes Scott College*
Simon Gilroy, *Pennsylvania State University*
Alan Gishlick, *National Center for Science Education*
John Glendinning, *Barnard College*
Sandra Gollnick, *State University of New York at Buffalo*
Robert Goodman, *University of Wisconsin-Madison*
Phyllis Griffard, *University of Houston-Downtown*
Joel Hage, *Radford University*
Jody Hall, *Brown University*
Douglas Hallett, *Northern Arizona University*
Sam Hammer, *Boston University*
Laszlo Hanzely, *Northern Illinois University*
Jeff Hardin, *University of Wisconsin-Madison*
Carla Hass, *Pennsylvania State University*
Chris Haufler, *University of Kansas*
Chris Haynes, *Shelton State Community College*
Blair Hedges, *Pennsylvania State University*
David Hein, *Tulane University*
Jean Heitz, *University of Wisconsin-Madison*
John D. Helmann, *Cornell University*
Michelle Henricks, *University of California, Los Angeles*
Mark Hens, *University of North Carolina at Greensboro*
Scott Herrick, *Missouri Western State College*
David Hibbett, *Clark University*
William Hillenius, *College of Charleston*
Robert Hinrichsen, *Indiana University of Pennsylvania*
William Hoese, *California State University, Fullerton*
A. Scott Holaday, *Texas Tech University*
Karl Holte, *Idaho State University*
Nancy Hopkins, *Tulane University*
Sandra Horikami, *Daytona Beach Community College*

Sandra Hsu, *Skyline College*
Cheryl Ingram-Smith, *Clemson University*
Stephen Johnson, *William Penn University*
Walter Judd, *University of Florida*
Thomas Kane, *University of Cincinnati*
Tamos Kapros, *University of Missouri*
Jennifer Katcher, *Pima Community College*
Norm Kenkel, *University of Manitoba*
Mark Kirk, *University of Missouri-Columbia*
Daniel Klionsky, *University of Michigan*
Ned Knight, *Linfield College*
David Kohl, *University of California, Santa Barbara*
David Kurijaka, *Ohio University*
Elaine La, *Brandeis University*
William L' Amoreaux, *College of Staten Island*
Dominic Lannutti, *El Paso Community College*
Janet Lanza, *University of Arkansas, Little Rock*
John Lepri, *University of North Carolina at Greensboro*
Donald Levin, *University of Texas, Austin*
Mike Levine, *University of California, Berkeley*
Clark Lindgren, *Grinnell College*
Mark Lyford, *University of Wyoming*
Steven Lynch, *Louisiana State University at Shreveport*
Philip M. Meneely, *Haverford College*
Richard Machemer Jr., *St. John Fisher College*
Elizabeth Machunis-Masuoka, *University of Virginia*
Linda Maier, *University of Alabama in Huntsville*
Jose Maldonado, *El Paso Community College*
Richard Malkin, *University of California, Berkeley*
William Margolin, *University of Texas Medical School*
Diane Marshall, *University of New Mexico*
Linda Martin-Morris, *University of Washington*
Lee McClenaghan, *San Diego State University*
Kerry McDonald, *University of Missouri-Columbia*
Neal McReynolds, *Texas A&M International*
Lisa Meffert, *Rice University*
Michael Meighan, *University of California, Berkeley*
Scott Meissner, *Cornell University*
John Merrill, *Michigan State University*
James Mickle, *North Carolina State University*
Alan Molumby, *University of Illinois at Chicago*
Nicholas Money, *Miami University*
Alex Motten, *Duke University*
Rita Moyes, *Texas A&M University*
Greg Nishiyama, *College of the Canyons*
Jane Noble-Harvey, *Delaware University*
Richard Norman, *University of Michigan-Dearborn*
Steven Norris, *California State University, Channel Islands*
Steve Nowicki, *Duke University*
Linda Ogren, *University of California, Santa Cruz*
Jeanette Oliver, *St. Louis Community College, Florissant Valley*
Laura J. Olsen, *University of Michigan*
John Oross, *University of California, Riverside*
Catherine Ortega, *Fort Lewis College*
Charissa Osborne, *Butler University*
Thomas Owens, *Cornell University*
Penny Padgett, *University of North Carolina at Chapel Hill*
Kevin Padian, *University of California, Berkeley*
Dianna Padilla, *State University of New York, Stony Brook*
Daniel Papaj, *University of Arizona*
Ronald Patterson, *Michigan State University*
Debra Pearce, *Northern Kentucky University*
Beverly Perry, *Houston Community College*
David Pfennig, *University of North Carolina at Chapel Hill*
Randall Phillis, *University of Massachusetts*
Daniel Potter, *University of California, Davis*
Andy Pratt, *University of Canterbury*
Mitch Price, *Pennsylvania State University*
Val Raghavan, *Ohio State University*
Talitha Rajah, *Indiana University Southeast*
Thomas Rand, *Saint Mary's University*
Ahnya Redman, *Pennsylvania State University*
Bruce Reid, *Kean University*
Douglas Rhoads, *University of Arkansas*
Carol Rivin, *Oregon State University*
Laurel Roberts, *University of Pittsburgh*
William Roosenburg, *Ohio University*
Neil Sabine, *Indiana University East*
Tyson Sacco, *Cornell University*
Fred Sack, *Ohio State University*
Rowan Sage, *University of Toronto*
K. Sathasivan, *University of Texas, Austin*
Gary Saunders, *University of New Brunswick*
David Schimpf, *University of Minnesota, Duluth*
Robert Schorr, *Colorado State University*
David Schwartz, *Houston Community College*
Christa Schwintzer, *University of Maine, Orono*
Shukdeb Sen, *Bethune-Cookman College*
Wendy Sera, *Seton Hill University*
Timothy Shannon, *Francis Marion University*
Joan Sharp, *Simon Fraser University*
Victoria C. Sharpe, *Blinn College*
Richard Sherwin, *University of Pittsburgh*
James Shinkle, *Trinity University*
Richard Showman, *University of South Carolina*
Anne Simon, *University of Maryland*
Alastair Simpson, *Dalhousie University*
Roger Sloboda, *Dartmouth University*
John Smarrelli, *Le Moyne College*
Kelly Smith, *University of North Florida*
Nancy Smith-Huerta, *Miami University, Ohio*
Amanda Starnes, *Emory University*
Margery Stinson, *Southwestern College*
James Stockand, *University of Texas Health Science Center, San Antonio*
Antony Stretton, *University of Wisconsin-Madison*
Mark Sturtevant, *University of Michigan-Flint*
Judith Sumner, *Assumption College*
Rong Sun Pu, *Kean University*
Marshall Sundberg, *Emporia State University*
Lucinda Swatzell, *Southeast Missouri State University*
Janice Swenson, *University of North Florida*
David Tauck, *Santa Clara University*
John Taylor, *University of California, Berkeley*
Thomas Terry, *University of Connecticut*
Cyril Thong, *Simon Fraser University*
Robert Thornton, *University of California, Davis*
Stephen Timme, *Pittsburg State University*
Leslie Towill, *Arizona State University*
James Traniello, *Boston University*
Constantine Tsoukas, *San Diego State University*
Marsha Turell, *Houston Community College*
Catherine Uekert, *Northern Arizona University*
Gerald Van Dyke, *North Carolina State University*
Brandi Van Roo, *Framingham State College*
Moira Van Staaden, *Bowling Green State University*
Neal Voelz, *St. Cloud State University*
Jyoti Wagle, *Houston Community College*
Edward Wagner, *University of California, Irvine*
D. Alexander Wait, *Southwest Missouri State University*
Beth Wee, *Tulane University*
Matt White, *Ohio University*
Elizabeth Willott, *University of Arizona*
Bill Wischusen, *Louisiana State University*
Clarence Wolfe, *Northern Virginia Community College*
Linda Yasui, *Northern Illinois University*
Robert Yost, *Indiana University/Purdue University, Indianapolis*
Edward Zalisko, *Blackburn College*
Zai Ming Zhao, *University of Texas, Austin*

Reviewers of Previous Editions

Kenneth Able (State University of New York, Albany), Martin Adamson (University of British Columbia), John Alcock (Arizona State University), Richard Almon (State University of New York, Buffalo), Katherine Anderson (University of California, Berkeley), Richard J. Andren (Montgomery County Community College), Estry Ang (University of Pittsburgh at Greensburg), J. David Archibald (Yale University), Howard J. Arnott (University of Texas at Arlington), Robert Atherton (University of Wyoming), Leigh Auleb (San Francisco State University), P. Stephen Baenziger (University of Nebraska), Katherine Baker (Millersville University), William Barklow (Framingham State College), Steven Barnhart (Santa Rosa Junior College), Ron Basmajian (Merced College), Tom Beatty (University of British Columbia), Wayne Becker (University of Wisconsin, Madison), Jane Beiswenger (University of Wyoming), Anne Bekoff (University of Colorado, Boulder), Marc Bekoff (University of Colorado, Boulder), Tania Beliz (College of San Mateo), Adrianne Bendich (Hoffman-La Roche, Inc.), Barbara Bentley (State University of New York, Stony Brook), Darwin Berg (University of California, San Diego), Werner Bergen (Michigan State University), Gerald Bergstrom (University of Wisconsin, Milwaukee), Anna W. Berkovitz (Purdue University), Dorothy Berner (Temple University), Annalisa Berta (San Diego State University), Paulette Bierzychudek (Pomona College), Charles Biggers (Memphis State University), Andrew R. Blaustein (Oregon State University), Judy Bluemer (Morton College), Robert Blystone (Trinity University), Robert Boley (University of Texas, Arlington), Eric Bonde (University of Colorado, Boulder), Richard Boohar (University of Nebraska, Omaha), Carey L. Booth (Reed College), James L. Botsford (New Mexico State University), J. Michael Bowes (Humboldt State University), Richard Bowker (Alma College), Barry Bowman (University of California, Santa Cruz), Deric Bownds (University of Wisconsin, Madison), Robert Boyd (Auburn University), Jerry Brand (University of Texas, Austin), Theodore A. Bremner (Howard University), James Brenneman (University of Evansville), Charles H. Brenner (Berkeley, California), Donald P. Briskin (University of Illinois, Urbana), Paul Broady (University of Canterbury), Danny Brower (University of Arizona), Carole Browne (Wake Forest University), Mark Browning (Purdue University), Herbert Bruneau (Oklahoma State University), Gary Brusca (Humboldt State University), Alan H. Brush (University of Connecticut, Storrs), Meg Burke (University of North Dakota), Edwin Burling (De Anza College), William Busa (Johns Hopkins University), John Bushnell (University of Colorado), Linda Butler (University of Texas, Austin), David Byres (Florida Community College, Jacksonville), Iain Campbell (University of Pittsburgh), Robert E. Cannon (University of North Carolina at Greensboro), Deborah Canington (University of California, Davis), Gregory Capelli (College of William and Mary), Richard Cardullo (University of California, Riverside), Nina Caris (Texas A & M University), Bruce Chase (University of Nebraska, Omaha), Doug Cheeseman (De Anza College), Shepley Chen (University of Illinois, Chicago), Joseph P. Chinnici (Virginia Commonwealth University), Henry Claman (University of Colorado Health Science Center), Ross C. Clark (Eastern Kentucky University), Lynwood Clemens (Michigan State University), William P. Coffman (University of Pittsburgh), J. John Cohen (University of Colorado Health Science Center), David Cone (Saint Mary's University), John Corliss (University of Maryland), James T. Costa (Western Carolina University), Stuart J. Coward (University of Georgia), Charles Creutz (University of Toledo), Bruce Criley (Illinois Wesleyan University), Norma Criley (Illinois Wesleyan University), Joe W. Crim (University of Georgia), Richard Cyr (Pennsylvania State University), W. Marshall Darley (University of Georgia), Marianne Dauwalder (University of Texas, Austin), Bonnie J. Davis (San Francisco State University), Jerry Davis (University of Wisconsin, La Crosse), Thomas Davis (University of New Hampshire), John Dearn (University of Canberra), James Dekloe (University of California, Santa Cruz), T. Delevoryas (University of Texas, Austin), Diane C. DeNagel (Northwestern University), Jean DeSaix (University of North Carolina), Michael Dini (Texas Tech University), Andrew Dobson (Princeton University), John Drees (Temple University School of Medicine), Charles Drewes (Iowa State University), Marvin Druger (Syracuse University), Susan Dunford (University of Cincinnati), Betsey Dyer (Wheaton College), Robert Eaton (University of Colorado), Robert S. Edgar (University of California, Santa Cruz), Betty J. Eidemiller (Lamar University), William D. Eldred (Boston University), Margaret T. Erskine (Lansing Community College), David Evans (University of Florida), Robert C. Evans (Rutgers University, Camden), Sharon Eversman (Montana State University), Lincoln Fairchild (Ohio State University), Peter Fajer (Florida State University), Bruce Fall (University of Minnesota), Lynn Fancher (College of DuPage), Larry Farrell (Idaho State University), Jerry F. Feldman (University of California, Santa Cruz), Eugene Fenster (Longview Community College), Russell Fernald (University of Oregon), Milton Fingerman (Tulane University), Barbara Finney (Regis College), David Fisher (University of Hawaii, Manoa), William Fixsen (Harvard University), Abraham Flexer (Manuscript Consultant, Boulder, Colorado), Kerry Foresman (University of Montana), Norma Fowler (University of Texas, Austin), Robert G. Fowler (San Jose State University), David Fox (University of Tennessee, Knoxville), Carl Frankel (Pennsylvania State University, Hazleton), Bill Freedman (Dalhousie University), Otto Friesen (University of Virginia), Virginia Fry (Monterey Peninsula College), Alice Fulton (University of Iowa), Sara Fultz (Stanford University), Berdell Funke (North Dakota State University), Anne Funkhouser (University of the Pacific), Arthur W. Galston (Yale University), Carl Gans (University of Michigan), John Gapter (University of Northern Colorado), Reginald Garrett (University of Virginia), Patricia Gensel (University of North Carolina), Chris George (California Polytechnic State University, San Luis Obispo), Robert George (University of Wyoming), Frank Gilliam (Marshall University), Simon Gilroy (Pennsylvania State University), Todd Gleeson (University of Colorado), David Glenn-Lewin (Wichita State University), William Glider (University of Nebraska), Elizabeth A. Godrick (Boston University), Lynda Goff (University of California, Santa Cruz), Elliott Goldstein (Arizona State University), Paul Goldstein (University of Texas, El Paso), Anne Good (University of California, Berkeley), Judith Goodenough (University of Massachusetts, Amherst), Wayne Goodey (University of British Columbia), Ester Goudsmit (Oakland University), Linda Graham (University of Wisconsin, Madison), Robert Grammer (Belmont University), Joseph Graves (Arizona State University), A. J. F. Griffiths (University of British Columbia), William Grimes (University of Arizona), Mark Gromko (Bowling Green State University), Serine Gropper (Auburn University), Katherine L. Gross (Ohio State University), Gary Gussin (University of Iowa), Mark Guyer (National Human Genome Research Institute), Ruth Levy Guyer(Bethesda, Maryland), R. Wayne Habermehl (Montgomery County Community College), Mac Hadley (University of Arizona), Jack P. Hailman (University of Wisconsin), Leah Haimo (University of California, Riverside), Rebecca Halyard (Clayton State College), Penny Hanchey-Bauer (Colorado State University), Laszlo Hanzely (Northern Illinois University), Jeff Hardin (University of Wisconsin, Madison), Richard Harrison (Cornell University), H. D. Heath (California State University, Hayward), George Hechtel (State University of New York, Stony Brook), Jean Heitz-Johnson (University of Wisconsin, Madison), Colin Henderson (University of Montana), Caroll Henry (Chicago State University), Frank Heppner (University of Rhode Island), Ira Herskowitz (University of California, San Francisco), Paul E. Hertz (Barnard College), R. James Hickey (Miami University), Ralph Hinegardner (University of California, Santa Cruz), William Hines (Foothill College), Helmut Hirsch (State University of New York, Albany), Tuanhua David Ho (Washington University), Carl Hoagstrom (Ohio Northern University), James Hoffman (University of Vermont), James Holland (Indiana State University, Bloomington), Charles Holliday (Lafayette College), Laura Hoopes (Occidental College), Nancy Hopkins (Massachusetts Institute of Technology), Kathy Hornberger (Widener University), Pius F. Horner (San Bernardino Valley College), Margaret Houk (Ripon College), Ronald R. Hoy (Cornell University), Donald Humphrey (Emory University School of Medicine), Robert J. Huskey (University of Virginia), Steven Hutcheson (University of Maryland, College Park), Bradley Hyman (University of California, Riverside), Mark Iked (San Bernardino Valley College), Alice Jacklet (State University of New York, Albany), John Jackson (North Hennepin Community College), John C. Jahoda (Bridgewater State College), Dan Johnson (East Tennessee State University), Randall Johnson (University of California, San Diego), Wayne Johnson (Ohio State University), Kenneth C. Jones (California State University, Northridge), Russell Jones (University of California, Berkeley), Alan Journet (Southeast Missouri State University), Thomas C. Kane (University of Cincinnati), E. L. Karlstrom (University of Puget Sound), George Khoury (National Cancer Institute), Robert Kitchin (University of Wyoming), Attila O. Klein (Brandeis University), Greg Kopf (University of Pennsylvania School of Medicine), Thomas Koppenheffer (Trinity University), Janis Kuby (San Francisco State University), J. A. Lackey (State University of New York, Oswego), Lynn Lamoreux (Texas A & M University), Carmine A. Lanciani (University of Florida), Kenneth Lang (Humboldt State University), Allan Larson (Washington University), Diane K. Lavett (State University of New York, Cortland, and Emory University), Charles Leavell (Fullerton College), C. S. Lee (University of Texas), Robert Leonard (University of California, Riverside), Joseph Levine (Boston College), Bill Lewis (Shoreline Community College), John Lewis (Loma Linda University), Lorraine Lica (California State University, Hayward), Harvey Liftin (Broward Community College), Harvey Lillywhite (University of Florida, Gainesville), Sam Loker (University of New Mexico), Jane Lubchenco (Oregon State University), Margaret A. Lynch (Tufts University), James MacMahon (Utah State University), Charles Mallery (University of Miami), Lynn Margulis (Boston University), Edith Marsh (Angelo State University), Karl Mattox (Miami University of Ohio), Joyce Maxwell (California State University, Northridge), Jeffrey D. May (Marshall University), Richard McCracken (Purdue University), Jacqueline McLaughlin (Pennsylvania State University, Lehigh Valley), Paul Melchior (North Hennepin Community College), Phillip Meneely (Haverford College), John Merrill (University of Washington), Brian Metscher (University of California, Irvine), Ralph Meyer (University of Cincinnati), Roger Milkman (University of Iowa), Helen Miller (Oklahoma State University), John Miller (University of California, Berkeley), Kenneth R. Miller (Brown University), John E. Minnich (University of Wisconsin, Milwaukee), Michael Misamore (Louisiana State University), Kenneth Mitchell (Tulane University School of Medicine), Russell Monson (University of Colorado, Boulder), Frank Moore (Oregon State University), Randy Moore (Wright State University), William Moore (Wayne State University), Carl Moos (Veterans Administration Hospital, Albany, New York), Michael Mote (Temple University), Deborah Mowshowitz (Columbia University), Darrel L. Murray (University of Illinois at Chicago), John Mutchmor (Iowa State University), Elliot Myerowitz (California Institute of Technology), Gavin Naylor (Iowa State University), John Neess (University of Wisconsin, Madison), Raymond Neubauer (University of Texas, Austin), Todd Newbury (University of California, Santa Cruz), Harvey Nichols (University of Colorado, Boulder), Deborah Nickerson (University of South Florida), Bette Nicotri (University of Washington), Caroline Niederman (Tomball College), Maria Nieto (California State University, Hayward), Charles R. Noback (College of Physicians and Surgeons, Columbia University), Mary C. Nolan (Irvine Valley College), Peter Nonacs (University of California, Los Angeles), David O. Norris (University of Colorado, Boulder), Cynthia Norton (University of Maine, Augusta), Steve Norton (East Carolina University), Bette H. Nybakken (Hartnell College), Brian O'Conner (University of Massachusetts, Amherst), Gerard O'Donovan (University of North Texas), Eugene Odum (University of Georgia), Patricia O'Hern (Emory University), Gary P. Olivetti (University of Vermont), John Olsen (Rhodes College), Sharman O'Neill (University of California, Davis), Wan Ooi (Houston Community College), Gay Ostarello (Diablo Valley College), Barry Palevitz (University of Georgia), Peter Pappas (County College of Morris), Bulah Parker (North Carolina State University), Stanton Parmeter (Chemeketa Community College), Robert Patterson (San Francisco State University), Crellin Pauling (San Francisco State University), Kay Pauling (Foothill Community College), Daniel Pavuk (Bowling Green State University), Debra Pearce (Northern Kentucky University), Patricia Pearson (Western Kentucky University), Shelley Penrod (North Harris College), Bob Pittman (Michigan State University), James Platt (University of Denver), Martin Poenie (University of Texas, Austin), Scott Poethig (University of Pennsylvania), Jeffrey Pommerville (Texas A & M University), Warren Porter (University of Wisconsin), Donald Potts (University of California, Santa Cruz), David Pratt (University of California, Davis), Halina Presley (University of Illinois, Chicago), Rebecca Pyles (East Tennessee State University), Scott Quackenbush (Florida International University), Ralph Quatrano (Oregon State University), Deanna Raineri (University of Illinois, Champaign-Urbana), Charles Ralph (Colorado State University), Kurt Redborg (Coe College), Brian Reeder (Morehead State University), C. Gary Reiness (Lewis & Clark College), Charles Remington (Yale University), David Reznick (University of California, Riverside), Fred Rhoades (Western Washington State University), David Reid (Blackburn College), Christopher Riegle (Irvine Valley College), Donna Ritch (Pennsylvania State University), Thomas Rodella (Merced College), Rodney Rogers (Drake University), Wayne Rosing (Middle Tennessee State University), Thomas Rost (University of California, Davis), Stephen I. Rothstein (University of California, Santa Barbara), John Ruben (Oregon State University), Albert Ruesink (Indiana University), Don Sakaguchi (Iowa State University), Walter Sakai (Santa Monica College), Mark F. Sanders (University of California, Davis), Ted Sargent (University of Massachusetts, Amherst), Gary Saunders (University of New Brunswick), Carl Schaefer (University of Connecticut), Lisa Shimeld (Crafton Hills College), David Schimpf (University of Minnesota, Duluth), William H. Schlesinger (Duke University), Erik P. Scully (Towson State University), Edna Seaman (Northeastern University), Elaine Shea (Loyola College, Maryland), Stephen Sheckler (Virginia Polytechnic Institute and State University), James Shinkle (Trinity University), Barbara Shipes (Hampton University), Peter Shugarman (University of Southern California), Alice Shuttey (DeKalb Community College), James Sidie (Ursinus College), Daniel Simberloff (Florida State University), Susan Singer (Carleton College), John Smarrelli (Loyola University), Andrew T. Smith (Arizona State University), John Smol (Queen's University), Andrew J. Snope (Essex Community College), Mitchell Sogin (Woods Hole Marine Biological Laboratory), Susan Sovonick-Dunford (University of Cincinnati), Frederick W. Spiegel (University of Arkansas), Karen Steudel (University of Wisconsin), Barbara Stewart (Swarthmore College), Cecil Still (Rutgers University, New Brunswick), John Stolz (California Institute of Technology), Richard D. Storey (Colorado College), Stephen Strand (University of California, Los Angeles), Eric Strauss (University of Massachusetts, Boston), Russell Stullken (Augusta College), John Sullivan (Southern Oregon State University), Gerald Summers (University of Missouri), Marshall D. Sundberg (Emporia State University), Daryl Sweeney (University of Illinois, Champaign-Urbana), Samuel S. Sweet (University of California, Santa Barbara), Lincoln Taiz (University of California, Santa Cruz), Samuel Tarsitano (Southwest Texas State University), David Tauck (Santa Clara University), James Taylor (University of New Hampshire), Martha R. Taylor (Cornell University), Roger Thibault (Bowling Green State University), William Thomas (Colby-Sawyer College), John Thornton (Oklahoma State University), Robert Thornton (University of California, Davis), James Traniello (Boston University), Robert Tuveson (University of Illinois, Urbana), Maura G. Tyrrell (Stonehill College), Gordon Uno (University of Oklahoma), Lisa A. Urry (Mills College), James W. Valentine (University of California, Santa Barbara), Joseph Vanable (Purdue University), Theodore Van Bruggen (University of South Dakota), Kathryn VandenBosch (Texas A & M University), Frank Visco (Orange Coast College), Laurie Vitt (University of California, Los Angeles), Thomas J. Volk (University of Wisconsin, La Crosse), Susan D. Waaland (University of Washington), William Wade (Dartmouth Medical College), John Waggoner (Loyola Marymount University), Dan Walker (San Jose State University), Robert L. Wallace (Ripon College), Jeffrey Walters (North Carolina State University), Margaret Waterman (University of Pittsburgh), Charles Webber (Loyola University of Chicago), Peter Webster (University of Massachusetts, Amherst), Terry Webster (University of Connecticut, Storrs), Peter Wejksnora (University of Wisconsin, Milwaukee), Kentwood Wells (University of Connecticut), David J. Westenberg (University of Missouri, Rolla), Stephen Williams (Glendale Community College), Christopher Wills (University of California, San Diego), Fred Wilt (University of California, Berkeley), Robert T. Woodland (University of Massachusetts Medical School), Joseph Woodring (Louisiana State University), Patrick Woolley (East Central College), Philip Yant (University of Michigan), Hideo Yonenaka (San Francisco State University), Edward Zalisko (Blackburn College), John Zimmerman (Kansas State University), Uko Zylstra (Calvin College).

Supplements

Supplements for the Student

Campbell BIOLOGY Student CD-ROM and Website (www.campbellbiology.com)

The CD-ROM and website that accompany each book include 230 interactive Activities, 85 Videos, and 55 Investigations. In addition, new graphing exercises (Graph It!) help students learn how to build and interpret graphs.

The CD-ROM and website are fully integrated with the text, reinforcing students' focus on the big ideas. The media organization mirrors that of the textbook, with all the Activities, Videos, and Investigations for a given chapter correlated to the key concepts in the Chapter Guide.

There are four quizzes per chapter: a Pre-Test, the Self-Quiz from the book, an Activities Quiz, and a Chapter Quiz. The new online Results Reporter and Gradebook automatically record students' quiz scores. The Cumulative Test, another new feature of the Seventh Edition, allows students to build a self-test with questions from more than one chapter. Feedback is provided to students on all quizzes and tests in the media, which have been upgraded in level of difficulty.

New Flashcards, Word Roots, and Key Terms linked to the Glossary help students master terminology. Students can also access Art from the book with and without labels, the Glossary with audio pronunciations, the Campbell *BIOLOGY* Interviews from previous editions, an E-Book, the Biology Tutor Center, Web Links, News, Further Readings, and Research Navigator.

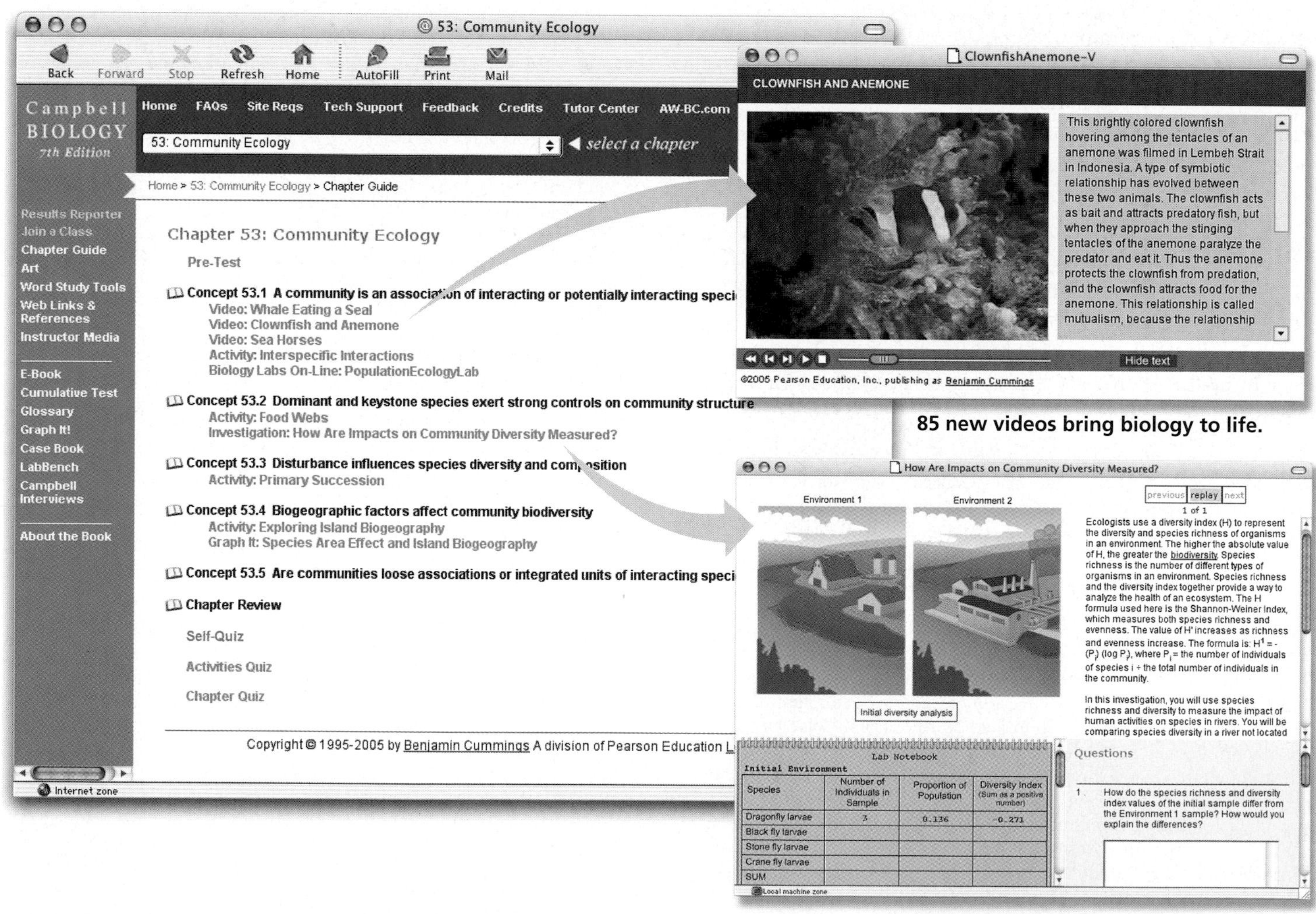

85 new videos bring biology to life.

55 Investigations help students develop scientific skills such as posing hypotheses, collecting data, and analyzing results.

NEW! Biological Inquiry: A Workbook of Investigative Cases (0-8053-7176-1)
Margaret Waterman, Southeast Missouri State University, and Ethel Stanley, Beloit College
This new workbook offers eight investigative cases, one for each unit of the textbook. In order to understand the science in each case, students will pose questions, analyze data, think critically, examine the relationship between evidence and conclusions, construct hypotheses, investigate options, graph data, interpret results, and communicate scientific arguments. Students will actively engage in the experimental nature of science as they gain new insight into how we know what we know. For example, in "The Donor's Dilemma" (the Unit 3 case) students explore the concepts of protein synthesis, viral genomes, and transmission pathways while investigating the case of a blood donor who may have been exposed to the West Nile virus. Web links and other online resources referred to in the investigative cases are provided on the Campbell *BIOLOGY* website.

Student Study Guide (0-8053-7155-9)
Martha R. Taylor, Cornell University
This popular study guide offers an interactive approach to learning, providing framework sections to orient students to the overall picture, concept maps to complete or create for most chapters, chapter summaries, word roots, chapter tests, and a variety of questions, including multiple choice, short-answer essay, art labeling, and interpreting graphs.

Practicing Biology: A Student Workbook, Second Edition (0-8053-7184-2)
Jean Heitz, University of Wisconsin, Madison
This workbook's hands-on activities emphasize key ideas, principles, and concepts that are basic to understanding biology. Suitable for group work in lecture, discussion sections, and/or lab, the workbook includes class-tested Process of Science activities, concept map development, drawing exercises, and modeling activities.

NEW! Kaplan MCAT®/GRE® Biology Test Preparation Guide for Campbell • Reece BIOLOGY, 7e (0-8053-7178-8)
Exclusively available with *BIOLOGY*, Seventh Edition, this new guide includes sample questions and answers from the Kaplan test preparation guides, correlated to specific pages in this edition of *BIOLOGY*.

Art Notebook (0-8053-7183-4)
This resource contains all the art from the text without labels, with plenty of room for students to take notes.

Study Card (0-8053-7175-3)
Useful as a quick reference guide, this fold-out card summarizes the basic concepts and content covered in *BIOLOGY*, Seventh Edition.

Spanish Glossary (0-8053-7182-6)
Laura P. Zanello

Biology Tutor Center (www.aw.com/tutorcenter)
This center provides one-to-one tutoring for college students via phone, fax, and email during evening hours and on weekends. Qualified college instructors are available to answer questions and provide instruction regarding self-quizzes and other content found in *BIOLOGY*, Seventh Edition. Visit the website for more information.

The Benjamin Cummings Special Topics Booklets
- **Understanding the Human Genome Project (0-8053-6774-8)**
- **Stem Cells and Cloning (0-8053-4864-6)**
- **Biological Terrorism (0-8053-4868-9)**
- **The Biology of Cancer (0-8053-4867-0)**

The Chemistry of Life CD-ROM, Second Edition (0-8053-3063-1)
Robert M. Thornton, University of California, Davis
This CD-ROM helps biology students grasp the essentials of chemistry with animations, interactive exercises, and quizzes with feedback.

An Introduction to Chemistry for Biology Students, Eighth Edition (0-8053-3970-1)
George I. Sackheim, University of Illinois, Chicago
This printed workbook helps students master all the basic facts, concepts, and terminology of chemistry that they need for their life science course.

Biomath: Problem Solving for Biology Students (0-8053-6524-9)
Robert W. Keck and Richard R. Patterson

A Short Guide to Writing about Biology, Fifth Edition (0-321-15981-0)
Jan A. Pechenik, Tufts University

Supplements for the Instructor

Campbell Media Manager (set of 8 CD-ROMs) (0-8053-7173-7)

The Campbell Media Manager combines all the instructor and student media for Campbell/Reece *BIOLOGY* into one chapter-by-chapter resource. It includes eight CD-ROMs, one for each unit in the text. Instructor media includes PowerPoint Lectures, PowerPoint TextEdit Art, PowerPoint Layered Art, PowerPoint Active Lecture Questions, the Image Library (1,600 photos, all art from the book with and without labels, selected art layered for step-by-step presentation, and tables), and the Media Library (85 videos and more than 100 animations). Also included are Lecture Outlines in Word format for each chapter, the Test Bank (in Word), and additional resources. (Most of these resources are also available in the Instructor Media section at *www.campbellbiology.com*.) The Campbell Media Manager also includes printed thumbnail-sized images for easy viewing of all the resources in the Image and Media Libraries and a convenient fold-out Quick Reference Guide.

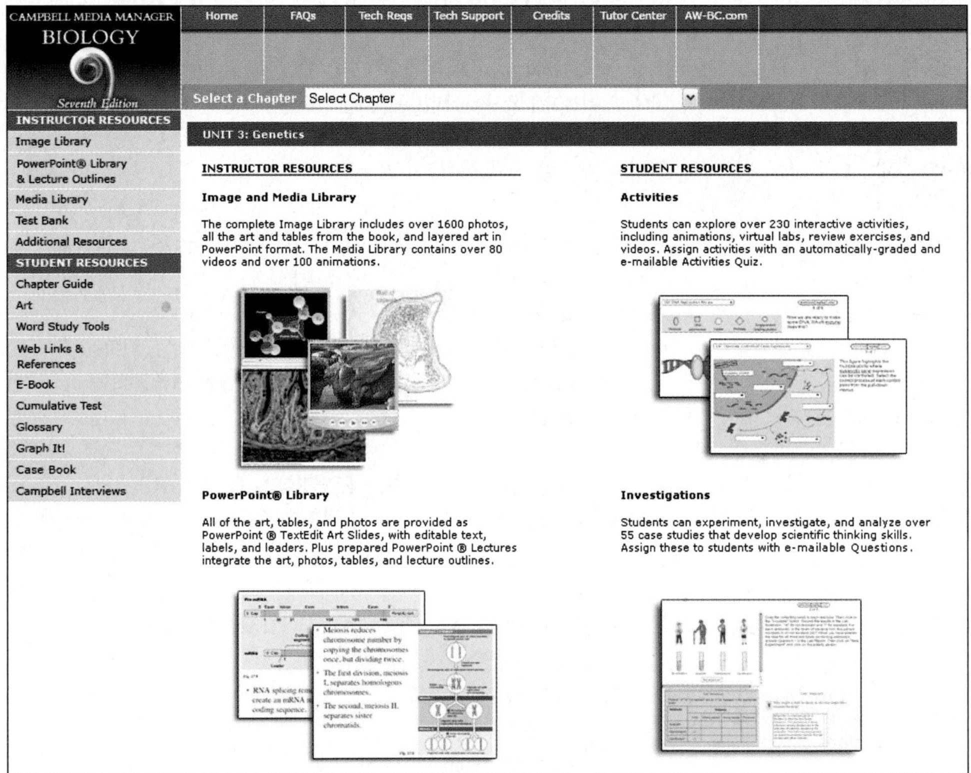

Instructor's Guide to Text and Media (0-8053-7148-6)

Joan Sharp, Simon Fraser University

This reference tool includes a chapter-by-chapter listing of all the text and media resources available to instructors and students organized by key concepts. It also includes objectives, key terms, and word roots. New to this edition are descriptions of frequent student misconceptions for each chapter and suggestions for how to address these concerns. The Instructor's Guide is available in print and in Word format on the Campbell Media Manager, at *www.campbellbiology.com*, and at *www.aw-bc.com*.

Lecture Outlines

Joan Sharp, Simon Fraser University

Chapter lecture outlines are available in Word format on the Campbell Media Manager and in the Instructor Media section at *www.campbellbiology.com*.

Transparency Acetates (0-8053-7149-4)

Approximately 1,000 acetates include all full-color illustrations and tables from the text, many of which incorporate photos. In addition, selected figures illustrating key concepts are broken down into layers for step-by-step lecture presentation.

Printed Test Bank (0-8053-7154-0)
Computerized Test Bank in TestGen (0-8053-7153-2)
Computerized Test Bank in ExamView (0-8053-7129-X)

Edited by William Barstow, University of Georgia

Thoroughly revised and updated, the Seventh Edition test bank also includes optional questions from the book's Self-Quizzes and questions related to the media to encourage students to use these resources. Available in print, on a cross-platform CD-ROM, on the Campbell Media Manager (in Word), at *www.aw-bc.com* (in Word, TestGen, and ExamView), and in the instructor section of CourseCompass™, Blackboard, and WebCT.

PowerPoint Lectures

Chris C. Romero, Front Range Community College, Larimer Campus

PowerPoint Lectures are provided for each chapter and include the art, photos, tables, and an editable lecture outline. The PowerPoint Lectures can be used as is or customized for your course with your own images and text and/or additional photos, videos, and animations from the Campbell Media Manager. The PowerPoint Lectures are available on the Campbell Media Manager and at *www.campbellbiology.com*.

PowerPoint TextEdit Art

All the art, photos, and tables are provided in PowerPoint, with selected figures layered for step-by-step presentation. The

PowerPoint TextEdit Art can be used as is or labels can be edited in PowerPoint. Available on the Campbell Media Manager and at *www.campbellbiology.com.*

NEW! Instructor Edition for Biological Inquiry: A Workbook of Investigative Cases (0-8053-7177-X)
Margaret Waterman, Southeast Missouri State University, and Ethel Stanley, Beloit College
The instructor version provides specific and detailed suggestions on how to use each case study effectively, outlining links to specific content in *BIOLOGY,* Seventh Edition, and other supplements, providing direction on how to facilitate problem-based learning, and listing suggested answers and opportunities for extended investigations.

Instructor's Guide to Practicing Biology: A Student Workbook, Second Edition
Jean Heitz, University of Wisconsin, Madison
The instructor's version is available online at:
www.campbellbiology.com and *www.aw-bc.com.*

Course Management Systems
The content from the Campbell *BIOLOGY* website is also available in these popular course management systems: CourseCompass, Blackboard, and WebCT. For more information, visit *http://cms.aw.com.*

NEW! Active Lecture Questions
Invigorate lectures with questions that can be used with classroom response systems (H-ITT and PRS). These multiple-choice questions are adapted from various sources, including the end-of-chapter questions, *Student Study Guide, Test Bank,* and *Biomath: Problem Solving for Biology* by Robert W. Keck and Richard R. Patterson. The questions are available in PowerPoint format on the Campbell Media Manager, at *www.campbellbiology.com*, at *www.aw-bc.com,* and preloaded on H-ITT and PRS. Sources and answers are located in the PowerPoint Notes field. Selected questions in each chapter include additional Discussion Points added as suggestions for the instructor. Visit *www.aw-bc.com/crs* for more information about classroom response systems.

Supplements for the Lab

Investigating Biology Laboratory Manual, Fifth Edition (0-8053-7179-6)
Judith Giles Morgan, Emory University, and M. Eloise Brown Carter, Oxford College of Emory University
With its distinctive investigative approach to learning, this laboratory manual encourages students to practice science. Students are invited to pose hypotheses, make predictions, conduct open-ended experiments, collect data, and then apply the results to new problems.

Annotated Instructor's Edition for Investigating Biology Laboratory Manual, Fifth Edition (0-8053-7180-X)
Teaching information, added to the original Student Edition text, includes margin notes with hints on lab procedures, additional art, and answers to in-text and end-of-chapter questions from the Student Edition. Also featured is a detailed Teaching Plan at the end of each lab with specific suggestions for organizing labs, including estimated time allotments and suggestions for encouraging independent thinking and collaborative discussion.

Preparation Guide for Investigating Biology Laboratory Manual, Fifth Edition (0-8053-7181-8)
Guides lab coordinators in ordering materials as well as in planning, setting up, and running labs.

NEW! New Designs for Bio-Explorations, Second Edition (0-8053-7229-6)
Janet Lanza, University of Arkansas at Little Rock
Eight inquiry-based laboratory exercises offer students creative control over the projects they undertake. Students are provided background information that enables them to design and conduct their own experiments.

NEW! Instructor's Guide to New Designs for Bio-Explorations, Second Edition (0-8053-7228-8)
The instructor's version is available online at:
www.campbellbiology.com and *www.aw-bc.com.*

Symbiosis Book Building Kit—Customized Lab Manuals (0-201-72142-2)
Build a customized lab manual, choosing the labs you want, importing artwork from our graphics library, and even adding your own material, and get a made-to-order black and white lab manual. Visit *http://www.pearsoncustom.com/database/symbiosis.html* for more information.

Biology Labs On-Line (www.biologylabsonline.com)
Twelve on-line labs enable students to expand their scientific horizons beyond the traditional wet lab setting and perform potentially dangerous, lengthy, or expensive experiments in an electronic environment. Each experiment can be repeated as often as necessary, employing a different set of variables each time. The labs are available for purchase individually or in a 12-pack with the printed Student Lab Manual.

Student Lab Manual for Biology Labs On-Line (0-8053-7017-X)

Instructor's Lab Manual for Biology Labs On-Line (0-8053-7018-8)

Featured Figures

Exploring Figures

1.3 Exploring Levels of Biological Organization *4*
1.14 Exploring Life's Three Domains *14*
4.10 Exploring Some Important Functional Groups of Organic Compounds *64*
5.20 Exploring Levels of Protein Structure *82*
6.9 Exploring Animal and Plant Cells *100*
6.30 Exploring Intercellular Junctions in Animal Tissues *121*
7.20 Exploring Endocytosis in Animal Cells *138*
11.7 Exploring Membrane Receptors *206*
12.6 Exploring the Mitotic Division of an Animal Cell *222*
13.8 Exploring the Meiotic Division of an Animal Cell *244*
21.2 Exploring Model Organisms for Genetic Studies of Development *412*
24.4 Exploring Reproductive Barriers *474*
27.13 Exploring Major Groups of Bacteria *542*
28.12 Exploring Structure and Function in the Ciliate *Paramecium caudatum* *557*
29.5 Exploring Derived Traits of Land Plants *576*
29.9 Exploring Bryophyte Diversity *582*
29.14 Exploring Seedless Vascular Plant Diversity *587*
30.4 Exploring Gymnosperm Diversity *594*
30.12 Exploring Angiosperm Diversity *602*
33.3 Exploring Invertebrate Diversity *639*
33.37 Exploring Insect Diversity *662*
34.36 Exploring Mammalian Diversity *698*
35.9 Exploring Examples of Differentiated Plant Cells *718*
37.13 Exploring Unusual Nutritional Adaptations in Plants *768*
38.3 Exploring Floral Variations *773*
40.5 Exploring Structure and Function in Animal Tissues *824*
41.2 Exploring Four Main Feeding Mechanisms of Animals *845*
42.4 Exploring Vertebrate Circulatory Systems *870*
44.18 Exploring Environmental Adaptations of the Vertebrate Kidney *939*
46.11 Exploring Human Oogenesis *974*
46.12 Exploring Human Spermatogenesis *975*
49.8 Exploring the Structure of the Human Ear *1051*
50.10 Exploring Global Climate Patterns *1088*
50.17 Exploring Aquatic Biomes *1094*
50.20 Exploring Terrestrial Biomes *1100*
54.17 Exploring Nutrient Cycles *1196*
55.22 Exploring Restoration Ecology Worldwide *1226*

Inquiry Figures

1.29 Does the presence of poisonous coral snakes affect predation rates on their mimics, king snakes? *23*
4.2 Could organic compounds have been synthesized abiotically on the early Earth? *59*
7.6 Do membrane proteins move? *127*
10.9 Which wavelengths of light are most effective in driving photosynthesis? *187*
12.8 During anaphase, do kinetochore microtubules shorten at their spindle pole ends or their kinetochore ends? *225*
12.13 Are there molecular signals in the cytoplasm that regulate the cell cycle? *228*
12.17 Does platelet-derived growth factor (PDGF) stimulate the division of human fibroblast cells in culture? *231*
14.3 When F_1 pea plants with purple flowers are allowed to self-pollinate, what flower color appears in the F_2 generation? *253*
14.8 Do the alleles for seed color and seed shape sort into gametes dependently (together) or independently? *257*
15.4 In a cross between a wild-type female fruit fly and a mutant white-eyed male, what color eyes will the F_1 and F_2 offspring have? *277*
15.5 Are the genes for body color and wing size in fruit flies located on the same chromosome or different chromosomes? *279*
16.2 Can the genetic trait of pathogenicity be transferred between bacteria? *294*
16.4 Is DNA or protein the genetic material of phage T2? *295*
16.11 Does DNA replication follow the conservative, semiconservative, or dispersive model? *300*
17.2 Do individual genes specify different enzymes in arginine biosynthesis? *311*
18.15 Can a bacterial cell acquire genes from another bacterial cell? *347*
21.5 Can a differentiated plant cell develop into a whole plant? *415*
21.6 Can the nucleus from a differentiated animal cell direct development of an organism? *416*
21.21 Which cell layers in the floral meristem determine the number of floral organs? *430*
22.12 Can predation pressure select for size and age at maturity in guppies? *447*

23.11 Does geographic variation in yarrow plants have a genetic component? *464*

24.7 Can divergence of allopatric fruit fly populations lead to reproductive isolation? *477*

24.10 Does sexual selection in cichlids result in reproductive isolation? *480*

26.2 Can organic molecules form in a reducing atmosphere? *513*

31.21 Does having mycorrhizae benefit a plant? *620*

36.19 What causes phloem sap to flow from source to sink? *753*

39.5 What part of a coleoptile senses light, and how is the signal transmitted? *792*

39.6 Does asymmetric distribution of a growth-promoting chemical cause a coleoptile to grow toward the light? *793*

39.7 What causes polar movement of auxin from shoot tip to base? *795*

39.13 How does ethylene concentration affect the triple response in seedlings? *799*

39.17 What wavelengths stimulate phototropic bending toward light? *803*

39.18 How does the order of red and far-red illumination affect seed germination? *803*

39.22 How does interrupting the dark period with a brief exposure to light affect flowering? *807*

39.23 Is phytochrome the pigment that measures the interruption of dark periods in photoperiodic response? *807*

39.24 Is there a flowering hormone? *808*

44.6 What role does fur play in water conservation by camels? *926*

47.4 What is the effect of sperm binding on on Ca^{2+} distribution in the egg? *990*

47.21 Does fibronectin promote cell migration? *1002*

47.22 Is cadherin required for development of the blastula? *1003*

47.24 How does distribution of the gray crescent at the first cleavage affect the potency of the two daughter cells? *1005*

47.25 Can the dorsal lip of the blastopore induce cells in another part of the amphibian embryo to change their developmental fate? *1006*

47.27 What role does the zone of polarizing activity (ZPA) play in limb pattern formation in vertebrates? *1008*

48.25 Are mammalian biological clocks influenced by external cues? *1031*

49.13 How do insects detect different tastes? *1055*

49.37 What are the energy costs of locomotion? *1074*

50.8 Does feeding by sea urchins and limpets affect seaweed distribution? *1086*

51.10 Are the different songs of closely related green lacewing species under genetic control? *1112*

51.12 How does dietary environment affect mate choice by female *Drosophila mojavensis*? *1113*

51.14 Does a digger wasp use landmarks to find her nest? *1115*

52.7 How does caring for offspring affect parental survival in kestrals? *1142*

52.18 How stable is the Isle Royale moose population? *1150*

53.2 Can a species' niche be influenced by interspecific competition? *1160*

53.28 How does species richness relate to area? *1178*

54.6 Which nutrient limits phytoplankton production along the coast of Long Island? *1189*

54.9 Is phosphorus or nitrogen the limiting nutrient in a Hudson Bay salt marsh? *1191*

55.10 What caused the drastic decline of the Illinois greater prairie chicken population? *1216*

Research Method Figures

2.5 Radioactive Tracers *35*

5.24 X-Ray Crystallography *86*

6.3 Light Microscopy *96*

6.4 Electron Microscopy *96*

6.5 Cell Fractionation *97*

7.4 Freeze Fracture *126*

10.8 Determining an Absorption Spectrum *187*

13.3 Preparing a Karyotype *240*

14.2 Crossing Pea Plants *252*

14.7 The Testcross *256*

15.7 Constructing a Linkage Map *281*

20.4 Cloning a human gene in a Bacterial Plasmid *387*

20.5 Nucleic Acid Probe Hybridization *389*

20.7 The Polymerase Chain Reaction (PCR) *391*

20.8 Gel Electrophoresis *393*

20.10 Southern Blotting of DNA Fragments *395*

20.12 Dideoxy Chain-Termination Method for Sequencing DNA *397*

20.14 DNA Microarray Assay of Gene Expression Levels *401*

20.19 Using the Ti Plasmid to Produce Transgenic Plants *407*

21.7 Reproductive Cloning of a Mammal by Nuclear Transplantation *417*

23.14 Using a Virtual Population to Study the Effects of Selection *467*

25.15 Applying Parsimony to a Problem in Molecular Systematics *502*

37.3 Hydroponic Culture *757*

48.9 Intracellular Recording *1016*

New to the Seventh Edition

The following list provides just a few highlights of what's new in *BIOLOGY*, Seventh Edition.

CHAPTER 1 Exploring Life

- Chapter 1 now includes a discussion of systems biology as one of the book's themes.
- The section on scientific inquiry is more robust and features a new case study of research on mimicry in snake populations.

UNIT ONE The Chemistry of Life

- At the suggestion of many instructors, the chapter on the basic principles of energy and metabolism, formerly Chapter 6, has been moved to Unit Two. In this edition, we provide a basic introduction to ATP in Chapter 4 and to enzymes in Chapter 5.

UNIT TWO The Cell

- The chapter "An Introduction to Metabolism" is now Chapter 8 in Unit Two, where it directly precedes the chapters on cellular respiration and photosynthesis. In addition to an improved presentation of the thermodynamic laws, Chapter 8 expands upon the introduction to ATP and enzymes given in Unit One.

UNIT THREE Genetics

- Chapter 19 has been updated throughout, including expanded coverage of histone modifications, DNA methylation, and epigenetic inheritance; a new discussion of regulation of gene expression by miRNAs and siRNAs; an updated discussion of the types of DNA sequences in the human genome; and a new section on genome evolution.
- In Chapter 20, new material ranges from a discussion of the current estimate of the number of human genes, to a more global view of gene interactions within a given genome, to comparisons of genomes of different species, all part of the current thrust to understand the biology of whole systems.
- Chapter 21 provides an expanded section on the evolution of development ("evo-devo"), including a new comparison of the genes involved in animal and plant development.

UNIT FOUR Mechanisms of Evolution

- Changes to this unit aim to combat misconceptions about evolutionary processes, as well as to eliminate hints of circular reasoning that are a target of anti-evolution arguments.
- New examples highlight vibrant research in evolutionary biology, including the continued impact of molecular systematics on phylogenetic studies and the use of virtual populations to model evolutionary processes.
- Chapter 25 has been revised to focus on the inquiry process involved in exploring phylogeny. Expanded coverage of genome evolution includes a new discussion of neutral theory.

UNIT FIVE The Evolutionary History of Biological Diversity

- Chapter 26, now titled "The Tree of Life: An Introduction to Biodiversity," is newly focused on placing life's diversity in the context of Earth's history, with emphasis on the major branchings in the tree of life.
- Updates in Chapter 27 (now titled "Prokaryotes") reflect new data regarding prokaryote classification and the growing evidence for cooperative relationships among prokaryotes.
- Chapters 28 (now titled "Protists"), 29, 30, and 31 include more information on the natural history, ecological roles, and human impact of various groups of protists, plants, and fungi. Updates include the implications of recent phylogenetic findings on classification, such as the recognition of a new fungal phylum (Glomeromycota).
- Chapters 32, 33, and 34 present a cohesive view of animal diversity, including an overview of hypotheses regarding animal phylogeny, an expanded survey of invertebrate phyla, added detail on natural history, updating of vertebrate classification, and recent findings relating to human origins.

UNIT SIX Plant Form and Function

- New examples highlight the role of biotechnology in agriculture, such as the development of genetically engineered "smart plants" that signal phosphorus deficiencies and the potential application of "terminator technology" to the problem of transgene escape from GM crops.
- New content in Chapter 39 focuses on the potential application of systems biology to the study of plant hormone interactions.

UNIT SEVEN Animal Form and Function

- New research examples highlight the physiology of diverse animals and relate physiological adaptations to the animals' ecological context.
- Content on thermoregulation has been moved from Chapter 44 (now titled "Osmoregulation and Excretion") to Chapter 40 ("Basic Principles of Animal Form and Function"), where it serves as an extended example of various animals' ability to maintain homeostasis.
- New sections on the vertebrate brain and on neurological disorders highlight recent discoveries as well as the tantalizing opportunities in these dynamic areas of research.

UNIT EIGHT Ecology

- A new Chapter 51 ("Behavioral Ecology") brings the subject into the 21st century, with expanded coverage of game theory, mate choice, and animal cognition.
- New examples throughout the unit highlight current research and applications, including facilitation and community structure, the FACTS-1 study of CO_2 impact on forests, and novel approaches in restoration ecology.

Interviews

UNIT ONE
The Chemistry of Life 30
Lydia Makhubu
University of Swaziland (retired)

UNIT TWO
The Cell 92
Peter Agre
Johns Hopkins University School of Medicine

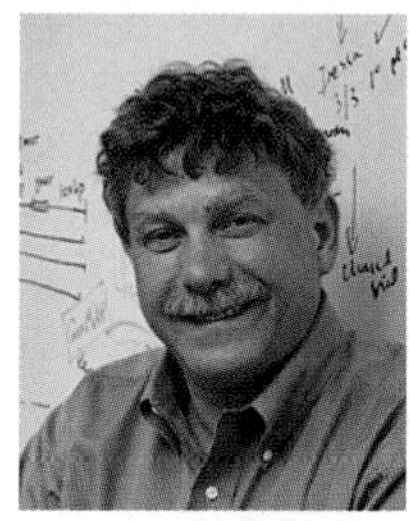

UNIT THREE
Genetics 236
Eric Lander
Massachusetts Institute of Technology

UNIT FOUR
Mechanisms of Evolution 436
Kenneth Kaneshiro
University of Hawaii, Manoa

UNIT FIVE
The Evolutionary History of Biological Diversity 510
Linda Graham
University of Wisconsin-Madison

UNIT SIX
Plant Form and Function 710
Natasha Raikhel
University of California, Riverside

UNIT SEVEN
Animal Form and Function 818
Erich Jarvis
Duke University Medical Center

UNIT EIGHT
Ecology 1078
Gene E. Likens
Institute of Ecosystem Studies

Brief Contents

1 Exploring Life 2

UNIT ONE The Chemistry of Life

2 The Chemical Context of Life 32
3 Water and the Fitness of the Environment 47
4 Carbon and the Molecular Diversity of Life 58
5 The Structure and Function of Macromolecules 68

UNIT TWO The Cell

6 A Tour of the Cell 94
7 Membrane Structure and Function 124
8 An Introduction to Metabolism 141
9 Cellular Respiration: Harvesting Chemical Energy 160
10 Photosynthesis 181
11 Cell Communication 201
12 The Cell Cycle 218

UNIT THREE Genetics

13 Meiosis and Sexual Life Cycles 238
14 Mendel and the Gene Idea 251
15 The Chromosomal Basis of Inheritance 274
16 The Molecular Basis of Inheritance 293
17 From Gene to Protein 309
18 The Genetics of Viruses and Bacteria 334
19 Eukaryotic Genomes: Organization, Regulation, and Evolution 359
20 DNA Technology and Genomics 384
21 The Genetic Basis of Development 411

UNIT FOUR Mechanisms of Evolution

22 Descent with Modification: A Darwinian View of Life 438
23 The Evolution of Populations 454
24 The Origin of Species 472
25 Phylogeny and Systematics 491

UNIT FIVE The Evolutionary History of Biological Diversity

26 The Tree of Life: An Introduction to Biological Diversity 512
27 Prokaryotes 534
28 Protists 549
29 Plant Diversity I: How Plants Colonized Land 573
30 Plant Diversity II: The Evolution of Seed Plants 591
31 Fungi 608
32 An Introduction to Animal Diversity 626
33 Invertebrates 638
34 Vertebrates 671

UNIT SIX Plant Form and Function

35 Plant Structure, Growth, and Development 712
36 Transport in Vascular Plants 738
37 Plant Nutrition 756
38 Angiosperm Reproduction and Biotechnology 771
39 Plant Responses to Internal and External Signals 788

UNIT SEVEN Animal Form and Function

40 Basic Principles of Animal Form and Function 820
41 Animal Nutrition 844
42 Circulation and Gas Exchange 867
43 The Immune System 898
44 Osmoregulation and Excretion 922
45 Hormones and the Endocrine System 943
46 Animal Reproduction 964
47 Animal Development 987
48 Nervous Systems 1011
49 Sensory and Motor Mechanisms 1045

UNIT EIGHT Ecology

50 An Introduction to Ecology and the Biosphere 1080
51 Behavioral Ecology 1106
52 Population Ecology 1136
53 Community Ecology 1159
54 Ecosystems 1184
55 Conservation Biology and Restoration Ecology 1209

Detailed Contents

1 Exploring Life 2

OVERVIEW: Biology's Most Exciting Era 2

CONCEPT 1.1 Biologists explore life from the microscopic to the global scale 2
- A Hierarchy of Biological Organization 3
- A Closer Look at Ecosystems 6
- A Closer Look at Cells 6

CONCEPT 1.2 Biological systems are much more than the sum of their parts 9
- The Emergent Properties of Systems 9
- The Power and Limitations of Reductionism 9
- Systems Biology 10
- Feedback Regulation in Biological Systems 11

CONCEPT 1.3 Biologists explore life across its great diversity of species 12
- Grouping Species: The Basic Idea 12
- The Three Domains of Life 13
- Unity in the Diversity of Life 14

CONCEPT 1.4 Evolution accounts for life's unity and diversity 15
- Natural Selection 16
- The Tree of Life 17

CONCEPT 1.5 Biologists use various forms of inquiry to explore life 19
- Discovery Science 19
- Hypothesis-Based Science 20
- A Case Study in Scientific Inquiry: Investigating Mimicry in Snake Populations 21
- Limitations of Science 24
- Theories in Science 24
- Model-Building in Science 24
- The Culture of Science 25
- Science, Technology, and Society 25

CONCEPT 1.6 A set of themes connects the concepts of biology 26

UNIT ONE

The Chemistry of Life

2 The Chemical Context of Life 32

OVERVIEW: Chemical Foundations of Biology 32

CONCEPT 2.1 Matter consists of chemical elements in pure form and in combinations called compounds 32
- Elements and Compounds 32
- Essential Elements of Life 33

CONCEPT 2.2 An element's properties depend on the structure of its atoms 34
- Subatomic Particles 34
- Atomic Number and Atomic Mass 34
- Isotopes 35
- The Energy Levels of Electrons 36
- Electron Configuration and Chemical Properties 37
- Electron Orbitals 38

CONCEPT 2.3 The formation and function of molecules depend on chemical bonding between atoms 39
- Covalent Bonds 39
- Ionic Bonds 41
- Weak Chemical Bonds 42
- Molecular Shape and Function 42

CONCEPT 2.4 Chemical reactions make and break chemical bonds 44

3 Water and the Fitness of the Environment 47

OVERVIEW: The Molecule That Supports All of Life 47

CONCEPT 3.1 The polarity of water molecules results in hydrogen bonding 47

CONCEPT 3.2 Four emergent properties of water contribute to Earth's fitness for life 48
- Cohesion 48
- Moderation of Temperatures 49
- Insulation of Bodies of Water by Floating Ice 50
- The Solvent of Life 51

CONCEPT 3.3 Dissociation of water molecules leads to acidic and basic conditions that affect living organisms 53
- Effects of Changes in pH 53
- The Threat of Acid Precipitation 55

4 Carbon and the Molecular Diversity of Life 58

OVERVIEW: Carbon—The Backbone of Biological Molecules 58

CONCEPT 4.1 Organic chemistry is the study of carbon compounds 58

CONCEPT 4.2 Carbon atoms can form diverse molecules by bonding to four other atoms 59
The Formation of Bonds with Carbon 59
Molecular Diversity Arising from Carbon Skeleton Variation 61

CONCEPT 4.3 Functional groups are the parts of molecules involved in chemical reactions 63
The Functional Groups Most Important in the Chemistry of Life 63
ATP: An Important Source of Energy for Cellular Processes 66
The Chemical Elements of Life: *A Review* 66

5 The Structure and Function of Macromolecules 68

OVERVIEW: The Molecules of Life 68

CONCEPT 5.1 Most macromolecules are polymers, built from monomers 68
The Synthesis and Breakdown of Polymers 68
Diversity of Polymers 69

CONCEPT 5.2 Carbohydrates serve as fuel and building material 69
Sugars 70
Polysaccharides 71

CONCEPT 5.3 Lipids are a diverse group of hydrophobic molecules 74
Fats 75
Phospholipids 76
Steroids 77

CONCEPT 5.4 Proteins have many structures, resulting in a wide range of functions 77
Polypeptides 78
Protein Conformation and Function 81

CONCEPT 5.5 Nucleic acids store and transmit hereditary information 86
The Roles of Nucleic Acids 86
The Structure of Nucleic Acids 87
The DNA Double Helix 88
DNA and Proteins as Tape Measures of Evolution 89

The Theme of Emergent Properties in the Chemistry of Life: *A Review* 89

UNIT TWO

The Cell 92

6 A Tour of the Cell 94

OVERVIEW: The Importance of Cells 94

CONCEPT 6.1 To study cells, biologists use microscopes and the tools of biochemistry 94
Microscopy 95
Isolating Organelles by Cell Fractionation 97

CONCEPT 6.2 Eukaryotic cells have internal membranes that compartmentalize their functions 98
Comparing Prokaryotic and Eukaryotic Cells 98
A Panoramic View of the Eukaryotic Cell 99

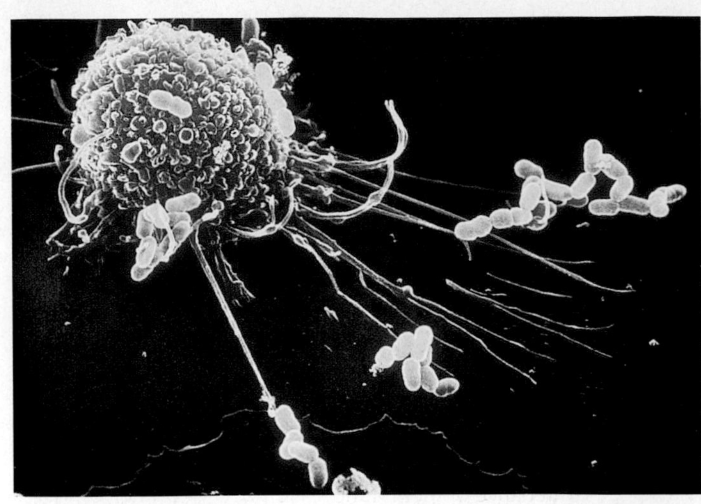

CONCEPT 6.3 The eukaryotic cell's genetic instructions are housed in the nucleus and carried out by the ribosomes 102
The Nucleus: Genetic Library of the Cell 102
Ribosomes: Protein Factories in the Cell 102

CONCEPT 6.4 The endomembrane system regulates protein traffic and performs metabolic functions in the cell 104
The Endoplasmic Reticulum: Biosynthetic Factory 104
The Golgi Apparatus: Shipping and Receiving Center 105
Lysosomes: Digestive Compartments 107
Vacuoles: Diverse Maintenance Compartments 108
The Endomembrane System: *A Review* 108

CONCEPT 6.5 Mitochondria and chloroplasts change energy from one form to another 109
Mitochondria: Chemical Energy Conversion 109
Chloroplasts: Capture of Light Energy 110
Peroxisomes: Oxidation 110

CONCEPT 6.6 The cytoskeleton is a network of fibers that organizes structures and activities in the cell 112
Roles of the Cytoskeleton: Support, Motility, and Regulation 112
Components of the Cytoskeleton 113

CONCEPT 6.7 Extracellular components and connections between cells help coordinate cellular activities 118
Cell Walls of Plants 118
The Extracellular Matrix (ECM) of Animal Cells 119
Intercellular Junctions 120

THE CELL: A Living Unit Greater Than the Sum of Its Parts 120

7 Membrane Structure and Function 124

OVERVIEW: Life at the Edge 124

CONCEPT 7.1 Cellular membranes are fluid mosaics of lipids and proteins 124
Membrane Models: *Scientific Inquiry* 125
The Fluidity of Membranes 126
Membrane Proteins and Their Functions 127

The Role of Membrane Carbohydrates in Cell-Cell Recognition 129
Synthesis and Sidedness of Membranes 129

CONCEPT 7.2 Membrane structure results in selective permeability 130
The Permeability of the Lipid Bilayer 130
Transport Proteins 130

CONCEPT 7.3 Passive transport is diffusion of a substance across a membrane with no energy investment 130
Effects of Osmosis on Water Balance 131
Facilitated Diffusion: Passive Transport Aided by Proteins 133

CONCEPT 7.4 Active transport uses energy to move solutes against their gradients 134
The Need for Energy in Active Transport 134
Maintenance of Membrane Potential by Ion Pumps 134
Cotransport: Coupled Transport by a Membrane Protein 136

CONCEPT 7.5 Bulk transport across the plasma membrane occurs by exocytosis and endocytosis 137
Exocytosis 137
Endocytosis 137

8 An Introduction to Metabolism 141

OVERVIEW: The Energy of Life 141

CONCEPT 8.1 An organism's metabolism transforms matter and energy, subject to the laws of thermodynamics 141
Organization of the Chemistry of Life into Metabolic Pathways 141
Forms of Energy 142
The Laws of Energy Transformation 143

CONCEPT 8.2 The free-energy change of a reaction tells us whether the reaction occurs spontaneously 145
Free-Energy Change, ΔG 145
Free Energy, Stability, and Equilibrium 145
Free Energy and Metabolism 146

CONCEPT 8.3 ATP powers cellular work by coupling exergonic reactions to endergonic reactions 148
The Structure and Hydrolysis of ATP 148
How ATP Performs Work 149
The Regeneration of ATP 150

CONCEPT 8.4 Enzymes speed up metabolic reactions by lowering energy barriers 150
The Activation Energy Barrier 150
How Enzymes Lower the E_A Barrier 152
Substrate Specificity of Enzymes 152
Catalysis in the Enzyme's Active Site 152
Effects of Local Conditions on Enzyme Activity 154

CONCEPT 8.5 Regulation of enzyme activity helps control metabolism 156
Allosteric Regulation of Enzymes 156
Specific Localization of Enzymes Within the Cell 157

9 Cellular Respiration: Harvesting Chemical Energy 160

OVERVIEW: Life Is Work 160

CONCEPT 9.1 Catabolic pathways yield energy by oxidizing organic fuels 161
Catabolic Pathways and Production of ATP 161
Redox Reactions: Oxidation and Reduction 161
The Stages of Cellular Respiration: *A Preview* 164

CONCEPT 9.2 Glycolysis harvests chemical energy by oxidizing glucose to pyruvate 165

CONCEPT 9.3 The citric acid cycle completes the energy-yielding oxidation of organic molecules 168

CONCEPT 9.4 During oxidative phosphorylation, chemiosmosis couples electron transport to ATP synthesis 170
The Pathway of Electron Transport 170
Chemiosmosis: The Energy-Coupling Mechanism 171
An Accounting of ATP Production by Cellular Respiration 173

CONCEPT 9.5 Fermentation enables some cells to produce ATP without the use of oxygen 174
Types of Fermentation 175
Fermentation and Cellular Respiration Compared 175
The Evolutionary Significance of Glycolysis 176

CONCEPT 9.6 Glycolysis and the citric acid cycle connect to many other metabolic pathways 176
The Versatility of Catabolism 176
Biosynthesis (Anabolic Pathways) 177
Regulation of Cellular Respiration via Feedback Mechanisms 177

10 Photosynthesis 181

OVERVIEW: The Process That Feeds the Biosphere 181

CONCEPT 10.1 Photosynthesis converts light energy to the chemical energy of food 182
Chloroplasts: The Sites of Photosynthesis in Plants 182
Tracking Atoms Through Photosynthesis: *Scientific Inquiry* 183
The Two Stages of Photosynthesis: *A Preview* 184

CONCEPT 10.2 The light reactions convert solar energy to the chemical energy of ATP and NADPH 186
The Nature of Sunlight 186
Photosynthetic Pigments: The Light Receptors 186
Excitation of Chlorophyll by Light 188
A Photosystem: A Reaction Center Associated with Light-Harvesting Complexes 189
Noncyclic Electron Flow 190
Cyclic Electron Flow 191
A Comparison of Chemiosmosis in Chloroplasts and Mitochondria 192

CONCEPT 10.3 The Calvin cycle uses ATP and NADPH to convert CO_2 to sugar 193

CONCEPT 10.4 Alternative mechanisms of carbon fixation have evolved in hot, arid climates 195
Photorespiration: An Evolutionary Relic? 195
C_4 Plants 196
CAM Plants 196
The Importance of Photosynthesis: *A Review* 197

11 Cell Communication 201

OVERVIEW: The Cellular Internet 201

CONCEPT 11.1 External signals are converted into responses within the cell 201

Evolution of Cell Signaling 201
Local and Long-Distance Signaling 202
The Three Stages of Cell Signaling: *A Preview* 203

CONCEPT 11.2 Reception: A signal molecule binds to a receptor protein, causing it to change shape 204

Intracellular Receptors 205
Receptors in the Plasma Membrane 205

CONCEPT 11.3 Transduction: Cascades of molecular interactions relay signals from receptors to target molecules in the cell 208

Signal Transduction Pathways 208
Protein Phosphorylation and Dephosphorylation 209
Small Molecules and Ions as Second Messengers 210

CONCEPT 11.4 Response: Cell signaling leads to regulation of cytoplasmic activities or transcription 212

Cytoplasmic and Nuclear Responses 212
Fine-Tuning of the Response 213

12 The Cell Cycle 218

OVERVIEW: The Key Roles of Cell Division 218

CONCEPT 12.1 Cell division results in genetically identical daughter cells 219

Cellular Organization of the Genetic Material 219
Distribution of Chromosomes During Cell Division 219

CONCEPT 12.2 The mitotic phase alternates with interphase in the cell cycle 221

Phases of the Cell Cycle 221
The Mitotic Spindle: *A Closer Look* 221
Cytokinesis: *A Closer Look* 224
Binary Fission 226
The Evolution of Mitosis 227

CONCEPT 12.3 The cell cycle is regulated by a molecular control system 228

Evidence for Cytoplasmic Signals 228
The Cell Cycle Control System 229
Loss of Cell Cycle Controls in Cancer Cells 232

UNIT THREE

Genetics 236

13 Meiosis and Sexual Life Cycles 238

OVERVIEW: Hereditary Similarity and Variation 238

CONCEPT 13.1 Offspring acquire genes from parents by inheriting chromosomes 238

Inheritance of Genes 239
Comparison of Asexual and Sexual Reproduction 239

CONCEPT 13.2 Fertilization and meiosis alternate in sexual life cycles 240

Sets of Chromosomes in Human Cells 240
Behavior of Chromosome Sets in the Human Life Cycle 241
The Variety of Sexual Life Cycles 242

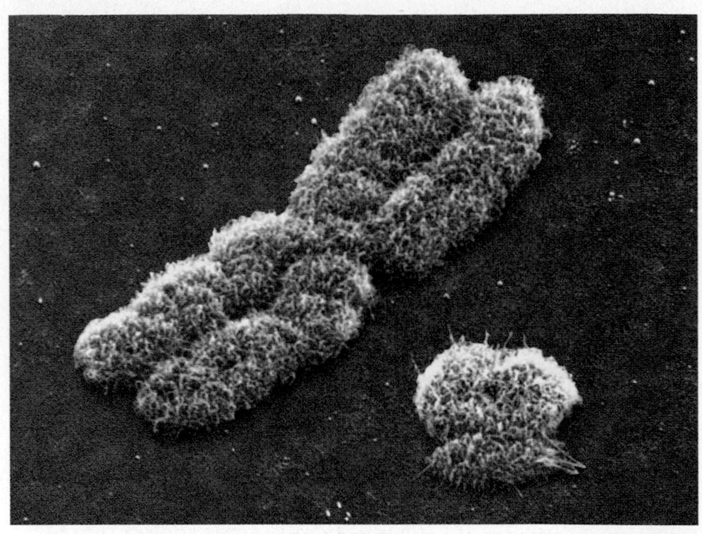

CONCEPT 13.3 Meiosis reduces the number of chromosome sets from diploid to haploid 243

The Stages of Meiosis 243
A Comparison of Mitosis and Meiosis 247

CONCEPT 13.4 Genetic variation produced in sexual life cycles contributes to evolution 247

Origins of Genetic Variation Among Offspring 247
Evolutionary Significance of Genetic Variation Within Populations 248

14 Mendel and the Gene Idea 251

OVERVIEW: Drawing from the Deck of Genes 251

CONCEPT 14.1 Mendel used the scientific approach to identify two laws of inheritance 251

Mendel's Experimental, Quantitative Approach 252
The Law of Segregation 253
The Law of Independent Assortment 256

CONCEPT 14.2 The laws of probability govern Mendelian inheritance 258

The Multiplication and Addition Rules Applied to Monohybrid Crosses 258
Solving Complex Genetics Problems with the Rules of Probability 259

CONCEPT 14.3 Inheritance patterns are often more complex than predicted by simple Mendelian genetics 260

Extending Mendelian Genetics for a Single Gene 260
Extending Mendelian Genetics for Two or More Genes 262
Nature and Nurture: The Environmental Impact on Phenotype 264
Integrating a Mendelian View of Heredity and Variation 264

CONCEPT 14.4 Many human traits follow Mendelian patterns of inheritance 265

Pedigree Analysis 265
Recessively Inherited Disorders 266
Dominantly Inherited Disorders 267
Multifactorial Disorders 268
Genetic Testing and Counseling 268

15 The Chromosomal Basis of Inheritance 274

OVERVIEW: Locating Genes on Chromosomes 274

CONCEPT 15.1 Mendelian inheritance has its physical basis in the behavior of chromosomes 274

Morgan's Experimental Evidence: *Scientific Inquiry* 276

CONCEPT 15.2 Linked genes tend to be inherited together because they are located near each other on the same chromosome 277

How Linkage Affects Inheritance: *Scientific Inquiry* 277
Genetic Recombination and Linkage 278
Linkage Mapping Using Recombination Data: *Scientific Inquiry* 279

CONCEPT 15.3 Sex-linked genes exhibit unique patterns of inheritance 282

The Chromosomal Basis of Sex 282
Inheritance of Sex-Linked Genes 283
X Inactivation in Female Mammals 284

CONCEPT 15.4 Alterations of chromosome number or structure cause some genetic disorders 285

Abnormal Chromosome Number 285
Alterations of Chromosome Structure 286
Human Disorders Due to Chromosomal Alterations 287

CONCEPT 15.5 Some inheritance patterns are exceptions to the standard chromosome theory 288

Genomic Imprinting 288
Inheritance of Organelle Genes 289

16 The Molecular Basis of Inheritance 293

OVERVIEW: Life's Operating Instructions 293

CONCEPT 16.1 DNA is the genetic material 293

The Search for the Genetic Material: *Scientific Inquiry* 293
Building a Structural Model of DNA: *Scientific Inquiry* 296

CONCEPT 16.2 Many proteins work together in DNA replication and repair 299

The Basic Principle: Base Pairing to a Template Strand 299
DNA Replication: *A Closer Look* 300
Proofreading and Repairing DNA 305
Replicating the Ends of DNA Molecules 306

17 From Gene to Protein 309

OVERVIEW: The Flow of Genetic Information 309

CONCEPT 17.1 Genes specify proteins via transcription and translation 309

Evidence from the Study of Metabolic Defects 309
Basic Principles of Transcription and Translation 311
The Genetic Code 312

CONCEPT 17.2 Transcription is the DNA-directed synthesis of RNA: *a closer look* 315

Molecular Components of Transcription 315
Synthesis of an RNA Transcript 316

CONCEPT 17.3 Eukaryotic cells modify RNA after transcription 317

Alteration of mRNA Ends 317
Split Genes and RNA Splicing 318

CONCEPT 17.4 Translation is the RNA-directed synthesis of a polypeptide: *a closer look* 320

Molecular Components of Translation 320
Building a Polypeptide 323
Completing and Targeting the Functional Protein 324

CONCEPT 17.5 RNA plays multiple roles in the cell: *a review* 327

CONCEPT 17.6 Comparing gene expression in prokaryotes and eukaryotes reveals key differences 327

CONCEPT 17.7 Point mutations can affect protein structure and function 328

Types of Point Mutations 328
Mutagens 329
What is a Gene? *Revisiting the Question* 330

18 The Genetics of Viruses and Bacteria 334

OVERVIEW: Microbial Model Systems 334

CONCEPT 18.1 A virus has a genome but can reproduce only within a host cell 334

The Discovery of Viruses: *Scientific Inquiry* 334
Structure of Viruses 335
General Features of Viral Reproductive Cycles 336
Reproductive Cycles of Phages 337
Reproductive Cycles of Animal Viruses 339
Evolution of Viruses 342

CONCEPT 18.2 Viruses, viroids, and prions are formidable pathogens in animals and plants 343

Viral Diseases in Animals 343
Emerging Viruses 344
Viral Diseases in Plants 345
Viroids and Prions: The Simplest Infectious Agents 345

CONCEPT 18.3 Rapid reproduction, mutation, and genetic recombination contribute to the genetic diversity of bacteria 346

The Bacterial Genome and Its Replication 346
Mutation and Genetic Recombination as Sources of Genetic Variation 346
Mechanisms of Gene Transfer and Genetic Recombination in Bacteria 348
Transposition of Genetic Elements 351

CONCEPT 18.4 Individual bacteria respond to environmental change by regulating their gene expression 352

Operons: The Basic Concept 353
Repressible and Inducible Operons: Two Types of Negative Gene Regulation 354
Positive Gene Regulation 356

19 Eukaryotic Genomes: Organization, Regulation, and Evolution 359

OVERVIEW: How Eukaryotic Genomes Work and Evolve 359

CONCEPT 19.1 Chromatin structure is based on successive levels of DNA packing 359

Nucleosomes, or "Beads on a String" 360
Higher Levels of DNA Packing 360

CONCEPT 19.2 **Gene expression can be regulated at any stage, but the key step is transcription 362**
Differential Gene Expression 362
Regulation of Chromatin Structure 363
Regulation of Transcription Initiation 364
Mechanisms of Post-Transcriptional Regulation 368

CONCEPT 19.3 **Cancer results from genetic changes that affect cell cycle control 370**
Types of Genes Associated with Cancer 370
Interference with Normal Cell-Signaling Pathways 371
The Multistep Model of Cancer Development 373
Inherited Predisposition to Cancer 374

CONCEPT 19.4 **Eukaryotic genomes can have many noncoding DNA sequences in addition to genes 374**
The Relationship Between Genomic Composition and Organismal Complexity 374
Transposable Elements and Related Sequences 375
Other Repetitive DNA, Including Simple Sequence DNA 376
Genes and Multigene Families 377

CONCEPT 19.5 **Duplications, rearrangements, and mutations of DNA contribute to genome evolution 378**
Duplication of Chromosome Sets 378
Duplication and Divergence of DNA Segments 378
Rearrangements of Parts of Genes: Exon Duplication and Exon Shuffling 380
How Transposable Elements Contribute to Genome Evolution 380

20 DNA Technology and Genomics 384

OVERVIEW: **Understanding and Manipulating Genomes 384**

CONCEPT 20.1 **DNA cloning permits production of multiple copies of a specific gene or other DNA segment 385**
DNA Cloning and Its Applications: *A Preview* 385
Using Restriction Enzymes to Make Recombinant DNA 386
Cloning a Eukaryotic Gene in a Bacterial Plasmid 386
Storing Cloned Genes in DNA Libraries 388
Cloning and Expressing Eukaryotic Genes 390
Amplifying DNA *in Vitro:* The Polymerase Chain Reaction (PCR) 391

CONCEPT 20.2 **Restriction fragment analysis detects DNA differences that affect restriction sites 392**
Gel Electrophoresis and Southern Blotting 392
Restriction Fragment Length Differences as Genetic Markers 394

CONCEPT 20.3 **Entire genomes can be mapped at the DNA level 394**
Genetic (Linkage) Mapping: Relative Ordering of Markers 396
Physical Mapping: Ordering DNA Fragments 396
DNA Sequencing 396

CONCEPT 20.4 **Genome sequences provide clues to important biological questions 398**
Identifying Protein-Coding Genes in DNA Sequences 399
Determining Gene Function 400
Studying Expression of Interacting Groups of Genes 400
Comparing Genomes of Different Species 400
Future Directions in Genomics 402

CONCEPT 20.5 **The practical applications of DNA technology affect our lives in many ways 402**
Medical Applications 402
Pharmaceutical Products 404
Forensic Evidence 404
Environmental Cleanup 405
Agricultural Applications 406
Safety and Ethical Questions Raised by DNA Technology 407

21 The Genetic Basis of Development 411

OVERVIEW: **From Single Cell to Multicellular Organism 411**

CONCEPT 21.1 **Embryonic development involves cell division, cell differentiation, and morphogenesis 412**

CONCEPT 21.2 **Different cell types result from differential gene expression in cells with the same DNA 415**
Evidence for Genomic Equivalence 415
Transcriptional Regulation of Gene Expression During Development 418
Cytoplasmic Determinants and Cell-Cell Signals in Cell Differentiation 420

CONCEPT 21.3 **Pattern formation in animals and plants results from similar genetic and cellular mechanisms 421**
Drosophila Development: A Cascade of Gene Activations 421
C. elegans: The Role of Cell Signaling 425
Plant Development: Cell Signaling and Transcriptional Regulation 429

CONCEPT 21.4 **Comparative studies help explain how the evolution of development leads to morphological diversity 431**
Widespread Conservation of Developmental Genes Among Animals 431
Comparison of Animal and Plant Development 433

UNIT FOUR

Mechanisms of Evolution 436

22 Descent with Modification: A Darwinian View of Life 438

OVERVIEW: **Darwin Introduces a Revolutionary Theory 438**

CONCEPT 22.1 **The Darwinian revolution challenged traditional views of a young Earth inhabited by unchanging species 438**
Resistance to the Idea of Evolution 439
Theories of Gradualism 440
Lamarck's Theory of Evolution 440

CONCEPT 22.2: **In *The Origin of Species*, Darwin proposed that species change through natural selection 441**
Darwin's Research 441
The Origin of Species 443

CONCEPT 22.3 **Darwin's theory explains a wide range of observations 446**
Natural Selection in Action 446
Homology, Biogeography, and the Fossil Record 448

What Is Theoretical about the Darwinian View of Life? 451

23 The Evolution of Populations 454

OVERVIEW: The Smallest Unit of Evolution 454

CONCEPT 23.1 Population genetics provides a foundation for studying evolution 454

The Modern Synthesis 455
Gene Pools and Allele Frequencies 455
The Hardy-Weinberg Theorem 456

CONCEPT 23.2 Mutation and sexual recombination produce the variation that makes evolution possible 459

Mutation 459
Sexual Recombination 460

CONCEPT 23.3 Natural selection, genetic drift, and gene flow can alter a population's genetic composition 460

Natural Selection 460
Genetic Drift 460
Gene Flow 462

CONCEPT 23.4 Natural selection is the primary mechanism of adaptive evolution 462

Genetic Variation 462
A Closer Look at Natural Selection 464
The Preservation of Genetic Variation 466
Sexual Selection 468
The Evolutionary Enigma of Sexual Reproduction 469
Why Natural Selection Cannot Fashion Perfect Organisms 469

24 The Origin of Species 472

OVERVIEW: That "Mystery of Mysteries" 472

CONCEPT 24.1 The biological species concept emphasizes reproductive isolation 473

The Biological Species Concept 473
Other Definitions of Species 476

CONCEPT 24.2 Speciation can take place with or without geographic separation 476

Allopatric ("Other Country") Speciation 477
Sympatric ("Same Country") Speciation 478
Allopatric and Sympatric Speciation: A Summary 480
Adaptive Radiation 480
Studying the Genetics of Speciation 481
The Tempo of Speciation 481

CONCEPT 24.3 Macroevolutionary changes can accumulate through many speciation events 482

Evolutionary Novelties 482
Evolution of the Genes That Control Development 484
Evolution Is Not Goal Oriented 486

25 Phylogeny and Systematics 491

OVERVIEW: Investigating the Tree of Life 491

CONCEPT 25.1 Phylogenies are based on common ancestries inferred from fossil, morphological, and molecular evidence 492

The Fossil Record 492
Morphological and Molecular Homologies 492

CONCEPT 25.2 Phylogenetic systematics connects classification with evolutionary history 495

Binomial Nomenclature 496
Hierarchical Classification 496
Linking Classification and Phylogeny 496

CONCEPT 25.3 Phylogenetic systematics informs the construction of phylogenetic trees based on shared characters 497

Cladistics 498
Phylogenetic Trees and Timing 499
Maximum Parsimony and Maximum Likelihood 501
Phylogenetic Trees as Hypotheses 501

CONCEPT 25.4 Much of an organism's evolutionary history is documented in its genome 504

Gene Duplications and Gene Families 505
Genome Evolution 505

CONCEPT 25.5 Molecular clocks help track evolutionary time 506

Molecular Clocks 506
The Universal Tree of Life 507

UNIT FIVE

The Evolutionary History of Biological Diversity 510

26 The Tree of Life: An Introduction to Biological Diversity 512

OVERVIEW: Changing Life on a Changing Earth 512

CONCEPT 26.1 Conditions on early Earth made the origin of life possible 513

Synthesis of Organic Compounds on Early Earth 513
Abiotic Synthesis of Polymers 514
Protobionts 515
The "RNA World" and the Dawn of Natural Selection 515

CONCEPT 26.2 The fossil record chronicles life on Earth 517

How Rocks and Fossils Are Dated 517
The Geologic Record 518
Mass Extinctions 518

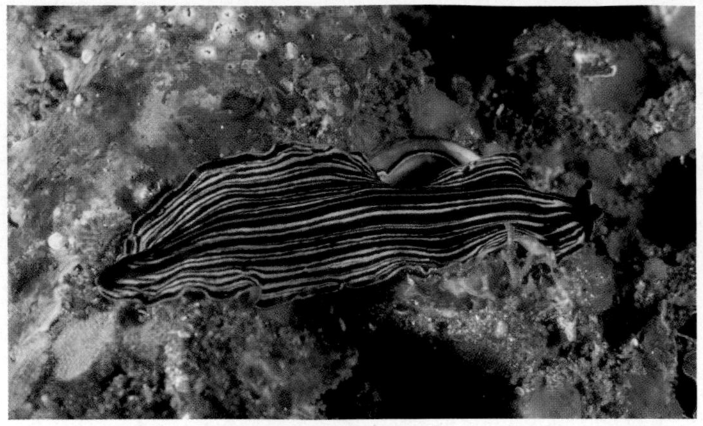

CONCEPT 26.3 **As prokaryotes evolved, they exploited and changed young Earth 521**
The First Prokaryotes 522
Electron Transport Systems 521
Photosynthesis and the Oxygen Revolution 522
CONCEPT 26.4 **Eukaryotic cells arose from symbioses and genetic exchanges between prokaryotes 523**
The First Eukaryotes 523
Endosymbiotic Origin of Mitochondria and Plastids 523
Eukaryotic Cells as Genetic Chimeras 524
CONCEPT 26.5 **Multicellularity evolved several times in eukaryotes 525**
The Earliest Multicellular Eukaryotes 525
The Colonial Connection 526
The "Cambrian Explosion" 526
Colonization of Land by Plants, Fungi, and Animals 527
Continental Drift 527
CONCEPT 26.6 **New information has revised our understanding of the tree of life 529**
Previous Taxonomic Systems 529
Reconstructing the Tree of Life: A Work in Progress 529

27 Prokaryotes 534
OVERVIEW: **They're (Almost) Everywhere! 534**
CONCEPT 27.1 **Structural, functional, and genetic adaptations contribute to prokaryote success 534**
Cell-Surface Structures 534
Motility 536
Internal and Genomic Organization 537
Reproduction and Adaptation 537
CONCEPT 27.2 **A great diversity of nutritional and metabolic adaptations have evolved in prokaryotes 538**
Metabolic Relationships to Oxygen 539
Nitrogen Metabolism 539
Metabolic Cooperation 539
CONCEPT 27.3 **Molecular systematics is illuminating prokaryotic phylogeny 540**
Lessons from Molecular Systematics 540
Bacteria 541
Archaea 541
CONCEPT 27.4 **Prokaryotes play crucial roles in the biosphere 544**
Chemical Recycling 544
Symbiotic Relationships 545
CONCEPT 27.5 **Prokaryotes have both harmful and beneficial impacts on humans 545**
Pathogenic Prokaryotes 545
Prokaryotes in Research and Technology 546

28 Protists 549
OVERVIEW: **A World in a Drop of Water 549**
CONCEPT 28.1 **Protists are an extremely diverse assortment of eukaryotes 549**
Endosymbiosis in Eukaryotic Evolution 550
CONCEPT 28.2 **Diplomonads and parabasalids have modified mitochondria 552**
Diplomonads 552
Parabasalids 553
CONCEPT 28.3 **Euglenozoans have flagella with a unique internal structure 553**
Kinetoplastids 553
Euglenids 554
CONCEPT 28.4 **Alveolates have sacs beneath the plasma membrane 555**
Dinoflagellates 555
Apicomplexans 555
Ciliates 556
CONCEPT 28.5 **Stramenopiles have "hairy" and smooth flagella 558**
Oomycetes (Water Molds and Their Relatives) 558
Diatoms 559
Golden Algae 560
Brown Algae 560
CONCEPT 28.6 **Cercozoans and radiolarians have threadlike pseudopodia 563**
Foraminiferans (Forams) 563
Radiolarians 563
CONCEPT 28.7 **Amoebozoans have lobe-shaped pseudopodia 564**
Gymnamoebas 564
Entamoebas 564
Slime Molds 564
CONCEPT 28.8 **Red algae and green algae are the closest relatives of land plants 567**
Red Algae 567
Green Algae 567

29 Plant Diversity I: How Plants Colonized Land 573
OVERVIEW: **The Greening of Earth 573**
CONCEPT 29.1 **Land plants evolved from green algae 573**
Morphological and Biochemical Evidence 573
Genetic Evidence 574
Adaptations Enabling the Move to Land 574
CONCEPT 29.2 **Land plants possess a set of derived terrestrial adaptations 575**
Defining the Plant Kingdom 575
Derived Traits of Plants 575

The Origin and Diversification of Plants 575

CONCEPT 29.3 **The life cycles of mosses and other bryophytes are dominated by the gametophyte stage 580**

Bryophyte Gametophytes 580
Bryophyte Sporophytes 580
The Ecological and Economic Importance of Mosses 583

CONCEPT 29.4 **Ferns and other seedless vascular plants formed the first forests 584**

Origins of Vascular Plants 584
Classification of Seedless Vascular Plants 586
The Significance of Seedless Vascular Plants 588

30 Plant Diversity II: The Evolution of Seed Plants 591

OVERVIEW: **Feeding the World 591**

CONCEPT 30.1 **The reduced gametophytes of seed plants are protected in ovules and pollen grains 591**

Advantages of Reduced Gametophytes 591
Heterospory: The Rule Among Seed Plants 592
Ovules and Production of Eggs 592
Pollen and Production of Sperm 592
The Evolutionary Advantage of Seeds 593

CONCEPT 30.2 **Gymnosperms bear "naked" seeds, typically on cones 593**

Gymnosperm Evolution 596
A Closer Look at the Life Cycle of a Pine 596

CONCEPT 30.3 **The reproductive adaptations of angiosperms include flowers and fruits 598**

Characteristics of Angiosperms 598
Angiosperm Evolution 601
Angiosperm Diversity 602
Evolutionary Links Between Angiosperms and Animals 604

CONCEPT 30.4 **Human welfare depends greatly on seed plants 605**

Products from Seed Plants 605
Threats to Plant Diversity 606

31 Fungi 608

OVERVIEW: **Mighty Mushrooms 608**

CONCEPT 31.1 **Fungi are heterotrophs that feed by absorption 608**

Nutrition and Fungal Lifestyles 608
Body Structure 609

CONCEPT 31.2 **Fungi produce spores through sexual or asexual life cycles 610**

Sexual Reproduction 610
Asexual Reproduction 611

CONCEPT 31.3 **Fungi descended from an aquatic, single-celled, flagellated protist 612**

The Origin of Fungi 612
The Move to Land 612

CONCEPT 31.4 **Fungi have radiated into a diverse set of lineages 612**

Chytrids 613
Zygomycetes 613
Glomeromycetes 615
Ascomycetes 616
Basidiomycetes 618

CONCEPT 31.5 **Fungi have a powerful impact on ecosystems and human welfare 620**

Decomposers 620
Symbionts 620
Pathogens 622
Practical Uses of Fungi 623

32 An Introduction to Animal Diversity 626

OVERVIEW: **Welcome to Your Kingdom 626**

CONCEPT 32.1 **Animals are multicellular, heterotrophic eukaryotes with tissues that develop from embryonic layers 626**

Nutritional Mode 626
Cell Structure and Specialization 626
Reproduction and Development 627

CONCEPT 32.2 **The history of animals may span more than a billion years 628**

Neoproterozoic Era (1 Billion–542 Million Years Ago) 628
Paleozoic Era (542–251 Million Years Ago) 629
Mesozoic Era (251–65.5 Million Years Ago) 629
Cenozoic Era (65.5 Million Years Ago to the Present) 629

CONCEPT 32.3 **Animals can be characterized by "body plans" 630**

Symmetry 630
Tissues 630
Body Cavities 631
Protostome and Deuterostome Development 631

CONCEPT 32.4 **Leading hypotheses agree on major features of the animal phylogenetic tree 633**

Points of Agreement 633
Disagreement over the Bilaterians 634
Future Directions in Animal Systematics 636

33 Invertebrates 638

OVERVIEW: **Life Without a Backbone 638**

CONCEPT 33.1 **Sponges are sessile and have a porous body and choanocytes 642**

CONCEPT 33.2 **Cnidarians have radial symmetry, a gastrovascular cavity, and cnidocytes 643**

Hydrozoans 644
Scyphozoans 644
Cubozoans 645
Anthozoans 645

CONCEPT 33.3 **Most animals have bilateral symmetry 646**

Flatworms 646
Rotifers 648
Lophophorates: Ectoprocts, Phoronids, and Brachiopods 649
Nemerteans 649

CONCEPT 33.4 **Molluscs have a muscular foot, a visceral mass, and a mantle 650**

Chitons 651
Gastropods 651

Bivalves 652
Cephalopods 652

CONCEPT 33.5 Annelids are segmented worms 653
Oligochaetes 653
Polychaetes 655
Leeches 655

CONCEPT 33.6 Nematodes are nonsegmented pseudocoelomates covered by a tough cuticle 655

CONCEPT 33.7 Arthropods are segmented coelomates that have an exoskeleton and jointed appendages 656
General Characteristics of Arthropods 656
Cheliceriforms 658
Myriapods 659
Insects 660
Crustaceans 664

CONCEPT 33.8 Echinoderms and chordates are deuterostomes 665
Echinoderms 665
Chordates 667

34 Vertebrates 671

OVERVIEW: Half a Billion Years of Backbones 671

CONCEPT 34.1 Chordates have a notochord and a dorsal, hollow nerve cord 671
Derived Characters of Chordates 673
Tunicates 673
Lanceletes 674
Early Chordate Evolution 674

CONCEPT 34.2 Craniates are chordates that have a head 675
Derived Characters of Craniates 676
The Origin of Craniates 676
Hagfishes 676

CONCEPT 34.3 Vertebrates are craniates that have a backbone 678
Derived Characters of Vertebrates 678
Lampreys 678
Fossils of Early Vertebrates 678
Origins of Bones and Teeth 679

CONCEPT 34.4 Gnathostomes are vertebrates that have jaws 679
Derived Characters of Gnathostomes 679
Fossil Gnathostomes 680
Chondricthyans (Sharks, Rays, and Their Relatives) 680
Ray-Finned Fishes and Lobe-Fins 682

CONCEPT 34.5 Tetrapods are gnathostomes that have limbs and feet 684
Derived Characters of Tetrapods 684
The Origin of Tetrapods 684
Amphibians 685

CONCEPT 34.6 Amniotes are tetrapods that have a terrestrially adapted egg 687
Derived Characters of Amniotes 688
Early Amniotes 688
Reptiles 688
Birds 691

CONCEPT 34.7 Mammals are amniotes that have hair and produce milk 694
Derived Characters of Mammals 694
Early Evolution of Mammals 694
Monotremes 695
Marsupials 695
Eutherians (Placental Mammals) 697

CONCEPT 34.8 Humans are bipedal hominids with a large brain 701
Derived Characters of Humans 701
The Earliest Hominids 702
Australopiths 703
Bipedalism 704
Tool Use 704
Early *Homo* 704
Neanderthals 705
Homo sapiens 705

UNIT SIX

Plant Form and Function 710

35 Plant Structure, Growth, and Development 712

OVERVIEW: No Two Plants Are Alike 712

CONCEPT 35.1 The plant body has a hierarchy of organs, tissues, and cells 712
The Three Basic Plant Organs: Roots, Stems, and Leaves 713
The Three Tissue Systems: Dermal, Vascular, and Ground 717
Common Types of Plant Cells 717

CONCEPT 35.2 Meristems generate cells for new organs 720

CONCEPT 35.3 Primary growth lengthens roots and shoots 721
Primary Growth of Roots 721
Primary Growth of Shoots 723

CONCEPT 35.4 Secondary growth adds girth to stems and roots in woody plants 725
The Vascular Cambium and Secondary Vascular Tissue 725
Cork Cambia and the Production of Periderm 728

CONCEPT 35.5 Growth, morphogenesis, and differentiation produce the plant body 728
Molecular Biology: Revolutionizing the Study of Plants 728
Growth: Cell Division and Cell Expansion 729
Morphogenesis and Pattern Formation 730
Gene Expression and Control of Cellular Differentiation 732
Location and a Cell's Developmental Fate 732
Shifts in Development: Phase Changes 733
Genetic Control of Flowering 734

36 Transport in Vascular Plants 738

OVERVIEW: Pathways for Survival 738

CONCEPT 36.1 Physical forces drive the transport of materials in plants over a range of distances 738
Selective Permeability of Membranes: *A Review* 738
The Central Role of Proton Pumps 739
Effects of Differences in Water Potential 740

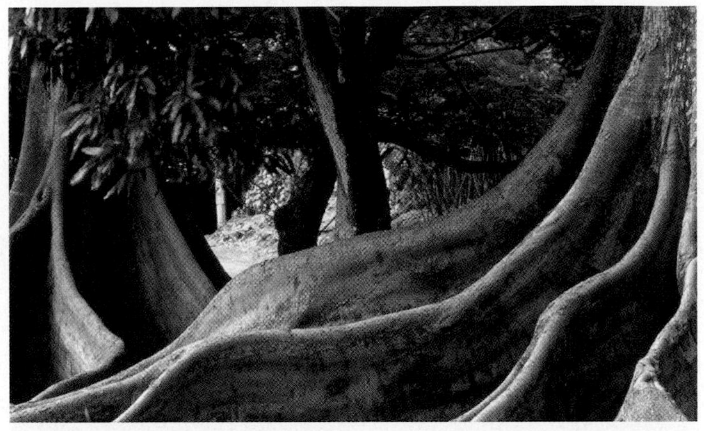

Three Major Compartments of Vacuolated Plant Cells 743
Functions of the Symplast and Apoplast in Transport 743
Bulk Flow in Long-Distance Transport 743

CONCEPT 36.2 Roots absorb water and minerals from the soil 744

The Roles of Root Hairs, Mycorrhizae, and Cortical Cells 744
The Endodermis: A Selective Sentry 744

CONCEPT 36.3 Water and minerals ascend from roots to shoots through the xylem 746

Factors Affecting the Ascent of Xylem Sap 746
Xylem Sap Ascent by Bulk Flow: *A Review* 748

CONCEPT 36.4 Stomata help regulate the rate of transpiration 749

Effects of Transpiration on Wilting and Leaf Temperature 749
Stomata: Major Pathways for Water Loss 750
Xerophyte Adaptations That Reduce Transpiration 751

CONCEPT 36.5 Organic nutrients are translocated through the phloem 751

Movement from Sugar Sources to Sugar Sinks 752
Pressure Flow: The Mechanism of Translocation in Angiosperms 753

37 Plant Nutrition 756

OVERVIEW: A Nutritional Network 756

CONCEPT 37.1 Plants require certain chemical elements to complete their life cycle 756

Macronutrients and Micronutrients 757
Symptoms of Mineral Deficiency 758

CONCEPT 37.2 Soil quality is a major determinant of plant distribution and growth 759

Texture and Composition of Soils 759
Soil Conservation and Sustainable Agriculture 760

CONCEPT 37.3 Nitrogen is often the mineral that has the greatest effect on plant growth 763

Soil Bacteria and Nitrogen Availability 763
Improving the Protein Yield of Crops 764

CONCEPT 37.4 Plant nutritional adaptations often involve relationships with other organisms 764

The Role of Bacteria in Symbiotic Nitrogen Fixation 764
Mycorrhizae and Plant Nutrition 766
Epiphytes, Parasitic Plants, and Carnivorous Plants 767

38 Angiosperm Reproduction and Biotechnology 771

OVERVIEW: To Seed or Not to Seed 771

CONCEPT 38.1 Pollination enables gametes to come together within a flower 771

Flower Structure 772
Gametophyte Development and Pollination 774
Mechanisms That Prevent Self-Fertilization 775

CONCEPT 38.2 After fertilization, ovules develop into seeds and ovaries into fruits 776

Double Fertilization 776
From Ovule to Seed 777
From Ovary to Fruit 778
Seed Germination 779

CONCEPT 38.3 Many flowering plants clone themselves by asexual reproduction 781

Mechanisms of Asexual Reproduction 781
Vegetative Propagation and Agriculture 781

CONCEPT 38.4 Plant biotechnology is transforming agriculture 783

Artificial Selection 783
Reducing World Hunger and Malnutrition 784
The Debate over Plant Biotechnology 784

39 Plant Responses to Internal and External Signals 788

OVERVIEW: Stimuli and a Stationary Life 788

CONCEPT 39.1 Signal transduction pathways link signal reception to response 788

Reception 789
Transduction 789
Response 790

CONCEPT 39.2 Plant hormones help coordinate growth, development, and responses to stimuli 791

The Discovery of Plant Hormones 792
A Survey of Plant Hormones 793
Systems Biology and Hormone Interactions 801

CONCEPT 39.3 Responses to light are critical for plant success 802

Blue-Light Photoreceptors 802
Phytochromes as Photoreceptors 802
Biological Clocks and Circadian Rhythms 805
The Effect of Light on the Biological Clock 806
Photoperiodism and Responses to Seasons 806

CONCEPT 39.4 Plants respond to a wide variety of stimuli other than light 808

Gravity 809
Mechanical Stimuli 809
Environmental Stresses 810

CONCEPT 39.5 Plants defend themselves against herbivores and pathogens 812

Defenses Against Herbivores 813
Defenses Against Pathogens 813

UNIT SEVEN

Animal Form and Function 818

40 Basic Principles of Animal Form and Function 820

OVERVIEW: Diverse Forms, Common Challenges 820

CONCEPT 40.1 Physical laws and the environment constrain animal size and shape 820
- Physical Laws and Animal Form 821
- Exchange with the Environment 821

CONCEPT 40.2 Animal form and function are correlated at all levels of organization 823
- Tissue Structure and Function 823
- Organs and Organ Systems 827

CONCEPT 40.3 Animals use the chemical energy in food to sustain form and function 828
- Bioenergetics 828
- Influences on Metabolic Rates 829
- Energy Budgets 830

CONCEPT 40.4 Many animals regulate their internal environment within relatively narrow limits 831
- Regulating and Conforming 832
- Mechanisms of Homeostasis 832

CONCEPT 40.5 Thermoregulation contributes to homeostasis and involves anatomy, physiology, and behavior 833
- Ectotherms and Endotherms 833
- Modes of Heat Exchange 834
- Balancing Heat Loss and Gain 834
- Feedback Mechanisms in Thermoregulation 838
- Adjustment to Changing Temperatures 839
- Torpor and Energy Conservation 840

41 Animal Nutrition 844

OVERVIEW: The Need to Feed 844

CONCEPT 41.1 Homeostatic mechanisms manage an animal's energy budget 844
- Glucose Regulation as an Example of Homeostasis 846
- Caloric Imbalance 846

CONCEPT 41.2 An animal's diet must supply carbon skeletons and essential nutrients 849
- Essential Amino Acids 849
- Essential Fatty Acids 850
- Vitamins 850
- Minerals 851

CONCEPT 41.3 The main stages of food processing are ingestion, digestion, absorption, and elimination 853
- Digestive Compartments 853

CONCEPT 41.4 Each organ of the mammalian digestive system has specialized food-processing functions 855
- The Oral Cavity, Pharynx, and Esophagus 856
- The Stomach 857
- The Small Intestine 858
- The Large Intestine 861

CONCEPT 41.5 Evolutionary adaptations of vertebrate digestive systems are often associated with diet 862
- Some Dental Adaptations 862

- Stomach and Intestinal Adaptations 863
- Symbiotic Adaptations 863

42 Circulation and Gas Exchange 867

OVERVIEW: Trading with the Environment 867

CONCEPT 42.1 Circulatory systems reflect phylogeny 867
- Invertebrate Circulation 868
- Survey of Vertebrate Circulation 869

CONCEPT 42.2 Double circulation in mammals depends on the anatomy and pumping cycle of the heart 871
- Mammalian Circulation: *The Pathway* 872
- The Mammalian Heart: *A Closer Look* 872
- Maintaining the Heart's Rhythmic Beat 873

CONCEPT 42.3 Physical principles govern blood circulation 874
- Blood Vessel Structure and Function 874
- Blood Flow Velocity 875
- Blood Pressure 876
- Capillary Function 877
- Fluid Return by the Lymphatic System 878

CONCEPT 42.4 Blood is a connective tissue with cells suspended in plasma 879
- Blood Composition and Function 879
- Cardiovascular Disease 882

CONCEPT 42.5 Gas exchange occurs across specialized respiratory surfaces 884
- Gills in Aquatic Animals 884
- Tracheal Systems in Insects 886
- Lungs 886

CONCEPT 42.6 Breathing ventilates the lungs 888
- How an Amphibian Breathes 888
- How a Bird Breathes 889
- Control of Breathing in Humans 890

CONCEPT 42.7 Respiratory pigments bind and transport gases 891
- The Role of Partial Pressure Gradients 891
- Respiratory Pigments 892
- Elite Animal Athletes 894

43 The Immune System 898

OVERVIEW: Reconnaissance, Recognition, and Response 898

CONCEPT 43.1 Innate immunity provides broad defenses against infection 898

External Defenses 899
Internal Cellular and Chemical Defenses 899
Invertebrate Immune Mechanisms 902

CONCEPT 43.2 In acquired immunity, lymphocytes provide specific defenses against infection 903

Antigen Recognition by Lymphocytes 903
Lymphocyte Development 905

CONCEPT 43.3 Humoral and cell-mediated immunity defend against different types of threats 908

Helper T Cells: A Response to Nearly All Antigens 909
Cytotoxic T Cells: A Response to Infected Cells and Cancer Cells 910
B Cells: A Response to Extracellular Pathogens 910
Active and Passive Immunization 914

CONCEPT 43.4 The immune system's ability to distinguish self from nonself limits tissue transplantation 915

Blood Groups and Transfusions 915
Tissue and Organ Transplants 916

CONCEPT 43.5 Exaggerated, self-directed, or diminished immune responses can cause disease 916

Allergies 916
Autoimmune Diseases 917
Immunodeficiency Diseases 917

44 Osmoregulation and Excretion 922

OVERVIEW: A Balancing Act 922

CONCEPT 44.1 Osmoregulation balances the uptake and loss of water and solutes 922

Osmosis 922
Osmotic Challenges 923
Transport Epithelia 926

CONCEPT 44.2 An animal's nitrogenous wastes reflect its phylogeny and habitat 927

Forms of Nitrogenous Waste 927
The Influence of Evolution and Environment on Nitrogenous Wastes 928

CONCEPT 44.3 Diverse excretory systems are variations on a tubular theme 928

Excretory Processes 929
Survey of Excretory Systems 929

CONCEPT 44.4 Nephrons and associated blood vessels are the functional units of the mammalian kidney 931

Structure and Function of the Nephron and Associated Structures 931
From Blood Filtrate to Urine: *A Closer Look* 933

CONCEPT 44.5 The mammalian kidney's ability to conserve water is a key terrestrial adaptation 934

Solute Gradients and Water Conservation 935
Regulation of Kidney Function 936

CONCEPT 44.6 Diverse adaptations of the vertebrate kidney have evolved in different environments 938

45 Hormones and the Endocrine System 943

OVERVIEW: The Body's Long-Distance Regulators 943

CONCEPT 45.1 The endocrine system and the nervous system act individually and together in regulating an animal's physiology 943

Overlap Between Endocrine and Neural Regulation 944
Control Pathways and Feedback Loops 944

CONCEPT 45.2 Hormones and other chemical signals bind to target cell receptors in target cells, initiating pathways that culminate in specific cell responses 945

Cell-Surface Receptors for Water-Soluble Hormones 946
Intracellular Receptors for Lipid-Soluble Hormones 947
Paracrine Signaling by Local Regulators 947

CONCEPT 45.3 The hypothalamus and pituitary integrate many functions of the vertebrate endocrine system 948

Relationship Between the Hypothalamus and Pituitary Gland 950
Posterior Pituitary Hormones 951
Anterior Pituitary Hormones 951

CONCEPT 45.4 Nonpituitary hormones help regulate metabolism, homeostasis, development, and behavior 953

Thyroid Hormones 953
Parathyroid Hormone and Calcitonin: Control of Blood Calcium 954
Insulin and Glucagon: Control of Blood Glucose 955
Adrenal Hormones: Response to Stress 956
Gonadal Sex Hormones 958
Melatonin and Biorhythms 959

CONCEPT 45.5 Invertebrate regulatory systems also involve endocrine and nervous system interactions 959

46 Animal Reproduction 964

OVERVIEW: Doubling Up for Sexual Reproduction 964

CONCEPT 46.1 Both asexual and sexual reproduction occur in the animal kingdom 964

Mechanisms of Asexual Reproduction 964
Reproductive Cycles and Patterns 965

CONCEPT 46.2 Fertilization depends on mechanisms that help sperm meet eggs of the same species 967

Ensuring the Survival of Offspring 967
Gamete Production and Delivery 968

CONCEPT 46.3 Reproductive organs produce and transport gametes: focus on humans 969

Female Reproductive Anatomy 969
Male Reproductive Anatomy 971
Human Sexual Response 972

CONCEPT 46.4 In humans and other mammals, a complex interplay of hormones regulates gametogenesis 973

The Reproductive Cycles of Females 973
Hormonal Control of the Male Reproductive System 977

CONCEPT 46.5 In humans and other placental mammals, an embryo grows into a newborn in the mother's uterus 978
Conception, Embryonic Development, and Birth 978
The Mother's Immune Tolerance of the Embryo and Fetus 982
Contraception and Abortion 982
Modern Reproductive Technology 983

47 Animal Development 987

OVERVIEW: A Body-Building Plan for Animals 987

CONCEPT 47.1 After fertilization, embryonic development proceeds through cleavage, gastrulation, and organogenesis 988
Fertilization 988
Cleavage 992
Gastrulation 994
Organogenesis 997
Developmental Adaptations of Amniotes 998
Mammalian Development 999

CONCEPT 47.2 Morphogenesis in animals involves specific changes in cell shape, position, and adhesion 1001
The Cytoskeleton, Cell Motility, and Convergent Extension 1001
Roles of the Extracellular Matrix and Cell Adhesion Molecules 1002

CONCEPT 47.3 The developmental fate of cells depends on their history and on inductive signals 1003
Fate Mapping 1004
Establishing Cellular Asymmetries 1004
Cell-Fate Determination and Pattern Formation by Inductive Signals 1006

48 Nervous Systems 1011

OVERVIEW: Command and Control Center 1011

CONCEPT 48.1 Nervous systems consist of circuits of neurons and supporting cells 1012
Organization of Nervous Systems 1012
Information Processing 1013
Neuron Structure 1013
Supporting Cells (Glia) 1014

CONCEPT 48.2 Ion pumps and ion channels maintain the resting potential of a neuron 1015
The Resting Potential 1016
Gated Ion Channels 1017

CONCEPT 48.3 Action potentials are the signals conducted by axons 1017
Production of Action Potentials 1018
Conduction of Action Potentials 1020

CONCEPT 48.4 Neurons communicate with other cells at synapses 1021
Direct Synaptic Transmission 1022
Indirect Synaptic Transmission 1023
Neurotransmitters 1024

CONCEPT 48.5 The vertebrate nervous system is regionally specialized 1026
The Peripheral Nervous System 1026
Embryonic Development of the Brain 1028
The Brainstem 1029
The Cerebellum 1030
The Diencephalon 1030
The Cerebrum 1031

CONCEPT 48.6 The cerebral cortex controls voluntary movement and cognitive functions 1032
Information Processing in the Cerebral Cortex 1032
Lateralization of Cortical Function 1033
Language and Speech 1034
Emotions 1034
Memory and Learning 1035
Consciousness 1036

CONCEPT 48.7 CNS injuries and diseases are the focus of much research 1037
Nerve Cell Development 1037
Neural Stem Cells 1038
Diseases and Disorders of the Nervous Systems 1039

49 Sensory and Motor Mechanisms 1045

OVERVIEW: Sensing and Acting 1045

CONCEPT 49.1 Sensory receptors transduce stimulus energy and transmit signals to the central nervous system 1046
Functions Performed by Sensory Receptors 1046
Types of Sensory Receptors 1048

CONCEPT 49.2 The mechanoreceptors involved with hearing and equilibrium detect settling particles or moving fluid 1050
Sensing Gravity and Sound in Invertebrates 1050
Hearing and Equilibrium in Mammals 1050
Hearing and Equilibrium in Other Vertebrates 1053

CONCEPT 49.3 The senses of taste and smell are closely related in most animals 1054
Taste in Humans 1055
Smell in Humans 1056

CONCEPT 49.4 Similar mechanisms underlie vision throughout the animal kingdom 1057
Vision in Invertebrates 1057
The Vertebrate Visual System 1058

CONCEPT 49.5 Animal skeletons function in support, protection, and movement 1063
Types of Skeletons 1063
Physical Support on Land 1064

CONCEPT 49.6 Muscles move skeletal parts by contracting 1066
Vertebrate Skeletal Muscle 1066
Other Types of Muscle 1072

CONCEPT 49.7 Locomotion requires energy to overcome friction and gravity 1073
Swimming 1073
Locomotion on Land 1073
Flying 1074
Comparing Costs of Locomotion 1074

UNIT EIGHT

Ecology 1078

50 An Introduction to Ecology and the Biosphere 1080

OVERVIEW: The Scope of Ecology 1080

CONCEPT 50.1 Ecology is the study of interactions between organisms and the environment 1080

- Ecology and Evolutionary Biology 1081
- Organisms and the Environment 1081
- Subfields of Ecology 1082
- Ecology and Environmental Issues 1083

CONCEPT 50.2 Interactions between organisms and the environment limit the distribution of species 1083

- Dispersal and Distribution 1084
- Behavior and Habitat Selection 1085
- Biotic Factors 1085
- Abiotic Factors 1086
- Climate 1087

CONCEPT 50.3 Abiotic and biotic factors influence the structure and dynamics of aquatic biomes 1092

CONCEPT 50.4 Climate largely determines the distribution and structure of terrestrial biomes 1098

- Climate and Terrestrial Biomes 1098
- General Features of Terrestrial Biomes 1098

51 Behavioral Ecology 1106

OVERVIEW: Studying Behavior 1106

CONCEPT 51.1 Behavioral ecologists distinguish between proximate and ultimate causes of behavior 1106

- What Is Behavior? 1107
- Proximate and Ultimate Questions 1107
- Ethology 1107

CONCEPT 51.2 Many behaviors have a strong genetic component 1109

- Directed Movements 1110
- Animal Signals and Communication 1111
- Genetic Influences on Mating and Parental Behavior 1112

CONCEPT 51.3 Environment, interacting with an animal's genetic makeup, influences the development of behaviors 1113

- Dietary Influence on Mate Choice Behavior 1113
- Social Environment and Aggressive Behavior 1114
- Learning 1114

CONCEPT 51.4 Behavioral traits can evolve by natural selection 1118

- Behavioral Variation in Natural Populations 1118
- Experimental Evidence for Behavioral Evolution 1120

CONCEPT 51.5 Natural selection favors behaviors that increase survival and reproductive success 1121

- Foraging Behavior 1122

- Mating Behavior and Mate Choice 1123
- Applying Game Theory 1127

CONCEPT 51.6 The concept of inclusive fitness can account for most altruistic social behavior 1128

- Altruism 1128
- Inclusive Fitness 1129
- Social Learning 1131
- Evolution and Human Culture 1132

52 Population Ecology 1136

OVERVIEW: Earth's Fluctuating Populations 1136

CONCEPT 52.1 Dynamic biological processes influence population density, dispersion, and demography 1136

- Density and Dispersion 1137
- Demography 1139

CONCEPT 52.2 Life history traits are products of natural selection 1141

- Life History Diversity 1141
- "Trade-offs" and Life Histories 1142

CONCEPT 52.3 The exponential model describes population growth in an idealized, unlimited environment 1143

- Per Capita Rate of Increase 1143
- Exponential Growth 1144

CONCEPT 52.4 The logistic growth model includes the concept of carrying capacity 1145

- The Logistic Growth Model 1145
- The Logistic Model and Real Populations 1146
- The Logistic Model and Life Histories 1147

CONCEPT 52.5 Populations are regulated by a complex interaction of biotic and abiotic influences 1148

- Population Change and Population Density 1148
- Density-Dependent Population Regulation 1148
- Population Dynamics 1150
- Population Cycles 1151

CONCEPT 52.6 Human population growth has slowed after centuries of exponential increase 1152

- The Global Human Population 1152
- Global Carrying Capacity 1155

53 Community Ecology 1159

OVERVIEW: What Is a Community? 1159

CONCEPT 53.1 A community's interactions include competition, predation, herbivory, symbiosis, and disease 1159

Competition 1160
Predation 1161
Herbivory 1163
Parasitism 1163
Disease 1163
Mutualism 1164
Commensalism 1164
Interspecific Interactions and Adaptation 1164

CONCEPT 53.2 Dominant and keystone species exert strong controls on community structure 1165

Species Diversity 1165
Trophic Structure 1166
Species with a Large Impact 1168
Bottom-up and Top-Down Controls 1170

CONCEPT 53.3 Disturbance influences species diversity and composition 1171

What Is Disturbance? 1172
Human Disturbance 1173
Ecological Succession 1173

CONCEPT 53.4 Biogeographic factors affect community biodiversity 1175

Equatorial-Polar Gradients 1176
Area Effects 1176
Island Equilibrium Model 1177

CONCEPT 53.5 Contrasting views of community structure are the subject of continuing debate 1178

Integrated and Individualistic Hypotheses 1178
Rivet and Redundancy Models 1180

54 Ecosystems 1184

OVERVIEW: Ecosystems, Energy, and Matter 1184

CONCEPT 54.1 Ecosystem ecology emphasizes energy flow and chemical cycling 1184

Ecosystems and Physical Laws 1185
Trophic Relationships 1185
Decomposition 1185

CONCEPT 54.2 Physical and chemical factors limit primary production in ecosystems 1186

Ecosystem Energy Budgets 1186
Primary Production in Marine and Freshwater Ecosystems 1188
Primary Production in Terrestrial and Wetland Ecosystems 1190

CONCEPT 54.3 Energy transfer between trophic levels is usually less than 20% efficient 1191

Production Efficiency 1191
The Green World Hypothesis 1193

CONCEPT 54.4 Biological and geochemical processes move nutrients between organic and inorganic parts of the ecosystem 1195

A General Model of Chemical Cycling 1195
Biogeochemical Cycles 1195
Decomposition and Nutrient Cycling Rates 1198
Vegetation and Nutrient Cycling: The Hubbard Brook Experimental Forest 1198

CONCEPT 54.5 The human population is disrupting chemical cycles throughout the biosphere 1200

Nutrient Enrichment 1200
Acid Precipitation 1201
Toxins in the Environment 1202
Atmospheric Carbon Dioxide 1203
Depletion of Atmospheric Ozone 1205

55 Conservation Biology and Restoration Ecology 1209

OVERVIEW: The Biodiversity Crisis 1209

CONCEPT 55.1 Human activities threaten Earth's biodiversity 1209

The Three Levels of Biodiversity 1210
Biodiversity and Human Welfare 1211
Four Major Threats to Biodiversity 1212

CONCEPT 55.2 Population conservation focuses on population size, genetic diversity, and critical habitat 1215

Small-Population Approach 1215
Declining-Population Approach 1218
Weighing Conflicting Demands 1219

CONCEPT 55.3 Landscape and regional conservation aim to sustain entire biotas 1220

Landscape Structure and Biodiversity 1220
Establishing Protected Areas 1222

CONCEPT 55.4 Restoration ecology attempts to restore degraded ecosystems to a more natural state 1224

Bioremediation 1225
Biological Augmentation 1225
Exploring Restoration 1225

CONCEPT 55.5 Sustainable development seeks to improve the human condition while conserving biodiversity 1228

Sustainable Biosphere Initiative 1228
Case Study: Sustainable Development in Costa Rica 1228
Biophilia and the Future of the Biosphere 1229

APPENDIX A Answers

APPENDIX B The Metric System

APPENDIX C A Comparison of the Light Microscope and the Electron Microscope

APPENDIX D Classification of Life

CREDITS

GLOSSARY

INDEX

Biology

1 Exploring Life

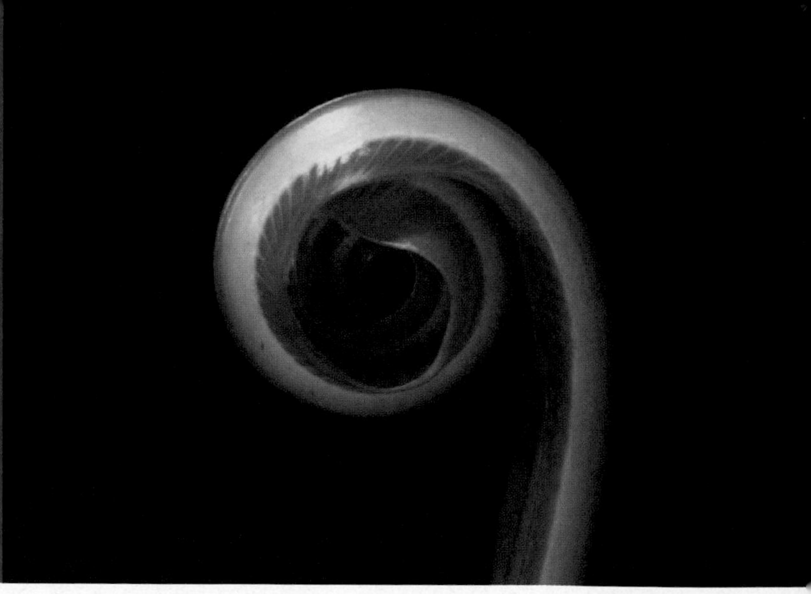

▲ **Figure 1.1 Biology is the science that focuses on life.**

Key Concepts

1.1 Biologists explore life from the microscopic to the global scale
1.2 Biological systems are much more than the sum of their parts
1.3 Biologists explore life across its great diversity of species
1.4 Evolution accounts for life's unity and diversity
1.5 Biologists use various forms of inquiry to explore life
1.6 A set of themes connects the concepts of biology

Overview

Biology's Most Exciting Era

Welcome to **biology,** the scientific study of life. You are becoming involved with biology during its most exciting era. The largest and best-equipped community of scientists in history is beginning to solve biological puzzles that once seemed unsolvable. We are moving ever closer to understanding how a single microscopic cell develops into a complex plant or animal; how plants convert solar energy to the chemical energy of food; how the human mind works; how various forms of life network in biological communities such as forests and coral reefs; and how the great diversity of life on Earth evolved from the first microbes. The more we learn about life, the more fascinating it becomes, as progress on one question leads to even more questions that will captivate curious minds for decades to come. More than anything else, biology is a quest, an ongoing inquiry about the nature of life.

Modern biology is as important as it is inspiring. Research breakthroughs in genetics and cell biology are transforming medicine and agriculture. Molecular biology is providing new tools for fields as diverse as anthropology and criminal science. Neuroscience and evolutionary biology are reshaping psychology and sociology. New models in ecology are helping societies evaluate environmental issues, such as the causes and biological consequences of global warming. These are just a few examples of how biology weaves into the fabric of our culture more than ever before. There has never been a better time to explore life.

The phenomenon we call life defies a simple, one-sentence definition. Yet almost any child perceives that a dog or a bug or a plant, such as the fern "fiddlehead" that graces the cover of this book **(Figure 1.1)**, is alive, while a rock is not. We recognize life by what living things do. **Figure 1.2** highlights some of the properties and processes we associate with life.

As we set off to explore life, it helps to have a panoramic view of the vast field of study before us. This opening chapter introduces the wide scope of biology, highlights the diversity of life, describes themes, such as evolution, that unify all of biology, and examines methods of inquiry that biologists use to explore life.

Concept 1.1

Biologists explore life from the microscopic to the global scale

The study of life extends from the microscopic scale of the molecules and cells that make up organisms to the global scale of the entire living planet. We can divide this enormous range into different levels of biological organization.

(a) Order. This close-up of a sunflower illustrates the highly ordered structure that characterizes life.

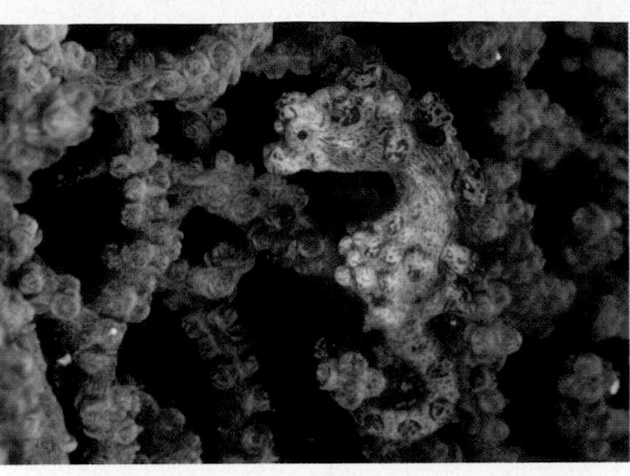

(b) Evolutionary adaptation. The appearance of this pygmy seahorse camouflages the animal in its environment. Such adaptations evolve over many generations by the reproductive success of those individuals with heritable traits that are best suited to their environments.

(c) Response to the environment. This Venus' flytrap closed its trap rapidly in response to the environmental stimulus of a damselfly landing on the open trap.

(d) Regulation. The regulation of blood flow through the blood vessels of this jackrabbit's ears helps maintain a constant body temperature by adjusting heat exchange with the surrounding air.

(e) Energy processing. This hummingbird obtains fuel in the form of nectar from flowers. The hummingbird will use the chemical energy stored in its food to power flight and other work.

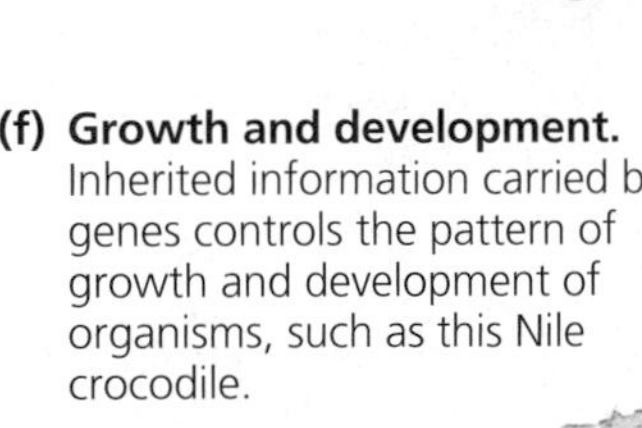

(f) Growth and development. Inherited information carried by genes controls the pattern of growth and development of organisms, such as this Nile crocodile.

(g) Reproduction. Organisms (living things) reproduce their own kind. Here an emperor penguin protects its baby.

▲ **Figure 1.2 Some properties of life.**

A Hierarchy of Biological Organization

Imagine zooming in from space to take a closer and closer look at life on Earth. Our destination is a forest in Ontario, Canada, where we will eventually use microscopes and other instruments to explore a maple leaf right down to the molecular level. **Figure 1.3** (on the next two pages) narrates this journey into life, with the circled numbers leading you through the levels of biological organization illustrated by the photographs.

Figure 1.3

Exploring Levels of Biological Organization

❶ **The biosphere.** As soon as we are near enough to Earth to make out its continents and oceans, we begin to see signs of life—in the green mosaic of the planet's forests, for example. This is our first view of the biosphere, which consists of all the environments on Earth that are inhabited by life. The biosphere includes most regions of land; most bodies of water, such as oceans, lakes, and rivers; and the atmosphere to an altitude of several kilometers.

❷ **Ecosystems.** As we approach Earth's surface for an imaginary landing in Ontario, we can begin to make out a forest with an abundance of deciduous trees (trees that lose their leaves in one season and grow new ones in another). Such a deciduous forest is an example of an ecosystem. Grasslands, deserts, and the ocean's coral reefs are other types of ecosystems. An ecosystem consists of all the living things in a particular area, along with all the nonliving components of the environment with which life interacts, such as soil, water, atmospheric gases, and light. All of Earth's ecosystems combined make up the biosphere.

❸ **Communities.** The entire array of organisms inhabiting a particular ecosystem is called a biological community. The community in our forest ecosystem includes many kinds of trees and other plants, a diversity of animals, various mushrooms and other fungi, and enormous numbers of diverse microorganisms, which are living forms such as bacteria that are too small to see without a microscope. Each of these forms of life is called a *species*.

❹ **Populations.** A population consists of all the individuals of a species living within the bounds of a specified area. For example, our Ontario forest includes a population of sugar maple trees and a population of American black bears. We can now refine our definition of a community as the set of populations that inhabit a particular area.

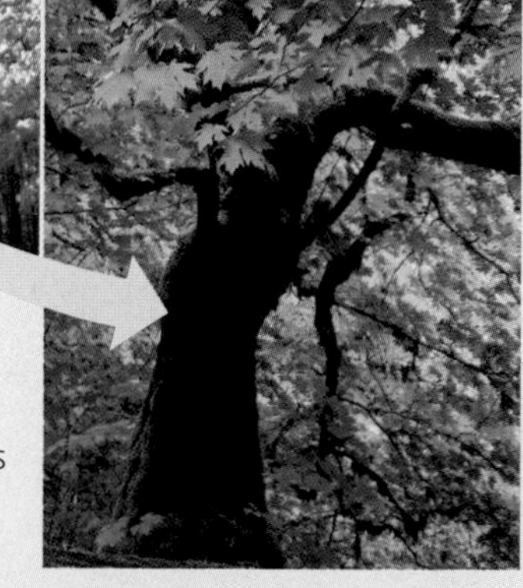

❺ **Organisms.** Individual living things are called organisms. Each of the maple trees and other plants in the forest is an organism, and so is each forest animal such as a frog, squirrel, bear, and insect. The soil teems with microorganisms such as bacteria.

8 Cells. The cell is life's fundamental unit of structure and function. Some organisms, such as amoebas and most bacteria, are single cells. Other organisms, including plants and animals, are multicellular. Instead of a single cell performing all the functions of life, a multicellular organism has a division of labor among specialized cells. A human body consists of trillions of microscopic cells of many different kinds, including muscle cells and nerve cells, which are organized into the various specialized tissues. For example, muscle tissue consists of bundles of muscle cells. And note again the cells of the tissue within a leaf's interior. Each of the cells you see is only about 25 μm (micrometers) across. It would take more than 700 of these cells to reach across a penny. As small as these cells are, you can see that each contains numerous green structures called chloroplasts, which are responsible for photosynthesis.

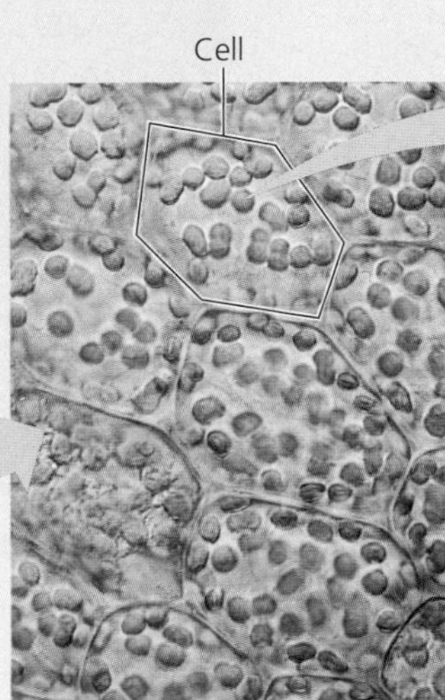

9 Organelles. Chloroplasts are examples of organelles, the various functional components that make up cells. In this figure, a very powerful tool called an electron microscope brings a single chloroplast into sharp focus.

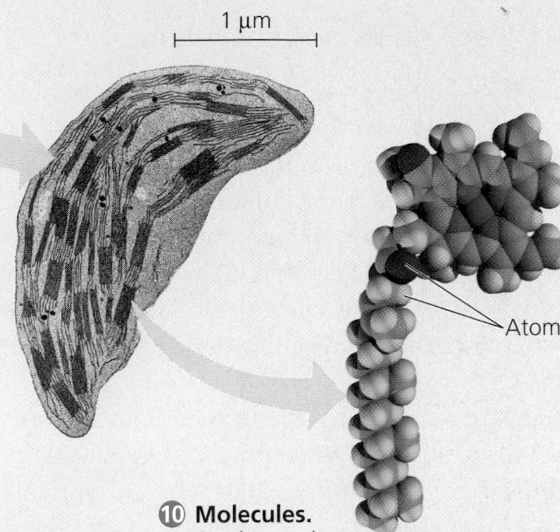

10 Molecules. Our last scale change vaults us into a chloroplast for a view of life at the molecular level. A molecule is a chemical structure consisting of two or more small chemical units called *atoms,* which are represented as balls in this computer graphic of a chlorophyll molecule. Chlorophyll is the pigment molecule that makes a maple leaf green. One of the most important molecules on Earth, chlorophyll absorbs sunlight during the first step of photosynthesis. Within each chloroplast, millions of chlorophylls and other molecules are organized into the equipment that converts light energy to the chemical energy of food.

50 μm

7 Tissues. Our next scale change to see a leaf's tissues requires a microscope. The leaf on the left has been cut on an angle. The honeycombed tissue in the interior of the leaf (upper half of photo) is the main location of photosynthesis, the process that converts light energy to the chemical energy of sugar and other food. We are viewing the sliced leaf from a perspective that also enables us to see the jigsaw puzzle-like tissue called epidermis, the "skin" on the surface of the leaf (lower half of photo). The pores through the epidermis allow the gas carbon dioxide, a raw material for sugar production, to reach the photosynthetic tissue in the interior of the leaf. At this scale, we can also see that each tissue has a cellular structure. In fact, each kind of tissue is a group of similar cells.

6 Organs and organ systems. The structural hierarchy of life continues to unfold as we explore the architecture of the more complex organisms. A maple leaf is an example of an organ, a body part consisting of two or more tissues (which we'll see upon our next scale change). Stems and roots are the other major organs of plants. Examples of human organs are the brain, heart, and kidney. The organs of humans and other complex animals are organized into organ systems, each a team of organs that cooperate in a specific function. For example, the human digestive system includes such organs as the tongue, stomach, and intestines.

In Figure 1.3, we traced the greenness that we first saw in our extraterrestrial view of the biosphere all the way down to the molecular level of chlorophyll. As a follow-up to this overview of life's structural hierarchy, we'll take a closer look at just two biological levels near opposite ends of the size scale: ecosystems and cells.

A Closer Look at Ecosystems

Life does not exist in a vacuum. Each organism interacts continuously with its environment, which includes other organisms as well as nonliving factors. The roots of a tree, for example, absorb water and minerals from the soil. The leaves take in carbon dioxide from the air. Solar energy absorbed by chlorophyll drives photosynthesis, which converts water and carbon dioxide to sugar and oxygen. The tree releases oxygen to the air, and its roots help form soil by breaking up rocks. Both organism and environment are affected by the interactions between them. The tree also interacts with other life, including soil microorganisms associated with its roots and animals that eat its leaves and fruit.

Ecosystem Dynamics

The dynamics of any ecosystem include two major processes. One process is the cycling of nutrients. For example, minerals acquired by plants will eventually be returned to the soil by microorganisms that decompose leaf litter, dead roots, and other organic debris. The second major process in an ecosystem is the flow of energy from sunlight to producers to consumers. **Producers** are plants and other photosynthetic organisms that convert light energy to chemical energy. **Consumers** are organisms, such as animals, that feed on producers and other consumers.

Energy Conversion

Moving, growing, reproducing, and other activities of life require organisms to perform work. And work depends on a source of energy. The exchange of energy between an organism and its surroundings often involves the transformation of one form of energy to another. For example, when a leaf produces sugar, it converts solar energy to chemical energy in sugar molecules. When an animal's muscle cells use sugar as fuel to power movements, they convert chemical energy to kinetic energy, the energy of motion. And in all these energy conversions, some of the available energy is converted to thermal energy, which working organisms dissipate to their surroundings as heat. In contrast to chemical nutrients, which recycle within an ecosystem, energy flows *through* an ecosystem, usually entering as light and exiting as heat **(Figure 1.4)**.

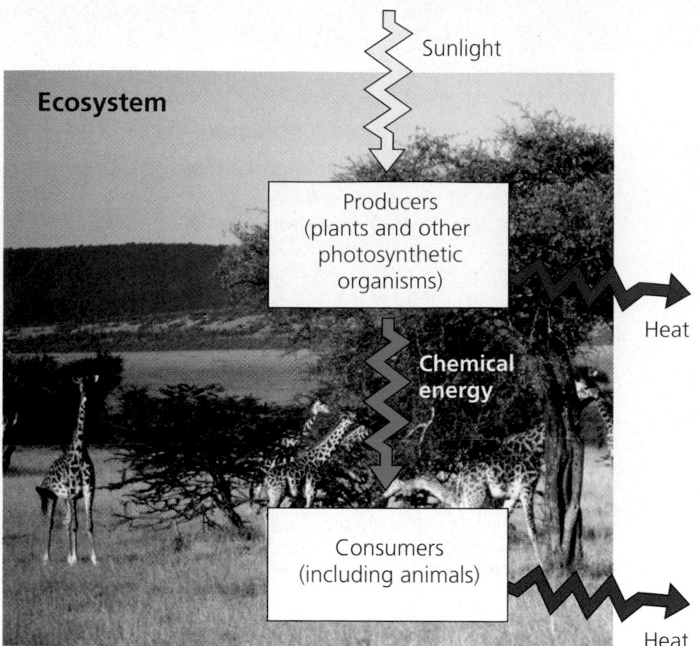

▲ **Figure 1.4 Basic scheme for energy flow through an ecosystem.**

A Closer Look at Cells

In life's structural hierarchy, the cell has a special place as the lowest level of organization that can perform *all* activities required for life. For example, the ability of cells to divide to form new cells is the basis for all reproduction and for the growth and repair of multicellular organisms **(Figure 1.5)**. Your every movement and thought are based on the activities of muscle cells and nerve cells. Even a global process such as the recycling of carbon, a chemical element essential to life, is the cumulative product of cellular activities, including the photosynthesis that occurs in the chloroplasts of leaf cells. Understanding how cells work is a major research focus of modern biology.

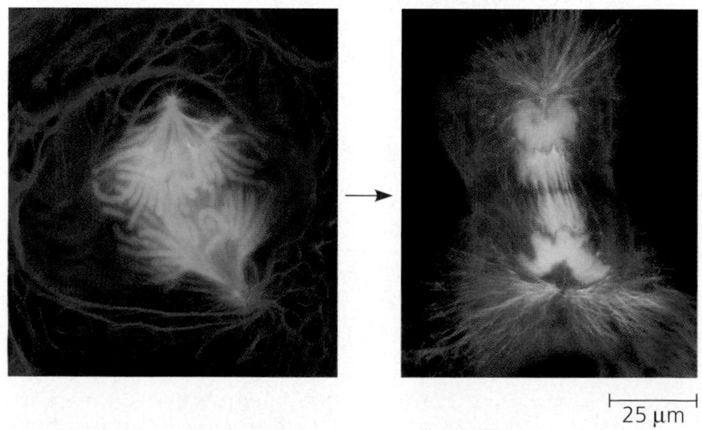

▲ **Figure 1.5 A lung cell from a newt divides into two smaller cells that will grow and divide again.**

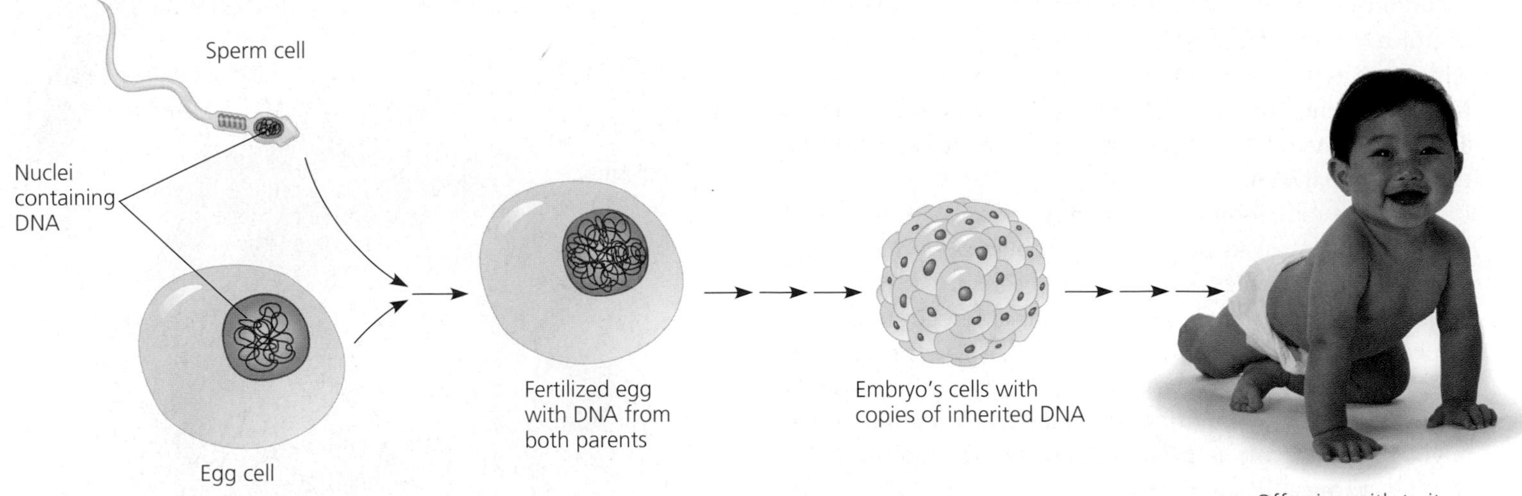

▲ **Figure 1.6 Inherited DNA directs development of an organism.**

The Cell's Heritable Information

Take another look at the dividing cell in Figure 1.5. Within the cells you can see structures called chromosomes, which are stained with a blue-glowing dye. The chromosomes are partly made of a substance called **deoxyribonucleic acid**, or **DNA** for short. DNA is the substance of **genes**, the units of inheritance that transmit information from parents to offspring. Your blood group (A, B, AB, or O), for example, is the result of certain genes that you inherited from your parents.

Each chromosome has one very long DNA molecule, with hundreds or thousands of genes arranged along its length. The DNA of chromosomes replicates as a cell prepares to divide; thus, each of the two cellular offspring inherits a complete set of genes.

Each of us began life as a single cell stocked with DNA inherited from our parents. Replication of that DNA transmitted those genes to our trillions of cells. In each cell, the genes along the length of DNA molecules encode the information for building the cell's other molecules. In this way, DNA directs the development and maintenance of the entire organism **(Figure 1.6)**.

The molecular structure of DNA accounts for its information-rich nature. Each DNA molecule is made up of two long chains arranged into what is called a double helix. Each link of a chain is one of four kinds of chemical building blocks called nucleotides **(Figure 1.7)**. The way DNA encodes a cell's information is analogous to the way we arrange the letters of the alphabet into precise sequences with specific meanings. The word *rat*, for example, conjures up an image of a rodent; the words *tar* and *art*, which contain the same letters, mean very different things. Libraries are filled with books containing information encoded in varying sequences of only 26 letters. We can think of nucleotides as the alphabet of

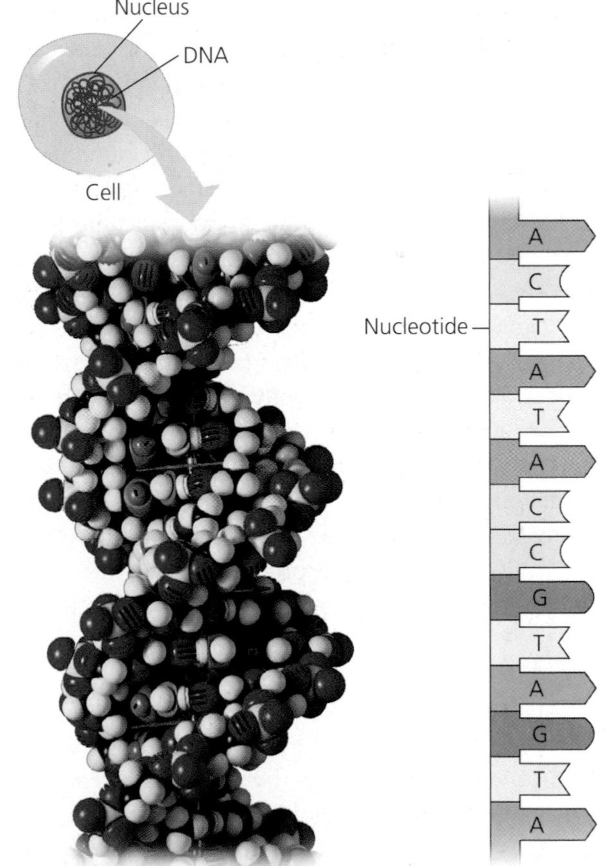

(a) DNA double helix. This model shows each atom in a segment of DNA. Made up of two long chains of building blocks called nucleotides, a DNA molecule takes the three-dimensional form of a double helix.

(b) Single strand of DNA. These geometric shapes and letters are simple symbols for the nucleotides in a small section of one chain of a DNA molecule. Genetic information is encoded in specific sequences of the four types of nucleotides (their names are abbreviated here as A, T, C, and G).

▲ **Figure 1.7 DNA: The genetic material.**

inheritance. Specific sequential arrangements of these four chemical letters encode the precise information in genes, which are typically hundreds or thousands of nucleotides long. One gene in a bacterial cell may be translated as "Build a purple pigment." A particular human gene may mean "Make the hormone insulin."

More generally, most genes program the cell's production of large molecules called proteins. The sequence of nucleotides along each gene codes for a specific protein with a unique shape and function in the cell. One protein might be part of the contractile apparatus of muscle cells. Another protein might be an antibody, part of the body's defense system against viruses and other disease agents. Still another protein might be an enzyme, a protein that catalyzes (speeds up) a specific chemical reaction in a cell. Almost all cellular activities involve the action of one or more proteins. DNA provides the heritable blueprints, but proteins are the tools that actually build and maintain the cell.

All forms of life employ essentially the same genetic code. A particular sequence of nucleotides says the same thing to one organism as it does to another. Differences between organisms reflect differences between their nucleotide sequences. But because the genetic code is universal, it is possible to engineer cells to produce proteins normally found only in some other organism. One of the first pharmaceutical products obtained using this technology was human insulin, produced by bacteria into which a gene for this human protein was inserted.

The entire "library" of genetic instructions that an organism inherits is called its **genome.** The chromosomes of each human cell pack a genome that is about 3 billion nucleotides long. If the one-letter symbols for this sequence of nucleotides were written in letters the size of those you are now reading, the genetic text would fill about 600 books the size of this one. Within this genomic library of nucleotide sequences are genes coding for the production of more than 75,000 different kinds of proteins, each with a specific function.

Two Main Forms of Cells

All cells share certain characteristics. For example, every cell is enclosed by a membrane that regulates the passage of materials between the cell and its surroundings. And every cell uses DNA as its genetic information.

We can distinguish two main forms of cells: prokaryotic cells and eukaryotic cells. The cells of two groups of microorganisms called bacteria and archaea are prokaryotic. All other forms of life, including plants and animals, are composed of eukaryotic cells.

A **eukaryotic cell** is subdivided by internal membranes into various membrane-enclosed organelles, including the chloroplasts of Figure 1.3. In most eukaryotic cells, the largest

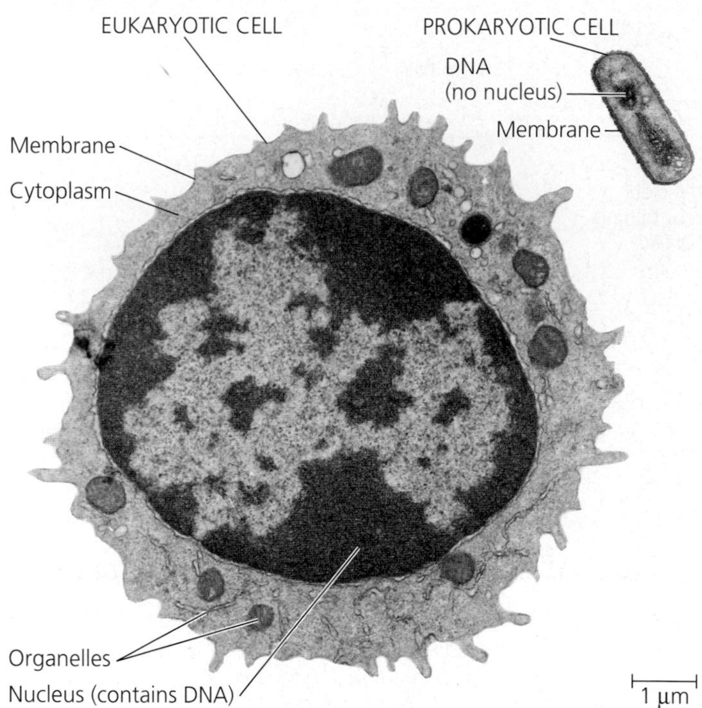

▲ **Figure 1.8 Contrasting eukaryotic and prokaryotic cells in size and complexity.**

organelle is the nucleus, which contains the cell's DNA (as chromosomal molecules). The other organelles are located in the cytoplasm, the entire region between the nucleus and outer membrane of the cell.

Prokaryotic cells are much simpler and generally smaller than eukaryotic cells **(Figure 1.8)**. In a **prokaryotic cell**, the DNA is not separated from the rest of the cell by enclosure in a membrane-bounded nucleus. Prokaryotic cells also lack the other kinds of membrane-enclosed organelles that characterize eukaryotic cells.

The prokaryote-eukaryote difference is an example of the biological diversity we will explore in Concept 1.3. But first, let's take one more look at the hierarchy of biological order, this time in the context of a research movement called systems biology.

Concept Check 1.1

1. For each biological level in Figure 1.3, write a sentence that includes the next "lower" level. Example: "A community consists of *populations* of the various species inhabiting a specific area."
2. What are the relationships between these three genetic terms: DNA, genes, and chromosomes?
3. Explain why, at the cellular level, plants have more in common with animals than they do with bacteria.

For suggested answers, see Appendix A.

Concept 1.2

Biological systems are much more than the sum of their parts

"The whole is greater than the sum of its parts." That familiar adage captures the important concept that a combination of components can form a more complex organization called a **system.** Examples of biological systems are cells, organisms, and ecosystems. To understand how such systems work, it is not enough to have a "parts list," even a complete one. The future of biology is in understanding the behavior of whole, integrated systems.

The Emergent Properties of Systems

Take another look at the levels of life in Figure 1.3. With each step upward in this hierarchy of biological order, novel properties emerge that are not present at the level just below. These **emergent properties** are due to the arrangement and interactions of parts as complexity increases. For example, a test-tube mixture of chlorophyll and all the other molecules found in a chloroplast cannot perform photosynthesis. The process of photosynthesis emerges from the very specific way in which the chlorophyll and other molecules are arranged in an intact chloroplast. To take another example, if a serious head injury disrupts the intricate architecture of a human brain, the mind may cease to function properly even though all of the brain parts are still present. Our thoughts and memories are emergent properties of a complex network of nerve cells. At an even higher level of biological organization—at the ecosystem level—the recycling of nutrients such as carbon depends on a network of diverse organisms interacting with each other and with the soil and air.

Emergent properties are neither supernatural nor unique to life. We can see the importance of arrangement in the distinction between a box of bicycle parts and a working bicycle. And while graphite and diamonds are both pure carbon, they have very different properties based on how their carbon atoms are arranged. But compared to such nonliving examples, the emergent properties of life are particularly challenging to study because of the unrivaled complexity of biological systems.

The Power and Limitations of Reductionism

Because the properties of life emerge from complex organization, scientists seeking to understand biological systems confront a dilemma. On the one hand, we cannot fully explain a higher level of order by breaking it down into its parts. A dissected animal no longer functions; a cell reduced to its chemical ingredients is no longer a cell. Disrupting a living system interferes with the meaningful understanding of its processes. On the other hand, something as complex as an organism or a cell cannot be analyzed without taking it apart.

Reductionism—reducing complex systems to simpler components that are more manageable to study—is a powerful strategy in biology. For example, by studying the molecular structure of DNA that had been extracted from cells, James Watson and Francis Crick inferred, in 1953, how this molecule could serve as the chemical basis of inheritance.

In 2001, almost half a century after the famous work of Watson and Crick, an international team of scientists published a "rough draft" of the sequence of 3 billion chemical letters in a human genome **(Figure 1.9)**. (Researchers have also sequenced the genomes of many other species.) The press and world leaders acclaimed the Human Genome Project as one of the greatest scientific achievements ever. But unlike past cultural zeniths, such as the moonwalk of Apollo astronauts, the sequencing of the human genome is more a

▶ **Figure 1.9 Modern biology as an information science.** Automatic DNA-sequencing machines and abundant computing power accelerated the Human Genome Project. This facility in Cambridge, United Kingdom, was one of many labs that collaborated in the international project.

commencement than a climax. As the quest continues, scientists are learning the functions of thousands of genes and their protein products. And research is now moving on to how the activities of these myriad molecules are coordinated in the development and maintenance of cells and whole organisms. At the cutting edge of this research is the approach called systems biology.

Systems Biology

Biology is turning in an exciting new direction as many researchers begin to complement reductionism with new strategies for understanding the emergent properties of life—how all the parts of biological systems such as cells are functionally integrated. This changing perspective is analogous to moving from ground level on a certain street corner to an aerial view above a city, where you can now see how variables such as time of day, construction projects, accidents, and traffic signal malfunctions affect traffic dynamics throughout the city.

The ultimate goal of **systems biology** is to model the dynamic behavior of whole biological systems. Accurate models will enable biologists to predict how a change in one or more variables will impact other components and the whole system. How, for example, will a slight increase in a muscle cell's calcium concentration affect the activities of the dozens of proteins that regulate muscle contraction? How will a drug that lowers blood pressure affect the function of organs throughout the human body and possibly cause harmful side effects? How will increasing the water supply to a crop impact key processes in the plants, such as the use of certain soil minerals and the storage of proteins essential for human nutrition? How will a gradual increase in atmospheric carbon dioxide alter ecosystems and the entire biosphere? The aim of systems biology is to make progress answering such big questions.

Systems biology is relevant to the study of life at all levels. Scientists investigating ecosystems pioneered the systems approach in the 1960s with elaborate models diagramming the network of interactions between species and nonliving components in salt marshes and other ecosystems. Even earlier, biologists studying the physiology (functioning) of humans and other organisms were integrating data on how multiple organs coordinate processes such as the regulation of sugar concentration in the blood. Such models of ecosystems and organisms have already been useful for predicting the responses of these systems to changing variables.

Systems biology is now taking hold in the study of life at the cellular and molecular levels, driven partly by the deluge of data from the sequencing of genomes and the growing catalog of known protein functions. In 2003, for example, a large research team published a network of protein interactions within the cell of a fruit fly, a popular research organism. The model is based on an extensive database of thousands of known proteins and their known interactions with other proteins. For example, protein A may bind to and alter the activities of proteins B, C, and D, which then go on to interact with still other proteins. **Figure 1.10** maps these protein partnerships to their cellular locales.

The basics of the systems strategy are straightforward enough. First, it is necessary to inventory as many parts as possible, such as all the known genes and proteins in a cell

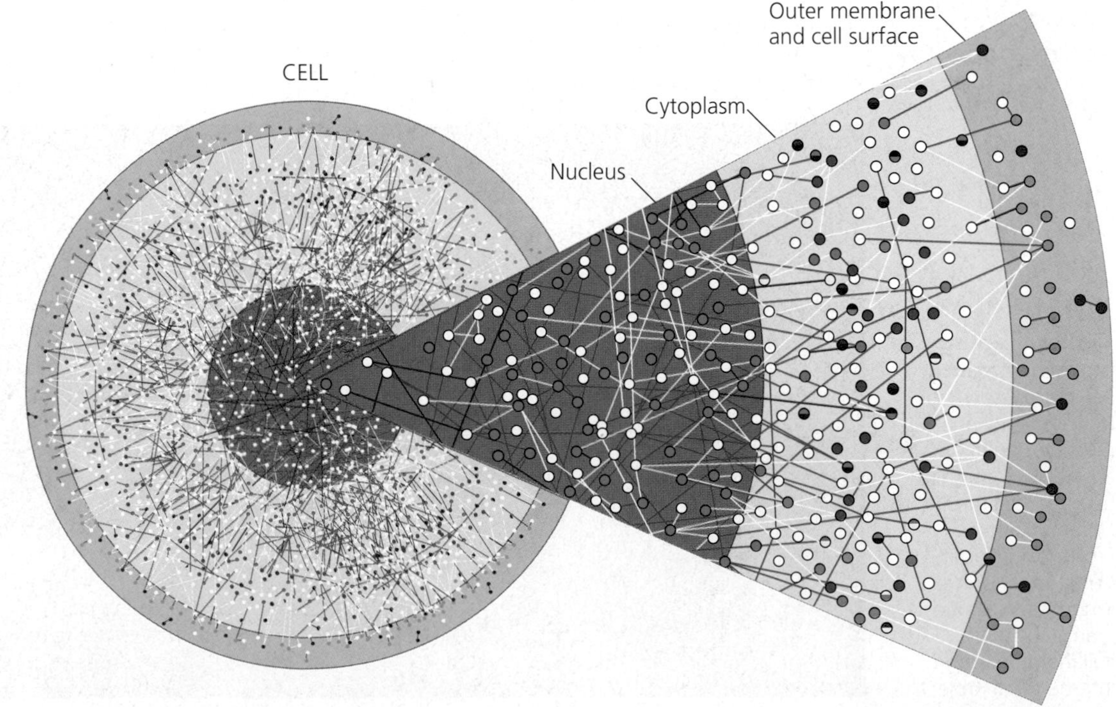

▶ **Figure 1.10 A systems map of interactions between proteins in a cell.** This diagram maps about 3,500 proteins (dots) and their network of interactions (lines connecting the proteins) in a fruit fly cell. Systems biologists develop such models from huge databases of information about molecules and their interactions in the cell. A major goal of this systems approach is to use such models to predict how one change, such as an increase in the activity of a particular protein, can ripple through the cell's molecular circuitry to cause other changes. One of the applications will be a more accurate prediction of the side effects of various drugs.

(reductionism). Then it is necessary to investigate how each part behaves in relation to others in the working system—all the protein-protein interactions, in the case of our fly cell example. Finally, with the help of computers and innovative software, it is possible to pool all the data from many research teams into the kind of system network modeled in Figure 1.10.

Though the basic idea of systems biology is simple, the practice is not, as you would expect from the complexity of biological systems. It has taken three key research developments to bring systems biology within reach:

- **High-throughput technology.** Systems biology depends on methods that can analyze biological materials very rapidly and produce enormous volumes of data. Such mega-data-collection methods are said to be "high-throughput." The automatic DNA-sequencing machines that made the Human Genome Project possible are examples of high-throughput devices (see Figure 1.9).
- **Bioinformatics.** The huge databases that result from high-throughput methods would be chaotic without the computing power, software, and mathematical models to process and integrate all this biological information. The new field of **bioinformatics** is extracting useful biological information from the enormous, ever-expanding data sets, such as DNA sequences and lists of protein interactions. The Internet is nurturing systems biology through dissemination of the digital data that feed bioinformatics.
- **Interdisciplinary research teams.** In 2003, Harvard Medical School formed a department of systems biology, its first new department in two decades. Nearby MIT is busy organizing over 80 faculty members from many departments into a new program for computational and systems biology. These and other systems biology start-ups are melting pots of diverse specialists, including engineers, medical scientists, chemists, physicists, mathematicians, computer scientists, and, of course, biologists from a variety of fields.

A number of prominent scientists are promoting systems biology with missionary zeal, but so far, the excitement exceeds the achievements. However, as systems biology gathers momentum, it is certain to have a growing impact on the questions biologists ask and the research they design. After all, scientists aspired to reach beyond reductionism to grasp how whole biological systems work long before new technology made modern systems biology possible. In fact, decades ago, biologists had already identified some of the key mechanisms that regulate the behavior of complex systems such as cells, organisms, and ecosystems.

Feedback Regulation in Biological Systems

A kind of supply-and-demand economy applies to some of the dynamics of biological systems. For example, when your muscle cells require more energy during exercise, they increase their consumption of the sugar molecules that provide fuel. In contrast, when you rest, a different set of chemical reactions converts surplus sugar to substances that store the fuel.

Like most of the cell's chemical processes, those that decompose or store sugar are accelerated, or catalyzed, by the specialized proteins called enzymes. Each type of enzyme catalyzes a specific chemical reaction. In many cases, these reactions are linked into chemical pathways, each step with its own enzyme. How does the cell coordinate its various chemical pathways? In our specific example of sugar management, how does the cell match fuel supply to demand by regulating its opposing pathways of sugar consumption and storage? The key is the ability of many biological processes to self-regulate by a mechanism called feedback.

In feedback regulation, the output, or product, of a process regulates that very process. In life, the most common form of regulation is **negative feedback**, in which accumulation of an end product of a process slows that process **(Figure 1.11)**. For example, the cell's breakdown of sugar generates chemical energy in the form of a substance called ATP. An excess accumulation of ATP "feeds back" and inhibits an enzyme near the beginning of the pathway.

Though less common than negative feedback, there are also many biological processes regulated by **positive feedback**, in which an end product *speeds up* its production. The clotting of your blood in response to injury is an example. When a blood vessel is damaged, structures in the blood called platelets begin to aggregate at the site. Positive feedback occurs as chemicals released by the platelets attract *more* platelets. The platelet pile then initiates a complex process that seals the

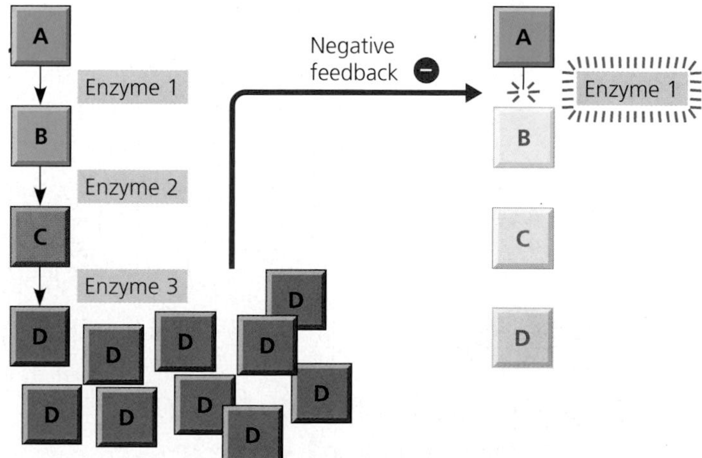

▲ **Figure 1.11 Negative feedback.** This three-step chemical pathway converts substance A to substance D. A specific enzyme catalyzes each chemical reaction. Accumulation of the final product (D) inhibits the first enzyme in the sequence, thus slowing down production of more D.

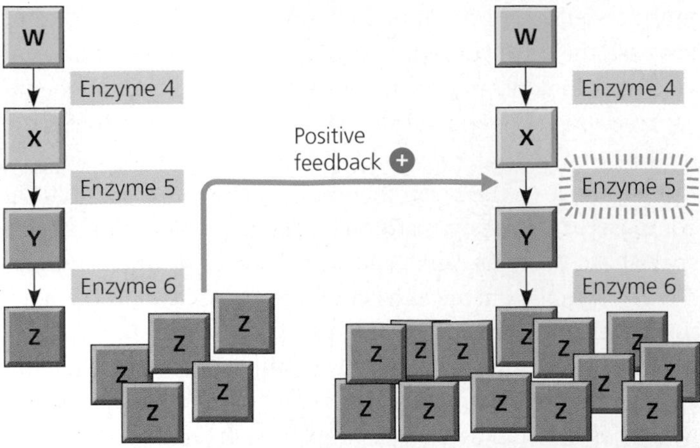

▲ **Figure 1.12 Positive feedback.** In positive feedback, a product stimulates an enzyme in the reaction sequence, increasing the rate of production of the product. Positive feedback is less common than negative feedback in living systems.

wound with a clot. **Figure 1.12** shows a simple model of positive feedback.

Feedback is a regulatory motif common to life at all levels, from the molecular level to the biosphere. Such regulation is an example of the integration that makes living systems much greater than the sum of their parts.

Concept Check 1.2

1. Apply the principle of emergent properties to explain the relationship of a sentence to the alphabet of letters from which that sentence is constructed.
2. How does high-throughput technology complement bioinformatics?
3. When you flush a toilet, water begins to fill the tank and lift a float attached to a lever. When the water level reaches a certain height, the lever shuts the water valve and prevents the tank from overflowing. What type of regulatory mechanism is at work in this nonliving system?

For suggested answers, see Appendix A.

Concept 1.3

Biologists explore life across its great diversity of species

We can think of biology's enormous scope as having two dimensions. The "vertical" dimension, which we examined in this chapter's first two concepts, is the size scale that reaches all the way from molecules to the biosphere. But biology's scope also has a "horizontal" dimension stretching across the great diversity of species, now and over life's long history.

▲ **Figure 1.13 Drawers of diversity.** This is just a small sample of the tens of thousands of species in the moth and butterfly collection at the National Museum of Natural History in Washington, D.C.

Diversity is a hallmark of life. Biologists have so far identified and named about 1.8 million species. This enormous diversity of life includes approximately 5,200 known species of prokaryotes, 100,000 fungi, 290,000 plants, 52,000 vertebrates (animals with backbones), and 1,000,000 insects (more than half of all known forms of life). Researchers identify thousands of additional species each year. Estimates of the total species count range from about 10 million to over 200 million. Whatever the actual number, the vast variety of life makes biology's scope very wide **(Figure 1.13)**.

Grouping Species: The Basic Idea

There seems to be a human tendency to group diverse items according to similarities. For instance, perhaps you organize your music collection according to artist. And then maybe you group the various artists into broader categories, such as dance music, party music, exercise music, and study-time music. In the same way, grouping species that are similar is natural for us. We may speak of squirrels and butterflies, though we recognize that many different species belong to each group. We may even sort groups into broader categories, such as rodents (which include squirrels) and insects (which include butterflies). Taxonomy, the branch of biology that names and classifies species, formalizes this ordering of species into a series of groups of increasing breadth **(Figure 1.14)**. You will learn more about this taxonomic scheme in Chapter 25. For now, we will focus on kingdoms and domains, the broadest units of classification.

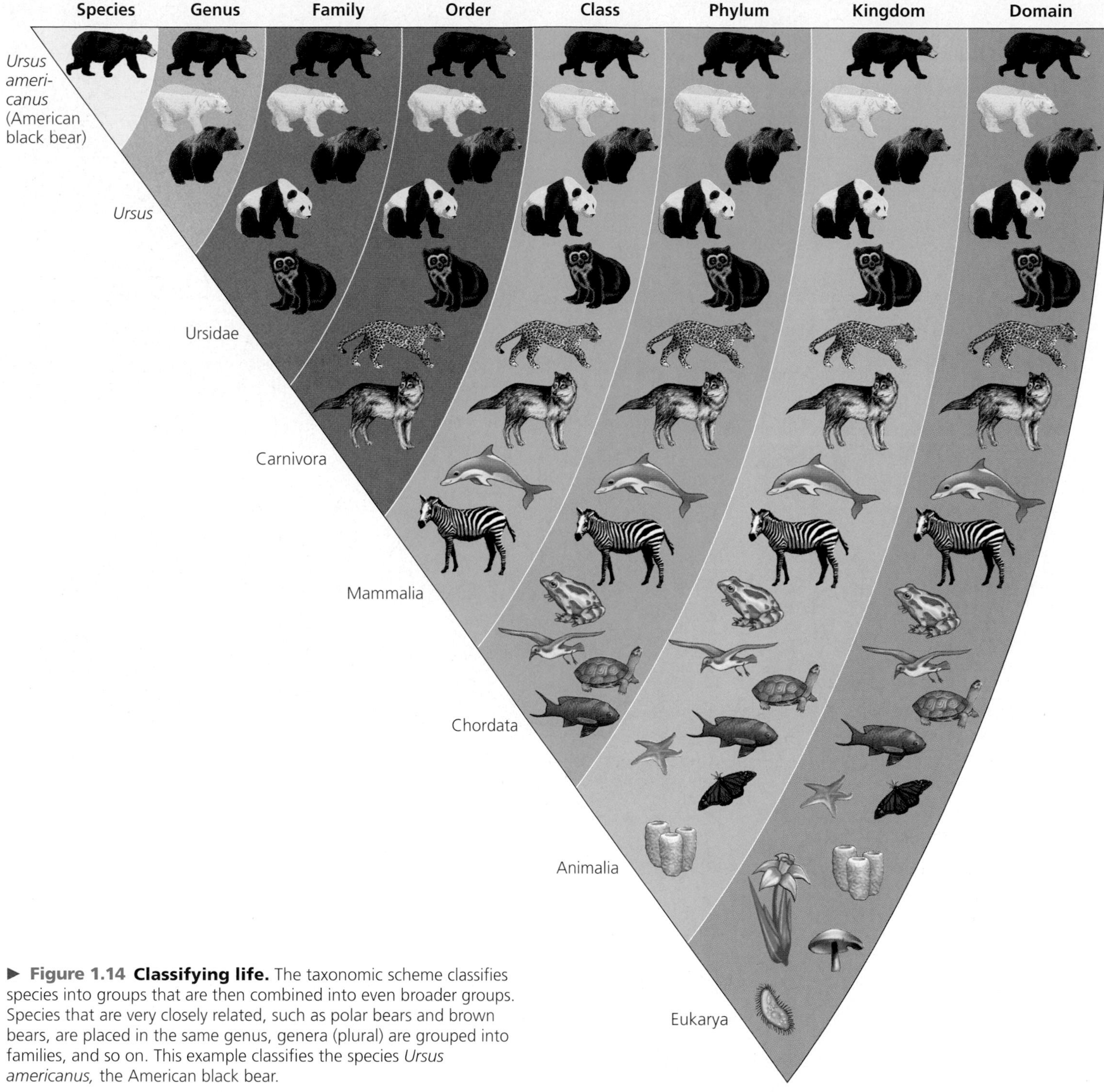

► **Figure 1.14 Classifying life.** The taxonomic scheme classifies species into groups that are then combined into even broader groups. Species that are very closely related, such as polar bears and brown bears, are placed in the same genus, genera (plural) are grouped into families, and so on. This example classifies the species *Ursus americanus,* the American black bear.

The Three Domains of Life

Until the last decade, most biologists adopted a taxonomic scheme that divided the diversity of life into five kingdoms, including the plant and animal kingdoms. But new methods, such as comparing the DNA sequences of diverse species, have led to an ongoing reevaluation of the number and boundaries of kingdoms. Different researchers have proposed anywhere from six kingdoms to dozens of kingdoms. But as debate continues at the kingdom level, there is more of a consensus that the kingdoms of life can now be grouped into three even higher levels of classification called domains. The three domains are named Bacteria, Archaea, and Eukarya.

The first two domains, **domain Bacteria** and **domain Archaea,** both consist of prokaryotes (organisms with prokaryotic cells). Most prokaryotes are unicellular and microscopic. In the five-kingdom system, bacteria and archaea were combined in a single kingdom, called kingdom Monera, because they shared the prokaryotic form of cell structure. But evidence now supports the view that bacteria and archaea represent two very

Figure 1.15

Exploring Life's Three Domains

DOMAIN BACTERIA

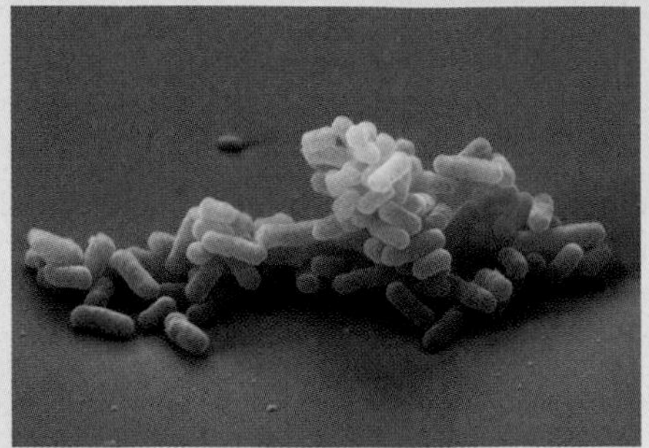

4 μm

Bacteria are the most diverse and widespread prokaryotes and are now divided among multiple kingdoms. Each of the rod-shaped structures in this photo is a bacterial cell.

DOMAIN ARCHAEA

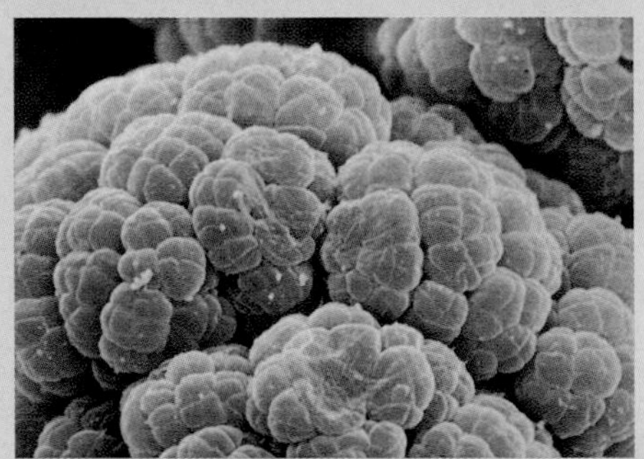

0.5 μm

Many of the prokaryotes known as **archaea** live in Earth's extreme environments, such as salty lakes and boiling hot springs. Domain Archaea includes multiple kingdoms. The photo shows a colony composed of many cells.

DOMAIN EUKARYA

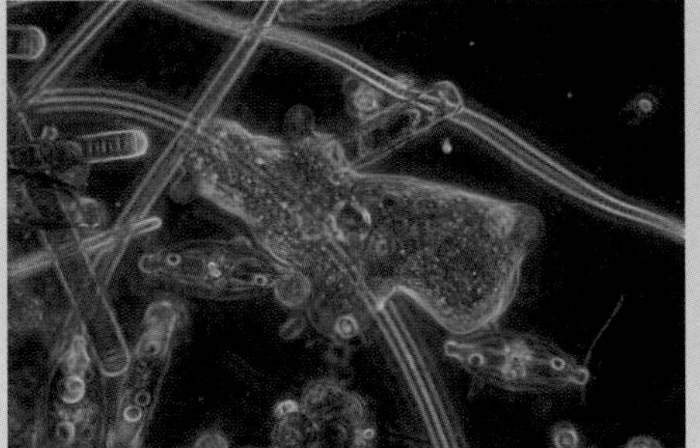

100 μm

Protists (multiple kingdoms) are unicellular eukaryotes and their relatively simple multicellular relatives. Pictured here is an assortment of protists inhabiting pond water. Scientists are currently debating how to split the protists into several kingdoms that better represent evolution and diversity.

Kingdom Plantae consists of multicellular eukaryotes that carry out photosynthesis, the conversion of light energy to food.

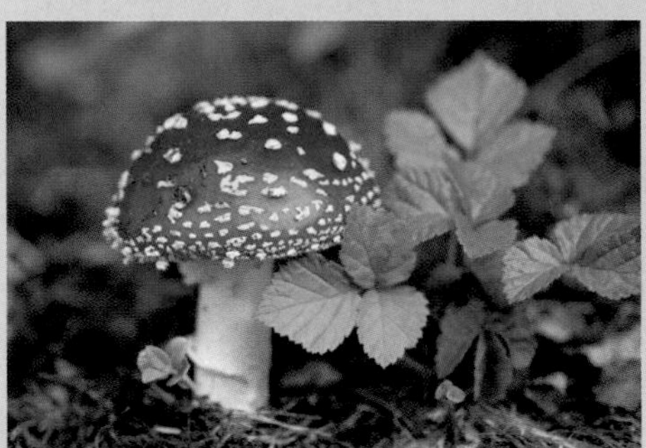

Kindom Fungi is defined in part by the nutritional mode of its members, such as this mushroom, which absorb nutrients after decomposing organic material.

Kindom Animalia consists of multicellular eukaryotes that ingest other organisms.

distinct branches of prokaryotic life, different in key ways that you'll learn about in Chapter 27. There is also molecular evidence that archaea are at least as closely related to eukaryotic organisms as they are to bacteria.

All the eukaryotes (organisms with eukaryotic cells) are now grouped into the various kingdoms of **domain Eukarya** **(Figure 1.15)**. In the era of the five-kingdom scheme, most of the single-celled eukaryotes, including the microorganisms known as protozoans, were placed in a single kingdom, the kingdom Protista. Many biologists extended the boundaries of the kingdom Protista to include some multicellular forms, such as seaweeds, that are closely related to certain unicellular protists. The recent taxonomic trend has been to split the protists into several kingdoms. In addition to these protistan kingdoms, the domain Eukarya includes three kingdoms of multicellular eukaryotes: the kingdoms Plantae, Fungi, and Animalia. These three kingdoms are distinguished partly by their modes of nutrition. Plants produce their own sugars and other foods by photosynthesis. Fungi are mostly decomposers that absorb nutrients by breaking down dead organisms and organic wastes, such as leaf litter and animal feces. Animals obtain food by ingestion, which is the eating and digesting of other organisms. It is, of course, the kingdom to which we belong.

Unity in the Diversity of Life

As diverse as life is, there is also evidence of remarkable unity, especially at the molecular and cellular levels. An example is the universal genetic language of DNA, which is common to organisms as different as bacteria and animals. And among eukaryotes, unity is evident in many features of cell structure **(Figure 1.16)**.

How can we account for life's dual nature of unity and diversity? The process of evolution, introduced in the next concept, illuminates both the similarities and differences among Earth's life.

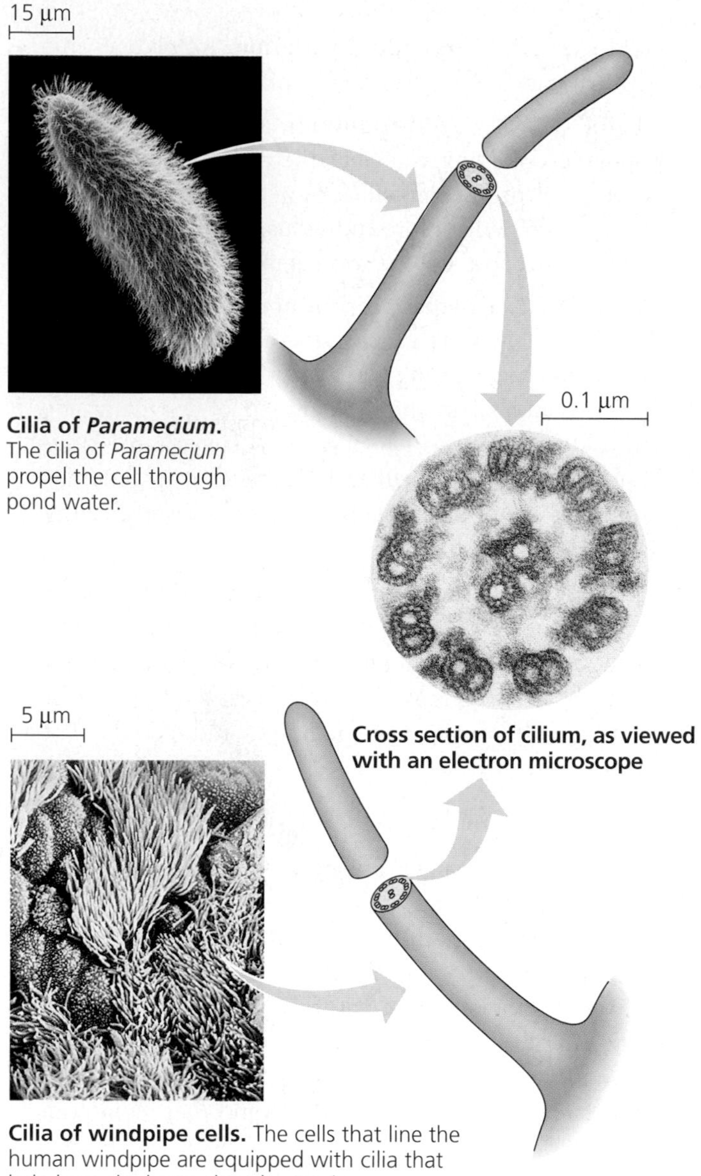

15 µm

Cilia of *Paramecium*. The cilia of *Paramecium* propel the cell through pond water.

0.1 µm

Cross section of cilium, as viewed with an electron microscope

5 µm

Cilia of windpipe cells. The cells that line the human windpipe are equipped with cilia that help keep the lungs clean by moving a film of debris-trapping mucus upward.

▲ **Figure 1.16 An example of unity underlying the diversity of life: the architecture of cilia in eukaryotes.** Cilia (singular, cilium) are extensions of cells that function in locomotion. They occur in eukaryotes as diverse as the single-celled *Paramecium* and humans. But even organisms so different share a common architecture for their cilia, which have an elaborate system of tubules that is revealed in cross-sectional views.

Concept Check 1.3

1. How is a mailing address analogous to biology's hierarchical taxonomic system?
2. What is the key difference that distinguishes organisms of domain Eukarya from the other two domains?

For suggested answers, see Appendix A.

▲ **Figure 1.17 Digging into the past.** Paleontologist Paul Sereno gingerly excavates the leg bones of a dinosaur fossil in Niger, Africa.

Concept 1.4

Evolution accounts for life's unity and diversity

The history of life, as documented by fossils and other evidence, is a saga of a changing Earth billions of years old, inhabited by an evolving cast of living forms **(Figure 1.17)**. This evolutionary view of life came into sharp focus in November 1859, when Charles Robert Darwin published one of the most important and controversial books ever written. Entitled *On the Origin of Species by Natural Selection,* Darwin's book was an immediate bestseller and soon made "Darwinism" almost synonymous with the concept of evolution **(Figure 1.18)**.

▲ **Figure 1.18 Charles Darwin in 1859, the year he published *The Origin of Species.***

The Origin of Species articulated two main points. First, Darwin presented evidence to support his view that contemporary species arose from

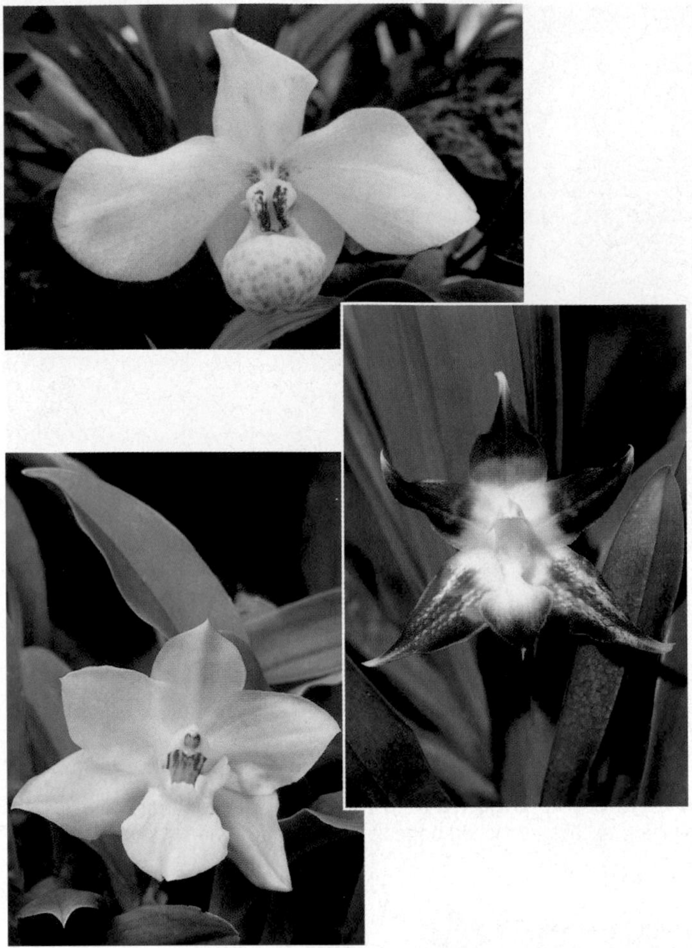

▲ **Figure 1.19 Unity and diversity in the orchid family.** These three rain forest orchids are variations on a common floral theme. For example, each of these flowers has a liplike petal that helps attract pollinating insects and provides a landing platform for the pollinators.

a succession of ancestors. (We will discuss the evidence for evolution in detail in Chapter 22.) Darwin called this evolutionary history of species "descent with modification." It was an insightful phrase, as it captured the duality of life's unity and diversity—unity in the kinship among species that descended from common ancestors; diversity in the modifications that evolved as species branched from their common ancestors **(Figure 1.19)**. Darwin's second main point was to propose a mechanism for descent with modification. He called this evolutionary mechanism natural selection.

Natural Selection

Darwin synthesized his theory of natural selection from observations that by themselves were neither new nor profound. Others had the pieces of the puzzle, but Darwin saw how they fit together. He inferred natural selection by connecting two readily observable features of life:

OBSERVATION: **Individual variation.** Individuals in a population of any species vary in many heritable traits.

OBSERVATION: **Overproduction and competition.** A population of any species has the potential to produce far more offspring than will survive to produce offspring of their own. With more individuals than the environment can support, competition is inevitable.

INFERENCE: **Unequal reproductive success.** From the observable facts of heritable variation and overproduction of offspring, Darwin inferred that individuals are unequal in their likelihood of surviving and reproducing. Those individuals with heritable traits best suited to the local environment will generally produce a disproportionately large number of healthy, fertile offspring.

INFERENCE: **Evolutionary adaptation.** This unequal reproductive success can adapt a population to its environment. Over the generations, heritable traits that enhance survival and reproductive success tend to increase in frequency among a population's individuals. The population evolves.

Darwin called this mechanism of evolutionary adaptation "natural selection" because the natural environment "selects" for the propagation of certain traits. **Figure 1.20** summarizes Darwin's theory of natural selection. The example in **Figure 1.21** illustrates the ability of natural selection to "edit" a population's heritable variations. We see the products of natural selection in the exquisite adaptations of organisms to the special circumstances of their way of life and their environment **(Figure 1.22)**.

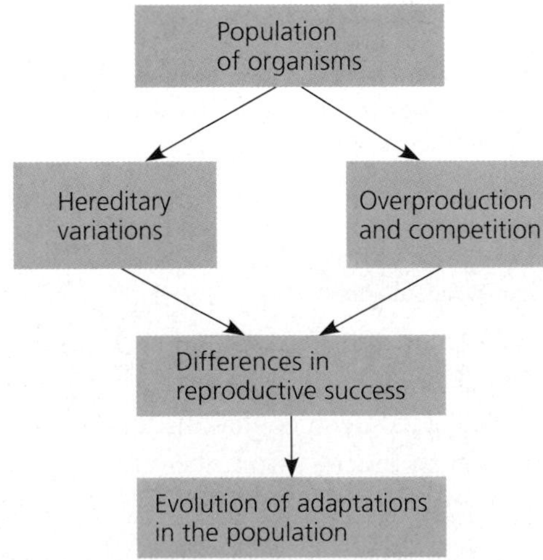

▲ **Figure 1.20 Summary of natural selection.**

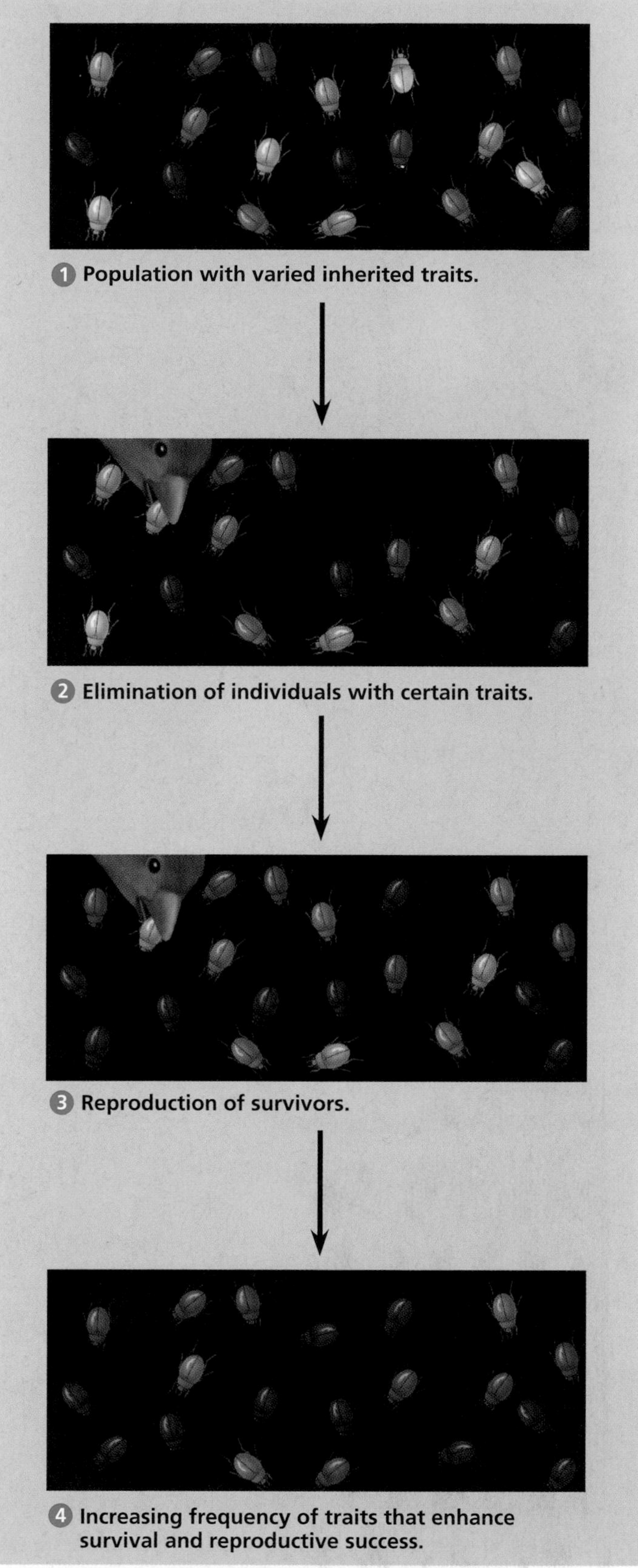

1 Population with varied inherited traits.
2 Elimination of individuals with certain traits.
3 Reproduction of survivors.
4 Increasing frequency of traits that enhance survival and reproductive success.

▲ **Figure 1.21 Natural selection.** This imaginary beetle population has colonized a locale where the soil has been blackened by a recent brush fire. Initially, the population varies extensively in the inherited coloration of the individuals, from very light gray to charcoal. For hungry birds that prey on the beetles, it is easiest to spot the beetles that are lightest in color.

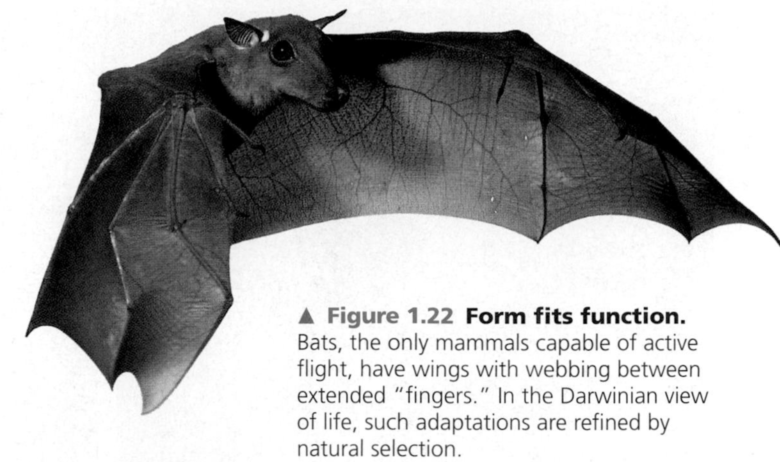

▲ **Figure 1.22 Form fits function.** Bats, the only mammals capable of active flight, have wings with webbing between extended "fingers." In the Darwinian view of life, such adaptations are refined by natural selection.

The Tree of Life

Take another look at the skeletal architecture of the bat's wings in Figure 1.22. These forelimbs, though adapted for flight, actually have all the same bones, joints, nerves, and blood vessels found in other limbs as diverse as the human arm, the horse's foreleg, and the whale's flipper. Indeed, all mammalian forelimbs are anatomical variations of a common architecture, much as the flowers in Figure 1.19 are variations on an underlying "orchid" theme. Such examples of kinship connect life's "unity in diversity" to the Darwinian concept of "descent with modification." In this view, the unity of mammalian limb anatomy reflects inheritance of that structure from a common ancestor—the "prototype" mammal from which all other mammals descended, their diverse forelimbs modified by natural selection operating over millions of generations in different environmental contexts. Fossils and other evidence corroborate anatomical unity in supporting this view of mammalian descent from a common ancestor.

Thus, Darwin proposed that natural selection, by its cumulative effects over vast spans of time, could enable an ancestral species to "split" into two or more descendant species. This would occur, for example, if one population fragmented into several subpopulations isolated in different environments. In these various arenas of natural selection, one species could gradually radiate into many species as the geographically isolated populations adapted over many generations to different sets of environmental factors.

The "family tree" of 14 finches in **Figure 1.23**, on the next page, illustrates a famous example of adaptive radiation of new species from a common ancestor. Darwin collected specimens of these birds during his 1835 visit to the remote Galápagos Islands, 900 kilometers (km) off the Pacific coast of South America. These relatively young, volcanic islands are home to many species of plants and animals found nowhere else in the world, though Galápagos organisms are clearly related to species on the South American mainland. After volcanism built the Galápagos several million years ago, finches probably diversified on the various islands from an ancestral finch

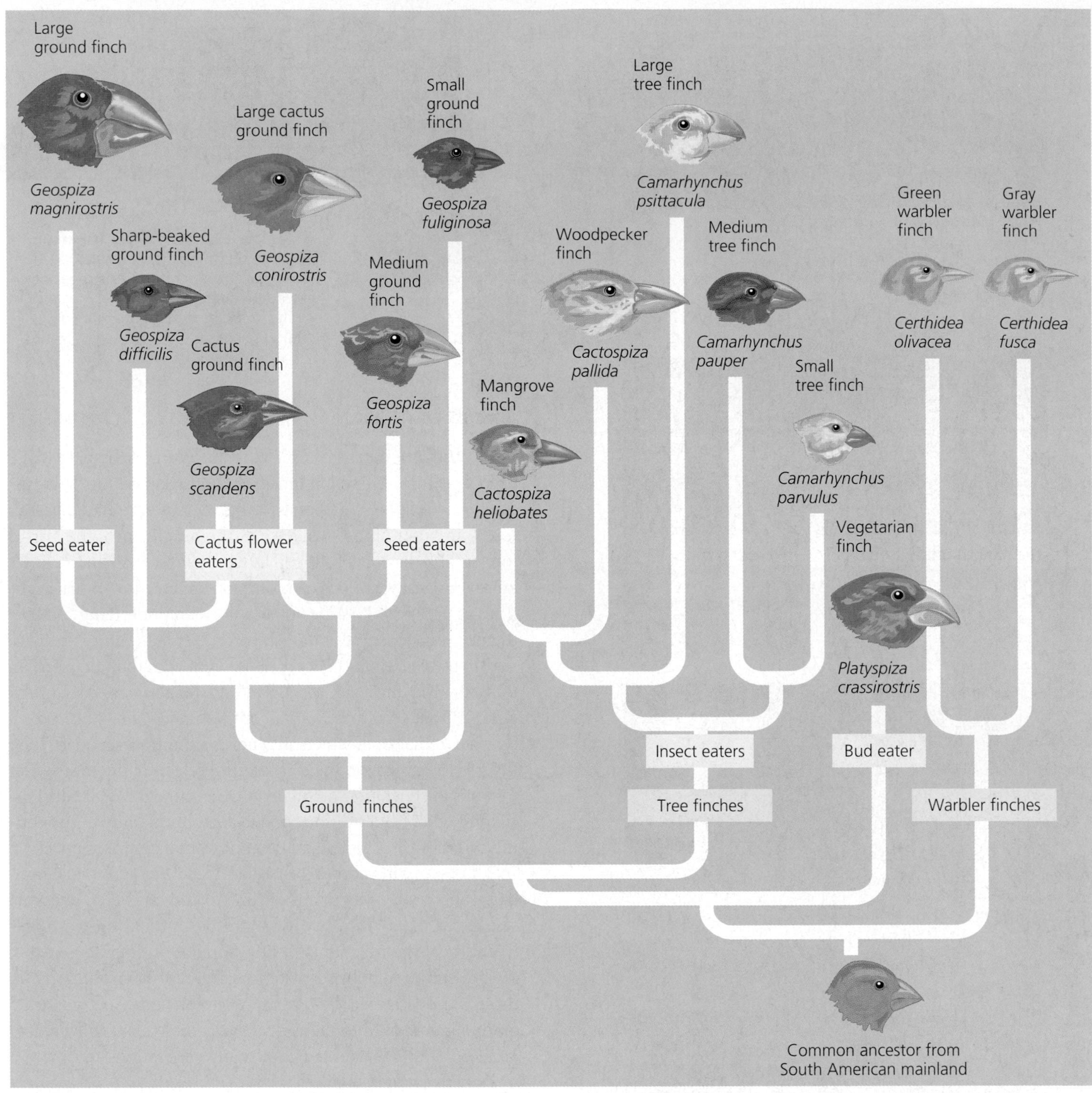

▲ **Figure 1.23 Descent with modification: adaptive radiation of finches on the Galápagos Islands.** Note the specialization of beaks, which are adapted to various food sources on the different islands.

species that by chance reached the archipelago from the mainland. Years after Darwin's collection of Galápagos finches, researchers began to sort out the relationships among the finch species, first from anatomical and geographic data and more recently with the help of DNA sequence comparisons.

Biologists' diagrams of evolutionary relationships generally take treelike forms, and for good reason. Just as an individual has a genealogy that can be diagrammed as a family tree, each species is one twig of a branching tree of life extending back in time through ancestral species more and more remote. Species that are very similar, such as the Galápagos finches, share a common ancestor at a relatively recent branch point on the tree of life. But through an ancestor that lived much farther back in time, finches are related to sparrows, hawks, penguins, and all other birds. And birds, mammals, and all other vertebrates (animals with backbones) share a common ancestor even more ancient. We find evidence of still broader relationships in such similarities as the matching machinery of all eukaryotic cilia (see Figure 1.16). Trace life back far enough, and there are only fossils of the primeval prokaryotes that inhabited Earth over 3.5 billion years ago. We can recognize their vestiges in our own cells—in the universal genetic code, for example. All of life is connected through its long evolutionary history.

Concept Check 1.4

1. Explain why "editing" is better than "creating" as a metaphor for how natural selection acts on a population's heritable variation.
2. The three domains you learned about in Concept 1.3 can be represented in the tree of life as the three main branches. On the eukaryotic branch, three of the subbranches are the kingdoms Plantae, Fungi, and Animalia. Evidence supports the view that fungi and animals are more closely related to each other than either of these kingdoms is to plants. Draw a simple branching pattern that symbolizes the relationship between these three eukaryotic kingdoms.

For suggested answers, see Appendix A.

Concept 1.5

Biologists use various forms of inquiry to explore life

The word *science* is derived from a Latin verb meaning "to know." Science is a way of knowing. It developed out of our curiosity about ourselves, other life-forms, the world, and the universe. Striving to understand seems to be one of our basic urges.

At the heart of science is **inquiry,** a search for information and explanation, often focusing on specific questions. Inquiry drove Darwin to seek answers in nature for how species adapt to their environments. And inquiry is driving the analyses of genomes that are helping us understand biological unity and diversity at the molecular level. In fact, the inquisitive mind is the engine that drives all progress in biology.

There is no formula for successful scientific inquiry, no single scientific method with a rule book that researchers must rigidly follow. As in all quests, science includes elements of challenge, adventure, and surprise, along with careful planning, reasoning, creativity, cooperation, competition, patience, and the persistence to overcome setbacks. Such diverse elements of inquiry make science far less structured than most people realize. That said, it is possible to distill certain characteristics that help to distinguish science from other ways of describing and explaining nature.

Biology blends two main processes of scientific inquiry: discovery science and hypothesis-based science. Discovery science is mostly about *describing* nature. Hypothesis-based science is mostly about *explaining* nature. Most scientific inquiries combine these two research approaches.

Discovery Science

Sometimes called descriptive science, **discovery science** describes natural structures and processes as accurately as possible through careful observation and analysis of data. For example, discovery science gradually built our understanding of cell structure, and it is discovery science that is expanding our databases of genomes of diverse species.

Types of Data

Observation is the use of the senses to gather information, either directly or indirectly with the help of tools such as microscopes that extend our senses. Recorded observations are called **data.** Put another way, data are items of information on which scientific inquiry is based.

The term *data* implies numbers to many people. But some data are *qualitative,* often in the form of recorded descriptions rather than numerical measurements. For example, Jane Goodall spent decades recording her observations of chimpanzee behavior during field research in a Gambian jungle **(Figure 1.24)**. She also documented her observations with

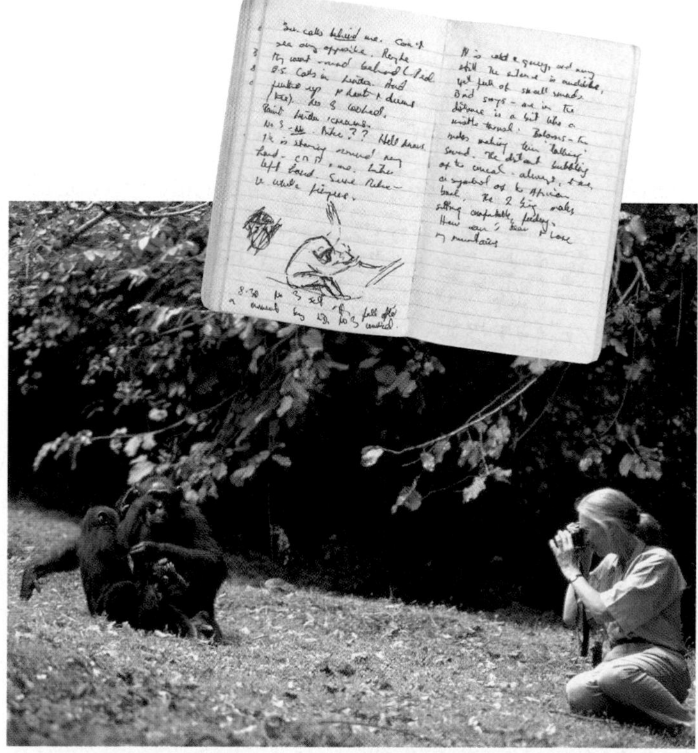

▲ **Figure 1.24 Jane Goodall collecting qualitative data on chimpanzee behavior.** Goodall recorded her observations in field notebooks, often with sketches of the animals' behavior.

photographs and movies. Along with these qualitative data, Goodall also enriched the field of animal behavior with volumes of *quantitative* data, which are generally recorded as measurements. Skim through any of the scientific journals in your college library, and you'll see many examples of quantitative data organized into tables and graphs.

Induction in Discovery Science

Discovery science can lead to important conclusions based on a type of logic called induction, or **inductive reasoning.** Through induction, we derive generalizations based on a large number of specific observations. "The sun always rises in the east" is an example. And so is "All organisms are made of cells." That generalization, part of the so-called cell theory, was based on two centuries of biologists discovering cells in the diverse biological specimens they observed with microscopes. The careful observations and data analyses of discovery science, along with the inductive generalizations they sometimes produce, are fundamental to our understanding of nature.

Hypothesis-Based Science

The observations and inductions of discovery science engage inquisitive minds to seek natural causes and explanations for those observations. What *caused* the diversification of finches on the Galápagos Islands? What *causes* the roots of a plant seedling to grow downward and the leaf-bearing shoot to grow upward? What *explains* the generalization that the sun always rises in the east? In science, such inquiry usually involves the proposing and testing of hypothetical explanations, or hypotheses.

The Role of Hypotheses in Inquiry

In science, a **hypothesis** is a tentative answer to a well-framed question—an explanation on trial. It is usually an educated postulate, based on past experience and the available data of discovery science. A scientific hypothesis makes predictions that can be tested by recording additional observations or by designing experiments.

We all use hypotheses in solving everyday problems. Let's say, for example, that your flashlight fails during a camp-out. That's an observation. The question is obvious: Why doesn't the flashlight work? Two reasonable hypotheses based on past experience are that (1) the batteries in the flashlight are dead or (2) the bulb is burnt out. Each of these alternative hypotheses makes predictions you can test with experiments. For example, the dead-battery hypothesis predicts that replacing the batteries will fix the problem. **Figure 1.25** diagrams this campground inquiry. Of course, we rarely dissect our thought processes this way when we are solving a problem using hypotheses, predictions, and experiments. But hypothesis-based science clearly has its origins in the human tendency to figure things out by tinkering.

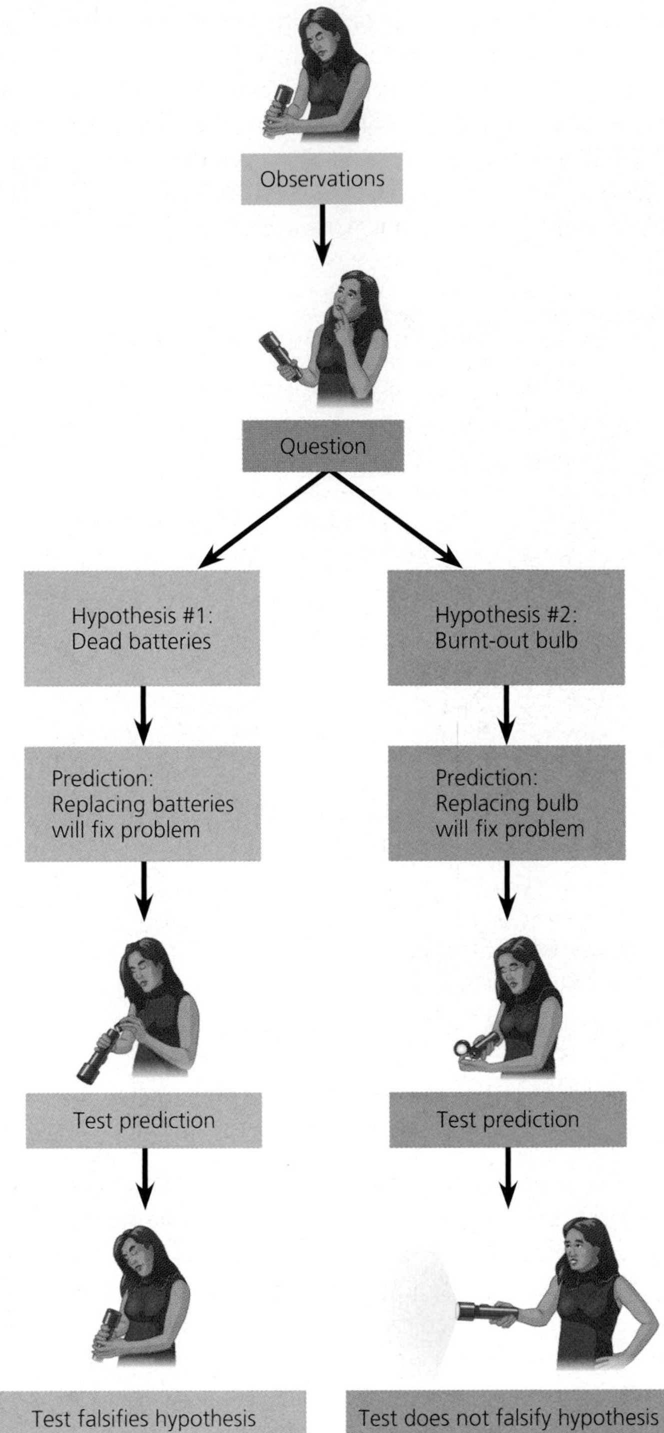

▲ **Figure 1.25 A campground example of hypothesis-based inquiry.**

Deduction: The "If . . . then" Logic of Hypothesis-Based Science

A type of logic called deduction is built into hypothesis-based science. Deduction contrasts with induction, which, remember, is reasoning from a set of specific observations to reach a general conclusion. In **deductive reasoning,** the logic flows in the opposite direction, from the general to the specific. From general premises, we extrapolate to the specific results we should expect if the premises are true. If all organisms are made of cells (premise 1), and humans are organisms (premise 2), then humans are composed of cells (deductive prediction about a specific case).

In hypothesis-based science, deduction usually takes the form of predictions about what outcomes of experiments or observations we should expect if a particular hypothesis (premise) is correct. We then test the hypothesis by performing the experiment to see whether or not the results are as predicted. This deductive testing takes the form of "*If . . . then*" logic. In the case of the flashlight example: *If* the dead-battery hypothesis is correct, and you replace the batteries with new ones, *then* the flashlight should work.

A Closer Look at Hypotheses in Scientific Inquiry

The flashlight example illustrates two important qualities of scientific hypotheses. First, a hypothesis must be *testable;* there must be some way to check the validity of the idea. Second, a hypothesis must be *falsifiable;* there must be some observation or experiment that *could* reveal if such an idea is actually *not* true. The hypothesis that dead batteries are the sole cause of the broken flashlight could be falsified by replacing the old batteries with new ones. But try to devise a test to falsify the hypothesis that invisible campground ghosts are fooling with your flashlight. Does restoring flashlight function by replacing the bulb falsify the ghost hypothesis? Not if the playful ghosts are continuing their mischief.

The flashlight inquiry illustrates another key point about hypothesis-based science. The ideal is to frame two or more alternative hypotheses and design experiments to falsify those candidate explanations. In addition to the two explanations tested in Figure 1.25, one of the many additional hypotheses is that *both* the batteries *and* the bulb are bad. What does this hypothesis predict about the outcome of the experiments in Figure 1.25? What additional experiment would you design to test this hypothesis of multiple malfunction?

We can mine the flashlight scenario for still one more important lesson about hypothesis-based science. Although the burnt-out bulb hypothesis stands up as the most likely explanation, notice that the testing supports that hypothesis *not* by proving that it is correct, but by not eliminating it through falsification. Perhaps the bulb was simply loose and the new bulb was inserted correctly. We could attempt to falsify the burnt-out bulb hypothesis by trying another experiment—removing the bulb and carefully reinstalling it. But no amount of experimental testing can *prove* a hypothesis beyond a shadow of doubt, because it is impossible to exhaust the testing of all alternative hypotheses. A hypothesis gains credibility by surviving various attempts to falsify it while testing eliminates (falsifies) alternative hypotheses.

The Myth of the Scientific Method

The steps in the flashlight example of Figure 1.25 trace an idealized process of inquiry called *the scientific method.* We can recognize the elements of this process in most of the research articles published by scientists, but rarely in such structured form. Very few scientific inquiries adhere rigidly to the sequence of steps prescribed by the "textbook" scientific method. For example, a scientist may start to design an experiment, but then backtrack upon realizing that more observations are necessary. In other cases, puzzling observations simply don't prompt well-defined questions until other research projects place those observations in a new context. For example, Darwin collected specimens of the Galápagos finches, but it wasn't until years later, as the idea of natural selection began to gel, that biologists began asking key questions about the history of those birds.

Moreover, scientists sometimes redirect their research when they realize they have been "barking up the wrong tree" by asking the wrong question. For example, in the early 20th century, much research on schizophrenia and manic-depressive disorder (now called bipolar disorder) got sidetracked by focusing too much on the question of how life experiences cause these serious maladies. Research on the causes and potential treatments became more productive when it was refocused on questions of how certain chemical imbalances in the brain contribute to mental illness. To be fair, we acknowledge that such twists and turns in scientific inquiry become more evident with the advantage of historical perspective.

There is still another reason that good science need not conform exactly to any one method of inquiry: Discovery science has contributed much to our understanding of nature without most of the steps of the so-called scientific method.

It is important for you to get some experience with the power of the scientific method—by using it for some of the laboratory inquiries in your biology course, for example. But it is also important to avoid stereotyping science as lock-step adherence to this method.

A Case Study in Scientific Inquiry: Investigating Mimicry in Snake Populations

Now that we have highlighted the key features of discovery science and hypothesis-based science, you should be able to recognize these forms of inquiry in a case study of actual scientific research.

The story begins with a set of observations and generalizations from discovery science. Many poisonous animals are brightly colored, often with distinctive patterns that stand out against the background. This is called warning coloration because it apparently signals "dangerous species" to potential predators. But there are also mimics. These imposters look like poisonous species, but are actually relatively harmless. An example is the flower fly, a nonstinging insect that mimics the appearance of a stinging honeybee (**Figure 1.26** on the next page).

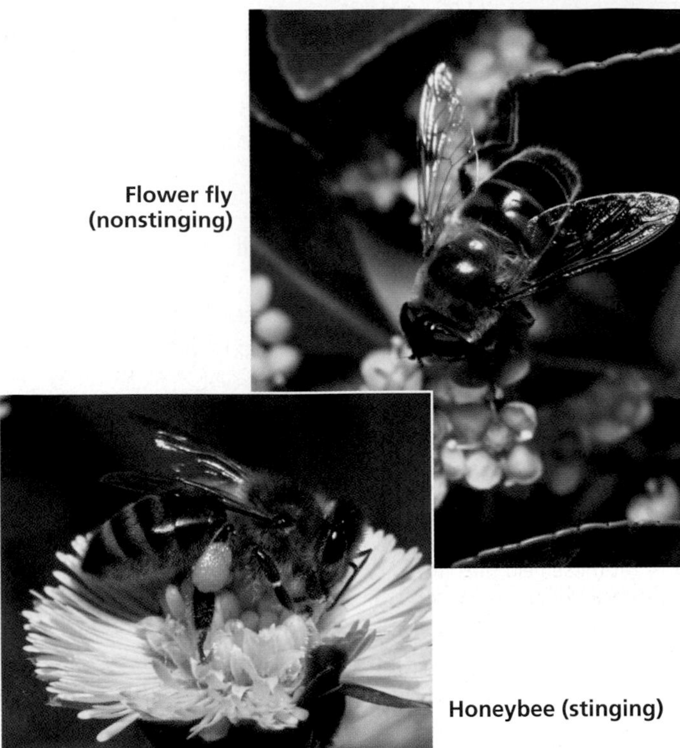

▲ **Figure 1.26 A stinging honeybee and its nonstinging mimic, a flower fly.**

What is the function of such mimicry? What advantage does it confer on the mimics? In 1862, British scientist Henry Bates proposed the reasonable hypothesis that mimics such as flower flies benefit when predators confuse them with the harmful species. In other words, the deception may be an evolutionary adaptation that evolved by reducing the mimic's risk of being eaten. As intuitive as this hypothesis may be, it has been relatively difficult to test, especially with field experiments. But then, in 2001, biologists David and Karin Pfennig, along with William Harcombe, an undergraduate at the University of North Carolina, designed a simple but elegant set of field experiments to test Bates's mimicry hypothesis.

The team investigated a case of mimicry among snakes that live in North and South Carolina. A poisonous snake called the eastern coral snake has warning coloration: bold, alternating rings of red, yellow, and black. Predators rarely attack these snakes. It is unlikely that predators *learn* this avoidance behavior, as a first strike by a coral snake is usually deadly. Natural selection may have increased the frequency of predators that have inherited an instinctive recognition and avoidance of the warning coloration of the coral snake.

A nonpoisonous snake named the scarlet king snake mimics the ringed coloration of the coral snake. Both king snakes and coral snakes live in the Carolinas, but the king snakes' geographic range extends farther north and west into regions where no coral snakes are found **(Figure 1.27)**.

The geographic distribution of the Carolina snakes made it possible to test the key prediction of the mimicry hypothesis. Mimicry should help protect king snakes from predators, but *only* in regions where coral snakes also live. The mimicry hypothesis predicts that predators in non-coral snake areas will attack king snakes more frequently than will predators that live where coral snakes are present.

Field Experiments with Artificial Snakes

To test the mimicry hypothesis, Harcombe made hundreds of artificial snakes out of wire covered with a claylike substance called plasticine. He fashioned two versions of fake snakes: an *experimental group* with the red, black, and yellow ring pattern of king snakes; and a *control group* of plain brown artificial snakes as a basis of comparison.

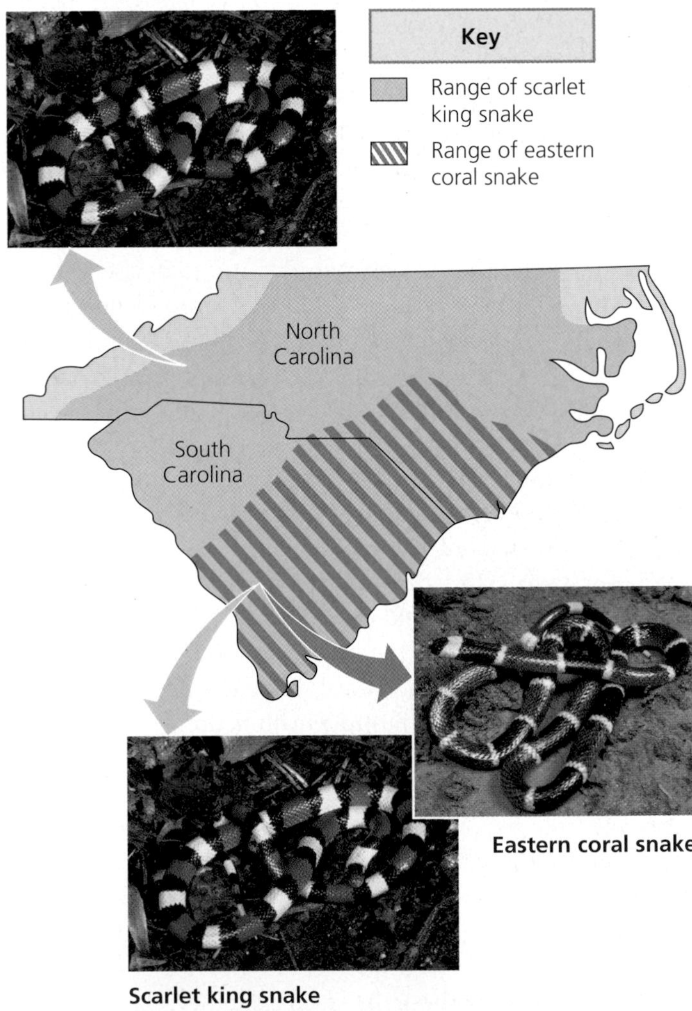

▲ **Figure 1.27 Geographic ranges of Carolina coral snakes and king snakes.** The scarlet king snake (*Lampropeltis triangulum*) mimics the warning coloration of the poisonous eastern coral snake (*Micrurus fulvius*). Though these two snake species cohabit many regions throughout North and South Carolina, the geographic range of the king snake extends north and west of the range of the coral snake.

The researchers placed equal numbers of the two types of artificial snakes in field sites throughout North and South Carolina, including the region where coral snakes are absent (see Figure 1.27). After four weeks, the scientists retrieved the fake snakes and recorded how many had been attacked by looking for bite or claw marks. The most common predators were foxes, coyotes, and raccoons, but black bears also attacked some of the artificial snakes **(Figure 1.28)**.

The data fit the key prediction of the mimicry hypothesis. Compared to the brown artificial snakes, the ringed snakes were attacked by predators less frequently *only* in field sites within the geographic range of the poisonous coral snakes. **Figure 1.29** summarizes the field experiments. This figure also introduces an illustration format we will use throughout the book to feature other examples of biological inquiry.

Designing Controlled Experiments

The snake mimicry experiment provides an example of how scientists design experiments to test the effect of one variable by canceling out the effects of any unwanted variables, such as the number of predators in this case. The design is called a **controlled experiment**, where an experimental group (the artificial king snakes, in this case) is compared with a control

(a) Artificial king snake

(b) Brown artificial snake that has been attacked

▲ **Figure 1.28 Artificial snakes used in field experiments to test the mimicry hypothesis.** You can see where a bear chomped on the brown artificial snake in (b).

Figure 1.29

Inquiry Does the presence of poisonous coral snakes affect predation rates on their mimics, king snakes?

EXPERIMENT David Pfennig and his colleagues made artificial snakes to test a prediction of the mimicry hypothesis: that king snakes benefit from mimicking the warning coloration of coral snakes *only* in regions where poisonous coral snakes are present. The Xs on the map below are field sites where the researchers placed equal numbers of artificial king snakes (experimental group) and brown artificial snakes (control group). The researchers recovered the artificial snakes after four weeks and tabulated predation data based on teeth and claw marks on the snakes (see Figure 1.28).

RESULTS In field sites where coral snakes were present, predators attacked far fewer artificial king snakes than brown artificial snakes. The warning coloration of the "king snakes" afforded no such protection where coral snakes were absent. In fact, at those field sites, the artificial king snakes were *more* likely to be attacked than the brown artificial snakes, perhaps because the bright pattern is particularly easy to spot against the background.

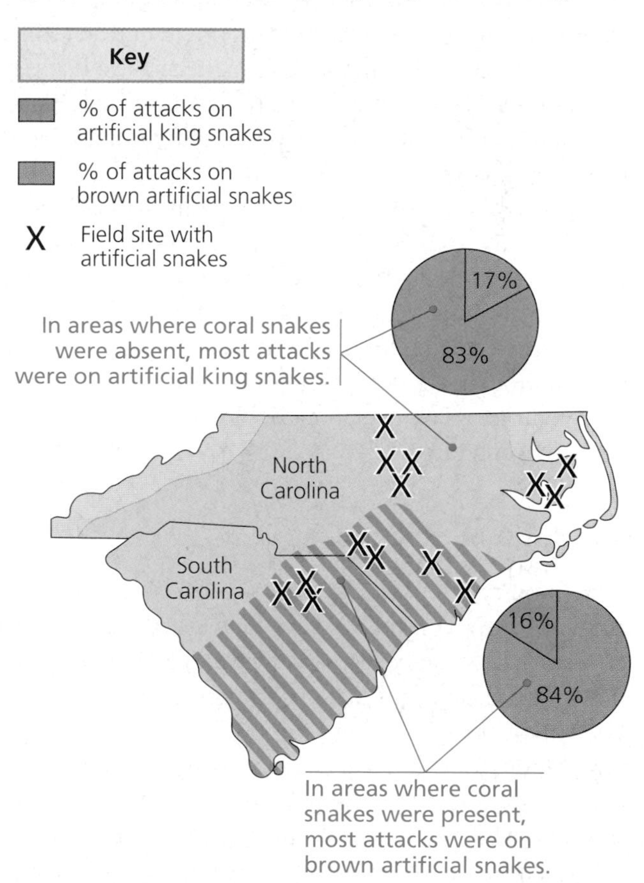

CONCLUSION The field experiments support the mimicry hypothesis by not falsifying the key prediction that imitation of coral snakes is only effective where coral snakes are present. The experiments also tested an alternative hypothesis that predators generally avoid all snakes with brightly colored rings, whether or not poisonous snakes with that coloration live in the environment. That alternative hypothesis was falsified by the data showing that the ringed coloration failed to repel predators where coral snakes were absent.

group (the brown artificial snakes). Ideally, the experimental and control groups differ only in the one factor the experiment is designed to test—in our example, the effect of the snakes' coloration on the behavior of predators.

What if the researchers had failed to control their experiment? Without the brown mock snakes as a control group, the number of attacks on the fake king snakes in different geographic regions would tell us nothing about the effect of snake coloration on predator behavior at the different field sites. Perhaps, for example, fewer predators attacked the artificial king snakes in the eastern and southern field sites simply because fewer predators live there. Or maybe warmer temperatures in those regions make predators less hungry. The brown artificial snakes enabled the scientists to rule out such variables as predator density and temperature because those factors would have had equal effects on the control group and experimental group. Yet predators in the eastern and southern field sites attacked more brown artificial snakes than "king snakes." The clever experimental design left coloration as the only factor that could account for the low predation rate on the artificial king snakes placed within the range of coral snakes. It was not the absolute number of attacks on the artificial king snakes that counted, but the difference between that number and the number of attacks on the brown snakes.

A common misconception is that the term *controlled experiment* means that scientists control the experimental environment to keep everything constant except the one variable being tested. But that's impossible in field research and not realistic even in highly regulated laboratory environments. Researchers usually "control" unwanted variables not by *eliminating* them through environmental regulation, but by *canceling* their effects by using control groups.

Limitations of Science

Scientific inquiry is a powerful way to know nature, but there are limitations to the kinds of questions it can answer. These limits are set by science's requirements that hypotheses be testable and falsifiable and that observations and experimental results be repeatable.

Observations that can't be verified may be interesting or even entertaining, but they cannot count as evidence in scientific inquiry. The headlines of supermarket tabloids would have you believe that humans are occasionally born with the head of a dog and that some of your classmates are extraterrestrials. The unconfirmed eyewitness accounts and the computer-rigged photos are amusing but unconvincing. In science, evidence from observations and experiments is only convincing if it stands up to the criterion of repeatability. The scientists who investigated snake mimicry in the Carolinas obtained similar data when they repeated their experiments with different species of coral snakes and king snakes in Arizona. And *you* should be able to obtain similar results if you were to repeat the snake experiments.

Ultimately, the limitations of science are imposed by its naturalism—its seeking of natural causes for natural phenomena. Science can neither support nor falsify hypotheses that angels, ghosts, or spirits, both benevolent and evil, cause storms, rainbows, illnesses, and cures. Such supernatural explanations are simply outside the bounds of science.

Theories in Science

"It's just a theory!" Our everyday use of the term *theory* often implies an untested speculation. But the term *theory* has a very different meaning in science. What is a scientific theory, and how is it different from a hypothesis or from mere speculation?

First, a scientific **theory** is much broader in scope than a hypothesis. *This* is a hypothesis: "Mimicking poisonous snakes is an adaptation that protects nonpoisonous snakes from predators." But *this* is a theory: "Evolutionary adaptations evolve by natural selection." Darwin's theory of natural selection accounts for an enormous diversity of adaptations, including mimicry.

Second, a theory is general enough to spin off many new, specific hypotheses that can be tested. For example, Peter and Rosemary Grant, of Princeton University, were motivated by the theory of natural selection to test the specific hypothesis that the beaks of the Galápagos finches evolve in response to changes in the types of available food.

And third, compared to any one hypothesis, a theory is generally supported by a much more massive body of evidence. Those theories that become widely adopted in science (such as the theory of natural selection) explain a great diversity of observations and are supported by an accumulation of evidence. In fact, scrutiny of general theories continues through testing of the specific, falsifiable hypotheses they spawn.

In spite of the body of evidence supporting a widely accepted theory, scientists must sometimes modify or even reject theories when new research methods produce results that don't fit. For example, the five-kingdom theory of biological diversity began to erode when new methods for comparing cells and molecules made it possible to test some of the hypothetical relationships between organisms that were based on the theory. If there is "truth" in science, it is conditional, based on the preponderance of available evidence.

Model Building in Science

You may work with many models in your biology course this year. Perhaps you'll model cell division by using pipe cleaners or other objects as chromosomes. Or maybe you'll practice using mathematical models to predict the growth of a bacterial population. Scientists often construct models as less abstract representations of ideas such as theories or

▶ **Figure 1.30 Modeling the pattern of blood flow through the four chambers of a human heart.**

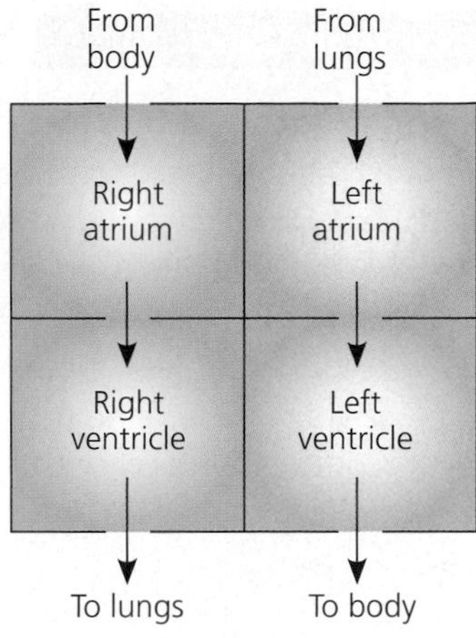

▶ **Figure 1.31 Science as a social process.** In her New York University laboratory, plant biologist Gloria Coruzzi mentors one of her students in the methods of molecular biology.

natural phenomena such as biological processes. Scientific **models** can take many forms, such as diagrams, graphs, three-dimensional objects, computer programs, or mathematical equations.

The choice of a model type depends on how it will be used to help explain and communicate the object, idea, or process it represents. Some models are meant to be as lifelike as possible. Other models are more useful if they are symbolic schematics. For example, the simple diagram in **Figure 1.30** does a good job of modeling blood flow through the chambers of a human heart without looking anything like a real heart. A heart model designed to help train a physician to perform heart surgery would look very different. Whatever its design, the test of a model is how well it fits the available data, how comfortably it accommodates new observations, how accurately it predicts the outcomes of new experiments, and how effectively it clarifies and communicates the idea or process it represents.

The Culture of Science

Movies and cartoons sometimes portray scientists as loners working in isolated labs. In reality, science is an intensely social activity. Most scientists work in teams, which often include both graduate and undergraduate students **(Figure 1.31)**. And to succeed in science, it helps to be a good communicator. Research results have no impact until shared with a community of peers through seminars, publications, and websites.

Both cooperation and competition characterize the scientific culture. Scientists working in the same research field often check one another's claims by attempting to confirm observations or repeat experiments. And when several scientists converge on the same research question, there is all the excitement of a race. Scientists enjoy the challenge of "getting there first" with an important discovery or key experiment.

The biology community is part of society at large, embedded in the cultural milieu of the times. For example, changing attitudes about career choices have increased the proportion of women in biology, which has in turn affected the emphasis in certain research fields. A few decades ago, for instance, biologists who studied the mating behavior of animals focused mostly on competition among males for access to females. More recent research, however, emphasizes the important role that females play in choosing mates. For example, in many bird species, females prefer the bright coloration that "advertises" a male's vigorous health, a behavior that enhances the female's probability of having healthy offspring.

Some philosophers of science argue that scientists are so influenced by cultural and political values that science is no more objective than other ways of "knowing nature." At the other extreme are people who speak of scientific theories as though they were natural laws instead of human interpretations of nature. The reality of science is probably somewhere in between—rarely perfectly objective, but continuously vetted through the expectation that observations and experiments be repeatable and hypotheses be testable and falsifiable.

Science, Technology, and Society

The relationship of science to society becomes clearer when we add technology to the picture. Though science and technology sometimes employ similar inquiry patterns, their basic goals differ. The goal of science is to understand natural phenomena. In contrast, **technology** generally *applies* scientific knowledge for some specific purpose. Biologists and other scientists often speak of "discoveries," while engineers and other technologists more often speak of "inventions." And the beneficiaries of those inventions include scientists, who put new technology to work in their research; the impact of information technology on systems biology is just one example. Thus, science and technology are interdependent.

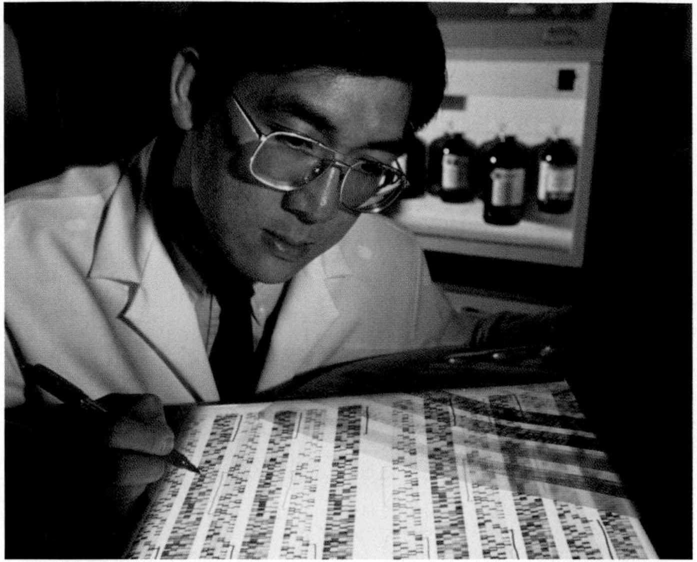

▲ **Figure 1.32 DNA technology and crime scene investigation.** Forensic technicians can use traces of DNA extracted from a blood sample or other body tissue collected at a crime scene to produce molecular fingerprints. The stained bands you see in this photograph represent fragments of DNA, and the pattern of bands varies from person to person.

The potent combination of science and technology has dramatic effects on society. For example, discovery of the structure of DNA by Watson and Crick 50 years ago and subsequent achievements in DNA science led to the many technologies of DNA engineering that are transforming a diversity of fields, including medicine, agriculture, and forensics **(Figure 1.32)**. Perhaps Watson and Crick envisioned that their discovery would someday produce important applications, but it is unlikely that they could have predicted exactly what those applications would be.

The directions that technology takes depend less on the curiosity that drives basic science than on the current needs and wants of people and on the social environment of the times. Debates about technology center more on "*should* we do it" than "*can* we do it." With advances in technology come difficult choices. For example, under what circumstances is it acceptable to use DNA technology to check if people have genes for hereditary diseases? Should such tests always be voluntary, or are there any circumstances when genetic testing should be mandatory? Should insurance companies or employers have access to the information, as they do for many other types of personal health data?

Such ethical issues have as much to do with politics, economics, and cultural values as with science and technology. But scientists and engineers have a responsibility to help educate politicians, bureaucrats, corporate leaders, and voters about how science works and about the potential benefits and risks of specific technologies. The crucial science-technology-society relationship is a theme that increases the significance of any biology course.

Concept Check 1.5

1. Contrast induction with deduction.
2. Critique this statement: "Scientists design controlled experiments to study a single variable by preventing all other factors from changing."
3. In the snake mimicry experiments, why did the researchers place some of their artificial snakes beyond the geographic range of coral snakes?
4. Contrast "theory" with "hypothesis."

For suggested answers, see Appendix A.

Concept 1.6

A set of themes connects the concepts of biology

In some ways, biology is the most demanding of all sciences, partly because living systems are so complex and partly because biology is an interdisciplinary science that requires knowledge of chemistry, physics, and mathematics. Modern biology is the decathlon of natural science. And of all the sciences, biology is the most connected to the humanities and social sciences. As biology students, you are definitely in the right place at the right time!

No matter what brings you to biology, you will find the study of life to be endlessly challenging and uplifting. But this ever-expanding subject can also be a bit intimidating, even to professional biologists. How, then, can beginning students develop a coherent view of life instead of hopelessly trying to memorize the details of a subject that is now far too big to memorize? One approach is to fit the many things you learn into a set of themes that pervade all of biology—ways of thinking about life that will still apply decades from now, when much of the specific information fossilized in any textbook will be obsolete. **Table 1.1** outlines a number of broad themes you will recognize from this first chapter of *Biology*. These unifying themes will reemerge throughout the book to provide touchstones as you explore life and begin asking important questions of your own.

Concept Check 1.6

1. Write a sentence relating the theme of "scientific inquiry" to the theme of "science, technology, and society."

For suggested answers, see Appendix A.

Table 1.1 Eleven Themes that Unify Biology

Theme	Description	Theme	Description
The cell 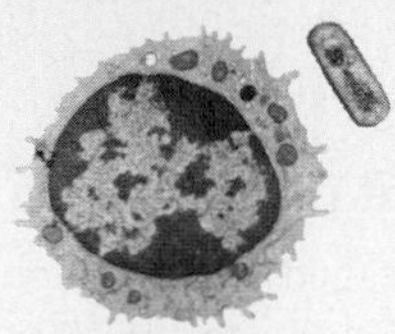	Cells are every organism's basic units of structure and function. The two main types of cells are prokaryotic cells (in bacteria and archaea) and eukaryotic cells (in protists, plants, fungi, and animals).	**Unity and diversity** 	Biologists group the diversity of life into three domains: Bacteria, Archaea, and Eukarya. As diverse as life is, we can also find unity, such as a universal genetic code. The more closely related two species are, the more characteristics they share.
Heritable information 	The continuity of life depends on the inheritance of biological information in the form of DNA molecules. This genetic information is encoded in the nucleotide sequences of the DNA.	**Evolution** 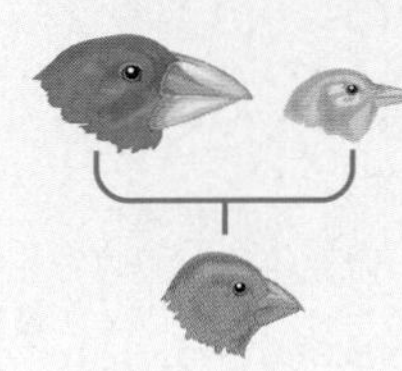	Evolution, biology's core theme, explains both the unity and diversity of life. The Darwinian theory of natural selection accounts for adaptation of populations to their environment through the differential reproductive success of varying individuals.
Emergent properties of biological systems 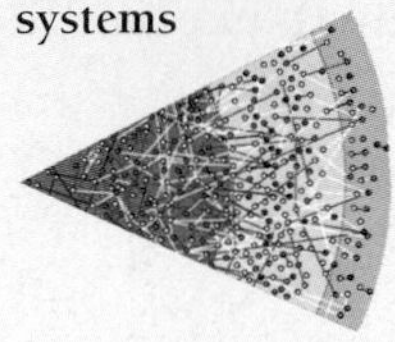	The living world has a hierarchical organization, extending from molecules to the biosphere. With each step upward in level, system properties emerge as a result of interactions among components at the lower levels.	**Structure and function** 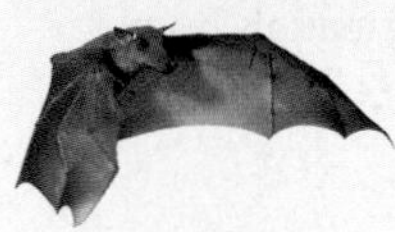	Form and function are correlated at all levels of biological organization.
Regulation 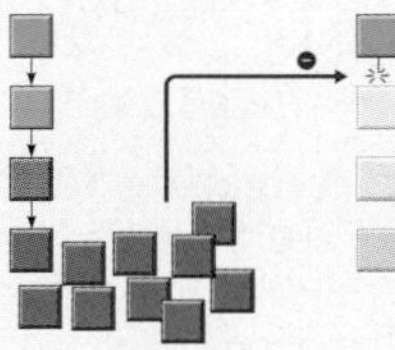	Feedback mechanisms regulate biological systems. In some cases, the regulation maintains a relatively steady state for internal factors such as body temperature.	**Scientific inquiry** 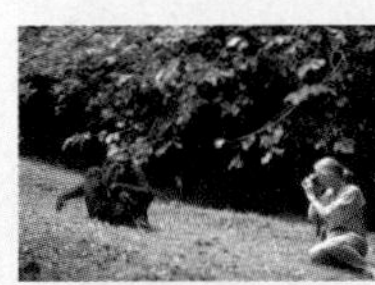	The process of science includes observation-based discovery and the testing of explanations through hypothesis-based inquiry. Scientific credibility depends on the repeatability of observations and experiments.
Interaction with the environment 	Organisms are open systems that exchange materials and energy with their surroundings. An organism's environment includes other organisms as well as nonliving factors.	**Science, technology, and society** 	Many technologies are goal-oriented applications of science. The relationships of science and technology to society are now more crucial to understand than ever before.
Energy and life 	All organisms must perform work, which requires energy. Energy flows from sunlight to producers to consumers.		

Chapter 1 Review

Go to www.campbellbiology.com or the student CD-ROM to explore Activities, Investigations, and other interactive study aids.

SUMMARY OF KEY CONCEPTS

Concept 1.1
Biologists explore life from the microscopic to the global scale

▶ **A Hierarchy of Biological Organization (pp. 3–6)** The hierarchy of life unfolds as follows: biosphere > ecosystem > community > population > organism > organ system > organ > tissue > cell > organelle > molecule > atom.
Activity *The Levels of Life Card Game*

▶ **A Closer Look at Ecosystems (p. 6)** Whereas chemical nutrients recycle within an ecosystem, energy flows through an ecosystem.
Activity *Energy Flow and Chemical Cycling*

▶ **A Closer Look at Cells (pp. 6–8)** The cell is the lowest level of organization that can perform all activities required for life. Cells contain DNA, the substance of genes, which program the cell's production of proteins and transmit information from parents to offspring. Eukaryotic cells contain membrane-enclosed organelles, including a DNA-containing nucleus. Prokaryotic cells lack such organelles.
Activity *Heritable Information: DNA*
Activity *Comparing Prokaryotic and Eukaryotic Cells*

Concept 1.2
Biological systems are much more than the sum of their parts

▶ **The Emergent Properties of Systems (p. 9)** Due to increasing complexity, new properties emerge with each step upward in the hierarchy of biological order.

▶ **The Power and Limitations of Reductionism (pp. 9–10)** Reductionism involves reducing complex systems to simpler components that are more manageable to study.

▶ **Systems Biology (pp. 10–11)** Systems biology seeks to create models of the dynamic behavior of whole biological systems. With such models, scientists will be able to predict how a change in one part of the system will affect the rest of the system.

▶ **Feedback Regulation in Biological Systems (pp. 11–12)** In negative feedback, accumulation of an end product slows the process that produces that product. In positive feedback, the end product speeds up its production.
Activity *Regulation: Negative and Positive Feedback*

Concept 1.3
Biologists explore life across its great diversity of species

▶ **Grouping Species: The Basic Idea (pp. 12–13)** Taxonomy is the branch of biology that names and classifies species according to a system of broader and broader groups.

▶ **The Three Domains of Life (pp. 13–14)** Domain Bacteria and Domain Archaea consist of prokaryotes. Domain Eukarya, the eukaryotes, includes the various protist kingdoms and the kingdoms Plantae, Fungi, and Animalia.
Activity *Classification Schemes*

▶ **Unity in the Diversity of Life (pp. 14–15)** As diverse as life is, there is also evidence of remarkable unity.

Concept 1.4
Evolution accounts for life's unity and diversity

▶ **Natural Selection (pp. 16–17)** Darwin called the evolutionary history of species "descent with modification." He proposed natural selection as the mechanism for evolutionary adaptation of populations to their environments. Natural selection is the evolutionary process that occurs when a population's heritable variations are exposed to environmental factors that favor the reproductive success of some individuals over others.
Investigation *How Do Environmental Changes Affect a Population?*
Activity *Form Fits Function: Cells*

▶ **The Tree of Life (pp. 17–19)** Each species is one twig of a branching tree of life extending back in time through ancestral species more and more remote. All of life is connected through its long evolutionary history.

Concept 1.5
Biologists use various forms of inquiry to explore life

▶ **Discovery Science (pp. 19–20)** In discovery science, scientists describe some aspect of the world and use inductive reasoning to draw general conclusions.
Graph It *An Introduction to Graphing*

▶ **Hypothesis-Based Science (pp. 20–21)** Based on observations, scientists propose hypotheses that lead to predictions and then test the hypotheses by seeing if the predictions come true. Deductive reasoning is used in testing hypotheses: *If* a hypothesis is correct, and we test it, *then* we can expect a particular outcome. Hypotheses must be testable and falsifiable.
Investigation *How Does Acid Precipitation Affect Trees?*

▶ **A Case Study in Scientific Inquiry: Investigating Mimicry in Snake Populations (pp. 21–24)** Experiments must be designed to test the effect of one variable by testing control groups and experimental groups that vary in only that one variable.

▶ **Limitations of Science (p. 24)** Science cannot address supernatural phenomena because hypotheses must be testable and falsifiable and observations and experimental results must be repeatable.

▶ **Theories in Science (p. 24)** A scientific theory is broad in scope, generates new hypotheses, and is supported by a large body of evidence.

▶ **Model Building in Science (pp. 24–25)** Models of ideas, structures, and processes help us understand scientific phenomena and make predictions.

▶ **The Culture of Science (p. 25)** Science is a social activity characterized by both cooperation and competition.

▶ **Science, Technology, and Society (pp. 25–26)** Technology applies scientific knowledge for some specific purpose.
Activity *Science, Technology, and Society: DDT*

Concept 1.6
A set of themes connects the concepts of biology

▶ Underlying themes provide a framework for understanding biology (pp. 26–27).

TESTING YOUR KNOWLEDGE

Self-Quiz

1. All the organisms on your campus make up
   a. an ecosystem.
   b. a community.
   c. a population.
   d. an experimental group.
   e. a taxonomic domain.

2. Which of the following is a correct sequence of levels in life's hierarchy, proceeding downward from an individual animal?
   a. brain, organ system, nerve cell, nervous tissue
   b. organ system, population of cells, nervous tissue, brain
   c. organism, organ system, tissue, cell, organ
   d. nervous system, brain, nervous tissue, nerve cell
   e. organ system, tissue, molecule, cell

3. Which of the following is *not* an observation or inference on which Darwin's theory of natural selection is based?
   a. Poorly adapted individuals never produce offspring.
   b. There is heritable variation among individuals.
   c. Because of overproduction of offspring, there is competition for limited resources.
   d. Individuals whose inherited characteristics best fit them to the environment will generally produce more offspring.
   e. A population can become adapted to its environment.

4. Systems biology is mainly an attempt to
   a. understand the integration of all levels of biological organization from molecules to the biosphere.
   b. simplify complex problems by reducing the system into smaller, less complex units.
   c. model one level of biological organization based on an understanding of the lower levels of organization.
   d. provide a systematic method for interpretation of large amounts of biological data.
   e. speed up the technological application of scientific knowledge.

5. Protists and bacteria are grouped into different domains because
   a. protists eat bacteria.
   b. bacteria are not made of cells.
   c. bacterial cells lack a nucleus.
   d. bacteria decompose protists.
   e. protists are photosynthetic.

6. Which of the following best demonstrates the unity among all organisms?
   a. matching DNA nucleotide sequences
   b. descent with modification
   c. the structure and function of DNA
   d. natural selection
   e. emergent properties

7. Which of the following is an example of qualitative data?
   a. The temperature decreased from 20°C to 15°C.
   b. The plant's height is 25 centimeters (cm).
   c. The fish swam in a zig-zag motion.
   d. The six pairs of robins hatched an average of three chicks.
   e. The contents of the stomach are mixed every 20 seconds.

8. Which of the following best describes the logic of hypothesis-based science?
   a. If I generate a testable hypothesis, tests and observations will support it.
   b. If my prediction is correct, it will lead to a testable hypothesis.
   c. If my observations are accurate, they will support my hypothesis.
   d. If my hypothesis is correct, I can expect certain test results.
   e. If my experiments are set up right, they will lead to a testable hypothesis.

9. A controlled experiment is one that
   a. proceeds slowly enough that a scientist can make careful records of the results.
   b. may include experimental groups and control groups tested in parallel.
   c. is repeated many times to make sure the results are accurate.
   d. keeps all environmental variables constant.
   e. is supervised by an experienced scientist.

10. Which of the following statements best distinguishes hypotheses from theories in science?
    a. Theories are hypotheses that have been proved.
    b. Hypotheses are guesses; theories are correct answers.
    c. Hypotheses usually are relatively narrow in scope; theories have broad explanatory power.
    d. Hypotheses and theories are essentially the same thing.
    e. Theories are proved true in all cases; hypotheses are usually falsified by tests.

For Self-Quiz answers, see Appendix A.

Go to the website or CD-ROM for more quiz questions.

Evolution Connection

A typical prokaryotic cell has about 3,000 genes in its DNA, while a human cell has about 25,000 genes. About 1,000 of these genes are present in both types of cells. Based on your understanding of evolution, explain how such different organisms could have the same subset of genes.

Scientific Inquiry

Based on the results of the snake mimicry case study, suggest another hypothesis researchers might investigate further.

Investigation ***How Do Environmental Changes Affect a Population?***
Investigation ***How Does Acid Precipitation Affect Trees?***

Science, Technology, and Society

The fruits of wild species of tomato are tiny compared to the giant beefsteak tomatoes available today. This difference in fruit size is almost entirely due to the larger number of cells in the domesticated fruits. Plant molecular biologists have recently discovered genes that are responsible for controlling cell division in tomatoes. Why would such a discovery be important to producers of other kinds of fruits and vegetables? To the study of human development and disease? To our basic understanding of biology?

AN INTERVIEW WITH
Lydia Makhubu

Until her recent retirement, Lydia Makhubu was the Vice Chancellor (in American terms, President) of the University of Swaziland, where she was also Professor of Chemistry. She received her higher education in Lesotho and at the Universities of Alberta and Toronto, where she earned a Ph.D. in Medicinal Chemistry. Building on her study of chemistry, Dr. Makhubu has had a distinguished career as a scientist in the area of health and traditional medicine, as a leader in higher education, and as a commentator on science, technology, and development in Africa and other developing regions. Among the many instances of her international service, she has been a consultant to several United Nations agencies and to the American Association for the Advancement of Science and has chaired the Association of Commonwealth Universities. Jane Reece met with her in Paris, where Dr. Makhubu was attending a meeting of the Executive Board of UNESCO, the United Nations Educational, Scientific and Cultural Organization.

Please tell us a little about Swaziland and its people.

Swaziland is a small landlocked country, only 17,400 square kilometers in area (smaller than New Jersey). It borders Mozambique on the east and South Africa on the west and south. The country ranges in altitude from high to low, and it has a great diversity of organisms, especially plants. The capital, Mbabane, is in the ecological zone called the highveld, where the altitude is close to 1,800 meters above sea level. The altitude drops to the middleveld, which has rich soils especially good for agriculture. As you go toward Mozambique and the sea, the land gets low and flat—the lowveld. The climate and the plant and animal species change as the altitude changes.

Swaziland is unusual these days because it is a kingdom, with a king who has executive authority. In the population of about 1 million, there is only one ethnic group, the Swazis, so we haven't had the conflicts that have afflicted some other African nations. Swaziland was a British colony, gaining independence in 1968. These days, the economy isn't as good as it used to be, in part because we've had a lot of drought and we're heavily dependent on agriculture.

What influenced you to become a medicinal chemist?

In the early days, my parents were teachers, but then my father took up a career in the Ministry of Health, becoming an orderly in a medical clinic. We lived at the clinic, and I could see him check people—since doctors were scarce, an orderly had a lot of responsibility. I wanted to be a medical doctor at that time. I ended up studying chemistry at college in Lesotho. From there, I went to Canada, where I did a master's degree and doctorate. I liked chemistry because it seemed to make sense: You mix this and that, and a product appears. I became interested in organic chemistry, and then, probably influenced by the importance of medicine in my society, I chose medicinal chemistry. I wanted to study the effects of drugs on the body.

What do you mean by "medicine in my society"?

The traditional medicine of my people. At college in Lesotho, we students used to argue about traditional medicine: Some believed it was absolute nonsense; others thought it worked. I was interested in this question. So when I came back from Canada, I immediately sought out traditional healers, including some of my relatives, and I was shown a few of the medicinal preparations they used. I started working in the laboratory to try to identify what was in those medicines.

In other countries of Southern Africa, traditional healers have organized themselves into associations and have even established some clinics. But in Swaziland, the British banned traditional medicine by the Witchcraft Act of 1901—and this law has not been repealed yet. Even people who had access to modern clinics, though, often continued to go to traditional healers, as well, and this continues today.

Tell us about the research on the plant *Phytolacca dodecandra* and its potential for preventing disease.

This plant, also called edod or soapberry, is a common bush in Africa. One day in 1964, an Ethiopian scientist, the late Aklilu Lemma of Addis Ababa University, was walking near a small stream, where he saw women washing clothes. He noticed a large pile of dead snails, of the type that transmits the disease schistosomiasis. He asked the women, "What are you using for soap?" and learned that they were using the berries of *Phytolacca*. He then took berries to the lab, along with some living snails, and found that extracts of the berries killed the snails.

What is schistosomiasis?

Also called bilharzia or snail fever, this is a debilitating disease that afflicts more than 200 million people worldwide. It is one of the greatest scourges in the developing world. The disease is caused by a parasitic flatworm (a fluke) that uses an aquatic snail as a host during part of its life cycle (see Figure 33.11). Fluke larvae released by the snails pierce the skin of people standing or swimming in the water, infecting them. You can control schistosomiasis by killing the parasite with a drug or by killing the snail with a synthetic molluscicide—but both are too expensive in Africa.

Is this where *Phytolacca* comes in?

Yes, *Phytolacca* berries are a better control method for schistosomiasis in Africa than synthetic chemicals of any sort because people can easily grow the plant and harvest the berries. Chemists have isolated the *Phytolacca* chemical that is lethal to snails, although it is not yet known exactly how it acts. Researchers have discovered that the chemical also kills some other parasites that live in African rivers, as well as the larvae of mosquitoes (which transmit malaria). And there are no bad environmental effects because the chemical readily decomposes.

At the University of Swaziland, we obtained seeds of *Phytolacca* from Ethiopia, grew the plants, and harvested the berries. The Ethiopians came to show us how to do everything; it was a true collaboration between Swazi and Ethiopian scientists, with help from some Zimbabwans and an American. Working in the lab, we discovered the concentration at which the berry extract killed the snails, and then we went into the field for further tests. Now we have selected an area in Swaziland where schistosomiasis is very prevalent, and we're working with the people there, teaching them how to grow and use *Phytolacca*. We hope that, in another year, the communities will be able to control the disease themselves.

What goes on at your university's institute of traditional medicine?

At this institute, officially the "Swaziland Institute of Research in Traditional Medicine, Medicinal and Indigenous Food Plants," multidisciplinary teams study all aspects of traditional medicine. Traditional healers are essential team members because they know the healing plants and how to use them. We have had several workshops with traditional healers, trying to convince them of the importance of sharing their knowledge with us —because they are going to die, like all of us, and the knowledge may soon be lost. However, the healers—even my relatives!—are reluctant to help. They think, "You with your white coats are going to make loads of money from my knowledge." Their belief system is another obstacle. The healers believe that they are given the power of healing by their ancestors, and they are supposed to pass on this knowledge only to their children. But mostly it is suspicion. You know, for a long time they were called witches, and quite a few of the older ones are still sore about that; they ask me, "When is that Witchcraft Act of *yours* going to be repealed?"—as if I had written it! But slowly we are managing to convince them. We want to involve them for the long term, not only to show us the plants and help us grow them but to come into the lab to teach us how they prepare the medicine, so that we can quantify everything. But it's not easy.

It is also important, I think, to study the spiritual beliefs of the healers because the whole system is based on those beliefs. They say they are shown the plants in a dream by their ancestors' spirits, and they make diagnoses by throwing bones and going into a trance, during which the spirits speak to them.

What is the state of the environment and biological diversity in Swaziland?

Not very good. I think the underlying problem is overreliance on the natural environment, especially plant resources. In many rural areas, people have chopped down trees for wood until the land is completely bare; they do not know how to replant. Their grazing animals, such as cattle, often eat whatever plants remain. And the healers may overharvest medicinal plants from the wild. Many plants are disappearing.

Preservation of diversity goes along with preservation of the environment. So, you find that, in parts of Swaziland, certain animals have disappeared because the plants they lived on are no longer there. Even the climate is affected. For instance, in the forested highveld of Swaziland, there used to be lots of rain. But as the plants are removed, the rainfall lessens.

Another issue is damage that can result from projects associated with economic development, such as mining and dam construction. It is only recently that companies carrying out these big projects are being required to take care of the environment.

What are the challenges that science education faces in Africa?

We don't have enough resources to build proper science facilities, and we don't have enough science teachers. Another serious problem is the underrepresentation of women in science; this is particularly bad in Africa. Women are left behind. Science, especially physical science, is not considered a field for women. Many people think that if women go too far, they won't get a husband. But the situation is starting to change.

You are the President of the Third World Organization for Women in Science (TWOWS). What does this organization do?

We provide fellowships for postgraduate study, enlisting support from organizational benefactors. The fellowship recipients are usually sent to good universities in developing countries, such as South Africa or Pakistan, where the available money can go a long way. TWOWS also promotes collaboration among women from developing countries who are already established scientists.

But it's crucial to start at the earliest level, primary school. Researchers have learned that once girls get started in science, they do well. But they need to be encouraged by their teachers. If there is equipment available, it is used by the boys; the girls' role may be simply recording the results! So we are working hard to encourage the involvement of women scientists at all levels of education, to show the teachers that girls can be scientists.

"You can control schistosomiasis by killing the parasite with a drug or by killing the snail with a synthetic molluscicide—but both are too expensive in Africa. . . . Phytolacca *berries are a better control method . . . because people can easily grow the plant."*

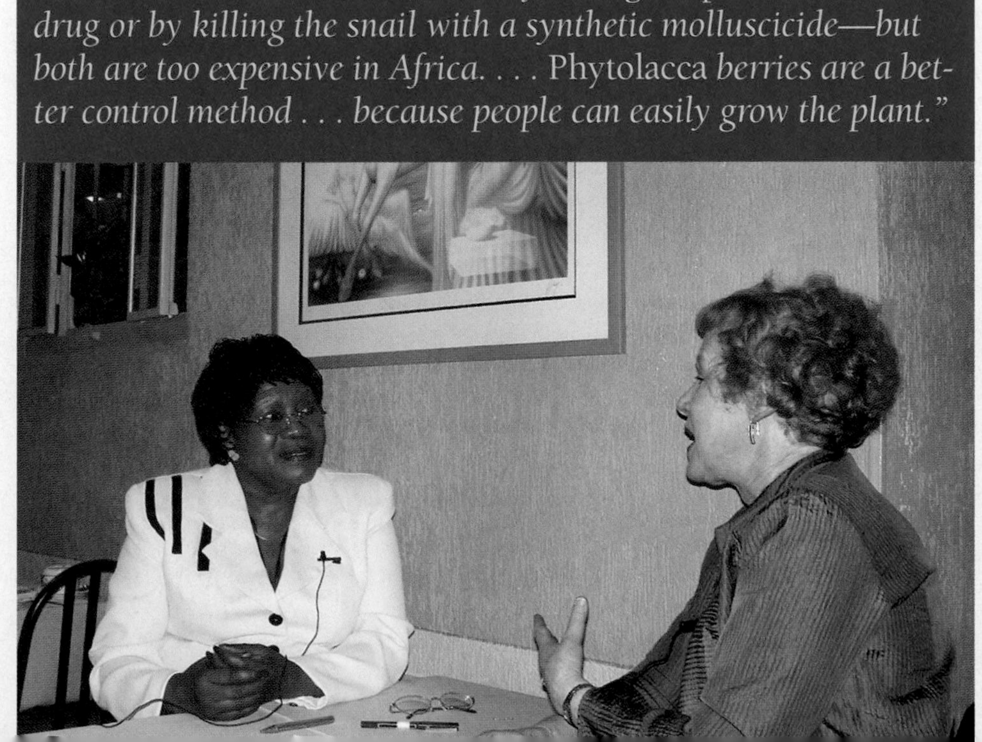

2 The Chemical Context of Life

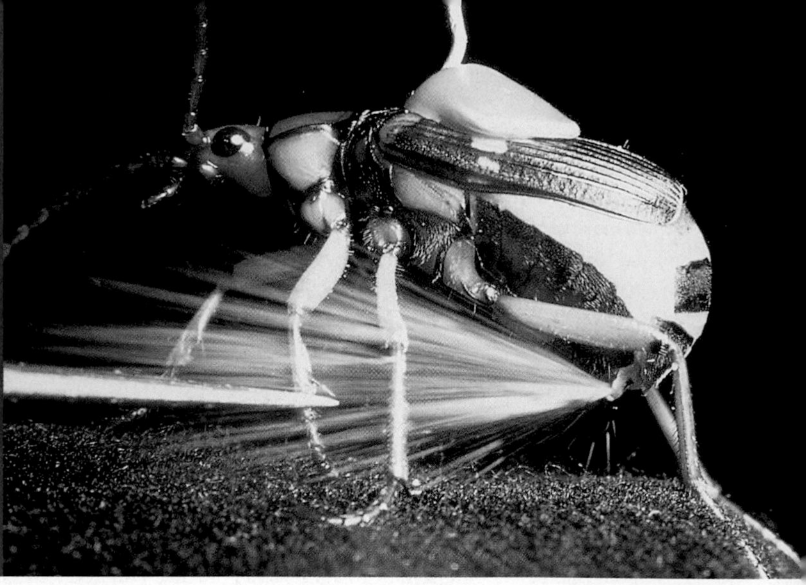

▲ **Figure 2.1 The bombardier beetle uses chemistry to defend itself.**

Key Concepts

2.1 Matter consists of chemical elements in pure form and in combinations called compounds
2.2 An element's properties depend on the structure of its atoms
2.3 The formation and function of molecules depend on chemical bonding between atoms
2.4 Chemical reactions make and break chemical bonds

Overview

Chemical Foundations of Biology

Like other animals, beetles have evolved structures and mechanisms that defend them from attack. The soil-dwelling bombardier beetle has a particularly effective mechanism for dealing with the ants that plague it. Upon detecting an ant on its body, this beetle ejects a spray of boiling hot liquid from glands in its abdomen, aiming the spray directly at the ant. (In **Figure 2.1**, the beetle aims its spray at a scientist's forceps.) The spray contains irritating chemicals that are generated at the moment of ejection by the explosive reaction of two sets of chemicals stored separately in the glands. The reaction produces heat and an audible pop.

Research on the bombardier beetle has involved chemistry, physics, and engineering, as well as biology. This is not surprising, for unlike a college catalog of courses, nature is not neatly packaged into the individual natural sciences. Biologists specialize in the study of life, but organisms and the world they live in are natural systems to which basic concepts of chemistry and physics apply. Biology is a multidisciplinary science.

This unit of chapters introduces key concepts of chemistry that will apply throughout our study of life. We will make many connections to the themes introduced in Chapter 1. One of those themes is the organization of life into a hierarchy of structural levels, with additional properties emerging at each successive level. In this unit, we will see how the theme of emergent properties applies to the lowest levels of biological organization—to the ordering of atoms into molecules and to the interactions of those molecules within cells. Somewhere in the transition from molecules to cells, we will cross the blurry boundary between nonlife and life. We begin by considering the chemical components that make up all matter. As Lydia Makhubu mentioned in the interview on pages 30 and 31, chemistry is an integral aspect of biology.

Concept 2.1

Matter consists of chemical elements in pure form and in combinations called compounds

Elements and Compounds

Organisms are composed of **matter,** which is anything that takes up space and has mass.* Matter exists in many diverse forms, each with its own characteristics. Rocks, metals, oils, gases, and humans are just a few examples of what seems an endless assortment of matter.

---

* Sometimes we substitute the term weight for mass, although the two are not identical. Mass is the amount of matter in an object, whereas the weight of an object is how strongly that mass is pulled by gravity. The weight of an astronaut walking on the moon is approximately 1/6 that on Earth, but his or her mass is the same. However, as long as we are earthbound, the weight of an object is a measure of its mass; in everyday language, therefore, we tend to use the terms interchangeably.

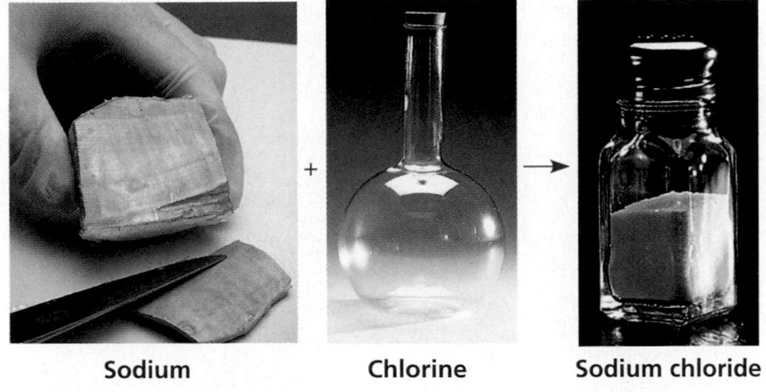

▲ **Figure 2.2 The emergent properties of a compound.** The metal sodium combines with the poisonous gas chlorine to form the edible compound sodium chloride, or table salt.

Matter is made up of elements. An **element** is a substance that cannot be broken down to other substances by chemical reactions. Today, chemists recognize 92 elements occurring in nature; gold, copper, carbon, and oxygen are examples. Each element has a symbol, usually the first letter or two of its name. Some of the symbols are derived from Latin or German names; for instance, the symbol for sodium is Na, from the Latin word *natrium*.

A **compound** is a substance consisting of two or more different elements combined in a fixed ratio. Table salt, for example, is sodium chloride (NaCl), a compound composed of the elements sodium (Na) and chlorine (Cl) in a 1:1 ratio. Pure sodium is a metal and pure chlorine is a poisonous gas. When chemically combined, however, sodium and chlorine form an edible compound. This is a simple example of organized matter having emergent properties: A compound has characteristics different from those of its elements **(Figure 2.2)**.

Essential Elements of Life

About 25 of the 92 natural elements are known to be essential to life. Just four of these—carbon (C), oxygen (O), hydrogen (H), and nitrogen (N)—make up 96% of living matter. Phosphorus (P), sulfur (S), calcium (Ca), potassium (K), and a few other elements account for most of the remaining 4% of an organism's weight. **Table 2.1** lists by percentage the elements that make up the human body; the percentages for other organisms are similar. **Figure 2.3a** illustrates the effect of a deficiency of nitrogen, an essential element, in a plant.

Table 2.1 Naturally Occurring Elements in the Human Body

Symbol	Element	Atomic Number (See p. 34)	Percentage of Human Body Weight
O	Oxygen	8	65.0
C	Carbon	6	18.5
H	Hydrogen	1	9.5
N	Nitrogen	7	3.3
Ca	Calcium	20	1.5
P	Phosphorus	15	1.0
K	Potassium	19	0.4
S	Sulfur	16	0.3
Na	Sodium	11	0.2
Cl	Chlorine	17	0.2
Mg	Magnesium	12	0.1

Trace elements (less than 0.01%): boron (B), chromium (Cr), cobalt (Co), copper (Cu), fluorine (F), iodine (I), iron (Fe), manganese (Mn), molybdenum (Mo), selenium (Se), silicon (Si), tin (Sn), vanadium (V), and zinc (Zn).

Trace elements are those required by an organism in only minute quantities. Some trace elements, such as iron (Fe), are needed by all forms of life; others are required only by certain species. For example, in vertebrates (animals with backbones), the element iodine (I) is an essential ingredient of a hormone produced by the thyroid gland. A daily intake of only 0.15 milligram (mg) of iodine is adequate for normal

(a) Nitrogen deficiency

(b) Iodine deficiency

► **Figure 2.3 The effects of essential-element deficiencies. (a)** This photo shows the effect of nitrogen deficiency in corn. In this controlled experiment, the plants on the left are growing in soil that was fertilized with compounds containing nitrogen, while the soil on the right is deficient in nitrogen. **(b)** Goiter, an enlarged thyroid gland, is the result of a deficiency of the trace element iodine. The goiter of this Malaysian woman can probably be reversed by iodine supplements.

activity of the human thyroid. An iodine deficiency in the diet causes the thyroid gland to grow to abnormal size, a condition called goiter **(Figure 2.3b)**. Where it is available, iodized salt has reduced the incidence of goiter.

Concept Check 2.1

1. Explain why table salt is a compound, while the oxygen we breathe is not.
2. What four chemical elements are most abundant in the food you ate yesterday?

For suggested answers, see Appendix A.

An element's properties depend on the structure of its atoms

Each element consists of a certain kind of atom that is different from the atoms of any other element. An **atom** is the smallest unit of matter that still retains the properties of an element. Atoms are so small that it would take about a million of them to stretch across the period printed at the end of this sentence. We symbolize atoms with the same abbreviation used for the element made up of those atoms; thus, C stands for both the element carbon and a single carbon atom.

Subatomic Particles

Although the atom is the smallest unit having the properties of its element, these tiny bits of matter are composed of even smaller parts, called *subatomic particles*. Physicists have split the atom into more than a hundred types of particles, but only three kinds of particles are stable enough to be of relevance here: **neutrons, protons,** and **electrons.** Neutrons and protons are packed together tightly to form a dense core, or **atomic nucleus,** at the center of the atom. The electrons, moving at nearly the speed of light, form a cloud around the nucleus. **Figure 2.4** shows two models of the structure of the helium atom as an example.

Electrons and protons are electrically charged. Each electron has one unit of negative charge, and each proton has one unit of positive charge. A neutron, as its name implies, is electrically neutral. Protons give the nucleus a positive charge, and it is the attraction between opposite charges that keeps the rapidly moving electrons in the vicinity of the nucleus.

The neutron and proton are almost identical in mass, each about 1.7×10^{-24} gram (g). Grams and other conventional units are not very useful for describing the mass of objects so

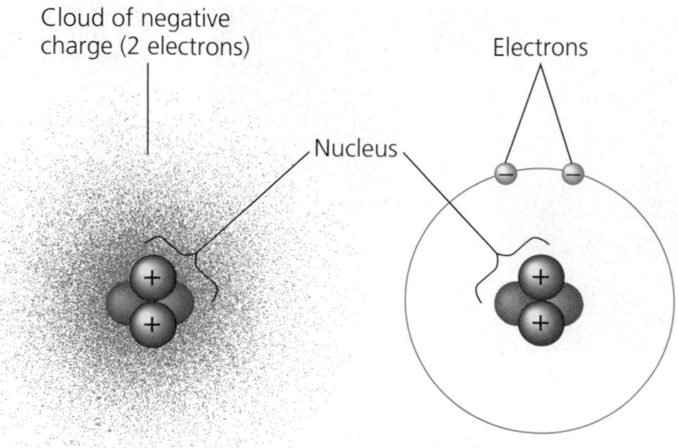

(a) This model represents the electrons as a cloud of negative charge, as if we had taken many snapshots of the 2 electrons over time, with each dot representing an electron's position at one point in time.

(b) In this even more simplified model, the electrons are shown as two small blue spheres on a circle around the nucleus.

▲ **Figure 2.4 Simplified models of a helium (He) atom.** The helium nucleus consists of 2 neutrons (brown) and 2 protons (pink). Two electrons (blue) move rapidly around the nucleus. These models are not to scale; they greatly overestimate the size of the nucleus in relation to the electron cloud.

minuscule. Thus, for atoms and subatomic particles (and for molecules as well), we use a unit of measurement called the **dalton,** in honor of John Dalton, the British scientist who helped develop atomic theory around 1800. (The dalton is the same as the *atomic mass unit,* or *amu,* a unit you may have encountered elsewhere.) Neutrons and protons have masses close to 1 dalton. Because the mass of an electron is only about $1/2{,}000$ that of a neutron or proton, we can ignore electrons when computing the total mass of an atom.

Atomic Number and Atomic Mass

Atoms of the various elements differ in their number of subatomic particles. All atoms of a particular element have the same number of protons in their nuclei. This number of protons, which is unique to that element, is called the **atomic number** and is written as a subscript to the left of the symbol for the element. The abbreviation $_2He$, for example, tells us that an atom of the element helium has 2 protons in its nucleus. Unless otherwise indicated, an atom is neutral in electrical charge, which means that its protons must be balanced by an equal number of electrons. Therefore, the atomic number tells us the number of protons and also the number of electrons in an electrically neutral atom.

We can deduce the number of neutrons from a second quantity, the **mass number,** which is the sum of protons plus neutrons in the nucleus of an atom. The mass number is written as a superscript to the left of an element's symbol. For example, we can use this shorthand to write an atom of helium as $^{4}_{2}He$. Because the atomic number indicates how many protons there are, we can determine the number of neutrons by subtracting the atomic number from the mass number: A $^{4}_{2}He$ atom has 2 neutrons. An atom of sodium, $^{23}_{11}Na$, has 11 protons, 11 electrons, and 12 neutrons. The simplest atom is hydrogen $^{1}_{1}H$, which has no neutrons; it consists of a lone proton with a single electron moving around it.

Almost all of an atom's mass is concentrated in its nucleus, because, as mentioned earlier, the contribution of electrons to mass is negligible. Because neutrons and protons each have a mass very close to 1 dalton, the mass number is an approximation of the total mass of an atom, called its **atomic mass.** So we might say that the atomic mass of sodium ($^{23}_{11}Na$) is 23 daltons, although more precisely it is 22.9898 daltons.

Isotopes

All atoms of a given element have the same number of protons, but some atoms have more neutrons than other atoms of the same element and therefore have greater mass. These different atomic forms are called **isotopes** of the element. In nature, an element occurs as a mixture of its isotopes. For example, consider the three isotopes of the element carbon, which has the atomic number 6. The most common isotope is carbon-12, $^{12}_{6}C$, which accounts for about 99% of the carbon in nature. It has 6 neutrons. Most of the remaining 1% of carbon consists of atoms of the isotope $^{13}_{6}C$, with 7 neutrons. A third, even rarer isotope, $^{14}_{6}C$, has 8 neutrons. Notice that all three isotopes of carbon have 6 protons—otherwise, they would not be carbon. Although isotopes of an element have slightly different masses, they behave identically in chemical reactions. (The number usually given as the atomic mass of an element, such as 22.9898 daltons for sodium, is actually an average of the atomic masses of all the element's naturally occurring isotopes.)

Both ^{12}C and ^{13}C are stable isotopes, meaning that their nuclei do not have a tendency to lose particles. The isotope ^{14}C, however, is unstable, or radioactive. A **radioactive isotope** is one in which the nucleus decays spontaneously, giving off particles and energy. When the decay leads to a change in the number of protons, it transforms the atom to an atom of a different element. For example, radioactive carbon decays to form nitrogen.

Radioactive isotopes have many useful applications in biology. In Chapter 26, you will learn how researchers use measurements of radioactivity in fossils to date those relics of past life. Radioactive isotopes are also useful as tracers to follow atoms through metabolism, the chemical processes of an organism. Cells use the radioactive atoms as they would nonradioactive isotopes of the same element, but the radioactive tracers can be readily detected. **Figure 2.5** presents an example of how biologists use radioactive tracers to monitor biological processes, in this case cells making copies of their DNA.

Figure 2.5
Research Method Radioactive Tracers

APPLICATION Scientists use radioactive isotopes to label certain chemical substances, creating tracers that can be used to follow a metabolic process or locate the substance within an organism. In this example, radioactive tracers are being used to determine the effect of temperature on the rate at which cells make copies of their DNA.

TECHNIQUE

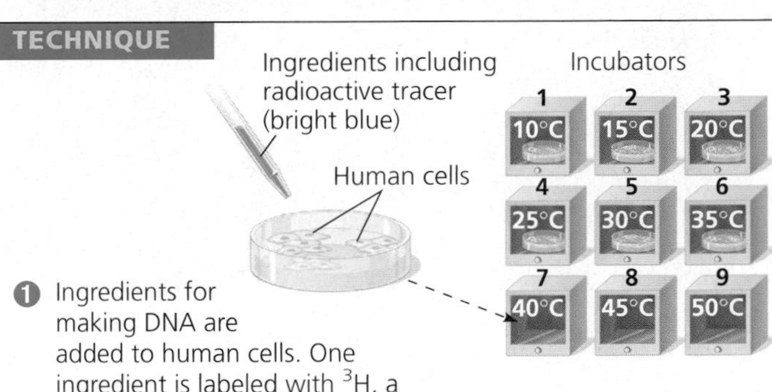

1 Ingredients for making DNA are added to human cells. One ingredient is labeled with ^{3}H, a radioactive isotope of hydrogen. Nine dishes of cells are incubated at different temperatures. The cells make new DNA, incorporating the radioactive tracer with ^{3}H.

2 The cells are placed in test tubes, their DNA is isolated, and unused ingredients are removed.

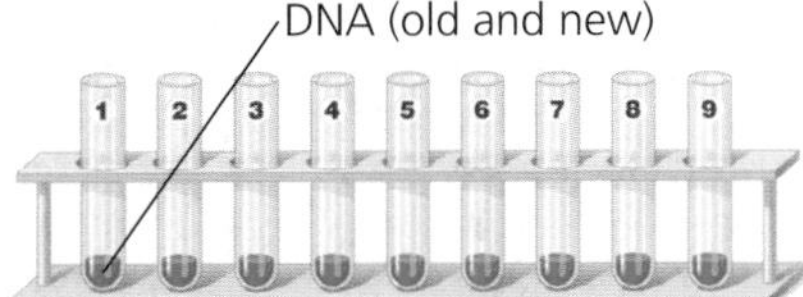

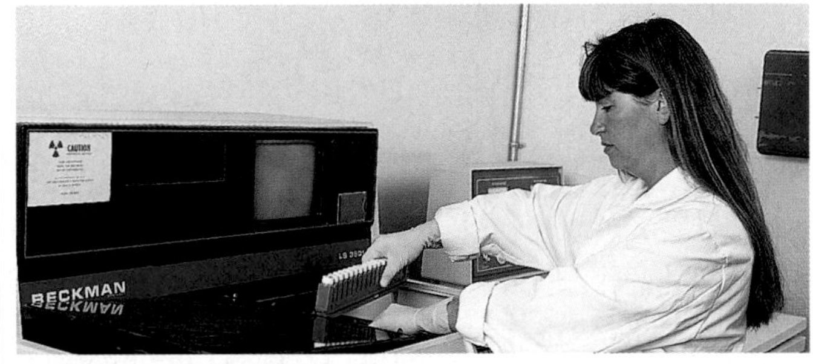

3 A solution called scintillation fluid is added to the test tubes and they are placed in a scintillation counter. As the ^{3}H in the newly made DNA decays, it emits radiation that excites chemicals in the scintillation fluid, causing them to give off light. Flashes of light are recorded by the scintillation counter.

RESULTS The frequency of flashes, which is recorded as counts per minute, is proportional to the amount of the radioactive tracer present, indicating the amount of new DNA. In this experiment, when the counts per minute are plotted against temperature, it is clear that temperature affects the rate of DNA synthesis—the most DNA was made at 35 °C.

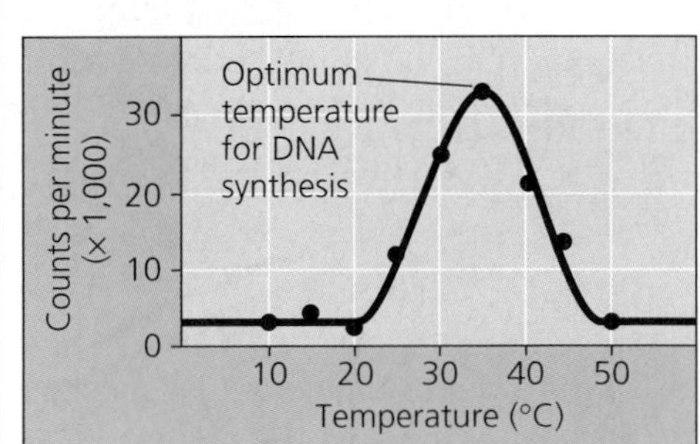

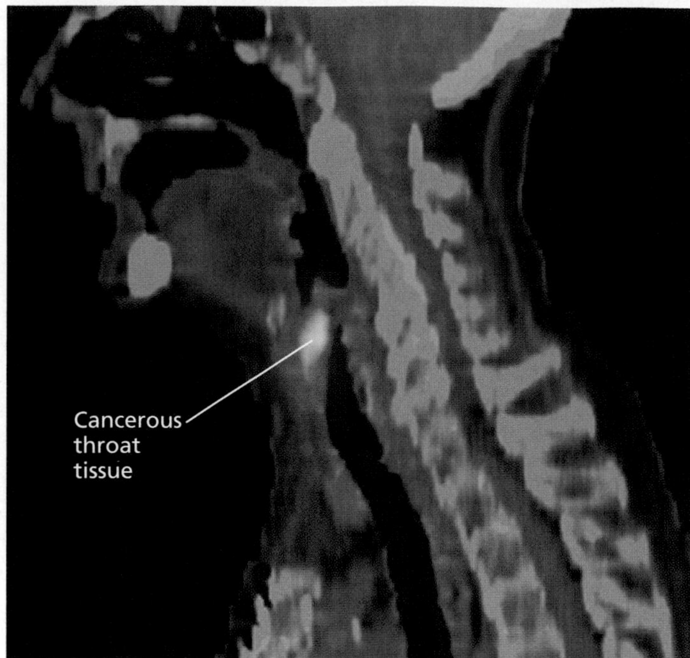

▲ **Figure 2.6 A PET scan, a medical use for radioactive isotopes.** PET, an acronym for positron-emission tomography, detects locations of intense chemical activity in the body. The patient is first injected with a nutrient such as glucose labeled with a radioactive isotope that emits subatomic particles. These particles collide with electrons made available by chemical reactions in the body. A PET scanner detects the energy released in these collisions and maps "hot spots," the regions of an organ that are most chemically active at the time. The color of the image varies with the amount of the isotope present, with the bright yellow color here identifying a hot spot of cancerous throat tissue.

Radioactive tracers are important diagnostic tools in medicine. For example, certain kidney disorders can be diagnosed by injecting small doses of substances containing radioactive isotopes into the blood and then measuring the amount of tracer excreted in the urine. Radioactive tracers are also used in combination with sophisticated imaging instruments, such as PET scanners, which can monitor chemical processes, such as those involved in cancerous growth, as they actually occur in the body **(Figure 2.6)**.

Although radioactive isotopes are very useful in biological research and medicine, radiation from decaying isotopes also poses a hazard to life by damaging cellular molecules. The severity of this damage depends on the type and amount of radiation an organism absorbs. One of the most serious environmental threats is radioactive fallout from nuclear accidents. The doses of most isotopes used in medical diagnosis, however, are relatively safe.

The Energy Levels of Electrons

The simplified models of the atom in Figure 2.4 greatly exaggerate the size of the nucleus relative to the volume of the whole atom. If an atom of helium were the size of Yankee Stadium, the nucleus would be only the size of a pencil eraser in the center of the field. Moreover, the electrons would be like two tiny gnats buzzing around the stadium. Atoms are mostly empty space.

When two atoms approach each other during a chemical reaction, their nuclei do not come close enough to interact. Of the three kinds of subatomic particles we have discussed, only electrons are directly involved in the chemical reactions between atoms.

An atom's electrons vary in the amount of energy they possess. **Energy** is defined as the capacity to cause change, for instance by doing work. **Potential energy** is the energy that matter possesses because of its location or structure. For example, because of its altitude, water in a reservoir on a hill has potential energy. When the gates of the reservoir's dam are opened and the water runs downhill, the energy can be used to do work, such as turning generators. Because energy has been expended, the water has less energy at the bottom of the hill than it did in the reservoir. Matter has a natural tendency to move to the lowest possible state of potential energy; in this example, water runs downhill. To restore the potential energy of a reservoir, work must be done to elevate the water against gravity.

The electrons of an atom also have potential energy because of how they are arranged in relation to the nucleus. The negatively charged electrons are attracted to the positively charged nucleus. It takes work to move an electron farther away from the nucleus, so the more distant the electrons are from the nucleus, the greater their potential energy. Unlike the continuous flow of water downhill, changes in the potential energy of electrons can occur only in steps of fixed amounts. An electron having a certain discrete amount of energy is something like a ball on a staircase **(Figure 2.7a)**. The ball can have different amounts of potential energy, depending on which step it is on, but it cannot spend much time between the steps. An electron cannot exist in between its fixed states of potential energy.

The different states of potential energy that electrons have in an atom are called **energy levels.** An electron's energy level is correlated with its average distance from the nucleus; these average distances are represented symbolically by **electron shells** **(Figure 2.7b)**. The first shell is closest to the nucleus, and electrons in this shell have the lowest potential energy. Electrons in the second shell have more energy, electrons in the third shell more energy still, and so on. An electron can change the shell it occupies, but only by absorbing or losing an amount of energy equal to the difference in potential energy between its position in the old shell and that in the new shell. When an electron absorbs energy, it moves to a shell farther out from the nucleus. For example, light energy can excite an electron to a higher energy level. (Indeed, this is the first step taken when plants harness the energy of sunlight for photosynthesis, the process that produces food from carbon dioxide and water.) When an electron loses energy, it "falls back" to a

(a) A ball bouncing down a flight of stairs provides an analogy for energy levels of electrons, because the ball can come to rest only on each step, not between steps.

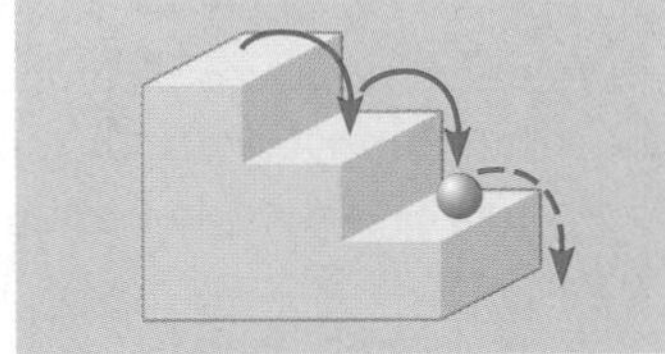

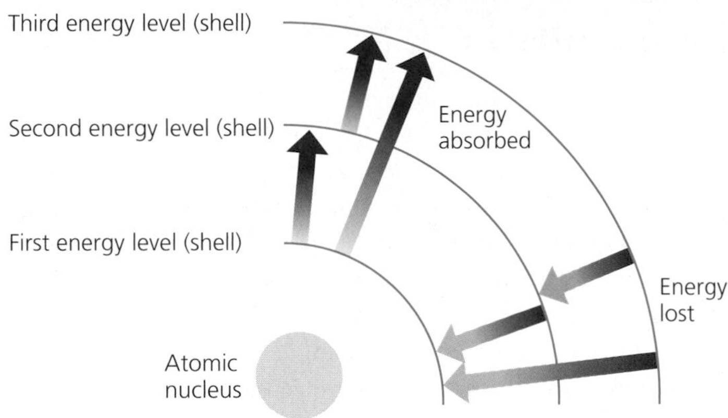

(b) An electron can move from one level to another only if the energy it gains or loses is exactly equal to the difference in energy between the two levels. Arrows indicate some of the step-wise changes in potential energy that are possible.

▲ **Figure 2.7 Energy levels of an atom's electrons.** Electrons exist only at fixed levels of potential energy, which are also called electron shells.

shell closer to the nucleus, and the lost energy is usually released to the environment in the form of heat. For example, sunlight excites electrons in the paint of a dark car to higher energy levels. When the electrons fall back to their original levels, the surface of the car heats up. This thermal energy can be transferred to the air or to your hand if you touch the car.

Electron Configuration and Chemical Properties

The chemical behavior of an atom is determined by its electron configuration—that is, the distribution of electrons in the atom's electron shells. Beginning with hydrogen, the simplest atom, we can imagine building the atoms of the other elements by adding 1 proton and 1 electron at a time (along with an appropriate number of neutrons). **Figure 2.8**, an abbreviated version of what is called the *periodic table of the elements*, shows this distribution of electrons for the first 18 elements, from hydrogen ($_1H$) to argon ($_{18}Ar$). The elements are arranged in three rows, or periods, corresponding to the number of electron shells in their atoms. The left-to-right sequence of elements in each row corresponds to the sequential addition of electrons (and protons).

Hydrogen's 1 electron and helium's 2 electrons are located in the first shell. Electrons, like all matter, tend to exist in the

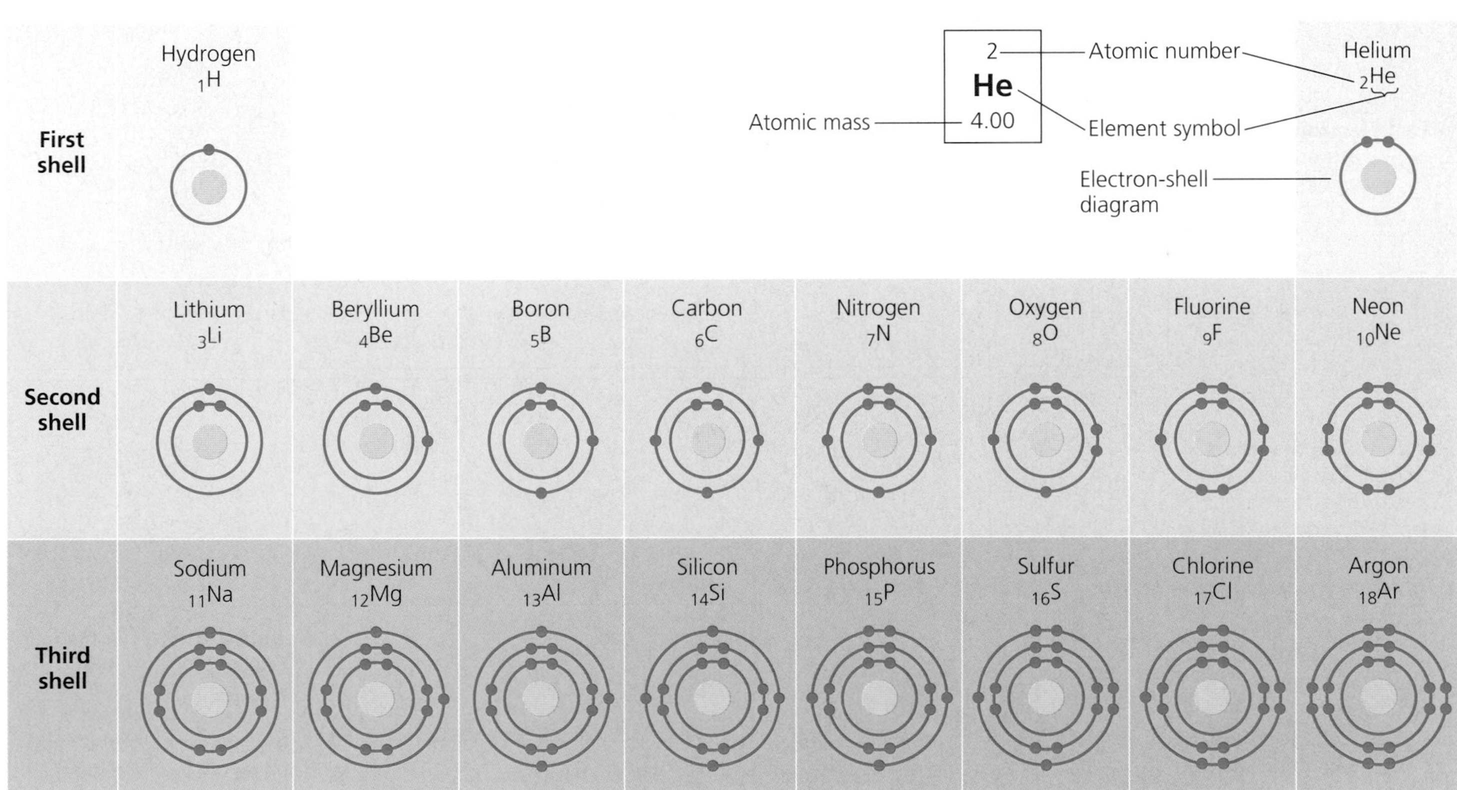

▲ **Figure 2.8 Electron-shell diagrams of the first 18 elements in the periodic table.** In a standard periodic table, information for each element is presented as shown for helium in the inset. In the diagrams in this modified table, electrons are shown as blue dots and electron shells (representing energy levels) as concentric rings. We are using these electron-shell diagrams as a convenient way to picture the distribution of an atom's electrons among its electron shells, but keep in mind that these are simplified models. The elements are arranged in rows, each representing the filling of an electron shell. As electrons are added, they occupy the lowest available shell.

lowest available state of potential energy, which they have in the first shell. However, the first shell can hold no more than 2 electrons; thus, there are only two elements (hydrogen and helium) in the first row of the table. An atom with more than 2 electrons must use higher shells because the first shell is full. The next element, lithium, has 3 electrons. Two of these electrons fill the first shell while the third electron occupies the second shell. The second shell holds a maximum of 8 electrons. Neon, at the end of the second row, has 8 electrons in the second shell, giving it a total of 10 electrons.

The chemical behavior of an atom depends mostly on the number of electrons in its *outermost* shell. We call those outer electrons **valence electrons** and the outermost electron shell the **valence shell.** In the case of lithium, there is only 1 valence electron, and the second shell is the valence shell. Atoms with the same number of electrons in their valence shells exhibit similar chemical behavior. For example, fluorine (F) and chlorine (Cl) both have 7 valence electrons, and both combine with the element sodium to form compounds (see Figure 2.2). An atom with a completed valence shell is unreactive; that is, it will not interact readily with other atoms it encounters. At the far right of the periodic table are helium, neon, and argon, the only three elements shown in Figure 2.8 that have full valence shells. These elements are said to be *inert*, meaning chemically unreactive. All the other atoms in Figure 2.8 are chemically reactive because they have incomplete valence shells.

Electron Orbitals

Early in the 20th century, the electron shells of an atom were visualized as concentric paths of electrons orbiting the nucleus, somewhat like planets orbiting the sun. It is still convenient to use two-dimensional concentric-circle diagrams to symbolize electron shells, as in Figure 2.8, if we bear in mind that an electron shell represents the *average* distance of an electron from the nucleus. This is only a model, however, and it does not give a real picture of an atom. In reality, we can never know the exact path of an electron. What we can do instead is describe the space in which an electron spends most of its time. The three-dimensional space where an electron is found 90% of the time is called an **orbital.**

Each electron shell consists of a specific number of orbitals of distinctive shapes and orientations **(Figure 2.9)**. You can

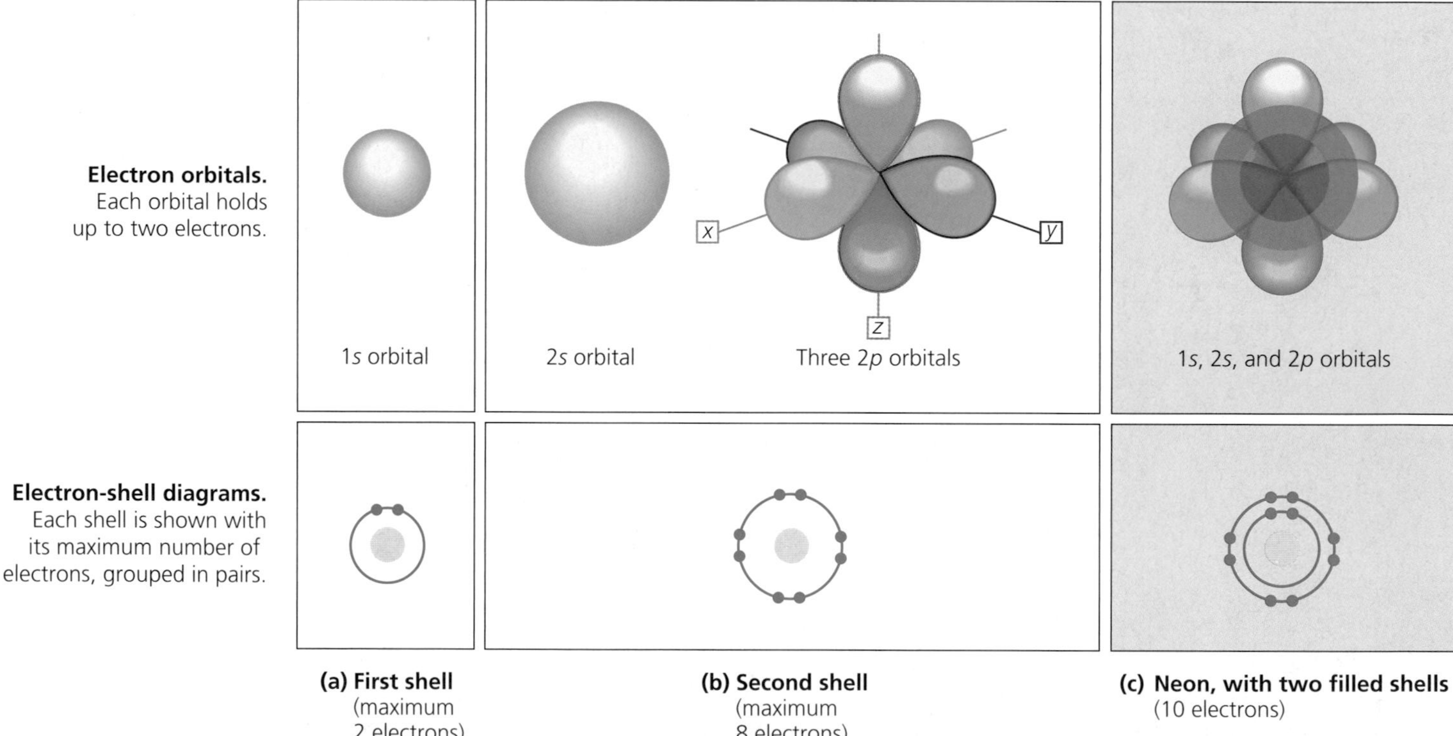

▲ **Figure 2.9 Electron orbitals.** The three-dimensional shapes in the top half of this figure represent electron orbitals—the volumes of space where the electrons of an atom are most likely to be found. Each orbital holds a maximum of 2 electrons. The bottom half of the figure shows the corresponding electron-shell diagrams. **(a)** The first electron shell has one spherical (*s*) orbital, designated 1*s*. **(b)** The second and all higher shells each have one larger *s* orbital (designated 2*s* for the second shell) plus three dumbbell-shaped orbitals called *p* orbitals (2*p* for the second shell). The three 2*p* orbitals lie at right angles to one another along imaginary *x*-, *y*-, and *z*-axes of the atom. Each 2*p* orbital is outlined here in a different color. **(c)** To symbolize the electron orbitals of the element neon, which has a total of 10 electrons, we superimpose the 1*s* orbital of the first shell and the 2*s* and three 2*p* orbitals of the second shell.

think of an orbital as a component of an electron shell. (Recall that an electron shell corresponds to a particular energy level.) The first electron shell has only one spherical *s* orbital (called 1*s*), but the second shell has four orbitals: one large spherical *s* orbital (called 2*s*) and three dumbbell-shaped *p* orbitals (called 2*p* orbitals). Each 2*p* orbital is oriented at right angles to the other two 2*p* orbitals (see Figure 2.9). (The third and higher electron shells also have *s* and *p* orbitals, as well as orbitals of more complex shapes.)

No more than 2 electrons can occupy a single orbital. The first electron shell can therefore accommodate a maximum of 2 electrons in its *s* orbital. The lone electron of a hydrogen atom occupies the 1*s* orbital, as do the 2 electrons of a helium atom. The four orbitals of the second electron shell can hold a maximum of 8 electrons. Electrons in each of the four orbitals have nearly the same energy, but they move in different volumes of space.

The reactivity of atoms arises from the presence of unpaired electrons in one or more orbitals of their valence shells. Notice that the electron configurations in Figure 2.8 build up with the addition of 1 electron at a time. For simplicity, we place 1 electron on each side of the outer shell until the shell is half full, and then pair up electrons until the shell is full. When atoms interact to complete their valence shells, it is the *unpaired* electrons that are involved.

Concept Check 2.2

1. A lithium atom has 3 protons and 4 neutrons. What is its atomic mass in daltons?
2. A nitrogen atom has 7 protons, and the most common isotope of nitrogen has 7 neutrons. A rarer isotope of nitrogen has 8 neutrons. What is the atomic number and mass number of the rarer isotope? Write as a chemical symbol with a subscript and superscript.
3. Look at Figure 2.8, and determine the atomic number of magnesium. How many protons and electrons does it have? How many electron shells? How many valence electrons are in the valence shell?
4. In an electron-shell diagram of phosphorus, in which shell do electrons have the most potential energy? In which shell do electrons have the least potential energy?
5. How many electrons does fluorine have? How many electron shells? Name the orbitals that are occupied. How many unpaired electrons does fluorine have?

For suggested answers, see Appendix A.

Concept 2.3

The formation and function of molecules depend on chemical bonding between atoms

Now that we have looked at the structure of atoms, we can move up the hierarchy of organization and see how atoms combine to form molecules and ionic compounds. Atoms with incomplete valence shells can interact with certain other atoms in such a way that each partner completes its valence shell: The atoms either share or transfer valence electrons. These interactions usually result in atoms staying close together, held by attractions called **chemical bonds**. The strongest kinds of chemical bonds are covalent bonds and ionic bonds.

Covalent Bonds

A **covalent bond** is the sharing of a pair of valence electrons by two atoms. For example, let's consider what happens when two hydrogen atoms approach each other. Recall that hydrogen has 1 valence electron in the first shell, but the shell's capacity is 2 electrons. When the two hydrogen atoms come close enough for their 1*s* orbitals to overlap, they share their electrons **(Figure 2.10)**. Each hydrogen atom now has 2 electrons associated with it in what amounts to a completed valence shell, shown in an electron-shell diagram in

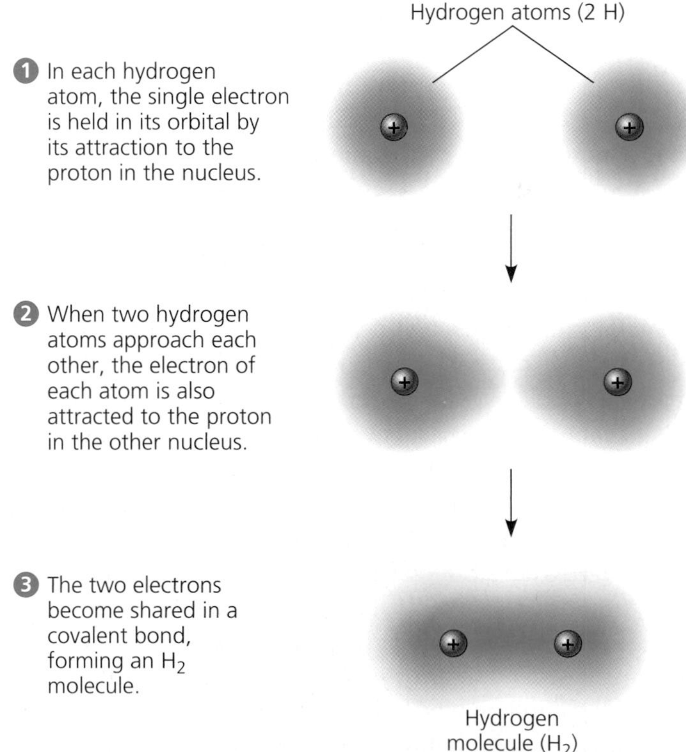
Hydrogen atoms (2 H)
1 In each hydrogen atom, the single electron is held in its orbital by its attraction to the proton in the nucleus.
2 When two hydrogen atoms approach each other, the electron of each atom is also attracted to the proton in the other nucleus.
3 The two electrons become shared in a covalent bond, forming an H_2 molecule.
Hydrogen molecule (H_2)

▲ **Figure 2.10 Formation of a covalent bond.**

Figure 2.11a. Two or more atoms held together by covalent bonds constitute a **molecule.** In this case, we have formed a hydrogen molecule. We can abbreviate the structure of this molecule as H—H, where the line represents a single covalent bond, or simply a **single bond**—that is, a pair of shared electrons. This notation, which represents both atoms and bonding, is called a **structural formula.** We can abbreviate even further by writing H_2, a **molecular formula** indicating simply that the molecule consists of two atoms of hydrogen.

With 6 electrons in its second electron shell, oxygen needs 2 more electrons to complete its valence shell. Two oxygen atoms form a molecule by sharing *two* pairs of valence electrons **(Figure 2.11b)**. The atoms are thus joined by what is called a double covalent bond, or simply a **double bond.**

Each atom that can share valence electrons has a bonding capacity corresponding to the number of covalent bonds the atom can form. When the bonds form, they give the atom a full complement of electrons in the valence shell. The bonding capacity of oxygen, for example, is 2. This bonding capacity is called the atom's **valence** and usually equals the number of unpaired electrons in the atom's outermost (valence) shell. See if you can determine the valences of hydrogen, oxygen, nitrogen, and carbon by studying the electron configurations in Figure 2.8. By counting the unpaired electrons, you can see that the valence of hydrogen is 1; oxygen, 2; nitrogen, 3; and carbon, 4. A more complicated case is phosphorus (P), another element important to life. Phosphorus can have a valence of 3, as we would predict from its 3 unpaired electrons. In biologically important molecules, however, we can consider it to have a valence of 5, forming three single bonds and one double bond.

The molecules H_2 and O_2 are pure elements, not compounds. (Recall that a compound is a combination of two or more *different* elements.) An example of a molecule that is a compound is water, with the molecular formula H_2O. It takes two atoms of hydrogen to satisfy the valence of one oxygen atom. **Figure 2.11c** shows the structure of a water molecule. Water is so important to life that Chapter 3 is devoted entirely to its structure and behavior.

Another molecule that is a compound is methane, the main component of natural gas, with the molecular formula CH_4 **(Figure 2.11d)**. It takes four hydrogen atoms, each with a valence of 1, to complement one atom of carbon, with its valence of 4. We will look at many other compounds of carbon in Chapter 4.

The attraction of a particular kind of atom for the electrons of a covalent bond is called its **electronegativity.** The more electronegative an atom, the more strongly it pulls shared electrons toward itself. In a covalent bond between two atoms of the same element, the outcome of the tug-of-war for common electrons is a standoff; the two atoms are equally electronegative. Such a bond, in which the electrons are shared equally, is a **nonpolar covalent bond.** For example, the covalent bond of H_2 is nonpolar, as is the double bond of O_2. In other compounds, however, where one atom is bonded to a more electronegative atom, the electrons of the bond are not shared equally. This sort of bond is called a **polar covalent bond.** Such bonds vary in their polarity, depending on the relative electronegativity of the two atoms. For example, the individual bonds of methane (CH_4) are slightly polar because carbon and hydrogen differ slightly in electronegativity. In a more extreme example, the bonds between the oxygen and hydrogen atoms of a water molecule are quite polar **(Figure 2.12)**. Oxygen is one of the most electronegative of the 92 elements, attracting shared electrons much more strongly than hydrogen does. In a covalent bond between oxygen and hydrogen, the electrons spend more time near the oxygen nucleus than they do near the hydrogen nucleus. Because electrons have a negative charge, the unequal sharing of electrons in water causes the oxygen atom to have a partial negative charge (indicated by the Greek letter δ before a minus sign, δ−, or "delta minus") and each hydrogen atom a partial positive charge (δ+, or "delta plus").

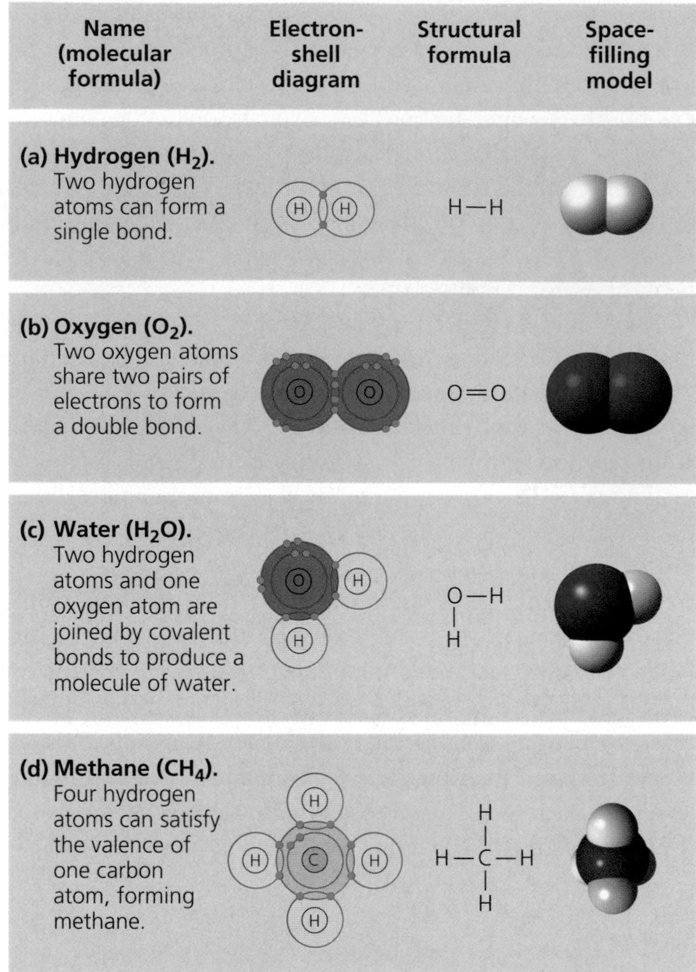

Name (molecular formula)
Electron-shell diagram
Structural formula
Space-filling model
(a) Hydrogen (H_2).
Two hydrogen atoms can form a single bond.
H—H
(b) Oxygen (O_2).
Two oxygen atoms share two pairs of electrons to form a double bond.
O=O
(c) Water (H_2O).
Two hydrogen atoms and one oxygen atom are joined by covalent bonds to produce a molecule of water.
O—H
H
(d) Methane (CH_4).
Four hydrogen atoms can satisfy the valence of one carbon atom, forming methane.
H
H—C—H
H

▲ **Figure 2.11 Covalent bonding in four molecules.** A single covalent bond consists of a pair of shared electrons. The number of electrons required to complete an atom's valence shell generally determines how many bonds that atom will form. Three ways of indicating bonds are shown; the space-filling model comes closest to representing the actual shape of the molecule (see also Figure 2.16).

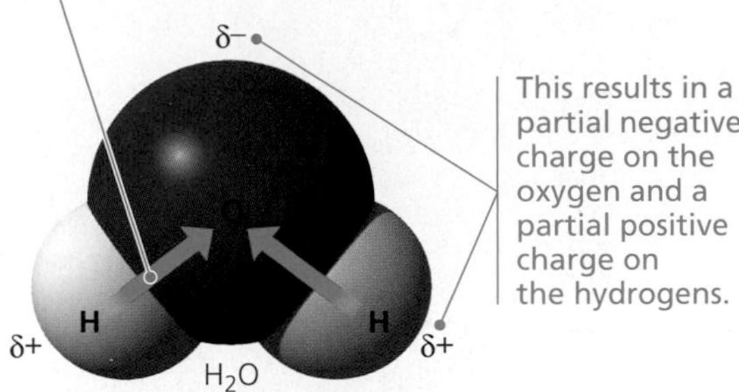
Because oxygen (O) is more electronegative than hydrogen (H), shared electrons are pulled more toward oxygen.
δ−
This results in a partial negative charge on the oxygen and a partial positive charge on the hydrogens.
H
H
δ+
δ+
H_2O

▲ **Figure 2.12 Polar covalent bonds in a water molecule.**

Ionic Bonds

In some cases, two atoms are so unequal in their attraction for valence electrons that the more electronegative atom strips an electron completely away from its partner. This is what happens when an atom of sodium ($_{11}Na$) encounters an atom of chlorine ($_{17}Cl$) **(Figure 2.13)**. A sodium atom has a total of 11 electrons, with its single valence electron in the third electron shell. A chlorine atom has a total of 17 electrons, with 7 electrons in its valence shell. When these two atoms meet, the lone valence electron of sodium is transferred to the chlorine atom, and both atoms end up with their valence shells complete. (Because sodium no longer has an electron in the third shell, the second shell is now the valence shell.)

The electron transfer between the two atoms moves one unit of negative charge from sodium to chlorine. Sodium, now with 11 protons but only 10 electrons, has a net electrical charge of 1+. A charged atom (or molecule) is called an **ion.** When the charge is positive, the ion is specifically called a **cation;** the sodium atom has become a cation. Conversely, the chlorine atom, having gained an extra electron, now has 17 protons and 18 electrons, giving it a net electrical charge of 1−. It has become a chloride ion—an **anion,** or negatively charged ion. Because of their opposite charges, cations and anions attract each other; this attraction is called an **ionic bond.** The transfer of an electron is not the formation of a bond; rather, it allows a bond to form because it results in two ions. Any two ions of opposite charge can form an ionic bond. The ions need not have acquired their charge by an electron transfer with each other.

Compounds formed by ionic bonds are called **ionic compounds,** or **salts.** We know the ionic compound sodium chloride (NaCl) as table salt **(Figure 2.14)**. Salts are often found in nature as crystals of various sizes and shapes, each an aggregate of vast numbers of cations and anions bonded by their electrical attraction and arranged in a three-dimensional lattice. A salt crystal does not consist of molecules in the sense that a covalent compound does, because a covalently bonded molecule has a definite size and number of atoms. The formula for an ionic compound, such as NaCl, indicates only the ratio of elements in a crystal of the salt. "NaCl" is not a molecule.

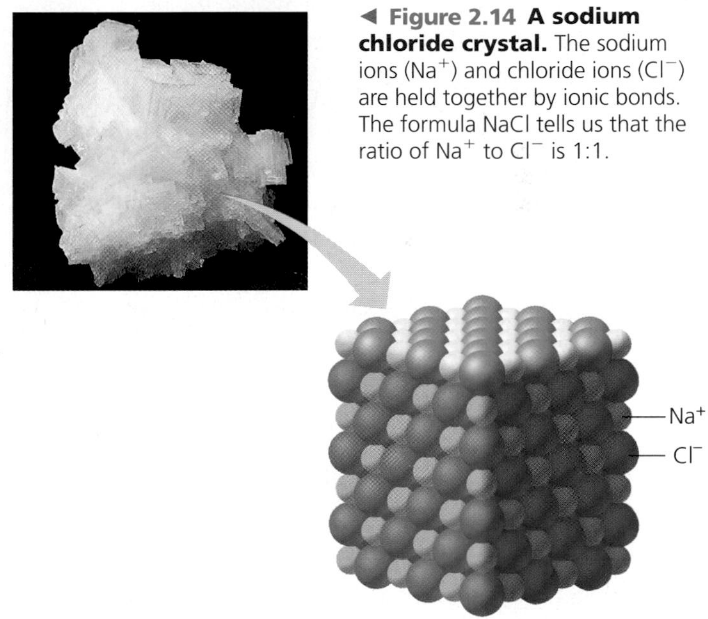
Na^+
Cl^-

◀ **Figure 2.14 A sodium chloride crystal.** The sodium ions (Na^+) and chloride ions (Cl^-) are held together by ionic bonds. The formula NaCl tells us that the ratio of Na^+ to Cl^- is 1:1.

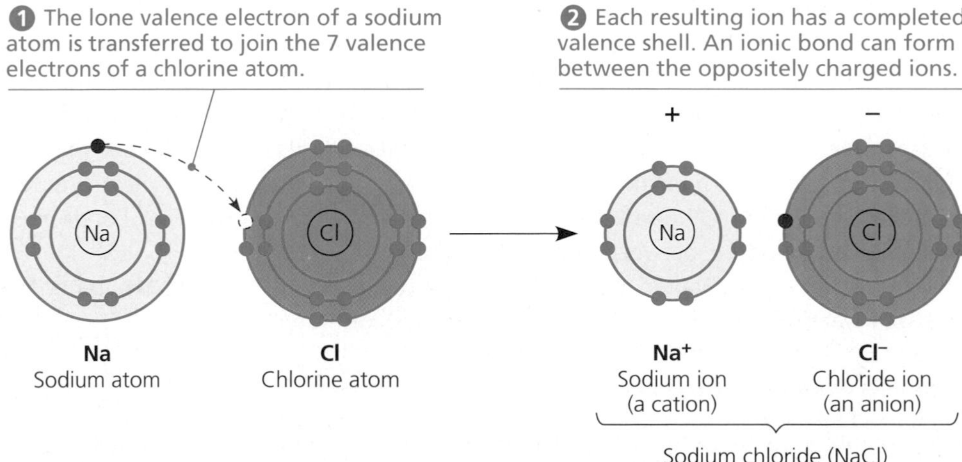
1 The lone valence electron of a sodium atom is transferred to join the 7 valence electrons of a chlorine atom.
2 Each resulting ion has a completed valence shell. An ionic bond can form between the oppositely charged ions.
Na
Cl
Na
Cl
+
−
Na
Sodium atom
Cl
Chlorine atom
Na^+
Sodium ion (a cation)
Cl^-
Chloride ion (an anion)
Sodium chloride (NaCl)

▶ **Figure 2.13 Electron transfer and ionic bonding.** The attraction between oppositely charged atoms, or ions, is an ionic bond. An ionic bond can form between any two oppositely charged ions, even if they have not been formed by transfer of an electron from one to the other.

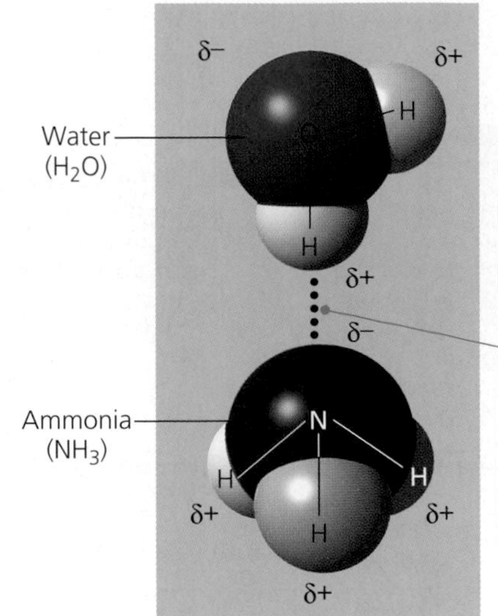

▲ **Figure 2.15 A hydrogen bond.**

Not all salts have equal numbers of cations and anions. For example, the ionic compound magnesium chloride ($MgCl_2$) has two chloride ions for each magnesium ion. Magnesium ($_{12}Mg$) must lose 2 outer electrons if the atom is to have a complete valence shell, so it tends to become a cation with a net charge of 2+ (Mg^{2+}). One magnesium cation can therefore form ionic bonds with two chloride anions.

The term *ion* also applies to entire molecules that are electrically charged. In the salt ammonium chloride (NH_4Cl), for instance, the anion is a single chloride ion (Cl^-), but the cation is ammonium (NH_4^+), a nitrogen atom with four covalently bonded hydrogen atoms. The whole ammonium ion has an electrical charge of 1+ because it is 1 electron short.

Environment affects the strength of ionic bonds. In a dry salt crystal, the bonds are so strong that it takes a hammer and chisel to break enough of them to crack the crystal in two. Place the same salt crystal in water, however, and the salt dissolves as the attractions between its ions decrease. In the next chapter, you will learn how water dissolves salts.

Weak Chemical Bonds

In living organisms, most of the strongest chemical bonds are covalent ones, which link atoms to form a cell's molecules. But weaker bonding within and between molecules is also indispensable in the cell, where the properties of life emerge from such interactions. The most important large biological molecules are held in their functional form by weak bonds. In addition, when two molecules in the cell make contact, they may adhere temporarily by weak bonds. The reversibility of weak bonding can be an advantage: Two molecules can come together, respond to one another in some way, and then separate.

Several types of weak chemical bonds are important in living organisms. One is the ionic bond, which we just discussed. Another type of weak bond, crucial to life, is known as a hydrogen bond.

Hydrogen Bonds

Among the various kinds of weak chemical bonds, hydrogen bonds are so important in the chemistry of life that they deserve special attention. A **hydrogen bond** forms when a hydrogen atom covalently bonded to one electronegative atom is also attracted to another electronegative atom. In living cells, the electronegative partners involved are usually oxygen or nitrogen atoms. Refer to **Figure 2.15** to examine the simple case of hydrogen bonding between water (H_2O) and ammonia (NH_3). In the next chapter, we'll see how hydrogen bonds between water molecules allow some insects to walk on water.

Van der Waals Interactions

Even a molecule with nonpolar covalent bonds may have positively and negatively charged regions. Because electrons are in constant motion, they are not always symmetrically distributed in the molecule; at any instant, they may accumulate by chance in one part of the molecule or another. The results are ever-changing "hot spots" of positive and negative charge that enable all atoms and molecules to stick to one another. These **van der Waals interactions** are weak and occur only when atoms and molecules are very close together. In spite of their weakness, van der Waals interactions were recently shown to be responsible for the ability of a gecko lizard (right) to walk up a wall. Each gecko toe has hundreds of thousands of tiny hairs, with multiple projections at the hair's tip that increase surface area. Apparently, the van der Waals interactions between the hair tip molecules and the molecules of the wall's surface are so numerous that in spite of their individual weakness, together they can support the gecko's body weight.

Van der Waals interactions, hydrogen bonds, ionic bonds, and other weak bonds may form not only between molecules but also between different regions of a single large molecule, such as a protein. Although these bonds are individually weak, their cumulative effect is to reinforce the three-dimensional shape of a large molecule. You will learn more about the very important biological roles of weak bonds in Chapter 5.

Molecular Shape and Function

A molecule has a characteristic size and shape. The precise shape of a molecule is usually very important to its function in the living cell.

A molecule consisting of two atoms, such as H_2 or O_2, is always linear, but molecules with more than two atoms have

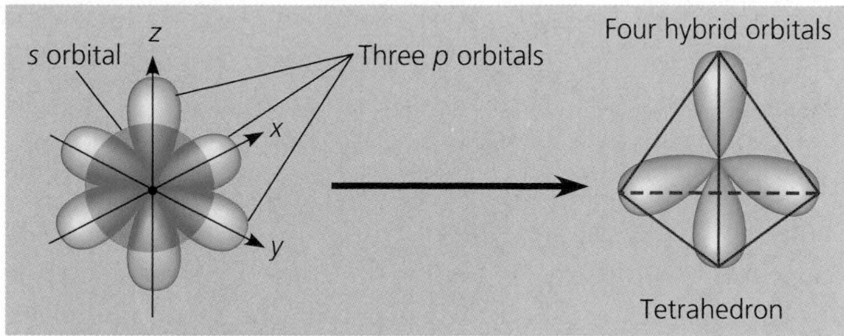

(a) Hybridization of orbitals. The single *s* and three *p* orbitals of a valence shell involved in covalent bonding combine to form four teardrop-shaped hybrid orbitals. These orbitals extend to the four corners of an imaginary tetrahedron (outlined in pink).

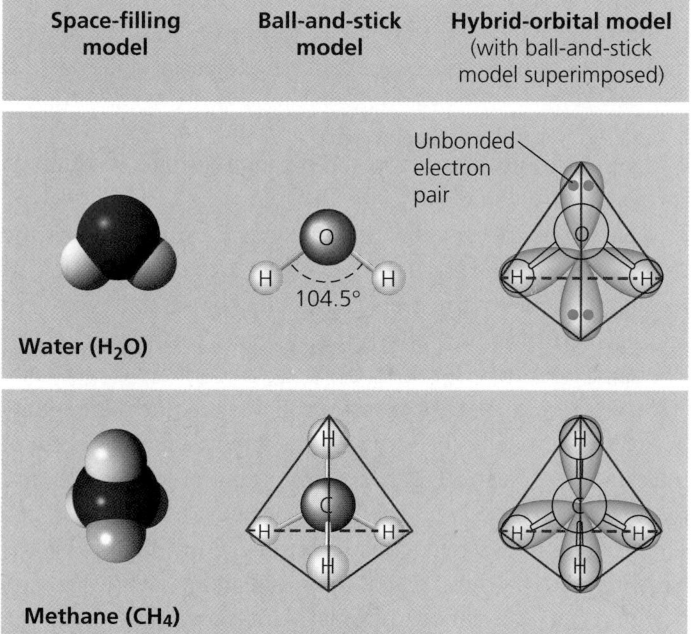

(b) Molecular shape models. Three models representing molecular shape are shown for two examples: water and methane. The positions of the hybrid orbitals determine the shapes of the molecules.

▲ **Figure 2.16 Molecular shapes due to hybrid orbitals.**

more complicated shapes. These shapes are determined by the positions of the atoms' orbitals. When an atom forms covalent bonds, the orbitals in its valence shell rearrange. For atoms with valence electrons in both *s* and *p* orbitals (review Figure 2.9), the single *s* and three *p* orbitals hybridize to form four new hybrid orbitals shaped like identical teardrops extending from the region of the atomic nucleus (**Figure 2.16a**). If we connect the larger ends of the teardrops with lines, we have the outline of a geometric shape called a tetrahedron, similar to a pyramid.

For the water molecule (H_2O), two of the hybrid orbitals in the oxygen atom's valence shell are shared with hydrogen atoms (**Figure 2.16b**). The result is a molecule shaped roughly like a V, its two covalent bonds spread apart at an angle of 104.5°.

The methane molecule (CH_4) has the shape of a completed tetrahedron because all four hybrid orbitals of carbon are shared with hydrogen atoms (see Figure 2.16b). The nucleus of the carbon is at the center, with its four covalent bonds radiating to hydrogen nuclei at the corners of the tetrahedron. Larger molecules containing multiple carbon atoms, including many of the molecules that make up living matter, have more complex overall shapes. However, the tetrahedral shape of a carbon atom bonded to four other atoms is often a repeating motif within such molecules.

Molecular shape is crucial in biology because it determines how biological molecules recognize and respond to one another with specificity. Only molecules with complementary shapes are able to bind to each other by weak bonds. An example of this specificity is provided by a mechanism of pain control. Natural signal molecules called endorphins bind to specific molecules, called receptors, on the surface of brain cells, producing euphoria and relieving pain. It turns out that molecules with shapes similar to endorphins have similar effects. Morphine, heroin, and other opiate drugs, for example, mimic endorphins by binding to endorphin receptors in the brain (**Figure 2.17**). The role of molecular shape in brain chemistry illustrates the relationship between structure and function, one of biology's unifying themes.

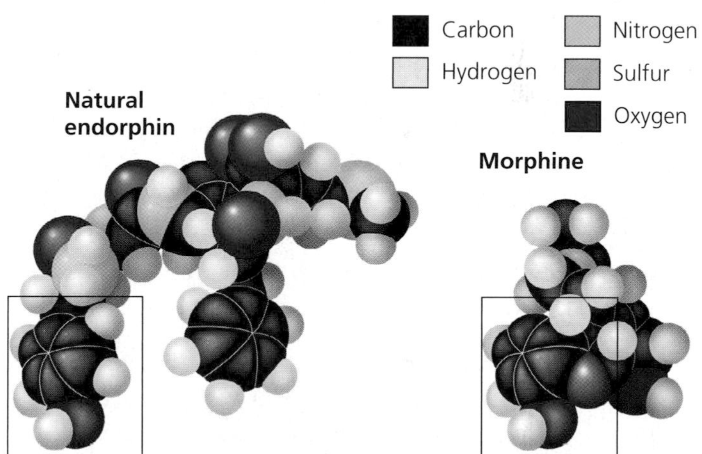

(a) Structures of endorphin and morphine. The boxed portion of the endorphin molecule (left) binds to receptor molecules on target cells in the brain. The boxed portion of the morphine molecule (right) is a close match.

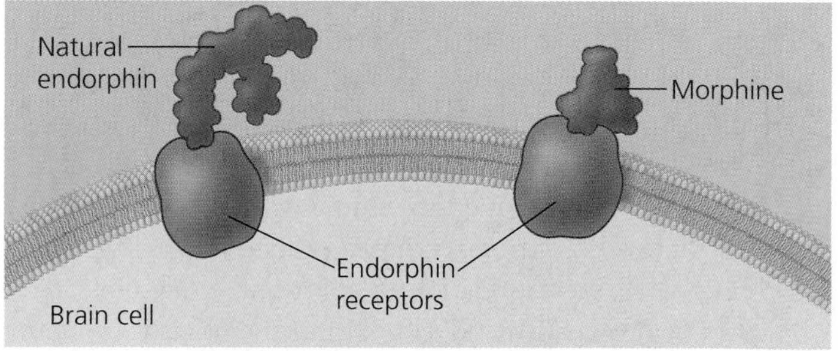

(b) Binding to endorphin receptors. Endorphin receptors on the surface of a brain cell can bind to both endorphin and morphine.

▲ **Figure 2.17 A molecular mimic.** Morphine affects pain perception and emotional state by mimicking the brain's natural endorphins.

Concept Check 2.3

1. Why does the following structure fail to make sense chemically?

$$H—C=C—H$$

2. Explain what holds together the atoms in a crystal of magnesium chloride ($MgCl_2$).

For suggested answers, see Appendix A.

Chemical reactions make and break chemical bonds

The making and breaking of chemical bonds, leading to changes in the composition of matter, are called **chemical reactions.** An example is the reaction between hydrogen and oxygen to form water:

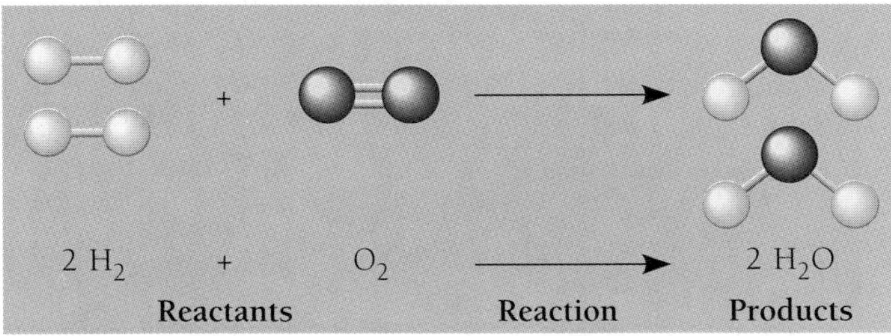

This reaction breaks the covalent bonds of H_2 and O_2 and forms the new bonds of H_2O. When we write a chemical reaction, we use an arrow to indicate the conversion of the starting materials, called the **reactants,** to the **products.** The coefficients indicate the number of molecules involved; for example, the coefficient 2 in front of the H_2 means that the reaction starts with two molecules of hydrogen. Notice that all atoms of the reactants must be accounted for in the products. Matter is conserved in a chemical reaction: Reactions cannot create or destroy matter but can only rearrange it.

Photosynthesis, which takes place within the cells of green plant tissues, is a particularly important example of how chemical reactions rearrange matter. Humans and other animals ultimately depend on photosynthesis for food and oxygen, and this process is at the foundation of almost all ecosystems. The following chemical shorthand summarizes the process of photosynthesis:

$$6\ CO_2 + 6\ H_2O \longrightarrow C_6H_{12}O_6 + 6\ O_2$$

The raw materials of photosynthesis are carbon dioxide (CO_2), which is taken from the air, and water (H_2O), which is absorbed from the soil. Within the plant cells, sunlight powers the conversion of these ingredients to a sugar called glucose ($C_6H_{12}O_6$) and oxygen molecules (O_2), a by-product that the plant releases into the surroundings **(Figure 2.18).** Although photosynthesis is actually a sequence of many chemical reactions, we still end up with the same number and kinds of atoms we had when we started. Matter has simply been rearranged, with an input of energy provided by sunlight.

Some chemical reactions go to completion; that is, all the reactants are converted to products. But most reactions are reversible, the products of the forward reaction becoming the reactants for the reverse reaction. For example, hydrogen and nitrogen molecules can combine to form ammonia, but ammonia can also decompose to regenerate hydrogen and nitrogen:

$$3\ H_2 + N_2 \rightleftharpoons 2\ NH_3$$

The opposite-headed arrows indicate that the reaction is reversible.

One of the factors affecting the rate of a reaction is the concentration of reactants. The greater the concentration of reactant molecules, the more frequently they collide with one another and have an opportunity to react to form products. The same holds true for the products. As products accumulate, collisions resulting in the reverse reaction become increasingly frequent. Eventually, the forward and reverse reactions occur at the same rate, and the relative concentrations of products and reactants stop changing. The point at which the reactions offset one another exactly is called **chemical equilibrium.** This is a dynamic equilibrium; reactions are still going on, but with no net effect on the concentrations of reactants and products. Equilibrium does *not* mean that the reactants and products are

▲ **Figure 2.18 Photosynthesis: a solar-powered rearrangement of matter.** *Elodea,* a freshwater plant, produces sugar by rearranging the atoms of carbon dioxide and water in the chemical process known as photosynthesis, which is powered by sunlight. Much of the sugar is then converted to other food molecules. Oxygen gas (O_2) is a by-product of photosynthesis; notice the bubbles of oxygen escaping from the leaves in the photo.

equal in concentration, but only that their concentrations have stabilized at a particular ratio. The reaction involving ammonia reaches equilibrium when ammonia decomposes as rapidly as it forms. In this case, there is far more ammonia than hydrogen and nitrogen at equilibrium.

We will return to the subject of chemical reactions after more detailed study of the various types of molecules that are important to life. In the next chapter, we focus on water, the substance in which all the chemical processes of living organisms occur.

Concept Check 2.4

1. Refer to the reaction between hydrogen and oxygen to form water, shown as a ball-and-stick model on page 44. Draw the electron-shell diagram representing this reaction.
2. Which occurs faster at equilibrium, the formation of products from reactants, or reactants from products?

For suggested answers, see Appendix A.

Chapter 2 Review

Go to www.campbellbiology.com or the student CD-ROM to explore Activities, Investigations, and other interactive study aids.

SUMMARY OF KEY CONCEPTS

Concept 2.1

Matter consists of chemical elements in pure form and in combinations called compounds

- **Elements and Compounds (pp. 32–33)** Elements cannot be broken down chemically to other substances. A compound contains two or more elements in a fixed ratio.
- **Essential Elements of Life (pp. 33–34)** Carbon, oxygen, hydrogen, and nitrogen make up approximately 96% of living matter.
  Investigation *How Are Space Rocks Analyzed for Signs of Life?*

Concept 2.2

An element's properties depend on the structure of its atoms

- **Subatomic Particles (p. 34)** An atom is the smallest unit of an element. An atom has a nucleus made up of positively charged protons and uncharged neutrons, as well as a surrounding cloud of negatively charged electrons.
  Activity *Structure of the Atomic Nucleus*
- **Atomic Number and Atomic Mass (pp. 34–35)** The number of electrons in an electrically neutral atom equals the number of protons.
- **Isotopes (pp. 35–36)** Most elements have two or more isotopes, different in neutron number and therefore mass. Some isotopes are unstable and give off particles and energy as radioactivity. Radioactive tracers help biologists monitor biological processes.
- **The Energy Levels of Electrons (pp. 36–37)** In an atom, electrons occupy specific energy levels, each of which can be represented by an electron shell of that atom.
- **Electron Configuration and Chemical Properties (pp. 37–38)** Electron configuration determines the chemical behavior of an atom. Chemical behavior depends on the number of valence electrons—electrons in the outermost shell. An atom with an incomplete valence shell is reactive.
  Activity *Electron Arrangement*
- **Electron Orbitals (pp. 38–39)** Electrons move within orbitals, three-dimensional spaces with specific shapes located within each successive shell.
  Activity *Build an Atom*

Concept 2.3

The formation and function of molecules depend on chemical bonding between atoms

- **Covalent Bonds (pp. 39–41)** Chemical bonds form when atoms interact and complete their valence shells. A single covalent bond is the sharing of a pair of valence electrons by two atoms; double bonds are the sharing of two pairs of electrons. Molecules consist of two or more covalently bonded atoms. Electrons of a polar covalent bond are pulled closer to the more electronegative atom. A covalent bond is nonpolar if both atoms are the same and therefore equally electronegative.
  Activity *Covalent Bonds*
  Activity *Nonpolar and Polar Molecules*
- **Ionic Bonds (pp. 41–42)** Two atoms may differ so much in electronegativity that one or more electrons are actually transferred from one atom to the other. The result is a negatively charged ion (anion) and a positively charged ion (cation). The attraction between two ions of opposite charge is called an ionic bond.
  Activity *Ionic Bonds*
- **Weak Chemical Bonds (p. 42)** A hydrogen bond is a weak attraction between one electronegative atom and a hydrogen atom that is covalently linked to another electronegative atom. Van der Waals interactions occur when transiently positive and negative regions of molecules attract each other. Weak bonds reinforce the shapes of large molecules and help molecules adhere to each other.
  Activity *Hydrogen Bonds*
- **Molecular Shape and Function (pp. 42–44)** A molecule's shape is determined by the positions of its atoms' valence orbitals. When covalent bonds form, the *s* and *p* orbitals in the valence shell of an atom may combine to form four hybrid orbitals that extend to the corners of an imaginary tetrahedron; such orbitals are responsible for the shapes of H_2O, CH_4, and many more complex biological molecules. Shape is usually the basis for the recognition of one biological molecule by another.

Concept 2.4

Chemical reactions make and break chemical bonds

- Chemical reactions change reactants into products while conserving matter. Most chemical reactions are reversible. Chemical equilibrium is reached when the forward and reverse reaction rates are equal (pp. 44–45).

TESTING YOUR KNOWLEDGE

Self-Quiz

1. An element is to a (an) _____ as an organ is to a (an) _____.
   a. atom; organism
   b. compound; organism
   c. molecule; cell
   d. atom; cell
   e. compound; organelle

2. In the term *trace element,* the modifier *trace* means
   a. the element is required in very small amounts.
   b. the element can be used as a label to trace atoms through an organism's metabolism.
   c. the element is very rare on Earth.
   d. the element enhances health but is not essential for the organism's long-term survival.
   e. the element passes rapidly through the organism.

3. Compared to ^{31}P, the radioactive isotope ^{32}P has
   a. a different atomic number.
   b. one more neutron.
   c. one more proton.
   d. one more electron.
   e. a different charge.

4. Atoms can be represented by simply listing the number of protons, neutrons, and electrons—for example, $2\,p^+$; $2\,n^0$; $2\,e^-$ for helium. Which atom represents the ^{18}O isotope of oxygen?
   a. $6\,p^+$; $8\,n^0$; $6\,e^-$
   b. $8\,p^+$; $10\,n^0$; $8\,e^-$
   c. $9\,p^+$; $9\,n^0$; $9\,e^-$
   d. $7\,p^+$; $2\,n^0$; $9\,e^-$
   e. $10\,p^+$; $8\,n^0$; $9\,e^-$

5. The atomic number of sulfur is 16. Sulfur combines with hydrogen by covalent bonding to form a compound, hydrogen sulfide. Based on the electron configuration of sulfur, we can predict that the molecular formula of the compound will be
   a. HS b. HS_2 c. H_2S d. H_3S_2 e. H_4S

6. Review the valences of carbon, oxygen, hydrogen, and nitrogen, and then determine which of the following molecules is most likely to exist.

a. O═C—H

b.
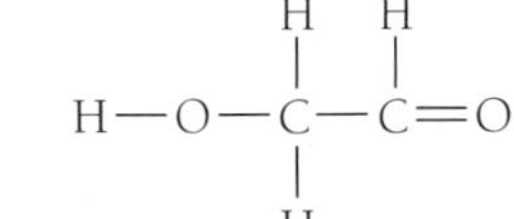

c.
```
   H       H
   |       |
H—C—H—C═O
   |
   H
```

d.
```
   O
   |
H—N═H
```

7. The reactivity of an atom arises from
   a. the average distance of the outermost electron shell from the nucleus.
   b. the existence of unpaired electrons in the valence shell.
   c. the sum of the potential energies of all the electron shells.
   d. the potential energy of the valence shell.
   e. the energy difference between the *s* and *p* orbitals.

8. Which of these statements is true of all anionic atoms?
   a. The atom has more electrons than protons.
   b. The atom has more protons than electrons.
   c. The atom has fewer protons than does a neutral atom of the same element.
   d. The atom has more neutrons than protons.
   e. The net charge is 1−.

9. What coefficients must be placed in the blanks so that all atoms are accounted for in the products?

$$C_6H_{12}O_6 \longrightarrow \_\_C_2H_6O + \_\_CO_2$$

   a. 1; 2 b. 2; 2 c. 1; 3 d. 1; 1 e. 3; 1

10. Which of the following statements correctly describes any chemical reaction that has reached equilibrium?
    a. The concentration of products equals the concentration of reactants.
    b. The rate of the forward reaction equals the rate of the reverse reaction.
    c. Both forward and reverse reactions have halted.
    d. The reaction is now irreversible.
    e. No reactants remain.

For Self-Quiz Answers, see Appendix A.

Go to the website or CD-ROM for more quiz questions.

Evolution Connection

The text states that the percentages of naturally occurring elements making up the human body (see Table 2.1) are similar to the percentages of these elements found in other organisms. How could you account for this similarity among organisms?

Scientific Inquiry

Female silkworm moths (*Bombyx mori*) attract males by emitting chemical signals that spread through the air. A male hundreds of meters away can detect these molecules and fly toward their source. The sensory organs responsible for this behavior are the comblike antennae visible in the photograph here. Each filament of an antenna is equipped with thousands of receptor cells that detect the sex attractant. Based on what you learned in this chapter, propose a hypothesis to account for the ability of the male moth to detect a specific molecule in the presence of many other molecules in the air. What predictions does your hypothesis make? Design an experiment to test one of these predictions.

Investigation *How Are Space Rocks Analyzed for Signs of Life?*

Science, Technology, and Society

While waiting at an airport, Neil Campbell once overheard this claim: "It's paranoid and ignorant to worry about industry or agriculture contaminating the environment with their chemical wastes. After all, this stuff is just made of the same atoms that were already present in our environment." How would you counter this argument?

3 Water and the Fitness of the Environment

▲ **Figure 3.1 A view of Earth from space, showing our planet's abundance of water.**

Key Concepts

3.1 **The polarity of water molecules results in hydrogen bonding**
3.2 **Four emergent properties of water contribute to Earth's fitness for life**
3.3 **Dissociation of water molecules leads to acidic and basic conditions that affect living organisms**

Overview

The Molecule That Supports All of Life

As astronomers study newly discovered planets orbiting distant stars, they hope to find evidence of water on these far-off celestial bodies, for water is the substance that makes possible life as we know it here on Earth. All organisms familiar to us are made mostly of water and live in an environment dominated by water. Water is the biological medium here on Earth, and possibly on other planets as well.

Life on Earth began in water and evolved there for 3 billion years before spreading onto land. Modern life, even terrestrial (land-dwelling) life, remains tied to water. All living organisms require water more than any other substance. Human beings, for example, can survive for quite a few weeks without food, but only a week or so without water. Molecules of water participate in many chemical reactions necessary to sustain life. Most cells are surrounded by water, and cells themselves are about 70–95% water. Three-quarters of Earth's surface is submerged in water (**Figure 3.1**). Although most of this water is in liquid form, water is also present on Earth as ice and vapor. Water is the only common substance to exist in the natural environment in all three physical states of matter: solid, liquid, and gas.

The abundance of water is a major reason Earth is habitable. In a classic book called *The Fitness of the Environment,* ecologist Lawrence Henderson highlights the importance of water to life. While acknowledging that life adapts to its environment through natural selection, Henderson emphasizes that for life to exist at all, the environment must first be a suitable abode. In this chapter, you will learn how the structure of a water molecule allows it to form weak chemical bonds with other molecules, including other water molecules. This ability leads to unique properties that support and maintain living systems on our planet. Your objective in this chapter is to develop a conceptual understanding of how water contributes to the fitness of Earth for life.

Concept 3.1

The polarity of water molecules results in hydrogen bonding

Water is so common that it is easy to overlook the fact that it is an exceptional substance with many extraordinary qualities. Following the theme of emergent properties, we can trace water's unique behavior to the structure and interactions of its molecules.

Studied in isolation, the water molecule is deceptively simple. Its two hydrogen atoms are joined to the oxygen atom by single covalent bonds. Because oxygen is more electronegative than hydrogen, the electrons of the polar bonds spend more time closer to the oxygen atom. In other words, the bonds that hold together the atoms in a water molecule are polar covalent bonds. The water molecule, shaped something like a wide V, is a **polar molecule,** meaning that opposite ends of the molecule have opposite charges: The oxygen region of the molecule has

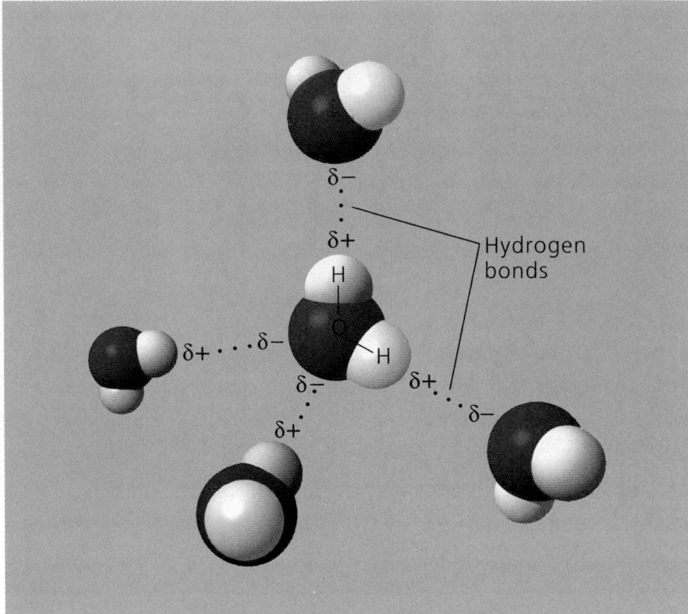

▲ **Figure 3.2 Hydrogen bonds between water molecules.** The charged regions of a polar water molecule are attracted to oppositely charged parts of neighboring molecules. Each molecule can hydrogen-bond to multiple partners, and these associations are constantly changing. At any instant in liquid water at 37°C (human body temperature), about 15% of the molecules are bonded to four partners in short-lived clusters.

a partial negative charge (δ−), and the hydrogens have a partial positive charge (δ+) (see Figure 2.12).

The anomalous properties of water arise from attractions between these polar molecules. The attraction is electrical; the slightly positive hydrogen of one molecule is attracted to the slightly negative oxygen of a nearby molecule. The two molecules are thus held together by a hydrogen bond **(Figure 3.2)**. Although the arrangement of molecules in a sample of liquid water is constantly changing, at any given moment, many of the molecules are linked by multiple hydrogen bonds. The extraordinary qualities of water are emergent properties resulting from the hydrogen bonding that orders molecules into a higher level of structural organization.

Concept Check 3.1

1. What is electronegativity and how does it affect interactions between water molecules?
2. Why is it unlikely that two neighboring water molecules would be arranged like this?

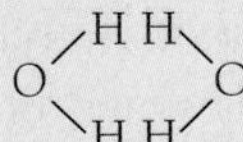

For suggested answers, see Appendix A.

Concept 3.2

Four emergent properties of water contribute to Earth's fitness for life

We will examine four of water's properties that contribute to the suitability of Earth as an environment for life. These are water's cohesive behavior, its ability to moderate temperature, its expansion upon freezing, and its versatility as a solvent.

Cohesion

Water molecules stay close to each other as a result of hydrogen bonding. When water is in its liquid form, its hydrogen bonds are very fragile, about one-twentieth as strong as covalent bonds. They form, break, and re-form with great frequency. Each hydrogen bond lasts only a few trillionths of a second, but the molecules are constantly forming new bonds with a succession of partners. Thus, at any instant, a substantial percentage of all the water molecules are bonded to their neighbors, making water more structured than most other liquids. Collectively, the hydrogen bonds hold the substance together, a phenomenon called **cohesion.**

Cohesion due to hydrogen bonding contributes to the transport of water and dissolved nutrients against gravity in plants **(Figure 3.3)**. Water from the roots reaches the leaves through

▲ **Figure 3.3 Water transport in plants.** Evaporation from leaves pulls water upward from the roots through water-conducting cells, in this case located in the trunk of a tree. Cohesion due to hydrogen bonding helps hold together the column of water within the cells. Adhesion of the water to cell walls helps resist the downward pull of gravity. Because of these properties, the tallest trees can transport water more than 100 meters (m) upward—approximately one-quarter the height of the Empire State Building in New York City.

▲ **Figure 3.4 Walking on water.** The high surface tension of water, resulting from the collective strength of its hydrogen bonds, allows the water strider to walk on the surface of a pond.

a network of water-conducting cells. As water evaporates from a leaf, hydrogen bonds cause water molecules leaving the veins to tug on molecules farther down, and the upward pull is transmitted through the water-conducting cells all the way down to the roots. **Adhesion,** the clinging of one substance to another, also plays a role. Adhesion of water to the walls of the cells helps counter the downward pull of gravity.

Related to cohesion is **surface tension,** a measure of how difficult it is to stretch or break the surface of a liquid. Water has a greater surface tension than most other liquids. At the interface between water and air is an ordered arrangement of water molecules, hydrogen-bonded to one another and to the water below. This makes the water behave as though coated with an invisible film. You can observe the surface tension of water by slightly overfilling a drinking glass; the water will stand above the rim. In a more biological example, some animals can stand, walk, or run on water without breaking the surface **(Figure 3.4)**.

Moderation of Temperature

Water moderates air temperature by absorbing heat from air that is warmer and releasing the stored heat to air that is cooler. Water is effective as a heat bank because it can absorb or release a relatively large amount of heat with only a slight change in its own temperature. To understand this capability of water, we must first look briefly at heat and temperature.

Heat and Temperature

Anything that moves has **kinetic energy,** the energy of motion. Atoms and molecules have kinetic energy because they are always moving, although not necessarily in any particular direction. The faster a molecule moves, the greater its kinetic energy. **Heat** is a measure of the *total* amount of kinetic energy due to molecular motion in a body of matter. **Temperature** measures the intensity of heat due to the *average* kinetic energy of the molecules. When the average speed of the molecules increases, a thermometer records this as a rise in temperature. Heat and temperature are related, but they are not the same. A swimmer crossing the English Channel has a higher temperature than the water, but the ocean contains far more heat because of its volume.

Whenever two objects of different temperature are brought together, heat passes from the warmer to the cooler object until the two are the same temperature. Molecules in the cooler object speed up at the expense of the kinetic energy of the warmer object. An ice cube cools a drink not by adding coldness to the liquid, but by absorbing heat from the liquid as the ice itself melts.

Throughout this book, we will use the **Celsius scale** to indicate temperature (Celsius degrees are abbreviated as °C). At sea level, water freezes at 0°C and boils at 100°C. The temperature of the human body averages 37°C, and comfortable room temperature is about 20–25°C.

One convenient unit of heat used in this book is the **calorie (cal).** A calorie is the amount of heat it takes to raise the temperature of 1 g of water by 1°C. Conversely, a calorie is also the amount of heat that 1 g of water releases when it cools by 1°C. A **kilocalorie (kcal),** 1,000 cal, is the quantity of heat required to raise the temperature of 1 kilogram (kg) of water by 1°C. (The "calories" on food packages are actually kilocalories.) Another energy unit used in this book is the **joule (J).** One joule equals 0.239 cal; one calorie equals 4.184 J.

Water's High Specific Heat

The ability of water to stabilize temperature stems from its relatively high specific heat. The **specific heat** of a substance is defined as the amount of heat that must be absorbed or lost for 1 g of that substance to change its temperature by 1°C. We already know water's specific heat because we have defined a calorie as the amount of heat that causes 1 g of water to change its temperature by 1°C. Therefore, the specific heat of water is 1 calorie per gram per degree Celsius, abbreviated as 1 cal/g/°C. Compared with most other substances, water has an unusually high specific heat. For example, ethyl alcohol, the type of alcohol in alcoholic beverages, has a specific heat of 0.6 cal/g/°C—that is, only 0.6 cal is required to raise the temperature of 1g of ethyl alcohol 1°C.

Because of the high specific heat of water relative to other materials, water will change its temperature less when it absorbs or loses a given amount of heat. The reason you can burn your fingers by touching the metal handle of a pot on the stove when the water in the pot is still lukewarm is that the specific heat of water is ten times greater than that of iron. In other words, it will take only 0.1 cal to raise the temperature of 1 g of iron 1°C. Specific heat can be thought of as a measure of how well a substance resists changing its temperature when it absorbs or releases heat. Water resists changing its temperature; when it does change its temperature, it absorbs or loses a relatively large quantity of heat for each degree of change.

We can trace water's high specific heat, like many of its other properties, to hydrogen bonding. Heat must be absorbed in order to break hydrogen bonds, and heat is released when hydrogen bonds form. A calorie of heat causes a relatively small change in the temperature of water because much of the heat is used to disrupt hydrogen bonds before the water molecules can begin moving faster. And when the temperature of water drops slightly, many additional hydrogen bonds form, releasing a considerable amount of energy in the form of heat.

What is the relevance of water's high specific heat to life on Earth? A large body of water can absorb and store a huge amount of heat from the sun in the daytime and during summer while warming up only a few degrees. And at night and during winter, the gradually cooling water can warm the air. This is the reason coastal areas generally have milder climates than inland regions. The high specific heat of water also tends to stabilize ocean temperatures, creating a favorable environment for marine life. Thus, because of its high specific heat, the water that covers most of Earth keeps temperature fluctuations on land and in water within limits that permit life. Also, because organisms are made primarily of water, they are more able to resist changes in their own temperature than if they were made of a liquid with a lower specific heat.

Evaporative Cooling

Molecules of any liquid stay close together because they are attracted to one another. Molecules moving fast enough to overcome these attractions can depart the liquid and enter the air as gas. This transformation from a liquid to a gas is called vaporization, or *evaporation*. Recall that the speed of molecular movement varies and that temperature is the *average* kinetic energy of molecules. Even at low temperatures, the speediest molecules can escape into the air. Some evaporation occurs at any temperature; a glass of water at room temperature, for example, will eventually evaporate. If a liquid is heated, the average kinetic energy of molecules increases and the liquid evaporates more rapidly.

Heat of vaporization is the quantity of heat a liquid must absorb for 1 g of it to be converted from the liquid to the gaseous state. For the same reason that water has a high specific heat, it also has a high heat of vaporization relative to most other liquids. To evaporate 1 g of water at 25°C, about 580 cal of heat is needed—nearly double the amount needed to vaporize a gram of alcohol or ammonia. Water's high heat of vaporization is another emergent property caused by hydrogen bonds, which must be broken before the molecules can make their exodus from the liquid.

Water's high heat of vaporization helps moderate Earth's climate. A considerable amount of solar heat absorbed by tropical seas is consumed during the evaporation of surface water. Then, as moist tropical air circulates poleward, it releases heat as it condenses to form rain.

As a liquid evaporates, the surface of the liquid that remains behind cools down. This **evaporative cooling** occurs because the "hottest" molecules, those with the greatest kinetic energy, are the most likely to leave as gas. It is as if the hundred fastest runners at a college transferred to another school; the average speed of the remaining students would decline.

Evaporative cooling of water contributes to the stability of temperature in lakes and ponds and also provides a mechanism that prevents terrestrial organisms from overheating. For example, evaporation of water from the leaves of a plant helps keep the tissues in the leaves from becoming too warm in the sunlight. Evaporation of sweat from human skin dissipates body heat and helps prevent overheating on a hot day or when excess heat is generated by strenuous activity. High humidity on a hot day increases discomfort because the high concentration of water vapor in the air inhibits the evaporation of sweat from the body.

Insulation of Bodies of Water by Floating Ice

Water is one of the few substances that are less dense as a solid than as a liquid. In other words, ice floats in liquid water. While other materials contract when they solidify, water expands. The cause of this exotic behavior is, once again, hydrogen bonding. At temperatures above 4°C, water behaves like other liquids, expanding as it warms and contracting as it cools. Water begins to freeze when its molecules are no longer moving vigorously enough to break their hydrogen bonds. As the temperature falls to 0°C, the water becomes locked into a crystalline lattice, each water molecule bonded to four partners (**Figure 3.5**). The hydrogen bonds keep the molecules at "arm's length," far enough apart to make ice about 10% less dense (10% fewer molecules for the same volume) than liquid water at 4°C. When ice absorbs enough heat for its temperature to rise above 0°C, hydrogen bonds between molecules are disrupted. As the crystal collapses, the ice melts, and molecules are free to slip closer together. Water reaches its greatest density at 4°C and then begins to expand as the molecules move faster. Keep in mind, however, that even in liquid water, many of the molecules are connected by hydrogen bonds, though only transiently: The hydrogen bonds are constantly breaking and re-forming.

The ability of ice to float because of the expansion of water as it solidifies is an important factor in the fitness of the environment. If ice sank, then eventually all ponds, lakes, and even oceans would freeze solid, making life as we know it impossible on Earth. During summer, only the upper few inches of the ocean would thaw. Instead, when a deep body of

Hydrogen bond
Ice
Hydrogen bonds are stable
Liquid water
Hydrogen bonds constantly break and re-form

▲ **Figure 3.5 Ice: crystalline structure and floating barrier.** In ice, each molecule is hydrogen-bonded to four neighbors in a three-dimensional crystal. Because the crystal is spacious, ice has fewer molecules than an equal volume of liquid water. In other words, ice is less dense than liquid water. Floating ice becomes a barrier that protects the liquid water below from the colder air. The marine organism shown here is called a euphausid shrimp; it was photographed beneath the antarctic ice.

water cools, the floating ice insulates the liquid water below, preventing it from freezing and allowing life to exist under the frozen surface, as shown in the photo in Figure 3.5.

The Solvent of Life

A sugar cube placed in a glass of water will dissolve. The glass will then contain a uniform mixture of sugar and water; the concentration of dissolved sugar will be the same everywhere in the mixture. A liquid that is a completely homogeneous mixture of two or more substances is called a **solution.** The dissolving agent of a solution is the **solvent,** and the substance that is dissolved is the **solute.** In this case, water is the solvent and sugar is the solute. An **aqueous solution** is one in which water is the solvent.

The medieval alchemists tried to find a universal solvent, one that would dissolve anything. They learned that nothing works better than water. However, water is not a universal solvent; if it were, it would dissolve any container in which it was stored, including our cells. But water is a very versatile solvent, a quality we can trace to the polarity of the water molecule.

Suppose, for example, that a crystal of the ionic compound sodium chloride (NaCl) is placed in water **(Figure 3.6)**. At the surface of the crystal, the sodium and chloride ions are exposed to the solvent. These ions and the water molecules have a mutual affinity through electrical attraction. The oxygen regions of the water molecules are negatively charged and cling to sodium cations. The hydrogen regions of the water molecules are positively charged and are attracted to chloride anions. As a result, water molecules surround the individual sodium and chloride ions, separating and shielding them from one another. The sphere of water molecules around each dissolved ion is called a **hydration shell.** Working inward from the surface of the salt crystal, water eventually dissolves all the ions. The result is a solution of two solutes, sodium cations and chloride anions, homogeneously mixed with water, the solvent. Other ionic compounds also dissolve in water. Seawater, for instance, contains a great variety of dissolved ions, as do living cells.

A compound does not need to be ionic to dissolve in water; compounds made up of nonionic polar molecules, such as

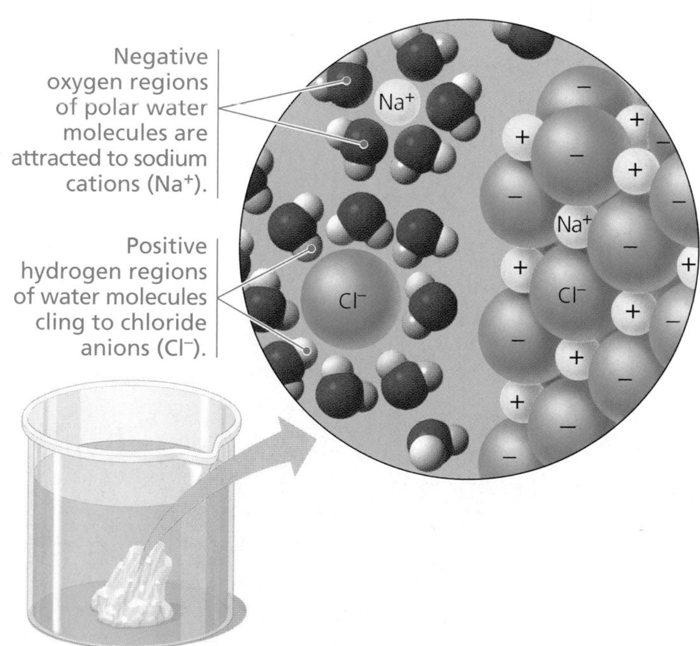
Negative oxygen regions of polar water molecules are attracted to sodium cations (Na^+).
Positive hydrogen regions of water molecules cling to chloride anions (Cl^-).
Na^+
Na^+
Cl^-
Cl^-

▲ **Figure 3.6 A crystal of table salt dissolving in water.** A sphere of water molecules, called a hydration shell, surrounds each solute ion.

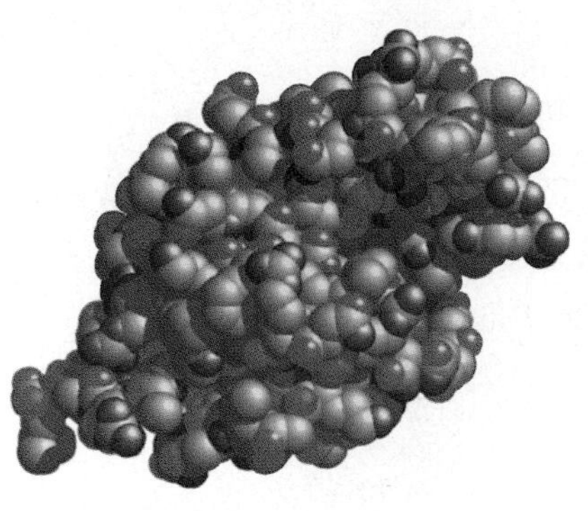

(a) Lysozyme molecule in a nonaqueous environment

(b) Lysozyme molecule (purple) in an aqueous environment such as tears or saliva

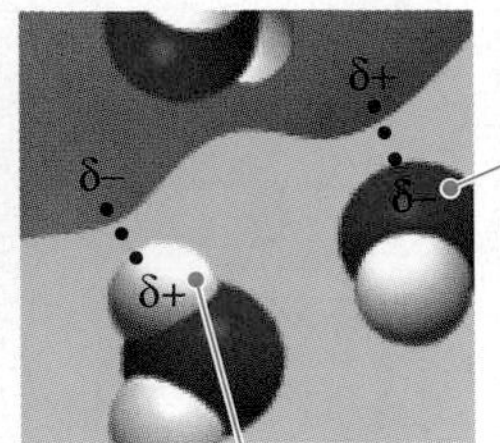

δ+
δ–
δ–
δ+
This oxygen is attracted to a slight positive charge on the lysozyme molecule.
This hydrogen is attracted to a slight negative charge on the lysozyme molecule.

(c) Ionic and polar regions on the protein's surface attract water molecules.

▲ **Figure 3.7 A water-soluble protein.** This figure shows human lysozyme, a protein found in tears and saliva that has antibacterial action.

sugars, are also water-soluble. Such compounds dissolve when water molecules surround each of the solute molecules. Even molecules as large as proteins can dissolve in water if they have ionic and polar regions on their surface **(Figure 3.7)**. Many different kinds of polar compounds are dissolved (along with ions) in the water of such biological fluids as blood, the sap of plants, and the liquid within all cells. Water is the solvent of life.

Hydrophilic and Hydrophobic Substances

Whether ionic or polar, any substance that has an affinity for water is said to be **hydrophilic** (from the Greek *hydro*, water, and *philios*, loving). In some cases, substances can be hydrophilic without actually dissolving. For example, some components in cells are such large molecules (or complexes of multiple molecules) that they do not dissolve. Instead, they remain suspended in the aqueous liquid of the cell. Such a mixture is an example of a **colloid**, a stable suspension of fine particles in a liquid. Another example of a hydrophilic substance that does not dissolve is cotton, a plant product. Cotton consists of giant molecules of cellulose, a compound with numerous regions of partial positive and partial negative charges associated with polar bonds. Water adheres to the cellulose fibers. Thus, a cotton towel does a great job of drying the body, yet does not dissolve in the washing machine. Cellulose is also present in the walls of water-conducting cells in a plant; you read earlier how the adhesion of water to these hydrophilic walls allows water transport to occur.

There are, of course, substances that do not have an affinity for water. Substances that are nonionic and nonpolar actually seem to repel water; these substances are said to be **hydrophobic** (from the Greek *phobos*, fearing). An example from the kitchen is vegetable oil, which, as you know, does not mix stably with water-based substances such as vinegar. The hydrophobic behavior of the oil molecules results from a prevalence of relatively nonpolar bonds, in this case bonds between carbon and hydrogen, which share electrons almost equally. Hydrophobic molecules related to oils are major ingredients of cell membranes. (Imagine what would happen to a cell if its membrane dissolved.)

Solute Concentration in Aqueous Solutions

Biological chemistry is "wet" chemistry. Most of the chemical reactions in organisms involve solutes dissolved in water. To understand chemical reactions, we need to know how many atoms and molecules are involved. Thus, it is important to learn how to calculate the concentration of solutes in an aqueous solution (the number of solute molecules in a volume of solution).

When carrying out experiments, we use mass to calculate the number of molecules. We know the mass of each atom in a given molecule, so we can calculate its **molecular mass**, which is simply the sum of the masses of all the atoms in a molecule. As an example, let's calculate the molecular mass of table sugar (sucrose), which has the molecular formula $C_{12}H_{22}O_{11}$. In round numbers of daltons, the mass of a carbon atom is 12, the mass of a hydrogen atom is 1, and the mass of an oxygen atom is 16. Thus, sucrose has a molecular mass of 342 daltons. Of course, weighing out small numbers of molecules is not practical. For this reason, we usually measure substances in units called moles. Just as a dozen always means 12 objects, a **mole (mol)** represents an exact number of objects—6.02×10^{23}, which is called Avogadro's number. Because of the way in which Avogadro's number and the unit *dalton* were originally defined, there are 6.02×10^{23} daltons in 1 gram. This is significant because once we determine the molecular mass of a molecule such as sucrose, we can use the same number (342), but with the unit *gram*, to represent the mass of 6.02×10^{23} molecules of sucrose, or one mole of sucrose (this is sometimes called the *molar mass*).

To obtain one mole of sucrose in the lab, therefore, we weigh out 342 g.

The practical advantage of measuring a quantity of chemicals in moles is that a mole of one substance has exactly the same number of molecules as a mole of any other substance. If the molecular mass of substance A is 342 daltons and that of substance B is 10 daltons, then 342 g of A will have the same number of molecules as 10 g of B. A mole of ethyl alcohol (C_2H_6O) also contains 6.02×10^{23} molecules, but its mass is only 46 g because the mass of a molecule of ethyl alcohol is less than that of a molecule of sucrose. Measuring in moles makes it convenient for scientists working in the laboratory to combine substances in fixed ratios of molecules.

How would we make a liter (L) of solution consisting of 1 mol of sucrose dissolved in water? We would measure out 342 g of sucrose and then gradually add water, while stirring, until the sugar was completely dissolved. We would then add enough water to bring the total volume of the solution up to 1 L. At that point, we would have a 1-molar (1 *M*) solution of sucrose. **Molarity**—the number of moles of solute per liter of solution—is the unit of concentration most often used by biologists for aqueous solutions.

Concept Check 3.2

1. Describe how properties of water contribute to the upward movement of water in a tree.
2. Explain the popular adage, "It's not the heat, it's the humidity."
3. How can the freezing of water crack boulders?
4. How would you make a 0.5-molar (0.5 *M*) solution of sodium chloride (NaCl)? (The atomic mass of Na is 23 daltons and that of Cl is 35.5 daltons.)

For suggested answers, see Appendix A.

Concept 3.3

Dissociation of water molecules leads to acidic and basic conditions that affect living organisms

Occasionally, a hydrogen atom participating in a hydrogen bond between two water molecules shifts from one molecule to the other. When this happens, the hydrogen atom leaves its electron behind, and what is actually transferred is a **hydrogen ion,** a single proton with a charge of 1+. The water molecule that lost a proton is now a **hydroxide ion** (OH^-), which has a charge of 1−. The proton binds to the other water molecule, making that molecule a hydronium ion (H_3O^+). We can picture the chemical reaction this way:

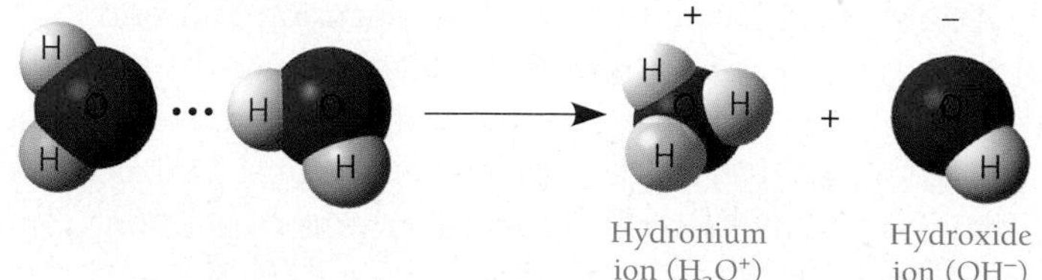

Although this is what actually happens, we can think of the process in a simplified way, as the dissociation (separation) of a water molecule into a hydrogen ion and a hydroxide ion:

$$H_2O \rightleftharpoons \underset{\text{Hydrogen ion}}{H^+} + \underset{\text{Hydroxide ion}}{OH^-}$$

As the double arrows indicate, this is a reversible reaction that will reach a state of dynamic equilibrium when water dissociates at the same rate that it is being re-formed from H^+ and OH^-. At this equilibrium point, the concentration of water molecules greatly exceeds the concentrations of H^+ and OH^-. In fact, in pure water, only one water molecule in every 554 million is dissociated. The concentration of each ion in pure water is 10^{-7} *M* (at 25°C). This means that there is only one ten-millionth of a mole of hydrogen ions per liter of pure water and an equal number of hydroxide ions.

Although the dissociation of water is reversible and statistically rare, it is exceedingly important in the chemistry of life. Hydrogen and hydroxide ions are very reactive. Changes in their concentrations can drastically affect a cell's proteins and other complex molecules. As we have seen, the concentrations of H^+ and OH^- are equal in pure water, but adding certain kinds of solutes, called acids and bases, disrupts this balance. Biologists use something called the pH scale to describe how acidic or basic (the opposite of acidic) a solution is. In the remainder of this chapter, you will learn about acids, bases, and pH and why changes in pH can adversely affect organisms.

Effects of Changes in pH

Before discussing the pH scale, let's see what acids and bases are and how they interact with water.

Acids and Bases

What would cause an aqueous solution to have an imbalance in its H^+ and OH^- concentrations? When the substances called acids dissolve in water, they donate additional H^+ to the solution. An **acid,** according to the definition often used by biologists, is a substance that increases the hydrogen ion concentration of a solution. For example, when hydrochloric acid (HCl) is added to water, hydrogen ions dissociate from chloride ions:

$$HCl \longrightarrow H^+ + Cl^-$$

This additional source of H^+ (dissociation of water is the other source) results in the solution having more H^+ than OH^-. Such a solution is known as an acidic solution.

A substance that reduces the hydrogen ion concentration of a solution is called a **base.** Some bases reduce the H^+ concentration directly by accepting hydrogen ions. Ammonia (NH_3), for instance, acts as a base when the unshared electron pair in nitrogen's valence shell attracts a hydrogen ion from the solution, resulting in an ammonium ion (NH_4^+):

$$NH_3 + H^+ \rightleftharpoons NH_4^+$$

Other bases reduce the H^+ concentration indirectly by dissociating to form hydroxide ions, which then combine with hydrogen ions in the solution to form water. One base that acts this way is sodium hydroxide (NaOH), which in water dissociates into its ions:

$$NaOH \longrightarrow Na^+ + OH^-$$

In either case, the base reduces the H^+ concentration. Solutions with a higher concentration of OH^- than H^+ are known as basic solutions. A solution in which the H^+ and OH^- concentrations are equal is said to be neutral.

Notice that single arrows were used in the reactions for HCl and NaOH. These compounds dissociate completely when mixed with water, and so hydrochloric acid is called a strong acid and sodium hydroxide a strong base. In contrast, ammonia is a relatively weak base. The double arrows in the reaction for ammonia indicate that the binding and release of hydrogen ions are reversible reactions, although at equilibrium there will be a fixed ratio of NH_4^+ to NH_3.

There are also weak acids, which reversibly release and accept back hydrogen ions. An example is carbonic acid, which has essential functions in many organisms:

$$\underset{\text{Carbonic acid}}{H_2CO_3} \rightleftharpoons \underset{\text{Bicarbonate ion}}{HCO_3^-} + \underset{\text{Hydrogen ion}}{H^+}$$

Here the equilibrium so favors the reaction in the left direction that when carbonic acid is added to water, only 1% of the molecules are dissociated at any particular time. Still, that is enough to shift the balance of H^+ and OH^- from neutrality.

The pH Scale

In any aqueous solution at 25°C, the *product* of the H^+ and OH^- concentrations is constant at 10^{-14}. This can be written

$$[H^+][OH^-] = 10^{-14}$$

In such an equation, brackets indicate molar concentration for the substance enclosed within them. In a neutral solution at room temperature (25°C), $[H^+] = 10^{-7}$ and $[OH^-] = 10^{-7}$, so in this case, 10^{-14} is the product of $10^{-7} \times 10^{-7}$. If enough acid is added to a solution to increase $[H^+]$ to 10^{-5} *M*, then $[OH^-]$ will decline by an equivalent amount to 10^{-9} *M* (note that $10^{-5} \times 10^{-9} = 10^{-14}$). This constant relationship expresses the behavior of acids and bases in an aqueous solution. An acid not only adds hydrogen ions to a solution, but also removes hydroxide ions because of the tendency for H^+ to combine with OH^- to form water. A base has the opposite effect, increasing OH^- concentration but also reducing H^+ concentration by the formation of water. If enough of a base is added to raise the OH^- concentration to 10^{-4} *M*, it will cause the H^+ concentration to drop to 10^{-10} *M*. Whenever we know the concentration of either H^+ or OH^- in an aqueous solution, we can deduce the concentration of the other ion.

Because the H^+ and OH^- concentrations of solutions can vary by a factor of 100 trillion or more, scientists have developed a way to express this variation more conveniently than in moles per liter. The pH scale **(Figure 3.8)** compresses the range of H^+ and OH^- concentrations by employing logarithms. The **pH** of a solution is defined as the negative logarithm (base 10) of the hydrogen ion concentration:

$$pH = -\log [H^+]$$

For a neutral aqueous solution, $[H^+]$ is 10^{-7} *M*, giving us

$$-\log 10^{-7} = -(-7) = 7$$

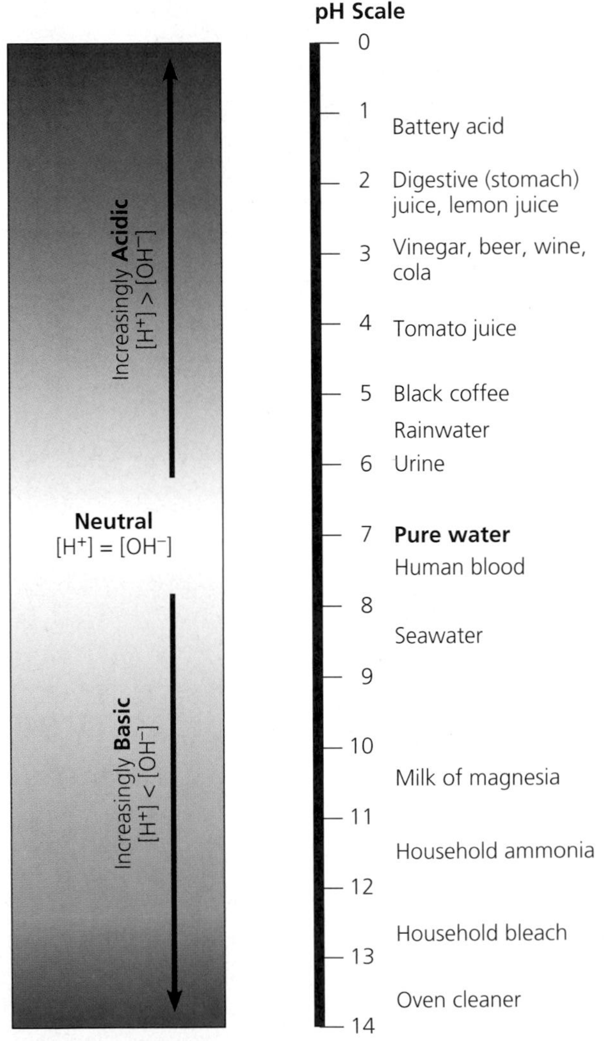

▲ **Figure 3.8 The pH scale and pH values of some aqueous solutions.**

Notice that pH *declines* as H^+ concentration *increases*. Notice, too, that although the pH scale is based on H^+ concentration, it also implies OH^- concentration. A solution of pH 10 has a hydrogen ion concentration of 10^{-10} *M* and a hydroxide ion concentration of 10^{-4} *M*.

The pH of a neutral aqueous solution at 25°C is 7, the midpoint of the scale. A pH value less than 7 denotes an acidic solution; the lower the number, the more acidic the solution. The pH for basic solutions is above 7. Most biological fluids are within the range pH 6–8. There are a few exceptions, however, including the strongly acidic digestive juice of the human stomach, which has a pH of about 2.

Remember that each pH unit represents a tenfold difference in H^+ and OH^- concentrations. It is this mathematical feature that makes the pH scale so compact. A solution of pH 3 is not twice as acidic as a solution of pH 6, but a thousand times more acidic. When the pH of a solution changes slightly, the actual concentrations of H^+ and OH^- in the solution change substantially.

Buffers

The internal pH of most living cells is close to 7. Even a slight change in pH can be harmful, because the chemical processes of the cell are very sensitive to the concentrations of hydrogen and hydroxide ions.

The presence of buffers in biological fluids allows for a relatively constant pH despite the addition of acids or bases. **Buffers** are substances that minimize changes in the concentrations of H^+ and OH^- in a solution. For example, buffers normally maintain the pH of human blood very close to 7.4, which is slightly basic. A person cannot survive for more than a few minutes if the blood pH drops to 7 or rises to 7.8. Under normal circumstances, the buffering capacity of the blood prevents such swings in pH.

A buffer works by accepting hydrogen ions from the solution when they are in excess and donating hydrogen ions to the solution when they have been depleted. Most buffer solutions contain a weak acid and its corresponding base, which combine reversibly with hydrogen ions. There are several buffers that contribute to pH stability in human blood and many other biological solutions. One of these is carbonic acid (H_2CO_3), which, as already mentioned, dissociates to yield a bicarbonate ion (HCO_3^-) and a hydrogen ion (H^+):

$$\underset{\substack{H^+\text{donor} \\ \text{(acid)}}}{H_2CO_3} \underset{\text{Response to a drop in pH}}{\overset{\text{Response to a rise in pH}}{\rightleftharpoons}} \underset{\substack{H^+\text{acceptor} \\ \text{(base)}}}{HCO_3^-} + \underset{\substack{\text{Hydrogen} \\ \text{ion}}}{H^+}$$

The chemical equilibrium between carbonic acid and bicarbonate acts as a pH regulator, the reaction shifting left or right as other processes in the solution add or remove hydrogen ions. If the H^+ concentration in blood begins to fall (that is, if pH rises), the reaction proceeds to the right and more carbonic acid dissociates, replenishing hydrogen ions. But when H^+ concentration in blood begins to rise (when pH drops), the reaction proceeds to the left, with HCO_3^- (the base) removing the hydrogen ions from the solution to form H_2CO_3. Thus, the carbonic acid–bicarbonate buffering system consists of an acid and a base in equilibrium with each other. Most other buffers are also acid-base pairs.

The Threat of Acid Precipitation

Considering the dependence of all life on water, contamination of rivers, lakes, and seas is a dire environmental problem. One of the most serious assaults on water quality is acid precipitation. Uncontaminated rain has a pH of about 5.6, slightly acidic, owing to the formation of carbonic acid from carbon dioxide and water. **Acid precipitation** refers to rain, snow, or fog with a pH lower or more acidic than pH 5.6.

Acid precipitation is caused primarily by the presence in the atmosphere of sulfur oxides and nitrogen oxides, gaseous compounds that react with water in the air to form strong acids, which fall to earth with rain or snow. A major source of these oxides is the burning of fossil fuels (coal, oil, and gas) in factories and automobiles. Electrical power plants that burn coal produce more of these pollutants than any other single source. Winds carry the pollutants away, and acid rain may fall hundreds of kilometers away from industrial centers. In certain sites in Pennsylvania and New York, the pH of rainfall in December 2001 averaged 4.3, about 20 times more acidic than normal rain. Acid precipitation falls on many other regions, including eastern Canada, the Cascade Mountains of the Pacific Northwest, and certain parts of Europe and Asia **(Figure 3.9)**.

Acid precipitation can damage life in lakes and streams. In addition, acid precipitation falling on land washes away certain mineral ions, such as calcium and magnesium ions, that

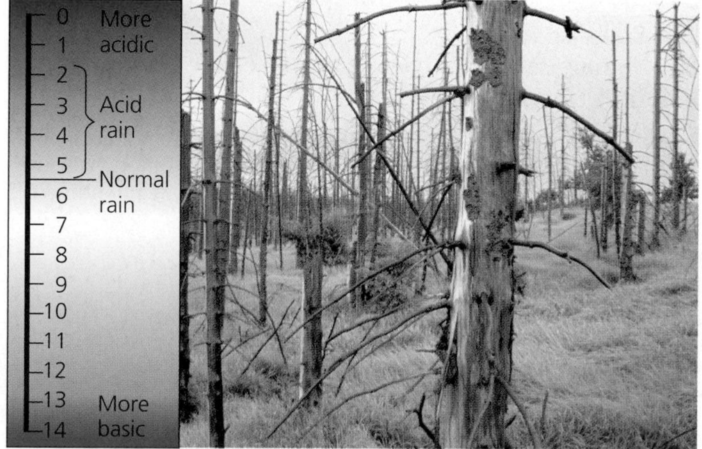

▲ **Figure 3.9 Acid precipitation and its effects on a forest.** Acid rain is thought to be responsible for killing trees in many forests, including the fir forest shown here in the Czech Republic.

ordinarily help buffer the soil solution and are essential nutrients for plant growth. At the same time, other minerals, such as aluminum, reach toxic concentrations when acidification increases their solubility. The effects of acid precipitation on soil chemistry have taken a toll on some North American forests and are contributing to the decline of European forests (see Figure 3.9). Nevertheless, studies indicate that the majority of North American forests are not currently suffering substantially from acid precipitation.

If there is reason for optimism about the future quality of water resources, it is that we have made progress in reducing acid precipitation (see Chapter 54). Continued progress can come only from the actions of people who are concerned about environmental quality. This requires understanding the crucial role that water plays in the environment's fitness for continued life on Earth.

Concept Check 3.3

1. Compared to a basic solution at pH 9, the same volume of an acidic solution at pH 4 has _____ times as many hydrogen ions (H^+).
2. HCl is a strong acid that dissociates completely in water: $HCl \rightarrow H^+ + Cl^-$. What is the pH of 0.01 *M* HCl?

For suggested answers, see Appendix A.

Chapter 3 Review

Go to www.campbellbiology.com or the student CD-ROM to explore Activities, Investigations, and other interactive study aids.

SUMMARY OF KEY CONCEPTS

Concept 3.1 The polarity of water molecules results in hydrogen bonding

- A hydrogen bond forms when the oxygen of one water molecule is electrically attracted to the hydrogen of a nearby molecule. Hydrogen bonding between water molecules is the basis for water's unusual properties (pp. 47–48).
  Activity *The Polarity of Water*

Concept 3.2 Four emergent properties of water contribute to Earth's fitness for life

- **Cohesion (pp. 48–49)** Hydrogen bonding keeps water molecules close to each other, and this cohesion helps pull water upward in the microscopic vessels of plants. Hydrogen bonding is also responsible for water's surface tension.
  Activity *Cohesion of Water*
- **Moderation of Temperature (pp. 49–50)** Hydrogen bonding gives water a high specific heat. Heat is absorbed when hydrogen bonds break and is released when hydrogen bonds form, helping minimize temperature fluctuations to within limits that permit life. Evaporative cooling is based on water's high heat of vaporization. Water molecules must have a relatively high kinetic energy to break hydrogen bonds. The evaporative loss of these energetic water molecules cools a surface.
- **Insulation of Bodies of Water by Floating Ice (pp. 50–51)** Ice is less dense than liquid water because its more organized hydrogen bonding causes expansion into a crystal formation. The lower density causes ice to float, which allows life to exist under the frozen surfaces of lakes and polar seas.
- **The Solvent of Life (pp. 51–53)** Water is an unusually versatile solvent because its polar molecules are attracted to charged and polar substances. Ions or polar substances surrounded by water molecules dissolve and are called solutes. Hydrophilic substances have an affinity for water; hydrophobic substances do not. Molarity, the number of moles of solute per liter of solution, is used as a measure of solute concentration in solutions. A mole is a certain number of molecules of a substance. The mass of a mole of the substance in grams is the same as the molecular mass in daltons.

Concept 3.3 Dissociation of water molecules leads to acidic and basic conditions that affect living organisms

- **Effects of Changes in pH (pp. 53–55)** Water can dissociate into H^+ and OH^-. The concentration of H^+ is expressed as pH, where $pH = -\log [H^+]$. Acids donate additional H^+ in aqueous solutions; bases donate OH^- or accept H^+. In a neutral solution at 25°C, $[H^+] = [OH^-] = 10^{-7}$ *M*, and pH = 7. In an acidic solution, $[H^+]$ is greater than $[OH^-]$, and the pH is less than 7. In a basic solution, $[H^+]$ is less than $[OH^-]$, and the pH is greater than 7. Buffers in biological fluids resist changes in pH. A buffer consists of an acid-base pair that combines reversibly with hydrogen ions.
  Activity *Dissociation of Water Molecules*
  Activity *Acids, Bases, and pH*
- **The Threat of Acid Precipitation (pp. 55–56)** Acid precipitation is rain, snow, or fog with a pH below 5.6. It often results from a reaction in the air between water vapor and sulfur oxides and nitrogen oxides produced by the combustion of fossil fuels.
  Investigation *How Does Acid Precipitation Affect Trees?*

TESTING YOUR KNOWLEDGE

Self-Quiz

1. What is the best explanation of the phrase "fitness of the environment," as used in this chapter?
   a. Earth's environment is constant.
   b. It is the physical environment, not life, that has changed.
   c. The environment of Earth has adapted to life.
   d. Life as we know it depends on certain environmental qualities on Earth.
   e. Water and other aspects of Earth's environment exist because they make the planet more suitable for life.

2. Many mammals control their body temperature by sweating. Which property of water is most directly responsible for the ability of sweat to lower body temperature?
   a. water's change in density when it condenses
   b. water's ability to dissolve molecules in the air
   c. the release of heat by the formation of hydrogen bonds
   d. the absorption of heat by the breaking of hydrogen bonds
   e. water's high surface tension

3. For two bodies of matter in contact, heat always flows from
   a. the body with greater heat to the one with less heat.
   b. the body of higher temperature to the one of lower temperature.
   c. the denser body to the less dense body.
   d. the body with more water to the one with less water.
   e. the larger body to the smaller body.

4. A slice of pizza has 500 kcal. If we could burn the pizza and use all the heat to warm a 50-L container of cold water, what would be the approximate increase in the temperature of the water? (*Note:* A liter of cold water weighs about 1 kg.)
   a. 50°C d. 100°C
   b. 5°C e. 1°C
   c. 10°C

5. The bonds that are broken when water vaporizes are
   a. ionic bonds.
   b. bonds between water molecules.
   c. bonds between atoms within individual water molecules.
   d. polar covalent bonds.
   e. nonpolar covalent bonds.

6. Which of the following is an example of a hydrophobic material?
   a. paper d. sugar
   b. table salt e. pasta
   c. wax

7. We can be sure that a mole of table sugar and a mole of vitamin C are equal in their
   a. mass in daltons. d. number of atoms.
   b. mass in grams. e. volume.
   c. number of molecules.

8. How many grams of acetic acid ($C_2H_4O_2$) would you use to make 10 L of a 0.1 *M* aqueous solution of acetic acid? (*Note:* The atomic masses, in daltons, are approximately 12 for carbon, 1 for hydrogen, and 16 for oxygen.)
   a. 10.0 g d. 60.0 g
   b. 0.1 g e. 0.6 g
   c. 6.0 g

9. Acid precipitation has lowered the pH of a particular lake to 4.0. What is the hydrogen ion concentration of the lake?
   a. 4.0 *M* d. 10^4 *M*
   b. 10^{-10} *M* e. 4%
   c. 10^{-4} *M*

10. What is the *hydroxide* ion concentration of the lake described in question 9?
   a. 10^{-7} *M* d. 10^{-14} *M*
   b. 10^{-4} *M* e. 10 *M*
   c. 10^{-10} *M*

For Self-Quiz answers, see Appendix A.

Go to the website or CD-ROM for more quiz questions.

Evolution Connection

The surface of the planet Mars has many landscape features reminiscent of those formed by flowing water on Earth, including what appear to be meandering channels and outwash areas. Recent probes sent to Mars have revealed strong evidence that liquid water was once present on its surface. Ice exists at the Martian poles today, and some scientists suspect a great deal more water may be present beneath the Martian surface. Why has there been so much interest in the presence of water on Mars? Does the presence of water make it more likely that life had evolved there? What other physical factors might also be important?

Scientific Inquiry

1. Design a controlled experiment to test the hypothesis that acid precipitation inhibits the growth of *Elodea,* a common freshwater plant.

2. In agricultural areas, farmers pay close attention to the weather forecast. Right before a predicted overnight freeze, farmers spray water on crops to protect the plants. Use the properties of water to explain how this works. Be sure to mention why hydrogen bonds are responsible for this phenomenon.

Investigation *How Does Acid Precipitation Affect Trees?*

Science, Technology, and Society

Agriculture, industry, and the growing populations of cities all compete, through political influence, for water. If you were in charge of water resources in an arid region, what would your priorities be for allocating the limited water supply for various uses? How would you try to build consensus among the different special-interest groups?

4 Carbon and the Molecular Diversity of Life

▲ **Figure 4.1 Life is based on carbon.**

Key Concepts

4.1 Organic chemistry is the study of carbon compounds
4.2 Carbon atoms can form diverse molecules by bonding to four other atoms
4.3 Functional groups are the parts of molecules involved in chemical reactions

Overview

Carbon—The Backbone of Biological Molecules

Although water is the universal medium for life on Earth, living organisms, including all the plants and the snail you see in **Figure 4.1**, are made up of chemicals based mostly on the element carbon. Carbon enters the biosphere through the action of plants, which use the sun's energy to transform CO_2 in the atmosphere into the molecules of life. These molecules are then passed along to animals that feed on plants, such as the snail in the photo. Of all chemical elements, carbon is unparalleled in its ability to form molecules that are large, complex, and diverse, and this molecular diversity has made possible the diversity of organisms that have evolved on Earth. Proteins, DNA, carbohydrates, and other molecules that distinguish living matter from inanimate material are all composed of carbon atoms bonded to one another and to atoms of other elements. Hydrogen (H), oxygen (O), nitrogen (N), sulfur (S), and phosphorus (P) are other common ingredients of these compounds, but it is carbon (C) that accounts for the large diversity of biological molecules.

Proteins and other very large molecules are the main focus of Chapter 5. In this chapter, we investigate the properties of smaller molecules, using them to illustrate a few concepts of molecular architecture that highlight carbon's importance to life and the theme that emergent properties arise from the organization of the matter of living organisms.

Concept 4.1

Organic chemistry is the study of carbon compounds

Compounds containing carbon are said to be organic, and the branch of chemistry that specializes in the study of carbon compounds is called **organic chemistry.** Organic compounds range from simple molecules, such as methane (CH_4), to colossal ones, such as proteins, with thousands of atoms and molecular masses in excess of 100,000 daltons. Most organic compounds contain hydrogen atoms in addition to carbon atoms.

The overall percentages of the major elements of life—C, H, O, N, S, and P—are quite uniform from one organism to another. Because of carbon's versatility, however, this limited assortment of atomic building blocks, taken in roughly the same proportions, can be used to build an inexhaustible variety of organic molecules. Different species of organisms, and different individuals within a species, are distinguished by variations in their organic molecules.

Since the dawn of human history, people have used other organisms as sources of valued substances—from foods to medicines and fabrics. The science of organic chemistry originated in attempts to purify and improve the yield of such products. By the early 19th century, chemists had learned to make many simple compounds in the laboratory by combining elements under the right conditions. Artificial synthesis of the complex molecules extracted from living matter seemed impossible, however. At that time, the Swedish chemist Jöns

Jakob Berzelius made the distinction between organic compounds, those that seemingly could arise only within living organisms, and inorganic compounds, those that were found in the nonliving world. The new discipline of organic chemistry was first built on a foundation of *vitalism*, the belief in a life force outside the jurisdiction of physical and chemical laws.

Chemists began to chip away at the foundation of vitalism when they learned to synthesize organic compounds in their laboratories. In 1828, Friedrich Wöhler, a German chemist who had studied with Berzelius, attempted to make an "inorganic" salt, ammonium cyanate, by mixing solutions of ammonium ions (NH_4^+) and cyanate ions (CNO^-). Wöhler was astonished to find that instead of the expected product, he had made urea, an organic compound present in the urine of animals. Wöhler challenged the vitalists when he wrote, "I must tell you that I can prepare urea without requiring a kidney or an animal, either man or dog." However, one of the ingredients used in the synthesis, the cyanate, had been extracted from animal blood, and the vitalists were not swayed by Wöhler's discovery. A few years later, however, Hermann Kolbe, a student of Wöhler's, made the organic compound acetic acid from inorganic substances that could themselves be prepared directly from pure elements.

Vitalism finally crumbled after several more decades of laboratory synthesis of increasingly complex organic compounds. In 1953, Stanley Miller, then a graduate student of Harold Urey at the University of Chicago, helped bring this abiotic (nonliving) synthesis of organic compounds into the context of evolution. Miller used a laboratory simulation of chemical conditions on the primitive Earth to demonstrate that the spontaneous synthesis of organic compounds could have been an early stage in the origin of life **(Figure 4.2)**.

The pioneers of organic chemistry helped shift the mainstream of biological thought from vitalism to *mechanism*, the view that all natural phenomena, including the processes of life, are governed by physical and chemical laws. Organic chemistry was redefined as the study of carbon compounds, regardless of their origin. Most naturally occurring organic compounds are produced by organisms, and these molecules represent a diversity and range of complexity unrivaled by inorganic compounds. However, the same rules of chemistry apply to inorganic and organic molecules alike. The foundation of organic chemistry is not some intangible life force, but the unique chemical versatility of the element carbon.

Concept Check 4.1

1. Stanley Miller found urea among the products of his experiment. What conclusion could be drawn from its presence?

For suggested answers, see Appendix A.

Figure 4.2

Inquiry Could organic compounds have been synthesized abiotically on the early Earth?

EXPERIMENT In 1953, Stanley Miller simulated what were thought to be environmental conditions on the lifeless, primordial Earth. As shown in this recreation, Miller used electrical discharges (simulated lightning) to trigger reactions in a primitive "atmosphere" of H_2O, H_2, NH_3 (ammonia), and CH_4 (methane)—some of the gases released by volcanoes.

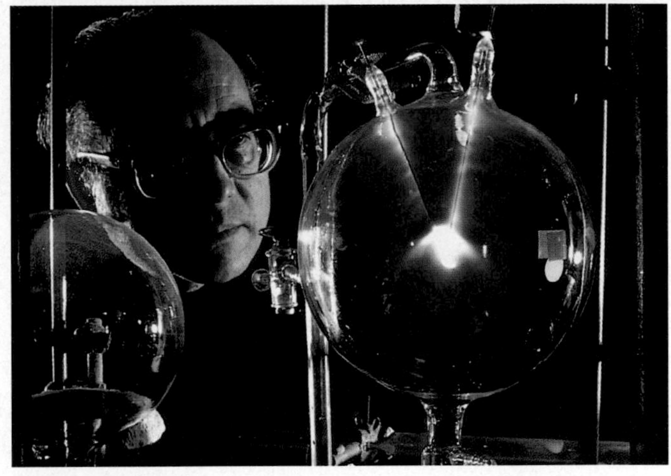

RESULTS A variety of organic compounds that play key roles in living cells were synthesized in Miller's apparatus.

CONCLUSION Organic compounds may have been synthesized abiotically on the early Earth, setting the stage for the origin of life. (We will explore this hypothesis in more detail in Chapter 26.)

Concept 4.2

Carbon atoms can form diverse molecules by bonding to four other atoms

The key to the chemical characteristics of an atom, as you learned in Chapter 2, is in its configuration of electrons. Electron configuration determines the kinds and number of bonds an atom will form with other atoms.

The Formation of Bonds with Carbon

Carbon has a total of 6 electrons, with 2 in the first electron shell and 4 in the second shell. Having 4 valence electrons in a shell that holds 8, carbon would have to donate or accept 4 electrons to complete its valence shell and become an ion. Instead, a carbon atom usually completes its valence shell by sharing its 4 electrons with other atoms in covalent bonds so that 8 electrons are present. Each carbon atom thus acts as an intersection point from which a molecule can branch off in up to

<table>
<tr><th>Name and Comment</th><th>Molecular Formula</th><th>Structural Formula</th><th>Ball-and-Stick Model</th><th>Space-Filling Model</th></tr>
<tr><td>(a) Methane. When a carbon atom has four single bonds to other atoms, the molecule is tetrahedral.</td><td>CH_4</td><td>H
|
H—C—H
|
H</td><td></td><td></td></tr>
<tr><td>(b) Ethane. A molecule may have more than one tetrahedral group of single-bonded atoms. (Ethane consists of two such groups.)</td><td>C_2H_6</td><td>H H
| |
H—C—C—H
| |
H H</td><td></td><td></td></tr>
<tr><td>(c) Ethene (ethylene). When two carbon atoms are joined by a double bond, all atoms attached to those carbons are in the same plane; the molecule is flat.</td><td>C_2H_4</td><td>H   H
C=C
H   H</td><td></td><td></td></tr>
</table>

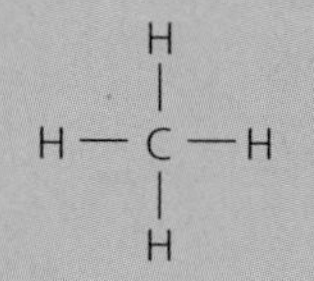

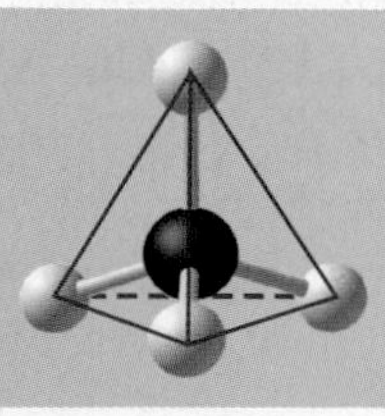

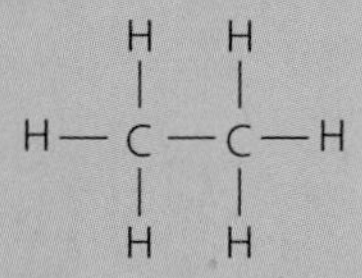

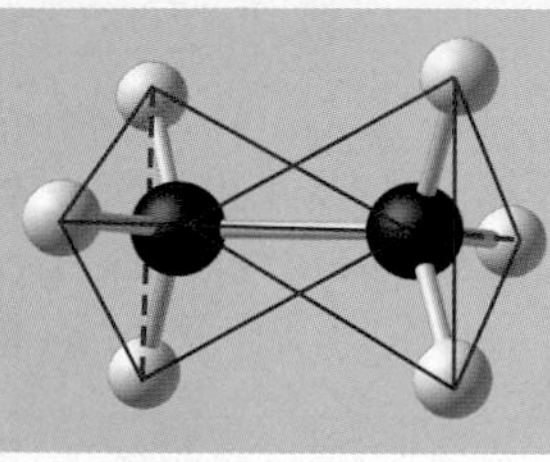

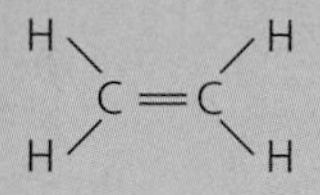

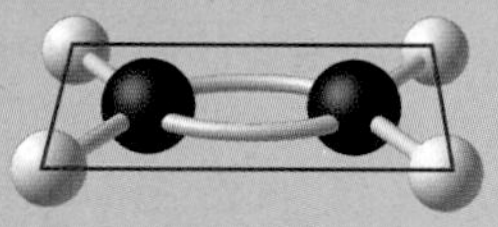

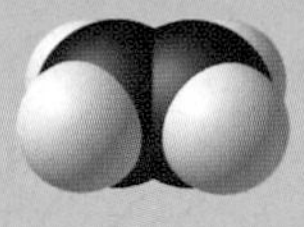

▲ **Figure 4.3 The shapes of three simple organic molecules.**

four directions. This *tetravalence* is one facet of carbon's versatility that makes large, complex molecules possible.

In Chapter 2, you also learned that when a carbon atom forms single covalent bonds, the arrangement of its four hybrid orbitals causes the bonds to angle toward the corners of an imaginary tetrahedron (see Figure 2.16b). The bond angles in methane (CH_4) are 109.5° **(Figure 4.3a)**, and they are approximately the same in any group of atoms where carbon has four single bonds. For example, ethane (C_2H_6) is shaped like two tetrahedrons overlapping at their apexes **(Figure 4.3b)**. In molecules with still more carbons, every grouping of a carbon bonded to four other atoms has a tetrahedral shape. But when two carbon atoms are joined by a double bond, all bonds around those carbons are in the same plane. For example, ethene (C_2H_4) is a flat molecule; its atoms all lie in the same plane **(Figure 4.3c)**. We find it convenient to write all structural formulas as though the molecules represented were flat, but keep in mind that molecules are three-dimensional and that the shape of a molecule often determines its function.

The electron configuration of carbon gives it covalent compatibility with many different elements. **Figure 4.4** shows electron-shell diagrams of the four major atomic components of organic molecules. As you may recall from Chapter 2, these models allow us to see the valences of carbon and its most frequent partners—oxygen, hydrogen, and nitrogen. We can think of these valences as the basis for the rules of covalent bonding in organic chemistry—the building code that governs the architecture of organic molecules.

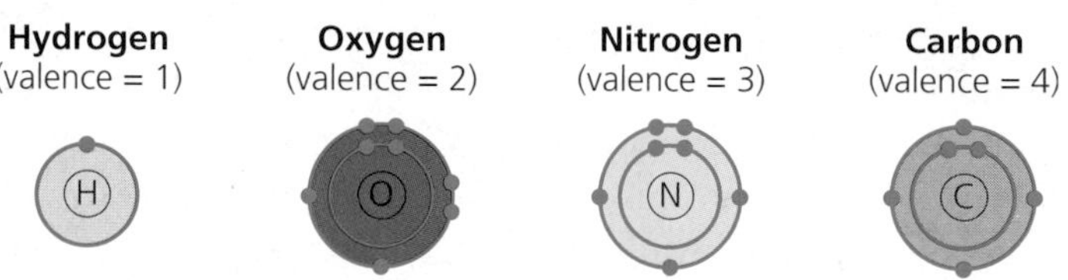

▲ **Figure 4.4 Electron-shell diagrams showing valences for the major elements of organic molecules.** Valence is the number of covalent bonds an atom can form. It is generally equal to the number of electrons required to complete the atom's outermost (valence) electron shell (see Figure 2.8).

A couple of additional examples will show how the rules of covalent bonding apply to carbon atoms with partners other than hydrogen. In the carbon dioxide molecule (CO_2), a single carbon atom is joined to two atoms of oxygen by double covalent bonds. The structural formula for CO_2 is shown here:

$$O{=}C{=}O$$

Each line in a structural formula represents a pair of shared electrons. Notice that the carbon atom in CO_2 is involved in two double bonds, the equivalent of four single covalent bonds. The arrangement completes the valence shells of all atoms in the molecule. Because carbon dioxide is a very simple molecule and lacks hydrogen, it is often considered inorganic, even though it contains carbon. Whether we call CO_2 organic or inorganic, there is no question about its importance to the living world. As previously mentioned, CO_2 is the source of carbon for all the organic molecules found in organisms.

Another relatively simple molecule is urea, $CO(NH_2)_2$. This is the organic compound found in urine that Wöhler

Ethane

Propane

(a) Length. Carbon skeletons vary in length.

Butane

2-methylpropane (commonly called isobutane)

(b) Branching. Skeletons may be unbranched or branched.

1-Butene

2-Butene

(c) Double bonds. The skeleton may have double bonds, which can vary in location.

Cyclohexane

Benzene

(d) Rings. Some carbon skeletons are arranged in rings. In the abbreviated structural formula for each compound (at the right), each corner represents a carbon and its attached hydrogens.

▲ **Figure 4.5 Variations in carbon skeletons.** Hydrocarbons, organic molecules consisting only of carbon and hydrogen, illustrate the diversity of the carbon skeletons of organic molecules.

learned to synthesize in the early 19th century. The structural formula for urea is shown at the right.

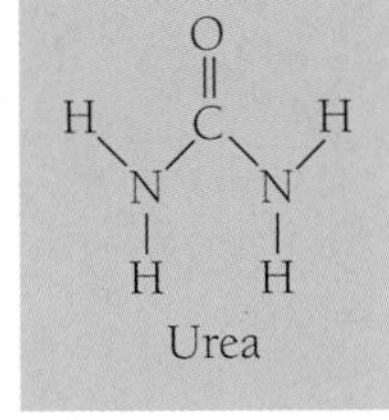

Again, each atom has the required number of covalent bonds. In this case, one carbon atom is involved in both single and double bonds.

Both urea and carbon dioxide are molecules with only one carbon atom. But as Figure 4.3 shows, a carbon atom can also use one or more of its valence electrons to form covalent bonds to other carbon atoms, making it possible to link the atoms into chains of seemingly infinite variety.

Molecular Diversity Arising from Carbon Skeleton Variation

Carbon chains form the skeletons of most organic molecules **(Figure 4.5)**. The skeletons vary in length and may be straight, branched, or arranged in closed rings. Some carbon skeletons have double bonds, which vary in number and location. Such variation in carbon skeletons is one important source of the molecular complexity and diversity that characterize living matter. In addition, atoms of other elements can be bonded to the skeletons at available sites.

Hydrocarbons

All the molecules shown in Figures 4.3 and 4.5 are **hydrocarbons**, organic molecules consisting only of carbon and hydrogen. Atoms of hydrogen are attached to the carbon skeleton wherever electrons are available for covalent bonding. Hydrocarbons are the major components of petroleum, which is called a fossil fuel because it consists of the partially decomposed remains of organisms that lived millions of years ago.

Although hydrocarbons are not prevalent in living organisms, many of a cell's organic molecules have regions consisting of only carbon and hydrogen. For example, the molecules known as fats have long hydrocarbon tails attached to a nonhydrocarbon component **(Figure 4.6)**. Neither petroleum nor fat dissolves in water; both are hydrophobic compounds

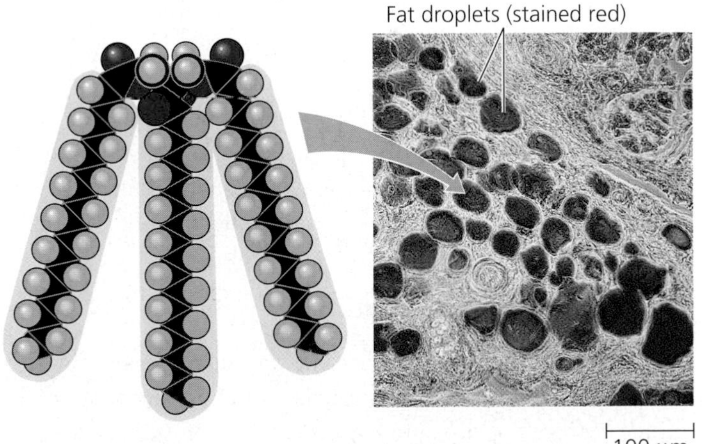

▲ **Figure 4.6 The role of hydrocarbons in fats. (a)** A fat molecule consists of a small, non-hydrocarbon component joined to three hydrocarbon tails. The tails can be broken down to provide energy. They also account for the hydrophobic behavior of fats. (Black = carbon; gray = hydrogen; red = oxygen.) **(b)** Mammalian adipose cells stockpile fat molecules as a fuel reserve. Each adipose cell in this micrograph is almost filled by a large fat droplet, which contains a huge number of fat molecules.

because the great majority of their bonds are nonpolar carbon-to-hydrogen linkages. Another characteristic of hydrocarbons is that they can undergo reactions that release a relatively large amount of energy. The gasoline that fuels a car consists of hydrocarbons, and the hydrocarbon tails of fat molecules serve as stored fuel for animal bodies.

Isomers

Variation in the architecture of organic molecules can be seen in **isomers,** compounds that have the same numbers of atoms of the same elements but different structures and hence different properties. Compare, for example, the two pentanes in **Figure 4.7a**. Both have the molecular formula C_5H_{12}, but they differ in the covalent arrangement of their carbon skeletons. The skeleton is straight in one form of pentane but branched in the other. We will examine three types of isomers: structural isomers, geometric isomers, and enantiomers.

Structural isomers differ in the covalent arrangements of their atoms. The number of possible isomers increases tremendously as carbon skeletons increase in size. There are only three pentanes (two are shown in Figure 4.7a), but there are 18 variations of C_8H_{18} and 366,319 possible structural isomers of $C_{20}H_{42}$. Structural isomers may also differ in the location of double bonds.

Geometric isomers have the same covalent partnerships, but they differ in their spatial arrangements. Geometric isomers arise from the inflexibility of double bonds, which, unlike single bonds, will not allow the atoms they join to rotate freely about the bond axis. If a double bond joins two carbon atoms, and each C also has two different atoms (or groups of atoms) attached to it, then two distinct geometric isomers are possible. Consider the simple example in **Figure 4.7b**. Each of the carbons has an H and an X attached to it, but one isomer has a *"cis"* arrangement, with two Xs on the same side relative to the double bond, and the other isomer has a *"trans"* arrangement, with the Xs on opposite sides. The subtle difference in shape between geometric isomers can dramatically affect the biological activities of organic molecules. For example, the biochemistry of vision involves a light-induced change of rhodopsin, a chemical compound in the eye, from the *cis* isomer to the *trans* isomer (see Chapter 49).

Enantiomers are molecules that are mirror images of each other. In the ball-and-stick models shown in **Figure 4.7c**, the middle carbon is called an *asymmetric carbon* because it is attached to four different atoms or groups of atoms. The four groups can be arranged in space about the asymmetric carbon in two different ways that are mirror images. They are, in a way, left-handed and right-handed versions of the molecule. A cell can distinguish these isomers based on their different shapes. Usually, one isomer is biologically active and the other is inactive.

The concept of enantiomers is important in the pharmaceutical industry because the two enantiomers of a drug may not be equally effective. For example, L-dopa is effective against Parkinson's disease, while its enantiomer, D-dopa, is not **(Figure 4.8)**. In some cases, one of the isomers may even produce harmful effects. This was the case with thalidomide, a drug prescribed for thousands of pregnant women in the late 1950s and early 1960s. The drug was a mixture of two enantiomers. One enantiomer reduced morning sickness, the desired effect, but the other caused severe birth defects. (Unfortunately, even if the "good" thalidomide enantiomer is used in purified form, some of it soon converts to the "bad" enantiomer in the patient's body.) The differing effects of enantiomers in the body demonstrate that organisms are sensitive to even

(a) Structural isomers differ in covalent partners, as shown in this example of two isomers of pentane.

cis isomer: The two Xs are on the same side.

trans isomer: The two Xs are on opposite sides.

(b) Geometric isomers differ in arrangement about a double bond. In these diagrams, X represents an atom or group of atoms attached to a double-bonded carbon.

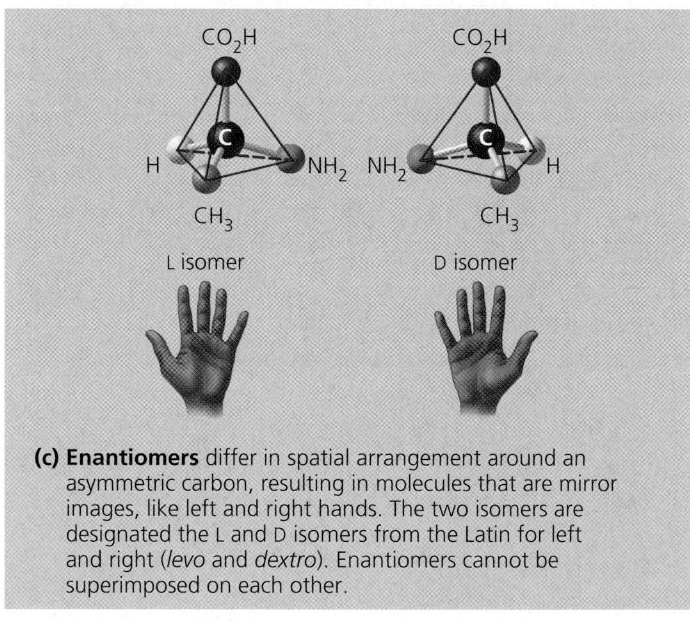

(c) Enantiomers differ in spatial arrangement around an asymmetric carbon, resulting in molecules that are mirror images, like left and right hands. The two isomers are designated the L and D isomers from the Latin for left and right (*levo* and *dextro*). Enantiomers cannot be superimposed on each other.

▲ **Figure 4.7 Three types of isomers.** Compounds with the same molecular formula but different structures, isomers are a source of diversity in organic molecules.

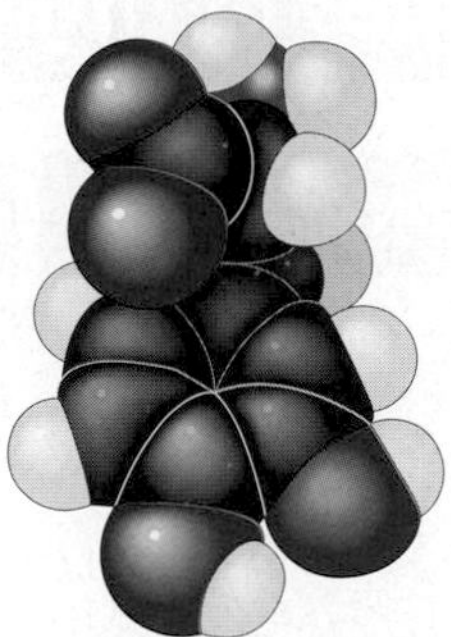

L-Dopa
(effective against Parkinson's disease)

D-Dopa
(biologically inactive)

▲ **Figure 4.8 The pharmacological importance of enantiomers.** L-Dopa is a drug used to treat Parkinson's disease, a disorder of the central nervous system. The drug's enantiomer, the mirror-image molecule designated D-dopa, has no effect on patients.

Female lion

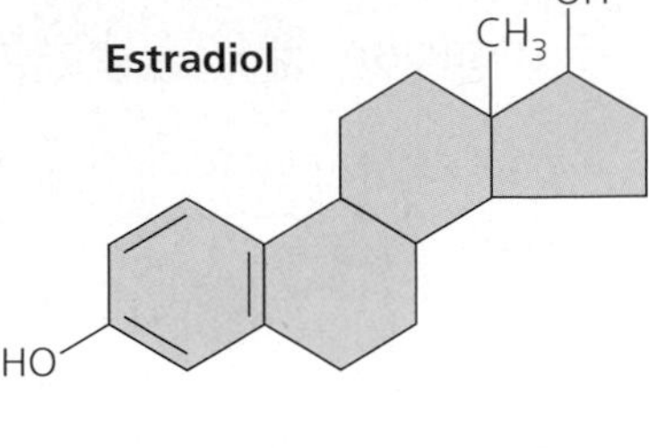

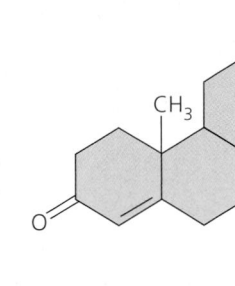

Male lion

▲ **Figure 4.9 A comparison of functional groups of female (estradiol) and male (testosterone) sex hormones.** The two molecules differ only in the functional groups attached to a common carbon skeleton of four fused rings, which is shown here in abbreviated form. These subtle variations in molecular architecture influence the development of the anatomical and physiological differences between female and male vertebrates.

the most subtle variations in molecular architecture. Once again, we see that molecules have emergent properties that depend on the specific arrangement of their atoms.

Concept Check 4.2

1. Draw a structural formula for C_2H_4.
2. Look at Figure 4.5, and determine which pair(s) of molecules is (are) isomers of each other, identifying the type(s) of isomer.
3. What is the chemical similarity between gasoline and fat?

For suggested answers, see Appendix A.

Functional groups are the parts of molecules involved in chemical reactions

The distinctive properties of an organic molecule depend not only on the arrangement of its carbon skeleton, but also on the molecular components attached to that skeleton. We will now examine certain groups of atoms that are frequently attached to the skeletons of organic molecules.

The Functional Groups Most Important in the Chemistry of Life

The components of organic molecules that are most commonly involved in chemical reactions are known as **functional groups.** If we think of hydrocarbons as the simplest organic molecules, we can view functional groups as attachments that replace one or more of the hydrogens bonded to the carbon skeleton of the hydrocarbon. (However, some functional groups include atoms of the carbon skeleton, as we will see.)

Each functional group behaves consistently from one organic molecule to another, and the number and arrangement of the groups help give each molecule its unique properties. Consider the differences between testosterone and estradiol (a type of estrogen). These compounds are male and female sex hormones, respectively, in humans and other vertebrates (**Figure 4.9**). Both are steroids, organic molecules with a common carbon skeleton in the form of four fused rings. These sex hormones differ only in the functional groups attached to the rings. The different actions of these two molecules on many targets throughout the body help produce the contrasting features of females and males. Thus, even our sexuality has its biological basis in variations of molecular architecture.

The six functional groups most important in the chemistry of life are the hydroxyl, carbonyl, carboxyl, amino, sulfhydryl, and phosphate groups. These groups are hydrophilic and thus increase the solubility of organic compounds in water. Before reading further, take time to familiarize yourself with the functional groups in **Figure 4.10** on the next two pages.

Figure 4.10

Exploring Some Important Functional Groups of Organic Compounds

FUNCTIONAL GROUP	HYDROXYL	CARBONYL	CARBOXYL
STRUCTURE	 —OH (may be written HO—) In a **hydroxyl group** (—OH), a hydrogen atom is bonded to an oxygen atom, which in turn is bonded to the carbon skeleton of the organic molecule. (Do not confuse this functional group with the hydroxide ion, OH^-.)	The **carbonyl group** (>CO) consists of a carbon atom joined to an oxygen atom by a double bond.	 When an oxygen atom is double-bonded to a carbon atom that is also bonded to a hydroxyl group, the entire assembly of atoms is called a **carboxyl group** (—COOH).
NAME OF COMPOUNDS	**Alcohols** (their specific names usually end in *-ol*)	**Ketones** if the carbonyl group is within a carbon skeleton **Aldehydes** if the carbonyl group is at the end of the carbon skeleton	**Carboxylic acids**, or organic acids
EXAMPLE	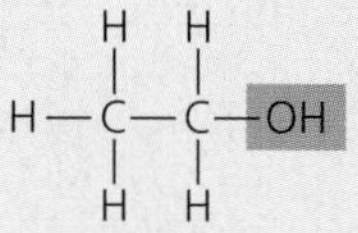 **Ethanol**, the alcohol present in alcoholic beverages	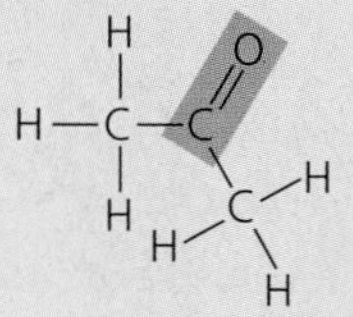 **Acetone**, the simplest ketone 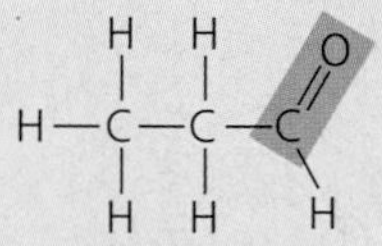 **Propanal**, an aldehyde	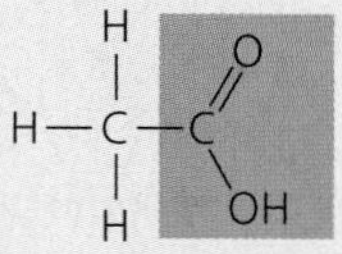 **Acetic acid**, which gives vinegar its sour taste
FUNCTIONAL PROPERTIES	▶ Is polar as a result of the electronegative oxygen atom drawing electrons toward itself. ▶ Attracts water molecules, helping dissolve organic compounds such as sugars (see Figure 5.3).	▶ A ketone and an aldehyde may be structural isomers with different properties, as is the case for acetone and propanal.	▶ Has acidic properties because it is a source of hydrogen ions. ▶ The covalent bond between oxygen and hydrogen is so polar that hydrogen ions (H^+) tend to dissociate reversibly; for example, 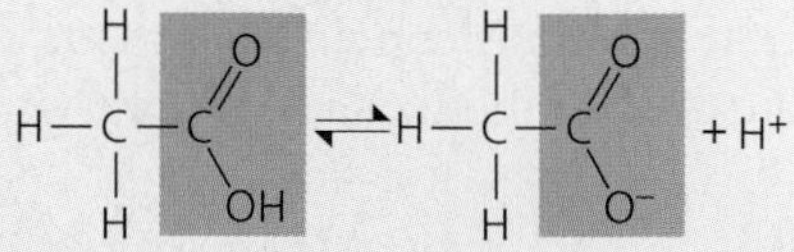 Acetic acid ⇌ Acetate ion + H^+ ▶ In cells, found in the ionic form, which is called a carboxylate group.

FUNCTIONAL GROUP	AMINO	SULFHYDRYL	PHOSPHATE
STRUCTURE	 The **amino group** (—NH_2) consists of a nitrogen atom bonded to two hydrogen atoms and to the carbon skeleton.	 (may be written HS —) The **sulfhydryl group** consists of a sulfur atom bonded to an atom of hydrogen; resembles a hydroxyl group in shape.	 In a **phosphate group**, a phosphorus atom is bonded to four oxygen atoms; one oxygen is bonded to the carbon skeleton; two oxygens carry negative charges; abbreviated Ⓟ. The phosphate group (—OPO_3^{2-}) is an ionized form of a phosphoric acid group (—OPO_3H_2; note the two hydrogens).
NAME OF COMPOUNDS	**Amines**	**Thiols**	**Organic phosphates**
EXAMPLE	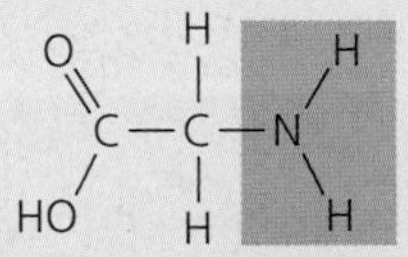 **Glycine** Because it also has a carboxyl group, glycine is both an amine and a carboxylic acid; compounds with both groups are called amino acids.	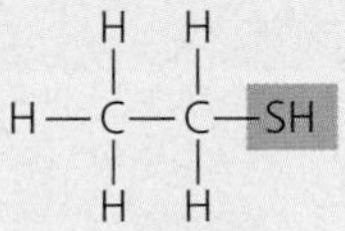 **Ethanethiol**	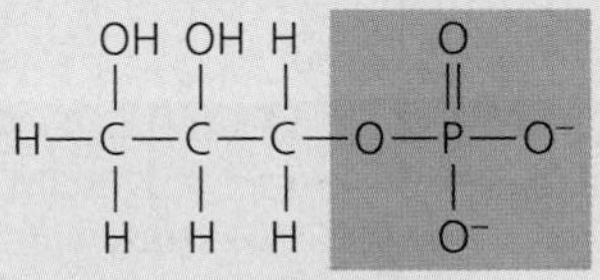 **Glycerol phosphate**
FUNCTIONAL PROPERTIES	▶ Acts as a base; can pick up a proton from the surrounding solution: 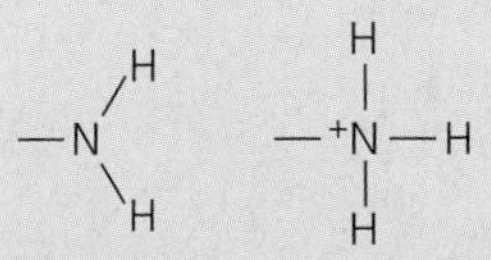 (nonionized) (ionized) ▶ Ionized, with a charge of 1+, under cellular conditions.	▶ Two sulfhydryl groups can interact to help stabilize protein structure (see Figure 5.20).	▶ Makes the molecule of which it is a part an anion (negatively charged ion). ▶ Can transfer energy between organic molecules.

ATP: An Important Source of Energy for Cellular Processes

The "Phosphate" column in Figure 4.10 shows a simple example of an organic phosphate molecule. A more complicated organic phosphate, **adenosine triphosphate**, or **ATP**, is worth mentioning because it is the primary energy-transferring molecule in the cell. ATP consists of an organic molecule called adenosine attached to a string of three phosphate groups:

$$^{-}O-\overset{\overset{O}{\|}}{\underset{\underset{O^{-}}{|}}{P}}-O-\overset{\overset{O}{\|}}{\underset{\underset{O^{-}}{|}}{P}}-O-\overset{\overset{O}{\|}}{\underset{\underset{O^{-}}{|}}{P}}-O-\boxed{\text{Adenosine}}$$

Where three phosphates are present in series, as in ATP, one phosphate may split off as an inorganic phosphate ion. This ion, $HOPO_3^{2-}$, is often abbreviated Ⓟ$_i$ in this book. Losing one phosphate, ATP becomes adenosine *di*phosphate, or ADP. The reaction releases energy that can be used by the cell, as you will learn in more detail in Chapter 8.

Ⓟ–Ⓟ–Ⓟ–[Adenosine] → Ⓟ$_i$ + Ⓟ–Ⓟ–[Adenosine] + Energy

ATP → Inorganic phosphate + ADP

Concept Check 4.3

1. What does the term "amino acid" signify about the structure of such a molecule?
2. What change usually occurs in ATP when it releases energy?

For suggested answers, see Appendix A.

The Chemical Elements of Life: *A Review*

Living matter, as you have learned, consists mainly of carbon, oxygen, hydrogen, and nitrogen, with smaller amounts of sulfur and phosphorus. These elements share the characteristic of forming strong covalent bonds, a quality that is essential in the architecture of complex organic molecules. Of all these elements, carbon is the virtuoso of the covalent bond. The versatility of carbon makes possible the great diversity of organic molecules, each with particular properties that emerge from the unique arrangement of its carbon skeleton and the functional groups appended to that skeleton. At the foundation of all biological diversity lies this variation at the molecular level.

Chapter 4 Review

Go to www.campbellbiology.com or the student CD-ROM to explore Activities, Investigations, and other interactive study aids.

SUMMARY OF KEY CONCEPTS

Concept 4.1

Organic chemistry is the study of carbon compounds

▶ Organic compounds were once thought to arise only within living organisms, but this idea (vitalism) was disproved when chemists were able to synthesize organic compounds in the laboratory (pp. 58–59).

Concept 4.2

Carbon atoms can form diverse molecules by bonding to four other atoms

▶ **The Formation of Bonds with Carbon (pp. 59–61)** A covalent-bonding capacity of four contributes to carbon's ability to form diverse molecules. Carbon can bond to a variety of atoms, including O, H, and N. Carbon atoms can also bond to other carbons, forming the carbon skeletons of organic compounds.

▶ **Molecular Diversity Arising from Carbon Skeleton Variation (pp. 61–63)** The carbon skeletons of organic molecules vary in length and shape and have bonding sites for atoms of other elements. Hydrocarbons consist only of carbon and hydrogen. Isomers are molecules with the same molecular formula but different structures and properties. Three types of isomers are structural isomers, geometric isomers, and enantiomers.
Activity *Diversity of Carbon-Based Molecules*
Activity *Isomers*
Investigation *What Factors Determine the Effectiveness of Drugs?*

Concept 4.3

Functional groups are the parts of molecules involved in chemical reactions

▶ **The Functional Groups Most Important in the Chemistry of Life (pp. 63–65)** Functional groups are chemically reactive groups of atoms within an organic molecule that give the molecule distinctive chemical properties. The hydroxyl group (—OH) is polar, thus helping compounds dissolve in water. The carbonyl group (>CO) can be either at the end of a carbon skeleton (aldehyde) or within the skeleton (ketone). The carboxyl group (—COOH) is found in carboxylic acids. The hydrogen of this group can dissociate, making such molecules acids. The amino group ($—NH_2$) can accept a proton (H^+), thereby acting as a base. The sulfhydryl group (—SH) helps stabilize the structure of some proteins. The phosphate group ($—OPO_3^{2-}$) has an important role in the transfer of energy.
Activity *Functional Groups*

▶ **ATP: An Important Source of Energy for Cellular Processes (p. 66)** When a phosphate group splits off from ATP, energy is released that can be used by the cell.

▶ **The Chemical Elements of Life: *A Review* (p. 66)** Living matter is made mostly of carbon, oxygen, hydrogen, and nitrogen, with some sulfur and phosphorus. Biological diversity has its molecular basis in carbon's ability to form a huge number of molecules with particular shapes and chemical properties.

TESTING YOUR KNOWLEDGE

Self-Quiz

1. Organic chemistry is currently defined as
   a. the study of compounds that can be made only by living cells.
   b. the study of carbon compounds.
   c. the study of vital forces.
   d. the study of natural (as opposed to synthetic) compounds.
   e. the study of hydrocarbons.

2. Choose the pair of terms that correctly completes this sentence: Hydroxyl is to ________ as ___________ is to aldehyde.
   a. carbonyl; ketone   d. amine; carboxyl
   b. oxygen; carbon   e. alcohol; ketone
   c. alcohol; carbonyl

3. Which of the following hydrocarbons has a double bond in its carbon skeleton?
   a. C_3H_8 d. C_2H_4
   b. C_2H_6 e. C_2H_2
   c. CH_4

4. The gasoline consumed by an automobile is a fossil fuel consisting mostly of
   a. aldehydes.   d. hydrocarbons.
   b. amino acids.   e. thiols.
   c. alcohols.

5. Choose the term that correctly describes the relationship between these two sugar molecules:

```
     H
     |                 H    O
 H—C—OH                 \  //
     |                    C
     C=O                  |
     |                H—C—OH
 H—C—OH                   |
     |                H—C—OH
     H                    |
                          H
```

   a. structural isomers   c. enantiomers
   b. geometric isomers   d. isotopes

6. Identify the asymmetric carbon in this molecule:

```
 O     OH H   H   H
  \\ a |b |c  |d  |e
    C—C—C—C—C—H
   /   |  |  |  |
  H    H  H  H  H
```

7. Which functional group is *not* present in this molecule?

```
   HO    O
     \  //
      C   H
      |   |
   H—C—C—OH
      |   |
      N   H
     / \
    H   H
```

   a. carboxyl   c. hydroxyl
   b. sulfhydryl   d. amino

8. Which action could produce a carbonyl group?
   a. the replacement of the hydroxyl of a carboxyl group with hydrogen
   b. the addition of a thiol to a hydroxyl
   c. the addition of a hydroxyl to a phosphate
   d. the replacement of the nitrogen of an amine with oxygen
   e. the addition of a sulfhydryl to a carboxyl

9. Which functional group is most likely to be responsible for an organic molecule behaving as a base?
   a. hydroxyl   c. carboxyl   e. phosphate
   b. carbonyl   d. amino

10. Given what you know about the electronegativity of oxygen, predict which of the following molecules would be the stronger acid. (*Hint:* Study Figure 4.10.) Explain your answer.

```
a.   H  H    O          b.   H  O    O
     |  |   //               |  ||  //
  H—C—C—C                 H—C—C—C
     |  |   \                |      \
     H  H    OH              H       OH
```

For Self-Quiz answers, see Appendix A.

Go to the website or CD-ROM for more quiz questions.

Evolution Connection

Some scientists believe that life elsewhere in the universe might be based on the element silicon, rather than on carbon, as on Earth. What properties does silicon share with carbon that would make silicon-based life more likely than, say, neon-based life or aluminum-based life? (See Figure 2.8.)

Scientific Inquiry

In 1918, an epidemic of sleeping sickness caused an unusual rigid paralysis in some survivors, similiar to symptoms of advanced Parkinson's disease. Years later, L-dopa, a chemical used to treat Parkinson's disease (see Figure 4.8), was given to some of these patients, as dramatized in the movie *Awakenings*. L-Dopa was remarkably effective at eliminating the paralysis, at least temporarily. However, its enantiomer, D-dopa, was subsequently shown to have no effect at all, as is the case for Parkinson's disease. Suggest a hypothesis to explain why, for *both* diseases, one enantiomer is effective and the other is not.

Investigation *What Factors Determine the Effectiveness of Drugs?*

Science, Technology, and Society

Thalidomide achieved notoriety 50 years ago because of a wave of birth defects among children born to women who took thalidomide during pregnancy as a treatment for morning sickness. However, in 1998 the U.S. Food and Drug Administration (FDA) approved this drug for the treatment of certain conditions associated with Hansen's disease (leprosy). In clinical trials, thalidomide also shows promise for use in treating patients suffering from AIDS, tuberculosis, and some types of cancer. Do you think approval of this drug is appropriate? If so, under what conditions? What criteria do you think the FDA should use in weighing a drug's benefits against its dangers?

5 The Structure and Function of Macromolecules

▲ **Figure 5.1 Scientists working with computer models of proteins.**

Key Concepts

5.1 Most macromolecules are polymers, built from monomers
5.2 Carbohydrates serve as fuel and building material
5.3 Lipids are a diverse group of hydrophobic molecules
5.4 Proteins have many structures, resulting in a wide range of functions
5.5 Nucleic acids store and transmit hereditary information

Overview

The Molecules of Life

We have seen how the concept of emergent properties applies to water and relatively simple organic molecules. Each type of small molecule has unique properties arising from the orderly arrangement of its atoms. Another level in the hierarchy of biological organization is reached when small organic molecules are joined inside cells, forming larger molecules. The four main classes of large biological molecules are carbohydrates, lipids, proteins, and nucleic acids. Many of these cellular molecules are, on the molecular scale, huge. For example, a protein may consist of thousands of covalently connected atoms that form a molecular colossus with a mass of over 100,000 daltons. Biologists use the term **macromolecule** for such giant molecules.

Considering the size and complexity of macromolecules, it is remarkable that biochemists have determined the detailed structures of so many of them **(Figure 5.1)**. The architecture of a macromolecule helps explain how that molecule works. Life's large molecules are the main subject of this chapter. For these molecules, as at all levels in the biological hierarchy, form and function are inseparable.

Concept 5.1

Most macromolecules are polymers, built from monomers

The large molecules in three of the four classes of life's organic compounds—carbohydrates, proteins, and nucleic acids—are chain-like molecules called polymers (from the Greek *polys*, many, and *meris*, part). A **polymer** is a long molecule consisting of many similar or identical building blocks linked by covalent bonds, much as a train consists of a chain of cars. The repeating units that serve as the building blocks of a polymer are small molecules called **monomers.** Some of the molecules that serve as monomers also have other functions of their own.

The Synthesis and Breakdown of Polymers

The classes of polymeric macromolecules differ in the nature of their monomers, but the chemical mechanisms by which cells make and break polymers are basically the same in all cases **(Figure 5.2)**. Monomers are connected by a reaction in which two molecules are covalently bonded to each other through loss of a water molecule; this is called a **condensation reaction,** specifically a **dehydration reaction,** because the molecule lost is water **(Figure 5.2a)**. When a bond forms between two monomers, each monomer contributes part of the water molecule that is lost: One molecule provides a hydroxyl group (—OH), while the other provides a hydrogen (—H). In making a polymer, this reaction is repeated as monomers are added to the chain one by one. The cell must expend energy to carry out these dehydration reactions, and the process oc-

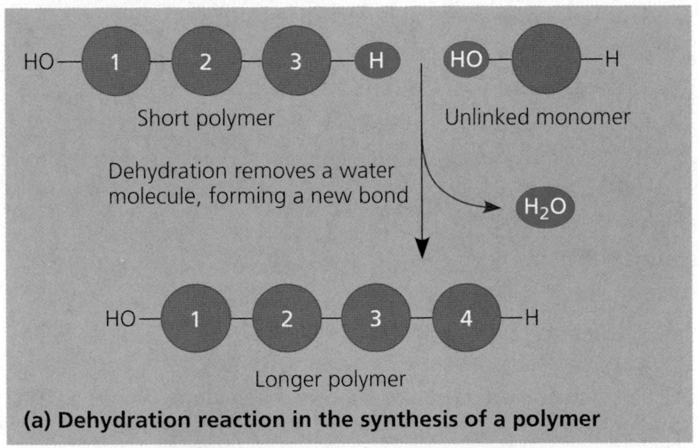

(a) Dehydration reaction in the synthesis of a polymer

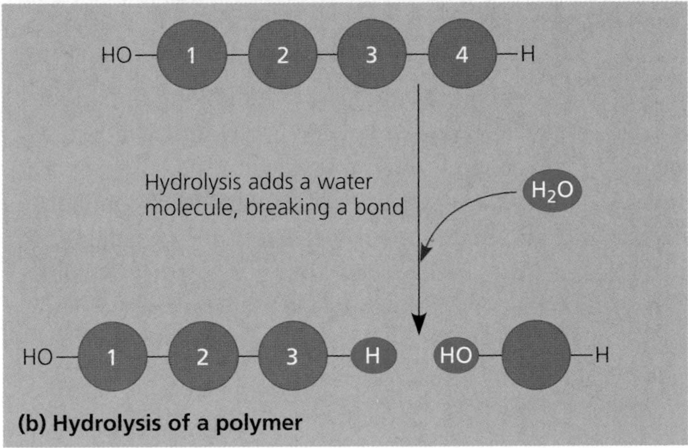

(b) Hydrolysis of a polymer

▲ **Figure 5.2 The synthesis and breakdown of polymers.**

curs only with the help of enzymes, specialized proteins that speed up chemical reactions in cells.

Polymers are disassembled to monomers by **hydrolysis**, a process that is essentially the reverse of the dehydration reaction **(Figure 5.2b)**. Hydrolysis means to break with water (from the Greek *hydro*, water, and *lysis*, break). Bonds between monomers are broken by the addition of water molecules, a hydrogen from the water attaching to one monomer and a hydroxyl group attaching to the adjacent monomer. An example of hydrolysis working in our bodies is the process of digestion. The bulk of the organic material in our food is in the form of polymers that are much too large to enter our cells. Within the digestive tract, various enzymes attack the polymers, speeding up hydrolysis. The released monomers are then absorbed into the bloodstream for distribution to all body cells. Those cells can then use dehydration reactions to assemble the monomers into new polymers that differ from the ones that were digested. The new polymers perform specific functions required by the cell.

The Diversity of Polymers

Each cell has thousands of different kinds of macromolecules; the collection varies from one type of cell to another even in the same organism. The inherent differences between human siblings reflect variations in polymers, particularly DNA and proteins. Molecular differences between unrelated individuals are more extensive and between species greater still. The diversity of macromolecules in the living world is vast, and the possible variety is effectively limitless.

What is the basis for such diversity in life's polymers? These molecules are constructed from only 40 to 50 common monomers and some others that occur rarely. Building an enormous variety of polymers from such a limited list of monomers is analogous to constructing hundreds of thousands of words from only 26 letters of the alphabet. The key is arrangement—variation in the linear sequence that the units follow. However, this analogy falls far short of describing the great diversity of macromolecules, because most biological polymers are much longer than the longest word. Proteins, for example, are built from 20 kinds of amino acids arranged in chains that are typically hundreds of amino acids long. The molecular logic of life is simple but elegant: Small molecules common to all organisms are ordered into unique macromolecules.

We are now ready to investigate the specific structures and functions of the four major classes of organic compounds found in cells. For each class, we will see that the large molecules have emergent properties not found in their individual building blocks.

Concept Check 5.1

1. What are the four main classes of large biological molecules?
2. How many molecules of water are needed to completely hydrolyze a polymer that is 10 monomers long?
3. After you eat a slice of apple, which reactions must occur for the amino acid monomers in the protein of the apple to be converted into proteins in your body?

For suggested answers, see Appendix A.

Concept 5.2

Carbohydrates serve as fuel and building material

Carbohydrates include both sugars and the polymers of sugars. The simplest carbohydrates are the monosaccharides, or single sugars, also known as simple sugars. Disaccharides are double sugars, consisting of two monosaccharides joined by a condensation reaction. The carbohydrates that are macromolecules are polysaccharides, polymers composed of many sugar building blocks.

Sugars

Monosaccharides (from the Greek *monos,* single, and *sacchar,* sugar) generally have molecular formulas that are some multiple of the unit CH_2O **(Figure 5.3)**. Glucose ($C_6H_{12}O_6$), the most common monosaccharide, is of central importance in the chemistry of life. In the structure of glucose, we can see the trademarks of a sugar: The molecule has a carbonyl group ($>C{=}O$) and multiple hydroxyl groups (—OH). Depending on the location of the carbonyl group, a sugar is either an aldose (aldehyde sugar) or a ketose (ketone sugar). Glucose, for example, is an aldose; fructose, a structural isomer of glucose, is a ketose. (Most names for sugars end in *-ose.*) Another criterion for classifying sugars is the size of the carbon skeleton, which ranges from three to seven carbons long. Glucose, fructose, and other sugars that have six carbons are called hexoses. Trioses (three-carbon sugars) and pentoses (five-carbon sugars) are also common.

Still another source of diversity for simple sugars is in the spatial arrangement of their parts around asymmetric carbons. (Recall from Chapter 4 that an asymmetric carbon is a carbon attached to four different kinds of partners.) Glucose and galactose, for example, differ only in the placement of parts around one asymmetric carbon (see the purple boxes in Figure 5.3). What seems like a small difference is significant enough to give the two sugars distinctive shapes and behaviors.

Although it is convenient to draw glucose with a linear carbon skeleton, this representation is not completely accurate. In aqueous solutions, glucose molecules, as well as most other sugars, form rings **(Figure 5.4)**.

Monosaccharides, particularly glucose, are major nutrients for cells. In the process known as cellular respiration, cells extract the energy stored in glucose molecules. Not only are simple sugar molecules a major fuel for cellular work, but their carbon skeletons serve as raw material for the synthesis of other types of small organic molecules, such as amino acids and fatty acids. Sugar molecules that are not immediately used in these ways are generally incorporated as monomers into disaccharides or polysaccharides.

A **disaccharide** consists of two monosaccharides joined by a **glycosidic linkage,** a covalent bond formed between two monosaccharides by a dehydration reaction. For example, maltose is a disaccharide formed by the linking of two molecules of glucose **(Figure 5.5a)**. Also known as malt sugar, maltose is an ingredient used in brewing beer. The most prevalent disaccharide is sucrose, which is table sugar. Its two monomers are glucose and fructose **(Figure 5.5b)**. Plants generally transport carbohydrates from leaves to roots and other nonphotosynthetic organs in the form of sucrose. Lactose, the sugar present in milk, is another disaccharide, in this case a glucose molecule joined to a galactose molecule.

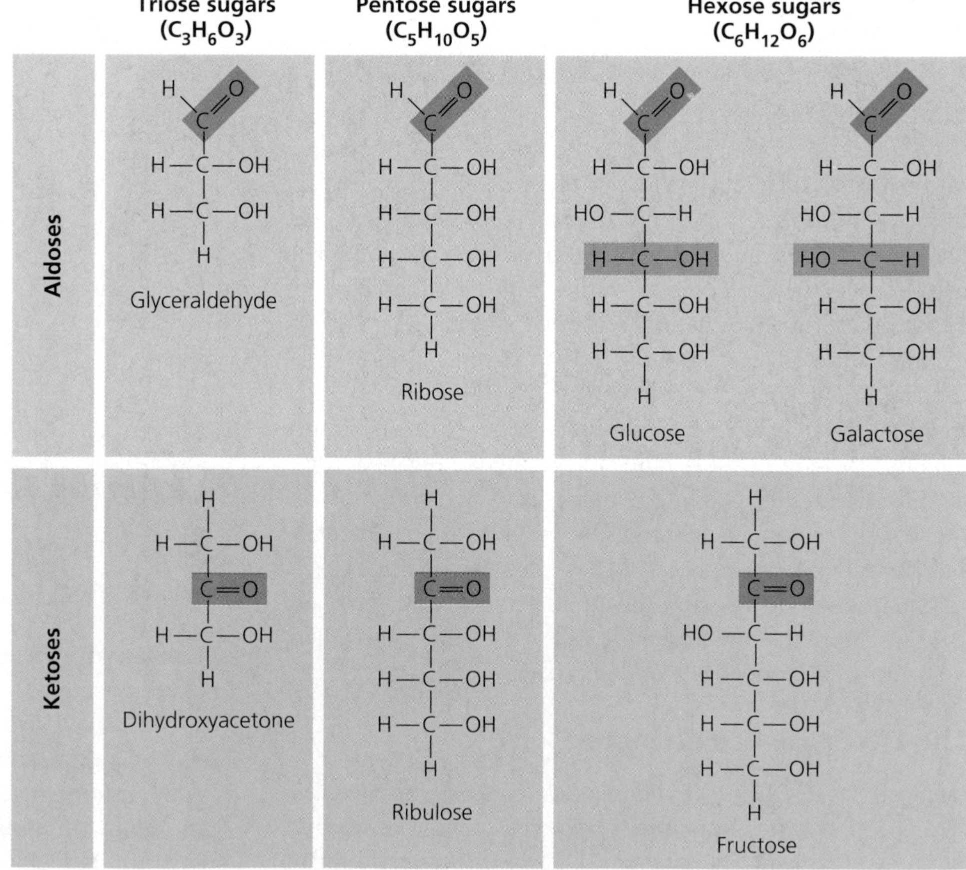

▶ **Figure 5.3 The structure and classification of some monosaccharides.** Sugars may be aldoses (aldehyde sugars, top row) or ketoses (ketone sugars, bottom row), depending on the location of the carbonyl group (dark orange). Sugars are also classified according to the length of their carbon skeletons. A third point of variation is the spatial arrangement around asymmetric carbons (compare, for example, the purple portions of glucose and galactose).

(a) **Linear and ring forms.** Chemical equilibrium between the linear and ring structures greatly favors the formation of rings. To form the glucose ring, carbon 1 bonds to the oxygen attached to carbon 5.

(b) **Abbreviated ring structure.** Each corner represents a carbon. The ring's thicker edge indicates that you are looking at the ring edge-on; the components attached to the ring lie above or below the plane of the ring.

▲ **Figure 5.4 Linear and ring forms of glucose.**

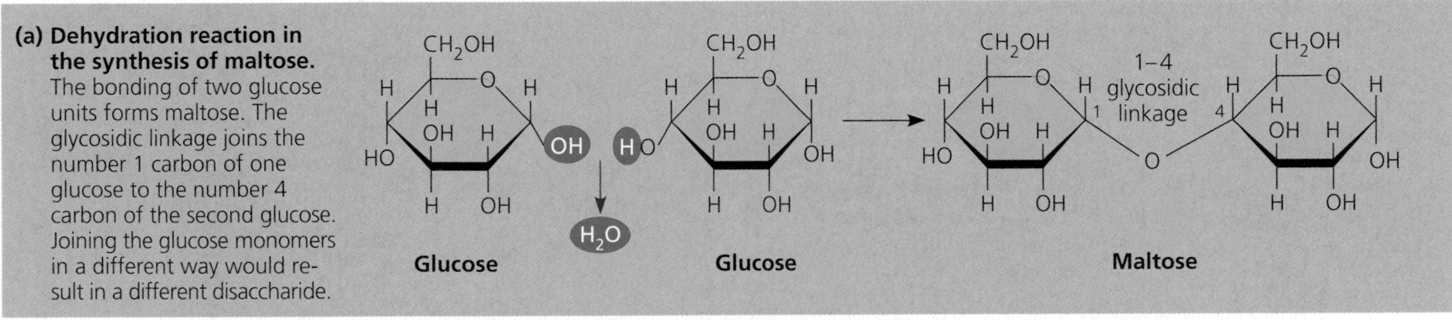

(a) **Dehydration reaction in the synthesis of maltose.** The bonding of two glucose units forms maltose. The glycosidic linkage joins the number 1 carbon of one glucose to the number 4 carbon of the second glucose. Joining the glucose monomers in a different way would result in a different disaccharide.

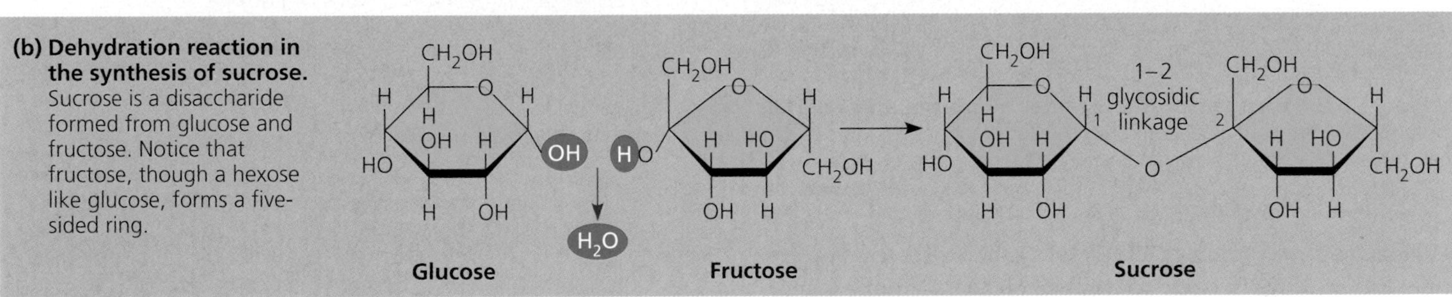

(b) **Dehydration reaction in the synthesis of sucrose.** Sucrose is a disaccharide formed from glucose and fructose. Notice that fructose, though a hexose like glucose, forms a five-sided ring.

▲ **Figure 5.5 Examples of disaccharide synthesis.**

Polysaccharides

Polysaccharides are macromolecules, polymers with a few hundred to a few thousand monosaccharides joined by glycosidic linkages. Some polysaccharides serve as storage material, hydrolyzed as needed to provide sugar for cells. Other polysaccharides serve as building material for structures that protect the cell or the whole organism. The architecture and function of a polysaccharide are determined by its sugar monomers and by the positions of its glycosidic linkages.

Storage Polysaccharides

Starch, a storage polysaccharide of plants, is a polymer consisting entirely of glucose monomers. Most of these monomers are joined by 1–4 linkages (number 1 carbon to number 4 carbon), like the glucose units in maltose (see Figure 5.5a). The angle of these bonds makes the polymer helical. The simplest form of starch, amylose, is unbranched. Amylopectin, a more complex form of starch, is a branched polymer with 1–6 linkages at the branch points.

Plants store starch as granules within cellular structures called plastids, which include chloroplasts **(Figure 5.6a)**. Synthesizing starch enables the plant to stockpile surplus glucose. Because glucose is a major cellular fuel, starch represents stored energy. The sugar can later be withdrawn from this carbohydrate "bank" by hydrolysis, which breaks the bonds between the glucose monomers. Most animals, including humans, also have enzymes that can hydrolyze plant starch, making glucose

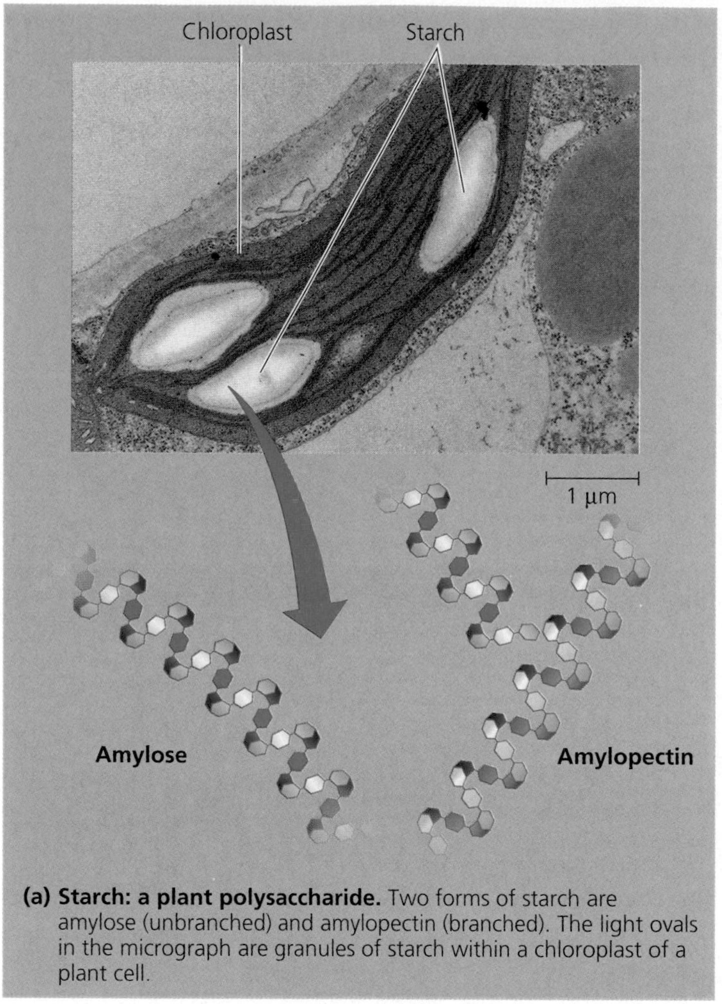

(a) Starch: a plant polysaccharide. Two forms of starch are amylose (unbranched) and amylopectin (branched). The light ovals in the micrograph are granules of starch within a chloroplast of a plant cell.

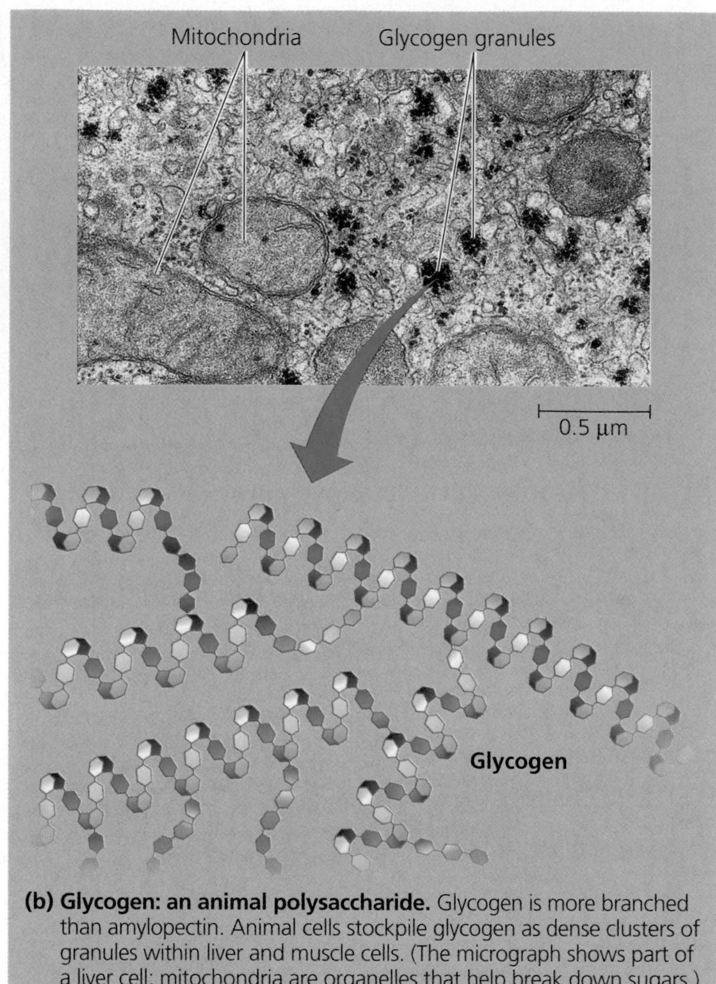

(b) Glycogen: an animal polysaccharide. Glycogen is more branched than amylopectin. Animal cells stockpile glycogen as dense clusters of granules within liver and muscle cells. (The micrograph shows part of a liver cell; mitochondria are organelles that help break down sugars.)

▲ **Figure 5.6 Storage polysaccharides of plants and animals.** These examples, starch and glycogen, are composed entirely of glucose monomers, represented here by hexagons. Due to their molecular structure, the polymer chains tend to form helices.

available as a nutrient for cells. Potato tubers and grains—the fruits of wheat, corn, rice, and other grasses—are the major sources of starch in the human diet.

Animals store a polysaccharide called **glycogen,** a polymer of glucose that is like amylopectin but more extensively branched **(Figure 5.6b)**. Humans and other vertebrates store glycogen mainly in liver and muscle cells. Hydrolysis of glycogen in these cells releases glucose when the demand for sugar increases. This stored fuel cannot sustain an animal for long, however. In humans, for example, glycogen stores are depleted in about a day unless they are replenished by consumption of food.

Structural Polysaccharides

Organisms build strong materials from structural polysaccharides. For example, the polysaccharide called **cellulose** is a major component of the tough walls that enclose plant cells. On a global scale, plants produce almost 10^{11} (100 billion) tons of cellulose per year; it is the most abundant organic compound on Earth. Like starch, cellulose is a polymer of glucose, but the glycosidic linkages in these two polymers differ. The difference is based on the fact that there are actually two slightly different ring structures for glucose **(Figure 5.7a)**. When glucose forms a ring, the hydroxyl group attached to the number 1 carbon is positioned either below or above the plane of the ring. These two ring forms for glucose are called alpha (α) and beta (β), respectively. In starch, all the glucose monomers are in the α configuration **(Figure 5.7b)**, the arrangement we saw in Figures 5.4 and 5.5. In contrast, the glucose monomers of cellulose are all in the β configuration, making every other glucose monomer upside down with respect to its neighbors **(Figure 5.7c)**.

The differing glycosidic linkages in starch and cellulose give the two molecules distinct three-dimensional shapes. Whereas a starch molecule is mostly helical, a cellulose molecule is straight (and never branched), and its hydroxyl groups are free to hydrogen-bond with the hydroxyls of other cellulose molecules lying parallel to it. In plant cell walls, parallel cellulose molecules held together in this way are grouped into units called microfibrils **(Figure 5.8)**. These cable-like microfibrils

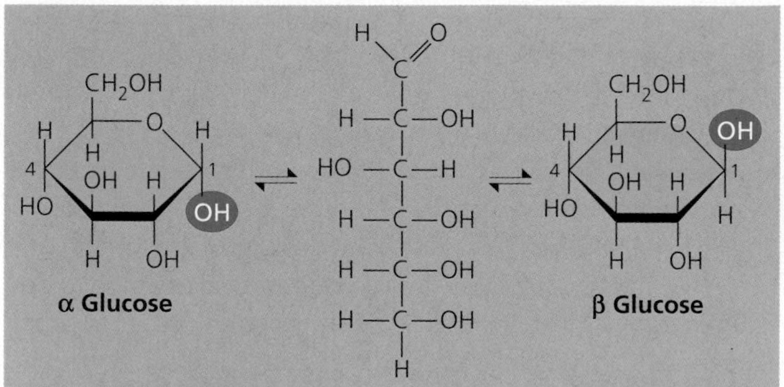

α Glucose

β Glucose

(a) α and β glucose ring structures. These two interconvertible forms of glucose differ in the placement of the hydroxyl group attached to the number 1 carbon.

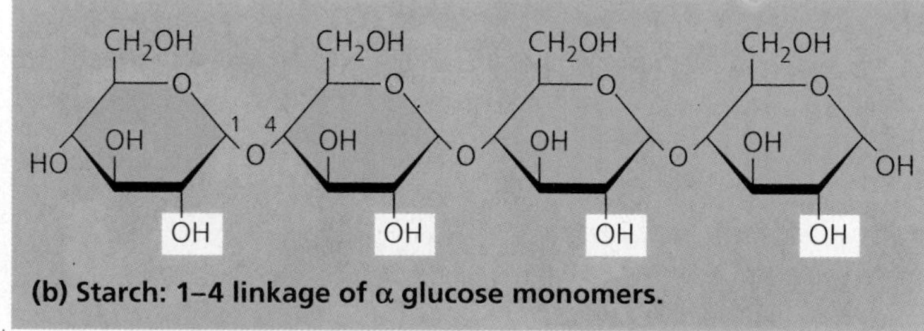

(b) Starch: 1–4 linkage of α glucose monomers.

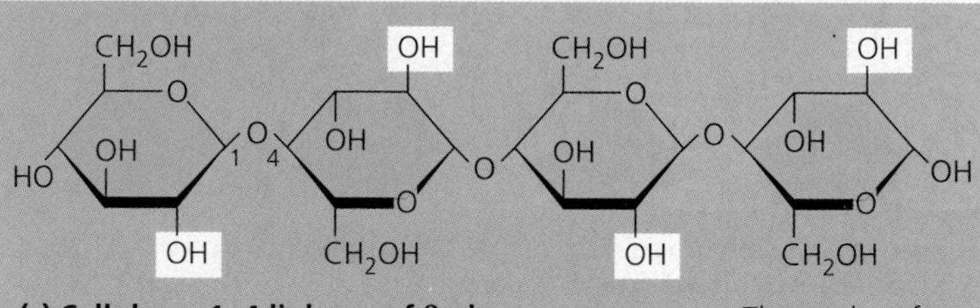

(c) Cellulose: 1–4 linkage of β glucose monomers. The angles of the bonds that link the rings make every other glucose monomer upside down with respect to its neighbors. Compare the positions of the highlighted –OH groups in (b) starch and (c) cellulose.

▲ **Figure 5.7 Starch and cellulose structures.**

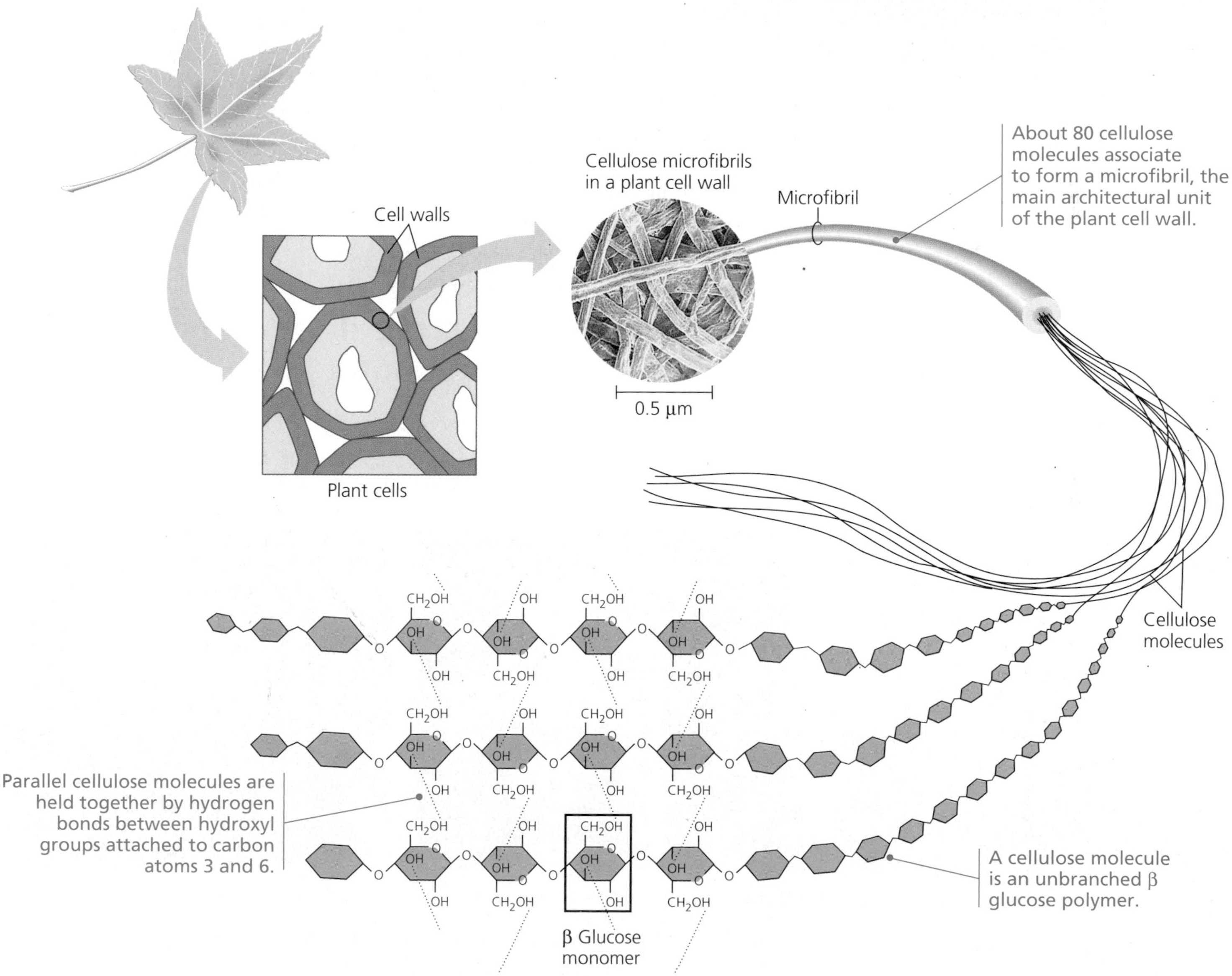

Cell walls

Plant cells

Cellulose microfibrils in a plant cell wall

0.5 μm

Microfibril

About 80 cellulose molecules associate to form a microfibril, the main architectural unit of the plant cell wall.

Cellulose molecules

Parallel cellulose molecules are held together by hydrogen bonds between hydroxyl groups attached to carbon atoms 3 and 6.

A cellulose molecule is an unbranched β glucose polymer.

β Glucose monomer

▲ **Figure 5.8 The arrangement of cellulose in plant cell walls.**

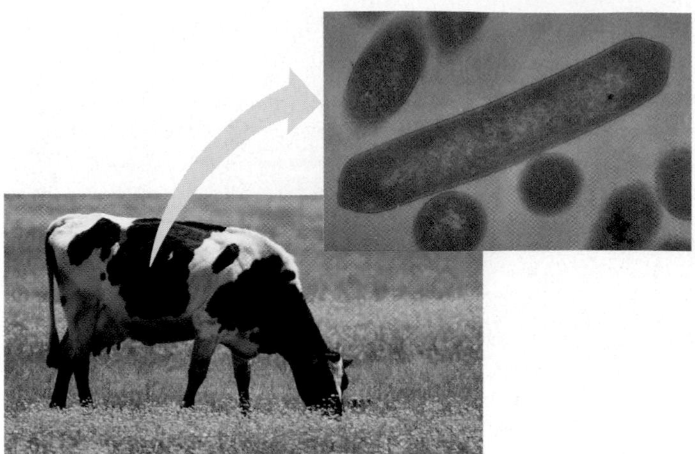

▲ **Figure 5.9 Cellulose-digesting bacteria are found in grazing animals such as this cow.**

are a strong building material for plants as well as for humans, who use wood, which is rich in cellulose, for lumber.

Enzymes that digest starch by hydrolyzing its α linkages are unable to hydrolyze the β linkages of cellulose because of the distinctly different shapes of these two molecules. In fact, few organisms possess enzymes that can digest cellulose. Humans do not; the cellulose in our food passes through the digestive tract and is eliminated with the feces. Along the way, the cellulose abrades the wall of the digestive tract and stimulates the lining to secrete mucus, which aids in the smooth passage of food through the tract. Thus, although cellulose is not a nutrient for humans, it is an important part of a healthful diet. Most fresh fruits, vegetables, and whole grains are rich in cellulose. On food packages, "insoluble fiber" refers mainly to cellulose.

Some microbes can digest cellulose, breaking it down to glucose monomers. A cow harbors cellulose-digesting bacteria in the rumen, the first compartment in its stomach **(Figure 5.9)**. The bacteria hydrolyze the cellulose of hay and grass and convert the glucose to other nutrients that nourish the cow. Similarly, a termite, which is unable to digest cellulose by itself, has microbes living in its gut that can make a meal of wood. Some fungi can also digest cellulose, thereby helping recycle chemical elements within Earth's ecosystems.

Another important structural polysaccharide is **chitin**, the carbohydrate used by arthropods (insects, spiders, crustaceans, and related animals) to build their exoskeletons **(Figure 5.10)**. An exoskeleton is a hard case that surrounds the soft parts of an animal. Pure chitin is leathery, but it becomes hardened when encrusted with calcium carbonate, a salt. Chitin is also found in many fungi, which use this polysaccharide rather than cellulose as the building material for their cell walls. Chitin is similar to cellulose, except that the glucose monomer of chitin has a nitrogen-containing appendage (see Figure 5.10a).

Concept Check 5.2

1. Write the formula for a monosaccharide that has three carbons.
2. A dehydration reaction joins two glucose molecules to form maltose. The formula for glucose is $C_6H_{12}O_6$. What is the formula for maltose?
3. Compare and contrast starch and cellulose.

For suggested answers, see Appendix A.

Concept 5.3

Lipids are a diverse group of hydrophobic molecules

Lipids are the one class of large biological molecules that does not consist of polymers. The compounds called **lipids** are grouped together because they share one important trait:

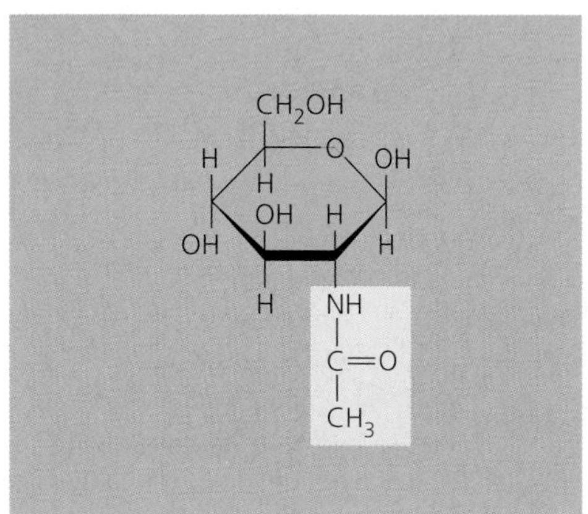

(a) The structure of the chitin monomer.

(b) Chitin forms the exoskeleton of arthropods. This cicada is molting, shedding its old exoskeleton and emerging in adult form.

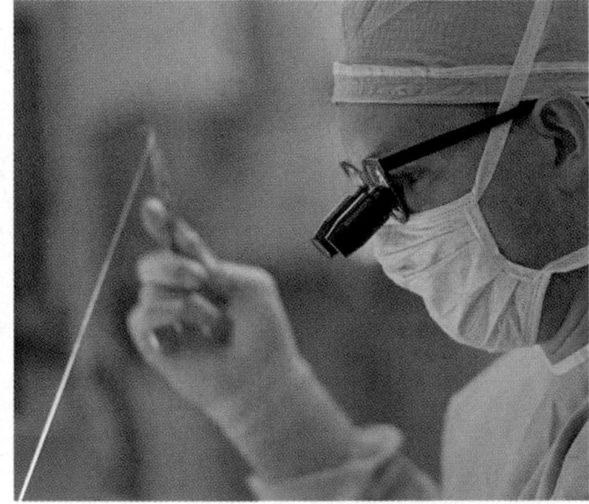

(c) Chitin is used to make a strong and flexible surgical thread that decomposes after the wound or incision heals.

▲ **Figure 5.10 Chitin, a structural polysaccharide.**

They have little or no affinity for water. The hydrophobic behavior of lipids is based on their molecular structure. Although they may have some polar bonds associated with oxygen, lipids consist mostly of hydrocarbons. Smaller than true (polymeric) macromolecules, lipids are a highly varied group in both form and function. Lipids include waxes and certain pigments, but we will focus on the most biologically important types of lipids: fats, phospholipids, and steroids.

Fats

Although fats are not polymers, they are large molecules, and they are assembled from smaller molecules by dehydration reactions. A **fat** is constructed from two kinds of smaller molecules: glycerol and fatty acids **(Figure 5.11a)**. Glycerol is an alcohol with three carbons, each bearing a hydroxyl group. A **fatty acid** has a long carbon skeleton, usually 16 or 18 carbon atoms in length. At one end of the fatty acid is a carboxyl group, the functional group that gives these molecules the name fatty *acid.* Attached to the carboxyl group is a long hydrocarbon chain. The nonpolar C—H bonds in the hydrocarbon chains of fatty acids are the reason fats are hydrophobic. Fats separate from water because the water molecules hydrogen-bond to one another and exclude the fats. A common example of this phenomenon is the separation of vegetable oil (a liquid fat) from the aqueous vinegar solution in a bottle of salad dressing.

In making a fat, three fatty acid molecules each join to glycerol by an ester linkage, a bond between a hydroxyl group and a carboxyl group. The resulting fat, also called a **triacylglycerol,** thus consists of three fatty acids linked to one glycerol molecule. (Still another name for a fat is *triglyceride,* a word often found in the list of ingredients on packaged foods.) The fatty acids in a fat can be the same, as in **Figure 5.11b**, or they can be of two or three different kinds.

Fatty acids vary in length and in the number and locations of double bonds. The terms *saturated fats* and *unsaturated fats* are commonly used in the context of nutrition **(Figure 5.12)**. These terms refer to the structure of the hydrocarbon chains of the fatty acids. If there are no double bonds between carbon atoms composing the chain, then as many hydrogen atoms as possible are bonded to the carbon skeleton. Such a structure is described as being *saturated* with hydrogen, so the resulting fatty acid is called a **saturated fatty acid (Figure 5.12a)**. An **unsaturated**

Fatty acid (palmitic acid)

H_2O

Glycerol

(a) Dehydration reaction in the synthesis of a fat

Ester linkage

(b) Fat molecule (triacylglycerol)

▲ **Figure 5.11 The synthesis and structure of a fat, or triacylglycerol.** The molecular building blocks of a fat are one molecule of glycerol and three molecules of fatty acids. **(a)** One water molecule is removed for each fatty acid joined to the glycerol. **(b)** A fat molecule with three identical fatty acid units. The carbons of the fatty acids are arranged zig-zag to suggest the actual orientations of the four single bonds extending from each carbon (see Figure 4.3a).

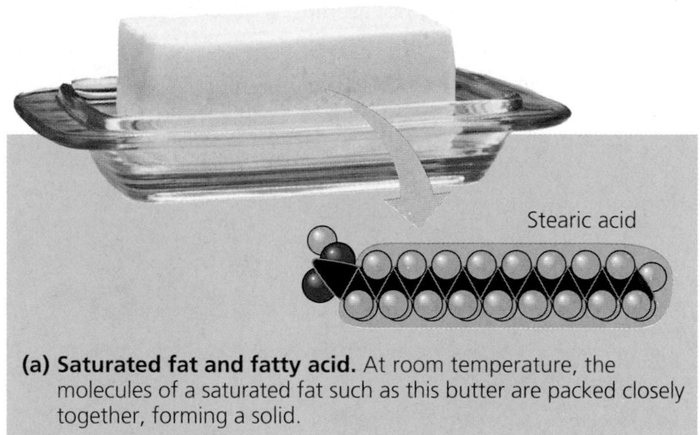

(a) Saturated fat and fatty acid. At room temperature, the molecules of a saturated fat such as this butter are packed closely together, forming a solid.

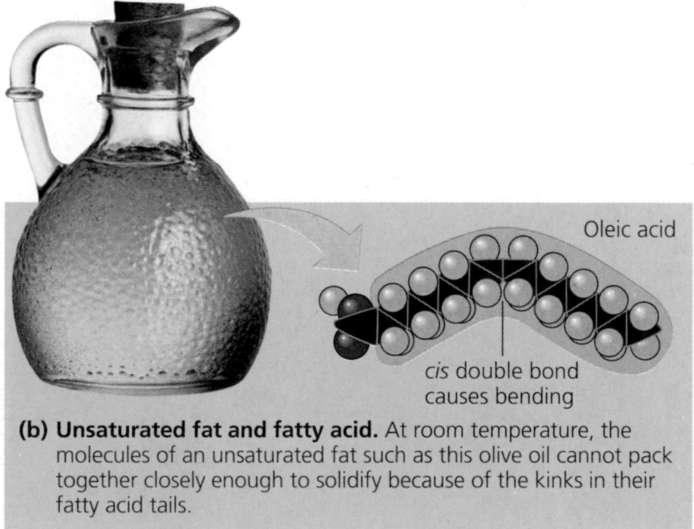

(b) Unsaturated fat and fatty acid. At room temperature, the molecules of an unsaturated fat such as this olive oil cannot pack together closely enough to solidify because of the kinks in their fatty acid tails.

▲ **Figure 5.12 Examples of saturated and unsaturated fats and fatty acids.**

fatty acid has one or more double bonds, formed by the removal of hydrogen atoms from the carbon skeleton. The fatty acid will have a kink in its hydrocarbon chain wherever a *cis* double bond occurs **(Figure 5.12b)**.

A fat made from saturated fatty acids is called a saturated fat. Most animal fats are saturated: The hydrocarbon chains of their fatty acids—the "tails" of the fat molecules—lack double bonds, and the molecules can pack tightly, side by side. Saturated animal fats—such as lard and butter—are solid at room temperature. In contrast, the fats of plants and fishes are generally unsaturated, meaning that they are built of one or more types of unsaturated fatty acids. Usually liquid at room temperature, plant and fish fats are referred to as oils—olive oil and cod liver oil are examples. The kinks where the *cis* double bonds are located prevent the molecules from packing together closely enough to solidify at room temperature. The phrase "hydrogenated vegetable oils" on food labels means that unsaturated fats have been synthetically converted to saturated fats by adding hydrogen. Peanut butter, margarine, and many other products are hydrogenated to prevent lipids from separating out in liquid (oil) form.

A diet rich in saturated fats is one of several factors that may contribute to the cardiovascular disease known as atherosclerosis. In this condition, deposits called plaques develop within the walls of blood vessels, causing inward bulges that impede blood flow and reduce the resilience of the vessels. Recent studies have shown that the process of hydrogenating vegetable oils produces not only saturated fats but also unsaturated fats with *trans* double bonds. These *trans* fat molecules may contribute more than saturated fats to atherosclerosis (see Chapter 42) and other problems.

Fat has come to have such a negative connotation in our culture that you might wonder whether fats serve any useful purpose. The major function of fats is energy storage. The hydrocarbon chains of fats are similar to gasoline molecules and just as rich in energy. A gram of fat stores more than twice as much energy as a gram of a polysaccharide, such as starch. Because plants are relatively immobile, they can function with bulky energy storage in the form of starch. (Vegetable oils are generally obtained from seeds, where more compact storage is an asset to the plant.) Animals, however, must carry their energy stores with them, so there is an advantage to having a more compact reservoir of fuel—fat. Humans and other mammals stock their long-term food reserves in adipose cells (see Figure 4.6b), which swell and shrink as fat is deposited and withdrawn from storage. In addition to storing energy, adipose tissue also cushions such vital organs as the kidneys, and a layer of fat beneath the skin insulates the body. This subcutaneous layer is especially thick in whales, seals, and most other marine mammals, protecting them from cold ocean water.

Phospholipids

A **phospholipid**, as shown in **Figure 5.13**, is similar to a fat, but has only two fatty acids attached to glycerol rather than three. The third hydroxyl group of glycerol is joined to a phosphate

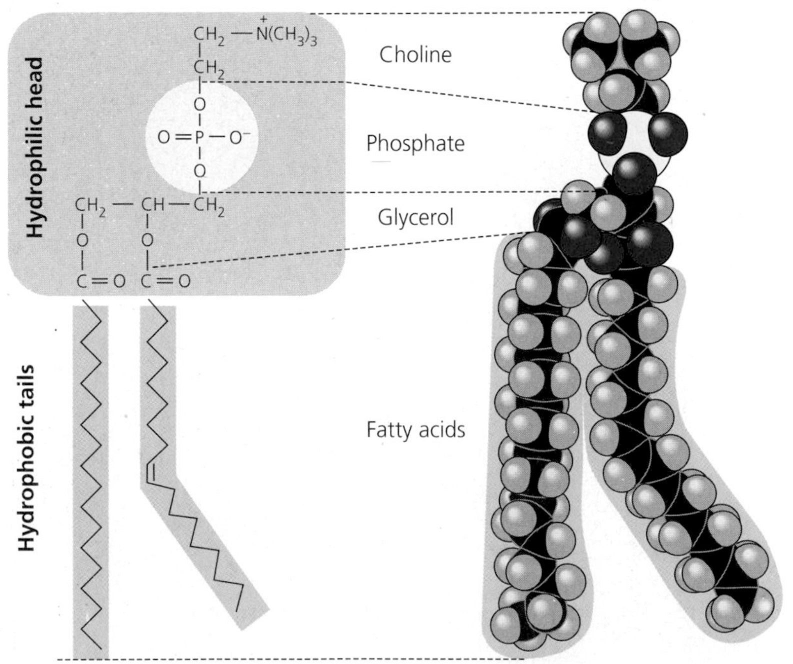

(a) Structural formula **(b) Space-filling model**

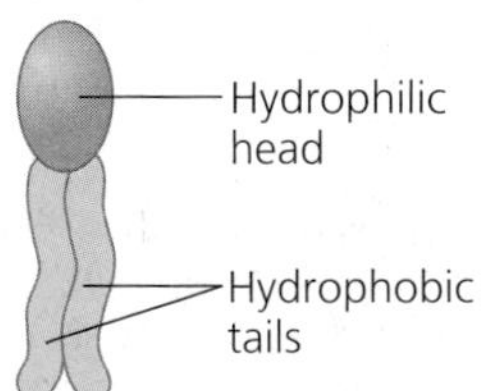

(c) Phospholipid symbol

◄ **Figure 5.13 The structure of a phospholipid.** A phospholipid has a hydrophilic (polar) head and two hydrophobic (nonpolar) tails. Phospholipid diversity is based on differences in the two fatty acids and in the groups attached to the phosphate group of the head. This particular phospholipid, called a phosphatidylcholine, has an attached choline group. The kink in one of its tails is due to a *cis* double bond. **(a)** The structural formula follows a common chemical convention of omitting the carbons and attached hydrogens of the hydrocarbon tails. **(b)** In the space-filling model, black = carbon, gray = hydrogen, red = oxygen, yellow = phosphorus, and blue = nitrogen. **(c)** This symbol for a phospholipid will appear throughout the book.

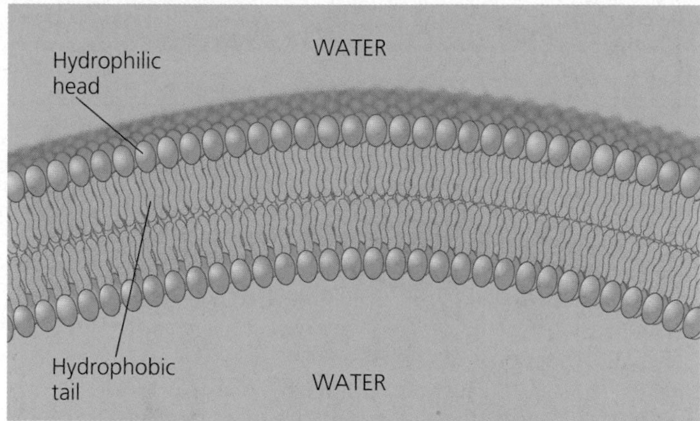

▲ **Figure 5.14 Bilayer structure formed by self-assembly of phospholipids in an aqueous environment.** The phospholipid bilayer shown here is the main fabric of biological membranes. Note that the hydrophilic heads of the phospholipids are in contact with water in this structure, whereas the hydrophobic tails are in contact with each other and remote from water.

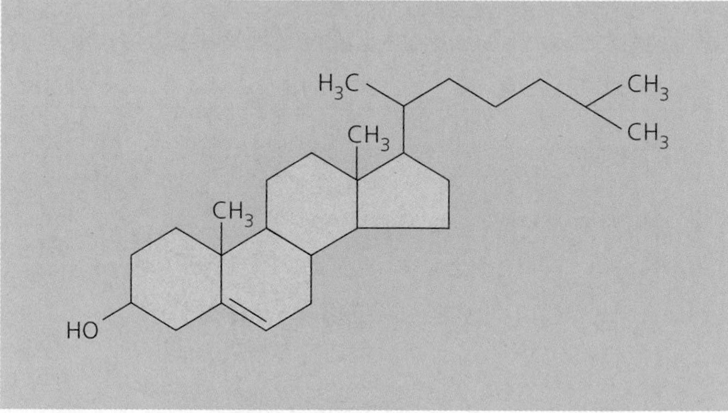

▲ **Figure 5.15 Cholesterol, a steroid.** Cholesterol is the molecule from which other steroids, including the sex hormones, are synthesized. Steroids vary in the functional groups attached to their four interconnected rings (shown in gold).

group, which has a negative electrical charge. Additional small molecules, usually charged or polar, can be linked to the phosphate group to form a variety of phospholipids.

Phospholipids show ambivalent behavior toward water. Their hydrocarbon tails are hydrophobic and are excluded from water. However, the phosphate group and its attachments form a hydrophilic head that has an affinity for water. When phospholipids are added to water, they self-assemble into double-layered aggregates—bilayers—that shield their hydrophobic portions from water **(Figure 5.14)**.

At the surface of a cell, phospholipids are arranged in a similar bilayer. The hydrophilic heads of the molecules are on the outside of the bilayer, in contact with the aqueous solutions inside and outside the cell. The hydrophobic tails point toward the interior of the bilayer, away from the water. The phospholipid bilayer forms a boundary between the cell and its external environment; in fact, phospholipids are major components of all cell membranes. This behavior provides another example of how form fits function at the molecular level.

Steroids

Steroids are lipids characterized by a carbon skeleton consisting of four fused rings **(Figure 5.15)**. Different steroids vary in the functional groups attached to this ensemble of rings. One steroid, **cholesterol,** is a common component of animal cell membranes and is also the precursor from which other steroids are synthesized. Many hormones, including vertebrate sex hormones, are steroids produced from cholesterol (see Figure 4.9). Thus, cholesterol is a crucial molecule in animals, although a high level of it in the blood may contribute to atherosclerosis. Both saturated fats and *trans* fats exert their negative impact on health by affecting cholesterol levels.

Concept Check 5.3

1. Compare the structure of a fat (triglyceride) with that of a phospholipid.
2. How do saturated fats differ from unsaturated fats, both in structure and in behavior?
3. Why are human sex hormones considered to be lipids?

For suggested answers, see Appendix A.

Concept 5.4

Proteins have many structures, resulting in a wide range of functions

The importance of proteins is implied by their name, which comes from the Greek word *proteios,* meaning "first place." Proteins account for more than 50% of the dry mass of most cells, and they are instrumental in almost everything organisms do. Some proteins speed up chemical reactions, while others play a role in structural support, storage, transport, cellular communications, movement, and defense against foreign substances (**Table 5.1**, on the next page).

The most important type of protein may be **enzymes.** Enzymatic proteins regulate metabolism by acting as **catalysts,** chemical agents that selectively speed up chemical reactions in

Table 5.1 An Overview of Protein Functions

Type of Protein	Function	Examples
Enzymatic proteins	Selective acceleration of chemical reactions	Digestive enzymes catalyze the hydrolysis of the polymers in food.
Structural proteins	Support	Insects and spiders use silk fibers to make their cocoons and webs, respectively. Collagen and elastin provide a fibrous framework in animal connective tissues. Keratin is the protein of hair, horns, feathers, and other skin appendages.
Storage proteins	Storage of amino acids	Ovalbumin is the protein of egg white, used as an amino acid source for the developing embryo. Casein, the protein of milk, is the major source of amino acids for baby mammals. Plants have storage proteins in their seeds.
Transport proteins	Transport of other substances	Hemoglobin, the iron-containing protein of vertebrate blood, transports oxygen from the lungs to other parts of the body. Other proteins transport molecules across cell membranes.
Hormonal proteins	Coordination of an organism's activities	Insulin, a hormone secreted by the pancreas, helps regulate the concentration of sugar in the blood of vertebrates.
Receptor proteins	Response of cell to chemical stimuli	Receptors built into the membrane of a nerve cell detect chemical signals released by other nerve cells.
Contractile and motor proteins	Movement	Actin and myosin are responsible for the movement of muscles. Other proteins are responsible for the undulations of the organelles called cilia and flagella.
Defensive proteins	Protection against disease	Antibodies combat bacteria and viruses.

the cell without being consumed by the reaction **(Figure 5.16)**. Because an enzyme can perform its function over and over again, these molecules can be thought of as workhorses that keep cells running by carrying out the processes of life.

A human has tens of thousands of different proteins, each with a specific structure and function; proteins, in fact, are the most structurally sophisticated molecules known. Consistent with their diverse functions, they vary extensively in structure, each type of protein having a unique three-dimensional shape, or conformation.

Polypeptides

Diverse as proteins are, they are all polymers constructed from the same set of 20 amino acids. Polymers of amino acids are called **polypeptides.** A **protein** consists of one or more polypeptides folded and coiled into specific conformations.

Amino Acid Monomers

Amino acids are organic molecules possessing both carboxyl and amino groups (see Chapter 4). The illustration at the right shows the general formula for an amino acid. At the center of the amino acid is an asymmetric carbon atom called the *alpha* (α) *carbon.* Its four different partners are an amino group, a carboxyl group, a hydrogen atom, and a variable group symbolized by R. The R group, also called the side chain, differs with each amino acid. **Figure 5.17** shows the 20 amino acids that cells use to

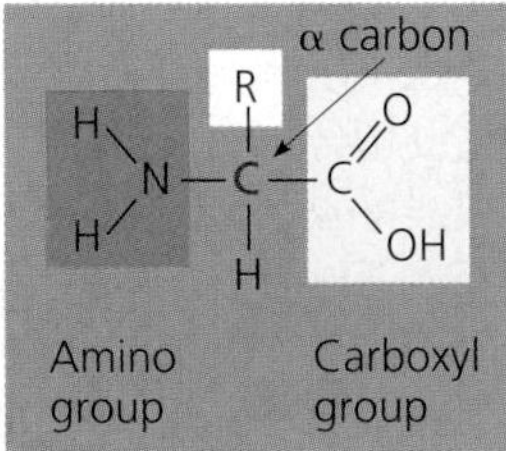

1 Active site is available for a molecule of substrate, the reactant on which the enzyme acts.
2 Substrate binds to enzyme.
Substrate (sucrose)
Glucose
OH
Enzyme (sucrase)
H_2O
Fructose
HO
4 Products are released.
3 Substrate is converted to products.

▲ **Figure 5.16 The catalytic cycle of an enzyme.** The enzyme sucrase accelerates hydrolysis of sucrose into glucose and fructose. Acting as a catalyst, the sucrase protein is not consumed during the cycle, but is available for further catalysis.

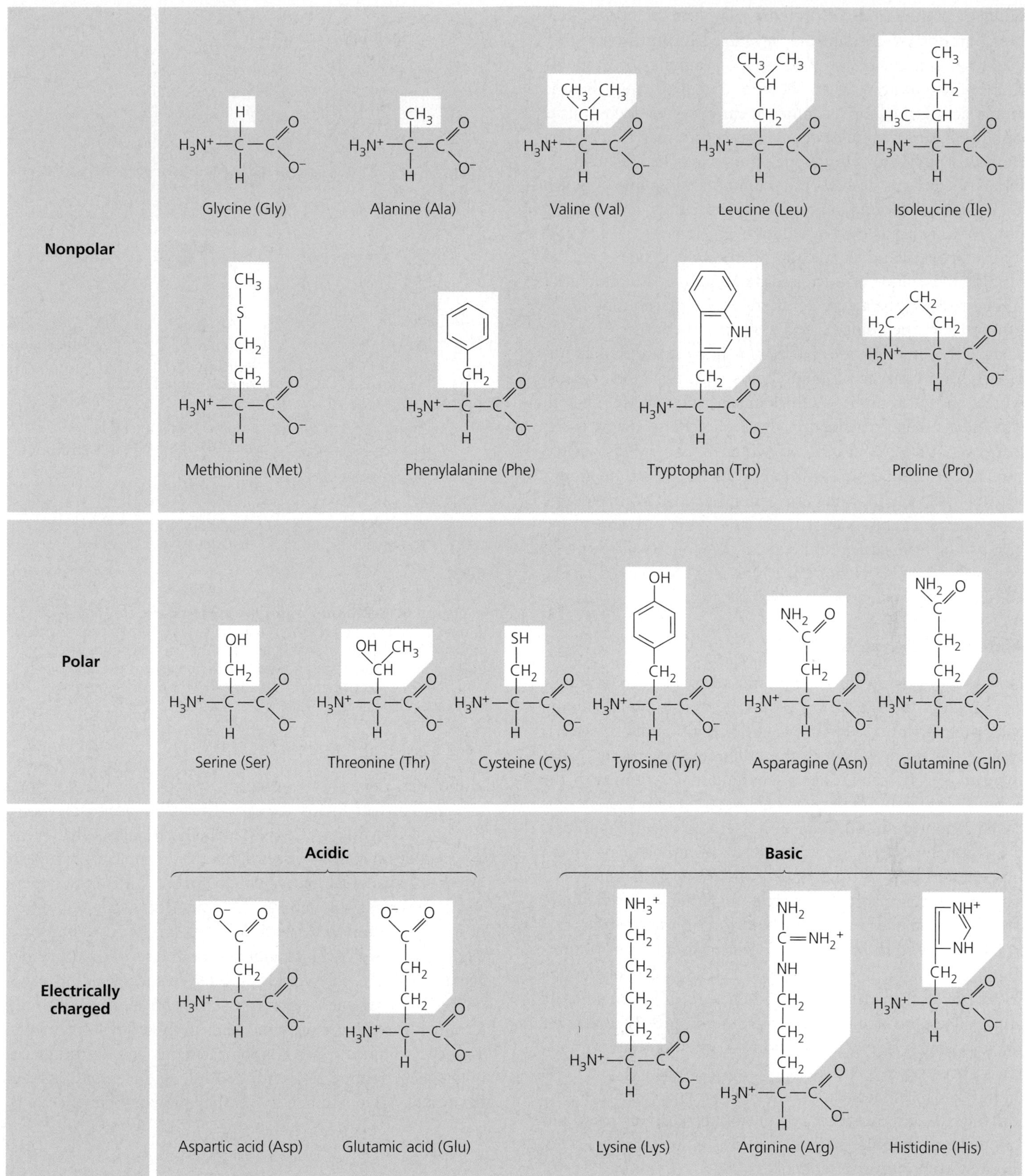

▲ **Figure 5.17 The 20 amino acids of proteins.** The amino acids are grouped here according to the properties of their side chains (R groups), highlighted in white. The amino acids are shown in their prevailing ionic forms at pH 7.2, the pH within a cell. The three-letter abbreviations for the amino acids are in parentheses. All the amino acids used in proteins are the same enantiomer, called the L form, as shown here (see Figure 4.7).

build their thousands of proteins. Here the amino and carboxyl groups are all depicted in ionized form, the way they usually exist at the pH in a cell. The R group may be as simple as a hydrogen atom, as in the amino acid glycine (the one amino acid lacking an asymmetric carbon, since two of its α carbon's partners are hydrogen atoms), or it may be a carbon skeleton with various functional groups attached, as in glutamine. (Organisms do have other amino acids, some of which are occasionally found in proteins. Because these are relatively rare, they are not shown in Figure 5.17.)

The physical and chemical properties of the side chain determine the unique characteristics of a particular amino acid. In Figure 5.17, the amino acids are grouped according to the properties of their side chains. One group consists of amino acids with nonpolar side chains, which are hydrophobic. Another group consists of amino acids with polar side chains, which are hydrophilic. Acidic amino acids are those with side chains that are generally negative in charge owing to the presence of a carboxyl group, which is usually dissociated (ionized) at cellular pH. Basic amino acids have amino groups in their side chains that are generally positive in charge. (Notice that *all* amino acids have carboxyl groups and amino groups; the terms *acidic* and *basic* in this context refer only to groups on the side chains.) Because they are charged, acidic and basic side chains are also hydrophilic.

Amino Acid Polymers

Now that we have examined amino acids, let's see how they are linked to form polymers **(Figure 5.18)**. When two amino acids are positioned so that the carboxyl group of one is adjacent to the amino group of the other, an enzyme can cause them to join by catalyzing a dehydration reaction, with the removal of a water molecule. The resulting covalent bond is called a **peptide bond.** Repeated over and over, this process yields a polypeptide, a polymer of many amino acids linked by peptide bonds. At one end of the polypeptide chain is a free amino group; at the opposite end is a free carboxyl group. Thus, the chain has an amino end (N-terminus) and a carboxyl end (C-terminus). The repeating sequence of atoms highlighted in purple in Figure 5.18b is called the polypeptide backbone. Attached to this backbone are different kinds of appendages, the side chains of the amino acids. Polypeptides range in length from a few monomers to a thousand or more. Each specific polypeptide has a unique linear sequence of amino acids. The immense variety of polypeptides in nature illustrates an important concept introduced earlier—that cells can make many different polymers by linking a limited set of monomers into diverse sequences.

Determining the Amino Acid Sequence of a Polypeptide

The pioneer in determining the amino acid sequence of proteins was Frederick Sanger, who, with his colleagues at Cambridge University in England, worked on the hormone insulin in the late 1940s and early 1950s. His approach was to use protein-digesting enzymes and other catalysts that break polypeptides at specific places rather than completely hydrolyzing the chains to amino acids. Treatment with one of these agents cleaves a polypeptide into fragments (each consisting of multiple amino acid subunits) that can be separated by a technique called chromatography. Hydrolysis with a different agent breaks the polypeptide at different sites, yielding a second group of fragments. Sanger used chemical methods to determine the sequence of amino acids in these small fragments. Then he searched for overlapping regions among the pieces obtained by hydrolyzing with the different agents. Consider, for instance, two fragments with the following sequences:

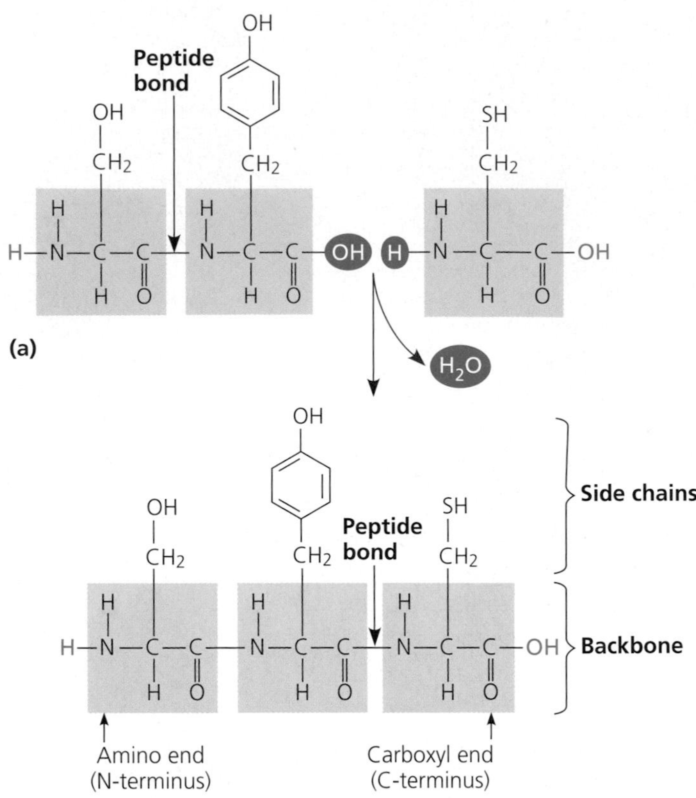

▲ **Figure 5.18 Making a polypeptide chain. (a)** Peptide bonds formed by dehydration reactions link the carboxyl group of one amino acid to the amino group of the next. **(b)** The peptide bonds are formed one at a time, starting with the amino acid at the amino end (N-terminus). The polypeptide has a repetitive backbone (purple) to which the amino acid side chains are attached.

Cys-Ser-Leu-Tyr-Gln-Leu
Tyr-Gln-Leu-Glu-Asn

We can deduce from the overlapping regions that the intact polypeptide contains in its primary structure the following segment:

Cys-Ser-Leu-Tyr-Gln-Leu-Glu-Asn

Just as we could reconstruct this sentence from a collection of fragments with overlapping sequences of letters, Sanger and his co-workers were able, after years of effort, to reconstruct the complete primary structure of insulin. Since then, most of the steps involved in sequencing a polypeptide have been automated.

Protein Conformation and Function

Once we have learned the amino acid sequence of a polypeptide, what can it tell us about protein conformation and function? The term *polypeptide* is not quite synonymous with the term *protein*. Even for a protein consisting of a single polypeptide, the relationship is somewhat analogous to that between a long strand of yarn and a sweater of particular size and shape that one can knit from the yarn. A functional protein is not *just* a polypeptide chain, but one or more polypeptides precisely twisted, folded, and coiled into a molecule of unique shape **(Figure 5.19)**. It is the amino acid sequence of a polypeptide that determines what three-dimensional conformation the protein will take.

When a cell synthesizes a polypeptide, the chain generally folds spontaneously, assuming the functional conformation for that protein. This folding is driven and reinforced by the formation of a variety of bonds between parts of the chain, which in turn depends on the sequence of amino acids. Many proteins are globular (roughly spherical), while others are fibrous in shape. Even within these broad categories, countless variations are possible.

A protein's specific conformation determines how it works. In almost every case, the function of a protein depends on its ability to recognize and bind to some other molecule. For instance, an antibody (a protein) binds to a particular foreign substance that has invaded the body, and an enzyme (another type of protein) recognizes and binds to its substrate, the substance the enzyme works on. In Chapter 2, you learned that natural signal molecules called endorphins bind to specific receptor proteins on the surface of brain cells in humans, producing euphoria and relieving pain. Morphine, heroin, and other opiate drugs are able to mimic endorphins because they all share a similar shape with endorphins and can thus fit into and bind to endorphin receptors in the brain. This fit is very specific, something like a lock and key (see Figure 2.17). Thus, the function of a protein—for instance, the ability of a receptor protein to identify and associate with a particular pain-relieving signal molecule—is an emergent property resulting from exquisite molecular order.

Four Levels of Protein Structure

In the complex architecture of a protein, we can recognize three superimposed levels of structure, known as primary, secondary, and tertiary structure. A fourth level, quaternary structure, arises when a protein consists of two or more polypeptide chains. **Figure 5.20**, on the following two pages, describes these four levels of protein structure. Be sure to study this figure thoroughly before going on to the next section.

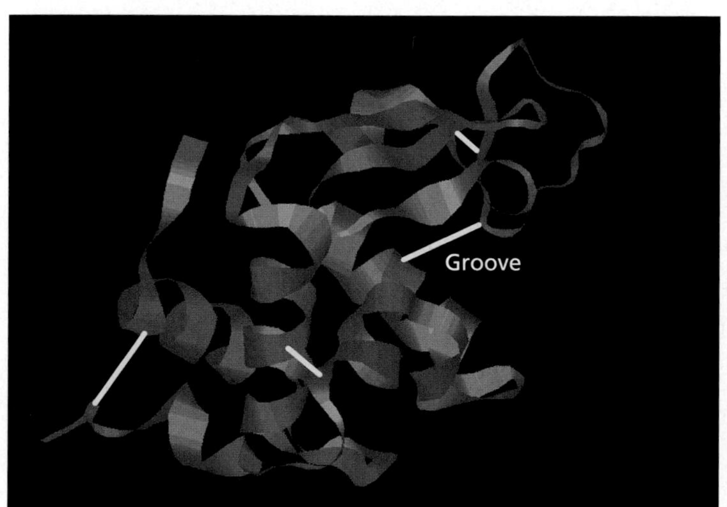

(a) A **ribbon model** shows how the single polypeptide chain folds and coils to form the functional protein. (The yellow lines represent one type of chemical bond that stabilizes the protein's shape.)

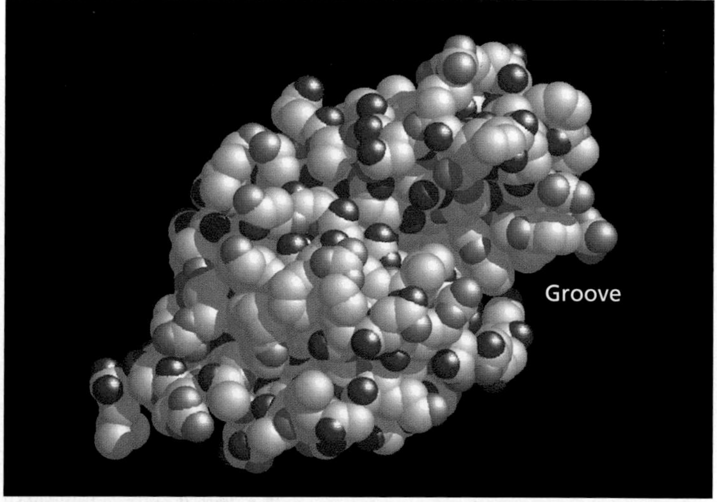

(b) A **space-filling model** shows more clearly the globular shape seen in many proteins, as well as the specific conformation unique to lysozyme.

▲ **Figure 5.19 Conformation of a protein, the enzyme lysozyme.** Present in our sweat, tears, and saliva, lysozyme is an enzyme that helps prevent infection by binding to and destroying specific molecules on the surface of many kinds of bacteria. The groove is the part of the protein that recognizes and binds to the target molecules on bacterial walls.

Figure 5.20

Exploring Levels of Protein Structure

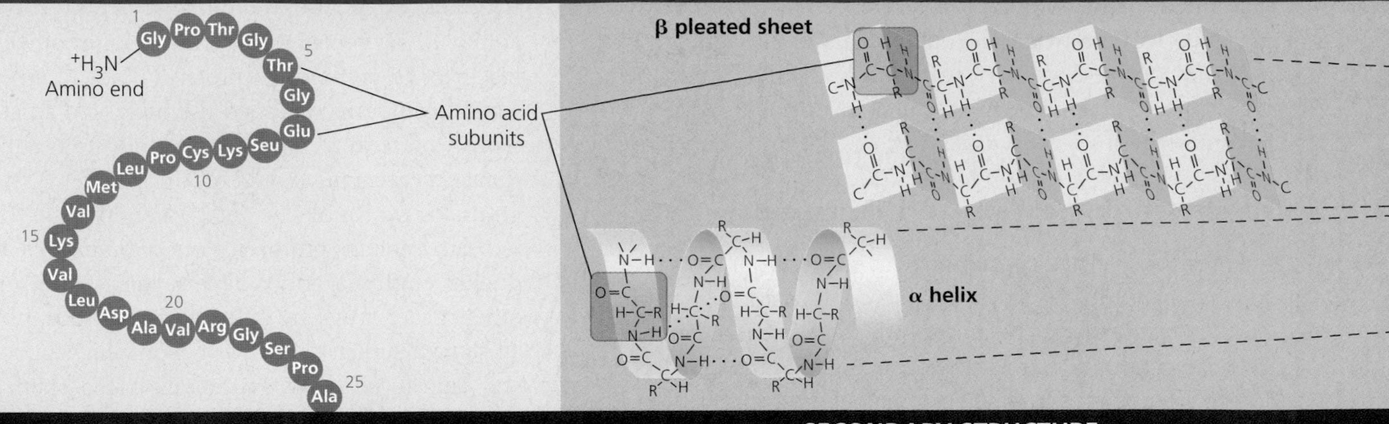

PRIMARY STRUCTURE

SECONDARY STRUCTURE

The **primary structure** of a protein is its unique sequence of amino acids. As an example, let's consider transthyretin, a globular protein found in the blood that transports vitamin A and a particular thyroid hormone throughout the body. Each of the four identical polypeptide chains that, together, make up transthyretin is composed of 127 amino acids. Shown here is one of these chains unraveled for a closer look at its primary structure. A specific one of the 20 amino acids, indicated here by its three-letter abbreviation, occupies each of the 127 positions along the chain. The primary structure is like the order of letters in a very long word. If left to chance, there would be 20^{127} different ways of making a polypeptide chain 127 amino acids long. However, the precise primary structure of a protein is determined not by the random linking of amino acids, but by inherited genetic information.

Glu Ile Asp Thr Lys Ser Tyr Trp Lys Ala Leu Gly Ile Ser Pro Phe His Glu His Ala Glu Val Val Phe Thr Ala Asn Asp Ser Gly Pro Arg Arg Tyr Thr Ile Ala Ala Leu Leu Ser Pro Tyr Ser Tyr Ser Thr Thr Ala Val Val Thr Asn Pro Lys Glu

Carboxyl end

Most proteins have segments of their polypeptide chains repeatedly coiled or folded in patterns that contribute to the protein's overall conformation. These coils and folds, collectively referred to as **secondary structure**, are the result of hydrogen bonds between the repeating constituents of the polypeptide backbone (not the amino acid side chains). Both the oxygen and the nitrogen atoms of the backbone are electronegative, with partial negative charges (see Figure 2.15). The weakly positive hydrogen atom attached to the nitrogen atom has an affinity for the oxygen atom of a nearby peptide bond. Individually, these hydrogen bonds are weak, but because they are repeated many times over a relatively long region of the polypeptide chain, they can support a particular shape for that part of the protein.

One such secondary structure is the **α helix**, a delicate coil held together by hydrogen bonding between every fourth amino acid, shown above for transthyretin. Although transthyretin has only one α helix region (see tertiary structure), other globular proteins have multiple stretches of α helix separated by nonhelical regions. Some fibrous proteins, such as α-keratin, the structural protein of hair, have the α helix formation over most of their length.

The other main type of secondary structure is the **β pleated sheet.** As shown above, in this structure two or more regions of the polypeptide chain lying side by side are connected by hydrogen bonds between parts of the two parallel polypeptide backbones. Pleated sheets make up the core of many globular proteins, as is the case for transthyretin, and dominate some fibrous proteins, including the silk protein of a spider's web. The teamwork of so many hydrogen bonds makes each spider silk fiber stronger than a steel strand of the same weight.

Abdominal glands of the spider secrete silk fibers that form the web.

The radiating strands, made of dry silk fibers, maintain the shape of the web.

The spiral strands (capture strands) are elastic, stretching in response to wind, rain, and the touch of insects.

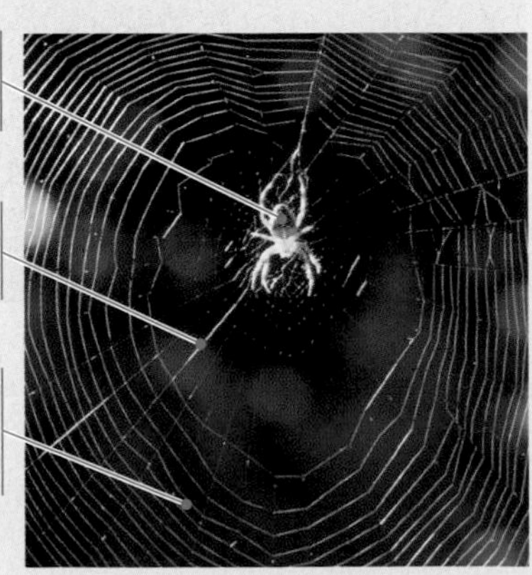

Spider silk: a structural protein containing β pleated sheets

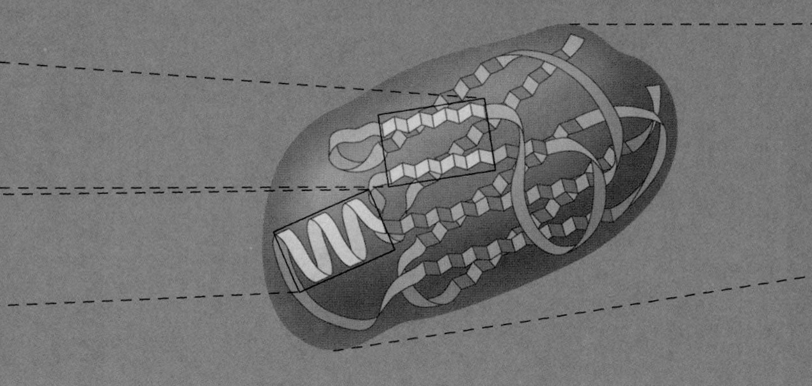

TERTIARY STRUCTURE

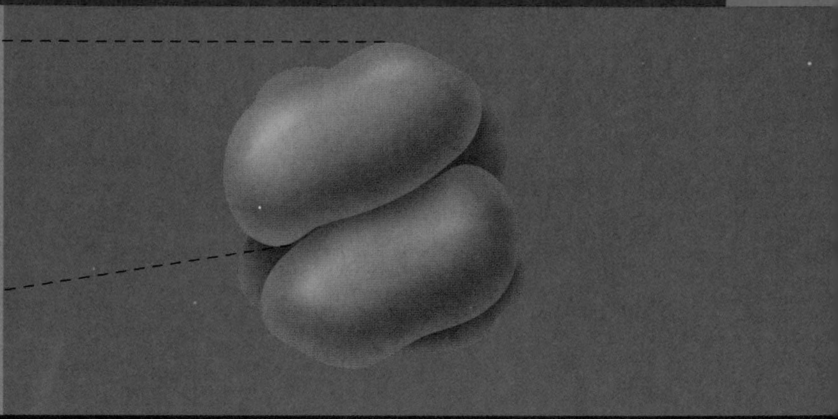

Superimposed on the patterns of secondary structure is a protein's **tertiary structure,** shown above for the transthyretin polypeptide. Rather than involving interactions between backbone constituents, tertiary structure is the overall shape of a polypeptide resulting from interactions between the side chains (R groups) of the various amino acids. One type of interaction that contributes to tertiary structure is—somewhat misleadingly—called a **hydrophobic interaction.** As a polypeptide folds into its functional conformation, amino acids with hydrophobic (nonpolar) side chains usually end up in clusters at the core of the protein, out of contact with water. Thus, what we call a hydrophobic interaction is actually caused by the action of water molecules, which exclude nonpolar substances as they form hydrogen bonds with each other and with hydrophilic parts of the protein. Once nonpolar amino acid side chains are close together, van der Waals interactions help hold them together. Meanwhile, hydrogen bonds between polar side chains and ionic bonds between positively and negatively charged side chains also help stabilize tertiary structure. These are all weak interactions, but their cumulative effect helps give the protein a unique shape.

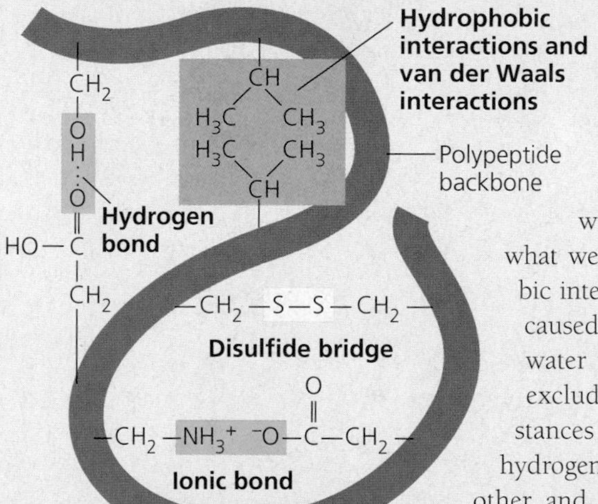

Hydrophobic interactions and van der Waals interactions

Polypeptide backbone

Hydrogen bond

$-CH_2-S-S-CH_2-$

Disulfide bridge

$-CH_2-NH_3^+ \; ^-O-C(=O)-CH_2-$

Ionic bond

The conformation of a protein may be reinforced further by covalent bonds called **disulfide bridges.** Disulfide bridges form where two cysteine monomers, amino acids with sulfhydryl groups (—SH) on their side chains, are brought close together by the folding of the protein. The sulfur of one cysteine bonds to the sulfur of the second, and the disulfide bridge (—S—S—) rivets parts of the protein together (see yellow lines in Figure 5.19a). All of these different kinds of bonds can occur in one protein, as shown above in a small part of a hypothetical protein.

QUATERNARY STRUCTURE

Some proteins consist of two or more polypeptide chains aggregated into one functional macromolecule. **Quaternary structure** is the overall protein structure that results from the aggregation of these polypeptide subunits. For example, shown above is the complete, globular transthyretin protein, made up of its four polypeptides. Another example is collagen, shown on the right, which is a fibrous protein that has helical subunits intertwined into a larger triple helix, giving the long fibers great strength. This suits collagen fibers to their function as the girders of connective tissue in skin, bone, tendons, ligaments, and other body parts (collagen accounts for 40% of the protein in a human body). Hemoglobin, the oxygen-binding protein of red blood cells shown below, is another example of a globular protein with quaternary structure. It consists of four polypeptide subunits, two of one kind (α chains) and two of another kind (β chains). Both α and β subunits consist primarily of α-helical secondary structure. Each subunit has a nonpolypeptide component, called heme, with an iron atom that binds oxygen.

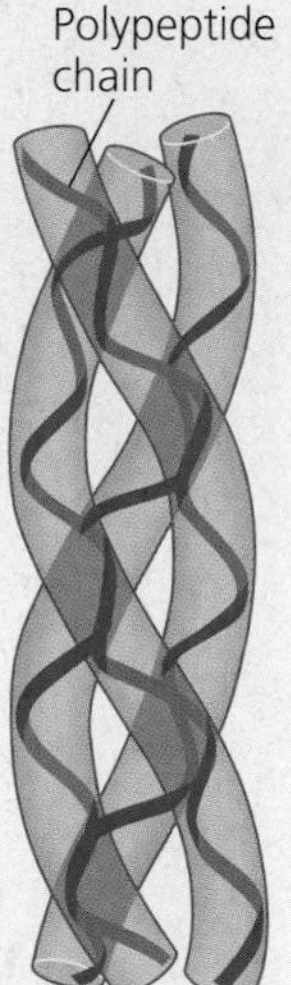

Collagen

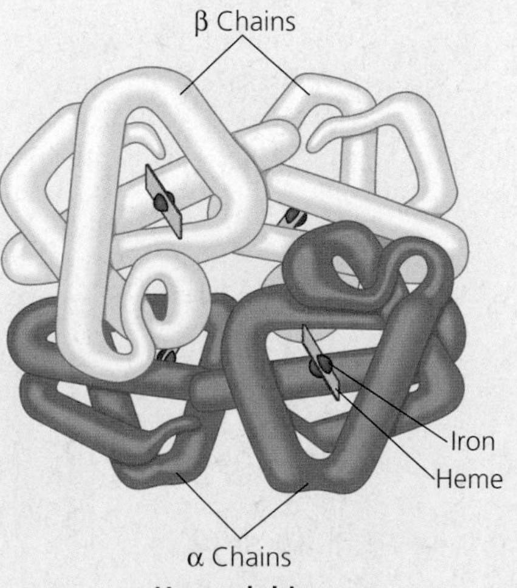

Hemoglobin

Sickle-Cell Disease: A Simple Change in Primary Structure

Even a slight change in primary structure can affect a protein's conformation and ability to function. For instance, the substitution of one amino acid (valine) for the normal one (glutamic acid) at a particular position in the primary structure of hemoglobin, the protein that carries oxygen in red blood cells, can cause *sickle-cell disease,* an inherited blood disorder. Normal red blood cells are disk-shaped, but in sickle-cell disease, the abnormal hemoglobin molecules tend to crystallize, deforming some of the cells into a sickle shape **(Figure 5.21)**. The life of someone with the disease is punctuated by "sickle-cell crises," which occur when the angular cells clog tiny blood vessels, impeding blood flow. The toll taken on such patients is a dramatic example of how a simple change in protein structure can have devastating effects on protein function.

What Determines Protein Conformation?

You've learned that a unique shape endows each protein with a specific function. But what are the key factors determining protein conformation? You already know most of the answer: A polypeptide chain of a given amino acid sequence can spontaneously arrange itself into a three-dimensional shape determined and maintained by the interactions responsible for secondary and tertiary structure. This folding normally occurs as the protein is being synthesized within the cell. However, protein conformation also depends on the physical and chemical conditions of the protein's environment. If the pH, salt concentration, temperature, or other aspects of its environment are altered, the protein may unravel and lose its native conformation, a change called **denaturation (Figure 5.22)**. Because it is misshapen, the denatured protein is biologically inactive.

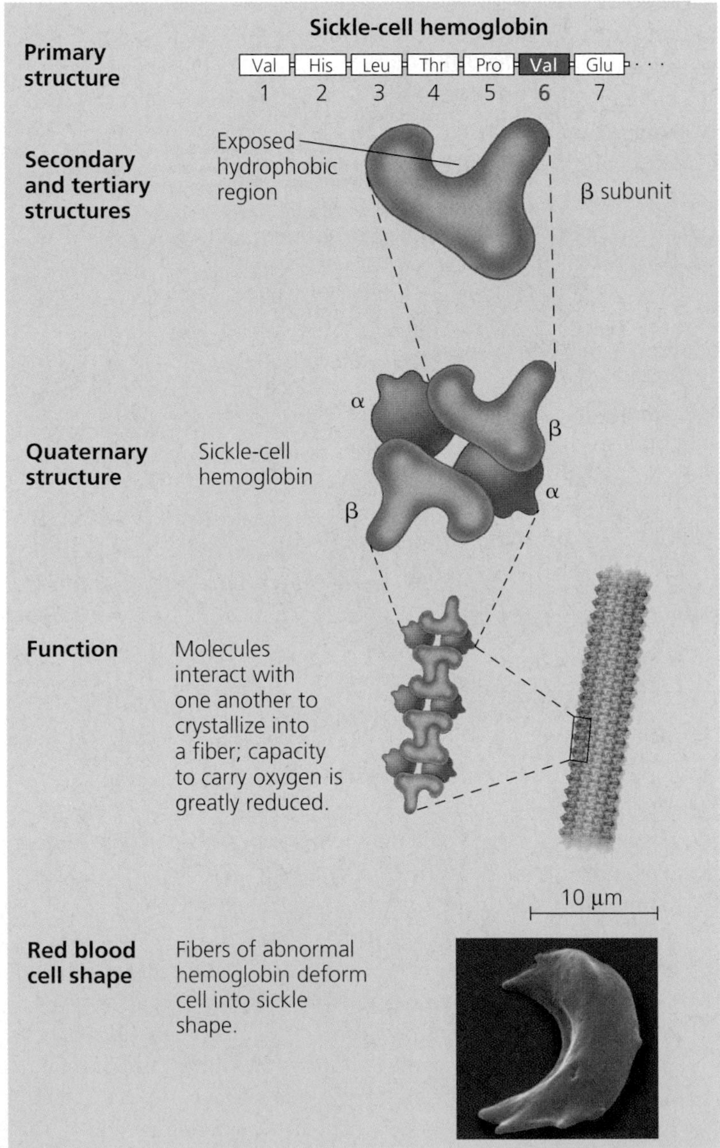

▲ **Figure 5.21 A single amino acid substitution in a protein causes sickle-cell disease.** To show fiber formation clearly, the orientation of the hemoglobin molecule here is different from that in Figure 5.20.

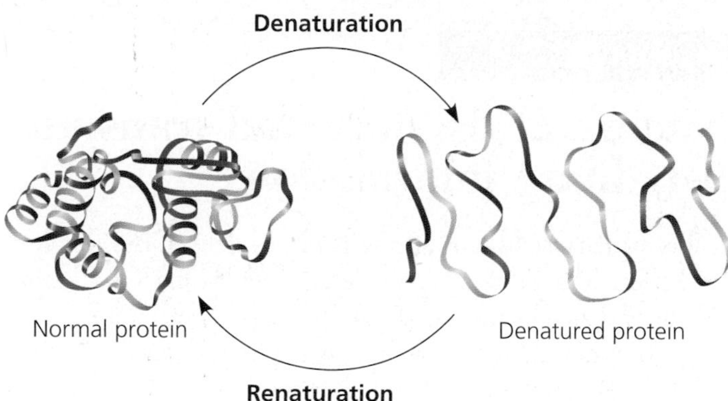
Denaturation
Normal protein
Denatured protein
Renaturation

▲ **Figure 5.22 Denaturation and renaturation of a protein.** High temperatures or various chemical treatments will denature a protein, causing it to lose its conformation and hence its ability to function. If the denatured protein remains dissolved, it can often renature when the chemical and physical aspects of its environment are restored to normal.

Most proteins become denatured if they are transferred from an aqueous environment to an organic solvent, such as ether or chloroform; the polypeptide chain refolds so that its hydrophobic regions face outward toward the solvent. Other denaturation agents include chemicals that disrupt the hydrogen bonds, ionic bonds, and disulfide bridges that maintain a protein's shape. Denaturation can also result from excessive heat, which agitates the polypeptide chain enough to overpower the weak interactions that stabilize conformation. The white of an egg becomes opaque during cooking because the denatured proteins are insoluble and solidify. This also explains why extremely high fevers can be fatal: Proteins in the blood become denatured by such high body temperatures.

When a protein in a test-tube solution has been denatured by heat or chemicals, it will often return to its functional shape when the denaturing agent is removed. We can conclude that the information for building specific shape is intrinsic to the protein's primary structure. The sequence of amino acids determines conformation—where an α helix can form, where β pleated sheets can occur, where disulfide bridges are located, where ionic bonds can form, and so on. However, in the crowded environment inside a cell, correct folding may be more of a problem than it is in a test tube.

The Protein-Folding Problem

Biochemists now know the amino acid sequences of more than 875,000 proteins and the three-dimensional shapes of about 7,000. One would think that by correlating the primary structures of many proteins with their conformations, it would be relatively easy to discover the rules of protein folding. Unfortunately, the protein-folding problem is not that simple. Most proteins probably go through several intermediate states on their way to a stable conformation, and looking at the mature conformation does not reveal the stages of folding required to achieve that form. However, biochemists have developed methods for tracking a protein through its intermediate stages of folding. Researchers have also discovered **chaperonins** (also called chaperone proteins), protein molecules that assist the proper folding of other proteins **(Figure 5.23)**. Chaperonins do not actually specify the correct final structure of a polypeptide. Instead, they work by keeping the new polypeptide segregated from "bad influences" in the cytoplasmic environment while it folds spontaneously. The well-studied chaperonin shown in Figure 5.23, from the bacterium *E. coli,* is a giant multiprotein complex shaped like a hollow cylinder. The cavity provides a shelter for folding polypeptides of various types.

Even when scientists have an actual protein in hand, determining its exact three-dimensional structure is not simple, for a single protein molecule is built of thousands of atoms. **X-ray crystallography** is an important method used to determine a protein's three-dimensional structure **(Figure 5.24)**. Another method that has recently been applied to this problem is nuclear magnetic resonance (NMR) spectroscopy, which does not require protein crystallization. These approaches have contributed greatly to our understanding of protein structure and have also given us valuable hints about protein function.

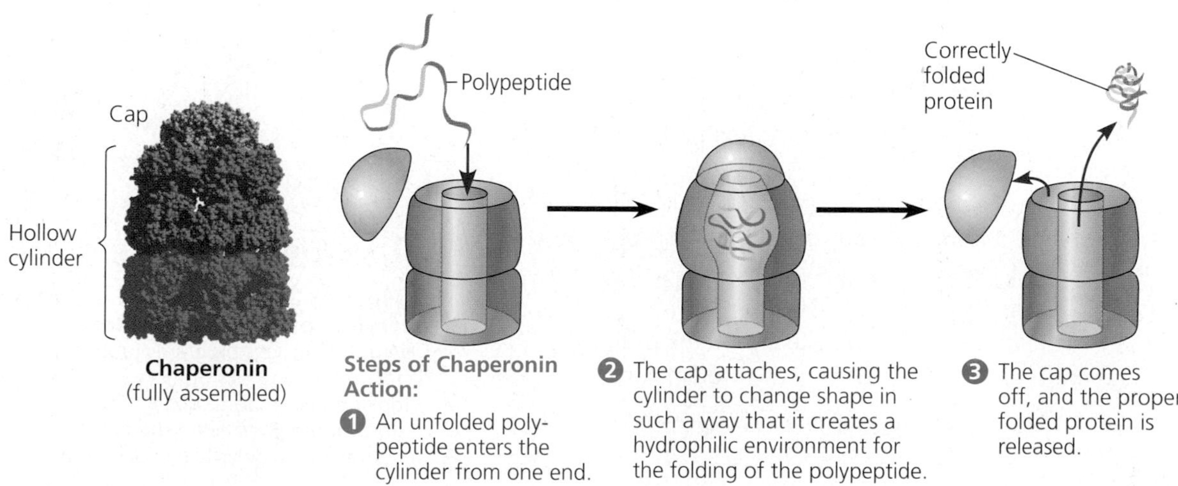
Cap
Hollow cylinder
Chaperonin
(fully assembled)
Polypeptide
Correctly folded protein
Steps of Chaperonin Action:
1 An unfolded polypeptide enters the cylinder from one end.
2 The cap attaches, causing the cylinder to change shape in such a way that it creates a hydrophilic environment for the folding of the polypeptide.
3 The cap comes off, and the properly folded protein is released.

▶ **Figure 5.23 A chaperonin in action.** The computer graphic (left) shows a large chaperonin protein complex with an interior space that provides a shelter for the proper folding of newly made polypeptides. The complex consists of two proteins: One protein is a hollow cylinder; the other is a cap that can fit on either end.

Figure 5.24
Research Method X-Ray Crystallography

APPLICATION Scientists use X-ray crystallography to determine the three-dimensional structure of macromolecules such as nucleic acids and proteins. In this figure we will examine how researchers at the University of California, Riverside, determined the structure of the protein ribonuclease, an enzyme whose function involves binding to a nucleic acid molecule.

TECHNIQUE Researchers aim an X-ray beam through the crystallized protein. The atoms of the crystal diffract (deflect) the X-rays into an orderly array. The diffracted X-rays expose photographic film, producing a pattern of spots known as an X-ray diffraction pattern.

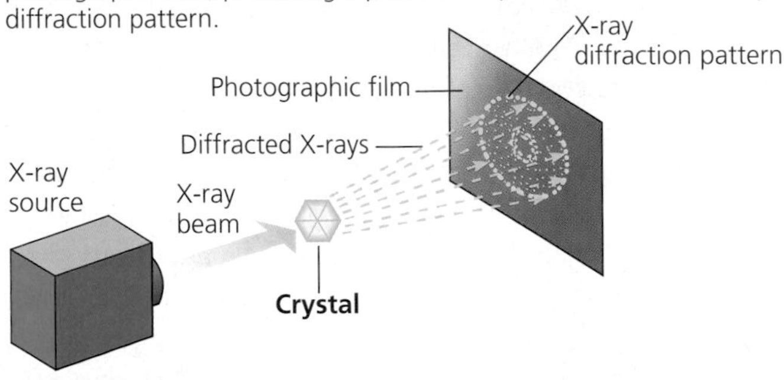

RESULTS Using data from X-ray diffraction patterns, as well as the amino acid sequence determined by chemical methods, scientists build a 3D computer model of the protein, such as this model of the protein ribonuclease (purple) bound to a short strand of nucleic acid (green).

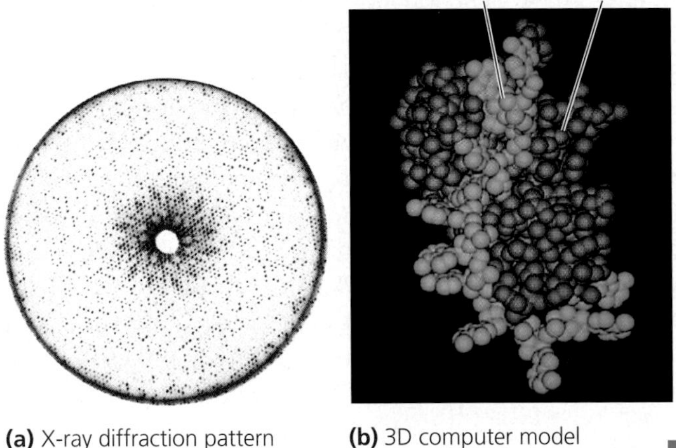

(a) X-ray diffraction pattern **(b)** 3D computer model

Concept Check 5.4

1. Why does a denatured protein no longer function normally?
2. Differentiate between secondary and tertiary structure by describing the parts of the polypeptide chain that participate in the bonds that hold together each level of structure.
3. A genetic mutation can change a protein's primary structure. How can this destroy the protein's function?

For suggested answers, see Appendix A.

Concept 5.5

Nucleic acids store and transmit hereditary information

If the primary structure of polypeptides determines the conformation of a protein, what determines primary structure? The amino acid sequence of a polypeptide is programmed by a unit of inheritance known as a **gene.** Genes consist of DNA, which is a polymer belonging to the class of compounds known as **nucleic acids.**

The Roles of Nucleic Acids

There are two types of nucleic acids: **deoxyribonucleic acid (DNA)** and **ribonucleic acid (RNA).** These are the molecules that enable living organisms to reproduce their complex components from one generation to the next. Unique among molecules, DNA provides directions for its own replication. DNA also directs RNA synthesis and, through RNA, controls protein synthesis **(Figure 5.25)**.

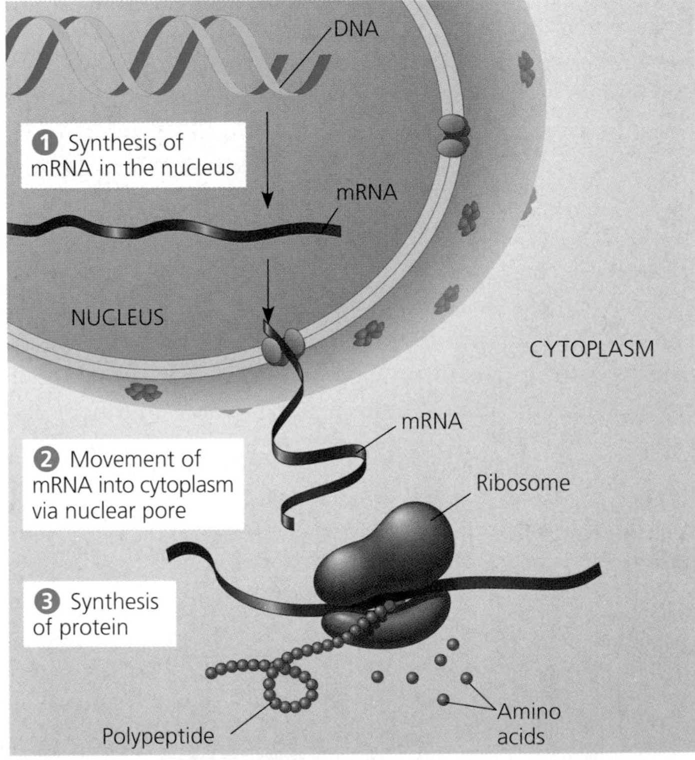

▲ **Figure 5.25 DNA → RNA → protein: a diagrammatic overview of information flow in a cell.** In a eukaryotic cell, DNA in the nucleus programs protein production in the cytoplasm by dictating the synthesis of messenger RNA (mRNA), which travels to the cytoplasm and binds to ribosomes. As a ribosome (greatly enlarged in this drawing) moves along the mRNA, the genetic message is translated into a polypeptide of specific amino acid sequence.

DNA is the genetic material that organisms inherit from their parents. Each chromosome contains one long DNA molecule, usually consisting of from several hundred to more than a thousand genes. When a cell reproduces itself by dividing, its DNA molecules are copied and passed along from one generation of cells to the next. Encoded in the structure of DNA is the information that programs all the cell's activities. The DNA, however, is not directly involved in running the operations of the cell, any more than computer software by itself can print a bank statement or read the bar code on a box of cereal. Just as a printer is needed to print out a statement and a scanner is needed to read a bar code, proteins are required to implement genetic programs. The molecular hardware of the cell—the tools for most biological functions—consists of proteins. For example, the oxygen carrier in the blood is the protein hemoglobin, not the DNA that specifies its structure.

How does RNA, the other type of nucleic acid, fit into the flow of genetic information from DNA to proteins? Each gene along the length of a DNA molecule directs the synthesis of a type of RNA called *messenger RNA* (mRNA). The mRNA molecule then interacts with the cell's protein-synthesizing machinery to direct the production of a polypeptide. We can summarize the flow of genetic information as DNA → RNA → protein (see Figure 5.25). The actual sites of protein synthesis are cellular structures called ribosomes. In a eukaryotic cell, ribosomes are located in the cytoplasm, but DNA resides in the nucleus. Messenger RNA conveys the genetic instructions for building proteins from the nucleus to the cytoplasm. Prokaryotic cells lack nuclei, but they still use RNA to send a message from the DNA to the ribosomes and other equipment of the cell that translate the coded information into amino acid sequences.

The Structure of Nucleic Acids

Nucleic acids are macromolecules that exist as polymers called **polynucleotides (Figure 5.26a)**. As indicated by the name, each polynucleotide consists of monomers called **nucleotides.** A nucleotide is itself composed of three parts: a nitrogenous base, a pentose (five-carbon sugar), and a phosphate group **(Figure 5.26b)**. The portion of this unit without the phosphate group is called a *nucleoside*.

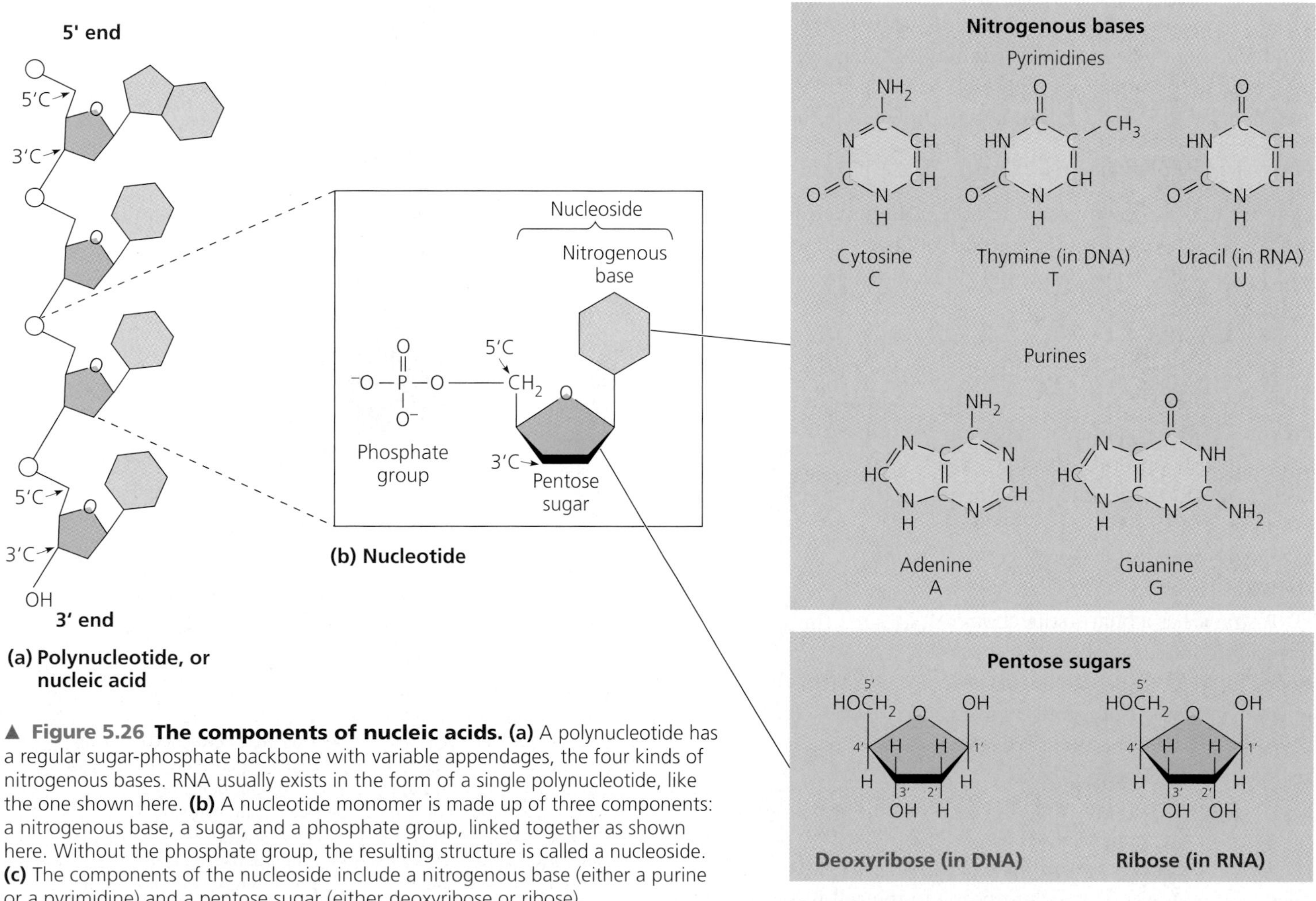

▲ **Figure 5.26 The components of nucleic acids. (a)** A polynucleotide has a regular sugar-phosphate backbone with variable appendages, the four kinds of nitrogenous bases. RNA usually exists in the form of a single polynucleotide, like the one shown here. **(b)** A nucleotide monomer is made up of three components: a nitrogenous base, a sugar, and a phosphate group, linked together as shown here. Without the phosphate group, the resulting structure is called a nucleoside. **(c)** The components of the nucleoside include a nitrogenous base (either a purine or a pyrimidine) and a pentose sugar (either deoxyribose or ribose).

Nucleotide Monomers

To build a nucleotide, let's first consider the two components of the nucleoside: the nitrogenous base and the sugar **(Figure 5.26c)**. There are two families of nitrogenous bases: pyrimidines and purines. A **pyrimidine** has a six-membered ring of carbon and nitrogen atoms. (The nitrogen atoms tend to take up H^+ from solution, which explains the term *nitrogenous base*.) The members of the pyrimidine family are cytosine (C), thymine (T), and uracil (U). **Purines** are larger, with a six-membered ring fused to a five-membered ring. The purines are adenine (A) and guanine (G). The specific pyrimidines and purines differ in the functional groups attached to the rings. Adenine, guanine, and cytosine are found in both types of nucleic acid; thymine is found only in DNA and uracil only in RNA.

The pentose connected to the nitrogenous base is **ribose** in the nucleotides of RNA and **deoxyribose** in DNA (see Figure 5.26c). The only difference between these two sugars is that deoxyribose lacks an oxygen atom on the second carbon in the ring; hence its name. Because the atoms in both the nitrogenous base and the sugar are numbered, the sugar atoms have a prime (′) after the number to distinguish them. Thus, the second carbon in the sugar ring is the 2′ ("2 prime") carbon, and the carbon that sticks up from the ring is called the 5′ carbon.

So far, we have built a nucleoside. To complete the construction of a nucleotide, we attach a phosphate group to the 5′ carbon of the sugar (see Figure 5.26b). The molecule is now a nucleoside monophosphate, better known as a nucleotide.

Nucleotide Polymers

Now we can see how these nucleotides are linked together to build a polynucleotide. Adjacent nucleotides are joined by covalent bonds called phosphodiester linkages between the —OH group on the 3′ carbon of one nucleotide and the phosphate on the 5′ carbon of the next. This bonding results in a backbone with a repeating pattern of sugar-phosphate units (see Figure 5.26a). The two free ends of the polymer are distinctly different from each other. One end has a phosphate attached to a 5′ carbon, and the other end has a hydroxyl group on a 3′ carbon; we refer to these as the 5′ end and the 3′ end, respectively. So we can say that the DNA strand has a built-in directionality along its sugar-phosphate backbone, from 5′ to 3′, somewhat like a one-way street. All along this sugar-phosphate backbone are appendages consisting of the nitrogenous bases.

The sequence of bases along a DNA (or mRNA) polymer is unique for each gene. Because genes are hundreds to thousands of nucleotides long, the number of possible base sequences is effectively limitless. A gene's meaning to the cell is encoded in its specific sequence of the four DNA bases. For example, the sequence AGGTAACTT means one thing, whereas the sequence CGCTTTAAC has a different meaning. (Real genes, of course, are much longer.) The linear order of bases in a gene specifies the amino acid sequence—the primary structure—of a protein, which in turn specifies that protein's three-dimensional conformation and function in the cell.

The DNA Double Helix

The RNA molecules of cells consist of a single polynucleotide chain like the one shown in Figure 5.26. In contrast, cellular DNA molecules have two polynucleotides that spiral around an imaginary axis, forming a **double helix (Figure 5.27)**. James Watson and Francis Crick, working at Cambridge University,

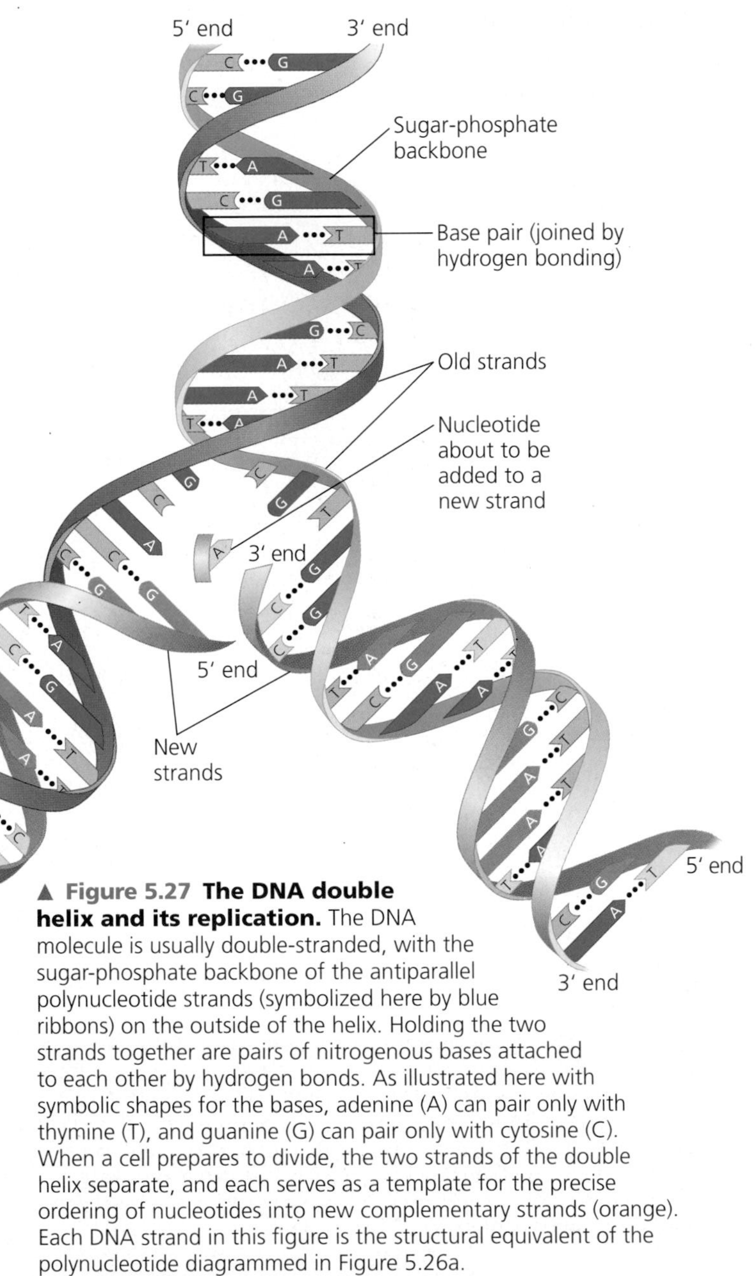

▲ **Figure 5.27 The DNA double helix and its replication.** The DNA molecule is usually double-stranded, with the sugar-phosphate backbone of the antiparallel polynucleotide strands (symbolized here by blue ribbons) on the outside of the helix. Holding the two strands together are pairs of nitrogenous bases attached to each other by hydrogen bonds. As illustrated here with symbolic shapes for the bases, adenine (A) can pair only with thymine (T), and guanine (G) can pair only with cytosine (C). When a cell prepares to divide, the two strands of the double helix separate, and each serves as a template for the precise ordering of nucleotides into new complementary strands (orange). Each DNA strand in this figure is the structural equivalent of the polynucleotide diagrammed in Figure 5.26a.

first proposed the double helix as the three-dimensional structure of DNA in 1953. The two sugar-phosphate backbones run in opposite 5′ → 3′ directions from each other, an arrangement referred to as **antiparallel,** somewhat like a divided highway. The sugar-phosphate backbones are on the outside of the helix, and the nitrogenous bases are paired in the interior of the helix. The two polynucleotides, or strands, as they are called, are held together by hydrogen bonds between the paired bases and by van der Waals interactions between the stacked bases. Most DNA molecules are very long, with thousands or even millions of base pairs connecting the two chains. One long DNA double helix includes many genes, each one a particular segment of the molecule.

Only certain bases in the double helix are compatible with each other. Adenine (A) always pairs with thymine (T), and guanine (G) always pairs with cytosine (C). If we were to read the sequence of bases along one strand as we traveled the length of the double helix, we would know the sequence of bases along the other strand. If a stretch of one strand has the base sequence 5′-AGGTCCG-3′, then the base-pairing rules tell us that the same stretch of the other strand must have the sequence 3′-TCCAGGC-5′. The two strands of the double helix are *complementary,* each the predictable counterpart of the other. It is this feature of DNA that makes possible the precise copying of genes that is responsible for inheritance (see Figure 5.27). In preparation for cell division, each of the two strands of a DNA molecule serves as a template to order nucleotides into a new complementary strand. The result is two identical copies of the original double-stranded DNA molecule, which are then distributed to the two daughter cells. Thus, the structure of DNA accounts for its function in transmitting genetic information whenever a cell reproduces.

DNA and Proteins as Tape Measures of Evolution

We are accustomed to thinking of shared traits, such as hair and milk production in mammals, as evidence of shared ancestors. Because we now understand that DNA carries heritable information in the form of genes, we can see that genes and their products (proteins) document the hereditary background of an organism. The linear sequences of nucleotides in DNA molecules are passed from parents to offspring; these sequences determine the amino acid sequences of proteins. Siblings have greater similarity in their DNA and proteins than do unrelated individuals of the same species. If the evolutionary view of life is valid, we should be able to extend this concept of "molecular genealogy" to relationships *between* species: We should expect two species that appear to be closely related based on fossil and anatomical evidence to also share a greater proportion of their DNA and protein sequences than do more distantly related species. In fact, that is the case. For example, if we compare a polypeptide chain of human hemoglobin with the corresponding hemoglobin polypeptide in five other vertebrates, we find the following. In this chain of 146 amino acids, humans and gorillas differ in just 1 amino acid, humans and gibbons differ in 2 amino acids, and humans and rhesus monkeys differ in 8 amino acids. More distantly related species have chains that are less similar. Humans and mice differ in 27 amino acids, and humans and frogs differ in 67 amino acids. Molecular biology has added a new tape measure to the toolkit biologists use to assess evolutionary kinship.

Concept Check 5.5

1. Go to Figure 5.26a and number all the carbons in the sugars for the top three nucleotides; circle the nitrogenous bases and star the phosphates.
2. In a DNA double helix, a region along one DNA strand has this sequence of nitrogenous bases: 5′-TAGGCCT-3′. List the base sequence along the other strand of the molecule, clearly indicating the 5′ and 3′ ends of this strand.

For suggested answers, see Appendix A.

The Theme of Emergent Properties in the Chemistry of Life: *A Review*

Recall that life is organized along a hierarchy of structural levels (see Figure 1.3). With each increasing level of order, new properties emerge in addition to those of the component parts. In Chapters 2–5, we have dissected the chemistry of life using the strategy of the reductionist. But we have also begun to develop a more integrated view of life as we have seen how properties emerge with increasing order.

We have seen that the unusual behavior of water, so essential to life on Earth, results from interactions of the water molecules, themselves an ordered arrangement of hydrogen and oxygen atoms. We reduced the great complexity and diversity of organic compounds to the chemical characteristics of carbon, but we also saw that the unique properties of organic compounds are related to the specific structural arrangements of carbon skeletons and their appended functional groups. We learned that small organic molecules are often assembled into giant molecules, but we also discovered that a macromolecule does not behave like a composite of its monomers but rather takes on additional properties owing to the interactions between those monomers.

By completing our overview of the molecular basis of life with an introduction to the important classes of macromolecules that build living cells, we have built a bridge to Unit Two, where we will study the cell's structure and function. We will maintain our balance between the need to reduce life to a conglomerate of simpler processes and the ultimate satisfaction of viewing those processes in their integrated context.

Chapter 5 Review

Go to www.campbellbiology.com or the student CD-ROM to explore Activities, Investigations, and other interactive study aids.

SUMMARY OF KEY CONCEPTS

Concept 5.1

Most macromolecules are polymers, built from monomers

▶ **The Synthesis and Breakdown of Polymers (pp. 68–69)** Carbohydrates, lipids, proteins, and nucleic acids are the four major classes of organic compounds in cells. Many of these compounds are very large molecules. Most macromolecules are polymers, chains of identical or similar monomers. Monomers form larger molecules by condensation reactions, in which water molecules are released (dehydration). Polymers can disassemble by the reverse process, hydrolysis.
Activity *Making and Breaking Polymers*

▶ **The Diversity of Polymers (p. 69)** Each class of polymer is formed from a specific set of monomers. Although organisms share the same limited number of monomer types, each organism is unique because of the specific arrangement of monomers into polymers. An immense variety of polymers can be built from a small set of monomers.

Concept 5.2

Carbohydrates serve as fuel and building material

▶ **Sugars (pp. 70–71)** Sugars, the smallest carbohydrates, serve as fuel and carbon sources. Monosaccharides are the simplest sugars. They are used for fuel, converted to other organic molecules, or combined into polymers. Disaccharides consist of two monosaccharides connected by a glycosidic linkage.
Activity *Models of Glucose*

▶ **Polysaccharides (pp. 71–74)** Polysaccharides, polymers of sugars, have storage and structural roles. The monomers of polysaccharides are connected by glycosidic linkages. Starch in plants and glycogen in animals are both storage polymers of glucose. Cellulose is an important structural polymer of glucose in plant cell walls. Starch, glycogen, and cellulose differ in the positions and orientations of their glycosidic linkages.
Activity *Carbohydrates*

Concept 5.3

Lipids are a diverse group of hydrophobic molecules

▶ **Fats (pp. 75–76)** Fats store large amounts of energy. Also known as triacylglycerols, fats are constructed by the joining of a glycerol molecule to three fatty acids by dehydration reactions. Saturated fatty acids have the maximum number of hydrogen atoms. Unsaturated fatty acids (present in oils) have one or more double bonds in their hydrocarbon chains.

▶ **Phospholipids (pp. 76–77)** Phospholipids, which are major components of cell membranes, consist of two fatty acids and a phosphate group linked to glycerol. Thus, the "head" of a phospholipid is hydrophilic and the "tail" hydrophobic.

▶ **Steroids (p. 77)** Steroids include cholesterol and certain hormones. Steroids have a basic structure of four fused rings of carbon atoms.
Activity *Lipids*

Concept 5.4

Proteins have many structures, resulting in a wide range of functions

▶ **Polypeptides (pp. 78–81)** A polypeptide is a polymer of amino acids connected in a specific sequence. A protein consists of one or more polypeptide chains folded into a specific three-dimensional conformation. Polypeptides are constructed from 20 different amino acids, each with a characteristic side chain (R group). The carboxyl and amino groups of adjacent amino acids link together in peptide bonds.

▶ **Protein Conformation and Function (pp. 81–86)** The primary structure of a protein is its unique sequence of amino acids. Secondary structure is the folding or coiling of the polypeptide into repeating configurations, mainly the α helix and the β pleated sheet, which result from hydrogen bonding between parts of the polypeptide backbone. Tertiary structure is the overall three-dimensional shape of a polypeptide and results from interactions between amino acid R groups. Proteins made of more than one polypeptide chain have a quaternary level of structure. Protein shape is ultimately determined by its primary structure, but the structure and function of a protein are sensitive to physical and chemical conditions.
Activity *Protein Functions*
Activity *Protein Structure*
Biology Labs On-Line *HemoglobinLab*

Concept 5.5

Nucleic acids store and transmit hereditary information

▶ **The Roles of Nucleic Acids (pp. 86–87)** DNA stores information for the synthesis of specific proteins. RNA (specifically, mRNA) carries this genetic information to the protein-synthesizing machinery.

▶ **The Structure of Nucleic Acids (pp. 87–88)** Each nucleotide monomer consists of a pentose covalently bonded to a phosphate group and to one of four different nitrogenous bases (A, G, C, and T or U). RNA has ribose as its pentose; DNA has deoxyribose. RNA has U; DNA, T. In a polynucleotide, nucleotides are joined to form a sugar-phosphate backbone from which the nitrogenous bases project. Each polynucleotide strand has polarity, with a 5′ end and a 3′ end. The sequence of bases along a gene specifies the amino acid sequence of a particular protein.

▶ **The DNA Double Helix (pp. 88–89)** DNA is a helical, double-stranded macromolecule with bases projecting into the interior of the molecule from the two antiparallel polynucleotide strands. Because A always hydrogen-bonds to T, and C to G, the nucleotide sequences of the two strands are complementary. One strand can serve as a template for the formation of the other. This unique feature of DNA provides a mechanism for the continuity of life.

▶ **DNA and Proteins as Tape Measures of Evolution (p. 89)** Molecular comparisons help biologists sort out the evolutionary connections among species.
Activity *Nucleic Acid Functions*
Activity *Nucleic Acid Structure*

• • •

▶ **The Theme of Emergent Properties in the Chemistry of Life: *A Review* (p. 89)** Higher levels of organization result in the emergence of new properties. Organization is the key to the chemistry of life.

TESTING YOUR KNOWLEDGE

Self-Quiz

1. Which term includes all others in the list?
   a. monosaccharide
   b. disaccharide
   c. starch
   d. carbohydrate
   e. polysaccharide

2. The molecular formula for glucose is $C_6H_{12}O_6$. What would be the molecular formula for a polymer made by linking ten glucose molecules together by dehydration reactions?
   a. $C_{60}H_{120}O_{60}$
   b. $C_6H_{12}O_6$
   c. $C_{60}H_{102}O_{51}$
   d. $C_{60}H_{100}O_{50}$
   e. $C_{60}H_{111}O_{51}$

3. The enzyme amylase can break glycosidic linkages between glucose monomers only if the monomers are the α form. Which of the following could amylase break down? (Choose all that apply.)
   a. cellulose
   b. chitin
   c. glycogen
   d. starch
   e. amylopectin

4. Choose the pair of terms that correctly completes this sentence: Nucleotides are to _______ as _________ are to proteins.
   a. nucleic acids; amino acids
   b. amino acids; polypeptides
   c. glycosidic linkages; polypeptide linkages
   d. genes; enzymes
   e. polymers; polypeptides

5. Which of the following statements concerning *unsaturated* fats is true?
   a. They are more common in animals than in plants.
   b. They have double bonds in the carbon chains of their fatty acids.
   c. They generally solidify at room temperature.
   d. They contain more hydrogen than saturated fats having the same number of carbon atoms.
   e. They have fewer fatty acid molecules per fat molecule.

6. The structural level of a protein least affected by a disruption in hydrogen bonding is the
   a. primary level.
   b. secondary level.
   c. tertiary level.
   d. quaternary level.
   e. All structural levels are equally affected.

7. Which of the following pairs of base sequences could form a short stretch of a normal double helix of DNA?
   a. 5′-purine-pyrimidine-purine-pyrimidine-3′ with 3′-purine-pyrimidine-purine-pyrimidine-5′
   b. 5′-A-G-C-T-3′ with 5′-T-C-G-A-3′
   c. 5′-G-C-G-C-3′ with 5′-T-A-T-A-3′
   d. 5′-A-T-G-C-3′ with 5′-G-C-A-T-3′
   e. a, b, and d are all correct

8. Enzymes that break down DNA catalyze the hydrolysis of the covalent bonds that join nucleotides together. What would happen to DNA molecules treated with these enzymes?
   a. The two strands of the double helix would separate.
   b. The phosphodiester bonds between deoxyribose sugars would be broken.
   c. The purines would be separated from the deoxyribose sugars.
   d. The pyrimidines would be separated from the deoxyribose sugars.
   e. All bases would be separated from the deoxyribose sugars.

9. Which of the following is *not* a protein?
   a. hemoglobin
   b. cholesterol
   c. an antibody
   d. an enzyme
   e. insulin

10. Which of the following statements about the 5′ end of a polynucleotide strand is correct?
    a. The 5′ end has a hydroxyl group.
    b. The 5′ end has a phosphate group.
    c. The 5′ end is identical to the 3′ end.
    d. The 5′ end is antiparallel to the 3′ end.
    e. The 5′ end is the fifth position on one of the nitrogenous bases.

For Self-Quiz answers, see Appendix A.

Go to the website or CD-ROM for more quiz questions.

Evolution Connection

Comparisons of the amino acid sequences of proteins or the nucleotide sequences of genes can shed light on the evolutionary divergence of related organisms. Would you expect all the proteins or genes of a given set of organisms living on Earth today to show the same degree of divergence? Why or why not?

Scientific Inquiry

During the Napoleonic Wars in the early 1800s, there was a sugar shortage in Europe because supply ships were blockaded from harbors. To create artificial sweeteners, German scientists hydrolyzed wheat starch. They did this by adding hydrochloric acid to heated starch solutions, which caused some of the glycosidic linkages between the glucose monomers to break. The process broke only about 50% of the glycosidic linkages, however, so the sweetener was less sweet than sugar. In addition, consumers complained of a slight bitterness resulting from by-products of the reaction. Sketch a glycosidic linkage in starch using Figures 5.5a and 5.7b for reference. Show how the acid was able to break this bond. Why do you think the acid broke only 50% of the linkages in the wheat starch?

Biological Inquiry: A Workbook of Investigative Cases *Explore macromolecules further with the case "Picture Perfect."*
Biology Labs On-Line *HemoglobinLab*

Science, Technology, and Society

Some amateur and professional athletes take anabolic steroids to help them "bulk up" or build strength. The health risks of this practice are extensively documented. Apart from health considerations, how do you feel about the use of chemicals to enhance athletic performance? Do you think an athlete who takes anabolic steroids is cheating, or is the use of such chemicals just part of the preparation required to succeed in a competitive sport? Defend your answer.

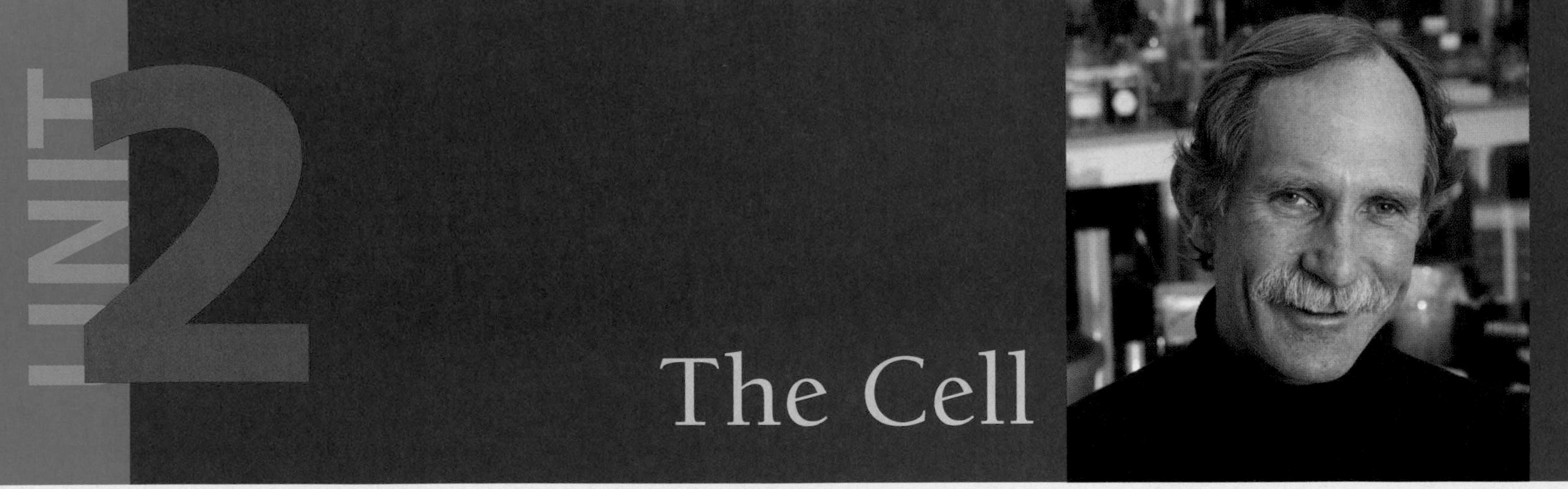

AN INTERVIEW WITH
Peter Agre

In 2003 Peter Agre received the Nobel Prize in Chemistry for the discovery of aquaporins, "water pore" proteins that allow water molecules to rapidly cross the cell membrane. (He shared the prize with Roderick MacKinnon of Rockefeller University, who works on the other main aspect of cellular "plumbing"—the transport of ions.) A medical doctor as well as a researcher in basic science, Dr. Agre is a professor in the Departments of Biological Chemistry and Medicine at the Johns Hopkins University School of Medicine.

How did you start in science?

As a little kid, really. I grew up in Northfield, Minnesota, where my dad was teaching chemistry at St. Olaf College. The kids in our family had an idyllic life, playing in apple orchards and roaming the campus. We lived right across the street from my dad's laboratory, a really nice lab with cheerful students who did summer research. I'd do little experiments with my father's help. We'd change the pH of some solution, and the indicator dye would turn pink or blue—it was amazing! As an adolescent I didn't want anything to do with any of that. But the first merit badge I ever earned as a Boy Scout was in chemistry. Then I became more interested in medicine.

What was the appeal of medicine?

In part, the people. The local doctors were really interesting and nice people, and it seemed like they did a lot of good things.

High school was a bit of a detour—I was kind of a hellraiser then. My friends and I started an underground newspaper (this was 1967, and I thought a revolution was about to occur). I didn't do a lot of school work. So I was invited to leave high school. I went to night school and then got into Augsberg College.

What turned you on to research?

At medical school, I spent time in a research lab, and that really fired me up. I got to work with a group of vibrant scientists from all over the world. I learned you could approach complex problems at a molecular level—reductionist biology. We purified a variant of cholera toxin produced by certain strains of the bacterium *E. coli*. Like the toxin of the true cholera bacterium (*Vibrio cholerae*), this can cause a terrible and sometimes fatal diarrhea.

If young people saw scientists the way I did in that lab, they would all want to be scientists. Too many people still stereotype scientists in a negative way—something like Doc from *Back to the Future*. I'm afraid we've lost good role models in science. And too often we scientists are so stressed out trying to get grants funded that our work doesn't look appealing to students.

People need to realize how much science is a social endeavor. Scientists have personalities and interests and fears and limitations. We're not cybertronic machines. And we collaborate. Unless you're at the level of Leonardo da Vinci, you get a lot of ideas from other people.

Tell us about combining medical practice with basic research.

Until 1993, I was an attending physician in Hematology here. I stopped in part because I was busy with starting a new graduate program in cellular and molecular medicine. (It provides Ph.D. training in basic science but with direct medical relevance.) Combining medical practice with basic research is a challenge because there are only so many hours in the day. I'm married with four children, and, as much as I liked my patients, my heart was more in the research. In practicing medicine, you can help individual patients, but with research, you might make a discovery that would help thousands of patients. As a student, I doubted I had what it took to succeed in basic science, but I wanted to try it.

I still have my medical license, which I use for Boy Scout camp physicals, ringside positions as medical doctor to amateur wrestlers, and inner city kids—I'm their doc. Anyway, my wife's not convinced that research is going to carry me to retirement. She may be right. I think it's better in life not to be overconfident.

Let's talk about the research that led to the discovery of aquaporins.

I'm a blood specialist (hematologist), and my particular interest has been proteins found in the plasma membrane of red blood cells. The plasma membrane is a phospholipid bilayer with inserted cholesterol molecules, "integral" proteins that span the bilayer, and membrane-associated proteins that bind to the membrane's inner surface. With red cells, it's easy to study plasma membrane in pure form, and there are interesting disorders associated with defects in the proteins.

When I joined the faculty at Hopkins, I began to study the Rh blood antigens. Blood group antigens are membrane glycoproteins of red cells that exist in the population in two or more forms. The differences can be in the attached carbohydrate, as with the antigens of the ABO blood groups (see Chapter 43), or they can be within the core protein. Rh is of medical importance because of Rh incompatibility, which occurs when Rh negative mothers have Rh positive babies and become sensitized to the Rh antigen. Unless the mother is treated, there's a significant risk that in subsequent pregnancies, her immune system will attack the baby's red blood cells. Surprisingly, as of the late 1980s, no one had yet been able to isolate the Rh antigen.

Membrane-spanning proteins like the Rh protein are really messy to work with. But we worked out a method to isolate and partially purify the Rh protein. Our sample seemed to consist of two proteins, but we were sure that the smaller one was just a breakdown product of the larger one. We were completely wrong, though: It was a water channel protein, which we later named aquaporin-1.

How did you figure that out?

It took a lot of convincing to make myself believe that this protein was even a new discovery. Using antibodies we made to the protein, we showed it to be one of the most abundant proteins in red cell membranes (200,000 copies per cell!) and even more abundant in certain renal (kidney) cells. I talked to 15 or so membrane biochemists: What could it be? Although many people study red cells, no one had ever seen this protein before because it doesn't stain with the usual protein stains. It was John Parker, a hematologist and superb red-cell physiologist, who finally said, "Maybe this protein is involved in water transport."

To test this hypothesis, we performed a simple experiment using frog egg cells: We injected the cells with mRNA that the cells translated into aquaporin and found that we could turn the cells from being almost watertight (without aquaporin) into highly water-permeable (with aquaporin). The cells with aquaporin would swell and explode in distilled water as a result of osmosis: With aquaporins, the cell membrane became permeable to water but not to things dissolved in it (solutes), so water molecules crossed the membrane from the side with less solute (distilled water) to the side with more (the cell interior).

You know, as a scientist, you can benefit from being a bit like Huckleberry Finn. You should explore your fancy to some degree. If something interesting shows up, it's good to go for it.

Why do cells have water channel proteins?

Not all cells do. Before our discovery, however, many physiologists thought that diffusion through the phospholipid bilayer was enough for getting water into and out of *all* cells. Others said this couldn't be enough, especially for cells whose water permeability needs to be very high or regulated—for example, cells forming the tubules of the kidneys, which control the concentration of the urine. A number of researchers had made observations indicating that diffusion through the phospholipid bilayer wasn't everything. But there was a lot of skepticism until our experiment with frog eggs ended the controversy.

Water transport is very important in our bodies. For example, our kidneys must filter and reabsorb many liters of water every day. If we don't reabsorb that water, we die of dehydration. Aquaporins allow us to reabsorb enough water, without reabsorbing ions.

How do aquaporins work?

The structure of an aquaporin molecule helps explain how it functions. The protein looks like an hourglass spanning the membrane. The two halves are symmetrical with opposite orientations. An hourglass works equally well if it's right side up or upside down, and aquaporins work equally well for the uptake or release of water. The driving force is not some kind of a pump action, but simply osmosis. An aquaporin allows osmosis to occur extremely rapidly. Nothing larger than water molecules can fit through the channel. In addition, the passage of ions is prevented—even small ones like the hydronium ions (H_3O^+) formed by the combining of H^+ with H_2O (see p. 53). They are repelled by positively charged amino acids at the narrowest part of the channel.

What happens when mutations disrupt aquaporin function?

Mutations in aquaporin genes can cause health problems. People whose kidney cells have defective aquaporin-2 molecules need to drink 20 liters of water a day to prevent dehydration. They can't concentrate their urine enough. Individuals who can't make aquaporin-1 generally do OK in modern life, but they get dehydrated much more easily than other people. In addition, some patients make too much aquaporin, causing them to retain too much fluid. Fluid retention in pregnant women is caused by the synthesis of too much aquaporin-2.

Knowledge of aquaporins may in the future contribute to the *solution* of medical problems. Specific aquaporins, such as those of the malaria parasite, might be useful targets for new drugs.

What has the study of aquaporins revealed about evolution?

Several hundred aquaporins are known so far. Plants have the most; rice has around 50, whereas mammals have only about 10. Apparently, aquaporins are even more important for plants than animals. The presence of aquaporins in almost all organisms and similarities among the molecules suggest that aquaporin arose very early in evolution.

What are your major goals now?

Of great importance to me right now is the success of the next generation of scientists. As a scientist, there are three things that count: original discoveries, the respect of your peers, and training future scientists—and the third is perhaps the most important in the long run. Almost everything we scientists do eventually gets revised. We usually pick up a story and carry it for a while until someone else takes it over. So the educating of young scientists is a big issue, as is scientific literacy for everyone.

"Science is a social endeavor. Scientists have personalities and interests and fears and limitations. We're not cybertronic machines. And we collaborate. Unless you're at the level of Leonardo da Vinci, you get a lot of ideas from other people."

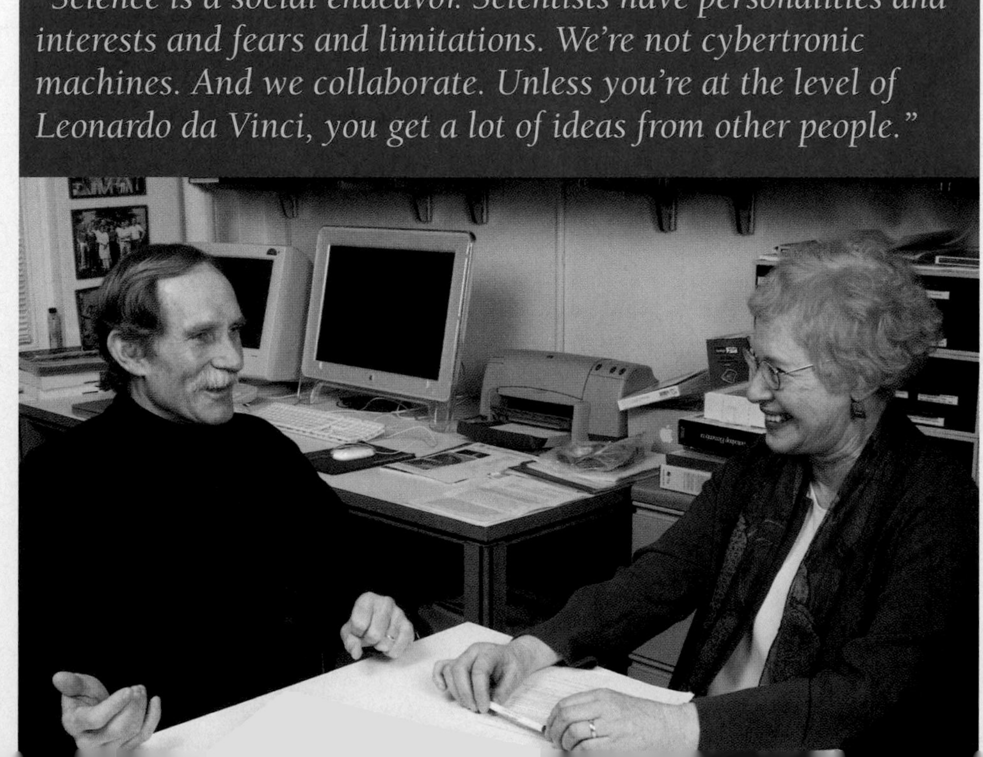

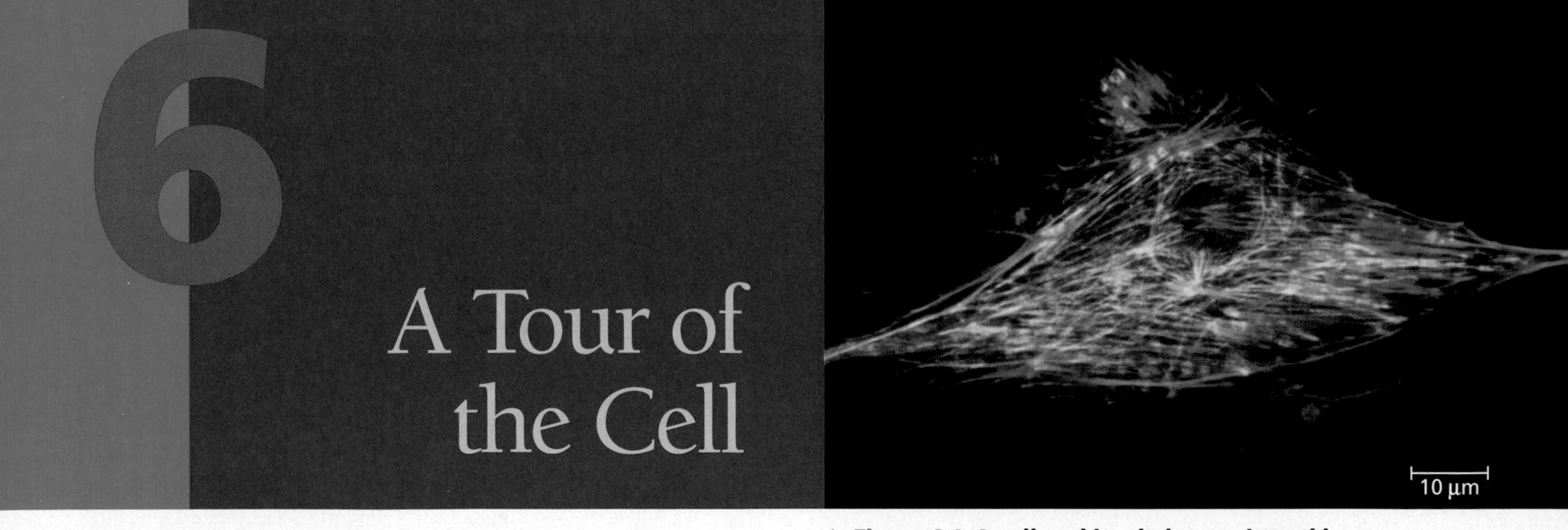

▲ **Figure 6.1 A cell and its skeleton viewed by fluorescence microscopy.**

Key Concepts

6.1 To study cells, biologists use microscopes and the tools of biochemistry
6.2 Eukaryotic cells have internal membranes that compartmentalize their functions
6.3 The eukaryotic cell's genetic instructions are housed in the nucleus and carried out by the ribosomes
6.4 The endomembrane system regulates protein traffic and performs metabolic functions in the cell
6.5 Mitochondria and chloroplasts change energy from one form to another
6.6 The cytoskeleton is a network of fibers that organizes structures and activities in the cell
6.7 Extracellular components and connections between cells help coordinate cellular activities

Overview

The Importance of Cells

The cell is as fundamental to biology as the atom is to chemistry: All organisms are made of cells. In the hierarchy of biological organization, the cell is the simplest collection of matter that can live. Indeed, there are diverse forms of life existing as single-celled organisms. More complex organisms, including plants and animals, are multicellular; their bodies are cooperatives of many kinds of specialized cells that could not survive for long on their own. However, even when they are arranged into higher levels of organization, such as tissues and organs, cells can be singled out as the organism's basic units of structure and function. The contraction of muscle cells moves your eyes as you read this sentence; when you decide to turn the next page, nerve cells will transmit that decision from your brain to the muscle cells of your hand. Everything an organism does occurs fundamentally at the cellular level.

The cell is a microcosm that demonstrates most of the themes introduced in Chapter 1. Life at the cellular level arises from structural order, reinforcing the themes of emergent properties and the correlation between structure and function. For example, the movement of an animal cell depends on an intricate interplay of the structures that make up a cellular skeleton (green and red in the micrograph in **Figure 6.1**). Another recurring theme in biology is the interaction of organisms with their environment. Cells sense and respond to environmental fluctuations. And keep in mind the one biological theme that unifies all others: evolution. All cells are related by their descent from earlier cells. However, they have been modified in many different ways during the long evolutionary history of life on Earth.

Although cells can differ substantially from each other, they share certain common characteristics. In this chapter, we'll first learn about the tools and experimental approaches that have allowed us to understand subcellular details; then we'll tour the cell and become acquainted with its components.

Concept 6.1

To study cells, biologists use microscopes and the tools of biochemistry

It can be difficult to understand how a cell, usually too small to be seen by the unaided eye, can be so complex. How can cell biologists possibly investigate the inner workings of such tiny entities? Before we actually tour the cell, it will be helpful to learn how cells are studied.

Microscopy

The advance of a scientific field often parallels the invention of instruments that extend human senses to new limits. The discovery and early study of cells progressed with the invention of microscopes in 1590 and their improvement in the 17th century. Microscopes of various types are still indispensable tools for the study of cells.

The microscopes first used by Renaissance scientists, as well as the microscopes you are likely to use in the laboratory, are all **light microscopes (LMs).** Visible light is passed through the specimen and then through glass lenses. The lenses refract (bend) the light in such a way that the image of the specimen is magnified as it is projected into the eye, onto photographic film or a digital sensor, or onto a video screen. (See the diagram of microscope structure in Appendix C.)

Two important parameters in microscopy are magnification and resolving power, or resolution. *Magnification* in microscopy is the ratio of an object's image size to its real size. *Resolution* is a measure of the clarity of the image; it is the minimum distance two points can be separated and still be distinguished as two points. For example, what appears to the unaided eye as one star in the sky may be resolved as twin stars with a telescope.

Just as the resolving power of the human eye is limited, the resolution of telescopes and microscopes is limited. Microscopes can be designed to magnify objects as much as desired, but the light microscope cannot resolve detail finer than about 0.2 micrometer (μm), or 200 nanometers (nm), the size of a small bacterium (**Figure 6.2**). This resolution is limited by the shortest wavelength of light used to illuminate the specimen. Light microscopes can magnify effectively to about 1,000 times the size of the actual specimen; at greater magnifications, the image becomes increasingly blurry. Most of the improvements in light microscopy since the beginning of the 20th century have involved new methods for enhancing contrast, which clarifies the details that can be resolved (**Figure 6.3**, next page). In addition, scientists have developed methods for staining or labeling particular cell components so that they stand out visually.

Although cells were discovered by Robert Hooke in 1665, the geography of the cell was largely uncharted until the 1950s. Most subcellular structures, or **organelles,** are too small to be resolved by the light microscope. Cell biology advanced rapidly in the 1950s with the introduction of the electron microscope. Instead of using light, the **electron microscope (EM)** focuses a beam of electrons through the specimen or onto its surface (see Appendix C). Resolution is inversely related to the wavelength of the radiation a microscope uses for imaging, and electron beams have wavelengths much shorter than the wavelengths of visible light. Modern electron microscopes can theoretically achieve a resolution of about 0.002 nm, but the practical limit for biological structures is generally only about 2 nm—still a hundredfold improvement over the light microscope. Biologists use the term *cell ultrastructure* to refer to a cell's anatomy as revealed by an electron microscope.

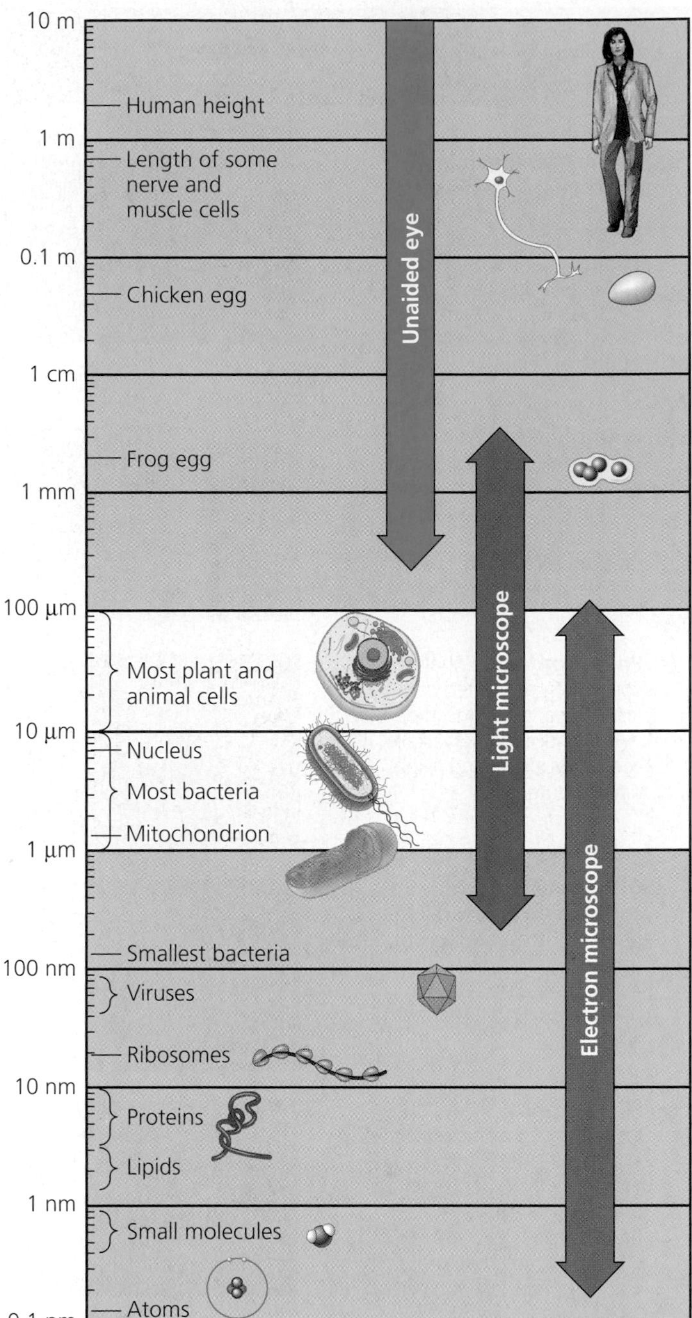

Measurements
1 centimeter (cm) = 10^{-2} meter (m) = 0.4 inch
1 millimeter (mm) = 10^{-3} m
1 micrometer (μm) = 10^{-3} mm = 10^{-6} m
1 nanometer (nm) = 10^{-3} μm = 10^{-9} m

▲ **Figure 6.2 The size range of cells.** Most cells are between 1 and 100 μm in diameter (yellow region of chart) and are therefore visible only under a microscope. Notice that the scale along the left side is logarithmic to accommodate the range of sizes shown. Starting at the top of the scale with 10 m and going down, each reference measurement marks a tenfold decrease in diameter or length. For a complete table of the metric system, see Appendix B.

Figure 6.3

Research Method Light Microscopy

TECHNIQUE | **RESULTS**

(a) Brightfield (unstained specimen). Passes light directly through specimen. Unless cell is naturally pigmented or artificially stained, image has little contrast. [Parts (a)–(d) show a human cheek epithelial cell.]

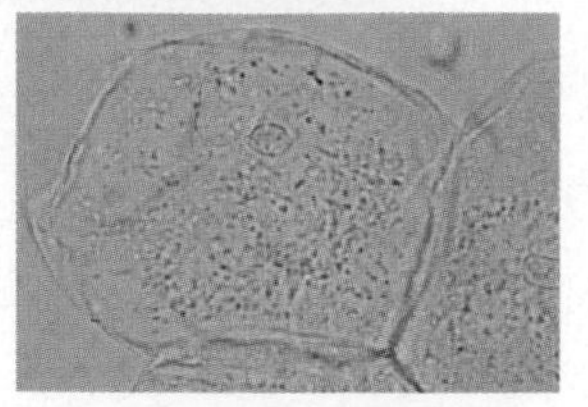

(b) Brightfield (stained specimen). Staining with various dyes enhances contrast, but most staining procedures require that cells be fixed (preserved).

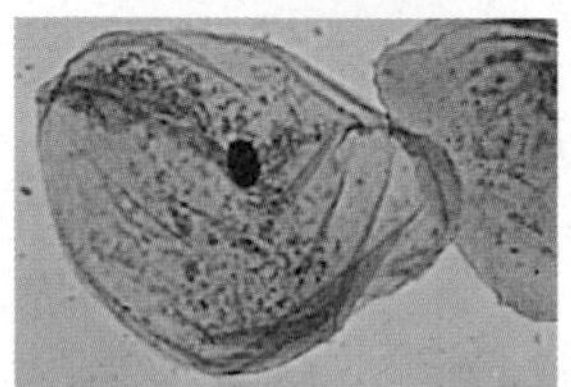

(c) Phase-contrast. Enhances contrast in unstained cells by amplifying variations in density within specimen; especially useful for examining living, unpigmented cells.

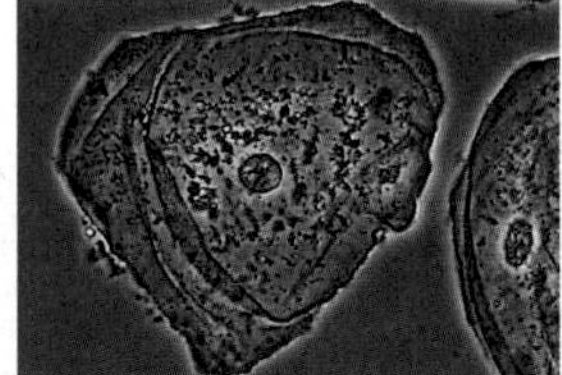

(d) Differential-interference-contrast (Nomarski). Like phase-contrast microscopy, it uses optical modifications to exaggerate differences in density, making the image appear almost 3D.

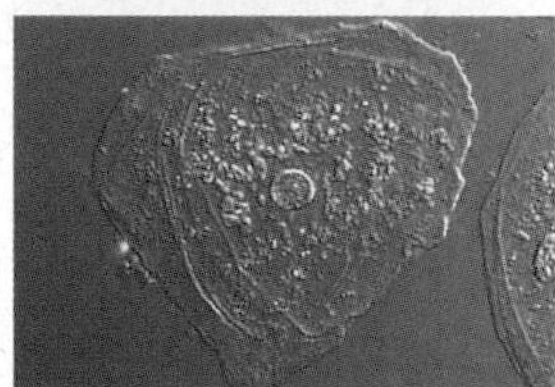

(e) Fluorescence. Shows the locations of specific molecules in the cell by tagging the molecules with fluorescent dyes or antibodies. These fluorescent substances absorb ultraviolet radiation and emit visible light, as shown here in a cell from an artery.

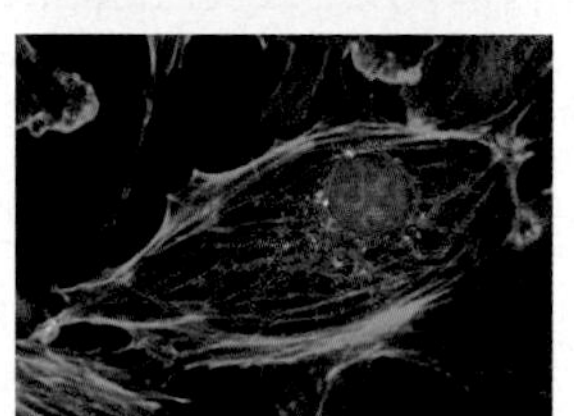

(f) Confocal. Uses lasers and special optics for "optical sectioning" of fluorescently-stained specimens. Only a single plane of focus is illuminated; out-of-focus fluorescence above and below the plane is subtracted by a computer. A sharp image results, as seen in stained nervous tissue (top), where nerve cells are green, support cells are red, and regions of overlap are yellow. A standard fluorescence micrograph (bottom) of this relatively thick tissue is blurry.

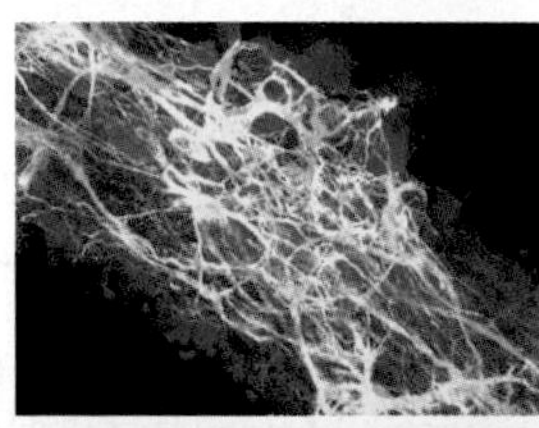

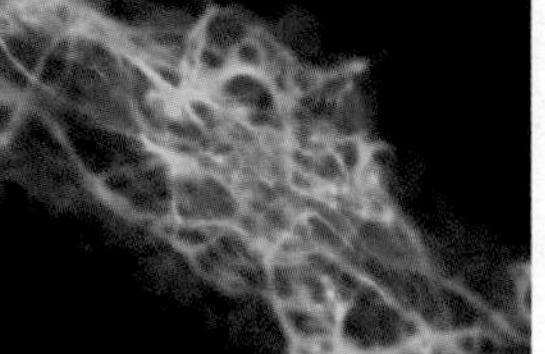

Figure 6.4

Research Method Electron Microscopy

TECHNIQUE | **RESULTS**

(a) Scanning electron microscopy (SEM). Micrographs taken with a scanning electron microscope show a 3D image of the surface of a specimen. This SEM shows the surface of a cell from a rabbit trachea (windpipe) covered with motile organelles called cilia. Beating of the cilia helps move inhaled debris upward toward the throat.

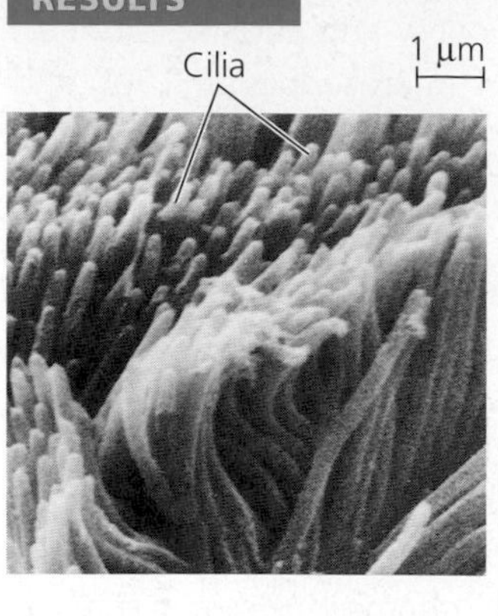

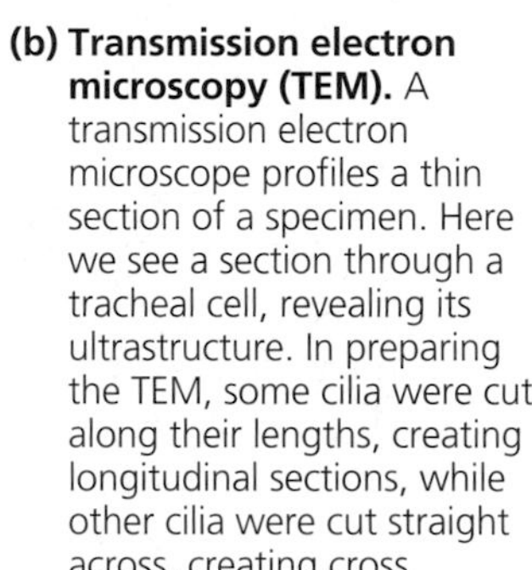

(b) Transmission electron microscopy (TEM). A transmission electron microscope profiles a thin section of a specimen. Here we see a section through a tracheal cell, revealing its ultrastructure. In preparing the TEM, some cilia were cut along their lengths, creating longitudinal sections, while other cilia were cut straight across, creating cross sections.

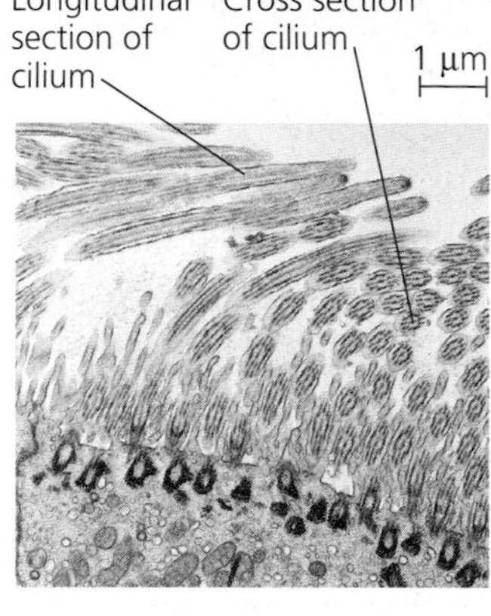

There are two basic types of electron microscopes: the **scanning electron microscope (SEM)** and the **transmission electron microscope (TEM).** The SEM is especially useful for detailed study of the surface of a specimen **(Figure 6.4a)**. The electron beam scans the surface of the sample, which is usually coated with a thin film of gold. The beam excites electrons on the sample's surface, and these secondary electrons are detected by a device that translates the pattern of electrons into an electronic signal to a video screen. The result is an image of the topography of the specimen. The SEM has great depth of field, which results in an image that appears three-dimensional.

Cell biologists use the TEM mainly to study the internal ultrastructure of cells **(Figure 6.4b)**. The TEM aims an electron beam through a very thin section of the specimen, similar to the way a light microscope transmits light through a slide. The specimen has been stained with atoms of heavy metals, which attach to certain cellular structures, thus enhancing the electron density of some parts of the cell more than others. The electrons passing through the specimen are scattered more in the denser regions, so fewer electrons are transmitted. The image is created by the pattern of transmitted electrons. Instead of using glass lenses, the TEM uses electromagnets as lenses to bend the paths of the electrons, ultimately focusing the image onto a screen for viewing or onto photographic film. Some microscopes are equipped with a digital camera to photograph the image on the screen; others are equipped with a digital detector in place of both screen and camera.

Electron microscopes reveal many organelles that are impossible to resolve with the light microscope. But the light microscope offers advantages, especially for the study of living cells. A disadvantage of electron microscopy is that the methods used to prepare the specimen kill the cells. Also, specimen preparation can introduce artifacts, structural features seen in micrographs that do not exist in the living cell (as is true for all microscopy techniques). From this point on in the book, micrographs are identified by the type of microscopy: LM for a light micrograph, SEM for a scanning electron micrograph, and TEM for a transmission electron micrograph.

Microscopes are the most important tools of *cytology,* the study of cell structure. But simply describing the diverse organelles within the cell reveals little about their function. Modern cell biology developed from an integration of cytology with *biochemistry,* the study of the molecules and chemical processes (metabolism) of cells. A biochemical approach called cell fractionation has been particularly important in cell biology.

Isolating Organelles by Cell Fractionation

The goal of **cell fractionation** is to take cells apart and separate the major organelles from one another **(Figure 6.5)**. The instrument used to fractionate cells is the centrifuge, which can spin test tubes holding mixtures of disrupted cells at various speeds. The resulting force separates the cell components by size and density. The most powerful machines, called **ultracentrifuges,** can spin as fast as 130,000 revolutions per minute (rpm) and apply forces on particles of more than 1 million times the force of gravity (1,000,000 *g*).

Cell fractionation enables the researcher to prepare specific components of cells in bulk quantity to study their composition and functions. By following this approach, biologists have been able to assign various functions of the cell to the different organelles, a task that would be far more difficult with intact cells. For example, one cellular fraction collected by centrifugation has enzymes that function in the metabolic process known as cellular respiration. The electron microscope reveals this fraction to be very rich in the organelles called mitochondria. This evidence helped cell biologists determine that mitochondria are the sites of cellular respiration. Cytology and biochemistry complement each other in correlating cellular structure and function.

Concept Check 6.1

1. Which type of microscope would you use to study (a) the changes in shape of a living white blood cell, (b) the details of surface texture of a hair, and (c) the detailed structure of an organelle?

For suggested answers, see Appendix A.

Figure 6.5
Research Method Cell Fractionation

APPLICATION Cell fractionation is used to isolate (fractionate) cell components, based on size and density.

TECHNIQUE First, cells are homogenized in a blender to break them up. The resulting mixture (cell homogenate) is then centrifuged at various speeds and durations to fractionate the cell components, forming a series of pellets.

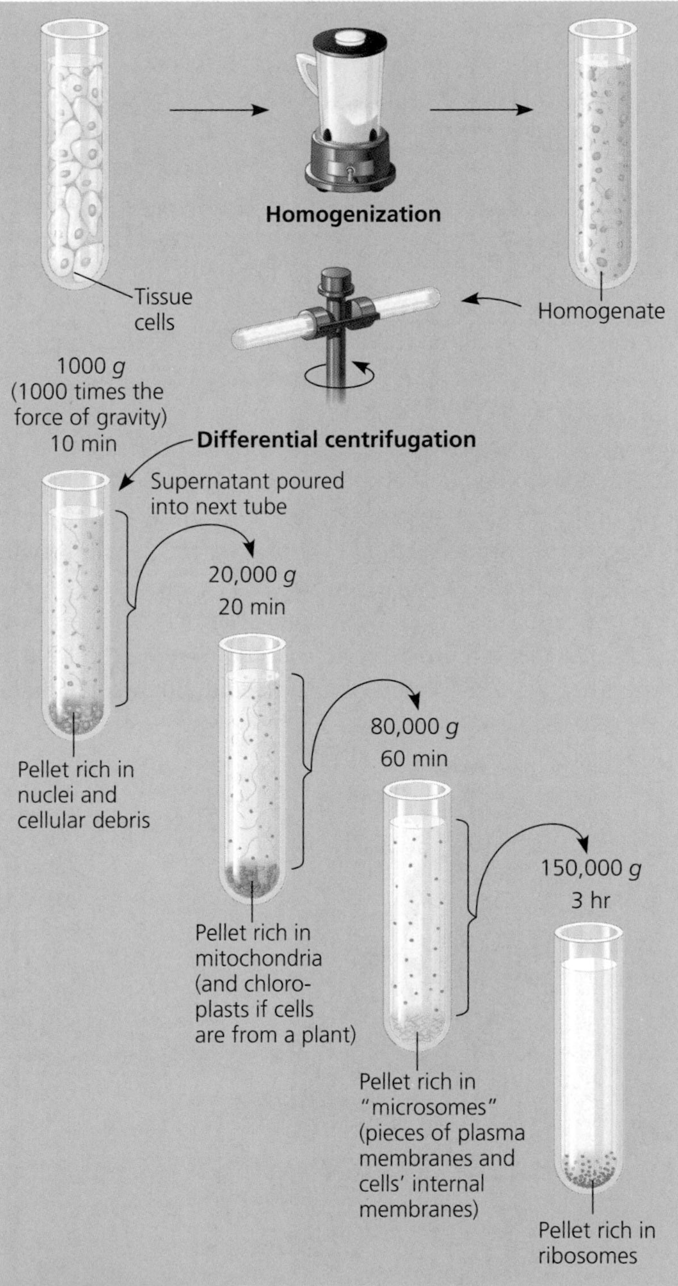

RESULTS In the original experiments, the researchers used microscopy to identify the organelles in each pellet, establishing a baseline for further experiments. In the next series of experiments, researchers used biochemical methods to determine the metabolic functions associated with each type of organelle. Researchers currently use cell fractionation to isolate particular organelles in order to study further details of their function.

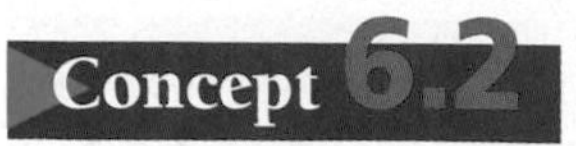

Eukaryotic cells have internal membranes that compartmentalize their functions

The basic structural and functional unit of every organism is one of two types of cells—prokaryotic or eukaryotic. Only organisms of the domains Bacteria and Archaea consist of prokaryotic cells. Protists, fungi, animals, and plants all consist of eukaryotic cells. This chapter focuses on generalized animal and plant cells, after first comparing them with prokaryotic cells.

Comparing Prokaryotic and Eukaryotic Cells

All cells have several basic features in common: They are all bounded by a membrane, called a *plasma membrane*. Within the membrane is a semifluid substance, **cytosol,** in which organelles are found. All cells contain *chromosomes,* carrying genes in the form of DNA. And all cells have *ribosomes,* tiny organelles that make proteins according to instructions from the genes.

A major difference between prokaryotic and eukaryotic cells, indicated by their names, is that the chromosomes of a eukaryotic cell are located in a membrane-enclosed organelle called the *nucleus.* The word *prokaryotic* is from the Greek *pro,* meaning "before," and *karyon,* meaning "kernel," referring here to the nucleus. In a **prokaryotic cell (Figure 6.6)**, the DNA is concentrated in a region called the **nucleoid,** but no membrane separates this region from the rest of the cell. In contrast, the **eukaryotic cell** (Greek *eu,* true, and *karyon*) has a true nucleus, bounded by a membranous nuclear envelope (see Figure 6.9, pp. 100–101). The entire region between the nucleus and the plasma membrane is called the **cytoplasm,** a term also used for the interior of a prokaryotic cell. Within the cytoplasm of a eukaryotic cell, suspended in cytosol, are a variety of membrane-bounded organelles of specialized form and function. These are absent in prokaryotic cells. Thus, the presence or absence of a true nucleus is just one example of the disparity in structural complexity between the two types of cells.

Eukaryotic cells are generally quite a bit bigger than prokaryotic cells (see Figure 6.2). Size is a general aspect of cell structure that relates to function. The logistics of carrying out cellular metabolism sets limits on cell size. At the lower limit, the smallest cells known are bacteria called mycoplasmas, which have diameters between 0.1 and 1.0 μm. These are perhaps the smallest packages with enough DNA to program metabolism and enough enzymes and other cellular equipment to carry out the activities necessary for a cell to sustain itself and reproduce. Most bacteria are 1–10 μm in diameter, a dimension about ten times greater than that of mycoplasmas. Eukaryotic cells are typically 10–100 μm in diameter.

Metabolic requirements also impose theoretical upper limits on the size that is practical for a single cell. As an object of a particular shape increases in size, its volume grows proportionately more than its surface area. (Area is proportional to a linear dimension squared, whereas volume is proportional to the linear dimension cubed.) Thus, the smaller the object, the greater its ratio of surface area to volume **(Figure 6.7)**.

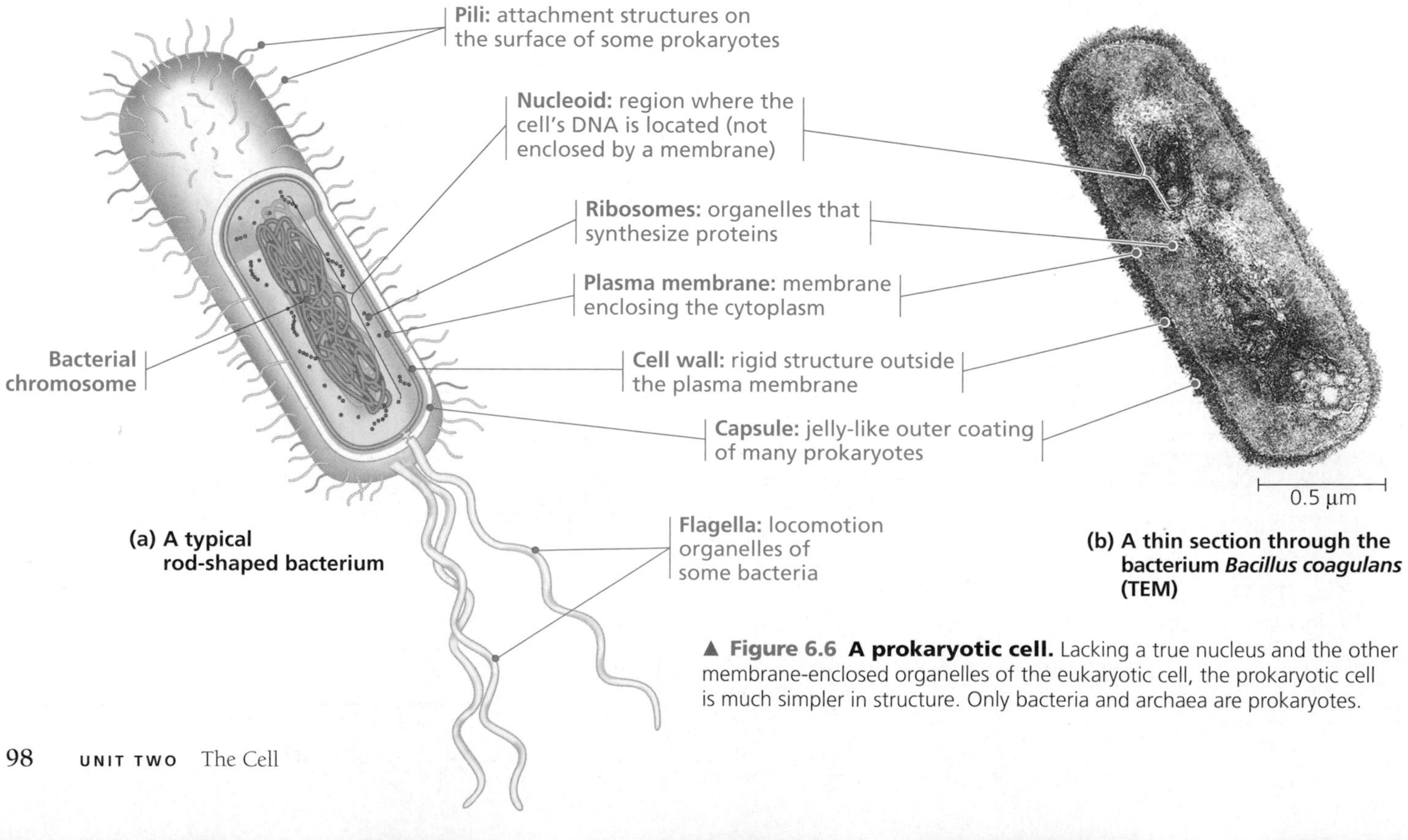

▲ **Figure 6.6 A prokaryotic cell.** Lacking a true nucleus and the other membrane-enclosed organelles of the eukaryotic cell, the prokaryotic cell is much simpler in structure. Only bacteria and archaea are prokaryotes.

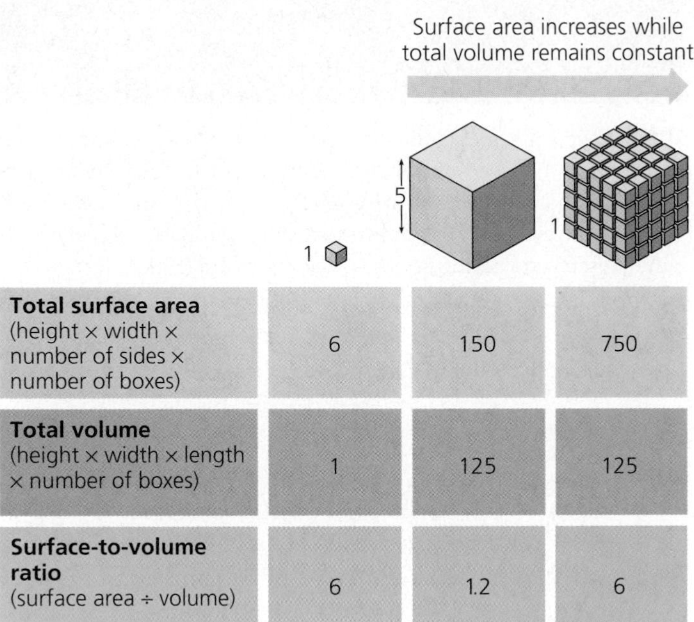

Total surface area (height × width × number of sides × number of boxes)	6	150	750
Total volume (height × width × length × number of boxes)	1	125	125
Surface-to-volume ratio (surface area ÷ volume)	6	1.2	6

▲ **Figure 6.7 Geometric relationships between surface area and volume.** In this diagram, cells are represented as boxes. Using arbitrary units of length, we can calculate the cell's surface area (in square units), volume (in cubic units), and ratio of surface area to volume. The smaller the cell, the higher the surface-to-volume ratio. A high surface-to-volume ratio facilitates the exchange of materials between a cell and its environment.

At the boundary of every cell, the **plasma membrane** functions as a selective barrier that allows sufficient passage of oxygen, nutrients, and wastes to service the entire volume of the cell **(Figure 6.8)**. For each square micrometer of membrane, only so much of a particular substance can cross per second. Rates of chemical exchange with the extracellular environment might be inadequate to maintain a cell with a very large cytoplasm. The need for a surface area sufficiently large to accommodate the volume helps explain the microscopic size of most cells. Larger organisms do not generally have *larger* cells than smaller organisms—simply *more* cells. A sufficiently high ratio of surface area to volume is especially important in cells that exchange a lot of material with their surroundings, such as intestinal cells. Such cells may have many long, thin projections from their surface called microvilli, which increase surface area without an appreciable increase in volume.

Prokaryotic cells will be described in detail in Chapters 18 and 27 (see Table 27.2 for a comparison of prokaryotes and eukaryotes), and the possible evolutionary relationships between prokaryotic and eukaryotic cells will be discussed in Chapter 26. Most of the discussion of cell structure that follows in this chapter applies to eukaryotic cells.

A Panoramic View of the Eukaryotic Cell

In addition to the plasma membrane at its outer surface, a eukaryotic cell has extensive and elaborately arranged internal membranes, which partition the cell into compartments—the membranous organelles mentioned earlier. These membranes also participate directly in the cell's metabolism, because many enzymes are built right into the membranes. Furthermore, the cell's compartments provide different local environments that facilitate specific metabolic functions, so incompatible processes can go on simultaneously inside the same cell.

Membranes of various kinds are fundamental to the organization of the cell. In general, biological membranes consist of a double layer of phospholipids and other lipids. Embedded in this lipid bilayer or attached to its surfaces are diverse proteins (see Figure 6.8). However, each type of membrane has a unique composition of lipids and proteins suited to that membrane's specific functions. For example, enzymes embedded in the membranes of the organelles called mitochondria function in cellular respiration.

Before continuing with this chapter, examine the overviews of eukaryotic cells in **Figure 6.9** on the next two pages. These generalized cell diagrams introduce the various organelles and provide a map of the cell for the detailed tour upon which we will now embark. Figure 6.9 also contrasts animal and plant cells. As eukaryotic cells, they have much more in common than either has with any prokaryotic cell. As you will see, however, there are important differences between animal and plant cells.

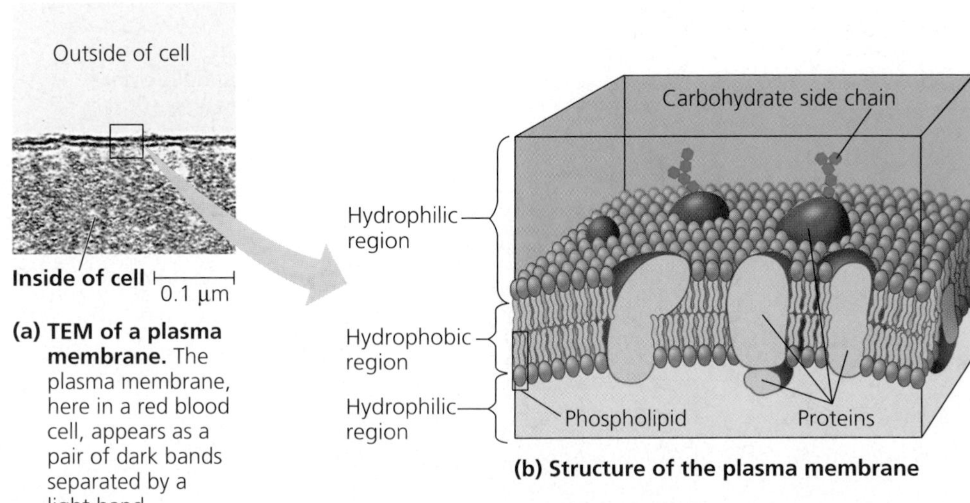

(a) TEM of a plasma membrane. The plasma membrane, here in a red blood cell, appears as a pair of dark bands separated by a light band.

(b) Structure of the plasma membrane

◀ **Figure 6.8 The plasma membrane.** The plasma membrane and the membranes of organelles consist of a double layer (bilayer) of phospholipids with various proteins attached to or embedded in it. The phospholipid tails in the interior of a membrane are hydrophobic; the interior portions of membrane proteins are also hydrophobic. The phospholipid heads, exterior proteins, exterior parts of proteins, and carbohydrate side chains are hydrophilic and in contact with the aqueous solution on either side of the membrane. Carbohydrate side chains are found only on the outer surface of the plasma membrane. The specific functions of a membrane depend on the kinds of phospholipids and proteins present.

Figure 6.9

Exploring Animal and Plant Cells

ANIMAL CELL

This drawing of a generalized animal cell incorporates the most common structures of animal cells (no cell actually looks just like this). As shown by this cutaway view, the cell has a variety of organelles ("little organs"), many of which are bounded by membranes. The most prominent organelle in an animal cell is usually the nucleus. Most of the cell's metabolic activities occur in the cytoplasm, the entire region between the nucleus and the plasma membrane. The cytoplasm contains many organelles suspended in a semifluid medium, the cytosol. Pervading much of the cytoplasm is a labyrinth of membranes called the endoplasmic reticulum (ER).

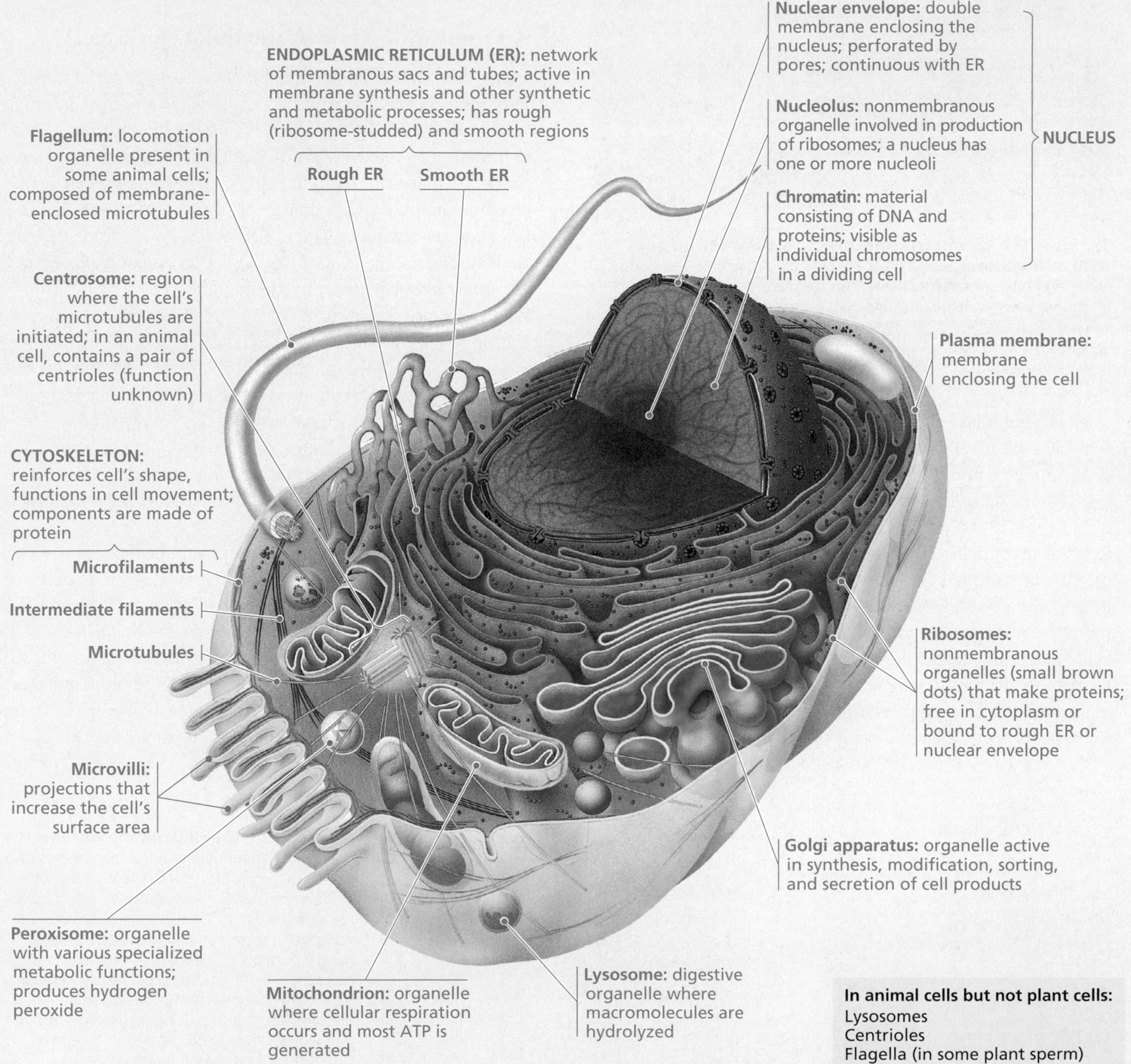

ENDOPLASMIC RETICULUM (ER): network of membranous sacs and tubes; active in membrane synthesis and other synthetic and metabolic processes; has rough (ribosome-studded) and smooth regions
Rough ER
Smooth ER
Nuclear envelope: double membrane enclosing the nucleus; perforated by pores; continuous with ER
Nucleolus: nonmembranous organelle involved in production of ribosomes; a nucleus has one or more nucleoli
Chromatin: material consisting of DNA and proteins; visible as individual chromosomes in a dividing cell
NUCLEUS
Flagellum: locomotion organelle present in some animal cells; composed of membrane-enclosed microtubules
Centrosome: region where the cell's microtubules are initiated; in an animal cell, contains a pair of centrioles (function unknown)
Plasma membrane: membrane enclosing the cell
CYTOSKELETON: reinforces cell's shape, functions in cell movement; components are made of protein
Microfilaments
Intermediate filaments
Microtubules
Ribosomes: nonmembranous organelles (small brown dots) that make proteins; free in cytoplasm or bound to rough ER or nuclear envelope
Microvilli: projections that increase the cell's surface area
Golgi apparatus: organelle active in synthesis, modification, sorting, and secretion of cell products
Peroxisome: organelle with various specialized metabolic functions; produces hydrogen peroxide
Mitochondrion: organelle where cellular respiration occurs and most ATP is generated
Lysosome: digestive organelle where macromolecules are hydrolyzed

In animal cells but not plant cells:
Lysosomes
Centrioles
Flagella (in some plant sperm)

PLANT CELL

This drawing of a generalized plant cell reveals the similarities and differences between an animal cell and a plant cell. In addition to most of the features seen in an animal cell, a plant cell has membrane-enclosed organelles called plastids. The most important type of plastid is the chloroplast, which carries out photosynthesis. Many plant cells have a large central vacuole; some may have one or more smaller vacuoles. Outside a plant cell's plasma membrane is a thick cell wall, perforated by channels called plasmodesmata.

If you preview the rest of the chapter now, you'll see Figure 6.9 repeated in miniature as orientation diagrams. In each case, a particular organelle is highlighted, color-coded to its appearance in Figure 6.9. As we take a closer look at individual organelles, the orientation diagrams will help you place those structures in the context of the whole cell.

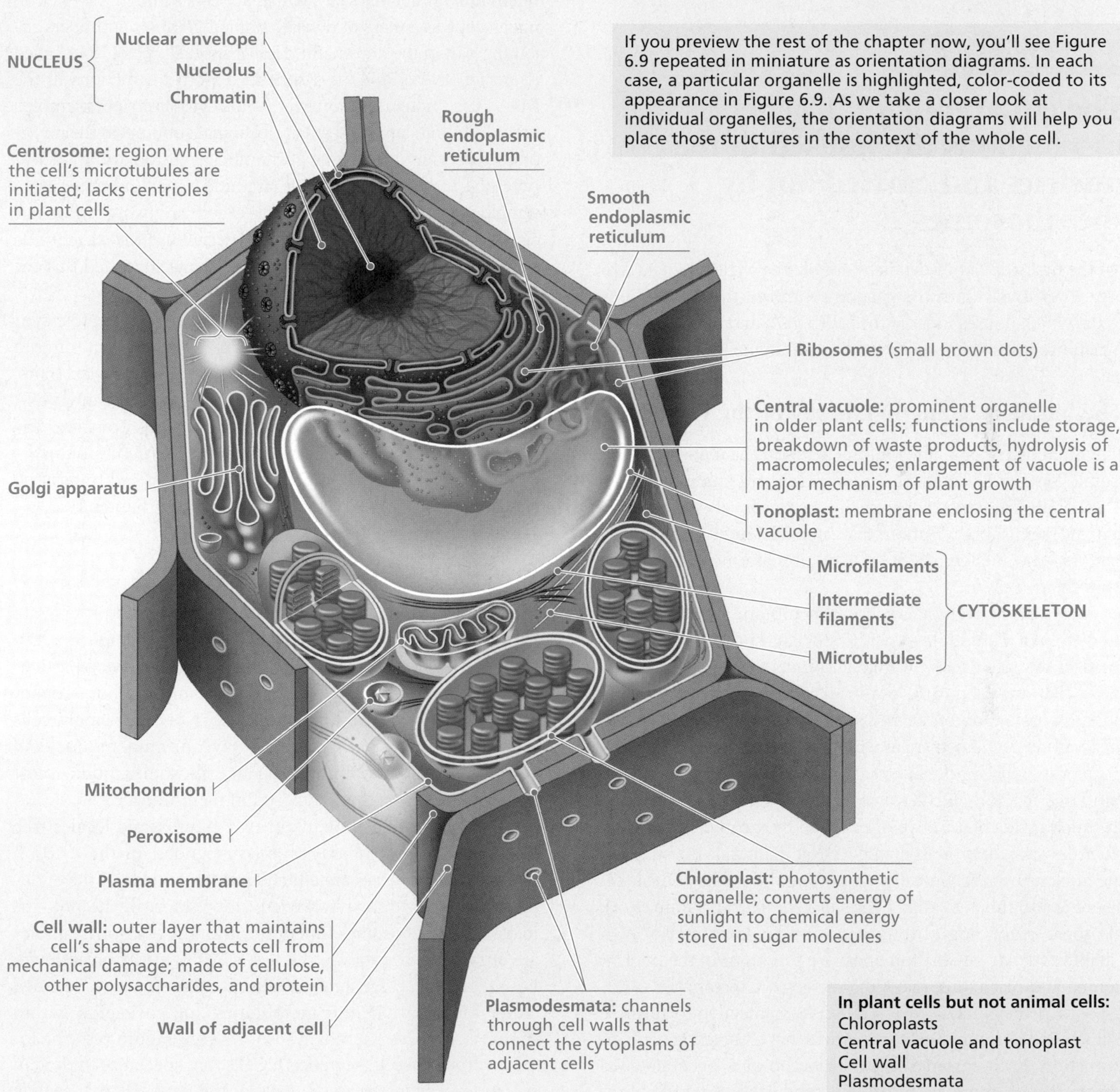

In plant cells but not animal cells:
Chloroplasts
Central vacuole and tonoplast
Cell wall
Plasmodesmata

Concept Check 6.2

1. After carefully reviewing Figure 6.9, briefly describe the structure and function of each of the following organelles: nucleus, mitochondrion, chloroplast, central vacuole, endoplasmic reticulum, and Golgi apparatus.

For suggested answers, see Appendix A.

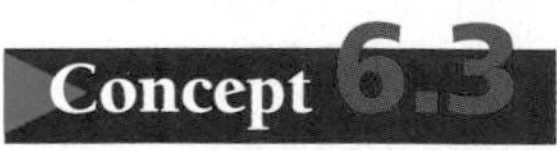

The eukaryotic cell's genetic instructions are housed in the nucleus and carried out by the ribosomes

On the first stop of our detailed tour of the cell, let's look at two organelles involved in the genetic control of the cell: the nucleus, which houses most of the cell's DNA, and the ribosomes, which use information from the DNA to make proteins.

The Nucleus: Genetic Library of the Cell

The **nucleus** contains most of the genes in the eukaryotic cell (some genes are located in mitochondria and chloroplasts). It is generally the most conspicuous organelle in a eukaryotic cell, averaging about 5 µm in diameter. The **nuclear envelope** encloses the nucleus **(Figure 6.10)**, separating its contents from the cytoplasm.

The nuclear envelope is a *double* membrane. The two membranes, each a lipid bilayer with associated proteins, are separated by a space of 20–40 nm. The envelope is perforated by pores that are about 100 nm in diameter. At the lip of each pore, the inner and outer membranes of the nuclear envelope are continuous. An intricate protein structure called a *pore complex* lines each pore and regulates the entry and exit of certain large macromolecules and particles. Except at the pores, the nuclear side of the envelope is lined by the **nuclear lamina,** a netlike array of protein filaments that maintains the shape of the nucleus by mechanically supporting the nuclear envelope. There is also much evidence for a *nuclear matrix,* a framework of fibers extending throughout the nuclear interior. (In Chapter 19, we will examine possible functions of the nuclear lamina and matrix in organizing the genetic material.)

Within the nucleus, the DNA is organized into discrete units called **chromosomes,** structures that carry the genetic information. Each chromosome is made up of a material called **chromatin,** a complex of proteins and DNA. Stained chromatin usually appears through both light microscopes and electron microscopes as a diffuse mass. As a cell prepares to divide, however, the thin chromatin fibers coil up (condense), becoming thick enough to be distinguished as the familiar separate structures we know as chromosomes. Each eukaryotic species has a characteristic number of chromosomes. A typical human cell, for example, has 46 chromosomes in its nucleus; the exceptions are the sex cells (eggs and sperm), which have only 23 chromosomes in humans. A fruit fly cell has 8 chromosomes in most cells, with 4 in the sex cells.

A prominent structure within the nondividing nucleus is the **nucleolus** (plural, *nucleoli*), which appears through the electron microscope as a mass of densely stained granules and fibers adjoining part of the chromatin. Here a special type of RNA called *ribosomal RNA* (rRNA) is synthesized from instructions in the DNA. Also, proteins imported from the cytoplasm are assembled with rRNA into large and small ribosomal subunits in the nucleolus. These subunits then exit the nucleus through the nuclear pores to the cytoplasm, where a large and a small subunit can assemble into a ribosome. Sometimes there are two or more nucleoli; the number depends on the species and the stage in the cell's reproductive cycle. Recent studies have suggested that the nucleolus may perform additional functions as well.

As we saw in Figure 5.25, the nucleus directs protein synthesis by synthesizing messenger RNA (mRNA) according to instructions provided by the DNA. The mRNA is then transported to the cytoplasm via the nuclear pores. Once an mRNA molecule reaches the cytoplasm, ribosomes translate the mRNA's genetic message into the primary structure of a specific polypeptide. This process of transcribing and translating genetic information is described in detail in Chapter 17.

Ribosomes: Protein Factories in the Cell

Ribosomes, particles made of ribosomal RNA and protein, are the organelles that carry out protein synthesis **(Figure 6.11)**. Cells that have high rates of protein synthesis have a particularly large number of ribosomes. For example, a human pancreas cell has a few million ribosomes. Not surprisingly, cells active in protein synthesis also have prominent nucleoli. (Keep in mind that both nucleoli and ribosomes, unlike most other organelles, are not enclosed in membrane.)

Ribosomes build proteins in two cytoplasmic locales (see Figure 6.11). *Free ribosomes* are suspended in the cytosol, while *bound ribosomes* are attached to the outside of the endoplasmic reticulum or nuclear envelope. Most of the proteins made on free ribosomes function within the cytosol; examples are enzymes that catalyze the first steps of sugar breakdown. Bound ribosomes generally make proteins that are destined either for insertion into membranes, for packaging within certain organelles such as lysosomes (see Figure 6.9), or for export from the cell (secretion). Cells that specialize in protein secretion—for instance, the cells of the pancreas that secrete digestive enzymes—frequently have a high proportion of bound ribosomes. Bound and free ribosomes are structurally

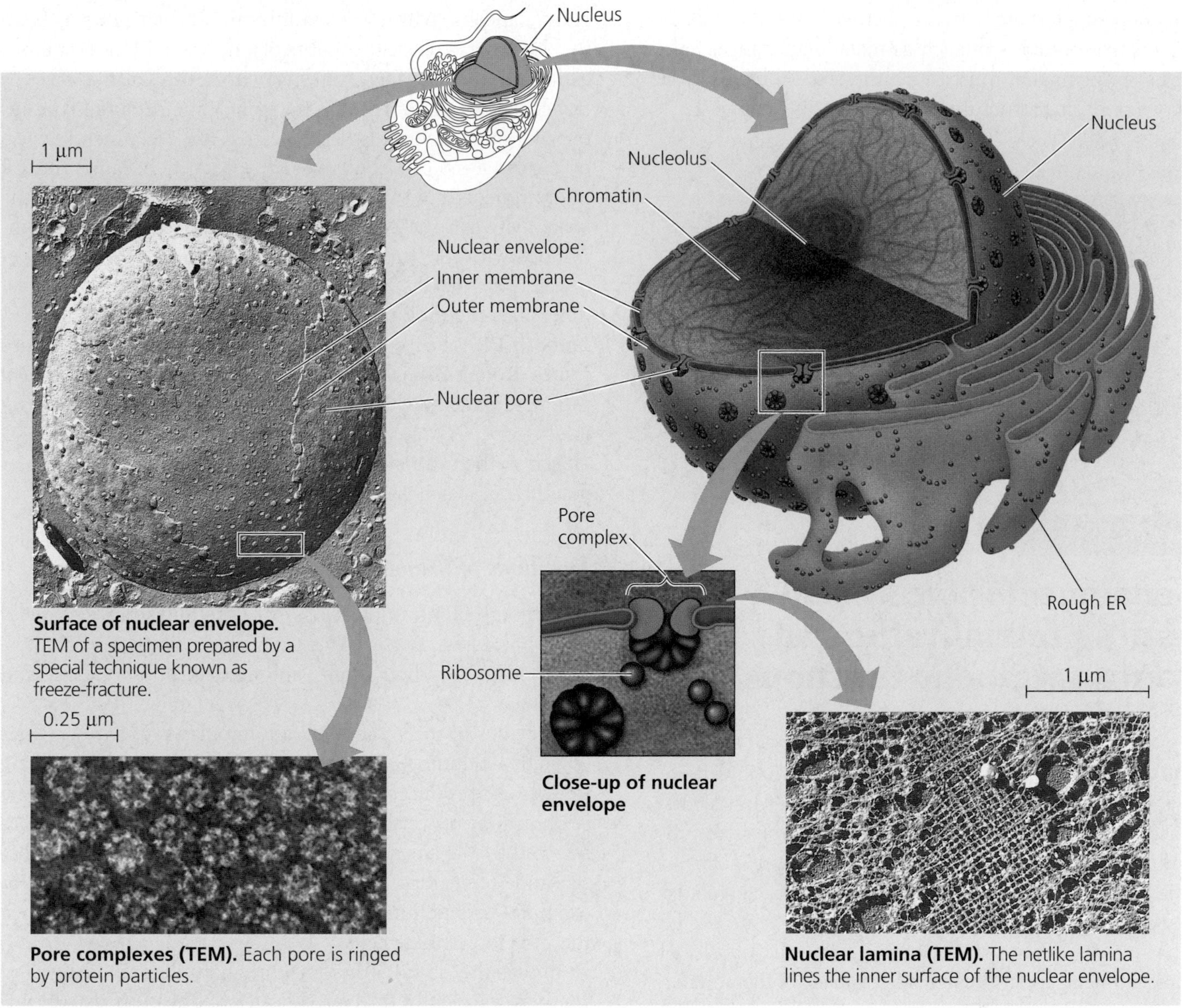

Nucleus

Nucleus

Nucleolus

Chromatin

Nuclear envelope:
Inner membrane
Outer membrane

Nuclear pore

1 µm

Surface of nuclear envelope. TEM of a specimen prepared by a special technique known as freeze-fracture.

Pore complex

Ribosome

Rough ER

Close-up of nuclear envelope

0.25 µm

Pore complexes (TEM). Each pore is ringed by protein particles.

1 µm

Nuclear lamina (TEM). The netlike lamina lines the inner surface of the nuclear envelope.

▲ **Figure 6.10 The nucleus and its envelope.** Within the nucleus are the chromosomes, which appear as a mass of chromatin (DNA and associated proteins), and one or more nucleoli (singular, nucleolus), which function in ribosome synthesis. The nuclear envelope, which consists of two membranes separated by a narrow space, is perforated with pores and lined by the nuclear lamina.

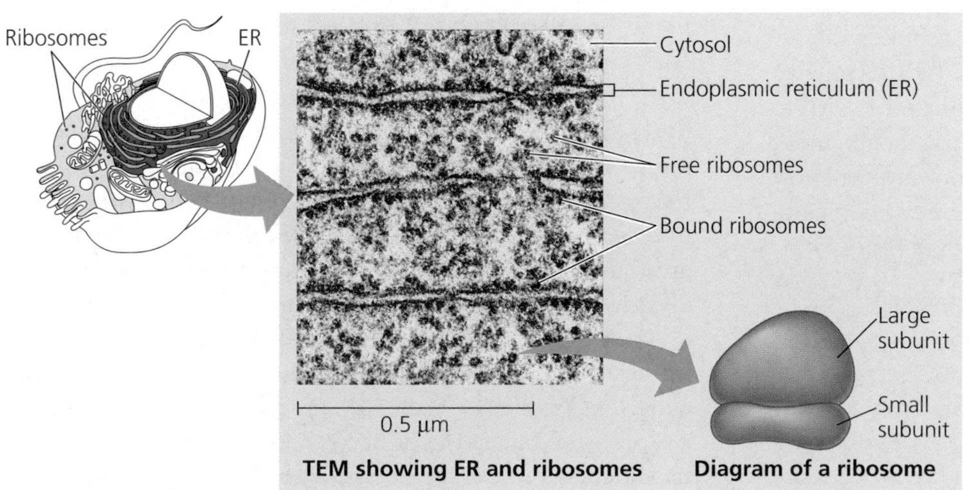

◀ **Figure 6.11 Ribosomes.** This electron micrograph of part of a pancreas cell shows many ribosomes, both free (in the cytosol) and bound (to the endoplasmic reticulum). The simplified diagram of a ribosome shows its two subunits.

identical and can alternate between the two roles; the cell adjusts the relative numbers of each as metabolic changes alter the types of proteins that must be synthesized. You will learn more about ribosome structure and function in Chapter 17.

Concept Check 6.3

1. What role do the ribosomes play in carrying out the genetic instructions?
2. Describe the composition of chromatin and of nucleoli and the function(s) of each.

For suggested answers, see Appendix A.

Concept 6.4

The endomembrane system regulates protein traffic and performs metabolic functions in the cell

Many of the different membranes of the eukaryotic cell are part of an **endomembrane system,** which carries out a variety of tasks in the cell. These tasks include synthesis of proteins and their transport into membranes and organelles or out of the cell, metabolism and movement of lipids, and detoxification of poisons. The membranes of this system are related either through direct physical continuity or by the transfer of membrane segments as tiny **vesicles** (sacs made of membrane). Despite these relationships, the various membranes are not identical in structure and function. Moreover, the thickness, molecular composition, and types of chemical reactions carried out by proteins in a given membrane are not fixed, but may be modified several times during the membrane's life. The endomembrane system includes the nuclear envelope, endoplasmic reticulum, Golgi apparatus, lysosomes, various kinds of vacuoles, and the plasma membrane (not actually an *endo*membrane in physical location, but nevertheless related to the endoplasmic reticulum and other internal membranes). We have already discussed the nuclear envelope and will now focus on the endoplasmic reticulum and the other endomembranes to which the endoplasmic reticulum gives rise.

The Endoplasmic Reticulum: Biosynthetic Factory

The **endoplasmic reticulum (ER)** is such an extensive network of membranes that it accounts for more than half the total membrane in many eukaryotic cells. (The word *endoplasmic* means "within the cytoplasm," and *reticulum* is Latin for "little net.") The ER consists of a network of membranous tubules and sacs called cisternae (from the Latin *cisterna,* a reservoir for a liquid). The ER membrane separates the internal compartment of the ER, called the ER lumen (cavity) or cisternal space, from the cytosol. And because the ER membrane is continuous with the nuclear envelope, the space between the two membranes of the envelope is continuous with the lumen of the ER **(Figure 6.12)**.

There are two distinct, though connected, regions of ER that differ in structure and function: smooth ER and rough ER. **Smooth ER** is so named because its outer surface lacks ribosomes. **Rough ER** has ribosomes that stud the outer surface of the membrane and thus appears rough through the electron microscope. As already mentioned, ribosomes are also attached to the cytoplasmic side of the nuclear envelope's outer membrane, which is continuous with rough ER.

Functions of Smooth ER

The smooth ER of various cell types functions in diverse metabolic processes. These processes include synthesis of lipids, metabolism of carbohydrates, and detoxification of drugs and poisons.

Enzymes of the smooth ER are important to the synthesis of lipids, including oils, phospholipids, and steroids. Among the steroids produced by the smooth ER in animal cells are the sex hormones of vertebrates and the various steroid hormones secreted by the adrenal glands. The cells that actually synthesize and secrete these hormones—in the testes and ovaries, for example—are rich in smooth ER, a structural feature that fits the function of these cells.

In the smooth ER, other enzymes help detoxify drugs and poisons, especially in liver cells. Detoxification usually involves adding hydroxyl groups to drugs, making them more soluble and easier to flush from the body. The sedative phenobarbital and other barbiturates are examples of drugs metabolized in this manner by smooth ER in liver cells. In fact, barbiturates, alcohol, and many other drugs induce the proliferation of smooth ER and its associated detoxification enzymes, thus increasing the rate of detoxification. This, in turn, increases tolerance to the drugs, meaning that higher doses are required to achieve a particular effect, such as sedation. Also, because some of the detoxification enzymes have relatively broad action, the proliferation of smooth ER in response to one drug can increase tolerance to other drugs as well. Barbiturate abuse, for example, may decrease the effectiveness of certain antibiotics and other useful drugs.

The smooth ER also stores calcium ions. In muscle cells, for example, a specialized smooth ER membrane pumps calcium ions from the cytosol into the ER lumen. When a muscle cell is stimulated by a nerve impulse, calcium ions rush back across the ER membrane into the cytosol and trigger contraction of the muscle cell. In other cell types, calcium ion release from the smooth ER can trigger different responses.

Functions of Rough ER

Many types of specialized cells secrete proteins produced by ribosomes attached to rough ER. For example, certain cells in the pancreas secrete the protein insulin, a hormone, into the bloodstream. As a polypeptide chain grows from a bound ribosome, it is threaded into the ER lumen through a pore formed by a protein complex in the ER membrane. As the new protein enters the ER lumen, it folds into its native conformation. Most secretory proteins are **glycoproteins,** proteins that have carbohydrates covalently bonded to them. The carbohydrate is attached to the protein in the ER by specialized molecules built into the ER membrane.

Once secretory proteins are formed, the ER membrane keeps them separate from the proteins, produced by free ribosomes, that will remain in the cytosol. Secretory proteins depart from the ER wrapped in the membranes of vesicles that bud like bubbles from a specialized region called transitional ER (see Figure 6.12). Vesicles in transit from one part of the cell to another are called **transport vesicles;** we will learn their fate in the next section.

In addition to making secretory proteins, rough ER is a membrane factory for the cell; it grows in place by adding membrane proteins and phospholipids to its own membrane. As polypeptides destined to be membrane proteins grow from the ribosomes, they are inserted into the ER membrane itself and are anchored there by their hydrophobic portions. The rough ER also makes its own membrane phospholipids; enzymes built into the ER membrane assemble phospholipids from precursors in the cytosol. The ER membrane expands and is transferred in the form of transport vesicles to other components of the endomembrane system.

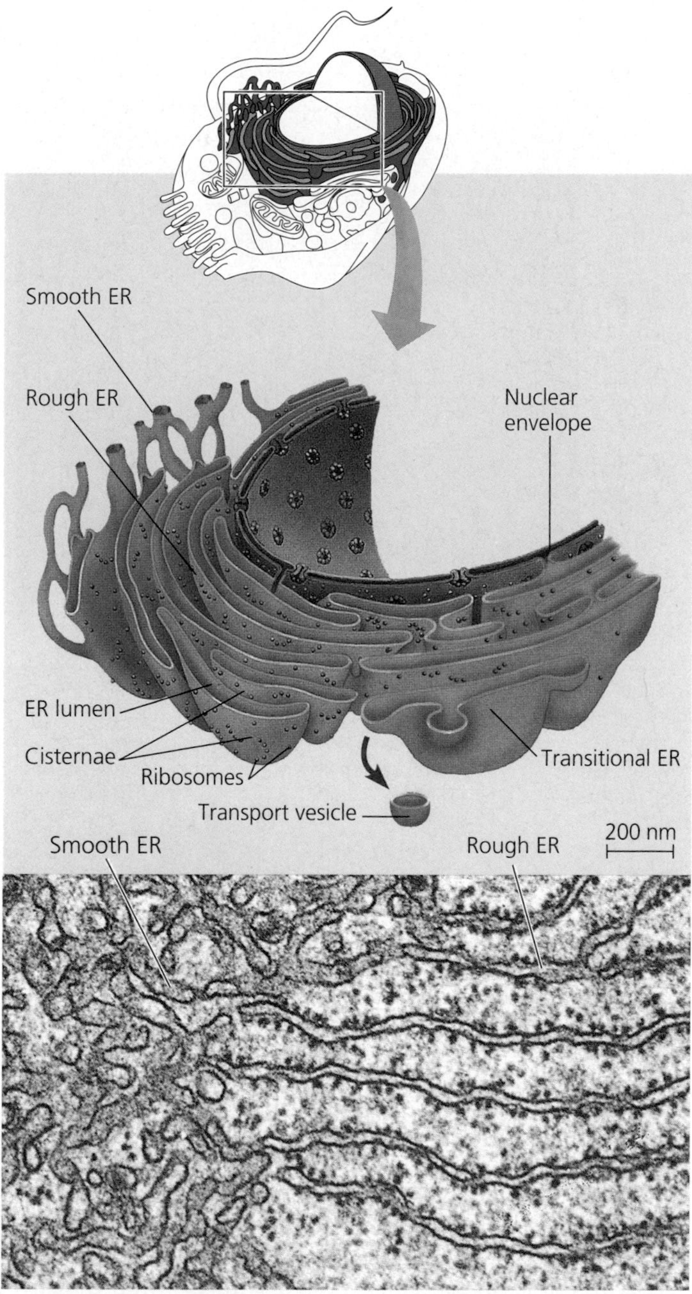

▲ **Figure 6.12 Endoplasmic reticulum (ER).** A membranous system of interconnected tubules and flattened sacs called cisternae, the ER is also continuous with the nuclear envelope. (The drawing is a cutaway view.) The membrane of the ER encloses a continuous compartment called the ER lumen (or cisternal space). Rough ER, which is studded on its outer surface with ribosomes, can be distinguished from smooth ER in the electron micrograph (TEM). Transport vesicles bud off from a region of the rough ER called transitional ER and travel to the Golgi apparatus and other destinations.

The Golgi Apparatus: Shipping and Receiving Center

After leaving the ER, many transport vesicles travel to the **Golgi apparatus.** We can think of the Golgi as a center of manufacturing, warehousing, sorting, and shipping. Here, products of the ER are modified and stored and then sent to other destinations. Not surprisingly, the Golgi apparatus is especially extensive in cells specialized for secretion.

The Golgi apparatus consists of flattened membranous sacs—cisternae—looking like a stack of pita bread (**Figure 6.13**, on the next page). A cell may have many or even hundreds of these stacks. The membrane of each cisterna in a stack separates its internal space from the cytosol. Vesicles concentrated in the vicinity of the Golgi apparatus are engaged in the transfer of material between the parts of the Golgi and other structures.

A Golgi stack has a distinct polarity, with the membranes of cisternae on opposite sides of the stack differing in thickness and molecular composition. The two poles of a Golgi stack are referred to as the *cis* face and the *trans* face; these act, respectively, as the receiving and shipping departments of the Golgi apparatus. The *cis* face is usually located near the ER. Transport vesicles move material from the ER to the Golgi apparatus. A vesicle that buds from the ER can add its membrane and the contents of its lumen to the *cis* face by fusing with a

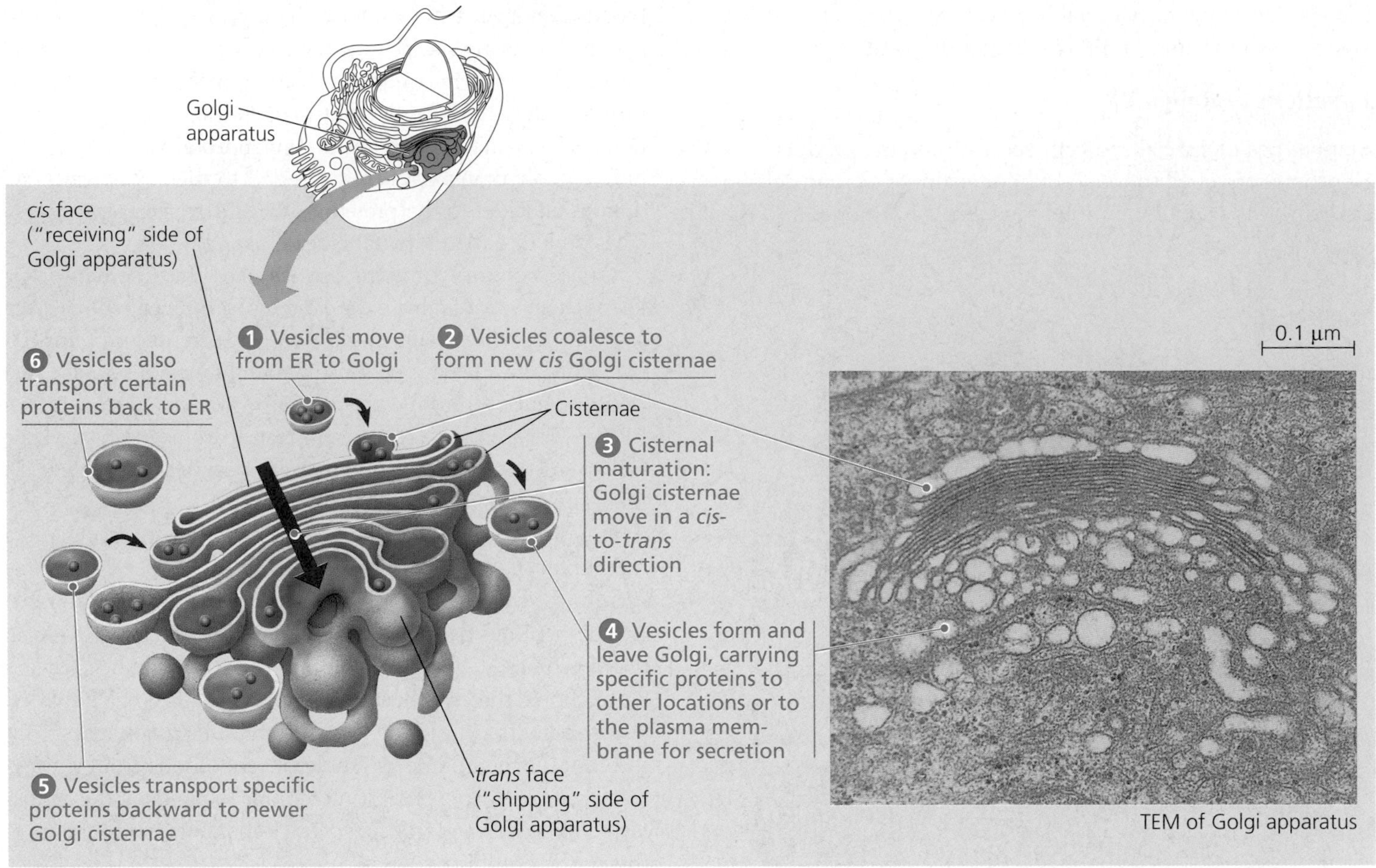

▲ **Figure 6.13 The Golgi apparatus.** The Golgi apparatus consists of stacks of flattened sacs, or cisternae, which, unlike ER cisternae, are not physically connected. (The drawing is a cutaway view.) A Golgi stack receives and dispatches transport vesicles and the products they contain. A Golgi stack has a structural and functional polarity, with a *cis* face that receives vesicles containing ER products and a *trans* face that dispatches vesicles. The cisternal maturation model suggests that the Golgi cisternae themselves appear to "mature," moving from the *cis* to the *trans* face while carrying some proteins along. In addition, some vesicles recycle enzymes that had been carried forward, in moving cisternae, "backward" to a newer region where their functions are needed (TEM).

Golgi membrane. The *trans* face gives rise to vesicles, which pinch off and travel to other sites.

Products of the ER are usually modified during their transit from the *cis* region to the *trans* region of the Golgi. Proteins and phospholipids of membranes may be altered. For example, various Golgi enzymes modify the carbohydrate portions of glycoproteins. Carbohydrates are first added to proteins in the rough ER, often during the process of polypeptide synthesis. The carbohydrate on the resulting glycoprotein is then modified as it passes through the rest of the ER and the Golgi. The Golgi removes some sugar monomers and substitutes others, producing a large variety of carbohydrates.

In addition to its finishing work, the Golgi apparatus manufactures certain macromolecules by itself. Many polysaccharides secreted by cells are Golgi products, including pectins and certain other non-cellulose polysaccharides made by plant cells and incorporated along with cellulose into their cell walls. (Cellulose is made by enzymes located within the plasma membrane, which directly deposit this polysaccharide on the outside surface.) Golgi products that will be secreted depart from the *trans* face of the Golgi inside transport vesicles that eventually fuse with the plasma membrane.

The Golgi manufactures and refines its products in stages, with different cisternae between the *cis* and *trans* regions containing unique teams of enzymes. Until recently, we viewed the Golgi as a static structure, with products in various stages of processing transferred from one cisterna to the next by vesicles. While this may occur, recent research has given rise to a new model of the Golgi as a more dynamic structure. According to the model called the *cisternal maturation model*, the cisternae of the Golgi actually progress forward from the *cis* to the *trans* face of the Golgi, carrying and modifying their protein cargo as they move. Figure 6.13 shows the details of this model.

Before a Golgi stack dispatches its products by budding vesicles from the *trans* face, it sorts these products and targets them for various parts of the cell. Molecular identification tags, such as phosphate groups that have been added to the Golgi products, aid in sorting by acting like ZIP codes on mailing labels. Finally, transport vesicles budded from the Golgi may have external molecules on their membranes that recognize "docking sites" on the surface of specific organelles or on the plasma membrane, thus targeting them appropriately.

Lysosomes: Digestive Compartments

A **lysosome** is a membranous sac of hydrolytic enzymes that an animal cell uses to digest all kinds of macromolecules. Lysosomal enzymes work best in the acidic environment found in lysosomes. If a lysosome breaks open or leaks its contents, the released enzymes are not very active, because the cytosol has a neutral pH. However, excessive leakage from a large number of lysosomes can destroy a cell by autodigestion.

Hydrolytic enzymes and lysosomal membrane are made by rough ER and then transferred to the Golgi apparatus for further processing. At least some lysosomes probably arise by budding from the *trans* face of the Golgi apparatus (see Figure 6.13). Proteins of the inner surface of the lysosomal membrane and the digestive enzymes themselves are thought to be spared from destruction by having three-dimensional conformations that protect vulnerable bonds from enzymatic attack.

Lysosomes carry out intracellular digestion in a variety of circumstances. Amoebas and many other protists eat by engulfing smaller organisms or other food particles, a process called **phagocytosis** (from the Greek *phagein*, to eat, and *kytos*, vessel, referring here to the cell). The *food vacuole* formed in this way then fuses with a lysosome, whose enzymes digest the food **(Figure 6.14a)**. Digestion products, including simple sugars, amino acids, and other monomers, pass into the cytosol and become nutrients for the cell. Some human cells also carry out phagocytosis. Among them are macrophages, a type of white blood cell that helps defend the body by engulfing and destroying bacteria and other invaders (see Figure 6.32).

Lysosomes also use their hydrolytic enzymes to recycle the cell's own organic material, a process called *autophagy*. During autophagy, a damaged organelle or small amount of cytosol

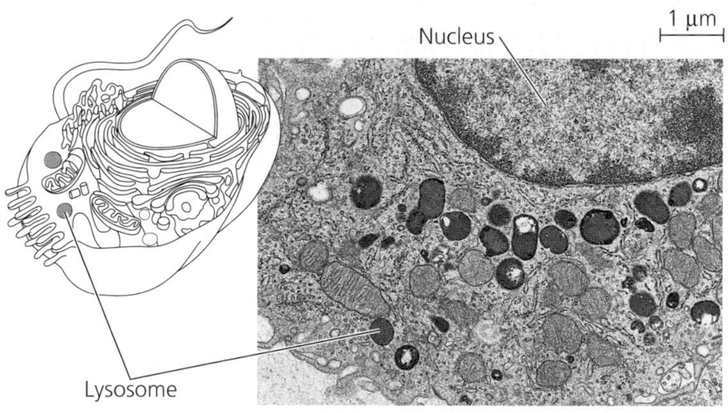

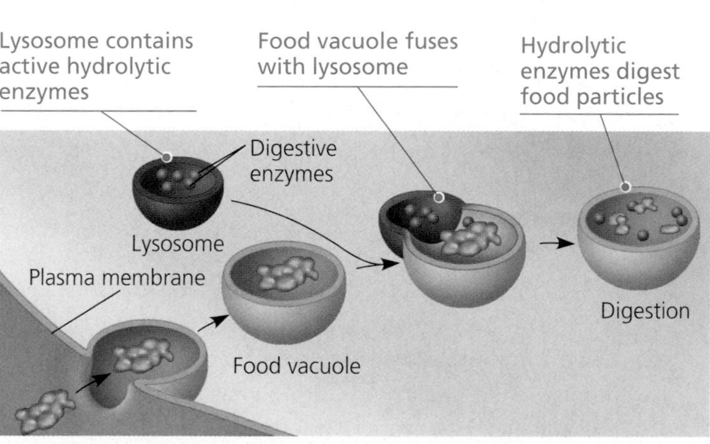

(a) Phagocytosis: lysosome digesting food

(b) Autophagy: lysosome breaking down damaged organelle

▲ **Figure 6.14 Lysosomes.** Lysosomes digest (hydrolyze) materials taken into the cell and recycle intracellular materials. **(a)** *Top* In this macrophage (a type of white blood cell) from a rat, the lysosomes are very dark because of a specific stain that reacts with one of the products of digestion within the lysosome (TEM). Macrophages ingest bacteria and viruses and destroy them using lysosomes. *Bottom* This diagram shows one lysosome fusing with a food vacuole during the process of phagocytosis. **(b)** *Top* In the cytoplasm of this rat liver cell, a lysosome has engulfed two disabled organelles, a mitochondrion and a peroxisome, in the process of autophagy (TEM). *Bottom* This diagram shows a lysosome fusing with a vesicle containing a damaged mitochondrion.

becomes surrounded by a membrane, and a lysosome fuses with this vesicle **(Figure 6.14b)**. The lysosomal enzymes dismantle the enclosed material, and the organic monomers are returned to the cytosol for reuse. With the help of lysosomes, the cell continually renews itself. A human liver cell, for example, recycles half of its macromolecules each week.

The cells of people with inherited lysosomal storage diseases lack a functioning hydrolytic enzyme normally present in lysosomes. The lysosomes become engorged with indigestible substrates, which begin to interfere with other cellular activities. In Tay-Sachs disease, for example, a lipid-digesting enzyme is missing or inactive, and the brain becomes impaired by an accumulation of lipids in the cells. Fortunately, lysosomal storage diseases are rare in the general population.

Vacuoles: Diverse Maintenance Compartments

A plant or fungal cell may have one or several vacuoles. While vacuoles carry out hydrolysis and are thus similar to lysosomes, they carry out other functions as well. **Food vacuoles**, formed by phagocytosis, have already been mentioned (see Figure 6.14a). Many freshwater protists have **contractile vacuoles** that pump excess water out of the cell, thereby maintaining the appropriate concentration of salts and other molecules (see Figure 7.14). Mature plant cells generally contain a large **central vacuole (Figure 6.15)** enclosed by a membrane called the **tonoplast.** The central vacuole develops by the coalescence of smaller vacuoles, themselves derived from the endoplasmic reticulum and Golgi apparatus. The vacuole is in this way an integral part of a plant cell's endomembrane system. Like all cellular membranes, the tonoplast is selective in transporting solutes; as a result, the solution inside the vacuole, called cell sap, differs in composition from the cytosol.

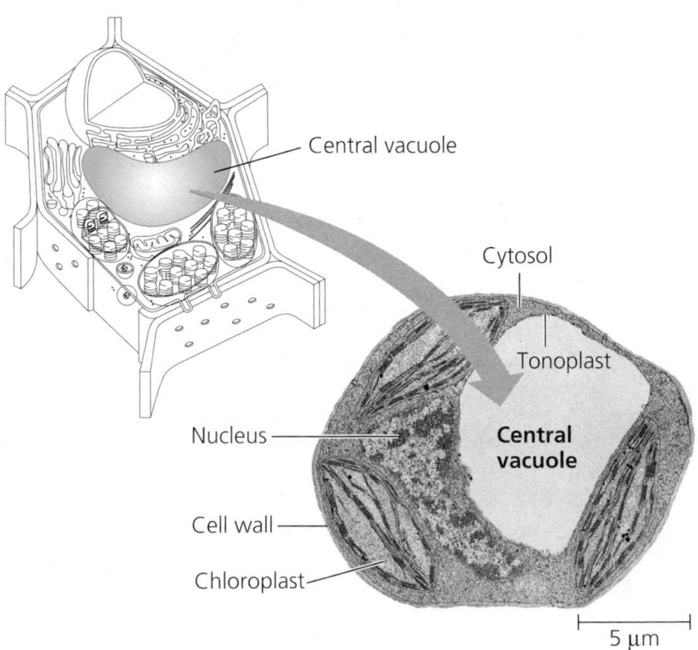

▲ **Figure 6.15 The plant cell vacuole.** The central vacuole is usually the largest compartment in a plant cell; the rest of the cytoplasm is generally confined to a narrow zone between the vacuolar membrane (tonoplast) and the plasma membrane (TEM).

The plant cell's central vacuole is a versatile compartment. It can hold reserves of important organic compounds, such as the proteins stockpiled in the vacuoles of storage cells in seeds. It is also the plant cell's main repository of inorganic ions, such as potassium and chloride. Many plant cells use their vacuoles as disposal sites for metabolic by-products that would endanger the cell if they accumulated in the cytosol. Some vacuoles contain pigments that color the cells, such as the red and blue pigments of petals that help attract pollinating insects to flowers. Vacuoles may also help protect the plant against predators by containing compounds that are poisonous or unpalatable to animals. The vacuole has a major role in the growth of plant cells, which enlarge as their vacuoles absorb water, enabling the cell to become larger with a minimal investment in new cytoplasm. And because of the large vacuole, the cytosol often occupies only a thin layer between the plasma membrane and the tonoplast, so the ratio of membrane surface to cytosolic volume is great, even for a large plant cell.

The Endomembrane System: *A Review*

Figure 6.16 reviews the endomembrane system, showing the flow of membrane lipids and proteins through the various organelles. As the membrane moves from the ER to the Golgi and then elsewhere, its molecular composition and metabolic functions are modified, along with those of its contents. The endomembrane system is a complex and dynamic player in the cell's compartmental organization.

We'll continue our tour of the cell with some membranous organelles that are *not* closely related to the endomembrane system, but play crucial roles in the energy transformations carried out by cells.

Concept Check 6.4

1. Describe the structural and functional distinctions between rough and smooth ER.
2. Imagine a protein that functions in the ER, but requires modification in the Golgi apparatus before it can achieve that function. Describe the protein's path through the cell, starting with the mRNA molecule that specifies the protein.
3. How do transport vesicles serve to integrate the endomembrane system?

For suggested answers, see Appendix A.

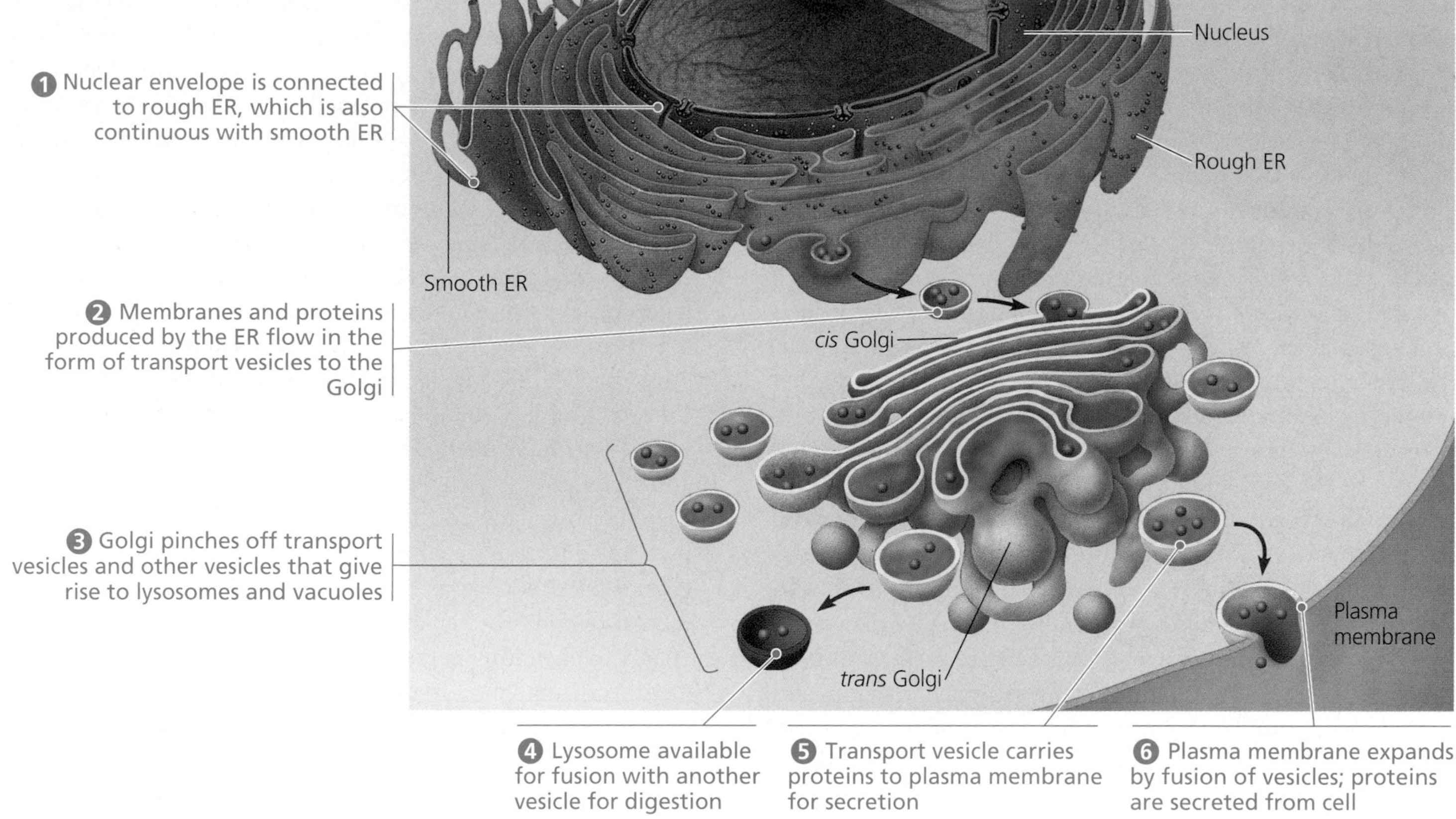
Nucleus
1 Nuclear envelope is connected to rough ER, which is also continuous with smooth ER
Rough ER
Smooth ER
2 Membranes and proteins produced by the ER flow in the form of transport vesicles to the Golgi
cis Golgi
3 Golgi pinches off transport vesicles and other vesicles that give rise to lysosomes and vacuoles
Plasma membrane
trans Golgi
4 Lysosome available for fusion with another vesicle for digestion
5 Transport vesicle carries proteins to plasma membrane for secretion
6 Plasma membrane expands by fusion of vesicles; proteins are secreted from cell

▲ **Figure 6.16 Review: relationships among organelles of the endomembrane system.** The red arrows show some of the migration pathways for membranes and the materials they enclose.

Concept 6.5

Mitochondria and chloroplasts change energy from one form to another

Organisms transform energy they acquire from their surroundings. In eukaryotic cells, mitochondria and chloroplasts are the organelles that convert energy to forms that cells can use for work. **Mitochondria** (singular, *mitochondrion*) are the sites of cellular respiration, the metabolic process that generates ATP by extracting energy from sugars, fats, and other fuels with the help of oxygen. **Chloroplasts,** found only in plants and algae, are the sites of photosynthesis. They convert solar energy to chemical energy by absorbing sunlight and using it to drive the synthesis of organic compounds such as sugars from carbon dioxide and water.

Although mitochondria and chloroplasts are enclosed by membranes, they are not part of the endomembrane system. In contrast to organelles of the endomembrane system, each of these organelles has at least two membranes separating the innermost space from the cytosol. Their membrane proteins are made not by the ER, but by free ribosomes in the cytosol and by ribosomes contained within these organelles themselves. Not only do these organelles have ribosomes, but they also contain a small amount of DNA. It is this DNA that programs the synthesis of the proteins made on the organelle's own ribosomes. (Proteins imported from the cytosol—constituting most of the organelle's proteins—are programmed by nuclear DNA.) Mitochondria and chloroplasts are semiautonomous organelles that grow and reproduce within the cell. In Chapters 9 and 10, we will focus on how mitochondria and chloroplasts function. We will consider the evolution of these organelles in Chapter 26. Here we are concerned mainly with the structure of these energy transformers.

In this section, we will also consider the **peroxisome,** an oxidative organelle that is not part of the endomembrane system. Like mitochondria and chloroplasts, the peroxisome imports its proteins primarily from the cytosol.

Mitochondria: Chemical Energy Conversion

Mitochondria are found in nearly all eukaryotic cells, including those of plants, animals, fungi, and protists. Some cells have a single large mitochondrion, but more often a cell has hundreds or even thousands of mitochondria; the number is correlated with the cell's level of metabolic activity. For example, motile or contractile cells have proportionally more mitochondria per volume than less active cells. Mitochondria are about 1–10 μm long. Time-lapse films of living cells reveal mitochondria moving around, changing their shapes, and dividing in two, unlike the static cylinders seen in electron micrographs of dead cells.

The mitochondrion is enclosed by two membranes, each a phospholipid bilayer with a unique collection of embedded proteins **(Figure 6.17)**. The outer membrane is smooth, but the inner membrane is convoluted, with infoldings called **cristae.** The inner membrane divides the mitochondrion into two internal compartments. The first is the intermembrane space, the narrow region between the inner and outer membranes. The second compartment, the **mitochondrial matrix,** is enclosed by the inner membrane. The matrix contains many different enzymes as well as the mitochondrial DNA and ribosomes. Some of the metabolic steps of cellular respiration are catalyzed by enzymes in the matrix. Other proteins that function in respiration, including the enzyme that makes ATP, are built into the inner membrane. As highly folded surfaces, the cristae give the inner mitochondrial membrane a large surface area for these proteins, thus enhancing the productivity of cellular respiration. This is another example of structure fitting function.

Chloroplasts: Capture of Light Energy

The chloroplast is a specialized member of a family of closely related plant organelles called **plastids.** *Amyloplasts* are colorless plastids that store starch (amylose), particularly in roots and tubers. *Chromoplasts* have pigments that give fruits and flowers their orange and yellow hues. *Chloroplasts* contain the green pigment chlorophyll, along with enzymes and other molecules that function in the photosynthetic production of sugar. These lens-shaped organelles, measuring about 2 μm by 5 μm, are found in leaves and other green organs of plants and in algae **(Figure 6.18)**.

The contents of a chloroplast are partitioned from the cytosol by an envelope consisting of at least two membranes separated by a very narrow intermembrane space. Inside the chloroplast is another membranous system in the form of flattened, interconnected sacs called **thylakoids.** In some regions, thylakoids are stacked like poker chips; each stack is called a **granum** (plural, *grana*). The fluid outside the thylakoids is the **stroma,** which contains the chloroplast DNA and ribosomes as well as many enzymes. The membranes of the chloroplast divide the chloroplast space into three compartments: the intermembrane space, the stroma, and the thylakoid space. In Chapter 10, you will learn how this compartmental organization enables the chloroplast to convert light energy to chemical energy during photosynthesis.

As with mitochondria, the static and rigid appearance of chloroplasts in micrographs or schematic diagrams is not true to their dynamic behavior in the living cell. Their shapes are changeable, and they grow and occasionally pinch in two, reproducing themselves. They are mobile and move around the cell with mitochondria and other organelles along tracks of the cytoskeleton, a structural network we will consider later in this chapter.

Peroxisomes: Oxidation

The peroxisome is a specialized metabolic compartment bounded by a single membrane **(Figure 6.19)**. Peroxisomes

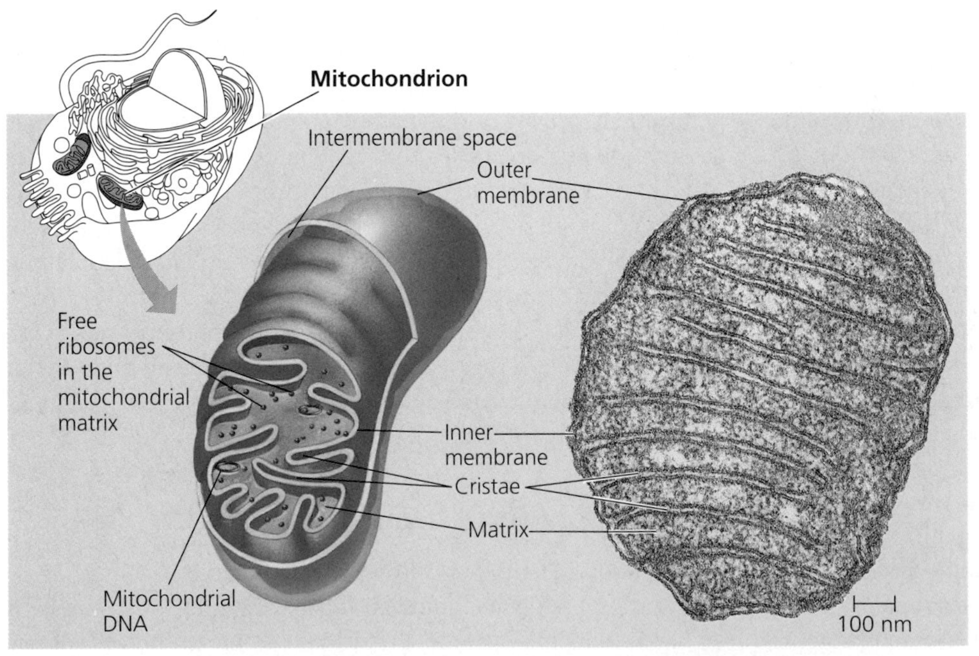

◀ **Figure 6.17 The mitochondrion, site of cellular respiration.** The inner and outer membranes of the mitochondrion are evident in the drawing and micrograph (TEM). The cristae are infoldings of the inner membrane. The cutaway drawing shows the two compartments bounded by the membranes: the intermembrane space and the mitochondrial matrix. Free ribosomes are seen in the matrix, along with one to several copies of the mitochondrial genome (DNA). The DNA molecules are usually circular and are attached to the inner mitochondrial membrane.

▼ **Figure 6.18 The chloroplast, site of photosynthesis.** A chloroplast is enclosed by at least two membranes separated by a narrow intermembrane space that constitutes an outer compartment. The inner membrane encloses a second compartment containing the fluid called stroma. Free ribosomes and copies of the chloroplast genome (DNA) are present in the stroma. The stroma surrounds a third compartment, the thylakoid space, delineated by the thylakoid membrane. Interconnected thylakoid sacs (thylakoids) are stacked to form structures called grana (singular, granum), which are further connected by thin tubules between individual thylakoids (TEM).

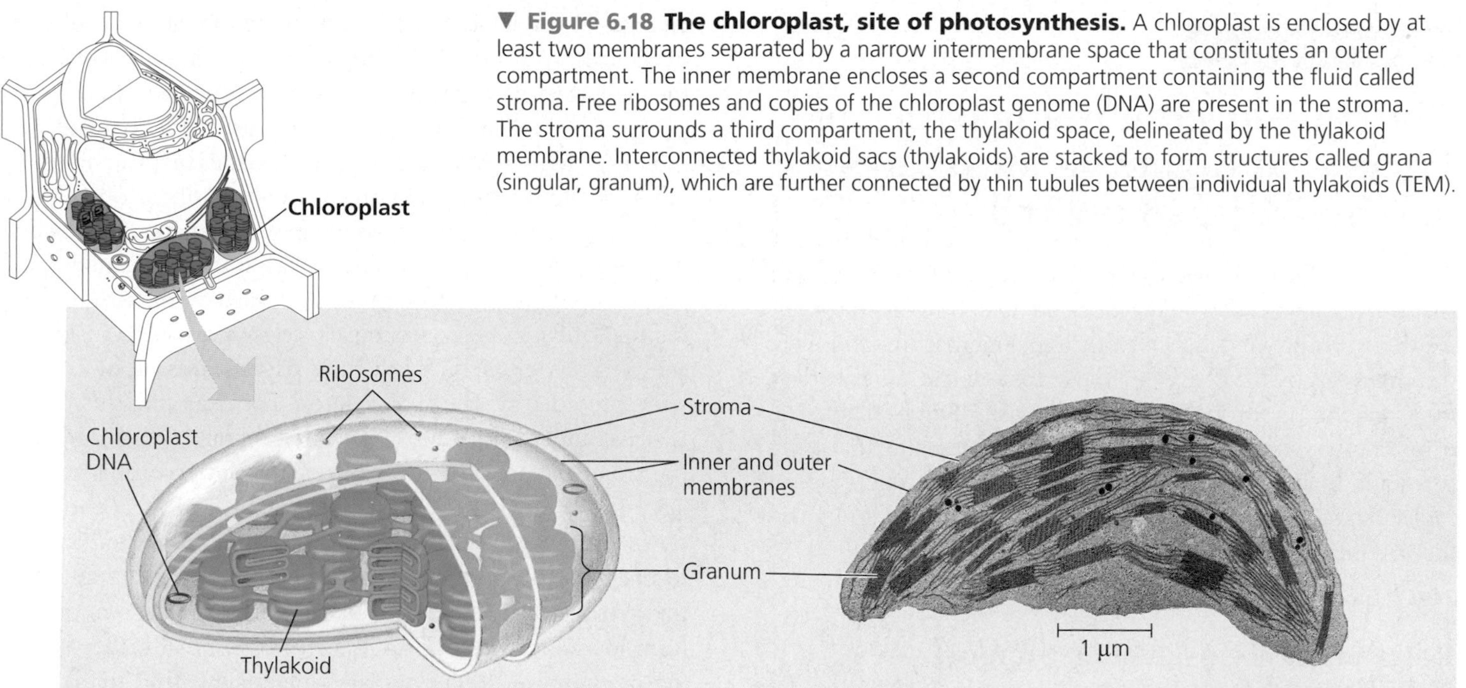

contain enzymes that transfer hydrogen from various substrates to oxygen, producing hydrogen peroxide (H_2O_2) as a by-product, from which the organelle derives its name. These reactions may have many different functions. Some peroxisomes use oxygen to break fatty acids down into smaller molecules that can then be transported to mitochondria, where they are used as fuel for cellular respiration. Peroxisomes in the liver detoxify alcohol and other harmful compounds by transferring hydrogen from the poisons to oxygen. The H_2O_2 formed by peroxisome metabolism is itself toxic, but the organelle contains an enzyme that converts the H_2O_2 to water. Enclosing in the same space both the enzymes that produce hydrogen peroxide and those that dispose of this toxic compound is another example of how the cell's compartmental structure is crucial to its functions.

Specialized peroxisomes called *glyoxysomes* are found in the fat-storing tissues of plant seeds. These organelles contain enzymes that initiate the conversion of fatty acids to sugar, which the emerging seedling can use as a source of energy and carbon until it is able to produce its own sugar by photosynthesis.

Unlike lysosomes, peroxisomes do not bud from the endomembrane system. They grow larger by incorporating proteins made primarily in the cytosol, lipids made in the ER, and lipids synthesized within the peroxisome itself. Peroxisomes may increase in number by splitting in two when they reach a certain size.

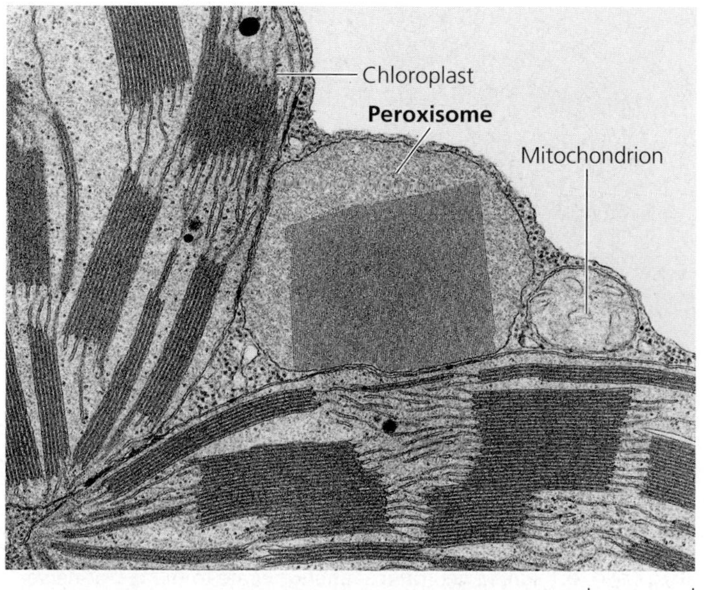

▲ **Figure 6.19 A peroxisome.** Peroxisomes are roughly spherical and often have a granular or crystalline core that is thought to be a dense collection of enzyme molecules. This peroxisome is in a leaf cell. Notice its proximity to two chloroplasts and a mitochondrion. These organelles cooperate with peroxisomes in certain metabolic functions (TEM).

Concept Check 6.5

1. Describe at least two common characteristics of chloroplasts and mitochondria.
2. Explain the characteristics of mitochondria and chloroplasts that place them in a separate category from organelles in the endomembrane system.

For suggested answers, see Appendix A.

Concept 6.6

The cytoskeleton is a network of fibers that organizes structures and activities in the cell

In the early days of electron microscopy, biologists thought that the organelles of a eukaryotic cell floated freely in the cytosol. But improvements in both light microscopy and electron microscopy have revealed the **cytoskeleton,** a network of fibers extending throughout the cytoplasm **(Figure 6.20)**. The cytoskeleton, which plays a major role in organizing the structures and activities of the cell, is composed of three types of molecular structures: microtubules, microfilaments, and intermediate filaments **(Table 6.1)**.

Roles of the Cytoskeleton: Support, Motility, and Regulation

The most obvious function of the cytoskeleton is to give mechanical support to the cell and maintain its shape. This is especially important for animal cells, which lack walls. The remarkable strength and resilience of the cytoskeleton as a whole is based on its architecture. Like a geodesic dome, the cytoskeleton is stabilized by a balance between opposing forces exerted by its elements. And just as the skeleton of an animal helps fix the positions of other body parts, the cytoskeleton provides anchorage for many organelles and even cytosolic enzyme molecules. The cytoskeleton is more dynamic than an animal skeleton, however. It can be quickly dismantled in one part of the cell and reassembled in a new location, changing the shape of the cell.

The cytoskeleton is also involved in several types of cell motility (movement). The term *cell motility* encompasses both changes in cell location and more limited movements of parts of the cell. Cell motility generally requires the interaction of the cytoskeleton with proteins called **motor proteins.** Examples of such cell motility abound. Cytoskeletal elements and motor proteins work together with plasma membrane molecules to allow whole cells to move along fibers outside the cell. Motor proteins bring about the movements of cilia and flagella by gripping microtubules within those organelles and propelling them past each other. A similar mechanism involving microfilaments causes muscle cells to contract. Inside the cell, vesicles often travel to their destinations along "monorails" provided by the cytoskeleton. For example, this is how vesicles containing neurotransmitter molecules migrate to the tips of axons, the long extensions of nerve cells that release these molecules as chemical signals to adjacent nerve cells **(Figure 6.21)**. The vesicles that bud off from the ER travel to the Golgi along tracks built of cytoskeletal elements. It is the cytoskeleton that manipulates the plasma membrane to form food vacuoles during phagocytosis. Finally, the streaming of cytoplasm that circulates materials within many large plant cells is yet another kind of cellular movement brought about by components of the cytoskeleton.

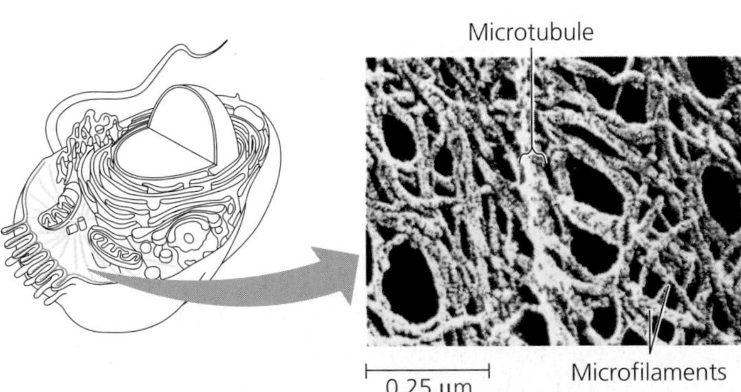

▲ **Figure 6.20 The cytoskeleton.** In this TEM, prepared by a method known as deep-etching, the thicker, hollow microtubules and the thinner, solid microfilaments are visible. A third component of the cytoskeleton, intermediate filaments, is not evident here.

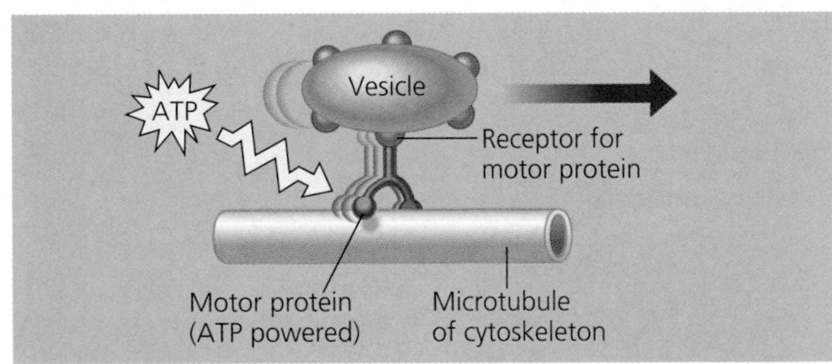

(a) Motor proteins that attach to receptors on organelles can "walk" the organelles along microtubules or, in some cases, microfilaments.

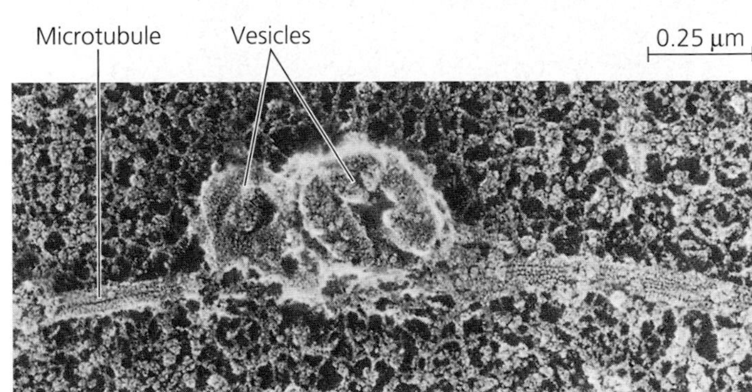

(b) Vesicles containing neurotransmitters migrate to the tips of nerve cell axons via the mechanism in (a). In this SEM of a squid giant axon, two vesicles can be seen moving along a microtubule. (A separate part of the experiment provided the evidence that they were in fact moving.)

▲ **Figure 6.21 Motor proteins and the cytoskeleton.**

Table 6.1 The Structure and Function of the Cytoskeleton

Property	Microtubules (Tubulin Polymers)	Microfilaments (Actin Filaments)	Intermediate Filaments
Structure	Hollow tubes; wall consists of 13 columns of tubulin molecules	Two intertwined strands of actin, each a polymer of actin subunits	Fibrous proteins supercoiled into thicker cables
Diameter	25 nm with 15-nm lumen	7 nm	8–12 nm
Protein subunits	Tubulin, consisting of α-tubulin and β-tubulin	Actin	One of several different proteins of the keratin family, depending on cell type
Main functions	Maintenance of cell shape (compression-resisting "girders") Cell motility (as in cilia or flagella) Chromosome movements in cell division Organelle movements	Maintenance of cell shape (tension-bearing elements) Changes in cell shape Muscle contraction Cytoplasmic streaming Cell motility (as in pseudopodia) Cell division (cleavage furrow formation)	Maintenance of cell shape (tension-bearing elements) Anchorage of nucleus and certain other organelles Formation of nuclear lamina
Micrographs of fibroblasts, a favorite cell type for cell biology studies. Each has been experimentally treated to fluorescently tag the structure of interest.			

The most recent addition to the list of possible cytoskeletal functions is the regulation of biochemical activities in the cell. Mounting evidence suggests that the cytoskeleton can transmit mechanical forces exerted by extracellular molecules via surface proteins of the cell to its interior—and even into the nucleus. In one experiment, investigators used a micromanipulation device to pull on certain plasma membrane proteins attached to the cytoskeleton. A video microscope captured almost instantaneous rearrangements of nucleoli and other structures in the nucleus. In this way, the transmission of naturally occurring mechanical signals by the cytoskeleton may help regulate cell function.

Components of the Cytoskeleton

Now let's look more closely at the three main types of fibers that make up the cytoskeleton (see Table 6.1). **Microtubules** are the thickest of the three types; **microfilaments** (also called actin filaments) are the thinnest; and **intermediate filaments** are fibers with diameters in a middle range.

Microtubules

Microtubules are found in the cytoplasm of all eukaryotic cells. They are hollow rods measuring about 25 nm in diameter and from 200 nm to 25 μm in length. The wall of the hollow tube is constructed from a globular protein called tubulin. Each tubulin molecule is a dimer consisting of two slightly different polypeptide subunits, α-tubulin and β-tubulin. A microtubule grows in length by adding tubulin dimers to its ends. Microtubules can be disassembled and their tubulin used to build microtubules elsewhere in the cell.

Microtubules shape and support the cell and also serve as tracks along which organelles equipped with motor proteins can move (see Figure 6.21). For example, microtubules guide secretory vesicles from the Golgi apparatus to the plasma membrane. Microtubules are also responsible for the separation of chromosomes during cell division (see Chapter 12).

Centrosomes and Centrioles. In many cells, microtubules grow out from a **centrosome,** a region often located near the nucleus that is considered to be a "microtubule-organizing center." These microtubules function as compression-resisting girders of the cytoskeleton. Within the centrosome of an animal cell are a pair of **centrioles,** each composed of nine sets of triplet microtubules arranged in a ring **(Figure 6.22)**. Before a cell divides, the centrioles replicate. Although centrioles may help organize microtubule assembly, they are not essential for this function in all eukaryotes; centrosomes of most plants lack centrioles, but have well-organized microtubules.

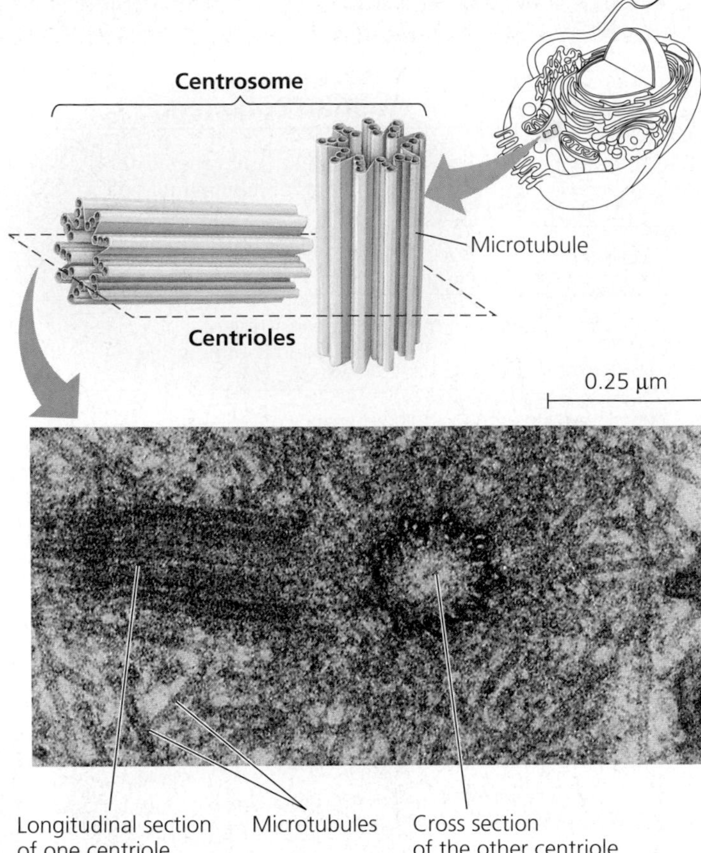

▲ **Figure 6.22 Centrosome containing a pair of centrioles.** An animal cell has a pair of centrioles within its centrosome, the region near the nucleus where the cell's microtubules are initiated. The centrioles, each about 250 nm (0.25 μm) in diameter, are found at right angles to each other, and each is made up of nine sets of three microtubules. The blue portions of the drawing represent nontubulin proteins that connect the microtubule triplets (TEM).

Cilia and Flagella. In eukaryotes, a specialized arrangement of microtubules is responsible for the beating of **flagella** (singular, *flagellum*) and **cilia** (singular, *cilium*), locomotor appendages that protrude from some cells. Many unicellular eukaryotic organisms are propelled through water by cilia or flagella, and the sperm of animals, algae, and some plants have flagella. When cilia or flagella extend from cells that are held in place as part of a tissue layer, they can move fluid over the surface of the tissue. For example, the ciliated lining of the windpipe sweeps mucus containing trapped debris out of the lungs (see Figure 6.4). In a woman's reproductive tract, the cilia lining the oviducts (fallopian tubes) help move an egg toward the uterus.

Cilia usually occur in large numbers on the cell surface. They are about 0.25 μm in diameter and about 2–20 μm in length. Flagella are the same diameter but longer than cilia, measuring 10–200 μm in length. Also, flagella are usually limited to just one or a few per cell.

Flagella and cilia differ in their beating patterns **(Figure 6.23)**. A flagellum has an undulating motion that generates force in the same direction as the flagellum's axis. In contrast, cilia work more like oars, with alternating power and recovery strokes generating force in a direction perpendicular to the cilium's axis.

Though different in length, number per cell, and beating pattern, cilia and flagella share a common ultrastructure. A cilium or flagellum has a core of microtubules sheathed in an extension of the plasma membrane **(Figure 6.24)**. Nine doublets of microtubules, the members of each pair sharing part of their walls, are arranged in a ring. In the center of the ring are two single microtubules. This arrangement, referred to as the "9 + 2" pattern, is found in nearly all eukaryotic flagella and cilia. (The flagella of motile prokaryotes, discussed in Chapter 27, do not contain microtubules.) Flexible "wagon wheels" of cross-linking proteins, evenly spaced along the length of the cilium or flagellum, connect the outer doublets to each other (the wheel rim) and to the two central microtubules (the wheel spokes). Each outer doublet also has pairs of side-arms spaced along its length and reaching toward the neighboring doublet; these are motor proteins. The microtubule assembly of a cilium or flagellum is anchored in the cell by a **basal body,** which is structurally identical to a centriole. In fact, in many animals (including humans), the basal body of the fertilizing sperm's flagellum enters the egg and becomes a centriole.

Each motor protein extending from one microtubule doublet to the next is a large protein called **dynein,** which is composed

(a) Motion of flagella. A flagellum usually undulates, its snakelike motion driving a cell in the same direction as the axis of the flagellum. Propulsion of a human sperm cell is an example of flagellate locomotion (LM).

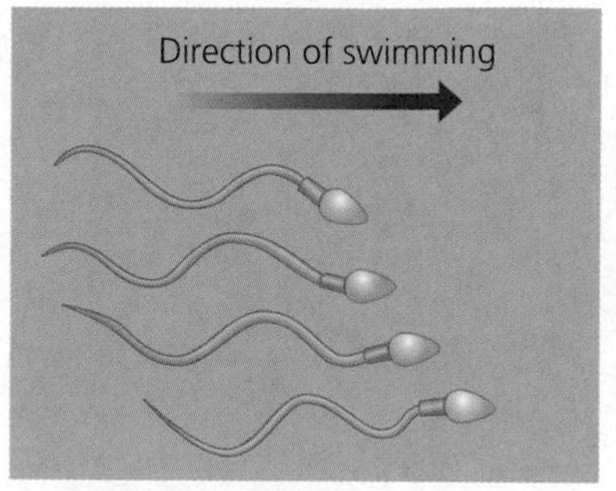

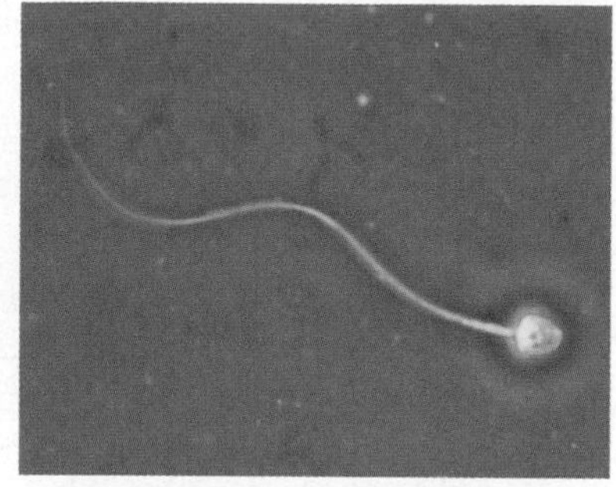

5 µm

(b) Motion of cilia. Cilia have a back-and-forth motion that moves the cell in a direction perpendicular to the axis of the cilium. A dense nap of cilia, beating at a rate of about 40 to 60 strokes a second, covers this *Colpidium*, a freshwater protozoan (colorized SEM).

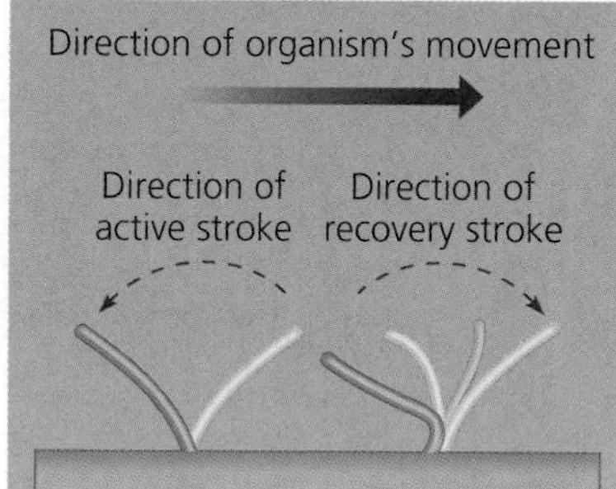

15 µm

◀ **Figure 6.23 A comparison of the beating of flagella and cilia.**

0.1 µm

Outer microtubule doublet

Dynein arms

Central microtubule

Cross-linking proteins inside outer doublets

Radial spoke

Plasma membrane

Microtubules

Plasma membrane

Basal body

0.5 µm

(a) A longitudinal section of a cilium shows microtubules running the length of the structure (TEM).

(b) A cross section through the cilium shows the "9 + 2" arrangement of microtubules (TEM). The outer microtubule doublets and the two central microtubules are held together by cross-linking proteins (purple in art), including the radial spokes. The doublets also have attached motor proteins, the dynein arms (red in art).

0.1 µm

Triplet

(c) Basal body: The nine outer doublets of a cilium or flagellum extend into the basal body, where each doublet joins another microtubule to form a ring of nine triplets. Each triplet is connected to the next by nontubulin proteins (blue). The two central microtubules terminate above the basal body (TEM).

Cross section of basal body

▲ **Figure 6.24 Ultrastructure of a eukaryotic flagellum or cilium.**

of several polypeptides. These dynein arms are responsible for the bending movements of cilia and flagella. A dynein arm performs a complex cycle of movements caused by changes in the conformation of the protein, with ATP providing the energy for these changes **(Figure 6.25)**.

The mechanics of dynein "walking" are reminiscent of a cat climbing a tree by attaching its claws, moving its legs, releasing its front claws, and grabbing again farther up the tree. Similarly, the dynein arms of one doublet attach to an adjacent doublet and pull so that the doublets slide past each other in opposite directions. The arms then release from the other doublet and reattach a little farther along its length. Without any restraints on the movement of the microtubule doublets, one doublet would continue to "walk" along and slide past the surface of the other, elongating the cilium or flagellum rather than bending it (see Figure 6.25a). For lateral movement of a cilium or flagellum, the dynein "walking" must have something to pull against, as when the muscles in your leg pull against your bones to move your knee. In cilia and flagella, the microtubule doublets seem to be held in place by the cross-linking proteins just inside the outer doublets and by the radial spokes and other structural elements. Thus, neighboring doublets cannot slide past each other very far. Instead, the forces exerted by the dynein arms cause the doublets to curve, bending the cilium or flagellum (see Figure 6.25b and c).

Microfilaments (Actin Filaments)

Microfilaments are solid rods about 7 nm in diameter. They are also called actin filaments, because they are built from molecules of **actin,** a globular protein. A microfilament is a twisted double chain of actin subunits (see Table 6.1). Besides occurring as linear filaments, microfilaments can form structural networks, due to the presence of proteins that bind along the side of an actin filament and allow a new filament to extend as a branch. Microfilaments seem to be present in all eukaryotic cells.

In contrast to the compression-resisting role of microtubules, the structural role of microfilaments in the cytoskeleton is to bear tension (pulling forces). The ability of microfilaments to form a three-dimensional network just inside the plasma membrane helps support the cell's shape. This network gives the cortex (outer cytoplasmic layer) of a cell the semisolid consistency of a gel, in contrast with the more fluid (sol) state of the interior cytoplasm. In animal cells specialized for transporting materials across the plasma membrane, such as intestinal cells, bundles of microfilaments make up the core of microvilli, the previously mentioned delicate projections that increase the cell surface area **(Figure 6.26)**.

Microfilaments are well known for their role in cell motility, particularly as part of the contractile apparatus of muscle

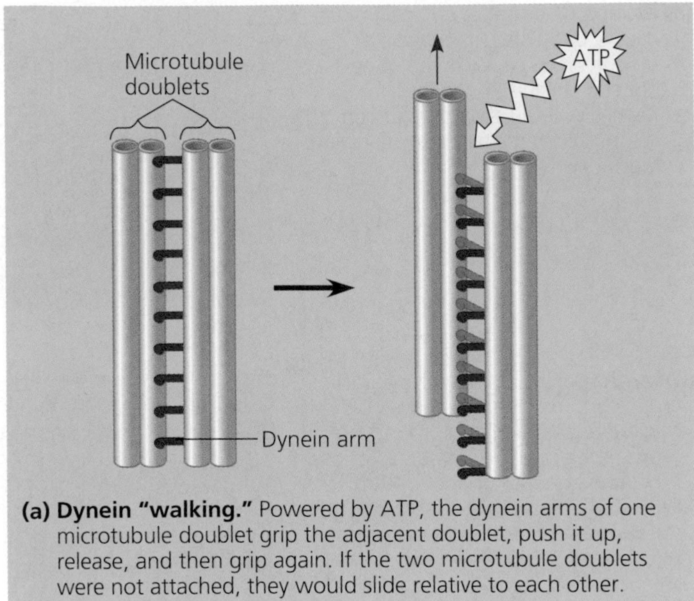

(a) Dynein "walking." Powered by ATP, the dynein arms of one microtubule doublet grip the adjacent doublet, push it up, release, and then grip again. If the two microtubule doublets were not attached, they would slide relative to each other.

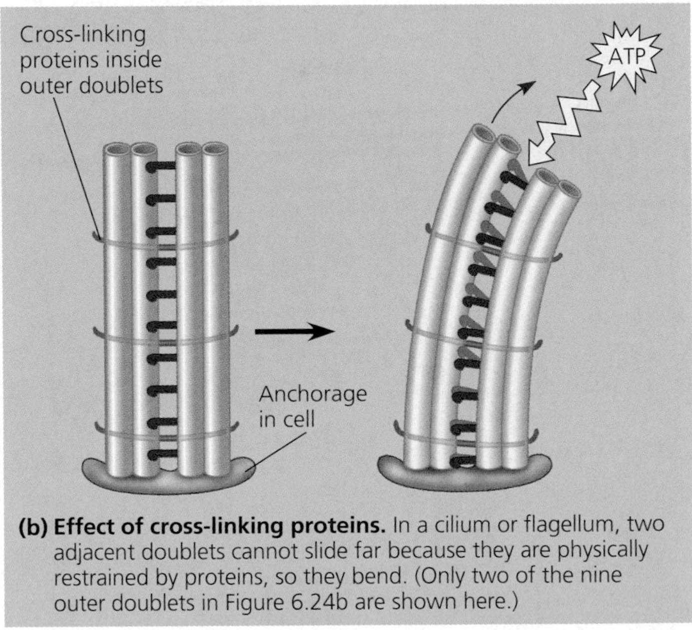

(b) Effect of cross-linking proteins. In a cilium or flagellum, two adjacent doublets cannot slide far because they are physically restrained by proteins, so they bend. (Only two of the nine outer doublets in Figure 6.24b are shown here.)

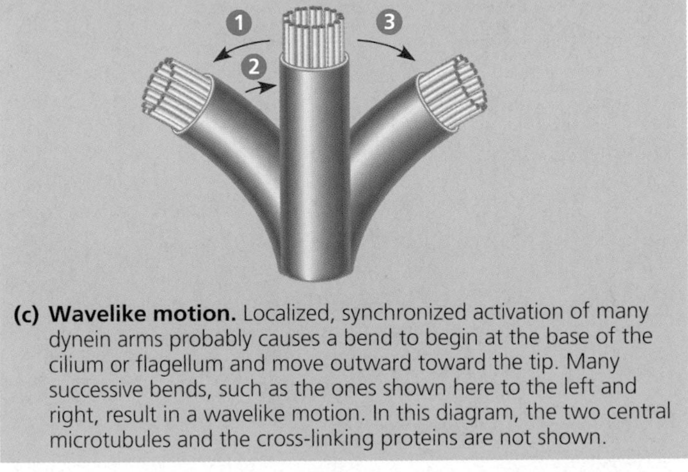

(c) Wavelike motion. Localized, synchronized activation of many dynein arms probably causes a bend to begin at the base of the cilium or flagellum and move outward toward the tip. Many successive bends, such as the ones shown here to the left and right, result in a wavelike motion. In this diagram, the two central microtubules and the cross-linking proteins are not shown.

▲ **Figure 6.25 How dynein "walking" moves flagella and cilia.**

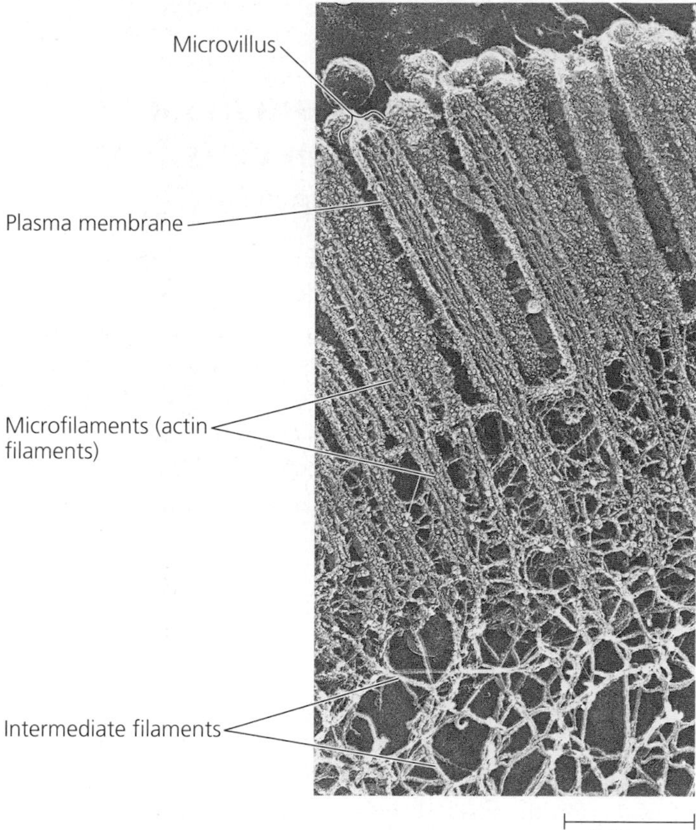

▲ **Figure 6.26 A structural role of microfilaments.** The surface area of this nutrient-absorbing intestinal cell is increased by its many microvilli (singular, microvillus), cellular extensions reinforced by bundles of microfilaments. These actin filaments are anchored to a network of intermediate filaments (TEM).

cells. Thousands of actin filaments are arranged parallel to one another along the length of a muscle cell, interdigitated with thicker filaments made of a protein called **myosin** **(Figure 6.27a)**. Myosin acts as a motor protein by means of projections (arms) that "walk" along the actin filaments. Contraction of the muscle cell results from the actin and myosin filaments sliding past one another in this way, shortening the cell. In other kinds of cells, actin filaments are associated with myosin in miniature and less elaborate versions of the arrangement in muscle cells. These actin-myosin aggregates are responsible for localized contractions of cells. For example, a contracting belt of microfilaments forms a cleavage furrow that pinches a dividing animal cell into two daughter cells.

Localized contraction brought about by actin and myosin also plays a role in amoeboid movement **(Figure 6.27b)**, in which a cell, such as an amoeba, for example, crawls along a surface by extending and flowing into cellular extensions called **pseudopodia** (from the Greek *pseudes*, false, and *pod*, foot). Pseudopodia extend and contract through the reversible assembly of actin subunits into microfilaments and of microfilaments into networks that convert cytoplasm from sol to gel. According to a widely accepted model, filaments near the cell's trailing end interact with myosin, causing contraction. Like squeezing on a toothpaste tube, this contraction forces the interior fluid into the pseudopodium, where the actin network has been weakened. The pseudopodium extends until

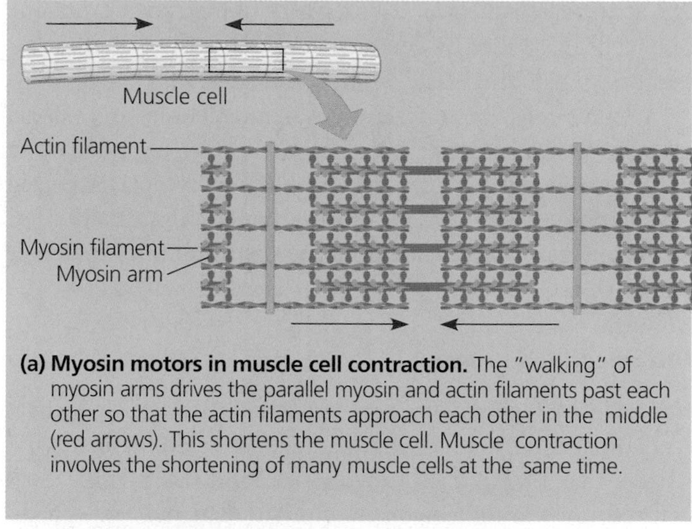

(a) Myosin motors in muscle cell contraction. The "walking" of myosin arms drives the parallel myosin and actin filaments past each other so that the actin filaments approach each other in the middle (red arrows). This shortens the muscle cell. Muscle contraction involves the shortening of many muscle cells at the same time.

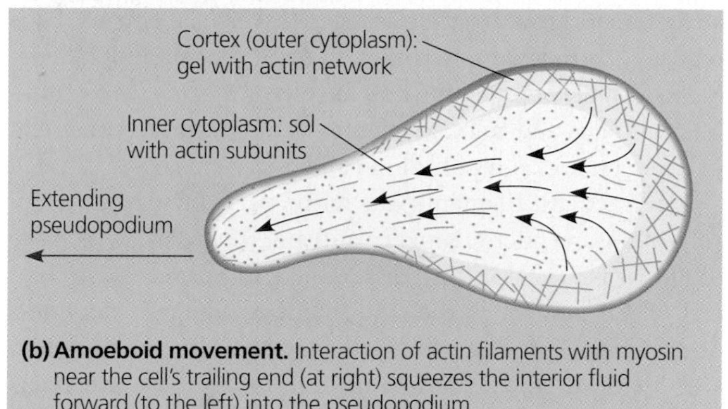

(b) Amoeboid movement. Interaction of actin filaments with myosin near the cell's trailing end (at right) squeezes the interior fluid forward (to the left) into the pseudopodium.

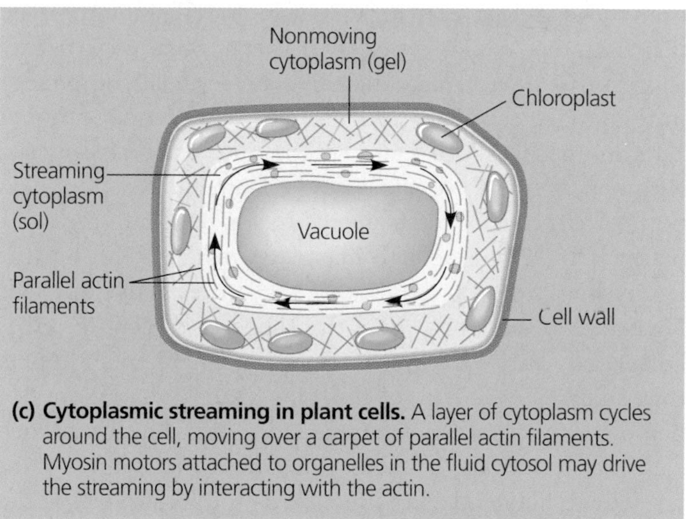

(c) Cytoplasmic streaming in plant cells. A layer of cytoplasm cycles around the cell, moving over a carpet of parallel actin filaments. Myosin motors attached to organelles in the fluid cytosol may drive the streaming by interacting with the actin.

▲ **Figure 6.27 Microfilaments and motility.** In the three examples shown in this figure, cell nuclei and most other organelles have been omitted for clarity.

the actin reassembles into a network. Amoebas are not the only cells that move by crawling; so do many cells in the animal body, including some white blood cells.

In plant cells, both actin-myosin interactions and sol-gel transformations brought about by actin may be involved in **cytoplasmic streaming,** a circular flow of cytoplasm within cells **(Figure 6.27c)**. This movement, which is especially common in large plant cells, speeds the distribution of materials within the cell.

Intermediate Filaments

Intermediate filaments are named for their diameter, which, at 8–12 nm, is larger than the diameter of microfilaments but smaller than that of microtubules (see Table 6.1, p. 113). Specialized for bearing tension (like microfilaments), intermediate filaments are a diverse class of cytoskeletal elements. Each type is constructed from a different molecular subunit belonging to a family of proteins whose members include the keratins. Microtubules and microfilaments, in contrast, are consistent in diameter and composition in all eukaryotic cells.

Intermediate filaments are more permanent fixtures of cells than are microfilaments and microtubules, which are often disassembled and reassembled in various parts of a cell. Even after cells die, intermediate filament networks often persist; for example, the outer layer of our skin consists of dead skin cells full of keratin proteins. Chemical treatments that remove microfilaments and microtubules from the cytoplasm of living cells leave a web of intermediate filaments that retains its original shape. Such experiments suggest that intermediate filaments are especially important in reinforcing the shape of a cell and fixing the position of certain organelles. For example, the nucleus commonly sits within a cage made of intermediate filaments, fixed in location by branches of the filaments that extend into the cytoplasm. Other intermediate filaments make up the nuclear lamina that lines the interior of the nuclear envelope (see Figure 6.10). In cases where the shape of the entire cell is correlated with function, intermediate filaments support that shape. For instance, the long extensions (axons) of nerve cells that transmit impulses are strengthened by one class of intermediate filament. Thus, the various kinds of intermediate filaments may function as the framework of the entire cytoskeleton.

Concept Check 6.6

1. Describe how the properties of microtubules, microfilaments, and intermediate filaments allow them to determine cell shape.
2. How do cilia and flagella bend?

For suggested answers, see Appendix A.

Concept 6.7

Extracellular components and connections between cells help coordinate cellular activities

Having crisscrossed the interior of the cell to explore various organelles, we complete our tour of the cell by returning to the surface of this microscopic world, where there are additional structures with important functions. The plasma membrane is usually regarded as the boundary of the living cell, but most cells synthesize and secrete materials of one kind or another that are external to the plasma membrane. Although they are outside the cell, the study of these extracellular structures is central to cell biology because they are involved in so many cellular functions.

Cell Walls of Plants

The **cell wall** is an extracellular structure of plant cells that distinguishes them from animal cells. The wall protects the plant cell, maintains its shape, and prevents excessive uptake of water. On the level of the whole plant, the strong walls of specialized cells hold the plant up against the force of gravity. Prokaryotes, fungi, and some protists also have cell walls, but we will postpone discussion of them until Unit Five.

Plant cell walls are much thicker than the plasma membrane, ranging from 0.1 μm to several micrometers. The exact chemical composition of the wall varies from species to species and even from one cell type to another in the same plant, but the basic design of the wall is consistent. Microfibrils made of the polysaccharide cellulose (see Figure 5.8) are embedded in a matrix of other polysaccharides and protein. This combination of materials, strong fibers in a "ground substance" (matrix), is the same basic architectural design found in steel-reinforced concrete and in fiberglass.

A young plant cell first secretes a relatively thin and flexible wall called the **primary cell wall (Figure 6.28)**. Between primary walls of adjacent cells is the **middle lamella,** a thin layer rich in sticky polysaccharides called pectins. The middle lamella glues adjacent cells together (pectin is used as a thickening agent in jams and jellies). When the cell matures and stops growing, it strengthens its wall. Some plant cells do this simply by secreting hardening substances into the primary wall. Other cells add a **secondary cell wall** between the plasma membrane and the primary wall. The secondary wall, often deposited in several laminated layers, has a strong and durable matrix that affords the cell protection and support. Wood, for example, consists mainly of secondary walls. Plant cell walls are commonly perforated by channels between adjacent cells called plasmodesmata (see Figure 6.28), which will be discussed shortly.

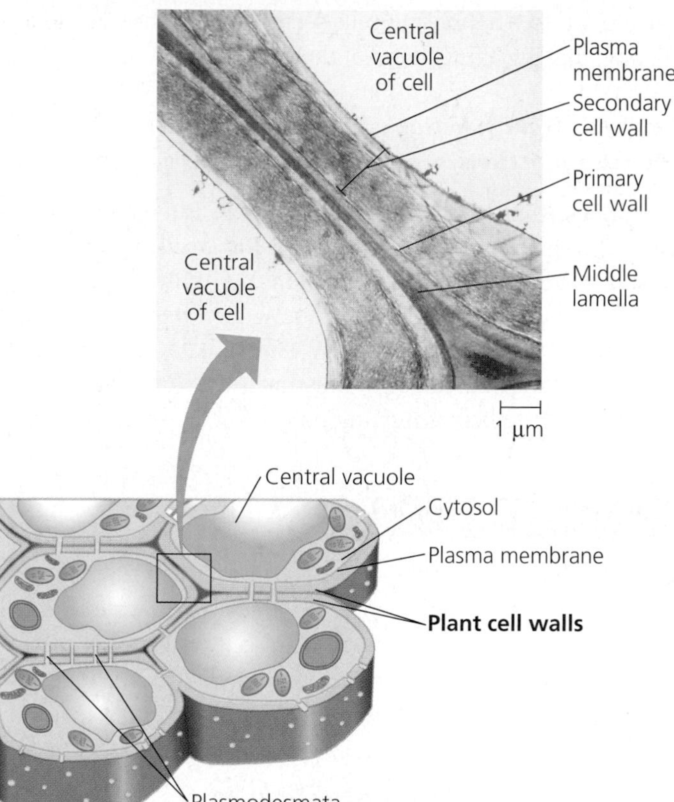
Central
vacuole
of cell
Plasma
membrane
Secondary
cell wall
Primary
cell wall
Central
vacuole
of cell
Middle
lamella
1 µm
Central vacuole
Cytosol
Plasma membrane
Plant cell walls
Plasmodesmata

▲ **Figure 6.28 Plant cell walls.** The orientation drawing shows several cells, each with a large vacuole, a nucleus, and several chloroplasts and mitochondria. The transmission electron micrograph (TEM) shows the cell walls where two cells come together. The multilayered partition between plant cells consists of adjoining walls individually secreted by the cells.

The Extracellular Matrix (ECM) of Animal Cells

Although animal cells lack walls akin to those of plant cells, they do have an elaborate **extracellular matrix (ECM)** **(Figure 6.29)**. The main ingredients of the ECM are glycoproteins secreted by the cells. (Recall that glycoproteins are proteins with covalently bonded carbohydrate, usually short chains of sugars.) The most abundant glycoprotein in the ECM of most animal cells is **collagen,** which forms strong fibers outside the cells. In fact, collagen accounts for about half of the total protein in the human body. The collagen fibers are embedded in a network woven from **proteoglycans,** which are glycoproteins of another class. A proteoglycan molecule consists of a small core protein with many carbohydrate chains covalently attached, so that it may be up to 95% carbohydrate. Large proteoglycan complexes can form when hundreds of proteoglycans become noncovalently attached to a single long polysaccharide molecule, as shown in Figure 6.29. Some cells are attached to the ECM by still other ECM glycoproteins, including **fibronectin.** Fibronectin and other ECM proteins bind to cell surface receptor proteins called **integrins** that are built into the plasma membrane. Integrins span the membrane and bind on their cytoplasmic side to associated proteins attached to microfilaments of the cytoskeleton. The name integrin is based on the word *integrate:* Integrins are in a position to transmit changes between the ECM and the cytoskeleton and thus to integrate changes occurring outside and inside the cell.

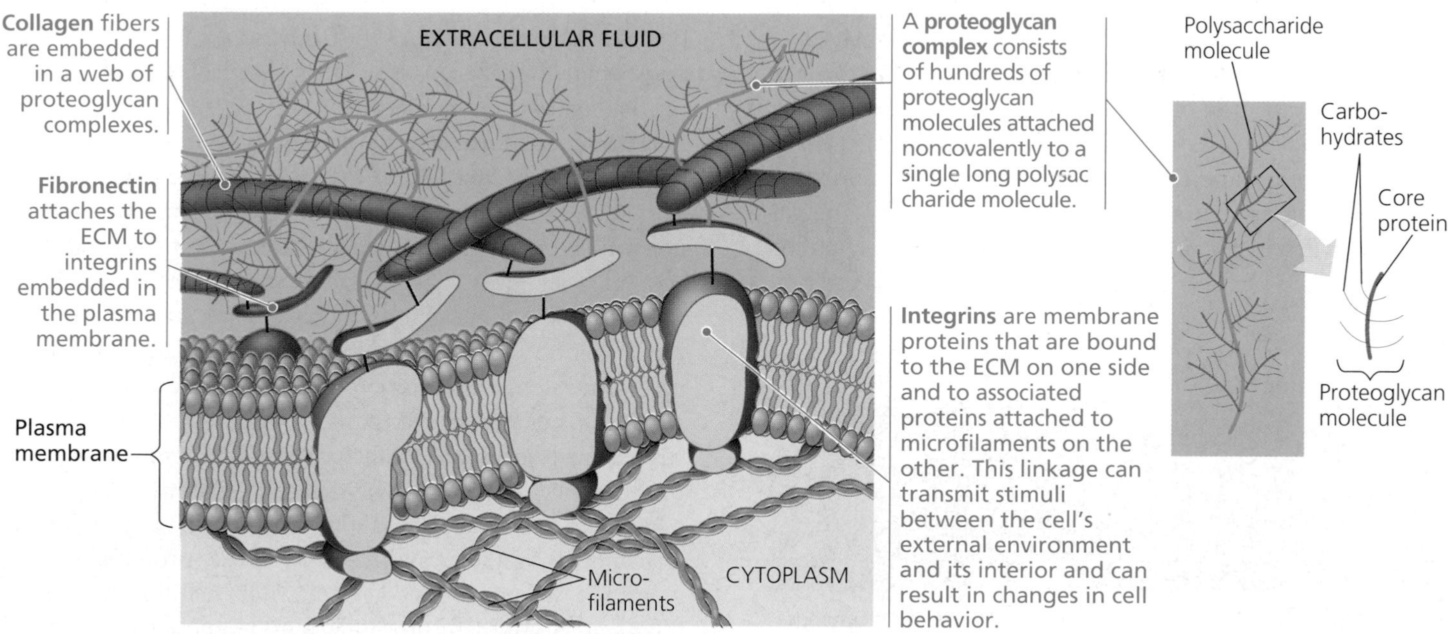
Collagen fibers are embedded in a web of proteoglycan complexes.
EXTRACELLULAR FLUID
A proteoglycan complex consists of hundreds of proteoglycan molecules attached noncovalently to a single long polysaccharide molecule.
Polysaccharide molecule
Carbo-hydrates
Core protein
Proteoglycan molecule
Fibronectin attaches the ECM to integrins embedded in the plasma membrane.
Plasma membrane
Integrins are membrane proteins that are bound to the ECM on one side and to associated proteins attached to microfilaments on the other. This linkage can transmit stimuli between the cell's external environment and its interior and can result in changes in cell behavior.
Micro-filaments
CYTOPLASM

▲ **Figure 6.29 Extracellular matrix (ECM) of an animal cell.** The molecular composition and structure of the ECM varies from one cell type to another. In this example, three different types of glycoproteins are present: proteoglycans, collagen, and fibronectin.

Current research on fibronectin, other ECM molecules, and integrins is revealing the influential role of the extracellular matrix in the lives of cells. By communicating with a cell through integrins, the ECM can regulate a cell's behavior. For example, some cells in a developing embryo migrate along specific pathways by matching the orientation of their microfilaments to the "grain" of fibers in the extracellular matrix. Researchers are also learning that the extracellular matrix around a cell can influence the activity of genes in the nucleus. Information about the ECM probably reaches the nucleus by a combination of mechanical and chemical signaling pathways. Mechanical signaling involves fibronectin, integrins, and microfilaments of the cytoskeleton. Changes in the cytoskeleton may in turn trigger chemical signaling pathways inside the cell, leading to changes in the set of proteins being made by the cell and therefore changes in the cell's function. In this way, the extracellular matrix of a particular tissue may help coordinate the behavior of all the cells within that tissue. Direct connections between cells also function in this coordination, as we discuss next.

Intercellular Junctions

The many cells of an animal or plant are organized into tissues, organs, and organ systems. Neighboring cells often adhere, interact, and communicate through special patches of direct physical contact.

Plants: Plasmodesmata

It might seem that the nonliving cell walls of plants would isolate cells from one another. But in fact, as shown in **Figure 6.30**, plant cell walls are perforated with channels called **plasmodesmata** (singular, *plasmodesma;* from the Greek *desmos,* to bind). Cytosol passes through the plasmodesmata and connects the chemical environments of adjacent cells. These connections unify most of the plant into one living continuum. The plasma membranes of adjacent cells line the channel of each plasmodesma and thus are continuous. Water and small solutes can pass freely from cell to cell, and recent experiments have shown that in certain circumstances, specific proteins and RNA molecules can also do this. The macromolecules to be transported to neighboring cells seem to reach the plasmodesmata by moving along fibers of the cytoskeleton.

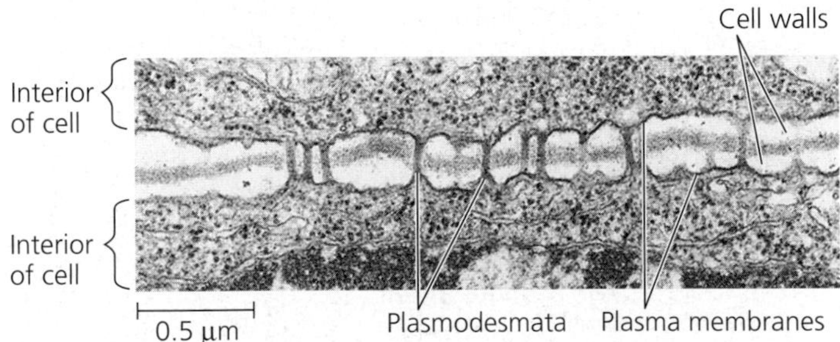

▲ **Figure 6.30 Plasmodesmata between plant cells.** The cytoplasm of one plant cell is continuous with the cytoplasm of its neighbors via plasmodesmata, channels through the cell walls (TEM).

Animals: Tight Junctions, Desmosomes, and Gap Junctions

In animals, there are three main types of intercellular junctions: *tight junctions, desmosomes,* and *gap junctions* (which are most like the plasmodesmata of plants). All three types are especially common in epithelial tissue, which lines the external and internal surfaces of the body. **Figure 6.31** uses epithelial cells of the intestinal lining to illustrate these junctions; please study this figure before moving on.

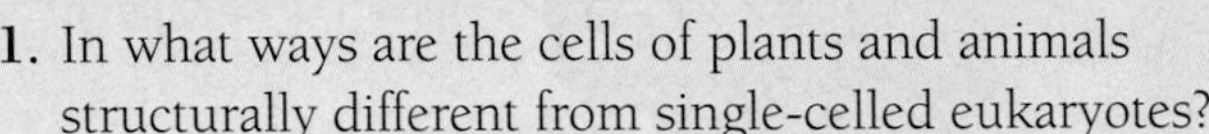

1. In what ways are the cells of plants and animals structurally different from single-celled eukaryotes?
2. What characteristics of the plant cell wall and animal cell extracellular matrix allow the cells to exchange matter and information with their external environment?

For suggested answers, see Appendix A.

The Cell: A Living Unit Greater Than the Sum of Its Parts

From our panoramic view of the cell's overall compartmental organization to our close-up inspection of each organelle's architecture, this tour of the cell has provided many opportunities to correlate structure with function. (This would be a good time to review cell structure by returning to Figure 6.9, pp. 100 and 101.) But even as we dissect the cell, remember that none of its organelles works alone. As an example of cellular integration, consider the microscopic scene in **Figure 6.32**. The large cell is a macrophage (see Figure 6.14a). It helps defend the body against infections by ingesting bacteria (the smaller cells) into phagocytic vesicles. The macrophage crawls along a surface and reaches out to the bacteria with thin pseudopodia (called filopodia). Actin filaments interact with other elements of the cytoskeleton in these movements. After the macrophage engulfs the bacteria, they are destroyed by lysosomes. The elaborate endomembrane system produces the lysosomes. The digestive enzymes of the lysosomes and the proteins of the cytoskeleton are all made on ribosomes. And the synthesis of these proteins is programmed by genetic messages dispatched from the DNA in the nucleus. All these processes require energy, which mitochondria supply in the form of ATP. Cellular functions arise from cellular order: The cell is a living unit greater than the sum of its parts.

Figure 6.31

Exploring Intercellular Junctions in Animal Tissues

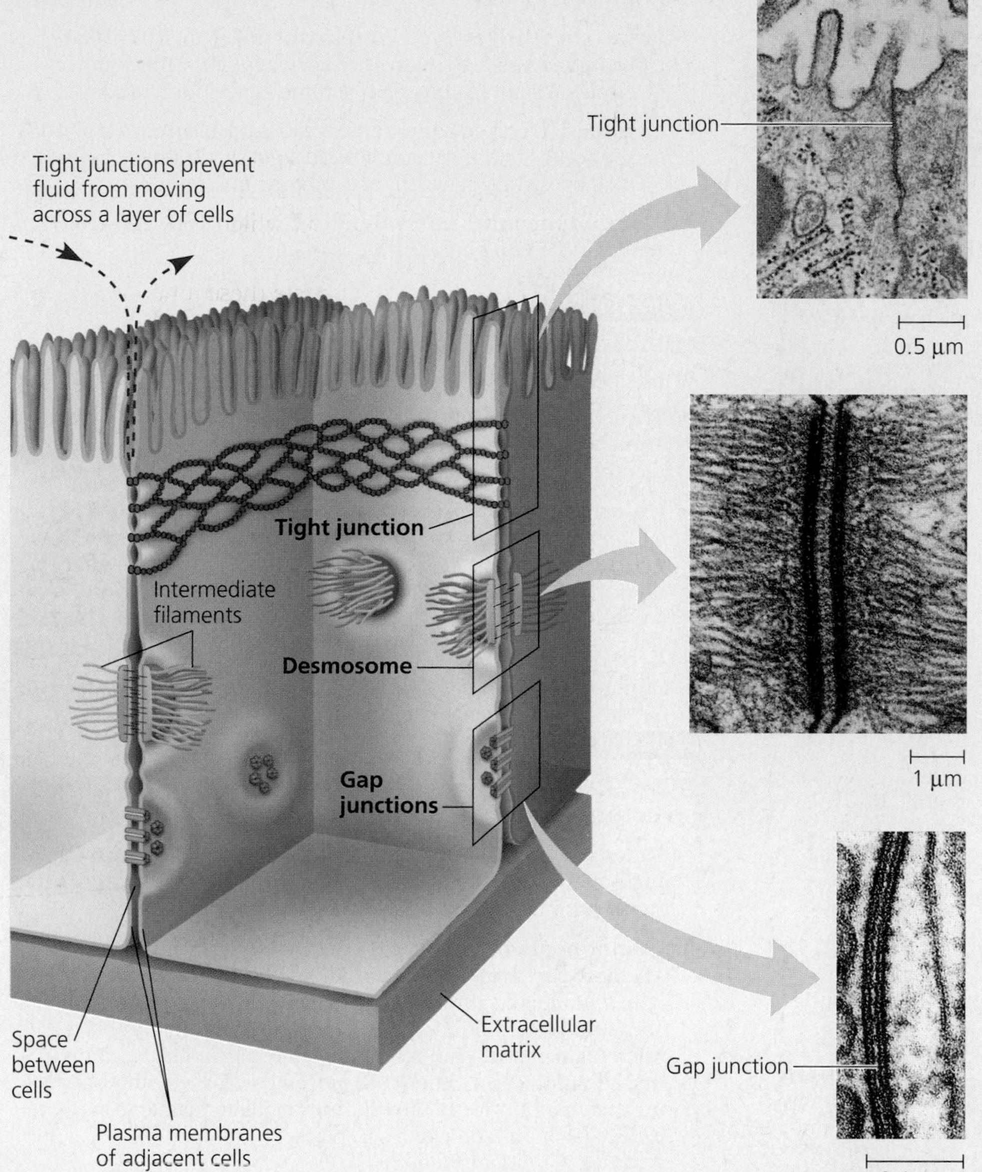

TIGHT JUNCTIONS

At **tight junctions**, the membranes of neighboring cells are very tightly pressed against each other, bound together by specific proteins (purple). Forming continuous seals around the cells, tight junctions prevent leakage of extracellular fluid across a layer of epithelial cells.

DESMOSOMES

Desmosomes (also called *anchoring junctions*) function like rivets, fastening cells together into strong sheets. Intermediate filaments made of sturdy keratin proteins anchor desmosomes in the cytoplasm.

GAP JUNCTIONS

Gap junctions (also called *communicating junctions*) provide cytoplasmic channels from one cell to an adjacent cell. Gap junctions consist of special membrane proteins that surround a pore through which ions, sugars, amino acids, and other small molecules may pass. Gap junctions are necessary for communication between cells in many types of tissues, including heart muscle, and in animal embryos.

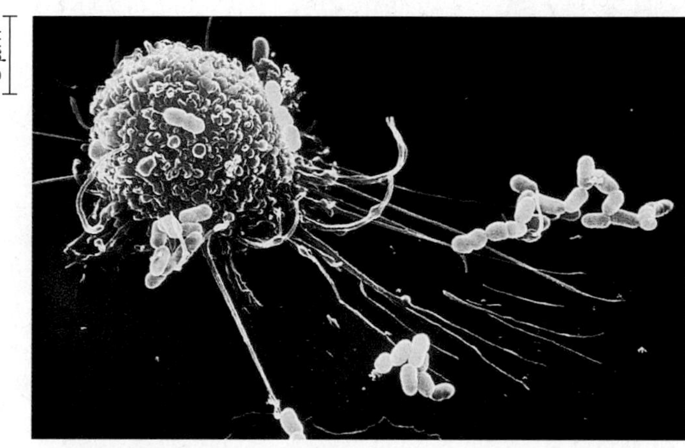

▶ **Figure 6.32 The emergence of cellular functions from the cooperation of many organelles.** The ability of this macrophage (brown) to recognize, apprehend, and destroy bacteria (yellow) is a coordinated activity of the whole cell. Its cytoskeleton, lysosomes, and plasma membrane are among the components that function in phagocytosis (colorized SEM).

Chapter 6 Review

Go to www.campbellbiology.com or the student CD-ROM to explore Activities, Investigations, and other interactive study aids.

SUMMARY OF KEY CONCEPTS

Concept 6.1

To study cells, biologists use microscopes and the tools of biochemistry

- **Microscopy** (pp. 95–97) Improvements in microscopy have catalyzed progress in the study of cell structure.
  Activity *Metric System Review*
  Investigation *What Is the Size and Scale of Our World?*
- **Isolating Organelles by Cell Fractionation** (p. 97) Cell biologists can obtain pellets enriched in specific organelles by centrifuging disrupted cells.

Concept 6.2

Eukaryotic cells have internal membranes that compartmentalize their functions

- **Comparing Prokaryotic and Eukaryotic Cells** (pp. 98–99) All cells are bounded by a plasma membrane. Unlike eukaryotic cells, prokaryotic cells lack nuclei and other membrane-enclosed organelles.
  Activity *Prokaryotic Cell Structure and Function*
  Activity *Comparing Prokaryotic and Eukaryotic Cells*
- **A Panoramic View of the Eukaryotic Cell** (pp. 99–101) Plant and animal cells have most of the same organelles.
  Activity *Build an Animal Cell and a Plant Cell*

Concept 6.3

The eukaryotic cell's genetic instructions are housed in the nucleus and carried out by the ribosomes

- **The Nucleus: Genetic Library of the Cell** (pp. 102–103) The nucleus houses DNA and nucleoli, where ribosomal subunits are made. Materials pass through pores in the nuclear envelope.
- **Ribosomes: Protein Factories in the Cell** (pp. 102–104) Free ribosomes in the cytosol and bound ribosomes on the outside of the ER and the nuclear envelope synthesize proteins.
  Activity *Role of the Nucleus and Ribosomes in Protein Synthesis*

Concept 6.4

The endomembrane system regulates protein traffic and performs metabolic functions in the cell

- The membranes of the endomembrane system are connected by physical continuity or through transport vesicles (p. 104).
- **The Endoplasmic Reticulum: Biosynthetic Factory** (pp. 104–105) Smooth ER synthesizes lipids, metabolizes carbohydrates, stores calcium, and detoxifies poisons. Rough ER has bound ribosomes and produces proteins and membranes, which are distributed by transport vesicles from the ER.
- **The Golgi Apparatus: Shipping and Receiving Center** (pp. 105–107) Proteins are transported from the ER to the Golgi, where they are modified, sorted, and released in transport vesicles.
- **Lysosomes: Digestive Compartments** (pp. 107–108) Lysosomes are sacs of hydrolytic enzymes. They break down ingested substances and cell macromolecules for recycling.
- **Vacuoles: Diverse Maintenance Compartments** (p. 108) A plant cell's central vacuole functions in digestion, storage, waste disposal, cell growth, and protection.
- **The Endomembrane System: *A Review*** (pp. 108–109)
  Activity *The Endomembrane System*

Concept 6.5

Mitochondria and chloroplasts change energy from one form to another

- **Mitochondria: Chemical Energy Conversion** (pp. 109–110) Mitochondria, the sites of cellular respiration, have an outer membrane and an inner membrane that is folded into cristae.
- **Chloroplasts: Capture of Light Energy** (pp. 110–111) Chloroplasts contain pigments that function in photosynthesis. At least two membranes surround the fluid stroma, which contains thylakoids stacked into grana.
  Activity *Build a Chloroplast and a Mitochondrion*
- **Peroxisomes: Oxidation** (pp. 110–111) Peroxisomes produce hydrogen peroxide (H_2O_2) and convert it to water.

Concept 6.6

The cytoskeleton is a network of fibers that organizes structures and activities in the cell

- **Roles of the Cytoskeleton: Support, Motility, and Regulation** (pp. 112–113) The cytoskeleton functions in structural support for the cell, motility, and signal transmission.
- **Components of the Cytoskeleton** (pp. 113–118) Microtubules shape the cell, guide movement of organelles, and help separate the chromosome copies in dividing cells. Cilia and flagella are motile appendages containing microtubules. Microfilaments are thin rods built from actin; they function in muscle contraction, amoeboid movement, cytoplasmic streaming, and support for microvilli. Intermediate filaments support cell shape and fix organelles in place.
  Activity *Cilia and Flagella*

Concept 6.7

Extracellular components and connections between cells help coordinate cellular activities

- **Cell Walls of Plants** (pp. 118–119) Plant cell walls are made of cellulose fibers embedded in other polysaccharides and protein.
- **The Extracellular Matrix (ECM) of Animal Cells** (pp. 119–120) Animal cells secrete glycoproteins that form the ECM, which functions in support, adhesion, movement, and regulation.
- **Intercellular Junctions** (pp. 120–121) Plants have plasmodesmata that pass through adjoining cell walls. Animal cells have tight junctions, desmosomes, and gap junctions.
  Activity *Cell Junctions*
- **The Cell: A Living Unit Greater Than the Sum of Its Parts** (pp. 120–121)
  Activity *Review: Animal Cell Structure and Function*
  Activity *Review: Plant Cell Structure and Function*

TESTING YOUR KNOWLEDGE

Self-Quiz

1. The symptoms of a certain inherited disorder in humans include breathing problems and, in males, sterility. Which of the following is a reasonable hypothesis for the molecular basis of this disorder?
   a. a defective enzyme in the mitochondria
   b. defective actin molecules in cellular microfilaments
   c. defective dynein molecules in cilia and flagella
   d. abnormal hydrolytic enzymes in the lysosomes
   e. defective ribosome assembly in the nucleolus

2. Choose the statement that correctly characterizes bound ribosomes.
   a. Bound ribosomes are enclosed in their own membrane.
   b. Bound and free ribosomes are structurally different.
   c. Bound ribosomes generally synthesize membrane proteins and secretory proteins.
   d. The most common location for bound ribosomes is the cytoplasmic surface of the plasma membrane.
   e. All of the above.

3. Which of the following is not considered part of the endomembrane system?
   a. nuclear envelope
   b. chloroplast
   c. Golgi apparatus
   d. plasma membrane
   e. ER

4. Cells of the pancreas will incorporate radioactively labeled amino acids into proteins. This "tagging" of newly synthesized proteins enables a researcher to track the location of these proteins in a cell. In this case, we are tracking an enzyme that is eventually secreted by pancreatic cells. Which of the following is the most likely pathway for movement of this protein in the cell?
   a. ER→Golgi→nucleus
   b. Golgi→ER→lysosome
   c. nucleus→ER→Golgi
   d. ER→Golgi→vesicles that fuse with plasma membrane
   e. ER→lysosomes→vesicles that fuse with plasma membrane

5. Which of the following structures is common to plant *and* animal cells?
   a. chloroplast
   b. wall made of cellulose
   c. tonoplast
   d. mitochondrion
   e. centriole

6. Which of the following is present in a prokaryotic cell?
   a. mitochondrion
   b. ribosome
   c. nuclear envelope
   d. chloroplast
   e. ER

7. Which type of cell would probably provide the best opportunity to study lysosomes?
   a. muscle cell
   b. nerve cell
   c. phagocytic white blood cell
   d. leaf cell of a plant
   e. bacterial cell

8. Which of the following statements is a correct distinction between prokaryotic and eukaryotic cells attributable to the absence of a prokaryotic cytoskeleton?
   a. Organelles are found only in eukaryotic cells.
   b. Cytoplasmic streaming is not observed in prokaryotes.
   c. Only eukaryotic cells are capable of movement.
   d. Prokaryotic cells have cell walls.
   e. Only the eukaryotic cell concentrates its genetic material in a region separate from the rest of the cell.

9. Which of the following structure-function pairs is *mismatched*?
   a. nucleolus; ribosome production
   b. lysosome; intracellular digestion
   c. ribosome; protein synthesis
   d. Golgi; protein trafficking
   e. microtubule; muscle contraction

10. Cyanide binds with at least one of the molecules involved in the production of ATP. Following exposure of a cell to cyanide, most of the cyanide could be expected to be found within the
   a. mitochondria.
   b. ribosomes.
   c. peroxisomes.
   d. lysosomes.
   e. endoplasmic reticulum.

For Self-Quiz Answers, see Appendix A.

Go to the website or CD-ROM for more quiz questions.

Evolution Connection

Although the similarities among cells reveal the evolutionary unity of life, cells can differ dramatically in structure. Which aspects of cell structure best reveal their evolutionary unity? What are some examples of specialized cellular modifications?

Scientific Inquiry

Imagine protein X, destined to go to the plasma membrane of a cell. Assume that the mRNA carrying the genetic message for protein X has already been translated by ribosomes in a cell culture. You collect the cells, break them open, and then fractionate their contents by differential centrifugation as shown in Figure 6.5. In the pellet of which fraction would you expect to find protein X? Explain your answer by describing the transit of protein X through the cell.

Investigation *What Is the Size and Scale of Our World?*

Science, Technology, and Society

Doctors at a California university removed a man's spleen, standard treatment for a type of leukemia. The disease did not recur. Researchers kept some of the spleen cells alive in a nutrient medium. They found that some of the cells produced a blood protein that showed promise for treating cancer and AIDS. The researchers patented the cells. The patient sued, claiming a share in profits from any products derived from his cells. The California Supreme Court ruled against the patient, stating that his suit "threatens to destroy the economic incentive to conduct important medical research." The U.S. Supreme Court agreed. Do you think the patient was treated fairly? What else would you like to know about this case that might help you make up your mind?

7 Membrane Structure and Function

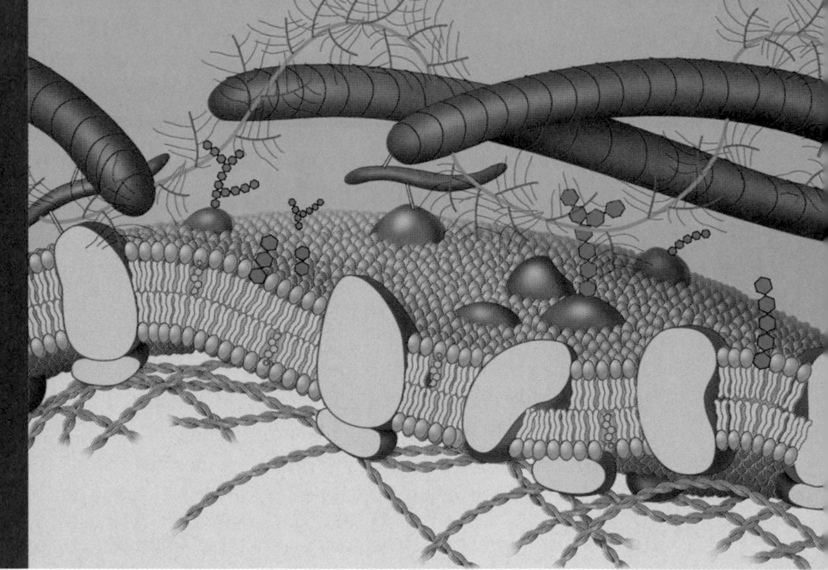

▲ **Figure 7.1 The plasma membrane.**

Key Concepts

7.1 Cellular membranes are fluid mosaics of lipids and proteins

7.2 Membrane structure results in selective permeability

7.3 Passive transport is diffusion of a substance across a membrane with no energy investment

7.4 Active transport uses energy to move solutes against their gradients

7.5 Bulk transport across the plasma membrane occurs by exocytosis and endocytosis

Overview

Life at the Edge

The plasma membrane is the edge of life, the boundary that separates the living cell from its nonliving surroundings. A remarkable film only about 8 nm thick—it would take over 8,000 to equal the thickness of this page—the plasma membrane controls traffic into and out of the cell it surrounds. Like all biological membranes, the plasma membrane exhibits **selective permeability;** that is, it allows some substances to cross it more easily than others. One of the earliest episodes in the evolution of life may have been the formation of a membrane that enclosed a solution different from the surrounding solution while still permitting the uptake of nutrients and elimination of waste products. This ability of the cell to discriminate in its chemical exchanges with its environment is fundamental to life, and it is the plasma membrane and its component molecules that make this selectivity possible.

In this chapter, you will learn how cellular membranes control the passage of substances. The importance of selective permeability was highlighted when the 2003 Nobel Prize in Chemistry was awarded to Peter Agre (see the interview on pp. 92–93) and Roderick MacKinnon, two scientists who worked out how water and specific ions are transported into and out of the cell. We will concentrate on the plasma membrane, the outermost membrane of the cell, represented in **Figure 7.1**. However, the same general principles of membrane traffic also apply to the many varieties of internal membranes that partition the eukaryotic cell. To understand how membranes work, we begin by examining their architecture.

Concept 7.1

Cellular membranes are fluid mosaics of lipids and proteins

Lipids and proteins are the staple ingredients of membranes, although carbohydrates are also important. The most abundant lipids in most membranes are phospholipids. The ability of phospholipids to form membranes is inherent in their molecular structure. A phospholipid is an **amphipathic molecule,** meaning it has both a hydrophilic region and a hydrophobic region (see Figure 5.13). Other types of membrane lipids are also amphipathic. Furthermore, most of the proteins of membranes have both hydrophobic and hydrophilic regions.

How are phospholipids and proteins arranged in the membranes of cells? You encountered the currently accepted model for the arrangement of these molecules in Chapter 6 (see Figure 6.8). In this **fluid mosaic model,** the membrane is a fluid structure with a "mosaic" of various proteins embedded in or attached to a double layer (bilayer) of phospholipids. We'll discuss this model in detail, starting with the story of how it was developed.

Membrane Models: *Scientific Inquiry*

Scientists began building molecular models of the membrane decades before membranes were first seen with the electron microscope in the 1950s. In 1915, membranes isolated from red blood cells were chemically analyzed and found to be composed of lipids and proteins. Ten years later, two Dutch scientists, E. Gorter and F. Grendel, reasoned that cell membranes must actually be phospholipid bilayers. Such a double layer of molecules could exist as a stable boundary between two aqueous compartments because the molecular arrangement shelters the hydrophobic tails of the phospholipids from water while exposing the hydrophilic heads to water **(Figure 7.2)**.

With the conclusion that a phospholipid bilayer was the main fabric of a membrane, the next question was where to place the proteins. Although the heads of phospholipids are hydrophilic, the surface of a membrane consisting of a pure phospholipid bilayer adheres less strongly to water than does the surface of a biological membrane. Given these data, in 1935, Hugh Davson and James Danielli suggested that this difference could be accounted for if the membrane were coated on both sides with hydrophilic proteins. They proposed a sandwich model: a phospholipid bilayer between two layers of proteins.

When researchers first used electron microscopes to study cells in the 1950s, the pictures seemed to support the Davson-Danielli model. By the 1960s, the Davson-Danielli sandwich had become widely accepted as the structure not only of the plasma membrane, but of all the internal membranes of the cell. By the end of that decade, however, many cell biologists recognized two problems with the model. First, the generalization that all membranes of the cell are identical was challenged. Whereas the plasma membrane is 7–8 nm thick and has a three-layered structure in electron micrographs, the inner membrane of the mitochondrion is only 6 nm thick and looks like a row of beads. Mitochondrial membranes also have a substantially greater percentage of proteins than do plasma membranes, and there are differences in the specific kinds of phospholipids and other lipids. In short, membranes with different functions differ in chemical composition and structure.

A second, more serious problem with the sandwich model was the placement of the proteins. Unlike proteins dissolved in the cytosol, membrane proteins are not very soluble in water. Membrane proteins have hydrophobic regions as well as hydrophilic regions (that is, they are amphipathic). If such proteins were layered on the surface of the membrane, their hydrophobic parts would be in an aqueous environment.

In 1972, S. J. Singer and G. Nicolson proposed that membrane proteins are dispersed and individually inserted into the phospholipid bilayer, with only their hydrophilic regions protruding far enough from the bilayer to be exposed to water **(Figure 7.3)**. This molecular arrangement would maximize contact of hydrophilic regions of proteins and phospholipids with water while providing their hydrophobic parts with a nonaqueous environment. According to this model, the membrane is a mosaic of protein molecules bobbing in a fluid bilayer of phospholipids.

A method of preparing cells for electron microscopy called freeze-fracture has demonstrated visually that proteins are indeed embedded in the phospholipid bilayer of the membrane. Freeze-fracture splits a membrane along the middle of the phospholipid bilayer. When the halves of the fractured membrane are viewed in the electron microscope, the interior of the bilayer appears cobblestoned, with protein particles interspersed in a smooth matrix, as in the fluid mosaic model (**Figure 7.4**, next page). Other kinds of evidence further support this arrangement.

Models are proposed by scientists as hypotheses, ways of organizing and explaining existing information. Replacing one model of membrane structure with another does not imply that the original model was worthless. The acceptance or rejection of a model depends on how well it fits observations and explains experimental results. A good model also makes predictions that shape future research. Models inspire experiments, and few models survive these tests without modification. New findings may make a model obsolete; even then, it may not be totally scrapped, but revised to incorporate the new observations. The fluid mosaic model is continually being refined and may one day undergo further revision.

Now let's take a closer look at membrane structure, beginning with the ability of lipids and proteins to drift laterally within the membrane.

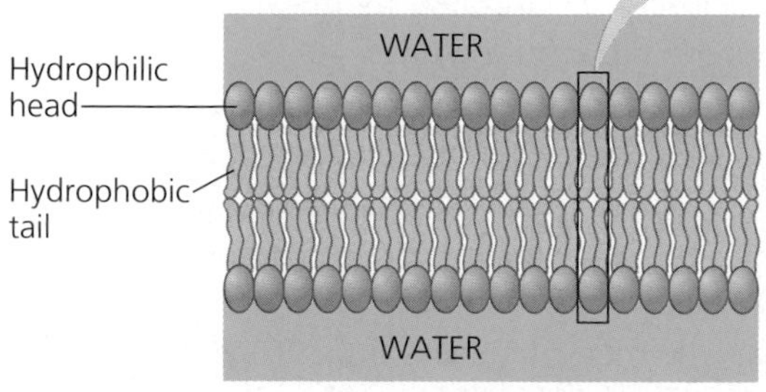

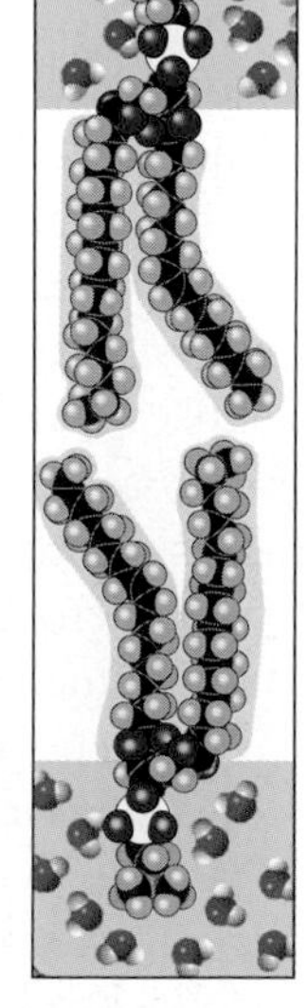

▲ **Figure 7.2 Phospholipid bilayer (cross section).**

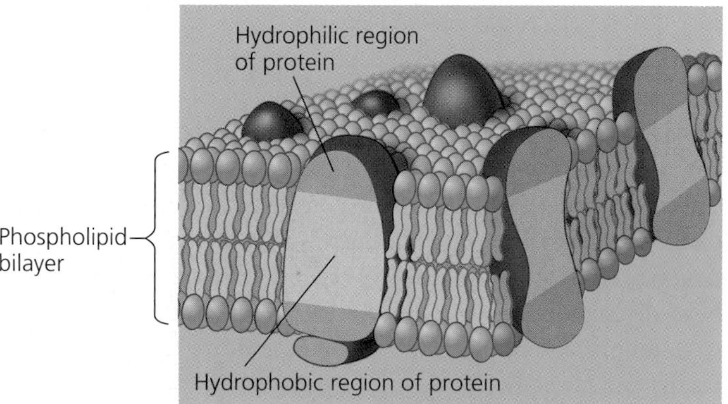

▲ **Figure 7.3 The fluid mosaic model for membranes.**

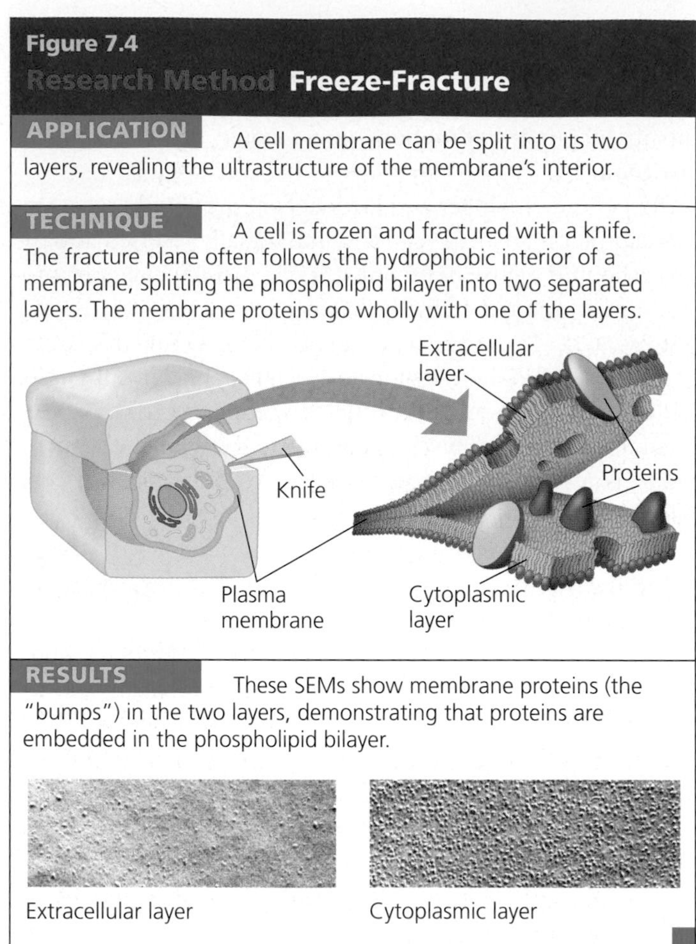

Figure 7.4
Research Method Freeze-Fracture
APPLICATION
A cell membrane can be split into its two layers, revealing the ultrastructure of the membrane's interior.
TECHNIQUE
A cell is frozen and fractured with a knife. The fracture plane often follows the hydrophobic interior of a membrane, splitting the phospholipid bilayer into two separated layers. The membrane proteins go wholly with one of the layers.
Extracellular layer
Knife
Proteins
Plasma membrane
Cytoplasmic layer
RESULTS
These SEMs show membrane proteins (the "bumps") in the two layers, demonstrating that proteins are embedded in the phospholipid bilayer.
Extracellular layer
Cytoplasmic layer

The Fluidity of Membranes

Membranes are not static sheets of molecules locked rigidly in place. A membrane is held together primarily by hydrophobic interactions, which are much weaker than covalent bonds (see Figure 5.20). Most of the lipids and some of the proteins can drift about laterally—that is, in the plane of the membrane **(Figure 7.5a)**. It is quite rare, however, for a molecule to flip-flop transversely across the membrane, switching from one phospholipid layer to the other; to do so, the hydrophilic part of the molecule must cross the hydrophobic core of the membrane.

The lateral movement of phospholipids within the membrane is rapid. Adjacent phospholipids switch positions about 10^7 times per second, which means that a phospholipid can travel about 2 μm—the length of a typical bacterial cell—in 1 second. Proteins are much larger than lipids and move more slowly, but some membrane proteins do, in fact, drift **(Figure 7.6)**. And some membrane proteins seem to move in a highly directed manner, perhaps driven along cytoskeletal fibers by motor proteins connected to the membrane proteins' cytoplasmic regions. However, many other membrane proteins seem to be held virtually immobile by their attachment to the cytoskeleton.

A membrane remains fluid as temperature decreases, until finally the phospholipids settle into a closely packed arrangement and the membrane solidifies, much as bacon grease forms lard when it cools. The temperature at which a membrane solidifies depends on the types of lipids it is made of. The membrane remains fluid to a lower temperature if it is rich in phospholipids with unsaturated hydrocarbon tails (see Figures 5.12 and 5.13). Because of kinks in the tails where double bonds are located, unsaturated hydrocarbons cannot pack together as closely as saturated hydrocarbons, and this makes the membrane more fluid **(Figure 7.5b)**.

The steroid cholesterol, which is wedged between phospholipid molecules in the plasma membranes of animal cells, has different effects on membrane fluidity at different temperatures **(Figure 7.5c)**. At relatively warm temperatures—at 37°C, the body temperature of humans, for example—cholesterol makes the membrane less fluid by restraining the movement of phospholipids. However, because cholesterol

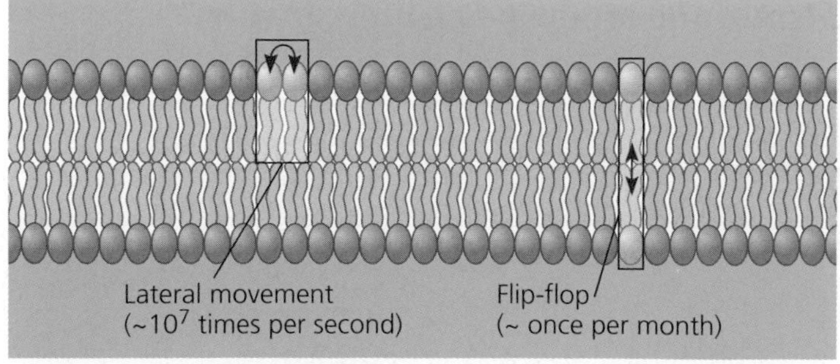

(a) Movement of phospholipids. Lipids move laterally in a membrane, but flip-flopping across the membrane is quite rare.

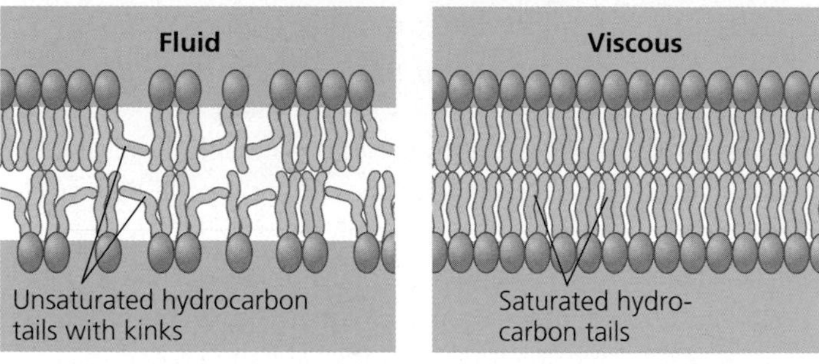

(b) Membrane fluidity. Unsaturated hydrocarbon tails of phospholipids have kinks that keep the molecules from packing together, enhancing membrane fluidity.

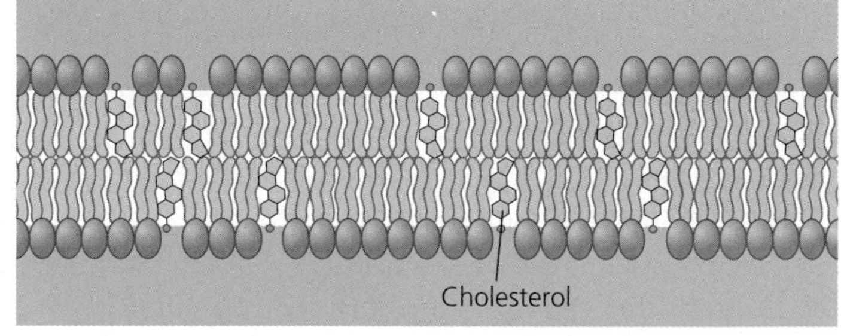

(c) Cholesterol within the animal cell membrane. Cholesterol reduces membrane fluidity at moderate temperatures by reducing phospholipid movement, but at low temperatures it hinders solidification by disrupting the regular packing of phospholipids.

▲ **Figure 7.5 The fluidity of membranes.**

Figure 7.6

Inquiry Do membrane proteins move?

EXPERIMENT Researchers labeled the plasma membrane proteins of a mouse cell and a human cell with two different markers and fused the cells. Using a microscope, they observed the markers on the hybrid cell.

RESULTS

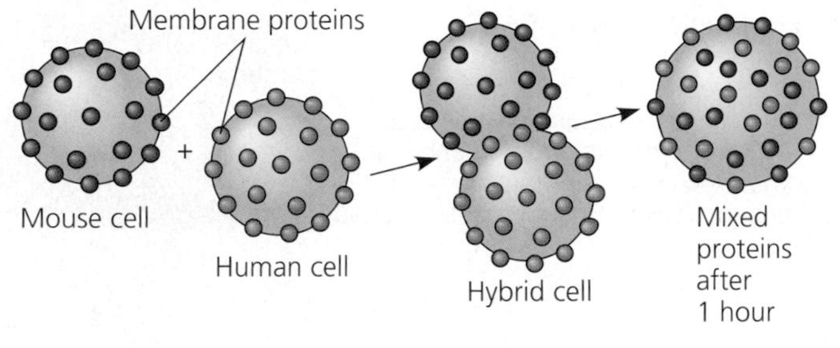

CONCLUSION The mixing of the mouse and human membrane proteins indicates that at least some membrane proteins move sideways within the plane of the plasma membrane.

also hinders the close packing of phospholipids, it lowers the temperature required for the membrane to solidify. Thus, cholesterol can be thought of as a "temperature buffer" for the membrane, resisting changes in membrane fluidity that can be caused by changes in temperature.

Membranes must be fluid to work properly; they are usually about as fluid as salad oil. When a membrane solidifies, its permeability changes, and enzymatic proteins in the membrane may become inactive—for example, if their activity requires them to be able to move laterally in the membrane. The lipid composition of cell membranes can change as an adjustment to changing temperature. For instance, in many plants that tolerate extreme cold, such as winter wheat, the percentage of unsaturated phospholipids increases in autumn, an adaptation that keeps the membranes from solidifying during winter.

Membrane Proteins and Their Functions

Now we come to the *mosaic* aspect of the fluid mosaic model. A membrane is a collage of different proteins embedded in the fluid matrix of the lipid bilayer **(Figure 7.7)**. More than 50 kinds

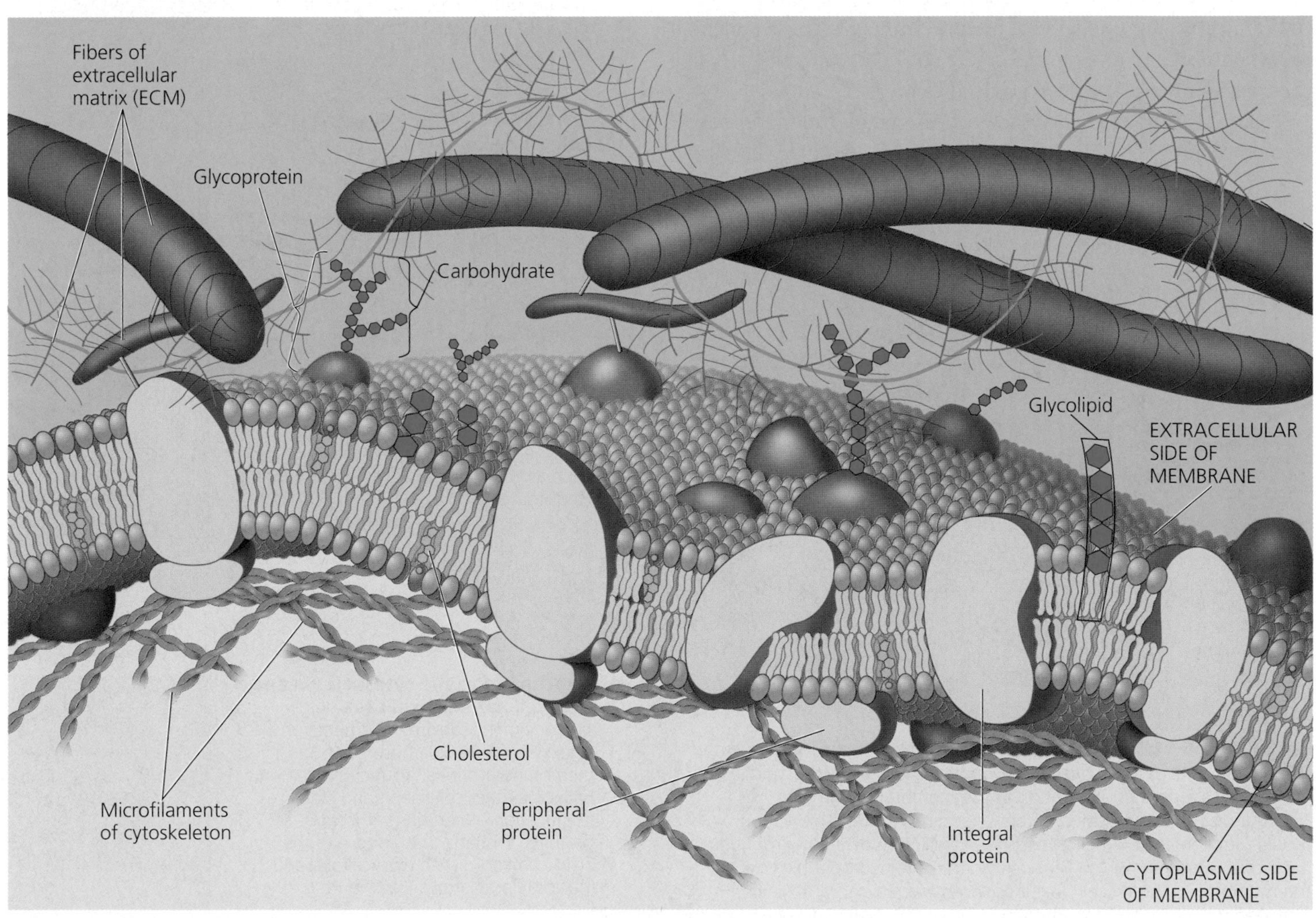

▲ **Figure 7.7 The detailed structure of an animal cell's plasma membrane, in cross section.**

of proteins have been found so far in the plasma membrane of red blood cells, for example. Phospholipids form the main fabric of the membrane, but proteins determine most of the membrane's specific functions. Different types of cells contain different sets of membrane proteins, and the various membranes within a cell each have a unique collection of proteins.

Notice in Figure 7.7 that there are two major populations of membrane proteins. **Integral proteins** penetrate the hydrophobic core of the lipid bilayer. Many are *transmembrane* proteins, which completely span the membrane. The hydrophobic regions of an integral protein consist of one or more stretches of nonpolar amino acids (see Figure 5.17), usually coiled into α helices (**Figure 7.8**). The hydrophilic parts of the molecule are exposed to the aqueous solutions on either side of the membrane. **Peripheral proteins** are not embedded in the lipid bilayer at all; they are appendages loosely bound to the surface of the membrane, often to the exposed parts of integral proteins (see Figure 7.7).

On the cytoplasmic side of the plasma membrane, some membrane proteins are held in place by attachment to the cytoskeleton. And on the exterior side, certain membrane proteins are attached to fibers of the extracellular matrix (see Figure 6.29; *integrins* are one type of integral protein). These attachments combine to give animal cells a stronger framework than the plasma membrane itself could provide.

Figure 7.9 gives an overview of six major functions performed by proteins of the plasma membrane. A single cell may

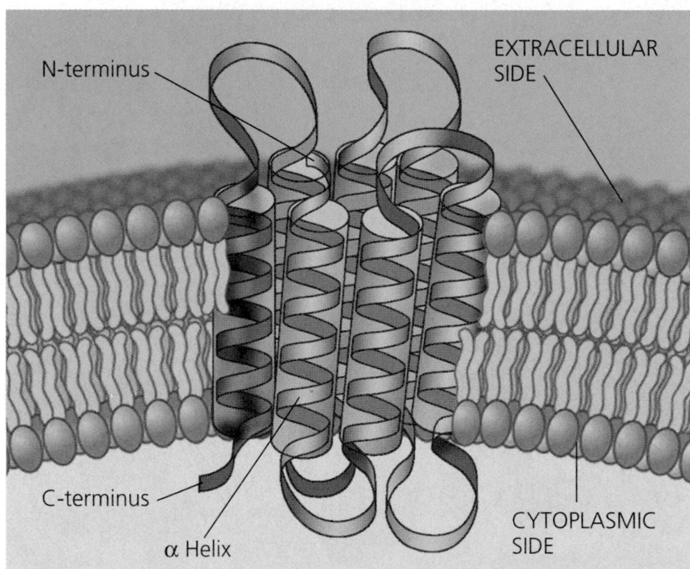

▲ **Figure 7.8 The structure of a transmembrane protein.** The protein shown here, bacteriorhodopsin (a bacterial transport protein), has a distinct orientation in the membrane, with the N-terminus outside the cell and the C-terminus inside. This ribbon model highlights the α-helical secondary structure of the hydrophobic parts of the protein, which lie mostly within the hydrophobic core of the membrane. The protein includes seven transmembrane helices (outlined with cylinders for emphasis). The nonhelical hydrophilic segments of the protein are in contact with the aqueous solutions on the extracellular and cytoplasmic sides of the membrane.

(a) Transport. (left) A protein that spans the membrane may provide a hydrophilic channel across the membrane that is selective for a particular solute. **(right)** Other transport proteins shuttle a substance from one side to the other by changing shape. Some of these proteins hydrolyze ATP as an energy source to actively pump substances across the membrane.

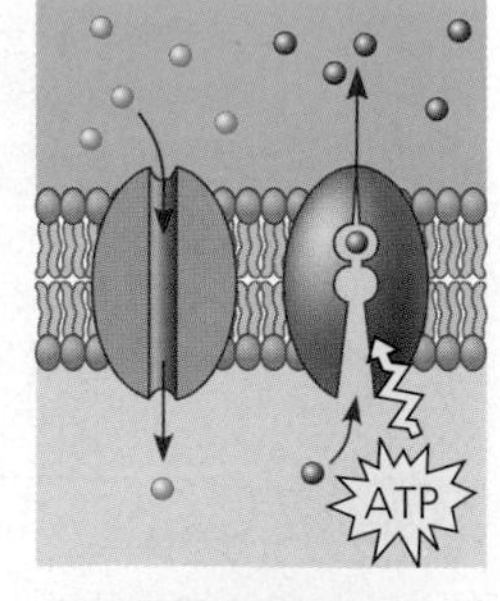

(b) Enzymatic activity. A protein built into the membrane may be an enzyme with its active site exposed to substances in the adjacent solution. In some cases, several enzymes in a membrane are organized as a team that carries out sequential steps of a metabolic pathway.

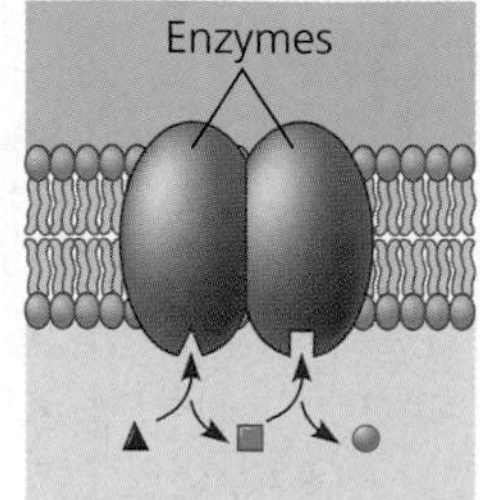

(c) Signal transduction. A membrane protein may have a binding site with a specific shape that fits the shape of a chemical messenger, such as a hormone. The external messenger (signal) may cause a conformational change in the protein (receptor) that relays the message to the inside of the cell.

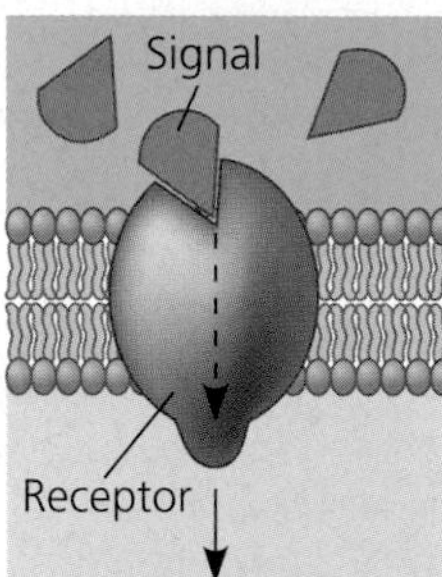

(d) Cell-cell recognition. Some glycoproteins serve as identification tags that are specifically recognized by other cells.

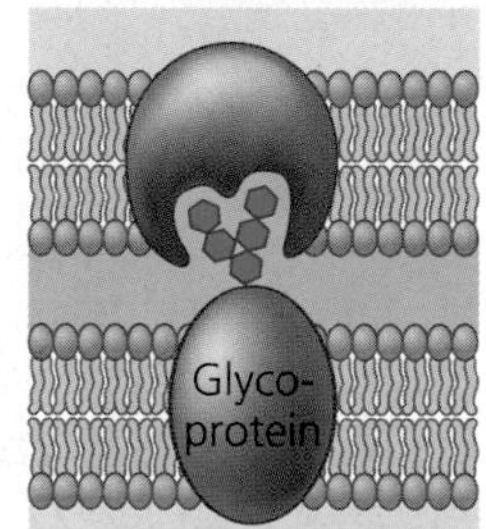

(e) Intercellular joining. Membrane proteins of adjacent cells may hook together in various kinds of junctions, such as gap junctions or tight junctions (see Figure 6.31).

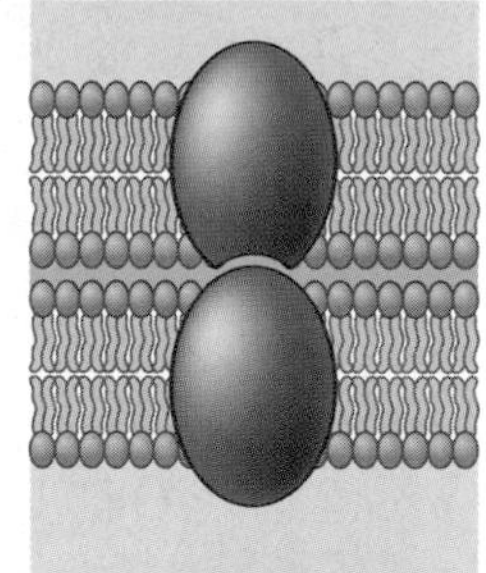

(f) Attachment to the cytoskeleton and extracellular matrix (ECM). Microfilaments or other elements of the cytoskeleton may be bonded to membrane proteins, a function that helps maintain cell shape and stabilizes the location of certain membrane proteins. Proteins that adhere to the ECM can coordinate extracellular and intracellular changes (see Figure 6.29).

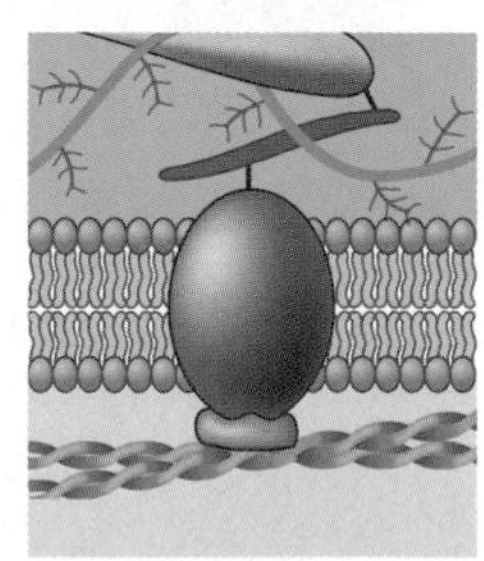

▲ **Figure 7.9 Some functions of membrane proteins.** In many cases, a single protein performs some combination of these tasks.

have membrane proteins carrying out several of these functions, and a single protein may have multiple functions. Thus, the membrane is a functional mosaic as well as a structural one.

The Role of Membrane Carbohydrates in Cell-Cell Recognition

Cell-cell recognition, a cell's ability to distinguish one type of neighboring cell from another, is crucial to the functioning of an organism. It is important, for example, in the sorting of cells into tissues and organs in an animal embryo. It is also the basis for the rejection of foreign cells (including those of transplanted organs) by the immune system, an important line of defense in vertebrate animals (see Chapter 43). The way cells recognize other cells is by binding to surface molecules, often carbohydrates, on the plasma membrane (see Figure 7.9d).

Membrane carbohydrates are usually short, branched chains of fewer than 15 sugar units. Some of these carbohydrates are covalently bonded to lipids, forming molecules called **glycolipids.** (Recall that *glyco* refers to the presence of carbohydrate.) Most, however, are covalently bonded to proteins, which are thereby **glycoproteins** (see Figure 7.7).

The carbohydrates on the external side of the plasma membrane vary from species to species, among individuals of the same species, and even from one cell type to another in a single individual. The diversity of the molecules and their location on the cell's surface enable membrane carbohydrates to function as markers that distinguish one cell from another. For example, the four human blood types designated A, B, AB, and O reflect variation in the carbohydrates on the surface of red blood cells.

Synthesis and Sidedness of Membranes

Membranes have distinct inside and outside faces. The two lipid layers may differ in specific lipid composition, and each protein has directional orientation in the membrane (see Figure 7.8). When a vesicle fuses with the plasma membrane, the outside layer of the vesicle becomes continuous with the cytoplasmic layer of the plasma membrane. Therefore, molecules that start out on the *inside* face of the ER end up on the *outside* face of the plasma membrane.

The process, shown in **Figure 7.10**, starts with ❶ the synthesis of membrane proteins and lipids in the endoplasmic reticulum. Carbohydrates (green) are added to the proteins (purple), making them glycoproteins. The carbohydrate portions may then be modified. ❷ Inside the Golgi apparatus, the glycoproteins undergo further carbohydrate modification, and lipids acquire carbohydrates, becoming glycolipids. ❸ Transmembrane proteins (purple dumbbells), membrane glycolipids, and secretory proteins (purple spheres) are transported in vesicles to the plasma membrane. ❹ There the vesicles fuse with the membrane, releasing secretory proteins from the cell. Vesicle fusion positions the carbohydrates of membrane glycoproteins and glycolipids on the outside of the plasma membrane. Thus, the asymmetrical distribution of proteins, lipids, and their associated carbohydrates in the plasma membrane is determined as the membrane is being built by the ER and Golgi apparatus.

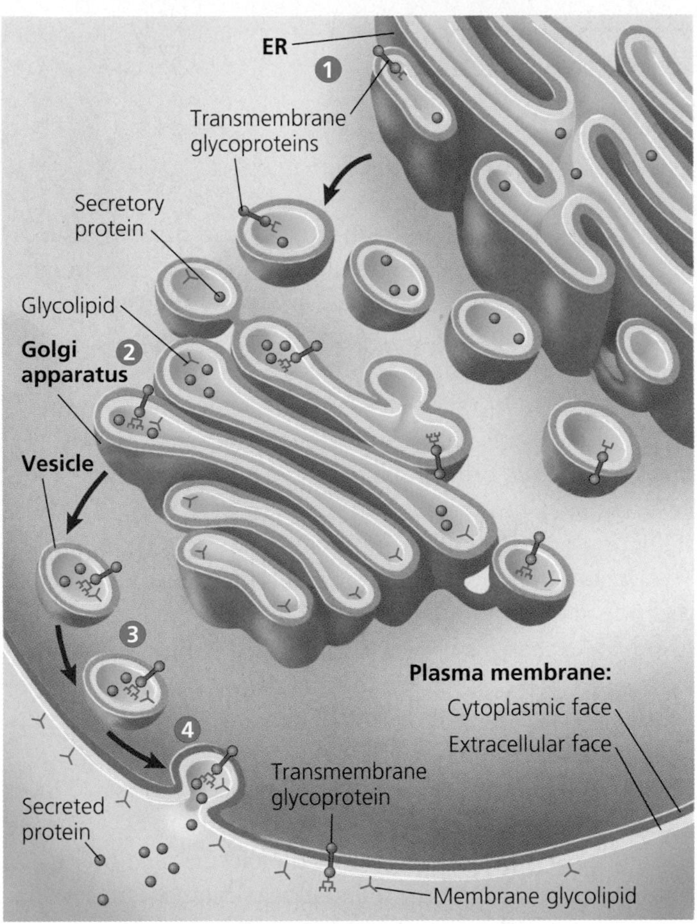

▲ **Figure 7.10 Synthesis of membrane components and their orientation on the resulting membrane.** The plasma membrane has distinct cytoplasmic and extracellular sides, or faces, with the extracellular face arising from the inside face of ER, Golgi, and vesicle membranes.

Concept Check 7.1

1. How would you expect the saturation levels of membrane fatty acids to differ in plants adapted to cold environments and plants adapted to hot environments?
2. The carbohydrates attached to some of the proteins and lipids of the plasma membrane are added as the membrane is made and refined in the ER and Golgi apparatus; the new membrane then forms transport vesicles that travel to the cell surface. On which side of the vesicle membrane are the carbohydrates?

For suggested answers, see Appendix A.

Concept 7.2

Membrane structure results in selective permeability

The biological membrane is an exquisite example of a supramolecular structure—many molecules ordered into a higher level of organization—with emergent properties beyond those of the individual molecules. The remainder of this chapter focuses on one of the most important of those properties: the ability to regulate transport across cellular boundaries, a function essential to the cell's existence. We will see once again that form fits function: The fluid mosaic model helps explain how membranes regulate the cell's molecular traffic.

A steady traffic of small molecules and ions moves across the plasma membrane in both directions. Consider the chemical exchanges between a muscle cell and the extracellular fluid that bathes it. Sugars, amino acids, and other nutrients enter the cell, and metabolic waste products leave it. The cell takes in oxygen for cellular respiration and expels carbon dioxide. It also regulates its concentrations of inorganic ions, such as Na^+, K^+, Ca^{2+}, and Cl^-, by shuttling them one way or the other across the plasma membrane. Although traffic through the membrane is extensive, cell membranes are selectively permeable, and substances do not cross the barrier indiscriminately. The cell is able to take up many varieties of small molecules and ions and exclude others. Moreover, substances that move through the membrane do so at different rates.

The Permeability of the Lipid Bilayer

Hydrophobic (nonpolar) molecules, such as hydrocarbons, carbon dioxide, and oxygen, can dissolve in the lipid bilayer of the membrane and cross it with ease, without the aid of membrane proteins. However, the hydrophobic core of the membrane impedes the direct passage of ions and polar molecules, which are hydrophilic, through the membrane. Polar molecules such as glucose and other sugars pass only slowly through a lipid bilayer, and even water, an extremely small polar molecule, does not cross very rapidly. A charged atom or molecule and its surrounding shell of water (see Figure 3.6) find the hydrophobic layer of the membrane even more difficult to penetrate. Fortunately, the lipid bilayer is only part of the story of a membrane's selective permeability. Proteins built into the membrane play key roles in regulating transport.

Transport Proteins

Cell membranes *are* permeable to specific ions and a variety of polar molecules. These hydrophilic substances can avoid contact with the lipid bilayer by passing through **transport proteins** that span the membrane. Some transport proteins, called *channel proteins*, function by having a hydrophilic channel that certain molecules or atomic ions use as a tunnel through the membrane (see Figure 7.9a, left). For example, the passage of water molecules through the membrane in certain cells is greatly facilitated by channel proteins known as **aquaporins.** (These were discovered in the laboratory of Peter Agre; see pp. 92–93.) Other transport proteins, called *carrier proteins*, hold onto their passengers and change shape in a way that shuttles them across the membrane (see Figure 7.9a, right). In both cases, the transport protein is specific for the substance it translocates (moves), allowing only a certain substance (or substances) to cross the membrane. For example, glucose carried in blood and needed by red blood cells for cellular activities enters these cells rapidly through specific transport proteins in the plasma membrane. This "glucose transporter" is so selective as a carrier protein that it even rejects fructose, a structural isomer of glucose.

Thus, the selective permeability of a membrane depends on both the discriminating barrier of the lipid bilayer and the specific transport proteins built into the membrane. But what determines the *direction* of traffic across a membrane? At a given time, will a particular substance enter or leave the cell? And what mechanisms actually drive molecules across membranes? We will address these questions next as we explore two modes of membrane traffic: passive transport and active transport.

Concept Check 7.2

1. Two molecules that can cross a lipid bilayer without help from membrane proteins are O_2 and CO_2. What properties allow this to occur?
2. Why would water molecules need a transport protein (aquaporin) to move rapidly and in large quantities across a membrane?

For suggested answers, see Appendix A.

Concept 7.3

Passive transport is diffusion of a substance across a membrane with no energy investment

Molecules have a type of energy called thermal motion (heat). One result of thermal motion is **diffusion,** the tendency for molecules of any substance to spread out evenly into the available space. Each molecule moves randomly, yet diffusion of a *population* of molecules may be directional. A good way to visualize this is to imagine a synthetic membrane separating pure water from a solution of a dye in water. Assume that this membrane has microscopic pores and is permeable to the dye molecules **(Figure 7.11a)**. Each dye molecule wanders randomly, but there will be a *net* movement of the dye molecules across the membrane to the side that began as pure water. The dye molecules will continue to spread across the membrane until both solutions have equal concentrations of the dye. Once that point is reached, there will be a dynamic equilibrium, with as many dye molecules crossing the membrane each second in one direction as in the other.

We can now state a simple rule of diffusion: In the absence of other forces, a substance will diffuse from where it is more concentrated to where it is less concentrated. Put another way, any substance will diffuse down its **concentration gradient.** No work must be done to make this happen; diffusion is a spontaneous process. Note that each substance diffuses down its *own* concentration gradient, unaffected by the concentration differences of other substances **(Figure 7.11b)**.

Much of the traffic across cell membranes occurs by diffusion. When a substance is more concentrated on one side of a membrane than on the other, there is a tendency for the substance to diffuse across the membrane down its concentration gradient (assuming that the membrane is permeable to that substance). One important example is the uptake of oxygen by a cell performing cellular respiration. Dissolved oxygen diffuses into the cell across the plasma membrane. As long as cellular respiration consumes the O_2 as it enters, diffusion into the cell will continue, because the concentration gradient favors movement in that direction.

The diffusion of a substance across a biological membrane is called **passive transport** because the cell does not have to expend energy to make it happen. The concentration gradient itself represents potential energy (see Chapter 2, p. 36) and drives diffusion. Remember, however, that membranes are selectively permeable and therefore have different effects on the rates of diffusion of various molecules. In the case of water, aquaporins allow water to diffuse very rapidly across the membranes of certain cells. The movement of water across the plasma membrane has important consequences for cells.

Effects of Osmosis on Water Balance

To see how two solutions with different solute concentrations interact, picture a U-shaped glass tube with a selectively permeable membrane separating two sugar solutions **(Figure 7.12)**. Pores in this synthetic membrane are too small for sugar molecules to pass through but large enough for water molecules. How does this affect the *water* concentration? It seems logical that the solution with the higher concentration of solute would have the lower concentration of water and that water would diffuse into it from the other side for that reason. However, for a dilute solution like most biological fluids,

(a) Diffusion of one solute. The membrane has pores large enough for molecules of dye to pass through. Random movement of dye molecules will cause some to pass through the pores; this will happen more often on the side with more molecules. The dye diffuses from where it is more concentrated to where it is less concentrated (called diffusing down a concentration gradient). This leads to a dynamic equilibrium: The solute molecules continue to cross the membrane, but at equal rates in both directions.

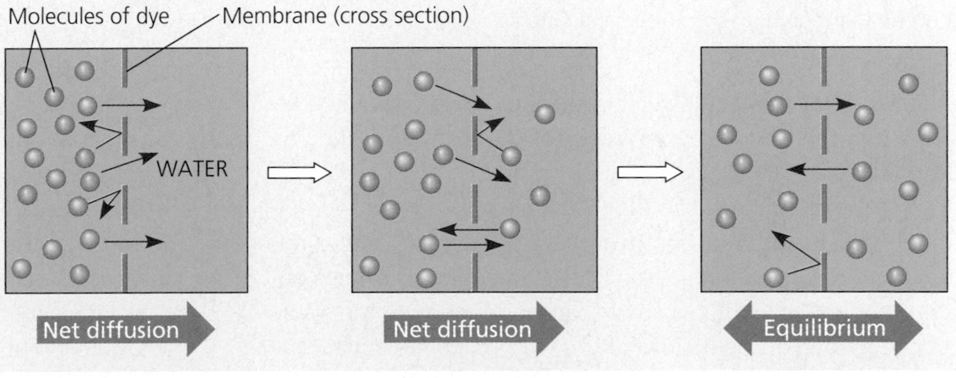

(b) Diffusion of two solutes. Solutions of two different dyes are separated by a membrane that is permeable to both. Each dye diffuses down its own concentration gradient. There will be a net diffusion of the purple dye toward the left, even though the *total* solute concentration was initially greater on the left side.

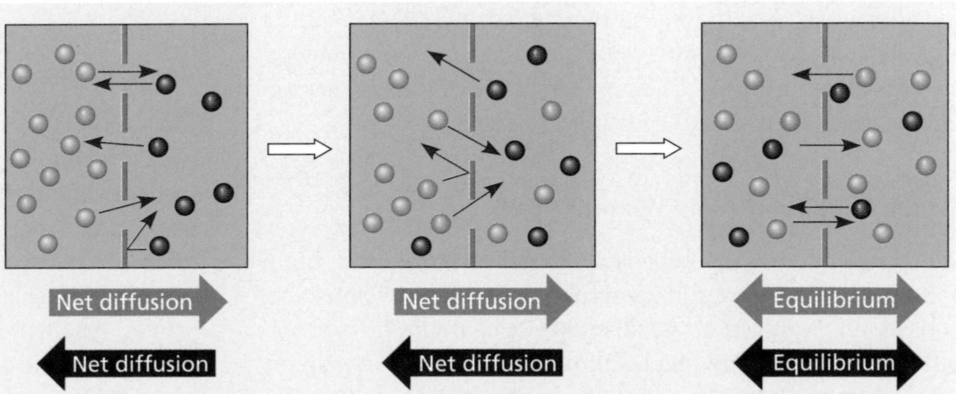

▲ **Figure 7.11 The diffusion of solutes across a membrane.** Each of the large arrows under the diagrams shows the net diffusion of the dye molecules of that color.

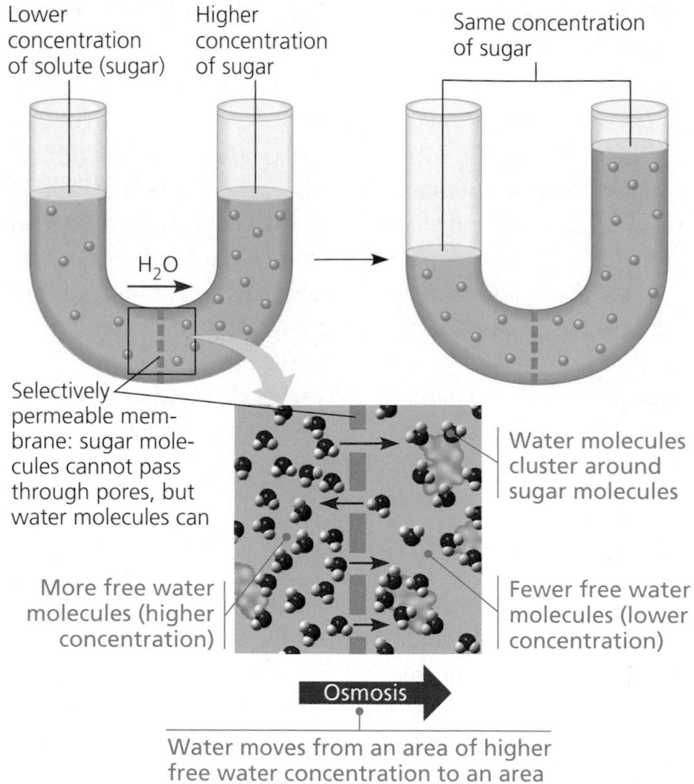

▲ **Figure 7.12 Osmosis.** Two sugar solutions of different concentrations are separated by a selectively permeable membrane, which the solvent (water) can pass through but the solute (sugar) cannot. Water molecules move randomly and may cross through the pores in either direction, but overall, water diffuses from the solution with less concentrated solute to that with more concentrated solute. This transport of water, or osmosis, eventually equalizes the sugar concentrations on both sides of the membrane.

solutes do not affect the water concentration significantly. Instead, tight clustering of water molecules around the hydrophilic solute molecules makes some of the water unavailable to cross the membrane. It is the difference in *free* water concentration that is imporant. But the effect is the same: Water diffuses across the membrane from the region of lower solute concentration to that of higher solute concentration until the solute concentrations on both sides of the membrane are equal. The diffusion of water across a selectively permeable membrane is called **osmosis.** The movement of water across cell membranes and the balance of water between the cell and its environment are crucial to organisms. Let's now apply to living cells what we have learned about osmosis in artificial systems.

Water Balance of Cells Without Walls

When considering the behavior of a cell in a solution, both solute concentration and membrane permeability must be considered. Both factors are taken into account in the concept of **tonicity,** the ability of a solution to cause a cell to gain or lose water. The tonicity of a solution depends in part on its concentration of solutes that cannot cross the membrane (nonpenetrating solutes), relative to that in the cell itself. If there are more nonpenetrating solutes in the surrounding solution, water will tend to leave the cell, and vice versa.

If a cell without a wall, such as an animal cell, is immersed in an environment that is **isotonic** to the cell (*iso* means "same"), there will be no *net* movement of water across the plasma membrane. Water flows across the membrane, but at the same rate in both directions. In an isotonic environment, the volume of an animal cell is stable **(Figure 7.13a)**.

Now let's transfer the cell to a solution that is **hypertonic** to the cell (*hyper* means "more," in this case more nonpenetrating solutes). The cell will lose water to its environment, shrivel, and probably die. This is one reason why an increase in the salinity (saltiness) of a lake can kill the animals there—if the lake water becomes hypertonic to the animals' cells, the cells might shrivel and die. However, taking up too much water can be just as hazardous to an animal cell as losing water. If we place the cell in a solution that is **hypotonic** to the cell (*hypo* means "less"), water will enter the cell faster than it leaves, and the cell will swell and lyse (burst) like an overfilled water balloon.

A cell without rigid walls can tolerate neither excessive uptake nor excessive loss of water. This problem of water balance is automatically solved if such a cell lives in isotonic surroundings. Seawater is isotonic to many marine invertebrates. The cells of most terrestrial (land-dwelling) animals are bathed in an extracellular fluid that is isotonic to the cells. Animals and other organisms without rigid cell walls living in hypertonic or hypotonic environments must have special adaptations for **osmoregulation,** the control of water balance. For example, the protist *Paramecium* lives in pond water, which is hypotonic to the cell. *Paramecium* has a plasma membrane that is much less permeable to water than the membranes of most other cells, but this only slows the uptake of water, which continually enters the cell. *Paramecium* doesn't burst because it is also equipped with a contractile vacuole, an organelle that functions as a bilge pump to force water out of the cell as fast as it enters by osmosis **(Figure 7.14)**. We will examine other evolutionary adaptations for osmoregulation in Chapter 44.

Water Balance of Cells with Walls

The cells of plants, prokaryotes, fungi, and some protists have walls. When such a cell is immersed in a hypotonic solution—bathed in rainwater, for example—the wall helps maintain the cell's water balance. Consider a plant cell. Like an animal cell, the plant cell swells as water enters by osmosis **(Figure 7.13b)**. However, the elastic wall will expand only so much before it exerts a back pressure on the cell that opposes further water uptake. At this point, the cell is **turgid** (very firm), which is the healthy state for most plant cells. Plants that are not

▶ **Figure 7.13 The water balance of living cells.** How living cells react to changes in the solute concentration of their environment depends on whether or not they have cell walls. **(a)** Animal cells such as this red blood cell do not have cell walls. **(b)** Plant cells do. (Arrows indicate net water movement since the cells were first placed in these solutions.)

(a) Animal cell. An animal cell fares best in an isotonic environment unless it has special adaptations to offset the osmotic uptake or loss of water.

(b) Plant cell. Plant cells are turgid (firm) and generally healthiest in a hypotonic environment, where the uptake of water is eventually balanced by the elastic wall pushing back on the cell.

Hypotonic solution

H_2O

Lysed

H_2O

Turgid (normal)

Isotonic solution

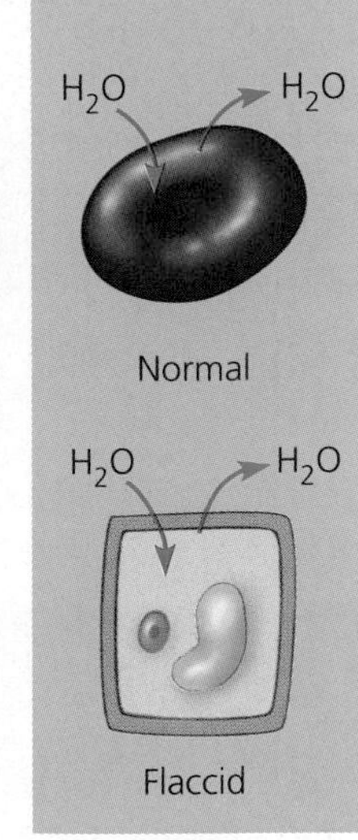

Hypertonic solution

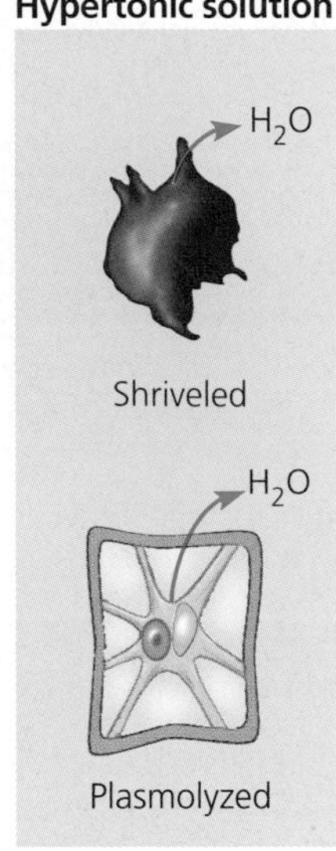

woody, such as most houseplants, depend for mechanical support on cells kept turgid by a surrounding hypotonic solution. If a plant's cells and their surroundings are isotonic, there is no net tendency for water to enter, and the cells become **flaccid** (limp).

However, a wall is of no advantage if the cell is immersed in a hypertonic environment. In this case, a plant cell, like an animal cell, will lose water to its surroundings and shrink. As the plant cell shrivels, its plasma membrane pulls away from the wall. This phenomenon, called **plasmolysis,** causes the plant to wilt and can be lethal. The walled cells of bacteria and fungi also plasmolyze in hypertonic environments.

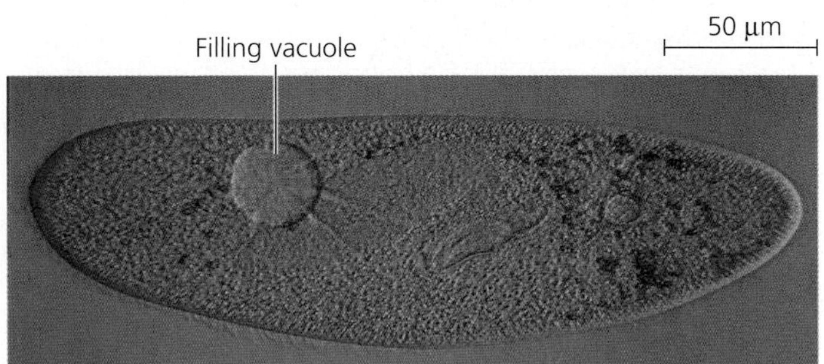

(a) A contractile vacuole fills with fluid that enters from a system of canals radiating throughout the cytoplasm.

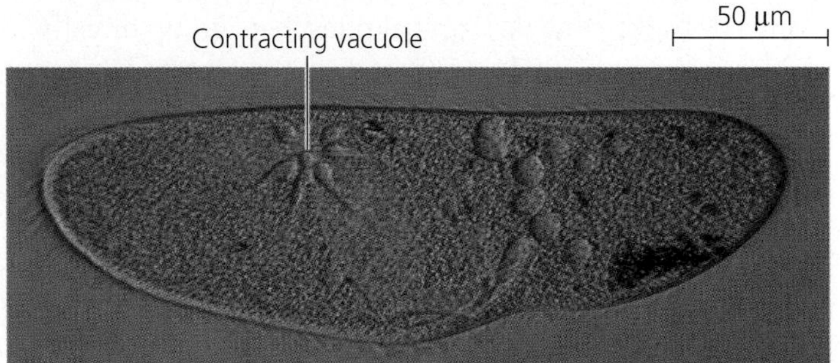

(b) When full, the vacuole and canals contract, expelling fluid from the cell.

▲ **Figure 7.14 The contractile vacuole of *Paramecium*: an evolutionary adaptation for osmoregulation.** The contractile vacuole of this freshwater protist offsets osmosis by bailing water out of the cell.

Facilitated Diffusion: Passive Transport Aided by Proteins

Let's look more closely at how water and certain hydrophilic solutes cross a membrane. As mentioned earlier, many polar molecules and ions impeded by the lipid bilayer of the membrane diffuse passively with the help of transport proteins that span the membrane. This phenomenon is called **facilitated diffusion.** Cell biologists are still trying to learn exactly how various transport proteins facilitate diffusion. Most transport proteins are very specific: They transport only particular substances but not others.

As described earlier, the two types of transport proteins are channel proteins and carrier proteins. Channel proteins simply provide corridors that allow a specific molecule or ion to cross the membrane **(Figure 7.15a)**. The hydrophilic passageways provided by these proteins allow water molecules or small ions to flow very quickly from one side of the membrane to the other. While water molecules are small enough to cross through the phospholipid bilayer, the rate of water movement by this route is relatively slow because of their polarity. Aquaporins, the water channel proteins, facilitate the massive amounts of diffusion that occur in plant cells and in animal cells such as red blood cells (see Figure 7.13). Another group of channels are **ion channels,** many of which function as **gated channels;** a stimulus causes them to open or close. The stimulus may be electrical or chemical; if chemical, the stimulus is a substance other than the one to be transported. For example, stimulation of a nerve cell by

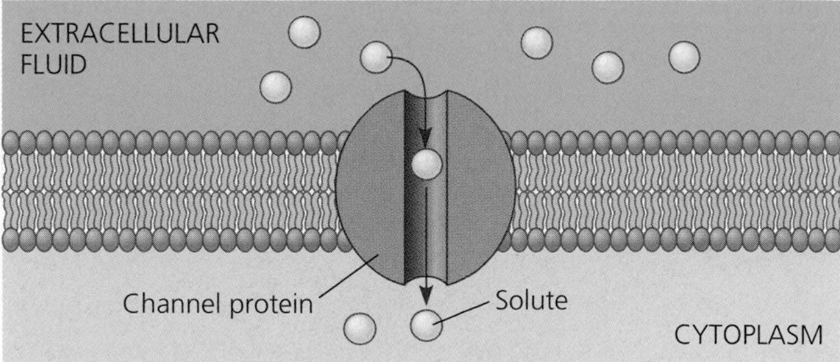

(a) A channel protein (purple) has a channel through which water molecules or a specific solute can pass.

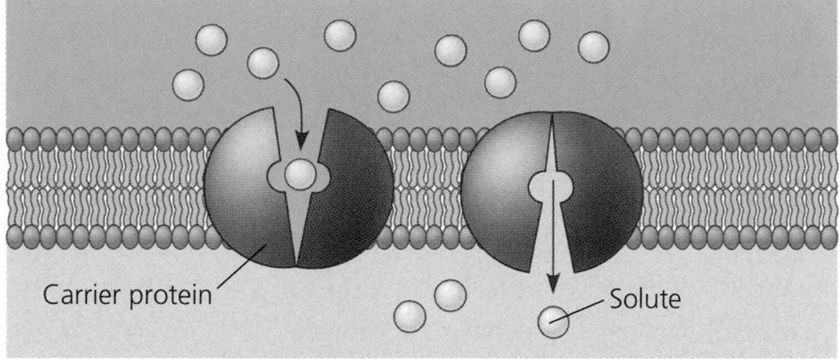

(b) A carrier protein alternates between two conformations, moving a solute across the membrane as the shape of the protein changes. The protein can transport the solute in either direction, with the net movement being down the concentration gradient of the solute.

▲ **Figure 7.15 Two types of transport proteins that carry out facilitated diffusion.** In both cases, the protein transports the solute down its concentration gradient.

certain neurotransmitter molecules opens gated channels that allow sodium ions into the cell.

Carrier proteins seem to undergo a subtle change in shape that somehow translocates the solute-binding site across the membrane **(Figure 7.15b)**. These changes in shape may be triggered by the binding and release of the transported molecule.

In certain inherited diseases, specific transport systems are either defective or missing altogether. An example is cystinuria, a human disease characterized by the absence of a protein that transports cysteine and some other amino acids across the membranes of kidney cells. Kidney cells normally reabsorb these amino acids from the urine and return them to the blood, but an individual afflicted with cystinuria develops painful stones from amino acids that accumulate and crystallize in the kidneys.

Concept Check 7.3

1. If a *Paramecium* were to swim from a hypotonic environment to an isotonic one, would the activity of its contractile vacuole increase or decrease? Why?

For suggested answers, see Appendix A.

Concept 7.4

Active transport uses energy to move solutes against their gradients

Despite the help of transport proteins, facilitated diffusion is still considered passive transport because the solute being transported is moving down its concentration gradient. Facilitated diffusion speeds the transport of a solute by providing an efficient passage through the membrane, but it does not alter the direction of transport. Some transport proteins, however, can move solutes against their concentration gradients, across the plasma membrane from the side where they are less concentrated to the side where they are more concentrated.

The Need for Energy in Active Transport

To pump a molecule across a membrane against its gradient requires work; the cell must expend energy. Therefore, this type of membrane traffic is called **active transport.** The transport proteins that move solutes against a concentration gradient are all carrier proteins, rather than channel proteins. This makes sense, because when channel proteins are open, they merely allow molecules to flow down their concentration gradient, rather than picking them up and transporting them against their gradient.

Active transport enables a cell to maintain internal concentrations of small molecules that differ from concentrations in its environment. For example, compared to its surroundings, an animal cell has a much higher concentration of potassium ions and a much lower concentration of sodium ions. The plasma membrane helps maintain these steep gradients by pumping sodium out of the cell and potassium into the cell.

As in other types of cellular work, ATP supplies the energy for most active transport. One way ATP can power active transport is by transferring its terminal phosphate group directly to the transport protein. This may induce the protein to change its conformation in a manner that translocates a solute bound to the protein across the membrane. One transport system that works this way is the **sodium-potassium pump**, which exchanges sodium (Na^+) for potassium (K^+) across the plasma membrane of animal cells **(Figure 7.16)**. **Figure 7.17** reviews the distinction between passive transport and active transport.

Maintenance of Membrane Potential by Ion Pumps

All cells have voltages across their plasma membranes. Voltage is electrical potential energy—a separation of opposite charges.

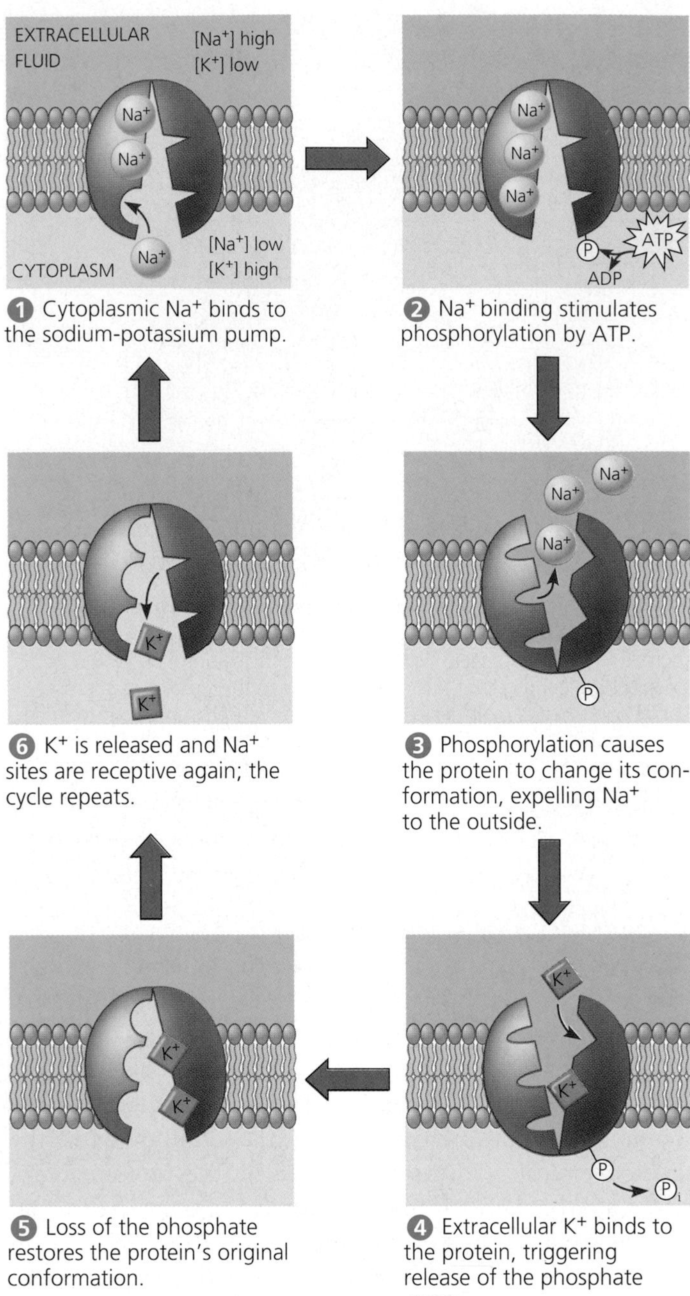

❶ Cytoplasmic Na^+ binds to the sodium-potassium pump.

❷ Na^+ binding stimulates phosphorylation by ATP.

❸ Phosphorylation causes the protein to change its conformation, expelling Na^+ to the outside.

❹ Extracellular K^+ binds to the protein, triggering release of the phosphate group.

❺ Loss of the phosphate restores the protein's original conformation.

❻ K^+ is released and Na^+ sites are receptive again; the cycle repeats.

▲ **Figure 7.16 The sodium-potassium pump: a specific case of active transport.** This transport system pumps ions against steep concentration gradients: Sodium ion concentration (represented as $[Na^+]$) is high outside the cell and low inside, while potassium ion concentration ($[K^+]$) is low outside the cell and high inside. The pump oscillates between two conformational states in a pumping cycle that translocates three sodium ions out of the cell for every two potassium ions pumped into the cell. ATP powers the changes in conformation by phosphorylating the transport protein (that is, by transferring a phosphate group to the protein).

The cytoplasm of a cell is negative in charge compared to the extracellular fluid because of an unequal distribution of anions and cations on opposite sides of the membrane. The voltage across a membrane, called a **membrane potential,** ranges from

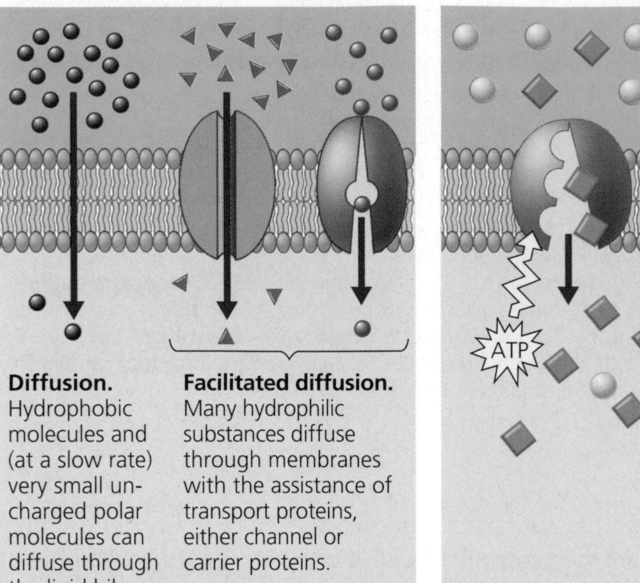

Passive transport. Substances diffuse spontaneously down their concentration gradients, crossing a membrane with no expenditure of energy by the cell. The rate of diffusion can be greatly increased by transport proteins in the membrane.

Active transport. Some transport proteins act as pumps, moving substances across a membrane against their concentration gradients. Energy for this work is usually supplied by ATP.

Diffusion. Hydrophobic molecules and (at a slow rate) very small uncharged polar molecules can diffuse through the lipid bilayer.

Facilitated diffusion. Many hydrophilic substances diffuse through membranes with the assistance of transport proteins, either channel or carrier proteins.

▲ **Figure 7.17 Review: passive and active transport compared.**

about −50 to −200 millivolts (mV). (The minus sign indicates that the inside of the cell is negative compared to the outside.)

The membrane potential acts like a battery, an energy source that affects the traffic of all charged substances across the membrane. Because the inside of the cell is negative compared to the outside, the membrane potential favors the passive transport of cations into the cell and anions out of the cell. Thus, *two* forces drive the diffusion of ions across a membrane: a chemical force (the ion's concentration gradient) and an electrical force (the effect of the membrane potential on the ion's movement). This combination of forces acting on an ion is called the **electrochemical gradient.** In the case of ions, we must refine our concept of passive transport: An ion does not simply diffuse down its *concentration* gradient, but diffuses down its *electrochemical* gradient. For example, the concentration of sodium ions (Na^+) inside a resting nerve cell is much lower than outside it. When the cell is stimulated, gated channels that facilitate Na^+ diffusion open. Sodium ions then "fall" down their electrochemical gradient, driven by the concentration gradient of Na^+ and by the attraction of cations to the negative side of the membrane.

Some membrane proteins that actively transport ions contribute to the membrane potential. An example is the sodium-potassium pump. Notice in Figure 7.16 that the pump does not

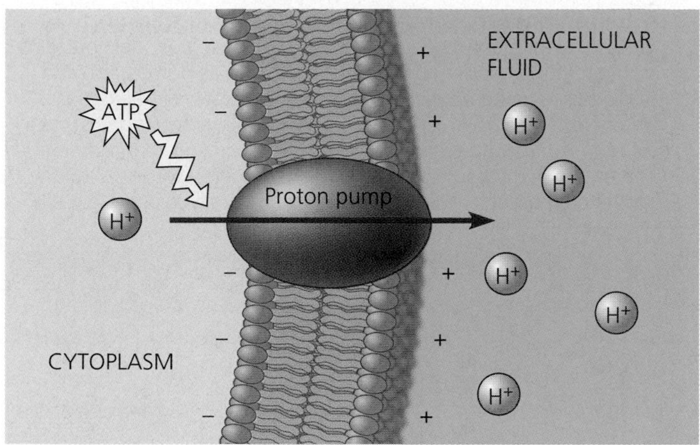

▲ **Figure 7.18 An electrogenic pump.** Proton pumps, the main electrogenic pumps of plants, fungi, and bacteria, are membrane proteins that store energy by generating voltage (charge separation) across membranes. Using ATP for power, a proton pump translocates positive charge in the form of hydrogen ions. The voltage and H^+ concentration gradient represent a dual energy source that can drive other processes, such as the uptake of nutrients.

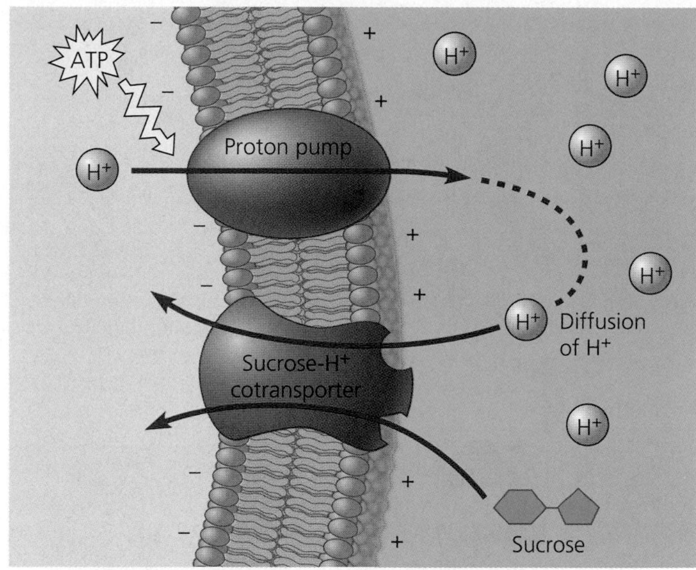

▲ **Figure 7.19 Cotransport: active transport driven by a concentration gradient.** A special carrier protein such as this sucrose-H^+ cotransporter is able to use the diffusion of H^+ down its electrochemical gradient into the cell to drive the uptake of sucrose. The H^+ gradient is maintained by an ATP-driven proton pump that concentrates H^+ outside the cell, thus storing potential energy that can be used for active transport, in this case of sucrose. Thus, ATP is indirectly providing the energy necessary for cotransport.

translocate Na^+ and K^+ one for one, but actually pumps three sodium ions out of the cell for every two potassium ions it pumps into the cell. With each "crank" of the pump, there is a net transfer of one positive charge from the cytoplasm to the extracellular fluid, a process that stores energy in the form of voltage. A transport protein that generates voltage across a membrane is called an **electrogenic pump.** The sodium-potassium pump seems to be the major electrogenic pump of animal cells. The main electrogenic pump of plants, fungi, and bacteria is a **proton pump,** which actively transports hydrogen ions (protons) out of the cell. The pumping of H^+ transfers positive charge from the cytoplasm to the extracellular solution **(Figure 7.18)**. By generating voltage across membranes, electrogenic pumps store energy that can be tapped for cellular work, including a type of membrane traffic called cotransport.

Cotransport: Coupled Transport by a Membrane Protein

A single ATP-powered pump that transports a specific solute can indirectly drive the active transport of several other solutes in a mechanism called **cotransport.** A substance that has been pumped across a membrane can do work as it moves back across the membrane by diffusion, analogous to water that has been pumped uphill and performs work as it flows back down. Another specialized transport protein, a cotransporter separate from the pump, can couple the "downhill" diffusion of this substance to the "uphill" transport of a second substance against its own concentration gradient. For example, a plant cell uses the gradient of hydrogen ions generated by its proton pumps to drive the active transport of amino acids, sugars, and several other nutrients into the cell. One specific transport protein couples the return of hydrogen ions to the transport of sucrose into the cell **(Figure 7.19)**. The protein can translocate sucrose into the cell against a concentration gradient, but only if the sucrose molecule travels in the company of a hydrogen ion. The hydrogen ion uses the common transport protein as an avenue to diffuse down the electrochemical gradient maintained by the proton pump. Plants use the mechanism of sucrose-H^+ cotransport to load sucrose produced by photosynthesis into specialized cells in the veins of leaves. The sugar can then be distributed by the vascular tissue of the plant to nonphotosynthetic organs, such as roots.

What we know about cotransport proteins, osmosis, and water balance in animal cells has helped us find more effective treatments for the dehydration resulting from diarrhea, a serious problem in developing countries where intestinal parasites are prevalent. Patients are given a solution to drink containing a high concentration of glucose and salt. The solutes are taken up by transport proteins on the intestinal cell surface and passed through the cells into the blood. The resulting osmotic pressure causes a flow of water from the intestine through the intestinal cells into the blood, rehydrating the patient. Because of the specific proteins involved, both glucose and the sodium ion from salt must be present. The same principle underlies athletes' consumption of solute-rich solutions after a demanding athletic event.

Concept Check 7.4

1. When nerve cells establish a voltage across their membrane with a sodium-potassium pump, does this pump use ATP or does it produce ATP? Why?
2. Explain why the sodium-potassium pump in Figure 7.16 would not be considered a cotransporter.

For suggested answers, see Appendix A.

Concept 7.5

Bulk transport across the plasma membrane occurs by exocytosis and endocytosis

Water and small solutes enter and leave the cell by passing through the lipid bilayer of the plasma membrane or by being pumped or carried across the membrane by transport proteins. However, large molecules, such as proteins and polysaccharides, as well as larger particles, generally cross the membrane by a different mechanism—one involving vesicles.

Exocytosis

As we described in Chapter 6, the cell secretes macromolecules by the fusion of vesicles with the plasma membrane; this is called **exocytosis.** A transport vesicle that has budded from the Golgi apparatus moves along microtubules of the cytoskeleton to the plasma membrane. When the vesicle membrane and plasma membrane come into contact, the lipid molecules of the two bilayers rearrange themselves so that the two membranes fuse. The contents of the vesicle then spill to the outside of the cell, and the vesicle membrane becomes part of the plasma membrane (see Figure 7.10).

Many secretory cells use exocytosis to export their products. For example, certain cells in the pancreas manufacture the hormone insulin and secrete it into the blood by exocytosis. Another example is the neuron, or nerve cell, which uses exocytosis to release neurotransmitters that signal other neurons or muscle cells. When plant cells are making walls, exocytosis delivers proteins and certain carbohydrates from Golgi vesicles to the outside of the cell.

Endocytosis

In **endocytosis,** the cell takes in macromolecules and particulate matter by forming new vesicles from the plasma membrane. Although the proteins involved in the processes are different, the events of endocytosis look like the reverse of exocytosis. A small area of the plasma membrane sinks inward to form a pocket. As the pocket deepens, it pinches in, forming a vesicle containing material that had been outside the cell. There are three types of endocytosis: phagocytosis ("cellular eating"), pinocytosis ("cellular drinking"), and receptor-mediated endocytosis. Please study **Figure 7.20** on page 138, which describes these processes, before going on.

Human cells use receptor-mediated endocytosis to take in cholesterol for use in the synthesis of membranes and as a precursor for the synthesis of other steroids. Cholesterol travels in the blood in particles called low-density lipoproteins (LDLs), complexes of lipids and proteins. These particles act as **ligands** (a general term for any molecule that binds specifically to a receptor site of another molecule) by binding to LDL receptors on membranes and then entering the cells by endocytosis. In humans with familial hypercholesterolemia, an inherited disease characterized by a very high level of cholesterol in the blood, the LDL receptor proteins are defective or missing, and the LDL particles cannot enter cells. Instead, cholesterol accumulates in the blood, where it contributes to early atherosclerosis, the buildup of lipid deposits within the walls of blood vessels, causing them to bulge inward and impede blood flow.

Vesicles not only transport substances between the cell and its surroundings but also provide a mechanism for rejuvenating or remodeling the plasma membrane. Endocytosis and exocytosis occur continually to some extent in most eukaryotic cells, and yet the amount of plasma membrane in a nongrowing cell remains fairly constant over the long run. Apparently, the addition of membrane by one process offsets the loss of membrane by the other.

Energy and cellular work have figured prominently in our study of membranes. We have seen, for example, that active transport is powered by ATP. In the next three chapters, you will learn more about how cells acquire chemical energy to do the work of life.

Concept Check 7.5

1. As a cell grows, its plasma membrane expands. Does this process involve endocytosis or exocytosis? Explain.
2. To send a signal, a neuron may carry out exocytosis of chemical signals that are recognized by a second neuron. In some cases, the first neuron ends the signal by taking up the signaling molecules by endocytosis. Would you expect this to occur by pinocytosis or by receptor-mediated endocytosis? Explain your reasoning.

For suggested answers, see Appendix A.

Figure 7.20

Exploring Endocytosis in Animal Cells

PHAGOCYTOSIS

In **phagocytosis**, a cell engulfs a particle by wrapping pseudopodia around it and packaging it within a membrane-enclosed sac large enough to be classified as a vacuole. The particle is digested after the vacuole fuses with a lysosome containing hydrolytic enzymes.

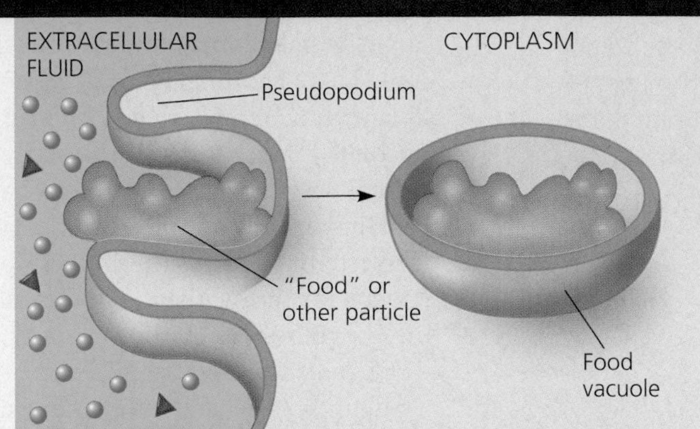

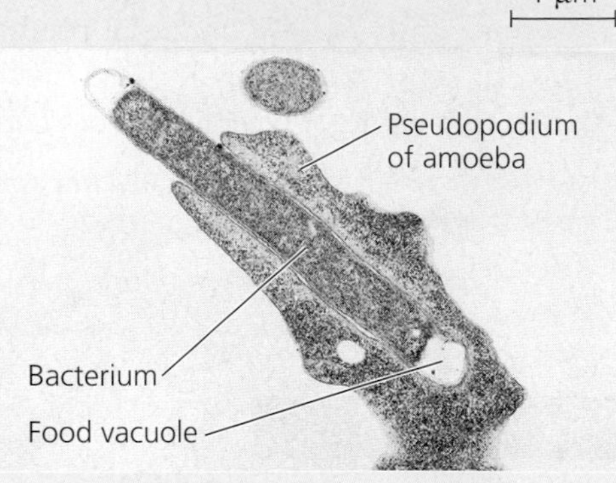

An amoeba engulfing a bacterium via phagocytosis (TEM).

PINOCYTOSIS

In **pinocytosis**, the cell "gulps" droplets of extracellular fluid into tiny vesicles. It is not the fluid itself that is needed by the cell, but the molecules dissolved in the droplet. Because any and all included solutes are taken into the cell, pinocytosis is nonspecific in the substances it transports.

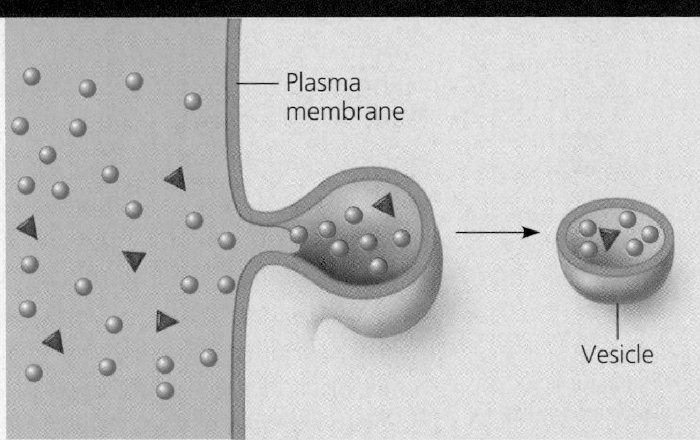

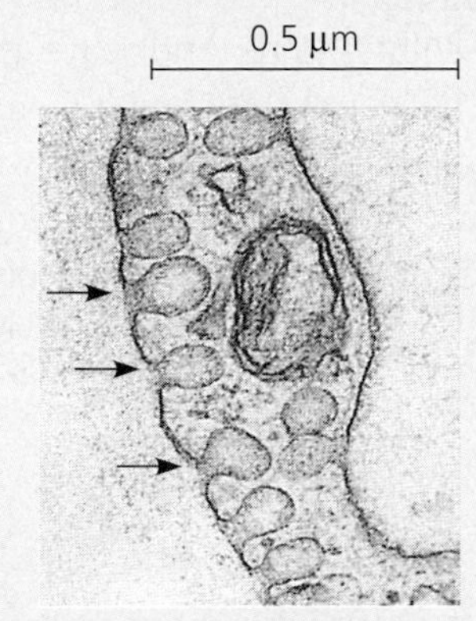

Pinocytosis vesicles forming (arrows) in a cell lining a small blood vessel (TEM).

RECEPTOR-MEDIATED ENDOCYTOSIS

Receptor-mediated endocytosis enables the cell to acquire bulk quantities of specific substances, even though those substances may not be very concentrated in the extracellular fluid. Embedded in the membrane are proteins with specific receptor sites exposed to the extracellular fluid. The receptor proteins are usually already clustered in regions of the membrane called coated pits, which are lined on their cytoplasmic side by a fuzzy layer of coat proteins. Extracellular substances (ligands) bind to these receptors. When binding occurs, the coated pit forms a vesicle containing the ligand molecules. Notice that there are relatively more bound molecules (purple) inside the vesicle, but other molecules (green) are also present. After this ingested material is liberated from the vesicle, the receptors are recycled to the plasma membrane by the same vesicle.

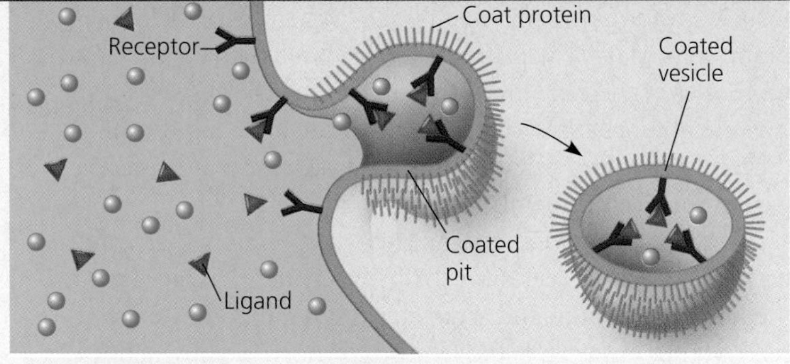

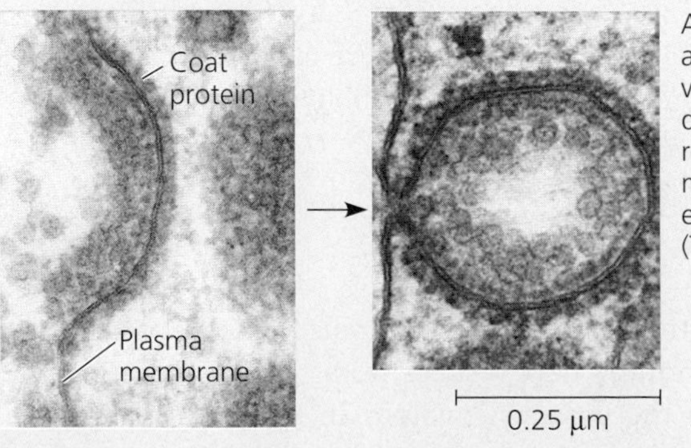

A coated pit and a coated vesicle formed during receptor-mediated endocytosis (TEMs).

Chapter 7 Review

Go to www.campbellbiology.com or the student CD-ROM to explore Activities, Investigations, and other interactive study aids.

SUMMARY OF KEY CONCEPTS

Concept 7.1

Cellular membranes are fluid mosaics of lipids and proteins

- **Membrane Models: *Scientific Inquiry*** (pp. 125–126) The Davson-Danielli sandwich model of the membrane has been replaced by the fluid mosaic model, in which amphipathic proteins are embedded in the phospholipid bilayer.
- **The Fluidity of Membranes** (pp. 126–127) Phospholipids and, to a lesser extent, proteins move laterally within the membrane. Cholesterol and unsaturated hydrocarbon tails in the phospholipids affect membrane fluidity.
- **Membrane Proteins and Their Functions** (pp. 127–129) Integral proteins are embedded in the lipid bilayer; peripheral proteins are attached to the surfaces. The functions of membrane proteins include transport, enzymatic activity, signal transduction, cell-cell recognition, intercellular joining, and attachment to the cytoskeleton and extracellular matrix.
- **The Role of Membrane Carbohydrates in Cell-Cell Recognition** (p. 129) Short chains of sugars are linked to proteins and lipids on the exterior side of the plasma membrane, where they can interact with the surface molecules of other cells.
- **Synthesis and Sidedness of Membranes** (p. 129) Membrane proteins and lipids are synthesized in the ER and modified in the ER and Golgi apparatus. The inside and outside faces of the membrane differ in composition.
  Activity *Membrane Structure*

Concept 7.2

Membrane structure results in selective permeability

- A cell must exchange small molecules and ions with its surroundings, a process controlled by the plasma membrane (p. 130).
- **The Permeability of the Lipid Bilayer** (p. 130) Hydrophobic substances are soluble in lipid and pass through membranes rapidly.
- **Transport Proteins** (p. 130) Polar molecules and ions generally require specific transport proteins to help them cross.
  Activity *Selective Permeability of Membranes*

Concept 7.3

Passive transport is diffusion of a substance across a membrane with no energy investment

- Diffusion is the spontaneous movement of a substance down its concentration gradient (pp. 130–131).
  Activity *Diffusion*
- **Effects of Osmosis on Water Balance** (pp. 131–133) Water flows across a membrane from the side where solute is less concentrated (hypotonic) to the side where solute is more concentrated (hypertonic). If the concentrations are equal (isotonic), no net osmosis occurs. Cell survival depends on balancing water uptake and loss. Cells lacking walls (as in animals and some protists) are isotonic with their environments or have adaptations for osmoregulation. Plants, prokaryotes, fungi, and some protists have elastic cell walls, so the cells don't burst in a hypotonic environment.
  Activity *Osmosis and Water Balance in Cells*
  Investigation *How Do Salt Concentrations Affect Cells?*
- **Facilitated Diffusion: Passive Transport Aided by Proteins** (pp. 133–134) In facilitated diffusion, a transport protein speeds the movement of water or a solute across a membrane down its concentration gradient.
  Activity *Facilitated Diffusion*

Concept 7.4

Active transport uses energy to move solutes against their gradients

- **The Need for Energy in Active Transport** (pp. 134–135) Specific membrane proteins use energy, usually in the form of ATP, to do the work of active transport.
  Activity *Active Transport*
- **Maintenance of Membrane Potential by Ion Pumps** (pp. 134–136) Ions can have both a concentration (chemical) gradient and an electrical gradient (voltage). These forces combine in the electrochemical gradient, which determines the net direction of ionic diffusion. Electrogenic pumps, such as sodium-potassium pumps and proton pumps, are transport proteins that contribute to electrochemical gradients.
- **Cotransport; Coupled Transport by a Membrane Protein** (p. 136) One solute's "downhill" diffusion drives the other's "uphill" transport.

Concept 7.5

Bulk transport across the plasma membrane occurs by exocytosis and endocytosis

- **Exocytosis** (p. 137) In exocytosis, transport vesicles migrate to the plasma membrane, fuse with it, and release their contents.
- **Endocytosis** (pp. 137–138) In endocytosis, molecules enter cells within vesicles that pinch inward from the plasma membrane. The three types of endocytosis are phagocytosis, pinocytosis, and receptor-mediated endocytosis.
  Activity *Exocytosis and Endocytosis*

TESTING YOUR KNOWLEDGE

Self-Quiz

1. In what way do the various membranes of a eukaryotic cell differ?
   a. Phospholipids are found only in certain membranes.
   b. Certain proteins are unique to each membrane.
   c. Only certain membranes of the cell are selectively permeable.
   d. Only certain membranes are constructed from amphipathic molecules.
   e. Some membranes have hydrophobic surfaces exposed to the cytoplasm, while others have hydrophilic surfaces facing the cytoplasm.

2. According to the fluid mosaic model of membrane structure, proteins of the membrane are mostly
   a. spread in a continuous layer over the inner and outer surfaces of the membrane.
   b. confined to the hydrophobic core of the membrane.
   c. embedded in a lipid bilayer.
   d. randomly oriented in the membrane, with no fixed inside-outside polarity.
   e. free to depart from the fluid membrane and dissolve in the surrounding solution.

3. Which of the following factors would tend to increase membrane fluidity?
   a. a greater proportion of unsaturated phospholipids
   b. a greater proportion of saturated phospholipids
   c. a lower temperature
   d. a relatively high protein content in the membrane
   e. a greater proportion of relatively large glycolipids compared to lipids having smaller molecular masses

4. Which of the following processes includes all others?
   a. osmosis
   b. diffusion of a solute across a membrane
   c. facilitated diffusion
   d. passive transport
   e. transport of an ion down its electrochemical gradient

5. Based on the model of sucrose uptake in Figure 7.19, which of the following experimental treatments would increase the rate of sucrose transport into the cell?
   a. decreasing extracellular sucrose concentration
   b. decreasing extracellular pH
   c. decreasing cytoplasmic pH
   d. adding an inhibitor that blocks the regeneration of ATP
   e. adding a substance that makes the membrane more permeable to hydrogen ions

Questions 6–10

An artificial cell consisting of an aqueous solution enclosed in a selectively permeable membrane has just been immersed in a beaker containing a different solution. The membrane is permeable to water and to the simple sugars glucose and fructose but completely impermeable to the disaccharide sucrose.

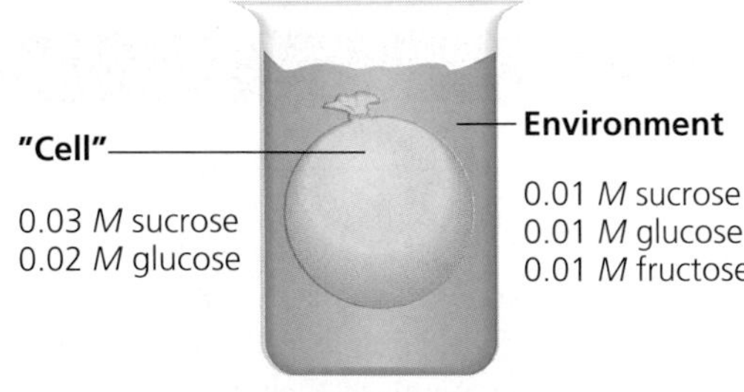

6. Which solute(s) will exhibit a net diffusion into the cell?

7. Which solute(s) will exhibit a net diffusion out of the cell?

8. Is the solution outside the cell isotonic, hypotonic, or hypertonic?

9. In which direction will there be a net osmotic movement of water?

10. After the cell is placed in the beaker, which of the following changes will occur? (Choose all that apply.)
    a. The artificial cell will become more flaccid.
    b. The artificial cell will become more turgid.
    c. Some water molecules will flow out of the cell, but the majority will flow into it.
    d. The membrane potential will decrease.
    e. In spite of the inability of sucrose to cross the membrane, eventually the two solutions will have the same solute concentrations.

For Self-Quiz answers, see Appendix A.

Go to the website or CD-ROM for more quiz questions.

Evolution Connection

Paramecium and other protists that live in hypotonic environments have cell membrane adaptations that slow osmotic water uptake, while those living in isotonic environments have more permeable cell membranes. What water regulation adaptations would you expect to have evolved in protists living in hypertonic habitats such as Great Salt Lake? How about those living in habitats where salt concentration fluctuates?

Scientific Inquiry

An experiment is designed to study the mechanism of sucrose uptake by plant cells. Cells are immersed in a sucrose solution, and the pH of the solution is monitored with a pH meter. Samples of the cells are taken at intervals, and the sucrose concentration in the sampled cells is measured. The measurements show that sucrose uptake by the cells correlates with a rise in the pH of the surrounding solution. The magnitude of the pH change is proportional to the starting concentration of sucrose in the extracellular solution. A metabolic poison known to block the ability of cells to regenerate ATP is found to inhibit the pH changes in the extracellular solution. Propose a hypothesis accounting for these results. Suggest an additional experiment to test your hypothesis.

Investigation *How Do Salt Concentrations Affect Cells?*

Science, Technology, and Society

Extensive irrigation in arid regions causes salts to accumulate in the soil. (The water contains low concentrations of salts, but when the water evaporates from the fields, the salts are left behind to concentrate in the soil.) Based on what you have learned about water balance in plant cells, explain why increasing soil salinity (saltiness) has an adverse effect on agriculture. Suggest some ways to minimize this damage. What costs are attached to your solutions?

8 An Introduction to Metabolism

▲ **Figure 8.1 Bioluminescence by a fungus.**

Key Concepts

8.1 An organism's metabolism transforms matter and energy, subject to the laws of thermodynamics
8.2 The free-energy change of a reaction tells us whether the reaction occurs spontaneously
8.3 ATP powers cellular work by coupling exergonic reactions to endergonic reactions
8.4 Enzymes speed up metabolic reactions by lowering energy barriers
8.5 Regulation of enzyme activity helps control metabolism

Overview

The Energy of Life

The living cell is a chemical factory in miniature, where thousands of reactions occur within a microscopic space. Sugars can be converted to amino acids that are linked together into proteins when needed, and proteins are dismantled into amino acids that can be converted to sugars when food is digested. Small molecules are assembled into polymers, which may be hydrolyzed later as the needs of the cell change. In multicellular organisms, many cells export chemical products that are used in other parts of the organism. The process known as cellular respiration drives the cellular economy by extracting the energy stored in sugars and other fuels. Cells apply this energy to perform various types of work, such as the transport of solutes across the plasma membrane, which we discussed in Chapter 7. In a more exotic example, cells of the fungus in **Figure 8.1** convert the energy stored in certain organic molecules to light, a process called bioluminescence. (The glow may attract insects that benefit the fungus by dispersing its spores.) Bioluminescence and all other metabolic activities carried out by a cell are precisely coordinated and controlled. In its complexity, its efficiency, its integration, and its responsiveness to subtle changes, the cell is peerless as a chemical factory. The concepts of metabolism that you learn in this chapter will help you understand how matter and energy flow during life's processes and how that flow is regulated.

Concept 8.1

An organism's metabolism transforms matter and energy, subject to the laws of thermodynamics

The totality of an organism's chemical reactions is called **metabolism** (from the Greek *metabole,* change). Metabolism is an emergent property of life that arises from interactions between molecules within the orderly environment of the cell.

Organization of the Chemistry of Life into Metabolic Pathways

We can picture a cell's metabolism as an elaborate road map of the thousands of chemical reactions that occur in a cell, arranged as intersecting metabolic pathways. A **metabolic pathway** begins with a specific molecule, which is then altered in a series of defined steps, resulting in a certain product. Each step of the pathway is catalyzed by a specific enzyme:

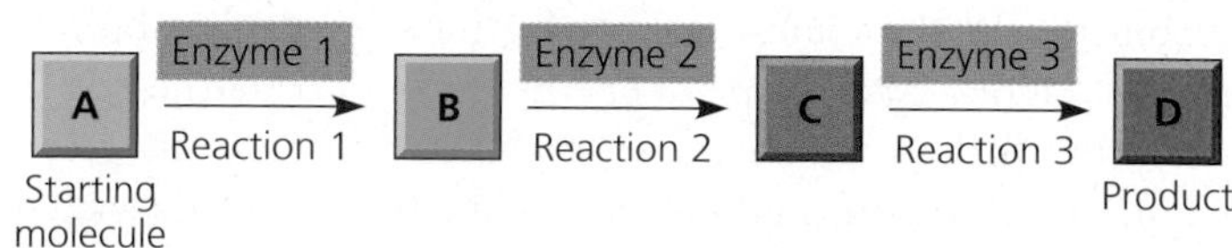

Analogous to the red, yellow, and green stoplights that control the flow of traffic, mechanisms that regulate enzymes balance metabolic supply and demand, averting deficits or surpluses of important cellular molecules.

Metabolism as a whole manages the material and energy resources of the cell. Some metabolic pathways release energy by breaking down complex molecules to simpler compounds. These degradative processes are called **catabolic pathways**, or breakdown pathways. A major pathway of catabolism is cellular respiration, in which the sugar glucose and other organic fuels are broken down in the presence of oxygen to carbon dioxide and water. (Pathways can have more than one starting molecule and/or product.) Energy that was stored in the organic molecules becomes available to do the work of the cell, such as ciliary beating or membrane transport. **Anabolic pathways**, in contrast, consume energy to build complicated molecules from simpler ones; they are sometimes called biosynthetic pathways. An example of anabolism is the synthesis of a protein from amino acids. Catabolic and anabolic pathways are the "downhill" and "uphill" avenues of the metabolic map. Energy released from the downhill reactions of catabolism can be stored and then used to drive the uphill reactions of the anabolic pathways.

In this chapter, we will focus on the mechanisms common to metabolic pathways. Because energy is fundamental to all metabolic processes, a basic knowledge of energy is necessary to understand how the living cell works. Although we will use some nonliving examples to study energy, keep in mind that the concepts demonstrated by these examples also apply to **bioenergetics**, the study of how organisms manage their energy resources.

Forms of Energy

Energy is the capacity to cause change. In everyday life, energy is important because some forms of energy can be used to do work—that is, to move matter against opposing forces, such as gravity and friction. Put another way, energy is the ability to rearrange a collection of matter. For example, you expend energy to turn the pages of this book, and your cells expend energy in transporting certain substances across membranes. Energy exists in various forms, and the work of life depends on the ability of cells to transform energy from one type into another.

Energy can be associated with the relative motion of objects; this energy is called **kinetic energy.** Moving objects can perform work by imparting motion to other matter: A pool player uses the motion of the cue stick to push the cue ball, which in turn moves the other balls; water gushing over a dam turns turbines; and the contraction of leg muscles pushes bicycle pedals. Light is also a type of energy that can be harnessed to perform work, such as powering photosynthesis in green plants. **Heat**, or **thermal energy**, is kinetic energy associated with the random movement of atoms or molecules.

An object not presently moving may still possess energy. Energy that is not kinetic is called **potential energy;** it is energy that matter possesses because of its location or structure. Water behind a dam, for instance, stores energy because of its altitude above sea level. Molecules store energy because of the arrangement of their atoms. **Chemical energy** is a term used by biologists to refer to the potential energy available for release in a chemical reaction. Recall that catabolic pathways release energy by breaking down complex molecules. Biologists say that these complex molecules, such as glucose, are high in chemical energy. During a catabolic reaction, atoms are rearranged and energy is released, resulting in lower-energy breakdown products. This transformation also occurs, for example, in the engine of a car when the hydrocarbons of gasoline react explosively with oxygen, releasing the energy that pushes the pistons and producing exhaust. Although less explosive, a similar reaction of food molecules with oxygen provides chemical energy in biological systems, producing carbon dioxide and water as waste products. It is the structures and biochemical pathways of cells that enable them to release chemical energy from food molecules, powering life processes.

How is energy converted from one form to another? Consider the divers in **Figure 8.2**. The young man climbing the

On the platform, a diver has more potential energy.
Diving converts potential energy to kinetic energy.
Climbing up converts kinetic energy of muscle movement to potential energy.
In the water, a diver has less potential energy.

▲ **Figure 8.2 Transformations between kinetic and potential energy.**

steps to the diving platform is releasing chemical energy from the food he ate for lunch and using some of that energy to perform the work of climbing. The kinetic energy of muscle movement is thus being transformed into potential energy owing to his increasing height above the water. The young man diving is converting his potential energy to kinetic energy, which is then transferred to the water as he enters it. A small amount of energy is lost as heat due to friction.

Now let's go back one step and consider the original source of the organic food molecules that provided the necessary chemical energy for the divers to climb the steps. This chemical energy was itself derived from light energy by plants during photosynthesis. Organisms are energy transformers.

The Laws of Energy Transformation

The study of the energy transformations that occur in a collection of matter is called **thermodynamics.** Scientists use the word *system* to denote the matter under study; they refer to the rest of the universe—everything outside the system—as the *surroundings.* A *closed system,* such as that approximated by liquid in a thermos bottle, is isolated from its surroundings. In an *open system,* energy (and often matter) can be transferred between the system and its surroundings. Organisms are open systems. They absorb energy—for instance, light energy or chemical energy in the form of organic molecules—and release heat and metabolic waste products, such as carbon dioxide, to the surroundings. Two laws of thermodynamics govern energy transformations in organisms and all other collections of matter.

The First Law of Thermodynamics

According to the **first law of thermodynamics,** the energy of the universe is constant. *Energy can be transferred and transformed, but it cannot be created or destroyed.* The first law is also known as the *principle of conservation of energy.* The electric company does not make energy, but merely converts it to a form that is convenient to use. By converting sunlight to chemical energy, a green plant acts as an energy transformer, not an energy producer.

The cheetah in **Figure 8.3a** will convert the chemical energy of the organic molecules in its food to kinetic and other forms of energy as it carries out biological processes. What happens to this energy after it has performed work? The second law helps to answer this question.

The Second Law of Thermodynamics

If energy cannot be destroyed, why can't organisms simply recycle their energy over and over again? It turns out that during every energy transfer or transformation, some energy becomes unusable energy, unavailable to do work. In most energy transformations, more usable forms of energy are at least partly converted to heat, which is the energy associated with the random motion of atoms or molecules. Only a small fraction of the chemical energy from the food in Figure 8.3a is transformed into the motion of the cheetah shown in **Figure 8.3b**; most is lost as heat, which dissipates rapidly through the surroundings.

In the process of carrying out chemical reactions that perform various kinds of work, living cells unavoidably convert organized forms of energy to heat. A system can put heat to work only when there is a temperature difference that results in the heat flowing from a warmer location to a cooler one. If temperature is uniform, as it is in a living cell, then the only use for heat energy generated during a chemical reaction is to warm a body of matter, such as the organism. (This can make a room crowded with people uncomfortably

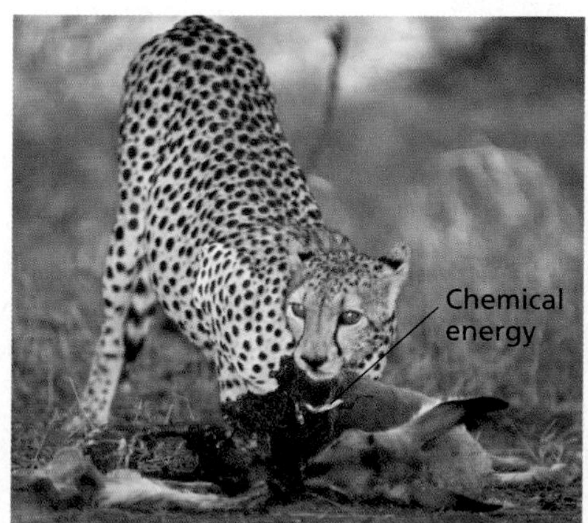

(a) First law of thermodynamics: Energy can be transferred or transformed but neither created nor destroyed. For example, the chemical (potential) energy in food will be converted to the kinetic energy of the cheetah's movement in (b).

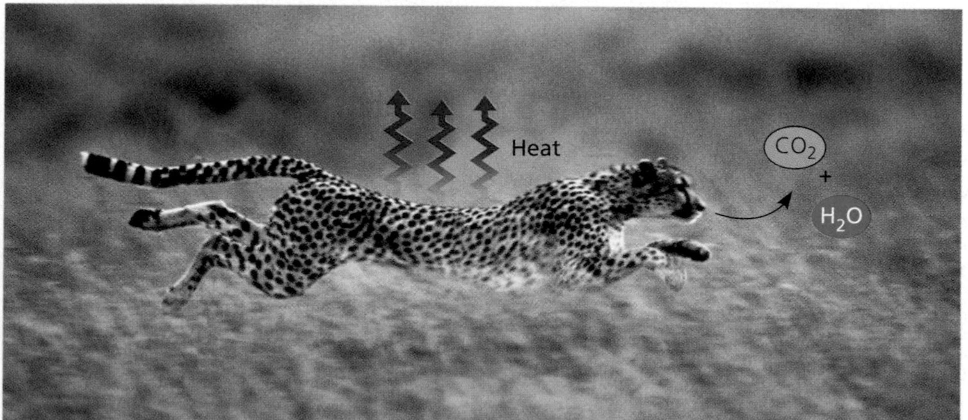

(b) Second law of thermodynamics: Every energy transfer or transformation increases the disorder (entropy) of the universe. For example, disorder is added to the cheetah's surroundings in the form of heat and the small molecules that are the by-products of metabolism.

▲ **Figure 8.3 The two laws of thermodynamics.**

warm, as each person is carrying out a multitude of chemical reactions!)

A logical consequence of the loss of usable energy during energy transfer or transformation is that each such event makes the universe more disordered. Scientists use a quantity called **entropy** as a measure of disorder, or randomness. The more randomly arranged a collection of matter is, the greater its entropy. We can now state the **second law of thermodynamics** as follows: *Every energy transfer or transformation increases the entropy of the universe.* Although order can increase locally, there is an unstoppable trend toward randomization of the universe as a whole.

In many cases, increased entropy is evident in the physical disintegration of a system's organized structure. For example, you can observe increasing entropy in the gradual decay of an unmaintained building. Much of the increasing entropy of the universe is less apparent, however, because it appears as increasing amounts of heat and less ordered forms of matter. As the cheetah in Figure 8.3b converts chemical energy to kinetic energy, it is also increasing the disorder of its surroundings by producing heat and the small molecules that are the breakdown products of food.

The concept of entropy helps us understand why certain processes occur. It turns out that for a process to occur on its own, without outside help (an input of energy), it must increase the entropy of the universe. Let's first agree to use the word *spontaneous* for a process that can occur without an input of energy. Note that as we're using it here, the word *spontaneous* does not imply that such a process would occur quickly. Some spontaneous processes may be virtually instantaneous, such as an explosion, while others may be much slower, such as the rusting of an old car over time. A process that cannot occur on its own is said to be nonspontaneous; it will happen only if energy is added to the system. We know from experience that certain events occur spontaneously and others do not. For instance, we know that water flows downhill spontaneously, but moves uphill only with an input of energy, for instance when a machine pumps the water against gravity. In fact, another way to state the second law is: *For a process to occur spontaneously, it must increase the entropy of the universe.*

Biological Order and Disorder

Living systems increase the entropy of their surroundings, as predicted by thermodynamic law. It is true that cells create ordered structures from less organized starting materials. For example, amino acids are ordered into the specific sequences of polypeptide chains. At the organismal level, **Figure 8.4** shows the extremely symmetrical anatomy of a plant's root, formed by biological processes from simpler starting materials. However, an organism also takes in organized forms of matter and energy from the surroundings and replaces them with less ordered forms. For example, an animal obtains starch, proteins, and other complex molecules from the food it eats. As catabolic pathways break these molecules down, the animal releases carbon dioxide and water—small molecules that store less chemical energy than the food did. The depletion of chemical energy is accounted for by heat generated during metabolism. On a larger scale, energy flows into an ecosystem in the form of light and leaves in the form of heat.

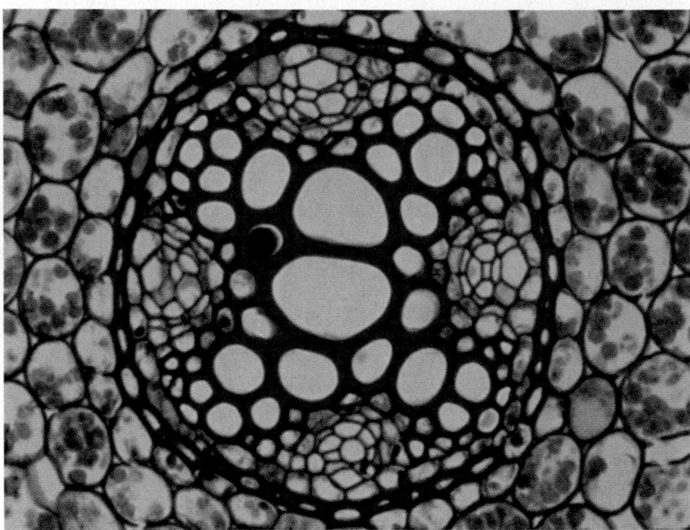

▲ **Figure 8.4 Order as a characteristic of life.** Order is evident in the detailed anatomy of this root tissue from a buttercup plant (LM, cross section). As open systems, organisms can increase their order as long as the order of their surroundings decreases.

During the early history of life, complex organisms evolved from simpler ancestors. For example, we can trace the ancestry of the plant kingdom to much simpler organisms called green algae. However, this increase in organization over time in no way violates the second law. The entropy of a particular system, such as an organism, may actually decrease, so long as the total entropy of the *universe*—the system plus its surroundings—increases. Thus, organisms are islands of low entropy in an increasingly random universe. The evolution of biological order is perfectly consistent with the laws of thermodynamics.

Concept Check 8.1

1. How does the second law of thermodynamics help explain the diffusion of a substance across a membrane?
2. What is the relationship between energy and work?
3. Describe the forms of energy found in an apple as it grows on a tree, then falls and is digested by someone who eats it.

For suggested answers, see Appendix A.

The free-energy change of a reaction tells us whether the reaction occurs spontaneously

The laws of thermodynamics that we've just discussed apply to the universe as a whole. As biologists, we want to understand the chemical reactions of life—for example, to know which reactions occur spontaneously and which ones require some input of energy from outside. But how can we know this without assessing the energy and entropy changes in the entire universe for each separate reaction?

Free-Energy Change, ΔG

Recall that the universe is really equivalent to "the system" plus "the surroundings." In 1878, J. Willard Gibbs, a professor at Yale, defined a very useful function called the Gibbs free energy of a system (without considering its surroundings), symbolized by the letter G. We'll refer to the Gibbs free energy simply as free energy. **Free energy** measures the portion of a system's energy that can perform work when temperature and pressure are uniform throughout the system, as in a living cell. Let's consider how we determine the free energy change that occurs when a system changes—for example, during a chemical reaction.

The change in free energy, ΔG, can be calculated for any specific chemical reaction with the following formula:

$$\Delta G = \Delta H - T\Delta S$$

This formula uses only properties of the system (the reaction) itself: ΔH symbolizes the change in the system's *enthalpy* (in biological systems, equivalent to total energy); ΔS is the change in the system's entropy; and T is the absolute temperature in Kelvin (K) units (K = °C + 273; see Appendix B).

Once we know the value of ΔG for a process, we can use it to predict whether the process will be spontaneous (that is, whether it will run without an outside input of energy). A century of experiments has shown that only processes with a negative ΔG are spontaneous. For a process to occur spontaneously, therefore, the system must either give up enthalpy (H must decrease), give up order (TS must increase), or both: When the changes in H and TS are tallied, ΔG must have a negative value ($\Delta G < 0$). This means that every spontaneous process decreases the system's free energy. Processes that have a positive or zero ΔG are never spontaneous.

This information is immensely interesting to biologists, for it gives us the power to predict which kinds of change can happen without help. Such spontaneous changes can be harnessed to perform work. This principle is very important in the study of metabolism, where a major goal is to determine which reactions can supply energy to do work in the living cell.

Free Energy, Stability, and Equilibrium

As we saw in the previous section, when a process occurs spontaneously in a system, we can be sure that ΔG is negative. Another way to think of ΔG is to realize that it represents the difference between the free energy of the final state and the free energy of the initial state:

$$\Delta G = G_{\text{final state}} - G_{\text{initial state}}$$

Thus, ΔG can only be negative when the process involves a loss of free energy during the change from initial state to final state. Because it has less free energy, the system in its final state is less likely to change and is therefore more stable than it was previously.

We can think of free energy as a measure of a system's instability—its tendency to change to a more stable state. Unstable systems (higher G) tend to change in such a way that they become more stable (lower G). For example, a diver on top of a platform is less stable than when floating in the water, a drop of concentrated dye is less stable than when the dye is spread randomly through the liquid, and a sugar molecule is less stable than the simpler molecules into which it can be broken (**Figure 8.5**, on the next page). Unless something prevents it, each of these systems will move toward greater stability: The diver falls, the solution becomes uniformly colored, and the sugar molecule is broken down.

Another term for a state of maximum stability is *equilibrium*, which you learned about in Chapter 2 in connection with chemical reactions. There is an important relationship between free energy and equilibrium, including chemical equilibrium. Recall that most chemical reactions are reversible and proceed to a point at which the forward and backward reactions occur at the same rate. The reaction is then said to be at chemical equilibrium, and there is no further net change in the relative concentration of products and reactants.

As a reaction proceeds toward equilibrium, the free energy of the mixture of reactants and products decreases. Free energy increases when a reaction is somehow pushed away from equilibrium, perhaps by removing some of the products (and thus changing their concentration relative to that of the reactants). For a system at equilibrium, G is at its lowest possible value in that system. We can think of the equilibrium state as an energy valley. Any small change from the equilibrium position will have a positive ΔG and will not be spontaneous. For this reason, systems never spontaneously move away from equilibrium. Because a system at equilibrium cannot spontaneously change, it can do no work. A process is spontaneous and can perform work only when it is moving toward equilibrium.

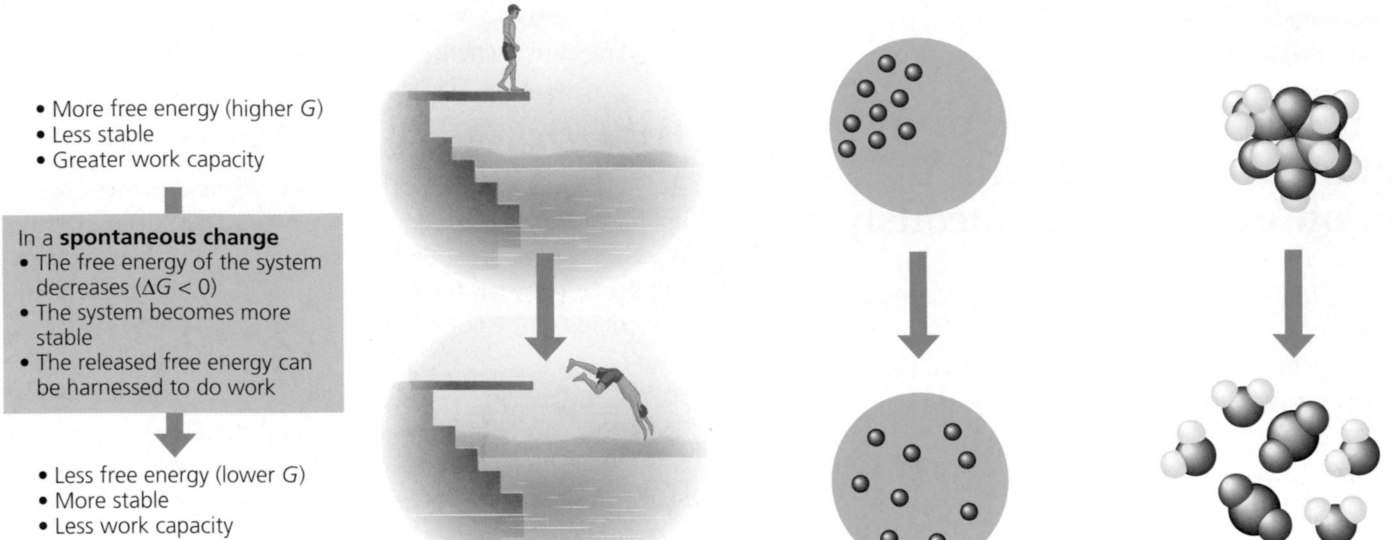

• More free energy (higher G)
• Less stable
• Greater work capacity
In a spontaneous change
• The free energy of the system decreases (ΔG < 0)
• The system becomes more stable
• The released free energy can be harnessed to do work
• Less free energy (lower G)
• More stable
• Less work capacity

(a) Gravitational motion. Objects move spontaneously from a higher altitude to a lower one.

(b) Diffusion. Molecules in a drop of dye diffuse until they are randomly dispersed.

(c) Chemical reaction. In a cell, a sugar molecule is broken down into simpler molecules.

▲ **Figure 8.5 The relationship of free energy to stability, work capacity, and spontaneous change.** Unstable systems (top diagrams) are rich in free energy, or *G*. They have a tendency to change spontaneously to a more stable state (bottom), and it is possible to harness this "downhill" change to perform work.

Free Energy and Metabolism

We can now apply the free-energy concept more specifically to the chemistry of life's processes.

Exergonic and Endergonic Reactions in Metabolism

Based on their free-energy changes, chemical reactions can be classified as either exergonic ("energy outward") or endergonic ("energy inward"). An **exergonic reaction** proceeds with a net release of free energy **(Figure 8.6a)**. Because the chemical mixture loses free energy (G decreases), ΔG is negative for an exergonic reaction. Using ΔG as a standard for spontaneity, exergonic reactions are those that occur spontaneously. (Remember, the word *spontaneous* does not imply that a reaction will occur instantaneously, or even rapidly.) The magnitude of ΔG for an exergonic reaction represents the maximum amount of work the reaction can perform.* The greater the decrease in free energy, the greater the amount of work that can be done.

We can use the overall reaction for cellular respiration as an example:

$$C_6H_{12}O_6 + O_2 \rightarrow 6\ CO_2 + 6\ H_2O$$

$$\Delta G = -686 \text{ kcal/mol } (-2{,}870 \text{ kJ/mol})$$

* The word *maximum* qualifies this statement, because some of the free energy is released as heat and cannot do work. Therefore, ΔG represents a theoretical upper limit of available energy.

For each mole (180 g) of glucose broken down by respiration under what are called "standard conditions" (1 *M* of each reactant and product, 25°C, pH 7), 686 kcal (2,870 kJ) of energy are made available for work. Because energy must be conserved, the chemical products of respiration store 686 kcal less free energy per mole than the reactants. The products are, in a sense, the spent exhaust of a process that tapped the free energy stored in the sugar molecules.

An **endergonic reaction** is one that absorbs free energy from its surroundings **(Figure 8.6b)**. Because this kind of reaction essentially *stores* free energy in molecules (G increases), ΔG is positive. Such reactions are nonspontaneous, and the magnitude of ΔG is the quantity of energy required to drive the reaction. If a chemical process is exergonic (downhill) in one direction, then the reverse process must be endergonic (uphill). A reversible process cannot be downhill in both directions. If $\Delta G = -686$ kcal/mol for respiration, which converts sugar and oxygen to carbon dioxide and water, then the reverse process—the conversion of carbon dioxide and water to sugar and oxygen—must be strongly endergonic, with $\Delta G = +686$ kcal/mol. Such a reaction would never happen by itself.

How, then, do plants make the sugar that organisms use for energy? They get the required energy—686 kcal to make a mole of sugar—from the environment by capturing light and converting its energy to chemical energy. Next, in a long series of exergonic steps, they gradually spend that chemical energy to assemble sugar molecules.

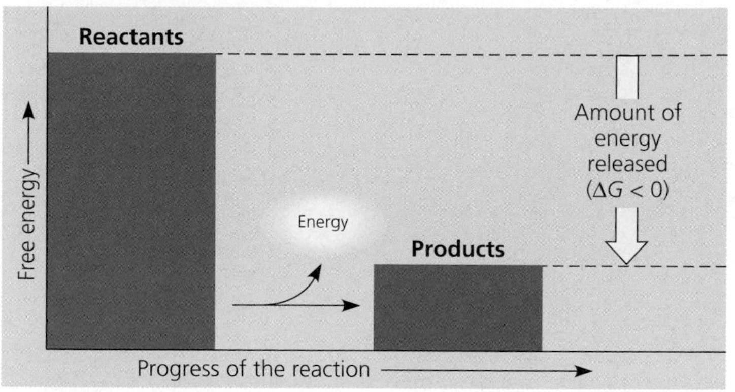

(a) Exergonic reaction: energy released

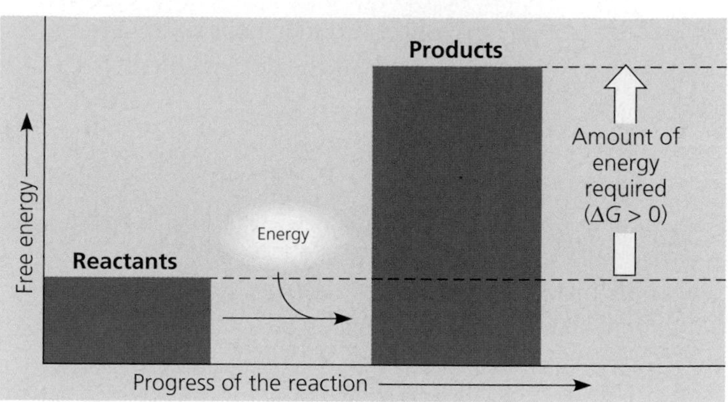

(b) Endergonic reaction: energy required

▲ **Figure 8.6 Free energy changes (ΔG) in exergonic and endergonic reactions.**

Equilibrium and Metabolism

Reactions in a closed system eventually reach equilibrium and can then do no work, as illustrated by the closed hydroelectric system in **Figure 8.7a**. The chemical reactions of metabolism are reversible, and they, too, would reach equilibrium if they occurred in the isolation of a test tube. Because systems at equilibrium are at a minimum of *G* and can do no work, a cell that has reached metabolic equilibrium is dead! The fact that metabolism as a whole is never at equilibrium is one of the defining features of life.

Like most systems, a cell in our body is not in equilibrium. The constant flow of materials in and out of the cell keeps the metabolic pathways from ever reaching equilibrium, and the cell continues to do work throughout its life. This principle is illustrated by the open (and more realistic) hydroelectric system in **Figure 8.7b**. However, unlike this simple single-step system, a catabolic pathway in a cell releases free energy in a series of reactions. An example is cellular respiration, illustrated by analogy in **Figure 8.7c**. Some of the reversible reactions of respiration are constantly "pulled" in one direction—that is, they are kept out of equilibrium. The key to maintaining this lack of equilibrium is that the product of one reaction does not accumulate, but instead becomes a reactant in the next step; finally, waste products are expelled from the cell. The overall sequence of reactions is kept going by the huge free-energy difference between glucose and oxygen at the top of the energy "hill" and carbon dioxide and water at the "downhill" end. As long as our cells have a steady supply of glucose or other fuels and oxygen and are able to expel waste products to the surroundings, their metabolic pathways never reach equilibrium and can continue to do the work of life.

We see once again how important it is to think of organisms as open systems. Sunlight provides a daily source of free energy for an ecosystem's plants and other photosynthetic organisms. Animals and other nonphotosynthetic organisms in an ecosystem must have a source of free energy in the form of the organic products of photosynthesis. Now that we have

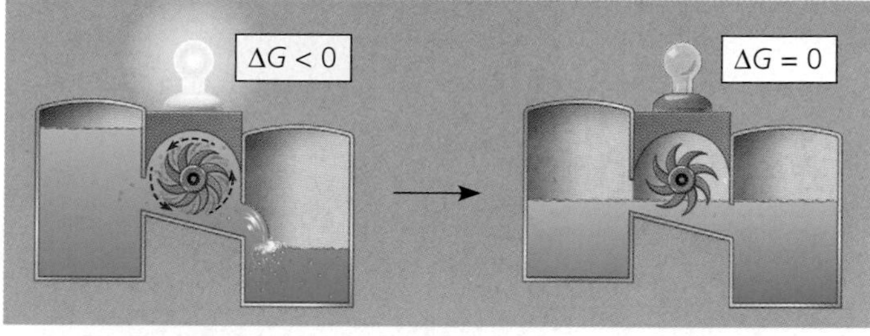

(a) A closed hydroelectric system. Water flowing downhill turns a turbine that drives a generator providing electricity to a light bulb, but only until the system reaches equilibrium.

(b) An open hydroelectric system. Flowing water keeps driving the generator because intake and outflow of water keep the system from reaching equilibrium.

(c) A multistep open hydroelectric system. Cellular respiration is analogous to this system: Glucose is broken down in a series of exergonic reactions that power the work of the cell. The product of each reaction becomes the reactant for the next, so no reaction reaches equilibrium.

▲ **Figure 8.7 Equilibrium and work in closed and open systems.**

applied the free-energy concept to metabolism, we are ready to see how a cell actually performs the work of life.

Concept Check 8.2

1. Cellular respiration uses glucose, which has a high level of free energy, and releases CO_2 and water, which have low levels of free energy. Is respiration spontaneous or not? Is it exergonic or endergonic? What happens to the energy released from glucose?
2. A key process in metabolism is the transport of H^+ ions across a membrane to create a concentration gradient. In some conditions, H^+ ions flow back across the membrane and come to equal concentrations on each side. In which conditions can the H^+ ions perform work in this system?

For suggested answers, see Appendix A.

ATP powers cellular work by coupling exergonic reactions to endergonic reactions

A cell does three main kinds of work:

- *Mechanical work,* such as the beating of cilia (see Chapter 6), the contraction of muscle cells, and the movement of chromosomes during cellular reproduction
- *Transport work,* the pumping of substances across membranes against the direction of spontaneous movement (see Chapter 7)
- *Chemical work,* the pushing of endergonic reactions, which would not occur spontaneously, such as the synthesis of polymers from monomers (the focus of this chapter, and Chapters 9 and 10)

A key feature in the way cells manage their energy resources to do this work is **energy coupling,** the use of an exergonic process to drive an endergonic one. ATP is responsible for mediating most energy coupling in cells, and in most cases it acts as the immediate source of energy that powers cellular work.

The Structure and Hydrolysis of ATP

ATP (adenosine triphosphate) was introduced in Chapter 4 when we discussed the phosphate group as a functional group. Here we will look more closely at the structure of this molecule. ATP contains the sugar ribose, with the nitrogenous base adenine and a chain of three phosphate groups bonded to it **(Figure 8.8)**.

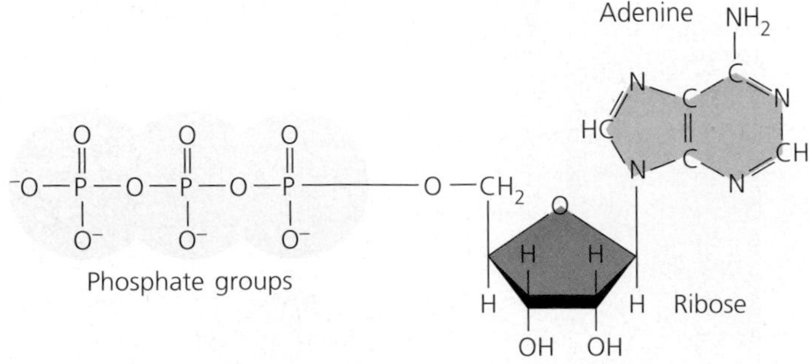

▲ **Figure 8.8 The structure of adenosine triphosphate (ATP).** In the cell, most hydroxyl groups of phosphates are ionized (—O^-).

The bonds between the phosphate groups of ATP's tail can be broken by hydrolysis. When the terminal phosphate bond is broken, a molecule of inorganic phosphate (abbreviated $Ⓟ_i$ throughout this book) leaves the ATP, which becomes adenosine diphosphate, or ADP **(Figure 8.9)**. The reaction is exergonic and under standard conditions releases 7.3 kcal of energy per mole of ATP hydrolyzed:

$$ATP + H_2O \rightarrow ADP + Ⓟ_i$$

$$\Delta G = -7.3 \text{ kcal/mol } (-30.5 \text{ kJ/mol}) \text{ (standard conditions)*}$$

The free-energy change for many different reactions has been measured in the laboratory under standard conditions. If the ΔG of an endergonic reaction is less than the amount of energy released by ATP hydrolysis, then the two reactions can be coupled so that, overall, the coupled reactions are exergonic **(Figure 8.10)**.

Because their hydrolysis releases energy, the phosphate bonds of ATP are sometimes referred to as high-energy phosphate

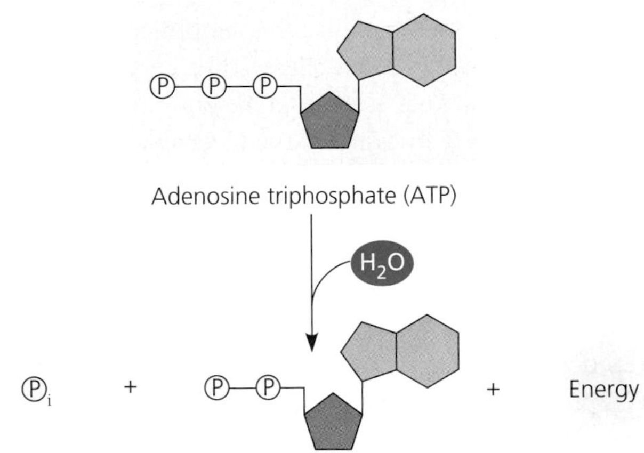

▲ **Figure 8.9 The hydrolysis of ATP.** The hydrolysis of ATP yields inorganic phosphate ($Ⓟ_i$) and ADP.

* In the cell, the conditions do not conform to standard conditions, owing in large part to reactant and product concentrations that differ from 1 *M*. For example, when ATP hydrolysis occurs under cellular conditions, the actual ΔG is about -13 kcal/mol, 78% greater than the energy released by ATP hydrolysis under standard conditions.

bonds, but the term is misleading. The phosphate bonds of ATP are not unusually strong bonds, as "high-energy" may imply; rather, the molecule itself has high energy in relation to that of the products (ADP and Ⓟ$_i$). The release of energy during the hydrolysis of ATP comes from the chemical change to a state of lower free energy, not from the phosphate bonds themselves.

ATP is useful to the cell because the energy it releases on hydrolyzing a phosphate group is somewhat greater than the energy most other molecules could deliver. But why does this hydrolysis release so much energy? If we reexamine the ATP molecule in Figure 8.8, we can see that all three phosphate groups are negatively charged. These like charges are crowded together, and their mutual repulsion contributes to the instability of this region of the ATP molecule. The triphosphate tail of ATP is the chemical equivalent of a compressed spring.

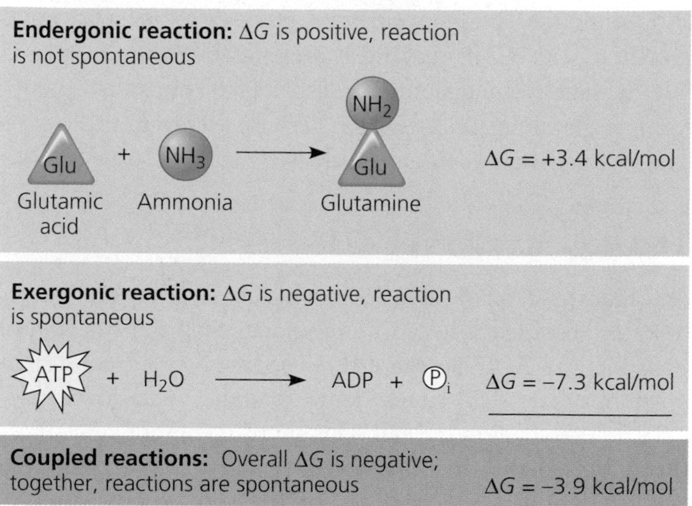

▲ **Figure 8.10 Energy coupling using ATP hydrolysis.** In this example, the exergonic process of ATP hydrolysis is used to drive an endergonic process—the synthesis of the amino acid glutamine from glutamic acid and ammonia.

How ATP Performs Work

When ATP is hydrolyzed in a test tube, the release of free energy merely heats the surrounding water. In an organism, this same generation of heat can sometimes be beneficial. For instance, the process of shivering uses ATP hydrolysis during muscle contraction to generate heat and warm the body. In most cases in the cell, however, the generation of heat alone would be an inefficient (and potentially dangerous) use of a valuable energy resource.

Instead, with the help of specific enzymes, the cell is able to couple the energy of ATP hydrolysis directly to endergonic processes by transferring a phosphate group from ATP to some other molecule, such as the reactant. The recipient of the phosphate group is then said to be **phosphorylated.** The key to coupling exergonic and endergonic reactions is the formation of this phosphorylated intermediate, which is more reactive (less stable) than the original unphosphorylated molecule.

The three types of cellular work—mechanical, transport, and chemical—are nearly always powered by the hydrolysis of ATP **(Figure 8.11)**. In each case, a phosphate group is transferred from ATP to some other molecule, and this phosphorylated molecule undergoes a change that performs work. An example is the synthesis of the amino acid glutamine from glutamic acid (another amino acid)

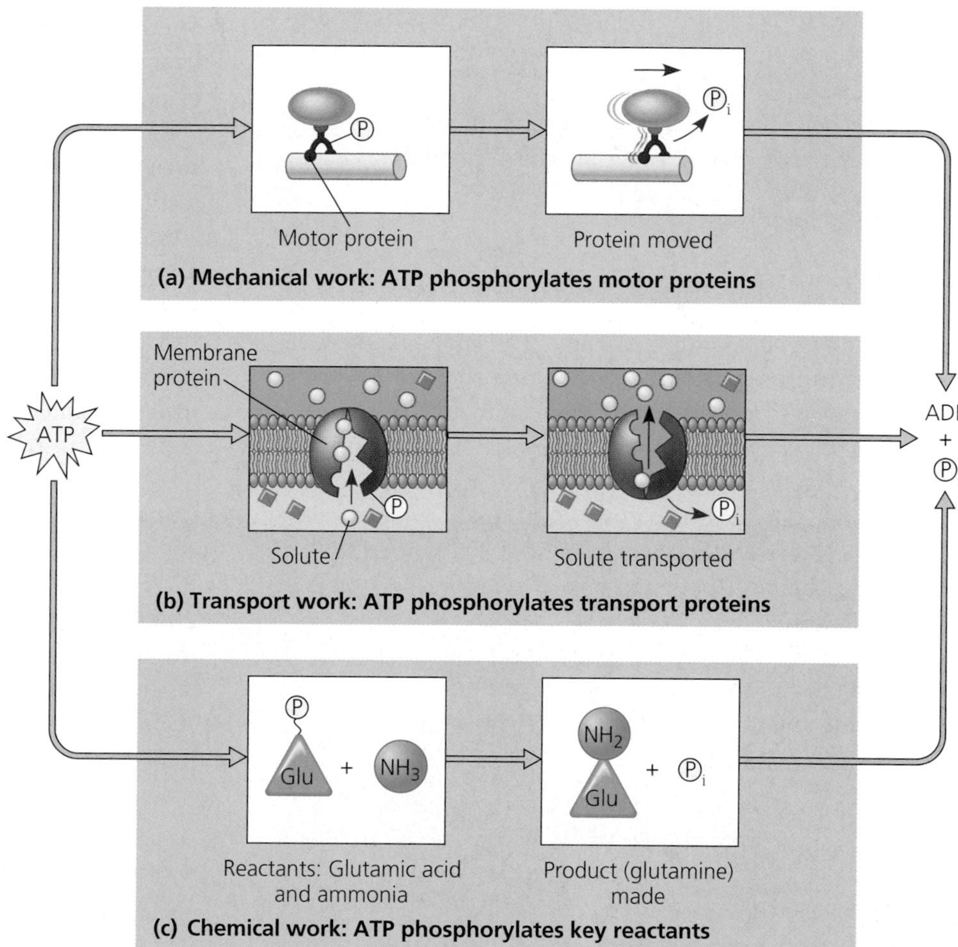

▲ **Figure 8.11 How ATP drives cellular work.** Phosphate group transfer is the mechanism responsible for most types of cellular work. For example, **(a)** ATP drives mechanical work by phosphorylating motor proteins, such as the ones that move organelles along cytoskeletal "tracks" in the cell. ATP also **(b)** drives active transport by phosphorylating certain membrane proteins. And ATP **(c)** drives chemical work by phosphorylating key reactants, in this case glutamic acid that is then converted to glutamine. The phosphorylated molecules lose the phosphate groups as work is performed, leaving ADP and inorganic phosphate (Ⓟ$_i$) as products. Cellular respiration replenishes the ATP supply by powering the phosphorylation of ADP, as we will see in the next chapter.

and ammonia (see Figure 8.11c). First, ATP phosphorylates glutamic acid (Glu), making it a less stable phosphorylated intermediate. Second, ammonia displaces the phosphate group, forming glutamine (Glu—NH_2). Because the overall process is exergonic, it occurs spontaneously (see Figure 8.10).

The Regeneration of ATP

An organism at work uses ATP continuously, but ATP is a renewable resource that can be regenerated by the addition of phosphate to ADP **(Figure 8.12)**. The free energy required to phosphorylate ADP comes from exergonic breakdown reactions (catabolism) in the cell. This shuttling of inorganic phosphate and energy is called the ATP cycle, and it couples the cell's energy-yielding (exergonic) processes to the energy-consuming (endergonic) ones. The ATP cycle moves at an astonishing pace. For example, a working muscle cell recycles its entire pool of ATP in less than a minute. That turnover represents 10 million molecules of ATP consumed and regenerated per second per cell. If ATP could not be regenerated by the phosphorylation of ADP, humans would use up nearly their body weight in ATP each day.

Because both directions of a reversible process cannot go downhill, the regeneration of ATP from ADP and Ⓟ$_i$ is necessarily endergonic:

$$ADP + Ⓟ_i \rightarrow ATP + H_2O$$

$$\Delta G = +7.3 \text{ kcal/mol } (+30.5 \text{ kJ/mol}) \text{ (standard conditions)}$$

Because ATP formation from ADP and Ⓟ$_i$ is not spontaneous, free energy must be spent to make it occur. Catabolic (exergonic) pathways, especially cellular respiration, provide the energy for the endergonic process of making ATP. Plants also use light energy to produce ATP.

Thus, the ATP cycle is a turnstile through which energy passes during its transfer from catabolic to anabolic pathways. In fact, the chemical potential energy temporarily stored in ATP drives most cellular work.

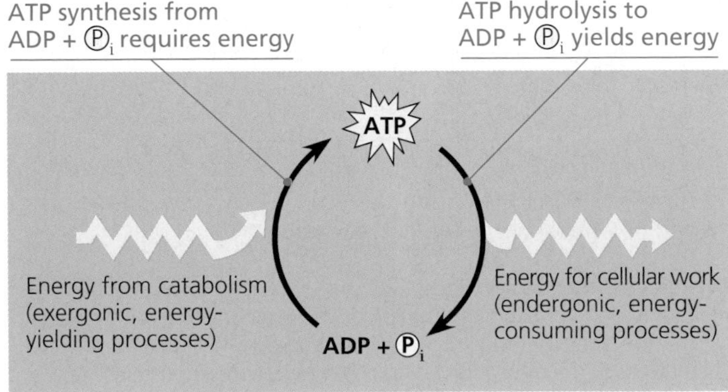

▲ **Figure 8.12 The ATP cycle.** Energy released by breakdown reactions (catabolism) in the cell is used to phosphorylate ADP, regenerating ATP. Energy stored in ATP drives most cellular work.

Concept Check 8.3

1. In most cases, how does ATP transfer energy from exergonic to endergonic processes in the cell?
2. Which of the following groups has more free energy: glutamic acid + ammonia + ATP, or glutamine + ADP + Ⓟ$_i$? Explain your answer.

For suggested answers, see Appendix A.

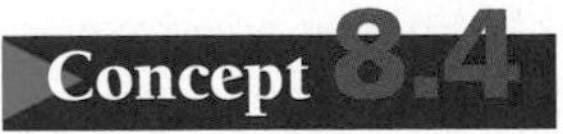

Concept 8.4

Enzymes speed up metabolic reactions by lowering energy barriers

The laws of thermodynamics tell us what will and will not happen under given conditions but say nothing about the rate of these processes. A spontaneous chemical reaction occurs without any requirement for outside energy, but it may occur so slowly that it is imperceptible. For example, even though the hydrolysis of sucrose (table sugar) to glucose and fructose is exergonic, occurring spontaneously with a release of free energy ($\Delta G = -7$ kcal/mol), a solution of sucrose dissolved in sterile water will sit for years at room temperature with no appreciable hydrolysis. However, if we add a small amount of a catalyst, such as the enzyme sucrase, to the solution, then all the sucrose may be hydrolyzed within seconds **(Figure 8.13)**. How does an enzyme do this?

A **catalyst** is a chemical agent that speeds up a reaction without being consumed by the reaction; an **enzyme** is a catalytic protein. (Another class of biological catalysts, made of RNA and called ribozymes, is discussed in Chapters 17 and 26.) In the absence of regulation by enzymes, chemical traffic through the pathways of metabolism would become hopelessly congested because many chemical reactions would take such a long time. In the next two sections, we will see what impedes a spontaneous reaction from occurring faster and how an enzyme changes the situation.

The Activation Energy Barrier

Every chemical reaction between molecules involves both bond breaking and bond forming. For example, the hydrolysis of sucrose involves breaking the bond between glucose and fructose and one of the bonds of a water molecule, and then forming two new bonds, as shown in Figure 8.13. Changing one molecule into another generally involves contorting the starting molecule into a highly unstable state before the reaction can proceed. This contortion can be compared to a metal

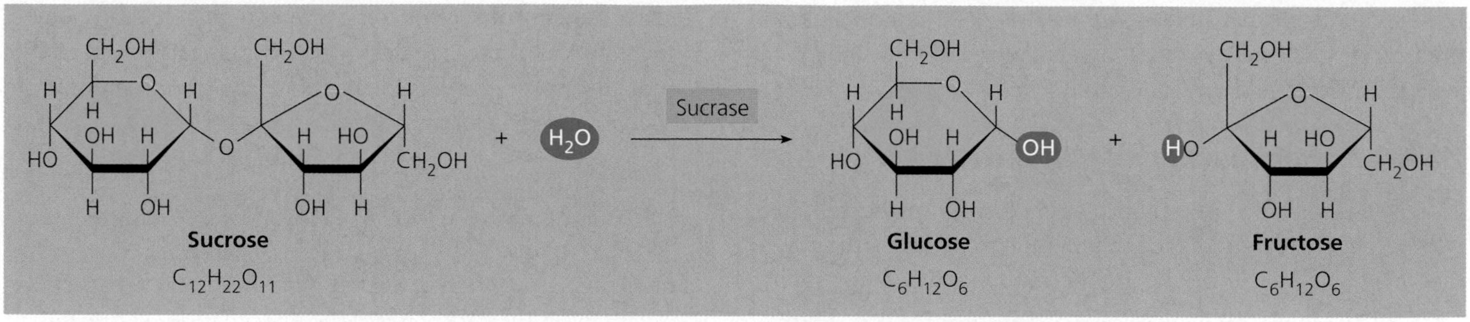

Sucrose $C_{12}H_{22}O_{11}$ + H_2O —Sucrase→ Glucose $C_6H_{12}O_6$ + Fructose $C_6H_{12}O_6$

▲ **Figure 8.13 Example of an enzyme-catalyzed reaction: hydrolysis of sucrose by sucrase.**

key ring when you bend it and pry it open to add a new key. The key ring is highly unstable in its opened form but returns to a stable state once the key is threaded all the way onto the ring. To reach the contorted state where bonds can change, reactant molecules must absorb energy from their surroundings. When the new bonds of the product molecules form, energy is released as heat, and the molecules return to stable shapes with lower energy.

The initial investment of energy for starting a reaction—the energy required to contort the reactant molecules so the bonds can change—is known as the **free energy of activation**, or **activation energy**, abbreviated E_A in this book. We can think of activation energy as the amount of energy needed to push the reactants over an energy barrier, or hill, so that the "downhill" part of the reaction can begin. **Figure 8.14** graphs the energy changes for a hypothetical exergonic reaction that swaps portions of two reactant molecules:

$$AB + CD \rightarrow AC + BD$$

The energizing, or activation, of the reactants is represented by the uphill portion of the graph, with the free-energy content of the reactant molecules increasing. At the summit, the reactants are in an unstable condition known as the *transition state:* They are activated, and the breaking and making of bonds can occur. The bond-forming phase of the reaction corresponds to the downhill part of the curve, which shows the loss of free energy by the molecules.

Activation energy is often supplied in the form of heat that the reactant molecules absorb from the surroundings. The bonds of the reactants break only when the molecules have absorbed enough energy to become unstable and are therefore more reactive (in the transition state at the peak of the curve in Figure 8.14). The absorption of thermal energy increases the speed of the reactant molecules, so they collide more often and more forcefully. Also, thermal agitation of the atoms in the molecules makes the bonds more likely to break. As the molecules settle into their new, more stable bonding arrangements, energy is released to the surroundings. If the reaction is exergonic, E_A will be repaid with dividends, as the formation of new bonds releases more energy than was invested in the breaking of old bonds.

The reactants AB and CD must absorb enough energy from the surroundings to reach the unstable transition state, where bonds can break.

Bonds break and new bonds form, releasing energy to the surroundings.

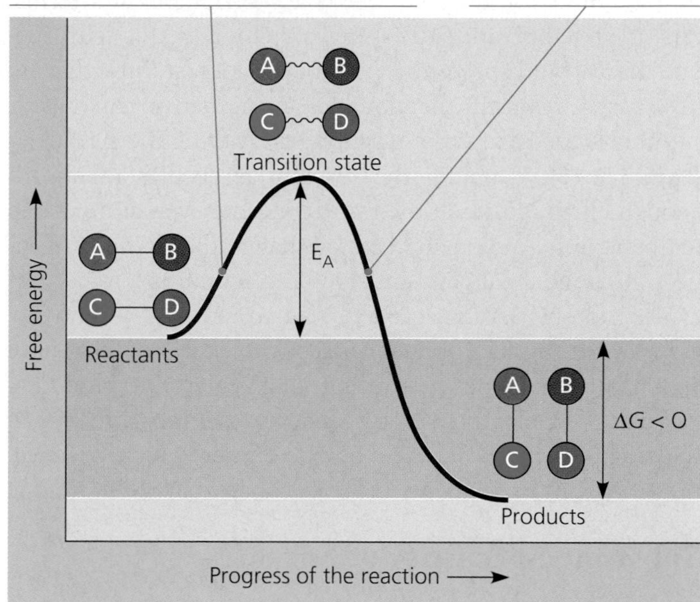

▲ **Figure 8.14 Energy profile of an exergonic reaction.** The "molecules" are hypothetical, with A, B, C, and D representing portions of the molecules. Thermodynamically, this is an exergonic reaction, with a negative ΔG, and the reaction occurs spontaneously. However, the activation energy (E_A) provides a barrier that determines the rate of the reaction.

The reaction shown in Figure 8.14 is exergonic and occurs spontaneously. However, the activation energy provides a barrier that determines the rate of the reaction. The reactants must absorb enough energy to reach the top of the activation energy barrier before the reaction can occur. For some reactions, E_A is modest enough that even at room temperature there is sufficient thermal energy for many of the reactants to reach the transition state in a short time. In most cases, however, E_A is so high and the transition state is reached so rarely that the reaction will hardly proceed at all. In these cases, the reaction will occur at a noticeable rate only if the reactants are heated. The spark plugs in an automobile engine energize the gasoline-oxygen mixture so that the molecules reach the transition state and react; only then can there be the explosive release of energy that pushes the pistons. Without a spark, a mixture of gasoline hydrocarbons and oxygen will not react because the E_A barrier is too high.

How Enzymes Lower the E_A Barrier

Proteins, DNA, and other complex molecules of the cell are rich in free energy and have the potential to decompose spontaneously; that is, the laws of thermodynamics favor their breakdown. These molecules persist only because at temperatures typical for cells, few molecules can make it over the hump of activation energy. However, the barriers for selected reactions must occasionally be surmounted for cells to carry out the processes necessary for life. Heat speeds a reaction by allowing reactants to attain the transition state more often, but this solution would be inappropriate for biological systems. First, high temperature denatures proteins and kills cells. Second, heat would speed up *all* reactions, not just those that are necessary. Organisms therefore use an alternative: catalysis.

An enzyme catalyzes a reaction by lowering the E_A barrier **(Figure 8.15)**, enabling the reactant molecules to absorb enough energy to reach the transition state even at moderate temperatures. An enzyme cannot change the ΔG for a reaction; it cannot make an endergonic reaction exergonic. Enzymes can only hasten reactions that would occur eventually anyway, but this function makes it possible for the cell to have a dynamic metabolism, routing chemical traffic smoothly through the cell. And because enzymes are very selective in the reactions they catalyze, they determine which chemical processes will be going on in the cell at any particular time.

Substrate Specificity of Enzymes

The reactant an enzyme acts on is referred to as the enzyme's **substrate.** The enzyme binds to its substrate (or substrates, when there are two or more reactants), forming an **enzyme-substrate complex.** While enzyme and substrate are joined, the catalytic action of the enzyme converts the substrate to the product (or products) of the reaction. The overall process can be summarized as follows:

Enzyme + Substrate(s)	⇌	Enzyme-substrate complex	⇌	Enzyme + Product(s)

For example, the enzyme sucrase (most enzyme names end in *-ase*) catalyzes the hydrolysis of the disaccharide sucrose into its two monosaccharides, glucose and fructose (see Figure 8.13):

Sucrase + Sucrose + H_2O	⇌	Sucrase-sucrose-H_2O complex	⇌	Sucrase + Glucose + Fructose

The reaction catalyzed by each enzyme is very specific; an enzyme can recognize its specific substrate even among closely related compounds, such as isomers. For instance, sucrase will act only on sucrose and will not bind to other disaccharides, such as maltose. What accounts for this molecular recognition? Recall that enzymes are proteins, and proteins are macromolecules with unique three-dimensional conformations. The specificity of an enzyme results from its shape, which is a consequence of its amino acid sequence.

Only a restricted region of the enzyme molecule actually binds to the substrate. This region, called the **active site,** is typically a pocket or groove on the surface of the protein **(Figure 8.16a)**. Usually, the active site is formed by only a few of the enzyme's amino acids, with the rest of the protein molecule providing a framework that determines the configuration of the active site. The specificity of an enzyme is attributed to a compatible fit between the shape of its active site and the shape of the substrate. The active site, however, is not a rigid receptacle for the substrate. As the substrate enters the active site, interactions between its chemical groups and those on the amino acids of the protein cause the enzyme to change its shape slightly so that the active site fits even more snugly around the substrate **(Figure 8.16b)**. This **induced fit** is like a clasping handshake. Induced fit brings chemical groups of the active site into positions that enhance their ability to catalyze the chemical reaction.

Catalysis in the Enzyme's Active Site

In an enzymatic reaction, the substrate binds to the active site **(Figure 8.17)**. In most cases, the substrate is held in the active site by weak interactions, such as hydrogen bonds and ionic bonds. Side chains (R groups) of a few of the amino acids that make up the active site catalyze the conversion of substrate to product, and the product departs from the active site. The enzyme is then free to take another substrate molecule into its active site. The entire cycle happens so fast that a single enzyme molecule typically acts on about a thousand substrate molecules per second. Some enzymes are much faster. Enzymes, like other catalysts, emerge from the reaction in their original form. There-

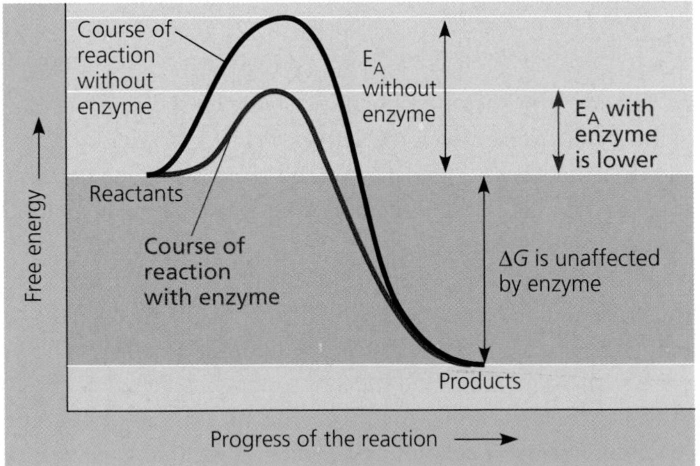

▲ **Figure 8.15 The effect of enzymes on reaction rate.** Without affecting the free-energy change (ΔG) for a reaction, an enzyme speeds the reaction by reducing its activation energy (E_A).

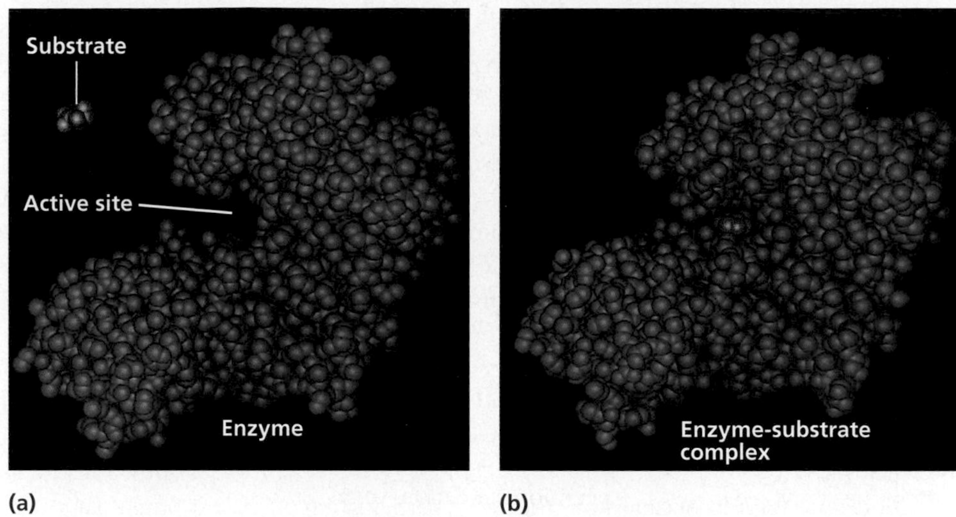

Substrate
Active site
Enzyme
Enzyme-substrate complex
(a)
(b)

▶ **Figure 8.16 Induced fit between an enzyme and its substrate. (a)** In this computer graphic model, the active site of this enzyme (hexokinase, shown in blue) forms a groove on its surface. Its substrate is glucose (red). **(b)** When the substrate enters the active site, it induces a change in the shape of the protein. This change allows more weak bonds to form, causing the active site to embrace the substrate and hold it in place.

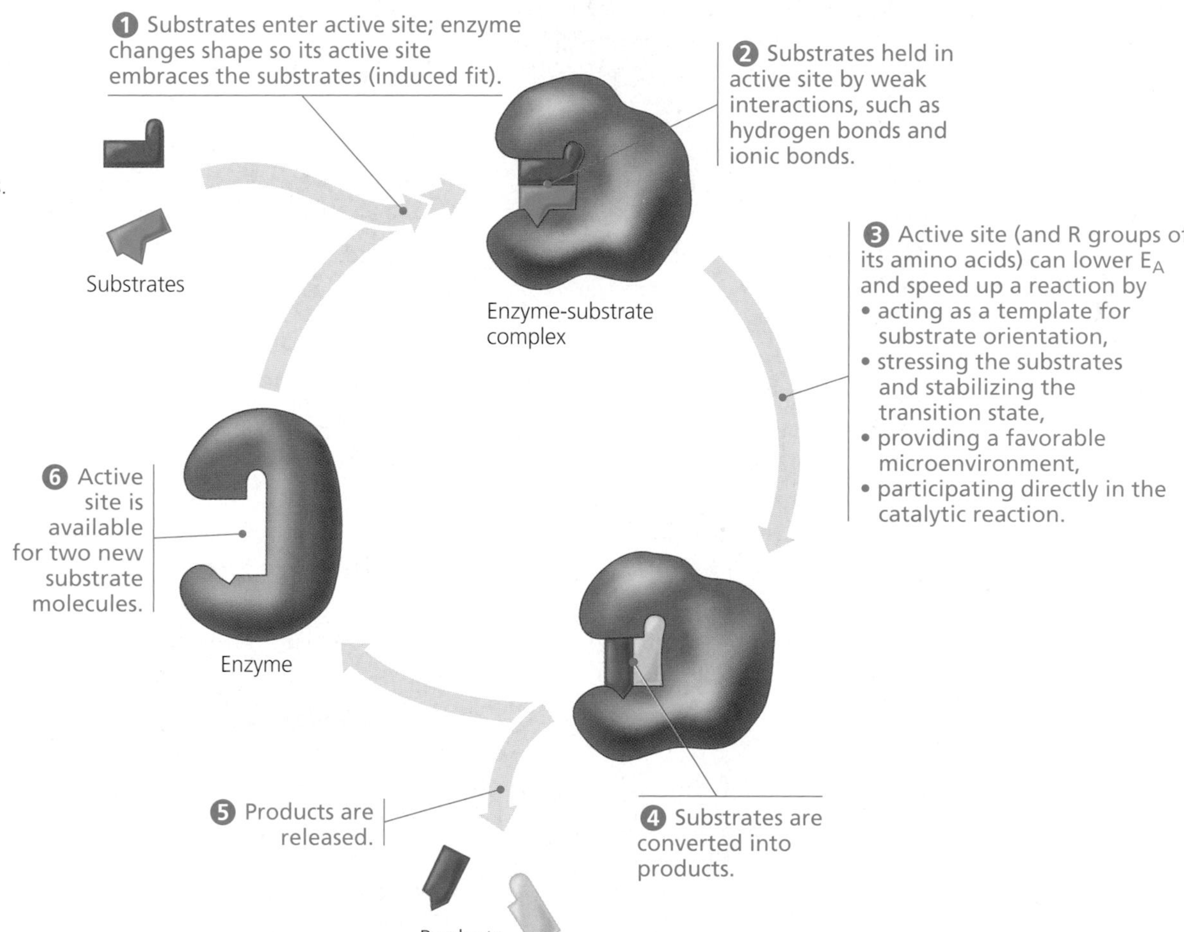

1 Substrates enter active site; enzyme changes shape so its active site embraces the substrates (induced fit).
2 Substrates held in active site by weak interactions, such as hydrogen bonds and ionic bonds.
Substrates
Enzyme-substrate complex
3 Active site (and R groups of its amino acids) can lower E_A and speed up a reaction by
• acting as a template for substrate orientation,
• stressing the substrates and stabilizing the transition state,
• providing a favorable microenvironment,
• participating directly in the catalytic reaction.
6 Active site is available for two new substrate molecules.
Enzyme
5 Products are released.
4 Substrates are converted into products.
Products

▶ **Figure 8.17 The active site and catalytic cycle of an enzyme.** An enzyme can convert one or more reactant molecules to one or more product molecules. The enzyme shown here converts two substrate molecules to two product molecules.

fore, very small amounts of enzyme can have a huge metabolic impact by functioning over and over again in catalytic cycles.

Most metabolic reactions are reversible, and an enzyme can catalyze both the forward and the reverse reactions. Which reaction prevails depends mainly on the relative concentrations of reactants and products. The enzyme always catalyzes the reaction in the direction of equilibrium.

Enzymes use a variety of mechanisms that lower activation energy and speed up a reaction (see Figure 8.17, step 3). First, in reactions involving two or more reactants, the active site provides a template for the substrates to come together in the proper orientation for a reaction to occur between them. Second, as the active site of an enzyme clutches the bound substrates, the enzyme may stretch the substrate molecules toward their transition-state conformation, stressing and bending critical chemical bonds that must be broken during the reaction. Because E_A is proportional to the difficulty of breaking the bonds, distorting the substrate makes it approach the transition state and thus reduces the amount of free energy that must be absorbed to achieve a transition state.

Third, the active site may also provide a microenvironment that is more conducive to a particular type of reaction than the solution itself would be without the enzyme. For example, if the active site has amino acids with acidic side chains (R groups), the active site may be a pocket of low pH in an otherwise neutral cell. In such cases, an acidic amino acid may facilitate H^+ transfer to the substrate as a key step in catalyzing the reaction.

A fourth mechanism of catalysis is the direct participation of the active site in the chemical reaction. Sometimes this process even involves brief covalent bonding between the substrate and a side chain of an amino acid of the enzyme. Subsequent steps of the reaction restore the side chains to their original states, so the active site is the same after the reaction as it was before.

The rate at which a particular amount of enzyme converts substrate to product is partly a function of the initial concentration of the substrate: The more substrate molecules are available, the more frequently they access the active sites of the enzyme molecules. However, there is a limit to how fast the reaction can be pushed by adding more substrate to a fixed concentration of enzyme. At some point, the concentration of substrate will be high enough that all enzyme molecules have their active sites engaged. As soon as the product exits an active site, another substrate molecule enters. At this substrate concentration, the enzyme is said to be *saturated,* and the rate of the reaction is determined by the speed at which the active site can convert substrate to product. When an enzyme population is saturated, the only way to increase the rate of product formation is to add more enzyme. Cells sometimes do this by making more enzyme molecules.

Effects of Local Conditions on Enzyme Activity

The activity of an enzyme—how efficiently the enzyme functions—is affected by general environmental factors, such as temperature and pH. It can also be affected by chemicals that specifically influence that enzyme.

Effects of Temperature and pH

Recall from Chapter 5 that the three-dimensional structures of proteins are sensitive to their environment. As a consequence, each enzyme works better under some conditions than under others, because these *optimal conditions* favor the most active conformation for the enzyme molecule.

Temperature and pH are environmental factors important in the activity of an enzyme. Up to a point, the rate of an enzymatic reaction increases with increasing temperature, partly because substrates collide with active sites more frequently when the molecules move rapidly. Above that temperature, however, the speed of the enzymatic reaction drops sharply. The thermal agitation of the enzyme molecule disrupts the hydrogen bonds, ionic bonds, and other weak interactions that stabilize the active conformation, and the protein molecule eventually denatures. Each enzyme has an optimal temperature at which its reaction rate is greatest. Without denaturing the enzyme, this temperature allows the greatest number of molecular collisions and the fastest conversion of the reactants to product molecules. Most human enzymes have optimal temperatures of about 35–40°C (close to human body temperature). Bacteria that live in hot springs contain enzymes with optimal temperatures of 70°C or higher **(Figure 8.18a)**.

Just as each enzyme has an optimal temperature, it also has a pH at which it is most active. The optimal pH values for most enzymes fall in the range of pH 6–8, but there are exceptions. For example, pepsin, a digestive enzyme in the stomach, works best at pH 2. Such an acidic environment denatures most enzymes, but the active conformation of pepsin is adapted to maintain its functional three-dimensional structure in the acidic environment of the stomach. In contrast, trypsin, a digestive enzyme residing in the alkaline environment of the intestine, has an optimal pH of 8 and would be denatured in the stomach **(Figure 8.18b)**.

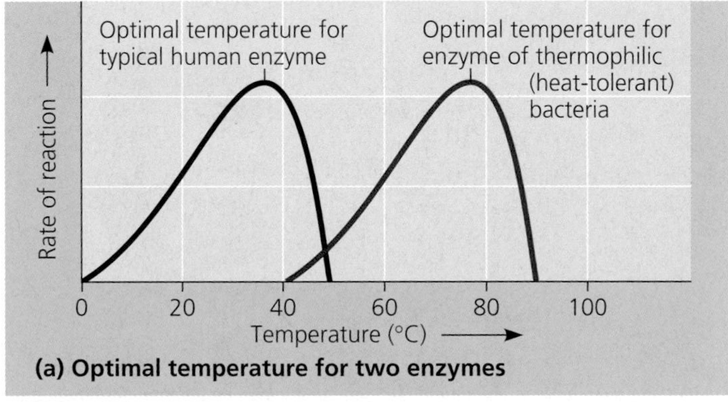

(a) Optimal temperature for two enzymes

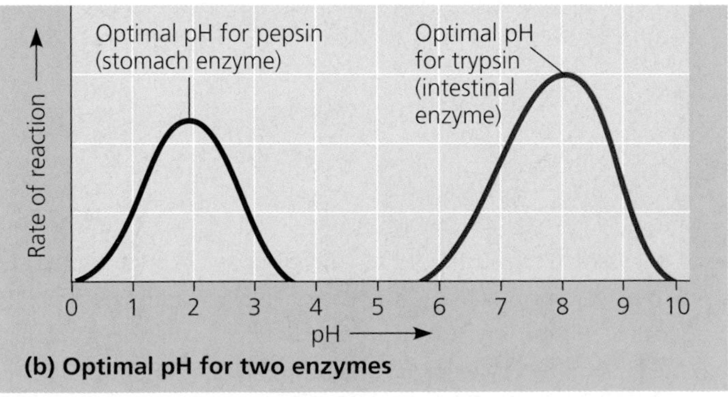

(b) Optimal pH for two enzymes

▲ **Figure 8.18 Environmental factors affecting enzyme activity.** Each enzyme has an optimal **(a)** temperature and **(b)** pH that favor the most active conformation of the protein molecule.

Cofactors

Many enzymes require nonprotein helpers for catalytic activity. These adjuncts, called **cofactors**, may be bound tightly to the enzyme as permanent residents, or they may bind loosely and reversibly along with the substrate. The cofactors of some enzymes are inorganic, such as the metal atoms zinc, iron, and copper in ionic form. If the cofactor is an organic molecule, it is more specifically called a **coenzyme.** Most vitamins are coenzymes or raw materials from which coenzymes are made. Cofactors function in various ways, but in all cases where they are used, they perform a crucial function in catalysis. You'll encounter examples of cofactors later in the book.

Enzyme Inhibitors

Certain chemicals selectively inhibit the action of specific enzymes, and we have learned a lot about enzyme function by studying the effects of these molecules. If the inhibitor attaches to the enzyme by covalent bonds, inhibition is usually irreversible.

Many enzyme inhibitors, however, bind to the enzyme by weak bonds, in which case inhibition is reversible. Some reversible inhibitors resemble the normal substrate molecule and compete for admission into the active site (**Figure 8.19a** and **b**). These mimics, called **competitive inhibitors,** reduce the productivity of enzymes by blocking substrates from entering active sites. This kind of inhibition can be overcome by increasing the concentration of substrate so that as active sites become available, more substrate molecules than inhibitor molecules are around to gain entry to the sites.

In contrast, **noncompetitive inhibitors** do not directly compete with the substrate to bind to the enzyme at the active site **(Figure 8.19c)**. Instead, they impede enzymatic reactions by binding to another part of the enzyme. This interaction causes the enzyme molecule to change its shape, rendering the active site less effective at catalyzing the conversion of substrate to product.

Toxins and poisons are often irreversible enzyme inhibitors. An example is sarin, a nerve gas that caused the death of several people and injury to many others when it was released by terrorists in the Tokyo subway in 1995. This small molecule binds covalently to the R group on the amino acid serine, which is found in the active site of acetylcholinesterase, an enzyme important in the nervous system. Other examples include the pesticides DDT and parathion, inhibitors of key enzymes in the nervous system. Finally, many antibiotics are inhibitors of specific enzymes in bacteria. For instance, penicillin blocks the active site of an enzyme that many bacteria use to make their cell walls.

Citing enzyme inhibitors that are metabolic poisons may give the impression that enzyme inhibition is generally abnormal and harmful. In fact, molecules naturally present in the cell often regulate enzyme activity by acting as inhibitors. Such regulation—selective inhibition—is essential to the control of cellular metabolism, as we discuss next.

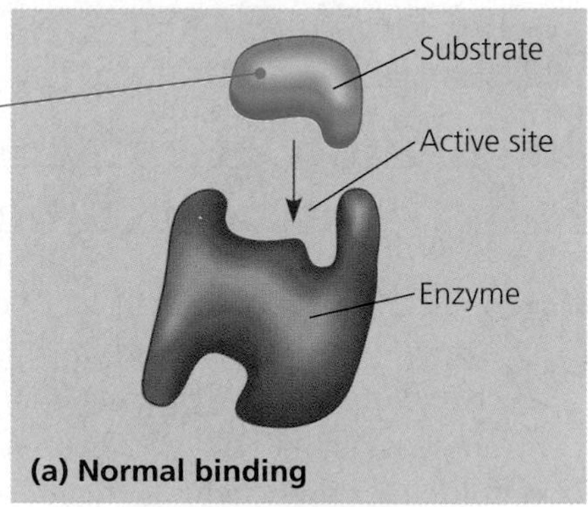
A substrate can bind normally to the active site of an enzyme.
Substrate
Active site
Enzyme
(a) Normal binding

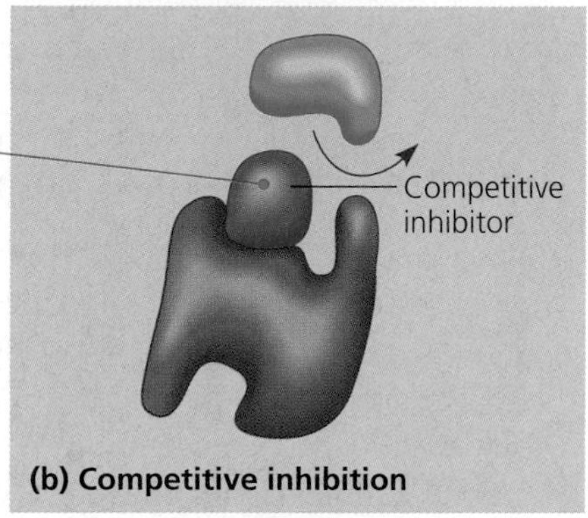
A competitive inhibitor mimics the substrate, competing for the active site.
Competitive inhibitor
(b) Competitive inhibition

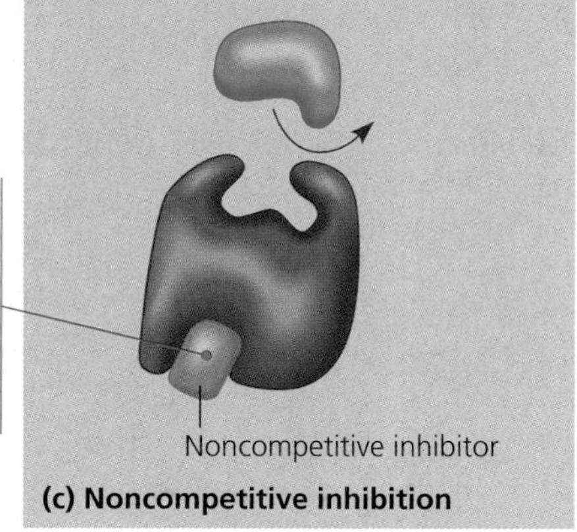
A noncompetitive inhibitor binds to the enzyme away from the active site, altering the conformation of the enzyme so that its active site no longer functions.
Noncompetitive inhibitor
(c) Noncompetitive inhibition

▲ Figure 8.19 **Inhibition of enzyme activity.**

Concept Check 8.4

1. Many spontaneous reactions occur very slowly. Why don't all spontaneous reactions occur instantly?
2. Describe why enzymes act only on very specific substrates.
3. Malonate is a competitive inhibitor of the enzyme succinate dehydrogenase. Describe how malonate would prevent the enzyme from acting on its normal substrate succinate.

For suggested answers, see Appendix A.

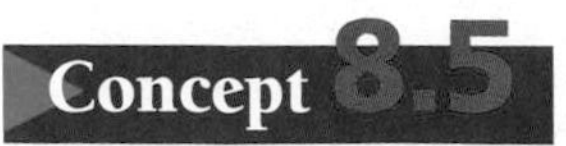

Regulation of enzyme activity helps control metabolism

Chemical chaos would result if all of a cell's metabolic pathways were operating simultaneously. Intrinsic to the process of life is a cell's ability to tightly regulate its metabolic pathways by controlling when and where its various enzymes are active. It does this either by switching on and off the genes that encode specific enzymes (as we will discuss in Unit Three) or, as we discuss here, by regulating the activity of enzymes once they are made.

Allosteric Regulation of Enzymes

In many cases, the molecules that naturally regulate enzyme activity in a cell behave something like reversible noncompetitive inhibitors (see Figure 8.19c): These regulatory molecules change an enzyme's shape and the functioning of its active site by binding to a site elsewhere on the molecule, via noncovalent bonds. **Allosteric regulation** is the term used to describe any case in which a protein's function at one site is affected by the binding of a regulatory molecule to a separate site. It may result in either inhibition or stimulation of an enzyme's activity.

Allosteric Activation and Inhibition

Most allosterically regulated enzymes are constructed from two or more polypeptide chains, or subunits **(Figure 8.20)**. Each subunit has its own active site. The entire complex oscillates between two conformational states, one catalytically active and the other inactive **(Figure 8.20a)**. In the simplest case of allosteric regulation, an activating or inhibiting regulatory molecule binds to a regulatory site (sometimes called an allosteric site), often located where subunits join. The binding of an *activator* to a regulatory site stabilizes the conformation that has functional active sites, whereas the binding of an *inhibitor* stabilizes the inactive form of the enzyme. The subunits of an allosteric enzyme fit together in such a way that a conformational change in one subunit is transmitted to all others. Through this interaction of subunits, a single activator or inhibitor molecule that binds to one regulatory site will affect the active sites of all subunits.

Fluctuating concentrations of regulators can cause a sophisticated pattern of response in the activity of cellular enzymes. The products of ATP hydrolysis (ADP and Ⓟ$_i$), for example, play a major role in balancing the flow of traffic between anabolic and catabolic pathways by their effects on key enzymes. For example, ATP binds to several catabolic enzymes allosterically, lowering their affinity for substrate and thus inhibiting their activity. ADP, however, functions as an activator of the same enzymes. This is logical because a major function of catabolism is to regenerate ATP. If ATP production lags behind its use, ADP accumulates and activates these key enzymes that speed up catabolism, producing more ATP. If the supply of ATP exceeds demand, then catabolism slows down as ATP molecules accumulate and bind these same enzymes, inhibiting them. ATP, ADP, and other related molecules also affect key enzymes in anabolic pathways. In this way, allosteric enzymes control the rates of key reactions in metabolic pathways.

In another kind of allosteric activation, a substrate molecule binding to one active site may stimulate the catalytic powers

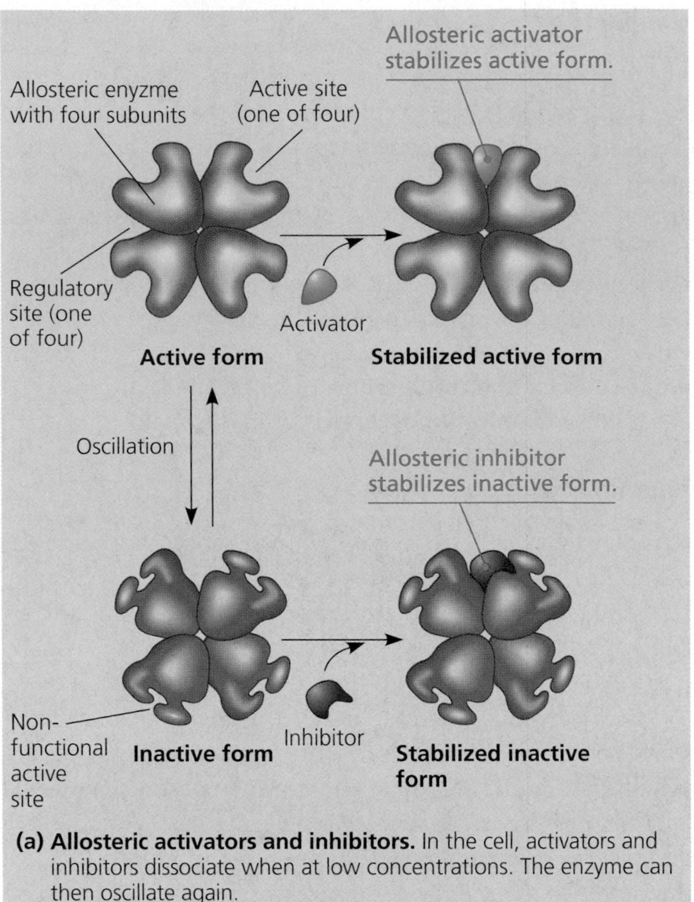

(a) Allosteric activators and inhibitors. In the cell, activators and inhibitors dissociate when at low concentrations. The enzyme can then oscillate again.

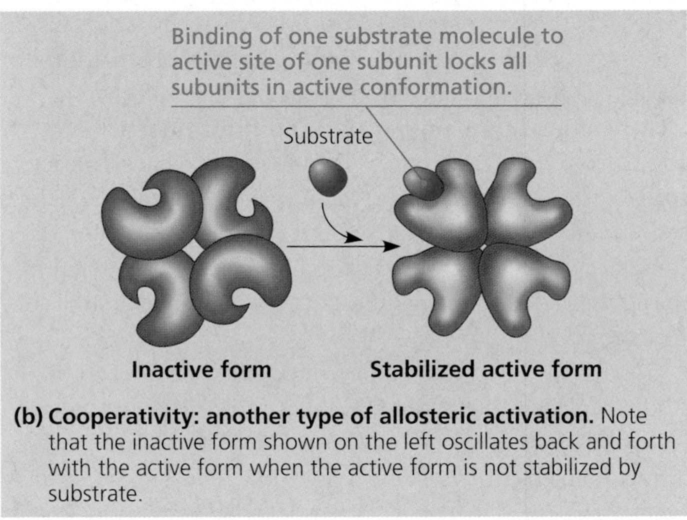

(b) Cooperativity: another type of allosteric activation. Note that the inactive form shown on the left oscillates back and forth with the active form when the active form is not stabilized by substrate.

▲ **Figure 8.20 Allosteric regulation of enzyme activity.**

of a multi-subunit enzyme by affecting the other active sites **(Figure 8.20b)**. If an enzyme has two or more subunits, a substrate molecule causing induced fit in one subunit can trigger the same favorable conformational change in all the other subunits of the enzyme. Called **cooperativity,** this mechanism amplifies the response of enzymes to substrates: One substrate molecule primes an enzyme to accept additional substrate molecules more readily.

Feedback Inhibition

When ATP allosterically inhibits an enzyme in an ATP-generating pathway, the result is feedback inhibition, a common method of metabolic control. In **feedback inhibition,** a metabolic pathway is switched off by the inhibitory binding of its end product to an enzyme that acts early in the pathway. **Figure 8.21** shows an example of this control mechanism operating on an anabolic pathway. Some cells use this pathway of five steps to synthesize the amino acid isoleucine from threonine, another amino acid. As isoleucine accumulates, it slows down its own synthesis by allosterically inhibiting the enzyme for the very first step of the pathway. Feedback inhibition thereby prevents the cell from wasting chemical resources by synthesizing more isoleucine than is necessary.

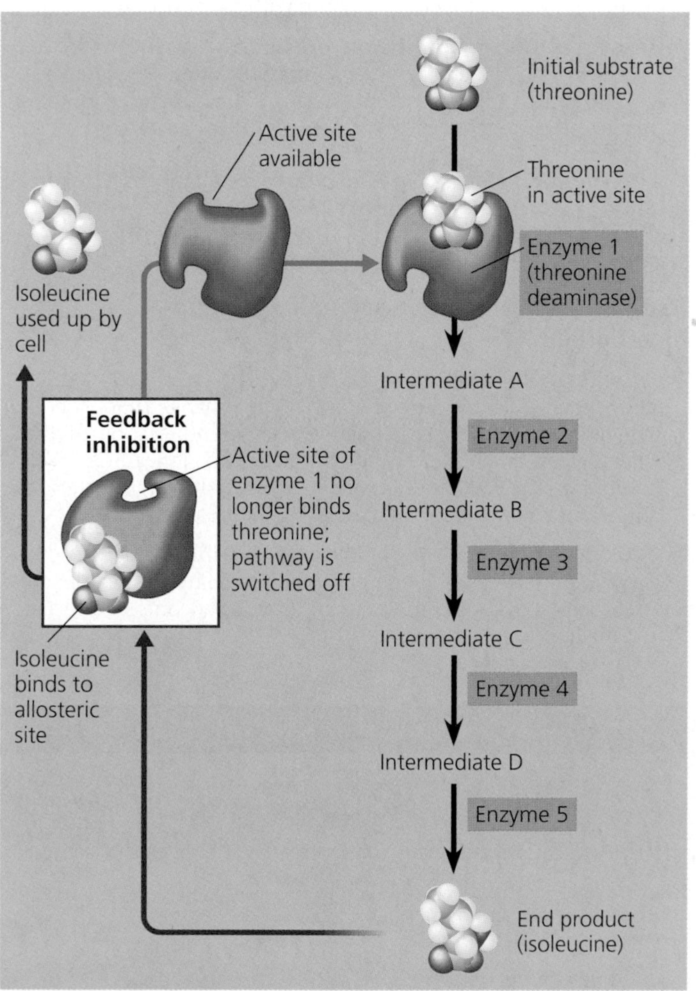

▲ **Figure 8.21 Feedback inhibition in isoleucine synthesis.**

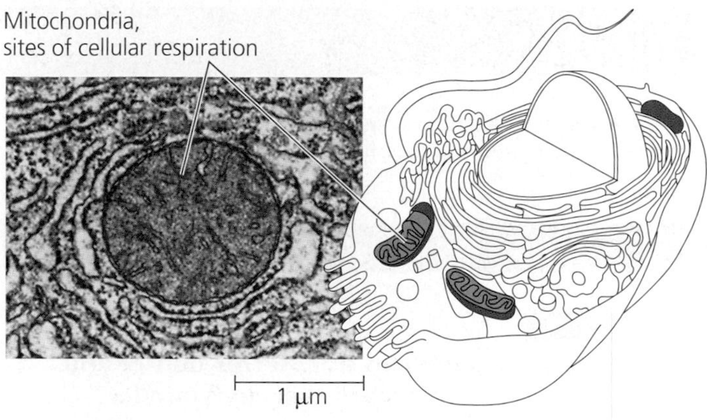

▲ **Figure 8.22 Organelles and structural order in metabolism.** Organelles such as these mitochondria (TEM) contain enzymes that carry out specific functions, in this case cellular respiration.

Specific Localization of Enzymes Within the Cell

The cell is not just a bag of chemicals with thousands of different kinds of enzymes and substrates in a random mix. Structures within the cell help bring order to metabolic pathways. In some cases, a team of enzymes for several steps of a metabolic pathway is assembled into a multienzyme complex. The arrangement controls and speeds up the sequence of reactions, as the product from the first enzyme becomes the substrate for an adjacent enzyme in the complex, and so on, until the end product is released. Some enzymes and enzyme complexes have fixed locations within the cell, and act as structural components of particular membranes. Others are in solution within specific membrane-enclosed eukaryotic organelles, each with its own internal chemical environment. For example, in eukaryotic cells, the enzymes for cellular respiration reside in specific locations within mitochondria. **(Figure 8.22)**.

In this chapter we have learned that metabolism, the intersecting set of chemical pathways characteristic of life, is a choreographed interplay of thousands of different kinds of cellular molecules. In the next chapter we explore cellular respiration, the major catabolic pathway that breaks down organic molecules, releasing energy for the crucial processes of life.

Concept Check 8.5

1. How can an activator and an inhibitor have different effects on an allosterically regulated enzyme?

For suggested answers, see Appendix A.

Chapter 8 Review

Go to www.campbellbiology.com or the student CD-ROM to explore Activities, Investigations, and other interactive study aids.

SUMMARY OF KEY CONCEPTS

Concept 8.1

An organism's metabolism transforms matter and energy, subject to the laws of thermodynamics

- **Organization of the Chemistry of Life into Metabolic Pathways (pp. 141–142)** Metabolism is the collection of chemical reactions that occur in an organism. Aided by enzymes, it follows intersecting pathways, which may be catabolic (breaking down molecules, releasing energy) or anabolic (building molecules, consuming energy).
- **Forms of Energy (pp. 142–143)** Energy is the capacity to cause change; some forms of energy do work by moving matter. Kinetic energy is associated with motion. Potential energy is stored in the location or structure of matter and includes chemical energy stored in molecular structure.
- **The Laws of Energy Transformation (pp. 143–144)** The first law, conservation of energy, states that energy cannot be created or destroyed, only transferred or transformed. The second law states that spontaneous changes, those requiring no outside input of energy, increase the entropy (disorder) of the universe.
  Activity *Energy Transformations*

Concept 8.2

The free-energy change of a reaction tells us whether the reaction occurs spontaneously

- **Free-Energy Change, ΔG (p. 145)** A living system's free energy is energy that can do work under cellular conditions. The change in free energy (ΔG) during a biological process is related directly to enthalpy change (ΔH) and to the change in entropy (ΔS): $\Delta G = \Delta H - T\Delta S$.
- **Free Energy, Stability, and Equilibrium (pp. 145–146)** Organisms live at the expense of free energy. During a spontaneous change, free energy decreases and the stability of a system increases. At maximum stability, the system is at equilibrium.
- **Free Energy and Metabolism (pp. 146–148)** In an exergonic (spontaneous) chemical reaction, the products have less free energy than the reactants ($-\Delta G$). Endergonic (nonspontaneous) reactions require an input of energy ($+\Delta G$). The addition of starting materials and the removal of end products prevent metabolism from reaching equilibrium.

Concept 8.3

ATP powers cellular work by coupling exergonic reactions to endergonic reactions

- **The Structure and Hydrolysis of ATP (pp. 148–149)** ATP is the cell's energy shuttle. Release of its terminal phosphate group produces ADP and phosphate and releases free energy.
  Activity *The Structure of ATP*
- **How ATP Performs Work (pp. 149–150)** ATP drives endergonic reactions by phosphorylation, the transfer of a phosphate group to specific reactants, making them more reactive. In this way, cells can carry out work, such as movement and anabolism.
  Activity *Chemical Reactions and ATP*
- **The Regeneration of ATP (p. 150)** Catabolic pathways drive the regeneration of ATP from ADP and phosphate.

Concept 8.4

Enzymes speed up metabolic reactions by lowering energy barriers

- **The Activation Energy Barrier (pp. 150–152)** In a chemical reaction, the energy necessary to break the bonds of the reactants is the activation energy, E_A.
- **How Enzymes Lower the E_A Barrier (p. 152)** Enzymes, which are proteins, are biological catalysts. They speed up reactions by lowering the activation energy barrier.
- **Substrate Specificity of Enzymes (p. 152)** Each type of enzyme has a unique active site that combines specifically with its substrate, the reactant molecule on which it acts. The enzyme changes shape slightly when it binds the substrate (induced fit).
  Activity *How Enzymes Work*
- **Catalysis in the Enzyme's Active Site (pp. 152–154)** The active site can lower an E_A barrier by orienting substrates correctly, straining their bonds, providing a favorable microenvironment, and even covalently bonding with the substrate.
- **Effects of Local Conditions on Enzyme Activity (pp. 154–155)** Each enzyme has an optimal temperature and pH. Inhibitors reduce enzyme function. A competitive inhibitor binds to the active site, while a noncompetitive inhibitor binds to a different site on the enzyme.
  Investigation *How Is the Rate of Enzyme Catalysis Measured?*
  Biology Labs On-Line *EnzymeLab*

Concept 8.5

Regulation of enzyme activity helps control metabolism

- **Allosteric Regulation of Enzymes (pp. 156–157)** Many enzymes are allosterically regulated: They change shape when regulatory molecules, either activators or inhibitors, bind to specific regulatory sites, affecting enzymatic function. In feedback inhibition, the end product of a metabolic pathway allosterically inhibits the enzyme for an earlier step in the pathway.
- **Specific Localization of Enzymes Within the Cell (p. 157)** Some enzymes are grouped into complexes, some are incorporated into membranes, and others are contained inside organelles.

TESTING YOUR KNOWLEDGE

Self-Quiz

1. Choose the pair of terms that correctly completes this sentence: Catabolism is to anabolism as ________ is to ________.
   a. exergonic; spontaneous
   b. exergonic; endergonic
   c. free energy; entropy
   d. work; energy
   e. entropy; enthalpy

2. Most cells cannot harness heat to perform work because
   a. heat is not a form of energy.
   b. cells do not have much heat; they are relatively cool.
   c. temperature is usually uniform throughout a cell.
   d. heat can never be used to do work.
   e. heat denatures enzymes.

3. According to the first law of thermodynamics,
   a. matter can be neither created nor destroyed.
   b. energy is conserved in all processes.
   c. all processes increase the order of the universe.
   d. systems rich in energy are intrinsically stable.
   e. the universe constantly loses energy because of friction.

4. Which of the following metabolic processes can occur without a net influx of energy from some other process?
   a. ADP + $Ⓟ_i \rightarrow$ ATP + H_2O
   b. $C_6H_{12}O_6 + 6\ O_2 \rightarrow 6\ CO_2 + 6\ H_2O$
   c. $6\ CO_2 + 6\ H_2O \rightarrow C_6H_{12}O_6 + 6\ O_2$
   d. amino acids $\rightarrow$ protein
   e. glucose + fructose $\rightarrow$ sucrose

5. If an enzyme has been inhibited noncompetitively,
   a. the ΔG for the reaction it catalyzes will always be negative.
   b. the active site will be occupied by the inhibitor molecule.
   c. raising substrate concentration will increase inhibition.
   d. more energy will be necessary to initiate the reaction.
   e. the inhibitor molecule may be chemically unrelated to the substrate.

6. If an enzyme solution is saturated with substrate, the most effective way to obtain an even faster yield of products is to
   a. add more of the enzyme.
   b. heat the solution to 90°C.
   c. add more substrate.
   d. add an allosteric inhibitor.
   e. add a noncompetitive inhibitor.

7. If an enzyme is added to a solution where its substrate and products are in equilibrium, what would occur?
   a. Additional product would be formed.
   b. Additional substrate would be formed.
   c. The reaction would change from endergonic to exergonic.
   d. The free energy of the system would change.
   e. Nothing; the reaction would stay at equilibrium.

8. Some bacteria are metabolically active in hot springs because
   a. they are able to maintain a cooler internal temperature.
   b. high temperatures make catalysis unnecessary.
   c. their enzymes have high optimal temperatures.
   d. their enzymes are completely insensitive to temperature.
   e. they use molecules other than proteins as their main catalysts.

9. Which of the following characteristics is not associated with allosteric regulation of an enzyme's activity?
   a. A mimic of the substrate competes for the active site.
   b. A naturally occurring molecule stabilizes a catalytically active conformation.
   c. Regulatory molecules bind to a site remote from the active site.
   d. Inhibitors and activators may compete with one another.
   e. The enzyme usually has a quaternary structure.

10. In this branched metabolic pathway, a red arrow with a minus sign symbolizes inhibition of a metabolic step by an end product:

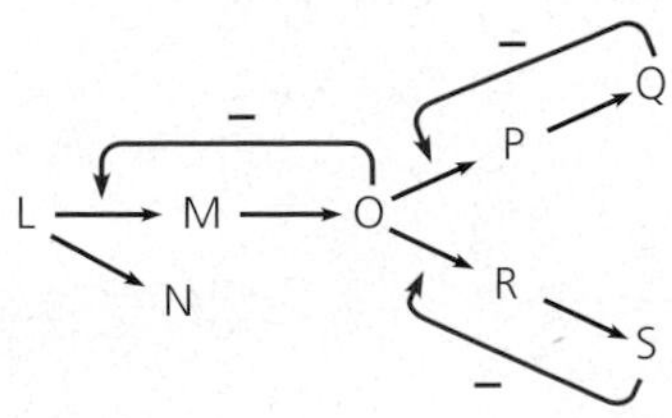

Which reaction would prevail if both Q and S were present in the cell in high concentrations?
   a. L → M
   b. M → O
   c. L → N
   d. O → P
   e. R → S

For Self-Quiz answers, see Appendix A.

Go to the website or CD-ROM for more quiz questions.

Evolution Connection

A recent revival of the anti-evolutionary "argument from design" holds that biochemical pathways are too complex to have evolved, because all intermediate steps in a given pathway must be present to produce the final product. Critique this argument. How could you use the existing diversity of metabolic pathways that produce the same or similar products to support your case?

Scientific Inquiry

A researcher has developed an assay to measure the activity of an important enzyme present in liver cells being grown in culture. She adds the enzyme's substrate to a dish of cells, then measures the appearance of reaction products. The results are graphed as the amount of product on the *y*-axis versus time on the *x*-axis. The researcher notes four sections of the graph. For a short period of time, no products appeared (section A). Then (section B) the reaction rate was quite high (the slope of the line was steep). After some time, the reaction slowed down considerably (section C), although products continued to appear (the line was not flat). Still later, the reaction resumed its original rapid rate (section D). Draw the graph, and propose a model to explain the molecular events underlying this interesting reaction profile.

Investigation *How Is the Rate of Enzyme Catalysis Measured?*
Biology Labs On-Line *EnzymeLab*

Science, Technology, and Society

The EPA is evaluating the safety of the most commonly used organophosphate insecticides (organic compounds containing phosphate groups). Organophosphates typically interfere with nerve transmission by inhibiting the enzymes that degrade the transmitter molecules diffusing from one neuron to another. Noxious insects are not uniquely susceptible; humans and other vertebrates can be affected as well. Thus, the use of organophosphate pesticides creates some health risks. As a consumer, what level of risk are you willing to accept in exchange for an abundant and affordable food supply? What other facts would you like to know before you defend your opinion?

9 Cellular Respiration

Harvesting Chemical Energy

▲ **Figure 9.1 This giant panda is consuming fuel to power the work of life.**

Key Concepts

9.1 Catabolic pathways yield energy by oxidizing organic fuels

9.2 Glycolysis harvests chemical energy by oxidizing glucose to pyruvate

9.3 The citric acid cycle completes the energy-yielding oxidation of organic molecules

9.4 During oxidative phosphorylation, chemiosmosis couples electron transport to ATP synthesis

9.5 Fermentation enables some cells to produce ATP without the use of oxygen

9.6 Glycolysis and the citric acid cycle connect to many other metabolic pathways

Overview

Life Is Work

Living cells require transfusions of energy from outside sources to perform their many tasks—for example, assembling polymers, pumping substances across membranes, moving, and reproducing. The giant panda in **Figure 9.1** obtains energy for its cells by eating plants; some animals feed on other organisms that eat plants. The energy stored in the organic molecules of food ultimately comes from the sun. Energy flows into an ecosystem as sunlight and leaves as heat **(Figure 9.2)**. In contrast, the chemical elements essential to life are recycled. Photosynthesis generates oxygen and organic molecules used by the mitochondria of eukaryotes (including plants and algae) as fuel for cellular respiration. Respiration breaks this fuel down, generating ATP. The waste products of respiration, carbon dioxide and water, are the raw materials for photosynthesis. In this chapter, we consider how cells harvest the chemical energy stored in organic molecules and use it to generate ATP, the molecule that drives most cellular work. After presenting some basics about respiration, we will focus on the three key pathways of respiration: glycolysis, the citric acid cycle, and oxidative phosphorylation.

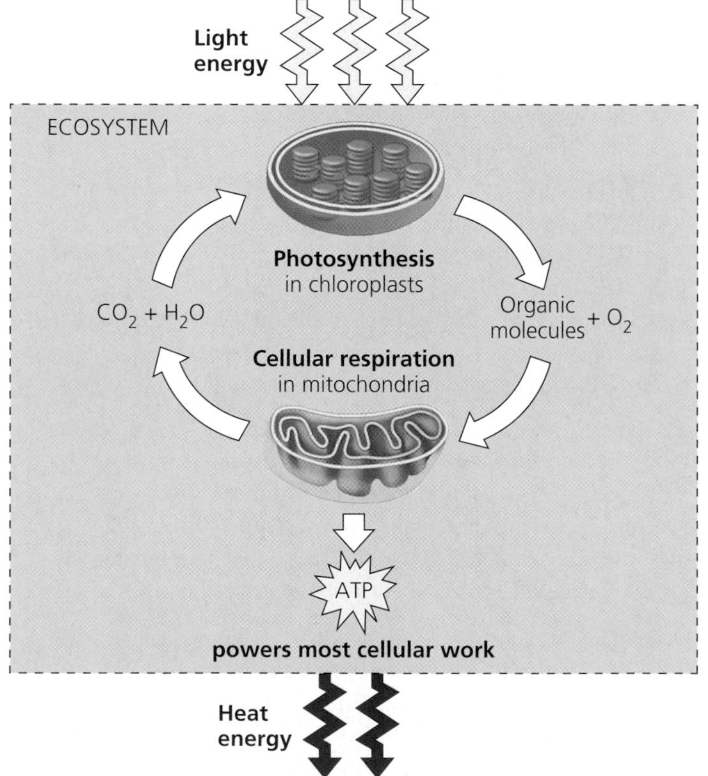

▲ **Figure 9.2 Energy flow and chemical recycling in ecosystems.** Energy flows into an ecosystem as sunlight and ultimately leaves as heat, while the chemical elements essential to life are recycled.

Catabolic pathways yield energy by oxidizing organic fuels

In this section, we consider several processes that are central to cellular respiration and related pathways.

Catabolic Pathways and Production of ATP

Organic compounds store energy in their arrangement of atoms. With the help of enzymes, a cell systematically degrades complex organic molecules that are rich in potential energy to simpler waste products that have less energy. Some of the energy taken out of chemical storage can be used to do work; the rest is dissipated as heat. As you learned in Chapter 8, metabolic pathways that release stored energy by breaking down complex molecules are called catabolic pathways. One catabolic process, **fermentation,** is a partial degradation of sugars that occurs without the use of oxygen. However, the most prevalent and efficient catabolic pathway is **cellular respiration,** in which oxygen is consumed as a reactant along with the organic fuel. In eukaryotic cells, mitochondria house most of the metabolic equipment for cellular respiration.

Although very different in mechanism, respiration is in principle similar to the combustion of gasoline in an automobile engine after oxygen is mixed with the fuel (hydrocarbons). Food provides the fuel for respiration, and the exhaust is carbon dioxide and water. The overall process can be summarized as follows:

$$\text{Organic compounds} + \text{Oxygen} \rightarrow \text{Carbon dioxide} + \text{Water} + \text{Energy}$$

Although carbohydrates, fats, and proteins can all be processed and consumed as fuel, it is helpful to learn the steps of cellular respiration by tracking the degradation of the sugar glucose ($C_6H_{12}O_6$), the fuel that cells most often use:

$$C_6H_{12}O_6 + 6\ O_2 \rightarrow 6\ CO_2 + 6\ H_2O + \text{Energy (ATP + heat)}$$

This breakdown of glucose is exergonic, having a free-energy change of −686 kcal (−2,870 kJ) per mole of glucose decomposed ($\Delta G = -686$ kcal/mol). Recall that a negative ΔG indicates that the products of the chemical process store less energy than the reactants and that the reaction can happen spontaneously—in other words, without an input of energy.

Catabolic pathways do not directly move flagella, pump solutes across membranes, polymerize monomers, or perform other cellular work. Catabolism is linked to work by a chemical drive shaft—ATP, which you learned about in Chapter 8. To keep working, the cell must regenerate its supply of ATP from ADP and Ⓟ$_i$ (see Figure 8.12). To understand how cellular respiration accomplishes this, let's examine the fundamental chemical processes known as oxidation and reduction.

Redox Reactions: Oxidation and Reduction

Why do the catabolic pathways that decompose glucose and other organic fuels yield energy? The answer is based on the transfer of electrons during the chemical reactions. The relocation of electrons releases energy stored in organic molecules, and this energy ultimately is used to synthesize ATP.

The Principle of Redox

In many chemical reactions, there is a transfer of one or more electrons (e^-) from one reactant to another. These electron transfers are called oxidation-reduction reactions, or **redox reactions** for short. In a redox reaction, the loss of electrons from one substance is called **oxidation,** and the addition of electrons to another substance is known as **reduction.** (Note that *adding* electrons is called *reduction;* negatively charged electrons added to an atom *reduce* the amount of positive charge of that atom.) To take a simple, nonbiological example, consider the reaction between the elements sodium (Na) and chlorine (Cl) that forms table salt:

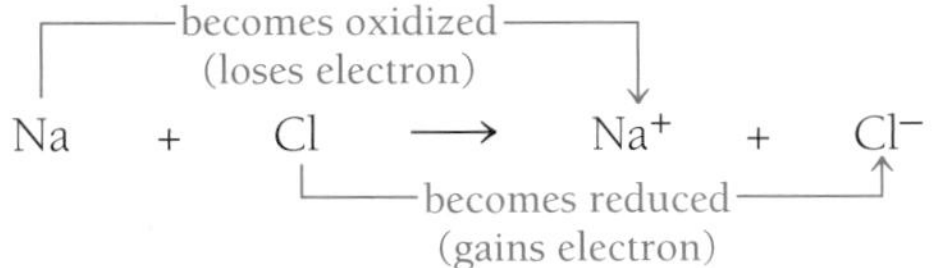

We could generalize a redox reaction this way:

becomes oxidized

$$Xe^- + Y \longrightarrow X + Ye^-$$

becomes reduced

In the generalized reaction, substance X, the electron donor, is called the **reducing agent;** it reduces Y, which accepts the donated electron. Substance Y, the electron acceptor, is the **oxidizing agent;** it oxidizes X by removing its electron. Because an electron transfer requires both a donor and an acceptor, oxidation and reduction always go together.

Not all redox reactions involve the complete transfer of electrons from one substance to another; some change the degree of electron sharing in covalent bonds. The reaction between methane and oxygen, shown in **Figure 9.3** on the next page, is an example. As explained in Chapter 2, the covalent electrons in methane are shared nearly equally between the bonded atoms because carbon and hydrogen have about the same affinity for valence electrons; they are about equally electronegative. But when methane reacts with oxygen, forming carbon dioxide, electrons end up farther away from the carbon atom and closer to their new covalent partners, the oxygen atoms, which are very electronegative. In effect, the carbon atom has partially "lost" its shared electrons; thus, methane has been oxidized.

Now let's examine the fate of the reactant O_2. The two atoms of the oxygen molecule (O_2) share their electrons equally. But when oxygen reacts with the hydrogen from methane, forming water, the electrons of the covalent bonds are drawn closer to

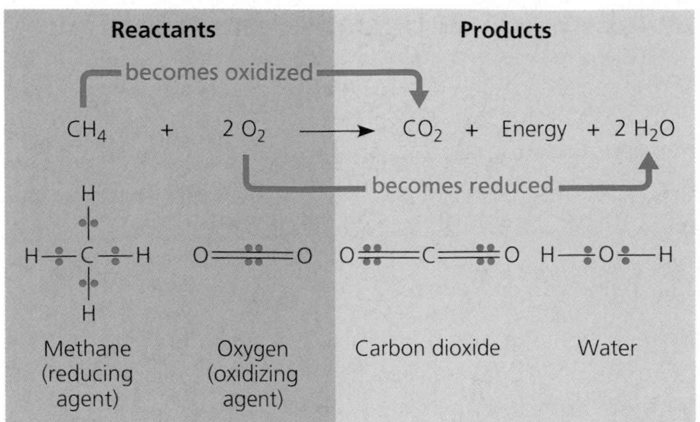

▲ **Figure 9.3 Methane combustion as an energy-yielding redox reaction.** The reaction releases energy to the surroundings because the electrons lose potential energy when they end up closer to electronegative atoms such as oxygen.

the oxygen (see Figure 9.3). In effect, each oxygen atom has partially "gained" electrons, and so the oxygen molecule has been reduced. Because oxygen is so electronegative, it is one of the most potent of all oxidizing agents.

Energy must be added to pull an electron away from an atom, just as energy is required to push a ball uphill. The more electronegative the atom (the stronger its pull on electrons), the more energy is required to take an electron away from it. An electron loses potential energy when it shifts from a less electronegative atom toward a more electronegative one, just as a ball loses potential energy when it rolls downhill. A redox reaction that relocates electrons closer to oxygen, such as the burning of methane, therefore releases chemical energy that can be put to work.

Oxidation of Organic Fuel Molecules During Cellular Respiration

The oxidation of methane by oxygen is the main combustion reaction that occurs at the burner of a gas stove. The combustion of gasoline in an automobile engine is also a redox reaction; the energy released pushes the pistons. But the energy-yielding redox process of greatest interest here is respiration: the oxidation of glucose and other molecules in food. Examine again the summary equation for cellular respiration, but this time think of it as a redox process:

$$\overbrace{C_6H_{12}O_6 + 6\,O_2 \longrightarrow 6\,CO_2}^{\text{becomes oxidized}} + \underbrace{6\,H_2O}_{\text{becomes reduced}} + \text{Energy}$$

As in the combustion of methane or gasoline, the fuel (glucose) is oxidized and oxygen is reduced. The electrons lose potential energy along the way, and energy is released.

In general, organic molecules that have an abundance of hydrogen are excellent fuels because their bonds are a source of "hilltop" electrons, whose energy may be released as these electrons "fall" down an energy gradient when they are transferred to oxygen. The summary equation for respiration indicates that hydrogen is transferred from glucose to oxygen. But the important point, not visible in the summary equation, is that the status of electrons changes as hydrogen is transferred to oxygen, liberating energy (ΔG is negative). By oxidizing glucose, respiration liberates stored energy from glucose and makes it available for ATP synthesis.

The main energy foods, carbohydrates and fats, are reservoirs of electrons associated with hydrogen. Only the barrier of activation energy holds back the flood of electrons to a lower energy state (see Figure 8.14). Without this barrier, a food substance like glucose would combine almost instantaneously with O_2. When we supply the activation energy by igniting glucose, it burns in air, releasing 686 kcal (2,870 kJ) of heat per mole of glucose (about 180 g). Body temperature is not high enough to initiate burning, of course. Instead, if you swallow some glucose, enzymes in your cells will lower the barrier of activation energy, allowing the sugar to be oxidized in a series of steps.

Stepwise Energy Harvest via NAD⁺ and the Electron Transport Chain

If energy is released from a fuel all at once, it cannot be harnessed efficiently for constructive work. For example, if a gasoline tank explodes, it cannot drive a car very far. Cellular respiration does not oxidize glucose in a single explosive step either. Rather, glucose and other organic fuels are broken down in a series of steps, each one catalyzed by an enzyme. At key steps, electrons are stripped from the glucose. As is often the case in oxidation reactions, each electron travels with a proton—thus, as a hydrogen atom. The hydrogen atoms are not transferred directly to oxygen, but instead are usually passed first to a coenzyme called **NAD^+** (nicotinamide adenine dinucleotide, a derivative of the vitamin niacin). As an electron acceptor, NAD^+ functions as an oxidizing agent during respiration.

How does NAD^+ trap electrons from glucose and other organic molecules? Enzymes called dehydrogenases remove a pair of hydrogen atoms (two electrons and two protons) from the substrate (a sugar, for example), thereby oxidizing it. The enzyme delivers the two electrons along with *one* proton to its coenzyme, NAD^+ **(Figure 9.4)**. The other proton is released as a hydrogen ion (H^+) into the surrounding solution:

$$\text{H}-\underset{|}{\overset{|}{\text{C}}}-\text{OH} + \text{NAD}^+ \xrightarrow{\text{Dehydrogenase}} \underset{|}{\overset{|}{\text{C}}}=\text{O} + \text{NADH} + \text{H}^+$$

By receiving two negatively charged electrons but only one positively charged proton, NAD^+ has its charge neutralized when it is reduced to NADH. The name NADH shows the hydrogen that has been received in the reaction. NAD^+ is the most versa-

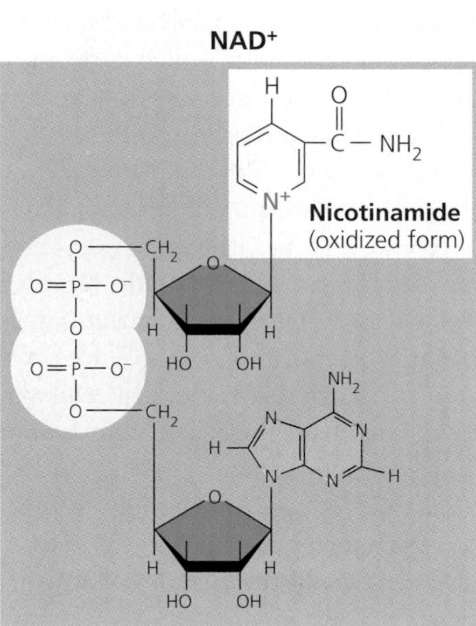

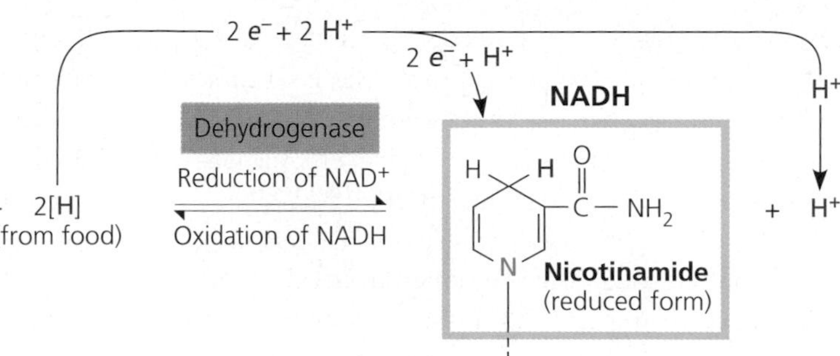

◀ **Figure 9.4 NAD$^+$ as an electron shuttle.** The full name for NAD$^+$, nicotinamide adenine dinucleotide, describes its structure; the molecule consists of two nucleotides joined together at their phosphate groups (shown in yellow). (Nicotinamide is a nitrogenous base, although not one that is present in DNA or RNA.) The enzymatic transfer of two electrons and one proton (H$^+$) from an organic molecule in food to NAD$^+$ reduces the NAD$^+$ to NADH; the second proton (H$^+$) is released. Most of the electrons removed from food are transferred initially to NAD$^+$.

tile electron acceptor in cellular respiration and functions in several of the redox steps during the breakdown of sugar.

Electrons lose very little of their potential energy when they are transferred from food to NAD$^+$. Each NADH molecule formed during respiration represents stored energy that can be tapped to make ATP when the electrons complete their "fall" down an energy gradient from NADH to oxygen.

How do electrons that are extracted from food and stored by NADH finally reach oxygen? It will help to compare the redox chemistry of cellular respiration to a much simpler reaction: the reaction between hydrogen and oxygen to form water **(Figure 9.5a)**. Mix H_2 and O_2, provide a spark for activation energy, and the gases combine explosively. The explosion represents a release of energy as the electrons of hydrogen fall closer to the electronegative oxygen atoms. Cellular respiration also brings hydrogen and oxygen together to form water, but there are two important differences. First, in cellular respiration, the hydrogen that reacts with oxygen is derived from organic molecules rather than H_2. Second, respiration uses an **electron transport chain** to break the fall of electrons to oxygen into several energy-releasing steps instead of one explosive reaction **(Figure 9.5b)**. The transport chain consists of a number of molecules, mostly proteins, built into the inner membrane of a mitochondrion. Electrons removed from food are shuttled

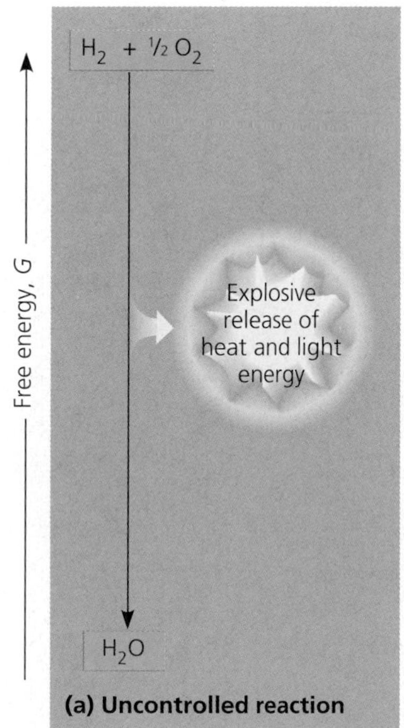

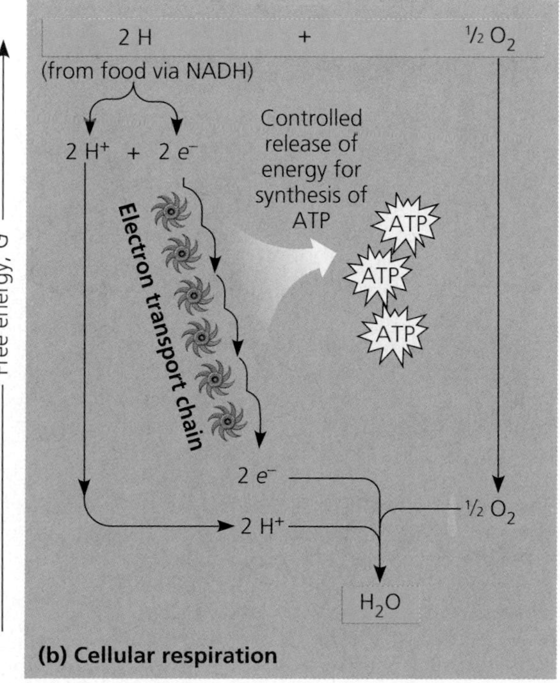

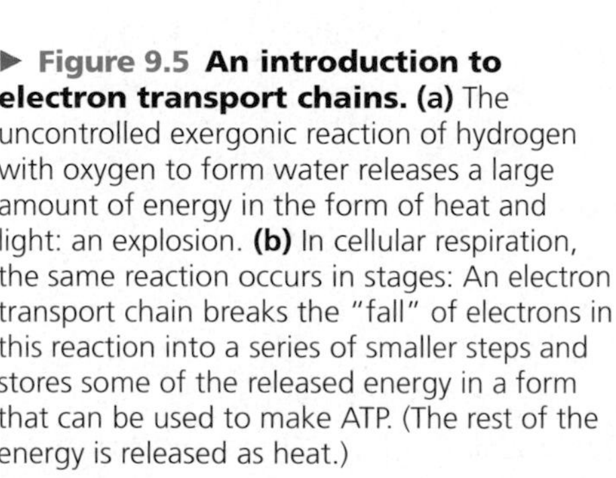
► **Figure 9.5 An introduction to electron transport chains. (a)** The uncontrolled exergonic reaction of hydrogen with oxygen to form water releases a large amount of energy in the form of heat and light: an explosion. **(b)** In cellular respiration, the same reaction occurs in stages: An electron transport chain breaks the "fall" of electrons in this reaction into a series of smaller steps and stores some of the released energy in a form that can be used to make ATP. (The rest of the energy is released as heat.)

by NADH to the "top," higher-energy end of the chain. At the "bottom," lower-energy end, oxygen captures these electrons along with hydrogen nuclei (H^+), forming water.

Electron transfer from NADH to oxygen is an exergonic reaction with a free-energy change of −53 kcal/mol (−222 kJ/mol). Instead of this energy being released and wasted in a single explosive step, electrons cascade down the chain from one carrier molecule to the next, losing a small amount of energy with each step until they finally reach oxygen, the terminal electron acceptor, which has a very great affinity for electrons. Each "downhill" carrier is more electronegative than its "uphill" neighbor, with oxygen at the bottom of the chain. Thus, the electrons removed from food by NAD^+ fall down an energy gradient in the electron transport chain to a far more stable location in the electronegative oxygen atom. Put another way, oxygen pulls electrons down the chain in an energy-yielding tumble analogous to gravity pulling objects downhill.

In summary, during cellular respiration, most electrons travel the following "downhill" route: food → NADH → electron transport chain → oxygen. Later in this chapter, you will learn more about how the cell uses the energy released from this exergonic electron fall to regenerate its supply of ATP.

Now that we have covered the basic redox mechanisms of cellular respiration, let's look at the entire process.

The Stages of Cellular Respiration: *A Preview*

Respiration is a cumulative function of three metabolic stages:

1. Glycolysis (color-coded teal throughout the chapter)
2. The citric acid cycle (color-coded salmon)
3. Oxidative phosphorylation: electron transport and chemiosmosis (color-coded violet)

Technically, cellular respiration is defined as including only the processes that require O_2: the citric acid cycle and oxidative phosphorylation. We include glycolysis, even though it doesn't require O_2, because most respiring cells deriving energy from glucose use this process to produce starting material for the citric acid cycle.

As diagrammed in **Figure 9.6**, the first two stages of cellular respiration, glycolysis and the citric acid cycle, are the catabolic pathways that decompose glucose and other organic fuels. **Glycolysis,** which occurs in the cytosol, begins the degradation process by breaking glucose into two molecules of a compound called pyruvate. The **citric acid cycle,** which takes place within the mitochondrial matrix, completes the breakdown of glucose by oxidizing a derivative of pyruvate to carbon dioxide. Thus, the carbon dioxide produced by respiration represents fragments of oxidized organic molecules.

Some of the steps of glycolysis and the citric acid cycle are redox reactions in which dehydrogenase enzymes transfer electrons from substrates to NAD^+, forming NADH. In the third stage of respiration, the electron transport chain accepts electrons from the breakdown products of the first two stages (most often via NADH) and passes these electrons from one molecule to another. At the end of the chain, the electrons are combined with molecular oxygen and hydrogen ions (H^+), forming water (see Figure 9.5b). The energy released at each step of the chain is stored in a form the mitochondrion can use to make ATP. This mode of ATP synthesis is called **oxidative phosphorylation** because it is powered by the redox reactions of the electron transport chain.

The inner membrane of the mitochondrion is the site of electron transport and chemiosmosis, the processes that together constitute oxidative phosphorylation. Oxidative phosphorylation accounts for almost 90% of the ATP generated by

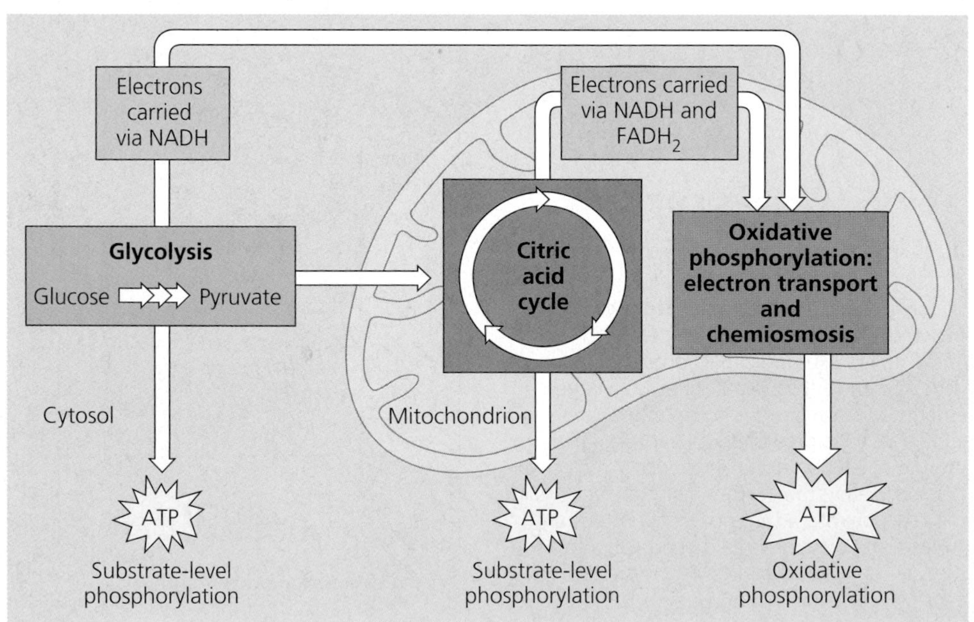

▶ **Figure 9.6 An overview of cellular respiration.** During glycolysis, each glucose molecule is broken down into two molecules of the compound pyruvate. The pyruvate enters the mitochondrion, where the citric acid cycle oxidizes it to carbon dioxide. NADH and a similar coenzyme called $FADH_2$ transfer electrons derived from glucose to electron transport chains, which are built into the inner mitochondrial membrane. During oxidative phosphorylation, electron transport chains convert the chemical energy to a form used for ATP synthesis in the process called chemiosmosis.

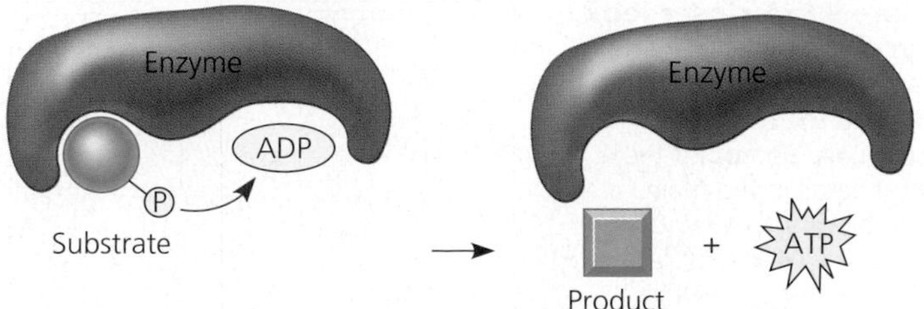

▲ **Figure 9.7 Substrate-level phosphorylation.** Some ATP is made by direct enzymatic transfer of a phosphate group from an organic substrate to ADP.

respiration. A smaller amount of ATP is formed directly in a few reactions of glycolysis and the citric acid cycle by a mechanism called **substrate-level phosphorylation** (**Figure 9.7**). This mode of ATP synthesis occurs when an enzyme transfers a phosphate group from a substrate molecule to ADP, rather than adding an inorganic phosphate to ADP as in oxidative phosphorylation. "Substrate molecule" here refers to an organic molecule generated during the catabolism of glucose.

For each molecule of glucose degraded to carbon dioxide and water by respiration, the cell makes up to about 38 molecules of ATP, each with 7.3 kcal/mol of free energy. Respiration cashes in the large denomination of energy banked in a single molecule of glucose (686 kcal/mol) for the small change of many molecules of ATP, which is more practical for the cell to spend on its work.

This preview has introduced how glycolysis, the citric acid cycle, and oxidative phosphorylation fit into the overall process of cellular respiration. We are now ready to take a closer look at each of these three stages of respiration.

Concept Check 9.1

1. In the following redox reaction, which compound is oxidized and which is reduced?

$$C_4H_6O_5 + NAD^+ \rightarrow C_4H_4O_5 + NADH + H^+$$

For suggested answers, see Appendix A.

Glycolysis harvests chemical energy by oxidizing glucose to pyruvate

The word *glycolysis* means "splitting of sugar," and that is exactly what happens during this pathway. Glucose, a six-carbon sugar, is split into two three-carbon sugars. These smaller sugars are then oxidized and their remaining atoms rearranged to form two molecules of pyruvate. (Pyruvate is the ionized form of pyruvic acid.)

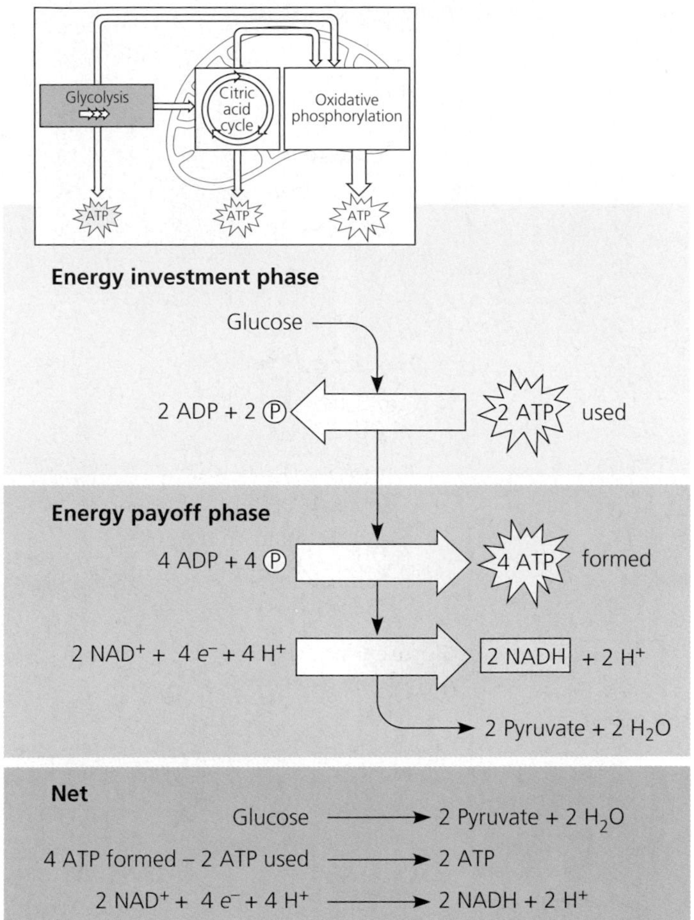

▲ **Figure 9.8 The energy input and output of glycolysis.**

As summarized in **Figure 9.8** and described in detail in **Figure 9.9**, on the next two pages, the pathway of glycolysis consists of ten steps, which can be divided into two phases. During the energy investment phase, the cell actually spends ATP. This investment is repaid with dividends during the energy payoff phase, when ATP is produced by substrate-level phosphorylation and NAD^+ is reduced to NADH by electrons released from the oxidation of the food (glucose in this example). The net energy yield from glycolysis, per glucose molecule, is 2 ATP plus 2 NADH.

Notice in Figure 9.9 that all of the carbon originally present in glucose is accounted for in the two molecules of pyruvate; no CO_2 is released during glycolysis. Glycolysis occurs whether or not O_2 is present. However, if O_2 *is* present, the chemical energy stored in pyruvate and NADH can be extracted by the citric acid cycle and oxidative phosphorylation, respectively.

Concept Check 9.2

1. During the redox reaction in glycolysis (step 6 in Figure 9.9), which molecule acts as the oxidizing agent? The reducing agent?

For suggested answers, see Appendix A.

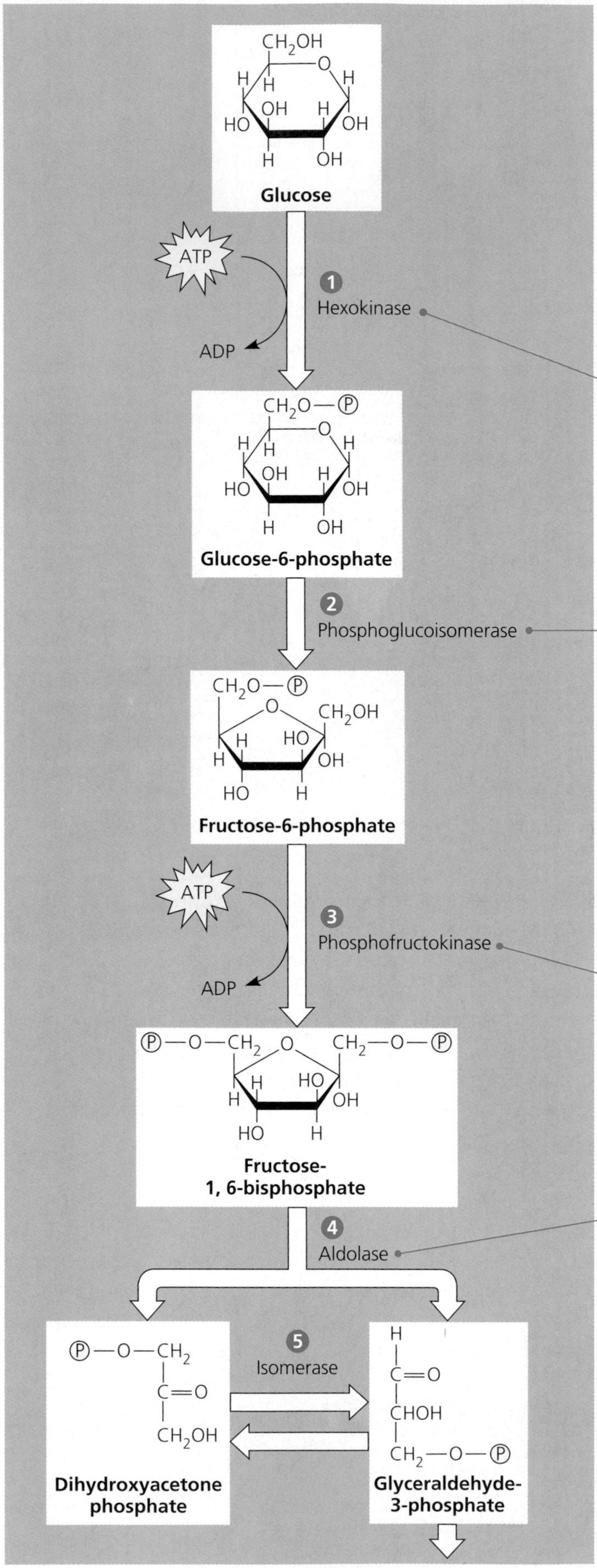

▼ **Figure 9.9 A closer look at glycolysis.** The orientation diagram at the right relates glycolysis to the whole process of respiration. Do not let the chemical detail in the main diagram block your view of glycolysis as a source of ATP and NADH.

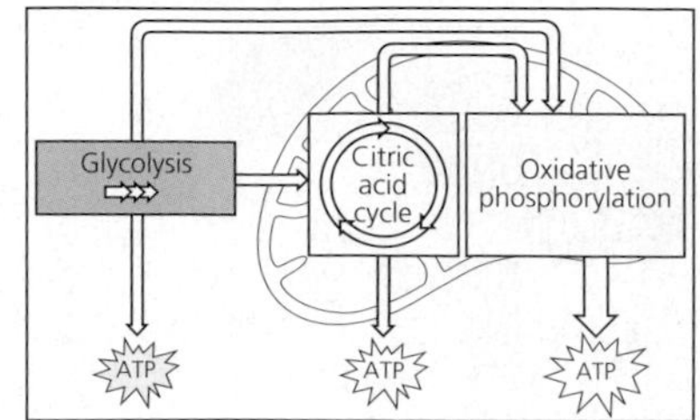

Glucose

ATP → ADP

1 Hexokinase

Glucose-6-phosphate

2 Phosphoglucoisomerase

Fructose-6-phosphate

ATP → ADP

3 Phosphofructokinase

Fructose-1, 6-bisphosphate

4 Aldolase

Dihydroxyacetone phosphate

5 Isomerase

Glyceraldehyde-3-phosphate

ENERGY INVESTMENT PHASE

1 Glucose enters the cell and is phosphorylated by the enzyme hexokinase, which transfers a phosphate group from ATP to the sugar. The charge of the phosphate group traps the sugar in the cell because the plasma membrane is impermeable to ions. Phosphorylation also makes glucose more chemically reactive. In this diagram, the transfer of a phosphate group or pair of electrons from one reactant to another is indicated by coupled arrows:

2 Glucose-6-phosphate is rearranged to convert it to its isomer, fructose-6-phosphate.

3 This enzyme transfers a phosphate group from ATP to the sugar, investing another molecule of ATP in glycolysis. So far, 2 ATP have been used. With phosphate groups on its opposite ends, the sugar is now ready to be split in half. This is a key step for regulation of glycolysis; phosphofructokinase is allosterically regulated by ATP and its products.

4 This is the reaction from which glycolysis gets its name. The enzyme cleaves the sugar molecule into two different three-carbon sugars: dihydroxyacetone phosphate and glyceraldehyde-3-phosphate. These two sugars are isomers of each other.

5 Isomerase catalyzes the reversible conversion between the two three-carbon sugars. This reaction never reaches equilibrium in the cell because the next enzyme in glycolysis uses only glyceraldehyde-3-phosphate as its substrate (and not dihydroxyacetone phosphate). This pulls the equilibrium in the direction of glyceraldehyde-3-phosphate, which is removed as fast as it forms. Thus, the net result of steps 4 and 5 is cleavage of a six-carbon sugar into two molecules of glyceraldehyde-3-phosphate; each will progress through the remaining steps of glycolysis.

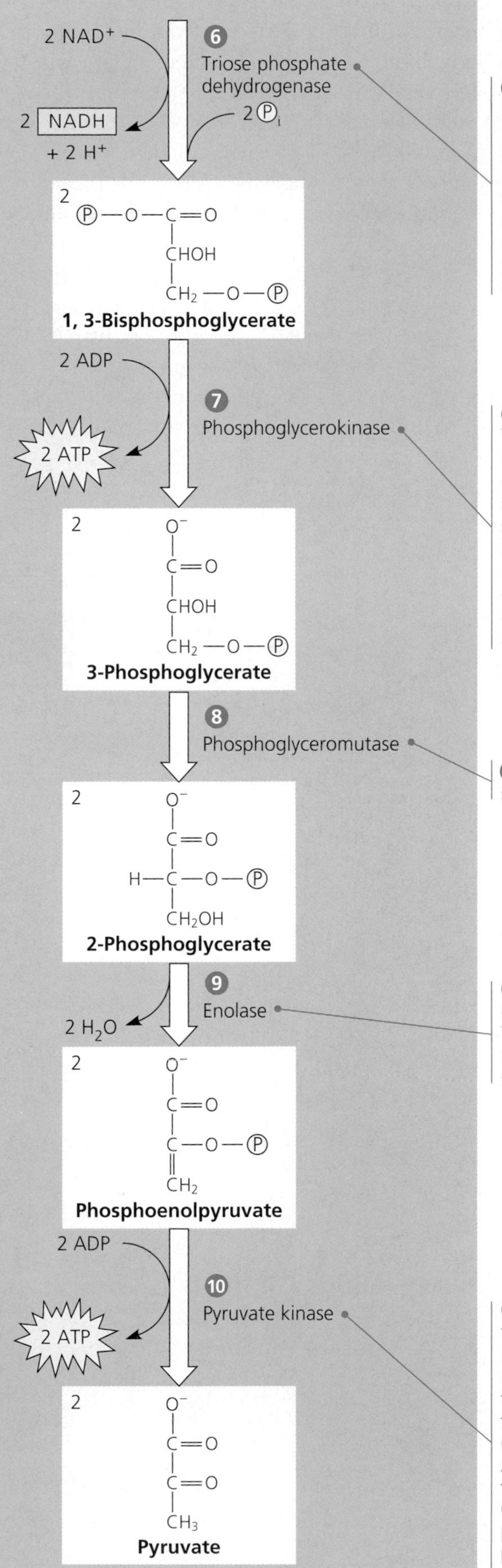

ENERGY PAYOFF PHASE

6 This enzyme catalyzes two sequential reactions while it holds glyceraldehyde-3-phosphate in its active site. First, the sugar is oxidized by the transfer of electrons and H^+ to NAD^+, forming NADH (a redox reaction). This reaction is very exergonic, and the enzyme uses the released energy to attach a phosphate group to the oxidized substrate, making a product of very high potential energy. The source of the phosphates is the pool of inorganic phosphate ions that are always present in the cytosol. Notice that the coefficient 2 precedes all molecules in the energy payoff phase; these steps occur after glucose is split into two three-carbon sugars (step 4).

7 Glycolysis produces some ATP by substrate-level phosphorylation. The phosphate group added in the previous step is transferred to ADP in an exergonic reaction. For each glucose molecule that began glycolysis, step 7 produces 2 ATP, since every product after the sugar-splitting step (step 4) is doubled. Recall that 2 ATP were invested to get sugar ready for splitting; this ATP debt has now been repaid. Glucose has been converted to two molecules of 3-phosphoglycerate, which is not a sugar. The carbonyl group that characterizes a sugar has been oxidized to a carboxyl group ($—COO^-$), the hallmark of an organic acid. The sugar was oxidized in step 6, and now the energy made available by that oxidation has been used to make ATP.

8 Next, this enzyme relocates the remaining phosphate group. This step prepares the substrate for the next reaction.

9 This enzyme causes a double bond to form in the substrate by extracting a water molecule, yielding phosphoenolpyruvate (PEP). The electrons of the substrate are rearranged in such a way that the remaining phosphate bond becomes very unstable, preparing the substrate for the next reaction.

10 The last reaction of glycolysis produces more ATP by transferring the phosphate group from PEP to ADP, a second example of substrate-level phosphorylation. Since this step occurs twice for each glucose molecule, 2 ATP are produced. Overall, glycolysis has used 2 ATP in the energy investment phase (steps 1 and 3) and produced 4 ATP in the energy payoff phase (steps 7 and 10), for a net gain of 2 ATP. Glycolysis has repaid the ATP investment with 100% interest. Additional energy was stored by step 6 in NADH, which can be used to make ATP by oxidative phosphorylation if oxygen is present. Glucose has been broken down and oxidized to two molecules of pyruvate, the end product of the glycolytic pathway. If oxygen is present, the chemical energy in pyruvate can be extracted by the citric acid cycle.

Concept 9.3

The citric acid cycle completes the energy-yielding oxidation of organic molecules

Glycolysis releases less than a quarter of the chemical energy stored in glucose; most of the energy remains stockpiled in the two molecules of pyruvate. If molecular oxygen is present, the pyruvate enters the mitochondrion, where the enzymes of the citric acid cycle complete the oxidation of the organic fuel.

Upon entering the mitochondrion via active transport, pyruvate is first converted to a compound called acetyl coenzyme A, or **acetyl CoA (Figure 9.10)**. This step, the junction between glycolysis and the citric acid cycle, is accomplished by a multienzyme complex that catalyzes three reactions: ❶ Pyruvate's carboxyl group (—COO^-), which is already fully oxidized and thus has little chemical energy, is removed and given off as a molecule of CO_2. (This is the first step in which CO_2 is released during respiration.) ❷ The remaining two-carbon fragment is oxidized, forming a compound named acetate (the ionized form of acetic acid). An enzyme transfers the extracted electrons to NAD^+, storing energy in the form of NADH. ❸ Finally, coenzyme A, a sulfur-containing compound derived from a B vitamin, is attached to the acetate by an unstable bond (the wavy line in Figure 9.10), making the acetyl group (the attached acetate) very reactive. The product of this chemical grooming, acetyl CoA, is now ready to feed its acetyl group into the citric acid cycle for further oxidation.

The citric acid cycle is also called the tricarboxylic acid cycle or the Krebs cycle, the latter honoring Hans Krebs, the German-British scientist who was largely responsible for elucidating the pathway in the 1930s. The cycle functions as a metabolic furnace that oxidizes organic fuel derived from pyruvate. **Figure 9.11** summarizes the inputs and outputs as pyruvate is broken down to 3 CO_2 molecules, including the molecule of CO_2 released during the conversion of pyruvate to acetyl CoA. The cycle generates 1 ATP per turn by substrate-level phosphorylation, but most of the chemical energy is transferred to NAD^+ and the related coenzyme FAD during the redox reactions. The reduced coenzymes, NADH and $FADH_2$, shuttle their cargo of high-energy electrons to the electron transport chain.

Now let's look at the citric acid cycle in more detail. The cycle has eight steps, each catalyzed by a specific enzyme. You can see in **Figure 9.12** that for each turn of the citric acid cycle, two carbons (red) enter in the relatively reduced form of an acetyl group (step 1), and two different carbons (blue) leave in the completely oxidized form of CO_2 (steps 3 and 4). The acetyl group of acetyl CoA joins the cycle by combining with the compound oxaloacetate, forming citrate (step 1).

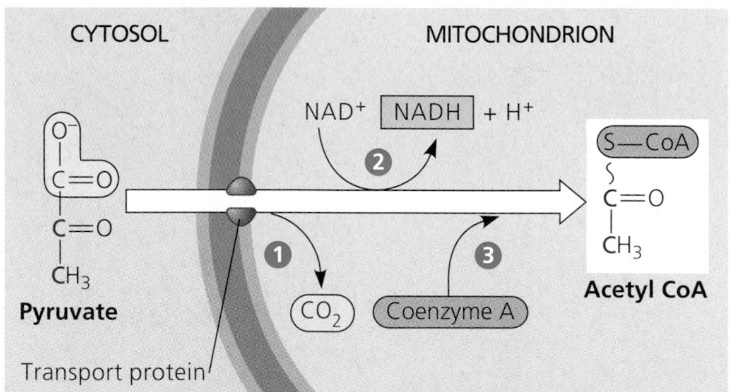

▲ Figure 9.10 Conversion of pyruvate to acetyl CoA, the junction between glycolysis and the citric acid cycle. Because pyruvate is a charged molecule, it must enter the mitochondrion via active transport, with the help of a transport protein. Next, a complex of several enzymes (the pyruvate dehydrogenase complex) catalyzes the three numbered steps, which are described in the text. The acetyl group of acetyl CoA will enter the citric acid cycle. The CO_2 molecule will diffuse out of the cell.

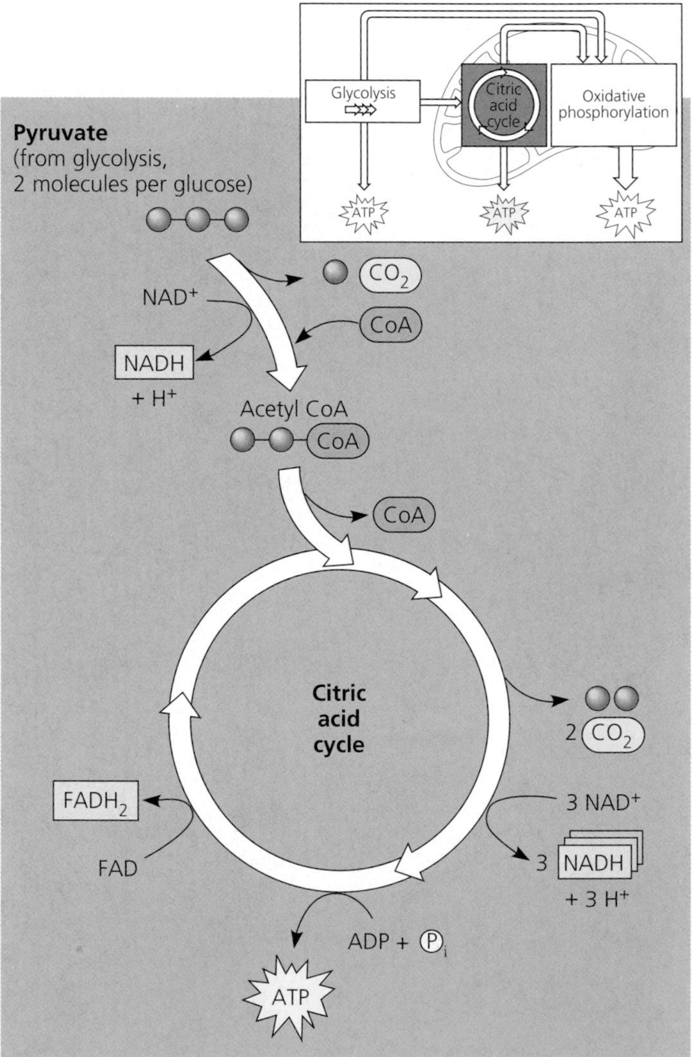

▲ Figure 9.11 An overview of the citric acid cycle. To calculate the inputs and outputs on a per-glucose basis, multiply by 2, because each glucose molecule is split during glycolysis into two pyruvate molecules.

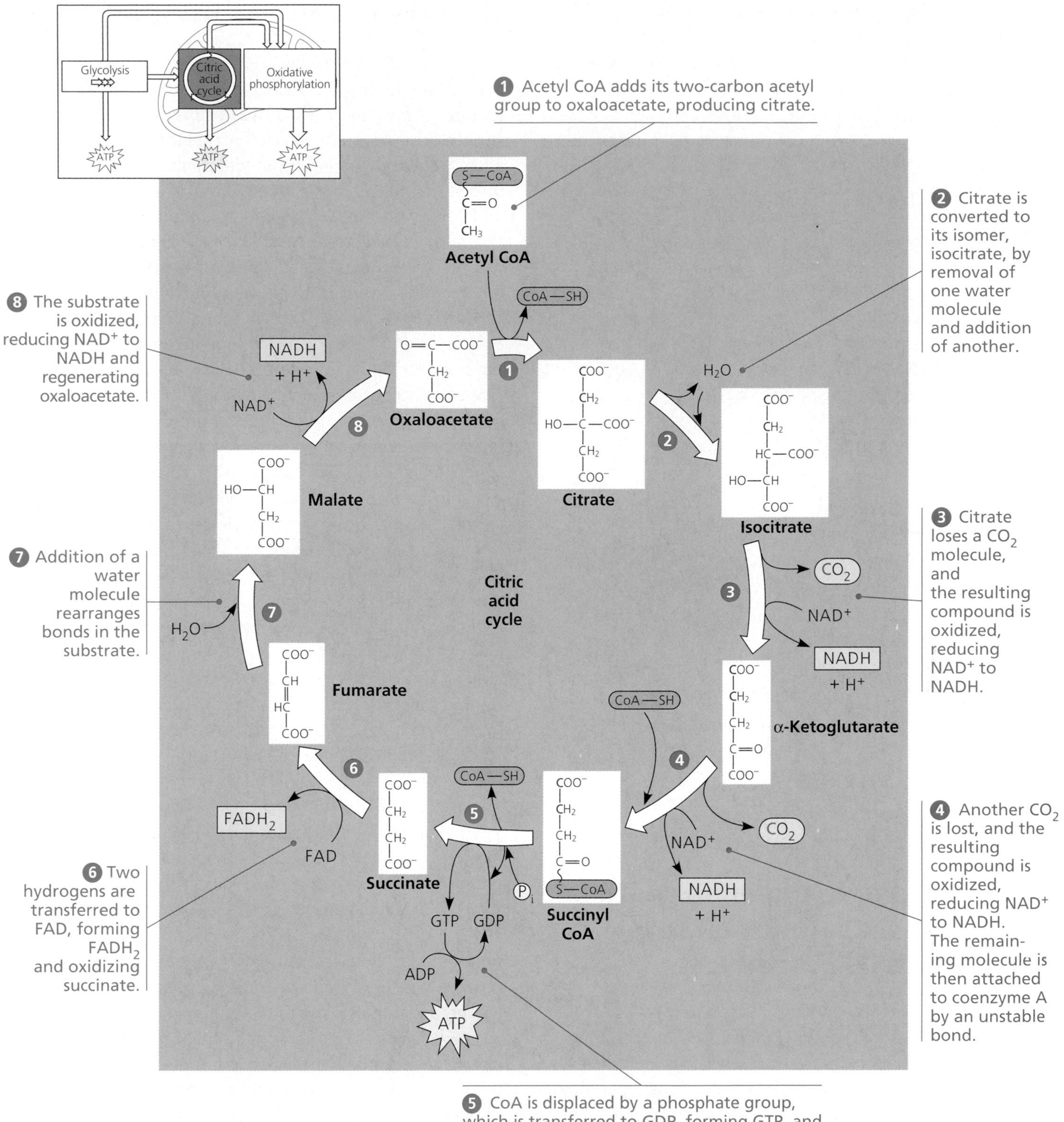

Glycolysis
Citric acid cycle
Oxidative phosphorylation
ATP
ATP
ATP
1 Acetyl CoA adds its two-carbon acetyl group to oxaloacetate, producing citrate.
Acetyl CoA
CoA—SH
Oxaloacetate
Citrate
H2O
Isocitrate
2 Citrate is converted to its isomer, isocitrate, by removal of one water molecule and addition of another.
3 Citrate loses a CO2 molecule, and the resulting compound is oxidized, reducing NAD+ to NADH.
CO2
NAD+
NADH
+ H+
α-Ketoglutarate
CoA—SH
CO2
NAD+
NADH
+ H+
4 Another CO2 is lost, and the resulting compound is oxidized, reducing NAD+ to NADH. The remaining molecule is then attached to coenzyme A by an unstable bond.
Succinyl CoA
CoA—SH
Pi
GTP
GDP
ADP
ATP
Succinate
5 CoA is displaced by a phosphate group, which is transferred to GDP, forming GTP, and then to ADP, forming ATP (substrate-level phosphorylation).
FADH2
FAD
6 Two hydrogens are transferred to FAD, forming FADH2 and oxidizing succinate.
Fumarate
H2O
7 Addition of a water molecule rearranges bonds in the substrate.
Malate
NADH
+ H+
NAD+
8 The substrate is oxidized, reducing NAD+ to NADH and regenerating oxaloacetate.
Citric acid cycle

▲ **Figure 9.12 A closer look at the citric acid cycle.** In the chemical structures, red type traces the fate of the two carbon atoms that enter the cycle via acetyl CoA (step 1), and blue type indicates the two carbons that exit the cycle as CO_2 in steps 3 and 4. (The red labeling only goes through step 5, but you can continue to trace the fate of those carbons.) Notice that the carbon atoms that enter the cycle from acetyl CoA do not leave the cycle in the same turn. They remain in the cycle, occupying a different location in the molecules on their next turn after another acetyl group is added. As a consequence, the oxaloacetate that is regenerated at step 8 is composed of different carbon atoms each time around. All the citric acid cycle enzymes are located in the mitochondrial matrix except for the enzyme that catalyzes step 6, which resides in the inner mitochondrial membrane. Carboxylic acids are represented in their ionized forms, as $—COO^-$, because the ionized forms prevail at the pH within the mitochondrion. For example, citrate is the ionized form of citric acid.

(Citrate is the ionized form of citric acid, for which the citric acid cycle is named.) The next seven steps decompose the citrate back to oxaloacetate. It is this regeneration of oxaloacetate that makes this process a *cycle*.

For each acetyl group that enters the cycle, 3 NAD^+ are reduced to NADH (steps 3, 4, and 8). In step 6, electrons are transferred not to NAD^+, but to a different electron acceptor, FAD (flavin adenine dinucleotide, derived from riboflavin, a B vitamin). Step 5 in the citric acid cycle forms a GTP molecule directly by substrate-level phosphorylation, similar to the ATP-generating steps of glycolysis. This GTP is then used to synthesize an ATP, the only ATP generated directly by the citric acid cycle. Most of the ATP output of respiration results from oxidative phosphorylation, when the NADH and $FADH_2$ produced by the citric acid cycle relay the electrons extracted from food to the electron transport chain. In the process, they supply the necessary energy for the phosphorylation of ADP to ATP. We will explore this process in the next section.

Concept Check 9.3

1. In which molecules is most of the energy from the citric acid cycle's redox reactions conserved? How will these molecules convert their energy to a form that can be used to make ATP?
2. What cellular processes produce the carbon dioxide that you exhale?

For suggested answers, see Appendix A.

Concept 9.4

During oxidative phosphorylation, chemiosmosis couples electron transport to ATP synthesis

Our main objective in this chapter is to learn how cells harvest the energy of food to make ATP. But the metabolic components of respiration we have dissected so far, glycolysis and the citric acid cycle, produce only 4 ATP molecules per glucose molecule, all by substrate-level phosphorylation: 2 net ATP from glycolysis and 2 ATP from the citric acid cycle. At this point, molecules of NADH (and $FADH_2$) account for most of the energy extracted from the food. These electron escorts link glycolysis and the citric acid cycle to the machinery of oxidative phosphorylation, which uses energy released by the electron transport chain to power ATP synthesis. In this section, you will learn first how the electron transport chain works, then how the inner membrane of the mitochondrion couples electron flow down the chain to ATP synthesis.

The Pathway of Electron Transport

The electron transport chain is a collection of molecules embedded in the inner membrane of the mitochondrion. The folding of the inner membrane to form cristae increases its surface area, providing space for thousands of copies of the chain in each mitochondrion. (Once again, we see that structure fits function.) Most components of the chain are proteins, which exist in multiprotein complexes numbered I through IV. Tightly bound to these proteins are *prosthetic groups,* nonprotein components essential for the catalytic functions of certain enzymes.

Figure 9.13 shows the sequence of electron carriers in the electron transport chain and the drop in free energy as electrons travel down the chain. During electron transport along the chain, electron carriers alternate between reduced and oxidized states as they accept and donate electrons. Each component of the chain becomes reduced when it accepts electrons from its "uphill" neighbor, which has a lower affinity for electrons (is less electronegative). It then returns to its oxidized form as it passes electrons to its "downhill," more electronegative neighbor.

Now let's take a closer look at the electron transport chain in Figure 9.13. Electrons removed from food by NAD^+, during glycolysis and the citric acid cycle, are transferred from NADH to the first molecule of the electron transport chain. This molecule is a flavoprotein, so named because it has a prosthetic group called flavin mononucleotide (FMN in complex I). In the next redox reaction, the flavoprotein returns to its oxidized form as it passes electrons to an iron-sulfur protein (Fe•S in complex I), one of a family of proteins with both iron and sulfur tightly bound. The iron-sulfur protein then passes the electrons to a compound called ubiquinone (Q in Figure 9.13). This electron carrier is a small hydrophobic molecule, the only member of the electron transport chain that is not a protein. Ubiquinone is mobile within the membrane rather than residing in a particular complex.

Most of the remaining electron carriers between ubiquinone and oxygen are proteins called **cytochromes.** Their prosthetic group, called a heme group, has an iron atom that accepts and donates electrons. (It is similar to the heme group in hemoglobin, the protein of red blood cells, except that the iron in hemoglobin carries oxygen, not electrons.) The electron transport chain has several types of cytochromes, each a different protein with a slightly different electron-carrying heme group. The last cytochrome of the chain, cyt a_3, passes its electrons to oxygen, which is *very* electronegative. Each oxygen atom also picks up a pair of hydrogen ions from the aqueous solution, forming water.

Another source of electrons for the transport chain is $FADH_2$, the other reduced product of the citric acid cycle. Notice in Figure 9.13 that $FADH_2$ adds its electrons to the electron transport chain at complex II, at a lower energy level than NADH does. Consequently, the electron transport chain provides about one-third less energy for ATP synthesis when the electron donor is $FADH_2$ rather than NADH.

The electron transport chain makes no ATP directly. Its function is to ease the fall of electrons from food to oxygen, breaking a large free-energy drop into a series of smaller steps that release energy in manageable amounts. How does the mitochondrion couple this electron transport and energy release to ATP synthesis? The answer is a mechanism called chemiosmosis.

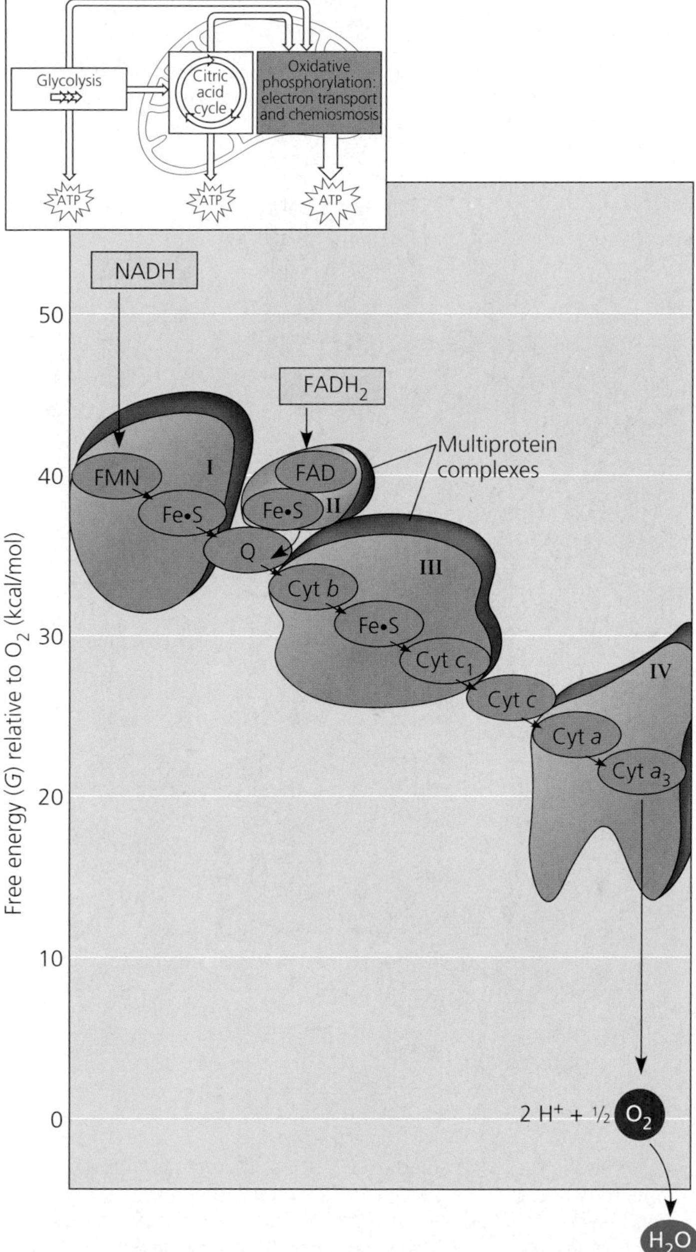

▲ **Figure 9.13 Free-energy change during electron transport.** The overall energy drop (ΔG) for electrons traveling from NADH to oxygen is 53 kcal/mol, but this "fall" is broken up into a series of smaller steps by the electron transport chain. (An oxygen atom is represented here as ½ O_2 to emphasize that the electron transport chain reduces molecular oxygen, O_2, not individual oxygen atoms. For every 2 NADH molecules, 1 O_2 molecule is reduced to 2 H_2O.)

Chemiosmosis: The Energy-Coupling Mechanism

Populating the inner membrane of the mitochondrion are many copies of a protein complex called **ATP synthase,** the enzyme that actually makes ATP from ADP and inorganic phosphate **(Figure 9.14)**. ATP synthase works like an ion pump running in reverse. Recall from Chapter 7 that ion pumps usually use ATP as an energy source to transport ions against their gradients. In the reverse of that process, ATP synthase uses the energy of an existing ion gradient to power ATP synthesis. The ion gradient that drives phosphorylation is a proton (hydrogen ion) gradient; that is, the power source for the ATP synthase is a difference in the concentration of H^+ on opposite sides of the inner mitochondrial membrane. (We can also think of this gradient as a difference in pH, since pH is a measure of H^+ concentration.) This process, in which energy stored in the form of a hydrogen ion gradient across a membrane is used to drive cellular work such as the synthesis of ATP, is called **chemiosmosis** (from the Greek *osmos*, push). We have previously used the word *osmosis* in discussing water transport, but here it refers to the flow of H^+ across a membrane.

From studying the structure of ATP synthase, scientists have learned how the flow of H^+ through this large enzyme powers ATP generation. ATP synthase is a multisubunit complex with

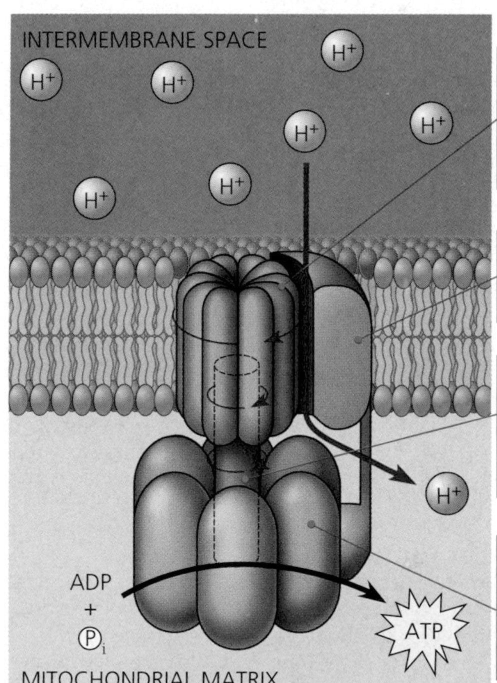

▲ **Figure 9.14 ATP synthase, a molecular mill.** The ATP synthase protein complex functions as a mill, powered by the flow of hydrogen ions. This complex resides in mitochondrial and chloroplast membranes of eukaryotes and in the plasma membranes of prokaryotes. Each of the four parts of ATP synthase consists of a number of polypeptide subunits.

four main parts, each made up of multiple polypeptides (see Figure 9.14): a rotor in the inner mitochondrial membrane; a knob that protrudes into the mitochondrial matrix; an internal rod extending from the rotor into the knob; and a stator, anchored next to the rotor, that holds the knob stationary. Hydrogen ions flow down a narrow space between the stator and rotor, causing the rotor and its attached rod to rotate, much as a rushing stream turns a waterwheel. The spinning rod causes conformational changes in the stationary knob, activating three catalytic sites in the subunits that make up the knob, such that ADP and inorganic phosphate combine to make ATP.

So how does the inner mitochondrial membrane generate and maintain the H^+ gradient that drives ATP synthesis in the ATP synthase protein complex? Creating the H^+ gradient is the function of the electron transport chain, which is shown in its mitochondrial location in **Figure 9.15**. The chain is an energy converter that uses the exergonic flow of electrons to pump H^+ across the membrane, from the mitochondrial matrix into the intermembrane space. The H^+ has a tendency to move back across the membrane, diffusing down its gradient. And the ATP synthases are the only sites on the membrane that are freely permeable to H^+. The ions pass through a channel in ATP synthase, which uses the exergonic flow of H^+ to drive the phosphorylation of ADP (see Figure 9.14). Thus, the energy stored in an H^+ gradient across a membrane couples the redox reactions of the electron transport chain to ATP synthesis, an example of chemiosmosis.

At this point, you may be wondering how the electron transport chain pumps hydrogen ions. Researchers have found that certain members of the electron transport chain

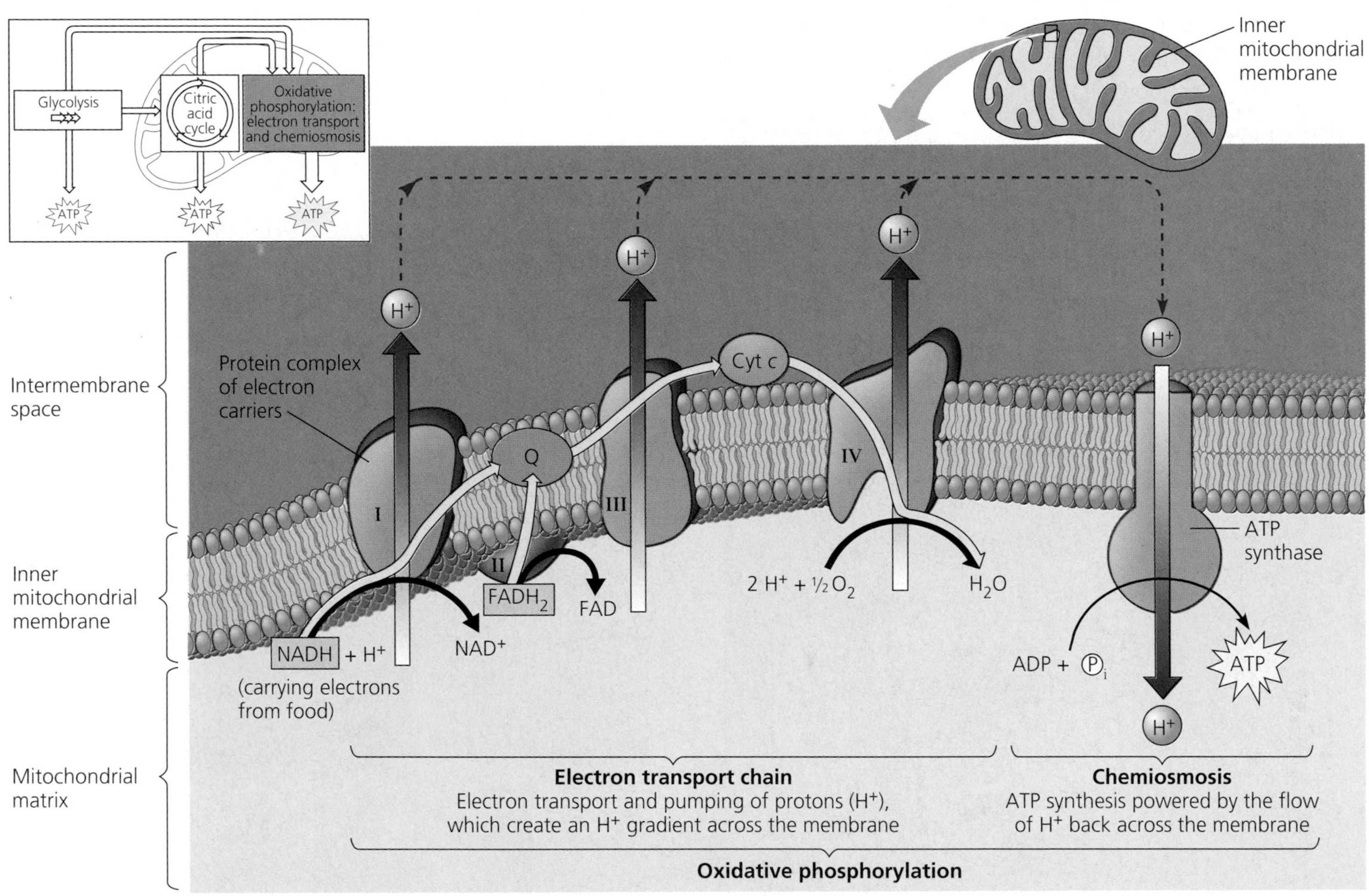

▲ **Figure 9.15 Chemiosmosis couples the electron transport chain to ATP synthesis.** NADH and $FADH_2$ shuttle high-energy electrons extracted from food during glycolysis and the citric acid cycle to an electron transport chain built into the inner mitochondrial membrane. The yellow arrow traces the transport of electrons, which finally pass to oxygen at the "downhill" end of the chain, forming water. As Figure 9.13 showed, most of the electron carriers of the chain are grouped into four complexes. Two mobile carriers, ubiquinone (Q) and cytochrome *c* (Cyt *c*), move rapidly along the membrane, ferrying electrons between the large complexes. As complexes I, III, and IV accept and then donate electrons, they pump hydrogen ions (protons) from the mitochondrial matrix into the intermembrane space. (Note that $FADH_2$ deposits its electrons via complex II and so results in fewer protons being pumped into the intermembrane space than NADH.) Chemical energy originally harvested from food is transformed into a proton-motive force, a gradient of H^+ across the membrane. The hydrogen ions flow back down their gradient through a channel in an ATP synthase, another protein complex built into the membrane. The ATP synthase harnesses the proton-motive force to phosphorylate ADP, forming ATP. The use of an H^+ gradient (proton-motive force) to transfer energy from redox reactions to cellular work (ATP synthesis, in this case) is called chemiosmosis. Together, electron transport and chemiosmosis compose oxidative phosphorylation.

accept and release protons (H^+) along with electrons. At certain steps along the chain, electron transfers cause H^+ to be taken up and released into the surrounding solution. The electron carriers are spatially arranged in the membrane in such a way that H^+ is accepted from the mitochondrial matrix and deposited in the intermembrane space (see Figure 9.15). The H^+ gradient that results is referred to as a **proton-motive force,** emphasizing the capacity of the gradient to perform work. The force drives H^+ back across the membrane through the specific H^+ channels provided by ATP synthases.

In general terms, *chemiosmosis is an energy-coupling mechanism that uses energy stored in the form of an H^+ gradient across a membrane to drive cellular work.* In mitochondria, the energy for gradient formation comes from exergonic redox reactions, and ATP synthesis is the work performed. But chemiosmosis also occurs elsewhere and in other variations. Chloroplasts use chemiosmosis to generate ATP during photosynthesis; in these organelles, light (rather than chemical energy) drives both electron flow down an electron transport chain and the resulting H^+ gradient formation. Prokaryotes, which lack both mitochondria and chloroplasts, generate H^+ gradients across their plasma membranes. They then tap the proton-motive force not only to make ATP but also to pump nutrients and waste products across the membrane and to rotate their flagella. Because of its central importance to energy conversions in prokaryotes and eukaryotes, chemiosmosis has helped unify the study of bioenergetics. Peter Mitchell was awarded the Nobel Prize in 1978 for originally proposing the chemiosmotic model.

An Accounting of ATP Production by Cellular Respiration

Now that we have looked more closely at the key processes of cellular respiration, let's return to its overall function: harvesting the energy of food for ATP synthesis.

During respiration, most energy flows in this sequence: glucose → NADH → electron transport chain → proton-motive force → ATP. We can do some bookkeeping to calculate the ATP profit when cellular respiration oxidizes a molecule of glucose to six molecules of carbon dioxide. The three main departments of this metabolic enterprise are glycolysis, the citric acid cycle, and the electron transport chain, which drives oxidative phosphorylation. **Figure 9.16** gives a detailed accounting of the ATP yield per glucose molecule oxidized. The tally adds the 4 ATP produced directly by substrate-level phosphorylation during glycolysis and the citric acid cycle to the many more molecules of ATP generated by oxidative phosphorylation. Each NADH that transfers a pair of electrons from food to the electron transport chain contributes enough to the proton-motive force to generate a maximum of about 3 ATP.

Why are the numbers in Figure 9.16 inexact? There are three reasons we cannot state an exact number of ATP molecules generated by the breakdown of one molecule of glucose. First, phos-

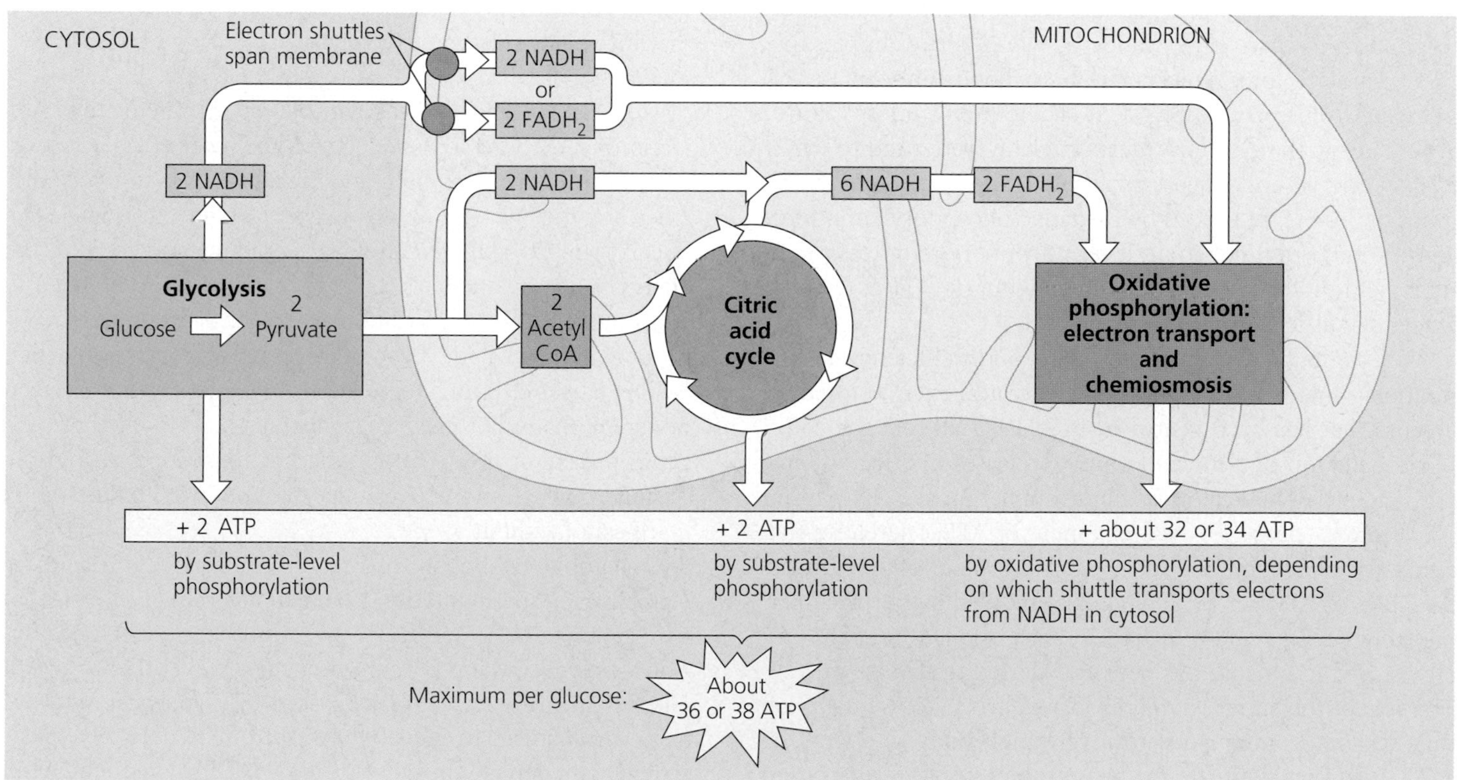

▲ **Figure 9.16 ATP yield per molecule of glucose at each stage of cellular respiration.**

phorylation and the redox reactions are not directly coupled to each other, so the ratio of number of NADH molecules to number of ATP molecules is not a whole number. We know that 1 NADH results in 10 H^+ being transported out across the inner mitochondrial membrane, and we also know that somewhere between 3 and 4 H^+ must reenter the mitochondrial matrix via ATP synthase to generate 1 ATP. Therefore, 1 NADH generates enough proton-motive force for synthesis of 2.5 to 3.3 ATP; generally, we round off and say that 1 NADH can generate about 3 ATP. The citric acid cycle also supplies electrons to the electron transport chain via $FADH_2$, but since it enters later in the chain, each molecule of this electron carrier is responsible for transport of only enough H^+ for the synthesis of 1.5 to 2 ATP.

Second, the ATP yield varies slightly depending on the type of shuttle used to transport electrons from the cytosol into the mitochondrion. The mitochondrial inner membrane is impermeable to NADH, so NADH in the cytosol is segregated from the machinery of oxidative phosphorylation. The two electrons of NADH captured in glycolysis must be conveyed into the mitochondrion by one of several electron shuttle systems. Depending on the type of shuttle in a particular cell type, the electrons are passed either to NAD^+ or to FAD. If the electrons are passed to FAD, as in brain cells, only about 2 ATP can result from each cytosolic NADH. If the electrons are passed to mitochondrial NAD^+, as in liver cells and heart cells, the yield is about 3 ATP.

A third variable that reduces the yield of ATP is the use of the proton-motive force generated by the redox reactions of respiration to drive other kinds of work. For example, the proton-motive force powers the mitochondrion's uptake of pyruvate from the cytosol. So, if *all* the proton-motive force generated by the electron transport chain were used to drive ATP synthesis, one glucose molecule could generate a maximum of 34 ATP produced by oxidative phosphorylation plus 4 ATP (net) from substrate-level phosphorylation to give a total yield of about 38 ATP (or only about 36 ATP if the less efficient shuttle were functioning).

We can now make a rough estimate of the efficiency of respiration—that is, the percentage of chemical energy stored in glucose that has been restocked in ATP. Recall that the complete oxidation of a mole of glucose releases 686 kcal of energy ($\Delta G = -686$ kcal/mol). Phosphorylation of ADP to form ATP stores at least 7.3 kcal per mole of ATP. Therefore, the efficiency of respiration is 7.3 kcal per mole of ATP times 38 moles of ATP per mole of glucose divided by 686 kcal per mole of glucose, which equals 0.4. Thus, about 40% of the energy stored in glucose has been transferred to storage in ATP. The rest of the stored energy is lost as heat. We use some of this heat to maintain our relatively high body temperature (37°C), and we dissipate the rest through sweating and other cooling mechanisms. Cellular respiration is remarkably efficient in its energy conversion. By comparison, the most efficient automobile converts only about 25% of the energy stored in gasoline to energy that moves the car.

Concept Check 9.4

1. What effect would an absence of O_2 have on the process shown in Figure 9.15?
2. In the absence of O_2, as above, what do you think would happen if you decreased the pH of the intermembrane space of the mitochondrion? Explain your answer.

For suggested answers, see Appendix A.

Fermentation enables some cells to produce ATP without the use of oxygen

Because most of the ATP generated by cellular respiration is the work of oxidative phosphorylation, our estimate of ATP yield from respiration is contingent upon an adequate supply of oxygen to the cell. Without the electronegative oxygen to pull electrons down the transport chain, oxidative phosphorylation ceases. However, fermentation provides a mechanism by which some cells can oxidize organic fuel and generate ATP *without* the use of oxygen.

How can food be oxidized without oxygen? Remember, oxidation refers to the loss of electrons to *any* electron acceptor, not just to oxygen. Glycolysis oxidizes glucose to two molecules of pyruvate. The oxidizing agent of glycolysis is NAD^+, not oxygen. Overall, glycolysis is exergonic, and some of the energy made available is used to produce 2 ATP (net) by substrate-level phosphorylation. If oxygen *is* present, then additional ATP is made by oxidative phosphorylation when NADH passes electrons removed from glucose to the electron transport chain. But glycolysis generates 2 ATP whether oxygen is present or not—that is, whether conditions are **aerobic** or **anaerobic** (from the Greek *aer*, air, and *bios*, life; the prefix *an-* means "without").

Anaerobic catabolism of organic nutrients can occur by fermentation. Fermentation is an extension of glycolysis that can generate ATP solely by substrate-level phosphorylation—as long as there is a sufficient supply of NAD^+ to accept electrons during the oxidation step of glycolysis. Without some mechanism to recycle NAD^+ from NADH, glycolysis would soon deplete the cell's pool of NAD^+ by reducing it all to NADH and shut itself down for lack of an oxidizing agent. Under aerobic conditions, NAD^+ is recycled productively from NADH by the transfer of electrons to the electron transport chain. The anaerobic alternative is to transfer electrons from NADH to pyruvate, the end product of glycolysis.

Types of Fermentation

Fermentation consists of glycolysis plus reactions that regenerate NAD^+ by transferring electrons from NADH to pyruvate or derivatives of pyruvate. The NAD^+ can then be reused to oxidize sugar by glycolysis, which nets two molecules of ATP by substrate-level phosphorylation. There are many types of fermentation, differing in the end products formed from pyruvate. Two common types are alcohol fermentation and lactic acid fermentation.

In **alcohol fermentation (Figure 9.17a)**, pyruvate is converted to ethanol (ethyl alcohol) in two steps. The first step releases carbon dioxide from the pyruvate, which is converted to the two-carbon compound acetaldehyde. In the second step, acetaldehyde is reduced by NADH to ethanol. This regenerates the supply of NAD^+ needed for the continuation of glycolysis. Many bacteria carry out alcohol fermentation under anaerobic conditions. Yeast (a fungus) also carries out alcohol fermentation. For thousands of years, humans have used yeast in brewing, winemaking, and baking. The CO_2 bubbles generated by baker's yeast allow bread to rise.

During **lactic acid fermentation (Figure 9.17b)**, pyruvate is reduced directly by NADH to form lactate as an end product, with no release of CO_2. (Lactate is the ionized form of lactic acid.) Lactic acid fermentation by certain fungi and bacteria is used in the dairy industry to make cheese and yogurt. Other types of microbial fermentation that are commercially important produce acetone and methanol (methyl alcohol).

Human muscle cells make ATP by lactic acid fermentation when oxygen is scarce. This occurs during the early stages of strenuous exercise, when sugar catabolism for ATP production outpaces the muscle's supply of oxygen from the blood. Under these conditions, the cells switch from aerobic respiration to fermentation. The lactate that accumulates may cause muscle fatigue and pain, but the lactate is gradually carried away by the blood to the liver. Lactate is converted back to pyruvate by liver cells.

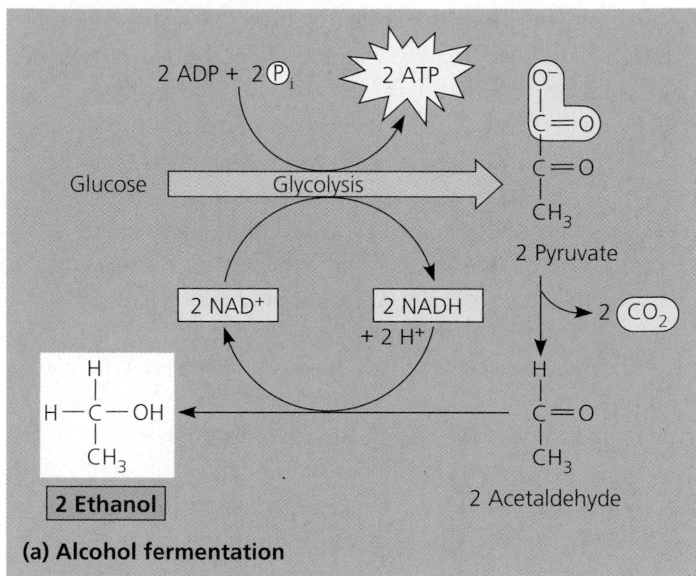

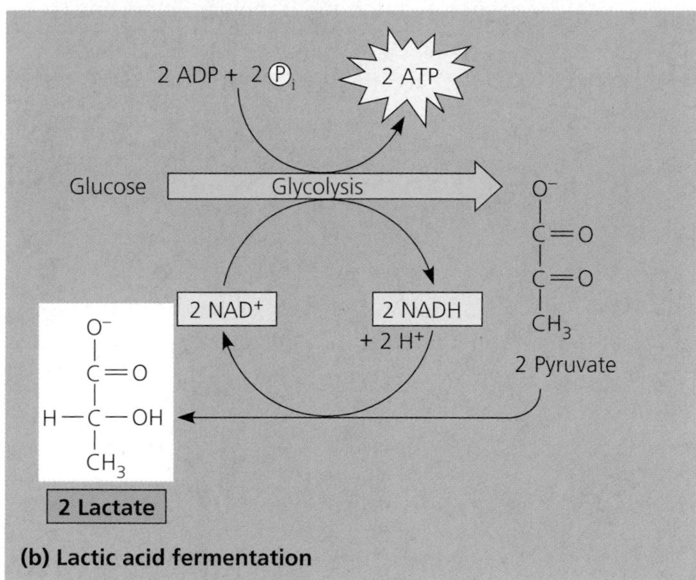

▲ **Figure 9.17 Fermentation.** In the absence of oxygen, many cells use fermentation to produce ATP by substrate-level phosphorylation. Pyruvate, the end product of glycolysis, serves as an electron acceptor for oxidizing NADH back to NAD^+, which can then be reused in glycolysis. Two of the common end products formed from fermentation are **(a)** ethanol and **(b)** lactate, the ionized form of lactic acid.

Fermentation and Cellular Respiration Compared

Fermentation and cellular respiration are anaerobic and aerobic alternatives, respectively, for producing ATP by harvesting the chemical energy of food. Both pathways use glycolysis to oxidize glucose and other organic fuels to pyruvate, with a net production of 2 ATP by substrate-level phosphorylation. And in both fermentation and respiration, NAD^+ is the oxidizing agent that accepts electrons from food during glycolysis. A key difference is the contrasting mechanisms for oxidizing NADH back to NAD^+, which is required to sustain glycolysis. In fermentation, the final electron acceptor is an organic molecule such as pyruvate (lactic acid fermentation) or acetaldehyde (alcohol fermentation). In respiration, by contrast, the final acceptor for electrons from NADH is oxygen. This not only regenerates the NAD^+ required for glycolysis but pays an ATP bonus when the stepwise electron transport from NADH to oxygen drives oxidative phosphorylation. An even bigger ATP payoff comes from the oxidation of pyruvate in the citric acid cycle, which is unique to respiration. Without oxygen, the energy still stored in pyruvate is unavailable to the cell. Thus, cellular respiration harvests much more energy from each sugar molecule than fermentation can. In fact, respiration yields as much as 19 times more ATP per glucose molecule than does fermentation—up to 38 ATP for respiration, compared to 2 ATP produced by substrate-level phosphorylation in fermentation.

Some organisms, including yeasts and many bacteria, can make enough ATP to survive using either fermentation or respiration. Such species are called **facultative anaerobes.** On the cellular level, our muscle cells behave as facultative anaerobes. In a facultative anaerobe, pyruvate is a fork in the metabolic road that leads to two alternative catabolic routes **(Figure 9.18)**. Under aerobic conditions, pyruvate can be converted to acetyl CoA, and oxidation continues in the citric acid cycle. Under anaerobic conditions, pyruvate is diverted from the citric acid cycle, serving instead as an electron acceptor to recycle NAD^+. To make the same amount of ATP, a facultative anaerobe would have to consume sugar at a much faster rate when fermenting than when respiring.

The Evolutionary Significance of Glycolysis

The role of glycolysis in both fermentation and respiration has an evolutionary basis. Ancient prokaryotes probably used glycolysis to make ATP long before oxygen was present in Earth's atmosphere. The oldest known fossils of bacteria date back 3.5 billion years, but appreciable quantities of oxygen probably did not begin to accumulate in the atmosphere until about 2.7 billion years ago. Cyanobacteria produced this O_2 as a by-product of photosynthesis. Therefore, early prokaryotes may have generated ATP exclusively from glycolysis, which does not require oxygen. In addition, glycolysis is the most widespread metabolic pathway, which suggests that it evolved very early in the history of life. The cytosolic location of glycolysis also implies great antiquity; the pathway does not require any of the membrane-bounded organelles of the eukaryotic cell, which evolved approximately 1 billion years after the prokaryotic cell. Glycolysis is a metabolic heirloom from early cells that continues to function in fermentation and as the first stage in the breakdown of organic molecules by respiration.

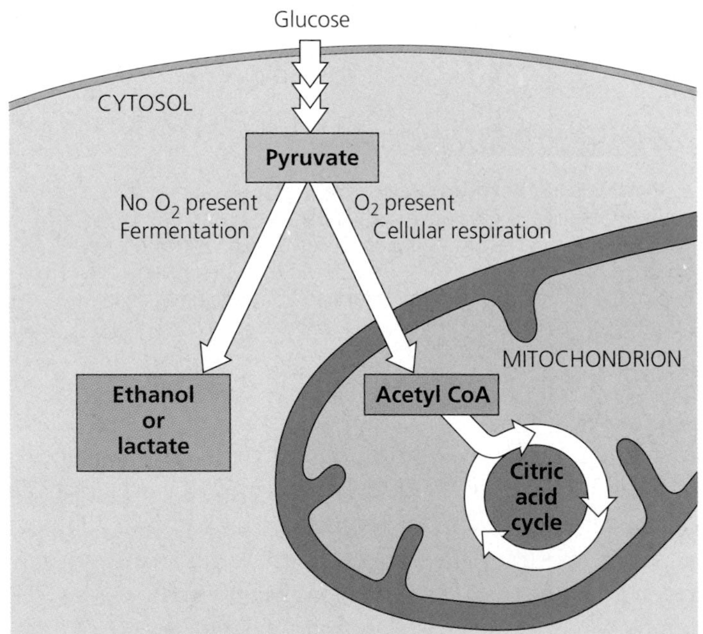

▲ **Figure 9.18 Pyruvate as a key juncture in catabolism.** Glycolysis is common to fermentation and cellular respiration. The end product of glycolysis, pyruvate, represents a fork in the catabolic pathways of glucose oxidation. In a cell capable of both cellular respiration and fermentation, pyruvate is committed to one of those two pathways, usually depending on whether or not oxygen is present.

Concept Check 9.5

1. Consider the NADH formed during glycolysis. What is the final acceptor for its electrons during fermentation? What is the final acceptor for its electrons during respiration?
2. A glucose-fed yeast cell is moved from an aerobic environment to an anaerobic one. For the cell to continue generating ATP at the same rate, how would its rate of glucose consumption need to change?

For suggested answers, see Appendix A.

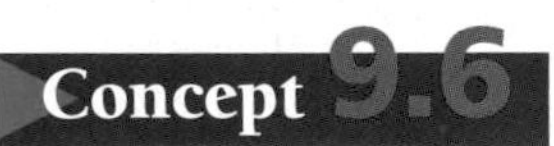

Glycolysis and the citric acid cycle connect to many other metabolic pathways

So far, we have treated the oxidative breakdown of glucose in isolation from the cell's overall metabolic economy. In this section, you will learn that glycolysis and the citric acid cycle are major intersections of various catabolic and anabolic (biosynthetic) pathways.

The Versatility of Catabolism

Throughout this chapter, we have used glucose as the fuel for cellular respiration. But free glucose molecules are not common in the diets of humans and other animals. We obtain most of our calories in the form of fats, proteins, sucrose and other disaccharides, and starch, a polysaccharide. All these organic molecules in food can be used by cellular respiration to make ATP **(Figure 9.19)**.

Glycolysis can accept a wide range of carbohydrates for catabolism. In the digestive tract, starch is hydrolyzed to glucose, which can then be broken down in the cells by glycolysis and the citric acid cycle. Similarly, glycogen, the polysaccharide that humans and many other animals store in their liver and muscle cells, can be hydrolyzed to glucose between meals as fuel for respiration. The digestion of disaccharides, including sucrose, provides glucose and other monosaccharides as fuel for respiration.

Proteins can also be used for fuel, but first they must be digested to their constituent amino acids. Many of the amino acids, of course, are used by the organism to build new proteins. Amino acids present in excess are converted by enzymes to intermediates of glycolysis and the citric acid cycle. Before amino acids can feed into glycolysis or the citric acid cycle, their amino groups must be removed, a process called deamination. The nitrogenous refuse is excreted from the animal in the form of ammonia, urea, or other waste products.

Catabolism can also harvest energy stored in fats obtained either from food or from storage cells in the body. After fats are digested to glycerol and fatty acids, the glycerol is converted to glyceraldehyde-3-phosphate, an intermediate of glycolysis. Most of the energy of a fat is stored in the fatty acids. A metabolic sequence called **beta oxidation** breaks the fatty acids down to two-carbon fragments, which enter the citric acid cycle as acetyl CoA. Fats make excellent fuel. A gram of fat oxidized by respiration produces more than twice as much ATP as a gram of carbohydrate. Unfortunately, this also means that a person who is trying to lose weight must work hard to use up fat stored in the body, because so many calories are stockpiled in each gram of fat.

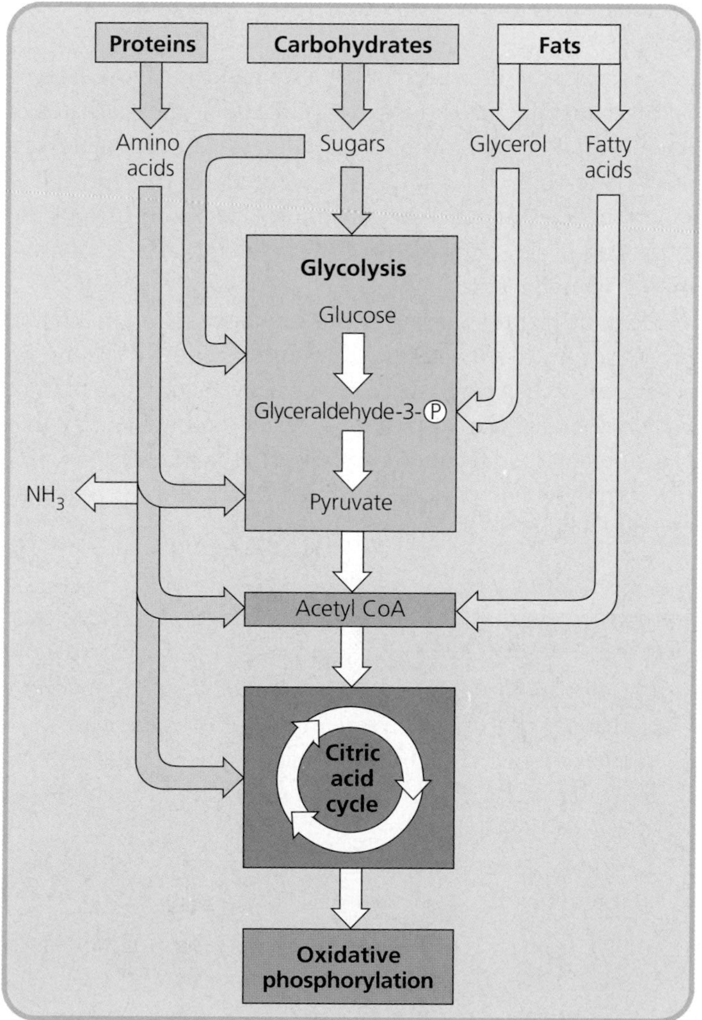

▲ **Figure 9.19 The catabolism of various molecules from food.** Carbohydrates, fats, and proteins can all be used as fuel for cellular respiration. Monomers of these molecules enter glycolysis or the citric acid cycle at various points. Glycolysis and the citric acid cycle are catabolic funnels through which electrons from all kinds of organic molecules flow on their exergonic fall to oxygen.

Biosynthesis (Anabolic Pathways)

Cells need substance as well as energy. Not all the organic molecules of food are destined to be oxidized as fuel to make ATP. In addition to calories, food must also provide the carbon skeletons that cells require to make their own molecules. Some organic monomers obtained from digestion can be used directly. For example, as previously mentioned, amino acids from the hydrolysis of proteins in food can be incorporated into the organism's own proteins. Often, however, the body needs specific molecules that are not present as such in food. Compounds formed as intermediates of glycolysis and the citric acid cycle can be diverted into anabolic pathways as precursors from which the cell can synthesize the molecules it requires. For example, humans can make about half of the 20 amino acids in proteins by modifying compounds siphoned away from the citric acid cycle. Also, glucose can be made from pyruvate, and fatty acids can be synthesized from acetyl CoA. Of course, these anabolic, or biosynthetic, pathways do not generate ATP, but instead consume it.

In addition, glycolysis and the citric acid cycle function as metabolic interchanges that enable our cells to convert some kinds of molecules to others as we need them. For example, an intermediate compound generated during glycolysis, dihydroxyacetone phosphate (see Figure 9.9, step 5), can be converted into one of the major precursors of fats. If we eat more food than we need, we store fat even if our diet is fat-free. Metabolism is remarkably versatile and adaptable.

Regulation of Cellular Respiration via Feedback Mechanisms

Basic principles of supply and demand regulate the metabolic economy. The cell does not waste energy making more of a particular substance than it needs. If there is a glut of a certain amino acid, for example, the anabolic pathway that synthesizes that amino acid from an intermediate of the citric acid cycle is switched off. The most common mechanism for this control is feedback inhibition: The end product of the anabolic pathway inhibits the enzyme that catalyzes an early step of the pathway (see Figure 8.21). This prevents the needless diversion of key metabolic intermediates from uses that are more urgent.

The cell also controls its catabolism. If the cell is working hard and its ATP concentration begins to drop, respiration speeds up. When there is plenty of ATP to meet demand, respiration slows down, sparing valuable organic molecules for other functions. Again, control is based mainly on regulating the activity of enzymes at strategic points in the catabolic pathway. One important switch is phosphofructokinase **(Figure 9.20)**, the enzyme that catalyzes step 3 of glycolysis (see Figure 9.9). That is the earliest step that commits substrate irreversibly to the glycolytic pathway. By controlling the rate of this step, the cell can speed up or slow down the entire catabolic process; phosphofructokinase can thus be considered the pacemaker of respiration.

Phosphofructokinase is an allosteric enzyme with receptor sites for specific inhibitors and activators. It is inhibited by ATP and stimulated by AMP (adenosine monophosphate), which the cell derives from ADP. As ATP accumulates, inhibition of the enzyme slows down glycolysis. The enzyme becomes active again as cellular work converts ATP to ADP (and AMP) faster than ATP is being regenerated. Phosphofructokinase is also sensitive to citrate, the first product of the citric acid cycle. If citrate accumulates in mitochondria, some of it passes into the cytosol and inhibits phosphofructokinase. This mechanism helps synchronize the rates of glycolysis and the citric acid cycle. As citrate accumulates, glycolysis slows down, and the supply of acetyl groups to the citric acid cycle decreases. If citrate consumption increases, either because of a demand for more ATP or because anabolic pathways are draining off intermediates of the citric acid cycle, glycolysis accelerates and meets the demand. Metabolic balance is augmented by the control of other enzymes at other key locations in glycolysis and the citric acid cycle. Cells are thrifty, expedient, and responsive in their metabolism.

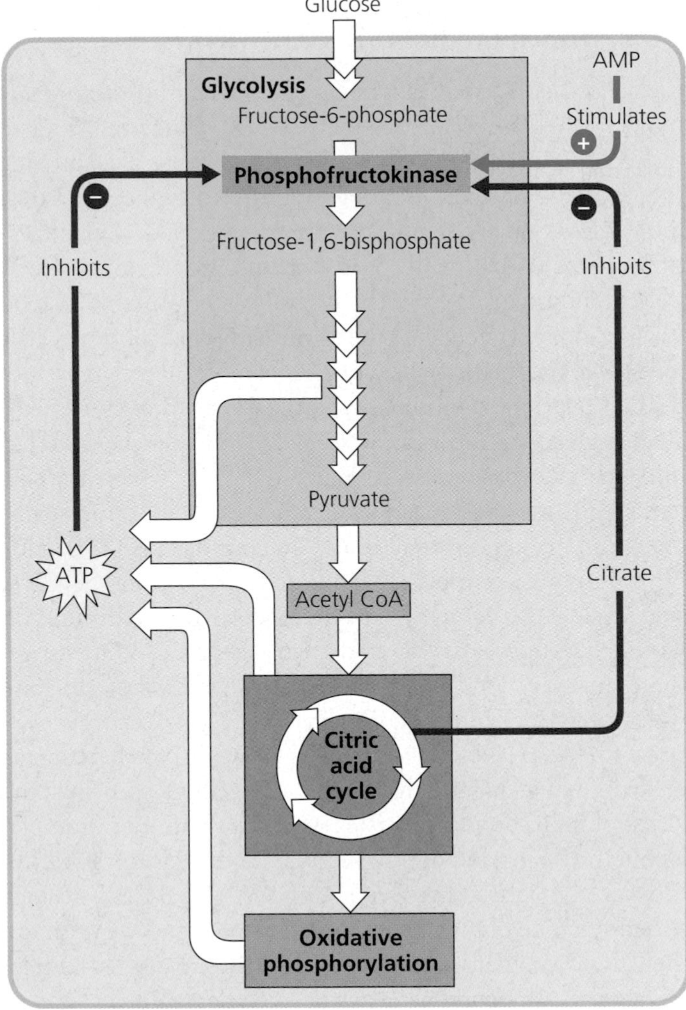

▲ **Figure 9.20 The control of cellular respiration.** Allosteric enzymes at certain points in the respiratory pathway respond to inhibitors and activators that help set the pace of glycolysis and the citric acid cycle. Phosphofructokinase, the enzyme that catalyzes step 3 of glycolysis (see Figure 9.9), is one such enzyme. It is stimulated by AMP (derived from ADP) but is inhibited by ATP and by citrate. This feedback regulation adjusts the rate of respiration as the cell's catabolic and anabolic demands change.

Examine Figure 9.2 again to put cellular respiration into the broader context of energy flow and chemical cycling in ecosystems. The energy that keeps us alive is *released*, but not *produced*, by cellular respiration. We are tapping energy that was stored in food by photosynthesis. In the next chapter, you will learn how photosynthesis captures light and converts it to chemical energy.

Concept Check 9.6

1. Compare the structure of a fat (see Figure 5.11) with that of a carbohydrate (see Figure 5.3). What features of their structures make fat a much better fuel?
2. Under what circumstances might your body synthesize fat molecules?
3. What will happen in a muscle cell that has used up its supply of oxygen and ATP? (See Figure 9.20.)

For suggested answers, see Appendix A.

Chapter 9 Review

Go to www.campbellbiology.com or the student CD-ROM to explore Activities, Investigations, and other interactive study aids.

SUMMARY OF KEY CONCEPTS

- Life's processes require energy that enters the ecosystem in the form of sunlight. Energy is used for work or dissipated as heat, while the essential chemical elements are recycled by respiration and photosynthesis (p. 160).
  Activity *Build a Chemical Cycling System*

Concept 9.1

Catabolic pathways yield energy by oxidizing organic fuels

- **Catabolic Pathways and Production of ATP (p. 161)** The breakdown of glucose and other organic fuels is exergonic. Starting with glucose or another organic molecule and using O_2, cellular respiration yields H_2O, CO_2, and energy in the form of ATP and heat. To keep working, a cell must regenerate ATP.
- **Redox Reactions: Oxidation and Reduction (pp. 161–164)** The cell taps the energy stored in food molecules through redox reactions, in which one substance partially or totally shifts electrons to another. The substance receiving electrons is reduced; the substance losing electrons is oxidized. During cellular respiration, glucose ($C_6H_{12}O_6$) is oxidized to CO_2, and O_2 is reduced to H_2O. Electrons lose potential energy during their transfer from organic compounds to oxygen. Electrons from organic compounds are usually passed first to NAD^+, reducing it to NADH. NADH passes the electrons to an electron transport chain, which conducts them to O_2 in energy-releasing steps. The energy released is used to make ATP.
- **The Stages of Cellular Respiration: *A Preview* (pp. 164–165)** Glycolysis and the citric acid cycle supply electrons (via NADH or $FADH_2$) to the electron transport chain, which drives oxidative phosphorylation. Oxidative phosphorylation generates ATP.
  Activity *Overview of Cellular Respiration*

Concept 9.2

Glycolysis harvests chemical energy by oxidizing glucose to pyruvate

- Glycolysis breaks down glucose into two pyruvate molecules and nets 2 ATP and 2 NADH per glucose molecule (pp. 165–167).
  Activity *Glycolysis*

Concept 9.3

The citric acid cycle completes the energy-yielding oxidation of organic molecules

- The import of pyruvate into the mitochondrion and its conversion to acetyl CoA links glycolysis to the citric acid cycle. The two-carbon acetyl group of acetyl CoA joins the four-carbon oxaloacetate, forming the six-carbon citrate, which is degraded back to oxaloacetate. The cycle releases 2 CO_2, forms 1 ATP, and passes electrons to NAD^+ and FAD, yielding 3 NADH and 1 $FADH_2$ per turn (pp. 168–170).
  Activity *The Citric Acid Cycle*

Concept 9.4

During oxidative phosphorylation, chemiosmosis couples electron transport to ATP synthesis

- NADH and $FADH_2$ donate electrons to the electron transport chain, which powers ATP synthesis via oxidative phosphorylation (p. 170).
- **The Pathway of Electron Transport (pp. 170–171)** In the electron transport chain, electrons from NADH and $FADH_2$ lose energy in several energy-releasing steps. At the end of the chain, electrons are passed to O_2, reducing it to H_2O.
- **Chemiosmosis: The Energy-Coupling Mechanism (pp. 171–173)** At certain steps along the electron transport chain, electron transfer causes protein complexes to move H^+ from the mitochondrial matrix to the intermembrane space, storing energy as a proton-motive force (H^+ gradient). As H^+ diffuses back into the matrix through ATP synthase, its passage drives the phosphorylation of ADP.
  Activity *Electron Transport*
- **An Accounting of ATP Production by Cellular Respiration (pp. 173–174)** About 40% of the energy stored in a glucose molecule is transferred to ATP during cellular respiration, producing a maximum of about 38 ATP.
  Biology Labs On-Line *MitochondrialLab*
  Investigation *How Is the Rate of Cellular Respiration Measured?*

Concept 9.5

Fermentation enables some cells to produce ATP without the use of oxygen

- **Types of Fermentation (p. 175)** Glycolysis nets two ATP by substrate-level phosphorylation whether oxygen is present or not. Under anaerobic conditions, the electrons from NADH are passed to pyruvate or a derivative of pyruvate, regenerating the NAD^+ required to oxidize more glucose. Two common types of fermentation are alcohol fermentation and lactic acid fermentation.
  Activity *Fermentation*
- **Fermentation and Cellular Respiration Compared (pp. 175–176)** Both use glycolysis to oxidize glucose, but differ in their final electron acceptor. Respiration yields more ATP.
- **The Evolutionary Significance of Glycolysis (p. 176)** Glycolysis occurs in nearly all organisms and probably evolved in ancient prokaryotes before there was O_2 in the atmosphere.

Concept 9.6

Glycolysis and the citric acid cycle connect to many other metabolic pathways

- **The Versatility of Catabolism (pp. 176–177)** Catabolic pathways funnel electrons from many kinds of organic molecules into cellular respiration.
- **Biosynthesis (Anabolic Pathways) (p. 177)** The body can use small molecules from food directly or use them to build other substances through glycolysis or the citric acid cycle.
- **Regulation of Cellular Respiration via Feedback Mechanisms (pp. 177–178)** Cellular respiration is controlled by allosteric enzymes at key points in glycolysis and the citric acid cycle.

TESTING YOUR KNOWLEDGE

Self-Quiz

1. What is the reducing agent in the following reaction?

$$\text{Pyruvate} + \text{NADH} + \text{H}^+ \rightarrow \text{Lactate} + \text{NAD}^+$$

a. oxygen b. NADH c. NAD^+
d. lactate e. pyruvate

2. The *immediate* energy source that drives ATP synthesis by ATP synthase during oxidative phosphorylation is
a. the oxidation of glucose and other organic compounds.
b. the flow of electrons down the electron transport chain.
c. the affinity of oxygen for electrons.
d. the H^+ concentration gradient across the inner mitochondrial membrane.
e. the transfer of phosphate to ADP.

3. Which metabolic pathway is common to both fermentation and cellular respiration?
a. the citric acid cycle
b. the electron transport chain
c. glycolysis
d. synthesis of acetyl CoA from pyruvate
e. reduction of pyruvate to lactate

4. In mitochondria, exergonic redox reactions
a. are the source of energy driving prokaryotic ATP synthesis.
b. are directly coupled to substrate-level phosphorylation.
c. provide the energy to establish the proton gradient.
d. reduce carbon atoms to carbon dioxide.
e. are coupled via phosphorylated intermediates to endergonic processes.

5. The final electron acceptor of the electron transport chain that functions in oxidative phosphorylation is
a. oxygen. b. water. c. NAD^+.
d. pyruvate. e. ADP.

6. When electrons flow along the electron transport chains of mitochondria, which of the following changes occurs?
a. The pH of the matrix increases.
b. ATP synthase pumps protons by active transport.
c. The electrons gain free energy.
d. The cytochromes phosphorylate ADP to form ATP.
e. NAD^+ is oxidized.

7. In the presence of a metabolic poison that specifically and completely inhibits all function of mitochondrial ATP synthase, which would you expect?
a. a decrease in the pH difference across the inner mitochondrial membrane
b. an increase in the pH difference across the inner mitochondrial membrane
c. increased synthesis of ATP
d. increased oxygen consumption
e. an accumulation of NAD^+

8. Cells do not catabolize carbon dioxide because
a. its double bonds are too stable to be broken.
b. CO_2 has fewer bonding electrons than other organic compounds.
c. CO_2 is already completely reduced.
d. CO_2 is already completely oxidized.
e. the molecule has too few atoms.

9. Which of the following is a true distinction between fermentation and cellular respiration?
a. Only respiration oxidizes glucose.
b. NADH is oxidized by the electron transport chain in respiration only.
c. Fermentation, but not respiration, is an example of a catabolic pathway.
d. Substrate-level phosphorylation is unique to fermentation.
e. NAD^+ functions as an oxidizing agent only in respiration.

10. Most CO_2 from catabolism is released during
a. glycolysis.
b. the citric acid cycle.
c. lactate fermentation.
d. electron transport.
e. oxidative phosphorylation.

For Self-Quiz Answers, see Appendix A.

Go to the website or CD-ROM for more quiz questions.

Evolution Connection

ATP synthase enzymes are found in the prokaryotic plasma membrane and in mitochondria and chloroplasts. What does this suggest about the evolutionary relationship of these eukaryotic organelles to prokaryotes? How might the amino acid sequences of the ATP synthases from the different sources support or refute your hypothesis?

Scientific Inquiry

In the 1940s, some physicians prescribed low doses of a drug called dinitrophenol (DNP) to help patients lose weight. This unsafe method was abandoned after a few patients died. DNP uncouples the chemiosmotic machinery by making the lipid bilayer of the inner mitochondrial membrane leaky to H^+. Explain how this causes weight loss.

Biology Labs On-Line ***MitochondriaLab***
Investigation ***How Is the Rate of Cellular Respiration Measured?***

Science, Technology, and Society

Nearly all human societies use fermentation to produce alcoholic drinks such as beer and wine. The practice dates back to the earliest days of agriculture. How do you suppose this use of fermentation was first discovered? Why did wine prove to be a more useful beverage, especially to a preindustrial culture, than the grape juice from which it was made?

Biological Inquiry: A Workbook of Investigative Cases ***Explore fermentation further in the case "Bean Brew."***

▲ **Figure 10.1 Sunlight consists of a spectrum of colors, visible here in a rainbow.**

Key Concepts

10.1 Photosynthesis converts light energy to the chemical energy of food
10.2 The light reactions convert solar energy to the chemical energy of ATP and NADPH
10.3 The Calvin cycle uses ATP and NADPH to convert CO_2 to sugar
10.4 Alternative mechanisms of carbon fixation have evolved in hot, arid climates

Overview

The Process That Feeds the Biosphere

Life on Earth is solar powered. The chloroplasts of plants capture light energy that has traveled 150 million kilometers from the sun and convert it to chemical energy stored in sugar and other organic molecules. This conversion process is called **photosynthesis.** Let's begin by placing photosynthesis in its ecological context.

Photosynthesis nourishes almost the entire living world directly or indirectly. An organism acquires the organic compounds it uses for energy and carbon skeletons by one of two major modes: autotrophic nutrition or heterotrophic nutrition. **Autotrophs** are "self-feeders" (*auto* means "self," and *trophos* means "feed"); they sustain themselves without eating anything derived from other organisms. Autotrophs produce their organic molecules from CO_2 and other inorganic raw materials obtained from the environment. They are the ultimate sources of organic compounds for all nonautotrophic organisms, and for this reason, biologists refer to autotrophs as the *producers* of the biosphere.

Almost all plants are autotrophs; the only nutrients they require are water and minerals from the soil and carbon dioxide from the air. Specifically, plants are *photo*autotrophs, organisms that use light as a source of energy to synthesize organic substances **(Figure 10.1)**. Photosynthesis also occurs in algae, certain other protists, and some prokaryotes (**Figure 10.2**, on the next page). In this chapter, our emphasis will be on plants; variations in autotrophic nutrition that occur in prokaryotes and algae will be discussed in Chapters 27 and 28.

Heterotrophs obtain their organic material by the second major mode of nutrition. Unable to make their own food, they live on compounds produced by other organisms (*hetero* means "other"). Heterotrophs are the biosphere's *consumers.* The most obvious form of this "other-feeding" occurs when an animal eats plants or other animals. But heterotrophic nutrition may be more subtle. Some heterotrophs consume the remains of dead organisms by decomposing and feeding on organic litter such as carcasses, feces, and fallen leaves; they are known as decomposers. Most fungi and many types of prokaryotes get their nourishment this way. Almost all heterotrophs, including humans, are completely dependent on photoautotrophs for food—and also for oxygen, a by-product of photosynthesis.

In this chapter, you will learn how photosynthesis works. After a discussion of the general principles of photosynthesis, we will consider the two stages of photosynthesis: the light reactions, in which solar energy is captured and transformed into chemical energy; and the Calvin cycle, in which the chemical energy is used to make organic molecules of food. Finally, we will consider photosynthesis from an evolutionary perspective.

▶ **Figure 10.2 Photoautotrophs.** These organisms use light energy to drive the synthesis of organic molecules from carbon dioxide and (in most cases) water. They feed not only themselves, but the entire living world. **(a)** On land, plants are the predominant producers of food. In aquatic environments, photosynthetic organisms include **(b)** multicellular algae, such as this kelp; **(c)** some unicellular protists, such as *Euglena;* **(d)** the prokaryotes called cyanobacteria; and **(e)** other photosynthetic prokaryotes, such as these purple sulfur bacteria, which produce sulfur (spherical globules) (c, d, e: LMs).

(a) Plants

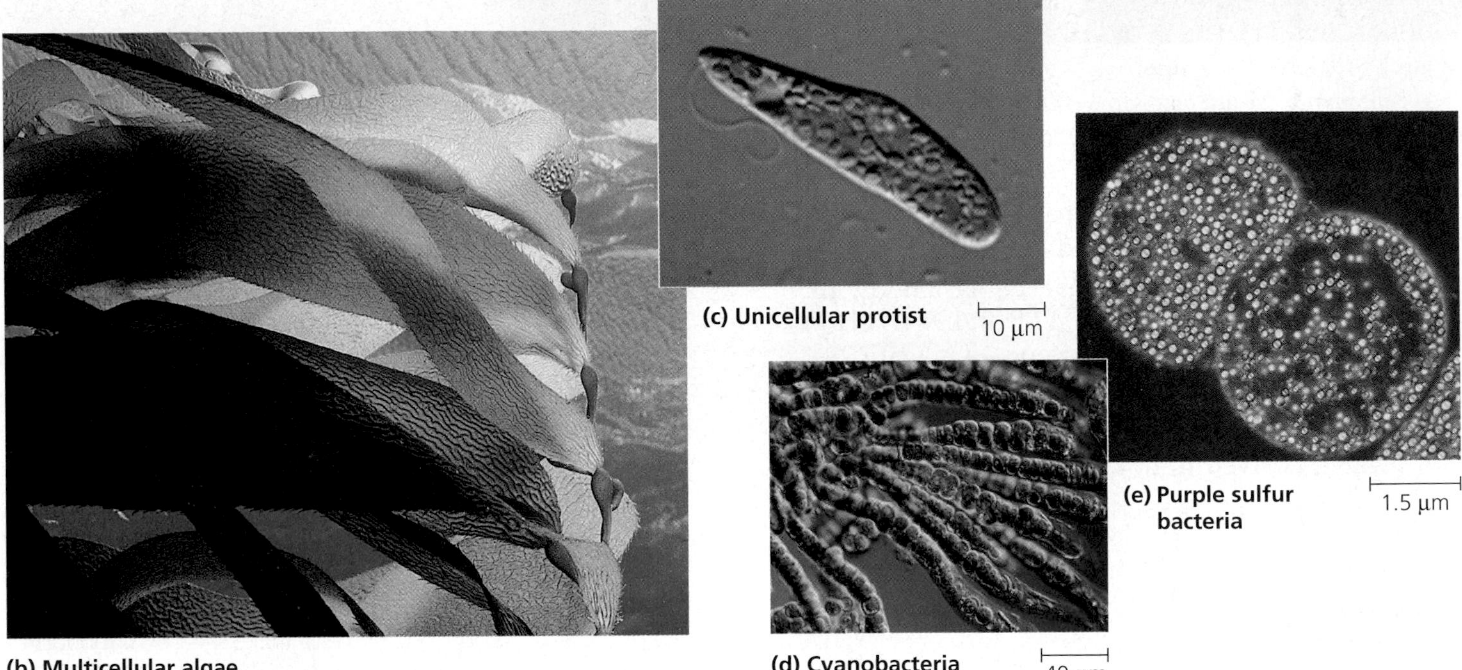

(b) Multicellular algae **(c) Unicellular protist** **(d) Cyanobacteria** **(e) Purple sulfur bacteria**

Concept 10.1

Photosynthesis converts light energy to the chemical energy of food

You were introduced to the chloroplast in Chapter 6. This remarkable organelle is responsible for feeding the vast majority of organisms on our planet. Chloroplasts are present in a variety of photosynthesizing organisms (see Figure 10.2), but here we will focus on plants.

Chloroplasts: The Sites of Photosynthesis in Plants

All green parts of a plant, including green stems and unripened fruit, have chloroplasts, but the leaves are the major sites of photosynthesis in most plants **(Figure 10.3)**. There are about half a million chloroplasts per square millimeter of leaf surface. The color of the leaf is from **chlorophyll**, the green pigment located within chloroplasts. It is the light energy absorbed by chlorophyll that drives the synthesis of organic molecules in the chloroplast. Chloroplasts are found mainly in the cells of the **mesophyll**, the tissue in the interior of the leaf. Carbon dioxide enters the leaf, and oxygen exits, by way of microscopic pores called **stomata** (singular, *stoma;* from the Greek, meaning "mouth"). Water absorbed by the roots is delivered to the leaves in veins. Leaves also use veins to export sugar to roots and other nonphotosynthetic parts of the plant.

A typical mesophyll cell has about 30 to 40 chloroplasts, each organelle measuring about 2–4 µm by 4–7 µm. An envelope of two membranes encloses the **stroma**, the dense fluid within the chloroplast. An elaborate system of interconnected membranous sacs called **thylakoids** segregates the stroma from another compartment, the interior of the thylakoids, or *thylakoid space*. In some places, thylakoid sacs are stacked in columns called *grana* (singular, granum). Chlorophyll resides

▶ **Figure 10.3 Focusing in on the location of photosynthesis in a plant.** Leaves are the major organs of photosynthesis in plants. These pictures take you into a leaf, then into a cell, and finally into a chloroplast, the organelle where photosynthesis occurs (middle, LM; bottom, TEM).

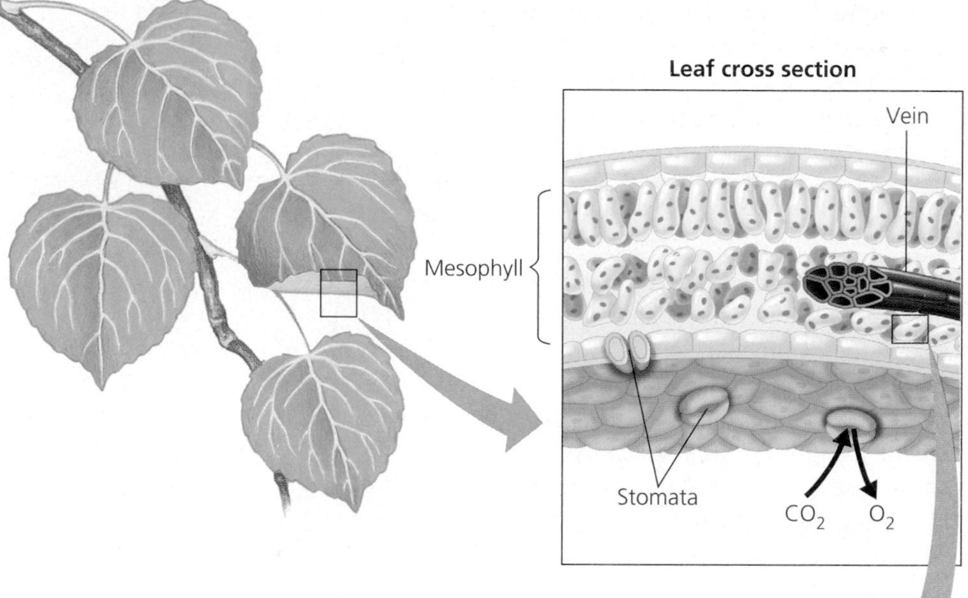

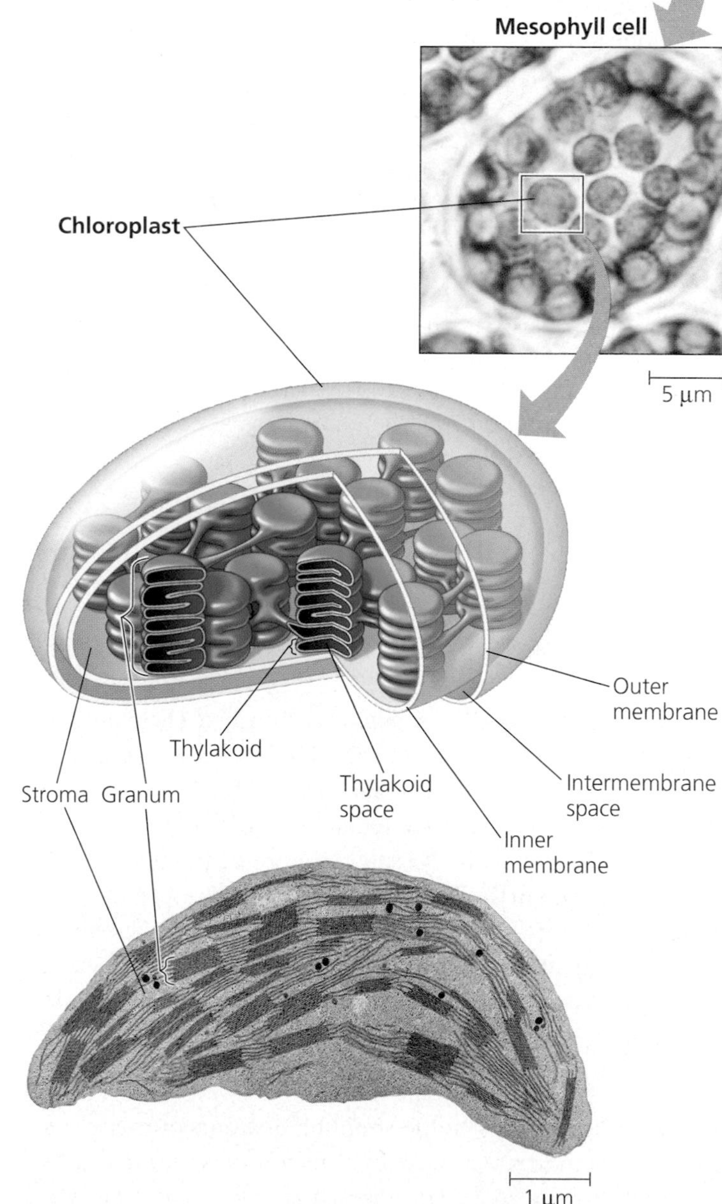

in the thylakoid membranes. (Photosynthetic prokaryotes lack chloroplasts, but they do have photosynthetic membranes arising from infolded regions of the plasma membrane that function in a manner similar to the thylakoid membranes of chloroplasts; see Figure 27.7b.) Now that we have looked at the sites of photosynthesis in plants, we are ready to look more closely at the process of photosynthesis.

Tracking Atoms Through Photosynthesis: *Scientific Inquiry*

Scientists have tried for centuries to piece together the process by which plants make food. Although some of the steps are still not completely understood, the overall photosynthetic equation has been known since the 1800s: In the presence of light, the green parts of plants produce organic compounds and oxygen from carbon dioxide and water. Using molecular formulas, we can summarize photosynthesis with this chemical equation:

$$6\ CO_2 + 12\ H_2O + \text{Light energy} \rightarrow C_6H_{12}O_6 + 6\ O_2 + 6\ H_2O$$

The carbohydrate $C_6H_{12}O_6$ is glucose.* Water appears on both sides of the equation because 12 molecules are consumed and 6 molecules are newly formed during photosynthesis. We can simplify the equation by indicating only the net consumption of water:

$$6\ CO_2 + 6\ H_2O + \text{Light energy} \rightarrow C_6H_{12}O_6 + 6\ O_2$$

Writing the equation in this form, we can see that the overall chemical change during photosynthesis is the reverse of the one that occurs during cellular respiration. Both of these metabolic processes occur in plant cells. However, as you will soon learn, plants do not make food by simply reversing the steps of respiration.

Now let's divide the photosynthetic equation by 6 to put it in its simplest possible form:

$$CO_2 + H_2O \rightarrow [CH_2O] + O_2$$

Here, the brackets indicate that CH_2O is not an actual sugar but represents the general formula for a carbohydrate. In

* The direct product of photosynthesis is actually a three-carbon sugar. Glucose is used here only to simplify the relationship between photosynthesis and respiration.

other words, we are imagining the synthesis of a sugar molecule one carbon at a time. Six repetitions would produce a glucose molecule. Let's now use this simplified formula to see how researchers tracked the chemical elements (C, H, and O) from the reactants of photosynthesis to the products.

The Splitting of Water

One of the first clues to the mechanism of photosynthesis came from the discovery that the oxygen given off by plants through their stomata is derived from water and not from carbon dioxide. The chloroplast splits water into hydrogen and oxygen. Before this discovery, the prevailing hypothesis was that photosynthesis split carbon dioxide ($CO_2 \rightarrow C + O_2$) and then added water to the carbon ($C + H_2O \rightarrow [CH_2O]$). This hypothesis predicted that the O_2 released during photosynthesis came from CO_2. This idea was challenged in the 1930s by C. B. van Niel of Stanford University. Van Niel was investigating photosynthesis in bacteria that make their carbohydrate from CO_2 but do not release O_2. Van Niel concluded that, at least in these bacteria, CO_2 is not split into carbon and oxygen. One group of bacteria used hydrogen sulfide (H_2S) rather than water for photosynthesis, forming yellow globules of sulfur as a waste product (these globules are visible in Figure 10.2e). Here is the chemical equation for photosynthesis in these sulfur bacteria:

$$CO_2 + 2\ H_2S \rightarrow [CH_2O] + H_2O + 2\ S$$

Van Niel reasoned that the bacteria split H_2S and used the hydrogen atoms to make sugar. He then generalized that idea, proposing that all photosynthetic organisms require a hydrogen source but that the source varies:

$$\text{Sulfur bacteria: } CO_2 + 2\ H_2S \rightarrow [CH_2O] + H_2O + 2\ S$$
$$\text{Plants: } CO_2 + 2\ H_2O \rightarrow [CH_2O] + H_2O + O_2$$
$$\text{General: } CO_2 + 2\ H_2X \rightarrow [CH_2O] + H_2O + 2\ X$$

Thus, van Niel hypothesized that plants split water as a source of electrons from hydrogen atoms, releasing oxygen as a by-product.

Nearly 20 years later, scientists confirmed van Niel's hypothesis by using oxygen-18 (^{18}O), a heavy isotope, as a radioactive tracer to follow the fate of oxygen atoms during photosynthesis. The experiments showed that the O_2 from plants was labeled with ^{18}O *only* if water was the source of the tracer (experiment 1). If the ^{18}O was introduced to the plant in the form of CO_2, the label did not turn up in the released O_2 (experiment 2). In the following summary, red denotes labeled atoms of oxygen (^{18}O):

$$\text{Experiment 1: } CO_2 + 2\ H_2O \rightarrow [CH_2O] + H_2O + O_2$$
$$\text{Experiment 2: } CO_2 + 2\ H_2O \rightarrow [CH_2O] + H_2O + O_2$$

A significant result of the shuffling of atoms during photosynthesis is the extraction of hydrogen from water and its incorporation into sugar. The waste product of photosynthesis, O_2, is released to the atmosphere. **Figure 10.4** shows the fates of all atoms in photosynthesis.

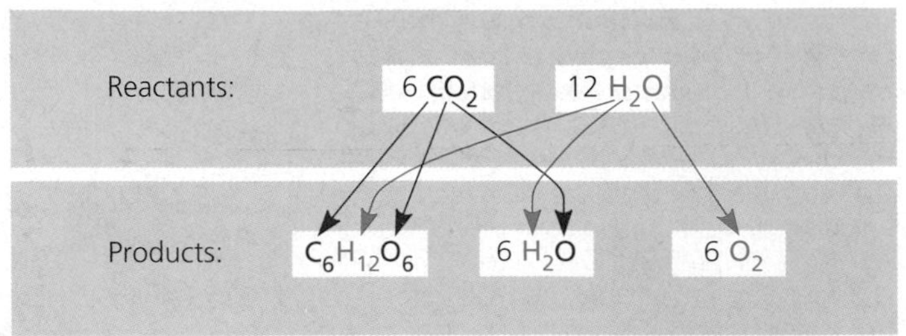

▲ **Figure 10.4 Tracking atoms through photosynthesis.**

Photosynthesis as a Redox Process

Let's briefly compare photosynthesis with cellular respiration. Both processes involve redox reactions. During cellular respiration, energy is released from sugar when electrons associated with hydrogen are transported by carriers to oxygen, forming water as a by-product. The electrons lose potential energy as they "fall" down the electron transport chain toward electronegative oxygen, and the mitochondrion harnesses that energy to synthesize ATP (see Figure 9.15). Photosynthesis reverses the direction of electron flow. Water is split, and electrons are transferred along with hydrogen ions from the water to carbon dioxide, reducing it to sugar. Because the electrons increase in potential energy as they move from water to sugar, this process requires energy. This energy boost is provided by light.

The Two Stages of Photosynthesis: *A Preview*

The equation for photosynthesis is a deceptively simple summary of a very complex process. Actually, photosynthesis is not a single process, but two processes, each with multiple steps. These two stages of photosynthesis are known as the **light reactions** (the *photo* part of photosynthesis) and the **Calvin cycle** (the *synthesis* part) **(Figure 10.5)**.

The light reactions are the steps of photosynthesis that convert solar energy to chemical energy. Light absorbed by chlorophyll drives a transfer of electrons and hydrogen from water to an acceptor called **$NADP^+$** (nicotinamide adenine dinucleotide phosphate), which temporarily stores the energized electrons. Water is split in the process, and thus it is the light reactions of photosynthesis that give off O_2 as a by-product. The electron acceptor of the light reactions, $NADP^+$, is first cousin to NAD^+, which functions as an electron carrier in cellular respiration; the two molecules differ only by the presence of an extra phosphate group in the $NADP^+$ molecule. The light reactions use solar power to reduce $NADP^+$ to NADPH by adding a pair of electrons along with a hydrogen nucleus, or H^+. The light reactions also generate ATP, using chemiosmosis to power the addition of a phosphate

▶ **Figure 10.5 An overview of photosynthesis: cooperation of the light reactions and the Calvin cycle.** In the chloroplast, the thylakoid membranes are the sites of the light reactions, whereas the Calvin cycle occurs in the stroma. The light reactions use solar energy to make ATP and NADPH, which function as chemical energy and reducing power, respectively, in the Calvin cycle. The Calvin cycle incorporates CO_2 into organic molecules, which are converted to sugar. (Recall from Chapter 5 that most simple sugars have formulas that are some multiple of $[CH_2O]$.)

A smaller version of this diagram will reappear in several subsequent figures as a reminder of whether the events being described occur in the light reactions or in the Calvin cycle.

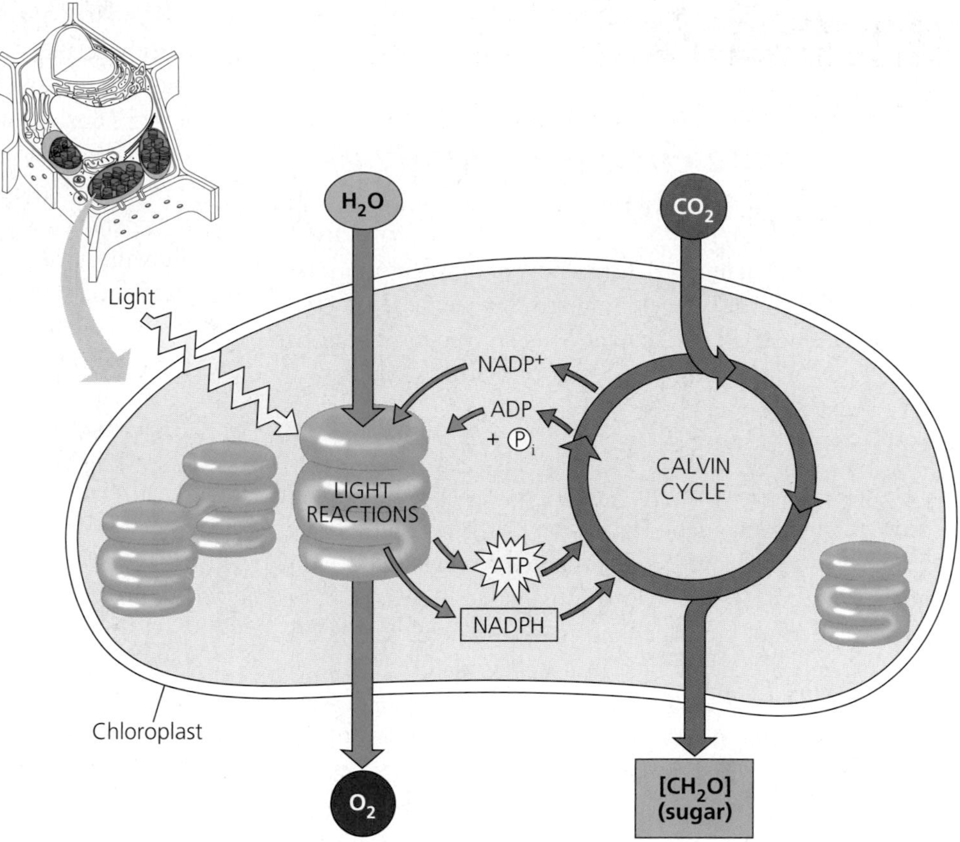

group to ADP, a process called **photophosphorylation.** Thus, light energy is initially converted to chemical energy in the form of two compounds: NADPH, a source of energized electrons ("reducing power"), and ATP, the versatile energy currency of cells. Notice that the light reactions produce no sugar; that happens in the second stage of photosynthesis, the Calvin cycle.

The Calvin cycle is named for Melvin Calvin, who, along with his colleagues, began to elucidate its steps in the late 1940s. The cycle begins by incorporating CO_2 from the air into organic molecules already present in the chloroplast. This initial incorporation of carbon into organic compounds is known as **carbon fixation.** The Calvin cycle then reduces the fixed carbon to carbohydrate by the addition of electrons. The reducing power is provided by NADPH, which acquired energized electrons in the light reactions. To convert CO_2 to carbohydrate, the Calvin cycle also requires chemical energy in the form of ATP, which is also generated by the light reactions. Thus, it is the Calvin cycle that makes sugar, but it can do so only with the help of the NADPH and ATP produced by the light reactions. The metabolic steps of the Calvin cycle are sometimes referred to as the dark reactions, or light-independent reactions, because none of the steps requires light *directly.* Nevertheless, the Calvin cycle in most plants occurs during daylight, for only then can the light reactions provide the NADPH and ATP that the Calvin cycle requires. In essence, the chloroplast uses light energy to make sugar by coordinating the two stages of photosynthesis.

As Figure 10.5 indicates, the thylakoids of the chloroplast are the sites of the light reactions, while the Calvin cycle occurs in the stroma. In the thylakoids, molecules of $NADP^+$ and ADP pick up electrons and phosphate, respectively, and then are released to the stroma, where they transfer their high-energy cargo to the Calvin cycle. The two stages of photosynthesis are treated in this figure as metabolic modules that take in ingredients and crank out products. Our next step toward understanding photosynthesis is to look more closely at how the two stages work, beginning with the light reactions.

Concept Check 10.1

1. How do the reactant molecules of photosynthesis reach the chloroplasts in leaves?
2. How did the use of an oxygen isotope help elucidate the chemistry of photosynthesis?
3. Describe how the two stages of photosynthesis are dependent on each other.

For suggested answers, see Appendix A.

Concept 10.2

The light reactions convert solar energy to the chemical energy of ATP and NADPH

Chloroplasts are chemical factories powered by the sun. Their thylakoids transform light energy into the chemical energy of ATP and NADPH. To understand this conversion better, we need to know about some important properties of light.

The Nature of Sunlight

Light is a form of energy known as electromagnetic energy, also called electromagnetic radiation. Electromagnetic energy travels in rhythmic waves analogous to those created by dropping a pebble into a pond. Electromagnetic waves, however, are disturbances of electrical and magnetic fields rather than disturbances of a material medium such as water.

The distance between the crests of electromagnetic waves is called the **wavelength.** Wavelengths range from less than a nanometer (for gamma rays) to more than a kilometer (for radio waves). This entire range of radiation is known as the **electromagnetic spectrum** (**Figure 10.6**). The segment most important to life is the narrow band from about 380 nm to 750 nm in wavelength. This radiation is known as **visible light** because it is detected as various colors by the human eye.

The model of light as waves explains many of light's properties, but in certain respects light behaves as though it consists of discrete particles, called **photons.** Photons are not tangible objects, but they act like objects in that each of them has a fixed quantity of energy. The amount of energy is inversely related to the wavelength of the light; the shorter the wavelength, the greater the energy of each photon of that light. Thus, a photon of violet light packs nearly twice as much energy as a photon of red light.

Although the sun radiates the full spectrum of electromagnetic energy, the atmosphere acts like a selective window, allowing visible light to pass through while screening out a substantial fraction of other radiation. The part of the spectrum we can see—visible light—is also the radiation that drives photosynthesis.

Photosynthetic Pigments: The Light Receptors

When light meets matter, it may be reflected, transmitted, or absorbed. Substances that absorb visible light are known as pigments. Different pigments absorb light of different wavelengths, and the wavelengths that are absorbed disappear. If a pigment is illuminated with white light, the color we see is the color most reflected or transmitted by the pigment. (If a pigment absorbs all wavelengths, it appears black.) We see green when we look at a leaf because chlorophyll absorbs violet-blue and red light while transmitting and reflecting green light (**Figure 10.7**). The ability of a pigment to absorb various wavelengths of light can be measured with an instrument called a **spectrophotometer.** This machine directs beams of light of different wavelengths through a solution of the pigment and measures the fraction of the light transmitted

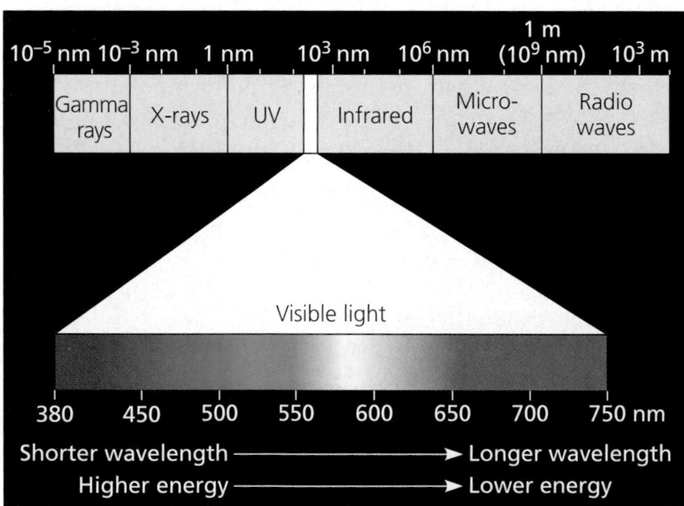

▲ **Figure 10.6 The electromagnetic spectrum.** White light is a mixture of all wavelengths of visible light. A prism can sort white light into its component colors by bending light of different wavelengths at different angles. (Droplets of water in the atmosphere can act as prisms, forming a rainbow; see Figure 10.1.) Visible light drives photosynthesis.

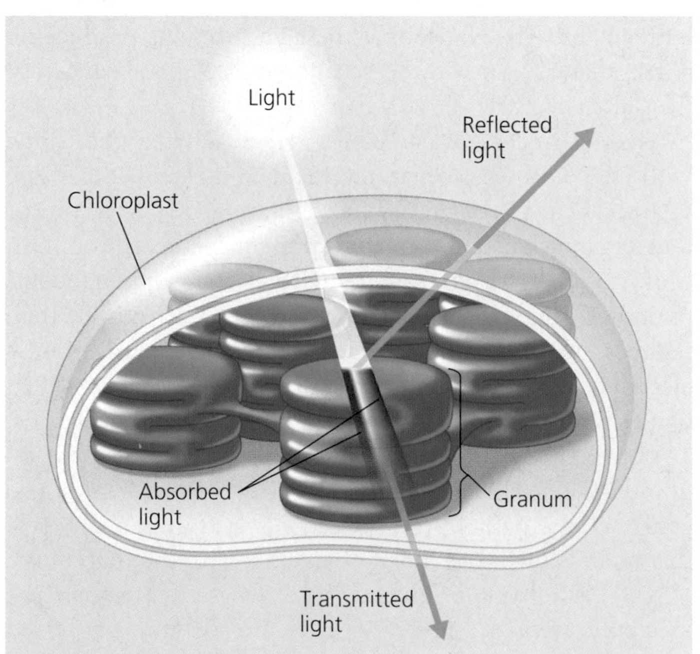

▲ **Figure 10.7 Why leaves are green: interaction of light with chloroplasts.** The chlorophyll molecules of chloroplasts absorb violet-blue and red light (the colors most effective in driving photosynthesis) and reflect or transmit green light. This is why leaves appear green.

at each wavelength **(Figure 10.8)**. A graph plotting a pigment's light absorption versus wavelength is called an **absorption spectrum**.

The absorption spectra of chloroplast pigments provide clues to the relative effectiveness of different wavelengths for driving photosynthesis, since light can perform work in chloroplasts only if it is absorbed. **Figure 10.9a** shows the absorption spectra of three types of pigments in chloroplasts. If we look first at the absorption spectrum of **chlorophyll *a***, it suggests that violet-blue and red light work best for photosynthesis, since they are

Figure 10.8

Research Method Determining an Absorption Spectrum

APPLICATION An absorption spectrum is a visual representation of how well a particular pigment absorbs different wavelengths of visible light. Absorption spectra of various chloroplast pigments help scientists decipher each pigment's role in a plant.

TECHNIQUE A spectrophotometer measures the relative amounts of light of different wavelengths absorbed and transmitted by a pigment solution.

1. White light is separated into colors (wavelengths) by a prism.
2. One by one, the different colors of light are passed through the sample (chlorophyll in this example). Green light and blue light are shown here.
3. The transmitted light strikes a photoelectric tube, which converts the light energy to electricity.
4. The electrical current is measured by a galvanometer. The meter indicates the fraction of light transmitted through the sample, from which we can determine the amount of light absorbed.

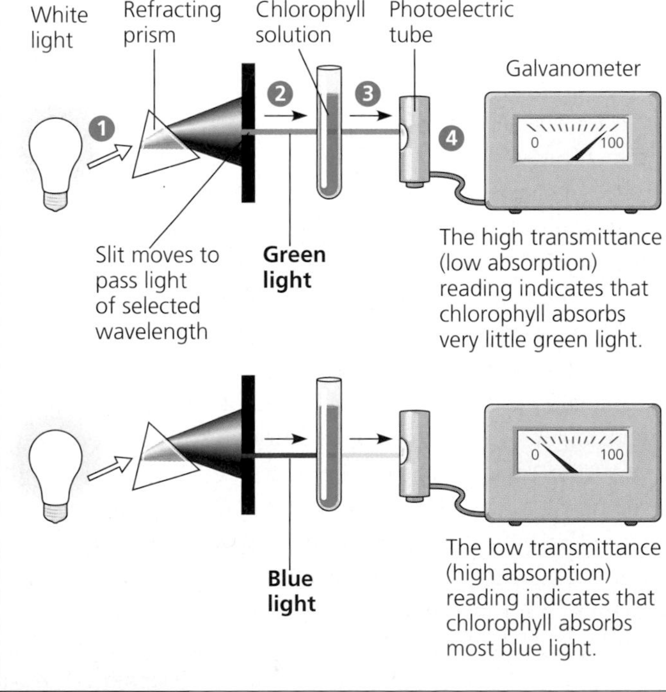

RESULTS See Figure 10.9a for absorption spectra of three types of chloroplast pigments.

Figure 10.9

Inquiry Which wavelengths of light are most effective in driving photosynthesis?

EXPERIMENT Three different experiments helped reveal which wavelengths of light are photosynthetically important. The results are shown below.

RESULTS

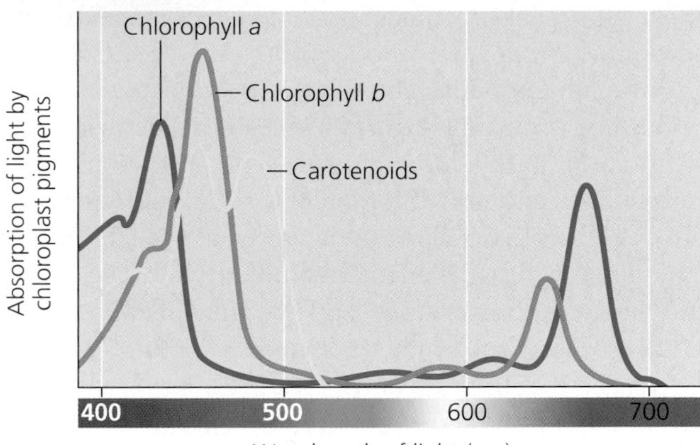

(a) Absorption spectra. The three curves show the wavelengths of light best absorbed by three types of chloroplast pigments.

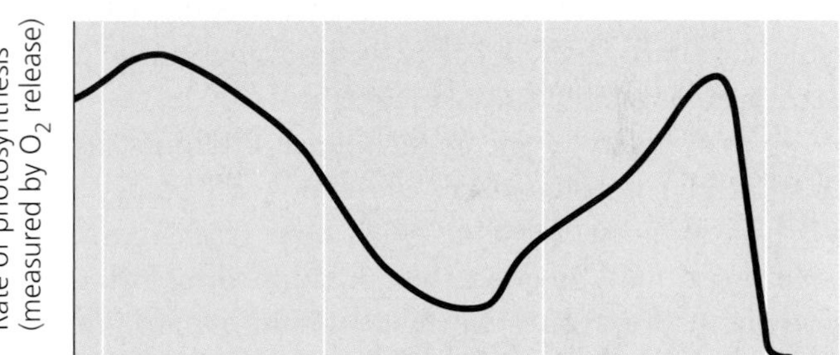

(b) Action spectrum. This graph plots the rate of photosynthesis versus wavelength. The resulting action spectrum resembles the absorption spectrum for chlorophyll *a* but does not match exactly (see part a). This is partly due to the absorption of light by accessory pigments such as chlorophyll *b* and carotenoids.

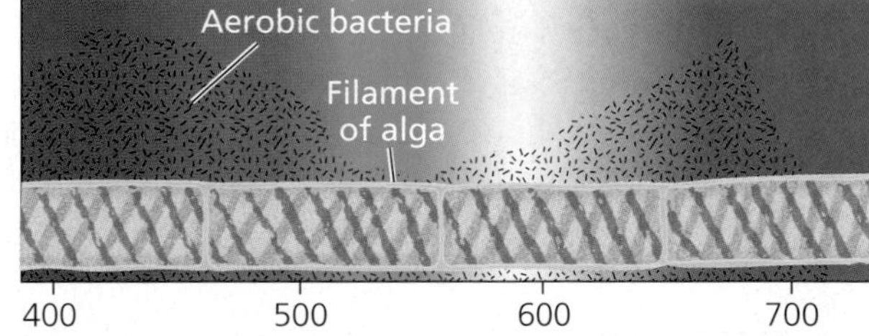

(c) Engelmann's experiment. In 1883, Theodor W. Engelmann illuminated a filamentous alga with light that had been passed through a prism, exposing different segments of the alga to different wavelengths. He used aerobic bacteria, which concentrate near an oxygen source, to determine which segments of the alga were releasing the most O_2 and thus photosynthesizing most. Bacteria congregated in greatest numbers around the parts of the alga illuminated with violet-blue or red light. Notice the close match of the bacterial distribution to the action spectrum in part b.

CONCLUSION Light in the violet-blue and red portions of the spectrum are most effective in driving photosynthesis.

absorbed, while green is the least effective color. This is confirmed by an **action spectrum** for photosynthesis **(Figure 10.9b)**, which profiles the relative effectiveness of different wavelengths of radiation in driving the process. An action spectrum is prepared by illuminating chloroplasts with light of different colors and then plotting wavelength against some measure of photosynthetic rate, such as CO_2 consumption or O_2 release. The action spectrum for photosynthesis was first demonstrated in 1883 in an elegant experiment performed by German botanist Theodor W. Engelmann, who used bacteria to measure rates of photosynthesis in filamentous algae **(Figure 10.9c)**.

Notice by comparing Figures 10.9a and 10.9b that the action spectrum for photosynthesis does not exactly match the absorption spectrum of chlorophyll *a*. The absorption spectrum of chlorophyll *a* alone underestimates the effectiveness of certain wavelengths in driving photosynthesis. This is partly because accessory pigments with different absorption spectra are also photosynthetically important in chloroplasts and broaden the spectrum of colors that can be used for photosynthesis. One of these accessory pigments is another form of chlorophyll, **chlorophyll *b***. Chlorophyll *b* is almost identical to chlorophyll *a*, but a slight structural difference between them **(Figure 10.10)** is enough to give the two pigments slightly different absorption spectra (see Figure 10.9a). As a result, they have different colors—chlorophyll *a* is blue-green, whereas chlorophyll *b* is yellow-green.

Other accessory pigments include **carotenoids**, hydrocarbons that are various shades of yellow and orange because they absorb violet and blue-green light (see Figure 10.9a). Carotenoids may broaden the spectrum of colors that can drive photosynthesis. However, a more important function of at least some carotenoids seems to be *photoprotection:* These compounds absorb and dissipate excessive light energy that would otherwise damage chlorophyll or interact with oxygen, forming reactive oxidative molecules that are dangerous to the cell. Interestingly, carotenoids similar to the photoprotective ones in chloroplasts have a photoprotective role in the human eye. These and other related molecules are highlighted in health food products as "phytochemicals" (from the Greek *phyton,* plant) that have antioxidant powers. Plants can synthesize all the antioxidants they require, whereas humans and other animals must obtain some of them from their diets.

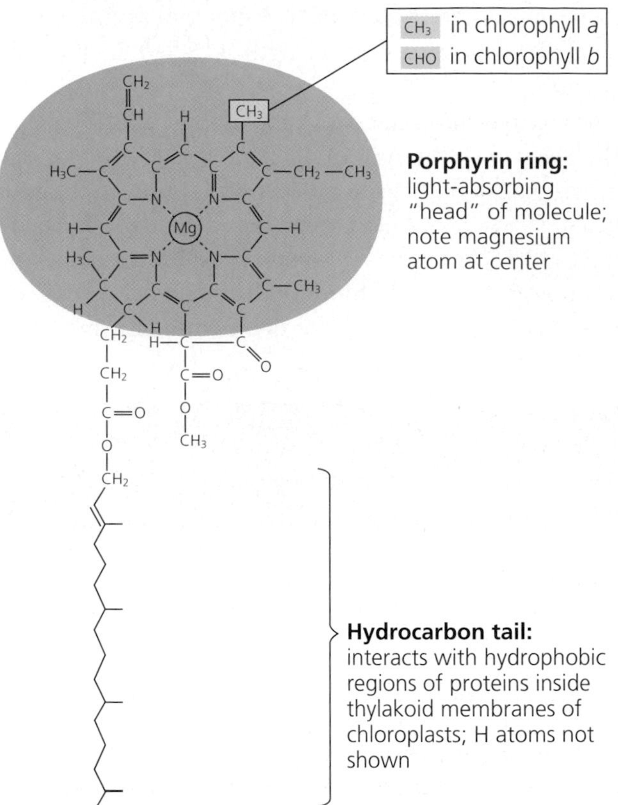

▲ **Figure 10.10 Structure of chlorophyll molecules in chloroplasts of plants.** Chlorophyll *a* and chlorophyll *b* differ only in one of the functional groups bonded to the porphyrin ring.

Excitation of Chlorophyll by Light

What exactly happens when chlorophyll and other pigments absorb light? The colors corresponding to the absorbed wavelengths disappear from the spectrum of the transmitted and reflected light, but energy cannot disappear. When a molecule absorbs a photon of light, one of the molecule's electrons is elevated to an orbital where it has more potential energy. When the electron is in its normal orbital, the pigment molecule is said to be in its ground state. Absorption of a photon boosts an electron to an orbital of higher energy, and the pigment molecule is then said to be in an excited state. The only photons absorbed are those whose energy is exactly equal to the energy difference between the ground state and an excited state, and this energy difference varies from one kind of atom or molecule to another. Thus, a particular compound absorbs only photons corresponding to specific wavelengths, which is why each pigment has a unique absorption spectrum.

Once absorption of a photon raises an electron from the ground state to an excited state, the electron cannot remain there long. The excited state, like all high-energy states, is unstable. Generally, when isolated pigment molecules absorb light, their excited electrons drop back down to the ground-state orbital in a billionth of a second, releasing their excess energy as heat. This conversion of light energy to heat is what makes the top of an automobile so hot on a sunny day. (White cars are coolest because their paint reflects all wavelengths of visible light, although it may absorb ultraviolet and other invisible radiation.) In isolation, some pigments, including chlorophyll, emit light as well as heat after absorbing photons. As excited electrons fall back to the ground state, photons are given off. This afterglow is called fluorescence. If a solution of chlorophyll isolated from chloroplasts is illuminated, it will fluoresce in the red-orange part of the spectrum and also give off heat **(Figure 10.11)**.

▶ **Figure 10.11 Excitation of isolated chlorophyll by light. (a)** Absorption of a photon causes a transition of the chlorophyll molecule from its ground state to its excited state. The photon boosts an electron to an orbital where it has more potential energy. If the illuminated molecule exists in isolation, its excited electron immediately drops back down to the ground-state orbital, and its excess energy is given off as heat and fluorescence (light). **(b)** A chlorophyll solution excited with ultraviolet light fluoresces with a red-orange glow.

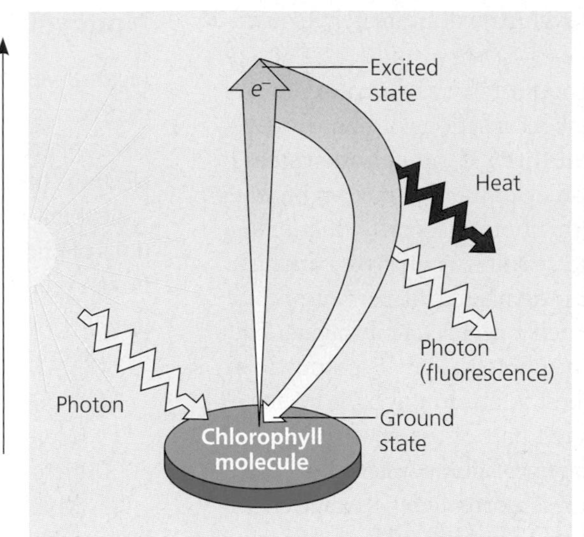

(a) Excitation of isolated chlorophyll molecule

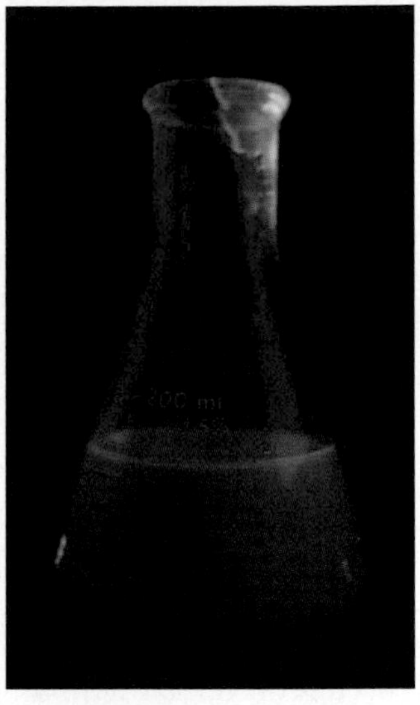

(b) Fluorescence

A Photosystem: A Reaction Center Associated with Light-Harvesting Complexes

Chlorophyll molecules excited by the absorption of light energy produce very different results in an intact chloroplast than they do in isolation (see Figure 10.11). In their native environment of the thylakoid membrane, chlorophyll molecules are organized along with other small organic molecules and proteins into photosystems.

A **photosystem** is composed of a reaction center surrounded by a number of light-harvesting complexes **(Figure 10.12)**. Each **light-harvesting complex** consists of pigment molecules (which may include chlorophyll *a*, chlorophyll *b*, and carotenoids) bound to particular proteins. The number and variety of pigment molecules enable a photosystem to harvest light over a larger surface and a larger portion of the spectrum than any single pigment molecule alone could. Together, these light-harvesting complexes act as an antenna for the reaction center. When a pigment molecule absorbs a photon, the energy is transferred from pigment molecule to pigment molecule within a light-harvesting complex until it is funneled into the reaction center. The **reaction center** is a protein complex that includes two special chlorophyll *a* molecules and a molecule called the **primary electron acceptor.** These chlorophyll *a* molecules are special because their molecular environment—their location and the other molecules with which they are associated—enables them to use the energy from light to boost one of their electrons to a higher energy level.

The solar-powered transfer of an electron from a special chlorophyll *a* molecule to the primary electron acceptor is the first step of the light reactions. As soon as the chlorophyll electron is excited to a higher energy level, the primary electron acceptor captures it; this is a redox reaction. Isolated chlorophyll fluoresces because there is no electron acceptor, so electrons of photoexcited chlorophyll drop right back to the ground state.

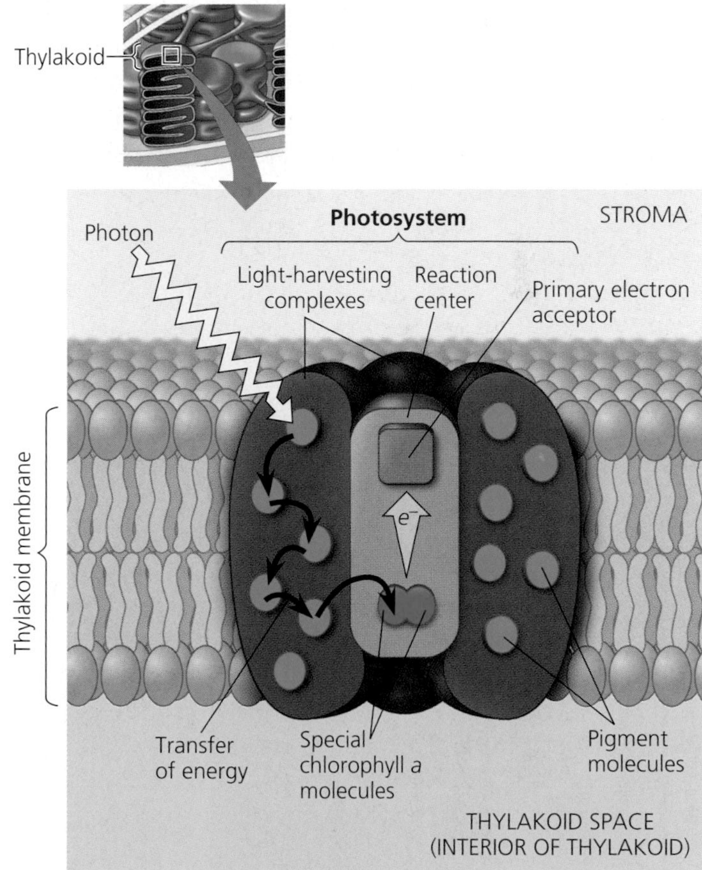

▲ **Figure 10.12 How a photosystem harvests light.** When a photon strikes a pigment molecule in a light-harvesting complex, the energy is passed from molecule to molecule until it reaches the reaction center. At the reaction center, an excited electron from one of the two special chlorophyll *a* molecules is captured by the primary electron acceptor.

In a chloroplast, this immediate plunge of high-energy electrons back to the ground state is prevented. Thus, each photosystem—a reaction center surrounded by light-harvesting complexes—functions in the chloroplast as a unit. It converts light energy to chemical energy, which will ultimately be used for the synthesis of sugar.

The thylakoid membrane is populated by two types of photosystems that cooperate in the light reactions of photosynthesis. They are called **photosystem II (PS II)** and **photosystem I (PS I).** (They were named in order of their discovery, but the two function sequentially, with photosystem II functioning first.) Each has a characteristic reaction center—a particular kind of primary electron acceptor next to a pair of special chlorophyll *a* molecules associated with specific proteins. The reaction-center chlorophyll *a* of photosystem II is known as P680 because this pigment is best at absorbing light having a wavelength of 680 nm (in the red part of the spectrum). The chlorophyll *a* at the reaction center of photosystem I is called P700 because it most effectively absorbs light of wavelength 700 nm (in the far red part of the spectrum). These two pigments, P680 and P700, are actually identical chlorophyll *a* molecules. However, their association with different proteins in the thylakoid membrane affects the electron distribution in the chlorophyll molecules and accounts for the slight differences in light-absorbing properties. Now let's see how the two photosystems work together in using light energy to generate ATP and NADPH, the two main products of the light reactions.

Noncyclic Electron Flow

Light drives the synthesis of NADPH and ATP by energizing the two photosystems embedded in the thylakoid membranes of chloroplasts. The key to this energy transformation is a flow of electrons through the photosystems and other molecular components built into the thylakoid membrane. During the light reactions of photosynthesis, there are two possible routes for electron flow: cyclic and noncyclic. **Noncyclic electron flow,** the predominant route, is shown in **Figure 10.13**. The numbers in the text description correspond to the numbered steps in the figure.

1. A photon of light strikes a pigment molecule in a light-harvesting complex and is relayed to other pigment molecules until it reaches one of the two P680 chlorophyll *a* molecules in the PS II reaction center. It excites one of the P680 electrons to a higher energy state.
2. This electron is captured by the primary electron acceptor.
3. An enzyme splits a water molecule into two electrons, two hydrogen ions, and an oxygen atom. The electrons are supplied one by one to the P680 molecules, each replacing an electron lost to the primary electron acceptor. (Missing an electron, P680 is the strongest biological oxidizing agent known; its electron hole must be filled.)

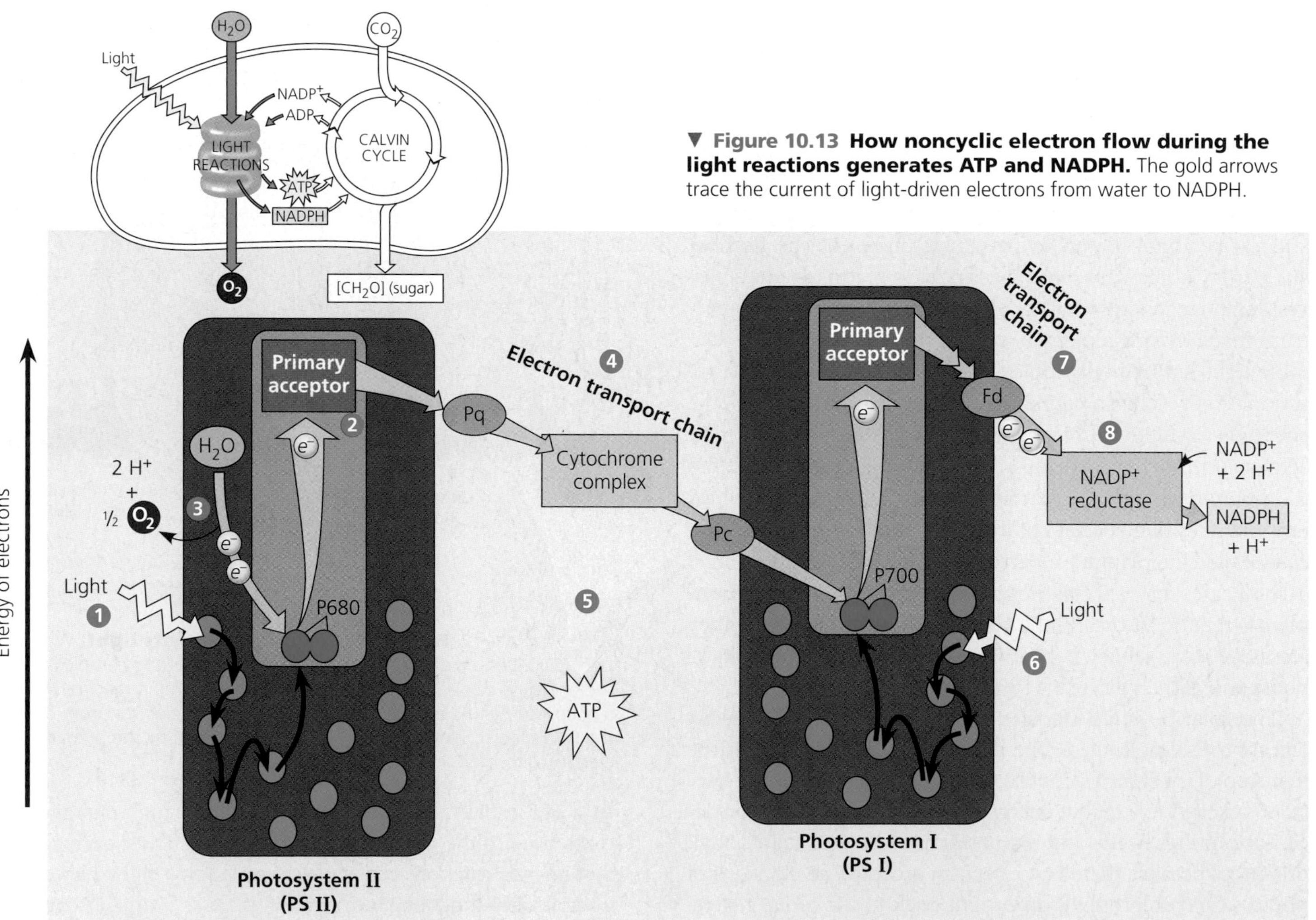

▼ **Figure 10.13 How noncyclic electron flow during the light reactions generates ATP and NADPH.** The gold arrows trace the current of light-driven electrons from water to NADPH.

The oxygen atom immediately combines with another oxygen atom, forming O_2.

4. Each photoexcited electron passes from the primary electron acceptor of PS II to PS I via an electron transport chain (similar to the electron transport chain that functions in cellular respiration). The electron transport chain between PS II and PS I is made up of the electron carrier plastoquinone (Pq), a cytochrome complex, and a protein called plastocyanin (Pc).
5. The exergonic "fall" of electrons to a lower energy level provides energy for the synthesis of ATP.
6. Meanwhile, light energy was transferred via a light-harvesting complex to the PS I reaction center, exciting an electron of one of the two P700 chlorophyll *a* molecules located there. The photoexcited electron was then captured by PS I's primary electron acceptor, creating an electron "hole" in the P700. The hole is filled by an electron that reaches the bottom of the electron transport chain from PS II.
7. Photoexcited electrons are passed from PS I's primary electron acceptor down a second electron transport chain through the protein ferredoxin (Fd).
8. The enzyme $NADP^+$ reductase transfers electrons from Fd to $NADP^+$. Two electrons are required for its reduction to NADPH.

As complicated as the scheme shown in Figure 10.13 is, do not lose track of its functions: The light reactions use solar power to generate ATP and NADPH, which provide chemical energy and reducing power, respectively, to the sugar-making reactions of the Calvin cycle. The energy changes of electrons as they flow through the light reactions are shown by analogy in **Figure 10.14**.

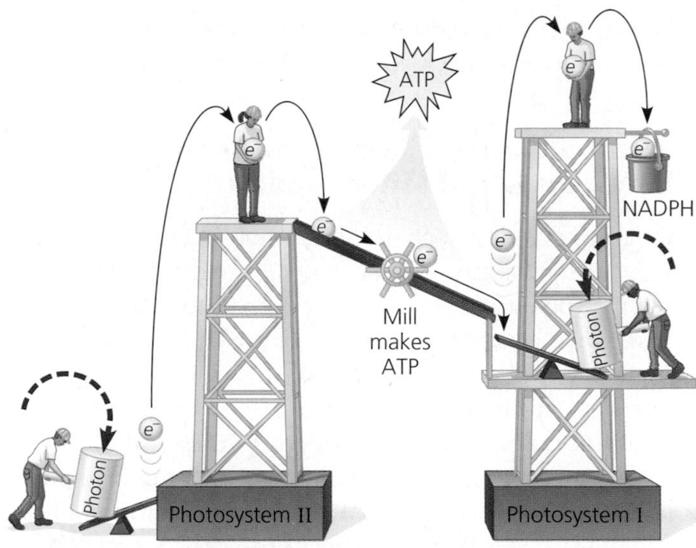

▲ **Figure 10.14 A mechanical analogy for the light reactions.**

Cyclic Electron Flow

Under certain conditions, photoexcited electrons take an alternative path called **cyclic electron flow,** which uses photosystem I but not photosystem II. You can see in **Figure 10.15** that cyclic flow is a short circuit: The electrons cycle back from ferredoxin (Fd) to the cytochrome complex and from there continue on to a P700 chlorophyll in the PS I reaction center. There is no production of NADPH and no release of oxygen. Cyclic flow does, however, generate ATP.

What is the function of cyclic electron flow? Noncyclic electron flow produces ATP and NADPH in roughly equal quantities, but the Calvin cycle consumes more ATP than NADPH. Cyclic electron flow makes up the difference, since it produces ATP but no NADPH. The concentration of NADPH in the chloroplast may help regulate which pathway, cyclic versus noncyclic, electrons take through the light reactions. If the chloroplast runs low on ATP for the Calvin cycle, NADPH will begin to accumulate as the Calvin cycle slows down. The rise in NADPH may stimulate a temporary shift from noncyclic to cyclic electron flow until ATP supply catches up with demand.

Whether ATP synthesis is driven by noncyclic or cyclic electron flow, the actual mechanism is the same. This is a good

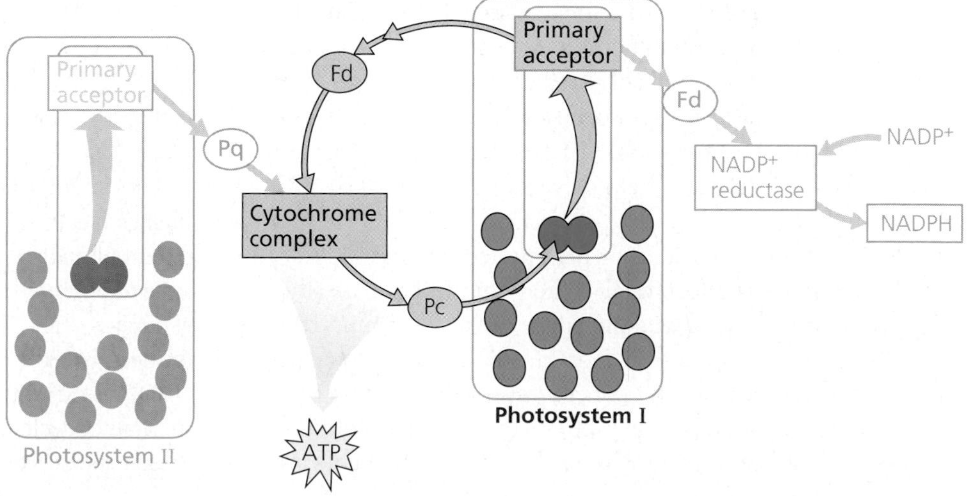

► **Figure 10.15 Cyclic electron flow.** Photoexcited electrons from PS I are occasionally shunted back from ferredoxin (Fd) to chlorophyll via the cytochrome complex and plastocyanin (Pc). This electron shunt supplements the supply of ATP (via chemiosmosis) but produces no NADPH. The "shadow" of noncyclic electron flow is included in the diagram for comparison with the cyclic route. The two ferredoxin molecules shown in this diagram are actually one and the same—the final electron carrier in the electron transport chain of PS I.

time to review chemiosmosis, the process that uses membranes to couple redox reactions to ATP production.

A Comparison of Chemiosmosis in Chloroplasts and Mitochondria

Chloroplasts and mitochondria generate ATP by the same basic mechanism: chemiosmosis. An electron transport chain assembled in a membrane pumps protons across the membrane as electrons are passed through a series of carriers that are progressively more electronegative. In this way, electron transport chains transform redox energy to a proton-motive force, potential energy stored in the form of an H^+ gradient across a membrane. Built into the same membrane is an ATP synthase complex that couples the diffusion of hydrogen ions down their gradient to the phosphorylation of ADP. Some of the electron carriers, including the iron-containing proteins called cytochromes, are very similar in chloroplasts and mitochondria. The ATP synthase complexes of the two organelles are also very much alike. But there are noteworthy differences between oxidative phosphorylation in mitochondria and photophosphorylation in chloroplasts. In mitochondria, the high-energy electrons dropped down the transport chain are extracted from organic molecules (which are thus oxidized). Chloroplasts do not need molecules from food to make ATP; their photosystems capture light energy and use it to drive electrons to the top of the transport chain. In other words, mitochondria transfer chemical energy from food molecules to ATP (and NADH), whereas chloroplasts transform light energy into chemical energy in ATP (and NADPH).

The spatial organization of chemiosmosis also differs in chloroplasts and mitochondria (**Figure 10.16**). The inner membrane of the mitochondrion pumps protons from the mitochondrial matrix out to the intermembrane space, which then serves as a reservoir of hydrogen ions that powers the ATP synthase. The thylakoid membrane of the chloroplast pumps protons from the stroma into the thylakoid space (interior of the thylakoid), which functions as the H^+ reservoir. The thylakoid membrane makes ATP as the hydrogen ions diffuse down their concentration gradient from the thylakoid space back to the stroma through ATP synthase complexes, whose catalytic knobs are on the stroma side of the membrane. Thus, ATP forms in the stroma, where it is used to help drive sugar synthesis during the Calvin cycle.

The proton (H^+) gradient, or pH gradient, across the thylakoid membrane is substantial. When chloroplasts are illuminated, the pH in the thylakoid space drops to about 5 (the H^+ concentration increases), and the pH in the stroma increases to about 8 (the H^+ concentration decreases). This gradient of three pH units corresponds to a thousandfold difference in H^+ concentration. If in the laboratory the lights are turned off, the pH gradient is abolished, but it can quickly be restored by turning the lights back on. Such experiments provide strong evidence in support of the chemiosmotic model.

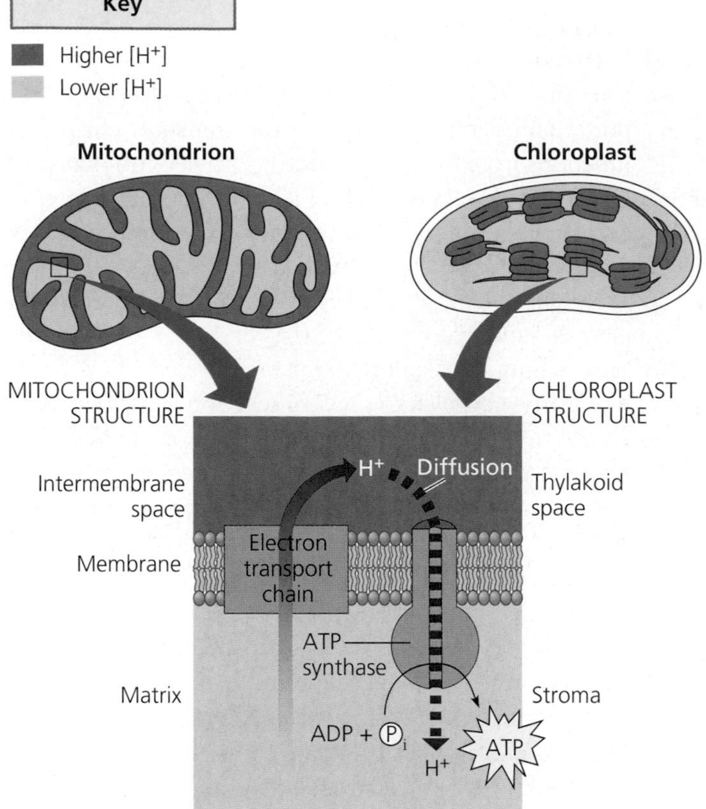

▲ **Figure 10.16 Comparison of chemiosmosis in mitochondria and chloroplasts.** In both kinds of organelles, electron transport chains pump protons (H^+) across a membrane from a region of low H^+ concentration (light gray in this diagram) to one of high H^+ concentration (dark gray). The protons then diffuse back across the membrane through ATP synthase, driving the synthesis of ATP.

Based on studies in several laboratories, **Figure 10.17** shows a current model for the organization of the light-reaction "machinery" within the thylakoid membrane. Each of the molecules and molecular complexes in the figure is present in numerous copies in each thylakoid. Notice that NADPH, like ATP, is produced on the side of the membrane facing the stroma, where the Calvin cycle reactions take place.

Let's summarize the light reactions. Noncyclic electron flow pushes electrons from water, where they are at a low state of potential energy, to NADPH, where they are stored at a high state of potential energy. The light-driven electron current also generates ATP. Thus, the equipment of the thylakoid membrane converts light energy to chemical energy stored in NADPH and ATP. (Oxygen is a by-product.) Let's now see how the Calvin cycle uses the products of the light reactions to synthesize sugar from CO_2.

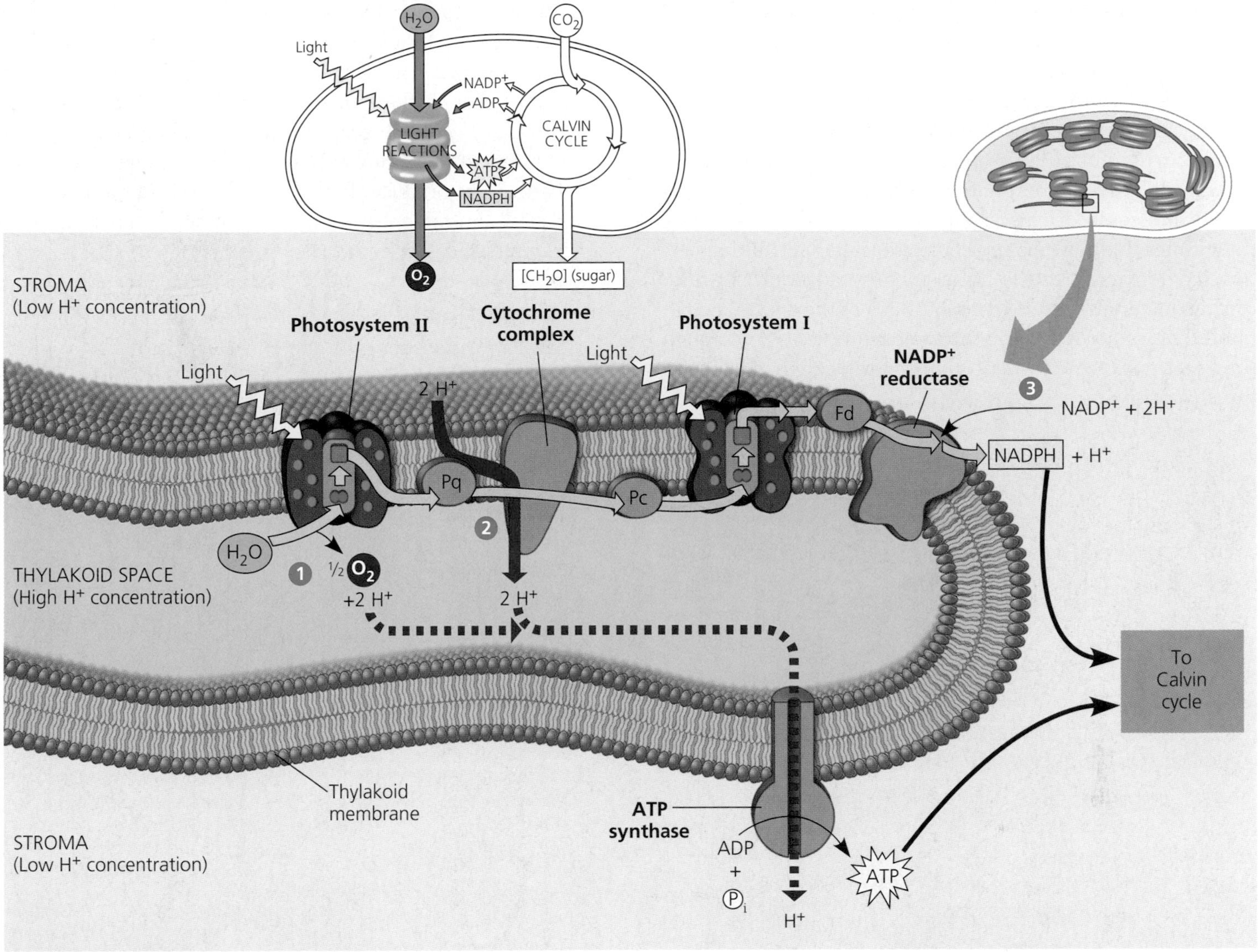

▲ **Figure 10.17 The light reactions and chemiosmosis: the organization of the thylakoid membrane.** This diagram shows a current model for the organization of the thylakoid membrane. The gold arrows track the noncyclic electron flow outlined in Figure 10.13. As electrons pass from carrier to carrier in redox reactions, hydrogen ions removed from the stroma are deposited in the thylakoid space, storing energy as a proton-motive force (H^+ gradient). At least three steps in the light reactions contribute to the proton gradient: ➊ Water is split by photosystem II on the side of the membrane facing the thylakoid space; ➋ as plastoquinone (Pq), a mobile carrier, transfers electrons to the cytochrome complex, protons are translocated across the membrane into the thylakoid space; and ➌ a hydrogen ion is removed from the stroma when it is taken up by $NADP^+$. Notice how, as in Figure 10.16, hydrogen ions are being pumped from the stroma into the thylakoid space. The diffusion of H^+ from the thylakoid space back to the stroma (along the H^+ concentration gradient) powers the ATP synthase. These light-driven reactions store chemical energy in NADPH and ATP, which shuttle the energy to the sugar-producing Calvin cycle.

Concept Check 10.2

1. What color of light is *least* effective in driving photosynthesis? Explain.
2. Compared to a solution of isolated chlorophyll, why do intact chloroplasts release less heat and fluorescence when illuminated?
3. In the light reactions, what is the electron donor? Where do the electrons end up?

For suggested answers, see Appendix A.

The Calvin cycle uses ATP and NADPH to convert CO_2 to sugar

The Calvin cycle is similar to the citric acid cycle in that a starting material is regenerated after molecules enter and leave the cycle. However, while the citric acid cycle is catabolic, oxidizing glucose and releasing energy, the Calvin cycle is anabolic, building sugar from smaller molecules and consuming energy. Carbon enters the Calvin cycle in the form of CO_2 and leaves in the form of sugar. The cycle spends ATP as an energy source and consumes NADPH as reducing power for adding high-energy electrons to make the sugar.

The carbohydrate produced directly from the Calvin cycle is actually not glucose, but a three-carbon sugar named **glyceraldehyde-3-phosphate (G3P).** For the net synthesis of one molecule of this sugar, the cycle must take place three times, fixing three molecules of CO_2. (Recall that carbon fixation refers to the initial incorporation of CO_2 into organic material.) As we trace the steps of the cycle, keep in mind that we are following three molecules of CO_2 through the reactions. **Figure 10.18** divides the Calvin cycle into three phases:

Phase 1: Carbon fixation. The Calvin cycle incorporates each CO_2 molecule, one at a time, by attaching it to a five-carbon sugar named ribulose bisphosphate (abbreviated RuBP). The enzyme that catalyzes this first step is RuBP carboxylase, or **rubisco.** (It is the most abundant protein in chloroplasts and probably the most abundant protein on Earth.) The product of the reaction is a six-carbon intermediate so unstable that it immediately splits in half, forming two molecules of 3-phosphoglycerate (for each CO_2).

Phase 2: Reduction. Each molecule of 3-phosphoglycerate receives an additional phosphate group from ATP, becoming 1,3-bisphosphoglycerate. Next, a pair of electrons donated from NADPH reduces 1,3-bisphosphoglycerate to G3P. Specifically, the electrons from NADPH reduce the

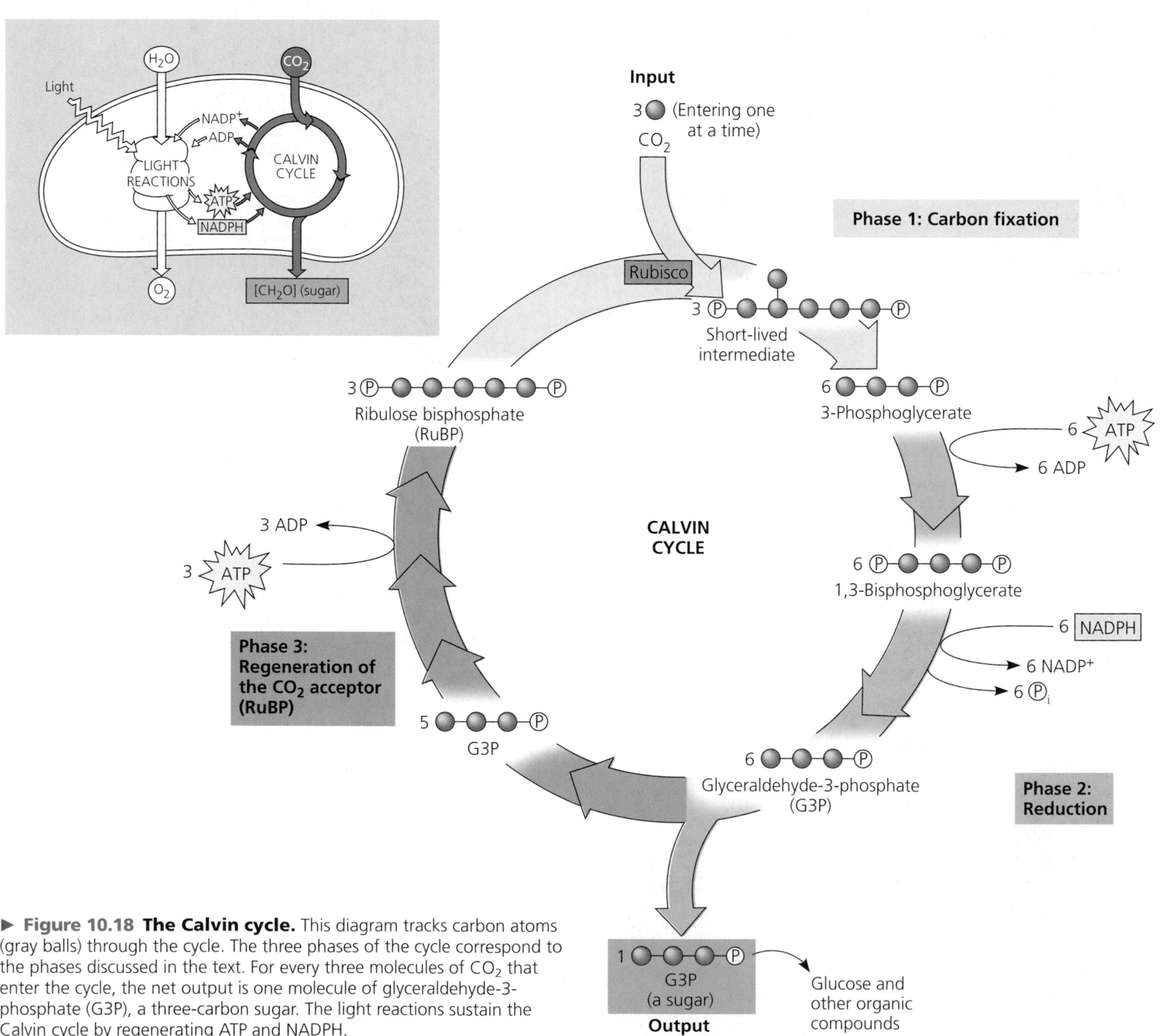

▶ **Figure 10.18 The Calvin cycle.** This diagram tracks carbon atoms (gray balls) through the cycle. The three phases of the cycle correspond to the phases discussed in the text. For every three molecules of CO_2 that enter the cycle, the net output is one molecule of glyceraldehyde-3-phosphate (G3P), a three-carbon sugar. The light reactions sustain the Calvin cycle by regenerating ATP and NADPH.

carboyxl group of 3-phosphoglycerate to the aldehyde group of G3P, which stores more potential energy. G3P is a sugar—the same three-carbon sugar formed in glycolysis by the splitting of glucose. Notice in Figure 10.18 that for every *three* molecules of CO_2, there are *six* molecules of G3P. But only one molecule of this three-carbon sugar can be counted as a net gain of carbohydrate. The cycle began with 15 carbons' worth of carbohydrate in the form of three molecules of the five-carbon sugar RuBP. Now there are 18 carbons' worth of carbohydrate in the form of six molecules of G3P. One molecule exits the cycle to be used by the plant cell, but the other five molecules must be recycled to regenerate the three molecules of RuBP.

Phase 3: Regeneration of the CO_2 acceptor (RuBP). In a complex series of reactions, the carbon skeletons of five molecules of G3P are rearranged by the last steps of the Calvin cycle into three molecules of RuBP. To accomplish this, the cycle spends three more molecules of ATP. The RuBP is now prepared to receive CO_2 again, and the cycle continues.

For the net synthesis of one G3P molecule, the Calvin cycle consumes a total of nine molecules of ATP and six molecules of NADPH. The light reactions regenerate the ATP and NADPH. The G3P spun off from the Calvin cycle becomes the starting material for metabolic pathways that synthesize other organic compounds, including glucose and other carbohydrates. Neither the light reactions nor the Calvin cycle alone can make sugar from CO_2. Photosynthesis is an emergent property of the intact chloroplast, which integrates the two stages of photosynthesis.

Concept Check 10.3

1. To synthesize one glucose molecule, the Calvin cycle uses ______ molecules of CO_2, ______ molecules of ATP, and ______ molecules of NADPH.
2. Explain why the high number of ATP and NADPH molecules used during the Calvin cycle is consistent with the high value of glucose as an energy source.
3. Explain why a poison that inhibits an enzyme of the Calvin cycle will also inhibit the light reactions.

For suggested answers, see Appendix A.

Concept 10.4

Alternative mechanisms of carbon fixation have evolved in hot, arid climates

Ever since plants first moved onto land about 475 million years ago, they have been adapting to the problems of terrestrial life, particularly the problem of dehydration. In Chapters 29 and 36, we will consider anatomical adaptations that help plants conserve water. Here we are concerned with metabolic adaptations. The solutions often involve trade-offs. An important example is the compromise between photosynthesis and the prevention of excessive water loss from the plant. The CO_2 required for photosynthesis enters a leaf via stomata, the pores through the leaf surface (see Figure 10.3). However, stomata are also the main avenues of transpiration, the evaporative loss of water from leaves. On a hot, dry day, most plants close their stomata, a response that conserves water. This response also reduces photosynthetic yield by limiting access to CO_2. With stomata even partially closed, CO_2 concentrations begin to decrease in the air spaces within the leaf, and the concentration of O_2 released from the light reactions begins to increase. These conditions within the leaf favor a seemingly wasteful process called photorespiration.

Photorespiration: An Evolutionary Relic?

In most plants, initial fixation of carbon occurs via rubisco, the Calvin cycle enzyme that adds CO_2 to ribulose bisphosphate. Such plants are called **C_3 plants** because the first organic product of carbon fixation is a three-carbon compound, 3-phosphoglycerate (see Figure 10.18). Rice, wheat, and soybeans are C_3 plants that are important in agriculture. When their stomata partially close on hot, dry days, C_3 plants produce less sugar because the declining level of CO_2 in the leaf starves the Calvin cycle. In addition, rubisco can bind O_2 in place of CO_2. As CO_2 becomes scarce within the air spaces of the leaf, rubisco adds O_2 to the Calvin cycle instead of CO_2. The product splits, and a two-carbon compound leaves the chloroplast. Peroxisomes and mitochondria rearrange and split this compound, releasing CO_2. The process is called **photorespiration** because it occurs in the light (*photo*) and consumes O_2 while producing CO_2 (*respiration*). However, unlike normal cellular respiration, photorespiration generates no ATP; in fact, photorespiration consumes ATP. And unlike photosynthesis, photorespiration produces no sugar. In fact, photorespiration *decreases* photosynthetic output by siphoning organic material from the Calvin cycle.

How can we explain the existence of a metabolic process that seems to be counterproductive for the plant? According to one hypothesis, photorespiration is evolutionary baggage—a metabolic relic from a much earlier time, when the atmosphere had less O_2 and more CO_2 than it does today. In the ancient atmosphere that prevailed when rubisco first evolved, the inability of the enzyme's active site to exclude O_2 would have made little difference. The hypothesis speculates that modern rubisco retains some of its chance affinity for O_2, which is now so concentrated in the atmosphere that a certain amount of photorespiration is inevitable.

It is not known whether photorespiration is beneficial to plants in any way. It *is* known that in many types of plants—

including crop plants—photorespiration drains away as much as 50% of the carbon fixed by the Calvin cycle. As heterotrophs that depend on carbon fixation in chloroplasts for our food, we naturally view photorespiration as wasteful. Indeed, if photorespiration could be reduced in certain plant species without otherwise affecting photosynthetic productivity, crop yields and food supplies might increase.

In certain plant species, alternate modes of carbon fixation have evolved that minimize photorespiration and optimize the Calvin cycle—even in hot, arid climates. The two most important of these photosynthetic adaptations are C_4 photosynthesis and CAM.

C_4 Plants

The **C_4 plants** are so named because they preface the Calvin cycle with an alternate mode of carbon fixation that forms a four-carbon compound as its first product. Several thousand species in at least 19 plant families use the C_4 pathway. Among the C_4 plants important to agriculture are sugarcane and corn, members of the grass family.

A unique leaf anatomy is correlated with the mechanism of C_4 photosynthesis (**Figure 10.19**; compare with Figure 10.3). In C_4 plants, there are two distinct types of photosynthetic cells: bundle-sheath cells and mesophyll cells. **Bundle-sheath cells** are arranged into tightly packed sheaths around the veins of the leaf. Between the bundle sheath and the leaf surface are the more loosely arranged **mesophyll cells.** The Calvin cycle is confined to the chloroplasts of the bundle sheath. However, the cycle is preceded by incorporation of CO_2 into organic compounds in the mesophyll. The first step, carried out by the enzyme **PEP carboxylase,** is the addition of CO_2 to phosphoenolpyruvate (PEP), forming the four-carbon product oxaloacetate. PEP carboxylase has a much higher affinity for CO_2 than rubisco and no affinity for O_2. Therefore, PEP carboxylase can fix carbon efficiently when rubisco cannot—that is, when it is hot and dry and stomata are partially closed, causing CO_2 concentration in the leaf to fall and O_2 concentration to rise. After the C_4 plant fixes carbon from CO_2, the mesophyll cells export their four-carbon products (malate in the example shown in Figure 10.19) to bundle-sheath cells through plasmodesmata (see Figure 6.30). Within the bundle-sheath cells, the four-carbon compounds release CO_2, which is reassimilated into organic material by rubisco and the Calvin cycle. Pyruvate is also regenerated for conversion to PEP in mesophyll cells.

In effect, the mesophyll cells of a C_4 plant pump CO_2 into the bundle sheath, keeping the CO_2 concentration in the bundle-sheath cells high enough for rubisco to bind carbon dioxide rather than oxygen. The cyclic series of reactions involving PEP carboxylase and the regeneration of PEP can be thought of as a CO_2-concentrating pump that is powered by ATP. In this way, C_4 photosynthesis minimizes photorespiration and enhances sugar production. This adaptation is especially advantageous in hot regions with intense sunlight, where stomata partially close during the day, and it is in such environments that C_4 plants evolved and thrive today.

CAM Plants

A second photosynthetic adaptation to arid conditions has evolved in succulent (water-storing) plants (including jade

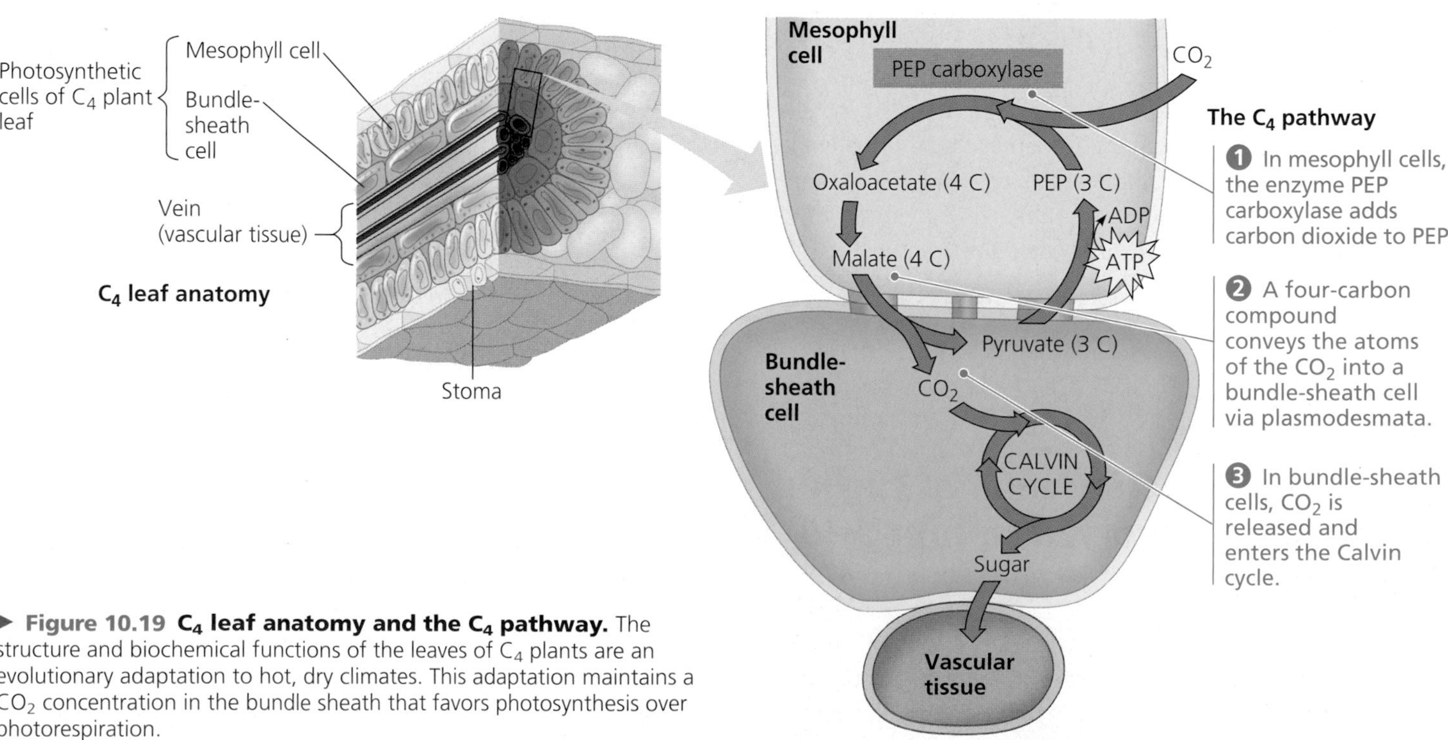

▶ **Figure 10.19 C_4 leaf anatomy and the C_4 pathway.** The structure and biochemical functions of the leaves of C_4 plants are an evolutionary adaptation to hot, dry climates. This adaptation maintains a CO_2 concentration in the bundle sheath that favors photosynthesis over photorespiration.

plants), many cacti, pineapples, and representatives of several other plant families. These plants open their stomata during the night and close them during the day, just the reverse of how other plants behave. Closing stomata during the day helps desert plants conserve water, but it also prevents CO_2 from entering the leaves. During the night, when their stomata are open, these plants take up CO_2 and incorporate it into a variety of organic acids. This mode of carbon fixation is called **crassulacean acid metabolism**, or **CAM**, after the plant family Crassulaceae, the succulents in which the process was first discovered. The mesophyll cells of **CAM plants** store the organic acids they make during the night in their vacuoles until morning, when the stomata close. During the day, when the light reactions can supply ATP and NADPH for the Calvin cycle, CO_2 is released from the organic acids made the night before to become incorporated into sugar in the chloroplasts.

Notice in **Figure 10.20** that the CAM pathway is similar to the C_4 pathway in that carbon dioxide is first incorporated into organic intermediates before it enters the Calvin cycle. The difference is that in C_4 plants, the initial steps of carbon fixation are separated structurally from the Calvin cycle, whereas in CAM plants, the two steps occur at separate times but within the same cell. (Keep in mind that CAM, C_4, and C_3 plants all eventually use the Calvin cycle to make sugar from carbon dioxide.)

Concept Check 10.4

1. Explain why photorespiration lowers photosynthetic output for plants.
2. How would you expect the relative abundance of C_3 versus C_4 and CAM species to change in a geographic region whose climate becomes much hotter and drier?

For suggested answers, see Appendix A.

The Importance of Photosynthesis: *A Review*

In this chapter, we have followed photosynthesis from photons to food. The light reactions capture solar energy and use

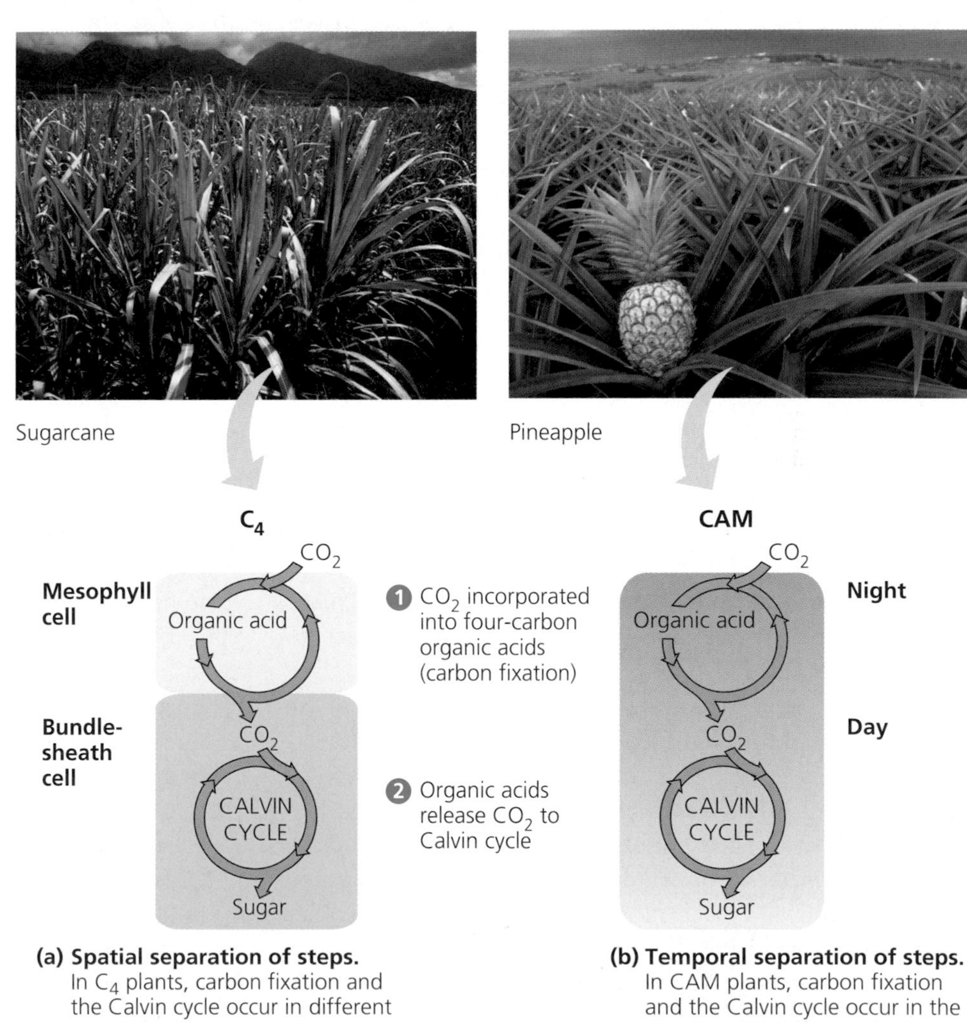

(a) Spatial separation of steps. In C_4 plants, carbon fixation and the Calvin cycle occur in different types of cells.

(b) Temporal separation of steps. In CAM plants, carbon fixation and the Calvin cycle occur in the same cells at different times.

▶ **Figure 10.20 C_4 and CAM photosynthesis compared.** Both adaptations are characterized by 1 preliminary incorporation of CO_2 into organic acids, followed by 2 transfer of CO_2 to the Calvin cycle. The C_4 and CAM pathways are two evolutionary solutions to the problem of maintaining photosynthesis with stomata partially or completely closed on hot, dry days.

▶ **Figure 10.21 A review of photosynthesis.** This diagram outlines the main reactants and products of the light reactions and the Calvin cycle as they occur in the chloroplasts of plant cells. The entire ordered operation depends on the structural integrity of the chloroplast and its membranes. Enzymes in the chloroplast and cytosol convert glyceraldehyde-3-phosphate (G3P), the direct product of the Calvin cycle, into many other organic compounds.

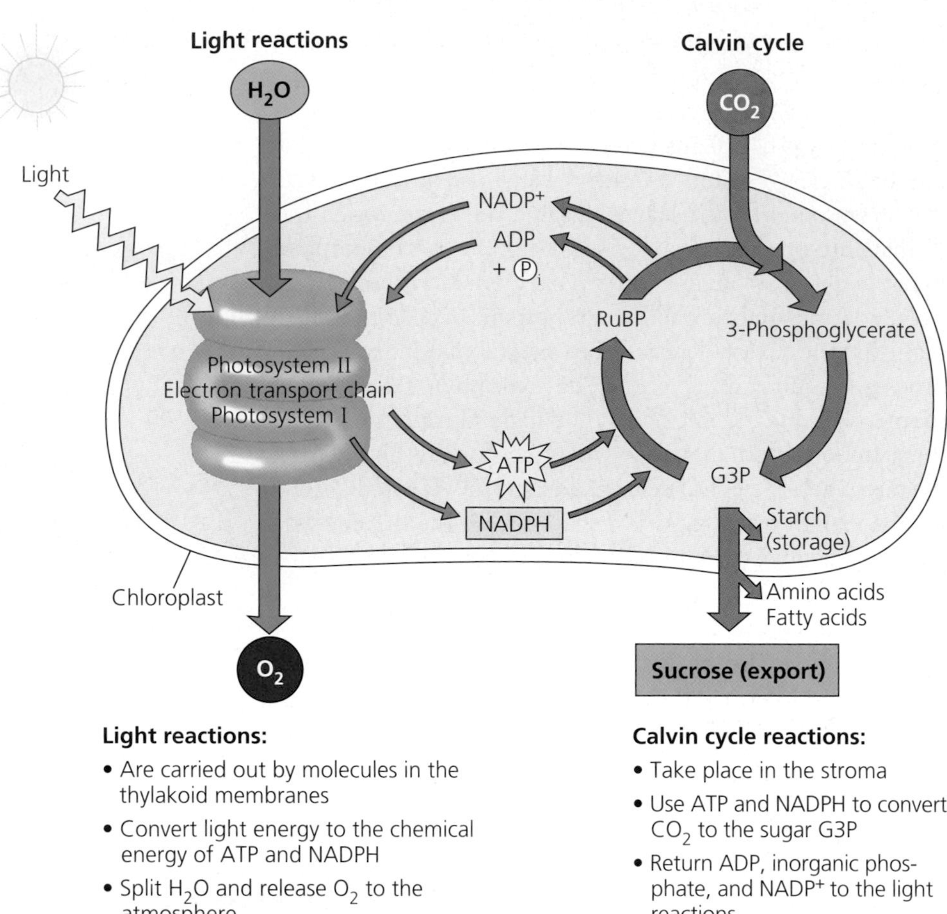

Light reactions:

- Are carried out by molecules in the thylakoid membranes
- Convert light energy to the chemical energy of ATP and NADPH
- Split H_2O and release O_2 to the atmosphere

Calvin cycle reactions:

- Take place in the stroma
- Use ATP and NADPH to convert CO_2 to the sugar G3P
- Return ADP, inorganic phosphate, and $NADP^+$ to the light reactions

it to make ATP and transfer electrons from water to $NADP^+$. The Calvin cycle uses the ATP and NADPH to produce sugar from carbon dioxide. The energy that enters the chloroplasts as sunlight becomes stored as chemical energy in organic compounds. See **Figure 10.21** for a review of the entire process.

What are the fates of photosynthetic products? The sugar made in the chloroplasts supplies the entire plant with chemical energy and carbon skeletons for the synthesis of all the major organic molecules of plant cells. About 50% of the organic material made by photosynthesis is consumed as fuel for cellular respiration in the mitochondria of the plant cells. Sometimes there is a loss of photosynthetic products to photorespiration.

Technically, green cells are the only autotrophic parts of the plant. The rest of the plant depends on organic molecules exported from leaves via veins. In most plants, carbohydrate is transported out of the leaves in the form of sucrose, a disaccharide. After arriving at nonphotosynthetic cells, the sucrose provides raw material for cellular respiration and a multitude of anabolic pathways that synthesize proteins, lipids, and other products. A considerable amount of sugar in the form of glucose is linked together to make the polysaccharide cellulose, especially in plant cells that are still growing and maturing. Cellulose, the main ingredient of cell walls, is the most abundant organic molecule in the plant—and probably on the surface of the planet.

Most plants manage to make more organic material each day than they need to use as respiratory fuel and precursors for biosynthesis. They stockpile the extra sugar by synthesizing starch, storing some in the chloroplasts themselves and some in storage cells of roots, tubers, seeds, and fruits. In accounting for the consumption of the food molecules produced by photosynthesis, let's not forget that most plants lose leaves, roots, stems, fruits, and sometimes their entire bodies to heterotrophs, including humans.

On a global scale, photosynthesis is the process that is responsible for the presence of oxygen in our atmosphere. Furthermore, in terms of food production, the collective productivity of the minute chloroplasts is prodigious; it is estimated that photosynthesis makes about 160 billion metric tons of carbohydrate per year (a metric ton is 1,000 kg, about 1.1 tons). That's organic matter equivalent to a stack of about 60 trillion copies of this textbook—17 stacks of books reaching from Earth to the sun! No other chemical process on the planet can match the output of photosynthesis. And no process is more important than photosynthesis to the welfare of life on Earth.

Chapter 10 Review

Go to www.campbellbiology.com or the student CD-ROM to explore Activities, Investigations, and other interactive study aids.

SUMMARY OF KEY CONCEPTS

▶ Plants and other autotrophs are the producers of the biosphere. Photoautotrophs use the energy of sunlight to make organic molecules from CO_2 and H_2O. Heterotrophs consume organic molecules from other organisms for energy and carbon (p. 181).

Concept 10.1

Photosynthesis converts light energy to the chemical energy of food

▶ **Chloroplasts: The Sites of Photosynthesis in Plants (pp. 182–183)** In autotrophic eukaryotes, photosynthesis occurs in chloroplasts, organelles containing thylakoids. Stacks of thylakoids form grana.
Activity *The Sites of Photosynthesis*

▶ **Tracking Atoms Through Photosynthesis: *Scientific Inquiry* (pp. 183–184)** Photosynthesis is summarized as

$$6\,CO_2 + 12\,H_2O + \text{Light energy} \rightarrow C_6H_{12}O_6 + 6\,O_2 + 6\,H_2O$$

Chloroplasts split water into hydrogen and oxygen, incorporating the electrons of hydrogen into sugar molecules. Photosynthesis is a redox process: H_2O is oxidized, CO_2 is reduced.

▶ **The Two Stages of Photosynthesis: *A Preview* (pp. 184–185)** The light reactions in the grana split water, releasing O_2, producing ATP, and forming NADPH. The Calvin cycle in the stroma forms sugar from CO_2, using ATP for energy and NADPH for reducing power.
Activity *Overview of Photosynthesis*

Concept 10.2

The light reactions convert solar energy to the chemical energy of ATP and NADPH

▶ **The Nature of Sunlight (p. 186)** Light is a form of electromagnetic energy. The colors we see as visible light include those wavelengths that drive photosynthesis.

▶ **Photosynthetic Pigments: The Light Receptors (pp. 186–188)** A pigment absorbs visible light of specific wavelengths. Chlorophyll *a* is the main photosynthetic pigment in plants. Accessory pigments absorb different wavelengths of light and pass the energy on to chlorophyll *a*.
Activity *Light Energy and Pigments*
Investigation *How Does Paper Chromatography Separate Plant Pigments?*

▶ **Excitation of Chlorophyll by Light (p. 188)** A pigment goes from a ground state to an excited state when a photon boosts one of its electrons to a higher-energy orbital. This excited state is unstable. Electrons from isolated pigments tend to fall back to the ground state, giving off heat and/or light.

▶ **A Photosystem: A Reaction Center Associated with Light-Harvesting Complexes (pp. 189–190)** A photosystem is composed of a reaction center surrounded by light-harvesting complexes that funnel the energy of photons to the reaction center. When a reaction-center chlorophyll *a* molecule absorbs energy, one of its electrons gets bumped up to the primary electron acceptor. Photosystem I contains P700 chlorophyll *a* molecules at the reaction center; photosystem II contains P680 molecules.

▶ **Noncyclic Electron Flow (pp. 190–191)** Noncyclic electron flow produces NADPH, ATP, and oxygen.

▶ **Cyclic Electron Flow (pp. 191–192)** Cyclic electron flow employs only photosystem I, producing ATP but no NADPH or O_2.

▶ **A Comparison of Chemiosmosis in Chloroplasts and Mitochondria (pp. 192–193)** In both organelles, the redox reactions of electron transport chains generate an H^+ gradient across a membrane. ATP synthase uses this proton-motive force to make ATP.
Activity *The Light Reactions*

Concept 10.3

The Calvin cycle uses ATP and NADPH to convert CO_2 to sugar

▶ The Calvin cycle occurs in the stroma and consists of carbon fixation, reduction, and regeneration of the CO_2 acceptor. Using electrons from NADPH and energy from ATP, the cycle synthesizes a three-carbon sugar (G3P). Most of the G3P is reused in the cycle, but some exits the cycle and is converted to glucose and other organic molecules (pp. 193–195).
Activity *The Calvin Cycle*
Investigation *How Is the Rate of Photosynthesis Measured?*
Biology Labs On-Line *LeafLab*

Concept 10.4

Alternative mechanisms of carbon fixation have evolved in hot, arid climates

▶ **Photorespiration: An Evolutionary Relic? (pp. 195–196)** On dry, hot days, plants close their stomata, conserving water. Oxygen from the light reactions builds up. In photorespiration, O_2 substitutes for CO_2 in the active site of rubisco. This process consumes organic fuel and releases CO_2 without producing ATP or sugar.

▶ **C_4 Plants (p. 196)** C_4 plants minimize the cost of photorespiration by incorporating CO_2 into four-carbon compounds in mesophyll cells. These compounds are exported to bundle-sheath cells, where they release carbon dioxide for use in the Calvin cycle.

▶ **CAM Plants (pp. 196–197)** CAM plants open their stomata at night, incorporating CO_2 into organic acids, which are stored in mesophyll cells. During the day the stomata close, and the CO_2 is released from the organic acids for use in the Calvin cycle.
Activity *Photosynthesis in Dry Climates*

▶ **The Importance of Photosynthesis: *A Review* (pp. 197–198)** Organic compounds produced by photosynthesis provide the energy and building material for ecosystems.

TESTING YOUR KNOWLEDGE

Self-Quiz

1. The light reactions of photosynthesis supply the Calvin cycle with
   a. light energy.
   b. CO_2 and ATP.
   c. H_2O and NADPH.
   d. ATP and NADPH.
   e. sugar and O_2.

2. Which of the following sequences correctly represents the flow of electrons during photosynthesis?
   a. NADPH → O_2 → CO_2
   b. H_2O → NADPH → Calvin cycle
   c. NADPH → chlorophyll → Calvin cycle
   d. H_2O → photosystem I → photosystem II
   e. NADPH → electron transport chain → O_2

3. Which of the following conclusions does *not* follow from studying the absorption spectrum for chlorophyll *a* and the action spectrum for photosynthesis (see Figure 10.9a and b)?
   a. Not all wavelengths are equally effective for photosynthesis.
   b. There must be accessory pigments that broaden the spectrum of light that contributes to photosynthesis.
   c. The red and blue areas of the spectrum are most effective in driving photosynthesis.
   d. Chlorophyll owes its color to the absorption of green light.
   e. Chlorophyll *a* has two absorption peaks.

4. Cooperation of the *two* photosystems is required for
   a. ATP synthesis.
   b. reduction of $NADP^+$.
   c. cyclic photophosphorylation.
   d. oxidation of the reaction center of photosystem I.
   e. generation of a proton-motive force.

5. In *mechanism,* photophosphorylation is most similar to
   a. substrate-level phosphorylation in glycolysis.
   b. oxidative phosphorylation in cellular respiration.
   c. the Calvin cycle.
   d. carbon fixation.
   e. reduction of $NADP^+$.

6. In what respect are the photosynthetic adaptations of C_4 plants and CAM plants similar?
   a. In both cases, only photosystem I is used.
   b. Both types of plants make sugar without the Calvin cycle.
   c. In both cases, an enzyme other than rubisco carries out the first step in carbon fixation.
   d. Both types of plants make most of their sugar in the dark.
   e. Neither C_4 plants nor CAM plants have thylakoids.

7. Which of the following processes is most directly driven by light energy?
   a. creation of a pH gradient by pumping protons across the thylakoid membrane
   b. carbon fixation in the stroma
   c. reduction of $NADP^+$ molecules
   d. removal of electrons from chlorophyll molecules
   e. ATP synthesis

8. Which of the following statements is a correct distinction between cyclic and noncyclic electron flow?
   a. Only noncyclic electron flow produces ATP.
   b. In addition to ATP, cyclic electron flow also produces O_2 and NADPH.
   c. Only cyclic electron flow utilizes light at 700 nm.
   d. Chemiosmosis is unique to noncyclic electron flow.
   e. Only cyclic electron flow can operate in the absence of photosystem II.

9. Which of the following statements is a correct distinction between autotrophs and heterotrophs?
   a. Only heterotrophs require chemical compounds from the environment.
   b. Cellular respiration is unique to heterotrophs.
   c. Only heterotrophs have mitochondria.
   d. Autotrophs, but not heterotrophs, can nourish themselves beginning with CO_2 and other nutrients that are inorganic.
   e. Only heterotrophs require oxygen.

10. Which of the following does *not* occur during the Calvin cycle?
    a. carbon fixation
    b. oxidation of NADPH
    c. release of oxygen
    d. regeneration of the CO_2 acceptor
    e. consumption of ATP

For Self-Quiz answers, see Appendix A.

Go to the website or CD-ROM for more quiz questions.

Evolution Connection

Photorespiration can substantially decrease soybeans' photosynthetic output by about 50%. Would you expect this figure to be higher or lower in wild relatives of soybeans? Why?

Scientific Inquiry

The diagram below represents an experiment with isolated chloroplasts. The chloroplasts were first made acidic by soaking them in a solution at pH 4. After the thylakoid space reached pH 4, the chloroplasts were transferred to a basic solution at pH 8. The chloroplasts then made ATP in the dark. Explain this result.

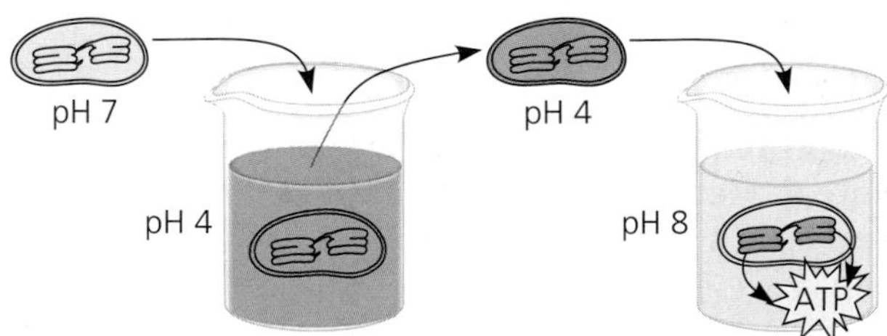

Investigation ***How Does Paper Chromatography Separate Plant Pigments?***
Investigation ***How Is the Rate of Photosynthesis Measured?***
Biology Labs On-Line ***LeafLab***

Science, Technology, and Society

CO_2 in the atmosphere traps heat and warms the air, just as clear glass does in a greenhouse. Scientific evidence indicates that the CO_2 added to the air by the burning of wood and fossil fuels is contributing to a rise in global temperature. Tropical rain forests are estimated to be responsible for more than 20% of global photosynthesis. It seems reasonable to expect that the rain forests would reduce global warming by consuming large amounts of CO_2, but many experts now think that rain forests make little or no *net* contribution to reduction of global warming. Why might this be? (*Hint:* What happens to the food produced by a rain forest tree when it is eaten by animals or the tree dies?)

11 Cell Communication

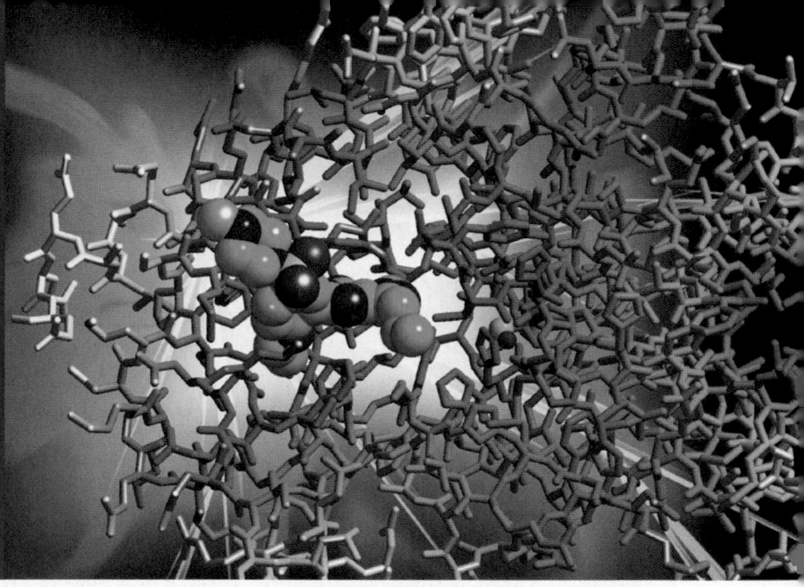

▲ **Figure 11.1 Viagra (multicolored) bound to an enzyme (purple) involved in a signaling pathway.**

Key Concepts

11.1 External signals are converted into responses within the cell
11.2 Reception: A signal molecule binds to a receptor protein, causing it to change shape
11.3 Transduction: Cascades of molecular interactions relay signals from receptors to target molecules in the cell
11.4 Response: Cell signaling leads to regulation of cytoplasmic activities or transcription

Overview

The Cellular Internet

A hiker slips and falls down a steep ravine, injuring her leg in the fall. Tragedy is averted when she is able to pull out a cell phone and call for help. Cell phones, the Internet, e-mail, instant messaging—no one would deny the importance of communication in our lives. The role of communication in life at the cellular level is equally critical. Cell-to-cell communication is absolutely essential for multicellular organisms such as humans and oak trees. The trillions of cells in a multicellular organism must communicate with each other to coordinate their activities in a way that enables the organism to develop from a fertilized egg, then survive and reproduce in turn. Communication between cells is also important for many unicellular organisms. Networks of communication between cells can be even more complicated than the World Wide Web.

In studying how cells signal to each other and how they interpret the signals they receive, biologists have discovered some universal mechanisms of cellular regulation, additional evidence for the evolutionary relatedness of all life. The same small set of cell-signaling mechanisms shows up again and again in many lines of biological research—from embryonic development to hormone action to cancer. In one example, a common cell-to-cell signaling pathway leads to dilation of blood vessels. Once the signal subsides, the response is shut down by the enzyme shown in purple in **Figure 11.1**. Also shown is a multicolored molecule that blocks the action of this enzyme and keeps blood vessels dilated. Enzyme-inhibiting compounds like this one are often prescribed for treatment of medical conditions. The action of the multicolored compound, known as Viagra, will be discussed later in the chapter. The signals received by cells, whether originating from other cells or from changes in the physical environment, take various forms, including light and touch. However, cells most often communicate with each other by chemical signals. In this chapter, we focus on the main mechanisms by which cells receive, process, and respond to chemical signals sent from other cells.

Concept 11.1

External signals are converted into responses within the cell

What does a "talking" cell say to a "listening" cell, and how does the latter cell respond to the message? Let's approach these questions by first looking at communication among microorganisms, for modern microbes are a window on the role of cell signaling in the evolution of life on Earth.

Evolution of Cell Signaling

One topic of cell "conversation" is sex—at least for the yeast *Saccharomyces cerevisiae,* which people have used for millennia to make bread, wine, and beer. Researchers have learned that cells of this yeast identify their mates by chemical signaling. There are two sexes, or mating types, called **a** and α (**Figure 11.2**). Cells of mating type **a** secrete a chemical signal called **a** factor, which can bind to specific receptor proteins on nearby α cells. At the same time, α cells secrete α factor, which binds to receptors on **a** cells. Without actually entering the cells, the two mating factors cause the cells to grow toward each other and bring about other cellular changes. The result is the fusion, or mating, of two cells of opposite type. The new **a**/α cell contains all the genes of both original cells, a combination of genetic resources that provides advantages to the cell's descendants, which arise by subsequent cell divisions.

How is the mating signal at the yeast cell surface changed, or *transduced*, into a form that brings about the cellular response of mating? The process by which a signal on a cell's surface is converted into a specific cellular response is a series of steps called a **signal transduction pathway.** Many such pathways have been extensively studied in both yeast and animal cells. Amazingly, the molecular details of signal transduction in yeast and mammals are strikingly similar, even though the last common ancestor of these two groups of organisms lived over a billion years ago. These similarities—and others more recently uncovered between signaling systems in bacteria and plants—suggest that early versions of the cell-signaling mechanisms used today evolved well before the first multicellular creatures appeared on Earth. Scientists think that signaling mechanisms evolved first in ancient prokaryotes and single-celled eukaryotes and were then adopted for new uses by their multicellular descendants.

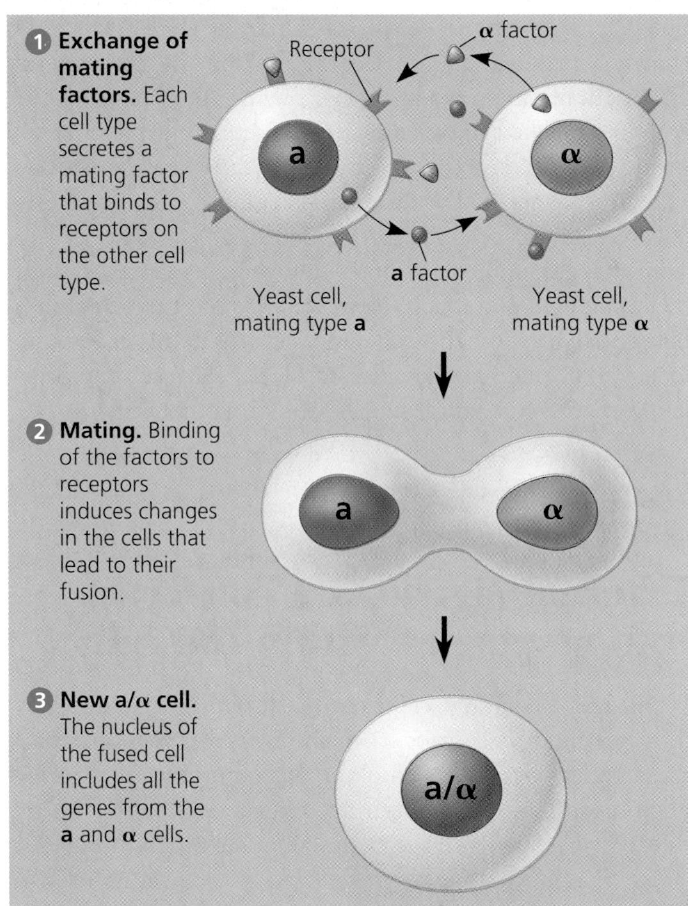

1 **Exchange of mating factors.** Each cell type secretes a mating factor that binds to receptors on the other cell type.
Receptor
α factor
a
α
a factor
Yeast cell, mating type **a**
Yeast cell, mating type α
2 **Mating.** Binding of the factors to receptors induces changes in the cells that lead to their fusion.
a
α
3 **New a/α cell.** The nucleus of the fused cell includes all the genes from the **a** and α cells.
a/α

▲ **Figure 11.2 Communication between mating yeast cells.** *Saccharomyces cerevisiae* cells use chemical signaling to identify cells of opposite mating type and initiate the mating process. The two mating types and their corresponding chemical signals, or mating factors, are called **a** and α.

Local and Long-Distance Signaling

Like yeast cells, cells in a multicellular organism usually communicate via chemical messengers targeted for cells that may or may not be immediately adjacent. Cells may communicate by direct contact, as we saw in Chapters 6 and 7. Both animals and plants have cell junctions that, where present, directly connect the cytoplasms of adjacent cells (**Figure 11.3a**). In these cases, signaling substances dissolved in the cytosol can pass freely between adjacent cells. Moreover, animal cells may communicate via direct contact between membrane-bound cell surface molecules (**Figure 11.3b**). This sort of signaling, called cell-cell recognition, is important in such processes as embryonic development and the immune response.

In many other cases, messenger molecules are secreted by the signaling cell. Some of these travel only short distances; such **local regulators** influence cells in the vicinity. One class of local regulators in animals, *growth factors*, are compounds

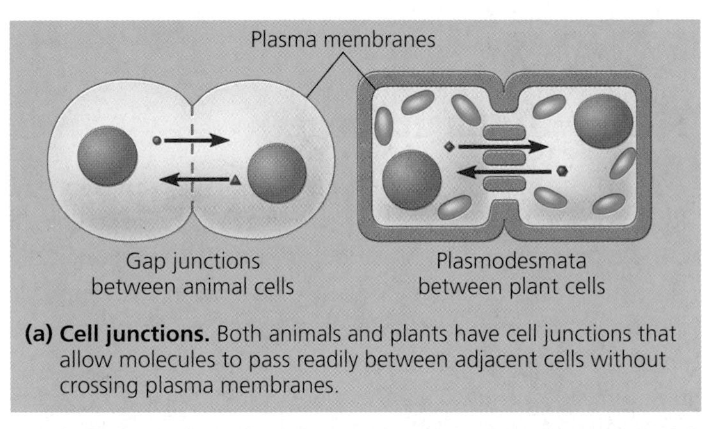

(a) Cell junctions. Both animals and plants have cell junctions that allow molecules to pass readily between adjacent cells without crossing plasma membranes.

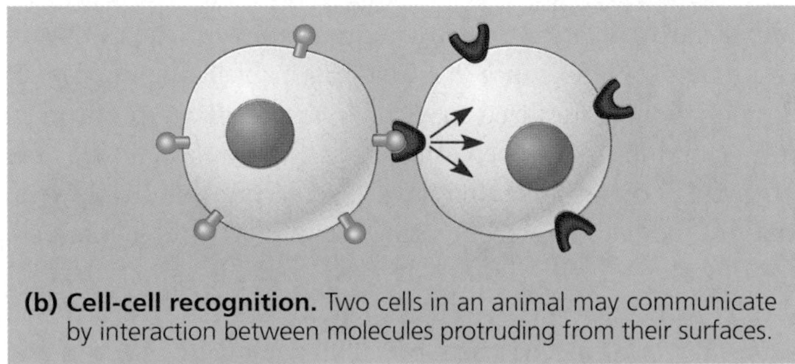

(b) Cell-cell recognition. Two cells in an animal may communicate by interaction between molecules protruding from their surfaces.

▲ **Figure 11.3 Communication by direct contact between cells.**

Local signaling

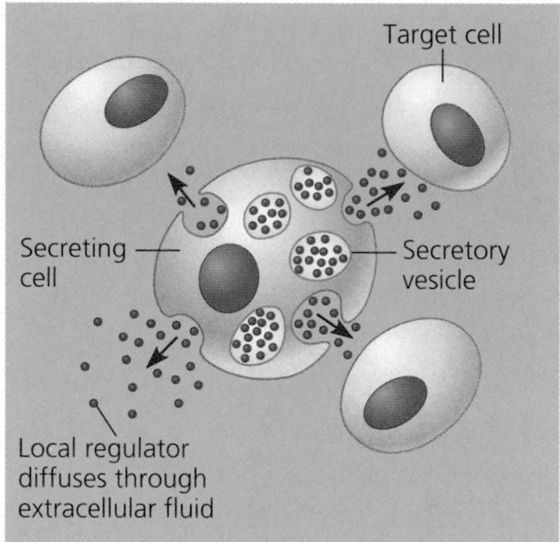

(a) Paracrine signaling. A secreting cell acts on nearby target cells by discharging molecules of a local regulator (a growth factor, for example) into the extracellular fluid.

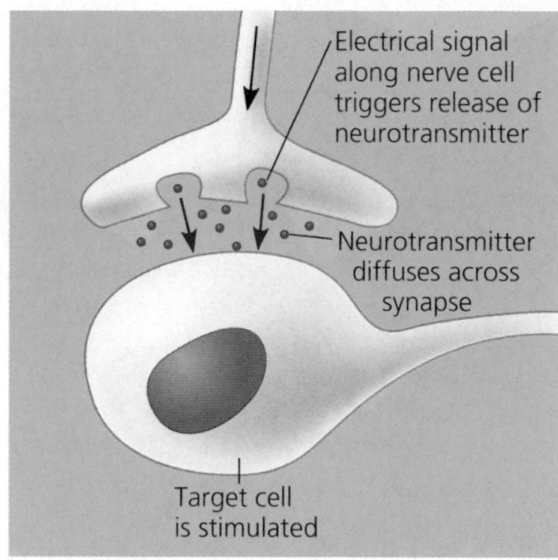

(b) Synaptic signaling. A nerve cell releases neurotransmitter molecules into a synapse, stimulating the target cell.

Long-distance signaling

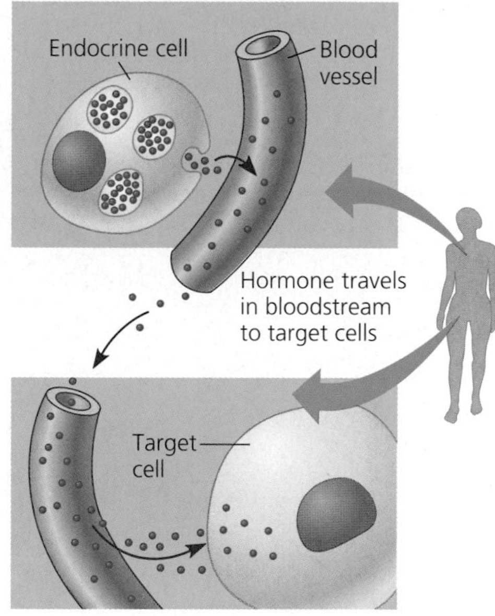

(c) Hormonal signaling. Specialized endocrine cells secrete hormones into body fluids, often the blood. Hormones may reach virtually all body cells.

▲ **Figure 11.4 Local and long-distance cell communication in animals.** In both local and long-distance signaling, only specific target cells recognize and respond to a given chemical signal.

that stimulate nearby target cells to grow and multiply. Numerous cells can simultaneously receive and respond to the molecules of growth factor produced by a single cell in their vicinity. This type of local signaling in animals is called *paracrine signaling* **(Figure 11.4a)**.

Another, more specialized type of local signaling called *synaptic signaling* occurs in the animal nervous system. An electrical signal along a nerve cell triggers the secretion of a chemical signal in the form of neurotransmitter molecules. These diffuse across the synapse, the narrow space between the nerve cell and its target cell (often another nerve cell). The neurotransmitter stimulates the target cell **(Figure 11.4b)**.

Local signaling in plants is not as well understood. Because of their cell walls, plants must use mechanisms somewhat different from those operating locally in animals.

Both animals and plants use chemicals called **hormones** for long-distance signaling. In hormonal signaling in animals, also known as endocrine signaling, specialized cells release hormone molecules into vessels of the circulatory system, by which they travel to target cells in other parts of the body **(Figure 11.4c)**. Plant hormones (often called *growth regulators*) sometimes travel in vessels but more often reach their targets by moving through cells (see Chapter 39) or by diffusion through the air as a gas. Hormones vary widely in molecular size and type, as do local regulators. For instance, the plant hormone ethylene, a gas that promotes fruit ripening and helps regulate growth, is a hydrocarbon of only six atoms (C_2H_4) that can pass through cell walls. In contrast, the mammalian hormone insulin, which regulates sugar levels in the blood, is a protein with thousands of atoms.

The transmission of a signal through the nervous system can also be considered an example of long-distance signaling. An electrical signal travels the length of a nerve cell and is then converted back to a chemical signal that crosses the synapse to another nerve cell. Here it is converted back into an electrical signal. In this way, a nerve signal can travel along a series of nerve cells. Since some nerve cells are quite long, the nerve signal can quickly travel great distances—from your brain to your big toe, for example. This type of long-distance signaling will be covered in detail in Chapter 48.

What happens when a cell encounters a signal? The signal must be recognized by a specific receptor molecule, and the information it carries must be changed into another form—transduced—inside the cell before the cell can respond. The remainder of the chapter discusses this process, primarily as it occurs in animal cells.

The Three Stages of Cell Signaling: *A Preview*

Our current understanding of how chemical messengers act via signal transduction pathways had its origins in the pioneering work of Earl W. Sutherland, whose research led to a Nobel Prize in 1971. Sutherland and his colleagues at Vanderbilt University were investigating how the animal hormone epinephrine stimulates the breakdown of the storage polysaccharide glycogen within liver cells and skeletal muscle cells. Glycogen

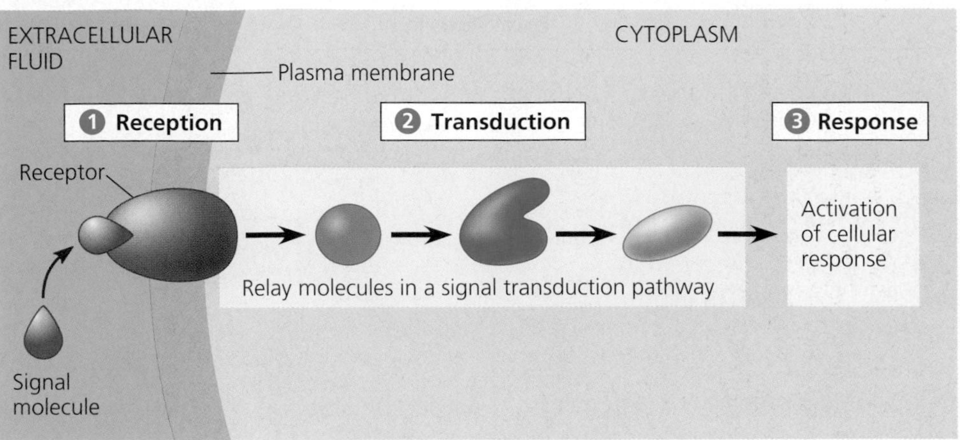

▶ **Figure 11.5 Overview of cell signaling.** From the perspective of the cell receiving the message, cell signaling can be divided into three stages: signal reception, signal transduction, and cellular response. When reception occurs at the plasma membrane, as shown here, the transduction stage is usually a pathway of several steps, with each molecule in the pathway bringing about a change in the next molecule. The last molecule in the pathway triggers the cell's response. The three stages are explained in the text.

breakdown releases the sugar glucose-1-phosphate, which the cell converts to glucose-6-phosphate. The cell (a liver cell, for example) can then use this compound, an early intermediate in glycolysis, for energy production. Alternatively, the compound can be stripped of phosphate and released from the liver cell into the blood as glucose, which can fuel cells throughout the body. Thus, one effect of epinephrine, which is secreted from the adrenal gland during times of physical or mental stress, is the mobilization of fuel reserves.

Sutherland's research team discovered that epinephrine stimulates glycogen breakdown by somehow activating a cytosolic enzyme, glycogen phosphorylase. However, when epinephrine was added to a test-tube mixture containing the enzyme and its substrate, glycogen, no breakdown occurred. Epinephrine could activate glycogen phosphorylase only when the hormone was added to a solution containing *intact* cells. This result told Sutherland two things. First, epinephrine does not interact directly with the enzyme responsible for glycogen breakdown; an intermediate step or series of steps must be occurring inside the cell. Second, the plasma membrane is somehow involved in transmitting the epinephrine signal.

Sutherland's early work suggested that the process going on at the receiving end of a cellular conversation can be dissected into three stages: reception, transduction, and response **(Figure 11.5)**:

1. **Reception.** Reception is the target cell's detection of a signal molecule coming from outside the cell. A chemical signal is "detected" when it binds to a receptor protein located at the cell's surface or inside the cell.
2. **Transduction.** The binding of the signal molecule changes the receptor protein in some way, initiating the process of transduction. The transduction stage converts the signal to a form that can bring about a specific cellular response. In Sutherland's system, the binding of epinephrine to a receptor protein in a liver cell's plasma membrane leads to activation of glycogen phosphorylase. Transduction sometimes occurs in a single step but more often requires a sequence of changes in a series of different molecules—a *signal transduction pathway.* The molecules in the pathway are often called relay molecules.
3. **Response.** In the third stage of cell signaling, the transduced signal finally triggers a specific cellular response. The response may be almost any imaginable cellular activity—such as catalysis by an enzyme (for example, glycogen phosphorylase), rearrangement of the cytoskeleton, or activation of specific genes in the nucleus. The cell-signaling process helps ensure that crucial activities like these occur in the right cells, at the right time, and in proper coordination with the other cells of the organism. We'll now explore the mechanisms of cell signaling in more detail.

Concept Check 11.1

1. Explain how nerve cells provide examples of both local and long-distance signaling.
2. When epinephrine is mixed with glycogen phosphorylase and glycogen in a test tube, is glucose-1-phosphate generated? Why or why not?

For suggested answers, see Appendix A.

Concept 11.2

Reception: A signal molecule binds to a receptor protein, causing it to change shape

When we speak to someone, others nearby may hear our message, sometimes with unfortunate consequences. However, errors of this kind rarely occur among cells. The signals emitted by an **a** yeast cell are "heard" only by its prospective mates, α cells. Similarly, although epinephrine encounters many types of cells as it circulates in the blood, only certain target cells detect and react to the hormone. A receptor protein on or in the target cell allows the cell to "hear" the signal and respond to it. The signal molecule is complementary in shape to a specific site on the receptor and attaches there, like a key in a lock or a substrate in the catalytic site of an enzyme. The signal molecule behaves as a **ligand**, the term for a molecule that specifically binds to another molecule, often a larger one. Ligand binding generally causes a receptor protein to undergo a change in conformation—that is, to change shape. For many receptors, this shape change directly activates the receptor, enabling it to interact with other cellular molecules. For other kinds of receptors, the immediate effect of ligand binding is to cause the aggregation of two or more receptor molecules, which leads to further molecular events inside the cell.

Most signal receptors are plasma membrane proteins. Their ligands are water-soluble and generally too large to pass freely through the plasma membrane. Other signal receptors, however, are located inside the cell. We discuss these next, before returning to membrane receptors.

Intracellular Receptors

Intracellular receptor proteins are found in either the cytoplasm or nucleus of target cells. To reach such a receptor, a chemical messenger passes through the target cell's plasma membrane. A number of important signaling molecules can do this because they are either hydrophobic enough or small enough to cross the phospholipid interior of the membrane. Such hydrophobic chemical messengers include the steroid hormones and thyroid hormones of animals. Another chemical signal with an intracellular receptor is nitric oxide (NO), a gas; its very small molecules readily pass between the membrane phospholipids.

The behavior of testosterone is representative of steroid hormones. Secreted by cells of the testis, the hormone travels through the blood and enters cells all over the body. In the cytoplasm of target cells, the only cells that contain receptor molecules for testosterone, the hormone binds to the receptor protein, activating it (**Figure 11.6**). With the hormone attached, the active form of the receptor protein then enters the nucleus and turns on specific genes that control male sex characteristics.

How does the activated hormone-receptor complex turn on genes? Recall that the genes in a cell's DNA function by being transcribed and processed into messenger RNA (mRNA), which leaves the nucleus and is translated into a specific protein by ribosomes in the cytoplasm (see Figure 5.25). Special proteins called *transcription factors* control which genes are turned on—that is, which genes are transcribed into mRNA—in a particular cell at a particular time. The testosterone receptor, when activated, acts as a transcription factor that turns on specific genes.

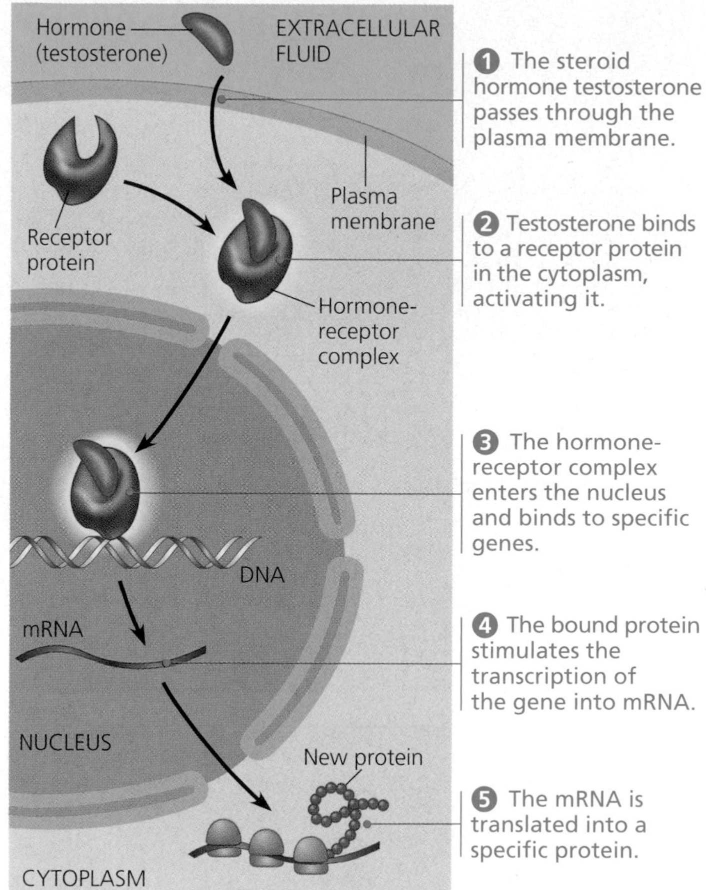
Hormone (testosterone)
EXTRACELLULAR FLUID
Receptor protein
Plasma membrane
Hormone-receptor complex
DNA
mRNA
NUCLEUS
New protein
CYTOPLASM
1 The steroid hormone testosterone passes through the plasma membrane.
2 Testosterone binds to a receptor protein in the cytoplasm, activating it.
3 The hormone-receptor complex enters the nucleus and binds to specific genes.
4 The bound protein stimulates the transcription of the gene into mRNA.
5 The mRNA is translated into a specific protein.

▲ **Figure 11.6 Steroid hormone interacting with an intracellular receptor.**

By acting as a transcription factor, the testosterone receptor itself carries out the complete transduction of the signal. Most other intracellular receptors function in the same way, although many of them are already in the nucleus before the signal molecule reaches them (an example is the thyroid hormone receptor). Interestingly, many of these intracellular receptor proteins are structurally similar, suggesting an evolutionary kinship. We will look more closely at hormones with intracellular receptors in Chapter 45.

Receptors in the Plasma Membrane

Most water-soluble signal molecules bind to specific sites on receptor proteins embedded in the cell's plasma membrane. Such a receptor transmits information from the extracellular environment to the inside of the cell by changing shape or aggregating when a specific ligand binds to it. We can see how membrane receptors work by looking at three major types: G-protein-linked receptors, receptor tyrosine kinases, and ion channel receptors. These receptors are discussed and illustrated in **Figure 11.7** on the next three pages; please study this figure before going on.

Figure 11.7

Exploring Membrane Receptors

G-PROTEIN-LINKED RECEPTORS

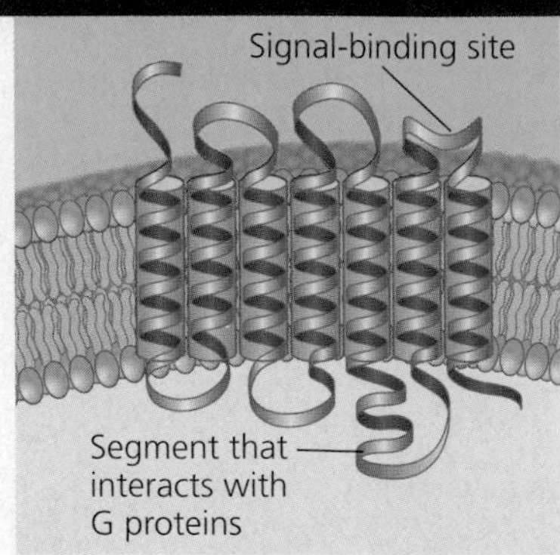

G-protein-linked receptor

A **G-protein-linked receptor** is a plasma membrane receptor that works with the help of a protein called a **G protein.** Many different signal molecules use G-protein-linked receptors, including yeast mating factors, epinephrine and many other hormones, and neurotransmitters. These receptors vary in their binding sites for recognizing signal molecules and for recognizing different G proteins inside the cell. Nevertheless, G-protein-linked receptor proteins are all remarkably similar in structure. They each have seven α helices spanning the membrane, as shown above.

A large family of eukaryotic receptor proteins has this secondary structure, where the single polypeptide, represented here as a ribbon, has seven transmembrane α helices, represented as cylinders and depicted in a row for clarity. Specific loops between the helices form binding sites for signal and G-protein molecules.

G-protein-linked receptor systems are extremely widespread and diverse in their functions, including roles in embryonic development and sensory reception. In humans, for example, both vision and smell depend on such proteins. Similarities in structure among G proteins and G-protein-linked receptors of modern organisms suggest that G proteins and associated receptors evolved very early.

G-protein systems are involved in many human diseases, including bacterial infections. The bacteria that cause cholera, pertussis (whooping cough), and botulism, among others, make their victims ill by producing toxins that interfere with G-protein function. Pharmacologists now realize that up to 60% of all medicines used today exert their effects by influencing G-protein pathways.

1 Loosely attached to the cytoplasmic side of the membrane, the G protein functions as a molecular switch that is either on or off, depending on which of two guanine nucleotides is attached, GDP or GTP—hence the term G protein. (GTP, or guanosine triphosphate, is similar to ATP.) When GDP is bound to the G protein, as shown above, the G protein is inactive. The receptor and G protein work together with another protein, usually an enzyme.

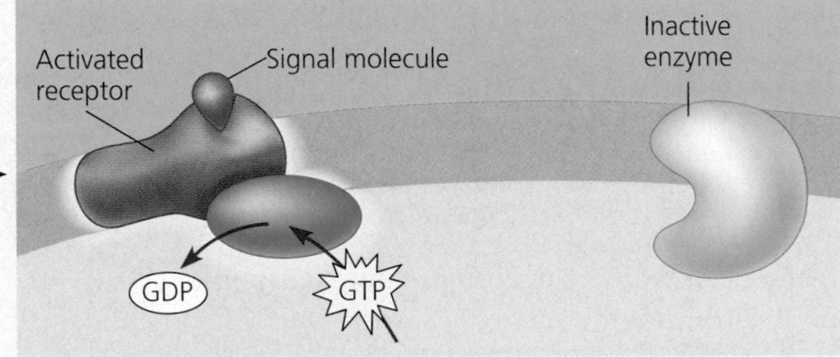

2 When the appropriate signal molecule binds to the extracellular side of the receptor, the receptor is activated and changes shape. Its cytoplasmic side then binds an inactive G protein, causing a GTP to displace the GDP. This activates the G protein.

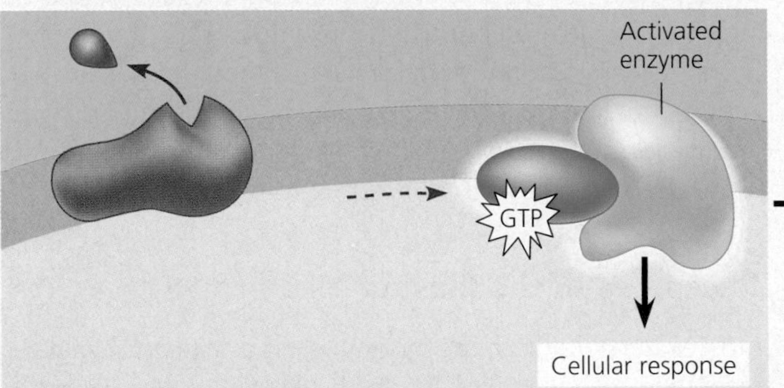

3 The activated G protein dissociates from the receptor and diffuses along the membrane, then binds to an enzyme and alters *its* activity. When the enzyme is activated, it can trigger the next step in a pathway leading to a cellular response.

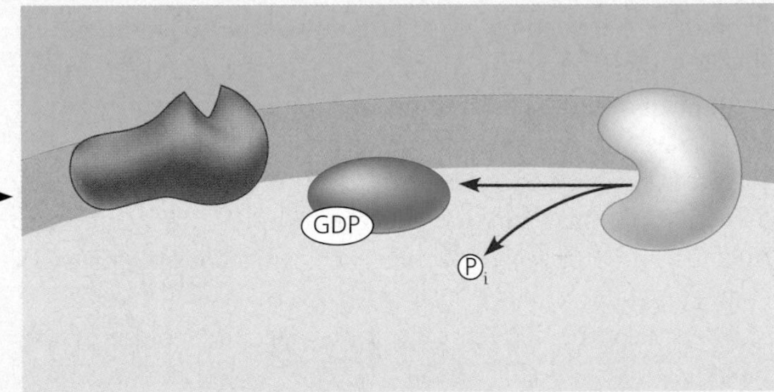

4 The changes in the enzyme and G protein are only temporary, because the G protein also functions as a GTPase enzyme and soon hydrolyzes its bound GTP to GDP. Now inactive again, the G protein leaves the enzyme, which returns to its original state. The G protein is now available for reuse. The GTPase function of the G protein allows the pathway to shut down rapidly when the signal molecule is no longer present.

Continued on next page

RECEPTOR TYROSINE KINASES

A **receptor tyrosine kinase** can trigger more than one signal transduction pathway at once, helping the cell regulate and coordinate many aspects of cell growth and cell reproduction. This receptor is one of a major class of plasma membrane receptors characterized by having enzymatic activity. A *kinase* is an enzyme that catalyzes the transfer of phosphate groups. The part of the receptor protein extending into the cytoplasm functions as a tyrosine kinase, an enzyme that catalyzes the transfer of a phosphate group from ATP to the amino acid tyrosine on a substrate protein. Thus, receptor tyrosine kinases are membrane receptors that attach phosphates to tyrosines.

One receptor tyrosine kinase complex may activate ten or more different transduction pathways and cellular responses. The ability of a single ligand-binding event to trigger so many pathways is a key difference between receptor tyrosine kinases and G-protein-linked receptors. Abnormal receptor tyrosine kinases that function even in the absence of signal molecules may contribute to some kinds of cancer.

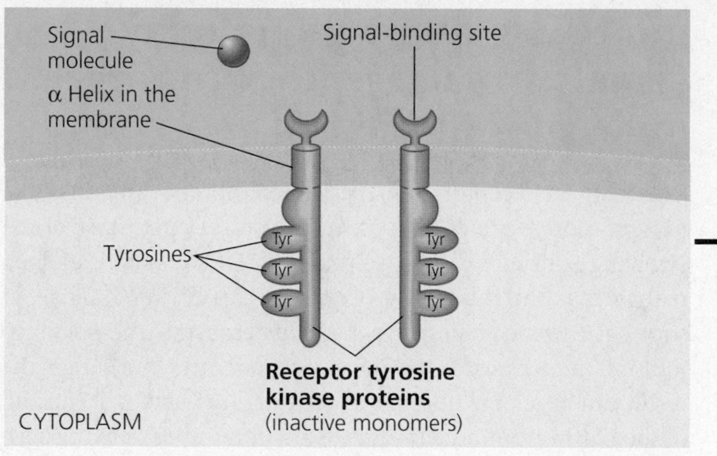

❶ Many receptor tyrosine kinases have the structure depicted schematically here. Before the signal molecule binds, the receptors exist as individual polypeptides. Notice that each has an extracellular signal-binding site, an α helix spanning the membrane, and an intracellular tail containing multiple tyrosines.

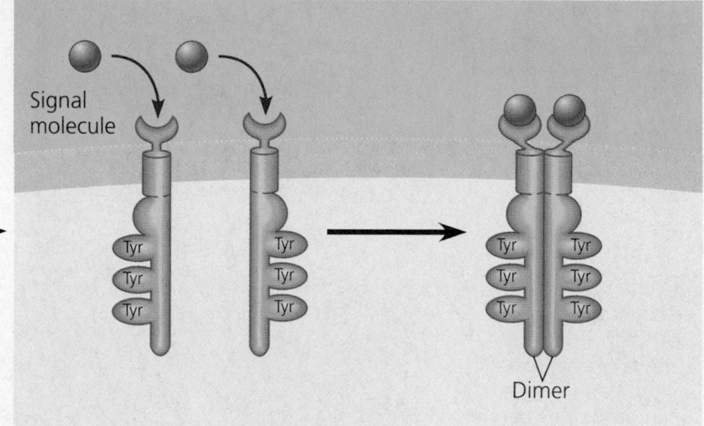

❷ The binding of a signal molecule (such as a growth factor) causes two receptor polypeptides to associate closely with each other, forming a dimer (dimerization).

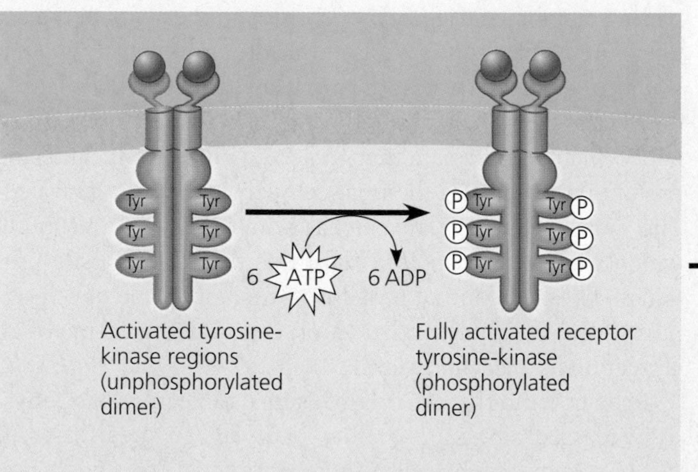

❸ Dimerization activates the tyrosine-kinase region of each polypeptide; each tyrosine kinase adds a phosphate from an ATP molecule to a tyrosine on the tail of the other polypeptide.

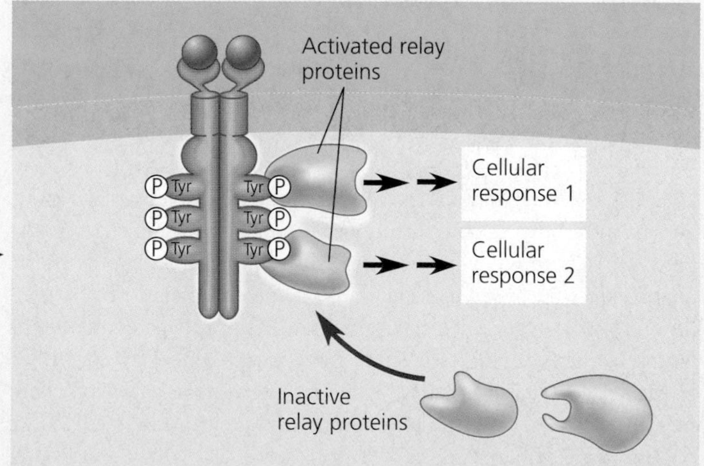

❹ Now that the receptor protein is fully activated, it is recognized by specific relay proteins inside the cell. Each such protein binds to a specific phosphorylated tyrosine, undergoing a resulting structural change that activates the bound protein. Each activated protein triggers a transduction pathway, leading to a cellular response.

Continued on next page

Figure 11.7 (continued)

Exploring Membrane Receptors

ION CHANNEL RECEPTORS

A **ligand-gated ion channel** is a type of membrane receptor, a region of which can act as a "gate" when the receptor changes shape. When a signal molecule binds as a ligand to the receptor protein, the gate opens or closes, allowing or blocking the flow of specific ions, such as Na^+ or Ca^{2+}, through a channel in the receptor. Like the other receptors we have discussed, these proteins bind the ligand at a specific site on their extracellular side.

1 Here we show a ligand-gated ion channel receptor in which the gate remains closed until a ligand binds to the receptor.

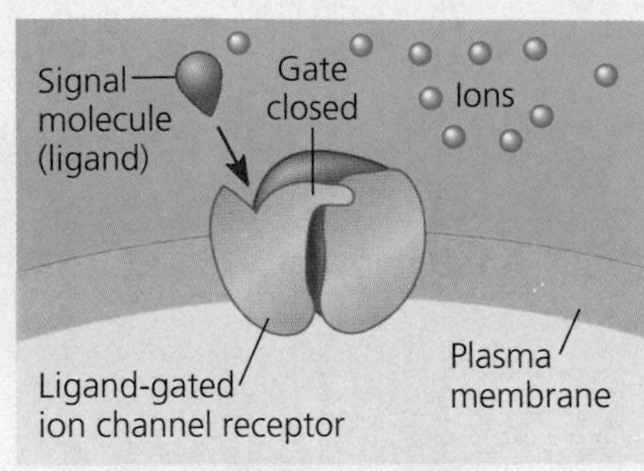

2 When the ligand binds to the receptor and the gate opens, specific ions can flow through the channel and rapidly change the concentration of that particular ion inside the cell. This change may directly affect the activity of the cell in some way.

Gate open

Cellular response

3 When the ligand dissociates from this receptor, the gate closes and ions no longer enter the cell.

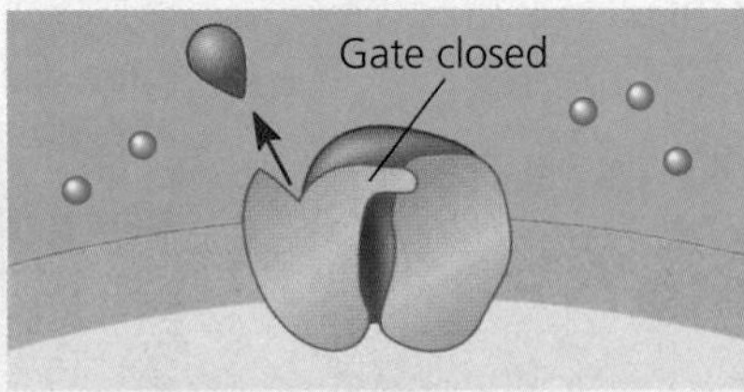

Ligand-gated ion channels are very important in the nervous system. For example, the neurotransmitter molecules released at a synapse between two nerve cells (see Figure 11.4b) bind as ligands to ion channels on the receiving cell, causing the channels to open. Ions flow in and trigger an electrical signal that propagates down the length of the receiving cell. Some gated ion channels are controlled by electrical signals instead of ligands; these voltage-gated ion channels are also crucial to the functioning of the nervous system, as we will discuss in Chapter 48.

Concept Check 11.2

1. Nerve growth factor (NGF) is a water-soluble signal molecule. Would you expect the receptor for NGF to be intracellular or in the plasma membrane? Why?

For suggested answers, see Appendix A.

Concept 11.3

Transduction: Cascades of molecular interactions relay signals from receptors to target molecules in the cell

When signal receptors are plasma membrane proteins, like most of those we have discussed, the transduction stage of cell signaling is usually a multistep pathway. One benefit of such pathways is the possibility of greatly amplifying a signal. If some of the molecules in a pathway transmit the signal to multiple molecules of the next component in the series, the result can be a large number of activated molecules at the end of the pathway. In other words, a small number of extracellular signal molecules can produce a large cellular response. Moreover, multistep pathways provide more opportunities for coordination and regulation than simpler systems do, as we'll discuss later.

Signal Transduction Pathways

The binding of a specific signal molecule to a receptor in the plasma membrane triggers the first step in the chain of molecular interactions—the signal transduction pathway—that leads to a particular response within the cell. Like falling dominoes, the signal-activated receptor activates another protein, which activates another molecule, and so on, until the protein that produces the final cellular response is activated. The molecules that relay a signal from receptor to response, which we call relay molecules in this book, are mostly proteins. The interaction of proteins is a major theme of cell signaling. Indeed, protein interaction is a unifying theme of all regulation at the cellular level.

Keep in mind that the original signal molecule is not physically passed along a signaling pathway; in most cases, it never even enters the cell. When we say that the signal is relayed along a pathway, we mean that certain information is passed on. At each step, the signal is transduced into a different form, commonly a conformational change in a protein. Very often, the conformational change is brought about by phosphorylation.

Protein Phosphorylation and Dephosphorylation

Previous chapters introduced the concept of activating a protein by adding one or more phosphate groups to it (see Figure 8.11). In Figure 11.7, we have already seen how phosphorylation is involved in the activation of receptor tyrosine kinases. In fact, the phosphorylation and dephosphorylation of proteins is a widespread cellular mechanism for regulating protein activity. The general name for an enzyme that transfers phosphate groups from ATP to a protein is **protein kinase.** Recall that receptor tyrosine kinases phosphorylate other receptor tyrosine kinase monomers. Most cytoplasmic protein kinases, however, act on proteins different from themselves. Another distinction is that most cytoplasmic protein kinases phosphorylate either the amino acid serine or threonine, rather than tyrosine. Such serine/threonine kinases are widely involved in signaling pathways in animals, plants, and fungi.

Many of the relay molecules in signal transduction pathways are protein kinases, and they often act on other protein kinases in the pathway. **Figure 11.8** depicts a hypothetical pathway containing three different protein kinases, which create a "phosphorylation cascade." The sequence shown is similar to many known pathways, including those triggered in yeast by mating factors and in animal cells by many growth factors. The signal is transmitted by a cascade of protein phosphorylations, each bringing with it a conformational change. Each shape change results from the interaction of the newly added phosphate groups with charged or polar amino acids (see Figure 5.17). The addition of phosphate groups often changes a protein from an inactive form to an active form (although in other cases phosphorylation *decreases* the activity of the protein).

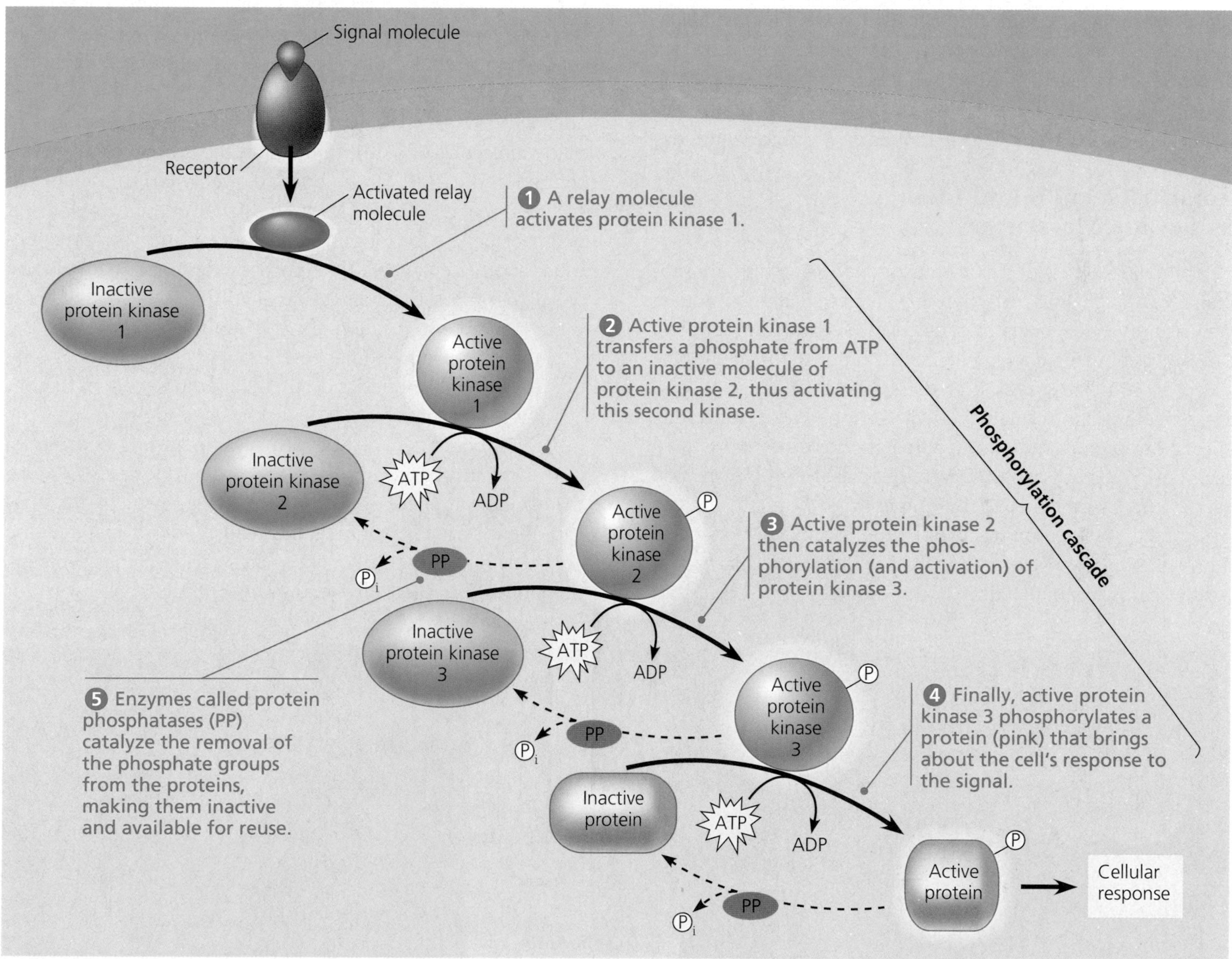

▲ **Figure 11.8 A phosphorylation cascade.** In a phosphorylation cascade, a series of different molecules in a pathway are phosphorylated in turn, each molecule adding a phosphate group to the next one in line. The active and inactive forms of each protein are represented by different shapes to remind you that activation is usually associated with a change in molecular conformation.

The importance of protein kinases can hardly be overstated. About 2% of our own genes are thought to code for protein kinases. A single cell may have hundreds of different kinds, each specific for a different substrate protein. Together, they probably regulate a large proportion of the thousands of proteins in a cell. Among these are most of the proteins that, in turn, regulate cell reproduction. Abnormal activity of such a kinase can cause abnormal cell growth and contribute to the development of cancer.

Equally important in the phosphorylation cascade are the **protein phosphatases,** enzymes that can rapidly remove phosphate groups from proteins, a process called dephosphorylation. By dephosphorylating and thus inactivating protein kinases, phosphatases provide the mechanism for turning off the signal transduction pathway when the initial signal is no longer present. Phosphatases also make the protein kinases available for reuse, enabling the cell to respond again to an extracellular signal. At any given moment, the activity of a protein regulated by phosphorylation depends on the balance in the cell between active kinase molecules and active phosphatase molecules. The phosphorylation/dephosphorylation system acts as a molecular switch in the cell, turning activities on or off as required.

Small Molecules and Ions as Second Messengers

Not all components of signal transduction pathways are proteins. Many signaling pathways also involve small, nonprotein, water-soluble molecules or ions called **second messengers.** (The extracellular signal molecule that binds to the membrane receptor is a pathway's "first messenger.") Because second messengers are both small and water-soluble, they can readily spread throughout the cell by diffusion. For example, as we'll see shortly, it is a second messenger called cyclic AMP that carries the signal initiated by epinephrine from the plasma membrane of a liver or muscle cell into the cell's interior, where it brings about glycogen breakdown. Second messengers participate in pathways initiated by both G-protein-linked receptors and receptor tyrosine kinases. The two most widely used second messengers are cyclic AMP and calcium ions, Ca^{2+}. A large variety of relay proteins are sensitive to the cytosolic concentration of one or the other of these second messengers.

Cyclic AMP

Once Earl Sutherland had established that epinephrine somehow causes glycogen breakdown without passing through the plasma membrane, the search began for the second messenger (he coined the term) that transmits the signal from the plasma membrane to the metabolic machinery in the cytoplasm.

Sutherland found that the binding of epinephrine to the plasma membrane of a liver cell elevates the cytosolic concentration of a compound called cyclic adenosine monophosphate, abbreviated **cyclic AMP** or **cAMP** (**Figure 11.9**). An enzyme embedded in the plasma membrane, **adenylyl cyclase,** converts ATP to cAMP in response to an extracellular signal—in this case, epinephrine. But the epinephrine doesn't stimulate the adenylyl cyclase directly. When epinephrine outside the cell binds to a specific receptor protein, the protein activates adenylyl cyclase, which in turn can catalyze the synthesis of many molecules of cAMP. In this way, the normal cellular concentration of cAMP can be boosted twentyfold in a matter of seconds. The cAMP broadcasts the signal to the cytoplasm. It does not persist for long in the absence of the hormone, because another enzyme, called phosphodiesterase, converts the cAMP to AMP. Another surge of epinephrine is needed to boost the cytosolic concentration of cAMP again.

Subsequent research has revealed that epinephrine is only one of many hormones and other signal molecules that trigger the formation of cAMP. It has also brought to light the other components of cAMP pathways, including G proteins, G-protein-linked receptors, and protein kinases (**Figure 11.10**). The immediate effect of cAMP is usually the activation of a serine/threonine kinase called *protein kinase A*. The activated kinase then phosphorylates various other proteins, depending on the cell type. (The complete pathway for

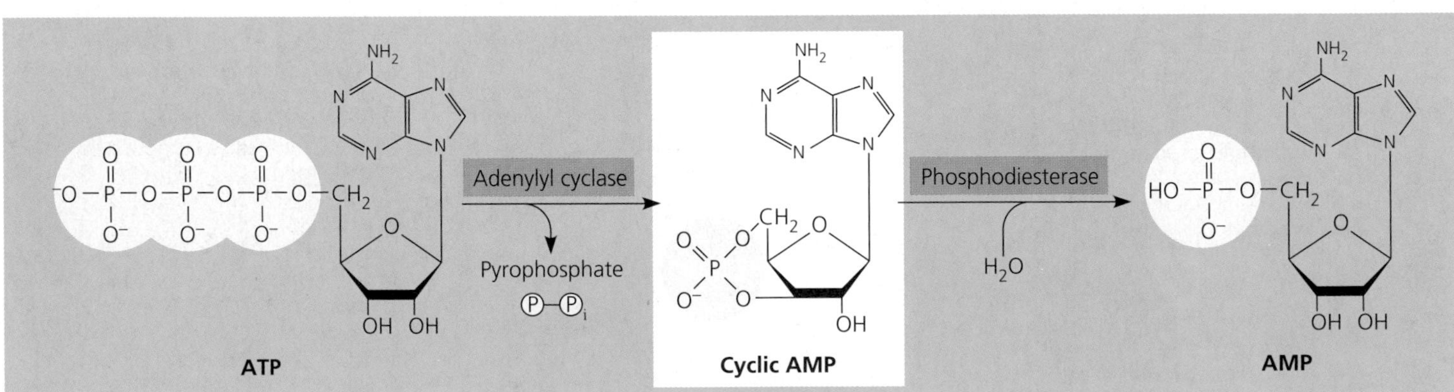

▲ **Figure 11.9 Cyclic AMP.** The second messenger cyclic AMP (cAMP) is made from ATP by adenylyl cyclase, an enzyme embedded in the plasma membrane. Cyclic AMP is inactivated by phosphodiesterase, an enzyme that converts it to AMP.

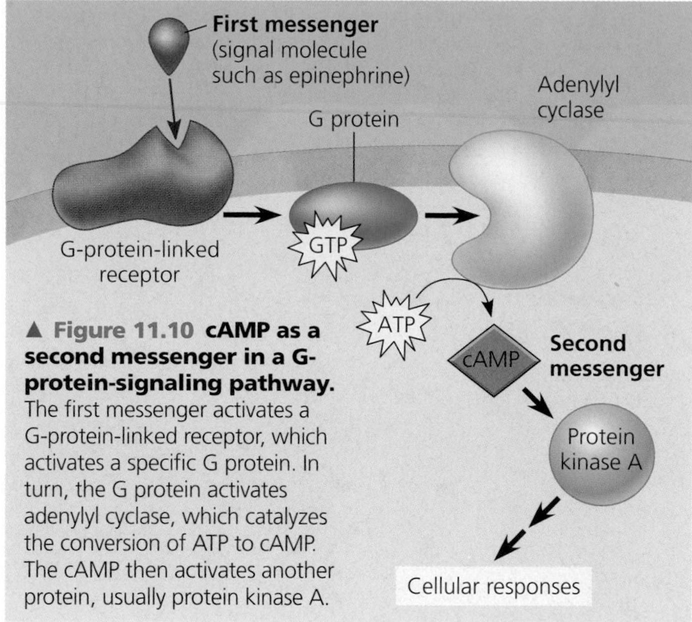

▲ **Figure 11.10 cAMP as a second messenger in a G-protein-signaling pathway.** The first messenger activates a G-protein-linked receptor, which activates a specific G protein. In turn, the G protein activates adenylyl cyclase, which catalyzes the conversion of ATP to cAMP. The cAMP then activates another protein, usually protein kinase A.

epinephrine's stimulation of glycogen breakdown is shown later, in Figure 11.13.)

Further regulation of cell metabolism is provided by other G-protein systems that *inhibit* adenylyl cyclase. In these systems, a different signal molecule activates a different receptor, which activates an *inhibitory* G protein.

Now that we know about the role of cAMP in G-protein-signaling pathways, we can explain in molecular detail how certain microbes cause disease. Consider cholera, a disease that is frequently epidemic in places where the water supply is contaminated with human feces. People acquire the cholera bacterium, *Vibrio cholerae,* by drinking contaminated water. The bacteria colonize the lining of the small intestine and produce a toxin. The cholera toxin is an enzyme that chemically modifies a G protein involved in regulating salt and water secretion. Because the modified G protein is unable to hydrolyze GTP to GDP, it remains stuck in its active form, continuously stimulating adenylyl cyclase to make cAMP. The resulting high concentration of cAMP causes the intestinal cells to secrete large amounts of water and salts into the intestines. An infected person quickly develops profuse diarrhea and if left untreated can soon die from the loss of water and salts.

Our understanding of signaling pathways involving cyclic AMP or related messengers has allowed us to develop treatments for certain conditions in humans. One such pathway uses *cyclic GMP,* or *cGMP,* as a signaling molecule; its effects include relaxation of smooth muscle cells in artery walls. A compound that inhibits the hydrolysis of cGMP to GMP, thus prolonging the signal, was originally prescribed for chest pains because it increased blood flow to the heart muscle. Under the trade name Viagra (see Figure 11.1), this compound is now widely used as a treatment for erectile dysfunction. Viagra causes dilation of blood vessels, which allows increased blood flow to the penis, optimizing physiological conditions for penile erections.

Calcium Ions and Inositol Trisphosphate (IP_3)

Many signal molecules in animals, including neurotransmitters, growth factors, and some hormones, induce responses in their target cells via signal transduction pathways that increase the cytosolic concentration of calcium ions (Ca^{2+}). Calcium is even more widely used than cAMP as a second messenger. Increasing the cytosolic concentration of Ca^{2+} causes many responses in animal cells, including muscle cell contraction, secretion of certain substances, and cell division. In plant cells, a wide range of hormonal and environmental stimuli can cause brief increases in cytosolic Ca^{2+} concentration, triggering various signaling pathways, such as the pathway for greening in response to light (see Figure 39.4). Cells use Ca^{2+} as a second messenger in both G-protein and receptor tyrosine kinase pathways.

Although cells always contain some Ca^{2+}, this ion can function as a second messenger because its concentration in the cytosol is normally much lower than the concentration outside the cell **(Figure 11.11)**. In fact, the level of Ca^{2+} in

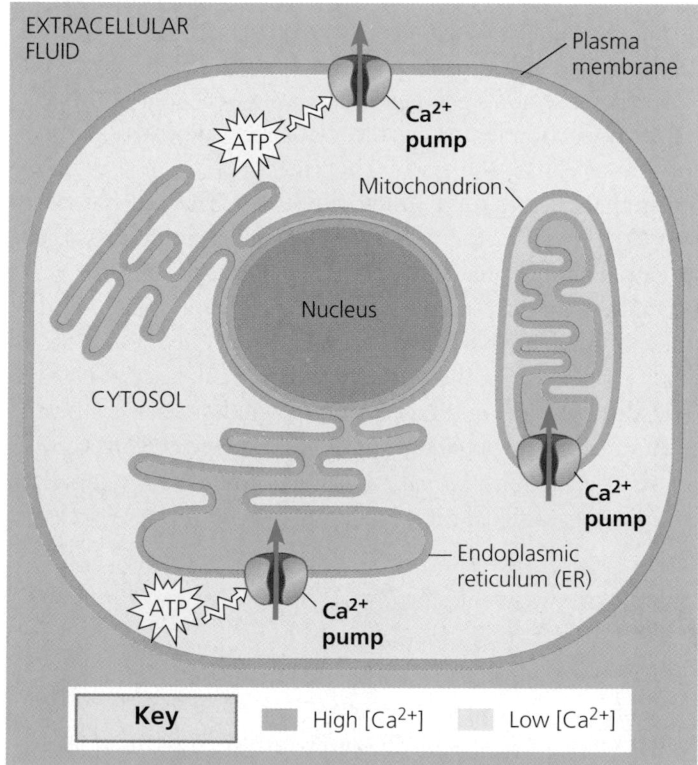

▲ **Figure 11.11 The maintenance of calcium ion concentrations in an animal cell.** The Ca^{2+} concentration in the cytosol is usually much lower (light blue) than in the extracellular fluid and ER (darker blue). Protein pumps in the plasma membrane and the ER membrane, driven by ATP, move Ca^{2+} from the cytosol into the extracellular fluid and into the lumen of the ER. Mitochondrial pumps, driven by chemiosmosis (see Chapter 9), move Ca^{2+} into mitochondria when the calcium level in the cytosol rises significantly.

▶ Figure 11.12 Calcium and IP_3 in signaling pathways. Calcium ions (Ca^{2+}) and inositol trisphosphate (IP_3) function as second messengers in many signal transduction pathways. In this figure, the process is initiated by the binding of a signal molecule to a G-protein-linked receptor. A receptor tyrosine kinase could also initiate this pathway by activating phospholipase C.

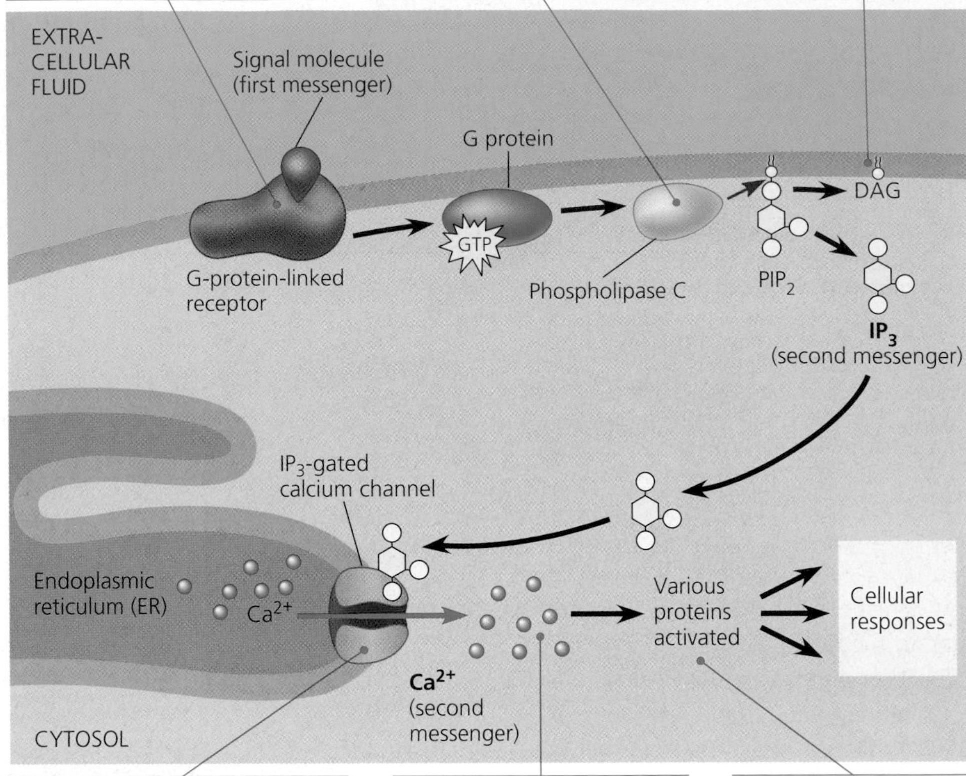

1 A signal molecule binds to a receptor, leading to activation of phospholipase C.
2 Phospholipase C cleaves a plasma membrane phospholipid called PIP2 into DAG and IP3.
3 DAG functions as a second messenger in other pathways.
EXTRA-CELLULAR FLUID
Signal molecule (first messenger)
G protein
DAG
GTP
G-protein-linked receptor
Phospholipase C
PIP2
IP3 (second messenger)
IP3-gated calcium channel
Endoplasmic reticulum (ER)
Ca2+
Various proteins activated
Cellular responses
Ca2+ (second messenger)
CYTOSOL
4 IP3 quickly diffuses through the cytosol and binds to an IP3-gated calcium channel in the ER membrane, causing it to open.
5 Calcium ions flow out of the ER (down their concentration gradient), raising the Ca2+ level in the cytosol.
6 The calcium ions activate the next protein in one or more signaling pathways.

the blood and extracellular fluid of an animal often exceeds that in the cytosol by more than 10,000 times. Calcium ions are actively transported out of the cell and are actively imported from the cytosol into the endoplasmic reticulum (and, under some conditions, into mitochondria and chloroplasts) by various protein pumps (see Figure 11.11). As a result, the calcium concentration in the ER is usually much higher than that in the cytosol. Because the cytosolic calcium level is low, a small change in absolute numbers of ions represents a relatively large percentage change in calcium concentration.

In response to a signal relayed by a signal transduction pathway, the cytosolic calcium level may rise, usually by a mechanism that releases Ca^{2+} from the cell's ER. The pathways leading to calcium release involve still other second messengers, **inositol trisphosphate (IP_3)** and **diacylglycerol (DAG)**. These two messengers are produced by cleavage of a certain kind of phospholipid in the plasma membrane. **Figure 11.12** shows how this occurs and how IP_3 stimulates the release of calcium from the ER. Because IP_3 acts before calcium in these pathways, calcium could be considered a "*third* messenger." However, scientists use the term *second messenger* for all small, nonprotein components of signal transduction pathways.

Concept Check 11.3

1. What is a protein kinase, and what is its role in a signal transduction pathway?
2. When a signal transduction pathway involves a phosphorylation cascade, how does the cell's response get turned off?
3. Upon activation of phospholipase C by ligand binding to a receptor, what effect does the IP_3-gated calcium channel have on Ca^{2+} concentration in the cytosol?

For suggested answers, see Appendix A.

Concept 11.4

Response: Cell signaling leads to regulation of cytoplasmic activities or transcription

We now take a closer look at the cell's subsequent response to an extracellular signal—what some researchers call the "output response." What is the nature of the final step in a signaling pathway?

Cytoplasmic and Nuclear Responses

Ultimately, a signal transduction pathway leads to the regulation of one or more cellular activities. The response may occur in the cytoplasm or may involve action in the nucleus.

In the cytoplasm, a signal may cause, for example, the opening or closing of an ion channel in the plasma membrane or a change in cell metabolism. As we have discussed already, the response of liver cells to signaling by the hormone epinephrine helps regulate cellular energy metabolism. The final step in the signaling pathway activates the enzyme that catalyzes the breakdown of glycogen. **Figure 11.13** shows the complete pathway leading to the release of glucose-1-phosphate from glycogen. Note that at each step the response is amplified, as we will discuss later.

Many other signaling pathways ultimately regulate not the *activity* of enzymes but the *synthesis* of enzymes or other proteins, usually by turning specific genes on or off in the nucleus. Like an activated steroid receptor (see Figure 11.6), the final activated molecule in a signaling pathway may function as a transcription factor. **Figure 11.14** shows an example in which a signaling pathway activates a transcription factor that turns a gene on: The response to the growth factor signal is the synthesis of mRNA, which will be translated in the cytoplasm into a specific protein. In other cases, the transcription factor might regulate a gene by turning it off. Often a transcription factor regulates several different genes.

All the different kinds of signal receptors and relay molecules introduced in this chapter participate in various gene-regulating pathways, as well as in pathways leading to other kinds of responses. The molecular messengers that produce gene regulation responses include growth factors and certain plant and animal hormones. Malfunctioning of growth factor pathways like the one in Figure 11.14 can contribute to the development of cancer, as we will see in Chapter 19.

Fine-Tuning of the Response

Why are there often so many steps between a signaling event at the cell surface and the cell's response? As mentioned earlier, signaling pathways with a multiplicity of steps have

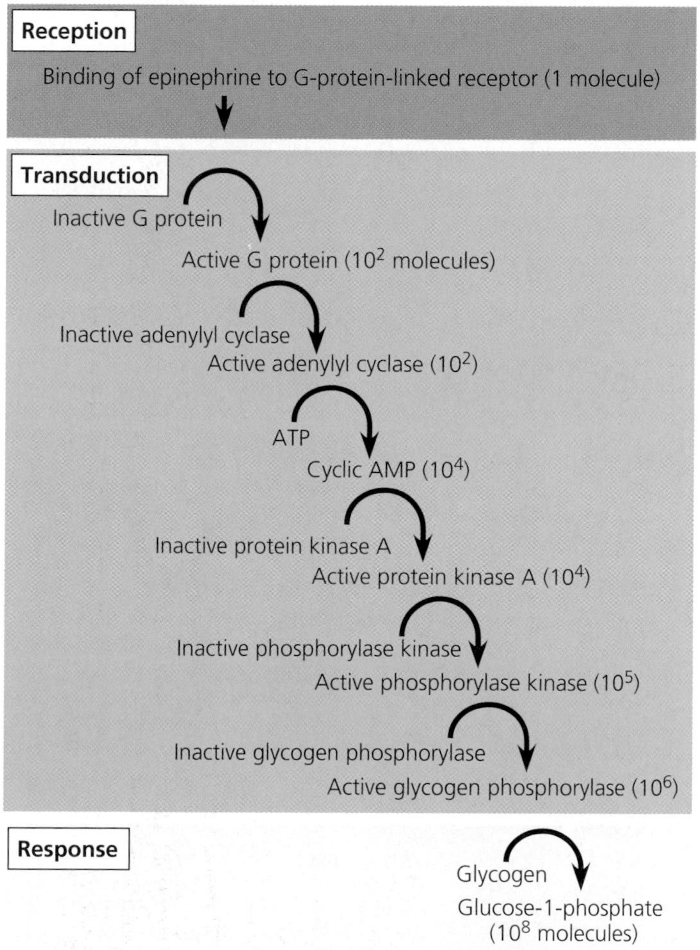

Reception

Binding of epinephrine to G-protein-linked receptor (1 molecule)

Transduction

Inactive G protein → Active G protein (10^2 molecules)

Inactive adenylyl cyclase → Active adenylyl cyclase (10^2)

ATP → Cyclic AMP (10^4)

Inactive protein kinase A → Active protein kinase A (10^4)

Inactive phosphorylase kinase → Active phosphorylase kinase (10^5)

Inactive glycogen phosphorylase → Active glycogen phosphorylase (10^6)

Response

Glycogen → Glucose-1-phosphate (10^8 molecules)

▲ **Figure 11.13 Cytoplasmic response to a signal: the stimulation of glycogen breakdown by epinephrine.** In this signaling system, the hormone epinephrine acts through a G-protein-linked receptor to activate a succession of relay molecules, including cAMP and two protein kinases (see also Figure 11.10). The final protein to be activated is the enzyme glycogen phosphorylase, which releases glucose-1-phosphate units from glycogen. This pathway amplifies the hormonal signal, because one receptor protein can activate about 100 molecules of G protein, and each enzyme in the pathway can act on many molecules of its substrate, the next molecule in the cascade. The number of activated molecules given for each step is approximate.

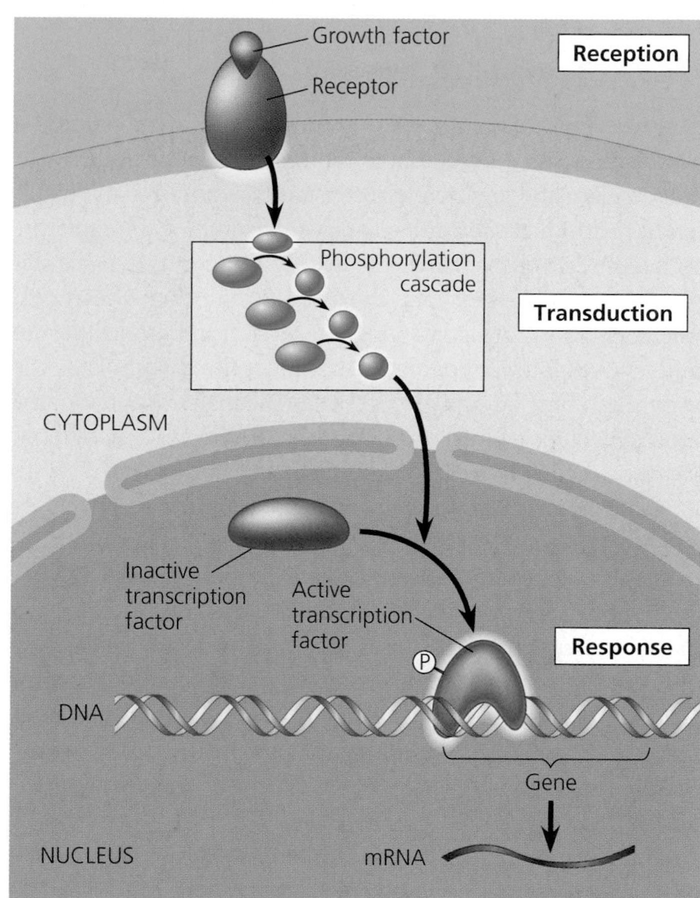

▲ **Figure 11.14 Nuclear responses to a signal: the activation of a specific gene by a growth factor.** This diagram is a simplified representation of a typical signaling pathway that leads to the regulation of gene activity in the cell nucleus. The initial signal molecule, a local regulator called a growth factor, triggers a phosphorylation cascade. (The ATP molecules that serve as sources of phosphate are not shown.) Once phosphorylated, the last kinase in the sequence enters the nucleus and there activates a gene-regulating protein, a transcription factor. This protein stimulates a specific gene so that an mRNA is synthesized, which then directs the synthesis of a particular protein in the cytoplasm.

two important benefits: They amplify the signal (and thus the response), and they contribute to the specificity of response.

Signal Amplification

Elaborate enzyme cascades amplify the cell's response to a signal. At each catalytic step in the cascade, the number of activated products is much greater than in the preceding step. For example, in the epinephrine-triggered pathway in Figure 11.13, each adenylyl cyclase molecule catalyzes the formation of many cAMP molecules, each molecule of protein kinase A phosphorylates many molecules of the next kinase in the pathway, and so on. The amplification effect stems from the fact that these proteins persist in the active form long enough to process numerous molecules of substrate before they become inactive again. As a result of the signal's amplification, a small number of epinephrine molecules binding to receptors on the surface of a liver cell or muscle cell can lead to the release of hundreds of millions of glucose molecules from glycogen.

The Specificity of Cell Signaling

Consider two different cells in your body—a liver cell and a heart muscle cell, for example. Both are in contact with your bloodstream and are therefore constantly exposed to many different hormone molecules, as well as to local regulators secreted by nearby cells. Yet the liver cell responds to some signals but ignores others, and the same is true for the heart cell. And some kinds of signals trigger responses in both cells—but different responses. For instance, epinephrine stimulates the liver cell to break down glycogen, but the main response of the heart cell to epinephrine is contraction, leading to a more rapid heartbeat. How do we account for this difference?

The explanation for the specificity exhibited in cellular responses to signals is the same as the basic explanation for virtually all differences between cells: *Different kinds of cells have different collections of proteins* **(Figure 11.15)**. The response of a particular cell to a signal depends on its particular collection of signal receptor proteins, relay proteins, and proteins needed to carry out the response. A liver cell, for example, is poised to respond appropriately to epinephrine by having the proteins listed in Figure 11.13 as well as those needed to manufacture glycogen.

Thus, two cells that respond differently to the same signal differ in one or more of the proteins that handle and respond to the signal. Notice in Figure 11.15 that different pathways may have some molecules in common. For example, cells A, B, and C all use the same receptor protein for the orange signal molecule; differences in other proteins account for their differing responses. In cell D, a different receptor protein is used for the same signal molecule, leading to yet another response. In cell B, a pathway that is triggered by a single kind of signal diverges to produce two responses; such branched pathways often involve receptor tyrosine kinases (which can activate multiple relay proteins) or second messengers (which can regulate numerous proteins). In cell C, two pathways triggered by separate signals converge to modulate a single response. Branching of pathways and "cross-talk" (interaction) between pathways are important in regulating and coordinating a cell's responses to information coming in from different sources in the body. Moreover, the use of some of the same proteins in more than one pathway allows the cell to economize on the number of different proteins it must make.

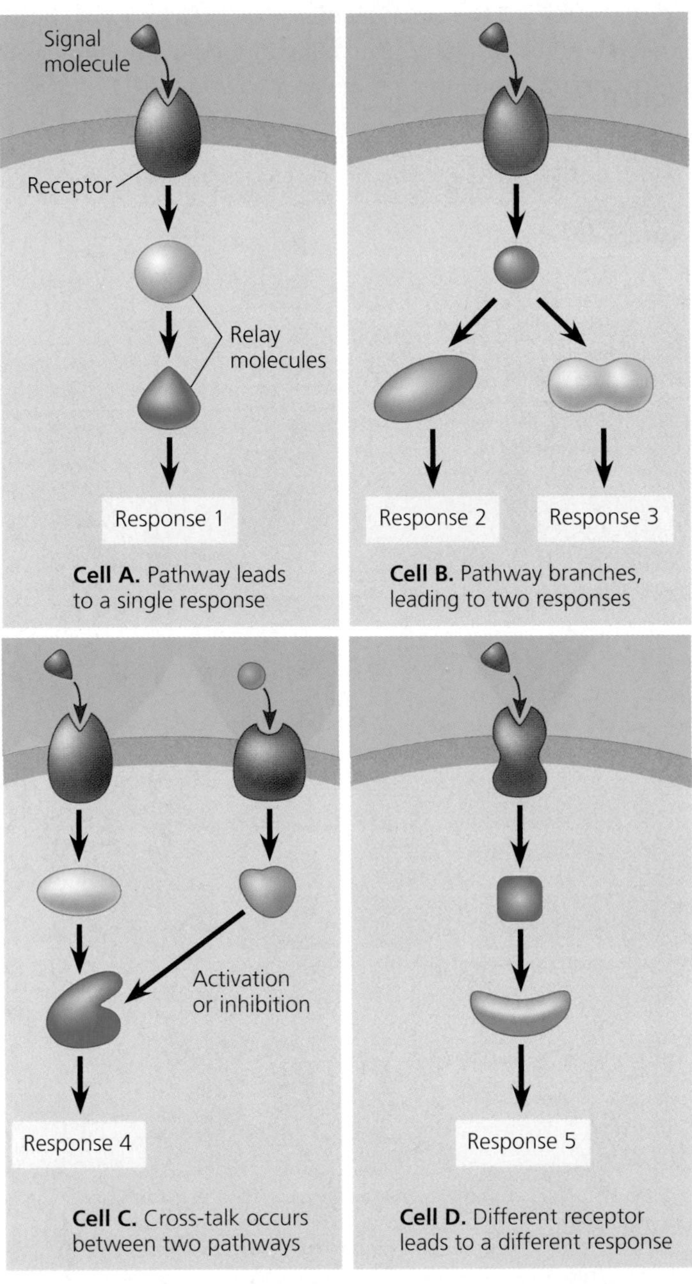

▲ **Figure 11.15 The specificity of cell signaling.** The particular proteins a cell possesses determine what signal molecules it responds to and the nature of the response. The four cells in these diagrams respond to the same signal molecule (orange) in different ways because each has a different set of proteins (purple and teal shapes). Note, however, that the same kinds of molecules can participate in more than one pathway.

Signaling Efficiency: Scaffolding Proteins and Signaling Complexes

The signaling pathways in Figure 11.15 (as well as some of the other pathway depictions in this chapter) are greatly simplified. The diagrams show only a few relay molecules and, for clarity's sake, display these molecules spread out in the cytosol. If this were true in the cell, signaling pathways would operate very inefficiently because most relay molecules are proteins, and proteins are too large to diffuse quickly through the viscous cytosol. How does a particular protein kinase, for instance, find its substrate?

Recent research suggests that the efficiency of signal transduction may in many cases be increased by the presence of **scaffolding proteins,** large relay proteins to which several other relay proteins are simultaneously attached. For example, one scaffolding protein isolated from mouse brain cells holds three protein kinases and carries these kinases with it when it binds to an appropriately activated membrane receptor; it thus facilitates a specific phosphorylation cascade **(Figure 11.16)**. In fact, researchers are finding scaffolding proteins in brain cells that *permanently* hold together networks of signaling-pathway proteins at synapses. This hardwiring enhances the speed and accuracy of signal transfer between cells.

When signaling pathways were first discovered, they were thought to be linear, independent pathways. Our understanding of the processes of cellular communication has benefited from the realization that things are not that simple. In fact, as seen in Figure 11.15, some proteins may participate in more than one pathway, either in different cell types or in the same cell at different times or under different conditions. This view underscores the importance of permanent or transient protein complexes in the functioning of a cell.

The importance of the relay proteins that serve as points of branching or intersection in signaling pathways is highlighted by the problems arising when these proteins are defective or missing. For instance, in an inherited disorder called Wiskott-Aldrich syndrome (WAS), the absence of a single relay protein leads to such diverse effects as abnormal bleeding, eczema, and a predisposition to infections and leukemia. These symptoms are thought to arise primarily from the absence of the protein in cells of the immune system. By studying normal cells, scientists found that the WAS protein is located just beneath the cell surface. The protein interacts both with microfilaments of the cytoskeleton and with several different components of signaling pathways that relay information from the cell surface, including pathways regulating immune cell proliferation. This multifunctional relay protein is thus both a branch point and an important intersection point in a complex signal transduction network that controls immune cell behavior. When the WAS protein is absent, the cytoskeleton is not properly organized and signaling pathways are disrupted, leading to the WAS symptoms.

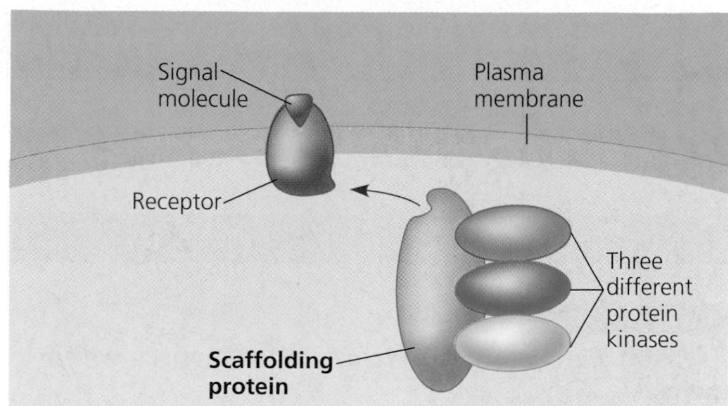

▲ **Figure 11.16 A scaffolding protein.** The scaffolding protein shown here (pink) simultaneously binds to a specific activated membrane receptor and three different protein kinases. This physical arrangement facilitates signal transduction by these molecules.

Termination of the Signal

To keep Figure 11.15 simple, we have not indicated the *inactivation* mechanisms that are an essential aspect of cell signaling. For a cell of a multicellular organism to remain alert and capable of responding to incoming signals, each molecular change in its signaling pathways must last only a short time. As we saw in the cholera example, if a signaling pathway component becomes locked into one state, whether active or inactive, dire consequences for the organism can result.

Thus, a key to a cell's continuing receptiveness to regulation is the reversibility of the changes that signals produce. The binding of signal molecules to receptors is reversible, with the result that the lower the concentration of signal molecules, the fewer will be bound at any given moment. When signal molecules leave the receptor, the receptor reverts to its inactive form. Then, by a variety of means, the relay molecules return to their inactive forms: The GTPase activity intrinsic to a G protein hydrolyzes its bound GTP; the enzyme phosphodiesterase converts cAMP to AMP; protein phosphatases inactivate phosphorylated kinases and other proteins; and so forth. As a result, the cell is soon ready to respond to a fresh signal.

This chapter has introduced you to many of the general mechanisms of cell communication, such as ligand binding, conformational changes, cascades of interactions, and protein phosphorylation. As you continue through the text, you will encounter numerous examples of cell signaling.

Concept Check 11.4

1. How can a target cell's response to a hormone be amplified more than a millionfold?
2. Explain how two cells with different scaffolding proteins could behave differently in response to the same signaling molecule.

For suggested answers, see Appendix A.

Chapter 11 Review

Go to www.campbellbiology.com or the student CD-ROM to explore Activities, Investigations, and other interactive study aids.

SUMMARY OF KEY CONCEPTS

Concept 11.1

External signals are converted into responses within the cell

- **Evolution of Cell Signaling (pp. 201–202)** Signaling in microbes has much in common with processes in multicellular organisms, suggesting an early origin of signaling mechanisms.
- **Local and Long-Distance Signaling (pp. 202–203)** In local signaling, animal cells may communicate by direct contact or by secreting local regulators, such as growth factors or neurotransmitters. For signaling over long distances, both animals and plants use hormones; animals also signal along nerve cells.
  Investigation *How Do Cells Communicate with Each Other?*
- **The Three Stages of Cell Signaling: *A Preview* (pp. 203–204)** Earl Sutherland discovered how the hormone epinephrine acts on cells. The signal molecule epinephrine binds to receptors on a cell's surface (reception), leading to a series of changes in the receptor and molecules inside the cell (transduction) and finally to the activation of an enzyme that breaks down glycogen (response).
  Activity *Overview of Cell Signaling*

Concept 11.2

Reception: A signal molecule binds to a receptor protein, causing it to change shape

- The binding between signal molecule (ligand) and receptor is highly specific. A conformational change in a receptor is often the initial transduction of the signal **(pp. 204–205).**
- **Intracellular Receptors (p. 205)** Intracellular receptors are cytoplasmic or nuclear proteins. Signal molecules that are small or hydrophobic and can readily cross the plasma membrane use these receptors.
- **Receptors in the Plasma Membrane (pp. 205–208)** A G-protein-linked receptor is a membrane receptor that works with the help of a cytoplasmic G protein. Ligand binding activates the receptor, which then activates a specific G protein, which activates yet another protein, thus propagating the signal along a signal transduction pathway.
  Receptor tyrosine kinases react to the binding of signal molecules by forming dimers and then adding phosphate groups to tyrosines on the cytoplasmic side of the other subunit of the dimer. Relay proteins in the cell can then be activated by binding to different phosphorylated tyrosines, allowing this receptor to trigger several pathways at once.
  Specific signal molecules cause ligand-gated ion channels in a membrane to open or close, regulating the flow of specific ions.
  Activity *Reception*

Concept 11.3

Transduction: Cascades of molecular interactions relay signals from receptors to target molecules in the cell

- **Signal Transduction Pathways (p. 208)** At each step in a pathway, the signal is transduced into a different form, commonly a conformational change in a protein.
- **Protein Phosphorylation and Dephosphorylation (pp. 209–210)** Many signal transduction pathways include phosphorylation cascades, in which a series of protein kinases each add a phosphate group to the next one in line, activating it. Phosphatase enzymes soon remove the phosphates.
- **Small Molecules and Ions as Second Messengers (pp. 210–212)** Second messengers, such as cyclic AMP (cAMP) and Ca^{2+}, diffuse readily through the cytosol and thus help broadcast signals quickly. Many G proteins activate adenylyl cyclase, which makes cAMP from ATP. Cells use Ca^{2+} as a second messenger in both G-protein and tyrosine kinase pathways. The tyrosine kinase pathways can also involve two other second messengers, DAG and IP_3. IP_3 can trigger a subsequent increase in Ca^{2+} levels.
  Activity *Signal Transduction Pathways*

Concept 11.4

Response: Cell signaling leads to regulation of cytoplasmic activities or transcription

- **Cytoplasmic and Nuclear Responses (pp. 212–213)** In the cytoplasm, signaling pathways regulate, for example, enzyme activity and cytoskeleton rearrangement. Other pathways regulate genes by activating transcription factors, proteins that turn specific genes on or off.
  Activity *Cellular Responses*
  Activity *Build a Signaling Pathway*
- **Fine-Tuning of the Response (pp. 213–215)** Each catalytic protein in a signaling pathway amplifies the signal by activating multiple copies of the next component of the pathway; for long pathways, the total amplification may be a millionfold or more. The particular combination of proteins in a cell gives the cell great specificity in both the signals it detects and the responses it carries out. Scaffolding proteins can increase signal transduction efficiency. Pathway branching and cross-talk further help the cell coordinate incoming signals. Signal response is terminated quickly by the reversal of ligand binding.

TESTING YOUR KNOWLEDGE

Self-Quiz

1. Phosphorylation cascades involving a series of protein kinases are useful for cellular signal transduction because
   a. they are species specific.
   b. they always lead to the same cellular response.
   c. they amplify the original signal manyfold.
   d. they counter the harmful effects of phosphatases.
   e. the number of molecules used is small and fixed.

2. Binding of a signal molecule to which type of receptor leads directly to a change in the distribution of anions and/or cations on opposite sides of the membrane?
   a. receptor tyrosine kinase
   b. G-protein-linked receptor
   c. phosphorylated receptor tyrosine kinase dimer
   d. ligand-gated ion channel
   e. intracellular receptor

3. The activation of receptor tyrosine kinases is always characterized by
   a. dimerization and phosphorylation.
   b. IP_3 binding.
   c. a phosphorylation cascade.
   d. GTP hydrolysis.
   e. channel protein conformational change.

4. Which of the following provides the best evidence that cell-signaling pathways evolved early in the history of life?
   a. They are seen in "primitive" cells such as yeast.
   b. Yeast cells signal each other for mating.
   c. Signal transduction molecules found in distantly related organisms are similar.
   d. Signals can be sent long distances by cells.
   e. Most signals are received by cell surface receptors.

5. Which observation suggested to Sutherland the involvement of a second messenger in epinephrine's effect on liver cells?
   a. Enzymatic activity was proportional to the amount of calcium added to a cell-free extract.
   b. Receptor studies indicated that epinephrine was a ligand.
   c. Glycogen breakdown was observed only when epinephrine was administered to intact cells.
   d. Glycogen breakdown was observed when epinephrine and glycogen phosphorylase were combined.
   e. Epinephrine was known to have different effects on different types of cells.

6. Protein phosphorylation is commonly involved with all of the following *except*
   a. regulation of transcription by extracellular signal molecules.
   b. enzyme activation.
   c. activation of G-protein-linked receptors.
   d. activation of receptor tyrosine kinases.
   e. activation of protein kinase molecules.

7. Amplification of a chemical signal occurs when
   a. a receptor in the plasma membrane activates several G-protein molecules while a signal molecule is bound to it.
   b. a cAMP molecule activates one protein kinase molecule before being converted to AMP.
   c. phosphorylase and phosphatase activities are balanced.
   d. receptor tyrosine kinases dimerize upon ligand binding.
   e. both a and d occur.

8. Lipid-soluble signal molecules, such as testosterone, cross the membranes of all cells but affect only target cells because
   a. only target cells retain the appropriate DNA segments.
   b. intracellular receptors are present only in target cells.
   c. most cells lack the Y chromosome required.
   d. only target cells possess the cytosolic enzymes that transduce the testosterone.
   e. only in target cells is testosterone able to initiate the phosphorylation cascade leading to activated transcription factor.

9. Signal transduction pathways benefit cells for all of the following reasons *except*
   a. they help cells respond to signal molecules that are too large or too polar to cross the plasma membrane.
   b. they enable different cells to respond appropriately to the same signal.
   c. they help cells use up phosphate generated by ATP breakdown.
   d. they can amplify a signal.
   e. variations in the signal transduction pathways can enhance response specificity.

10. Consider this pathway: epinephrine → G-protein-linked receptor → G protein → adenylyl cyclase → cAMP. Identify the second messenger.
   a. cAMP
   b. G protein
   c. GTP
   d. adenylyl cyclase
   e. G-protein-linked receptor

For Self-Quiz answers, see Appendix A.

Go to the website or CD-ROM for more quiz questions.

Evolution Connection

You learned in this chapter that cell-to-cell signaling is thought to have arisen early in the history of life, because the same mechanisms of signaling are found in distantly related organisms. But why hasn't some "better" mechanism arisen? Is the evolution of wholly new signaling mechanisms too difficult, or are existing mechanisms simply adequate and therefore maintained? Put another way, does natural selection favor the evolution of superior signaling mechanisms if existing mechanisms are adequate and effective? Why or why not?

Scientific Inquiry

Epinephrine initiates a signal transduction pathway that involves production of cyclic AMP (cAMP) and leads to the breakdown of glycogen to glucose, a major energy source for cells. But glycogen breakdown is actually only part of a "fight-or-flight response" that epinephrine brings about; the overall effect on the body includes increased heart rate and alertness, as well as a burst of energy. Given that caffeine blocks the activity of cAMP phosphodiesterase, propose a mechanism by which caffeine ingestion leads to heightened alertness and sleeplessness.

Investigation *How Do Cells Communicate with Each Other?*

Science, Technology, and Society

The aging process is thought to be initiated at the cellular level. Among the changes that can occur after a certain number of cell divisions is the loss of a cell's ability to respond to growth factors and other chemical signals. Much research into aging is aimed at understanding such losses, with the ultimate goal of significantly extending the human life span. Not everyone, however, agrees that this is a desirable goal. If life expectancy were greatly increased, what might be the social and ecological consequences? How might we cope with them?

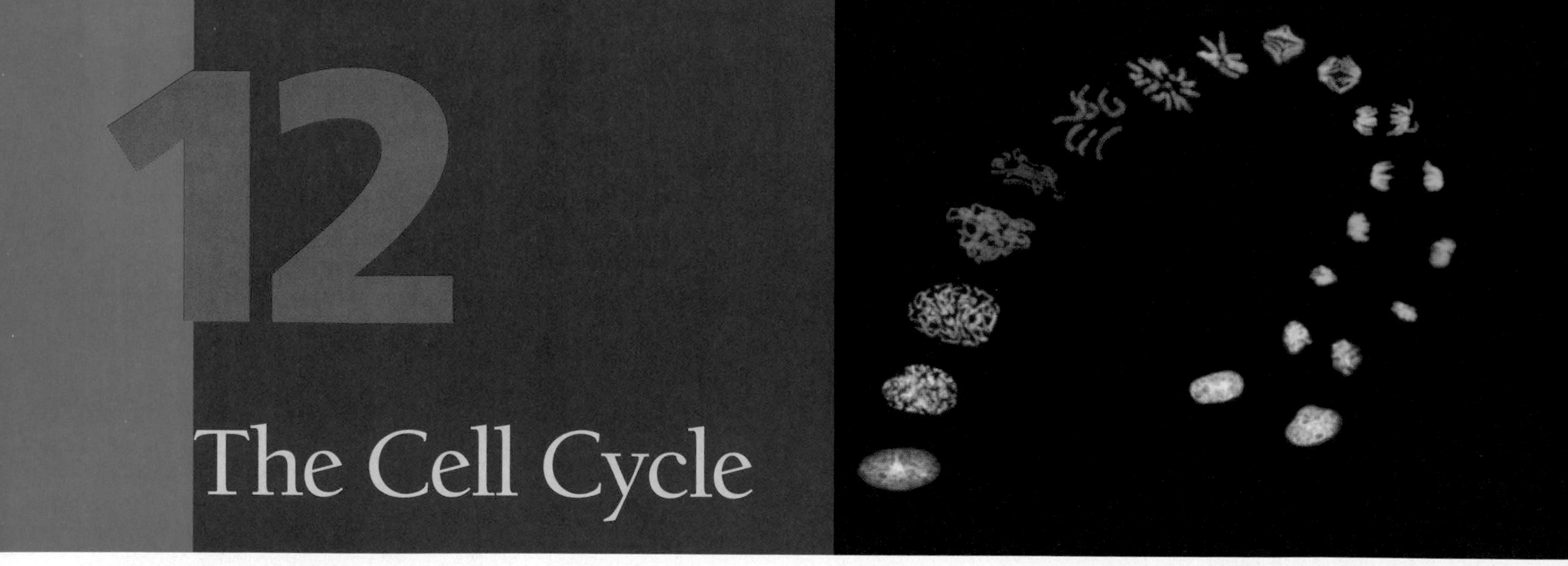

12 The Cell Cycle

▲ **Figure 12.1 Chromosomes in a dividing cell.**

Key Concepts

12.1 Cell division results in genetically identical daughter cells

12.2 The mitotic phase alternates with interphase in the cell cycle

12.3 The cell cycle is regulated by a molecular control system

Overview

The Key Roles of Cell Division

The ability of organisms to reproduce their own kind is the one characteristic that best distinguishes living things from nonliving matter. This unique capacity to procreate, like all biological functions, has a cellular basis. Rudolf Virchow, a German physician, put it this way in 1855: "Where a cell exists, there must have been a preexisting cell, just as the animal arises only from an animal and the plant only from a plant." He summarized this concept with the Latin axiom *"Omnis cellula e cellula,"* meaning "Every cell from a cell." The continuity of life is based on the reproduction of cells, or **cell division.** The series of fluorescence micrographs in **Figure 12.1** follows an animal cell's chromosomes, from lower left to lower right, as one cell divides into two.

Cell division plays several important roles in the life of an organism. When a unicellular organism, such as an amoeba, divides and forms duplicate offspring, the division of one cell reproduces an entire organism **(Figure 12.2a)**. Cell division on a larger scale can produce progeny from some multicellular organisms (such as plants that grow from cuttings). Cell division also enables sexually reproducing organisms to develop from a single cell—the fertilized egg, or zygote **(Figure 12.2b)**. And after an organism is fully grown, cell division continues to function in renewal and repair, replacing cells that die from normal wear and tear or accidents. For example, dividing cells in your bone marrow continuously make new blood cells **(Figure 12.2c)**.

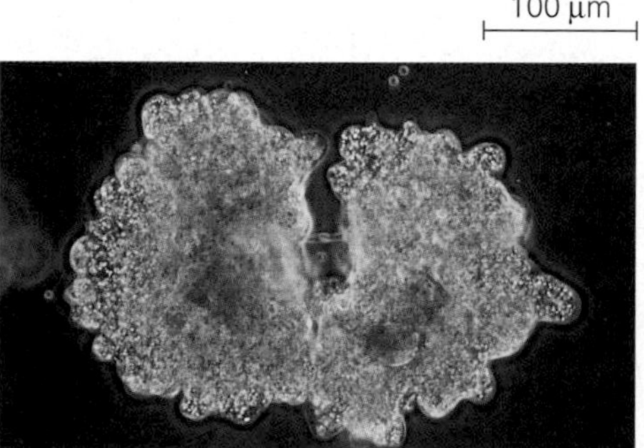

(a) Reproduction. An amoeba, a single-celled eukaryote, is dividing into two cells. Each new cell will be an individual organism (LM).

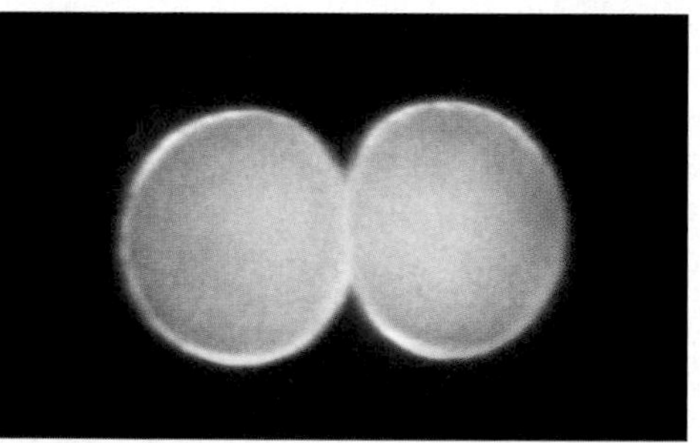

(b) Growth and development. This micrograph shows a sand dollar embryo shortly after the fertilized egg divided, forming two cells (LM).

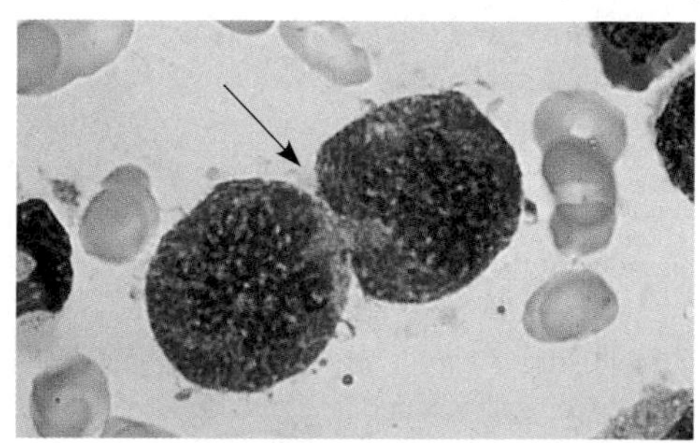

(c) Tissue renewal. These dividing bone marrow cells (arrow) will give rise to new blood cells (LM).

▲ **Figure 12.2 The functions of cell division.**

The cell division process is an integral part of the **cell cycle**, the life of a cell from the time it is first formed from a dividing parent cell until its own division into two cells. Passing identical genetic material to cellular offspring is a crucial function of cell division. In this chapter, you will learn how cell division distributes identical genetic material to daughter cells.* After studying the cellular mechanics of cell division, you will learn about the molecular control system that regulates progress through the cell cycle and what happens when the control system malfunctions. Because cell cycle regulation, or a lack thereof, plays a major role in cancer development, this aspect of cell biology is an active area of research.

Concept 12.1

Cell division results in genetically identical daughter cells

The reproduction of an ensemble as complex as a cell cannot occur by a mere pinching in half; a cell is not like a soap bubble that simply enlarges and splits in two. Cell division involves the distribution of identical genetic material—DNA—to two daughter cells. What is most remarkable about cell division is the fidelity with which the DNA is passed along from one generation of cells to the next. A dividing cell duplicates its DNA, allocates the two copies to opposite ends of the cell, and only then splits into daughter cells.

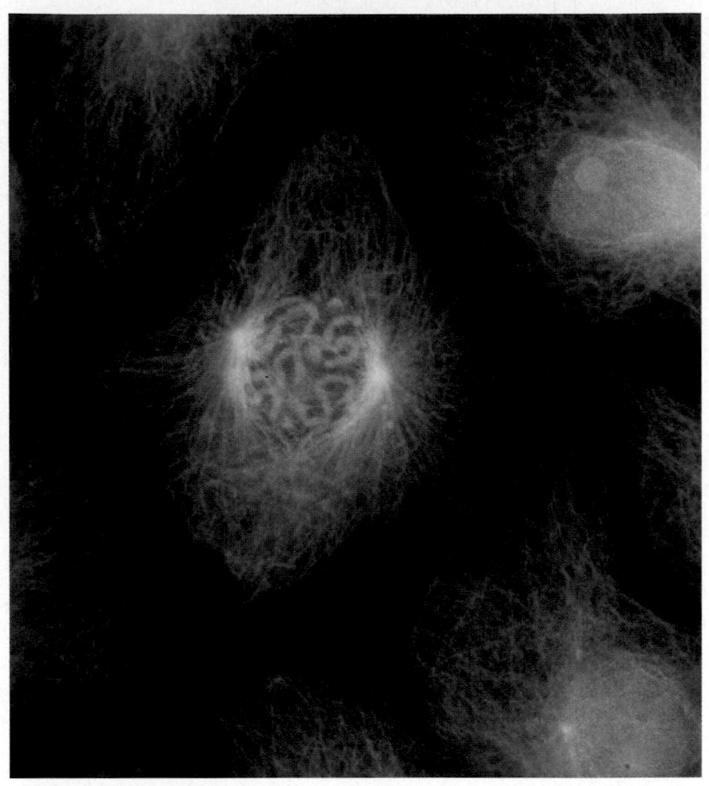

▲ **Figure 12.3 Eukaryotic chromosomes.** Chromosomes (stained orange) are visible within the nucleus of the kangaroo rat epithelial cell in the center of this micrograph. The cell is preparing to divide (LM).

Cellular Organization of the Genetic Material

A cell's endowment of DNA, its genetic information, is called its **genome.** Although a prokaryotic genome is often a single long DNA molecule, eukaryotic genomes usually consist of a number of DNA molecules. The overall length of DNA in a eukaryotic cell is enormous. A typical human cell, for example, has about 2 m of DNA—a length about 250,000 times greater than the cell's diameter. Yet before the cell can divide, all of this DNA must be copied and then the two copies separated so that each daughter cell ends up with a complete genome.

The replication and distribution of so much DNA is manageable because the DNA molecules are packaged into **chromosomes**, so named because they take up certain dyes used in microscopy (from the Greek *chroma*, color, and *soma*, body) **(Figure 12.3)**. Every eukaryotic species has a characteristic number of chromosomes in each cell nucleus. For example, the nuclei of human **somatic cells** (all body cells except the reproductive cells) each contain 46 chromosomes made up of two sets of 23, one set inherited from each parent. Reproductive cells, or **gametes**—sperm cells and egg cells—have half as many chromosomes as somatic cells, or one set of 23 chromosomes in humans.

Eukaryotic chromosomes are made of **chromatin**, a complex of DNA and associated protein molecules. Each single chromosome contains one very long, linear DNA molecule that carries several hundred to a few thousand genes, the units that specify an organism's inherited traits. The associated proteins maintain the structure of the chromosome and help control the activity of the genes.

Distribution of Chromosomes During Cell Division

When a cell is not dividing, and even as it duplicates its DNA in preparation for cell division, each chromosome is in the form of a long, thin chromatin fiber. After DNA duplication, however, the chromosomes condense: Each chromatin fiber becomes densely coiled and folded, making the chromosomes much shorter and so thick that we can see them with a light microscope.

Each duplicated chromosome has two **sister chromatids.** The two chromatids, each containing an identical DNA molecule, are initially attached by adhesive proteins all along their lengths. In its condensed form, the duplicated chromosome

* Although the terms *daughter cells* and *sister chromatids* (a term you will encounter later in the chapter) are traditional and will be used throughout this book, the structures they refer to have no gender.

▶ **Figure 12.4 Chromosome duplication and distribution during cell division.** A eukaryotic cell preparing to divide duplicates each of its chromosomes. The micrograph shows a duplicated human chromosome (SEM). The copies of each chromosome are then distributed to two daughter cells during cell division. (Chromosomes normally exist in the highly condensed state shown here only during the process of cell division; the chromosomes in the top and bottom cells are shown in condensed form for illustration purposes only.)

A eukaryotic cell has multiple chromosomes, one of which is represented here. Before duplication, each chromosome has a single DNA molecule.
Chromosome duplication (including DNA synthesis)
Once duplicated, a chromosome consists of two sister chromatids connected at the centromere. Each chromatid contains a copy of the DNA molecule.
Centromere
Sister chromatids
Separation of sister chromatids
Mechanical processes separate the sister chromatids into two chromosomes and distribute them to two daughter cells.
Centromeres
0.5 μm
Sister chromatids

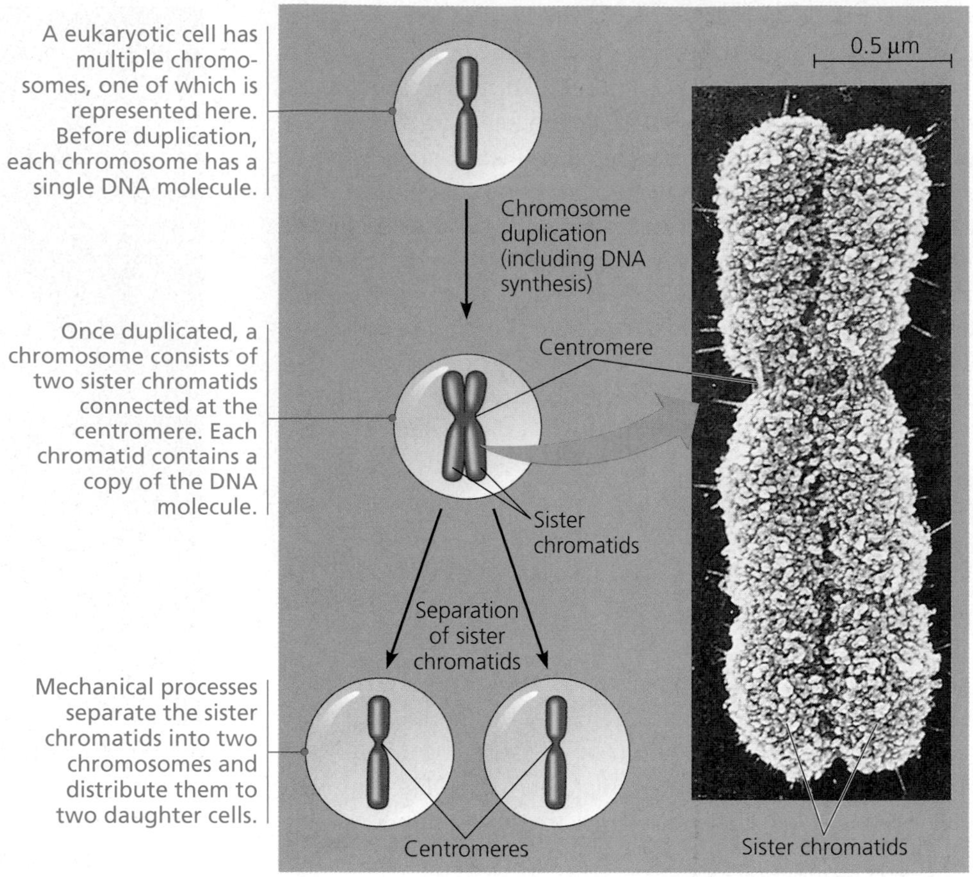

has a narrow "waist" at a specialized region called the **centromere** where the two chromatids are most closely attached **(Figure 12.4)**. Later in the cell division process, the two sister chromatids of each duplicated chromosome separate and move into two new nuclei, one at each end of the cell. Once the sister chromatids separate, they are considered individual chromosomes. Thus, each new nucleus receives a group of chromosomes identical to the original group in the parent cell. **Mitosis,** the division of the nucleus, is usually followed immediately by **cytokinesis,** the division of the cytoplasm. Where there was one cell, there are now two, each the genetic equivalent of the parent cell.

What happens to chromosome number as we follow the human life cycle through the generations? You inherited 46 chromosomes, one set of 23 from each parent. They were combined in the nucleus of a single cell when a sperm cell from your father united with an egg cell from your mother, forming a fertilized egg, or zygote. Mitosis and cytokinesis produced the 200 trillion somatic cells that now make up your body, and the same processes continue to generate new cells to replace dead and damaged ones. In contrast, you produce gametes—eggs or sperm cells—by a variation of cell division called **meiosis,** which yields nonidentical daughter cells that have only one set of chromosomes, thus half as many chromosomes as the parent cell. Meiosis occurs only in your gonads (ovaries or testes). In each generation of humans, meiosis reduces the chromosome number from 46 (two sets of chromosomes) to 23 (one set). Fertilization fuses two gametes together and returns the chromosome number to 46, and mitosis conserves that number in every somatic cell nucleus of the new individual. In Chapter 13, we will examine the role of meiosis in reproduction and inheritance in more detail. In the remainder of this chapter, we focus on mitosis and the rest of the cell cycle.

Concept Check 12.1

1. Starting with a fertilized egg (zygote), a series of five cell divisions would produce an early embryo with how many cells?
2. How many chromatids are in a duplicated chromosome?
3. A chicken has 78 chromosomes in its somatic cells; how many chromosomes did the chicken inherit from each parent? How many chromosomes are in each of the chicken's gametes? How many chromosomes will be in each somatic cell of the chicken's offspring? How many chromosomes are in a "set"?

For suggested answers, see Appendix A.

Concept 12.2

The mitotic phase alternates with interphase in the cell cycle

In 1882, a German anatomist named Walther Flemming developed dyes that allowed him to observe, for the first time, the behavior of chromosomes during mitosis and cytokinesis. (In fact, Flemming coined the terms *mitosis* and *chromatin.*) During the period between one cell division and the next, it appeared to Flemming that the cell was simply growing larger. But we now know that many critical events occur during this stage in the life of a cell.

Phases of the Cell Cycle

Mitosis is just one part of the cell cycle **(Figure 12.5)**. In fact, the **mitotic (M) phase**, which includes both mitosis and cytokinesis, is usually the shortest part of the cell cycle. Mitotic cell division alternates with a much longer stage called **interphase**, which often accounts for about 90% of the cycle. It is during interphase that the cell grows and copies its chromosomes in preparation for cell division. Interphase can be divided into subphases: the **G_1 phase** ("first gap"), the **S phase** ("synthesis"), and the **G_2 phase** ("second gap"). During all three subphases, the cell grows by producing proteins and cytoplasmic organelles such as mitochondria and endoplasmic reticulum. However, chromosomes are duplicated only during the S phase (we discuss synthesis of DNA in Chapter 16). Thus, a cell grows (G_1), continues to grow as it copies its chromosomes (S), grows more as it completes preparations for cell division (G_2), and divides (M). The daughter cells may then repeat the cycle.

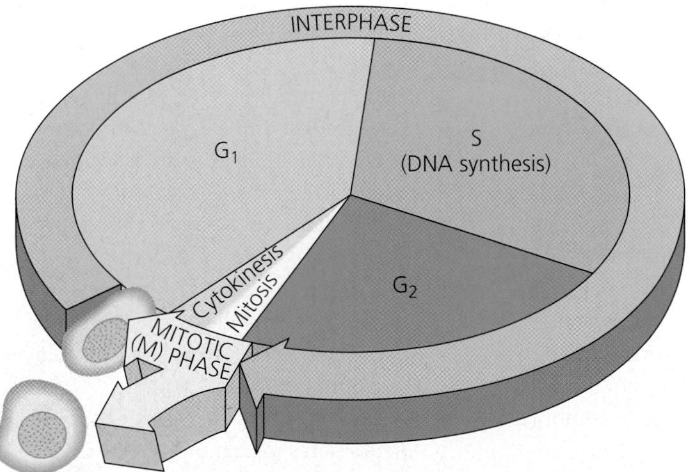

▲ **Figure 12.5 The cell cycle.** In a dividing cell, the mitotic (M) phase alternates with interphase, a growth period. The first part of interphase, called G_1, is followed by the S phase, when the chromosomes replicate; the last part of interphase is called G_2. In the M phase, mitosis divides the nucleus and distributes its chromosomes to the daughter nuclei, and cytokinesis divides the cytoplasm, producing two daughter cells.

A typical human cell might undergo one division in 24 hours. Of this time, the M phase would occupy less than 1 hour, while the S phase might occupy about 10–12 hours, or about half the cycle. The rest of the time would be apportioned between the G_1 and G_2 phases. The G_2 phase usually takes 4–6 hours; in our example, G_1 would occupy about 5–6 hours. G_1 is the most variable in length in different types of cells.

Time-lapse films of living, dividing cells reveal the dynamics of mitosis as a continuum of changes. For purposes of description, however, mitosis is conventionally broken down into five stages: **prophase**, **prometaphase**, **metaphase**, **anaphase**, and **telophase.** Overlapping with the latter stages of mitosis, cytokinesis completes the mitotic phase. **Figure 12.6**, on the next two pages, describes these stages in an animal cell. Be sure to study this figure thoroughly before progressing to the next two sections, which examine mitosis and cytokinesis more closely.

The Mitotic Spindle: *A Closer Look*

Many of the events of mitosis depend on the **mitotic spindle**, which begins to form in the cytoplasm during prophase. This structure consists of fibers made of microtubules and associated proteins. While the mitotic spindle assembles, the other microtubules of the cytoskeleton partially disassemble, probably providing the material used to construct the spindle. The spindle microtubules elongate by incorporating more subunits of the protein tubulin (see Table 6.1).

The assembly of spindle microtubules starts at the **centrosome**, a nonmembranous organelle that functions throughout the cell cycle to organize the cell's microtubules (it is also called the *microtubule-organizing center*). In animal cells, a pair of centrioles is located at the center of the centrosome, but the centrioles are not essential for cell division. In fact, the centrosomes of most plants lack centrioles, and if the centrioles of an animal cell are destroyed with a laser microbeam, a spindle nevertheless forms during mitosis.

During interphase, the single centrosome replicates, forming two centrosomes, which remain together near the nucleus (see Figure 12.6). The two centrosomes move apart from each other during prophase and prometaphase of mitosis, as spindle microtubules grow out from them. By the end of prometaphase, the two centrosomes, one at each pole of the spindle, are at opposite ends of the cell. An **aster,** a radial array of short microtubules, extends from each centrosome. The spindle includes the centrosomes, the spindle microtubules, and the asters.

Each of the two sister chromatids of a chromosome has a **kinetochore,** a structure of proteins associated with specific sections of chromosomal DNA at the centromere. The chromosome's two kinetochores face in opposite directions. During prometaphase, some of the spindle microtubules attach to the kinetochores; these are called kinetochore microtubules. (The number of microtubules attached to a kinetochore varies

Figure 12.6
Exploring The Mitotic Division of an Animal Cell

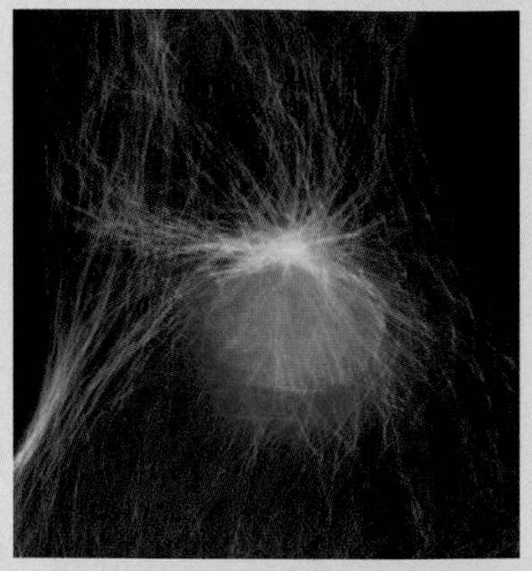

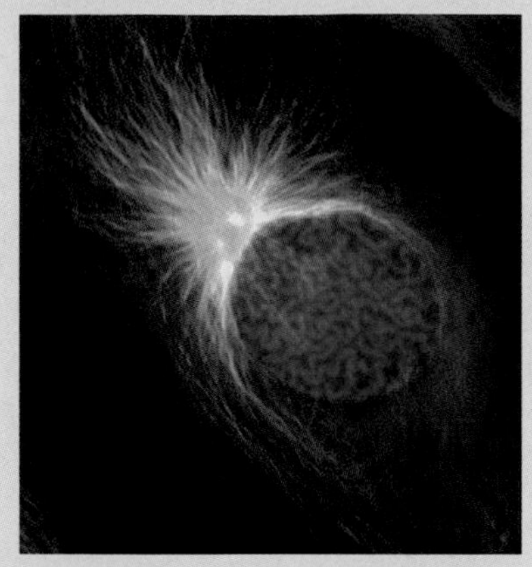

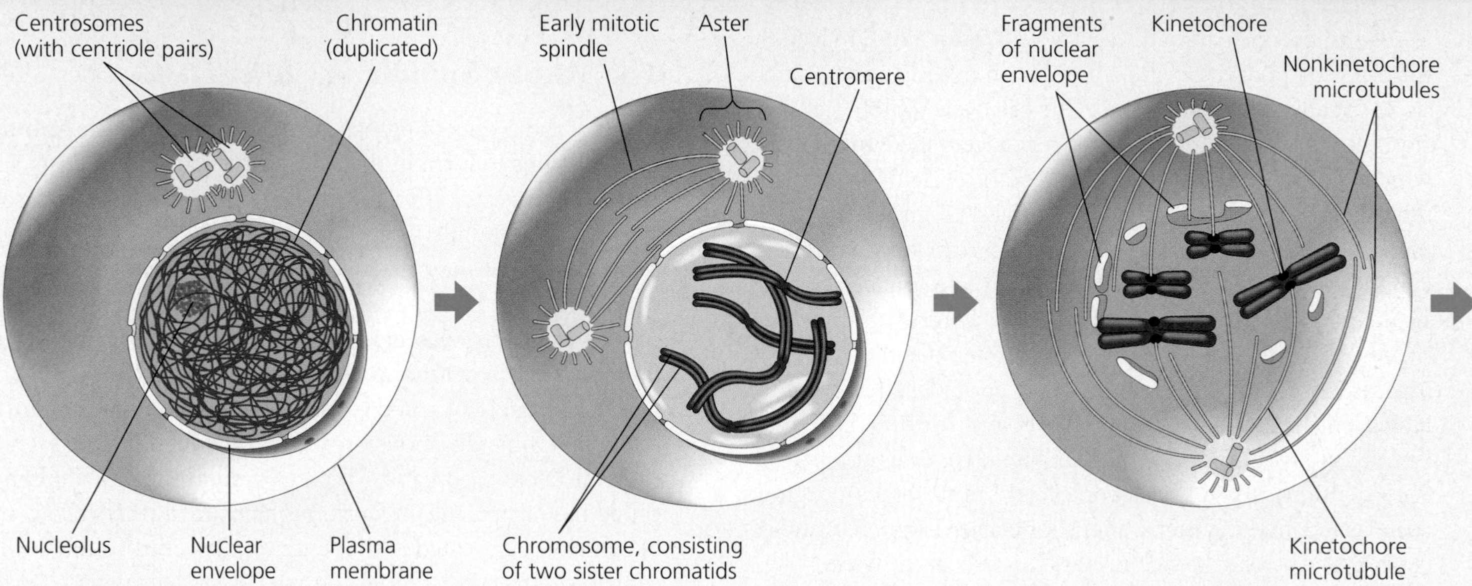

G$_2$ of Interphase

- A nuclear envelope bounds the nucleus.
- The nucleus contains one or more nucleoli (singular, *nucleolus*).
- Two centrosomes have formed by replication of a single centrosome.
- In animal cells, each centrosome features two centrioles.
- Chromosomes, duplicated during S phase, cannot be seen individually because they have not yet condensed.

The light micrographs show dividing lung cells from a newt, which has 22 chromosomes in its somatic cells (chromosomes appear blue, microtubules green, intermediate filaments red). For simplicity, the drawings show only four chromosomes.

Prophase

- The chromatin fibers become more tightly coiled, condensing into discrete chromosomes observable with a light microscope.
- The nucleoli disappear.
- Each duplicated chromosome appears as two identical sister chromatids joined together.
- The mitotic spindle begins to form. It is composed of the centrosomes and the microtubules that extend from them. The radial arrays of shorter microtubules that extend from the centrosomes are called asters ("stars").
- The centrosomes move away from each other, apparently propelled by the lengthening microtubules between them.

Prometaphase

- The nuclear envelope fragments.
- The microtubules of the spindle can now invade the nuclear area and interact with the chromosomes, which have become even more condensed.
- Microtubules extend from each centrosome toward the middle of the cell.
- Each of the two chromatids of a chromosome now has a kinetochore, a specialized protein structure located at the centromere.
- Some of the microtubules attach to the kinetochores, becoming "kinetochore microtubules"; these jerk the chromosomes back and forth.
- Nonkinetochore microtubules interact with those from the opposite pole of the spindle.

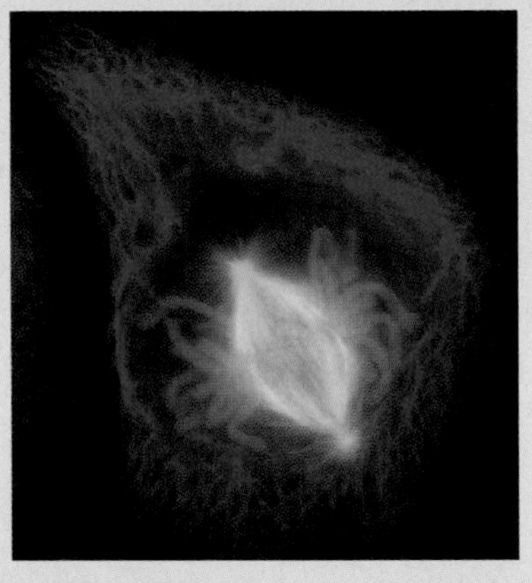

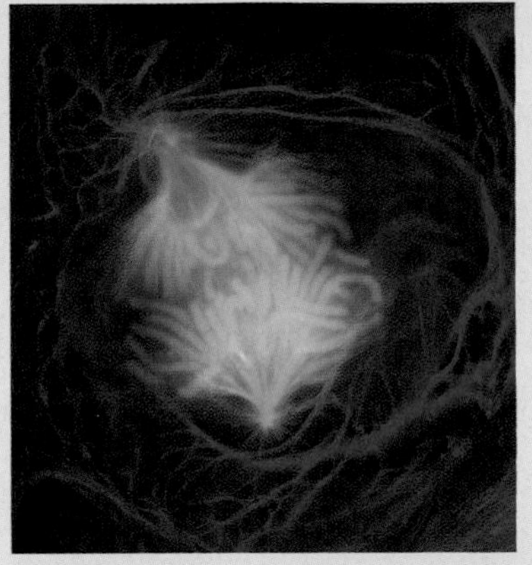

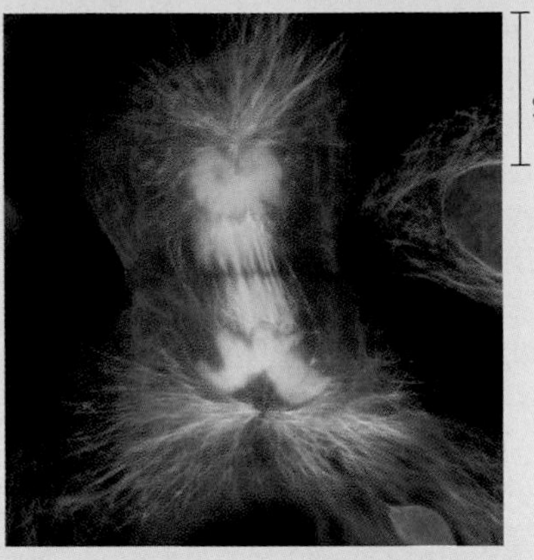

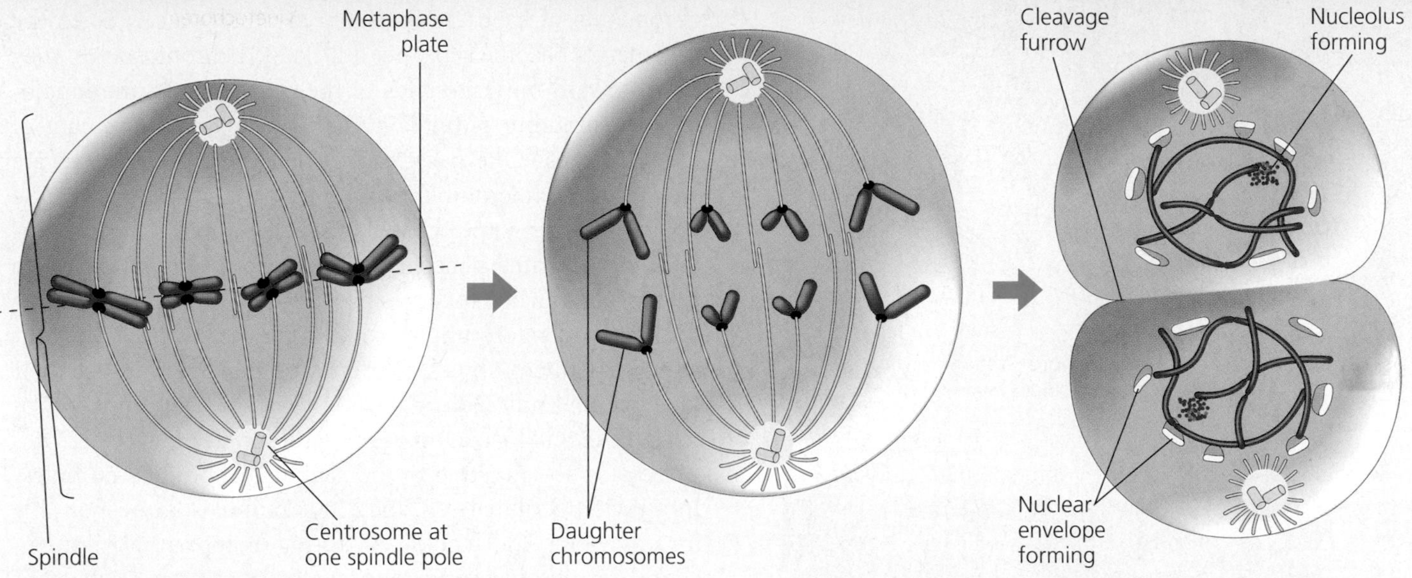

Metaphase

- Metaphase is the longest stage of mitosis, lasting about 20 minutes.
- The centrosomes are now at opposite ends of the cell.
- The chromosomes convene on the metaphase plate, an imaginary plane that is equidistant between the spindle's two poles. The chromosomes' centromeres lie on the metaphase plate.
- For each chromosome, the kinetochores of the sister chromatids are attached to kinetochore microtubules coming from opposite poles.
- The entire apparatus of microtubules is called the spindle because of its shape.

Anaphase

- Anaphase is the shortest stage of mitosis, lasting only a few minutes.
- Anaphase begins when the two sister chromatids of each pair suddenly part. Each chromatid thus becomes a full-fledged chromosome.
- The two liberated chromosomes begin moving toward opposite ends of the cell, as their kinetochore microtubules shorten. Because these microtubules are attached at the centromere region, the chromosomes move centromere first (at about 1 μm/min).
- The cell elongates as the nonkinetochore microtubules lengthen.
- By the end of anaphase, the two ends of the cell have equivalent—and complete—collections of chromosomes.

Telophase

- Two daughter nuclei begin to form in the cell.
- Nuclear envelopes arise from the fragments of the parent cell's nuclear envelope and other portions of the endomembrane system.
- The chromosomes become less condensed.
- Mitosis, the division of one nucleus into two genetically identical nuclei, is now complete.

Cytokinesis

- The division of the cytoplasm is usually well underway by late telophase, so the two daughter cells appear shortly after the end of mitosis.
- In animal cells, cytokinesis involves the formation of a cleavage furrow, which pinches the cell in two.

among species, from one microtubule in yeast cells to 40 or so in some mammalian cells.) When one of a chromosome's kinetochores is "captured" by microtubules, the chromosome begins to move toward the pole from which those microtubules extend. However, this movement is checked as soon as microtubules from the opposite pole attach to the other kinetochore.

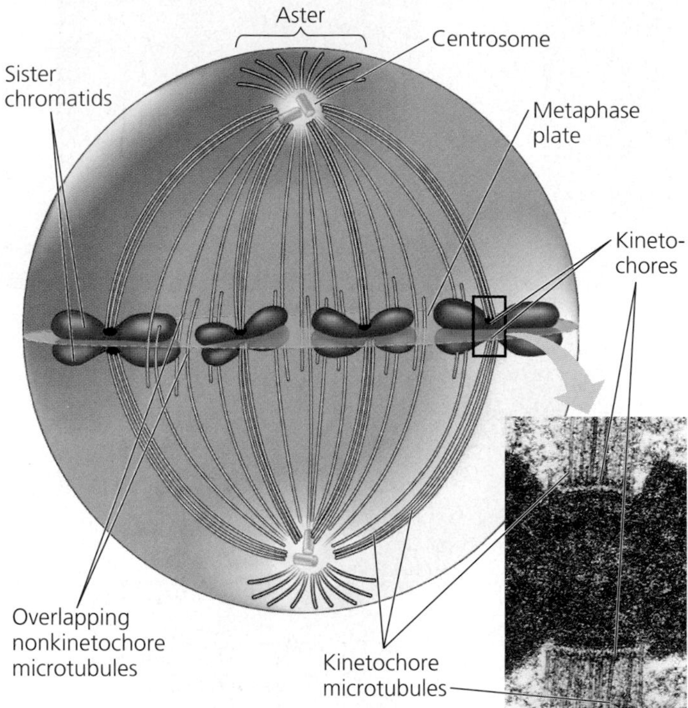

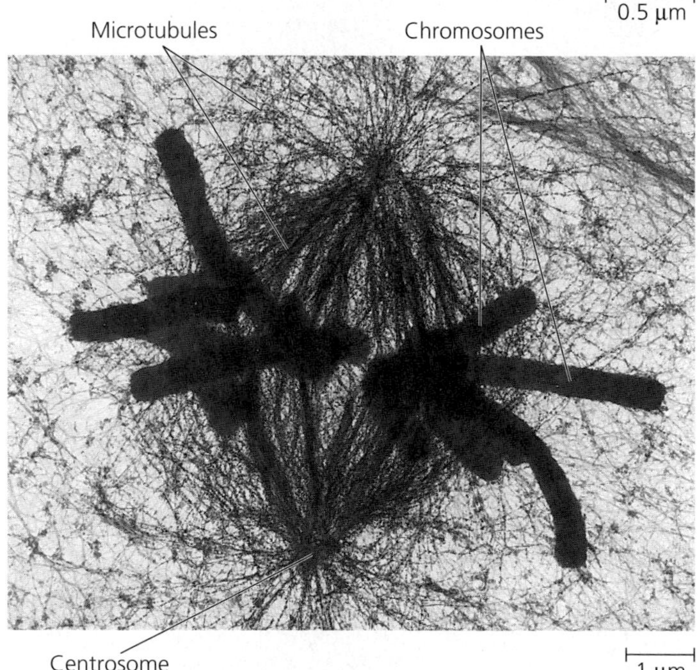

▲ **Figure 12.7 The mitotic spindle at metaphase.** The kinetochores of a chromosome's two sister chromatids face in opposite directions. Here, each kinetochore is actually attached to a *cluster* of kinetochore microtubules extending from the nearest centrosome. Nonkinetochore microtubules overlap at the metaphase plate (TEMs).

What happens next is like a tug-of-war that ends in a draw. The chromosome moves first in one direction, then the other, back and forth, finally settling midway between the two ends of the cell. At metaphase, the centromeres of all the duplicated chromosomes are on a plane midway between the spindle's two poles. This imaginary plane is called the **metaphase plate** of the cell **(Figure 12.7)**. Meanwhile, microtubules that do not attach to kinetochores have been growing, and by metaphase they overlap and interact with other nonkinetochore microtubules from the opposite pole of the spindle. (These are sometimes called "polar" microtubules.) By metaphase, the microtubules of the asters have also grown and are in contact with the plasma membrane. The spindle is now complete.

Let's now see how the structure of the completed spindle correlates with its function during anaphase. Anaphase commences suddenly when proteins holding together the sister chromatids of each chromosome are inactivated. Once the chromatids become separate, full-fledged chromosomes, they move toward opposite ends of the cell. How do the kinetochore microtubules function in this poleward movement of chromosomes? One possibility is that the chromosomes are "reeled in" by microtubules that are shortening at the spindle poles. However, experimental evidence supports the hypothesis that the primary mechanism of movement involves motor proteins on the kinetochores that "walk" a chromosome along the attached microtubules toward the nearest pole. Meanwhile, the microtubules shorten by depolymerizing at their kinetochore ends **(Figure 12.8)**. (To review how motor proteins move an object along a microtubule, see Figure 6.21.)

What is the function of the *non*kinetochore microtubules? In a dividing animal cell, these microtubules are responsible for elongating the whole cell during anaphase. Nonkinetochore microtubules from opposite poles overlap each other extensively during metaphase (see Figure 12.7). During anaphase, the region of overlap is reduced as motor proteins attached to the microtubules walk them away from one another, using energy from ATP. As the microtubules push apart from each other, their spindle poles are pushed apart, elongating the cell. At the same time, the microtubules lengthen somewhat by the addition of tubulin subunits to their overlapping ends. As a result, the microtubules continue to overlap.

At the end of anaphase, duplicate groups of chromosomes have arrived at opposite ends of the elongated parent cell. Nuclei re-form during telophase. Cytokinesis generally begins during these later stages of mitosis, and the spindle eventually disassembles.

Cytokinesis: *A Closer Look*

In animal cells, cytokinesis occurs by a process known as **cleavage.** The first sign of cleavage is the appearance of a **cleavage furrow,** a shallow groove in the cell surface near the old metaphase plate **(Figure 12.9a)**. On the cytoplasmic side of

Figure 12.8

Inquiry During anaphase, do kinetochore microtubules shorten at their spindle pole ends or their kinetochore ends?

EXPERIMENT

1 The microtubules of a cell in early anaphase were labeled with a fluorescent dye that glows in the microscope (yellow).

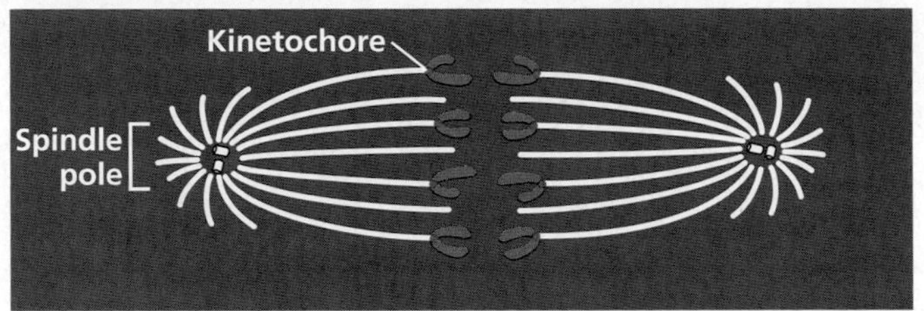

2 A laser was used to mark the kinetochore microtubules by eliminating the fluorescence in a region between one spindle pole and the chromosomes. As anaphase proceeded, researchers monitored the changes in the lengths of the microtubules on either side of the mark.

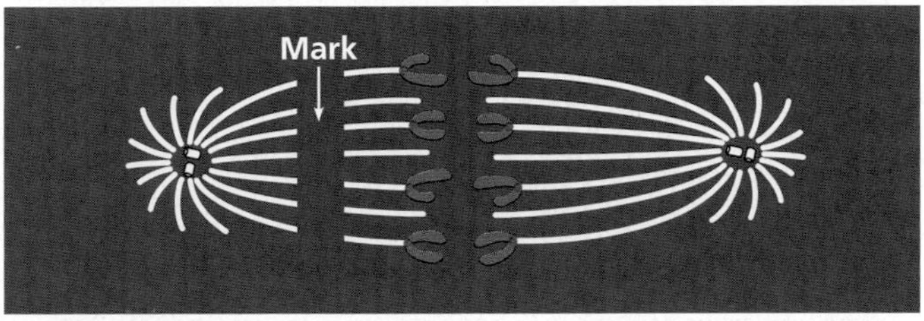

RESULTS As the chromosomes moved toward the poles, the microtubule segments on the kinetochore side of the laser mark shortened, while those on the spindle pole side stayed the same length.

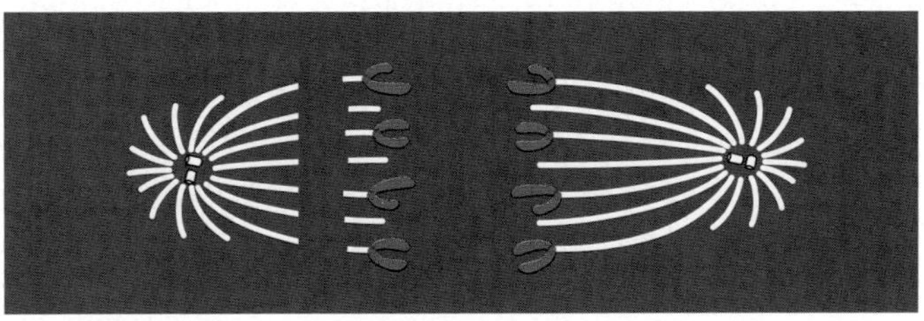

CONCLUSION This experiment demonstrated that during anaphase, kinetochore microtubules shorten at their kinetochore ends, not at their spindle pole ends. This is just one of the experiments supporting the hypothesis that during anaphase, a chromosome tracks along a microtubule as the microtubule depolymerizes at its kinetochore end, releasing tubulin subunits.

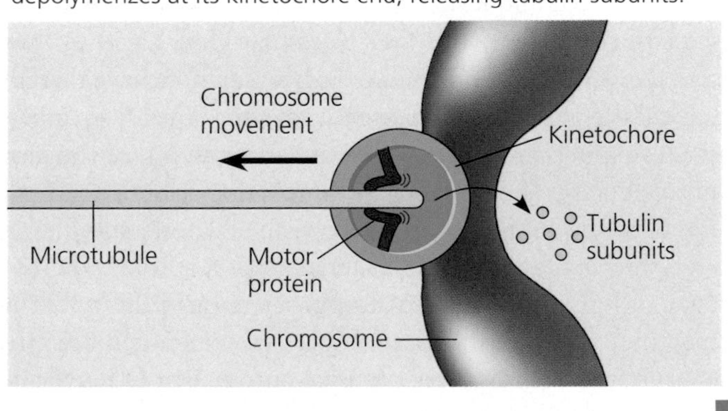

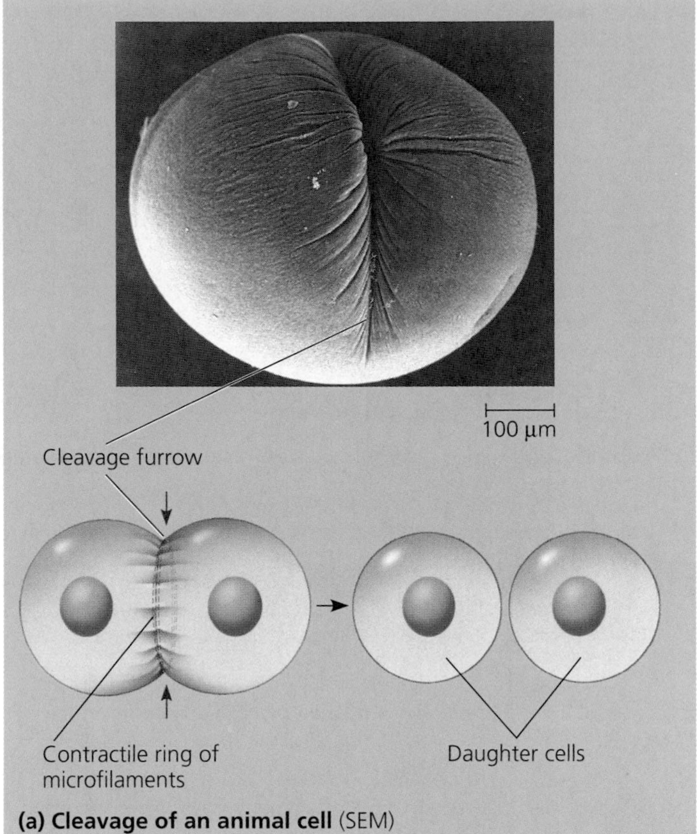

(a) Cleavage of an animal cell (SEM)

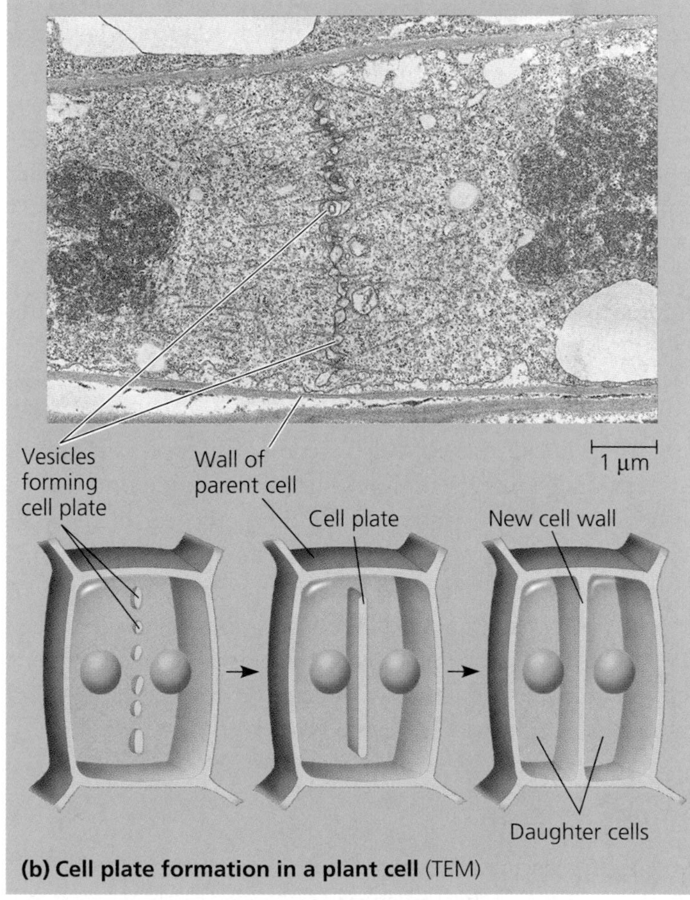

(b) Cell plate formation in a plant cell (TEM)

▲ **Figure 12.9 Cytokinesis in animal and plant cells.**

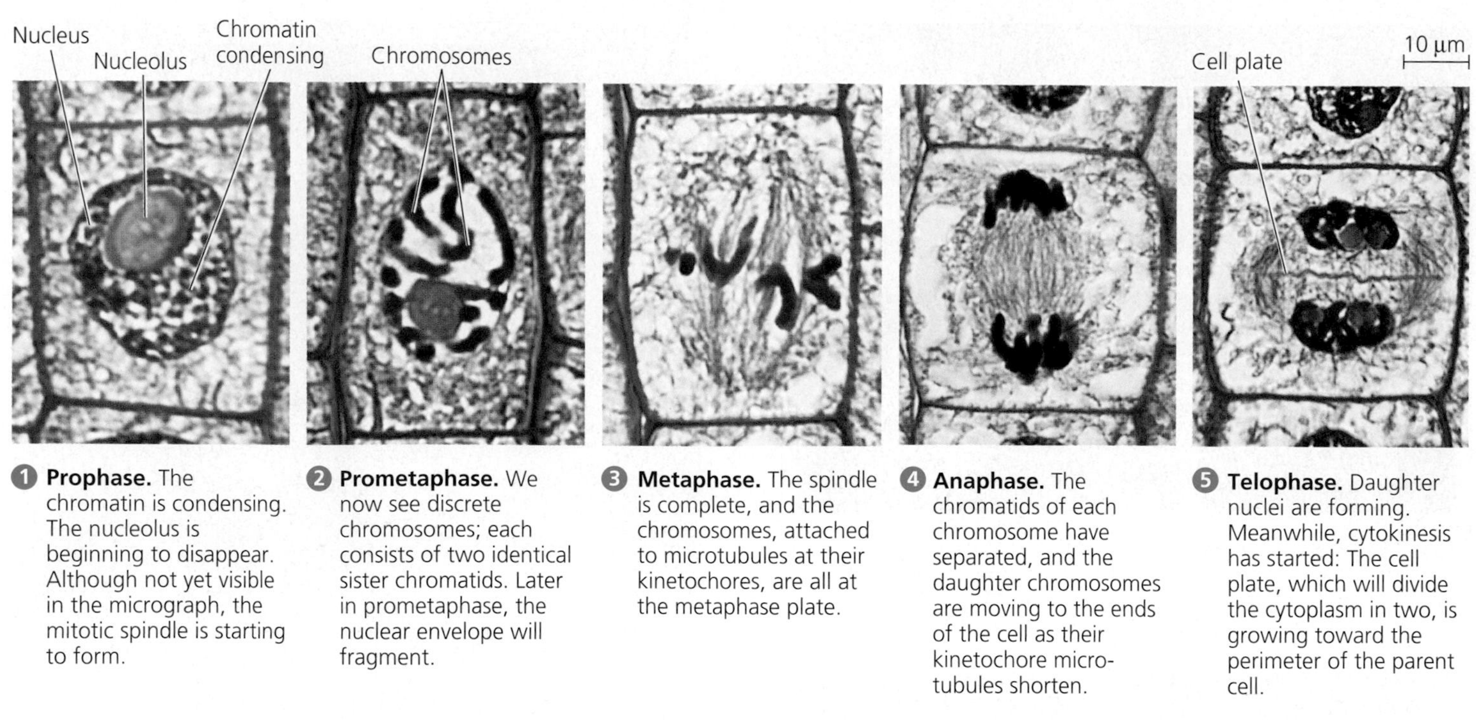

❶ **Prophase.** The chromatin is condensing. The nucleolus is beginning to disappear. Although not yet visible in the micrograph, the mitotic spindle is starting to form.

❷ **Prometaphase.** We now see discrete chromosomes; each consists of two identical sister chromatids. Later in prometaphase, the nuclear envelope will fragment.

❸ **Metaphase.** The spindle is complete, and the chromosomes, attached to microtubules at their kinetochores, are all at the metaphase plate.

❹ **Anaphase.** The chromatids of each chromosome have separated, and the daughter chromosomes are moving to the ends of the cell as their kinetochore microtubules shorten.

❺ **Telophase.** Daughter nuclei are forming. Meanwhile, cytokinesis has started: The cell plate, which will divide the cytoplasm in two, is growing toward the perimeter of the parent cell.

▲ **Figure 12.10 Mitosis in a plant cell.** These light micrographs show mitosis in cells of an onion root.

the furrow is a contractile ring of actin microfilaments associated with molecules of the protein myosin. (Actin and myosin are the same proteins that are responsible for muscle contraction as well as many other kinds of cell movement.) The actin microfilaments interact with the myosin molecules, causing the ring to contract. The contraction of the dividing cell's ring of microfilaments is like the pulling of drawstrings. The cleavage furrow deepens until the parent cell is pinched in two, producing two completely separated cells, each with its own nucleus and share of cytosol and organelles.

Cytokinesis in plant cells, which have cell walls, is markedly different. There is no cleavage furrow. Instead, during telophase, vesicles derived from the Golgi apparatus move along microtubules to the middle of the cell, where they coalesce, producing a **cell plate** **(Figure 12.9b)**. Cell wall materials carried in the vesicles collect in the cell plate as it grows. The cell plate enlarges until its surrounding membrane fuses with the plasma membrane along the perimeter of the cell. Two daughter cells result, each with its own plasma membrane. Meanwhile, a new cell wall arising from the contents of the cell plate has formed between the daughter cells.

Figure 12.10 is a series of micrographs of a dividing plant cell. Examining this figure will help you review mitosis and cytokinesis.

Binary Fission

Prokaryotes (bacteria and archaea) reproduce by a type of cell division called **binary fission,** meaning "division in half." In bacteria, most genes are carried on a single *bacterial chromosome* that consists of a circular DNA molecule and associated proteins. Although bacteria are smaller and simpler than eukaryotic cells, the problem of replicating their genomes in an orderly fashion and distributing the copies equally to two daughter cells is still formidable. The chromosome of the bacterium *Escherichia coli,* for example, when it is fully stretched out, is about 500 times longer than the length of the cell. Clearly, such a long chromosome must be highly coiled and folded within the cell—and it is.

In *E. coli,* the process of cell division begins when the DNA of the bacterial chromosome begins to replicate at a specific place on the chromosome called the **origin of replication,** producing two origins. As the chromosome continues to replicate, one origin moves rapidly toward the opposite end of the cell **(Figure 12.11)**. While the chromosome is replicating, the cell elongates. When replication is complete and the bacterium has reached about twice its initial size, its plasma membrane grows inward, dividing the parent *E. coli* cell into two daughter cells. Each cell inherits a complete genome.

Using the techniques of modern DNA technology to tag the origins of replication with molecules that glow green in fluorescence microscopy (see Figure 6.3), researchers have directly observed the movement of bacterial chromosomes. This movement is reminiscent of the poleward movements of the centromere regions of eukaryotic chromosomes during anaphase of mitosis, but bacteria don't have visible mitotic spindles or even microtubules. In most bacterial species studied, the two origins of replication end up at opposite ends of the cell or in some other very specific location, possibly anchored there by one or more proteins. How bacterial chromosomes move and how their specific location is established and maintained are

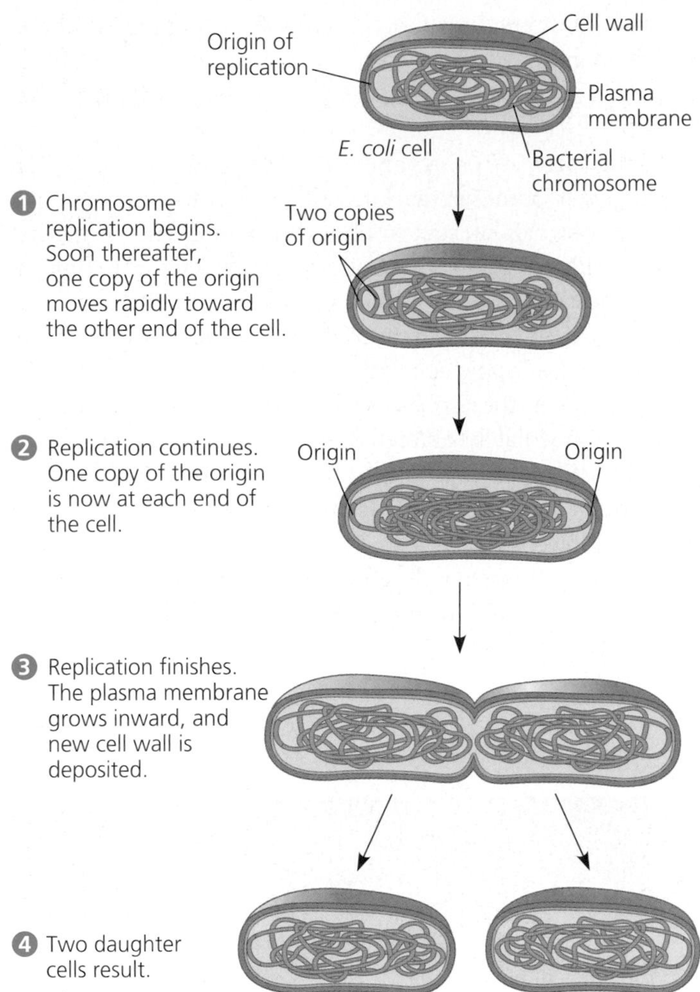
Cell wall
Origin of replication
Plasma membrane
E. coli cell
Bacterial chromosome
1 Chromosome replication begins. Soon thereafter, one copy of the origin moves rapidly toward the other end of the cell.
Two copies of origin
2 Replication continues. One copy of the origin is now at each end of the cell.
Origin
Origin
3 Replication finishes. The plasma membrane grows inward, and new cell wall is deposited.
4 Two daughter cells result.

▲ **Figure 12.11 Bacterial cell division (binary fission).** The example shown here is the bacterium *E. coli*. The single, circular chromosome replicates, and the two copies move apart by an unknown mechanism, so that the two origins of replication (green) end up at opposite ends of the cell. Meanwhile, the cell elongates. When chromosomal replication is complete, the plasma membrane grows inward, dividing the cell in two as a new cell wall is deposited between the daughter cells.

becoming clearer but are still not fully understood. Several proteins have been identified that play important roles.

The Evolution of Mitosis

How did mitosis evolve? Given that prokaryotes preceded eukaryotes on Earth by more than a billion years, we might hypothesize that mitosis had its origins in simpler prokaryotic mechanisms of cell reproduction. In fact, some of the proteins involved in bacterial binary fission are related to eukaryotic proteins, strengthening the case for the evolution of mitosis from bacterial cell division. Intriguingly, recent work has shown that two of the proteins involved in binary fission are related to eukaryotic tubulin and actin proteins.

As eukaryotes evolved, along with their larger genomes and nuclear envelopes, the ancestral process of binary fission somehow gave rise to mitosis. **Figure 12.12** traces a

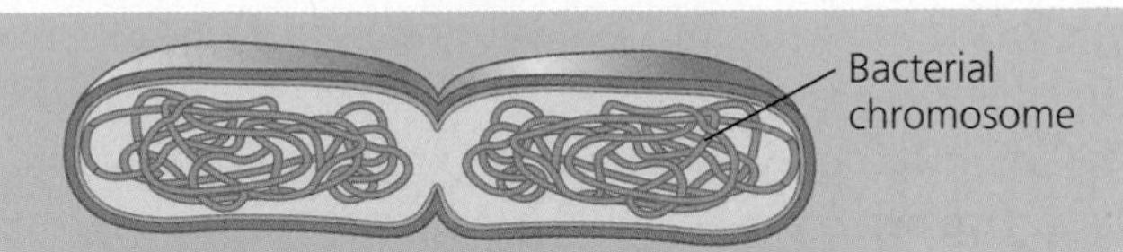
Bacterial chromosome

(a) Prokaryotes. During binary fission, the origins of the daughter chromosomes move to opposite ends of the cell. The mechanism is not fully understood, but proteins may anchor the daughter chromosomes to specific sites on the plasma membrane.

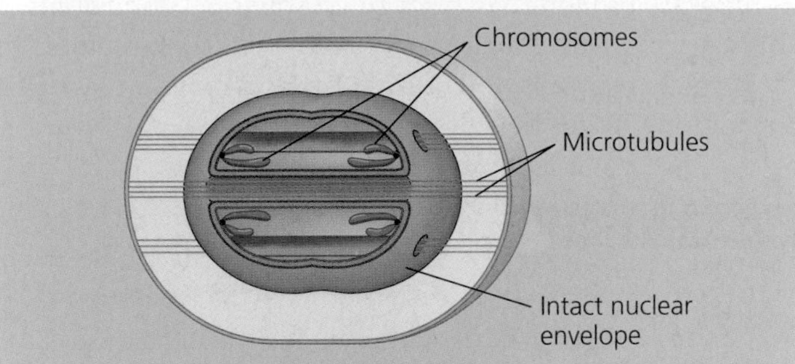
Chromosomes
Microtubules
Intact nuclear envelope

(b) Dinoflagellates. In unicellular protists called dinoflagellates, the nuclear envelope remains intact during cell division, and the chromosomes attach to the nuclear envelope. Microtubules pass through the nucleus inside cytoplasmic tunnels, reinforcing the spatial orientation of the nucleus, which then divides in a fission process reminiscent of bacterial division.

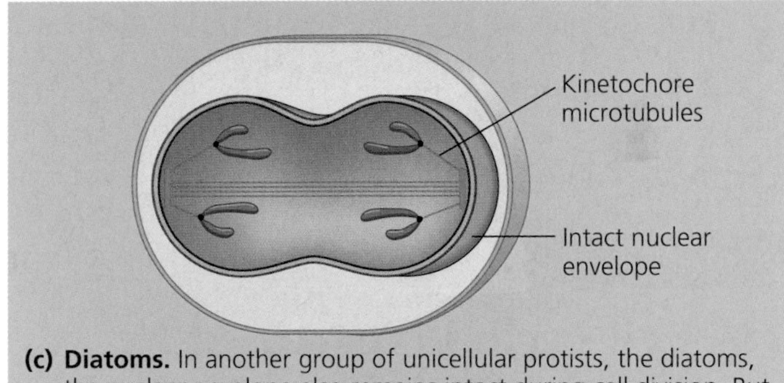
Kinetochore microtubules
Intact nuclear envelope

(c) Diatoms. In another group of unicellular protists, the diatoms, the nuclear envelope also remains intact during cell division. But in these organisms, the microtubules form a spindle *within* the nucleus. Microtubules separate the chromosomes, and the nucleus splits into two daughter nuclei.

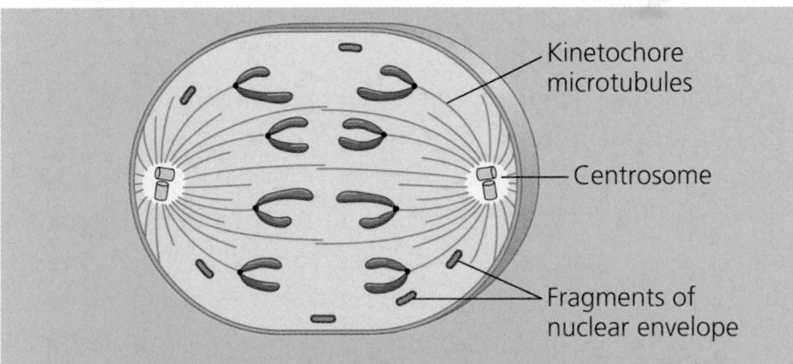
Kinetochore microtubules
Centrosome
Fragments of nuclear envelope

(d) Most eukaryotes. In most other eukaryotes, including plants and animals, the spindle forms outside the nucleus, and the nuclear envelope breaks down during mitosis. Microtubules separate the chromosomes, and the nuclear envelope then re-forms.

▲ **Figure 12.12 A hypothetical sequence for the evolution of mitosis.** In modern organisms, researchers have observed mechanisms of cell division that appear to be intermediate between the binary fission of bacteria (a) and mitosis as it occurs in most eukaryotes (d). Except for (a), these schematic diagrams do not show cell walls.

hypothesis for the stepwise evolution of mitosis. Possible intermediate stages are represented by two unusual types of nuclear division found in certain modern unicellular protists. These two examples of nuclear division are thought to be cases where ancestral mechanisms have remained relatively unchanged over evolutionary time. In both types, the nuclear envelope remains intact. In dinoflagellates, replicated chromosomes are attached to the nuclear envelope and separate as the nucleus elongates prior to cell division. In diatoms, a spindle within the nucleus separates the chromosomes. In most eukaryotic cells, the nuclear envelope breaks down and a spindle separates the chromosomes.

Concept Check 12.2

1. During which stages of a cell cycle would a chromosome consist of two identical chromatids?
2. How many chromosomes are shown in the Figure 12.7 diagram? How many chromatids are shown?
3. Compare cytokinesis in animal cells and plant cells.
4. What is a function of nonkinetochore microtubules?
5. Identify three similarities between bacterial chromosomes and eukaryotic chromosomes, considering both structure and behavior during cell division.

For suggested answers, see Appendix A.

Concept 12.3

The cell cycle is regulated by a molecular control system

The timing and rate of cell division in different parts of a plant or animal are crucial to normal growth, development, and maintenance. The frequency of cell division varies with the type of cell. For example, human skin cells divide frequently throughout life, whereas liver cells maintain the ability to divide but keep it in reserve until an appropriate need arises—say, to repair a wound. Some of the most specialized cells, such as mature, fully formed nerve cells and muscle cells, do not divide at all in a mature human. These cell cycle differences result from regulation at the molecular level. The mechanisms of this regulation are of intense interest, not only for understanding the life cycles of normal cells but also for understanding how cancer cells manage to escape the usual controls.

Evidence for Cytoplasmic Signals

What drives the cell cycle? One reasonable hypothesis might be that each event in the cycle triggers the next. According to this hypothesis, for example, the replication of chromosomes in the S phase might cause cell growth during the G_2 phase, which might in turn directly trigger the onset of mitosis. However, this apparently logical hypothesis is not in fact correct.

In the early 1970s, a variety of experiments suggested an alternative hypothesis: that the cell cycle is driven by specific molecular signals present in the cytoplasm. Some of the first strong evidence for this hypothesis came from experiments with mammalian cells grown in culture. In these experiments, two cells in different phases of the cell cycle were fused to form a single cell with two nuclei. If one of the original cells was in the S phase and the other was in G_1, the G_1 nucleus immediately entered the S phase, as though stimulated by chemicals present in the cytoplasm of the first cell. Similarly, if a cell undergoing mitosis (M phase) was fused with another cell in any stage of its cell cycle, even G_1, the second nucleus immediately entered mitosis, with condensation of the chromatin and formation of a mitotic spindle **(Figure 12.13)**.

Figure 12.13

Inquiry Are there molecular signals in the cytoplasm that regulate the cell cycle?

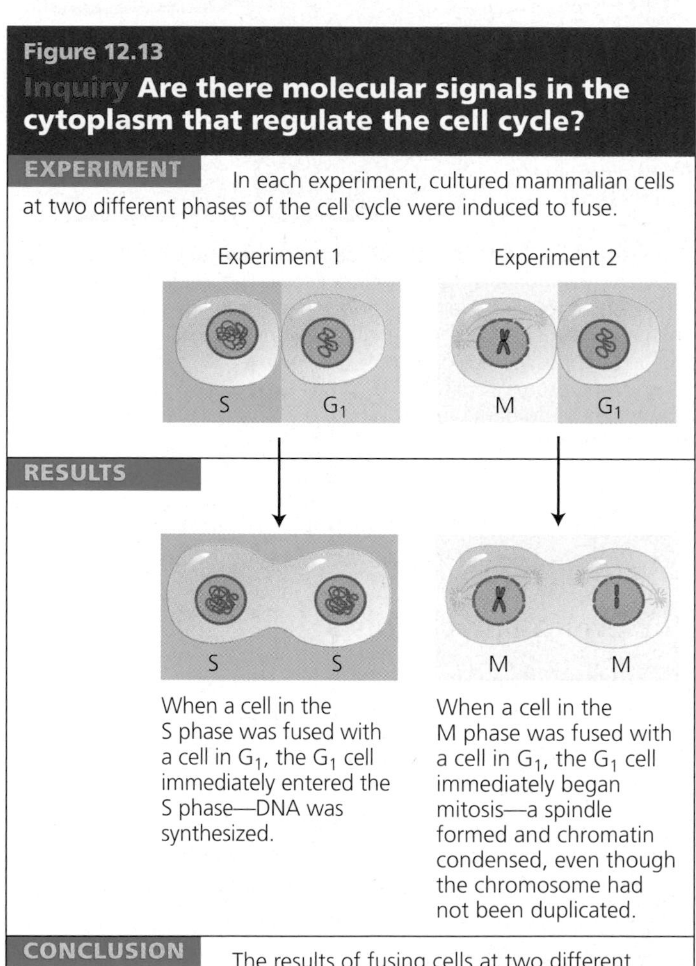

EXPERIMENT
In each experiment, cultured mammalian cells at two different phases of the cell cycle were induced to fuse.
Experiment 1
Experiment 2
S
G1
M
G1
RESULTS
S
S
M
M
When a cell in the S phase was fused with a cell in G1, the G1 cell immediately entered the S phase—DNA was synthesized.
When a cell in the M phase was fused with a cell in G1, the G1 cell immediately began mitosis—a spindle formed and chromatin condensed, even though the chromosome had not been duplicated.
CONCLUSION
The results of fusing cells at two different phases of the cell cycle suggest that molecules present in the cytoplasm of cells in the S or M phase control the progression of phases.

The Cell Cycle Control System

The experiment shown in Figure 12.13 and other experiments demonstrated that the sequential events of the cell cycle are directed by a distinct **cell cycle control system,** a cyclically operating set of molecules in the cell that both triggers and coordinates key events in the cell cycle. The cell cycle control system has been compared to the control device of an automatic washing machine **(Figure 12.14)**. Like the washer's timing device, the cell cycle control system proceeds on its own, driven by a built-in clock. However, just as a washer's cycle is subject to both internal control (such as the sensor that detects when the tub is filled with water) and external adjustment (such as activation of the start mechanism), the cell cycle is regulated at certain checkpoints by both internal and external controls.

A **checkpoint** in the cell cycle is a critical control point where stop and go-ahead signals can regulate the cycle. (The signals are transmitted within the cell by the kinds of signal transduction pathways discussed in Chapter 11.) Animal cells generally have built-in stop signals that halt the cell cycle at checkpoints until overridden by go-ahead signals. Many signals registered at checkpoints come from cellular surveillance mechanisms inside the cell; the signals report whether crucial cellular processes up to that point have been completed correctly and thus whether or not the cell cycle should proceed. Checkpoints also register signals from outside the cell, as we will discuss later. Three major checkpoints are found in the G_1, G_2, and M phases (see Figure 12.14).

For many cells, the G_1 checkpoint—dubbed the "restriction point" in mammalian cells—seems to be the most important. If a cell receives a go-ahead signal at the G_1 checkpoint, it will usually complete the S, G_2, and M phases and divide. Alternatively, if it does not receive a go-ahead signal at that point, it will exit the cycle, switching into a nondividing state called the **G_0 phase** **(Figure 12.15)**. Most cells of the human body are actually in the G_0 phase. As mentioned earlier, fully formed, mature nerve cells and muscle cells never divide. Other cells, such as liver cells, can be "called back" from the G_0 phase to the cell cycle by certain external cues, such as growth factors released during injury.

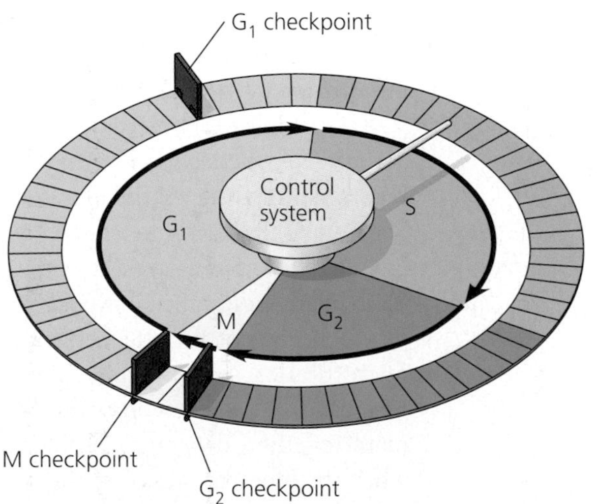

▲ **Figure 12.14 Mechanical analogy for the cell cycle control system.** In this diagram of the cell cycle, the flat "stepping stones" around the perimeter represent sequential events. Like the control device of an automatic washer, the cell cycle control system proceeds on its own, driven by a built-in clock. However, the system is subject to regulation at various checkpoints, of which three are shown (red).

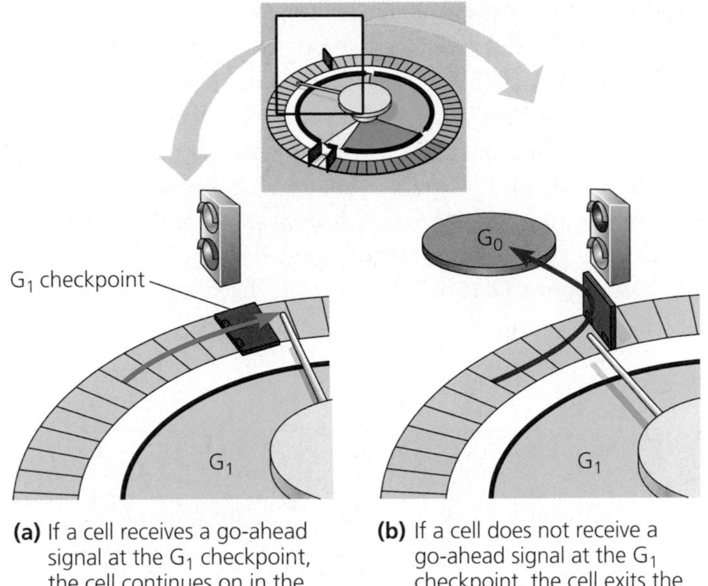

(a) If a cell receives a go-ahead signal at the G_1 checkpoint, the cell continues on in the cell cycle.

(b) If a cell does not receive a go-ahead signal at the G_1 checkpoint, the cell exits the cell cycle and goes into G_0, a nondividing state.

▲ **Figure 12.15 The G_1 checkpoint.**

To understand how cell cycle checkpoints work, we first need to see what kinds of molecules make up the cell cycle control system (the molecular basis for the cell cycle clock) and how a cell progresses forward through the cycle. Then we will consider the internal and external checkpoint signals that can make the clock pause or continue.

The Cell Cycle Clock: Cyclins and Cyclin-Dependent Kinases

Rhythmic fluctuations in the abundance and activity of cell cycle control molecules pace the sequential events of the cell cycle. These regulatory molecules are proteins of two main types: kinases and cyclins. Protein kinases are enzymes that activate or inactivate other proteins by phosphorylating them (see Chapter 11). Particular protein kinases give the go-ahead signals at the G_1 and G_2 checkpoints.

The kinases that drive the cell cycle are actually present at a constant concentration in the growing cell, but much of the time they are in an inactive form. To be active, such a kinase must be attached to a **cyclin,** a protein that gets its name from its cyclically fluctuating concentration in the cell. Because of this requirement, these kinases are called **cyclin-dependent kinases,** or **Cdks.** The activity of a Cdk rises and falls with changes in the concentration of its cyclin partner.

Figure 12.16a shows the fluctuating activity of the cyclin-Cdk complex that was discovered first, called **MPF.** Note that the peaks of MPF activity correspond to the peaks of cyclin concentration. The cyclin level rises during the S and G_2 phases, then falls abruptly during mitosis (M).

The initials MPF stand for "maturation-promoting factor," but we can think of MPF as "M-phase-promoting factor" because it triggers the cell's passage past the G_2 checkpoint into M phase **(Figure 12.16b)**. When cyclins that accumulate during G_2 associate with Cdk molecules, the resulting MPF complex initiates mitosis, phosphorylating a variety of proteins. MPF acts both directly as a kinase and indirectly by activating other kinases. For example, MPF causes phosphorylation of various proteins of the nuclear lamina (see Figure 6.10), which promotes fragmentation of the nuclear envelope during prometaphase of mitosis. There is also evidence that MPF contributes to molecular events required for chromosome condensation and spindle formation during prophase.

During anaphase, MPF helps switch itself off by initiating a process that leads to the destruction of its own cyclin. The noncyclin part of MPF, the Cdk, persists in the cell in inactive form until it associates with new cyclin molecules synthesized during the S and G_2 phases of the next round of the cycle.

What about the G_1 checkpoint? Recent research suggests the involvement of at least three Cdk proteins and several different cyclins at this checkpoint. The fluctuating activities of different cyclin-Cdk complexes seem to control all the stages of the cell cycle.

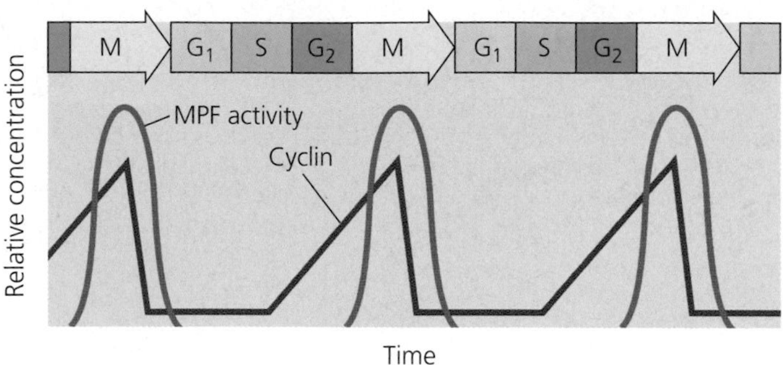
M
G1
S
G2
M
G1
S
G2
M
MPF activity
Cyclin
Relative concentration
Time

(a) Fluctuation of MPF activity and cyclin concentration during the cell cycle

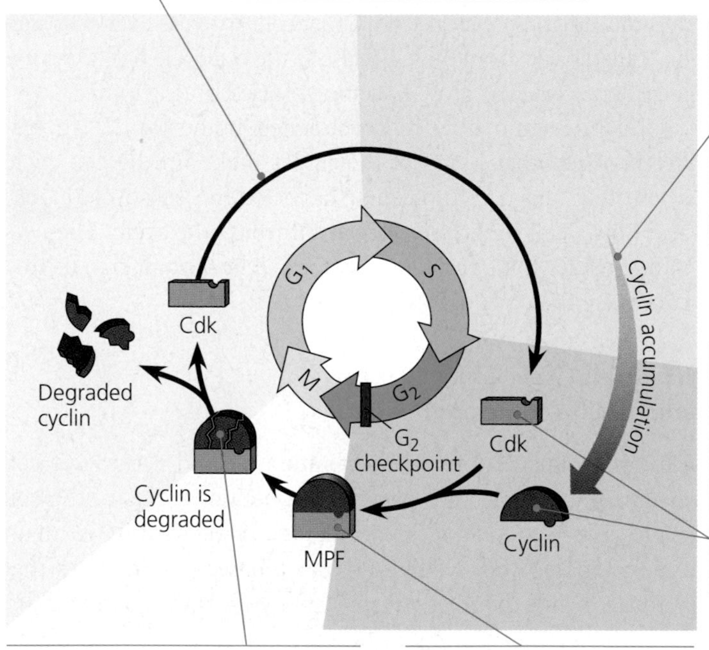
5 During G1, conditions in the cell favor degradation of cyclin, and the Cdk component of MPF is recycled.
1 Synthesis of cyclin begins in late S phase and continues through G2. Because cyclin is protected from degradation during this stage, it accumulates.
G1
S
M
G2
Cdk
Degraded cyclin
G2 checkpoint
Cdk
Cyclin accumulation
Cyclin is degraded
MPF
Cyclin
2 Accumulated cyclin molecules combine with recycled Cdk molecules, producing enough molecules of MPF to pass the G2 checkpoint and initiate the events of mitosis.
4 During anaphase, the cyclin component of MPF is degraded, terminating the M phase. The cell enters the G1 phase.
3 MPF promotes mitosis by phosphorylating various proteins. MPF's activity peaks during metaphase.

(b) Molecular mechanisms that help regulate the cell cycle

▲ **Figure 12.16 Molecular control of the cell cycle at the G_2 checkpoint.** The steps of the cell cycle are timed by rhythmic fluctuations in the activity of cyclin-dependent kinases (Cdks). Here we focus on a cyclin-Cdk complex called MPF, which acts at the G_2 checkpoint as a go-ahead signal, triggering the events of mitosis.

Stop and Go Signs: Internal and External Signals at the Checkpoints

Research scientists are only in the early stages of working out the signaling pathways that link cyclin-dependent kinases to other molecules and events inside and outside the cell. For example, they know that in general, active Cdks function by phosphorylating substrate proteins that affect particular steps in the cell cycle. In many cases, though, scientists don't yet know what the various Cdks actually do. However, they have identified some steps of the signaling pathways that convey information to the cell cycle machinery.

An example of an internal signal occurs at the M phase checkpoint. Anaphase, the separation of sister chromatids, does not begin until all the chromosomes are properly attached to the spindle at the metaphase plate. Researchers have learned that kinetochores not yet attached to spindle microtubules send a molecular signal that causes the sister chromatids to remain together, delaying anaphase. Only when the kinetochores of all the chromosomes are attached to the spindle will the sister chromatids separate (owing to inactivation of the proteins holding them together). This mechanism ensures that daughter cells do not end up with missing or extra chromosomes.

By growing animal cells in culture, researchers have been able to identify many external factors, both chemical and physical, that can influence cell division. For example, cells fail to divide if an essential nutrient is left out of the culture medium. (This is analogous to trying to run an automatic washing machine without the water supply hooked up.) And even if all other conditions are favorable, most types of mammalian cells divide in culture only if the growth medium includes specific growth factors. As mentioned in Chapter 11, a **growth factor** is a protein released by certain cells that stimulates other cells to divide. While called a growth factor for historical reasons, a protein that promotes mitosis is sometimes more narrowly called a mitogen.

One such growth factor is *platelet-derived growth factor (PDGF)*, which is made by blood cells called platelets. The experiment illustrated in **Figure 12.17** demonstrates that PDGF is required for the division of fibroblasts in culture. Fibroblasts, a type of connective tissue cell, have PDGF receptors on their plasma membranes. The binding of PDGF molecules to these receptors (which are receptor tyrosine kinases; see Chapter 11) triggers a signal transduction pathway that allows the cells to pass the G_1 checkpoint and divide. PDGF stimulates fibroblast division not only in the artificial conditions of cell culture, but in an animal's body as well. When an injury occurs, platelets release PDGF in the vicinity. The resulting proliferation of fibroblasts helps heal the wound. Researchers have discovered at least 50 different growth factors that can trigger cells to divide. Different cell types respond specifically to a certain growth factor or combination of growth factors.

The effect of an external physical factor on cell division is clearly seen in **density-dependent inhibition,** a phenomenon in which crowded cells stop dividing (**Figure 12.18a**, on the next page). As first observed many years ago, cultured cells normally divide until they form a single layer of cells on the inner surface of the culture container, at which point the cells stop dividing. If some cells are removed, those bordering the open space begin dividing again and continue until the vacancy is filled. It was originally thought that a cell's physical contact with neighboring cells signaled it to stop dividing. However, while physical contact may have some influence, it turns out that the amount of required growth factors and nutrients available to each cell has a more important effect: Apparently, when a cell population reaches a certain density, the availability of nutrients becomes insufficient to allow continued cell growth and division.

Most animal cells also exhibit **anchorage dependence** (see Figure 12.18a). To divide, they must be attached to a substratum, such as the inside of a culture jar or the extracellular matrix of a tissue. Experiments suggest that anchorage is signaled to the cell cycle control system via pathways involving plasma membrane proteins and elements of the cytoskeleton linked to them.

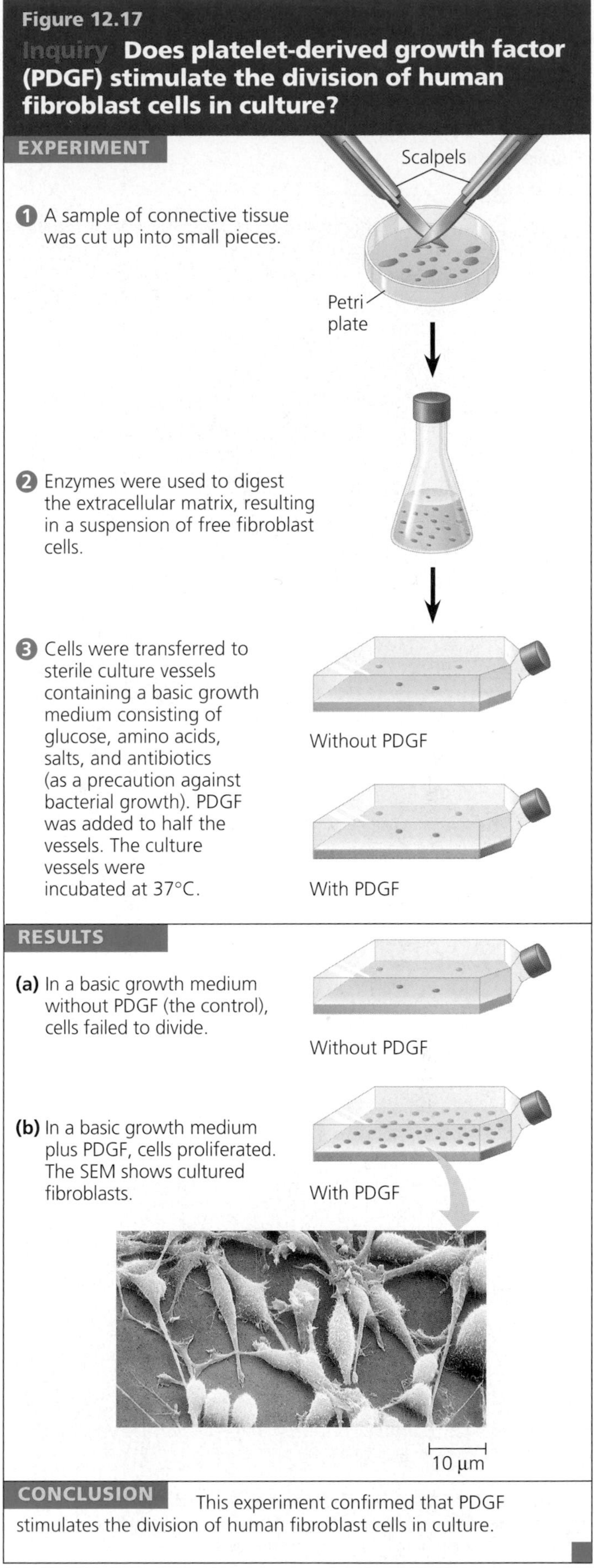

Figure 12.17
Inquiry Does platelet-derived growth factor (PDGF) stimulate the division of human fibroblast cells in culture?
EXPERIMENT
Scalpels
1 A sample of connective tissue was cut up into small pieces.
Petri plate
2 Enzymes were used to digest the extracellular matrix, resulting in a suspension of free fibroblast cells.
3 Cells were transferred to sterile culture vessels containing a basic growth medium consisting of glucose, amino acids, salts, and antibiotics (as a precaution against bacterial growth). PDGF was added to half the vessels. The culture vessels were incubated at 37°C.
Without PDGF
With PDGF
RESULTS
(a) In a basic growth medium without PDGF (the control), cells failed to divide.
Without PDGF
(b) In a basic growth medium plus PDGF, cells proliferated. The SEM shows cultured fibroblasts.
With PDGF
10 µm
CONCLUSION This experiment confirmed that PDGF stimulates the division of human fibroblast cells in culture.

Density-dependent inhibition and anchorage dependence appear to function in the body's tissues as well as in cell culture, checking the growth of cells at some optimal density and location. Cancer cells, which we discuss next, exhibit neither density-dependent inhibition nor anchorage dependence **(Figure 12.18b)**.

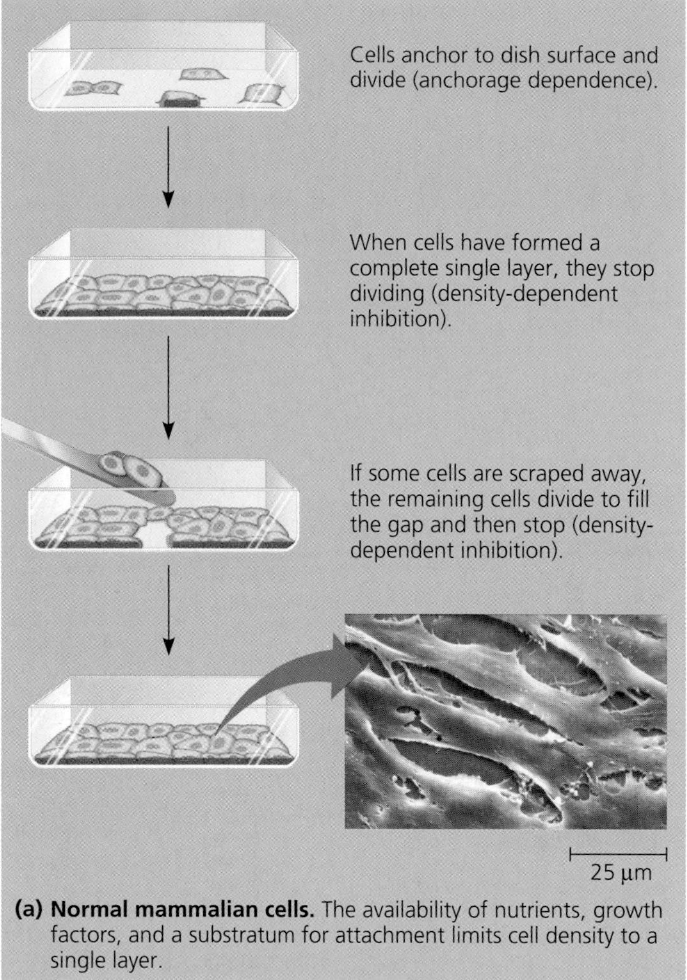

Cells anchor to dish surface and divide (anchorage dependence).
When cells have formed a complete single layer, they stop dividing (density-dependent inhibition).
If some cells are scraped away, the remaining cells divide to fill the gap and then stop (density-dependent inhibition).
25 µm

(a) Normal mammalian cells. The availability of nutrients, growth factors, and a substratum for attachment limits cell density to a single layer.

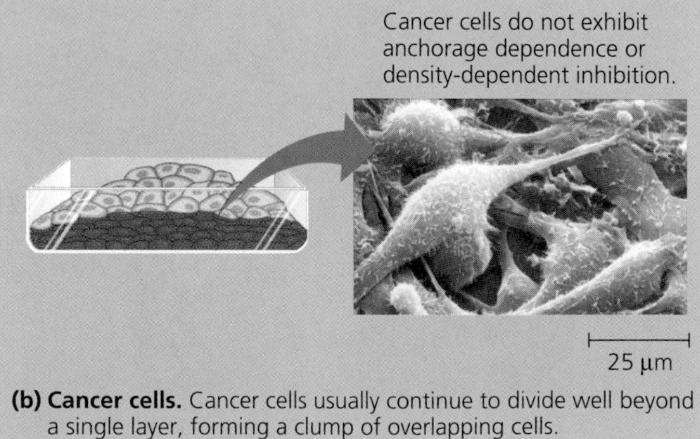

Cancer cells do not exhibit anchorage dependence or density-dependent inhibition.
25 µm

(b) Cancer cells. Cancer cells usually continue to divide well beyond a single layer, forming a clump of overlapping cells.

▲ **Figure 12.18 Density-dependent inhibition and anchorage dependence of cell division.** Individual cells are shown disproportionately large in the drawings.

Loss of Cell Cycle Controls in Cancer Cells

Cancer cells do not respond normally to the body's control mechanisms. They divide excessively and invade other tissues. If unchecked, they can kill the organism.

By studying cells growing in culture, researchers have learned that cancer cells do not heed the normal signals that regulate the cell cycle. For example, as Figure 12.18b shows, cancer cells do not exhibit density-dependent inhibition when growing in culture; they do not stop dividing when growth factors are depleted. A logical hypothesis to explain this behavior is that cancer cells do not need growth factors in their culture medium in order to grow and divide. They may make a required growth factor themselves, or they may have an abnormality in the signaling pathway that conveys the growth factor's signal to the cell cycle control system even in the absence of that factor. Another possibility is an abnormal cell cycle control system. In fact, as you will learn in Chapter 19, these are all conditions that may lead to cancer.

There are other important differences between normal cells and cancer cells that reflect derangements of the cell cycle. If and when they stop dividing, cancer cells do so at random points in the cycle, rather than at the normal checkpoints. Moreover, in culture, cancer cells can go on dividing indefinitely if they are given a continual supply of nutrients; they are said to be "immortal." A striking example is a cell line that has been reproducing in culture since 1951. Cells of this line are called HeLa cells because their original source was a tumor removed from a woman named Henrietta Lacks. By contrast, nearly all normal mammalian cells growing in culture divide only about 20 to 50 times before they stop dividing, age, and die. (We'll see a possible reason for this phenomenon when we discuss chromosome replication in Chapter 16.)

The abnormal behavior of cancer cells can be catastrophic when it occurs in the body. The problem begins when a single cell in a tissue undergoes **transformation,** the process that converts a normal cell to a cancer cell. The body's immune system normally recognizes a transformed cell as an insurgent and destroys it. However, if the cell evades destruction, it may proliferate and form a tumor, a mass of abnormal cells within otherwise normal tissue. If the abnormal cells remain at the original site, the lump is called a **benign tumor.** Most benign tumors do not cause serious problems and can be completely removed by surgery. In contrast, a **malignant tumor** becomes invasive enough to impair the functions of one or more organs **(Figure 12.19)**. An individual with a malignant tumor is said to have cancer.

The cells of malignant tumors are abnormal in many ways besides their excessive proliferation. They may have unusual numbers of chromosomes (whether this is a cause or an effect of transformation is a current topic of debate). Their metabolism may be disabled, and they may cease to function in any constructive way. Also, owing to abnormal changes on the

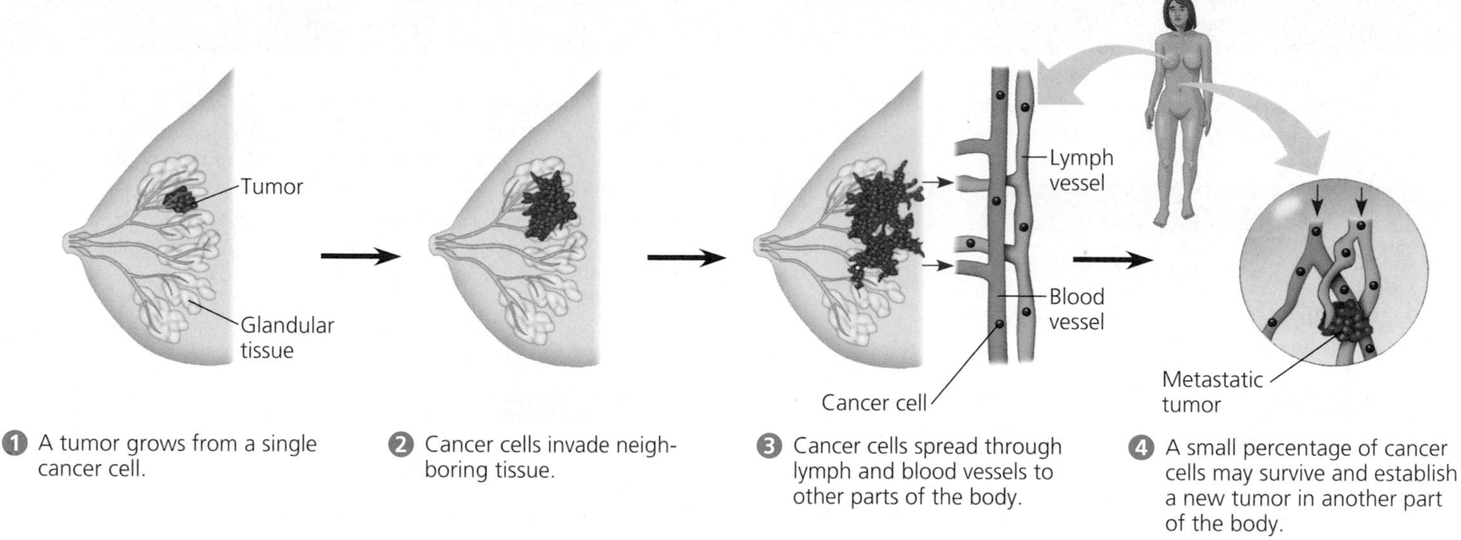

▲ **Figure 12.19 The growth and metastasis of a malignant breast tumor.** The cells of malignant (cancerous) tumors grow in an uncontrolled way and can spread to neighboring tissues and, via lymph and blood vessels, to other parts of the body. The spread of cancer cells beyond their original site is called metastasis.

cells' surfaces, they lose or destroy their attachments to neighboring cells and the extracellular matrix and can spread into nearby tissues. Cancer cells may also secrete signal molecules that cause blood vessels to grow toward the tumor. A few tumor cells may separate from the original tumor, enter blood vessels and lymph vessels, and travel to other parts of the body. There, they may proliferate and form a new tumor. This spread of cancer cells to locations distant from their original site is called **metastasis** (see Figure 12.19).

Tumor
Glandular tissue
Lymph vessel
Blood vessel
Cancer cell
Metastatic tumor
1 A tumor grows from a single cancer cell.
2 Cancer cells invade neighboring tissue.
3 Cancer cells spread through lymph and blood vessels to other parts of the body.
4 A small percentage of cancer cells may survive and establish a new tumor in another part of the body.

A tumor that appears to be localized may be treated with high-energy radiation, which damages DNA in cancer cells much more than it does in normal cells, apparently because cancer cells have lost the ability to repair such damage. To treat known or suspected metastatic tumors, chemotherapy is used, in which drugs that are toxic to actively dividing cells are administered through the circulatory system. As you might expect, chemotherapeutic drugs interfere with specific steps in the cell cycle. For example, the drug Taxol freezes the mitotic spindle by preventing microtubule depolymerization, which stops actively dividing cells from proceeding past metaphase. The side effects of chemotherapy are due to the drugs' effects on normal cells. For example, nausea results from chemotherapy's effects on intestinal cells, hair loss from effects on hair follicle cells, and susceptibility to infection from effects on immune system cells.

Researchers are beginning to understand how a normal cell is transformed into a cancer cell. You will learn more about the molecular biology of cancer in Chapter 19. Though the causes of cancer are diverse, cellular transformation always involves the alteration of genes that somehow influence the cell cycle control system. Our knowledge of how changes in the genome lead to the various abnormalities of cancer cells remains rudimentary, however.

Perhaps the reason we have so many unanswered questions about cancer cells is that there is still so much to learn about how normal cells function. The cell, life's basic unit of structure and function, holds enough secrets to engage researchers well into the future.

Concept Check 12.3

1. A researcher treats cells with a chemical that prevents DNA synthesis. This treatment traps the cells in which part of the cell cycle?
2. In Figure 12.13, why do the nuclei resulting from experiment 2 contain different amounts of DNA?
3. What is the go-ahead signal for a cell to pass the G_2 phase checkpoint and enter mitosis? (See Figure 12.16.)
4. What would happen if you performed the experiment in Figure 12.17 with cancer cells?
5. What phase of the cell cycle are most of your body cells in?
6. Compare and contrast a benign tumor and a malignant tumor.

For suggested answers, see Appendix A.

Chapter 12 Review

Go to www.campbellbiology.com or the student CD-ROM to explore Activities, Investigations, and other interactive study aids.

SUMMARY OF KEY CONCEPTS

▶ Unicellular organisms reproduce by cell division. Multicellular organisms depend on it for development from a fertilized egg, growth, and repair (pp. 218–219).
Activity ***Roles of Cell Division***

Concept 12.1

Cell division results in genetically identical daughter cells

▶ Cells duplicate their genetic material before they divide, ensuring that each daughter cell receives an exact copy of the genetic material, DNA (p. 219).

▶ **Cellular Organization of the Genetic Material (p. 219)** DNA is partitioned among chromosomes. Eukaryotic chromosomes consist of chromatin, a complex of DNA and protein that condenses during mitosis. In animals, gametes have one set of chromosomes and somatic cells have two sets.

▶ **Distribution of Chromosomes During Cell Division (pp. 219–220)** In preparation for cell division, chromosomes replicate, each one then consisting of two identical sister chromatids. The chromatids separate during cell division, becoming the chromosomes of the new daughter cells. Eukaryotic cell division consists of mitosis (division of the nucleus) and cytokinesis (division of the cytoplasm).

Concept 12.2

The mitotic phase alternates with interphase in the cell cycle

▶ **Phases of the Cell Cycle (pp. 221–223)** Between divisions, cells are in interphase: the G_1, S, and G_2 phases. The cell grows throughout interphase, but DNA is replicated only during the synthesis (S) phase. Mitosis and cytokinesis make up the mitotic (M) phase of the cell cycle. Mitosis is a continuous process, often described as occurring in five stages: prophase, prometaphase, metaphase, anaphase, and telophase.
Activity ***The Cell Cycle***

▶ **The Mitotic Spindle: *A Closer Look* (pp. 221–225)** The mitotic spindle is an apparatus of microtubules that controls chromosome movement during mitosis. The spindle arises from the centrosomes and includes spindle microtubules and asters. Some spindle microtubules attach to the kinetochores of chromsomes and move the chromosomes to the metaphase plate. In anaphase, sister chromatids separate and move along the kinetochore microtubules toward opposite ends of the cell. Meanwhile, nonkinetochore microtubules from opposite poles overlap and push against each other, elongating the cell. In telophase, genetically identical daughter nuclei form at opposite ends of the cell.

▶ **Cytokinesis: *A Closer Look* (pp. 224–226)** Mitosis is usually followed by cytokinesis. Animal cells carry out cytokinesis by cleavage, and plant cells form a cell plate.
Activity ***Mitosis and Cytokinesis Animation***
Activity ***Mitosis and Cytokinesis Video***
Investigation ***How Much Time Do Cells Spend in Each Phase of Mitosis?***

▶ **Binary Fission (pp. 226–227)** During binary fission, the bacterial chromosome replicates and the two daughter chromosomes actively move apart. The specific proteins involved in this movement are a subject of current research.

▶ **The Evolution of Mitosis (pp. 227–228)** Since prokaryotes preceded eukaryotes by more than a billion years, it is likely that mitosis evolved from prokaryotic cell division. Certain protists exhibit types of cell division that seem intermediate between bacterial binary fission and the process of mitosis carried out by most eukaryotic cells.

Concept 12.3

The cell cycle is regulated by a molecular control system

▶ **Evidence for Cytoplasmic Signals (p. 228)** Molecules present in the cytoplasm regulate progress through the cell cycle.

▶ **The Cell Cycle Control System (pp. 229–232)** Cyclic changes in regulatory proteins work as a cell cycle clock. The clock has specific checkpoints where the cell cycle stops until a go-ahead signal is received. The key molecules are cyclins and cyclin-dependent kinases (Cdks). Cell culture has enabled researchers to study the molecular details of cell division. Both internal signals and external signals control the cell cycle checkpoints via signal transduction pathways. Most cells exhibit density-dependent inhibition of cell division as well as anchorage dependence.

▶ **Loss of Cell Cycle Controls in Cancer Cells (pp. 232–233)** Cancer cells elude normal regulation and divide out of control, forming tumors. Malignant tumors invade surrounding tissues and can metastasize, exporting cancer cells to other parts of the body, where they may form secondary tumors.
Activity ***Causes of Cancer***

TESTING YOUR KNOWLEDGE

Self-Quiz

1. Increases in the enzymatic activity of some protein kinases important for the regulation of the cell cycle are due to
   a. kinase synthesis by ribosomes.
   b. activation of inactive kinases by binding to cyclins.
   c. conversion of inactive cyclins to active kinases by means of phosphorylation.
   d. cleavage of the inactive kinase molecules by cytoplasmic proteases.
   e. a decline in external growth factors to a concentration below the inhibitory threshold.

2. Through a microscope, you can see a cell plate beginning to develop across the middle of the cell and nuclei re-forming on either side of the cell plate. This cell is most likely
   a. an animal cell in the process of cytokinesis.
   b. a plant cell in the process of cytokinesis.
   c. an animal cell in the S phase of the cell cycle.
   d. a bacterial cell dividing.
   e. a plant cell in metaphase.

3. Vinblastine is a standard chemotherapeutic drug used to treat cancer. Because it interferes with the assembly of microtubules, its effectiveness must be related to
   a. disruption of mitotic spindle formation.
   b. inhibition of regulatory protein phosphorylation.
   c. suppression of cyclin production.
   d. myosin denaturation and inhibition of cleavage furrow formation.
   e. inhibition of DNA synthesis.

4. A particular cell has half as much DNA as some of the other cells in a mitotically active tissue. The cell in question is most likely in
   a. G_1.
   b. G_2.
   c. prophase.
   d. metaphase.
   e. anaphase.

5. One difference between a cancer cell and a normal cell is that
   a. the cancer cell is unable to synthesize DNA.
   b. the cell cycle of the cancer cell is arrested at the S phase.
   c. cancer cells continue to divide even when they are tightly packed together.
   d. cancer cells cannot function properly because they suffer from density-dependent inhibition.
   e. cancer cells are always in the M phase of the cell cycle.

6. The decline of MPF activity at the end of mitosis is caused by
   a. the destruction of the protein kinase (Cdk).
   b. decreased synthesis of cyclin.
   c. the degradation of cyclin.
   d. synthesis of DNA.
   e. an increase in the cell's volume-to-genome ratio.

7. A red blood cell (RBC) has a 120-day life span. If an average adult has 5 L of blood, and each microliter (μL) contains 5 million RBCs, how many new cells must be produced each *second* to replace the entire RBC population? ($1\ \mu L = 10^{-6}\ L$)
   a. 30,000
   b. 2,400
   c. 2,400,000
   d. 18,000
   e. 30,000,000

8. The drug cytochalasin B blocks the function of actin. Which of the following aspects of the cell cycle would be most disrupted by cytochalasin B?
   a. spindle formation
   b. spindle attachment to kinetochores
   c. DNA synthesis
   d. cell elongation during anaphase
   e. cleavage furrow formation

9. In some organisms, mitosis occurs without cytokinesis occurring. This will result in
   a. cells with more than one nucleus.
   b. cells that are unusually small.
   c. cells lacking nuclei.
   d. destruction of chromosomes.
   e. cell cycles lacking an S phase.

10. Which of the following does *not* occur during mitosis?
   a. condensation of the chromosomes
   b. replication of the DNA
   c. separation of sister chromatids
   d. spindle formation
   e. separation of the centrosomes

11. In the light micrograph below of dividing cells near the tip of an onion root, identify a cell in each of the following stages: interphase, prophase, metaphase, and anaphase. Describe the major events occurring at each stage.

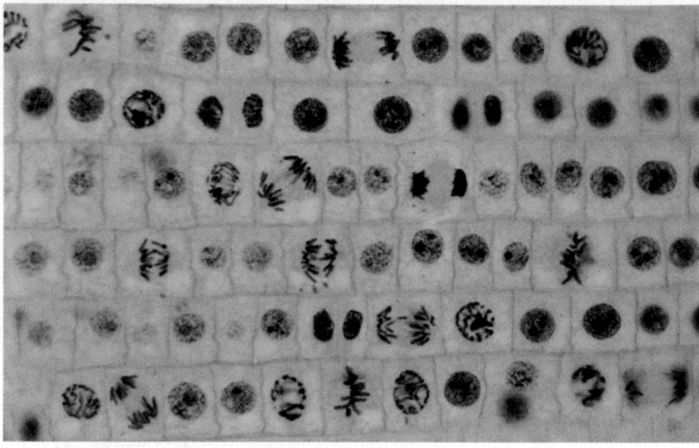

For Self-Quiz answers, see Appendix A.

Go to the website or CD-ROM for more quiz questions.

Evolution Connection

The result of mitosis is that the daughter cells end up with the same number of chromosomes as the parent cell had. Another way to maintain the number of chromosomes would be to carry out cell division first and then duplicate the chromosomes in each daughter cell. What would be the problems with this alternative? Or do you think it would be an equally good way of organizing the cell cycle?

Scientific Inquiry

Microtubules are polar structures in that one end (called the + end) polymerizes and depolymerizes at a much higher rate than the other end (the − end). The experiment shown in Figure 12.8 clearly identifies these two ends.

a. From the results, identify the + end and explain your reasoning.
b. If the opposite end were the + end, what would the results be? Make a sketch.
c. Redesign the model in the conclusion of Figure 12.8 to reflect your new version of the results.

Investigation *How Much Time Do Cells Spend in Each Phase of Mitosis?*

Science, Technology, and Society

Hundreds of millions of dollars are spent each year in the search for effective treatments for cancer; far less money is spent preventing cancer. Why do you think this is true? What kinds of lifestyle changes could we make to help prevent cancer? What kinds of prevention programs could be initiated or strengthened to encourage these changes? What factors might impede such changes and programs?

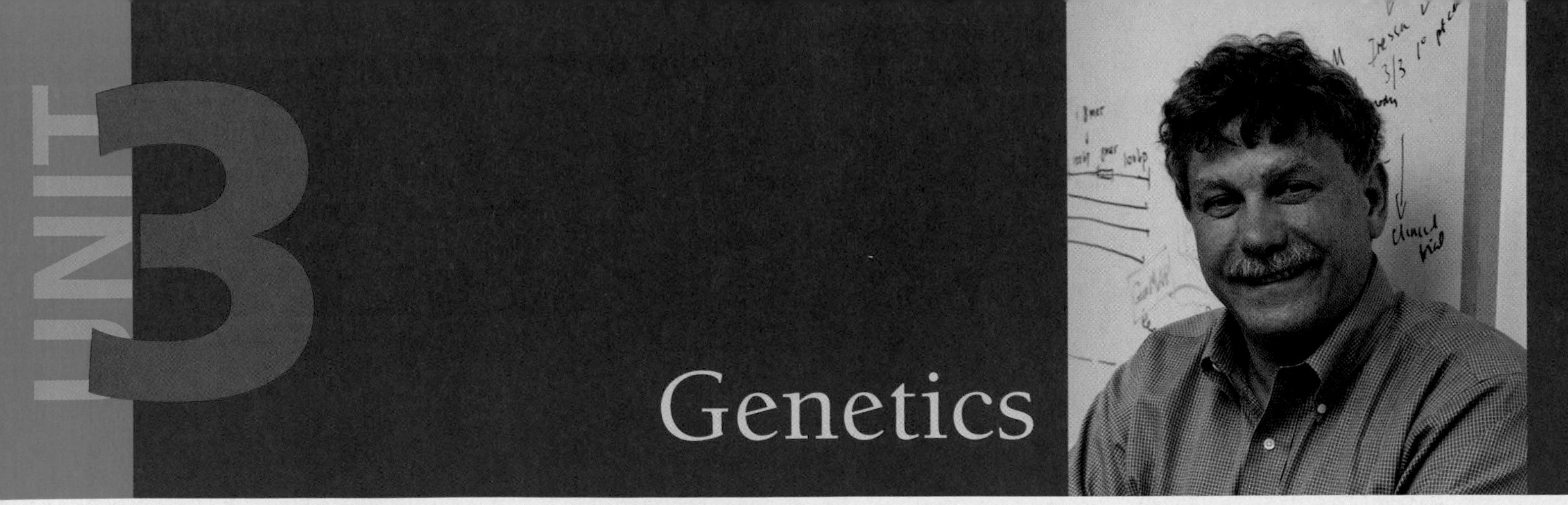

UNIT 3 Genetics

AN INTERVIEW WITH Eric Lander

Genomics, the study of all the genes in an organism and how they function together, is a new field that is bringing about a revolution in biology. Having played a major role in the Human Genome Project, Eric Lander is a leader of this revolution. Dr. Lander is the founding director of the Broad Institute of MIT and Harvard, which uses genomics to develop new tools and approaches to understanding and treating disease. The institute includes the former Whitehead Center for Genome Research, which played a leading role in the sequencing of the human genome.

A graduate of Princeton, Dr. Lander earned a doctorate in mathematics at Oxford University, where he was a Rhodes Scholar. He then taught managerial economics at Harvard Business School until turning full time to biology in 1990. Among his many honors have been a MacArthur Foundation Fellowship and election to the U.S. National Academy of Sciences. Dr. Lander is a professor at both MIT and Harvard, and he has taught MIT's core introductory biology course for a decade.

How did you get into genetics?

In high school, I took biology, but I loved math, and I was a math major in college. I went on to get my Ph.D. in mathematics but decided I didn't want to be a pure mathematician. One day, my brother suggested I might be interested in the coding theory of the brain and sent me some papers on mathematical neurobiology. I realized that, to understand them, I had to learn something about cellular neurobiology. This required me to study cell biology. Next came molecular biology, and finally, I really had to know genetics. So one thing led to another—and here I am still learning genetics!

What was the main purpose of the Human Genome Project?

The ultimate purpose of the Human Genome Project was to read out and make freely available the complete DNA sequence of the human being. This information is fundamental to our biology. It contains the parts list in terms of which all our biological structures and processes must eventually be described.

You can study the detailed properties of individual genes, as biologists did before the Human Genome Project and still do, or you can study how all the components of the system interact. Important discoveries are made at both levels, but there are some things you only see when you step back. Imagine looking at a pointillist painting. Up close, the dots are interesting, but when you step back you can see patterns that weren't evident at first. Until the Human Genome Project, it hadn't been possible to step back and get the big picture of the human genome.

Analysis of the genome is unveiling a comprehensive picture of how the genes are turned on and off in different tissues and at different times, so we can see how genes collaborate in modules or circuits. It's an ensemble picture that we've just never had before.

What were the main challenges of the Human Genome Project?

The biggest challenge was the necessity of a 10,000-fold improvement in our methods for mapping and sequencing DNA—the project was that much more ambitious than anything ever done before! In any realistic sense, the scientific community was crazy to propose it. But realism is much overrated. Once we recognized how important the sequence would be to thousands of scientists, we began to ask why, exactly, we couldn't do it. Then we took on the barriers one at a time. We set intermediate goals, both to obtain some information that would be immediately useful and to show ourselves that we were on the right track. And one after another, we were able to reach our goals: first, genetic maps we could use to trace the inheritance of diseases, then physical maps of the chromosomal DNA, and, finally, the nucleotide-by-nucleotide sequence of the whole genome.

It was a great experience. Everybody involved in the Human Genome Project knew we were working on something that would still be fundamental to science a hundred years from now. We felt it would be important to our children because the medicine of 50 years from now will be based on this work. And it was the work of no one individual, no one center, no one country. Overall, the project involved several thousand people at 20 centers all over the world—in the United States, the United Kingdom, France, Germany, Japan, and China. It was science at its best, an international collaboration of people working together for something bigger than themselves.

To achieve our goals, we constantly had to invent new methodologies, and we had to figure out how to automate as much as possible. Then we had to figure out how to analyze the data. My own background in mathematics actually turned out to be useful.

The genome sequences are just long chains of A's, C's, T's, and G's. How do you know which sections are genes?

Knowing the three billion letters of the human genome is still a far cry from understanding what they say. It's not easy to identify the genes within a very long sequence. This is particularly so in humans and other multicellular eukaryotes, which generally have huge amounts of noncoding DNA and gene-coding sequences split up into small segments (exons) interrupted by stretches of noncoding DNA (introns). In searching for human genes, we're looking for small signals in a sea of noise.

Computer programs can do a passable job of picking out gene sequences, but only passable. The sequences you can look for include ATG, which starts the coding sequence of all genes, but ATG is so common in the genome that it's not much of a signal. You can also look for places where that start signal is followed by a fairly long stretch of nucleotides before one of the three-nucleotide stop signals appears, but that can easily occur by chance. So, using a computer to look for those kinds of signals alone does a pretty crummy job of identifying genes.

Happily, we have other tricks up our sleeves. One way to identify genes is to compare the genome to copies of messenger RNA molecules from cells (see Figure 5.25). Another powerful approach is to compare the human sequence to those of related species. Because many genes are crucial to an organism's survival, they tend to be conserved (not change very much) during evolution. So if we look at the portions of the human genome that are very similar in other mammals, they tend to contain genes. The task of finding all the genes is a fascinating puzzle that involves computer scientists, evolutionary biologists, molecular biologists, and others.

How many human genes are there?

My best guess today is that there are 20,000–25,000 protein-coding genes. One of the most surprising findings of the Human Genome Project was a gene count much lower than people had expected based on the total size of the human genome; not so many years ago, textbooks gave 100,000 as the likely number of human genes. When we got the rough draft of the sequence, we thought there might be as many as 40,000 genes, but we soon learned that many of these were actually pseudogenes, which are defective, nonfunctional copies of true genes. The gene count has been falling and falling. Of course, genes for functional RNA molecules, such as rRNA, that are not translated into protein are also present in the genome, but they don't add up to a lot of territory, probably only a few thousand genes—although we're not certain.

How close are we to understanding the human genome?

The extent of our ignorance became clear from comparing the human genome with the mouse genome. When we lined up the two, we found that about 5% of the human genome showed strong similarity to that of the mouse, indicating strong evolutionary conservation since the last common ancestor of mouse and human. Thus, the sequences in this DNA must be functionally very important. But only about a third of the 5% could be accounted for by known genes and regulatory sequences, leaving a lot more that evolution "cares about" than we can explain today.

That's what I love about genomics. We're learning that there are vast tracts of biology we have missed. It's as if we suddenly could look at the whole Earth and see that, golly, there are several continents we hadn't known about. Genomics is revealing huge territories for the next generation of young scientists to explore.

"The Human Genome Project . . . was science at its best, an international collaboration of people working together for something bigger than themselves."

How is genomics affecting the study of evolution?

Evolutionary biology is being transformed by the availability of genomic data. Now we can sit down with the complete sequences of a number of organisms and begin to reconstruct their evolutionary relationships and histories in exquisite detail. We can see the forces at work in different lineages. For instance, we can see that in humans the most actively evolving genes are those involved in reproduction and the immune system. Scientists have long thought that genomes carried such information about evolution, but we're finally getting to see it laid out clearly. Evolution is, so to speak, an experimentalist that has been running experiments for three and a half billion years, since the origin of life on Earth. And, wonderfully, the genomes of today's organisms retain the lab notes of these experiments, so that we can go back and reconstruct the events that took place.

Let's talk about the application of genomics to medicine. What goes on at the Broad Institute?

The key challenge ahead is to understand the molecular basis of common human diseases such as diabetes, lung cancer, psychiatric diseases, and many others. Understanding molecular mechanisms is key to creating therapies. The sequence of the human genome has given us a great foundation for systematic understanding of disease—but it's just a start. We now need a full understanding of all the functional elements encoded in the human genome sequence and how they regulate genes; of all the common genetic variants in the human population and how they correlate with risk of disease; of the gene and protein expression patterns that reflect the activation of cellular pathways; and of all the genetic mutations that underlie cancers. In parallel, we need to develop ways to use this comprehensive picture to identify the right targets for drug development and other therapies.

The Broad Institute's mission is to help drive this next stage of biomedicine. The institute includes many kinds of scientists—biologists, chemists, physicians, mathematicians, and others—from across MIT and Harvard, including the Harvard hospitals. Together, we're developing new ways to generate biological information and new ways to apply it in medical studies. We also feel strongly about making all the data publicly available, so that anyone can use it to ask new questions.

13 Meiosis and Sexual Life Cycles

▲ **Figure 13.1** **Francis Ford Coppola and his family.**

Key Concepts

13.1 Offspring acquire genes from parents by inheriting chromosomes

13.2 Fertilization and meiosis alternate in sexual life cycles

13.3 Meiosis reduces the number of chromosome sets from diploid to haploid

13.4 Genetic variation produced in sexual life cycles contributes to evolution

Overview

Hereditary Similarity and Variation

Living organisms are distinguished by their ability to reproduce their own kind. Only oak trees produce oaks, and only elephants can make more elephants. Furthermore, offspring resemble their parents more than they do less closely related individuals of the same species. The transmission of traits from one generation to the next is called inheritance, or **heredity** (from the Latin *heres,* heir). Along with inherited similarity, there is also **variation:** Offspring differ somewhat in appearance from parents and siblings. Farmers have exploited these observations for thousands of years, breeding plants and animals for desired traits. Just as ancient is a curiosity about genetic similarities and differences among people, including family members **(Figure 13.1)**. The mechanisms of heredity and variation, however, eluded biologists until the development of genetics in the 20th century.

Genetics is the scientific study of heredity and hereditary variation. In this unit, you will learn about genetics at the levels of organism, cell, and molecule. On the practical side, you will learn how modern genetics is revolutionizing medicine and agriculture, and you will be asked to consider some social and ethical questions raised by our ability to manipulate DNA, the genetic material. At the end of the unit, you will learn about the contributions geneticists have made to solving the mystery of how multicellular animals and plants arise from a single cell, the fertilized egg. In fact, genetic methods and discoveries are catalyzing progress in all areas of biology—from cell biology to physiology, evolutionary biology, ecology, and even behavior.

We begin our study of genetics in this chapter by examining how chromosomes pass from parents to offspring in sexually reproducing organisms. The processes of meiosis (a special type of cell division) and fertilization (the fusion of sperm and egg) maintain a species' chromosome count during the sexual life cycle. We will describe the cellular mechanics of meiosis and how this process differs from mitosis. Finally, we will consider how both meiosis and fertilization contribute to genetic variation, such as the variation obvious in the Coppola family (see Figure 13.1).

Concept 13.1

Offspring acquire genes from parents by inheriting chromosomes

Family friends may tell you that you have your mother's freckles or your father's eyes. However, parents do not, in any literal sense, give their children freckles, eyes, hair, or any other traits. What, then, *is* actually inherited?

Inheritance of Genes

Parents endow their offspring with coded information in the form of hereditary units called **genes.** The tens of thousands of genes we inherit from our mothers and fathers constitute our genome. This genetic link to our parents accounts for family resemblances such as shared eye color or freckles. Our genes program the specific traits that emerge as we develop from fertilized eggs into adults.

Genes are segments of DNA. You learned in Chapters 1 and 5 that DNA is a polymer of four different kinds of monomers called nucleotides. Inherited information is passed on in the form of each gene's specific sequence of nucleotides, much as printed information is communicated in the form of meaningful sequences of letters. Language is symbolic. The brain translates words and sentences into mental images and ideas; for example, the object you imagine when you read "apple" looks nothing like the word itself. Analogously, cells translate genetic "sentences" into freckles and other features that bear no resemblance to genes. Most genes program cells to synthesize specific enzymes and other proteins whose cumulative action produces an organism's inherited traits. The programming of these traits in the form of DNA is one of the unifying themes of biology.

The transmission of hereditary traits has its molecular basis in the precise replication of DNA, which produces copies of genes that can be passed along from parents to offspring. In animals and plants, reproductive cells called **gametes** are the vehicles that transmit genes from one generation to the next. During fertilization, male and female gametes (sperm and eggs) unite, thereby passing on the genes of both parents to their offspring.

Except for tiny amounts of DNA in mitochondria and chloroplasts, the DNA of a eukaryotic cell is subdivided into chromosomes within the nucleus. Every living species has a characteristic number of chromosomes. For example, humans have 46 chromosomes in almost all of their cells. Each chromosome consists of a single long DNA molecule elaborately coiled in association with various proteins. One chromosome includes several hundred to a few thousand genes, each of which is a specific sequence of nucleotides within the DNA molecule. A gene's specific location along the length of a chromosome is called the gene's **locus** (plural, *loci*). Our genetic endowment consists of the genes carried on the chromosomes we inherited from our parents.

Comparison of Asexual and Sexual Reproduction

Only organisms that reproduce asexually produce offspring that are exact copies of themselves. In **asexual reproduction,** a single individual is the sole parent and passes copies of all its genes to its offspring. For example, single-celled eukaryotic organisms can reproduce asexually by mitotic cell division, in which DNA is copied and allocated equally to two daughter cells. The genomes of the offspring are virtually exact copies of the parent's genome. Some multicellular organisms are also capable of reproducing asexually. A hydra, which is related to the jellies (jellyfishes), can reproduce by budding **(Figure 13.2)**. Because the cells of a bud are derived by mitosis in the parent, the "chip off the old block" is usually genetically identical to its parent. An individual that reproduces asexually gives rise to a **clone,** a group of genetically identical individuals. Genetic differences occasionally arise in asexually reproducing organisms as a result of changes in the DNA called mutations, which we will discuss in Chapter 17.

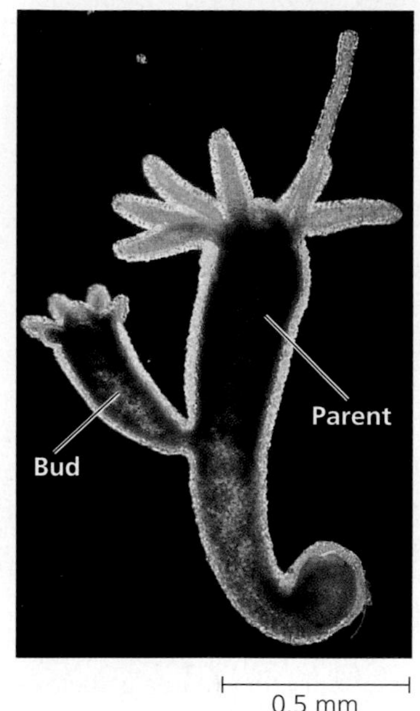

◄ **Figure 13.2 The asexual reproduction of a hydra.** This relatively simple multicellular animal reproduces by budding. The bud, a localized mass of mitotically dividing cells, develops into a small hydra, which detaches from the parent (LM).

In **sexual reproduction,** two parents give rise to offspring that have unique combinations of genes inherited from the two parents. In contrast to a clone, offspring of sexual reproduction vary genetically from their siblings and both parents: They are variations on a common theme of family resemblance, not exact replicas. Genetic variation like that shown in Figure 13.1 is an important consequence of sexual reproduction. What mechanisms generate this genetic variation? The key is the behavior of chromosomes during the sexual life cycle.

Concept Check 13.1

1. How are the traits of parents (such as hair color) transmitted to their offspring?
2. In the absence of mutation, asexually reproducing organisms produce offspring that are genetically identical to each other and to their parents. Explain.
3. In organisms that reproduce sexually, how similar are the offspring to their parents? Explain.

For suggested answers, see Appendix A.

Concept 13.2

Fertilization and meiosis alternate in sexual life cycles

A **life cycle** is the generation-to-generation sequence of stages in the reproductive history of an organism, from conception to production of its own offspring. In this section, we use humans as an example to track the behavior of chromosomes through sexual life cycles. We begin by considering the chromosome count in human somatic cells and gametes; we will then explore how the behavior of chromosomes relates to the human life cycle and other types of sexual life cycles.

Sets of Chromosomes in Human Cells

In humans, each **somatic cell**—any cell other than a gamete—has 46 chromosomes. During mitosis, the chromosomes become condensed enough to be visible in a light microscope. Because chromosomes differ in size, in the positions of their centromeres, and in the pattern of colored bands produced by certain stains, they can be distinguished from one another by microscopic examination when sufficiently condensed.

Careful examination of a micrograph of the 46 human chromosomes from a single cell in mitosis reveals that there are two chromosomes of each type. This becomes clear when images of the chromosomes are arranged in pairs, starting with the longest chromosomes. The resulting ordered display is called a **karyotype** (**Figure 13.3**). The two chromosomes composing a pair have the same length, centromere position, and staining pattern: These are called **homologous chromosomes,** or homologues. Both chromosomes of each pair carry genes controlling the same inherited characters. For example, if a gene for eye color is situated at a particular locus on a certain chromosome, then the homologue of that chromosome will also have a gene specifying eye color at the equivalent locus.

The two distinct chromosomes referred to as X and Y are an important exception to the general pattern of homologous chromosomes in human somatic cells. Human females have a homologous pair of X chromosomes (XX), but males have one X and one Y chromosome (XY). Only small parts of the X and Y are homologous. Most of the genes carried on the X chromosome do not have counterparts on the tiny Y, and the Y chromosome has genes lacking on the X. Because they determine an individual's sex, the X and Y chromosomes are called **sex chromosomes.** The other chromosomes are called **autosomes.**

The occurrence of homologous pairs of chromosomes in each human somatic cell is a consequence of our sexual origins. We inherit one chromosome of each pair from each parent. So the 46 chromosomes in our somatic cells are actually two sets of 23 chromosomes—a maternal set (from our mother) and a paternal set (from our father). The number of chromosomes in a single set is represented by *n*. Any cell with two chromosome

Figure 13.3
Research Method Preparing a Karyotype

APPLICATION A karyotype is a display of condensed chromosomes arranged in pairs. Karyotyping can be used to screen for abnormal numbers of chromosomes or defective chromosomes associated with certain congenital disorders, such as Down syndrome.

TECHNIQUE Karyotypes are prepared from isolated somatic cells, which are treated with a drug to stimulate mitosis and then grown in culture for several days. A slide of cells arrested in metaphase is stained and then viewed with a microscope equipped with a digital camera. A digital photograph of the chromosomes is entered into a computer, and the chromosomes are electronically rearranged into pairs according to size and shape.

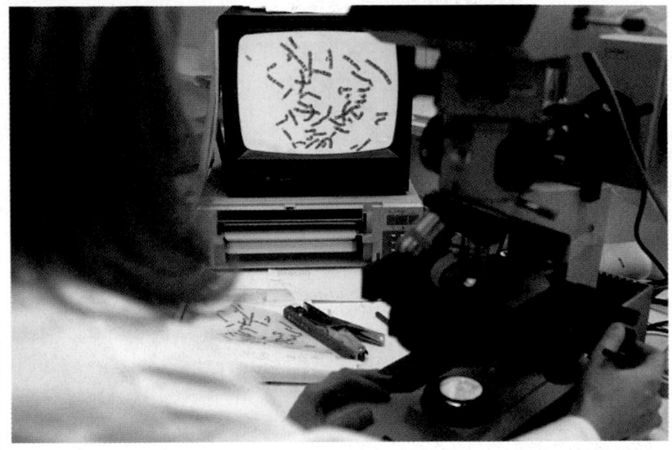

RESULTS This karyotype shows the chromosomes from a normal human male. The patterns of stained bands help identify specific chromosomes and parts of chromosomes. Although difficult to discern in the karyotype, each metaphase chromosome consists of two, closely attached sister chromatids (see diagram).

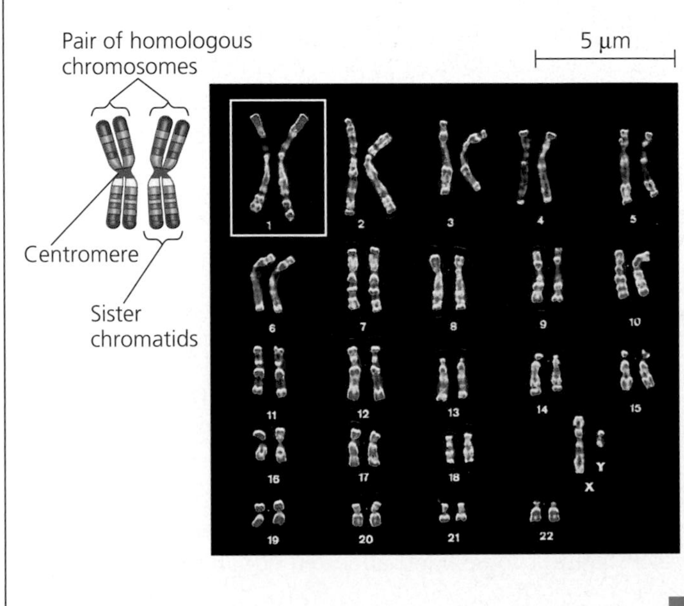

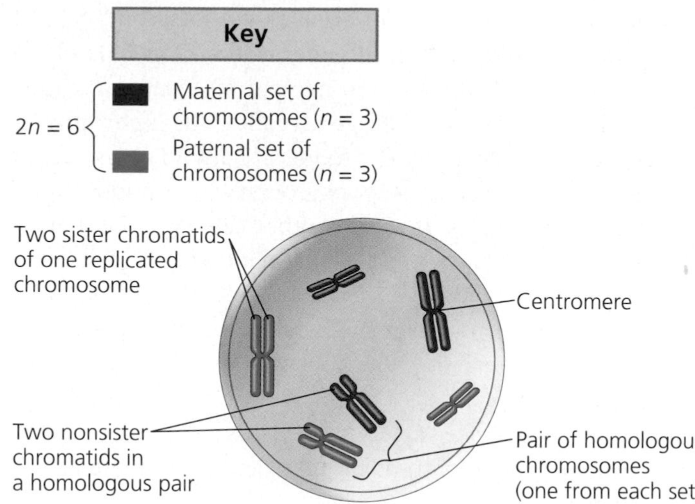

▲ **Figure 13.4 Describing chromosomes.** A cell with a diploid number of 6 ($2n = 6$) is depicted here in G_2 of interphase, following chromosome replication. (The chromosomes have been artificially condensed.) Each of the six duplicated chromosomes consists of two sister chromatids joined at the centromere. Each homologous pair is composed of one chromosome from the maternal set (red) and one from the paternal set (blue). Each set is made up of three chromosomes. Nonsister chromatids are any two chromatids in a pair of homologous chromosomes that are not sister chromatids.

sets is called a **diploid cell** and has a diploid number of chromosomes, abbreviated $2n$. For humans, the diploid number is 46 ($2n = 46$), the number of chromosomes in our somatic cells. In a cell in which DNA synthesis has occurred, all the chromosomes are duplicated and thus each consists of two identical sister chromatids. **Figure 13.4** helps clarify the various terms that we use in describing duplicated chromosomes in a diploid cell. Study this figure so that you understand the differences between homologous chromosomes, sister chromatids, nonsister chromatids, and chromosome sets.

Unlike somatic cells, gametes (sperm and egg cells) contain a single chromosome set. Such cells are called **haploid cells**, and each has a haploid number of chromosomes (n). For humans, the haploid number is 23 ($n = 23$), the number of chromosomes found in a gamete. The set of 23 consists of the 22 autosomes plus a single sex chromosome. An unfertilized egg cell (also called an ovum) contains an X chromosome, but a sperm cell may contain an X or a Y chromosome.

Note that each sexually reproducing species has a characteristic haploid number and diploid number. These may be higher than, lower than, or the same as the values for humans. Now let's extend the concepts of haploid and diploid to understand chromosome behavior during the human life cycle.

Behavior of Chromosome Sets in the Human Life Cycle

The human life cycle begins when a haploid sperm cell from the father fuses with a haploid ovum from the mother. This union of gametes, culminating in fusion of their nuclei, is called **fertilization.** The resulting fertilized egg, or **zygote**, is diploid because it contains two haploid sets of chromosomes bearing genes representing the maternal and paternal family lines. As a human develops from a zygote to a sexually mature adult, mitosis generates all the somatic cells of the body. Both chromosome sets in the zygote and all the genes they carry are passed with precision to our somatic cells.

The only cells of the human body *not* produced by mitosis are the gametes, which develop in the gonads—ovaries in females and testes in males **(Figure 13.5)**. Imagine what would happen if human gametes were made by mitosis: They would be diploid like the somatic cells. At the next round of fertilization, when two gametes fused, the normal chromosome number of 46 would double to 92, and each subsequent generation would double the number of chromosomes yet again. This hypothetical situation of constantly increasing chromosome number in sexually reproducing organisms is avoided through the

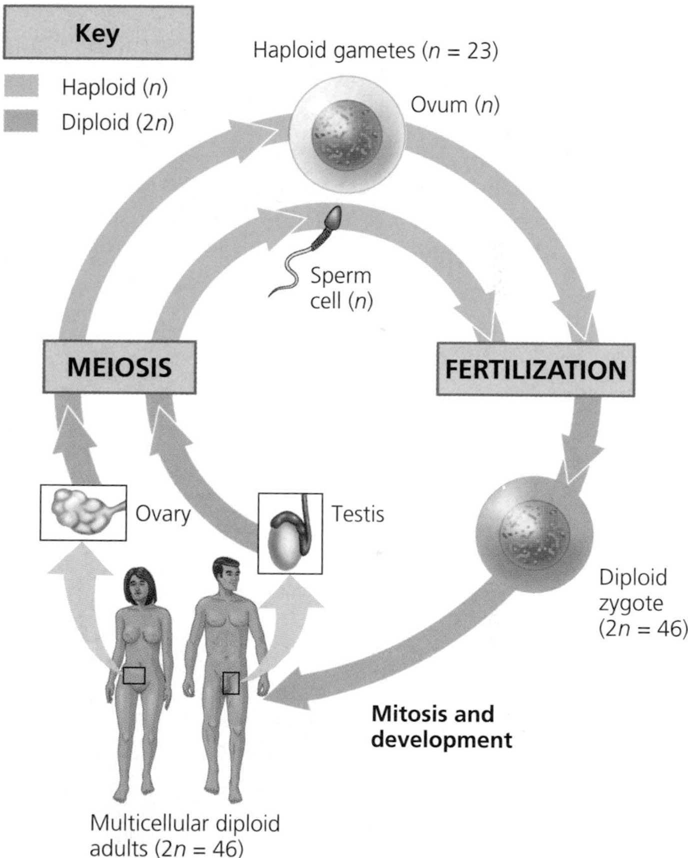

▲ **Figure 13.5 The human life cycle.** In each generation, the doubling of the number of chromosome sets that results from fertilization is offset by the halving of the number of sets that results from meiosis. For humans, the number of chromosomes in a haploid cell is 23, consisting of one set ($n = 23$); the number of chromosomes in the diploid zygote and all somatic cells arising from it is 46, consisting of two sets ($2n = 46$).

This figure introduces a color code that will be used for other life cycles later in this book. The teal-colored arrows highlight haploid stages of a life cycle, and the beige-colored arrows highlight diploid stages.

process of **meiosis.** This type of cell division reduces the number of sets of chromosomes from two to one in the gametes, compensating for the doubling that occurs at fertilization. In animals, meiosis occurs only in the ovaries or testes. As a result, each human sperm and ovum is haploid ($n = 23$). Fertilization restores the diploid condition by combining two haploid sets of chromosomes, and the human life cycle is repeated, generation after generation (see Figure 13.5). You will learn more about the production of sperm and ova in Chapter 46.

In general, the steps of the human life cycle are typical of many animals. Indeed, the processes of fertilization and meiosis are the unique trademarks of sexual reproduction. Fertilization and meiosis alternate in sexual life cycles, offsetting each other's effects on the chromosome number and thus perpetuating a species' chromosome count.

The Variety of Sexual Life Cycles

Although the alternation of meiosis and fertilization is common to all organisms that reproduce sexually, the timing of these two events in the life cycle varies, depending on the species. These variations can be grouped into three main types of life cycles. In the type of life cycle that occurs in humans and most other animals, gametes are the only haploid cells. Meiosis occurs during the production of gametes, which undergo no further cell division prior to fertilization. The diploid zygote divides by mitosis, producing a multicellular organism that is diploid **(Figure 13.6a)**.

Plants and some species of algae exhibit a second type of life cycle called **alternation of generations.** This type of life cycle includes both diploid and haploid multicellular stages. The multicellular diploid stage is called the **sporophyte.** Meiosis in the sporophyte produces haploid cells called **spores.** Unlike a gamete, a spore gives rise to a multicellular individual without fusing with another cell. A spore divides mitotically to generate a multicellular haploid stage called the **gametophyte.** The haploid gametophyte makes gametes by mitosis. Fertilization among the haploid gametes results in a diploid zygote, which develops into the next sporophyte generation. Therefore, in this type of life cycle, the sporophyte generation produces a gametophyte as its offspring, and the gametophyte generation produces the next sporophyte generation **(Figure 13.6b)**.

A third type of life cycle occurs in most fungi and some protists, including some algae. After gametes fuse and form a diploid zygote, meiosis occurs without a diploid offspring developing. Meiosis produces not gametes but haploid cells that then divide by mitosis and give rise to a haploid multicellular adult organism. Subsequently, the haploid organism carries out mitosis, producing the cells that develop into gametes. The only diploid stage in these species is the single-celled zygote **(Figure 13.6c)**. (Note that *either* haploid or diploid cells can divide by mitosis, depending on the type of life cycle. Only diploid cells, however, can undergo meiosis.)

Though the three types of sexual life cycles differ in the timing of meiosis and fertilization, they share a fundamental

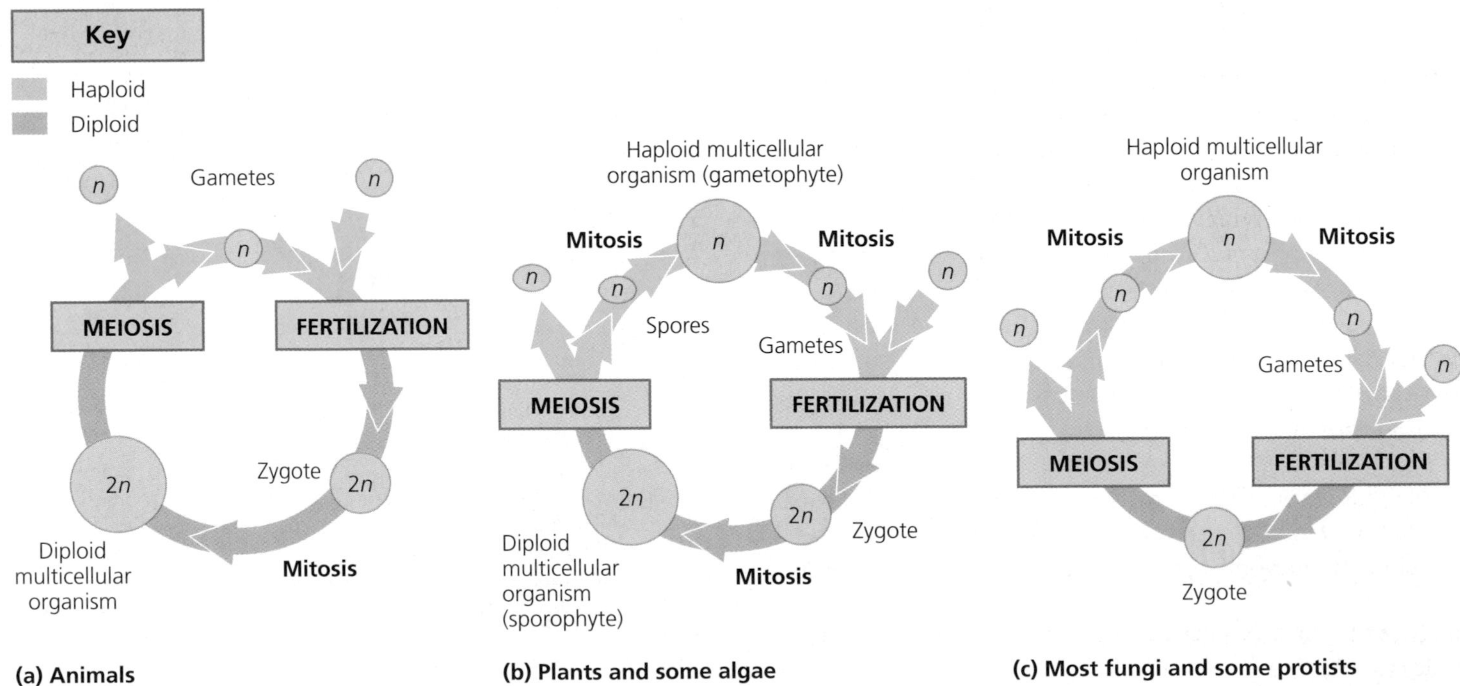

▲ **Figure 13.6 Three types of sexual life cycles.** The common feature of all three cycles is the alternation of meiosis and fertilization, key events that contribute to genetic variation among offspring. The cycles differ in the timing of these two key events.

result: Each cycle of chromosome halving and doubling contributes to genetic variation among offspring. A closer look at meiosis will reveal the sources of this variation.

Concept Check 13.2

1. How does the karyotype of a human female differ from that of a human male?
2. How does the alternation of meiosis and fertilization in the life cycles of sexually reproducing organisms maintain the normal chromosome count for each species?
3. Dog sperm contain 39 chromosomes. What are the haploid number and diploid number for dogs?
4. What process (meiosis or mitosis) is more directly involved in the production of gametes in animals? In plants and most fungi?

For suggested answers, see Appendix A.

Concept 13.3 Meiosis reduces the number of chromosome sets from diploid to haploid

Many of the steps of meiosis closely resemble corresponding steps in mitosis. Meiosis, like mitosis, is preceded by the replication of chromosomes. However, this single replication is followed by two consecutive cell divisions, called **meiosis I** and **meiosis II.** These divisions result in four daughter cells (rather than the two daughter cells of mitosis), each with only half as many chromosomes as the parent cell.

The Stages of Meiosis

The overview of meiosis in **Figure 13.7** shows how both members of a single homologous pair of chromosomes in a diploid cell are replicated and the copies then sorted into four haploid daughter cells. Recall that sister chromatids are two copies of *one* chromosome, attached at the centromere; together they make up one duplicated chromosome (see Figure 13.4). In contrast, the two chromosomes of a homologous pair are individual chromosomes that were inherited from different parents; they are not usually connected to each other. Homologues appear alike in the microscope, but they may have different versions of genes at corresponding loci (for example, a gene for freckles on one chromosome and a gene for the absence of freckles at the same locus on the homologue).

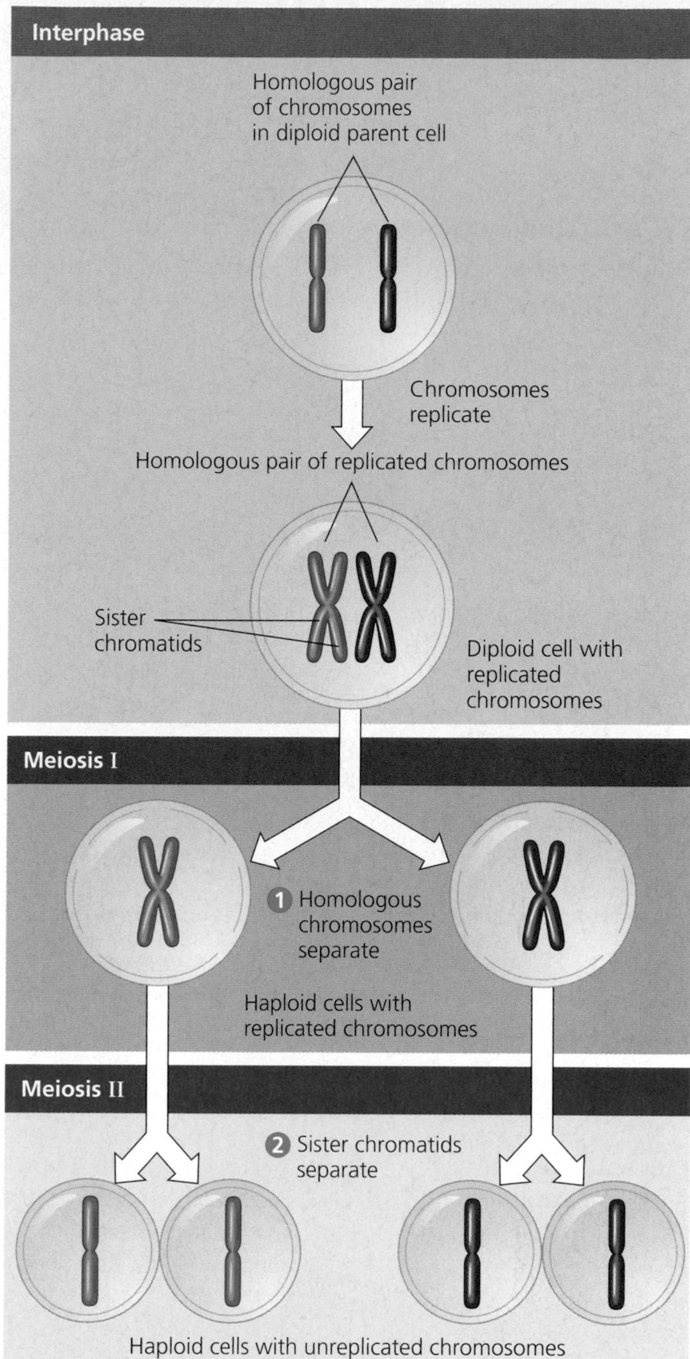

▲ **Figure 13.7 Overview of meiosis: how meiosis reduces chromosome number.** After the chromosomes replicate in interphase, the diploid cell divides *twice*, yielding four haploid daughter cells. This overview tracks just one pair of homologous chromosomes, which for the sake of simplicity are drawn in the condensed state throughout (they would not normally be condensed during interphase). The red chromosome was inherited from the female parent, the blue chromosome from the male parent.

Figure 13.8, on the next two pages, describes in detail the stages of the two divisions of meiosis for an animal cell whose diploid number is 6. Meiosis halves the total number of chromosomes in a very specific way, reducing the number of sets from two to one, with each daughter cell receiving one set of chromosomes. Study Figure 13.8 thoroughly before going on to the next section.

Figure 13.8

Exploring The Meiotic Division of an Animal Cell

INTERPHASE

MEIOSIS I: Separates homologous chromosomes

PROPHASE I | METAPHASE I | ANAPHASE I

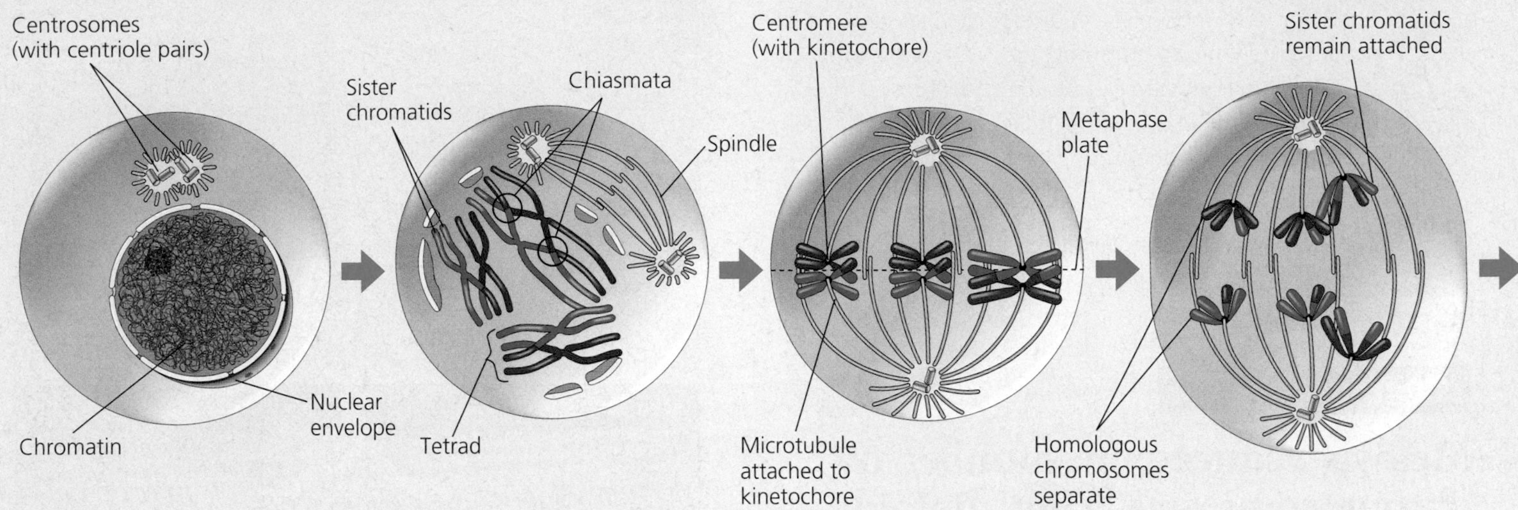

Chromosomes duplicate

Homologous chromosomes (red and blue) pair and exchange segments; $2n = 6$ in this example

Tetrads line up

Pairs of homologous chromosomes split up

Interphase

- ▶ Chromosomes replicate during S phase but remain uncondensed.
- ▶ Each replicated chromosome consists of two genetically identical sister chromatids connected at the centromere.
- ▶ The centrosome replicates, forming two centrosomes.

Prophase I

- ▶ This phase typically occupies more than 90% of the time required for meiosis.
- ▶ Chromosomes begin to condense.
- ▶ Homologous chromosomes loosely pair along their lengths, precisely aligned gene by gene.
- ▶ In crossing over, the DNA molecules in nonsister chromatids break at corresponding places and then rejoin to the other's DNA.
- ▶ In synapsis, a protein structure called the synaptonemal complex forms between homologues, holding them tightly together along their lengths.
- ▶ The synaptonemal complex disassembles in late prophase, and each chromosome pair becomes visible in the microscope as a tetrad, a group of four chromatids.
- ▶ Each tetrad has one or more chiasmata, criss-crossed regions where crossing over has occurred; these hold the homologues together until anaphase I.
- ▶ The movement of centrosomes, formation of spindle microtubules, breakdown of the nuclear envelope, and dispersal of nucleoli occur as in mitosis.
- ▶ In late prophase I (not shown here), the kinetochores of each homologue attach to microtubules from one pole or the other. The homologous pairs then move toward the metaphase plate.

Metaphase I

- ▶ The pairs of homologous chromosomes, in the form of tetrads, are now arranged on the metaphase plate, with one chromosome of each pair facing each pole.
- ▶ Both chromatids of a homologue are attached to kinetochore microtubules from one pole; those of the other homologue are attached to microtubules from the opposite pole.

Anaphase I

- ▶ The chromosomes move toward the poles, guided by the spindle apparatus.
- ▶ Sister chromatids remain attached at the centromere and move as a single unit toward the same pole.
- ▶ Homologous chromosomes, each composed of two sister chromatids, move toward opposite poles.

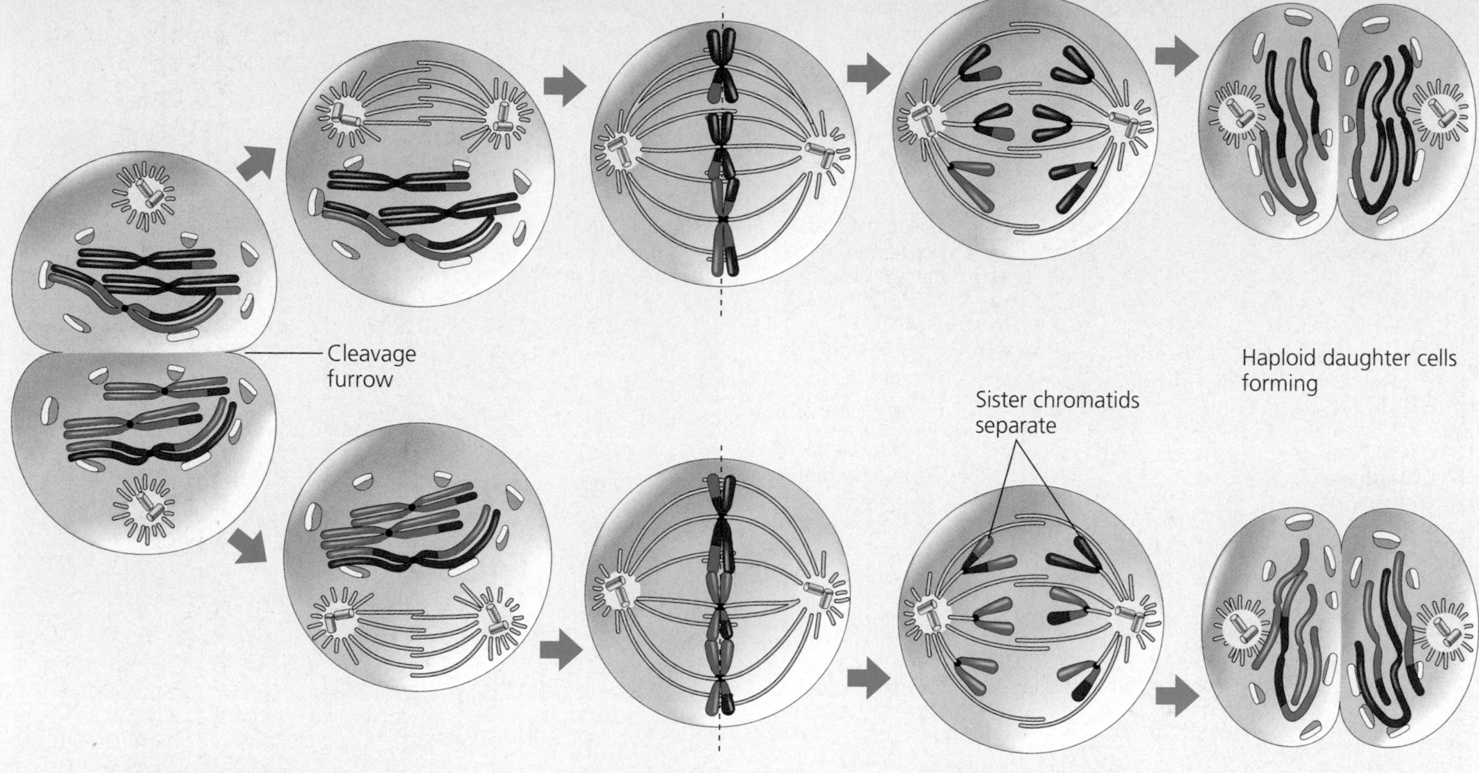
MEIOSIS II: Separates sister chromatids
TELOPHASE I AND CYTOKINESIS
PROPHASE II
METAPHASE II
ANAPHASE II
TELOPHASE II AND CYTOKINESIS
Cleavage furrow
Haploid daughter cells forming
Sister chromatids separate
Two haploid cells form; chromosomes are still double
During another round of cell division, the sister chromatids finally separate; four haploid daughter cells result, containing single chromosomes

Telophase I and Cytokinesis

- ▶ At the beginning of telophase I, each half of the cell has a complete haploid set of chromosomes, but each chromosome is still composed of two sister chromatids.
- ▶ Cytokinesis (division of the cytoplasm) usually occurs simultaneously with telophase I, forming two haploid daughter cells.
- ▶ In animal cells, a cleavage furrow forms. (In plant cells, a cell plate forms.)
- ▶ In some but not all species, the chromosomes decondense and the nuclear envelope and nucleoli re-form.
- ▶ No chromosome replication occurs between the end of meiosis I and the beginning of meiosis II, as the chromosomes are already replicated.

Prophase II

- ▶ A spindle apparatus forms.
- ▶ In late prophase II (not shown here), chromosomes, each still composed of two chromatids, move toward the metaphase II plate.

Metaphase II

- ▶ The chromosomes are positioned on the metaphase plate as in mitosis.
- ▶ Because of crossing over in meiosis I, the two sister chromatids of each chromosome are *not* genetically identical.
- ▶ The kinetochores of sister chromatids are attached to microtubules extending from opposite poles.

Anaphase II

- ▶ The centromeres of each chromosome finally separate, and the sister chromatids come apart.
- ▶ The sister chromatids of each chromosome now move as two individual chromosomes toward opposite poles.

Telophase II and Cytokinesis

- ▶ Nuclei form, the chromosomes begin decondensing, and cytokinesis occurs.
- ▶ The meiotic division of one parent cell produces four daughter cells, each with a haploid set of (unreplicated) chromosomes.
- ▶ Each of the four daughter cells is genetically distinct from the other daughter cells and from the parent cell.

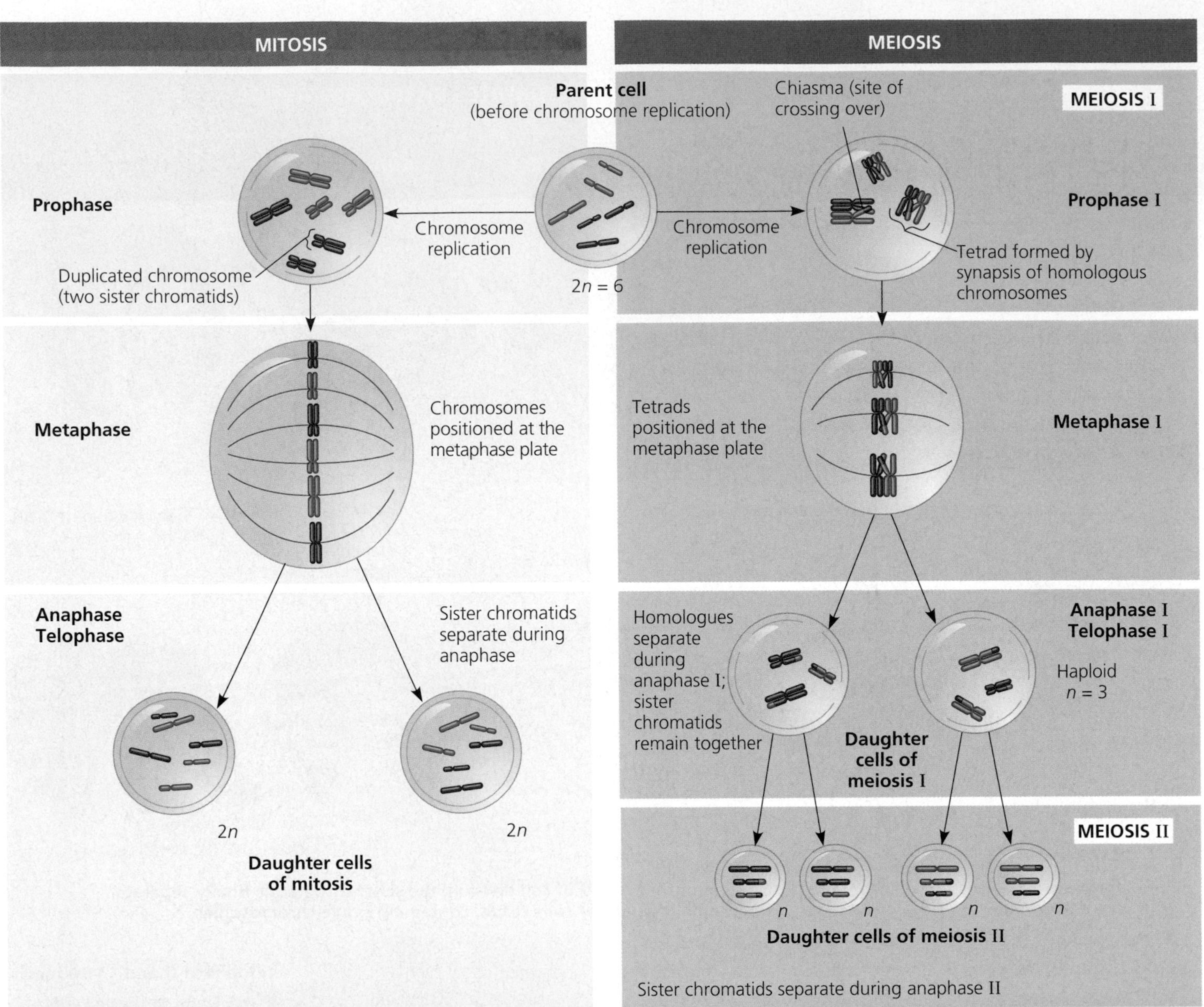

SUMMARY

Property	Mitosis	Meiosis
DNA replication	Occurs during interphase before mitosis begins	Occurs during interphase before meiosis I begins
Number of divisions	One, including prophase, metaphase, anaphase, and telophase	Two, each including prophase, metaphase, anaphase, and telophase
Synapsis of homologous chromosomes	Does not occur	Occurs during prophase I, forming tetrads (groups of four chromatids); is associated with crossing over between non-sister chromatids
Number of daughter cells and genetic composition	Two, each diploid ($2n$) and genetically identical to the parent cell	Four, each haploid (n), containing half as many chromosomes as the parent cell; genetically different from the parent cell and from each other
Role in the animal body	Enables multicellular adult to arise from zygote; produces cells for growth and tissue repair	Produces gametes; reduces number of chromosomes by half and introduces genetic variability among the gametes

▲ **Figure 13.9 A comparison of mitosis and meiosis.**

A Comparison of Mitosis and Meiosis

Now let's summarize the key differences between meiosis and mitosis. Meiosis reduces the number of chromosome sets from two (diploid) to one (haploid), whereas mitosis conserves the number of chromosome sets. Therefore, mitosis produces daughter cells genetically identical to their parent cell and to each other, whereas meiosis produces cells that differ genetically from their parent cell and from each other.

Figure 13.9 compares mitosis and meiosis. Three events are unique to meiosis, and all three occur during meiosis I:

1. **Synapsis and crossing over.** During prophase I, duplicated homologous chromosomes line up and become physically connected along their lengths by a zipper-like protein structure, the *synaptonemal complex;* this process is called **synapsis.** Genetic rearrangement between nonsister chromatids, known as **crossing over,** also occurs during prophase I. Following disassembly of the synaptonemal complex in late prophase, the four chromatids of a homologous pair are visible in the light microscope as a **tetrad.** Each tetrad normally contains at least one X-shaped region called a **chiasma** (plural, *chiasmata*), the physical manifestation of crossing over. Synapsis and crossing over normally do not occur during mitosis.
2. **Tetrads on the metaphase plate.** At metaphase I of meiosis, paired homologous chromosomes (tetrads) are positioned on the metaphase plate, rather than individual replicated chromosomes, as in mitosis.
3. **Separation of homologues.** At anaphase I of meiosis, the duplicated chromosomes of each homologous pair move toward opposite poles, but the sister chromatids of each duplicated chromosome remain attached. In mitosis, sister chromatids separate.

Meiosis I is called the *reductional division* because it halves the number of chromosome sets per cell—a reduction from two sets (the diploid state) to one set (the haploid state). The sister chromatids then separate during the second meiotic division, meiosis II, producing haploid daughter cells. The mechanism for separating sister chromatids is virtually identical in meiosis II and mitosis.

Concept Check 13.3

1. Using the concept of chromosome sets, explain briefly how mitosis conserves chromosome number, whereas meiosis reduces the number of chromosomes by half.
2. How are the chromosomes in a cell at metaphase of mitosis similar to and different from the chromosomes in a cell at metaphase of meiosis II?

For suggested answers, see Appendix A.

Concept 13.4

Genetic variation produced in sexual life cycles contributes to evolution

How do we account for the genetic variation illustrated in Figure 13.1? As you will learn in later chapters, mutations are the original source of genetic diversity. These changes in an organism's DNA create different versions of genes. Once these differences arise, reshuffling of the versions during sexual reproduction produces the variation that results in each member of a species having its own unique combination of traits.

Origins of Genetic Variation Among Offspring

In species that reproduce sexually, the behavior of chromosomes during meiosis and fertilization is responsible for most of the variation that arises each generation. Let's examine three mechanisms that contribute to the genetic variation arising from sexual reproduction: independent assortment of chromosomes, crossing over, and random fertilization.

Independent Assortment of Chromosomes

One aspect of sexual reproduction that generates genetic variation is the random orientation of homologous pairs of chromosomes at metaphase of meiosis I. At metaphase I, the homologous pairs, each consisting of one maternal and one paternal chromosome, are situated on the metaphase plate. (Note that the terms *maternal* and *paternal* refer, respectively, to the mother and father of the individual whose cells are undergoing meiosis.) Each pair may orient with either its maternal or paternal homologue closer to a given pole—its orientation is as random as the flip of a coin. Thus, there is a 50% chance that a particular daughter cell of meiosis I will get the maternal chromosome of a certain homologous pair and a 50% chance that it will receive the paternal chromosome.

Because each homologous pair of chromosomes is positioned independently of the other pairs at metaphase I, the first meiotic division results in each pair sorting its maternal and paternal homologues into daughter cells independently of every other pair. This is called *independent assortment.* Each daughter cell represents one outcome of all possible combinations of maternal and paternal chromosomes. As shown in **Figure 13.10** on the next page, the number of combinations possible for daughter cells formed by meiosis of a diploid cell with two homologous pairs of chromosomes ($2n = 4$) is four. Note that only two of the four combinations of daughter cells shown in the figure would result from meiosis of a *single* diploid cell, because a single parent cell would have one or the other possible chromosomal arrangement at metaphase I, but not both. However, the population of daughter cells resulting from meiosis of a large number of diploid cells contains all four types in approximately equal numbers. In the case of $n = 3$, eight combinations of chromosomes are possible for daughter cells. More generally, the number of possible combinations when chromosomes sort independently during meiosis is 2^n, where n is the haploid number of the organism.

In the case of humans, the haploid number (n) in the formula is 23. Thus, the number of possible combinations of maternal and paternal chromosomes in the resulting gametes is 2^{23}, or about 8 million. Each gamete that you produce in your lifetime contains one of roughly 8 million possible combinations of chromosomes inherited from your mother and father.

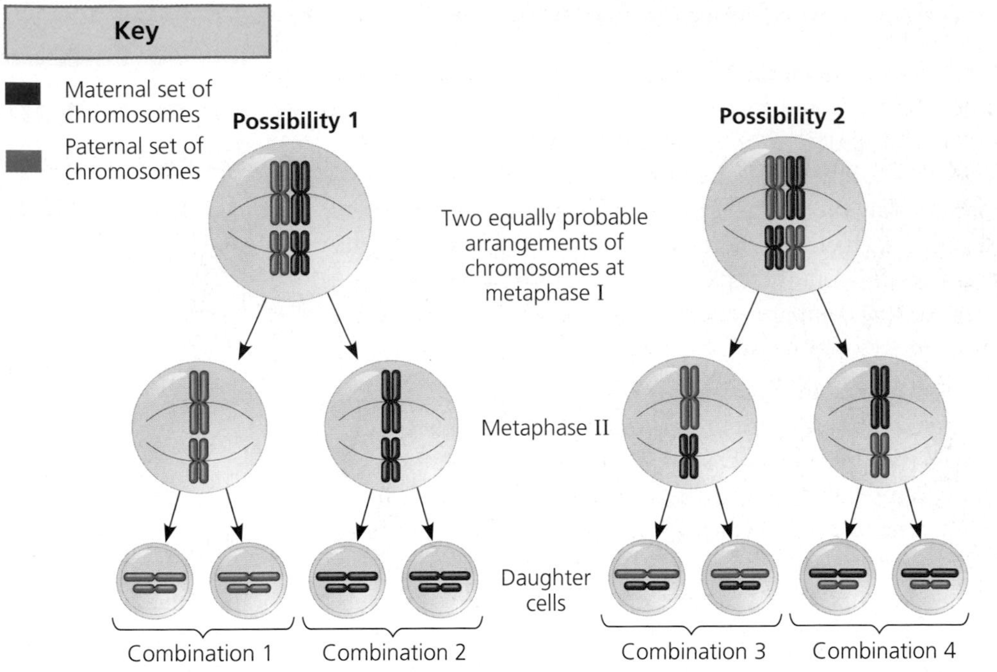

▲ **Figure 13.10 The independent assortment of homologous chromosomes in meiosis.**

Crossing Over

As a consequence of the independent assortment of chromosomes during meiosis, each of us produces a collection of gametes differing greatly in their combinations of the chromosomes we inherited from our two parents. Figure 13.10 suggests that each individual chromosome in a gamete is exclusively maternal or paternal in origin. In fact, this is *not* the case, because crossing over produces **recombinant chromosomes**, individual chromosomes that carry genes (DNA) derived from two different parents **(Figure 13.11)**.

Crossing over begins very early in prophase I, as homologous chromosomes pair loosely along their lengths. Each gene on one homologue is aligned precisely with the corresponding gene on the other homologue. In a single crossover event, the DNA molecules of two *nonsister* chromatids—one maternal and one paternal chromatid of a homologous pair—are broken at the same place and then rejoined to each other's DNA. That is, the segment of each sister chromatid from the break point to the end is joined to the rest of the other chromatid. In effect, two homologous segments trade places, or cross over, producing chromosomes with new combinations of maternal and paternal genes (see Figure 13.11).

In humans, an average of one to three crossover events occur per chromosome pair, depending on the size of the chromosomes and the position of their centromeres. Recent research indicates that, in some species, crossing over may be essential for synapsis and the proper assortment of chromosomes in meiosis I. However, the exact relationship between crossing over and synapsis is not yet fully understood and seems to vary among species.

At metaphase II, chromosomes that contain one or more recombinant chromatids can be oriented in two alternative, nonequivalent ways with respect to other chromosomes, because their sister chromatids are no longer identical twins. The independent assortment of these nonidentical sister chromatids during meiosis II increases even more the number of genetic types of daughter cells that can result from meiosis.

You will learn more about crossing over in Chapter 15. The important point for now is that crossing over, by combining DNA inherited from two parents into a single chromosome, is an important source of genetic variation in sexual life cycles.

Random Fertilization

The random nature of fertilization adds to the genetic variation arising from meiosis. In humans, for instance, each male and female gamete represents one of approximately 8 million possible chromosome combinations due to independent assortment during meiosis. The fusion of a single male gamete with a single female gamete during fertilization will produce a zygote with any of about 64 trillion (8 million × 8 million) diploid combinations. (If you calculate $2^{23} \times 2^{23}$ exactly, you will find that the total is actually over 70 trillion.) Adding in the variation brought about by crossing over, the number of possibilities is truly astronomical. No wonder brothers and sisters can be so different. You really *are* unique.

Evolutionary Significance of Genetic Variation Within Populations

Now that you've learned how new combinations of genes arise among offspring in a sexually reproducing population, let's see how the genetic variation in a population relates to evolution. Darwin recognized that a population evolves through the differential reproductive success of its variant members. On

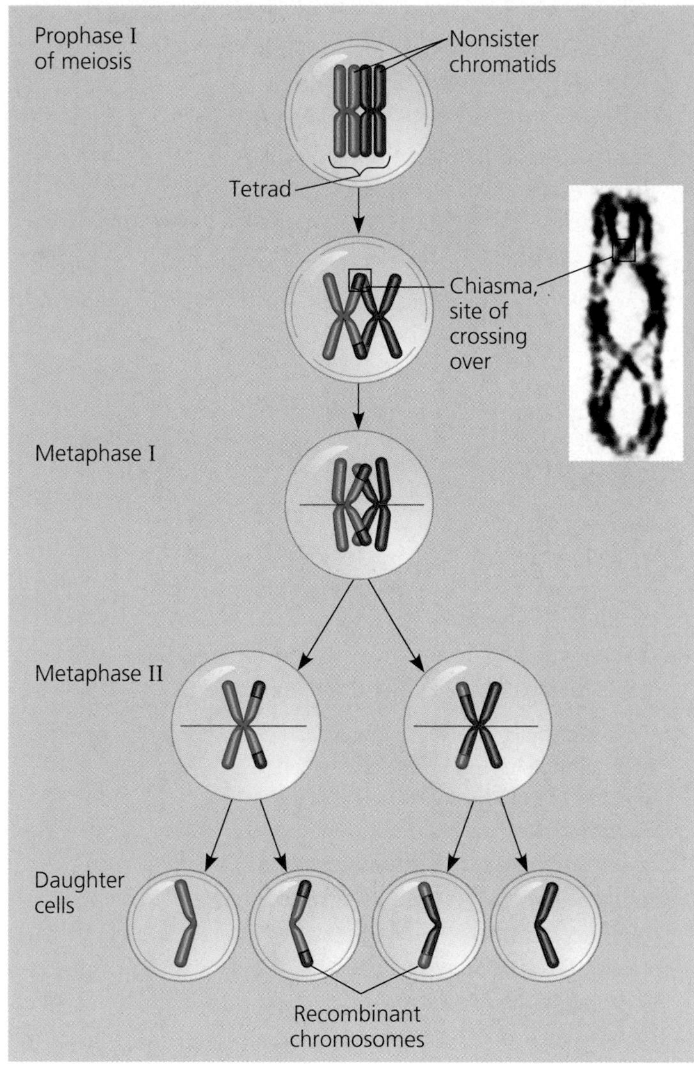

▲ **Figure 13.11 The results of crossing over during meiosis.**

average, those individuals best suited to the local environment leave the most offspring, thus transmitting their genes. This natural selection results in the accumulation of those genetic variations favored by the environment. As the environment changes, the population may survive if, in each generation, at least some of its members can cope effectively with the new conditions. Different genetic variations may work better than those that previously prevailed. In this chapter, we have seen how sexual reproduction contributes to the genetic variation present in a population, which ultimately results from mutations.

Although Darwin realized that heritable variation is what makes evolution possible, he could not explain why offspring resemble—but are not identical to—their parents. Ironically, Gregor Mendel, a contemporary of Darwin, published a theory of inheritance that helps explain genetic variation, but his discoveries had no impact on biologists until 1900, more than 15 years after Darwin (1809–1882) and Mendel (1822–1884) had died. In the next chapter, you will learn how Mendel discovered the basic rules governing the inheritance of specific traits.

Concept Check 13.4

1. Fruit flies have a diploid number of 8, and honeybees have a diploid number of 32. Assuming no crossing over, is the genetic variation among offspring from the same two parents likely to be greater in fruit flies or honeybees? Explain.
2. Under what circumstances would crossing over during meiosis *not* contribute to genetic variation among daughter cells?

For suggested answers, see Appendix A.

Chapter 13 Review

Go to www.campbellbiology.com or the student CD-ROM to explore Activities, Investigations, and other interactive study aids.

SUMMARY OF KEY CONCEPTS

Concept 13.1

Offspring acquire genes from parents by inheriting chromosomes

▶ **Inheritance of Genes (p. 239)** Each gene in an organism's DNA has a specific locus on a certain chromosome. We inherit one set of chromosomes from our mother and one set from our father.

▶ **Comparison of Asexual and Sexual Reproduction (p. 239)** In asexual reproduction, one parent produces genetically identical offspring by mitosis. Sexual reproduction combines sets of genes from two different parents, forming genetically diverse offspring.
Activity *Asexual and Sexual Life Cycles*

Concept 13.2

Fertilization and meiosis alternate in sexual life cycles

▶ **Sets of Chromosomes in Human Cells (pp. 240–241)** Normal human somatic cells have 46 chromosomes made up of two sets—one set of 23 chromosomes derived from each parent. In diploid cells ($2n = 46$), each of the 22 maternal autosomes has a homologous paternal chromosome. The 23rd pair, the sex chromosomes, determines whether the person is female (XX) or male (XY).

▶ **Behavior of Chromosome Sets in the Human Life Cycle (pp. 241–242)** At sexual maturity, ovaries and testes (the gonads) produce haploid gametes by meiosis, each gamete containing a single set of 23 chromosomes. During fertilization, an ovum and sperm unite, forming a diploid ($2n$) single-celled zygote, which develops into a multicellular organism by mitosis.

▶ **The Variety of Sexual Life Cycles (pp. 242–243)** Sexual life cycles differ in the timing of meiosis in relation to fertilization. Multicellular organisms may be diploid or haploid or may alternate between haploid and diploid generations.

Concept 13.3

Meiosis reduces the number of chromosome sets from diploid to haploid

▶ **The Stages of Meiosis (pp. 243–245)** The two cell divisions of meiosis produce four haploid daughter cells. The number of chromosome sets is reduced from diploid to haploid during meiosis I, the reductional division.

▶ **A Comparison of Mitosis and Meiosis (pp. 246–247)** Meiosis is distinguished from mitosis by three events of meiosis I: synapsis, which is associated with crossing over; positioning of paired homologous chromosomes (tetrads) on the metaphase plate; and movement of the two chromosomes of each homologous pair (not the sister chromatids) to opposite poles during anaphase I. Meiosis II separates the sister chromatids.
Activity *Meiosis Animation*

Genetic variation produced in sexual life cycles contributes to evolution

▶ **Origins of Genetic Variation Among Offspring (pp. 247–249)** The events of sexual reproduction that contribute to genetic variation in a population are independent assortment of chromosomes during meiosis, crossing over during meiosis I, and random fertilization of egg cells by sperm.
Activity *Origins of Genetic Variation*
Investigation *How Can the Frequency of Crossing Over Be Estimated?*

▶ **Evolutionary Significance of Genetic Variation Within Populations (pp. 248–249)** Genetic variation is the raw material for evolution by natural selection. Mutations are the original source of this variation; the production of new combinations of variant genes in sexual reproduction generates additional genetic diversity.

TESTING YOUR KNOWLEDGE

Self-Quiz

1. A human cell containing 22 autosomes and a Y chromosome is
   a. a somatic cell of a male.
   b. a zygote.
   c. a somatic cell of a female.
   d. a sperm cell.
   e. an ovum.

2. Homologous chromosomes move toward opposite poles of a dividing cell during
   a. mitosis.
   b. meiosis I.
   c. meiosis II.
   d. fertilization.
   e. binary fission.

3. Meiosis II is similar to mitosis in that
   a. homologous chromosomes synapse.
   b. DNA replicates before the division.
   c. the daughter cells are diploid.
   d. sister chromatids separate during anaphase.
   e. the chromosome number is reduced.

4. If the DNA content of a diploid cell in the G_1 phase of the cell cycle is x, then the DNA content of the same cell at metaphase of meiosis I would be
   a. $0.25x$. b. $0.5x$. c. x. d. $2x$. e. $4x$.

5. If we continued to follow the cell lineage from question 4, then the DNA content at metaphase of meiosis II would be
   a. $0.25x$. b. $0.5x$. c. x. d. $2x$. e. $4x$.

6. How many different combinations of maternal and paternal chromosomes can be packaged in gametes made by an organism with a diploid number of 8 ($2n = 8$)?
   a. 2 b. 4 c. 8 d. 16 e. 32

7. The immediate product of meiosis in a plant is a
   a. spore. c. sporophyte. e. zygote.
   b. gamete. d. gametophyte.

8. Multicellular haploid organisms
   a. are typically called sporophytes.
   b. produce new cells for growth by meiosis.
   c. produce gametes by mitosis.
   d. are found only in aquatic environments.
   e. are the direct result of fertilization.

9. Crossing over usually contributes to genetic variation by exchanging chromosomal segments between
   a. sister chromatids of a chromosome.
   b. chromatids of nonhomologues.
   c. nonsister chromatids of homologues.
   d. nonhomologous loci of the genome.
   e. autosomes and sex chromosomes.

10. In comparing the typical life cycles of plants and animals, a stage found in plants but not in animals is a
    a. gamete. c. multicellular diploid.
    b. zygote. d. multicellular haploid.

For Self-Quiz answers, see Appendix A.

Go to the website or CD-ROM for more quiz questions.

Evolution Connection

Many species can reproduce either asexually or sexually. Speculate about the evolutionary significance of the switch from asexual to sexual reproduction that occurs in some organisms when the environment becomes unfavorable.

Scientific Inquiry

You prepare a karyotype of an animal you are studying and discover that its somatic cells each have three homologous sets of chromosomes, a condition called triploidy. What might have happened?

Investigation *How Can the Frequency of Crossing Over Be Estimated?*

Science, Technology, and Society

Starting with short pieces of needles from straight, fast-growing pine trees, we can grow thousands of genetically identical trees that are superior producers of lumber. What are the short-term and long-term benefits and drawbacks of this approach?

14 Mendel and the Gene Idea

▲ **Figure 14.1 Gregor Mendel and his garden peas.**

Key Concepts

14.1 Mendel used the scientific approach to identify two laws of inheritance
14.2 The laws of probability govern Mendelian inheritance
14.3 Inheritance patterns are often more complex than predicted by simple Mendelian genetics
14.4 Many human traits follow Mendelian patterns of inheritance

Overview

Drawing from the Deck of Genes

Eyes of brown, blue, green, or gray; hair of black, brown, blond, or red—these are just a few examples of heritable variations that we may observe among individuals in a population. What genetic principles account for the transmission of such traits from parents to offspring?

One possible explanation of heredity is a "blending" hypothesis, the idea that genetic material contributed by the two parents mixes in a manner analogous to the way blue and yellow paints blend to make green. This hypothesis predicts that over many generations, a freely mating population will give rise to a uniform population of individuals. However, our everyday observations and the results of breeding experiments with animals and plants contradict such a prediction. The blending hypothesis also fails to explain other phenomena of inheritance, such as traits reappearing after skipping a generation.

An alternative to the blending model is a "particulate" hypothesis of inheritance: the gene idea. According to this model, parents pass on discrete heritable units—genes—that retain their separate identities in offspring. An organism's collection of genes is more like a deck of cards or a bucket of marbles than a pail of paint. Like cards and marbles, genes can be sorted and passed along, generation after generation, in undiluted form.

Modern genetics had its genesis in an abbey garden, where a monk named Gregor Mendel documented a particulate mechanism of inheritance. The painting in **Figure 14.1** depicts Mendel working with his experimental organism, garden peas. Mendel developed his theory of inheritance several decades before the behavior of chromosomes was observed in the microscope and their significance understood. So in this chapter, we digress from the study of chromosomes to recount how Mendel arrived at his theory. We will also explore how to predict the inheritance of certain characteristics and consider inheritance patterns more complex than those Mendel observed in garden peas. Finally, we will see how the Mendelian model applies to the inheritance of human variations, including hereditary disorders such as sickle-cell disease.

Concept 14.1

Mendel used the scientific approach to identify two laws of inheritance

Mendel discovered the basic principles of heredity by breeding garden peas in carefully planned experiments. As we retrace his work, the key elements of the scientific process that were introduced in Chapter 1 will be evident.

Mendel's Experimental, Quantitative Approach

Mendel grew up on his parents' small farm in a region of Austria that is now part of the Czech Republic. At school in this agricultural area, Mendel and the other children received agricultural training along with basic education. Later, Mendel overcame financial hardship and illness to excel in high school and at the Olmutz Philosophical Institute.

In 1843, at the age of 21, Mendel entered an Augustinian monastery. After failing an examination to become a teacher, he went to the University of Vienna, where he studied from 1851 to 1853. These were very important years for Mendel's development as a scientist. Two professors were especially influential. One was the physicist Christian Doppler, who encouraged his students to learn science through experimentation and trained Mendel to use mathematics to help explain natural phenomena. The second was a botanist named Franz Unger, who aroused Mendel's interest in the causes of variation in plants. These influences came together in Mendel's subsequent experiments with garden peas.

After attending the university, Mendel was assigned to teach at a school where several other teachers shared his enthusiasm for scientific research. What's more, many university professors and researchers lived at the monastery with Mendel. Most important, the monks had a long-standing interest in the breeding of plants. Around 1857, Mendel began breeding garden peas in the abbey garden in order to study inheritance. In itself, this does not seem extraordinary. What *was* extraordinary was Mendel's fresh approach to very old questions about heredity.

Mendel most likely chose to work with peas because they are available in many varieties. For example, one variety has purple flowers, while another variety has white flowers. A **character** is a heritable feature, such as flower color, that varies among individuals. Each variant for a character, such as purple or white color for flowers, is called a **trait.** (Some geneticists use the terms *character* and *trait* synonymously, but in this book we distinguish between them.)

Another advantage in using peas was that Mendel could strictly control which plants mated with which. The reproductive organs of a pea plant are in its flowers, and each pea flower has both pollen-producing organs (stamens) and an egg-bearing organ (carpel). In nature, pea plants usually self-fertilize: Pollen grains released from the stamens land on the carpel of the same flower, and sperm from the pollen fertilize eggs in the carpel. To achieve cross-pollination (fertilization between different plants), Mendel removed the immature stamens of a plant before they produced pollen and then dusted pollen from another plant onto the altered flowers **(Figure 14.2)**. Each resulting zygote then developed into a plant embryo encased in a seed (pea). Whether ensuring self-pollination or executing artificial cross-pollination, Mendel could always be sure of the parentage of new seeds.

Mendel chose to track only those characters that varied in an "either-or" manner rather than a "more-or-less" manner. For example, his plants had either purple flowers or white flowers; there was nothing intermediate between these two varieties. Had Mendel focused instead on characters that varied in a continuum among individuals—seed weight, for example—he would not have discovered the particulate nature of inheritance (you'll learn why later).

Mendel also made sure that he started his experiments with varieties that were **true-breeding.** When true-breeding plants self-pollinate, all their offspring are of the same variety. For example, a plant with purple flowers is true-breeding if

Figure 14.2 Research Method **Crossing Pea Plants**

APPLICATION By crossing (mating) two true-breeding varieties of an organism, scientists can study patterns of inheritance. In this example, Mendel crossed pea plants that varied in flower color.

TECHNIQUE

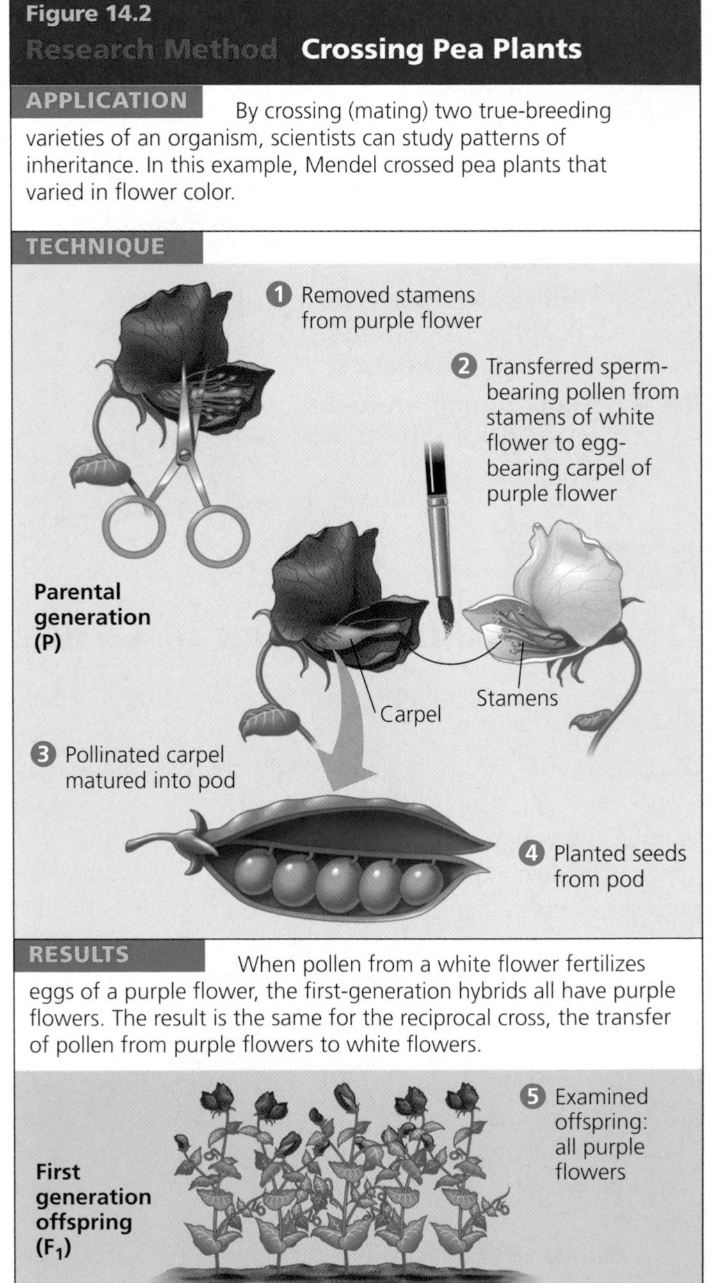
1 Removed stamens from purple flower
2 Transferred sperm-bearing pollen from stamens of white flower to egg-bearing carpel of purple flower
Parental generation (P)
Carpel
Stamens
3 Pollinated carpel matured into pod
4 Planted seeds from pod
5 Examined offspring: all purple flowers
First generation offspring (F_1)

RESULTS When pollen from a white flower fertilizes eggs of a purple flower, the first-generation hybrids all have purple flowers. The result is the same for the reciprocal cross, the transfer of pollen from purple flowers to white flowers.

the seeds produced by self-pollination all give rise to plants that also have purple flowers.

In a typical breeding experiment, Mendel cross-pollinated two contrasting, true-breeding pea varieties—for example, purple-flowered plants and white-flowered plants (see Figure 14.2). This mating, or crossing, of two true-breeding varieties is called **hybridization.** The true-breeding parents are referred to as the **P generation** (parental generation), and their hybrid offspring are the **F_1 generation** (first filial generation, the word *filial* from the Latin word for "son"). Allowing these F_1 hybrids to self-pollinate produces an **F_2 generation** (second filial generation). Mendel usually followed traits for at least the P, F_1, and F_2 generations. Had Mendel stopped his experiments with the F_1 generation, the basic patterns of inheritance would have eluded him. It was mainly Mendel's quantitative analysis of F_2 plants that revealed the two fundamental principles of heredity that are now known as the law of segregation and the law of independent assortment.

The Law of Segregation

If the blending model of inheritance were correct, the F_1 hybrids from a cross between purple-flowered and white-flowered pea plants would have pale purple flowers, intermediate between the two varieties of the P generation. Notice in Figure 14.2 that the experiment produced a very different result: All the F_1 offspring had flowers just as purple as the purple-flowered parents. What happened to the white-flowered plants' genetic contribution to the hybrids? If it were lost, then the F_1 plants could produce only purple-flowered offspring in the F_2 generation. But when Mendel allowed the F_1 plants to self-pollinate and planted their seeds, the white-flower trait reappeared in the F_2 generation.

Mendel used very large sample sizes and kept accurate records of his results: 705 of the F_2 plants had purple flowers, and 224 had white flowers. These data fit a ratio of approximately three purple to one white **(Figure 14.3)**. Mendel reasoned that the heritable factor for white flowers did not disappear in the F_1 plants, but only the purple-flower factor was affecting flower color in these hybrids. In Mendel's terminology, purple flower color is a *dominant* trait and white flower color is a *recessive* trait. The reappearance of white-flowered plants in the F_2 generation was evidence that the heritable factor causing that recessive trait had not been diluted by coexisting with the purple-flower factor in the F_1 hybrids.

Mendel observed the same pattern of inheritance in six other characters, each represented by two different traits (**Table 14.1**, on the next page). For example, the parental pea seeds either had a smooth, round shape or were wrinkled. When Mendel crossed his two true-breeding varieties, all the F_1 hybrids produced round seeds; this is the dominant trait. In the F_2 generation, 75% of the seeds were round and 25% were wrinkled—a 3:1 ratio, as in Figure 14.3. Now let's see how Mendel deduced the law of segregation from his experimental results. In our discussion, we will use modern terms instead of some of the terms used by Mendel (for example, we'll use "gene" instead of Mendel's "heritable factor").

Mendel's Model

Mendel developed a hypothesis, or model, to explain the 3:1 inheritance pattern that he consistently observed among the F_2 offspring in his pea experiments. We describe four related concepts making up this model, the fourth of which is the law of segregation.

First, *alternative versions of genes account for variations in inherited characters.* The gene for flower color in pea plants, for example, exists in two versions, one for purple flowers and the other for white. These alternative versions of a gene are

Figure 14.3
Inquiry When F_1 pea plants with purple flowers are allowed to self-pollinate, what flower color appears in the F_2 generation?

EXPERIMENT True-breeding purple-flowered pea plants and white-flowered pea plants were crossed (symbolized by ×). The resulting F_1 hybrids were allowed to self-pollinate or were cross-pollinated with other F_1 hybrids. Flower color was then observed in the F_2 generation.

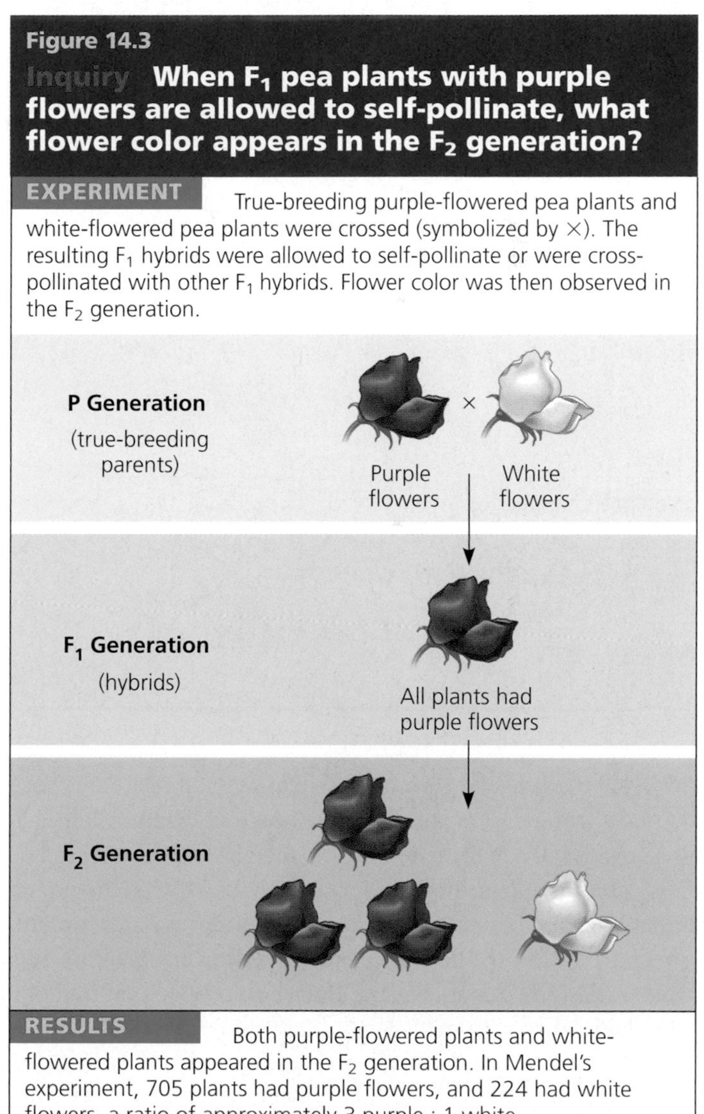

RESULTS Both purple-flowered plants and white-flowered plants appeared in the F_2 generation. In Mendel's experiment, 705 plants had purple flowers, and 224 had white flowers, a ratio of approximately 3 purple : 1 white.

Table 14.1 The Results of Mendel's F_1 Crosses for Seven Characters in Pea Plants

Character	Dominant Trait	×	Recessive Trait	F_2 Generation Dominant:Recessive	Ratio
Flower color	Purple	×	White	705:224	3.15:1
Flower position	Axial	×	Terminal	651:207	3.14:1
Seed color	Yellow	×	Green	6022:2001	3.01:1
Seed shape	Round	×	Wrinkled	5474:1850	2.96:1
Pod shape	Inflated	×	Constricted	882:299	2.95:1
Pod color	Green	×	Yellow	428:152	2.82:1
Stem length	Tall	×	Dwarf	787:277	2.84:1

now called **alleles (Figure 14.4)**. Today, we can relate this concept to chromosomes and DNA. As noted in Chapter 13, each gene resides at a specific locus on a specific chromosome. The DNA at that locus, however, can vary somewhat in its sequence of nucleotides and hence in its information content. The purple-flower allele and the white-flower allele are two DNA variations possible at the flower-color locus on one of a pea plant's chromosomes.

Second, *for each character, an organism inherits two alleles, one from each parent.* Remarkably, Mendel made this deduction without knowing about the role of chromosomes. Recall from Chapter 13 that each somatic cell in a diploid organism has two sets of chromosomes, one set inherited from each parent. Thus, a genetic locus is actually represented twice in a diploid cell. The two alleles at a particular locus may be identical, as in the true-breeding plants of Mendel's P generation. Or the alleles may differ, as in the F_1 hybrids (see Figure 14.4).

Third, *if the two alleles at a locus differ, then one, the* ***dominant allele,*** *determines the organism's appearance; the other, the* ***recessive allele,*** *has no noticeable effect on the organism's appearance.* Thus, Mendel's F_1 plants had purple flowers because the allele for that trait is dominant and the allele for white flowers is recessive.

The fourth and final part of Mendel's model, now known as the **law of segregation,** states that *the two alleles for a heritable character separate (segregate) during gamete formation and end up in different gametes.* Thus, an egg or a sperm gets only one of the two alleles that are present in the somatic cells of the organism. In terms of chromosomes, this segregation corresponds to the distribution of homologous chromosomes to different gametes in meiosis (see Figure 13.7). Note that if an organism has identical alleles for a particular character—that is, the organism is true-breeding for that character—then that allele is present in all gametes. But if different alleles are present, as in the F_1 hybrids, then 50% of the gametes receive the dominant allele and 50% receive the recessive allele.

Does Mendel's segregation model account for the 3:1 ratio he observed in the F_2 generation of his numerous crosses? For the flower-color character, the model predicts that the two different alleles present in an F_1 individual will segregate into gametes such that half the gametes will have the purple-flower allele and half will have the white-flower allele. During self-pollination, gametes of each class unite randomly. An egg with a purple-flower allele has an equal chance of being fertilized by a sperm with a purple-flower allele or one with a white-flower allele. Since the same is true for an egg with a white-flower allele, there are a total of four equally likely combinations of sperm and egg. **Figure 14.5** illustrates these combinations using a **Punnett square**, a handy diagrammatic device for predicting the allele composition of offspring from a cross between individuals of known genetic makeup. Notice that we use a capital letter to symbolize a dominant allele and a lowercase letter for a recessive allele. In our example, *P* is the purple-flower allele, and *p* is the white-flower allele.

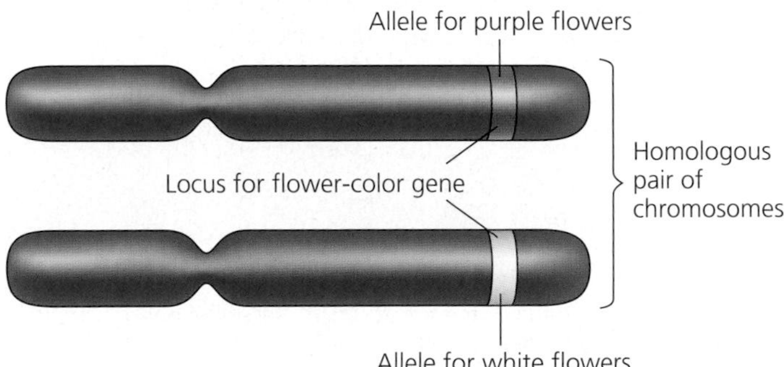

▲ **Figure 14.4 Alleles, alternative versions of a gene.** A somatic cell has two copies of each chromosome (forming a homologous pair) and thus two alleles of each gene, which may be identical or different. This figure depicts a homologous pair of chromosomes in an F_1 pea hybrid. The chromosome with an allele for purple flowers was inherited from one parent, and an allele for white flowers was inherited from the other parent.

What will be the physical appearance of these F_2 offspring? One-fourth of the plants have inherited two purple-flower alleles; clearly, these plants will have purple flowers. One-half of the F_2 offspring have inherited one purple-flower allele and one white-flower allele; these plants will also have purple flowers, the dominant trait. Finally, one-fourth of the F_2 plants have inherited two white-flower alleles and will, in fact, express the recessive trait. Thus, Mendel's model accounts for the 3:1 ratio of traits that he observed in the F_2 generation.

Useful Genetic Vocabulary

An organism having a pair of identical alleles for a character is said to be **homozygous** for the gene controlling that character. A pea plant that is true-breeding for purple flowers (*PP*) is an example. Pea plants with white flowers are also homozygous, but for the recessive allele (*pp*). If we cross dominant homozygotes with recessive homozygotes, as in the parental (P generation) cross of Figure 14.5, every offspring will have two different alleles—*Pp* in the case of the F_1 hybrids of our flower-color experiment. An organism that has two different alleles for a gene is said to be **heterozygous** for that gene. Unlike

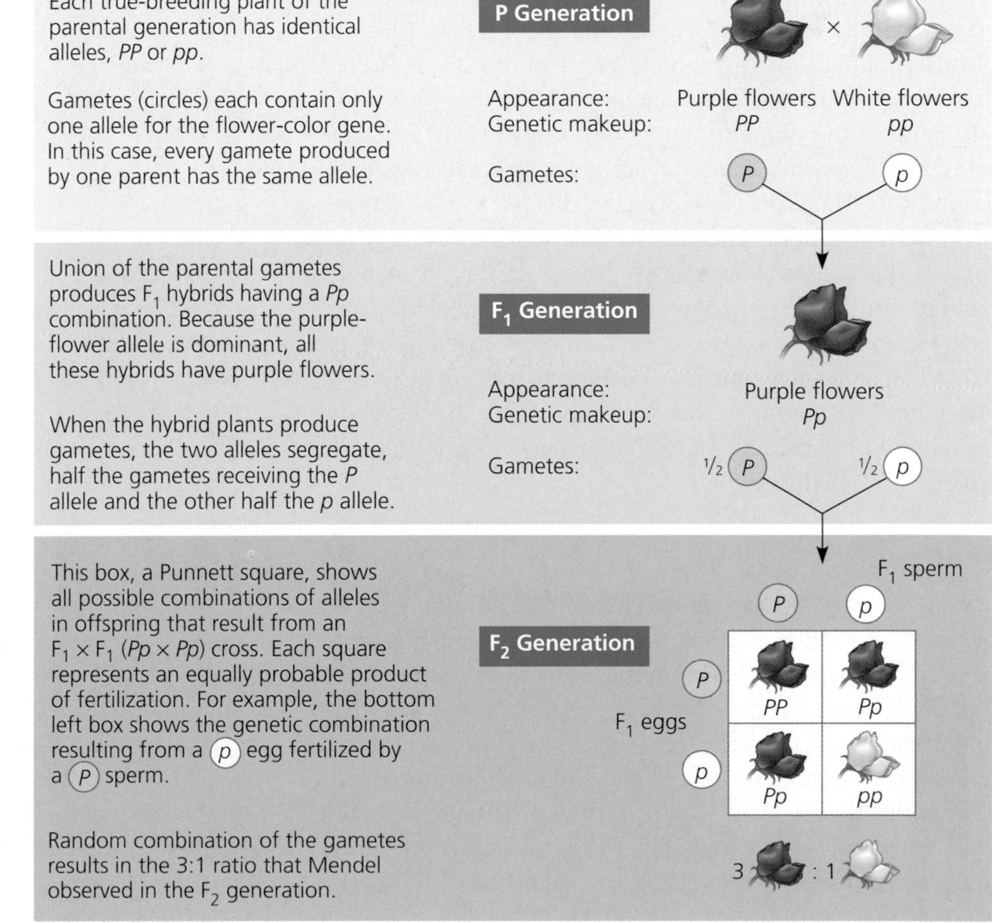

Each true-breeding plant of the parental generation has identical alleles, *PP* or *pp*.
Gametes (circles) each contain only one allele for the flower-color gene. In this case, every gamete produced by one parent has the same allele.
P Generation
×
Appearance: Purple flowers White flowers
Genetic makeup: *PP* *pp*
Gametes: *P* *p*
Union of the parental gametes produces F_1 hybrids having a *Pp* combination. Because the purple-flower allele is dominant, all these hybrids have purple flowers.
When the hybrid plants produce gametes, the two alleles segregate, half the gametes receiving the *P* allele and the other half the *p* allele.
F_1 Generation
Appearance: Purple flowers
Genetic makeup: *Pp*
Gametes: 1/2 *P* 1/2 *p*
This box, a Punnett square, shows all possible combinations of alleles in offspring that result from an $F_1 \times F_1$ ($Pp \times Pp$) cross. Each square represents an equally probable product of fertilization. For example, the bottom left box shows the genetic combination resulting from a *p* egg fertilized by a *P* sperm.
F_2 Generation
F_1 sperm
F_1 eggs
P *p*
PP *Pp*
Pp *pp*
Random combination of the gametes results in the 3:1 ratio that Mendel observed in the F_2 generation.
3 : 1

► **Figure 14.5 Mendel's law of segregation.** This diagram shows the genetic makeup of the generations in Figure 14.3. It illustrates Mendel's model for inheritance of the alleles of a single gene. Each plant has two alleles for the gene controlling flower color, one allele inherited from each parent. To construct a Punnett square, list all the possible female gametes along one side of the square and all the possible male gametes along an adjacent side. The boxes represent the offspring resulting from all the possible unions of male and female gametes.

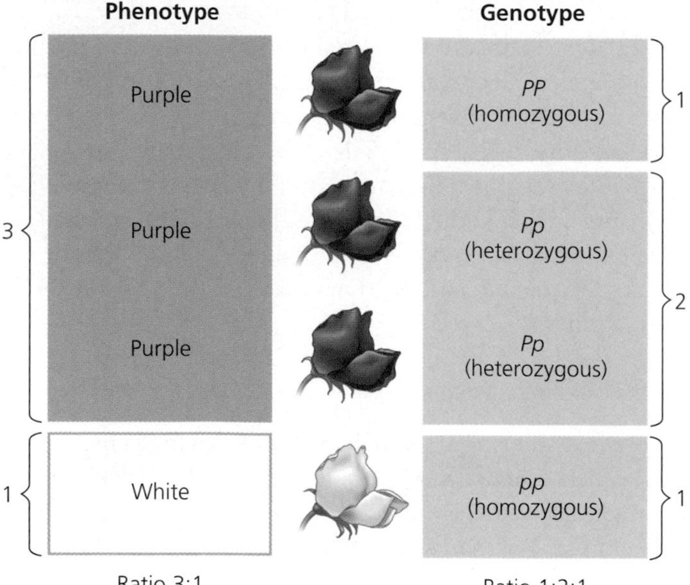

▲ **Figure 14.6 Phenotype versus genotype.** Grouping F_2 offspring from a cross for flower color according to phenotype results in the typical 3:1 phenotypic ratio. In terms of genotype, however, there are actually two categories of purple-flowered plants, *PP* (homozygous) and *Pp* (heterozygous), giving a 1:2:1 genotypic ratio.

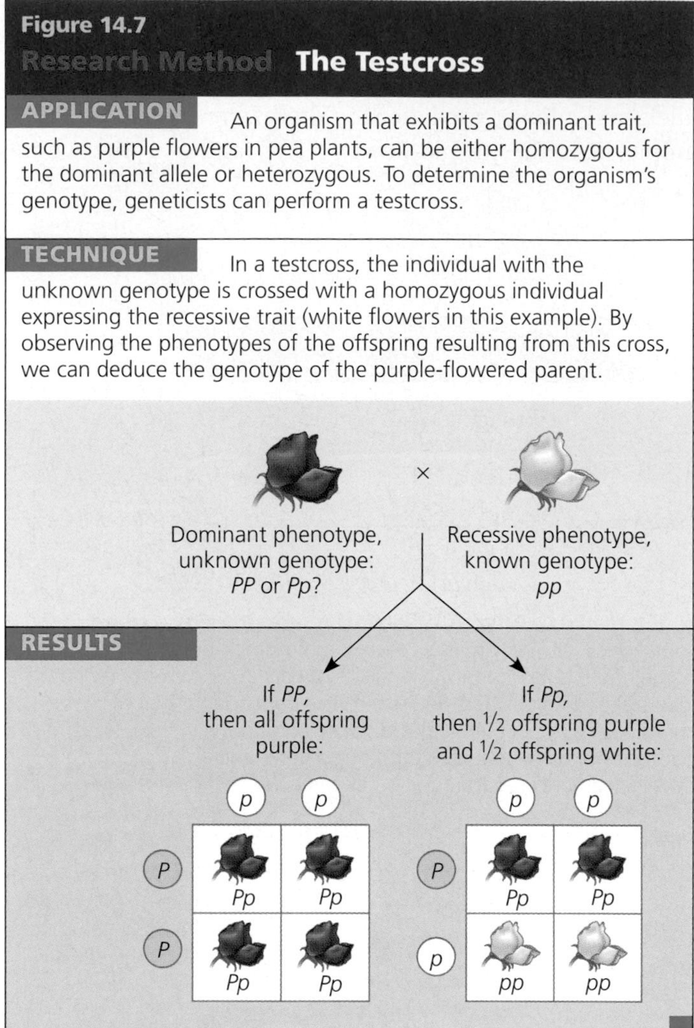

homozygotes, heterozygotes are not true-breeding because they produce gametes with different alleles—for example, *P* and *p* in the F_1 hybrids of Figure 14.5. As a result, those F_1 hybrids produce both purple-flowered and white-flowered offspring when they self-pollinate.

Figure 14.7
Research Method The Testcross
APPLICATION
An organism that exhibits a dominant trait, such as purple flowers in pea plants, can be either homozygous for the dominant allele or heterozygous. To determine the organism's genotype, geneticists can perform a testcross.
TECHNIQUE
In a testcross, the individual with the unknown genotype is crossed with a homozygous individual expressing the recessive trait (white flowers in this example). By observing the phenotypes of the offspring resulting from this cross, we can deduce the genotype of the purple-flowered parent.
×
Dominant phenotype, unknown genotype: PP or Pp?
Recessive phenotype, known genotype: pp
RESULTS
If PP, then all offspring purple:
If Pp, then 1/2 offspring purple and 1/2 offspring white:
p
p
P
P
Pp
Pp
Pp
Pp
p
p
P
p
Pp
Pp
pp
pp

Because of the different effects of dominant and recessive alleles, an organism's traits do not always reveal its genetic composition. Therefore, we distinguish between an organism's traits, called its **phenotype,** and its genetic makeup, its **genotype.** In the case of flower color in pea plants, *PP* and *Pp* plants have the same phenotype (purple) but different genotypes. **Figure 14.6** reviews these terms. Note that phenotype refers to physiological traits as well as traits relating directly to appearance. For example, there is a pea variety that lacks the normal trait of being able to self-pollinate. This physiological variation is a phenotype.

The Testcross

Suppose we have a pea plant that has purple flowers. We cannot tell from its flower color if this plant is homozygous or heterozygous because the genotypes *PP* and *Pp* result in the same phenotype. But if we cross this pea plant with one having white flowers, the appearance of the offspring will reveal the genotype of the purple-flowered parent **(Figure 14.7)**. Because white flowers is a recessive trait, the white-flowered parent must be homozygous (*pp*). If all the offspring of the cross have purple flowers, then the other parent must be homozygous for the dominant allele, because a *PP* × *pp* cross produces all *Pp* offspring. But if both the purple and the white phenotypes appear among the offspring, then the purple-flowered parent must be heterozygous. The offspring of a *Pp* × *pp* cross will have a 1:1 phenotypic ratio. This breeding of a recessive homozygote with an organism of dominant phenotype but unknown genotype is called a **testcross.** It was devised by Mendel and continues to be an important tool of geneticists.

The Law of Independent Assortment

Mendel derived the law of segregation by performing breeding experiments in which he followed only a *single* character, such as flower color. All the F_1 progeny produced in his crosses of true-breeding parents were **monohybrids,** meaning that they were heterozygous for one character. We refer to a cross between such heterozygotes as a *monohybrid cross.*

Mendel identified his second law of inheritance by following *two* characters at the same time. For instance, two of the seven characters Mendel studied were seed color and seed shape. Seeds may be either yellow or green. They also may be either round (smooth) or wrinkled. From single-character crosses, Mendel knew that the allele for yellow seeds is dominant (*Y*) and that the allele for green seeds is recessive (*y*). For the seed-shape character, the allele for round is dominant (*R*), and the allele for wrinkled is recessive (*r*).

Imagine crossing two true-breeding pea varieties differing in *both* of these characters—a parental cross between a plant with yellow-round seeds (*YYRR*) and a plant with green-wrinkled seeds (*yyrr*). The F_1 plants will be **dihybrids**, heterozygous for both characters (*YyRr*). But are these two characters, seed color and seed shape, transmitted from parents to offspring as a package? Put another way, will the *Y* and *R* alleles always stay together, generation after generation? Or are seed color and seed shape inherited independently of each other? **Figure 14.8** illustrates how a *dihybrid cross,* a cross between F_1 dihybrids, can determine which of these two hypotheses is correct.

The F_1 plants, of genotype *YyRr,* exhibit both dominant phenotypes, yellow seeds with round shapes, no matter which hypothesis is correct. The key step in the experiment is to see what happens when F_1 plants self-pollinate and produce F_2 offspring. If the hybrids must transmit their alleles in the same combinations in which they were inherited from the P generation, then there will only be two classes of gametes: *YR* and *yr.* This hypothesis predicts that the phenotypic ratio of the F_2 generation will be 3:1, just as in a monohybrid cross (see Figure 14.8).

The alternative hypothesis is that the two pairs of alleles segregate independently of each other. In other words, genes are packaged into gametes in all possible allelic combinations, as long as each gamete has one allele for each gene. In our example, four classes of gametes would be produced by an F_1 plant in equal quantities: *YR, Yr, yR,* and *yr.* If sperm of the four classes are mixed with eggs of the four classes, there will be 16 (4 × 4) equally probable ways in which the alleles can combine in the F_2 generation, as shown in the Punnett square on the right in Figure 14.8. These combinations make up four phenotypic categories with a ratio of 9:3:3:1 (nine yellow-round to three green-round to three yellow-wrinkled to one green-wrinkled). When Mendel did the experiment and "scored" (classified) the F_2 offspring, his results were close to the predicted 9:3:3:1 phenotypic ratio, supporting the hypothesis that each character—seed color or seed shape—is inherited independently of the other character.

Mendel tested his seven pea characters in various dihybrid combinations and always observed a 9:3:3:1 phenotypic ratio in the F_2 generation. Notice in Figure 14.8, however, that, if you consider the two characters separately, there is a 3:1 phenotypic ratio for each: three yellow to one green; three round to one wrinkled. As far as a single character is concerned, the alleles segregate as if this were a monohybrid cross. The results of Mendel's dihybrid experiments are the basis for what we

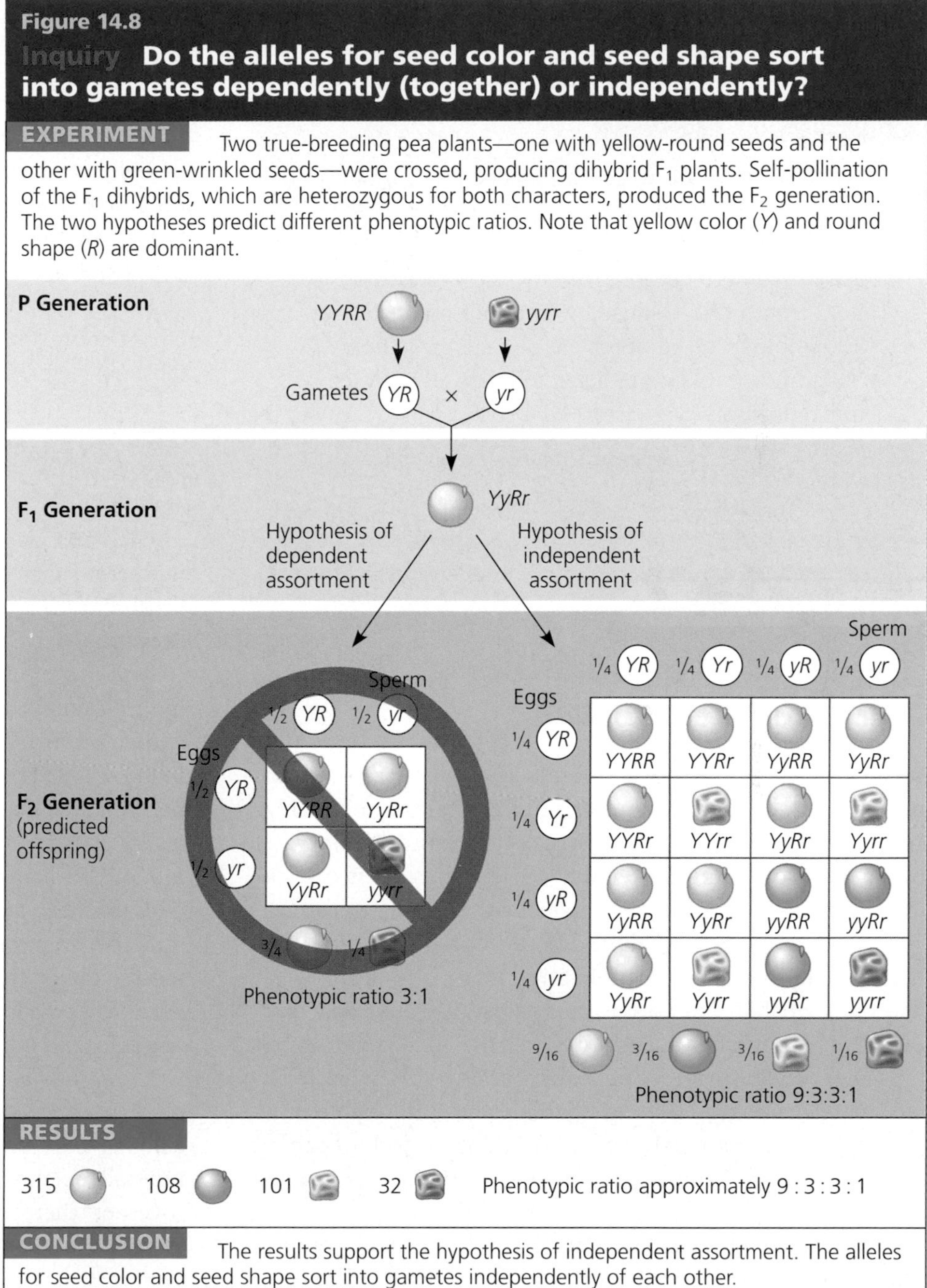

now call the **law of independent assortment,** which states that *each pair of alleles segregates independently of other pairs of alleles during gamete formation.*

Strictly speaking, this law applies only to genes (allele pairs) located on different chromosomes—that is, on chromosomes that are not homologous. Genes located near each other on the same chromosome tend to be inherited together and have more complex inheritance patterns than predicted by the law of independent assortment. We will describe such inheritance patterns in Chapter 15. All the pea characters studied by Mendel were controlled by genes on different chromosomes (or behaved as though they were); this fortuitous situation greatly simplified interpretation of his multi-character pea crosses. All the examples we consider in the rest of this chapter involve genes located on different chromosomes.

Concept Check 14.1

1. Briefly describe how the experimental results shown in Figure 14.3 support Mendel's "particulate" hypothesis of inheritance.
2. Pea plants heterozygous for flower position and stem length (*AaTt*) are allowed to self-pollinate, and 400 of the resulting seeds are planted. How many offspring would be predicted to be dwarf with terminal flowers? (See Table 14.1.)

For suggested answers, see Appendix A.

Concept 14.2

The laws of probability govern Mendelian inheritance

Mendel's laws of segregation and independent assortment reflect the same rules of probability that apply to tossing coins, rolling dice, and drawing cards from a deck. The probability scale ranges from 0 to 1. An event that is certain to occur has a probability of 1, while an event that is certain *not* to occur has a probability of 0. With a coin that has heads on both sides, the probability of tossing heads is 1, and the probability of tossing tails is 0. With a normal coin, the chance of tossing heads is $1/2$, and the chance of tossing tails is $1/2$. The probability of drawing the ace of spades from a 52-card deck is $1/52$. The probabilities of all possible outcomes for an event must add up to 1. With a deck of cards, the chance of picking a card other than the ace of spades is $51/52$.

Tossing a coin illustrates an important lesson about probability. For every toss, the probability of heads is $1/2$. The outcome of any particular toss is unaffected by what has happened on previous trials. We refer to phenomena such as coin tosses as independent events. Each toss of a coin, whether done sequentially with one coin or simultaneously with many, is independent of every other toss. And like two separate coin tosses, the alleles of one gene segregate into gametes independently of another gene's alleles (the law of independent assortment). Two basic rules of probability can help us predict the outcome of the fusion of such gametes in simple monohybrid crosses and more complicated crosses.

The Multiplication and Addition Rules Applied to Monohybrid Crosses

How do we determine the probability that two or more independent events will occur together in some specific combination? For example, what is the chance that two coins tossed simultaneously will both land heads up? The *multiplication rule* states that to determine this probability, we multiply the probability of one event (one coin coming up heads) by the probability of the other event (the other coin coming up heads). By the multiplication rule, then, the probability that both coins will land heads up is $1/2 \times 1/2 = 1/4$.

We can apply the same reasoning to an F_1 monohybrid cross **(Figure 14.9)**. With seed shape in pea plants as the heritable character, the genotype of F_1 plants is *Rr.* Segregation in a heterozygous plant is like flipping a coin: Each egg produced has a $1/2$ chance of carrying the dominant allele (*R*) and a $1/2$ chance of carrying the recessive allele (*r*). The same odds apply to each sperm cell produced. For a particular F_2 plant to have wrinkled seeds, the recessive trait, both the egg and the sperm that come together must carry the *r* allele. The probability that two *r* alleles will be present in gametes at fertilization is $1/2$ (the probability that the egg will have an *r*) $\times$ $1/2$ (the probability that the sperm will have an *r*). Thus, the multiplication rule tells us that the probability of an F_2 plant with wrinkled seeds (*rr*) is $1/4$ (see the Punnett square in Figure 14.9). Likewise, the probability of an F_2 plant carrying both dominant alleles for seed shape (*RR*) is $1/4$.

To figure out the probability that an F_2 plant from a monohybrid cross will be heterozygous rather than homozygous, we need to invoke a second rule. Notice in Figure 14.9 that the dominant allele can come from the egg and the recessive allele from the sperm, or vice versa. That is, F_1 gametes can combine to produce *Rr* offspring in two independent and mutually exclusive ways: For any particular heterozygous F_2 plant, the dominant allele can come from the egg *or* the sperm, but not from both. According to the *addition rule,* the probability that any one of two or more mutually exclusive events will occur is calculated by adding together their individual probabilities. As we have just seen, the multiplication rule gives us the individual probabilities to add together. The probability for one possible way of obtaining an F_2 heterozygote—the dominant allele from

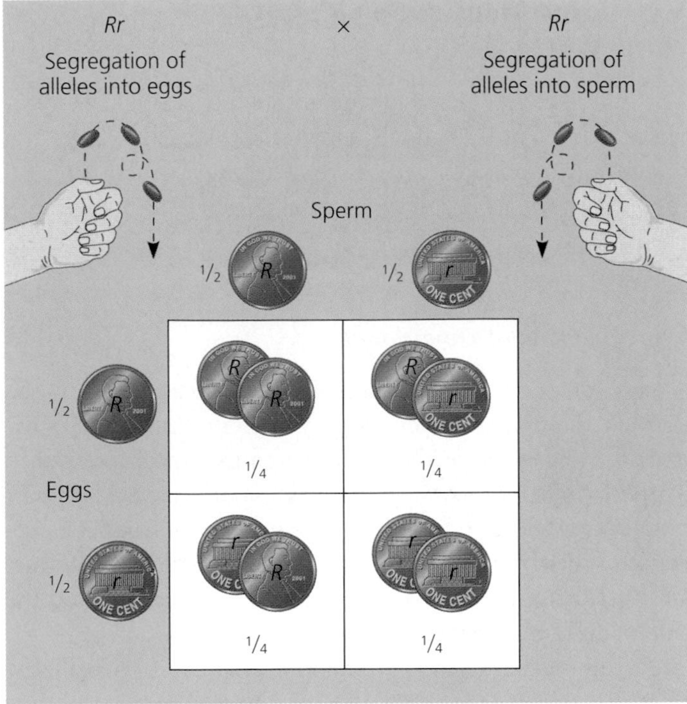

▲ **Figure 14.9 Segregation of alleles and fertilization as chance events.** When a heterozygote (*Rr*) forms gametes, segregation of alleles is like the toss of a coin. We can determine the probability for any genotype among the offspring of two heterozygotes by multiplying together the individual probabilities of an egg and sperm having a particular allele (*R* or *r* in this example).

the egg and the recessive allele from the sperm—is 1/4. The probability for the other possible way—the recessive allele from the egg and the dominant allele from the sperm—is also 1/4 (see Figure 14.9). Using the rule of addition, then, we can calculate the probability of an F_2 heterozygote as $1/4 + 1/4 = 1/2$.

Solving Complex Genetics Problems with the Rules of Probability

We can also apply the rules of probability to predict the outcome of crosses involving multiple characters. Recall that each allelic pair segregates independently during gamete formation (the law of independent assortment). Thus, a dihybrid or other multi-character cross is equivalent to two or more independent monohybrid crosses occurring simultaneously. By applying what we have learned about monohybrid crosses, we can determine the probability of specific genotypes occurring in the F_2 generation without having to construct unwieldy Punnett squares.

Consider the dihybrid cross between *YyRr* heterozygotes shown in Figure 14.8. We will focus first on the seed-color character. For a monohybrid cross of *Yy* plants, the probabilities of the offspring genotypes are 1/4 for *YY*, 1/2 for *Yy*, and 1/4 for *yy*. The same probabilities apply to the offspring genotypes for seed shape: 1/4 *RR*, 1/2 *Rr*, and 1/4 *rr*. Knowing these probabilities, we can simply use the multiplication rule to determine the probability of each of the genotypes in the F_2 generation. For example, the probability of an F_2 plant having the *YYRR* genotype is $1/4 \times 1/4 = 1/16$. This corresponds to the upper left box in the Punnett square on the right in Figure 14.8. To give another example, the probability of an F_2 plant with the *YyRR* genotype is $1/2$ (*Yy*) $\times$ $1/4$ (*RR*) $= 1/8$. If you look closely at the Punnett square on the right in Figure 14.8, you will see that 2 of the 16 boxes (1/8) correspond to the *YyRR* genotype.

Now let's see how we can combine the multiplication and addition rules to solve even more complex problems in Mendelian genetics. For instance, imagine a cross of two pea varieties in which we track the inheritance of three characters. Suppose we cross a trihybrid with purple flowers and yellow, round seeds (heterozygous for all three genes) with a plant with purple flowers and green, wrinkled seeds (heterozygous for flower color but homozygous recessive for the other two characters). Using Mendelian symbols, our cross is *PpYyRr* × *Ppyyrr*. What fraction of offspring from this cross would be predicted to exhibit the recessive phenotypes for *at least two* of the three characters?

To answer this question, we can start by listing all genotypes that fulfill this condition: *ppyyRr*, *ppYyrr*, *Ppyyrr*, *PPyyrr*, and *ppyyrr*. (Because the condition is *at least two* recessive traits, the last genotype, which produces all three recessive phenotypes, counts.) Next, we calculate the probability for each of these genotypes resulting from our *PpYyRr* × *Ppyyrr* cross by multiplying together the individual probabilities for the allele pairs, just as we did in our dihybrid example. Note that in a cross involving heterozygous and homozygous allele pairs (for example, *Yy* × *yy*), the probability of heterozygous offspring is 1/2 and the probability of homozygous offspring is 1/2. Finally, we use the addition rule to add together the probabilities for all the different genotypes that fulfill the condition of at least two recessive traits, as shown below.

ppyyRr	1/4 (probability of *pp*) × 1/2 (*yy*) × 1/2 (*Rr*)	= 1/16
ppYyrr	1/4 × 1/2 × 1/2	= 1/16
Ppyyrr	1/2 × 1/2 × 1/2	= 2/16
PPyyrr	1/4 × 1/2 × 1/2	= 1/16
ppyyrr	1/4 × 1/2 × 1/2	= 1/16
Chance of *at least two* recessive traits		= 6/16 or 3/8

With practice, you'll be able to solve genetics problems faster by using the rules of probability than by filling in Punnett squares.

We cannot predict with certainty the exact numbers of progeny of different genotypes resulting from a genetic cross. But the rules of probability give us the *chance* of various outcomes. Usually, the larger the sample size, the closer the results will conform to our predictions. The reason Mendel counted

so many offspring from his crosses is that he understood this statistical feature of inheritance and had a keen sense of the rules of chance.

Concept Check 14.2

1. For any gene with a dominant allele *C* and recessive allele *c*, what proportions of the offspring from a *CC* × *Cc* cross are expected to be homozygous dominant, homozygous recessive, and heterozygous?
2. An organism with the genotype *BbDD* is mated to one with the genotype *BBDd*. Assuming independent assortment of these two genes, write the genotypes of all possible offspring from this cross and calculate the chance of each genotype occurring using the rules of probability.
3. What is the probability that an offspring from the cross in question 2 will exhibit either of the two recessive traits coded by the *b* and *d* alleles? Explain.

For suggested answers, see Appendix A.

Concept 14.3

Inheritance patterns are often more complex than predicted by simple Mendelian genetics

In the 20th century, geneticists extended Mendelian principles not only to diverse organisms, but also to patterns of inheritance more complex than Mendel actually described. It was brilliant (and lucky) that Mendel chose pea plant characters that turned out to have a relatively simple genetic basis: Each character he studied is determined by one gene, for which there are only two alleles, one completely dominant to the other.* But these conditions are not met by all heritable characters, even in garden peas. The relationship between genotype and phenotype is rarely so simple. This does not diminish the utility of Mendelian genetics (also called Mendelism), however, because the basic principles of segregation and independent assortment apply even to more complex patterns of inheritance. In this section, we will extend Mendelian genetics to hereditary patterns that were not reported by Mendel.

* There is one exception: Geneticists have found that Mendel's flower-position character is actually determined by two genes.

Extending Mendelian Genetics for a Single Gene

The inheritance of characters determined by a single gene deviates from simple Mendelian patterns when alleles are not completely dominant or recessive, when a particular gene has more than two alleles, or when a single gene produces multiple phenotypes. We will describe examples of each of these situations in this section.

The Spectrum of Dominance

Alleles can show different degrees of dominance and recessiveness in relation to each other. We refer to this range as the *spectrum of dominance*. One extreme on this spectrum is seen in the F_1 offspring of Mendel's classic pea crosses. These F_1 plants always looked like one of the two parental varieties because of the **complete dominance** of one allele over another. In this situation, the phenotypes of the heterozygote and the dominant homozygote are indistinguishable.

At the other extreme is the **codominance** of both alleles; that is, the two alleles both affect the phenotype in separate, distinguishable ways. For example, the human MN blood group is determined by codominant alleles for two specific molecules located on the surface of red blood cells, the M and N molecules. A single gene locus, at which two allelic variations are possible, determines the phenotype of this blood group. Individuals homozygous for the *M* allele (*MM*) have red blood cells with only M molecules; individuals homozygous for the *N* allele (*NN*) have red blood cells with only N molecules. But *both* M and N molecules are present on the red blood cells of individuals heterozygous for the *M* and *N* alleles (*MN*). Note that the MN phenotype is *not* intermediate between the M and N phenotypes. Rather, both the M and N phenotypes are exhibited by heterozygotes, since both molecules are present.

The alleles for some characters fall in the middle of the spectrum of dominance. In this case, the F_1 hybrids have a phenotype somewhere in between the phenotypes of the two parental varieties. This phenomenon, called the **incomplete dominance** of either allele, is seen when red snapdragons are crossed with white snapdragons: All the F_1 hybrids have pink flowers **(Figure 14.10)**. This third phenotype results from flowers of the heterozygotes having less red pigment than the red homozygotes (unlike the situation in Mendel's pea plants, where the *Pp* heterozygotes make enough pigment for the flowers to be a purple color indistinguishable from those of *PP* plants).

At first glance, incomplete dominance of either allele seems to provide evidence for the blending hypothesis of inheritance, which would predict that the red or white trait could never be retrieved from the pink hybrids. In fact, interbreeding F_1 hybrids produces F_2 offspring with a phenotypic ratio of one red to two pink to one white. (Because

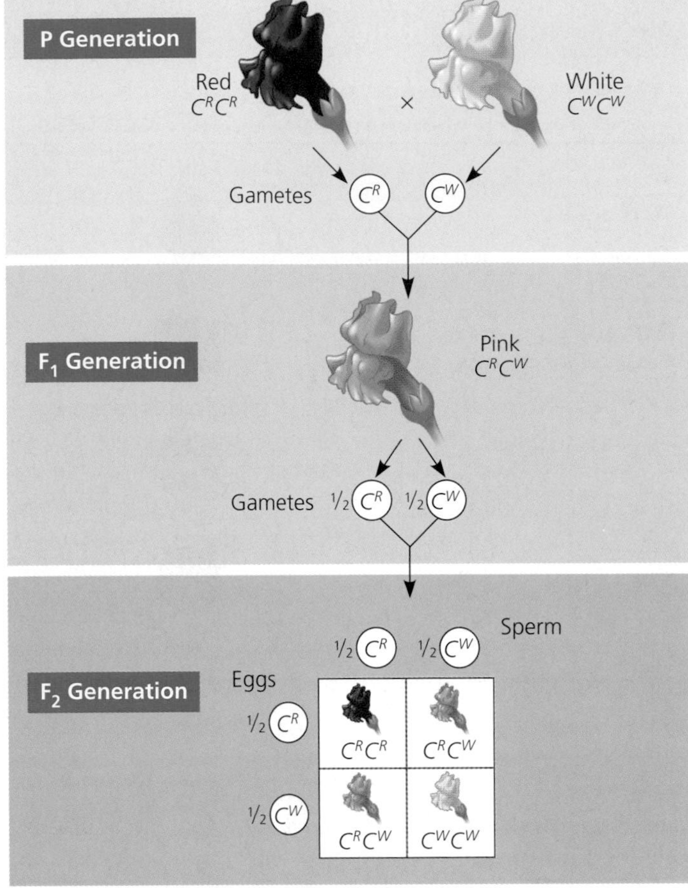

▲ **Figure 14.10 Incomplete dominance in snapdragon color.** When red snapdragons are crossed with white ones, the F_1 hybrids have pink flowers. Segregation of alleles into gametes of the F_1 plants results in an F_2 generation with a 1:2:1 ratio for both genotype and phenotype. Superscripts indicate alleles for flower color: C^R for red and C^W for white.

heterozygotes have a separate phenotype, the genotypic and phenotypic ratios for the F_2 generation are the same, 1:2:1.) The segregation of the red-flower and white-flower alleles in the gametes produced by the pink-flowered plants confirms that the alleles for flower color are heritable factors that maintain their identity in the hybrids; that is, inheritance is particulate.

The Relation Between Dominance and Phenotype. We've now seen that the relative effects of two alleles range from complete dominance of one allele, through incomplete dominance of either allele, to codominance of both alleles. It is important to understand that an allele is not termed *dominant* because it somehow subdues a recessive allele. Recall that alleles are simply variations in a gene's nucleotide sequence. When a dominant allele coexists with a recessive allele in a heterozygote, they do not actually interact at all. It is in the pathway from genotype to phenotype that dominance and recessiveness come into play.

To illustrate the relation between dominance and phenotype, we can use one of Mendel's characters—round versus wrinkled pea seed shape. The dominant allele (round) codes for the synthesis of an enzyme that helps convert sugar to starch in the seed. The recessive allele (wrinkled) codes for a defective form of this enzyme. Thus, in a recessive homozygote, sugar accumulates in the seed because it is not converted to starch. As the seed develops, the high sugar concentration causes the osmotic uptake of water, and the seed swells. Then when the mature seed dries, it develops wrinkles. In contrast, if a dominant allele is present, sugar is converted to starch, the seeds do not take up excess water, and so the seeds do not wrinkle when they dry. One dominant allele results in enough of the enzyme to convert sugar to starch, and thus dominant homozygotes and heterozygotes have the same phenotype: round seeds.

A closer look at the relation between dominance and phenotype reveals an intriguing fact: For any character, the observed dominance/recessiveness relationship of alleles depends on the level at which we examine phenotype. **Tay-Sachs disease,** an inherited disorder in humans, provides an example. The brain cells of a baby with Tay-Sachs disease are unable to metabolize certain lipids because a crucial enzyme does not work properly. As these lipids accumulate in brain cells, an infant begins to suffer seizures, blindness, and degeneration of motor and mental performance. An affected child dies within a few years.

Only children who inherit two copies of the Tay-Sachs allele (homozygotes) have the disease. Thus, at the *organismal* level, the Tay-Sachs allele qualifies as recessive. However, the activity level of the lipid-metabolizing enzyme in heterozygotes is intermediate between that in individuals homozygous for the normal allele and that in individuals with Tay-Sachs disease. The intermediate phenotype observed at the *biochemical* level is characteristic of incomplete dominance of either allele. Fortunately, the heterozygote condition does not lead to disease symptoms, apparently because half the normal enzyme activity is sufficient to prevent lipid accumulation in the brain. Extending our analysis to yet another level, we find that heterozygous individuals produce equal numbers of normal and dysfunctional enzyme molecules. Thus, at the *molecular* level, the normal allele and the Tay-Sachs allele are codominant. As you can see, whether alleles appear to be completely dominant, incompletely dominant, or codominant relative to each other depends on which phenotypic trait is considered.

Frequency of Dominant Alleles. Although you might assume that the dominant allele for a particular character would be more common in a population than the recessive allele for that character, this is not necessarily the case. For example, about one baby out of 400 in the United States is born with extra fingers or toes, a condition known as

polydactyly. The allele for the unusual trait of polydactyly is dominant to the allele for the more common trait of five digits per appendage. In other words, 399 out of every 400 people are recessive homozygotes for this character; the recessive allele is far more prevalent than the dominant allele in the population. In Chapter 23, you will learn how the relative frequencies of alleles in a population are affected by natural selection.

Multiple Alleles

Only two alleles exist for the pea characters that Mendel studied, but most genes actually exist in populations in more than two allelic forms. The ABO blood group in humans, for instance, is determined by multiple alleles of a single gene. There are four possible phenotypes for this character: A person's blood group may be either A, B, AB, or O. These letters refer to two carbohydrates—A and B—that may be found on the surface of red blood cells. A person's blood cells may have carbohydrate A (type A blood), carbohydrate B (type B), both (type AB), or neither (type O), as shown schematically in **Table 14.2**.

The four blood groups result from various combinations of three different alleles for the enzyme (I) that attaches the A or B carbohydrate to red blood cells. The enzyme encoded by the I^A allele adds the A carbohydrate, whereas the enzyme encoded by I^B adds the B carbohydrate (the superscripts indicate the carbohydrate). The enzyme encoded by the *i* allele adds neither A nor B. Because each person carries two alleles, six genotypes are possible, resulting in four phenotypes (see Table 14.2). Both the I^A and the I^B alleles are dominant to the *i* allele. Thus, I^AI^A and I^Ai individuals have type A blood, and I^BI^B and I^Bi individuals have type B blood. Recessive homozygotes, *ii*, have type O blood, because their red blood cells have neither the A nor the B carbohydrate. The I^A and I^B alleles are codominant; both are expressed in the phenotype of I^AI^B heterozygotes, who have type AB blood.

Matching compatible blood groups is critical for safe blood transfusions. For example, if a type A person receives blood from a type B or type AB donor, the recipient's immune system recognizes the "foreign" B substance on the donated blood cells and attacks them. This response causes the donated blood cells to clump together, potentially killing the recipient (see Chapter 43).

Pleiotropy

So far, we have treated Mendelian inheritance as though each gene affects one phenotypic character. Most genes, however, have multiple phenotypic effects, a property called **pleiotropy** (from the Greek *pleion*, more). For example, pleiotropic alleles are responsible for the multiple symptoms associated with certain hereditary diseases in humans, such as cystic fibrosis and sickle-cell disease, discussed later in this chapter. Considering the intricate molecular and cellular interactions responsible for an organism's development and physiology, it is not surprising that a single gene can affect a number of characteristics in an organism.

Table 14.2 Determination of ABO Blood Group by Multiple Alleles

Genotype	Phenotype (Blood Group)	Red Blood Cells
I^AI^A or I^Ai	A	
I^BI^B or I^Bi	B	
I^AI^B	AB	
ii	O	

Extending Mendelian Genetics for Two or More Genes

Dominance relationships, multiple alleles, and pleiotropy all have to do with the effects of the alleles of a single gene. We now consider two situations in which two or more genes are involved in determining a particular phenotype.

Epistasis

In **epistasis** (from the Greek for "stopping"), a gene at one locus alters the phenotypic expression of a gene at a second locus. An example will help clarify this concept. In mice and many other mammals, black coat color is dominant to brown. Let's designate *B* and *b* as the two alleles for this character. For a mouse to have brown fur, its genotype must be *bb*. But there is more to the story. A second gene determines whether or not pigment will be deposited in the hair. The dominant allele, symbolized by *C* (for color), results in the deposition of either black or brown pigment, depending on the genotype at the first locus. But if the mouse is homozygous recessive for the second locus (*cc*), then the coat is white (albino), regardless of the genotype at the black/brown locus. The gene for pigment deposition is said to be epistatic to the gene that codes for black or brown pigment.

What happens if we mate black mice that are heterozygous for both genes (*BbCc*)? Although the two genes affect the same phenotypic character (coat color), they follow the law of independent assortment. Thus, our breeding experiment represents an F_1 dihybrid cross, like those that produced a 9:3:3:1 ratio in Mendel's experiments. We can use a Punnett square to represent the genotypes of the F_2 offspring **(Figure 14.11)**. As a result of epistasis, the phenotypic ratio among the F_2 offspring is 9 black to 3 brown to 4 (3 + 1) white. Other types of epistatic interactions produce different ratios, but all are modified versions of 9:3:3:1.

Polygenic Inheritance

Mendel studied characters that could be classified on an either-or basis, such as purple versus white flower color. But for many characters, such as human skin color and height, an either-or classification is impossible because the characters vary in the population along a continuum (in gradations). These are called **quantitative characters.** Quantitative variation usually indicates **polygenic inheritance**, an additive effect of two or more genes on a single phenotypic character (the converse of pleiotropy, where a single gene affects several phenotypic characters).

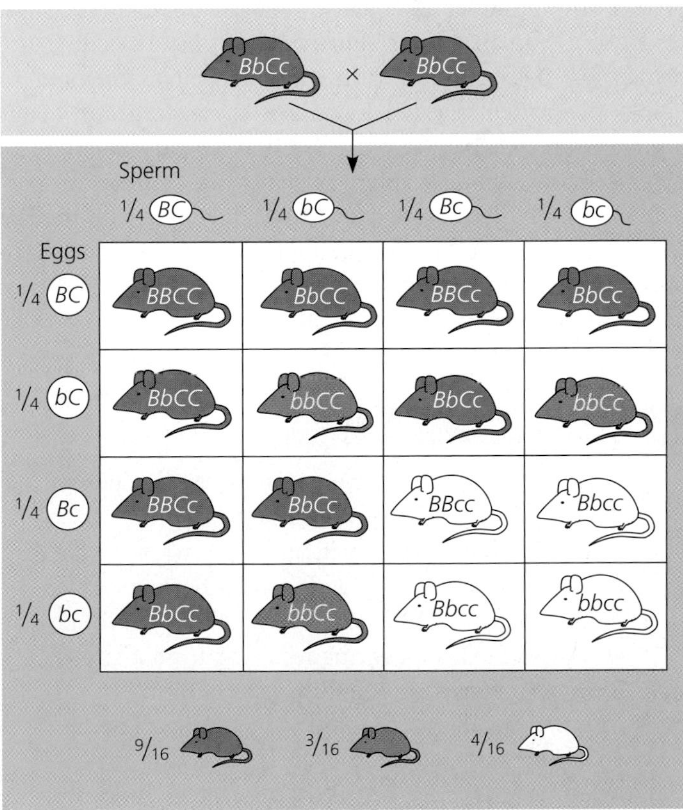

▲ **Figure 14.11 An example of epistasis.** This Punnett square illustrates the genotypes and phenotypes predicted for offspring of matings between two black mice of genotype *BbCc*. The *C/c* gene, which is epistatic to the *B/b* gene, controls whether or not pigment of any color will be deposited in the hair.

There is evidence, for instance, that skin pigmentation in humans is controlled by at least three separately inherited genes (probably more, but we will simplify). Let's consider three genes, with a dark-skin allele for each gene (*A*, *B*, or *C*) contributing one "unit" of darkness to the phenotype and being incompletely dominant to the other allele (*a*, *b*, or *c*). An *AABBCC* person would be very dark, while an *aabbcc* individual would be very light. An *AaBbCc* person would have skin of an intermediate shade. Because the alleles have a cumulative effect, the genotypes *AaBbCc* and *AABbcc* would make the same genetic contribution (three units) to skin darkness. **Figure 14.12** shows how this polygenic inheritance could result in a bell-shaped curve, called a normal distribution, for skin darkness among the progeny of hypothetical matings between individuals heterozygous for all three genes. (You are probably familiar with the concept of a normal distribution for class curves of test scores.) Environmental factors, such as exposure to the sun, also affect the skin-color phenotype and help make the graph a smooth curve rather than a stair-like histogram.

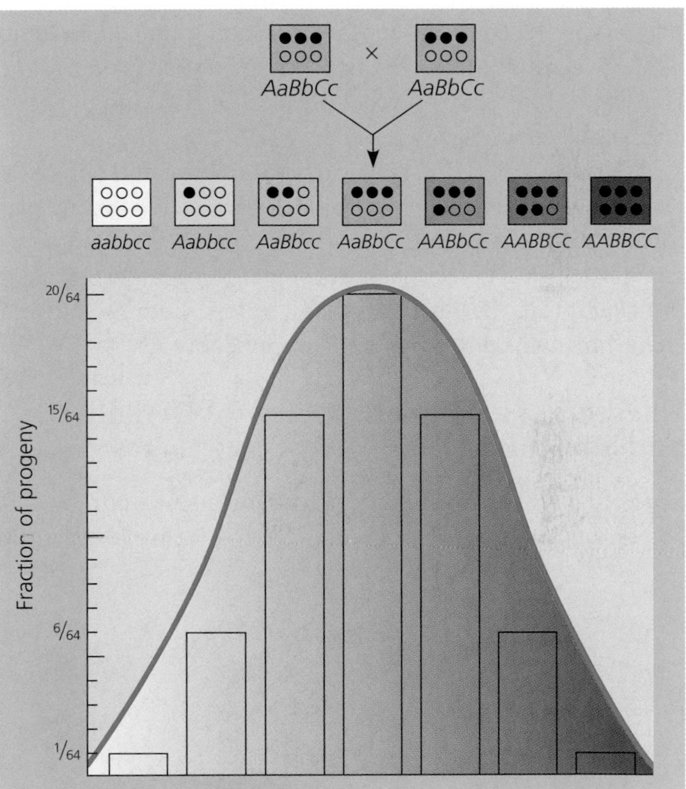

▲ **Figure 14.12 A simplified model for polygenic inheritance of skin color.** According to this model, three separately inherited genes affect the darkness of skin. The heterozygous individuals (*AaBbCc*), represented by the two rectangles at the top of this figure, each carry three dark-skin alleles (black circles) and three light-skin alleles (open circles). The variations in genotype and skin color that can occur among offspring from a large number of hypothetical matings between these heterozygotes are shown above the graph. The *y*-axis represents the fraction of progeny with each skin color. The resulting histogram is smoothed into a bell-shaped curve by environmental factors that affect skin color.

Nature and Nurture: The Environmental Impact on Phenotype

Another departure from simple Mendelian genetics arises when the phenotype for a character depends on environment as well as on genotype. A single tree, locked into its inherited genotype, has leaves that vary in size, shape, and greenness, depending on exposure to wind and sun. For humans, nutrition influences height, exercise alters build, sun-tanning darkens the skin, and experience improves performance on intelligence tests. Even identical twins, who are genetic equals, accumulate phenotypic differences as a result of their unique experiences.

Whether human characteristics are more influenced by genes or the environment—nature or nurture—is a very old and hotly contested debate that we will not attempt to settle here. We can say, however, that a genotype generally is not associated with a rigidly defined phenotype, but rather with a range of phenotypic possibilities due to environmental influences. This phenotypic range is called the **norm of reaction** for a genotype **(Figure 14.13)**. For some characters, such as the ABO blood group, the norm of reaction has no breadth whatsoever; that is, a given genotype mandates a very specific phenotype. In contrast, a person's blood count of red and white cells varies quite a bit, depending on such factors as the altitude, the customary level of physical activity, and the presence of infectious agents.

Generally, norms of reaction are broadest for polygenic characters. Environment contributes to the quantitative nature of these characters, as we have seen in the continuous variation of skin color. Geneticists refer to such characters as **multifactorial,** meaning that many factors, both genetic and environmental, collectively influence phenotype.

Integrating a Mendelian View of Heredity and Variation

Over the past several pages, we have broadened our view of Mendelian inheritance by exploring the spectrum of dominance as well as multiple alleles, pleiotropy, epistasis, polygenic inheritance, and the phenotypic impact of the environment. How can we integrate these refinements into a comprehensive theory of Mendelian genetics? The key is to make the transition from the reductionist emphasis on single genes and phenotypic characters to the emergent properties of the organism as a whole, one of the themes of this book.

The term *phenotype* can refer not only to specific characters, such as flower color and blood group, but also to an organism in its entirety—*all* aspects of its physical appearance, internal anatomy, physiology, and behavior. Similarly, the term *genotype* can refer to an organism's entire genetic makeup, not just its alleles for a single genetic locus. In most cases, a gene's impact on phenotype is affected by other genes and by the environment. In this integrated view of heredity and variation, an organism's phenotype reflects its overall genotype and unique environmental history.

Considering all that can occur in the pathway from genotype to phenotype, it is indeed impressive that Mendel could uncover the fundamental principles governing the transmission of individual genes from parents to offspring. Mendel's two laws, segregation and independent assortment, explain heritable variations in terms of alternative forms of genes (hereditary "particles") that are passed along, generation after generation, according to simple rules of probability. This theory of inheritance is equally valid for peas, flies, fishes, birds, and human beings. Furthermore, by extending the principles of segregation and independent assortment to help explain such hereditary patterns as epistasis and quantitative characters, we begin to see how broadly Mendelism applies. From Mendel's abbey garden came a theory of particulate inheritance that anchors modern genetics. In the last section of this chapter, we will apply Mendelian genetics to human inheritance, with emphasis on the transmission of hereditary diseases.

▲ **Figure 14.13 The effect of environment on phenotype.** The outcome of a genotype lies within its norm of reaction, a phenotypic range that depends on the environment in which the genotype is expressed. For example, hydrangea flowers of the same genetic variety range in color from blue-violet to pink, depending on the acidity of the soil.

Concept Check 14.3

1. A rooster with gray feathers is mated with a hen of the same phenotype. Among their offspring, 15 chicks are gray, 6 are black, and 8 are white. What is the simplest explanation for the inheritance of these colors in chickens? What phenotypes would you expect in the offspring resulting from a cross between a gray rooster and a black hen?
2. In humans, tall parents tend to have tall children, and short parents tend to have short children. Adult heights, however, vary in the population over a wide range, following a normal bell-shaped curve. Explain these observations.

For suggested answers, see Appendix A.

Concept 14.4

Many human traits follow Mendelian patterns of inheritance

Whereas peas are convenient subjects for genetic research, humans are not. The human generation span is about 20 years, and human parents produce relatively few offspring compared to peas and most other species. Furthermore, breeding experiments like the ones Mendel performed are unacceptable with humans. In spite of these difficulties, the study of human genetics continues to advance, spurred on by the desire to understand our own inheritance. New techniques in molecular biology have led to many breakthrough discoveries, as we will see in Chapter 20, but basic Mendelism endures as the foundation of human genetics.

Pedigree Analysis

Unable to manipulate the mating patterns of people, geneticists must analyze the results of matings that have already occurred. They do so by collecting information about a family's history for a particular trait and assembling this information into a family tree describing the interrelationships of parents and children across the generations—the family **pedigree.**

Figure 14.14a shows a three-generation pedigree that traces the occurrence of a pointed contour of the hairline on the forehead. This trait, called a widow's peak, is due to a dominant allele, *W.* Because the widow's-peak allele is dominant, all individuals who lack a widow's peak must be homozygous recessive (*ww*). The two grandparents with widow's peaks must have the *Ww* genotype, since some of their offspring are homozygous recessive. The offspring in the second generation who *do* have widow's peaks must also be heterozygous, because they are the products of *Ww* × *ww* matings. The third generation in this pedigree consists of two sisters. The one who has a widow's peak could be either homozygous (*WW*) or heterozygous (*Ww*), given what we know about the genotypes of her parents (both *Ww*).

Figure 14.14b is a pedigree of the same family, but this time we focus on a recessive trait, attached earlobes. We'll use *f* for the recessive allele and *F* for the dominant allele, which results in free earlobes. As you work your way through the pedigree, notice once again that you can apply what you have learned about Mendelian inheritance to fill in the genotypes for most individuals.

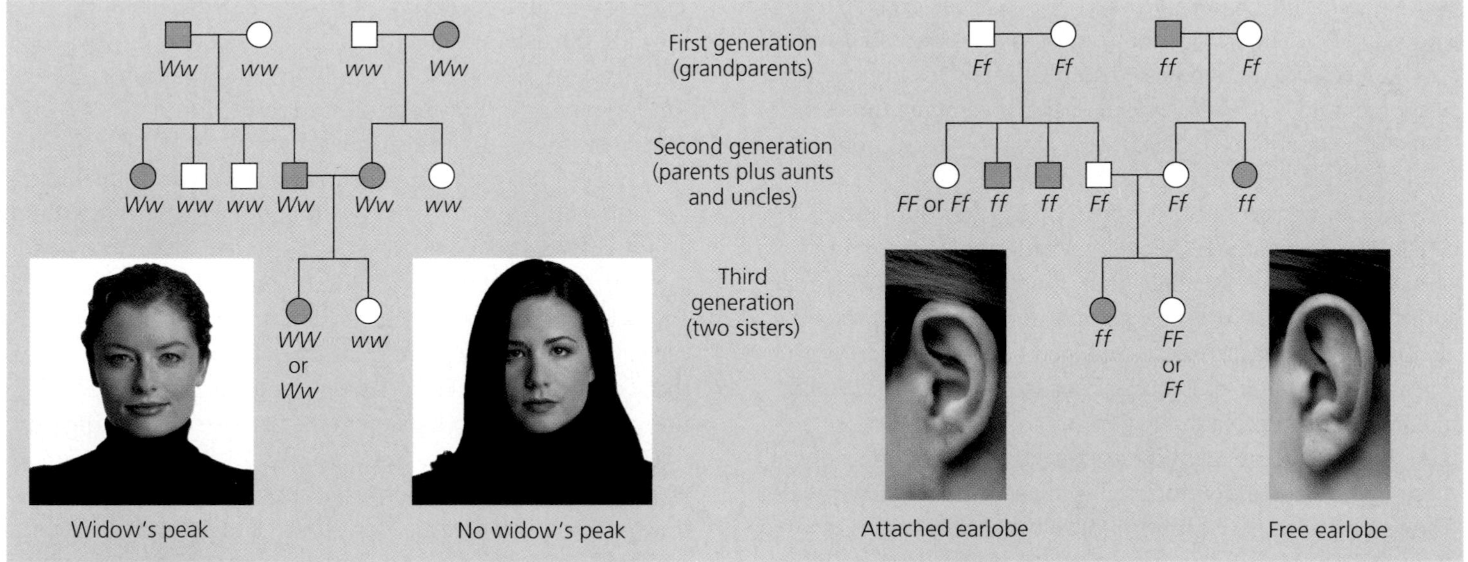

(a) Dominant trait (widow's peak). This pedigree traces the trait called widow's peak through three generations of a family. Notice in the third generation that the second-born daughter lacks a widow's peak, although both of her parents had the trait. Such a pattern of inheritance supports the hypothesis that the trait is due to a dominant allele. If the trait were due to a *recessive* allele, and both parents had the recessive phenotype, then *all* of their offspring would also have the recessive phenotype.

(b) Recessive trait (attached earlobe). This is the same family, but in this case we are tracing the inheritance of a recessive trait, attached earlobes. Notice that the first-born daughter in the third generation has attached earlobes, although both of her parents lack that trait (they have free earlobes). Such a pattern is easily explained if the attached-lobe phenotype is due to a recessive allele. If it were due to a *dominant* allele, then at least one parent would also have had the trait.

▲ **Figure 14.14 Pedigree analysis.** In these family trees, squares represent males and circles represent females. A horizontal line connecting a male and female (□—O) indicates a mating, with offspring listed below in their order of birth from left to right. Shaded squares and circles represent individuals who exhibit the trait being traced.

An important application of a pedigree is to help us predict the future. Suppose that the couple represented in the second generation of Figure 14.14 decide to have one more child. What is the probability that the child will have a widow's peak? This is equivalent to a Mendelian F_1 monohybrid cross (*Ww* × *Ww*), and thus the probability that a child will inherit a dominant allele and have a widow's peak is $\frac{3}{4}$ ($\frac{1}{4}$ *WW* + $\frac{1}{2}$ *Ww*). What is the probability that the child will have attached earlobes? Again, we can treat this as a monohybrid cross (*Ff* × *Ff*), but this time we want to know the chance that the offspring will be homozygous recessive (*ff*). That probability is $\frac{1}{4}$. Finally, what is the chance that the child will have a widow's peak *and* attached earlobes? Assuming that the genes for these two characters are on different chromosomes, the two pairs of alleles will assort independently in this dihybrid cross (*WwFf* × *WwFf*). Thus, we can use the multiplication rule: $\frac{3}{4}$ (chance of widow's peak) × $\frac{1}{4}$ (chance of attached earlobes) = $\frac{3}{16}$ (chance of widow's peak and attached earlobes).

Pedigrees are a more serious matter when the alleles in question cause disabling or deadly hereditary diseases instead of innocuous human variations such as hairline or earlobe configuration. However, for disorders inherited as simple Mendelian traits, the same techniques of pedigree analysis apply.

Recessively Inherited Disorders

Thousands of genetic disorders are known to be inherited as simple recessive traits. These disorders range in severity from relatively mild, such as albinism (lack of pigmentation, which results in susceptibility to skin cancers and vision problems), to life-threatening, such as cystic fibrosis.

How can we account for the recessive behavior of the alleles causing these disorders? Recall that genes code for proteins of specific function. An allele that causes a genetic disorder codes either for a malfunctional protein or for no protein at all. In the case of disorders classified as recessive, heterozygotes are normal in phenotype because one copy of the normal allele produces a sufficient amount of the specific protein. Thus, a recessively inherited disorder shows up only in the homozygous individuals who inherit one recessive allele from each parent. We can symbolize the genotype of such individuals as *aa* and the genotypes of individuals with the normal phenotype as either *AA* or *Aa*. Although phenotypically normal with regard to the disorder, heterozygotes (*Aa*) may transmit the recessive allele to their offspring and thus are called **carriers.**

Most people who have recessive disorders are born to parents who are carriers of the disorder but themselves have a normal phenotype. A mating between two carriers corresponds to a Mendelian F_1 monohybrid cross (*Aa* × *Aa*); the genotypic ratio for the offspring is 1 *AA* : 2 *Aa* : 1 *aa*. Thus, each child has a $\frac{1}{4}$ chance of inheriting a double dose of the recessive allele and being affected by the disorder. From the genotypic ratio, we also can see that out of three offspring with the *normal* phenotype (one *AA* plus two *Aa*), two are predicted to be heterozygous carriers, a $\frac{2}{3}$ chance. Recessive homozygotes could also result from *Aa* × *aa* and *aa* × *aa* matings, but if the disorder is lethal before reproductive age or results in sterility, no *aa* individuals will reproduce. Even if recessive homozygotes are able to reproduce, such individuals will still account for a much smaller percentage of the population than heterozygous carriers (for reasons we will examine in Chapter 23).

In general, a genetic disorder is not evenly distributed among all groups of humans. For example, the incidence of Tay-Sachs disease, which we described earlier in this chapter, is disproportionately high among Ashkenazic Jews, Jewish people whose ancestors lived in central Europe. In that population, Tay-Sachs disease occurs in one out of 3,600 births, about 100 times greater than the incidence among non-Jews or Mediterranean (Sephardic) Jews. Such an uneven distribution results from the different genetic histories of the world's peoples during less technological times, when populations were more geographically (and hence genetically) isolated. We will now examine two other recessively inherited diseases, which are also more prevalent in some groups than others.

Cystic Fibrosis

The most common lethal genetic disease in the United States is **cystic fibrosis,** which strikes one out of every 2,500 people of European descent but is much rarer in other groups. Among people of European descent, one out of 25 (4%) is a carrier of the cystic fibrosis allele. The normal allele for this gene codes for a membrane protein that functions in chloride ion transport between certain cells and the extracellular fluid. These chloride transport channels are defective or absent in the plasma membranes of children who inherit two recessive alleles for cystic fibrosis. The result is an abnormally high concentration of extracellular chloride, which causes the mucus that coats certain cells to become thicker and stickier than normal. The mucus builds up in the pancreas, lungs, digestive tract, and other organs, leading to multiple (pleiotropic) effects, including poor absorption of nutrients from the intestines, chronic bronchitis, foul stools, and recurrent bacterial infections. Recent research indicates that the extracellular chloride also contributes to infection by disabling a natural antibiotic made by some body cells. When immune cells come to the rescue, their remains add to the mucus, creating a vicious cycle.

If untreated, most children with cystic fibrosis die before their fifth birthday. Gentle pounding on the chest to clear mucus from clogged airways, daily doses of antibiotics to prevent infection, and other preventive treatments can prolong life. In the United States, more than half of the people with cystic fibrosis now survive into their late 20s or even 30s and beyond.

Sickle-Cell Disease

The most common inherited disorder among people of African descent is **sickle-cell disease,** which affects one out of 400 African-Americans. Sickle-cell disease is caused by the substitution of a single amino acid in the hemoglobin protein of red blood cells. When the oxygen content of an affected individual's blood is low (at high altitudes or under physical stress, for instance), the sickle-cell hemoglobin molecules aggregate into long rods that deform the red cells into a sickle shape (see Figure 5.21). Sickled cells may clump and clog small blood vessels, often leading to other symptoms throughout the body, including physical weakness, pain, organ damage, and even paralysis. The multiple effects of a double dose of the sickle-cell allele are another example of pleiotropy. Regular blood transfusions can ward off brain damage in children with sickle-cell disease, and new drugs can help prevent or treat other problems, but there is no cure.

Although two sickle-cell alleles are necessary for an individual to manifest full-blown sickle-cell disease, the presence of one sickle-cell allele can affect the phenotype. Thus, at the organismal level, the normal allele is incompletely dominant to the sickle-cell allele. Heterozygotes, said to have *sickle-cell trait,* are usually healthy, but they may suffer some sickle-cell symptoms during prolonged periods of reduced blood oxygen. At the molecular level, the two alleles are codominant; both normal and abnormal (sickle-cell) hemoglobins are made in heterozygotes.

About one out of ten African-Americans has sickle-cell trait, an unusually high frequency of heterozygotes for an allele with severe detrimental effects in homozygotes. One explanation for this is that a single copy of the sickle-cell allele reduces the frequency and severity of malaria attacks, especially among young children. The malaria parasite spends part of its life cycle in red blood cells (see Figure 28.11), and the presence of even heterozygous amounts of sickle-cell hemoglobin results in lower parasite densities and hence reduced malaria symptoms. Thus, in tropical Africa where infection with the malaria parasite is common, the sickle-cell allele is both boon and bane. The relatively high frequency of African-Americans with sickle-cell trait is a vestige of their African roots.

Mating of Close Relatives

When a disease-causing recessive allele is rare, it is relatively unlikely that two carriers of the same harmful allele will meet and mate. However, if the man and woman are close relatives (for example, siblings or first cousins), the probability of passing on recessive traits increases greatly. These are called consanguineous ("same blood") matings, and they are indicated in pedigrees by double lines. Because people with recent common ancestors are more likely to carry the same recessive alleles than are unrelated people, it is more likely that a mating of close relatives will produce offspring homozygous for recessive traits—including harmful ones. Such effects can be observed in many types of domesticated and zoo animals that have become inbred.

There is debate among geneticists about the extent to which human consanguinity increases the risk of inherited diseases. Many deleterious alleles have such severe effects that a homozygous embryo spontaneously aborts long before birth. Still, most societies and cultures have laws or taboos forbidding marriages between close relatives. These rules may have evolved out of empirical observation that in most populations, stillbirths and birth defects are more common when parents are closely related. Social and economic factors have also influenced the development of customs and laws against consanguineous marriages.

Dominantly Inherited Disorders

Although many harmful alleles are recessive, a number of human disorders are due to dominant alleles. One example is *achondroplasia,* a form of dwarfism with a prevalence of one among every 25,000 people. Heterozygous individuals have the dwarf phenotype **(Figure 14.15)**. Therefore, all people who are not achondroplastic dwarfs—99.99% of the population—are homozygous for the recessive allele. Like the presence of extra fingers or toes mentioned earlier, achondroplasia is a trait for which the recessive allele is much more prevalent than the corresponding dominant allele.

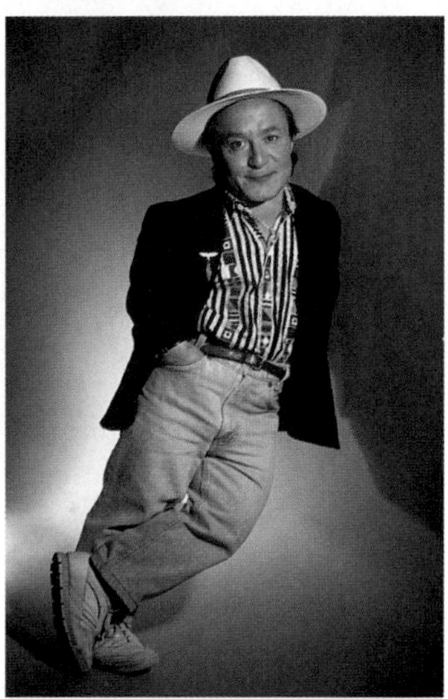

▲ **Figure 14.15 Achondroplasia.** The late David Rappaport, an actor, had achondroplasia, a form of dwarfism that is caused by a dominant allele.

Dominant alleles that cause a lethal disease are much less common than recessive alleles that do so. All such lethal alleles arise by mutations (changes to the DNA) in a sperm or egg; presumably, such mutations occur equally often whether the mutant allele is dominant or recessive. However, if a lethal dominant allele causes the death of offspring before they mature and can reproduce, the allele will not be passed on to future generations. In contrast, a lethal recessive allele can be perpetuated from generation to generation by heterozygous carriers who have normal phenotypes. These carriers can reproduce and pass on the recessive allele. Only homozygous recessive offspring will have the lethal disease.

A lethal dominant allele can escape elimination if it causes death only at a relatively advanced age. By the time the symptoms become evident, the individual may have already transmitted the lethal allele to his or her children. For example, **Huntington's disease**, a degenerative disease of the nervous system, is caused by a lethal dominant allele that has no obvious phenotypic effect until the individual is about 35 to 45 years old. Once the deterioration of the nervous system begins, it is irreversible and inevitably fatal. Any child born to a parent who has the allele for Huntington's disease has a 50% chance of inheriting the allele and the disorder. (The mating can be symbolized as *Aa* × *aa*, with *A* being the dominant allele that causes Huntington's disease.) In the United States, this devastating disease afflicts about one in 10,000 people.

Until relatively recently, the onset of symptoms was the only way to know if a person had inherited the Huntington's allele. This is no longer the case. By analyzing DNA samples from a large family with a high incidence of the disorder, geneticists tracked the Huntington's allele to a locus near the tip of chromosome 4 (**Figure 14.16**). This information led to development of a test that can detect the presence of the Huntington's allele in an individual's genome. (The methods that make such tests possible are discussed in Chapter 20.) For those with a family history of Huntington's disease, the availability of this test poses an agonizing dilemma: Under what circumstances is it beneficial for a presently healthy person to find out whether he or she has inherited a fatal and not yet curable disease? Some individuals may want to be tested for the disease before planning a family.

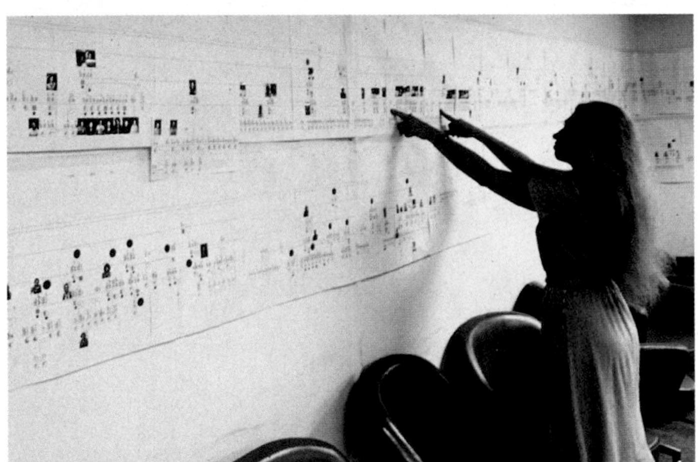

▲ **Figure 14.16 Large families as excellent case studies of human genetics.** Here, Nancy Wexler, of Columbia University and the Hereditary Disease Foundation, studies a huge pedigree that traces Huntington's disease through several generations of one large family in Venezuela. Classical Mendelian analysis of this family, coupled with the techniques of molecular biology, enabled scientists to develop a test for the presence of the dominant allele that causes Huntington's disease—a test that can be used before symptoms appear. Dr. Wexler's mother died of Huntington's disease, and thus there is a 50% chance that Dr. Wexler inherited the dominant allele that causes the disease. To date she has shown no symptoms.

Multifactorial Disorders

The hereditary diseases we have discussed so far are sometimes described as simple Mendelian disorders because they result from abnormality of one or both alleles at a single genetic locus. Many more people are susceptible to diseases that have a multifactorial basis—a genetic component plus a significant environmental influence. Heart disease, diabetes, cancer, alcoholism, certain mental illnesses such as schizophrenia and bipolar disorder, and many other diseases are multifactorial. In many cases, the hereditary component is polygenic. For example, many genes affect cardiovascular health, making some of us more prone than others to heart attacks and strokes. But our lifestyle intervenes tremendously between genotype and phenotype for cardiovascular health and other multifactorial characters. Exercise, a healthful diet, abstinence from smoking, and an ability to handle stressful situations all reduce our risk of heart disease and some types of cancer.

At present, so little is understood about the genetic contributions to most multifactorial diseases that the best public health strategy is to educate people about the importance of environmental factors and to promote healthful behavior.

Genetic Testing and Counseling

A preventive approach to simple Mendelian disorders is possible when the risk of a particular genetic disorder can be assessed before a child is conceived or during the early stages of the pregnancy. Many hospitals have genetic counselors who can provide information to prospective parents concerned about a family history for a specific disease.

Counseling Based on Mendelian Genetics and Probability Rules

Consider the case of a hypothetical couple, John and Carol. Both had a brother who died from the same recessively inherited lethal disease. Before conceiving their first child, John and Carol seek genetic counseling to determine the risk of having a child with the disease. From the information about their brothers, we know that both parents of John and both parents of Carol must have been carriers of the recessive allele. Thus, John and Carol are both products of *Aa* × *Aa* crosses, where *a* symbolizes the allele that causes this particular disease. We also know that John and Carol are not homozygous recessive (*aa*), because they do not have the disease. Therefore, their genotypes are either *AA* or *Aa*.

Given a genotypic ratio of 1*AA*:2*Aa*:1*aa* for offspring of an *Aa* × *Aa* cross, John and Carol each have a $2/3$ chance of being carriers (*Aa*). According to the rule of multiplication, the overall probability of their firstborn having the disorder is $2/3$ (the chance that John is a carrier) multiplied by $2/3$ (the chance that Carol is a carrier) multiplied by $1/4$ (the chance of two carriers having a child with the disease), which equals $1/9$.

Suppose that Carol and John decide to have a child—after all, there is an $8/9$ chance that their baby will not have the disorder. If, despite these odds, their child is born with the disease, then we would know that *both* John and Carol are, in fact, carriers (*Aa* genotype). If both John and Carol are carriers, there is a $1/4$ chance that any subsequent child this couple has will have the disease.

When we use Mendel's laws to predict possible outcomes of matings, it is important to remember that each child represents an independent event in the sense that its genotype is unaffected by the genotypes of older siblings. Suppose that John and Carol have three more children, and *all three* have the hypothetical hereditary disease. There is only one chance in 64 ($1/4 \times 1/4 \times 1/4$) that such an outcome will occur. Despite this run of misfortune, the chance that still another child of this couple will have the disease remains $1/4$.

Tests for Identifying Carriers

Because most children with recessive disorders are born to parents with normal phenotypes, the key to assessing more accurately the genetic risk for a particular disease is determining whether the prospective parents are heterozygous carriers of the recessive allele. For an increasing number of heritable disorders, tests are available that can distinguish individuals of normal phenotype who are dominant homozygotes from those who are heterozygotes. There are now tests that can identify carriers of the alleles for Tay-Sachs disease, sickle-cell disease, and the most common form of cystic fibrosis.

These tests for identifying carriers enable people with family histories of genetic disorders to make informed decisions about having children. But these new methods for genetic screening pose potential problems. If confidentiality is breached, will carriers be stigmatized? Will they be denied health or life insurance, even though they themselves are healthy? Will misinformed employers equate "carrier" with disease? And will sufficient genetic counseling be available to help a large number of individuals understand their test results? New biotechnology offers possibilities for reducing human suffering, but not before key ethical issues are resolved. The dilemmas posed by human genetics reinforce one of this book's themes: the immense social implications of biology.

Fetal Testing

Suppose a couple learns that they are both Tay-Sachs carriers, but they decide to have a child anyway. Tests performed in conjunction with a technique known as **amniocentesis** can determine, beginning at the 14th to 16th week of pregnancy, whether the developing fetus has Tay-Sachs disease (**Figure 14.17a**, on the next page). To perform this procedure, a physician inserts a needle into the uterus and extracts about 10 milliliters of amniotic fluid, the liquid that bathes the fetus. Some genetic disorders can be detected from the presence of certain chemicals in the amniotic fluid itself. Tests for other disorders, including Tay-Sachs disease, are performed on cells grown in the laboratory, descendants of the fetal cells sloughed off into the amniotic fluid. These cultured cells can also be used for karyotyping to identify certain chromosomal defects (see Figure 13.3).

In an alternative technique called **chorionic villus sampling (CVS),** a physician inserts a narrow tube through the cervix into the uterus and suctions out a tiny sample of tissue from the placenta, the organ that transmits nutrients and fetal wastes between the fetus and the mother (**Figure 14.17b**). The cells of the chorionic villi of the placenta, the portion sampled, are derived from the fetus and have the same genotype as the new individual. These cells are proliferating rapidly enough to allow karyotyping to be carried out immediately. This rapid analysis is an advantage over amniocentesis, in which the cells must be cultured for several weeks before karyotyping. Another advantage of CVS is that it can be performed as early as the eighth to tenth week of pregnancy. However, CVS is not suitable for tests requiring amniotic fluid, and it is less widely available than amniocentesis. Recently, medical scientists have developed methods for isolating fetal cells that have escaped into the mother's blood. Although very few in number, these cells can be cultured and then tested.

Imaging techniques allow a physician to examine a fetus directly for major anatomical abnormalities. In the *ultrasound* technique, sound waves are used to produce an image of the fetus by a simple noninvasive procedure. In *fetoscopy,* a needle-thin tube containing a viewing scope and fiber optics (to transmit light) is inserted into the uterus.

Ultrasound has no known risk to either mother or fetus, but amniocentesis and fetoscopy cause complications, such as maternal bleeding or even fetal death, in about 1% of cases. For this reason, these techniques generally are used only when the chance of a genetic disorder or other type of birth defect is relatively great. If the fetal tests reveal a serious disorder, the parents face the difficult choice of terminating the pregnancy or preparing to care for a child with a genetic disorder.

Newborn Screening

Some genetic disorders can be detected at birth by simple tests that are now routinely performed in most hospitals in the United States. One common screening program is for phenylketonuria (PKU), a recessively inherited disorder that occurs in about one out of every 10,000 to 15,000 births in the United States. Children with this disease cannot properly break down the amino acid phenylalanine. This compound and its by-product, phenylpyruvate, can accumulate to toxic levels in the blood, causing mental retardation. However, if the deficiency is detected in the newborn, a special diet low in phenylalanine can usually promote normal development and

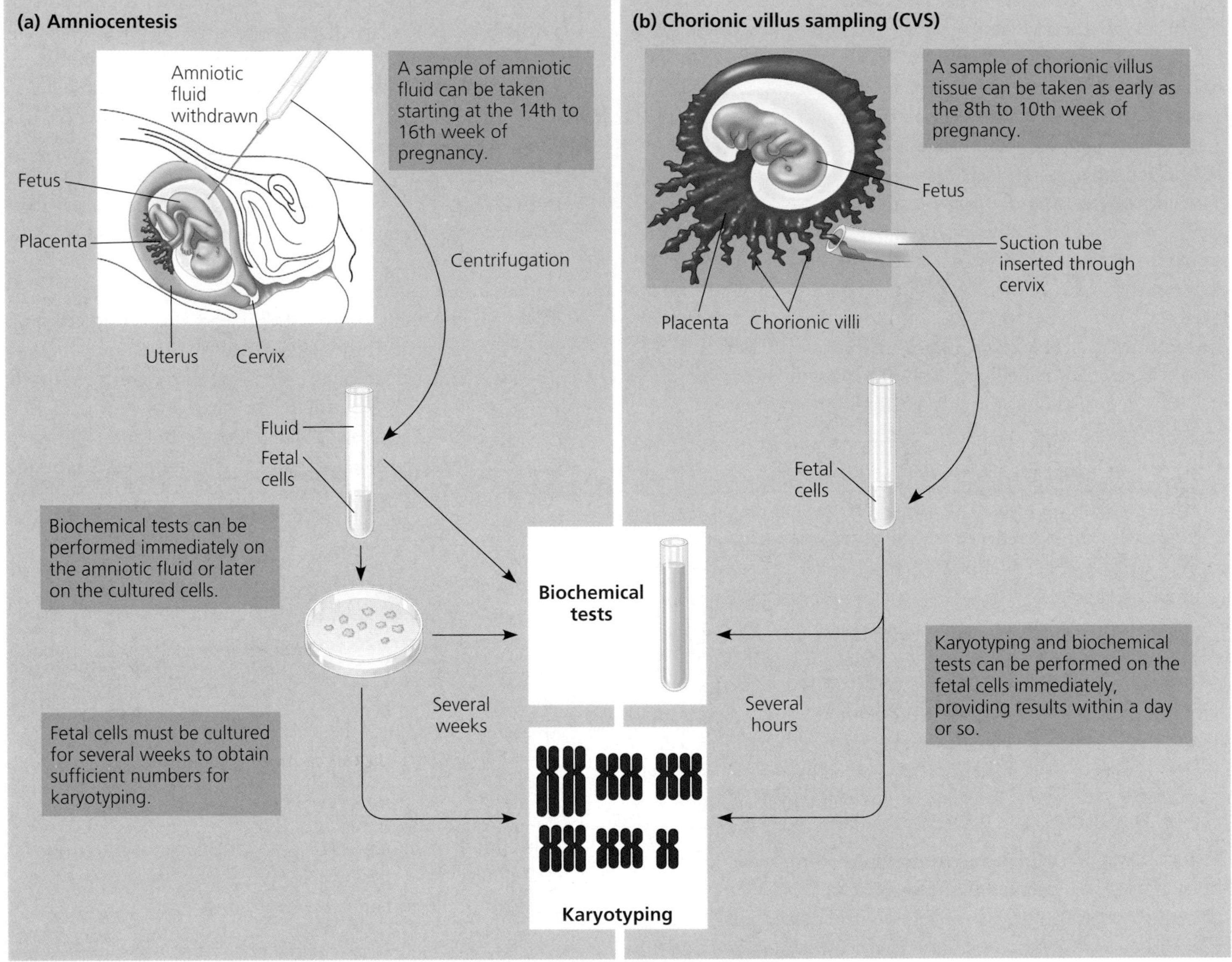
(a) Amniocentesis
Amniotic fluid withdrawn
Fetus
Placenta
Uterus
Cervix
A sample of amniotic fluid can be taken starting at the 14th to 16th week of pregnancy.
Centrifugation
Fluid
Fetal cells
Biochemical tests can be performed immediately on the amniotic fluid or later on the cultured cells.
Fetal cells must be cultured for several weeks to obtain sufficient numbers for karyotyping.
Several weeks
Biochemical tests
Karyotyping
(b) Chorionic villus sampling (CVS)
A sample of chorionic villus tissue can be taken as early as the 8th to 10th week of pregnancy.
Fetus
Suction tube inserted through cervix
Placenta
Chorionic villi
Fetal cells
Several hours
Karyotyping and biochemical tests can be performed on the fetal cells immediately, providing results within a day or so.

▲ **Figure 14.17 Testing a fetus for genetic disorders.** Biochemical tests may detect substances associated with particular disorders. Karyotyping shows whether the chromosomes of the fetus are normal in number and appearance.

prevent retardation. Unfortunately, very few other genetic disorders are treatable at the present time.

Screening of newborns and fetuses for serious inherited diseases, tests for identifying carriers, and genetic counseling—all these tools of modern medicine rely on the Mendelian model of inheritance. We owe the "gene idea"—the concept of particulate heritable factors transmitted according to simple rules of chance—to the elegant quantitative experiments of Gregor Mendel. The importance of his discoveries was overlooked by most biologists until early in the 20th century, several decades after his findings were reported. In the next chapter, you will learn how Mendel's laws have their physical basis in the behavior of chromosomes during sexual life cycles and how the synthesis of Mendelism and a chromosome theory of inheritance catalyzed progress in genetics.

Concept Check 14.4

1. Beth and Tom each have a sibling with cystic fibrosis, but neither Beth nor Tom nor any of their parents have the disease. Calculate the probability that if this couple has a child, the child will have cystic fibrosis. What would be the probability if a test revealed that Tom is a carrier but Beth is not?
2. Joan was born with six toes on each foot, a dominant trait called polydactyly. Two of her five siblings and her mother, but not her father, also have extra digits. What is Joan's genotype for the number-of-digits character? Explain your answer. Use *D* and *d* to symbolize the alleles for this character.

For suggested answers, see Appendix A.

Chapter 14 Review

Go to www.campbellbiology.com or the student CD-ROM to explore Activities, Investigations, and other interactive study aids.

SUMMARY OF KEY CONCEPTS

Concept 14.1

Mendel used the scientific approach to identify two laws of inheritance

▶ **Mendel's Experimental, Quantitative Approach (pp. 252–253)** Gregor Mendel formulated a particulate theory of inheritance based on experiments with garden peas, carried out in the 1860s. He showed that parents pass on to their offspring discrete genes that retain their identity through the generations.

▶ **The Law of Segregation (pp. 253–256)** This law states that the two alleles of a gene separate (segregate) during gamete formation, so that a sperm or an egg carries only one allele of each pair. Mendel proposed this law to explain the 3:1 ratio of F_2 phenotypes he observed when monohybrids self-pollinated. According to Mendel's model, genes have alternative forms (alleles), and each organism inherits one allele for each gene from each parent. If the two alleles of a gene are different, expression of one (the dominant allele) masks the phenotypic effect of the other (the recessive allele). Homozygous individuals have identical alleles of a given gene and are true-breeding. Heterozygous individuals have two different alleles of a given gene.
Activity *Monohybrid Cross*

▶ **The Law of Independent Assortment (pp. 256–258)** This law states that each pair of alleles segregates into gametes independently of other pairs. Mendel proposed this law based on dihybrid crosses between plants heterozygous for two genes. Alleles of each gene segregate into gametes independently of alleles of other genes. The offspring of a dihybrid cross (the F_2 generation) have four phenotypes in a 9:3:3:1 ratio.
Activity *Dihybrid Cross*

Concept 14.2

The laws of probability govern Mendelian inheritance

▶ **The Multiplication and Addition Rules Applied to Monohybrid Crosses (pp. 258–259)** The multiplication rule states that the probability of a compound event is equal to the product of the individual probabilities of the independent single events. The addition rule states that the probability of an event that can occur in two or more independent, mutually exclusive ways is the sum of the individual probabilities.
Activity *Gregor's Garden*

▶ **Solving Complex Genetics Problems with the Rules of Probability (pp. 259–260)** A dihybrid or other multi-character cross is equivalent to two or more independent monohybrid crosses occurring simultaneously. In calculating the chances for the various offspring genotypes from such crosses, each character first is considered separately and then the individual probabilities are multiplied together.

Concept 14.3

Inheritance patterns are often more complex than predicted by simple Mendelian genetics

▶ **Extending Mendelian Genetics for a Single Gene (pp. 260–262)** For a gene with complete dominance of one allele, the heterozygous phenotype is the same as that for the homozygous dominant phenotype. For a gene with codominance of both alleles, both phenotypes are expressed in heterozygotes. For a gene with incomplete dominance of either allele, the heterozygous phenotype is intermediate between the two homozygous phenotypes. Many genes exist in multiple (more than two) alleles in a population. Pleiotropy is the ability of a single gene to affect multiple phenotypic characters.
Activity *Incomplete Dominance*

▶ **Extending Mendelian Genetics for Two or More Genes (pp. 262–263)** In epistasis, one gene affects the expression of another gene. In polygenic inheritance, a single phenotypic character is affected by two or more genes. Characters influenced by multiple genes are often quantitative, meaning that they vary continuously.

▶ **Nature and Nurture: The Environmental Impact on Phenotype (p. 264)** The expression of a genotype can be affected by environmental influences. The phenotypic range of a particular genotype is called its norm of reaction. Polygenic characters that are also influenced by the environment are called multifactorial characters.

▶ **Integrating a Mendelian View of Heredity and Variation (p. 264)** An organism's overall phenotype, including its physical appearance, internal anatomy, physiology, and behavior, reflects its overall genotype and unique environmental history. Even in more complex inheritance patterns, Mendel's fundamental laws of segregation and independent assortment still apply.

Concept 14.4

Many human traits follow Mendelian patterns of inheritance

▶ **Pedigree Analysis (pp. 265–266)** Family pedigrees can be used to deduce the possible genotypes of individuals and make predictions about future offspring. Predictions are usually statistical probabilities rather than certainties.

▶ **Recessively Inherited Disorders (pp. 266–267)** Tay-Sachs disease, cystic fibrosis, sickle-cell disease, and many other genetic disorders are inherited as simple recessive traits. Most affected individuals (with the homozygous recessive genotype) are children of phenotypically normal, heterozygous carriers.

▶ **Dominantly Inherited Disorders (pp. 267–268)** Lethal dominant alleles are eliminated from the population if affected people die before reproducing. Nonlethal dominant alleles and lethal ones that strike relatively late in life, such as the allele that causes Huntington's disease, are inherited in a Mendelian way.

▶ **Multifactorial Disorders (p. 268)** Many human diseases, such as most forms of cancer and heart disease, have both genetic and environmental components. These diseases do not follow simple Mendelian inheritance patterns.

▶ **Genetic Testing and Counseling (pp. 268–270)** Using family histories, genetic counselors help couples determine the odds that their children will have genetic disorders. For a growing number of diseases, tests that identify carriers define the odds more accurately. Once a child is conceived, amniocentesis and chorionic villus sampling can help determine whether a suspected genetic disorder is present. Further genetic tests can be performed after a child is born.
Investigation *How Do You Diagnose a Genetic Disorder?*

TESTING YOUR KNOWLEDGE

Genetics Problems

1. In some plants, a true-breeding, red-flowered strain gives all pink flowers when crossed with a white-flowered strain: C^RC^R (red) × C^WC^W (white) → C^RC^W (pink). If flower position (axial or terminal) is inherited as it is in peas (see Table 14.1), what will be the ratios of genotypes and phenotypes of the F_1 generation resulting from the following cross: axial-red (true-breeding) × terminal-white? What will be the ratios in the F_2 generation?

2. Flower position, stem length, and seed shape were three characters that Mendel studied. Each is controlled by an independently assorting gene and has dominant and recessive expression as follows:

Character	*Dominant*	*Recessive*
Flower position	Axial (*A*)	Terminal (*a*)
Stem length	Tall (*T*)	Dwarf (*t*)
Seed shape	Round (*R*)	Wrinkled (*r*)

If a plant that is heterozygous for all three characters is allowed to self-fertilize, what proportion of the offspring would you expect to be as follows? (*Note:* Use the rules of probability instead of a huge Punnett square.)
a. homozygous for the three dominant traits
b. homozygous for the three recessive traits
c. heterozygous for all three characters
d. homozygous for axial and tall, heterozygous for seed shape

3. A black guinea pig crossed with an albino guinea pig produces 12 black offspring. When the albino is crossed with a second black one, 7 blacks and 5 albinos are obtained. What is the best explanation for this genetic situation? Write genotypes for the parents, gametes, and offspring.

4. In sesame plants, the one-pod condition (*P*) is dominant to the three-pod condition (*p*), and normal leaf (*L*) is dominant to wrinkled leaf (*l*). Pod type and leaf type are inherited independently. Determine the genotypes for the two parents for all possible matings producing the following offspring:
a. 318 one-pod, normal leaf : 98 one-pod, wrinkled leaf
b. 323 three-pod, normal leaf : 106 three-pod, wrinkled leaf
c. 401 one-pod, normal leaf
d. 150 one-pod, normal leaf : 147 one-pod, wrinkled leaf : 51 three-pod, normal leaf : 48 three-pod, wrinkled leaf
e. 223 one-pod, normal leaf : 72 one-pod, wrinkled leaf : 76 three-pod, normal leaf : 27 three-pod, wrinkled leaf

5. A man with type A blood marries a woman with type B blood. Their child has type O blood. What are the genotypes of these individuals? What other genotypes, and in what frequencies, would you expect in offspring from this marriage?

6. Phenylketonuria (PKU) is an inherited disease caused by a recessive allele. If a woman and her husband, who are both carriers, have three children, what is the probability of each of the following?
a. All three children are of normal phenotype.
b. One or more of the three children have the disease.
c. All three children have the disease.
d. At least one child is phenotypically normal.

(*Note:* Remember that the probabilities of all possible outcomes always add up to 1.)

7. The genotype of F_1 individuals in a tetrahybrid cross is *AaBbCcDd*. Assuming independent assortment of these four genes, what are the probabilities that F_2 offspring will have the following genotypes?
a. *aabbccdd*
b. *AaBbCcDd*
c. *AABBCCDD*
d. *AaBBccDd*
e. *AaBBCCdd*

8. What is the probability that each of the following pairs of parents will produce the indicated offspring? (Assume independent assortment of all gene pairs.)
a. *AABBCC* × *aabbcc* → *AaBbCc*
b. *AABbCc* × *AaBbCc* → *AAbbCC*
c. *AaBbCc* × *AaBbCc* → *AaBbCc*
d. *aaBbCC* × *AABbcc* → *AaBbCc*

9. Karen and Steve each have a sibling with sickle-cell disease. Neither Karen nor Steve nor any of their parents have the disease, and none of them have been tested to reveal sickle-cell trait. Based on this incomplete information, calculate the probability that if this couple has a child, the child will have sickle-cell disease.

10. In 1981, a stray black cat with unusual rounded, curled-back ears was adopted by a family in California. Hundreds of descendants of the cat have since been born, and cat fanciers hope to develop the curl cat into a show breed. Suppose you owned the first curl cat and wanted to develop a true-breeding variety. How would you determine whether the curl allele is dominant or recessive? How would you obtain true-breeding curl cats? How could you be sure they are true-breeding?

11. Imagine that a newly discovered, recessively inherited disease is expressed only in individuals with type O blood, although the disease and blood group are independently inherited. A normal man with type A blood and a normal woman with type B blood have already had one child with the disease. The woman is now pregnant for a second time. What is the probability that the second child will also have the disease? Assume that both parents are heterozygous for the gene that causes the disease.

12. In tigers, a recessive allele causes an absence of fur pigmentation (a white tiger) and a cross-eyed condition. If two phenotypically normal tigers that are heterozygous at this locus are mated, what percentage of their offspring will be cross-eyed? What percentage will be white?

13. In corn plants, a dominant allele *I* inhibits kernel color, while the recessive allele *i* permits color when homozygous.

At a different locus, the dominant allele *P* causes purple kernel color, while the homozygous recessive genotype *pp* causes red kernels. If plants heterozygous at both loci are crossed, what will be the phenotypic ratio of the offspring?

14. The pedigree below traces the inheritance of alkaptonuria, a biochemical disorder. Affected individuals, indicated here by the colored circles and squares, are unable to break down a substance called alkapton, which colors the urine and stains body tissues. Does alkaptonuria appear to be caused by a dominant allele or by a recessive allele? Fill in the genotypes of the individuals whose genotypes can be deduced. What genotypes are possible for each of the other individuals?

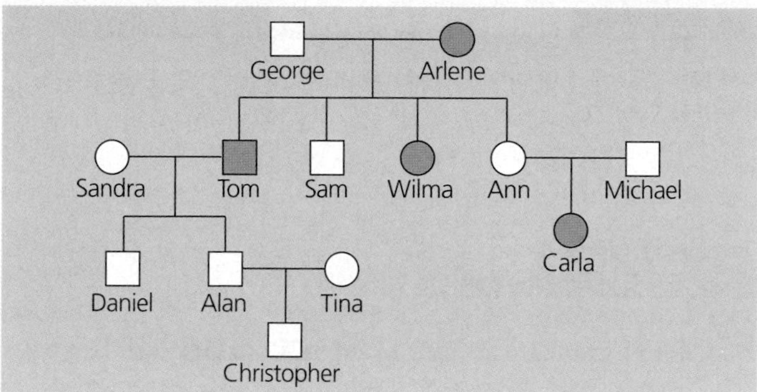

15. A man has six fingers on each hand and six toes on each foot. His wife and their daughter have the normal number of digits. Extra digits is a dominant trait. What fraction of this couple's children would be expected to have extra digits?

16. Imagine that you are a genetic counselor, and a couple planning to start a family come to you for information. Charles was married once before, and he and his first wife had a child with cystic fibrosis. The brother of his current wife Elaine died of cystic fibrosis. What is the probability that Charles and Elaine will have a baby with cystic fibrosis? (Neither Charles nor Elaine has cystic fibrosis.)

17. In mice, black color (*B*) is dominant to white (*b*). At a different locus, a dominant allele (*A*) produces a band of yellow just below the tip of each hair in mice with black fur. This gives a frosted appearance known as agouti. Expression of the recessive allele (*a*) results in a solid coat color. If mice that are heterozygous at both loci are crossed, what is the expected phenotypic ratio of their offspring?

For Genetics Problems answers, see Appendix A.

Go to the website or CD-ROM for more quiz questions.

Evolution Connection

Over the past half century, there has been a trend in the United States and other developed countries for people to marry and start families later in life than did their parents and grandparents. Speculate on the effects this trend may have on the incidence (frequency) of late-acting dominant lethal alleles in the population.

Scientific Inquiry

You are handed a mystery pea plant with long stems and axial flowers and asked to determine its genotype as quickly as possible. You know the allele for tall stems (*T*) is dominant to that for dwarf stems (*t*) and that the allele for axial flowers (*A*) is dominant to that for terminal flowers (*a*).

a. What are *all* the possible genotypes for your mystery plant?
b. Describe the *one* cross you would do, out in your garden, to determine the exact genotype of your mystery plant.
c. While waiting for the results of your cross, you predict the results for each possible genotype listed in part a. How do you do this?
d. Make your predictions using the following format: If the genotype of my mystery plant is ______, the plants resulting from my cross will be ______.
e. If ½ of your offspring plants have tall stems with axial flowers and ½ have tall stems with terminal flowers, what must be the genotype of your mystery plant?
f. Explain why the activities you performed in parts c and d were not "doing a cross."

Investigation ***How Do You Diagnose a Genetic Disorder?***

Science, Technology, and Society

Imagine that one of your parents had Huntington's disease. What is the probability that you, too, will someday manifest the disease? There is no cure for Huntington's. Would you want to be tested for the Huntington's allele? Why or why not?

15 The Chromosomal Basis of Inheritance

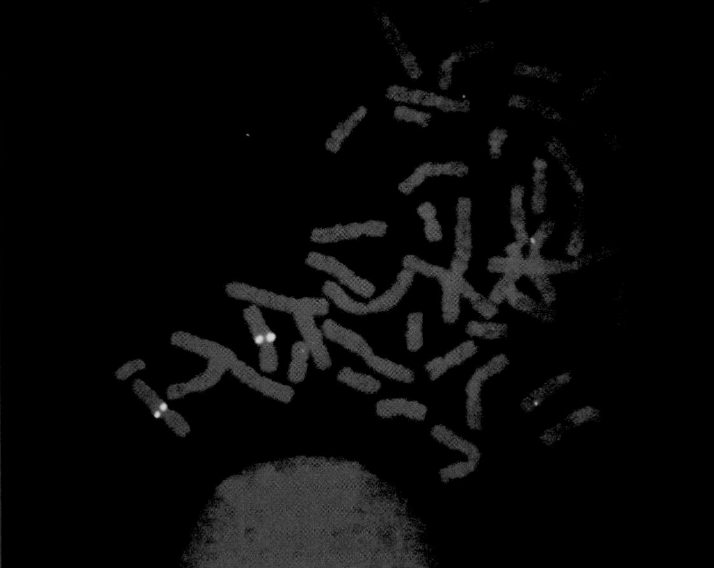

▲ **Figure 15.1 Chromosomes tagged to reveal a specific gene (yellow).**

Key Concepts

15.1 Mendelian inheritance has its physical basis in the behavior of chromosomes
15.2 Linked genes tend to be inherited together because they are located near each other on the same chromosome
15.3 Sex-linked genes exhibit unique patterns of inheritance
15.4 Alterations of chromosome number or structure cause some genetic disorders
15.5 Some inheritance patterns are exceptions to the standard chromosome theory

Overview

Locating Genes on Chromosomes

Today, we can show that genes—Gregor Mendel's "hereditary factors"—are located on chromosomes. We can see the location of a particular gene by tagging isolated chromosomes with a fluorescent dye that highlights that gene. For example, the yellow dots in **Figure 15.1** mark the locus of a specific gene on a homologous pair of human chromosomes. (Because the chromosomes in this light micrograph have already replicated, we see two dots per chromosome, one on each sister chromatid.) A century or so ago, however, the relation of genes and chromosomes was not immediately obvious. Many biologists remained skeptical about Mendel's laws of segregation and independent assortment until evidence accumulated that these principles of heredity had a physical basis in the behavior of chromosomes. In this chapter, which integrates and extends what you learned in the past two chapters, we describe the chromosomal basis for the transmission of genes from parents to offspring, along with some important exceptions.

Concept 15.1

Mendelian inheritance has its physical basis in the behavior of chromosomes

Using improved techniques of microscopy, cytologists worked out the process of mitosis in 1875 and meiosis in the 1890s. Then, around 1900, cytology and genetics converged as biologists began to see parallels between the behavior of chromosomes and the behavior of Mendel's "factors" during sexual life cycles: Chromosomes and genes are both present in pairs in diploid cells; homologous chromosomes separate and alleles segregate during the process of meiosis; and fertilization restores the paired condition for both chromosomes and genes. Around 1902, Walter S. Sutton, Theodor Boveri, and others independently noted these parallels, and the **chromosome theory of inheritance** began to take form. According to this theory, Mendelian genes have specific loci (positions) on chromosomes, and it is the chromosomes that undergo segregation and independent assortment.

Figure 15.2 shows that the behavior of homologous chromosomes during meiosis can account for the segregation of the alleles at each genetic locus to different gametes. The figure also shows that the behavior of nonhomologous chromosomes can account for the independent assortment of the alleles for two or more genes located on different chromosomes. By carefully studying this figure, which traces the same dihybrid pea cross you learned about in Figure 14.8, you can see how the behavior of chromosomes during meiosis in the F_1 generation and subsequent random fertilization gives rise to the F_2 phenotypic ratio observed by Mendel.

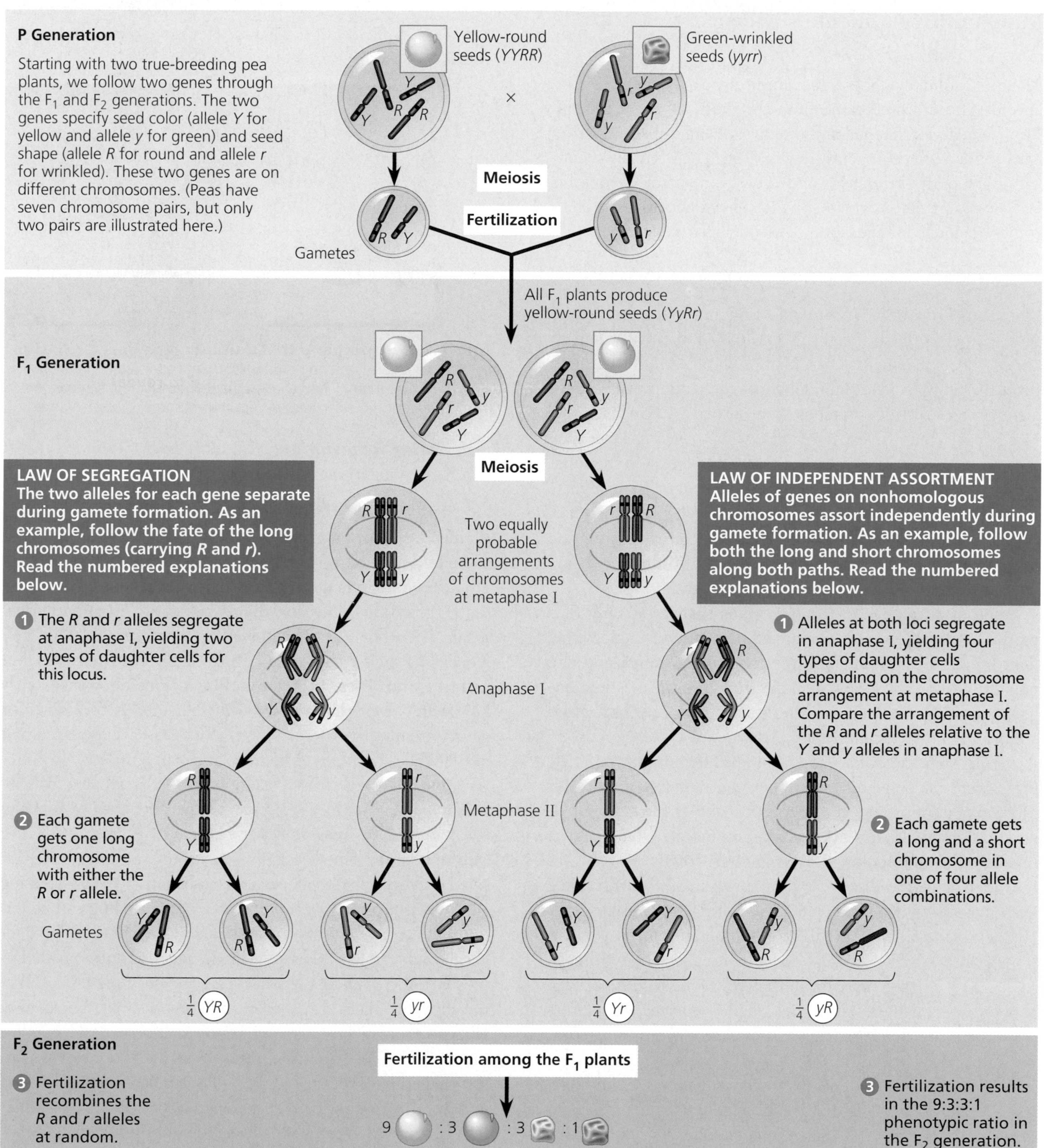

P Generation
Starting with two true-breeding pea plants, we follow two genes through the F1 and F2 generations. The two genes specify seed color (allele Y for yellow and allele y for green) and seed shape (allele R for round and allele r for wrinkled). These two genes are on different chromosomes. (Peas have seven chromosome pairs, but only two pairs are illustrated here.)
Yellow-round seeds (YYRR)
Green-wrinkled seeds (yyrr)
×
Meiosis
Fertilization
Gametes
All F1 plants produce yellow-round seeds (YyRr)
F1 Generation
Meiosis
LAW OF SEGREGATION
The two alleles for each gene separate during gamete formation. As an example, follow the fate of the long chromosomes (carrying R and r). Read the numbered explanations below.
LAW OF INDEPENDENT ASSORTMENT
Alleles of genes on nonhomologous chromosomes assort independently during gamete formation. As an example, follow both the long and short chromosomes along both paths. Read the numbered explanations below.
Two equally probable arrangements of chromosomes at metaphase I
1 The R and r alleles segregate at anaphase I, yielding two types of daughter cells for this locus.
1 Alleles at both loci segregate in anaphase I, yielding four types of daughter cells depending on the chromosome arrangement at metaphase I. Compare the arrangement of the R and r alleles relative to the Y and y alleles in anaphase I.
Anaphase I
Metaphase II
2 Each gamete gets one long chromosome with either the R or r allele.
2 Each gamete gets a long and a short chromosome in one of four allele combinations.
Gametes
1/4 YR
1/4 yr
1/4 Yr
1/4 yR
F2 Generation
Fertilization among the F1 plants
3 Fertilization recombines the R and r alleles at random.
9 : 3 : 3 : 1
3 Fertilization results in the 9:3:3:1 phenotypic ratio in the F2 generation.

▲ **Figure 15.2 The chromosomal basis of Mendel's laws.** Here we correlate the results of one of Mendel's dihybrid crosses (see Figure 14.8) with the behavior of chromosomes during meiosis (see Figure 13.8). The arrangement of chromosomes at metaphase I of meiosis and their movement during anaphase I account for the segregation and independent assortment of the alleles for seed color and shape. Each cell that undergoes meiosis in an F_1 plant produces two kinds of gametes. Overall, however, F_1 plants produce equal numbers of all four kinds of gametes because the alternative chromosome arrangements at metaphase I are equally likely.

Morgan's Experimental Evidence: *Scientific Inquiry*

The first solid evidence associating a specific gene with a specific chromosome came from the work of Thomas Hunt Morgan, an experimental embryologist at Columbia University early in the 20th century. Although Morgan was initially skeptical about both Mendelism and the chromosome theory, his early experiments provided convincing evidence that chromosomes are indeed the location of Mendel's heritable factors.

Morgan's Choice of Experimental Organism

Many times in the history of biology, important discoveries have come to those insightful enough or lucky enough to choose an experimental organism suitable for the research problem being tackled. Mendel chose the garden pea because a number of distinct varieties were available. For his work, Morgan selected a species of fruit fly, *Drosophila melanogaster,* a common, generally innocuous insect that feeds on the fungi growing on fruit. Fruit flies are prolific breeders; a single mating will produce hundreds of offspring, and a new generation can be bred every two weeks. These characteristics make the fruit fly a convenient organism for genetic studies. Morgan's laboratory soon became known as "the fly room."

Another advantage of the fruit fly is that it has only four pairs of chromosomes, which are easily distinguishable with a light microscope. There are three pairs of autosomes and one pair of sex chromosomes. Female fruit flies have a homologous pair of X chromosomes, and males have one X chromosome and one Y chromosome.

While Mendel could readily obtain different pea varieties, there were no convenient suppliers of fruit fly varieties for Morgan to employ. Indeed, he was probably the first person to want different varieties of this common insect. After a year of breeding flies and looking for variant individuals, Morgan was rewarded with the discovery of a single male fly with white eyes instead of the usual red. The normal phenotype for a character (the phenotype most common in natural populations), such as red eyes in *Drosophila,* is called the **wild type** (**Figure 15.3**). Traits that are alternatives to the wild type, such as white eyes in *Drosophila,* are called *mutant phenotypes* because they are due to alleles assumed to have originated as changes, or mutations, in the wild-type allele.

Morgan and his students invented a notation for symbolizing alleles in *Drosophila* that is still widely used for fruit flies. For a given character in flies, the gene takes its symbol from the first mutant (non–wild type) discovered. Thus, the allele for white eyes in *Drosophila* is symbolized by *w.* A superscript + identifies the allele for the wild-type trait—w^+ for the allele for red eyes, for example. Over the years, different gene notation systems have been developed for different organisms. For example, human genes are usually written in all capitals, such as *HD* for the allele for Huntington's disease.

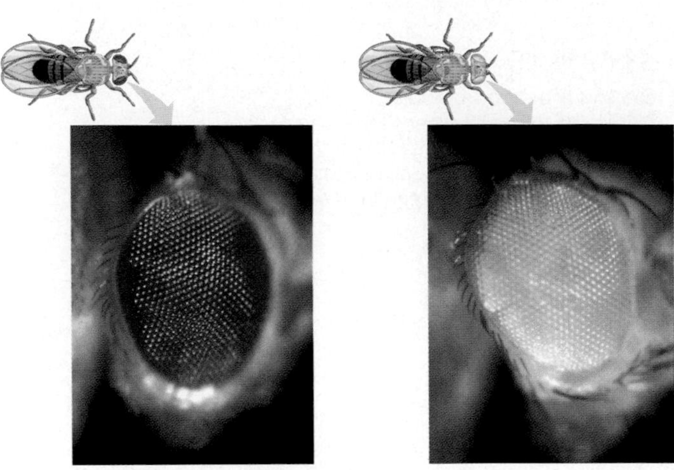

▲ **Figure 15.3 Morgan's first mutant.** Wild-type *Drosophila* flies have red eyes (left). Among his flies, Morgan discovered a mutant male with white eyes (right). This variation made it possible for Morgan to trace a gene for eye color to a specific chromosome (LMs).

Correlating Behavior of a Gene's Alleles with Behavior of a Chromosome Pair

Morgan mated his white-eyed male fly with a red-eyed female. All the F_1 offspring had red eyes, suggesting that the wild-type allele is dominant. When Morgan bred the F_1 flies to each other, he observed the classical 3:1 phenotypic ratio among the F_2 offspring. However, there was a surprising additional result: The white-eye trait showed up only in males. All the F_2 females had red eyes, while half the males had red eyes and half had white eyes. Therefore, Morgan concluded that somehow a fly's eye color was linked to its sex. (If the eye-color gene were unrelated to gender, one would have expected half of the white-eyed flies to be male and half female.)

A female fly has two X chromosomes (XX), while a male fly has an X and a Y (XY). The correlation between the trait of white eye color and the male sex of the affected F_2 flies suggested to Morgan that the gene affected in his white-eyed mutant was located exclusively on the X chromosome, with no corresponding allele present on the Y chromosome. His reasoning can be followed in **Figure 15.4**. For a male, a single copy of the mutant allele would confer white eyes; since a male has only one X chromosome, there can be no wild-type allele (w^+) present to offset the recessive allele. On the other hand, a female could have white eyes only if both her X chromosomes carried the recessive mutant allele (*w*). This was impossible for the F_2 females in Morgan's experiment because all the F_1 fathers had red eyes.

Morgan's finding of the correlation between a particular trait and an individual's sex provided support for the chromosome theory of inheritance: namely, that a specific gene is carried on a specific chromosome (in this case, the eye-color gene on the X chromosome). In addition, Morgan's work indicated that genes located on a sex chromosome exhibit unique inheritance patterns, which we will discuss later in this chapter. Recognizing the importance of Morgan's early work, many bright students were attracted to his fly room.

Figure 15.4

Inquiry In a cross between a wild-type female fruit fly and a mutant white-eyed male, what color eyes will the F_1 and F_2 offspring have?

EXPERIMENT Morgan mated a wild-type (red-eyed) female with a mutant white-eyed male. The F_1 offspring all had red eyes.

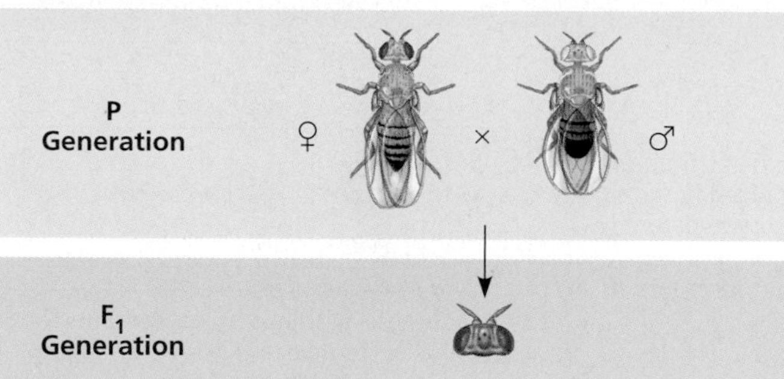

Morgan then bred an F_1 red-eyed female to an F_1 red-eyed male to produce the F_2 generation.

RESULTS The F_2 generation showed a typical Mendelian 3:1 ratio of red eyes to white eyes. However, no females displayed the white-eye trait; they all had red eyes. Half the males had white eyes, and half had red eyes.

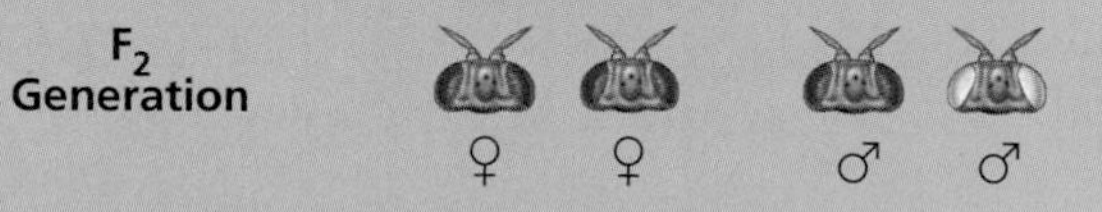

CONCLUSION Since all F_1 offspring had red eyes, the mutant white-eye trait (w) must be recessive to the wild-type red-eye trait (w^+). Since the recessive trait—white eyes—was expressed only in males in the F_2 generation, Morgan hypothesized that the eye-color gene is located on the X chromosome and that there is no corresponding locus on the Y chromosome, as diagrammed here.

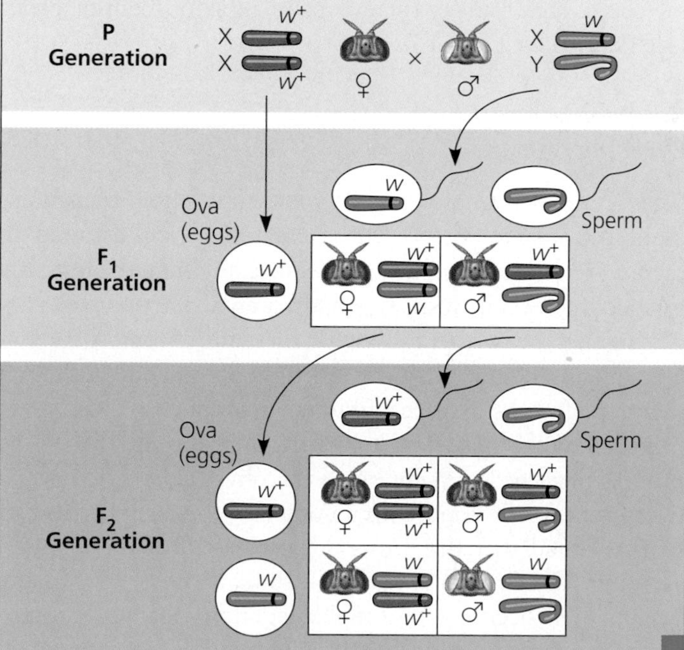

Concept Check 15.1

1. Which one of Mendel's laws relates to the inheritance of alleles for a single character? Which law relates to the inheritance of alleles for two characters in a dihybrid cross? What is the physical basis of these laws?
2. If the eye-color locus in *Drosophila* were located on an autosome, what would be the sex and phenotype of all the F_2 offspring produced by the crosses in Figure 15.4?

For suggested answers, see Appendix A.

Concept 15.2

Linked genes tend to be inherited together because they are located near each other on the same chromosome

The number of genes in a cell is far greater than the number of chromosomes; in fact, each chromosome has hundreds or thousands of genes. Genes located on the same chromosome that tend to be inherited together in genetic crosses are said to be **linked genes.** When geneticists follow linked genes in breeding experiments, the results deviate from those expected from Mendel's law of independent assortment.

How Linkage Affects Inheritance: *Scientific Inquiry*

To see how linkage between genes affects the inheritance of two different characters, let's examine another of Morgan's *Drosophila* experiments. In this case, the characters are body color and wing size, each with two different phenotypes. Wild-type flies have gray bodies and normal-sized wings. In addition to these flies, Morgan had doubly mutant flies with black bodies and vestigial wings (much smaller than normal wings). The alleles for these traits are represented by the following symbols: b^+ = gray, b = black; vg^+ = normal wings, vg = vestigial wings. The mutant alleles are recessive to the wild-type alleles, and neither gene is on a sex chromosome.

In studying these two genes, Morgan carried out the crosses shown in **Figure 15.5** (p. 279). He first mated true-breeding wild-type flies ($b^+\ b^+\ vg^+\ vg^+$) with black, vestigial-winged ones ($b\ b\ vg\ vg$) to produce heterozygous F_1 dihybrids ($b^+\ b\ vg^+\ vg$), all of which were wild-type in appearance. He then crossed female dihybrids with true-breeding males of the double-mutant phenotype ($b\ b\ vg\ vg$). In this second cross, which corresponds to a Mendelian testcross, we know the genotype of the female parent ($b^+\ b\ vg^+\ vg$), and we also know which allele combinations are "parental," meaning derived from the parents in the P generation: b^+ with vg^+ and b with vg. We don't know, however, whether the two genes are located on the same or different chromosomes. In the testcross, all the sperm will donate recessive alleles (b and vg); so the phenotypes of the offspring will depend on the ova's alleles. Therefore, from the phenotypes of the offspring, we can determine whether or not the parental allele combinations, b^+ with vg^+ and b with vg, stayed together during formation of the F_1 female's ova.

When Morgan "scored" (classified according to phenotype) 2,300 offspring from the testcross matings, he observed a much higher proportion of parental phenotypes than would be expected if the two genes assorted independently (see Figure 15.5). Based on these results, Morgan reasoned that body color and wing size are usually inherited together in specific combinations (the parental combinations) because the genes for these characters are on the same chromosome:

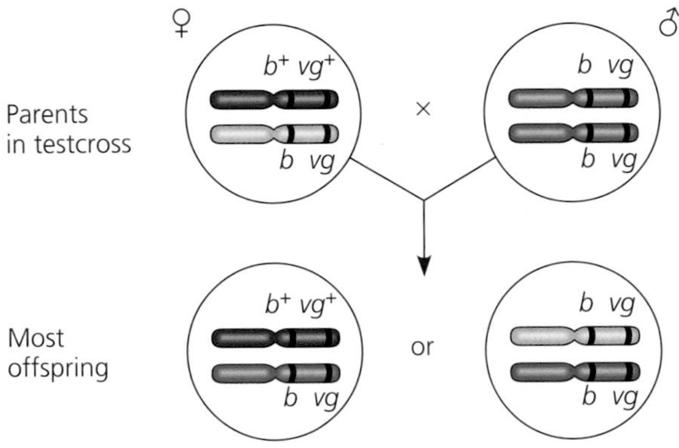

However, if the genes for body color and wing size were always inherited together in those parental combinations, no nonparental phenotypes would have been observed among the offspring of Morgan's testcross. In fact, both of the nonparental phenotypes *were* produced in Morgan's experiments (see Figure 15.5), suggesting that the body-color and wing-size genes are only partially linked genetically. To understand this result, we need to further explore **genetic recombination,** the production of offspring with combinations of traits differing from those found in either parent.

Genetic Recombination and Linkage

In Chapter 13, you learned that meiosis and random fertilization generate genetic variation among offspring of sexually reproducing organisms. Here we will examine the chromosomal basis of recombination in relation to the genetic findings of Mendel and Morgan.

Recombination of Unlinked Genes: Independent Assortment of Chromosomes

Mendel learned from crosses in which he followed two characters that some offspring have combinations of traits that do not match those of either parent. For example, we can represent the cross between a pea plant with yellow-round seeds that is heterozygous for both seed color and seed shape (*YyRr*) and a plant with green-wrinkled seeds (homozygous for both recessive alleles, *yyrr*) by the following Punnett square:

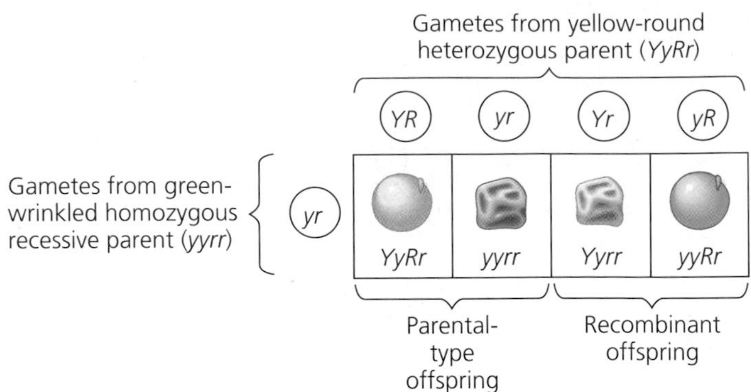

Notice in this Punnett square that one-half of the offspring are expected to inherit a phenotype that matches one of the parental phenotypes. These offspring are called **parental types.** But two nonparental phenotypes are also found among the offspring. Because these offspring have new combinations of seed shape and color, they are called **recombinant types,** or **recombinants** for short. When 50% of all offspring are recombinants, as in this example, geneticists say that there is a 50% frequency of recombination. The predicted phenotypic ratios among the offspring are similar to what Mendel actually found in *YyRr* × *yyrr* crosses.

A 50% frequency of recombination is observed for any two genes that are located on different chromosomes. The physical basis of recombination between unlinked genes is the random orientation of homologous chromosomes at metaphase I of meiosis, which leads to the independent assortment of alleles (see Figures 15.2 and 13.10).

Recombination of Linked Genes: Crossing Over

Now let's return to Morgan's fly room to see how we can explain the results of the *Drosophila* testcross illustrated in Figure 15.5. Recall that most of the offspring from the testcross for body color and wing size had parental phenotypes, suggesting that the two genes were on the same chromosome, but a small number of offspring were recombinants. Although there was linkage, it appeared to be incomplete.

Faced with these results, Morgan proposed that some process must occasionally break the physical connection between genes on the same chromosome. Subsequent experiments demonstrated that this process, now called **crossing over,** accounts for the recombination of linked genes. In crossing over, which occurs while replicated homologous chromosomes are paired during prophase of meiosis I, one maternal

Figure 15.5

Inquiry Are the genes for body color and wing size in fruit flies located on the same chromosome or different chromosomes?

EXPERIMENT Morgan first mated true-breeding wild-type flies with black, vestigial-winged flies to produce heterozygous F_1 dihybrids, all of which are wild-type in appearance. He then mated wild-type F_1 dihybrid females with black, vestigial-winged males, producing 2,300 F_2 offspring, which he classified according to phenotype.

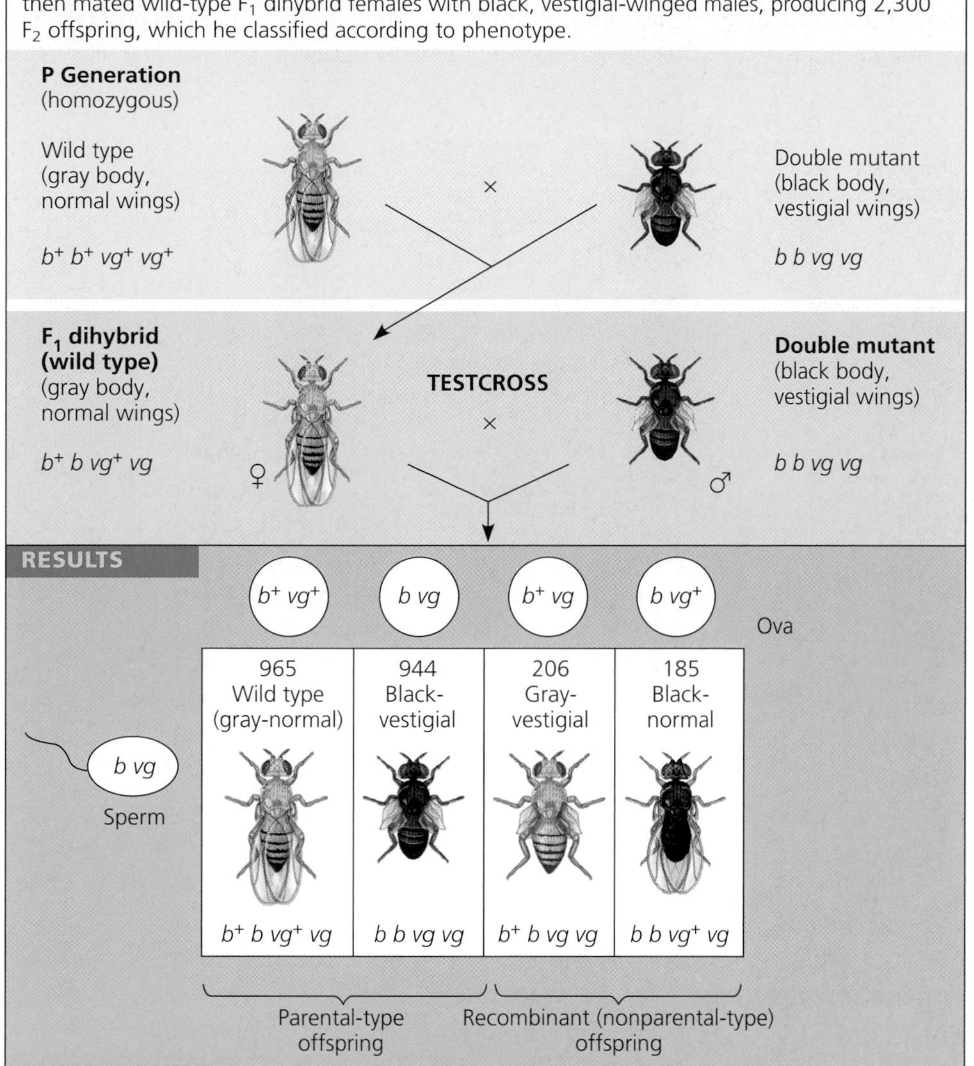

CONCLUSION If these two genes were on different chromosomes, the alleles from the F_1 dihybrid would sort into gametes independently, and we would expect to see equal numbers of the four types of offspring. If these two genes were on the same chromosome, we would expect each allele combination, b^+ vg^+ and *b vg,* to stay together as gametes formed. In this case, only offspring with parental phenotypes would be produced. Since most offspring had a parental phenotype, Morgan concluded that the genes for body color and wing size are located on the same chromosome. However, the production of a small number of offspring with nonparental phenotypes indicated that some mechanism occasionally breaks the linkage between genes on the same chromosome.

and one paternal chromatid break at corresponding points and then are rejoined to each other (see Figure 13.11). In effect, the end portions of two nonsister chromatids trade places each time a crossover occurs.

The recombinant chromosomes resulting from crossing over may bring alleles together in new combinations, and the subsequent events of meiosis distribute the recombinant chromosomes to gametes. **Figure 15.6** (p. 280) shows how crossing over in a dihybrid female fly resulted in recombinant ova and ultimately recombinant offspring in Morgan's testcross. Most of the ova had a chromosome with either the b^+ vg^+ or b vg parental genotype for body color and wing size, but some ova had a recombinant chromosome (b^+ vg or b vg^+). Fertilization of these various classes of ova by homozygous recessive sperm (b vg) produced an offspring population in which 17% exhibited a nonparental, recombinant phenotype (see Figure 15.6). As we discuss next, the percentage of recombinant offspring, the *recombination frequency,* is related to the distance between linked genes.

Linkage Mapping Using Recombination Data: *Scientific Inquiry*

The discovery of linked genes and recombination due to crossing over led one of Morgan's students, Alfred H. Sturtevant, to a method for constructing a **genetic map,** an ordered list of the genetic loci along a particular chromosome.

Sturtevant hypothesized that recombination frequencies calculated from experiments like the one in Figures 15.5 and 15.6 depend on the distances between genes on a chromosome. He assumed that crossing over is a random event, and thus the chance of crossing over is approximately equal at all points along a chromosome. Based on these assumptions, Sturtevant predicted that *the farther apart two genes are, the higher the probability that a crossover will occur between them and therefore the higher the recombination frequency.* His reasoning was simple: The greater the distance between two genes, the more points there are between them where crossing over can occur. Using recombination data from various fruit fly crosses, Sturtevant proceeded to assign relative positions to genes on the same chromosomes—that is, to *map* genes.

A genetic map based on recombination frequencies is specifically called a **linkage map. Figure 15.7** (p. 281) shows Sturtevant's linkage map of three genes: the body-color (b) and wing-size (vg) genes depicted in Figure 15.6 and a third gene, called cinnabar (cn). Cinnabar is one of many *Drosophila* genes affecting eye color. Cinnabar eyes, a mutant phenotype, are a brighter red than the wild-type color. The recombination frequency between *cn* and *b* is 9%; that between *cn* and *vg*, 9.5%; and that between *b* and *vg*, 17%. In other words, crossovers between *cn* and *b* and between *cn* and *vg* are about half as frequent as crossovers between *b* and *vg*. Only a map that locates *cn* about midway between *b* and *vg* is consistent with these data, as you can prove to yourself by drawing alternative maps. Sturtevant expressed the distances between genes in **map units,** defining one map unit as equivalent to a 1% recombination frequency. Today, map units often are called *centimorgans* in honor of Morgan.

In practice, the interpretation of recombination data is more complicated than this example suggests. For example, some genes on a chromosome are so far from each other that a crossover between them is virtually certain. The observed

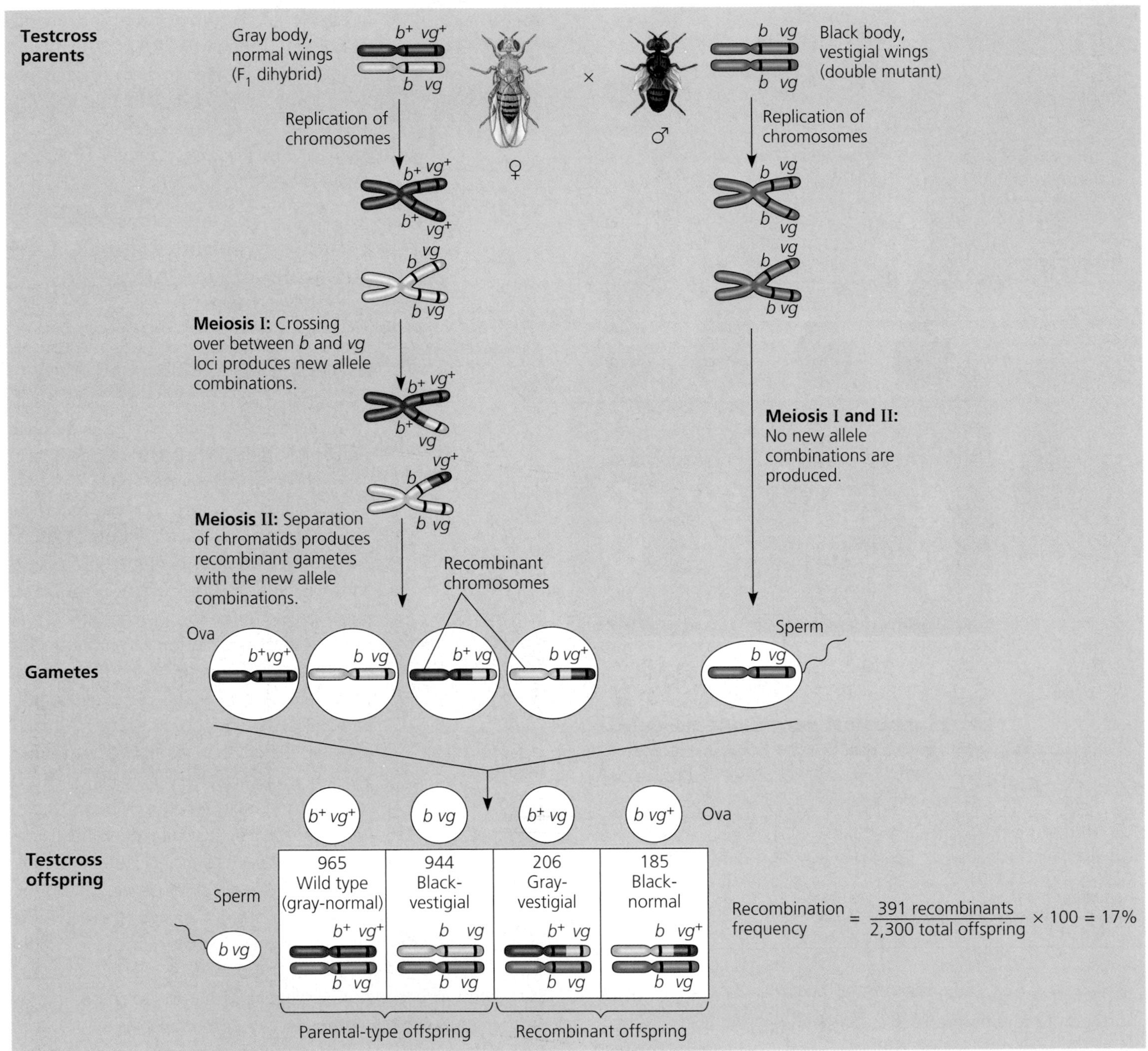

▲ **Figure 15.6 Chromosomal basis for recombination of linked genes.** In these diagrams re-creating the testcross in Figure 15.5, we track chromosomes as well as genes. The maternal chromosomes are color-coded to distinguish one homologue from the other. Because crossing over between the *b* and *vg* loci occurs in some, but not all, ovum-producing cells, more ova with parental-type chromosomes than with recombinant ones are produced in the mating females. Fertilization of the ova by sperm of genotype *b vg* gives rise to some recombinant offspring. The recombination frequency is the percentage of recombinant flies in the total pool of offspring.

Figure 15.7

Research Method Constructing a Linkage Map

APPLICATION A linkage map shows the relative locations of genes along a chromosome.

TECHNIQUE A linkage map is based on the assumption that the probability of a crossover between two genetic loci is proportional to the distance separating the loci. The recombination frequencies used to construct a linkage map for a particular chromosome are obtained from experimental crosses, such as the cross depicted in Figure 15.6. The distances between genes are expressed as map units (centimorgans), with one map unit equivalent to a 1% recombination frequency. Genes are arranged on the chromosome in the order that best fits the data.

RESULTS In this example, the observed recombination frequencies between three *Drosophila* gene pairs (*b–cn* 9%, *cn–vg* 9.5%, and *b–vg* 17%) best fit a linear order in which *cn* is positioned about halfway between the other two genes:

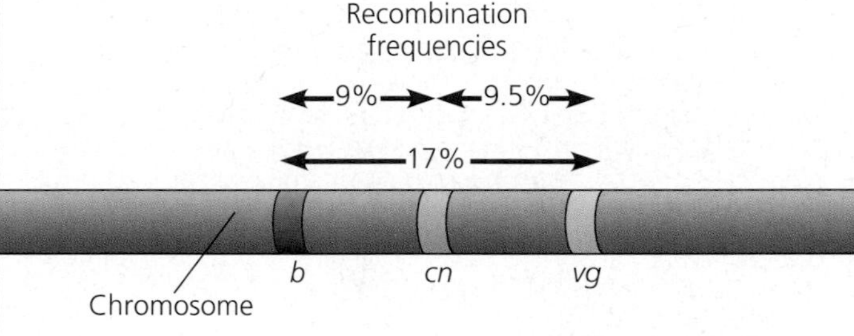

The *b–vg* recombination frequency is slightly less than the sum of the *b–cn* and *cn–vg* frequencies because double crossovers are fairly likely to occur between *b* and *vg* in matings tracking these two genes. A second crossover would "cancel out" the first and thus reduce the observed *b–vg* recombination frequency.

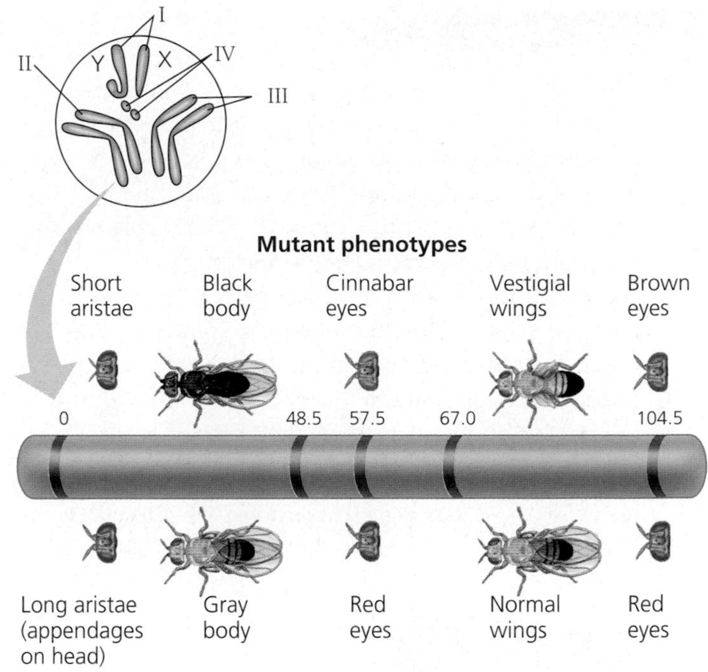

▲ **Figure 15.8 A partial genetic (linkage) map of a *Drosophila* chromosome.** This simplified map shows just a few of the genes that have been mapped on *Drosophila* chromosome II. The number at each gene locus indicates the number of map units between that locus and the locus for aristae length (left). Notice that more than one gene can affect a given phenotypic characteristic, such as eye color. Also, note that in contrast to the homologous autosomes (II–IV), the X and Y sex chromosomes (I) have distinct shapes.

frequency of recombination in crosses involving two such genes can have a maximum value of 50%, a result indistinguishable from that for genes on different chromosomes. In this case, the physical connection between genes on the same chromosome is not reflected in the results of genetic crosses. Despite being on the same chromosome and thus being *physically linked,* the genes are *genetically unlinked;* alleles of such genes assort independently as if they were on different chromosomes. In fact, the genes for two of the pea characters that Mendel studied—seed color and flower color—are now known to be on the same chromosome, but the distance between them is so great that linkage is not observed in genetic crosses. Genes located far apart on a chromosome are mapped by adding the recombination frequencies from crosses involving each of the distant genes and a number of genes lying between them.

Using recombination data, Sturtevant and his colleagues were able to map numerous *Drosophila* genes in linear arrays. They found that the genes clustered into four groups of linked genes. Because microscopists had found four pairs of chromosomes in *Drosophila* cells, this clustering of genes was additional evidence that genes are located on chromosomes. Each chromosome has a linear array of specific gene loci **(Figure 15.8)**.

Because a linkage map is based on recombination frequencies, it gives only an approximate picture of a chromosome. The frequency of crossing over is not actually uniform over the length of a chromosome, as Sturtevant assumed, and therefore map units do not correspond to actual physical distances (in nanometers, for instance). A linkage map does portray the order of genes along a chromosome, but it does not accurately portray the precise locations of those genes. Other methods enable geneticists to construct **cytogenetic maps** of chromosomes, which locate genes with respect to chromosomal features, such as stained bands, that can be seen in the microscope. The ultimate maps, which we will discuss in Chapter 20, show the physical distances between gene loci in DNA nucleotides. Comparing a linkage map with such a physical map or with a cytogenetic map of the same chromosome, we find that the linear order of genes is identical in all the maps, but the spacing between genes is not.

Concept Check 15.2

1. When two genes are located on the same chromosome, what is the physical basis for the production of recombinant offspring in a testcross between a dihybrid parent and a double-mutant parent?
2. For each type of offspring in Figure 15.5, explain the relationship between its phenotype and the alleles contributed by the female parent.
3. Genes *A*, *B*, and *C* are located on the same chromosome. Testcrosses show that the recombination frequency between *A* and *B* is 28% and between *A* and *C* is 12%. Can you determine the linear order of these genes?

For suggested answers, see Appendix A.

Concept 15.3

Sex-linked genes exhibit unique patterns of inheritance

As you learned earlier, Morgan's discovery of a trait (white eyes) that correlated with the sex of flies was a key episode in the development of the chromosome theory of inheritance. In this section, we consider the role of sex chromosomes in inheritance in more detail. We begin by reviewing the chromosomal basis of sex determination in humans and some other animals.

The Chromosomal Basis of Sex

Whether we are male or female is one of our more obvious phenotypic characters. Although the anatomical and physiological differences between women and men are numerous, the chromosomal basis for determining sex is rather simple. In humans and other mammals, there are two varieties of sex chromosomes, designated X and Y. A person who inherits two X chromosomes, one from each parent, usually develops as a female. A male develops from a zygote containing one X chromosome and one Y chromosome **(Figure 15.9a)**. The Y chromosome is much smaller than the X chromosome (see the micrograph to the left), and only relatively short segments at either end of the Y chromosome are homologous with corresponding regions of the X. These homologous regions allow the X and Y chromosomes in males to pair and behave like homologous chromosomes during meiosis in the testes.

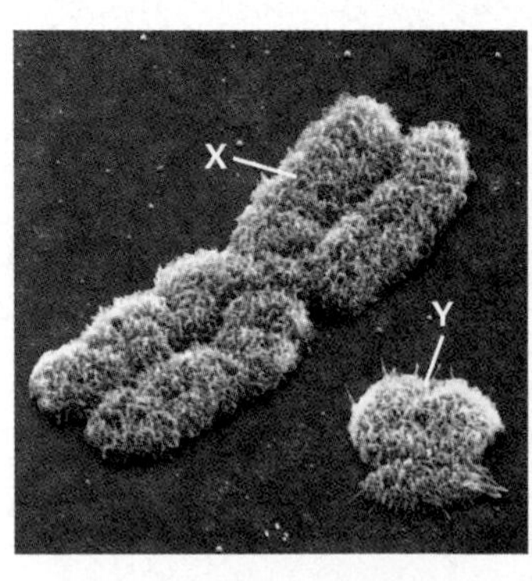

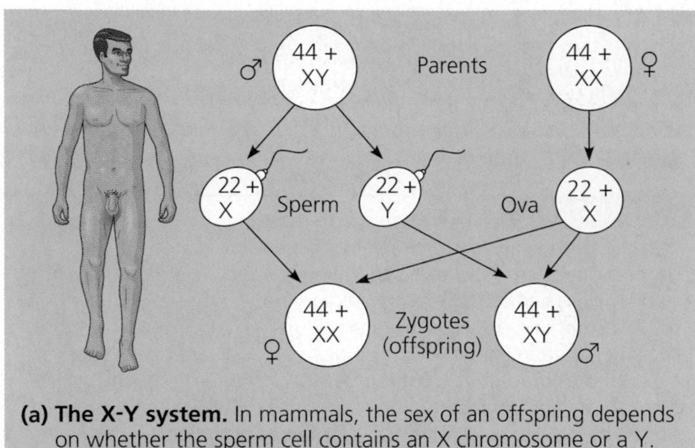

(a) The X-Y system. In mammals, the sex of an offspring depends on whether the sperm cell contains an X chromosome or a Y.

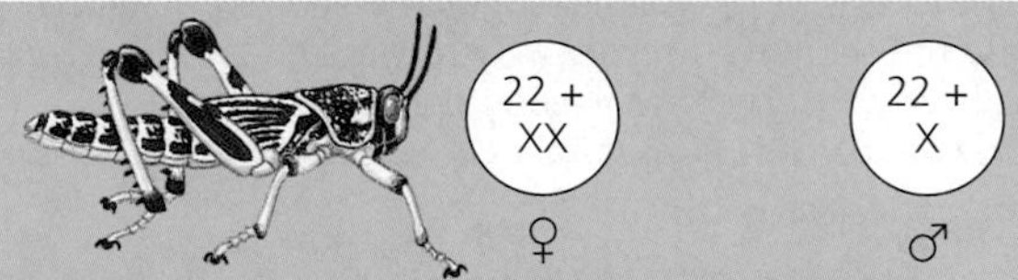

(b) The X-0 system. In grasshoppers, cockroaches, and some other insects, there is only one type of sex chromosome, the X. Females are XX; males have only one sex chromosome (X0). Sex of the offspring is determined by whether the sperm cell contains an X chromosome or no sex chromosome.

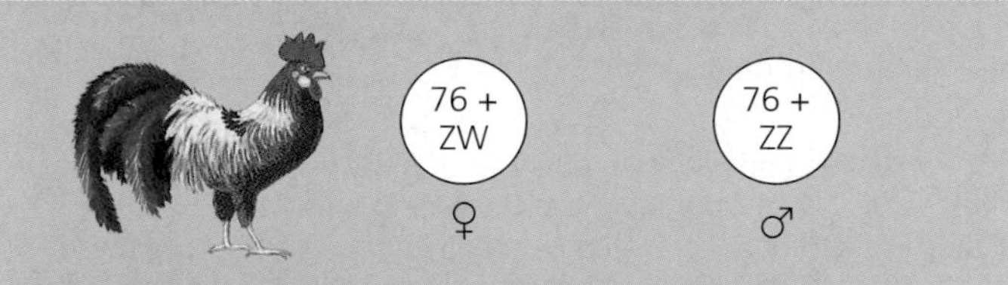

(c) The Z-W system. In birds, some fishes, and some insects, the sex chromosome present in the ovum (not the sperm) determines the sex of offspring. The sex chromosomes are designated Z and W. Females are ZW and males are ZZ.

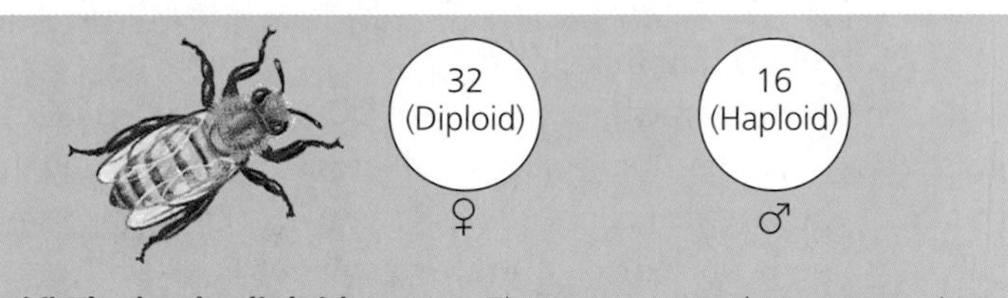

(d) The haplo-diploid system. There are no sex chromosomes in most species of bees and ants. Females develop from fertilized ova and are thus diploid. Males develop from unfertilized ova and are haploid; they have no fathers.

▲ **Figure 15.9 Some chromosomal systems of sex determination.** Numerals indicate the number of autosomes. In *Drosophila*, males are XY, but sex depends on the ratio between the number of X chromosomes and the number of autosome sets, not simply on the presence of a Y chromosome.

In both testes and ovaries, the two sex chromosomes segregate during meiosis, and each gamete receives one. Each ovum contains one X chromosome. In contrast, sperm fall into two categories: Half the sperm cells a male produces contain an X chromosome, and half contain a Y chromosome. We can trace the sex of each offspring to the moment of conception: If a sperm cell bearing an X chromosome happens to fertilize an ovum, the zygote is XX, a female; if a sperm cell containing a Y chromosome fertilizes an ovum, the zygote is XY, a male (see Figure 15.9a). Thus sex determination is a matter of chance—a fifty-fifty chance. Besides the mammalian X-Y system, three other chromosomal systems for determining sex are shown in Figure 15.9, in parts b–d.

In humans, the anatomical signs of sex begin to emerge when the embryo is about two months old. Before then, the rudiments of the gonads are generic—they can develop into either ovaries or testes, depending on hormonal conditions within the embryo. Which of these two possibilities occurs depends on whether or not a Y chromosome is present. In 1990, a British research team identified a gene on the Y chromosome required for the development of testes. They named the gene *SRY*, for sex-determining region of Y. In the absence of *SRY*, the gonads develop into ovaries. The researchers emphasized that the presence (or absence) of *SRY* is just a trigger. The biochemical, physiological, and anatomical features that distinguish males and females are complex, and many genes are involved in their development. *SRY* codes for a protein that regulates other genes. Researchers have subsequently identified a number of additional genes on the Y chromosome that are required for normal testis functioning. In the absence of these genes, an XY individual is male but does not produce normal sperm.

Inheritance of Sex-Linked Genes

In addition to their role in determining sex, the sex chromosomes, especially X chromosomes, have genes for many characters unrelated to sex. A gene located on either sex chromosome is called a **sex-linked gene**, although in humans the term has historically referred specifically to a gene on the X chromosome. (Note the distinction between the terms *sex-linked gene*, referring to a gene on a sex chromosome, and *linked genes*, referring to genes on the same chromosome that tend to be inherited together.) Sex-linked genes in humans follow the same pattern of inheritance that Morgan observed for the eye-color locus in *Drosophila* (see Figure 15.4). Fathers pass sex-linked alleles to all of their daughters but to none of their sons. In contrast, mothers can pass sex-linked alleles to both sons and daughters **(Figure 15.10)**.

If a sex-linked trait is due to a recessive allele, a female will express the phenotype only if she is a homozygote. Because males have only one locus, the terms *homozygous* and *heterozygous* lack meaning for describing their sex-linked genes (the term *hemizygous* is used in such cases). Any male receiving the recessive allele from his mother will express the trait. For this reason, far more males than females have sex-linked recessive disorders. However, even though the chance of a female inheriting a double dose of the mutant allele is much less than the probability of a male inheriting a single dose, there *are* females with sex-linked disorders. For instance, color blindness is a mild disorder inherited as a sex-linked trait. A color-blind daughter may be born to a color-blind father whose mate is a carrier (see Figure 15.10c). However, because the sex-linked allele for color blindness is relatively rare, the probability that such a man and woman will mate is low.

A number of human sex-linked disorders are much more serious than color blindness. An example is **Duchenne muscular dystrophy**, which affects about one out of every 3,500 males

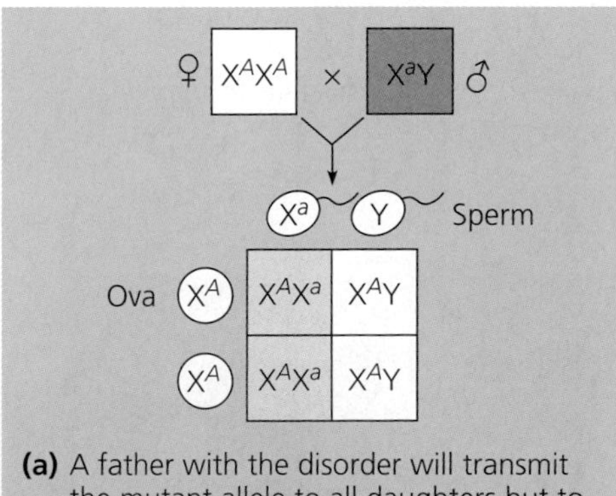

(a) A father with the disorder will transmit the mutant allele to all daughters but to no sons. When the mother is a dominant homozygote, the daughters will have the normal phenotype but will be carriers of the mutation.

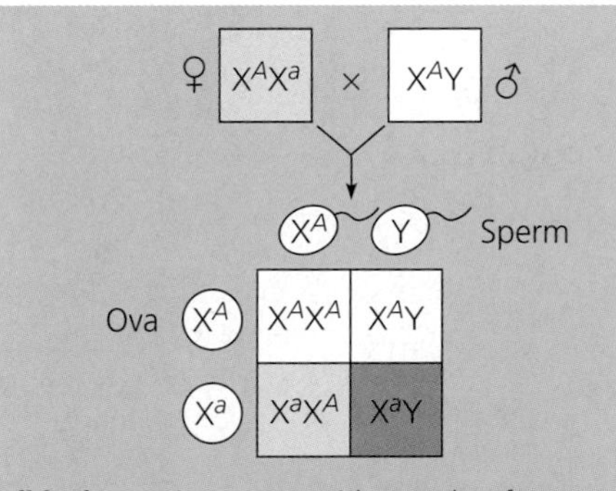

(b) If a carrier mates with a male of normal phenotype, there is a 50% chance that each daughter will be a carrier like her mother, and a 50% chance that each son will have the disorder.

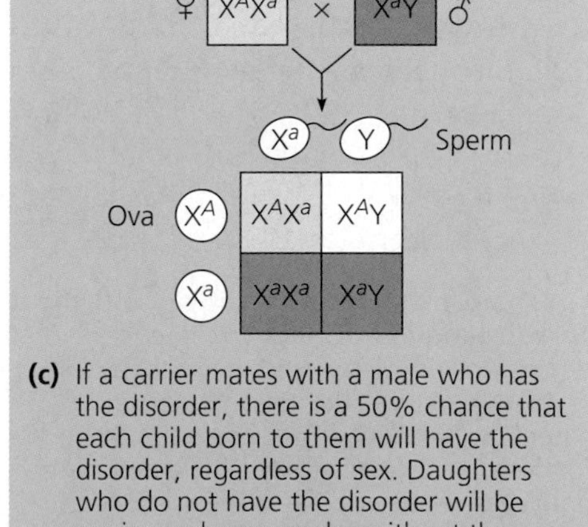

(c) If a carrier mates with a male who has the disorder, there is a 50% chance that each child born to them will have the disorder, regardless of sex. Daughters who do not have the disorder will be carriers, whereas males without the disorder will be completely free of the recessive allele.

▲ **Figure 15.10 The transmission of sex-linked recessive traits.** In this diagram, the superscript *A* represents a dominant allele carried on the X chromosome, and the superscript *a* represents a recessive allele. Imagine that this recessive allele is a mutation that causes a sex-linked disorder, such as color blindness. White boxes indicate unaffected individuals, light-colored boxes indicate carriers, and dark-colored boxes indicate individuals with the sex-linked disorder.

born in the United States. The disease is characterized by a progressive weakening of the muscles and loss of coordination. Affected individuals rarely live past their early 20s. Researchers have traced the disorder to the absence of a key muscle protein called dystrophin and have mapped the gene for this protein to a specific locus on the X chromosome.

Hemophilia is a sex-linked recessive disorder defined by the absence of one or more of the proteins required for blood clotting. When a person with hemophilia is injured, bleeding is prolonged because a firm clot is slow to form. Small cuts in the skin are usually not a problem, but bleeding in the muscles or joints can be painful and can lead to serious damage. Today, people with hemophilia are treated as needed with intravenous injections of the missing protein.

X Inactivation in Female Mammals

Although female mammals, including humans, inherit two X chromosomes, one X chromosome in each cell becomes almost completely inactivated during embryonic development. As a result, the cells of females and males have the same effective dose (one copy) of genes with loci on the X chromosome. The inactive X in each cell of a female condenses into a compact object called a **Barr body,** which lies along the inside of the nuclear envelope. Most of the genes of the X chromosome that forms the Barr body are not expressed. In the ovaries, Barr-body chromosomes are reactivated in the cells that give rise to ova, so every female gamete has an active X.

British geneticist Mary Lyon demonstrated that selection of which X chromosome will form the Barr body occurs randomly and independently in each embryonic cell present at the time of X inactivation. As a consequence, females consist of a *mosaic* of two types of cells: those with the active X derived from the father and those with the active X derived from the mother. After an X chromosome is inactivated in a particular cell, all mitotic descendants of that cell have the same inactive X. Thus, if a female is heterozygous for a sex-linked trait, about half her cells will express one allele, while the others will express the alternate allele. **Figure 15.11** shows how this mosaicism results in the mottled coloration of a tortoiseshell cat. In humans, mosaicism can be observed in a recessive X-linked mutation that prevents the development of sweat glands. A woman who is heterozygous for this trait has patches of normal skin and patches of skin lacking sweat glands.

Inactivation of an X chromosome involves modification of the DNA, such as attachment of methyl groups ($—CH_3$) to one of the nitrogenous bases of DNA nucleotides. (The regulatory role of DNA methylation is discussed further in Chapter 19.) Researchers also have discovered a gene called *XIST* (for X-inactive specific transcript) that is active *only* on the Barr-body chromosome. Multiple copies of the RNA molecule produced from this gene apparently attach to the X chromosome on which they are made, eventually almost covering it. Interaction of this RNA with the chromosome seems to initiate X inactivation. Our understanding of X inactivation is still rudimentary, however.

Concept Check 15.3

1. A white-eyed female *Drosophila* is mated with a red-eyed (wild-type) male, the reciprocal cross of that shown in Figure 15.4. What phenotypes and genotypes do you predict for the offspring?
2. Neither Tim nor Rhoda has Duchenne muscular dystrophy, but their firstborn son does have it. What is the probability that a second child of this couple will have the disease?

For suggested answers, see Appendix A.

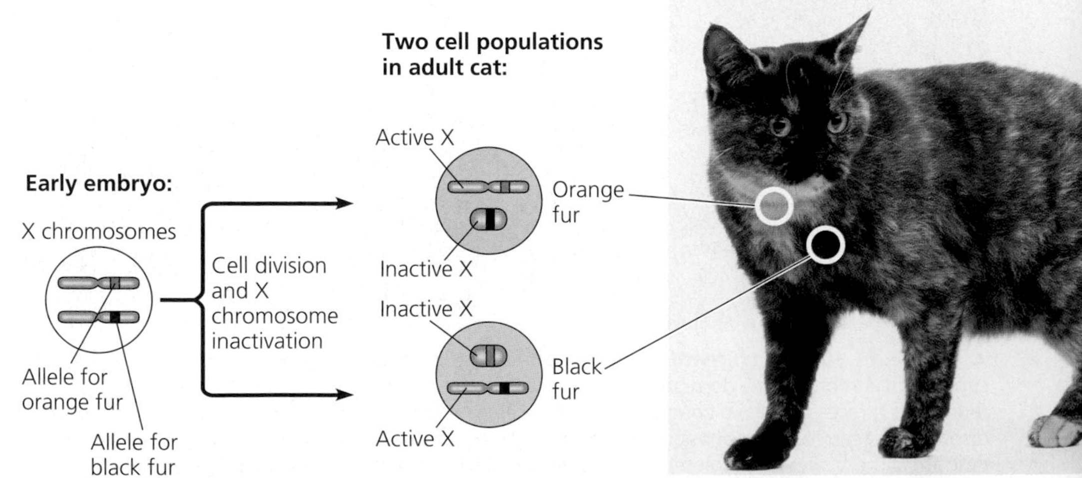

▶ **Figure 15.11 X inactivation and the tortoiseshell cat.** The tortoiseshell gene is on the X chromosome, and the tortoiseshell phenotype requires the presence of two different alleles, one for orange fur and one for black fur. Normally, only females can have both alleles, because only they have two X chromosomes. If a female is heterozygous for the tortoiseshell gene, she is tortoiseshell. Orange patches are formed by populations of cells in which the X chromosome with the orange allele is active; black patches have cells in which the X chromosome with the black allele is active. ("Calico" cats also have white areas, which are determined by yet another gene.)

Concept 15.4

Alterations of chromosome number or structure cause some genetic disorders

Sex-linked traits are not the only notable deviation from the inheritance patterns observed by Mendel, and the gene mutations that generate new alleles are not the only kind of changes to the genome that can affect phenotype. Physical and chemical disturbances, as well as errors during meiosis, can damage chromosomes in major ways or alter their number in a cell. Large-scale chromosomal alterations often lead to spontaneous abortion (miscarriage) of a fetus, and individuals born with these types of genetic defects commonly exhibit various developmental disorders. In plants, such genetic defects may be tolerated to a greater extent than in animals.

Abnormal Chromosome Number

Ideally, the meiotic spindle distributes chromosomes to daughter cells without error. But there is an occasional mishap, called a **nondisjunction**, in which the members of a pair of homologous chromosomes do not move apart properly during meiosis I or sister chromatids fail to separate during meiosis II. In these cases, one gamete receives two of the same type of chromosome and another gamete receives no copy **(Figure 15.12)**. The other chromosomes are usually distributed normally. If either of the aberrant gametes unites with a normal one at fertilization, the offspring will have an abnormal number of a particular chromosome, a condition known as **aneuploidy.**

If a chromosome is present in triplicate in the fertilized egg (so that the cell has a total of $2n + 1$ chromosomes), the aneuploid cell is said to be **trisomic** for that chromosome. If a chromosome is missing (so that the cell has $2n - 1$ chromosomes), the aneuploid cell is **monosomic** for that chromosome. Mitosis will subsequently transmit the anomaly to all embryonic cells. If the organism survives, it usually has a set of symptoms caused by the abnormal dose of the genes associated with the extra or missing chromosome. Nondisjunction can also occur during mitosis. If such an error takes place early in embryonic development, then the aneuploid condition is passed along by mitosis to a large number of cells and is likely to have a substantial effect on the organism.

Some organisms have more than two complete chromosome sets. The general term for this chromosomal alteration is **polyploidy**, with the specific terms *triploidy* ($3n$) and *tetraploidy* ($4n$) indicating three or four chromosomal sets, respectively. One way a triploid cell may be produced is by the fertilization of an abnormal diploid egg produced by nondisjunction of all its chromosomes. An example of an accident that would result in tetraploidy is the failure of a $2n$ zygote to divide after replicating its chromosomes. Subsequent normal mitotic divisions would then produce a $4n$ embryo.

Polyploidy is fairly common in the plant kingdom. As we will see in Chapter 24, the spontaneous origin of polyploid individuals plays an important role in the evolution of plants. In the animal kingdom, polyploid species are much less common, although they are known to occur among the fishes and amphibians. Researchers in Chile were the first to identify a

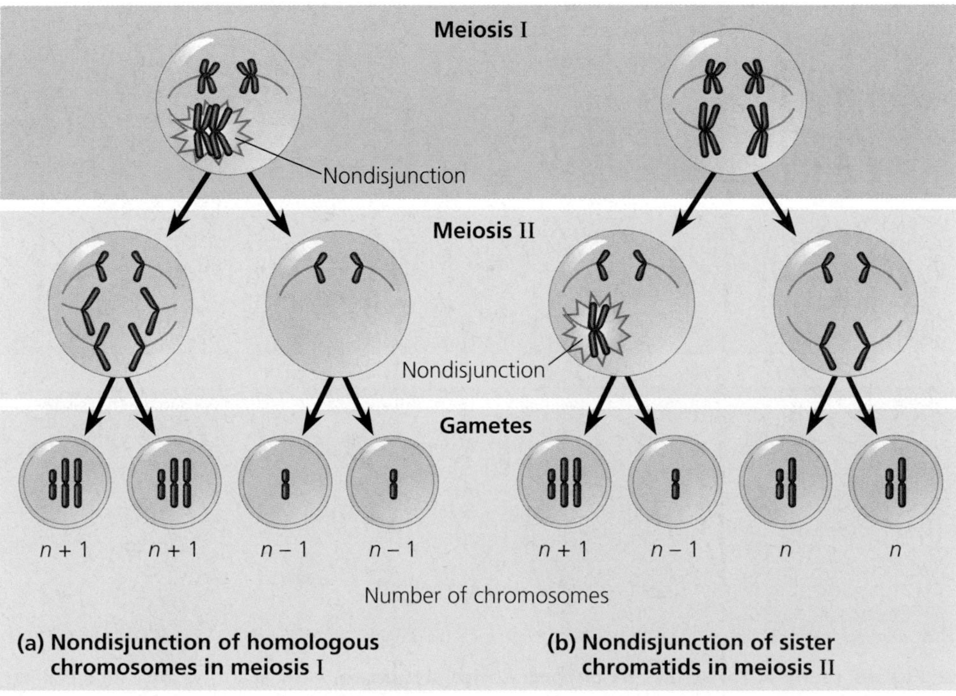

▶ **Figure 15.12 Meiotic nondisjunction.** Gametes with an abnormal chromosome number can arise by nondisjunction in either meiosis I or meiosis II.

▲ **Figure 15.13 A tetraploid mammal.** The somatic cells of this burrowing rodent, *Tympanoctomys barrerae*, have about twice as many chromosomes as those of closely related species. Interestingly, its sperm's head is unusually large, presumably a necessity for holding all that genetic material. Scientists think that this tetraploid species may have arisen when an ancestor doubled its chromosome number, presumably by errors in mitosis or meiosis within the animal's reproductive organs.

polyploid mammal, a rodent whose cells are tetraploid **(Figure 15.13)**. Additional research has found that a closely related species also appears to be tetraploid. In general, polyploids are more nearly normal in appearance than aneuploids. One extra (or missing) chromosome apparently disrupts genetic balance more than does an entire extra set of chromosomes.

Alterations of Chromosome Structure

Breakage of a chromosome can lead to four types of changes in chromosome structure, depicted in **Figure 15.14**. A **deletion** occurs when a chromosomal fragment lacking a centromere is lost. The affected chromosome is then missing certain genes. In some cases, if meiosis is in progress, such a "deleted" fragment may become attached as an extra segment to a sister chromatid, producing a **duplication.** Alternatively, a detached fragment could attach to a nonsister chromatid of a homologous chromosome. In that case, though, the "duplicated" segments might not be identical because the homologues could carry different alleles of certain genes. A chromosomal fragment may also reattach to the original chromosome but in the reverse orientation, producing an **inversion.** A fourth possible result of chromosomal breakage is for the fragment to join a nonhomologous chromosome, a rearrangement called a **translocation.**

Deletions and duplications are especially likely to occur during meiosis. In crossing over, nonsister chromatids sometimes break and rejoin at "incorrect" places, so that one partner gives up more genes than it receives. The products of such a *nonreciprocal* crossover are one chromosome with a deletion and one chromosome with a duplication.

A diploid embryo that is homozygous for a large deletion (or has a single X chromosome with a large deletion, in a male) is usually missing a number of essential genes, a condition that is ordinarily lethal. Duplications and translocations also tend to have harmful effects. In reciprocal translocations, in which segments are exchanged between nonhomologous chromosomes, and in inversions, the balance of genes is not abnormal—all genes are present in their normal doses. Never-

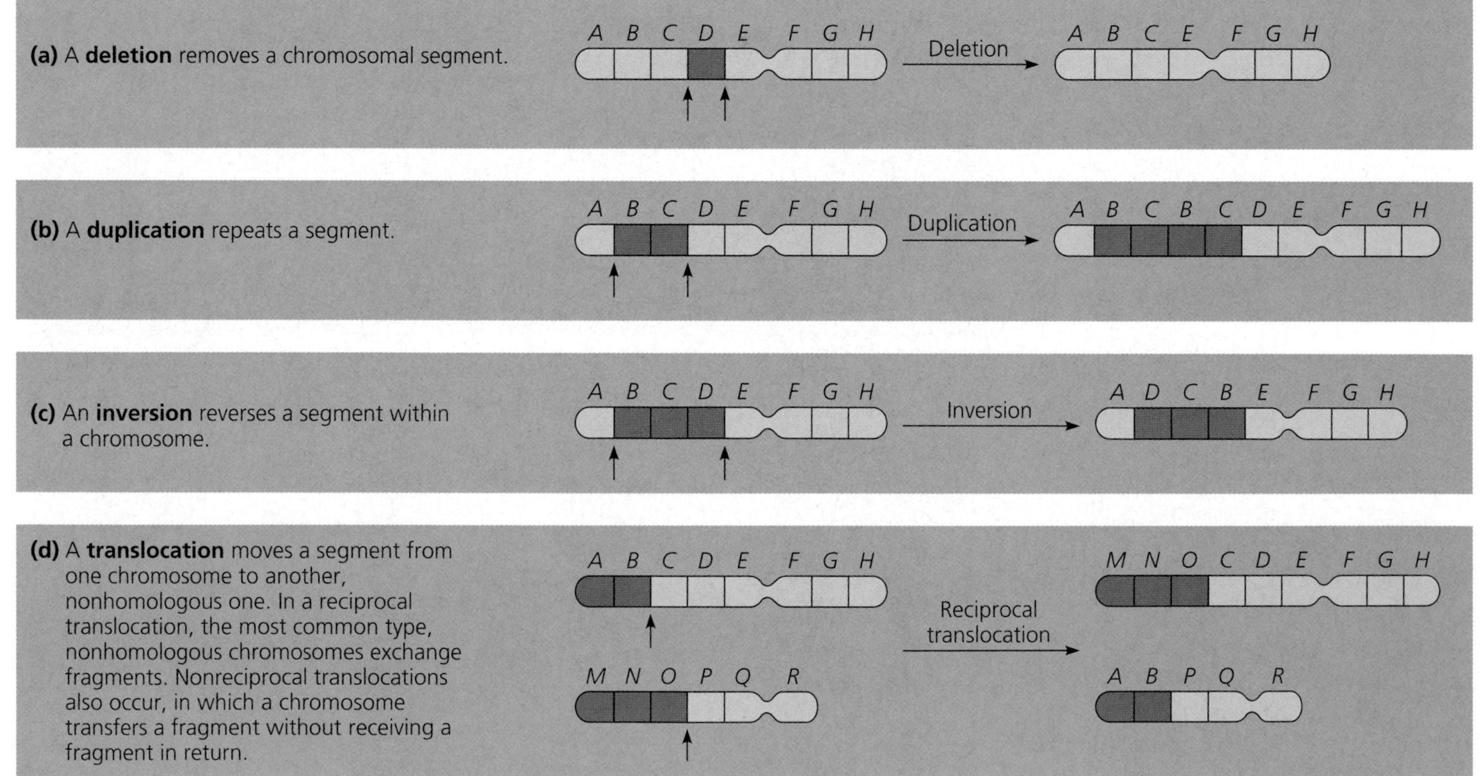

▲ **Figure 15.14 Alterations of chromosome structure.** Vertical arrows indicate breakage points. Dark purple highlights the chromosomal parts affected by the rearrangements.

theless, translocations and inversions can alter phenotype because a gene's expression can be influenced by its location among neighboring genes.

Human Disorders Due to Chromosomal Alterations

Alterations of chromosome number and structure are associated with a number of serious human disorders. Nondisjunction in meiosis results in aneuploid gametes. If an aneuploid gamete combines with a normal haploid gamete during fertilization, the result is an aneuploid zygote. Although the frequency of aneuploid zygotes may be quite high in humans, most of these chromosomal alterations are so disastrous to development that the embryos are spontaneously aborted long before birth. However, some types of aneuploidy appear to upset the genetic balance less than others, with the result that individuals with certain aneuploid conditions can survive to birth and beyond. These individuals have a set of symptoms—a *syndrome*—characteristic of the type of aneuploidy. Genetic disorders caused by aneuploidy can be diagnosed before birth by fetal testing (see Figure 14.17).

Down Syndrome (Trisomy 21)

One aneuploid condition, **Down syndrome**, affects approximately one out of every 700 children born in the United States **(Figure 15.15)**. Down syndrome is usually the result of an extra chromosome 21, so that each body cell has a total of 47 chromosomes. Because the cells are trisomic for chromosome 21, Down syndrome is often called *trisomy 21*. Down syndrome includes characteristic facial features, short stature, heart defects, susceptibility to respiratory infection, and mental retardation. Furthermore, individuals with Down syndrome are prone to developing leukemia and Alzheimer's disease. Although people with Down syndrome, on average, have a life span shorter than normal, some live to middle age or beyond. Most are sexually underdeveloped and sterile.

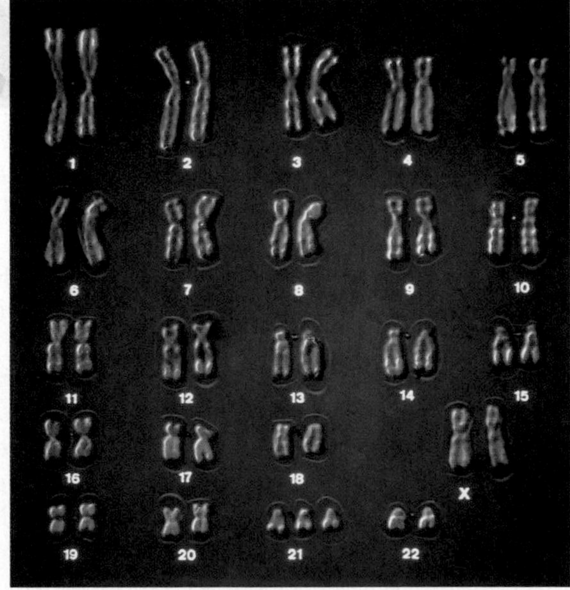

▲ **Figure 15.15 Down syndrome.** The child exhibits the facial features characteristic of Down syndrome. The karyotype shows trisomy 21, the most common cause of this disorder.

The frequency of Down syndrome increases with the age of the mother. While the disorder occurs in just 0.04% of children born to women under age 30, the risk climbs to 1.25% for mothers in their early 30s and is even higher for older mothers. Because of this relatively high risk, pregnant women over 35 are candidates for fetal testing to check for trisomy 21 in the embryo. The correlation of Down syndrome with maternal age has not yet been explained. Most cases result from nondisjunction during meiosis I, and some research points to an age-dependent abnormality in a meiosis checkpoint that normally delays anaphase until all the kinetochores are attached to the spindle (like the M phase checkpoint of the mitotic cell cycle; see Chapter 12). Trisomies of some other chromosomes also increase in incidence with maternal age, although infants with these autosomal trisomies rarely survive for long.

Aneuploidy of Sex Chromosomes

Nondisjunction of sex chromosomes produces a variety of aneuploid conditions. Most of these conditions appear to upset genetic balance less than aneuploid conditions involving autosomes. This may be because the Y chromosome carries relatively few genes and because extra copies of the X chromosome become inactivated as Barr bodies in somatic cells.

An extra X chromosome in a male, producing XXY, occurs approximately once in every 2,000 live births. People with this disorder, called *Klinefelter syndrome,* have male sex organs, but the testes are abnormally small and the man is sterile. Even though the extra X is inactivated, some breast enlargement and other female body characteristics are common. The affected individual is usually of normal intelligence. Males with an extra Y chromosome (XYY) do not exhibit any well-defined syndrome, but they tend to be somewhat taller than average.

Females with trisomy X (XXX), which occurs once in approximately 1,000 live births, are healthy and cannot be distinguished from XX females except by karyotype. Monosomy X, called *Turner syndrome,* occurs about once in every 5,000 births and is the only known viable monosomy in humans. Although these X0 individuals are phenotypically female, they are sterile because their sex organs do not mature. When provided with estrogen replacement therapy, girls with Turner syndrome do develop secondary sex characteristics. Most have normal intelligence.

▶ **Figure 15.16 Translocation associated with chronic myelogenous leukemia (CML).** The cancerous cells in nearly all CML patients contain an abnormally short chromosome 22, the so-called Philadelphia chromosome, and an abnormally long chromosome 9. These altered chromosomes result from the translocation shown here.

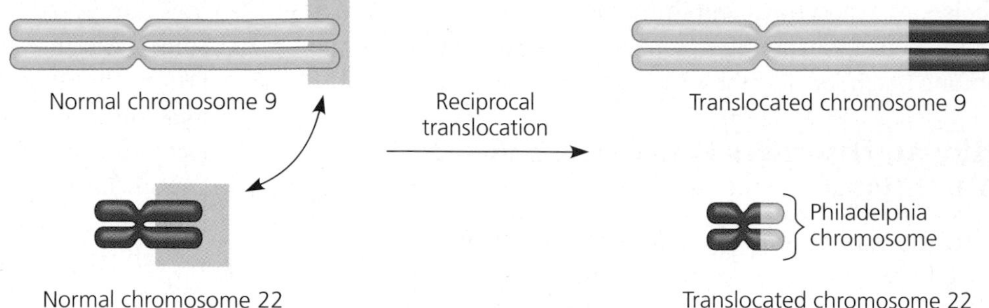

Disorders Caused by Structurally Altered Chromosomes

Many deletions in human chromosomes, even in a heterozygous state, cause severe problems. One such syndrome, known as *cri du chat* ("cry of the cat"), results from a specific deletion in chromosome 5. A child born with this deletion is mentally retarded, has a small head with unusual facial features, and has a cry that sounds like the mewing of a distressed cat. Such individuals usually die in infancy or early childhood.

Another type of chromosomal structural alteration associated with human disorders is translocation, the attachment of a fragment from one chromosome to another, nonhomologous chromosome. Chromosomal translocations have been implicated in certain cancers, including *chronic myelogenous leukemia (CML)*. Leukemia is a cancer affecting the cells that give rise to white blood cells, and in the cancerous cells of CML patients, a reciprocal translocation has occurred. In these cells, the exchange of a large portion of chromosome 22 with a small fragment from a tip of chromosome 9 produces a much shortened, easily recognized chromosome 22, called the *Philadelphia chromosome* (**Figure 15.16**). We will discuss how such an exchange might cause cancer in Chapter 19.

Concept Check 15.4

1. More common than completely polyploid animals are mosaic polyploids, animals that are diploid except for patches of polyploid cells. How might a mosaic tetraploid—an animal with some cells containing four sets of chromosomes—arise?
2. About 5% of individuals with Down syndrome have a chromosomal translocation in which one copy of chromosome 21 is attached to chromosome 14. How could this translocation in a parent's gonad lead to Down syndrome in a child?
3. Explain how a male cat could have the tortoiseshell phenotype.

For suggested answers, see Appendix A.

Concept 15.5

Some inheritance patterns are exceptions to the standard chromosome theory

In the previous section, you learned about abnormal deviations from the usual patterns of chromosomal inheritance. We conclude this chapter by describing two *normal* exceptions to Mendelian genetics, one involving genes located in the nucleus and the other involving genes located outside of the nucleus.

Genomic Imprinting

Throughout our discussions of Mendelian genetics and the chromosomal basis of inheritance, we have assumed that a specific allele will have the same effect regardless of whether it was inherited from the mother or the father. This is probably a safe assumption most of the time. For example, when Mendel crossed purple-flowered pea plants with white-flowered pea plants, he observed the same results regardless of whether the purple-flowered parent supplied the ova or the pollen. In recent years, however, geneticists have identified two to three dozen traits in mammals that depend on which parent passed along the alleles for those traits. Such variation in phenotype depending on whether an allele is inherited from the male or female parent is called **genomic imprinting.** (Note that the issue here is not sex linkage; most imprinted genes are on autosomes.)

Genomic imprinting occurs during the formation of gametes and results in the silencing of one allele of certain genes. Because these genes are imprinted differently in sperm and ova, a zygote expresses only one allele of an imprinted gene, either the allele inherited from the female parent or the allele inherited from the male parent. The imprints are transmitted to all the body cells during development, so the same allele of a given gene—either the maternally inherited allele or the paternally inherited allele—is expressed in all cells of that organism. In each generation, the old imprints are "erased" in gamete-producing cells, and the chromosomes of the developing gametes are newly imprinted according to the sex of the individual. In a given species, the imprinted genes are always imprinted in the same way. For instance, a gene imprinted for maternal allele expression is always imprinted for maternal allele expression, generation after generation.

Consider, for example, the gene for insulin-like growth factor 2 (*Igf2*), one of the first imprinted genes to be identified. Although this growth factor is required for normal prenatal growth, only the paternal allele is expressed **(Figure 15.17a)**. Evidence that the *Igf2* gene is imprinted initially came from crosses between wild-type mice and dwarf mice homozygous for a recessive mutation in the *Igf2* gene. The phenotypes of heterozygous offspring (one normal allele and one mutant) differed, depending on whether the mutant allele came from the mother or the father **(Figure 15.17b)**.

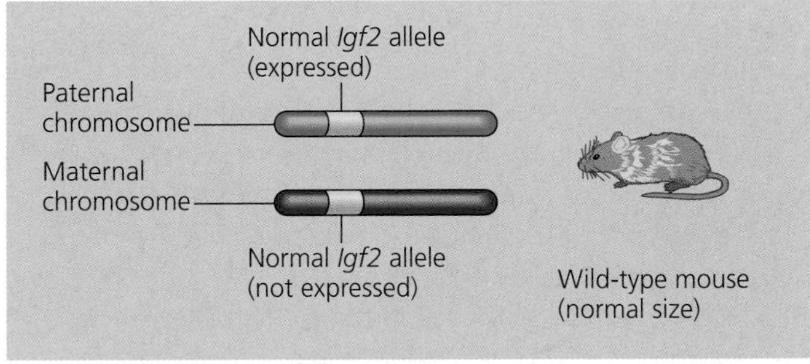

(a) A wild-type mouse is homozygous for the normal *Igf2* allele.

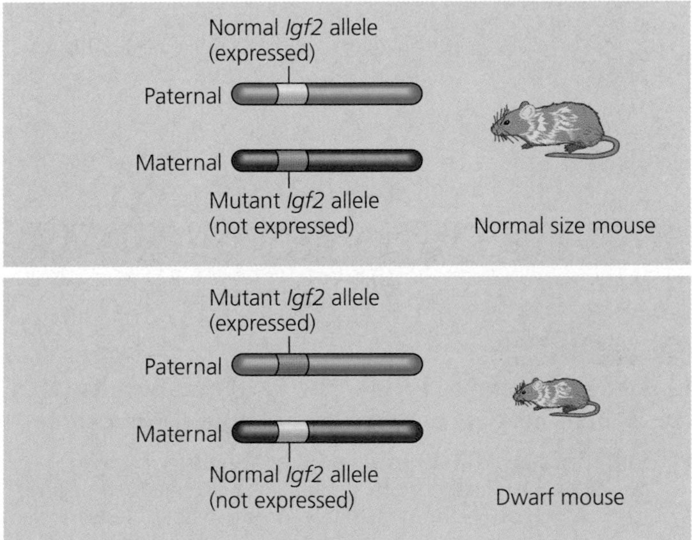

(b) When a normal *Igf2* allele is inherited from the father, heterozygous mice grow to normal size. But when a mutant allele is inherited from the father, heterozygous mice have the dwarf phenotype.

▲ **Figure 15.17 Genomic imprinting of the mouse *Igf2* gene. (a)** In mice, the paternal *Igf2* allele is expressed and the maternal allele is not. **(b)** Matings between wild-type mice and those homozygous for the recessive mutant *Igf2* allele produce heterozygous offspring that can be either normal size or dwarf, depending on which parent passes on the mutant allele.

What exactly is a genomic imprint? In many cases, it seems to consist of methyl (—CH_3) groups that are added to cytosine nucleotides of one of the alleles. Such methylation may directly silence the allele, an effect consistent with evidence that heavily methylated genes are usually inactive (see Chapter 19). However, for a few genes, methylation has been shown to *activate* expression of the allele. This is the case for the *Igf2* gene: Methylation of a certain DNA sequence on the paternal chromosome leads to expression of the paternal *Igf2* allele.

Genomic imprinting is thought to affect only a small fraction of the genes in mammalian genomes, but most of the known imprinted genes are critical for embryonic development. In experiments with mice, for example, embryos engineered to inherit both copies of certain chromosomes from the same parent inevitably die before birth, whether that parent is male or female. Normal development apparently requires that embryonic cells have exactly one active copy—not zero, not two—of certain genes. The association of aberrant imprinting with abnormal development and certain cancers is stimulating numerous studies on how different genes are imprinted.

Inheritance of Organelle Genes

Although our focus in this chapter has been on the chromosomal basis of inheritance, we end with an important amendment: Not all of a eukaryotic cell's genes are located on nuclear chromosomes, or even in the nucleus. Some genes are located in organelles in the cytoplasm; these genes are sometimes called *extranuclear genes*. Mitochondria, as well as chloroplasts and other plant plastids, contain small circular DNA molecules that carry genes coding for proteins and RNA. These organelles reproduce themselves and transmit their genes to daughter organelles. Because organelle genes are not distributed to offspring according to the same rules that direct the distribution of nuclear chromosomes during meiosis, they do not display Mendelian inheritance.

The first hint that extranuclear genes exist came from studies by Karl Correns on the inheritance of yellow or white patches on the leaves of an otherwise green plant. In 1909, he observed that the coloration of the offspring was determined only by the maternal parent (the source of seeds that germinate to give rise to the offspring) and not by the paternal parent (the pollen source). Subsequent research showed that such coloration patterns, or variegation, are due to mutations in plastid genes that control pigmentation. In most plants, a zygote receives all its plastids from the cytoplasm of the egg and none from pollen, which contributes little more than a haploid set of chromosomes. As the zygote develops, plastids containing wild-type or mutant pigment genes are distributed randomly to daughter cells. The pattern of leaf coloration exhibited by a

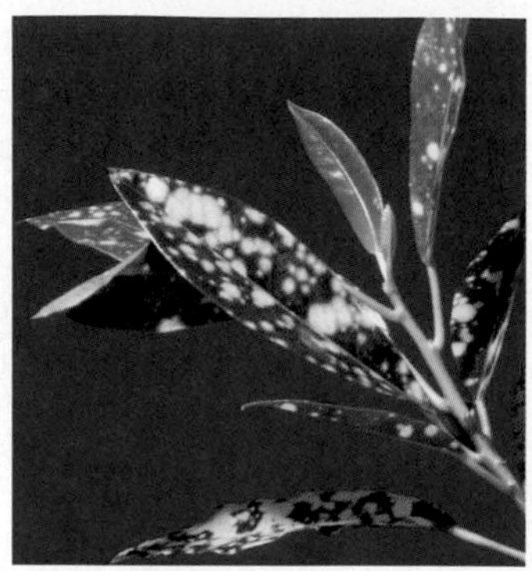

▶ **Figure 15.18 Variegated leaves from *Croton dioicus*.** Variegated (striped or spotted) leaves result from mutations in pigment genes located in plastids, which generally are inherited from the maternal parent.

plant depends on the ratio of wild-type to mutant plastids in its various tissues **(Figure 15.18)**.

Similar maternal inheritance is also the rule for mitochondrial genes in most animals and plants, because almost all the mitochondria passed on to a zygote come from the cytoplasm of the egg. The products of most mitochondrial genes help make up the protein complexes of the electron transport chain and ATP synthase (see Chapter 9). Defects in one or more of these proteins, therefore, reduce the amount of ATP the cell can make and have been shown to cause a number of rare human disorders. Because the parts of the body most susceptible to energy deprivation are the nervous system and the muscles, most mitochondrial diseases primarily affect these systems. For example, a person with the disease called *mitochondrial myopathy* suffers from weakness, intolerance of exercise, and muscle deterioration.

In addition to the rare diseases clearly caused by defects in mitochondrial DNA, mitochondrial mutations inherited from a person's mother may contribute to at least some cases of diabetes and heart disease, as well as to other disorders that commonly debilitate the elderly, such as Alzheimer's disease. In the course of a lifetime, new mutations gradually accumulate in our mitochondrial DNA, and some researchers think that these mutations play a role in the normal aging process.

Wherever genes are located in the cell—in the nucleus or in cytoplasmic organelles—their inheritance depends on the precise replication of DNA, the genetic material. In the next chapter, you will learn how this molecular reproduction occurs.

Concept Check 15.5

1. Gene dosage, the number of active copies of a gene, is important to proper development. Identify and describe two processes that help establish the proper dosage of certain genes.
2. Reciprocal crosses between two primrose varieties, A and B, produced the following results: A female × B male → offspring with all green (nonvariegated) leaves. B female × A male → offspring with spotted (variegated) leaves. Explain these results.
3. Mitochondrial genes are critical to the energy metabolism of cells, but mitochondrial disorders caused by mutations in these genes are generally not lethal. Why not?

For suggested answers, see Appendix A.

Chapter 15 Review

Go to www.campbellbiology.com or the student CD-ROM to explore Activities, Investigations, and other interactive study aids.

SUMMARY OF KEY CONCEPTS

Concept 15.1
Mendelian inheritance has its physical basis in the behavior of chromosomes

- ▶ In the early 1900s, several researchers proposed that genes are located on chromosomes and that the behavior of chromosomes during meiosis accounts for Mendel's laws of segregation and independent assortment (pp. 274–275).
- ▶ **Morgan's Experimental Evidence: *Scientific Inquiry* (pp. 276–277)** Morgan's discovery that transmission of the X chromosome in *Drosophila* correlates with inheritance of the eye-color trait was the first solid evidence indicating that a specific gene is associated with a specific chromosome.

Concept 15.2
Linked genes tend to be inherited together because they are located near each other on the same chromosome

- ▶ **How Linkage Affects Inheritance: *Scientific Inquiry* (pp. 277–278)** Each chromosome has hundreds or thousands of genes. Genes on the same chromosome whose alleles are so close together that they do not assort independently are said to be linked. The alleles of unlinked genes are either on separate chromosomes or so far apart on the same chromosome that they assort independently.
- ▶ **Genetic Recombination and Linkage (pp. 278–280)** Recombinant offspring exhibit new combinations of traits inherited from two parents. Because of the independent assortment of chromosomes and random fertilization, unlinked genes exhibit a 50% frequency of recombination. Even with crossing over between nonsister chromatids during the first meiotic division, linked genes exhibit recombination frequencies less than 50%.

▶ **Linkage Mapping Using Recombination Data: *Scientific Inquiry*** **(pp. 279–281)** Geneticists can deduce the order of genes on a chromosome and the relative distances between them from recombination frequencies observed in genetic crosses. In general, the farther apart genes are on a chromosome, the more likely they are to be separated during crossing over.
Activity *Linked Genes and Crossing Over*

Concept 15.3

Sex-linked genes exhibit unique patterns of inheritance

▶ **The Chromosomal Basis of Sex (pp. 282–283)** An organism's sex is an inherited phenotypic character usually determined by the presence or absence of certain chromosomes. Humans and other mammals have an X-Y system in which sex normally is determined by the presence or absence of a Y chromosome. Different systems of sex determination are found in birds, fishes, and insects.

▶ **Inheritance of Sex-Linked Genes (pp. 283–284)** The sex chromosomes carry certain genes for traits that are unrelated to maleness or femaleness. For instance, recessive alleles causing color blindness, hemophilia, and Duchenne muscular dystrophy are carried on the X chromosome. Fathers transmit such sex-linked alleles to all daughters but to no sons. Any male who inherits a single sex-linked recessive allele from his mother will express the trait.
Activity *Sex-Linked Genes*
Investigation *What Can Fruit Flies Reveal About Inheritance?*
Biology Labs On-Line *FlyLab*
Biology Labs On-Line *PedigreeLab*

▶ **X Inactivation in Female Mammals (p. 284)** In mammalian females, one of the two X chromosomes in each cell is randomly inactivated during early embryonic development. If a female is heterozygous for a particular gene located on the X chromosome, she will be mosaic for that character, with about half her cells expressing the maternal allele and about half expressing the paternal allele.

Concept 15.4

Alterations of chromosome number or structure cause some genetic disorders

▶ **Abnormal Chromosome Number (pp. 285–286)** Aneuploidy can arise when a normal gamete unites with one containing two copies or no copies of a particular chromosome as a result of nondisjunction during meiosis. The cells of the resulting zygote have either one extra copy of that chromosome (trisomy) or are missing a copy (monosomy). Polyploidy, in which there are more than two complete sets of chromosomes, can result from complete nondisjunction during gamete formation.
Activity *Polyploid Plants*

▶ **Alterations of Chromosome Structure (pp. 286–287)** Chromosome breakage can result in various rearrangements. A lost fragment leaves one chromosome with a deletion; the deleted fragment may reattach to the same chromosome in a different orientation, producing an inversion. Or the fragment may attach to a homologous chromosome, producing a duplication, or to a nonhomologous chromosome, producing a translocation.

▶ **Human Disorders Due to Chromosomal Alterations (pp. 287–288)** Changes in the number of chromosomes per cell or in the structure of individual chromosomes can affect phenotype. Such alterations cause Down syndrome (usually due to trisomy of chromosome 21), certain cancers associated with chromosomal translocations, and various other human disorders.

Some inheritance patterns are exceptions to the standard chromosome theory

▶ **Genomic Imprinting (pp. 288–289)** In mammals, the phenotypic effects of certain genes depend on which allele is inherited from the mother and which is inherited from the father. Imprints are formed during gamete production, with the result that one allele (either maternal or paternal) is not expressed in offspring. Most imprinted genes now known play a role in embryonic development.

▶ **Inheritance of Organelle Genes (pp. 289–290)** The inheritance of traits controlled by the genes present in mitochondria and chloroplasts depends solely on the maternal parent because the zygote's cytoplasm comes from the egg. Some diseases affecting the nervous and muscular systems are caused by defects in mitochondrial genes that prevent cells from making enough ATP.

TESTING YOUR KNOWLEDGE

Genetics Problems

1. A man with hemophilia (a recessive, sex-linked condition) has a daughter of normal phenotype. She marries a man who is normal for the trait. What is the probability that a daughter of this mating will be a hemophiliac? That a son will be a hemophiliac? If the couple has four sons, what is the probability that all four will be born with hemophilia?

2. Pseudohypertrophic muscular dystrophy is an inherited disorder that causes gradual deterioration of the muscles. It is seen almost exclusively in boys born to apparently normal parents and usually results in death in the early teens. Is this disorder caused by a dominant or a recessive allele? Is its inheritance sex-linked or autosomal? How do you know? Explain why this disorder is almost never seen in girls.

3. Red-green color blindness is caused by a sex-linked recessive allele. A color-blind man marries a woman with normal vision whose father was color-blind. What is the probability that they will have a color-blind daughter? What is the probability that their first son will be color-blind? (*Note:* The two questions are worded a bit differently.)

4. A wild-type fruit fly (heterozygous for gray body color and normal wings) is mated with a black fly with vestigial wings. The offspring have the following phenotypic distribution: wild type, 778; black-vestigial, 785; black-normal, 158; gray-vestigial, 162. What is the recombination frequency between these genes for body color and wing size?

5. In another cross, a wild-type fruit fly (heterozygous for gray body color and red eyes) is mated with a black fruit fly with purple eyes. The offspring are as follows: wild type, 721; black-purple, 751; gray-purple, 49; black-red, 45. What is the recombination frequency between these genes for body color and eye color? Using information from problem 4, what fruit flies (genotypes and phenotypes) would you mate to determine the sequence of the body-color, wing-size, and eye-color genes on the chromosome?

6. What pattern of inheritance would lead a geneticist to suspect that an inherited disorder of cell metabolism is due to a defective mitochondrial gene?

7. Women born with an extra X chromosome (XXX) are healthy and phenotypically indistinguishable from normal XX women. What is a likely explanation for this finding? How could you test this explanation?

8. Determine the sequence of genes along a chromosome based on the following recombination frequencies: *A–B*, 8 map units; *A–C*, 28 map units; *A–D*, 25 map units; *B–C*, 20 map units; *B–D*, 33 map units.

9. Assume that genes *A* and *B* are linked and are 50 map units apart. An animal heterozygous at both loci is crossed with one that is homozygous recessive at both loci. What percentage of the offspring will show phenotypes resulting from crossovers? If you did not know that genes *A* and *B* were linked, how would you interpret the results of this cross?

10. A space probe discovers a planet inhabited by creatures who reproduce with the same hereditary patterns seen in humans. Three phenotypic characters are height (T = tall, t = dwarf), head appendages (A = antennae, a = no antennae), and nose morphology (S = upturned snout, s = downturned snout). Since the creatures are not "intelligent," Earth scientists are able to do some controlled breeding experiments, using various heterozygotes in testcrosses. For tall heterozygotes with antennae, the offspring are: tall-antennae, 46; dwarf-antennae, 7; dwarf-no antennae, 42; tall-no antennae, 5. For heterozygotes with antennae and an upturned snout, the offspring are: antennae-upturned snout, 47; antennae-downturned snout, 2; no antennae-downturned snout, 48; no antennae-upturned snout, 3. Calculate the recombination frequencies for both experiments.

11. Using the information from problem 10, scientists do a further testcross using a heterozygote for height and nose morphology. The offspring are: tall-upturned snout, 40; dwarf-upturned snout, 9; dwarf-downturned snout, 42; tall-downturned snout, 9. Calculate the recombination frequency from these data; then use your answer from problem 10 to determine the correct sequence of the three linked genes.

12. The ABO blood type locus has been mapped on chromosome 9. A father who has blood type AB and a mother who has blood type O have a child with trisomy 9 and blood type A. Using this information, can you tell in which parent the nondisjunction occurred? Explain your answer.

13. Two genes of a flower, one controlling blue (*B*) versus white (*b*) petals and the other controlling round (*R*) versus oval (*r*) stamens, are linked and are 10 map units apart. You cross a homozygous blue-oval plant with a homozygous white-round plant. The resulting F_1 progeny are crossed with homozygous white-oval plants, and 1,000 F_2 progeny are obtained. How many F_2 plants of each of the four phenotypes do you expect?

14. You design *Drosophila* crosses to provide recombination data for gene *a*, which is located on the same chromosome shown in Figure 15.8. Gene *a* has recombination frequencies of 14% with the vestigial-wing locus and 26% with the brown-eye locus. Where is *a* located on the chromosome?

For Genetics Problems answers, see Appendix A.

Go to the website or CD-ROM for more quiz questions.

Evolution Connection

You have seen that crossing over, or recombination, is thought to be evolutionarily advantageous because this process continually shuffles genetic alleles into novel combinations. Some organisms, however, have apparently lost the recombination mechanism, while in others, certain chromosomes do not recombine. What factors do you think may favor reduced levels of recombination?

Scientific Inquiry

Consider Figure 15.5, in which the F_1 dihybrid females resulted from a cross between parental (P) flies with genotypes $b^+ \, b^+ \, vg^+ \, vg^+$ and $b \, b \, vg \, vg$. Now, imagine you make F_1 females by crossing two different P generation flies: $b^+ \, b^+ \, vg \, vg \times b \, b \, vg^+ \, vg^+$.

a. What will be the genotype of your F_1 females? Is this the same as that for the F_1 females in Figure 15.5?

b. Draw the chromosomes for the F_1 females, indicating the position of each allele. Are these the same as for the F_1 females in Figure 15.5?

c. Knowing that the distance between these two genes is 17 map units, predict the phenotypic ratios you will get from this cross. Will they be the same as in Figure 15.5?

d. Draw the chromosomes of the P, F_1, and F_2 generations (as is done in Figure 15.6 for the cross in Figure 15.5), showing how this arrangement of alleles in the P generation leads, via F_1 gametes, to the phenotypic ratios seen in the F_2 flies.

Investigation *What Can Fruit Flies Reveal About Inheritance?*
Biology Labs On-Line *FlyLab*
Biology Labs On-Line *PedigreeLab*

Science, Technology, and Society

About one in every 1,500 boys and one in every 2,500 girls are born with a fragile X chromosome, the tip of which hangs on to the rest of the chromosome by a thin thread of DNA. This abnormality causes mental retardation. Opinions differ about whether children with learning disorders should be tested by karyotyping for the presence of a fragile X chromosome. Some argue that it's always better to know the cause of the problem so that education specialized for that disorder can be prescribed. Others counter that attaching a specific biological cause to a learning disability stigmatizes a child and limits his or her opportunities. What is your evaluation of these arguments?

16 The Molecular Basis of Inheritance

▲ Figure 16.1 **Watson and Crick with their DNA model.**

Key Concepts

16.1 DNA is the genetic material
16.2 Many proteins work together in DNA replication and repair

Overview

Life's Operating Instructions

In April 1953, James Watson and Francis Crick shook the scientific world with an elegant double-helical model for the structure of deoxyribonucleic acid, or DNA. **Figure 16.1** shows Watson and Crick admiring their DNA model, which they built from tin and wire. Over the past 50 years, their model has evolved from a novel proposition to an icon of modern biology. DNA, the substance of inheritance, is the most celebrated molecule of our time. Mendel's heritable factors and Morgan's genes on chromosomes are, in fact, composed of DNA. Chemically speaking, your genetic endowment is the DNA contained in the 46 chromosomes you inherited from your parents.

Of all nature's molecules, nucleic acids are unique in their ability to direct their own replication from monomers. Indeed, the resemblance of offspring to their parents has its basis in the precise replication of DNA and its transmission from one generation to the next. Hereditary information is encoded in the chemical language of DNA and reproduced in all the cells of your body. It is this DNA program that directs the development of your biochemical, anatomical, physiological, and, to some extent, behavioral traits. In this chapter, you will learn how biologists deduced that DNA is the genetic material, how Watson and Crick discovered its structure, and how cells replicate and repair their DNA—the molecular basis of inheritance.

Concept 16.1

DNA is the genetic material

Today, even schoolchildren have heard of DNA, and scientists routinely manipulate DNA in the laboratory and use it to change the heritable characteristics of cells. Early in the 20th century, however, the identification of the molecules of inheritance loomed as a major challenge to biologists.

The Search for the Genetic Material: *Scientific Inquiry*

Once T. H. Morgan's group showed that genes are located on chromosomes (described in Chapter 15), the two chemical components of chromosomes—DNA and protein—became the candidates for the genetic material. Until the 1940s, the case for proteins seemed stronger, especially since biochemists had identified them as a class of macromolecules with great heterogeneity and specificity of function, essential requirements for the hereditary material. Moreover, little was known about nucleic acids, whose physical and chemical properties seemed far too uniform to account for the multitude of specific inherited traits exhibited by every organism. This view gradually changed as experiments with microorganisms yielded unexpected results. As with the work of Mendel and Morgan, a key factor in determining the identity of the genetic material was the choice of appropriate experimental organisms. The role of DNA in heredity was first worked out by studying bacteria and the viruses that infect them, which are far simpler than pea plants, fruit flies, or humans. In this section, we will trace the search for the genetic material in some detail as a case study in scientific inquiry.

Evidence That DNA Can Transform Bacteria

We can trace the discovery of the genetic role of DNA back to 1928. Frederick Griffith, a British medical officer, was studying *Streptococcus pneumoniae,* a bacterium that causes pneumonia in mammals. Griffith had two strains (varieties) of the bacterium, a pathogenic (disease-causing) one and a nonpathogenic (harmless) strain. He was surprised to find that when he killed the pathogenic bacteria with heat and then mixed the cell remains with living bacteria of the nonpathogenic strain, some of the living cells became pathogenic **(Figure 16.2)**. Furthermore, this new trait of pathogenicity was inherited by all the descendants of the transformed bacteria. Clearly, some chemical component of the dead pathogenic cells caused this heritable change, although the identity of the substance was not known. Griffith called the phenomenon **transformation,** now defined as a change in genotype and phenotype due to the assimilation of external DNA by a cell. (This use of the word *transformation* should not be confused with the conversion of a normal animal cell to a cancerous one, discussed in Chapter 12.)

Figure 16.2

Inquiry **Can the genetic trait of pathogenicity be transferred between bacteria?**

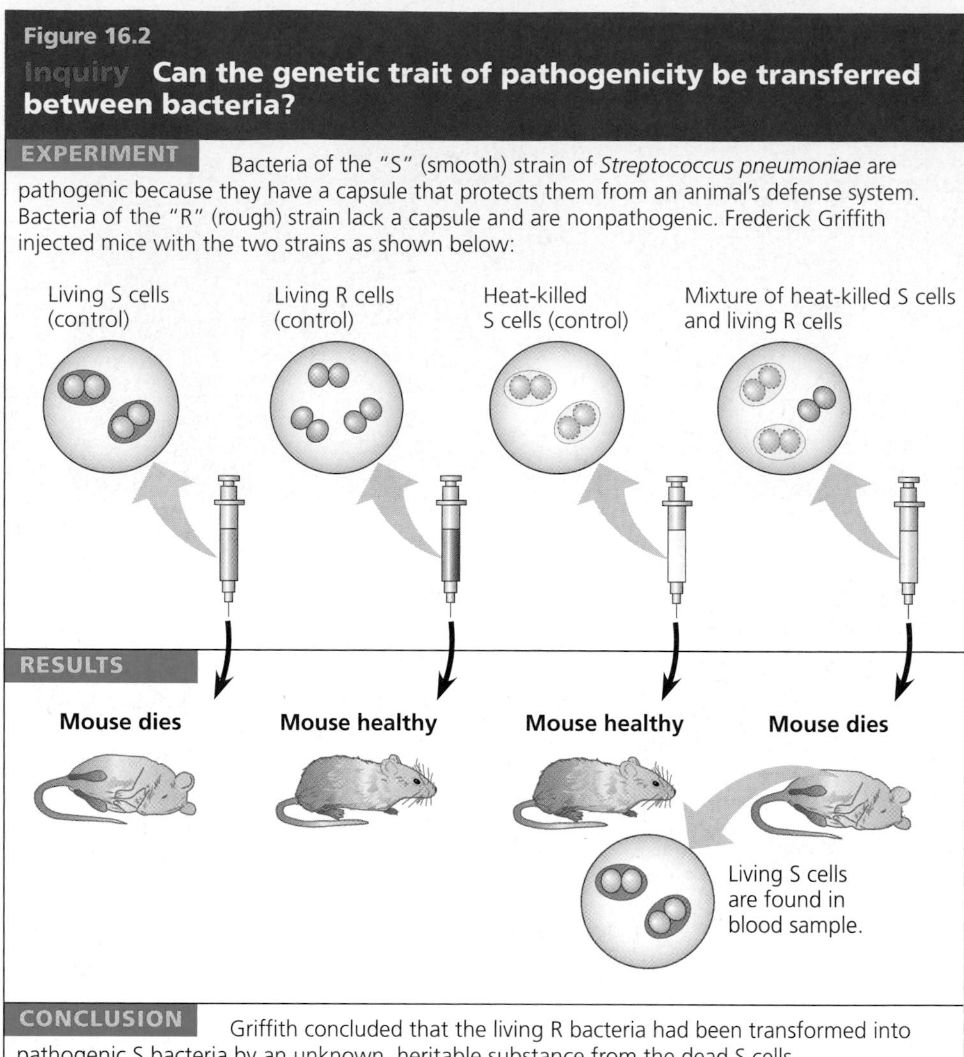

EXPERIMENT
Bacteria of the "S" (smooth) strain of *Streptococcus pneumoniae* are pathogenic because they have a capsule that protects them from an animal's defense system. Bacteria of the "R" (rough) strain lack a capsule and are nonpathogenic. Frederick Griffith injected mice with the two strains as shown below:
Living S cells (control)
Living R cells (control)
Heat-killed S cells (control)
Mixture of heat-killed S cells and living R cells
RESULTS
Mouse dies
Mouse healthy
Mouse healthy
Mouse dies
Living S cells are found in blood sample.
CONCLUSION
Griffith concluded that the living R bacteria had been transformed into pathogenic S bacteria by an unknown, heritable substance from the dead S cells.

Griffith's work set the stage for a 14-year search for the identity of the transforming substance by American bacteriologist Oswald Avery. Avery purified various types of molecules from the heat-killed pathogenic bacteria, then tried to transform live nonpathogenic bacteria with each type. Only DNA worked. Finally, in 1944, Avery and his colleagues Maclyn McCarty and Colin MacLeod announced that the transforming agent was DNA. Their discovery was greeted with interest but considerable skepticism, in part because of the lingering belief that proteins were better candidates for the genetic material. Moreover, many biologists were not convinced that the genes of bacteria would be similar in composition and function to those of more complex organisms. But the major reason for the continued doubt was that so little was known about DNA.

Evidence That Viral DNA Can Program Cells

Additional evidence for DNA as the genetic material came from studies of a virus that infects bacteria. Viruses are much simpler than cells. A virus is little more than DNA (or sometimes RNA) enclosed by a protective coat, which is often simply protein. To reproduce, a virus must infect a cell and take over the cell's metabolic machinery.

Viruses that infect bacteria are widely used as tools by researchers in molecular genetics. These viruses are called **bacteriophages** (meaning "bacteria-eaters"), or just **phages** **(Figure 16.3)**. In 1952, Alfred Hershey and Martha Chase performed experiments showing that DNA is the genetic material of a phage known as T2. This is one of many phages that infect *Escherichia coli (E. coli),* a bacterium that normally lives in the intestines of mammals. At that time, biologists already knew that T2, like many other viruses, was composed almost entirely of DNA and protein. They also knew that the T2 phage could quickly turn an *E. coli* cell into a T2-producing factory that released many copies when the cell ruptured. Somehow, T2 could reprogram its host cell to produce viruses. But which viral component—protein or DNA—was responsible?

Hershey and Chase answered this question by devising an experiment showing that only one of the two components of T2 actually enters the *E. coli* cell during infection **(Figure 16.4)**. In preparation for their experiment, they used different radioactive isotopes to tag phage DNA and protein. First, they grew T2 with *E. coli* in the presence of radioactive sulfur.

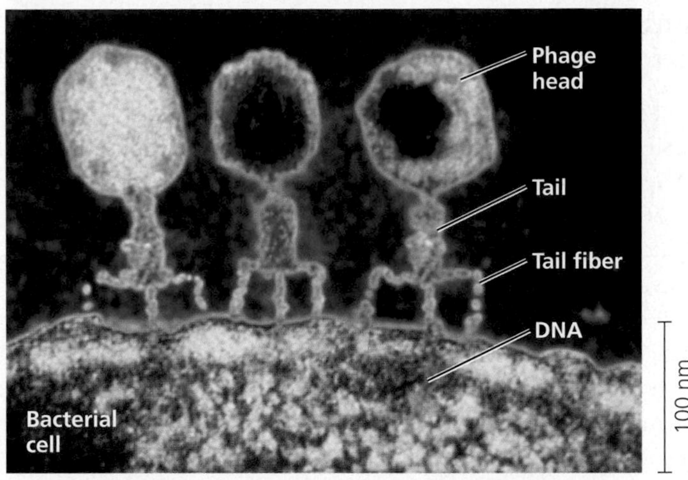

▲ **Figure 16.3 Viruses infecting a bacterial cell.** T2 and related phages attach to the host cell and inject their genetic material (colorized TEM).

Because protein, but not DNA, contains sulfur, the radioactive atoms were incorporated only into the protein of the phage. Next, in a similar way, the DNA of a separate batch of T2 was labeled with atoms of radioactive phosphorus; because nearly all the phage's phosphorus is in its DNA, this procedure left the phage protein unlabeled. In the experiment, the protein-labeled and DNA-labeled batches of T2 were each allowed to infect separate samples of nonradioactive *E. coli* cells. Shortly after the onset of infection, the cultures were whirled in a blender to shake loose any parts of the phages that remained outside the bacterial cells. The mixtures were then spun in a centrifuge, forcing the bacterial cells to form a pellet at the bottom of the centrifuge tubes, but allowing free phages and parts of phages, which are lighter, to remain suspended in the liquid, or supernatant. The scientists then measured the radioactivity in the pellet and in the supernatant.

Figure 16.4

Inquiry Is DNA or protein the genetic material of phage T2?

EXPERIMENT In their famous 1952 experiment, Alfred Hershey and Martha Chase used radioactive sulfur and phosphorus to trace the fates of the protein and DNA, respectively, of T2 phages that infected bacterial cells.

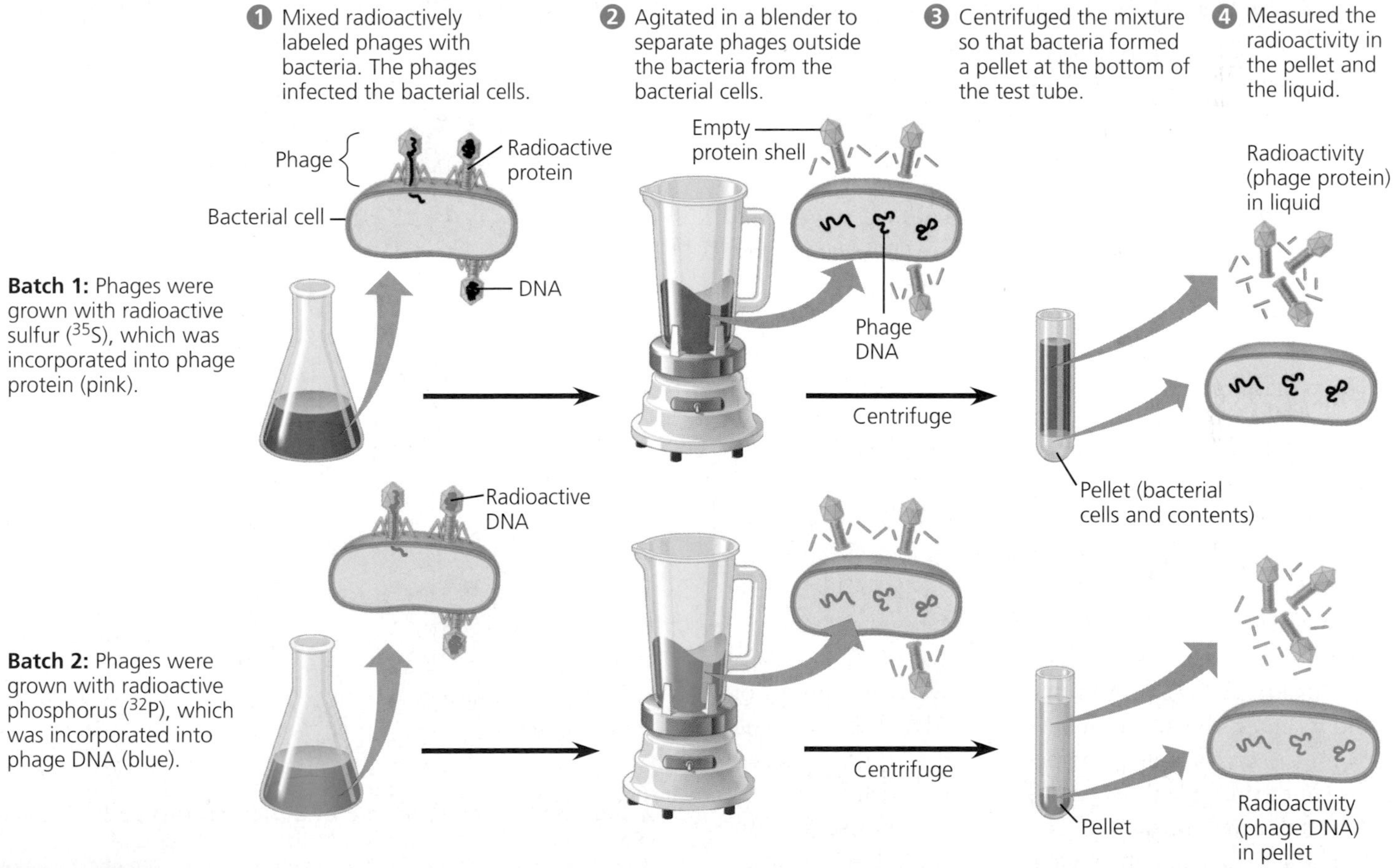

1. Mixed radioactively labeled phages with bacteria. The phages infected the bacterial cells.
2. Agitated in a blender to separate phages outside the bacteria from the bacterial cells.
3. Centrifuged the mixture so that bacteria formed a pellet at the bottom of the test tube.
4. Measured the radioactivity in the pellet and the liquid.

Phage
Radioactive protein
Bacterial cell
DNA
Empty protein shell
Phage DNA
Centrifuge
Radioactivity (phage protein) in liquid
Pellet (bacterial cells and contents)

Batch 1: Phages were grown with radioactive sulfur (^{35}S), which was incorporated into phage protein (pink).

Radioactive DNA
Centrifuge
Pellet
Radioactivity (phage DNA) in pellet

Batch 2: Phages were grown with radioactive phosphorus (^{32}P), which was incorporated into phage DNA (blue).

RESULTS Phage proteins remained outside the bacterial cells during infection, while phage DNA entered the cells. When cultured, bacterial cells with radioactive phage DNA released new phages with some radioactive phosphorus.

CONCLUSION Hershey and Chase concluded that DNA, not protein, functions as the T2 phage's genetic material.

Hershey and Chase found that when the bacteria had been infected with the T2 phage containing radioactively labeled proteins, most of the radioactivity was found in the supernatant, which contained phage particles (but not bacteria). This result suggested that the protein of the phage did not enter the host cells. But when the bacteria had been infected with the T2 phage containing radioactively labeled DNA, most of the radioactivity was found in the pellet, which contained the host bacteria. This result suggested that the phage DNA entered the host cells. Moreover, when these bacteria were returned to a culture medium, the infection ran its course, and the *E. coli* released phages that contained some radioactive phosphorus.

Hershey and Chase concluded that the DNA of the virus is injected into the host cell during infection, leaving the protein outside. The injected DNA provides genetic information that makes the cells produce new viral DNA and proteins, which assemble into new viruses. Thus, the Hershey-Chase experiment provided powerful evidence that nucleic acids, rather than proteins, are the hereditary material, at least for viruses.

Additional Evidence That DNA Is the Genetic Material

Further evidence that DNA is the genetic material came from the laboratory of biochemist Erwin Chargaff. It was already known that DNA is a polymer of nucleotides, each consisting of three components: a nitrogenous (nitrogen-containing) base, a pentose sugar called deoxyribose, and a phosphate group **(Figure 16.5)**. The base can be adenine (A), thymine (T), guanine (G), or cytosine (C). Chargaff analyzed the base composition of DNA from a number of different organisms. In 1947, he reported that DNA composition varies from one species to another. For example, 30.3% of human DNA nucleotides have the base A, whereas DNA from the bacterium *E. coli* has only 26.0% A. This evidence of molecular diversity among species, which had been presumed absent from DNA, made DNA a more credible candidate for the genetic material.

Chargaff also found a peculiar regularity in the ratios of nucleotide bases within a single species. In the DNA of each species he studied, the number of adenines approximately equaled the number of thymines, and the number of guanines approximately equaled the number of cytosines. In human DNA, for example, the four bases are present in these percentages: A = 30.3% and T = 30.3%; G = 19.5% and C = 19.9%. The equivalences for any given species between the number of A and T bases and the number of G and C bases became known as *Chargaff's rules*. The basis for these rules remained unexplained until the discovery of the double helix.

Additional circumstantial evidence was consistent with DNA being the genetic material in eukaryotes. Prior to mitosis, a eukaryotic cell exactly doubles its DNA content, and during mitosis, this DNA is distributed equally to the two daughter cells. Also, in a given species, a diploid set of chromosomes has twice as much DNA as the haploid set.

Building a Structural Model of DNA: *Scientific Inquiry*

Once most biologists were convinced that DNA was the genetic material, the challenge was to determine how the structure of DNA could account for its role in inheritance. By the early 1950s, the arrangement of covalent bonds in a nucleic acid polymer was well established (see Figure 16.5), and researchers focused on discovering the three-dimensional structure of DNA. Among the scientists working on the problem were Linus Pauling, in California, and Maurice Wilkins and Rosalind Franklin, in London. First to come up with the correct answer, however, were two scientists who were relatively unknown at the time—the American James Watson and the Englishman Francis Crick.

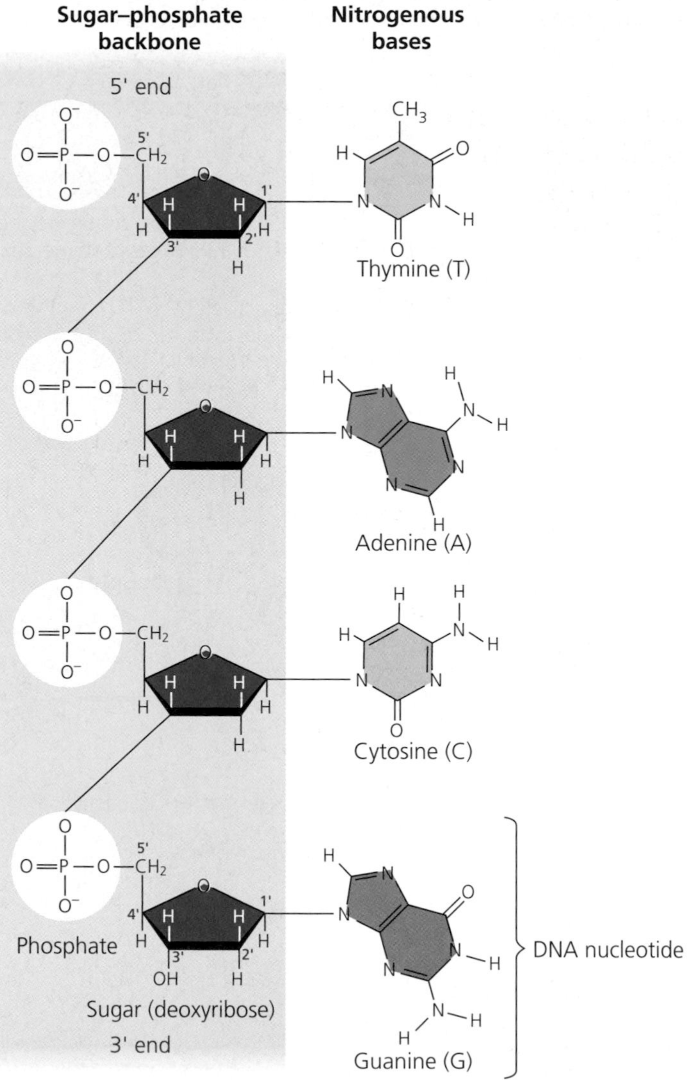

▲ **Figure 16.5 The structure of a DNA strand.** Each nucleotide (monomer) consists of a nitrogenous base (T, A, C, or G), the sugar deoxyribose (blue), and a phosphate group (yellow). The phosphate of one nucleotide is attached to the sugar of the next, resulting in a "backbone" of alternating phosphates and sugars from which the bases project. The polynucleotide strand has directionality, from the 5′ end (with the phosphate group) to the 3′ end (with the —OH group). 5′ and 3′ refer to the numbers assigned to the carbons in the sugar ring.

Concept 16.2

Many proteins work together in DNA replication and repair

The relationship between structure and function is manifest in the double helix. The idea that there is specific pairing of nitrogenous bases in DNA was the flash of inspiration that led Watson and Crick to the correct double helix. At the same time, they saw the functional significance of the base-pairing rules. They ended their classic paper with this wry statement: "It has not escaped our notice that the specific pairing we have postulated immediately suggests a possible copying mechanism for the genetic material." In this section, you will learn about the basic principle of DNA replication, as well as some important details of the process.

The Basic Principle: Base Pairing to a Template Strand

In a second paper, Watson and Crick stated their hypothesis for how DNA replicates:

> Now our model for deoxyribonucleic acid is, in effect, a pair of templates, each of which is complementary to the other. We imagine that prior to duplication the hydrogen bonds are broken, and the two chains unwind and separate. Each chain then acts as a template for the formation onto itself of a new companion chain, so that eventually we shall have two pairs of chains, where we only had one before. Moreover, the sequence of the pairs of bases will have been duplicated exactly.*

Figure 16.9 illustrates Watson and Crick's basic idea. To make it easier to follow, we show only a short section of double helix in untwisted form. Notice that if you cover one of the two DNA strands of Figure 16.9a, you can still determine its linear sequence of bases by referring to the uncovered strand and applying the base-pairing rules. The two strands are complementary; each stores the information necessary to reconstruct the other. When a cell copies a DNA molecule, each strand serves as a template for ordering nucleotides into a new, complementary strand. As nucleotides line up along the template strand according to the base-pairing rules, they are linked to form the new strands. Where there was one double-stranded DNA molecule at the beginning of the process, there are soon two, each an exact replica of the "parent" molecule. The copying mechanism is analogous to using a photographic negative to make a positive image, which can in turn be used to make another negative, and so on.

This model of DNA replication remained untested for several years following publication of the DNA structure. The requisite experiments were simple in concept but difficult to perform. Watson and Crick's model predicts that when a double helix replicates, each of the two daughter molecules will have one old strand, derived from the parent molecule, and one newly made strand. This **semiconservative model** can be distinguished from a conservative model of replication, in which the parent molecule somehow re-forms after the process (that is, it is conserved). In yet a third model, called the dispersive model, all four strands of DNA following replication have a mixture of old and new DNA (**Figure 16.10**, on the next page). Although mechanisms for conservative or dispersive DNA replication are not easy to devise, these models remained possibilities until they could be ruled out. Finally, in the late 1950s, Matthew Meselson and Franklin Stahl devised experiments that tested the three hypotheses. Their experiments

*F. H. C. Crick and J. D. Watson, "The Complementary Structure of Deoxyribonucleic Acid." *Proc. Roy. Soc.* (A) 223 (1954): 80.

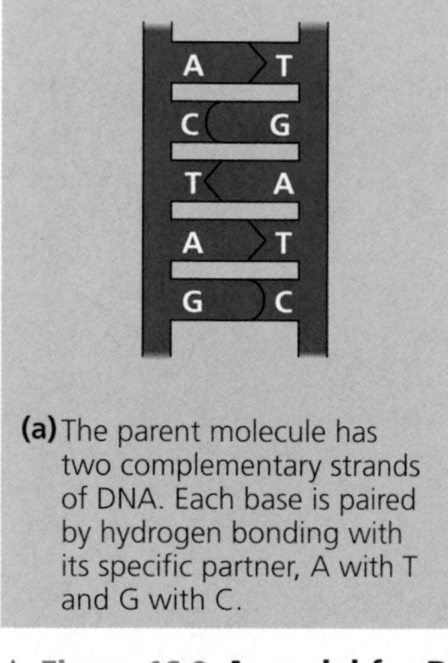

(a) The parent molecule has two complementary strands of DNA. Each base is paired by hydrogen bonding with its specific partner, A with T and G with C.

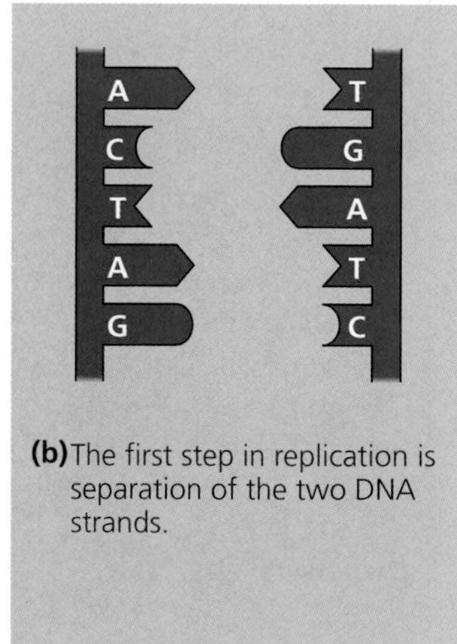

(b) The first step in replication is separation of the two DNA strands.

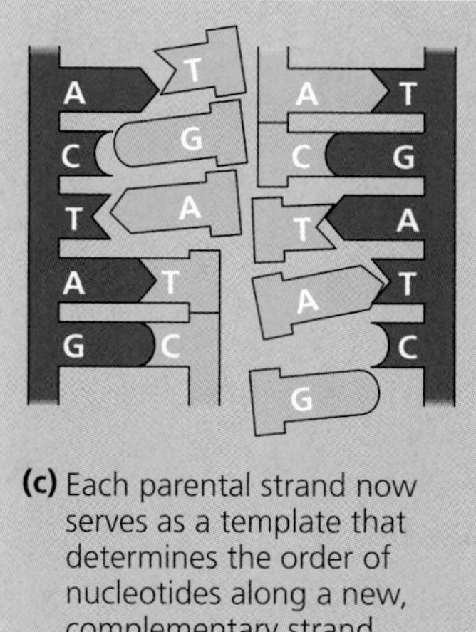

(c) Each parental strand now serves as a template that determines the order of nucleotides along a new, complementary strand.

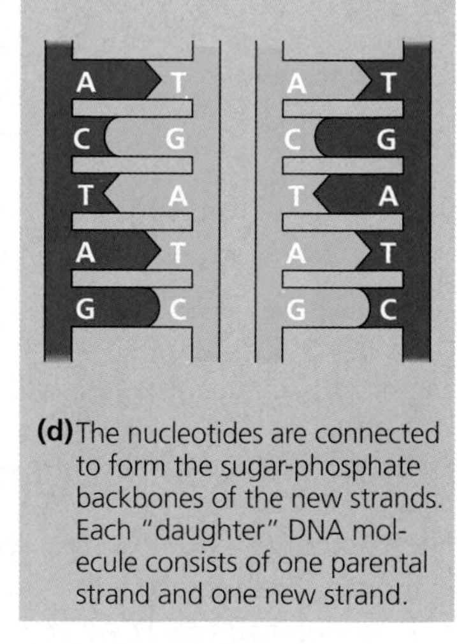

(d) The nucleotides are connected to form the sugar-phosphate backbones of the new strands. Each "daughter" DNA molecule consists of one parental strand and one new strand.

▲ **Figure 16.9 A model for DNA replication: the basic concept.** In this simplification, a short segment of DNA has been untwisted into a structure that resembles a ladder. The rails of the ladder are the sugar-phosphate backbones of the two DNA strands; the rungs are the pairs of nitrogenous bases. Simple shapes symbolize the four kinds of bases. Dark blue represents DNA strands present in the parent molecule; light blue represents free nucleotides and newly synthesized DNA.

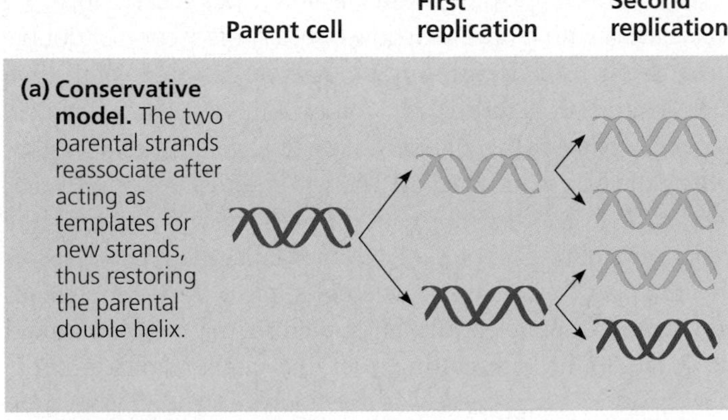

Parent cell
First replication
Second replication
(a) Conservative model. The two parental strands reassociate after acting as templates for new strands, thus restoring the parental double helix.

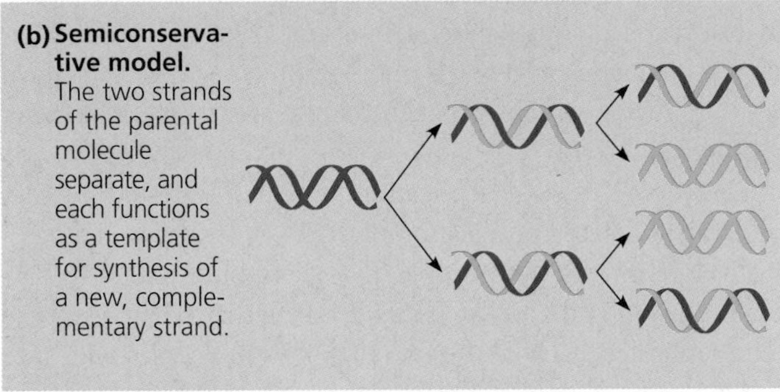

(b) Semiconservative model. The two strands of the parental molecule separate, and each functions as a template for synthesis of a new, complementary strand.

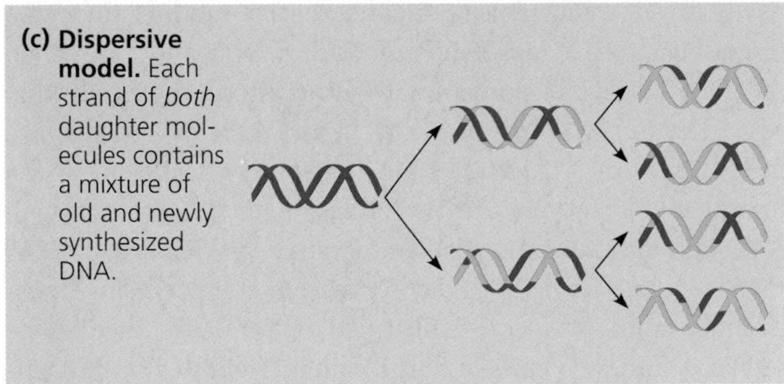

(c) Dispersive model. Each strand of *both* daughter molecules contains a mixture of old and newly synthesized DNA.

▲ **Figure 16.10 Three alternative models of DNA replication.** The short segments of double helix here symbolize the DNA within a cell. Beginning with a parent cell, we follow the DNA for two generations of cells—two rounds of DNA replication. Newly made DNA is light blue.

supported the semiconservative model of DNA replication, as predicted by Watson and Crick **(Figure 16.11)**.

The basic principle of DNA replication is elegantly simple. However, the actual process involves some complicated biochemical gymnastics, as we will now see.

DNA Replication: *A Closer Look*

The bacterium *E. coli* has a single chromosome of about 4.6 million nucleotide pairs. In a favorable environment, an *E. coli* cell can copy all this DNA and divide to form two genetically identical daughter cells in less than an hour. Each of *your* cells has 46 DNA molecules in its nucleus, one long double helical molecule per chromosome. In all, that represents about

Figure 16.11

Inquiry Does DNA replication follow the conservative, semiconservative, or dispersive model?

EXPERIMENT Matthew Meselson and Franklin Stahl cultured *E. coli* bacteria for several generations on a medium containing nucleotide precursors labeled with a heavy isotope of nitrogen, ^{15}N. The bacteria incorporated the heavy nitrogen into their DNA. The scientists then transferred the bacteria to a medium with only ^{14}N, the lighter, more common isotope of nitrogen. Any new DNA that the bacteria synthesized would be lighter than the parental DNA made in the ^{15}N medium. Meselson and Stahl could distinguish DNA of different densities by centrifuging DNA extracted from the bacteria.

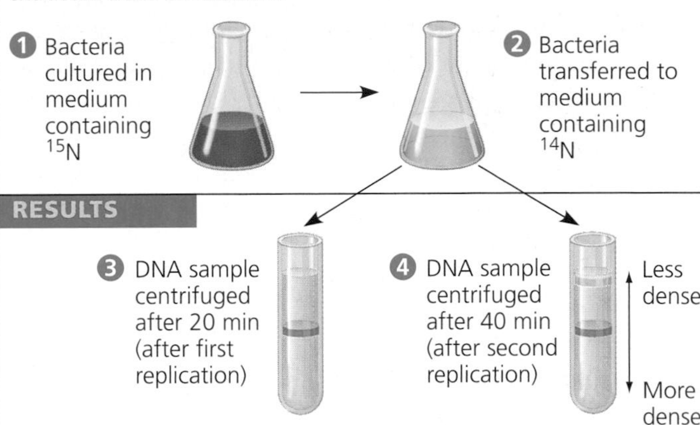

The bands in these two centrifuge tubes represent the results of centrifuging two DNA samples from the flask in step 2, one sample taken after 20 minutes and one after 40 minutes.

CONCLUSION Meselson and Stahl concluded that DNA replication follows the semiconservative model by comparing their result to the results predicted by each of the three models in Figure 16.10. The first replication in the ^{14}N medium produced a band of hybrid (^{15}N–^{14}N) DNA. This result eliminated the conservative model. A second replication produced both light and hybrid DNA, a result that eliminated the dispersive model and supported the semiconservative model.

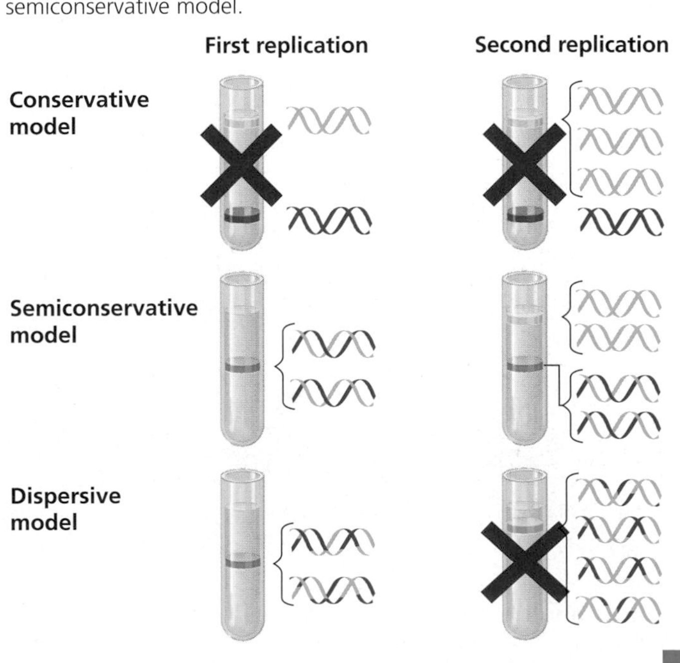

6 billion base pairs, or over a thousand times more DNA than is found in a bacterial cell. If we were to print the one-letter symbols for these bases (A, G, C, and T) the size of the letters you are now reading, the 6 billion bases of information in a diploid human cell would fill about 1,200 books as thick as this text. Yet it takes a cell just a few hours to copy all this DNA. This replication of an enormous amount of genetic information is achieved with very few errors—only about one per 10 billion nucleotides. The copying of DNA is remarkable in its speed and accuracy.

More than a dozen enzymes and other proteins participate in DNA replication. Much more is known about how this "replication machine" works in bacteria than in eukaryotes, and we will describe the basic steps of the process for *E. coli,* except where otherwise noted. What scientists have learned about eukaryotic DNA replication suggests, however, that most of the process is fundamentally similar for prokaryotes and eukaryotes.

Getting Started: Origins of Replication

The replication of a DNA molecule begins at special sites called **origins of replication.** The bacterial chromosome, which is circular, has a single origin, a stretch of DNA having a specific sequence of nucleotides. Proteins that initiate DNA replication recognize this sequence and attach to the DNA, separating the two strands and opening up a replication "bubble." Replication of DNA then proceeds in both directions until the entire molecule is copied (see Figure 18.14). In contrast to a bacterial chromosome, a eukaryotic chromosome may have hundreds or even thousands of replication origins. Multiple replication bubbles form and eventually fuse, thus speeding up the copying of the very long DNA molecules **(Figure 16.12)**. As in bacteria, eukaryotic DNA replication proceeds in both directions from each origin. At each end of a replication bubble is a **replication fork,** a Y-shaped region where the new strands of DNA are elongating.

Elongating a New DNA Strand

Elongation of new DNA at a replication fork is catalyzed by enzymes called **DNA polymerases.** As individual nucleotides align with complementary nucleotides along a template strand of DNA, DNA polymerase adds them, one by one, to the growing end of the new DNA strand. The rate of elongation is about 500 nucleotides per second in bacteria and 50 per second in human cells. In *E. coli,* two different DNA polymerases are involved in replication: DNA polymerase III and DNA polymerase I. The situation in eukaryotes is more complicated, with at least 11 different DNA polymerases discovered so far; however, the general principles are the same.

Each nucleotide that is added to a growing DNA strand is actually a nucleoside triphosphate, which is a nucleoside (a sugar and a base) with three phosphate groups. You have already encountered such a molecule—ATP (adenosine triphosphate; see Figure 8.8). The only difference between the ATP of energy metabolism and the nucleoside triphosphate that supplies adenine to DNA is the sugar component, which is deoxyribose in the building block of DNA, but ribose in ATP. Like ATP, the triphosphate monomers used for DNA synthesis are chemically reactive, partly because their triphosphate tails have an unstable cluster of negative charge. As each monomer joins the growing end of a DNA strand, it loses two phosphate groups as a molecule of pyrophosphate Ⓟ–Ⓟ$_i$. Subsequent hydrolysis of the pyrophosphate to two molecules of inorganic

❶ Replication begins at specific sites where the two parental strands separate and form replication bubbles.

❷ The bubbles expand laterally, as DNA replication proceeds in both directions.

❸ Eventually, the replication bubbles fuse, and synthesis of the daughter strands is complete.

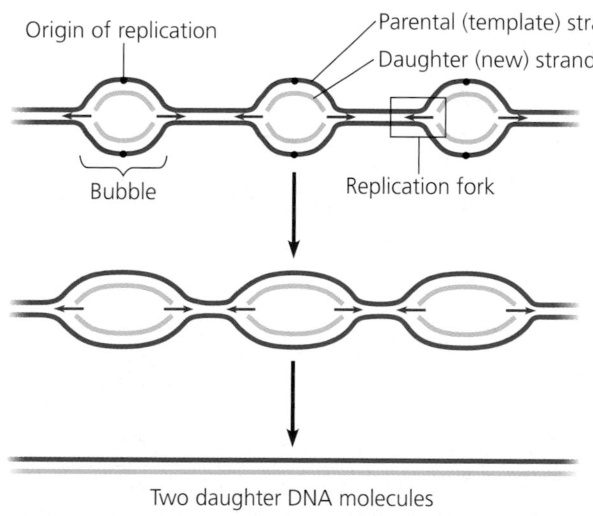

(a) In eukaryotes, DNA replication begins at many sites along the giant DNA molecule of each chromosome.

0.25 μm

(b) In this micrograph, three replication bubbles are visible along the DNA of a cultured Chinese hamster cell (TEM).

▲ **Figure 16.12 Origins of replication in eukaryotes.** The red arrows indicate the movement of the replication forks and thus the overall directions of DNA replication within each bubble.

▶ **Figure 16.13 Incorporation of a nucleotide into a DNA strand.** DNA polymerase catalyzes the addition of a nucleoside triphosphate to the 3′ end of a growing DNA strand.

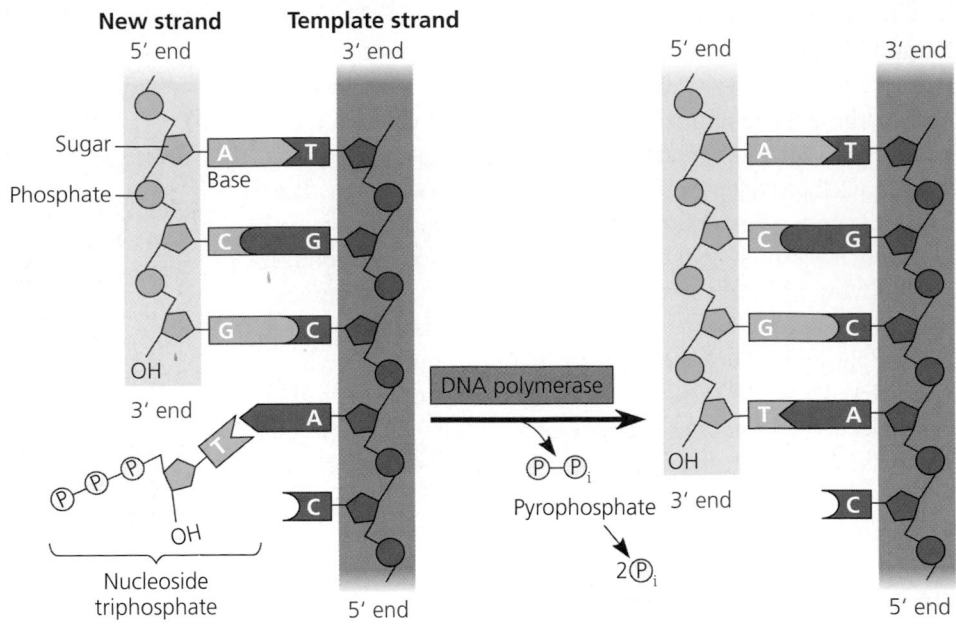

phosphate Ⓟ$_i$ is the exergonic reaction that drives the polymerization reaction **(Figure 16.13)**.

Antiparallel Elongation

As we have noted throughout this chapter, the two ends of a DNA strand are different (see Figure 16.5). In addition, the two strands of DNA in a double helix are antiparallel, meaning that they are oriented in opposite directions to each other (see Figure 16.13). Clearly, the two new strands formed during DNA replication must also be antiparallel to their template strands.

How does the antiparallel structure of the double helix affect replication? DNA polymerases add nucleotides only to the free 3′ end of a growing DNA strand, never to the 5′ end (see Figure 16.13). Thus, a new DNA strand can elongate only in the 5′→3′ direction. With this in mind, let's examine a replication fork **(Figure 16.14)**. Along one template strand, DNA polymerase III (abbreviated DNA pol III) can synthesize a complementary strand continuously by elongating the new DNA in the mandatory 5′→3′ direction. DNA pol III simply nestles in the replication fork on that template strand and continuously adds nucleotides to the complementary strand as the fork progresses. The DNA strand made by this mechanism is called the **leading strand.**

To elongate the other new strand of DNA in the mandatory 5′→3′ direction, DNA pol III must work along the other template strand in the direction *away from* the replication fork. The DNA strand synthesized in this direction is called the **lagging strand.*** In contrast to the leading strand, which elongates continuously, the lagging strand is synthesized as a series of segments. Once a replication bubble opens far enough, a DNA pol III molecule attaches to the lagging strand's template and moves away from the replication fork, synthesizing a short segment of DNA. As the bubble grows, another segment of the lagging strand can be made in a similar way. These segments of the lagging strand are called **Okazaki fragments,** after the Japanese scientist who

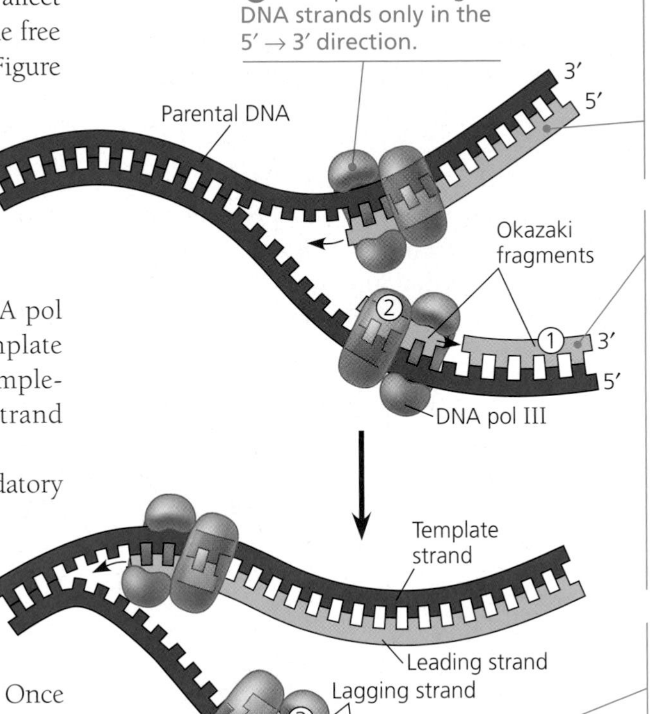

❷ One new strand, the leading strand, can elongate continuously 5′ → 3′ as the replication fork progresses.

❸ The other new strand, the lagging strand, must grow in an overall 3′ → 5′ direction by addition of short segments, Okazaki fragments, that grow 5′ → 3′ (numbered here in the order they were made).

❹ DNA ligase joins Okazaki fragments by forming a bond between their free ends. This results in a continuous strand.

▲ **Figure 16.14 Synthesis of leading and lagging strands during DNA replication.** DNA polymerase III (DNA pol III) is closely associated with a protein that encircles the newly synthesized double helix like a doughnut. Note that Okazaki fragments are actually much longer than the ones shown here. In this figure, we depict only five bases per fragment for simplicity.

*Synthesis of the leading strand and synthesis of the lagging strand occur concurrently and at the same rate. The lagging strand is so named because its synthesis is slightly delayed relative to synthesis of the leading strand; each new fragment cannot be started until enough template has been exposed at the replication fork.

discovered them. The fragments are about 1,000 to 2,000 nucleotides long in *E. coli* and 100 to 200 nucleotides long in eukaryotes. Another enzyme, **DNA ligase**, eventually joins (ligates) the sugar-phosphate backbones of the Okazaki fragments, forming a single new DNA strand.

Priming DNA Synthesis

DNA polymerases cannot *initiate* the synthesis of a polynucleotide; they can only add nucleotides to the 3′ end of an already existing chain that is base-paired with the template strand (see Figure 16.13). The initial nucleotide chain is a short one called a **primer.** Primers may consist of either DNA or RNA (the other class of nucleic acid), and in initiating the replication of cellular DNA, the primer is a short stretch of RNA with an available 3′ end. An enzyme called **primase** can start an RNA chain from scratch. Primase joins RNA nucleotides together one at a time, making a primer complementary to the template strand at the location where initiation of the new DNA strand will occur. (Primers are generally 5 to 10 nucleotides long.) DNA pol III then adds a DNA nucleotide to the 3′ end of the RNA primer and continues adding DNA nucleotides to the growing DNA strand according to the base-pairing rules.

Only one primer is required for DNA pol III to begin synthesizing the leading strand. For synthesis of the lagging strand, however, each Okazaki fragment must be primed separately **(Figure 16.15)**. Another DNA polymerase, DNA polymerase I (DNA pol I), replaces the RNA nucleotides of the primers with DNA versions, adding them one by one onto the 3′ end of the adjacent Okazaki fragment (fragment 2 in Figure 16.15). But DNA pol I cannot join the final nucleotide of this replacement DNA segment to the first DNA nucleotide of the Okazaki fragment whose primer was just replaced (fragment 1 in Figure 16.15). DNA ligase accomplishes this task, joining the sugar-phosphate backbones of all the Okazaki fragments into a continuous DNA strand.

Other Proteins That Assist DNA Replication

You have learned about three kinds of proteins that function in DNA synthesis: DNA polymerases, ligase, and primase. Other kinds of proteins also participate, including helicase, topoisomerase, and single-strand binding proteins. **Helicase** is an enzyme that untwists the double helix at the replication forks, separating the two parental strands and making them available as template strands. This untwisting causes tighter twisting and strain ahead of the replication fork, and **topoisomerase** helps relieve this strain. After helicase separates the two parental strands, molecules of **single-strand binding protein** then bind to the unpaired DNA strands, stabilizing them until they serve as templates for the synthesis of new complementary strands.

Table 16.1 and **Figure 16.16**, on the next page, summarize DNA replication. Study them carefully before proceeding.

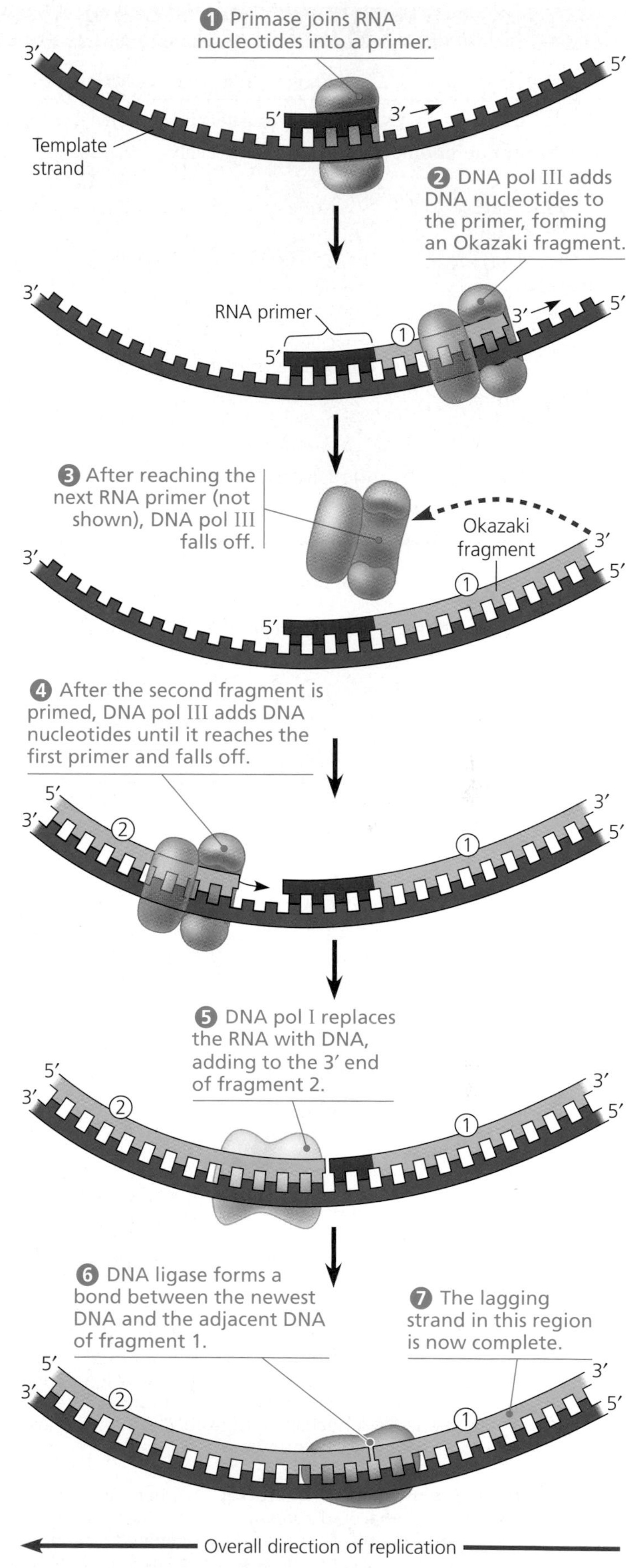
1 Primase joins RNA nucleotides into a primer.
3′
5′
5′
3′
Template strand
2 DNA pol III adds DNA nucleotides to the primer, forming an Okazaki fragment.
RNA primer
1
3 After reaching the next RNA primer (not shown), DNA pol III falls off.
Okazaki fragment
4 After the second fragment is primed, DNA pol III adds DNA nucleotides until it reaches the first primer and falls off.
2
5 DNA pol I replaces the RNA with DNA, adding to the 3′ end of fragment 2.
6 DNA ligase forms a bond between the newest DNA and the adjacent DNA of fragment 1.
7 The lagging strand in this region is now complete.
Overall direction of replication

▲ **Figure 16.15 Synthesis of the lagging strand.**

Table 16.1 Bacterial DNA replication proteins and their functions

Protein	Function for Leading and Lagging Strands	
Helicase	Unwinds parental double helix at replication forks	
Single-strand binding protein	Binds to and stabilizes single-stranded DNA until it can be used as a template	
Topoisomerase	Corrects "overwinding" ahead of replication forks by breaking, swiveling, and rejoining DNA strands	
	Function for Leading Strand	**Function for Lagging Strand**
Primase	Synthesizes a single RNA primer at the 5′ end of the leading strand	Synthesizes an RNA primer at the 5′ end of each Okazaki fragment
DNA pol III	Continuously synthesizes the leading strand, adding on to the primer	Elongates each Okazaki fragment, adding on to its primer
DNA pol I	Removes primer from the 5′ end of leading strand and replaces it with DNA, adding on to the adjacent 3′ end	Removes the primer from the 5′ end of each fragment and replaces it with DNA, adding on to the 3′ end of the adjacent fragment
DNA Ligase	Joins the 3′ end of the DNA that replaces the primer to the rest of the leading strand	Joins the Okazaki fragments

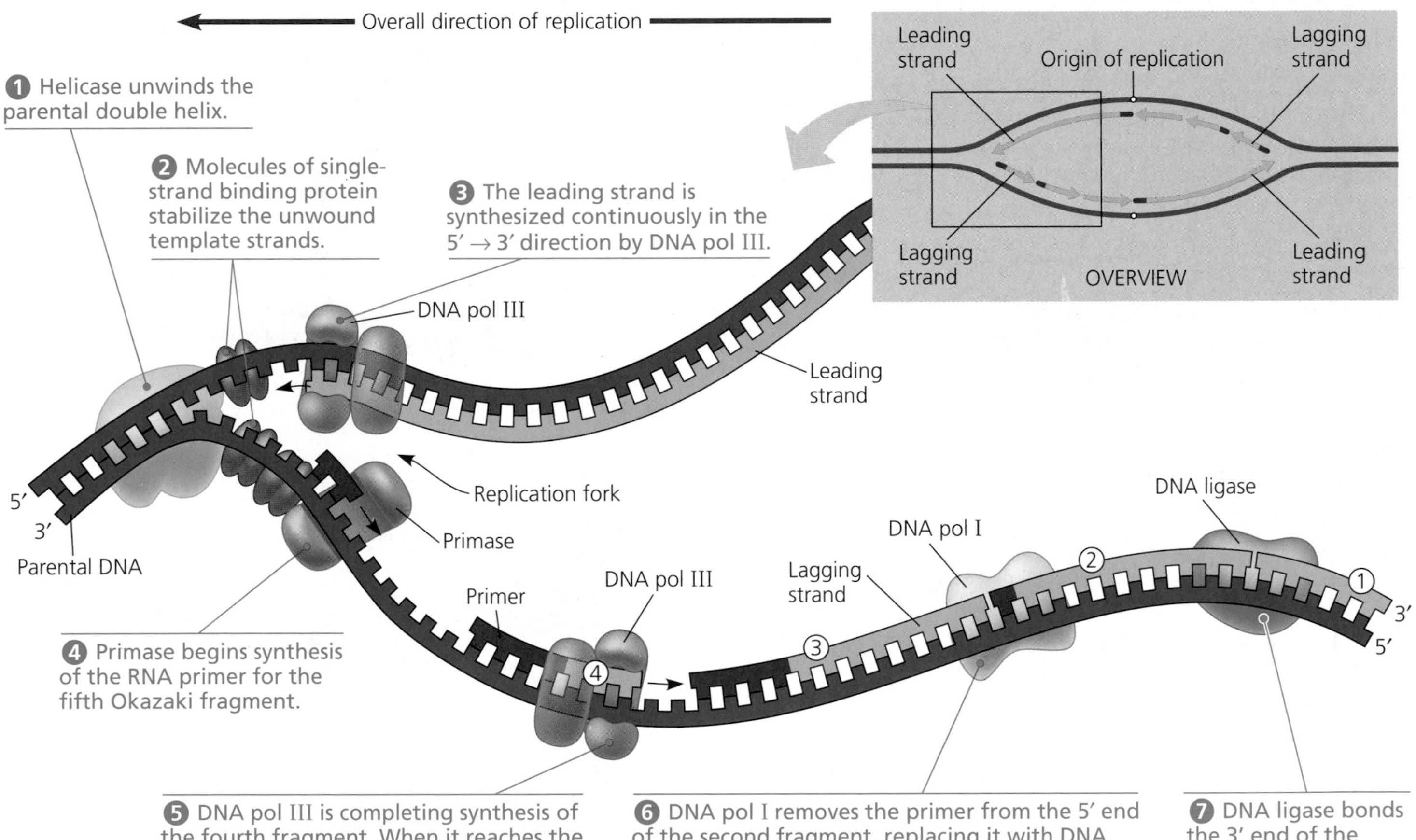

Overall direction of replication
Leading strand
Origin of replication
Lagging strand
Lagging strand
OVERVIEW
Leading strand
1 Helicase unwinds the parental double helix.
2 Molecules of single-strand binding protein stabilize the unwound template strands.
3 The leading strand is synthesized continuously in the 5′ → 3′ direction by DNA pol III.
DNA pol III
Leading strand
Replication fork
Primase
5′
3′
Parental DNA
DNA ligase
DNA pol I
Lagging strand
DNA pol III
Primer
3′
5′
4 Primase begins synthesis of the RNA primer for the fifth Okazaki fragment.
5 DNA pol III is completing synthesis of the fourth fragment. When it reaches the RNA primer on the third fragment, it will dissociate, move to the replication fork, and add DNA nucleotides to the 3′ end of the fifth fragment primer.
6 DNA pol I removes the primer from the 5′ end of the second fragment, replacing it with DNA nucleotides that it adds one by one to the 3′ end of the third fragment. The replacement of the last RNA nucleotide with DNA leaves the sugar-phosphate backbone with a free 3′ end.
7 DNA ligase bonds the 3′ end of the second fragment to the 5′ end of the first fragment.

▲ **Figure 16.16 A summary of bacterial DNA replication.** The detailed diagram shows one replication fork, but as indicated in the overview diagram, replication usually occurs simultaneously at two forks, one at either end of a replication bubble. Notice in the overview diagram that a leading strand is initiated by an RNA primer (red), as is each Okazaki fragment in a lagging strand. Viewing each daughter strand in its entirety in the overview, you can see that half of it is made continuously as a leading strand, while the other half (on the other side of the origin) is synthesized in fragments as a lagging strand.

The DNA Replication Machine as a Stationary Complex

It is traditional—and convenient—to represent DNA polymerase molecules as locomotives moving along a DNA "railroad track," but such a model is inaccurate in two important ways. First, the various proteins that participate in DNA replication actually form a single large complex, a DNA replication "machine." Many protein-protein interactions facilitate the efficiency of this machine; for example, helicase works much more rapidly when it is in contact with primase. Second, the DNA replication machine is probably stationary during the replication process. In eukaryotic cells, multiple copies of the machine, perhaps grouped into "factories," may anchor to the nuclear matrix, a framework of fibers extending through the interior of the nucleus. Recent studies support a model in which DNA polymerase molecules "reel in" the parental DNA and extrude newly made daughter DNA molecules. Additional evidence suggests that the lagging strand is looped through the complex, so that when a DNA polymerase completes synthesis of an Okazaki fragment and dissociates, it doesn't have far to travel to reach the primer for the next fragment, near the replication fork. This looping of the lagging strand enables more Okazaki fragments to be synthesized in less time.

Proofreading and Repairing DNA

We cannot attribute the accuracy of DNA replication solely to the specificity of base pairing. Although errors in the completed DNA molecule amount to only one in 10 billion nucleotides, initial pairing errors between incoming nucleotides and those in the template strand are 100,000 times more common—an error rate of one in 100,000 base pairs. During DNA replication, DNA polymerases proofread each nucleotide against its template as soon as it is added to the growing strand. Upon finding an incorrectly paired nucleotide, the polymerase removes the nucleotide and then resumes synthesis. (This action is similar to fixing a typing error by using the "delete" key and then entering the correct letter.)

Mismatched nucleotides sometimes evade proofreading by a DNA polymerase or arise after DNA synthesis is completed—by damage to an existing nucleotide base, for instance. In **mismatch repair**, cells use special enzymes to fix incorrectly paired nucleotides. Researchers spotlighted the importance of such enzymes when they found that a hereditary defect in one of them is associated with a form of colon cancer. Apparently, this defect allows cancer-causing errors to accumulate in the DNA at a faster rate than normal.

Maintenance of the genetic information encoded in DNA requires frequent repair of various kinds of damage to existing DNA. DNA molecules are constantly subjected to potentially harmful chemical and physical agents, as we'll discuss in Chapter 17. Reactive chemicals (in the environment and occurring naturally in cells), radioactive emissions, X-rays, and ultraviolet light can change nucleotides in ways that can affect encoded genetic information, usually adversely. In addition, DNA bases often undergo spontaneous chemical changes under normal cellular conditions. However, changes in DNA are usually corrected before they become self-perpetuating mutations. Each cell continuously monitors and repairs its genetic material. Because repair of damaged DNA is so important to the survival of an organism, it is no surprise that many different DNA repair enzymes have evolved. Almost 100 are known in *E. coli,* and about 130 have been identified so far in humans.

Most mechanisms for repairing DNA damage take advantage of the base-paired structure of DNA. Usually, a segment of the strand containing the damage is cut out (excised) by a DNA-cutting enzyme—a **nuclease**—and the resulting gap is filled in with nucleotides properly paired with the nucleotides in the undamaged strand. The enzymes involved in filling the gap are a DNA polymerase and ligase. DNA repair of this type is called **nucleotide excision repair (Figure 16.17)**.

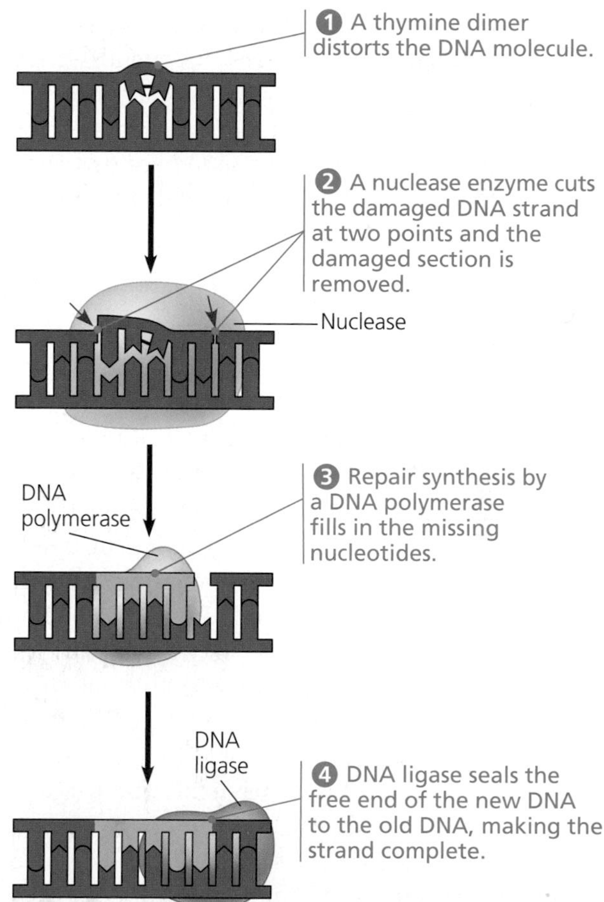

1 A thymine dimer distorts the DNA molecule.
2 A nuclease enzyme cuts the damaged DNA strand at two points and the damaged section is removed.
Nuclease
DNA polymerase
3 Repair synthesis by a DNA polymerase fills in the missing nucleotides.
DNA ligase
4 DNA ligase seals the free end of the new DNA to the old DNA, making the strand complete.

▲ **Figure 16.17 Nucleotide excision repair of DNA damage.** A team of enzymes detects and repairs damaged DNA. This figure shows DNA containing a thymine dimer, a type of damage often caused by ultraviolet radiation. A nuclease enzyme cuts out the damaged region of DNA, and a DNA polymerase (in bacteria, DNA pol I) replaces it with a normal DNA segment. Ligase completes the process by closing the remaining break in the sugar-phosphate backbone.

One function of the DNA repair enzymes in our skin cells is to repair genetic damage caused by the ultraviolet rays of sunlight. One type of damage, the type shown in Figure 16.17, is the covalent linking of thymine bases that are adjacent on a DNA strand. Such thymine dimers cause the DNA to buckle and interfere with DNA replication. The importance of repairing this kind of damage is underscored by the disorder xeroderma pigmentosum, which in most cases is caused by an inherited defect in a nucleotide excision repair enzyme. Individuals with this disorder are hypersensitive to sunlight; mutations in their skin cells caused by ultraviolet light are left uncorrected and cause skin cancer.

Replicating the Ends of DNA Molecules

In spite of the major role played by DNA polymerases in DNA replication and repair, it turns out that there is a small portion of the cell's DNA that DNA polymerases cannot replicate or repair. For linear DNA, such as the DNA of eukaryotic chromosomes, the fact that a DNA polymerase can only add nucleotides to the 3′ end of a preexisting polynucleotide leads to a problem. The usual replication machinery provides no way to complete the 5′ ends of daughter DNA strands. Even if an Okazaki fragment can be started with an RNA primer bound to the very end of the template strand, once that primer is removed, it cannot be replaced with DNA, because there is no 3′ end onto which DNA polymerase can add DNA nucleotides **(Figure 16.18)**. As a result, repeated rounds of replication produce shorter and shorter DNA molecules.

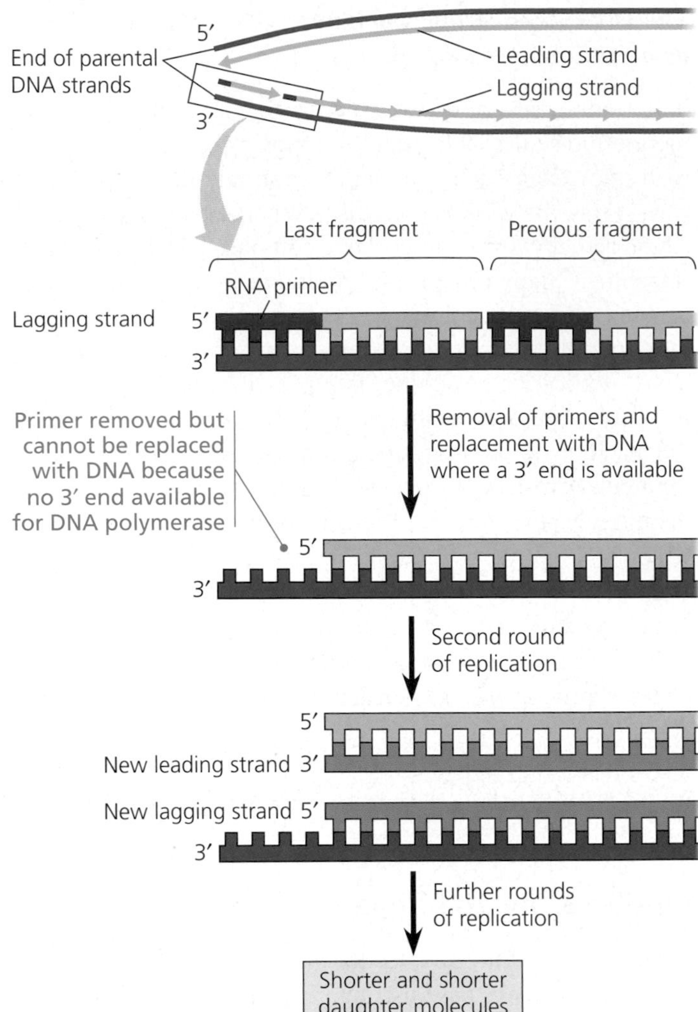

▲ **Figure 16.18 Shortening of the ends of linear DNA molecules.** Here we follow the end of one strand of a DNA molecule through two rounds of replication. After the first round, the new lagging strand is shorter than its template. After a second round, both the leading and lagging strands have become shorter than the original parental DNA. Although not shown here, the other ends of these DNA molecules also become shorter.

Prokaryotes do not have this problem because their DNA is circular (with no ends), but what about eukaryotes? Eukaryotic chromosomal DNA molecules have nucleotide sequences called **telomeres** at their ends **(Figure 16.19)**. Telomeres do not contain genes; instead, the DNA typically consists of multiple repetitions of one short nucleotide sequence. The repeated unit in human telomeres, for example, is the six-nucleotide sequence TTAGGG. The number of repetitions in a telomere varies from about 100 to 1,000. Telomeric DNA protects the organism's genes from being eroded through successive rounds of DNA replication. In addition, telomeric DNA and specific proteins associated with it somehow prevent the staggered ends of the daughter molecule from activating the cell's systems for monitoring DNA damage. (The end of a DNA molecule that is "seen" as a double-strand break may otherwise trigger signal transduction pathways leading to cell cycle arrest or cell death.)

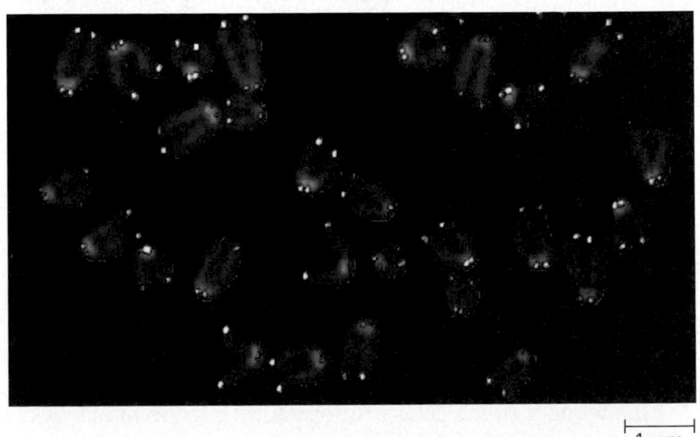

▲ **Figure 16.19 Telomeres.** Eukaryotes have repetitive, noncoding sequences called telomeres at the ends of their DNA, marked in these mouse chromosomes by a bright orange stain (LM).

Telomeres do not prevent the shortening of DNA molecules due to successive rounds of replication; they just postpone the erosion of genes near the ends of DNA molecules. As shown in Figure 16.18, telomeres become shorter during every round of replication. As we would expect, telomeric DNA does tend to be shorter in dividing somatic cells of older individuals and in cultured cells that have divided many times. It has been proposed that shortening of telomeres is somehow connected to the aging process of certain tissues and even to aging of the organism as a whole.

But what about the cells whose genomes persist unchanged from an organism to its offspring over many generations? If the chromosomes of germ cells (which give rise to gametes) became shorter in every cell cycle, essential genes would eventually be missing from the gametes they produce. However, this does not occur: An enzyme called **telomerase** catalyzes the lengthening of telomeres in eukaryotic germ cells, thus restoring their original length and compensating for the shortening that occurs during DNA replication. The lengthening process is made possible by the presence, in telomerase, of a short molecule of RNA that serves as a template for new telomere segments. Telomerase is not active in most somatic cells, but its activity in germ cells results in telomeres of maximum length in the zygote.

Normal shortening of telomeres may protect organisms from cancer by limiting the number of divisions that somatic cells can undergo. Cells from large tumors often have unusually short telomeres, as one would expect for cells that have undergone many cell divisions. Further shortening would presumably lead to self-destruction of the cancer. Intriguingly, researchers have found telomerase activity in cancerous somatic cells, suggesting that its ability to stabilize telomere length may allow these cancer cells to persist. Many cancer cells do seem capable of unlimited cell division, as do immortal strains of cultured cells (see Chapter 12). If telomerase is indeed an important factor in many cancers, it may provide a useful target for both cancer diagnosis and chemotherapy.

In this chapter, you have learned how DNA replication provides the copies of genes that parents pass to offspring. However, it is not enough that genes be copied and transmitted; they must also be expressed. In the next chapter, we will examine how the cell translates genetic information encoded in DNA.

Concept Check 16.2

1. What role does complementary base pairing play in the replication of DNA?
2. Identify two major functions of DNA pol III in DNA replication.
3. Why is DNA pol I necessary to complete synthesis of a leading strand? Point out in the overview box in Figure 16.16 where DNA pol I would function on the top leading strand.
4. How are telomeres important for preserving eukaryotic genes?

For suggested answers, see Appendix A.

Chapter 16 Review

Go to www.campbellbiology.com or the student CD-ROM to explore Activities, Investigations, and other interactive study aids.

SUMMARY OF KEY CONCEPTS

Concept 16.1

DNA is the genetic material

► **The Search for the Genetic Material: *Scientific Inquiry* (pp. 293–296)** Experiments with bacteria and with phages provided the first strong evidence that the genetic material is DNA.
Activity ***The Hershey-Chase Experiment***

► **Building a Structural Model of DNA: *Scientific Inquiry* (pp. 296–298)** Watson and Crick deduced that DNA is a double helix. Two antiparallel sugar-phosphate chains wind around the outside of the molecule; the nitrogenous bases project into the interior, where they hydrogen-bond in specific pairs, A with T and G with C.
Activity ***DNA and RNA Structure***
Activity ***DNA Double Helix***

Many proteins work together in DNA replication and repair

► **The Basic Principle: Base Pairing to a Template Strand (pp. 299–300)** DNA replication is semiconservative: The parent molecule unwinds, and each strand then serves as a template for the synthesis of a new strand according to base-pairing rules.
Activity ***DNA Replication: An Overview***
Investigation ***What Is the Correct Model for DNA Replication?***

► **DNA Replication: *A Closer Look* (pp. 300–305)** DNA replication begins at origins of replication. Y-shaped replication forks form at opposite ends of a replication bubble, where the two DNA strands separate. DNA synthesis starts at the 3′ end of an RNA primer, a short polynucleotide complementary to the template strand. DNA polymerases catalyze the synthesis of new DNA strands, working in the 5′→3′ direction. The leading strand is synthesized continuously, and the lagging strand is synthesized in short segments, called Okazaki fragments. The fragments are joined together by DNA ligase.
Activity ***DNA Replication: A Closer Look***
Activity ***DNA Replication: A Review***

► **Proofreading and Repairing DNA (pp. 305–306)** DNA polymerases proofread newly made DNA, replacing any incorrect nucleotides. In mismatch repair of DNA, repair enzymes correct errors in base pairing. In nucleotide excision repair, enzymes cut out and replace damaged stretches of DNA.

► **Replicating the Ends of DNA Molecules (pp. 306–307)** The ends of eukaryotic chromosomal DNA get shorter with each round of replication. The presence of telomeres, repetitive sequences at the ends of linear DNA molecules, postpones the erosion of genes. Telomerase catalyzes the lengthening of telomeres in germ cells.

TESTING YOUR KNOWLEDGE

Self-Quiz

1. In his work with pneumonia-causing bacteria and mice, Griffith found that
   a. the protein coat from pathogenic cells was able to transform nonpathogenic cells.
   b. heat-killed pathogenic cells caused pneumonia.
   c. some substance from pathogenic cells was transferred to nonpathogenic cells, making them pathogenic.
   d. the polysaccharide coat of bacteria caused pneumonia.
   e. bacteriophages injected DNA into bacteria.

2. *E. coli* cells grown on ^{15}N medium are transferred to ^{14}N medium and allowed to grow for two more generations (two rounds of DNA replication). DNA extracted from these cells is centrifuged. What density distribution of DNA would you expect in this experiment?
   a. one high-density and one low-density band
   b. one intermediate-density band
   c. one high-density and one intermediate-density band
   d. one low-density and one intermediate-density band
   e. one low-density band

3. A biochemist isolates and purifies molecules needed for DNA replication. When she adds some DNA, replication occurs, but each DNA molecule consists of a normal strand paired with numerous segments of DNA a few hundred nucleotides long. What has she probably left out of the mixture?
   a. DNA polymerase
   b. DNA ligase
   c. nucleotides
   d. Okazaki fragments
   e. primase

4. What is the basis for the difference in how the leading and lagging strands of DNA molecules are synthesized?
   a. The origins of replication occur only at the 5′ end.
   b. Helicases and single-strand binding proteins work at the 5′ end.
   c. DNA polymerase can join new nucleotides only to the 3′ end of a growing strand.
   d. DNA ligase works only in the 3′→5′ direction.
   e. Polymerase can work on only one strand at a time.

5. In analyzing the number of different bases in a DNA sample, which result would be consistent with the base-pairing rules?
   a. $A = G$
   b. $A + G = C + T$
   c. $A + T = G + T$
   d. $A = C$
   e. $G = T$

6. Synthesis of a new DNA strand usually begins with
   a. an RNA primer.
   b. a DNA primer.
   c. an Okazaki fragment.
   d. DNA ligase.
   e. a thymine dimer.

7. A eukaryotic cell lacking active telomerase would
   a. be unable to take up DNA from the surrounding solution.
   b. be unable to identify and correct mismatched nucleotides.
   c. experience a gradual reduction of chromosome length with each replication cycle.
   d. have a greater potential to become cancerous.
   e. be unable to connect Okazaki fragments.

8. The elongation of the leading strand during DNA synthesis
   a. progresses away from the replication fork.
   b. occurs in the 3′→5′ direction.
   c. produces Okazaki fragments.
   d. depends on the action of DNA polymerase.
   e. does not require a template strand.

9. The spontaneous loss of amino groups from adenine results in hypoxanthine, an unnatural base, opposite thymine in DNA. What combination of molecules could repair such damage?
   a. nuclease, DNA polymerase, DNA ligase
   b. telomerase, primase, DNA polymerase
   c. telomerase, helicase, single-strand binding protein
   d. DNA ligase, replication fork proteins, adenylyl cyclase
   e. nuclease, telomerase, primase

10. The most reasonable inference from the observation that defects in DNA repair enzymes contribute to some cancers is that
    a. cancer is generally inherited.
    b. uncorrected changes in DNA can lead to cancer.
    c. cancer cannot occur when repair enzymes work properly.
    d. mutations generally lead to cancer.
    e. cancer is caused by environmental factors that damage DNA repair enzymes.

For Self-Quiz answers, see Appendix A.

Go to the website or CD-ROM for more quiz questions.

Evolution Connection

Many bacteria may be able to respond to environmental stress by increasing the rate at which mutations occur during cell division. How might this be accomplished, and what might be an evolutionary advantage of this ability?

Scientific Inquiry

Demonstrate your understanding of the Meselson-Stahl experiment by answering the following questions.

a. Describe in your own words exactly what each of the centrifugation bands pictured in Figure 16.11 represents.
b. Imagine that the experiment is done as follows: Bacteria are first grown for several generations in a medium containing the *lighter* isotope of nitrogen, ^{14}N, then switched into a medium containing ^{15}N. The rest of the experiment is identical. Redraw Figure 16.11 to reflect this experiment, predicting what band positions you would expect after one generation and after two generations if each of the three models shown in Figure 16.10 were true.

Investigation ***What Is the Correct Model for DNA Replication?***

Science, Technology, and Society

Cooperation and competition are both common in science. What roles did these two social behaviors play in Watson and Crick's discovery of the double helix? How might competition between scientists accelerate progress? How might it slow progress?

17 From Gene to Protein

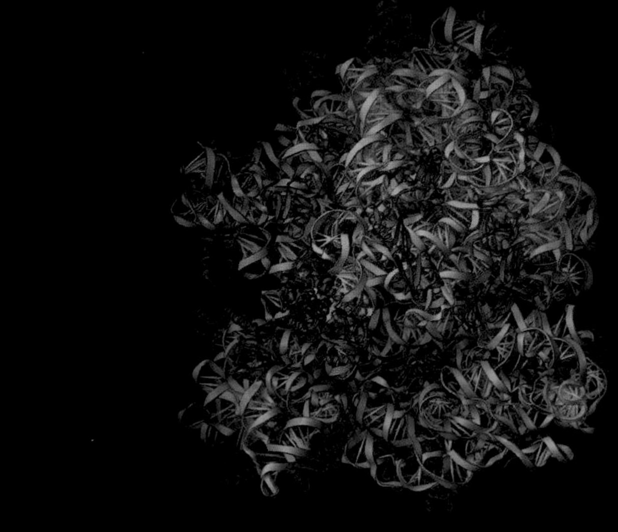

▲ **Figure 17.1 A computer model of a ribosome, part of the protein synthesis machinery.**

Key Concepts

17.1 Genes specify proteins via transcription and translation

17.2 Transcription is the DNA-directed synthesis of RNA: *a closer look*

17.3 Eukaryotic cells modify RNA after transcription

17.4 Translation is the RNA-directed synthesis of a polypeptide: *a closer look*

17.5 RNA plays multiple roles in the cell: *a review*

17.6 Comparing gene expression in prokaryotes and eukaryotes reveals key differences

17.7 Point mutations can affect protein structure and function

Overview

The Flow of Genetic Information

The information content of DNA, the genetic material, is in the form of specific sequences of nucleotides along the DNA strands. But how does this information determine an organism's traits? Put another way, what does a gene actually say? And how is its message translated by cells into a specific trait, such as brown hair or type A blood?

Consider, once again, Mendel's peas. One of the characters Mendel studied was stem length (see Table 14.1). Mendel did not know the physiological basis for the difference between the tall and dwarf varieties of pea plants, but plant scientists have since worked out the explanation: Dwarf peas lack growth hormones called gibberellins, which stimulate the normal elongation of stems. A dwarf plant treated with gibberellins from an external source grows to normal height. Why do dwarf peas fail to make their own gibberellins? They are missing a key protein, an enzyme required for gibberellin synthesis. And they are missing that protein because they do not have a properly functioning gene for that protein.

This example illustrates the main point of this chapter: The DNA inherited by an organism leads to specific traits by dictating the synthesis of proteins. In other words, proteins are the links between genotype and phenotype. The process by which DNA directs protein synthesis, *gene expression,* includes two stages, called transcription and translation. In **Figure 17.1**, you can see a computer model of a ribosome, which is part of the cellular machinery for translation—polypeptide synthesis. This chapter describes the flow of information from gene to protein in detail. By the end, you will understand how genetic mutations, such as the one causing the dwarf trait in pea plants, affect organisms through their proteins.

Concept 17.1

Genes specify proteins via transcription and translation

Before going into the details of how genes direct protein synthesis, let's step back and examine how the fundamental relationship between genes and proteins was discovered.

Evidence from the Study of Metabolic Defects

In 1909, British physician Archibald Garrod was the first to suggest that genes dictate phenotypes through enzymes that catalyze specific chemical reactions in the cell. Garrod postulated that the symptoms of an inherited disease reflect a person's inability to make a particular enzyme. He referred to such diseases as "inborn errors of metabolism." Garrod gave as one example the hereditary condition called alkaptonuria, in which the urine is black because it contains the chemical alkapton, which darkens upon exposure to air. Garrod reasoned that most people have an enzyme that breaks down alkapton, whereas people with alkaptonuria have inherited an inability to make the enzyme that metabolizes alkapton.

Garrod's idea was ahead of its time, but research conducted several decades later supported his hypothesis that a gene dictates the production of a specific enzyme. Biochemists accumulated much evidence that cells synthesize and degrade most organic molecules via metabolic pathways, in which each chemical reaction in a sequence is catalyzed by a specific enzyme. Such metabolic pathways lead, for instance, to the synthesis of the pigments that give fruit flies (*Drosophila*) their eye color (see Figure 15.3). In the 1930s, George Beadle and Boris Ephrussi speculated that in *Drosophila,* each of the various mutations affecting eye color blocks pigment synthesis at a specific step by preventing production of the enzyme that catalyzes that step. However, neither the chemical reactions nor the enzymes that catalyze them were known at the time.

Nutritional Mutants in Neurospora: *Scientific Inquiry*

A breakthrough in demonstrating the relationship between genes and enzymes came a few years later, when Beadle and Edward Tatum began working with a bread mold, *Neurospora crassa.* They bombarded *Neurospora* with X-rays and then looked among the survivors for mutants that differed in their nutritional needs from the wild-type mold. Wild-type *Neurospora* has modest food requirements. It can survive in the laboratory on agar (a moist support medium) mixed only with inorganic salts, glucose, and the vitamin biotin. From this *minimal medium,* the mold uses its metabolic pathways to produce all the other molecules it needs. Beadle and Tatum identified mutants that could not survive on minimal medium, apparently because they were unable to synthesize certain essential molecules from the minimal ingredients. However, most such nutritional mutants *can* survive on a *complete growth medium,* minimal medium supplemented with all 20 amino acids and a few other nutrients.

To characterize the metabolic defect in each nutritional mutant, Beadle and Tatum took samples from the mutant growing on complete medium and distributed them to a number of different vials. Each vial contained minimal medium plus a single additional nutrient. The particular supplement that allowed growth indicated the metabolic defect. For example, if the only supplemented vial that supported growth of the mutant was the one fortified with the amino acid arginine, the researchers could conclude that the mutant was defective in the biochemical pathway that wild-type cells use to synthesize arginine.

Beadle and Tatum went on to pin down each mutant's defect more specifically. Their work with arginine-requiring mutants was especially instructive. Using genetic crosses, they determined that their mutants fell into three classes, each mutated in a different gene. The researchers then showed that they could distinguish among the classes of mutants nutritionally by additional tests of their growth requirements (**Figure 17.2**). In the synthetic pathway leading to arginine, they suspected, a precursor nutrient is converted to ornithine, which is converted to citrulline, which is converted to arginine. When they tested their arginine mutants for growth on ornithine and citrulline, they found that one class could grow on either compound (or arginine), the second class only on citrulline (or arginine), and the third on neither—it absolutely required arginine. The three classes of mutants, the researchers reasoned, must be blocked at different steps in the pathway that synthesizes arginine, with each mutant class lacking the enzyme that catalyzes the blocked step.

Because each mutant was defective in a single gene, Beadle and Tatum's results provided strong support for the *one gene–one enzyme hypothesis,* as they dubbed it, which states that the function of a gene is to dictate the production of a specific enzyme. The researchers also showed how a combination of genetics and biochemistry could be used to work out the steps in a metabolic pathway. Further support for the one gene–one enzyme hypothesis came with experiments that identified the specific enzymes lacking in the mutants.

The Products of Gene Expression: A Developing Story

As researchers learned more about proteins, they made minor revisions to the one gene–one enzyme hypothesis. First of all, not all proteins are enzymes. Keratin, the structural protein of animal hair, and the hormone insulin are two examples of nonenzyme proteins. Because proteins that are not enzymes are nevertheless gene products, molecular biologists began to think in terms of one gene–one protein. However, many proteins are constructed from two or more different polypeptide chains, and each polypeptide is specified by its own gene. For example, hemoglobin, the oxygen-transporting protein of vertebrate red blood cells, is built from two kinds of polypeptides, and thus two genes code for this protein (see Figure 5.20). Beadle and Tatum's idea has therefore been restated as the **one gene–one polypeptide hypothesis.** Even this statement is not entirely accurate, though. As you will learn later in this chapter, some genes code for RNA molecules that have important functions in cells even though they are never translated into protein. But for now, we will focus on genes that code for polypeptides. (Note that it is common to refer to proteins, rather than polypeptides, as the gene products, a practice you will encounter in this book.)

Figure 17.2

Inquiry Do individual genes specify different enzymes in arginine biosynthesis?

EXPERIMENT Working with the mold *Neurospora crassa*, George Beadle and Edward Tatum had isolated mutants requiring arginine in their growth medium and had shown genetically that these mutants fell into three classes, each defective in a different gene. From other considerations, they suspected that the metabolic pathway of arginine biosynthesis included the precursors ornithine and citrulline. Their most famous experiment, shown here, tested both their one gene–one enzyme hypothesis and their postulated arginine pathway. In this experiment, they grew their three classes of mutants under the four different conditions shown in the Results section below.

RESULTS The wild-type strain required only the minimal medium for growth. The three classes of mutants had different growth requirements.

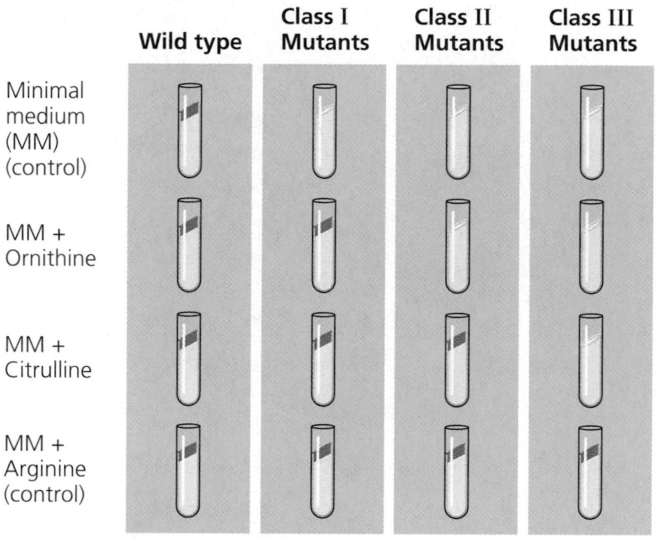

CONCLUSION From the growth patterns of the mutants, Beadle and Tatum deduced that each mutant was unable to carry out one step in the pathway for synthesizing arginine, presumably because it lacked the necessary enzyme. Because each of their mutants was mutated in a single gene, they concluded that each mutated gene must normally dictate the production of one enzyme. Their results supported the one gene–one enzyme hypothesis and also confirmed the arginine pathway. (Notice that a mutant can grow only if supplied with a compound made *after* the defective step.)

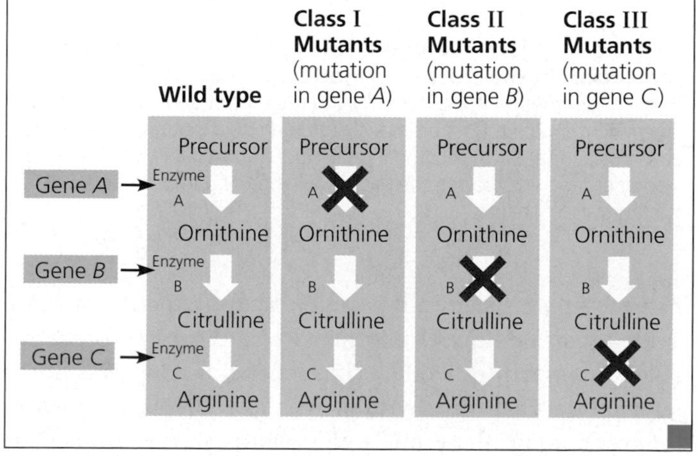

Basic Principles of Transcription and Translation

Genes provide the instructions for making specific proteins. But a gene does not build a protein directly. The bridge between DNA and protein synthesis is the nucleic acid RNA. You learned in Chapter 5 that RNA is chemically similar to DNA, except that it contains ribose instead of deoxyribose as its sugar and has the nitrogenous base uracil rather than thymine (see Figure 5.26). Thus, each nucleotide along a DNA strand has A, G, C, or T as its base, and each nucleotide along an RNA strand has A, G, C, or U as its base. An RNA molecule usually consists of a single strand.

It is customary to describe the flow of information from gene to protein in linguistic terms because both nucleic acids and proteins are polymers with specific sequences of monomers that convey information, much as specific sequences of letters communicate information in a language like English. In DNA or RNA, the monomers are the four types of nucleotides, which differ in their nitrogenous bases. Genes are typically hundreds or thousands of nucleotides long, each gene having a specific sequence of bases. Each polypeptide of a protein also has monomers arranged in a particular linear order (the protein's primary structure), but its monomers are the 20 amino acids. Thus, nucleic acids and proteins contain information written in two different chemical languages. Getting from DNA to protein requires two major stages, transcription and translation.

Transcription is the synthesis of RNA under the direction of DNA. Both nucleic acids use the same language, and the information is simply transcribed, or copied, from one molecule to the other. Just as a DNA strand provides a template for the synthesis of a new complementary strand during DNA replication, it provides a template for assembling a sequence of RNA nucleotides. The resulting RNA molecule is a faithful transcript of the gene's protein-building instructions. In discussing protein-coding genes, this type of RNA molecule is called **messenger RNA (mRNA)**, because it carries a genetic message from the DNA to the protein-synthesizing machinery of the cell. (Transcription is the general term for the synthesis of *any* kind of RNA on a DNA template. Later in this chapter, you will learn about other types of RNA produced by transcription.)

Translation is the actual synthesis of a polypeptide, which occurs under the direction of mRNA. During this stage, there is a change in language: The cell must translate the base sequence of an mRNA molecule into the amino acid sequence of a polypeptide. The sites of translation are **ribosomes**, complex particles that facilitate the orderly linking of amino acids into polypeptide chains.

You might wonder why proteins couldn't simply be translated directly from DNA. There are evolutionary reasons for using an RNA intermediate. First, it provides protection for the DNA and its genetic information. As an analogy, when an architect designs a house, the original specifications (analogous to DNA) are not what the construction workers use at the site. Instead they use *copies* of the originals (analogous to mRNA), keeping the originals pristine and undamaged. Second, using an RNA intermediate allows more copies of a protein to be made simultaneously, since many RNA transcripts can be made from one gene. Also, each RNA transcript can be translated repeatedly.

Although the basic mechanics of transcription and translation are similar for prokaryotes and eukaryotes, there is an important difference in the flow of genetic information within the cells. Because bacteria lack nuclei, their DNA is not segregated from ribosomes and the other protein-synthesizing equipment **(Figure 17.3a)**. As you will see later, this allows translation of an mRNA to begin while its transcription is still in progress (see Figure 17.22). In a eukaryotic cell, by contrast, the nuclear envelope separates transcription from translation in space and time **(Figure 17.3b)**. Transcription occurs in the nucleus, and mRNA is transported to the cytoplasm, where translation occurs. But before they can leave the nucleus, eukaryotic RNA transcripts are modified in various ways to produce the final, functional mRNA. The transcription of a protein-coding eukaryotic gene results in *pre-mRNA,* and **RNA processing** yields the finished mRNA. The initial RNA transcript from any gene, including those coding for RNA that is not translated into protein, is more generally called a **primary transcript.**

Let's summarize: Genes program protein synthesis via genetic messages in the form of messenger RNA. Put another way, cells are governed by a molecular chain of command: DNA → RNA → protein. In the next section, we discuss how the instructions for assembling amino acids into a specific order are encoded in nucleic acids.

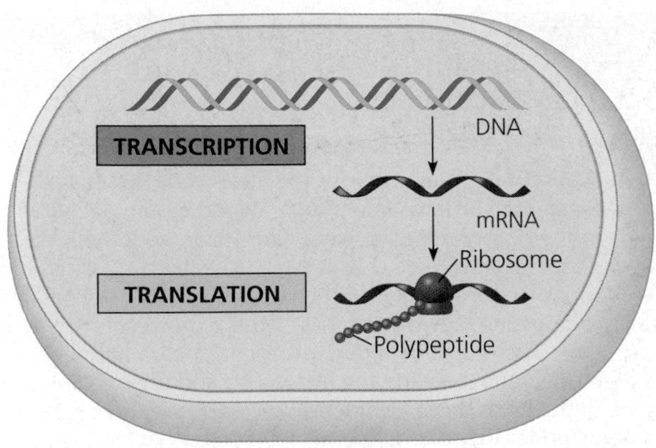

(a) Prokaryotic cell. In a cell lacking a nucleus, mRNA produced by transcription is immediately translated without additional processing.

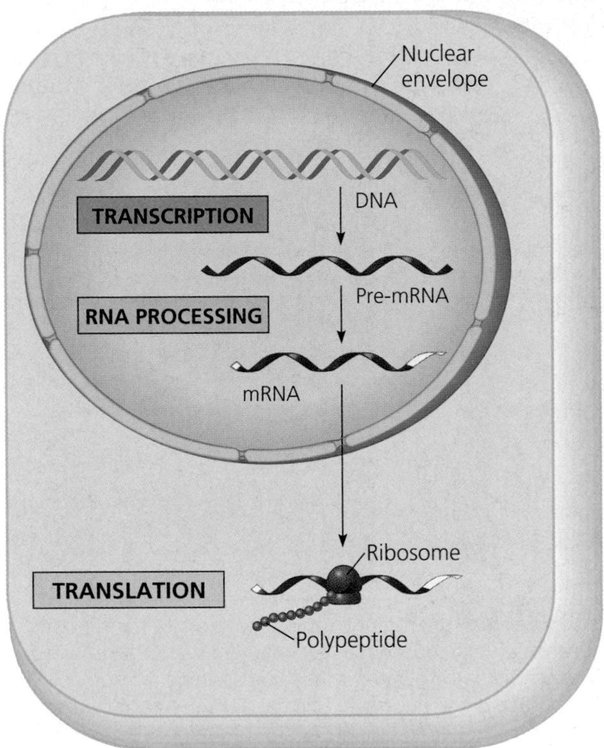

(b) Eukaryotic cell. The nucleus provides a separate compartment for transcription. The original RNA transcript, called pre-mRNA, is processed in various ways before leaving the nucleus as mRNA.

▲ **Figure 17.3 Overview: the roles of transcription and translation in the flow of genetic information.** In a cell, inherited information flows from DNA to RNA to protein. The two main stages of information flow are transcription and translation. A miniature version of part (a) or (b) accompanies several figures later in the chapter as an orientation diagram to help you see where a particular figure fits into the overall scheme.

The Genetic Code

When biologists began to suspect that the instructions for protein synthesis were encoded in DNA, they recognized a problem: There are only four nucleotide bases to specify 20 amino acids. Thus, the genetic code cannot be a language like Chinese, where each written symbol corresponds to a single word. How many bases, then, correspond to an amino acid?

Codons: Triplets of Bases

If each nucleotide base were translated into an amino acid, only 4 of the 20 amino acids could be specified. Would a language of two-letter code words suffice? The base sequence AG, for example, could specify one amino acid, and GT could specify another. Since there are four bases, this would give us 16 (that is, 4^2) possible arrangements—still not enough to code for all 20 amino acids.

Triplets of nucleotide bases are the smallest units of uniform length that can code for all the amino acids. If each arrangement of three consecutive bases specifies an amino acid, there can be 64 (that is, 4^3) possible code words—more than enough to specify all the amino acids. Experiments have verified that the flow of information from gene to protein is based on a **triplet code:** The genetic instructions for a polypeptide chain are written in the DNA as a series of nonoverlapping, three-nucleotide words. For example, the base triplet AGT at a particular position along a DNA strand results in the placement of the amino acid serine at the corresponding position of the polypeptide to be produced.

During transcription, the gene determines the sequence of bases along the length of an mRNA molecule **(Figure 17.4)**. For each gene, only one of the two DNA strands is transcribed. This strand is called the **template strand** because it provides the template for ordering the sequence of nucleotides in an RNA transcript. A given DNA strand can be the template strand for some genes along a DNA molecule, while for other genes in other regions, the complementary strand may function as the template. Note, however, that for a given gene, the same strand is used as the template every time it is transcribed.

An mRNA molecule is complementary rather than identical to its DNA template because RNA bases are assembled on the template according to base-pairing rules. The pairs are similar to those that form during DNA replication, except that U, the RNA substitute for T, pairs with A and the mRNA nucleotides contain ribose instead of deoxyribose. Like a new strand of DNA, the RNA molecule is synthesized in an antiparallel direction to the template strand of DNA. (To review what is meant by "antiparallel" and the 5′ and 3′ ends of a nucleic acid chain, see Figure 16.7.) For example, the base triplet ACC along the DNA (written as 3′-ACC-5′) provides a template for 5′-UGG-3′ in

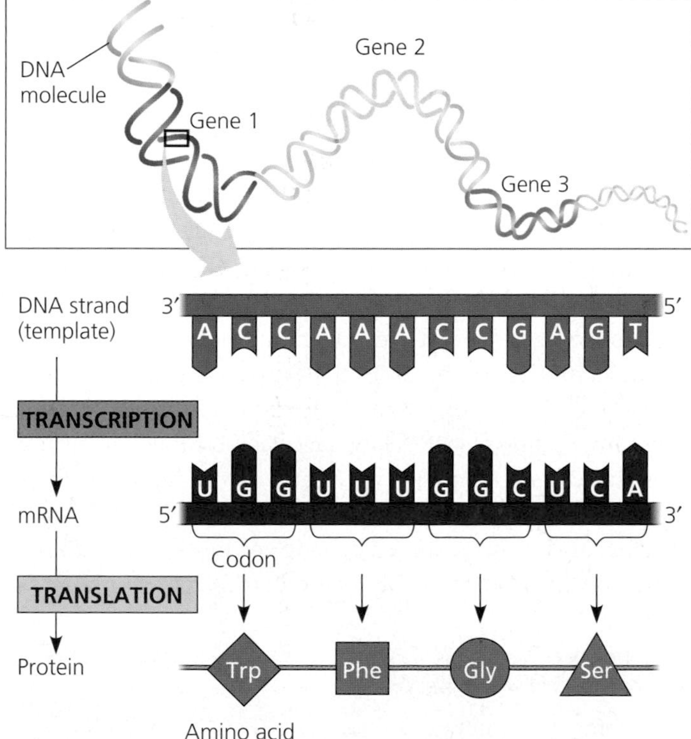

▲ **Figure 17.4 The triplet code.** For each gene, one DNA strand functions as a template for transcription. The base-pairing rules for DNA synthesis also guide transcription, but uracil (U) takes the place of thymine (T) in RNA. During translation, the mRNA is read as a sequence of base triplets, called codons. Each codon specifies an amino acid to be added to the growing polypeptide chain. The mRNA is read in the 5′ → 3′ direction.

the mRNA molecule. The mRNA base triplets are called **codons**, and they are customarily written in the 5′ → 3′ direction. In our example, UGG is the codon for the amino acid tryptophan (abbreviated Trp). The term *codon* is also sometimes used for the DNA base triplets along the *nontemplate* strand. These codons are complementary to the template strand and thus identical in sequence to the mRNA except that they have T instead of U. (For this reason, the nontemplate DNA strand is sometimes called the "coding strand.")

During translation, the sequence of codons along an mRNA molecule is decoded, or translated, into a sequence of amino acids making up a polypeptide chain. The codons are read by the translation machinery in the 5′ → 3′ direction along the mRNA. Each codon specifies which one of the 20 amino acids will be incorporated at the corresponding position along a polypeptide. Because codons are base triplets, the number of nucleotides making up a genetic message must be three times the number of amino acids making up the protein product. For example, it takes 300 nucleotides along an mRNA strand to code for a polypeptide that is 100 amino acids long.

Cracking the Code

Molecular biologists cracked the code of life in the early 1960s, when a series of elegant experiments disclosed the amino acid translations of each of the RNA codons. The first codon was deciphered in 1961 by Marshall Nirenberg, of the National Institutes of Health, and his colleagues. Nirenberg synthesized an artificial mRNA by linking identical RNA nucleotides containing uracil as their base. No matter where this message started or stopped, it could contain only one codon in repetition: UUU. Nirenberg added this "poly-U" to a test-tube mixture containing amino acids, ribosomes, and the other components required for protein synthesis. His artificial system translated the poly-U into a polypeptide containing a single amino acid, phenylalanine (Phe), strung together as a long polyphenylalanine chain. Thus, Nirenberg determined that the mRNA codon UUU specifies the amino acid phenylalanine. Soon, the amino acids specified by the codons AAA, GGG, and CCC were also determined.

Although more elaborate techniques were required to decode mixed triplets such as AUA and CGA, all 64 codons were deciphered by the mid-1960s. As **Figure 17.5** on the next page shows, 61 of the 64 triplets code for amino acids. The three codons that do not designate amino acids are "stop" signals, or termination codons, marking the end of translation. Notice that the codon AUG has a dual function: It codes for the amino acid methionine (Met) and also functions as a "start" signal, or initiation codon. Genetic messages begin with the mRNA codon AUG, which signals the protein-synthesizing machinery to begin translating the mRNA at that location. (Because AUG also stands for methionine, polypeptide chains begin with methionine when they are synthesized. However, an enzyme may subsequently remove this starter amino acid from the chain.)

First mRNA base (5′ end)	Second mRNA base: U	C	A	G	Third mRNA base (3′ end)
U	UUU Phe	UCU Ser	UAU Tyr	UGU Cys	U
U	UUC Phe	UCC Ser	UAC Tyr	UGC Cys	C
U	UUA Leu	UCA Ser	UAA Stop	UGA Stop	A
U	UUG Leu	UCG Ser	UAG Stop	UGG Trp	G
C	CUU Leu	CCU Pro	CAU His	CGU Arg	U
C	CUC Leu	CCC Pro	CAC His	CGC Arg	C
C	CUA Leu	CCA Pro	CAA Gln	CGA Arg	A
C	CUG Leu	CCG Pro	CAG Gln	CGG Arg	G
A	AUU Ile	ACU Thr	AAU Asn	AGU Ser	U
A	AUC Ile	ACC Thr	AAC Asn	AGC Ser	C
A	AUA Ile	ACA Thr	AAA Lys	AGA Arg	A
A	AUG Met or start	ACG Thr	AAG Lys	AGG Arg	G
G	GUU Val	GCU Ala	GAU Asp	GGU Gly	U
G	GUC Val	GCC Ala	GAC Asp	GGC Gly	C
G	GUA Val	GCA Ala	GAA Glu	GGA Gly	A
G	GUG Val	GCG Ala	GAG Glu	GGG Gly	G

▲ **Figure 17.5 The dictionary of the genetic code.** The three bases of an mRNA codon are designated here as the first, second, and third bases, reading in the 5′ → 3′ direction along the mRNA. (Practice using this dictionary by finding the codons in Figure 17.4.) The codon AUG not only stands for the amino acid methionine (Met) but also functions as a "start" signal for ribosomes to begin translating the mRNA at that point. Three of the 64 codons function as "stop" signals, marking the end of a genetic message.

Notice in Figure 17.5 that there is *redundancy* in the genetic code, but no ambiguity. For example, although codons GAA and GAG both specify glutamic acid (redundancy), neither of them ever specifies any other amino acid (no ambiguity). The redundancy in the code is not altogether random. In many cases, codons that are synonyms for a particular amino acid differ only in the third base of the triplet. We will consider a possible benefit for this redundancy later in the chapter.

Our ability to extract the intended message from a written language depends on reading the symbols in the correct groupings—that is, in the correct **reading frame.** Consider this statement: "The red dog ate the cat." Group the letters incorrectly by starting at the wrong point, and the result will probably be gibberish: for example, "her edd oga tet hec at." The reading frame is also important in the molecular language of cells. The short stretch of polypeptide shown in Figure 17.4, for instance, will only be made correctly if the mRNA nucleotides are read from left to right (5′ → 3′) in the groups of three shown in the figure: UGG UUU GGC UCA. Although a genetic message is written with no spaces between the codons, the cell's protein-synthesizing machinery reads the message as a series of nonoverlapping three-letter words. The message is *not* read as a series of overlapping words—UGGUUU, and so on—which would convey a very different message.

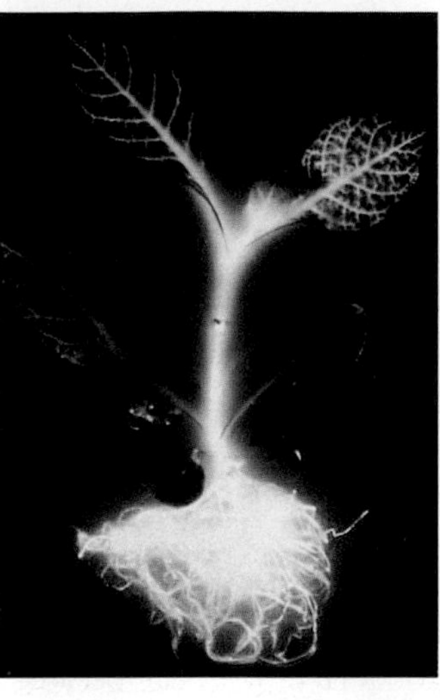

▶ **Figure 17.6 A tobacco plant expressing a firefly gene.** Because diverse forms of life share a common genetic code, it is possible to program one species to produce proteins characteristic of another species by transplanting DNA. In this experiment, researchers were able to incorporate a gene from a firefly into the DNA of a tobacco plant. The firefly gene codes for an enzyme that catalyzes a chemical reaction that releases light energy.

Evolution of the Genetic Code

The genetic code is nearly universal, shared by organisms from the simplest bacteria to the most complex animals. The RNA codon CCG, for instance, is translated as the amino acid proline in all organisms whose genetic code has been examined. In laboratory experiments, genes can be transcribed and translated after being transplanted from one species to another **(Figure 17.6)**. Bacteria can be programmed by the insertion of human genes to synthesize certain human proteins for medical use. Such applications have produced many exciting developments in biotechnology (see Chapter 20).

Exceptions to the universality of the genetic code include translation systems where a few codons differ from the standard ones. Slight variations in the genetic code exist in certain unicellular eukaryotes and in the organelle genes of some species. Some prokaryotes can translate stop codons into one of two amino acids not found in most organisms. Despite these exceptions, the evolutionary significance of the code's *near* universality is clear. A language shared by all living things must have been operating very early in the history of life—early enough to be present in the common ancestors of all modern organisms. A shared genetic vocabulary is a reminder of the kinship that bonds all life on Earth.

Concept Check 17.1

1. Draw the nontemplate strand of DNA for the template shown in Figure 17.4. Compare and contrast its base sequence with the mRNA molecule.
2. What protein product would you expect from a poly-G mRNA that is 30 nucleotides long?

For suggested answers, see Appendix A.

Concept 17.2

Transcription is the DNA-directed synthesis of RNA: *a closer look*

Now that we have considered the linguistic logic and evolutionary significance of the genetic code, we are ready to reexamine transcription, the first stage of gene expression, in more detail.

Molecular Components of Transcription

Messenger RNA, the carrier of information from DNA to the cell's protein-synthesizing machinery, is transcribed from the template strand of a gene. An enzyme called an **RNA polymerase** pries the two strands of DNA apart and hooks together the RNA nucleotides as they base-pair along the DNA template **(Figure 17.7)**. Like the DNA polymerases that function in DNA replication, RNA polymerases can only assemble a polynucleotide in its $5' \rightarrow 3'$ direction. Unlike DNA polymerases, however, RNA polymerases are able to start a chain from scratch; they don't need a primer.

Specific sequences of nucleotides along the DNA mark where transcription of a gene begins and ends. The DNA sequence where RNA polymerase attaches and initiates transcription is known as the **promoter;** in prokaryotes, the sequence that signals the end of transcription is called the **terminator.** (The termination mechanism is different in eukaryotes, which we'll describe later.) Molecular biologists refer to the direction of transcription as "downstream" and the other direction as "upstream." These terms are also used to describe the positions of nucleotide sequences within the DNA or RNA. Thus, the promoter sequence in DNA is said to be upstream from the terminator. The stretch of DNA that is transcribed into an RNA molecule is called a **transcription unit.**

Bacteria have a single type of RNA polymerase that synthesizes not only mRNA but also other types of RNA that function in

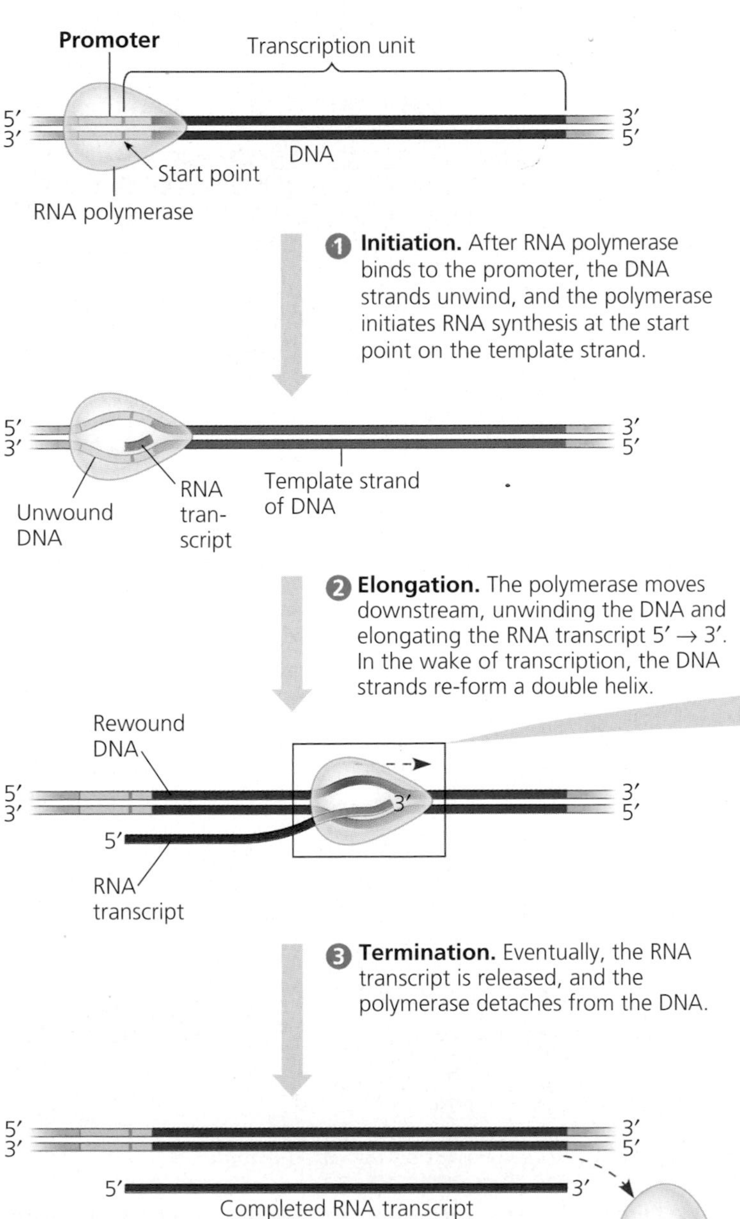

Promoter
Transcription unit
5′
3′
3′
5′
Start point
DNA
RNA polymerase
1 Initiation. After RNA polymerase binds to the promoter, the DNA strands unwind, and the polymerase initiates RNA synthesis at the start point on the template strand.
Unwound DNA
RNA transcript
Template strand of DNA
2 Elongation. The polymerase moves downstream, unwinding the DNA and elongating the RNA transcript 5′ → 3′. In the wake of transcription, the DNA strands re-form a double helix.
Rewound DNA
RNA transcript
3 Termination. Eventually, the RNA transcript is released, and the polymerase detaches from the DNA.
Completed RNA transcript

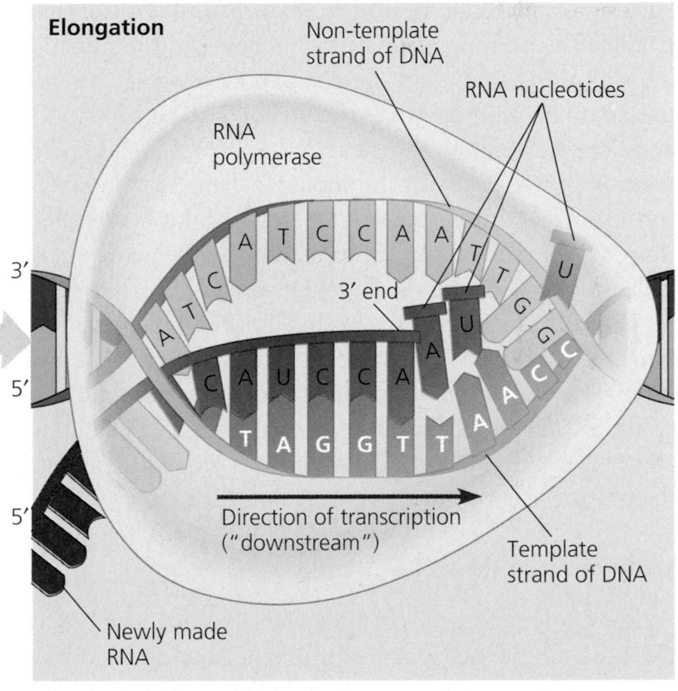

Elongation
Non-template strand of DNA
RNA nucleotides
RNA polymerase
3′ end
3′
5′
5′
Direction of transcription ("downstream")
Template strand of DNA
Newly made RNA

◀ **Figure 17.7 The stages of transcription: initiation, elongation, and termination.** This general depiction of transcription applies to both prokaryotes and eukaryotes, but the details of termination vary for prokaryotes and eukaryotes, as described in the text. Also, in a prokaryote, the RNA transcript is immediately usable as mRNA, whereas in a eukaryote, it must first undergo processing to become mRNA.

protein synthesis. In contrast, eukaryotes have three types of RNA polymerase in their nuclei, numbered I, II, and III. The one used for mRNA synthesis is RNA polymerase II. The other two RNA polymerases transcribe RNA molecules that are not translated into protein. In the discussion of transcription that follows, we start with the features of mRNA synthesis common to both prokaryotes and eukaryotes and then describe some key differences.

Synthesis of an RNA Transcript

The three stages of transcription, as shown in Figure 17.7 and described next, are initiation, elongation, and termination of the RNA chain. Study Figure 17.7 to familiarize yourself with the stages and the terms used to describe them.

RNA Polymerase Binding and Initiation of Transcription

The promoter of a gene includes within it the transcription start point (the nucleotide where RNA synthesis actually begins) and typically extends several dozen nucleotide pairs "upstream" from the start point. In addition to serving as a binding site for RNA polymerase and determining where transcription starts, the promoter determines which of the two strands of the DNA helix is used as the template.

Certain sections of a promoter are especially important for binding RNA polymerase. In prokaryotes, the RNA polymerase itself specifically recognizes and binds to the promoter. In eukaryotes, a collection of proteins called **transcription factors** mediate the binding of RNA polymerase and the initiation of transcription. Only after certain transcription factors are attached to the promoter does RNA polymerase II bind to it. The completed assembly of transcription factors and RNA polymerase II bound to the promoter is called a **transcription initiation complex.** **Figure 17.8** shows the role of transcription factors and a crucial promoter DNA sequence called a **TATA box** in forming the initiation complex in eukaryotes.

The interaction between eukaryotic RNA polymerase II and transcription factors is an example of the importance of protein-protein interactions in controlling eukaryotic transcription (as we will discuss further in Chapter 19). Once the polymerase is firmly attached to the promoter DNA, the two DNA strands unwind there, and the enzyme starts transcribing the template strand.

Elongation of the RNA Strand

As RNA polymerase moves along the DNA, it continues to untwist the double helix, exposing about 10 to 20 DNA bases at a time for pairing with RNA nucleotides (see Figure 17.7). The enzyme adds nucleotides to the 3′ end of the growing RNA molecule as it continues along the double helix. In the wake of this advancing wave of RNA synthesis, the new RNA molecule peels away from its DNA template and the DNA double helix re-forms. Transcription progresses at a rate of about 60 nucleotides per second in eukaryotes.

A single gene can be transcribed simultaneously by several molecules of RNA polymerase following each other like trucks in a convoy. A growing strand of RNA trails off from each polymerase, with the length of each new strand reflecting how far along the template the enzyme has traveled from the start point (see Figure 17.22). The congregation of many polymerase molecules simultaneously transcribing a single gene increases the amount of mRNA transcribed from it, which helps the cell make the encoded protein in large amounts.

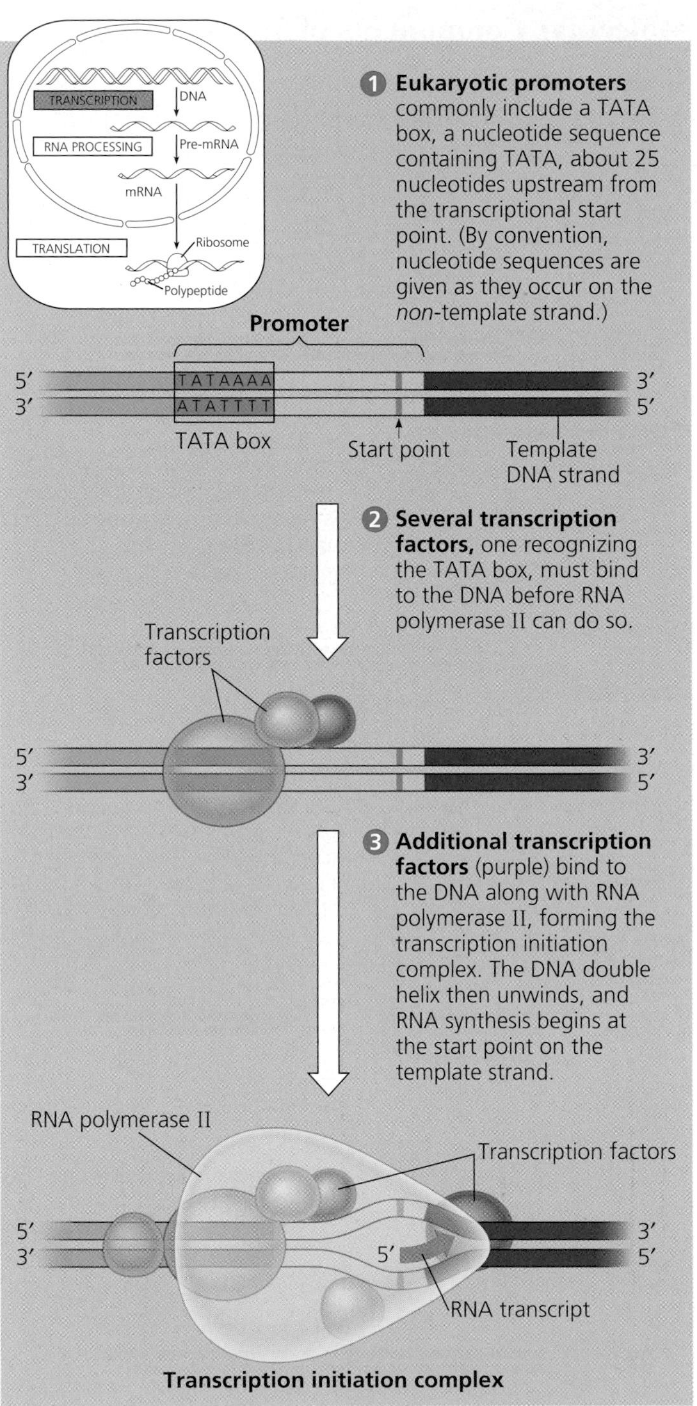

TRANSCRIPTION
DNA
RNA PROCESSING
Pre-mRNA
mRNA
TRANSLATION
Ribosome
Polypeptide
1 Eukaryotic promoters commonly include a TATA box, a nucleotide sequence containing TATA, about 25 nucleotides upstream from the transcriptional start point. (By convention, nucleotide sequences are given as they occur on the non-template strand.)
Promoter
5′
3′
TATAAAA
ATATTTT
3′
5′
TATA box
Start point
Template DNA strand
2 Several transcription factors, one recognizing the TATA box, must bind to the DNA before RNA polymerase II can do so.
Transcription factors
3 Additional transcription factors (purple) bind to the DNA along with RNA polymerase II, forming the transcription initiation complex. The DNA double helix then unwinds, and RNA synthesis begins at the start point on the template strand.
RNA polymerase II
Transcription factors
RNA transcript
Transcription initiation complex

▲ **Figure 17.8 The initiation of transcription at a eukaryotic promoter.** In eukaryotic cells, proteins called transcription factors mediate the initiation of transcription by RNA polymerase II.

Termination of Transcription

The mechanism of termination differs between prokaryotes and eukaryotes. In prokaryotes, transcription proceeds through a terminator sequence in the DNA. The transcribed terminator (an RNA sequence) functions as the termination signal, causing the polymerase to detach from the DNA and release the transcript, which is available for immediate use as mRNA. In eukaryotes, however, the pre-mRNA is cleaved from the growing RNA chain while RNA polymerase II continues to transcribe the DNA. Specifically, the polymerase transcribes a sequence on the DNA called the polyadenylation signal sequence, which codes for a polyadenylation signal (AAUAAA) in the pre-mRNA. Then, at a point about 10 to 35 nucleotides downstream from the AAUAAA signal, proteins associated with the growing RNA transcript cut it free from the polymerase, releasing the pre-mRNA. The polymerase continues transcribing for hundreds of nucleotides past the site where the pre-mRNA was released. Transcription is terminated when the polymerase eventually falls off the DNA (by a mechanism that is not fully understood). Once the pre-mRNA has been made, it is modified during RNA processing, the topic of our next section.

Concept Check 17.2

1. Compare and contrast the functioning of DNA polymerase and RNA polymerase.
2. Is the promoter at the upstream or downstream end of a transcription unit?
3. In a prokaryote, how does RNA polymerase "know" where to start transcribing a gene? In a eukaryote?
4. How is the primary transcript produced by a prokaryotic cell different from that produced by a eukaryotic cell?

For suggested answers, see Appendix A.

Eukaryotic cells modify RNA after transcription

Enzymes in the eukaryotic nucleus modify pre-mRNA in specific ways before the genetic messages are dispatched to the cytoplasm. During this RNA processing, both ends of the primary transcript are usually altered. Also, in most cases, certain interior sections of the molecule are cut out and the remaining parts spliced together. These modifications help form an mRNA molecule that is ready to be translated.

Alteration of mRNA Ends

Each end of a pre-mRNA molecule is modified in a particular way **(Figure 17.9)**. The 5′ end, the end transcribed first, is capped off with a modified form of a guanine (G) nucleotide after transcription of the first 20 to 40 nucleotides, forming a **5′ cap.** The 3′ end of the pre-mRNA molecule is also modified before the mRNA exits the nucleus. Recall that the pre-mRNA is released soon after the polyadenylation signal, AAUAAA, is transcribed. At the 3′ end, an enzyme adds 50 to 250 adenine (A) nucleotides, forming a **poly-A tail.** The 5′ cap and poly-A tail share several important functions. First, they seem to facilitate the export of the mature mRNA from the nucleus. Second, they help protect the mRNA from degradation by hydrolytic enzymes. And third, once the mRNA reaches the cytoplasm, both structures help ribosomes attach to the 5′ end of the mRNA. Figure 17.9 shows a diagram of a eukaryotic mRNA molecule with cap and tail. The figure also shows the untranslated regions (UTRs) at the 5′ and 3′ ends of the mRNA (referred to as the *5′ UTR* and *3′ UTR*). The UTRs are parts of the mRNA that will not be translated into protein, but they have other functions, such as ribosome binding.

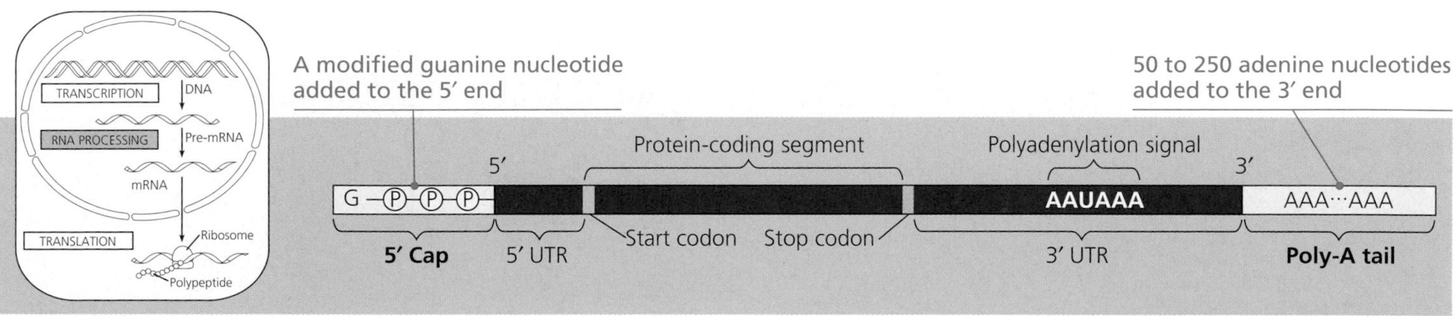

▲ **Figure 17.9 RNA processing: addition of the 5′ cap and poly-A tail.** Enzymes modify the two ends of a eukaryotic pre-mRNA molecule. The modified ends may promote the export of mRNA from the nucleus and help protect the mRNA from degradation. When the mRNA reaches the cytoplasm, the modified ends, in conjunction with certain cytoplasmic proteins, facilitate ribosome attachment. The 5′ cap and poly-A tail are not translated into protein, nor are the regions called the 5′ untranslated region (5′ UTR) and 3′ untranslated region (3′ UTR).

Split Genes and RNA Splicing

The most remarkable stage of RNA processing in the eukaryotic nucleus is the removal of a large portion of the RNA molecule that is initially synthesized—a cut-and-paste job called **RNA splicing** **(Figure 17.10)**. The average length of a transcription unit along a eukaryotic DNA molecule is about 8,000 nucleotides, so the primary RNA transcript is also that long. But it takes only about 1,200 nucleotides to code for an average-sized protein of 400 amino acids. (Remember, each amino acid is encoded by a *triplet* of nucleotides.) This means that most eukaryotic genes and their RNA transcripts have long noncoding stretches of nucleotides, regions that are not translated. Even more surprising is that most of these noncoding sequences are interspersed between coding segments of the gene and thus between coding segments of the pre-mRNA. In other words, the sequence of DNA nucleotides that codes for a eukaryotic polypeptide is usually not continuous; it is split into segments. The noncoding segments of nucleic acid that lie between coding regions are called intervening sequences, or **introns** for short. The other regions are called **exons,** because they are eventually expressed, usually by being translated into amino acid sequences. (Exceptions include the UTRs of the exons at the ends of the RNA, which make up part of the mRNA but are not translated into protein. Because of these exceptions, you may find it helpful to think of exons as sequences of RNA that *exit* the nucleus.) The terms *intron* and *exon* are used for both RNA sequences and the DNA sequences that encode them.

In making a primary transcript from a gene, RNA polymerase II transcribes both introns and exons from the DNA, but the mRNA molecule that enters the cytoplasm is an abridged version. The introns are cut out from the molecule and the exons joined together, forming an mRNA molecule with a continuous coding sequence. This is the process of RNA splicing.

How is pre-mRNA splicing carried out? Researchers have learned that the signal for RNA splicing is a short nucleotide sequence at each end of an intron. Particles called *small nuclear ribonucleoproteins,* abbreviated *snRNPs* (pronounced "snurps"), recognize these splice sites. As the name implies, snRNPs are located in the cell nucleus and are composed of RNA and protein molecules. The RNA in a snRNP particle is called a *small nuclear RNA (snRNA);* each molecule is about 150 nucleotides long. Several different snRNPs join with additional proteins to form an even larger assembly called a **spliceosome,** which is almost as big as a ribosome. The spliceosome interacts with certain sites along an intron, releasing the intron and joining together the two exons that flanked the intron **(Figure 17.11)**. There is strong evidence that snRNAs play a major role in these catalytic processes, as well as in spliceosome assembly and splice site recognition.

Ribozymes

The idea of a catalytic role for snRNA arose from the discovery of **ribozymes,** RNA molecules that function as enzymes. In some organisms, RNA splicing can occur without proteins or additional RNA molecules: The intron RNA functions as a ribozyme and catalyzes its own excision! For example, in the protozoan *Tetrahymena,* self-splicing occurs in the production of ribosomal RNA (rRNA), a component of the organism's ribosomes. The pre-rRNA actually removes its own introns.

The fact that RNA is single-stranded plays an important role in allowing certain RNA molecules to function as ribozymes. A region of an RNA molecule may base-pair with a complementary region elsewhere in the same molecule, thus imparting specific structure to the RNA molecule as a whole. Also, some of the bases contain functional groups that may participate in catalysis. Just as the specific shape of an enzymatic protein and the functional groups on its amino acid side chains allow the protein to function as a catalyst, the structure of some RNA molecules allows them to function as catalysts, too. The discovery of ribozymes rendered obsolete the belief that all biological catalysts were proteins.

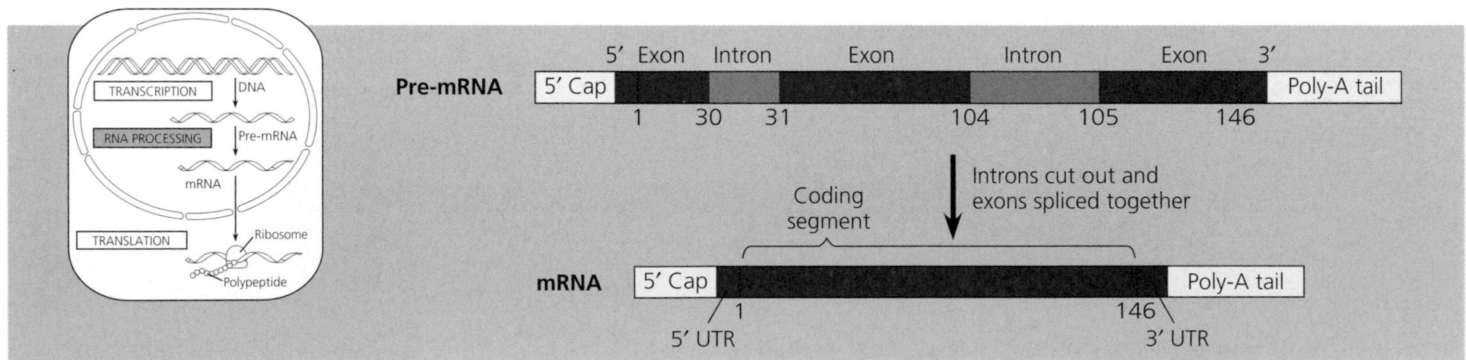

▲ Figure 17.10 RNA processing: RNA splicing. The RNA molecule shown here codes for β-globin, one of the polypeptides of hemoglobin. The numbers under the RNA refer to codons; β-globin is 146 amino acids long. The β-globin gene and its pre-mRNA transcript have three exons, corresponding to sequences that will leave the nucleus as mRNA. (The 5′ UTR and 3′ UTR are parts of exons because they are included in the mRNA; however, they do not code for protein.) During RNA processing, the introns are cut out and the exons spliced together.

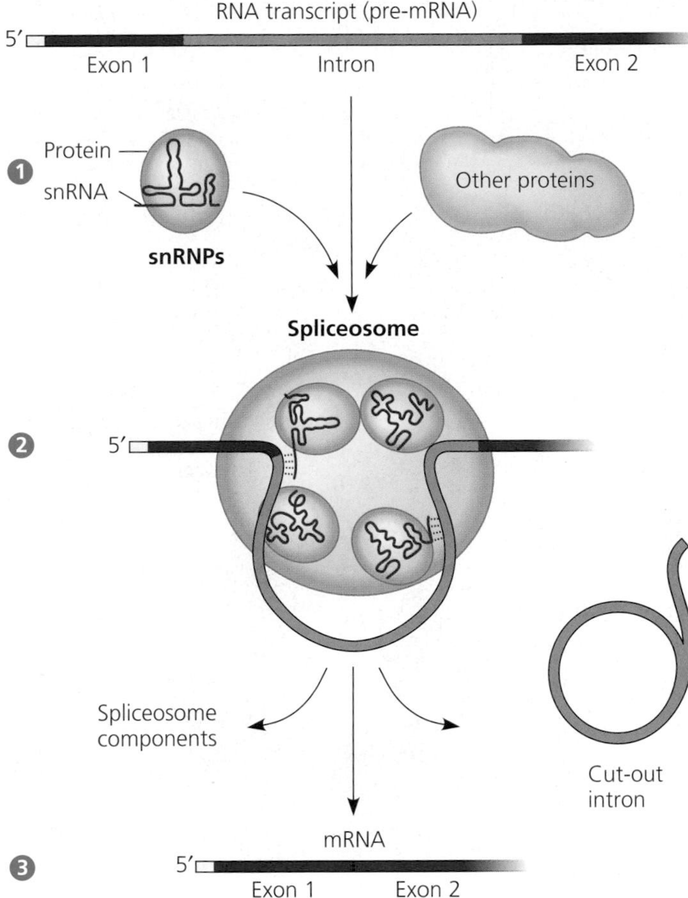

▲ **Figure 17.11 The roles of snRNPs and spliceosomes in pre-mRNA splicing.** The diagram shows only a portion of the pre-mRNA transcript; additional introns and exons lie downstream from the ones pictured here. ❶ Small nuclear ribonucleoproteins (snRNPs) and other proteins form a molecular complex called a spliceosome on a pre-mRNA containing exons and introns. ❷ Within the spliceosome, snRNA base-pairs with nucleotides at specific sites along the intron. ❸ The RNA transcript is cut, releasing the intron and at the same time splicing the exons together. The spliceosome then comes apart, releasing spliced mRNA, which now contains only exons.

The Functional and Evolutionary Importance of Introns

What are the biological functions of introns and RNA splicing? One idea is that introns play regulatory roles in the cell; at least some introns contain sequences that control gene activity in some way. And the splicing process itself is necessary for the passage of mRNA from the nucleus to the cytoplasm.

One consequence of the presence of exons and introns in genes is that a single gene can encode more than one kind of polypeptide. A number of genes are known to give rise to two or more different polypeptides, depending on which segments are treated as exons during RNA processing; this is called **alternative RNA splicing** (see Figure 19.8). For example, sex differences in fruit flies are largely due to differences in how males and females splice the RNA transcribed from certain genes. Early results from the Human Genome Project (discussed in Chapter 20) suggest that alternative RNA splicing may be one reason humans can get along with a relatively small number of genes—not quite twice as many as a fruit fly. Because of alternative splicing, the number of different protein products an organism can produce is much greater than its number of genes.

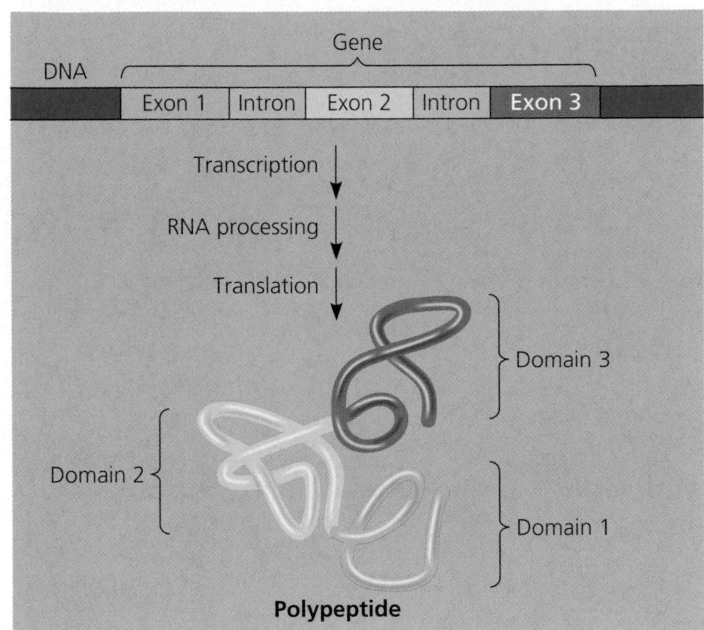

▲ **Figure 17.12 Correspondence between exons and protein domains.** In a large number of genes, different exons encode separate domains of the protein product.

Proteins often have a modular architecture consisting of discrete structural and functional regions called **domains.** One domain of an enzymatic protein, for instance, might include the active site, while another might attach the protein to a cellular membrane. In many cases, different exons code for the different domains of a protein **(Figure 17.12)**. The presence of introns in a gene may facilitate the evolution of new and potentially useful proteins as a result of a process known as *exon shuffling*. Introns increase the probability of potentially beneficial crossing over between the exons of alleles—simply by providing more terrain for crossovers without interrupting coding sequences. We can also imagine the occasional mixing and matching of exons between completely different (nonallelic) genes. Exon shuffling of either sort could lead to new proteins with novel combinations of functions. While most of the shuffling would result in nonbeneficial changes, occasionally a beneficial variant might arise.

Concept Check 17.3

1. How does alteration of the 5′ and 3′ ends of pre-mRNA affect the mRNA that exits the nucleus?
2. Describe the role of snRNPs in RNA splicing.
3. How can alternative RNA splicing generate a greater number of polypeptide products than there are genes?

For suggested answers, see Appendix A.

Concept 17.4

Translation is the RNA-directed synthesis of a polypeptide: *a closer look*

We will now examine in greater detail how genetic information flows from mRNA to protein—the process of translation. As we did for transcription, we'll concentrate on the basic steps of translation that occur in both prokaryotes and eukaryotes while pointing out key differences.

Molecular Components of Translation

In the process of translation, a cell interprets a genetic message and builds a polypeptide accordingly. The message is a series of codons along an mRNA molecule, and the interpreter is called **transfer RNA (tRNA)**. The function of tRNA is to transfer amino acids from the cytoplasmic pool of amino acids to a ribosome. A cell keeps its cytoplasm stocked with all 20 amino acids, either by synthesizing them from other compounds or by taking them up from the surrounding solution. The ribosome adds each amino acid brought to it by tRNA to the growing end of a polypeptide chain (**Figure 17.13**).

Molecules of tRNA are not all identical. The key to translating a genetic message into a specific amino acid sequence is that each type of tRNA molecule translates a particular mRNA codon into a particular amino acid. As a tRNA molecule arrives at a ribosome, it bears a specific amino acid at one end. At the other end of the tRNA is a nucleotide triplet called an **anticodon**, which base-pairs with a complementary codon on mRNA. For example, consider the mRNA codon UUU, which is translated as the amino acid phenylalanine. The tRNA that base-pairs with this codon by hydrogen bonding has AAA as its anticodon and carries phenylalanine at its other end (see the middle tRNA in Figure 17.13). As an mRNA molecule is moved through a ribosome, phenylalanine will be added to the polypeptide chain whenever the codon UUU is presented for translation. Codon by codon, the genetic message is translated as tRNAs deposit amino acids in the order prescribed, and the ribosome joins the amino acids into a chain. The tRNA molecule is a translator because it can read a nucleic acid word (the mRNA codon) and interpret it as a protein word (the amino acid).

Translation is simple in principle but complex in its biochemistry and mechanics, especially in the eukaryotic cell. In dissecting translation, we'll concentrate on the slightly less complicated version of the process that occurs in prokaryotes. Let's first look at some of the major players in this cellular drama, then see how they act together to make a polypeptide.

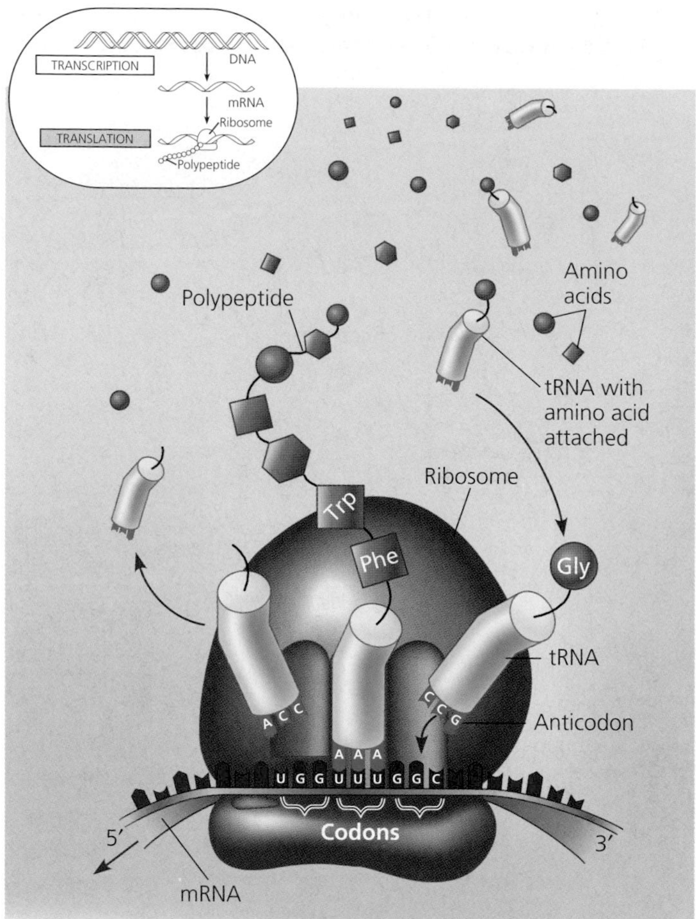

▲ **Figure 17.13 Translation: the basic concept.** As a molecule of mRNA is moved through a ribosome, codons are translated into amino acids, one by one. The interpreters are tRNA molecules, each type with a specific anticodon at one end and a corresponding amino acid at the other end. A tRNA adds its amino acid cargo to a growing polypeptide chain when the anticodon bonds to a complementary codon on the mRNA. The figures that follow show some of the details of translation in the prokaryotic cell.

The Structure and Function of Transfer RNA

Like mRNA and other types of cellular RNA, transfer RNA molecules are transcribed from DNA templates. In a eukaryotic cell, tRNA, like mRNA, is made in the nucleus and must travel from the nucleus to the cytoplasm, where translation occurs. In both prokaryotic and eukaryotic cells, each tRNA molecule is used repeatedly, picking up its designated amino acid in the cytosol, depositing this cargo at the ribosome, and then leaving the ribosome to pick up another amino acid.

As illustrated in **Figure 17.14**, a tRNA molecule consists of a single RNA strand that is only about 80 nucleotides long (compared to hundreds of nucleotides for most mRNA molecules). Because of the presence of complementary stretches of bases that can hydrogen-bond to each other, this single strand can fold back upon itself, forming a molecule with a three-dimensional structure. Flattened into one plane to reveal this base pairing, a tRNA molecule looks like a cloverleaf (**Figure 17.14a**). The tRNA actually twists and folds into a

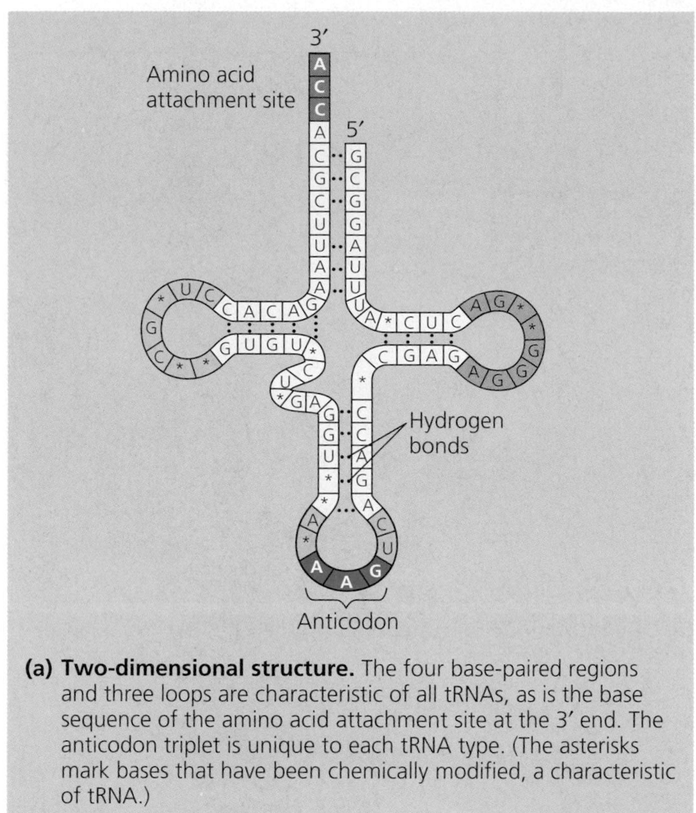

(a) Two-dimensional structure. The four base-paired regions and three loops are characteristic of all tRNAs, as is the base sequence of the amino acid attachment site at the 3′ end. The anticodon triplet is unique to each tRNA type. (The asterisks mark bases that have been chemically modified, a characteristic of tRNA.)

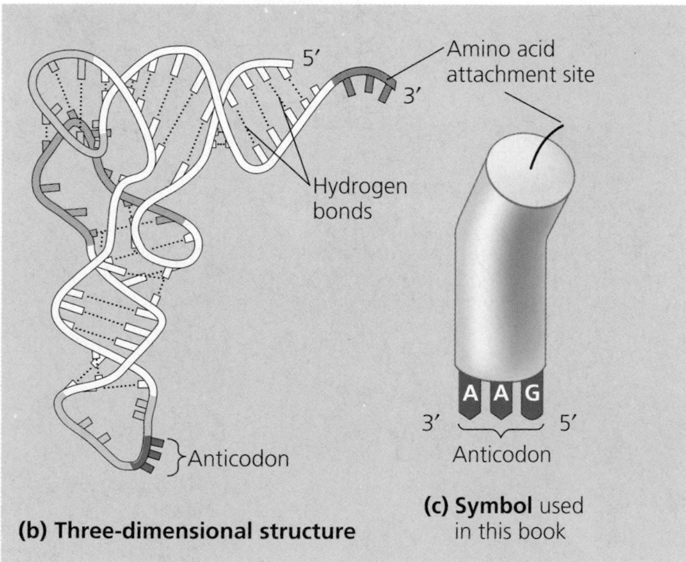

(b) Three-dimensional structure

(c) Symbol used in this book

▲ **Figure 17.14 The structure of transfer RNA (tRNA).** Anticodons are conventionally written 3′ → 5′ to align properly with codons written 5′ → 3′ (see Figure 17.13). For base pairing, RNA strands must be antiparallel, like DNA. For example, anticodon 3′-AAG-5′ pairs with mRNA codon 5′-UUC-3′.

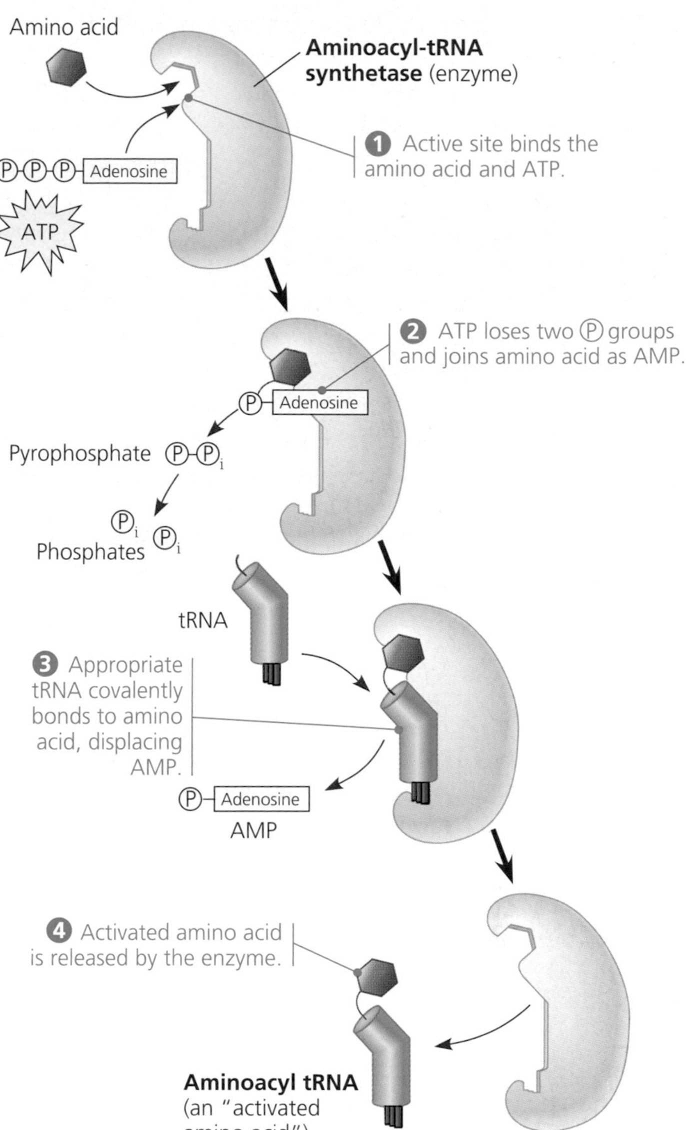

▲ **Figure 17.15 An aminoacyl-tRNA synthetase joins a specific amino acid to a tRNA.** Linkage of the tRNA and amino acid is an endergonic process that occurs at the expense of ATP. The ATP loses two phosphate groups, becoming AMP (adenosine monophosphate).

compact three-dimensional structure that is roughly L-shaped **(Figure 17.14b)**. The loop protruding from one end of the L includes the anticodon, the special base triplet that binds to a specific mRNA codon. From the other end of the L-shaped tRNA molecule protrudes its 3′ end, which is the attachment site for an amino acid. Thus, the structure of a tRNA molecule fits its function.

The accurate translation of a genetic message requires two recognition steps. First, there must be a correct match between a tRNA and an amino acid. A tRNA that binds to an mRNA codon specifying a particular amino acid must carry *only* that amino acid to the ribosome. Each amino acid is joined to the correct tRNA by a specific enzyme called an **aminoacyl-tRNA synthetase (Figure 17.15)**. The active site of each type of aminoacyl-tRNA synthetase fits only a specific combination of amino acid and tRNA. There are 20 different synthetases, one for each amino acid; each synthetase is able to bind all the different tRNAs that code for its specific amino acid. The synthetase catalyzes the covalent attachment of the amino acid to its tRNA in a process driven by the hydrolysis of ATP. The resulting aminoacyl tRNA, also called an activated amino acid, is released from the enzyme and delivers its amino acid to a growing polypeptide chain on a ribosome.

The second recognition step involves a correct match between the tRNA anticodon and an mRNA codon. If one tRNA variety existed for each of the mRNA codons that specifies an amino acid, there would be 61 tRNAs (see Figure 17.5). In fact, there are only about 45, signifying that some tRNAs must be able to bind to more than one codon. Such versatility is possible because the rules for base pairing between the third base of a codon and the corresponding base of a tRNA anticodon are not as strict as those for DNA and mRNA codons. For example, the base U at the 5′ end of a tRNA anticodon can pair with either A or G in the third position (at the 3′ end) of an mRNA codon. This relaxation of the base-pairing rules is called **wobble.** Wobble explains why the synonymous codons for a given amino acid can differ in their third base, but usually not in their other bases.

Ribosomes

Ribosomes facilitate the specific coupling of tRNA anticodons with mRNA codons during protein synthesis. A ribosome, which is large enough to be seen with an electron microscope, is made up of two subunits, called the large and small subunits **(Figure 17.16)**. The ribosomal subunits are constructed of proteins and RNA molecules named **ribosomal RNA,** or **rRNA.** In eukaryotes, the subunits are made in the nucleolus. Ribosomal RNA genes on the chromosomal DNA are transcribed, and the RNA is processed and assembled with proteins imported from the cytoplasm. The resulting ribosomal subunits are then exported via nuclear pores to the cytoplasm. In both prokaryotes and eukaryotes, large and small subunits join to form a functional ribosome only when they attach to an mRNA molecule. About two-thirds of the mass of a ribosome is rRNA. Because most cells contain thousands of ribosomes, rRNA is the most abundant type of RNA.

Although the ribosomes of prokaryotes and eukaryotes are very similar in structure and function, those of eukaryotes are slightly larger and differ somewhat from prokaryotic ribosomes in their molecular composition. The differences are medically significant. Certain antibiotic drugs can inactivate prokaryotic ribosomes without inhibiting the ability of eukaryotic ribosomes to make proteins. These drugs, including tetracycline and streptomycin, are used to combat bacterial infections.

The structure of a ribosome reflects its function of bringing mRNA together with amino acid-bearing tRNAs. In addition to a binding site for mRNA, each ribosome has three binding sites for tRNA (see Figure 17.16). The **P site** (peptidyl-tRNA site) holds the tRNA carrying the growing polypeptide chain, while the **A site** (aminoacyl-tRNA site) holds the tRNA carrying the next amino acid to be added to the chain. Discharged tRNAs leave the ribosome from the **E site** (exit site). The ribosome holds the tRNA and mRNA in close proximity and

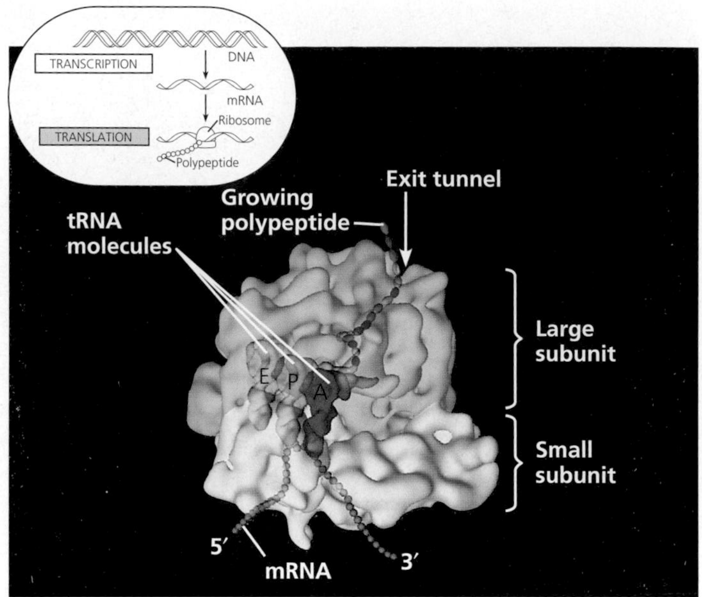

(a) Computer model of functioning ribosome. This is a model of a bacterial ribosome, showing its overall shape. The eukaryotic ribosome is roughly similar. A ribosomal subunit is an aggregate of ribosomal RNA molecules and proteins.

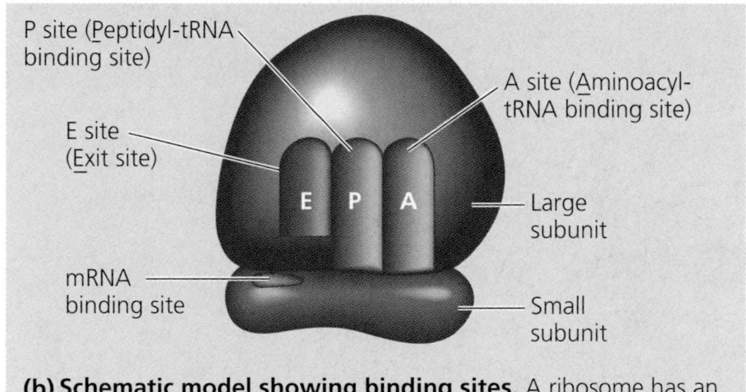

(b) Schematic model showing binding sites. A ribosome has an mRNA binding site and three tRNA binding sites, known as the A, P, and E sites. This schematic ribosome will appear in later diagrams.

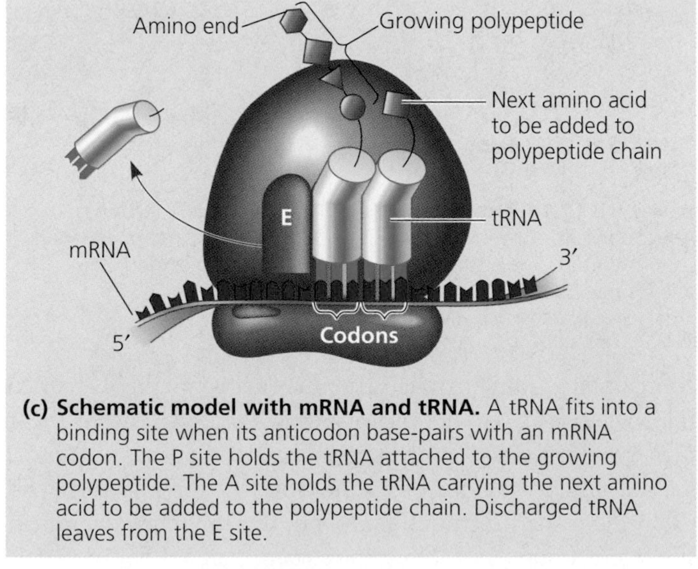

(c) Schematic model with mRNA and tRNA. A tRNA fits into a binding site when its anticodon base-pairs with an mRNA codon. The P site holds the tRNA attached to the growing polypeptide. The A site holds the tRNA carrying the next amino acid to be added to the polypeptide chain. Discharged tRNA leaves from the E site.

▲ **Figure 17.16 The anatomy of a functioning ribosome.**

positions the new amino acid for addition to the carboxyl end of the growing polypeptide. It then catalyzes the formation of the peptide bond. As the polypeptide becomes longer, it passes through an *exit tunnel* in the ribosome's large subunit. When the polypeptide is complete, it is released to the cytosol through the exit tunnel.

Four decades of genetic and biochemical research on ribosome structure have culminated in the detailed structure of the bacterial ribosome, which appears as a ribbon model in Figure 17.1. Recent research strongly supports the hypothesis that rRNA, not protein, is primarily responsible for both structure and function of the ribosome. The proteins, which are largely on the exterior, support the conformational changes of the rRNA molecules as they carry out catalysis during translation. RNA is the main constituent of the interface between the two subunits and of the A and P sites, and it is the catalyst of peptide bond formation. Thus, a ribosome can be regarded as one colossal ribozyme!

Building a Polypeptide

We can divide translation, the synthesis of a polypeptide chain, into three stages (analogous to those of transcription): initiation, elongation, and termination. All three stages require protein "factors" that aid mRNA, tRNA, and ribosomes in the translation process. For certain aspects of chain initiation and elongation, energy is also required. It is provided by the hydrolysis of GTP (guanosine triphosphate), a molecule closely related to ATP.

Ribosome Association and Initiation of Translation

The initiation stage of translation brings together mRNA, a tRNA bearing the first amino acid of the polypeptide, and the two subunits of a ribosome **(Figure 17.17)**. First, a small ribosomal subunit binds to both mRNA and a specific initiator tRNA, which carries the amino acid methionine. The small subunit then moves, or *scans,* downstream along the mRNA until it reaches the start codon, AUG, which signals the start of translation; this is important because it establishes the codon reading frame for the mRNA. The initiator tRNA, already associated with the complex, then hydrogen-bonds with the start codon.

The union of mRNA, initiator tRNA, and a small ribosomal subunit is followed by the attachment of a large ribosomal subunit, completing a translation initiation complex. Proteins called *initiation factors* are required to bring all these components together. The cell also spends energy in the form of a GTP molecule to form the initiation complex. At the completion of the initiation process, the initiator tRNA sits in the P site of the ribosome, and the vacant A site is ready for the next aminoacyl tRNA. Note that a polypeptide is always synthesized in one direction, from the initial methionine at the amino end, also called the N-terminus, toward the final amino acid at the carboxyl end, also called the C-terminus (see Figure 5.18).

Elongation of the Polypeptide Chain

In the elongation stage of translation, amino acids are added one by one to the preceding amino acid. Each addition involves the participation of several proteins called *elongation factors* and occurs in a three-step cycle described in **Figure 17.18** on the next page. Energy expenditure occurs in the first and third steps. Codon recognition requires hydrolysis of two molecules of GTP, which increases the accuracy and efficiency of this step. One more GTP is hydrolyzed to provide energy for the translocation step.

The mRNA is moved through the ribosome in one direction only, 5′ end first; this is equivalent to the ribosome moving 5′ → 3′ on the mRNA. The important point is that the ribosome and the mRNA move relative to each other, unidirectionally, codon by codon. The elongation cycle takes less than a tenth of a second in prokaryotes and is repeated as each amino acid is added to the chain until the polypeptide is completed.

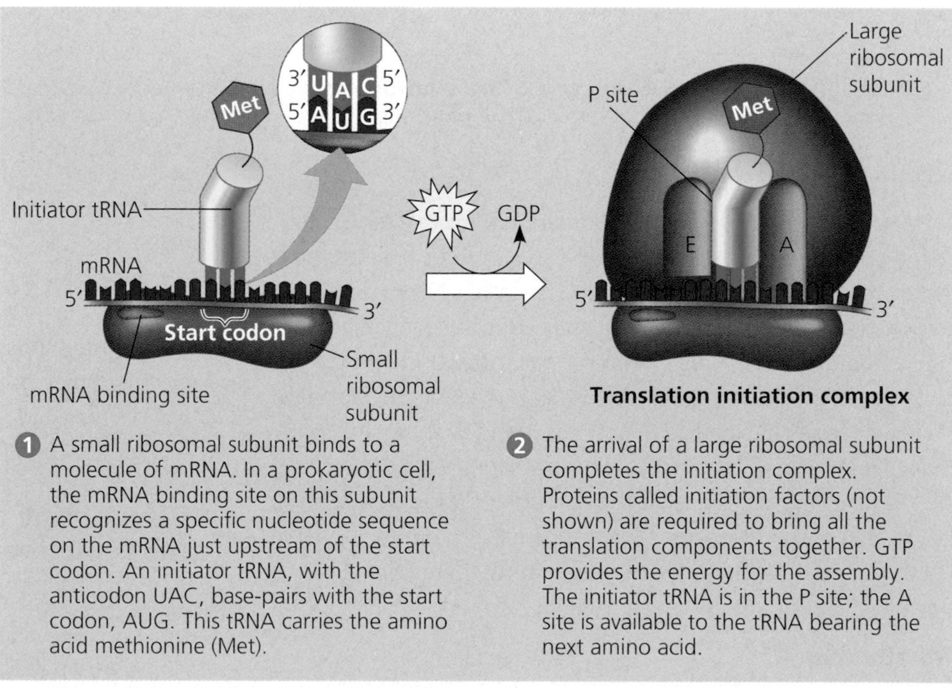

3′ U A C 5′
5′ A U G 3′
Met
Initiator tRNA
mRNA
5′
3′
Start codon
mRNA binding site
Small ribosomal subunit
GTP
GDP
P site
Large ribosomal subunit
Met
E
A
5′
3′
Translation initiation complex
1 A small ribosomal subunit binds to a molecule of mRNA. In a prokaryotic cell, the mRNA binding site on this subunit recognizes a specific nucleotide sequence on the mRNA just upstream of the start codon. An initiator tRNA, with the anticodon UAC, base-pairs with the start codon, AUG. This tRNA carries the amino acid methionine (Met).
2 The arrival of a large ribosomal subunit completes the initiation complex. Proteins called initiation factors (not shown) are required to bring all the translation components together. GTP provides the energy for the assembly. The initiator tRNA is in the P site; the A site is available to the tRNA bearing the next amino acid.

▲ **Figure 17.17 The initiation of translation.**

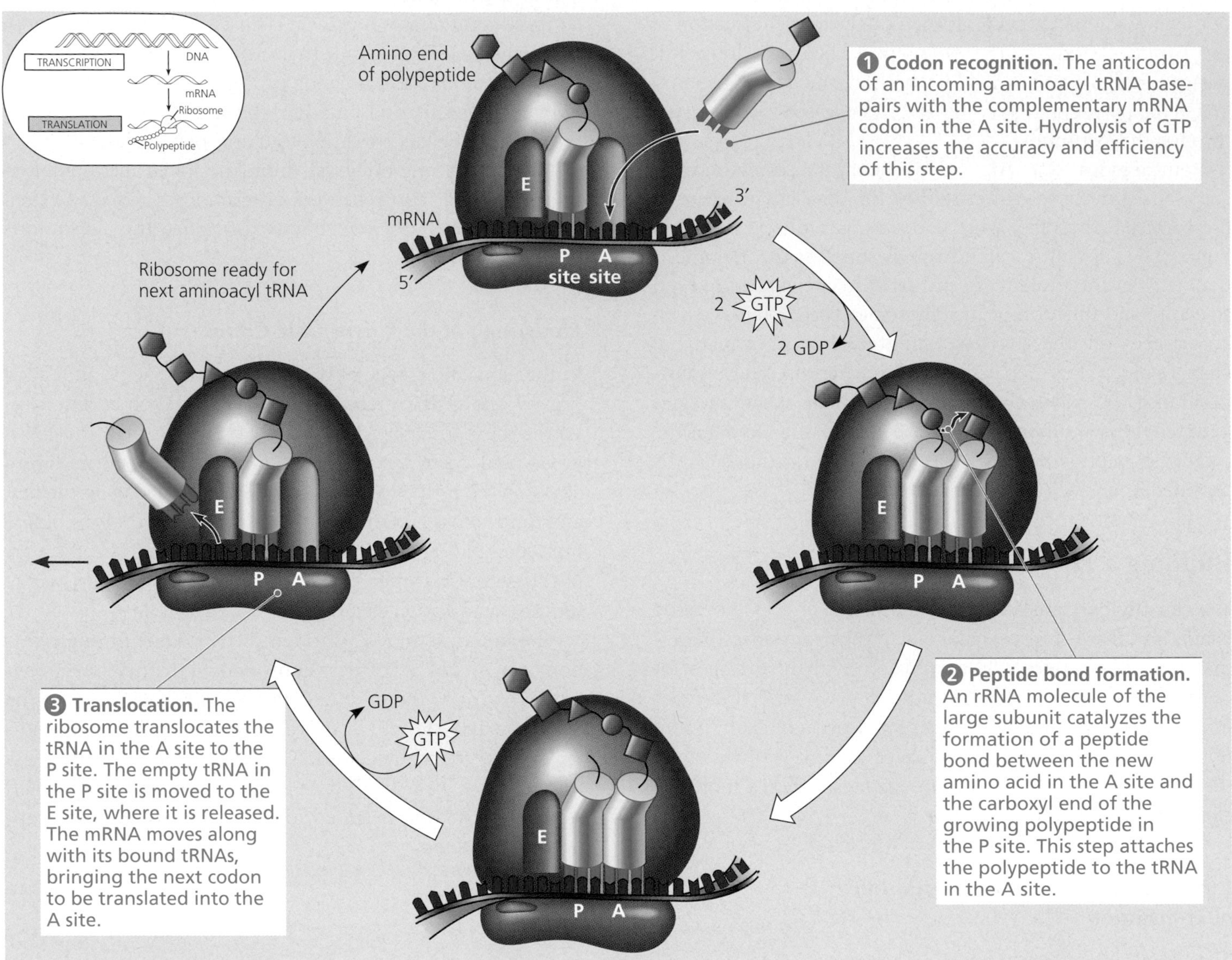

TRANSCRIPTION
DNA
mRNA
Ribosome
TRANSLATION
Polypeptide
Amino end
of polypeptide
1 Codon recognition. The anticodon of an incoming aminoacyl tRNA base-pairs with the complementary mRNA codon in the A site. Hydrolysis of GTP increases the accuracy and efficiency of this step.
E
3′
mRNA
5′
P site
A site
Ribosome ready for
next aminoacyl tRNA
2 GTP
2 GDP
E
P A
E
P A
2 Peptide bond formation. An rRNA molecule of the large subunit catalyzes the formation of a peptide bond between the new amino acid in the A site and the carboxyl end of the growing polypeptide in the P site. This step attaches the polypeptide to the tRNA in the A site.
3 Translocation. The ribosome translocates the tRNA in the A site to the P site. The empty tRNA in the P site is moved to the E site, where it is released. The mRNA moves along with its bound tRNAs, bringing the next codon to be translated into the A site.
GDP
GTP
E
P A

▲ **Figure 17.18 The elongation cycle of translation.** Not shown in this diagram are the proteins called elongation factors. The hydrolysis of GTP plays an important role in the elongation process.

Termination of Translation

The final stage of translation is termination (**Figure 17.19**). Elongation continues until a stop codon in the mRNA reaches the A site of the ribosome. The base triplets UAG, UAA, and UGA do not code for amino acids but instead act as signals to stop translation. A protein called a *release factor* binds directly to the stop codon in the A site. The release factor causes the addition of a water molecule instead of an amino acid to the polypeptide chain. This reaction hydrolyzes the completed polypeptide from the tRNA in the P site, releasing the polypeptide through the exit tunnel of the ribosome's large subunit (see Figure 17.16a). The remainder of the translation assembly then comes apart.

Polyribosomes

A single ribosome can make an average-sized polypeptide in less than a minute. Typically, however, a single mRNA is used to make many copies of a polypeptide simultaneously because several ribosomes can translate the message from one mRNA at the same time. Once a ribosome moves past the start codon, a second ribosome can attach to the mRNA; thus, a number of ribosomes may trail along one mRNA. Such strings of ribosomes, called **polyribosomes** (or **polysomes**), can be seen with an electron microscope (**Figure 17.20**). Polyribosomes are found in both prokaryotic and eukaryotic cells. They enable a cell to make many copies of a polypeptide very quickly.

Completing and Targeting the Functional Protein

The process of translation is often not sufficient to make a functional protein. In this section, you will learn about modifications that polypeptide chains undergo after the translation process as well as some of the mechanisms used to target completed proteins to specific sites in the cell.

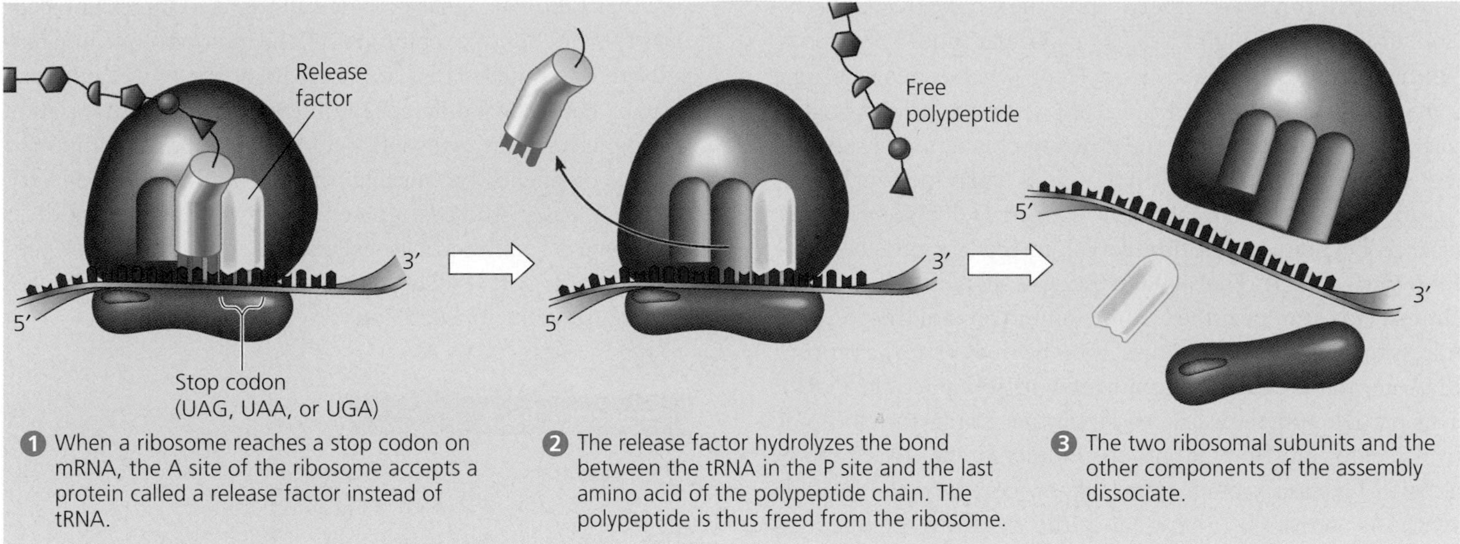

Release factor
Free polypeptide
Stop codon (UAG, UAA, or UGA)
5′
3′
5′
3′
5′
3′
1 When a ribosome reaches a stop codon on mRNA, the A site of the ribosome accepts a protein called a release factor instead of tRNA.
2 The release factor hydrolyzes the bond between the tRNA in the P site and the last amino acid of the polypeptide chain. The polypeptide is thus freed from the ribosome.
3 The two ribosomal subunits and the other components of the assembly dissociate.

▲ **Figure 17.19 The termination of translation.**

Protein Folding and Post-Translational Modifications

During its synthesis, a polypeptide chain begins to coil and fold spontaneously, forming a functional protein of specific conformation: a three-dimensional molecule with secondary and tertiary structure (see Figure 5.20). A gene determines primary structure, and primary structure in turn determines conformation. In many cases, a chaperone protein (chaperonin) helps the polypeptide fold correctly (see Figure 5.23).

Additional steps—*post-translational modifications*—may be required before the protein can begin doing its particular job in the cell. Certain amino acids may be chemically modified by the attachment of sugars, lipids, phosphate groups, or other additions. Enzymes may remove one or more amino acids from the leading (amino) end of the polypeptide chain. In some cases, a single polypeptide chain may be enzymatically cleaved into two or more pieces. For example, the protein insulin is first synthesized as a single polypeptide chain but becomes active only after an enzyme cuts out a central part of the chain, leaving a protein made up of two polypeptide chains connected by disulfide bridges. In other cases, two or more polypeptides that are synthesized separately may come together, becoming the subunits of a protein that has quaternary structure.

Targeting Polypeptides to Specific Locations

In electron micrographs of eukaryotic cells active in protein synthesis, two populations of ribosomes (and polyribosomes) are evident: free and bound (see Figure 6.11). Free ribosomes are suspended in the cytosol and mostly synthesize proteins that dissolve in the cytosol and function there. In contrast, bound ribosomes are attached to the cytosolic side of the endoplasmic reticulum (ER), or to the nuclear envelope. Bound ribosomes make proteins of the endomembrane system (the nuclear envelope, ER, Golgi apparatus, lysosomes, vacuoles, and plasma membrane) as well as proteins secreted from the cell, such as insulin. The ribosomes themselves are identical and can switch their status from free to bound.

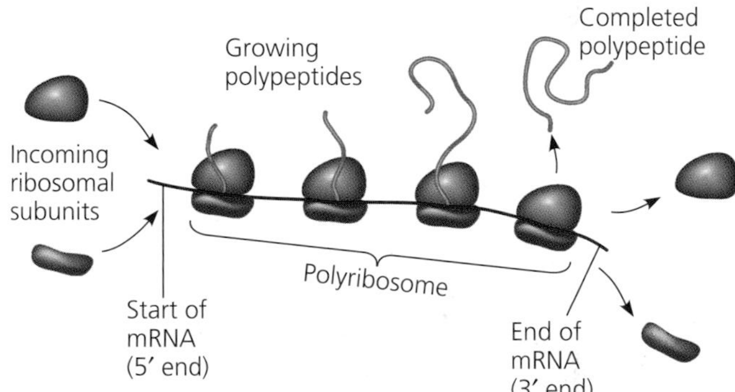

(a) An mRNA molecule is generally translated simultaneously by several ribosomes in clusters called polyribosomes.

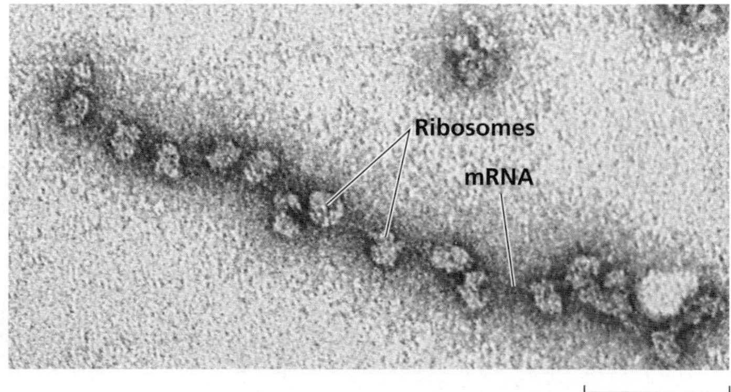

(b) This micrograph shows a large polyribosome in a prokaryotic cell (TEM).

▲ **Figure 17.20 Polyribosomes.**

What determines whether a ribosome will be free in the cytosol or bound to rough ER at any particular time? Polypeptide synthesis always begins in the cytosol, when a free ribosome starts to translate an mRNA molecule. There the process continues to completion—*unless* the growing polypeptide itself cues the ribosome to attach to the ER. The polypeptides of proteins destined for the endomembrane system or for secretion are marked by a **signal peptide,** which targets the protein to the ER **(Figure 17.21)**. The signal peptide, a sequence of about 20 amino acids at or near the leading (amino) end of the polypeptide, is recognized as it emerges from the ribosome by a protein-RNA complex called a **signal-recognition particle (SRP).** This particle functions as an adapter that brings the ribosome to a receptor protein built into the ER membrane. This receptor is part of a multiprotein translocation complex. Polypeptide synthesis continues there, and the growing polypeptide snakes across the membrane into the ER lumen via a protein pore. The signal peptide is usually removed by an enzyme. The rest of the completed polypeptide, if it is to be a secretory protein, is released into solution within the ER lumen (as in Figure 17.21). Alternatively, if the polypeptide is to be a membrane protein, it remains partially embedded in the ER membrane.

Other kinds of signal peptides are used to target polypeptides to mitochondria, chloroplasts, the interior of the nucleus, and other organelles that are not part of the endomembrane system. The critical difference in these cases is that translation is completed in the cytosol before the polypeptide is imported into the organelle. The mechanisms of translocation also vary, but in all cases studied to date, the "zip codes" that address proteins for secretion or to cellular locations are signal peptides of some sort. Prokaryotes also employ signal sequences to target proteins for secretion.

Concept Check 17.4

1. Which two processes ensure that the correct amino acid is added to a growing polypeptide chain?
2. Describe how the formation of polyribosomes can benefit the cell.
3. Describe how a polypeptide to be secreted is transported to the endomembrane system.

For suggested answers, see Appendix A.

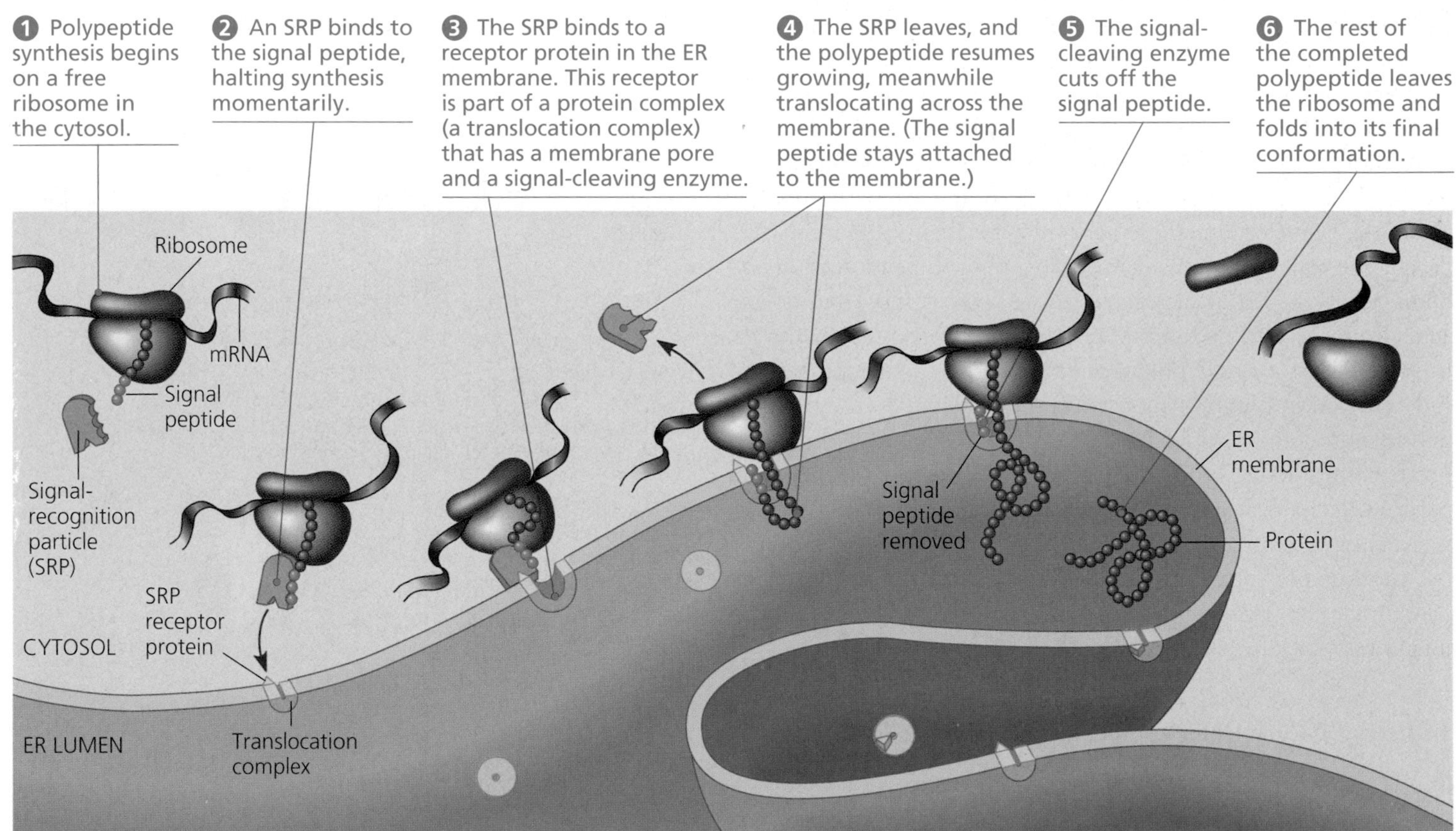

❶ Polypeptide synthesis begins on a free ribosome in the cytosol.
❷ An SRP binds to the signal peptide, halting synthesis momentarily.
❸ The SRP binds to a receptor protein in the ER membrane. This receptor is part of a protein complex (a translocation complex) that has a membrane pore and a signal-cleaving enzyme.
❹ The SRP leaves, and the polypeptide resumes growing, meanwhile translocating across the membrane. (The signal peptide stays attached to the membrane.)
❺ The signal-cleaving enzyme cuts off the signal peptide.
❻ The rest of the completed polypeptide leaves the ribosome and folds into its final conformation.
Ribosome
mRNA
Signal peptide
Signal-recognition particle (SRP)
SRP receptor protein
CYTOSOL
ER LUMEN
Translocation complex
Signal peptide removed
ER membrane
Protein

▲ **Figure 17.21 The signal mechanism for targeting proteins to the ER.** A polypeptide destined for the endomembrane system or for secretion from the cell begins with a signal peptide, a series of amino acids that targets it for the ER. This figure shows the synthesis of a secretory protein and its simultaneous import into the ER. In the ER and then in the Golgi, the protein is further processed. Finally, a transport vesicle conveys it to the plasma membrane for release from the cell (see Figure 7.10).

Concept 17.5

RNA plays multiple roles in the cell: *a review*

As we have seen, the cellular machinery of protein synthesis (and ER targeting) is dominated by RNA of various kinds. In addition to mRNA, these include tRNA, rRNA, and, in eukaryotes, snRNA and SRP RNA **(Table 17.1)**. A type of RNA called *small nucleolar RNA* (*snoRNA*) aids in processing pre-rRNA transcripts in the nucleolus, a process necessary for ribosome formation. The diverse functions of these small RNA molecules range from structural to informational to catalytic. Recent research has also revealed the presence of small, single- and double-stranded RNA molecules that play unexpectedly important roles in regulating which genes get expressed. These types of RNA are called *small interfering RNA (siRNA)* and *microRNA (miRNA)* (see Chapter 19).

The ability of RNA to perform so many different functions is based on three properties. First, RNA can hydrogen-bond to other nucleic acid molecules (DNA or RNA). Second, it can assume a specific three-dimensional shape by forming hydrogen bonds between bases in different parts of its own polynucleotide chain (as seen in tRNA; see Figure 17.14). Third, it has functional groups that allow it to act as a catalyst (ribozyme). These three properties make RNA quite multifunctional.

DNA may be the genetic material of all living cells, but RNA is much more versatile. You will learn in Chapter 18 that many viruses use RNA rather than DNA as their genetic material. In the past few years, scientists have begun to appreciate the diverse functions carried out by RNA molecules. In fact, the journal *Science* bestowed its 2002 "Breakthrough of the Year" award on the discovery of the small regulatory RNA molecules siRNA and miRNA.

Table 17.1 Types of RNA in a Eukaryotic Cell

Type of RNA	Functions
Messenger RNA (mRNA)	Carries information specifying amino acid sequences of proteins from DNA to ribosomes.
Transfer RNA (tRNA)	Serves as adapter molecule in protein synthesis; translates mRNA codons into amino acids.
Ribosomal RNA (rRNA)	Plays catalytic (ribozyme) roles and structural roles in ribosomes.
Primary transcript	Serves as a precursor to mRNA, rRNA, or tRNA, before being processed by splicing or cleavage. Some intron RNA acts as a ribozyme, catalyzing its own splicing.
Small nuclear RNA (snRNA)	Plays structural and catalytic roles in spliceosomes, the complexes of protein and RNA that splice pre-mRNA.
SRP RNA	Is a component of the signal-recognition particle (SRP), the protein-RNA complex that recognizes the signal peptides of polypeptides targeted to the ER.
Small nucleolar RNA (snoRNA)	Aids in processing of pre-rRNA transcripts for ribosome subunit formation in the nucleolus.
Small interfering RNA (siRNA) and microRNA (miRNA)	Are involved in regulation of gene expression.

Concept Check 17.5

1. Describe three properties of RNA that allow it to perform diverse roles in the cell.

For suggested answers, see Appendix A.

Concept 17.6

Comparing gene expression in prokaryotes and eukaryotes reveals key differences

Although prokaryotes and eukaryotes carry out transcription and translation in very similar ways, we have noted certain differences in cellular machinery and in details of the processes. Prokaryotic and eukaryotic RNA polymerases are different, and those of eukaryotes depend on a complex set of transcription factors. Transcription is terminated differently in the two kinds of cells. Also, prokaryotic and eukaryotic ribosomes are slightly different. The most important differences, however, arise from the eukaryotic cell's compartmental organization. Like a one-room workshop, a prokaryotic cell ensures a streamlined operation. In the absence of a nucleus, it can simultaneously transcribe and translate the same gene (**Figure 17.22**, on the next page), and the newly made protein can quickly diffuse to its site of function. In contrast, the eukaryotic cell's nuclear envelope segregates transcription from translation and provides a compartment for extensive RNA processing. This processing stage provides additional steps whose regulation can help coordinate the eukaryotic cell's elaborate activities (see Chapter 19). Finally, eukaryotic cells have complicated mechanisms for targeting proteins to the appropriate cellular compartment (organelle).

Where did eukaryotes and prokaryotes get the genes that encode the huge diversity of proteins they synthesize? For the past few billion years, the ultimate source of new genes has been the mutation of preexisting genes, the topic of the next section.

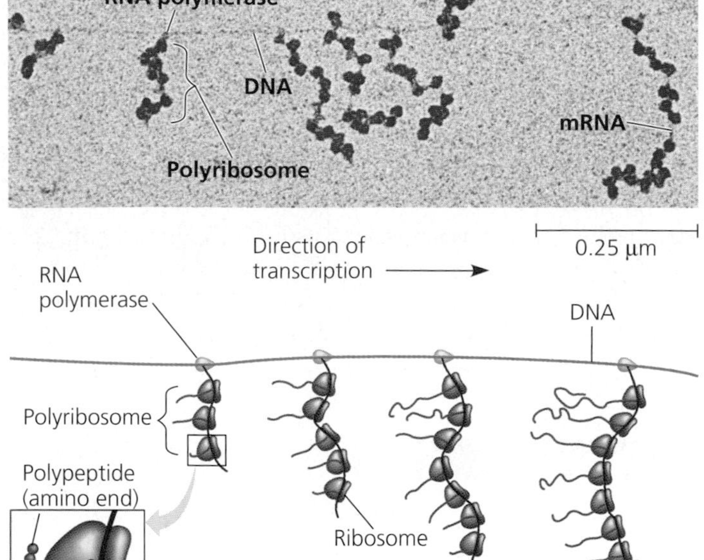

▲ **Figure 17.22 Coupled transcription and translation in bacteria.** In prokaryotic cells, the translation of mRNA can begin as soon as the leading (5′) end of the mRNA molecule peels away from the DNA template. The micrograph (TEM) shows a strand of *E. coli* DNA being transcribed by RNA polymerase molecules. Attached to each RNA polymerase molecule is a growing strand of mRNA, which is already being translated by ribosomes. The newly synthesized polypeptides are not visible in the micrograph but are shown in the diagram.

Concept Check 17.6

1. In Figure 17.22, number the RNA polymerases in order of their initiation of transcription. Then number each mRNA's ribosomes in order of their initiation of translation.
2. Would the arrangement shown in Figure 17.22 be found in a eukaryotic cell? Explain.

For suggested answers, see Appendix A.

Concept 17.7

Point mutations can affect protein structure and function

Mutations are changes in the genetic material of a cell (or virus). In Figure 15.14, we considered large-scale mutations, chromosomal rearrangements that affect long segments of DNA. Now we can examine **point mutations**, chemical changes in just one base pair of a gene.

If a point mutation occurs in a gamete or in a cell that gives rise to gametes, it may be transmitted to offspring and to a succession of future generations. If the mutation has an adverse effect on the phenotype of an organism, the mutant condition is referred to as a genetic disorder, or hereditary disease. For example, we can trace the genetic basis of sickle-cell disease to a mutation of a single base pair in the gene that codes for one of the polypeptides of hemoglobin. The change of a single nucleotide in the DNA's template strand leads to the production of an abnormal protein (**Figure 17.23**, and see Figure 5.21). In individuals who are homozygous for the mutant allele, the sickling of red blood cells caused by the altered hemoglobin produces the multiple symptoms associated with sickle-cell disease (see Chapter 14). Let's see how different types of point mutations translate into altered proteins.

Types of Point Mutations

Point mutations within a gene can be divided into two general categories: base-pair substitutions and base-pair insertions or deletions. While reading about how these mutations affect proteins, refer to **Figures 17.24** and **17.25**, on the next two pages.

Substitutions

A **base-pair substitution** is the replacement of one nucleotide and its partner with another pair of nucleotides. Some substitutions are called *silent mutations* because, owing to the redundancy of the genetic code, they have no effect on the encoded protein. In other words, a change in a base pair may transform one codon into another that is translated into the same amino acid. For example, if 3′-CCG-5′ on the template strand mutated to 3′-CCA-5′, the mRNA codon that used to be GGC would become GGU, and a glycine would still be inserted at the proper location in the protein (see Figure 17.5). Other substitutions may change an amino acid but have little effect on the protein. The new amino acid may have properties similar to those of the amino acid it replaces, or it may be in a region of the protein where the exact sequence of amino acids is not essential to the protein's function.

However, the base-pair substitutions of greatest interest are those that cause a readily detectable change in a protein. The alteration of a single amino acid in a crucial area of a protein—in the active site of an enzyme, for example—will significantly alter protein activity. Occasionally, such a mutation leads to an improved protein or one with novel capabilities, but much more often such mutations are detrimental, leading to a useless or less active protein that impairs cellular function.

Substitution mutations are usually **missense mutations**; that is, the altered codon still codes for an amino acid and thus makes sense, although not necessarily the *right* sense. But a point mutation can also change a codon for an amino acid into a stop codon. This is called a **nonsense mutation**, and it causes translation to be terminated prematurely; the resulting polypeptide will be shorter than the polypeptide encoded by the normal gene (see Figure 17.24). Nearly all nonsense mutations lead to nonfunctional proteins.

▶ **Figure 17.23 The molecular basis of sickle-cell disease: a point mutation.** The allele that causes sickle-cell disease differs from the wild-type (normal) allele by a single DNA base pair.

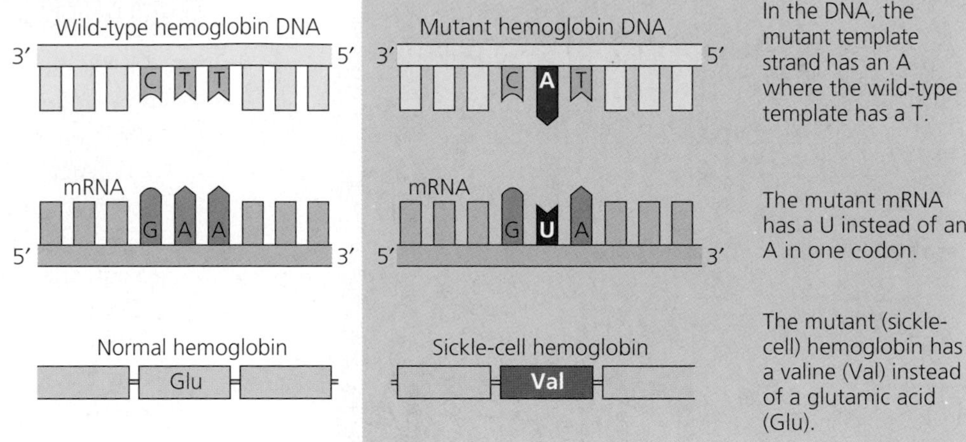

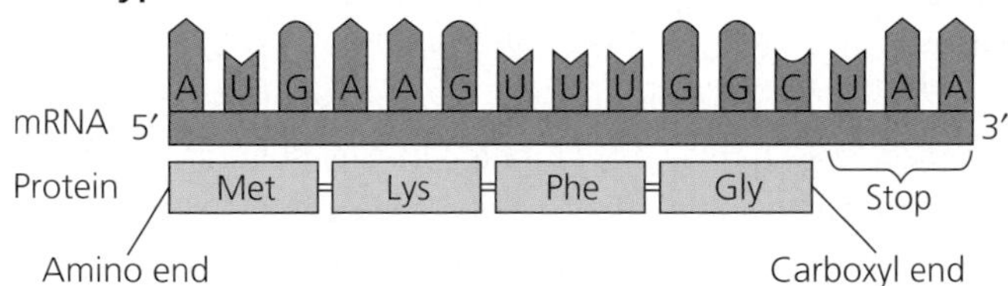

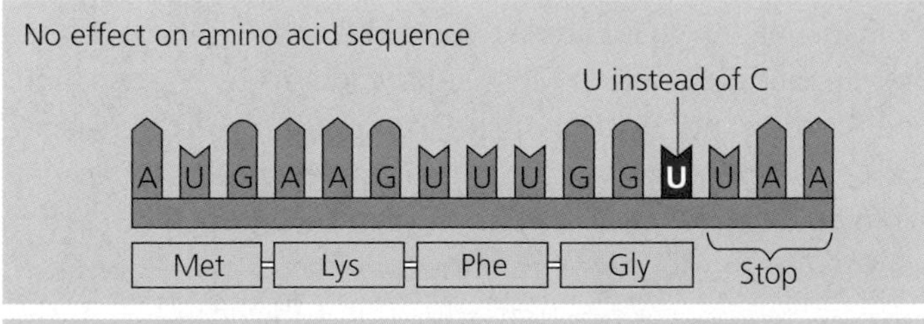

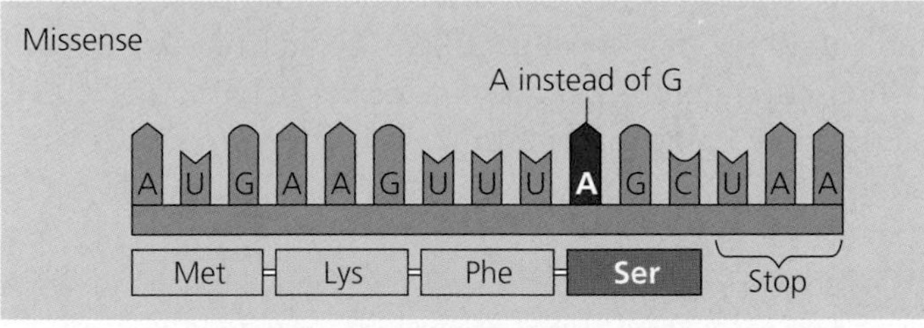

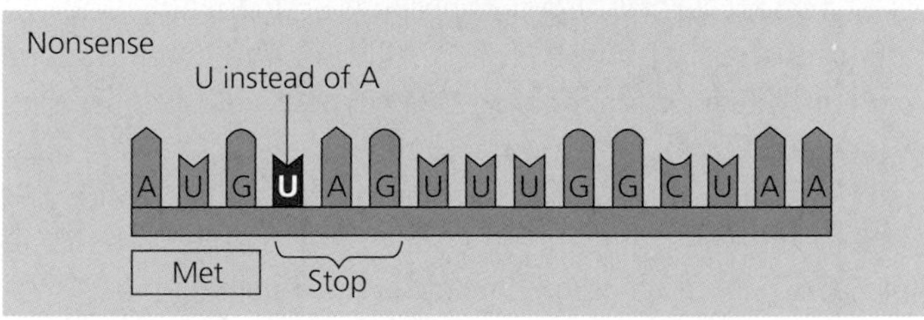

▲ **Figure 17.24 Base-pair substitution.** Mutations are changes in DNA, but they are represented here as they are reflected in mRNA and its protein product. Base-pair substitutions may lead to silent, missense, or nonsense mutations.

Insertions and Deletions

Insertions and **deletions** are additions or losses of nucleotide pairs in a gene. These mutations have a disastrous effect on the resulting protein more often than substitutions do. Because mRNA is read as a series of nucleotide triplets during translation, the insertion or deletion of nucleotides may alter the reading frame (triplet grouping) of the genetic message. Such a mutation, called a **frameshift mutation,** will occur whenever the number of nucleotides inserted or deleted is not a multiple of three (see Figure 17.25, on the next page). All the nucleotides that are downstream of the deletion or insertion will be improperly grouped into codons, and the result will be extensive missense probably ending sooner or later in nonsense and premature termination. Unless the frameshift is very near the end of the gene, it will produce a protein that is almost certain to be nonfunctional.

Mutagens

Mutations can arise in a number of ways. Errors during DNA replication, repair, or recombination can lead to base-pair substitutions, insertions, or deletions, as well as to mutations affecting longer stretches of DNA. Mutations resulting from such errors are called *spontaneous mutations.* It is difficult to calculate the rate at which such mutations occur. Rough estimates have been made of the rate of mutation during DNA replication for both *E. coli* and eukaryotes, and the numbers are similar: About 1 nucleotide in every 10^{10} is altered and passed on to the next generation of cells.

A number of physical and chemical agents, called **mutagens,** interact with DNA in ways that cause mutations. In the 1920s, Hermann Muller discovered that X-rays caused genetic changes in fruit flies. With X-rays, he was able to make *Drosophila* mutants that he could use in his genetic studies. But he also recognized an alarming implication of his discovery: X-rays and other forms of high-energy radiation pose hazards to the genetic material of people as well as laboratory

Wild type

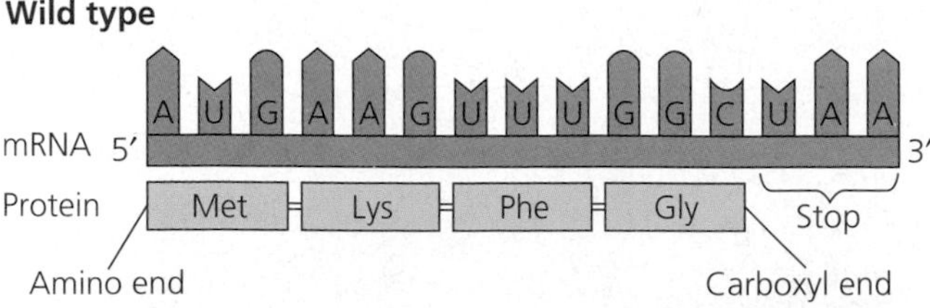

Base-pair insertion or deletion

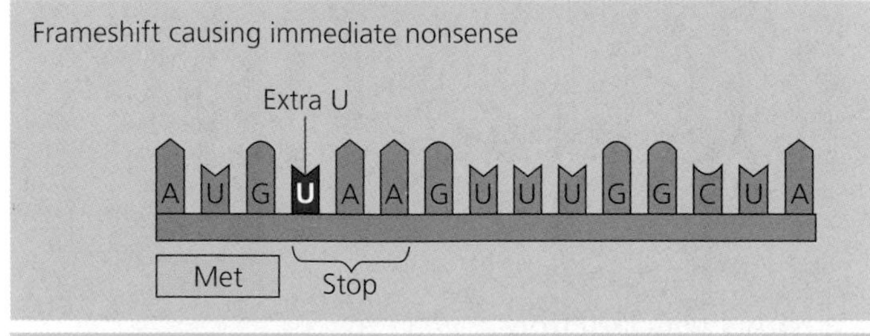

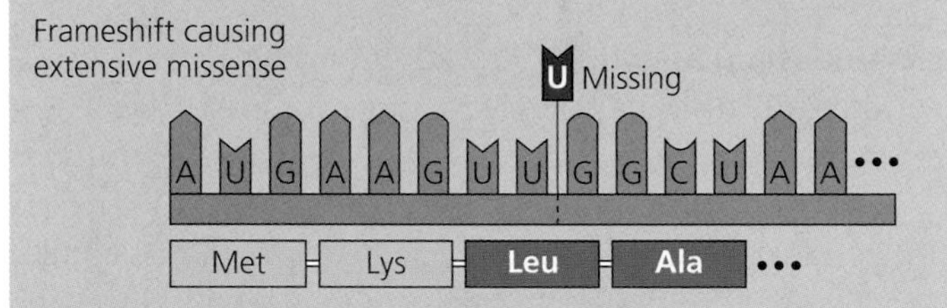

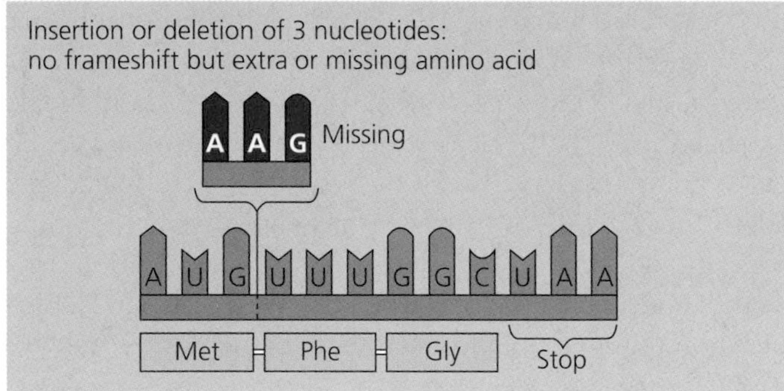

▲ **Figure 17.25 Base-pair insertion or deletion.** Strictly speaking, the example at the bottom is not a point mutation because it involves insertion or deletion of more than one nucleotide.

organisms. Mutagenic radiation, a physical mutagen, includes ultraviolet (UV) light, which can cause disruptive thymine dimers in DNA (see Figure 16.17).

Chemical mutagens fall into several categories. Base analogs are chemicals that are similar to normal DNA bases but that pair incorrectly during DNA replication. Some other chemical mutagens interfere with correct DNA replication by inserting themselves into the DNA and distorting the double helix. Still other mutagens cause chemical changes in bases that change their pairing properties.

Researchers have developed various methods to test the mutagenic activity of different chemicals. A major application of these tests is the preliminary screening of chemicals to identify those that may cause cancer. This approach makes sense because most carcinogens (cancer-causing chemicals) are mutagenic and, conversely, most mutagens are carcinogenic.

Concept Check 17.7

1. What happens when one nucleotide pair is lost from the middle of the coding sequence of a gene?
2. The template strand of a gene contains the sequence 3′-TACTTGTCCGATATC-5′. Draw the double strand of DNA and the resulting mRNA, labeling all 5′ and 3′ ends. Determine the amino acid sequence. Then show the same after a mutation changes the template DNA sequence to 3′-TACTTGTCCAATATC-5′. What is the effect on the amino acid sequence?

For suggested answers, see Appendix A.

What is a gene? *revisiting the question*

Our definition of a gene has evolved over the past few chapters, as it has through the history of genetics. We began with the Mendelian concept of a gene as a discrete unit of inheritance that affects a phenotypic character (Chapter 14). We saw that Morgan and his colleagues assigned such genes to specific loci on chromosomes (Chapter 15). We went on to view a gene as a region of specific nucleotide sequence along the length of a DNA molecule (Chapter 16). Finally, in this chapter, we have considered a functional definition of a gene as a DNA sequence coding for a specific polypeptide chain. All these definitions are useful, depending on the context in which genes are being studied. (**Figure 17.26** summarizes the path from gene to polypeptide in a eukaryotic cell.)

Even the one gene–one polypeptide model must be refined and applied selectively. Most eukaryotic genes contain noncoding segments (introns), so large portions of these genes have no corresponding segments in polypeptides. Molecular biologists also often include promoters and certain other regulatory regions of DNA within the boundaries of a gene. These DNA sequences are not transcribed, but they can be considered part of the functional gene because they must be present for transcription to occur. Our molecular definition of a gene must also be broad enough to include the DNA that is transcribed into rRNA, tRNA, and other RNAs that are not translated. These genes have no polypeptide products. Thus, we arrive at the following definition: *A gene is a region of DNA whose final product is either a polypeptide or an RNA molecule.*

For most genes, however, it is still useful to retain the one gene–one polypeptide idea. In this chapter, you have learned in molecular terms how a typical gene is expressed—by transcription into RNA and then translation into a polypeptide that forms a protein of specific structure and function. Proteins, in turn, bring about an organism's observable phenotype.

Genes are regulated. We will explore the regulation of gene expression in eukaryotes in Chapters 19 and 21. In the next chapter, we begin our discussion of gene regulation by focusing on the simpler molecular biology of bacteria and viruses.

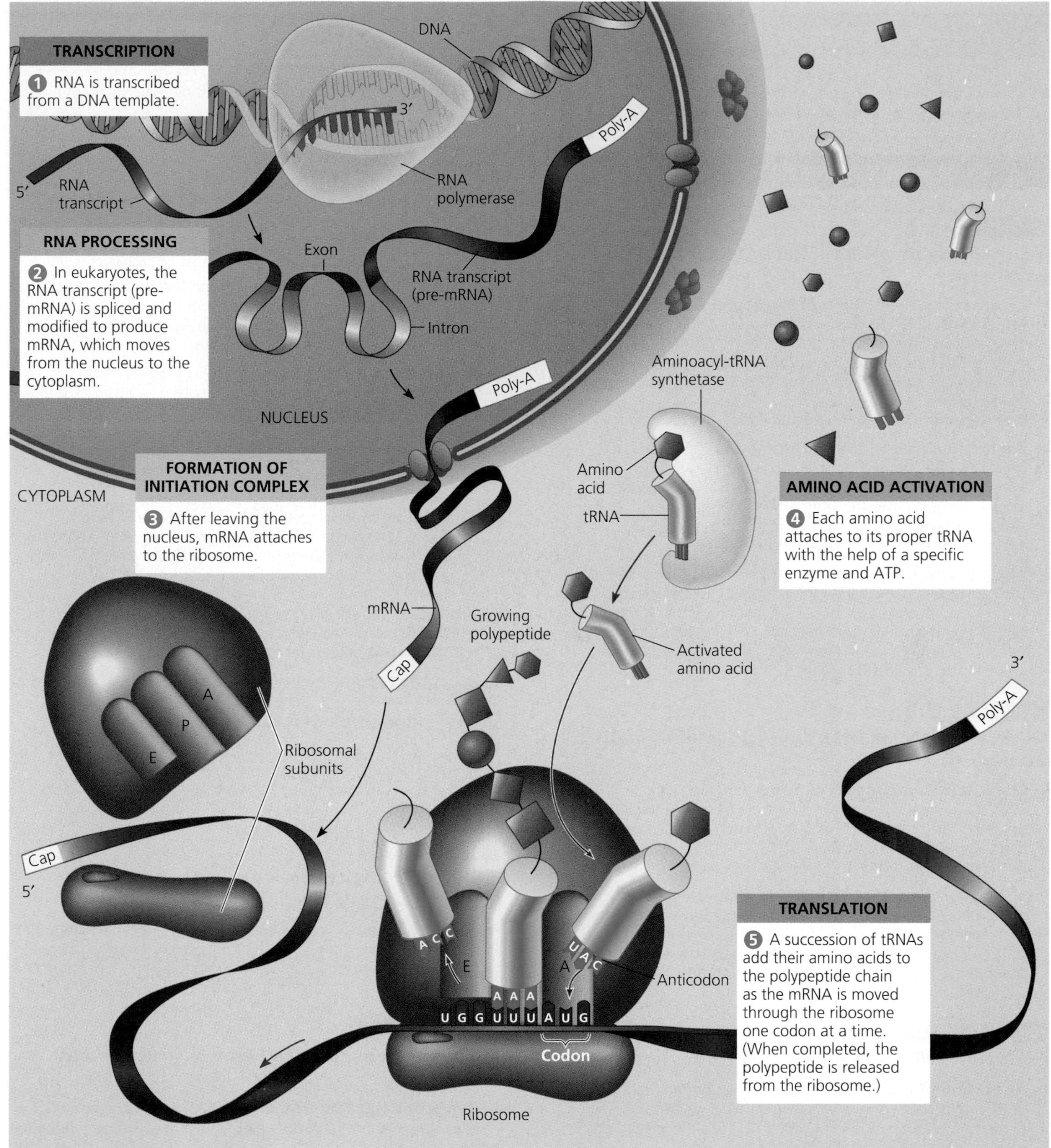
DNA
TRANSCRIPTION
1 RNA is transcribed from a DNA template.
3′
5′
RNA transcript
RNA polymerase
Poly-A
RNA PROCESSING
2 In eukaryotes, the RNA transcript (pre-mRNA) is spliced and modified to produce mRNA, which moves from the nucleus to the cytoplasm.
Exon
RNA transcript (pre-mRNA)
Intron
Poly-A
NUCLEUS
CYTOPLASM
FORMATION OF INITIATION COMPLEX
3 After leaving the nucleus, mRNA attaches to the ribosome.
Aminoacyl-tRNA synthetase
Amino acid
tRNA
AMINO ACID ACTIVATION
4 Each amino acid attaches to its proper tRNA with the help of a specific enzyme and ATP.
mRNA
Growing polypeptide
Activated amino acid
Cap
A
P
E
Ribosomal subunits
3′
Poly-A
Cap
5′
E
A
A C C
U A C
A A A
U G G U U U A U G
Anticodon
Codon
Ribosome
TRANSLATION
5 A succession of tRNAs add their amino acids to the polypeptide chain as the mRNA is moved through the ribosome one codon at a time. (When completed, the polypeptide is released from the ribosome.)

▲ **Figure 17.26 A summary of transcription and translation in a eukaryotic cell.** This diagram shows the path from one gene to one polypeptide. Keep in mind that each gene in the DNA can be transcribed repeatedly into many RNA molecules, and that each mRNA can be translated repeatedly to yield many polypeptide molecules. (Also, remember that the final products of some genes are not polypeptides but RNA molecules, including tRNA and rRNA.) In general, the steps of transcription and translation are similar in prokaryotic and eukaryotic cells. The major difference is the occurrence of RNA processing in the eukaryotic nucleus. Other significant differences are found in the initiation stages of both transcription and translation and in the termination of transcription.

Chapter 17 Review

Go to www.campbellbiology.com or the student CD-ROM to explore Activities, Investigations, and other interactive study aids.

SUMMARY OF KEY CONCEPTS

Concept 17.1

Genes specify proteins via transcription and translation

▶ **Evidence from the Study of Metabolic Defects (pp. 309–311)** DNA controls metabolism by directing cells to make specific enzymes and other proteins. Beadle and Tatum's experiments with mutant strains of *Neurospora* supported the one gene–one enzyme hypothesis. Genes code for polypeptide chains or for RNA molecules.
Investigation *How Is a Metabolic Pathway Analyzed?*

▶ **Basic Principles of Transcription and Translation (pp. 311–312)** Transcription is the nucleotide-to-nucleotide transfer of information from DNA to RNA, while translation is the informational transfer from nucleotide sequence in RNA to amino acid sequence in a polypeptide.
Activity *Overview of Protein Synthesis*

▶ **The Genetic Code (pp. 312–314)** Genetic information is encoded as a sequence of nonoverlapping base triplets, or codons. A codon in messenger RNA (mRNA) either is translated into an amino acid (61 codons) or serves as a translational stop signal (3 codons). Codons must be read in the correct reading frame for the specified polypeptide to be produced.

Concept 17.2

Transcription is the DNA-directed synthesis of RNA: *a closer look*

▶ **Molecular Components of Transcription (pp. 315–316)** RNA synthesis is catalyzed by RNA polymerase. It follows the same base-pairing rules as DNA replication, except that in RNA, uracil substitutes for thymine.
Activity *Transcription*

▶ **Synthesis of an RNA Transcript (pp. 316–317)** The three stages of transcription are initiation, elongation, and termination. Promoters signal the initiation of RNA synthesis. Transcription factors help eukaryotic RNA polymerase recognize promoter sequences. The mechanisms of termination are different in prokaryotes and eukaryotes.

Concept 17.3

Eukaryotic cells modify RNA after transcription

▶ **Alteration of mRNA Ends (p. 317)** Eukaryotic mRNA molecules are processed before leaving the nucleus by modification of their ends and by RNA splicing. The 5′ end receives a modified nucleotide cap, and the 3′ end a poly-A tail.
Activity *RNA Processing*

▶ **Split Genes and RNA Splicing (pp. 318–319)** Most eukaryotic genes have introns interspersed among the coding regions, the exons. In RNA splicing, introns are removed and exons joined. RNA splicing is carried out by spliceosomes, but in some cases, RNA alone catalyzes splicing. Catalytic RNA molecules are called ribozymes. The presence of introns allows for alternative RNA splicing.

Concept 17.4

Translation is the RNA-directed synthesis of a polypeptide: *a closer look*

▶ **Molecular Components of Translation (pp. 320–323)** A cell translates an mRNA message into protein with the help of transfer RNA (tRNA). After binding specific amino acids, tRNA molecules line up by means of their anticodons at complementary codons on mRNA. Ribosomes help facilitate this coupling with binding sites for mRNA and tRNA.

▶ **Building a Polypeptide (pp. 323–325)** Ribosomes coordinate the three stages of translation: initiation, elongation, and termination. The formation of peptide bonds between amino acids is catalyzed by rRNA. A number of ribosomes can translate a single mRNA molecule simultaneously, forming a polyribosome.
Activity *Translation*
Biology Labs On-Line *TranslationLab*

▶ **Completing and Targeting the Functional Protein (pp. 324–326)** After translation, proteins may be modified in ways that affect their three-dimensional shape. Free ribosomes in the cytosol initiate the synthesis of all proteins, but proteins destined for the endomembrane system or for secretion must be transported into the ER. Such proteins have signal peptides to which a signal-recognition particle (SRP) binds, enabling the translating ribosome to bind to the ER.

Concept 17.5

RNA plays multiple roles in the cell: *a review*

▶ RNA can hydrogen-bond to other nucleic acid molecules (DNA or RNA). It can assume a specific three-dimensional shape. And it has functional groups that allow it to act as a catalyst, a ribozyme (p. 327).

Concept 17.6

Comparing gene expression in prokaryotes and eukaryotes reveals key differences

▶ Because prokaryotic cells lack a nuclear envelope, translation can begin while transcription is still in progress. In a eukaryotic cell, the nuclear envelope separates transcription from translation, and extensive RNA processing occurs in the nucleus (pp. 327–328).

Concept 17.7

Point mutations can affect protein structure and function

▶ **Types of Point Mutations (pp. 328–330)** A point mutation is a change in one DNA base pair, which may lead to production of a nonfunctional protein or no protein at all. Base-pair substitutions can cause missense or nonsense mutations. Base-pair insertions or deletions may produce frameshift mutations.

▶ **Mutagens (pp. 329–330)** Spontaneous mutations can occur during DNA replication, recombination, or repair. Chemical and physical mutagens can also alter genes.

TESTING YOUR KNOWLEDGE

Self-Quiz

1. Base-pair substitutions involving the third base of a codon are unlikely to result in an error in the polypeptide. This is because
   a. substitutions are corrected before transcription begins.
   b. substitutions are restricted to introns.
   c. the base-pairing rules are less strict for the third base of codons and anticodons.
   d. a signal-recognition particle corrects coding errors.
   e. transcribed errors attract snRNPs, which then stimulate splicing and correction.

2. In eukaryotic cells, transcription cannot begin until
   a. the two DNA strands have completely separated and exposed the promoter.
   b. several transcription factors have bound to the promoter.
   c. the 5′ caps are removed from the mRNA.
   d. the DNA introns are removed from the template.
   e. DNA nucleases have isolated the transcription unit.

3. Which of the following is *not* true of a codon?
   a. It consists of three nucleotides.
   b. It may code for the same amino acid as another codon.
   c. It never codes for more than one amino acid.
   d. It extends from one end of a tRNA molecule.
   e. It is the basic unit of the genetic code.

4. The metabolic pathway of arginine synthesis is as follows:

   Precursor $\xrightarrow{A}$ Ornithine $\xrightarrow{B}$ Citrulline $\xrightarrow{C}$ Arginine

   Beadle and Tatum discovered several classes of *Neurospora* mutants that were able to grow on minimal medium with arginine added (see Figure 17.2). They were able to conclude that
   a. one gene codes for the entire metabolic pathway.
   b. the genetic code of DNA is a triplet code.
   c. class I mutants have their mutations later in the nucleotide chain than do class II mutants.
   d. class I mutants have a nonfunctional enzyme at step A, and class II mutants have one at step B.
   e. class III mutants have nonfunctional enzymes for all three steps.

5. The anticodon of a particular tRNA molecule is
   a. complementary to the corresponding mRNA codon.
   b. complementary to the corresponding triplet in rRNA.
   c. the part of tRNA that bonds to a specific amino acid.
   d. changeable, depending on the amino acid that attaches to the tRNA.
   e. catalytic, making the tRNA a ribozyme.

6. Which of the following is *not* true of RNA processing?
   a. Exons are cut out before mRNA leaves the nucleus.
   b. Nucleotides may be added at both ends of the RNA.
   c. Ribozymes may function in RNA splicing.
   d. RNA splicing can be catalyzed by spliceosomes.
   e. A primary transcript is often much longer than the final RNA molecule that leaves the nucleus.

7. Which of the following is true of translation in both prokaryotes and eukaryotes?
   a. Translation is coupled to transcription.
   b. The product of transcription is immediately ready for translation.
   c. The codon UUU codes for phenylalanine.
   d. Ribosomes are affected by streptomycin.
   e. The signal-recognition particle (SRP) binds to the first 20 amino acids of certain polypeptides.

8. Using Figure 17.5, identify a 5′ → 3′ sequence of nucleotides in the DNA template strand for an mRNA coding for the polypeptide sequence Phe-Pro-Lys.
   a. UUU-GGG-AAA
   b. GAA-CCC-CTT
   c. AAA-ACC-TTT
   d. CTT-CGG-GAA
   e. AAA-CCC-UUU

9. Which of the following mutations would be *most* likely to have a harmful effect on an organism?
   a. a base-pair substitution
   b. a deletion of three nucleotides near the middle of a gene
   c. a single nucleotide deletion in the middle of an intron
   d. a single nucleotide deletion near the end of the coding sequence
   e. a single nucleotide insertion downstream of, and close to, the start of the coding sequence

10. Which component is *not* directly involved in translation?
   a. mRNA b. DNA c. tRNA d. ribosomes e. GTP

For Self-Quiz answers, see Appendix A.

Go to the website or CD-ROM for more quiz questions.

Evolution Connection

The genetic code (see Figure 17.5) is rich with evolutionary implications. For instance, notice that the 20 amino acids are not randomly scattered; most amino acids are coded for by a similar set of codons. What evolutionary explanations can be given for this pattern? (*Hint:* There is one explanation relating to historical ancestry, and some less obvious ones of a "form-fits-function" type.)

Scientific Inquiry

A biologist inserts a gene from a human liver cell into the chromosome of a bacterium. The bacterium then transcribes and translates this gene. The protein produced is useless and is found to contain many more amino acids than does the protein made by the eukaryotic cell. Explain why.

Investigation ***How Is a Metabolic Pathway Analyzed?***
Biology Labs On-Line ***TranslationLab***

Science, Technology, and Society

Our civilization generates many potentially mutagenic chemicals (pesticides, for example) and modifies the environment in ways that increase exposure to other mutagens, notably UV radiation. What role should government play in identifying mutagens and regulating their release to the environment?

18 The Genetics of Viruses and Bacteria

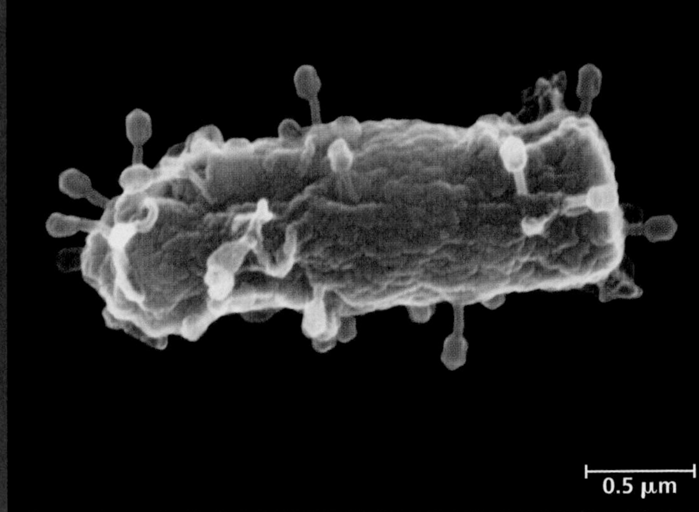

▲ **Figure 18.1 T4 bacteriophages infecting an *E. coli* cell.**

Key Concepts

18.1 A virus has a genome but can reproduce only within a host cell
18.2 Viruses, viroids, and prions are formidable pathogens in animals and plants
18.3 Rapid reproduction, mutation, and genetic recombination contribute to the genetic diversity of bacteria
18.4 Individual bacteria respond to environmental change by regulating their gene expression

Overview

Microbial Model Systems

The photo in **Figure 18.1** shows a remarkable event: the attack of a bacterial cell by numerous structures that resemble miniature lollipops. These structures, a type of virus called T4 bacteriophage, are seen infecting the bacterium *Escherichia coli* in this colorized SEM. By injecting its DNA into the cell, the virus sets in motion a genetic takeover of the bacterium. Molecular biology was born in the laboratories of microbiologists studying viruses and bacteria. Microbes such as *E. coli* and its viruses are called *model systems* because of their frequent use by researchers in studies that reveal broad biological principles. Experiments with viruses and bacteria provided most of the evidence that genes are made of DNA, and they were critical in working out the molecular mechanisms of the fundamental processes of DNA replication, transcription, and translation.

Beyond their value as model systems, viruses and bacteria have unique genetic mechanisms that are interesting in their own right. These specialized mechanisms have important applications for understanding how viruses and bacteria cause disease. In addition, techniques enabling scientists to manipulate genes and transfer them from one organism to another have emerged from the study of microbes. These techniques are having an important impact on both basic research and biotechnology (see Chapter 20).

In this chapter, we explore the genetics of viruses and bacteria. Recall that bacteria are prokaryotes, with cells much smaller and more simply organized than those of eukaryotes, such as plants and animals. Viruses are smaller and simpler still (**Figure 18.2**). Lacking the structures and metabolic machinery found in cells, most viruses are little more than genes packaged in protein coats. We will begin with the structure of these simplest of all genetic systems and their role as disease-causing agents, or pathogens. Then we will discuss the genetics of bacteria and regulation of their gene expression.

Concept 18.1

A virus has a genome but can reproduce only within a host cell

Scientists were able to detect viruses indirectly long before they were actually able to see them. The story of how viruses were discovered begins near the end of the 19th century.

The Discovery of Viruses: *Scientific Inquiry*

Tobacco mosaic disease stunts the growth of tobacco plants and gives their leaves a mottled, or mosaic, coloration (**Figure 18.3**). In 1883, Adolf Mayer, a German scientist, discovered that he could transmit the disease from plant to plant by rubbing sap extracted from diseased leaves onto healthy plants. After

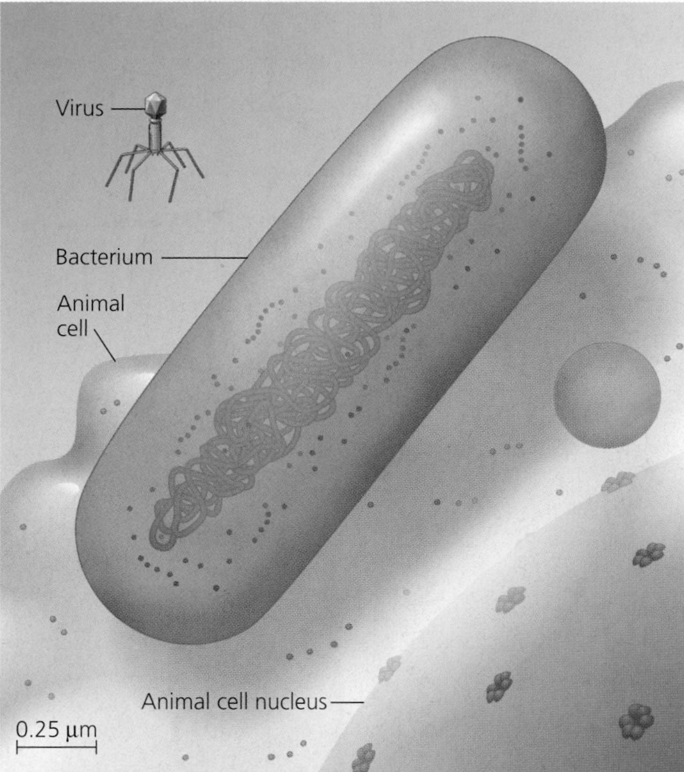

▲ **Figure 18.2 Comparing the size of a virus, a bacterium, and an animal cell.** Only a portion of a typical animal cell is shown. Its diameter is about ten times greater than the length of *E. coli*.

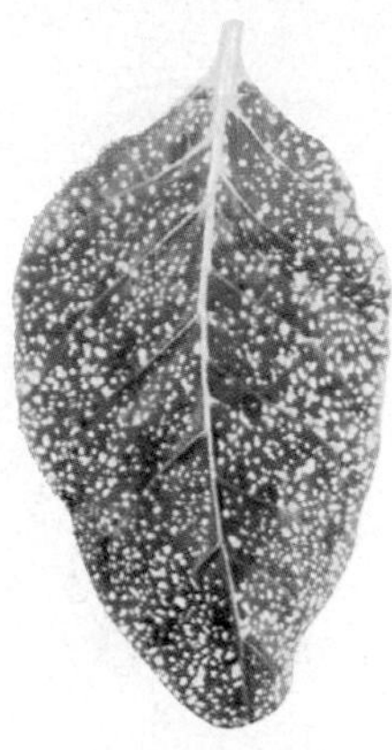

▲ **Figure 18.3 Infection by tobacco mosaic virus (TMV).** A healthy, uninfected tobacco leaf (left) compared with a leaf experimentally infected with TMV (right).

an unsuccessful search for an infectious microbe in the sap, Mayer concluded that the disease was caused by unusually small bacteria that could not be seen with the microscope. This hypothesis was tested a decade later by Dimitri Ivanowsky, a Russian who passed sap from infected tobacco leaves through a filter designed to remove bacteria. After filtering, the sap still produced mosaic disease.

Ivanowsky clung to the hypothesis that bacteria caused tobacco mosaic disease. Perhaps, he reasoned, the bacteria were so small that they passed through the filter or made a filterable toxin that caused the disease. This latter possibility was ruled out when the Dutch botanist Martinus Beijerinck discovered that the infectious agent in the filtered sap could reproduce. He rubbed plants with filtered sap, and after these plants developed mosaic disease, he used their sap to infect more plants, continuing this process through a series of infections. The pathogen must have been reproducing, for its ability to cause disease was undiluted after several transfers from plant to plant.

In fact, the pathogen reproduced only within the host it infected. Unlike bacteria, the mysterious agent of mosaic disease could not be cultivated on nutrient media in test tubes or petri dishes. Beijerinck imagined a reproducing particle much smaller and simpler than bacteria. His suspicions were confirmed in 1935 when the American scientist Wendell Stanley crystallized the infectious particle, now known as tobacco mosaic virus (TMV). Subsequently, TMV and many other viruses were actually seen with the help of the electron microscope.

Structure of Viruses

The tiniest viruses are only 20 nm in diameter—smaller than a ribosome. Millions could easily fit on a pinhead. Even the largest viruses are barely visible in the light microscope. Stanley's discovery that some viruses could be crystallized was exciting and puzzling news. Not even the simplest of cells can aggregate into regular crystals. But if viruses are not cells, then what are they? They are infectious particles consisting of nucleic acid enclosed in a protein coat and, in some cases, a membranous envelope. Let's examine the structure of viruses more closely and then how they reproduce.

Viral Genomes

We usually think of genes as being made of double-stranded DNA—the conventional double helix—but many viruses defy this convention. Their genomes may consist of double-stranded DNA, single-stranded DNA, double-stranded RNA, or single-stranded RNA, depending on the kind of virus. A virus is called a DNA virus or an RNA virus, according to the kind of nucleic acid that makes up its genome. In either case, the genome is usually organized as a single linear or circular molecule of nucleic acid. The smallest viruses have only four genes, while the largest have several hundred.

Capsids and Envelopes

The protein shell enclosing the viral genome is called a **capsid.** Depending on the type of virus, the capsid may be rod-shaped, polyhedral, or more complex in shape (like T4). Capsids are built from a large number of protein subunits called *capsomeres*, but the number of different *kinds* of proteins is usually small. Tobacco mosaic virus has a rigid, rod-shaped capsid

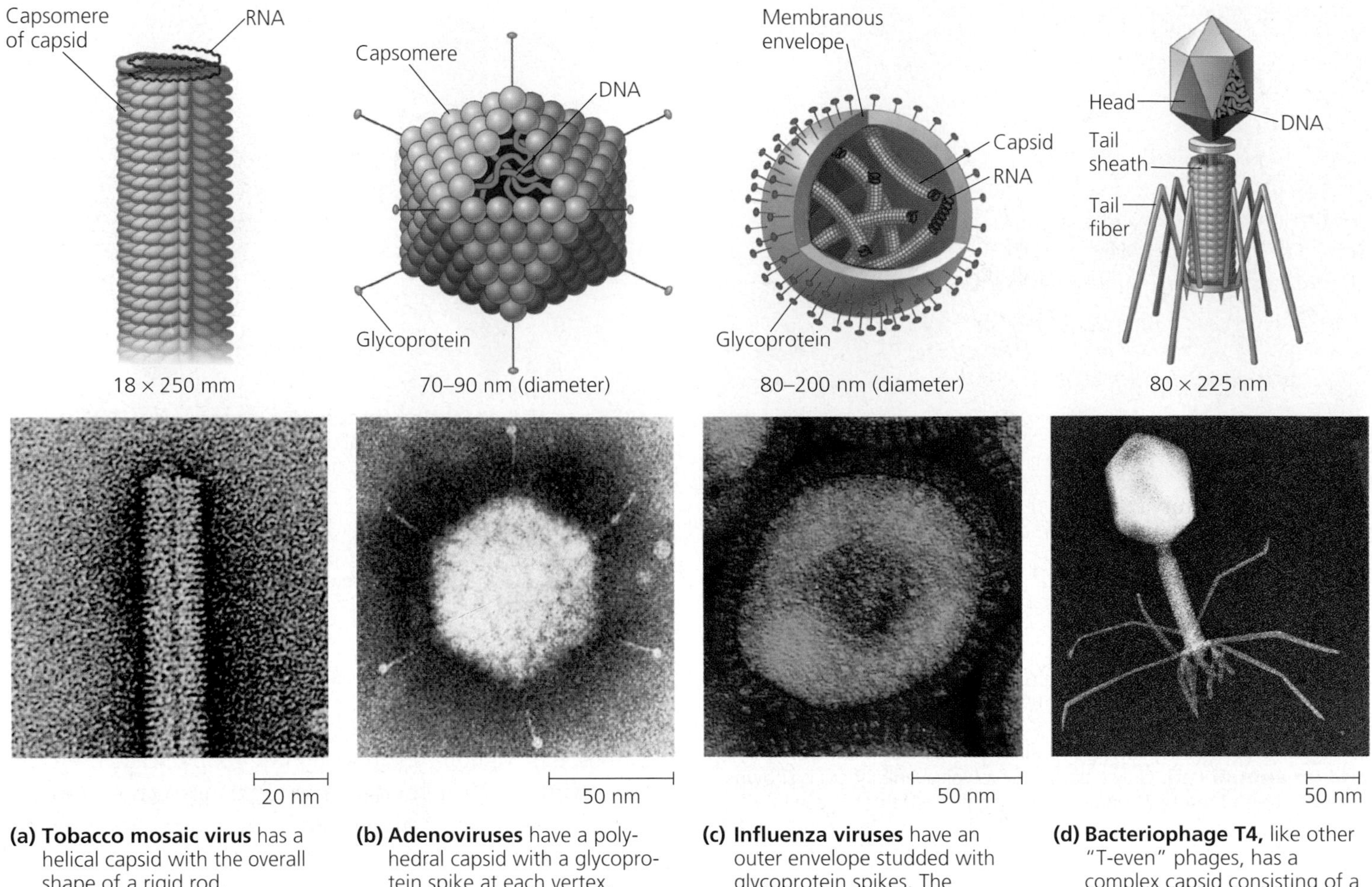

(a) Tobacco mosaic virus has a helical capsid with the overall shape of a rigid rod.

(b) Adenoviruses have a polyhedral capsid with a glycoprotein spike at each vertex.

(c) Influenza viruses have an outer envelope studded with glycoprotein spikes. The genome consists of eight different RNA molecules, each wrapped in a helical capsid.

(d) Bacteriophage T4, like other "T-even" phages, has a complex capsid consisting of a polyhedral head and a tail apparatus.

▲ **Figure 18.4 Viral structure.** Viruses are made up of nucleic acid (DNA or RNA) enclosed in a protein coat (the capsid) and sometimes further wrapped in a membranous envelope. The individual protein subunits making up the capsid are called capsomeres. Although diverse in size and shape, viruses have common structural features, most of which appear in the four examples shown here. (All the micrographs are colorized TEMs.)

made from over a thousand molecules of a single type of protein arranged in a helix **(Figure 18.4a)**. Adenoviruses, which infect the respiratory tracts of animals, have 252 identical protein molecules arranged in a polyhedral capsid with 20 triangular facets—an icosahedron **(Figure 18.4b)**.

Some viruses have accessory structures that help them infect their hosts. For instance, a membranous envelope surrounds the capsids of influenza viruses and many other viruses found in animals **(Figure 18.4c)**. These **viral envelopes,** which are derived from the membrane of the host cell, contain host cell phospholipids and membrane proteins. They also contain proteins and glycoproteins of viral origin (glycoproteins are proteins with carbohydrate covalently attached). Some viruses carry a few viral enzyme molecules within their capsids.

The most complex capsids are found among viruses that infect bacteria, called **bacteriophages,** or simply **phages.** The first phages studied included seven that infect *E. coli*. These seven phages were named type 1 (T1), type 2 (T2), and so forth, in the order of their discovery. The three T-even phages (T2, T4, and T6) turned out to be very similar in structure. Their capsids have elongated icosahedral heads enclosing their DNA. Attached to the head is a protein tail piece with fibers that the phages use to attach to a bacterium **(Figure 18.4d)**.

General Features of Viral Reproductive Cycles

Viruses are obligate intracellular parasites: They can reproduce only within a host cell. An isolated virus is unable to reproduce or do anything else except infect an appropriate host cell. Viruses lack metabolic enzymes, ribosomes, and other equipment for making proteins. Thus, isolated viruses are merely packaged sets of genes in transit from one host cell to another.

Each type of virus can infect only a limited range of host cells, called its **host range.** This host specificity results from the evolution of recognition systems by the virus. Viruses identify their host cells by a "lock-and-key" fit between proteins on the outside of the virus and specific receptor molecules on the surface of cells. (Presumably, the receptors first evolved because they carried out functions of benefit to the organism.) Some viruses have broad host ranges. West Nile virus, for example, can infect mosquitoes, birds, and humans, and equine encephalitis virus can infect mosquitoes, birds, horses, and humans. Other viruses have host ranges so narrow that they infect only a single species. Measles virus and poliovirus, for instance, can infect only humans. Furthermore, infection by viruses of multicellular eukaryotes is usually limited to particular tissues. Human cold viruses infect only the cells lining the upper respiratory tract, and the AIDS virus binds to specific receptors on certain types of white blood cells.

A viral infection begins when the genome of a virus makes its way into a host cell **(Figure 18.5)**. The mechanism by which this nucleic acid enters the cell varies, depending on the type of virus and the type of host cell. For example, the T-even phages use their elaborate tail apparatus to inject DNA into a bacterium (see Figure 18.4d). Once inside, the viral genome can commandeer its host, reprogramming the cell to copy the viral nucleic acid and manufacture viral proteins. The host provides the nucleotides for making viral nucleic acids, as well as enzymes, ribosomes, tRNAs, amino acids, ATP, and other components needed for making the viral proteins dictated by viral genes. Most DNA viruses use the DNA polymerases of the host cell to synthesize new genomes along the templates provided by the viral DNA. In contrast, to replicate their genomes, RNA viruses use special virus-encoded polymerases that can use RNA as a template. (Uninfected cells generally make no enzymes for carrying out this latter process.)

After the viral nucleic acid molecules and capsomeres are produced, their assembly into new viruses is often a spontaneous process of self-assembly. In fact, the RNA and capsomeres of TMV can be separated in the laboratory and then reassembled to form complete viruses simply by mixing the components together under the right conditions. The simplest type of viral reproductive cycle ends with the exit of hundreds or thousands of viruses from the infected host cell, a process that often damages or destroys the cell. Such cellular damage and death, as well as the body's responses to this destruction, cause some of the symptoms associated with viral infections. The viral progeny that exit a cell have the potential to infect additional cells, spreading the viral infection.

There are many variations on the simplified viral reproductive cycle we have traced in this overview. We will now take a closer look at some of these variations in bacterial viruses (phages) and animal viruses; later in the chapter, we will consider plant viruses.

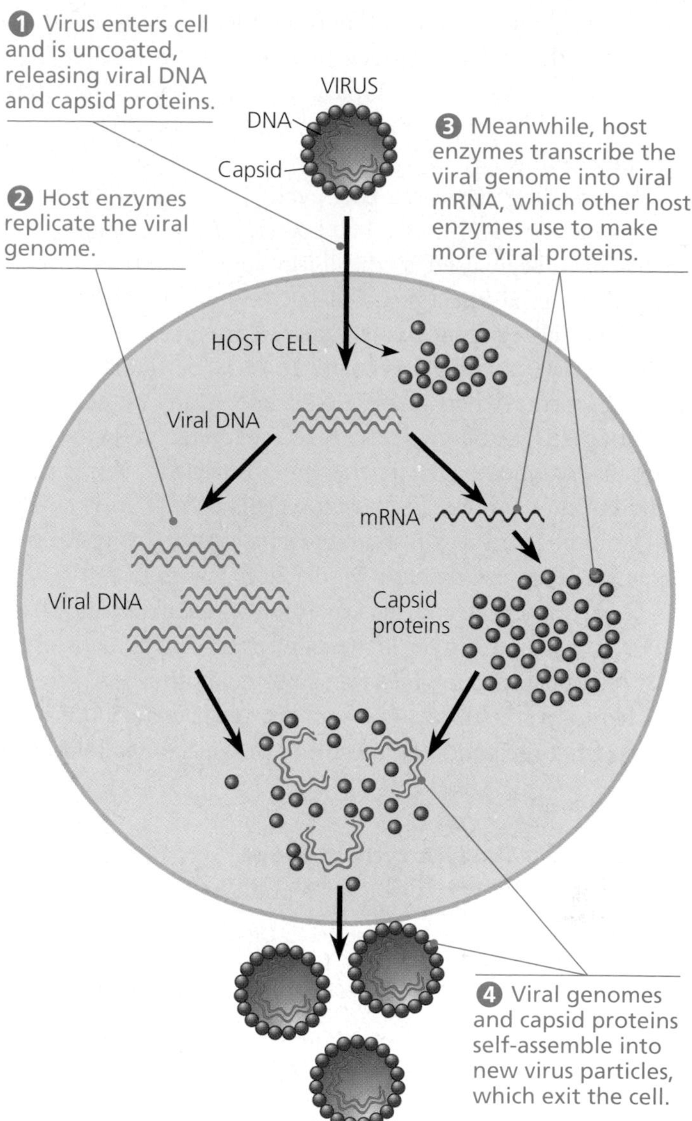

1 Virus enters cell and is uncoated, releasing viral DNA and capsid proteins.
VIRUS
DNA
Capsid
2 Host enzymes replicate the viral genome.
3 Meanwhile, host enzymes transcribe the viral genome into viral mRNA, which other host enzymes use to make more viral proteins.
HOST CELL
Viral DNA
mRNA
Viral DNA
Capsid proteins
4 Viral genomes and capsid proteins self-assemble into new virus particles, which exit the cell.

▲ **Figure 18.5 A simplified viral reproductive cycle.** A virus is an obligate intracellular parasite that uses the equipment and small precursors of its host cell to reproduce. In this simplest of viral cycles, the parasite is a DNA virus with a capsid consisting of a single type of protein.

Reproductive Cycles of Phages

Phages are the best understood of all viruses, although some of them are also among the most complex. Research on phages led to the discovery that some double-stranded DNA viruses can reproduce by two alternative mechanisms: the lytic cycle and the lysogenic cycle.

The Lytic Cycle

A phage reproductive cycle that culminates in death of the host cell is known as a **lytic cycle.** The term refers to the last stage of infection, during which the bacterium lyses (breaks open) and releases the phages that were produced within the cell. Each of these phages can then infect a healthy cell, and a few successive lytic cycles can destroy an entire bacterial population in just a few hours. A phage that reproduces only by a lytic cycle is a **virulent phage.** **Figure 18.6** illustrates the major steps in the lytic cycle of T4, a typical virulent phage. The figure and legend describe the process, which you should study before proceeding.

After reading about the lytic cycle, you may wonder why phages haven't exterminated all bacteria. In fact, phage treatments have been used medically in some countries to help control bacterial infections. But bacteria are not defenseless. First, natural selection favors bacterial mutants with receptor sites that are no longer recognized by a particular type of phage. Second, when phage DNA successfully enters a bacterium, the DNA often is recognized as foreign and cut up by cellular enzymes called *restriction endonucleases,* or simply **restriction enzymes.** The bacterial cell's own DNA is chemically modified in a way that prevents attack by restriction enzymes. But just as natural selection favors bacteria with effective restriction enzymes, natural selection favors phage mutants that are resistant to these enzymes. Thus, the parasite-host relationship is in constant evolutionary flux.

There is yet a third important reason bacteria have been spared from extinction as a result of phage activity. Instead of lysing their host cells, many phages coexist with them in what is called the lysogenic cycle.

The Lysogenic Cycle

In contrast to the lytic cycle, which kills the host cell, the **lysogenic cycle** replicates the phage genome without destroying the host. Phages capable of using both modes of reproducing within a bacterium are called **temperate phages.** A temperate phage called lambda, written with the Greek letter λ, is widely used in biological research. Phage λ resembles T4, but its tail has only one short tail fiber.

Infection of an *E. coli* cell by phage λ begins when the phage binds to the surface of the cell and injects its DNA **(Figure 18.7)**. Within the host, the λ DNA molecule forms a circle. What happens next depends on the reproductive mode: lytic cycle or lysogenic cycle. During a lytic cycle, the viral genes immediately turn the host cell into a λ-producing factory, and the cell soon lyses and releases its viral products. During a lysogenic cycle, however, the λ DNA molecule is incorporated by genetic recombination (crossing over) into a specific site on the host cell's chromosome. When integrated into the bacterial chromosome in this way, the viral DNA is

▶ **Figure 18.6 The lytic cycle of phage T4, a virulent phage.** Phage T4 has about 100 genes, which are transcribed and translated using the host cell's machinery. One of the first phage genes translated after the viral DNA enters the host cell codes for an enzyme that degrades the host cell's DNA (step 2); the phage DNA is protected from breakdown because it contains a modified form of cytosine that is not recognized by the enzyme. The entire lytic cycle, from the phage's first contact with the cell surface to cell lysis, takes only 20–30 minutes at 37°C.

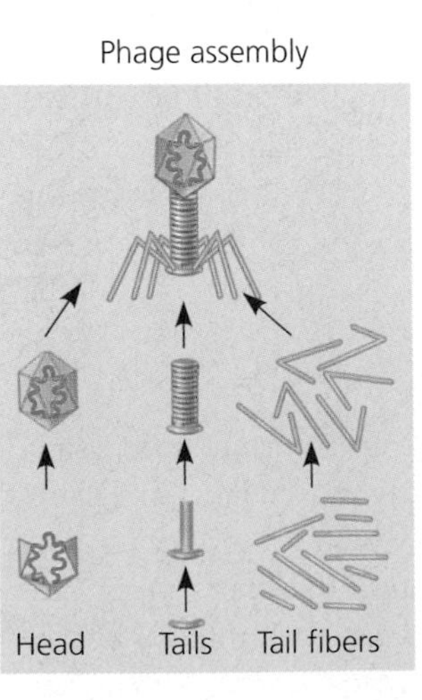

Phage assembly
Head
Tails
Tail fibers

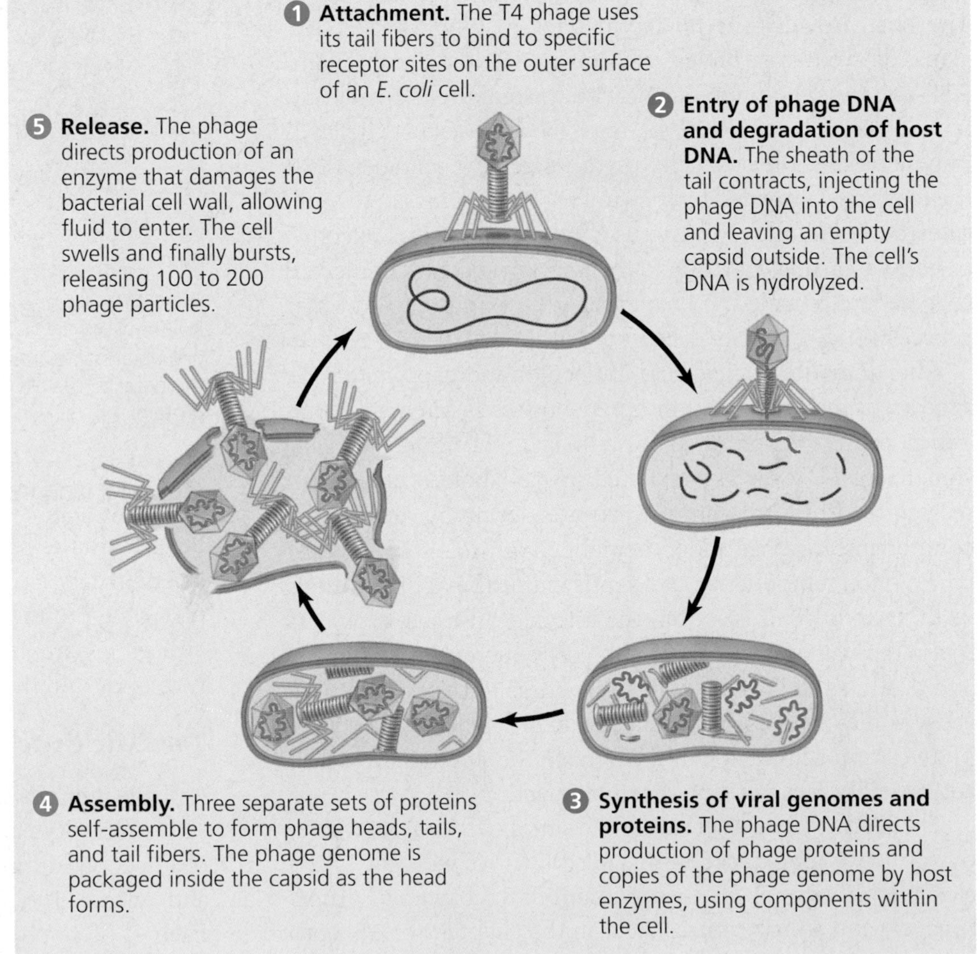

1 Attachment. The T4 phage uses its tail fibers to bind to specific receptor sites on the outer surface of an E. coli cell.
2 Entry of phage DNA and degradation of host DNA. The sheath of the tail contracts, injecting the phage DNA into the cell and leaving an empty capsid outside. The cell's DNA is hydrolyzed.
3 Synthesis of viral genomes and proteins. The phage DNA directs production of phage proteins and copies of the phage genome by host enzymes, using components within the cell.
4 Assembly. Three separate sets of proteins self-assemble to form phage heads, tails, and tail fibers. The phage genome is packaged inside the capsid as the head forms.
5 Release. The phage directs production of an enzyme that damages the bacterial cell wall, allowing fluid to enter. The cell swells and finally bursts, releasing 100 to 200 phage particles.

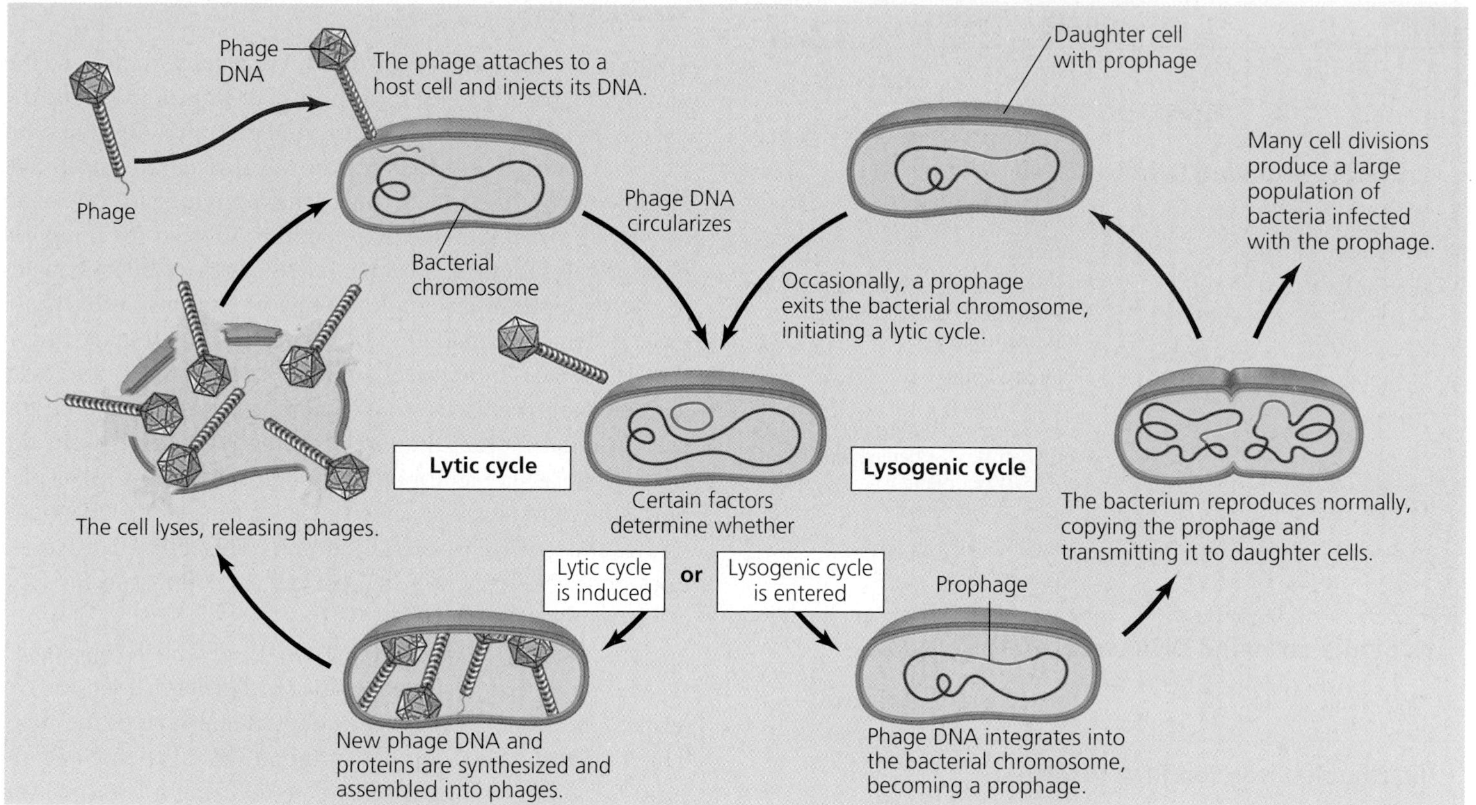

▲ **Figure 18.7 The lytic and lysogenic cycles of phage λ, a temperate phage.** After entering the bacterial cell and circularizing, the λ DNA can immediately initiate the production of a large number of progeny phages (lytic cycle) or integrate into the bacterial chromosome (lysogenic cycle). In most cases, phage λ follows the lytic pathway, which is similar to that detailed in Figure 18.6. However, once a lysogenic cycle begins, the prophage may be carried in the host cell's chromosome for many generations. Phage λ has one main tail fiber, which is short.

known as a **prophage.** One prophage gene codes for a protein that prevents transcription of most of the other prophage genes. Thus, the phage genome is mostly silent within the bacterium. Every time the *E. coli* cell prepares to divide, it replicates the phage DNA along with its own and passes the copies on to daughter cells. A single infected cell can quickly give rise to a large population of bacteria carrying the virus in prophage form. This mechanism enables viruses to propagate without killing the host cells on which they depend.

The term *lysogenic* implies that prophages are capable of giving rise to active phages that lyse their host cells. This occurs when the λ genome exits the bacterial chromosome and initiates a lytic cycle. What triggers the switchover from the lysogenic to the lytic mode is usually an environmental signal, such as radiation or the presence of certain chemicals.

In addition to the gene for the transcription-preventing protein, a few other prophage genes may also be expressed during lysogenic cycles. Expression of these genes may alter the host's phenotype, a phenomenon that can have important medical significance. For example, the bacteria that cause the human diseases diphtheria, botulism, and scarlet fever would be harmless to humans if it were not for certain prophage genes that cause the host bacteria to make toxins.

Reproductive Cycles of Animal Viruses

Everyone has suffered from viral infections, whether cold sores, influenza, or the common cold. Like all viruses, those that cause illness in humans and other animals can reproduce only inside host cells. Many variations on the basic scheme of viral infection and reproduction are represented among the animal viruses. One key variable is the nature of the viral genome: Is it composed of DNA or RNA? Is it double-stranded or single-stranded? The nature of the genome is the basis for the common classification of viruses shown in **Table 18.1** on the next page. Single-stranded RNA viruses are further classified into three classes (IV–VI) according to how the RNA genome functions in a host cell.

Another important characteristic of a virus is the presence or absence of a membranous envelope derived from host cell membrane. Rather than consider all the mechanisms of viral infection and reproduction, we will focus on the roles of viral envelopes and on the functioning of RNA as the genetic material of many animal viruses. Whereas few bacteriophages have an envelope or RNA genome, nearly all the animal viruses with RNA genomes have an envelope, as do some with DNA genomes (see Table 18.1).

Table 18.1 Classes of Animal Viruses

Class/ Family	Envelope	Examples/ Disease
I. Double-stranded DNA (dsDNA)		
Adenovirus (see Figure 18.4b)	No	Respiratory diseases; animal tumors
Papovavirus	No	Papillomavirus (warts, cervical cancer); polyomavirus (animal tumors)
Herpesvirus	Yes	Herpes simplex I and II (cold sores, genital sores); varicella zoster (shingles, chicken pox); Epstein-Barr virus (mononucleosis, Burkitt's lymphoma)
Poxvirus	Yes	Smallpox virus; cowpox virus
II. Single-stranded DNA (ssDNA)		
Parvovirus	No	B19 parvovirus (mild rash)
III. Double-stranded RNA (dsRNA)		
Reovirus	No	Rotavirus (diarrhea); Colorado tick fever virus
IV. Single-stranded RNA (ssRNA); serves as mRNA		
Picornavirus	No	Rhinovirus (common cold); poliovirus, hepatitis A virus, and other enteric (intestinal) viruses
Coronavirus (see Figure 18.11b)	Yes	Severe acute respiratory syndrome (SARS)
Flavivirus	Yes	Yellow fever virus; West Nile virus; hepatitis C virus
Togavirus	Yes	Rubella virus; equine encephalitis viruses
V. ssRNA; template for mRNA synthesis		
Filovirus	Yes	Ebola virus (hemorrhagic fever)
Orthomyxovirus (see Figure 18.4c)	Yes	Influenza virus
Paramyxovirus	Yes	Measles virus; mumps virus
Rhabdovirus	Yes	Rabies virus
VI. ssRNA; template for DNA synthesis		
Retrovirus (see Figure 18.9)	Yes	HIV, human immunodeficiency virus (AIDS); RNA tumor viruses (leukemia)

Viral Envelopes

An animal virus equipped with an outer membrane, or viral envelope, uses it to enter the host cell. Protruding from the outer surface of this envelope are viral glycoproteins that bind to specific receptor molecules on the surface of a host cell. **Figure 18.8** outlines the events in the reproductive cycle of an enveloped virus with an RNA genome. You can see that viral glycoproteins for new envelopes are made by cellular enzymes in the endoplasmic reticulum (ER) of the host cell. These glycoproteins, embedded in ER membrane, are transported to the cell surface. In a process much like exocytosis, new virus capsids wrap themselves in membrane as they bud from the cell. In other words, the viral envelope is derived from the host cell's plasma membrane, although some of the molecules of this membrane are specified by viral genes. The enveloped viruses are now free to infect other cells. This reproductive cycle does not necessarily kill the host cell, in contrast to the lytic cycles of phages.

Some viruses have envelopes that are not derived from plasma membrane. The envelopes of herpesviruses, for example, are derived from the nuclear membrane of the host. These viruses have a double-stranded DNA genome and reproduce within the host cell nucleus, using a combination of viral and cellular enzymes to replicate and transcribe their DNA. In some cases, copies of the herpesvirus DNA remain behind as minichromosomes in the nuclei of certain nerve cells. There they remain latent until some sort of physical or emotional stress triggers a new round of active virus production. The infection of other cells by these new viruses causes the blisters characteristic of herpes, such as cold sores or genital sores. Once someone acquires a herpesvirus infection, flare-ups may recur throughout the person's life.

RNA as Viral Genetic Material

Although some phages and most plant viruses are RNA viruses, the broadest variety of RNA genomes is found among the viruses that infect animals. Among the three types of single-stranded RNA genomes found in animal viruses, the genome of class IV viruses can directly serve as mRNA and thus can be translated into viral protein immediately after infection. Figure 18.8 shows a virus of class V, in which the RNA genome serves as a *template* for mRNA synthesis. The RNA genome is transcribed into complementary RNA strands, which function both as mRNA and as templates for the synthesis of additional copies of genome RNA. Like all viruses that require RNA → RNA synthesis to make mRNA, this one uses a viral enzyme that is packaged with the genome inside the capsid.

The RNA animal viruses with the most complicated reproductive cycles are the **retroviruses** (class VI). These viruses are

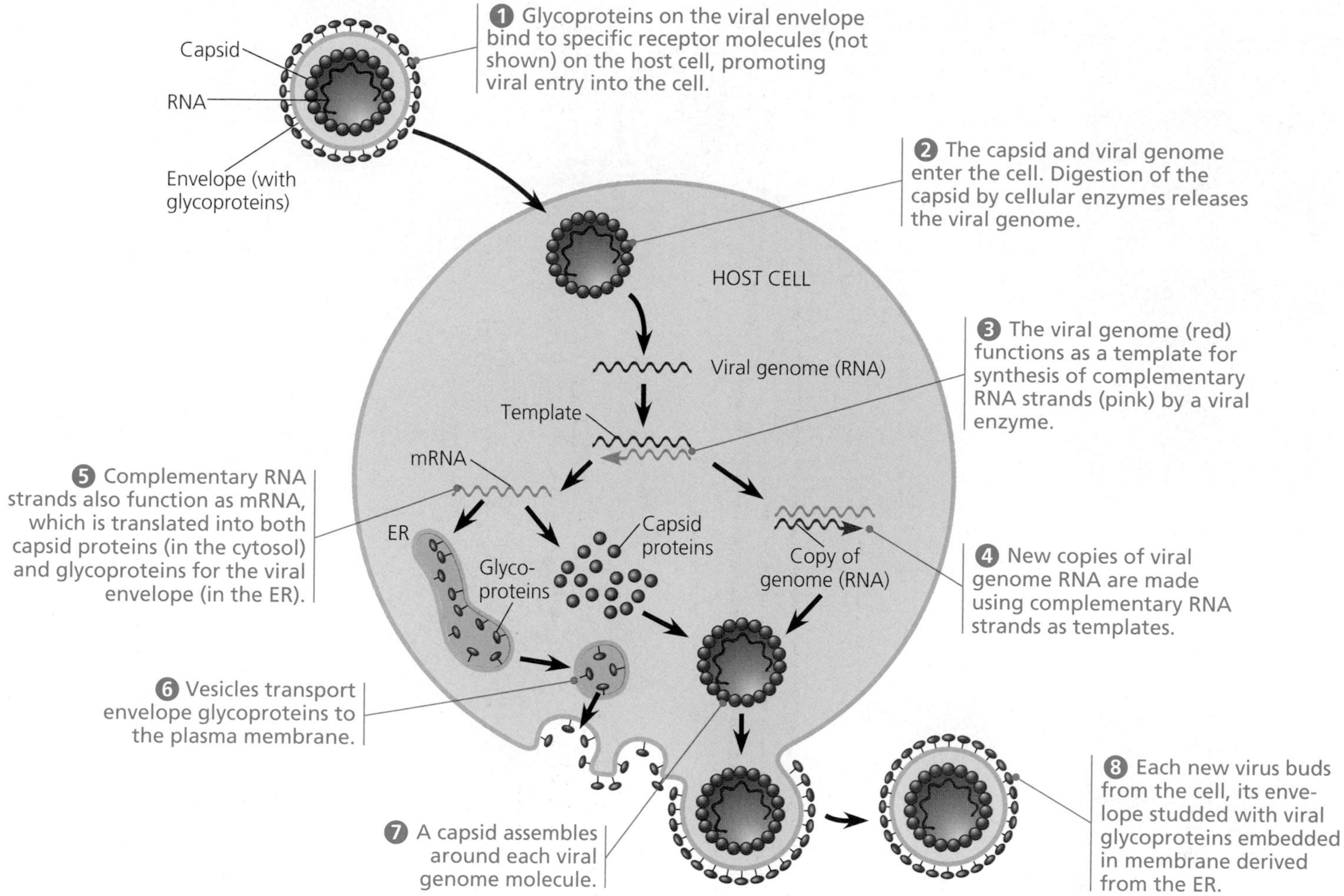

Capsid
RNA
Envelope (with glycoproteins)

1 Glycoproteins on the viral envelope bind to specific receptor molecules (not shown) on the host cell, promoting viral entry into the cell.

2 The capsid and viral genome enter the cell. Digestion of the capsid by cellular enzymes releases the viral genome.

HOST CELL

Viral genome (RNA)

Template

3 The viral genome (red) functions as a template for synthesis of complementary RNA strands (pink) by a viral enzyme.

mRNA

5 Complementary RNA strands also function as mRNA, which is translated into both capsid proteins (in the cytosol) and glycoproteins for the viral envelope (in the ER).

ER

Glyco-proteins

Capsid proteins

Copy of genome (RNA)

4 New copies of viral genome RNA are made using complementary RNA strands as templates.

6 Vesicles transport envelope glycoproteins to the plasma membrane.

7 A capsid assembles around each viral genome molecule.

8 Each new virus buds from the cell, its envelope studded with viral glycoproteins embedded in membrane derived from the ER.

▲ **Figure 18.8 The reproductive cycle of an enveloped RNA virus.** Shown here is a virus with a single-stranded RNA genome that functions as a template for synthesis of mRNA. Some enveloped viruses enter the host cell by fusion of the envelope with the cell's plasma membrane; others enter by endocytosis. For all enveloped RNA viruses, the formation of new envelopes for progeny viruses occurs by the mechanism depicted in this figure.

equipped with an enzyme called **reverse transcriptase**, which transcribes an RNA template into DNA, providing an RNA → DNA information flow, the opposite of the usual direction. This unusual phenomenon gave rise to the name retroviruses (*retro* means "backward"). Of particular medical importance is **HIV (human immunodeficiency virus)**, the retrovirus that causes **AIDS (acquired immunodeficiency syndrome)**. HIV and other retroviruses are enveloped viruses that contain two identical molecules of single-stranded RNA and two molecules of reverse transcriptase **(Figure 18.9)**.

After HIV enters a host cell, its reverse transcriptase molecules are released into the cytoplasm and catalyze synthesis of viral DNA. The newly made viral DNA then enters the cell's nucleus and integrates into the DNA of a chromosome. The integrated viral DNA, called a **provirus**, never leaves the host's genome, remaining a permanent resident of the cell. (Unlike a prophage, a provirus never leaves.) The host's RNA polymerase transcribes the proviral DNA into RNA molecules, which can

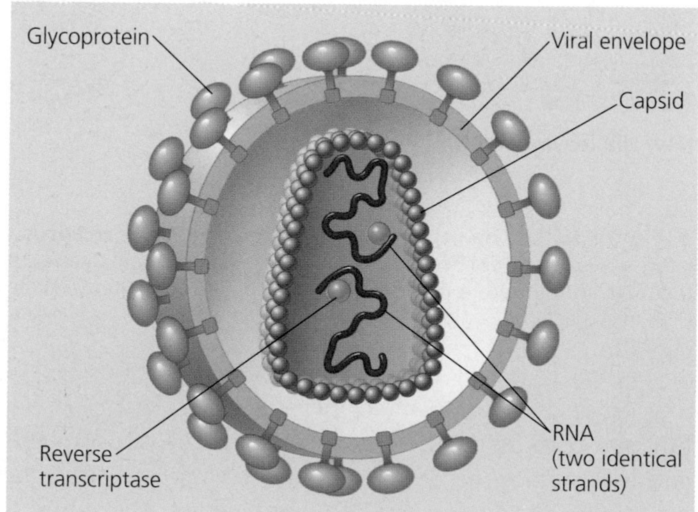

▲ **Figure 18.9 The structure of HIV, the retrovirus that causes AIDS.** The envelope glycoproteins enable the virus to bind to specific receptors on certain white blood cells.

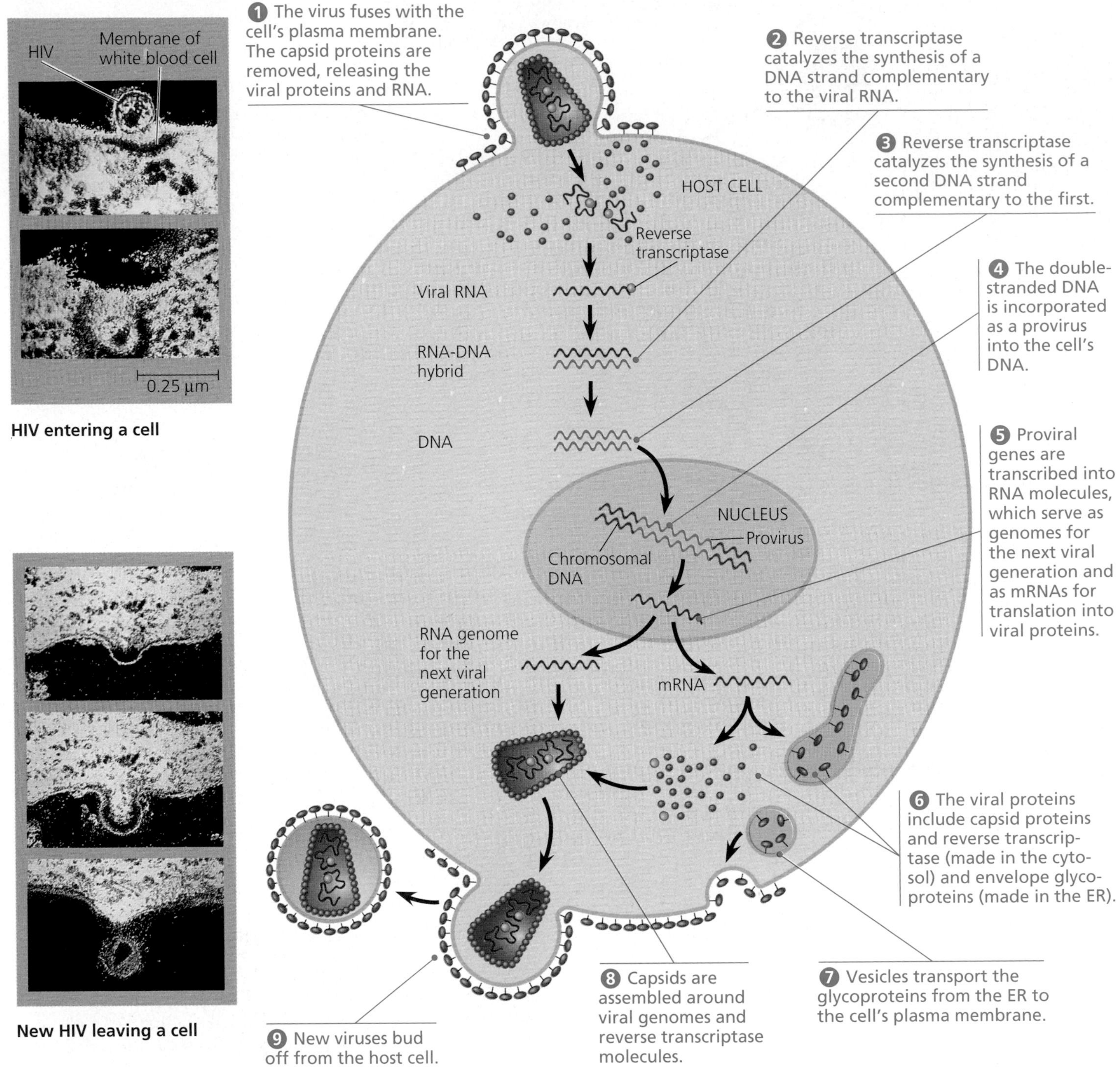
HIV
Membrane of white blood cell
0.25 µm
HIV entering a cell
New HIV leaving a cell
1 The virus fuses with the cell's plasma membrane. The capsid proteins are removed, releasing the viral proteins and RNA.
2 Reverse transcriptase catalyzes the synthesis of a DNA strand complementary to the viral RNA.
3 Reverse transcriptase catalyzes the synthesis of a second DNA strand complementary to the first.
4 The double-stranded DNA is incorporated as a provirus into the cell's DNA.
5 Proviral genes are transcribed into RNA molecules, which serve as genomes for the next viral generation and as mRNAs for translation into viral proteins.
6 The viral proteins include capsid proteins and reverse transcriptase (made in the cytosol) and envelope glycoproteins (made in the ER).
7 Vesicles transport the glycoproteins from the ER to the cell's plasma membrane.
8 Capsids are assembled around viral genomes and reverse transcriptase molecules.
9 New viruses bud off from the host cell.
HOST CELL
Reverse transcriptase
Viral RNA
RNA-DNA hybrid
DNA
NUCLEUS
Provirus
Chromosomal DNA
RNA genome for the next viral generation
mRNA

▲ **Figure 18.10 The reproductive cycle of HIV, a retrovirus.** The photos on the left (artificially colored TEMs) reveal HIV entering and leaving a human white blood cell. Note in step 4 that DNA synthesized from the viral RNA genome is integrated into the host cell chromosomal DNA, a characteristic unique to retroviruses.

function both as mRNA for the synthesis of viral proteins and as genomes for new virus particles released from the cell. **Figure 18.10** traces the HIV reproductive cycle, which is typical of a retrovirus. In Chapter 43, we will describe how HIV causes the deterioration of the immune system that occurs in AIDS.

Evolution of Viruses

Viruses do not really fit our definition of living organisms. An isolated virus is biologically inert, unable to replicate its genes or regenerate its own supply of ATP. Yet it has a genetic program written in the universal language of life. Do we think of viruses as nature's most complex molecules or as the simplest forms of life? Either way, we must bend our usual definitions. Although viruses are obligate intracellular parasites that cannot reproduce independently, their use of the genetic code makes it hard to deny their evolutionary connection to the living world.

How did viruses originate? Because they depend on cells for their own propagation, viruses most likely are not the descendants of precellular forms of life, but evolved *after* the first cells appeared, possibly multiple times. Most molecular biologists favor the hypothesis that viruses originated from fragments of cellular nucleic acids that could move from one cell to another. Consistent with this idea is the observation that a viral genome usually has more in common with the genome of its host than with the genomes of viruses infecting other hosts. Indeed, some viral genes are essentially identical to genes of the host. On the other hand, recent sequencing of many viral genomes has found that the genetic sequences of some viruses are quite similar to those of seemingly distantly related viruses (such as an animal virus and a plant virus that share similar sequences). This genetic similarity may reflect the persistence of groups of viral genes that were evolutionarily successful during the early evolution of viruses and the eukaryotic cells serving as their hosts. The origin of viruses is still a topic of much debate.

Perhaps the earliest viruses were naked bits of nucleic acid that made it from one cell to another via injured cell surfaces. The evolution of genes coding for capsid proteins may have facilitated the infection of undamaged cells. Candidates for the original sources of viral genomes include plasmids and transposons, genetic elements that we will discuss in more detail later in the chapter. Plasmids are small, circular DNA molecules, found in bacteria and also in yeasts, which are unicellular eukaryotes. Plasmids exist apart from the cell's genome, can replicate independently of the genome, and are occasionally transferred between cells. Transposons are DNA segments that can move from one location to another within a cell's genome. Thus, plasmids, transposons, and viruses all share an important feature: They are mobile genetic elements.

The ongoing evolutionary relationship between viruses and the genomes of their host cells is an association that makes viruses very useful model systems in molecular biology. Knowledge about viruses also has many practical applications, since viruses have a tremendous impact on all living organisms through their ability to cause disease.

Concept Check 18.1

1. Compare the effect on the host cell of a lytic (virulent) phage and a lysogenic (temperate) phage.
2. How do some viruses reproduce without possessing or ever synthesizing DNA?
3. Why is HIV called a retrovirus?

For suggested answers, see Appendix A.

Concept 18.2

Viruses, viroids, and prions are formidable pathogens in animals and plants

Diseases caused by viral infections afflict humans, agricultural crops, and livestock worldwide. Other smaller, less complex entities known as viroids and prions also cause disease in plants and animals.

Viral Diseases in Animals

The link between a viral infection and the symptoms it produces is often obscure. Viruses may damage or kill cells by causing the release of hydrolytic enzymes from lysosomes. Some viruses cause infected cells to produce toxins that lead to disease symptoms, and some have molecular components that are toxic, such as envelope proteins. How much damage a virus causes depends partly on the ability of the infected tissue to regenerate by cell division. People usually recover completely from colds because the epithelium of the respiratory tract, which the viruses infect, can efficiently repair itself. In contrast, damage inflicted by poliovirus to mature nerve cells is permanent, because these cells do not divide and usually cannot be replaced. Many of the temporary symptoms associated with viral infections, such as fever and aches, actually result from the body's own efforts at defending itself against infection.

The immune system is a complex and critical part of the body's natural defenses (see Chapter 43). The immune system is also the basis for the major medical tool for preventing viral infections—vaccines. **Vaccines** are harmless variants or derivatives of pathogenic microbes that stimulate the immune system to mount defenses against the actual pathogen. Vaccination has eradicated smallpox, at one time a devastating scourge in many parts of the world. The viruses that cause smallpox, polio, and measles infect only humans. This very narrow host range was critical to the successful effort of the World Health Organization to eradicate smallpox; similar worldwide vaccination campaigns currently are under way to eradicate the other two viruses as well. Effective vaccines are also available against rubella, mumps, hepatitis B, and a number of other viral diseases.

Although vaccines can prevent certain viral illnesses, medical technology can do little, at present, to cure most viral infections once they occur. The antibiotics that help us recover from bacterial infections are powerless against viruses. Antibiotics kill bacteria by inhibiting enzyme-catalyzed processes specific to the pathogens, but viruses have few or no enzymes of their own. However, a few drugs effectively combat certain viruses. Most antiviral drugs resemble nucleosides and as a result interfere with viral nucleic acid synthesis. One such drug is acyclovir, which impedes herpesvirus reproduction by inhibiting the viral polymerase that synthesizes viral DNA. Similarly, azidothymidine (AZT) curbs HIV reproduction by interfering with the synthesis of DNA by reverse transcriptase. In the past ten years, much effort has gone into developing drugs against HIV. Currently, multidrug treatments, sometimes called "cocktails," have been found to be most effective. Such a regimen commonly includes a combination of two nucleoside mimics and a protease inhibitor, which interferes with an enzyme required for assembly of virus particles.

Emerging Viruses

Viruses that appear suddenly or that suddenly come to the attention of medical scientists are often referred to as *emerging viruses*. HIV, the AIDS virus, is a classic example: This virus appeared in San Francisco in the early 1980s, seemingly out of nowhere. The deadly Ebola virus, recognized initially in 1976 in central Africa, is one of several emerging viruses that cause *hemorrhagic fever,* an often fatal syndrome characterized by fever, vomiting, massive bleeding, and circulatory system collapse. A number of other dangerous new viruses cause encephalitis, inflammation of the brain. One example is the West Nile virus, which appeared for the first time in North America in 1999 and has spread to all 48 contiguous states in the U.S.

An even more recent viral disease to emerge is *severe acute respiratory syndrome (SARS),* which first appeared in southern China in November 2002 **(Figure 18.11a)**. During a global outbreak from November 2002 to July 2003, about 8,000 people were known to be infected, of whom more than 700 subsequently died. Researchers quickly identified the agent causing SARS as a *coronavirus,* a virus with a single-stranded RNA genome (class IV) that was not previously known to cause disease in humans **(Figure 18.11b)**.

From where and how do such viruses burst on the human scene, giving rise to previously rare or unknown diseases? Three processes contribute to the emergence of viral diseases. First, the mutation of existing viruses is a major source of these new diseases. RNA viruses tend to have an unusually high rate of mutation because errors in replicating their RNA genomes are not corrected by proofreading. Some mutations enable existing viruses to evolve into new genetic varieties (strains) that can cause disease in individuals who had developed immunity to the ancestral virus. Flu epidemics, for instance, are caused by new strains of influenza virus genetically different enough from earlier strains that people have little immunity to them.

Another source of new viral diseases is the spread of existing viruses from one host species to another. Scientists estimate that about three-quarters of new human diseases originate in other animals. For example, hantavirus is common in rodents, especially deer mice. The population of deer mice in the southwestern United States exploded in 1993 after unusually wet weather increased the rodents' food supply. Many people who inhaled dust containing traces of urine and feces from infected mice became infected with hantavirus, and dozens died. The source of the SARS-causing virus was still undetermined as of spring 2004, although candidates include the exotic animals found in food markets in China. And early 2004 brought reports of the first cases of people in southeast Asia infected with a flu virus previously seen only in birds. If this virus evolves so that it can spread easily from person to person, the potential for a major human outbreak is significant. Indeed, evidence is strong that the flu pandemic of 1918–1919, which killed about 40 million people, originated in birds.

(a) Young ballet students in Hong Kong wear face masks to protect themselves from the virus causing SARS.

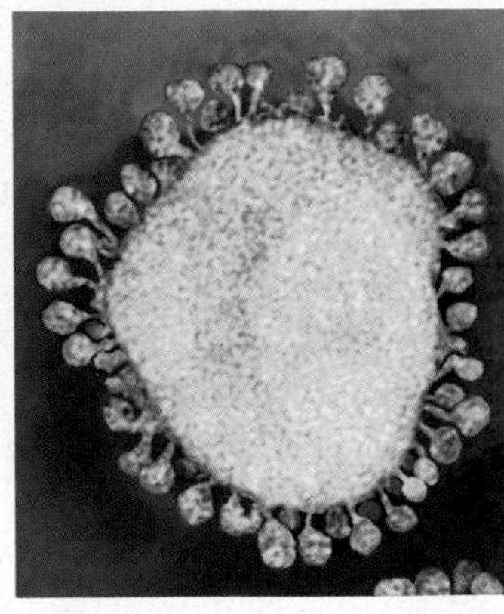

(b) The SARS-causing agent is a coronavirus like this one (colorized TEM), so named for the "corona" of glycoprotein spikes protruding from the envelope.

▲ **Figure 18.11 SARS (severe acute respiratory syndrome), a recently emerging viral disease.**

Finally, the dissemination of a viral disease from a small, isolated population can lead to widespread epidemics. For instance, AIDS went unnamed and virtually unnoticed for decades before it began to spread around the world. In this case, technological and social factors, including affordable international travel, blood transfusions, sexual promiscuity, and the abuse of intravenous drugs, allowed a previously rare human disease to become a global scourge.

Thus, emerging viruses are generally not new; rather, they are existing viruses that mutate, spread to new host species, or disseminate more widely in the current host species. Changes in host behavior or environmental changes can increase the viral traffic responsible for emerging diseases. For example, new roads through remote areas can allow viruses to spread between previously isolated human populations. Another problem is the destruction of forests to expand cropland, an environmental disturbance that brings humans into contact with other animals that may host viruses capable of infecting humans.

the host cell lyses. Occasionally, a small piece of the host cell's degraded DNA is accidentally packaged within a phage capsid in place of the phage genome. Such a virus is defective because it lacks its own genetic material. However, after its release from the lysed host, the phage can attach to another bacterium (the recipient) and inject the piece of bacterial DNA acquired from the first cell (the donor). Some of this DNA can subsequently replace the homologous region of the recipient cell's chromosome, if a crossover takes place at each end of the piece. In this case, the recipient cell's chromosome becomes a combination of DNA derived from two cells; genetic recombination has occurred.

Temperate phages, those able to integrate their genome into the bacterial chromosome as a prophage (see Figure 18.7), can carry out *specialized transduction*. In this process, a prophage picks up just a few adjacent bacterial genes as it exits the chromosome and transfers them to a new host cell. This process can result in efficient transfer, but only of genes adjacent to the prophage site.

Conjugation and Plasmids

Sometimes referred to as bacterial "sex," **conjugation** is the direct transfer of genetic material between two bacterial cells that are temporarily joined. The DNA transfer is one-way: One cell donates DNA, and its "mate" receives the DNA. The donor, sometimes called the "male," uses appendages called sex pili (singular, sex pilus) to attach to the recipient, sometimes called the "female" **(Figure 18.17)**. After contacting a recipient cell, a sex pilus retracts, pulling the two cells together, much like a grappling hook. A temporary cytoplasmic *mating bridge* then forms between the two cells, providing an avenue for DNA transfer.

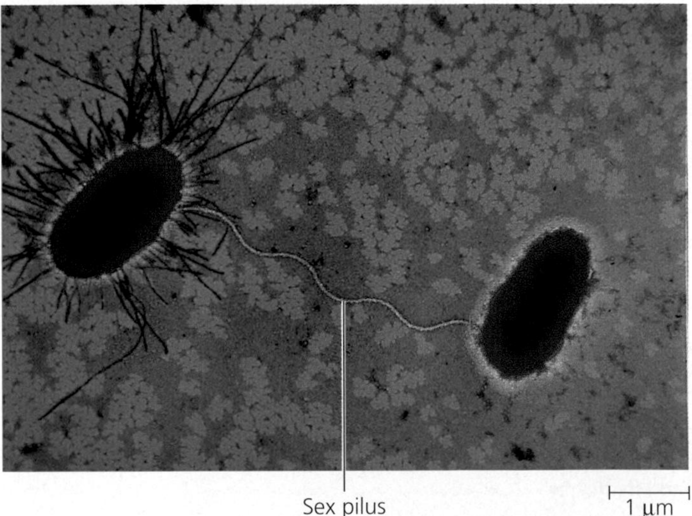

▲ **Figure 18.17 Bacterial conjugation.** The *E. coli* donor cell (left) extends sex pili, one of which is attached to a recipient cell. The two cells will be drawn close together, allowing a cytoplasmic mating bridge to form between them. Through this bridge, the donor will transfer DNA to the recipient (colorized TEM).

In most cases, the ability to form sex pili and donate DNA during conjugation results from the presence of a special piece of DNA called an **F factor** (F for fertility). An F factor can exist either as a segment of DNA within the bacterial chromosome or as a plasmid. A **plasmid** is a small, circular, self-replicating DNA molecule separate from the bacterial chromosome. Certain plasmids, such as F plasmids, can undergo reversible integration into the cell's chromosome. A genetic element that can replicate either as part of the bacterial chromosome or independently of it is called an **episome.** In addition to some plasmids, temperate viruses, such as phage λ, qualify as episomes.

A plasmid has only a small number of genes, and these genes are not required for the survival and reproduction of the bacterium under normal conditions. However, the genes of plasmids can confer advantages on bacteria living in stressful environments. For example, the F plasmid facilitates genetic recombination, which may be advantageous in a changing environment that no longer favors existing strains in a bacterial population.

The F Plasmid and Conjugation. The F factor and its plasmid form, the **F plasmid,** consist of about 25 genes, most required for the production of sex pili. Cells containing the F plasmid, designated F^+ cells, function as DNA donors during conjugation. The F plasmid replicates in synchrony with the chromosomal DNA, and division of an F^+cell usually gives rise to two offspring that are both F^+. Cells lacking the F factor in either form, designated F^-, function as DNA recipients during conjugation. The F^+ condition is transferable in the sense that an F^+ cell converts an F^- cell to F^+ when the two cells conjugate, as shown in **Figure 18.18a**, on the next page. The original cell remains F^+ because the process of transfer involves a special type of DNA replication: One parental strand of F factor DNA is transferred across the mating bridge, and each parental strand acts as a template for synthesis of the second strand in its respective cell. In a mating of F^+ and F^- cells, only F plasmid DNA is transferred.

Chromosomal genes can be transferred during conjugation when the donor cell's F factor is integrated into the chromosome **(Figure 18.18b, top)**. A cell with the F factor built into its chromosome is called an *Hfr cell* (for *H*igh *f*requency of *r*ecombination). Like an F^+ cell, an Hfr cell functions as a donor during conjugation: DNA replication is initiated at a specific point on the integrated F factor DNA; from that point, a single strand of the F factor DNA moves into the F^- partner, dragging along adjacent chromosomal DNA **(Figure 18.18b, center)**. Random movements of the bacteria almost always disrupt conjugation long before an entire strand of the Hfr chromosome can be passed to the F^- cell. The single strand in each cell serves as a template for synthesis of a second strand. Thus, the Hfr cell's DNA remains the same, while the F^- cell acquires new DNA, some of it chromosomal. Temporarily, the recipient cell is a partial diploid, containing its own complete

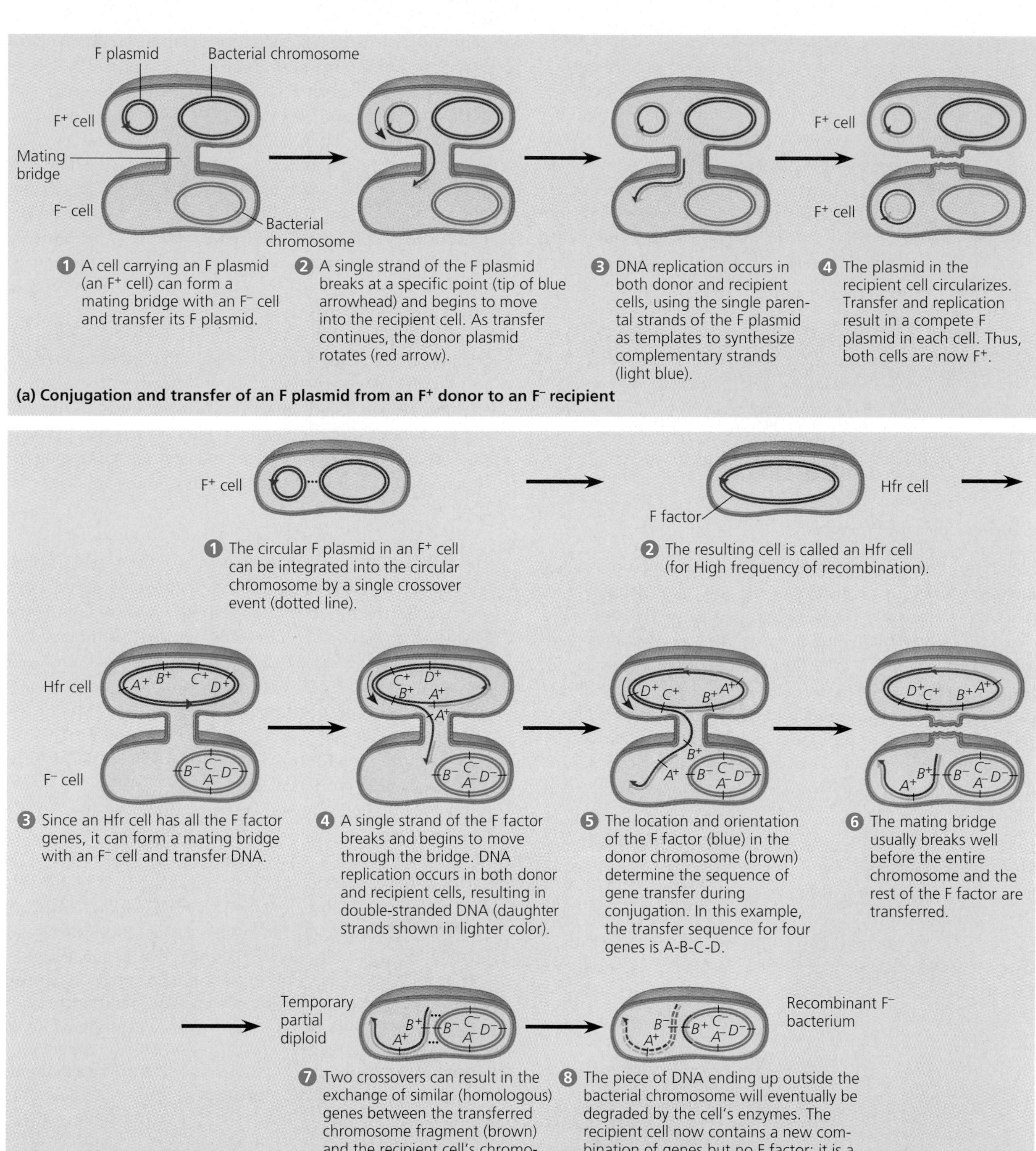

▲ **Figure 18.18 Conjugation and recombination in *E. coli*.** The DNA replication that accompanies transfer of an F plasmid or part of an Hfr bacterial chromosome is called *rolling circle replication*. This is sometimes referred to as the "toilet paper" model because of the way the single strand rolls off the donor cell DNA and moves into the recipient cell.

F plasmid
Bacterial chromosome
F⁺ cell
Mating bridge
F⁻ cell
Bacterial chromosome
F⁺ cell
F⁺ cell
❶ A cell carrying an F plasmid (an F⁺ cell) can form a mating bridge with an F⁻ cell and transfer its F plasmid.
❷ A single strand of the F plasmid breaks at a specific point (tip of blue arrowhead) and begins to move into the recipient cell. As transfer continues, the donor plasmid rotates (red arrow).
❸ DNA replication occurs in both donor and recipient cells, using the single parental strands of the F plasmid as templates to synthesize complementary strands (light blue).
❹ The plasmid in the recipient cell circularizes. Transfer and replication result in a compete F plasmid in each cell. Thus, both cells are now F⁺.
(a) Conjugation and transfer of an F plasmid from an F⁺ donor to an F⁻ recipient
F⁺ cell
F factor
Hfr cell
❶ The circular F plasmid in an F⁺ cell can be integrated into the circular chromosome by a single crossover event (dotted line).
❷ The resulting cell is called an Hfr cell (for High frequency of recombination).
Hfr cell
F⁻ cell
A⁺ B⁺ C⁺ D⁺
A⁻ B⁻ C⁻ D⁻
❸ Since an Hfr cell has all the F factor genes, it can form a mating bridge with an F⁻ cell and transfer DNA.
❹ A single strand of the F factor breaks and begins to move through the bridge. DNA replication occurs in both donor and recipient cells, resulting in double-stranded DNA (daughter strands shown in lighter color).
❺ The location and orientation of the F factor (blue) in the donor chromosome (brown) determine the sequence of gene transfer during conjugation. In this example, the transfer sequence for four genes is A-B-C-D.
❻ The mating bridge usually breaks well before the entire chromosome and the rest of the F factor are transferred.
Temporary partial diploid
Recombinant F⁻ bacterium
❼ Two crossovers can result in the exchange of similar (homologous) genes between the transferred chromosome fragment (brown) and the recipient cell's chromosome (green).
❽ The piece of DNA ending up outside the bacterial chromosome will eventually be degraded by the cell's enzymes. The recipient cell now contains a new combination of genes but no F factor; it is a recombinant F⁻ cell.
(b) Conjugation and transfer of part of the bacterial chromosome from an Hfr donor to an F⁻ recipient, resulting in recombination

F^- chromosome plus transferred chromosomal DNA from the Hfr donor. If part of the newly acquired DNA aligns with the homologous region of the F^- chromosome, segments of DNA can be exchanged **(Figure 18.18b, bottom).** Reproduction of this cell gives rise to a population of recombinant bacteria with genes derived from two different cells. This process of conjugation and recombination accounts for the results of the experiment shown in Figure 18.15, in which one of the bacterial strains was Hfr and the other F^-.

R Plasmids and Antibiotic Resistance. In the 1950s, Japanese physicians began to notice that some hospital patients suffering from bacterial dysentery, which produces severe diarrhea, did not respond to antibiotics that had generally been effective in the past. Apparently, resistance to these antibiotics had evolved in certain strains of *Shigella,* the pathogen. Eventually, researchers began to identify the specific genes that confer antibiotic resistance in *Shigella* and other pathogenic bacteria. Sometimes, mutation in a chromosomal gene of the pathogen can cause resistance. For example, a mutation in one gene may reduce the pathogen's ability to transport a particular antibiotic into the cell. Mutation in a different gene may alter the intracellular target protein for an antibiotic molecule, reducing its inhibitory effect. Some bacteria have resistance genes coding for enzymes that specifically destroy certain antibiotics, such as tetracycline or ampicillin. Genes conferring this type of resistance are generally carried by plasmids known as **R plasmids** (R for resistance).

Exposure of a bacterial population to a specific antibiotic, whether in a laboratory culture or within a host organism, will kill antibiotic-sensitive bacteria but not those that happen to have R plasmids with genes that counter the antibiotic. The theory of natural selection predicts that under these circumstances, the fraction of the bacterial population carrying genes for antibiotic resistance will increase, and that is exactly what happens. The medical consequences are also predictable: Resistant strains of pathogens are becoming more common, making the treatment of certain bacterial infections more difficult. The problem is compounded by the fact that many R plasmids, like F plasmids, have genes that encode sex pili and enable plasmid transfer from one bacterial cell to another by conjugation. Making the problem still worse, some R plasmids carry as many as ten genes for resistance to that many antibiotics. How do so many antibiotic resistance genes become part of a single plasmid? The answer involves another type of mobile genetic element, which we investigate next.

Transposition of Genetic Elements

In the previous section, you learned how DNA from one bacterial cell can be transferred to another cell and recombined into the genome of the recipient. The DNA of a single cell can also undergo recombination owing to movement of so-called *transposable genetic elements,* or simply **transposable elements,** within the cell's genome. Unlike a plasmid or prophage, transposable elements never exist independently but are always part of chromosomal or plasmid DNA. During the movement of these elements, called *transposition,* the transposable element moves from one site in a cell's DNA to another site—a target site—by a type of recombination process. In a bacterial cell, a transposable element may move within the chromosome, from a plasmid to the chromosome (or vice versa), or from one plasmid to another.

Transposable elements are sometimes called "jumping genes," but the phrase is misleading because they never completely detach from the cell's DNA. (The original and new DNA sites are brought together by DNA folding.) Some transposable elements move from one DNA location to another by a "cut-and-paste" mechanism. Others move by a "copy-and-paste" mechanism, in which the transposable element replicates at its original site, and a copy inserts elsewhere. In other words, the transposable element is added at a new site without being lost from the old site.

Although transposable elements vary in their selectivity for target sites, most can move to many alternative locations in the DNA. This ability to scatter certain genes throughout the genome makes transposition fundamentally different from other mechanisms of genetic shuffling. During bacterial transformation, generalized transduction, and conjugation (and during meiosis in eukaryotes as well), recombination occurs between homologous regions of DNA, regions of identical or very similar base sequence that can undergo base pairing. In contrast, the insertion of a transposable element in a new site does not depend on complementary base sequences. A transposable element can move genes to a site where genes of that sort have never before existed.

Insertion Sequences

The simplest transposable elements, called **insertion sequences,** exist only in bacteria. An insertion sequence contains a single gene, which codes for transposase, an enzyme that catalyzes movement of the insertion sequence from one site to another within the genome. The transposase gene is bracketed by a pair of noncoding DNA sequences about 20 to 40 nucleotides long. These sequences are called *inverted repeats* because the base sequence at one end of the insertion sequence is repeated upside down and backward (inverted) at the other end (**Figure 18.19a**, on the next page). Transposase recognizes these inverted repeats as the boundaries of the insertion sequence. During transposition, molecules of the enzyme bind to the inverted repeats and to a target site elsewhere in the genome and catalyze the necessary DNA cutting and resealing.

An insertion sequence can cause mutations if it transposes into the coding sequence of a gene or into a DNA region that regulates gene expression. This mechanism of mutation is intrinsic to the cell, in contrast to mutagenesis by extrinsic factors such as environmental radiation and chemicals. Insertion sequences account for about 1.5% of the *E. coli* genome.

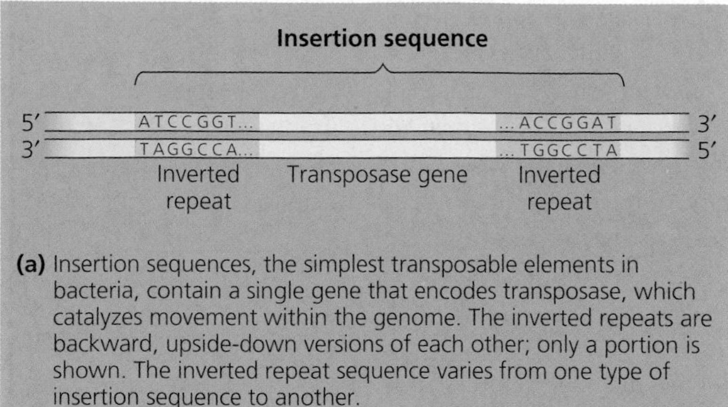

(a) Insertion sequences, the simplest transposable elements in bacteria, contain a single gene that encodes transposase, which catalyzes movement within the genome. The inverted repeats are backward, upside-down versions of each other; only a portion is shown. The inverted repeat sequence varies from one type of insertion sequence to another.

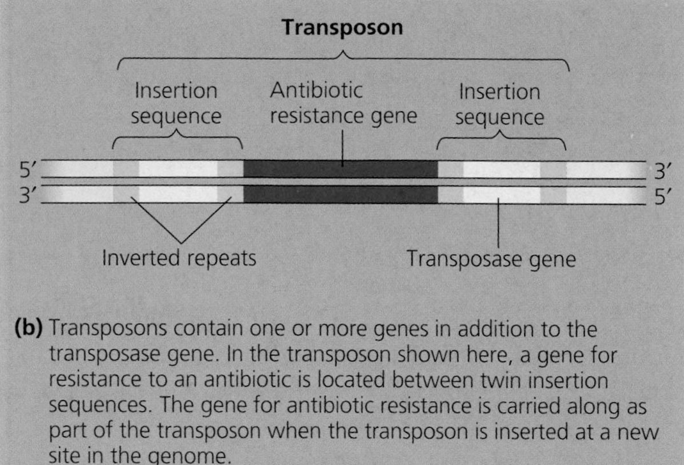

(b) Transposons contain one or more genes in addition to the transposase gene. In the transposon shown here, a gene for resistance to an antibiotic is located between twin insertion sequences. The gene for antibiotic resistance is carried along as part of the transposon when the transposon is inserted at a new site in the genome.

▲ **Figure 18.19 Transposable genetic elements in bacteria.** These diagrams are not to scale; most transposons are considerably longer than insertion sequences.

However, mutation of a given gene by transposition occurs only rarely—about once in every 10 million generations. This is about the same as the spontaneous mutation rate due to other factors.

Transposons

Transposable elements longer and more complex than insertion sequences, called **transposons,** also move about in the bacterial genome. In addition to the DNA required for transposition, transposons include extra genes that go along for the ride, such as genes for antibiotic resistance. In some bacterial transposons, the extra genes are sandwiched between two insertion sequences **(Figure 18.19b)**. It is as though two insertion sequences happened to land relatively close together in the genome and now travel together, along with all the DNA between them, as a single transposable element. Other bacterial transposons do not contain insertion sequences; these have different inverted repeats at their ends.

In contrast to insertion sequences, which are not known to benefit bacteria in any specific way, transposons may help bacteria adapt to new environments. We mentioned earlier that a single R plasmid can carry several genes for resistance to different antibiotics. This is explained by transposons, which can add a gene for antibiotic resistance to a plasmid already carrying genes for resistance to other antibiotics. The transmission of this composite plasmid to other bacterial cells by cell division or conjugation can then spread resistance to a variety of antibiotics throughout a bacterial population. In an antibiotic-rich environment, natural selection favors bacteria that have built up R plasmids with multiple antibiotic resistance genes through a series of transpositions.

Transposons are not unique to bacteria and are important components of eukaryotic genomes as well. You will learn about transposable elements in eukaryotes in Chapter 19.

Concept Check 18.3

1. Distinguish between the three mechanisms of transferring DNA from one bacterial cell to another.
2. What are the similarities and differences between lysogenic phage DNA and a plasmid?
3. Explain why the process of conjugation can lead to genetic recombination of chromosomal DNA in an Hfr × F^- mating, but not in an F^+ × F^- mating.

For suggested answers, see Appendix A.

Concept 18.4

Individual bacteria respond to environmental change by regulating their gene expression

Mutations and various types of gene transfer generate the genetic variation that makes natural selection possible. And natural selection, acting over many generations, can increase the proportion of individuals in a bacterial population that are adapted to some new environmental condition. But how can an individual bacterium, locked into the genome it has inherited, cope with environmental fluctuation?

Consider, for instance, an individual *E. coli* cell living in the erratic environment of a human colon, dependent for its nutrients on the whimsical eating habits of its host. If the environment is lacking in the amino acid tryptophan, which the bacterium needs to survive, the cell responds by activating a metabolic pathway that makes tryptophan from another compound. Later, if the human host eats a tryptophan-rich meal, the bacterial cell stops producing tryptophan, thus saving itself from squandering its resources to produce a substance that is available from the surrounding solution in prefabricated form. This is just one example of how bacteria tune their metabolism to changing environments.

Metabolic control occurs on two levels **(Figure 18.20)**. First, cells can adjust the activity of enzymes already present. This is a fairly fast response, which relies on the sensitivity of many enzymes to chemical cues that increase or decrease their catalytic activity (see Chapter 8). For example, the activity of the first enzyme in the tryptophan synthesis pathway is inhibited by the pathway's end product. Thus, if tryptophan accumulates in a cell, it shuts down the synthesis of more tryptophan by inhibiting enzyme activity. Such *feedback inhibition*, typical of anabolic (biosynthetic) pathways, allows a cell to adapt to short-term fluctuations in the supply of a substance it needs.

Second, cells can adjust the amount being made of certain enzymes; that is, they can regulate the expression of the genes encoding the enzymes. If, in our example, the environment continues to provide all the tryptophan the cell needs, the cell stops making the enzymes that work in the tryptophan pathway. This control of enzyme production occurs at the level of transcription, the synthesis of messenger RNA coding for these enzymes. More generally, many genes of the bacterial genome are switched on or off by changes in the metabolic status of the cell. The basic mechanism for this control of gene expression in bacteria, described as the *operon model,* was discovered in 1961 by François Jacob and Jacques Monod at the Pasteur Institute in Paris. Let's see what an operon is and how it works, using the control of tryptophan synthesis as our first example.

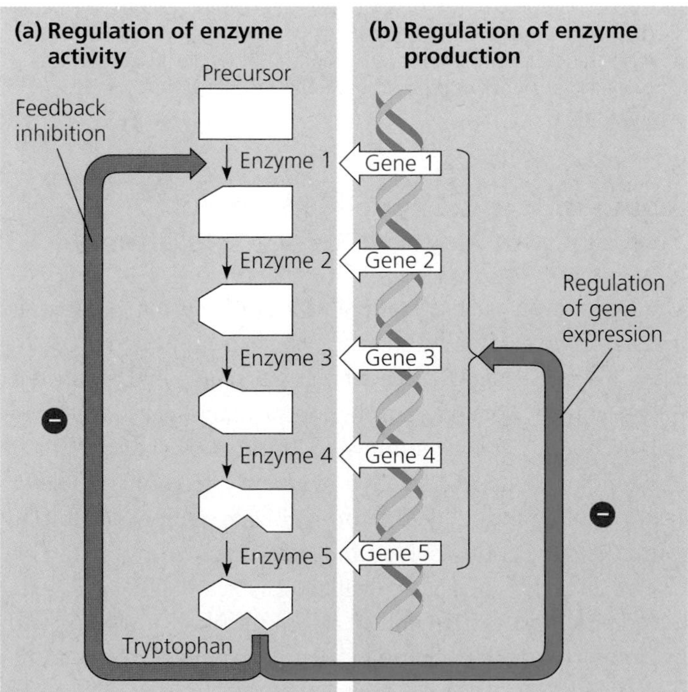

▲ **Figure 18.20 Regulation of a metabolic pathway.** In the pathway for tryptophan synthesis, an abundance of tryptophan can both **(a)** inhibit the activity of the first enzyme in the pathway (feedback inhibition), a rapid response, and **(b)** repress expression of the genes for all the enzymes needed for the pathway, a longer-term response. The ⊖ symbol stands for inhibition.

Operons: The Basic Concept

E. coli synthesizes tryptophan from a precursor molecule in a series of steps, each reaction catalyzed by a specific enzyme (see Figure 18.20). The five genes coding for these enzymes are clustered together on the bacterial chromosome. A single promoter serves all five genes, which constitute a transcription unit. (Recall from Chapter 17 that a promoter is a site where RNA polymerase can bind to DNA and begin transcription.) Thus, transcription gives rise to one long mRNA molecule that codes for all five enzymes in the tryptophan pathway. The cell can translate this mRNA into five separate polypeptides because the mRNA is punctuated with start and stop codons that signal where the coding sequence for each polypeptide begins and ends.

A key advantage of grouping genes of related function into one transcription unit is that a single on-off "switch" can control the whole cluster of functionally related genes. When an *E. coli* cell must make tryptophan for itself because the nutrient medium lacks this amino acid, all the enzymes for the metabolic pathway are synthesized at one time. The switch is a segment of DNA called an **operator.** Both its location and name suit its function: Positioned within the promoter or between the promoter and the enzyme-coding genes, the operator controls the access of RNA polymerase to the genes. All together, the operator, the promoter, and the genes they control—the entire stretch of DNA required for enzyme production for the tryptophan pathway—constitute an **operon.** Here we are dissecting the *trp* operon (*trp* for tryptophan), one of many operons in the *E. coli* genome (**Figure 18.21**, on the next page).

If the operator is the switch for controlling transcription, how does this switch work? By itself, the *trp* operon is turned on; that is, RNA polymerase can bind to the promoter and transcribe the genes of the operon. The operon can be switched off by a protein called the *trp* **repressor.** The repressor binds to the operator and blocks attachment of RNA polymerase to the promoter, preventing transcription of the genes. A repressor protein is specific; that is, it recognizes and binds only to the operator of a particular operon. The repressor that switches off the *trp* operon by binding to the *trp* operator has no effect on other operons in the *E. coli* genome.

The *trp* repressor is the product of a **regulatory gene** called *trpR,* which is located some distance away from the operon it controls and has its own promoter. Regulatory genes are expressed continuously, although at a low rate, and a few *trp* repressor molecules are always present in *E. coli* cells. Why, then, is the *trp* operon not switched off permanently? First, the binding of repressors to operators is reversible. An operator vacillates between two states: one

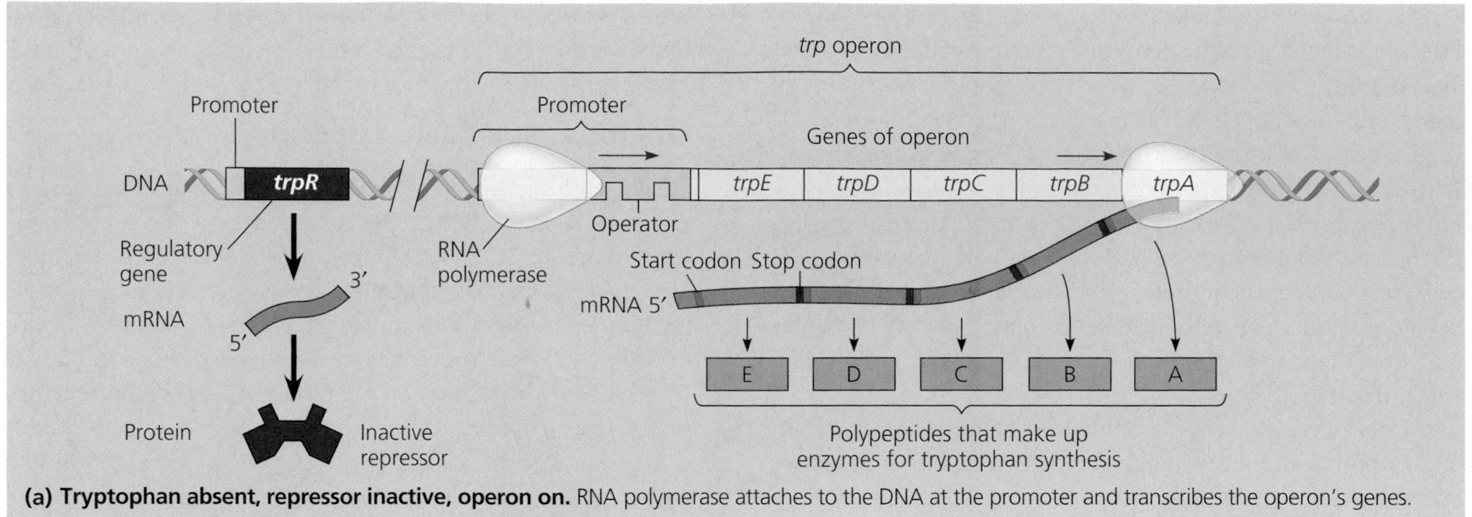

(a) Tryptophan absent, repressor inactive, operon on. RNA polymerase attaches to the DNA at the promoter and transcribes the operon's genes.

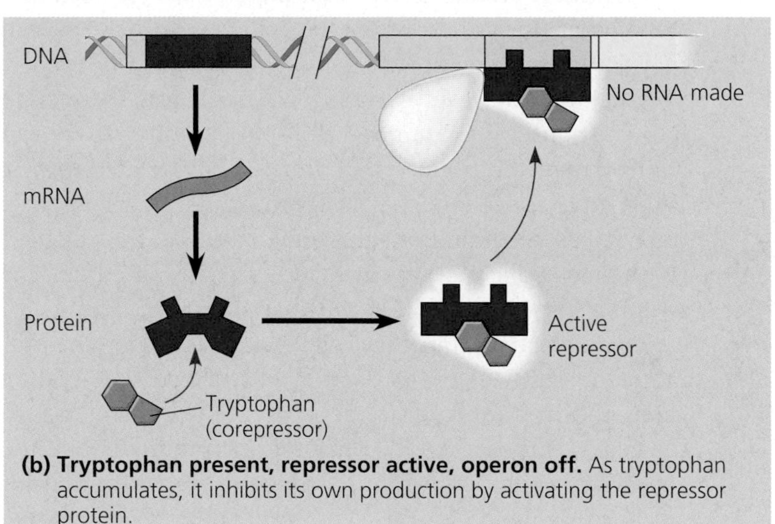

(b) Tryptophan present, repressor active, operon off. As tryptophan accumulates, it inhibits its own production by activating the repressor protein.

◀ **Figure 18.21 The *trp* operon: regulated synthesis of repressible enzymes.** Tryptophan is an amino acid produced by an anabolic pathway catalyzed by repressible enzymes. **(a)** Five genes encoding the polypeptides that make up the enzymes of this pathway are grouped, along with a promoter and an operator, into the *trp* operon. The *trp* operator is located within the *trp* promoter. **(b)** Accumulation of tryptophan, the end product of the pathway, represses transcription of the *trp* operon, thus blocking synthesis of all the enzymes in the pathway. The mechanism in *E. coli* is shown here.

without the repressor bound and one with the repressor bound. The relative duration of each state depends on the number of active repressor molecules around. Second, the *trp* repressor, like most regulatory proteins, is an allosteric protein, with two alternative shapes, active and inactive (see Figure 8.20). The *trp* repressor is synthesized in an inactive form with little affinity for the *trp* operator. Only if tryptophan binds to the *trp* repressor at an allosteric site does the repressor protein change to the active form that can attach to the operator, turning the operon off.

Tryptophan functions in this system as a **corepressor**, a small molecule that cooperates with a repressor protein to switch an operon off. As tryptophan accumulates, more tryptophan molecules associate with *trp* repressor molecules, which can then bind to the *trp* operator and shut down production of the tryptophan pathway enzymes. If the cell's tryptophan level drops, transcription of the operon's genes resumes. This is one example of how gene expression responds rapidly to changes in the cell's internal and external environment.

Repressible and Inducible Operons: Two Types of Negative Gene Regulation

The *trp* operon is said to be a *repressible operon* because its transcription is usually on but can be inhibited (repressed) when a specific small molecule (tryptophan) binds allosterically to a regulatory protein. In contrast, an *inducible operon* is usually off but can be stimulated (induced) when a specific small molecule interacts with a regulatory protein. The classic example of an inducible operon is the *lac* operon (*lac* for lactose), the subject of Jacob and Monod's pioneering research.

The disaccharide lactose (milk sugar) is available to *E. coli* if the human host drinks milk. Lactose metabolism begins with hydrolysis of the disaccharide into its component monosaccharides, glucose and galactose, a reaction catalyzed by the enzyme β-galactosidase. Only a few molecules of this enzyme are present in an *E. coli* cell growing in the absence of lactose. But if lactose is added to the bacterium's environment, the number of β-galactosidase molecules in the cell increases a thousandfold within about 15 minutes.

The gene for β-galactosidase is part of the *lac* operon, which includes two other genes coding for enzymes that function in lactose metabolism. The entire transcription unit is under the command of a single operator and promoter. The regulatory gene, *lacI,* located outside the operon, codes for an allosteric repressor protein that can switch off the *lac* operon by binding to the operator. So far, this sounds just like regulation of the *trp* operon, but there is one important difference. Recall that the *trp* repressor is inactive by itself and requires tryptophan as a corepressor in order to bind to the operator. The *lac* repressor, in contrast, is active all by itself, binding to the operator and switching the *lac* operon off. In this case, a specific small molecule, called an **inducer,** *inactivates* the repressor.

For the *lac* operon, the inducer is allolactose, an isomer of lactose formed in small amounts from lactose that enters the cell. In the absence of lactose (and hence allolactose), the *lac* repressor is in its active configuration, and the genes of the *lac* operon are silenced **(Figure 18.22a)**. If lactose is added to the cell's surroundings, allolactose binds to the *lac* repressor and alters its conformation, nullifying the repressor's ability to attach to the operator. Without bound repressor, the *lac* operon is transcribed into mRNA for the lactose-utilizing enzymes **(Figure 18.22b)**.

In the context of gene regulation, the enzymes of the lactose pathway are referred to as *inducible enzymes,* because their synthesis is induced by a chemical signal (allolactose, in this case). Analogously, the enzymes for tryptophan synthesis are said to be repressible. *Repressible enzymes* generally function in anabolic pathways, which synthesize essential end products from raw materials (precursors). By suspending production of an end product when it is already present in sufficient quantity, the cell can allocate its organic precursors and energy for other uses. In contrast, inducible enzymes usually function in catabolic pathways, which break down a nutrient to simpler molecules. By producing the appropriate enzymes only when the nutrient is available, the cell avoids wasting energy and precursors making proteins that are not needed.

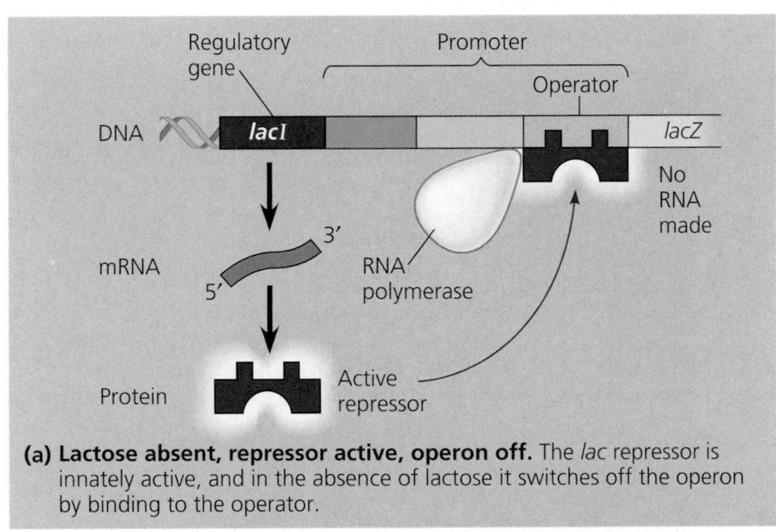

(a) Lactose absent, repressor active, operon off. The *lac* repressor is innately active, and in the absence of lactose it switches off the operon by binding to the operator.

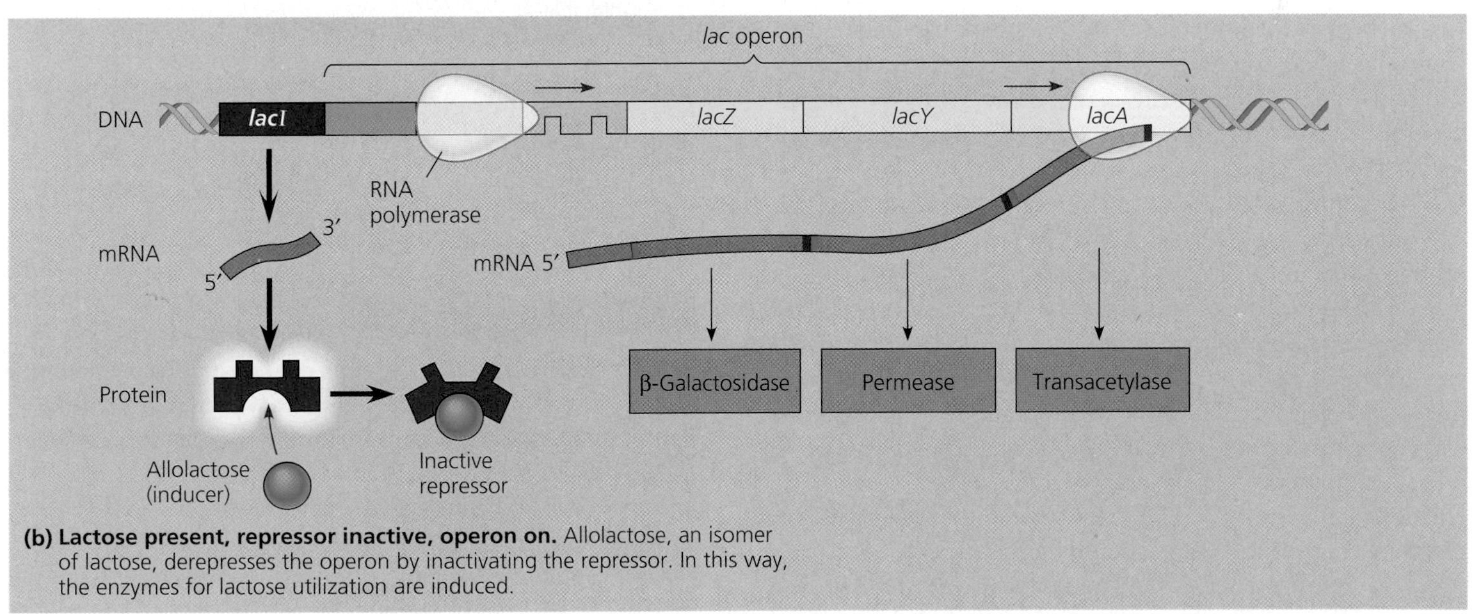

(b) Lactose present, repressor inactive, operon on. Allolactose, an isomer of lactose, derepresses the operon by inactivating the repressor. In this way, the enzymes for lactose utilization are induced.

◀ **Figure 18.22 The *lac* operon: regulated synthesis of inducible enzymes.** *E. coli* uses three enzymes to take up and metabolize lactose. The genes for these three enzymes are clustered in the *lac* operon. One gene, *lacZ,* codes for β-galactosidase, which hydrolyzes lactose to glucose and galactose. The second gene, *lacY,* codes for a permease, the membrane protein that transports lactose into the cell. The third gene, *lacA,* codes for an enzyme called transacetylase, whose function in lactose metabolism is still unclear. The gene for the *lac* repressor, *lacI,* happens to be adjacent to the *lac* operon, an unusual situation. The function of the darker green region at the upstream (left) end of the promoter is revealed in Figure 18.23.

Regulation of both the *trp* and *lac* operons involves the *negative* control of genes, because the operons are switched off by the active form of the repressor protein. It may be easier to see this for the *trp* operon, but it is also true for the *lac* operon. Allolactose induces enzyme synthesis not by acting directly on the genome, but by freeing the *lac* operon from the negative effect of the repressor. Gene regulation is said to be *positive* only when a regulatory protein interacts directly with the genome to switch transcription on. Let's look at an example of positive control of genes, again involving the *lac* operon.

Positive Gene Regulation

For the lactose-utilizing enzymes to be synthesized in appreciable quantity, it is not sufficient for lactose to be present in the bacterial cell. The other requirement is that the simple sugar glucose be in short supply. Given a choice of substrates for glycolysis and other catabolic pathways, *E. coli* preferentially uses glucose. The enzymes for glucose breakdown (glycolysis; see Figure 9.9) are continually present.

How does the *E. coli* cell sense the glucose concentration, and how is this information relayed to the genome? Again, the mechanism depends on the interaction of an allosteric regulatory protein with a small organic molecule, in this case **cyclic AMP (cAMP),** which accumulates when glucose is scarce (see Figure 11.9 for the structure of cAMP). The regulatory protein, called *catabolite activator protein (CAP),* is an **activator** of transcription. When cAMP binds to CAP, the regulatory protein (CAP) assumes its active shape and can bind to a specific site at the upstream end of the *lac* promoter **(Figure 18.23a)**. The attachment of CAP to the promoter directly stimulates gene expression. Thus, this mechanism qualifies as positive regulation.

If the amount of glucose in the cell increases, the cAMP concentration falls, and without it, CAP detaches from the operon. Because CAP is inactive, transcription of the *lac* operon proceeds at only a low level, even in the presence of lactose **(Figure 18.23b)**. Thus, the *lac* operon is under dual control: negative control by the *lac* repressor and positive control by CAP. The state of the *lac* repressor (with or without bound allolactose) determines whether or not transcription of the *lac* operon's genes occurs at all; the state of CAP (with or without bound cAMP) controls the *rate* of transcription if the operon is repressor-free. It is as though the operon has both an on-off switch and a volume control.

In addition to the *lac* operon, CAP helps regulate several other operons that encode enzymes used in catabolic pathways. When glucose is plentiful and CAP is inactive, the synthesis of enzymes that catabolize compounds other than glucose generally slows down. The cell's ability to catabolize other compounds, such as lactose, enables a cell deprived of glucose to survive. The specific compounds present at the moment determine which operons are switched on. These multiple contingency mechanisms suit an organism that cannot control what its host eats. Bacteria are remarkable in their ability to adapt—over the long term by evolutionary changes in their genetic makeup and over the short term by the control of gene expression in individual cells. Of course, the various control mechanisms are also evolutionary products that exist because they have been favored by natural selection.

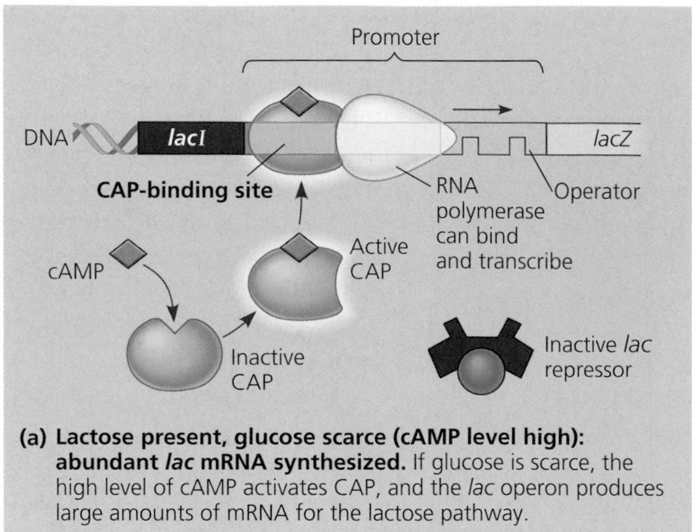

(a) Lactose present, glucose scarce (cAMP level high): abundant *lac* mRNA synthesized. If glucose is scarce, the high level of cAMP activates CAP, and the *lac* operon produces large amounts of mRNA for the lactose pathway.

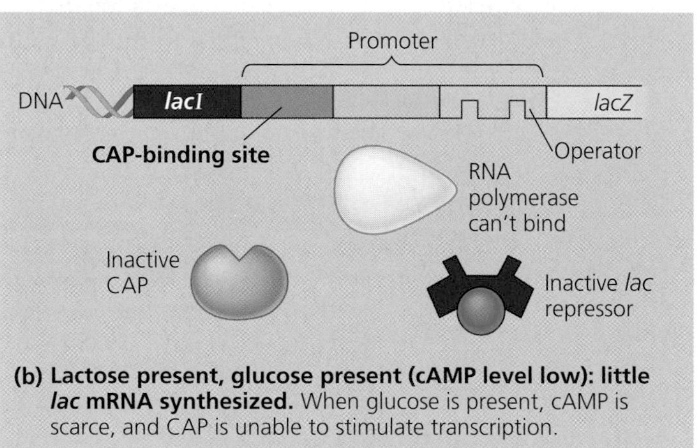

(b) Lactose present, glucose present (cAMP level low): little *lac* mRNA synthesized. When glucose is present, cAMP is scarce, and CAP is unable to stimulate transcription.

▲ **Figure 18.23 Positive control of the *lac* operon by catabolite activator protein (CAP).** RNA polymerase has high affinity for the *lac* promoter only when catabolite activator protein (CAP) is bound to a DNA site at the upstream end of the promoter. CAP attaches to its DNA site only when associated with cyclic AMP (cAMP), whose concentration in the cell rises when the glucose concentration falls. Thus, when glucose is present, even if lactose also is available, the cell preferentially catabolizes glucose and does not make the lactose-utilizing enzymes.

Concept Check 18.4

1. A certain mutation in *E. coli* changes the *lac* operator so that the active repressor cannot bind. How would this affect the cell's production of β-galactosidase?
2. How does binding of the *trp* corepressor and the *lac* inducer to their respective repressor proteins alter repressor function and transcription in each case?

For suggested answers, see Appendix A.

Chapter 18 Review

Go to www.campbellbiology.com or the student CD-ROM to explore Activities, Investigations, and other interactive study aids.

SUMMARY OF KEY CONCEPTS

A virus has a genome but can reproduce only within a host cell

- **The Discovery of Viruses: *Scientific Inquiry*** **(pp. 334–335)** Researchers discovered viruses in the late 1800s by studying a plant disease, tobacco mosaic disease.
- **Structure of Viruses (pp. 335–336)** A virus is a small nucleic acid genome enclosed in a protein capsid and sometimes a membranous envelope containing viral proteins that help viruses enter cells. The genome may be single- or double-stranded DNA or RNA.
- **General Features of Viral Reproductive Cycles (pp. 336–337)** Viruses use enzymes, ribosomes, and small molecules of host cells to synthesize progeny viruses. Each type of virus has a characteristic host range.
  Activity *Simplified Viral Reproductive Cycle*
- **Reproductive Cycles of Phages (pp. 337–339)** In the lytic cycle, entry of the viral genome into a bacterium programs destruction of host DNA, production of new phages, and digestion of the host's cell wall, releasing the progeny phages. In the lysogenic cycle, the genome of a temperate phage inserts into the bacterial chromosome as a prophage, which is passed on to host daughter cells until it is induced to leave the chromosome and initiate a lytic cycle.
  Activity *Phage Lytic Cycle*
  Activity *Phage Lysogenic and Lytic Cycles*
- **Reproductive Cycles of Animal Viruses (pp. 339–342)** Many animal viruses have an envelope. Retroviruses (such as HIV) use the enzyme reverse transcriptase to copy their RNA genome into DNA, which can be integrated into the host genome as a provirus.
  Activity *Retrovirus (HIV) Reproductive Cycle*
- **Evolution of Viruses (pp. 342–343)** Since viruses can reproduce only within cells, they probably evolved after the first cells appeared, perhaps as packaged fragments of cellular nucleic acid.

Viruses, viroids, and prions are formidable pathogens in animals and plants

- **Viral Diseases in Animals (pp. 343–344)** Symptoms may be caused by direct viral harm to cells or by the body's immune response. Vaccines stimulate the immune system to defend the host against specific viruses.
- **Emerging Viruses (p. 344)** Outbreaks of "new" viral diseases in humans are usually caused by existing viruses that expand their host territory.
  Investigation *What Causes Infections in AIDS Patients?*
  Investigation *Why Do AIDS Rates Differ Across the U.S.?*
- **Viral Diseases in Plants (p. 345)** Viruses enter plant cells through damaged cell walls (horizontal transmission) or are inherited from a parent (vertical transmission).
- **Viroids and Prions: The Simplest Infectious Agents (pp. 345–346)** Viroids are naked RNA molecules that infect plants and disrupt their growth. Prions are slow-acting, virtually indestructible infectious proteins that cause brain diseases in mammals.

Concept 18.3

Rapid reproduction, mutation, and genetic recombination contribute to the genetic diversity of bacteria

- **The Bacterial Genome and Its Replication (p. 346)** The bacterial chromosome is usually a circular DNA molecule with few associated proteins. Plasmids are smaller circular DNA molecules that can replicate independently of the chromosome.
- **Mutation and Genetic Recombination as Sources of Genetic Variation (pp. 346–348)** Because bacteria can proliferate rapidly, new mutations can quickly increase a population's genetic variation. Further diversity can arise by recombination of the DNA from two different bacterial cells.
- **Mechanisms of Gene Transfer and Genetic Recombination in Bacteria (pp. 348–351)** New bacterial strains can arise by the transfer of DNA from one cell to another cell. In transformation, naked DNA enters the cell from the surroundings. In transduction, bacterial DNA is carried from one cell to another by phages. In conjugation, an F^+ donor cell, which contains the F plasmid, transfers plasmid DNA to an F^- recipient cell. The F factor of an Hfr cell, which is integrated into the bacterial chromosome, brings some chromosomal DNA along with it when it is transferred to an F^- cell. R plasmids confer resistance to various antibiotics.
  Investigation *What Are the Patterns of Antibiotic Resistance?*
- **Transposition of Genetic Elements (pp. 351–352)** DNA segments that can insert at multiple sites in a cell's DNA contribute to genetic shuffling in bacteria. Insertion sequences, the simplest bacterial transposable elements, consist of inverted repeats of DNA flanking a gene for transposase. Bacterial transposons have additional genes, such as those for antibiotic resistance.

Concept 18.4

Individual bacteria respond to environmental change by regulating their gene expression

- **Operons: The Basic Concept (pp. 353–354)** Cells control metabolism by regulating enzyme activity or the expression of genes coding for enzymes. In bacteria, genes are often clustered into operons, with one promoter serving several adjacent genes. An operator site on the DNA switches the operon on or off.
- **Repressible and Inducible Operons: Two Types of Negative Gene Regulation (pp. 354–356)** In a repressible operon, binding of a specific repressor protein to the operator shuts off transcription. The repressor is active when bound to a corepressor, usually the end product of an anabolic pathway. In an inducible operon, binding of an inducer to an innately active repressor inactivates the repressor and turns on transcription. Inducible enzymes usually function in catabolic pathways.
- **Positive Gene Regulation (p. 356)** Some operons are also subject to positive control via a stimulatory activator protein, such as catabolite activator protein (CAP), which promotes transcription when bound to a site within the promoter.
  Activity *The* lac *Operon in* E. coli

TESTING YOUR KNOWLEDGE

Self-Quiz

1. A bacterium is infected with an experimentally constructed bacteriophage composed of the T2 phage protein coat and T4 phage DNA. The new phages produced would have
   a. T2 protein and T4 DNA.
   b. T2 protein and T2 DNA.
   c. a mixture of the DNA and proteins of both phages.
   d. T4 protein and T4 DNA.
   e. T4 protein and T2 DNA.

2. RNA viruses require their own supply of certain enzymes because
   a. host cells rapidly destroy the viruses.
   b. host cells lack enzymes that can replicate the viral genome.
   c. these enzymes translate viral mRNA into proteins.
   d. these enzymes penetrate host cell membranes.
   e. these enzymes cannot be made in host cells.

3. Which of the following is descriptive of an R plasmid?
   a. Its transfer converts an F^- cell into an F^+ cell.
   b. It has genes for antibiotic resistance and maybe for sex pili.
   c. It is transferred between bacteria by transduction.
   d. It is a good example of a composite transposon.
   e. It makes bacteria resistant to phage.

4. Transposition differs from other mechanisms of genetic recombination because it
   a. occurs only in bacteria.
   b. moves genes between homologous regions of the DNA.
   c. plays little or no role in evolution.
   d. occurs only in eukaryotes.
   e. scatters genes to new loci in the genome.

5. If a particular operon encodes enzymes for making an essential amino acid and is regulated like the *trp* operon, then
   a. the amino acid inactivates the repressor.
   b. the enzymes produced are called inducible enzymes.
   c. the repressor is active in the absence of the amino acid.
   d. the amino acid acts as a corepressor.
   e. the amino acid turns on transcription of the operon.

6. What would occur if the repressor of an inducible operon were mutated so it could not bind the operator?
   a. continuous transcription of the operon's genes
   b. reduced transcription of the operon's genes
   c. buildup of a substrate for the pathway controlled by the operon
   d. irreversible binding of the repressor to the promoter
   e. overproduction of catabolite activator protein (CAP)

7. During conjugation between an Hfr cell and an F^- cell,
   a. the F^- cell becomes an F^+ cell.
   b. the F^- cell becomes an Hfr cell.
   c. the chromosome of the F^- cell is degraded.
   d. genes from the Hfr cell may replace genes of the F^- cell by recombination.
   e. DNA from the F^- cell transfers to the Hfr cell, and DNA from the Hfr cell transfers to the F^- cell.

8. Genetic variation in bacterial populations never results from
   a. transduction.
   b. transformation.
   c. conjugation.
   d. mutation.
   e. meiosis.

9. Which of the following characteristics or processes is common to *both* bacteria and viruses?
   a. binary fission
   b. ribosomes
   c. genetic material of nucleic acid
   d. mitosis
   e. conjugation

10. Emerging viruses arise by
    a. mutation of existing viruses.
    b. the spread of existing viruses to new host species.
    c. the spread of existing viruses more widely within their host species.
    d. all of the above
    e. none of the above

For Self-Quiz answers, see Appendix A.

Go to the website or CD-ROM for more quiz questions.

Evolution Connection

The success of some viruses lies in their ability to evolve within the host. Such a virus evades the host's defenses by rapidly mutating and producing many altered progeny viruses before the body can mount an attack. Thus, the viruses present late in infection differ from those that initially infected the body. Discuss this as an example of evolution in microcosm. Which viral lineages tend to survive?

Scientific Inquiry

When bacteria infect an animal, the number of bacteria in the body increases in an exponential fashion (graph A). After infection by a virulent animal virus with a lytic reproductive cycle, there is no evidence of infection for a while. Then, the number of viruses rises suddenly and subsequently increases in a series of steps (graph B). Explain the difference in the growth curves.

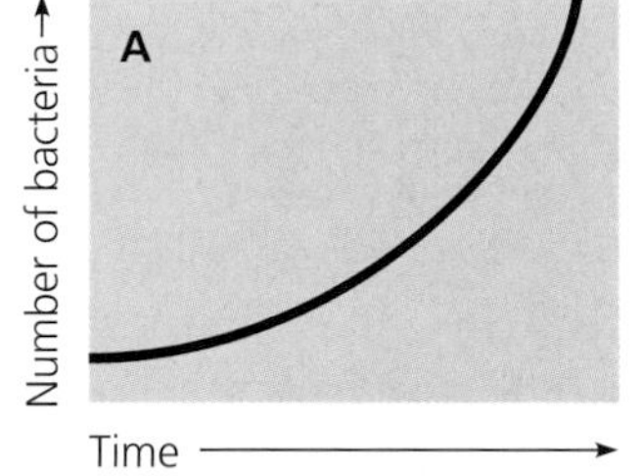

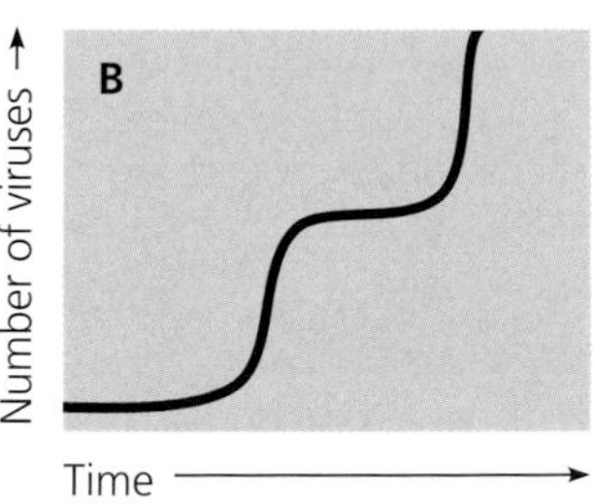

Biological Inquiry: A Workbook of Investigative Cases
Explore West Nile virus in the case "The Donor's Dilemma."
Investigation *What Causes Infections in AIDS Patients?*
Investigation *Why Do AIDS Rates Differ Across the U.S.?*
Investigation *What Are the Patterns of Antibiotic Resistance?*

Science, Technology, and Society

Explain how the excessive or inappropriate use of antibiotics poses a health hazard for a human population.

19 Eukaryotic Genomes
Organization, Regulation, and Evolution

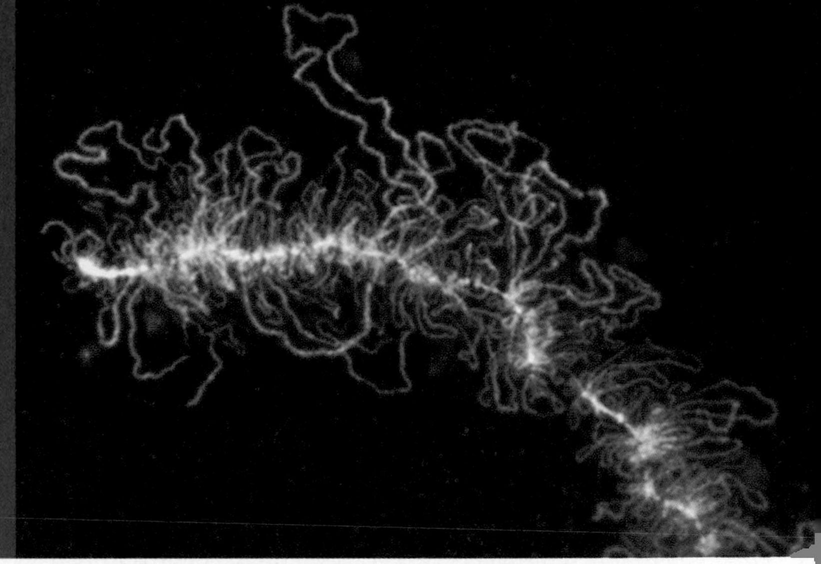

▲ **Figure 19.1 DNA in a eukaryotic chromosome from a developing salamander egg.**

Key Concepts

19.1 Chromatin structure is based on successive levels of DNA packing

19.2 Gene expression can be regulated at any stage, but the key step is transcription

19.3 Cancer results from genetic changes that affect cell cycle control

19.4 Eukaryotic genomes can have many noncoding DNA sequences in addition to genes

19.5 Duplications, rearrangements, and mutations of DNA contribute to genome evolution

Overview

How Eukaryotic Genomes Work and Evolve

Eukaryotic cells face the same challenges as prokaryotic cells in expressing their genes. However, the typical eukaryotic genome is much larger, and, in multicellular eukaryotes, cell specialization is crucial. These two features present a formidable information-processing task for the eukaryotic cell.

The human genome, for instance, has an estimated 25,000 genes—more than five times that of a typical prokaryote. It also includes an enormous amount of DNA that does not code for RNA or protein. Managing so much DNA requires that the genome be elaborately organized. In all organisms, DNA associates with proteins that condense it. In eukaryotes, the DNA-protein complex, called **chromatin,** is ordered into higher structural levels than the DNA-protein complex in prokaryotes. The fluorescence micrograph in **Figure 19.1** gives a sense of the complex organization of the chromatin in a eukaryotic chromosome. Part of the chromatin is packed into the main axis (white) of the chromosome, while those parts that are being actively transcribed are spread out in loops (red).

Both prokaryotes and eukaryotes must alter their patterns of gene expression in response to changes in environmental conditions. Multicellular eukaryotes, in addition, must develop and maintain multiple cell types. Each of these cell types contains the same genome but expresses a different subset of genes, a significant challenge in gene regulation.

In this chapter, you will first learn about the structure of chromatin and how changes in chromatin structure affect gene expression. Then we will discuss other mechanisms for regulating expression of eukaryotic genes. As in prokaryotes, eukaryotic gene expression is most often regulated at the stage of transcription. Next, we will describe how disruptions in gene regulation can lead to cancer. In the remainder of the chapter, we will consider the various types of nucleotide sequences in eukaryotic genomes and explain how they arose and changed during genome evolution. Being aware of the forces that have shaped—and continue to shape—genomes will help you understand how biological diversity has evolved.

Concept 19.1

Chromatin structure is based on successive levels of DNA packing

Eukaryotic DNA is precisely combined with a large amount of protein, and the resulting chromatin undergoes striking changes in the course of the cell cycle (see Figure 12.6). In interphase cells stained for light microscopy, the chromatin usually appears as a diffuse mass within the nucleus, suggesting that the chromatin is highly extended. As a cell prepares for mitosis, its chromatin coils and folds up (condenses), eventually forming a characteristic number of short, thick chromosomes that are distinguishable from each other with the light microscope.

Eukaryotic chromosomes contain an enormous amount of DNA relative to their condensed length. Each chromosome contains a single linear DNA double helix that, in humans, averages about 1.5×10^8 nucleotide pairs. If completely stretched out, such a DNA molecule would be about 4 cm long, thousands of times longer than the diameter of a cell nucleus. All this DNA—as well as the DNA of the other 45 human chromosomes—fits into the nucleus through the elaborate, multilevel system of DNA packing outlined in **Figure 19.2**.

Nucleosomes, or "Beads on a String"

Proteins called **histones** are responsible for the first level of DNA packing in chromatin. The mass of histone in chromatin is approximately equal to the mass of DNA. Histones have a high proportion of positively charged amino acids (lysine and arginine), and they bind tightly to the negatively charged DNA. (Recall that the phosphate groups of DNA give it a negative charge all along its length; see Figure 16.7.) Histones are very similar from one eukaryote to another, and similar proteins are found even in prokaryotes. The apparent conservation of histone genes during evolution probably reflects the pivotal role of histones in organizing DNA within cells.

In electron micrographs, unfolded chromatin has the appearance of beads on a string, as shown in **Figure 19.2a**. In this configuration, a chromatin fiber is 10 nm in diameter (the *10-nm fiber*). Each "bead" is a **nucleosome,** the basic unit of DNA packing; the "string" between the beads is called *linker DNA*. A nucleosome consists of DNA wound around a protein core composed of two molecules each of four types of histone: H2A, H2B, H3, and H4. The amino end (N-terminus) of each histone protein (the *histone tail*) extends outward from the nucleosome. A molecule of a fifth histone, called H1, attaches to the DNA near the nucleosome when a 10-nm chromatin fiber undergoes the next level of packing.

The association of DNA and histones in nucleosomes seems to remain essentially intact throughout the cell cycle. The histones leave the DNA only transiently during DNA replication, and, with very few exceptions, they stay with the DNA during transcription. How can DNA be transcribed when it is wrapped around histones in a nucleosome? Researchers have learned that changes in the shapes and positions of nucleosomes can allow RNA-synthesizing polymerases to move along the DNA. Later in this chapter, we'll discuss some recent discoveries about the roles of histone tails and nucleosomes in the regulation of gene expression.

Higher Levels of DNA Packing

The next level of packing is due to interactions between the histone tails of one nucleosome and the linker DNA and nucleosomes to either side. With the aid of histone H1, these interactions cause the extended 10-nm fiber to coil or fold, forming a chromatin fiber roughly 30 nm in thickness, the *30-nm fiber* (**Figure 19.2b**). The 30-nm fiber, in turn, forms loops called *looped domains* attached to a chromosome scaffold made of nonhistone proteins, thus making up a *300-nm fiber* (**Figure 19.2c**). In a mitotic chromosome, the looped domains themselves coil and fold, further compacting all the chromatin to produce the characteristic metaphase chromosome shown in the micrograph at the bottom of **Figure 19.2d**. Particular genes always end up located at the same places in metaphase chromosomes, indicating that the packing steps are highly specific and precise.

Though interphase chromatin is generally much less condensed than the chromatin of mitotic chromosomes, it shows several of the same levels of higher-order packing. Much of the chromatin comprising a chromosome is present as a 10-nm fiber, but some is compacted into a 30-nm fiber, which in some regions is further folded into looped domains. Although an interphase chromosome lacks an obvious scaffold, its looped domains seem to be attached to the nuclear lamina, on the inside of the nuclear envelope, and perhaps also to fibers of the nuclear matrix. These attachments may help organize regions of active transcription. The chromatin of each chromosome occupies a specific restricted area within the interphase nucleus, and the chromatin fibers of different chromosomes do not become entangled.

Even during interphase, the centromeres and telomeres of chromosomes, as well as other chromosomal regions in some cells, exist in a highly condensed state similar to that seen in a metaphase chromosome. This type of interphase chromatin, visible as irregular clumps with a light microscope, is called **heterochromatin,** to distinguish it from the less compacted **euchromatin** ("true chromatin"). Because of its compaction, heterochromatin DNA is largely inaccessible to transcription enzymes and thus generally is not transcribed. In contrast, the looser packing of euchromatin makes its DNA accessible to enzymes and available for transcription. In the next section, we will consider how changes in chromatin and other mechanisms allow a cell to regulate which of its genes are expressed.

Concept Check 19.1

1. Describe the structure of a nucleosome, the basic unit of DNA packing in eukaryotic cells.
2. What chemical properties of histones and DNA enable these molecules to bind tightly together?
3. In general, how does dense packing of DNA in chromosomes prevent gene expression?

For suggested answers, see Appendix A.

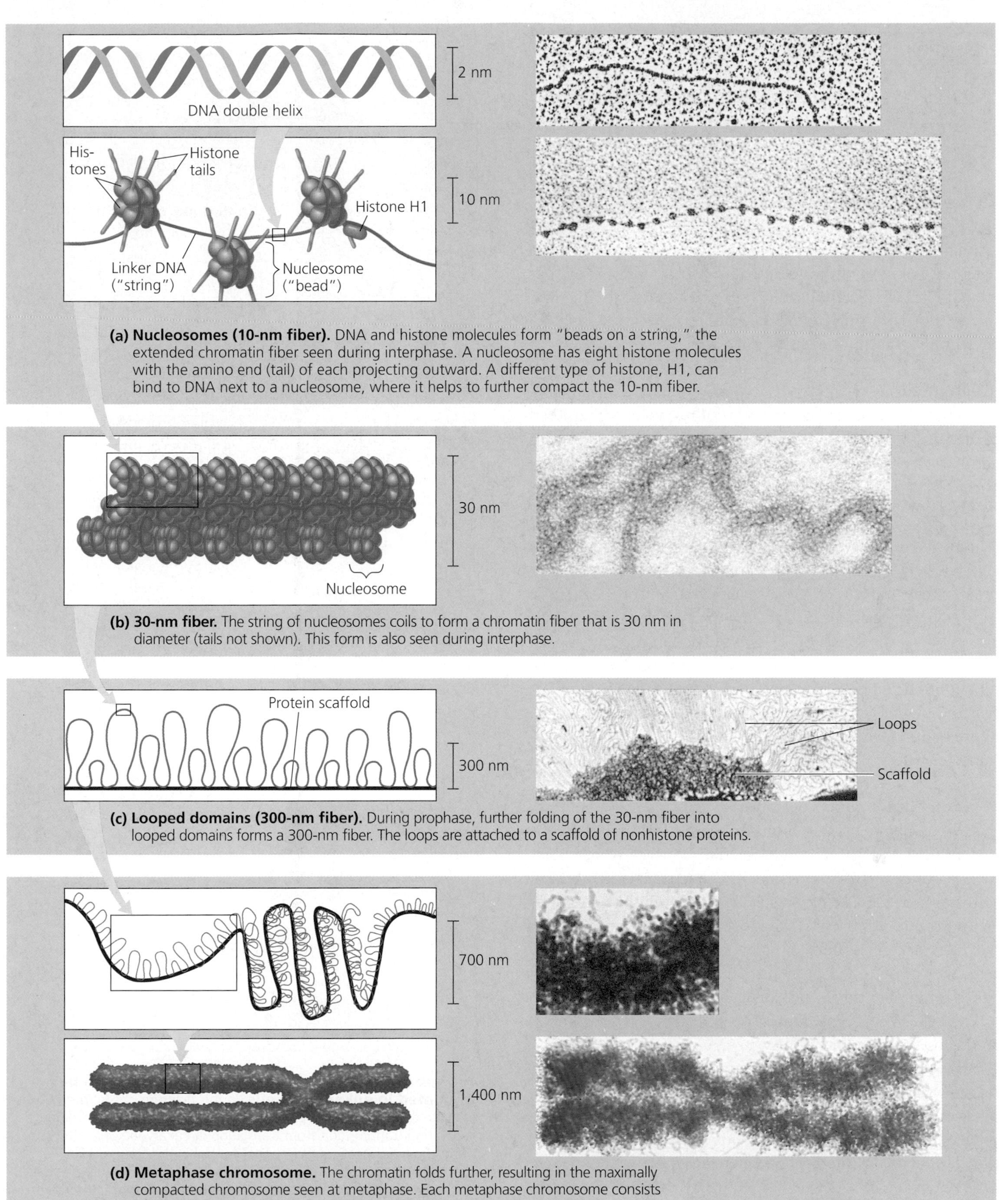

(a) Nucleosomes (10-nm fiber). DNA and histone molecules form "beads on a string," the extended chromatin fiber seen during interphase. A nucleosome has eight histone molecules with the amino end (tail) of each projecting outward. A different type of histone, H1, can bind to DNA next to a nucleosome, where it helps to further compact the 10-nm fiber.

(b) 30-nm fiber. The string of nucleosomes coils to form a chromatin fiber that is 30 nm in diameter (tails not shown). This form is also seen during interphase.

(c) Looped domains (300-nm fiber). During prophase, further folding of the 30-nm fiber into looped domains forms a 300-nm fiber. The loops are attached to a scaffold of nonhistone proteins.

(d) Metaphase chromosome. The chromatin folds further, resulting in the maximally compacted chromosome seen at metaphase. Each metaphase chromosome consists of two chromatids.

▲ **Figure 19.2 Levels of chromatin packing.** This series of diagrams and transmission electron micrographs depicts a current model for the progressive stages of DNA coiling and folding.

Concept 19.2

Gene expression can be regulated at any stage, but the key step is transcription

All organisms must regulate which genes are expressed at any given time. Both unicellular organisms and the cells of multicellular organisms must continually turn genes on and off in response to signals from their external and internal environments. The cells of a multicellular organism must also regulate their gene expression on a more long-term basis. During development of a multicellular organism, its cells undergo a process of specialization in form and function called **cell differentiation**, resulting in several or many differentiated cell types. The mature human body, for instance, is composed of about 200 different cell types. Examples are muscle cells and nerve cells.

Differential Gene Expression

A typical human cell probably expresses about 20% of its genes at any given time. Highly differentiated cells, such as muscle cells, express an even smaller fraction of their genes. Although almost all the cells in an organism contain an identical genome,* the subset of genes expressed in the cells of each type is unique, allowing these cells to carry out their specific function. The differences between cell types, therefore, are due not to different genes being present, but to **differential gene expression**, the expression of different genes by cells with the same genome.

The genomes of eukaryotes may contain tens of thousands of genes, but for quite a few species, only a small amount of the DNA—about 1.5% in humans—codes for protein. Of the remaining DNA, a very small fraction consists of genes for RNA products such as ribosomal RNA and transfer RNA. Most of the rest of the DNA seems to be noncoding, although recently researchers have learned that a significant amount of it may be transcribed into RNAs of unknown function. In any case, the transcription proteins of a cell must locate the right genes at the right time, a task on a par with finding a needle in a haystack. When gene expression goes awry, serious imbalances and diseases, including cancer, can arise.

Figure 19.3 summarizes the entire process of gene expression in a eukaryotic cell, highlighting key stages in the expression of a protein-coding gene. Each stage depicted in Figure 19.3 is a potential control point at which gene expression can be turned on or off, accelerated, or slowed down.

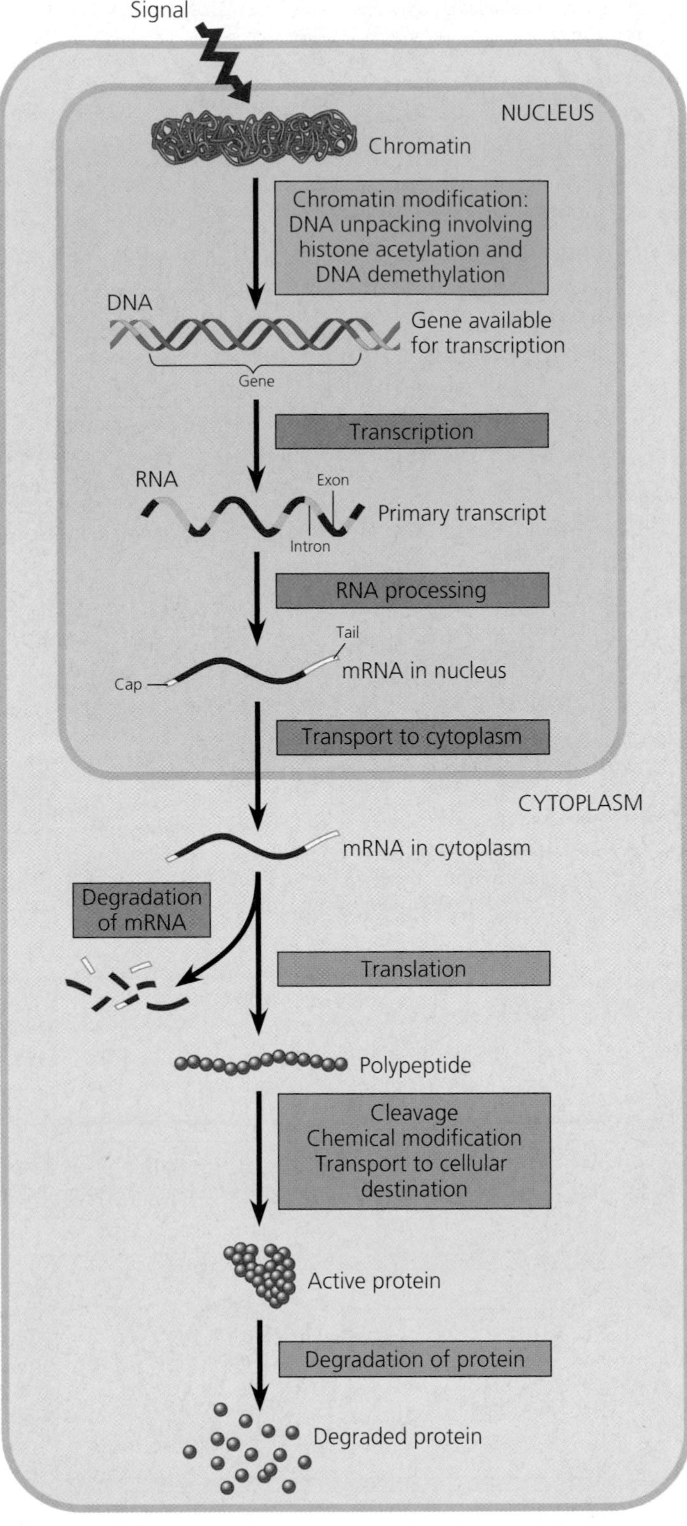

▲ **Figure 19.3 Stages in gene expression that can be regulated in eukaryotic cells.** In this diagram, the colored boxes indicate the processes most often regulated. The nuclear envelope separating transcription from translation in eukaryotic cells offers an opportunity for post-transcriptional control in the form of RNA processing that is absent in prokaryotes. In addition, eukaryotes have a greater variety of control mechanisms operating before transcription and after translation. The expression of any given gene, however, does not necessarily involve every stage shown; for example, not every polypeptide is cleaved. As in prokaryotes, transcription initiation is the most important control point.

*Cells of the immune system are an exception. During their differentiation, rearrangement of the immunoglobulin genes results in a change in the genome, which will be discussed in Chapter 43.

Only 40 years ago, an understanding of the mechanisms that control gene expression in eukaryotes seemed almost hopelessly out of reach. Since then, new research methods, including advances in DNA technology (see Chapter 20), have empowered molecular biologists to uncover many of the details of eukaryotic gene regulation. In all organisms, the expression of specific genes is most commonly regulated at transcription, often in response to signals coming from outside the cell. For that reason, the term *gene expression* is often equated with transcription for both prokaryotes and eukaryotes. However, the greater complexity of eukaryotic cell structure and function provides opportunities for regulating gene expression at additional stages. In the following three sections, we'll examine some of the important control points of eukaryotic gene expression more closely.

Regulation of Chromatin Structure

As mentioned earlier, the structural organization of chromatin not only packs a cell's DNA into a compact form that fits inside the nucleus but also is important in helping regulate gene expression. Genes within heterochromatin, which is highly condensed, are usually not expressed. The repressive effect of heterochromatin has been seen in experiments in which a transcriptionally active gene was inserted into a region of heterochromatin in yeast cells; the inserted gene was no longer expressed. In addition, the location of a gene's promoter relative to nucleosomes and to the sites where the DNA attaches to the chromosome scaffold or nuclear lamina can also affect whether it is transcribed. A flurry of recent research indicates that certain chemical modifications to the histones and DNA of chromatin influence both chromatin structure and gene expression. Here we examine the effects of these modifications, which are catalyzed by specific enzymes.

Histone Modifications

There is mounting evidence that chemical modifications to histones play a direct role in the regulation of gene transcription. The N-terminus of each histone molecule in a nucleosome protrudes outward from the nucleosome **(Figure 19.4a)**. These histone tails are accessible to various modifying enzymes, which catalyze the addition or removal of specific chemical groups.

In **histone acetylation**, acetyl groups ($—COCH_3$) are attached to positively charged lysines in histone tails; deacetylation is the removal of acetyl groups. When the histone tails of a nucleosome are acetylated, their positive charges are neutralized and they no longer bind to neighboring nucleosomes **(Figure 19.4b)**. Recall that such binding promotes the folding of chromatin into a more compact structure; when this binding does not occur, chromatin has a looser structure. As a result, transcription proteins have easier access to genes in an acetylated region. Researchers have shown that some enzymes that acetylate or deacetylate histones are closely associated with or even components of the transcription factors that bind to promoters (see Figure 17.8). In other words, histone acetylation enzymes may promote the initiation of transcription not only by modifying chromatin structure, but also by binding to, and thus "recruiting," components of the transcription machinery.

Several other chemical groups can be reversibly attached to amino acids in histone tails. For example, the addition of methyl groups ($—CH_3$) to histone tails (methylation) can lead

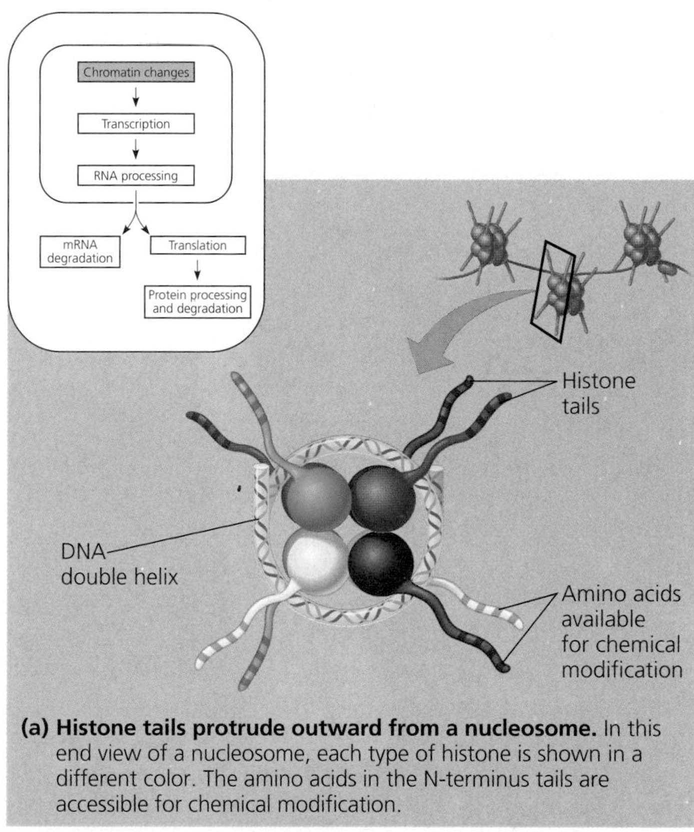

(a) Histone tails protrude outward from a nucleosome. In this end view of a nucleosome, each type of histone is shown in a different color. The amino acids in the N-terminus tails are accessible for chemical modification.

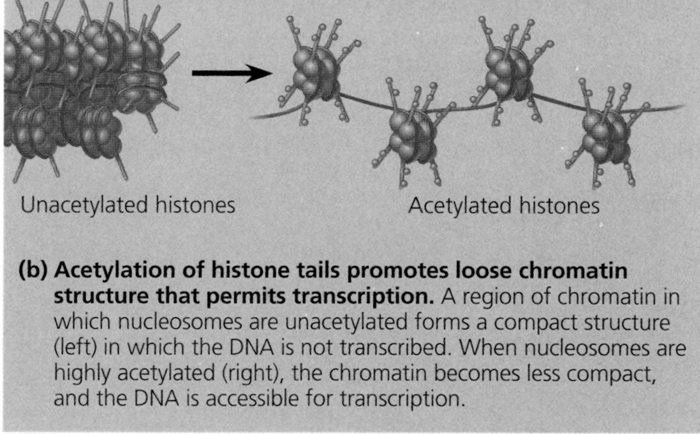

(b) Acetylation of histone tails promotes loose chromatin structure that permits transcription. A region of chromatin in which nucleosomes are unacetylated forms a compact structure (left) in which the DNA is not transcribed. When nucleosomes are highly acetylated (right), the chromatin becomes less compact, and the DNA is accessible for transcription.

▲ **Figure 19.4 A simple model of histone tails and the effect of histone acetylation.** In addition to acetylation, histones can undergo several other types of modifications that also help determine the chromatin configuration in a region.

to condensation of the chromatin. The recent discovery that this and many other modifications to histone tails can affect chromatin structure and gene expression has led to the *histone code hypothesis*. According to this model, specific combinations of modifications, rather than the overall level of histone acetylation, help determine the chromatin configuration, which in turn influences transcription.

DNA Methylation

Distinct from methylation of histone tails is the addition of methyl groups to certain bases in DNA after DNA is synthesized. In fact, the DNA of most plants and animals has methylated bases, usually cytosine. Inactive DNA, such as that of inactivated mammalian X chromosomes (see Figure 15.11), is generally highly methylated compared with DNA that is actively transcribed, although there are exceptions.

Comparison of the same genes in different tissues shows that the genes are usually more heavily methylated in cells in which they are not expressed. Removal of the extra methyl groups can turn on certain of these genes. Moreover, researchers have discovered that certain proteins that bind to methylated DNA recruit histone deacetylation enzymes. Thus, a dual mechanism, involving both DNA methylation and histone deacetylation, can repress transcription.

At least in some species, DNA methylation seems to be essential for the long-term inactivation of certain genes that occurs during normal cell differentiation in the embryo. For instance, experiments have shown that deficient DNA methylation due to lack of a methylating enzyme leads to abnormal embryonic development in organisms as different as mice and *Arabidopsis* (a plant). Once methylated, genes usually stay that way through successive cell divisions. At DNA sites where one strand is already methylated, methylation enzymes correctly methylate the daughter strand after each round of DNA replication. Methylation patterns are thus passed on, and cells forming specialized tissues keep a chemical record of what occurred during embryonic development. A methylation pattern maintained in this way also accounts for **genomic imprinting** in mammals, where methylation permanently regulates expression of either the maternal or paternal allele of certain genes at the start of development (see Chapter 15).

Epigenetic Inheritance

The chromatin modifications that we have just discussed do not involve a change in the DNA sequence, and yet they may be passed along to future generations of cells. Inheritance of traits transmitted by mechanisms not directly involving the nucleotide sequence is called **epigenetic inheritance.** Researchers are amassing more and more evidence for the importance of epigenetic information in regulation of gene expression. Clearly, enzymes that modify chromatin structure appear to be integral parts of the cell's machinery for regulating transcription.

Regulation of Transcription Initiation

Chromatin-modifying enzymes provide initial control of gene expression by making a region of DNA either more or less able to bind the transcription machinery. Once a gene is optimally modified for expression, the initiation of transcription is the most important and universally used stage at which gene expression is regulated. Before looking at how cells control their transcription, let's review the structure of a typical eukaryotic gene and its transcript.

Organization of a Typical Eukaryotic Gene

A eukaryotic gene and the DNA elements (segments) that control it are typically organized as shown in **Figure 19.5**, which extends what you learned about eukaryotic genes in Chapter 17. Recall that a cluster of proteins called a *transcription initiation complex* assembles on the promoter sequence at the "upstream" end of the gene. One of these proteins, RNA polymerase II, then proceeds to transcribe the gene, synthesizing a primary RNA transcript (pre-mRNA). RNA processing includes enzymatic addition of a 5′ cap and a poly-A tail, as well as splicing out of introns, to yield a mature mRNA. Associated with most eukaryotic genes are multiple **control elements,** segments of noncoding DNA that help regulate transcription by binding certain proteins. These control elements and the proteins they bind are critical to the precise regulation of gene expression seen in different cell types.

The Roles of Transcription Factors

To initiate transcription, eukaryotic RNA polymerase requires the assistance of proteins called **transcription factors** (see Figure 17.8). Because the transcription factors mentioned in Chapter 17 are essential for the transcription of *all* protein-coding genes, they are sometimes called *general* transcription factors. Only a few general transcription factors independently bind a DNA sequence, such as the TATA box within the promoter; the others primarily bind proteins, including each other and RNA polymerase II. Protein-protein interactions are crucial to the initiation of eukaryotic transcription. Only when the complete initiation complex has assembled can the polymerase begin to move along the DNA template strand, producing a complementary strand of RNA.

The interaction of general transcription factors and RNA polymerase II with a promoter usually leads to only a low rate of initiation and production of few RNA transcripts. In eukaryotes, high levels of transcription of particular genes at the appropriate time and place depend on the interaction of control elements with other proteins that can be thought of as *specific* transcription factors.

Enhancers and Specific Transcription Factors. As you can see in Figure 19.5, some control elements, named *proximal*

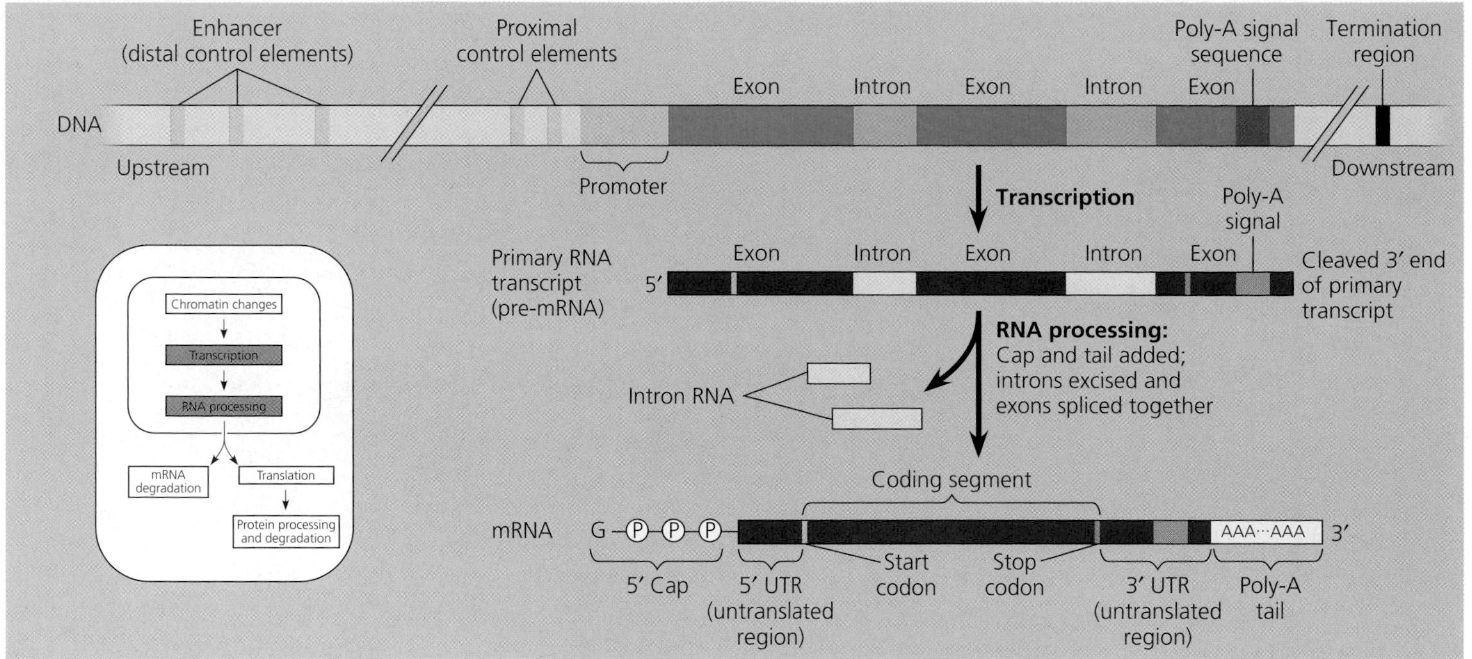

▲ **Figure 19.5 A eukaryotic gene and its transcript.** Each eukaryotic gene has a promoter, a DNA sequence where RNA polymerase binds and starts transcription, proceeding "downstream." A number of control elements (gold) are involved in regulating the initiation of transcription; these are DNA sequences located near (proximal to) or far from (distal to) the promoter. Distal control elements can be grouped together as enhancers. A polyadenylation (poly-A) signal in the last exon of the gene is transcribed into an RNA sequence that determines where the transcript is cleaved and the poly-A tail added. Transcription may continue for hundreds of nucleotides beyond the poly-A signal before terminating. RNA processing of the primary transcript into a functional mRNA involves three steps: addition of the 5′ cap, addition of the poly-A tail, and splicing. In the cell, the 5′ cap is added soon after transcription is initiated; splicing and poly-A tail addition may also occur while transcription is still under way (see Figure 17.9).

control elements, are located close to the promoter. (Although some biologists consider proximal control elements part of the promoter, we do not.) The more distant *distal control elements,* groups of which are called **enhancers,** may be thousands of nucleotides upstream or downstream of a gene or even within an intron. A given gene may have multiple enhancers, each active at a different time or in a different cell type or location in the organism.

The interactions between enhancers and specific transcription factors called activators or repressors are particularly important in controlling gene expression. An **activator** is a protein that binds to an enhancer and stimulates transcription of a gene. **Figure 19.6**, on the next page, shows a current model for how binding of activators to enhancers located far from the promoter can influence transcription. Protein-mediated bending of the DNA is thought to bring the bound activators in contact with a group of so-called *mediator proteins,* which in turn interact with proteins at the promoter. These multiple protein-protein interactions help assemble and position the initiation complex on the promoter.

Hundreds of transcription activators have been discovered in eukaryotes. Researchers have identified two common structural elements in a large number of activator proteins: a DNA-binding domain—a part of the protein's three-dimensional structure that binds to DNA—and one or more activation domains. Activation domains bind other regulatory proteins or components of the transcription machinery, allowing a sequence of protein-protein interactions that result in transcription of a given gene.

Some specific transcription factors function as **repressors** to inhibit expression of a particular gene. Eukaryotic repressors can cause inhibition of gene expression in several different ways. Certain repressors block the binding of activators either to their control elements or to components of the transcription machinery. Other repressors bind directly to their own control elements in an enhancer and act to turn off transcription even in the presence of activators.

In addition to affecting assembly of the transcription machinery directly, some activators and repressors act indirectly by influencing chromatin structure. Recall that a gene present in a region of chromatin with high levels of histone acetylation is able to bind the transcription machinery, whereas a gene in a region of chromatin with low levels of histone acetylation is not (see Figure 19.4). Studies in yeast and mammals show that some activators recruit proteins that acetylate histones near the promoters of specific genes, thus promoting transcription. In contrast, some repressors recruit proteins that deacetylate histones, leading to reduced transcription, a phenomenon referred to as *silencing*. Indeed, recruitment of chromatin-modifying proteins seems to be the most common mechanism of repression in eukaryotes.

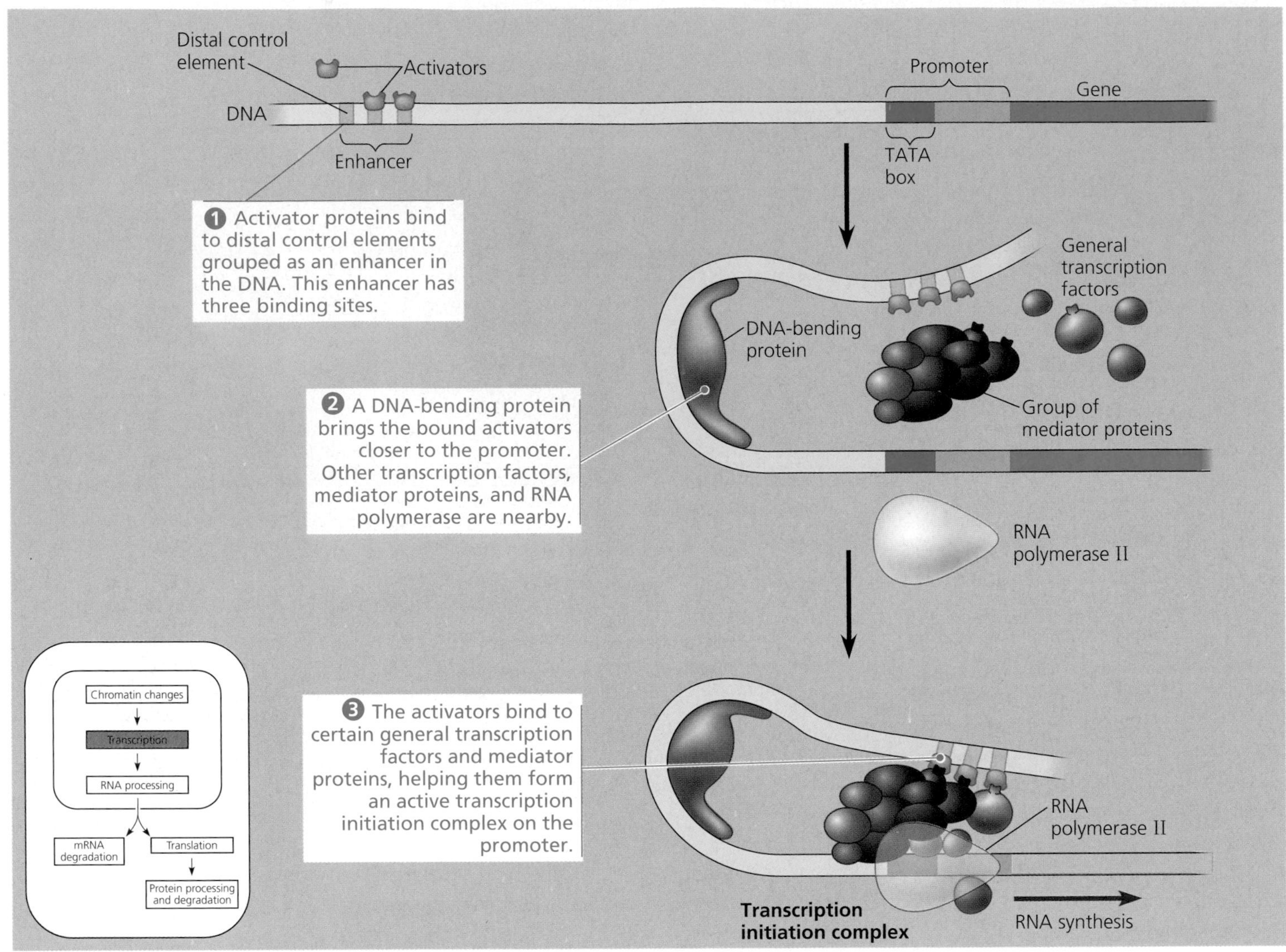

▲ **Figure 19.6 A model for the action of enhancers and transcription activators.** Bending of the DNA by a protein enables enhancers to influence a promoter hundreds or even thousands of nucleotides away. Specific transcription factors called activators bind to the enhancer DNA sequences and then to a group of mediator proteins, which in turn bind to general transcription factors assembling the transcription initiation complex. These protein-protein interactions facilitate the correct positioning of the complex on the promoter and the initiation of RNA synthesis. Only one enhancer is shown in this figure, but a gene may have several that act at different times or in different cell types.

Combinatorial Control of Gene Activation. In eukaryotes, the precise control of transcription depends largely on the binding of activators to DNA control elements. Considering the very large number of genes that must be regulated in a typical animal or plant cell, the number of completely different nucleotide sequences found in control elements is surprisingly small. A dozen or so short nucleotide sequences appear again and again in the control elements for different genes. On average, each enhancer is composed of about ten control elements, each of which can bind only one or two specific transcription factors. The particular *combination* of control elements in an enhancer associated with a gene turns out to be more important than the presence of a single unique control element in regulating transcription of the gene.

Even with only a dozen control element sequences available, a large number of combinations are possible. A particular combination of control elements will be able to activate transcription only when the appropriate activator proteins are present, such as at a precise time during development or in a particular cell type. The example in **Figure 19.7** illustrates how the use of different combinations of control elements to activate transcription allows exquisite regulation of transcription with a small set of control elements.

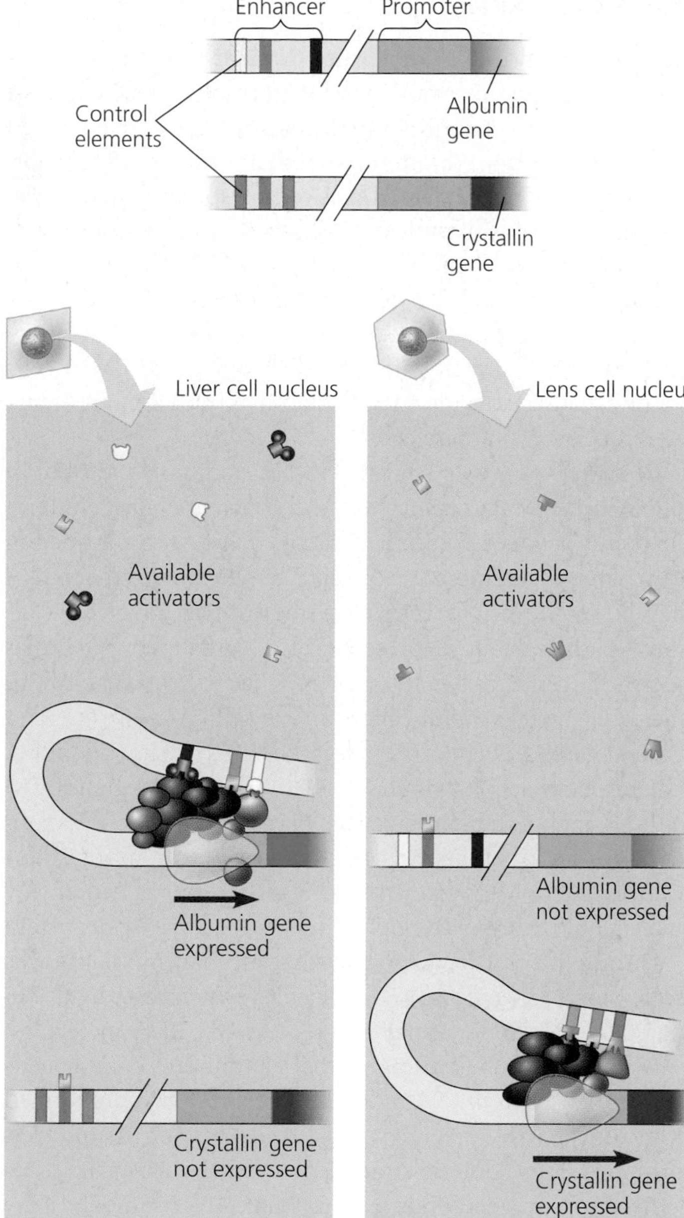

(a) Liver cell. The albumin gene is expressed, and the crystallin gene is not.

(b) Lens cell. The crystallin gene is expressed, and the albumin gene is not.

▲ **Figure 19.7 Cell type–specific transcription.** Both liver cells and lens cells have the genes for making the proteins albumin and crystallin, but only liver cells make albumin (a blood protein) and only lens cells make crystallin (the main component of the lens of the eye). The specific transcription factors (activators and repressors) made in a particular type of cell determine which genes are expressed. In this example, the genes for albumin and crystallin are shown at the top, each with an enhancer made up of three different control elements. Although the enhancers for the two genes share one control element, each enhancer has a unique combination of elements. All the activators required for high-level expression of the albumin gene are present only in liver cells (a), whereas the activators needed for expression of the crystallin gene are present only in lens cells (b). For simplicity, we consider only the role of activators here, although the presence or absence of repressors may also influence transcription in certain cell types.

Coordinately Controlled Genes

How does the eukaryotic cell deal with genes of related function that need to be turned on or off at the same time? In Chapter 18, you learned that in prokaryotes, such coordinately controlled genes are often clustered into an operon, which is regulated by a single promoter and transcribed into a single mRNA molecule. Thus, the genes are expressed together, and the encoded proteins are produced concurrently. With rare exceptions, operons that work in this way have not been found in eukaryotic cells.

Recent studies of the genomes of several eukaryotic species have found that some co-expressed genes are clustered near one another on the same chromosome. Examples include certain genes in the testis of the fruit fly and muscle-related genes in a small worm called a nematode. Unlike genes in prokaryotic operons, however, each eukaryotic gene in these clusters has its own promoter and is individually transcribed. The coordinate regulation of clustered genes in eukaryotic cells is thought to involve changes in the chromatin structure that make the entire group of genes either available or unavailable for transcription.

More commonly, co-expressed eukaryotic genes, such as genes coding for the enzymes of a metabolic pathway, are found scattered over different chromosomes. In these cases, coordinate gene expression seems to depend on the association of a specific control element or combination of elements with every gene of a dispersed group. Copies of the activators that recognize these control elements bind to them, promoting simultaneous transcription of the genes, no matter where they are in the genome.

Coordinate control of dispersed genes in a eukaryotic cell often occurs in response to external chemical signals. A steroid hormone, for example, enters a cell and binds to a specific intracellular receptor protein, forming a hormone-receptor complex that serves as a transcription activator (see Figure 11.6). Every gene whose transcription is stimulated by a particular steroid hormone, regardless of its chromosomal location, has a control element recognized by that hormone-receptor complex.

Many signal molecules, such as nonsteroid hormones and growth factors, bind to receptors on a cell's surface and never actually enter the cell. This kind of signal can control gene expression indirectly by triggering signal transduction pathways that lead to activation of particular transcription activators or repressors (see Figure 11.14). The principle of coordinate regulation is the same as in the case of steroid hormones: Genes with the same control elements are activated by the same chemical signals. Systems for coordinating gene regulation probably arose early in evolutionary history and evolved by the duplication and distribution of control elements within the genome.

Mechanisms of Post-Transcriptional Regulation

Transcription alone does not constitute gene expression. The expression of a protein-coding gene is ultimately measured by the amount of functional protein a cell makes, and much happens between the synthesis of the RNA transcript and the activity of the protein in the cell. An increasing number of examples are being found of regulatory mechanisms that operate at various stages after transcription (see Figure 19.3). These mechanisms allow a cell to fine-tune gene expression rapidly in response to environmental changes without altering its transcription patterns. Here we discuss how cells can regulate gene expression once a gene has been transcribed.

RNA Processing

RNA processing in the nucleus and the export of mature RNA to the cytoplasm provide several opportunities for regulating gene expression that are not available in prokaryotes. One example of regulation at the RNA-processing level is **alternative RNA splicing,** in which different mRNA molecules are produced from the same primary transcript, depending on which RNA segments are treated as exons and which as introns **(Figure 19.8)**. Regulatory proteins specific to a cell type control intron-exon choices by binding to regulatory sequences within the primary transcript.

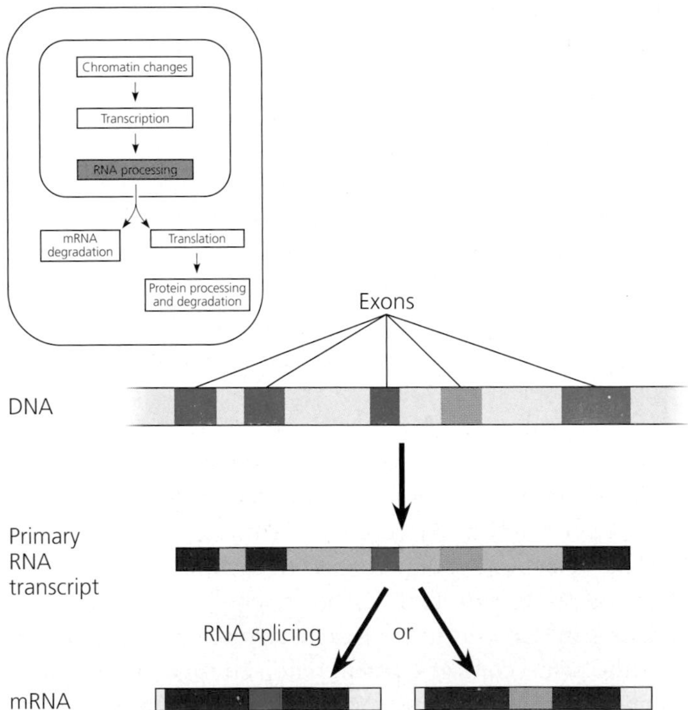

▲ **Figure 19.8 Alternative RNA splicing.** The primary transcripts of some genes can be spliced in more than one way, generating different mRNA molecules. Notice in this example that one mRNA molecule has ended up with the green exon and the other with the purple exon. With alternative splicing, an organism can produce more than one type of polypeptide from a single gene.

mRNA Degradation

The life span of mRNA molecules in the cytoplasm is an important factor in determining the pattern of protein synthesis in a cell. Prokaryotic mRNA molecules typically are degraded by enzymes within a few minutes of their synthesis. This short life span of prokaryotic mRNAs is one reason prokaryotes can vary their patterns of protein synthesis so quickly in response to environmental changes. In contrast, mRNAs in multicellular eukaryotes typically survive for hours, days, or even weeks. For instance, the mRNAs for the hemoglobin polypeptides (α-globin and β-globin) in developing red blood cells are unusually stable, and these long-lived mRNAs are translated repeatedly in these cells.

Research on yeasts suggests that a common pathway of mRNA breakdown begins with the enzymatic shortening of the poly-A tail (see Figure 19.5). This helps trigger the action of enzymes that remove the 5′ cap (the two ends of the mRNA may be briefly held together by the proteins involved). Removal of the cap, a critical step, is also regulated by particular nucleotide sequences in the mRNA. Once the cap is removed, nuclease enzymes rapidly chew up the mRNA.

Nucleotide sequences that affect the length of time an mRNA remains intact are often found in the untranslated region (UTR) at the 3′ end of the molecule (see Figure 19.5). In one experiment, researchers transferred such a sequence from the short-lived mRNA for a growth factor to the 3′ end of a normally stable globin mRNA. The globin mRNA was quickly degraded.

During the past few years, another mechanism that blocks expression of specific mRNA molecules has come to light. Researchers have found small single-stranded RNA molecules, called **microRNAs (miRNAs),** that can bind to complementary sequences in mRNA molecules. The miRNAs are formed from longer RNA precursors that fold back on themselves, forming a long, double-stranded hairpin structure held together by hydrogen bonds **(Figure 19.9)**. An enzyme, fittingly called Dicer, then cuts the double-stranded RNA molecule into short fragments. One of the two strands is degraded, and the other strand (miRNA) associates with a large protein complex and allows the complex to bind to any mRNA molecules that have the complementary sequence. Depending on various factors, the miRNA-protein complex then either degrades the target mRNA or blocks its translation.

Inhibition of gene expression by RNA molecules was first observed by biologists who noticed that injecting double-stranded RNA molecules into a cell somehow turned off a gene with the same sequence. They called this experimental phenomenon **RNA interference** (or **RNAi**). It was later shown to be due to **small interfering RNAs (siRNAs),** RNAs of similar size and function as miRNAs. In fact, subsequent research showed that the cellular machinery that generates siRNAs is the very same as that responsible for producing miRNAs naturally in the cell. The mechanisms by which these small RNAs function appear to be similar as well.

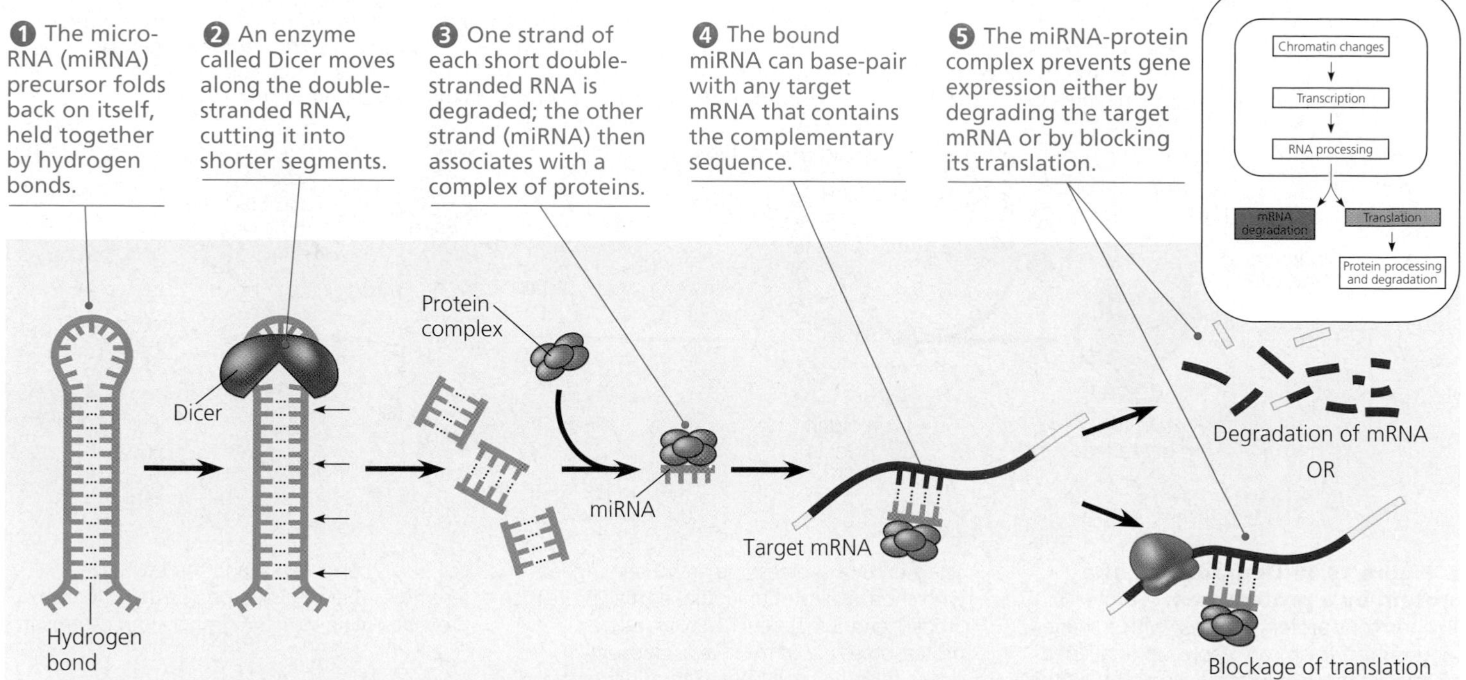

1 The micro-RNA (miRNA) precursor folds back on itself, held together by hydrogen bonds.
2 An enzyme called Dicer moves along the double-stranded RNA, cutting it into shorter segments.
3 One strand of each short double-stranded RNA is degraded; the other strand (miRNA) then associates with a complex of proteins.
4 The bound miRNA can base-pair with any target mRNA that contains the complementary sequence.
5 The miRNA-protein complex prevents gene expression either by degrading the target mRNA or by blocking its translation.
Chromatin changes
Transcription
RNA processing
mRNA degradation
Translation
Protein processing and degradation
Hydrogen bond
Dicer
Protein complex
miRNA
Target mRNA
Degradation of mRNA
OR
Blockage of translation

▲ **Figure 19.9 Regulation of gene expression by microRNAs (miRNAs).** RNA transcripts from miRNA-encoding genes are processed into miRNAs, which prevent expression of complementary mRNAs (siRNAs are believed to be generated and to function in a similar way).

Because the cellular RNAi pathway can lead to the destruction of RNAs with sequences complementary to those found in double-stranded RNA molecules, it is commonly believed to have originated as a natural defense against infection by RNA viruses. However, the fact that the RNAi pathway can affect expression of cellular genes also supports alternative models. In any case, it is clear that RNAi plays an important role in regulating gene expression in the cell.

Initiation of Translation

Translation presents another opportunity for regulating gene expression; such regulation occurs most commonly at the initiation stage (see Figure 17.17). The initiation of translation of selected mRNAs can be blocked by regulatory proteins that bind to specific sequences or structures within the untranslated region at the 5′ end (5′ UTR) of the mRNA, preventing the attachment of ribosomes. A different mechanism for blocking translation is seen in a variety of mRNAs present in the egg cells of many organisms: These stored mRNAs lack poly-A tails of sufficient size to allow translation initiation. At the appropriate time during embryonic development, a cytoplasmic enzyme adds more A residues, allowing translation to begin.

Alternatively, translation of *all* the mRNAs in a cell may be regulated simultaneously. In a eukaryotic cell, such "global" control usually involves the activation or inactivation of one or more of the protein factors required to initiate translation. This mechanism also plays a role in starting translation of mRNAs that are stored in egg cells. Just after fertilization, translation is triggered by the sudden activation of translation initiation factors. The response is a burst of synthesis of the proteins encoded by the stored mRNAs. Some plants and algae store mRNAs during periods of darkness; light then triggers the reactivation of the translational apparatus.

Protein Processing and Degradation

The final opportunities for controlling gene expression occur after translation. Often, eukaryotic polypeptides must be processed to yield functional protein molecules. For instance, cleavage of the initial insulin polypeptide (pro-insulin) forms the active hormone. In addition, many proteins undergo chemical modifications that make them functional. Regulatory proteins are commonly activated or inactivated by the reversible addition of phosphate groups, and proteins destined for the surface of animal cells acquire sugars. Cell-surface proteins and many others must also be transported to target destinations in the cell in order to function. Regulation might occur at any of the steps involved in modifying or transporting a protein.

Finally, the length of time each protein functions in the cell is strictly regulated by means of selective degradation. Many proteins, such as the cyclins involved in regulating the cell cycle, must be relatively short-lived if the cell is to function appropriately (see Figure 12.16). To mark a particular protein for destruction, the cell commonly attaches molecules of a small protein called ubiquitin to the protein. Giant protein complexes called **proteasomes** then recognize the ubiquitin-tagged

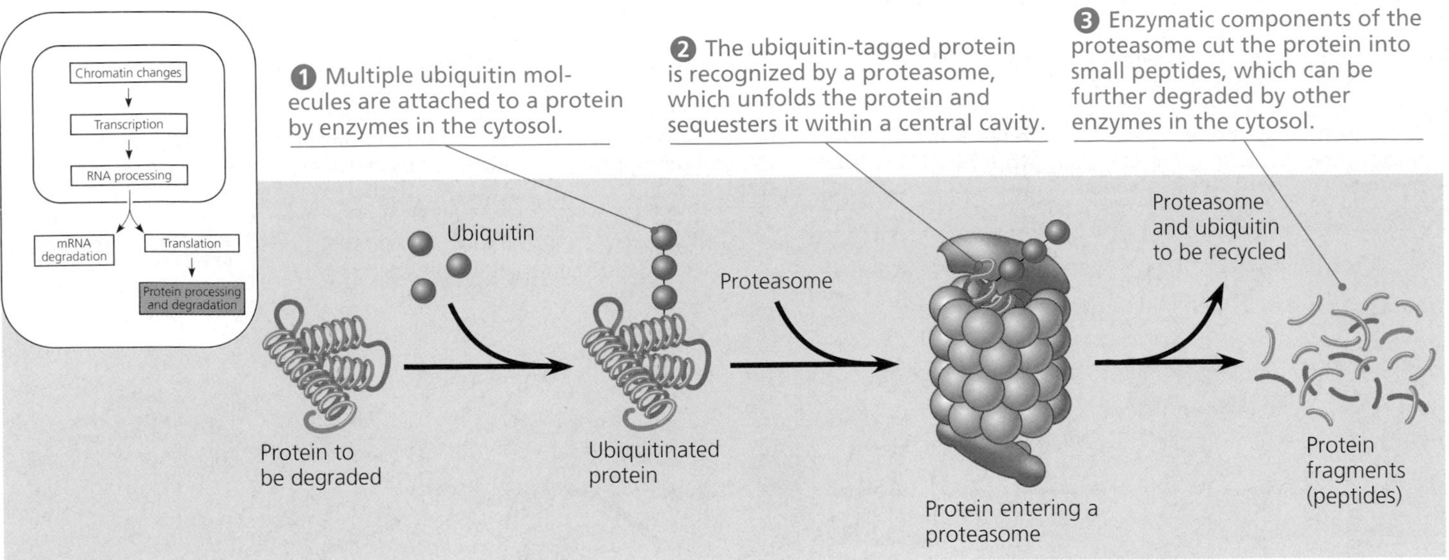

Chromatin changes
Transcription
RNA processing
mRNA degradation
Translation
Protein processing and degradation
1 Multiple ubiquitin molecules are attached to a protein by enzymes in the cytosol.
2 The ubiquitin-tagged protein is recognized by a proteasome, which unfolds the protein and sequesters it within a central cavity.
3 Enzymatic components of the proteasome cut the protein into small peptides, which can be further degraded by other enzymes in the cytosol.
Ubiquitin
Proteasome
Proteasome and ubiquitin to be recycled
Protein to be degraded
Ubiquitinated protein
Protein entering a proteasome
Protein fragments (peptides)

▲ **Figure 19.10 Degradation of a protein by a proteasome.** A proteasome, an enormous protein complex with a shape suggesting a trash can, chops up unneeded proteins in the cell. In most cases, the proteins attacked by a proteasome have been tagged with short chains of ubiquitin, a small protein. Steps 1 and 3 require ATP. Eukaryotic proteasomes are as massive as ribosomal subunits and are distributed throughout the cell. Their barrel-like shape somewhat resembles that of chaperone proteins, which protect protein structure rather than destroy it (see Figure 5.23).

protein molecules and degrade them **(Figure 19.10)**. The importance of proteasomes is underscored by the finding that mutations making cell cycle proteins impervious to proteasome degradation can lead to cancer.

Concept Check 19.2

1. In general, what is the effect of histone acetylation and DNA methylation on gene expression?
2. Compare the roles of general and specific transcription factors in regulating gene expression.
3. If you compared the nucleotide sequences of the distal control elements in the enhancers of three coordinately regulated genes, what would you expect to find? Why?
4. Once mRNA encoding a particular protein reaches the cytoplasm, what are four mechanisms that can regulate the amount of the active protein in the cell?

For suggested answers, see Appendix A.

Concept 19.3

Cancer results from genetic changes that affect cell cycle control

In Chapter 12, we considered cancer as a set of diseases in which cells escape from the control mechanisms normally limiting their growth. Now that we have discussed the molecular basis of gene expression and its regulation, we are ready to look at cancer more closely. The gene regulation systems that go wrong during cancer turn out to be the very same systems that play important roles in embryonic development, the immune response, and many other biological processes. Thus, research into the molecular basis of cancer has both benefited from and informed many other fields of biology.

Types of Genes Associated with Cancer

The genes that normally regulate cell growth and division during the cell cycle include genes for growth factors, their receptors, and the intracellular molecules of signaling pathways. (To review the cell cycle, see Chapter 12.) Mutations that alter any of these genes in somatic cells can lead to cancer. The agent of such change can be random spontaneous mutation. However, it is likely that many cancer-causing mutations result from environmental influences, such as chemical carcinogens, X-rays, and certain viruses.

An early breakthrough in understanding cancer came in 1911, when Peyton Rous discovered a virus that causes cancer in chickens. Since then, scientists have recognized a number of *tumor viruses* that cause cancer in various animals, including humans (see Table 18.1). The Epstein-Barr virus, a herpesvirus that causes infectious mononucleosis, has been linked to several types of cancer, notably Burkitt's lymphoma. Papilloma viruses (of the papovavirus group) are associated with cancer of the cervix. Among the retroviruses, one called HTLV-1 causes a type of adult leukemia. All tumor viruses transform cells into cancer cells through the integration of viral nucleic acid into host cell DNA.

Oncogenes and Proto-Oncogenes

Research on tumor viruses led to the discovery of cancer-causing genes called **oncogenes** (from the Greek *onco*, tumor) in certain retroviruses. Subsequently, close counterparts of these oncogenes were found in the genomes of humans and other animals. The normal cellular genes, called **proto-oncogenes,** code for proteins that stimulate normal cell growth and division.

How might a proto-oncogene—a gene that has an essential function in normal cells—become an oncogene, a cancer-causing gene? In general, an oncogene arises from a genetic change that leads to an increase either in the amount of the proto-oncogene's protein product or in the intrinsic activity of each protein molecule. The genetic changes that convert proto-oncogenes to oncogenes fall into three main categories: movement of DNA within the genome, amplification of a proto-oncogene, and point mutations in a control element or in the proto-oncogene itself **(Figure 19.11)**.

Cancer cells are frequently found to contain chromosomes that have broken and rejoined incorrectly, translocating fragments from one chromosome to another (see Figure 15.14). If a translocated proto-oncogene ends up near an especially active promoter (or other control element), its transcription may increase, making it an oncogene. Movement of transposable elements also may place a more active promoter near a proto-oncogene, increasing its expression. (Eukaryotic transposable elements are described later in the chapter.) The second main type of genetic change, amplification, increases the number of copies of the proto-oncogene in the cell. The third possibility is a point mutation either (1) in the promoter or an enhancer that controls a proto-oncogene, causing an increase in its expression, or (2) in the coding sequence, changing the gene's product to a protein that is more active or more resistant to degradation than the normal protein. All these mechanisms can lead to abnormal stimulation of the cell cycle and put the cell on the path to malignancy.

Tumor-Suppressor Genes

In addition to genes whose products normally promote cell division, cells contain genes whose normal products *inhibit* cell division. Such genes are called **tumor-suppressor genes** because the proteins they encode help prevent uncontrolled cell growth. Any mutation that decreases the normal activity of a tumor-suppressor protein may contribute to the onset of cancer, in effect stimulating growth through the absence of suppression.

The protein products of tumor-suppressor genes have various functions. Some tumor-suppressor proteins normally repair damaged DNA, a function that prevents the cell from accumulating cancer-causing mutations. Other tumor-suppressor proteins control the adhesion of cells to each other or to the extracellular matrix; proper cell anchorage is crucial in normal tissues—and often absent in cancers. Still other tumor-suppressor proteins are components of cell-signaling pathways that inhibit the cell cycle.

Interference with Normal Cell-Signaling Pathways

The proteins encoded by many proto-oncogenes and tumor-suppressor genes are components of cell-signaling pathways. Let's take a closer look at how such proteins function in normal cells and what goes wrong with their function in cancer cells. We will focus on the products of two key genes, the *ras* proto-oncogene and the *p53* tumor-suppressor gene. Mutations in *ras* occur in about 30% of human cancers; mutations in *p53* in more than 50%.

The Ras protein, encoded by the ***ras* gene,** is a G protein that relays a signal from a growth factor receptor on the plasma membrane to a cascade of protein kinases. The cellular response at the end of the pathway is the synthesis of a protein that stimulates the cell cycle (**Figure 19.12a**, on the next page). Normally, such a pathway will not operate unless triggered by

▼ Figure 19.11 Genetic changes that can turn proto-oncogenes into oncogenes.

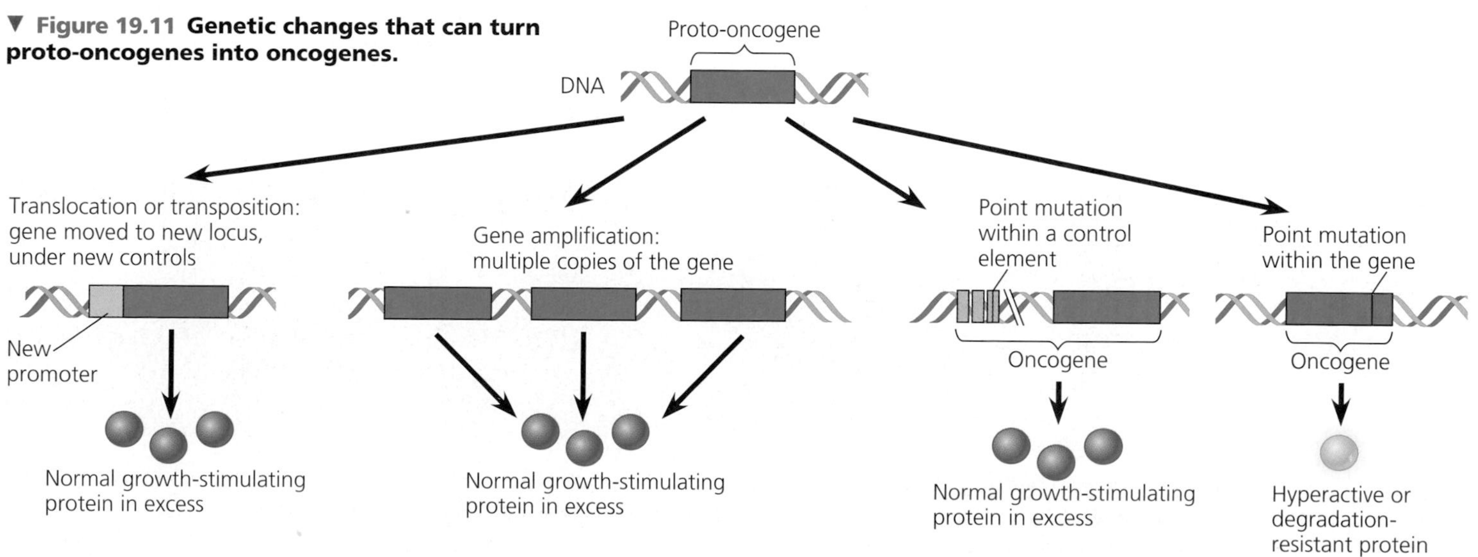

(a) Cell cycle–stimulating pathway. This pathway is triggered by ❶ a growth factor that binds to ❷ its receptor in the plasma membrane. The signal is relayed to ❸ a G protein called Ras. Like all G proteins, Ras is active when GTP is bound to it. Ras passes the signal to ❹ a series of protein kinases. The last kinase activates ❺ a transcription activator that turns on one or more genes for proteins that stimulate the cell cycle. If a mutation makes Ras or any other pathway component abnormally active, excessive cell division and cancer may result.

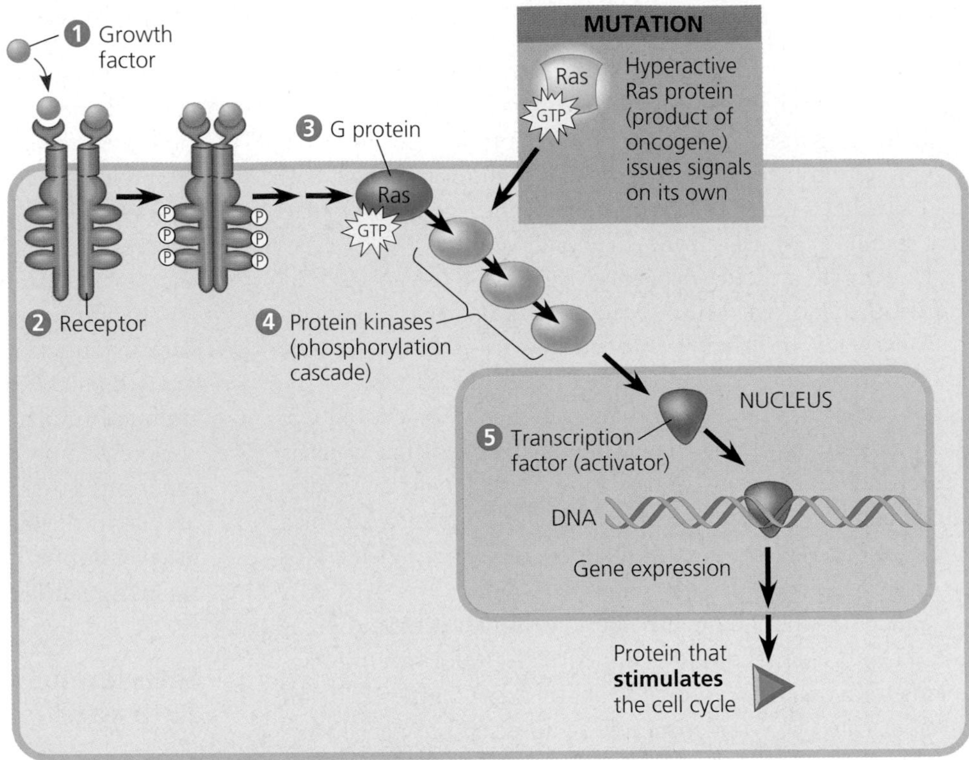

(b) Cell cycle–inhibiting pathway. In this pathway, ❶ DNA damage is an intracellular signal that is passed via ❷ protein kinases and leads to activation of ❸ p53. Activated p53 promotes transcription of the gene for a protein that inhibits the cell cycle. The resulting suppression of cell division ensures that the damaged DNA is not replicated. Mutations causing deficiencies in any pathway component can contribute to the development of cancer.

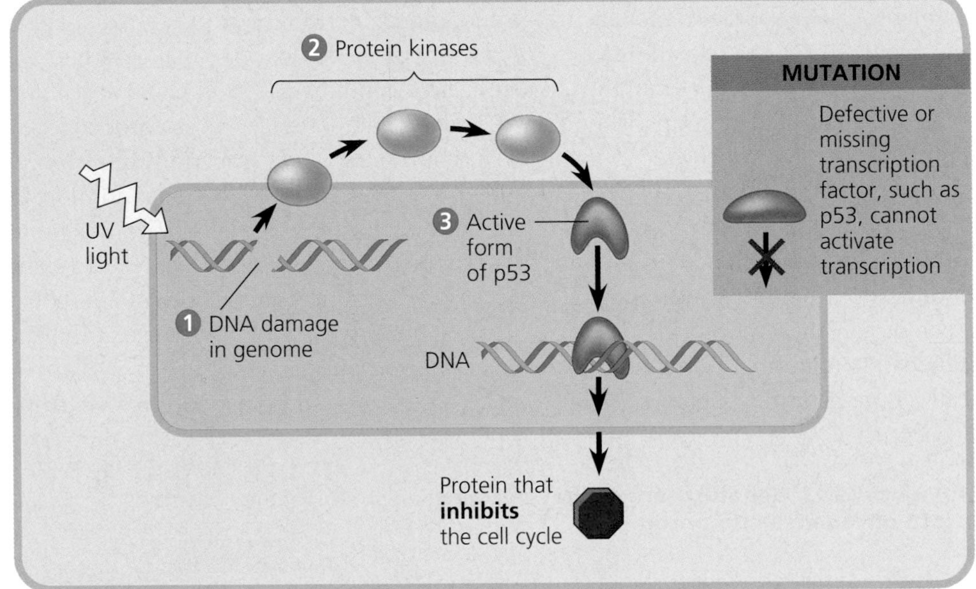

(c) Effects of mutations. Increased cell division, possibly leading to cancer, can result if the cell cycle is overstimulated, as in (a), or not inhibited when it normally would be, as in (b).

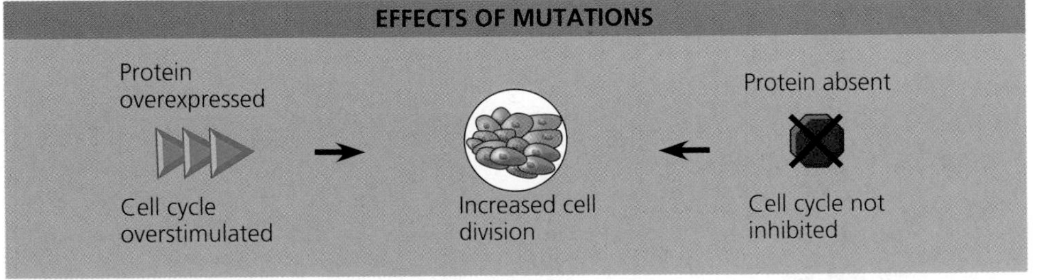

▲ **Figure 19.12 Signaling pathways that regulate cell division.** Both stimulatory and inhibitory pathways regulate the cell cycle, commonly by influencing transcription. Cancer can result from aberrations in such pathways, in combination with other factors.

the appropriate growth factor. But certain mutations in the *ras* gene can lead to production of a hyperactive Ras protein that triggers the kinase cascade, resulting in increased cell division even in the absence of growth factor. In fact, hyperactive versions or excess amounts of any of the pathway's components can have the same outcome: excessive cell division.

Figure 19.12b shows a pathway in which a signal leads to the synthesis of a protein that suppresses the cell cycle. In this case, the signal is damage to the cell's DNA, perhaps as the result of exposure to ultraviolet light. Operation of this signaling pathway blocks the cell cycle until the damage has been repaired. Otherwise, the damage might contribute to tumor formation by causing mutations or chromosomal abnormalities. Thus, the genes for the components of the pathway act as tumor-suppressor genes. The tumor-suppressor protein encoded by the wild-type *p53* gene is a specific transcription factor that promotes the synthesis of cell cycle–inhibiting proteins. That is why a mutation knocking out the *p53* gene, like a mutation that leads to a hyperactive Ras protein, can lead to excessive cell growth and cancer **(Figure 19.12c)**.

The ***p53* gene,** named for the 53,000-dalton molecular weight of its protein product, has been called the "guardian angel of the genome." Once activated, for example by DNA damage, the p53 protein functions as an activator for several genes. Often it activates a gene called *p21,* whose product halts the cell cycle by binding to cyclin-dependent kinases, allowing time for the cell to repair the DNA; the p53 protein can also turn on genes directly involved in DNA repair. When DNA damage is irreparable, p53 activates "suicide" genes, whose protein products cause cell death by a process called apoptosis (see Figure 21.18). Thus, in at least three ways, p53 prevents a cell from passing on mutations due to DNA damage. If mutations do accumulate and the cell survives through many divisions—as is more likely if the *p53* tumor-suppressor gene is defective or missing—cancer may ensue.

The Multistep Model of Cancer Development

More than one somatic mutation is generally needed to produce all the changes characteristic of a full-fledged cancer cell. This may help explain why the incidence of cancer increases greatly with age. If cancer results from an accumulation of mutations and if mutations occur throughout life, then the longer we live, the more likely we are to develop cancer.

The model of a multistep path to cancer is well supported by studies of one of the best understood types of human cancer, colorectal cancer. About 135,000 new cases of colorectal cancer are diagnosed each year in the United States, and the disease causes 60,000 deaths per year. Like most cancers, colorectal cancer develops gradually **(Figure 19.13)**. The first sign is often a polyp, a small, benign growth in the colon lining. The cells of the polyp look normal, although they divide unusually frequently. The tumor grows and may eventually become malignant, invading other tissues. The development of a malignant tumor is paralleled by a gradual accumulation of mutations that convert proto-oncogenes to oncogenes and knock out tumor-suppressor genes. A *ras* oncogene and a mutated *p53* tumor-suppressor gene are often involved.

About a half dozen changes must occur at the DNA level for a cell to become fully cancerous. These usually include the appearance of at least one active oncogene and the mutation or loss of several tumor-suppressor genes. Furthermore, since mutant tumor-suppressor alleles are usually recessive, in most cases mutations must knock out *both* alleles in a cell's genome to block tumor suppression. (Most oncogenes, on the other hand, behave as dominant alleles.) Finally, in many malignant tumors, the gene for telomerase is activated. This enzyme prevents the shortening of chromosome ends during DNA replication (see Figure 16.18). Production of telomerase in cancer cells removes a natural limit on the number of times the cells can divide.

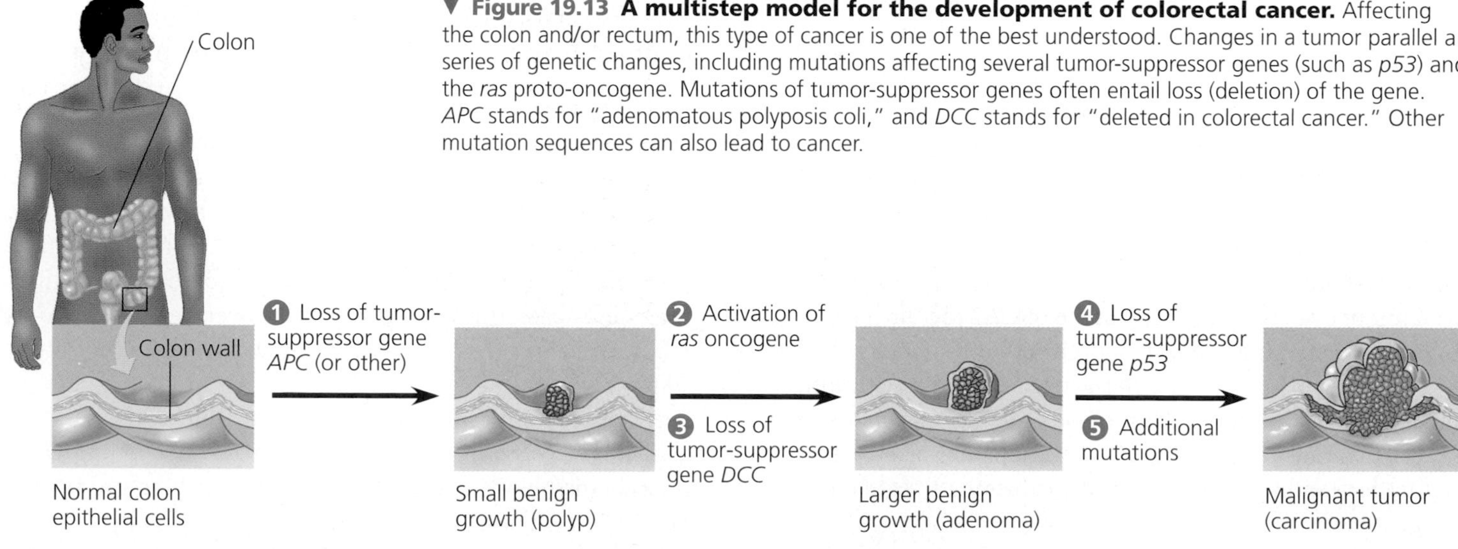

▼ **Figure 19.13 A multistep model for the development of colorectal cancer.** Affecting the colon and/or rectum, this type of cancer is one of the best understood. Changes in a tumor parallel a series of genetic changes, including mutations affecting several tumor-suppressor genes (such as *p53*) and the *ras* proto-oncogene. Mutations of tumor-suppressor genes often entail loss (deletion) of the gene. *APC* stands for "adenomatous polyposis coli," and *DCC* stands for "deleted in colorectal cancer." Other mutation sequences can also lead to cancer.

Viruses seem to play a role in about 15% of human cancer cases worldwide. Viruses contribute to cancer development by integrating their genetic material into the DNA of infected cells. By this process, a retrovirus may donate an oncogene to the cell. Alternatively, integrated viral DNA may disrupt a tumor-suppressor gene or convert a proto-oncogene to an oncogene. Finally, some viruses produce proteins that inactivate p53 and other tumor-suppressor proteins, thus making the cell more prone to becoming cancerous.

Inherited Predisposition to Cancer

The fact that multiple genetic changes are required to produce a cancer cell helps explain the observation that certain cancers run in some families. An individual inheriting an oncogene or a mutant allele of a tumor-suppressor gene is one step closer to accumulating the necessary mutations for cancer to develop than is an individual without any such mutations.

Geneticists are devoting much effort to identifying inherited cancer alleles so that predisposition to certain cancers can be detected early in life. About 15% of colorectal cancers, for example, involve inherited mutations. Many of these affect the tumor-suppressor gene called *adenomatous polyposis coli,* or *APC* (see Figure 19.13). This gene has multiple functions in the cell, including regulation of cell migration and adhesion. Even in patients with no family history of the disease, the *APC* gene is mutated in 60% of colorectal cancers. In these individuals, new mutations must occur in both *APC* alleles before the gene's function is lost. Since only 15% of colorectal cancers are associated with known inherited mutations, researchers continue in their efforts to identify "markers" that could predict the risk of developing this type of cancer.

There is evidence of a strong inherited predisposition in 5–10% of patients with breast cancer. This is the second most common type of cancer in the United States, striking over 180,000 women (and some men) annually and killing 40,000 each year. Mutations in the *BRCA1* or *BRCA2* gene are found in at least half of inherited breast cancers (*BRCA* stands for BReast CAncer). A woman who inherits one mutant *BRCA1* allele has a 60% probability of developing breast cancer before the age of 50, compared with only a 2% probability for an individual homozygous for the normal allele. Both *BRCA1* and *BRCA2* are considered to be tumor-suppressor genes because their wild-type alleles protect against breast cancer and their mutant alleles are recessive. Researchers have not yet determined what the normal products of these genes actually do in the cell. However, recent evidence suggests that the BRCA2 protein is directly involved in repairing breaks that occur in both strands of DNA.

The study of these and other genes associated with inherited cancer may lead to new methods for early diagnosis and treatment of all cancers. Studying these genes also increases our understanding of the normal processes of regulation of the genome.

Concept Check 19.3

1. Compare the usual functions of proteins encoded by proto-oncogenes with those encoded by tumor-suppressor genes.
2. Explain how the types of mutations that lead to cancer are different for a proto-oncogene and a tumor-suppressor gene.
3. Under what circumstances do we consider cancer to have a hereditary component?

For suggested answers, see Appendix A.

Eukaryotic genomes can have many noncoding DNA sequences in addition to genes

We have spent most of this chapter, and indeed this unit, focusing on genes that code for proteins. Yet the coding regions of these genes and the genes for RNA products such as rRNA and tRNA make up only a tiny portion of the genomes of most multicellular eukaryotes. The bulk of most eukaryotic genomes consists of noncoding DNA sequences, often described in the past as "junk DNA." However, much evidence is accumulating that noncoding DNA plays important roles in the cell, an idea supported by its persistence in diverse genomes over many hundreds of generations. In this section, we examine how genes and noncoding DNA sequences are organized within eukaryotic genomes, using the human genome as our prime example. Genome organization tells us much about how genomes have evolved and continue to evolve, the subject of the final section of the chapter.

The Relationship Between Genomic Composition and Organismal Complexity

Several trends are evident when we compare the genomes of prokaryotes and those of eukaryotes, including more complex groups such as mammals. Although there are exceptions, we find a general progression from smaller to larger genomes, but with fewer genes in a given length of DNA. For example, humans have 500 to 1,500 times as many base pairs in their genome as most prokaryotes, but on average only 5 to 15 times as many genes—thus many fewer genes in any given length of DNA.

In prokaryotic genomes, most of the DNA codes for protein, tRNA, or rRNA; the small amount of noncoding DNA consists mainly of regulatory sequences, such as promoters. The coding sequence of nucleotides along a prokaryotic gene proceeds from start to finish without interruption by noncoding sequences (introns). In eukaryotic genomes, by contrast, most of the DNA does *not* encode protein or RNA, and it includes more complex regulatory sequences. In fact, humans have 10,000 times as much noncoding DNA as prokaryotes. Some of the noncoding DNA in multicellular eukaryotes is present as introns within genes. Indeed, introns account for most of the difference in average length between human genes (27,000 base pairs) and prokaryotic genes (1,000 base pairs).

Now that the complete sequence of the human genome is available, we know what makes up most of the 98.5% that does not code for proteins, rRNAs, or tRNAs **(Figure 19.14)**. Gene-related regulatory sequences and introns account for 24% of the human genome. The remaining sequences, located between functional genes, include some unique noncoding DNA, such as gene fragments and mutated genes that are nonfunctional. Most intergenic DNA, however, is **repetitive DNA,** sequences that are present in multiple copies in the genome. Somewhat surprisingly, about three-fourths of this repetitive DNA (44% of the entire human genome) is made up of transposable elements and sequences related to them.

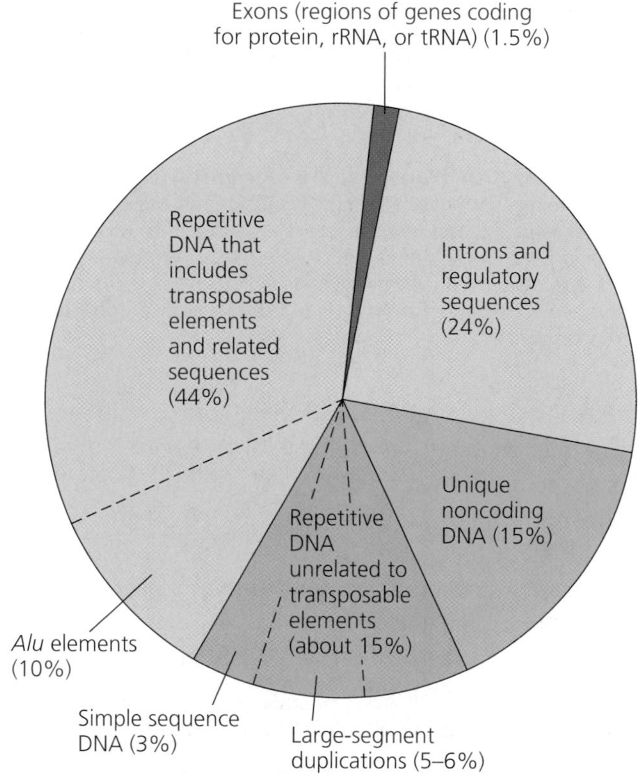

▲ **Figure 19.14 Types of DNA sequences in the human genome.** The coding sequences in genes (dark purple) make up only about 1.5% of the human genome, while introns and regulatory sequences associated with genes (light purple) make up about a quarter. The vast majority of the human genome does not code for human proteins or RNAs, and much of it is repetitive DNA (dark and light green). Because repetitive DNA is the most difficult to sequence and analyze, classification of some portions is tentative, and the percentages given here may shift slightly as genome analysis continues.

Transposable Elements and Related Sequences

All organisms seem to have stretches of DNA that can move from one location to another within the genome. In Chapter 18, we described transposable elements in prokaryotes, which may have been an evolutionary source of viruses. However, the first evidence for wandering DNA segments did not come from experiments with prokaryotes, but from American geneticist Barbara McClintock's breeding experiments with Indian corn (maize) in the 1940s and 1950s **(Figure 19.15)**. McClintock identified changes in the color of corn kernels that made sense only if she postulated the existence of genetic elements capable of moving from other locations in the genome into the genes for kernel color. McClintock's discovery received little attention until transposable elements were discovered in bacteria many years later and microbial geneticists learned more about the molecular basis of transposition.

Movement of Transposons and Retrotransposons

Eukaryotic transposable elements are of two types: **transposons**, which move within a genome by means of a DNA intermediate, and **retrotransposons,** which move by means of an RNA intermediate, a transcript of the retrotransposon DNA. Transposons can move by a "cut-and-paste" mechanism, which removes the element from the original site, or by

▲ **Figure 19.15 The effect of transposable elements on corn kernel color.** Barbara McClintock first proposed the idea of mobile genetic elements after observing variegations in corn kernel color. Although her idea was met with skepticism when she proposed it in the 1940s, it was later validated. She received the Nobel Prize in 1983, at the age of 81, for her pioneering research.

a "copy-and-paste" mechanism, which leaves a copy behind **(Figure 19.16a)**.

Retrotransposons always leave a copy at the original site during transposition, since they are initially transcribed into an RNA intermediate **(Figure 19.16b)**. To insert at another site, the RNA intermediate is first converted back to DNA by reverse transcriptase, an enzyme encoded in the retrotransposon itself. Thus, reverse transcriptase can be present in cells not infected with retroviruses. (In fact, retroviruses may have evolved from retrotransposons.) A cellular enzyme catalyzes insertion of the reverse-transcribed DNA at a new site. Most transposable elements in eukaryotic genomes are retrotransposons.

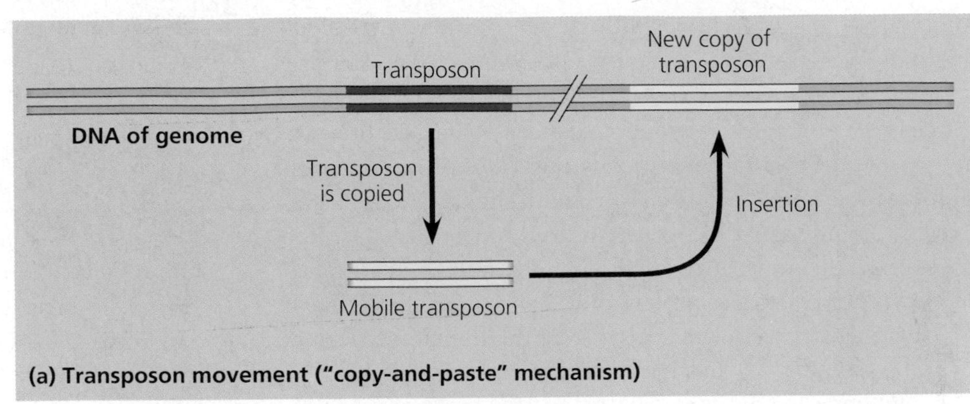

(a) Transposon movement ("copy-and-paste" mechanism)

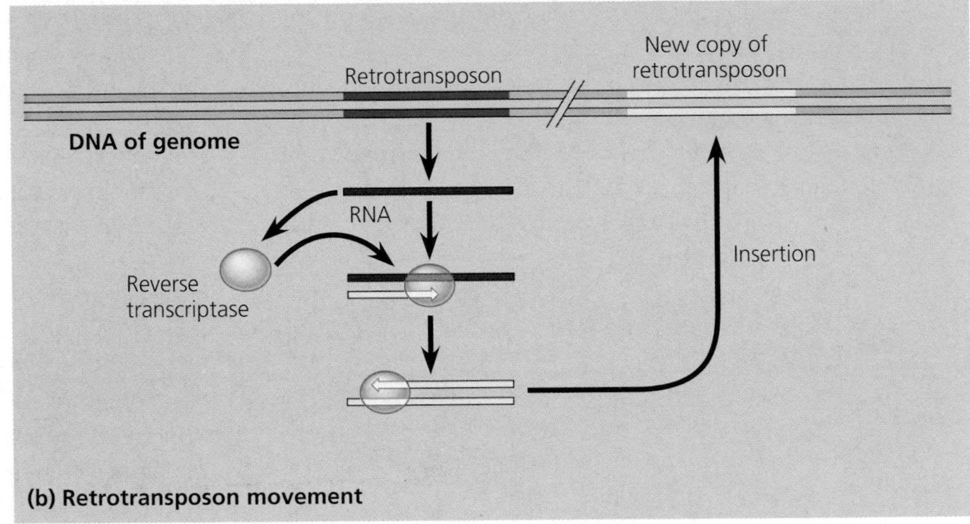

(b) Retrotransposon movement

▲ **Figure 19.16 Movement of eukaryotic transposable elements. (a)** Movement of transposons by either the cut-and-paste mechanism or the copy-and-paste mechanism (shown here) involves a double-stranded DNA intermediate that is inserted into the genome. **(b)** Movement of retrotransposons begins with formation of a single-stranded RNA intermediate. The remaining steps are essentially identical to part of the retrovirus reproductive cycle (see Figure 18.10). In movement of transposons by the copy-and-paste mechanism and in movement of retrotransposons, the DNA sequence remains in the original site as well as appearing in a new site.

Sequences Related to Transposable Elements

Multiple copies of transposable elements and sequences related to them are scattered throughout eukaryotic genomes. A single unit is usually hundreds to thousands of base pairs long, and the dispersed "copies" are similar but usually not identical to each other. Some of these are transposable elements that move using enzymes encoded either by themselves or by other transposable elements, and some are related sequences that have lost the ability to move altogether. Transposable elements and related sequences make up 25–50% of most mammalian genomes and even higher percentages in amphibians and many plants (see Figure 19.14).

In humans and other primates, a large portion of transposable element–related DNA consists of a family of similar sequences called *Alu elements.* These sequences alone account for approximately 10% of the human genome. *Alu* elements are about 300 nucleotides long, much shorter than most functional transposable elements, and they do not code for any protein. However, many *Alu* elements are transcribed into RNA molecules; their cellular function, if any, is unknown.

Although many transposable elements encode proteins, these proteins do not carry out normal cellular functions. Thus, these elements often are described as "noncoding" DNA, along with other repetitive sequences.

Other Repetitive DNA, Including Simple Sequence DNA

Repetitive DNA that is not related to transposable elements probably arose by mistakes that occurred during DNA replication or recombination. It accounts for about 15% of the human genome (see Figure 19.14). About a third of this (5% of the human genome) consists of large-segment duplications, in which a long stretch of DNA, between 10,000 and 300,000 nucleotide pairs, seems to have been copied from one chromosomal location to another, on the same or a different chromosome.

In contrast to single duplications of long sequences, *simple sequence DNA* contains many copies of tandemly repeated short sequences, as in the following example (showing one DNA strand only):

. . . GTTACGTTACGTTACGTTACGTTACGTTAC . . .

In this case, the repeated unit consists of five nucleotides (GTTAC). Repeated units often contain fewer than 15 nucleotides, but they may include as many as 500 nucleotides. The number of repeated units at a particular site in the genome varies as well. For example, there could be as many as several hundred thousand repetitions of the GTTAC unit at one site. Altogether, simple sequence DNA makes up 3% of the human genome.

The nucleotide composition of simple sequence DNA is often different enough from the rest of the cell's DNA to have an intrinsically different density. If genomic DNA is cut into pieces and centrifuged at high speed, segments of different density migrate to different positions in the centrifuge tube. Repetitive DNA isolated in this way was originally called *satellite DNA* because it appeared as a "satellite" band in the centrifuge tube, separate from the rest of the DNA. Now the term is often used interchangeably with *simple sequence DNA*.

Much of a genome's simple sequence DNA is located at chromosomal telomeres and centromeres, suggesting that this DNA plays a structural role for chromosomes. The DNA at centromeres is essential for the separation of chromatids in cell division (see Chapter 12). Centromeric DNA, along with simple sequence DNA located elsewhere, also may help organize the chromatin within the interphase nucleus. The simple sequence DNA located at telomeres, at the tips of chromosomes, prevents genes from being lost as the DNA shortens with each round of replication (see Chapter 16). Telomeric DNA also binds proteins that protect the ends of a chromosome from degradation and from joining to other chromosomes.

Genes and Multigene Families

We finish our discussion of the various types of DNA sequences in eukaryotic genomes with a closer look at genes. Recall that sequences coding for proteins and structural RNAs compose a mere 1.5% of the human genome (see Figure 19.14). If we include introns and regulatory sequences associated with genes, the total amount of gene-related DNA—coding and noncoding—constitutes about 25% of the human genome.

As in prokaryotes, most eukaryotic genes are present as unique sequences, with only one copy per haploid set of chromosomes. But in the human genome, such solitary genes make up only about half of the total coding DNA. The rest occurs in **multigene families,** collections of identical or very similar genes.

Some multigene families consist of *identical* DNA sequences, usually clustered tandemly. With the notable exception of the genes for histone proteins, multigene families of identical genes code for RNA products. An example is the family of identical sequences encoding the three largest ribosomal RNA (rRNA) molecules **(Figure 19.17a)**. These rRNA molecules are encoded in a single transcription unit that is repeated tandemly hundreds to thousands of times in one or several clusters in the genome of a multicellular eukaryote. The many copies of this rRNA transcription unit help cells to quickly make the millions of ribosomes needed for active protein synthesis. The primary transcript is cleaved to yield the three rRNA molecules. These are then

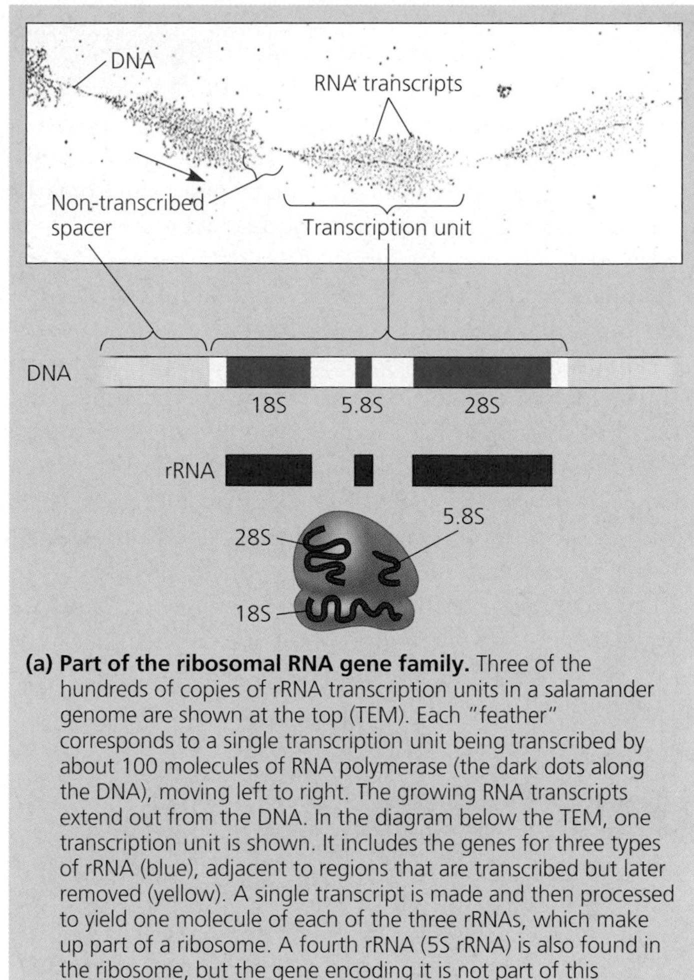

(a) Part of the ribosomal RNA gene family. Three of the hundreds of copies of rRNA transcription units in a salamander genome are shown at the top (TEM). Each "feather" corresponds to a single transcription unit being transcribed by about 100 molecules of RNA polymerase (the dark dots along the DNA), moving left to right. The growing RNA transcripts extend out from the DNA. In the diagram below the TEM, one transcription unit is shown. It includes the genes for three types of rRNA (blue), adjacent to regions that are transcribed but later removed (yellow). A single transcript is made and then processed to yield one molecule of each of the three rRNAs, which make up part of a ribosome. A fourth rRNA (5S rRNA) is also found in the ribosome, but the gene encoding it is not part of this transcription unit.

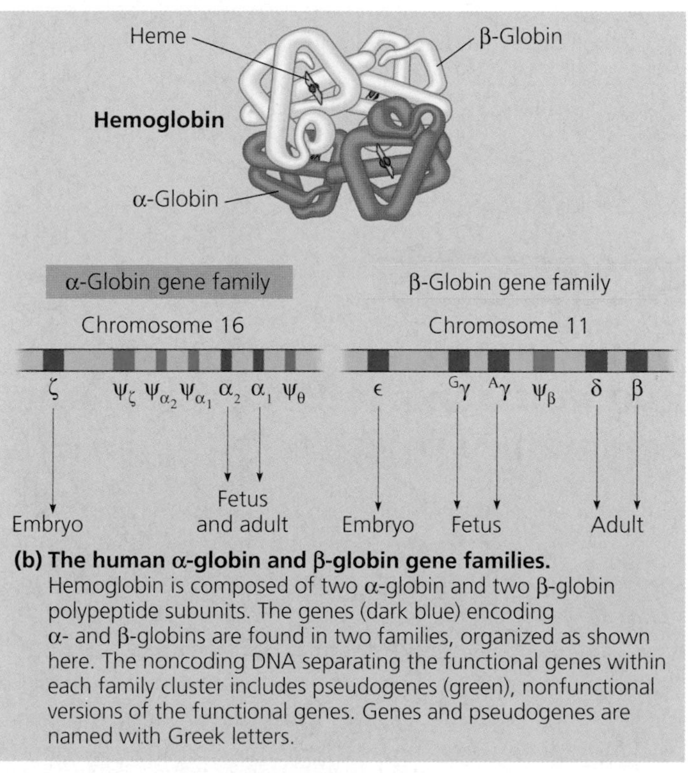

(b) The human α-globin and β-globin gene families. Hemoglobin is composed of two α-globin and two β-globin polypeptide subunits. The genes (dark blue) encoding α- and β-globins are found in two families, organized as shown here. The noncoding DNA separating the functional genes within each family cluster includes pseudogenes (green), nonfunctional versions of the functional genes. Genes and pseudogenes are named with Greek letters.

▲ **Figure 19.17 Gene families.**

combined with proteins and one other kind of rRNA (5S rRNA) to form ribosomal subunits.

The classic examples of multigene families of *nonidentical* genes are two related families of genes that encode globins, a group of proteins that include the α and β polypeptide subunits of hemoglobin. One family, located on chromosome 16 in humans, encodes various forms of α-globin; the other, on chromosome 11, encodes forms of β-globin **(Figure 19.17b)**. The different forms of each globin subunit are expressed at different times in development, allowing hemoglobin to function effectively in the changing environment of the developing animal. In humans, for example, the embryonic and fetal forms of hemoglobin have a higher affinity for oxygen than the adult forms, ensuring the efficient transfer of oxygen from mother to developing fetus. Also found in the globin gene family clusters are several **pseudogenes,** nonfunctional nucleotide sequences quite similar to the functional genes.

The arrangement of the genes in gene families has provided insight into the evolution of genomes. We will consider some of the processes that have shaped the genomes of different species over evolutionary time in the next section.

Concept Check 19.4

1. Discuss the characteristics that make mammalian genomes larger than prokaryotic genomes.
2. How do introns, transposable elements, and simple sequence DNA differ in their distribution in the genome?
3. Discuss the differences in the organization of the rRNA gene family and the globin gene families. How do these gene families benefit the organism?

For suggested answers, see Appendix A.

Concept 19.5

Duplications, rearrangements, and mutations of DNA contribute to genome evolution

The basis of change at the genomic level is mutation, which underlies much of genome evolution. It seems likely that the earliest forms of life had a minimal number of genes—those necessary for survival and reproduction. If this was indeed the case, one aspect of evolution must have been an increase in the size of the genome, with the extra genetic material providing the raw material for gene diversification. In this section, we will first describe how extra copies of all or part of a genome can arise and then consider subsequent processes that can lead to the evolution of proteins (or RNA products) with related or entirely new functions.

Duplication of Chromosome Sets

An accident in meiosis can result in one or more extra sets of chromosomes, a condition known as polyploidy. In a polyploid organism, one complete set of genes can provide essential functions for the organism. The genes in the one or more extra sets can diverge by accumulating mutations; these variations may persist if the organism carrying them survives and reproduces. In this way, genes with novel functions can evolve. As long as one copy of a crucial gene is expressed, the divergence of another copy can lead to its encoded protein acting in a novel way, thereby changing the organism's phenotype. The accumulation of mutations in many (or even a few) genes may lead to the branching off of a new species, as happens often in plants (see Chapter 24). Although polyploid animals exist, they are rare.

Duplication and Divergence of DNA Segments

Errors during meiosis can also lead to the duplication of individual genes. Unequal crossing over during prophase I of meiosis, for instance, can result in one chromosome with a deletion and another with a duplication of a particular region. As illustrated in **Figure 19.18**, transposable elements in the

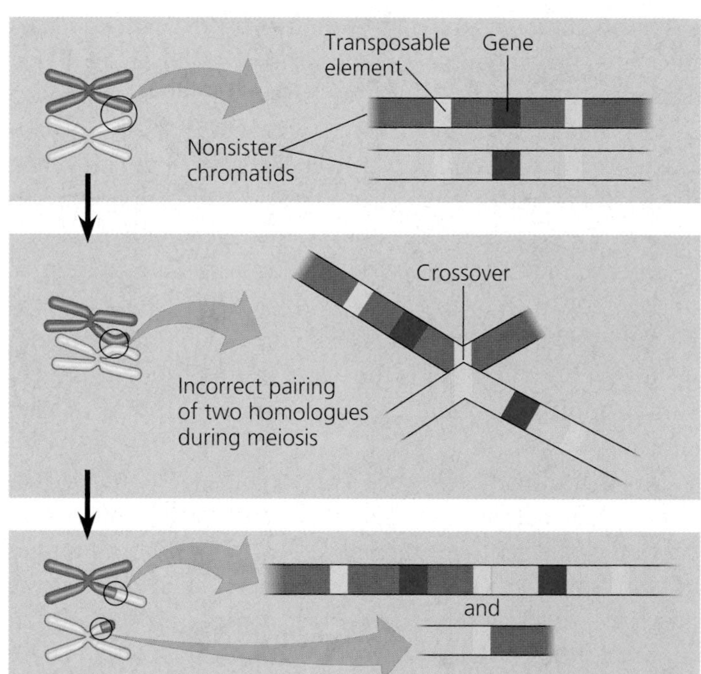

▲ **Figure 19.18 Gene duplication due to unequal crossing over.** One mechanism by which a gene (or other DNA segment) can be duplicated is recombination during meiosis between copies of a transposable element flanking the gene. Such recombination between misaligned nonsister chromatids of homologous chromosomes produces one chromatid with two copies of the gene and one chromatid with no copy.

genome can provide sites where nonsister chromatids can cross over, even when their homologous gene sequences are not correctly aligned.

Also, slippage can occur during DNA replication, such that the template shifts with respect to the new complementary strand, and one region of the template strand is either not copied or copied twice. As a result, a region of DNA is deleted or duplicated. It is easy to imagine how such errors could occur in regions of repeats, such as the simple sequence DNA described previously. The variability in numbers of repeated units of simple sequence DNA at the same site is probably due to errors like these. Evidence that molecular events such as unequal crossing over and slippage lead to duplication of genes is found in the existence of multigene families.

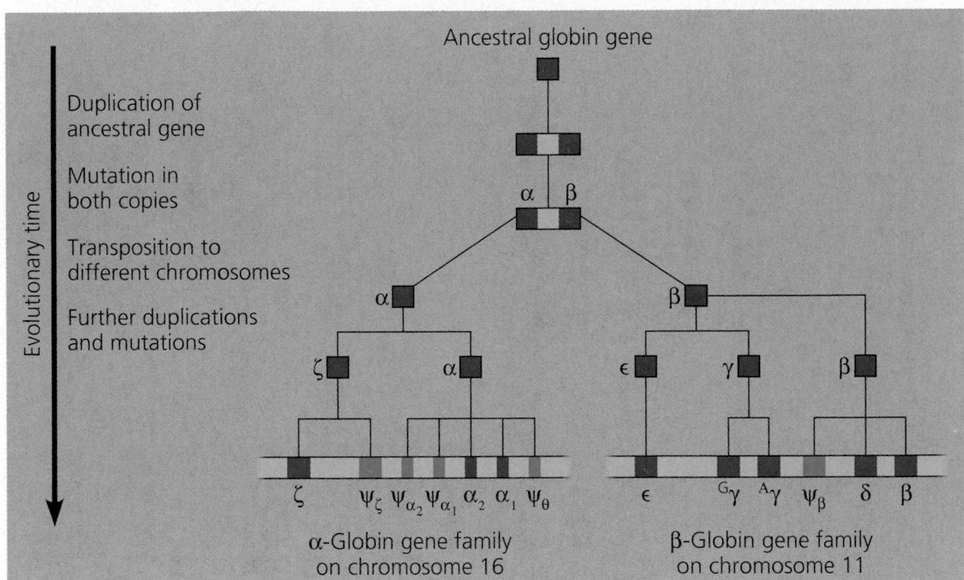

▲ **Figure 19.19 Evolution of the human α-globin and β-globin gene families.** Shown here is a model for evolution of the modern α-globin and β-globin gene families from a single ancestral globin gene.

Evolution of Genes with Related Functions: The Human Globin Genes

Duplication events can lead to the evolution of genes with related functions, such as those of the α-globin and β-globin gene families (see Figure 19.17b). A comparison of gene sequences within a multigene family can suggest the order in which the genes arose. This approach to re-creating the evolutionary history of the various globin genes indicates that they all evolved from one common ancestral globin gene, which was duplicated and diverged into α-globin and β-globin ancestral genes about 450–500 million years ago **(Figure 19.19)**. Each of these genes was later duplicated several times, and the copies then diverged from each other in sequence, yielding the current family members. In fact, the common ancestral globin gene also gave rise to the oxygen-binding muscle protein myoglobin and to the plant protein leghemoglobin. The latter two proteins function as monomers, and their genes are included in a "globin superfamily."

After the duplication events, the differences between the genes in the globin families undoubtedly arose from mutations that accumulated in the gene copies over many generations. The current model is that the necessary function provided by an α-globin protein, for example, was fulfilled by one gene, while other copies of the α-globin gene accumulated random mutations. Some mutations may have had an adverse affect on the organism and some may have had no effect, but some mutations may have altered the function of the protein product in a way that was advantageous to the organism at a particular life stage without substantially changing its oxygen-carrying function. Presumably, natural selection acted on these altered genes, maintaining them in the population and leading to production of alternative forms of α-globin protein.

The similarity in the amino acid sequences of the various α-globin and β-globin proteins supports this model of gene duplication and mutation **(Table 19.1)**. The amino acid sequences of the β-globins, for instance, are much more similar to each other than to the α-globin sequences. The existence of several pseudogenes among the functional globin genes provides additional evidence for this model (see Figure 19.17b). That is, random mutations in these "genes" over evolutionary time have destroyed their function.

Table 19.1 Percentage of Similarity in Amino Acid Sequence Between Human Globin Proteins

		α-Globins		β-Globins		
		α	ζ	β	γ	ε
α-Globins	α	100	58	42	39	37
	ζ	58	100	34	38	37
β-Globins	β	42	34	100	73	75
	γ	39	38	73	100	80
	ε	37	37	75	80	100

Evolution of Genes with Novel Functions

In the evolution of the globin gene families, gene duplication and subsequent divergence produced family members whose protein products performed related functions. Alternatively, one copy of a duplicated gene can undergo alterations that lead to a completely new function for the protein product. The genes for lysozyme and α-lactalbumin are good examples.

Lysozyme is an enzyme that helps prevent infection by hydrolyzing the cell walls of bacteria; α-lactalbumin is a nonenzymatic protein that plays a role in milk production in mammals. The two proteins are quite similar in their amino acid sequences and three-dimensional structures. Both genes are found in mammals, whereas only the lysozyme gene is present in birds. These findings suggest that at some time after the lineages leading to mammals and birds had separated, the lysozyme gene underwent a duplication event in the mammalian lineage but not in the avian lineage. Subsequently, one copy of the duplicated lysozyme gene evolved into a gene encoding α-lactalbumin, a protein with a completely different function.

Rearrangements of Parts of Genes: Exon Duplication and Exon Shuffling

Rearrangement of existing DNA sequences has also contributed to genome evolution. The presence of introns in most eukaryotic genes may have promoted the evolution of new and potentially useful proteins by facilitating the duplication or repositioning of exons in the genome. Recall from Chapter 17 that an exon often codes for a domain, a distinct structural or functional region of a protein.

We've already seen that unequal crossing over during meiosis can lead to duplication of a gene on one chromosome and its loss from the homologous chromosome (see Figure 19.18). By a similar process, a particular exon within a gene could be duplicated on one chromosome and deleted from the homologous chromosome. The gene with the duplicated exon would code for a protein containing a second copy of the encoded domain. This change in the protein's structure could augment its function by increasing its stability, enhancing its ability to bind a particular ligand, or altering some other property. Quite a few protein-coding genes have multiple copies of related exons, which presumably arose by duplication and then diverged. The gene encoding the extracellular matrix protein collagen is a good example. Collagen is a structural protein with a highly repetitive amino acid sequence, which is reflected in the repetitive pattern of exons in the collagen gene.

Alternatively, we can imagine the occasional mixing and matching of different exons either within a gene or between two nonallelic genes owing to errors in meiotic recombination. This process, termed *exon shuffling*, could lead to new proteins with novel combinations of functions. As an example, let's consider the gene for tissue plasminogen activator (TPA). The TPA protein is an extracellular protein involved in limiting blood clotting. It has four domains of three types, each encoded by an exon; one exon is present in two copies. Because each type of exon is also found in other proteins, the gene for TPA is believed to have arisen by several instances of exon shuffling and duplication **(Figure 19.20)**. The TPA protein slows the clotting reaction and therefore limits the damage that can result from heart attacks and some types of stroke, as long as it is administered immediately to victims.

How Transposable Elements Contribute to Genome Evolution

The persistence of transposable elements as a large fraction of some eukaryotic genomes is consistent with the idea that they can play an important role in shaping a genome over evolutionary time. These elements can contribute to the evolution of the genome in several ways. They can promote recombination, disrupt cellular genes or control elements, and carry entire genes or individual exons to new locations.

The presence of homologous transposable element sequences scattered throughout the genome allows recombination to take place between different chromosomes. Most such alterations are probably detrimental, causing chromosomal translocations and other changes in the genome that may be lethal to the organism. But over the course of evolutionary time, an occasional recombination like this may be advantageous to the organism.

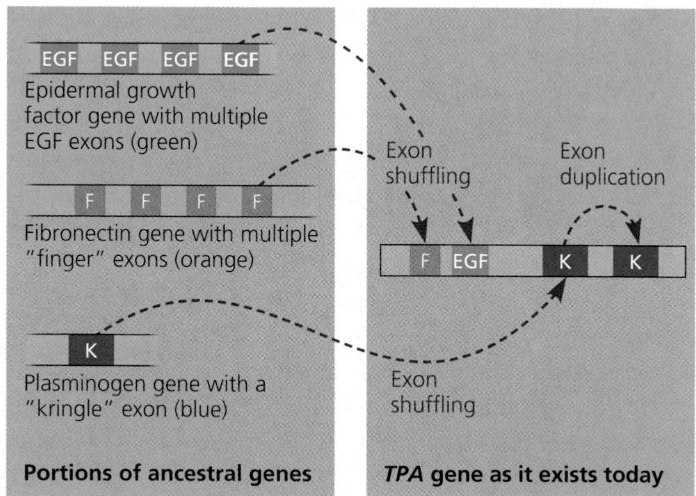

▲ **Figure 19.20 Evolution of a new gene by exon shuffling.** Exon shuffling could have moved exons from ancestral forms of the genes for epidermal growth factor, fibronectin, and plasminogen (left) into the evolving gene for tissue plasminogen activator, TPA (right). The order in which these events might have occurred is unknown. Duplication of the "kringle" exon from plasminogen after its movement could account for the two copies of this exon in the *TPA* gene. Each type of exon encodes a particular domain in the TPA protein.

The movement of transposable elements around the genome can have several direct consequences. For instance, if a transposable element "jumps" into the middle of a coding sequence of a protein-coding gene, it prevents the normal functioning of the interrupted gene. If a transposable element inserts within a regulatory sequence, the transposition may lead to increased or decreased production of one or more proteins. Transposition caused both types of effects on the genes coding for pigment-synthesizing enzymes in McClintock's corn kernels. Again, while such changes may usually be harmful, in the long run some may prove beneficial.

During transposition, a transposable element may carry along a gene or group of genes to a new position in the genome. This mechanism probably accounts for the location of the α-globin and β-globin gene families on different human chromosomes, as well as the dispersion of the genes of certain other gene families. By a similar tag-along process, an exon from one gene may be inserted into another gene in a mechanism similar to that of exon shuffling during recombination. For example, an exon may be inserted by transposition into the intron of a protein-coding gene. If the inserted exon is retained in the RNA transcript during RNA splicing, the protein that is synthesized will have an additional domain, which may confer a new function on the protein.

Recent research reveals yet another way that transposable elements can lead to new coding sequences. This work shows that an *Alu* element may hop into introns in a way that creates a weak alternative splice site in the RNA transcript. During processing of the transcript, the regular splice sites are used more often, so that the original protein is made. On occasion, however, splicing occurs at the new weak site, with the result that some of the *Alu* element ends up in the mRNA, coding for a new portion of the protein. In this way, alternative genetic combinations can be "tried out" while the function of the original gene product is retained.

Clearly, these processes produce either no effect or harmful effects in most individual cases. However, over long periods of time, the generation of genetic diversity provides more raw material for natural selection to work on during evolution. Recent advances in DNA technology have allowed researchers to sequence and compare the genomes of many different species, increasing our understanding of how genomes evolve. You will learn more about these topics in the next chapter.

Concept Check 19.5

1. Describe three examples of errors in cellular processes that lead to DNA duplications.
2. What processes are thought to have led to the evolution of the globin gene families?
3. Look at the portions of the fibronectin and EGF genes shown in Figure 19.20 (left). How might they have arisen?
4. What are three ways transposable elements are thought to contribute to the evolution of the genome?

For suggested answers, see Appendix A.

Chapter 19 Review

Go to www.campbellbiology.com or the student CD-ROM to explore Activities, Investigations, and other interactive study aids.

SUMMARY OF KEY CONCEPTS

Concept 19.1

Chromatin structure is based on successive levels of DNA packing

▶ **Nucleosomes, or "Beads on a String" (pp. 360–361)** Eukaryotic chromatin is composed mostly of DNA and histone proteins that bind to each other and to the DNA to form nucleosomes, the most basic units of DNA packing. Histone tails extend outward from each bead-like nucleosome core.

▶ **Higher Levels of DNA Packing (pp. 360–361)** Additional folding leads ultimately to the highly condensed chromatin of the metaphase chromosome. In interphase cells, most chromatin is less compacted (euchromatin), but some remains highly condensed (heterochromatin).
Activity *DNA Packing*

Concept 19.2

Gene expression can be regulated at any stage, but the key step is transcription

▶ **Differential Gene Expression (pp. 362–363)** Each cell of a multicellular eukaryote expresses only a fraction of its genes. In each type of differentiated cell, a unique subset of genes is expressed. Key stages at which gene expression may be regulated include changes in chromatin structure, initiation of transcription, RNA processing, mRNA degradation, translation, and protein processing and degradation.
Activity *Overview: Control of Gene Expression*

▶ **Regulation of Chromatin Structure (pp. 363–364)** Genes in highly compacted chromatin are generally not transcribed. Chemical modification of histone tails can affect the configuration of chromatin and thus gene expression. Histone acetylation seems to loosen chromatin structure and thereby enhance transcription. DNA methylation is associated with reduced transcription.

▶ **Regulation of Transcription Initiation (pp. 364–367)** Multiple DNA control elements distant from the promoter (in one or more enhancers) bind specific transcription factors

(activators or repressors) that regulate transcription initiation for specific genes within the genome. Bending of DNA enables activators bound to enhancers to contact proteins at the promoter. Unlike the genes of a prokaryotic operon, coordinately controlled eukaryotic genes each have a promoter and control elements. The same regulatory sequences are common to all the genes of a group, enabling recognition by the same specific transcription factors.
Activity *Control of Transcription*
Investigation *How Do You Design a Gene Expression System?*

- **Mechanisms of Post-Transcriptional Regulation (pp. 368–370)** Regulation at the RNA-processing level is exemplified by alternative RNA splicing. Also, each mRNA has a characteristic life span, determined in part by sequences in the leader and trailer regions. RNA interference by single-stranded micro-RNAs can lead to degradation of an mRNA or block its translation. The initiation of translation can be controlled via regulation of initiation factors. After translation, various types of protein processing (such as cleavage and the addition of chemical groups) are subject to control, as is the degradation of proteins by proteasomes.
Activity *Post-Transcriptional Control Mechanisms*
Activity *Review: Control of Gene Expression*

Cancer results from genetic changes that affect cell cycle control

- **Types of Genes Associated with Cancer (pp. 370–371)** The products of proto-oncogenes and tumor-suppressor genes control cell division. A DNA change that makes a proto-oncogene excessively active converts it to an oncogene, which may promote excessive cell division and cancer. A tumor-suppressor gene encodes a protein that inhibits abnormal cell division. A mutation in such a gene that reduces the activity of its protein product may also lead to excessive cell division and possibly to cancer.
- **Interference with Normal Cell-Signaling Pathways (pp. 371–373)** Many proto-oncogenes and tumor-suppressor genes encode components of growth-stimulating and growth-inhibiting signaling pathways, respectively. A hyperactive version of a protein in a stimulatory pathway, such as Ras (a G protein), functions as an oncogene protein. A defective version of a protein in an inhibitory pathway, such as p53 (a transcription activator), fails to function as a tumor suppressor.
- **The Multistep Model of Cancer Development (pp. 373–374)** Normal cells are converted to cancer cells by the accumulation of multiple mutations affecting proto-oncogenes and tumor-suppressor genes. Certain viruses promote cancer by integration of viral DNA into a cell's genome.
- **Inherited Predisposition to Cancer (p. 374)** Individuals who inherit a mutant oncogene or tumor-suppressor allele have an increased risk of developing certain types of cancer.
Activity *Causes of Cancer*

Eukaryotic genomes can have many noncoding DNA sequences in addition to genes

- **The Relationship Between Genomic Composition and Organismal Complexity (pp. 374–375)** Compared with prokaryotic genomes, the genomes of eukaryotes generally are larger, have longer genes, and contain a much greater amount of noncoding DNA both associated with genes (introns, regulatory sequences) and between genes (much of it repetitive sequences).
- **Transposable Elements and Related Sequences (pp. 375–376)** The most abundant type of repetitive DNA in multicellular eukaryotes consists of transposable elements and related sequences. Two types of transposable elements occur in eukaryotes: transposons, which move via a DNA intermediate, and retrotransposons, which are the most prevalent and move via an RNA intermediate. Each element may be hundreds or thousands of base pairs long, and similar but usually not identical copies are dispersed throughout the genome.
- **Other Repetitive DNA, Including Simple Sequence DNA (pp. 376–377)** Short noncoding sequences that are tandemly repeated thousands of times (simple sequence DNA) are especially prominent in centromeres and telomeres, where they probably play structural roles in the chromosome.
- **Genes and Multigene Families (pp. 377–378)** Most eukaryotic genes are present in one copy per haploid set of chromosomes. However, the transcription unit encoding the three largest rRNAs is tandemly repeated hundreds to thousands of times at one or several chromosomal sites, enabling the cell to make the rRNA for millions of ribosomes quickly. The multiple, slightly different genes in the two globin gene families encode polypeptides used at different developmental stages of an animal.

Duplications, rearrangements, and mutations of DNA contribute to genome evolution

- **Duplication of Chromosome Sets (p. 378)** Accidents in cell division can lead to extra copies of all or part of a genome, which may then diverge if one set accumulates sequence changes.
- **Duplication and Divergence of DNA Segments (pp. 378–380)** The genes encoding the various globin proteins evolved from one common ancestral globin gene, which duplicated and diverged into α-globin and β-globin ancestral genes. Subsequent duplications of these genes and random mutations gave rise to the present globin genes, all of which code for oxygen-binding proteins. The copies of some duplicated genes have diverged so much during evolutionary time that the functions of their encoded proteins are now substantially different.
- **Rearrangements of Parts of Genes: Exon Duplication and Exon Shuffling (p. 380)** Rearrangement of exons within and between genes during evolution has led to genes containing multiple copies of similar exons and/or several different exons derived from other genes.
- **How Transposable Elements Contribute to Genome Evolution (pp. 380–381)** Movement of transposable elements or recombination between copies of the same element occasionally generates new sequence combinations that are beneficial to the organism. Such mechanisms can alter the functions of genes or their patterns of expression and regulation.

TESTING YOUR KNOWLEDGE

Self-Quiz

1. In a nucleosome, the DNA is wrapped around
   a. polymerase molecules.
   b. ribosomes.
   c. histones.
   d. the nucleolus.
   e. satellite DNA.

2. Muscle cells differ from nerve cells mainly because
   a. they express different genes.
   b. they contain different genes.
   c. they use different genetic codes.
   d. they have unique ribosomes.
   e. they have different chromosomes.

3. One of the characteristics of retrotransposons is that
   a. they code for an enzyme that synthesizes DNA using an RNA template.
   b. they are found only in animal cells.
   c. they generally move by a cut-and-paste mechanism.
   d. they contribute a significant portion of the genetic variability seen within a population of gametes.
   e. their amplification is dependent on a retrovirus.

4. The functioning of enhancers is an example of
   a. transcriptional control of gene expression.
   b. a post-transcriptional mechanism for editing mRNA.
   c. the stimulation of translation by initiation factors.
   d. post-translational control that activates certain proteins.
   e. a eukaryotic equivalent of prokaryotic promoter functioning.

5. Multigene families are
   a. groups of enhancers that control transcription.
   b. usually clustered at the telomeres.
   c. equivalent to the operons of prokaryotes.
   d. sets of genes that are coordinately controlled.
   e. identical or similar genes that have evolved by gene duplication.

6. Which of the following statements about the DNA in one of your brain cells is true?
   a. Some DNA sequences are present in multiple copies.
   b. Most of the DNA codes for protein.
   c. The majority of genes are likely to be transcribed.
   d. Each gene lies immediately adjacent to an enhancer.
   e. Many genes are grouped into operon-like clusters.

7. Two eukaryotic proteins have one domain in common but are otherwise very different. Which of the following processes is most likely to have contributed to this phenomenon?
   a. gene duplication
   b. RNA splicing
   c. exon shuffling
   d. histone modification
   e. random point mutations

8. Which of the following is an example of a possible step in the post-transcriptional control of gene expression?
   a. the addition of methyl groups to cytosine bases of DNA
   b. the binding of transcription factors to a promoter
   c. the removal of introns and splicing together of exons
   d. gene amplification during a stage in development
   e. the folding of DNA to form heterochromatin

9. Within a cell, the amount of protein made using a given mRNA molecule depends partly on
   a. the degree of DNA methylation.
   b. the rate at which the mRNA is degraded.
   c. the presence of certain transcription factors.
   d. the number of introns present in the mRNA.
   e. the types of ribosomes present in the cytoplasm.

10. Proto-oncogenes can change into oncogenes that cause cancer. Which of the following best explains the presence of these potential time bombs in eukaryotic cells?
   a. Proto-oncogenes first arose from viral infections.
   b. Proto-oncogenes normally help regulate cell division.
   c. Proto-oncogenes are genetic "junk."
   d. Proto-oncogenes are mutant versions of normal genes.
   e. Cells produce proto-oncogenes as they age.

For Self-Quiz answers, see Appendix A.

Go to the website or CD-ROM for more quiz questions.

Evolution Connection

One of the revelations of the human genome sequence was the presence of relict prokaryotic sequences—genes of prokaryotes incorporated into our genome but now defunct molecular fossils. What may have occurred to maroon prokaryotic genes in our genome?

Scientific Inquiry

Prostate cells usually require testosterone and other androgens to survive. But some prostate cancer cells thrive despite treatments that eliminate androgens. One hypothesis is that estrogen, often considered a female hormone, may be activating genes normally controlled by an androgen in these cancer cells. Describe one or more experiments to test this hypothesis. (See Figure 11.6 to review the action of these steroid hormones.)

Investigation *How Do You Design a Gene Expression System?*

Science, Technology, and Society

Trace amounts of dioxin were present in Agent Orange, a defoliant sprayed on vegetation during the Vietnam War. Animal tests suggest that dioxin can cause birth defects, cancer, liver and thymus damage, and immune system suppression, sometimes leading to death. But the animal tests are equivocal; a hamster is not affected by a dose that can kill a guinea pig. Dioxin acts somewhat like a steroid hormone, entering a cell and binding to a receptor protein that then attaches to the cell's DNA. How might this mechanism help explain the variety of dioxin's effects on different body systems and in different animals? How might you determine whether a type of illness is related to dioxin exposure? How might you determine whether a particular individual became ill as a result of exposure to dioxin? Which would be more difficult to demonstrate? Why?

20 DNA Technology and Genomics

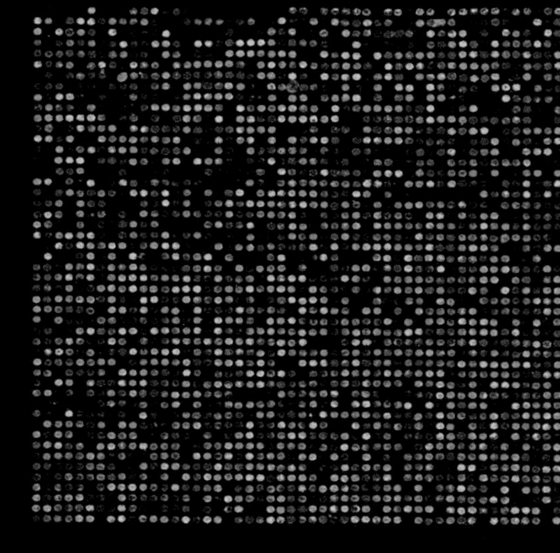

▲ **Figure 20.1 DNA microarray that reveals expression levels of 2,400 human genes (enlarged photo).**

Key Concepts

20.1 DNA cloning permits production of multiple copies of a specific gene or other DNA segment

20.2 Restriction fragment analysis detects DNA differences that affect restriction sites

20.3 Entire genomes can be mapped at the DNA level

20.4 Genome sequences provide clues to important biological questions

20.5 The practical applications of DNA technology affect our lives in many ways

Overview

Understanding and Manipulating Genomes

One of the great achievements of modern science has been the sequencing of the human genome, which was largely completed by 2003. The sequencing of the first complete genome, that of a bacterium, had been carried out a mere eight years previously. During the intervening years, researchers accelerated the pace of DNA sequencing, while working on other genomes, aided by the development of faster and faster sequencing machines. These sequencing accomplishments have all depended on advances in DNA technology, starting with the invention of methods for making **recombinant DNA.** This is DNA in which nucleotide sequences from two different sources—often different species—are combined *in vitro* into the same DNA molecule.

The methods for making recombinant DNA are central to **genetic engineering,** the direct manipulation of genes for practical purposes. Applications of genetic engineering include the manufacture of hundreds of protein products, such as hormones and blood-clotting factors. Using DNA technology, scientists can make recombinant DNA and then introduce it into cultured cells that replicate the DNA and express its genes, yielding a desired protein.

DNA technology has launched a revolution in the area of **biotechnology,** the manipulation of organisms or their components to make useful products. Practices that go back centuries are forms of biotechnology: for example, the use of microbes to make wine and cheese and the selective breeding of livestock, which exploits naturally occurring mutations and genetic recombination. Modern biotechnology based on the manipulation of DNA *in vitro* differs from earlier practices by enabling scientists to modify specific genes and move them between organisms as distinct as bacteria, plants, and animals.

DNA technology is now applied in areas ranging from agriculture to criminal law. More important, its use allows researchers in virtually all fields of biology to tackle age-old questions in a more comprehensive way. For instance, the level of expression of thousands of different genes can now be measured at the same time, as shown in the DNA microarray in **Figure 20.1**. In the photograph, the color of each spot represents the relative expression of one of 2,400 human genes in a particular tissue. With this technique, researchers can compare gene expression in particular tissues or under different conditions. The knowledge gained from such global expression studies was largely inaccessible only a few decades ago.

In this chapter, we first describe the main techniques for manipulating DNA and then discuss how genomes are analyzed and compared at the DNA level. In the last section, we survey the practical applications of DNA technology, concluding the chapter by considering some of the social and ethical issues that arise as DNA technology becomes more pervasive in our lives.

Concept 20.1

DNA cloning permits production of multiple copies of a specific gene or other DNA segment

The molecular biologist studying a particular gene faces a challenge. Naturally occurring DNA molecules are very long, and a single molecule usually carries many genes. Moreover, genes may occupy only a small proportion of the chromosomal DNA, the rest being noncoding nucleotide sequences. A single human gene, for example, might constitute only 1/100,000 of a chromosomal DNA molecule. As a further complication, the distinctions between a gene and the surrounding DNA are subtle, consisting only of differences in nucleotide sequence. To work directly with specific genes, scientists have developed methods for preparing well-defined, gene-sized pieces of DNA in multiple identical copies, a process called **gene cloning.**

DNA Cloning and Its Applications: *A Preview*

Most methods for cloning pieces of DNA in the laboratory share certain general features. One common approach uses bacteria (most often, *Escherichia coli*) and their plasmids. Recall from Chapter 18 that bacterial plasmids are relatively small, circular DNA molecules that replicate separate from a bacterial chromosome. For cloning genes or other pieces of DNA in the laboratory, a plasmid is first isolated from a bacterial cell, and then the foreign DNA is inserted into it **(Figure 20.2)**. The resulting plasmid is now a recombinant DNA molecule, combining DNA from two sources. The plasmid is returned to a bacterial cell, producing a *recombinant bacterium,* which reproduces to form a **clone** of identical cells. Because the dividing bacteria replicate the recombinant plasmid and pass it on to their descendants, the foreign gene is "cloned" at the same time; that is, the clone of cells contains multiple copies of the gene.

Cloned genes are useful for two basic purposes: to make many copies of a particular gene and to produce a protein product. Researchers can isolate copies of a cloned gene from bacteria for use in basic research or to endow an organism with a new metabolic capability, such as pest resistance. For example, a resistance gene present in one crop species might be cloned and transferred into plants of another species. Alternatively, a protein with medical uses, such as human growth hormone, can be harvested in large quantities from bacterial cultures carrying the cloned gene for the protein.

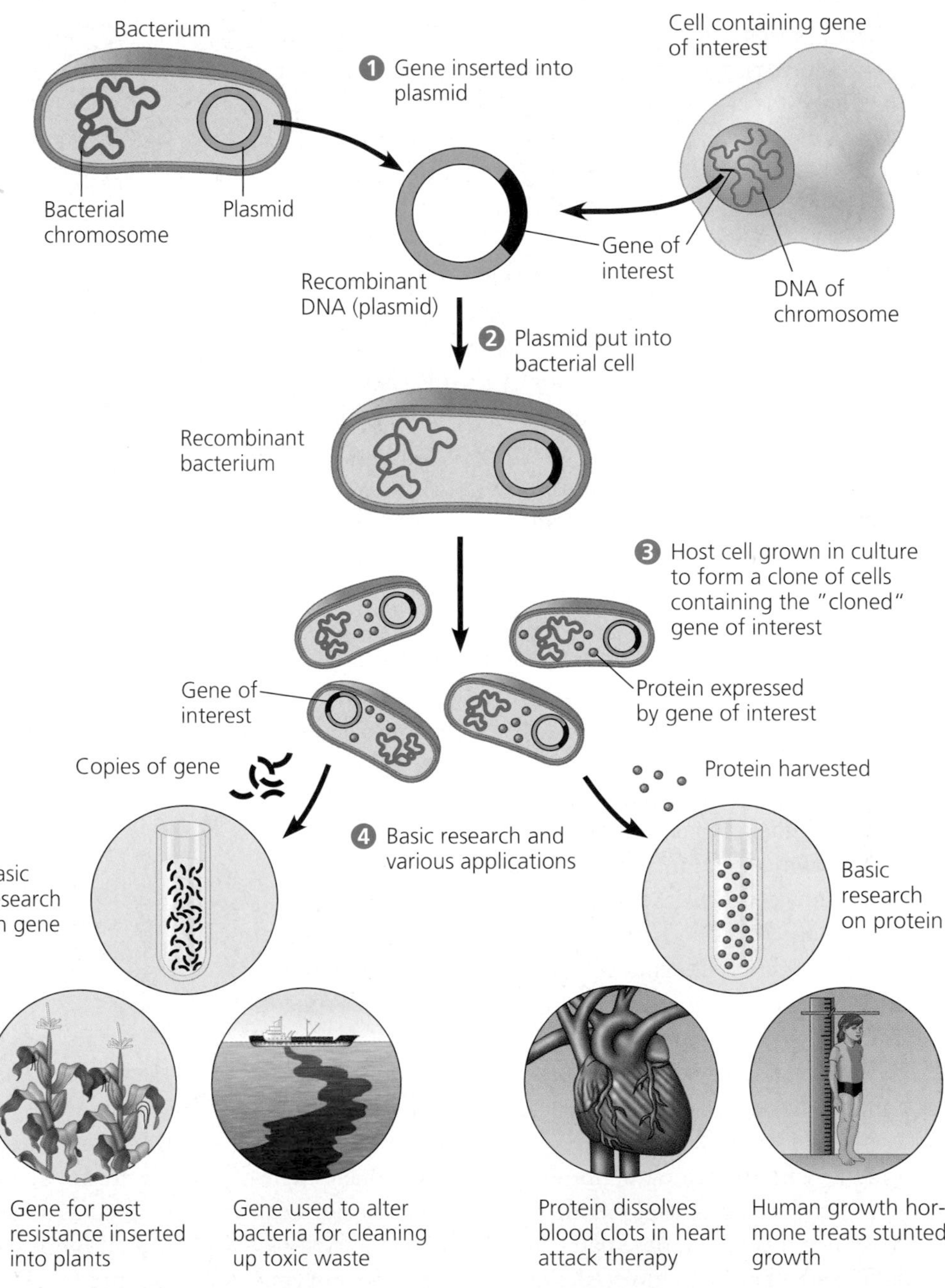

▲ **Figure 20.2 Overview of gene cloning with a bacterial plasmid, showing various uses of cloned genes.** In this simplified diagram of gene cloning in the laboratory, we start with a plasmid isolated from a bacterial cell and a gene of interest from another organism. Only one copy of the plasmid and one copy of the gene of interest are shown at the top of the figure, but the starting materials would include many copies of each.

Most protein-coding genes exist in only one copy per genome—something on the order of one part per million of DNA—so the ability to clone such rare DNA fragments is extremely valuable. In the remainder of this chapter, you will learn more about the techniques outlined in Figure 20.2 and related methods.

Using Restriction Enzymes to Make Recombinant DNA

Gene cloning and genetic engineering were made possible by the discovery of enzymes that cut DNA molecules at a limited number of specific locations. These enzymes, called restriction endonucleases, or **restriction enzymes**, were discovered in the late 1960s by researchers studying bacteria. In nature, these enzymes protect the bacterial cell against intruding DNA from other organisms, such as phages or other species of bacteria. They work by cutting up the foreign DNA, a process called *restriction.*

Hundreds of different restriction enzymes have been identified and isolated. Each restriction enzyme is very specific, recognizing a particular short DNA sequence, or **restriction site,** and cutting both DNA strands at specific points within this restriction site. The DNA of a bacterial cell is protected from the cell's own restriction enzymes by the addition of methyl groups (—CH_3) to adenines or cytosines within the sequences recognized by the enzymes.

The top of **Figure 20.3** illustrates a restriction site recognized by a particular restriction enzyme from *E. coli.* As shown in this example, most restriction sites are symmetrical: That is, the sequence of nucleotides is the same on both strands when read in the $5' \rightarrow 3'$ direction. Most restriction enzymes recognize sequences containing four to eight nucleotides. Because any sequence this short usually occurs (by chance) many times in a long DNA molecule, a restriction enzyme will make many cuts in a DNA molecule, yielding a set of **restriction fragments.** All copies of a particular DNA molecule always yield the same set of restriction fragments when exposed to the same restriction enzyme. In other words, a restriction enzyme cuts a DNA molecule in a reproducible way. (Later you will learn how the different fragments can be separated.)

The most useful restriction enzymes cleave the sugar-phosphate backbones in both DNA strands in a staggered way, as indicated in Figure 20.3. The resulting double-stranded restriction fragments have at least one single-stranded end, called a **sticky end.** These short extensions can form hydrogen-bonded base pairs with complementary sticky ends on any other DNA molecules cut with the same enzyme. The associations formed in this way are only temporary, but the associations between fragments can be made permanent by the enzyme **DNA ligase.** This enzyme catalyzes the formation of covalent bonds that close up the sugar-phosphate backbones. As you can see at the bottom of Figure 20.3, the ligase-catalyzed joining of DNA from two different sources produces a stable recombinant DNA molecule.

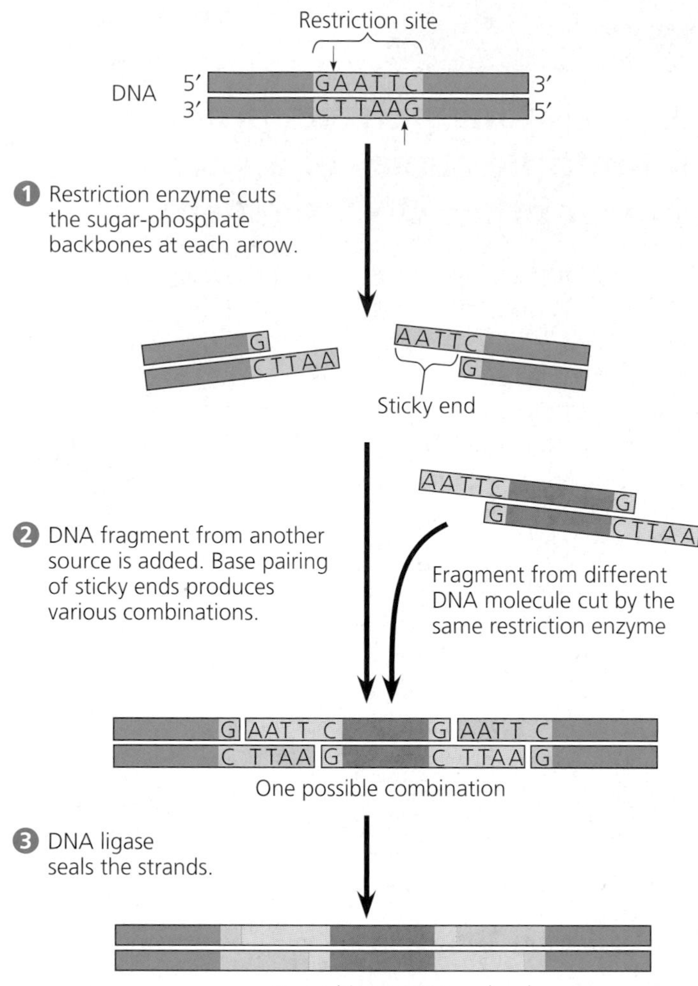

▲ **Figure 20.3 Using a restriction enzyme and DNA ligase to make recombinant DNA.** The restriction enzyme in this example (called *Eco*RI) recognizes a specific six-base-pair sequence, the restriction site, and makes staggered cuts in the sugar-phosphate backbones within this sequence, producing fragments with sticky ends. Any fragments with complementary sticky ends can base-pair; if they come from different DNA molecules, recombinant DNA is the product.

Cloning a Eukaryotic Gene in a Bacterial Plasmid

Now that you've learned about restriction enzymes and DNA ligase, we can take a closer look at how genes are cloned in plasmids. The original plasmid is called a **cloning vector,** defined as a DNA molecule that can carry foreign DNA into a cell and replicate there. Bacterial plasmids are widely used as cloning vectors for several reasons. They can be easily isolated from bacteria, manipulated to form recombinant plasmids by insertion of foreign DNA *in vitro,* and then reintroduced into bacterial cells. Moreover, bacterial cells reproduce rapidly and in the process multiply any foreign DNA they carry.

Producing Clones of Cells

Figure 20.4 details one method for cloning a particular gene from humans or other eukaryotic species using a bacterial plasmid as the cloning vector. The step numbers in the text that follows correspond to those in the figure.

❶ We begin by isolating the bacterial plasmid from *E. coli* cells and DNA containing the gene of interest from human cells grown in laboratory culture. The plasmid has been engineered to carry two genes that will later prove useful: *amp*^{R}, which makes *E. coli* cells resistant to the antibiotic ampicillin, and *lacZ*, which encodes β-galactosidase. This enzyme hydrolyzes the sugar lactose, as well as a synthetic molecular mimic called X-gal. Within the *lacZ* gene is a single copy of the restriction site recognized by the restriction enzyme used in the next step.

❷ Both the plasmid and the human DNA are digested with the same restriction enzyme, one that produces sticky ends. The enzyme cuts the plasmid DNA at its single restriction site within the *lacZ* gene, but cuts the human DNA at multiple sites, generating many thousands of fragments. One of the human DNA fragments carries the gene of interest.

❸ Next we mix the human DNA fragments with the cut plasmids, allowing base pairing between their complementary sticky ends. We then add DNA ligase, which permanently joins each base-paired plasmid and human DNA fragment. Some of the resulting recombinant plasmids contain human DNA fragments like the three shown in Figure 20.4. This step may also generate other products, such as a plasmid containing several human DNA fragments, a combination of two plasmids, or a rejoined, nonrecombinant version of the original plasmid.

❹ The DNA prepared in step 3 is mixed with bacteria that have a mutation in their own *lacZ* gene, making them unable to hydrolyze lactose. Under

Figure 20.4
Research Method **Cloning a Human Gene in a Bacterial Plasmid**

APPLICATION Cloning is used to prepare many copies of a gene of interest for use in sequencing the gene, in producing its encoded protein, in gene therapy, or in basic research.

TECHNIQUE In this example, a human gene is inserted into a plasmid from *E. coli*. The plasmid contains the *amp*^{R} gene, which makes *E. coli* cells resistant to the antibiotic ampicillin. It also contains the *lacZ* gene, which encodes β-galactosidase. This enzyme hydrolyzes a molecular mimic of lactose (X-gal) to form a blue product. Only three plasmids and three human DNA fragments are shown, but millions of copies of the plasmid and a mixture of millions of different human DNA fragments would be present in the samples.

❶ Isolate plasmid DNA from bacterial cells and DNA from human cells containing the gene of interest.

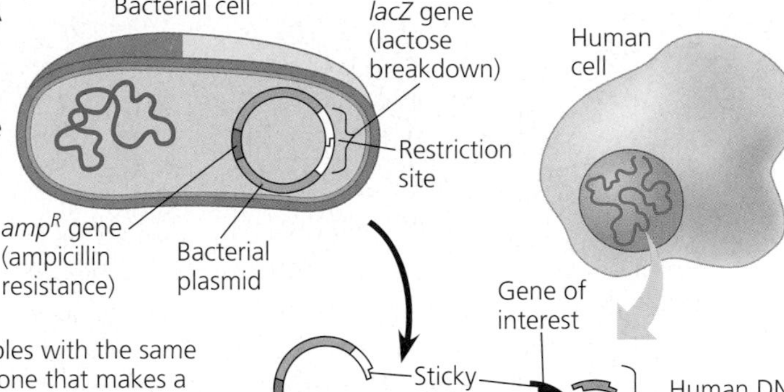

❷ Cut both DNA samples with the same restriction enzyme, one that makes a single cut within the *lacZ* gene and many cuts within the human DNA.

Sticky ends
Human DNA fragments

❸ Mix the cut plasmids and DNA fragments. Some join by base pairing; add DNA ligase to seal them together. The products are recombinant plasmids (shown here) and many nonrecombinant plasmids.

Recombinant DNA plasmids

❹ Introduce the DNA into bacterial cells that have a mutation in their own *lacZ* gene. Under suitable conditions, some cells will take up a recombinant plasmid or other DNA molecule by transformation.

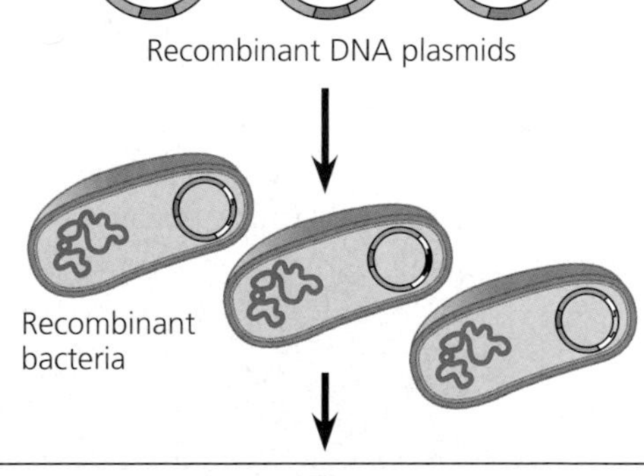

❺ Plate the bacteria on agar containing ampicillin and X-gal. Incubate until colonies grow.

RESULTS Only a cell that took up a plasmid, which has the *ampR* gene, will reproduce and form a colony. Colonies with nonrecombinant plasmids will be blue, because they can hydrolyze X-gal. Colonies with recombinant plasmids, in which *lacZ* is disrupted, will be white, because they cannot hydrolyze X-gal. By screening the white colonies with a nucleic acid probe (see Figure 20.5), researchers can identify clones of bacterial cells carrying the gene of interest.

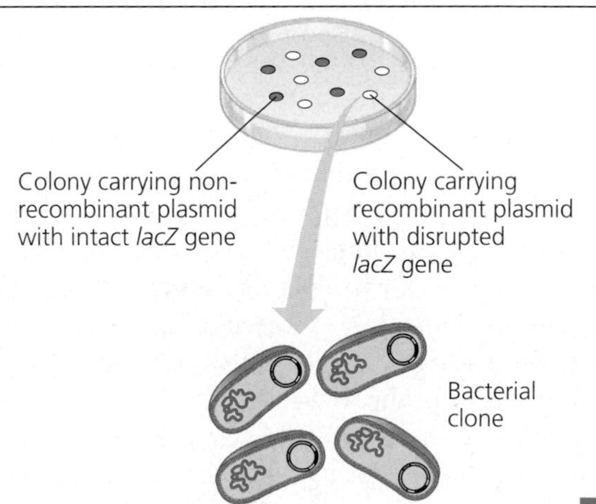

suitable experimental conditions, the cells take up foreign DNA by transformation (see p. 348). Some cells acquire a recombinant plasmid carrying the gene of interest. Many other cells, however, take up a recombinant plasmid carrying a different gene, a nonrecombinant plasmid, or a human DNA fragment. These different possibilities will be sorted out later.

5. In the actual cloning step, the bacteria are plated out on solid nutrient medium (agar) containing ampicillin and X-gal, the molecular mimic of lactose. Use of this medium allows us to identify clones of cells transformed with a recombinant plasmid.

How do we recognize the cell clones carrying recombinant plasmids? First, only cells with a plasmid will reproduce, for only they have the amp^R gene conferring resistance to the ampicillin in the medium. Each reproducing bacterium forms a clone by repeated cell divisions, generating a large group of cells that have all descended from the parent cell. Once the clone contains about 10^5 cells, it is visible as a mass, or *colony*, on the agar. As cells reproduce, any foreign genes carried by recombinant plasmids also are copied, or cloned.

Second, the color of the colonies allows us to distinguish colonies of bacteria with recombinant plasmids from those with nonrecombinant plasmids. Colonies containing nonrecombinant plasmids with the intact *lacZ* gene will be blue because they produce functional β-galactosidase, which hydrolyzes the X-gal in the medium, forming a blue product. In contrast, no functional β-galactosidase is produced in colonies containing recombinant plasmids with foreign DNA inserted into the *lacZ* gene; thus, these colonies will be white.

The procedure to this point will have cloned many different human DNA fragments, not just the one that interests us. The final and most difficult part of cloning a particular gene is identifying a colony containing that gene among the many thousands of colonies carrying other pieces of human DNA.

Identifying Clones Carrying a Gene of Interest

To screen all the colonies with recombinant plasmids (the white colonies in the above method) for a clone of cells containing a gene of interest, we can look either for the gene itself or for its protein product. In the first approach, which we describe here, the DNA of the gene is detected by its ability to base-pair with a complementary sequence on another nucleic acid molecule, a process called **nucleic acid hybridization.** The complementary molecule, a short, single-stranded nucleic acid that can be either RNA or DNA, is called a **nucleic acid probe.** If we know at least part of the nucleotide sequence of the gene of interest (perhaps from knowing the protein it encodes or its sequence in the genome of a closely related species), we can synthesize a probe complementary to it. For example, if part of the sequence on one strand of the desired gene is

5′ ···GGCTAACTTAGC··· 3′

then we would synthesize this probe:

3′ CCGATTGAATCG 5′

Each probe molecule, which will hydrogen-bond specifically to a complementary strand in the desired gene, is labeled with a radioactive isotope or a fluorescent tag so we can track it.

For example, we could transfer a few cells from each white colony in Figure 20.4 (step 5) to a spot on a new agar plate and allow each to grow into a new colony. **Figure 20.5** shows how a number of such bacterial clones can be simultaneously screened for the presence of DNA complementary to a DNA probe. An essential step in this method is the **denaturation** of the cells' DNA—that is, the separation of its two strands. As with protein denaturation, DNA denaturation is routinely accomplished with chemicals or heat.

Once we've identified the location of a colony carrying the desired gene, we can grow some cells from that colony in liquid culture in a large tank and then easily isolate large amounts of the gene. Also, we can use the cloned gene itself as a probe to identify similar or identical genes in DNA from other sources, such as DNA from other species.

Storing Cloned Genes in DNA Libraries

The cloning procedure in Figure 20.4, which starts with a mixture of fragments from the entire genome of an organism, is called a "shotgun" approach; no single gene is targeted for cloning. Thousands of different recombinant plasmids are produced in step 3, and a clone of each ends up as a (white) colony in step 5. The complete set of plasmid clones, each carrying copies of a particular segment from the initial genome, is referred to as a **genomic library (Figure 20.6a)**. Scientists often obtain such libraries (or even particular cloned genes) from another researcher or a commercial source (sometimes referred to as "cloning by phone"!).

Certain bacteriophages are also common cloning vectors for making genomic libraries. Fragments of foreign DNA can be spliced into a phage genome, as into a plasmid, by using a restriction enzyme and DNA ligase. An advantage of using phages as vectors is that a phage can carry a larger DNA insert than a bacterial plasmid. The recombinant phage DNA is packaged into capsids *in vitro* and introduced into a bacterial cell through the normal infection process. Inside the cell, the phage DNA replicates and produces new phage particles, each carrying the foreign DNA. A genomic library made using phage is stored as a collection of phage clones **(Figure 20.6b)**. Since restriction enzymes do not recognize gene boundaries,

Figure 20.5
Research Method **Nucleic Acid Probe Hybridization**

APPLICATION Hybridization with a complementary nucleic acid probe detects a specific DNA within a mixture of DNA molecules. In this example, a collection of bacterial clones (colonies) are screened to identify those carrying a plasmid with a gene of interest.

TECHNIQUE Cells from each colony known to contain recombinant plasmids (white colonies in Figure 20.4) are transferred to separate locations on a new agar plate and allowed to grow into visible colonies. This collection of bacterial colonies is the master plate.

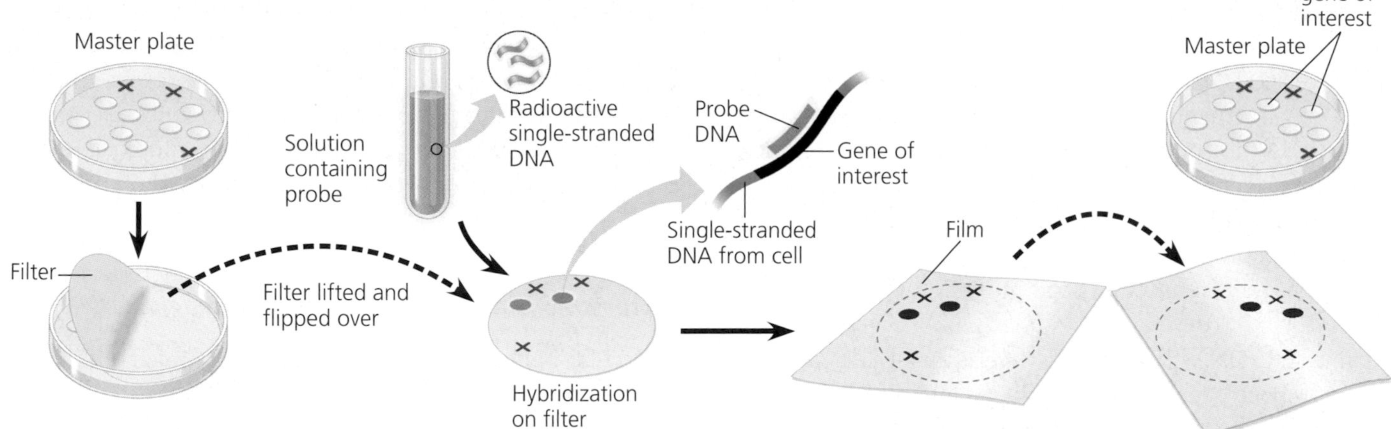

❶ A special filter paper is pressed against the master plate, transferring cells to the bottom side of the filter. Marks (Xs) are placed on the filter paper and agar to establish the position of each individual colony in reference to the marks.

❷ The filter is treated to break open the cells and denature their DNA; the resulting single-stranded DNA molecules are treated so that they stick to the filter. Radioactive probe molecules complementary to part of the gene of interest are incubated with the filter. The single-stranded probe base-pairs with any complementary DNA on the filter; excess DNA is rinsed off. (Colonies with radioactive probe–DNA hybrids are shown here in orange but would not be visible yet.)

❸ The filter is laid under photographic film, allowing any radioactive areas to expose the film (autoradiography). Black spots on the film correspond to the locations on the filter of DNA that has hybridized to the probe.

❹ After the developed film is flipped over, the reference marks on the film and master plate are aligned, so that colonies carrying the gene of interest can be located.

RESULTS Colonies of cells containing the gene of interest have been identified by nucleic acid hybridization. Cells from colonies tagged with the probe can be grown in large tanks of liquid growth medium. Large amounts of the DNA containing the gene of interest can be isolated from these cultures. By using probes with different nucleotide sequences, the collection of bacterial clones can be screened for different genes.

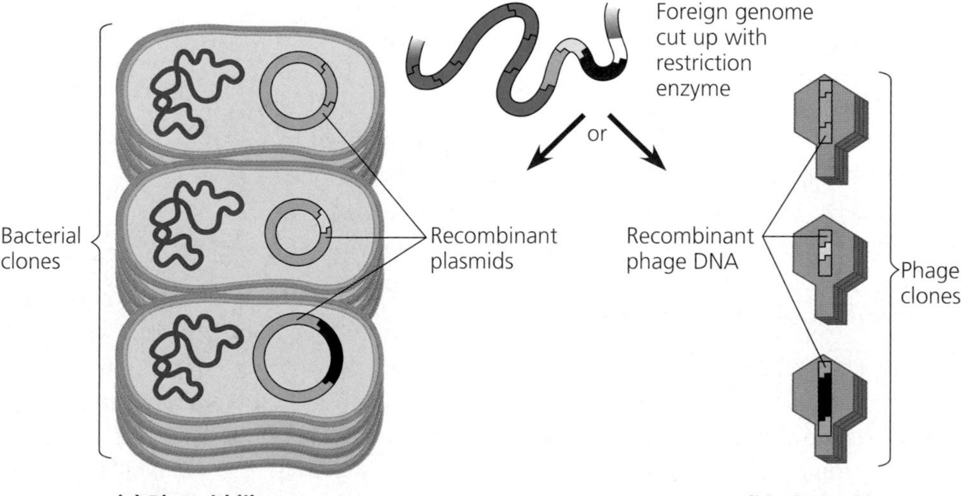

▶ **Figure 20.6 Genomic libraries.** A genomic library is a collection of many bacterial or phage clones, each containing copies of a particular DNA segment from a foreign genome. In a complete genomic library, the foreign DNA segments cover the entire genome of an organism. **(a)** Shown are three of the thousands of "books" in a plasmid library. Each "book" is a clone of bacterial cells, each containing copies of a particular foreign genome fragment (pink, yellow, black segments) in its recombinant plasmid. **(b)** The same three foreign genome segments are shown in three "books" of a phage library.

some genes in either type of genomic library will be cut and divided up among two or more clones.

Researchers can make another kind of DNA library starting with mRNA extracted from cells. The enzyme reverse transcriptase (obtained from retroviruses) is used *in vitro* to make single-stranded DNA transcripts of the mRNA molecules. Following enzymatic degradation of the mRNA, a second DNA strand, complementary to the first, is synthesized by DNA polymerase. This double-stranded DNA, called **complementary DNA (cDNA),** is then modified by addition of restriction enzyme recognition sequences at each end. Finally, the cDNA is inserted into vector DNA in a manner similar to the insertion of genomic DNA fragments. The isolated mRNA was a mixture of all the mRNA molecules in the cells used, transcribed from many different genes. Therefore, the cDNAs that are cloned make up a library containing a collection of genes. However, a **cDNA library** represents only part of the genome—only the subset of genes that were transcribed into mRNA in the original cells.

Genomic and cDNA libraries each have advantages, depending on what is being studied. If you want to clone a gene but are unsure in what cell type it is expressed or are unable to obtain that cell type, a genomic library is almost certain to contain the gene. Also, if you are interested in the regulatory sequences or introns associated with a gene, a genomic library is required because these sequences are absent from the fully processed mRNAs used in making a cDNA library. For this very reason, if you are only interested in the coding sequence of a gene, you can obtain a stripped-down version of the gene from a cDNA library. A cDNA library is also useful for studying genes responsible for the specialized functions of a particular cell type, such as brain or liver cells. Finally, changes in patterns of gene expression during development can be traced by making cDNA from cells of the same type at different times in the life of an organism.

Cloning and Expressing Eukaryotic Genes

As an alternative to screening a DNA library for a particular nucleotide sequence, the clones can sometimes be screened for a desired gene based on detection of its encoded protein. For example, if the protein is an enzyme, its activity can be measured; alternatively, the protein can be detected using antibodies that bind specifically to it. And once a particular gene has been cloned in host cells, its protein product can be produced in larger amounts for research purposes or valuable practical applications.

Bacterial Expression Systems

Getting a cloned eukaryotic gene to function in bacterial host cells can be difficult because certain aspects of gene expression are different in eukaryotes and prokaryotes. To overcome differences in promoters and other DNA control sequences, scientists usually employ an **expression vector,** a cloning vector that contains a highly active prokaryotic promoter just upstream of a restriction site where the eukaryotic gene can be inserted in the correct reading frame. The bacterial host cell will recognize the promoter and proceed to express the foreign gene now linked to that promoter. Such expression vectors allow the synthesis of many eukaryotic proteins in bacterial cells.

Another problem with expressing cloned eukaryotic genes in bacteria is the presence of noncoding regions (introns) in most eukaryotic genes. Introns can make a eukaryotic gene very long and unwieldy, and they prevent correct expression of the gene by bacterial cells, which do not have RNA-splicing machinery. This problem can be surmounted by using a cDNA form of the gene, which includes only the exons. Bacteria can express a eukaryotic cDNA gene if the vector contains a bacterial promoter and any other control elements necessary for the gene's transcription and translation.

Eukaryotic Cloning and Expression Systems

Molecular biologists can avoid eukaryotic-prokaryotic incompatibility by using eukaryotic cells such as yeasts, rather than bacteria, as hosts for cloning and/or expressing eukaryotic genes of interest. Yeast cells, single-celled fungi, offer two advantages: They are as easy to grow as bacteria, and they have plasmids, a rarity among eukaryotes. Scientists have even constructed recombinant plasmids that combine yeast and bacterial DNA and can replicate in either type of cell. Another useful tool for cloning eukaryotic genes are **yeast artificial chromosomes (YACs),** which combine the essentials of a eukaryotic chromosome—an origin for DNA replication, a centromere, and two telomeres—with foreign DNA. These chromosome-like vectors behave normally in mitosis, cloning the foreign DNA as the yeast cell divides. Because a YAC can carry a much longer DNA segment than can a plasmid vector, a cloned fragment is more likely to contain an entire gene rather than just a portion of it.

Another reason to use eukaryotic host cells for expressing a cloned eukaryotic gene is that many eukaryotic proteins will not function unless they are modified after translation, for example by the addition of carbohydrate or lipid groups. Bacterial cells cannot carry out these modifications, and if the gene product requiring such processing is from a mammal, even yeast cells will not be able to modify the protein correctly. The use of host cells from an animal cell culture may therefore be necessary.

Scientists have developed a variety of methods for introducing recombinant DNA into eukaryotic cells. In **electroporation,** a brief electrical pulse applied to a solution containing cells creates temporary holes in their plasma membranes, through which DNA can enter. (This technique is now commonly used for bacteria as well.) Alternatively, scientists can inject DNA directly into single eukaryotic cells using microscopically thin needles. To get DNA into plant cells, the soil bacterium *Agrobacterium* can be used, as you will learn later. If the introduced DNA is incorporated into a cell's genome by genetic recombination, then it may be expressed by the cell.

Amplifying DNA *in Vitro:* The Polymerase Chain Reaction (PCR)

DNA cloning in cells remains the best method for preparing large quantities of a particular gene or other DNA sequence. However, when the source of DNA is scanty or impure, the **polymerase chain reaction,** or **PCR,** is quicker and more selective. In this technique, any specific target segment within one or many DNA molecules can be quickly amplified (copied many times) in a test tube. With automation, PCR can make billions of copies of a target segment of DNA in a few hours, significantly faster than the days it would take to obtain the same number of copies by screening a DNA library for a clone with the desired gene and letting it replicate within host cells.

In the PCR procedure **(Figure 20.7)**, a three-step cycle brings about a chain reaction that produces an exponentially growing population of identical DNA molecules. During each cycle, the reaction mixture is heated to denature (separate) the DNA strands and then cooled to allow annealing (hydrogen bonding) of short, single-stranded DNA primers complementary to sequences on opposite strands at each end of the target sequence; finally, a heat-stable DNA polymerase extends the primers in the $5' \rightarrow 3'$ direction. If a standard DNA polymerase were used, the protein would be denatured along with the DNA during the first heating step and would have to be replaced after each cycle. The key to automating PCR was the discovery of an unusual heat-stable DNA polymerase, first isolated from prokaryotes living in hot springs, that could withstand the heat at the start of each cycle.

Just as impressive as the speed of PCR is its specificity. Only minute amounts of DNA need be present in the starting material, and this DNA can be in a partially degraded state. The key to this high

Figure 20.7
Research Method The Polymerase Chain Reaction (PCR)

APPLICATION With PCR, any specific segment—the target sequence—within a DNA sample can be copied many times (amplified) completely *in vitro.*

TECHNIQUE The starting materials for PCR are double-stranded DNA containing the target nucleotide sequence to be copied, a heat-resistant DNA polymerase, all four nucleotides, and two short, single-stranded DNA molecules that serve as primers. One primer is complementary to one strand at one end of the target sequence; the second is complementary to the other strand at the other end of the sequence.

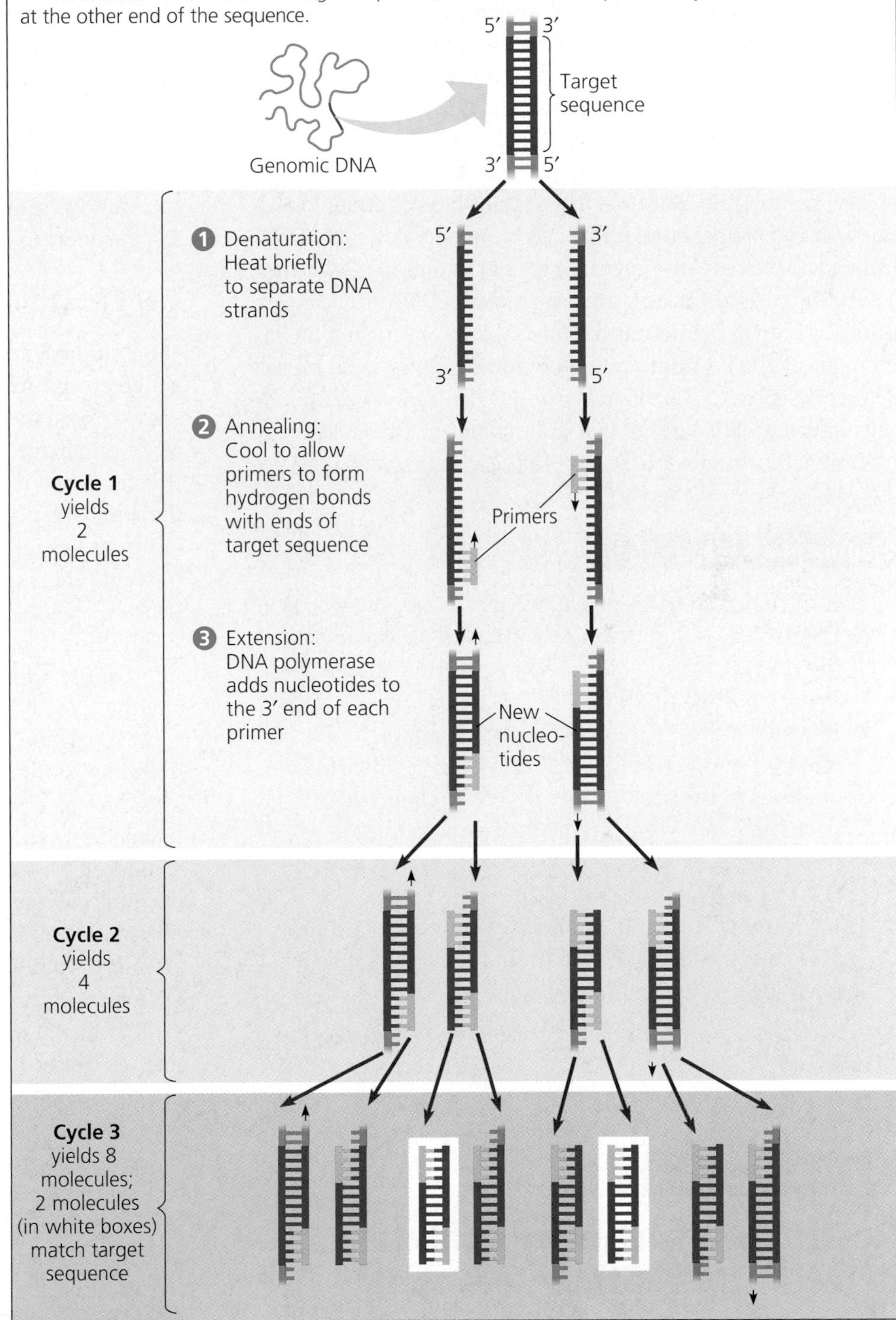

5′
3′
Target sequence
Genomic DNA
3′
5′
Cycle 1 yields 2 molecules
1 Denaturation: Heat briefly to separate DNA strands
5′
3′
3′
5′
2 Annealing: Cool to allow primers to form hydrogen bonds with ends of target sequence
Primers
3 Extension: DNA polymerase adds nucleotides to the 3′ end of each primer
New nucleo-tides
Cycle 2 yields 4 molecules
Cycle 3 yields 8 molecules; 2 molecules (in white boxes) match target sequence

RESULTS During each PCR cycle, the target DNA sequence is doubled. By the end of the third cycle, one-fourth of the molecules correspond exactly to the target sequence, with both strands of the correct length (see white boxes above). After 20 or so cycles, the target sequence molecules outnumber all others by a billionfold or more.

specificity is the primers, which hydrogen-bond *only* to sequences at opposite ends of the target segment. By the end of the third cycle, one-fourth of the molecules are identical to the target segment, with both strands the appropriate length. With each successive cycle, the number of target segment molecules of the correct length doubles, soon greatly outnumbering all other DNA molecules in the reaction.

Despite its speed and specificity, PCR amplification cannot substitute for gene cloning in cells when large amounts of a gene are desired. Occasional errors during PCR replication impose limits on the number of good copies that can be made by this method. Increasingly, however, PCR is used to make enough of a specific DNA fragment to clone it merely by inserting it into a vector.

Devised in 1985, PCR has had a major impact on biological research and biotechnology. PCR has been used to amplify DNA from a wide variety of sources: fragments of ancient DNA from a 40,000-year-old frozen woolly mammoth; DNA from fingerprints or from tiny amounts of blood, tissue, or semen found at crime scenes; DNA from single embryonic cells for rapid prenatal diagnosis of genetic disorders; and DNA of viral genes from cells infected with viruses that are difficult to detect, such as HIV. We'll return to applications of PCR later in the chapter.

Concept Check 20.1

1. If the medium used for plating cells in step 5 of Figure 20.4 did not contain ampicillin, cells containing no plasmid would be able to grow into colonies. What color would those colonies be and why?
2. Imagine you want to study human β-globin, a protein present in red blood cells. To obtain sufficient amounts of the protein, you decide to clone the β-globin gene. Would you construct a genomic library or a cDNA library? What material would you use as a source of DNA or RNA?
3. What are two potential difficulties in using plasmid vectors and bacterial host cells for production of large quantities of human proteins from cloned genes?

For suggested answers, see Appendix A.

Concept 20.2

Restriction fragment analysis detects DNA differences that affect restriction sites

With techniques available for making homogeneous preparations of large numbers of identical DNA segments, we can begin to tackle interesting questions about specific genes and their functions. Does a particular gene differ from person to person, and are there certain alleles associated with a hereditary disorder? Where in the body and when is the gene expressed? Where is the gene located within the genome? Is expression of the gene related to expression of other genes? We can also ask how the gene differs from species to species and begin to unravel its evolutionary history.

To answer such questions, we need to know the complete nucleotide sequence of the gene and its counterparts in various individuals and species, as well as its expression pattern. These topics are covered in the next two sections. In this section, we discuss a more indirect approach, called *restriction fragment analysis,* that detects certain differences in the nucleotide sequences of DNA molecules. This type of analysis can rapidly provide useful comparative information about DNA sequences.

Gel Electrophoresis and Southern Blotting

Many approaches for studying DNA molecules make use of **gel electrophoresis.** This technique uses a gel as a molecular sieve to separate nucleic acids or proteins on the basis of size, electrical charge, and other physical properties **(Figure 20.8)**. Because nucleic acid molecules carry negative charges on their phosphate groups, they all travel toward the positive electrode in an electric field. As they move, the thicket of polymer fibers impedes longer molecules more than it does shorter ones, separating them by length. Thus, gel electrophoresis separates a mixture of linear DNA molecules into bands, each consisting of DNA molecules of the same length.

In restriction fragment analysis, the DNA fragments produced by restriction enzyme digestion of a DNA molecule are sorted by gel electrophoresis. When the mixture of restriction fragments from a particular DNA molecule undergoes electrophoresis, it yields a band pattern characteristic of the starting molecule and the restriction enzyme used. In fact, the relatively small DNA molecules of viruses and plasmids can be identified simply by their restriction fragment patterns. (Larger DNA molecules, such as those of eukaryotic chromosomes, yield so many fragments that they appear as a smear, rather than as distinct bands.) Because DNA can be recovered undamaged from gels, the procedure also provides a way to prepare pure samples of individual fragments.

Restriction fragment analysis is also useful for comparing two different DNA molecules—for example, two alleles of a gene. A restriction enzyme recognizes a specific sequence of nucleotides, and a change in even one base pair will prevent it from cutting at a particular site. Thus, if the nucleotide differences between the alleles occur within a restriction enzyme recognition sequence, digestion with that enzyme will produce a different mixture of fragments from each allele. And each mixture will give its own band pattern in gel electrophoresis. For example, sickle-cell disease is caused by a mutation in a

Figure 20.8
Research Method **Gel Electrophoresis**

APPLICATION Gel electrophoresis is used for separating nucleic acids or proteins that differ in size, electrical charge, or other physical properties. DNA molecules are separated by gel electrophoresis in restriction fragment analysis of both cloned genes (see Figure 20.9) and genomic DNA (see Figure 20.10).

TECHNIQUE Gel electrophoresis separates macromolecules on the basis of their rate of movement through a gel in an electric field. How far a DNA molecule travels while the current is on is inversely proportional to its length. A mixture of DNA molecules, usually fragments produced by restriction enzyme digestion, is separated into "bands"; each band contains thousands of molecules of the same length.

1 Each sample, a mixture of DNA molecules, is placed in a separate well near one end of a thin slab of gel. The gel is supported by glass plates, bathed in an aqueous solution, and has electrodes attached to each end.

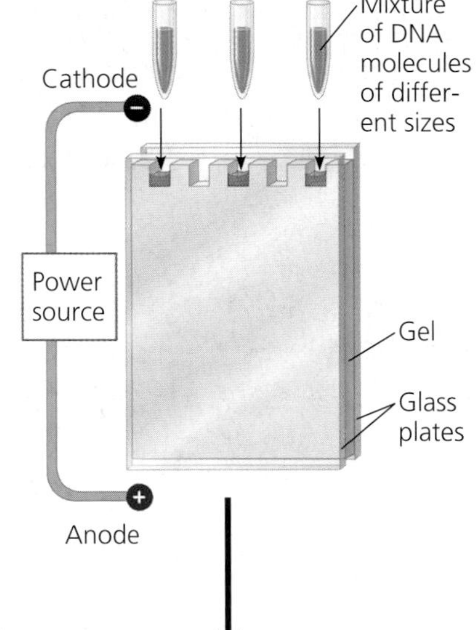

2 When the current is turned on, the negatively charged DNA molecules move toward the positive electrode, with shorter molecules moving faster than longer ones. Bands are shown here in blue, but on an actual gel, DNA bands are not visible until a DNA-binding dye is added. The shortest molecules, having traveled farthest, end up in bands at the bottom of the gel.

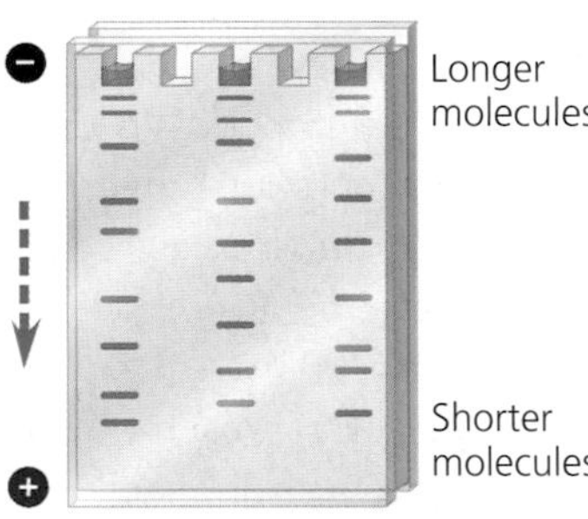

RESULTS After the current is turned off, a DNA-binding dye is added. This dye fluoresces pink in ultraviolet light, revealing the separated bands to which it binds. In this actual gel, the pink bands correspond to DNA fragments of different lengths separated by electrophoresis. If all the samples were initially cut with the same restriction enzyme, then the different band patterns indicate that they came from different sources.

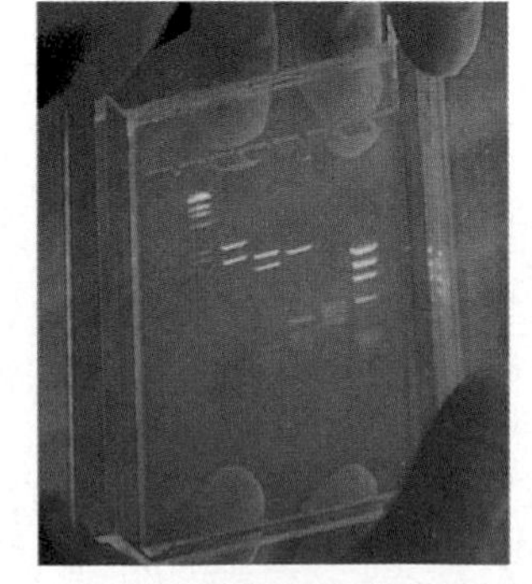

single nucleotide that is located within a restriction sequence in the β-globin gene (see Figure 17.23 and p. 267). As shown in **Figure 20.9**, restriction fragment analysis by electrophoresis can distinguish the normal and sickle-cell alleles of the β-globin gene.

The starting materials in Figure 20.9 are samples of the cloned and purified β-globin alleles. Suppose, however, we wanted to compare genomic DNA samples from three individuals: a person homozygous for the normal β-globin allele; a person with sickle-cell disease, homozygous for the mutant allele; and a heterozygous carrier. Electrophoresis of genomic DNA digested with a restriction enzyme yields too many bands to distinguish them

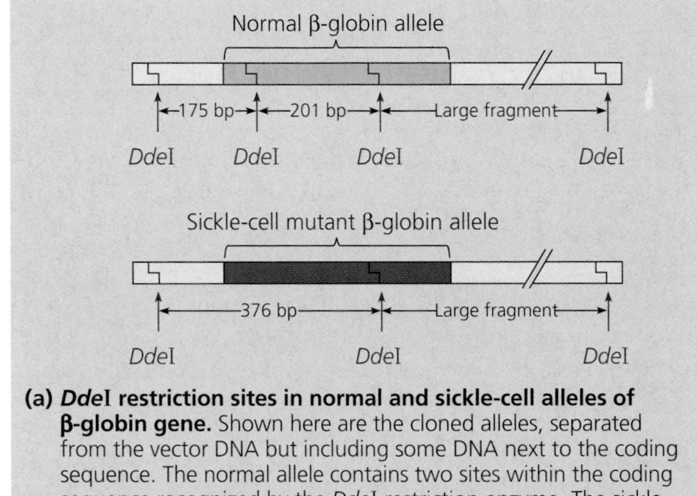

(a) *Dde*I restriction sites in normal and sickle-cell alleles of β-globin gene. Shown here are the cloned alleles, separated from the vector DNA but including some DNA next to the coding sequence. The normal allele contains two sites within the coding sequence recognized by the *Dde*I restriction enzyme. The sickle-cell allele lacks one of these sites.

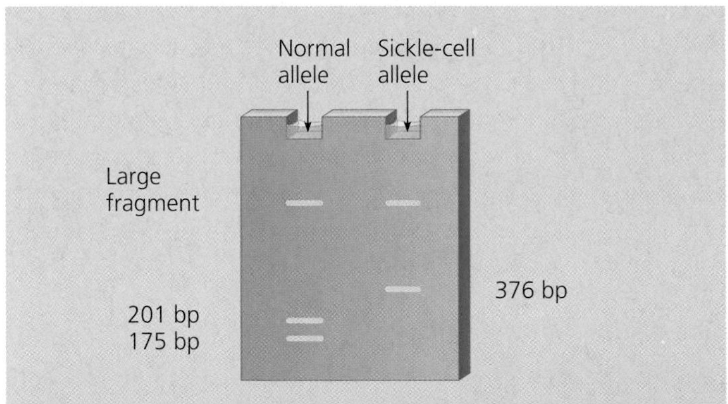

(b) Electrophoresis of restriction fragments from normal and sickle-cell alleles. Samples of each purified allele were cut with the *Dde*I enzyme and then subjected to gel electrophoresis, resulting in three bands for the normal allele and two bands for the sickle-cell allele. (The tiny fragments on the ends of both initial DNA molecules are identical and are not seen here.)

▲ **Figure 20.9 Using restriction fragment analysis to distinguish the normal and sickle-cell alleles of the β-globin gene. (a)** The sickle-cell mutation destroys one of the *Dde*I restriction sites within the β-globin gene. **(b)** As a result, digestion with the *Dde*I enzyme generates different fragments from the normal and sickle-cell alleles.

individually. But with a method called **Southern blotting**, which combines gel electrophoresis and nucleic acid hybridization, we can detect just those bands that include parts of the β-globin gene. The principle is the same as in nucleic acid hybridization for screening bacterial clones (see Figure 20.5). In this case, the probe is a radioactive single-stranded DNA molecule that is complementary to the β-globin gene. **Figure 20.10** outlines the entire procedure and demonstrates how it can be used to compare DNA samples from the three individuals mentioned above. Southern blotting reveals not only whether a particular sequence is present in a sample of DNA but also the size of the restriction fragments that contain the sequence. One of its many applications, as in our β-globin example, is to identify heterozygote carriers of mutant alleles associated with genetic diseases.

Restriction Fragment Length Differences as Genetic Markers

Restriction fragment analysis proved invaluable when biologists turned their attention to *noncoding* DNA, which comprises most of the DNA of animal and plant genomes (see Figure 19.14). When researchers subjected cloned segments of noncoding DNA from different individuals to procedures like that shown in Figure 20.8, they were excited to discover many differences in band patterns. Like different alleles of a gene, noncoding DNA sequences on homologous chromosomes may exhibit small nucleotide differences.

Differences in the restriction sites on homologous chromosomes that result in different restriction fragment patterns are called **restriction fragment length polymorphisms** (**RFLPs,** pronounced "Rif-lips"). RFLPs are scattered abundantly throughout genomes, including the human genome. This type of sequence difference in noncoding DNA is conceptually the same as a difference in coding sequence. Analogous to the single base-pair difference that identifies the sickle-cell allele, a RFLP can serve as a genetic marker for a particular location (locus) in the genome. A given RFLP may occur in numerous variants in a population. (The word *polymorphisms* comes from the Greek for "many forms.")

RFLPs are detected and analyzed by Southern blotting, with the probe complementary to the sequence under consideration. The example shown in Figure 20.10 could as easily represent the detection of a RFLP in noncoding DNA as one in the coding sequences of two alleles. Because of the sensitivity of DNA hybridization, the entire genome can be used as the DNA starting material. (Samples of human DNA are typically obtained from white blood cells.)

Because RFLP markers are inherited in a Mendelian fashion, they can serve as genetic markers for making linkage maps. The geneticist uses the same reasoning illustrated in Figure 15.6: The frequency with which two RFLP markers—or a RFLP marker and a certain allele for a gene—are inherited together is a measure of the closeness of the two loci on a chromosome. The discovery of RFLPs greatly increased the number of markers available for mapping the human genome. No longer were geneticists limited to genetic variations that lead to obvious phenotypic differences (such as genetic diseases) or even to differences in protein products.

1. Suppose you carry out electrophoresis on a sample of genomic DNA isolated from an individual and treated with a restriction enzyme. After staining the gel with a DNA-binding dye, what would you see? Explain.
2. Explain why restriction fragment length polymorphisms (RFLPs) can serve as genetic markers even though they produce no visible phenotypic differences.

For suggested answers, see Appendix A.

Concept 20.3

Entire genomes can be mapped at the DNA level

As early as 1980, molecular biologist David Botstein and colleagues proposed that the DNA variations reflected in RFLPs could serve as the basis for an extremely detailed map of the entire human genome. Since then, researchers have used such markers in conjunction with the tools and techniques of DNA technology to develop more and more detailed maps of the genomes of a number of species.

The most ambitious mapping project to date has been sequencing of the human genome, officially begun as the **Human Genome Project** in 1990. This effort was largely completed in 2003 when the nucleotide sequence of the vast majority of DNA in each human chromosome (the 22 autosomes and the pair of sex chromosomes) was obtained. Organized by an international, publicly funded consortium of researchers at universities and research institutes, the project proceeded through three stages that provided progressively more detailed views of the human genome: genetic (or linkage) mapping, physical mapping, and DNA sequencing. (The interview with Eric Lander on pp. 236–237 gives a personal view of the project.)

In addition to mapping human DNA, researchers with the Human Genome Project are also working on the genomes of other species important in biological research. They have completed sequences for *E. coli* and numerous other prokaryotes, *Saccharomyces cerevisiae* (yeast), *Caenorhabditis elegans* (nematode), *Drosophila melanogaster* (fruit fly), *Mus musculus* (mouse), and quite a few others. These genomes are of great

Figure 20.10
Research Method **Southern Blotting of DNA Fragments**

APPLICATION Researchers can detect specific nucleotide sequences within a DNA sample with this method. In particular, Southern blotting is useful for comparing the restriction fragments produced from different samples of genomic DNA.

TECHNIQUE In this example, we compare genomic DNA samples from three individuals: a homozygote for the normal β-globin allele (I), a homozygote for the mutant sickle-cell allele (II), and a heterozygote (III).

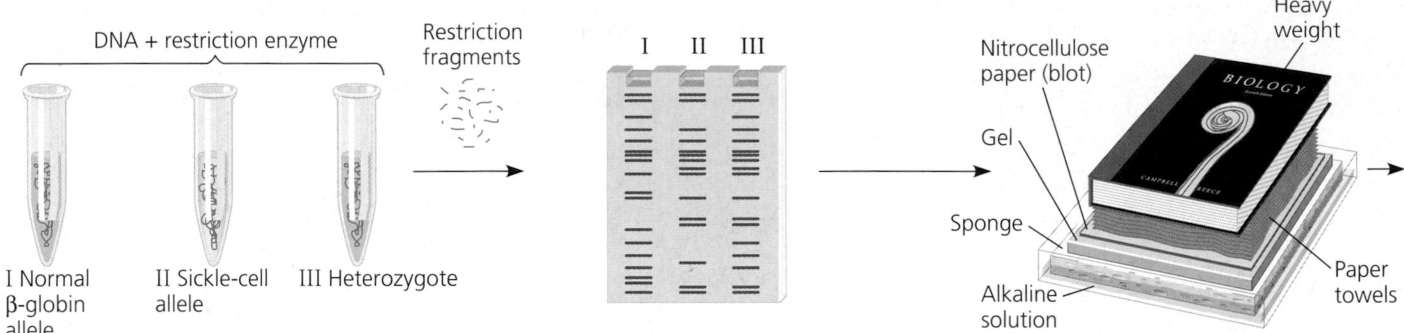

❶ **Preparation of restriction fragments.** Each DNA sample is mixed with the same restriction enzyme, in this case *Dde*I. Digestion of each sample yields a mixture of thousands of restriction fragments.

❷ **Gel electrophoresis.** The restriction fragments in each sample are separated by electrophoresis, forming a characteristic pattern of bands. (In reality, there would be many more bands than shown here, and they would be invisible unless stained.)

❸ **Blotting.** With the gel arranged as shown above, capillary action pulls the alkaline solution upward through the gel, transferring the DNA to a sheet of nitrocellulose paper (the blot) and denaturing it in the process. The single strands of DNA stuck to the paper blot are positioned in bands corresponding to those on the gel.

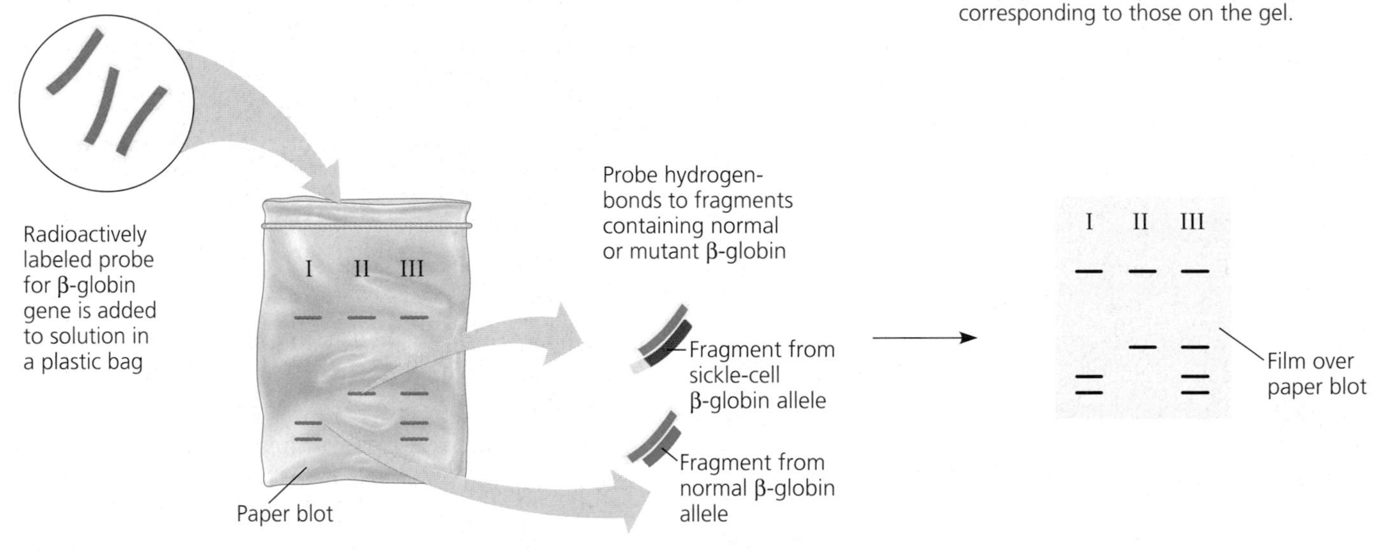

❹ **Hybridization with radioactive probe.** The paper blot is exposed to a solution containing a radioactively labeled probe. In this example, the probe is single-stranded DNA complementary to the β-globin gene. Probe molecules attach by base-pairing to any restriction fragments containing a part of the β-globin gene. (The bands would not be visible yet.)

❺ **Autoradiography.** A sheet of photographic film is laid over the paper blot. The radioactivity in the bound probe exposes the film to form an image corresponding to those bands containing DNA that base-pairs with the probe.

RESULTS Because the band patterns for the three samples are clearly different, this method can be used to identify heterozygous carriers of the sickle-cell allele (III), as well as those with the disease, who have two mutant alleles (II), and unaffected individuals, who have two normal alleles (I). The band patterns for samples I and II resemble those observed for the purified normal and mutant alleles, respectively, seen in Figure 20.9b. The band pattern for the sample from the heterozygote (III) is a combination of the patterns for the two homozygotes (I and II).

interest in their own right and are also providing important insights of general biological significance, as we'll discuss later. In addition, the early mapping efforts on these genomes were useful for developing the strategies, methods, and new technologies necessary for deciphering the human genome, which is much larger.

Genetic (Linkage) Mapping: Relative Ordering of Markers

Even before the Human Genome Project began, earlier research had painted a rough picture of the organization of the genomes of many organisms. For instance, the karyotype of a species reveals the number of chromosomes and their overall banding pattern (see Figure 13.3). And some genes had already been located on a particular region of a whole chromosome by fluorescence *in situ* hybridization (FISH), a method in which fluorescently labeled probes are allowed to hybridize to an immobilized array of whole chromosomes (see Figure 15.1). Cytogenetic maps based on this type of information provided the starting point for more detailed mapping.

With cytogenetic maps of the chromosomes in hand, the initial stage in mapping a large genome is to construct a **linkage map** of several thousand genetic markers spaced throughout each of the chromosomes (**Figure 20.11**, stage ❶). The order of the markers and the relative distances between them on such a map are based on recombination frequencies (see Chapter 15). The markers can be genes or any other identifiable sequences in the DNA, such as RFLPs or the simple sequence DNA discussed in Chapter 19. Relying primarily on simple sequence DNA, which is abundant in the human genome and has various "alleles" differing in length, researchers compiled a human genetic map with some 5,000 markers. Such a map enabled them to locate other markers, including genes, by testing for genetic linkage to the known markers. It was also valuable as a framework for organizing more detailed maps of particular regions.

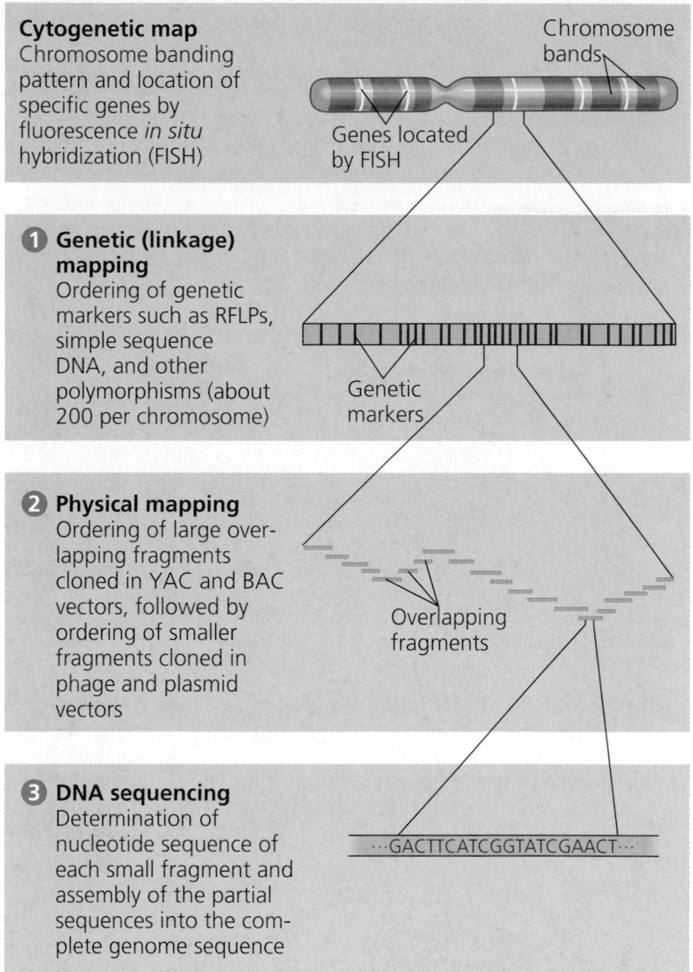

Cytogenetic map
Chromosome banding pattern and location of specific genes by fluorescence in situ hybridization (FISH)
Chromosome bands
Genes located by FISH
❶ Genetic (linkage) mapping
Ordering of genetic markers such as RFLPs, simple sequence DNA, and other polymorphisms (about 200 per chromosome)
Genetic markers
❷ Physical mapping
Ordering of large overlapping fragments cloned in YAC and BAC vectors, followed by ordering of smaller fragments cloned in phage and plasmid vectors
Overlapping fragments
❸ DNA sequencing
Determination of nucleotide sequence of each small fragment and assembly of the partial sequences into the complete genome sequence
···GACTTCATCGGTATCGAACT···

▲ **Figure 20.11 Three-stage approach to mapping an entire genome.** Starting with a cytogenetic map of each chromosome, researchers with the Human Genome Project proceeded through three stages of mapping to reach the ultimate goal, the nearly complete nucleotide sequence of every chromosome.

Physical Mapping: Ordering DNA Fragments

In a **physical map**, the distances between markers are expressed in some physical measure, usually the number of base pairs along the DNA. For whole-genome mapping, a physical map is made by cutting the DNA of each chromosome into a number of restriction fragments and then determining the original order of the fragments in the chromosomal DNA. The key is to make fragments that overlap and then use probes or automated nucleotide sequencing of the ends to find the overlaps (Figure 20.11, stage ❷). In this way, more and more fragments can be assigned to a sequential order that corresponds to their order in a chromosome.

Supplies of the DNA fragments used for physical mapping are prepared by cloning. In working with large genomes, researchers carry out several rounds of DNA cutting, cloning, and physical mapping. The first cloning vector is often a yeast artificial chromosome (YAC), which can carry inserted fragments a million base pairs long, or a **bacterial artificial chromosome (BAC),** an artificial version of a bacterial chromosome that can carry inserts of 100,000 to 500,000 base pairs. After such long fragments are ordered, each fragment is cut into smaller pieces, which are cloned in plasmids or phages, ordered in turn, and finally sequenced.

DNA Sequencing

The ultimate goal in mapping a genome is to determine the complete nucleotide sequence of each chromosome (Figure 20.11, stage ❸). If a pure preparation of many copies of a DNA fragment up to about 800 base pairs in length is available, the sequence of the fragment can be determined by a sequencing machine. The usual sequencing technique, described in **Figure 20.12**, was developed by British scientist Frederick Sanger; it is

Figure 20.12

Research Method Dideoxy Chain-Termination Method for Sequencing DNA

APPLICATION The sequence of nucleotides in any cloned DNA fragment up to about 800 base pairs in length can be determined rapidly with specialized machines that carry out sequencing reactions and separate the labeled reaction products by length.

TECHNIQUE This method synthesizes a nested set of DNA strands complementary to the original DNA fragment. Each strand starts with the same primer and ends with a dideoxyribonucleotide (ddNTP), a modified nucleotide. Incorporation of a ddNTP terminates a growing DNA strand because it lacks a 3′ —OH group, the site for attachment of the next nucleotide (see Figure 16.13). In the set of strands synthesized, each nucleotide position along the original sequence is represented by strands ending at that point with the complementary ddNTP. Because each type of ddNTP is tagged with a distinct fluorescent label, the identity of the ending nucleotides of the new strands, and ultimately the entire original sequence, can be determined.

1 The fragment of DNA to be sequenced is denatured into single strands and incubated in a test tube with the necessary ingredients for DNA synthesis: a primer designed to base pair with the known 3′ end of the template strand, DNA polymerase, the four deoxyribonucleotides, and the four dideoxyribonucleotides, each tagged with a specific fluorescent molecule.

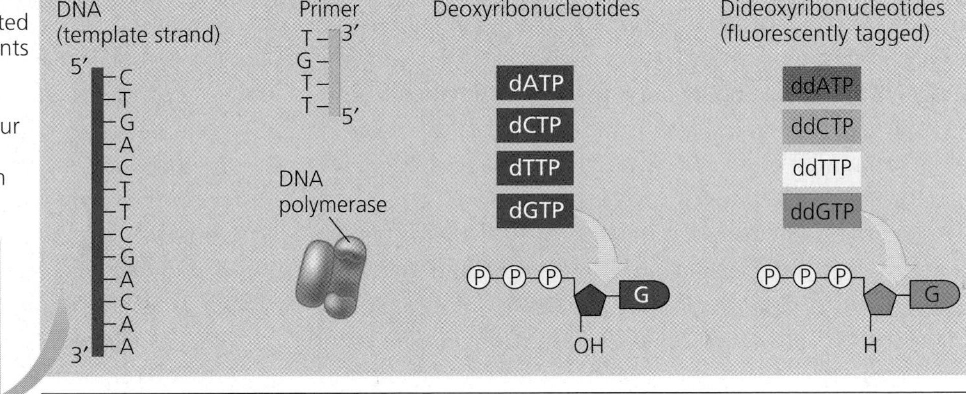

2 Synthesis of each new strand starts at the 3′ end of the primer and continues until a dideoxyribonucleotide is inserted, at random, instead of the normal equivalent deoxyribonucleotide. This prevents further elongation of the strand. Eventually, a set of labeled strands of various lengths is generated, with the color of the tag representing the last nucleotide in the sequence.

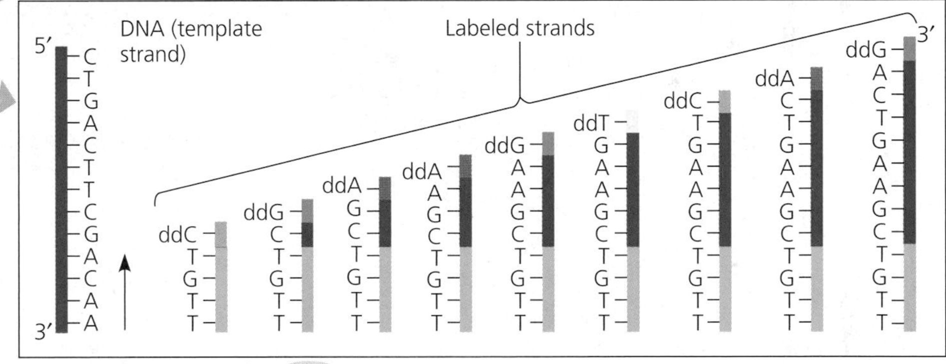

3 The labeled strands in the mixture are separated by passage through a polyacrylamide gel in a capillary tube, with shorter strands moving through faster. A fluorescence detector senses the color of each fluorescent tag as the strands come through. Strands differing by as little as one nucleotide in length can be distinguished.

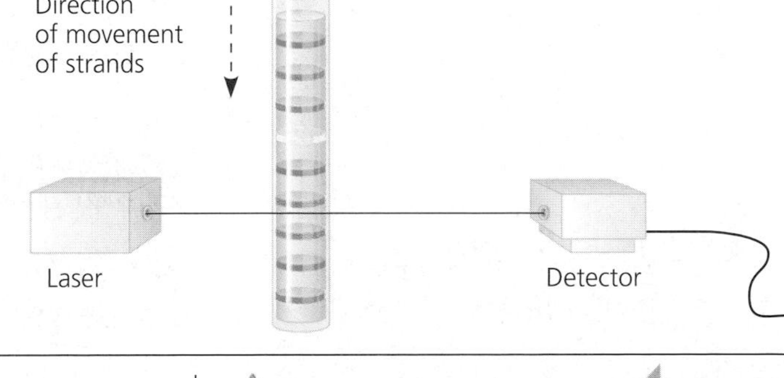

RESULTS The color of the fluorescent tag on each strand indicates the identity of the nucleotide at its end. The results can be printed out as a spectrogram, and the sequence, which is complementary to the template strand, can then be read from bottom to top. (Notice that the sequence here begins after the primer.)

G
A
C
T
G
A
A
G
C

often called the *dideoxyribonucleotide* (or *dideoxy,* for short) *chain-termination method.* Even with automation, the sequencing of all 2.9 billion base pairs in a haploid set of human chromosomes presented a formidable challenge. In fact, as discussed in the interview on pp. 236–237, a major thrust of the Human Genome Project was the development of technology for faster sequencing, along with more sophisticated computer software for analyzing and assembling the partial sequences.

In practice, the three stages shown in Figure 20.11 overlap in a way that our simplified version does not portray, but they accurately represent the overarching strategy employed in the Human Genome Project. In 1992, emboldened by advances in sequencing and computer technology, molecular biologist J. Craig Venter devised an alternative approach to the sequencing of whole genomes. His idea was essentially to skip the genetic mapping and physical mapping stages and start directly with the sequencing of random DNA fragments. Powerful computer programs would then assemble the resulting very large number of overlapping short sequences into a single continuous sequence **(Figure 20.13)**.

Despite the skepticism of many scientists, the value of Venter's approach became clear in 1995 when he and colleagues reported the first complete genome sequence of an organism, the bacterium *Hemophilus influenzae*. In May 1998, he set up a company, Celera Genomics, and promised a completed human sequence in three years. His whole-genome shotgun approach was further validated in March 2000 with completion of the *Drosophila melanogaster* genome sequence. As promised, in February 2001, Celera announced the sequencing of over 90% of the human genome simultaneously with a similar announcement by the public Human Genome Project.

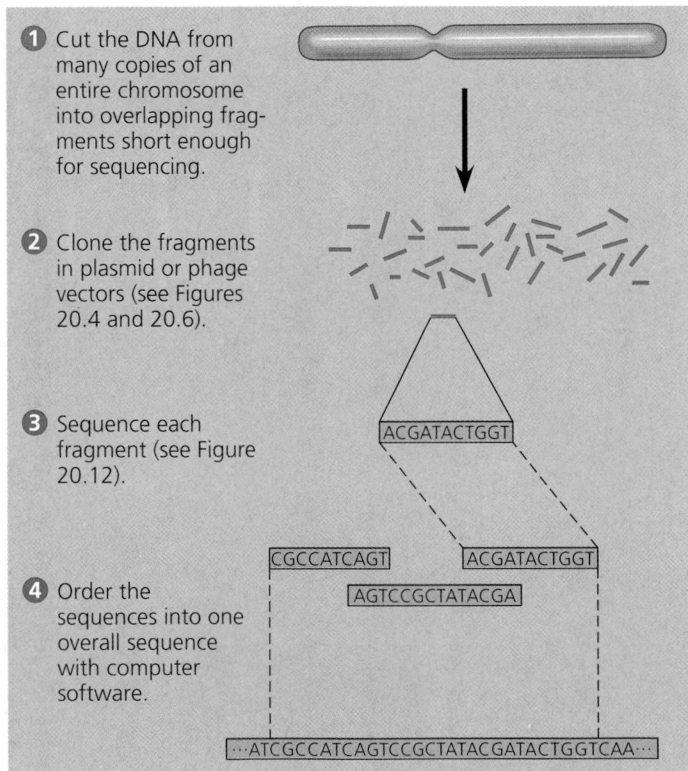

1 Cut the DNA from many copies of an entire chromosome into overlapping fragments short enough for sequencing.
2 Clone the fragments in plasmid or phage vectors (see Figures 20.4 and 20.6).
3 Sequence each fragment (see Figure 20.12).
ACGATACTGGT
4 Order the sequences into one overall sequence with computer software.
CGCCATCAGT
ACGATACTGGT
AGTCCGCTATACGA
···ATCGCCATCAGTCCGCTATACGATACTGGTCAA···

▲ **Figure 20.13 Whole-genome shotgun approach to sequencing.** In this approach, developed by Celera Genomics, random DNA fragments are sequenced and then ordered relative to each other. Compare this approach with the hierarchical, three-stage approach shown in Figure 20.11.

Representatives of the public consortium point out that Celera made much use of the consortium's maps and sequence data, which are immediately made freely available to all researchers, unlike Celera's data. They further assert that the infrastructure established by their approach greatly facilitated Celera's efforts. Venter, on the other hand, argues for the efficiency and economy of Celera's methods, and indeed, the public consortium has made some use of them. Clearly, both approaches are valuable and have contributed to the rapid completion of genome sequencing for quite a few species.

Sequencing of the human genome is now virtually complete, although some gaps remain to be mapped. Because of the presence of repetitive DNA and for other poorly understood reasons, certain parts of the chromosomes of multicellular organisms resist detailed mapping by the usual methods.

On one level, genome sequences of humans and other organisms are simply dry lists of nucleotide bases—millions of A's, T's, C's, and G's in mind-numbing succession. But on another level, analyses of these sequences for various species and comparisons between species are leading to exciting discoveries, which we discuss next.

Concept Check 20.3

1. What is the major difference between a genetic (linkage) map and a physical map of a chromosome?
2. In general, how does the approach to genome mapping used in the Human Genome Project differ from the shotgun approach?

For suggested answers, see Appendix A.

Concept 20.4

Genome sequences provide clues to important biological questions

Now that the sequences of entire genomes are available, scientists can study whole sets of genes and their interactions, an approach called **genomics.** Genomics is yielding new insights into fundamental questions about genome organization, regulation of gene expression, growth and development, and evolution. Using the methods of DNA technology, geneticists can study genes directly, without having to infer genotype from phenotype, as in classical genetics. But the newer approach poses the opposite problem, determining the phenotype from the genotype. Starting with a long DNA sequence, how does one recognize genes and determine their function?

Identifying Protein-Coding Genes in DNA Sequences

DNA sequences are collected in computer data banks available via the Internet to researchers all over the world. To identify as-yet unknown protein-coding genes, scientists use software to scan these stored sequences for transcriptional and translational start and stop signals, for RNA-splicing sites, and for other telltale signs of protein-coding genes. The software also looks for certain short coding sequences similar to those present in known genes. Thousands of such sequences, called *expressed sequence tags,* or *ESTs,* are cataloged in computer databases. This type of analysis identifies sequences that may be "new" protein-coding genes, so-called putative genes or gene candidates.

Although genome size generally increases from prokaryotes to eukaryotes, it does not always correlate with biological complexity among eukaryotes. For instance, the genome of *Fritillaria assyriaca,* a flowering plant, contains 120×10^9 base pairs, about 40 times the size of the human genome. Moreover, the number of genes an organism has is often lower than expected from the size of its genome. In particular, the current estimated number of human genes—about 25,000 or less—is much lower than the 50,000 to 100,000 expected and only about one and a half times the number found in the fruit fly and the nematode worm **(Table 20.1)**. This initially seemed surprising, given the larger diversity of cell types in humans and other vertebrates and their generally greater biological complexity. Relative to the other organisms studied so far, genes account for a much smaller fraction of the human genome. Much of the enormous amount of noncoding DNA in the human genome is repetitive DNA, but unusually long introns also contribute significantly.

So what makes humans and other vertebrates apparently more complex than flies or worms? For one thing, gene expression is regulated in more subtle and complicated ways in vertebrates than in other organisms. Some of the large amount of noncoding DNA in vertebrates may function in these regulatory mechanisms. Also, vertebrate genomes tend to "get more bang for the buck" from their coding sequences because of alternative splicing of RNA transcripts. Recall that this process generates more than one functional protein from a single gene (see Figure 19.8). For instance, nearly all human genes contain multiple exons, and an estimated 75% of these multi-exon genes are alternatively spliced. If we assume that each alternatively spliced human gene on average specifies three different polypeptides, then the total number of different human polypeptides would be about 75,000. Additional polypeptide diversity could result from variations in post-translational cleavage or addition of carbohydrate groups in different cell types or at different developmental stages. Another likely contribution to the biological complexity of vertebrates is the much larger number of possible interactions between gene products that result from greater polypeptide diversity. Later we will examine experimental methods for uncovering these interactions.

The identities of about half of the human genes were known before the Human Genome Project began. But what about the others, the new genes revealed by analysis of DNA sequences? Clues about their identities can come from comparing the sequences of new gene candidates with those of known genes from various organisms. In some cases, a newly identified gene sequence will match, at least partially, the sequence of a gene whose function is well known. For example, part of a new gene may match a known gene that encodes a protein kinase, suggesting that the new gene does, too. In other cases, however, the new gene sequence will be similar to a previously encountered sequence whose function is still unknown. In still other cases, the sequence may be entirely unlike anything ever seen before. In the organisms that have been sequenced so far, many of the gene candidate sequences are entirely new to science. For example, about a third of the genes of *E. coli,* the best studied of research organisms, are new to us.

Table 20.1 Genome Sizes and Estimated Numbers of Genes*

Organism	Haploid Genome Size (Mb)	Number of Genes	Genes per Mb
Hemophilus influenzae (bacterium)	1.8	1,700	940
Escherichia coli (bacterium)	4.6	4,400	950
Saccharomyces cerevisiae (yeast)	12	5,800	480
Caenorhabditis elegans (nematode)	97	19,000	200
Arabidopsis thaliana (plant)	118	25,500	215
Drosophila melanogaster (fruit fly)	180	13,700	76
Oryza sativa (rice)	430	60,000	140
Danio rerio (zebrafish)	1,700	22,000	13
Mus musculus (house mouse)	2,600	25,000	11
Homo sapiens (human)	2,900	25,000	10
Fritillaria assyriaca (plant)	120,000	ND	ND

*Strictly defined, "genome" refers to the *haploid* genome of an organism. Some values given here are likely to be revised as genome analysis continues. Mb = million base pairs. ND = not determined.

Determining Gene Function

So how do scientists determine the function of a new gene identified by genome sequencing and comparative analysis? Perhaps the most common approach is to disable the gene and then observe the consequences in the cell or organism. In one application of this approach, called ***in vitro* mutagenesis**, specific mutations are introduced into the sequence of a cloned gene, after which the mutated gene is returned to a cell. If the introduced mutations alter or destroy the function of the gene product, the phenotype of the mutant cell may help reveal the function of the missing normal protein. Researchers can even put such a mutated gene into cells from the early embryo of a multicellular organism (such as a mouse) to study the role of the gene in the development and functioning of the whole organism.

A simpler and faster method for silencing expression of selected genes exploits the phenomenon of **RNA interference (RNAi)**, described in Chapter 19. This experimental approach uses synthetic double-stranded RNA molecules matching the sequence of a particular gene to trigger breakdown or to block translation of the gene's messenger RNA. To date, the RNAi technique has had some limited success in mammalian cells, including human cells in culture. But in other organisms, such as the nematode and the fruit fly, RNAi is already proving valuable for analyzing the functions of genes on a large scale. In one study, RNAi was used to prevent expression of 86% of the genes in early nematode embryos, one gene at a time. Analysis of the phenotypes of the worms that developed from these embryos allowed the researchers to group most of the genes into a small number of functional groups. This type of genome-wide analysis of gene function is sure to become more common as research focuses on the importance of interactions between genes in the system as a whole—the basis of systems biology (see Chapter 1).

Studying Expression of Interacting Groups of Genes

A major goal of genomics is to learn how genes act together to produce and maintain a functioning organism. As mentioned earlier, part of the explanation for how humans get along with so few genes probably lies in the complexity of networks of interactions among genes and their products. Once the sequences of entire genomes of several organisms neared completion, some researchers began using these sequences to investigate which genes are transcribed in different situations, such as in different tissues or at different stages of development. They also consider whether groups of genes are expressed in a coordinated manner, with the aim of identifying global patterns or networks of expression. The results of such studies will begin to reveal how genes act together as a functional network in an organism.

The basic strategy in global expression studies is to isolate the mRNAs made in particular cells, use these molecules as templates for making the corresponding cDNAs by reverse transcription, and then compare this set of cDNAs with collections of genomic DNA fragments. DNA technology makes such studies possible; with automation, they are easily performed on a large scale. Scientists can now measure the expression of thousands of genes at one time.

Currently, the main approach for genome-wide expression studies uses **DNA microarray assays.** A DNA microarray consists of tiny amounts of a large number of single-stranded DNA fragments representing different genes fixed to a glass slide in a tightly spaced array (grid). (The array is also called a *DNA chip* by analogy to a computer chip.) Ideally, these fragments represent all the genes of an organism, as is already possible for organisms whose genomes have been completely sequenced. **Figure 20.14** outlines how the DNA fragments on a microarray are tested for hybridization with samples of cDNA molecules prepared from the mRNAs in particular cells of interest and labeled with fluorescent dyes.

For example, in one study, researchers performed microarray assays of more than 90% of the genes of *C. elegans* during every stage of its life cycle. The results showed that expression of nearly 60% of the genes changed dramatically during development, and many were expressed in a sex-specific pattern. Such studies illustrate the value of DNA microarrays to reveal general profiles of gene expression over the lifetime of an organism.

In addition to uncovering gene interactions and providing clues to gene function, DNA microarray assays may contribute to a better understanding of certain diseases and suggest new diagnostic techniques or therapies. For example, comparing patterns of gene expression in breast cancer tumors and noncancerous breast tissue has already resulted in more informed and effective treatment protocols. Ultimately, information from DNA microarray assays should provide us a grander view—of how ensembles of genes interact to form a living organism.

Comparing Genomes of Different Species

The genomes of about 150 species had been completely or almost completely sequenced by the spring of 2004, with many more in progress. Of these, the vast majority are genomes of prokaryotes, including about 20 archaean genomes. Among the 20 or so eukaryotic species in the group are vertebrates, invertebrates, and plants. The first eukaryotic genome to be completed was that of the yeast *Saccharomyces cerevisiae*, a single-celled organism; the nematode *Caenorhabditis elegans*, a simple worm, was the first multicellular organism whose genome was sequenced. The plant *Arabidopsis thaliana*, another important research organism, has also been completed. Other species whose genomes have been or are currently being sequenced include the honey bee, the dog, the rat, the chicken, and the frog.

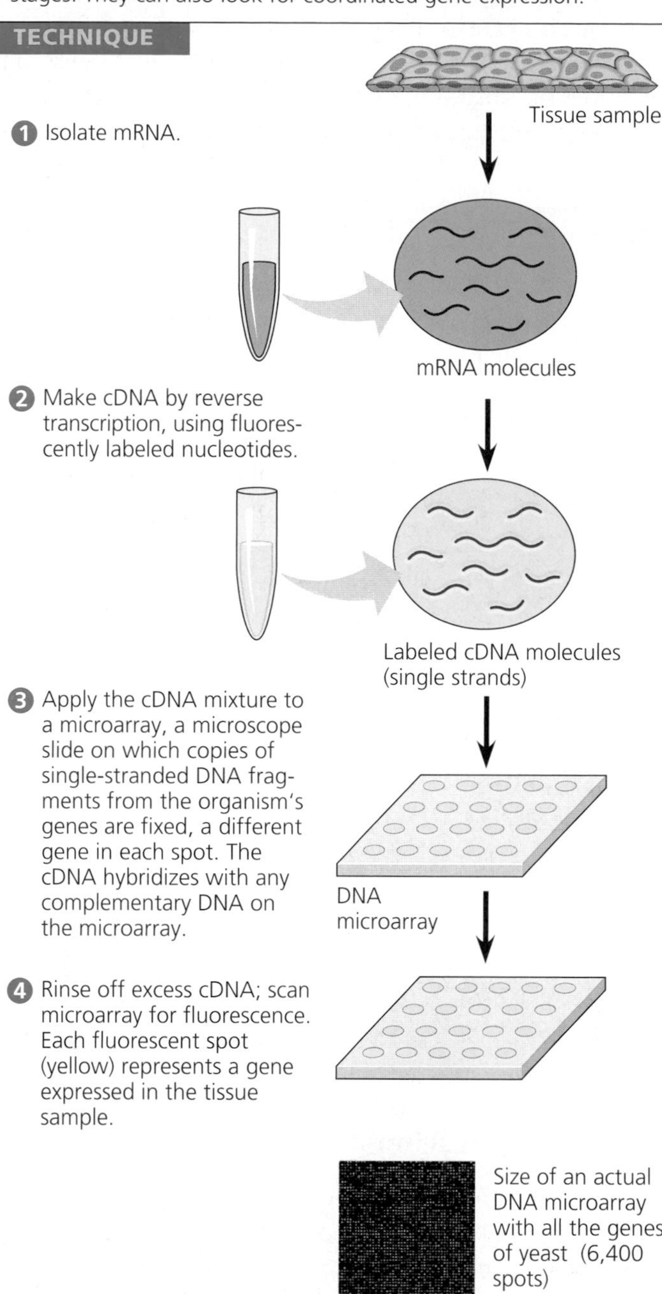

Figure 20.14
Research Method DNA Microarray Assay of Gene Expression Levels
APPLICATION
With this method, researchers can test thousands of genes simultaneously to determine which ones are expressed in a particular tissue, under different environmental conditions in various disease states, or at different developmental stages. They can also look for coordinated gene expression.
TECHNIQUE
Tissue sample
1 Isolate mRNA.
mRNA molecules
2 Make cDNA by reverse transcription, using fluorescently labeled nucleotides.
Labeled cDNA molecules (single strands)
3 Apply the cDNA mixture to a microarray, a microscope slide on which copies of single-stranded DNA fragments from the organism's genes are fixed, a different gene in each spot. The cDNA hybridizes with any complementary DNA on the microarray.
DNA microarray
4 Rinse off excess cDNA; scan microarray for fluorescence. Each fluorescent spot (yellow) represents a gene expressed in the tissue sample.
Size of an actual DNA microarray with all the genes of yeast (6,400 spots)
RESULTS
The intensity of fluorescence at each spot is a measure of the expression of the gene represented by that spot in the tissue sample. Commonly, two different samples are tested together by labeling the cDNAs prepared from each sample with a differently colored fluorescence label. The resulting color at a spot reveals the relative levels of expression of a particular gene in the two samples, which may be from different tissues or the same tissue under different conditions.

Comparisons of genome sequences from different species allow us to determine evolutionary relationships between those species. The more similar in sequence a gene is in two species, the more closely related those species are in their evolutionary history. Likewise, comparing multiple genes between species can shed light on higher groupings of species, which reflect their evolutionary history. Indeed, comparisons of the complete genome sequences of bacteria, archaea, and eukarya strongly support the theory that these are the three fundamental domains of life.

In addition to their value in evolutionary biology, comparative genome studies confirm the relevance of research on simpler organisms to our understanding of biology in general and human biology in particular. The similarities between genes of disparate organisms can be surprising, to the point that one researcher now views fruit flies as "little people with wings." The yeast genome also is proving quite useful in helping us to understand the human genome. For example, the large amount of noncoding DNA in the human genome initially hindered the search for regulatory control elements. But comparisons of noncoding sequences in the human genome with those in the much smaller yeast genome revealed regions with highly conserved sequences; these turned out to be important regulatory sequences in both organisms. In another example, several yeast protein-coding genes are so similar to certain human disease genes that researchers have figured out the functions of the disease genes by studying their normal yeast counterparts.

Comparing the genomes of two closely related species can also be quite useful, because their genomes are likely to be organized similarly. Once the sequence and organization of one genome is known, it can serve as a scaffold for organizing the DNA sequences from a closely related species as they are determined, greatly accelerating mapping of the second genome. For instance, the mouse genome, which is similar in size to the human genome, was mapped at a rapid pace, with the human genome sequence serving as a guide. This approach is particularly helpful when one of two related species has a much shorter genome than the other. An example is the tsetse fly, *Glossina palpalis*, which transmits the parasite causing African sleeping sickness. The tsetse fly genome contains 7×10^9 base pairs (more than twice the size of the human genome), but the genome of a closely related fly is only one-tenth as large. Researchers are sequencing the smaller genome first. Then they will set their sights on the much larger tsetse fly genome, focusing on the coding sequences the two species are expected to share.

The small number of gene differences between closely related species also makes it easier to correlate phenotypic differences between the species with particular genetic differences. For example, one gene that is clearly different in humans and chimpanzees appears to function in speech, a characteristic that obviously distinguishes the two species. And the genetic similarity between mice and humans, which share 80% of their genes, can be exploited in studying certain human genetic diseases. If researchers know or can hypothesize the organ or tissue in which a defective gene causes a particular disease, they can look for genes that are expressed in these locations in experiments with mice. This approach has revealed several human genes of interest, including one that may be involved in Down syndrome.

Other research efforts are under way to extend genomic studies to many more microbial species and to neglected species from diverse branches of the tree of life. These studies will advance our understanding of all aspects of biology, including health, ecology, and evolution.

Future Directions in Genomics

The success in sequencing genomes and studying entire sets of genes is encouraging scientists to attempt similar systematic study of the full protein sets *(proteomes)* encoded by genomes, an approach called **proteomics.** For reasons already mentioned, the number of proteins in humans and our close relatives undoubtedly exceeds the number of genes. Because proteins, not genes, actually carry out the activities of the cell, we must study when and where they are produced in an organism, and also how they interact, if we are to understand the functioning of cells and organisms. Assembling and analyzing proteomes pose many experimental challenges, but ongoing technical advances are providing the necessary tools to meet those challenges.

Genomics and proteomics are enabling biologists to approach the study of life from an increasingly global perspective. Biologists are now in a position to compile catalogs of genes and proteins—a listing of all the "parts" that contribute to the operation of cells, tissues, and organisms. With such catalogs in hand, researchers are shifting their attention from the individual parts to their functional integration in biological systems. A first step in this systems biology approach is defining gene circuits and protein interaction networks (see Figure 1.10). With the use of computer science and mathematics to process and integrate vast amounts of biological data, researchers can detect and quantify the many combinations of interactions.

Another exciting prospect is our increasing understanding of the spectrum of genetic variation in humans. Because the history of the human species is so short, the amount of DNA variation among humans is small compared to that of many other species. Most of our diversity seems to be in the form of **single nucleotide polymorphisms (SNPs,** pronounced "snips"), which are single base-pair variations in the genome, usually detected by sequencing. In the human genome, SNPs occur on average about once in 1,000 base pairs. In other words, if you could compare your personal DNA sequence with that of a person of the same gender—either sitting next to you or on the other side of the world—you would find them to be 99.9% identical.

Scientists are already well on their way to identifying the locations of the several million SNP sites in the human genome. These will be useful genetic markers for studying human evolution, the differences between human populations, and the migratory routes of human populations throughout history. SNPs and other polymorphisms in noncoding (and coding) DNA will also be valuable markers for identifying disease genes and genes that affect our health in more subtle ways. This is likely to change the practice of medicine later in the 21st century. However, applications of DNA research and technology are already affecting our lives in many ways, as we discuss in the final section of the chapter.

Concept Check 20.4

1. Current estimates are that the human genome contains about 25,000 genes, but there is evidence for many more different human polypeptides. What processes might explain this discrepancy?
2. What is the major value of DNA microarray analysis for studying gene expression?
3. Why is the genetic variation among people so much less than it is among individuals of many other species?

For suggested answers, see Appendix A.

Concept 20.5

The practical applications of DNA technology affect our lives in many ways

DNA technology is in the news almost every day. Most often, the topic of the story is a new and promising application in medicine, but this is just one of numerous fields benefiting from DNA technology and genetic engineering.

Medical Applications

One obvious benefit of DNA technology is the identification of human genes whose mutation plays a role in genetic diseases. These discoveries may lead to ways of diagnosing, treating, and even preventing such conditions. DNA technology is also contributing to our understanding of "nongenetic" diseases, from arthritis to AIDS, since a person's genes influence susceptibility to these diseases. Furthermore, diseases of all sorts involve changes in gene expression within the affected cells and often within the patient's immune system. By using DNA microarray assays or other techniques to compare gene expression in healthy and diseased tissues, researchers hope to find many of the genes that are turned on or off in particular diseases. These genes and their products are potential targets for prevention or therapy.

Diagnosis of Diseases

A new chapter in the diagnosis of infectious diseases has been opened by DNA technology, in particular the use of PCR and labeled nucleic acid probes to track down certain pathogens. For example, because the sequence of the HIV genetic material (RNA) is known, PCR can be used to amplify, and thus detect, HIV RNA in blood or tissue samples. RNA cannot be directly amplified by PCR, but the RNA genome is first converted to double-stranded cDNA with reverse transcriptase (RT). PCR is then performed on the cDNA, using a probe specific for one of the HIV genes. This technique, called *RT-PCR,* is often the best way to detect an otherwise elusive infection.

Medical scientists can now diagnose hundreds of human genetic disorders by using PCR and primers corresponding to cloned disease genes, then sequencing the amplified product to look for the disease-causing mutation. Among the genes for human diseases that have been cloned are those for sickle-cell disease, hemophilia, cystic fibrosis, Huntington's disease, and Duchenne muscular dystrophy. Affected individuals with such diseases often can be identified before the onset of symptoms, even before birth. It is also possible to identify symptomless carriers of potentially harmful recessive alleles (see Figure 20.10).

Even when a disease gene has not yet been cloned, the presence of an abnormal allele can be diagnosed with reasonable accuracy if a closely linked RFLP marker has been found **(Figure 20.15)**. Alleles for Huntington's disease and a number of other genetic diseases were first detected in this indirect way. If the marker and the gene itself are close enough, crossing over between the marker and the gene is very unlikely to occur during gamete formation. Therefore, the marker and gene will almost always be inherited together, even though the RFLP marker is not part of the gene. The same principle applies to all kinds of markers, including SNPs.

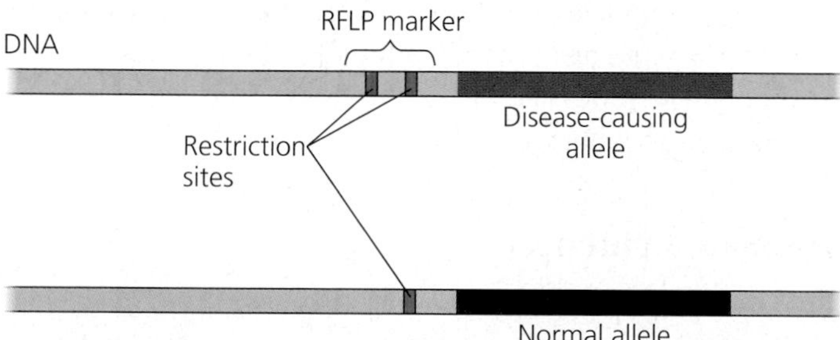
DNA
RFLP marker
Restriction sites
Disease-causing allele
Normal allele

▲ **Figure 20.15 RFLPs as markers for disease-causing alleles.** This diagram depicts homologous segments of DNA from a family in which some members have a genetic disease. In this family, different versions of a RFLP marker are found in unaffected family members and in those who exhibit the disease. If a family member has inherited the version of the RFLP marker with two restriction sites near the gene (rather than one), there is a high probability that the individual has also inherited the disease-causing allele.

Human Gene Therapy

Gene therapy—the alteration of an afflicted individual's genes—holds great potential for treating disorders traceable to a single defective gene. In theory, a normal allele of the defective gene could be inserted into the somatic cells of the tissue affected by the disorder.

For gene therapy of somatic cells to be permanent, the cells that receive the normal allele must be ones that multiply throughout the patient's life. Bone marrow cells, which include the stem cells that give rise to all the cells of the blood and immune system, are prime candidates. **Figure 20.16** outlines one possible procedure for gene therapy of an individual

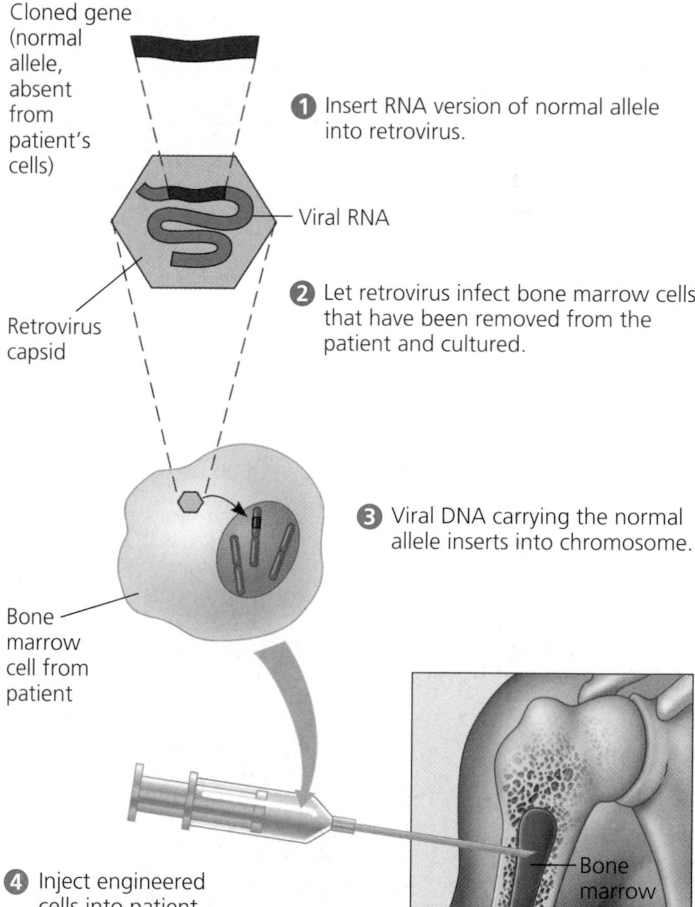
Cloned gene (normal allele, absent from patient's cells)
1 Insert RNA version of normal allele into retrovirus.
Viral RNA
Retrovirus capsid
2 Let retrovirus infect bone marrow cells that have been removed from the patient and cultured.
3 Viral DNA carrying the normal allele inserts into chromosome.
Bone marrow cell from patient
4 Inject engineered cells into patient.
Bone marrow

▲ **Figure 20.16 Gene therapy using a retroviral vector.** A retrovirus that has been rendered harmless is used as a vector in this procedure, which exploits the ability of a retrovirus to insert a DNA transcript of its RNA genome into the chromosomal DNA of its host cell (see Figure 18.10). If the foreign gene carried by the retroviral vector is expressed, the cell and its descendants will possess the gene product, and the patient may be cured. Cells that reproduce throughout life, such as bone marrow cells, are ideal candidates for gene therapy.

whose bone marrow cells do not produce a vital enzyme because of a single defective gene. One type of severe combined immunodeficiency (SCID) is caused by just this kind of defect. If the treatment is successful, the patient's bone marrow cells will begin producing the missing protein, and the patient will be cured.

The procedure shown in Figure 20.16 was used in the first gene therapy trial for SCID, which began in 1990. But the clinical results from this and succeeding studies during the 1990s did not convincingly demonstrate the effectiveness of the treatment. In another trial that started in 2000, ten young children with SCID were treated by the same procedure. Nine of these patients showed significant, definitive improvement after two years, the first indisputable success of gene therapy. However, two of the patients subsequently developed leukemia, a type of blood cell cancer. Researchers discovered that in both cases, the retroviral vector used to carry the normal allele into bone marrow cells had inserted near a gene involved in the proliferation and development of blood cells, causing leukemia. This and other trials of retrovirus-based gene therapy were temporarily suspended in several countries. By learning more about the behavior of retroviruses, researchers may be able to control the insertion of retroviral vectors to a location that might avoid such problems.

Several other technical questions are also posed by gene therapy. For example, how can the activity of the transferred gene be controlled so that cells make appropriate amounts of the gene product at the right time and in the right place? How can we be sure that the insertion of the therapeutic gene does not harm some other necessary cell function? As more is learned about DNA control elements and gene interactions, researchers may be able to answer such questions.

In addition to technical challenges, gene therapy raises difficult ethical questions. Some critics suggest that tampering with human genes in any way will inevitably lead to the practice of eugenics, a deliberate effort to control the genetic makeup of human populations. Other observers see no fundamental difference between the transplantation of genes into somatic cells and the transplantation of organs.

Treatment of human germ-line cells in the hope of correcting a defect in future generations raises further ethical issues. Such genetic engineering is routinely done in laboratory mice, and the technical problems relating to similar genetic engineering in humans will eventually be solved. Under what circumstances, if any, should we alter the genomes of human germ lines or embryos? In a way, this could be thought of as interfering with evolution. From a biological perspective, the elimination of unwanted alleles from the gene pool could backfire. Genetic variation is a necessary ingredient for the survival of a species as environmental conditions change with time. Genes that are damaging under some conditions may be advantageous under other conditions (one example is the sickle-cell allele, discussed in Chapter 14). Are we willing to risk making genetic changes that could be detrimental to the survival of our species in the future? We may have to face this question soon.

Pharmaceutical Products

You learned earlier in the chapter about DNA cloning and expression systems for producing large quantities of proteins that are present naturally in only minute amounts. The host cells used in such expression systems can even be engineered to secrete a protein as it is made, thereby simplifying the task of purifying it by traditional biochemical methods.

Among the first pharmaceutical products "manufactured" in this way were human insulin and human growth hormone (HGH). Some 2 million people with diabetes in the United States depend on insulin treatment to control their disease. Human growth hormone has been a boon to children born with a form of dwarfism caused by inadequate amounts of HGH. Another important pharmaceutical product produced by genetic engineering is tissue plasminogen activator (TPA). If administered shortly after a heart attack, this protein helps dissolve blood clots and reduces the risk of subsequent heart attacks.

The most recent developments in pharmaceutical products involve truly novel ways to fight certain diseases that do not respond to traditional drug treatments. One approach is the use of genetically engineered proteins that either block or mimic surface receptors on cell membranes. One such experimental drug mimics a receptor protein that HIV binds to while entering white blood cells. The HIV binds to the drug molecules instead and fails to enter the blood cells.

DNA technology also can be used to produce vaccines, which stimulate the immune system to defend against specific pathogens (see Chapter 43). Traditional vaccines are of two types: inactivated (killed) microbes and viable but weakened (attenuated) microbes that generally do not cause illness. Most pathogens have one or more specific proteins on their surface that trigger an immune response against it. This type of protein, produced by recombinant DNA techniques, can be used as a vaccine against the pathogen. Alternatively, genetic engineering methods can be used to modify the genome of the pathogen to attenuate it.

Forensic Evidence

In violent crimes, body fluids or small pieces of tissue may be left at the scene or on the clothes or other possessions of the victim or assailant. If enough blood, semen, or tissue is available, forensic laboratories can determine the blood type or tissue type by using antibodies to detect specific cell-surface proteins. However, such tests require fairly fresh samples in relatively large amounts. Also, because many people have the same blood or tissue type, this approach can only exclude a suspect; it cannot provide strong evidence of guilt.

DNA testing, on the other hand, can identify the guilty individual with a high degree of certainty, because the DNA sequence of every person is unique (except for identical twins). RFLP analysis by Southern blotting is a powerful method for detecting similarities and differences in DNA samples and requires only tiny amounts of blood or other tissue (about 1,000 cells). In a murder case, for example, this method can be used to compare DNA samples from the suspect, the victim, and a small amount of blood found at the crime scene. The forensic scientist usually tests for about five RFLP markers; in other words, only a few selected portions of the DNA are tested. However, even such a small set of markers from an individual can provide a **DNA fingerprint**, or specific pattern of bands, that is of forensic use, because the probability that two people (who are not identical twins) would have the exact same set of RFLP markers is very small. The autoradiograph in **Figure 20.17** resembles the type of evidence presented to juries in murder trials.

DNA fingerprinting can also be used to establish paternity. A comparison of the DNA of a mother, her child, and the purported father can conclusively settle a question of paternity. Sometimes paternity is of historical interest: DNA fingerprinting has provided strong evidence that Thomas Jefferson or one of his close male relatives fathered at least one of the children of his slave Sally Hemings.

Today, instead of RFLPs, variations in the lengths of certain repeated base sequences in simple sequence DNA within the genome are increasingly used as markers for DNA fingerprinting. These repetitive DNA sequences are highly variable from person to person, providing even more markers than RFLPs. For example, one individual may have the repeat unit ACA repeated 65 times at one genome locus, 118 times at a second locus, and so on, whereas another individual is likely to have different numbers of repeats at these loci. Such polymorphic genetic loci are sometimes called *simple tandem repeats (STRs)*. The greater the number of markers examined in a DNA sample, the more likely it is that the DNA fingerprint is unique to one individual. PCR is often used to amplify particular STRs or other markers before electrophoresis. PCR is especially valuable when the DNA is in poor condition or available only in minute quantities. A tissue sample as small as 20 cells can be sufficient for PCR amplification.

Just how reliable is DNA fingerprinting? In most forensic cases, the probability of two people having identical DNA fingerprints is between one chance in 100,000 and one in a billion. The exact figure depends on the number of markers compared and on the frequency of those markers in the general population. Information on how common various markers are in different ethnic groups is critical because these marker frequencies may vary considerably among ethnic groups and between a particular ethnic group and the population as a whole. With the increasing availability of frequency data, forensic scientists can make extremely accurate statistical calculations. Thus, despite problems that can still arise from insufficient data, human error, or flawed evidence, DNA fingerprints are now accepted as compelling evidence by legal experts and scientists alike. In fact, DNA analysis on stored forensic samples has provided the evidence needed to solve many "cold cases" in recent years.

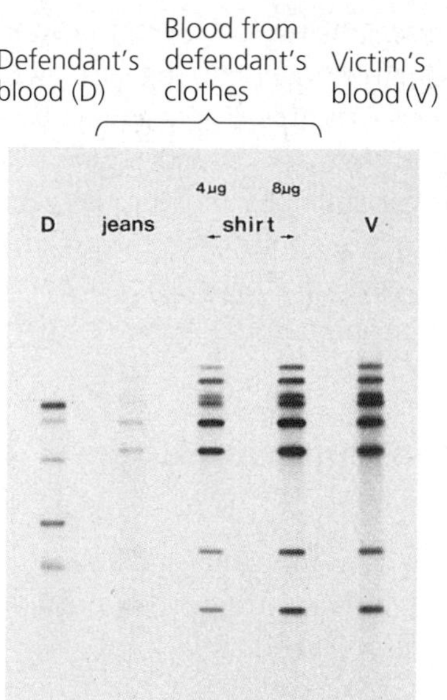

▲ **Figure 20.17 DNA fingerprints from a murder case.** This autoradiograph shows that DNA in blood from the defendant's clothes matches the DNA fingerprint of the victim but differs from the DNA fingerprint of the defendant. This is evidence that the blood on the defendant's clothes came from the victim, not the defendant. The three DNA samples were subjected to Southern blotting using radioactive probes (see Figure 20.10). The DNA bands resulting from electrophoresis were exposed to probes for several different RFLP markers in succession, with the previous probe washed off before the next one was applied.

Environmental Cleanup

Increasingly, the remarkable ability of certain microorganisms to transform chemicals is being exploited for environmental cleanup. Scientists are now engineering these metabolic capabilities into other microorganisms, which are then used to help treat some environmental problems. For example, many bacteria can extract heavy metals, such as copper, lead, and nickel, from their environments and incorporate the metals into compounds such as copper sulfate or lead sulfate, which are readily recoverable. Genetically engineered microbes may become important in both mining minerals (especially as ore reserves are depleted) and cleaning up highly toxic mining wastes. Biotechnologists are also trying to engineer microbes that can degrade chlorinated hydrocarbons and other harmful compounds. These microbes could be used in wastewater treatment plants or by manufacturers before the compounds are ever released into the environment.

A related research area is the identification and engineering of microbes capable of detoxifying specific toxic wastes found in spills and waste dumps. For example, bacterial strains have been developed that can degrade some of the chemicals released during oil spills. By moving the genes responsible for these transformations into different organisms, bioengineers may be able to develop strains that can survive the harsh conditions of environmental disasters and help detoxify the wastes.

Agricultural Applications

Scientists are working to learn more about the genomes of agriculturally important plants and animals, and for a number of years they have been using DNA technology in an effort to improve agricultural productivity.

Animal Husbandry and "Pharm" Animals

DNA technology is now routinely used to make vaccines and growth hormones for treating farm animals. On a still largely experimental basis, scientists can also introduce a gene from one animal into the genome of another animal, which is then called a **transgenic** animal. To do this, they first remove egg cells from a female and fertilize them *in vitro*. Meanwhile, they have cloned the desired gene from another organism. They then inject the cloned DNA directly into the nuclei of the fertilized eggs. Some of the cells integrate the foreign DNA, the *transgene,* into their genomes and are able to express the foreign gene. The engineered embryos are then surgically implanted in a surrogate mother. If an embryo develops successfully, the result is a transgenic animal, containing a gene from a third "parent" that may even be of another species.

The goals of creating a transgenic animal are often the same as the goals of traditional breeding—for instance, to make a sheep with better quality wool, a pig with leaner meat, or a cow that will mature in a shorter time. Scientists might, for example, identify and clone a gene that causes the development of larger muscles (muscles make up most of the meat we eat) in one variety of cattle and transfer it to other cattle or even to sheep.

Transgenic animals also have been engineered to be pharmaceutical "factories"—producers of a large amount of an otherwise rare biological substance for medical use. For example, a transgene for a desired human protein, such as a hormone or blood-clotting factor, can be inserted into the genome of a farm mammal in such a way that the transgene's product is secreted in the animal's milk **(Figure 20.18)**. The protein can then be purified from the milk, usually more easily than from a cell culture. Recently, researchers have engineered transgenic chickens that express large amounts of the transgene's product in eggs. Their success suggests that transgenic chickens may emerge as relatively inexpensive pharmaceutical factories in the near future.

▲ **Figure 20.18 "Pharm" animals.** These transgenic sheep carry a gene for a human blood protein, which they secrete in their milk. This protein inhibits an enzyme that contributes to lung damage in patients with cystic fibrosis and some other chronic respiratory diseases. Easily purified from the sheep's milk, the protein is currently under evaluation as a treatment for cystic fibrosis.

Human proteins produced by farm animals may differ in some ways from the corresponding natural human proteins. Thus, these proteins must be tested very carefully to ensure that they will not cause allergic reactions or other adverse effects in patients receiving them. Also, the health and welfare of farm animals carrying genes from humans and other foreign species are important issues; problems such as low fertility or increased susceptibility to disease are not uncommon.

Genetic Engineering in Plants

Agricultural scientists have already endowed a number of crop plants with genes for desirable traits, such as delayed ripening and resistance to spoilage and disease. In one striking way, plants are easier to genetically engineer than most animals. For many plant species, a single tissue cell grown in culture can give rise to an adult plant (see Figure 21.5). Thus, genetic manipulations can be performed on a single cell and the cell then used to generate an organism with new traits.

The most commonly used vector for introducing new genes into plant cells is a plasmid, called the **Ti plasmid**, from the soil bacterium *Agrobacterium tumefaciens.* This plasmid integrates a segment of its DNA, known as T DNA, into the chromosomal DNA of its host plant cells. For vector purposes, researchers work with a version of the plasmid that does not cause disease, as the wild-type version does. **Figure 20.19** outlines one method for using the Ti plasmid to produce transgenic plants. Scientists can introduce recombinant Ti plasmids into plant cells by electroporation. Alternatively, the recombinant plasmid can be put back into *Agrobacterium;* susceptible plants or plant cells growing in culture are then infected with bacteria that contain the recombinant plasmid.

Figure 20.19

Research Method **Using the Ti Plasmid to Produce Transgenic Plants**

APPLICATION Genes conferring useful traits, such as pest resistance, herbicide resistance, delayed ripening, and increased nutritional value, can be transferred from one plant variety or species to another using the Ti plasmid as a vector.

TECHNIQUE

1. The Ti plasmid is isolated from the bacterium *Agrobacterium tumefaciens*. The segment of the plasmid that integrates into the genome of host cells is called T DNA.

2. Isolated plasmids and foreign DNA containing a gene of interest are incubated with a restriction enzyme that cuts in the middle of T DNA. After base pairing occurs between the sticky ends of the plasmids and foreign DNA fragments, DNA ligase is added. Some of the resulting stable recombinant plasmids contain the gene of interest.

3. Recombinant plasmids can be introduced into cultured plant cells by electroporation. Or plasmids can be returned to *Agrobacterium*, which is then applied as a liquid suspension to the leaves of susceptible plants, infecting them. Once a plasmid is taken into a plant cell, its T DNA integrates into the cell's chromosomal DNA.

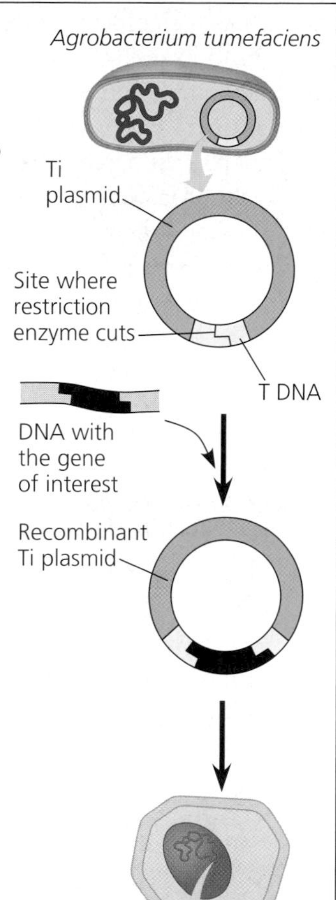

RESULTS Transformed cells carrying the transgene of interest can regenerate complete plants that exhibit the new trait conferred by the transgene.

Genetic engineering is rapidly replacing traditional plant-breeding programs, especially for useful traits, such as herbicide or pest resistance, determined by one or a few genes. For example, crops engineered with a bacterial gene making the plants resistant to herbicides can grow while weeds are destroyed. Similarly, genetically engineered crops that can resist destructive microbes and insects have reduced the need for chemical insecticides.

Genetic engineering also has great potential for improving the nutritional value of crop plants. For instance, scientists have developed transgenic rice plants that produce yellow rice grains containing beta-carotene, which our body uses to make vitamin A (see Figure 38.16). This "golden" rice could help prevent vitamin A deficiency in the half of the world's population that depends on rice as a staple food. Currently, large numbers of young children in Southeast Asia suffer from vitamin A deficiency, which leads to vision impairment and increases susceptibility to disease.

In a novel twist, the pharmaceutical industry is beginning to develop "pharm" plants, analogous to "pharm" animals. Although natural plants have long been sources of drugs, researchers are now creating plants that make human proteins for medical use and viral proteins for use as vaccines. Several such products are being tested in clinical trials, including vaccines for hepatitis B and an antibody produced in transgenic tobacco plants that interferes with the bacteria that cause tooth decay. Large amounts of these proteins might be produced more economically by plants than by cultured cells.

Safety and Ethical Questions Raised by DNA Technology

Early concerns about potential dangers associated with recombinant DNA technology focused on the possibility that hazardous new pathogens might be created. What might happen, for instance, if cancer cell genes were transferred into bacteria or viruses? To guard against such rogue microbes, scientists developed a set of guidelines that were adopted as formal government regulations in the United States and some other countries. One safety measure is a set of strict laboratory procedures designed to protect researchers from infection by engineered microbes and to prevent the microbes from accidentally leaving the laboratory. In addition, strains of microorganisms to be used in recombinant DNA experiments are genetically crippled to ensure that they cannot survive outside the laboratory. Finally, certain obviously dangerous experiments have been banned.

Today, most public concern about possible hazards centers not on recombinant microbes but on **genetically modified (GM) organisms** used as food. In common language, a GM organism is one that has acquired by artificial means one or more genes from the same or another species. Salmon, for example, have been genetically modified by addition of a more active salmon growth hormone gene. However, the majority of the GM organisms that contribute to our food supply are not animals, but crop plants.

Some countries have been wary of the GM revolution, with the safety of GM foods and possible environmental consequences of growing GM plants being the major concerns. In

1999, for instance, the European Union suspended the introduction of new GM crops pending new legislation. Early in 2000, negotiators from 130 countries (including the United States) agreed on a Biosafety Protocol that requires exporters to identify GM organisms present in bulk food shipments and allows importing countries to decide whether the products pose environmental or health risks.

Advocates of a cautious approach toward GM crops fear that transgenic plants might pass their new genes to close relatives in nearby wild areas. We know that lawn and crop grasses, for example, commonly exchange genes with wild relatives via pollen transfer. If crop plants carrying genes for resistance to herbicides, diseases, or insect pests pollinated wild ones, the offspring might become "super weeds" that are very difficult to control. Another possible hazard, suggested by one laboratory-based study, is that a transgene encoding a pesticide-type protein might cause plants to produce pollen toxic to butterflies. However, scientists with the Agricultural Research Service concluded from a two-year study that butterflies were unlikely to be exposed to toxic levels of pollen.

As for the risks to human health from GM foods, some people fear that the protein products of transgenes might lead to allergic reactions. Although there is some evidence that this could happen, advocates claim that these proteins could be tested for their ability to cause allergic reactions.

Today, governments and regulatory agencies throughout the world are grappling with how to facilitate the use of biotechnology in agriculture, industry, and medicine while ensuring that new products and procedures are safe. In the United States, such applications of biotechnology are evaluated for potential risks by various regulatory agencies, including the Food and Drug Administration, Environmental Protection Agency, National Institutes of Health, and Department of Agriculture. These agencies are under increasing pressure from some consumer groups. Meanwhile, these same agencies and the public must consider the ethical implications of biotechnology.

Completion of the mapping of the human genome, for instance, raises significant ethical questions. Who should have the right to examine someone else's genes? How should that information be used? Should a person's genome be a factor in suitability for a job or eligibility for insurance? Ethical considerations, as well as concerns about potential environmental and health hazards, will likely slow some applications of biotechnology. There is always a danger that too much regulation will stifle basic research and its potential benefits. However, the power of DNA technology and genetic engineering—our ability to profoundly and rapidly alter species that have been evolving for millennia—demands that we proceed with humility and caution.

Concept Check 20.5

1. What is the advantage of using stem cells for gene therapy?
2. List at least three different properties that have been acquired by crop plants via genetic engineering.

For suggested answers, see Appendix A.

Chapter 20 Review

Go to www.campbellbiology.com or the student CD-ROM to explore Activities, Investigations, and other interactive study aids.

SUMMARY OF KEY CONCEPTS

Concept 20.1

DNA cloning permits production of multiple copies of a specific gene or other DNA segment

- **DNA Cloning and Its Applications: *A Preview* (pp. 385–386)** DNA cloning and other techniques, collectively termed DNA technology, can be used to manipulate and analyze DNA and to produce useful new products and organisms.
  Activity *Applications of DNA Technology*

- **Using Restriction Enzymes to Make Recombinant DNA (p. 386)** Bacterial restriction enzymes cut DNA molecules within short, specific nucleotide sequences to yield a set of double-stranded DNA fragments with single-stranded sticky ends. The sticky ends on fragments from one DNA source can base-pair with complementary sticky ends on fragments from other DNA molecules; sealing of the base-paired fragments with DNA ligase produces recombinant DNA molecules.
  Activity *Restriction Enzymes*

- **Cloning a Eukaryotic Gene in a Bacterial Plasmid (pp. 386–388)** A recombinant plasmid is made by inserting restriction fragments from DNA containing a gene of interest into a plasmid vector that has been cut open by the same enzyme. Gene cloning results when the recombinant plasmid is introduced into a host bacterial cell and the foreign genes are replicated along with the bacterial chromosome as the host cell reproduces. A clone of cells carrying the gene of interest can be identified with a radioactively labeled nucleic acid probe that has a sequence complementary to the gene.
  Activity *Cloning a Gene in Bacteria*
  Investigation *How Can Antibiotic-Resistant Plasmids Transform* E. coli*?*

- **Storing Cloned Genes in DNA Libraries (pp. 388–390)** A genomic library is the collection of recombinant vector clones produced by cloning DNA fragments derived from an entire genome. A cDNA (complementary DNA) library is made by cloning DNA made *in vitro* by reverse transcription of all the mRNA produced by a particular kind of cell.

▶ **Cloning and Expressing Eukaryotic Genes** **(pp. 390–391)** Several technical difficulties hinder the expression of cloned eukaryotic genes in bacterial host cells. The use of cultured eukaryotic cells as host cells and yeast artificial chromosomes (YACs) as vectors helps avoid these problems.

▶ **Amplifying DNA *in Vitro:* The Polymerase Chain Reaction (PCR)** **(pp. 391–392)** PCR can produce many copies of a specific target segment of DNA, using primers that bracket the desired sequence and a heat-resistant DNA polymerase.

Concept 20.2

Restriction fragment analysis detects DNA differences that affect restriction sites

▶ **Gel Electrophoresis and Southern Blotting** **(pp. 392–394)** DNA restriction fragments of different lengths can be separated by gel electrophoresis. Specific fragments can be identified by Southern blotting, using labeled probes that hybridize to the DNA immobilized on a "blot" of the gel.
Activity *Gel Electrophoresis of DNA*

▶ **Restriction Fragment Length Differences as Genetic Markers** **(p. 394)** Restriction fragment length polymorphisms (RFLPs) are differences in DNA sequence on homologous chromosomes that result in restriction fragments of different lengths, which can be detected by Southern blotting. The thousands of RFLPs present throughout eukaryotic DNA can serve as genetic markers.
Activity *Analyzing DNA Fragments Using Gel Electrophoresis*
Investigation *How Can Gel Electrophoresis Be Used to Analyze DNA?*

Concept 20.3

Entire genomes can be mapped at the DNA level

▶ **Genetic (Linkage) Mapping: Relative Ordering of Markers** **(p. 396)** The order of genes and other inherited markers in the genome and the relative distances between them can be determined from recombination frequencies.

▶ **Physical Mapping: Ordering DNA Fragments** **(p. 396)** A physical map is constructed by cutting a DNA molecule into many short fragments and arranging them in order by identifying overlaps. A physical map gives the actual distance in base pairs between markers.

▶ **DNA Sequencing** **(pp. 396–398)** Relatively short DNA fragments can be sequenced by the dideoxy chain-termination method, which can be performed in automated sequencing machines.
Activity *The Human Genome Project: Genes on Human Chromosome 17*

Concept 20.4

Genome sequences provide clues to important biological questions

▶ **Identifying Protein-Coding Genes in DNA Sequences** **(p. 399)** Computer analysis of genome sequences helps researchers identify sequences that are likely to encode proteins. Current estimates are that the human genome contains about 25,000 genes, but the number of human proteins is much larger. Comparison of the sequences of "new" genes with those of known genes in other species may help identify new genes.

▶ **Determining Gene Function** **(p. 400)** For a gene of unknown function, experimental inactivation of the gene and observation of the resulting phenotypic effects can provide clues to its function.

▶ **Studying Expression of Interacting Groups of Genes** **(p. 400)** DNA microarray assays allow researchers to compare patterns of gene expression in different tissues, at different times, or under different conditions.

▶ **Comparing Genomes of Different Species** **(pp. 400–402)** Comparative studies of genomes from related and widely divergent species are providing valuable information in many fields of biology.

▶ **Future Directions in Genomics** **(p. 402)** Genomics is the systematic study of entire genomes; proteomics is the systematic study of all the proteins encoded by a genome. Single nucleotide polymorphisms (SNPs) provide useful markers for studying human genetic variation.

Concept 20.5

The practical applications of DNA technology affect our lives in many ways

▶ **Medical Applications** **(pp. 402–404)** DNA technology is increasingly being used in the diagnosis of genetic and other diseases and offers potential for better treatment of certain genetic disorders or even permanent cures.

▶ **Pharmaceutical Products** **(p. 404)** Large-scale production of human hormones and other proteins with therapeutic uses, including safer vaccines, are possible with DNA technology.

▶ **Forensic Evidence** **(pp. 404–405)** DNA "fingerprints" obtained by analysis of tissue or body fluids found at crime scenes can provide definitive evidence that a suspect is guilty or not. Such fingerprints are also useful in parenthood disputes.
Activity *DNA Fingerprinting*

▶ **Environmental Cleanup** **(pp. 405–406)** Genetic engineering can be used to modify the metabolism of microorganisms so that they can be used to extract minerals from the environment or degrade various types of potentially toxic waste materials.

▶ **Agricultural Applications** **(pp. 406–407)** The aim of developing transgenic plants and animals is to improve agricultural productivity and food quality.

▶ **Safety and Ethical Questions Raised by DNA Technology** **(pp. 407–408)** The potential benefits of genetic engineering must be carefully weighed against the potential hazards of creating products or developing procedures that are harmful to humans or the environment.
Activity *Making Decisions About DNA Technology: Golden Rice*

TESTING YOUR KNOWLEDGE

Self-Quiz

1. Which of the following tools of recombinant DNA technology is *incorrectly* paired with its use?
   a. restriction enzyme—production of RFLPs
   b. DNA ligase—enzyme that cuts DNA, creating the sticky ends of restriction fragments
   c. DNA polymerase—used in a polymerase chain reaction to amplify sections of DNA
   d. reverse transcriptase—production of cDNA from mRNA
   e. electrophoresis—separation of DNA fragments

2. Which of the following would *not* be true of cDNA produced using human brain tissue as the starting material?
   a. It could be amplified by the polymerase chain reaction.
   b. It could be used to create a complete genomic library.
   c. It is produced from mRNA using reverse transcriptase.
   d. It could be used as a probe to locate genes expressed in the brain.
   e. It lacks the introns of the human genes and thus can probably be introduced into phage vectors.

3. Plants are more readily manipulated by genetic engineering than are animals because
   a. plant genes do not contain introns.
   b. more vectors are available for transferring recombinant DNA into plant cells.
   c. a somatic plant cell can often give rise to a complete plant.
   d. genes can be inserted into plant cells by microinjection.
   e. plant cells have larger nuclei.

4. A paleontologist has recovered a bit of tissue from the 400-year-old preserved skin of an extinct dodo (a bird). The researcher would like to compare DNA from the sample with DNA from living birds. Which of the following would be most useful for increasing the amount of dodo DNA available for testing?
   a. RFLP analysis
   b. polymerase chain reaction (PCR)
   c. electroporation
   d. gel electrophoresis
   e. Southern hybridization

5. Expression of a cloned eukaryotic gene in a prokaryotic cell involves many difficulties. The use of mRNA and reverse transcriptase is part of a strategy to solve the problem of
   a. post-transcriptional processing.
   b. electroporation.
   c. post-translational processing.
   d. nucleic acid hybridization.
   e. restriction fragment ligation.

6. DNA technology has many medical applications. Which of the following is *not* done routinely at present?
   a. production of hormones for treating diabetes and dwarfism
   b. production of viral subunits for vaccines
   c. introduction of genetically engineered genes into human gametes
   d. prenatal identification of genetic disease genes
   e. genetic testing for carriers of harmful alleles

7. Which of the following has the largest genome size and the smallest number of genes per million base pairs?
   a. *Hemophilus influenzae* (bacterium)
   b. *Saccharomyces cerevisiae* (yeast)
   c. *Arabidopsis thaliana* (plant)
   d. *Drosophila melanogaster* (fruit fly)
   e. *Homo sapiens* (human)

8. Which of the following sequences in double-stranded DNA is most likely to be recognized as a cutting site for a restriction enzyme?

a. AAGG	b. AGTC	c. GGCC	d. ACCA	e. AAAA
TTCC	TCAG	CCGG	TGGT	TTTT

9. In recombinant DNA methods, the term *vector* can refer to
   a. the enzyme that cuts DNA into restriction fragments.
   b. the sticky end of a DNA fragment.
   c. a RFLP marker.
   d. a plasmid used to transfer DNA into a living cell.
   e. a DNA probe used to identify a particular gene.

10. When using the shotgun approach to genome mapping, researchers carry out
   a. linkage mapping of each chromosome.
   b. extensive physical mapping of each chromosome, starting with large chromosomal fragments.
   c. DNA sequencing of small fragments and then ordering of the fragments to determine the overall nucleotide sequence.
   d. a and b.
   e. a, b, and c.

For Self-Quiz answers, see Appendix A.

Go to the website or CD-ROM for more quiz questions.

Evolution Connection

If DNA-based technologies become widely used, how might they change the way evolution proceeds, as compared with the natural evolutionary mechanisms of the past 4 billion years?

Scientific Inquiry

You hope to study a gene that codes for a neurotransmitter protein in human brain cells. You know the amino acid sequence of the protein. Explain how you might (a) identify the genes expressed in a specific type of brain cell, (b) identify the gene for the neurotransmitter, (c) produce multiple copies of the gene for study, and (d) produce a quantity of the neurotransmitter for evaluation as a potential medication.

Investigation ***How Can Antibiotic-Resistant Plasmids Transform E. coli?***
Investigation ***How Can Gel Electrophoresis Be Used to Analyze DNA?***

Science, Technology, and Society

Is there danger of discrimination based on testing for "harmful" genes? What policies can you suggest that would prevent such abuses?

21 The Genetic Basis of Development

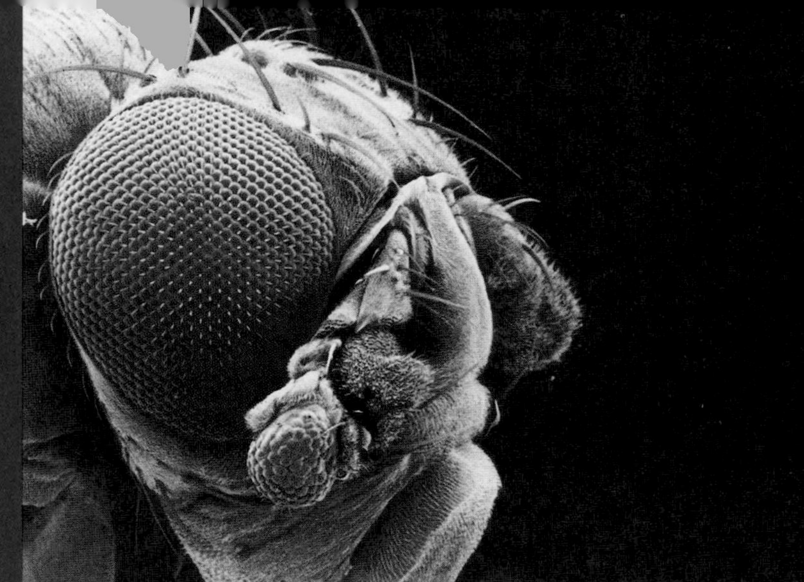

▲ **Figure 21.1 Mutant *Drosophila* with an extra small eye on its antenna.**

Key Concepts

21.1 Embryonic development involves cell division, cell differentiation, and morphogenesis

21.2 Different cell types result from differential gene expression in cells with the same DNA

21.3 Pattern formation in animals and plants results from similar genetic and cellular mechanisms

21.4 Comparative studies help explain how the evolution of development leads to morphological diversity

Overview

From Single Cell to Multicellular Organism

This chapter applies much of what you've learned about molecules, cells, and genes to one of biology's most important questions—how a complex multicellular organism develops from a single cell. The application of genetic analysis and DNA technology to the study of development has revolutionized the field. In much the same way that researchers have used mutations to deduce pathways of cellular metabolism, they now use mutations to dissect developmental pathways. In one striking example, Swiss researchers demonstrated in 1995 that a particular gene functions as a master switch that triggers development of the eye in *Drosophila*. The scanning electron micrograph in **Figure 21.1** shows part of the head of an abnormal fly that has a small extra eye on each antenna. In this fly, the master gene triggering eye development was expressed in an abnormal body location, causing extra eyes. A similar gene activates eye development in mice and other mammals. In fact, developmental biologists are discovering remarkable similarities in the mechanisms that form diverse organisms.

The scientific study of development got under way about 130 years ago, around the same time as genetics. But for decades, the two disciplines proceeded along mostly separate paths. Developmental biologists focused on embryology, the study of the stages of development leading from a fertilized egg to a fully formed organism. They studied animals that lay their eggs in water, including marine invertebrates and freshwater vertebrate amphibians, such as frogs. By studying these and other animals, as well as plants, biologists worked out a description of animal development (see Chapter 47) and plant development (see Chapter 35) at the macroscopic and microscopic levels.

In recent years, scientists have applied the concepts and tools of molecular genetics to the study of developmental biology with remarkably fruitful results. In this chapter, we introduce some of the basic mechanisms that control development in animals and plants, focusing on what has been learned from molecular and genetic studies. After introducing the basic cellular processes underlying development, we consider how cells become different from each other and the factors that establish the spatial pattern of these different types of cells in the embryo. Next, we examine in more detail the molecular basis of several specific developmental phenomena as examples of some general principles of development. Finally, we discuss what researchers can learn about evolution from comparing developmental processes in different species.

Figure 21.2

Exploring Model Organisms for Genetic Studies of Development

DROSOPHILA MELANOGASTER (FRUIT FLY)

Researchers can draw on a wealth of information about the fruit fly *Drosophila melanogaster* (often referred to simply as *Drosophila*), one of the most important model organisms in developmental genetics. First chosen as a model organism by the pioneering geneticist T. H. Morgan in the early 20th century, *Drosophila* has been studied intensively by generations of geneticists since then. Small and easily grown in the laboratory, *Drosophila* has a generation time of only two weeks. It produces many offspring and the embryos develop outside the mother's body—both assets for developmental studies. Sequencing of the *Drosophila* genome was completed in 2000; it has 180×10^6 base pairs (180 million bases, Mb) and contains about 13,700 genes. Although early development of fruit flies is at least superficially quite different from that of many other animals, research on *Drosophila* development has yielded deep insights into basic principles of animal development.

CAENORHABDITIS ELEGANS (NEMATODE)

The soil-dwelling nematode *Caenorhabditis elegans* (or *C. elegans*) is easily grown in the laboratory in petri dishes. It is about 1 mm long, has a simple, transparent body with only a few types of cells, and grows from zygote to mature adult in only three and a half days. The nematode genome is 97 Mb long and contains an estimated 19,000 genes. Most individuals are hermaphrodites, producing both eggs and sperm. Hermaphrodites are convenient for genetic studies because recessive mutations are easy to detect. If a worm with the wild-type phenotype self-fertilizes and one-fourth of its offspring have a mutant phenotype (homozygous for a recessive allele), then the parent must be heterozygous for the recessive mutant allele. Even if homozygotes with the recessive mutation do not reproduce, the mutation can be maintained in the heterozygotes. Another advantage of *C. elegans* is that every adult hermaphrodite has exactly 959 somatic cells, which arise from the zygote in virtually the same way for every individual. Using a microscope to follow all the cell divisions starting immediately after a zygote forms, biologists have been able to reconstruct the entire ancestry of every cell in the adult body.

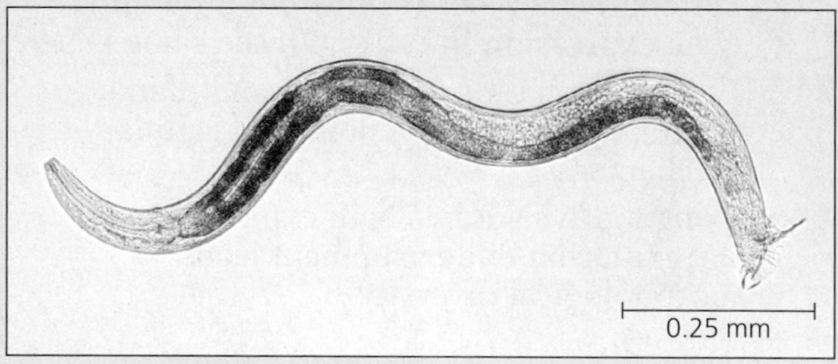

When the primary research goal is to understand broad biological principles, the organism chosen for study is called a **model organism.** Researchers select model organisms that lend themselves to the study of a particular question, are representative of a larger group, and are easy to grow in the lab. For research aimed at uncovering the connections between genes and development, biologists have turned to organisms which have relatively short generation times and small genomes and about which much is already known. Organisms with these attributes are particularly convenient for genetic analysis. Among the favorite model organisms in developmental genetics are the fruit fly *Drosophila melanogaster,* the nematode *Caenorhabditis elegans,* the mouse *Mus musculus,* the zebrafish *Danio rerio,* and the plant *Arabidopsis thaliana.* Before proceeding, acquaint yourself with these model organisms in **Figure 21.2**. Developmental principles we have learned from studying these organisms will be presented throughout this chapter.

Embryonic development involves cell division, cell differentiation, and morphogenesis

In the embryonic development of most organisms, a single-celled zygote (fertilized egg) gives rise to cells of many different types, each type with a different structure and corresponding function. For example, an animal will have muscle cells that enable it to move and nerve cells that transmit signals to the muscle cells; a plant will have mesophyll cells that carry out photosynthesis and guard cells around stomata (pores) that regulate the passage of gases into and out of leaves. Within a multicellular organism, cells of different types are organized into tissues, tissues into organs, organs into organ systems, and

MUS MUSCULUS (MOUSE)

Among vertebrates, two in particular lend themselves to the genetic analysis of development, the mouse and the zebrafish. The mouse *Mus musculus* has a long history as a mammalian model, and much is known about its biology. The mouse genome is about 2,600 Mb long with about 25,000 genes, roughly the same as the human genome. Researchers are now adept at manipulating mouse genes to make transgenic mice and mice in which particular genes are "knocked out" by mutation. However, mice have a generation time of about nine weeks, and their embryos develop in the mother's uterus, hidden from view, both disadvantages for developmental studies.

DANIO RERIO (ZEBRAFISH)

Many of the disadvantages of the mouse as a vertebrate model are absent in the zebrafish *Danio rerio*. These small fish (2–4 cm long) are easy to breed in the laboratory, and the transparent embryos develop outside the mother's body. Although the generation time is relatively long (two to four months), early development proceeds quickly: By 24 hours after fertilization, most of the tissues and rudiments of the organs have formed, and by two days, a tiny fish hatches out of the egg case. The zebrafish genome (estimated to be 1,700 Mb long) is still being mapped and sequenced, but researchers have already identified many genes involved in this animal's development.

ARABIDOPSIS THALIANA (COMMON WALL CRESS)

For studying the molecular genetics of plant development, researchers often use a small flowering plant in the mustard family called *Arabidopsis thaliana* (or simply *Arabidopsis*). One of these plants can grow in a test tube and produce thousands of progeny after eight to ten weeks; as in Mendel's pea plants, each flower makes both eggs and sperm. For research on gene function, scientists can make transgenic *Arabidopsis* plants (see Figure 20.19). Compared with some other plant species, *Arabidopsis* has a relatively small genome, about 118 Mb, which contains an estimated 25,500 genes.

organ systems into the whole organism. Thus, the process of embryonic development must give rise not only to cells of different types but to higher-level structures arranged in a particular way in three dimensions.

The photos in **Figure 21.3** illustrate the dramatic transformation of a zygote into an organism. This transformation results from three interrelated processes: cell division, cell differentiation, and morphogenesis. Through a succession of mitotic cell divisions, the zygote gives rise to a large number of cells. Cell division alone, however, would produce only a great ball of identical cells, nothing like an animal or plant. During embryonic development, cells not only increase in number, but also undergo **cell differentiation**, the process by which cells become specialized in structure and function. Moreover, the different kinds of cells are not randomly distributed but are organized into tissues and organs. The physical processes that give an organism its shape constitute **morphogenesis**, meaning "creation of form."

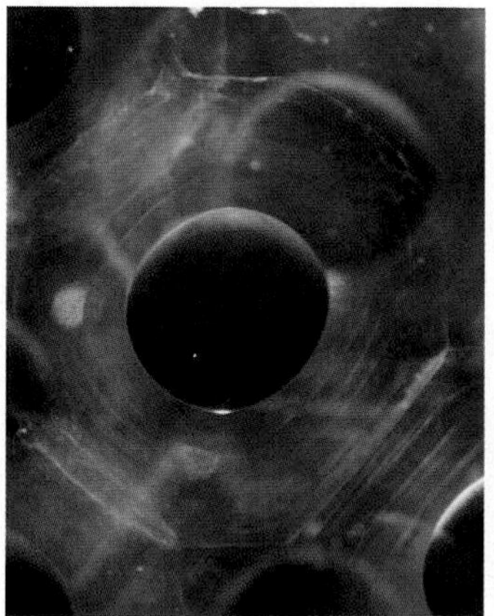

(a) Fertilized eggs of a frog

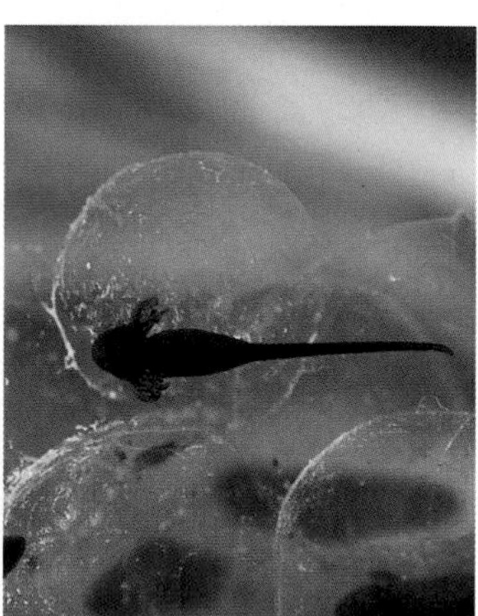

(b) Tadpole hatching from egg

▲ **Figure 21.3 From fertilized egg to animal: what a difference a week makes.** It takes just one week for cell division, differentiation, and morphogenesis to transform each of the fertilized frog eggs shown in (a) into a hatching tadpole like the one in (b). A protective jelly coat surrounds the eggs and tadpole.

The processes of cell division, differentiation, and morphogenesis overlap in time **(Figure 21.4)**. Morphogenetic events lay out the basic body plan very early in embryonic development, establishing, for example, which end of an animal embryo will be the head or which end of a plant embryo will become the roots. These early events determine the body axes of the organism, such as the anterior-posterior (head-to-tail) axis and the dorsal-ventral (back-to-belly) axis. Later morphogenetic events establish relative locations of structures within smaller regions of the embryo, such as the appendages on a fly's body, the fins on a fish, or the digits on a vertebrate limb—and then within regions still smaller.

Cell division and differentiation play important roles in morphogenesis in all multicellular organisms, as does the appropriately timed programmed death of certain cells. However, the overall schemes of morphogenesis in animals and plants exhibit significant differences. In addition to many shared developmental mechanisms, the development of animals and plants differs in two major ways:

- In animals, but not in plants, *movements* of cells and tissues are necessary to transform the early embryo into the characteristic three-dimensional form of the organism.
- In plants, but not in animals, morphogenesis and growth in overall size are not limited to embryonic and juvenile periods but occur throughout the life of the plant.

The structures responsible for a plant's continual growth and formation of new organs are **apical meristems**, perpetually embryonic regions in the tips of shoots and roots. In animals, ongoing development in adults is normally restricted to the generation of cells that must be continually replenished throughout the animal's lifetime. Examples are blood cells, skin cells, and the cells lining the intestines.

During differentiation and morphogenesis, embryonic cells behave and function in different ways from each other, even though all of them have arisen from the same cell—the zygote. In the next section, you will learn about the principal way in which this occurs.

(a) Animal development. Most animals go through some variation of the blastula and gastrula stages. The blastula is a sphere of cells surrounding a fluid-filled cavity. The gastrula forms when a region of the blastula folds inward, creating a tube—a rudimentary gut. Once the animal is mature, differentiation occurs in only a limited way—for the replacement of damaged or lost cells.

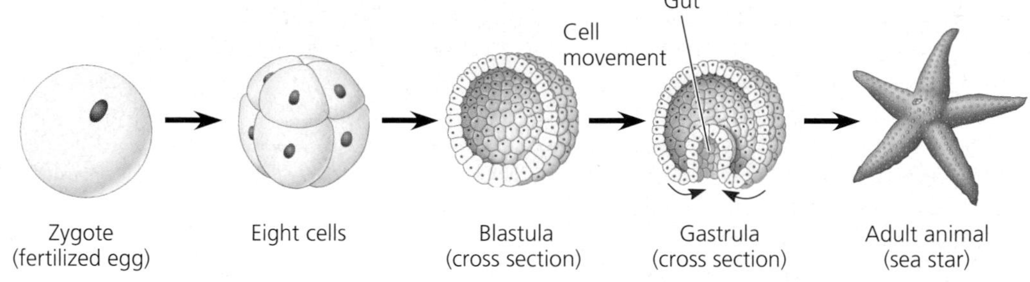

(b) Plant development. In plants with seeds, a complete embryo develops within the seed. Morphogenesis, which involves cell division and cell wall expansion rather than cell or tissue movement, occurs throughout the plant's lifetime. Apical meristems (purple) continuously arise and develop into the various plant organs as the plant grows to an indeterminate size.

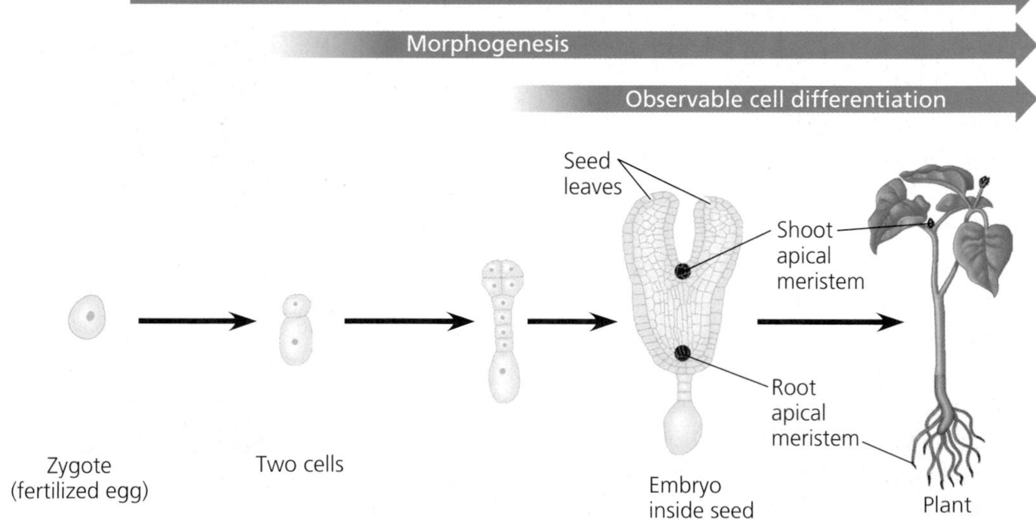

▲ **Figure 21.4 Some key stages of development in animals and plants.** Cell division, morphogenesis, and cell differentiation occur in both animal development and plant development. Molecular events leading to cell differentiation begin as early as the two-cell stage, but observable differences among cells are not evident until much later.

Concept Check 21.1

1. As you learned in Chapter 12, mitosis gives rise to two daughter cells that are genetically identical to the parent cell. Yet you, the product of many mitotic divisions, are not just a ball of identical cells. Why?
2. What are the fundamental differences between plants and animals in their mechanisms of development?

For suggested answers, see Appendix A.

Concept 21.2

Different cell types result from differential gene expression in cells with the same DNA

In earlier chapters, we have stated that differences between cells in a multicellular organism come almost entirely from differences in gene *expression*, not from differences in the cells' genomes. (There are a few exceptions, such as antibody-producing cells; see Figure 43.11.) Furthermore, we have mentioned that these differences arise during development as regulatory mechanisms turn specific genes on and off. Let's now look at some of the evidence for this assertion.

Evidence for Genomic Equivalence

The results of many experiments support the conclusion that nearly all the cells of an organism have *genomic equivalence*—that is, they all have the same genes. What happens to these genes as a cell begins to differentiate? We can shed some light on this question by asking whether genes are irreversibly inactivated during differentiation. For example, does an epidermal cell in your finger contain a functional gene specifying eye color, or has the eye-color gene been destroyed or permanently inactivated there?

Totipotency in Plants

One experimental approach for testing genomic equivalence is to see whether a differentiated cell can generate a whole organism. Such experiments were performed during the 1950s by F. C. Steward and his students at Cornell University, working with carrot plants **(Figure 21.5)**. They found that differentiated cells taken from the root (the carrot) and placed in culture medium could grow into normal adult plants, each genetically identical to the "parent" plant. These results show that differentiation does not necessarily involve irreversible changes in the DNA. In plants, at least, mature cells can dedifferentiate and then give rise to all the specialized cell types of the mature organism. Any cell with this potential is said to be **totipotent.**

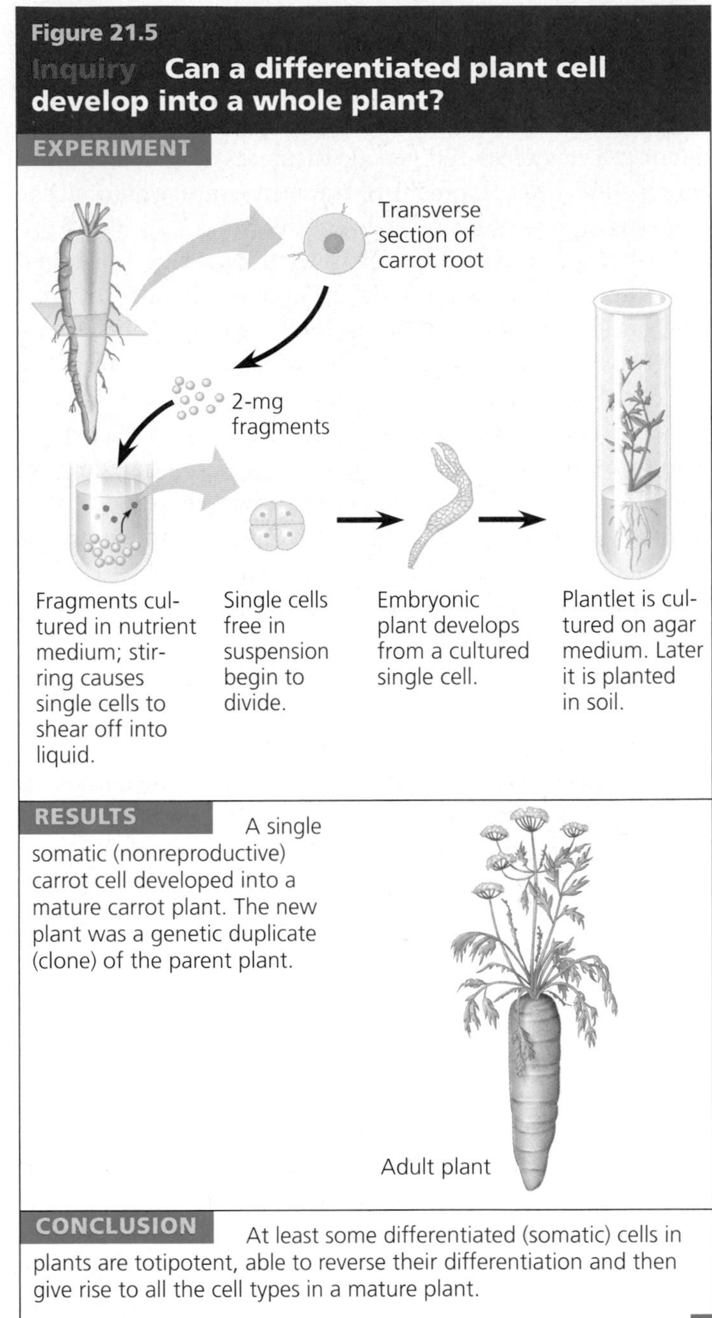

Figure 21.5
Inquiry Can a differentiated plant cell develop into a whole plant?
EXPERIMENT
Transverse section of carrot root
2-mg fragments
Fragments cultured in nutrient medium; stirring causes single cells to shear off into liquid.
Single cells free in suspension begin to divide.
Embryonic plant develops from a cultured single cell.
Plantlet is cultured on agar medium. Later it is planted in soil.
RESULTS A single somatic (nonreproductive) carrot cell developed into a mature carrot plant. The new plant was a genetic duplicate (clone) of the parent plant.
Adult plant
CONCLUSION At least some differentiated (somatic) cells in plants are totipotent, able to reverse their differentiation and then give rise to all the cell types in a mature plant.

Using one or more somatic cells from a multicellular organism to make another genetically identical individual is called **cloning,** and each new individual made in this way can be called a **clone** (from the Greek *klon,* twig). Plant cloning is now used extensively in agriculture. Indeed, if you have ever grown a new plant from a cutting, you have practiced cloning.

Nuclear Transplantation in Animals

Differentiated cells from animals generally do not divide in culture, much less develop into the multiple cell types of a new organism. Therefore, animal researchers had to use a different approach to the question of whether differentiated animal cells can be totipotent. Their approach was to remove the nucleus of an unfertilized egg cell or zygote and replace it with the nucleus of a differentiated cell, a method called *nuclear transplantation*. If the nucleus from the differentiated donor cell retains its full genetic capability, then it should be able to direct development of the recipient egg into all the proper tissues and organs of an organism.

Such experiments were conducted on frogs by Robert Briggs and Thomas King in the 1950s and extended by John Gurdon in the 1980s. These researchers transplanted a nucleus from an embryonic or tadpole cell into an enucleated egg of the same species. The transplanted nucleus was often able to support normal development of the egg into a tadpole **(Figure 21.6)**. However, the "potency" of transplanted nuclei in directing normal development was inversely related to the age of the donor: the older the donor nucleus, the lower the percentage of normally developing tadpoles.

From these results, we can conclude that something in the nucleus *does* change as animal cells differentiate. In frogs and most other animals, nuclear potency tends to be restricted more and more as embryonic development and cell differentiation progress. Research has shown that although the base sequence of the DNA usually does not change, the chromatin structure is altered in specific ways, usually involving chemical modifications of histones or DNA methylation (see Chapter 19). However, these chromatin changes are sometimes reversible, and biologists agree that the nuclei of most differentiated animal cells have all the genes required for making the entire organism. In other words, the various cell types in the body of an animal differ in structure and function not because they contain different genes, but because they express different sets of genes from a common genome.

Reproductive Cloning of Mammals. Evidence that all cells in an organism have the same DNA also comes from experiments with mammals. Researchers have long been able to clone mammals using nuclei or cells from a variety of early embryos. But it was not known whether a nucleus from a fully differentiated cell could be "reprogrammed" to be totipotent. However, in 1997, Scottish researchers captured newspaper headlines when they announced the birth of Dolly, a lamb cloned from an adult sheep by nuclear transplantation from a differentiated cell **(Figure 21.7)**. These researchers achieved the necessary dedifferentiation of donor nuclei by culturing mammary cells in nutrient-poor medium. The researchers then fused these cells with sheep egg cells whose nuclei had been removed. The resulting diploid cells divided to form early embryos, which were implanted into surrogate mothers. One of several hundred implanted embryos successfully completed normal development, and Dolly was born.

Later analyses showed that Dolly's chromosomal DNA was indeed identical to that of the nucleus donor. (Her mitochondrial DNA came from the egg cell donor, as expected.)

Figure 21.6

Inquiry Can the nucleus from a differentiated animal cell direct development of an organism?

EXPERIMENT Researchers enucleated frog egg cells by exposing them to ultraviolet light, which destroyed the nucleus. Nuclei from cells of embryos up to the tadpole stage were transplanted into the enucleated egg cells.

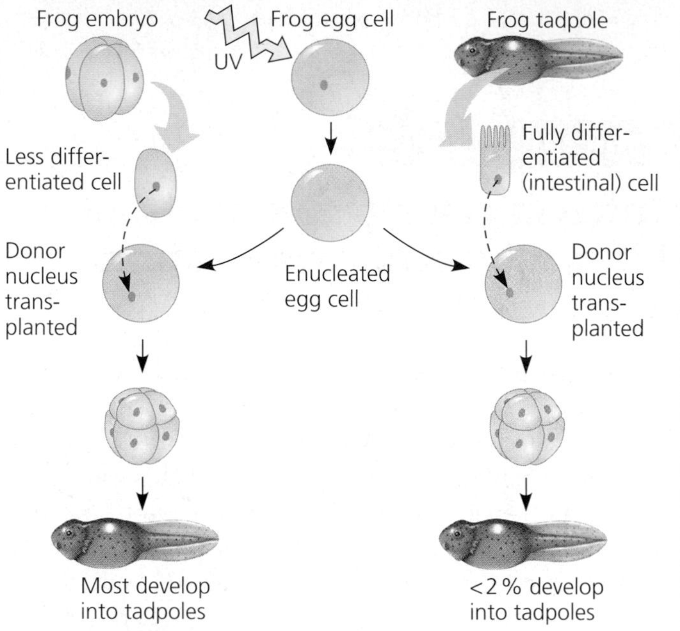

Frog embryo
UV
Frog egg cell
Frog tadpole
Less differentiated cell
Fully differentiated (intestinal) cell
Donor nucleus transplanted
Enucleated egg cell
Donor nucleus transplanted
Most develop into tadpoles
<2% develop into tadpoles

RESULTS Most of the recipient eggs developed into tadpoles when the transplanted nuclei came from cells of an early embryo, which are relatively undifferentiated cells. But with nuclei from the fully differentiated intestinal cells of a tadpole, fewer than 2% of the eggs developed into normal tadpoles, and most of the embryos died at a much earlier developmental stage.

CONCLUSION The nucleus from a differentiated frog cell can direct development of a tadpole. However, its ability to do so decreases as the donor cell becomes more differentiated, presumably because of changes in the nucleus.

In 2003, at age 6, Dolly suffered complications from a lung disease usually seen in much older sheep and was euthanized. Dolly's premature death, as well as her arthritic condition, led to speculation that her cells were "older" than those of a normal sheep, possibly reflecting incomplete reprogramming of the original transplanted nucleus.

Since 1997, cloning has also been demonstrated in numerous other mammals, including mice, cats, cows, horses, and pigs. In most cases, the goal has been to produce new individuals; this is known as *reproductive cloning*. We have already learned much of interest from such experiments. For example, cloned animals of the same species do *not* always look or behave identically. In a herd of cows cloned from the same cell line, certain cows are dominant and others are more submissive. Another example is the first cloned

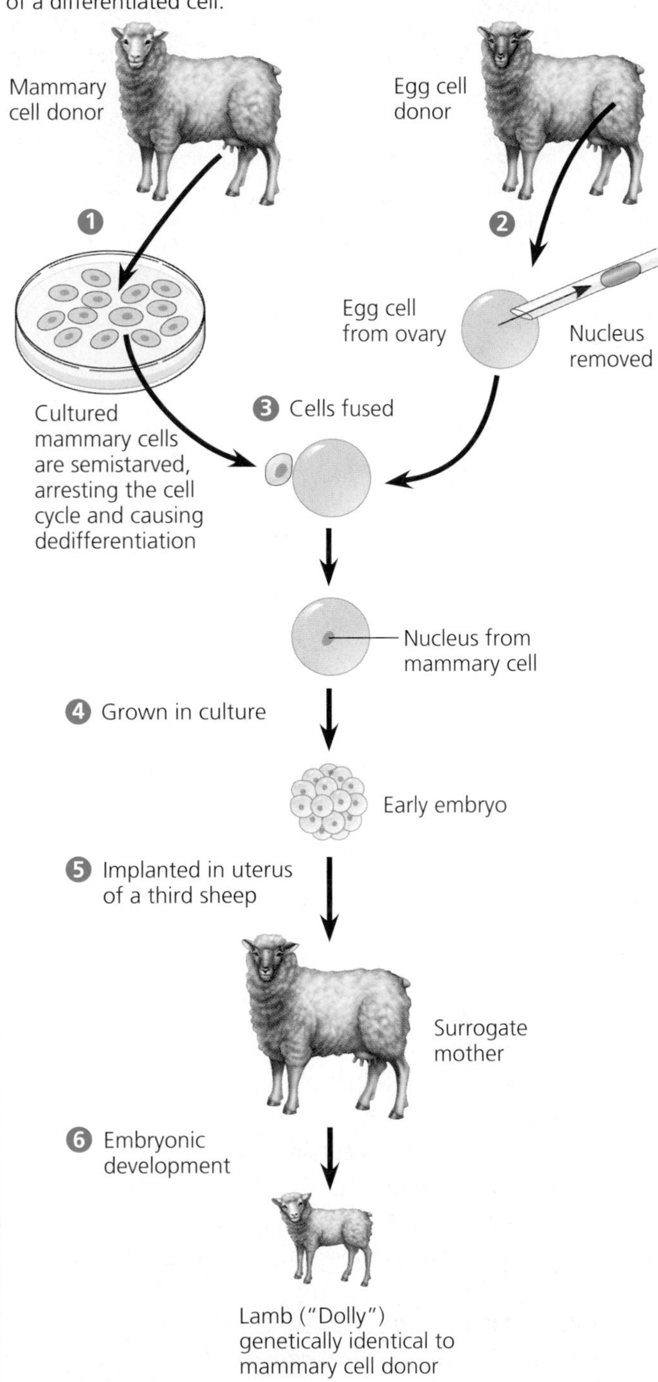
Figure 21.7
Research Method Reproductive Cloning of a Mammal by Nuclear Transplantation
APPLICATION This method is used to produce cloned animals whose nuclear genes are identical to the donor animal supplying the nucleus.
TECHNIQUE Shown here is the procedure used to produce Dolly, the first reported case of a mammal cloned using the nucleus of a differentiated cell.
Mammary cell donor
Egg cell donor
1
2
Egg cell from ovary
Nucleus removed
Cultured mammary cells are semistarved, arresting the cell cycle and causing dedifferentiation
3 Cells fused
Nucleus from mammary cell
4 Grown in culture
Early embryo
5 Implanted in uterus of a third sheep
Surrogate mother
6 Embryonic development
Lamb ("Dolly") genetically identical to mammary cell donor
RESULTS The cloned animal is identical in appearance and genetic makeup to the donor animal supplying the nucleus, but differs from the egg cell donor and surrogate mother.

cat, named Copy Cat **(Figure 21.8)**. She has a calico coat, like her single female parent, but the color and pattern are different due to random X chromosome inactivation, which is a normal occurrence during embryonic development (see Figure 15.11). Clearly, environmental influences and random phenomena can play a significant role during development.

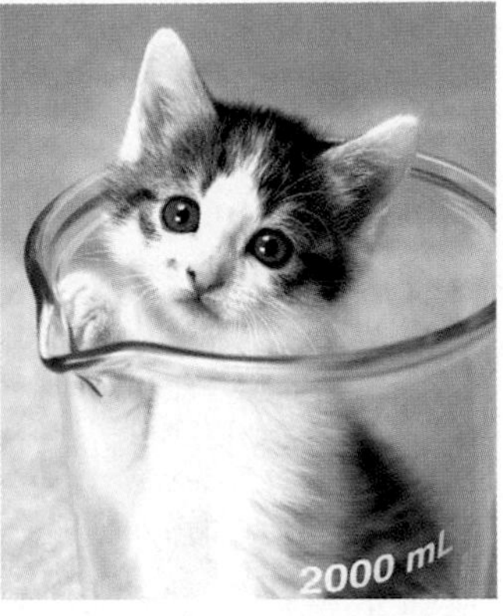
2000 mL

▲ **Figure 21.8 Copy Cat, the first cloned cat.**

The successful cloning of various mammals has heightened speculation about the cloning of humans. In early 2004, South Korean researchers reported success in the first step of reproductive cloning of humans. In this work, nuclei from differentiated human cells were transplanted into unfertilized eggs from which the nuclei had been removed. These eggs were stimulated to divide, and some reached the blastocyst stage, an early embryonic stage similar to the blastula stage in Figure 21.4. Although the embryos were not allowed to develop beyond the blastocyst stage, the work of these researchers brings us one step closer to the possibility of human reproductive cloning, which raises unprecedented ethical issues. However, problems associated with the cloning process have bought us a little more time for thought.

Problems Associated with Animal Cloning. In most nuclear transplantation studies thus far, only a small percentage of cloned embryos develop normally to birth. And like Dolly, many cloned animals exhibit various defects. Cloned mice, for instance, are prone to obesity, pneumonia, liver failure, and premature death. Scientists believe that even cloned animals that appear normal are likely to have subtle defects.

In recent years, we have begun to learn possible reasons underlying the low efficiency of cloning and the high incidence of abnormalities. In the nuclei of fully differentiated cells, a small subset of genes is turned on and expression of the rest is repressed. This regulation often is the result of epigenetic changes in chromatin, such as acetylation of histones or methylation of DNA (see Figure 19.4). Many of these changes must be reversed in the nucleus from a donor animal in order for genes to be expressed or repressed appropriately for early stages of development. Researchers have found that the DNA in embryonic cells from cloned embryos, like that of differentiated cells, often has more methyl groups than does the DNA in equivalent cells from uncloned embryos of the same species. This finding suggests that the reprogramming of donor nuclei is not always complete. Because DNA methylation helps regulate gene expression, misplaced methyl groups in the DNA of donor nuclei may interfere with the pattern of gene expression necessary for normal embryonic development.

The Stem Cells of Animals

Further support for the idea that cells become different by expressing different sets of genes from the same genome comes from the study of a group of cells called stem cells. A **stem cell** is a relatively unspecialized cell that can both reproduce itself indefinitely and, under appropriate conditions, differentiate into specialized cells of one or more types. Thus, stem cells are able both to replenish their own population and to generate cells that travel down various differentiation pathways.

Many early animal embryos contain totipotent stem cells, which can give rise to differentiated cells of any type. Stem cells can be isolated from early embryos at the blastula stage or its human equivalent, the blastocyst stage **(Figure 21.9)**. In culture, these *embryonic stem cells* reproduce indefinitely; and depending on culture conditions, they can differentiate into various specialized cells, including even eggs and sperm.

The adult body also has a variety of stem cells, which serve to replace nonreproducing specialized cells as needed. In contrast to totipotent embryonic stem cells, adult stem cells are said to be **pluripotent,** able to give rise to multiple but not all cell types. For example, stem cells in the bone marrow give rise to all the different kinds of blood cells (see Figure 21.9), and those in the intestinal wall regenerate the various cells forming the lining of the intestine. To the surprise of many, we have recently discovered that the adult brain contains stem cells that continue to produce certain kinds of nerve cells there. Although adult animals have only tiny numbers of stem cells, scientists are learning to identify and isolate these cells from various tissues and, in some cases, to grow them in culture. Taking this research further, scientists have found that with the right culture conditions (for instance, the addition of specific growth factors), cultured stem cells from adult animals can differentiate into multiple types of specialized cells.

Research with embryonic or adult stem cells is providing valuable information about differentiation and has enormous potential for medical applications. The ultimate aim is to supply cells for the repair of damaged or diseased organs: for example, insulin-producing pancreatic cells for people with diabetes or certain kinds of brain cells for people with Parkinson's disease or Huntington's disease. Currently, embryonic stem cells are more promising than adult stem cells for such applications, but because the cells are derived from human embryos, their use raises ethical and political issues.

Embryonic stem cells are currently obtained from embryos donated by patients undergoing infertility treatment or from long-term cell cultures originally established with cells isolated from donated embryos. With the recent cloning of human embryos to the blastocyst stage, scientists might be able to use such clones as the source of embryonic stem cells in the future. When the major aim of cloning is to produce embryonic stem cells to treat disease, the process is called *therapeutic cloning*. Although most people believe that reproductive cloning of humans is unethical, opinions vary about the morality of therapeutic cloning. Some believe that it is wrong to create embryos that will be destroyed, whereas others, in the words of the researcher who created Dolly, believe that "cloning promises such great benefits that it would be immoral not to do it."

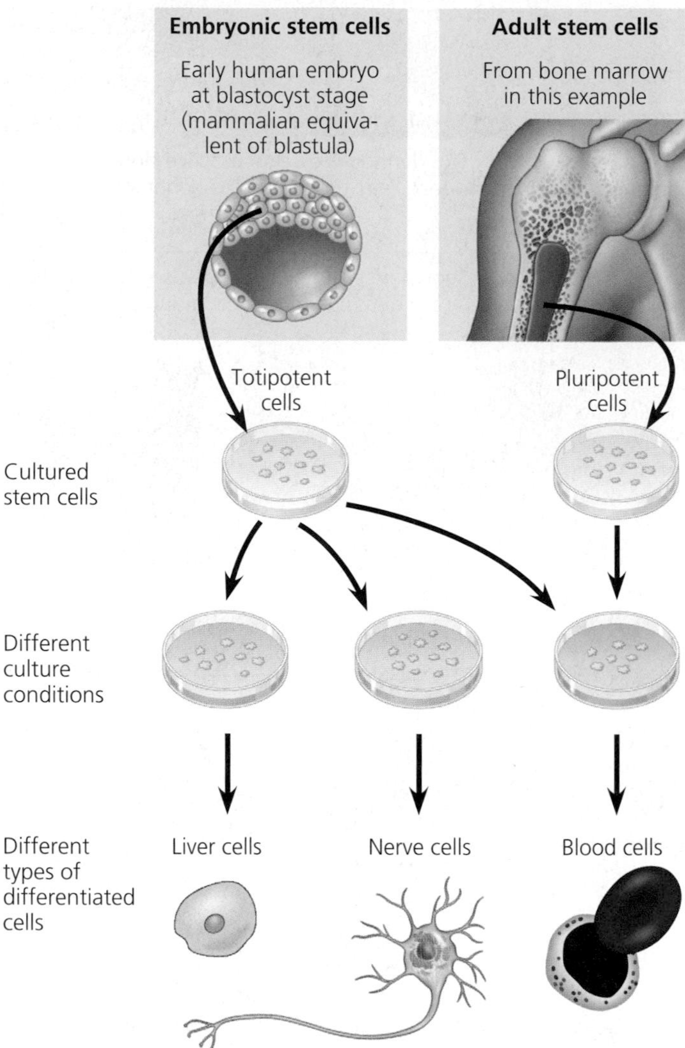

▲ **Figure 21.9 Working with stem cells.** Animal stem cells, which can be isolated from early embryos or adult tissues and grown in culture, are self-perpetuating, relatively undifferentiated cells. Embryonic stem cells are easier to grow than adult stem cells and can theoretically give rise to *all* types of cells. The range of cell types that can arise from adult stem cells is not yet fully understood.

The evidence we've discussed shows that nearly all the cells in a plant or animal contain the same set of genes. Next we look at the main processes that give rise to different cell types—that is, the molecular basis of cell differentiation.

Transcriptional Regulation of Gene Expression During Development

As the tissues and organs of an embryo take shape, the cells become visibly different in structure and function. These observable changes are actually the outcome of a cell's developmental history extending back to the first mitotic divisions of the zygote. However, the earliest changes that set a cell on a path to specialization are subtle ones, showing up only at the molecular level. Before biologists knew much about the molecular changes occurring in embryos, they coined the term **determination** to refer to the events that lead to the observable differentiation of a cell. At the end of this process, an embryonic cell is irreversibly committed to its final fate, and it is said to be determined. If a committed cell is experimentally placed in another location in the embryo, it will still differentiate into the cell type that is its normal fate.

Today we understand determination in terms of molecular changes. The outcome of determination—observable cell differentiation—is marked by the expression of genes for *tissue-specific proteins*. These proteins are found only in a specific cell type and give the cell its characteristic structure and function. The first evidence of differentiation is the appearance of mRNAs for these proteins. Eventually, differentiation is observable with a microscope as changes in cellular structure. In most cases, the pattern of gene expression in a differentiated cell is controlled at the level of transcription.

Differentiated cells are specialists at making tissue-specific proteins. For example, as the result of transcriptional regulation, liver cells specialize in making albumin, and lens cells specialize in making crystallins (see Figure 19.7). Indeed, lens cells devote 80% of their capacity for protein synthesis to making crystallin proteins, which enable the lens to transmit and focus light. Skeletal muscle cells are another instructive example. The "cells" of skeletal muscle are long fibers containing many nuclei within a single plasma membrane. They contain high concentrations of muscle-specific versions of the contractile proteins myosin and actin, as well as membrane receptor proteins that detect signals from nerve cells.

Muscle cells develop from embryonic precursor cells that have the potential to develop into a number of alternative cell types, including cartilage cells or fat cells, but particular conditions commit them to becoming muscle cells. Although the committed cells appear unchanged under the microscope, determination has occurred, and they are now *myoblasts*. Eventually, myoblasts start to churn out large amounts of muscle-specific proteins and fuse to form mature, elongated, multinucleate skeletal muscle cells (**Figure 21.10**, left).

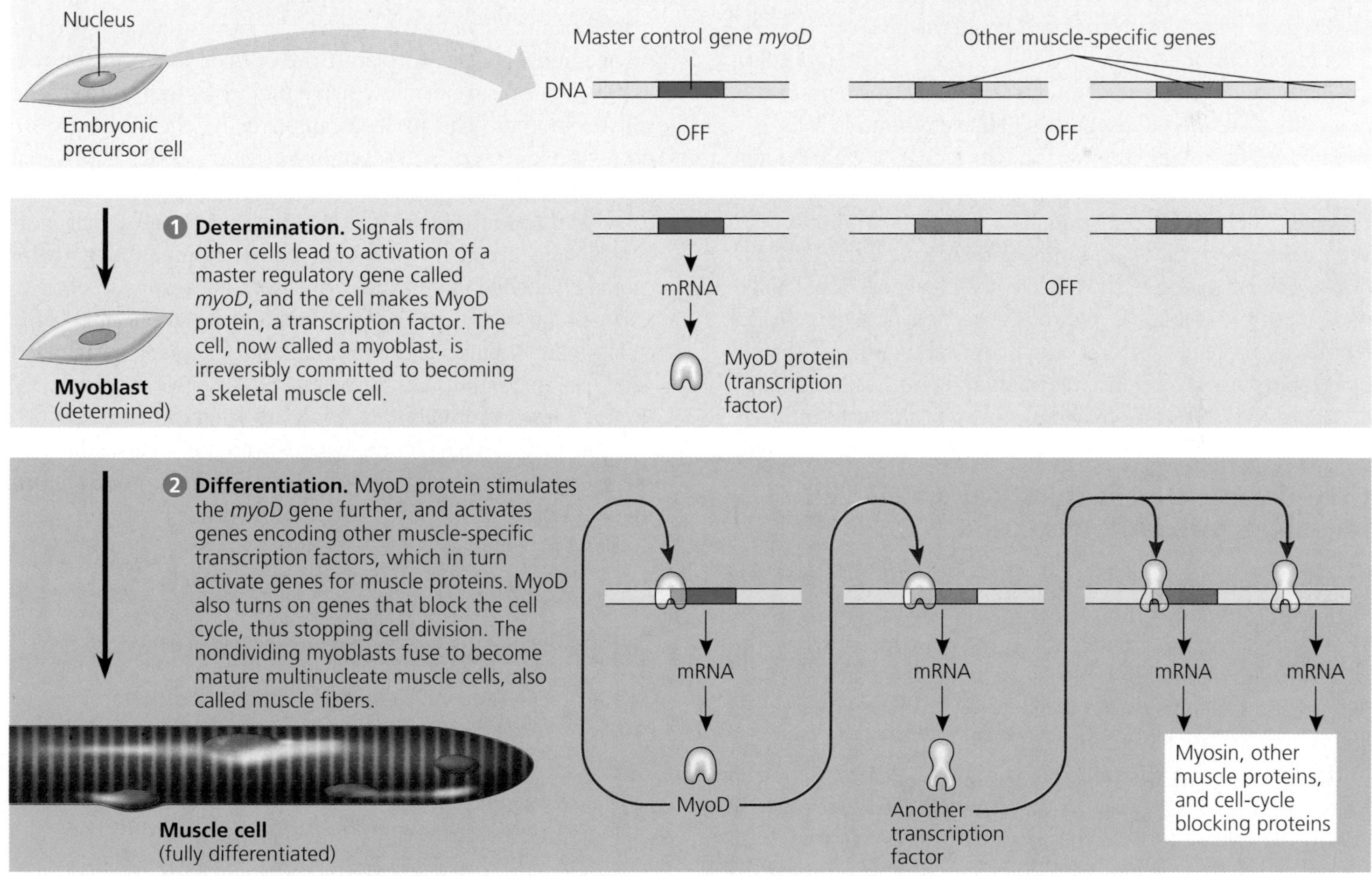

▲ **Figure 21.10 Determination and differentiation of muscle cells.** This figure depicts a simplified version of how skeletal muscle cells arise from embryonic cells. These precursor cells resemble fibroblasts (see the photo in Figure 12.17).

Researchers have worked out what happens at the molecular level during muscle cell determination by growing myoblasts in culture and applying some of the techniques you learned about in Chapter 20. They first produced a cDNA library containing all the genes that are expressed in cultured myoblasts. The researchers then inserted each of the cloned genes into a separate embryonic precursor cell and looked for differentiation into myoblasts and muscle cells. In this way, they identified several so-called "master regulatory genes" whose protein products commit the cells to becoming skeletal muscle. Thus, in the case of muscle cells, the molecular basis of determination is the expression of one or more of these master regulatory genes.

To understand more about how commitment occurs in muscle cell differentiation, let's focus on the master regulatory gene called *myoD* (see Figure 21.10, right). This gene encodes MyoD protein, a transcription factor that binds to specific control elements in the enhancers of various target genes and stimulates their expression (see Figure 19.6). Some target genes for MyoD encode still other muscle-specific transcription factors. MyoD also stimulates expression of the *myoD* gene itself, thus perpetuating its effect in maintaining the cell's differentiated state. Presumably, all the genes activated by MyoD have enhancers recognized by MyoD and are thus coordinately controlled. Finally, the secondary transcription factors activate the genes for proteins, such as myosin and actin, that confer the unique properties of skeletal muscle cells.

The MyoD protein is powerful. Researchers have been able to use it to change some kinds of fully differentiated nonmuscle cells, such as fat cells and liver cells, into muscle cells. Why doesn't it work on *all* kinds of cells? One likely explanation is that activation of the muscle-specific genes is not solely dependent on MyoD but requires a particular *combination* of regulatory proteins, some of which are lacking in cells that do not respond to MyoD. The determination and differentiation of other kinds of tissues may play out in a similar fashion.

Cytoplasmic Determinants and Cell-Cell Signals in Cell Differentiation

Explaining the role of *myoD* in muscle cell differentiation is a long way from explaining the development of an organism. The *myoD* story immediately raises the question of what triggers the expression of *that* gene and then raises a series of similar questions leading back to the zygote. What generates the *first* differences that arise among the cells in an early embryo? And what controls morphogenesis and the differentiation of all the different cell types as development proceeds? As we saw in the case of muscle cells, this question comes down to which genes are transcribed in the cells of a developing organism. Two sources of information, used to varying extents in different species, "tell" a cell which genes to express at any given time during embryonic development.

One important source of information early in development is the egg cell's cytoplasm, which contains both RNA and protein molecules encoded by the mother's DNA. The cytoplasm of an unfertilized egg cell is not homogeneous. Messenger RNA, proteins, other substances, and organelles are distributed unevenly in the unfertilized egg; and this heterogeneity has a profound impact on the development of the future embryo in many species. Maternal substances in the egg that influence the course of early development are called **cytoplasmic determinants.** After fertilization, early mitotic divisions distribute the zygote's cytoplasm into separate cells. The nuclei of many of these cells are thus exposed to different cytoplasmic determinants, depending on which portions of the zygotic cytoplasm a cell received **(Figure 21.11a)**. The set of cytoplasmic determinants a particular cell receives helps determine its developmental fate by regulating expression of the cell's genes during the course of cell differentiation.

The other important source of developmental information, which becomes increasingly important as the number of embryonic cells increases, is the environment around a particular cell. Most important are the signals impinging on an embryonic cell from other embryonic cells in the vicinity. In animals, these include contact with cell-surface molecules on neighboring cells and the binding of growth factors secreted by neighboring cells. In plants, the cell-cell junctions known as plasmodesmata can allow signal molecules to pass from one cell to another. The molecules conveying these signals are proteins expressed by the embryo's own genes. The signal molecules cause changes in nearby target cells, a process called **induction (Figure 21.11b)**. In general, the signal molecules send a cell down a specific developmental path by causing a change in its gene expression that results in observable cellular changes. Thus, interactions between embryonic cells eventually induce differentiation of the many specialized cell types making up a new organism.

You'll learn more about cytoplasmic determinants and induction in the next section, where we take a closer look at some important genetic and cellular mechanisms of development in three model organisms: *Drosophila, C. elegans,* and *Arabidopsis.*

Concept Check 21.2

1. Why can't a single embryonic stem cell develop into an embryo?
2. If you clone a carrot, will all the progeny plants ("clones") look identical? Why or why not?
3. The signal molecules released by an embryonic cell can induce changes in a neighboring cell without entering the cell. How?

For suggested answers, see Appendix A.

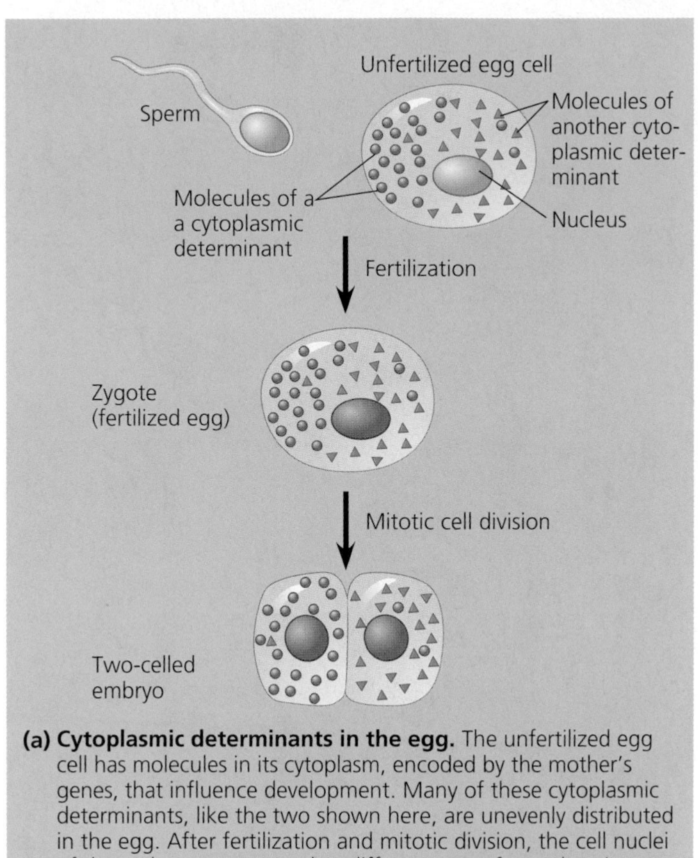

(a) Cytoplasmic determinants in the egg. The unfertilized egg cell has molecules in its cytoplasm, encoded by the mother's genes, that influence development. Many of these cytoplasmic determinants, like the two shown here, are unevenly distributed in the egg. After fertilization and mitotic division, the cell nuclei of the embryo are exposed to different sets of cytoplasmic determinants and, as a result, express different genes.

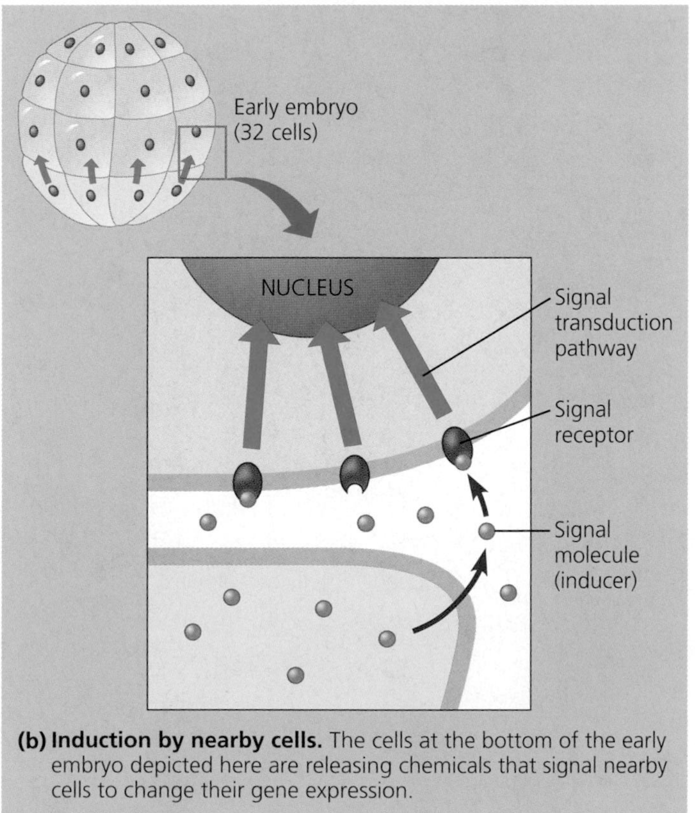

(b) Induction by nearby cells. The cells at the bottom of the early embryo depicted here are releasing chemicals that signal nearby cells to change their gene expression.

▲ **Figure 21.11 Sources of developmental information for the early embryo.**

Pattern formation in animals and plants results from similar genetic and cellular mechanisms

Before morphogenesis can shape an animal or plant, the organism's *body plan*—its overall three-dimensional arrangement—must be established. Cytoplasmic determinants and inductive signals contribute to this process, but what roles do they play? We'll explore this question in the context of **pattern formation**, the development of a spatial organization in which the tissues and organs of an organism are all in their characteristic places. In the life of a plant, pattern formation occurs continually in the apical meristems (see Figure 21.4b). In animals, pattern formation is mostly limited to embryos and juveniles, except in those species where lost parts can be regenerated.

Pattern formation in animals begins in the early embryo, when the major axes of an animal are established. Before construction begins on a new building, the location of the front, back, and sides are determined. In the same way, before specialized tissues or organs of a bilaterally symmetrical animal appear, the relative positions of the animal's head and tail, right and left sides, and back and front are set up, thus establishing the three major body axes. In plants, the root-shoot axis is determined at a similarly early stage of development. The molecular cues that control pattern formation, collectively called **positional information**, are provided by cytoplasmic determinants and inductive signals (see Figure 21.11). These cues tell a cell its location relative to the body axes and to neighboring cells and determine how the cell and its progeny will respond to future molecular signals.

Drosophila Development: A Cascade of Gene Activations

Pattern formation has been most extensively studied in *Drosophila melanogaster,* where genetic approaches have had spectacular success. These studies have established that genes control development and have led to an understanding of the key roles that specific molecules play in defining position and directing differentiation. Combining anatomical, genetic, and biochemical approaches to the study of *Drosophila* development, researchers have discovered developmental principles common to many other species, including humans.

The Life Cycle of Drosophila

Fruit flies and other arthropods have a modular construction, an ordered series of segments. These segments make up the body's three major parts: the head, the thorax (midbody, from which the wings and legs extend), and the abdomen. Like other bilaterally symmetrical animals, *Drosophila* has an anterior-posterior (head-to-tail) axis and a dorsal-ventral (back-to-belly) axis. In *Drosophila,* cytoplasmic determinants that are localized in the unfertilized egg provide positional information for the placement of these two axes even before fertilization. After fertilization, positional information operating on a finer and finer scale establishes a specific number of correctly oriented segments and finally triggers the formation of each segment's characteristic structures.

The *Drosophila* egg cell develops in the female's ovary, surrounded by ovarian cells called nurse cells and follicle cells (**Figure 21.12**, top). These supply the egg cell with nutrients, mRNAs, and other substances needed for development and make the egg shell. After fertilization and laying of the egg, the following events ensue (see key numbers in Figure 21.12):

1. The first ten mitotic divisions have two notable features. First, these early divisions, which occur very quickly, consist of S and M phases only, with no growth, so the amount of cytoplasm does not change. Second, cytokinesis does not occur. As a result, the early *Drosophila* embryo is one big multinucleate cell (in contrast to the embryos of many other animals; see Figure 21.4).
2. At the tenth nuclear division, the nuclei begin to migrate to the periphery of the embryo, forming an early blastula-like stage called the blastoderm.
3. At division 13, plasma membranes finally partition the 6,000 or so nuclei into separate cells of the late blastoderm. Although not yet apparent under the microscope, the basic body plan—including body axes and segment boundaries—has already been determined by this time. A centrally placed yolk nourishes the embryo, and the egg shell continues to protect it.
4. Subsequent events in the embryo create clearly visible segments, which at first look very much alike.
5. Then, some cells move to new positions, organs form, and a wormlike larva (juvenile form) hatches out of the shell. *Drosophila* goes through three larval stages, during which the larva eats, grows, and molts (sheds its tough outer layer).
6. The third larval stage becomes enclosed in a case, forming a pupa.
7. Metamorphosis, the change from larva to adult fly, occurs inside the pupa, and the fly emerges.

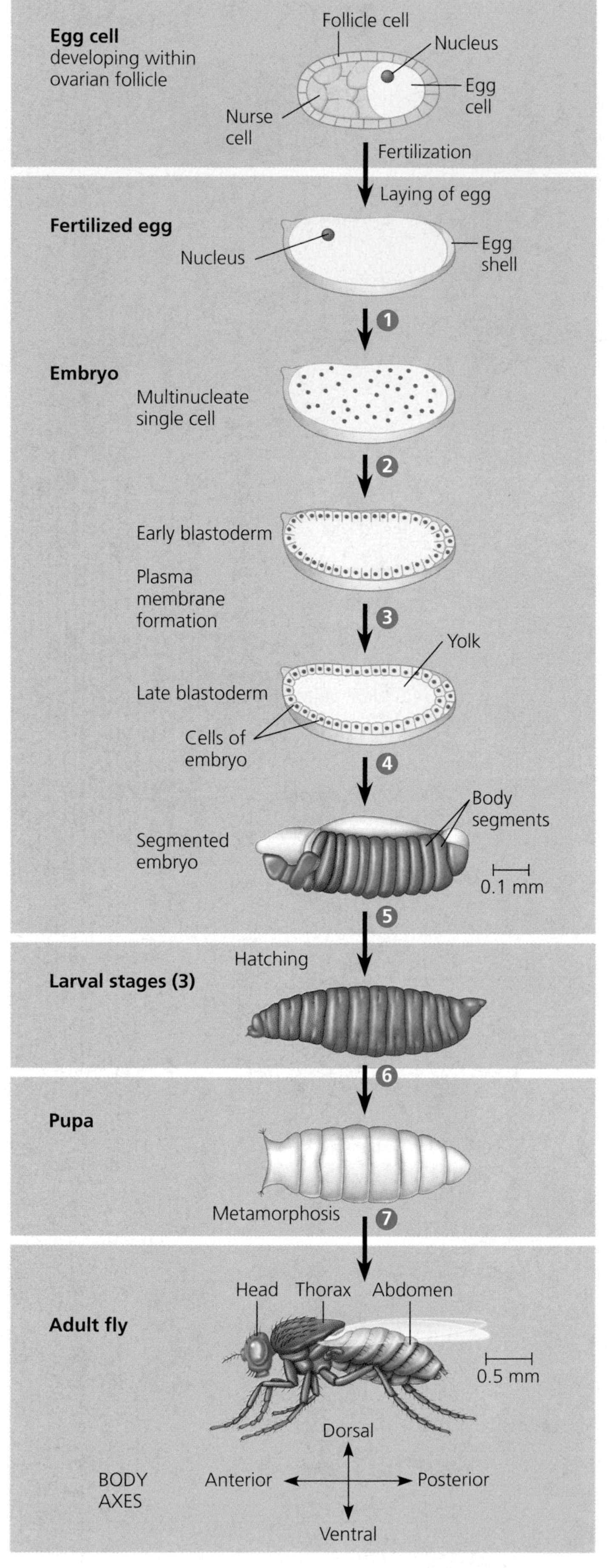

▶ **Figure 21.12 Key developmental events in the life cycle of *Drosophila*.** The yellow egg cell (top) is surrounded by other cells, which form a structure called the follicle within one of the mother's ovaries. The nurse cells shrink and eventually disappear; the egg cell grows and matures, eventually filling the egg shell that is secreted by the follicle cells. The egg is fertilized within the mother and then laid. See text for description of steps 1–7, which generate the segmented adult fly, each segment bearing a characteristic appendage.

In the adult fly, each segment is anatomically distinct, with characteristic appendages (see Figure 21.12, bottom). For example, the first thoracic segment bears a pair of legs; the second thoracic segment has a pair of legs plus a pair of wings; and the third thoracic segment bears a pair of legs plus a pair of balancing organs called halteres.

Genetic Analysis of Early Development: Scientific Inquiry

During the first half of the 20th century, classical embryologists made detailed anatomical observations of embryonic development in a number of species and performed experiments in which they manipulated embryonic tissues. Although this research laid the groundwork for understanding the mechanisms of development, it did not reveal the specific molecules that guide development or determine how patterns are established. Then, in the 1940s, a visionary American biologist, Edward B. Lewis, showed that a genetic approach—the study of mutants—could be used to investigate *Drosophila* development.

Lewis studied bizarre mutant flies with developmental defects that led to extra wings or legs in the wrong places **(Figure 21.13)**. He located the mutations on the fly's genetic map, thus connecting the developmental abnormalities to specific genes. This research supplied the first concrete evidence that genes somehow direct the developmental processes studied by embryologists. The genes Lewis discovered control pattern formation in the late embryo; we'll return to them shortly.

Insight into pattern formation during early development did not come for another 30 years or so, when two researchers in Germany, Christiane Nüsslein-Volhard and Eric Wieschaus, set out to identify *all* the genes that affect segment formation in *Drosophila*. The project was daunting for three reasons. The first was the sheer number of *Drosophila* genes, now known to total about 13,700. The genes affecting segmentation might be just a few needles in a haystack or might be so numerous and varied that the scientists would be unable to make sense of them. Second, mutations affecting a process as fundamental as segmentation would surely be **embryonic lethals,** mutations with phenotypes causing death at the embryonic or larval stage. Because organisms with embryonic lethal mutations never reproduce, they cannot be bred for study. Third, cytoplasmic determinants in the egg were known to play a role in axis formation, and thus the researchers knew they would have to study the mother's genes as well as those of the embryo.

To get around the problem of embryonic lethality, Nüsslein-Volhard and Wieschaus looked for recessive mutations, which can be propagated in heterozygous flies. Their basic strategy was to expose flies to a mutagenic chemical to create mutations in the flies' gametes and then look among the flies' descendants for dead embryos (or larvae) with abnormal segmentation. By doing the appropriate crosses, they would be able to identify living heterozygotes carrying an embryonic lethal mutant allele and a normal allele of the same gene. The researchers hoped that the abnormalities visible in the dead homozygous embryos would suggest how the affected genes normally functioned.

Nüsslein-Volhard and Wieschaus eventually identified about 1,200 genes essential for embryonic development. Of these, about 120 were essential for pattern formation leading to normal segmentation. Over several years, the researchers were able to group these segmentation genes by general function, to map them, and to clone many of them. The result was a detailed molecular understanding of the early steps in pattern formation in *Drosophila*.

When the results of Nüsslein-Volhard and Wieschaus were combined with Lewis's earlier work, a coherent picture of *Drosophila* development emerged. In recognition of their discoveries, the three researchers were awarded a Nobel Prize in 1995. Before we discuss how the segmentation genes function, we need to back up and look at the cytoplasmic determinants deposited in the egg by the mother, for these control the expression of the segmentation genes.

Axis Establishment

As previously mentioned, cytoplasmic determinants in the egg are the substances that initially establish the axes of the *Drosophila* body. These substances are encoded by genes of the

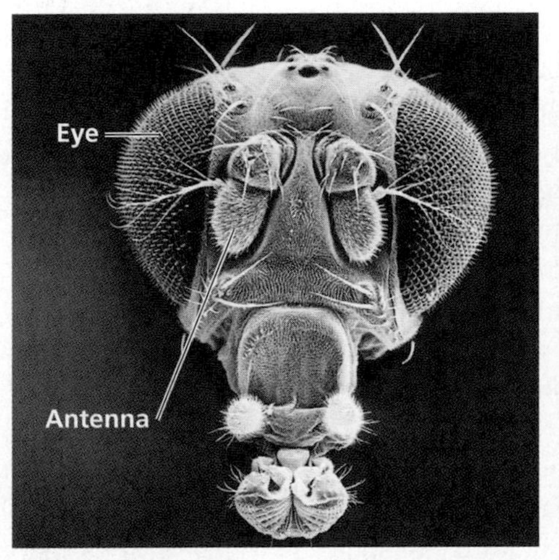

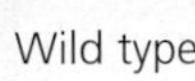

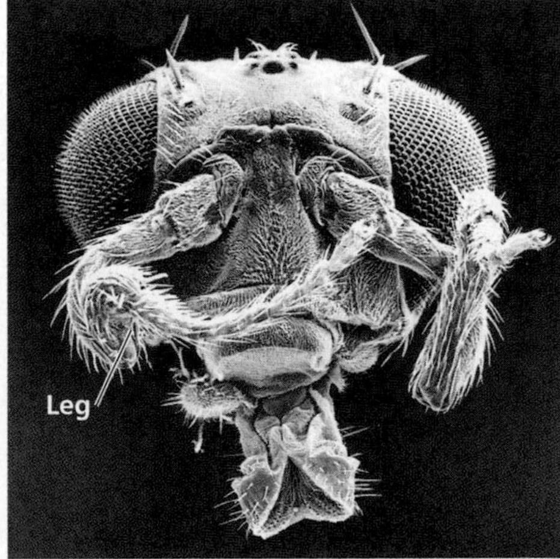

▲ **Figure 21.13 Abnormal pattern formation in *Drosophila*.** Mutations in certain genes, called homeotic mutations, cause a misplacement of structures in an animal. These micrographs contrast the head of a wild-type fly, bearing a pair of small antennae, with that of a homeotic mutant, bearing a pair of legs in place of antennae (SEMs).

mother, fittingly called maternal effect genes. A **maternal effect gene** is a gene that, when mutant in the mother, results in a mutant phenotype in the offspring, regardless of their own genotype. In fruit fly development, the mRNA or protein products of maternal effect genes are placed in the egg while it is still in the mother's ovary. When the mother has a mutation in such a gene, she makes a defective gene product (or none at all), and her eggs are defective; when these eggs are fertilized, they fail to develop properly.

Because they control the orientation (polarity) of the egg and consequently of the fly, maternal effect genes are also called **egg-polarity genes.** One group of these genes sets up the anterior-posterior axis of the embryo, while a second group establishes the dorsal-ventral axis. Like mutations in segmentation genes, mutations in maternal effect genes are generally embryonic lethals.

To see how maternal effect genes determine the body axes of the offspring, we focus on one such gene called *bicoid,* a term meaning "two-tailed." An embryo whose mother has a mutant *bicoid* gene lacks the front half of its body and has posterior structures at both ends **(Figure 21.14a)**. This phenotype suggested to researchers that the product of the mother's *bicoid* gene is essential for setting up the anterior end of the fly and might be concentrated at the future anterior end. This hypothesis is a specific example of the gradient hypothesis first proposed by embryologists a century ago. According to this idea, gradients of substances called **morphogens** establish an embryo's axes and other features of its form.

DNA technology and other modern biochemical methods enabled researchers to test whether the *bicoid* product is in fact a morphogen that determines the anterior end of the fly. The researchers cloned the *bicoid* gene and used a nucleic acid probe derived from the gene to reveal the location of *bicoid* mRNA in the eggs produced by wild-type female flies. As predicted by the hypothesis, the *bicoid* mRNA is highly concentrated at the extreme anterior end of the mature egg cell **(Figure 21.14b)**. After the egg is fertilized, the mRNA is translated into protein. The Bicoid protein then diffuses from the anterior end toward the posterior, resulting in a gradient of protein within the early embryo, with the highest concentration at the anterior end. These results are consistent with the hypothesis that Bicoid protein is responsible for specifying the fly's anterior end. To test the hypothesis more specifically, scientists injected pure *bicoid* mRNA into various regions of early embryos. The protein that resulted from its translation caused anterior structures to form at the injection sites.

The *bicoid* research is important for several reasons. First, it led to the identification of a specific protein required for some of the earliest steps in pattern formation. Second, it increased our understanding of the mother's critical role in the initial phases of embryonic development. (As one developmental biologist has put it, "Mom tells Junior which way is up.") Finally, the principle that a gradient of molecules can determine

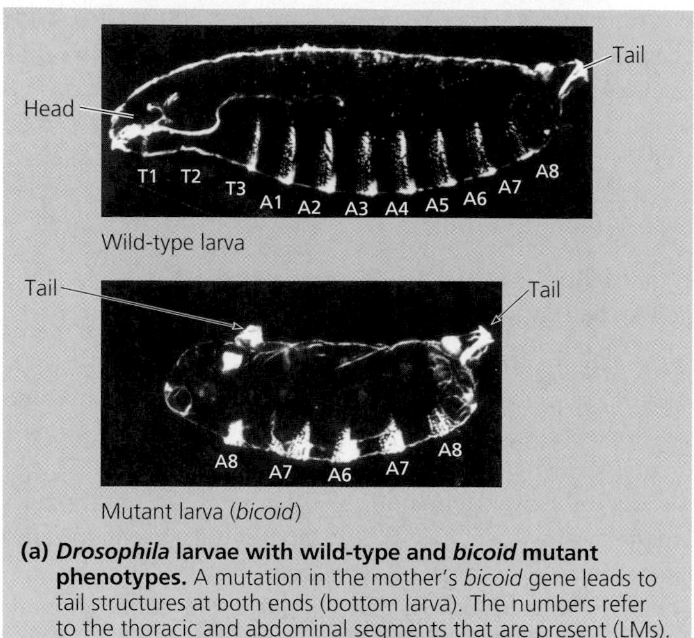

(a) *Drosophila* larvae with wild-type and *bicoid* mutant phenotypes. A mutation in the mother's *bicoid* gene leads to tail structures at both ends (bottom larva). The numbers refer to the thoracic and abdominal segments that are present (LMs).

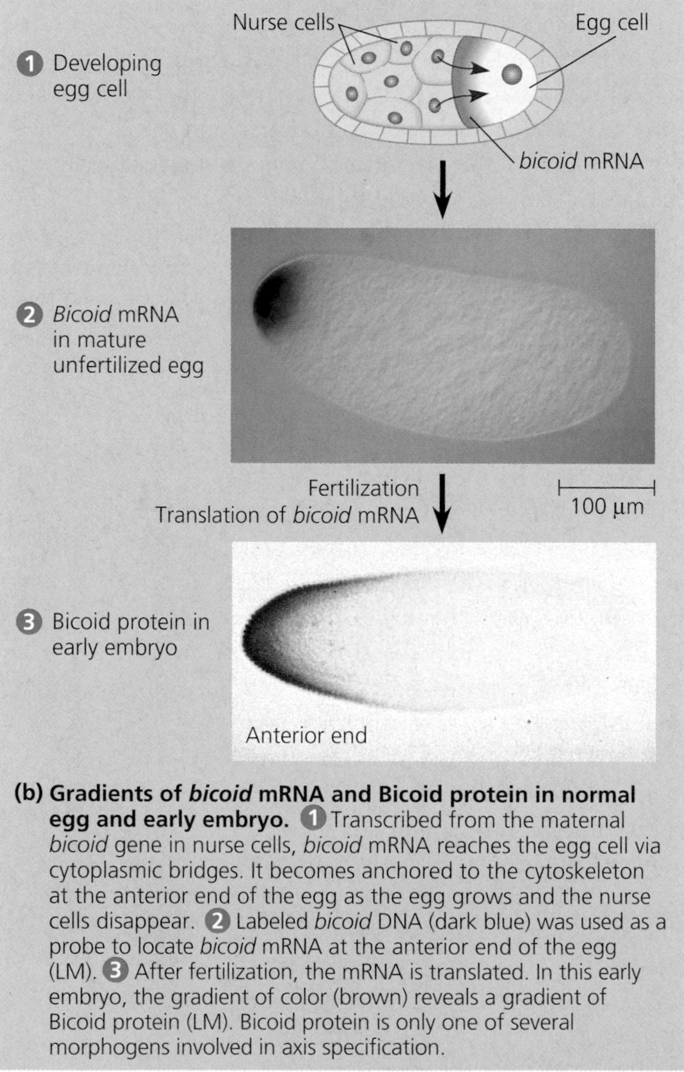

(b) Gradients of *bicoid* mRNA and Bicoid protein in normal egg and early embryo. 1 Transcribed from the maternal *bicoid* gene in nurse cells, *bicoid* mRNA reaches the egg cell via cytoplasmic bridges. It becomes anchored to the cytoskeleton at the anterior end of the egg as the egg grows and the nurse cells disappear. 2 Labeled *bicoid* DNA (dark blue) was used as a probe to locate *bicoid* mRNA at the anterior end of the egg (LM). 3 After fertilization, the mRNA is translated. In this early embryo, the gradient of color (brown) reveals a gradient of Bicoid protein (LM). Bicoid protein is only one of several morphogens involved in axis specification.

▲ **Figure 21.14 The effect of the *bicoid* gene, a maternal effect (egg-polarity) gene in *Drosophila*.**

polarity and position has proved to be a key developmental concept for a number of species, just as early embryologists had thought. In *Drosophila,* gradients of specific proteins determine the posterior end as well as the anterior and also are responsible for establishing the dorsal-ventral axis.

Segmentation Pattern

The Bicoid protein and other proteins encoded by egg-polarity genes regulate the expression of some of the embryo's own genes. Gradients of these proteins bring about regional differences in the expression of **segmentation genes,** the genes of the embryo whose products direct formation of segments after the embryo's major body axes are defined.

In a cascade of gene activations, sequential activation of three sets of segmentation genes provides the positional information for increasingly fine details of the animal's modular body plan. The three sets are called the *gap genes,* the *pair-rule genes,* and the *segment polarity genes.*

The products of many segmentation genes, like those of egg-polarity genes, are transcription factors that directly activate the next set of genes in the hierarchical scheme of pattern formation. Other segmentation genes operate more indirectly, supporting the functioning of the transcription factors in various ways. For example, some are components of cell-signaling pathways, including signal molecules used in cell-cell communication and the membrane receptors that recognize them (see Chapter 11). Cell-signaling molecules are critically important once plasma membranes have divided the embryo into separate cellular compartments.

Working together, the products of egg-polarity genes like *bicoid* regulate the regional expression of gap genes, which control the localized expression of pair-rule genes, which in turn activate specific segment polarity genes in different parts of each segment. The boundaries and axes of the segments are now set. In the hierarchy of gene activations responsible for pattern formation, the next genes to be expressed determine the specific anatomy of each segment along the length of the embryo.

Identity of Body Parts

In a normal fly, structures such as antennae, legs, and wings develop on the appropriate segments. The anatomical identity of the segments is set by master regulatory genes called **homeotic genes.** These are the genes discovered by Edward Lewis. Once the segmentation genes have staked out the fly's segments, homeotic genes specify the types of appendages and other structures that each segment will form. Mutations in homeotic genes can cause an entire structure characteristic of a particular segment of the animal to arise in the wrong segment, as Lewis observed in flies (see Figure 21.13).

Like many of the egg-polarity and segmentation genes, the homeotic genes encode specific transcription factors. These regulatory proteins are gene activators or repressors, controlling expression of genes responsible for particular anatomical structures. For example, a homeotic protein made in the cells of a particular head segment specifies antenna development. In contrast, a homeotic protein active in a certain thoracic segment selectively activates genes that bring about leg development. A mutant version of the gene encoding the thoracic homeotic protein causes the protein to be expressed also in the head segment. There it overrides the normal antennal gene-activating protein, labeling the segment as "thoracic" instead of "head" and causing legs to develop in place of antennae.

Scientists are now busy identifying the genes activated by the homeotic proteins—the genes specifying the proteins that actually build the fly structures. The following flowchart summarizes the cascade of gene activity in the *Drosophila* embryo:

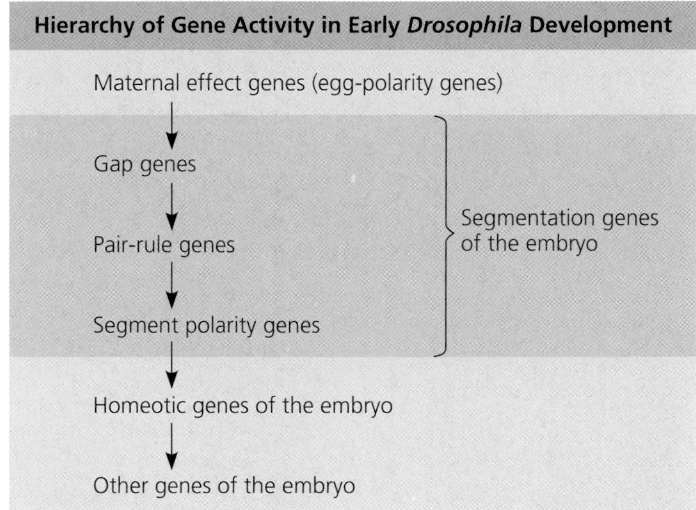

Although this simplified summary suggests a strictly sequential series of gene actions, the reality is more complicated. For instance, the genes in each set not only activate the next set of genes but also maintain their own expression in most cases.

Amazingly, many of the molecules and mechanisms revealed by research on fly pattern formation have turned out to have close counterparts throughout the animal kingdom. The homeotic genes and their products exhibit these similarities in a most striking way. We will return to this point later in the chapter when we consider the evolution of development.

C. elegans: The Role of Cell Signaling

The development of a multicellular organism requires close communication between cells. Indeed, even before fertilization in *Drosophila,* molecules made in neighboring nurse cells cause localization of *bicoid* mRNA at one end of the egg, thus helping to establish the anterior end of the future embryo. Once the embryo is truly multicellular, with membranes enclosing each individual nucleus and accompanying cytoplasm, inductive signaling among the embryo's own cells becomes increasingly important. As we've seen, the ultimate basis for the differences between cells is transcriptional regulation—the turning on and off of specific genes. It is induction, signaling from one group of cells to an adjacent group, that brings about differentiation. In some cases, cell signaling also leads to the programmed death of specific cells, a phenomenon that is also crucial to normal embryonic development.

The nematode *C. elegans* has proved to be a very useful model organism for investigating the roles of cell signaling, induction, and programmed cell death in development (see Figure 21.2). Researchers know the entire ancestry of every cell in the body of an adult *C. elegans,* the organism's complete **cell lineage.** This information can be represented in a cell lineage diagram, somewhat like a pedigree, that shows the fates of all the cells in the developing embryo **(Figure 21.15)**.

Because the lineage of each *C. elegans* cell is so reproducible, scientists at first thought that it must be determined from the start, suggesting that cytoplasmic determinants are the most important means of establishing cell fate in nematodes. However, although cytoplasmic determinants do play a key role very early in *C. elegans* development, a combination of genetic, biochemical, and embryological approaches have revealed important contributions of inductive events as well.

Induction

As early as the four-cell stage in *C. elegans,* cell signaling helps direct daughter cells down the appropriate pathways. For instance, as shown in **Figure 21.16a**, a signal from cell 4 acts on cell 3 so that one of the daughter cells of cell 3 eventually gives rise to the intestine. The signal is a cell-surface protein, made by cell 4, that can be recognized and bound by a cell-surface receptor protein on cell 3. This interaction triggers events inside cell 3 that result in one end of the cell (the posterior end) becoming different from the other. When cell 3 divides, the posterior daughter cell will go on to make the intestine, whereas the anterior daughter cell has a different fate. If cell 4 is experimentally removed early in the four-cell stage, no intestine forms, but if an isolated cell 3 and cell 4 are recombined, the intestine develops as normal. These results helped researchers recognize the role of induction in early nematode development.

Induction is also critical later in nematode development as the embryo passes through three larval stages prior to becoming an adult. The vulva, the tiny opening through which a worm lays its eggs, arises from six cells that are present on the ventral surface at the second larval stage **(Figure 21.16b)**. A single cell of the embryonic gonad, the *anchor cell,* initiates a cascade of inductive signals that establishes the fates of the six

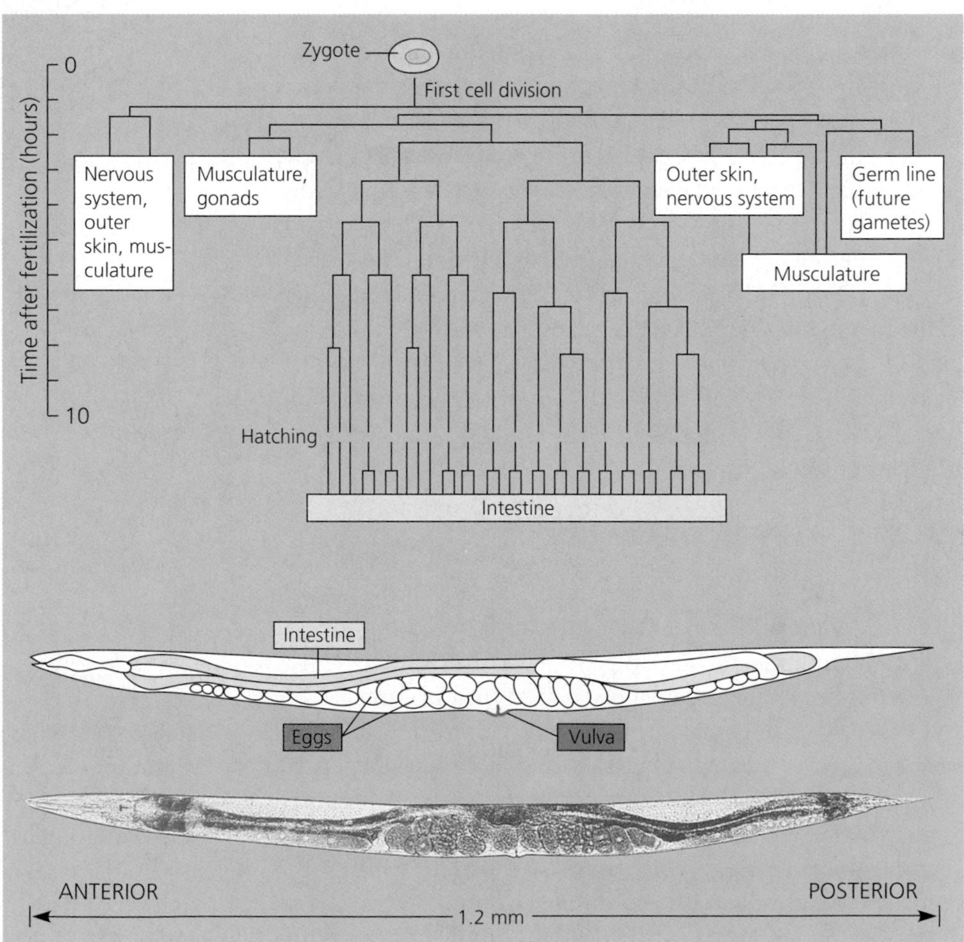

▶ **Figure 21.15 Cell lineage in *C. elegans.*** The *Caenorhabditis elegans* embryo is transparent, making it possible for researchers to trace the lineage of every cell, from the zygote to the adult worm (LM). The diagram shows a detailed lineage only for the intestine (gold), which is derived exclusively from one of the first four cells formed from the zygote. The intestinal cell lineage does not happen to include any programmed cell death, an important aspect of the lineages for some other parts of the animal. The large white cells are eggs, which will be fertilized internally and released through the vulva.

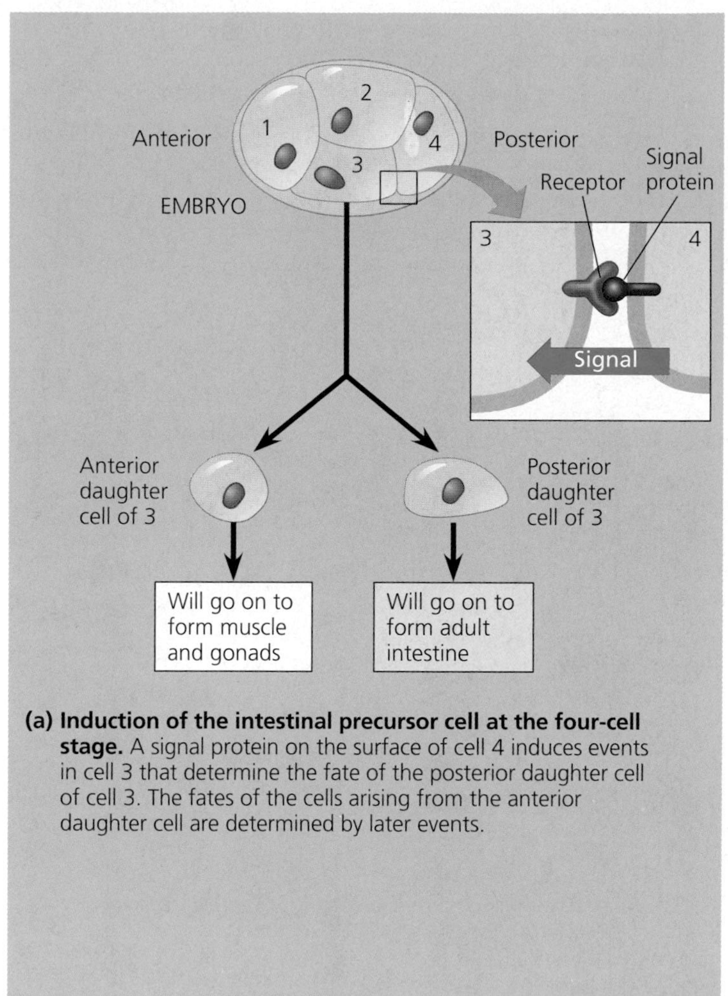

(a) Induction of the intestinal precursor cell at the four-cell stage. A signal protein on the surface of cell 4 induces events in cell 3 that determine the fate of the posterior daughter cell of cell 3. The fates of the cells arising from the anterior daughter cell are determined by later events.

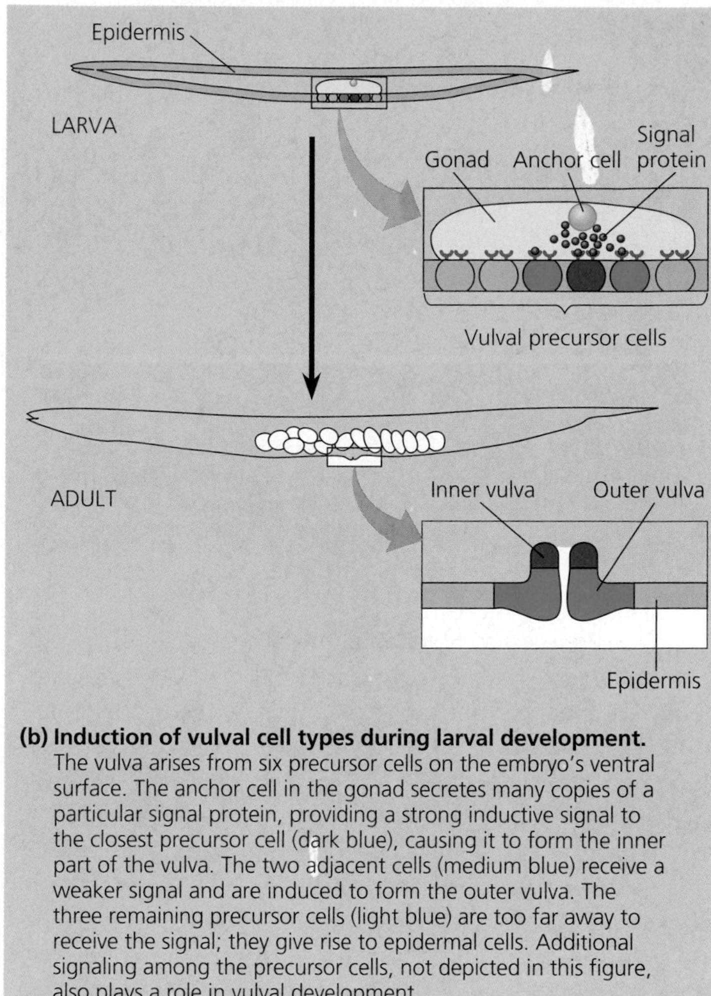

(b) Induction of vulval cell types during larval development. The vulva arises from six precursor cells on the embryo's ventral surface. The anchor cell in the gonad secretes many copies of a particular signal protein, providing a strong inductive signal to the closest precursor cell (dark blue), causing it to form the inner part of the vulva. The two adjacent cells (medium blue) receive a weaker signal and are induced to form the outer vulva. The three remaining precursor cells (light blue) are too far away to receive the signal; they give rise to epidermal cells. Additional signaling among the precursor cells, not depicted in this figure, also plays a role in vulval development.

▲ **Figure 21.16 Cell signaling and induction during development of the nematode.** In both examples, a protein on the surface of or secreted from one cell signals one or more nearby target cells, inducing differentiation of the target cells.

vulval precursor cells. If an experimenter destroys the anchor cell with a laser beam, the vulva fails to form, and the precursor cells simply become part of the worm's epidermis.

The signaling mechanisms in both of these examples are similar to those discussed in Chapter 11. Secreted growth factors or cell-surface proteins bind to a receptor on the recipient cell, initiating intracellular signal transduction pathways. Transcriptional regulation and differential gene expression in the induced cell are the usual results.

These two examples of induction during nematode development illustrate a number of important concepts that apply elsewhere in the development of *C. elegans* and many other animals:

- In the developing embryo, sequential inductions drive the formation of organs.
- The effect of an inducer can depend on its concentration (just as we saw with cytoplasmic determinants in *Drosophila*).
- Inducers produce their effects via signal transduction pathways similar to those operating in adult cells.
- The induced cell's response is often the activation (or inactivation) of genes—transcriptional regulation—which in turn establishes the pattern of gene activity characteristic of a particular kind of differentiated cell.

Programmed Cell Death (Apoptosis)

Lineage analysis of *C. elegans* has underscored another outcome of cell signaling that is crucial in animal development: programmed cell death, or **apoptosis.** The timely suicide of cells occurs exactly 131 times in the course of normal development in *C. elegans*, at precisely the same points in the cell lineage of each worm. In worms and other species, apoptosis is triggered by signals that activate a cascade of "suicide" proteins in the cells destined to die. During apoptosis, a cell shrinks and becomes lobed (called "blebbing"), the nucleus condenses, and the DNA is fragmented (**Figure 21.17**, p. 428). Neighboring cells quickly engulf and digest the membrane-bounded remains, leaving no trace.

Genetic screening of *C. elegans* has revealed two key apoptosis genes, *ced-3* and *ced-4* (*ced* stands for "cell death"), which

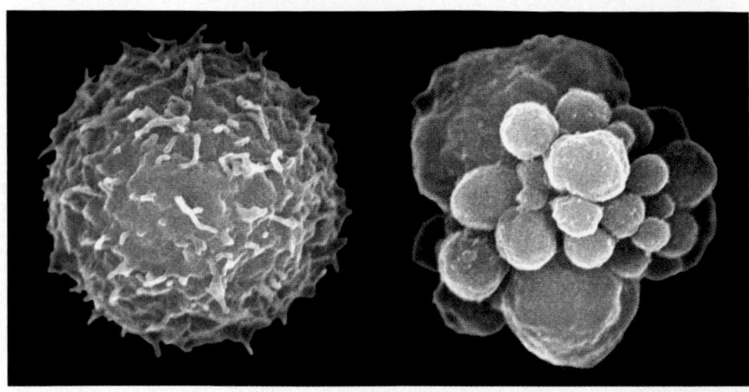

▲ **Figure 21.17 Apoptosis of human white blood cells.** A normal white blood cell (left) is compared with a white blood cell undergoing apoptosis (right). The apoptotic cell is shrinking and forming lobes ("blebs"), which eventually are shed as membrane-bound cell fragments.

encode proteins essential for apoptosis. The proteins are called Ced-3 and Ced-4, respectively. These and most other proteins involved in apoptosis are continually present in cells, but in inactive form; thus, protein *activity* is regulated in this case, not transcription or translation. In *C. elegans*, a protein in the outer mitochondrial membrane, called Ced-9 (the product of gene *ced-9*), serves as a master regulator of apoptosis, acting as a brake in the absence of a signal promoting apoptosis **(Figure 21.18)**. When a death signal is received by the cell, the apoptosis pathway activates proteases and nucleases, enzymes that cut up the proteins and DNA of the cell. The main proteases of apoptosis are called *caspases;* in the nematode, the chief caspase is Ced-3.

In humans and other mammals, several different pathways, involving about 15 different caspases, can lead to apoptosis. The pathway that is used depends on the type of cell and on the particular signal that triggers apoptosis. One important pathway involves mitochondrial proteins. Apoptosis pathway proteins or other signals somehow cause the mitochondrial outer membrane to leak, releasing proteins that promote apoptosis. Surprisingly, these include cytochrome *c*, which functions in mitochondrial electron transport in healthy cells (see Figure 9.15), but acts as a cell death factor when released from mitochondria. The mitochondrial apoptosis of mammals uses proteins homologous to the worm proteins Ced-3, Ced-4, and Ced-9. Mammalian cells make life-or-death "decisions" by somehow integrating the signals they receive, both "death" signals and "life" signals such as growth factors.

A built-in cell suicide mechanism is essential to development in all animals. The similarities between apoptosis genes in nematodes and mammals, as well as the observation that apoptosis occurs in multicellular fungi and single-cell yeasts, indicate that the basic mechanism evolved early in animal evolution. In vertebrates, apoptosis is essential for normal development of the nervous system, for normal operation of the immune system, and for normal morphogenesis of hands and feet in humans and paws in other mammals **(Figure 21.19)**. A lower level of apoptosis in developing limbs accounts for the webbed feet of ducks and other water birds, in contrast to chickens and other land birds with nonwebbed feet. In the case of humans, the failure of appropriate apoptosis can result in webbed fingers and toes. Also, researchers are investigating the possibility that certain degenerative diseases of the nervous system result from the inappropriate activation of apoptosis genes and that some

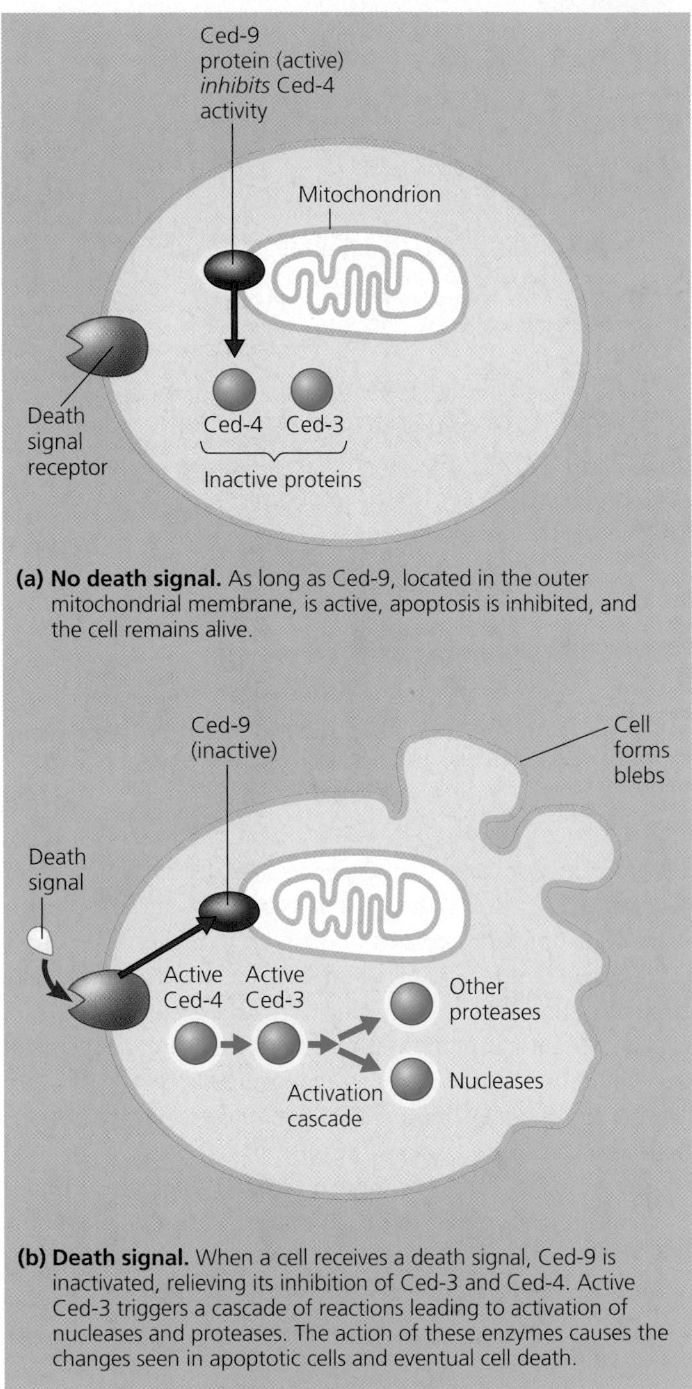

(a) No death signal. As long as Ced-9, located in the outer mitochondrial membrane, is active, apoptosis is inhibited, and the cell remains alive.

(b) Death signal. When a cell receives a death signal, Ced-9 is inactivated, relieving its inhibition of Ced-3 and Ced-4. Active Ced-3 triggers a cascade of reactions leading to activation of nucleases and proteases. The action of these enzymes causes the changes seen in apoptotic cells and eventual cell death.

▲ **Figure 21.18 Molecular basis of apoptosis in *C. elegans*.** Three proteins, Ced-3, Ced-4, and Ced-9, are critical to apoptosis and its regulation in the nematode. Apoptosis is more complicated in mammals but involves proteins similar to those in the nematode.

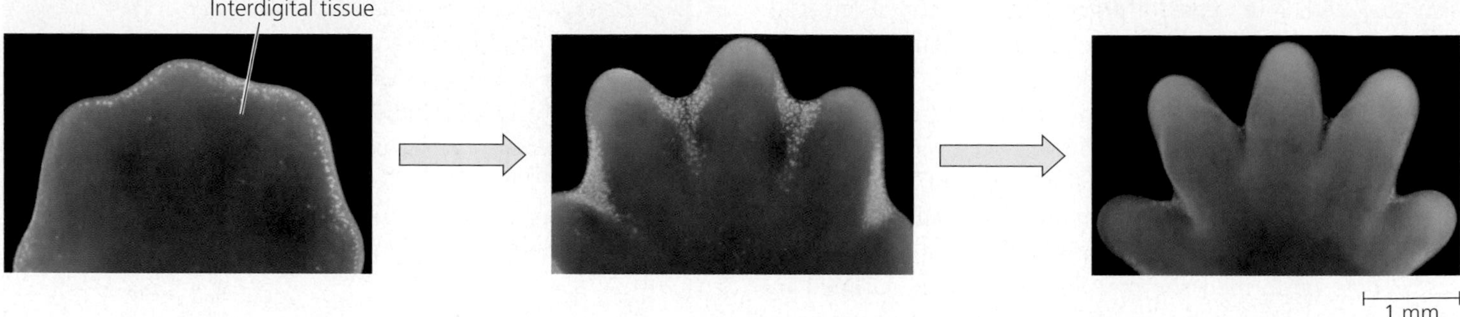

▲ **Figure 21.19 Effect of apoptosis during paw development in the mouse.** In mice, humans, and other mammals, as well as land birds, the embryonic region that develops into feet or hands initially has a solid, platelike structure. Apoptosis eliminates the cells in the interdigital regions, thus forming the digits. The embryonic mouse paws shown in these light micrographs are stained so that cells undergoing apoptosis appear bright yellow. Apoptosis of cells begins at the margin of each interdigital region (left), peaks as the tissue in these regions is reduced (middle), and is no longer visible when the interdigital tissue has been eliminated.

cancers result from a failure of cell suicide. Cells that have suffered irreparable damage, including DNA damage that could lead to cancer, normally generate *internal* signals that trigger apoptosis.

Studies on the roles of induction and apoptosis during development of *C. elegans* were begun less than 30 years ago by Sydney Brenner, John E. Sulston, and H. Robert Horvitz. The importance of their studies was highlighted in 2003, when the Nobel Prize for Medicine was awarded to these researchers for significantly advancing our understanding of how genes regulate organ growth (such as the nematode vulva) and the process of programmed cell death.

Plant Development: Cell Signaling and Transcriptional Regulation

The genetic analysis of plant development, using model organisms such as *Arabidopsis* (see Figure 21.2), has lagged behind that of animal models simply because there are fewer researchers working on plants. For example, in 2000, when the *Arabidopsis* DNA sequence was completed, fewer than 5% of its genes had been defined by mutational analysis, whereas over 25% of the genes in both *Drosophila* and *C. elegans* had been identified in that way. We are just beginning to understand the molecular basis of plant development in detail. Thanks to DNA technology and clues from animal research, plant research is now progressing rapidly.

Mechanisms of Plant Development

In general, cell lineage is much less important for pattern formation in plants than in animals. As mentioned previously, many plant cells are totipotent, and their fates depend more on positional information than on cell lineage. Therefore, the major mechanisms regulating development are cell-signaling (induction) and transcriptional regulation.

The embryonic development of most plants occurs inside the seed and thus is relatively inaccessible to study (a mature seed already contains a fully formed embryo). However, other important aspects of plant development are observable throughout a plant's life in its meristems, particularly the apical meristems at the tips of shoots. It is there that cell division, morphogenesis, and differentiation give rise to new organs, such as leaves or the petals of flowers. We'll discuss two aspects of pattern formation in floral meristems, the apical meristems that produce flowers.

Pattern Formation in Flowers

Environmental signals, such as day length and temperature, trigger signal transduction pathways that convert ordinary shoot meristems to floral meristems, causing a plant to flower. Researchers have combined a genetic approach with tissue transplantation to study induction in the development of tomato flowers. As shown in **Figure 21.20**, a floral meristem is a bump consisting of three layers of cells (L1–L3). All three layers participate in the formation of a flower, a reproductive

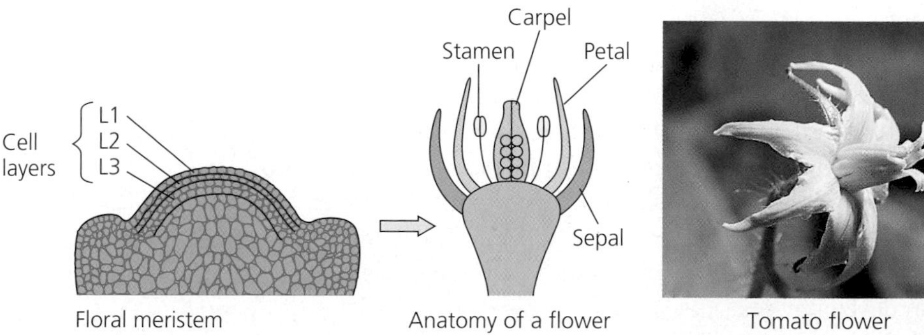

▲ **Figure 21.20 Flower development.** A flower develops from three cell layers (L1–L3) in a floral meristem. A specific pattern of cell division, differentiation, and enlargement produces a flower. The four types of organs (carpels, stamens, petals, and sepals) that make up a flower are arranged in concentric circles (whorls). Each species has a characteristic number of organs in each whorl. The tomato has six sepals, six petals, six stamens, and four carpels.

structure with four types of organs: *carpels* (containing egg cells), *stamens* (containing sperm-bearing pollen), *petals*, and *sepals* (leaflike structures outside the petals). In a mature plant, the four types of organs are arranged radially, rather than linearly like the body structures in *Drosophila*.

Tomato plants homozygous for a mutant allele called *fasciated* (*f*) produce flowers with an abnormally large number of organs. To study what controls the number of organs, researchers performed the grafting experiment outlined in **Figure 21.21**. They grafted stems from fasciated plants onto those of wild-type plants (*FF*, homozygous for normal allele) and then grew new plants from the shoots that sprouted near the graft sites. Many of the new plants were **chimeras,** organisms with a mixture of genetically different cells. Some chimeras produced floral meristems in which the three cell layers did not all come from the same "parent." The researchers identified the parental sources of the meristem layers by monitoring other genetic markers, such as an unrelated mutation causing yellow leaves. The results showed that whether the number of floral organs was normal or abnormally high depended on whether the L3 layer arose from wild-type or mutant cells. Thus, the L3 cell layer induces the overlying L2 and L1 layers to form a particular number of organs. The mechanism of cell-cell signaling leading to this induction is not yet known, but is currently under study.

In contrast to genes controlling organ number in flowers are genes controlling organ identity. An **organ identity gene** determines the type of structure that will grow from a meristem—for instance, whether a particular outgrowth from a floral meristem becomes a petal or a stamen. Most of what we know about such organ identity genes comes from research on flower development in *Arabidopsis*.

Organ identity genes are analogous to homeotic genes in animals and are often referred to as plant homeotic genes. Just as a mutation in a fruit fly homeotic gene can cause legs to grow in place of antennae, a mutation in an organ identity gene can cause carpels to grow in place of sepals. By collecting and studying mutants with abnormal flowers, researchers have been able to identify and clone a number of floral organ identity genes. In plants with a "homeotic" mutation, specific organs are missing or repeated **(Figure 21.22)**. Some of these mutant phenotypes are reminiscent of those caused by *bicoid* or other pattern formation mutations in *Drosophila*. Like the homeotic genes of animals, the organ identity genes of plants encode transcription factors that regulate specific target genes by binding to their enhancers in the DNA. In Chapter 35, you will learn about a current model of how these genes control organ development.

Clearly, the developmental mechanisms used by plants are similar to those used by the two animal species we discussed earlier. In the next section, we will see what can be learned from comparing developmental strategies and molecular mechanisms across all multicellular organisms.

Figure 21.21

Inquiry Which cell layers in the floral meristem determine the number of floral organs?

EXPERIMENT Tomato plants with the *fasciated (ff)* mutation develop extra floral organs.

Sepal
Petal
Carpel
Stamen
Wild-type, normal
Fasciated *(ff)*, extra organs

Researchers grafted stems from mutant plants onto wild-type plants. They then planted the shoots that emerged near the graft site, many of which were chimeras.

Graft
Chimeras

For each chimera, researchers recorded the flower phenotype: wild-type or fasciated. Analysis using other genetic markers identified the parental source for each of the three cell layers of the floral meristem (L1–L3) in the chimeras.

RESULTS The flowers of the chimeric plants had the fasciated phenotype only when the L3 layer came from the fasciated parent.

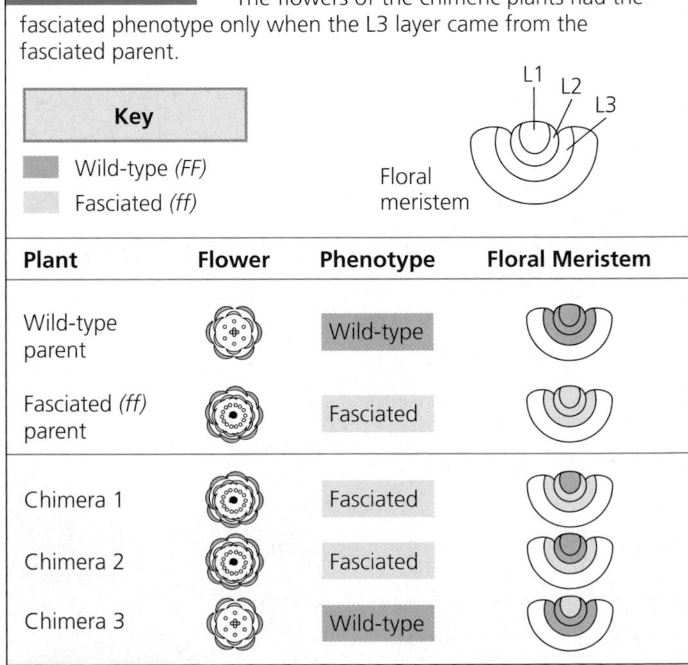

Plant	Flower	Phenotype	Floral Meristem
Wild-type parent		Wild-type	
Fasciated *(ff)* parent		Fasciated	
Chimera 1		Fasciated	
Chimera 2		Fasciated	
Chimera 3		Wild-type	

CONCLUSION Cells in the L3 layer induce the L1 and L2 layers to form flowers with a particular number of organs. (The nature of the inductive signal from L3 is not entirely understood.)

▲ **Figure 21.22 Mutations in floral organ identity genes.** Wild-type *Arabidopsis* has four sepals, four petals, six stamens, and two carpels. If an organ identity gene called *apetala2* is mutated, the identities of the organs in the four whorls are carpels, stamens, stamens, and carpels (there are no petals or sepals).

Concept Check 21.3

1. Why are fruit fly maternal effect genes also called egg-polarity genes?
2. If a researcher removes the anchor cell from a *C. elegans* embryo, the vulva does not form, even though all the cells that would have made the vulva are present. Explain why.
3. Explain why cutting and rooting a shoot from a plant, then planting it successfully, provides evidence that plant cells are totipotent.

For suggested answers, see Appendix A.

Concept 21.4

Comparative studies help explain how the evolution of development leads to morphological diversity

Biologists in the field of evolutionary developmental biology, or "evo-devo" as it is often called, compare developmental processes of different multicellular organisms. Their aim is to understand how developmental processes have evolved and how changes in these processes can modify existing organismal features or lead to new ones. With the advent of molecular techniques and the recent flood of genomic information, we are beginning to realize that the genomes of related species with strikingly different forms may have only minor differences in gene sequence or regulation. Discovering the molecular basis underlying these differences, in turn, helps us understand how the myriad of diverse forms that cohabit this planet have arisen, thus informing the study of evolution.

Widespread Conservation of Developmental Genes Among Animals

Molecular analysis of the homeotic genes in *Drosophila* has shown that they all include a 180-nucleotide sequence called a **homeobox**, which specifies a 60-amino-acid *homeodomain* in the protein. An identical or very similar nucleotide sequence has been discovered in the homeotic genes of many invertebrates and vertebrates. In fact, the vertebrate genes homologous to the homeotic genes of fruit flies have even kept their chromosomal arrangement **(Figure 21.23)**. (Homeotic genes in

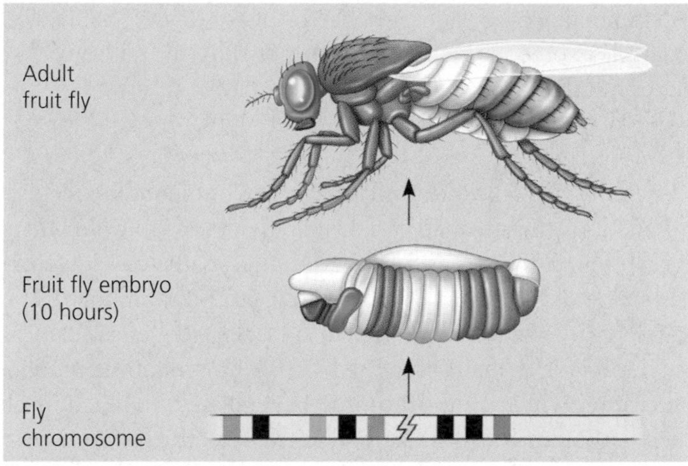

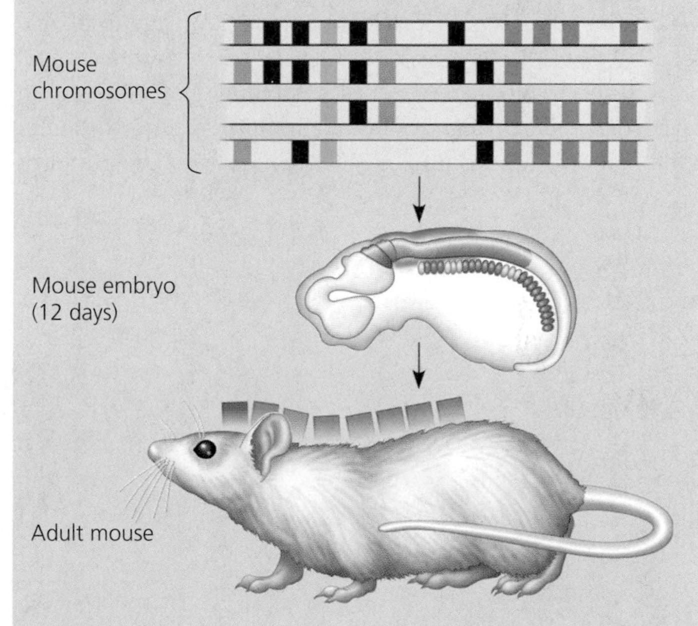

▲ **Figure 21.23 Conservation of homeotic genes in a fruit fly and a mouse.** Homeotic genes that control the form of anterior and posterior structures of the body occur in the same linear sequence on chromosomes in *Drosophila* and mice. Each colored band on the chromosomes shown here represents a homeotic gene. In fruit flies, all homeotic genes are found on one chromosome. The mouse and other mammals have the same or similar sets of genes on four chromosomes. The color code indicates the parts of the embryos in which these genes are expressed and the adult body regions that result. All of these genes are essentially identical in flies and mice, except for those represented by black bands, which are less similar in the two animals.

animals are often called *Hox* genes.) Furthermore, related sequences have been found in regulatory genes of much more distantly related eukaryotes, including plants and yeasts, and even in prokaryotes. From these similarities, we can deduce that the homeobox DNA sequence evolved very early in the history of life and was sufficiently valuable to organisms to have been conserved in animals and plants virtually unchanged for hundreds of millions of years.

Not all homeobox-containing genes are homeotic genes; that is, some do not directly control the identity of body parts. However, most of these genes, in animals at least, are associated with development, suggesting their ancient and fundamental importance in that process. In *Drosophila,* for example, homeoboxes are present not only in the homeotic genes but also in the egg-polarity gene *bicoid,* in several of the segmentation genes, and in the master regulatory gene for eye development.

Researchers have found that the homeobox-encoded homeodomain is the part of a protein that binds to DNA when the protein functions as a transcriptional regulator. However, the shape of the homeodomain allows it to bind to any DNA segment; by itself it cannot select a specific sequence. Rather, more variable domains in a homeodomain-containing protein determine which genes the protein regulates. Interaction of these latter domains with still other transcription factors helps a homeodomain-containing protein recognize specific enhancers in the DNA. Proteins with homeodomains probably regulate development by coordinating the transcription of batteries of developmental genes, switching them on or off. In embryos of *Drosophila* and other animal species, different combinations of homeobox genes are active in different parts of the embryo. This selective expression of regulatory genes, varying over time and space, is central to pattern formation.

Developmental biologists have found that in addition to homeotic genes, many other genes involved in development are highly conserved from species to species. These include numerous genes encoding components of signaling pathways. The extraordinary similarity among particular developmental genes in different animal species raises the question, How can the same genes be involved in the development of animals whose forms are so very different from each other?

Current studies are suggesting likely answers to this question. In some cases, small changes in regulatory sequences of particular genes can lead to major changes in body form. For example, the differing patterns of expression of the *Hox* genes along the body axis in insects and crustaceans can explain the different number of leg-bearing segments among these segmented animals **(Figure 21.24)**. In other cases, similar genes direct different developmental processes in different organisms, resulting in different body shapes. Several *Hox* genes, for instance, are expressed in the embryonic and larval stages of the sea urchin, a nonsegmented animal that has a body plan quite different from those of insects and mice. Sea urchin adults make the pincushion-shaped shells you may have seen on the beach. They are among the organisms long used in classical embryological studies (see Chapter 47).

Sequencing of the *Arabidopsis* genome has revealed that plants do have some homeobox-containing genes. However, these apparently do not function as master regulatory switches as do the homeobox-containing homeotic genes in animals. Other genes appear to carry out basic processes of pattern formation in plants.

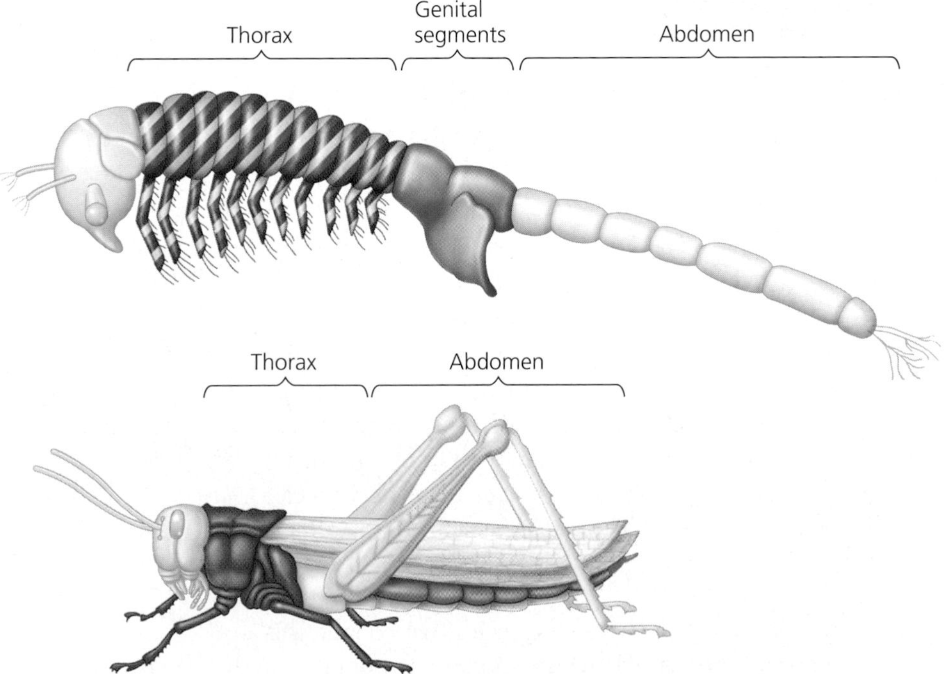

▶ **Figure 21.24 Effect of differences in *Hox* gene expression during development in crustaceans and insects.** Changes in the expression patterns of four *Hox* genes have occurred over evolutionary time. These changes account in part for the different body plans of the brine shrimp *Artemia,* a crustacean (top), and the grasshopper, an insect. Shown here are regions of the adult body color-coded for expression of the *Hox* genes that determine formation of particular body parts during embryonic development.

Comparison of Animal and Plant Development

The last common ancestor of plants and animals was probably a single-celled microbe that lived hundreds of millions of years ago, so the processes of development must have evolved independently in the two lineages of organisms. Plants evolved with rigid cell walls that make the movement of cells and tissue layers virtually impossible, ruling out the morphogenetic movements of cells and tissues that are important in animals. Instead, morphogenesis in plants relies more heavily on differing planes of cell division and on selective cell enlargement. (You will learn about these processes in Chapter 35.) But despite the differences between plants and animals, there are some basic similarities in the actual mechanisms of development—legacies of their shared cellular origins.

In both plants and animals, development relies on a cascade of transcriptional regulators turning on or turning off genes in a finely tuned series—for example, setting up the head-to-tail axis in *Drosophila* and establishing the organ identities in a radial pattern in the *Arabidopsis* flower. But the genes that direct these processes differ considerably in plants and animals. While quite a few of the master regulatory switches in *Drosophila* are homeobox-containing *Hox* genes, those in *Arabidopsis* belong to a completely different family of genes, called the *Mads-box* genes. And although homeobox-containing genes can be found in plants and *Mads-box* genes in animals, in neither case do they perform the same major roles in development that they do in the other group.

In this final chapter of the genetics unit, you have learned how genetic studies can reveal much about the molecular and cellular mechanisms underlying development. The unity of life is reflected in the similarity of biological mechanisms used to establish body pattern, although the genes directing development may differ among organisms. The similarities reflect the common ancestry of life on Earth. But the differences are also crucial, for they have created the huge diversity of organisms that have evolved. In the remainder of the book, we expand our perspective beyond the level of molecules, cells, and genes to explore this diversity on the organismal level.

Concept Check 21.4

1. The DNA sequences called homeoboxes, which help homeotic genes in animals direct development, are common to flies and mice. Given this similarity, explain why these animals are so different.

For suggested answers, see Appendix A.

Chapter 21 Review

Go to www.campbellbiology.com or the student CD-ROM to explore Activities, Investigations, and other interactive study aids.

SUMMARY OF KEY CONCEPTS

Concept 21.1

Embryonic development involves cell division, cell differentiation, and morphogenesis

- In addition to mitosis, embryonic cells undergo differentiation, becoming specialized in structure and function. Morphogenesis encompasses the processes that give shape to the organism and its various parts. Several model organisms are commonly used to study different aspects of the genetic basis of development (pp. 412–415).

Concept 21.2

Different cell types result from differential gene expression in cells with the same DNA

- **Evidence for Genomic Equivalence (pp. 415–418)** Cells differ in structure and function not because they contain different genes but because they express different portions of a common genome; they have genomic equivalence. Differentiated cells from mature plants are often totipotent, capable of generating a complete new plant. The nucleus from a differentiated animal cell can sometimes give rise to a new animal if transplanted to an enucleated egg cell. Pluripotent stem cells from animal embryos or adult tissues can reproduce and differentiate *in vitro* as well as *in vivo,* offering the potential for medical use.
- **Transcriptional Regulation of Gene Expression During Development (pp. 418–420)** Differentiation is heralded by the appearance of tissue-specific proteins. These proteins enable differentiated cells to carry out their specialized roles.
- **Cytoplasmic Determinants and Cell-Cell Signals in Cell Differentiation (p. 420)** Cytoplasmic determinants in the cytoplasm of the unfertilized egg regulate the expression of genes in the zygote that affect the developmental fate of embryonic cells. In the process called induction, signal molecules from embryonic cells cause transcriptional changes in nearby target cells.
  Activity ***Signal Transduction Pathways***

Concept 21.3

Pattern formation in animals and plants results from similar genetic and cellular mechanisms

- Pattern formation, the development of a spatial organization of tissues and organs, occurs continually in plants but is mostly limited to embryos and juveniles in animals. Positional information, the molecular cues that control pattern formation, tell a cell its location relative to the body's axes and to other cells (p. 421).
- ***Drosophila* Development: A Cascade of Gene Activations (pp. 421–425)** After fertilization, positional information on an increasingly fine scale specifies the segments in *Drosophila* and

finally triggers the formation of each segment's characteristic structures. Gradients of morphogens encoded by maternal effect genes, such as *bicoid,* produce regional differences in the sequential expression of three sets of segmentation genes, the products of which direct the actual formation of segments. Finally, master regulatory genes, called homeotic genes, specify the type of appendages and other structures that form on each segment. Transcription factors encoded by the homeotic genes are regulatory proteins that control the expression of genes responsible for specific anatomical structures.
Activity *Role of* bicoid *Gene in* Drosophila *Development*
Investigation *How Do* bicoid *Mutations Alter Development?*

▶ ***C. elegans:* The Role of Cell Signaling** (pp. 425–429) The complete lineage of each cell in *C. elegans* is known. Cell signaling and induction are critical in determining worm cell fates, including apoptosis (programmed cell death). An inducing signal produced by one cell in the embryo can initiate a chain of inductions that results in the formation of a particular organ, such as the intestine or vulva. In apoptosis, precisely timed signals trigger the activation of a cascade of "suicide" proteins in the cells destined to die.

▶ **Plant Development: Cell Signaling and Transcriptional Regulation** (pp. 429–431) Induction by cell-cell signaling helps determine the numbers of floral organs that develop from a floral meristem. Organ identity genes determine the type of structure (stamen, carpal, sepal, or petal) that grows from each whorl of a floral meristem. The organ identity genes apparently act as master regulatory genes, each controlling the activity of other genes that more directly bring about an organ's structure and function.

Concept 21.4

Comparative studies help explain how the evolution of development leads to morphological diversity

▶ **Widespread Conservation of Developmental Genes Among Animals** (pp. 431–432) Homeotic genes and some other genes associated with animal development contain a homeobox region, whose sequence is identical or similar in diverse species. Related sequences are present in the genes of yeasts, plants, and even prokaryotes. Other developmental genes also are highly conserved among animal species. In many cases, genes with conserved sequences play different roles in the development of different species. In plants, for instance, homeobox-containing genes do not function in pattern formation as they do in many animals.

▶ **Comparison of Animal and Plant Development** (p. 433) During embryonic development in both plants and animals, a cascade of transcription regulators turns genes on or off in a carefully regulated sequence. But the genes that direct analogous developmental processes differ considerably in sequence in plants and animals, as a result of their remote ancestry.

TESTING YOUR KNOWLEDGE

Self-Quiz

1. Which of the following processes is most directly responsible for the lack of webbing between the fingers of most humans?
   a. pattern formation.
   b. transcriptional regulation.
   c. apoptosis.
   d. cell division.
   e. induction.

2. The criteria for a good model organism for studying development would probably include all of the following *except*
   a. observable embryonic development.
   b. short generation time.
   c. a relatively small genome.
   d. preexisting knowledge of the organism's life history.
   e. a rare pattern of development when compared to most organisms.

3. Totipotency is demonstrated when
   a. mutations in homeotic genes result in the development of misplaced appendages.
   b. a cell isolated from a plant leaf grows into a normal adult plant.
   c. an embryonic cell divides and differentiates.
   d. replacing the nucleus of an unfertilized egg with that of an intestinal cell converts the egg to an intestinal cell.
   e. segment-specific organs develop along the anterior-posterior axis of a *Drosophila* embryo.

4. Cell differentiation always involves
   a. the production of tissue-specific proteins, such as muscle actin.
   b. the movement of cells.
   c. the transcription of the *myoD* gene.
   d. the selective loss of certain genes from the genome.
   e. the cell's sensitivity to environmental cues such as light or heat.

5. The development of *Drosophila* is somewhat unusual in that
   a. the early mitotic divisions proceed without cytokinesis.
   b. metamorphosis occurs during the larval stage rather than the pupal stage, as with other insects.
   c. homeotic genes are mutated.
   d. cell migration within the embryo does not occur.
   e. the initial cell divisions have lengthy G_1 phases.

6. In *Drosophila,* which genes initiate a cascade of gene activation that includes all other genes in the list?
   a. homeotic genes
   b. gap genes
   c. pair-rule genes
   d. egg-polarity genes
   e. segment polarity genes

7. Absence of *bicoid* mRNA from a *Drosophila* egg leads to the absence of anterior larval body parts and mirror-image duplication of posterior parts. This is evidence that the product of the *bicoid* gene
   a. is transcribed in the early embryo.
   b. normally leads to formation of tail structures.
   c. normally leads to formation of head structures.
   d. is a protein present in all head structures.
   e. leads to programmed cell death.

8. Homeotic genes
   a. encode transcription factors that control the expression of genes responsible for specific anatomical structures.
   b. are found only in *Drosophila* and other arthropods.
   c. specify the anterior-posterior axis for each fruit fly segment.
   d. create the basic subdivisions of the anterior-posterior axis of the fly embryo.
   e. are responsible for the programmed cell death occurring during morphogenesis.

9. The embryonic development of *C. elegans* illustrates all of the following developmental concepts *except*:
   a. An inducer's effect can depend on its concentration gradient.
   b. The response of an induced cell involves the establishment of a unique pattern of gene activity.
   c. The signal transduction pathways activated by inducers are unique to embryonic cells.
   d. Sequential inductions direct the formation of complex structures in the developing embryo.
   e. Inducers bring about their effects via the activation or inactivation of genes that code for transcriptional regulators.

10. Although quite different in structure, plants and animals share some basic similarities in their development, such as
    a. the importance of cell and tissue movements.
    b. the importance of selective cell enlargement.
    c. the importance of homeobox-containing homeotic genes.
    d. the retention of meristematic tissues in the adult.
    e. master regulatory genes that encode DNA-binding proteins.

For Self-Quiz answers, see Appendix A.

Go to the website or CD-ROM for more quiz questions.

Evolution Connection

Genes important in the embryonic development of animals, such as homeobox-containing genes, have been relatively well conserved during evolution; that is, they are more similar among different species than are many other genes. Why is this?

Scientific Inquiry

Stem cells in an adult organism can divide to form two daughter stem cells, thus maintaining a population of relatively undifferentiated cells. Alternatively, a given mitotic division may yield one daughter cell that remains a stem cell and a second daughter cell that initiates a differentiation pathway. Propose one or more hypotheses to explain how this can happen. (*Note:* There is no easy answer to this question, but it is worth considering. For a hint, look at Figure 21.16a.)

Investigation *How Do* bicoid *Mutations Alter Development?*

Science, Technology, and Society

Government funding of embryonic stem cell research has been a contentious political issue. Why has this debate been so heated? Summarize the arguments for and against embryonic stem cell research, and explain your own position on the issue.

UNIT 4

Mechanisms of Evolution

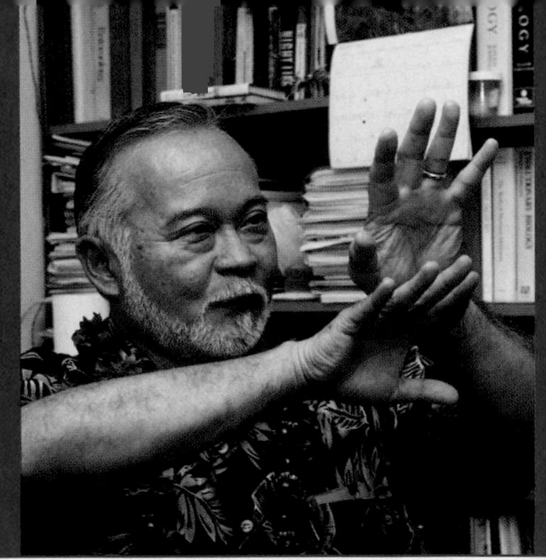

AN INTERVIEW WITH

Kenneth Kaneshiro

The Hawaiian Islands are one of Earth's greatest natural laboratories for understanding mechanisms of evolution. Dr. Kenneth Kaneshiro has contributed much to that understanding through his research on the diverse species of Hawaiian *Drosophila* flies. Professor Kaneshiro is the director of the Hawaiian Evolutionary Biology Program at the University of Hawaii, Manoa, where he is also the director of the Center for Conservation Research and Training. I first met Dr. Kaneshiro in 2003, when I was a visiting scholar at Iolani School in Honolulu, where Kaneshiro began his education. It was a joy returning to Hawaii a year later to conduct this interview. I love this job!

How did the Hawaiian Islands form?

Geologically, the islands are very young, with the oldest island, Kauai, being only about 5 to 6 million years old. The islands popped up in chronological fashion as the Pacific Plate moved northwest over a volcanic hotspot on the seafloor. The youngest island, the Big Island of Hawaii, is over the hotspot now, and there is a new island, Loihi, beginning to form as an undersea mountain to the southeast of the Big Island.

And what makes the Hawaiian Islands such a compelling place to study evolution?

First, they're the most isolated landmass in the world, sitting out in the middle of the Pacific nearly 2,000 miles away from continents in any direction. So for any species of organism that arrived here—by blowing across the ocean on the wind, for example—the founding population would have then been isolated from gene flow from other populations of that species. And some of the organisms that made it here radiated profusely into new species by colonizing other islands. Since the islands formed in single file, from Kauai to the Big Island, you also have a chronological sequence in the origin of species as founders went from older islands to newer ones. In the case of *Drosophila*, the evidence points to a single founder (a fertilized female) that arrived several million years ago and whose progeny eventually radiated into the more than 500 described species of Hawaiian *Drosophila* flies. That's about a quarter of all known *Drosophila* species in the world.

Obviously, that's far more *Drosophila* species than there are islands. Did environmental variation on each island contribute to that radiation of species?

Yes, differences in elevation, rainfall, and other factors make each island very environmentally diverse. And each island also has what are called *kipukas,* "islands" of vegetation surrounded by lava. So, there are islands within islands, with the lava forming barriers between the kipukas. For example, we have studied two kipukas on the Big Island that used to be connected before a lava flow separated them just 100 years ago. And we're already detecting significant genetic differences between *Drosophila* populations that live in these two kipukas. They're still the same species, but they've started to diverge. Speciation is still a very dynamic process on the Hawaiian Islands.

How did your interest in biology develop?

It probably started when I was a child growing up here in Hawaii. When my dad took us fishing around Oahu, it was very scientific, though it didn't seem that way to me at the time. My dad considered the tides and wind, and he matched the colors of his fishing lures to the kinds of baitfish that might be out there. When I fish today, I still apply those lessons from my dad. I think it was that sort of scientific approach to fishing that first got me interested in biology.

When did that interest turn to evolutionary biology?

When I was a freshman here at the University of Hawaii, I wanted to go into marine biology. But because of family pressures, I enrolled in the premed program instead. To help pay my way, I took a job with The Hawaiian *Drosophila* Project, starting as a dishwasher in the lab and learning how to prepare the nutrient media for breeding the flies. Within a few months, I was involved in actual research, dissecting genitalia and looking at other morphological characters that give us clues about the evolutionary history of the Hawaiian *Drosophila* species. I also had the opportunity to go into the field. So ecology also became part of my undergraduate education, not so much from the classroom, but by being involved in the *Drosophila* research. Every summer, 10 or 12 top scientists with different research specialties would visit to work on various *Drosophila* projects. Looking over the shoulders of these eminent scientists when I was an undergraduate really got me hooked. After I graduated, I switched to entomology for my graduate work so I could continue studying the Hawaiian *Drosophila*.

And that work included your research on mating behavior in *Drosophila*. I think many students will be surprised to learn that these flies court their mates.

The courtship between males and females is very elaborate. And the mating behavior also includes competition between males. A male will defend a territory to which females are attracted for mating. In one *Drosophila* species, for example, the males have very wide heads, and two males will butt their heads together—like rams—and then push each other back and forth as they joust for territory. In another species, the competition is more like sumo wrestling; the males get up "tippy-toe" on their hind legs, grappling with their middle and forelegs, and lock heads. But being a good fighter doesn't mean that you're also a good lover. For a male fly, being able to fight off other males and defend a territory gives him the most opportunity to encounter females, but he still has to be able to perform the very complex behavior that satisfies the courtship requirements of a female.

Such as?

In one species, the male hoists his abdomen up and over his head, in a scorpion-like pose. This displays to the female a row of specialized bristles that are on the underside of the abdomen. Each bristle is flattened like a fan. Then the male vibrates his abdomen, and the bristles waft a sex-attractant vapor called a pheromone secreted from an abdominal gland. At the same time, the male spreads his wings and rocks back and forth, emitting a sound. While dancing and singing, the male extends his mouthparts from a very white face. In response, the female actually kisses the male. Mating only occurs if the male can perform this elaborate display.

According to a model now known as the "Kaneshiro Hypothesis," changes in such mating behavior played a key role in the origin of Hawaiian *Drosophila* species, especially in the early stages of speciation. What's the basic idea?

Shifts in mating behavior can occur in a small population after a founding event. Say you have a population of flies on Kauai, the oldest island. Then Oahu pops up, and a fertilized female happens to make it there. She may found a new population on Oahu by producing a few hundred offspring. The males will vary in their ability to perform the species' original courtship rituals. But in such a small population, females who are very choosy will have less opportunity to reproduce than less choosy females, who will encounter more mates they are willing to accept. So selection favors fresh combinations of genes that combine adaptations to the new environment along with less rigid mating behavior than in the "parent" species back on Kauai. That would explain why mating behavior is typically the most complex in the oldest *Drosophila* species. I think such shifts in mating behavior have been very important in the evolution of the Hawaiian *Drosophila*, and probably in many other groups of organisms as well.

At the same time that biologists are studying the evolution of such a diversity of species on the Hawaiian islands, the islands have been designated a biodiversity hotspot, meaning that many species are endangered. What are the biggest threats to biodiversity in Hawaii?

There is the destruction of habitat, but the impact of invasive species—nonnative species that are accidentally or purposefully brought to the islands—is probably the biggest threat. And ants may be the number one problem. Hawaii doesn't have any native ants; they're all alien. Their foraging in native forest ecosystems has severely impacted the native arthropod fauna. Invasive rats threaten the native birds by getting into nests and eating the eggs and young. Wild pigs are also a serious problem. They descended from hybridization between pigs the native Hawaiians brought to the islands and the pigs brought by Europeans. The wild pigs root in the forests, which creates puddles in which mosquitoes breed. The mosquitoes carry pathogens that cause malaria in birds. Also, invasive plants are crowding out many of the native plant species. Unfortunately, in spite of its small size, Hawaii is the extinction capital of the United States, if not the world.

One of your many hats is your role as director of the Center for Conservation Research and Training here at the University of Hawaii. What kind of work does the Center conduct?

Our research interest in conservation biology mainly takes an ecosystem and ecoregional approach: What happens on the mountaintops affects ecosystems all the way down to the coral reefs and beyond. I think we have to understand such relationships between ecosystems in order to protect the islands' biodiversity. The Center also has a major commitment to education. We have a National Science Foundation grant that supports outreach programs where our graduate students work with schoolchildren and mentor their teachers. We get these K-12 students out into the field participating in research: They are collecting valid scientific data, discovering new species, and helping us understand how to eradicate some of the alien species. The sheer numbers of these young scientists enable us to make certain kinds of measurements that would be impossible otherwise. For example, researchers would usually monitor contaminants in a stream by collecting water samples at a few points along the stream. But our graduate students worked with 320 seventh graders to sample the water all the way from a mountain waterfall to where the stream drains into the ocean. To me, this kind of environmental education, beginning at a young age, will eventually increase public awareness and make us all more effective in protecting our water resources and our native ecosystems.

Aloha and mahalo, Dr. Kaneshiro!

In the case of Drosophila, *the evidence points to a single founder (a fertilized female) that arrived several million years ago and whose progeny eventually radiated into the more than 500 described species of Hawaiian* Drosophila *flies.*

22 Descent with Modification

A Darwinian View of Life

▲ **Figure 22.1 An iguana of the Galápagos Islands, well-suited to its rocky habitat.**

Key Concepts

22.1 **The Darwinian revolution challenged traditional views of a young Earth inhabited by unchanging species**

22.2 **In *The Origin of Species*, Darwin proposed that species change through natural selection**

22.3 **Darwin's theory explains a wide range of observations**

Overview

Darwin Introduces a Revolutionary Theory

A new era of biology began on November 24, 1859, the day Charles Darwin published *On the Origin of Species by Means of Natural Selection*. Darwin's book drew a cohesive picture of life by connecting the dots among what had once seemed a bewildering array of unrelated observations. *The Origin of Species* focused biologists' attention on the great diversity of organisms—their origins and relationships, their similarities and differences, their geographic distribution, and their adaptations to surrounding environments **(Figure 22.1)**.

Darwin made two major points in *The Origin of Species*. First, he presented evidence that the many species of organisms presently inhabiting Earth are descendants of ancestral species that were different from the modern species. Second, he proposed a mechanism for this evolutionary process, which he termed **natural selection.** The basic idea of natural selection is that a population can change over generations if individuals that possess certain heritable traits leave more offspring than other individuals. The result of natural selection is **evolutionary adaptation,** an accumulation of inherited characteristics that enhance organisms' ability to survive and reproduce in specific environments. In modern terms, we can define **evolution** as a change over time in the genetic composition of a population. Eventually, a population may accumulate enough change that it constitutes a new species—a new life-form. Thus we can also use the term *evolution* on a grand scale to mean the gradual appearance of all of biological diversity, from the earliest microbes to the enormous variety of organisms alive today.

Evolution is such a fundamental concept that its study illuminates biology at every level from molecules to ecosystems, and it continues to transform medicine, agriculture, biotechnology, and conservation biology. You have already encountered evolution as the main thematic thread woven throughout this book. In this chapter, you will learn about the historical development of the Darwinian view of life.

Concept 22.1

The Darwinian revolution challenged traditional views of a young Earth inhabited by unchanging species

The impact of an intellectual revolution such as Darwinism depends on timing as well as logic. To understand why Darwin's ideas were revolutionary, we need to examine his views in the context of other Western ideas about Earth and its life **(Figure 22.2)**.

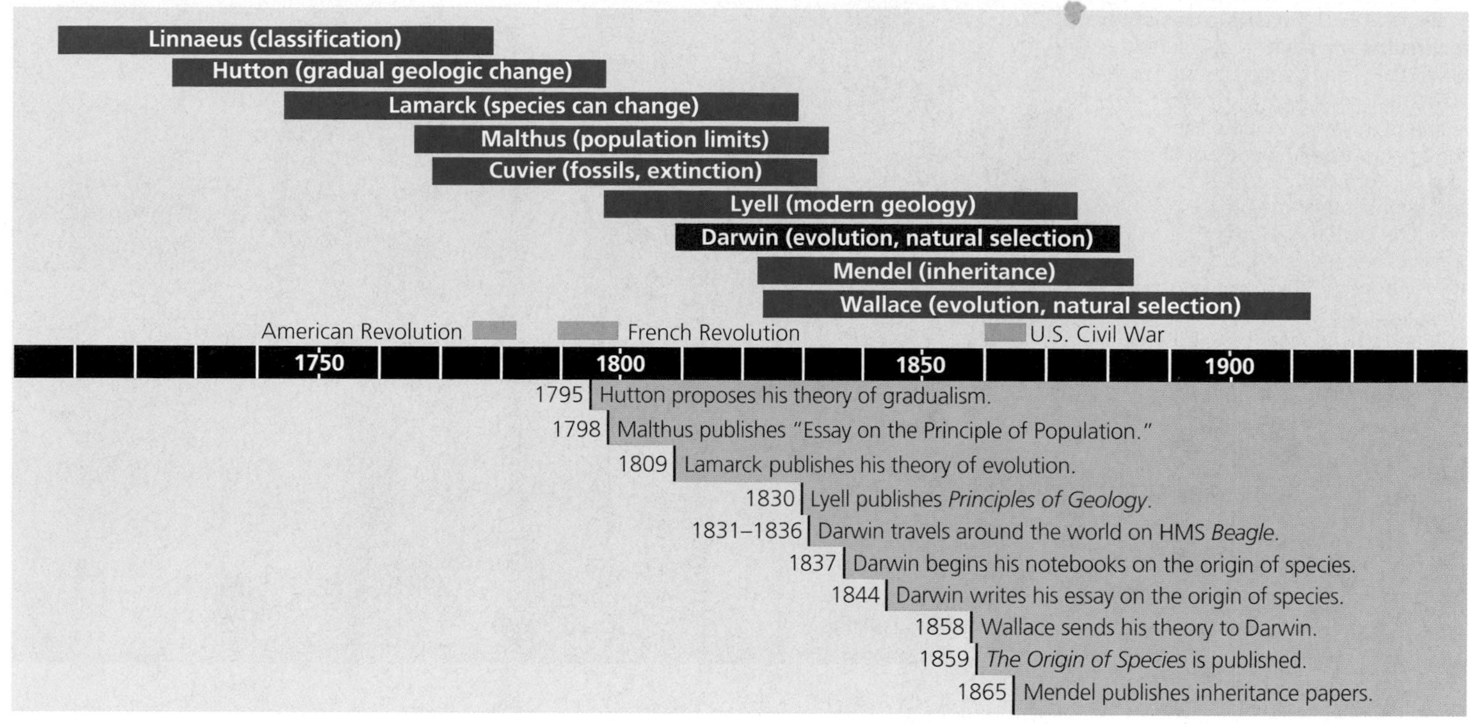

▲ **Figure 22.2 The historical context of Darwin's life and ideas.** The dark blue bars above the timeline represent the lives of some individuals whose ideas have contributed to our modern understanding of evolution.

Resistance to the Idea of Evolution

The Origin of Species not only challenged prevailing scientific views but also shook the deepest roots of Western culture. Darwin's view of life contrasted sharply with traditional beliefs of an Earth only a few thousand years old, populated by forms of life that had been created at the beginning and remained unchanged ever since. Darwin's book challenged a worldview that had been prevalent for centuries.

The Scale of Nature and Classification of Species

Although several Greek philosophers suggested that life might have evolved gradually, one philosopher who greatly influenced early Western science, Aristotle (384–322 B.C.), viewed species as fixed (unchanging). Through his observations of nature, Aristotle recognized certain "affinities" among living things, leading him to conclude that life-forms could be arranged on a ladder, or scale, of increasing complexity, later called the *scala naturae* ("scale of nature"). Each form of life, perfect and permanent, had its allotted rung on this ladder.

These ideas coincided with the Old Testament account of creation, which holds that species were individually designed by God and therefore perfect. In the 1700s, many scientists interpreted the superb adaptations of organisms to their environments as evidence that the Creator had designed each species for a particular purpose.

One such scientist was Carolus Linnaeus (1707–1778), a Swedish physician and botanist who sought to classify life's diversity "for the greater glory of God." Linnaeus was a founder of **taxonomy,** the branch of biology concerned with naming and classifying organisms. He developed the two-part, or binomial, system of naming organisms according to genus and species that is still used today. In contrast to the linear hierarchy of the *scala naturae,* Linnaeus adopted a nested classification system, grouping similar species into increasingly general categories. For example, similar species are grouped in the same genus, similar genera (plural of genus) are grouped in the same family, and so on (see Figure 1.14).

To Linnaeus, the observation that some species resemble each other did not imply evolutionary kinship, but rather the pattern of their creation. However, a century later his taxonomic system would play a role in Darwin's arguments for evolution.

Fossils, Cuvier, and Catastrophism

The study of fossils also helped to lay the groundwork for Darwin's ideas. **Fossils** are remains or traces of organisms from the past. Most fossils are found in **sedimentary rocks** formed from the sand and mud that settle to the bottom of seas, lakes, and marshes. New layers of sediment cover older ones and compress them into superimposed layers of rock called strata (singular, *stratum*). Later, erosion may carve

▶ **Figure 22.3 Fossils from strata of sedimentary rock.** The Colorado River carved the Grand Canyon through more than 2,000 m of rock, exposing sedimentary layers that are like huge pages from the book of life. Each stratum entombs fossils that represent some of the organisms from that period of Earth's history. The seed fern leaf fossil (top) is from the shallower Hermit Shale layer (265 million years old), and the trilobite fossil (bottom) is from the deeper Bright Angel Shale layer (530 million years old).

through upper (younger) strata and reveal older strata that had been buried. Fossils in each layer provide a glimpse of some of the organisms that populated Earth at the time that layer formed **(Figure 22.3)**.

Paleontology, the study of fossils, was largely developed by French scientist Georges Cuvier (1769–1832). In examining rock layers in the region around Paris, Cuvier noted that the deeper (older) the strata, the more dissimilar the fossils are from current life. He also observed that from one stratum to the next, some new species appear while others disappear. He inferred that extinctions must have been a common occurrence in the history of life. Yet Cuvier staunchly opposed the idea of gradual evolutionary change. Instead, he advocated **catastrophism**, speculating that each boundary between strata represents a catastrophe, such as a flood or drought, that destroyed many of the species living at that time. He proposed that these periodic catastrophes were usually confined to local geographic regions, which were repopulated by species immigrating from other areas.

Theories of Gradualism

In contrast to catastrophism, the work of other scientists promoted the concept of **gradualism**—the idea that profound change can take place through the cumulative effect of slow but continuous processes. In 1795, Scottish geologist James Hutton (1726–1797) proposed that Earth's geologic features could be explained by gradual mechanisms *currently* operating in the world. He suggested that valleys were formed by rivers wearing through rocks and that sedimentary rocks containing marine fossils were formed from particles that had eroded from the land and been carried by rivers to the sea.

The leading geologist of Darwin's time, Charles Lyell (1797–1875), incorporated Hutton's thinking into a more comprehensive theory known as **uniformitarianism.** Lyell proposed that the same geologic processes are operating today as in the past, and at the same rate.

Hutton and Lyell's ideas exerted a strong influence on Darwin's thinking. Darwin agreed that if geologic change results from slow, continuous actions rather than sudden events, then Earth must be much older than the 6,000 years that theologians estimated. He later reasoned that perhaps similarly slow and subtle processes could act on living organisms over a long period of time, producing substantial change. Darwin was not the first to apply the principle of gradualism to biological evolution, however.

Lamarck's Theory of Evolution

During the 18th century, several naturalists (including Darwin's grandfather, Erasmus Darwin) suggested that life evolves as environments change. But only one of Charles Darwin's predecessors developed a comprehensive model for *how* life evolves: French biologist Jean-Baptiste de Lamarck (1744–1829). Alas, Lamarck is primarily remembered today not for his visionary recognition that evolutionary change explains the fossil record and organisms' adaptations to their environments, but for the incorrect mechanism he proposed to explain how evolution occurs.

Lamarck published his theory in 1809, the year Darwin was born. By comparing current species with fossil forms, Lamarck had found what appeared to be several lines of descent, each a chronological series of older to younger fossils leading to a living species. He explained this with two principles that

▲ **Figure 22.4 Acquired traits cannot be inherited.** This bonsai tree was "trained" to grow as a dwarf by pruning and shaping. However, seeds from this tree would produce offspring of normal size.

were commonly accepted at that time. The first was *use and disuse,* the idea that parts of the body that are used extensively become larger and stronger, while those that are not used deteriorate. As an example, he cited a giraffe stretching its neck to reach leaves on high branches. The second principle, *inheritance of acquired characteristics,* stated that an organism could pass these modifications to its offspring. Lamarck reasoned that the long, muscular neck of the living giraffe had evolved over many generations as giraffes stretched their necks ever higher. Lamarck also thought that evolution happens because organisms have an innate drive to become more complex. Darwin rejected this idea in favor of natural selection, but he too thought that variation was introduced into the evolutionary process through inheritance of acquired characteristics. However, our modern understanding of genetics refutes this principle; there is no evidence that acquired characteristics can be inherited **(Figure 22.4)**.

Lamarck was vilified in his own time, especially by Cuvier, who denied that species ever evolve. In retrospect, however, Lamarck deserves credit for his insightful observations of nature and recognition of gradual evolutionary change as the best explanation for those observations.

Concept Check 22.1

1. Which of the individuals discussed in this section viewed species as fixed, and which viewed species as being able to change over time?
2. What was Lamarck's theory of evolution? Explain its significance.

For suggested answers, see Appendix A.

Concept 22.2

In *The Origin of Species,* Darwin proposed that species change through natural selection

As the 19th century dawned, it was generally believed that species had remained unchanged since their creation. A few clouds of doubt about the permanence of species were beginning to gather, but no one could have forecast the thundering storm just over the horizon.

Darwin's Research

Charles Darwin (1809–1882) was born in Shrewsbury in western England. Even as a boy, he had a consuming interest in nature. When he was not reading nature books, he was fishing, hunting, and collecting insects. Darwin's father, an eminent physician, could see no future for his 16-year-old son as a naturalist and sent him to the University of Edinburgh to study medicine. But Charles found medical school boring and surgery before the days of anesthesia horrifying. He left Edinburgh without a degree and enrolled at Cambridge University with the intention of becoming a clergyman. At that time in England, many scholars of science belonged to the clergy.

At Cambridge, Darwin became the protégé of the Reverend John Henslow, a professor of botany. Soon after Darwin received his B.A. degree, Henslow recommended the young graduate to Captain Robert FitzRoy, who was preparing the survey ship HMS *Beagle* for a voyage around the world. FitzRoy accepted Darwin on board because of his education and because of his age and social class, which were similar to FitzRoy's.

The Voyage of the Beagle

In 1831, the 22-year-old Darwin left England aboard the *Beagle*. The primary mission of the voyage was to chart poorly known stretches of the South American coastline. While the ship's crew surveyed the coast, Darwin spent most of his time on shore, observing and collecting thousands of South American plants and animals. He observed the various adaptations of plants and animals that inhabited such diverse environments as the Brazilian jungles, the expansive grasslands of the Argentine pampas, the desolate lands of Tierra del Fuego near Antarctica, and the towering heights of the Andes Mountains.

Darwin noted that the plants and animals in temperate regions of South America more closely resembled species living in the South American tropics than species in temperate regions of Europe. Furthermore, the fossils he found, though clearly different from living species, were distinctly South American in their resemblance to the living organisms of that continent.

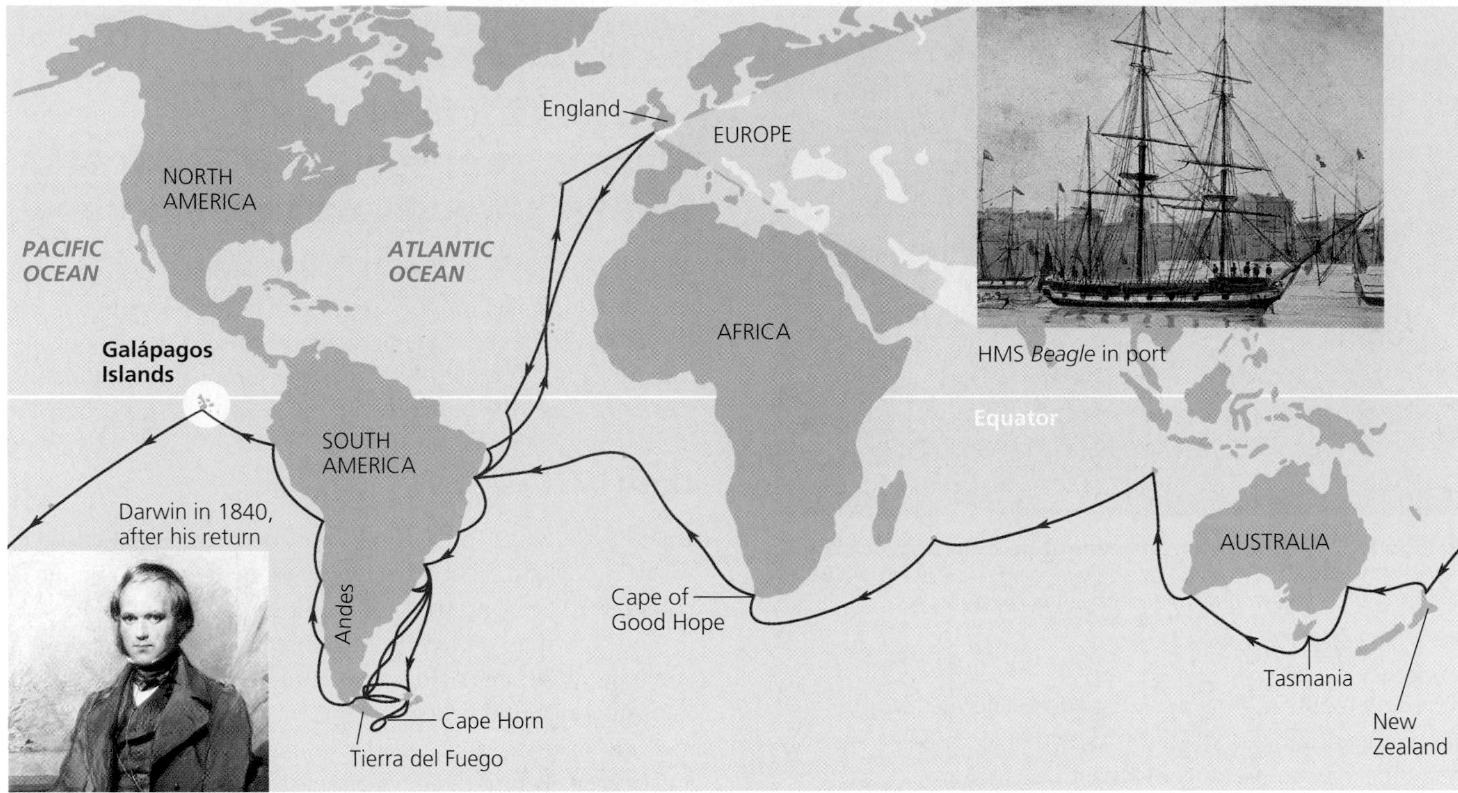

▲ **Figure 22.5 The voyage of HMS *Beagle*.**

Geologic observations also impressed Darwin during the voyage. Despite bouts of seasickness, he read Lyell's *Principles of Geology* while aboard the *Beagle*. He experienced geologic change firsthand when a violent earthquake rocked the coast of Chile, and he observed afterward that the coastline had risen by several feet. Finding fossils of ocean organisms high in the Andes Mountains, he inferred that the rocks containing the fossils must have been raised there by a long series of similar earthquakes. These observations reinforced what he had learned from Lyell—the physical evidence did not support the traditional view of a static Earth only a few thousand years old.

Darwin's interest in the geographic distribution of species was further kindled by the *Beagle*'s stop at the Galápagos, a group of geologically young volcanic islands located near the equator about 900 km (540 miles) west of South America **(Figure 22.5)**. Darwin was fascinated by the unusual organisms he found there. Among the birds he collected on the Galápagos were several kinds of finches that, although quite similar, seemed to be different species. Some were unique to individual islands, while others were distributed on two or more adjacent islands. But Darwin did not fully grasp the significance of these observations until after his return to England in 1836. He and others found that although the animals on the Galápagos resemble species living on the South American mainland, most of the Galápagos animals live nowhere else in the world. He hypothesized that the Galápagos had been colonized by organisms that had strayed from South America and had then diversified on the various islands.

Darwin's Focus on Adaptation

As Darwin reassessed all that he had observed during the voyage, he began to perceive adaptation to the environment and the origin of new species as closely related processes. Could a new species arise from an ancestral form by the gradual accumulation of adaptations to a different environment? From studies made years after Darwin's voyage, biologists have concluded that this is indeed what happened to the Galápagos finches. Their beaks and behaviors are adapted to the specific foods available on their home islands **(Figure 22.6)**. Darwin realized that an explanation for such adaptations was essential to understanding evolution.

By the early 1840s, Darwin had worked out the major features of his theory of natural selection as the mechanism of evolution. However, he had not yet published his ideas. He was in poor health, and he rarely left his home near London. Despite his reclusiveness, Darwin was not isolated. Already famous as a naturalist because of the letters and specimens he had sent to England during the voyage of the *Beagle,* Darwin corresponded extensively with scientists around the world and was often visited by Lyell, Henslow, and others.

In 1844, Darwin wrote a long essay on the origin of species and natural selection. However, he was reluctant to introduce

▶ **Figure 22.6 Beak variation in Galápagos finches.** The Galápagos Islands are home to more than a dozen species of closely related finches, some found only on a single island. The most striking differences among them are their beaks, which are adapted for specific diets.

(a) Cactus eater. The long, sharp beak of the cactus ground finch (*Geospiza scandens*) helps it tear and eat cactus flowers and pulp.

(b) Insect eater. The green warbler finch (*Certhidea olivacea*) uses its narrow, pointed beak to grasp insects.

(c) Seed eater. The large ground finch (*Geospiza magnirostris*) has a large beak adapted for cracking seeds that fall from plants to the ground.

his theory publicly, apparently because he anticipated the uproar it would cause. Darwin asked his wife to publish his essay if he were to die before finishing a more thorough dissertation. Even as he procrastinated, he continued to compile evidence in support of his theory. Lyell, not yet himself convinced of evolution, nevertheless urged Darwin to publish on the subject before someone else came to the same conclusions and published first.

In June 1858, Lyell's prediction came true. Darwin received a manuscript from Alfred Russel Wallace (1823–1913), a young British naturalist working in the East Indies who had developed a theory of natural selection similar to Darwin's. Wallace asked Darwin to evaluate his paper and forward it to Lyell if it merited publication. Darwin complied, writing to Lyell: "Your words have come true with a vengeance. . . . I never saw a more striking coincidence . . . so all my originality, whatever it may amount to, will be smashed." Lyell and a colleague then presented Wallace's paper, along with extracts from Darwin's unpublished 1844 essay, to the Linnean Society of London on July 1, 1858. Darwin quickly finished *The Origin of Species* and published it the next year. Although Wallace had written up his ideas for publication first, he was a great admirer of Darwin and agreed that Darwin had developed the theory of natural selection so extensively that he should be known as its main architect.

Within a decade, Darwin's book and its proponents had convinced most biologists that biological diversity was the product of evolution. Darwin succeeded where previous evolutionists had failed, mainly because he presented his reasoning with immaculate logic and an avalanche of supporting evidence. He soon followed up his first book with other pioneering work, in particular an exploration of the type of natural selection known as sexual selection (see Chapter 23).

The Origin of Species

In publishing his theory, Darwin developed two main ideas: that evolution explains life's unity and diversity and that natural selection is a cause of adaptive evolution.

Descent with Modification

In the first edition of *The Origin of Species,* Darwin did not use the word *evolution* until the very end. Instead, he referred to **descent with modification,** a phrase that summarized his view of life. Darwin perceived unity in life, with all organisms related through descent from an ancestor that lived in the remote past. As the descendants of that ancestral organism spilled into various habitats over millions of years, they accumulated diverse modifications, or adaptations, that fit them to specific ways of life.

In the Darwinian view, the history of life is like a tree, with multiple branchings from a common trunk to the tips of the youngest twigs that represent the diversity of living organisms. Each fork of the tree represents an ancestor of all the lines of evolution that subsequently branched from that point. Closely related species, such as the Asian elephant and African elephants, are very similar because they shared the same line of descent until a relatively recent divergence

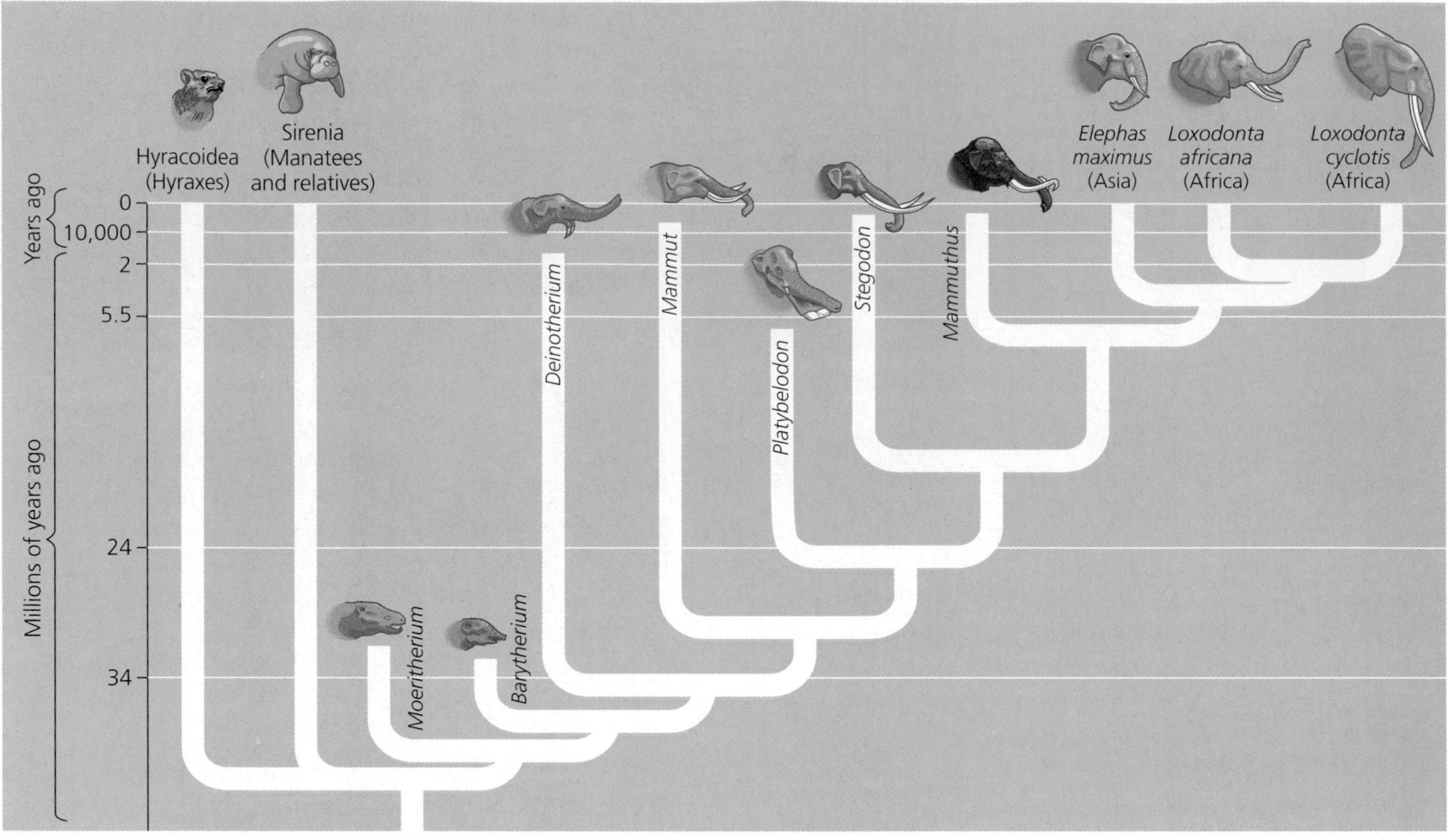

▲ **Figure 22.7 Descent with modification.** This evolutionary tree of the elephants is based mainly on fossils—their anatomy, order of appearance in strata, and geographic distribution. Note that most branches of descent ended in extinction. Despite their very different appearances, the manatees and hyraxes are the elephants' closest living relatives. (Timeline not to scale.)

from their common ancestor **(Figure 22.7)**. Most branches of evolution, even some major ones, are dead ends; about 99% of all species that have ever lived are now extinct. Thus, there are no living animals that fill the gap between the elephants and their nearest relatives today, the manatees and hyraxes, though some fossils have been found.

Linnaeus had realized that some organisms resemble each other more closely than others, but he had not linked these resemblances to evolution. Nonetheless, because he had recognized that the great diversity of organisms could be organized into "groups subordinate to groups" (Darwin's phrase), his taxonomic scheme largely fit with Darwin's theory. To Darwin, the Linnaean hierarchy reflected the branching history of the tree of life, with organisms at the various taxonomic levels related through descent from common ancestors.

Natural Selection and Adaptation

How does natural selection work, and how does it explain adaptation? Evolutionary biologist Ernst Mayr has dissected the logic of Darwin's theory of natural selection into three inferences based on five observations:*

OBSERVATION #1: For any species, population sizes would increase exponentially if all individuals that are born reproduced successfully **(Figure 22.8)**.

OBSERVATION #2: Nonetheless, populations tend to remain stable in size, except for seasonal fluctuations.

OBSERVATION #3: Resources are limited.

INFERENCE #1: Production of more individuals than the environment can support leads to a struggle for existence among individuals of a population, with only a fraction of their offspring surviving each generation.

OBSERVATION #4: Members of a population vary extensively in their characteristics; no two individuals are exactly alike **(Figure 22.9)**.

OBSERVATION #5: Much of this variation is heritable.

INFERENCE #2: Survival depends in part on inherited traits. Individuals whose inherited traits give them a high probability of surviving and reproducing in a given environment have higher fitness and are likely to leave more offspring than less fit individuals.

INFERENCE #3: This unequal ability of individuals to survive and reproduce will lead to a gradual change in a population, with favorable characteristics accumulating over generations.

---

* Adapted from E. Mayr, *The Growth of Biological Thought: Diversity, Evolution and Inheritance* (Cambridge, MA: Harvard University Press, 1982).

▶ **Figure 22.8 Overproduction of offspring.** Just one branch of a maple tree bears dozens of winged seeds. If all the tree's offspring survived, we would quickly be overwhelmed by maple forests.

▲ **Figure 22.9 Variation in a population.** To the extent that the variation in color and dot patterns among the members of this population of ladybird beetles is heritable, it can be acted on by natural selection.

Darwin perceived an important connection between natural selection, which results from what he called the struggle for existence, and the capacity of organisms to "overreproduce." Apparently, this insight came after he read a 1798 essay on population growth by Thomas Malthus. Malthus contended that much of human suffering—disease, famine, homelessness, and war—was the inescapable consequence of the human population's potential to increase faster than food and other resources. The capacity to overreproduce seems to be characteristic of all species (see Figure 22.8). Of the many eggs laid, young born, and seeds spread, only a tiny fraction complete their development and leave offspring of their own. The rest are eaten, starved, diseased, unmated, or unable to reproduce for some other reason.

In each generation, environmental factors filter heritable variations, favoring some over others. Organisms with traits favored by the environment tend to produce more offspring than do organisms without those traits. This differential reproductive success results in favored traits being disproportionately represented in the next generation. Increases in the frequencies of favored traits in a population, which are always taking place regardless of whether the environment is changing, are an important source of evolutionary modification.

Artificial Selection. Darwin derived another piece of his theory from the many familiar examples of selective breeding of domesticated plants and animals. Humans have modified other species over many generations by selecting and breeding individuals that possess desired traits—a process called **artificial selection.** As a result of artificial selection, crop plants and animals bred as livestock or pets often bear little resemblance to their wild ancestors **(Figure 22.10)**.

If artificial selection can achieve so much change in a relatively short period of time, Darwin reasoned, then what he termed "natural selection" should be capable of considerable modification of species over hundreds or thousands of generations. Even if the advantages of some heritable traits over others are slight, the advantageous variations will gradually accumulate in the population, and less favorable variations will diminish.

Summary of Natural Selection. Let's once more state the main ideas of natural selection:

- Natural selection is the differential success in reproduction among individuals that vary in their heritable traits. These reproductive differences emerge as each individual interacts with its environment.

▲ **Figure 22.10 Artificial selection.** These vegetables have all been selected from one species of wild mustard. By selecting variations in different parts of the plant, breeders have obtained these divergent results.

(a) A flower mantid in Malaysia

(b) A stick mantid in Africa

▲ **Figure 22.11 Camouflage as an example of evolutionary adaptation.** Related species of the insects called mantids have diverse shapes and colors that evolved in different environments.

- Over time, natural selection can increase the adaptation of organisms to their environment **(Figure 22.11)**.
- If an environment changes over time, or if individuals of a particular species move to a new environment, natural selection may result in adaptation to these new conditions, sometimes giving rise to new species in the process.

Before continuing, we need to emphasize three subtle but important points about evolution by natural selection. The first is that although natural selection occurs through interactions between individual organisms and their environment, individuals do not evolve. A population is the smallest unit that can evolve. (For now, we will define a population as a group of interbreeding individuals belonging to a particular species and sharing a common geographic area.) Evolution can be measured only as changes in relative proportions of heritable variations in a population over a succession of generations.

Another key point is that natural selection can amplify or diminish *only* heritable traits—that is, traits that are passed from organisms to their offspring. Though an organism may become modified through its own interactions with the environment during its lifetime, and these acquired characteristics may even adapt the organism to its environment, there is no evidence that such acquired characteristics can be inherited by offspring. We must distinguish between adaptations that an organism acquires during its lifetime and inherited adaptations that accumulate in a population over many generations as a result of natural selection.

Third, remember that environmental factors vary from place to place and from time to time. A trait that is favorable in one situation may be useless—or even detrimental—in different circumstances. Natural selection is always operating, but which traits are favored depends on the environment.

Thus, Darwin envisioned life as evolving through this gradual accumulation of small changes. He postulated that natural selection, operating in varying contexts over vast spans of time as revealed by the emerging science of geology, could account for the entire diversity of life.

Concept Check 22.2

1. Describe how the following concepts relate to Darwin's theory of evolution by natural selection: overreproduction of populations, limited resources, and heritable variation.
2. Explain why an individual organism cannot be said to evolve.

For suggested answers, see Appendix A.

Darwin's theory explains a wide range of observations

Scientific inquiry seeks *natural* causes for natural phenomena (see Chapter 1). The power of evolution as a unifying theory is its versatility as a natural explanation for a diversity of data from biology's many subfields. And like all general theories in science, Darwin's theory of evolution continues to be tested by how effectively it can account for additional observations and experimental outcomes. Perhaps the most direct validations of Darwin's theory are provided by the many cases where scientists can actually observe natural selection at work.

Natural Selection in Action

Let's look at two examples of natural selection as a mechanism of evolution in populations.

Differential Predation and Guppy Populations

Guppies (*Poecilia reticulata*) are small freshwater fish that you may know as aquarium pets. For many years, John Endler, of the University of California, Santa Barbara, and David Reznick, of the University of California, Riverside, have studied wild guppy populations living in pools in the Aripo River system on the Caribbean island of Trinidad.

The researchers observed significant differences between populations in the average age and size at which the guppies reached sexual maturity. These variations were correlated with the type of predator most active on that population. In some pools, the main predator is the small killifish, which preys predominantly on juvenile guppies. In other pools, the major predator is the pike-cichlid, a larger fish that primarily eats sexually mature individuals. Guppies in populations preyed on by pike-cichlids begin reproducing at a younger age and are smaller at maturity, on average, than guppies that are preyed on by killifish.

But although correlation with type of predator is suggestive, it does not necessarily indicate cause and effect. To test whether these differences are due to natural selection, Reznick and Endler introduced guppies from "pike-cichlid pools" to new pools that contained killifish but no guppies **(Figure 22.12)**. Over the next 11 years, the researchers compared the age and size at maturity of the transplanted guppies with the original population. After periods of 30 to 60 generations, the transplanted guppies were an average of 14% heavier at maturity than the nontransplanted individuals. Their average age at maturity had also risen. This result supports the hypothesis that natural selection caused the difference in the populations. The conclusion is that because pike-cichlids prey mainly on reproductively mature adults, the chance that a guppy will survive to reproduce several times is relatively low. The guppies with greatest reproductive success in these ponds are those that mature at a young age and small size, enabling them to produce at least one brood before growing to a size preferred by the pike-cichlids. But in ponds with killifish, the guppies that survive early predation can grow slowly and still produce many broods of young. This research by Reznick and Endler is one of many documented examples of evolution in natural settings over relatively short time periods.

Figure 22.12

Inquiry Can predation pressure select for size and age at maturity in guppies?

EXPERIMENT Reznick and Endler transplanted guppies from pike-cichlid pools to killifish pools and measured the average age and size of guppies at maturity over an 11-year period (30 to 60 generations).

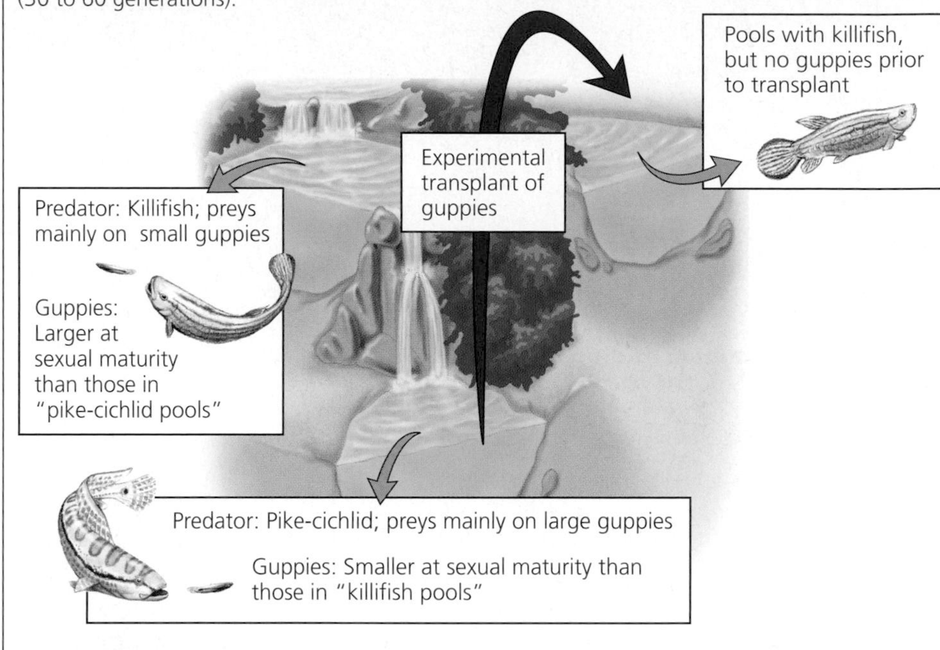

RESULTS After 11 years, the average size and age at maturity of guppies in the transplanted populations increased compared to those of guppies in control populations.

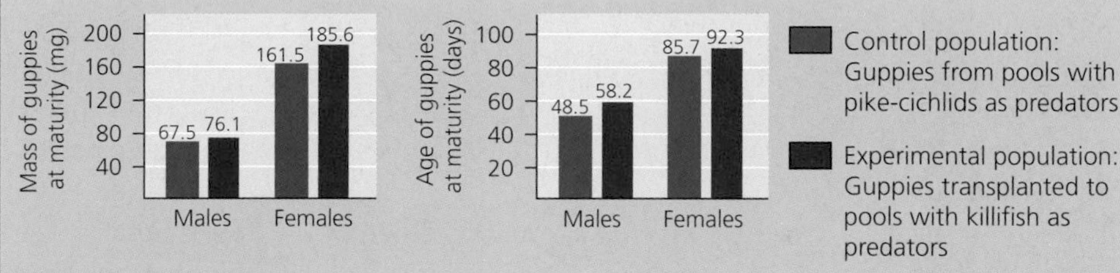

CONCLUSION Reznick and Endler concluded that the change in predator resulted in different variations in the population (larger size and slower maturation) being favored. Over a relatively short time, this altered selection pressure resulted in an observable evolutionary change in the experimental population.

The Evolution of Drug-Resistant HIV

An unsettling example of ongoing natural selection that directly affects our own lives is the evolution of drug-resistant pathogens. This is a particular problem in bacteria and viruses with rapid rates of reproduction, because a variation that makes individuals resistant to a particular drug can increase in frequency very quickly in the population.

Consider the example of HIV (human immunodeficiency virus), which causes AIDS (see Chapter 18 and Chapter 43). Researchers have developed numerous drugs to combat this pathogen, but using these medications selects for viruses resistant to the drugs. A few drug-resistant viruses may be present by chance at the beginning of treatment. Those that survive the early doses pass on the genes that enable them to

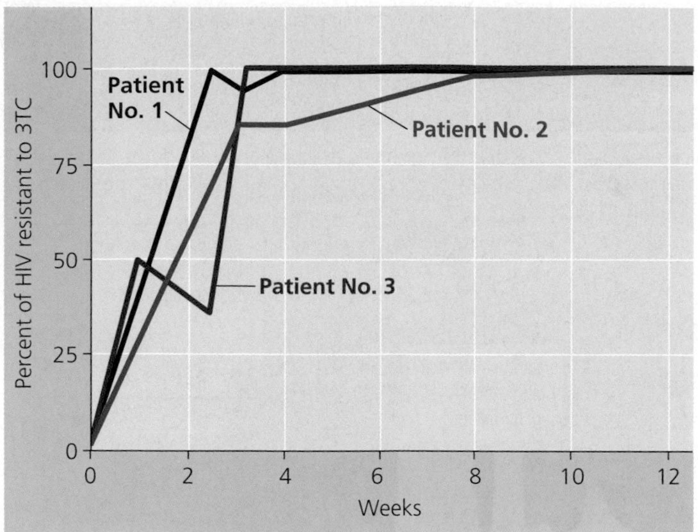

▲ **Figure 22.13 Evolution of drug resistance in HIV.** Rare resistant viruses multiply quickly when each of these patients is treated with the anti-HIV drug 3TC. Within just a few weeks, 3TC-resistant organisms comprise 100% of the virus population in each case.

resist the drug to their progeny, rapidly increasing the frequency of resistant viruses.

Figure 22.13 illustrates the evolution of HIV resistance to the drug 3TC. Scientists designed 3TC to interfere with reverse transcriptase, the enzyme HIV uses to copy its RNA genome into the DNA of the human host cell (see Figure 18.10). Because the 3TC molecule is similar in shape to the cytosine (C)-bearing nucleotide of DNA, HIV's reverse transcriptase picks up a 3TC molecule instead of a C-bearing nucleotide and inserts the 3TC into a growing DNA chain. This error terminates further elongation of the DNA and thus blocks reproduction of HIV.

The 3TC-resistant varieties of HIV carry slightly different versions of reverse transcriptase that are able to discriminate between the drug and the normal C-bearing nucleotide. The viruses that carry these genes have no advantage in the absence of 3TC; in fact, they replicate more slowly than those that carry the more common form. But once 3TC is added to their environment it becomes a powerful selecting force, favoring reproduction of resistant individuals.

These examples highlight two key points about natural selection. First, natural selection is more a process of editing than a creative mechanism. A drug does not *create* resistant pathogens; it *selects for* resistant individuals that were already present in the population. Second, natural selection depends on time and place. It favors those characteristics in a genetically variable population that increase fitness in the current, local environment. What is adaptive in one situation may be useless or even harmful in another. In the guppy example, individuals that mature at an early age and small size are at an advantage in a pool with pike-cichlids, but at a disadvantage in a pool with killifish.

Homology, Biogeography, and the Fossil Record

The two examples you have just read about show that natural selection can bring about change rapidly enough to be observed directly. In addition, Darwin's theory provides a cohesive explanation for observations made by researchers in the fields of anatomy, embryology, molecular biology, biogeography, and paleontology.

Homology

Darwin's concept of descent with modification can explain why certain characteristics in related species have an underlying similarity even though they may have very different functions. Such similarity resulting from common ancestry is known as **homology.**

Anatomical Homologies. The view of evolution as a remodeling process accounts for the findings of comparative anatomy, the comparison of body structures between species. The forelimbs of all mammals, including humans, cats, whales, and bats, show the same arrangement of bones from the shoulder to the tips of the digits, even though these appendages can have very different functions—lifting, walking, swimming, and flying **(Figure 22.14)**. Such striking anatomical resemblances would not exist if these structures had arisen anew in each species. The arms, forelegs, flippers, and wings of different mammals are **homologous structures** that represent variations on a structural theme that was present in their common ancestor.

Comparative embryology, the comparison of early stages of animal development, reveals additional anatomical homologies not visible in adult organisms. For example, at some point in their development, all vertebrate embryos have a tail posterior to the anus, as well as structures called pharyngeal (throat) pouches **(Figure 22.15)**. These embryonic structures develop into homologous structures with very different functions, such as gills in fishes and parts of the ears and throat in humans.

Some of the most intriguing homologous structures are **vestigial organs,** structures of marginal, if any, importance to the organism. Vestigial organs are remnants of structures that served important functions in the organism's ancestors. For instance, the skeletons of some snakes retain vestiges of the pelvis and leg bones of walking ancestors. Because limbs are a hindrance to a snake's way of life, natural selection has favored snake ancestors with successively smaller limbs. We would not expect to see these vestigial structures if snakes had an origin separate from other vertebrate animals.

Because evolution can modify only existing structures and functions, it often produces less than perfect results. For example, the human knee joint and spine derive from ancestral structures that supported four-legged mammals. Few people

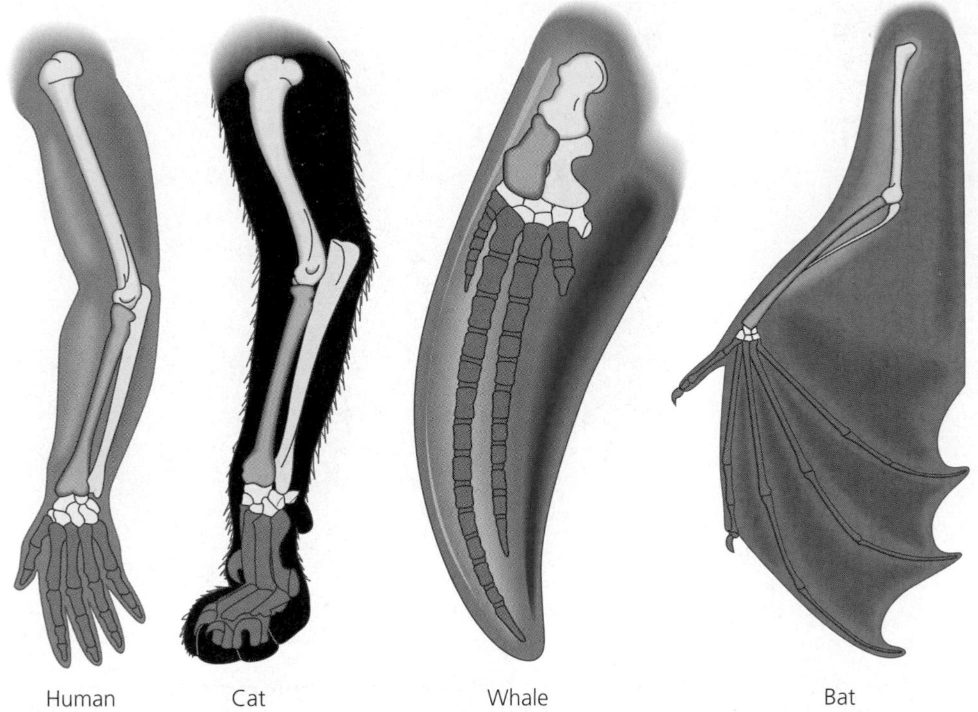

▲ **Figure 22.14 Mammalian forelimbs: homologous structures.** Even though they have become adapted for different functions, the forelimbs of all mammals are constructed from the same basic skeletal elements: one large bone (gray), attached to two smaller bones (orange and beige), attached to several small bones (yellow), attached to approximately five digits, or phalanges (brown).

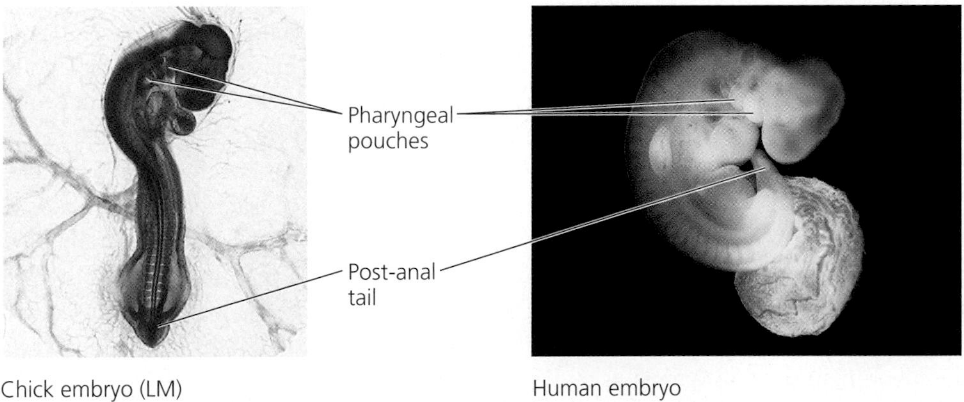

▲ **Figure 22.15 Anatomical similarities in vertebrate embryos.** At some stage in their embryonic development, all vertebrates have a tail located posterior to the anus, as well as pharyngeal (throat) pouches. Descent from a common ancestor can explain such similarities.

reach old age without experiencing knee or back problems. If these structures had originally evolved in animals that were bipedal, we would expect these parts of the body to be less subject to injury. The anatomical remodeling that enabled our ancestors to stand upright was apparently constrained by our evolutionary history.

Molecular Homologies. Biologists also observe similarities among organisms at the molecular level. All forms of life use the same genetic machinery of DNA and RNA, and the genetic code is essentially universal (see Chapter 17). Because the genetic code is shared by all organisms, it is likely that all species descended from a common ancestor. Molecular homologies also go beyond a shared code. Organisms as dissimilar as humans and bacteria share many genes that have been inherited from a distant common ancestor. Like the forelimbs of humans and whales, these genes have often acquired different functions.

Homologies and the Tree of Life. The Darwinian concept of an evolutionary tree of life can explain the homologies that researchers have observed. Some homologies, such as the genetic code, are shared by all species because they date to the deep ancestral past. Homologies that evolved more recently are shared only within smaller branches of the tree. For example, all tetrapods (from the Greek *tetra*, four, and *pod*, foot), the vertebrate branch consisting of amphibians, reptiles (including birds), and mammals, possess the same basic five-digit limb structure illustrated in Figure 22.14. Thus, homologies result in a nested pattern, with all life sharing the deepest layer and each smaller group adding homologies to those they share with larger groups. This nested pattern is exactly what we would expect if life evolved and diversified from a common ancestor.

Anatomical resemblances among species are generally reflected in their molecules—in their genes (DNA) and gene products (proteins). **Figure 22.16**, on the next page, compares the amino acid sequence of human hemoglobin, the protein that transports oxygen in blood, with the hemoglobin sequences of other vertebrates. The data show the same pattern of evolutionary relationships that researchers find when they compare other proteins or assess relationships based on nonmolecular methods such as comparative anatomy. The Darwinian view of life predicts that different kinds of homologies in a group of organisms will all tend to show the same branching pattern through their evolutionary history.

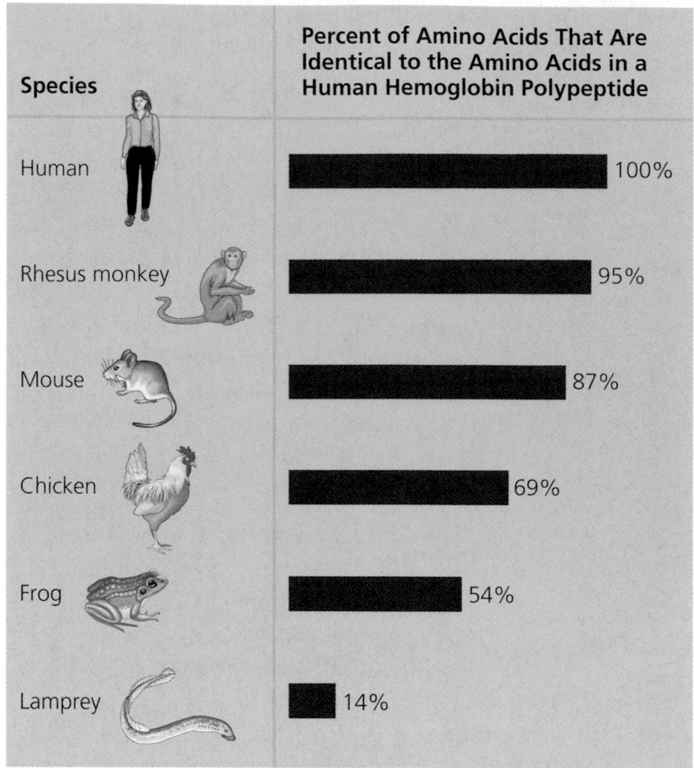

▲ **Figure 22.16 Comparison of a protein found in diverse vertebrates.**

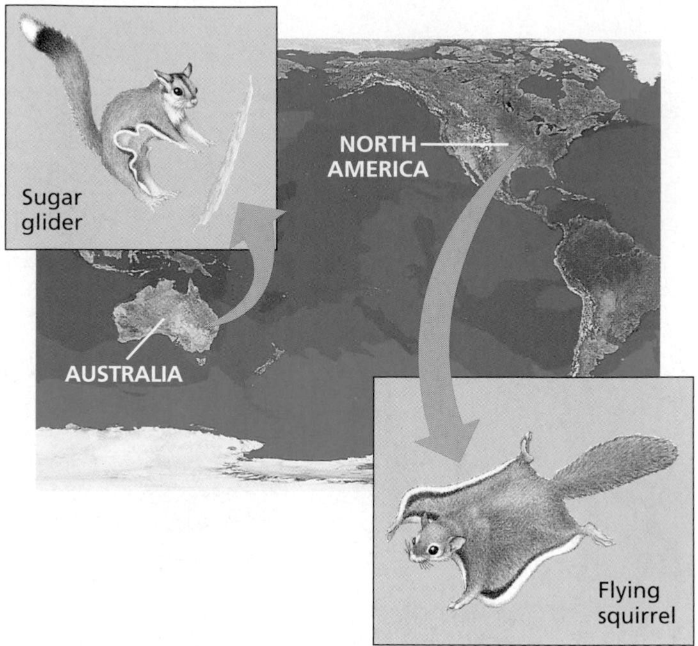

▲ **Figure 22.17 Different geographic regions, different mammalian "brands."** The sugar glider is an example of the diverse marsupial life that evolved in isolation on the island continent of Australia. While sugar gliders superficially resemble the eutherian flying squirrels of North America, the ability to glide through the air evolved independently in these two distantly related groups of mammals.

Biogeography

Darwin's observations of the geographic distribution of species—or **biogeography**—formed an important part of his theory of evolution. Closely related species tend to be found in the same geographic region, whereas the same ecological niches in distant regions are occupied by very different (though sometimes similar-looking) species. Consider Australia, which is home to a group of mammals—the marsupials—that are distinct from another group of mammals—the eutherians—that live elsewhere on Earth. (Eutherians are mammals that complete their embryonic development in the uterus, whereas marsupials are born as embryos and complete their development in an external pouch.) Some Australian marsupials have eutherian look-alikes with similar adaptations living on other continents. For instance, a forest-dwelling Australian marsupial called the sugar glider is superficially very similar to "flying" squirrels, gliding eutherians that live in North American forests **(Figure 22.17)**. But the sugar glider has many other characteristics that define it as distinctly marsupial, much more closely related to kangaroos and other Australian marsupials than to flying squirrels or other eutherians. Once again Darwin's theory can explain these observations. Although these two mammals have adapted to similar environments in similar ways, they evolved independently from different ancestors. The sugar glider is a marsupial not because that is a requirement for its gliding lifestyle, but simply because its ancestors were marsupials. The unique fauna of Australia diversified after that island continent became isolated from the landmasses on which eutherians diversified. The resemblance between sugar gliders and flying squirrels is not homologous; rather, it is an example of what biologists call convergent evolution (we'll take a closer look at convergence in Chapter 25).

It's no wonder that Darwin's experiences in the Galápagos were so pivotal to his thinking, since islands are showcases of the influence of geography on evolution. They generally have many species of plants and animals that are **endemic,** which means they are found nowhere else in the world. Yet, as Darwin observed when he reassessed his collections from the voyage, most island species are closely related to species from the nearest mainland or neighboring island. This explains why two islands with similar environments in different parts of the world are populated not by closely related species but rather by species that resemble those of the nearest mainland, where the environment is often quite different. Island chains like the Galápagos, or archipelagos, are especially interesting in their biogeography. If a species that disperses from a mainland to an island succeeds in its new environment, it may give rise to several new species as populations spread to other islands in the archipelago. The finches of the Galápagos archipelago are an example of this process (see Figure 1.23), as are the hundreds of Hawaiian *Drosophila* species studied by Kenneth Kaneshiro (see interview on pages 436–437).

The Fossil Record

The succession of forms observed in the fossil record is consistent with other inferences about the major branches of descent in the tree of life. For instance, comparative data from biochemistry, molecular biology, and cell biology suggest that prokaryotes are the ancestors of all life and predict that prokaryotes should precede all eukaryotic life in the fossil record. Indeed, the oldest known fossils are prokaryotes (see Chapter 26).

The Darwinian view of life predicts that evolutionary transitions should leave signs in the fossil record. Paleontologists have discovered fossils of many such transitional forms that link ancient organisms to modern species. For example, researchers have found fossil evidence that birds descended from one branch of dinosaurs, and they have also found fossilized whales linking these aquatic mammals to their terrestrial ancestors **(Figure 22.18)**.

Darwin's theory endures in biology because it explains so many different kinds of observations: anatomical and molecular homologies that match patterns in space (biogeography) and time (the fossil record). Natural selection can also explain how similar adaptations can evolve independently among distantly related species, such as sugar gliders and flying squirrels.

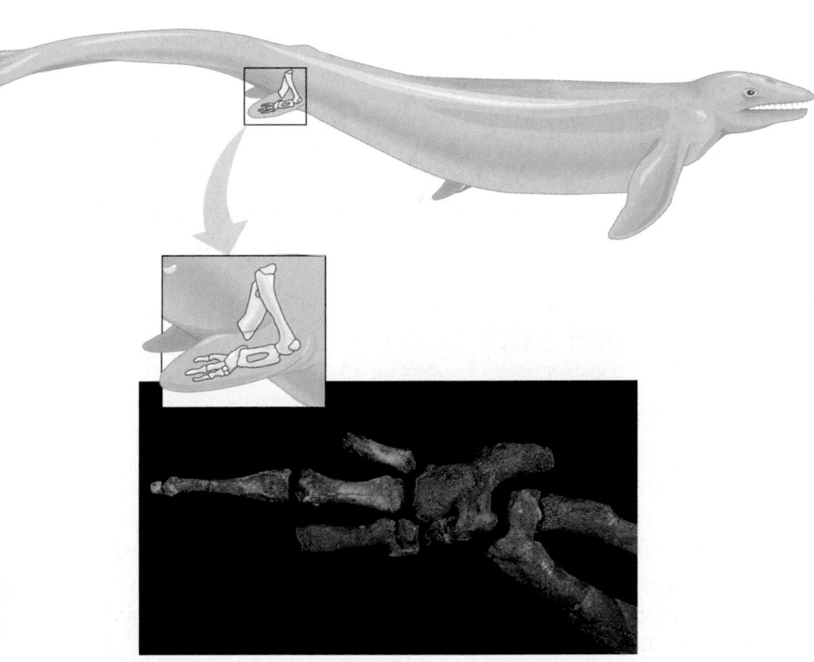

▲ **Figure 22.18 A transitional fossil linking past and present.** The hypothesis that whales evolved from terrestrial (land-dwelling) ancestors predicts a four-limbed beginning for whales. Paleontologists digging in Egypt and Pakistan have identified extinct whales that had hind limbs. Shown here are the fossilized leg bones of *Basilosaurus,* one of those ancient whales. Other fossil whales have been found with larger or smaller limbs, and all are related to still older animals that spent only part of their time in the water.

What Is Theoretical about the Darwinian View of Life?

Some people dismiss the Darwinian view of life as "just a theory." However, as we have seen, Darwin's explanation makes sense of massive amounts of data, and the effects of natural selection can be observed in nature.

What, then, is theoretical about evolution? The term *theory* has very different meanings in science and in everyday use. The colloquial use of the word *theory* comes close to what scientists mean by a hypothesis. In science, a theory is more comprehensive than a hypothesis. A theory, such as Newton's theory of gravitation or Darwin's theory of evolution by natural selection, accounts for many observations and data and attempts to explain and integrate a great variety of phenomena. Such a unifying theory does not become widely accepted unless its predictions stand up to thorough and continual testing by experiment and additional observation (see Chapter 1). As the next three chapters demonstrate, this has certainly been the case with the theory of evolution by natural selection.

The skepticism of scientists as they continue to test theories prevents these ideas from becoming dogma. For example, many evolutionary biologists now question whether natural selection is the only evolutionary mechanism responsible for the evolutionary history inferred from the fossil and molecular data. As we'll explore further in this unit, other factors have also played an important role, particularly in the evolution of genes and proteins. The study of evolution is livelier than ever as scientists find more ways to test the predictions of Darwin's theory.

By attributing the diversity of life to natural processes, Darwin gave biology a sound, scientific basis. Nevertheless, the diverse products of evolution remain elegant and inspiring. As Darwin said in the closing paragraph of *The Origin of Species,* "There is grandeur in this view of life. . . ."

Concept Check 22.3

1. Explain why the following statement is inaccurate: "Anti-HIV drugs have created drug resistance in the virus."
2. How does Darwin's theory account for both the similar mammalian forelimbs with different functions shown in Figure 22.14 and the similar lifestyle of the two distantly related mammals shown in Figure 22.17?
3. Explain how the fossil record can be used to test predictions of evolutionary theory.

For suggested answers, see Appendix A.

Chapter 22 Review

Go to www.campbellbiology.com or the student CD-ROM to explore Activities, Investigations, and other interactive study aids.

SUMMARY OF KEY CONCEPTS

The Darwinian revolution challenged traditional views of a young Earth inhabited by unchanging species

- **Resistance to the Idea of Evolution (pp. 439–440)** Darwin's theory that life's diversity has arisen from fewer ancestral species through natural selection was a radical departure from the prevailing views of Western culture.
- **Theories of Gradualism (p. 440)** Geologists Hutton and Lyell perceived that changes in Earth's surface can result from slow, continuous actions still operating at the present time.
- **Lamarck's Theory of Evolution (pp. 440–441)** Lamarck hypothesized that species evolve, but the mechanisms he proposed are unsupported by evidence.

Concept 22.2

In *The Origin of Species*, Darwin proposed that species change through natural selection

- **Darwin's Research (pp. 441–443)** Darwin's experiences during the voyage of HMS *Beagle* provided much of the background for his idea that new species originate from ancestral forms through the gradual accumulation of adaptations. After returning to England, he refined his theory and finally published it in 1859 after learning that Wallace had the same idea.
  Activity *Darwin and the Galápagos Islands*
  Activity *The Voyage of the* Beagle: *Darwin's Trip Around the World*
  Biology Labs On-Line *EvolutionLab*
- ***The Origin of Species* (pp. 443–446)** In his book describing his theory of descent with modification, Darwin recognized that there are heritable variations in populations and that some of those variations are better suited than others to a particular environment. Because organisms tend to produce many more offspring than the environment can support, there is competition for resources, and those that are better suited to the environment tend to have greater success in surviving and reproducing. Thus, these better-suited individuals leave more offspring than other individuals. Over time, this process of natural selection can result in adaptation of organisms to their environment.

Darwin's theory explains a wide range of observations

- **Natural Selection in Action (pp. 446–448)** Researchers have observed natural selection leading to adaptive evolution in wild guppy populations. In humans, the use of drugs selects for pathogens that through chance mutations are resistant to the drugs' effects. The ability of bacteria and viruses to evolve rapidly poses a challenge to our society.
  Investigation *How Do Environmental Changes Affect a Population?*
  Investigation *What Are the Patterns of Antibiotic Resistance?*
- **Homology, Biogeography, and the Fossil Record (pp. 448–451)** Evolutionary theory explains many kinds of observations, including structural and molecular similarities, geographic distribution of organisms, and the fossil record.
  Activity *Reconstructing Forelimbs*
- **What Is Theoretical about the Darwinian View of Life? (p. 451)** Darwin's theory of evolution by natural selection integrates diverse areas of biological study and stimulates many new research questions.

TESTING YOUR KNOWLEDGE

Self-Quiz

1. Which of the following statements reflects aspects of Hutton and Lyell's ideas of gradualism that were incorporated into Darwin's theory of evolution?
   a. There is a struggle in populations for survival and reproduction.
   b. Natural selection acts on heritable variation.
   c. Small changes accumulated over vast spans of time can produce dramatic results.
   d. Characteristics acquired gradually over an organism's lifetime can lead to changes in the characteristics of the next generation.
   e. Homologous structures are found in organisms with a common ancestor.

2. Which of the following is *not* an observation or inference on which natural selection is based?
   a. There is heritable variation among individuals.
   b. Poorly adapted individuals never produce offspring.
   c. There is a struggle for limited resources, and only a fraction of offspring survive.
   d. Individuals whose characteristics are best suited to the environment generally leave more offspring than those whose characteristics are less well suited.
   e. Organisms interact with their environments.

3. Analysis of forelimb anatomy of humans, bats, and whales shows that humans and bats have fairly similar skeletal structures, while whales have diverged considerably in the shapes and proportions of their bones. However, analysis of several genes in these species suggests that all three diverged from a common ancestor at about the same time. Which of the following is the best explanation for these data?
   a. Humans and bats evolved by natural selection, and whales evolved by Lamarckian mechanisms.
   b. Evolution of human and bat forelimbs was adaptive, but not for whales.
   c. Natural selection in an aquatic environment resulted in significant changes to whale forelimb anatomy.
   d. Genes mutate more rapidly in whales than in humans or bats.
   e. Whales are not properly defined as mammals.

4. Which of the following observations helped Darwin shape his concept of descent with modification?
   a. Species diversity declines as distance from the equator increases.
   b. Fewer species are found living on islands than on the nearest continents.
   c. Birds can be found on islands that are farther from the mainland than the birds' maximum nonstop flight distance.
   d. South American temperate plants are more similar to the tropical plants of South America than to the temperate plants of Europe.
   e. Earthquakes reshape life by causing mass extinctions.

5. Darwin synthesized information from several sources in developing his theory of evolution by natural selection. Which of the following did *not* influence his thinking?
   a. Linnaeus' hierarchical classification of species
   b. Lyell's *Principles of Geology*
   c. observations of molecular homologies
   d. examples of major changes in domesticated species produced by artificial selection
   e. the distribution of species that he observed on the Galápagos Islands and during his journey around South America

6. In science, the term *theory* generally applies to an idea that
   a. is a speculation lacking supportive observations or experiments.
   b. attempts to explain many related phenomena.
   c. is synonymous with what biologists mean by a hypothesis.
   d. is considered a law of nature.
   e. all of the above.

7. Within a few weeks of treatment with the drug 3TC, a patient's HIV population consists entirely of 3TC-resistant viruses. How can this result best be explained?
   a. HIV has the ability to change its surface proteins and resist vaccines.
   b. The patient must have become reinfected with 3TC-resistant viruses.
   c. HIV began making drug-resistant versions of reverse transcriptase in response to the drug.
   d. A few drug-resistant viruses were present at the start of treatment, and natural selection increased their frequency.
   e. The drug caused the HIV RNA to change.

8. The smallest biological unit that can evolve over time is
   a. a cell.
   b. an individual organism.
   c. a population.
   d. a species.
   e. an ecosystem.

9. Which of the following ideas is common to both Darwin's and Lamarck's theories of evolution?
   a. Adaptation results from differential reproductive success.
   b. Evolution drives organisms to greater and greater complexity.
   c. Evolutionary adaptation results from interactions between organisms and their environments.
   d. Adaptation results from the use and disuse of anatomical structures.
   e. The fossil record supports the view that species are fixed.

10. Which of the following pairs of structures is *least* likely to represent homology?
   a. the wings of a bat and the forelimbs of a human
   b. the hemoglobin of a baboon and that of a gorilla
   c. the mitochondria of a plant and those of an animal
   d. the wings of a bird and those of an insect
   e. the brain of a cat and that of a dog

For Self-Quiz answers, see Appendix A.

Go to the website or CD-ROM for more quiz questions.

Evolution Connection

Explain why anatomical and molecular homologies generally fit a similar nested pattern.

Scientific Inquiry

Darwin's argument for the idea that evolution has occurred is largely inductive, while his argument for the mechanism of natural selection is essentially deductive. Summarize in your own words the inductive and deductive components of Darwin's theory. (You can review induction and deduction in Chapter 1.)

Investigation *How Do Environmental Changes Affect a Population?*
Investigation *What Are the Patterns of Antibiotic Resistance?*
Biology Labs On-Line *EvolutionLab*

Science, Technology, and Society

Is the concept of natural selection relevant in an economic or political context? In other words, if a particular corporation or nation achieves success or dominance, does this mean that it is fitter than its competitors and that dominance is justified? Why or why not?

23 The Evolution of Populations

▲ **Figure 23.1 Variation in a natural population.**

Key Concepts

23.1 Population genetics provides a foundation for studying evolution
23.2 Mutation and sexual recombination produce the variation that makes evolution possible
23.3 Natural selection, genetic drift, and gene flow can alter a population's genetic composition
23.4 Natural selection is the primary mechanism of adaptive evolution

Overview

The Smallest Unit of Evolution

One common misconception about evolution is that individual organisms evolve, in the Darwinian sense, during their lifetimes. It is true that natural selection acts on individuals; each organism's combination of traits affects its survival and reproductive success compared to other individuals. But the evolutionary impact of natural selection is only apparent in the changes in a population of organisms over time. Consider, for example, the collection of shells from a population of Cuban tree snails (*Polymita picta*) shown in **Figure 23.1**. Their different patterns and colors are largely the result of genetic differences among the individuals. Suppose that predators are less likely to feed on snails with a particular coloration, perhaps because these snails are better camouflaged against their surroundings. The proportion of snails of that color will tend to increase from one generation to the next. Thus, the population, not its individual members, evolves; some traits become more common within the population, while other traits become less common.

Concept 23.1

Population genetics provides a foundation for studying evolution

Today we can define evolutionary change on its smallest scale, or **microevolution**, as change in the genetic makeup of a population from generation to generation **(Figure 23.2)**. But Darwin did not define evolution in this way. In natural selection, Darwin had found a mechanism for change in species

▲ **Figure 23.2 Individuals are selected; populations evolve.** The bent grass (*Agrostis tenuis*) in the foreground is growing on the tailings of an abandoned mine. These plants tolerate concentrations of heavy metals that are toxic to other plants of the same species in the pasture beyond the fence. Many seeds from the pasture drift onto the tailings, but only those with genes that enable them to tolerate metallic soil survive and reproduce.

over time. However, he did not have a satisfactory explanation for how the heritable variations required for natural selection appear in populations or how organisms transmit these variations to their offspring. Ideas about inheritance from Darwin's time could not explain how inherited variations are maintained in populations. For instance, the widely accepted blending hypothesis proposed that the traits of parents are blended in the offspring. But Darwin and others realized that over time, blending would eliminate the differences between individuals. Although breeding experiments and observations of nature refuted this prediction, Darwin lacked an alternate model of inheritance that could support his hypothesis.

Just a few years after Darwin published *On the Origin of Species,* Gregor Mendel proposed just such a model: the particulate hypothesis of inheritance, which stated that parents pass on discrete heritable units (genes) that retain their identities in offspring. But Darwin never saw Mendel's paper, and its implications were not understood by the few scientists who read it at the time. Mendel's contribution to evolutionary theory was not appreciated until half a century later.

The Modern Synthesis

Ironically, when Mendel's work was rediscovered and extended in the early 20th century, many geneticists thought that his laws of inheritance were at odds with Darwin's theory. Darwin considered the raw material for natural selection to be "quantitative" characters—those characteristics in a population that vary along a continuum, such as fur length in mammals or the running speed of animals fleeing from predators. But Mendel and other early geneticists worked only with discrete "either-or" traits, such as purple or white flowers in pea plants. It was therefore not obvious that there was a genetic basis to the more subtle variations that were central to Darwin's theory. Within a few decades, however, geneticists determined that quantitative characters are influenced by multiple genetic loci and that the alleles at each of these loci follow Mendelian patterns of inheritance (see Chapter 14). These discoveries helped reconcile Darwin's and Mendel's ideas and led to the formal founding of **population genetics,** the study of how populations change genetically over time.

By the mid-20th century, population genetics also gave rise to what is called the **modern synthesis,** a comprehensive theory of evolution that integrated ideas from many other fields. The first architects of the modern synthesis included statistician R. A. Fisher (1890–1962), who demonstrated the rules by which Mendelian characters are inherited, and biologist J. B. S. Haldane (1892–1964), who studied the rules of natural selection. Later contributors included geneticists Theodosius Dobzhansky (1900–1975) and Sewall Wright (1889–1988), biogeographer Ernst Mayr (1904–2005), paleontologist George Gaylord Simpson (1902–1984), and botanist G. Ledyard Stebbins (1906–2000).

Of course, scientific paradigms rarely endure without modification. The modern synthesis is still expanding to integrate, for example, the discovery of genetic changes in populations caused by mechanisms other than natural selection. Although the modern synthesis itself continues to evolve, its focus on populations has shaped much of our current thinking about evolutionary processes.

Gene Pools and Allele Frequencies

Before continuing our discussion of population genetics, we need to define what we mean by a population. A **population** is a localized group of individuals that are capable of interbreeding and producing fertile offspring. Populations of the same species may be isolated from one another, thus exchanging genetic material only rarely. Such isolation is common for populations confined to different, widely separated islands or lakes. But not all populations are isolated, nor must they have sharp boundaries **(Figure 23.3)**. Still, individuals near the population center are more likely to breed with members of their own population than other populations and thus on average are more closely related to one another than to members of other populations.

The aggregate of genes in a population at any one time is called the population's **gene pool.** It consists of all alleles

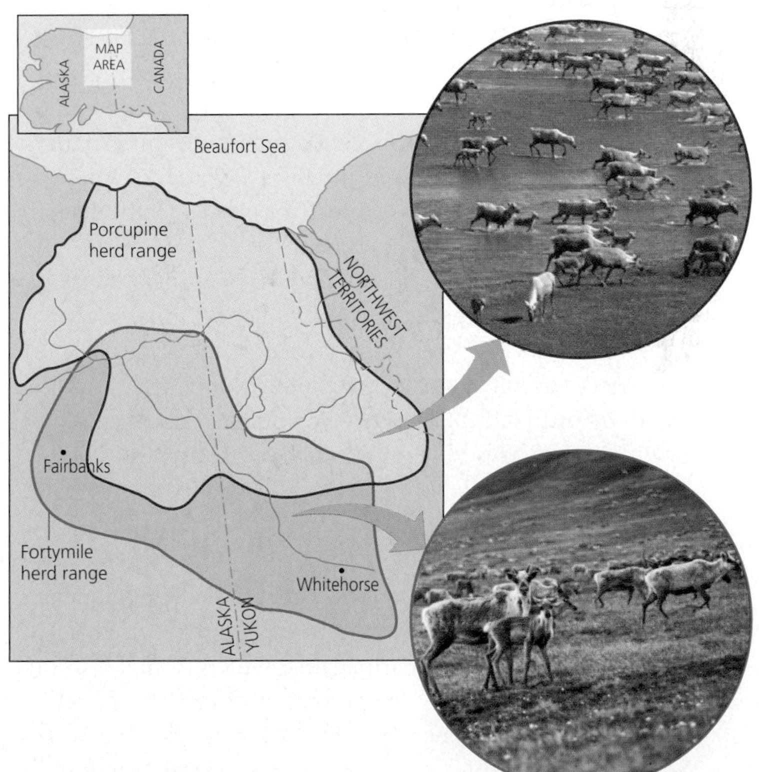

▲ **Figure 23.3 One species, two populations.** These two caribou populations in the Yukon are not totally isolated—they sometimes share the same area. Nonetheless, members of either population are more likely to breed with members of their own population than with members of the other population.

at all gene loci in all individuals of the population. If only one allele exists at a particular locus in a population, that allele is said to be *fixed* in the gene pool, and all individuals are homozygous for that allele. (Recall that homozygous individuals have two identical alleles for a given locus, while heterozygous individuals have two different alleles at that locus.) But if there are two or more alleles for a particular locus in a population, individuals may be either homozygous or heterozygous.

Each allele has a frequency (proportion) in the population. For example, imagine a population of 500 wildflower plants with two alleles, C^R and C^W, at a locus that codes for flower pigment. Plants homozygous for the C^R allele (C^RC^R) produce red pigment and have red flowers; plants homozygous for the C^W allele (C^WC^W) produce no red pigment and have white flowers; and heterozygotes (C^RC^W) produce some red pigment and have pink flowers. In our population there are 320 plants with red flowers, 160 with pink flowers, and 20 with white flowers. These alleles show incomplete dominance (see Chapter 14). Because these are diploid organisms, there are a total of 1,000 copies of genes for flower color in the population of 500 individuals. The C^R allele accounts for 800 of these genes ($320 \times 2 = 640$ for C^RC^R plants, plus $160 \times 1 = 160$ for C^RC^W plants).

When there are two alleles at a particular locus, the convention is to use p to represent the frequency of one allele and q to represent the frequency of the other allele. Thus, p, the frequency of the C^R allele in the gene pool of this population, is $800/1{,}000 = 0.8 = 80\%$. And because there are only two allelic forms of this gene, then the frequency of the C^W allele, represented by q, must be 0.2, or 20%. At loci that have more than two alleles, the sum of all allele frequencies must still equal 1 (100%).

We can easily measure genetic variability at this flower-color locus because each genotype has a distinct phenotype. Although many loci in gene pools have more than one allele, this variability is usually less easy to quantify than in our flower example, because one allele may be completely dominant or the alleles may not have obvious effects on phenotypes.

The Hardy-Weinberg Theorem

In a moment we will explore how genes and allele frequencies can change over time, resulting in evolutionary change. But first, to provide a benchmark against which to measure such changes, we'll investigate the properties of gene pools that are *not* evolving. Such gene pools are described by the **Hardy-Weinberg theorem,** named for the two scientists who independently derived the principle in 1908. The theorem states that the frequencies of alleles and genotypes in a population's gene pool remain constant from generation to generation, provided that only Mendelian segregation and recombination of alleles are at work **(Figure 23.4)**.

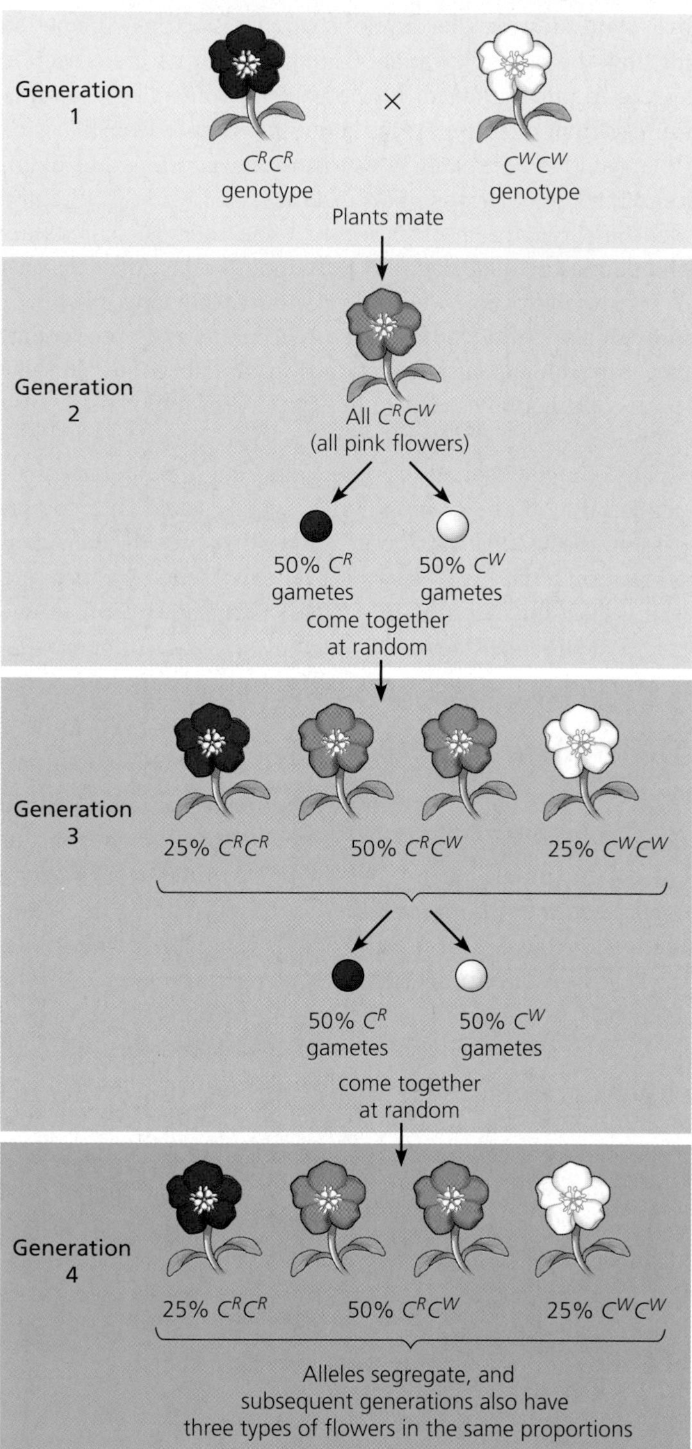

▲ **Figure 23.4 Mendelian inheritance preserves genetic variation from one generation to the next.**

Preservation of Allele Frequencies

The Hardy-Weinberg theorem describes how Mendelian inheritance preserves genetic variation from one generation to the next in populations that are not evolving. But it does more than this. We will see that the theorem lays the groundwork for understanding long-term evolutionary changes that Darwin, lacking knowledge of genetics, could not have envisioned. The preservation of genetic variation provides the opportunity for natural selection to act over many generations.

Let's apply the Hardy-Weinberg theorem to our hypothetical wildflower population. Recall that 80% (0.8) of the flower-color alleles in the gene pool are C^R and 20% (0.2) are C^W. Because each gamete produced by the flowers is haploid, it carries only one allele for flower color. The chance that a gamete will carry a C^R allele is 0.8, and the chance that it will carry a C^W allele is 0.2. The allele frequencies in all the gametes produced by the population will be the same as in the original population. Provided that gametes are contributed to the next generation at random, allele frequencies will be unchanged.

Hardy-Weinberg Equilibrium

Suppose that the individuals in a population not only donate gametes to the next generation at random but also mate at random—that is, all male-female matings are equally likely. Not only will this population have the same allele frequencies from one generation to the next, but its genotype frequencies can be predicted from the allele frequencies. Such populations are in a state of **Hardy-Weinberg equilibrium.**

It turns out that our wildflower population is in Hardy-Weinberg equilibrium **(Figure 23.5)**. Using the rule of multiplication (see Chapter 14), we can calculate the frequencies of the three possible genotypes assuming random unions of sperm and ova. The probability that two C^R alleles will come together is $0.8 \times 0.8 = p \times p = p^2 = 0.64$. Thus, about 64% of the plants in the next generation will have the genotype C^RC^R. The frequency of C^WC^W individuals will be about $0.2 \times 0.2 = q^2 = 0.04$, or 4%. C^RC^W heterozygotes can arise in two different ways. If the sperm provides the C^R allele and the ovum provides the C^W allele, the resulting heterozygotes will be $0.8 \times 0.2 = 16\%$ of the total. If the ovum provides the C^R allele and the sperm the C^W allele, the heterozygous offspring will make up another $0.8 \times 0.2 = 16\%$. We can summarize these unions of gametes in an algebraic equation:

$(p + q)$	$\times$	$(p + q)$	$=$	$p^2 + 2pq + q^2$
Allele frequencies of male gametes		Allele frequencies of female gametes		Genotype frequencies in the next generation

As with allele frequencies, all of these genotype frequencies sum to 1. Thus, the equation for Hardy-Weinberg equilibrium states that at a locus with two alleles, the three genotypes will appear in the following proportions:

$$p^2 + 2pq + q^2 = 1$$

If a population were in Hardy-Weinberg equilibrium and its members continued to mate randomly generation after generation, allele and genotype frequencies would remain constant. The system operates somewhat like a deck of cards: No matter how many times the deck is reshuffled to deal out new hands, the deck itself remains the same. Aces do not grow more numerous than jacks. And the repeated shuffling of a population's gene pool over the generations cannot, in itself, change the frequency of one allele relative to another.

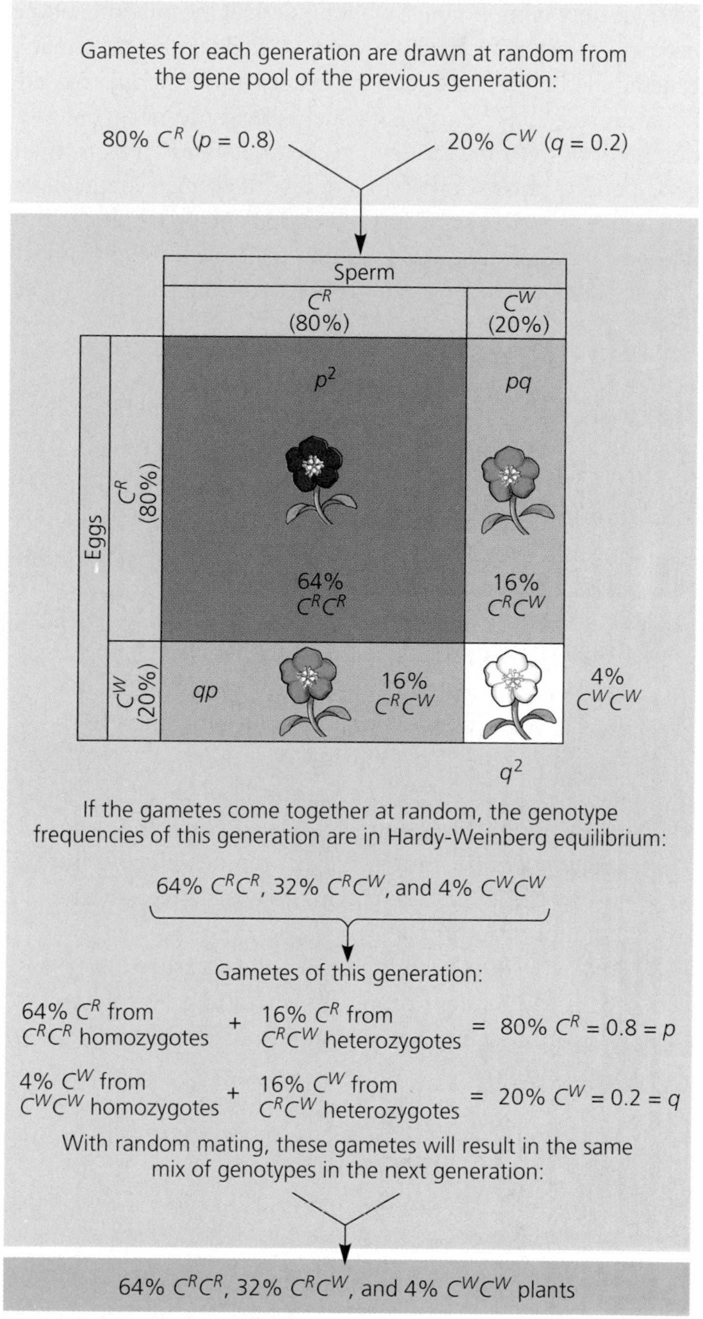

Note that a population does not need to be in Hardy-Weinberg equilibrium for its allele frequencies to remain constant. Many species, such as the peas that Mendel used in his experiments, do not mate randomly. Since pea pollen normally matures before the flower opens, it fertilizes the same flower that contains it. (Mendel could cross his pea plants only by artificial fertilization.) Because of self-fertilization, pea populations are far from Hardy-Weinberg equilibrium. All the homozygous plants

▲ **Figure 23.5 The Hardy-Weinberg theorem.** In our wildflower population, the gene pool remains constant from one generation to the next. Mendelian processes alone do not alter frequencies of alleles or genotypes.

produce only homozygotes, while about half the offspring of the heterozygous plants are homozygous. It does not take many generations before almost all the plants in the population are homozygous. But even though such species do not mate randomly, they still produce their gametes at random from their gene pools. Without other influences, their allele frequencies will not change over time. Just as in populations that do mate at random, genetic variation—the raw material for evolutionary change—is preserved even in the self-fertilizing peas.

Conditions for Hardy-Weinberg Equilibrium

The Hardy-Weinberg theorem describes a hypothetical population that is not evolving. But in real populations, allele and genotype frequencies *do* change over time. This is because the following five conditions for non-evolving populations are rarely met for long in nature:

1. *Extremely large population size.* The smaller the population, the greater the role played by chance fluctuations in allele frequencies from one generation to the next, known as genetic drift.
2. *No gene flow.* Gene flow, the transfer of alleles between populations, can alter allele frequencies.
3. *No mutations.* By introducing or removing genes from chromosomes or by changing one allele into another, mutations modify the gene pool.
4. *Random mating.* If individuals preferentially choose mates with certain genotypes, including close relatives (inbreeding), random mixing of gametes does not occur.
5. *No natural selection.* Differential survival and reproductive success of individuals carrying different genotypes will alter allele frequencies.

Departure from these conditions usually results in evolution. While natural populations rarely if ever are in true Hardy-Weinberg equilibrium, in many populations the rate of evolutionary change is so slow that these populations *appear* to be close to equilibrium. This enables us to obtain approximate estimates of allele and genotype frequencies, as the following example shows.

Population Genetics and Human Health

The Hardy-Weinberg equation can be used to estimate the percentage of the population carrying the allele for an inherited disease. For example, about one out of every 10,000 babies in the United States is born with phenylketonuria (PKU), a metabolic disorder that results from homozygosity for a recessive allele. Left untreated, PKU results in mental retardation and other problems. (Newborns are now tested for PKU, and their symptoms can be lessened with a phenylalanine-free diet.)

To use the Hardy-Weinberg equation, we must assume that people do not choose their mates on the basis of whether or not they carry this gene and do not generally mate with close relatives (inbreeding). We must also neglect any effects of gene flow from other populations into the United States, introduction of new PKU mutations, and differential survival and reproductive success among PKU genotypes. These assumptions are reasonable, since inbreeding in the United States is not common, populations outside the United States have PKU allele frequencies similar to those seen in the United States, the mutation rate for the PKU gene is low, and selection occurs only against the rare homozygotes.

If all these assumptions hold, then the frequency of individuals in the population born with PKU will correspond to q^2 in the Hardy-Weinberg equation (q^2 = frequency of the homozygotes for this allele). Because the allele is recessive, we must estimate the number of heterozygotes rather than counting them directly as we did with the pink flowers. Since we know there is one PKU occurrence per 10,000 births ($q^2 = 0.0001$), the frequency of the recessive allele for PKU is

$$q = \sqrt{0.0001} = 0.01$$

and the frequency of the dominant allele is

$$p = 1 - q = 1 - 0.01 = 0.99$$

The frequency of carriers, heterozygous people who do not have PKU but may pass the PKU allele to offspring, is

$$2pq = 2 \times 0.99 \times 0.01 = 0.0198$$ (approximately 2% of the U.S. population)

Remember, the assumption of Hardy-Weinberg equilibrium results in only an approximation; the real number of carriers may be somewhat different. Nevertheless, we can conclude that many harmful recessive alleles at this and other loci are concealed in the human population because they are carried by healthy heterozygotes.

Concept Check 23.1

1. What did Mendel's findings about genetics add to Darwin's theory of evolution by natural selection?
2. Suppose a population of organisms with 500 gene loci is fixed at half of these loci, and has two alleles at each of the other loci. How many alleles are found in its gene pool? Explain.
3. Which parts of the Hardy-Weinberg equation ($p^2 + 2pq + q^2 = 1$) correspond to the frequency of individuals that have at least one PKU allele?

For suggested answers, see Appendix A.

Concept 23.2

Mutation and sexual recombination produce the variation that makes evolution possible

As you have learned, natural selection operates on the heritable differences, often slight ones, among individuals in a population. Two processes, mutation and sexual recombination, produce the variation in gene pools that contributes to these individual differences.

Mutation

New genes and new alleles originate only by **mutations** **(Figure 23.6)**, which are changes in the nucleotide sequence of DNA. A mutation is like a shot in the dark—it is not possible to predict how it will alter DNA and what its effects will be. Most mutations occur in somatic cells and are lost when the individual dies. Only mutations in cell lines that produce gametes can be passed to offspring, and only a small fraction of these spread through populations.

Point Mutations

A change of as little as one base in a gene—a "point mutation"—can have a significant impact on phenotype, as in sickle-cell disease (see Figure 5.21). But most point mutations are probably harmless. One reason is that much of the DNA in eukaryotic genomes does not code for protein products. And because the genetic code is redundant, even point mutations in genes that code for protein may have little effect because they do not alter the protein's amino acid composition (see Figure 17.24). However, some noncoding regions of DNA *do* regulate the expression of genes. Changes in these regulatory regions of DNA can have profound effects.

▲ **Figure 23.6 Mutations are the source of all heritable variation.** The diverse coat colors of these wild mustangs are the products of past mutations over many generations.

Organisms reflect thousands of generations of past selection, and a single mutational change is about as likely to improve the genome as blindly firing a gunshot through the hood of a car is likely to improve engine performance. On rare occasions, however, a mutant allele may actually make its bearer better suited to the environment, enhancing reproductive success. This is more likely when the environment is changing and mutations that were once selected against become favorable. For example, as you read in Chapter 22, certain mutations that make HIV resistant to particular drugs also slow its reproductive rate. Only after the drugs were introduced were these mutant alleles favored, and natural selection increased their frequency as a result.

Mutations that Alter Gene Number or Sequence

Chromosomal mutations that delete, disrupt, or rearrange many loci at once are almost certain to be harmful. However, when such mutations leave genes intact, their effects on organisms may be neutral. In rare cases, chromosomal rearrangements may even be beneficial. For example, the translocation of part of one chromosome to a different chromosome could link genes that together have some heightened positive effect.

Gene **duplication** is an important source of variation. Duplications of chromosome segments, like other chromosomal mutations, are nearly always harmful. But smaller pieces of DNA are often introduced into a genome through the activity of transposable elements (see Chapter 19). If such a duplicated segment does not have severe effects, it can persist over generations, providing an expanded genome with new loci that may take on new functions by further mutations and subsequent selection. New genes may also arise when the coding portions of genes (exons) are shuffled within the genome.

Such beneficial increases in gene number appear to have played a major role in evolution. For example, the remote ancestors of mammals carried a single gene for detecting odors that has been duplicated through a variety of mutational mechanisms. As a result, humans today have about 1,000 olfactory receptor genes, and mice have 1,300. About 60% of human olfactory receptor genes have been inactivated by subsequent mutations, whereas mice have lost only 20% of theirs—a remarkable demonstration that a versatile sense of smell is more important to mice than it is to us!

Mutation Rates

Mutation rates tend to be low in animals and plants, averaging about one mutation in every 100,000 genes per generation. But in microorganisms and viruses with short generation spans, mutations can rapidly generate genetic variation. For instance, HIV has a generation span of about two days and an RNA genome, which has a much higher mutation rate than a typical DNA genome. For this reason, single-drug treatments will probably never be effective against HIV—mutant forms of the virus that are resistant to a particular drug are likely to proliferate in relatively short order. Even double-drug treatments rarely work for long, because viruses with double mutations conferring resistance to *both* drugs arise daily. This explains why the most effective AIDS treatments at this time are drug "cocktails" combining several medications. It is less likely that multiple mutations conferring resistance to *all* the drugs will occur in a short time period.

Sexual Recombination

In sexually reproducing populations, sexual recombination is far more important than mutation on a generation-to-generation time scale in producing the variations that make adaptation possible. Nearly all phenotypic variations based on genetic differences result from recombinational shuffling of the existing alleles in the gene pool. (Of course, this allelic variation originated from past mutations.) A population contains myriad possible mating combinations, and fertilization brings together the gametes of individuals that are likely to have different genetic backgrounds. Sexual reproduction rearranges alleles into fresh combinations every generation.

Bacteria and many viruses can also undergo recombination, but they do so less regularly than animals and plants and often in ways that allow them to cross species barriers (see Chapter 18). Many of the genes of the *Escherichia coli* bacterium O157:H7, a frequent culprit in food poisoning cases, are actually "mosaics" of genes from different types of bacteria. The ability of pathogens to evolve rapidly through such extensive recombination, combined with their high mutation rates, makes them especially dangerous adversaries.

Concept Check 23.2

1. Of all the mutations that occur, why do only a small fraction become widespread in a gene pool?
2. How does sexual recombination produce variation?

For suggested answers, see Appendix A.

Concept 23.3

Natural selection, genetic drift, and gene flow can alter a population's genetic composition

Note again on page 458 the five conditions required for a population to be in Hardy-Weinberg equilibrium. A deviation from any of these conditions is a potential cause of evolution. However, although new mutations can modify allele frequencies, the change from one generation to the next is likely to be small. Recombination reshuffles alleles, but does not change their frequencies. Nonrandom mating can affect relative frequencies of homozygous and heterozygous genotypes but by itself usually has no effect on allele frequencies. The three major factors that alter allele frequencies and cause most evolutionary change are natural selection, genetic drift, and gene flow.

Natural Selection

As you read in Chapter 22, Darwin's concept of natural selection is based on differential success in reproduction: Individuals in a population exhibit variations in their heritable traits, and those with variations that are better suited to their environment tend to produce more offspring than those with variations that are less well suited.

We now know that selection results in alleles being passed to the next generation in proportions different from their relative frequencies in the present generation. For example, in our imaginary wildflower population, white flowers ($C^W C^W$) might be more visible to plant-eating insects, resulting in more white flowers being eaten. And perhaps red flowers ($C^R C^R$) are more attractive to pollinators, increasing the opportunity for red flowers to produce offspring. These differences in survival and reproductive success would disturb Hardy-Weinberg equilibrium: The frequency of the C^W allele would decline in the gene pool, and the frequency of the C^R allele would increase. Later in the chapter, we'll examine the process of natural selection more closely.

Genetic Drift

Flip a coin 1,000 times, and a result of 700 heads and 300 tails would make you suspicious about that coin. But flip a coin 10 times, and an outcome of 7 heads and 3 tails would not be surprising. The smaller the sample, the greater the chance of deviation from the predicted result—in this case, an equal number of heads and tails.

Similar deviations from the expected result—which occur because real populations are finite in size rather than infinite—explain how allele frequencies can fluctuate unpredictably from one generation to the next. Such fluctuations are called **genetic drift.** Consider the example in **Figure 23.7**. Note that in this example, one of the alleles has been lost by an indiscriminate event; that is, it was a matter of chance that the C^W allele rather than the C^R allele was lost. Over time, drift tends to reduce genetic variation through such losses of alleles from the gene pool.

Two situations that can increase the likelihood that genetic drift will have a large impact on a population are referred to as the bottleneck effect and the founder effect.

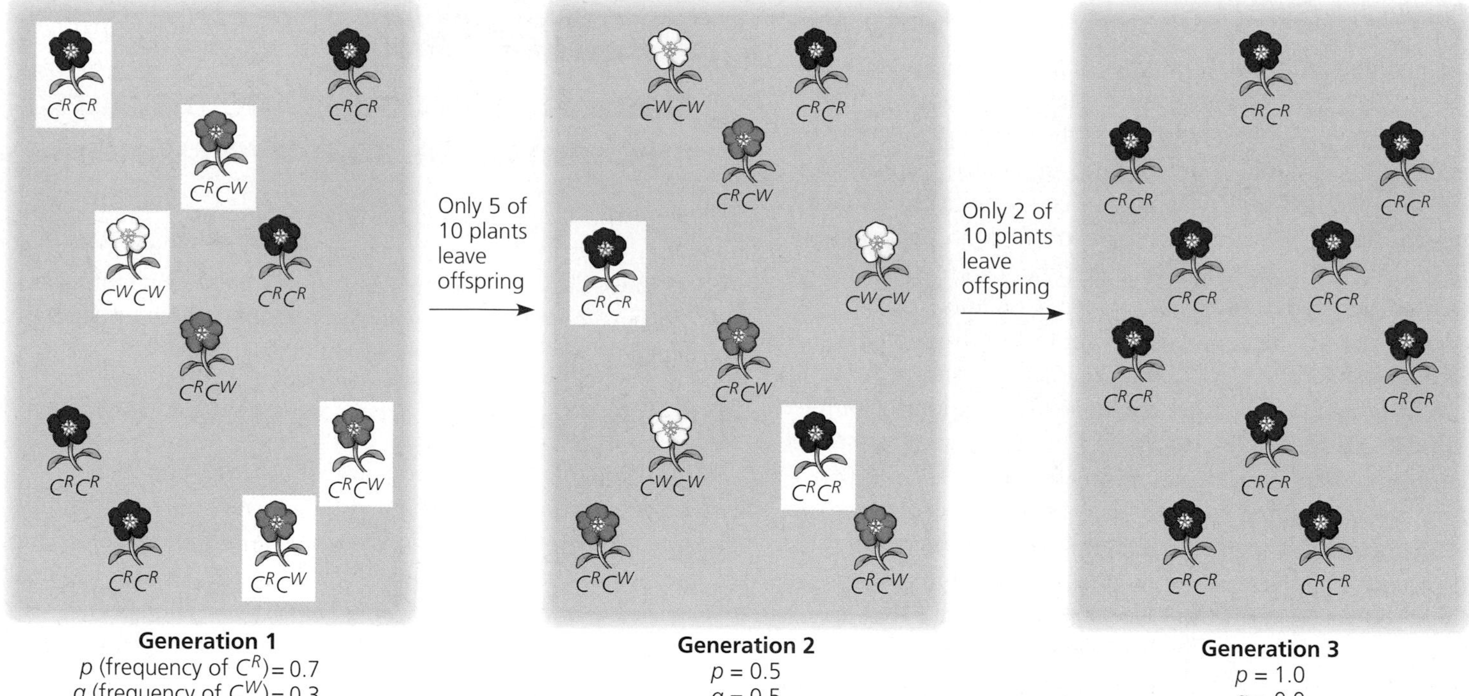

▲ **Figure 23.7 Genetic drift.** This small wildflower population has a stable size of ten plants. Only five plants (those in white boxes) of generation 1 produce fertile offspring. By chance, only two plants of generation 2 manage to leave fertile offspring. The C^W allele first increases in generation 2, then falls to zero in generation 3.

The Bottleneck Effect

A sudden change in the environment, such as a fire or flood, may drastically reduce the size of a population. In effect, the survivors have passed through a restrictive "bottleneck," and their gene pool may no longer be reflective of the original population's gene pool; this is called the **bottleneck effect** **(Figure 23.8a)**. By chance, certain alleles may be overrepresented among the survivors, others may be underrepresented, and some may be eliminated altogether. Genetic drift may continue to have substantial effects on the gene pool for many generations until the population is large enough that chance fluctuations have less effect.

One reason it is important to understand the bottleneck effect is that human actions can sometimes create severe bottlenecks for other species. For example, in the 1890s, hunters reduced the population of northern elephant seals in California to about 20 individuals. Since then, this mammal has been protected, and the population has rebounded to over 30,000 **(Figure 23.8b)**. However, in examining 24 gene loci in a representative sample of the seals, researchers found *no* variation: For each of the 24 genes, there was only one allele. In contrast, populations of the southern elephant seal, a closely related species that was not so severely reduced, show considerable genetic variation at these loci.

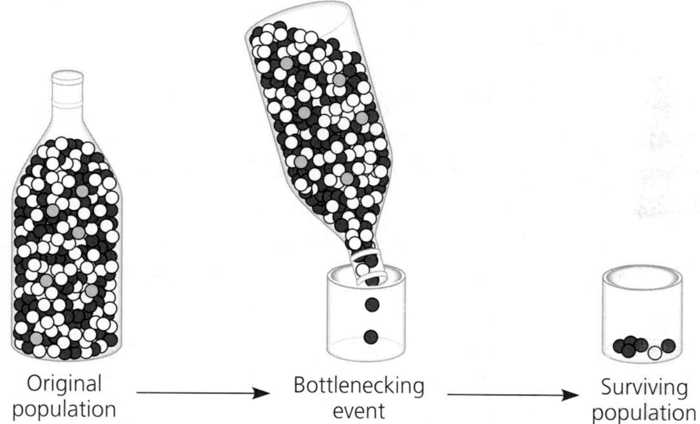

(a) Shaking just a few marbles through the narrow neck of a bottle is analogous to a drastic reduction in the size of a population after some environmental disaster. By chance, blue marbles are overrepresented in the new population and gold marbles are absent.

(b) Similarly, bottlenecking a population of organisms tends to reduce genetic variation, as in these northern elephant seals in California that were once hunted nearly to extinction.

▲ **Figure 23.8 The bottleneck effect.**

The Founder Effect

When a few individuals become isolated from a larger population, this smaller group may establish a new population whose gene pool is not reflective of the source population; this is called the **founder effect.** For example, the founder effect can occur when a few members of a population colonize a new location. These founders pass through an "isolation bottleneck" and represent a distinct gene pool with different allele frequencies from those of the parent population.

The founder effect probably accounts for the relatively high frequency of certain inherited disorders among isolated human populations. For example, in 1814, 15 colonists founded a British settlement on Tristan da Cunha, a group of small islands in the Atlantic Ocean midway between Africa and South America. Apparently, one of the colonists carried a recessive allele for retinitis pigmentosa, a progressive form of blindness that afflicts homozygous individuals. Of the 240 descendants living on the island in the late 1960s, four had the disease. The frequency of this allele remains ten times higher on Tristan da Cunha than in the populations from which the founders came.

Gene Flow

A population may gain or lose alleles by **gene flow,** genetic additions to and/or subtractions from a population resulting from the movement of fertile individuals or gametes. Suppose, for example, that near our original hypothetical wildflower population there is a newly established wildflower population consisting primarily of white-flowered individuals (C^WC^W). Insects carrying pollen from these plants may fly to and pollinate plants in our original population. The introduced C^W alleles will modify our original population's allele frequencies in the next generation.

Gene flow tends to reduce differences between populations. If it is extensive enough, gene flow can amalgamate neighboring populations into a single population with a common gene pool. For example, humans today move much more freely about the world than in the past, and gene flow has become an important agent of evolutionary change in human populations that were previously quite isolated.

Concept Check 23.3

1. In what sense is natural selection more "predictable" than genetic drift?
2. Distinguish genetic drift and gene flow in terms of (a) how they occur and (b) their implications for future genetic variation in a population.

For suggested answers, see Appendix A.

Concept 23.4

Natural selection is the primary mechanism of adaptive evolution

Of all the factors that can change a gene pool, only natural selection is likely to adapt a population to its environment. Natural selection accumulates and maintains favorable genotypes in a population. As you have read, the process of selection depends on the existence of genetic variation.

Genetic Variation

You probably have no trouble recognizing your friends in a crowd. Each person has a unique genome, reflected in individual phenotypic variations such as appearance, voice, and temperament. Individual variation occurs in populations of all species. Although most people are very conscious of human diversity, we are generally less sensitive to individuality in other organisms. But variations are always present, and as Darwin realized, variations that are heritable are the raw material for natural selection. In addition to the differences that we can see or hear, populations have extensive genetic variation that can only be observed at the molecular level. For example, you cannot identify a person's blood group (A, B, AB, or O) from his or her appearance.

Not all phenotypic variation is heritable **(Figure 23.9)**. Phenotype is the cumulative product of an inherited genotype and a multitude of environmental influences. For example, bodybuilders alter their phenotypes dramatically but do not pass their huge muscles on to the next generation. It is important to remember that only the genetic component of variation can have evolutionary consequences as a result of natural selection.

(a) Map butterflies that emerge in spring: orange and brown

(b) Map butterflies that emerge in late summer: black and white

▲ **Figure 23.9 Nonheritable variation within a population.** These European map butterflies are seasonal forms of the same species (*Araschnia levana*). Owing to seasonal differences in hormones, **(a)** individuals that emerge in the spring are orange and brown, while **(b)** individuals that emerge in the late summer are black and white. These two forms are genetically identical at the loci for coloration. Therefore, if these two forms differ in reproductive success, that in itself would not lead to any change in the ability of the butterflies to develop in these two different ways.

Variation Within a Population

Both discrete and quantitative characters contribute to variation *within* a population. *Discrete characters,* such as the red, pink, and white colors of our hypothetical wildflower population, can be classified on an either-or basis (each plant has flowers that are all either red or pink or white). Discrete characters often are determined by a single gene locus with different alleles that produce distinct phenotypes. However, as discussed in Chapter 22, most heritable variation consists of *quantitative characters* that vary along a continuum within a population. Heritable quantitative variation results from the influence of two or more genes on a single phenotypic character.

Polymorphism. When individuals differ in a discrete character, the different forms are called *morphs.* A population is said to display **phenotypic polymorphism** for a character if two or more distinct morphs are each represented in high enough frequencies to be readily noticeable. (Obviously, the definition of "readily noticeable" is somewhat subjective, but a population is not considered polymorphic if it consists primarily of a single morph and other morphs are extremely rare.)

In contrast, height variation in the human population does not show phenotypic polymorphism because it does not consist of distinct and separate morphs—heights vary along a continuum. Nonetheless, polymorphisms play a role in such characters at the genetic level. The heritable component of height is the result of such **genetic polymorphisms** for alleles at the several loci that influence height.

Measuring Genetic Variation. Population geneticists measure the number of polymorphisms in a population by determining the amount of heterozygosity at both the level of whole genes (gene variability) and the molecular level of DNA (nucleotide variability). To see how this works, consider a population of fruit flies (*Drosophila*). The genome of a fruit fly has about 13,700 loci. The **average heterozygosity** of *Drosophila* is measured as the average percent of these loci that are heterozygous. On average, a fruit fly is heterozygous (has two different alleles) at about 14% of its loci. You can therefore say that the fly population has an average heterozygosity of 14%, meaning that a typical fruit fly is heterozygous at about 1,920 of its 13,700 gene loci and homozygous at all the rest.

Nucleotide variability is measured by comparing the nucleotide sequences of DNA samples from two individuals and then averaging the data from many such comparisons. The fruit fly genome has about 180 million nucleotides, and the sequences of any two flies differ on average by approximately 1%.

Why does average heterozygosity tend to be greater than nucleotide variability? This is because a gene can consist of thousands of bases of DNA. A difference at only one of these bases is sufficient to make two alleles of that gene different and increase average heterozygosity.

Based on measurements of nucleotide variability, humans have relatively little genetic variation compared to most species. Two humans differ by only about 0.1% of their bases, a tenth of the nucleotide variability found within *Drosophila* populations. Clearly, we humans are all far more genetically alike than we are different. Still, this 0.1% nucleotide variability encompasses the entire heritable component of the many different ways that people look, sound, and act, along with their biochemical differences, such as their ABO blood group, that are not outwardly visible.

Variation Between Populations

Most species exhibit **geographic variation,** differences between the gene pools of separate populations or population subgroups. **Figure 23.10** illustrates an example of geographic variation observed in isolated populations of house mice (*Mus musculus*) that were inadvertently introduced to the Atlantic island of Madeira by Portuguese settlers in the 15th century.

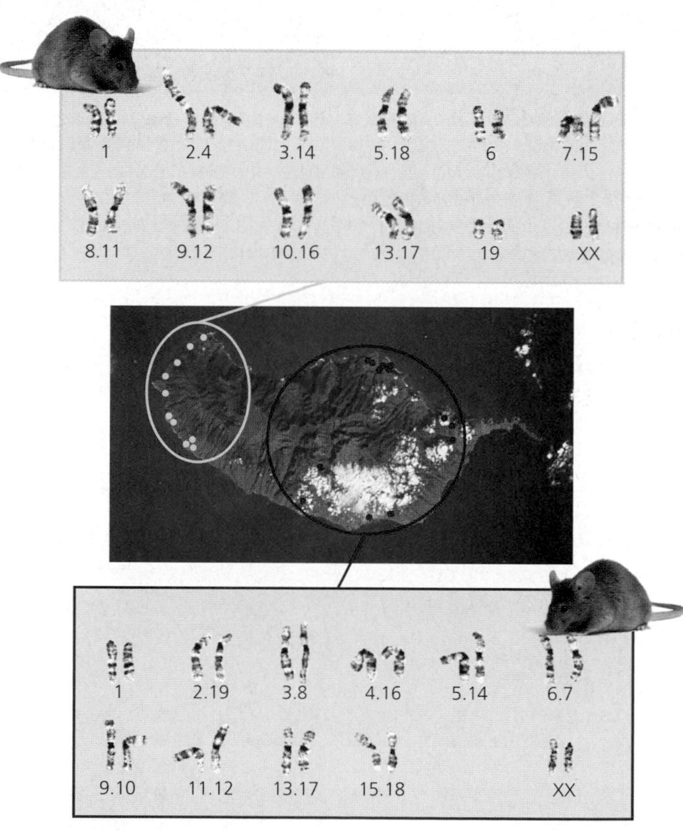

▲ **Figure 23.10 Geographic variation in chromosomal mutations.** Separated by mountains, several populations of house mice on the island of Madeira have evolved in isolation from one another. Researchers have observed differences in the karyotypes (chromosome sets) of these isolated populations. In some of the populations, the original chromosomes have become fused. For example, "2.4" indicates fusion of chromosome 2 and chromosome 4. However, the patterns of fused chromosomes differ from one mouse population to another. Mice in the areas indicated by the gold dots have the set of fused chromosomes in the gold box; mice in the locales with the red dots have the different pattern of fusions in the red box. Because these mutations leave genes intact, their effects on the mice appear to be neutral.

Because environmental factors are likely to differ from one place to another, natural selection can contribute to geographic variation. For example, one population of our hypothetical wildflower species might have a higher frequency of C^W alleles than other populations because local pollinators prefer white flowers. Genetic drift can also produce allele frequency differences between populations through the cumulative effect of random fluctuations in frequencies rather than natural selection.

Some examples of geographic variation occur as a **cline,** a graded change in a trait along a geographic axis. In some cases, a cline may represent a graded region of overlap where individuals of neighboring populations are interbreeding. In other cases, a gradation in some environmental variable may produce a cline. For example, the average body size of many North American species of birds and mammals increases gradually with increasing latitude. Presumably, the reduced ratio of surface area to volume that accompanies larger size is an adaptation that helps animals conserve body heat in cold environments. Experimental studies of many clines have confirmed that both genetic variation and environment play a role in the geographic differences of phenotype **(Figure 23.11)**.

A Closer Look at Natural Selection

From the range of available variations in a population, natural selection increases the frequencies of certain genotypes, fitting organisms to their environment over generations. This section takes a closer look at natural selection as a mechanism of evolutionary adaptation.

Evolutionary Fitness

The phrases "struggle for existence" and "survival of the fittest" are commonly used to describe natural selection, yet these expressions are misleading if taken to mean direct competitive contests among individuals. There *are* animal species in which individuals, usually the males, lock horns or otherwise do combat to determine mating privilege. But reproductive success is generally subtler and depends on many factors besides outright battles for mates. For example, a barnacle may produce more eggs than its neighbors because it is more efficient at collecting food from the water. A moth may have more offspring than other moths in the same population because its body colors more effectively conceal it from predators. Wildflowers may differ in reproductive success because some attract more pollinators owing to slight variations in color, shape, or fragrance. These are all examples of adaptive advantage, and they are all components of what we call **fitness:** the contribution an individual makes to the gene pool of the next generation, relative to the contributions of other individuals.

In a more quantitative approach to natural selection, population geneticists define **relative fitness** as the contribution of a genotype to the next generation compared to the contributions of alternative genotypes for the same locus. Consider our wildflower population once again. Let's assume that, on average, individuals with red flowers produce fewer offspring than those with white or pink flowers, which produce equal numbers of offspring. The relative fitness of the most reproductively successful variants is set at 1 as a basis for comparison, so the relative fitness of a white-flowered or pink-flowered plant is 1. If plants with red flowers average only 80% as many offspring, their relative fitness is 0.8.

Although population geneticists often refer to the relative fitness of a genotype, it is important to remember that natural selection acts on phenotypes, not genotypes. The entity that is subjected to natural selection is the whole organism. Thus, the relative fitness of a particular allele depends on the entire genetic and environmental context in which it is expressed. For example, alleles that enhance the growth of a tree's trunk and limbs may be useless or even detrimental in the absence of alleles at other loci that enhance the growth of roots necessary

Figure 23.11

Inquiry Does geographic variation in yarrow plants have a genetic component?

EXPERIMENT Researchers observed that the average size of yarrow plants *(Achillea)* growing on the slopes of the Sierra Nevada mountains gradually decreases with increasing elevation. To eliminate the effect of environmental differences at different elevations, researchers collected seeds from various altitudes and planted them in a common garden. They then measured the heights of the resulting plants.

RESULTS The average plant sizes in the common garden were inversely correlated with the altitudes at which the seeds were collected, although the height differences were less than in the plants' natural environments.

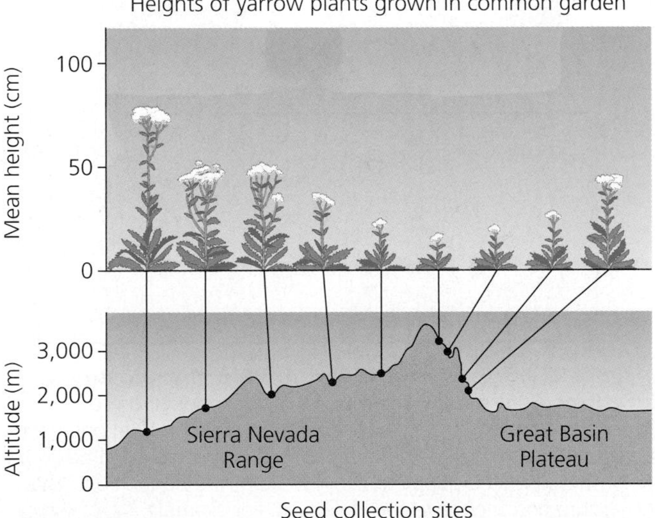

CONCLUSION The lesser but still measurable clinal variation in yarrow plants grown at a common elevation demonstrates the role of genetic as well as environmental differences.

to support the tree. On the other hand, alleles or other pieces of DNA that contribute nothing to an organism's success (or that may even be slightly maladaptive) may be perpetuated in individuals whose overall fitness is high. In this way, thousands of pieces of old retroviruses and transposable elements have accumulated and continue to accumulate in our genomes. These functionless bits of DNA are carried along from one generation to the next because they do no harm.

Survival alone does not guarantee reproductive success. Relative fitness is zero for a sterile plant or animal, even if it is robust and outlives other members of the population. But survival is necessary for reproduction, and longevity increases fitness if it means that long-lived individuals leave more descendants than those that die sooner. On the other hand, an individual that matures quickly and becomes fertile at an early age, even if it only lives for a short time, may have more offspring than an individual that lives longer but matures late. Thus, many factors affect both survival and fertility and contribute to evolutionary fitness.

Directional, Disruptive, and Stabilizing Selection

Natural selection can alter the frequency distribution of heritable traits in three ways, depending on which phenotypes in a population are favored. These three modes of selection are called directional selection, disruptive selection, and stabilizing selection.

Directional selection (**Figure 23.12a**) is most common when a population's environment changes or when members of a population migrate to a new habitat with different environmental conditions than their former one. Directional selection shifts the frequency curve for some phenotypic character in one direction or the other by favoring individuals that deviate from the average. For instance, fossil evidence indicates that the average size of black bears in Europe increased with each glacial period, only to decrease again during the warmer interglacial periods. Larger bears, with a smaller surface-to-volume ratio, are better at conserving body heat and surviving periods of extreme cold.

▼ **Figure 23.12 Modes of selection.** These cases describe three ways in which a hypothetical deer mouse population with heritable variation in fur coloration from light to dark might evolve. The graphs show how the frequencies of individuals with different fur colors change over time. The large white arrows symbolize natural selection working against certain phenotypes.

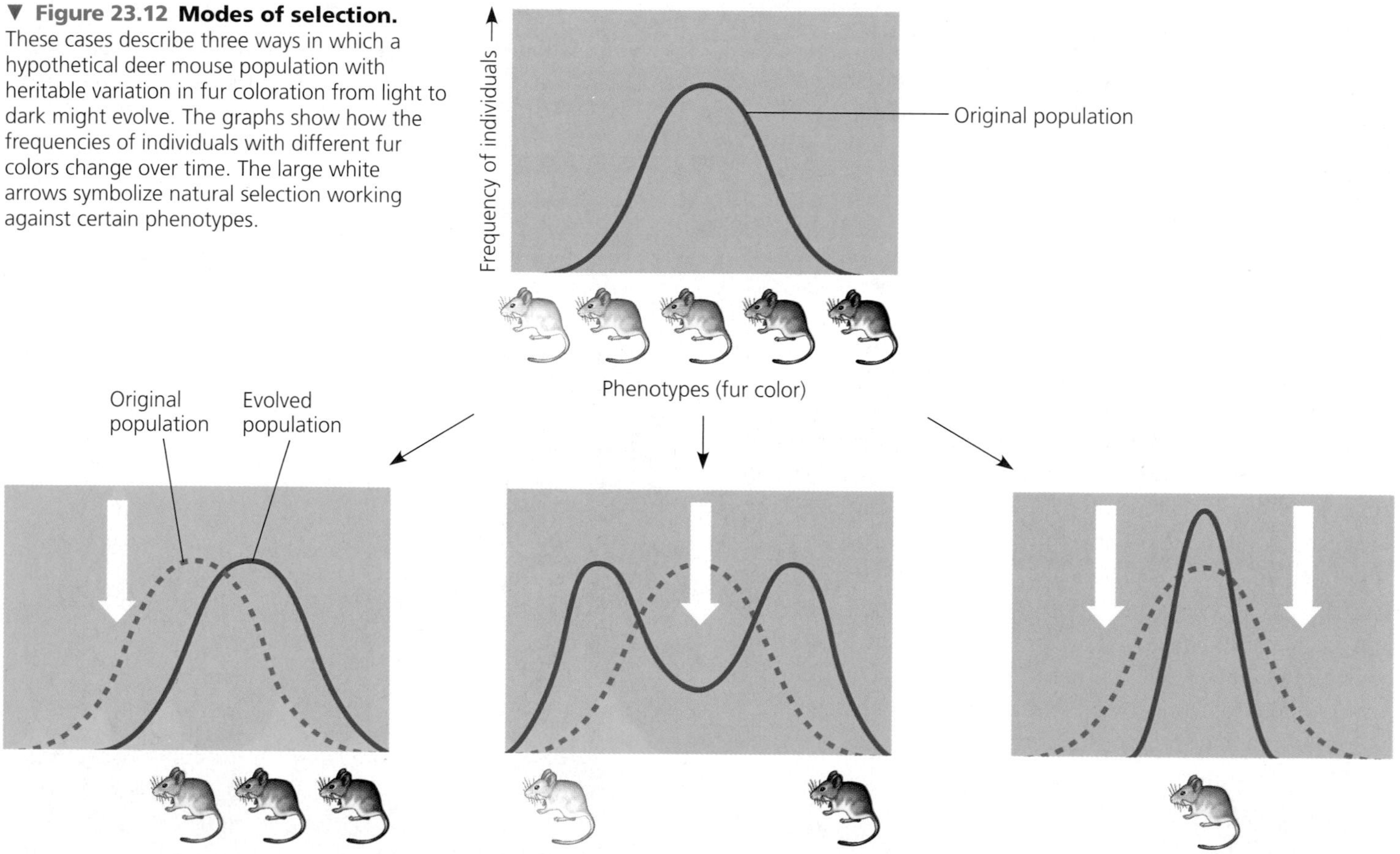

(a) Directional selection shifts the overall makeup of the population by favoring variants at one extreme of the distribution. In this case, darker mice are favored because they live among dark rocks, and a darker fur color conceals them from predators.

(b) Disruptive selection favors variants at both ends of the distribution. These mice have colonized a patchy habitat made up of light and dark rocks, with the result that mice of an intermediate color are at a disadvantage.

(c) Stabilizing selection removes extreme variants from the population and preserves intermediate types. If the environment consists of rocks of an intermediate color, both light and dark mice will be selected against.

Disruptive selection **(Figure 23.12b)** occurs when conditions favor individuals on both extremes of a phenotypic range over individuals with intermediate phenotypes. For example, a population of black-bellied seedcracker finches in Cameroon displays two distinctly different beak sizes. Small-billed birds feed mainly on soft seeds, whereas large-billed birds specialize in cracking hard seeds. It appears that birds with intermediate-sized bills are relatively inefficient at cracking both types of seeds, and thus have lower relative fitness. As you will read in the next chapter, disruptive selection can be important in the early stages of speciation.

Stabilizing selection **(Figure 23.12c)** acts against extreme phenotypes and favors intermediate variants. This mode of selection reduces variation and maintains the status quo for a particular phenotypic character. For example, the birth weights of most human babies lie in the range of 3–4 kg; babies who are much smaller or larger suffer higher rates of mortality.

Regardless of the mode of selection, however, the basic mechanism remains the same. Selection favors certain heritable traits through differential reproductive success.

The Preservation of Genetic Variation

What prevents natural selection from reducing genetic variation by culling all unfavorable genotypes? The tendency for directional and stabilizing selection to reduce variation is countered by mechanisms that preserve or restore it.

Diploidy

Because most eukaryotes are diploid, a considerable amount of genetic variation is hidden from selection in the form of recessive alleles. Recessive alleles that are less favorable than their dominant counterparts, or even harmful in the current environment, can persist because they are propagated in heterozygous individuals. This latent variation is exposed to natural selection only when both parents carry the same recessive allele and combine two copies in one zygote. This happens only rarely if the frequency of the recessive allele is very low. The rarer the recessive allele, the greater the degree of protection from natural selection. Heterozygote protection maintains a huge pool of alleles that might not be favored under present conditions but some of which could bring new benefits when the environment changes.

Balancing Selection

Selection itself may preserve variation at some gene loci. **Balancing selection** occurs when natural selection maintains stable frequencies of two or more phenotypic forms in a population, a state called **balanced polymorphism.** This type of selection includes heterozygote advantage and frequency-dependent selection.

Heterozygote Advantage. If individuals who are heterozygous at a particular gene locus have greater fitness than the homozygotes, natural selection will tend to maintain two or more alleles at that locus. An example of this **heterozygote advantage** occurs at the locus in humans that codes for one of the peptide subunits in hemoglobin, the oxygen-carrying protein of red blood cells. In homozygous individuals, a recessive allele at that locus causes sickle-cell disease (see Figures 5.21 and 17.23). However, heterozygotes are protected against the severest effects of malaria (although they are not resistant to malarial infection). This protection is an important advantage in tropical regions where malaria is a major killer. The environment in these areas favors heterozygotes over homozygous dominant individuals, who are more susceptible to malaria, and also over homozygous recessive individuals, who develop sickle-cell disease. The frequency of the sickle-cell allele in Africa is generally highest in areas where the malaria parasite is most common **(Figure 23.13)**. In some tribes, it accounts for 20% of the hemoglobin alleles in the gene pool, a very high frequency for such a harmful allele. But even at this high frequency ($q = 0.2$), only 4% of the population suffers from sickle-cell disease ($q^2 = 0.04$). There are far more heterozygotes who are resistant to malaria ($2pq = 2 \times 0.8 \times 0.2 = 0.32$). Although the fitness advantage to the heterozygotes is much smaller than the disadvantage to the homozygotes, there are so many more heterozygotes than homozygotes that the aggregate benefit of the allele in the population *balances* its aggregate harm.

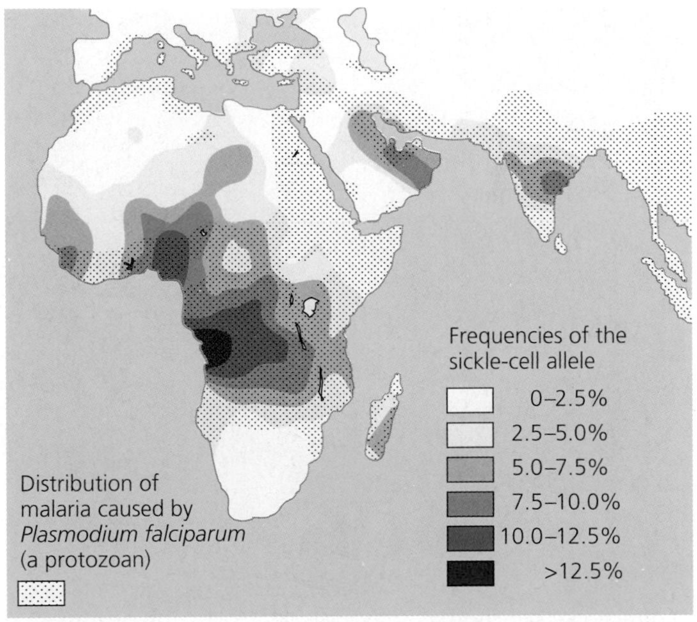

▲ **Figure 23.13 Mapping malaria and the sickle-cell allele.** The sickle-cell allele is most common in Africa, but it is not the only balanced polymorphism that confers protection against malaria. Different balanced polymorphisms are found in populations in other parts of the world, such as around the Mediterranean Sea and in Southeast Asia, where malaria is widespread.

Figure 23.14

Research Method **Using a Virtual Population to Study the Effects of Selection**

APPLICATION Using virtual organisms that can "reproduce" and pass on characteristics to their offspring, biologists can model the effects of selection over many generations in a compressed period of time. This method also enables researchers to isolate the effect of the study variable by controlling for other influences on the population's evolution—a nearly impossible goal in nature.

TECHNIQUE Studies involving virtual populations are most meaningful when based on a natural system. For example, Alan Bond and Alan Kamil of the University of Nebraska created a "virtual ecology" based on the well-studied predator-prey relationship between the North American blue jay and the woodland moth. First, the researchers converted gray-scale photographs of moths into digital moths. The wing patterns were determined by a complex computerized genome based on real moth genetics, including polygenic inheritance, crossing-over events, and mutation. These moths can be "mated" and the genotypes and phenotypes of the offspring determined by various mathematical models.

Bond and Kamil trained captive blue jays to "hunt" digital moths displayed against patterned backgrounds, as the birds would hunt moths hidden on tree trunks in the wild. To model the effects of blue jay predation on the moth population, the researchers presented each moth to a jay and measured the detection time (or lack of detection). The researchers used a statistical model to calculate each moth's likelihood of breeding, producing the next generation of potential prey. They repeated this process for 100 generations each for the experimental group (exposed to frequency-dependent selection by jays) and two control groups. In one control (no selection by jays), the researchers randomly recombined the genotypes in each generation. In the other control (frequency-independent selection), jays selected moths, but a computer program eliminated the "search image" effect.

On pecking a moth image the blue jay receives a food reward. If the bird does not detect a moth on either screen, it pecks the green circle to continue to a new set of images (a new feeding opportunity).

RESULTS Using a virtual prey population enabled the researchers to model the long-term evolutionary impact of selection by real predators. In this study, the experimental moth population became more difficult to detect against the patterned backgrounds compared to the nonselected control (as did the frequency-independent control). But as shown in the graph below, in three trials, the experimental group (colored lines) showed much greater phenotypic variation than the frequency-independent control group (black line). This result supports Bond and Kamil's hypothesis that frequency-dependent selection by visual predators promotes polymorphism in a prey population.

Parental population sample

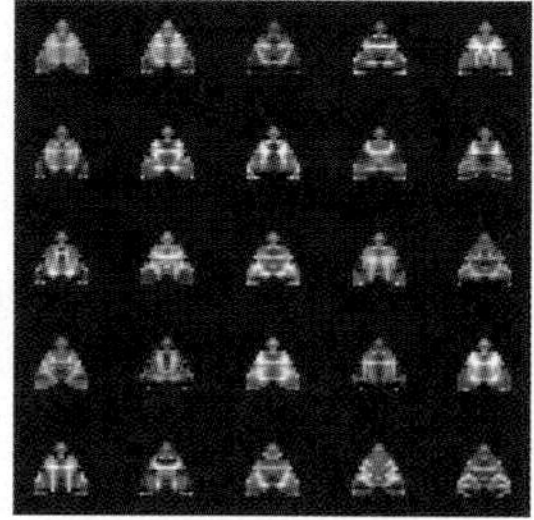

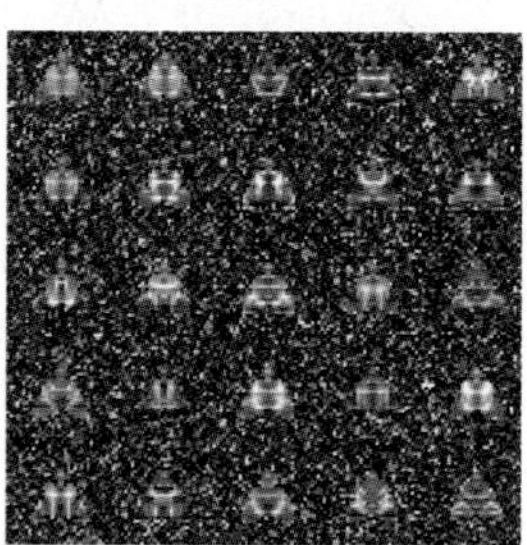

Experimental group sample

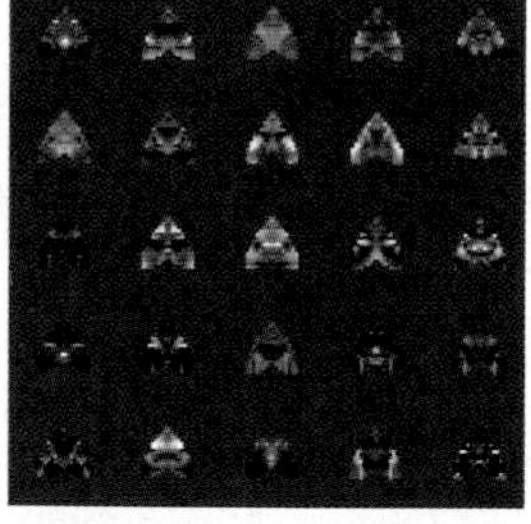

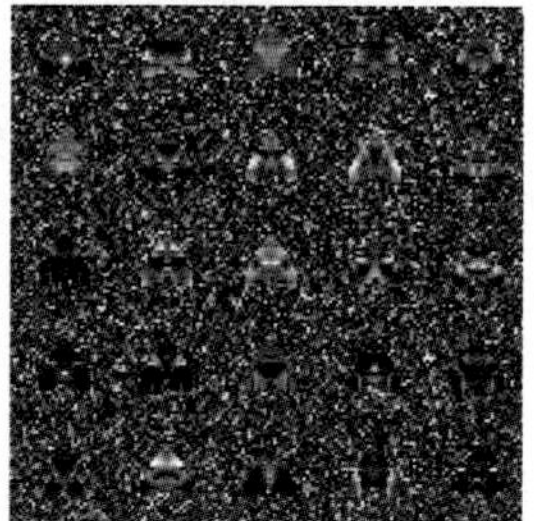

Plain background

Patterned background

Frequency-Dependent Selection. In **frequency-dependent selection,** the fitness of any one morph declines if it becomes too common in the population. **Figure 23.14** illustrates an experiment that simulates frequency-dependent selection. In the wild, blue jays locate and eat moths that sit motionless on tree trunks during the day. Captive jays can be trained to find virtual prey on a computer screen. These jays were presented with a series of screens that showed a speckled background, sometimes including an image of a moth. When a jay pecked at a moth image, it was rewarded with a small piece of food. The jays soon learned a "search image"—a quick way to recognize the most common type of moth. As their skill grew, the computer program adjusted the frequencies of the virtual prey, making the common type rare and formerly rare types common, just as natural selection would do. The researchers also programmed "mutations" that introduced new moth patterns.

Whenever a type of moth became common, the jays rapidly learned a new search image and targeted it. Frequency-dependent selection preserved polymorphism in this population: Rare moths of any pattern were at an advantage and common moths of any pattern were at a disadvantage because the jays learned to find the common moths more easily than the rare ones. Similar frequency-dependent selection has been observed in a number of predator-prey interactions in the wild.

Neutral Variation

Some of the genetic variation in populations probably has little or no impact on reproductive success, and thus natural selection does not affect these alleles. For example, most of the DNA base differences between humans that are found in untranslated parts of the genome appear to confer no selective advantage and therefore are considered **neutral variation.** And in **pseudogenes,** genes that have become inactivated by mutations, genetic "noise" is free to accumulate in all parts of the gene. The relative frequencies of neutral variations are not affected by natural selection; over time, some neutral alleles increase in frequency and others decrease through genetic drift. Even mutational changes that alter proteins can be neutral. For example, data from *Drosophila* suggest that roughly half of the amino-acid-changing mutations that arise and subsequently become fixed have no selective effect—that is, they do not affect protein function.

Of course, it is possible that genetic differences that appear to be neutral may actually influence survival and reproductive success in ways that are difficult to measure. Furthermore, a variant allele may be neutral in one environment but not in another. The debate over the extent of neutral variation continues (as you will read further in Chapter 25), but one thing is clear: Even if only a fraction of the extensive variation in a gene pool significantly affects fitness, that is still an enormous reservoir of raw material for natural selection and adaptive evolution.

Sexual Selection

Charles Darwin was the first to explore the implications of **sexual selection,** which is natural selection for mating success. This type of selection can result in **sexual dimorphism,** marked differences between the sexes in secondary sexual characteristics, which are not directly associated with reproduction **(Figure 23.15)**. These distinctions include differences in size, color, and ornamentation. In vertebrates, males are usually the showier sex.

It is important to distinguish between *intra*sexual and *inter*sexual selection. **Intrasexual selection,** meaning selection "within the same sex," is a direct competition among individuals of one sex for mates of the opposite sex. Intrasexual selection is usually most obvious in males. For example, in many species a single male patrols a group of females and prevents other males from mating with them. The patrolling male may defend his status by defeating smaller, weaker, or less fierce males in combat; more often, he is the psychological victor in ritualized displays that discourage would-be competitors but do not risk injury that would reduce his own fitness (see Chapter 51). But evidence is growing that intrasexual selection can take place between females as well. In ring-tailed lemurs, for example, females dominate males and also establish hierarchies of dominance among themselves.

▲ **Figure 23.15 Sexual dimorphism and sexual selection.** Peacocks and peahens show extreme sexual dimorphism. There is intrasexual selection between competing males followed by intersexual selection when the females choose among the showiest males.

In **intersexual selection,** also called mate choice, individuals of one sex (usually females) are choosy in selecting their mates from the other sex. In many cases, the female's choice depends on the showiness of the male's appearance or behavior (see Figure 23.15). However, because females have fewer chances to mate than males, a female only gains an advantage over other females if she chooses a mate that enables her to produce more fit offspring.

What intrigued Darwin about mate choice is that some of the male showiness does not seem to be adaptive in any other way and may in fact pose some risk. For example, bright plumage may make male birds more visible to predators. But if such secondary sexual characteristics help a male gain a mate, and this benefit outweighs the risk from predation, then both the bright plumage and the female preference for it will be reinforced for the most Darwinian of reasons—because they enhance reproductive success. Every time a female chooses a mate based on a certain appearance or behavior, she perpetuates the alleles that influenced her to make that choice, allowing a male with an especially showy phenotype to pass on his alleles to offspring.

How do female preferences for certain male characteristics evolve? Could there be fitness benefits to showy traits? Several researchers are testing the hypothesis that these sexual advertisements reflect overall health. For example, male birds with serious parasitic infections may have dull, disheveled plumage. They don't usually win many females. If the female chooses a showy male, she may be choosing a healthy one as well, and the benefit will be a greater chance of healthy offspring.

The Evolutionary Enigma of Sexual Reproduction

Even as biologists study the mechanisms of sexual selection, they continue to wonder about the evolution of sex itself. As a mechanism of rapid population expansion, it is far inferior to asexual reproduction. Consider, for example, a population of insects in which half the females reproduce only sexually and half reproduce only asexually. Even if both types of females produced the same number of offspring each generation, the asexual condition would increase in frequency because all of the females' offspring would be daughters that are able to give birth to more reproductive daughters in turn. In contrast, half of the sexual females' offspring would be males, which would be required for reproduction but would themselves be unable to give birth. **Figure 23.16** diagrams this "reproductive handicap" of sex.

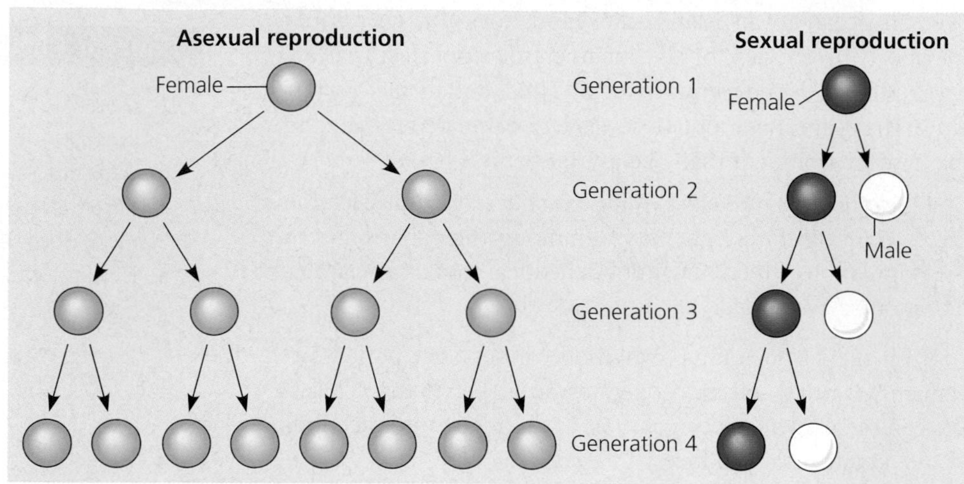

▲ **Figure 23.16 The "reproductive handicap" of sex.** These diagrams contrast the reproductive output of females (blue circles) over four generations for asexual versus sexual reproduction. The scenarios assume two surviving offspring per female. The asexual population rapidly outgrows the sexual one.

Nonetheless, sex is maintained in the vast majority of eukaryotic species, even in those that can also reproduce asexually. It must enhance reproductive success somehow, for otherwise natural selection would act in favor of alleles that promote asexual reproduction. But what advantage does sex provide?

One explanation is that the processes of meiotic recombination and fertilization generate the genetic variation on which natural selection acts. The assumption is that natural selection sustains sex, in spite of its reproductive drawbacks, because genetic variation enables *future* adaptation to an ever-changing environment. But this assumption is difficult to defend. Natural selection always acts in the here and now, favoring the success of those individuals that reproduce best in the current environment.

So instead, consider how the genetic variation that results from sex might benefit sexually reproducing individuals over the short term—on a generation-to-generation time scale. One hypothesis emphasizes the importance of genetic variation in resistance to disease. Many bacterial and viral pathogens recognize and infect a host by attaching to receptor molecules on the host's cells. There should be an advantage in producing offspring that vary in their resistance to different diseases. For example, one offspring may have cellular markers that make it resistant to virus A, while another may be resistant to virus B. This hypothesis predicts that gene loci that code for the receptors to which pathogens attach should have many alleles. And that seems to be the case. In humans, for example, there are hundreds of alleles for each of two gene loci coding for the proteins that give cell surfaces their molecular fingerprints (you'll read more about these cellular markers in Chapter 43). Of course, because most pathogens evolve rapidly in their ability to key on specific host receptors, resistance of a particular genotype to a given disease is not permanent. But sex provides a mechanism for "changing the locks" and varying them among offspring. Such an example of coevolution, in which host and parasite must evolve quickly to keep up with each other, is sometimes described as a "Red Queen race," after the Red Queen in Lewis Carroll's *Through the Looking Glass,* who advised Alice to run as fast as she could just to stay in the same place.

Why Natural Selection Cannot Fashion Perfect Organisms

As we noted in Chapter 22, nature abounds with examples of organisms that seem to be less than ideally "engineered" for their lifestyles, for several reasons.

1. *Evolution is limited by historical constraints.* Each species has a legacy of descent with modification from a long line of ancestral forms. Evolution does not scrap ancestral anatomy and build each new complex structure from scratch; it co-opts existing structures and adapts them to new situations. For example, we could imagine how some bird species might benefit by having wings for flight as well as *four* legs for speedy, maneuverable running, instead of just two legs. But birds descended from reptiles that had only two pairs of limbs, and co-opting the forelimbs for flight left only the two hind limbs for moving around on the ground.

2. *Adaptations are often compromises.* Each organism must do many different things. A seal spends part of its time on rocks; it could probably walk better if it had legs instead of flippers, but it would not swim nearly as well. We humans owe much of our versatility and athleticism to our prehensile hands and flexible limbs, which also make us prone to sprains, torn ligaments, and dislocations. Structural reinforcement has been compromised for agility.

3. *Chance and natural selection interact.* Chance events affect the subsequent evolutionary history of populations. For instance, when a storm blows insects or birds hundreds of miles

over an ocean to an island, the wind does not necessarily transport the species, or the members of a species, that are best suited to the new environment. Thus, not all alleles present in the founding population's gene pool are better suited to the new environment than alleles "left behind."

4. *Selection can edit only existing variations.* Natural selection favors only the fittest phenotypes among those currently in the population, which may not be the ideal traits. New alleles do not arise on demand.

With these constraints, evolution cannot craft perfect organisms. Natural selection operates on a "better than" basis. We can see evidence for evolution in the many imperfections of the organisms it produces.

Concept Check 23.4

1. Does nucleotide variability in a population always correspond to phenotypic polymorphism? Why or why not?
2. What is the relative fitness of a sterile mule? Explain.
3. How does sexual selection lead to sexual dimorphism?
4. Explain what is meant by the "reproductive handicap" of sex.

For suggested answers, see Appendix A.

Chapter 23 Review

Go to www.campbellbiology.com or the student CD-ROM to explore Activities, Investigations, and other interactive study aids.

SUMMARY OF KEY CONCEPTS

Concept 23.1

Population genetics provides a foundation for studying evolution

- **The Modern Synthesis** (p. 455) The modern synthesis integrates Mendelian genetics with the Darwinian theory of evolution by natural selection and focuses on populations as the basic unit of evolution.
- **Gene Pools and Allele Frequencies** (pp. 455–456) A population, a localized group of organisms that all belong to the same species, is united by its gene pool, the aggregate of all alleles in the population.
- **The Hardy-Weinberg Theorem** (pp. 456–458) The Hardy-Weinberg theorem states that the frequencies of alleles and genotypes in a population will remain constant if Mendelian segregation and random mating are the only processes that affect the gene pool. If p and q represent the relative frequencies of the only two possible alleles at a particular locus, then $p^2 + 2pq + q^2 = 1$, where p^2 and q^2 are the frequencies of the homozygous genotypes and $2pq$ is the frequency of the heterozygous genotype. Although many populations approximate Hardy-Weinberg equilibrium, the equilibrium in its strictest sense applies only if the population is large, mating is random, mutation is negligible, there is no gene flow from other populations, and all individuals have equal reproductive success.
  Investigation *How Can Frequency of Alleles Be Calculated?*

Concept 23.2

Mutation and sexual recombination produce the variation that makes evolution possible

- **Mutation** (pp. 459–460) New genes and new alleles originate only by mutation. Most mutations have no effect or are harmful, but a few increase adaptation.
- **Sexual Recombination** (p. 460) Genetic recombination between sexually reproducing organisms produces most of the variation in traits that makes adaptation possible.
  Activity *Genetic Variation from Sexual Recombination*

Concept 23.3

Natural selection, genetic drift, and gene flow can alter a population's genetic composition

- **Natural Selection** (p. 460) Differential success in reproduction results in certain alleles being passed to the next generation in greater proportions than others.
- **Genetic Drift** (pp. 460–462) Chance fluctuations in allele frequencies from generation to generation tend to reduce genetic variation in populations.
- **Gene Flow** (p. 462) Genetic exchange between populations tends to reduce differences between populations over time.
  Biology Labs On-Line *PopulationGeneticsLab*
  Activity *Causes of Evolutionary Change*

Concept 23.4

Natural selection is the primary mechanism of adaptive evolution

- **Genetic Variation** (pp. 462–464) Genetic variation includes variation among individuals within a population in discrete and quantitative characters, as well as geographic variation between populations.
- **A Closer Look at Natural Selection** (pp. 464–466) One organism has a greater relative fitness than another if it leaves more descendants. Selection favors certain genotypes in a population by acting on the phenotypes of individual organisms. Natural selection can favor relatively rare individuals at one end of the phenotypic range (directional selection), can favor individuals at both extremes of the range rather than intermediate phenotypes (disruptive selection), or can act against extreme phenotypes (stabilizing selection).
- **The Preservation of Genetic Variation** (pp. 466–468) Diploidy maintains a reservoir of concealed recessive variation in heterozygotes. Balanced polymorphism may maintain variation at some gene loci as a result of heterozygote advantage or frequency-dependent selection.
- **Sexual Selection** (p. 468) Sexual selection leads to the evolution of secondary sex characteristics, which can give individuals an advantage in mating.
- **The Evolutionary Enigma of Sexual Reproduction** (p. 469) Enhanced disease resistance based on genetic variation is one possible explanation for the persistence of sexual reproduction despite its lesser reproductive output compared to asexual reproduction.

▶ **Why Natural Selection Cannot Fashion Perfect Organisms** (pp. 469–470) Structures result from modified ancestral anatomy; adaptations are often compromises; the gene pool can be affected by genetic drift; and natural selection can act only on available variation.
Biology Labs On-Line *EvolutionLab*

TESTING YOUR KNOWLEDGE

Self-Quiz

1. In the gene pool of a population with 100 individuals, a fixed allele for a particular gene locus has a frequency of
 a. 0. b. 0.5. c. 1. d. 100.
 e. cannot be calculated based on this information

2. Researchers examining a particular gene in a fruit fly population discovered that the gene can have either of two slightly different sequences, designated A1 and A2. Further tests showed that 70% of the gametes produced in the population contained the A1 sequence. If the population is at Hardy-Weinberg equilibrium, what proportion of the flies carries both A1 and A2?
 a. 0.7 b. 0.49 c. 0.21 d. 0.42 e. 0.09

3. At a locus with a dominant and a recessive allele in Hardy-Weinberg equilibrium, 16% of the individuals are homozygous for the recessive allele. What is the frequency of the dominant allele in the population?
 a. 0.84 b. 0.36 c. 0.6 d. 0.4 e. 0.48

4. The average length of jackrabbit ears decreases gradually with increasing latitude. This variation is an example of
 a. directional selection. d. genetic drift.
 b. discrete variation. e. disruptive selection.
 c. polymorphism.

5. Which of the following is a polymorphic trait in humans?
 a. variation in height
 b. variation in intelligence
 c. free versus attached earlobes (see Figure 14.14)
 d. variation in the number of fingers
 e. variation in fingerprints

6. Natural selection changes allele frequencies in populations because some ________ survive and reproduce more successfully than others.
 a. alleles d. gene loci
 b. individual organisms e. species
 c. gene pools

7. Longer tails of male barn swallows evolve because female barn swallows prefer to mate with the males that have the longest tails. This process is best described as
 a. genetic drift that changes the frequencies of the alleles for tail length.
 b. natural selection for sexual reproduction that maintains variation in the genes that influence tail length.
 c. intersexual selection for traits, such as long tails, that help males attract mates.
 d. intrasexual selection for traits, such as long tails, that help males win contests for access to females.
 e. directional selection for traits, such as long tails, that improve males' ability to fly strongly and forage for food over large areas.

8. No two human individuals are alike, except for identical twins. The chief cause of the variation among individuals is
 a. new mutations that occurred in the preceding generation.
 b. sexual recombination.
 c. genetic drift due to the small size of the population.
 d. geographic variation within the population.
 e. environmental effects.

9. Road construction has isolated a small portion of a beetle population from the main population. After a few generations, this new population exhibits dramatic genetic differences from the old one, most likely because
 a. mutations are more common in the new environment.
 b. allele frequencies among the stranded beetles differed by chance from those in the parent population's gene pool and subsequent genetic drift caused even more divergence from the original gene pool.
 c. the new environment is different from the old, favoring directional selection.
 d. gene flow increases in a new environment.
 e. members of a small population tend to migrate, removing alleles from the gene pool.

10. Sparrows with average-sized wings survive severe storms better than those with longer or shorter wings, illustrating
 a. the bottleneck effect. d. neutral variation.
 b. stabilizing selection. e. disruptive selection.
 c. frequency-dependent selection.

For Self-Quiz answers, see Appendix A.

Go to the website or CD-ROM for more quiz questions.

Evolution Connection

How is the process of evolution revealed by the imperfections of living organisms?

Scientific Inquiry

In the population of wildflowers we used to test the Hardy-Weinberg theorem, the frequency of C^R is 0.8 and the frequency of C^W is 0.2. In a different population of these flowers, the frequencies of genotypes do not conform to Hardy-Weinberg equilibrium: 60% of the plants are C^RC^R and 40% are C^WC^W. Assuming that all Hardy-Weinberg conditions are met, show that genotypes will reach equilibrium in the next generation. Suppose instead that the plants mate only with themselves (self-fertilize). What will the allele and genotype frequencies be in the next generation?

Investigation *How Can Frequency of Alleles Be Calculated?*
Biology Labs On-Line *PopulationGeneticsLab*
Biology Labs On-Line *EvolutionLab*

Science, Technology, and Society

To what extent are humans living in a technological society exempt from natural selection? Justify your answer.

24 The Origin of Species

▲ **Figure 24.1 The flightless cormorant *(Nannopterum harrisi)*, one of many new species that have originated on the isolated Galápagos Islands.**

Key Concepts

24.1 The biological species concept emphasizes reproductive isolation

24.2 Speciation can take place with or without geographic separation

24.3 Macroevolutionary changes can accumulate through many speciation events

Overview

That "Mystery of Mysteries"

Darwin came to the Galápagos Islands eager to explore landforms newly emerged from the sea. He noted that these volcanic islands, despite their geologic youth, were filled with plants and animals known nowhere else in the world, and later realized that these species, like the islands, were new **(Figure 24.1)**. He wrote in his diary: "Both in space and time, we seem to be brought somewhat near to that great fact—that mystery of mysteries—the first appearance of new beings on this Earth."

The origin of new species, or **speciation,** is at the focal point of evolutionary theory, because the appearance of new species is the source of biological diversity. It is not enough to explain how adaptations evolve in a population (the topic covered in Chapter 23; such changes, confined to a single gene pool, are described as **microevolution**). Evolutionary theory must also explain how new species originate and develop through the subdivision and subsequent divergence of gene pools. The fossil record reveals the cumulative effects of speciation over vast tracts of time. The term **macroevolution** refers to such evolutionary change above the species level—for example, the appearance of feathers during the evolution of birds from one group of dinosaurs, and other such "evolutionary novelties" that can be used to define higher taxa.

We can distinguish two basic patterns of evolutionary change: anagenesis and cladogenesis **(Figure 24.2)**. Anagenesis (from the Greek *ana*, new, and *genos*, race), also called phyletic evolution, is the accumulation of changes that gradually transform a given species into a species with different characteristics. Cladogenesis (from the Greek *klados*, branch), also called branching evolution, is the splitting of a gene pool into two or more separate pools, which each give rise to one or more new species. Only cladogenesis can promote biological diversity by increasing the number of species.

▲ **Figure 24.2 Two patterns of evolutionary change.**
(a) Anagenesis is the accumulation of heritable changes, altering the characteristics of a species. **(b)** Cladogenesis is branching evolution, in which a new species arises from a population that buds from a parent species. (Note that the "parent" species may change as well.) Cladogenesis is the basis for biological diversity.

In this chapter, we will explore the mechanisms by which species arise. We will also examine the possible origins of some novel features that define higher taxonomic groups. But first, let's consider what we really mean when we refer to a "species."

Concept 24.1

The biological species concept emphasizes reproductive isolation

Species is a Latin word meaning "kind" or "appearance." We distinguish between various "kinds" of plants or animals—between dogs and cats, for instance—from differences in their appearance. But are organisms truly divided into the discrete units we call species, or is this classification an arbitrary attempt to impose order on the natural world? To answer this question, biologists have compared not only the morphology (body form) of different groups of organisms but also less obvious differences in physiology, biochemistry, and DNA sequences. The results generally confirm that morphologically distinct species are indeed discrete groups, with many differences in addition to morphological ones.

The Biological Species Concept

The primary definition of species used most frequently in this textbook was proposed in 1942 by biologist Ernst Mayr. This **biological species concept** defines a species as a population or group of populations whose members have the potential to interbreed in nature and produce viable, fertile offspring, but are unable to produce viable, fertile offspring with members of other populations **(Figure 24.3)**. In other words, the members of a biological species are united by being reproductively compatible, at least potentially. All humans, for example, belong to the same biological species. A businesswoman in Manhattan may be unlikely to meet a dairy farmer in Mongolia, but if the two should happen to get together, they could have viable babies that develop into fertile adults. In contrast, humans and chimpanzees remain distinct biological species even where they share territory, because many factors keep them from interbreeding and producing fertile offspring.

(a) Similarity between different species. The eastern meadowlark (*Sturnella magna*, left) and the western meadowlark (*Sturnella neglecta*, right) have similar body shapes and colorations. Nevertheless, they are distinct biological species because their songs and other behaviors are different enough to prevent interbreeding should they meet in the wild.

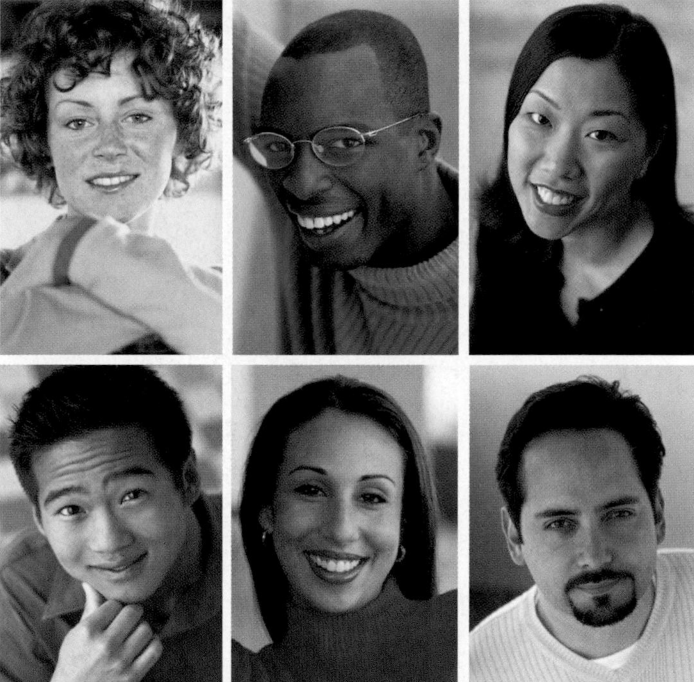

(b) Diversity within a species. As diverse as we may be in appearance, all humans belong to a single biological species (*Homo sapiens*), defined by our capacity to interbreed.

▲ **Figure 24.3 The biological species concept is based on the potential to interbreed rather than on physical similarity.**

Reproductive Isolation

Because biological species are distinguished based on reproductive incompatibility, the concept hinges on **reproductive isolation**—the existence of biological factors (barriers) that impede members of two species from producing viable, fertile hybrids. Although a single barrier may not block all genetic exchange between species, a combination of several barriers can effectively isolate a species' gene pool.

Clearly, a fly cannot mate with a frog or a fern, but the reproductive barriers between more closely related species are not so obvious. These barriers can be classified according to whether they contribute to reproductive isolation before or after fertilization. **Prezygotic barriers** ("before the zygote") impede mating between species or hinder the fertilization of ova if members of different species attempt to mate. If a sperm cell from one species does overcome prezygotic barriers and fertilizes an ovum from another species, **postzygotic barriers** ("after the zygote") often prevent the hybrid zygote from developing into a viable, fertile adult. **Figure 24.4**, on the next two pages, describes various prezygotic and postzygotic barriers.

Figure 24.4

Exploring Reproductive Barriers

Prezygotic barriers impede mating or hinder fertilization if mating does occur

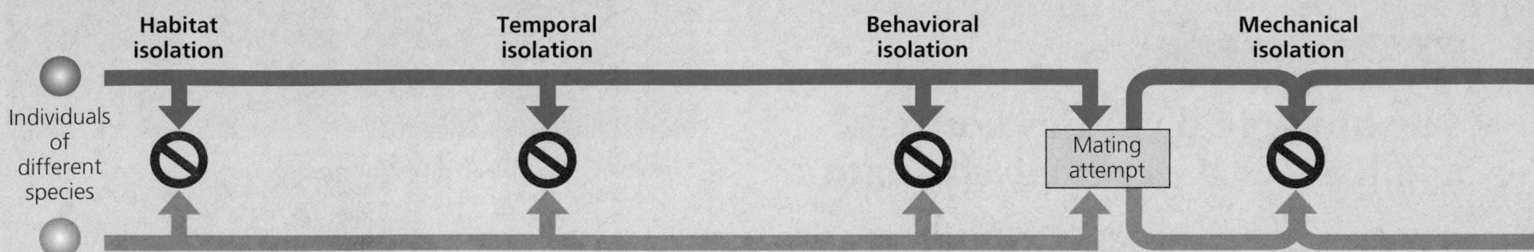

HABITAT ISOLATION

Two species that occupy different habitats within the same area may encounter each other rarely, if at all, even though they are not isolated by obvious physical barriers such as mountain ranges.

Example: Two species of garter snakes in the genus *Thamnophis* occur in the same geographic areas, but one lives mainly in water (a) while the other is primarily terrestrial (b).

TEMPORAL ISOLATION

Species that breed during different times of the day, different seasons, or different years cannot mix their gametes.

Example: In North America, the geographic ranges of the eastern spotted skunk (*Spilogale putorius*) (c) and the western spotted skunk (*Spilogale gracilis*) (d) overlap, but *S. putorius* mates in late winter and *S. gracilis* mates in late summer.

BEHAVIORAL ISOLATION

Courtship rituals that attract mates and other behaviors unique to a species are effective reproductive barriers, even between closely related species.

Example: Blue-footed boobies, inhabitants of the Galápagos, mate only after a courtship display unique to their species. Part of the "script" calls for the male to high-step, a behavior that calls the female's attention to his bright blue feet (e).

MECHANICAL ISOLATION

Morphological differences can prevent successful mating.

Example: Even in closely related species of plants, the flowers often have distinct appearances that attract different pollinators. These two species of monkey flower (*Mimulus*) differ greatly in the shapes and colors of their blossoms (f, g). Thus, cross-pollination between the plants does not occur.

(b) (a)

(e)

(d) (c)

(g) (f)

Postzygotic barriers prevent a hybrid zygote from developing into a viable, fertile adult

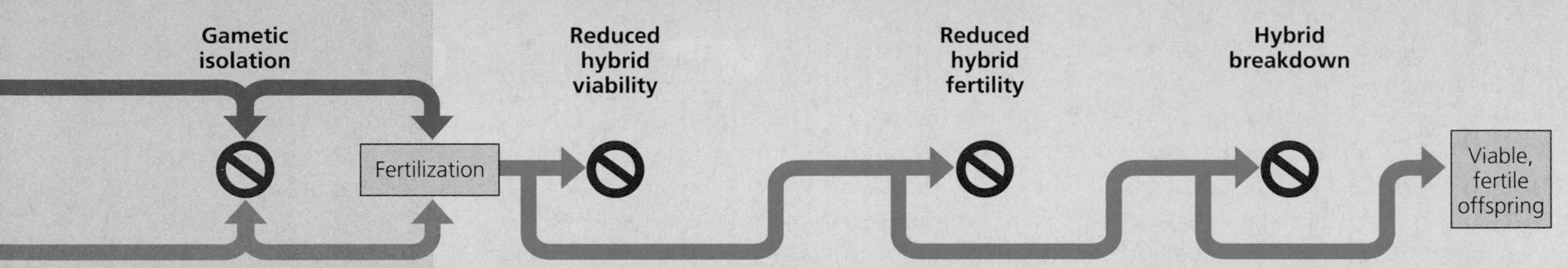

GAMETIC ISOLATION

Sperm of one species may not be able to fertilize the eggs of another species. Many mechanisms can produce this isolation. For instance, sperm may not be able to survive in the reproductive tract of females of the other species, or biochemical mechanisms may prevent the sperm from penetrating the membrane surrounding the other species' eggs.

Example: Gametic isolation separates certain closely related species of aquatic animals such as sea urchins (h). The sea urchins release their sperm and eggs into the surrounding water, where they fuse and form zygotes. Gametes of different species, like the red and purple urchins shown here, are unable to fuse.

(h)

REDUCED HYBRID VIABILITY

The genes of different parent species may interact and impair the hybrid's development.

Example: Some salamander subspecies of the genus *Ensatina* live in the same regions and habitats, where they may occasionally hybridize. But most of the hybrids do not complete development, and those that do are frail (i).

(i)

REDUCED HYBRID FERTILITY

Even if hybrids are vigorous, they may be sterile. If chromosomes of the two parent species differ in number or structure, meiosis in the hybrids may fail to produce normal gametes. Since the infertile hybrids cannot produce offspring when they mate with either parental species, genes cannot flow freely between the species.

Example: The hybrid offspring of a donkey (j) and a horse (k), a mule (l), is robust but sterile.

(k) (j)

(l)

HYBRID BREAKDOWN

Some first-generation hybrids are viable and fertile, but when they mate with one another or with either parent species, offspring of the next generation are feeble or sterile.

Example: Strains of cultivated rice have accumulated different mutant recessive alleles at two loci in the course of their divergence from a common ancestor. Hybrids between them are vigorous and fertile (m, left and right), but plants in the next generation that carry too many of these recessive alleles are small and sterile (m, center). Although these rice strains are not yet considered different species, they have already begun to be separated by postzygotic barriers.

(m)

Limitations of the Biological Species Concept

While the biological species concept's emphasis on reproductive isolation has greatly influenced evolutionary theory, the number of species to which this concept can be usefully applied is limited. For example, there is no way to evaluate the reproductive isolation of fossils or asexual organisms such as prokaryotes. (Many prokaryotes do transfer genes by conjugation and other processes—see Chapter 18—but this transfer is different from sexual recombination. Furthermore, genes are often transferred between distantly related prokaryotes.) It is also difficult to apply the biological species concept to the many sexual organisms about which little is known regarding their ability to mate with different kinds of organisms. For such reasons, alternative species concepts are useful in certain situations.

Other Definitions of Species

While the biological species concept emphasizes the *separateness* of species from one another due to reproductive barriers, several other definitions emphasize the *unity* within a species. For example, the **morphological species concept** characterizes a species by its body shape, size, and other structural features. The morphological species concept has advantages: It can be applied to asexual and sexual organisms, and it can be useful even without information on the extent of gene flow. In practice, this is how scientists distinguish most species. One disadvantage, however, is that this definition relies on subjective criteria; researchers may disagree on which structural features distinguish a species.

The **paleontological species concept** focuses on morphologically discrete species known only from the fossil record. We are forced to distinguish many species in this way because there is little or no information about their mating capability.

The **ecological species concept** views a species in terms of its ecological niche, its role in a biological community (see Chapter 53). For example, two species of Galápagos finches may be similar in appearance but distinguishable based on what they eat. Unlike the biological species concept, this definition can accommodate asexual as well as sexual species.

The **phylogenetic species concept** defines a species as a set of organisms with a unique genetic history—that is, as one branch on the tree of life. Biologists trace the phylogenetic history of a species by comparing its physical characteristics or its molecular sequences with those of other organisms. Such analysis can distinguish groups of individuals that are sufficiently different to be considered separate species. (Of course, the difficulty is in determining the degree of difference required to indicate separate species.) Phylogenetic information sometimes reveals the existence of "sibling species": species that appear so similar that they cannot be distinguished on morphological grounds. Scientists can then apply the biological species concept to determine if the phylogenetic distinction is confirmed by reproductive incompatibility.

The usefulness of each of these definitions depends on the situation and the questions we are asking. The biological species concept, with its focus on reproductive barriers, is particularly valuable for studying how species originate.

Concept Check 24.1

1. Two bird species in a forest are not known to interbreed. One species feeds and mates in the treetops and the other on the ground. But in captivity, the two species can interbreed and produce viable, fertile offspring. What type of reproductive barrier most likely keeps these species separate? Explain.
2. **a.** Which species concept can be used for both asexual and sexual species? **b.** Which can only be applied to sexual species? **c.** Which would be most useful for identifying species in the field?

For suggested answers, see Appendix A.

Concept 24.2

Speciation can take place with or without geographic separation

Speciation can occur in two main ways, depending on how gene flow between the populations is interrupted **(Figure 24.5)**.

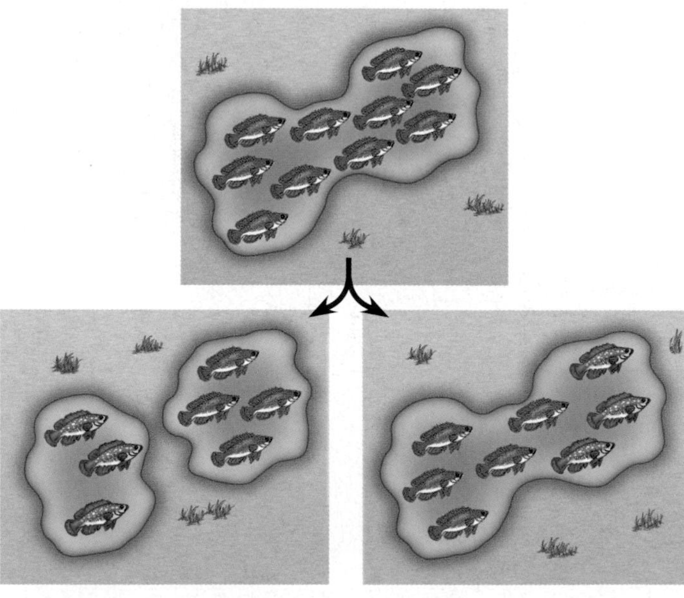

(a) Allopatric speciation. A population forms a new species while geographically isolated from its parent population.

(b) Sympatric speciation. A small population becomes a new species without geographic separation.

▲ **Figure 24.5 Two main modes of speciation.**

Allopatric ("Other Country") Speciation

In **allopatric speciation** (from the Greek *allos*, other, and *patra*, homeland), gene flow is interrupted when a population is divided into geographically isolated subpopulations. For example, the water level in a lake may subside, resulting in smaller lakes that are home to separated populations (see Figure 24.5). A river may change course and split a population of animals that cannot cross it. Allopatric speciation can also occur without geologic remodeling, such as when individuals colonize a remote area, and their descendants become geographically isolated from the parent population. An example is the speciation that occurred on the Galápagos Islands following colonization by mainland organisms.

How formidable must a geographic barrier be to keep allopatric populations apart? The answer depends on the ability of the organisms to move about. Birds, mountain lions, and coyotes can cross hills, rivers, and canyons. Nor do such barriers hinder the windblown pollen of pine trees or the seeds of many flowering plants. In contrast, small rodents may find a deep canyon or a wide river a formidable barrier **(Figure 24.6)**.

Once geographic separation has occurred, the separated gene pools diverge through any or all of the mechanisms described in Chapter 23: different mutations arise, sexual selection takes a different course in the respective populations, other selective pressures act differently on the separated organisms, and genetic drift alters allele frequencies. Because small, isolated populations are more likely than large populations to undergo a significant change in their gene pool in a relatively short time due to selection and drift, they are also more likely to experience allopatric speciation. In less than 2 million years, the few animals and plants from the South American mainland that colonized the Galápagos Islands gave rise to all the new species now found there. But for each small, isolated population that becomes a new species, many more perish in their new environment.

▲ **Figure 24.6 Allopatric speciation of antelope squirrels on opposite rims of the Grand Canyon.** Harris's antelope squirrel (*Ammospermophilus harrisi*) inhabits the canyon's south rim (left). Just a few miles away on the north rim (right) lives the closely related white-tailed antelope squirrel (*Ammospermophilus leucurus*). In contrast, birds and other organisms that can disperse easily across the canyon have not diverged into different species on opposite rims.

To confirm a case of allopatric speciation, it is necessary to determine whether the allopatric populations have changed enough that they no longer have the potential to interbreed and produce fertile offspring. In some cases, researchers evaluate whether speciation has occurred by bringing together members of separated populations in a laboratory setting **(Figure 24.7)**.

Figure 24.7

Inquiry Can divergence of allopatric fruit fly populations lead to reproductive isolation?

EXPERIMENT Diane Dodd, of Yale University, divided a fruit-fly population, raising some populations on a starch medium and others on a maltose medium. After many generations, natural selection resulted in divergent evolution: Populations raised on starch digested starch more efficiently, while those raised on maltose digested maltose more efficiently. Dodd then put flies from the same or different populations in mating cages and measured mating frequencies.

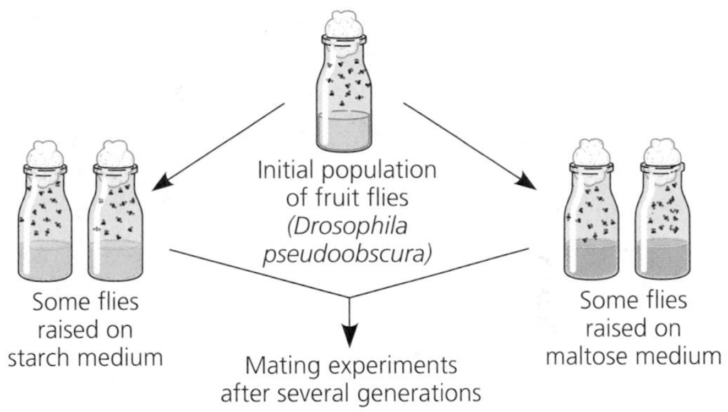

RESULTS When flies from "starch populations" were mixed with flies from "maltose populations," the flies tended to mate with like partners. In the control group, flies taken from different populations that were adapted to the same medium were about as likely to mate with each other as with flies from their own populations.

Male \ Female	Starch	Maltose
Starch	22	9
Maltose	8	20

Mating frequencies in experimental group

Male \ Female	Same population	Different populations
Same population	18	15
Different populations	12	15

Mating frequencies in control group

CONCLUSION The strong preference of "starch flies" and "maltose flies" to mate with like-adapted flies, even if they were from different populations, indicates that a reproductive barrier is forming between the divergent populations of flies. The barrier is not absolute (some mating between starch flies and maltose flies did occur) but appears to be under way after several generations of divergence resulting from the separation of these allopatric populations into different environments.

Biologists can also assess allopatric speciation in the wild. For example, females of the Galápagos ground finch *Geospiza difficilis* respond to the song of males from the same island, but ignore the songs of males of the same species from other islands (allopatric populations) that they encounter. This finding indicates that different behavioral (prezygotic) barriers have developed in these allopatric *G. difficilis* populations, which may eventually become separate species.

We need to emphasize that geographic isolation, though obviously preventing interbreeding between allopatric populations, is not in itself a biological isolating mechanism. Isolating mechanisms—which are intrinsic to the organisms themselves—prevent interbreeding even in the absence of geographic isolation.

Next, let's turn to mechanisms that can produce a new species *without* geographic isolation from the parent population.

Sympatric ("Same Country") Speciation

In **sympatric speciation** (from the Greek *syn*, together), speciation takes place in geographically overlapping populations. How can reproductive barriers between sympatric populations evolve when the members remain in contact with each other? Mechanisms of sympatric speciation include chromosomal changes and nonrandom mating that reduces gene flow.

Polyploidy

Some plant species have their origins in accidents during cell division that result in extra sets of chromosomes, a mutational change that results in the condition called **polyploidy.** An **autopolyploid** (from the Greek *autos*, self) is an individual that has more than two chromosome sets, all derived from a single species. For example, a failure of cell division can double a cell's chromosome number from the diploid number ($2n$) to a tetraploid number ($4n$) **(Figure 24.8)**. This mutation prevents a tetraploid from successfully interbreeding with diploid plants of the original population—the triploid ($3n$) offspring of such unions are sterile because their unpaired chromosomes result in abnormal meiosis. However, the tetraploid plants can still produce fertile tetraploid offspring by self-pollinating or mating with other tetraploids. Thus, in just one generation, autopolyploidy can generate reproductive isolation without any geographic separation.

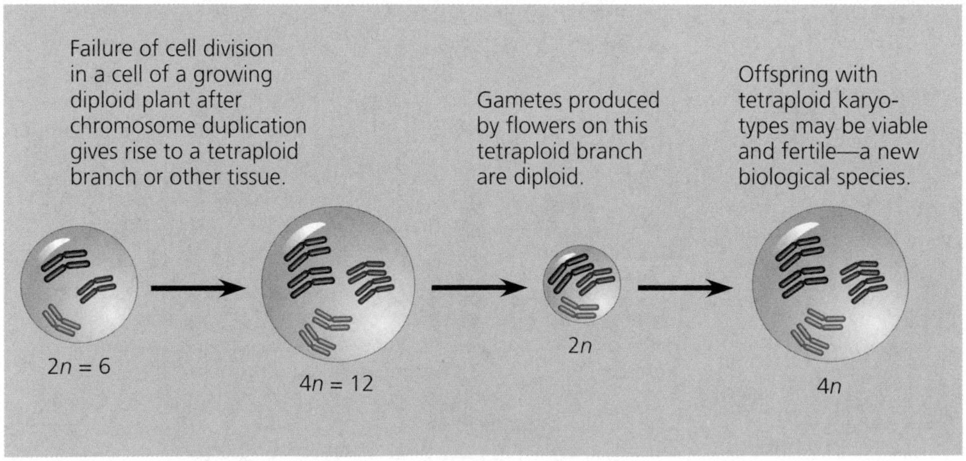

A much more common form of polyploidy can occur when two different species interbreed and produce a hybrid. Interspecific hybrids are often sterile because the set of chromosomes from one species cannot pair during meiosis with the set of chromosomes from the other species. However, though infertile, the hybrid may be able to propagate itself asexually (as many plants can do). In subsequent generations, various mechanisms can change a sterile hybrid into a fertile polyploid known as an **allopolyploid** (**Figure 24.9** illustrates one such mechanism). The allopolyploids are fertile with each other but cannot interbreed with either parental species—thus they represent a new biological species.

The origin of new polyploid plant species is common enough and rapid enough that scientists have documented several such speciations. For example, two new species of goatsbeard plants (genus *Tragopogon*) originated in the Pacific Northwest in the mid-1900s. Although the goatsbeard genus is native to Europe, humans had introduced three species to the Americas in the early 1900s. These species, *T. dubius*, *T. pratensis*, and *T. porrifolius*, are now common weeds in abandoned parking lots and other urban wastelands. In the 1950s, botanists identified two new *Tragopogon* species in regions of Idaho and Washington where all three European species are also found. One new species, *T. miscellus*, is a tetraploid hybrid of *T. dubius* and *T. pratensis*; the other new species, *T. mirus*, is also an allopolyploid, but its ancestors are *T. dubius* and *T. porrifolius*. While the *T. mirus* population grows mainly by reproduction of its own members, additional episodes of hybridization between the ancestral species are continuing to add to the *T. mirus* population—just one example of an ongoing speciation process that can be observed.

Many important agricultural crops—such as oats, cotton, potatoes, tobacco, and wheat—are polyploids. The wheat used for bread, *Triticum aestivum*, is an allohexaploid (six sets of chromosomes, two sets from each of three different species). The first of the polyploidy events that eventually led to modern wheat probably occurred about 8,000 years ago in the Middle East as a spontaneous hybrid of an early cultivated wheat and a wild grass. Today, plant geneticists create new polyploids in the laboratory by using chemicals that induce meiotic and mitotic errors. By harnessing the evolutionary process,

Failure of cell division in a cell of a growing diploid plant after chromosome duplication gives rise to a tetraploid branch or other tissue.
Gametes produced by flowers on this tetraploid branch are diploid.
Offspring with tetraploid karyotypes may be viable and fertile—a new biological species.
2n = 6
4n = 12
2n
4n

▲ **Figure 24.8 Sympatric speciation by autopolyploidy in plants.**

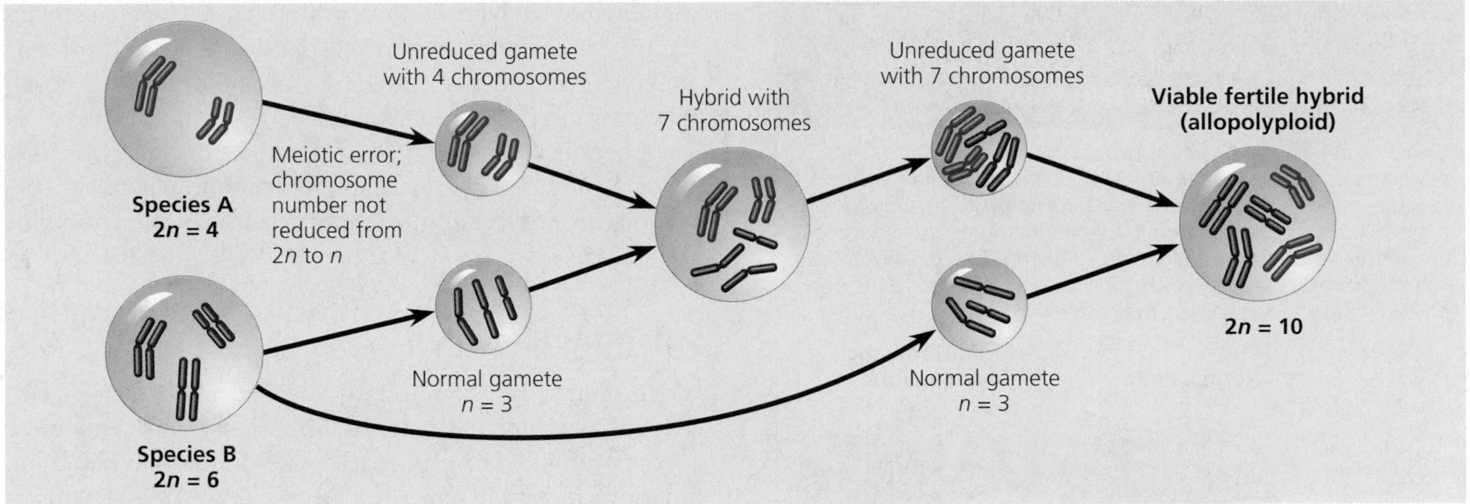

▲ **Figure 24.9 One mechanism for allopolyploid speciation in plants.** A hybrid of two different species is usually sterile because its chromosomes are not homologous and cannot pair during meiosis. However, such a hybrid may be able to reproduce asexually. This diagram traces one mechanism that can produce fertile hybrids (allopolyploids) as new species. The new species has a diploid chromosome number equal to the sum of the diploid chromosome numbers of the two parent species.

researchers can produce new hybrid species with desired qualities, such as a hybrid combining the high yield of wheat with the hardiness of rye.

Habitat Differentiation and Sexual Selection

Polyploid speciation also occurs in animals, although it is less common than in plants. Other mechanisms can also lead to sympatric speciation in both animals and plants. For example, reproductive isolation can occur when genetic factors enable a subpopulation to exploit a resource not used by the parent population. Such is the case with the North American apple maggot fly, *Rhagoletis pomonella*. The fly's original habitat was native hawthorn trees, but about 200 years ago, some populations colonized apple trees introduced by European settlers. Apples mature more quickly than hawthorn fruit, so the apple-feeding flies have been selected for rapid development. These apple-feeding populations now show temporal isolation from the hawthorn-feeding *R. pomonella*. Although the two groups are still classified as subspecies rather than separate species, speciation appears to be well under way.

One of Earth's hot spots of animal speciation is Lake Victoria in eastern Africa. This vast, shallow lake has filled and dried up repeatedly in response to climate changes. The current lake, which is only 12,000 years old, is home to more than 500 species of cichlid fishes. The species are so genetically similar that it is very likely that many have arisen since the lake last filled. The subdivision of the original fish population into groups adapted to exploiting different food sources was one factor contributing to this rapid speciation. But researchers from the University of Leiden in the Netherlands have shown that an additional factor may have been nonrandom mating (sexual selection), in which females select males based on their appearance.

These researchers studied two closely related sympatric species of cichlids that differ mainly in coloration: *Pundamilia pundamilia* has a blue-tinged back, and *Pundamilia nyererei* has a red-tinged back. It is a reasonable hypothesis that a preference for mates of like coloration functions as a behavioral barrier to interbreeding. In an aquarium with natural light, females of each species mated only with males of their own species. But in an aquarium illuminated with a monochromatic orange lamp, which made the two cichlid species appear identical, females of each species mated indiscriminately with males of both species (**Figure 24.10**, on the next page). The hybrids from the *P. pundamilia* × *P. nyererei* matings were viable and fertile.

From the results of these experiments, we can infer that mate choice based on coloration is the main reproductive barrier that normally keeps the gene pools of these two species separate. And we can also infer from their ability to interbreed in the laboratory that, like the apple maggot flies, these species have only begun to diverge. It seems likely that the ancestral population was polymorphic for color and that divergence began with the appearance of two ecological niches that divided the fish into subpopulations. Genetic drift resulted in chance differences in their genetic makeup, with the result that females in one subpopulation favored red in their mates, while females in the other subpopulation preferred blue. Sexual selection then reinforced the color difference as females mated preferentially with males having genes for the least mistakable preferred coloration (see Chapter 23 to review sexual selection). Pollution is now clouding Lake Victoria's waters, so perhaps the cichlids' divergence

Figure 24.10

Inquiry Does sexual selection in cichlids result in reproductive isolation?

EXPERIMENT Researchers from the University of Leiden placed males and females of *Pundamilia pundamilia* and *P. nyererei* together in two aquarium tanks, one with natural light and one with a monochromatic orange lamp. Under normal light, the two species are noticeably different in coloration; under monochromatic orange light, the two species appear identical in color. The researchers then observed the mating choices of the fish in each tank.

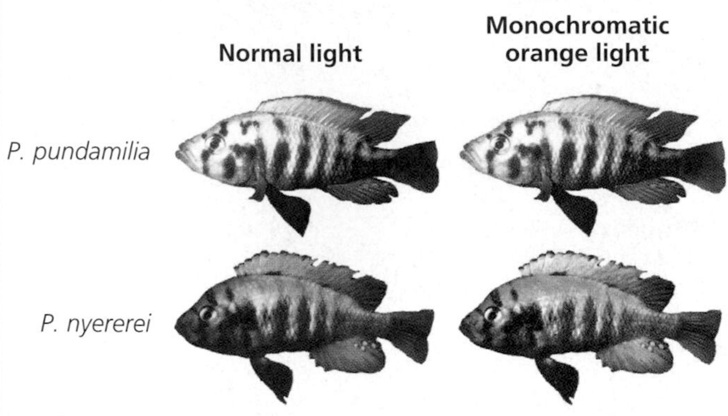

RESULTS Under normal light, females of each species mated only with males of their own species. But under orange light, females of each species mated indiscriminately with males of both species. The resulting hybrids were viable and fertile.

CONCLUSION The researchers concluded that mate choice by females based on coloration is the main reproductive barrier that normally keeps the gene pools of these two species separate. Since the species can still interbreed when this prezygotic behavioral barrier is breached in the laboratory, the genetic divergence between the species is likely to be small. This suggests that speciation in nature has occurred relatively recently.

will be reversed. As it becomes more difficult for female cichlids to visually distinguish between the males, the gene pools of *P. pundamilia* and *P. nyererei* may again blend together.

Allopatric and Sympatric Speciation: *A Summary*

Before continuing our discussion, let's recap the two main modes by which new species form. In allopatric speciation, a new species forms while geographically isolated from its parent population. As the isolated population evolves by natural selection and genetic drift, reproductive isolation from the ancestral species may arise as a by-product of the genetic change. Such reproductive barriers prevent interbreeding with the parent population, even if the populations come back into contact.

Sympatric speciation, in contrast, requires the emergence of a reproductive barrier that isolates a subset of a population without geographic separation from the parent population. A common mechanism, especially in plants, is allopolyploidy: hybridization between closely related species coupled with errors during cell division that lead to fertile polyploid individuals. Sympatric speciation can also result when a subset of a population becomes reproductively isolated because of a switch to a habitat, food source, or other resource not used by the parent population. In animals, speciation can also result from sexual selection in a polymorphic population.

Adaptive Radiation

The evolution of many diversely adapted species from a common ancestor upon introduction to various new environmental opportunities and challenges is called **adaptive radiation.** Adaptive radiation typically occurs when a few organisms make their way to new, often distant areas **(Figure 24.11)** or when environmental changes cause numerous extinctions, opening up ecological niches for the survivors. For example, fossil evidence indicates that mammals underwent a dramatic adaptive radiation after the mass extinctions of dinosaurs 65 million years ago.

The Hawaiian archipelago is one of the world's great showcases of adaptive radiation, as Kenneth Kaneshiro explained in the interview on pages 436–437. Located about 3,500 km from the nearest continent, the volcanic islands are progressively younger as one follows the chain toward the southeast; the youngest island, Hawaii, is less than a million years old and still has active volcanoes. Each island was born "naked" and was gradually populated by stray organisms that rode the ocean currents and winds either from far-distant land areas or from older islands of the archipelago itself. The physical diversity of each island—including immense ranges of altitudes and rainfall—provides many opportunities for evolutionary divergence by natural selection **(Figure 24.12)**. Multiple

▲ **Figure 24.11 Long-distance dispersal.** Seeds of the plant *Pisonia* cling like Velcro to this migratory black noddy tern off the Australian coast. (In fact, studying how such seeds bind tightly but reversibly to feathers and fur inspired the inventors of Velcro.) This mechanism enables the seeds to be transported over long distances.

▲ **Figure 24.12 Adaptive radiation.** Molecular analysis indicates that these remarkably varied Hawaiian plants, known collectively as the "silversword alliance," are all descended from an ancestral tarweed that arrived on the islands about 5 million years ago from North America. Members of the silversword alliance have since spread into many different habitats on the islands, evolving into distinct forms through allopatric and sympatric speciation.

invasions and allopatric and sympatric speciation events have ignited an explosion of adaptive radiation—most of the thousands of species that inhabit the islands are found nowhere else on Earth.

Studying the Genetics of Speciation

Rapid strides in genetics are enabling researchers to pinpoint genes that play a key role in particular cases of speciation. In one such study, Douglas Schemske and his colleagues at Michigan State University examined two species of monkey flower (*Mimulus*; see Figure 24.4f and g). *Mimulus lewisii* and *Mimulus cardinalis* are pollinated by bees and hummingbirds, respectively, keeping their gene pools separate through prezygotic isolation. But the species so far show little postzygotic isolation; they can be mated readily in the greenhouse. The resulting hybrids produce offspring with flowers in a wide variety of colors and shapes. By observing which pollinators visit which flowers and then investigating the genetic differences among the plants, the researchers concluded that two gene loci are largely responsible for pollinator choice. One locus influences flower color and the other affects the amount of nectar that the flowers make. Thus, by determining the attractiveness of the flowers to different pollinators, allelic diversity at these loci has ultimately resulted in speciation.

The Tempo of Speciation

The fossil record includes many episodes in which new species appear suddenly in a geologic stratum, persist essentially unchanged through several strata, and then apparently disappear. Paleontologists Niles Eldredge of the American Museum of Natural History and Stephen Jay Gould of Harvard University

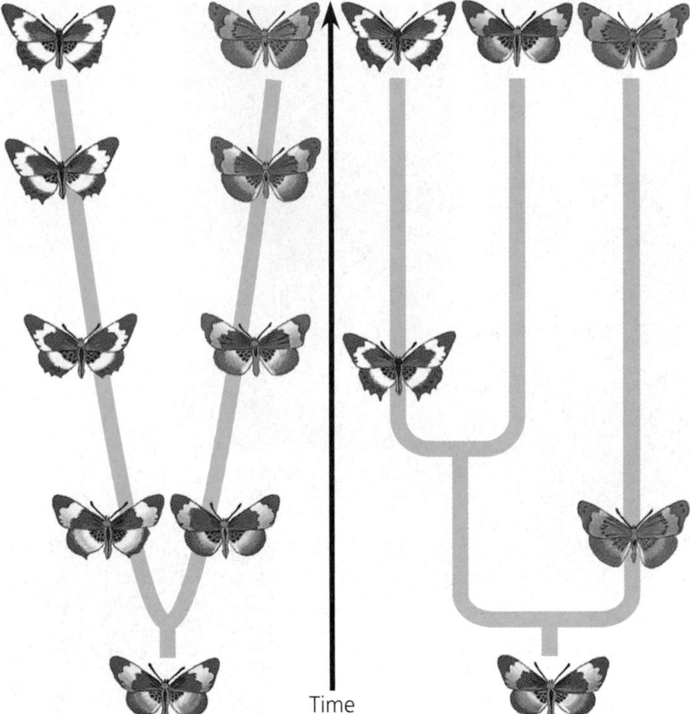

(a) Gradualism model. Species descended from a common ancestor gradually diverge more and more in their morphology as they acquire unique adaptations.

(b) Punctuated equilibrium model. A new species changes most as it buds from a parent species and then changes little for the rest of its existence.

▲ **Figure 24.13 Two models for the tempo of speciation.**

coined the term **punctuated equilibrium** to describe these periods of apparent stasis punctuated by sudden change.

Some scientists suggest that these patterns require an explanation outside the Darwinian model of gradual descent with modification. However, this is not necessarily the case. For one thing, such punctuations may have been less abrupt than they appear from the fossil record. Suppose that a species survived for 5 million years, but most of its morphological alterations occurred during the first 50,000 years of its existence—just 1% of its total lifetime. Because time periods this short often cannot be distinguished in fossil strata, the species would seem to have appeared suddenly and then lingered with little or no change before becoming extinct. Even though the emergence of this species actually took tens of thousands of years, this period of change left no fossil record **(Figure 24.13)**.

In fact, Darwin himself noted this pattern in the fossil record. He anticipated the concept of punctuated equilibrium when he wrote, "Although each species must have passed through numerous transitional stages, it is probable that the periods during which each underwent modification, though many and long as measured by years, have been short in comparison with the periods during which each remained in an unchanged condition."

Stasis, too, can be explained. All species continue to adapt after they come into existence, but often in ways that cannot be detected from fossils, such as small modifications in their biochemistry. By necessity, paleontologists base hypotheses about descent almost entirely on external anatomy and skeletons. During periods of apparent equilibrium, changes in behavior, internal anatomy, and physiology may go undetected. If the environment changes, however, the stasis will be broken by punctuations that leave visible traces in the fossil record.

Concept Check 24.2

1. Explain why allopatric speciation would be less likely to occur on an island close to a mainland than on a more isolated island of the same size.
2. Normal watermelon plants are diploid ($2n = 22$), but breeders have produced tetraploid ($4n = 44$) watermelons. If tetraploid plants are hybridized with their diploid relatives, they produce triploid ($3n = 33$) seeds. These offspring can produce triploid seedless watermelons and can be further propagated by cuttings. Are the diploid and tetraploid watermelon plants different species? Explain.
3. In the fossil record, transitional fossils linking newer species to older ones are relatively rare. Suggest an explanation for this observation.

For suggested answers, see Appendix A.

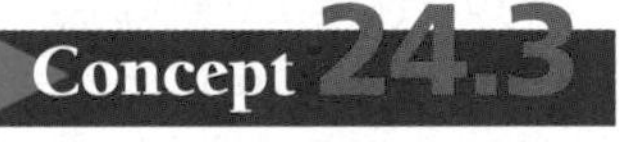

Macroevolutionary changes can accumulate through many speciation events

Speciation can result from differences as seemingly small as the color on a cichlid's back. However, as species diverge and speciate again and again, these differences can accumulate and become more pronounced. Thus, speciation constitutes the beginning of macroevolutionary change. Macroevolutionary transformations, like the microevolutionary changes that take place within a single gene pool, accumulate through the processes that we examined in Chapter 23—natural selection, mutation, genetic drift, and gene flow. It is the cumulative change during thousands of small speciation episodes that accounts for sweeping evolutionary changes. We will now look at some of the ways in which these major transformations have taken place.

Evolutionary Novelties

The Darwinian concept of descent with modification can be extended to account for even major morphological transformations. In most cases, complex structures have evolved in increments from much simpler versions that performed the same basic function.

For example, consider the human eye, an intricate organ constructed from numerous parts that work together in forming an image and transmitting it to the brain. How could the human eye have evolved in gradual increments? If the eye needs all its components to function, how could a partial eye have been of any use to our ancestors?

The flaw in this argument, as Darwin himself noted, lies in the assumption that only complicated eyes are useful. In fact, many animals depend on eyes that are far less complex than our own **(Figure 24.14)**. The simplest eyes that we know of are patches of light-sensitive photoreceptor cells. These simple eyes appear to have had a single evolutionary origin and are now found in a variety of animals, including limpets, small molluscs (members of the phylum Mollusca). These eyes have no lenses or other equipment for focusing images, but they do enable the animal to distinguish light from dark. Limpets cling more tightly to their rock when a shadow falls on them—a behavioral adaptation that reduces the risk of being eaten. Since limpets have had a long evolutionary history, we can conclude that their "simple" eyes are quite adequate to support their survival and reproduction.

In the animal kingdom, complex eyes of various types evolved independently from such basic structures many times. Some molluscs, including the squids and octopuses, have eyes as complex as those of humans and other vertebrates (see Figure 24.14). Although complex mollusc eyes evolved independently of complex vertebrate eyes, both evolved from an ancestral simple cluster of photoreceptor cells. This divergence took place through a series of incremental modifications that benefited the eyes' owners at every stage. Evidence for this evolutionary divergence comes from phylogenetic analysis of the genes that act as "master regulators" of eye development and are shared by all animals with eyes.

Throughout their evolutionary history, eyes retained their basic function of vision. But evolutionary novelty can also arise when structures that originally played one role gradually acquire a different one. Structures that evolve in one context but become co-opted for another function are sometimes called *exaptations* to distinguish them from the adaptive origin of the original structure.

Note that the concept of exaptation does not imply that a structure somehow evolves in anticipation of future use. Natural selection cannot predict the future; it can only improve a structure in the context of its *current* utility. For example, the honeycombed bones of birds are homologous to the bones of birds' earthbound ancestors. However, honeycombed bones could not have evolved in the ancestors as an adaptation for future flight—they must have had another benefit. Fossils of some close relatives of birds have been found in China. These small animals were unable to fly but had feet that enabled them to climb and perch in trees. It is possible that winglike forelimbs and feathers, which would have increased the surface area of these small animals' forelimbs, were co-opted for flight after functioning in some other capacity, such as courtship, thermoregulation, or camouflage (functions that feathers still serve today). The first flights might have been only short glides. Indeed, the first feathers to leave traces in the fossil record were short and downy; feathers adapted for flight appeared later. Over time, natural selection screening additional mutations remodeled feathers and wings in ways that better fit their new function.

The concept of exaptation offers one explanation for how novel features can arise gradually through a series of intermediate stages, each of which has some function in the organism's current context. In the words of Harvard University zoologist Karel Liem, "Evolution is like modifying a machine while it's running."

(a) Patch of pigmented cells. The limpet *Patella* has a simple patch of photoreceptors.

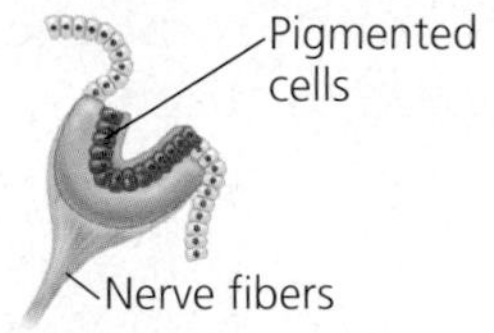

(b) Eyecup. The slit shell mollusc *Pleurotomaria* has an eyecup.

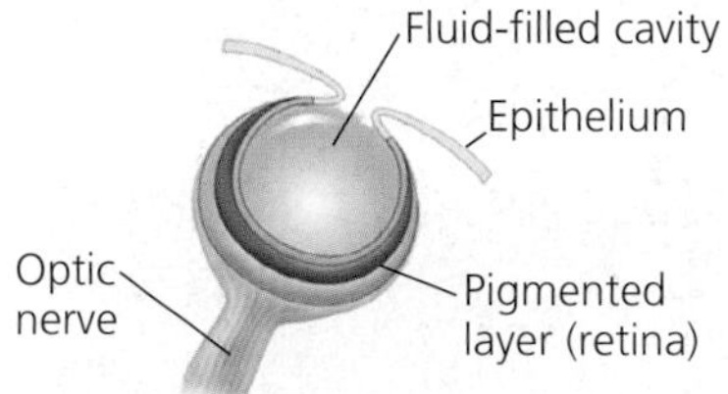

(c) Pinhole camera-type eye. The *Nautilus* eye functions like a pinhole camera (an early type of camera lacking a lens).

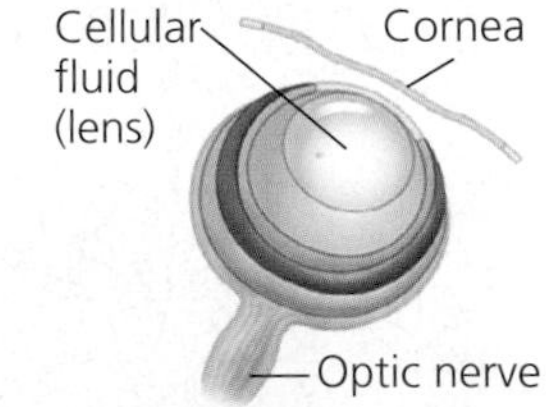

(d) Eye with primitive lens. The marine snail *Murex* has a primitive lens consisting of a mass of crystal-like cells. The cornea is a transparent region of epithelium (outer skin) that protects the eye and helps focus light.

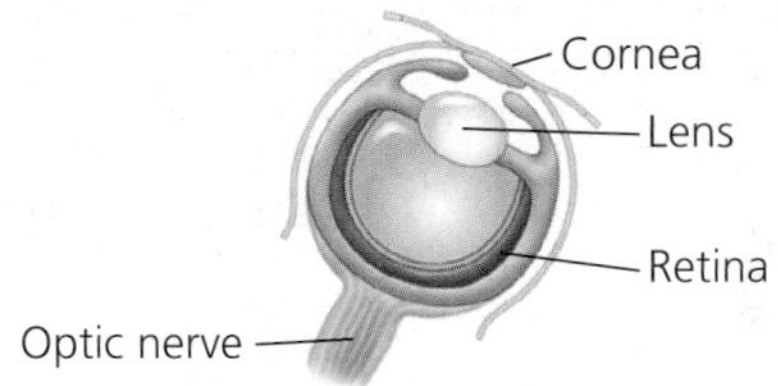

(e) Complex camera-type eye. The squid *Loligo* has a complex eye whose features (cornea, lens, and retina), though similar to those of vertebrate eyes, evolved independently.

▲ **Figure 24.14 A range of eye complexity among molluscs.**

Evolution of the Genes That Control Development

As you read in Chapter 21, "evo-devo"—the interface between evolutionary biology and developmental biology—is illuminating how slight genetic divergences can be magnified into major morphological differences between species. Genes that program development control the rate, timing, and spatial pattern of changes in an organism's form as it develops from a zygote into an adult.

Changes in Rate and Timing

Many striking evolutionary transformations are the result of **heterochrony** (from the Greek *hetero*, different, and *chronos*, time) an evolutionary change in the rate or timing of developmental events. For example, an organism's shape depends in part on the relative growth rates of different body parts during development. This proportioning that helps give a body its specific form is called **allometric growth** (from the Greek *allos*, other, and *metron*, measure). **Figure 24.15a** tracks how allometric growth alters human body proportions during development. Changing these relative rates of growth even slightly changes the adult form substantially. For example, different allometric patterns contribute to the contrasting shapes of human and chimpanzee skulls **(Figure 24.15b)**.

Heterochrony has also played a role in the evolution of salamander feet **(Figure 24.16)**. Most salamanders live on the ground, but some species live in trees. The feet of the tree-dwelling species are adapted for climbing vertically rather

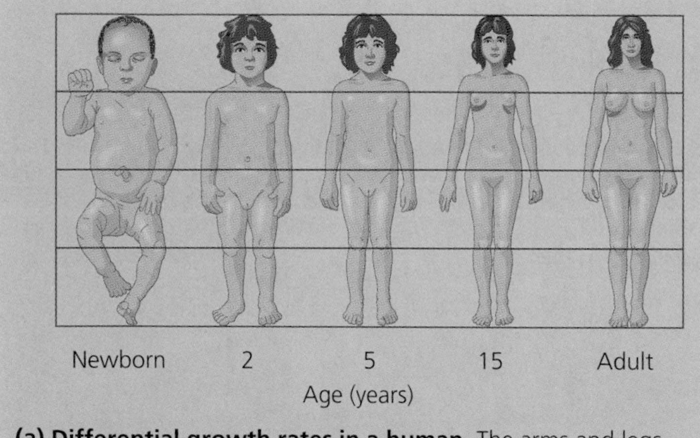

(a) Differential growth rates in a human. The arms and legs lengthen more during growth than the head and trunk, as can be seen in this conceptualization of an individual at different ages all rescaled to the same height.

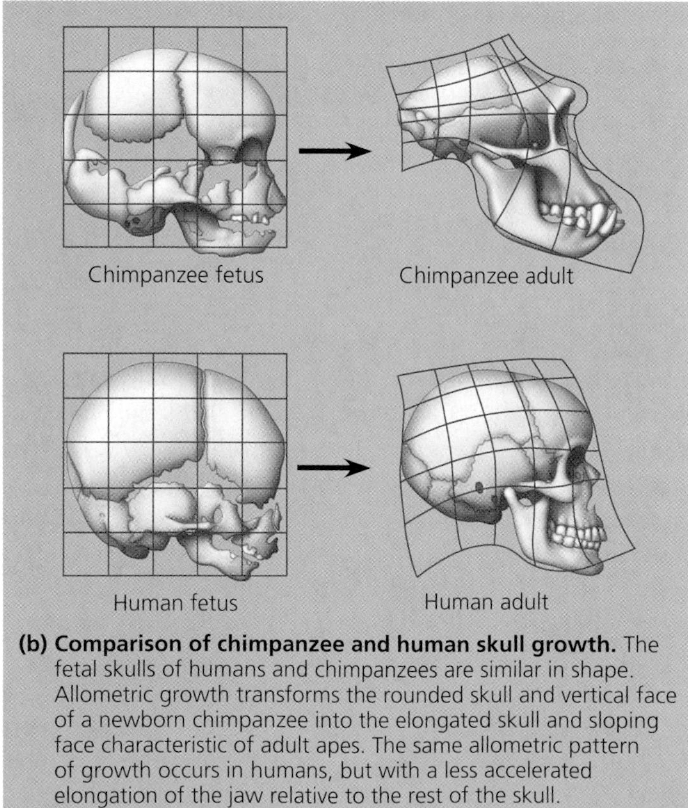

(b) Comparison of chimpanzee and human skull growth. The fetal skulls of humans and chimpanzees are similar in shape. Allometric growth transforms the rounded skull and vertical face of a newborn chimpanzee into the elongated skull and sloping face characteristic of adult apes. The same allometric pattern of growth occurs in humans, but with a less accelerated elongation of the jaw relative to the rest of the skull.

▲ **Figure 24.15 Allometric growth.** Different growth rates for different parts of the body determine body proportions.

(a) Ground-dwelling salamander. A longer time period for foot growth results in longer digits and less webbing.

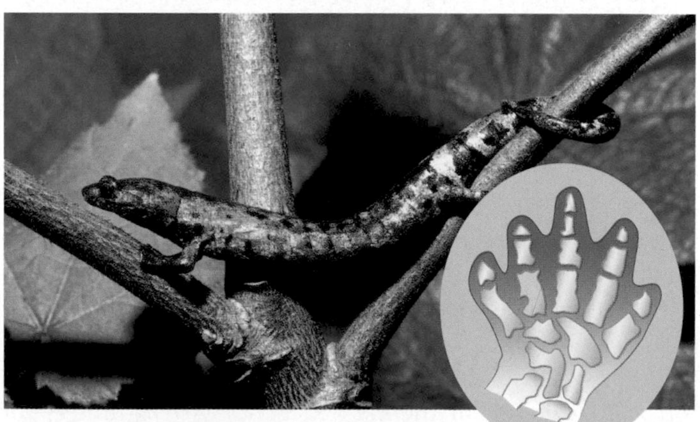

(b) Tree-dwelling salamander. Foot growth ends sooner. This evolutionary timing change accounts for the shorter digits and more extensive webbing, which help the salamander climb vertically on tree branches.

▲ **Figure 24.16 Heterochrony and the evolution of salamander feet in closely related species.**

than walking on the ground; for example, their shorter digits and more extensive webbing provide better traction. The basis for this adaptation was probably selection for alleles of genes that control the timing of foot development. According to this hypothesis, the foot of the ancestral salamander grew until the products of certain regulatory genes switched off growth, resulting in a foot of a certain size. A mutation in one or more of these regulatory genes could have switched off foot growth sooner, resulting in the stunted feet of tree-dwelling salamanders. In this way, a relatively small amount of genetic change can be amplified into substantial morphological change.

Heterochrony can also alter the timing of reproductive development relative to the development of somatic (nonreproductive) organs. If reproductive development accelerates compared to somatic development, the sexually mature stage of a species may retain body features that were juvenile structures in an ancestral species—a condition called **paedomorphosis** (from the Greek *paedos*, of a child, and *morphosis*, formation). For example, most salamander species have a larval stage that undergoes metamorphosis in becoming an adult. But some species grow to adult size and become sexually mature while retaining gills and other larval features **(Figure 24.17)**. Such an evolutionary alteration of developmental timing can produce animals that appear very different from their ancestors, even though the overall genetic change may be small. Indeed, recent evidence indicates that a genetic change at a single locus was probably sufficient to bring about paedomorphosis in the axolotl salamander, although other genes may have contributed as well.

In summary, heterochrony affects the evolution of morphology by altering the rates at which various body parts develop or by changing the timing of onset or completion of a particular part's development.

▲ **Figure 24.17 Paedomorphosis.** Some species retain as adults features that were juvenile in ancestors. This salamander is an axolotl, which grows to full size, becomes sexually mature, and reproduces while retaining certain larval (tadpole) characteristics, including gills.

Changes in Spatial Pattern

Substantial evolutionary changes can also result from alterations in genes that control the placement and spatial organization of body parts. For example, as described in Chapter 21, **homeotic genes** determine such basic features as where a pair of wings and a pair of legs will develop on a bird or how a plant's flower parts are arranged.

The products of one class of homeotic genes called *Hox* genes provide positional information in an animal embryo. This information prompts cells to develop into structures appropriate for a particular location. Changes in *Hox* genes and in the genes that regulate them in turn can have a profound impact on morphology. Consider, for example, the evolution of tetrapods (the terrestrial vertebrates including amphibians, birds and other reptiles, and mammals) from an ancestral aquatic vertebrate. During this process, four of the ancestral vertebrate's fins evolved into limbs. Among the adaptations of the tetrapod limb are digits (fingers and toes in humans) that extend skeletal support to the tip of the limb.

In fish embryos, a *Hox* gene is expressed in a band of cells running along one edge of the fin bud **(Figure 24.18)**. In

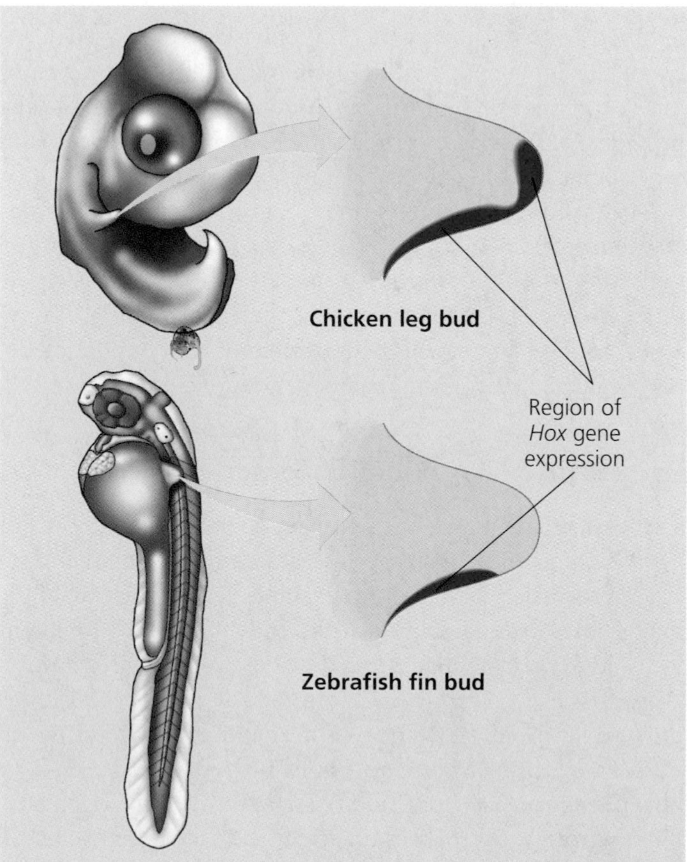

▲ **Figure 24.18 *Hox* genes and the evolution of tetrapod limbs.** The red zones indicate regions where a *Hox* gene involved in skeleton development is expressed.

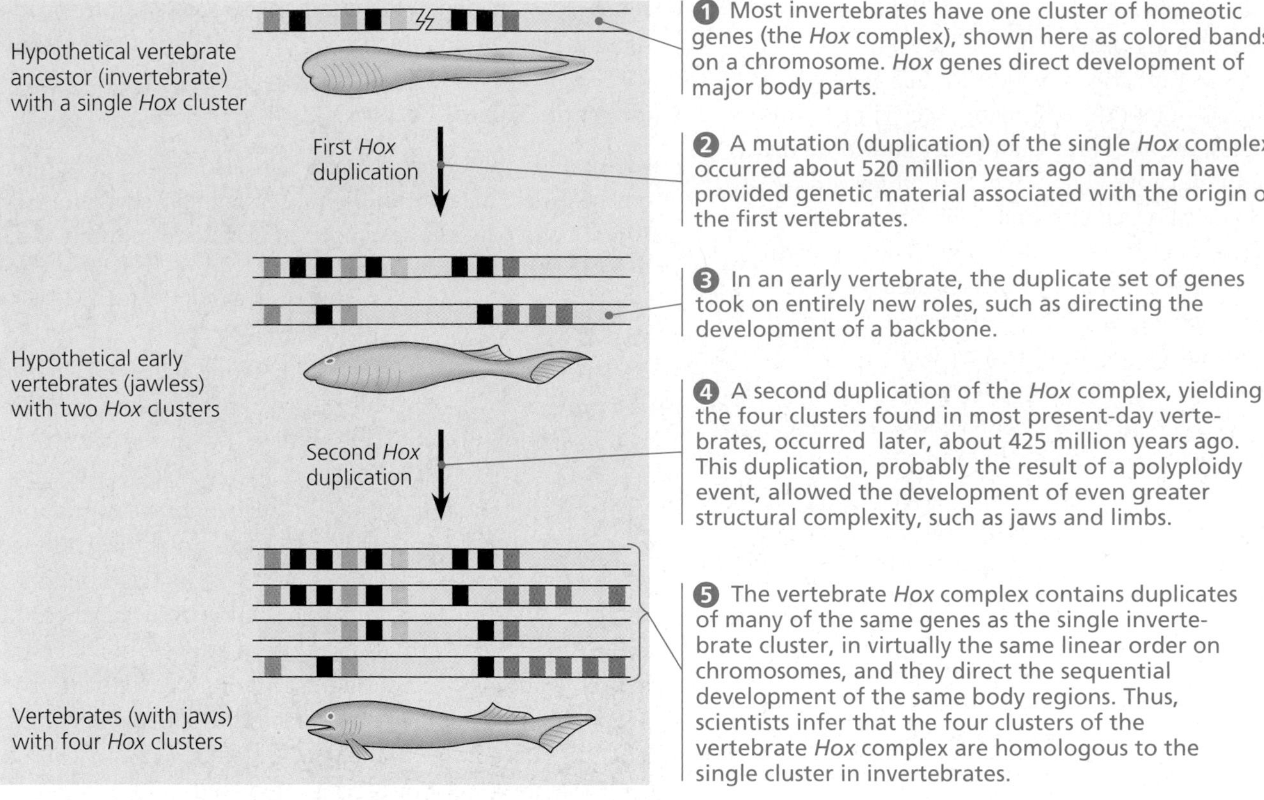

Hypothetical vertebrate ancestor (invertebrate) with a single Hox cluster
First Hox duplication
Hypothetical early vertebrates (jawless) with two Hox clusters
Second Hox duplication
Vertebrates (with jaws) with four Hox clusters
1 Most invertebrates have one cluster of homeotic genes (the Hox complex), shown here as colored bands on a chromosome. Hox genes direct development of major body parts.
2 A mutation (duplication) of the single Hox complex occurred about 520 million years ago and may have provided genetic material associated with the origin of the first vertebrates.
3 In an early vertebrate, the duplicate set of genes took on entirely new roles, such as directing the development of a backbone.
4 A second duplication of the Hox complex, yielding the four clusters found in most present-day vertebrates, occurred later, about 425 million years ago. This duplication, probably the result of a polyploidy event, allowed the development of even greater structural complexity, such as jaws and limbs.
5 The vertebrate Hox complex contains duplicates of many of the same genes as the single invertebrate cluster, in virtually the same linear order on chromosomes, and they direct the sequential development of the same body regions. Thus, scientists infer that the four clusters of the vertebrate Hox complex are homologous to the single cluster in invertebrates.

▲ **Figure 24.19** ***Hox* mutations and the origin of vertebrates.**

tetrapod embryos, this same *Hox* gene is also expressed at the tip of the limb bud, the embryonic structure that develops into a forelimb or hind limb. The product of this *Hox* gene provides positional information about how far digits and other bones should extend from the limb.

The evolution of vertebrates from invertebrate animals was an even larger evolutionary change, and it, too, was associated with alterations in *Hox* genes and the genes that regulate them **(Figure 24.19)**. In producing such morphological novelties, changes in developmental dynamics have undoubtedly played important roles in macroevolution.

Evolution Is Not Goal Oriented

The fossil record often shows apparent trends in evolution. For instance, some evolutionary lineages exhibit a trend toward larger or smaller body size. An example is the evolution of the living horse (genus *Equus*), a descendant of the 40-million-year-old *Hyracotherium* **(Figure 24.20)**. About the size of a large dog, *Hyracotherium* had four toes on its front feet, three toes on its hind feet, and teeth adapted for browsing on bushes and trees. In comparison, living horses are larger, have only one toe on each foot, and possess teeth modified for grazing on grasses. Do these changes represent real trends, and if so, how do we account for them?

Extracting a single evolutionary progression from the fossil record can be misleading; it is like describing a bush as growing toward a single point by tracing only the branches that lead to that twig. For instance, by selecting certain species from the available fossils, it is possible to arrange a succession of animals intermediate between *Hyracotherium* and living horses that shows trends toward increased size, reduced number of toes, and modification of teeth for grazing (yellow line in Figure 24.20). However, if we include all fossil horses known today, this apparent trend vanishes. The genus *Equus* did not evolve in a straight line; it is the only surviving twig of an evolutionary tree that is so branched that it is more like a bush. *Equus* actually descended through a series of speciation episodes that included several adaptive radiations, not all of which led to large, one-toed, grazing horses. For instance, notice in Figure 24.20 that only those lineages derived from *Parahippus* include grazers; other lineages derived from *Miohippus,* all of which are now extinct, remained multi-toed browsers for 35 million years.

Branching evolution can result in an evolutionary trend even if some new species counter the trend. For example, one model of long-term trends proposed by Steven Stanley of Johns Hopkins University considers species to be analogous to individuals: Speciation is their birth, extinction is their death, and new species that diverge from them are their offspring. In this model, Stanley suggests that just as individual organisms undergo natural selection, species undergo **species selection.** The species that endure the longest and generate the most new offspring species determine the direction of major

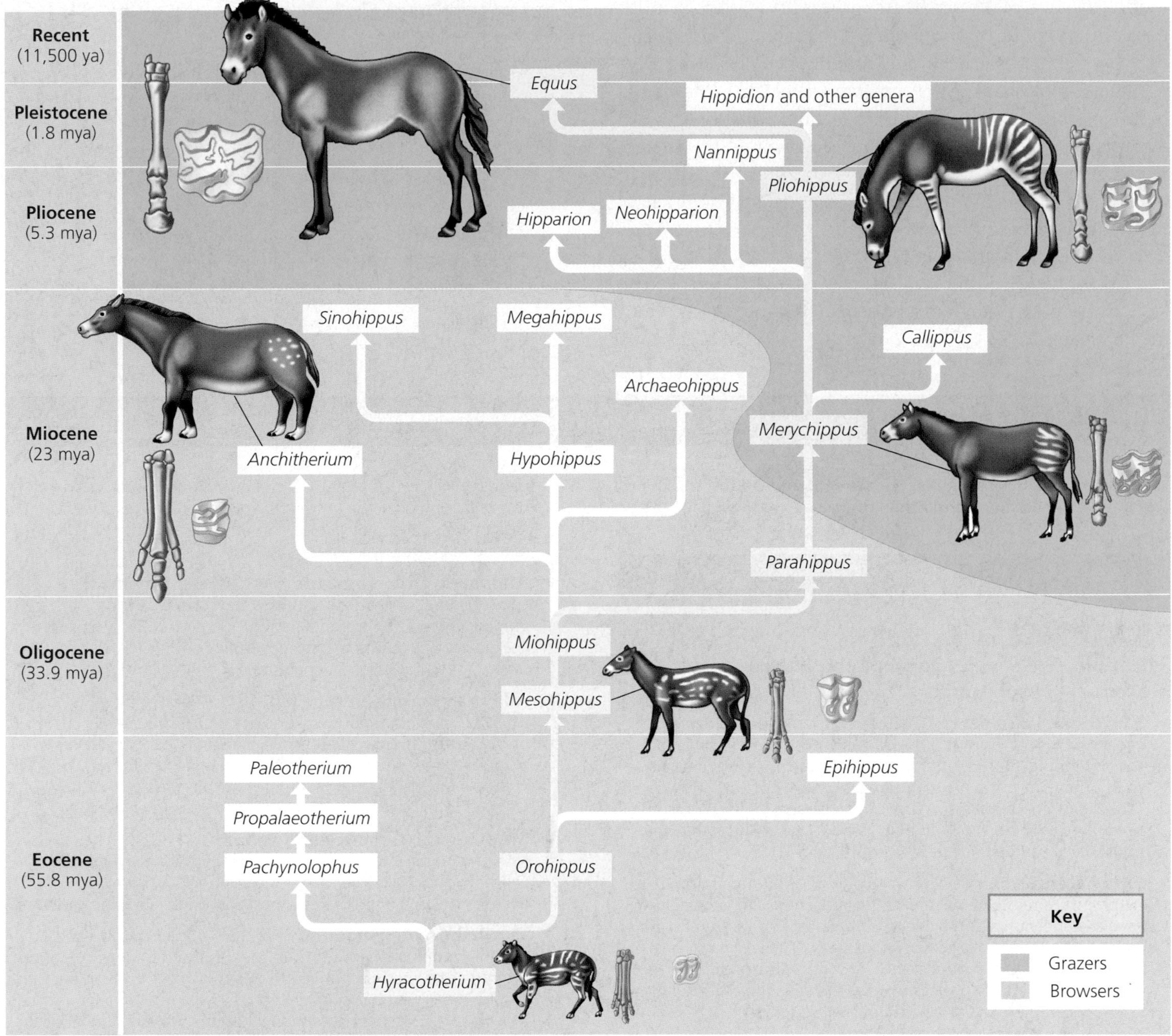

▲ **Figure 24.20 The branched evolution of horses.** Using a yellow highlighter to trace the sequence of fossil horses that are intermediate in form between the living horse (*Equus*) and its Eocene ancestor *Hyracotherium* creates the illusion of a progressive trend toward larger size, reduced number of toes, and teeth modified for grazing. In fact, the living horse is the only surviving twig of an evolutionary bush with many divergent trends.

evolutionary trends. The species selection model suggests that "differential speciation success" plays a role in macroevolution similar to the role of differential reproductive success in microevolution.

To the extent that speciation rates and species longevity reflect success, the analogy to natural selection is even stronger. It is possible that qualities unrelated to the overall success of organisms in specific environments may become important when entire species are selected. For example, if a species is able to disperse easily to new locations, it may give rise to a large number of "daughter species."

Whether or not the species selection model is valid, there are other possible sources of trends observed in the fossil record. For example, sexual selection in the Irish elk of northern Europe drove an immense increase in the size of their antlers—though this did not prevent the elk from becoming extinct because of changing vegetation and hunting by humans. And when horse ancestors invaded the grasslands that spread during the Age of Mammals, there was strong selection for grazers that could escape predators by running faster. This trend would not have occurred without open grasslands.

Whatever its cause, the appearance of an evolutionary trend does not imply that there is some intrinsic drive toward a particular phenotype. Evolution is the result of the interactions between organisms and their current environments. If environmental conditions change, an apparent evolutionary trend may cease or even reverse itself.

In the next chapter, we will continue our study of speciation and the further divergence of organisms, with a more detailed look at the fossil record and the effects of significant environmental change.

Concept Check 24.3

1. How can the Darwinian concept of descent with modification explain the evolution of such complex structures as the vertebrate eye or heart?
2. Explain why the concept of exaptation does not mean that a structure evolves in anticipation of some future environmental change.
3. How can heterochrony cause the evolution of different body forms?

For suggested answers, see Appendix A.

Chapter 24 Review

Go to www.campbellbiology.com or the student CD-ROM to explore Activities, Investigations, and other interactive study aids.

SUMMARY OF KEY CONCEPTS

Concept 24.1
The biological species concept emphasizes reproductive isolation

▶ **The Biological Species Concept (pp. 473–476)** A biological species is a group of populations whose individuals have the potential to interbreed and produce fertile offspring with each other but not with members of other species. The biological species concept emphasizes reproductive isolation through prezygotic and postzygotic barriers that can result in separating the gene pools of different populations.

▶ **Other Definitions of Species (p. 476)** Although particularly helpful in thinking about speciation processes, the biological species concept has some major limitations. For instance, it cannot be applied to organisms that are known only as fossils or to organisms that reproduce only asexually. Thus, scientists maintain alternative species concepts, such as the morphological species concept, that are useful in various contexts.
Activity *Overview of Macroevolution*

Concept 24.2
Speciation can take place with or without geographic separation

▶ **Allopatric ("Other Country") Speciation (pp. 477–478)** Allopatric speciation may occur when two populations of one species become geographically separated from each other. One or both populations may undergo evolutionary change during the period of separation. Should they come into contact once more, they may be separated by the prezygotic and postzygotic isolating mechanisms that have accumulated.

▶ **Sympatric ("Same Country") Speciation (pp. 478–480)** A new species can originate while remaining in a geographically overlapping area with the parent species. In particular, many plant species have evolved sympatrically through polyploidy (multiplications of the chromosome number). Autopolyploids are species derived this way from one ancestral species. Allopolyploids are species with multiple sets of chromosomes derived from different species. Sympatric speciation can also result from the appearance of new ecological niches and from nonrandom mating in polymorphic populations.

▶ **Allopatric and Sympatric Speciation: *A Summary* (p. 480)** In allopatric speciation, a new species forms while geographically isolated from its parent population. In sympatric speciation, a reproductive barrier isolates a subset of a population without geographic separation.

▶ **Adaptive Radiation (pp. 480–481)** Adaptive radiation can occur when a population encounters a multiplicity of new or newly available ecological niches. This may happen during colonization of a new environment, such as newly formed volcanic islands, or after an environmental change that has resulted in mass extinctions of other species in an area.

▶ **Studying the Genetics of Speciation (p. 481)** The explosion of genomics is enabling researchers to identify specific genes involved in some cases of speciation.
Investigation *How Do New Species Arise by Genetic Isolation?*

▶ **The Tempo of Speciation (pp. 481–482)** Eldredge and Gould's punctuated equilibrium model draws on fossil evidence showing that species change most as they arise from an ancestral species, after which they undergo relatively little change for the rest of their existence. This model contrasts with a model of gradual change throughout a species' existence.

Concept 24.3
Macroevolutionary changes can accumulate through many speciation events

▶ **Evolutionary Novelties (pp. 482–483)** Most novel biological structures evolve in many stages from previously existing structures. Some complex structures, such as the eye, have had similar functions during all stages of their evolution. The most important functions of others, such as feathers, have changed.

▶ **Evolution of the Genes That Control Development (pp. 484–486)** Many large evolutionary changes may have been associated with mutations in genes that regulate development. Such changes can affect the timing of developmental events (heterochrony) or the spatial organization of body parts. Some of these changes result from mutational changes in homeotic genes and in the genes that regulate them.
Activity *Allometric Growth*

▶ **Evolution Is Not Goal Oriented (pp. 486–488)** Long-term evolutionary trends may arise because of adaptation to a changing environment. In addition, according to the species selection model, trends may result when species with certain characteristics endure longer and speciate more often than those with other characteristics.

TESTING YOUR KNOWLEDGE

Self-Quiz

1. The biological species concept is *not* useful for organisms known only from fossils because
   a. fossils are rarely preserved well enough to distinguish species based on morphology.
   b. it is not possible to test reproductive isolation of fossil forms.
   c. it is not possible to infer the types of habitats occupied by the fossil forms before their extinction.
   d. in examining fossil organisms, it is not possible to distinguish males from females.
   e. the fossil record can only be used for studying anagenesis, but not cladogenesis.

2. The *largest* unit within which gene flow can readily occur is a
   a. population.
   b. species.
   c. genus.
   d. hybrid.
   e. phylum.

3. Bird guides once listed the myrtle warbler and Audubon's warbler as distinct species. Recently, these birds have been classified as eastern and western forms of a single species, the yellow-rumped warbler. Which of the following pieces of evidence, if true, would be cause for this reclassification?
   a. The two forms are observed to interbreed successfully where their habitats overlap.
   b. The two forms live in similar habitats.
   c. The two forms have many genes in common.
   d. The two forms have similar food requirements.
   e. The two forms are very similar in coloration.

4. Males of different species of the fruit fly *Drosophila* that live in the same parts of the Hawaiian islands have different elaborate courtship rituals that involve fighting other males and stylized movements that attract females. What type of reproductive isolation does this represent?
   a. habitat isolation
   b. temporal isolation
   c. behavioral isolation
   d. gametic isolation
   e. postzygotic barriers

5. Which of the following factors would *not* contribute to allopatric speciation?
   a. A population becomes geographically isolated from the parent population.
   b. The separated population is small, and genetic drift occurs.
   c. The isolated population is exposed to different selection pressures than the ancestral population.
   d. Different mutations begin to distinguish the gene pools of the separated populations.
   e. Gene flow between the two populations is extensive.

6. Plant species A has a diploid number of 12. Plant species B has a diploid number of 16. A new species, C, arises as an allopolyploid from A and B. The likely diploid number for species C would probably be
   a. 12.
   b. 14.
   c. 16.
   d. 28.
   e. 56.

7. The speciation episode described in question 6 is most likely a case of
   a. allopatric speciation.
   b. sympatric speciation.
   c. speciation based on sexual selection.
   d. adaptive radiation.
   e. anagenesis.

8. *Mimulus lewisii* and *M. cardinalis* are plants that do not hybridize in nature but can be readily crossed in the laboratory to produce fertile offspring. Which of the following is *least* likely to keep the gene pools of these two plants separate in nature?
   a. gametic incompatibility
   b. different attractiveness to pollinators
   c. different ecological niches
   d. different geographic ranges
   e. seasonal differences in flowering

9. According to the punctuated equilibrium model,
   a. natural selection is unimportant as a mechanism of evolution.
   b. given enough time, most existing species will branch gradually into new species.
   c. most new species accumulate their unique features as they come into existence, then change little for the rest of their duration as a species.
   d. most evolution is goal oriented.
   e. speciation is usually due to a single mutation.

10. A genetic change that caused a certain *Hox* gene to be expressed along the tip of a vertebrate limb bud instead of farther back made possible the evolution of the tetrapod limb. This type of change is illustrative of
    a. the influence of environment on an individual's development.
    b. paedomorphosis, or retention of ancestral juvenile structures in an adult organism.
    c. a change in a developmental gene or in its regulation that altered the spatial organization of body parts.
    d. punctuated equilibrium.
    e. the origin of a new species due to allopolyploidy.

For Self-Quiz answers, see Appendix A.

Go to the website or CD-ROM for more quiz questions.

Evolution Connection

In the margin of one of his notebooks, Darwin scrawled a note to remind himself never to apply the terms *higher* or *lower* to species. It was, and still is, very common for people to think of some species or species groups as more or less evolved than others. This probably stems from a notion of evolutionary "progress." Is there such a thing as evolutionary progress? Why or why not? Defend your position as if you were debating someone holding the opposite view.

Scientific Inquiry

Cultivated American cotton plants have a total of 52 chromosomes ($2n = 52$). In each cell, 13 pairs of chromosomes (26 chromosomes) are smaller than the other 13 pairs. Old World cotton plants have a total of 26 chromosomes ($2n = 26$), all large. Wild American cotton plants have 26 chromosomes, all small. Propose a hypothesis to explain how cultivated American cotton may have originated. How could you test your hypothesis?

Investigation ***How Do New Species Arise by Genetic Isolation?***

Science, Technology, and Society

What is the biological basis for assigning all human populations to a single species? Can you think of a scenario in which a second human species could arise in the future by cladogenesis?

25 Phylogeny and Systematics

▲ **Figure 25.1 A dragonfly fossil from Brazil, more than 100 million years old.**

Key Concepts

25.1 Phylogenies are based on common ancestries inferred from fossil, morphological, and molecular evidence

25.2 Phylogenetic systematics connects classification with evolutionary history

25.3 Phylogenetic systematics informs the construction of phylogenetic trees based on shared characters

25.4 Much of an organism's evolutionary history is documented in its genome

25.5 Molecular clocks help track evolutionary time

Overview

Investigating the Tree of Life

Evolutionary biology is about both process and history. In this unit, we have already examined the processes of evolution—natural selection and other mechanisms that change the genetic composition of populations (see Chapter 23) and that can lead to the origin of new species (see Chapter 24). But evolutionary biologists also seek to reconstruct the long-term results of these processes—the entire history of life on Earth.

This chapter describes how biologists trace **phylogeny** (from the Greek *phylon*, tribe, and *genesis*, origin), the evolutionary history of a species or group of species. In constructing phylogenies, biologists draw on the fossil record, which provides information about ancient organisms **(Figure 25.1)**. They also utilize **systematics**, an analytical approach to understanding the diversity and relationships of organisms, both present-day and extinct. Systematists have traditionally studied morphological and biochemical resemblances among organisms as a basis for inferring evolutionary relationships. In recent decades, systematists have gained a powerful new tool: **molecular systematics**, which uses comparisons of DNA, RNA, and other molecules to infer evolutionary relationships between individual genes and even between entire genomes **(Figure 25.2)**. This information explosion is enabling evolutionary biologists to construct a universal tree of all life, which will continue to be refined as the database of DNA and RNA sequences grows.

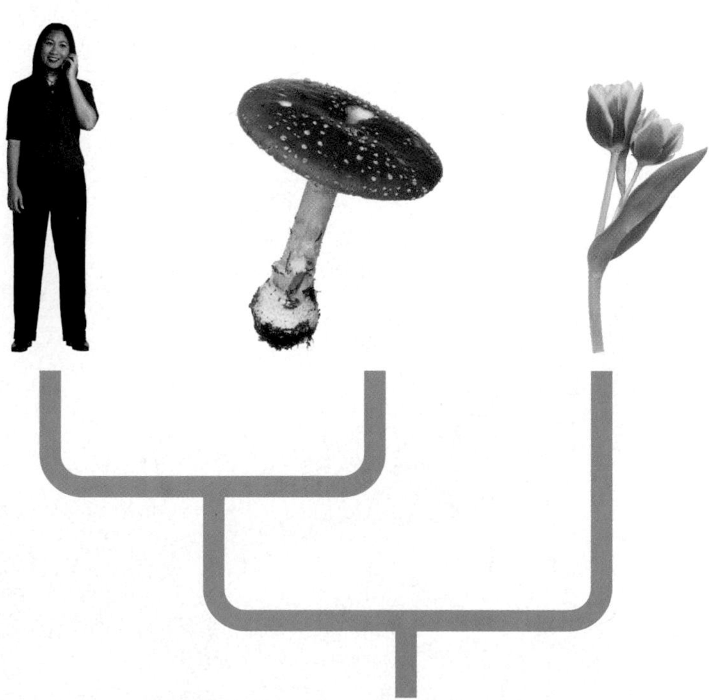

▲ **Figure 25.2 An unexpected family tree.** What are the evolutionary relationships among a human, a mushroom, and a tulip? Molecular systematics has revealed that—despite appearances—animals, including humans, and fungi, such as mushrooms, are more closely related to each other than either are to plants.

Concept 25.1

Phylogenies are based on common ancestries inferred from fossil, morphological, and molecular evidence

In order to infer phylogenies, we must gather as much information as we can about the morphologies, development, and biochemistry of living organisms. But it is also essential to study fossils, the preserved remnants or impressions left by organisms that lived in the past. Fossils can help establish relationships between living organisms because they reveal ancestral characteristics that may have been lost over time in certain lineages.

The Fossil Record

Sedimentary rocks are the richest source of fossils. Sand and silt eroded from the land are carried by rivers to seas and swamps, where the minerals settle to the bottom along with the remains of organisms. Over millions of years, deposits pile up and compress the older sediments below into layers called strata **(Figure 25.3)**. The **fossil record** is based on the sequence in which fossils have accumulated in such strata.

Though sedimentary fossils are the most common, paleontologists also study other types of fossils **(Figure 25.4)**. But fossils inform phylogeny only if we can determine their ages, clarifying the order in which various characteristics appeared and disappeared. In Chapter 26, we will discuss methods for dating fossils and for assigning the divisions of the geologic record. For now, keep in mind that the fossil record is a substantial, but incomplete, chronicle of evolutionary change. A large number of Earth's species probably did not die in the right place at the right time to be preserved as fossils; of those fossils that were formed, many were probably destroyed by later geologic processes; and only a fraction of existing fossils have been discovered. Rather than giving a true indication of the diversity of past life, the fossil record is biased in favor of species that existed for a long time, were abundant and widespread, and had hard shells, skeletons, or other parts that facilitated their fossilization. Even with its limitations, however, the fossil record is a remarkably detailed account of biological change over the vast scale of geologic time.

Morphological and Molecular Homologies

In addition to fossil organisms, certain morphological and molecular similarities among living organisms can indicate phylogenetic history. Recall that similarities due to shared ancestry are called homologies. For example, the similarity in the number and arrangement of bones in the forelimbs of mammals is due to their descent from a common ancestor with the same bone structure; this is an example of a morphological homology (see Figure 22.14). In the same way, genes or other DNA sequences are homologous if the nature of their similarity suggests that they are descended from the sequences carried by a common ancestor.

In general, organisms that share very similar morphologies or similar DNA sequences are likely to be more closely related than organisms with vastly different structures or sequences. In some cases, however, the morphological divergence between related species can be great and their genetic divergence small (or vice versa). For example, consider the Hawaiian silversword plants discussed in Chapter 24. These species vary dramatically in appearance throughout the islands: Some

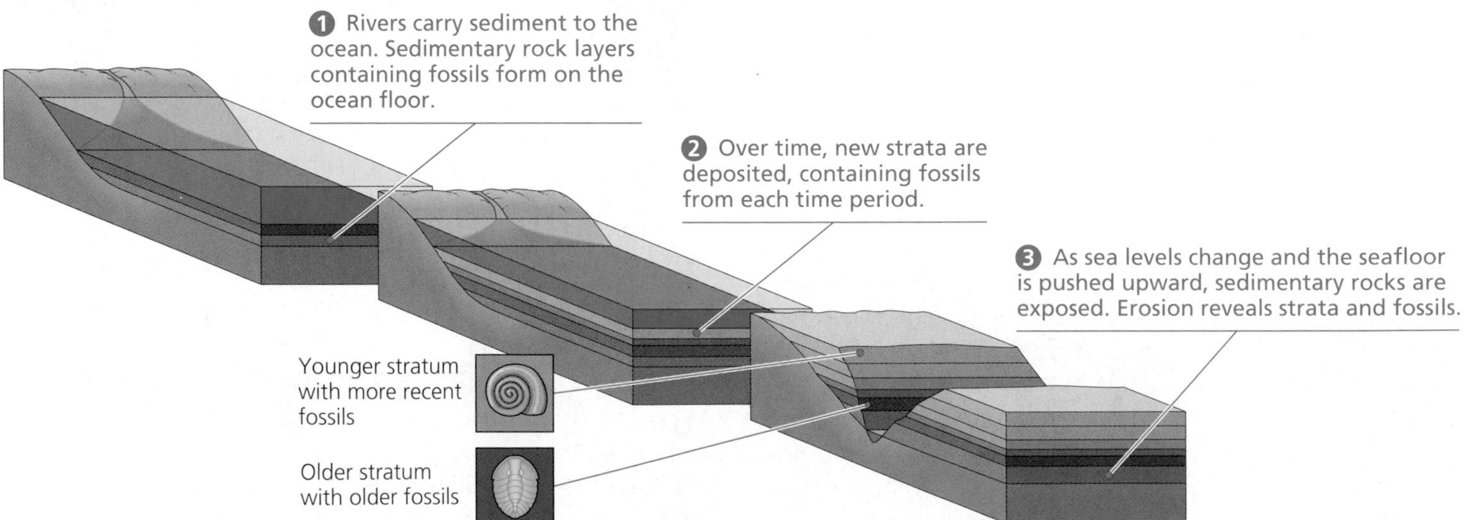

1 Rivers carry sediment to the ocean. Sedimentary rock layers containing fossils form on the ocean floor.
2 Over time, new strata are deposited, containing fossils from each time period.
3 As sea levels change and the seafloor is pushed upward, sedimentary rocks are exposed. Erosion reveals strata and fossils.
Younger stratum with more recent fossils
Older stratum with older fossils

▲ **Figure 25.3 Formation of sedimentary strata containing fossils.**

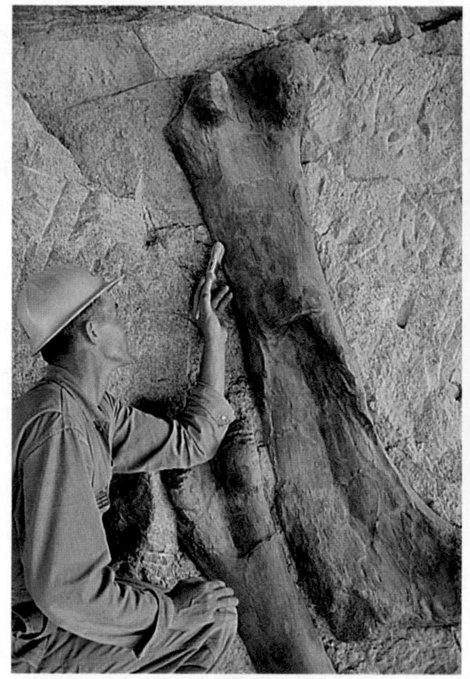

(a) Dinosaur bones being excavated from sandstone

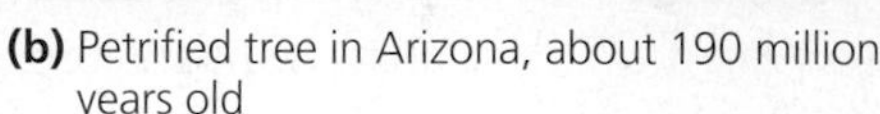

(b) Petrified tree in Arizona, about 190 million years old

(c) Leaf fossil, about 40 million years old

(d) Casts of ammonites, about 375 million years old

(f) Insects preserved whole in amber

(g) Tusks of a 23,000-year-old mammoth, frozen whole in Siberian ice

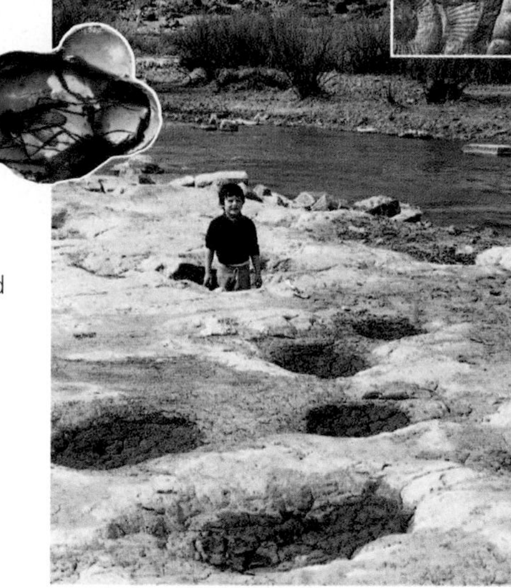

(e) Boy standing in a 150-million-year-old dinosaur track in Colorado

▲ **Figure 25.4 A gallery of fossil types. (a)** An organism's mineral-containing hard parts, such as bones, shells, or teeth, are most likely to be preserved as fossils. **(b)** Some fossils form when minerals seep into and replace organic matter. **(c)** Some sedimentary fossils retain organic material from which scientists can extract molecules for analysis. **(d)** Buried organisms that decay may leave impressions that are filled by water containing dissolved minerals. The casts that form when the minerals harden are replicas of the organisms. **(e)** Trace fossils are footprints, burrows, and other remnants of an ancient organism's activities. **(f)** Entire organisms are sometimes found preserved in amber (hardened resin from a tree). **(g)** Rarely, ice or an acid bog preserves the body of a very large organism.

are tall, twiggy trees, and others are dense, ground-hugging shrubs (see Figure 24.12). But in spite of these striking phenotypic differences, the silverswords' genes are very similar. Based on these small molecular divergences, scientists estimate that the silversword group began to diverge 5 million years ago, which is also about the time when the oldest of the current islands formed. We can infer that the morphological diversity of the silverswords is controlled by relatively few genetic differences.

Sorting Homology from Analogy

A potential red herring in constructing a phylogeny is similarity due to convergent evolution—called **analogy**—rather than to shared ancestry (homology). As you read in Chapter 22, convergent evolution occurs when similar environmental pressures and natural selection produce similar (analogous) adaptations in organisms from different evolutionary lineages (see Figure 22.17). For example, Australian and North American burrowing moles are very similar in appearance **(Figure 25.5)**. However, their reproductive systems are very dissimilar: Australian moles are marsupials (their young complete their embryonic development in a pouch outside the mother's body), whereas North American moles are eutherians (their young complete their embryonic development in a uterus within the mother's body). Indeed, genetic comparisons and the fossil record provide evidence that the moles' common ancestor lived 120 million years ago, about the time the marsupial and eutherian mammals diverged. This ancestor and most of its descendants were not mole-like, but similar characteristics evolved independently in these two mole lineages as they adapted to similar lifestyles.

Distinguishing between homology and analogy is critical in reconstructing phylogenies. For example, both bats and birds have adaptations that enable them to fly. This superficial resemblance might imply that bats are more closely related to birds than they are to cats, which cannot fly. But a closer examination reveals that the complex structure of the bat's flight apparatus is far more similar to the forelimbs of cats and other mammals than to a bird's wing. Fossil evidence also documents that bat forelimbs and bird wings arose independently from walking forelimbs of different ancestors. Thus, we can state that the bat's forelimb is *homologous* to those of other mammals, but *analogous* in function to a bird's wing. Analogous structures that have evolved independently, such as bat forelimbs and bird wings, are also sometimes called **homoplasies** (from the Greek for "to mold in the same way").

Besides searching for corroborating similarities or fossil evidence, another clue to distinguishing between homology and analogy is to consider the complexity of the characters being compared. The more points of resemblance that two complex structures have, the less likely it is that they evolved independently. For instance, the skulls of a human and a chimpanzee do not consist of a single bone but rather of many bones fused together. The compositions of the two skulls match almost perfectly, bone for bone. It is highly improbable that such complex structures, matching in so many details, have separate origins. More likely, the genes involved in the development of both skulls were inherited from a common ancestor. The same argument applies to comparing genes themselves, which are sequences of thousands of nucleotides. Each nucleotide position along a stretch of DNA or RNA represents an inherited character in the form of one of the four DNA bases: A (adenine), G (guanine), C (cytosine), or T (thymine). Thus, comparable regions of DNA that are 1,000 nucleotides long provide 1,000 points of similarity or difference between two species. Systematists compare long stretches of DNA and even entire genomes to assess relationships between species. If genes in two organisms share many portions of their nucleotide sequences, it is highly likely that the genes are homologous.

▲ **Figure 25.5 Convergent evolution of analogous burrowing characteristics.** An elongated body, enlarged front paws, small eyes, and a pad of thickened skin that protects a tapered nose all evolved independently in the marsupial Australian mole (top) and eutherian North American mole (bottom).

Evaluating Molecular Homologies

Molecular comparisons of nucleic acids often pose technical challenges. The first step is to align comparable nucleic acid sequences from the two species being studied. If the species are very closely related, the sequences likely differ at only one or a few sites. In contrast, comparable nucleic acid sequences in distantly related species usually have different bases at many sites and may even have different lengths. This is because, over longer periods of time, insertions and deletions accumulate, altering the lengths of the gene sequences (see Chapter 23). Suppose, for example, that certain noncoding DNA sequences in a particular gene in two species are very similar, but a deletion mutation has eliminated the first base of the sequence in one of the species. The effect is that the remaining sequence shifts back one notch. A comparison of the two sequences that does not take this deletion into account would overlook what in fact is a very good match. To avoid this problem, systematists use computer programs to analyze comparable DNA segments of differing lengths and realign them appropriately **(Figure 25.6)**.

Such molecular comparisons reveal that a large number of base substitutions and other differences have accumulated between the comparable genes of the Australian and North American moles, indicating that their lineages have diverged greatly since their common ancestor; thus, we say that the living species are not closely related. In contrast, the high degree of gene-sequence similarity among the silverswords supports the hypothesis that they are all very closely related, in spite of their considerable morphological differences.

The fact that molecules have diverged between species does not in itself tell us how long ago their common ancestor lived.

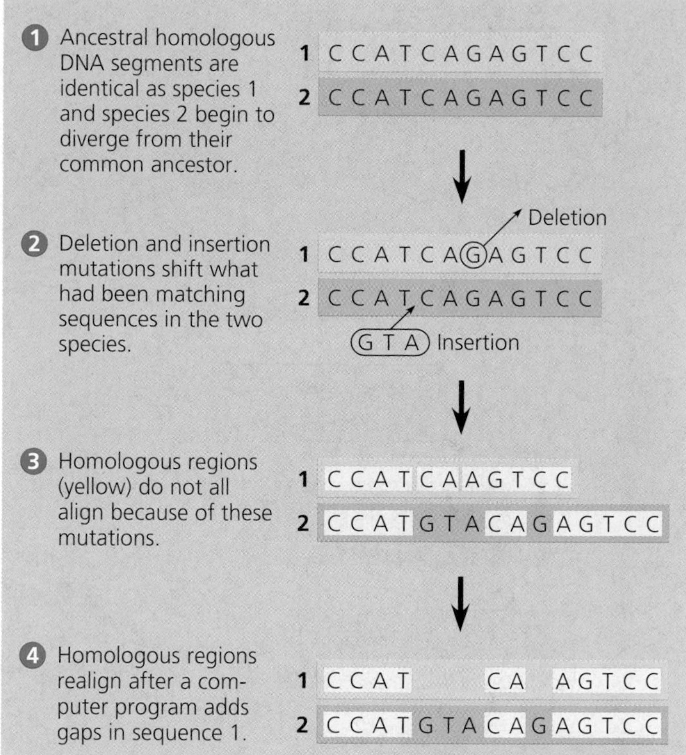

1 Ancestral homologous DNA segments are identical as species 1 and species 2 begin to diverge from their common ancestor.
1 CCATCAGAGTCC
2 CCATCAGAGTCC
2 Deletion and insertion mutations shift what had been matching sequences in the two species.
Deletion
1 CCATCAGAGTCC
2 CCATCAGAGTCC
GTA Insertion
3 Homologous regions (yellow) do not all align because of these mutations.
1 CCATCAAGTCC
2 CCATGTACAGAGTCC
4 Homologous regions realign after a computer program adds gaps in sequence 1.
1 CCAT CA AGTCC
2 CCATGTACAGAGTCC

▲ **Figure 25.6 Aligning segments of DNA.** Systematists use computer software to find and realign similar sequences along DNA segments from two species. (In this example, no bases have changed and the comparable sequences are still identical once the length is adjusted.)

Sometimes, as in the case of the moles, the fossil record provides data about when their common ancestor probably lived. But in the case of the silverswords, few fossils have been found. For such species, researchers may be able to compare their molecular divergence with that found in other plant lineages that have more complete fossil records. These values can serve as a sort of molecular yardstick to measure the approximate time span of various degrees of divergence. (This is how researchers calculated that the silverswords' common ancestor lived approximately 5 million years ago, as we discussed earlier.) Just as with morphological characters, it is necessary to distinguish homology from analogy to determine the usefulness of molecular similarities for evolutionary studies. Two sequences that resemble each other at many points along their length are most likely homologous (see Figure 25.6). But in organisms that do not appear to be closely related, the bases that their otherwise very different sequences happen to share may simply be coincidental matches, or molecular homoplasies **(Figure 25.7)**. Scientists have developed mathematical tools that can distinguish "distant" homologies from such coincidental matches in extremely divergent sequences. For example, such molecular analysis has provided evidence that, despite our lack of morphological similarity, we humans do indeed share a distant common ancestor with bacteria.

ACGGATAGTCCACTAGGCACTA

TCACCGACAGGTCTTTGACTAG

▲ **Figure 25.7 A molecular homoplasy.** These two DNA sequences from organisms that are not closely related coincidentally share 25% of their bases. Many tools have been developed to determine whether DNA sequences that share higher proportions of bases do so because they are homologous.

Scientists have so far sequenced more than 20 billion bases' worth of nucleic acid data from thousands of species. This enormous collection of data has fed a boom in the study of phylogeny, clarifying many evolutionary relationships, such as those between the Australian and North American moles and those between the various silverswords. In the rest of this chapter and the next unit, you will see many examples of the tremendous impact of molecular systematics.

Concept Check 25.1

1. Suggest whether each of the following pairs of structures more likely represents analogy or homology, and explain your reasoning: (a) a porcupine's quills and a cactus's spines; (b) a cat's paw and a human's hand; (c) an owl's wing and a hornet's wing.
2. Which of the following are more likely to be closely related: two species with similar appearances but very divergent gene sequences, or two species with very different appearances but nearly identical genes? Explain.

For suggested answers, see Appendix A.

Concept 25.2

Phylogenetic systematics connects classification with evolutionary history

The discipline of systematics dates to the 18th century. In 1748, Swedish botanist and anatomist Carolus Linnaeus published *Systema naturae* ("System of Nature"), his taxonomic classification of all plants and animals known at the time. **Taxonomy** is an ordered division of organisms into categories based on a set of characteristics used to assess similarities and differences. Although Linneaus' classification was not based on evolutionary relationships but simply on resemblances, many features of his system remain useful in phylogenetic systematics. Two of these are binomial designations for species and hierarchical classification.

Binomial Nomenclature

Common names for organisms—such as monkey, finch, and lilac—convey meaning in casual usage, but they can also cause confusion. Each of these names, for example, refers to more than one species. Moreover, some common names do not accurately reflect the type of organism. Consider these three "fishes": jellyfish (a cnidarian), crayfish (a small lobster-like crustacean), and silverfish (an insect). And of course, different languages have different words for various organisms.

To avoid ambiguity when communicating about organisms, biologists refer to each species by a scientific name. The two-part format of the scientific name, called a **binomial,** was instituted by Linnaeus. The first part of a species' name is the **genus** (plural, *genera*) to which the species belongs. The second part is the **specific epithet,** which is unique for each species within the genus. An example of a binomial is *Panthera pardus,* the scientific name for the large cat commonly called the leopard. Notice that the first letter of the genus is capitalized and the entire binomial is italicized. (Scientific names are also "latinized"; you can name an insect you discover after a friend, but you must add an appropriate Latin ending.) Many of the more than 11,000 binomials assigned by Linnaeus are still used today, including the optimistic name he designated for our own species—*Homo sapiens,* meaning "wise man."

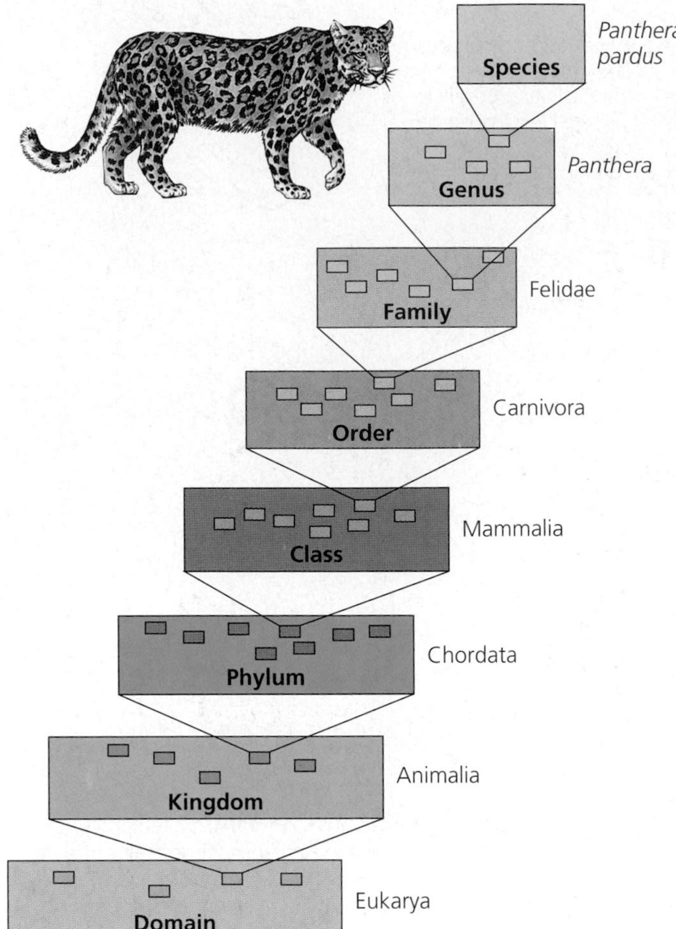

▲ **Figure 25.8 Hierarchical classification.** Species are placed into groups belonging to more comprehensive groups.

Hierarchical Classification

In addition to naming species, Linnaeus also grouped them into a hierarchy of increasingly broad categories. The first grouping is built into the binomial; species that appear to be closely related are grouped into the same genus. For example, the leopard (*Panthera pardus*) belongs to a genus that also includes the African lion (*Panthera leo*), the tiger (*Panthera tigris*), and the jaguar (*Panthera onca*). Beyond genera, systematists employ progressively more comprehensive categories of classification **(Figure 25.8)**. They place related genera in the same **family,** group families into **orders,** orders into **classes,** classes into **phyla** (singular, *phylum*), phyla into **kingdoms,** and, more recently, kingdoms into **domains.** The named taxonomic unit at any level is called a **taxon** (plural, *taxa*). For example, *Panthera* is a taxon at the genus level, and Mammalia is a taxon at the class level that includes all the many orders of mammals. Note that taxa broader than the genus level are not italicized, though they are capitalized. The resulting biological classification of a particular organism is somewhat like a postal address identifying a person in a particular apartment, in a building with many apartments, on a street with many apartment buildings, in a city with many streets, in a state with many cities, and so on.

Classifying species seems to come naturally to humans—it is a way to structure our view of the world. We lump together several species of tree to which we give the common name of oaks and distinguish them from other species of trees we call chestnuts. Taxonomists have decided that oaks and chestnuts are different enough that they should belong to separate genera. Yet oaks and chestnuts are deemed similar enough to be grouped into the same family, the Fagaceae. This decision was ultimately arbitrary, because higher classification levels are generally defined by various morphological characters chosen by taxonomists rather than by some quantitative measurement applicable to all organisms. For this reason, the larger categories are often not comparable between lineages; that is, an order of snails does not necessarily exhibit the same degree of morphological or genetic diversity as an order of mammals.

Linking Classification and Phylogeny

We saw earlier how systematists explore phylogeny by examining various characteristics in living and fossil organisms. They use branching diagrams called **phylogenetic trees** to depict their hypotheses about evolutionary relationships. The

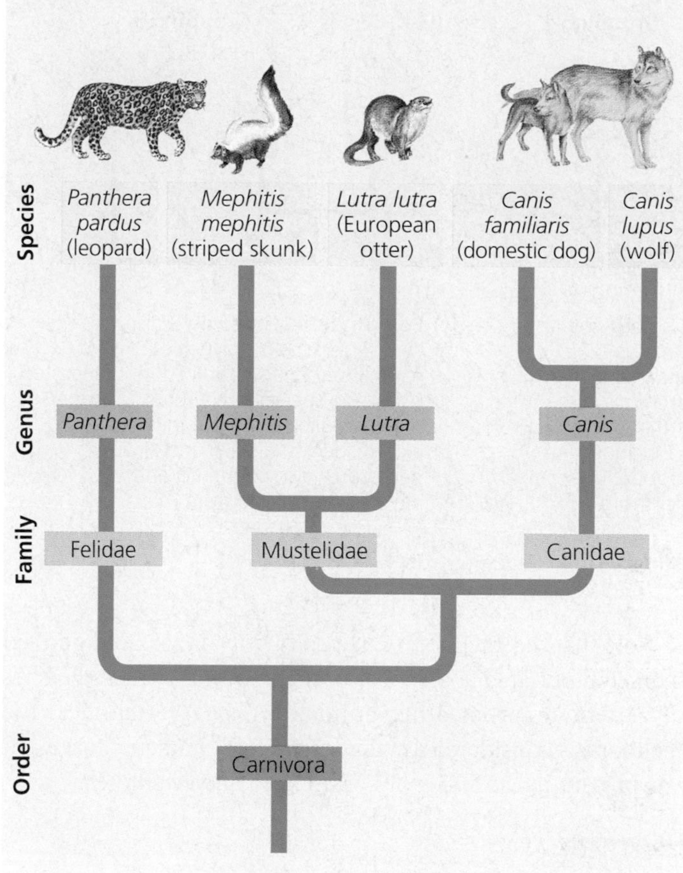

▲ **Figure 25.9 The connection between classification and phylogeny.** Hierarchical classification is reflected in the progressively finer branching of phylogenetic trees. This tree traces possible evolutionary relationships between some of the taxa within the order Carnivora, itself a branch of the class Mammalia.

branching of such trees reflects the hierarchical classification of groups nested within more inclusive groups **(Figure 25.9)**. A phylogenetic tree is often constructed from a series of dichotomies, or two-way branch points; each branch point represents the divergence of two species from a common ancestor. For example, we could represent a branch point within the cat family this way:

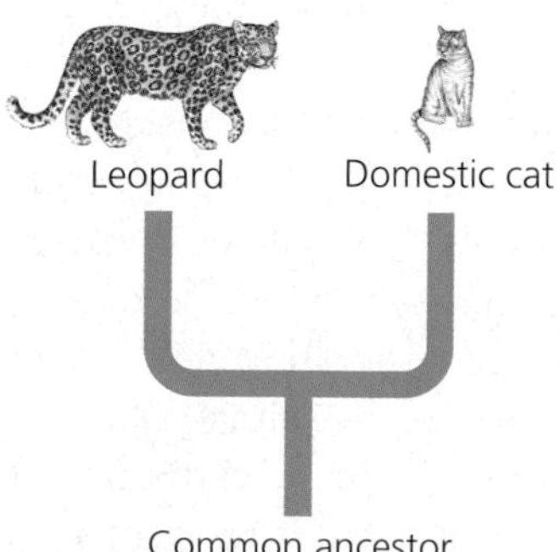

As in Figure 25.9, we can also diagram dichotomous branching of taxa that are more inclusive than species, such as families and orders. Each of the "deeper" branch points represents progressively greater amounts of divergence:

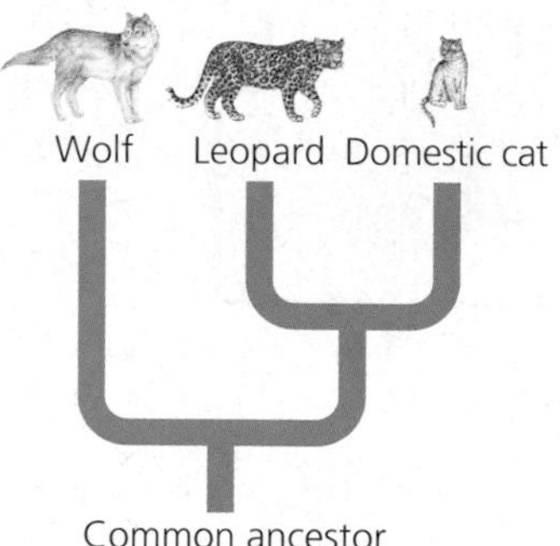

Do not confuse the sequence of branching in a tree with the vintages (actual ages) of the particular species. The tree in Figure 25.9 does *not* indicate that the wolf evolved more recently than the European otter, but only that their common ancestor preceded the last common ancestor of the wolf and the domestic dog.

Methods for tracing phylogeny began with Darwin, who, unlike Linnaeus, realized the evolutionary implications of hierarchical classification. Darwin introduced phylogenetic systematics in *The Origin of Species,* writing: "Our classifications will come to be, as far as they can be so made, genealogies."

Concept Check 25.2

1. Which levels of the classification in Figure 25.8 do humans share with the leopard?
2. What does the phylogenetic tree in Figure 25.9 indicate about the evolutionary relationships of the leopard, striped skunk, and wolf?

For suggested answers, see Appendix A.

Concept 25.3

Phylogenetic systematics informs the construction of phylogenetic trees based on shared characters

Patterns of shared characteristics can be depicted in a diagram called a **cladogram** (see Figure 25.11b on p. 499). A cladogram by itself does not imply evolutionary history. But if the shared characteristics are due to common ancestry (that is, if they are homologous), then the cladogram forms the basis of a phylogenetic tree. Within the tree, a **clade** (from the Greek *klados,* branch) is defined as a group of species that includes an ancestral species and all its descendants. The analysis of how species may be grouped into clades is called **cladistics.**

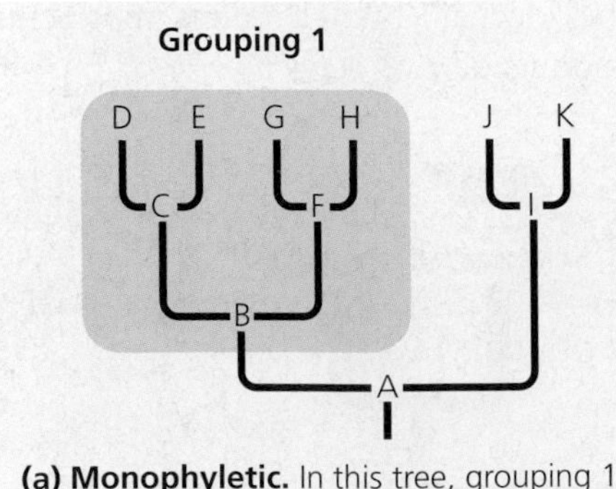

(a) Monophyletic. In this tree, grouping 1, consisting of the seven species B–H, is a monophyletic group, or clade. A monophyletic group is made up of an ancestral species (species B in this case) and *all* of its descendant species. Only monophyletic groups qualify as legitimate taxa derived from cladistics.

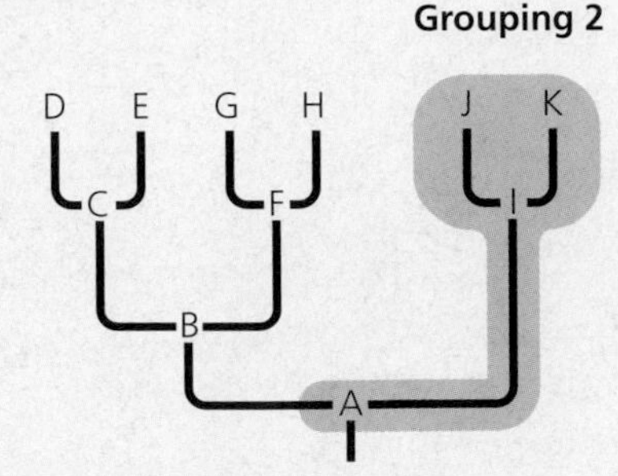

(b) Paraphyletic. Grouping 2 does not meet the cladistic criterion: It is paraphyletic, which means that it consists of an ancestor (A in this case) and *some*, but not all, of that ancestor's descendants. (Grouping 2 includes the descendants I, J, and K, but excludes B–H, which also descended from A.)

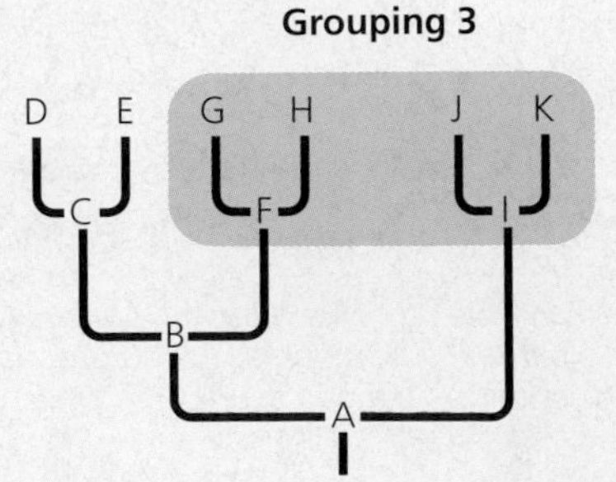

(c) Polyphyletic. Grouping 3 also fails the cladistic test. It is polyphyletic, which means that it lacks the common ancestor (A) of the species in the group. Furthermore, a valid taxon that includes the extant species G, H, J, and K would necessarily also contain D and E, which are also descended from A.

▲ **Figure 25.10 Monophyletic, paraphyletic, and polyphyletic groupings.**

Cladistics

Clades, like taxonomic ranks, can be nested within larger clades. For example, the cat family represents a clade within a larger clade that also includes the dog family. But not all groupings of organisms qualify as clades. A valid clade is **monophyletic** (meaning "single tribe"), signifying that it consists of the ancestral species and all its descendants **(Figure 25.10a)**. When we lack information about some members of a clade, the result is a **paraphyletic** grouping that consists of an ancestral species and some, but not all, of the descendants **(Figure 25.10b)**. Or the result may be a **polyphyletic** grouping of several species that lack a common ancestor **(Figure 25.10c)**. Such situations call for further reconstruction to uncover species that tie together these groupings into monophyletic clades.

Shared Primitive and Shared Derived Characters

After systematists have separated homologous from analogous similarities, they must sort through the homologies to distinguish between shared primitive and shared derived characters. "Character" here refers to any feature that a particular taxon possesses. The characters that are relevant to phylogeny, of course, are the homologous ones. For example, all mammals share the homologous character of a backbone. However, the presence of a backbone does not distinguish mammals from other vertebrates because nonmammalian vertebrates such as fishes and reptiles also have backbones. The backbone is a homologous structure that predates the branching of the mammalian clade from the other vertebrates; it is a **shared primitive character**, a character that is shared beyond the taxon we are trying to define. In contrast, hair, a character shared by all mammals but not found in nonmammalian vertebrates, is a **shared derived character**, an evolutionary novelty unique to a particular clade—in this case, the mammalian clade.

Note that the backbone can also qualify as a shared derived character, but only at a deeper branch point that distinguishes *all* vertebrates from other animals. *Among* vertebrates, the backbone is considered a shared primitive character because it was present in the ancestor common to all vertebrates.

Outgroups

Systematists use outgroup comparison to differentiate between shared derived characters and shared primitive characters. To demonstrate this method, let's arrange five vertebrates—a leopard, turtle, salamander, tuna, and lamprey (a jawless aquatic vertebrate)—into a cladogram. As a basis of comparison, we need to designate an **outgroup**, which is a species or group of species that is closely related to the **ingroup**, the various species we are studying; the outgroup is known to be less closely related than any of the ingroup members are to each other based on other evidence (such as paleontology, embryonic development, and gene sequences). A good choice of an outgroup for our example is a lancelet, a small animal that lives in mudflats and (like vertebrates) is a member of the phylum Chordata, but does not have a backbone. We can begin building our cladogram by comparing the ingroup with the outgroup.

Outgroup comparison is based on the assumption that homologies present in both the outgroup and ingroup must be primitive characters that predate the divergence of both groups from a common ancestor. In our study case, an example of such a character is a structure called a notochord, a flexible rod running the length of the animal. Lancelets have notochords throughout their lives, but in vertebrates the notochord is present only in embryos and is replaced later in development by the backbone. The species making up the ingroup display a mixture of shared primitive and shared derived characters. The outgroup comparison enables us to focus on just those characters that were derived at the various branch points of

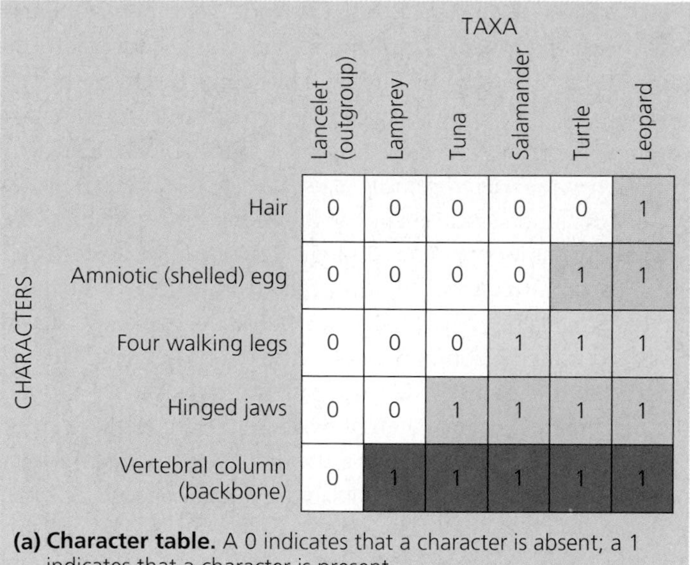

CHARACTERS \ TAXA	Lancelet (outgroup)	Lamprey	Tuna	Salamander	Turtle	Leopard
Hair	0	0	0	0	0	1
Amniotic (shelled) egg	0	0	0	0	1	1
Four walking legs	0	0	0	1	1	1
Hinged jaws	0	0	1	1	1	1
Vertebral column (backbone)	0	1	1	1	1	1

(a) Character table. A 0 indicates that a character is absent; a 1 indicates that a character is present.

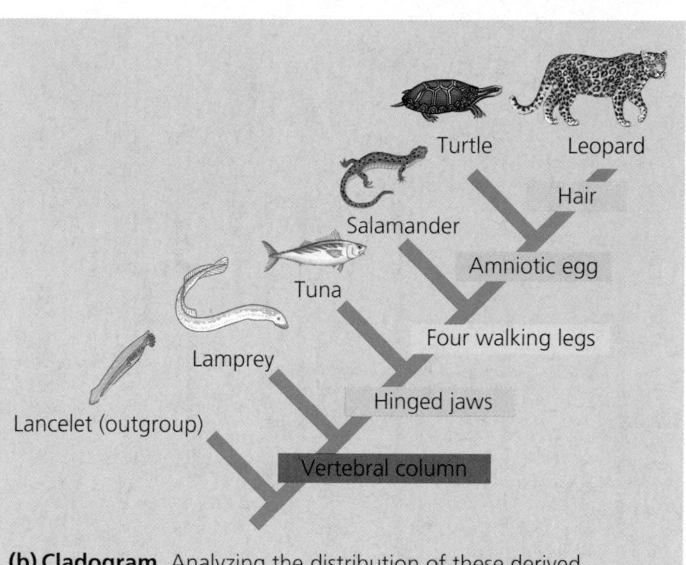

(b) Cladogram. Analyzing the distribution of these derived characters can provide insight into vertebrate phylogeny.

▲ **Figure 25.11 Constructing a cladogram.**

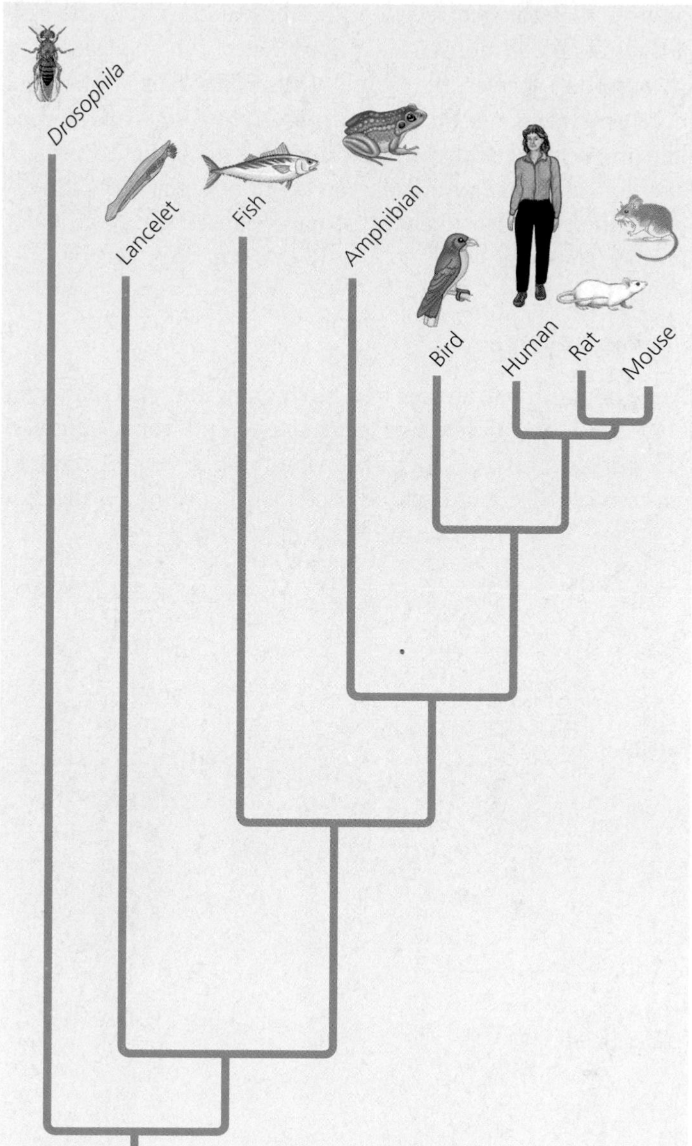

▲ **Figure 25.12 Phylogram.** This phylogram was constructed by comparing homologous *hedgehog* genes, employing the *Drosophila* gene as an outgroup. The *hedgehog* gene is important in development. The varying lengths of the branches indicate that the gene has evolved at slightly different rates in the different lineages.

vertebrate evolution. **Figure 25.11a** tabulates examples of these characters. Note that *all* the vertebrates in the ingroup have backbones; this is a shared primitive character that was present in the ancestral vertebrate, though not in the outgroup. Now note that hinged jaws are a character absent in lampreys but present in other members of the ingroup; this character helps us identify an early branch point in the vertebrate clade. **Figure 25.11b** illustrates how the data in our table of homologies can be translated into a cladogram.

Note that the cladogram in Figure 25.11b is *not* a phylogenetic tree. We may suspect that it tells an evolutionary story, but to convert it to a phylogenetic tree we would need more information—for example, from fossils, which can indicate when, and in which groups, the characters first appeared.

Phylogenetic Trees and Timing

Any chronology represented by the branching pattern of a phylogenetic tree is relative (earlier versus later) rather than absolute (so many millions of years ago). But some kinds of tree diagrams can be used to present more specific information about timing: Phylograms present information about the sequence of events relative to one another, and ultrametric trees present information about the actual time that given events occurred.

Phylograms

In a **phylogram**, the length of a branch reflects the number of changes that have taken place in a particular DNA sequence in that lineage **(Figure 25.12)**. Note that in the phylogram in

Figure 25.12, the *total length* of the vertical lines from the base of the tree to the mouse is less than that of the line leading to the outgroup species, the fruit fly *Drosophila*. This implies that more genetic changes have occurred in the *Drosophila* lineage than in the bird and mammal lineages since they diverged. Later in this chapter you will read about how scientists can estimate the time that likely must have elapsed for a particular number of changes in a DNA or RNA sequence to occur.

Ultrametric Trees

Even though the branches in a phylogram may have different lengths, all the different lineages that descend from a common ancestor have survived for the same number of years. To take an extreme example, humans and bacteria had a common ancestor that lived over 3 billion years ago. Evidence indicates that this ancestor was a single-celled prokaryote, and thus it must have been more like a modern bacterium than like a human. Even though bacteria have apparently changed little in their structure since that common ancestor, there have nonetheless been 3 billion years of evolution in the bacterial lineage, just as there have been 3 billion years of evolution in the eukaryote lineage that includes humans. These equal amounts of chronological time can be represented in an **ultrametric tree.** In an ultrametric tree, the branching pattern is the same as in a phylogram, but all the branches that can be traced from the common ancestor to the present are of equal length **(Figure 25.13)**. While ultrametric trees do not contain the information about different evolutionary rates that can be found in phylograms, they can draw on data from the fossil record to place certain branch points in the context of geologic time.

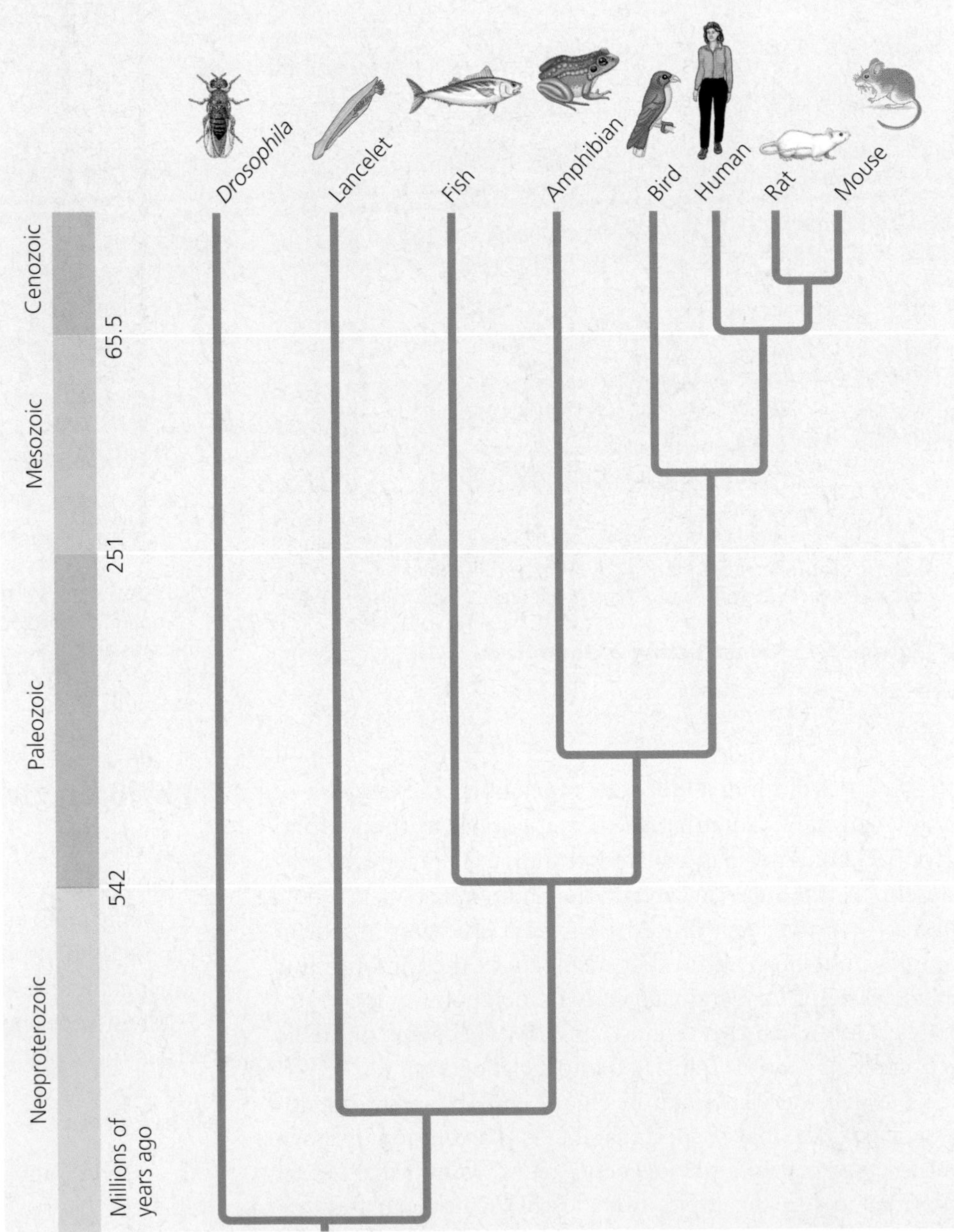

▶ **Figure 25.13 Ultrametric tree.** This ultrametric tree was constructed from the same molecular data as the phylogram in Figure 25.12. These data were then fit to known branching events in the fossil record. In this ultrametric tree, all the branches now have the same total length from the base of the tree through the various vertical segments leading to the labels at the top. This reflects the evidence that all of these lineages have diverged from the common ancestor at the base for equal amounts of time. Note also that as you move up the tree, each branch point sprouts two lineages of equal length, representing the equal times of divergence from that common ancestor. Because the root of the tree is located at a time before a substantial fossil record begins, its date is less certain.

Maximum Parsimony and Maximum Likelihood

As available data about DNA sequences increase and it becomes possible to link more and more species, the difficulty of building the phylogenetic tree that best describes evolutionary history also grows. What if you are analyzing data for 50 species? There are 3×10^{76} different ways to arrange 50 species into a tree! And which tree in this huge forest reflects the true phylogeny? Systematists can never be sure of finding the single best tree in such a large data set, but they can narrow the possibilities by applying the principles of maximum parsimony and maximum likelihood.

According to the principle of **maximum parsimony**, we should first investigate the simplest explanation that is consistent with the facts. (The parsimony principle is also called "Occam's Razor" after William of Occam, a 14th-century English philosopher who advocated this minimalist problem-solving approach of "shaving away" unnecessary complications.) In the case of trees based on morphological characters, the most parsimonious tree is the one that requires the fewest evolutionary events to have occurred in the form of shared derived characters. For phylograms based on DNA sequences, the most parsimonious tree requires the fewest base changes.

The principle of **maximum likelihood** states that, given certain rules about how DNA changes over time, a tree can be found that reflects the most likely sequence of evolutionary events. Maximum likelihood methods are complex and incorporate as much information as possible, but as a simple illustration of more likely and less likely trees, let us return to the phylogenetic relationships between a human, a mushroom, and a tulip. **Figure 25.14** shows two possible, equally parsimonious trees for this trio. In tree 1, the human is more closely related to the mushroom, whereas in tree 2, the human is more closely related to the tulip. Tree 1 is more likely if we assume that DNA changes have occurred at equal rates along all the branches of the tree from the common ancestor. Tree 2 is also possible, but it requires assuming that the rates of evolution slowed greatly in the mushroom clade and sped up greatly in the tulip clade. Thus, assuming that equal rates are more common than unequal rates, tree 1 is more likely. We will soon see that many genes do evolve at approximately equal rates in different lineages. But note that if we find new evidence of unequal rates, tree 2 might be more likely! The likelihood of a tree depends on the assumptions on which it is based.

Many computer programs have been developed to search for trees that are parsimonious and likely. These include the following approaches:

1. "Distance" methods minimize the total of all the percentage differences among all the sequences.
2. More complex "character-state" methods minimize the total number of base changes or search for the most likely pattern of base changes among all the sequences.

	Human	Mushroom	Tulip
Human	0	30%	40%
Mushroom		0	40%
Tulip			0

(a) Percentage differences between sequences

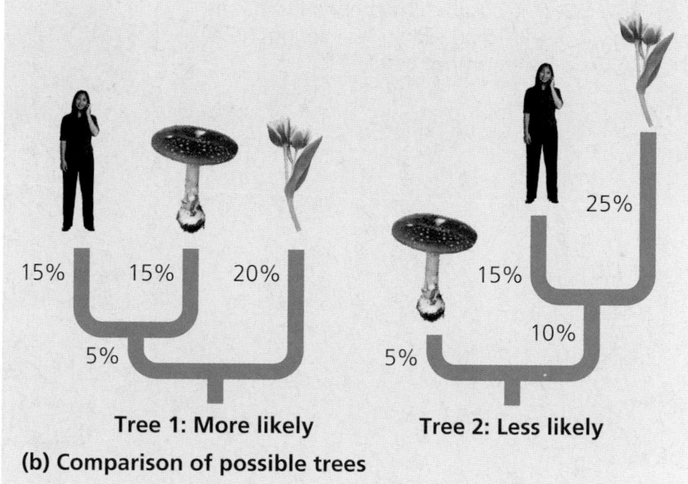

(b) Comparison of possible trees

▲ **Figure 25.14 A simple illustration of trees having different likelihoods.** Based on percentage differences between genes carried by a human, a mushroom, and a tulip **(a)**, we can construct two possible phylograms with the same total branch length **(b).** The sum of the percentages from a point of divergence in a tree equals the percentage differences as listed in (a). For example, in tree 1, the human–tulip divergence is 15% + 5% + 20% = 40%. In tree 2, this divergence also equals 40% (15% + 25%). Assuming that the genes have evolved at the same rate in the different branches, tree 1 is more likely than tree 2.

Although researchers can never be certain precisely which tree truly reflects phylogeny, if they have collected a large amount of accurate data, the various methods usually yield similar trees. As an example of one method, **Figure 25.15**, on the next two pages, walks you through the process of identifying the most parsimonious molecular tree for a four-species problem.

Phylogenetic Trees as Hypotheses

This is a good point at which to reiterate that any phylogenetic tree represents a hypothesis about how the various organisms in the tree are related to one another. The best hypothesis is the one that best fits all the available data. A phylogenetic hypothesis may be modified when new evidence compels systematists to revise their trees. Indeed, many older phylogenetic hypotheses have been changed or rejected since the introduction of molecular methods for comparing species and tracing phylogenies.

Figure 25.15

Research Method **Applying Parsimony to a Problem in Molecular Systematics**

APPLICATION In considering possible phylogenies for a group of species, systematists compare molecular data for the species. The most efficient way to study the various phylogenetic hypotheses is to begin by first considering the most parsimonious—that is, which hypothesis requires the fewest total evolutionary events (molecular changes) to have occurred.

TECHNIQUE Follow the numbered steps as we apply the principle of parsimony to a hypothetical phylogenetic problem involving four closely related bird species.

❶ First, draw the possible phylogenies for the species (only 3 of the 15 possible trees relating these four species are shown here).

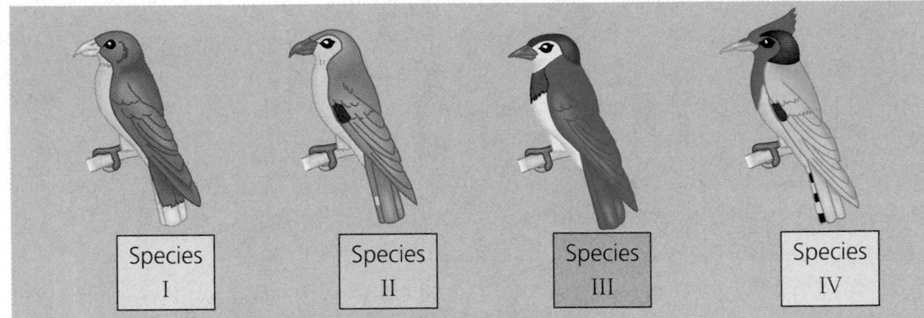

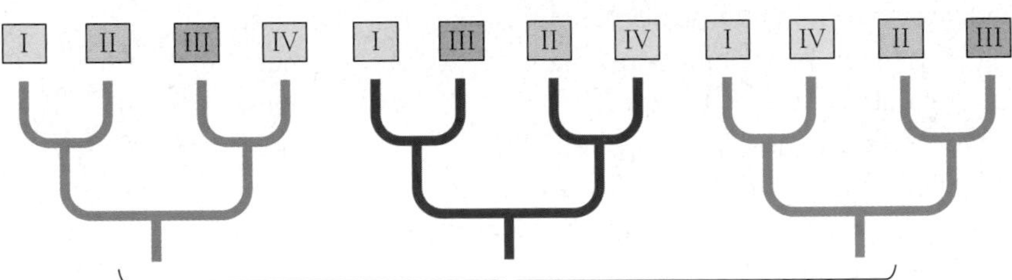

Three possible phylogenetic hypotheses

❷ Tabulate the molecular data for the species (in this simplified example, the data represent a DNA sequence consisting of just seven nucleotide bases).

Sites in DNA sequence

Species	1	2	3	4	5	6	7
I	A	G	G	G	G	G	T
II	G	G	G	A	G	G	G
III	G	A	G	G	A	A	T
IV	G	G	A	G	A	A	G

❸ Now focus on site 1 in the DNA sequence. A single base-change event, marked by the crossbar in the branch leading to species I, is sufficient to account for the site 1 data.

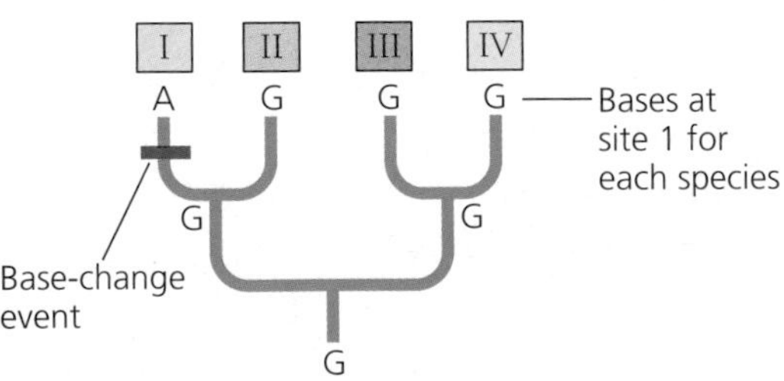

❹ Continuing the comparison of bases at sites 2, 3, and 4 reveals that each of these possible trees requires a total of four base-change events (marked again by crossbars). Thus, the first four sites in this DNA sequence do not help us identify the most parsimonious tree.

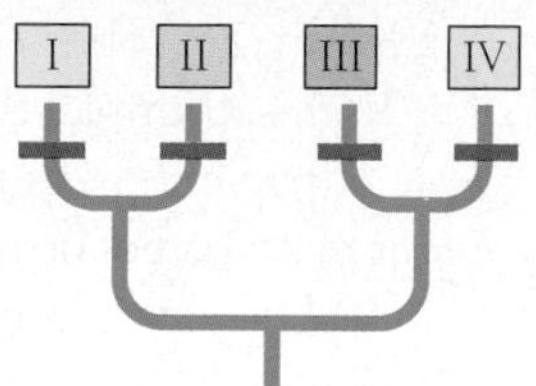

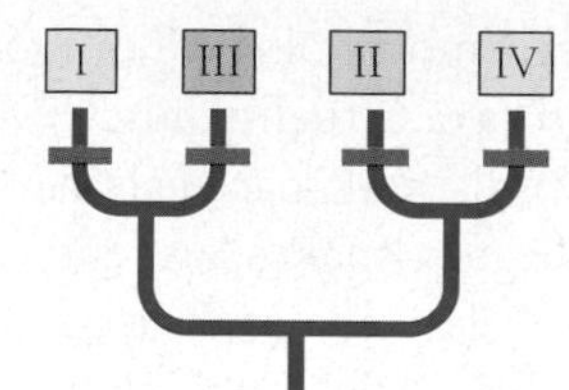

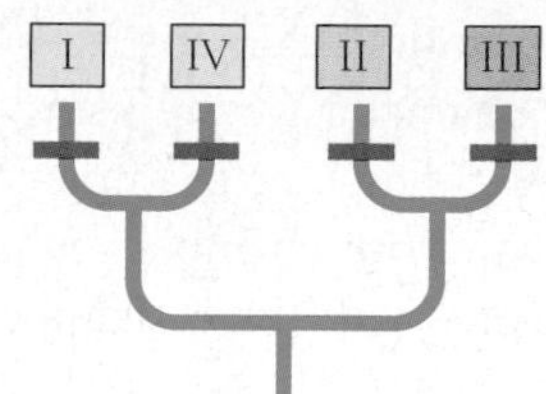

❺ After analyzing sites 5 and 6, we find that the first tree requires fewer evolutionary events than the other two trees (two base changes versus four). Note that in these diagrams, we assume that the common ancestor had GG at sites 5 and 6. But even if we started with an AA ancestor, the first tree still would require only two changes, while four changes would be required to make the other hypotheses work. Keep in mind that parsimony only considers the total number of events, not the particular nature of the events (how likely the particular base changes are to occur).

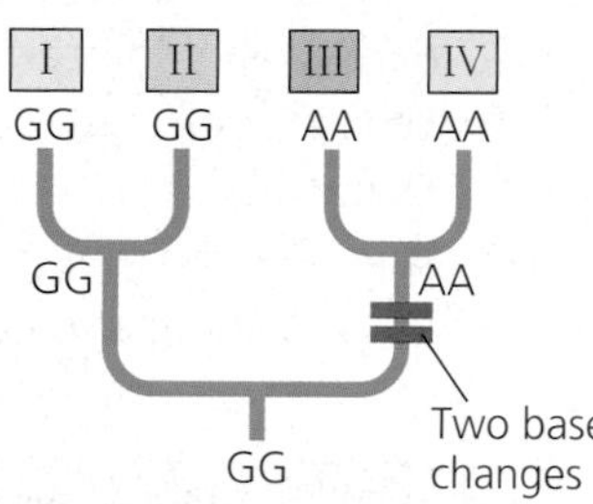

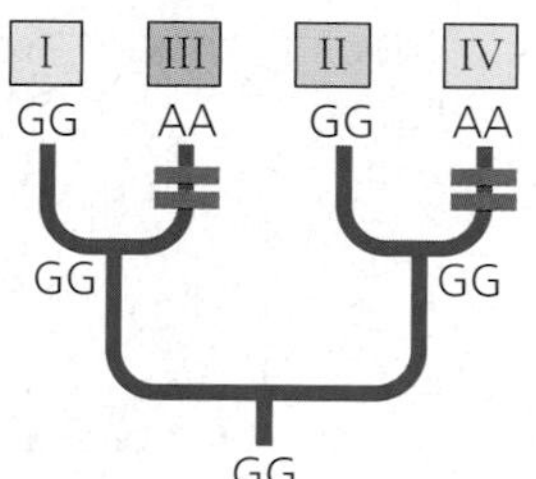

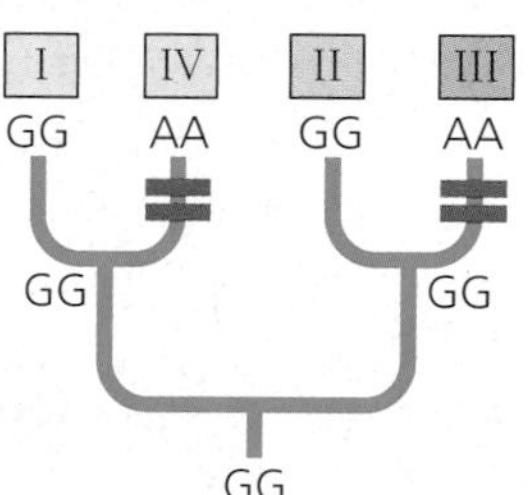

❻ At site 7, the three trees also differ in the number of evolutionary events required to explain the DNA data.

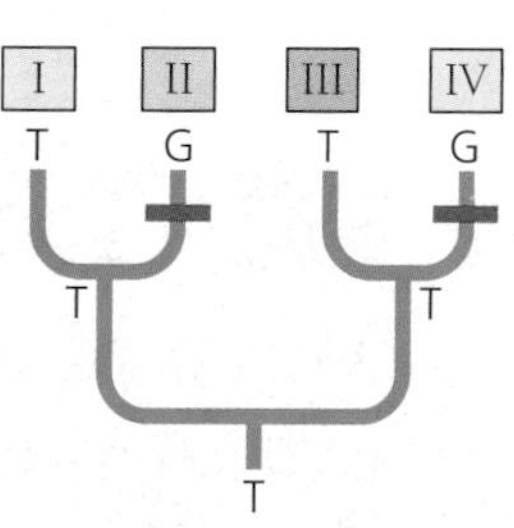

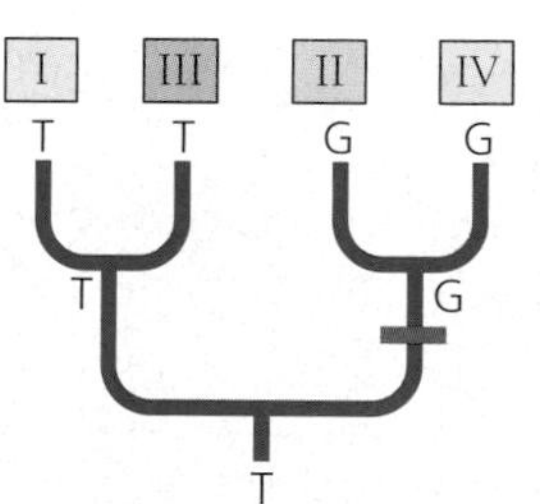

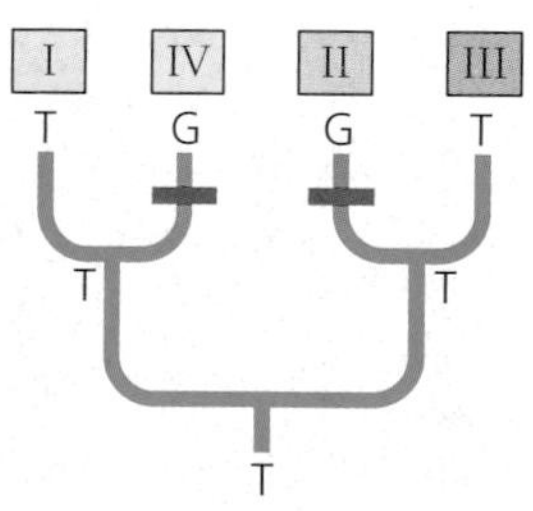

RESULTS To identify the most parsimonious tree, we total all the base-change events noted in steps 3–6 (don't forget to include the changes for site 1, on the facing page). We conclude that the first tree is the most parsimonious of these three possible phylogenies. (But now we must complete our search by investigating the 12 other possible trees.)

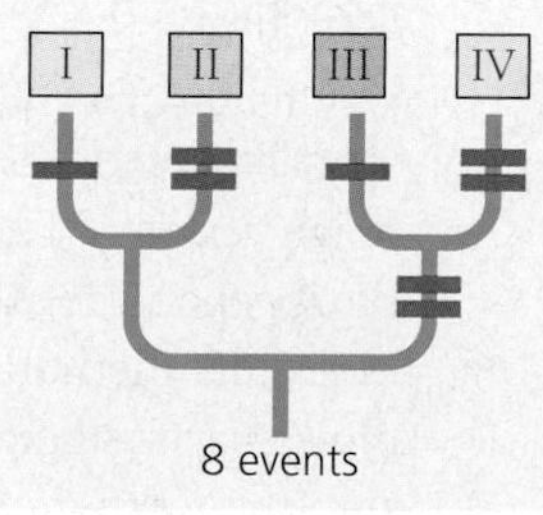

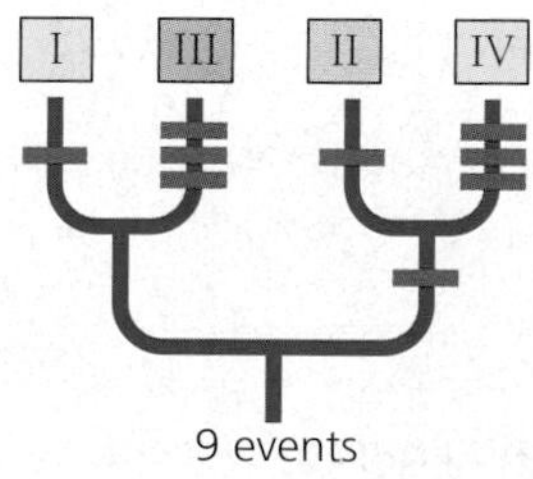

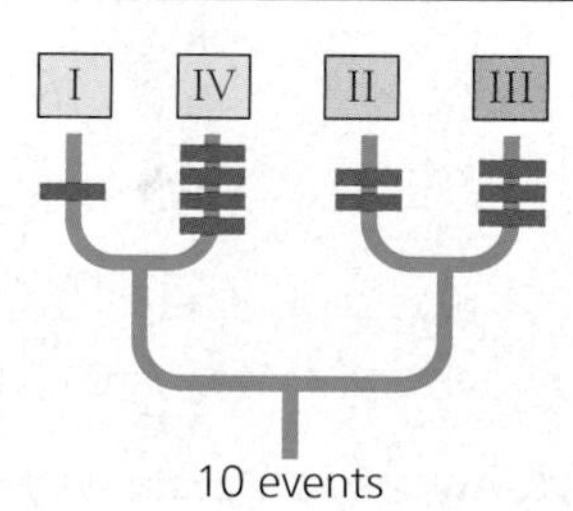

Often, in the absence of conflicting information, the most parsimonious tree is also the most likely. But sometimes there is compelling evidence that the best hypothesis is *not* the most parsimonious **(Figure 25.16)**. Nature does not always take the simplest course. Perhaps the particular morphological or molecular character we are using to sort taxa actually did evolve multiple times. For example, both birds and mammals have hearts with four chambers, whereas lizards, snakes, and turtles have hearts with three chambers (see Chapter 42). The parsimonious assumption would be that the four-chambered heart evolved once and was present in an ancestor common to birds and mammals but not to lizards, snakes, and turtles. However, abundant evidence indicates that birds are more closely related to lizards, snakes, and turtles than they are to mammals. Thus, the four-chambered heart appears to have evolved independently in birds and mammals. Indeed, studies have shown that the four-chambered hearts of birds and mammals develop differently, which supports the hypothesis that they evolved independently. The apparently parsimonious tree in Figure 25.16a is not consistent with these facts, while the tree in Figure 25.16b is consistent with the additional facts.

In this example, the problem is not so much with the principle of parsimony as it is with the analogy-homology issue. The four-chambered hearts of birds and mammals turn out to be analogous, not homologous. Matching changes of bases in DNA sequences in two species can also occur independently, but the more bases that are involved, the less likely it is that the matching changes are coincidental. Applying parsimony in molecular systematics is more reliable for a data set of many long DNA sequences than for a smaller data set. Similarly, occasionally misjudging an analogous similarity in morphology as a shared derived (homologous) character is less likely to distort a phylogenetic tree if each clade in the tree is defined by several derived characters. The strongest phylogenetic hypotheses are those supported by multiple lines of molecular and morphological evidence as well as by fossil evidence.

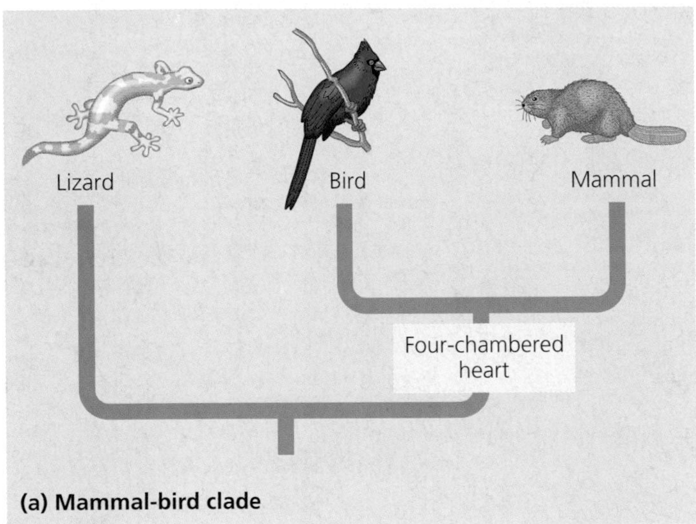

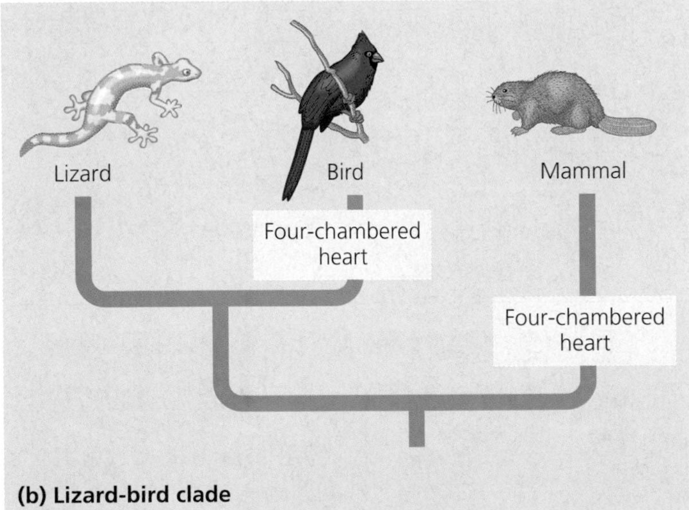

▲ **Figure 25.16 Parsimony and the analogy-versus-homology pitfall.** If we interpret the four-chambered hearts of birds and mammals as homologous instead of analogous and use no other information, the tree in (a) appears to be the more parsimonious tree. In fact, abundant evidence supports the hypothesis that birds and lizards are more closely related than birds and mammals are and that four-chambered hearts evolved more than once, supporting the tree in (b).

Concept Check 25.3

1. To distinguish a particular clade of mammals within the larger clade that corresponds to class Mammalia, would hair be a useful character? Why or why not?
2. Why might the most parsimonious tree not necessarily be the most accurate in representing evolutionary relationships among a particular group of species?

For suggested answers, see Appendix A.

Concept 25.4

Much of an organism's evolutionary history is documented in its genome

You have seen throughout this chapter that molecular systematics—comparing nucleic acids or other molecules to infer relatedness—is a valuable tool for tracing organisms' evolutionary history. The molecular approach helps us to understand phylogenetic relationships that cannot be measured by nonmolecular methods such as comparative anatomy. For example, molecular systematics helps us uncover evolutionary relationships between groups that have little common ground for morphological comparison, such as mammals and bacteria. It is possible to reconstruct phylogenies among groups of present-day bacteria and other microorganisms for which we have no fossil record at all. And molecular systematics enables scientists to compare genetic divergence within a species. Molecular biology has helped to extend systematics to evolutionary relationships far above and below the species level, ranging from the major branches of the tree of life to its finest twigs. Still, its findings are often inconclusive, as in cases where taxa diverged at nearly the same time in the distant past. The differences may be apparent, but not the order of their appearance.

The ability of molecular trees to encompass both short and long periods of time is based on the fact that different genes evolve at different rates, even in the same evolutionary lineage. For example, the DNA that codes for ribosomal RNA (rRNA) changes relatively slowly, so comparisons of DNA sequences in these genes are useful for investigating relationships between taxa that diverged hundreds of millions of years ago. Studies of rRNA sequences, for example, indicate that fungi are more closely related to animals than to green plants (see Figure 25.2). In contrast, the DNA in mitochondria (mtDNA) evolves relatively rapidly and can be used to explore recent evolutionary events. One research team has traced the relationships among Native American groups through their mtDNA sequences. The molecular findings corroborate other evidence that the Pima of Arizona, the Maya of Mexico, and the Yanomami of Venezuela are closely related, probably descending from the first of three waves of immigrants that crossed the Bering Land Bridge from Asia to the Americas about 13,000 years ago.

Gene Duplications and Gene Families

Gene duplication is one of the most important types of mutation in evolution because it increases the number of genes in the genome, providing opportunities for further evolutionary changes. The molecular phylogenies of gene duplications and the influence of these duplications on genome evolution can now be followed in detail. These phylogenies must account for repeated duplications that have resulted in gene families, which are groups of related genes within an organism's genome (see Figure 19.17). Like homologous genes in different species, these duplicated genes have a common ancestor. We distinguish these types of homologous genes by different names: orthologous genes and paralogous genes.

The term **orthologous genes** (from the Greek *orthos*, straight) refers to homologous genes that are passed in a straight line from one generation to the next but have ended up in different gene pools because of speciation **(Figure 25.17a)**. The β hemoglobin genes in humans and in mice are orthologous.

Paralogous genes (from the Greek *para*, at the side of) result from gene duplication, so they are found in more than one copy in the same genome **(Figure 25.17b)**. In Chapter 23 you encountered an example: the olfactory receptor genes, which have undergone many gene duplications in vertebrate animals. Humans and mice each have huge families of more than 1,000 of these paralogous genes.

It is possible to describe most of the genes that make up genomes as representing one of these two types of homology.

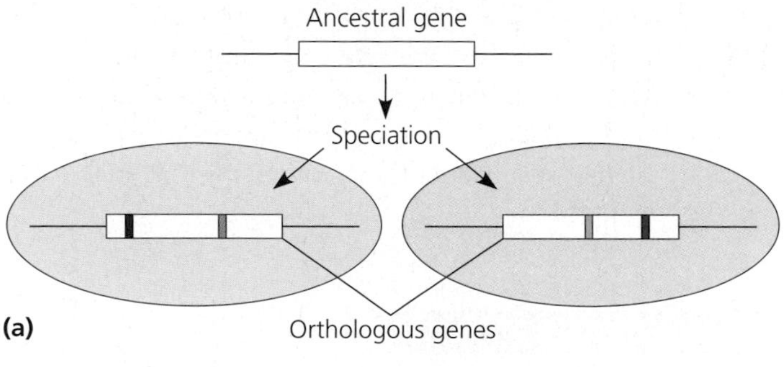

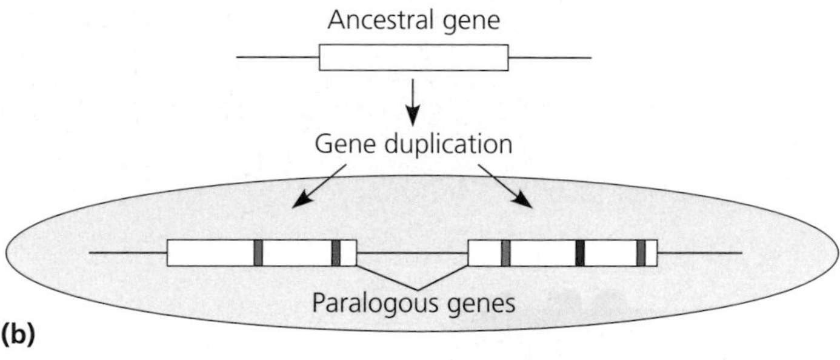

▲ **Figure 25.17 Two types of homologous genes.** The colored bands mark regions of the genes where differences in base sequences have accumulated. Orthologous genes can only diverge after speciation, whereas paralogous genes can diverge within the same evolutionary lineage.

Note that orthologous genes can only diverge after speciation has taken place, with the result that the genes are found in separate gene pools. For example, humans and mice each have one functioning β hemoglobin gene. These genes serve similar functions, but their sequences have diverged since the time that humans and mice had a common ancestor. Paralogous genes, on the other hand, can diverge even while they are in the same gene pool, because they are present in more than one copy in the genome. The paralogous genes comprising the olfactory receptor gene family in humans have diverged from each other during our long evolutionary history. They now specify proteins that confer sensitivity to a wide variety of odors, ranging from various foods to sex pheromones.

Genome Evolution

Now that we can compare the entire genomes of different organisms, including our own, two remarkable facts have emerged. First, orthologous genes are widespread and can extend over huge evolutionary distances. Ninety-nine percent of the genes of humans and mice are detectably orthologous, and 50% of our genes are orthologous with those of yeast. This remarkable commonality demonstrates that all living organisms share many biochemical and developmental pathways.

And second, the number of genes seems not to have increased through duplication at the same rate as phenotypic complexity. Humans have only about five times as many genes as yeast, a simple single-celled eukaryote, even though—unlike yeast cells—we have a large, complex brain and a body that contains more than 200 different types of tissues. Evidence is emerging that many human genes are more versatile than those of yeast and are able to carry out a wide variety of tasks in various body tissues. Before us lies a huge and exciting scientific challenge: unraveling the mechanisms that enable this genomic versatility.

Concept Check 25.4

1. Explain how comparisons between the proteins of two species can yield data about their evolutionary relationship.
2. Contrast orthologous genes with paralogous genes.

For suggested answers, see Appendix A.

Concept 25.5

Molecular clocks help track evolutionary time

As we stated at the beginning of this chapter, one of the goals of evolutionary biology is to understand the relationships between all living organisms, including those for which there are no fossils. When we extend molecular phylogenies beyond the fossil record, however, we must rely on an important assumption about how change occurs at the molecular level.

Molecular Clocks

Recall our statement earlier that researchers have estimated that the common ancestor of Hawaiian silverswords lived about 5 million years ago. How did they make this estimate? It relies on the concept of a **molecular clock,** a yardstick for measuring the absolute time of evolutionary change based on the observation that some genes and other regions of genomes appear to evolve at constant rates. The assumption underlying the molecular clock is that the number of nucleotide substitutions in orthologous genes is proportional to the time that has elapsed since the species branched from their common ancestor. In the case of paralogous genes, the number of substitutions is proportional to the time since the genes became duplicated.

We can calibrate the molecular clock of a gene that has a reliable average rate of evolution by graphing the number of nucleotide differences against the times of a series of evolutionary branch points that are known from the fossil record. The graph line representing the evolution rate of this molecular clock can then be used to estimate the date of evolutionary episodes that cannot be discerned from the fossil record, such as the origin of the silverswords.

Of course, no genes mark time with complete accuracy. In fact, some portions of the genome appear to have evolved in irregular fits and starts that are not at all clocklike. And even those genes that seem to have reliable molecular clocks are accurate only in the statistical sense of showing a fairly smooth *average* rate of change; over time, there may still be chance deviations above and below that average rate. Finally, even among genes that are clocklike, the rate of the clock may vary greatly from one gene to another; some genes evolve a million times faster than others.

Neutral Theory

The regularity of change that enables us to use some genes as molecular clocks raises the possibility that many of the changes in these sequences result from genetic drift, and the changes are mostly neutral—neither adaptive nor detrimental. In the 1960s, Jack King and Thomas Jukes, at the University of California, Berkeley, and Matoo Kimura, of the Japanese National Institute of Genetics, published papers supporting this **neutral theory**—that is, that much evolutionary change in genes and proteins has no effect on fitness and therefore is not influenced by Darwinian selection. Kimura pointed out that many new mutations are harmful and are removed quickly. But if most of the rest are neutral and have little or no effect on fitness, then the rate of molecular change should indeed be regular like a clock. Differences in the rate of the clock in different genes are a function of how important each gene is. If the exact sequence of amino acids a gene specifies is essential to survival, most of the mutational changes will be harmful and only a few will be neutral. As a result, such genes change only slowly. But if the exact sequence of amino acids is less critical, fewer of the new mutations will be harmful and more will be neutral. Such genes change more quickly.

Difficulties with Molecular Clocks

In fact, the molecular clock does not run as smoothly as neutral theory predicts. Many irregularities are likely to be the result of natural selection in which some DNA changes are favored over others. Consequently, some scientists question the utility of molecular clocks for timing evolution. Their skepticism is part of a broader debate about the extent to which neutral genetic variation can account for DNA diversity. Indeed, new evidence suggests that almost half of the amino acid differences in proteins of two *Drosophila* species, *D. simulans* and *D. yakuba,* are not neutral but have resulted from directional natural selection. Nonetheless, it is likely that over extremely long periods of time, fluctuations in evolutionary rate due to selective pressure will average out, so that even genes with irregular clocks can serve as approximate markers of elapsed time.

Another question arises when researchers attempt to extend molecular clocks beyond the time span documented by the fossil record. Although some fossils date back as far as 3 billion years, these are very rare. An abundant fossil record extends back only about 550 million years, but molecular clocks have been used to date evolutionary divergences that occurred a billion or more years ago. These estimates assume that the clocks have been constant for all that time. Thus, such estimates are likely to have a high degree of uncertainty.

Applying a Molecular Clock: The Origin of HIV

Recently, researchers at Los Alamos National Laboratory in New Mexico used a molecular clock to date the origin of HIV infection in humans. Phylogenetic analysis shows that HIV, the virus that causes AIDS, is descended from viruses that infect chimpanzees and other primates. (The viruses do not cause any AIDS-like diseases in nonhuman hosts.) When did HIV jump to humans? There is no simple answer, because the virus has spread to humans more than once. The multiple origins of HIV are reflected in the variety of strains (genetic types) of the virus. HIV's genetic material is made of RNA, and like other RNA viruses, it evolves quickly.

The most widespread strain in humans is HIV-1 M. To pinpoint the earliest HIV-1 M infection, the researchers compared samples of the virus from various times during the epidemic, including one partial viral sequence from 1959. The samples showed that the virus has evolved in a remarkably clocklike fashion since 1959. By extrapolating from their molecular clock, the researchers concluded that the first HIV-1 M invasion into humans took place during the 1930s.

The Universal Tree of Life

In the 1960s, researchers deciphering the genetic code found that it is universal in all forms of life. They inferred that all present-day organisms must therefore have a common ancestor. Today, researchers are applying systematics to link all organisms in a tree of life **(Figure 25.18)**.

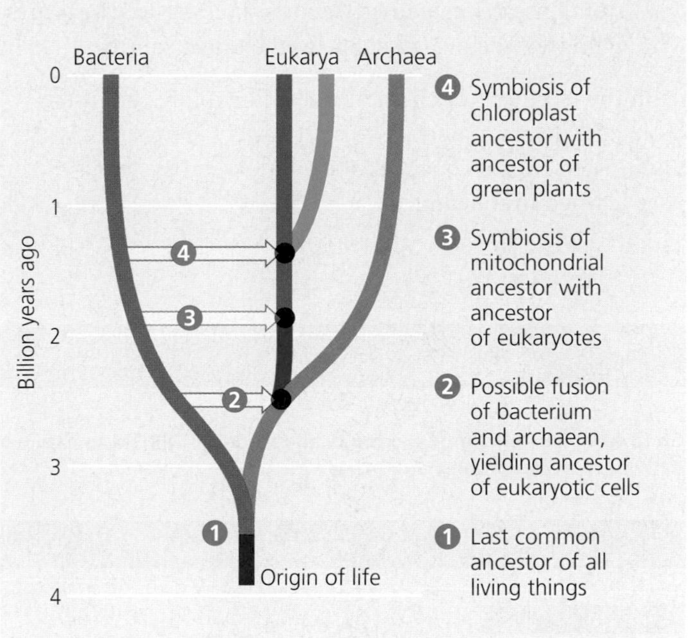

▲ **Figure 25.18 The universal tree of life.** All living organisms are divided into three domains: Bacteria, Archaea, and Eukarya. The details of this tree will change as further information emerges, but its general outlines appear firm.

Investigators use two criteria to identify regions of DNA molecules that can demonstrate the branching pattern of this tree: The regions must be able to be sequenced, and they must have evolved so slowly that homologies between even distantly related organisms can still be detected. The rRNA genes, which code for the RNA parts of ribosomes, fit these requirements. Because rRNA genes are fundamental to the workings of the cell, their molecular clock runs so slowly that they can serve as the basis for a universal tree of life. We will examine this tree in greater detail in later chapters, but for the moment let us make two points:

1. **The tree of life consists of three great domains: Bacteria, Archaea, and Eukarya.** The domain Bacteria includes most of the currently known prokaryotes, including the bacteria closely related to chloroplasts and mitochondria (we'll examine the bacterial origins of these organelles in Chapter 26). The second domain, Archaea, consists of a diverse group of prokaryotic organisms that inhabit a wide variety of environments. Some archaea can use hydrogen as an energy source, and some were the chief source of the deposits of natural gas that are found throughout Earth's crust. Chapter 27 will explore bacteria and archaea in more detail. The third domain, Eukarya, consists of all the organisms that have cells containing true nuclei. This domain includes many groups of single-celled organisms as well as multicellular plants, fungi, and animals, and is the topic of Chapters 28–34.
2. **The early history of these domains is not yet clear.** Comparisons of complete genomes from the three domains show that, especially during the early history of life, there have been substantial interchanges of genes between organisms in the different domains. These took place through **horizontal gene transfer,** in which genes are transferred from one genome to another through mechanisms such as transposable elements, and perhaps through fusions of different organisms. The first eukaryote may have arisen through such a fusion between an ancestral bacterium and an ancestral archaean. Because phylogenetic trees are based on the assumption that genes are passed vertically from one generation to the next, the occurrence of such horizontal events means that universal trees built from different genes often give inconsistent results, particularly near the root of the tree. As a result, details of the tree are continually being revised. However, the division into three great

domains has remained secure since systematists first glimpsed the overall shape of the universal tree.

In the next unit of chapters, we will explore the history of biological diversity and its current variety. Our study will be illuminated by the concepts of evolution and systematics you have learned in this unit.

Concept Check 25.5

1. What is a molecular clock? What assumption underlies the use of a molecular clock?
2. Explain how numerous base changes could occur in DNA, yet have no effect on an organism's fitness.

For suggested answers, see Appendix A.

Chapter 25 Review

Go to www.campbellbiology.com or the student CD-ROM to explore Activities, Investigations, and other interactive study aids.

SUMMARY OF KEY CONCEPTS

Phylogenies are based on common ancestries inferred from fossil, morphological, and molecular evidence

- **The Fossil Record** (pp. 492–493) The fossil record is based on fossil organisms preserved in geologic strata of different ages and reveals ancestral characteristics that may have been lost.
Activity *A Scrolling Geologic Record*
- **Morphological and Molecular Homologies** (pp. 492–495) Organisms that share very similar morphologies or DNA sequences are likely to be more closely related than organisms with very different structures and genetic sequences. But homology (similarity due to shared ancestry) must be sorted from analogy (similarity due to convergent evolution).

Concept 25.2

Phylogenetic systematics connects classification with evolutionary history

- **Binomial Nomenclature** (p. 496) Linnaeus's system gives organisms two-part names: a genus plus a specific epithet.
- **Hierarchical Classification** (p. 496) Linnaeus introduced a system for grouping species in increasingly broad categories.
- **Linking Classification and Phylogeny** (pp. 496–497) Systematists depict evolutionary relationships as branching phylogenetic trees, which may be based on various kinds of evidence.
Activity *Classification Schemes*

Concept 25.3

Phylogenetic systematics informs the construction of phylogenetic trees based on shared characters

- **Cladistics** (pp. 498–499) A clade is a monophyletic grouping of species that includes an ancestral species and all its descendants. In cladistic analysis, clades are defined by their evolutionary novelties, or shared derived characters. These are identified by comparing ingroup species with an outgroup species that does not have the shared derived character.
- **Phylogenetic Trees and Timing** (pp. 499–500) In phylograms, the length of a branch reflects the number of evolutionary changes in that lineage. Ultrametric trees place evolutionary branch points in the context of geologic time.
Investigation *How Is Phylogeny Determined by Comparing Proteins?*
- **Maximum Parsimony and Maximum Likelihood** (pp. 501–503) Among phylogenetic hypotheses, the most parsimonious tree is the one that requires the fewest evolutionary changes, and the most likely tree is the one based on the most likely pattern of changes.
- **Phylogenetic Trees as Hypotheses** (pp. 501–504) The best phylogenetic hypotheses are those that are consistent with the most data: morphological, molecular, and fossil.

Concept 25.4

Much of an organism's evolutionary history is documented in its genome

- **Gene Duplications and Gene Families** (p. 505) Orthologous genes, found in a single copy in the genome, can diverge only after speciation has taken place. Paralogous genes arise through duplication within a genome and can diverge within a clade, often adding new functions.
- **Genome Evolution** (pp. 505–506) Orthologous genes are often shared by distantly related species. The relatively small variation in total gene number in organisms of varying complexity indicates that genes in complex organisms are extremely versatile and that each gene can perform many functions.

Concept 25.5

Molecular clocks help track evolutionary time

- **Molecular Clocks** (pp. 506–507) The base sequences of some regions of DNA change at a rate consistent enough to allow dating of episodes in past evolution. These molecular clocks may result from the fixation of neutral mutations, but even when selection plays a role, many genes have tended to change in a clocklike fashion over long periods of time. Researchers have measured some molecular clocks and shown that they are remarkably constant. Other genes, however, change in a less predictable fashion.
- **The Universal Tree of Life** (pp. 507–508) The tree of life has three great clades (domains): Bacteria, Archaea, and Eukarya.

TESTING YOUR KNOWLEDGE

Self-Quiz

1. If humans and pandas belong to the same class, then they must also belong to the same
   a. order.
   b. phylum.
   c. family.
   d. genus.
   e. species.

2. Three living species X, Y, and Z share a common ancestor T, as do extinct species U and V. A grouping of species T, X, Y, and Z make up
   a. a valid taxon.
   b. a monophyletic clade.
   c. a paraphyletic grouping.
   d. a polyphyletic grouping.
   e. an ingroup for comparison with species U as the outgroup.

3. In a comparison of birds with mammals, having four appendages is
   a. a shared primitive character.
   b. a shared derived character.
   c. a character useful for distinguishing the birds from mammals.
   d. an example of analogy rather than homology.
   e. a character useful for sorting bird species.

4. How would one apply the principle of parsimony to the construction of a phylogenetic tree?
   a. Choose the tree that assumes all evolutionary changes are equally probable.
   b. Choose the tree in which the branch points are based on as many shared derived characters as possible.
   c. Base phylogenetic trees only on the fossil record, as this provides the simplest explanation for evolution.
   d. Choose the tree that represents the fewest evolutionary changes, either in DNA sequences or morphology.
   e. Choose the tree with the fewest branch points.

5. What would be the best source of data for determining phylogenetic relationships of lineages of protists that diverged hundreds of millions of years ago?
   a. fossils from the Proterozoic eon
   b. morphological characters that are shared and derived
   c. amino acid sequences of their various chlorophyll molecules
   d. mtDNA sequences
   e. rRNA gene sequences

6. If you were using cladistic analysis to build a phylogenetic tree of cats, which of the following would make the best outgroup?
   a. lion
   b. domestic cat
   c. wolf
   d. leopard
   e. tiger

7. Which of the following would be most useful for constructing a phylogenetic tree for several fish species?
   a. several analogous characteristics shared by all the fishes
   b. a single homologous characteristic shared by all the fishes
   c. the total degree of morphological similarity among various fish species
   d. several characteristics thought to have evolved after different fishes diverged from one another
   e. a single characteristic that is different in all the fishes

8. The relative lengths of the amphibian and mouse branches in the phylogram in Figure 25.12 indicate that
   a. amphibians evolved before mice.
   b. mice evolved before amphibians.
   c. the genes of amphibians and mice have only coincidental homoplasies.
   d. the homologous gene has evolved more rapidly in amphibians.
   e. the homologous gene has evolved more rapidly in mice.

9. Choose the pair of orthologous genes from the following list:
   a. human α hemoglobin and chimpanzee α hemoglobin genes
   b. two alleles of the human α hemoglobin gene
   c. mouse insulin gene and yeast mating-type gene
   d. two different rat olfactory receptor genes
   e. the multiple copies of rRNA genes in a eukaryotic genome

10. The recent estimate that HIV-1 M first jumped from chimpanzees to humans in the 1930s is based on
   a. the first clinical evidence of AIDS recorded in local village records in Africa.
   b. a molecular clock that used changes in sequences of an HIV gene sampled from patients over the past 40 years to project backward to an estimated origin.
   c. a comparison of homologous genes in HIV found in chimpanzees and in humans.
   d. a parsimonious explanation of the evolutionary relationships between the various strains of the virus found in humans at the present time.
   e. the discovery of HIV in a blood sample from the 1930s.

For Self-Quiz answers, see Appendix A.

Go to the website or CD-ROM for more quiz questions.

Evolution Connection

Darwin suggested looking at a species' close relatives to learn what its ancestors may have been like. How does his suggestion anticipate the use of outgroups in cladistic analysis?

Biological Inquiry: A Workbook of Investigative Cases ***Explore the use of outgroups and other methods of modern cladistic analysis in "Tree Thinking."***

Scientific Inquiry

Some nucleotide changes cause amino acid substitutions in the encoded protein (nonsynonymous changes), and others do not (synonymous changes). In a comparison of rodent and human genes, rodents were found to accumulate synonymous changes 2.0 times faster than humans, and nonsynonymous substitutions 1.3 times faster. What factors could explain this difference? How do such data complicate the use of molecular clocks to date evolutionary events?

Investigation ***How Is Phylogeny Determined by Comparing Proteins?***

Science, Technology, and Society

The ability to compare genomes has opened up new avenues in medical research. Because humans and mice share so many orthologous genes, it is possible to infer gene function in humans by "knocking out" the corresponding orthologous genes in mice. What medical applications might such research lead to? What might be the consequences of these discoveries to society?

UNIT 5

The Evolutionary History of Biological Diversity

AN INTERVIEW WITH Linda Graham

Professor Linda Graham is a time traveler. By peering through microscopes at fossils of the earliest plants and their closest living relatives, Dr. Graham looks back a half-billion years to investigate a major breakthrough in the history of life: the origin of land plants from their aquatic algal ancestor. Those early plants were the first macroscopic creatures on land. In addition to her research on the origin and early diversification of land plants, Graham and her students study how plants, especially mosses, continue to have an enormous impact on the biosphere today. Dr. Graham, who is a professor of botany and environmental studies at the University of Wisconsin-Madison, is also a gifted teacher, and the interview that follows offers some sage advice for why the study of biological diversity is an important part of your education.

What is the ancestor of land plants?

Plants originated from a particular group of green algae known as charophyceans. The evidence is very strong that these ancestors had already acquired some degree of complexity in terms of being able to branch. They also had some reproductive complexities that were probably inherited by the earliest land plants as well.

How long ago did this origin of land plants from algae occur?

The origin of land plants is still somewhat controversial in terms of timeline. The consensus among the paleobotanical community is that there were plants on the terrestrial surface at about 475 million years ago. This conclusion is based on evidence from fossil spores and other types of fossilized plant material. However, some of my colleagues have found fossil spores that they believe to have originated from land plants that lived as far back as 500 million years ago, in the mid-Cambrian period. I predict that in the future we will extend the origin of land plants back in time to at least the mid-Cambrian, perhaps even further back.

Which living plants do you think are most similar to the earliest land plants?

The molecular evidence indicates that the bryophytes—liverworts, hornworts, and mosses—are the oldest branches of the plant kingdom. And yet no one had found complete fossils of plants similar to bryophytes that were older than fossils of early vascular plants—plants with veins that transport water and nutrients—which the molecular data indicate diverged *after* bryophytes. That just didn't make sense. The standard explanation for the absence of early bryophyte fossils was that they just didn't preserve well. However, I knew that bryophytes produced spores and suspected that they produced other resistant materials as well. So, we wanted to test the idea that bryophytes have materials that could fossilize.

How did you do that?

We used two techniques to try to mimic the effects of degradation that would occur when a plant dies and falls into a water body and becomes partially degraded by microbial action. First, we treated living bryophyte material with an extreme technique known as acetolysis that combines high heat with strong acids. This is the same technique that the paleobotanists use to extract spores from rocks. We hypothesized that any plant material that would survive such an extreme treatment was fossilizable and should appear in a fossil record. Some of our colleagues argued that we might actually be generating resistant materials by such extreme treatment. So we added rotting techniques to our repertoire. We would leave our living bryophyte material in moist soil for months, and then retrieve the material and see what was left. Amazingly, the same types of resistant bryophyte materials that stood up to acetolysis also survived the more gentle process of rotting. And amazingly, those bits and pieces that survived rotting and acetolysis looked like some scrappy fossils of very ancient origin that people simply didn't recognize as being plant material.

And some of those fossil fragments are older than vascular plants?

Yes. Those scrappy fossils consisting of spores, tubelike structures, and bits of cellular sheets were much older than the vascular plant fossils. So we think that our work helps cement the idea that bryophyte-like plants were indeed present prior to the origin of vascular plants, as is supported by the molecular data.

How did the spread of these early land plants change the biosphere?

First, they helped produce early soils. Even the earliest plants had some organic materials that weren't easily degraded by microbes. So these materials built up as an organic layer in the soil. Second, by this photosynthetic conversion of carbon dioxide to resistant organic materials, early plants began to lower the amount of CO_2 in the atmosphere. That started a trend that culminated in the lowest historic CO_2 level as a result of the early woody plants of the coal swamps during the Carboniferous period. In addition, by producing organic acids, early plants probably released phosphate from the soil, and the runoff of phosphate would have stimulated growth of photosynthetic microorganisms in marine and fresh water ecosystems. Finally, early land plants established terrestrial ecosystems that eventually had sufficient organic productivity to support early land animals through food chains.

In addition to your interest in the origin of land plants, you also study the ecology of peat bogs. What are peat bogs, and why are they important?

Peat bogs are wetlands in which the dominant plant is *Sphagnum*, or "peat moss." This is a particularly important moss because it is an ecological engineer. The *Sphagnum* of peat bogs absorbs massive amounts of CO_2 from the atmosphere and stores it in organic materials that are not easily broken down by microorganisms—much like the early bryophytes we just talked about. Peat bogs are very extensive across the Northern Hemisphere—far more than most people realize because the bogs are located in northern regions that are not heavily populated. Very large areas of North America, Europe, and Asia are covered by vast peat lands that store an enormous amount of carbon. By helping to regulate atmospheric CO_2, which is a greenhouse gas, peat bogs function as a global thermostat. The moss helps stabilize the climate. If the temperature rises just a little bit, then that will facilitate moss growth, which will pull more CO_2 out of the atmosphere and help cool the planet. If it gets too cool, the moss won't grow as much, and there is a net release of CO_2 due to microbial degradation, which helps to warm the climate. So, we should be thankful for the vast peat lands that perform this thermostat function for our planet.

And how are we treating this important peat bog ecosystem?

Ecologists who study peat bogs are concerned about the disruption of peat bogs for mining or for agricultural uses such as cranberry production. By reducing the area of peat bogs, which play such an important role in regulating climate, we may be accelerating global warming.

How did you get started in science, Dr. Graham, and how did that interest turn to plants?

I had some wonderful schoolteachers when I was in elementary school, middle school, and high school—teachers who stimulated my interest in science. In particular, I had a female chemistry teacher who was a wonderful role model. She was intelligent and confident, and I got the idea from being in her class that I could be a scientist too. I have always been attracted to microscopes and being able to see the wonders of intricate structures. Plants have very interesting internal organization, and their relatives, the algae, are also very beautiful under a microscope. So, my specific interest in plants mainly came from my fascination with microscopic structure.

And how did that fascination become focused on the origin of land plants?

That was a pivotal event that occurred while studying for a botany final exam when I was an undergraduate at Washington University in St. Louis. One of the topics on the exam was the life cycles of plants, which have an alternation of multicellular haploid and diploid generations. The puzzle of how this complex life cycle originated stimulated my curiosity about the evolution of plants from their algal ancestors.

By helping to regulate atmospheric CO_2, which is a greenhouse gas, peat bogs function as a global thermostat. The moss helps stabilize the climate.

Why do you think it's important for first-year biology students to learn about the diversity of life, including plants, even if they plan to specialize in cellular or molecular biology or plan to go to medical school?

One reason we include biological diversity in our curriculum here at the University of Wisconsin is that we recognize that this may be the only point in the education of biology students that they will be exposed to the variety of organisms. And we recognize that biological diversity is important to all citizens because of its impact on human health. As students learn about prokaryotes, protists, plants, fungi, and animals, including invertebrates, they begin to see how these diverse organisms perform essential roles in ecosystems. And ultimately, our own health depends on the health of these ecosystems, which sustain humans with such services as clean water and clean air.

What other main points do you emphasize in your first-year courses?

I think that one of the most interesting aspects of biology is the linkage that occurs between the different hierarchical approaches. For example, understanding molecular processes and structures tells us a lot about processes at the organismal level and also at the ecological level. I think that what I can contribute to beginning biologists is an appreciation for integrative thinking and thinking across hierarchical levels. So in my classes, I try to point out that biological knowledge is not compartmented, but rather each topic is linked very tightly to other areas of biological inquiry. I also encourage first-year students to think about large issues and big questions—even if those big questions can't be answered by the application of a single experiment or a single set of observations. There are many questions in my personal research area, the origin of plants and extending back into the origin of life, that seem to some people to be unanswerable because they occurred so long ago that we can't perform direct observations. But I would encourage students to expect that we can answer such questions by deduction and integrative thinking. Thinking large and not focusing too much on the details of particular systems will prove useful in understanding all of biology.

26 The Tree of Life
An Introduction to Biological Diversity

▲ **Figure 26.1 An artist's conception of Earth 3 billion years ago.**

Key Concepts

26.1 Conditions on early Earth made the origin of life possible
26.2 The fossil record chronicles life on Earth
26.3 As prokaryotes evolved, they exploited and changed young Earth
26.4 Eukaryotic cells arose from symbioses and genetic exchanges between prokaryotes
26.5 Multicellularity evolved several times in eukaryotes
26.6 New information has revised our understanding of the tree of life

Overview

Changing Life on a Changing Earth

Life is a continuum extending from the earliest organisms to the great variety of species that exist today. In this unit, we will survey the diversity of life and trace the evolution of this diversity.

One of this book's themes is the interaction between organisms and their environments (see Chapter 1). Throughout this unit, we will see examples of the connection between biological history and geologic history. Geologic events that alter environments change the course of biological evolution. When a large lake subsides and forms several small lakes, for instance, some populations of organisms in the lakes become isolated and may evolve into different species (see Chapter 24). Conversely, living things change the planet they inhabit. For example, the evolution of photosynthetic organisms that released oxygen into the air had a dramatic impact on Earth's atmosphere. (The first of these organisms included prokaryotes similar to those in the dense mats that resemble stepping stones in **Figure 26.1**.) Another example of life changing Earth occurred when plants moved onto land, as Linda Graham explained in the interview on pages 510–511. Much more recently, the emergence of *Homo sapiens* has changed the land, water, and air on a scale and at a rate unprecedented for a single species. The histories of Earth and its life are inseparable.

The chapters in this unit also emphasize key junctures in evolution that have punctuated the history of biological diversity. Geologic history and biological history have been episodic, marked by what were in essence revolutions that opened many new ways of life.

Historical study of any sort is an inexact discipline that depends on the preservation, reliability, and interpretation of past records. The fossil record of past life is generally less and less complete the farther into the past we delve. However, each organism alive today carries traces of its evolutionary history in its molecules, metabolism, and anatomy. As we saw in Unit Four, such traces are clues to the past that augment the fossil record. Still, the evolutionary episodes of greatest antiquity are generally the most obscure.

We will begin this chapter by discussing the origin of life. That discussion is the most speculative in the entire unit, for no fossil evidence of that seminal episode exists. We will then turn to the fossil record and the connection between biological events and the physical history of Earth. Next, we will present an overview of major milestones in the 3.8-billion-year story of life on Earth. Finally, we will consider how biologists now understand the tree of life, a prelude to the survey of biological diversity presented in Chapters 27-34.

Concept 26.1

Conditions on early Earth made the origin of life possible

Scientific evidence is accumulating that chemical and physical processes on early Earth, aided by the emerging force of selection, produced very simple cells through a sequence of four main stages: (1) the abiotic (nonliving) synthesis of small organic molecules, such as amino acids and nucleotides; (2) the joining of these small molecules (monomers) into polymers, including proteins and nucleic acids; (3) the packaging of these molecules into "protobionts," droplets with membranes that maintained an internal chemistry different from that of their surroundings; and (4) the origin of self-replicating molecules that eventually made inheritance possible. This scenario has many uncertainties, but it does lead to predictions that can be tested in the laboratory. In this section, we will examine some of the evidence for each of these four stages.

Synthesis of Organic Compounds on Early Earth

Earth and the other planets of the solar system formed about 4.6 billion years ago, condensing from a vast cloud of dust and rocks that surrounded the young sun. It is unlikely that life could have originated or survived on Earth for the first few hundred million years because the planet was still being bombarded by huge chunks of rock and ice left over from the formation of the solar system. The collisions generated enough heat to vaporize all the available water and prevent seas from forming. This phase likely ended about 3.9 billion years ago. The oldest known rocks on Earth's surface, located at a site called Issua in Greenland, are 3.8 billion years old. Although certain chemical data from these rocks suggest that life may have existed then, this evidence is open to different interpretations, and no one has yet found fossils that old.

As the bombardment of early Earth slowed, conditions on the planet were extremely different from those of today. The first atmosphere was probably thick with water vapor, along with various compounds released by volcanic eruptions, including nitrogen and its oxides, carbon dioxide, methane, ammonia, hydrogen, and hydrogen sulfide. As Earth cooled, the water vapor condensed into oceans, and much of the hydrogen quickly escaped into space.

In the 1920s, Russian chemist A. I. Oparin and British scientist J. B. S. Haldane independently postulated that Earth's early atmosphere had been a reducing (electron-adding) environment, in which organic compounds could have formed from simple molecules. The energy for this organic synthesis could have come from lightning and intense UV radiation. Haldane suggested that the early oceans were a solution of organic molecules, a "primitive soup" from which life arose. In 1953, Stanley Miller and Harold Urey, of the University of Chicago, tested the Oparin-Haldane hypothesis by creating laboratory conditions comparable to those that scientists at the time thought existed on early Earth. Their apparatus yielded a variety of amino acids found in organisms today, along with other organic compounds (**Figure 26.2**; also see Figure 4.2). Many laboratories have since repeated the experiment using different

Figure 26.2
Inquiry Can organic molecules form in a reducing atmosphere?

EXPERIMENT Miller and Urey set up a closed system in their laboratory to simulate conditions thought to have existed on early Earth. A warmed flask of water simulated the primeval sea. The strongly reducing "atmosphere" in the system consisted of hydrogen (H_2), methane (CH_4), ammonia (NH_3), and water vapor. Sparks were discharged in the synthetic atmosphere to mimic lightning. A condenser cooled the atmosphere, raining water and any dissolved compounds into the miniature sea.

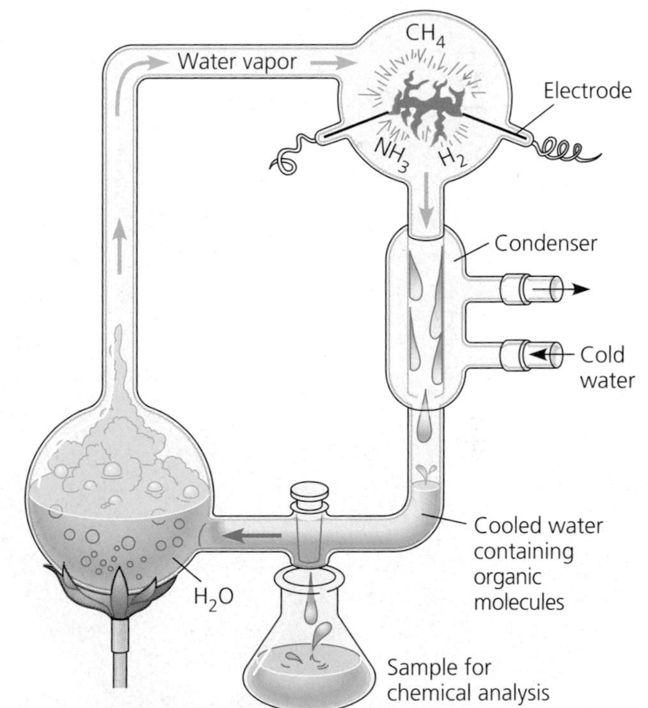

RESULTS As material circulated through the apparatus, Miller and Urey periodically collected samples for analysis. They identified a variety of organic molecules, including amino acids such as alanine and glutamic acid that are common in the proteins of organisms. They also found many other amino acids and complex, oily hydrocarbons.

CONCLUSION Organic molecules, a first step in the origin of life, can form in a strongly reducing atmosphere.

recipes for the atmosphere. Organic compounds were also produced in some of these modified models.

However, it is unclear whether young Earth's atmosphere contained enough methane and ammonia to be reducing. Growing evidence suggests that the early atmosphere was made up primarily of nitrogen and carbon dioxide and was neither reducing nor oxidizing (electron-removing). Miller-Urey-type experiments using such atmospheres have not produced organic molecules. Still, it is likely that small "pockets" of the early atmosphere—perhaps near volcanic openings—were reducing.

Instead of forming in the atmosphere, the first organic compounds on Earth may have been synthesized near submerged volcanoes and deep-sea vents—weak points in Earth's crust where hot water and minerals gush into the ocean **(Figure 26.3)**. These regions are also rich in inorganic sulfur and iron compounds, which are important in ATP synthesis by present-day organisms.

Extraterrestrial Sources of Organic Compounds

Some of the organic compounds from which the first life on Earth arose may have come from space. Among the meteorites that land on Earth are carbonaceous chondrites, rocks that are 1–2% carbon compounds by mass. Fragments of a 4.5-billion-year-old chondrite collected in southern Australia in 1969 contain more than 80 amino acids, some in large amounts. Remarkably, the proportions of these amino acids are similar to those produced in the Miller-Urey experiment. The chondrite amino acids cannot be contaminants from Earth because they consist of an equal mix of D and L isomers (see Chapter 4); organisms make and use only the L isomers, with a few rare exceptions. Amino acids that reached early Earth aboard chondrites could have added to the primitive soup, but recent calculations suggest that this contribution was likely to have been small.

▲ **Figure 26.3 A window to early life?** An instrument mounted to a robotic arm on the research submarine *Alvin* samples the water around a hydrothermal vent in the Sea of Cortés. More than 1.5 km below the surface, the vent releases hydrogen sulfide and iron sulfide, which react and produce pyrite (fool's gold) and hydrogen gas. Prokaryotes that live near the vent use the hydrogen as an energy source. Such environments are among the most extreme in which life exists today, and some researchers favor the hypothesis that life may have begun in similar regions of early Earth.

Looking Outside Earth for Clues About the Origin of Life

The possibility that life is not restricted to Earth is becoming more accessible to scientific testing. Mars is a good place to test hypotheses about the chemistry on Earth before life existed (prebiotic chemistry). The surface of Mars is now a cold, dry, and apparently lifeless desert, but evidence is growing that billions of years ago it was relatively warm for a brief period, with liquid water and a carbon dioxide-rich atmosphere. During that period, prebiotic chemistry similar to that on early Earth may have occurred on Mars. Did life evolve there and then die out, or was prebiotic chemistry terminated by dropping temperatures and a thinning atmosphere before any life-forms developed? Robot explorers are collecting data that may answer these questions in the next decade.

Measurements by the spacecraft Galileo indicate that liquid water lies beneath the ice-covered surface of Europa, one of Jupiter's moons, raising the possibility that Europa's hidden ocean may support prokaryotic life. Outside our solar system, more than a hundred planets orbiting other stars have been found. All these planets are at least as big as Jupiter, but space-based observatories planned for coming decades should have more sensitive telescopes and be able to detect smaller planets and moons of larger planets. If any of them has a strong spectroscopic signal of free oxygen in its atmosphere, that would provide evidence for the presence of oxygen-producing photosynthetic organisms. At present, however, it is unclear what atmospheric properties might indicate the presence of other forms of life.

Abiotic Synthesis of Polymers

Of course, living cells are more than a collection of amino acids. Every cell has a vast assortment of macromolecules, including proteins and the nucleic acids that are essential for self-replication, and it is difficult to imagine the emergence of life in an environment that did not contain similar macromolecules. Researchers have produced amino acid polymers by dripping solutions of amino acids onto hot sand, clay, or rock. The polymers formed spontaneously, without the help of enzymes or ribosomes. But unlike proteins, these polymers are a complex mix of linked and cross-linked amino acids, and each polymer

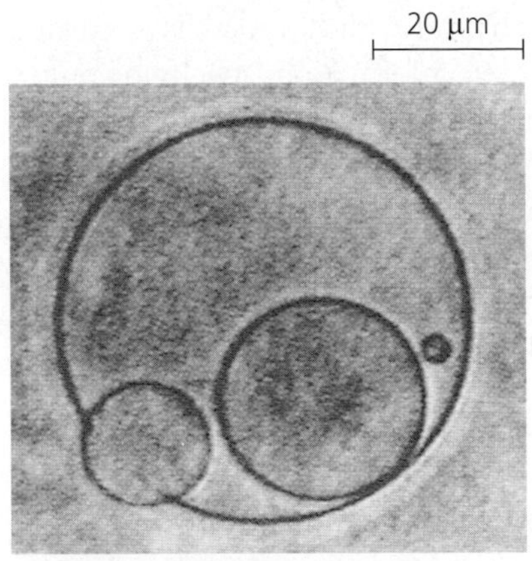

(a) **Simple reproduction.** This liposome is "giving birth" to smaller liposomes (LM).

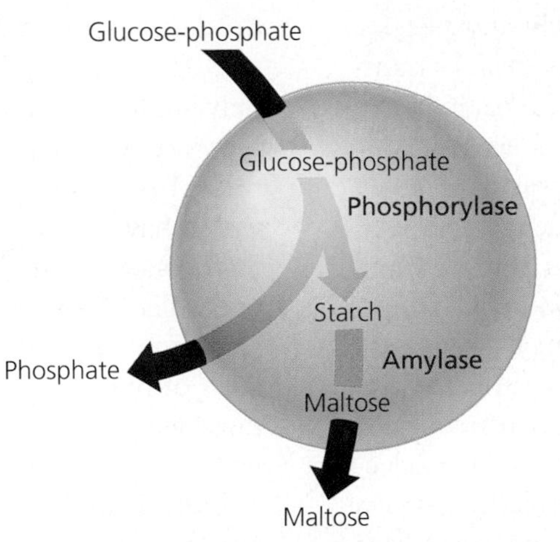

(b) **Simple metabolism.** If enzymes—in this case, phosphorylase and amylase—are included in the solution from which the droplets self-assemble, some liposomes can carry out simple metabolic reactions and export the products.

▲ **Figure 26.4 Laboratory versions of protobionts.**

is different. Nevertheless, such macromolecules may have acted as weak catalysts for a variety of reactions on early Earth.

Protobionts

Life is defined partly by two properties: accurate replication and metabolism. Neither property can exist without the other. DNA molecules carry genetic information, including the instructions needed to replicate themselves accurately. But the replication of DNA requires an elaborate enzymatic machinery, along with a copious supply of nucleotide building blocks that must be provided by the cell's metabolism (see Chapter 16). While Miller-Urey-type experiments have yielded some of the nitrogenous bases of DNA and RNA, they have not produced anything like nucleotides. Therefore, the building blocks of nucleic acids probably were not part of the early organic soup. Self-replicating molecules and a metabolism-like source of the building blocks must have appeared together. How did that happen?

The necessary conditions may have been met by **protobionts,** aggregates of abiotically produced molecules surrounded by a membrane or membrane-like structure. Protobionts exhibit some of the properties associated with life, including simple reproduction and metabolism, as well as the maintenance of an internal chemical environment different from that of their surroundings.

Laboratory experiments demonstrate that protobionts could have formed spontaneously from abiotically produced organic compounds. For example, small membrane-bounded droplets called liposomes can form when lipids or other organic molecules are added to water **(Figure 26.4)**. The hydrophobic molecules in the mixture organize into a bilayer at the surface of the droplet, much like the lipid bilayer of a plasma membrane. Because the liposome bilayer is selectively permeable, liposomes undergo osmotic swelling or shrinking when placed in solutions of different solute concentrations. Some liposomes store energy in the form of a membrane potential, a voltage across the surface. Such liposomes can discharge the voltage in nerve cell-like fashion; such excitability is characteristic of all life (which is not to say that liposomes are alive, but only that they display some of the properties of life.) If similar droplets forming in ponds on early Earth incorporated random polymers of linked amino acids into their membranes, and if some of those polymers made the membranes permeable to certain organic molecules, then those droplets could have selectively taken up organic molecules from their environment.

The "RNA World" and the Dawn of Natural Selection

The first genetic material was probably RNA, not DNA. Thomas Cech, of the University of Colorado, and Sidney Altman, of Yale University, found that RNA, which plays a central role in protein synthesis, can also carry out a number of enzyme-like catalytic functions. Cech called these RNA catalysts **ribozymes.** Some ribozymes can make complementary copies of short pieces of RNA, provided that they are supplied with nucleotide building blocks **(Figure 26.5)**. Others can

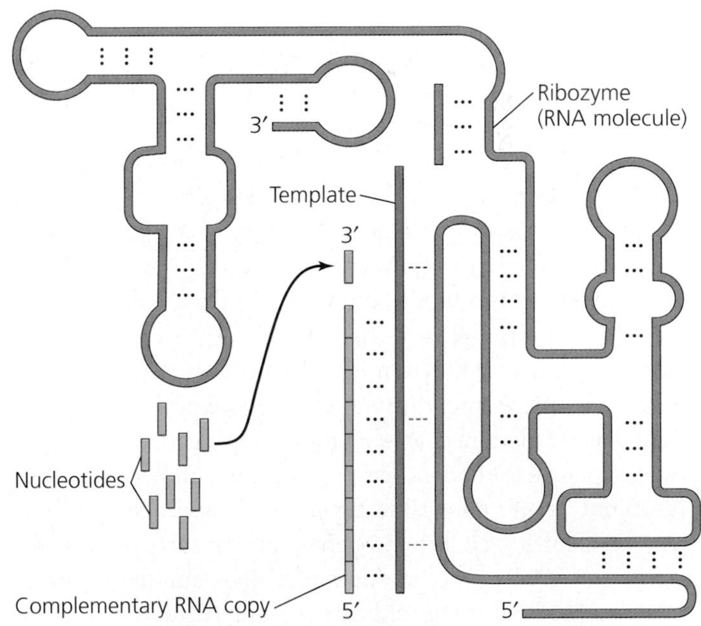

▲ **Figure 26.5 A ribozyme capable of replicating RNA.** This RNA molecule can make a complementary copy of another piece of RNA (a template) containing up to 14 nucleotides.

remove segments of themselves (self-splicing introns; see Chapter 17), or can act on different molecules, such as transfer RNA, excising pieces of these molecules and making them fully functional. Ribozyme-catalyzed reactions are relatively slow, but proteins normally associated with ribozymes can increase the reaction rate more than a thousandfold.

Natural selection on the molecular level has been observed operating on RNA populations in the laboratory. Unlike double-stranded DNA, which takes the form of a uniform helix, single-stranded RNA molecules assume a variety of specific three-dimensional shapes mandated by their nucleotide sequences. The molecule thus has both a genotype (its nucleotide sequence) and a phenotype (its conformation, which interacts with surrounding molecules in specific ways). In a particular environment, RNA molecules with certain base sequences are more stable and replicate faster and with fewer errors than other sequences. Beginning with a diversity of RNA molecules, the molecule whose sequence is best suited to the temperature, salt concentration, and other features of the surrounding solution and has the greatest autocatalytic activity will replicate most often. Its descendants will not be a single RNA species but will be a family of closely related sequences because of copying errors. Selection will screen mutations in the original sequence, and occasionally a copying error will result in a molecule that folds into a shape that is even more stable or more adept at self-replication than the ancestral sequence. Similar selection events may have occurred on early Earth. Thus, the molecular biology of today may have been preceded by an "RNA world," in which small RNA molecules that carried genetic information were able to replicate and to store information about the protobionts that carried them.

Physicist Freeman Dyson, of Princeton University, has suggested that the first RNA molecules may have been short, virus-like sequences and that these sequences were aided in their replication by random amino acid polymers that had rudimentary catalytic capabilities. Perhaps this early replication took place inside protobionts that had accumulated large numbers of organic monomers. If some of these RNA molecules could also bind specific amino acids to bases along part of their length, this binding could have held a few amino acids together long enough for them to be linked in peptides. (Indeed, this is one function of rRNA in modern ribosomes, the cellular sites of protein synthesis.) Some of these peptides might have functioned as enzymes, helping the RNA molecules replicate. Others might have become embedded in the protobiont membrane, allowing it to use high-energy inorganic molecules such as hydrogen sulfide to carry out organic reactions, including the production of RNA building blocks.

A protobiont with self-replicating, catalytic RNA would differ from its many neighbors that did not carry RNA or that carried RNA without such capabilities. If that protobiont could grow, split, and pass its RNA molecules to its daughters, the daughters would have some of the properties of their parent. Although the first such protobionts must have carried only limited amounts of genetic information, specifying only a few properties, they had inheritance and could be acted on by natural selection. The most successful of these protobionts would have increased in number, because they could exploit their resources effectively and pass their abilities on to subsequent generations. The emergence of such protobionts may seem unlikely, but remember that there could have been trillions of protobionts in bodies of water on early Earth. Even those with only a limited capacity for inheritance would have had a huge advantage over the rest.

Once RNA sequences that carried genetic information appeared in protobionts, many further changes were possible. For example, RNA could have provided the template on which DNA nucleotides were assembled. Double-stranded DNA is a much more stable repository for genetic information than the more fragile single-stranded RNA, and it can be replicated more accurately. Accurate replication was a necessity as genomes grew larger through gene duplication and other processes, and as more properties of the protobionts became coded in genetic information. After DNA appeared, perhaps RNA molecules began to take on their modern roles as intermediates in the translation of genetic programs, and the "RNA world" gave way to a "DNA world." The stage was now set for an explosion of life forms, driven by natural selection, that has continued to the present day.

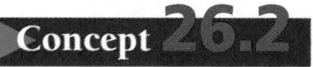

Concept Check 26.1

1. What hypothesis did Miller and Urey test in their experiment?
2. Why was the appearance of protobionts surrounded by membranes likely a key step in the origin of life?
3. What is a ribozyme?

For suggested answers, see Appendix A.

Concept 26.2

The fossil record chronicles life on Earth

Questions about the earliest stages in the origin of life on Earth may never be fully answered because, as far as we know, there is no record of these ancient events. Many later events, however, are well documented in the fossil record. Careful study of fossils opens a window into the lives of organisms that existed long ago and provides information about the evolution of life over billions of years.

How Rocks and Fossils Are Dated

Recall from Chapter 25 that most fossils are found in sedimentary rocks. The trapping of dead organisms in sediments freezes fossils in time. Thus, the fossils in each stratum of sedimentary rock are a local sample of the organisms that existed when the sediment that formed the stratum was deposited. Because younger sediments are superimposed on older ones, this book of sedimentary pages tells the relative ages of fossils (see Figure 25.3).

The strata at one location can often be correlated with strata at another location by the presence of similar fossils, known as index fossils. The best index fossils for correlating strata that are far apart are the shells of marine organisms that were widespread **(Figure 26.6)**. Index fossils reveal that at any particular location there are likely to be gaps in the sequence. That location may have been above sea level during certain periods, and thus no sedimentation occurred; or some of the sedimentary layers that were deposited when the location was submerged may have been eroded away.

The relative sequence of fossils in rock strata tells us the order in which the fossils were laid down, but it does not tell us their ages. Examining the relative positions of fossils in strata is like peeling layers of wallpaper from a very old house that has had many occupants. You can determine the sequence in which the layers were applied but not the year each layer was added.

Geologists have developed several methods for obtaining absolute dates for fossils. (Note that "absolute" dating does not mean errorless dating, but only that age is given in years rather than relative terms such as *before* and *after*.) One of the most common techniques is **radiometric dating**, which is based on the decay of radioactive isotopes. Each radioactive isotope has a fixed rate of decay. An isotope's **half-life**, the number of years it takes for 50% of the original sample to decay, is unaffected by temperature, pressure, and other environmental variables **(Figure 26.7)**.

Fossils contain isotopes of elements that accumulated in the organisms when they were alive. For example, the carbon in a living organism includes the most common carbon isotope, carbon-12, as well as a radioactive isotope, carbon-14. When the organism dies, it stops accumulating carbon, and the carbon-14 that it contained at the time of death slowly decays and becomes another element, nitrogen-14. Thus, by measuring the ratio of carbon-14 to total carbon or to nitrogen-14 in a fossil, we can determine the fossil's age. With a half-life of 5,730 years, carbon-14 is useful for dating fossils up to about 75,000 years old. Fossils older than that contain too little carbon-14 to be detected with current techniques. (After 13 half-lives—74,490 years—a fossil contains only $1/2^{13}$, or 0.012%, of the carbon-14 that was present when the fossil formed.) Radioactive isotopes with longer half-lives are used to date older fossils.

Paleontologists can often determine the age of fossils sandwiched between layers of volcanic rocks by measuring the amount of the radioactive isotope potassium-40 in those layers. Potassium-40 decays to the chemically unreactive gas argon-40, which is trapped in the rock. When the rock is heated during a volcanic eruption, the argon is driven off but the potassium remains. This resets the clock for potassium-40 decay to zero. The current ratio of potassium-40 to argon-40 in a layer of volcanic rock gives an estimate of when that layer was formed. Thus, if fossils are found in a sedimentary layer between two volcanic layers that are 530 and 520 million years old, we can infer that the fossils came from organisms that lived about 525 million years ago.

▲ **Figure 26.6 Index fossils.** Shelled animals called brachiopods were extremely abundant in the ancient seas. Their fossils are useful indicators of the relative ages of rock strata in different locations.

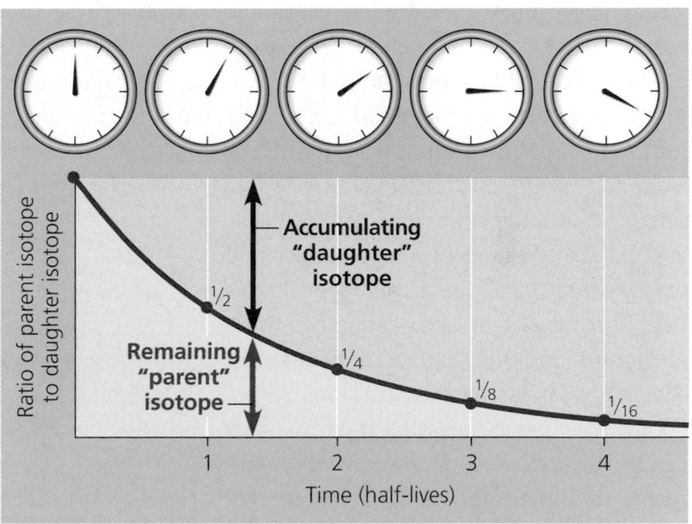

▲ **Figure 26.7 Radiometric dating.** A radioactive "parent" isotope decays to a "daughter" isotope at a constant rate. The rate of decay is expressed by the half-life, the time required for 50% of the remaining parent isotope to decay (in this diagram, each division of the clock face represents a half-life). Each type of radioactive isotope has a characteristic half-life. For example, carbon-14 decays relatively quickly and has a half-life of 5,730 years; uranium-238 decays slowly and has a half-life of 4.5 billion years.

The magnetism of rocks can also provide dating information. During the formation of volcanic or sedimentary rock, iron particles in the rock align themselves with Earth's magnetic field. When the rock hardens, their orientation is frozen in time. By measuring the magnetism of rocks with a device called a magnetometer, geologists have determined that Earth's north and south magnetic poles have reversed repeatedly in the past. Because these **magnetic reversals** affect the entire planet, they have left their record on rocks throughout the world. Therefore, patterns of magnetic reversals in one location can be matched with corresponding patterns elsewhere, allowing rocks to be dated when other methods are not available.

The Geologic Record

Geologists have established a **geologic record** of Earth's history, which is divided into three eons (**Table 26.1**, on the facing page). The first two eons—the Archaean and the Proterozoic—lasted approximately 4 billion years. Collectively, these two eons are often referred to as the Precambrian. The Phanerozoic eon, roughly the last half billion years, encompasses most of the time that multicellular eukaryotic life has existed on Earth, and it is divided in turn into three eras: the Paleozoic, Mesozoic, and Cenozoic. Each era represents a distinct age in the history of Earth and its life. For example, the Mesozoic era is sometimes called the "age of reptiles" because of its abundance of reptilian fossils, including those of dinosaurs. The boundaries between the eras correspond to times of mass extinctions seen clearly in the fossil record, when many forms of life disappeared and were replaced by forms that evolved from the survivors.

Lesser extinctions also mark the boundaries of many of the periods into which each era has been divided. Periods are further subdivided into intervals called epochs. Table 26.1 lists the epochs of the current era, the Cenozoic.

Mass Extinctions

A species may become extinct for many reasons. Its habitat may have been destroyed, or its environment may have changed in a direction unfavorable to the species. If ocean temperatures fall by even a few degrees, many species that are otherwise well adapted will perish. Even if physical factors in the environment remain stable, biological factors may change; the environment in which a species lives includes the other organisms that live there, and evolutionary change in one species is likely to impact others.

The fossil record chronicles a number of occasions when global environmental changes were so rapid and disruptive that a majority of species were swept away **(Figure 26.8)**. Such mass extinctions are known primarily from the decimation of hard-bodied animals that lived in shallow seas, the organisms for which the fossil record is most complete. Two mass extinctions—the Permian and the Cretaceous—have received the most attention. The Permian mass extinction, which defines the boundary between the Paleozoic and Mesozoic eras, claimed about 96% of marine animal species. Terrestrial life was also affected. For example, 8 out of 27 orders of insects were wiped out. This mass extinction occurred in less than 5 million years, possibly much less—an instant in the context of geologic time. The Cretaceous mass extinction of 65 million years ago, which marks the boundary between the Mesozoic and Cenozoic eras, doomed more than half of all marine species and exterminated many families of terrestrial plants and animals, including most of the dinosaurs.

The Permian mass extinction occurred at the time of enormous volcanic eruptions in what is now Siberia, constituting the most extreme episode of volcanism during the past half billion

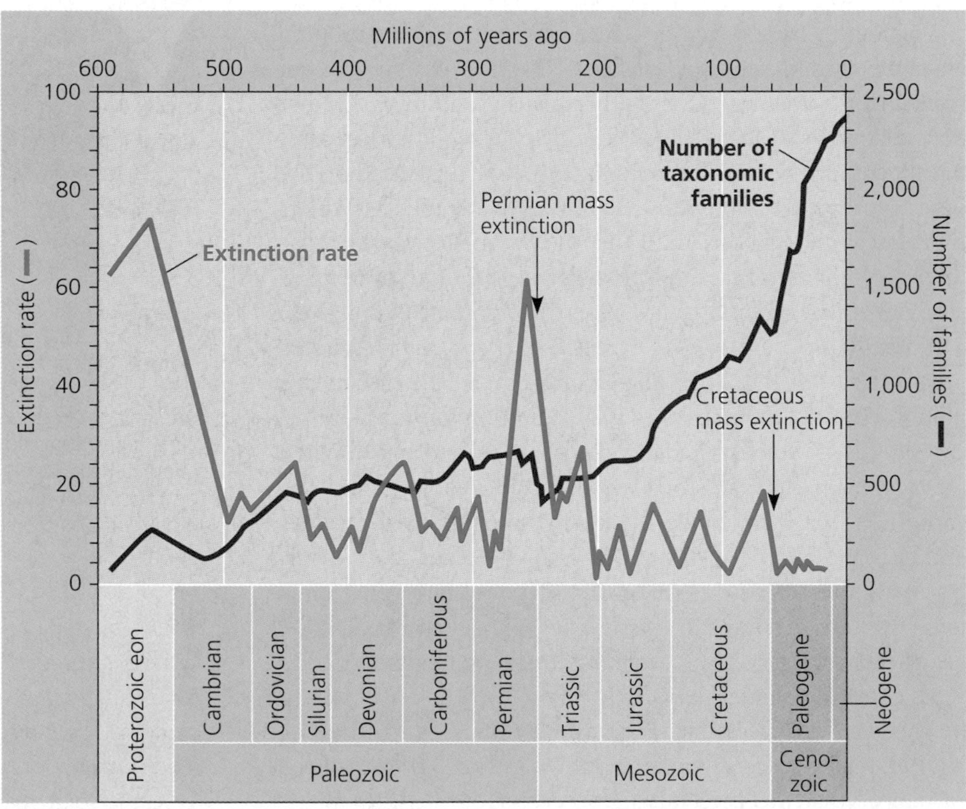

▲ **Figure 26.8 Diversity of life and periods of mass extinction.** The fossil record of terrestrial and marine organisms reveals a general increase in the diversity of organisms over time (red line and right vertical axis). Mass extinctions, represented by peaks in the extinction rate (blue line and left vertical axis) interrupted the buildup of diversity. The extinction rate is the estimated percentage of extant taxonomic families that died out in each period of geologic time.

Table 26.1 The Geologic Record

Relative Duration of Eons	Era	Period	Epoch	Age (Millions of Years Ago)	Some Important Events in the History of Life
Phanerozoic	Cenozoic	Neogene	Holocene	0.01	Historical time
			Pleistocene	1.8	Ice ages; humans appear
			Pliocene	5.3	Origin of genus *Homo*
			Miocene	23	Continued radiation of mammals and angiosperms; apelike ancestors of humans appear
		Paleogene	Oligocene	33.9	Origins of many primate groups, including apes
			Eocene	55.8	Angiosperm dominance increases; continued radiation of most modern mammalian orders
			Paleocene	65.5	Major radiation of mammals, birds, and pollinating insects
	Mesozoic	Cretaceous		145.5	Flowering plants (angiosperms) appear; many groups of organisms, including dinosaurs, become extinct at end of period (Cretaceous extinctions)
		Jurassic		199.6	Gymnosperms continue as dominant plants; dinosaurs abundant and diverse
		Triassic		251	Cone-bearing plants (gymnosperms) dominate landscape; radiation of dinosaurs; origin of mammal-like reptiles
	Paleozoic	Permian		299	Radiation of reptiles; origin of most present-day orders of insects; extinction of many marine and terrestrial organisms at end of period
		Carboniferous		359.2	Extensive forests of vascular plants; first seed plants; origin of reptiles; amphibians dominant
		Devonian		416	Diversification of bony fishes; first tetrapods and insects
		Silurian		443.7	Diversification of early vascular plants
		Ordovician		488.3	Marine algae abundant; colonization of land by plants and arthropods
		Cambrian		542	Sudden increase in diversity of many animal phyla (Cambrian explosion)
Proterozoic				600	Diverse algae and soft-bodied invertebrate animals
				2,200	Oldest fossils of eukaryotic cells
				2,500	
Archaean				2,700	Concentration of atmospheric oxygen begins to increase
				3,500	Oldest fossils of cells (prokaryotes)
				3,800	Oldest known rocks on Earth's surface
				Approx. 4,600	Origin of Earth

years. Besides spewing lava and ash into the atmosphere, the eruptions may have produced enough carbon dioxide to warm the global climate. Reduced temperature differences between the equator and the poles would have slowed the mixing of ocean water, which in turn would have reduced the amount of oxygen available to marine organisms. An oxygen deficit in the oceans may have played a large role in the Permian extinction.

One clue to a possible cause of the Cretaceous mass extinction is a thin layer of clay enriched in iridium that separates sediments from the Mesozoic and Cenozoic eras. Iridium is an element very rare on Earth but common in many of the meteorites and other extraterrestrial objects that occasionally fall to Earth. Walter and Luis Alvarez and their colleagues at the University of California proposed that this clay is fallout from a huge cloud of debris that billowed into the atmosphere when an asteroid or a large comet collided with Earth. This cloud would have blocked sunlight and severely disturbed the global climate for several months.

Where did the asteroid or comet hit? Research has focused on the Chicxulub crater, a 65-million-year-old scar beneath sediments off the Yucatán coast of Mexico **(Figure 26.9)**. About 180 km in diameter, the crater is the right size to have been caused by an object with a diameter of 10 km.

Critical evaluation of various hypotheses for these and other mass extinctions continues. At the time of the Cretaceous mass extinction, a spike in volcanic activity took place in what is now India. Was this spike triggered by the Chicxulub impact? Some scientists hypothesize that the Siberian volcanism that coincided with the Permian mass extinction also may have had an extraterrestrial trigger. Recent evidence suggests that there may have been a huge meteorite impact off the coast of Australia at that time. However, this interpretation is controversial, in part because no iridium-rich layer has been detected in Permian rocks. Alternatively, a near miss or a glancing blow by an asteroid or other body may have been enough to accelerate volcanic activity.

Much remains to be learned about the causes of mass extinctions, but it is clear that they provided life with unparalleled opportunities for adaptive radiations into newly vacated ecological niches. In the rest of this chapter, we will consider some of the major events in the history of life. **Figure 26.10** uses the analogy of a clock to place these events in the context of the geological record. This clock will reappear at various points in the chapter as a quick reminder of when the events we are discussing took place.

Concept Check 26.2

1. Your measurements indicate that a fossilized skull you unearthed has a carbon-14/carbon-12 ratio about 1/16 that of the skulls of present-day animals. What is the approximate age of the fossilized skull?
2. Based on Table 26.1, how long did prokaryotes inhabit Earth before eukaryotes evolved?

For suggested answers, see Appendix A.

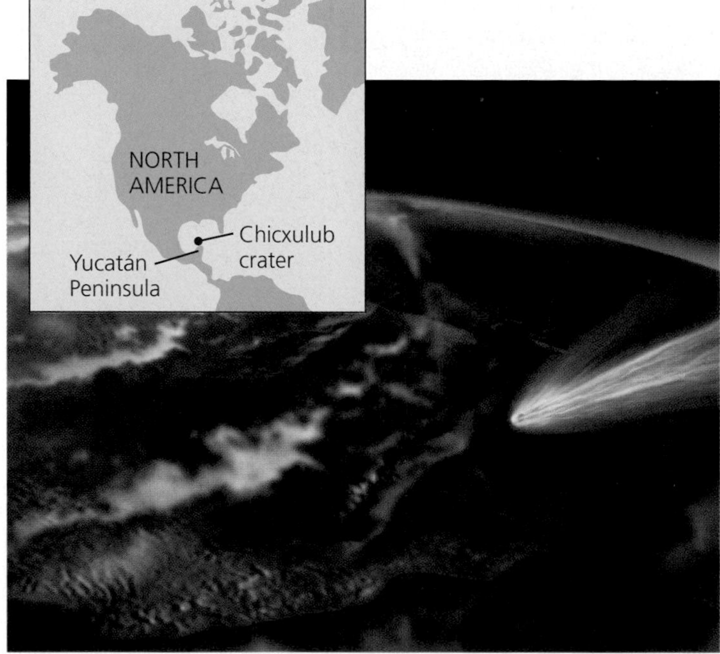

▲ **Figure 26.9 Trauma for Earth and its Cretaceous life.** The 65-million-year-old Chicxulub impact crater is located in the Caribbean Sea near the Yucatán Peninsula of Mexico. The horseshoe shape of the crater and the pattern of debris in sedimentary rocks indicate that an asteroid or comet struck at a low angle from the southeast. This artist's interpretation represents the impact and its immediate effect—a cloud of hot vapor and debris that could have killed most of the plants and animals in North America within hours.

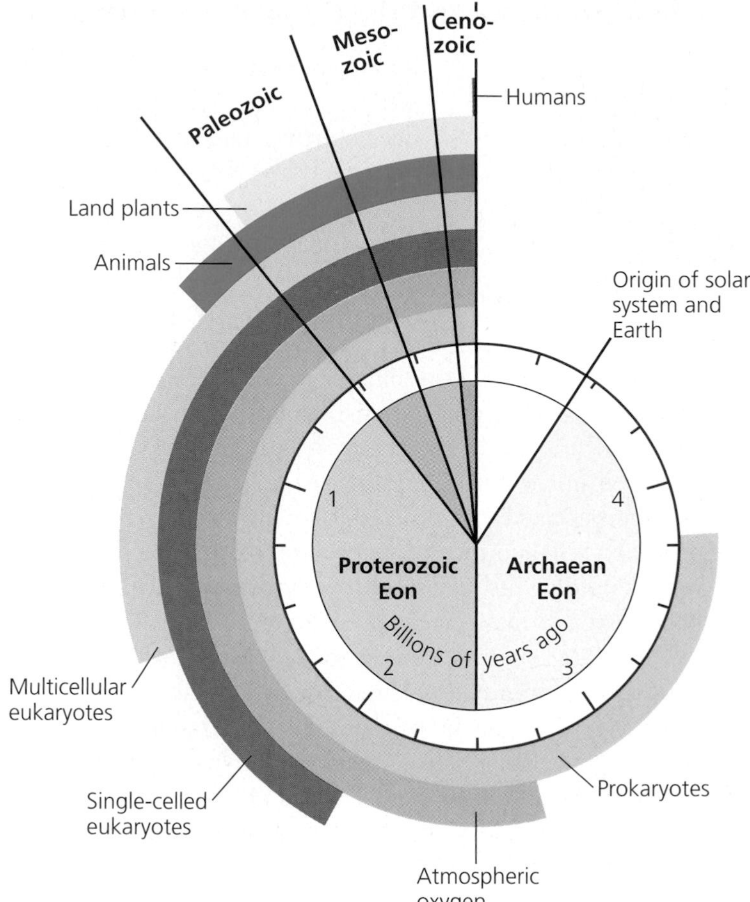

▲ **Figure 26.10 Clock analogy for some key events in Earth's history.** The clock ticks down from the origin of Earth 4.6 billion years ago to the present.

Concept 26.3

As prokaryotes evolved, they exploited and changed young Earth

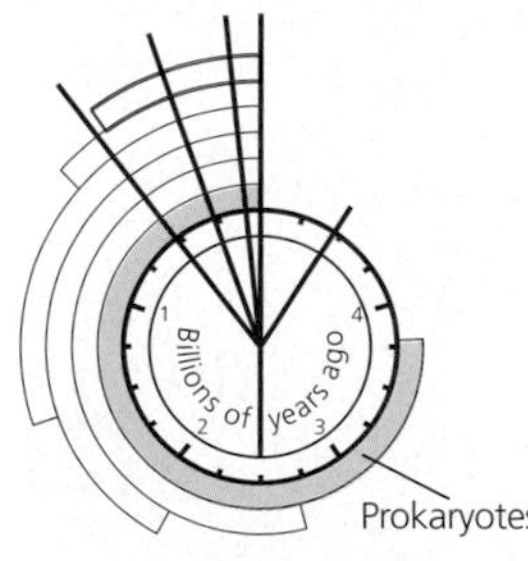

The oldest known fossils, dating from 3.5 billion years ago, are fossils of **stromatolites**, which are rocklike structures composed of many layers of bacteria and sediment **(Figure 26.11)**. Present-day stromatolites are found in a few warm, shallow, salty bays. If bacterial communities so complex existed 3.5 billion years ago, it is a reasonable hypothesis that life originated much earlier, perhaps as early as 3.9 billion years ago, when Earth began to cool to a temperature at which liquid water could exist. It is clear that prokaryotic life was already flourishing when Earth was still relatively young. Fairly early

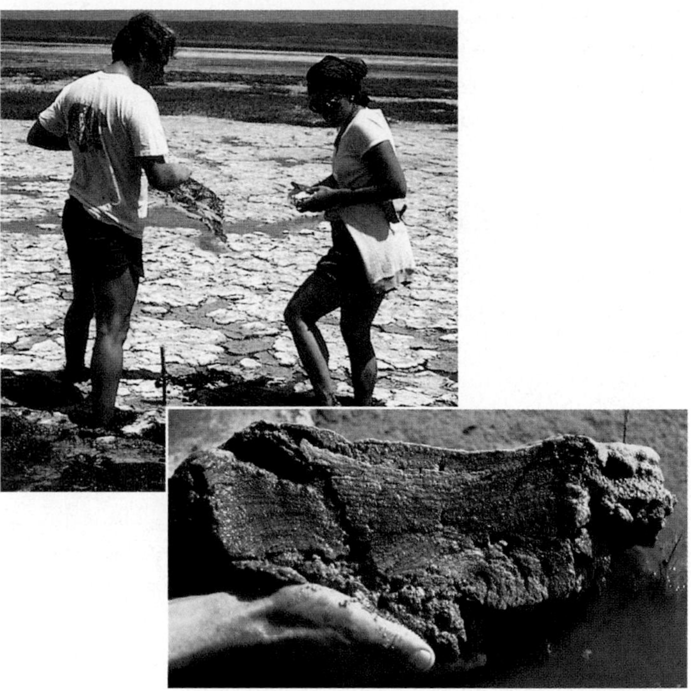

(a) Lynn Margulis (top right), of the University of Massachussetts, and Kenneth Nealson, of the University of Southern California, are shown collecting bacterial mats in a Baja California lagoon. The mats are produced by colonies of bacteria that live in environments inhospitable to most other life. A section through a mat (inset) shows layers of sediment that adhere to the sticky bacteria as the bacteria migrate upward.

(b) Some bacterial mats form rocklike structures called stromatolites, such as these in Shark Bay, Western Australia. The Shark Bay stromatolites began forming about 3,000 years ago. The inset shows a section through a fossilized stromatolite that is about 3.5 billion years old.

▲ **Figure 26.11 Bacterial mats and stromatolites.**

in prokaryotic history, two main evolutionary branches, the bacteria and the archaea, diverged. As you will read in Chapter 27, many species in these two lineages continue to thrive in a wide variety of environments today.

The First Prokaryotes

The early protobionts that had both self-replicating and metabolic capabilities must have used molecules for their growth and replication that were already present in the primitive soup. However, even the earliest successful protobionts probably had to make at least some of the molecules they required. Eventually, these protobionts were replaced by organisms that could produce all their needed compounds from molecules in their environment. These protobionts diversified into a rich variety of autotrophs, some of which could use light energy. The diversification of autotrophs likely encouraged the emergence of heterotrophs, which could live on products that the autotrophs excreted or on the autotrophs themselves.

These autotrophs and heterotrophs were the first prokaryotes, and they were Earth's sole inhabitants from at least 3.5 to about 2 billion years ago. As we will see, these organisms transformed the biosphere of our planet.

Electron Transport Systems

The chemiosmotic mechanism of ATP synthesis, in which a complex set of membrane-bound proteins pass electrons to reducible electron acceptors with the generation of ATP from ADP (see Chapter 9), is common to all three domains of life—Bacteria, Archaea, and Eukarya. There is strong evidence that this electron transport mechanism actually originated in organisms that lived before the last common ancestor of all present-day life. The earliest of these electron transport systems likely evolved before there was any free oxygen in the environment and before the appearance of photosynthesis; the organisms that used it would have required a plentiful supply of energy-rich compounds such as molecular hydrogen, methane, and hydrogen sulfide. A great challenge facing scientists studying the origin of life is to determine the steps by which this electron transport mechansim originated, and how important early versions of it might have been in the emergence of the first cells.

We'll explore the metabolically diverse prokaryotes that use different types of electron transport in detail in Chapter 27. For now, the important point is that considerable metabolic diversity among prokaryotes living in various environments had already evolved more than 3 billion years ago. Most subsequent evolution has been more structural than metabolic.

Photosynthesis and the Oxygen Revolution

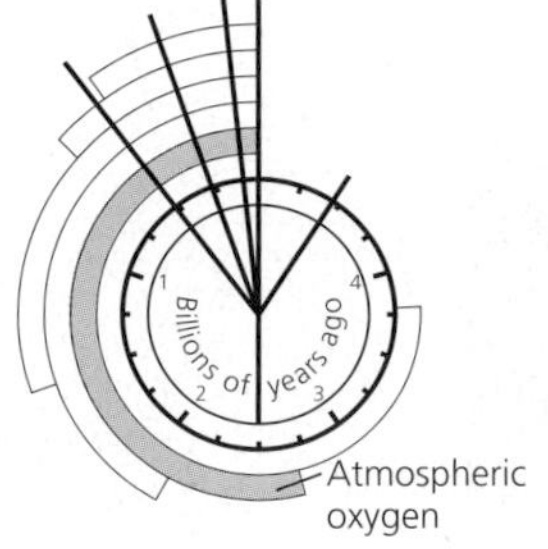

Photosynthesis probably evolved very early in prokaryotic history, but in metabolic versions that did not split water and liberate oxygen. In Chapter 27, we'll see examples of such nonoxygenic photosynthesis among living prokaryotes. The only living photosynthetic prokaryotes that generate O_2 are the cyanobacteria.

Most atmospheric O_2 is of biological origin, from the water-splitting step of photosynthesis. When this oxygenic photosynthesis first evolved, the free O_2 it produced probably dissolved in the surrounding water until the seas and lakes became saturated with O_2. Additional O_2 then reacted with dissolved iron and precipitated as iron oxide, which accumulated as sediments. These sediments were compressed into banded iron formations, red layers of rock containing iron oxide that are a source of iron ore today **(Figure 26.12)**. Once all the dissolved iron had precipitated, additional O_2 finally began to "gas out" of the seas and lakes and enter the atmosphere. This change left its mark in the rusting of iron-rich terrestrial rocks, a process that began about 2.7 billion years ago. This chronology implies that cyanobacteria may have originated as early as 3.5 billion years ago, when the microbial mats that left fossilized stromatolites began forming.

The amount of atmospheric O_2 increased gradually from about 2.7 to 2.2 billion years ago, but then shot up relatively rapidly to more than 10% of its present level. This "oxygen revolution" had an enormous impact on life. In its free molecular and ionized forms and in compounds such as hydrogen peroxide, oxygen attacks chemical bonds and can inhibit enzymes and damage cells. Its increasing atmospheric

▲ **Figure 26.12 Banded iron formations: evidence of oxygenic photosynthesis.** The reddish streaks in this sedimentary rock are bands of iron oxide.

concentration likely doomed many prokaryotic groups. Some species survived in habitats that remained anaerobic, where we find their descendants still living today as obligate anaerobes (see Chapter 27). Among other survivors, a diversity of adaptations to the changing atmosphere evolved, including cellular respiration, which uses oxygen in the process of harvesting the energy stored in organic molecules.

The early, gradual rise in atmospheric O_2 levels was associated with photosynthesis by ancient cyanobacteria. But what caused the accelerated rise in O_2 a few hundred million years later? One hypothesis is that it followed the evolution of eukaryotic cells containing chloroplasts, as we will discuss in the next section.

Concept Check 26.3

1. What do fossilized stromatolites suggest about the evolution of prokaryotes?
2. The first appearance of free oxygen in the atmosphere must have triggered a massive wave of extinctions among the prokaryotes of the time. Why?

For suggested answers, see Appendix A.

Concept 26.4

Eukaryotic cells arose from symbioses and genetic exchanges between prokaryotes

Eukaryotic cells differ in many respects from the generally smaller cells of bacteria and archaea (see Chapter 6). Even the smallest single-celled eukaryote is far more complex in structure than any prokaryote. Among the most fundamental questions in biology is how these complex eukaryotic cells evolved from much simpler prokaryotic cells.

The First Eukaryotes

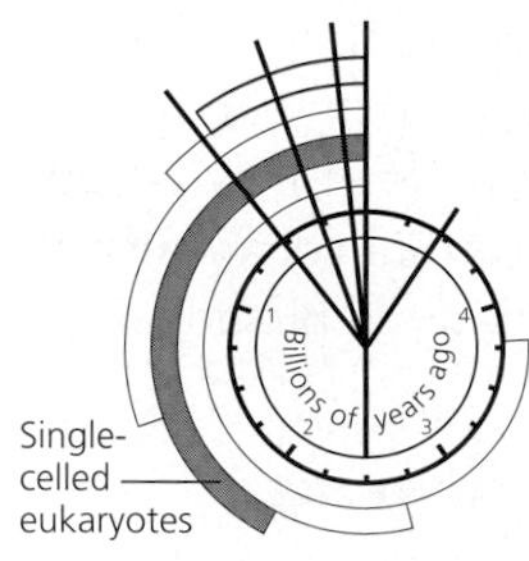

The oldest fossils that most researchers agree are eukaryotic are about 2.1 billion years old. Other fossils, of corkscrew-shaped organisms that resembled simple, single-celled algae, are slightly older (2.2 billion years), but their eukaryotic nature is less certain. However, some researchers postulate a much earlier eukaryotic origin based on traces of molecules similar to cholesterol found in rocks dating back 2.7 billion years. Such molecules are made only by eukaryotic cells that can respire aerobically. If confirmed, these findings could mean that eukaryotes evolved when the oxygen revolution was beginning to transform Earth's environments dramatically.

Endosymbiotic Origin of Mitochondria and Plastids

Prokaryotes lack many internal structures, such as the nuclear envelope, endoplasmic reticulum, and Golgi apparatus, that are characteristic of eukaryotic cells. Prokaryotes also have no cytoskeleton, so they are generally unable to change the shape of their cells. In contrast, eukaryotic cells have a cytoskeleton and can change shape, enabling them to surround and engulf other cells. Indeed, the first eukaryotes may have been predators of other cells. A cytoskeleton also enables a eukaryotic cell to shift its internal structures from one part of the cell to another, and it facilitates the regular movement of chromosomes in mitosis and meiosis (see Chapters 12 and 13). Mitosis made it possible to reproduce the large genomes of eukaryotes, and the closely related mechanism of meiosis became an essential part of sexual recombination of genes in eukaryotes.

How did the more complex organization of the eukaryotic cell evolve from the simpler prokaryotic condition? A process called endosymbiosis probably led to mitochondria and plastids (the general term for chloroplasts and related organelles, both photosynthetic and nonphotosynthetic). The theory of endosymbiosis proposes that mitochondria and plastids were formerly small prokaryotes living within larger cells. The term *endosymbiont* refers to such a cell that lives within another cell, which is called the *host cell*. The proposed ancestors of mitochondria were aerobic heterotrophic prokaryotes that became endosymbionts; the proposed ancestors of plastids were photosynthetic prokaryotes that became endosymbionts.

The prokaryotic ancestors of mitochondria and plastids probably gained entry to the host cell as undigested prey or internal parasites. Such a mechanism suggests an earlier evolution of an endomembrane system and a cytoskeleton that made it possible for the larger cell to engulf the smaller prokaryotes and package them within vesicles. By whatever means the relationships began, it is not hard to imagine the symbiosis eventually becoming mutually beneficial. A heterotrophic host could use nutrients released from photosynthetic endosymbionts. And in a world that was becoming increasingly aerobic, a cell that was itself an anaerobe would have benefited from aerobic endosymbionts that turned the oxygen to advantage. In the process of becoming more interdependent, the host and endosymbionts would have become a single organism, its parts inseparable. All eukaryotes, whether heterotrophic or autotrophic, have mitochondria or genetic

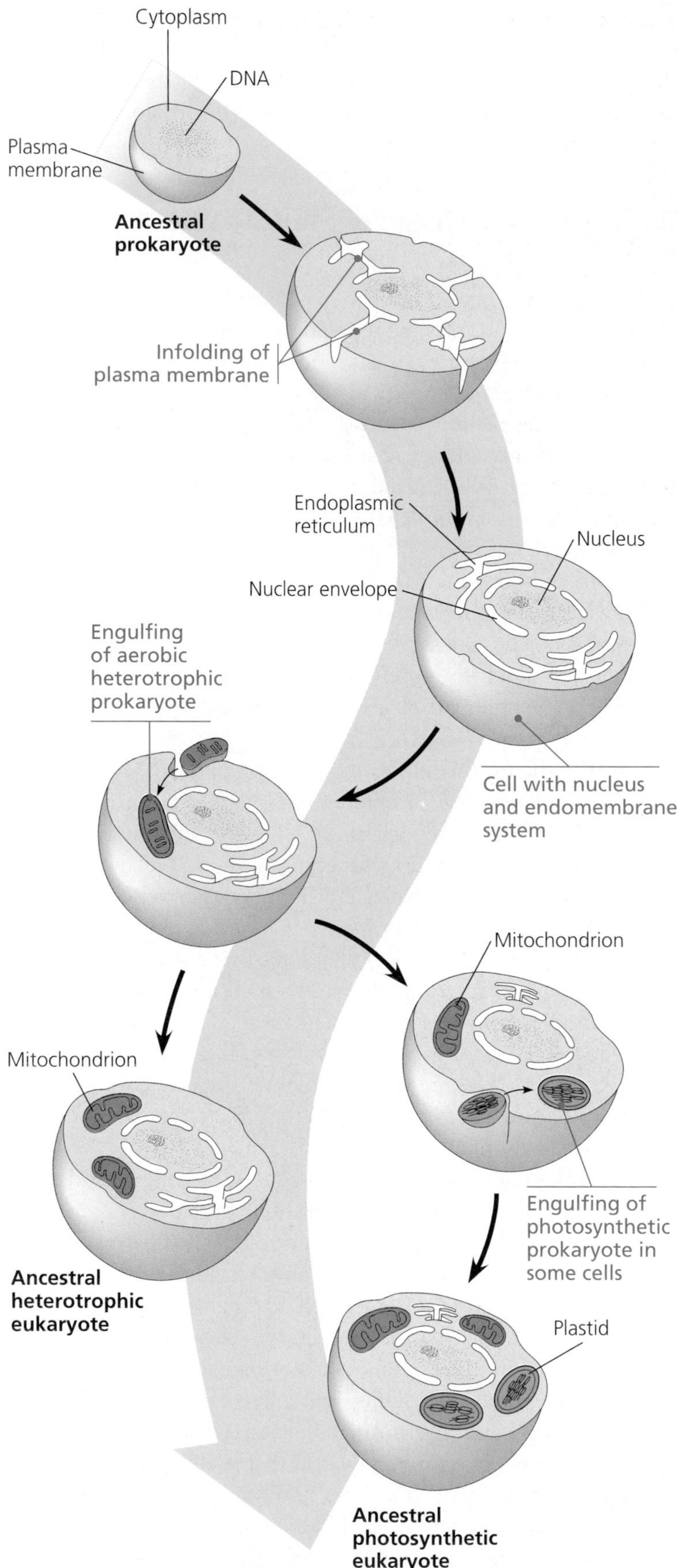

▲ **Figure 26.13 A model of the origin of eukaryotes through serial endosymbiosis.**

remnants of these organelles. However, not all eukaryotes have plastids. Thus, the hypothesis of **serial endosymbiosis** (a sequence of endosymbiotic events) supposes that mitochondria evolved before plastids **(Figure 26.13)**.

The evidence supporting an endosymbiotic origin of plastids and mitochondria is overwhelming. The inner membranes of both organelles have enzymes and transport systems that are homologous to those found in the plasma membranes of living prokaryotes. Mitochondria and plastids replicate by a splitting process reminiscent of binary fission in certain prokaryotes. Each organelle contains a single, circular DNA molecule that, like the chromosomes of bacteria, is not associated with histones or other proteins. These organelles contain the transfer RNAs, ribosomes, and other molecules needed to transcribe and translate their DNA into proteins. In terms of size, nucleotide sequence, and sensitivity to certain antibiotics, the ribosomes of mitochondria and plastids are more similar to prokaryotic ribosomes than they are to the cytoplasmic ribosomes of eukaryotic cells.

Which prokaryotic lineages gave rise to mitochondria and plastids? To answer this question, systematists have focused on the nucleotide sequence of the RNA in one of the ribosomal subunits. The gene for this small-subunit ribosomal RNA (SSU-rRNA) is present in all organisms, making it a good choice for studying the deepest branches of the tree of life (see Chapter 25). Comparisons of SSU-rRNA from mitochondria, plastids, and various living prokaryotes indicate that a group of bacteria called the alpha proteobacteria are the closest relatives of mitochondria, and that cyanobacteria are the closest relatives of plastids.

Over time, some of the genes originally present in mitochondria and plastids were transferred to the nucleus, a process that may have been accomplished by transposable elements (see Chapter 18). As a result, some mitochondrial and plastid proteins are encoded by the organelles' own DNA, whereas others are encoded by genes in the nucleus. Still other proteins are combinations of polypeptides encoded by genes in both locations; an example is the mitochondrial ATP synthase, the protein complex that generates ATP during cellular respiration (see Chapter 9). The transfer of genes to the nucleus has advanced further in some eukaryotes than in others: Mitochondria in the freshwater protist *Reclinomonas americana* have 97 genes, including several genes that in other eukaryotes have been transferred to the nucleus; vertebrate mitochondria have only 34 genes. It is fair to say that eukaryotic cells now contain only *one* genome, which is mostly nuclear but complemented by DNA that has remained in mitochondria and plastids.

Eukaryotic Cells as Genetic Chimeras

In Greek mythology, the chimera was a monster that was part goat, part lion, and part serpent. The eukaryotic cell is a chimera of prokaryotic parts, its mitochondria derived from one type of bacteria, its plastids from another, and its nuclear genome from parts of these endosymbionts' genomes and from at least one other cell, the cell that hosted the endosymbionts. Mitochondria and plastids have provided clues to their origins because they still carry small DNA molecules that have genes orthologous to prokaryotic genes (see Figure 25.17). Clues to other aspects of the origin of eukaryotic cells are far more numerous and confusing. For example, some researchers have proposed that the nucleus evolved from an endosymbiont belonging to domain Archaea. Genes with close relatives in both bacteria and archaea have been found in eukaryotic nuclei, but the picture is cloudy because different nuclei have different combinations of such genes.

The genome of eukaryotic cells may be the product of **genetic annealing,** in which horizontal gene transfers occurred between many different bacterial and archaean lineages. Carl Woese, of the University of Illinois, has suggested that most of these transfers took place during the early evolution of life, but new evidence indicates that horizontal gene transfers have happened repeatedly up to the present time. Ford Doolittle, of Dalhousie University, has used this evidence of repeated transfers to suggest a variant of genetic annealing: the "you are what you eat" hypothesis, in which evolving eukaryotes consumed various bacteria and archaea and occasionally incorporated some of the engulfed prokaryotes' genes into their nuclei.

The origin of other eukaryotic structures is the subject of active research. The Golgi apparatus and the endoplasmic reticulum may have originated from infoldings of the plasma membrane. Proteins that appear to be homologous to the cytoskeletal proteins actin and tubulin have been found in bacteria, where they are involved in the "pinching off" of bacterial cells during cell division. Further investigation of these bacterial proteins and their activities may provide information about the origin of the eukaryotic cytoskeleton. Some investigators have speculated that eukaryotic flagella and cilia evolved from symbiotic bacteria, an idea supported by the observation that bacteria have entered into more recent symbiotic relationships with some protozoans **(Figure 26.14)**. However, the 9 + 2 microtubule apparatus of eukaryotic flagella and cilia (see Chapter 6) has not been found in any prokaryote.

An orchestra can play a greater variety of musical compositions than a violin soloist can. Put simply, increased complexity makes more variations possible. The origin of the eukaryotic cell catalyzed the evolution of much more structural diversity than was possible for the simpler prokaryotic cells. This built on the first great adaptive radiation, the metabolic diversification of the prokaryotes. A third wave of diversification followed the origin of multicellular bodies in several eukaryotic lineages.

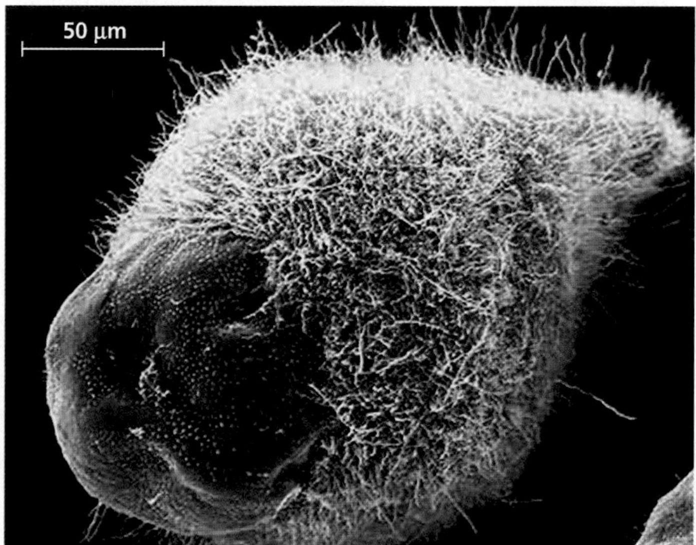

▲ **Figure 26.14 A complex symbiosis.** The remarkable protist *Mixotricha paradoxa,* which lives in the gut of termites, has three types of bacteria attached to its surface that provide it with motility (SEM). A fourth type of bacterium lives inside *Mixotricha* and digests wood fragments eaten by the termite.

Concept Check 26.4

1. What evidence supports the hypothesis that mitochondria preceded plastids in the evolution of eukaryotic cells?
2. How is a eukaryotic cell like a chimera?

For suggested answers, see Appendix A.

Concept 26.5

Multicellularity evolved several times in eukaryotes

After the first eukaryotes appeared, a great range of unicellular forms evolved, giving rise to the diversity of single-celled eukaryotes that continue to flourish today. But multicellular forms also evolved. Their descendants include a variety of algae, plants, fungi, and animals.

The Earliest Multicellular Eukaryotes

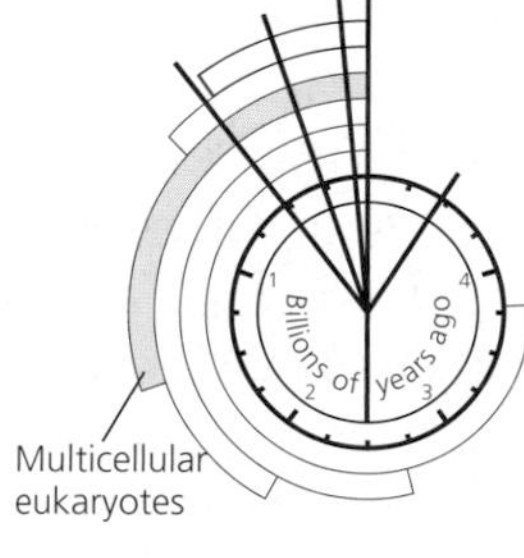

Molecular clocks date the common ancestor of multicellular eukaryotes to 1.5 billion years ago. However, the oldest known fossils of multicellular eukaryotes are of relatively small algae that lived about 1.2 billion years ago. Larger organisms do not appear in the fossil record until several hundred million years later, in the late Proterozoic. Chinese paleontologists recently described a particularly rich site that contains

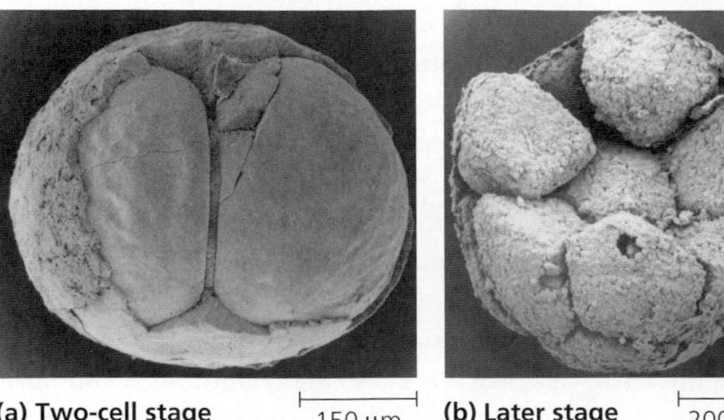

▲ **Figure 26.15 Fossils of Proterozoic animal embryos (SEM).**

570-million-year-old fossils of a diversity of algae and animals, including beautifully preserved structures that are probably animal embryos **(Figure 26.15)**.

Why were multicellular eukaryotes relatively limited in size, diversity, and distribution until the late Proterozoic? Geologic evidence indicates that a severe ice age occurred 750 to 570 million years ago. During this period, glaciers covered the planet's landmasses from pole to pole, and the seas were iced over. According to the **snowball Earth hypothesis**, most life would have been confined to areas near deep-sea vents and hot springs or to those sparse regions of the ocean where enough ice had melted for sunlight to penetrate the surface waters. The fossil record of the first major diversification of multicellular eukaryotes corresponds to the time when snowball Earth thawed.

The Colonial Connection

The first multicellular organisms were **colonies**, collections of autonomously replicating cells **(Figure 26.16)**. Some cells in the colonies became specialized for different functions. The first such specializations had already appeared in the prokaryotic world. For example, certain cells of the filamentous cyanobacterium *Nostoc* differentiate into nitrogen-fixing cells called heterocysts, which lose their ability to replicate. The undifferentiated cells are photosynthetic and capable of replicating, with the result that the filaments grow longer and eventually break into smaller filaments. *Nostoc* has more than 7,000 genes, twice as many as its close relatives that are single-celled, such as *Synechocystis,* and many of the additional genes are involved in the regulation of developmental pathways that can lead to cellular differentiation.

The evolution of colonies and of cellular specialization was carried much further in eukaryotes. A multicellular eukaryote, such as an animal, generally develops from a single cell—the fertilized egg, or zygote, in the case of sexual reproduction (see Figure 13.5). Cell division and cell differentiation help transform the single cell into a multicellular organism with many types of specialized cells. Increasing cell specialization made it possible for multicellular organisms to divide particular life functions, such as obtaining nutrients and sensing the environment, among specific groups of cells. This division of function eventually led to the evolution of tissues, organs, and organ systems in many eukaryotes.

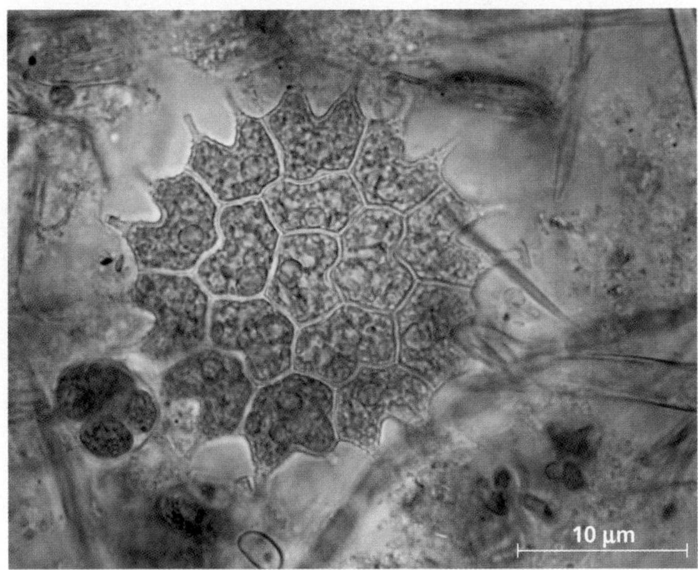

▲ **Figure 26.16 A colonial eukaryote.** *Pediastrum* is a photosynthetic eukaryote that forms flat colonies (LM).

Just as the evolution of unicellular eukaryotes increased the structural complexity of cells, the evolution of multicellular eukaryotes broke through another threshold in structural organization and provided the stock for new waves of diversification. Multicellularity evolved several times among early eukaryotes, giving rise to several lineages of algae as well as to plants, fungi, and animals.

The "Cambrian Explosion"

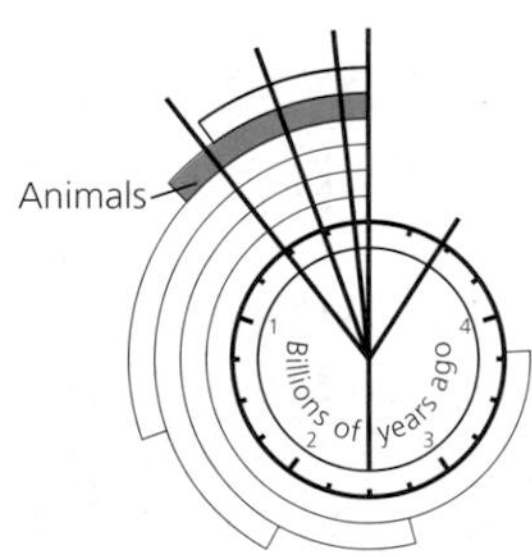

Most of the major phyla of animals appear suddenly in the fossil record that was laid down during the first 20 million years of the Cambrian period, a phenomenon referred to as the "Cambrian explosion." Fossils of two animal phyla—Cnidaria (sea anemones and their relatives) and Porifera (sponges)—are somewhat older, dating from the late Proterozoic **(Figure 26.17)**. However, molecular clocks suggest that many animal phyla originated and began to diverge much earlier, between 1 billion and 700 million years ago. In the memorable phrase of Cambridge University paleontologist Simon Conway Morris, "the Cambrian explosion had a long fuse." At the beginning of the Cambrian, for reasons not yet understood, these already divergent phyla expanded in diversity relatively suddenly and simultaneously. The "explosion" in diversity included many large animals with hard shells and exoskeletons that fossilized easily. We will evaluate possible explanations for this diversification in Chapter 32.

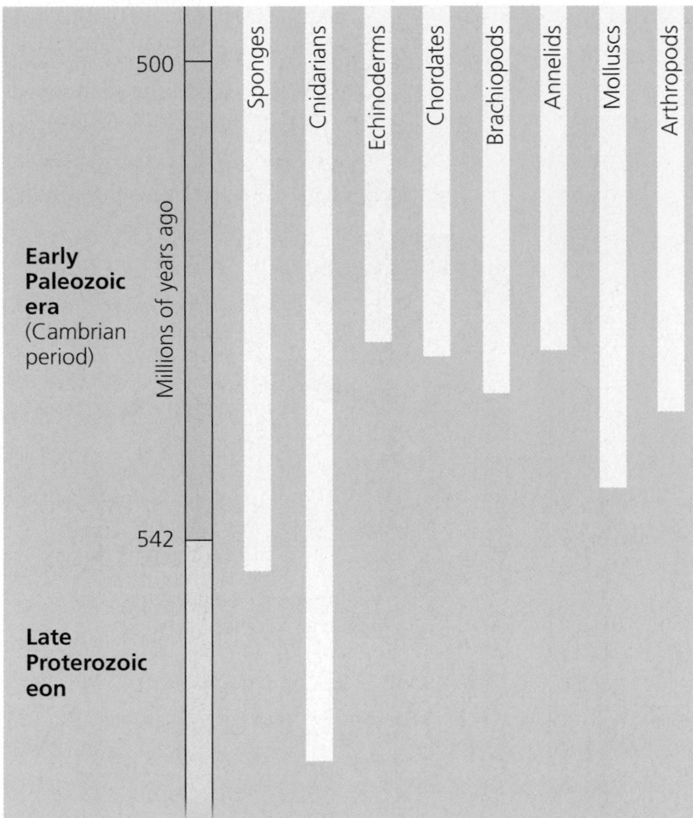

▲ **Figure 26.17 The Cambrian radiation of animals.** The bars in this diagram show the earliest appearance in the fossil record of several animal phyla. However, molecular evidence suggests that these phyla may have originated much earlier.

Colonization of Land by Plants, Fungi, and Animals

The colonization of land was one of the pivotal milestones in the history of life. There is fossil evidence that cyanobacteria and other photosynthetic prokaryotes coated damp terrestrial surfaces well over a billion years ago. However, macroscopic life in the form of plants, fungi, and animals did not colonize land until about 500 million years ago, during the early Paleozoic era. This gradual evolutionary venture out of ancestral aquatic environments was associated with adaptations that helped prevent dehydration and that made it possible to reproduce on land. For example, plants, which evolved from green algae, have a waterproof coating of wax on their leaves that slows the loss of water to the air.

Plants colonized land in the company of fungi. Even today, the roots of most plants are associated with fungi that aid in the absorption of water and minerals from the soil (see Chapter 31). These root fungi, in turn, obtain their organic nutrients from the plants. Such symbiotic associations of plants and fungi are evident in some of the oldest fossilized roots, dating this relationship back to the early spread of life onto land.

Although many animal groups are represented in terrestrial environments, the most widespread and diverse land animals are arthropods (particularly insects and spiders) and vertebrates (chiefly amphibians; reptiles, including birds; and mammals). Terrestrial vertebrates are called tetrapods, a reference to the four limbs that distinguish them from the various aquatic vertebrates. Tetrapods include humans, but we are late arrivals on the scene—the human lineage diverged from other hominoids (apes) only within the last 6 to 7 million years. If the clock of Earth's history were rescaled to represent an hour, humans appeared only one second ago.

Continental Drift

As we close this overview of the evolution of life on Earth, note that evolution has dimension in space as well as in time. Indeed, it was biogeography that first nudged Darwin and Wallace toward an evolutionary view of life. On a global scale, the major geographic factor that correlates with the spatial distribution of life is continental drift. This term refers to the fact that Earth's continents are not fixed; they drift across our planet's surface on great plates of crust that float on the hot, underlying mantle. In many cases, these plates pull away from or push against other plates where they meet each other **(Figure 26.18)**. For example, North America and Europe are presently drifting apart at a rate of about 2 cm per year, and California's infamous San Andreas Fault is part of a border where two plates slide past each other. Plate movements rearrange geography slowly, but their cumulative effects are dramatic. Many important geologic processes, including the

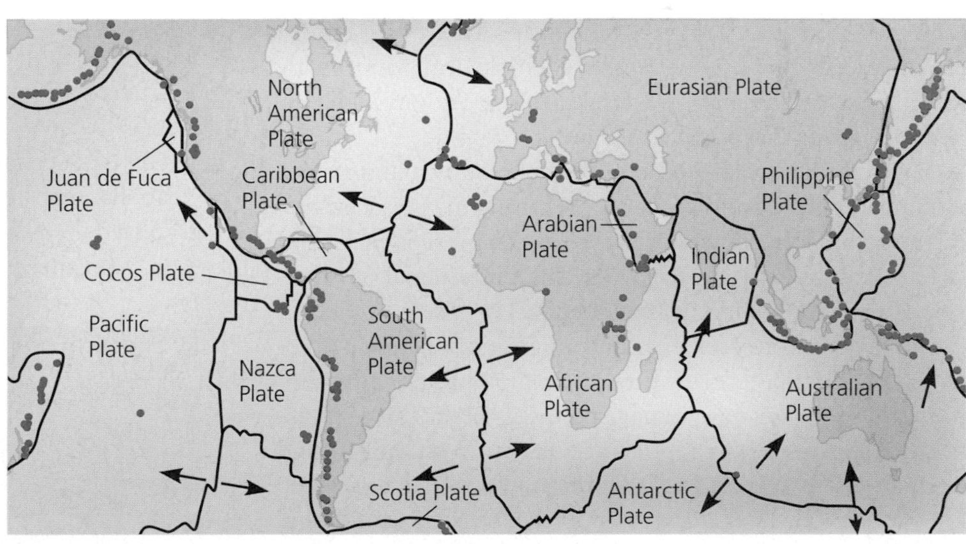

▲ **Figure 26.18 Earth's major crustal plates.** The plates change constantly; for example, a new plate boundary (not shown here) is forming between East Africa and the rest of the continent. The red dots indicate zones of violent tectonic events; many of these are subduction zones (see Figure 26.19).

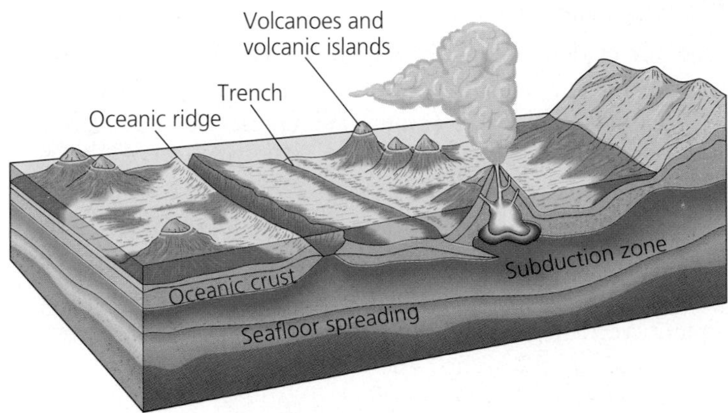
Volcanoes and volcanic islands
Trench
Oceanic ridge
Subduction zone
Oceanic crust
Seafloor spreading

▲ **Figure 26.19 Events at plate boundaries.** At some plate boundaries, such as oceanic ridges (indicated by paired, opposing arrows in Figure 26.18), the plates separate, and molten rock wells up in the gap. The rock solidifies and adds crust symmetrically to both plates, causing seafloor spreading. At subduction zones, where plates move toward each other, the more dense plate dives below the less dense one, forming a trench. Sudden release of accumulating tension at subduction zones causes earthquakes. When continents carried by different plates collide, continental and ocean floor material piles up, forming mountain ranges.

building of mountains and islands, occur at plate boundaries or at weak points in the plates themselves **(Figure 26.19)**.

Two chapters in the continuing saga of continental drift had especially strong influences on life. About 250 million years ago, near the end of the Paleozoic era, plate movements brought all the previously separated landmasses together into a supercontinent that has been named **Pangaea**, meaning "all land" **(Figure 26.20)**. Ocean basins became deeper, which lowered sea level and drained shallow coastal seas. At that time, as now, most marine species inhabited shallow waters, and the formation of Pangaea destroyed a considerable amount of that habitat. The interior of the vast continent was cold and dry, probably an even more severe environment than that of central Asia today. The formation of Pangaea had a tremendous environmental impact that reshaped biological diversity by causing extinctions and providing new opportunities for taxonomic groups that survived the crisis.

The second chapter in the history of continental drift was written about 180 million years ago, during the Mesozoic era. As the continents drifted apart, each became a separate evolutionary arena, with lineages of plants and animals that diverged from those on other continents.

This pattern of continental mergings and separations provides the solution to many biogeographic puzzles. For example, paleontologists have discovered fossils of the same species of Permian freshwater reptiles in Ghana (West Africa) and Brazil. These two parts of the world, now separated by 3,000 km of ocean, were joined together during the late Paleozoic and early Mesozoic eras. Continental drift also explains much about the current distribution of organisms, such as why Australian fauna and flora contrast so sharply with those of the rest of the world. Marsupial mammals fill ecological roles in Australia

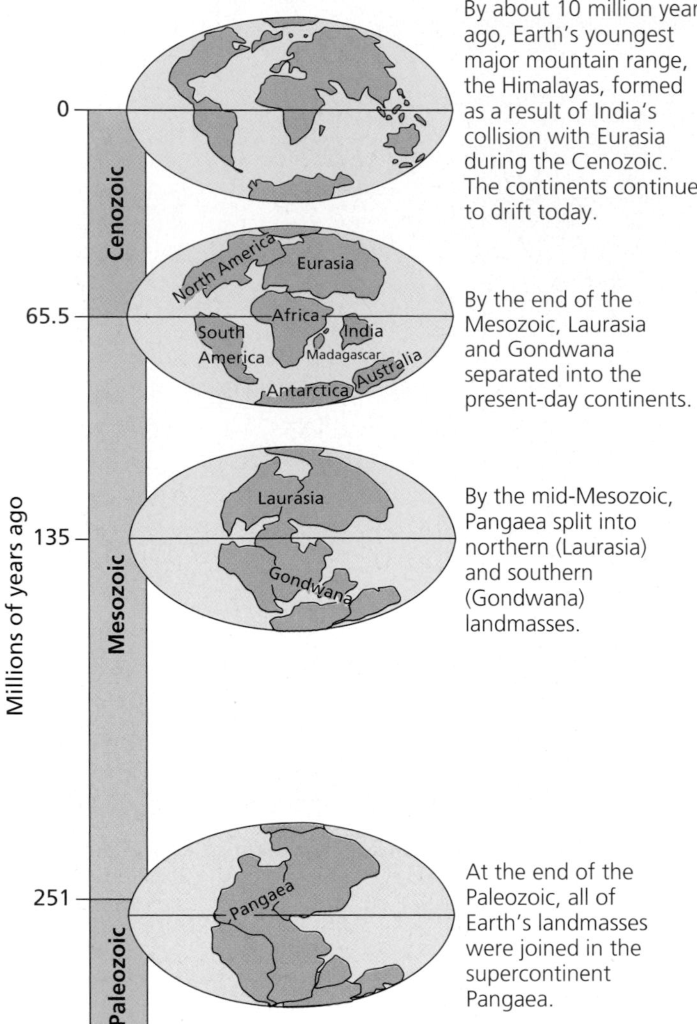
Millions of years ago
0
65.5
135
251
Cenozoic
Mesozoic
Paleozoic
By about 10 million years ago, Earth's youngest major mountain range, the Himalayas, formed as a result of India's collision with Eurasia during the Cenozoic. The continents continue to drift today.
North America
Eurasia
Africa
South America
India
Madagascar
Australia
Antarctica
By the end of the Mesozoic, Laurasia and Gondwana separated into the present-day continents.
Laurasia
Gondwana
By the mid-Mesozoic, Pangaea split into northern (Laurasia) and southern (Gondwana) landmasses.
Pangaea
At the end of the Paleozoic, all of Earth's landmasses were joined in the supercontinent Pangaea.

▲ **Figure 26.20 The history of continental drift during the Phanerozoic.**

analogous to those filled by eutherians (placental mammals) on other continents (see Figure 25.5). Marsupials probably evolved first in what is now North America and reached Australia via South America and Antarctica while the continents were still joined. The subsequent breakup of the southern continents set Australia "afloat" like a great ark of marsupials. In Australia, marsupials diversified, and the few early eutherians that lived there became extinct; on other continents, most marsupials became extinct, and the eutherians diversified.

Concept Check 26.5

1. How does the division of function differ for single-celled and multicellular organisms?
2. In what way is "Cambrian explosion" a good description of the early part of the fossil record of animal history? What is meant by the metaphor of a "long fuse" for the Cambrian explosion?

For suggested answers, see Appendix A.

Concept 26.6

New information has revised our understanding of the tree of life

As we discussed in Chapter 25, systematics is an analytical approach to reconstructing the evolutionary relationships of life's diverse forms. In recent decades, molecular data have provided new insights regarding the deepest branches of the tree of life, and ongoing research spurs biologists to continually revise the finer divisions of these branches.

Previous Taxonomic Systems

Many of us grew up with the notion that there are only two kingdoms of life—plants and animals—because we see a macroscopic, terrestrial realm and rarely notice those organisms that do not fit neatly into a plant-animal dichotomy. The two-kingdom scheme also had a long tradition in formal taxonomy; Linnaeus divided all known forms of life between the plant and animal kingdoms.

Even with the discovery of the diverse microbial world, the two-kingdom system persisted. Taxonomists placed bacteria in the plant kingdom, citing the rigid cell walls of bacteria as justification. Eukaryotic unicellular organisms with chloroplasts were also considered plants. Fungi, too, were classified as plants, partly because most fungi, like most plants, are unable to move about, even though fungi are not photosynthetic and have little in common structurally with plants. In the two-kingdom system, unicellular organisms that move and ingest food—protozoans—were called animals. Microbes such as *Euglena* that move and are photosynthetic were claimed by both botanists and zoologists and showed up in both kingdoms.

Taxonomic schemes with more than two kingdoms did not became popular with the majority of biologists until 1969, when Cornell University's Robert H. Whittaker argued effectively for a system consisting of five kingdoms: Monera, Protista, Plantae, Fungi, and Animalia **(Figure 26.21)**. Whittaker's system recognized the two fundamentally different types of cells, prokaryotic and eukaryotic, and set the prokaryotes apart from all eukaryotes by placing them in their own kingdom, Monera.

Whittaker distinguished three kingdoms of multicellular eukaryotes—Plantae, Fungi, and Animalia—partly on the criterion of nutrition. Plants are autotrophs, making their food by photosynthesis. Fungi and animals are heterotrophs. Most fungi are decomposers that invade their food source, secreting digestive enzymes and absorbing the small organic molecules produced by digestion. Most animals ingest (eat) food and digest it within a specialized body cavity.

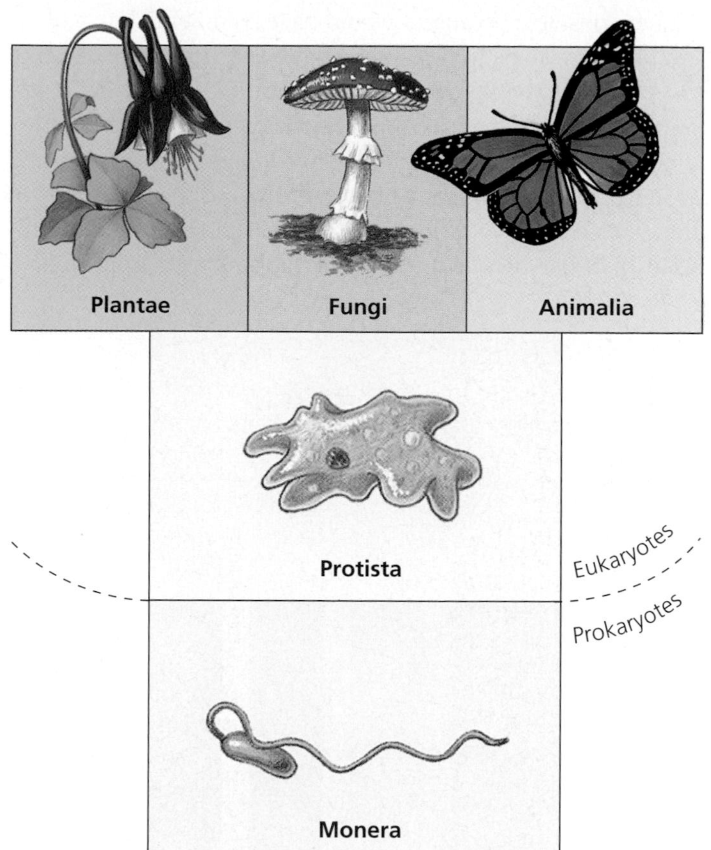

▲ **Figure 26.21 Whittaker's five-kingdom system.**

The kingdom Protista was not as clearly defined in Whittaker's five-kingdom system. Protista consisted of all eukaryotes that did not fit the definitions of plants, fungi, or animals. Most protists are unicellular, but the boundaries of Protista were expanded to include some multicellular organisms, such as seaweeds, because of their relationships to certain unicellular protists. With such refinements, the five-kingdom system prevailed in biology for more than 20 years.

Reconstructing the Tree of Life: A Work in Progress

Taxonomic systems are human constructs—attempts to order the diversity of life in a scheme that is useful and that reflects phylogenetic relationships. During the past three decades, systematists have applied cladistic analysis to taxonomy, including the construction of cladograms based on molecular data. We saw in Chapter 25 that these data have led biologists to adopt a **three-domain system.** The domains Bacteria, Archaea, and Eukarya are essentially superkingdoms, a taxonomic level higher than the kingdom level. Bacteria differ from archaea in many key structural, biochemical, and physiological characteristics, which will be highlighted in Chapter 27. These differences justify placing bacteria and archaea in separate domains. Note that the three-domain system makes the kingdom

Monera obsolete because it would have members in two different domains. In fact, many microbiologists now divide each of the two prokaryotic domains into multiple kingdoms based on phylogenetic analysis of molecular data **(Figure 26.22)**.

Another major challenge to the five-kingdom system is being mounted by systematists who are sorting out the phylogeny of the diverse eukaryotes formerly grouped in the kingdom Protista. Biologists who study these organisms now split most of them into five or more newly designated kingdoms, which have a common ancestor near the base of the eukaryotic branch of the tree. Debate continues over whether some single-celled protists should be transferred to Plantae, Fungi, or Animalia.

Much more research is needed before we can arrive at a consensus about how the three domains of life are related and about how many kingdoms each domain should include. As we explore newly discovered microbial communities, such as

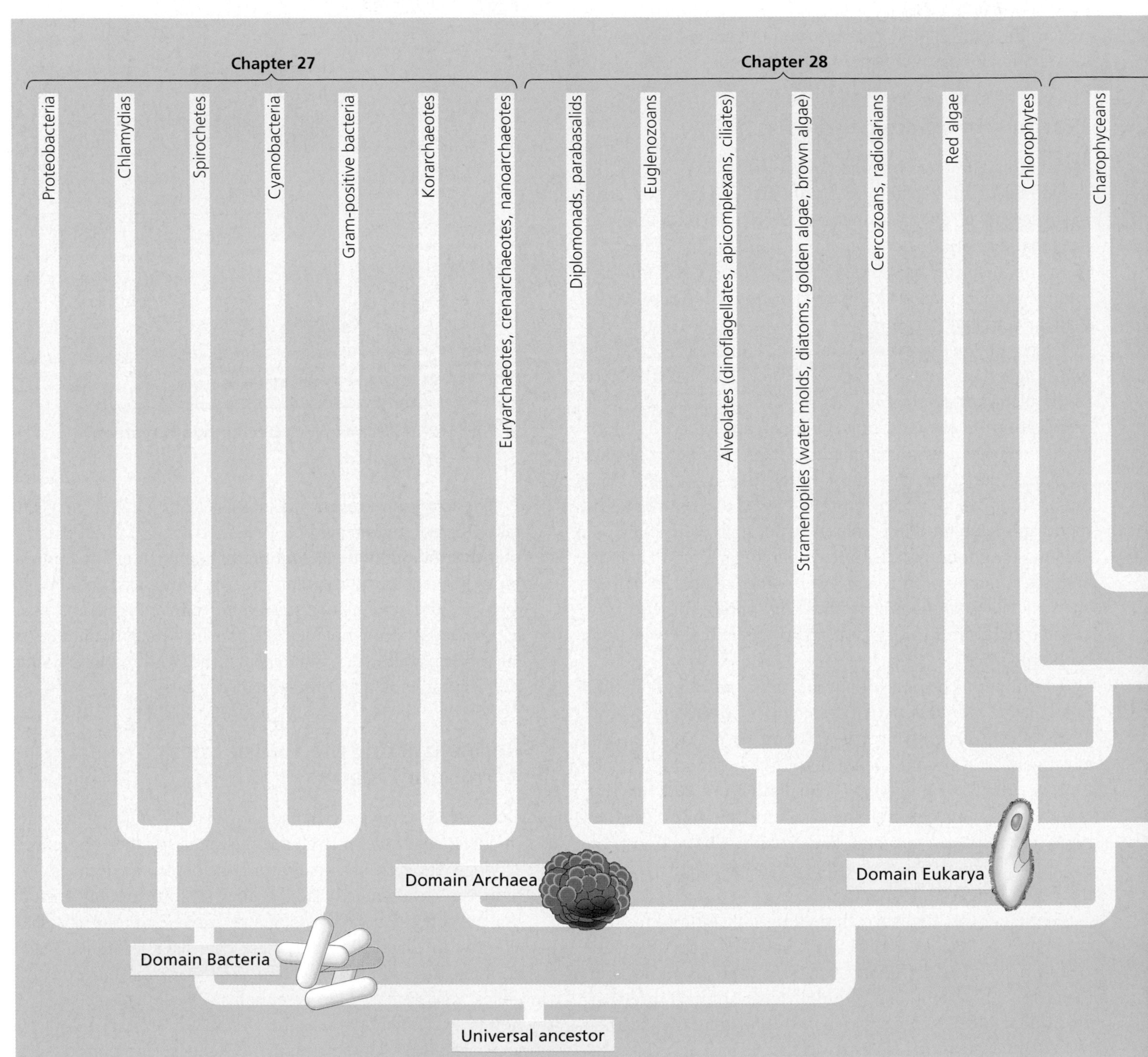

▲ **Figure 26.22 One current view of biological diversity.** This tree summarizes the diversification of life over evolutionary time. There is lively debate about some of the relationships shown here, as you will read in the chapters listed across the top of the branches.

those that live deep underground, and as we learn how to culture more of the organisms in those communities, we will no doubt discover new groups that will lead to further taxonomic remodeling. As you survey the diversity of life in Chapters 27–34, keep in mind that phylogenetic trees and taxonomic groupings are hypotheses that fit the best available data. It is the continuing scrutiny of testable hypotheses that validates evolutionary biology as a natural science.

Concept Check 26.6

1. Which kingdoms in Whittaker's five-kingdom system include organisms now in the domain Eukarya?
2. Based on Figure 26.22, explain why the kingdom Monera is no longer considered a valid taxon.

For suggested answers, see Appendix A.

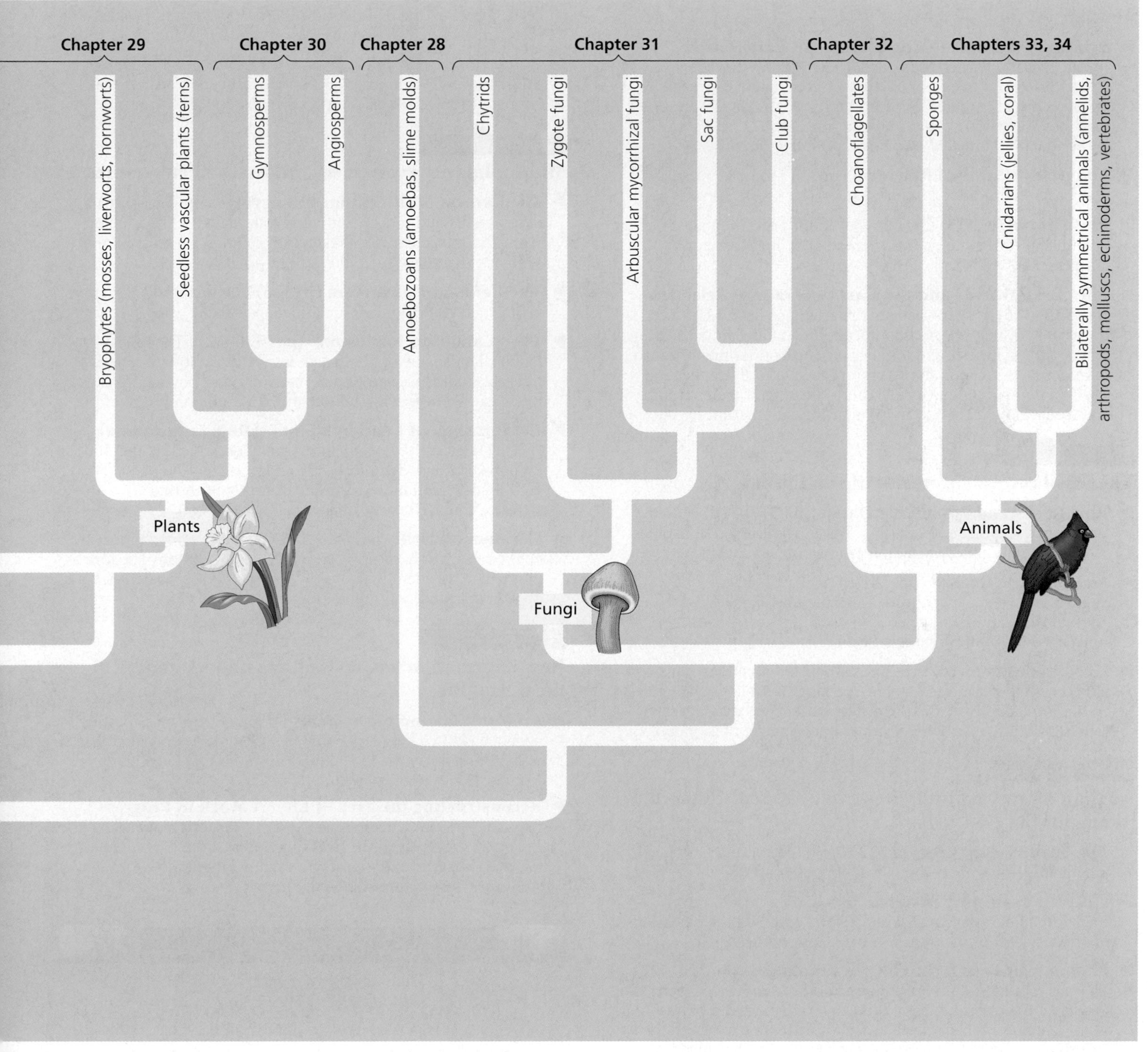

Chapter 26 Review

Go to www.campbellbiology.com or the student CD-ROM to explore Activities, Investigations, and other interactive study aids.

SUMMARY OF KEY CONCEPTS

Concept 26.1

Conditions on early Earth made the origin of life possible

▶ **Synthesis of Organic Compounds on Early Earth (pp. 513–514)** Earth formed about 4.6 billion years ago. Laboratory experiments simulating a reducing atmosphere have produced organic molecules from inorganic precursors. Amino acids have also been found in meteorites.
Investigation *How Did Life Begin on Early Earth?*

▶ **Abiotic Synthesis of Polymers (pp. 514–515)** Amino acids polymerize when added to hot sand, clay, or rock.

▶ **Protobionts (p. 515)** Organic compounds can spontaneously assemble in the laboratory into protobionts, lipid-bounded droplets that have some of the properties of cells.

▶ **The "RNA World" and the Dawn of Natural Selection (pp. 515–516)** The first genetic material may have been short pieces of RNA that served as templates for aligning amino acids in polypeptide synthesis and for aligning nucleotides in a primitive form of self-replication. Early protobionts with self-replicating, catalytic RNA would have been more effective at using resources and would have increased in number through natural selection.

Concept 26.2

The fossil record chronicles life on Earth

▶ **How Rocks and Fossils Are Dated (pp. 517–518)** Sedimentary strata reveal the relative ages of fossils. The absolute ages of fossils can be determined by radiometric dating and other methods.

▶ **The Geologic Record (pp. 518–519)** Earth's history is divided into geologic eons, eras, and periods, many of which mark major changes in the composition of fossil species.
Activity *A Scrolling Geologic Record*

▶ **Mass Extinctions (p. 518, pp. 520–521)** Evolutionary history has been punctuated by several mass extinctions followed by grand episodes of adaptive radiation. These extinctions may have been caused by volcanic activity or impacts from meteorites or comets.

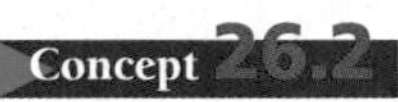

Concept 26.3

As prokaryotes evolved, they exploited and changed young Earth

▶ **The First Prokaryotes (p. 522)** Prokaryotes were Earth's sole inhabitants from 3.5 to about 2 billion years ago.

▶ **Electron Transport Systems (p. 522)** The first electron transport chains may have saved ATP by coupling the oxidation of organic acids to the transport of H^+ out of the cell.

▶ **Photosynthesis and the Oxygen Revolution (pp. 522–523)** The earliest types of photosynthesis did not produce oxygen. Oxygenic photosynthesis probably evolved about 3.5 billion years ago in cyanobacteria. The accumulation of oxygen in the atmosphere about 2.7 billion years ago posed a challenge for life, but it also selected for certain adaptations such as cellular respiration using oxygen.

Concept 26.4

Eukaryotic cells arose from symbioses and genetic exchanges between prokaryotes

▶ **The First Eukaryotes (p. 523)** The oldest fossils of eukaryotic cells date back 2.1 billion years.

▶ **Endosymbiotic Origin of Mitochondria and Plastids (pp. 523–524)** Eukaryotes may have evolved as predators of other cells. Mitochondria and plastids likely evolved from prokaryotes that were ingested by and lived symbiotically inside larger cells.

▶ **Eukaryotic Cells as Genetic Chimeras (pp. 524–525)** Additional endosymbioses and horizontal gene transfers may have contributed to the complex structures of eukaryotic cells.

Concept 26.5

Multicellularity evolved several times in eukaryotes

▶ **The Earliest Multicellular Eukaryotes (pp. 525–526)** The oldest fossils of multicellular eukaryotes date back 1.2 billion years, but molecular clocks date the common ancestor of multicellular eukaryotes to 1.5 billion years ago.

▶ **The Colonial Connection (p. 526)** The first multicellular organisms were colonies containing specialized cells.

▶ **The "Cambrian Explosion" (pp. 526–527)** The oldest fossils of most animal phyla date from the Cambrian period, but molecular evidence shows that many of these phyla originated much earlier, between 1 billion and 700 million years ago.

▶ **Colonization of Land by Plants, Fungi, and Animals (p. 527)** Plants, fungi, and animals colonized land about 500 million years ago. Symbiotic relationships between plant roots and fungi are common today and date from this time.
Activity *The History of Life*

▶ **Continental Drift (pp. 527–528)** By rearranging landmasses, continental drift has significantly affected evolution. The formation and subsequent breakup of the supercontinent Pangaea explain many biogeographic puzzles.

Concept 26.6

New information has revised our understanding of the tree of life

▶ **Previous Taxonomic Systems (p. 529)** Early classification systems had two kingdoms: plants and animals. A system that was proposed later had five kingdoms: Monera, Protista, Plantae, Fungi, and Animalia.

▶ **Reconstructing the Tree of Life: A Work in Progress (pp. 529–531)** A three-domain system (Bacteria, Archaea, and Eukarya) has replaced the five-kingdom system. Each domain has been split by taxonomists into many kingdoms.
Activity *Classification Schemes*

TESTING YOUR KNOWLEDGE

Self-Quiz

1. One current view of the origin of life proposes that, rather than forming in the atmosphere, the first organic compounds on Earth may have formed
   a. on dry land.
   b. near deep-sea vents.

c. from viruses.
d. in northern Africa.
e. when chunks that broke off from the moon bombarded Earth.

2. Which of the following steps has *not* yet been accomplished by scientists studying the origin of life?
   a. synthesis of small RNA polymers by ribozymes
   b. abiotic synthesis of polypeptides
   c. formation of molecular aggregates with selectively permeable membranes
   d. formation of protobionts that use DNA to direct the polymerization of amino acids
   e. abiotic synthesis of organic molecules

3. Populations of protobionts could begin to undergo evolutionary change by natural selection only when
   a. they were first able to catalyze chemical reactions.
   b. some kind of hereditary mechanism developed.
   c. they were able to grow and split in two.
   d. photosynthesis evolved.
   e. DNA first appeared.

4. Which statement does *not* support the hypothesis that RNA functioned as the first genetic material of early protobionts?
   a. Short RNA sequences can add limited numbers of complementary bases in the presence of nucleotide monomers.
   b. Catalytic activity has been demonstrated for RNA in living cells.
   c. Variations in base sequences produce molecules with variable stabilities in different environments.
   d. Present-day cells synthesize proteins using an RNA template.
   e. In present-day cells, RNA provides the template on which DNA nucleotides are assembled.

5. When a certain rock formed, it contained 12 mg of potassium-40. The rock now contains 3 mg of potassium-40. The half-life of potassium-40 is 1.3 billion years. How old is the rock?
   a. 0.4 billion years
   b. 0.3 billion years
   c. 1.3 billion years
   d. 2.6 billion years
   e. 5.2 billion years

6. Fossilized stromatolites
   a. all date from 2.7 billion years ago.
   b. formed around deep-sea vents.
   c. resemble bacterial communities that are found today in some warm, shallow, salty bays.
   d. provide evidence that plants moved onto land in the company of fungi around 500 million years ago.
   e. contain the first undisputed fossils of eukaryotes and date from 2.1 billion years ago.

7. The oxygen revolution changed Earth's environment dramatically. Which of the following adaptations took advantage of the presence of free oxygen in the oceans and atmosphere?
   a. the evolution of chloroplasts after early protists engulfed photosynthetic cyanobacteria
   b. the persistence of some animal groups in anaerobic habitats
   c. the evolution of photosynthetic pigments that protected early algae from the corrosive effects of oxygen
   d. the evolution of cellular respiration, which used oxygen to help harvest energy from organic molecules
   e. the evolution of multicellular eukaryotic colonies from symbiotic communities of prokaryotes

8. Which of the following represents a probable sequence of events in the biological history of Earth?
   a. metabolism before mitosis
   b. an oxidizing atmosphere before a reducing atmosphere
   c. eukaryotes before prokaryotes
   d. DNA genes before RNA genes
   e. animals before algae

9. The animals and plants of India show large differences from species in nearby Southeast Asia. Why might this be true?
   a. The species have become separated by convergent evolution.
   b. The climates of the two regions are completely different.
   c. India is in the process of separating from the rest of Asia.
   d. Life in India was wiped out by ancient volcanic eruptions.
   e. India was a separate continent until relatively recently.

10. Current debates about the number and boundaries of the kingdoms of life center *mainly* on which groups of organisms?
   a. plants and animals
   b. plants and fungi
   c. prokaryotes and single-celled eukaryotes
   d. fungi and animals
   e. amphibians and reptiles

For Self-Quiz answers, see Appendix A.

Go to the website or CD-ROM for more quiz questions.

Evolution Connection

Describe the minimum structural, metabolic, and genetic equipment for a protobiont to be considered a true cell.

Scientific Inquiry

If life were discovered elsewhere in the solar system, an obvious question would be whether terrestrial and extraterrestrial life had independent origins. If the physical and chemical attributes of such new life could be studied, what kinds of evidence would support a single origin? What scientific questions would be raised by evidence of a single origin? Of multiple origins?

Investigation *How Did Life Begin on Early Earth?*

Science, Technology, and Society

Experts estimate that human activities cause the extinction of hundreds of species every year. In contrast, the natural rate of extinction is thought to average only a few species per year. As we continue to alter the global environment, especially by destroying tropical rain forests and altering Earth's climate, the resulting wave of extinctions will probably rival those at the end of the Cretaceous period. Considering that life has endured numerous mass extinctions, should we be concerned about the present mass extinction? How does it differ from previous extinctions? What might be some of the consequences for the surviving species, including ourselves?

Biology Inquiry: A Workbook of Investigative Cases *Life on Earth has gone through many cycles of mass extinction and diversification. Explore the latter with the case "Unveiling the Carboniferous."*

27 Prokaryotes

▲ **Figure 27.1 Orange and yellow colonies of "heat-loving" prokaryotes in the hot water of a Nevada geyser.**

Key Concepts

27.1 Structural, functional, and genetic adaptations contribute to prokaryotic success
27.2 A great diversity of nutritional and metabolic adaptations have evolved in prokaryotes
27.3 Molecular systematics is illuminating prokaryotic phylogeny
27.4 Prokaryotes play crucial roles in the biosphere
27.5 Prokaryotes have both harmful and beneficial impacts on humans

Overview

They're (Almost) Everywhere!

Most prokaryotes are microscopic, but what they lack in size they more than make up for in numbers. Their collective biological mass (biomass) is at least ten times that of all eukaryotes. The number of prokaryotes in a single handful of fertile soil is greater than the number of people who have ever lived. What has enabled these tiny organisms to dominate the biosphere throughout their history? One reason for their success is a wealth of adaptations that enable various prokaryotes to inhabit diverse environments. Prokaryotes thrive almost everywhere, including places too acidic, too salty, too cold, or too hot for most other organisms **(Figure 27.1)**. They have even been discovered in rocks 2 miles below Earth's surface. In reconstructing the evolutionary history underlying the varied lifestyles of prokaryotes, biologists are discovering that these organisms have an astonishing genetic diversity. For example, comparing ribosomal RNA reveals that two strains of the bacterial species *Escherichia coli* are genetically more different than a human and a platypus.

As you read in Chapter 26, prokaryotes are classified into two domains, Bacteria and Archaea, which differ in many structural, physiological, and biochemical characteristics. In this chapter, you will read about the remarkable adaptations of prokaryotes, as well as some of the essential ecological services they perform, such as the recycling of chemicals. You will also read about the minority of prokaryotic species that cause serious illness in humans. Finally, you will learn how humans depend on benign prokaryotes for our very survival, and how biotechnology is beginning to harness the metabolic powers of these pervasive organisms.

Concept 27.1

Structural, functional, and genetic adaptations contribute to prokaryotic success

Most prokaryotes are unicellular, although some species aggregate transiently or permanently in colonies. Prokaryotic cells typically have diameters in the range of 1–5 µm, much smaller than the 10–100 µm diameter of many eukaryotic cells. (One notable exception is the giant prokaryote *Thiomargarita namibiensis,* which is about 750 µm in diameter—just visible to the unaided eye.) Prokaryotic cells have a variety of shapes, the three most common of which are spheres (cocci), rods (bacilli), and spirals **(Figure 27.2)**.

Cell-Surface Structures

One of the most important features of nearly all prokaryotic cells is their cell wall, which maintains cell shape, provides physical protection, and prevents the cell from bursting in a hypotonic environment (see Chapter 7). In a hypertonic environment,

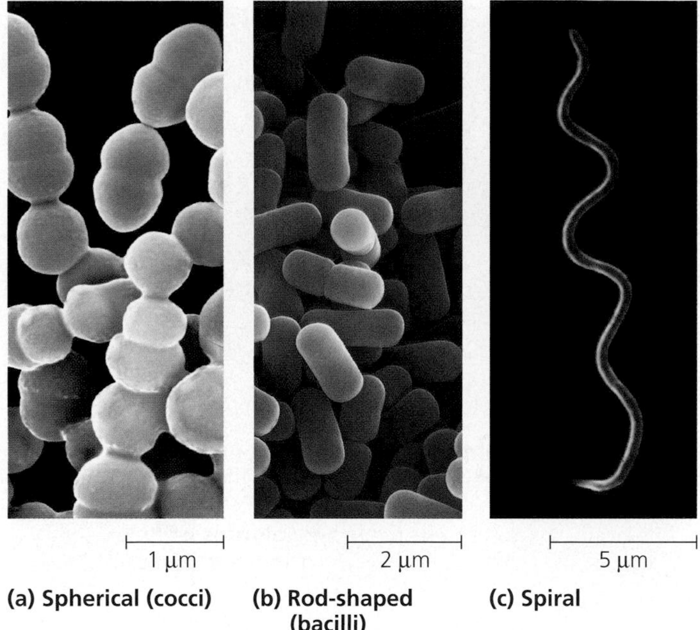

▲ **Figure 27.2 The most common shapes of prokaryotes.** **(a)** Cocci (singular, *coccus*) are spherical prokaryotes. They occur singly, in pairs (diplococci), in chains of many cells (streptococci, shown here), and in clusters resembling bunches of grapes (staphylococci). **(b)** Bacilli (singular, *bacillus*) are rod-shaped prokaryotes. They are usually solitary, but in some forms the rods are arranged in chains (streptobacilli). **(c)** Spiral prokaryotes include spirilla, which range from comma-like shapes to long coils, and spirochetes (shown here), which are corkscrew-shaped (colorized SEMs).

most prokaryotes lose water and shrink away from their wall (plasmolyze), like other walled cells. Severe water loss inhibits the reproduction of prokaryotes, which explains why salt can be used to preserve certain foods, such as pork and fish.

The cell walls of prokaryotes differ in molecular composition and construction from those of eukaryotes. As you read in Chapter 5, eukaryotic cell walls are usually made of cellulose or chitin. In contrast, most bacterial cell walls contain **peptidoglycan,** a network of modified-sugar polymers cross-linked by short polypeptides. This molecular fabric encloses the entire bacterium and anchors other molecules that extend from its surface. Archaeal cell walls contain a variety of polysaccharides and proteins but lack peptidoglycan.

Using a technique called the **Gram stain,** developed by the nineteenth-century Danish physician Hans Christian Gram, scientists can classify many bacterial species into two groups based on differences in cell wall composition. **Gram-positive** bacteria have simpler walls with a relatively large amount of peptidoglycan **(Figure 27.3a)**. **Gram-negative** bacteria have less peptidoglycan and are structurally more complex, with an outer membrane that contains lipopolysaccharides (carbohydrates bonded to lipids) **(Figure 27.3b)**.

Gram staining is a particularly valuable identification tool in medicine. Among pathogenic, or disease-causing, bacteria, gram-negative species are generally more threatening than gram-positive species. The lipopolysaccharides on the walls of gram-negative bacteria are often toxic, and the outer membrane helps protect these bacteria against the body's defenses. Furthermore, gram-negative bacteria are commonly more resistant than gram-positive species to antibiotics because the outer membrane impedes entry of the drugs.

The effectiveness of certain antibiotics, including penicillin, derives from their inhibition of the peptidoglycan cross-linking, thus preventing the formation of a functional cell wall, particularly in gram-positive bacteria. Such drugs destroy many

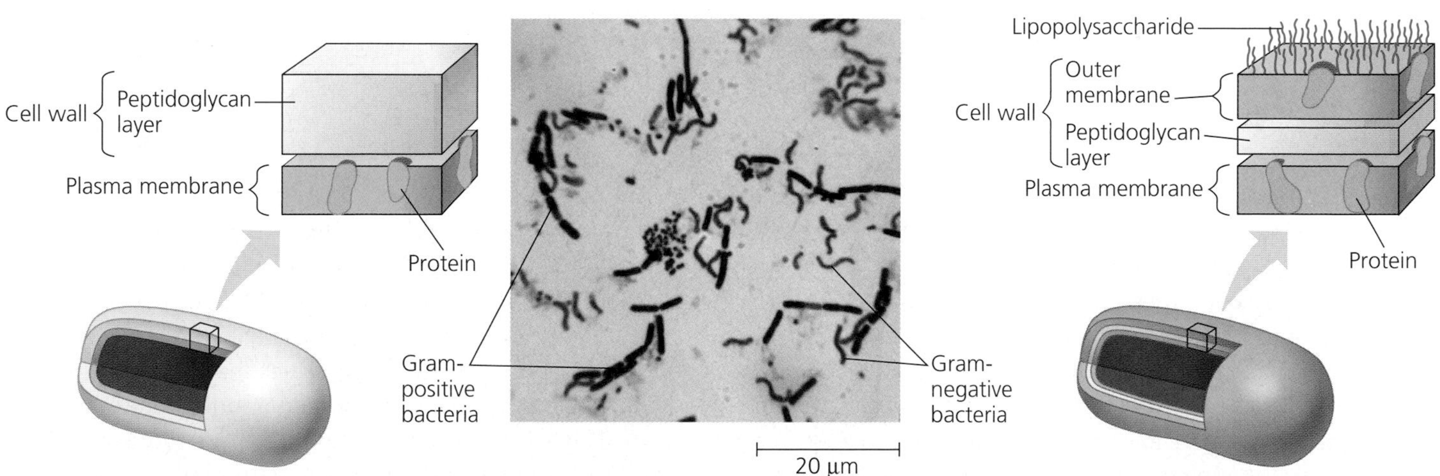

(a) Gram-positive. Gram-positive bacteria have a cell wall with a large amount of peptidoglycan that traps the violet dye in the cytoplasm. The alcohol rinse does not remove the violet dye, which masks the added red dye.

(b) Gram-negative. Gram-negative bacteria have less peptidoglycan, and it is located in a layer between the plasma membrane and an outer membrane. The violet dye is easily rinsed from the cytoplasm, and the cell appears pink or red.

▲ **Figure 27.3 Gram staining.** Bacteria are stained with a violet dye and iodine, rinsed in alcohol, and then stained with a red dye. The structure of the cell wall determines the staining response (LM).

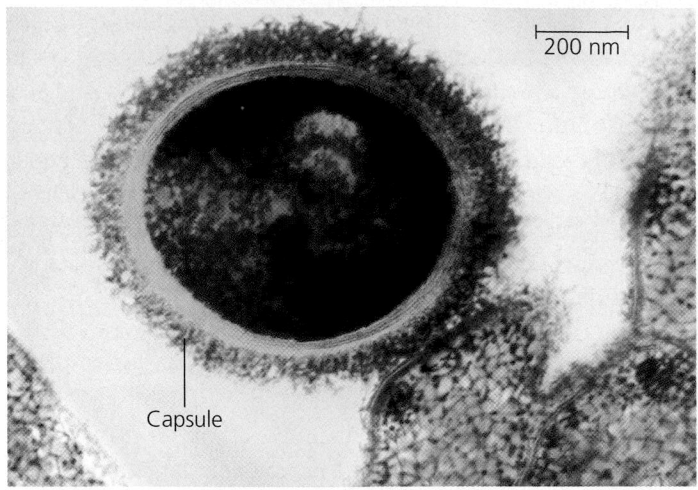

▲ **Figure 27.4 Capsule.** The polysaccharide capsule surrounding this *Streptococcus* bacterium enables the pathogenic prokaryote to attach to cells that line the human respiratory tract—in this image, a tonsil cell (colorized TEM).

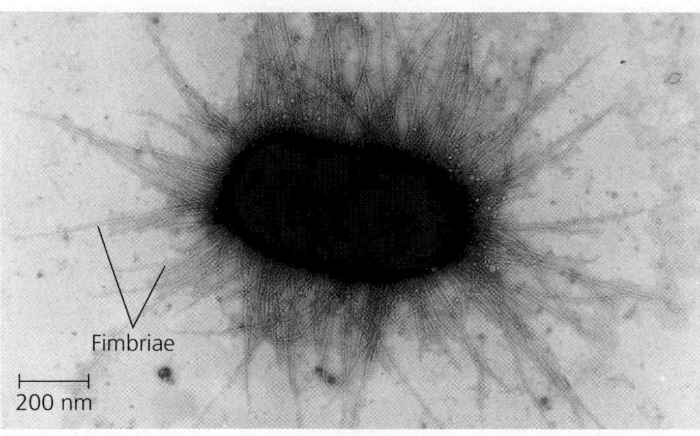

▲ **Figure 27.5 Fimbriae.** These numerous appendages enable some prokaryotes to attach to surfaces or to other prokaryotes (colorized TEM).

▶ **Figure 27.6 Prokaryotic flagellum.** The motor of the prokaryotic flagellum is the basal apparatus, a system of rings embedded in the cell wall and plasma membrane (TEM). ATP-driven pumps transport protons out of the cell, and the diffusion of protons back into the cell powers the basal apparatus, which turns a curved hook. The hook is attached to a filament composed of chains of flagellin, a globular protein. (This diagram shows flagellar structures characteristic of gram-negative bacteria.)

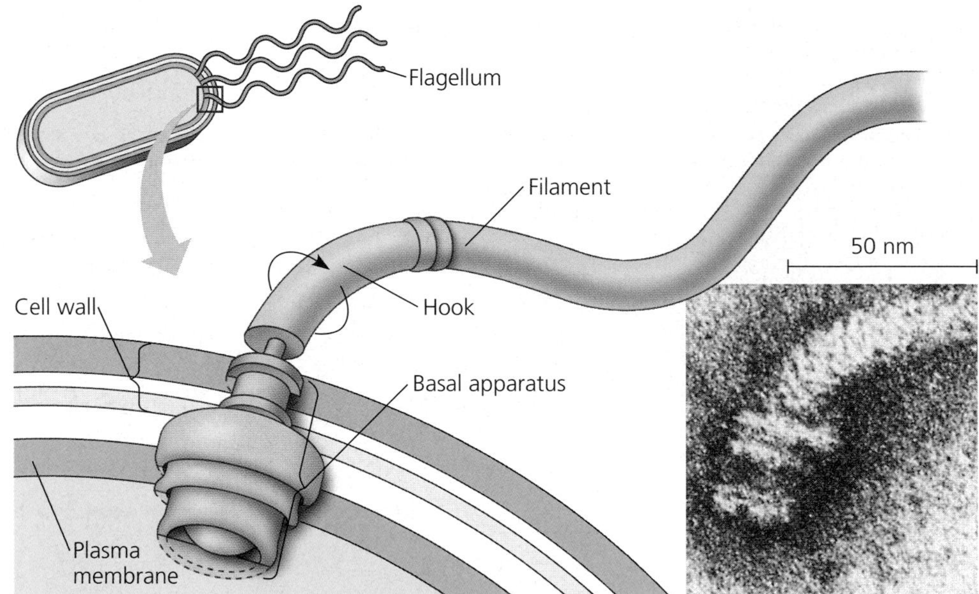

species of pathogenic bacteria without adversely affecting human cells, which do not contain peptidoglycan.

The cell wall of many prokaryotes is covered by a **capsule**, a sticky layer of polysaccharide or protein **(Figure 27.4)**. The capsule enables prokaryotes to adhere to their substrate or to other individuals in a colony. Capsules can also shield pathogenic prokaryotes from attacks by their host's immune system.

Some prokaryotes stick to their substrate or to one another by means of hairlike appendages called **fimbriae** (singular, *fimbria*) and **pili** (singular, *pilus*). Fimbriae are usually more numerous and shorter than pili **(Figure 27.5)**. *Neisseria gonorrhoeae,* the bacterium that causes gonorrhea, uses fimbriae to fasten itself to the mucous membranes of its host. Specialized pili, called sex pili, link prokaryotes during conjugation, a process in which one cell transfers DNA to another cell (see Figure 18.17).

Motility

About half of all prokaryotes are capable of directional movement. Some species can move at speeds exceeding 50 μm/sec—up to 50 times their body length per second.

Of the various structures that enable prokaryotes to move, the most common are flagella, which may be scattered over the entire cell surface or concentrated at one or both ends of the cell. The flagella of prokaryotes differ from those of eukaryotes in both structure and mechanism of propulsion **(Figure 27.6)**. Prokaryotic flagella are one-tenth the width of eukaryotic flagella and are not covered by an extension of the plasma membrane (see Figures 6.24 and 6.25 to review eukaryotic flagella).

In a relatively uniform environment, flagellated prokaryotes may move randomly. In a heterogeneous environment, however, many prokaryotes exhibit **taxis**, movement toward or away from a stimulus (from the Greek *taxis*, to arrange). For example, prokaryotes that exhibit chemotaxis respond to chemicals by changing their movement pattern. They may move *toward* nutrients or oxygen (positive chemotaxis) or *away from* a toxic substance (negative chemotaxis). In 2003, scientists at Princeton University and the Institut Curie in Paris demonstrated that solitary *E. coli* cells exhibit positive chemotaxis toward other members of their species, enabling the formation of colonies.

Internal and Genomic Organization

The cells of prokaryotes are simpler than those of eukaryotes, in both their internal structure and their genomic organization. Prokaryotic cells lack the complex compartmentalization found in eukaryotic cells (see Figure 6.6). However, some prokaryotic cells do have specialized membranes that perform metabolic functions **(Figure 27.7)**. These membranes are usually infoldings of the plasma membrane.

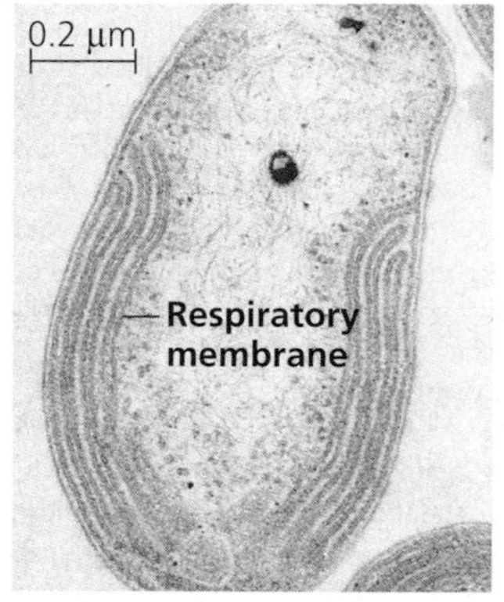

(a) Aerobic prokaryote

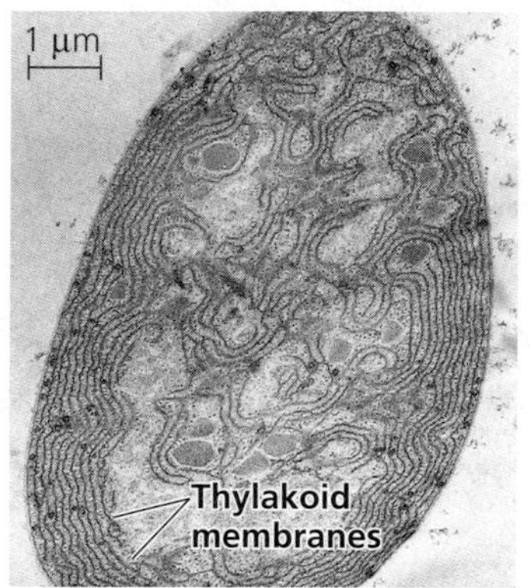

(b) Photosynthetic prokaryote

▲ **Figure 27.7 Specialized membranes of prokaryotes.** **(a)** Infoldings of the plasma membrane, reminiscent of the cristae of mitochondria, function in cellular respiration in some aerobic prokaryotes (TEM). **(b)** Photosynthetic prokaryotes called cyanobacteria have thylakoid membranes, much like those in chloroplasts (TEM).

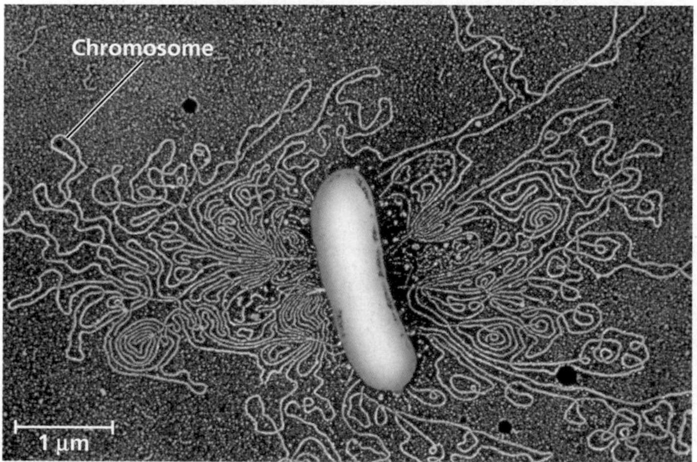

▲ **Figure 27.8 A prokaryotic chromosome.** The thin, tangled loops surrounding this ruptured *E. coli* cell are parts of a single ring of DNA (colorized TEM).

The genome of a prokaryote is structurally very different from a eukaryotic genome and has on average only about one-thousandth as much DNA. In the majority of prokaryotes, most of the genome consists of a ring of DNA that has relatively few proteins associated with it. This ring of genetic material is usually called the prokaryotic chromosome **(Figure 27.8)**. Unlike eukaryotic chromosomes, which are contained within the nucleus, the prokaryotic chromosome is located in a **nucleoid region,** a part of the cytoplasm that appears lighter than the surrounding cytoplasm in electron micrographs.

In addition to its single chromosome, a typical prokaryotic cell may also have much smaller rings of DNA called **plasmids,** most consisting of only a few genes. The plasmid genes provide resistance to antibiotics, direct the metabolism of rarely encountered nutrients, or have other such "contingency" functions. In most environments, the prokaryotic cell can survive without its plasmids, since all essential functions are encoded by the chromosome. But in certain circumstances, such as when antibiotics are used to treat an infection, the presence of a plasmid can significantly increase a prokaryote's chance of survival. Plasmids replicate independently of the main chromosome, and many can be readily transferred between partners when prokaryotes conjugate (see Figure 18.18).

As explained in Chapters 16 and 17, DNA replication, transcription, and translation are fundamentally similar in prokaryotes and eukaryotes, although there are some differences. For example, prokaryotic ribosomes are slightly smaller than eukaryotic ribosomes and differ in their protein and RNA content. These differences are great enough that certain antibiotics, such as erythromycin and tetracycline, bind to ribosomes and block protein synthesis in prokaryotes but not in eukaryotes. As a result, we can use these antibiotics to kill bacteria without harming ourselves.

Reproduction and Adaptation

Prokaryotes are highly successful in part because of their potential to reproduce quickly in a favorable environment. Dividing by binary fission (see Figure 12.11), a single prokaryotic cell becomes 2 cells, which then become 4, 8, 16, and so on. While most prokaryotes can divide every 1–3 hours, some species can produce a new generation in only 20 minutes under optimal conditions. If reproduction continued unchecked at this rate, a single prokaryote could give rise to a colony outweighing Earth in only three days! In reality, of course, prokaryotic reproduction is limited, as the cells eventually exhaust their nutrient supply, poison themselves with metabolic wastes, or are consumed by other organisms. Prokaryotes in nature also face competition from other microorganisms, many of which produce antibiotic chemicals that slow prokaryotic reproduction.

The ability of some prokaryotes to withstand harsh conditions also contributes to their success. Certain bacteria, for example, can form resistant cells called **endospores** when an

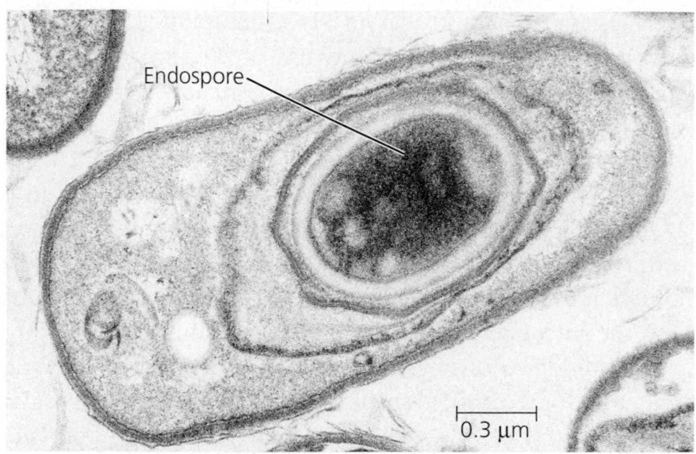

▲ **Figure 27.9 An endospore.** *Bacillus anthracis,* the bacterium that causes the deadly disease anthrax, produces endospores (TEM). An endospore's thick, protective coat helps it survive in the soil for years.

essential nutrient is lacking in the environment **(Figure 27.9)**. The original cell produces a copy of its chromosome and surrounds it with a tough wall, forming the endospore. Water is removed from the endospore, and metabolism inside it comes to a halt. The rest of the original cell then disintegrates, leaving the endospore behind. Most endospores are so durable that they can survive in boiling water. To kill endospores, microbiologists must heat their lab equipment with steam at 121°C under high pressure. In less hostile environments, endospores can remain dormant but viable for centuries, able to rehydrate and resume metabolism when they receive cues that their environment has become more benign.

Prokaryotes can adapt quickly to changes in their environment through evolution by natural selection. Because of prokaryotes' rapid reproduction, mutations that confer greater fitness can swiftly become more common in a population. For this reason, prokaryotes are important model organisms for scientists who study evolution in the laboratory. At Michigan State University, for example, Richard Lenski and his team have maintained colonies of *E. coli* through more than 20,000 generations since 1988. The researchers regularly freeze samples of the colonies and later thaw them to compare their characteristics with those of later generations. Such comparisons have revealed that the colonies today can grow 60% faster than the 1988 colonies under the same environmental conditions. Lenski's team is exploring the genetic changes underlying the colonies' evolutionary adaptation to their environment. In 2003, they reported that two colonies showed parallel changes in expression for the same 59 genes compared to the original colonies. The rapid reproduction of *E. coli* enabled the scientists to document this example of adaptive evolution.

Horizontal gene transfer (see Chapter 25) also facilitates rapid evolution in prokaryotes. For example, conjugation can permit the exchange of a plasmid containing a few genes or even large groups of genes. Once the transferred genes are incorporated into a prokaryote's genome, they are subject to natural selection during subsequent rounds of binary fission. Horizontal gene transfer is a major force in the long-term evolution of pathogenic bacteria, a topic explored later in this chapter.

Concept Check 27.1

1. Identify and explain at least two examples of adaptations that enable prokaryotes to survive in environments too harsh for other organisms.
2. Contrast the cellular and genomic organization of prokaryotes and eukaryotes.
3. Explain how rapid reproduction allows prokaryotes to adapt to changing environments.

For suggested answers, see Appendix A.

Concept 27.2

A great diversity of nutritional and metabolic adaptations have evolved in prokaryotes

All organisms can be categorized by nutrition—how they obtain energy and carbon used in building the organic molecules that make up cells. Nutritional diversity is greater among prokaryotes than among all eukaryotes: Every type of nutrition observed in eukaryotes is represented among prokaryotes, along with some nutritional modes unique to prokaryotes.

Organisms that obtain energy from light are called *phototrophs,* and those that obtain energy from chemicals are called *chemotrophs.* Organisms that need only the inorganic compound CO_2 as a carbon source are called *autotrophs.* In contrast, *heterotrophs* require at least one organic nutrient—such as glucose—to make other organic compounds. Combining these possibilities for energy sources and carbon sources results in four major modes of nutrition, described here and summarized in **Table 27.1**.

1. **Photoautotrophs** are photosynthetic organisms that capture light energy and use it to drive the synthesis of organic compounds from CO_2. Cyanobacteria and many other groups of prokaryotes are photoautotrophs, as are plants and algae.
2. **Chemoautotrophs** also need only CO_2 as a carbon source. However, instead of using light for energy, they oxidize inorganic substances, such as hydrogen sulfide (H_2S), ammonia (NH_3), or ferrous ions (Fe^{2+}). This mode of nutrition is unique to certain prokaryotes.

Table 27.1 Major Nutritional Modes

Mode of Nutrition	Energy Source	Carbon Source	Types of Organisms
Autotroph			
Photoautotroph	Light	CO_2	Photosynthetic prokaryotes (for example, cyanobacteria); plants; certain protists (algae)
Chemoautotroph	Inorganic chemicals	CO_2	Certain prokaryotes (for example, *Sulfolobus*)
Heterotroph			
Photoheterotroph	Light	Organic compounds	Certain prokaryotes (for example, *Rhodobacter, Chloroflexus*)
Chemoheterotroph	Organic compounds	Organic compounds	Many prokaryotes (for example, *Clostridium*) and protists; fungi; animals; some plants

3. **Photoheterotrophs** use light for energy but must obtain their carbon in organic form. A number of marine prokaryotes use this mode of nutrition.
4. **Chemoheterotrophs** must consume organic molecules for both energy and carbon. This nutritional mode is found widely among prokaryotes as well as protists, fungi, animals, and even some parasitic plants.

Metabolic Relationships to Oxygen

Prokaryotic metabolism also varies with respect to oxygen (see Chapter 9). **Obligate aerobes** use O_2 for cellular respiration and cannot grow without it. **Facultative anaerobes** use O_2 if it is present but can also grow by fermentation in an anaerobic environment. **Obligate anaerobes** are poisoned by O_2. Some obligate anaerobes live exclusively by fermentation; others extract chemical energy by **anaerobic respiration,** in which substances other than O_2, such as nitrate ions (NO_3^-) or sulfate ions (SO_4^{2-}), accept electrons at the "downhill" end of electron transport chains.

Nitrogen Metabolism

Nitrogen is essential for the production of amino acids and nucleic acids in all organisms. While eukaryotes are limited in the nitrogenous compounds they can use, prokaryotes can metabolize nitrogen in a wide variety of forms. For example, certain prokaryotes, including some cyanobacteria, convert atmospheric nitrogen (N_2) to ammonia (NH_3), a process called **nitrogen fixation.** The cells can then incorporate this "fixed" nitrogen into amino acids and other organic molecules. In terms of their nutrition, nitrogen-fixing cyanobacteria are the most self-sufficient of all organisms. They require only light, CO_2, N_2, water, and some minerals to grow. Chapter 54 discusses the essential roles that prokaryotes play in the nitrogen cycles of ecosystems.

Metabolic Cooperation

Cooperation between prokaryotes allows them to use environmental resources they could not use as individual cells. In some cases, this cooperation takes place between specialized cells of a colony. For instance, the cyanobacterium *Anabaena* has genes encoding proteins for photosynthesis and for nitrogen fixation, but a single cell cannot carry out both processes at the same time. The reason is that photosynthesis produces O_2, which inactivates the enzymes involved in nitrogen fixation. Instead of living as isolated cells, *Anabaena* forms filamentous colonies **(Figure 27.10)**. Most cells in a filament carry out only photosynthesis, while a few specialized cells called *heterocytes* (formerly *heterocysts*) carry out only nitrogen fixation. Heterocytes are surrounded by a thickened cell wall that restricts entry of O_2 produced by neighboring photosynthetic cells. Intercellular connections allow heterocytes to transport fixed nitrogen to neighboring cells in exchange for carbohydrates.

In some prokaryotic species, metabolic cooperation occurs in surface-coating colonies known as **biofilms (Figure 27.11)**.

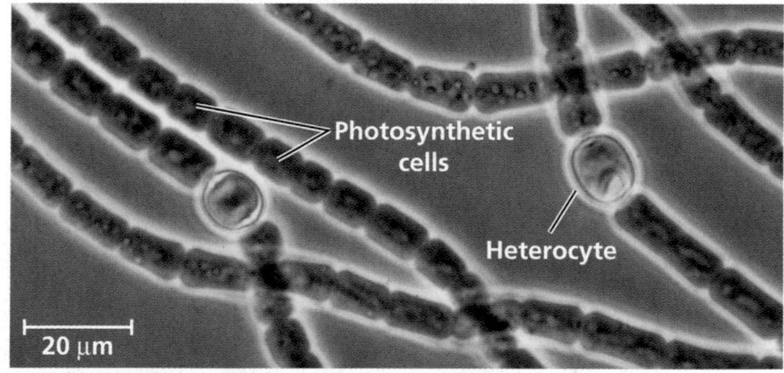

▲ **Figure 27.10 Metabolic cooperation in a colonial prokaryote.** In the filamentous cyanobacterium *Anabaena,* cells known as heterocytes fix nitrogen, while the other cells carry out photosynthesis (LM). *Anabaena* is found in many freshwater lakes.

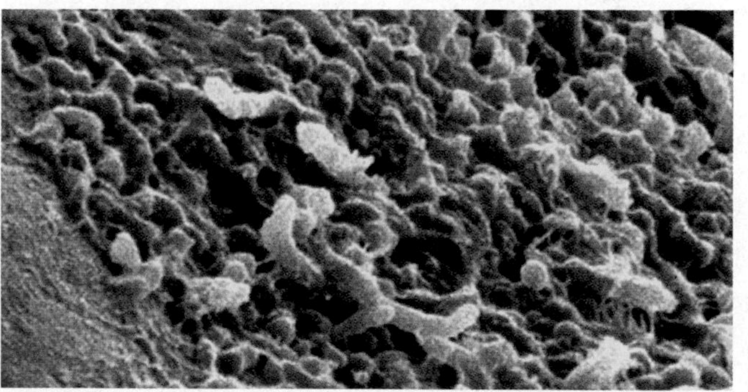

▲ **Figure 27.11 A biofilm.** The yellow mass in this colorized SEM is dental plaque, a biofilm that forms on tooth surfaces.

Cells in a colony secrete signaling molecules that recruit nearby cells, causing the colony to grow. The cells also produce proteins that adhere the cells to the substrate and to one another. Channels in the biofilm allow nutrients to reach cells in the interior and wastes to be expelled.

Prokaryotes belonging to different species also cooperate. For example, sulfate-consuming bacteria and methane-consuming archaea coexist in ball-shaped aggregates on the ocean floor. The bacteria appear to use the archaea's waste products, such as organic compounds and hydrogen. In turn, the bacteria produce compounds that facilitate methane consumption by the archaea. This partnership has global ramifications: Each year, these archaea consume an estimated 300 billion kg of methane, a major contributor to the greenhouse effect (see Chapter 54).

Concept Check 27.2

1. A bacterium requires only the amino acid methionine as an organic nutrient and lives in lightless caves. What mode of nutrition does it employ? Explain.
2. What are the sources of carbon and nitrogen for the cyanobacterium *Anabaena*?

For suggested answers, see Appendix A.

Concept 27.3

Molecular systematics is illuminating prokaryotic phylogeny

Until the late 20th century, systematists based prokaryotic taxonomy on phenotypic criteria such as shape, motility, nutritional mode, and response to Gram staining. These criteria are still valuable in certain contexts, such as the rapid identification of pathogenic bacteria cultured from a patient's blood. But when it comes to prokaryotic phylogeny, comparing these characteristics does not reveal a clear history. Applying molecular systematics to the investigation of prokaryotic phylogeny, however, has produced dramatic results.

Lessons from Molecular Systematics

As discussed in Chapter 25, microbiologists began comparing the sequences of prokaryotic genes in the 1970s. Using small-subunit ribosomal RNA (SSU-rRNA) as a marker for evolutionary relationships, Carl Woese and his colleagues concluded that many prokaryotes once classified as bacteria are actually

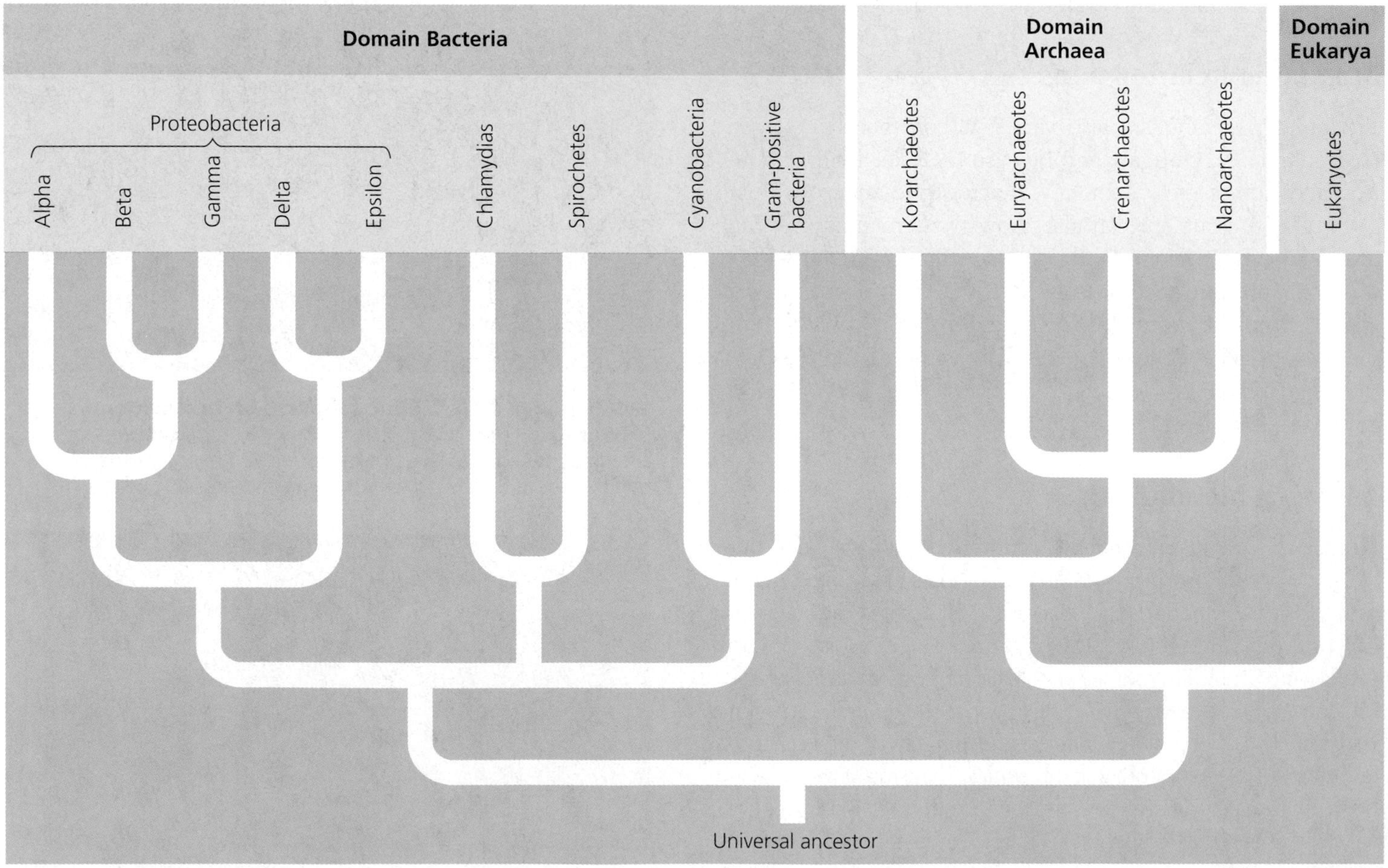

▲ **Figure 27.12 A simplified phylogeny of prokaryotes.** This phylogenetic tree based on molecular data highlights the relationships between the major prokaryotic groups discussed in this chapter.

more closely related to eukaryotes and belong in a domain of their own—Archaea. Microbiologists have since analyzed larger amounts of genetic data—in some cases, entire genomes—and have concluded that a few traditional taxonomic groups, such as cyanobacteria, are in fact monophyletic. However, other groups, such as gram-negative bacteria, are scattered throughout several lineages. **Figure 27.12** shows a tentative phylogeny of some of the major taxa of prokaryotes based on molecular systematics.

Prokaryotic phylogeny is still a work in progress, but two important lessons have already emerged. One is that the genetic diversity of prokaryotes is immense. When researchers began to sequence the genes of prokaryotes, they could investigate only those species that could be cultured in the laboratory—a small minority of all prokaryotic species. Methods pioneered by Norman Pace of the University of Colorado in the 1980s now allow researchers to sample genetic material directly from the environment. Every year this "genetic prospecting" adds major new branches to the tree of life. (Some researchers suggest that certain branches represent entire new kingdoms.) While only 4,500 prokaryotic species have been fully characterized, a single handful of soil could contain 10,000 prokaryotic species, according to some estimates. You can see why taking full stock of this diversity will require many years of research.

Another important lesson is the apparent significance of horizontal gene transfer in the evolution of prokaryotes. Over hundreds of millions of years, prokaryotes have acquired genes from distantly related species, and they continue to do so today. As a result, significant portions of the genomes of many prokaryotes are actually mosaics of genes imported from other species.

Note again in Figure 27.12 a key inference based on molecular systematics: the very early divergence of prokaryotes into two main lineages, bacteria and archaea.

Bacteria

Bacteria include the vast majority of prokaryotes that most people are aware of, from the pathogenic species that cause strep throat and other diseases to the beneficial species used to make Swiss cheese. Every major mode of nutrition and metabolism is represented among bacteria, and even a small taxonomic group of bacteria may contain species exhibiting many different nutritional modes. Examine **Figure 27.13** on the following pages for a closer look at several major groups of bacteria.

Archaea

Archaea share certain traits with bacteria and other traits with eukaryotes **(Table 27.2)**. However, archaea also have many unique characteristics, as we would expect for a taxon that has followed a separate evolutionary path for so long.

Table 27.2 A Comparison of the Three Domains of Life

CHARACTERISTIC	DOMAIN		
	Bacteria	**Archaea**	**Eukarya**
Nuclear envelope	Absent	Absent	Present
Membrane-enclosed organelles	Absent	Absent	Present
Peptidoglycan in cell wall	Present	Absent	Absent
Membrane lipids	Unbranched hydrocarbons	Some branched hydrocarbons	Unbranched hydrocarbons
RNA polymerase	One kind	Several kinds	Several kinds
Initiator amino acid for protein synthesis	Formyl-methionine	Methionine	Methionine
Introns (noncoding parts of genes)	Rare	Present in some genes	Present
Response to the antibiotics streptomycin and chloramphenicol	Growth inhibited	Growth not inhibited	Growth not inhibited
Histones associated with DNA	Absent	Present	Present
Circular chromosome	Present	Present	Absent
Ability to grow at temperatures > 100°C	No	Some species	No

The first prokaryotes that were classified in domain Archaea are species that live in environments so extreme that few other organisms can survive there. Such organisms are known as **extremophiles**, meaning "lovers" of extreme conditions (from the Greek *philos*, lover). Extremophiles include extreme thermophiles, extreme halophiles, and methanogens.

Extreme thermophiles (from the Greek *thermos*, hot) thrive in very hot environments (see Figure 27.1). For example, archaea in the genus *Sulfolobus* live in sulfur-rich volcanic springs at temperatures up to 90°C. *Pyrolobus fumarii*, an extreme thermophile found around deep-sea hydrothermal vents on the Mid-Atlantic Ridge, can survive at temperatures as high as 113°C. Another extreme thermophile, *Pyrococcus furiosus*, is used in biotechnology as a source of DNA polymerase for the polymerase chain reaction (PCR) technique (see Chapter 20).

Extreme halophiles (from the Greek *halo*, salt) live in highly saline environments, such as the Great Salt Lake and the Dead Sea. Some species merely tolerate salinity, while others require an environment that is several times saltier than seawater. Colonies of certain extreme halophiles form a purple-red scum that owes its color to bacteriorhodopsin, a

Figure 27.13

Exploring Major Groups of Bacteria

GROUP/DESCRIPTION	EXAMPLE

PROTEOBACTERIA

This large and diverse clade of gram-negative bacteria includes photoautotrophs, chemoautotrophs, and heterotrophs. Some proteobacteria are anaerobic; others are aerobic. Molecular systematists currently recognize five subgroups of proteobacteria.

SUBGROUP: ALPHA PROTEOBACTERIA

Many of the species in this subgroup are closely associated with eukaryotic hosts. For example, *Rhizobium* species live in nodules within the roots of legumes (plants of the pea/bean family), where the bacteria convert atmospheric N_2 to compounds the host plant can use to make proteins. Species in the genus *Agrobacterium* produce tumors in plants; genetic engineers use these bacteria to carry foreign DNA into the genomes of crop plants (see Figure 20.19). As explained in Chapter 26, scientists hypothesize that mitochondria evolved from aerobic alpha proteobacteria through endosymbiosis.

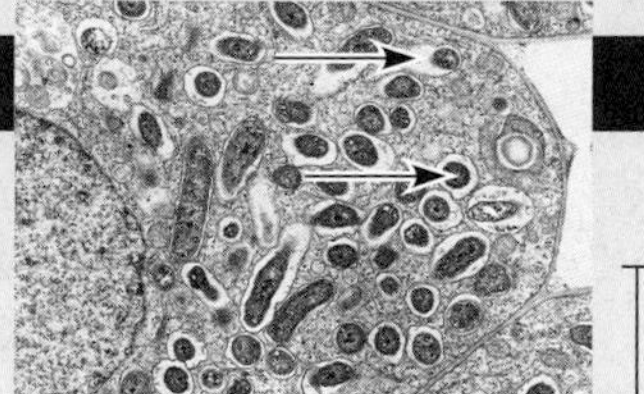

Rhizobium (arrows) inside a root cell of a legume (TEM)

SUBGROUP: BETA PROTEOBACTERIA

This nutritionally diverse subgroup includes *Nitrosomonas,* a genus of soil bacteria that play an important role in nitrogen recycling by oxidizing ammonium (NH_4^+), producing nitrite (NO_2^-) as a waste product.

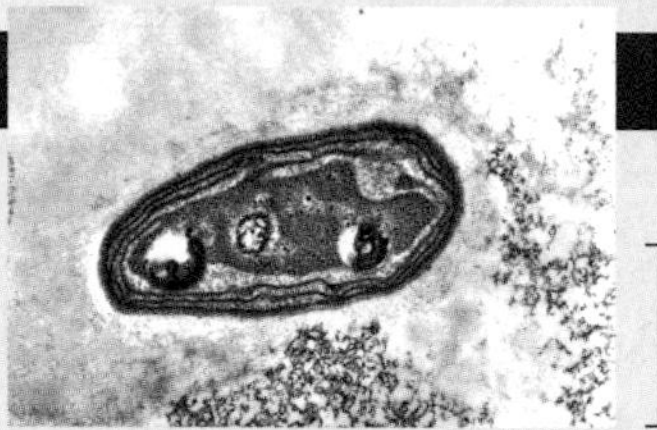

Nitrosomonas (colorized TEM)

SUBGROUP: GAMMA PROTEOBACTERIA

This subgroup's photosynthetic members include sulfur bacteria such as *Chromatium,* which obtain energy by oxidizing H_2S, producing sulfur as a waste. Some heterotrophic gamma proteobacteria are pathogens; for example, *Legionella* causes Legionnaires' disease, *Salmonella* is responsible for some cases of food poisoning, and *Vibrio cholerae* causes cholera. *Escherichia coli,* a common resident of the intestines of humans and other mammals, normally is not pathogenic.

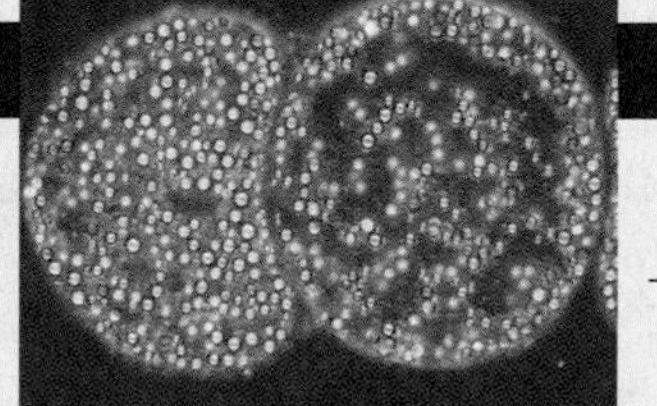

Chromatium; the small globules are sulfur wastes (LM)

SUBGROUP: DELTA PROTEOBACTERIA

This subgroup includes the slime-secreting myxobacteria, which form elaborate colonies. When the soil dries out or food is scarce, the cells congregate into a fruiting body that releases resistant spores. The spores become active and found new colonies in favorable environments. Bdellovibrios are delta proteobacteria that attack other bacteria. They charge their prey at up to 100 µm/sec (comparable to a human running 600 km/hr) and bore into the prey by spinning at 100 revolutions per second.

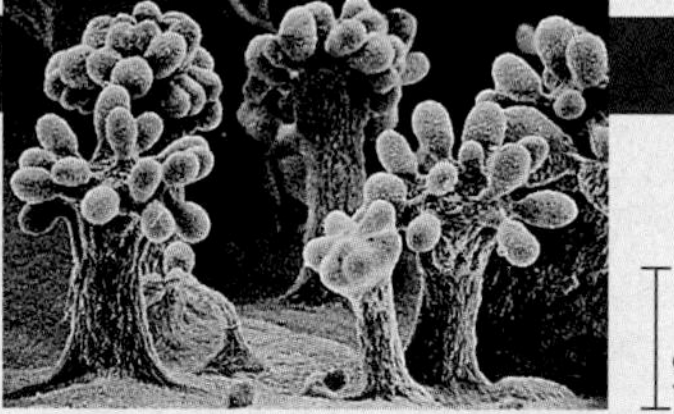

Fruiting bodies of *Chondromyces crocatus,* a myxobacterium (SEM)

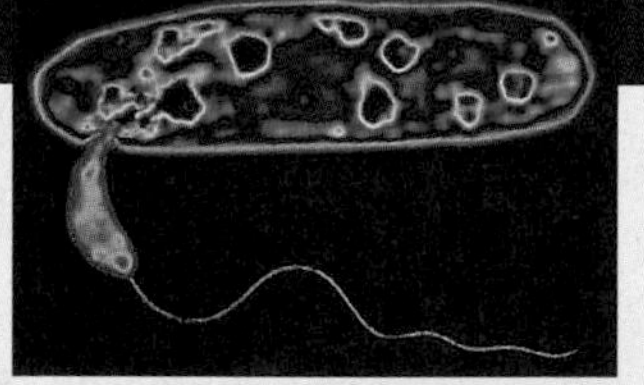

Bdellovibrio bacteriophorus attacking a larger bacterium (colorized TEM)

SUBGROUP: EPSILON PROTEOBACTERIA

Most species in this subgroup are pathogenic to humans or other animals. Epsilon proteobacteria include *Campylobacter,* which causes blood poisoning and intestinal inflammation, and *Helicobacter pylori,* which causes stomach ulcers.

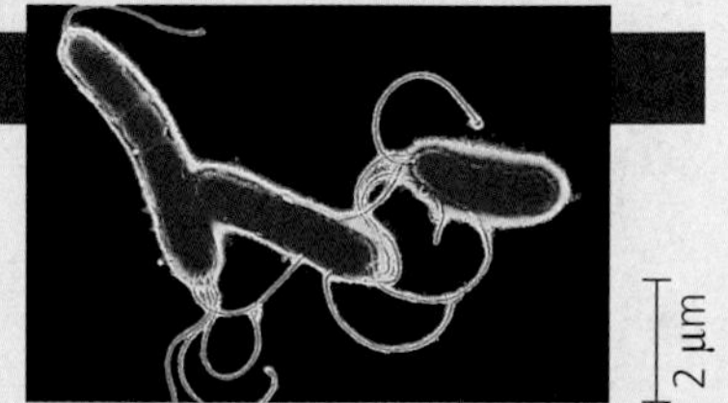

Helicobacter pylori (colorized TEM)

GROUP/DESCRIPTION	EXAMPLE

CHLAMYDIAS

These parasites can survive only within animal cells, depending on their hosts for resources as basic as ATP. The gram-negative walls of chlamydias are unusual in that they lack peptidoglycan. One species, *Chlamydia trachomatis,* is the most common cause of blindness in the world and also causes nongonococcal urethritis, the most common sexually transmitted disease in the United States.

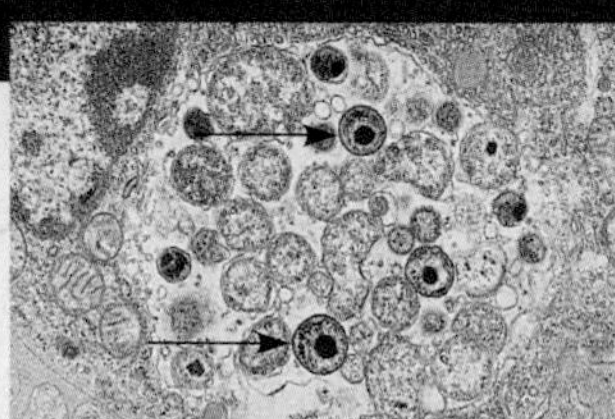

Chlamydia (arrows) inside an animal cell (colorized TEM)

SPIROCHETES

These helical heterotrophs spiral through their environment by means of rotating, internal, flagellum-like filaments. Many spirochetes are free-living, but others are notorious pathogenic parasites: *Treponema pallidum* causes syphilis, and *Borrelia burgdorferi* causes Lyme disease.

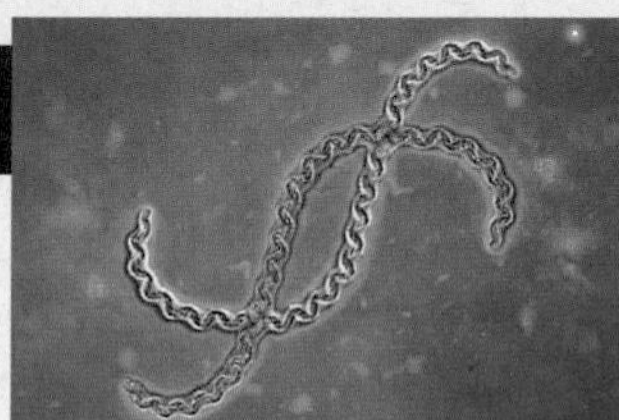

Leptospira, a spirochete (colorized TEM)

GRAM-POSITIVE BACTERIA

Gram-positive bacteria rival the proteobacteria in diversity. Species in one subgroup, the actinomycetes (from the Greek *mykes,* fungus, for which these bacteria were once mistaken), form colonies containing branched chains of cells. Two species of actinomycetes cause tuberculosis and leprosy. However, most actinomycetes are free-living species that help decompose the organic matter in soil; their secretions are partly responsible for the "earthy" odor of rich soil. Soil-dwelling species in the genus *Streptomyces* are cultured by pharmaceutical companies as a source of many antibiotics, including streptomycin.

In addition to the colonial actinomycetes, gram-positive bacteria include many solitary species, such as *Bacillus anthracis* (see Figure 27.9), which causes anthrax, and *Clostridium botulinum*, which causes botulism. Various species of *Staphylococcus* and *Streptococcus* are also gram-positive bacteria.

Mycoplasmas are the only bacteria known to lack cell walls. They are also the tiniest of all known cells, with diameters as small as 0.1 μm, only about five times as large as a ribosome. Mycoplasmas have remarkably small genomes—*Mycoplasma genitalium* has only 517 genes, for example. Many mycoplasmas are free-living soil bacteria, but others are pathogens, including a species that causes "walking pneumonia" in humans.

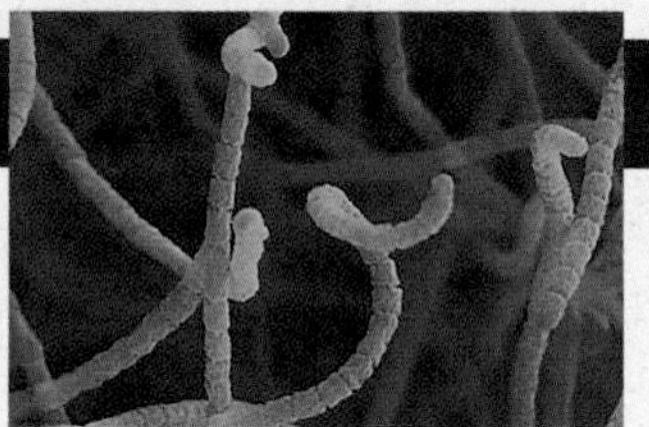

Streptomyces, the source of many antibiotics (colorized SEM)

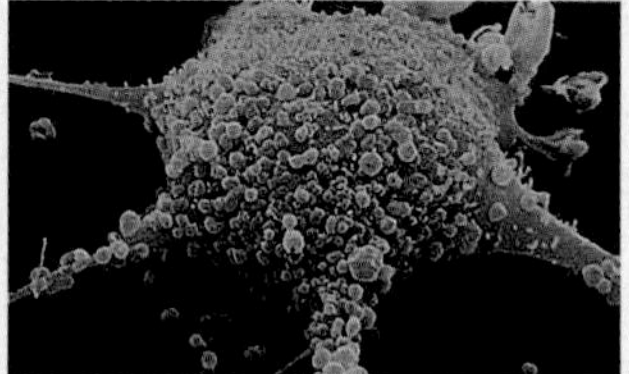

Hundreds of mycoplasmas covering a human fibroblast cell (colorized SEM)

CYANOBACTERIA

These photoautotrophs are the only prokaryotes with plantlike, oxygen-generating photosynthesis. (In fact, chloroplasts likely evolved from an endosymbiotic cyanobacterium; see Chapter 26). Both solitary and colonial cyanobacteria are abundant wherever there is water, providing an enormous amount of food for freshwater and marine ecosystems. Some filamentous colonies have cells specialized for nitrogen fixation, the process that converts atmospheric N_2 to compounds that can be incorporated into proteins and other organic molecules (see Figure 27.10).

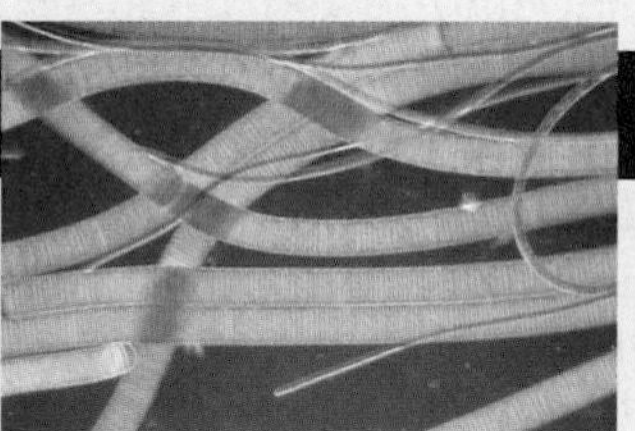

Two species of *Oscillatoria,* filamentous cyanobacteria (LM)

▲ **Figure 27.14 Extreme halophiles.** Colorful "salt-loving" archaea thrive in these ponds near San Francisco. Used for commercial salt production, the ponds contain water that is five to six times as salty as seawater.

photosynthetic pigment very similar to the visual pigments in the vertebrate retina **(Figure 27.14)**.

Methanogens are named for the unique way they obtain energy: They use CO_2 to oxidize H_2, releasing methane as a waste product. Among the strictest of anaerobes, methanogens are poisoned by O_2. Some species live in swamps and marshes where other microorganisms have consumed all the O_2. The "marsh gas" found in such environments is the methane produced by these archaea. Other species of methanogens inhabit the anaerobic environment within the guts of cattle, termites, and other herbivores, playing an essential role in the nutrition of these animals. Methanogens are also important decomposers in sewage treatment facilities.

All known extreme halophiles and methanogens are members of a clade called Euryarchaeota (from the Greek *eurys*, broad, a reference to the habitat range of these prokaryotes). Euryarchaeota also includes some extreme thermophiles, though most thermophilic species belong to a second clade, Crenarchaeota (*cren* means "spring," as in hydrothermal springs). Genetic prospecting has revealed that both Euryarchaeota and Crenarchaeota also include many species of archaea that are not extremophiles. These species exist in habitats ranging from farm soils to lake sediments to the surface waters of the open ocean.

New findings continue to update the picture of archaeal phylogeny. In 1996, researchers sampling a hot spring in Yellowstone National Park discovered archaea that do not appear to belong to either Euryarchaeota or Crenarchaeota. They placed these archaea in a new clade, Korarchaeota (from the Greek *koron,* young man). The oldest lineage in the domain Archaea, Korarchaeota may offer clues to the early evolution of life on Earth. In 2002, researchers exploring hydrothermal vents off the coast of Iceland discovered archaeal cells only 0.4 μm in diameter attached to a much larger crenarchaeote. The genome of the tiny archaean is one of the smallest known of any organism, containing only 500,000 base pairs. Analysis of the genome indicates that this prokaryote belongs to a fourth archaeal clade, now called Nanoarchaeota (from the Greek *nanos,* dwarf). Within a year after this clade was named, three other DNA sequences from nanoarchaeote species were isolated: one from Yellowstone's hot springs, one from hot springs in Siberia, and one from a hydrothermal vent in the Pacific. As prospecting continues, it seems likely that the tree in Figure 27.12 will undergo further change in years to come.

Concept Check 27.3

1. Explain how molecular systematics has greatly increased our understanding of prokaryotic phylogeny.
2. What do syphilis and Lyme disease have in common?
3. What characteristics enable some species of archaea to live in extreme environments?

For suggested answers, see Appendix A.

Concept 27.4

Prokaryotes play crucial roles in the biosphere

If humans were to disappear from the planet tomorrow, life on Earth would go on for most other species. But prokaryotes are so important to the biosphere that if they were to disappear, the prospects for any other life surviving would be dim.

Chemical Recycling

The atoms that make up the organic molecules in all living things were at one time part of inorganic compounds in the soil, air, and water. Sooner or later, that is where those atoms will return. Ecosystems depend on the continual recycling of chemical elements between the living and nonliving components of the environment, and prokaryotes play a major role in this process. For example, chemoheterotrophic prokaryotes function as **decomposers,** breaking down corpses, dead vegetation, and waste products, thereby unlocking supplies of carbon, nitrogen, and other elements. (See Chapter 54 for a detailed discussion of chemical cycles.)

Prokaryotes also convert inorganic compounds into forms that can be taken up by other organisms. Autotrophic prokaryotes, for example, use CO_2 to make organic compounds, which are then passed up through food chains. Cyanobacteria produce atmospheric O_2, and some species also fix nitrogen into a form that other organisms can use to make proteins.

Symbiotic Relationships

Just as certain species of prokaryotes have beneficial associations with other prokaryotes (metabolic cooperation), some prokaryotes form similarly intimate relationships with eukaryotes. An ecological relationship between organisms of different species that are in direct contact is called **symbiosis** (from a Greek word meaning "living together"). If one of the symbiotic organisms is much larger than the other, the larger is known as the **host** and the smaller is known as the **symbiont.** Symbiotic relationships are categorized as either mutualism, commensalism, or parasitism. In **mutualism**, both symbiotic organisms benefit **(Figure 27.15)**. In **commensalism**, one organism benefits while neither harming nor helping the other in any significant way. (Commensalism is rare in nature, as further discussed in Chapter 53.) In **parasitism**, one organism, called a **parasite**, benefits at the expense of the host.

The well-being of many eukaryotes—yourself included—depends on mutualistic prokaryotes. For example, human intestines are home to an estimated 500 to 1,000 species of bacteria; their cells outnumber all human cells in the body by as much as ten times. Many of these species are mutualists, digesting food that our own intestines cannot break down. In 2003, scientists at Washington University in St. Louis published the first complete genome of one of these gut mutualists, *Bacteroides thetaiotaomicron.* The genome includes a large array of genes involved in synthesizing carbohydrates, vitamins, and other nutrients needed by humans. Signals from the bacterium activate human genes that build the network of intestinal blood vessels necessary to absorb food. Other signals induce human cells to produce antimicrobial compounds to which *B. thetaiotaomicron* is not susceptible. Keeping other competing bacteria out of the intestines benefits *B. thetaiotaomicron* as well as its human host.

▲ **Figure 27.15 Mutualism: bacterial "headlights."** The glowing oval below the eye of the flashlight fish (*Photoblepharon palpebratus*) is an organ harboring bioluminescent bacteria. The fish uses the light to attract prey and to signal potential mates. The bacteria receive nutrients from the fish.

Concept Check 27.4

1. Although individual prokaryotes may be tiny, they are giants in their collective impact on Earth and its life. Explain.
2. Explain how the relationship between humans and *B. thetaiotaomicron* is an example of mutualism.

For suggested answers, see Appendix A.

Concept 27.5

Prokaryotes have both harmful and beneficial impacts on humans

While the best-known prokaryotes tend to be those that cause illness in humans, these pathogens represent only a small fraction of prokaryotic species. Many other prokaryotes have positive interactions with humans, even serving as essential tools in agriculture and industry.

Pathogenic Prokaryotes

The prokaryotic species that are human parasites deserve their negative reputation. All told, prokaryotes cause about half of all human diseases. Between 2 and 3 million people a year die of the lung disease tuberculosis, which is caused by the bacillus *Mycobacterium tuberculosis*, while another 2 million die from various diarrheal diseases caused by other prokaryotes. In the United States, the most widespread pest-carried disease is Lyme disease **(Figure 27.16)**. Caused by a bacterium carried

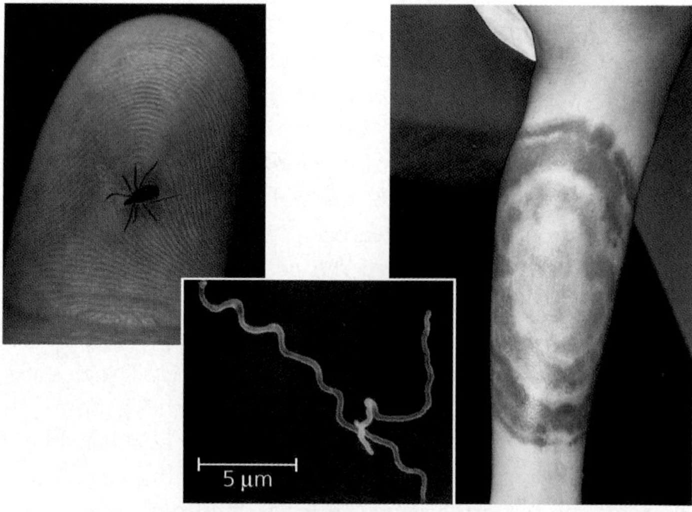

▲ **Figure 27.16 Lyme disease.** Ticks in the genus *Ixodes* spread the disease by transmitting the spirochete *Borrelia burgdorferi* (colorized SEM). A large, ring-shaped rash may develop at the site of the tick's bite, as shown in the photograph of a person's lower leg.

by ticks that live on deer and field mice, Lyme disease can produce debilitating arthritis, heart disease, and nervous disorders if untreated.

Pathogenic prokaryotes usually cause illness by producing poisons, which are classified as exotoxins or endotoxins. **Exotoxins** are proteins secreted by prokaryotes. Cholera, a dangerous diarrheal disease, is caused by an exotoxin released by the proteobacterium *Vibrio cholerae.* The exotoxin stimulates intestinal cells to release chloride ions into the gut, and water follows by osmosis. Exotoxins can produce disease even if the prokaryotes that manufacture them are not present. For example, the fatal disease botulism is caused by botulinum toxin, an exotoxin secreted by the gram-positive bacterium *Clostridium botulinum* as it ferments improperly canned foods.

Endotoxins are lipopolysaccharide components of the outer membrane of gram-negative bacteria. In contrast to exotoxins, endotoxins are released only when the bacteria die and their cell walls break down. Examples of endotoxin-producing bacteria include nearly all species in the genus *Salmonella,* which are not normally present in healthy animals. *Salmonella typhi* causes typhoid fever, and several other *Salmonella* species, some of which are frequently found in poultry, cause food poisoning.

Since the 19th century, improvements in sanitation in the developed world have greatly reduced the threat of pathogenic prokaryotes. Antibiotics have saved a great many lives and reduced the incidence of disease. However, resistance to antibiotics is currently evolving in many strains of prokaryotes. As you read earlier, the rapid reproduction of prokaryotes enables genes conferring resistance to multiply quickly throughout prokaryotic populations as a result of natural selection, and these genes can spread to other species by horizontal gene transfer.

Horizontal gene transfer can also spread genes associated with virulence, turning normally harmless prokaryotes into fatal pathogens. *E. coli,* for instance, is ordinarily a harmless symbiont in the human intestines, but pathogenic strains that cause bloody diarrhea have emerged. One of the most dangerous strains, called O157:H7, first came to the attention of microbiologists in 1982. Today it is a global threat; in the United States alone there are 75,000 cases of O157:H7 infection per year, often from contaminated beef. In 2001, an international team of scientists sequenced the genome of O157:H7 and compared it with the genome of a harmless strain of *E. coli* called K-12. They discovered that 1,387 out of the 5,416 genes in O157:H7 have no counterpart in K-12. These 1,387 genes must have been incorporated into the genome of O157:H7 through horizontal gene transfer, most likely through the action of bacteriophages (see Figure 18.16). Many of the imported genes are associated with the pathogenic bacterium's invasion of its host. For example, some genes code for exotoxins that enable O157:H7 to attach itself to the intestinal wall and extract nutrients.

Pathogenic prokaryotes pose a potential threat as weapons of bioterrorism. In October 2001, endospores of *Bacillus anthracis,* the bacterium that causes anthrax, were found in envelopes mailed to members of news media and the U.S. Senate. Eighteen people developed cases of anthrax, and five died. Other prokaryotes that could be candidates as weapons include *C. botulinum* and *Yersinia pestis,* which causes plague. The threat has stimulated intense research on pathogenic prokaryotic species. In May 2003, scientists at the Institute for Genomic Research in Maryland published the complete genome of the strain of *B. anthracis* that had been used in the October 2001 attack, in the hope of developing new vaccines and antibiotics.

Prokaryotes in Research and Technology

On a positive note, we reap many benefits from the metabolic capabilities of prokaryotes. For example, humans have long used bacteria to convert milk to cheese and yogurt. In recent years, our greater understanding of prokaryotes has led to an explosion of new applications in biotechnology; the use of *E. coli* in gene cloning and of *Agrobacterium tumefaciens* in producing transgenic plants are two examples (see Chapter 20).

Prokaryotes are the principal agents in **bioremediation**, the use of organisms to remove pollutants from soil, air, or water. For example, anaerobic bacteria and archaea decompose the organic matter in sewage, converting it to material that can be used as landfill or fertilizer after chemical sterilization. Other bioremediation applications include breaking down radioactive waste and cleaning up oil spills **(Figure 27.17)**.

In the mining industry, prokaryotes help recover metals from ores. Bacteria assist in extracting over 30 billion kg of copper from copper sulfides each year. Harnessing other prokaryotes that can extract gold from ore, one factory in the

▲ **Figure 27.17 Bioremediation of an oil spill.** A worker sprays fertilizers on an oil-soaked beach in Alaska. The fertilizers stimulate growth of native bacteria that initiate the breakdown of the oil—in some cases, speeding the natural breakdown process fivefold.

African nation of Ghana processes 1 million kg of gold concentrate a day—about half of Ghana's foreign exchange.

Through genetic engineering, humans can now modify prokaryotes to produce vitamins, antibiotics, hormones, and other products (see Chapter 20). One of the most radical ideas for modifying prokaryotes has come from Craig Venter (one of the leaders of the Human Genome Project), who has announced that he and his colleagues are attempting to build "synthetic chromosomes" for prokaryotes—in effect, producing entirely new species from scratch. Venter hopes to "design" prokaryotes that can perform specific tasks, such as producing large amounts of hydrogen to reduce dependence on fossil fuels.

The usefulness of prokaryotes largely derives from their diverse forms of nutrition and metabolism. All this metabolic versatility evolved prior to the appearance of the structural novelties that heralded the evolution of eukaryotic organisms, the topic of the remainder of this unit.

Concept Check 27.5

1. Contrast exotoxins and endotoxins.
2. What features of prokaryotes make them a potential bioterrorism threat?
3. Identify at least two ways that prokaryotes have affected you positively today.

For suggested answers, see Appendix A.

Chapter 27 Review

Go to www.campbellbiology.com or the student CD-ROM to explore Activities, Investigations, and other interactive study aids.

SUMMARY OF KEY CONCEPTS

Concept 27.1

Structural, functional, and genetic adaptations contribute to prokaryotic success

- **Cell-Surface Structures (pp. 534–536)** Nearly all prokaryotes have a cell wall. Gram-positive and gram-negative bacteria differ in the structure of their walls. Many species have a capsule, fimbriae, and pili outside the cell wall, which help the cells adhere to one another or to a substrate.
  Activity *Prokaryotic Cell Structure and Function*
- **Motility (pp. 536–537)** Most motile bacteria propel themselves by flagella, which are structurally and functionally different from eukaryotic flagella. In heterogeneous environments, many prokaryotes can move toward or away from certain stimuli.
- **Internal and Genomic Organization (p. 537)** Prokaryotic cells usually lack complex compartmentalization. The typical prokaryotic genome is a ring of DNA that is not surrounded by a membrane. Some species also have smaller rings of DNA called plasmids.
- **Reproduction and Adaptation (pp. 537–538)** Prokaryotes reproduce quickly by binary fission. Many form endospores, which can remain viable in harsh conditions for centuries. Rapid reproduction and horizontal gene transfer facilitate the evolution of prokaryotes in changing environments.

Concept 27.2

A great diversity of nutritional and metabolic adaptations have evolved in prokaryotes

- Examples of all four modes of nutrition—photoautotrophy, chemoautotrophy, photoheterotrophy, and chemoheterotrophy—are found among prokaryotes (pp. 538–539).
- **Metabolic Relationships to Oxygen (p. 539)** Obligate aerobes require O_2, obligate anaerobes are poisoned by O_2, and facultative anaerobes can survive with or without O_2.
- **Nitrogen Metabolism (p. 539)** Prokaryotes can metabolize a wide variety of nitrogenous compounds. Some can convert atmospheric nitrogen to ammonia in a process called nitrogen fixation.
- **Metabolic Cooperation (pp. 539–540)** Many prokaryotes depend on the metabolic activities of other prokaryotes. In the cyanobacterium *Anabaena,* photosynthetic cells and nitrogen-fixing cells exchange metabolic products. Some prokaryotes can form surface-coating colonies called biofilms, which may include different species.
  Investigation *What Are the Modes of Nutrition in Prokaryotes?*

Concept 27.3

Molecular systematics is illuminating prokaryotic phylogeny

- **Lessons from Molecular Systematics (pp. 540–541)** Molecular systematics is leading to a phylogenetic classification of prokaryotes, allowing systematists to identify major new clades.
  Activity *Classification of Prokaryotes*
- **Bacteria (pp. 541–543)** Diverse nutritional types are scattered among the major groups of bacteria. The two largest groups are the proteobacteria and the gram-positive bacteria.
- **Archaea (pp. 541–544)** Archaea share certain traits with bacteria and other traits with eukaryotes. Some archaea live in extreme environments; they include extreme thermophiles, extreme halophiles, and methanogens.

Concept 27.4

Prokaryotes play crucial roles in the biosphere

- **Chemical Recycling (p. 544)** Decomposition by heterotrophic prokaryotes and the synthetic activities of autotrophic and nitrogen-fixing prokaryotes contribute to the recycling of elements in ecosystems.
- **Symbiotic Relationships (p. 545)** Many prokaryotes live with other organisms in symbiotic relationships: mutualism, commensalism, or parasitism.

Prokaryotes have both harmful and beneficial impacts on humans

- **Pathogenic Prokaryotes (pp. 545–546)** Pathogenic prokaryotes typically cause disease by releasing exotoxins or endotoxins and are potential weapons of bioterrorism. Horizontal gene transfer can spread genes associated with virulence to harmless strains.
- **Prokaryotes in Research and Technology (pp. 546–547)** Experiments involving prokaryotes such as *E. coli* and *A. tumefaciens* have led to important advances in DNA technology. Prokaryotes are major tools in bioremediation, mining, and the synthesis of vitamins, antibiotics, and other products.

TESTING YOUR KNOWLEDGE

Self-Quiz

1. Which of the following is *not* true of peptidoglycan?
   a. It is composed of modified-sugar polymers.
   b. It anchors other molecules on the surface of a bacterium.
   c. Gram-positive bacteria have a relatively large amount of it.
   d. It is found in the cell walls of all prokaryotes.
   e. It is located outside the plasma membrane of most bacteria.
2. Photoautotrophs use
   a. light as an energy source and CO_2 as a carbon source.
   b. light as an energy source and methane as a carbon source.
   c. N_2 as an energy source and CO_2 as a carbon source.
   d. CO_2 as both an energy source and a carbon source.
   e. H_2S as an energy source and CO_2 as a carbon source.
3. Which of the following statements is *not* true?
   a. Archaea and bacteria have different membrane lipids.
   b. Both archaea and bacteria generally lack membrane-enclosed organelles.
   c. The cell walls of archaea lack peptidoglycan.
   d. Only bacteria have histones associated with DNA.
   e. Bacteria include the spirochetes.
4. Which of the following features of prokaryotic biology involves metabolic cooperation among cells?
   a. binary fission
   b. endospore formation
   c. endotoxin release
   d. biofilms
   e. photoautotrophy
5. Which of the following statements about archaea is *not* true?
   a. Archaea include euryarchaeotes and crenarchaeotes.
   b. Archaea include methanogens and extreme thermophiles.
   c. Archaea are not found in terrestrial habitats.
   d. Archaea share features with both bacteria and eukaryotes.
   e. Archaea are distinguished from bacteria based partly on molecular systematics.
6. Which prokaryotic group is mismatched with its members?
   a. Proteobacteria—diverse gram-negative bacteria
   b. Chlamydias—intracellular parasites
   c. Spirochetes—helical heterotrophs
   d. Gram-positive bacteria—symbionts in legume root nodules
   e. Cyanobacteria—solitary and colonial photoautotrophs
7. What kind of relationship exists between the Lyme disease-causing bacterium *B. burgdorferi* and humans?
   a. mutualism
   b. commensalism
   c. parasitism
   d. metabolic cooperation
   e. none
8. Erythromycin functions as an antibiotic mainly by inhibiting the ability of some prokaryotes to
   a. form spores.
   b. replicate DNA.
   c. synthesize normal cell walls.
   d. synthesize protein on ribosomes.
   e. synthesize ATP.
9. Plant-like photosynthesis that releases O_2 occurs in
   a. cyanobacteria.
   b. chlamydias.
   c. archaea.
   d. actinomycetes.
   e. chemoautotrophic bacteria.
10. An example of bioremediation is
    a. the use of prokaryotes to treat sewage or clean up oil spills.
    b. the production of antibiotics by cultured prokaryotes.
    c. the application of bacteria to produce transgenic plants.
    d. the introduction of parasitic bacteria to kill other bacteria.
    e. all of the above.

For Self-Quiz answers, see Appendix A.

Go to the website or CD-ROM for more quiz questions.

Evolution Connection

Health officials worldwide are concerned about a resurgence of diseases caused by bacteria that are resistant to standard antibiotics. For instance, antibiotic-resistant bacteria are causing an epidemic of tuberculosis (TB), a lung disease spread by airborne droplets. Drugs can relieve TB symptoms in a few weeks, but it takes much longer to halt the infection, and patients are likely to discontinue treatment while bacteria are still present. Why can prokaryotes quickly reinfect a patient if they are not wiped out? How might this result in the evolution of drug-resistant pathogens?

Scientific Inquiry

You read that some scientists are investigating the possibility of engineering entirely new prokaryotic species. What are the risks and potential benefits of such a project? What insights into this research can be found in the natural history and evolutionary biology of prokaryotes?

Investigation *What Are the Modes of Nutrition in Prokaryotes?*

Science, Technology, and Society

Many local newspapers regularly publish a list of restaurants that have been cited by inspectors for poor sanitation. Locate such a report and highlight the cases that are likely associated with potential food contamination by pathogenic prokaryotes.

28 Protists

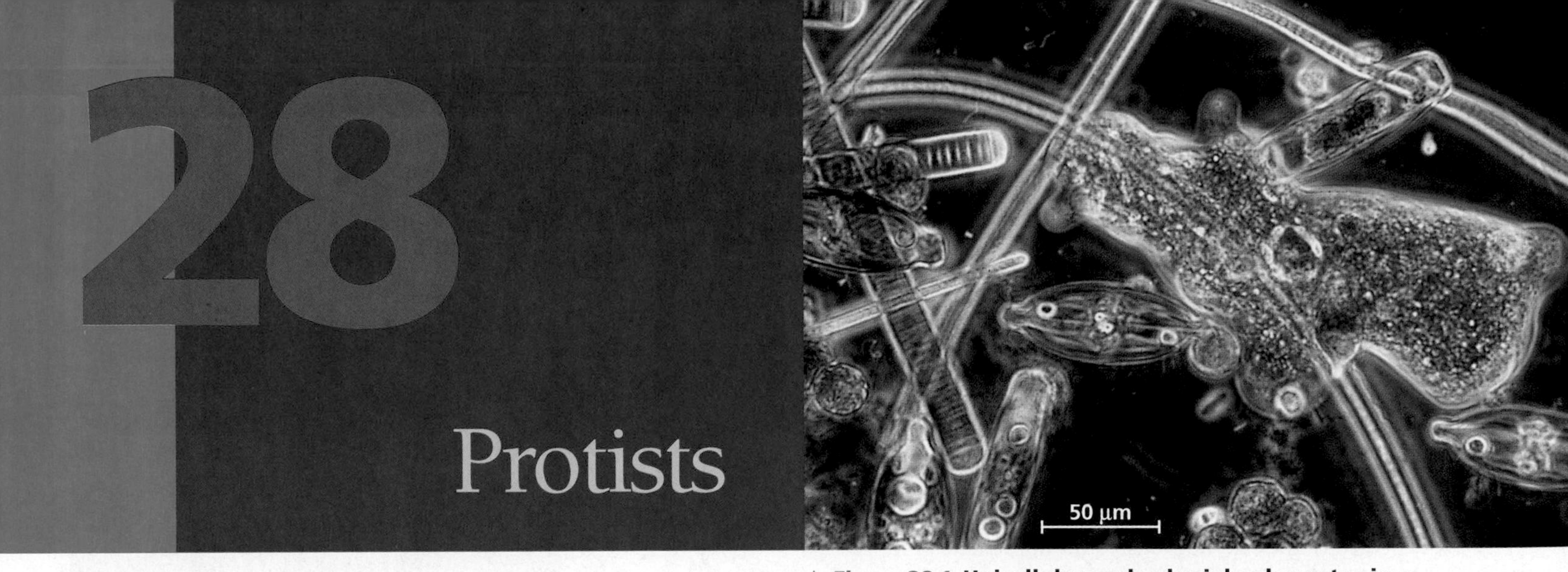

▲ **Figure 28.1 Unicellular and colonial eukaryotes in a drop of pond water (LM).**

Key Concepts

28.1 Protists are an extremely diverse assortment of eukaryotes
28.2 Diplomonads and parabasalids have modified mitochondria
28.3 Euglenozoans have flagella with a unique internal structure
28.4 Alveolates have sacs beneath the plasma membrane
28.5 Stramenopiles have "hairy" and smooth flagella
28.6 Cercozoans and radiolarians have threadlike pseudopodia
28.7 Amoebozoans have lobe-shaped pseudopodia
28.8 Red algae and green algae are the closest relatives of land plants

Overview

A World in a Drop of Water

Even a low-power light microscope can reveal an astonishing menagerie of organisms in a drop of pond water (**Figure 28.1**). Some of these tiny organisms propel themselves with whipping flagella, while others creep along by means of blob-like appendages. Some resemble miniature jewelry; others look like tumbling green globes. These beautiful creatures belong to the many diverse kingdoms of mostly unicellular eukaryotes informally known as **protists.** They have been intriguing scientists for more than 300 years, ever since the Dutch microscopist Antoni van Leeuwenhoek first laid eyes on them. Recalling his discovery of these organisms, he wrote, "No more pleasant sight has met my eye than this."

In the past, taxonomists classified all protists in a single kingdom, Protista. However, advances in eukaryotic systematics have caused the kingdom to crumble. It has become clear that Protista is in fact paraphyletic (see Figure 25.10): Some protists are more closely related to plants, fungi, or animals than they are to other protists. As a result, the kingdom Protista has been abandoned, and various lineages of protists are now recognized as kingdoms in their own right by some biologists. Most biologists still use the term *protist,* but only as a convenient way to refer to eukaryotes that are neither plants, animals, nor fungi.

In this chapter, you will become acquainted with some of the most significant groups of protists. You will learn about their structural and biochemical adaptations as well as their enormous impact on ecosystems, industry, and human health.

Concept 28.1

Protists are an extremely diverse assortment of eukaryotes

Given the paraphyletic nature of the group once called Protista, it isn't surprising that few general characteristics of protists can be cited without exceptions. In fact, protists exhibit more structural and functional diversity than any other group of organisms.

Most protists are unicellular, although there are some colonial and multicellular species. Unicellular protists are justifiably considered the simplest eukaryotes, but at the cellular level, many protists are exceedingly complex—the most elaborate of all cells. We should expect this of organisms that must carry out within the boundaries of a single cell the basic functions performed by all of the specialized cells in a multicellular organism.

Protists are the most nutritionally diverse of all eukaryotes. Some protists are photoautotrophs, containing chloroplasts. Some are heterotrophs, absorbing organic molecules or ingesting larger food particles. Still others, called **mixotrophs,** combine photosynthesis *and* heterotrophic nutrition. Photoautotrophy, heterotrophy, and mixotrophy have all arisen independently in many protist lineages. Distinguishing these nutritional modes helps us to understand the roles of protists in biological communities. In such an ecological context, we can divide protists into three categories: photosynthetic (plant-like) protists, or algae (singular, *alga*); ingestive (animal-like) protists, or protozoans; and absorptive (fungus-like) protists, which have no other general name. Bear in mind, however, that although the terms *alga* and *protozoan* are useful in discussing protist ecology, they do not refer to monophyletic groups.

Protist habitats, too, are diverse **(Figure 28.2)**. Most protists are aquatic, and they are found almost anywhere there is water, including moist terrestrial habitats such as damp soil and leaf litter. In oceans, ponds, and lakes, many protists are bottom-dwellers that attach themselves to rocks and other substrates or creep through the sand and silt. Protists are also important constituents of plankton (from the Greek *planktos*, wandering), the communities of organisms that drift passively near the water's surface. Phytoplankton (planktonic algae and cyanobacteria) form the foundation of most marine and freshwater food webs. In addition to these free-living species, many protists live as symbionts in other organisms.

Reproduction and life cycles are highly varied among protists. Some protists are exclusively asexual; others can also reproduce sexually or at least employ the sexual processes of meiosis and syngamy. All three basic types of sexual life cycles (see Figure 13.6) are represented among protists, along with some variations that do not quite fit any of these types. We will investigate the life cycles of several protist groups in this chapter.

Endosymbiosis in Eukaryotic Evolution

What gave rise to the enormous diversity of protists that exist today? There is now considerable evidence that much of protist diversity has its origins in endosymbiosis, a process in which certain unicellular organisms engulfed other cells, which became endosymbionts and ultimately organelles in the host cell. For example, as we discussed in Chapter 26, the earliest eukaryotes probably acquired mitochondria by engulfing alpha proteobacteria. The early origin of mitochondria is supported by the fact that all eukaryotes studied so far either have mitochondria or show signs that they had them in the past.

Biologists postulate that later in eukaryotic history, one lineage of heterotrophic eukaryotes acquired an additional endosymbiont—a photosynthetic cyanobacterium—that then evolved into plastids. In the model illustrated in **Figure 28.3**, this plastid-bearing lineage eventually gave rise to red algae

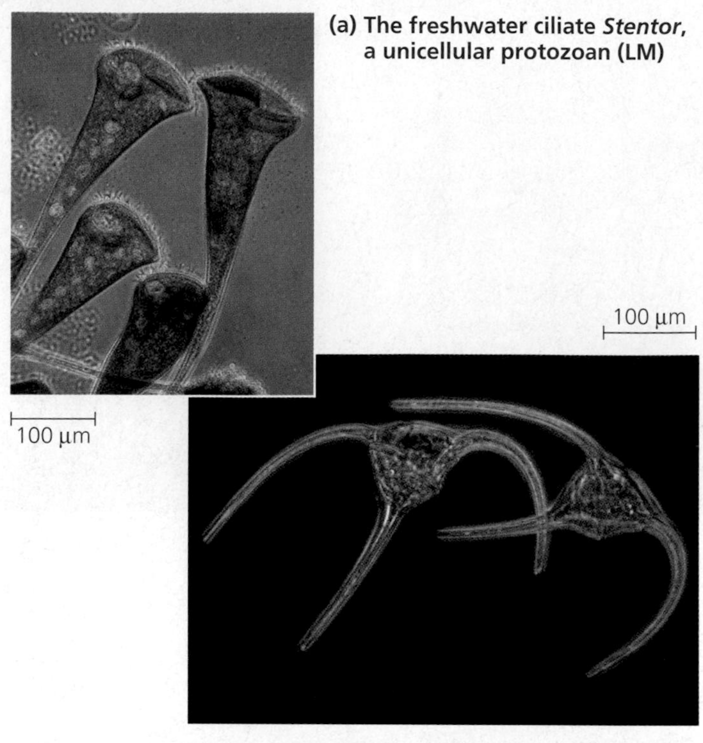

(a) The freshwater ciliate *Stentor*, a unicellular protozoan (LM)

(b) *Ceratium tripos*, a unicellular marine dinoflagellate (LM)

(c) *Delesseria sanguinea*, a multicellular marine red alga

(d) *Spirogyra*, a filamentous freshwater green alga (inset LM)

▲ **Figure 28.2 A small sample of protist diversity.**

▼ **Figure 28.3 Diversity of plastids produced by secondary endosymbiosis.** Studies of plastid-bearing eukaryotes suggest that all plastids evolved from a gram-negative cyanobacterium that was engulfed by an ancestral heterotrophic eukaryote (primary endosymbiosis). That ancestral eukaryote diversified into red algae and green algae, some of which were subsequently engulfed by other eukaryotes (secondary endosymbiosis).

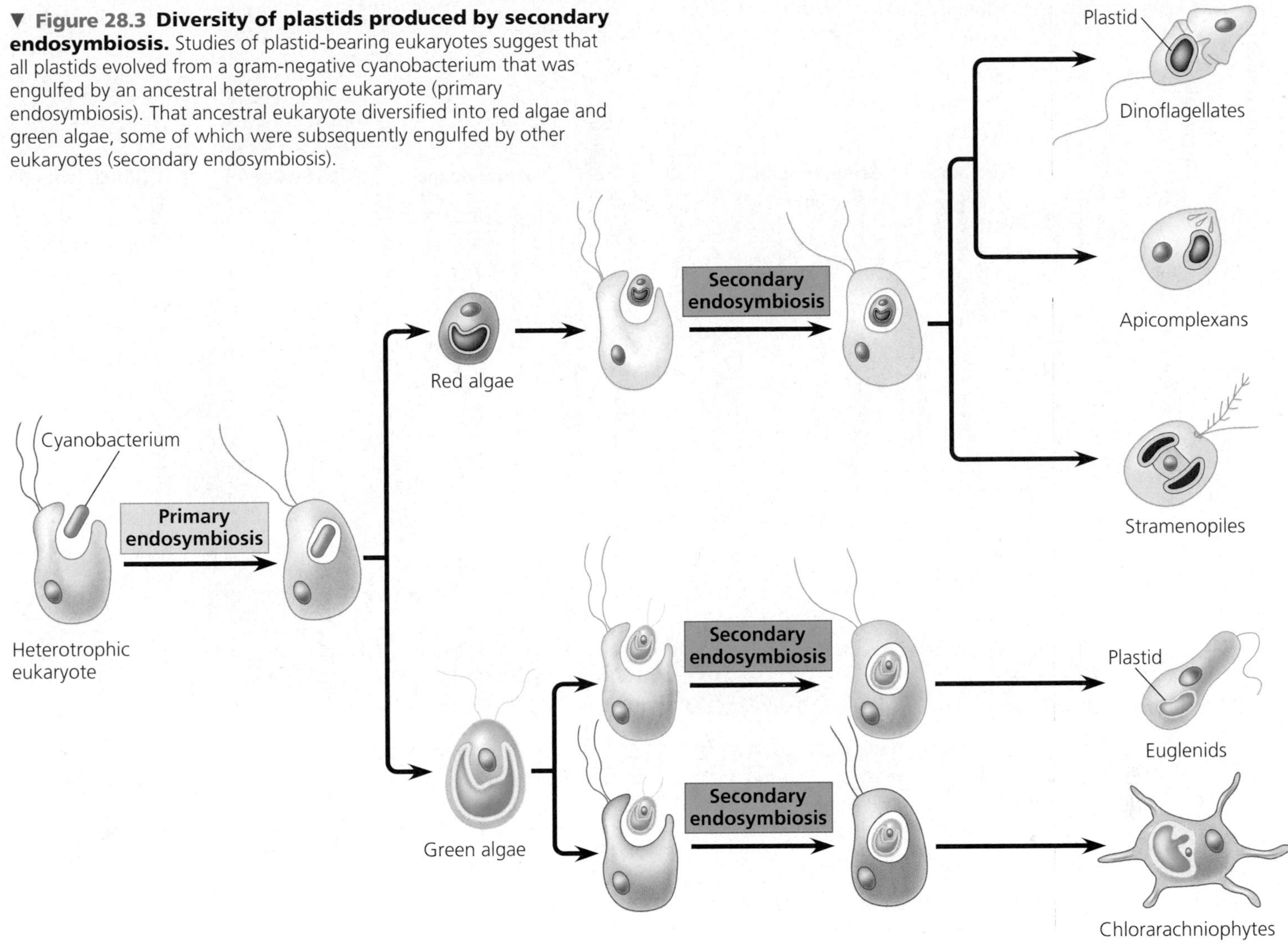

and green algae. These hypotheses are supported by the observation that the DNA of plastid genes in red algae and green algae closely resembles the DNA of cyanobacteria. In addition, plastids in red algae and green algae are surrounded by two membranes, which correspond to the inner and outer membranes of the gram-negative cyanobacterial endosymbionts.

On several occasions during eukaryotic evolution, red algae and green algae underwent **secondary endosymbiosis:** They were ingested in the food vacuole of a heterotrophic eukaryote and became endosymbionts themselves. For example, algae known as chlorarachniophytes evolved when a heterotrophic eukaryote engulfed a green alga. This process likely occurred relatively recently in evolutionary time because the engulfed alga still carries out photosynthesis with its plastids and contains a tiny, vestigial nucleus of its own, called a *nucleomorph.* Consistent with the hypothesis that chlorarachniophytes evolved from a eukaryote that engulfed another eukaryote, their plastids are surrounded by *four* membranes. The two inner membranes originated as the inner and outer membranes of the ancient cyanobacterium. The third membrane is derived from the engulfed alga's plasma membrane, and the outermost membrane is derived from the heterotrophic eukaryote's food vacuole.

Among lineages in which secondary endosymbiosis occurred in the more distant past, many parts of the engulfed alga have become reduced or have been lost completely. In most of these lineages, for example, the nucleomorph has disappeared.

Concept Check 28.1

1. Cite at least four examples of structural and functional diversity among protists.
2. Immediately after a eukaryotic cell ingests a gram-negative cyanobacterium during primary endosymbiosis, how many membranes separate the cytoplasm of the bacterium from the fluid outside the eukaryotic cell? Identify each membrane.

For suggested answers, see Appendix A.

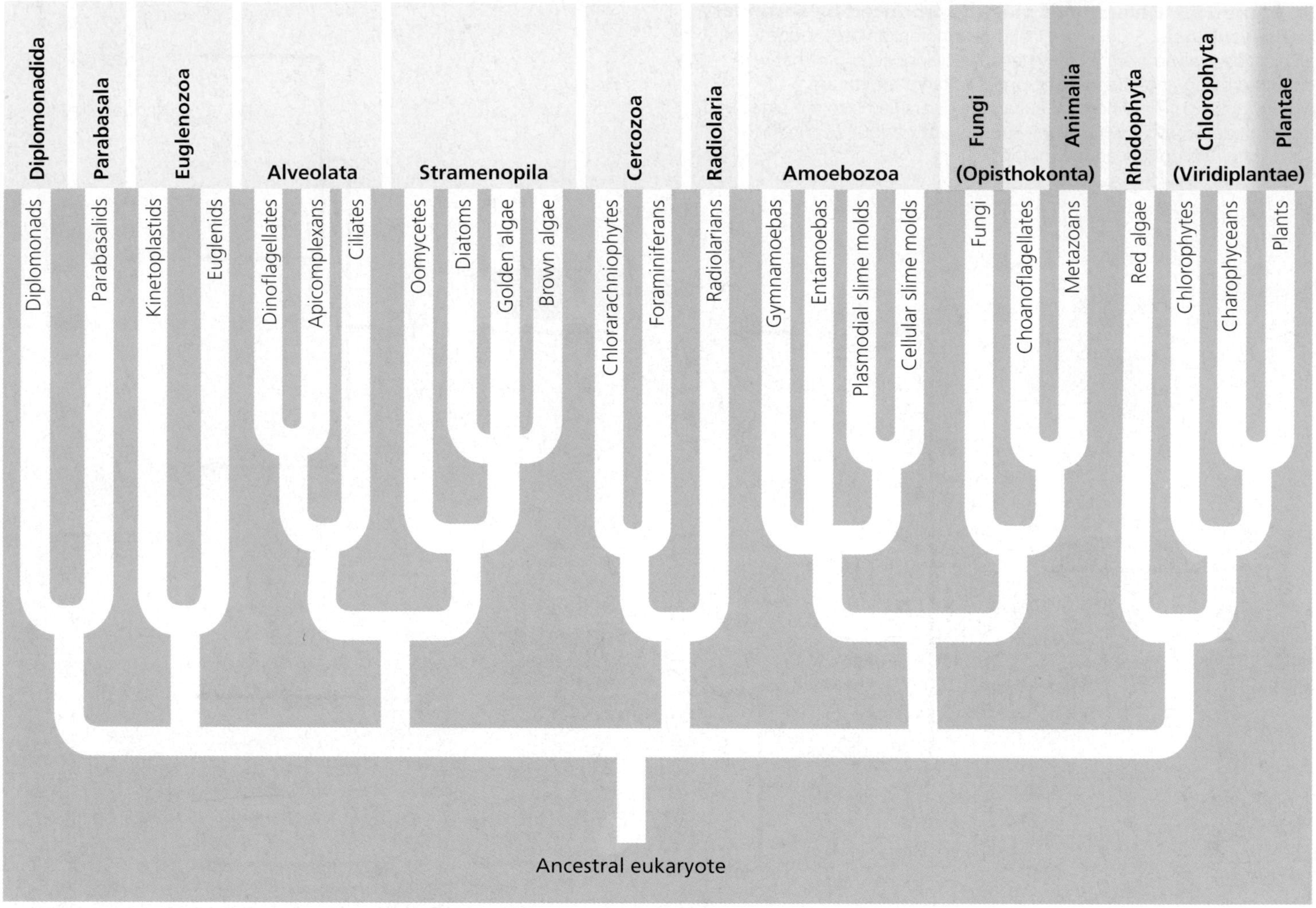

Diplomonadida
Parabasala
Euglenozoa
Alveolata
Stramenopila
Cercozoa
Radiolaria
Amoebozoa
Fungi
Animalia
(Opisthokonta)
Rhodophyta
Chlorophyta
Plantae
(Viridiplantae)

Diplomonads
Parabasalids
Kinetoplastids
Euglenids
Dinoflagellates
Apicomplexans
Ciliates
Oomycetes
Diatoms
Golden algae
Brown algae
Chlorarachniophytes
Foraminiferans
Radiolarians
Gymnamoebas
Entamoebas
Plasmodial slime molds
Cellular slime molds
Fungi
Choanoflagellates
Metazoans
Red algae
Chlorophytes
Charophyceans
Plants

Ancestral eukaryote

▲ **Figure 28.4 A tentative phylogeny of eukaryotes.** Eukaryotes labeled on the branches are grouped into larger clades, which are named at the top of the tree. The kingdoms Fungi, Animalia, and Plantae have survived from the five-kingdom system of classification, although their boundaries have changed somewhat. Clades that used to be included in the kingdom Protista are color-coded yellow.

Diplomonads and parabasalids have modified mitochondria

Now that we have examined some of the broad patterns in eukaryotic evolution, we will look more closely at several of the main clades of protists **(Figure 28.4)**.

We begin this tour with Diplomonadida (the diplomonads) and Parabasala (the parabasalids). Protists in these two clades lack plastids, and their mitochondria do not have DNA, electron transport chains, or enzymes that are normally needed for the citric acid cycle. In some species, the mitochondria are very small and produce cofactors for enzymes involved in ATP production in the cytosol. Most diplomonads and parabasalids are found in anaerobic environments.

Diplomonads

Diplomonads have two equal-sized nuclei and multiple flagella. Recall that eukaryotic flagella are extensions of the cytoplasm, consisting of bundles of microtubules covered by the cell's plasma membrane (see Figure 6.24). They are quite different from prokaryotic flagella, which are filaments composed of the globular protein flagellin attached to the cell surface (see Figure 27.6).

An infamous example of a diplomonad is *Giardia intestinalis* **(Figure 28.5a)**, a parasite that inhabits the intestine of mammals. People most often pick up *Giardia* by drinking water contaminated with feces containing the parasite in a dormant cyst stage. Drinking such contaminated water from a seemingly pristine stream or river can cause severe diarrhea and ruin a camping trip. Boiling the water before drinking it kills the cysts.

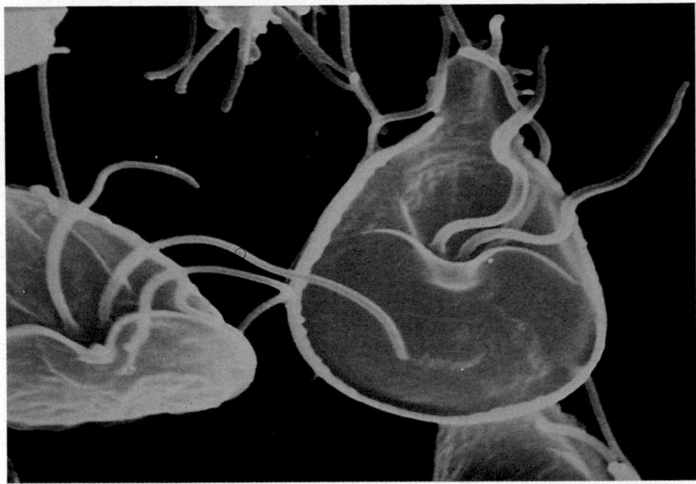

(a) *Giardia intestinalis*, a diplomonad (colorized SEM)

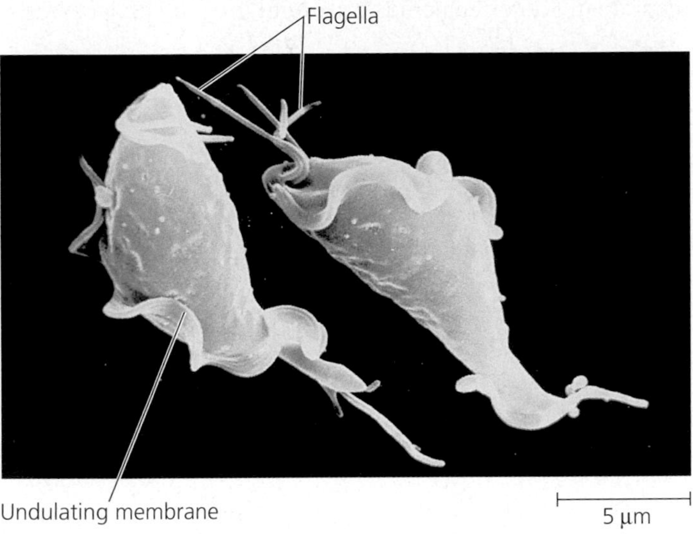

(b) *Trichomonas vaginalis*, a parabasalid (colorized SEM)

▲ **Figure 28.5 Diplomonads and parabasalids.**

Parabasalids

Parabasalids include the protists called trichomonads. The most well-known species is *Trichomonas vaginalis,* a common inhabitant of the vagina of human females **(Figure 28.5b)**. *T. vaginalis* travels along the mucus-coated lining of the reproductive and urinary tracts of its host by moving its flagella and by undulating part of its plasma membrane. If the normal acidity of the vagina is disturbed, *T. vaginalis* can outcompete beneficial microbes and infect the vaginal lining. Such infections, which can be sexually transmitted, can also occur in the urethra of males, though often without symptoms. Genetic studies of *T. vaginalis* suggest that the species became pathogenic when some of these parabasalids acquired a particular gene through horizontal gene transfer from bacteria that also dwell in the vagina. The gene allows *T. vaginalis* to feed on epithelial cells, resulting in infection.

Concept Check 28.2

1. Why do some biologists describe the mitochondria of diplomonads and parabasalids as "highly reduced"?
2. How is the structure of *Trichomonas vaginalis* well suited to its parasitic lifestyle inside its host's reproductive and urinary tracts?

For suggested answers, see Appendix A.

Concept 28.3

Euglenozoans have flagella with a unique internal structure

Euglenozoa (the euglenozoans) is a diverse clade that includes predatory heterotrophs, photosynthetic autotrophs, and pathogenic parasites. The main feature that distinguishes protists in this clade is the presence of a spiral or crystalline rod of unknown function inside their flagella **(Figure 28.6)**. Most euglenozoans also have disk-shaped mitochondrial cristae. The two best-studied groups of euglenozoans are the kinetoplastids and the euglenids.

Kinetoplastids

Kinetoplastids have a single, large mitochondrion that contains an organized mass of DNA called a kinetoplast. These protists include free-living consumers of prokaryotes in freshwater, marine, and moist terrestrial ecosystems, as well as species that parasitize animals, plants, and other protists. For example,

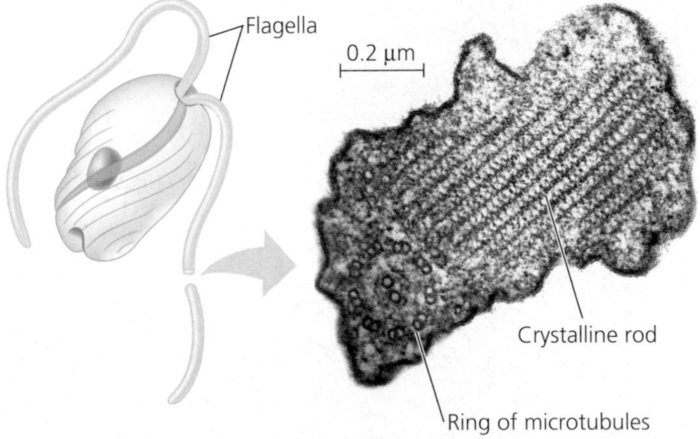

▲ **Figure 28.6 Euglenozoan flagellum.** Most euglenozoans have a crystalline rod inside one of their flagella (TEM). The rod lies alongside the 9 + 2 ring of microtubules found in all eukaryotic flagella (compare with Figure 6.24).

kinetoplastids in the genus *Trypanosoma* cause sleeping sickness in humans, a disease spread by the African tsetse fly that is invariably fatal if left untreated **(Figure 28.7)**. Trypanosomes also cause Chagas' disease, which is transmitted by bloodsucking insects and can lead to congestive heart failure.

Trypanosomes evade immune detection with an effective "bait-and-switch" defense. The surface of a trypanosome is coated with millions of copies of a single protein. However, before the host's immune system can recognize the protein and mount an attack, new generations of the parasite switch to another surface protein with a slightly different molecular structure. Frequent changes in the structure of the surface protein prevent the host from developing immunity. As much as a third of *Trypanosoma*'s genome is dedicated to the production of these surface proteins.

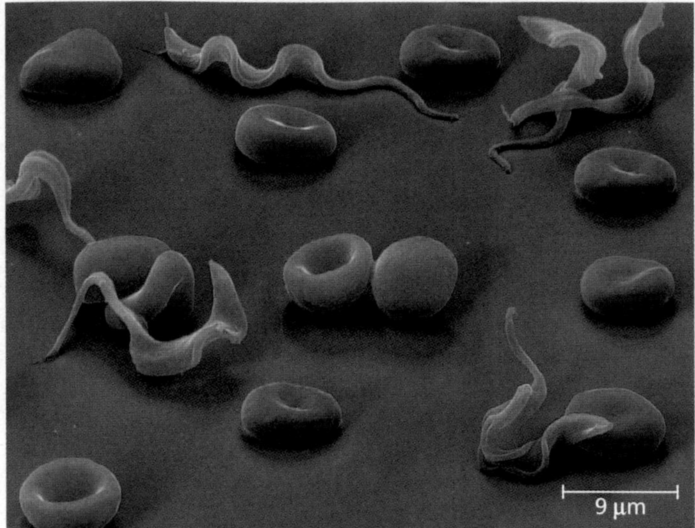
9 μm

▲ **Figure 28.7 *Trypanosoma*, the kinetoplastid that causes sleeping sickness.** The squiggles among these red blood cells are the trypanosomes (colorized SEM).

Euglenids

Euglenids have a pocket at one end of the cell from which one or two flagella emerge **(Figure 28.8)**. Paramylon, a glucose polymer that functions as a storage molecule, is also characteristic of euglenids. Many species of the euglenid *Euglena* are autotrophic, but when sunlight is unavailable, they can become heterotrophic, absorbing organic nutrients from their environment. Many other euglenids can engulf prey by phagocytosis.

Concept Check 28.3

1. How is *Trypanosoma*'s ability to produce an array of cell-surface proteins advantageous to its survival?
2. Is *Euglena* an alga? Explain your answer.

For suggested answers, see Appendix A.

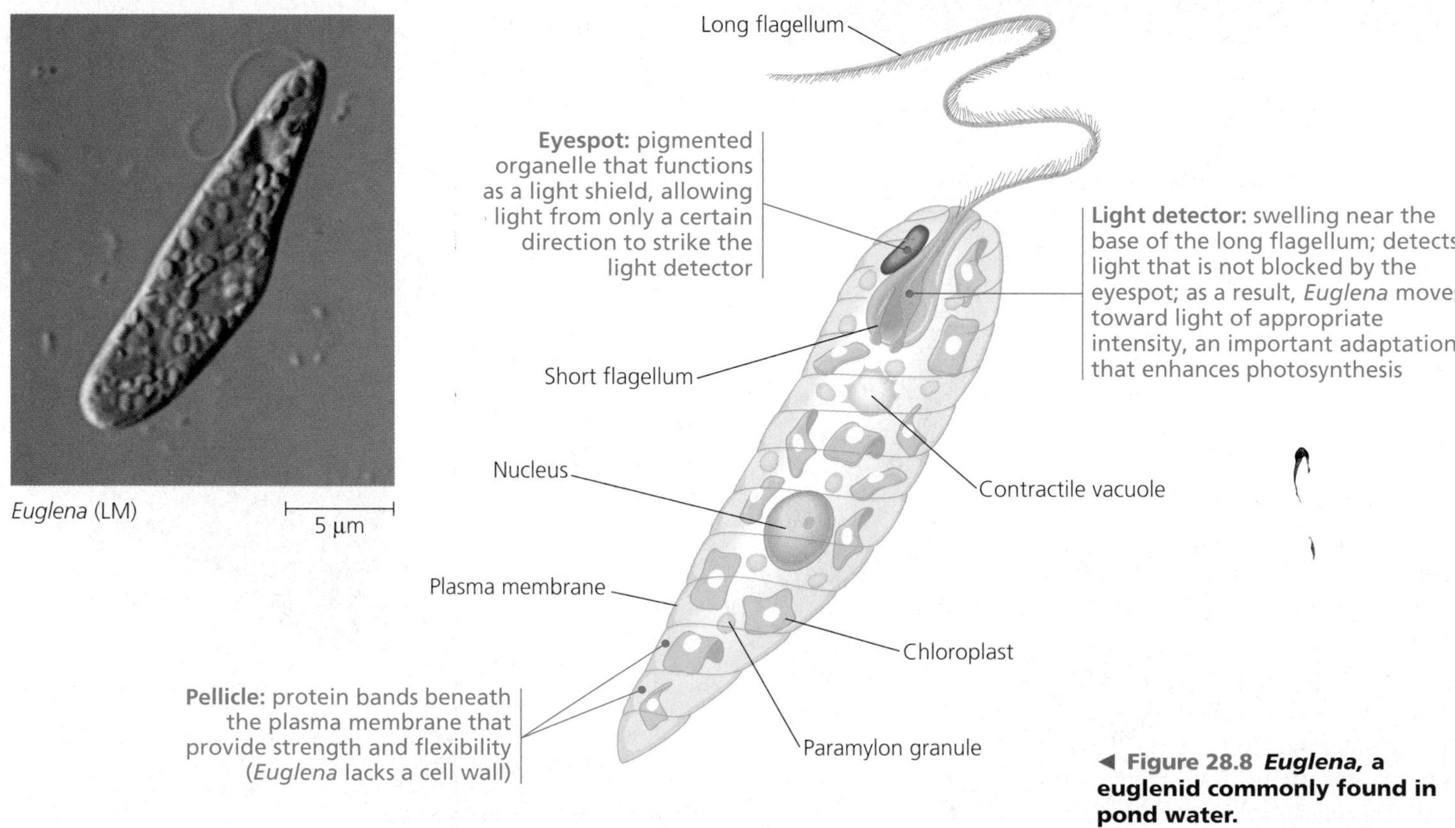
Euglena (LM)
5 μm
Long flagellum
Eyespot: pigmented organelle that functions as a light shield, allowing light from only a certain direction to strike the light detector
Light detector: swelling near the base of the long flagellum; detects light that is not blocked by the eyespot; as a result, Euglena moves toward light of appropriate intensity, an important adaptation that enhances photosynthesis
Short flagellum
Contractile vacuole
Nucleus
Plasma membrane
Chloroplast
Pellicle: protein bands beneath the plasma membrane that provide strength and flexibility (Euglena lacks a cell wall)
Paramylon granule

◀ **Figure 28.8 *Euglena*, a euglenid commonly found in pond water.**

Concept 28.4

Alveolates have sacs beneath the plasma membrane

Another clade of protists whose identity is emerging from molecular systematics, Alveolata (the alveolates), is characterized by membrane-bounded sacs (alveoli) just under the plasma membrane **(Figure 28.9)**. The function of the alveoli is unknown; researchers hypothesize that they may help stabilize the cell surface or regulate the cell's water and ion content.

Alveolata includes three groups: a group of flagellates (dinoflagellates), a group of parasites (apicomplexans), and a group of protists that move by means of cilia (ciliates).

Dinoflagellates

Dinoflagellates are abundant components of both marine and freshwater phytoplankton. There are also heterotrophic dinoflagellates. Of the several thousand known dinoflagellate species, most are unicellular, but some are colonial. Each has a characteristic shape that in many species is reinforced by internal plates of cellulose. Two flagella located in perpendicular grooves in this "armor" make dinoflagellates (from the Greek *dinos*, whirling) spin as they move through the water **(Figure 28.10)**.

Dinoflagellate blooms—episodes of explosive population growth—can cause a phenomenon called "red tide" in coastal waters. The blooms appear brownish red or pinkish orange because of the presence of carotenoids, the most common pigments in dinoflagellate plastids. Toxins produced by certain dinoflagellates can cause massive kills of invertebrates and fishes. Humans who consume molluscs that have accumulated the toxins are affected as well, sometimes fatally.

Some dinoflagellates are spectacularly bioluminescent: An ATP-driven chemical reaction creates an eerie glow at night when waves, boats, or swimmers agitate seawater with dense populations of the dinoflagellates. One possible function of this bioluminescence is that when the water is disturbed by organisms that feed on dinoflagellates, the light attracts fishes that eat those predators.

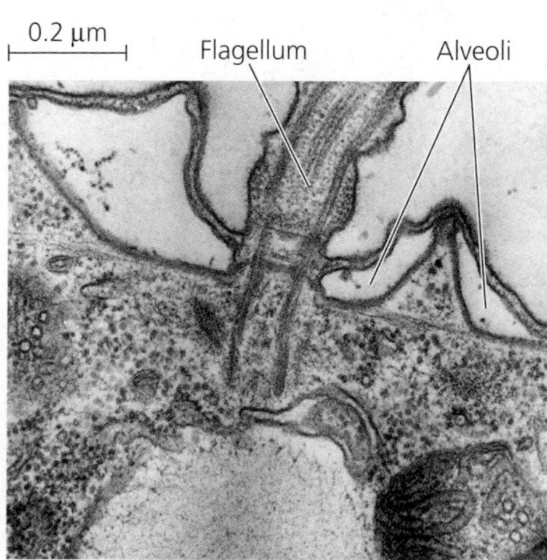

◄ **Figure 28.9 Alveoli.** These sacs under the plasma membrane are a defining characteristic of alveolates (TEM).

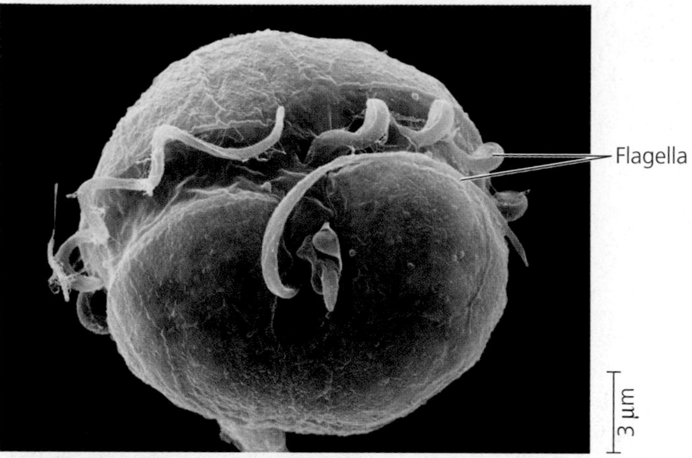

▲ **Figure 28.10 *Pfiesteria shumwayae*, a dinoflagellate.** Beating of the spiral flagellum, which lies in a groove that encircles the cell, makes this alveolate spin (colorized SEM).

Some dinoflagellates are mutualistic symbionts of coral polyps, animals that build coral reefs. The dinoflagellates' photosynthetic output is the main food for reef communities.

Apicomplexans

All **apicomplexans** are parasites of animals, and some cause serious human diseases. The parasites spread through their host as tiny infectious cells called **sporozoites.** Apicomplexans are so named because one end (the *apex*) of the sporozoite cell contains a *complex* of organelles specialized for penetrating host cells and tissues. Apicomplexans also have a nonphotosynthetic plastid, called the apicoplast. Although apicomplexans are not photosynthetic, their apicoplast has vital functions, such as the synthesis of fatty acids.

Most apicomplexans have intricate life cycles with both sexual and asexual stages. Those life cycles often require two or more different host species for completion. For example, *Plasmodium*, the parasite that causes malaria, lives in both mosquitoes and humans (**Figure 28.11**, on the next page).

The incidence of malaria was greatly diminished in the 1960s by the use of insecticides that reduced populations of *Anopheles* mosquitoes, which spread the disease, and by drugs that killed *Plasmodium* in humans. But the emergence of resistant varieties of both *Anopheles* and *Plasmodium* has led to a resurgence of malaria. About 300 million people in the tropics are now infected, and up to 2 million die each year from the disease.

The search for malarial vaccines has been hampered by the fact that *Plasmodium* spends most of its time inside human cells, hidden from the host's immune system. And, like trypanosomes, *Plasmodium* continually changes its surface proteins. The urgent need for new treatments for malaria inspired an ambitious effort to sequence *Plasmodium*'s genome. By 2003, researchers

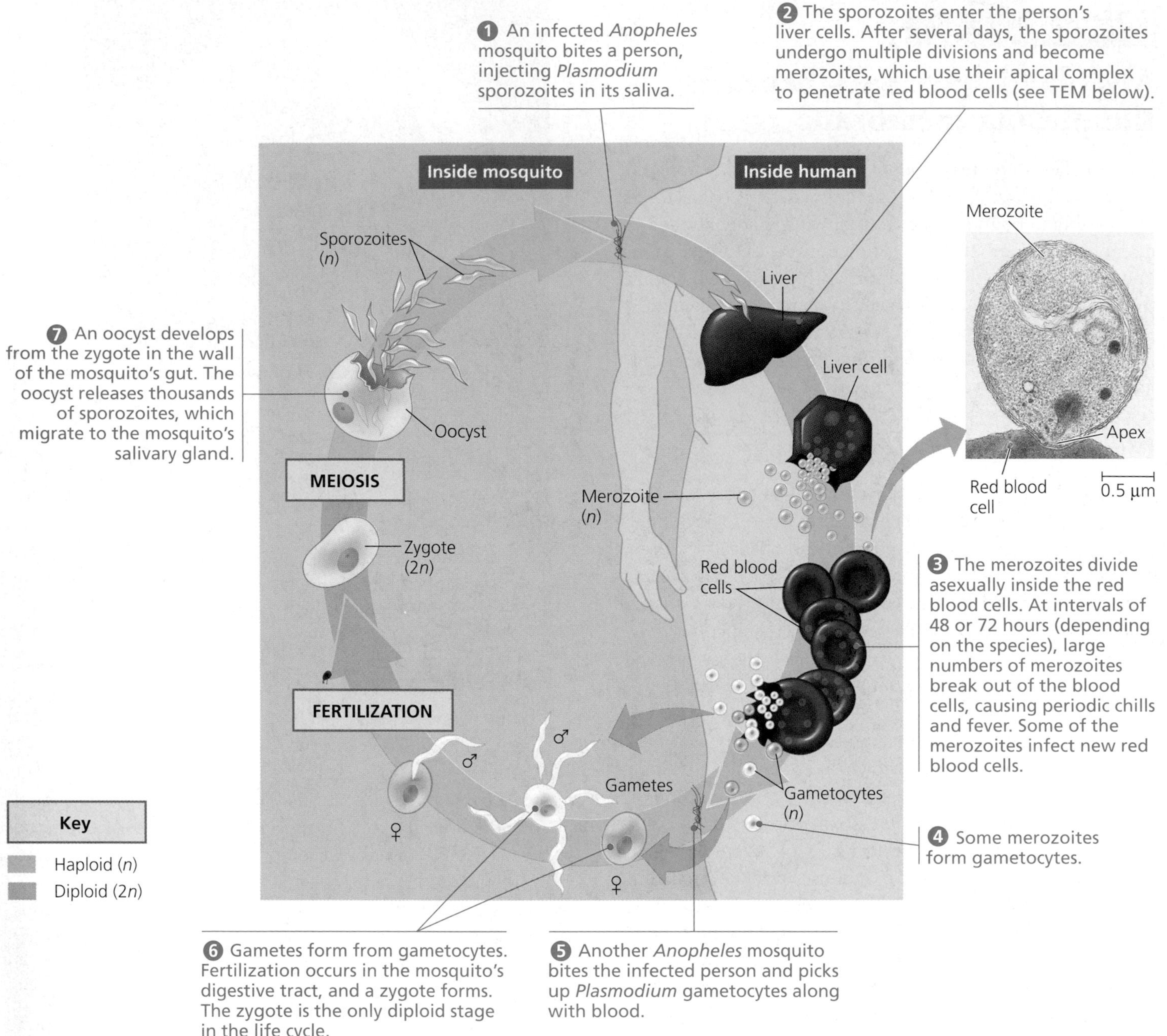

1 An infected *Anopheles* mosquito bites a person, injecting *Plasmodium* sporozoites in its saliva.
2 The sporozoites enter the person's liver cells. After several days, the sporozoites undergo multiple divisions and become merozoites, which use their apical complex to penetrate red blood cells (see TEM below).
Inside mosquito
Inside human
Merozoite
Sporozoites (*n*)
Liver
7 An oocyst develops from the zygote in the wall of the mosquito's gut. The oocyst releases thousands of sporozoites, which migrate to the mosquito's salivary gland.
Liver cell
Oocyst
Apex
MEIOSIS
Merozoite (*n*)
Red blood cell
0.5 µm
Zygote (2*n*)
Red blood cells
3 The merozoites divide asexually inside the red blood cells. At intervals of 48 or 72 hours (depending on the species), large numbers of merozoites break out of the blood cells, causing periodic chills and fever. Some of the merozoites infect new red blood cells.
FERTILIZATION
♂
♂
Gametes
Key
Haploid (*n*)
Diploid (2*n*)
♀
Gametocytes (*n*)
4 Some merozoites form gametocytes.
♀
6 Gametes form from gametocytes. Fertilization occurs in the mosquito's digestive tract, and a zygote forms. The zygote is the only diploid stage in the life cycle.
5 Another *Anopheles* mosquito bites the infected person and picks up *Plasmodium* gametocytes along with blood.

▲ **Figure 28.11 The two-host life cycle of *Plasmodium,* the apicomplexan that causes malaria.** (Colors are not true to life.)

had tracked the expression of most of the parasite's genes at numerous points in its life cycle. This research could help scientists identify potential new targets for vaccines.

Ciliates

Ciliates are a large, varied group of protists named for their use of cilia to move and feed. The cilia may completely cover the cell surface or may be clustered in a few rows or tufts. In certain species, such as *Stentor* (see Figure 28.2a), rows of tightly packed cilia function collectively in locomotion. Other ciliates scurry about on leglike structures constructed from many cilia bonded together. A submembrane system of microtubules coordinates ciliary movements.

A distinctive feature of ciliates is the presence of two types of nuclei: large macronuclei and tiny micronuclei. A cell may have one or more nucleus of each type. Each macronucleus typically contains dozens of copies of the ciliate's genome. The genes are not organized in chromosomes but instead are packaged in smaller units, each bearing many duplicates of just a few genes. Macronuclear genes control the everyday functions of the cell, such as feeding, waste removal, and maintaining water balance. The upper portion of **Figure 28.12** explores these functions in the ciliate *Paramecium.*

Figure 28.12
Exploring Structure and Function in the Ciliate *Paramecium caudatum*

FEEDING, WASTE REMOVAL, AND WATER BALANCE

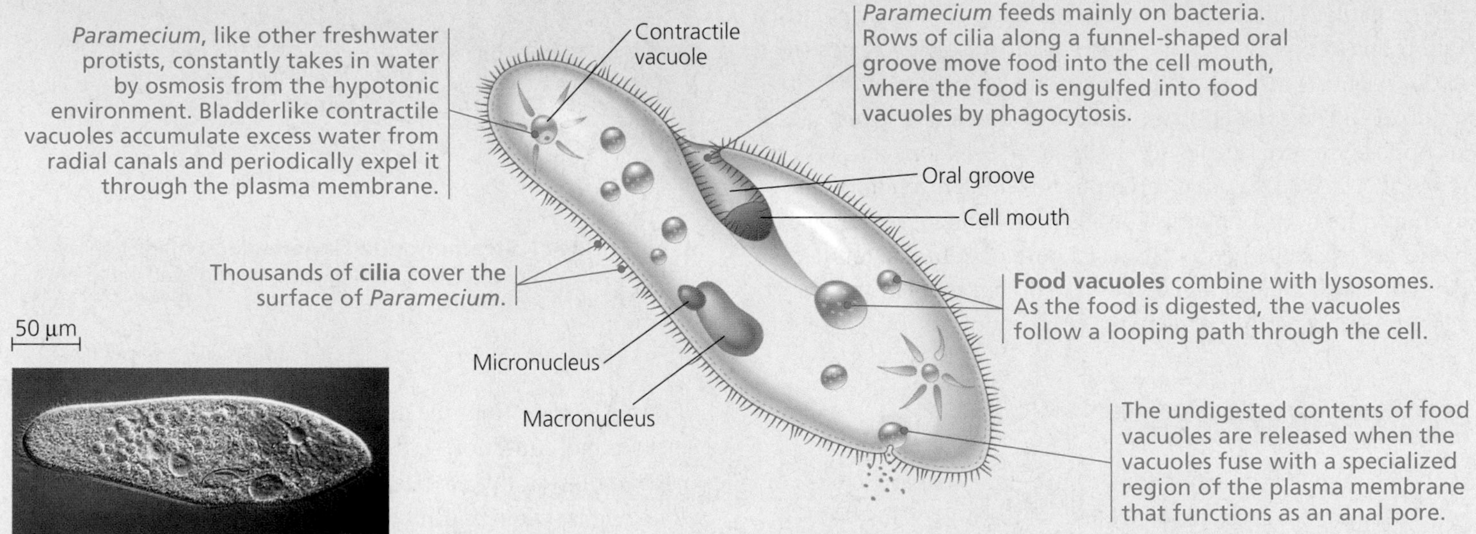

Paramecium, like other freshwater protists, constantly takes in water by osmosis from the hypotonic environment. Bladderlike contractile vacuoles accumulate excess water from radial canals and periodically expel it through the plasma membrane.
Contractile vacuole
Paramecium feeds mainly on bacteria. Rows of cilia along a funnel-shaped oral groove move food into the cell mouth, where the food is engulfed into food vacuoles by phagocytosis.
Oral groove
Cell mouth
Thousands of **cilia** cover the surface of *Paramecium*.
50 μm
Food vacuoles combine with lysosomes. As the food is digested, the vacuoles follow a looping path through the cell.
Micronucleus
Macronucleus
The undigested contents of food vacuoles are released when the vacuoles fuse with a specialized region of the plasma membrane that functions as an anal pore.

CONJUGATION AND REPRODUCTION

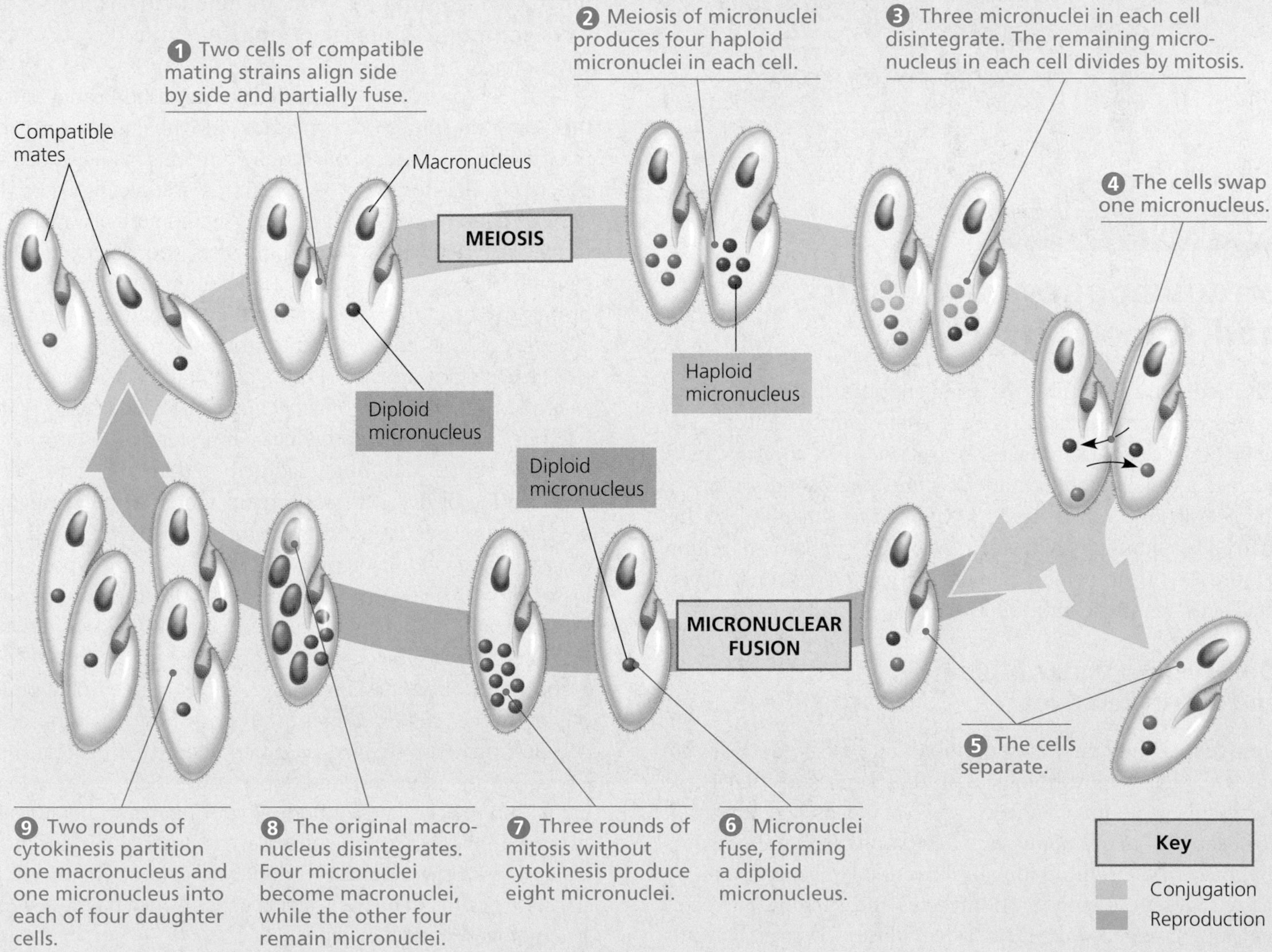

Compatible mates

1. Two cells of compatible mating strains align side by side and partially fuse.

Macronucleus

MEIOSIS

Diploid micronucleus

2. Meiosis of micronuclei produces four haploid micronuclei in each cell.

Haploid micronucleus

3. Three micronuclei in each cell disintegrate. The remaining micronucleus in each cell divides by mitosis.

4. The cells swap one micronucleus.

5. The cells separate.

MICRONUCLEAR FUSION

Diploid micronucleus

6. Micronuclei fuse, forming a diploid micronucleus.

7. Three rounds of mitosis without cytokinesis produce eight micronuclei.

8. The original macronucleus disintegrates. Four micronuclei become macronuclei, while the other four remain micronuclei.

9. Two rounds of cytokinesis partition one macronucleus and one micronucleus into each of four daughter cells.

Key
- Conjugation
- Reproduction

Ciliates generally reproduce asexually by binary fission, during which the macronucleus elongates and splits, rather than undergoing mitotic division. Genetic variation results from **conjugation**, a sexual process in which two individuals exchange haploid micronuclei. Notice in Figure 28.12 that conjugation and reproduction are separate processes in ciliates. In 2003, scientists at the Fred Hutchinson Cancer Research Center in Seattle discovered that sexual reproduction also provides an opportunity for ciliates to eliminate transposons and other types of "selfish" DNA that can replicate within a genome. During reproduction following conjugation, foreign genetic elements are excised when micronuclei develop into macronuclei. Up to 15% of a ciliate's genome may be removed in this way every time it reproduces following conjugation.

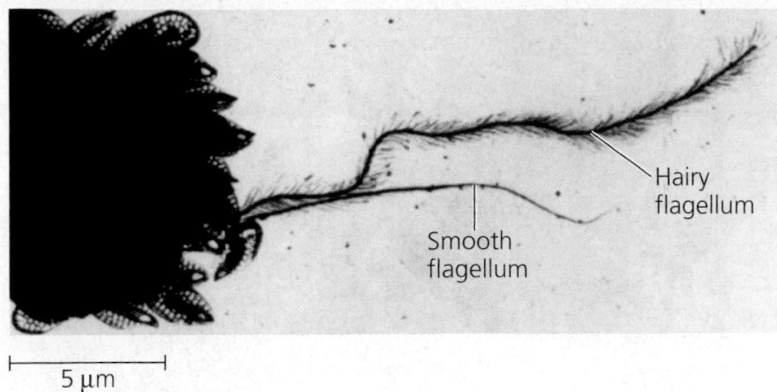

▲ **Figure 28.13 Stramenopile flagella.** Most stramenopiles, such as *Synura petersenii,* have two flagella: one covered with fine, stiff hairs and one smooth.

Concept Check 28.4

1. What morphological feature supports molecular data that suggest dinoflagellates, apicomplexans, and ciliates are members of a single clade?
2. Why is a "red tide" a cause for concern to people who eat locally caught seafood?
3. Why is it incorrect to refer to conjugation in ciliates as a form of reproduction?

For suggested answers, see Appendix A.

Concept 28.5

Stramenopiles have "hairy" and smooth flagella

The clade Stramenopila (the stramenopiles) includes several groups of heterotrophs as well as certain groups of algae. The name of the clade (from the Latin *stramen,* straw, and *pilos,* hair) refers to a flagellum with numerous fine, hairlike projections, a characteristic of these protists. In most stramenopiles, this "hairy" flagellum is paired with a "smooth" (nonhairy) flagellum **(Figure 28.13)**. In some stramenopile groups, the only flagellated cells are motile reproductive cells.

Oomycetes (Water Molds and Their Relatives)

Oomycetes include water molds, white rusts, and downy mildews. Early morphological studies suggested that these organisms were fungi (in fact, *oomycete* means "egg fungus"). For example, many oomycetes have multinucleate filaments (hyphae) that resemble fungal filaments. However, there are many differences between oomycetes and fungi. Oomycetes typically have cell walls made of cellulose, whereas the walls of fungi consist mainly of another polysaccharide, chitin. The diploid condition, which is reduced in fungi, predominates in most oomycete life cycles. Oomycetes also have flagellated cells, whereas almost all fungi lack flagella. Molecular systematics has confirmed that oomycetes are not closely related to fungi. Their superficial similarity is a case of convergent evolution (see Chapter 25). In both oomycetes and fungi, the high surface-to-volume ratio of filamentous structures enhances the uptake of nutrients from the organisms' environments.

Although oomycetes descended from plastid-bearing ancestors, they no longer have plastids and no longer carry out photosynthesis. Instead, they acquire nutrients mainly as decomposers or parasites. Most water molds are decomposers that grow as cottony masses on dead algae and animals, mainly in fresh water **(Figure 28.14)**. White rusts and downy mildews generally live on land as parasites of plants. They are dispersed primarily by windblown spores, although they also form flagellated zoospores at some point during their life cycles.

The ecological impact of oomycetes can be significant. For example, the oomycete *Phytophthora infestans* causes potato late blight, which turns the stalk and stem of potato plants to black slime. Late blight contributed to the devastating Irish famine of the 19th century, in which a million people died and at least that many were forced to leave Ireland. The disease remains a major problem today, causing typical crop losses of 15% in North America and 70% in some parts of Russia where pesticides are unavailable. To understand this pathogen better, molecular biologists have isolated DNA from a specimen of *P. infestans* preserved from the Irish potato blight of the 1840s. Genetic studies show that in recent decades, the oomycete has acquired genes that make it more aggressive and more resistant to pesticides. Scientists are looking within the genomes of both *Phytophthora* and potatoes for new weapons against the disease. In 2003, for example, a team of researchers produced domestic potatoes that were resistant to late blight by transferring genes from a blight-resistant strain of wild potato.

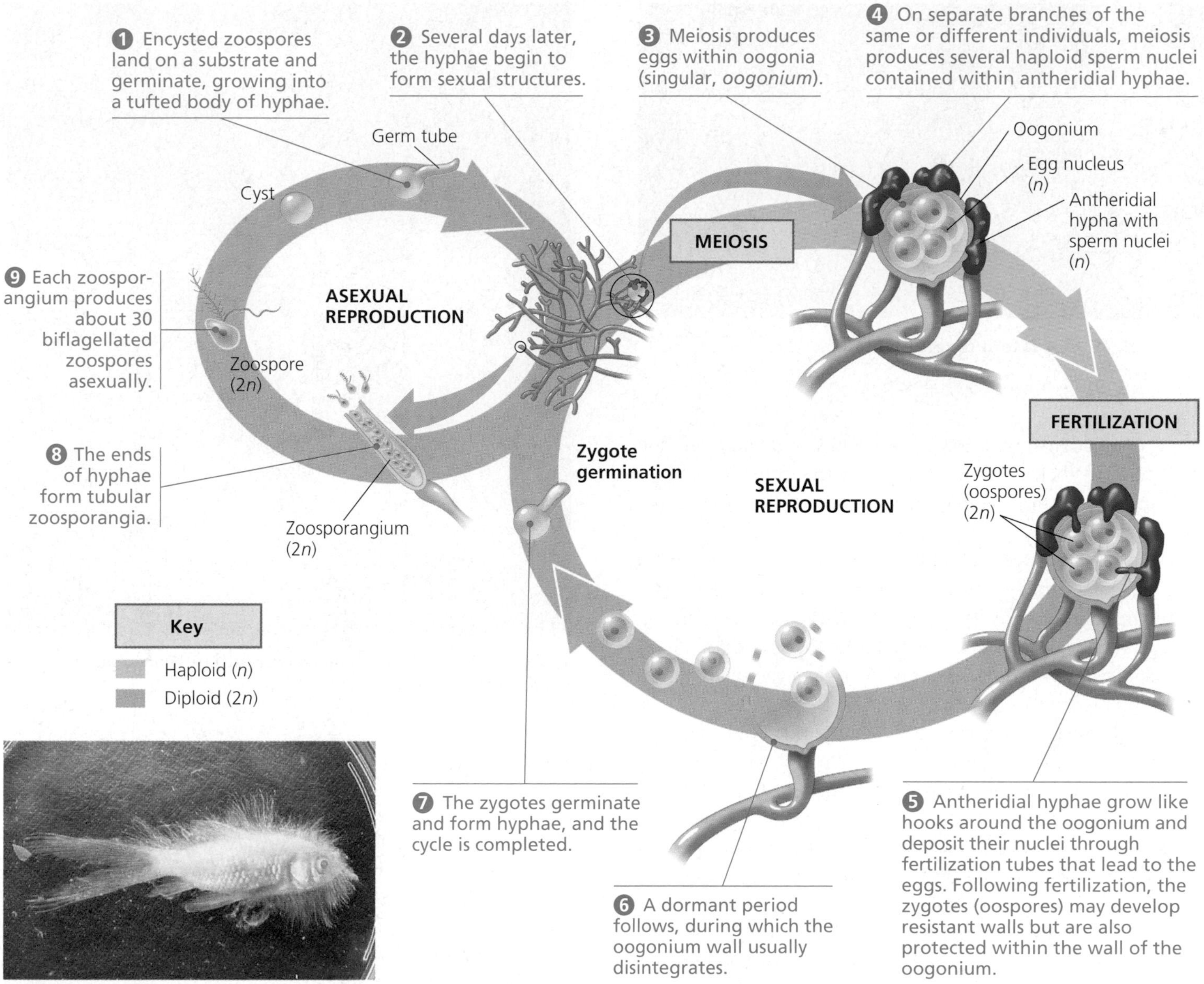

❶ Encysted zoospores land on a substrate and germinate, growing into a tufted body of hyphae.

❷ Several days later, the hyphae begin to form sexual structures.

❸ Meiosis produces eggs within oogonia (singular, *oogonium*).

❹ On separate branches of the same or different individuals, meiosis produces several haploid sperm nuclei contained within antheridial hyphae.

Germ tube

Cyst

Oogonium

Egg nucleus (n)

Antheridial hypha with sperm nuclei (n)

MEIOSIS

❾ Each zoosporangium produces about 30 biflagellated zoospores asexually.

ASEXUAL REPRODUCTION

Zoospore ($2n$)

FERTILIZATION

❽ The ends of hyphae form tubular zoosporangia.

Zygote germination

SEXUAL REPRODUCTION

Zygotes (oospores) ($2n$)

Zoosporangium ($2n$)

Key

Haploid (n)

Diploid ($2n$)

❼ The zygotes germinate and form hyphae, and the cycle is completed.

❺ Antheridial hyphae grow like hooks around the oogonium and deposit their nuclei through fertilization tubes that lead to the eggs. Following fertilization, the zygotes (oospores) may develop resistant walls but are also protected within the wall of the oogonium.

❻ A dormant period follows, during which the oogonium wall usually disintegrates.

▲ **Figure 28.14 The life cycle of a water mold.** Water molds help decompose dead insects, fishes, and other animals in fresh water. (Note the hyphal mass on the goldfish in the inset.)

Diatoms

Diatoms (also called bacillariophytes) are unicellular algae that have a unique glass-like wall made of hydrated silica embedded in an organic matrix. The wall consists of two parts that overlap like a shoe box and its lid **(Figure 28.15)**. These walls provide effective protection from the crushing jaws of predators: In 2003, German researchers discovered that live diatoms can withstand pressures of up to 1.4 million kg/m^2, equal to the pressure under each leg of a table supporting an elephant! Much of the strength of diatoms comes from the delicate lacework of holes and grooves in their walls—if the walls were smooth, it would take 60% less force to crush them.

Most of the year, diatoms reproduce asexually by mitosis, with each daughter cell receiving half of the parental cell wall and generating a new half that fits inside it. Some species form

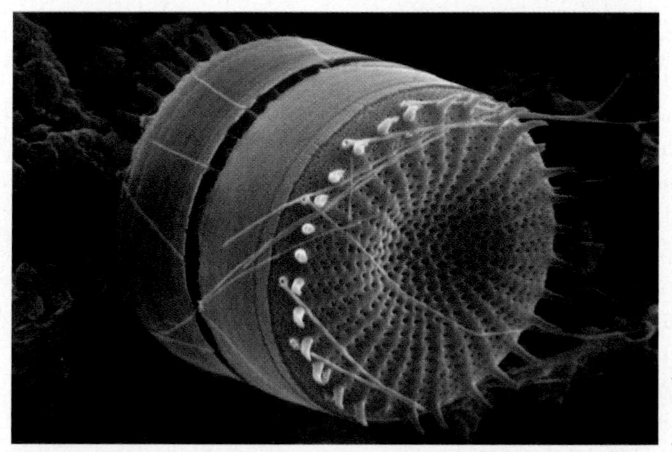

▲ **Figure 28.15 A freshwater diatom (colorized SEM).**

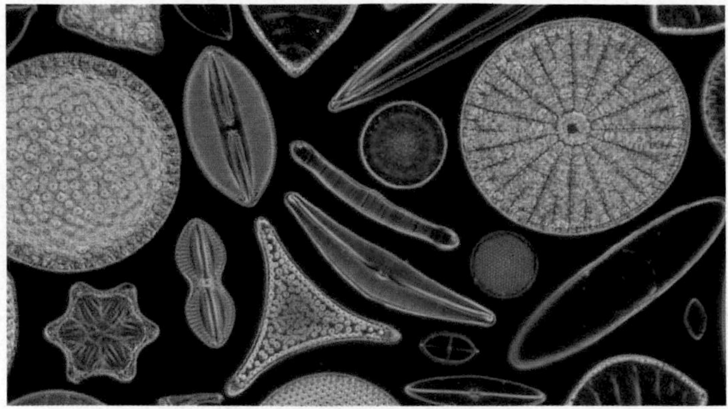

▲ **Figure 28.16 Diatom diversity (LM).**

cysts as resistant stages. Sexual reproduction is not common in diatoms. When it occurs, it involves the formation of eggs and sperm; sperm cells may be amoeboid or flagellated, depending on the species.

With an estimated 100,000 living species, diatoms are a highly diverse group of protists **(Figure 28.16)**. They are a major component of phytoplankton both in the ocean and in lakes: One bucket of water scooped from the surface of the sea may contain millions of these microscopic algae. Like golden algae and brown algae, diatoms store their food reserves in the form of a glucose polymer called laminarin. Some diatoms also store food as oil.

Massive accumulations of fossilized diatom walls are major constituents of the sediments known as diatomaceous earth, which is mined for its quality as a filtering medium and for many other uses. Diatoms also have an interesting application in the field of nanotechnology—the fashioning of microscopic devices. In building their shells, diatoms perform an intricate, three-dimensional self-assembly of microscopic components. Nano-engineers are studying this process as a model for the manufacture of miniature motors, lasers, and medicine-delivery systems. They are even exploring the idea of modifying diatom DNA to produce specific structures for humans.

Golden Algae

Golden algae, or chrysophytes (from the Greek *chrysos*, golden), are named for their color, which results from their yellow and brown carotenoids. The cells of golden algae are typically biflagellated, with both flagella attached near one end of the cell. Many golden algae are components of freshwater and marine plankton. While all golden algae are photosynthetic, some species are mixotrophic, and can also absorb dissolved organic compounds or ingest food particles and prokaryotes by phagocytosis. Most species are unicellular, but some, such as the freshwater genus *Dinobryon*, are colonial **(Figure 28.17)**. If cell density exceeds a certain level, many species form resistant cysts that can survive for decades.

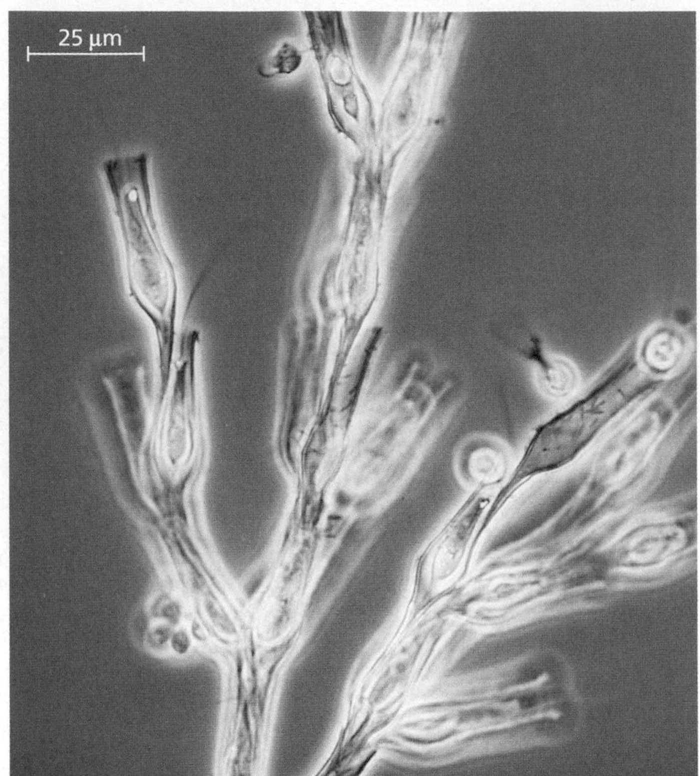

▲ **Figure 28.17 *Dinobryon*, a colonial golden alga found in fresh water (LM).**

Brown Algae

The largest and most complex algae are **brown algae**, or phaeophytes (from the Greek *phaios*, dusky, brown). All are multicellular, and most are marine. Brown algae are especially common along temperate coasts, where the water is cool. They owe their characteristic brown or olive color to the carotenoids in their plastids, which are homologous to the plastids of golden algae and diatoms.

Brown algae include many of the species commonly called seaweeds. (Some large, multicellular species of red and green algae are also referred to as seaweeds. We will examine them later in this chapter.) Seaweeds have the most complex multicellular anatomy of all algae. Some even have specialized tissues and organs that resemble those in plants. But various evidence indicates that the similarities evolved independently in the algal and plant lineages and are thus analogous, not homologous.

The term **thallus** (plural, *thalli;* from the Greek *thallos*, sprout) refers to a seaweed body that is plant-like. Unlike the body of a plant, however, a thallus lacks true roots, stems, and leaves. A typical seaweed thallus consists of a rootlike **holdfast**, which anchors the alga, and a stemlike **stipe**, which supports leaflike **blades (Figure 28.18)**. The blades provide most of the surface for photosynthesis. Some brown algae are equipped with floats, which keep the blades near the water surface. Beyond the intertidal zone in deeper waters live giant seaweeds

▲ **Figure 28.18 Seaweeds: adapted to life at the ocean's margins.** The sea palm (*Postelsia*) lives on rocks along the coast of the northwestern United States and western Canada. The thallus of this brown alga is well adapted to maintaining a firm foothold despite the crashing surf.

▲ **Figure 28.19 A kelp forest.** The great kelp beds of temperate coastal waters provide habitat and food for a variety of organisms, including many fish species caught by humans. *Macrocystis,* a kelp common along the Pacific coast of the United States, can grow more than 60 m in a single season, the fastest linear growth recorded in any organism.

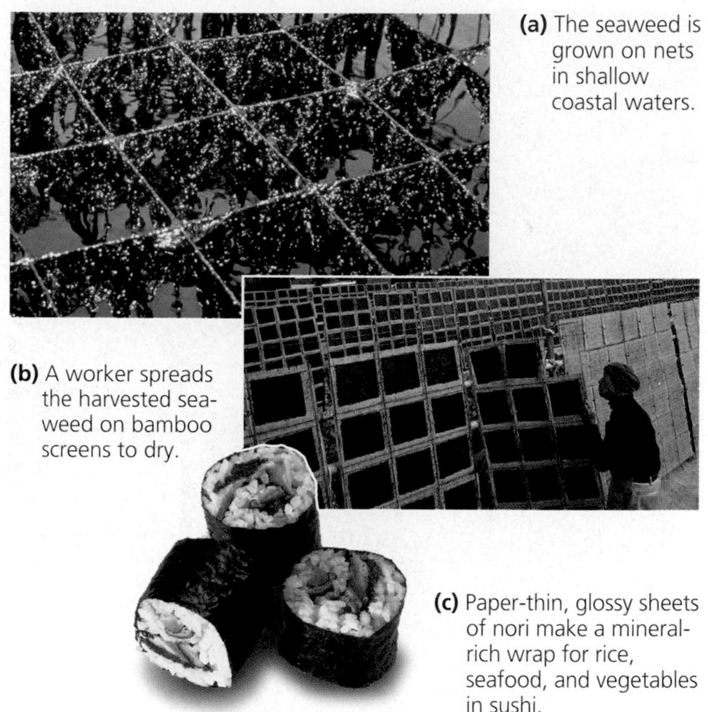

(a) The seaweed is grown on nets in shallow coastal waters.

(b) A worker spreads the harvested seaweed on bamboo screens to dry.

(c) Paper-thin, glossy sheets of nori make a mineral-rich wrap for rice, seafood, and vegetables in sushi.

▲ **Figure 28.20 Edible seaweed.** Nori is a traditional Japanese food made from the seaweed *Porphyra* (a red alga, which we discuss in more detail later in the chapter).

known as kelps **(Figure 28.19)**. The stipes of these brown algae may be as long as 60 m.

Seaweeds that inhabit the intertidal zone must cope with water churned by waves and wind, along with twice-daily low tides that expose the algae to the drying atmosphere and intense rays of the sun. Unique adaptations enable these seaweeds to survive. For example, their cell walls are composed of cellulose and gel-forming polysaccharides that help cushion the thalli from waves and reduce drying when the algae are exposed.

Human Uses of Seaweeds

Seaweeds—brown algae as well as large, multicellular red and green algae, which we discuss later in the chapter—are important commodities for humans. Many seaweeds are harvested for food. For example, the brown alga *Laminaria* (Japanese "kombu") is used in soups, and the red alga *Porphyra* (Japanese "nori") is eaten as crispy sheets or used to wrap sushi **(Figure 28.20)**. The gel-forming substances in the cell walls of algae (algin in brown algae, agar and carrageenan in red algae) are also used to thicken many processed foods, such as pudding, ice cream, and salad dressing.

Alternation of Generations

A variety of life cycles have evolved among the multicellular algae. The most complex life cycles include an **alternation of generations,** the alternation of multicellular haploid and diploid forms. Although haploid and diploid conditions

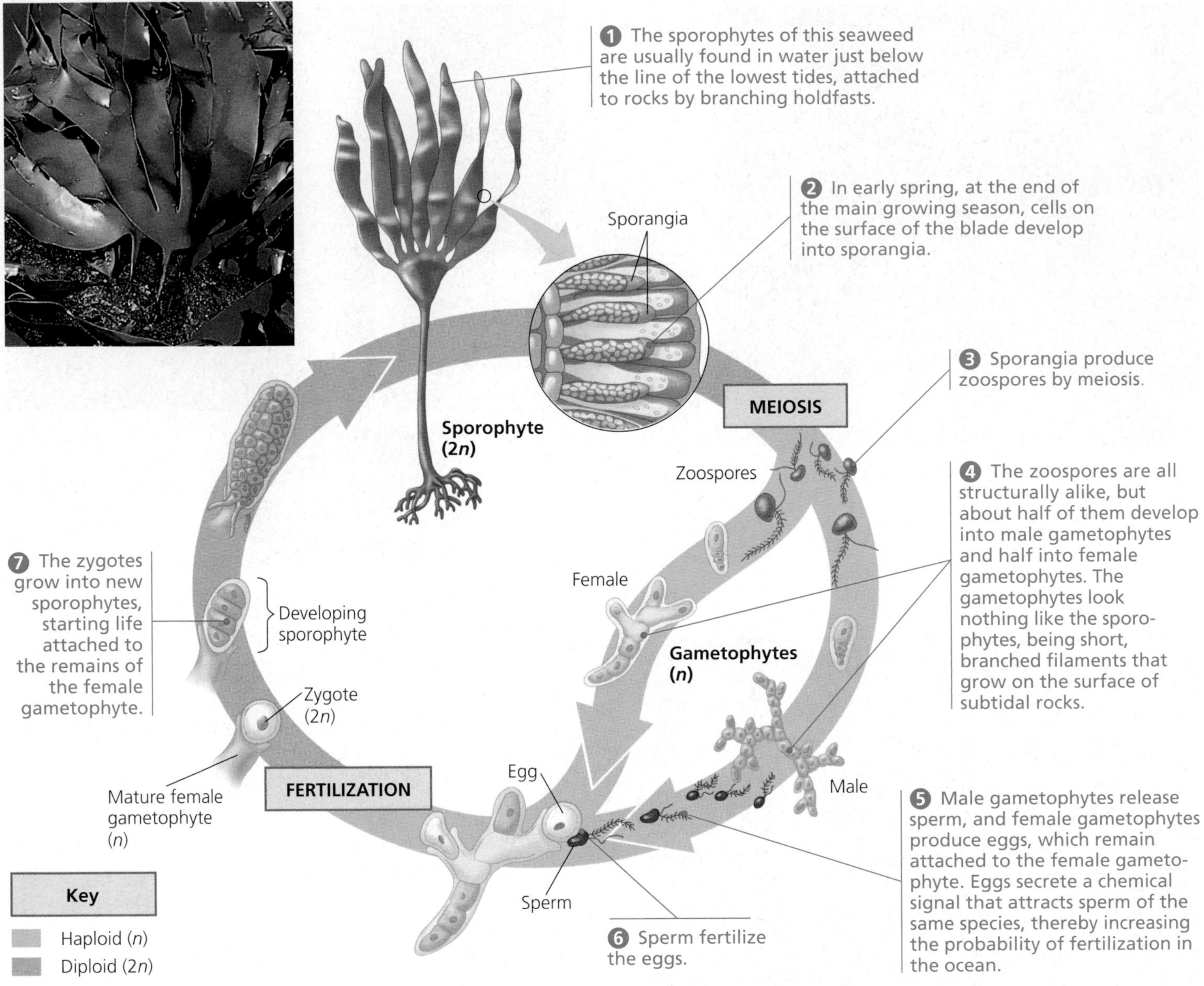

1 The sporophytes of this seaweed are usually found in water just below the line of the lowest tides, attached to rocks by branching holdfasts.
Sporangia
2 In early spring, at the end of the main growing season, cells on the surface of the blade develop into sporangia.
3 Sporangia produce zoospores by meiosis.
MEIOSIS
Sporophyte (2n)
Zoospores
4 The zoospores are all structurally alike, but about half of them develop into male gametophytes and half into female gametophytes. The gametophytes look nothing like the sporophytes, being short, branched filaments that grow on the surface of subtidal rocks.
Female
Gametophytes (n)
7 The zygotes grow into new sporophytes, starting life attached to the remains of the female gametophyte.
Developing sporophyte
Zygote (2n)
Male
Egg
Mature female gametophyte (n)
FERTILIZATION
5 Male gametophytes release sperm, and female gametophytes produce eggs, which remain attached to the female gametophyte. Eggs secrete a chemical signal that attracts sperm of the same species, thereby increasing the probability of fertilization in the ocean.
Key
Haploid (n)
Diploid (2n)
Sperm
6 Sperm fertilize the eggs.

▲ **Figure 28.21 The life cycle of *Laminaria*: an example of alternation of generations.**

alternate in *all* sexual life cycles—human gametes, for example, are haploid—the term alternation of generations applies only to life cycles in which both haploid and diploid stages are multicellular. As you will read in Chapter 29, alternation of generations also evolved in the life cycles of all plants.

The complex life cycle of the brown alga *Laminaria* is an example of alternation of generations **(Figure 28.21)**. The diploid individual is called the sporophyte because it produces reproductive cells called zoospores. The zoospores develop into haploid male and female gametophytes, which produce gametes. The union of two gametes (fertilization, or syngamy) results in a diploid zygote, which gives rise to a new sporophyte.

In *Laminaria*, the two generations are **heteromorphic**, meaning that the sporophytes and gametophytes are structurally different. Other algal life cycles have an alternation of **isomorphic** generations, in which the sporophytes and gametophytes look similar to each other, although they differ in chromosome number.

Concept Check 28.5

1. What unique cellular feature is common to all stramenopiles?
2. Compare the nutrition of oomycetes with that of golden algae.
3. How is the structure of a brown alga such as *Laminaria* well suited to its intertidal zone habitat?

For suggested answers, see Appendix A.

Concept 28.6

Cercozoans and radiolarians have threadlike pseudopodia

A newly recognized clade, Cercozoa (the cercozoans), contains a diversity of species that are among the organisms referred to as **amoebas.** Amoebas were formerly defined as protists that move and feed by means of **pseudopodia,** extensions that may bulge from virtually anywhere on the cell surface. When an amoeba moves, it extends a pseudopodium and anchors the tip, and then more cytoplasm streams into the pseudopodium. However, based on molecular systematics, it is now clear that amoebas do not constitute a monophyletic group but are dispersed across many distantly related eukaryotic taxa. Those that belong to the clade Cercozoa are distinguished morphologically from most other amoebas by their threadlike pseudopodia. Cercozoans include chlorarachniophytes (mentioned earlier in this chapter in the discussion of secondary endosymbiosis) and foraminiferans. Protists in another clade, Radiolaria (the radiolarians), also have threadlike pseudopodia and are closely related to cercozoans.

Foraminiferans (Forams)

Foraminiferans (from the Latin *foramen,* little hole, and *ferre,* to bear), or **forams,** are named for their porous shells, called **tests** (**Figure 28.22**). Foram tests are generally multichambered and consist of organic material hardened with calcium carbonate. The pseudopodia that extend through the pores function in swimming, test formation, and feeding. Many forams also derive nourishment from the photosynthesis of symbiotic algae that live within the tests.

▲ **Figure 28.22** ***Globigerina,*** **a foram with a snail-like test.** Threadlike pseudopodia extend through pores in the test (LM). The inset SEM shows a calcareous foram test.

Forams are found in both the ocean and fresh water. Most species live in the sand or attach themselves to rocks or algae, but some are abundant in plankton. The largest forams, though single-celled, grow to a diameter of several centimeters.

Ninety percent of all identified species of forams are known from fossils. Along with the calcareous remains of other protists, the fossilized tests of forams are components of marine sediments, including sedimentary rocks that are now land formations. Foram fossils are excellent markers for correlating the ages of sedimentary rocks in different parts of the world.

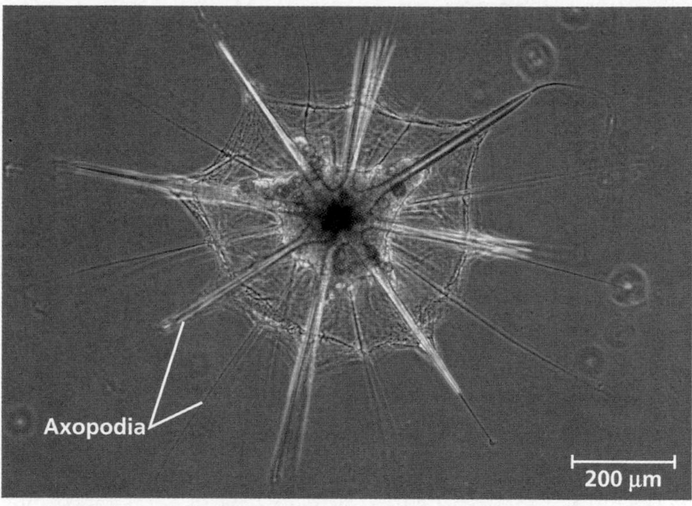

▲ **Figure 28.23 A radiolarian.** Numerous threadlike axopodia radiate from the central body of this radiolarian, which is found in the Red Sea (LM).

Radiolarians

Radiolarians are mostly marine protists whose tests are fused into one delicate piece, which is generally made of silica. The pseudopodia of radiolarians, known as axopodia, radiate from the central body and are reinforced by bundles of microtubules (**Figure 28.23**). The microtubules are covered by a thin layer of cytoplasm, which surrounds through phagocytosis smaller microorganisms that become attached to the axopodia. Cytoplasmic streaming then carries the engulfed prey into the main part of the cell. After radiolarians die, their tests settle to the seafloor, where they have accumulated as an ooze that is hundreds of meters thick in some locations.

Concept Check 28.6

1. Why do forams have such a well-preserved fossil record?
2. Compare feeding in forams and radiolarians.

For suggested answers, see Appendix A.

Concept 28.7

Amoebozoans have lobe-shaped pseudopodia

Many species of amoebas that have lobe-shaped, rather than threadlike, pseudopodia belong to the clade Amoebozoa (the amoebozoans). Amoebozoans include gymnamoebas, entamoebas, and slime molds.

Gymnamoebas

Gymnamoebas constitute a large and varied group of amoebozoans. These unicellular protists are ubiquitous in soil as well as freshwater and marine environments. Most are heterotrophs that actively seek and consume bacteria and other protists **(Figure 28.24)**. Some gymnamoebas also feed on detritus (nonliving organic matter).

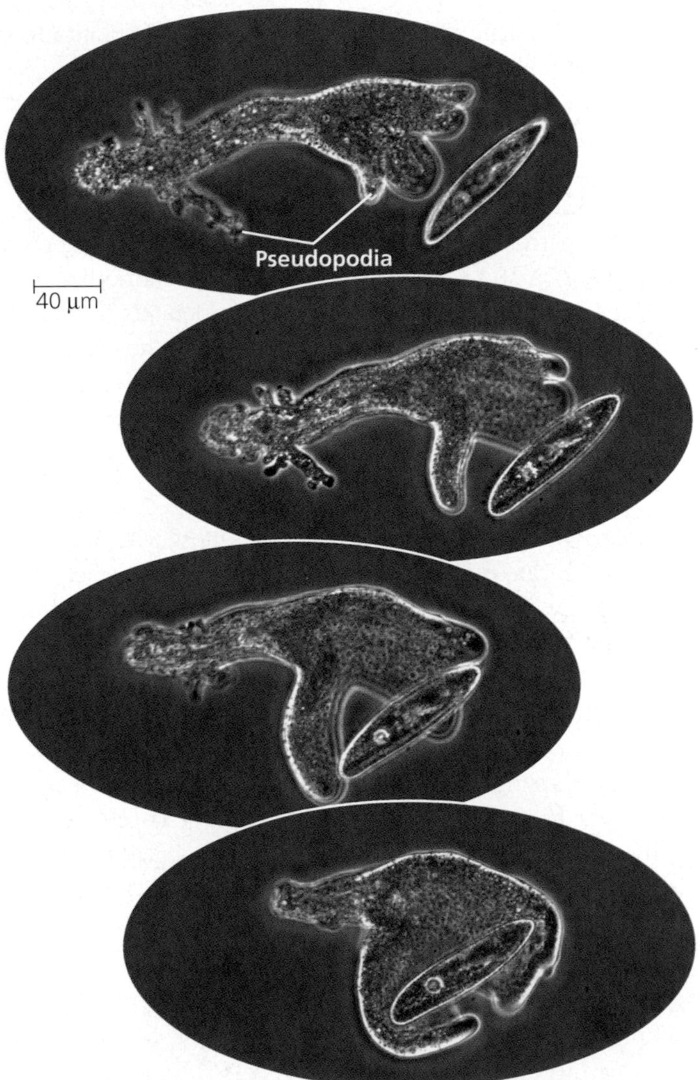

▲ **Figure 28.24 A gymnamoeba feeding.** In this series of images selected from a video, a gymnamoeba (*Amoeba* sp.) uses its lobe-shaped pseudopodia to engulf its prey, a ciliate.

Entamoebas

Whereas most amoebozoans are free-living, those that belong to the genus *Entamoeba* are parasites that infect all classes of vertebrates as well as some invertebrates. Humans host at least six species of *Entamoeba,* but only one, *E. histolytica,* is known to be pathogenic. *E. histolytica* causes amebic dysentery and is spread via contaminated drinking water, food, or eating utensils. Responsible for up to 100,000 deaths worldwide every year, the disease is the third-leading cause of death due to parasites, after malaria and schistosomiasis (see Chapter 33).

Slime Molds

Slime molds, or mycetozoans (from the Latin meaning "fungus animals"), were once thought to be fungi because, like fungi, they produce fruiting bodies that aid in spore dispersal. However, the resemblance between mycetozoans and fungi appears to be another example of evolutionary convergence. Molecular systematics places slime molds in the clade Amoebozoa and suggests that they descended from unicellular, gymnamoeba-like ancestors. Slime molds have diverged into two main branches, plasmodial slime molds and cellular slime molds, which are distinguished in part by their distinctive life cycles.

Plasmodial Slime Molds

Many species of **plasmodial slime molds** are brightly pigmented, usually yellow or orange **(Figure 28.25)**. At one stage in their life cycle, they form a mass called a **plasmodium,** which may grow to a diameter of many centimeters **(Figure 28.26)**. (Don't confuse a slime mold's plasmodium with the apicomplexan parasite *Plasmodium.*) Despite its size, the plasmodium is not multicellular; it is a single mass of cytoplasm that is undivided by membranes and that contains many diploid nuclei. This "supercell" is the product of mitotic nuclear divisions that are not followed by cytokinesis, the division of cytoplasm. In most species, the mitotic divisions are synchronous, meaning that

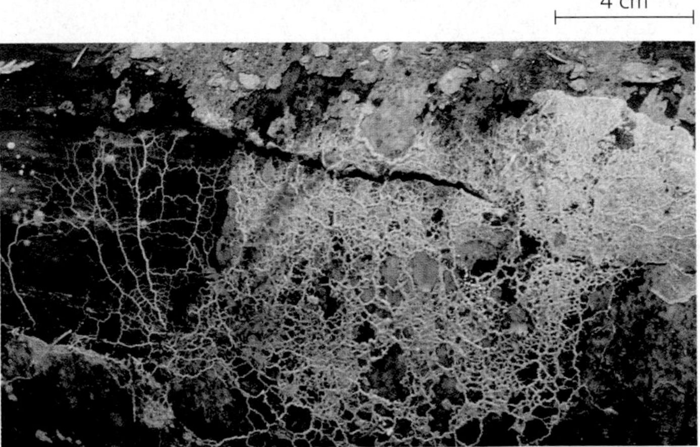

▲ **Figure 28.25 *Physarum polycephalum,* a plasmodial slime mold.**

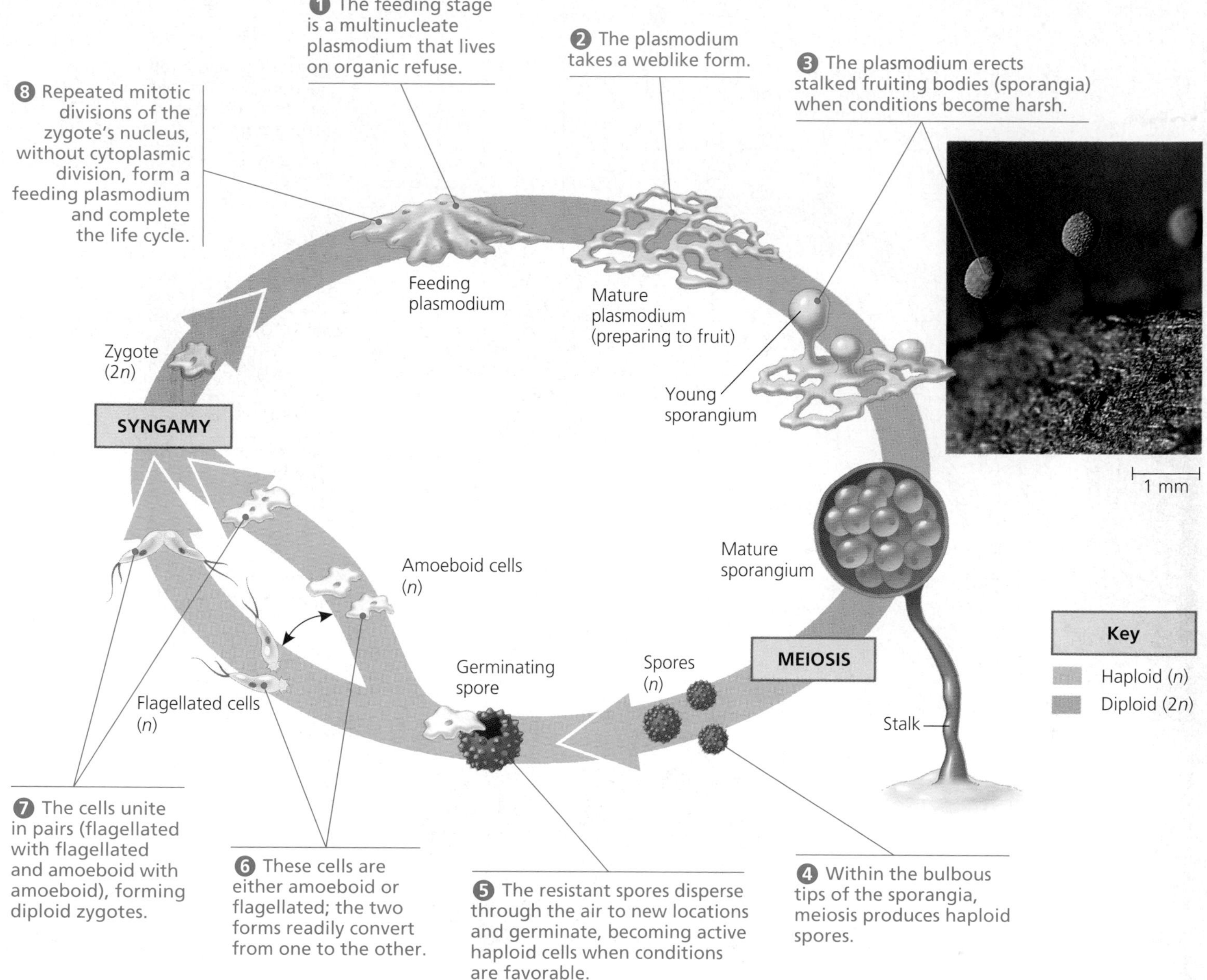

1 The feeding stage is a multinucleate plasmodium that lives on organic refuse.
2 The plasmodium takes a weblike form.
3 The plasmodium erects stalked fruiting bodies (sporangia) when conditions become harsh.
8 Repeated mitotic divisions of the zygote's nucleus, without cytoplasmic division, form a feeding plasmodium and complete the life cycle.
Feeding plasmodium
Mature plasmodium (preparing to fruit)
Young sporangium
Zygote (2n)
SYNGAMY
1 mm
Mature sporangium
Amoeboid cells (n)
Key
Haploid (n)
Diploid (2n)
MEIOSIS
Germinating spore
Spores (n)
Flagellated cells (n)
Stalk
7 The cells unite in pairs (flagellated with flagellated and amoeboid with amoeboid), forming diploid zygotes.
6 These cells are either amoeboid or flagellated; the two forms readily convert from one to the other.
5 The resistant spores disperse through the air to new locations and germinate, becoming active haploid cells when conditions are favorable.
4 Within the bulbous tips of the sporangia, meiosis produces haploid spores.

▲ **Figure 28.26 The life cycle of a plasmodial slime mold.**

thousands of nuclei go through each phase of mitosis at the same time. Because of this characteristic, plasmodial slime molds have been used to study the molecular details of the cell cycle.

Within the fine channels of the plasmodium, cytoplasm streams first one way, then the other, in pulsing flows that are beautiful to watch through a microscope. This cytoplasmic streaming apparently helps distribute nutrients and oxygen. The plasmodium extends pseudopodia through moist soil, leaf mulch, or rotting logs, engulfing food particles by phagocytosis as it grows. If the habitat begins to dry up or there is no food left, the plasmodium stops growing and differentiates into a stage of the life cycle that produces fruiting bodies, which function in sexual reproduction. In most plasmodial slime molds, the diploid condition is the predominant part of the life cycle.

Cellular Slime Molds

Cellular slime molds pose a semantic question about what it means to be an individual organism. The feeding stage of the life cycle consists of solitary cells that function individually, but when food is depleted the cells form an aggregate that functions as a unit (**Figure 28.27**, on the next page). Although this mass of cells superficially resembles a plasmodial slime mold, the cells remain separated by their membranes.

Cellular slime molds also differ from plasmodial slime molds in being haploid organisms (only the zygote is diploid) and in having fruiting bodies that function in asexual rather than sexual reproduction. Furthermore, cellular slime molds have no flagellated stages.

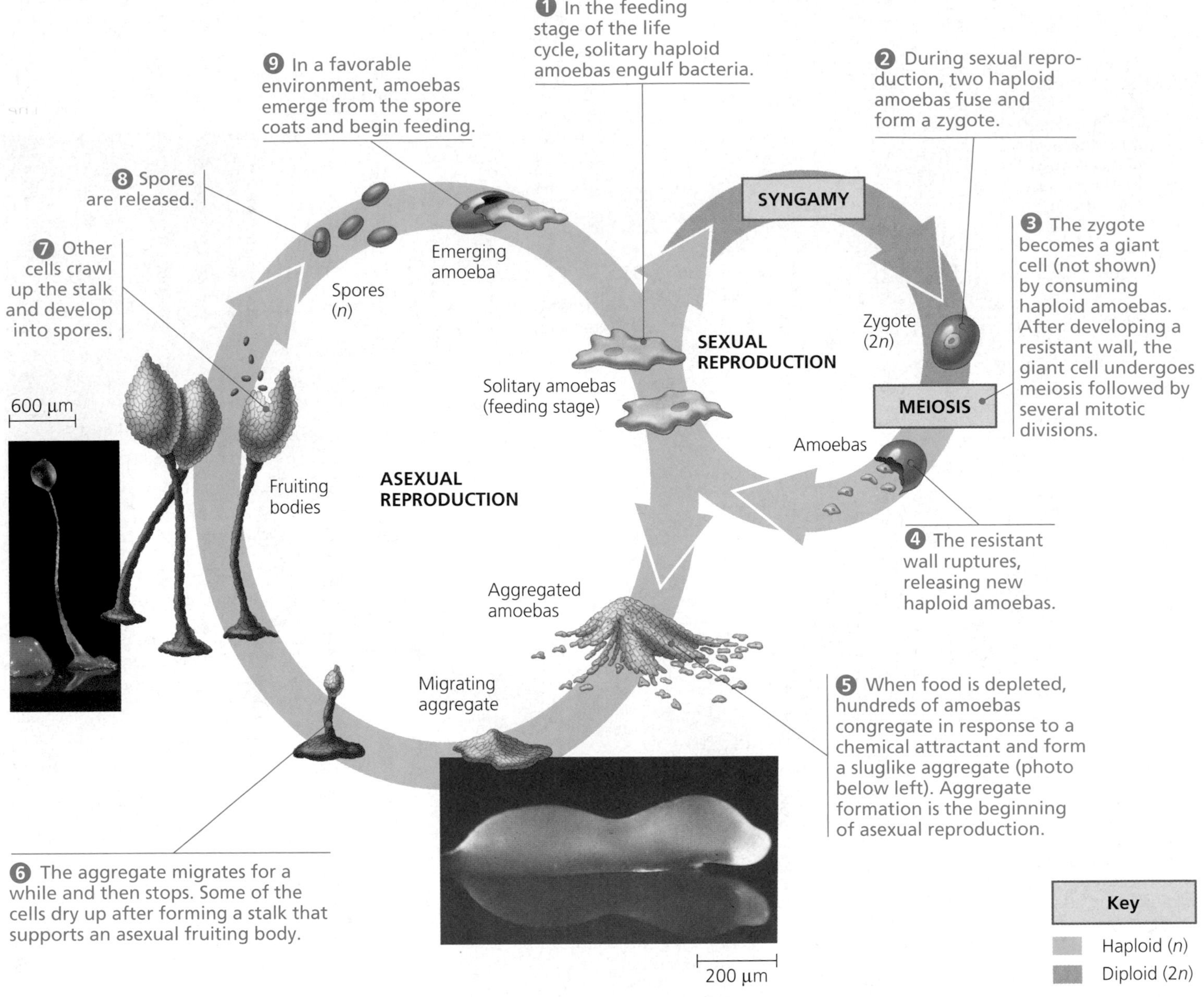

1 In the feeding stage of the life cycle, solitary haploid amoebas engulf bacteria.
2 During sexual reproduction, two haploid amoebas fuse and form a zygote.
3 The zygote becomes a giant cell (not shown) by consuming haploid amoebas. After developing a resistant wall, the giant cell undergoes meiosis followed by several mitotic divisions.
4 The resistant wall ruptures, releasing new haploid amoebas.
5 When food is depleted, hundreds of amoebas congregate in response to a chemical attractant and form a sluglike aggregate (photo below left). Aggregate formation is the beginning of asexual reproduction.
6 The aggregate migrates for a while and then stops. Some of the cells dry up after forming a stalk that supports an asexual fruiting body.
7 Other cells crawl up the stalk and develop into spores.
8 Spores are released.
9 In a favorable environment, amoebas emerge from the spore coats and begin feeding.
SYNGAMY
MEIOSIS
SEXUAL REPRODUCTION
ASEXUAL REPRODUCTION
Zygote (2n)
Amoebas
Solitary amoebas (feeding stage)
Emerging amoeba
Spores (n)
Fruiting bodies
Aggregated amoebas
Migrating aggregate
600 µm
200 µm
Key
Haploid (n)
Diploid (2n)

▲ **Figure 28.27 The life cycle of *Dictyostelium*, a cellular slime mold.**

Dictyostelium discoideum, a common cellular slime mold on forest floors, has become a model organism for addressing the evolution of multicellularity. One line of research has focused on the fruiting body stage. During this stage, the cells that form the stalk die as they dry out, while the spore cells at the top survive and have the potential to reproduce. Scientists have found that mutations to a single gene can turn individual *Dictyostelium* cells into "cheaters" that never become part of the stalk. Since these mutants gain a strong reproductive advantage over noncheaters, it is puzzling that all *Dictyostelium* cells are not cheaters.

In 2003, scientists at Rice University and the University of Turin, Italy, discovered why most of the cells do not cheat. Cheating mutants lack a protein on their cell surface, and noncheating cells can recognize this difference. Noncheaters preferentially aggregate with other noncheaters, thus depriving cheaters of the opportunity to exploit them. Such a recognition system may have been important in the evolution of multicellular eukaryotes such as animals and plants.

Concept Check 28.7

1. Contrast the pseudopodia of amoebozoans and forams.
2. In what sense is "fungus animal" a fitting description of a slime mold? In what sense is it not a fitting description?
3. Does cooperation between cells exist in amoebozoans? Explain.

For suggested answers, see Appendix A.

Concept 28.8

Red algae and green algae are the closest relatives of land plants

As we described in Chapter 26, molecular systematics and studies of cell structure support this phylogenetic scenario: More than a billion years ago, a heterotrophic protist acquired a cyanobacterial endosymbiont, and the photosynthetic descendants of this ancient protist evolved into red algae and green algae (see Figure 28.3). At least 475 million years ago, the lineage that produced green algae gave rise to land plants. We will examine land plants in Chapters 29 and 30; here we will look at the diversity of their closest algal relatives, red algae and green algae.

Red Algae

Many of the 6,000 known species of **red algae** (rhodophytes, from the Greek *rhodos,* red) are reddish, owing to an accessory pigment called phycoerythrin, which masks the green of chlorophyll. However, species adapted to shallower water have less phycoerythrin. As a result, red algae species may be almost black in deep water, bright red at more moderate depths, and greenish in very shallow water. Some species lack pigmentation altogether and function heterotrophically as parasites on other red algae.

Red algae are the most abundant large algae in the warm coastal waters of tropical oceans. Their accessory pigments allow them to absorb blue and green light, which penetrate relatively far into the water. A species of red alga has recently been discovered living near the Bahamas at a depth of more than 260 m. There are also some freshwater and terrestrial species.

Most red algae are multicellular, and the largest are included in the informal designation "seaweeds" **(Figure 28.28)**, although no red algae are as big as the giant brown kelps. The thalli of many red algae are filamentous, often branched and interwoven in lacy patterns. The base of the thallus is usually differentiated as a simple holdfast.

Red algae have especially diverse life cycles, and alternation of generations is common. But unlike other algae, they have no flagellated stages in their life cycle and depend on water currents to bring gametes together for fertilization.

Green Algae

Green algae are named for their grass-green chloroplasts. In their ultrastructure and pigment composition, these chloroplasts are much like those of organisms we traditionally call plants. Molecular systematics and cellular morphology leave little doubt that green algae and land plants are closely related. In fact, some systematists now advocate the inclusion of green algae in an expanded "plant" kingdom, Viridiplantae (from the Latin *viridis,* green).

Green algae are divided into two main groups, chlorophytes (from the Greek *chloros,* green) and charophyceans (see Figure 28.4). More than 7,000 species of chlorophytes have been identified. Most live in fresh water, but there are also many marine species. The simplest chlorophytes are biflagellated unicellular organisms such as *Chlamydomonas,* which resemble the gametes and zoospores of more complex chlorophytes. Various species of unicellular chlorophytes exist as

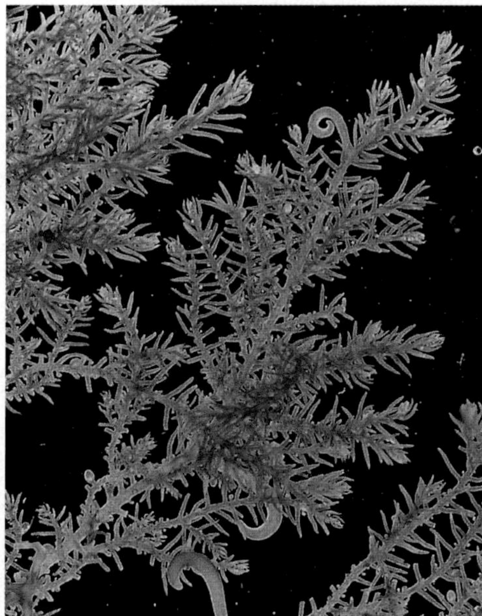

(a) ***Bonnemaisonia hamifera.*** This red alga has a filamentous form.

(b) Dulse *(Palmaria palmata).* This edible species has a "leafy" form.

(c) A coralline alga. The cell walls of coralline algae are hardened by calcium carbonate. Some coralline algae are members of the biological communities around coral reefs.

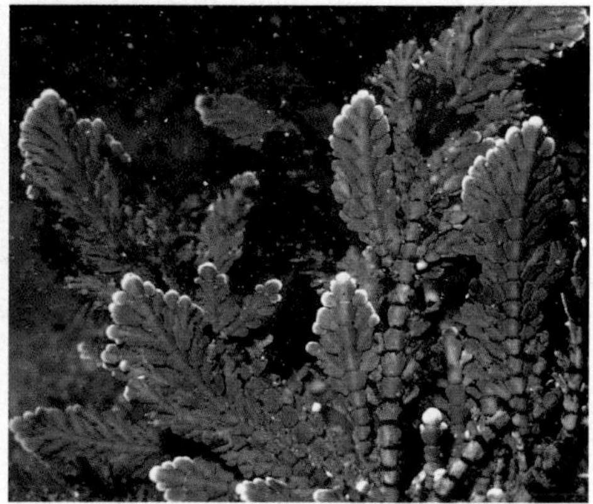

▲ **Figure 28.28 Red algae.**

plankton or inhabit damp soil; some live symbiotically within other eukaryotes, contributing part of their photosynthetic output to the food supply of their hosts. Chlorophytes are among the algae that live symbiotically with fungi in the associations called lichens (see Figure 31.24). Some chlorophytes have even adapted to one of the last places on Earth you might expect to find them: snow. For example, *Chlamydomonas nivalis* can form dense algal blooms on high-altitude glaciers and snowfields, where its reddish pigments produce an effect known as "watermelon snow" **(Figure 28.29)**. These chlorophytes carry out photosynthesis despite subfreezing temperatures and intense visible and ultraviolet radiation. They are protected by radiation-blocking compounds in their cytoplasm and by the snow itself, which acts as a shield.

Larger size and greater complexity evolved in chlorophytes by three different mechanisms: (1) the formation of colonies of individual cells, as seen in *Volvox* **(Figure 28.30a)** and in filamentous forms that contribute to the stringy masses known as pond scum; (2) the repeated division of nuclei with no

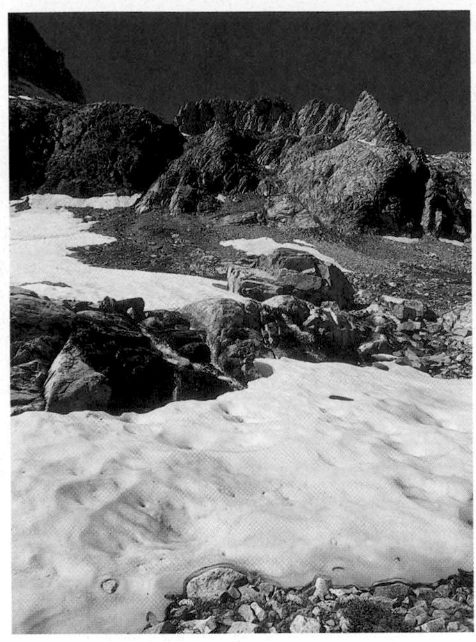

▶ **Figure 28.29 Watermelon snow.** Carotenoids in some snow-dwelling chlorophytes, such as *Chlamydomonas nivalis,* turn the snow red.

▼ Figure 28.30 **Colonial and multicellular chlorophytes.**

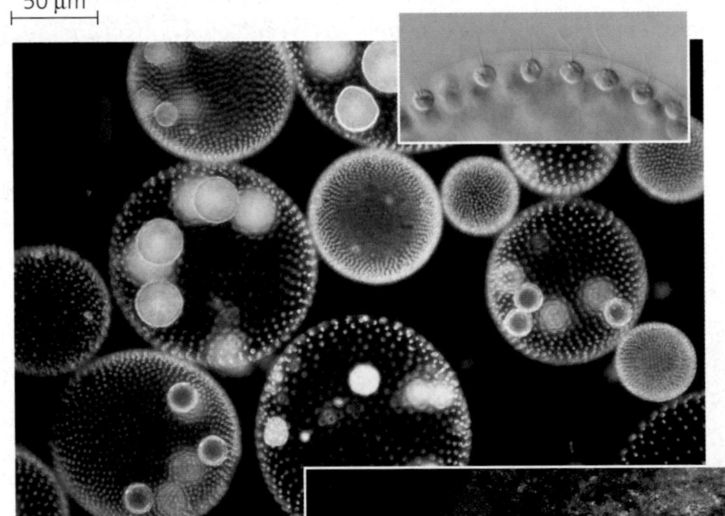

(a) ***Volvox*, a colonial freshwater chlorophyte.** The colony is a hollow ball whose wall is composed of hundreds or thousands of biflagellated cells (see inset LM) embedded in a gelatinous matrix. The cells are usually connected by strands of cytoplasm; if isolated, these cells cannot reproduce. The large colonies seen here will eventually release the small "daughter" colonies within them (LM).

(b) ***Caulerpa*, an intertidal chlorophyte.** The branched filaments lack cross-walls and thus are multinucleate. In effect, the thallus is one huge "supercell."

(c) ***Ulva*, or sea lettuce.** This edible seaweed has a multicellular thallus differentiated into leaflike blades and a rootlike holdfast that anchors the alga against turbulent waves and tides.

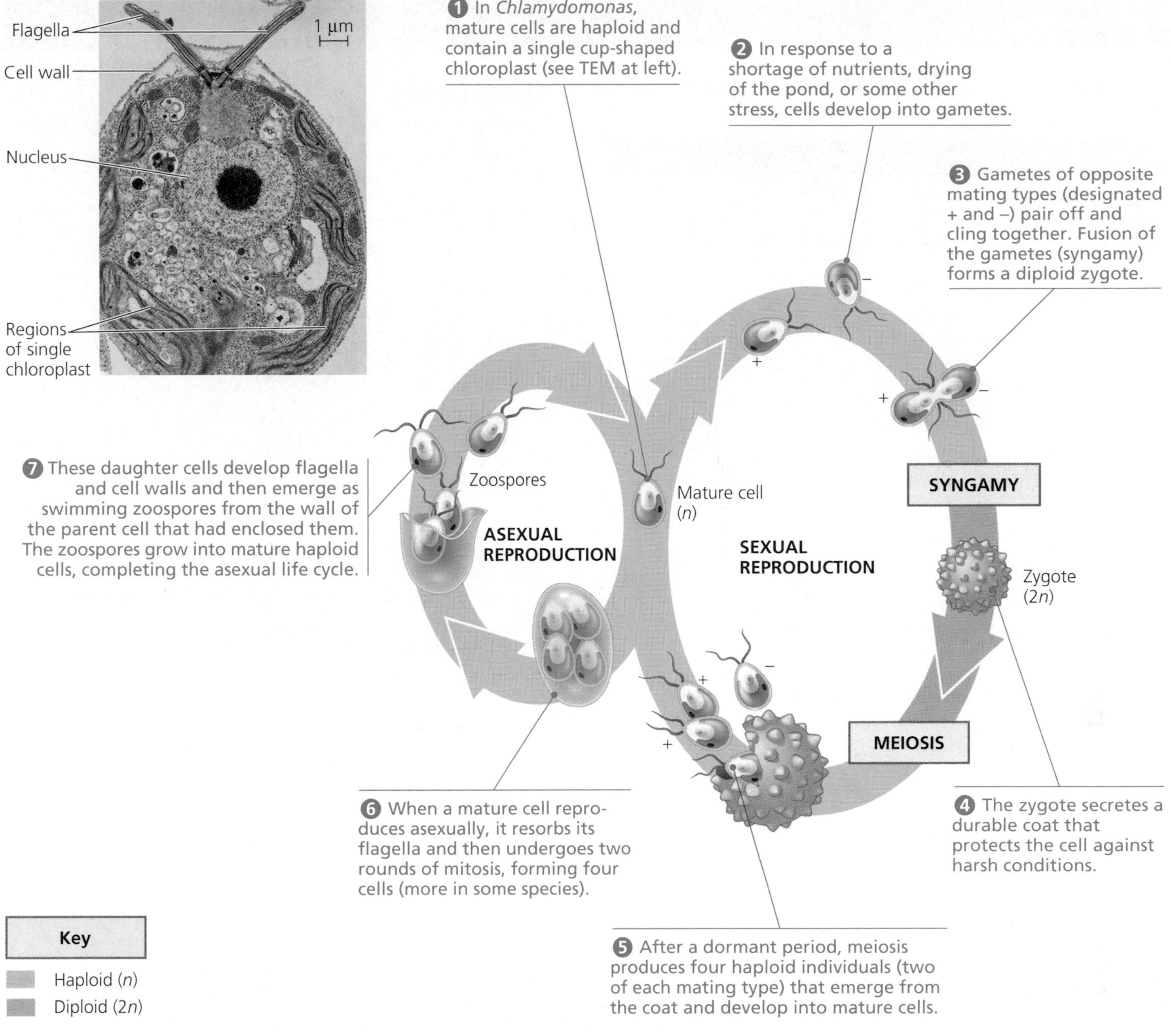

Flagella
Cell wall
1 µm
Nucleus
Regions of single chloroplast
1 In *Chlamydomonas*, mature cells are haploid and contain a single cup-shaped chloroplast (see TEM at left).
2 In response to a shortage of nutrients, drying of the pond, or some other stress, cells develop into gametes.
3 Gametes of opposite mating types (designated + and –) pair off and cling together. Fusion of the gametes (syngamy) forms a diploid zygote.
+
–
SYNGAMY
Zygote (2n)
Mature cell (n)
Zoospores
ASEXUAL REPRODUCTION
SEXUAL REPRODUCTION
MEIOSIS
7 These daughter cells develop flagella and cell walls and then emerge as swimming zoospores from the wall of the parent cell that had enclosed them. The zoospores grow into mature haploid cells, completing the asexual life cycle.
6 When a mature cell reproduces asexually, it resorbs its flagella and then undergoes two rounds of mitosis, forming four cells (more in some species).
4 The zygote secretes a durable coat that protects the cell against harsh conditions.
5 After a dormant period, meiosis produces four haploid individuals (two of each mating type) that emerge from the coat and develop into mature cells.
Key
Haploid (n)
Diploid (2n)

▲ **Figure 28.31 The life cycle of *Chlamydomonas,* a unicellular chlorophyte.**

cytoplasmic division, as seen in the multinucleate filaments of *Caulerpa* **(Figure 28.30b)**; and (3) the formation of true multicellular forms by cell division and cell differentiation, as in *Ulva* **(Figure 28.30c)**. Some multicellular marine chlorophytes are large and complex enough to qualify as seaweeds.

Most chlorophytes have complex life cycles, with both sexual and asexual reproductive stages. Nearly all reproduce sexually by means of biflagellated gametes that have cup-shaped chloroplasts **(Figure 28.31)**. The exceptions are the conjugating algae, such as *Spirogyra* (see Figure 28.2d), which produce amoeboid gametes. Alternation of generations evolved in the life cycles of some green algae, including *Ulva,* in which alternate generations are isomorphic.

The other main group of green algae, the charophyceans, are the most closely related to land plants. For that reason, we will discuss them along with plants in Chapter 29.

Concept Check 28.8

1. Identify two ways in which red algae are different from brown algae.
2. Why is it accurate to say that *Ulva* has true multicellularity but *Caulerpa* does not?

For suggested answers, see Appendix A.

Chapter 28 Review

Go to www.campbellbiology.com or the student CD-ROM to explore Activities, Investigations, and other interactive study aids.

SUMMARY OF KEY CONCEPTS

Concept 28.1

Protists are an extremely diverse assortment of eukaryotes

- Protists are more diverse than all other eukaryotes and are no longer classified in a single kingdom. Most protists are unicellular; some are colonial or multicellular. Protists include photoautotrophs, heterotrophs, and mixotrophs. Most species are aquatic, but others are found in moist terrestrial habitats. Some species are exclusively asexual; others reproduce sexually. **Table 28.1** reviews the groups of protists surveyed in this chapter (pp. 549–550).
- **Endosymbiosis in Eukaryotic Evolution (pp. 550–551)** Some biologists hypothesize that mitochondria and plastids are descendants of alpha proteobacteria and cyanobacteria, respectively, that were engulfed by other cells and became endosymbionts. The plastid-bearing lineage eventually evolved into red algae and green algae. Other protist groups evolved from secondary endosymbiotic events in which red algae or green algae were themselves engulfed.
  Activity *Tentative Phylogeny of Eukaryotes*

Concept 28.2

Diplomonads and parabasalids have modified mitochondria

- Diplomonads and parabasalids are adapted to anaerobic environments. They lack plastids, and their mitochondria do not contain DNA, an electron transport chain, or citric-acid cycle enzymes (p. 552).
- **Diplomonads (p. 552)** Diplomonads have multiple flagella and two nuclei.
- **Parabasalids (p. 553)** Parabasalids include trichomonads, which move by means of flagella and an undulating part of the plasma membrane.

Concept 28.3

Euglenozoans have flagella with a unique internal structure

- Euglenozoans have a spiral or crystalline rod inside their flagella, and most have disk-shaped mitochondrial cristae. They include autotrophs, predatory heterotrophs, and parasites (p. 553).
- **Kinetoplastids (pp. 553–554)** Kinetoplastids have a kinetoplast, an organized mass of DNA within a large mitochondrion. Parasitic kinetoplastids cause sleeping sickness and Chagas' disease.
- **Euglenids (p. 554)** Euglenids have one or two flagella that emerge from a pocket at one end of the cell. These protists store the glucose polymer paramylon. Some species of *Euglena* can switch from autotrophy to heterotrophy when sunlight is unavailable.

Concept 28.4

Alveolates have sacs beneath the plasma membrane

- Alveoli, membrane-bounded sacs beneath the plasma membrane, distinguish alveolates from other protists (p. 555).
- **Dinoflagellates (p. 555)** Dinoflagellates are a diverse group of aquatic photoautotrophs and heterotrophs. Their characteristic spinning movements are produced by two flagella that lie in perpendicular grooves in their cell surface. Rapid growth of some dinoflagellate populations causes red tides.
- **Apicomplexans (pp. 555–556)** Apicomplexans are parasites that have an apical complex of organelles specialized for invading host cells. They also have a nonphotosynthetic plastid, the apicoplast. The apicomplexan *Plasmodium* causes malaria.
- **Ciliates (pp. 556–558)** Ciliates use cilia to move and feed. They have large macronuclei and small micronuclei. The micronuclei function during conjugation, a sexual process that produces genetic variation. Conjugation is separate from reproduction, which generally occurs by binary fission.

Concept 28.5

Stramenopiles have "hairy" and smooth flagella

- **Oomycetes (Water Molds and Their Relatives) (pp. 558–559)** Most oomycetes are decomposers or parasites and have filaments (hyphae) that facilitate nutrient uptake. The oomycete *Phytophthora infestans* causes potato late blight.
- **Diatoms (pp. 559–560)** Diatoms are surrounded by a two-part glass-like wall and are a major component of phytoplankton. Accumulations of fossilized diatom walls comprise much of the sediments known as diatomaceous earth.
- **Golden Algae (p. 560)** Golden algae typically have two flagella attached near one end of the cell. Many species are planktonic. Their color results from the carotenoids they contain.
- **Brown Algae (pp. 560–562)** Brown algae are multicellular, mostly marine protists. They include some of the most complex algae commonly known as seaweeds, many of which are commercially important to humans. Like some red algae and green algae and all plants, some brown algae have a life cycle that features alternation of generations: A multicellular diploid form alternates with a multicellular haploid form.

Concept 28.6

Cercozoans and radiolarians have threadlike pseudopodia

- **Foraminiferans (Forams) (p. 563)** Forams are marine and freshwater amoebas with porous, generally multichambered shells (tests) made of organic material and calcium carbonate. Pseudopodia extend through the pores. Foram tests in marine sediments form an extensive fossil record.
- **Radiolarians (p. 563)** Radiolarians have fused tests usually made of silica. They ingest microorganisms by phagocytosis using their pseudopodia, which radiate from their central body.

Table 28.1 A Sample of Protist Diversity

Major Clade	Key Characteristics	Examples from Chapter
Diplomonadida (diplomonads)	Two equal-sized nuclei; modified mitochondria	*Giardia*
Parabasala (parabasalids)	Undulating membrane; modified mitochondria	*Trichomonas*
Euglenozoa (euglenozoans)	Spiral or crystalline rod inside flagella	
Kinetoplastida (kinetoplastids)	Kinetoplast (DNA in mitochondrion)	*Trypanosoma*
Euglenophyta (euglenids)	Paramylon as storage molecule	*Euglena*
Alveolata (alveolates)	Alveoli beneath plasma membrane	
Dinoflagellata (dinoflagellates)	Armor of cellulose plates	*Ceratium, Pfiesteria*
Apicomplexa (apicomplexans)	Apical complex of organelles	*Plasmodium*
Ciliophora (ciliates)	Cilia used in movement and feeding; macro- and micronuclei	*Paramecium, Stentor*
Stramenopila (stramenopiles)	Hairy and smooth flagella	
Oomycota (oomycetes)	Hyphae that absorb nutrients	Water molds, white rusts, downy mildews
Bacillariophyta (diatoms)	Glassy, two-part wall	
Chrysophyta (golden algae)	Flagella attached near one end of cell	*Dinobryon*
Phaeophyta (brown algae)	All multicellular, some with alternation of generations	*Laminaria, Macrocystis, Postelsia*
Cercozoa (cercozoans) and Radiolaria (radiolarians)	Amoebas with threadlike pseudopodia	
Foraminfera (forams)	Porous shell	*Globigerina*
Radiolaria (radiolarians)	Pseudopodia radiating from central body	
Amoebozoa (amoebozoans)	Amoebas with lobe-shaped pseudopodia	
Gymnamoeba (gymnamoebas)	Soil-dwelling, freshwater, or marine	*Amoeba*
Entamoeba (entamoebas)	Parasites	*Entamoeba*
Myxogastrida (plasmodial slime molds)	Multinucleate plasmodium; fruiting bodies that function in sexual reproduction	*Physarum*
Dictyostelida (cellular slime molds)	Multicellular aggregate that forms asexual fruiting bodies	*Dictyostelium*
Rhodophyta (red algae)	Phycoerythrin (accessory pigment); no flagellated stages	*Bonnemaisonia, Delesseria, Palmaria*
Chlorophyta (one group of green algae)	Plant-type chloroplasts	*Caulerpa, Chlamydomonas, Spirogyra, Ulva, Volvox*

Concept 28.7

Amoebozoans have lobe-shaped pseudopodia

- **Gymnamoebas (p. 564)** Gymnamoebas are common unicellular amoebozoans in soil as well as freshwater and marine environments. Most are heterotrophs.
- **Entamoebas (p. 564)** Entamoebas are parasites of vertebrates and some invertebrates. *Entamoeba histolytica* causes amebic dysentery in humans.
- **Slime Molds (pp. 564–566)** Plasmodial slime molds aggregate into a plasmodium, a multinucleate mass of cytoplasm undivided by membranes. The plasmodium extends pseudopodia through decomposing material, engulfing food by phagocytosis. Cellular slime molds form multicellular aggregates in which the cells remain separated by their membranes. The cellular slime mold *Dictyostelium discoideum* has become an experimental model for studying the evolution of multicellularity.

Concept 28.8

Red algae and green algae are the closest relatives of land plants

- **Red Algae (p. 567)** Red algae have colors ranging from green to black, owing to varying amounts of the accessory pigment phycoerythrin. Most red algae are multicellular; the largest are seaweeds. They are the most abundant large algae in coastal waters of the tropics.
- **Green Algae (pp. 567–569)** Green algae (chlorophytes and charophyceans) are closely related to land plants. Most chlorophytes live in fresh water, although many are marine; others live in damp soil, in snow, or as symbionts in lichens. Chlorophytes include unicellular, colonial, and multicellular forms. Most have complex life cycles.
  Investigation ***What Kinds of Protists Do Various Habitats Support?***

TESTING YOUR KNOWLEDGE

Self-Quiz

1. Plastids that are surrounded by more than two membranes are evidence of
   a. evolution from mitochondria.
   b. fusion of plastids.
   c. origin of the plastids from archaea.
   d. secondary endosymbiosis.
   e. budding of the plastids from the nuclear envelope.

2. Biologists suspect that endosymbiosis gave rise to mitochondria before plastids partly because
   a. the products of photosynthesis could not be metabolized without mitochondrial enzymes.
   b. almost all eukaryotes have mitochondria, whereas only autotrophic eukaryotes generally have plastids.
   c. mitochondrial DNA is less similar to prokaryotic DNA than is plastid DNA.
   d. without mitochondrial CO_2 production, photosynthesis could not occur.
   e. mitochondrial proteins are synthesized on cytosolic ribosomes, whereas plastids utilize their own ribosomes.

3. Which of the following eukaryotes have mitochondria that lack an electron transport chain?
   a. golden algae
   b. diplomonads
   c. apicomplexans
   d. kinetoplastids
   e. diatoms

4. Which organism is *incorrectly* paired with its description?
   a. cercozoans—amoebas with threadlike pseudopodia
   b. euglenids—protists that store paramylon
   c. forams—ciliated algae with numerous micronuclei
   d. apicomplexans—parasites with intricate life cycles
   e. diplomonads—protists with modified mitochondria

5. Dinoflagellates, apicomplexans, and ciliates are placed in the clade Alveolata because they all
   a. have flagella or cilia.
   b. are parasites of animals.
   c. are found exclusively in freshwater or marine habitats.
   d. have mitochondria.
   e. have membrane-bounded sacs under their plasma membrane.

6. In ciliates, the process that produces genetic variation through the exchange of nuclei is
   a. mixotrophy.
   b. endosymbiosis.
   c. meiosis.
   d. conjugation.
   e. binary fission.

7. The protist that contributed to the Irish potato famine was
   a. a foram.
   b. a ciliate.
   c. an oomycete.
   d. a plasmodial slime mold.
   e. a cellular slime mold.

8. Which algal group is *mismatched* with its description?
   a. dinoflagellates—glassy, two-part shells
   b. green algae—closest relatives of land plants
   c. red algae—no flagellated stages in life cycle
   d. brown algae—include the largest seaweeds
   e. diatoms—a major component of phytoplankton

9. In life cycles with an alternation of generations, multicellular haploid forms alternate with
   a. unicellular haploid forms.
   b. unicellular diploid forms.
   c. multicellular haploid forms.
   d. multicellular diploid forms.
   e. multicellular polyploid forms.

10. Which protists form colorful, multinucleate masses?
    a. euglenids
    b. plasmodial slime molds
    c. cellular slime molds
    d. forams
    e. water molds

For Self-Quiz answers, see Appendix A.

Go to the website or CD-ROM for more quiz questions.

Evolution Connection

Explain why systematists consider the kingdom Protista to be an obsolete taxon.

Scientific Inquiry

Applying the "If . . . then" logic of science (see Chapter 1), what are a few of the predictions that arise from the hypothesis that plants evolved from green algae? Put another way, how could you test this hypothesis?

Investigation ***What Kinds of Protists Do Various Habitats Support?***

Science, Technology, and Society

The ability of the pathogen *Plasmodium* to evade the human immune system is one reason developing a malaria vaccine is so difficult. Another reason is that less money is spent on malaria research than on research into diseases that affect far fewer people, such as cystic fibrosis. What are the possible reasons for this imbalance in research effort?

29 Plant Diversity I
How Plants Colonized Land

▲ **Figure 29.1 Tree ferns and a moss-covered log.**

Key Concepts

29.1 Land plants evolved from green algae
29.2 Land plants possess a set of derived terrestrial adaptations
29.3 The life cycles of mosses and other bryophytes are dominated by the gametophyte stage
29.4 Ferns and other seedless vascular plants formed the first forests

Overview

The Greening of Earth

Looking at a lush landscape, such as the forest scene in **Figure 29.1**, it is difficult to imagine the land without any plants or other organisms. Yet for more than the first 3 billion years of Earth's history, the terrestrial surface was lifeless. Geochemical evidence suggests that thin coatings of cyanobacteria existed on land about 1.2 billion years ago. But it was only about 500 million years ago that plants, fungi, and animals joined them ashore.

In this chapter, we will focus on land plants and how they evolved from aquatic green algae. Although some plant species, such as sea grasses, returned to aquatic habitats during their evolution, most plants live in terrestrial environments. Therefore, we will refer to all plants as *land* plants, even those that are now aquatic, to distinguish them from algae, which are photosynthetic protists. Since colonizing land, plants have diversified into roughly 290,000 living species inhabiting all but the harshest environments, such as some mountaintops and some desert and polar regions. The presence of plants has enabled other life-forms—including humans—to survive on land. Plant roots have created habitats for other organisms by stabilizing landscapes. More importantly, plants supply oxygen and are the ultimate provider of most of the food eaten by land animals.

In this chapter, we will trace the first 100 million years of plant evolution, including the emergence of seedless plants such as mosses and ferns. In Chapter 30, we will examine the later evolution of seed plants.

Concept 29.1

Land plants evolved from green algae

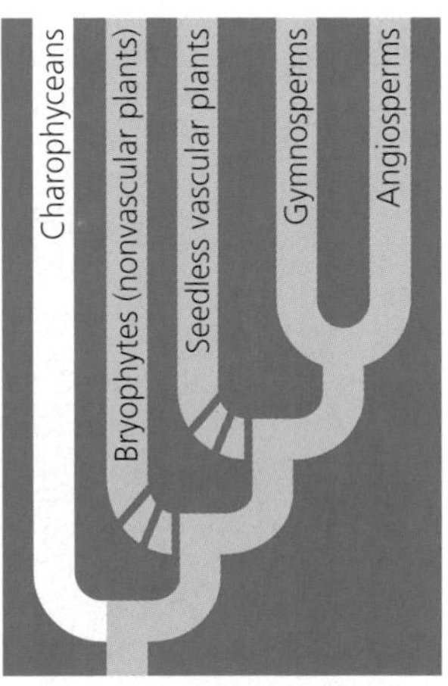

As you read in Chapter 28, researchers have identified green algae called charophyceans as the closest relatives of land plants. Here we will examine the evidence of this relationship and consider what it suggests about how the algal ancestors of land plants might have been well suited to make the move to land.

Morphological and Biochemical Evidence

Many key characteristics of land plants also appear in a variety of protists, primarily algae. For example, plants are multicellular, eukaryotic, photosynthetic autotrophs, as are brown, red, and certain green algae (see Chapter 28). Plants have cell walls made of cellulose, and so do green algae, dinoflagellates, and brown algae. And chloroplasts with chlorophylls *a* and *b* are present in green algae, euglenids, and a few dinoflagellates, as well as in plants.

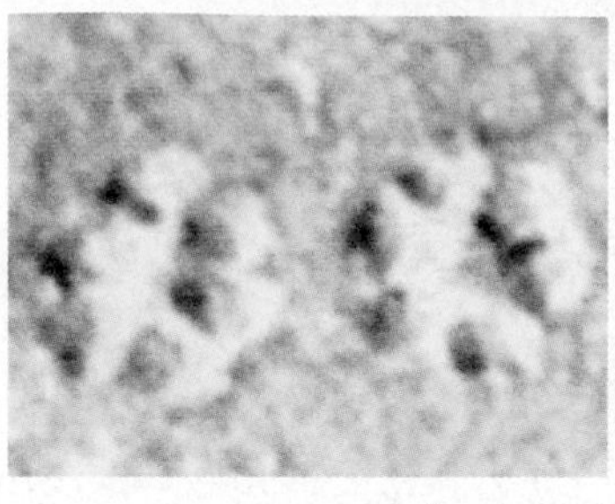

▶ **Figure 29.2 Rosette cellulose-synthesizing complexes.** These distinctively rose-shaped arrays of proteins are found only in land plants and charophycean algae, suggesting their close kinship (SEM).

There are four key traits, however, that land plants share only with the charophyceans, strongly suggesting a close relationship between the two groups:

- **Rose-shaped complexes for cellulose synthesis.** The cells of both land plants and charophyceans have **rosette cellulose-synthesizing complexes.** These are rose-shaped arrays of proteins in the plasma membrane that synthesize the cellulose microfibrils of the cell walls **(Figure 29.2)**. In contrast, linear arrays of proteins synthesize cellulose in noncharophycean algae. Also, the cell walls of plants and charophyceans contain a higher percentage of cellulose than the cell walls of noncharophycean algae. These differences indicate that the cellulose walls in plants and charophyceans evolved independently of those in other algae.
- **Peroxisome enzymes.** The peroxisomes (see Figure 6.19) of both land plants and charophyceans contain enzymes that help minimize the loss of organic products as a result of photorespiration (see Chapter 10). The peroxisomes of other algae lack these enzymes.
- **Structure of flagellated sperm.** In species of land plants that have flagellated sperm, the structure of the sperm closely resembles that of charophycean sperm.
- **Formation of a phragmoplast.** Certain details of cell division occur only in land plants and certain charophyceans, including the genera *Chara* and *Coleochaete*. For example, the synthesis of new cross-walls (cell plates) during cell division involves the formation of a **phragmoplast,** an alignment of cytoskeletal elements and Golgi-derived vesicles across the midline of the dividing cell (see Figure 12.10).

Genetic Evidence

Over the past decade, researchers involved in an international initiative called "Deep Green" have conducted a large-scale study of major transitions in plant evolution, analyzing genes from a wide range of plant and algal species. Comparisons of both nuclear and chloroplast genes agree with the morphological and biochemical data in pointing to charophyceans—particularly *Chara* and *Coleochaete*—as the closest living relatives of land plants **(Figure 29.3)**. Note that this does not mean that these *living* algae are the ancestors of plants; however, they do offer a glimpse of what those ancestors might have been like.

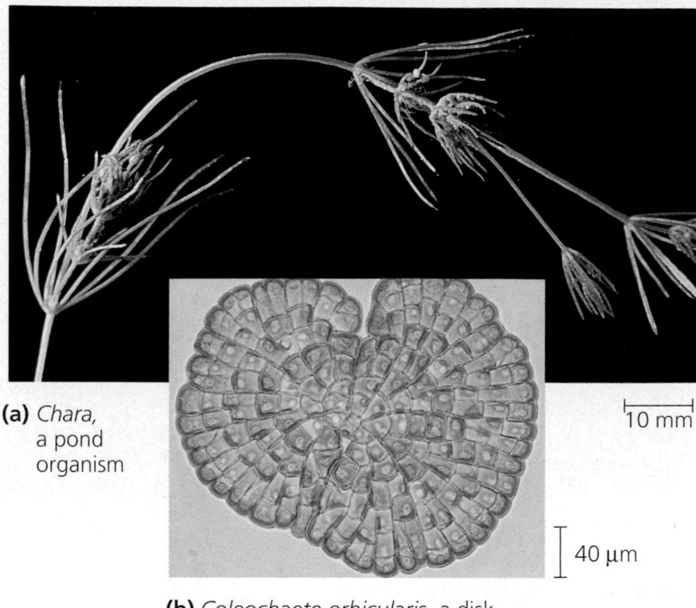

(a) *Chara,* a pond organism

(b) *Coleochaete orbicularis,* a disk-shaped charophycean (LM)

▲ **Figure 29.3 Examples of charophyceans, the closest algal relatives of land plants.**

Adaptations Enabling the Move to Land

Many species of charophycean algae inhabit shallow waters around the edges of ponds and lakes, where they are subject to occasional drying. In such environments, natural selection favors individual algae that can survive periods when they are not submerged in water. In charophyceans, a layer of a durable polymer called **sporopollenin** prevents exposed zygotes from drying out. An ancestral form of this chemical adaptation may have also been the precursor to the tough sporopollenin walls that encase plant spores.

It is likely that the accumulation of such traits by at least one population of charophycean ancestors enabled their descendants—the first land plants—to live permanently above the waterline. These evolutionary novelties opened an expanse of terrestrial habitat, a new frontier that offered enormous benefits. The bright sunlight was unfiltered by water and plankton; the atmosphere had an abundance of CO_2; the soil was rich in mineral nutrients; and initially there were relatively few herbivores and pathogens. Benefiting from these environmental opportunities became possible as adaptations evolved in plants that allowed them to survive and reproduce on land.

Concept Check 29.1

1. Describe the evidence linking plants to a charophycean ancestry.

For suggested answers, see Appendix A.

Land plants possess a set of derived terrestrial adaptations

Many of the adaptations that emerged after land plants diverged from their charophycean relatives facilitated survival and reproduction on dry land. We will explore the most important derived traits of plants. We will then examine some fossil evidence of the divergence of land plants from charophyceans and survey main groups within the plant kingdom.

Defining the Plant Kingdom

Where exactly is the line dividing land plants from algae? Systematists are currently debating the boundaries of the plant kingdom (**Figure 29.4**). The traditional scheme equates the kingdom Plantae with embryophytes (plants with embryos; *phyte* is from the Greek word for plant). Some plant biologists now propose that the boundaries of the plant kingdom should be expanded to include the green algae most closely related to plants—the charophyceans and a few related groups—and named kingdom Streptophyta. Others suggest an even broader definition of plants that also includes chlorophytes (the non-charophycean green algae) in a kingdom Viridiplantae (see Chapter 28). Since the debate is ongoing, this text retains the embryophyte definition of the plant kingdom and uses kingdom Plantae as the formal name for the taxon.

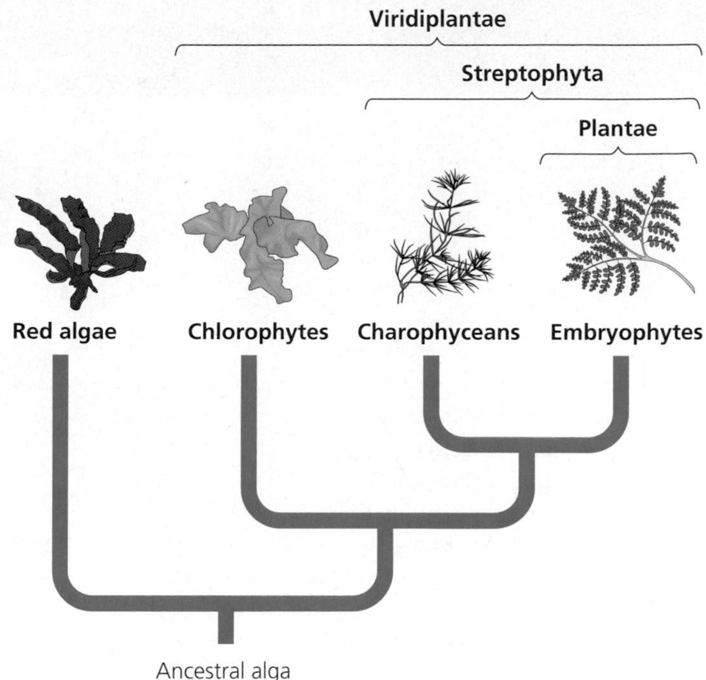

▲ **Figure 29.4 Three clades that are candidates for designation as the plant kingdom.** This textbook adopts the embryophyte definition of plants and uses the name Plantae for the kingdom.

Derived Traits of Plants

Five key traits appear in nearly all land plants but are absent in the charophyceans: apical meristems; alternation of generations; walled spores produced in sporangia; multicellular gametangia; and multicellular, dependent embryos. **Figure 29.5**, on the next two pages, depicts these traits. We can infer that these traits were absent in the ancestor common to land plants and charophyceans but instead evolved independently as derived traits of land plants. Some of these traits are not unique to plants, having evolved separately in other lineages, and some traits have been lost in certain lineages of plants. However, these key traits set the first land plants apart from their closest algal relatives.

Additional derived traits that relate to terrestrial life have evolved in many plant species. Permanently exposed to the air, land plants run a far greater risk of desiccation (drying out) than their algal ancestors. The epidermis in many species has a covering, known as a **cuticle,** that consists of polymers called polyesters and waxes. The cuticle acts as waterproofing, helping prevent excessive water loss from the above-ground plant organs, while also providing some protection from microbial attack.

Many land plants produce molecules called *secondary compounds,* so named because they are products of secondary metabolic pathways—side branches off the primary metabolic pathways that produce the lipids, carbohydrates, amino acids, and other compounds common to all organisms. Secondary compounds include alkaloids, terpenes, tannins, and phenolics such as flavonoids. Various alkaloids, terpenes, and tannins have a bitter taste, strong odor, or toxic effect that helps defend against herbivores and parasites. Flavonoids absorb harmful UV radiation and may act as signals in symbiotic relationships with beneficial soil microbes. Some phenolics deter attack by pathogenic microbes. As you survey the major groups of land plants, note how secondary compounds aid their survival. Humans also benefit from secondary compounds; just one example is use of the alkaloid quinine's antimicrobial properties to fight malaria.

The Origin and Diversification of Plants

Paleobotanists seeking the evolutionary origin of plants have long debated what constitutes the oldest fossil evidence of land plants. In the 1970s, researchers found fossil spores dating to the Ordovician period, up to 475 million years old. Although the fossil spores resemble those of living plants, they also have some striking differences. For example, spores of living plants are typically dispersed as single grains, but the fossil spores are fused together in groups of two or four. This difference raises the possibility that the fossil spores were not produced by plants, but by some extinct algal relative. Furthermore, the oldest known fragments of plant tissue are 50 million years younger than the puzzling spores.

Figure 29.5

Exploring Derived Traits of Land Plants

APICAL MERISTEMS

In terrestrial habitats, a photosynthetic organism finds essential resources in two very different places. Light and CO_2 are mainly available above-ground; water and mineral nutrients are found mainly in the soil. Though plants cannot move from place to place, their roots and shoots can elongate, increasing exposure to environmental resources. This growth in length is sustained throughout the plant's life by the activity of **apical meristems**, localized regions of cell division at the tips of shoots and roots. Cells produced by apical meristems differentiate into various tissues, including a surface epidermis that protects the body and several types of internal tissues. Shoot apical meristems also generate leaves in most plants. Thus, the complex bodies of plants show structural specialization for subterranean and aerial organs—roots and leaf-bearing shoots, respectively, in most plants.

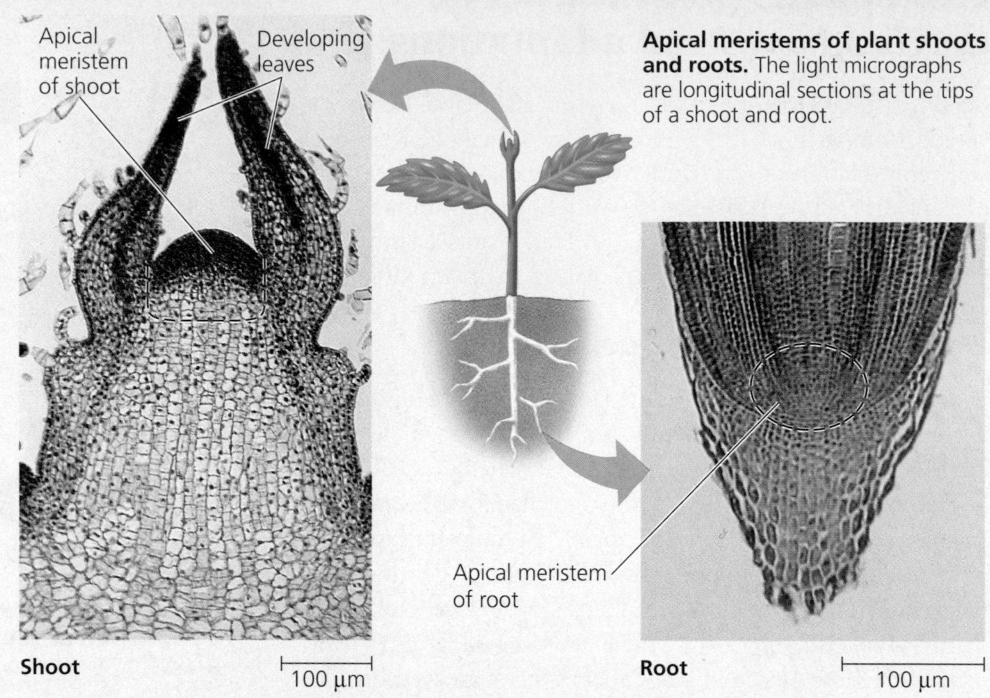

Apical meristems of plant shoots and roots. The light micrographs are longitudinal sections at the tips of a shoot and root.

ALTERNATION OF GENERATIONS

The life cycles of all land plants alternate between two different multicellular bodies, with each form producing the other. This type of reproductive cycle, called **alternation of generations**, also evolved in various groups of algae but does not occur in the charophyceans, the algae most closely related to land plants. We can infer that alternation of generations is a derived characteristic of land plants—it was not present in the ancestor common to land plants and charophyceans.

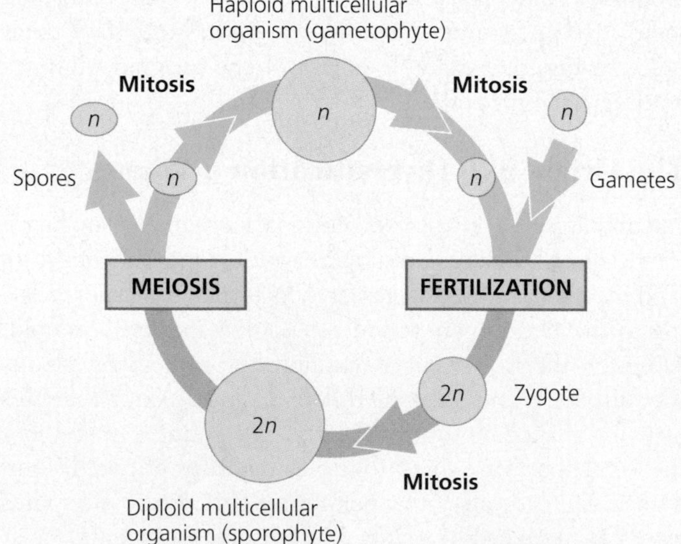

Alternation of generations: a generalized scheme

Take care not to confuse the alternation of generations in plants with the haploid and diploid stages in the life cycles of *all* sexually reproducing organisms (see Figure 13.6). In humans, for example, meiosis in the gonads (ovaries and testes) produces haploid gametes that unite, forming diploid zygotes that divide and become multicellular. The haploid stage is represented only by single-celled gametes. In contrast, alternation of generations is distinguished by the fact that there are both multicellular haploid and multicellular diploid stages in the life cycle.

The two multicellular body forms that alternate in the life cycles of land plants are the gametophyte and sporophyte generations. The cells of the **gametophyte** are haploid, meaning they have a single set of chromosomes. The gametophyte is named for its production by mitosis of haploid gametes—eggs and sperm—that fuse during fertilization, forming diploid zygotes. Mitotic division of the zygote produces the multicellular **sporophyte**, the spore-producing generation. Thus, the cells of the sporophyte are diploid, having two sets of chromosomes—one from each gamete. Meiosis in a mature sporophyte produces haploid **spores**, reproductive cells that can develop into a new organism without fusing with another cell. In contrast, gametes cannot develop directly into a multicellular organism but instead must fuse and form a zygote. Mitotic division of a plant spore produces a new multicellular gametophyte. And so the alternation of generations continues, with sporophytes producing spores that develop into gametophytes, and gametophytes producing gametes that unite, forming the zygotes that develop into sporophytes.

WALLED SPORES PRODUCED IN SPORANGIA

Plant spores are haploid reproductive cells that have the potential to grow into multicellular, haploid gametophytes by mitosis. The polymer sporopollenin makes the walls of plant spores very tough and resistant to harsh environments. This chemical adaptation makes it possible for spores to be dispersed through dry air without harm.

The sporophyte has multicellular organs called **sporangia** (singular, *sporangium*) that produce plant spores. Within a sporangium, diploid cells called **sporocytes**, also known as spore mother cells, undergo meiosis and generate the haploid spores. The outer tissues of the sporangium protect the developing spores until they are released into the air.

Multicellular sporangia that produce spores with sporopollenin-enriched walls are key terrestrial adaptations of land plants. Although charophyceans produce spores, these algae lack multicellular sporangia, and their flagellated, water-dispersed spores lack sporopollenin.

Sporophyte and sporangium of *Sphagnum* (a moss)

MULTICELLULAR GAMETANGIA

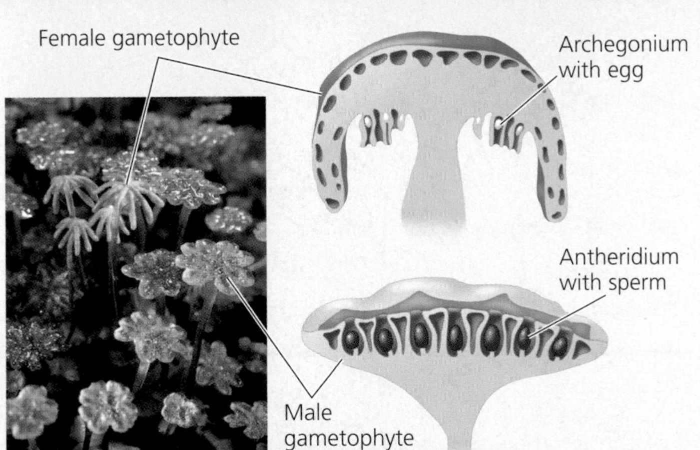

Archegonia and antheridia of *Marchantia* (a liverwort)

Another feature distinguishing early land plants from their algal ancestors was the production of gametes within multicellular organs called **gametangia**. The female gametangia are called **archegonia** (singular, *archegonium*), each of which is a vase-shaped organ that produces a single egg retained within the base of the organ. Male gametangia, called **antheridia** (singular, *antheridium*), produce and release sperm into the environment. In many major groups of living plants, the sperm have flagella and swim to the eggs through water droplets or water films. Each egg is fertilized within an archegonium, where the zygote develops into an embryo. As you will see in Chapter 30, the gametophytes of seed plants are so reduced in size that the archegonia and antheridia have been lost in some lineages.

MULTICELLULAR, DEPENDENT EMBRYOS

Multicellular plant embryos develop from zygotes that are retained within tissues of the female parent. The parental tissues provide the developing embryo with nutrients, such as sugars and amino acids. The embryo has specialized **placental transfer cells**, sometimes present in the adjacent maternal tissue as well, which enhance the transfer of nutrients from parent to embryo through elaborate ingrowths of the wall surface (plasma membrane and cell wall). This interface is analogous to the nutrient-transferring embryo-mother interface of eutherian (placental) mammals. The multicellular, dependent embryo of land plants is such a significant derived trait that land plants are also known as **embryophytes**.

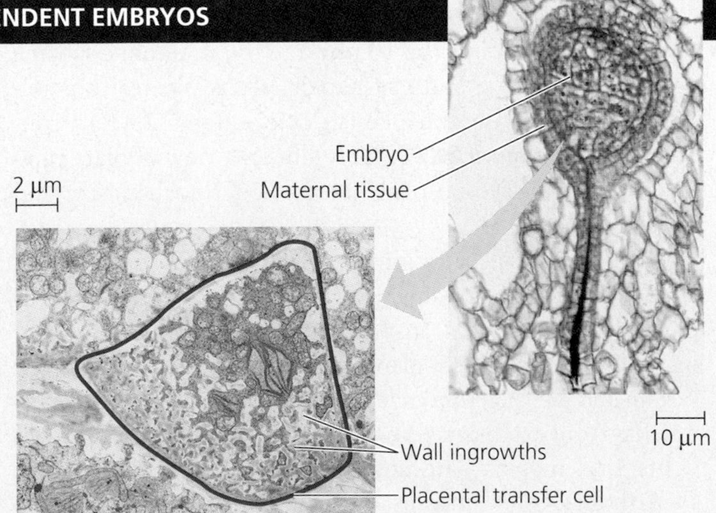

Embryo (LM) and placental transfer cell (TEM) of *Marchantia*

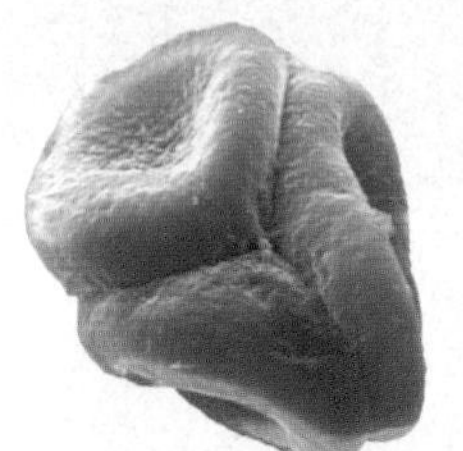
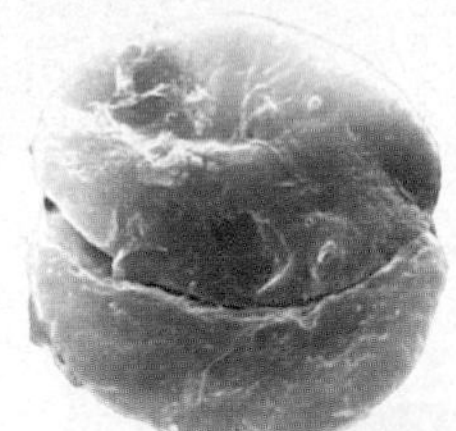

(a) Fossilized spores. Unlike the spores of most living plants, which are single grains, these spores found in Oman are in groups of four (left; one hidden) and two (right).

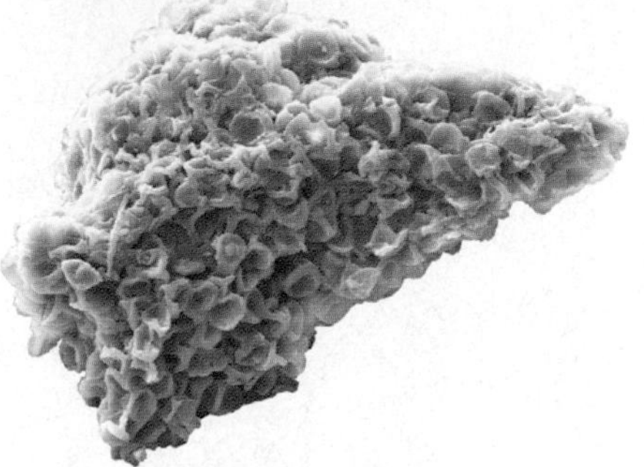

(b) Fossilized sporophyte tissue. The spores were embedded in tissue that appears to be from plants.

▲ **Figure 29.6 Ancient plant spores and tissue.** (colorized SEMs)

In 2003, scientists from Britain and the Middle Eastern country of Oman shed some light on this mystery when they extracted spores from 475-million-year-old rocks from Oman **(Figure 29.6a)**. Unlike previously discovered spores of this age, these were embedded in plant cuticle material that is similar to spore-bearing tissue in living plants **(Figure 29.6b)**. After finding other small fragments of tissue that clearly belonged to plants, the scientists concluded that the spores from Oman represent fossil plants rather than algae.

A 2001 "molecular clock" study (see Chapter 25) of plants suggested that the common ancestor of living plants existed 700 million years ago. If this is true, then the fossil record is missing the first 225 million years of plant evolution. In 2003, however, Michael Sanderson of the University of California reported an estimate based on molecular data of 490 to 425 million years, roughly the age of the spores discovered in Oman.

Whatever the precise age of the first land plants, those ancestral species gave rise to the vast diversity of living plants. **Table 29.1** summarizes the 10 phyla in the taxonomic scheme used in this chapter and in Chapter 30. As you read the following overview of land plants, look at Table 29.1 in conjunction with **Figure 29.7**, which reflects a view of plant phylogeny that is based on plant morphology, biochemistry, and genetics.

Land plants can be informally grouped based on the presence or absence of an extensive system of **vascular tissue,** cells joined into tubes that transport water and nutrients throughout the plant body. Most plants have a complex vascular tissue system and are therefore called **vascular plants.** Plants that do not have an extensive transport system—liverworts, hornworts, and mosses—are described as "nonvascular" plants, even though some mosses do have simple vascular tissue. Nonvascular plants are often informally called **bryophytes** (from the Greek *bryon,* moss, and *phyton,* plant).

Table 29.1 Ten Phyla of Extant Plants

	Common Name	Approximate Number of Extant Species
Bryophytes (nonvascular plants)		
Phylum Hepatophyta	Liverworts	9,000
Phylum Anthocerophyta	Hornworts	100
Phylum Bryophyta	Mosses	15,000
Vascular Plants		
Seedless Vascular Plants		
Phylum Lycophyta	Lycophytes (club mosses, spike mosses, and quillworts)	1,200
Phylum Pterophyta	Pterophytes (ferns, horsetails, and whisk ferns)	12,000
Seed Plants		
Gymnosperms		
Phylum Ginkgophyta	Ginkgo	1
Phylum Cycadophyta	Cycads	130
Phylum Gnetophyta	Gnetophytes (*Gnetum, Ephedra,* and *Welwitschia*)	75
Phylum Coniferophyta	Conifers	600
Angiosperms		
Phylum Anthophyta	Flowering plants	250,000

Although the term *bryophyte* is still commonly used to refer to all nonvascular plants, debate continues over the relationships of liverworts, hornworts, and mosses to each other and to vascular plants. Whereas some molecular studies have concluded that bryophytes are not monophyletic, a recent analysis of amino acid sequences in chloroplasts asserts that bryophytes form a clade. The broken lines in Figure 29.7 reflect the current uncertainty regarding bryophyte phylogeny. Whether or not bryophytes are monophyletic, they share some derived traits with vascular plants, such as multicellular embryos and apical meristems, while lacking many innovations of vascular plants, such as roots and true leaves.

Vascular plants form a clade, consisting of about 93% of all plant species. Vascular plants can be categorized further into three smaller clades. Two of these clades are the **lycophytes** (club mosses and their relatives) and the **pterophytes** (ferns and their relatives). Each of these clades lacks seeds, which is why collectively the two clades are often informally called **seedless vascular plants.** However, notice in Figure 29.7 that seedless vascular plants are not monophyletic. The third clade

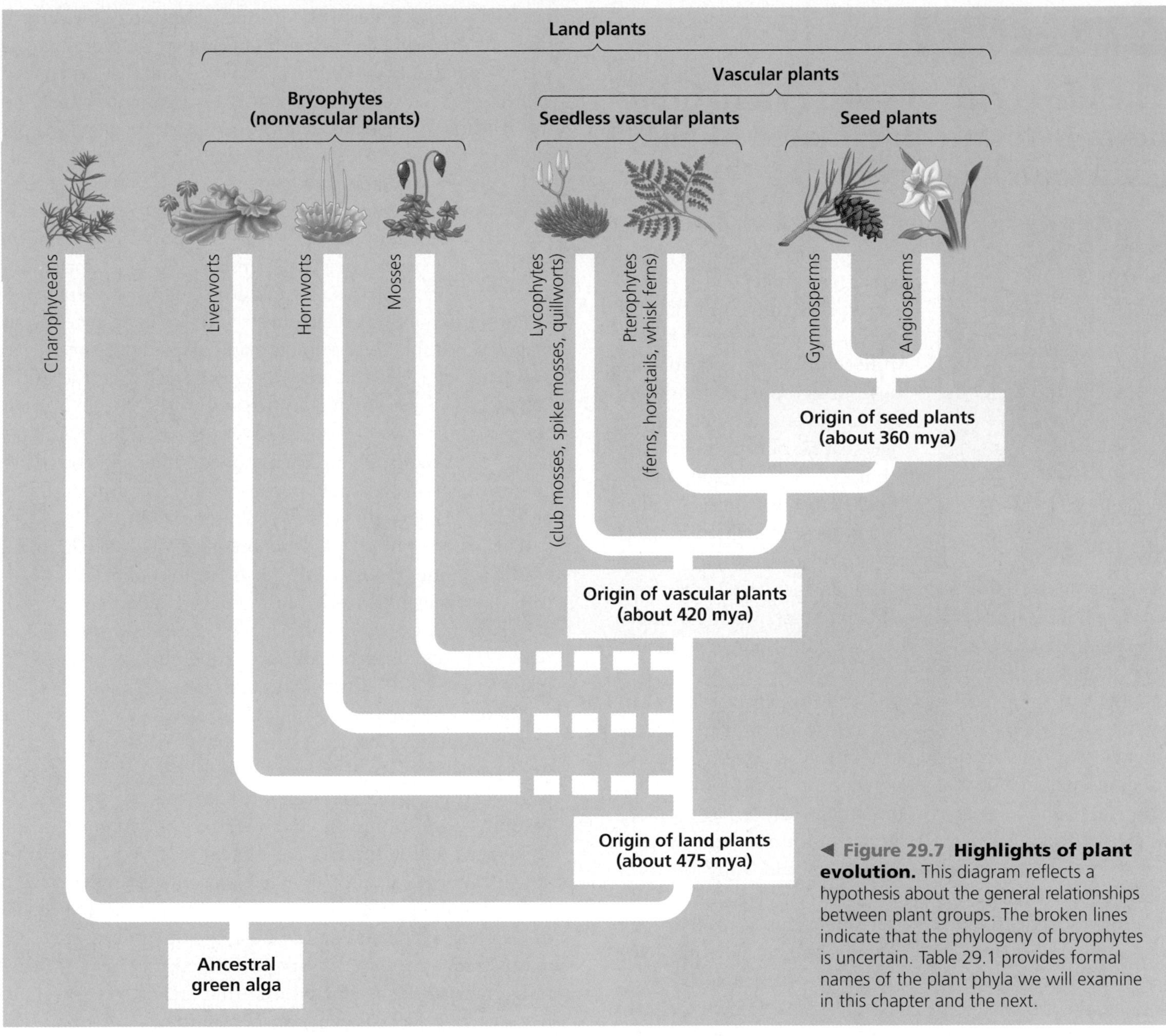

◀ **Figure 29.7 Highlights of plant evolution.** This diagram reflects a hypothesis about the general relationships between plant groups. The broken lines indicate that the phylogeny of bryophytes is uncertain. Table 29.1 provides formal names of the plant phyla we will examine in this chapter and the next.

of vascular plants consists of seed plants, the vast majority of living plant species. A **seed** is an embryo packaged with a supply of nutrients inside a protective coat. Seed plants can be divided into two groups, gymnosperms and angiosperms, based on the absence or presence of enclosed chambers in which seeds mature. **Gymnosperms** (from the Greek *gymnos,* naked, and *sperm,* seed) are grouped together as "naked seed" plants because their seeds are not enclosed in chambers. Surviving gymnosperm species, which consist primarily of conifers, probably form a clade. **Angiosperms** (from the Greek *angion,* container) are a huge clade consisting of all flowering plants. Angiosperm seeds develop inside chambers called ovaries, which originate within flowers and mature into fruits.

Note that the phylogeny depicted in Figure 29.7 focuses only on the relationships between *extant* plant lineages—those that have surviving members in addition to extinct members. Paleobotanists have also discovered fossils belonging to extinct plant lineages. Many of these fossils reveal the intermediate steps leading to the distinctive plant groups found on Earth today.

Concept Check 29.2

1. Identify three derived traits that distinguish plants from charophyceans *and* facilitate life on land. Explain.
2. Identify each structure as either haploid or diploid: (a) sporophyte; (b) spore; (c) gametophyte; (d) zygote; (e) sperm; (f) egg.

For suggested answers, see Appendix A.

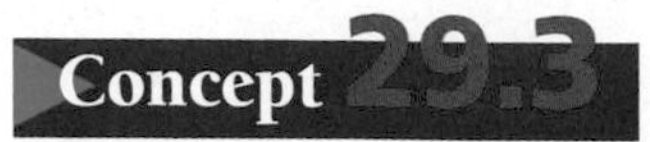

The life cycles of mosses and other bryophytes are dominated by the gametophyte stage

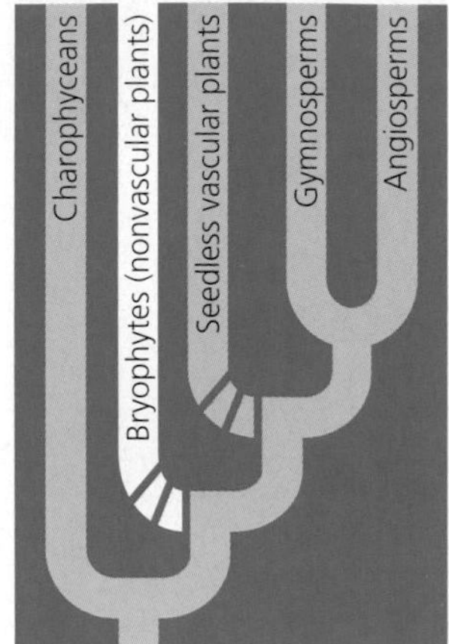

Bryophytes are represented today by three phyla of small herbaceous (nonwoody) plants: **liverworts** (phylum Hepatophyta), **hornworts** (phylum Anthocerophyta), and **mosses** (phylum Bryophyta). Liverworts and hornworts are named for their shapes, plus the suffix *wort* (from the Anglo-Saxon, herb). Mosses are the most familiar bryophytes, although some plants commonly called "mosses" are not really mosses at all. These include Irish moss (a red seaweed), reindeer moss (a lichen), club mosses (seedless vascular plants), and Spanish mosses (lichens in some regions and flowering plants in others).

Note that the terms *Bryophyta* and *bryophyte* are not synonymous. Bryophyta is the formal taxonomic name for the phylum that consists solely of mosses. The term *bryophyte* is used informally to refer to *all* nonvascular plants—liverworts, hornworts, and mosses. As noted previously, it has not been established whether these three groups form a clade. Systematists also continue to debate the sequence in which the three phyla of bryophytes evolved.

Bryophytes acquired many unique adaptations after their evolutionary split from the ancestors they share with living vascular plants. Nevertheless, living bryophytes likely reflect some traits of the earliest plants. The oldest known fossils of plant fragments, for example, include tissues that look very much like the interior of liverworts. Researchers are eager to discover more parts of these ancient plants to see if this resemblance is reflected more broadly.

Bryophyte Gametophytes

Unlike vascular plants, in all three bryophyte phyla the gametophytes are larger and longer-living than sporophytes, as shown in the moss life cycle in **Figure 29.8**. Sporophytes are typically present only part of the time.

If bryophyte spores are dispersed to a favorable habitat, such as moist soil or tree bark, they may germinate and grow into gametophytes. Germinating moss spores, for example, characteristically produce a mass of green, branched, one-cell-thick filaments known as a **protonema** (plural, *protonemata;* from the Greek *proto,* first, and *nema,* threads). A protonema has a large surface area that enhances absorption of water and minerals. In favorable conditions, a protonema produces one or more "buds," each with an apical meristem that generates a gamete-producing structure known as a **gametophore** ("gamete bearer"). Together, a protonema and a gametophore make up the body of a moss gametophyte.

Bryophyte gametophytes generally form ground-hugging carpets and are at most only a few cells thick. Such thin body parts could not support a tall plant. A second constraint on the height of bryophytes is the absence, in most species, of vascular tissue, which is required for long-distance transport of water and nutrients. The thin structure of bryophyte organs makes it possible to distribute materials without specialized vascular tissue. Some mosses, however, including the genus *Polytrichum,* have conducting tissues in the center of their "stems," and a few of these mosses can grow as tall as 2 m as a result. Plant biologists are trying to determine whether these conducting tissues are homologous with vascular plant tissues or are the result of convergent evolution.

The gametophytes are anchored by delicate **rhizoids,** which are long, tubular single cells (in liverworts and hornworts) or filaments of cells (in mosses). Unlike roots, which are characteristic of vascular plants, rhizoids are not composed of tissues, lack specialized conducting cells, and do not play a primary role in water and mineral absorption.

Mature gametophytes produce gametes in gametangia covered by protective tissue. A gametophyte may have multiple gametangia. Eggs are produced singly in vase-shaped archegonia, whereas antheridia each produce many sperm. Some bryophyte gametophytes are bisexual, but in mosses the archegonia and antheridia are typically carried on separate female and male gametophytes. Flagellated sperm swim through a film of water toward eggs, entering the archegonia in response to chemical attractants. Eggs are not released but instead remain within the bases of archegonia. After fertilization, embryos are retained within the archegonia. Layers of placental transfer cells help transport nutrients to the embryos as they develop into sporophytes.

Bryophyte Sporophytes

Although bryophyte sporophytes are usually green and photosynthetic when young, they cannot live independently. They remain attached to their parental gametophytes, from which they absorb sugars, amino acids, minerals, and water.

Bryophytes have the smallest and simplest sporophytes of all extant plant groups, consistent with the hypothesis that larger and more complex sporophytes evolved only later in vascular plants. A typical sporophyte consists of a foot, a seta, and a sporangium. Embedded in the archegonium, the **foot** absorbs nutrients from the gametophyte. The **seta** (plural, *setae*), or stalk, conducts these materials to the sporangium, also called a **capsule,** which uses them to produce spores by meiosis. One capsule can generate up to 50 million spores.

▼ **Figure 29.8 The life cycle of a *Polytrichum* moss.**

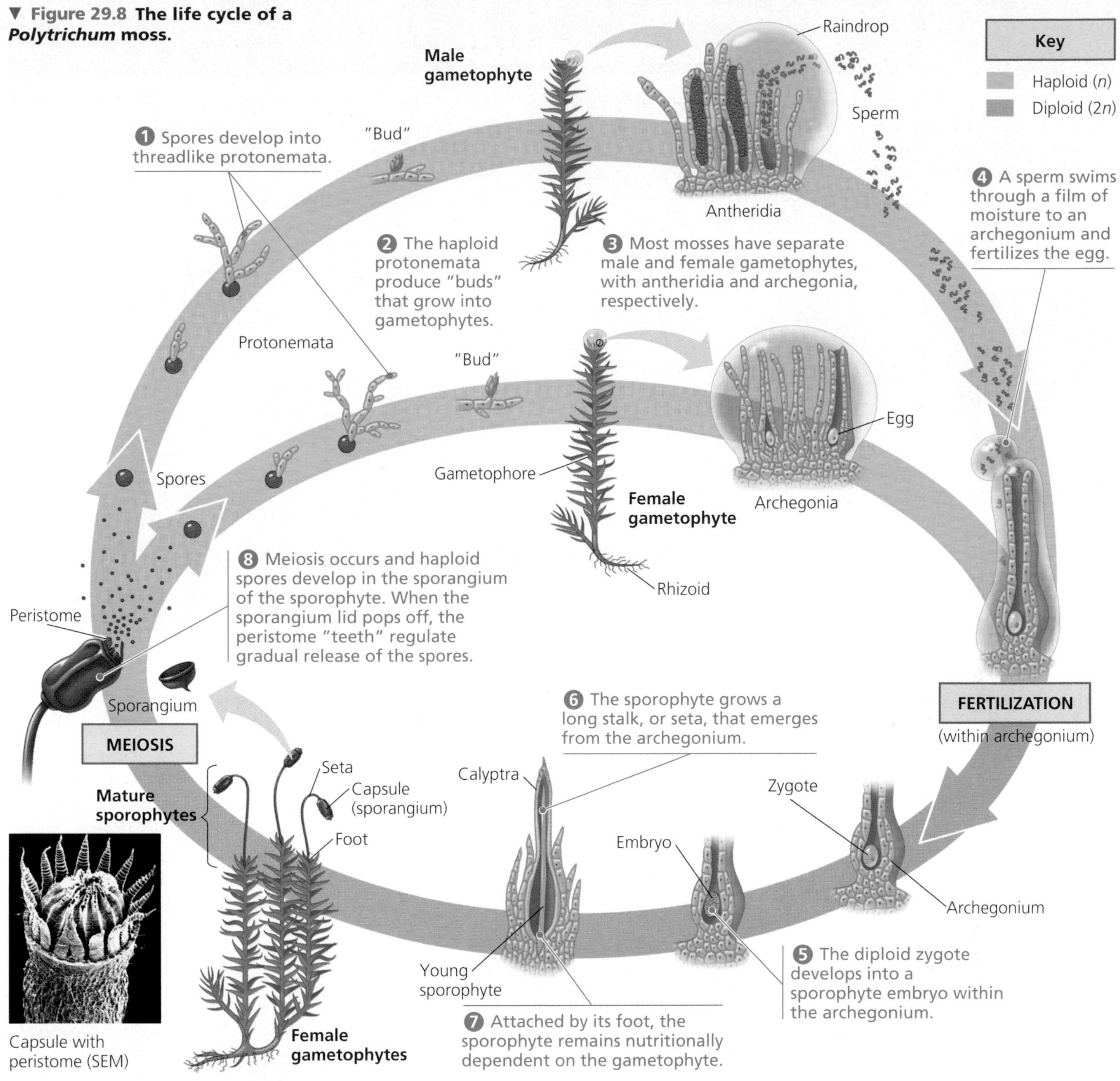

Key
Haploid (n)
Diploid (2n)
Raindrop
Male gametophyte
Sperm
Antheridia
"Bud"
1 Spores develop into threadlike protonemata.
2 The haploid protonemata produce "buds" that grow into gametophytes.
3 Most mosses have separate male and female gametophytes, with antheridia and archegonia, respectively.
4 A sperm swims through a film of moisture to an archegonium and fertilizes the egg.
Protonemata
"Bud"
Egg
Spores
Gametophore
Female gametophyte
Archegonia
Rhizoid
8 Meiosis occurs and haploid spores develop in the sporangium of the sporophyte. When the sporangium lid pops off, the peristome "teeth" regulate gradual release of the spores.
Peristome
Sporangium
MEIOSIS
FERTILIZATION
(within archegonium)
6 The sporophyte grows a long stalk, or seta, that emerges from the archegonium.
Seta
Capsule (sporangium)
Foot
Calyptra
Zygote
Mature sporophytes
Embryo
Archegonium
Young sporophyte
5 The diploid zygote develops into a sporophyte embryo within the archegonium.
7 Attached by its foot, the sporophyte remains nutritionally dependent on the gametophyte.
Capsule with peristome (SEM)
Female gametophytes

In most mosses, the seta becomes elongated, enhancing spore dispersal by elevating the capsule. An immature capsule has a protective cap of gametophyte tissue called the **calyptra**, which is shed when the capsule is mature. In most moss species, the upper part of the capsule features a ring of toothlike structures known as the **peristome** (see Figure 29.8). The peristome is specialized for gradual spore discharge, taking advantage of periodic wind gusts that can carry spores long distances.

Hornwort and moss sporophytes are larger and more complex than those of liverworts. Both hornwort and moss sporophytes also have specialized pores called **stomata** (singular, *stoma*), which are also found in all vascular plants. These pores support photosynthesis by allowing the exchange of CO_2 and O_2 between the outside air and the sporophyte interior (see Figure 10.3). Stomata are the main avenues by which water evaporates from the sporophyte. In hot, dry conditions, the stomata can close, minimizing water loss.

The fact that stomata are present in mosses and hornworts but absent in liverworts suggests three possible hypotheses for their evolution. If liverworts are the deepest-branching lineage

Figure 29.9

Exploring Bryophyte Diversity

LIVERWORTS (PHYLUM HEPATOPHYTA)

The common name and scientific name (from the Latin *hepaticus*, liver) refer to the liver-shaped gametophytes of *Marchantia*, a genus common in the Northern Hemisphere. In medieval times, their shape was thought to be a sign that the plants could help treat liver diseases. Some liverworts, such as *Marchantia*, are described as "thalloid" because of the flattened shape of their gametophytes. (Recall from Chapter 28 that the body of a multicellular alga is called a thallus.) *Marchantia* gametangia are elevated on gametophores that look like miniature trees (see also Figure 29.5). You would need a magnifying glass to see the sporophytes, which have a short seta (stalk) with a round sporangium. Some liverworts are called "leafy" because their gametophytes have stemlike structures with many leaflike appendages. Typically found in tropical and subtropical regions, leafy liverworts are much more common than thalloid species.

Gametophore of female gametophyte

Marchantia polymorpha, a "thalloid" liverwort

Foot

Seta

Sporangium

500 µm

***Marchantia* sporophyte (LM)**

***Plagiochila deltoidea,* a "leafy" liverwort**

HORNWORTS (PHYLUM ANTHOCEROPHYTA)

The common name and scientific name (from the Greek *keras*, horn) refer to the shape of the sporophyte, which also resembles a small grass blade. A typical sporophyte can grow up to about 5 cm high. A sporangium extends along its length and splits open to release mature spores, starting at the tip of the horn. Gametophytes, which are usually 1 to 2 cm in diameter, grow mostly horizontally and often have multiple sporophytes attached.

An *Anthoceros* hornwort species

Sporophyte

Gametophyte

MOSSES (PHYLUM BRYOPHYTA)

In contrast with liverworts and hornworts, moss gametophytes typically grow more vertically than horizontally. Heights range from less than a millimeter to more than 50 cm but are less than 15 cm in most species. The gametophytes are the structures that primarily comprise a carpet of moss. Their "leaves" are usually only one cell thick, but more complex "leaves" with ridges coated with cuticle can be found on the common hairy-cap moss and its close relatives. Moss sporophytes are typically elongated and visible to the naked eye, with sizes ranging up to about 20 cm. Though green and photosynthetic when young, they turn tan or brownish red when ready to release spores.

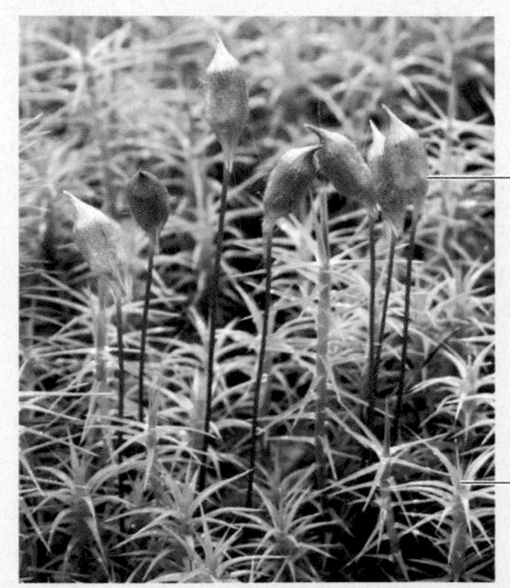

***Polytrichum commune,* hairy-cap moss**

of land plants, then stomata evolved once in the ancestor of hornworts, mosses, and vascular plants. If hornworts are the deepest-branching lineage, then stomata may have evolved once and then been lost in the liverwort lineage. Or perhaps hornworts acquired stomata independently of mosses and vascular plants. This question is important to understanding plant evolution because stomata play a crucial role in the success of vascular plants, as you will see in Chapter 36.

Figure 29.9, on the facing page, illustrates examples of gametophytes and sporophytes in the three bryophyte phyla.

Ecological and Economic Importance of Mosses

Wind dispersal of lightweight spores has distributed mosses around the world. These plants are particularly common and diverse in moist forests and wetlands, where they form habitats for tiny animals. Some moss species even inhabit such extreme environments as mountaintops, tundra, and deserts (see Chapter 50). Many mosses exist in very cold or dry habitats because they can survive the loss of most of their body water, then rehydrate when moisture is available. Few vascular plants can survive the same degree of desiccation. Moreover, phenolic compounds in moss cell walls absorb damaging levels of radiation present in deserts or at high altitudes and latitudes.

One wetland moss genus, *Sphagnum*, or "peat moss," is especially widespread, forming extensive deposits of partially decayed organic material known as **peat (Figure 29.10)**. Boggy regions dominated by this moss are called peat bogs. Because of the resistant phenolic compounds embedded in its cell walls, *Sphagnum* does not decay readily. In addition, it secretes compounds that may reduce bacterial activity. Low temperatures and nutrient levels in peat bogs also inhibit decay. As a result, peat bogs can preserve mummified corpses for thousands of years.

Worldwide, an estimated 400 billion tons of organic carbon are stored in peat. These carbon reservoirs help stabilize global atmospheric CO_2 concentrations (see Chapter 54).

Peat has long been a fuel source in Europe and Asia, and it is still harvested for fuel today, notably in Ireland and Canada. Because peat moss has large dead cells that can absorb some 20 times the moss's weight in water, it also serves as a soil conditioner and is used for packing plant roots during shipment. Current overharvesting of *Sphagnum* may reduce its beneficial ecological effects.

Concept Check 29.3

1. How do bryophytes differ from other plants?
2. Give three examples of how structure fits function in bryophytes.

For suggested answers, see Appendix A.

(a) Peat being harvested from a peat bog

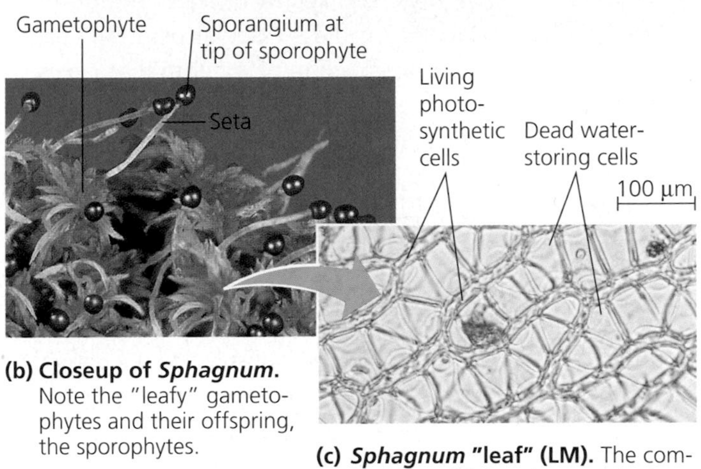

(b) Closeup of *Sphagnum*. Note the "leafy" gametophytes and their offspring, the sporophytes.

(c) *Sphagnum* "leaf" (LM). The combination of living photosynthetic cells and dead water-storing cells gives the moss its spongy quality.

(d) "Tolland Man," a bog mummy dating from 405–100 B.C. The acidic, oxygen-poor conditions produced by *Sphagnum* can preserve human or other animal bodies for thousands of years.

▲ **Figure 29.10 *Sphagnum*, or peat moss: a bryophyte with economic, ecological, and archaeological significance.**

Concept 29.4

Ferns and other seedless vascular plants formed the first forests

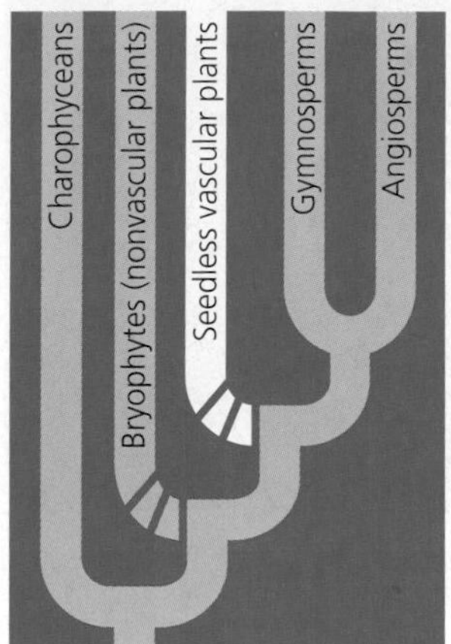

Whereas bryophytes or bryophyte-like plants were the prevalent vegetation during the first 100 million years of plant evolution, vascular plants dominate most landscapes today. Living seedless vascular plants provide insights into plant evolution during the Carboniferous period, when vascular plants began to diversify but most groups of seed plants had not yet evolved. The sperm of ferns and all other seedless vascular plants are flagellated and must swim through a film of water to reach eggs, as in bryophytes. Due to these swimming sperm, as well as their fragile gametophytes, living seedless vascular plants are most common in damp environments. Thus it is likely that before the emergence of seed plants, most plant life on Earth was limited to relatively damp habitats.

Origins and Traits of Vascular Plants

Fossils of the forerunners of today's vascular plants date back about 420 million years. Unlike bryophytes, these species had branched sporophytes that were not dependent on gametophytes for growth **(Figure 29.11)**. Although these plants grew no taller than about 50 cm, their branching made possible more complex bodies with multiple sporangia. This evolutionary development facilitated greater production of spores and increased survival despite herbivory, for even if some sporangia were eaten, others might survive.

The ancestors of vascular plants already displayed some derived traits of living vascular plants, but they lacked other key adaptations that evolved later. This section describes the main traits that characterize living vascular plants: life cycles with dominant sporophytes, transport in vascular tissues called xylem and phloem, and the presence of roots and leaves, including spore-bearing leaves called sporophylls.

Life Cycles with Dominant Sporophytes

Fossils suggest that the ancestors of vascular plants had life cycles characterized by gametophytes and sporophytes that were about equal in size. Among extant vascular plants, however, the sporophyte (diploid) generation is the larger and more complex plant in the alternation of generations. In ferns, for example, the familiar leafy plants are the sporophytes. You would have to get down on your hands and knees and search the ground carefully to find fern gametophytes, which are tiny structures that grow on or just below the soil surface. Until you have a chance to do that, you can study the sporophyte-dominant life cycle of seedless vascular plants in **Figure 29.12**, which uses a fern as an example. Then, for review, compare this life cycle with Figure 29.8, which represents a typical gametophyte-dominated life cycle of mosses and other nonvascular plants. In Chapter 30, you will see that gametophytes became even more reduced during the evolution of seed plants.

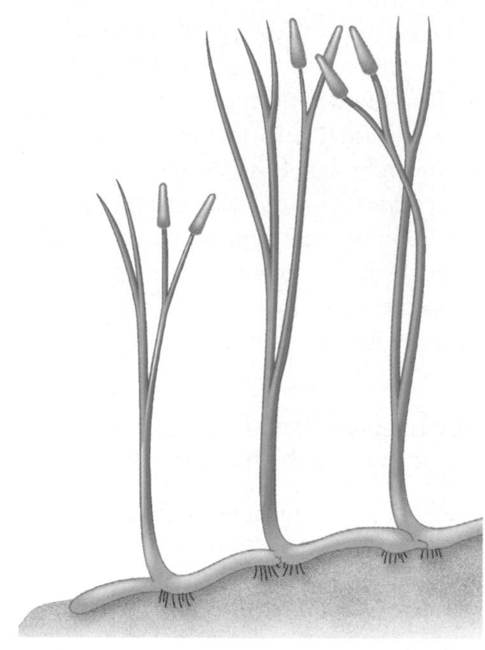

▶ **Figure 29.11** ***Aglaophyton major,* an ancient relative of modern vascular plants.** This reconstruction from fossils dating to about 420 million years ago exhibits dichotomous (Y-shaped) branching and terminal sporangia. These traits characterize living vascular plants but are lacking in bryophytes (nonvascular plants).

Transport in Xylem and Phloem

Vascular plants have two types of vascular tissue: xylem and phloem. **Xylem** conducts most of the water and minerals. The xylem of all vascular plants includes **tracheids,** tube-shaped cells that carry water and minerals up from roots (see Figure 35.9). Because nonvascular plants lack tracheids, vascular plants are sometimes referred to as tracheophytes. Tracheids are actually dead cells—only their walls remain and provide a system of microscopic water pipes. The water-conducting cells in vascular plants are *lignified;* that is, their cell walls are strengthened by the phenolic polymer **lignin.** The tissue called **phloem** includes living sugar-conducting cells arranged into tubes that distribute sugars, amino acids, and other organic products (see Figure 35.9).

Lignified vascular tissue permitted vascular plants to grow to greater heights than bryophytes. Their stems became strong enough to withstand drooping, and they could transport water and mineral nutrients high above the ground.

Evolution of Roots

Lignified vascular tissue also provides benefits below-ground. Instead of the rhizoids seen in bryophytes, roots evolved in almost all vascular plants. **Roots** are organs that anchor vascular plants and enable them to absorb water and nutrients from the soil. Roots also allow the shoot system to grow taller.

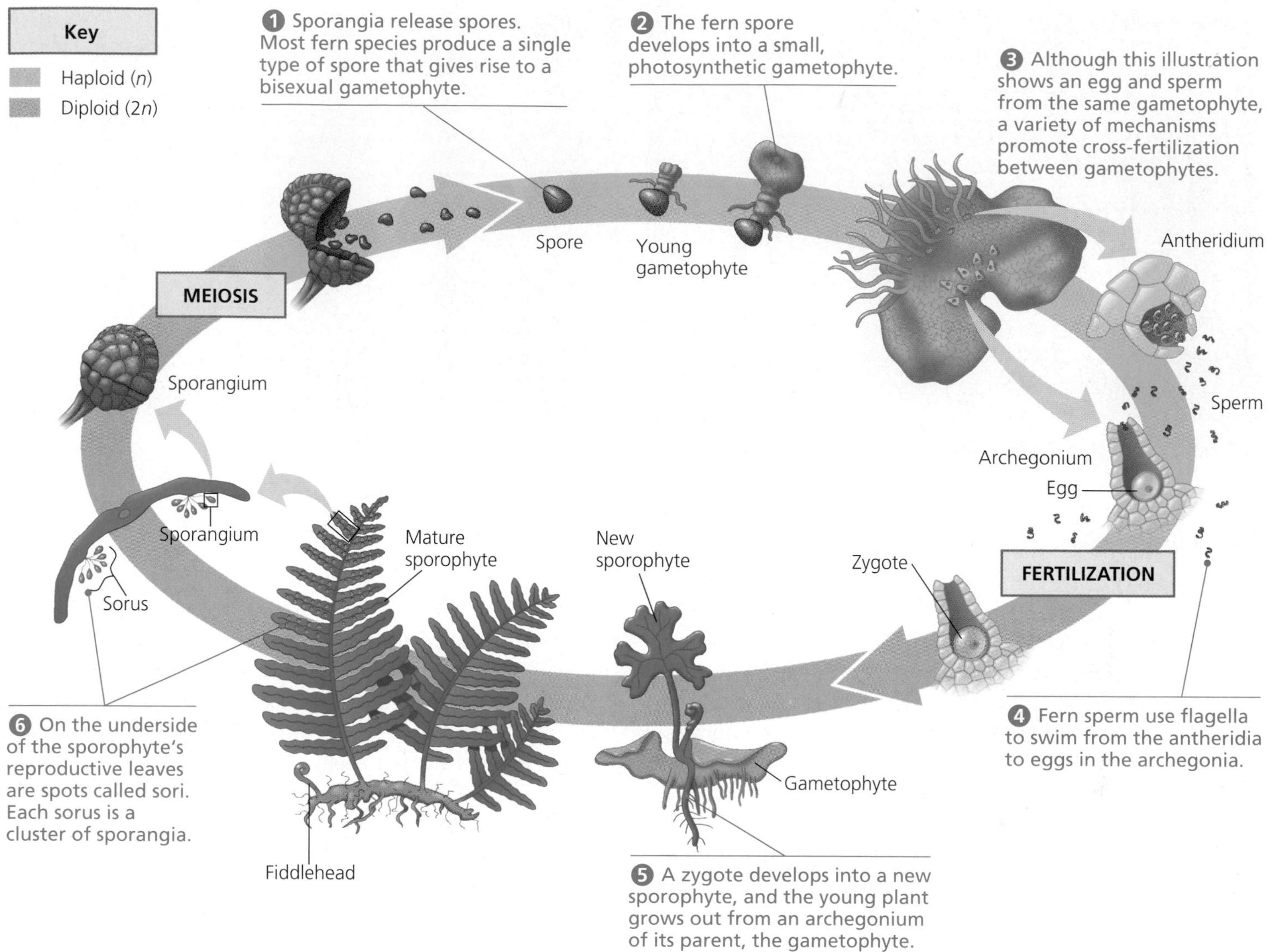
Key
Haploid (n)
Diploid (2n)
1 Sporangia release spores. Most fern species produce a single type of spore that gives rise to a bisexual gametophyte.
2 The fern spore develops into a small, photosynthetic gametophyte.
3 Although this illustration shows an egg and sperm from the same gametophyte, a variety of mechanisms promote cross-fertilization between gametophytes.
Spore
Young gametophyte
Antheridium
MEIOSIS
Sporangium
Sperm
Archegonium
Egg
Sporangium
Mature sporophyte
New sporophyte
Zygote
FERTILIZATION
Sorus
6 On the underside of the sporophyte's reproductive leaves are spots called sori. Each sorus is a cluster of sporangia.
Gametophyte
4 Fern sperm use flagella to swim from the antheridia to eggs in the archegonia.
Fiddlehead
5 A zygote develops into a new sporophyte, and the young plant grows out from an archegonium of its parent, the gametophyte.

▲ **Figure 29.12 The life cycle of a fern.**

Root tissues of living plants closely resemble stem tissues of early vascular plants preserved in fossils. This suggests that roots may have evolved from the lowermost, subterranean portions of stems in ancient vascular plants. It is not clear whether roots evolved only once in the common ancestor of all vascular plants or independently in different lineages. Although the roots of living members of these lineages of vascular plants share many similarities, the fossil evidence hints at convergent evolution. The oldest lycophyte fossils, for example, already displayed simple roots 400 million years ago, when the ancestors of ferns and seed plants still had none. Studying genes that control root development in different vascular plant species may help resolve this controversy.

Evolution of Leaves

Leaves are organs that increase the surface area of vascular plants, thereby capturing more solar energy for photosynthesis. In terms of size and complexity, leaves can be classified as either microphylls or megaphylls. All lycophytes (the oldest lineage of modern vascular plants) have **microphylls,** small, usually spine-shaped leaves with a single vein. Almost all other vascular plants have **megaphylls,** leaves with a highly branched vascular system. So named because they are typically larger than microphylls, megaphylls support greater photosynthetic productivity than microphylls as a result of their greater surface area served by a network of veins. Microphylls first appear in the fossil record 410 million years ago, but megaphylls do not emerge until about 370 million years ago, near the end of the Devonian period.

According to one model of leaf evolution, microphylls originated as small outgrowths of stems. These outgrowths were supported by single strands of vascular tissue **(Figure 29.13a)**. Megaphylls, by contrast, may have evolved from a series of branches lying close together on a stem, which became flattened and developed webbing that joined branches together **(Figure 29.13b)**. To better understand the origin of leaves, scientists are exploring genetic control of leaf development.

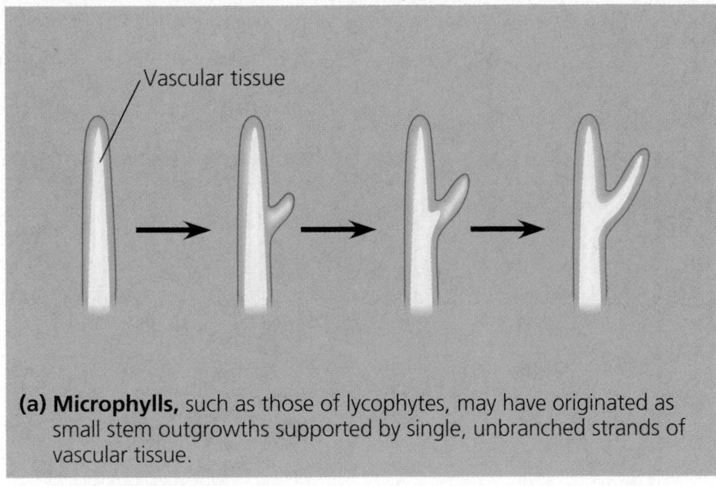

(a) Microphylls, such as those of lycophytes, may have originated as small stem outgrowths supported by single, unbranched strands of vascular tissue.

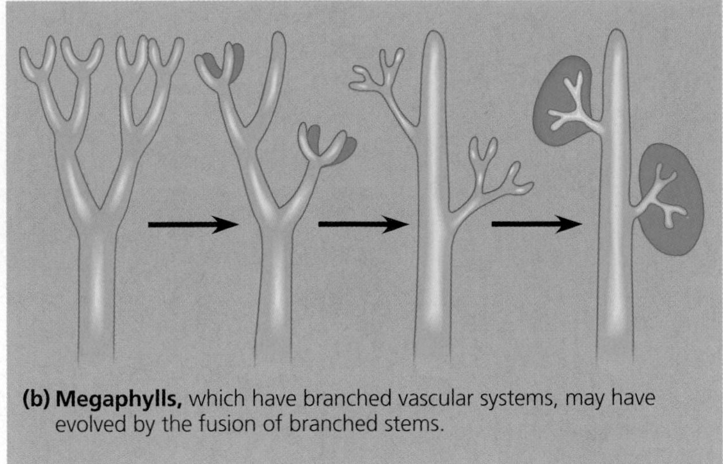

(b) Megaphylls, which have branched vascular systems, may have evolved by the fusion of branched stems.

▲ **Figure 29.13 Hypotheses for the evolution of leaves.**

Sporophylls and Spore Variations

One evolutionary milestone was the emergence of **sporophylls,** modified leaves that bear sporangia. Sporophylls vary greatly in structure. For example, fern sporophylls produce clusters of sporangia known as **sori** (singular, *sorus*), usually on the undersides of the sporophylls (see Figure 29.12). In many lycophytes and in most gymnosperms, groups of sporophylls form cones, technically called **strobili** (from the Greek *strobilos,* cone). In Chapter 30, you will see how sporophylls form gymnosperm strobili and parts of flowers.

Most seedless vascular plant species are **homosporous,** having one type of sporophyll producing one type of spore that typically develops into a bisexual gametophyte, as in most ferns. In contrast, a **heterosporous** species has two types of sporophylls and produces two kinds of spores. Megasporangia in megasporophylls produce **megaspores,** which develop into female gametophytes. Microsporangia in microsporophylls produce **microspores,** which develop into male gametophytes. All seed plants and a few seedless vascular plants are heterosporous. The following diagram compares the two conditions:

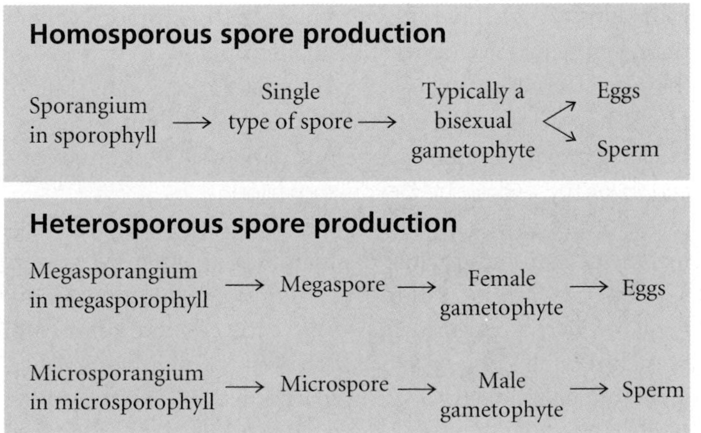

Classification of Seedless Vascular Plants

As noted earlier, living seedless vascular plants form two clades: lycophytes and pterophytes. Lycophytes (phylum Lycophyta) include club mosses, spike mosses, and quillworts. Pterophytes (phylum Pterophyta) include ferns, horsetails, and whisk ferns and their relatives. Because they differ greatly in appearance, ferns, horsetails, and whisk ferns have long been considered separate phyla: phylum Pterophyta (ferns), phylum Sphenophyta (horsetails, also called sphenophytes), and phylum Psilophyta (whisk ferns and a related genus, together known as psilophytes). However, recent molecular comparisons provide convincing evidence that all three groups form a clade. Accordingly, many systematists now classify them together as the phylum Pterophyta, as we do in this chapter. Others refer to them as three separate phyla within a clade.

Figure 29.14, on the facing page, describes the general groups of seedless vascular plants.

Phylum Lycophyta: Club Mosses, Spike Mosses, and Quillworts

Modern species of lycophytes—the most primitive group of vascular plants—are relicts of a far more eminent past. By the Carboniferous period, there were two evolutionary lines—one composed of small herbaceous plants and another line of giant woody trees with diameters of more than 2 m and heights of more than 40 m. The giant lycophytes thrived for millions of years in moist swamps, but they became extinct when the climate became drier at the end of the Carboniferous period. The small lycophytes survived, represented today by about 1,200 species. Though some are commonly called club mosses and spike mosses, they are not true mosses, which are bryophytes.

Exploring Seedless Vascular Plant Diversity

LYCOPHYTES (PHYLUM LYCOPHYTA)

Many lycophytes grow on tropical trees as epiphytes, plants that use other plants as a substrate but are not parasites. Other species grow on temperate forest floors. In some species, the tiny gametophytes live aboveground and are photosynthetic. Others live underground, nurtured by symbiotic fungi.

Sporophytes have upright stems with many small leaves, as well as ground-hugging stems that produce dichotomously branching roots. In club mosses, sporophylls are clustered into club-shaped cones (strobili). Spike mosses are usually smaller and often grow horizontally. Quillworts, named for their leaf shape, are a single genus that lives in marshy areas. Club mosses are all homosporous, whereas spike mosses and quillworts are all heterosporous. Spores are released in clouds and are so rich in oil that magicians and photographers once ignited them to create smoke or flashes of light.

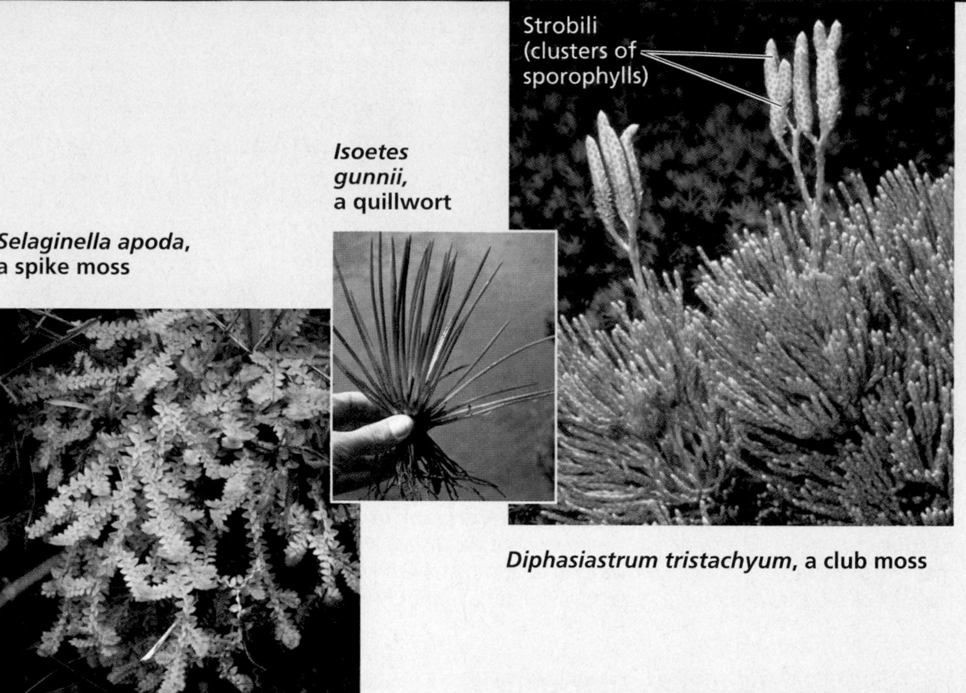

Selaginella apoda, a spike moss

Isoetes gunnii, a quillwort

Diphasiastrum tristachyum, a club moss

PTEROPHYTES (PHYLUM PTEROPHYTA)

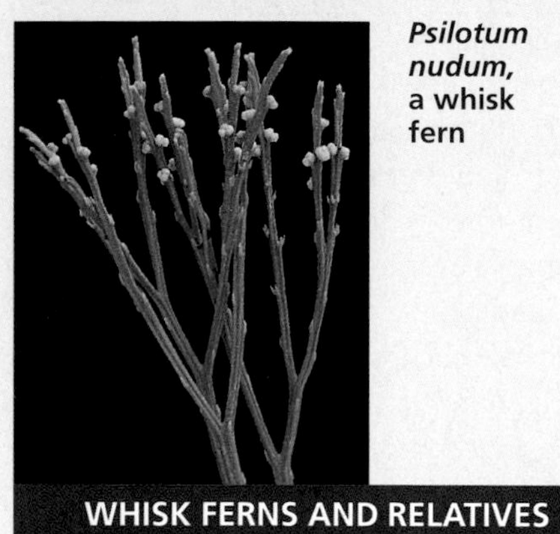

Psilotum nudum, a whisk fern

WHISK FERNS AND RELATIVES

Like primitive vascular plant fossils, the sporophytes of whisk ferns (*Psilotum*) have dichotomously branching stems but no roots. Stems have scale-like outgrowths that lack vascular tissue and may have evolved as very reduced leaves. Each yellow knob on a stem consists of three fused sporangia. Species of the genus *Tmesipteris*, closely related to whisk ferns and found only in the South Pacific, also lack roots but have small, leaflike outgrowths in their stems, giving them a vinelike appearance. Both genera are homosporous, with spores giving rise to bisexual gametophytes that grow underground and are only about a centimeter long.

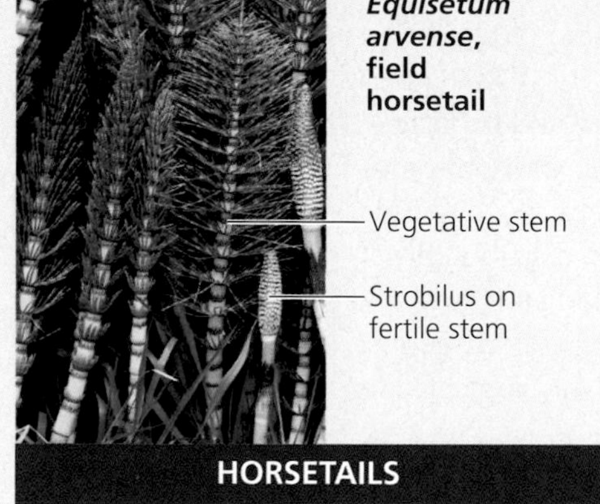

Equisetum arvense, field horsetail

HORSETAILS

The name refers to the brushy appearance of the stems, which have a gritty texture that made them historically useful as "scouring rushes" to scrub pots and pans. Some species have separate fertile (cone-bearing) and vegetative stems. Horsetails are homosporous, with cones releasing spores that give rise to tiny male or bisexual gametophytes.

Horsetails are also called arthrophytes ("jointed plants") because their stems have joints. Rings of small leaves or branches emerge from each joint, but the stem is the main photosynthetic organ. Large air canals carry oxygen to the roots, which often grow in waterlogged soil.

Athyrium filix-femina, lady fern

FERNS

Unlike most other seedless vascular plants, ferns have megaphylls (see Figure 29.13b). Sporophytes typically have horizontal stems that give rise to large leaves called fronds, often divided into leaflets. A frond grows as its coiled tip, the fiddlehead, unfurls (see the cover of this textbook).

Almost all species are homosporous. The gametophyte shrivels and dies after the young sporophyte detaches itself. In most species, sporophytes have stalked sporangia with spring-like devices that catapult spores several meters. Airborne spores can be carried far from their origin. Some species produce more than a trillion spores in a plant's lifetime.

▲ **Figure 29.15 Artist's conception of a Carboniferous forest based on fossil evidence.** Most of the large trees with straight trunks are lycophytes. The tree on the left with feathery branches is a horsetail. Tree ferns, not shown here, were also abundant in the "coal forests" of the Carboniferous. Animals, including giant dragonflies, also thrived.

Phylum Pterophyta: Ferns, Horsetails, and Whisk Ferns and Relatives

Ferns radiated extensively from their Devonian origins and grew alongside tree lycophytes and horsetails in the great Carboniferous swamp forests. Today ferns are by far the most widespread seedless vascular plants, numbering more than 12,000 species. Though most diverse in the tropics, many ferns thrive in temperate forests, and some species are even adapted to arid habitats.

Horsetails were very diverse during the Carboniferous period, some growing as tall as 15 m. However, today only about 15 species survive as a single, widely distributed genus, *Equisetum,* found in marshy places or along streams.

Psilotum (whisk ferns) and a closely related genus, *Tmesipteris,* form a clade consisting mainly of tropical epiphytes. Whisk ferns, the only vascular plants lacking true roots and leaves, are called "living fossils" because of their striking resemblance to fossils of ancient relatives of living vascular plants (see Figures 29.11 and 29.14). However, much evidence, including analyses of DNA sequences and sperm structure, indicates that *Psilotum* and *Tmesipteris* are closely related to ferns. This hypothesis suggests that their ancestor's true roots and leaves were lost during evolution.

The Significance of Seedless Vascular Plants

The ancestors of living lycophytes, horsetails, and ferns, along with their seedless vascular relatives, grew to great heights during the Carboniferous, forming the first forests **(Figure 29.15)**. With the evolution of vascular tissue, roots, and leaves, these plants accelerated their rate of photosynthesis, dramatically increasing the removal of CO_2 from the atmosphere. Scientists estimate that CO_2 levels dropped by as much as a factor of five during the Carboniferous, causing global cooling that resulted in widespread glacier formation.

The seedless vascular plants that formed the first forests eventually became coal. In the stagnant waters of Carboniferous swamps, dead plants did not completely decay. This organic material turned to thick layers of peat, later covered by the sea. Marine sediments piled on top, and over millions of years, heat and pressure converted the peat to coal. In fact, Carboniferous coal deposits are the most extensive ever formed. Coal was crucial to the Industrial Revolution, and people worldwide still burn 6 billion tons a year. It is ironic that coal, formed from plants that contributed to a global cooling, now contributes to global warming by returning carbon to the atmosphere (see Chapter 54).

Growing along with the seedless plants in Carboniferous swamps were primitive seed plants. Though seed plants were not dominant at that time, they rose to prominence after the swamps began to dry up at the end of the Carboniferous period. The next chapter traces the origin and diversification of seed plants, continuing our theme of terrestrial adaptation.

Concept Check 29.4

1. What are a few key differences between seedless vascular plants and bryophytes?
2. What is the major difference between most lycophytes and most ferns and their relatives?

For suggested answers, see Appendix A.

Chapter 29 Review

Go to www.campbellbiology.com or the student CD-ROM to explore Activities, Investigations, and other interactive study aids.

SUMMARY OF KEY CONCEPTS

Concept 29.1

Land plants evolved from green algae

- **Morphological and Biochemical Evidence (pp. 573–574)** The traits shared by plants and charophyceans include rosette cellulose-synthesizing complexes, peroxisome enzymes, similarities in flagellated sperm structure, and formation of phragmoplasts during cell division.
- **Genetic Evidence (p. 574)** Similarities in nuclear and chloroplast genes suggest that charophyceans are the closest living relatives of land plants.
- **Adaptations Enabling the Move to Land (p. 574)** Traits such as sporopollenin allow charophyceans to withstand occasional drying along the edges of ponds and lakes. These traits may have enabled the algal ancestors of plants to survive in terrestrial conditions, opening the way to the colonization of dry land.

Land plants possess a set of derived terrestrial adaptations

- **Defining the Plant Kingdom (p. 575)** Some biologists think that the plant kingdom should be expanded to include some or all green algae. Until this phylogenetic debate is resolved, this textbook retains the embryophyte definition of kingdom Plantae.
- **Derived Traits of Plants (pp. 575–577)** Derived traits that distinguish the clade of land plants from charophyceans, their closest algal relatives, include apical meristems, alternation of generations, walled spores in sporangia, multicellular gametangia, and multicellular, dependent embryos. Additional derived traits, such as the cuticle and secondary compounds, evolved in many plant species.
  Activity *Terrestrial Adaptations of Plants*
- **The Origin and Diversification of Plants (pp. 575, 578–579)** Fossil evidence indicates that plants were on land at least 475 million years ago. Subsequently, plants diverged into several major groups, including bryophytes (nonvascular plants); seedless vascular plants, such as lycophytes and ferns; and the two groups of seed plants: gymnosperms and angiosperms (flowering plants). Most systematists divide plants into ten phyla.
  Activity *Highlights of Plant Phylogeny*

Concept 29.3

The life cycles of mosses and other bryophytes are dominated by the gametophyte stage

- The three phyla of bryophytes—liverworts, hornworts, and mosses—may not form a clade. Debate continues over the sequence of their evolution (p. 580).
- **Bryophyte Gametophytes (p. 580)** Liverwort and hornwort gametophytes grow more horizontally, whereas those of mosses are more vertical. Gametophytes are the dominant generation and are typically most visible, such as a mat of moss. Rhizoids anchor gametophytes to the surface. The flagellated sperm produced by antheridia require a film of water to travel to the eggs in archegonia.
- **Bryophyte Sporophytes (pp. 580–583)** Sporophytes grow out of archegonia and are dependent on the haploid gametophytes for nourishment. Smaller and simpler than vascular plant sporophytes, they typically consist of a foot, seta (stalk), and sporangium. Hornwort and moss sporophytes have stomata.
  Activity *Moss Life Cycle*
- **Ecological and Economic Importance of Mosses (p. 583)** *Sphagnum* covers great expanses of land as peat bogs, playing an important role in the carbon cycle.

Concept 29.4

Ferns and other seedless vascular plants formed the first forests

- **Origins and Traits of Vascular Plants (pp. 584–586)** Fossils of the forerunners of today's vascular plants date back about 420 million years ago and show that these tiny plants had independent, branching sporophytes but lacked other derived traits of vascular plants, such as xylem and phloem, roots, and leaves.

  Life Cycles with Dominant Sporophytes. In contrast with bryophytes, sporophytes of seedless vascular plants are the larger generation, as in the example of the familiar leafy fern plant. The gametophytes are tiny plants that grow on or below the surface.
  Activity *Fern Life Cycle*
  Investigation *What Are the Different Stages of a Fern Life Cycle?*

  Transport in Xylem and Phloem. Vascular plants have two vascular tissues: xylem and phloem. Xylem conducts most of the water and minerals. Xylem of all vascular plants includes dead cells called tracheids. The lignin in xylem enables most vascular plants to grow taller than bryophytes. Phloem, a living tissue, conducts sugars and other organic nutrients.

  Evolution of Roots. Unlike the rhizoids of bryophytes, roots play an important role in absorbing water and nutrients. Roots may have evolved from subterranean stems. It is unclear whether roots evolved independently in different lineages.

  Evolution of Leaves. In terms of evolution, leaves are categorized into two types: microphylls and megaphylls. Microphylls, leaves with a single vein, evolved first and are typical of lycophytes. Almost all other vascular plants have megaphylls, leaves with a highly branched vascular system. Megaphylls are usually larger, with more photosynthetic productivity.

  Sporophylls and Spore Variations. Sporophylls are modified leaves with sporangia. Most seedless vascular plant species are homosporous, producing one type of spore, which usually develops into a bisexual gametophyte. All seed plants and some seedless vascular plant species are heterosporous, having two types of spores that give rise to male and female gametophytes.
- **Classification of Seedless Vascular Plants (pp. 586–588)** Seedless vascular plants include the phylum Lycophyta (club mosses, spike mosses, and quillworts) and the phylum Pterophyta (ferns, horsetails, and whisk ferns and relatives). Ancient lycophytes included woody and herbaceous plants that dominated the first forests. Modern lycophytes are small herbaceous plants. Lycophyte sporophytes have upright stems bearing many microphylls and horizontal stems that grow along the surface. Ferns are the most diverse seedless vascular plants. Most fern species are homosporous and produce clusters of sporangia known as sori. Horsetails and whisk ferns are close relatives of ferns.

▶ **The Significance of Seedless Vascular Plants** (p. 588)
Seedless vascular plants dominated the earliest forests. Their growth may have helped produce the major global cooling that characterized the end of the Carboniferous period. The decaying remnants of the first forests eventually became coal.

TESTING YOUR KNOWLEDGE

Self-Quiz

1. Which of the following is *not* evidence that charophyceans are the closest algal relatives of plants?
   a. similar sperm structure
   b. similar cell wall structure
   c. similarities in cell wall formation during cell division
   d. genetic similarities in chloroplasts
   e. similarities in proteins that synthesize cellulose

2. Which of the following characteristics of plants is absent in their closest relatives, the charophycean algae?
   a. chlorophyll *b*
   b. cellulose in cell walls
   c. alternation of multicellular generations
   d. sexual reproduction
   e. formation of a cell plate during cytokinesis

3. Which of the following is a clade (monophyletic group)?
   a. bryophytes and seedless vascular plants
   b. lycophytes and pterophytes
   c. liverworts, hornworts, and mosses
   d. seedless vascular plants and seed plants
   e. charophyceans and bryophytes

4. Which of the following characteristics do mosses, liverworts, and hornworts share?
   a. reproductive cells in gametangia; embryos
   b. branched sporophytes
   c. vascular tissues, true leaves, and a waxy cuticle
   d. seeds
   e. lignified walls

5. Which of the following is *not* common to all phyla of vascular plants?
   a. the development of seeds
   b. alternation of generations
   c. dominance of the diploid generation
   d. xylem and phloem
   e. the addition of lignin to cell walls

6. A heterosporous plant species is one that
   a. produces a gametophyte that bears both antheridia and archegonia.
   b. produces microspores and megaspores, which give rise to male and female gametophytes.
   c. produces spores all year long instead of during just one season.
   d. produces two kinds of spores, one asexually by mitosis and the other sexually by meiosis.
   e. reproduces only sexually.

7. Which of the following is diploid?
   a. the archegonia of a liverwort
   b. a nonreproductive cell in the gametangia of a moss
   c. a cell that is part of the stalk (seta) of a moss sporophyte
   d. a spore produced by a fern sporophyte
   e. a subterranean gametophyte of a lycophyte

8. Microphylls are characteristic of which types of plants?
   a. mosses
   b. liverworts
   c. lycophytes
   d. ferns
   e. hornworts

9. During the Carboniferous period, the dominant plants were
   a. giant lycophytes, horsetails, and ferns.
   b. conifers.
   c. angiosperms.
   d. charophyceans.
   e. early seed plants.

10. Which of the following is a land plant that produces flagellated sperm and has a sporophyte-dominant life cycle?
    a. fern
    b. moss
    c. liverwort
    d. charophycean
    e. hornwort

For Self-Quiz answers, see Appendix A.

Go to the website or CD-ROM for more quiz questions.

Evolution Connection

Draw a cladogram that includes a moss, a fern, and a gymnosperm. Use a charophycean alga as the outgroup. (See Chapter 25 to review cladistics.) Label each branch of the cladogram with at least one derived characteristic unique to that clade.

Scientific Inquiry

In April 1986, an accident at a nuclear power plant in Chernobyl, Ukraine, scattered radioactive fallout for hundreds of miles. In assessing the biological effects of the radiation, researchers found mosses to be especially valuable as organisms for monitoring the damage. Radiation damages organisms by causing mutations. Explain why the genetic effects of radiation can be observed sooner in bryophytes than in plants from other groups. Imagine that you are conducting tests shortly after a nuclear accident. Using potted moss plants as your experimental organisms, design an experiment to test the hypothesis that the frequency of mutations decreases with the organism's distance from the source of radiation.

Investigation *What Are the Different Stages of a Fern Life Cycle?*

Science, Technology, and Society

Bryophytes (nonvascular plants) and seedless vascular plants are common, and several of them have important economic and ecological uses. Nevertheless, very few are important agriculturally. Why? What attributes do they lack that would make a plant useful in agriculture? What attributes limit their agricultural utility?

30 Plant Diversity II
The Evolution of Seed Plants

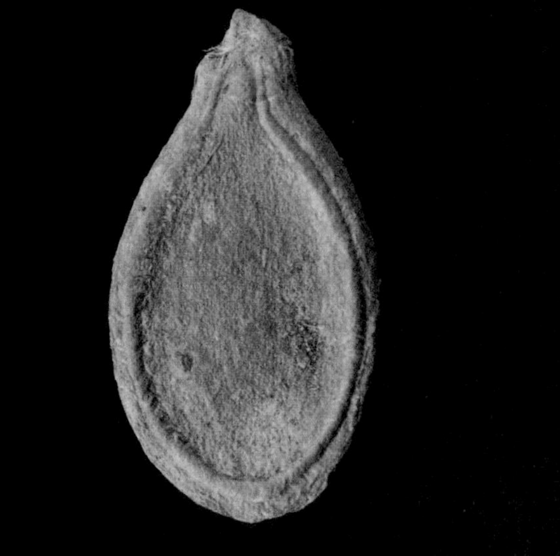

▲ **Figure 30.1 An ancient squash seed.**

Key Concepts

30.1 The reduced gametophytes of seed plants are protected in ovules and pollen grains
30.2 Gymnosperms bear "naked" seeds, typically on cones
30.3 The reproductive adaptations of angiosperms include flowers and fruits
30.4 Human welfare depends greatly on seed plants

Overview

Feeding the World

Continuing the saga of how plants transformed Earth, this chapter follows the emergence and diversification of seed plants. Fossils and comparative studies of living plants offer clues about the origin of seed plants some 360 million years ago. Seeds changed the course of plant evolution, enabling their bearers to become the dominant producers in most of the terrestrial ecosystems and to make up the vast majority of plant biodiversity.

Seed plants have also had an enormous impact on human society. Starting about 13,000 years ago, humans began to domesticate wheat, maize (commonly called corn in the United States), bananas, and other wild seed plants. This practice emerged separately in various regions of the world, including the Near East, East Asia, Africa, and the Americas. One piece of evidence, the well-preserved squash seed in **Figure 30.1**, was found in a cave in Mexico and dates from between 8,000 and 10,000 years ago. This seed differs from wild squash seeds, suggesting that squash was being cultivated by that time. The domestication of seed plants, particularly angiosperms, produced the most important cultural change in human history, transforming most human societies from roving bands of hunter-gatherers to permanent settlements anchored by agriculture.

In this chapter, we will first examine the general characteristics of seed plants. Then we will look at the distinguishing features and evolution of gymnosperms and angiosperms.

Concept 30.1

The reduced gametophytes of seed plants are protected in ovules and pollen grains

We begin with an overview of key terrestrial adaptations that seed plants added to those already present in bryophytes and seedless vascular plants (see Chapter 29). In addition to seeds, the following are common to all seed plants: reduced gametophytes, heterospory, ovules, and pollen. Seedless plants lack these adaptations, except for a few species that exhibit heterospory.

Advantages of Reduced Gametophytes

Mosses and other bryophytes have life cycles dominated by gametophytes, whereas ferns and other seedless vascular plants have sporophyte-dominated life cycles. The evolutionary trend of gametophyte reduction continued further in the vascular plant lineage that led to seed plants. While the gametophytes of seedless vascular plants are visible to the naked eye, the gametophytes of seed plants are mostly microscopic.

This miniaturization allowed for an important evolutionary innovation in seed plants: Their tiny gametophytes can develop from spores retained within the sporangia of the parental sporophyte. This arrangement protects the delicate female (egg-containing) gametophytes from environmental stresses. The moist reproductive tissues of the sporophyte shield the gametophytes from drought conditions and from UV radiation. This relationship also enables the dependent gametophytes to obtain nutrients from the sporophyte. In contrast, the free-living gametophytes of seedless plants must fend for themselves. **Figure 30.2** contrasts the sporophyte-gametophyte relationships in bryophytes, seedless vascular plants, and seed plants.

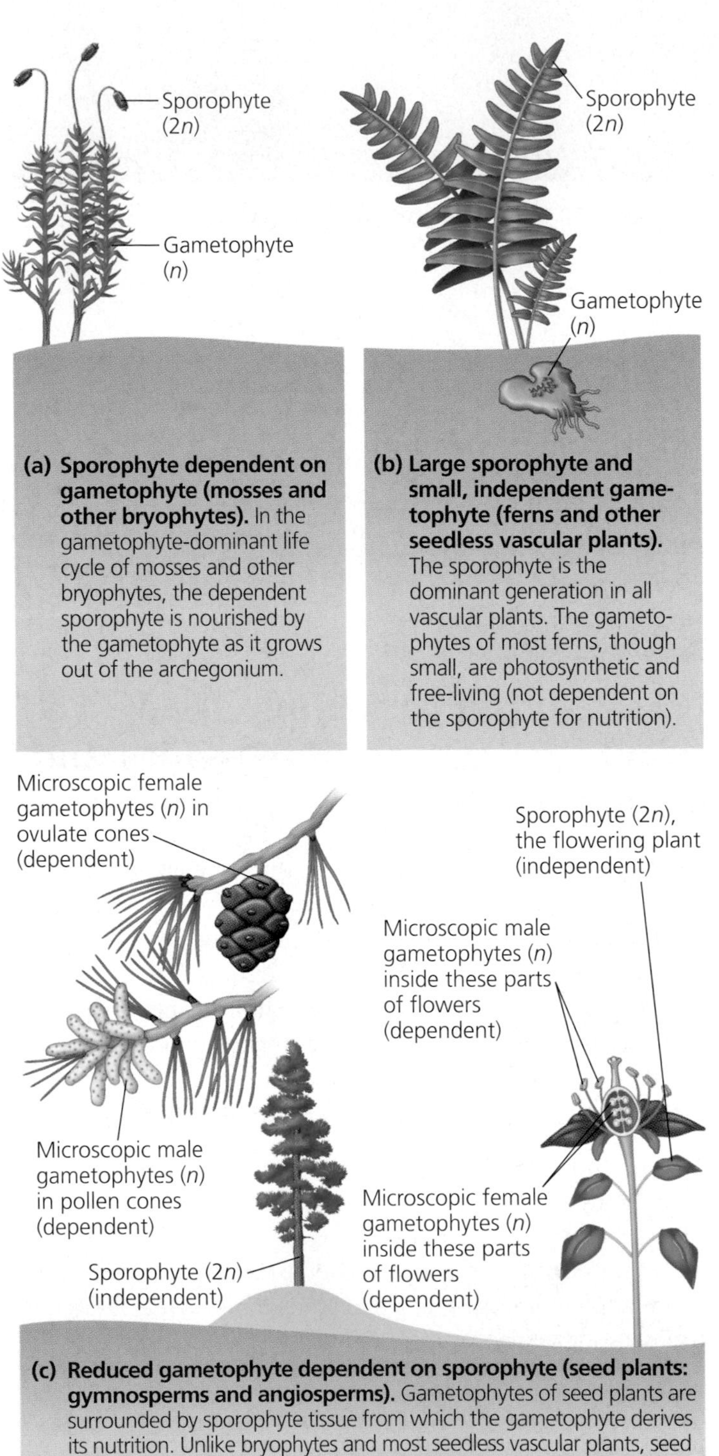

Sporophyte (2*n*)
Gametophyte (*n*)
Sporophyte (2*n*)
Gametophyte (*n*)
(a) Sporophyte dependent on gametophyte (mosses and other bryophytes). In the gametophyte-dominant life cycle of mosses and other bryophytes, the dependent sporophyte is nourished by the gametophyte as it grows out of the archegonium.
(b) Large sporophyte and small, independent gametophyte (ferns and other seedless vascular plants). The sporophyte is the dominant generation in all vascular plants. The gametophytes of most ferns, though small, are photosynthetic and free-living (not dependent on the sporophyte for nutrition).
Microscopic female gametophytes (*n*) in ovulate cones (dependent)
Sporophyte (2*n*), the flowering plant (independent)
Microscopic male gametophytes (*n*) inside these parts of flowers (dependent)
Microscopic male gametophytes (*n*) in pollen cones (dependent)
Microscopic female gametophytes (*n*) inside these parts of flowers (dependent)
Sporophyte (2*n*) (independent)
(c) Reduced gametophyte dependent on sporophyte (seed plants: gymnosperms and angiosperms). Gametophytes of seed plants are surrounded by sporophyte tissue from which the gametophyte derives its nutrition. Unlike bryophytes and most seedless vascular plants, seed plants typically have microscopic gametophytes.

▲ **Figure 30.2 Gametophyte/sporophyte relationships.**

Heterospory: The Rule Among Seed Plants

You read in Chapter 29 that nearly all seedless plants are homosporous—they produce one kind of spore, which usually gives rise to a bisexual gametophyte. The closest relatives of seed plants are all homosporous, suggesting that seed plants also had homosporous ancestors. At some point, seed plants or their ancestors became heterosporous: Megasporangia in megasporophylls produce megaspores that give rise to female gametophytes, and microsporangia in microsporophylls produce microspores that give rise to male gametophytes. Each megasporangium has a single functional megaspore, whereas each microsporangium contains vast numbers of microspores.

As we noted previously, the miniaturization of seed plant gametophytes contributed to the great success of this clade. Next we will look at the development of the female gametophyte within an ovule and the development of the male gametophyte in a pollen grain. Then we will follow the transformation of the ovule into a seed after fertilization.

Ovules and Production of Eggs

Although a few species of seedless plants are heterosporous, seed plants are unique in retaining the megaspore within the parent sporophyte (see Figure 30.2c). Layers of sporophyte tissue called **integuments** envelop and protect the megasporangium. Gymnosperm megaspores are surrounded by one integument, whereas those in angiosperms usually have two integuments. The whole structure—megasporangium, megaspore, and their integument(s)—is called an **ovule (Figure 30.3a)**. Inside each ovule (from the Latin *ovulum*, little egg), a female gametophyte develops from a megaspore and produces one or more egg cells.

Pollen and Production of Sperm

Microspores develop into **pollen grains,** which contain the male gametophytes of seed plants. Protected by a tough coat containing the polymer sporopollenin, pollen grains can be carried away from their parent plant by wind or by hitchhiking on the body of an animal that visits the plant to feed. The transfer of pollen to the part of a seed plant containing the ovules is called **pollination.** If a pollen grain germinates (begins growing), it gives rise to a pollen tube that discharges two sperm into the female gametophyte within the ovule, as shown in **Figure 30.3b**.

Recall that in bryophytes and seedless vascular plants such as ferns, free-living gametophytes release flagellated sperm that must swim through a film of water to reach egg cells. The distance for this sperm transport rarely exceeds a few centimeters. In seed plants, by contrast, the female gametophyte never leaves the sporophyte ovule, and the male gametophytes in pollen grains are durable travelers that can be carried long distances by the wind or by pollinators, depending on the species. Living gymnosperms provide evidence of this evolutionary transition. The sperm of some gymnosperm

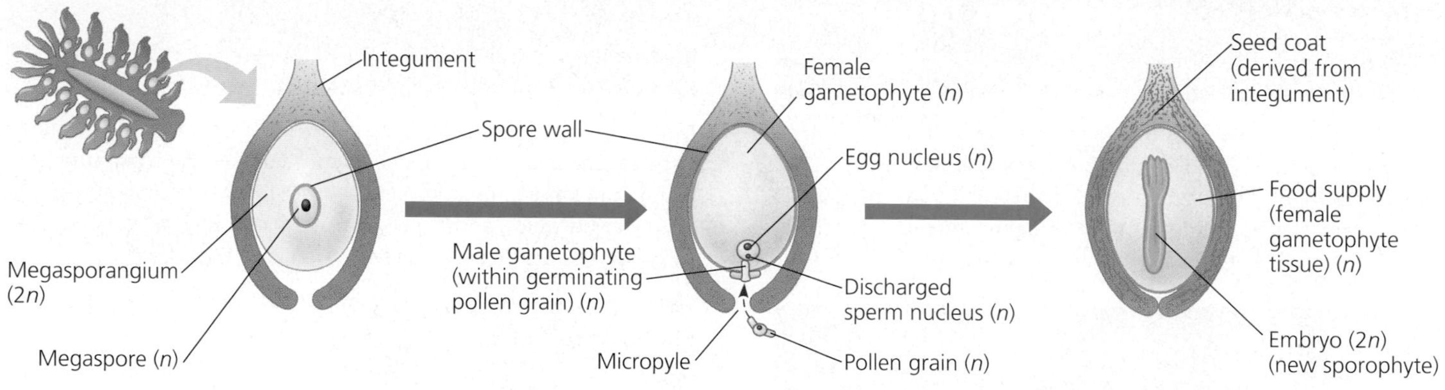

(a) Unfertilized ovule. In this sectional view through the ovule of a pine (a gymnosperm), a fleshy megasporangium is surrounded by a protective layer of tissue called an integument. (Angiosperms have two integuments.)

(b) Fertilized ovule. A megaspore develops into a multicellular female gametophyte. The micropyle, the only opening through the integument, allows entry of a pollen grain. The pollen grain contains a male gametophyte, which develops a pollen tube that discharges sperm.

(c) Gymnosperm seed. Fertilization initiates the transformation of the ovule into a seed, which consists of a sporophyte embryo, a food supply, and a protective seed coat derived from the integument.

▲ **Figure 30.3 From ovule to seed.**

species retain the ancient flagellated condition, but flagella have been lost in the sperm of most gymnosperms and all angiosperms. The sperm of seed plants do not require water or motility because the pollen is carried to the pollen tube, which then conveys the sperm to the ovule.

The Evolutionary Advantage of Seeds

We have been discussing characteristics of seed plants, but what exactly is a seed? If a sperm fertilizes an egg of a seed plant, the zygote grows into a sporophyte embryo. As shown in **Figure 30.3c**, the whole ovule develops into a **seed**, which consists of the embryo, along with a food supply, packaged within a protective coat derived from the integument(s).

The evolution of seeds enabled plants bearing them to better resist harsh environments and to disperse offspring more widely. Until the advent of seeds, the spore was the only protective stage in any plant life cycle. For example, moss spores may survive even if the local environment becomes too cold, too hot, or too dry for the mosses themselves to live. Their tiny size enables the moss spores to be dispersed in a dormant state to a new area, where they can germinate and give rise to new moss gametophytes if and when the environment is favorable enough for them to break dormancy. Spores were the main way that mosses and other seedless plants spread over Earth for the first 100 million years of plant life on land.

In contrast to a spore, which is single-celled, a seed is a multicellular structure that is much more resistant and complex. Its protective coat is derived from the integument(s) of the ovule. After being released from the parent plant, a seed may remain dormant for days, months, or even years. Under favorable conditions, it can then germinate, with the sporophyte embryo emerging as a seedling. Some seeds drop close to their parent sporophyte plant; others are carried far by the wind or animals.

Concept Check 30.1

1. Contrast sperm delivery in seedless vascular plants with sperm delivery in seed plants.
2. What additional features of seed plants, not present in seedless plants, contributed to the enormous success of seed plants on land?

For suggested answers, see Appendix A.

Concept 30.2

Gymnosperms bear "naked" seeds, typically on cones

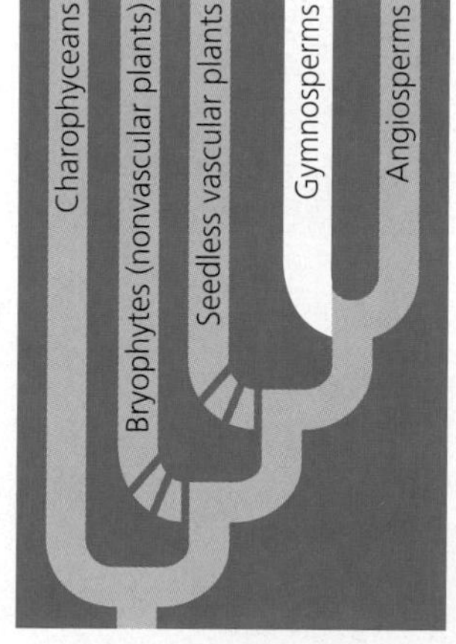

Recall from Chapter 29 that gymnosperms are plants that have "naked" seeds that are not enclosed in ovaries. Their seeds are exposed on modified leaves that usually form cones (strobili). In contrast, angiosperm seeds are enclosed in fruits, which are mature ovaries. Among the gymnosperms are cone-bearing plants called **conifers**, which include such trees as pines, firs and redwoods. Of the ten plant phyla in the taxonomic scheme adopted by this textbook (see Table 29.1), four are gymnosperms: Cycadophyta, Ginkgophyta, Gnetophyta, and Coniferophyta. The relationships of these four phyla to each other are uncertain. **Figure 30.4**, on the next two pages, surveys the diversity of extant gymnosperms.

Figure 30.4
Exploring Gymnosperm Diversity

PHYLUM CYCADOPHYTA

Cycads are the next largest group of gymnosperms after the conifers. They have large cones and palmlike leaves (true palm species are angiosperms). Only about 130 species survive today, but cycads thrived during the Mesozoic era, known as the "Age of Cycads" as well as the "Age of Dinosaurs."

Cycas revoluta

PHYLUM GINKGOPHYTA

Ginkgo biloba is the only extant species of this phylum. Also known as the maidenhair tree, it has deciduous fanlike leaves that turn gold in autumn. It is a popular ornamental tree in cities because it tolerates air pollution well. Landscapers usually plant only pollen-producing trees because the fleshy seeds smell rancid as they decay.

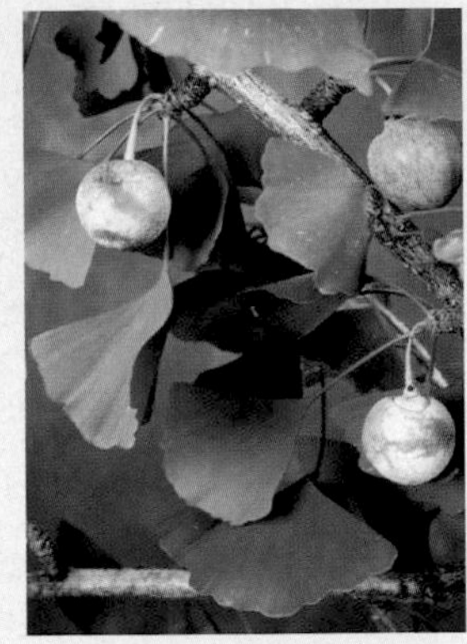

PHYLUM GNETOPHYTA

Plants in the phylum Gnetophyta, called gnetophytes, consist of three genera: *Gnetum*, *Ephedra*, and *Welwitschia*. Some species are tropical, whereas others live in deserts. Although very different in appearance, the genera are grouped together based on molecular data.

Welwitschia. This genus consists of one species, *Welwitschia mirabilis*, a plant that lives only in the deserts of southwestern Africa. Its strap-like leaves are among the largest leaves known.

Gnetum. This genus includes about 35 species of tropical trees, shrubs, and vines, mainly native to Africa and Asia. Their leaves look similar to those of flowering plants, and their seeds look somewhat like fruits.

Ephedra. This genus includes about 40 species that inhabit arid regions worldwide. These desert shrubs, commonly called "Mormon tea," produce the compound ephedrine, which is used medicinally as a decongestant.

PHYLUM CONIFEROPHYTA

Phylum Coniferophyta is by far the largest of the gymnosperm phyla, consisting of about 600 species of conifers (from the Latin *conus*, cone, and *ferre*, to carry). Many are large trees, such as cypresses and redwoods. A few conifer species dominate vast forested regions of the Northern Hemisphere, where the growing season is relatively short because of latitude or altitude.

Most conifers are evergreens; they retain their leaves throughout the year. Even during winter, a limited amount of photosynthesis occurs on sunny days. When spring comes, conifers already have fully developed leaves that can take advantage of the sunnier, warmer days. Some conifers, such as the dawn redwood, tamarack, and larch, are deciduous trees that lose leaves each autumn.

Douglas fir. "Doug fir" *(Pseudotsuga menziesii)* provides more timber than any other North American tree species. Some uses include house framing, plywood, pulpwood for paper, railroad ties, and boxes and crates.

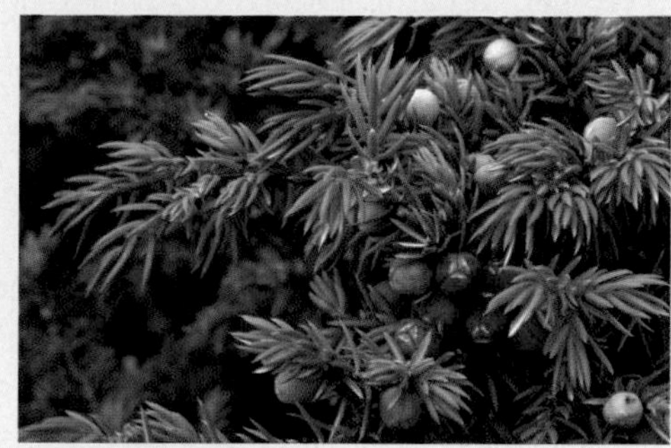

Common juniper. The "berries" of the common juniper (*Juniperus communis*) are actually ovule-producing cones consisting of fleshy sporophylls.

Pacific yew. The bark of Pacific yew (*Taxa brevifolia*) is a source of taxol, a compound used to treat women with ovarian cancer. The leaves of a European yew species produce a similar compound, which can be harvested without destroying the plants. Pharmaceutical companies are continuing to develop techniques for synthesizing drugs with taxol-like properties.

Wollemia pine. Survivors of a conifer group once known only from fossils, living Wollemia pines (*Wollemia nobilis*) were discovered in 1994 in a national park only 150 kilometers from Sydney, Australia. The species consists of just 40 known individuals in two small groves. The inset photo compares the leaves of this "living fossil" with actual fossils.

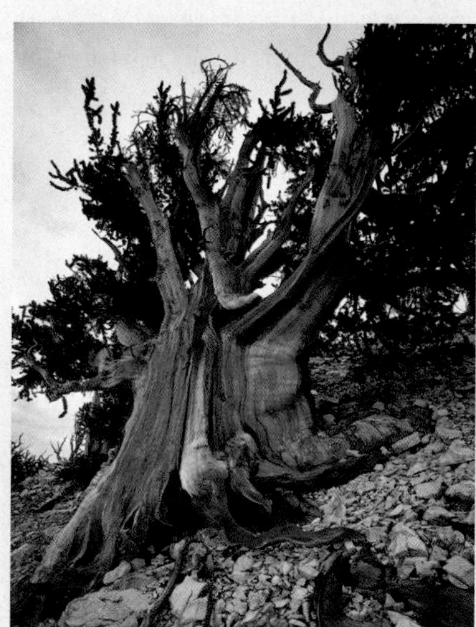

Bristlecone pine. This species (*Pinus longaeva*), which is found in the White Mountains of California, includes some of the oldest living organisms, reaching ages of more than 4,600 years. One tree (not shown here) is called Methuselah because it may be the world's oldest living tree. In order to protect the tree, scientists keep its location a secret.

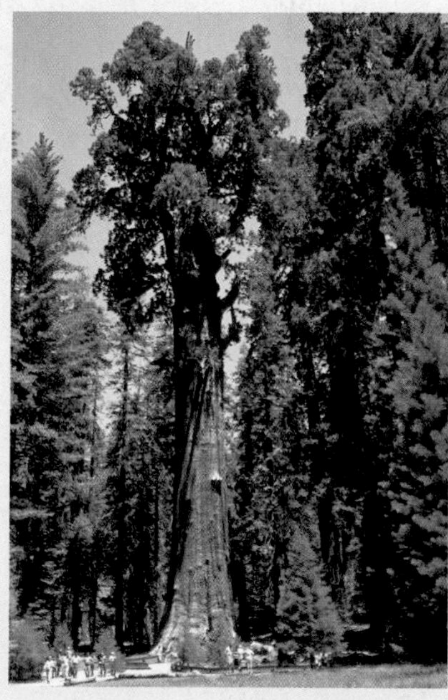

Sequoia. This giant sequoia (*Sequoiadendron giganteum*) in California's Sequoia National Park weighs about 2,500 metric tons, equivalent to about 24 blue whales (the largest animals), or 40,000 people. The giant sequoia is one of the largest living organisms and also among the most ancient, with some individuals estimated to be between 1,800 and 2,700 years old. Their cousins, the coast redwoods (*Sequoia sempervirens*), grow to heights of more than 110 meters (taller than the Statue of Liberty) and are found only in a narrow coastal strip of northern California and southern Oregon.

Gymnosperm Evolution

Fossil evidence reveals that by the late Devonian period, some plants had begun to acquire adaptations that characterize seed plants. For example, *Archaeopteris* was a heterosporous tree that produced wood **(Figure 30.5)**. It did not, however, bear seeds. Such transitional species of seedless vascular plants are sometimes called **progymnosperms.**

The first seed-bearing plants to appear in the fossil record were gymnosperms dating from around 360 million years ago, more than 200 million years before the first angiosperm fossils. These early gymnosperm species became extinct, along with several later gymnosperm lineages. Although the relationships between extinct and surviving lineages of seed plants remain uncertain, morphological and molecular evidence places surviving lineages of seed plants into two clades: the gymnosperms and the angiosperms (see Figure 29.7).

Early gymnosperms lived in Carboniferous ecosystems still dominated by lycophytes, horsetails, ferns, and other seedless vascular plants. As the Carboniferous period gave way to the Permian, markedly drier climatic conditions favored the spread of gymnosperms. The flora and fauna changed dramatically, as many groups of organisms disappeared and others became prominent (see Chapter 26). Though most pronounced in the seas, the changeover also affected terrestrial life. For example, in the animal kingdom, amphibians decreased in diversity and were replaced by reptiles, which were especially well adapted to the arid conditions. Similarly, the lycophytes, horsetails, and ferns that dominated the Carboniferous swamps were largely replaced by gymnosperms, which were more suited to the drier climate. In such gymnosperms as pines and firs, among the adaptations to arid conditions are their needle-shaped leaves, which have thick cuticles and relatively small surface areas.

▶ **Figure 30.5 A progymnosperm.** *Archaeopteris*, which lived 360 million years ago, produced wood and was heterosporous, but it did not produce seeds. Growing up to 20 m tall, it had fernlike leaves.

Geologists consider the end of the Permian period, about 251 million years ago, to be the boundary between the Paleozoic ("old life") and Mesozoic ("middle life") eras. Life changed profoundly as gymnosperms dominated terrestrial ecosystems throughout the Mesozoic, serving as the food supply for giant herbivorous dinosaurs. The Mesozoic era ended with mass extinctions of most of the dinosaurs and many other groups, and the planet gradually cooled. Although angiosperms now dominate most terrestrial ecosystems, many gymnosperms remain an important part of Earth's flora.

A Closer Look at the Life Cycle of a Pine

As you read earlier in the chapter, the evolution of seed plants included three key reproductive adaptations: the increasing dominance of the sporophyte generation; the advent of the seed as a resistant, dispersible stage in the life cycle; and the evolution of pollen as an airborne agent that brings gametes together. To reinforce your understanding of these adaptations, study **Figure 30.6**, which shows the life cycle of a pine, one of the most familiar gymnosperms.

The pine tree is the sporophyte; its sporangia are located on scalelike structures packed densely in cones. Like all seed plants, conifers are heterosporous. In conifers, the two types of spores are produced by separate cones: small pollen cones and large ovulate cones. In most pine species, each tree has both types of cones. In pollen cones, microsporocytes (microspore mother cells) undergo meiosis and produce haploid microspores. Each microspore develops into a pollen grain containing a male gametophyte. In pines and other conifers, the yellow pollen is released in large amounts and carried by the wind, dusting everything in its path. Meanwhile, in ovulate cones, megasporocytes (megaspore mother cells) undergo meiosis and produce haploid megaspores. Surviving megaspores develop into female gametophytes, which are retained within the sporangia. From the time young pollen and ovulate cones appear on the tree, it takes nearly three years for the male and female gametophytes to be produced and brought together, and for mature seeds to form from the fertilized ovules. The scales of each ovulate cone then separate, and the seeds travel on the wind. A seed that lands in a habitable place germinates, its embryo emerging as a pine seedling.

Concept Check 30.2

1. Based on Figure 30.4, explain why the various types of gymnosperms can be described as being similar yet distinctive.
2. Explain how the pine life cycle (see Figure 30.6) reflects basic characteristics of seed plants.

For suggested answers, see Appendix A.

▼ **Figure 30.6 The life cycle of a pine.**

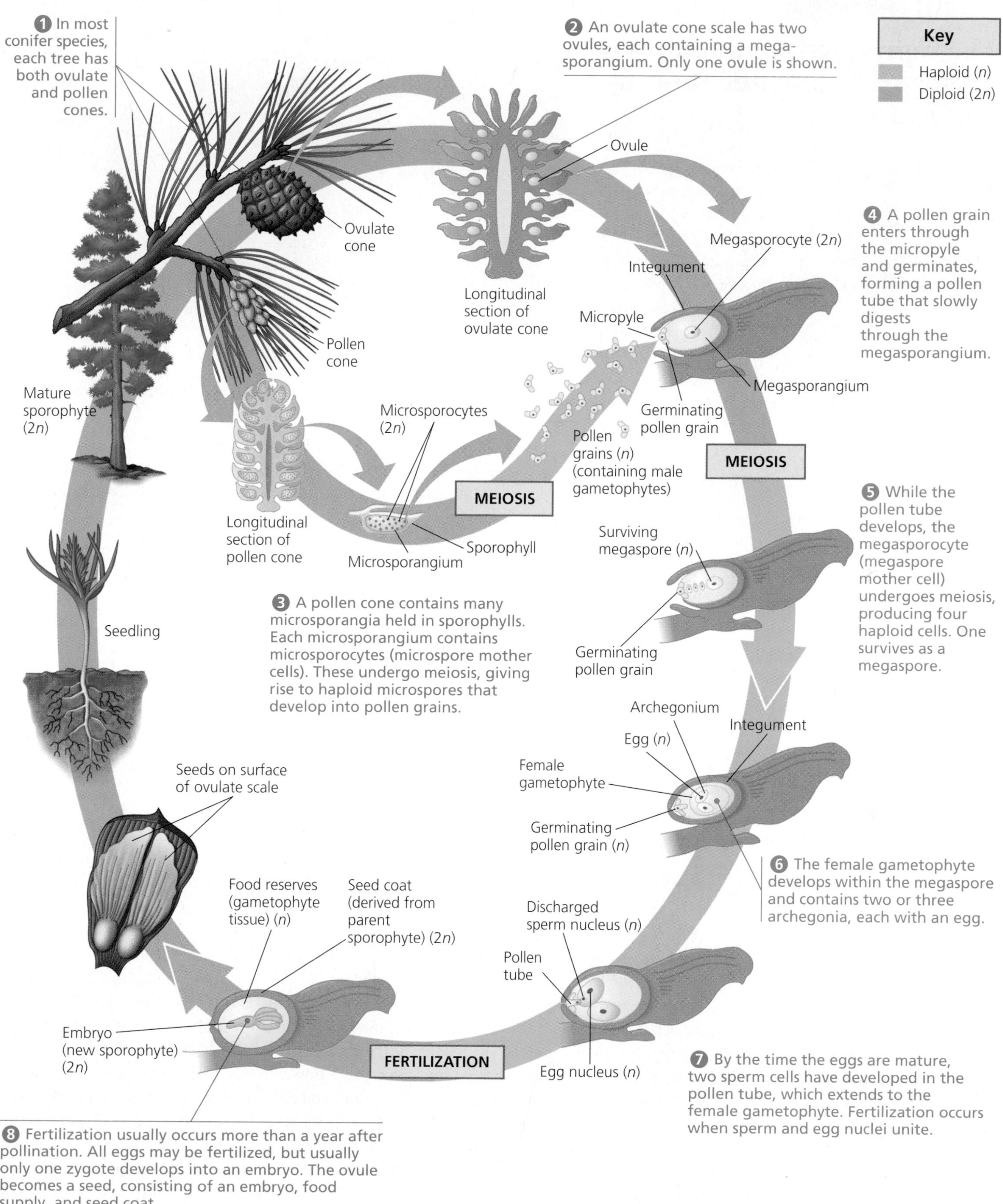

1 In most conifer species, each tree has both ovulate and pollen cones.
2 An ovulate cone scale has two ovules, each containing a megasporangium. Only one ovule is shown.
Key
Haploid (n)
Diploid (2n)
Ovule
Ovulate cone
Longitudinal section of ovulate cone
Megasporocyte (2n)
Integument
Micropyle
4 A pollen grain enters through the micropyle and germinates, forming a pollen tube that slowly digests through the megasporangium.
Pollen cone
Megasporangium
Mature sporophyte (2n)
Microsporocytes (2n)
Germinating pollen grain
Pollen grains (n) (containing male gametophytes)
MEIOSIS
MEIOSIS
Longitudinal section of pollen cone
Microsporangium
Sporophyll
Surviving megaspore (n)
5 While the pollen tube develops, the megasporocyte (megaspore mother cell) undergoes meiosis, producing four haploid cells. One survives as a megaspore.
3 A pollen cone contains many microsporangia held in sporophylls. Each microsporangium contains microsporocytes (microspore mother cells). These undergo meiosis, giving rise to haploid microspores that develop into pollen grains.
Seedling
Germinating pollen grain
Archegonium
Integument
Egg (n)
Female gametophyte
Seeds on surface of ovulate scale
Germinating pollen grain (n)
6 The female gametophyte develops within the megaspore and contains two or three archegonia, each with an egg.
Food reserves (gametophyte tissue) (n)
Seed coat (derived from parent sporophyte) (2n)
Discharged sperm nucleus (n)
Pollen tube
Embryo (new sporophyte) (2n)
FERTILIZATION
Egg nucleus (n)
7 By the time the eggs are mature, two sperm cells have developed in the pollen tube, which extends to the female gametophyte. Fertilization occurs when sperm and egg nuclei unite.
8 Fertilization usually occurs more than a year after pollination. All eggs may be fertilized, but usually only one zygote develops into an embryo. The ovule becomes a seed, consisting of an embryo, food supply, and seed coat.

Concept 30.3

The reproductive adaptations of angiosperms include flowers and fruits

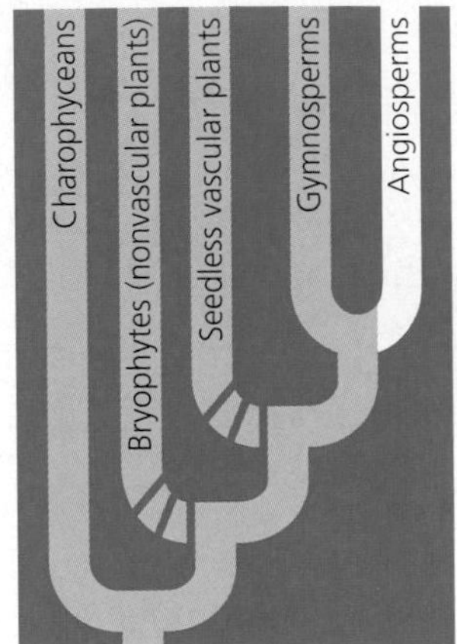

Commonly known as flowering plants, angiosperms are seed plants that produce the reproductive structures called flowers and fruits. The name *angiosperm* (from the Greek *angion,* container) refers to seeds contained in fruits, the mature ovaries. Today, angiosperms are the most diverse and widespread of all plants, with more than 250,000 species (about 90% of all plant species).

Characteristics of Angiosperms

All angiosperms are classified in a single phylum, Anthophyta (from the Greek *anthos,* flower). Before considering the evolution of angiosperms, we will examine their key adaptations—flowers and fruits—and the roles they play in the angiosperm life cycle.

Flowers

The **flower** is an angiosperm structure specialized for sexual reproduction. In many angiosperm species, insects or other animals transfer pollen from one flower to the female sex organs on another flower, which makes pollination more directed than the wind-dependent pollination of most gymnosperms. However, some angiosperms *are* wind-pollinated, particularly those species that occur in dense populations, such as grasses and tree species in temperate forests.

A flower is a specialized shoot that can have up to four rings of modified leaves called floral organs: sepals, petals, stamens, and carpels **(Figure 30.7)**. Starting at the base of the flower are the **sepals,** which are usually green and enclose the flower before it opens (think of a rosebud). Above the sepals are the **petals,** which are brightly colored in most flowers and aid in attracting pollinators. Flowers that are wind-pollinated, however, generally lack brightly colored parts. The sepals and petals are sterile floral organs, meaning that they are not directly involved in reproduction. Within the whorl of petals are the fertile sporophylls, floral organs that produce spores. The two whorls of sporophylls are the stamens and carpels. **Stamens,** the microsporophylls, produce microspores that give rise to pollen grains containing male gametophytes. A stamen consists of a stalk called the **filament** and a terminal sac, the **anther,** where pollen is produced. **Carpels** are the megasporophylls, which make

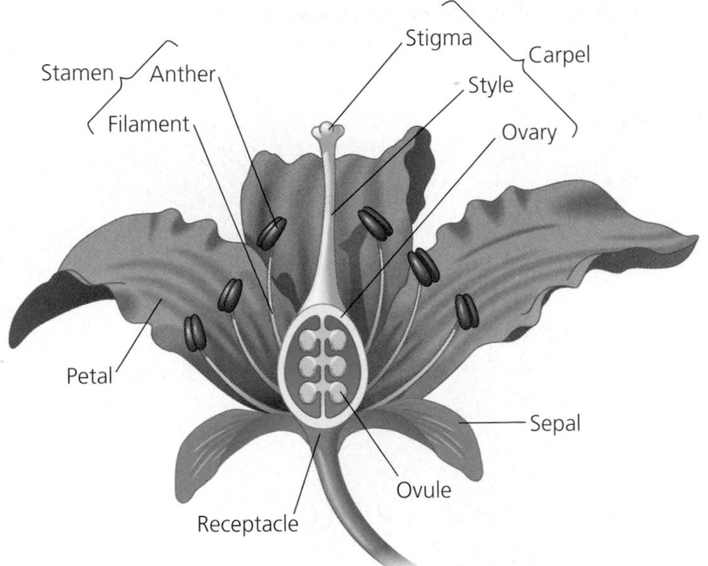

▲ **Figure 30.7 The structure of an idealized flower.**

megaspores and their products, female gametophytes. Many angiosperms have flowers with multiple carpels. At the tip of the carpel is a sticky **stigma** that receives pollen. A **style** leads to the **ovary** at the base of the carpel, which contains one or more ovules. If fertilized, an ovule develops into a seed. The whorls of floral organs—sepals, petals, stamens, and carpels—are attached to a part of the stem called the **receptacle.**

Some angiosperms, such as garden peas, have flowers with a single carpel. Others, such as magnolias, have several separate carpels. Still other species, such as lilies, have two or more fused carpels, usually forming an ovary with multiple ovule-containing chambers. A single carpel or a group of fused carpels is sometimes called a **pistil.**

Fruits

A **fruit** typically consists of a mature ovary, although it can include other flower parts as well. As seeds develop from ovules after fertilization, the wall of the ovary thickens. A pea pod is an example of a fruit, with seeds (mature ovules, the peas) encased in the ripened ovary (the pod). Fruits protect dormant seeds and aid in their dispersal.

The fruit begins to develop after pollination triggers hormonal changes that cause the ovary to grow. The wall of the ovary becomes the **pericarp,** the thickened wall of the fruit. As the ovary grows, the other flower parts wither away in many plants. If a flower is not pollinated, the fruit usually does not develop, and the entire flower withers and falls away.

Mature fruits can be either fleshy or dry **(Figure 30.8)**. Oranges, strawberries, and grapes are examples of fleshy fruits, in which one or more pericarp layers become soft during ripening. Dry fruits include beans, nuts, and grains. The dry, wind-dispersed fruits of grasses, harvested while on the

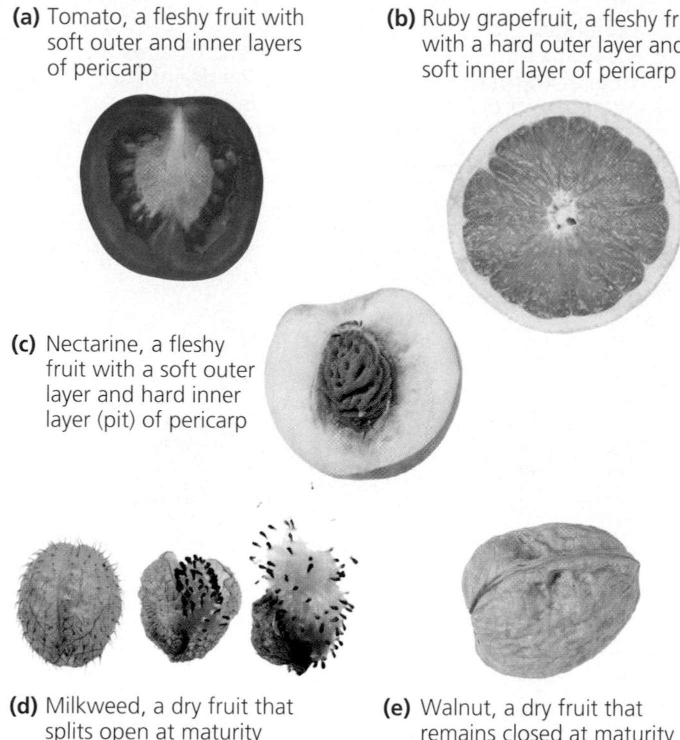

(a) Tomato, a fleshy fruit with soft outer and inner layers of pericarp
(b) Ruby grapefruit, a fleshy fruit with a hard outer layer and soft inner layer of pericarp
(c) Nectarine, a fleshy fruit with a soft outer layer and hard inner layer (pit) of pericarp
(d) Milkweed, a dry fruit that splits open at maturity
(e) Walnut, a dry fruit that remains closed at maturity

▲ **Figure 30.8 Some variations in fruit structure.**

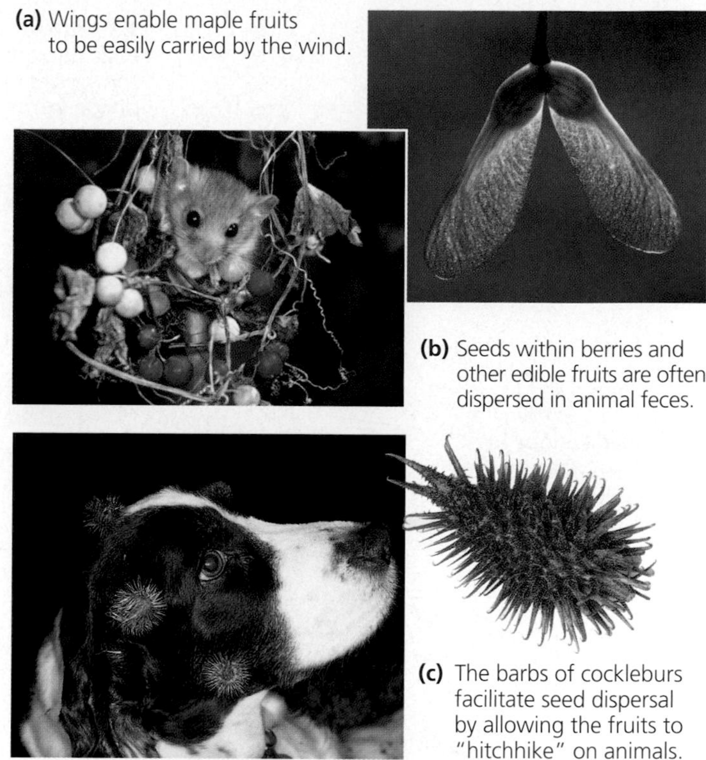

(a) Wings enable maple fruits to be easily carried by the wind.
(b) Seeds within berries and other edible fruits are often dispersed in animal feces.
(c) The barbs of cockleburs facilitate seed dispersal by allowing the fruits to "hitchhike" on animals.

▲ **Figure 30.9 Fruit adaptations that enhance seed dispersal.**

plant, are major staple foods for humans. The cereal grains of wheat, rice, maize, and other grasses are easily mistaken for seeds, but each is actually a fruit with a dry pericarp that adheres to the seed coat of the seed within. The ripening of a dry fruit involves the aging and drying out of the fruit tissues.

Fruits can also be categorized according to whether they develop from a single ovary, from multiple ovaries, or even from more than one flower. In Chapter 38, we will examine the transition from flower to fruit in more detail.

Various fruit adaptations help disperse seeds (**Figure 30.9**). The seeds of some flowering plants, such as dandelions and maples, are contained within fruits that function like kites or propellers, adaptations that enhance dispersal by wind. Some, such as coconuts, are adapted to dispersal by water. And many angiosperms rely on animals to carry seeds. Some of these plants have fruits modified as burrs that cling to animal fur (or the clothes of humans). Other angiosperms produce edible fruits, which are usually nutritious, sweet tasting, and vividly colored, advertising their ripeness. When an animal eats the fruit, it digests the fruit's fleshy part, but the tough seeds usually pass unharmed through the animal's digestive tract. Animals may deposit the seeds, along with a supply of fertilizer, miles from where the fruit was eaten.

The Angiosperm Life Cycle

A typical angiosperm life cycle is shown in **Figure 30.10**, on the next page. The flower of the sporophyte produces microspores that form male gametophytes and megaspores that form female gametophytes. The male gametophytes are in the pollen grains, which develop within microsporangia in the anthers. Each pollen grain has two haploid cells: a *generative cell* that divides, forming two sperm, and a *tube cell* that produces a pollen tube. Each ovule, which develops in an ovary, contains a female gametophyte, also known as an **embryo sac.** The embryo sac consists of only a few cells, one of which is the egg. Chapter 38 will discuss gametophyte development in more detail.

After its release from the anther, the pollen is carried to the sticky stigma at the tip of a carpel. Although some flowers self-pollinate, most have mechanisms that ensure **cross-pollination,** which in angiosperms is the transfer of pollen from an anther of a flower on one plant to the stigma of a flower on another plant of the same species. Cross-pollination enhances genetic variability. In some cases, stamens and carpels of a single flower may mature at different times, or they may be so arranged that self-pollination is unlikely.

The pollen grain germinates after it adheres to the stigma of a carpel. Now a mature male gametophyte, it extends a pollen tube that grows down within the style of the carpel. After reaching the ovary, the pollen tube penetrates through the **micropyle,** a pore in the integuments of the ovule, and discharges two sperm cells into the female gametophyte (embryo sac). One sperm fertilizes the egg, forming a diploid zygote. The other sperm fuses with the two nuclei in the large central cell of the female gametophyte. This type of **double fertilization** is unique to angiosperms.

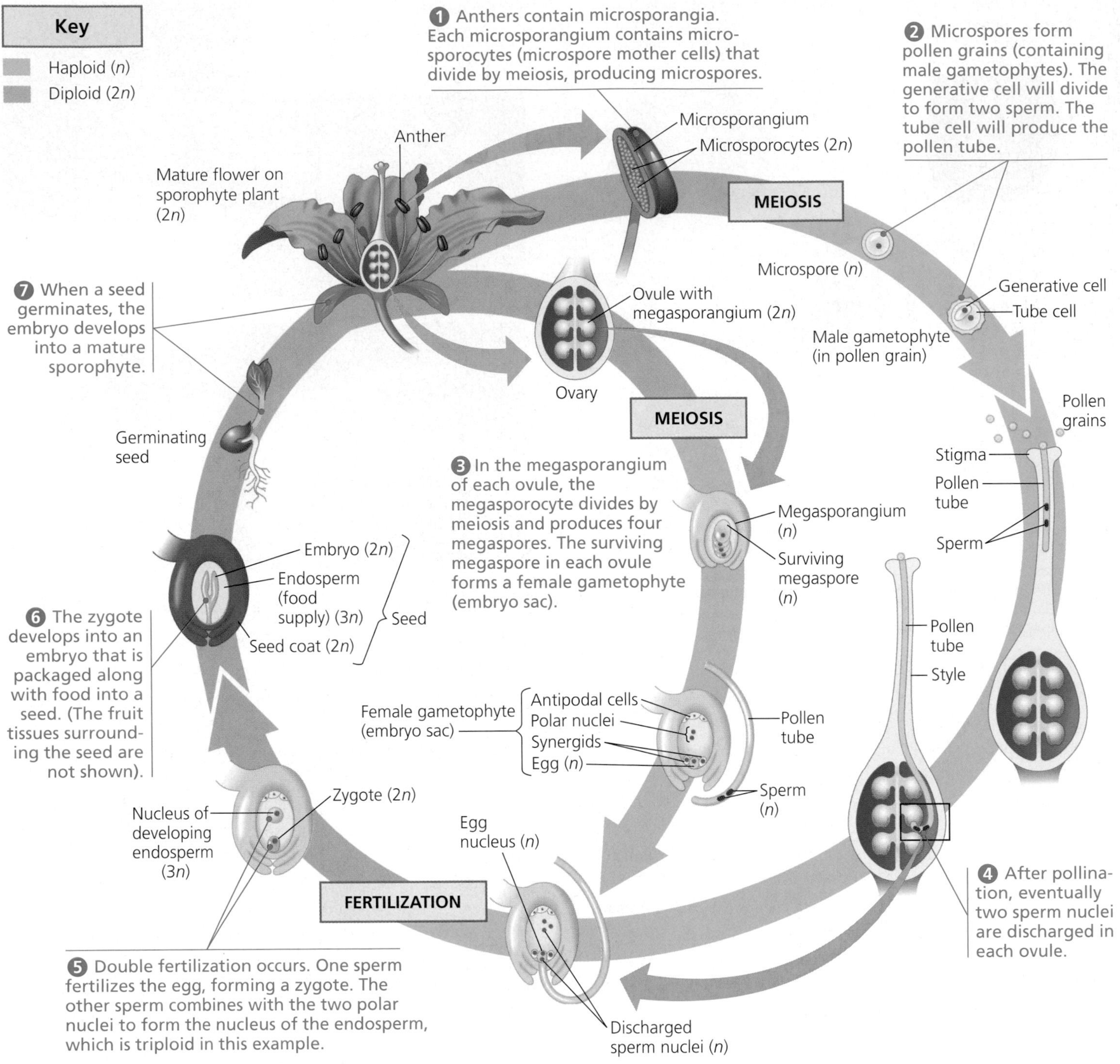
Key
Haploid (n)
Diploid (2n)
1 Anthers contain microsporangia. Each microsporangium contains microsporocytes (microspore mother cells) that divide by meiosis, producing microspores.
2 Microspores form pollen grains (containing male gametophytes). The generative cell will divide to form two sperm. The tube cell will produce the pollen tube.
Microsporangium
Microsporocytes (2n)
Anther
Mature flower on sporophyte plant (2n)
MEIOSIS
Microspore (n)
Generative cell
Tube cell
7 When a seed germinates, the embryo develops into a mature sporophyte.
Ovule with megasporangium (2n)
Male gametophyte (in pollen grain)
Ovary
Pollen grains
MEIOSIS
Germinating seed
Stigma
Pollen tube
3 In the megasporangium of each ovule, the megasporocyte divides by meiosis and produces four megaspores. The surviving megaspore in each ovule forms a female gametophyte (embryo sac).
Megasporangium (n)
Sperm
Surviving megaspore (n)
Embryo (2n)
Endosperm (food supply) (3n)
Seed
Seed coat (2n)
6 The zygote develops into an embryo that is packaged along with food into a seed. (The fruit tissues surrounding the seed are not shown).
Pollen tube
Style
Female gametophyte (embryo sac)
Antipodal cells
Polar nuclei
Synergids
Egg (n)
Pollen tube
Sperm (n)
Zygote (2n)
Nucleus of developing endosperm (3n)
Egg nucleus (n)
FERTILIZATION
4 After pollination, eventually two sperm nuclei are discharged in each ovule.
5 Double fertilization occurs. One sperm fertilizes the egg, forming a zygote. The other sperm combines with the two polar nuclei to form the nucleus of the endosperm, which is triploid in this example.
Discharged sperm nuclei (n)

▲ **Figure 30.10 The life cycle of an angiosperm.**

After double fertilization, the ovule matures into a seed. The zygote develops into a sporophyte embryo with a rudimentary root and one or two seed leaves called **cotyledons.** The nucleus of the central cell of the female gametophyte divides repeatedly and develops into **endosperm**, tissue rich in starch and other food reserves.

What is the function of double fertilization in angiosperms? One hypothesis is that double fertilization synchronizes the development of food storage in the seed with the development of the embryo. If a particular flower is not pollinated or sperm cells are not discharged into the embryo sac, fertilization does not occur, and neither endosperm nor embryo forms. So perhaps double fertilization is an adaptation that prevents flowering plants from squandering nutrients on infertile ovules.

Another type of double fertilization occurs in some gymnosperm species belonging to phylum Gnetophyta. However, in these species double fertilization gives rise to two embryos rather than to an embryo and endosperm. This difference indicates that double fertilization evolved independently in angiosperms and gymnosperms.

As you read earlier, the seed consists of the embryo, endosperm, remnants of the sporangium, and a seed coat derived from the integuments. An ovary develops into a fruit as its ovules become seeds. After being dispersed, a seed may germinate if environmental conditions are favorable. The coat ruptures and the embryo emerges as a seedling, using food stored in the endosperm and cotyledons.

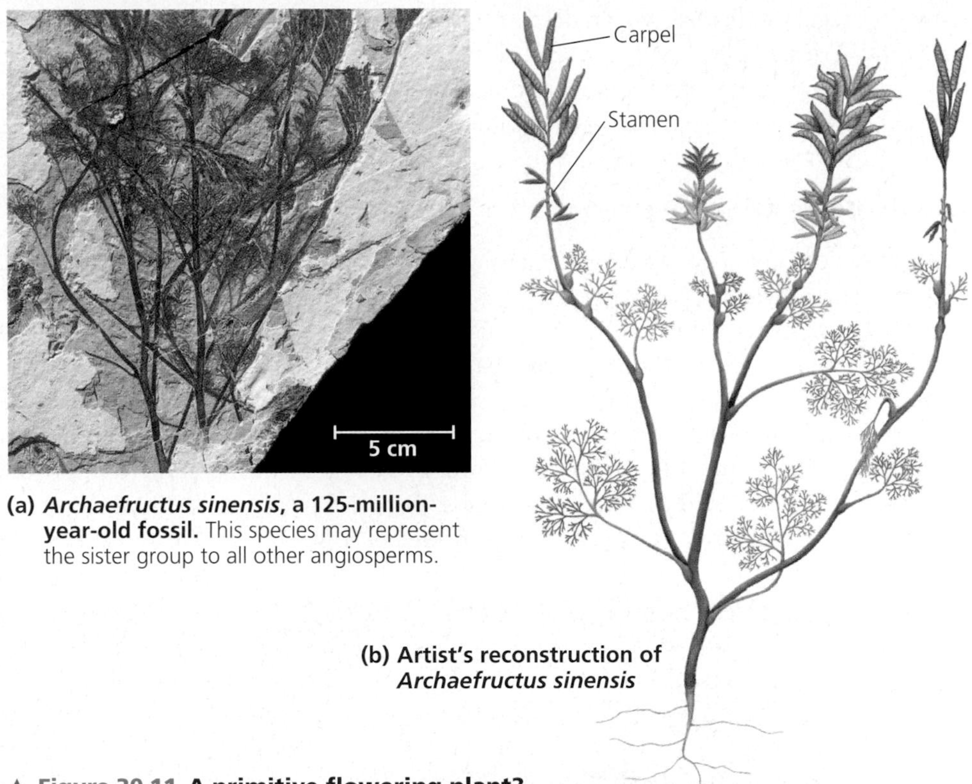

(a) *Archaefructus sinensis*, a 125-million-year-old fossil. This species may represent the sister group to all other angiosperms.

(b) Artist's reconstruction of *Archaefructus sinensis*

▲ **Figure 30.11 A primitive flowering plant?**

Angiosperm Evolution

Clarifying the origin and diversification of angiosperms—what Charles Darwin once called an "abominable mystery"—poses fascinating challenges to evolutionary biologists. Angiosperms originated at least 140 million years ago, and during the late Mesozoic, the major branches of the clade diverged from their common ancestor. But it was not until the end of the Mesozoic that angiosperms began to dominate many terrestrial ecosystems. Landscapes changed dramatically as conifers, cycads, and other gymnosperms gave way to flowering plants in many parts of the world.

With their flowers and fruits, angiosperms are profoundly different from living or fossil gymnosperms, which makes their origins puzzling. To understand how the angiosperm body plan emerged, scientists are studying newly discovered fossils and developmental genes that underlie flowers and other angiosperm innovations (see Chapter 35).

Fossil Angiosperms

In the late 1990s, scientists in China discovered some intriguing fossils of 125-million-year-old angiosperms. These fossils, now named *Archaefructus liaoningensis* and *Archaefructus sinensis* (**Figure 30.11**), display both derived and primitive traits. *A. sinensis*, for example, has anthers and has seeds inside closed carpels but lacks petals and sepals. In 2002, scientists completed a phylogenetic comparison of *A. sinensis* with 173 living plants. (*A. liaoningensis* was not included because its fossils are not as well preserved.) The researchers concluded that among all known plant fossils, *Archaefructus* is the most closely related to all living angiosperms.

If *Archaefructus* is indeed a "proto-angiosperm," that would suggest that the ancestors of flowering plants were herbaceous rather than woody. *Archaefructus* was discovered along with fish fossils and has bulbous structures that might be aquatic adaptations, implying that perhaps angiosperms originated as aquatic plants. The discoverers of *Archaefructus* have suggested that fast-growing herbaceous plants could have returned to water and thrived there, escaping competition with other seed plants and then later taking over the land.

More recently, however, some paleobotantists have disputed this interpretation. More derived lineages of angiosperms also became aquatic, and along the way their flowers became simpler, resembling the "primitive" flowers of *Archaefructus*. Such debates, typical whenever a seemingly transitional fossil is discovered, may be resolved when more fossils and other types of evidence emerge.

An "Evo-Devo" Hypothesis of Flower Origins

Chapter 24 discussed how studying developmental genes informs hypotheses about animal evolution. The same holds true for flowering plants. Botanist Michael Frohlich of the Natural History Museum in London has used the "evo-devo" approach—the synthesis of evolutionary and developmental biology—in hypothesizing how pollen-producing and ovule-producing structures were combined into a single flower. He proposes that the ancestor of angiosperms had separate pollen-producing and ovule-producing structures. Then, as a result of a mutation, ovules developed on some microsporophylls, which evolved into carpels.

Because Frohlich asserts that the flower evolved mainly from the pollen-producing ("male") reproductive structure of a gymnosperm ancestor, his view is known as the "mostly male" hypothesis. Supporting evidence includes comparisons of genes that give rise to flowers and cones. Flower-development genes are usually related to pollen-producing gymnosperm genes. Also, certain mutations cause flowering plants to grow ovules on sepals and petals, which demonstrates that the position of ovules can easily be changed. By comparing angiosperm and gymnosperm genes, botanists are testing Frohlich's hypothesis and other evo-devo models for the origin of flowers.

Angiosperm Diversity

From humble beginnings in the Mesozoic, angiosperms diversified into more than 250,000 species that now dominate most terrestrial ecosystems. Until the late 1990s, most systematists divided flowering plants into two groups, based partly on the number of cotyledons, or seed leaves, in the embryo. Species with one cotyledon were called **monocots,** and those with two were called **dicots.** Other features, such as flower and leaf structure, were also used to define the two groups. For example, monocots typically have parallel leaf veins (think of a grass blade), while the veins of most dicots have a netlike pattern (think of an oak leaf). Some examples of monocots are orchids, palms, and grain crops such as maize, wheat, and rice. Some examples of dicots are roses, peas, sunflowers, and maples.

Recent DNA studies, however, indicate that the monocot-dicot distinction likely does not completely reflect evolutionary relationships. Current research supports the view that monocots form a clade but reveals that the remaining angiosperm species are not monophyletic. The vast majority of species traditionally called dicots do form a clade, now known as **eudicots** ("true" dicots), but the others are now divided into several small lineages. Three of these lineages are informally called **basal angiosperms** because they appear to include the flowering plants belonging to the oldest lineages. Another lineage, known as the **magnoliids,** evolved later. **Figure 30.12** provides an overview of angiosperm diversity.

Figure 30.12

Exploring Angiosperm Diversity

BASAL ANGIOSPERMS

Surviving basal angiosperms are currently thought to consist of three small lineages, including only about 100 species. The oldest lineage seems to be represented by a single species, *Amborella trichopoda.* Two other surviving lineages diverged later: a clade that includes water lilies and then a clade consisting of the star anise and its relatives.

Amborella trichopoda. This small shrub, found only on the South Pacific island of New Caledonia, may be the sole survivor of a branch at the very base of the angiosperm tree. *Amborella* lacks vessels, which are present in more derived angiosperms. Consisting of xylem cells arranged in continuous tubes, vessels transport water more efficiently than tracheids. Their absence in *Amborella* indicates they evolved after the lineage that gave rise to *Amborella* branched off.

Water lily (*Nymphaea "Rene Gerard"*). Water lilies are living members of a clade that may be predated only by the *Amborella* lineage.

Star anise (*Illicium floridanum*). This species represents a third surviving lineage of basal angiosperms.

HYPOTHETICAL TREE OF FLOWERING PLANTS

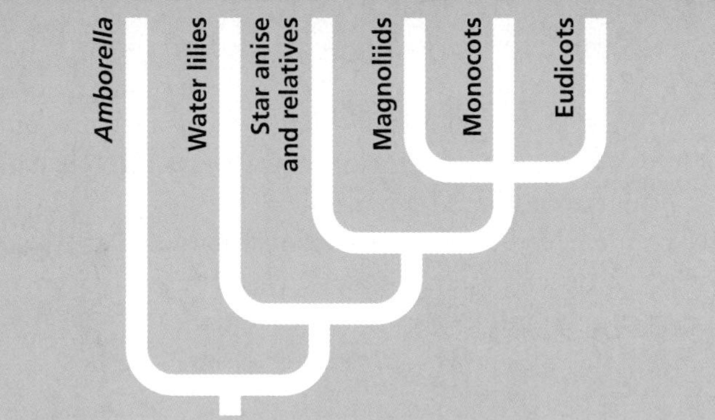

MAGNOLIIDS

Magnoliids consist of about 8,000 species, most notably magnolias, laurels, and black pepper plants. They include both woody and herbaceous species. Although they share some primitive traits with basal angiosperms, such as a typically spiral rather than whorled arrangement of flower organs, magnoliids are actually more closely related to monocots and eudicots.

Southern magnolia (*Magnolia grandiflora*). This member of the magnolia family is a woody magnoliid. This variety of southern magnolia, called "Goliath," has flowers that measure up to about a foot across.

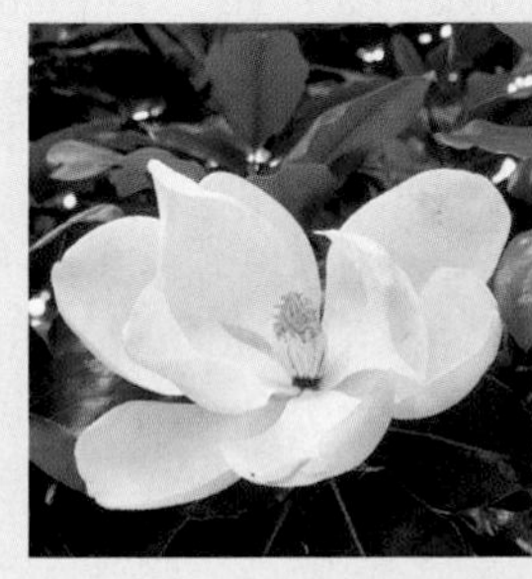

MONOCOTS

More than one-quarter of angiosperm species are monocots—about 70,000 species. These examples represent some of the largest families.

Orchid *(Lemboglossum rossii)*

Pygmy date palm (*Phoenix roebelenii*)

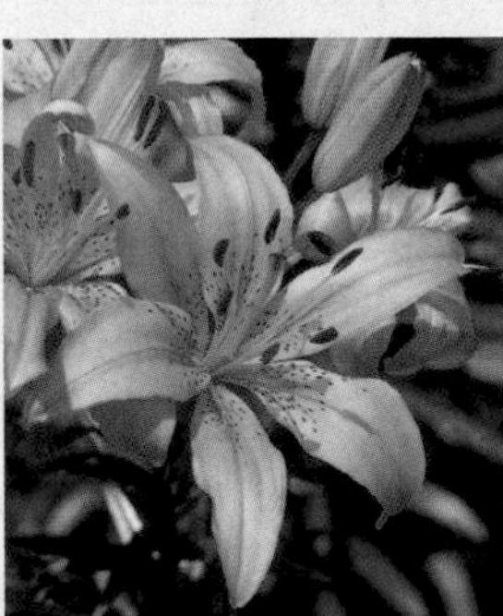

Lily (*Lilium* "Enchantment")

Barley (*Hordeum vulgare*), a grass

EUDICOTS

More than two-thirds of angiosperm species are eudicots—roughly 170,000 species. Below is a sampling of floral diversity.

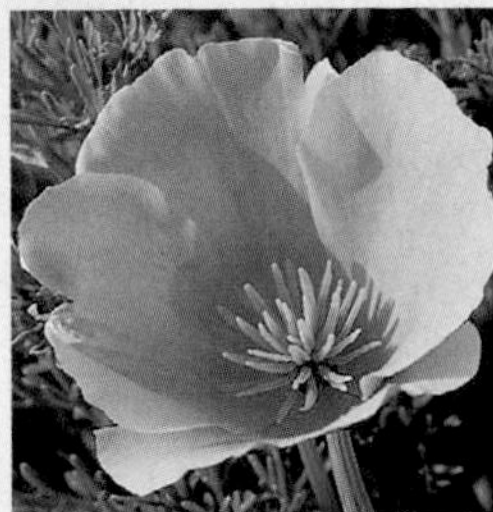

California poppy (*Eschscholzia californica*)

Pyrenean oak (*Quercus pyrenaica*)

Dog rose (*Rosa canina*), a wild rose

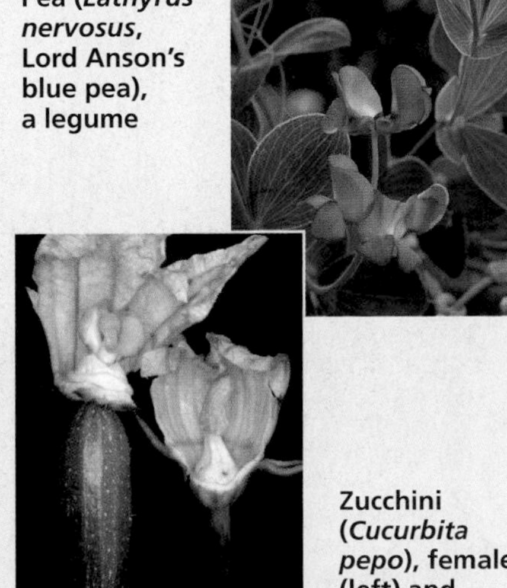

Pea (*Lathyrus nervosus*, Lord Anson's blue pea), a legume

Zucchini (*Cucurbita pepo*), female (left) and male flowers

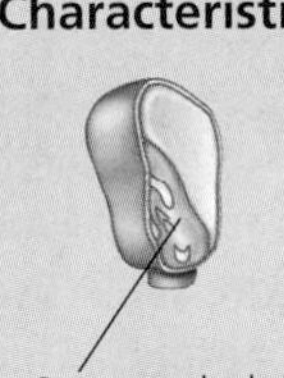

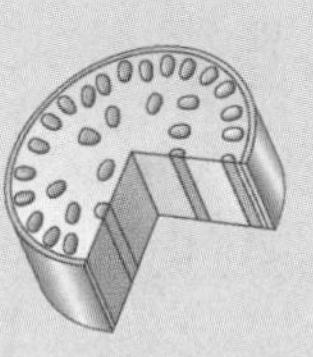

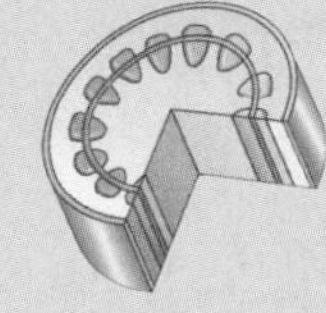

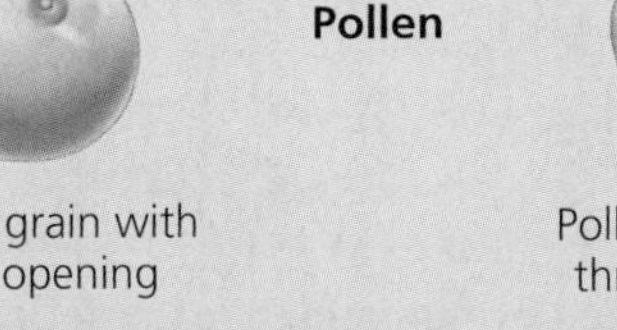

Monocot Characteristics		Eudicot Characteristics
One cotyledon	**Embryos**	Two cotyledons
Veins usually parallel	**Leaf venation**	Veins usually netlike
Vascular tissue scattered	**Stems**	Vascular tissue usually arranged in ring
Root system usually fibrous (no main root)	**Roots**	Taproot (main root) usually present
Pollen grain with one opening	**Pollen**	Pollen grain with three openings
Floral organs usually in multiples of three	**Flowers**	Floral organs usually in multiples of four or five

Evolutionary Links Between Angiosperms and Animals

Ever since they colonized land, animals have influenced the evolution of terrestrial plants, and vice versa. Animals crawling on the forest floor created a selective pressure favoring plants that kept their spores and gametophytes off the ground, out of easy reach of most animals. This trend in plants may in turn have favored the evolution of flight in insects.

This type of mutual evolutionary influence between species is often advantageous for both partners. For example, as flowers and fruits evolved in angiosperms, some animals helped plants by dispersing their pollen and seeds. Meanwhile, the animals received a benefit as they ate the plants' nectars, seeds, and fruits.

Plant-pollinator relationships have likely played a role in increasing angiosperm and animal diversity **(Figure 30.13)**. The most extreme of these relationships is between an individual species of flower that can be pollinated only by a single animal species. In Madagascar, for example, one species of orchid has an 11-inch-long nectary, so the nectar can be consumed only by a moth with an 11-inch-long proboscis! Such linked adaptations, involving reciprocal genetic change in two species, are described as *coevolution,* as we will discuss further in Chapter 53.

Most relationships are less specific, however. For example, the flowers of a particular species may attract insects rather than birds, but many different insect species may serve as pollinators. Conversely, a single animal species—a honeybee species, for example—may pollinate many plant species. But even in these less specific relationships, flower color, fragrance, and structure often reflect specialization for a particular taxonomic *group* of pollinators, such as diverse species of bees or hummingbirds. Other influences, such as animals that eat flowers while providing no pollination, can also help to shape flower diversity.

Angiosperm evolution has contributed particularly to the variety of insect species. But the expansion of grasslands during the past 65 million years has also increased the diversity of grazing mammals such as horses. Grasses that are C_4 photosynthesizers spread because a decline in the atmospheric CO_2 levels gave a selective advantage to C_4 photosynthesis (see Chapter 10). And as you will read in Chapter 34, the shift from forests to grasslands in Africa between 10 and 2 million years ago may have been a factor in the evolution of our own lineage.

Concept Check 30.3

1. It has been said that an oak is an acorn's way of making more acorns. Write an explanation that includes these terms: sporophyte, gametophyte, ovule, seed, ovary, and fruit.
2. Compare and contrast a pine cone and a flower in terms of structure and function.
3. Explain the use of the terms *monocot, dicot,* and *eudicot.*

For suggested answers, see Appendix A.

(a) A flower pollinated by honeybees. This honeybee is harvesting pollen and nectar (a sugary solution secreted by flower glands) from a Scottish broom flower. The flower has a tripping mechanism that arches the stamens over the bee and dusts it with pollen, some of which will rub off onto the stigma of the next flower the bee visits.

(b) A flower pollinated by hummingbirds. The long, thin beak and tongue of this rufous hummingbird enable the animal to probe flowers that secrete nectar deep within floral tubes. As the bird feeds, anthers dust its beak and head feathers with pollen. Many flowers that are pollinated by birds are red or pink, colors to which bird eyes are especially sensitive.

(c) A flower pollinated by nocturnal animals. Some angiosperms, such as this cactus, depend mainly on nocturnal pollinators, including bats. Common adaptations of such plants include large, light-colored, highly fragrant flowers that nighttime pollinators can locate.

▲ **Figure 30.13 Flower-pollinator relationships.**

Concept 30.4

Human welfare depends greatly on seed plants

Throughout this unit, we have been highlighting the important ways in which humans depend on various groups of organisms. No group is more important to our survival than seed plants. In forests and on farms, seed plants are key sources of food, fuel, wood products, and medicine. Our reliance on them makes the preservation of plant diversity critical.

Products from Seed Plants

Most of our food comes from angiosperms. Just six crops—wheat, rice, maize, potatoes, cassava, and sweet potatoes—yield 80% of all the calories consumed by humans. We also depend on angiosperms to feed livestock: It takes 4.8 kilograms of grain to produce a kilogram of grain-fed beef.

Modern crops are the products of a relatively recent burst of genetic change, resulting from artificial selection after humans began domesticating plants approximately 13,000 years ago. To appreciate the scale of the transformation, contrast the sizes of domesticated plants and their smaller wild relatives, as in the case of maize and teosinte (see Figure 38.14). At the genetic level, scientists can also glean information about domestication by comparing the genes of crops with those of wild relatives. With maize, dramatic changes, such as increased cob size and removal of the hard coating around teosinte kernels, may have been initiated by as few as five gene mutations.

How did wild plants change so dramatically in such relatively little time? For thousands of years, farmers have selected the seeds of plants with desirable traits (large fruits, for example) to plant for the next year's crops. But humans may also be domesticating certain other plants unconsciously, as in the case of wild almonds. Almonds contain a bitter compound called amygdalin that repels birds and other animals. Amygdalin breaks down into cyanide, so eating a large number of wild almonds can be fatal. But mutations can reduce the level of amygdalin, making almonds sweet rather than bitter. Wild birds eat almonds from trees with such mutations. According to one hypothesis, humans may have observed birds eating the almonds and then eaten the almonds themselves, passing some seeds in their feces. The seeds germinated around human settlements and grew into plants with sweeter, less dangerous almonds.

In addition to basic crops, flowering plants provide non-staple foods. Two of the world's most popular beverages come from tea leaves and coffee beans, and you can thank the tropical cacao tree for cocoa and chocolate. Spices are derived from various plant parts, such as flowers (cloves, saffron), fruits and seeds (vanilla, black pepper, mustard, cumin), leaves (basil, mint, sage), and even bark (cinnamon).

Many seed plants, both gymnosperms and angiosperms, are sources of wood, which is absent in all living seedless plants. Wood consists of an accumulation of tough-walled xylem cells (see Chapter 35). Wood is the primary source of fuel for much of the world, and wood pulp, typically derived from conifers such as fir and pine, is used to make paper. Worldwide, wood also remains the most widely used construction material.

For centuries, humans have also depended on seed plants for medicines. Many cultures have a long tradition of using herbal remedies, and scientific research has identified the relevant secondary compounds in many of these plants, leading to the synthesis of medicines. Willow leaves and bark, for instance, have been used since ancient times in pain-relieving remedies, including prescriptions by the Greek physician Hippocrates. In the 1800s, scientists traced the willow's medicinal property to the chemical salicin. A synthesized derivative, acetylsalicylic acid, is what we call aspirin. Although modern chemistry facilitates laboratory synthesis, plants remain an important direct source of medicinal compounds. In the United States, for example, about 25% of prescription drugs contain one or more active ingredients extracted or derived from plants, typically from seed plants. Other ingredients were first discovered in seed plants and then synthesized artificially. **Table 30.1** lists some medicinal uses of secondary compounds of seed plants.

Table 30.1 A Sampling of Medicines Derived from Seed Plants

Compound	Example of Source	Example of Use
Atropine	Belladonna plant	Pupil dilator in eye exams
Digitalin	Foxglove	Heart medication
Menthol	Eucalyptus tree	Ingredient in cough medicines
Morphine	Opium poppy	Pain reliever
Quinine	Cinchona tree (see below)	Malaria preventative
Taxol	Pacific yew	Ovarian cancer drug
Tubocurarine	Curare tree	Muscle relaxant during surgery
Vinblastine	Periwinkle	Leukemia drug

Cinchona bark, source of quinine

Threats to Plant Diversity

Although plants may be a renewable resource, plant diversity is not. The exploding human population and its demand for space and resources are extinguishing plant species at an unprecedented rate. The problem is especially severe in the tropics, where more than half the human population lives and where population growth is fastest. Fifty million acres of tropical rain forest, an area about the size of the state of Washington, are cleared each year, a rate that would completely eliminate Earth's tropical forests within 25 years. The most common cause of this destruction is slash-and-burn clearing of forests for agricultural use (see Chapter 55). As forests disappear, so do thousands of plant species. Once a species becomes extinct, it can never return.

The loss of plant species is often accompanied by the loss of insects and other rain forest animals. Researchers estimate that habitat destruction in rain forests and other ecosystems is claiming hundreds of species each year. This rate is faster than in any other period, even during the extensive Permian and Cretaceous extinctions. While the toll is greatest in the tropics, the threat is global.

Many people have ethical concerns about contributing to the extinction of living forms. But there are also practical reasons to be concerned about the loss of plant diversity. So far, we have explored the potential uses of only a tiny fraction of the more than 290,000 known plant species. For example, almost all our food is based on the cultivation of only about two dozen species of seed plants. And fewer than 5,000 plant species have been studied as potential sources of medicines. The tropical rain forest may be a medicine chest of healing plants that could be extinct before we even know they exist. If we begin to view rain forests and other ecosystems as living treasures that can regenerate only slowly, we may learn to harvest their products at sustainable rates. What else can we do to preserve plant diversity? Few questions are as important.

Concept Check 30.4

1. Explain why it is accurate to consider plant diversity to be a nonrenewable resource.

For suggested answers, see Appendix A.

Chapter 30 Review

Go to www.campbellbiology.com or the student CD-ROM to explore Activities, Investigations, and other interactive study aids.

SUMMARY OF KEY CONCEPTS

The reduced gametophytes of seed plants are protected in ovules and pollen grains

- **Advantages of Reduced Gametophytes (pp. 591–592)** The gametophytes of seed plants develop within the walls of spores retained within tissues of the parent sporophyte, which provides protection and nutrients.
- **Heterospory: The Rule Among Seed Plants (p. 592)** Seed plants evolved from plants that had megasporangia (which produce megaspores that give rise to female gametophytes) and microsporangia (which produce microspores that give rise to male gametophytes).
- **Ovules and Production of Eggs (p. 592)** An ovule consists of a megasporangium, megaspore, and protective integument(s). The female gametophyte develops from the megaspore and produces one or more eggs.
- **Pollen and Production of Sperm (pp. 592–593)** Pollen, which can be dispersed by air or animals, eliminated the water requirement for fertilization.
- **The Evolutionary Advantage of Seeds (p. 593)** A seed is a sporophyte embryo, along with its food supply, packaged in a protective coat. Seeds are more resistant than spores and can be distributed widely by wind or animals.

Concept 30.2

Gymnosperms bear "naked" seeds, typically on cones

- Extant gymnosperms include cycads, *Ginkgo biloba*, gnetophytes, and conifers (pp. 593–595).
- **Gymnosperm Evolution (p. 596)** Gymnosperms appear early in the plant fossil record and dominated Mesozoic terrestrial ecosystems. Living seed plants can be divided into two groups: gymnosperms and angiosperms.
- **A Closer Look at the Life Cycle of a Pine (pp. 596–597)** Dominance of the sporophyte generation, the development of seeds from fertilized ovules, and the role of pollen in transferring sperm to ovules are key features of a typical gymnosperm life cycle.
  Activity *Pine Life Cycle*

Concept 30.3

The reproductive adaptations of angiosperms include flowers and fruits

- **Characteristics of Angiosperms (pp. 598–601)** Flowers generally consist of four whorls of modified leaves: sepals, petals, stamens (which produce pollen), and carpels (which produce ovules). Ovaries ripen into fruits, which are often carried by wind, water, or animals to new locations. In the angiosperm life cycle, double fertilization occurs when a pollen tube discharges two sperm into the female gametophyte (embryo sac) within an ovule. One sperm fertilizes the egg, while the other combines with two nuclei in the center cell of the female gametophyte and initiates development of food-storing endosperm tissue. The endosperm nourishes the developing embryo.
  Activity *Angiosperm Life Cycle*

- **Angiosperm Evolution (pp. 601–602)** An adaptive radiation of angiosperms occurred during the late Mesozoic period. Fossils and evo-devo studies offer insights into the origin of flowers.
- **Angiosperm Diversity (pp. 602–603)** The two main groups of angiosperms are monocots and eudicots. Basal angiosperms are less derived. Magnoliids share some traits with basal angiosperms but are more closely related to monocots and eudicots.
  Investigation *How Are Trees Identified by Their Leaves?*
- **Evolutionary Links Between Angiosperms and Animals (p. 604)** Pollination of flowers by animals and transport of seeds by animals are two important relationships in terrestrial ecosystems.

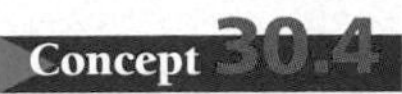

Human welfare depends greatly on seed plants

- **Products from Seed Plants (p. 605)** Humans depend on seed plants for food, wood, and many medicines.
- **Threats to Plant Diversity (p. 606)** Destruction of habitat is causing extinction of many plant species and the animal species they support.

TESTING YOUR KNOWLEDGE

Self-Quiz

1. Where in an angiosperm would you find a megasporangium?
   a. in the style of a flower
   b. inside the tip of a pollen tube
   c. enclosed in the stigma of a flower
   d. within an ovule contained within an ovary of a flower
   e. packed into pollen sacs within the anthers found on a stamen

2. A fruit is most commonly
   a. a mature ovary.
   b. a thickened style.
   c. an enlarged ovule.
   d. a modified root.
   e. a mature female gametophyte.

3. With respect to angiosperms, which of the following is *incorrectly* paired with its chromosome count?
   a. egg cell—n
   b. megaspore—$2n$
   c. microspore—n
   d. zygote—$2n$
   e. sperm—n

4. Which of the following is *not* a characteristic that distinguishes gymnosperms and angiosperms from other plants?
   a. alternation of generations
   b. ovules
   c. integuments
   d. pollen
   e. dependent gametophytes

5. Which of the following traits is *not* shared by most angiosperms?
   a. two cotyledons
   b. vessel elements
   c. taproot
   d. pollen grain with three openings
   e. parallel leaf venation

6. Gymnosperms and angiosperms have the following in common *except*
   a. seeds.
   b. pollen.
   c. vascular tissue.
   d. ovaries.
   e. ovules.

7–10. Match the derived characters listed below with the correct branch points in the diagram:
   a. flowers
   b. embryos
   c. seeds
   d. vascular tissue

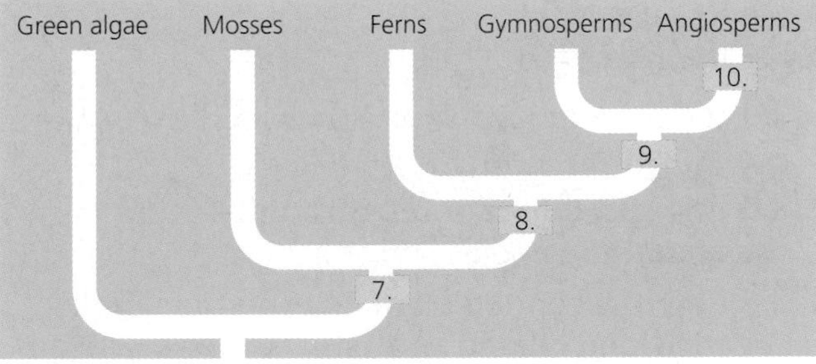

For Self-Quiz answers, see Appendix A.

Go to the website or CD-ROM for more quiz questions.

Evolution Connection

The history of life has been punctuated by several mass extinctions. For example, the impact of a meteorite may have wiped out most of the dinosaurs and many forms of marine life at the end of the Cretaceous period (see Chapter 26). Fossils indicate that plants were much less severely affected by this and other mass extinctions. What adaptations may have enabled plants to withstand these disasters better than animals?

Scientific Inquiry

Suggest a way to test the hypothesis that a particular angiosperm species is pollinated exclusively by beetles.

Investigation *How Are Trees Identified by Their Leaves?*

Science, Technology, and Society

Why are tropical rain forests being destroyed at such an alarming rate? What kinds of social, technological, and economic factors are responsible? Most forests in developed Northern Hemisphere countries have already been cut. Do the developed nations have a right to ask the developing nations in the Southern Hemisphere to slow or stop the destruction of their forests? Defend your answer. What kinds of benefits, incentives, or programs might slow the assault on the rain forests?

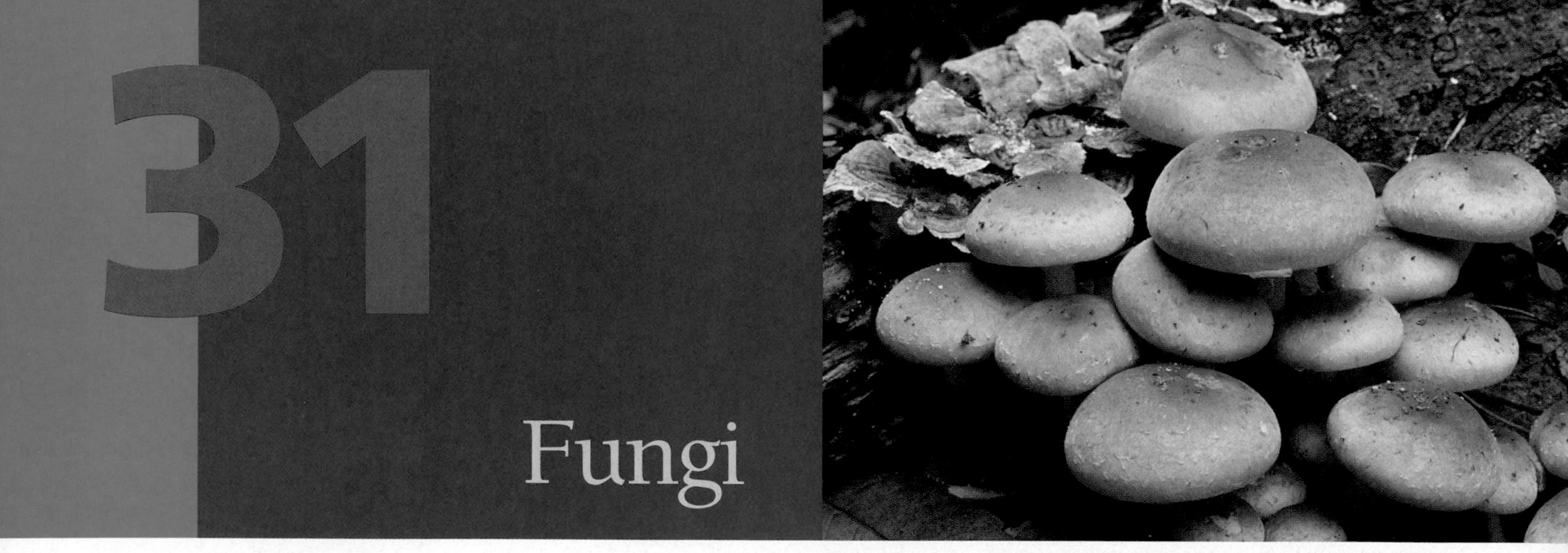

31 Fungi

▲ **Figure 31.1 Two species of fungi decomposing a log.**

Key Concepts

31.1 Fungi are heterotrophs that feed by absorption
31.2 Fungi produce spores through sexual or asexual life cycles
31.3 Fungi descended from an aquatic, single-celled, flagellated protist
31.4 Fungi have radiated into a diverse set of lineages
31.5 Fungi have a powerful impact on ecosystems and human welfare

Overview

Mighty Mushrooms

If you were to walk through the Malheur National Forest in eastern Oregon, you might notice a few clusters of honey mushrooms (*Armillaria ostoyae*) scattered here and there beneath the towering trees. Though the trees might appear to dwarf the mushrooms, as strange as it sounds, the reverse is actually true. These mushrooms are only the above-ground portion of a single enormous fungus. Its subterranean network of filaments spreads through 890 hectares (2,200 acres) of the forest—more than the area of 1,600 football fields. Based on its current growth rate, scientists estimate that this fungus, which weighs hundreds of tons, has been growing for 2,600 years.

The innocuous honey mushrooms on the forest floor are a fitting symbol of the neglected grandeur of the kingdom Fungi. Most of us are barely aware of these eukaryotes beyond the occasional brush with athlete's foot or spoiled food. Yet fungi are a huge and important component of the biosphere. Their diversity alone is staggering: Although 100,000 species have been described, it is estimated that there are actually as many as 1.5 million species of fungi. Some fungi are single-celled, but most form complex multicellular bodies, which in many cases include the above-ground structures we know as mushrooms. This diversity has enabled fungi to colonize just about every imaginable terrestrial habitat; airborne spores have even been found 160 km above the ground.

Fungi are not only diverse and widespread, but they are also essential for the well-being of most terrestrial ecosystems. They break down organic material and recycle nutrients **(Figure 31.1)**, allowing other organisms to assimilate essential chemical elements. Almost all plants depend on a symbiotic relationship with fungi that helps their roots absorb minerals from the soil. Humans also benefit from fungi's services to agriculture and forestry as well as their essential role in making products ranging from bread to antibiotics. But it is also true that a tiny fraction of fungi causes diseases in plants and animals.

In this chapter, we will investigate the structure of fungi, survey the members of the kingdom Fungi, and discuss their ecological and commercial significance.

Concept 31.1

Fungi are heterotrophs that feed by absorption

Despite their vast diversity, fungi share some key traits, most importantly the way in which they derive nutrition.

Nutrition and Fungal Lifestyles

Like animals, fungi are heterotrophs—they cannot make their own food as plants and algae can. But unlike animals, fungi do not ingest (eat) their food. Instead, they digest their food while it is still in the environment by secreting powerful hydrolytic enzymes, called **exoenzymes,** into their surroundings. Exoenzymes break down complex molecules to smaller organic compounds that the fungi can absorb into their bodies and use.

This absorptive mode of nutrition is related to the diverse lifestyles exhibited by fungi. Among the fungi are species that live as decomposers (also called saprobes), parasites, and mutualistic symbionts. Saprobic fungi break down and absorb nutrients from nonliving organic material, such as fallen logs, animal corpses, and the wastes of living organisms. Parasitic fungi absorb nutrients from the cells of living hosts. Some parasitic fungi are pathogenic, including species that infect human lungs and other species that are responsible for about 80% of plant diseases. Mutualistic fungi also absorb nutrients from a host organism, but the mutualistic fungi reciprocate with functions beneficial to the host in some way, such as aiding a plant in the uptake of minerals from the soil.

Body Structure

Some fungi exist as single cells and are known as yeasts. However, most species are multicellular. The morphology of multicellular fungi enhances their ability to absorb nutrients from their surroundings **(Figure 31.2)**. The bodies of these fungi typically form a network of tiny filaments called **hyphae** (singular, *hypha*). Hyphae are composed of tubular cell walls surrounding the plasma membrane and cytoplasm of the cells. Unlike the cellulose walls of plants, fungal cell walls contain **chitin,** a strong but flexible nitrogen-containing polysaccharide that is also found in the external skeletons of insects and other arthropods.

Fungal hyphae form an interwoven mass called a **mycelium** (plural, *mycelia*) that surrounds and infiltrates the material on which the fungus feeds. The structure of a mycelium maximizes the ratio of its surface area to its volume, making feeding more efficient. Just 1 cm^3 of rich organic soil may contain as much as 1 km of hyphae with a total surface area of more than 300 cm^2 interfacing with the soil. A fungal mycelium grows rapidly, as proteins and other materials synthesized by the fungus are channeled through cytoplasmic streaming to the tips of the extending hyphae. The fungus concentrates its energy and resources on adding hyphal length and thus overall absorptive surface area, rather than on increasing hyphal girth. Fungal mycelia are nonmotile; they cannot run, swim, or fly in search of food or mates. But the mycelium makes up for its lack of mobility by swiftly extending the tips of its hyphae into new territory.

In most fungi, the hyphae are divided into cells by crosswalls, or **septa** (singular, *septum*). Septa generally have pores large enough to allow ribosomes, mitochondria, and even nuclei to flow from cell to cell **(Figure 31.3a)**. Some fungi lack septa. Known as **coenocytic** fungi, these organisms consist of a continuous cytoplasmic mass containing hundreds or thousands of nuclei **(Figure 31.3b)**. The coenocytic condition results from the repeated division of nuclei without cytoplasmic division. This description may remind you of the plasmodial slime molds you read about in Chapter 28, which also consist of cytoplasmic masses containing many nuclei. This similarity was one reason that slime molds were formerly

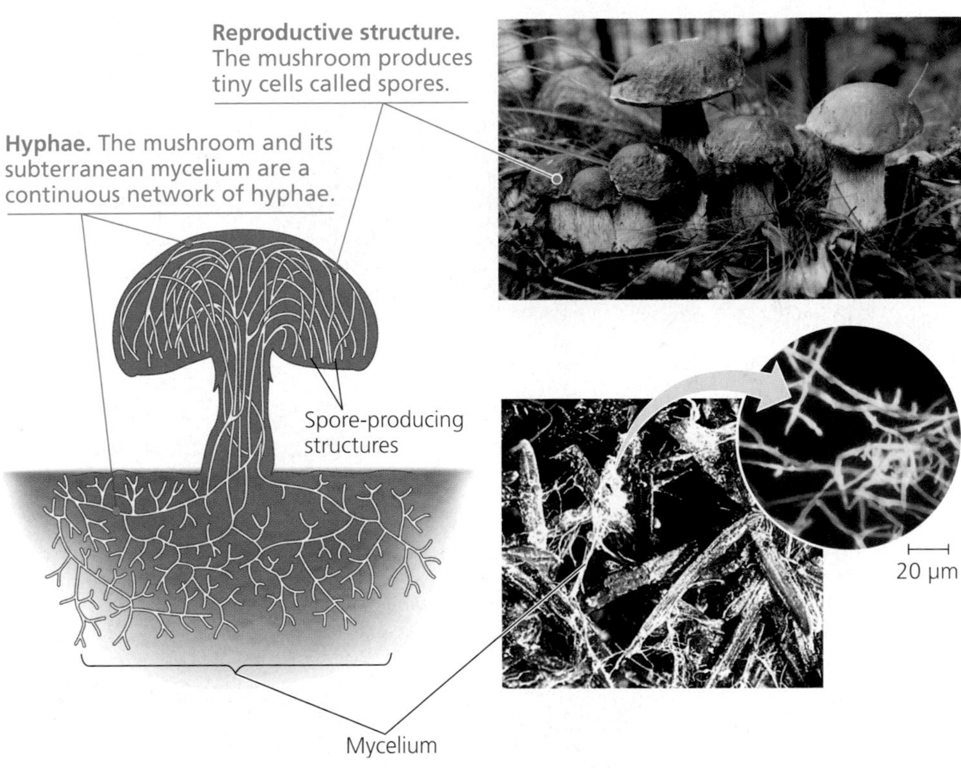

Reproductive structure. The mushroom produces tiny cells called spores.
Hyphae. The mushroom and its subterranean mycelium are a continuous network of hyphae.
Spore-producing structures
20 μm
Mycelium

▲ **Figure 31.2 Structure of a multicellular fungus.** The top photograph shows the sexual structures, called fruiting bodies, of the penny bun fungus *(Boletus edulis)*. The bottom left photograph shows a macroscopic view of a mycelium growing on fallen conifer needles, and the bottom right photograph shows a microscopic view of a mycelium (LM).

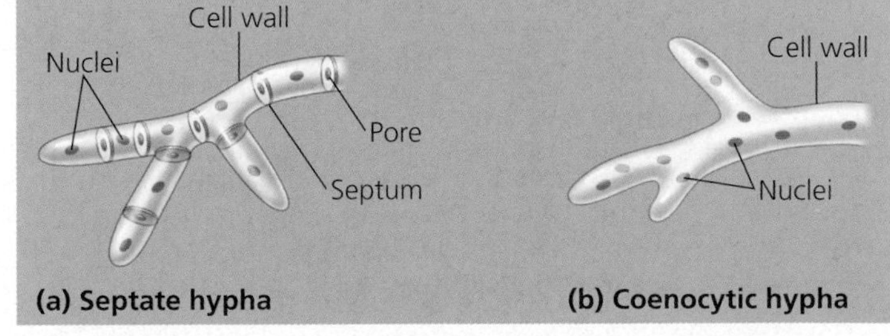

▶ **Figure 31.3 Structure of hyphae.**

classified as fungi; molecular comparisons have since confirmed that in fact the two clades are not closely related.

Some fungi have specialized hyphae that allow them to feed on living animals **(Figure 31.4a)**. Other species have specialized hyphae called **haustoria** that enable them to penetrate the tissue of their hosts **(Figure 31.4b)**. Mutually beneficial relationships between such fungi and plant roots are called **mycorrhizae.** (*Mycorrhizae* means "fungus roots.") Mycorrhizal fungi (fungi that form mycorrhizae) can deliver phosphate ions and other minerals to plants, which the plants themselves cannot acquire on their own. In exchange, the plants supply the fungi with organic nutrients. There are several different types of mycorrhizal fungi. **Ectomycorrhizal fungi** (from the Greek *ektos*, out) form sheaths of hyphae over the surface of a root and also grow into the extracellular spaces of the root cortex (see Figure 37.12a). **Endomycorrhizal fungi** (from the Greek *entos*, in) extend their hyphae through the root cell wall and into tubes formed by invagination (pushing inward) of the root cell membrane (see Figure 37.12b). You will read more about these types of mycorrhizal fungi later in this chapter.

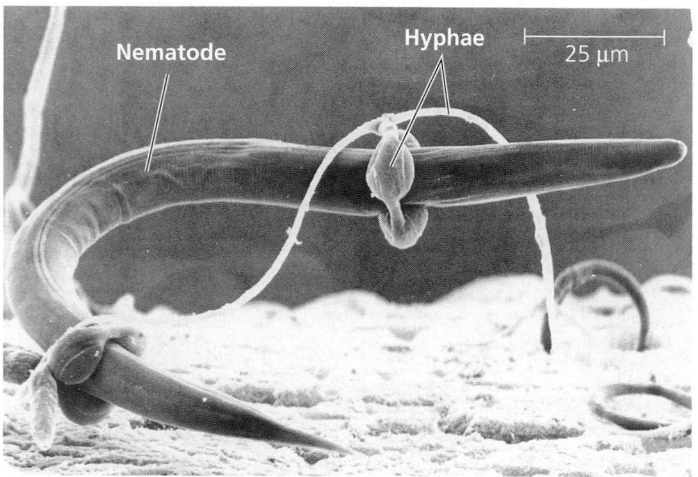

(a) Hyphae adapted for trapping and killing prey. In *Arthrobotrys*, a soil fungus, portions of the hyphae are modified as hoops that can constrict around a nematode (roundworm) in less than a second. The fungus then penetrates its prey with hyphae and digests the prey's inner tissues (SEM).

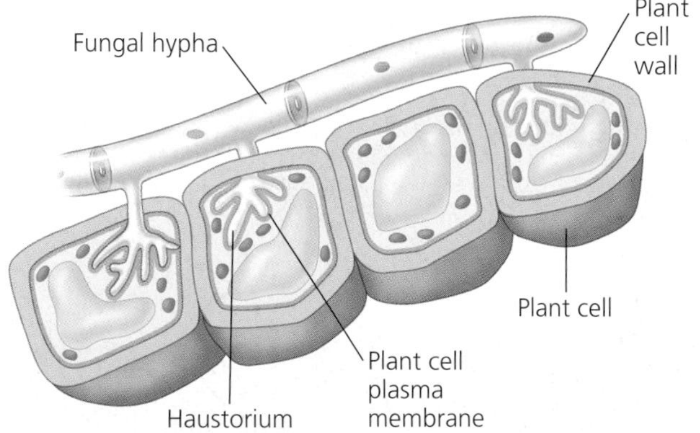

(b) Haustoria. Mutualistic and parasitic fungi grow specialized hyphae called haustoria that can penetrate the cell walls of plants. Haustoria remain separated from a plant cell's cytoplasm by the plasma membrane of the plant cell (dark gold).

▲ **Figure 31.4 Specialized hyphae.**

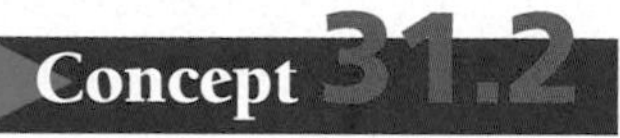

Concept Check 31.1

1. Compare and contrast the nutritional mode of a fungus with your own nutritional mode.
2. Describe how the structure of a fungus is adapted to its nutritional mode.

For suggested answers, see Appendix A.

Concept 31.2

Fungi produce spores through sexual or asexual life cycles

Fungi propagate themselves by producing vast numbers of spores, either sexually or asexually. For example, puffballs, the reproductive structures of certain fungi, may release trillions of spores in cloud-like bursts (see Figure 31.18d). Spores can be carried long distances by wind or water, and if they land in a moist place where there is food, they germinate, producing new mycelia. To appreciate how effective spores are at dispersing fungi, try leaving a slice of melon exposed to the air. Within a week or so, you should see fuzzy mycelia growing from the microscopic spores constantly falling onto the melon slice.

Figure 31.5 generalizes the many different life cycles that can produce fungal spores. In this section, we will survey general aspects of the sexual and asexual life cycles of fungi. Later in the chapter we will examine more closely the life cycles of specific phyla of fungi.

Sexual Reproduction

The nuclei of fungal hyphae and spores of most species are haploid, except for transient diploid stages that form during sexual life cycles. Generally, sexual reproduction in fungi begins when hyphae from two distinct mycelia release sexual signaling molecules called **pheromones.** If the mycelia are of different mating types, the pheromones from each partner bind to receptors on the surface of the other, and the hyphae extend toward the source of the pheromones. When the hyphae meet, they fuse. This "compatibility test" contributes to genetic variation by preventing hyphae from fusing with other hyphae from the same mycelium or another genetically identical mycelium.

The union of the cytoplasm of the two parent mycelia is known as **plasmogamy.** In many fungi, the haploid nuclei contributed by each parent do not fuse right away. Instead, parts of the mycelia contain coexisting, genetically different

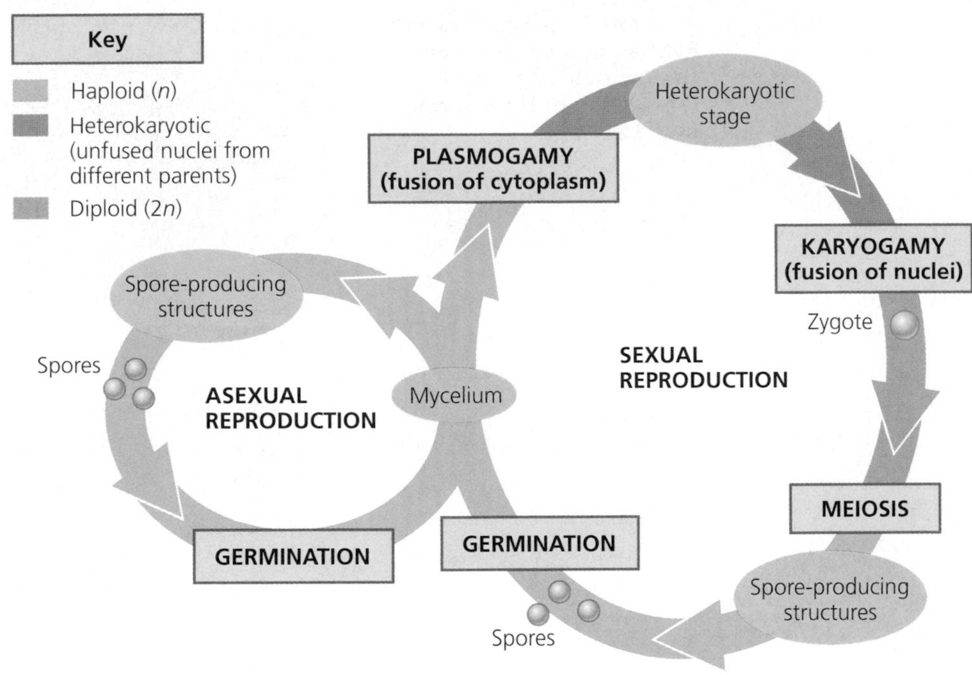

▲ **Figure 31.5 Generalized life cycle of fungi.** Many—but not all—fungi reproduce both sexually and asexually. Some reproduce only sexually, others only asexually.

nuclei. Such a mycelium is said to be a **heterokaryon** (meaning "different nuclei"). In some species, heterokaryotic mycelia become mosaics, with different nuclei restricted to different parts of the network. In other species, the different nuclei mingle and may even exchange chromosomes and genes in a process similar to crossing over (see Chapter 13).

In some fungi, the haploid nuclei pair off two to a cell, one from each parent. Such a mycelium is said to be **dikaryotic** (meaning "two nuclei"). As a dikaryotic mycelium grows, the two nuclei in each cell divide in tandem without fusing.

Hours, days, or even centuries may pass between plasmogamy and the next stage in the sexual cycle, **karyogamy.** During karyogamy, the haploid nuclei contributed by the two parents fuse, producing diploid cells. Zygotes and other transient structures form during karyogamy, the only stage in which diploidy exists in most fungi. Meiosis restores the haploid condition, and the mycelium then produces specialized reproductive structures that produce and disperse the spores.

The sexual processes of karyogamy and meiosis generate extensive genetic variation, an important prerequisite for adaptive evolution. (See Chapters 13 and 23 to review how sex can increase genetic diversity within a population.) The heterokaryotic condition also offers some of the advantages of diploidy, in that one haploid genome may be able to compensate for harmful mutations in the other.

Asexual Reproduction

In addition to sexual reproduction, many fungi can reproduce asexually. Clones are produced by mitotic production of spores, which can be spread by air or water. Some species only reproduce asexually. As with sexual reproduction, the processes of asexual reproduction vary widely among fungi.

Some fungi that can reproduce asexually grow as **mold.** Mold is a familiar sight in the kitchen, where it forms furry carpets on fruit, bread, and other foods **(Figure 31.6)**. Molds grow rapidly as mycelia and produce spores. Many species that can grow as molds can also reproduce sexually if they come in contact with other mating types.

Other asexual fungi are **yeasts**. Yeasts inhabit liquid or moist habitats, including plant sap and animal tissues. Instead of producing spores, yeasts reproduce asexually by simple cell division or by the pinching of small "bud cells" off a parent cell **(Figure 31.7)**. Some yeast species can also grow as filamentous mycelia, depending on the availability of nutrients. Yeasts sometimes can also reproduce sexually.

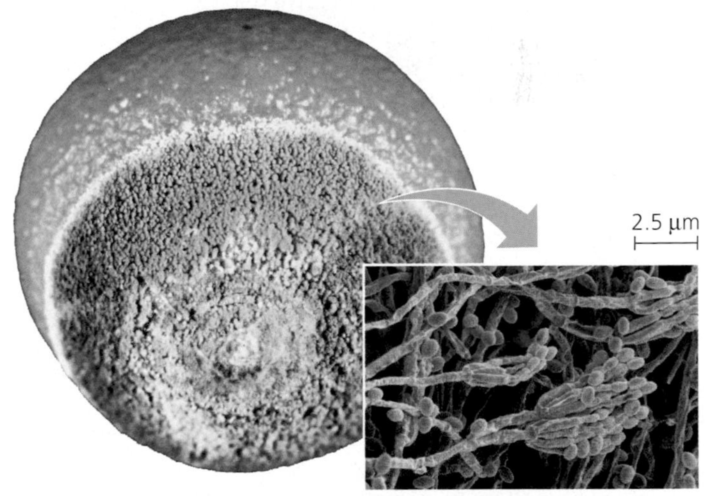

▲ **Figure 31.6 *Penicillium*, a mold commonly encountered as a saprobe on food.** The clusters of bead-like structures in the colorized SEM are conidia, involved in asexual reproduction.

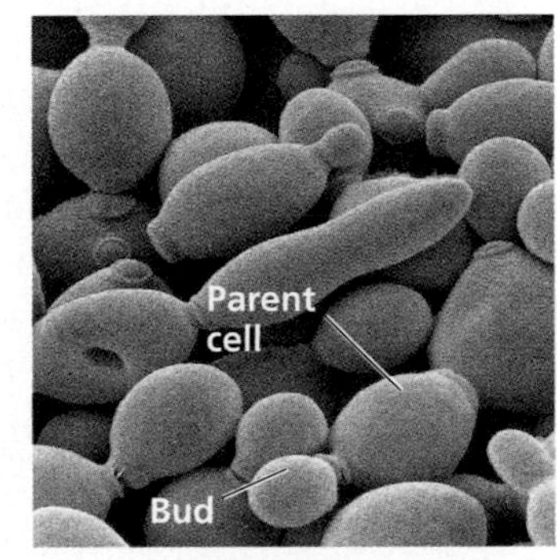

► **Figure 31.7 The yeast *Saccharomyces cerevisiae* in various stages of budding (SEM).**

Many molds and yeasts have no known sexual stage. Biologists who study fungi, or mycologists, have traditionally called such fungi **deuteromycetes** (the suffix *-mycete* means "fungus") or **imperfect fungi** (from the botanical use of the term *perfect* to refer to the sexual stages of life cycles). Whenever a sexual stage of a so-called deuteromycete is discovered, the species is classified in a particular phylum, depending on the type of sexual structures it forms. In addition to searching for little-known sexual stages of such unassigned fungi, mycologists also can now avail themselves of genetic techniques to identify taxonomic status.

Concept Check 31.2

1. In terms of haploidy versus diploidy, how do the life cycles of humans and fungi differ?
2. Suppose that you sample the DNA of two mushrooms on opposite sides of your yard and find that they are identical. What are two hypotheses that could reasonably account for this result?

For suggested answers, see Appendix A.

Concept 31.3

Fungi descended from an aquatic, single-celled, flagellated protist

Data from both paleontology and molecular systematics offer insights into the early evolution of fungi. Systematists now recognize Fungi and Animalia as sister kingdoms. In other words, fungi and animals are more closely related to each other than they are to plants or other eukaryotes.

The Origin of Fungi

Phylogenetic systematics suggests that fungi evolved from a flagellated ancestor. While the majority of fungi lack flagella, the lineages of fungi thought to be the earliest to diverge (the chytrids, discussed later in this chapter) do have flagella. Moreover, most of the protists that share a close common ancestor with animals and fungi also have flagella. These three groups of eukaryotes—the fungi, animals, and their protistan relatives—are called **opisthokonts**, members of the clade Opisthokonta. This name refers to the posterior (*opistho*) location of the flagellum in these organisms.

Phylogenetic evidence also suggests that the ancestor of fungi was unicellular. The fact that other evidence indicates that animals are more closely related to some unicellular opisthokonts than to fungi suggests that animals and fungi must have evolved multicellularity independently from different single-celled ancestors.

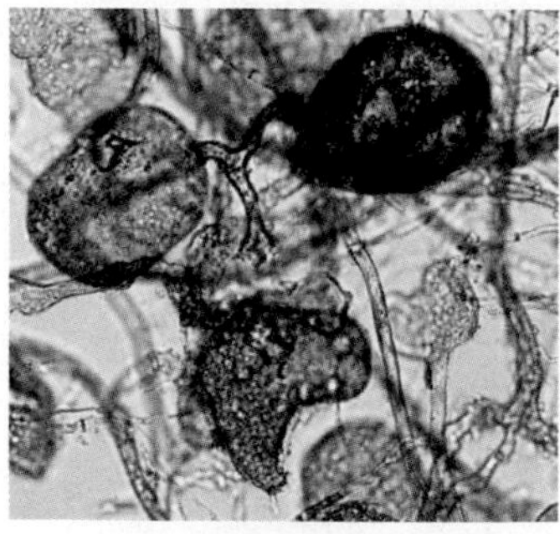

▶ **Figure 31.8 Fossil fungal hyphae and spores from the Ordovician period (about 460 million years ago; LM).**

Based on molecular clock analysis (see Chapter 25), scientists have estimated that the ancestors of animals and fungi diverged into separate lineages 1.5 billion years ago. However, the oldest undisputed fossils of fungi are only about 460 million years old (**Figure 31.8**). One possible explanation for this discrepancy is that the microscopic ancestors of terrestrial fungi fossilized poorly.

The Move to Land

Much of the fungal diversity we observe today may have had its phylogenetic origin in an adaptive radiation when life began to colonize land. Fossils of the earliest known vascular plants from the late Silurian period contain evidence of mycorrhizae, the symbiotic relationships between plants and subterranean fungi we discussed earlier. Plants probably existed in this symbiotic relationship from the earliest periods of colonization of land.

Concept Check 31.3

1. Why are fungi classified as opisthokonts when most fungi lack flagella?
2. Explain the evolutionary significance of the presence of mycorrhizae in the earliest vascular plants.

For suggested answers, see Appendix A.

Concept 31.4

Fungi have radiated into a diverse set of lineages

The phylogeny of fungi is currently the subject of much research. In the past decade, molecular analysis has helped clarify the evolutionary relationships between fungal groups, although there are still areas of uncertainty. **Figure 31.9** presents a simplified version of a current hypothesis of fungal phylogeny. In this section, we will survey each of the major fungal groups identified in this phylogenetic tree.

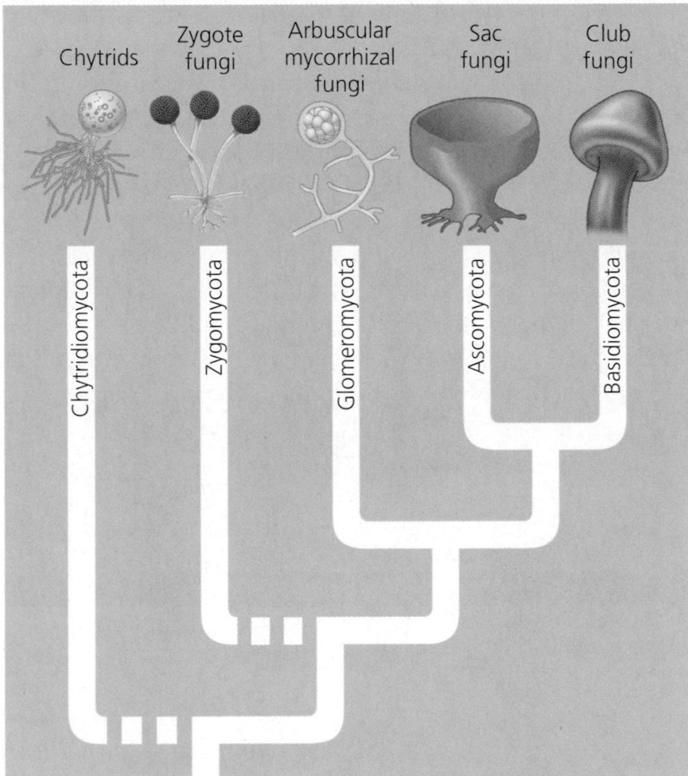

▲ **Figure 31.9 Phylogeny of fungi.** Most mycologists currently recognize five phyla of fungi. Molecular systematics is providing information on the relationships between these phyla; the dashed branches indicate groups thought to be paraphyletic (see Figure 31.11).

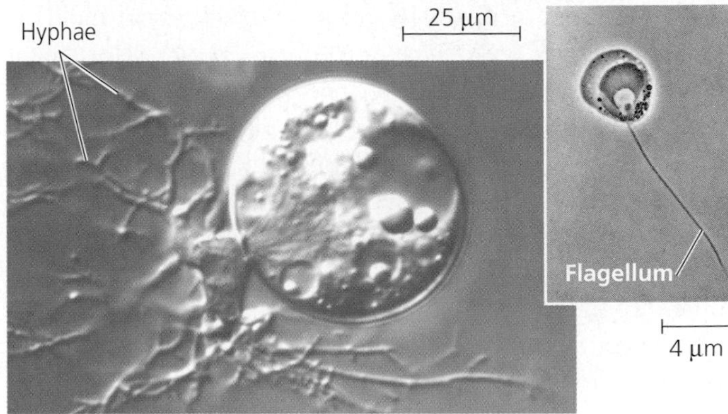

▲ **Figure 31.10 Chytrids.** The globular fruiting body of *Chytridium* sprouts branched hyphae (LM). Inset: Chytrids have a flagellated stage called a zoospore (TEM).

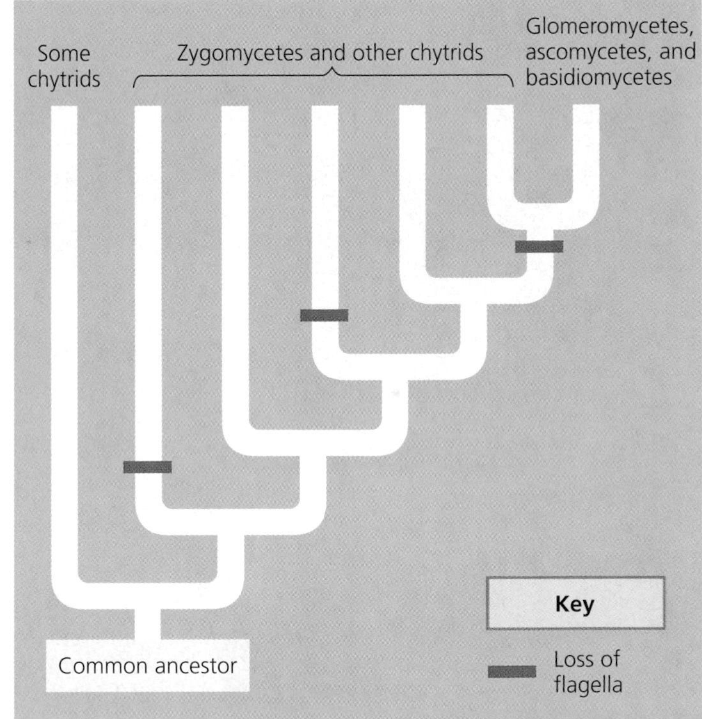

▲ **Figure 31.11 Multiple evolutionary losses of flagella.** Phylogenetic studies indicate that the common ancestor of fungi had flagella, which have been lost independently in several lineages.

Chytrids

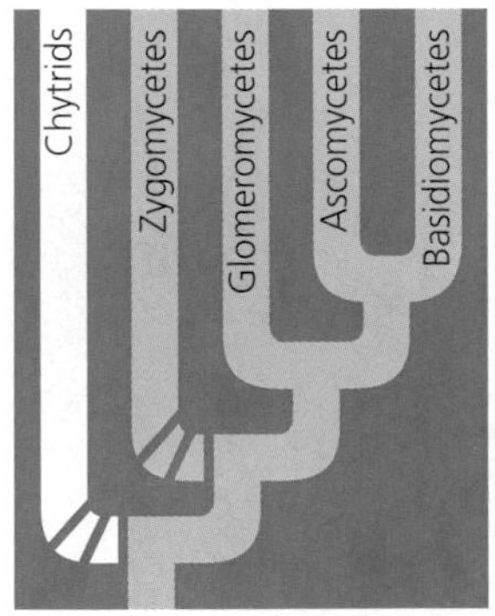

Fungi classified in the phylum Chytridiomycota, called **chytrids,** are ubiquitous in lakes and soil. Some are saprobes; others parasitize protists, plants, or animals.

Molecular evidence supports the hypothesis that chytrids diverged earliest in fungal evolution. Like other fungi, chytrids have cell walls made of chitin, and they also share certain key enzymes and metabolic pathways with other fungal groups. Some chytrids form colonies with hyphae, while others exist as single spherical cells. But chytrids are unique among fungi in having flagellated spores, called **zoospores (Figure 31.10)**.

Until recently, systematists thought that fungi lost flagella only once in their history, after chytrids had diverged from other lineages. However, molecular data indicate that some "chytrids" are actually more closely related to another fungal group, the zygomycetes. If this is true, then flagella were lost on more than one occasion during the evolution of fungi **(Figure 31.11)**. For this reason, many systematists consider the phyla Chytridiomycota and Zygomycota to be paraphyletic (compare Figures 31.9 and 31.11).

Zygomycetes

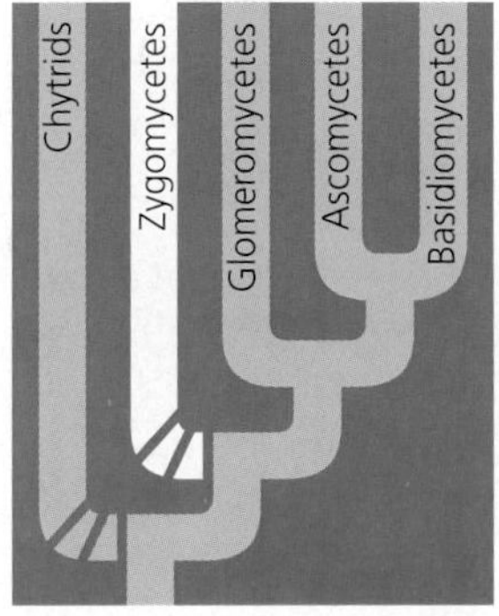

The approximately 1,000 known species of **zygomycetes** (fungi in the phylum Zygomycota) exhibit a considerable diversity of life histories. This phylum includes fast-growing molds responsible for rotting produce such as peaches, strawberries, and sweet potatoes during storage. Other zygomycetes live as parasites or as commensal (neutral) symbionts of animals.

The life cycle of *Rhizopus stolonifer* (black bread mold) is fairly typical of zygomycetes **(Figure 31.12)**. Horizontal hyphae spread out over the food, penetrate it, and absorb nutrients. The hyphae are coenocytic, with septa found only where reproductive cells are formed. In the asexual phase, bulbous black sporangia develop at the tips of upright hyphae. Within each sporangium, hundreds of haploid spores develop and are dispersed through the air. Spores that happen to land on moist food germinate, growing into new mycelia. Some zygomycetes, such as *Pilobolus*, can actually "aim" their sporangia toward conditions associated with good food sources **(Figure 31.13)**.

If environmental conditions deteriorate—for instance, if the mold consumes all its food—*Rhizopus* may reproduce sexually. The parents in a sexual union are mycelia of different mating types that possess different chemical markers, although they may appear identical. Plasmogamy produces a sturdy structure called a **zygosporangium**, in which karyogamy and then meiosis occur. Note that while a zygosporangium represents the zygote ($2n$) stage in the life cycle, it is not a zygote in the usual sense (that is, a cell with one diploid nucleus). Rather, a zygosporangium is a multinucleate structure, first heterokaryotic with many haploid nuclei from the two parents, then with many diploid nuclei after karyogamy.

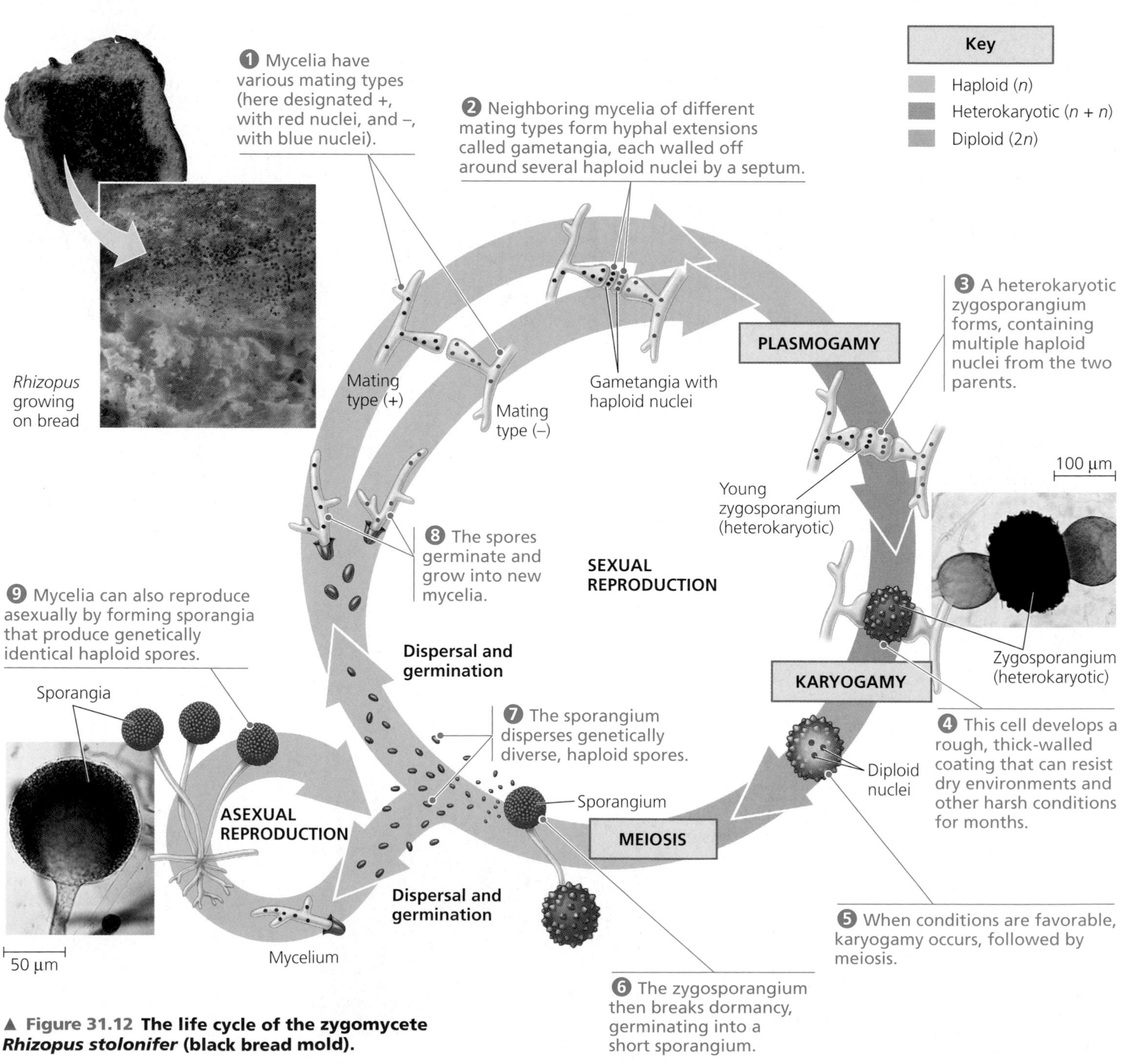
Key
Haploid (n)
Heterokaryotic (n + n)
Diploid (2n)
1 Mycelia have various mating types (here designated +, with red nuclei, and –, with blue nuclei).
2 Neighboring mycelia of different mating types form hyphal extensions called gametangia, each walled off around several haploid nuclei by a septum.
Rhizopus growing on bread
Mating type (+)
Mating type (–)
Gametangia with haploid nuclei
PLASMOGAMY
3 A heterokaryotic zygosporangium forms, containing multiple haploid nuclei from the two parents.
Young zygosporangium (heterokaryotic)
100 µm
8 The spores germinate and grow into new mycelia.
SEXUAL REPRODUCTION
9 Mycelia can also reproduce asexually by forming sporangia that produce genetically identical haploid spores.
Dispersal and germination
Zygosporangium (heterokaryotic)
KARYOGAMY
Sporangia
7 The sporangium disperses genetically diverse, haploid spores.
4 This cell develops a rough, thick-walled coating that can resist dry environments and other harsh conditions for months.
Diploid nuclei
Sporangium
ASEXUAL REPRODUCTION
MEIOSIS
Dispersal and germination
Mycelium
50 µm
5 When conditions are favorable, karyogamy occurs, followed by meiosis.
6 The zygosporangium then breaks dormancy, germinating into a short sporangium.

▲ **Figure 31.12 The life cycle of the zygomycete *Rhizopus stolonifer* (black bread mold).**

▲ **Figure 31.13 *Pilobolus* aiming its sporangia.** This zygomycete decomposes animal dung. The mycelium bends its spore-bearing hyphae toward bright light, where grass is likely to be growing. The fungus then shoots its sporangia like cannonballs as far as 2 m. Grazing animals such as cows ingest the fungi with the grass and then scatter the spores in feces.

Zygosporangia are resistant to freezing and drying and are metabolically inactive. When conditions improve, a zygosporangium undergoes meiosis, germinates into a sporangium, and releases genetically diverse haploid spores that may colonize a new substrate.

Microsporidia

Microsporidia are unicellular parasites of animals and protists **(Figure 31.14)**. They are often used in the biological control of insect pests. While microsporidia do not normally infect humans, they do pose a risk to people with AIDS and other immune-compromised conditions.

In many ways, microsporidia are unlike most other eukaryotes. They do not have conventional mitochondria, for example. As a result, microsporidia have been something of a taxonomic mystery, thought by some researchers to be an ancient, deep-branching lineage of eukaryotes. In recent years, however, it has become clear that microsporidia are not primitive eukaryotes, but rather are highly derived parasites. In 2002, Bryony Williams and colleagues at the Natural History Museum in London discovered that microsporidia actually have tiny organelles derived from mitochondria. Meanwhile, molecular comparisons have revealed that microsporidia are more closely related to fungi than to any other eukaryotes. A 2003 analysis provided some data suggesting that microsporidia should be classified as zygomycetes.

Microsporidia demonstrate the extraordinary changes that can occur as organisms adapt to a parasitic lifestyle. In their case, they lost almost all resemblance to their fungal relatives.

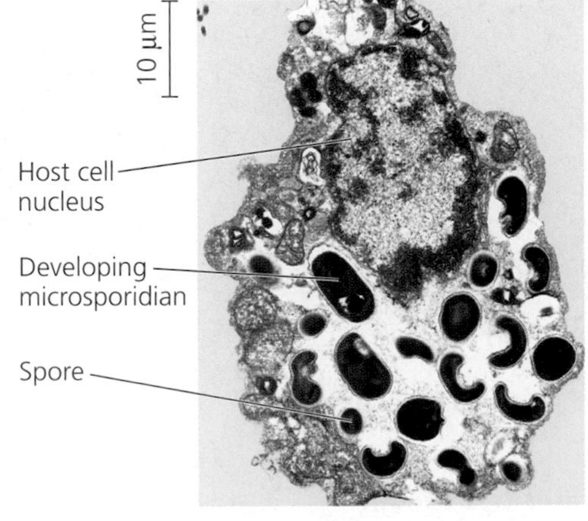

► **Figure 31.14 A eukaryotic cell infected by microsporidia.** A large vacuole inside this host eukaryotic cell contains spores and developing forms of the parasite *Encephalitozoon intestinalis* (TEM).

Glomeromycetes

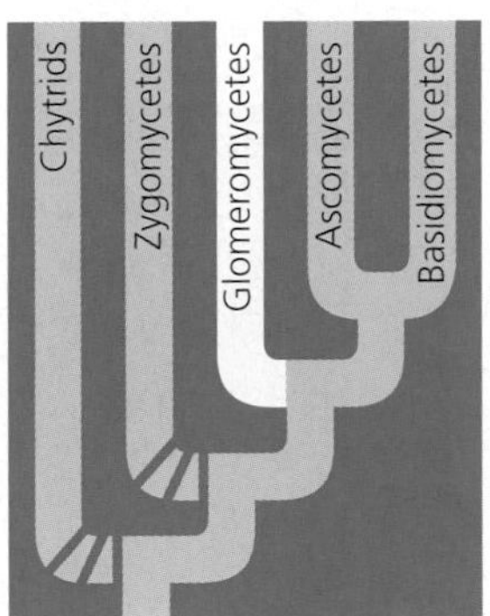

The **glomeromycetes**, fungi assigned to the phylum Glomeromycota, were formerly thought to be zygomycetes. But analyses of hundreds of fungal genomes indicate that glomeromycetes form a separate clade (monophyletic group). In spite of their small numbers—only 160 species have been identified to date—the glomeromycetes are an ecologically significant group. All glomeromycetes form a distinct type of endomycorrhizae called **arbuscular mycorrhizae (Figure 31.15)**. The tips of the hyphae that push into plant root cells branch into tiny treelike structures known as arbuscules. About 90% of all plants have such symbiotic partnerships with glomeromycetes.

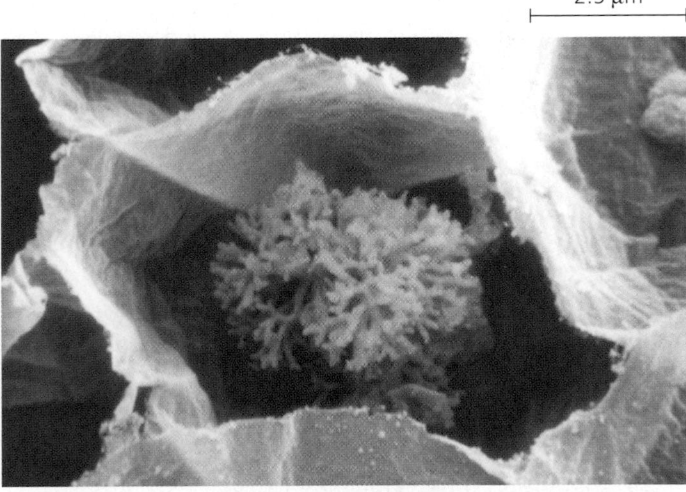

▲ **Figure 31.15 Arbuscular mycorrhizae.** Glomeromycetes form endomycorrhizae with plant roots, supplying minerals and other nutrients to the roots. This SEM depicts the branched hyphae—an arbuscule—of *Glomus mosseae* bulging into a root cell by pushing in the membrane (the root has been treated to remove the cytoplasm).

Ascomycetes

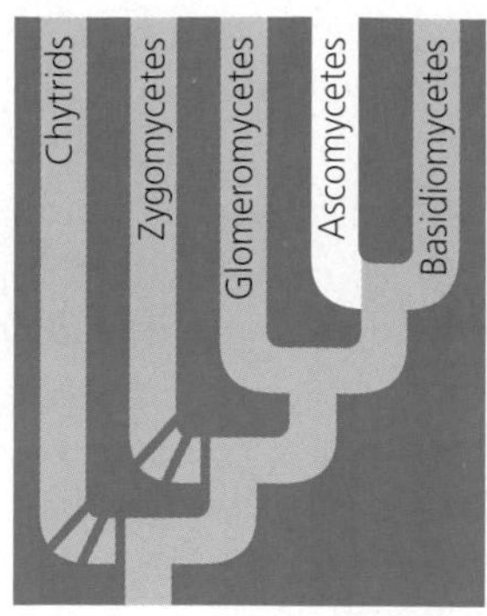

Mycologists have described more than 32,000 species of **ascomycetes** (fungi in the phylum Ascomycota) from a variety of marine, freshwater, and terrestrial habitats. The defining feature of ascomycetes is the production of sexual spores in saclike **asci** (singular, *ascus*); thus, they are commonly called **sac fungi.** Unlike zygomycetes, most ascomycetes bear their sexual stages in fruiting bodies, or **ascocarps**, which range in size from microscopic to macroscopic. The spore-forming asci are found in the ascocarps.

Ascomycetes vary in size and complexity from unicellular yeasts to elaborate cup fungi and morels **(Figure 31.16)**. They include some of the most devastating plant pathogens, which will be discussed later in the chapter. However, many ascomycetes are important saprobes, particularly of plant material. More than 40% of all ascomycete species live with green algae or cyanobacteria in symbiotic associations called lichens. Some ascomycetes form mycorrhizae with plants. Still others live between mesophyll cells in leaves, where they release toxic compounds that apparently help protect the plant from insects.

One of the best-studied ascomycetes is *Neurospora crassa,* a bread mold (see Figure 31.16d). Though *Neurospora* grows readily on bread, in the wild it is found growing on burned vegetation. As discussed in Chapter 17, biologists in the 1930s used *Neurospora* to formulate the one gene–one enzyme hypothesis. Today, this ascomycete continues to serve as a model organism; in 2003 its entire genome was published. With 10,000 genes, the genome of this tiny fungus is three-fourths the size of the *Drosophila* genome and one-third the size of the human genome! However, the *Neurospora* genome is relatively compact, having few of the stretches of noncoding DNA that occupy much space in the genomes of humans and many other eukaryotes. Biologists have found evidence suggesting that *Neurospora* has a genomic defense system that prevents "junk DNA" from accumulating to harmful levels.

Although the life cycles of various ascomycete groups differ in the details of their reproductive structures and processes, there are some common elements. Ascomycetes reproduce asexually by producing enormous numbers of asexual spores called **conidia (Figure 31.17).** Conidia are not formed inside sporangia, as are the asexual spores of most zygomycetes. Rather, they are produced externally at the tips of specialized hyphae called conidiophores, often in clusters or long chains, from which they may be dispersed by the wind.

Conidia may also be involved in sexual reproduction, fusing with hyphae from a mycelium of a different mating type, as occurs in *Neurospora* (see Figure 31.17). Fusion of two different mating types is followed by plasmogamy, resulting in a heterokaryotic nonseptated bulge, or ascogonium. The

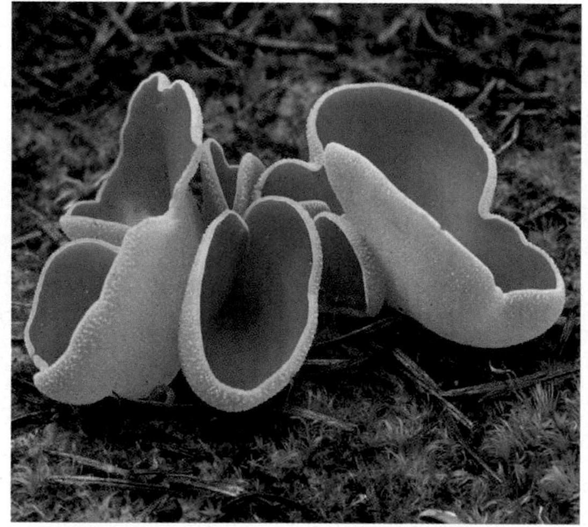

(a) The cup-shaped ascocarps (fruiting bodies) of *Aleuria aurantia* give this species its common name: orange peel fungus.

(b) The edible ascocarp of *Morchella esculenta*, the succulent morel, is often found under trees in orchards.

(c) *Tuber melanosporum* is a truffle, an ascocarp that grows underground and emits strong odors. These ascocarps have been dug up and the middle one sliced open.

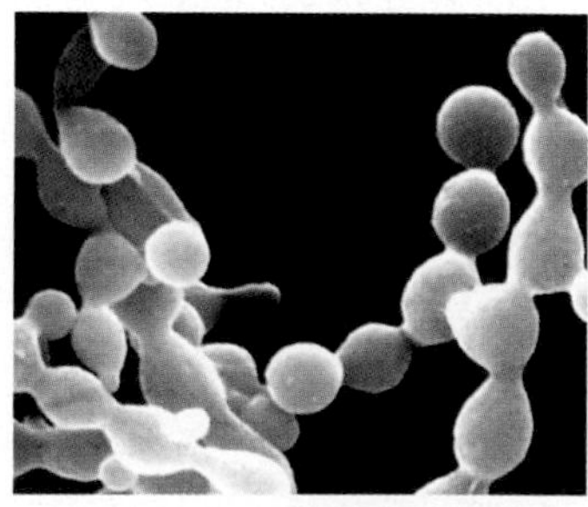

(d) *Neurospora crassa* feeds as a mold on bread and other food (SEM).

▲ **Figure 31.16 Ascomycetes (sac fungi).**

coenocytic ascogonium extends hyphae that are partitioned by septa, forming dikaryotic cells, each with two haploid nuclei representing the two parents. The cells at the tips of these dikaryotic hyphae develop into asci. Within an ascus, karyogamy combines the two parental genomes, and then meiosis forms four genetically different nuclei. This is followed by a mitotic division, forming eight ascospores. In many asci, the eight ascospores are lined up in a row in the order in which they formed from a single zygotic nucleus. The ascospores develop in and are eventually discharged from the ascocarp.

In contrast to the life cycle of zygomycetes, the extended dikaryotic stage of ascomycetes (and also basidiomycetes) provides increased opportunity for genetic recombination. For example, in some ascomycetes, certain dikaryotic cells fuse repeatedly, recombining genomes and resulting in a multitude of genetically different offspring from one mating event.

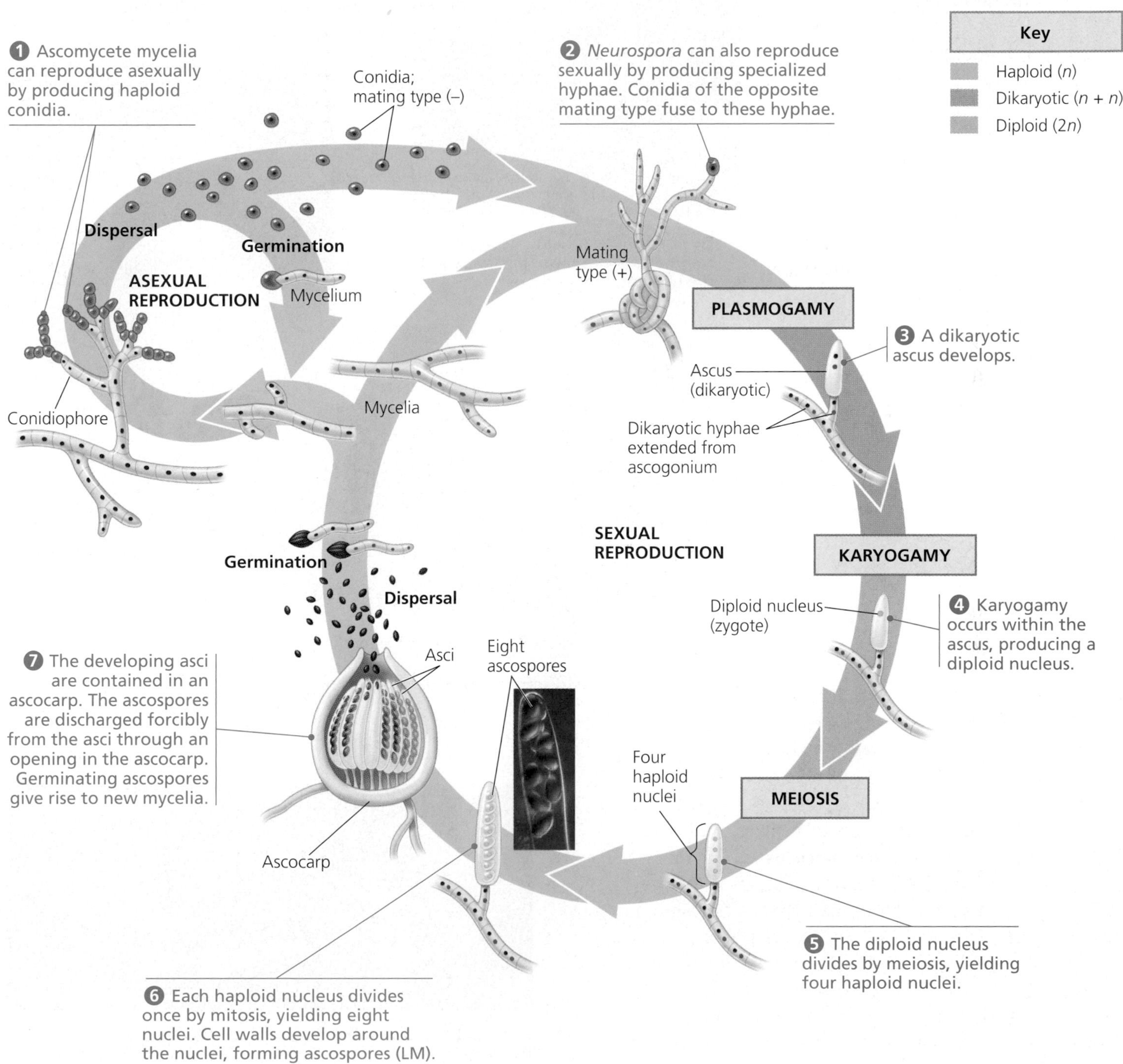

Key
Haploid (n)
Dikaryotic (n + n)
Diploid (2n)
1 Ascomycete mycelia can reproduce asexually by producing haploid conidia.
Conidia; mating type (–)
2 *Neurospora* can also reproduce sexually by producing specialized hyphae. Conidia of the opposite mating type fuse to these hyphae.
Dispersal
Germination
ASEXUAL REPRODUCTION
Mycelium
Conidiophore
Mycelia
Mating type (+)
PLASMOGAMY
3 A dikaryotic ascus develops.
Ascus (dikaryotic)
Dikaryotic hyphae extended from ascogonium
SEXUAL REPRODUCTION
KARYOGAMY
Diploid nucleus (zygote)
4 Karyogamy occurs within the ascus, producing a diploid nucleus.
Germination
Dispersal
Asci
Eight ascospores
7 The developing asci are contained in an ascocarp. The ascospores are discharged forcibly from the asci through an opening in the ascocarp. Germinating ascospores give rise to new mycelia.
Ascocarp
Four haploid nuclei
MEIOSIS
5 The diploid nucleus divides by meiosis, yielding four haploid nuclei.
6 Each haploid nucleus divides once by mitosis, yielding eight nuclei. Cell walls develop around the nuclei, forming ascospores (LM).

▲ **Figure 31.17 The life cycle of *Neurospora crassa*, an ascomycete.**

Basidiomycetes

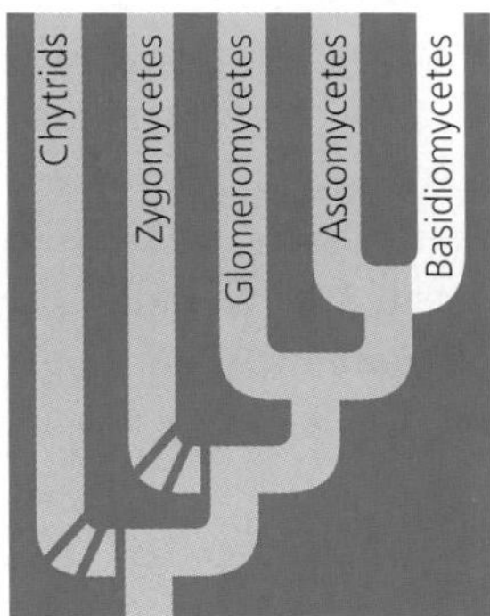

Approximately 30,000 fungi, including mushrooms and shelf fungi, are called **basidiomycetes** and are classified in the phylum Basidiomycota **(Figure 31.18)**. The phylum also includes molds, mutualists that form mycorrhizae, and two groups of particularly destructive plant parasites, the rusts and smuts. The name of the phylum derives from the **basidium** (Latin for "little pedestal"), a cell in which a transient diploid stage occurs during the fungal life cycle. The club-like shape of the basidium also gives rise to the common name **club fungus.**

Basidiomycetes are important decomposers of wood and other plant material. Of all the fungi, the saprobic basidiomycetes are the best at decomposing the complex polymer lignin, an abundant component of wood. Many shelf fungi break down the wood of weak or damaged trees and continue to decompose the wood after the tree dies.

(a) Fly agaric (*Amanita muscaria*), a common species in conifer forests in the northern hemisphere

(b) Maiden veil fungus (*Dictyphora*), a fungus with an odor like rotting meat

(c) Shelf fungi, important decomposers of wood

(d) Puffballs emitting spores

▲ **Figure 31.18 Basidiomycetes (club fungi).**

The life cycle of a basidiomycete usually includes a long-lived dikaryotic mycelium. As in ascomycetes, this extended dikaryotic stage provides opportunities for many genetic recombinations to occur, in effect multiplying the result of a single mating. Periodically, in response to environmental stimuli, the mycelium reproduces sexually by producing elaborate fruiting bodies called **basidiocarps** (see Figure 31.20). A mushroom is a familiar example of a basidiocarp.

By concentrating growth in the hyphae of mushrooms, a basidiomycete mycelium can quickly erect its fruiting structures in just a few hours; a mushroom pops up as it absorbs water and as cytoplasm streams in from the dikaryotic mycelium. By this process, a ring of mushrooms, popularly called a "fairy ring," may appear on a lawn literally overnight **(Figure 31.19)**. The mycelium below the fairy ring expands outward at a rate of about 30 cm per year, decomposing

▲ **Figure 31.19 A fairy ring.** According to legend, these mushrooms spring up where fairies have danced in a ring on a moonlit night. (The text provides a biological explanation of how fairy rings form.)

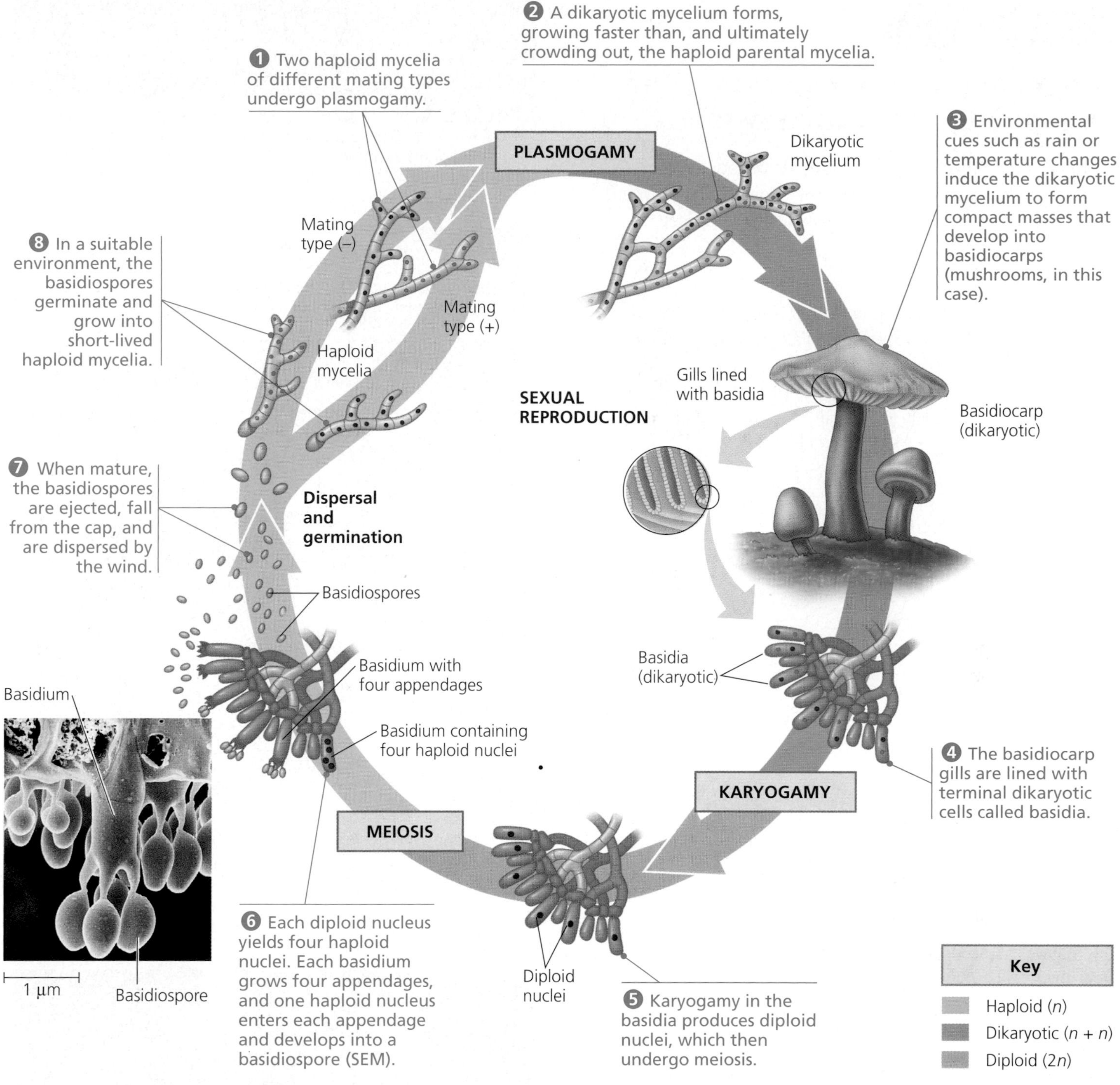

1 Two haploid mycelia of different mating types undergo plasmogamy.
2 A dikaryotic mycelium forms, growing faster than, and ultimately crowding out, the haploid parental mycelia.
PLASMOGAMY
Dikaryotic mycelium
3 Environmental cues such as rain or temperature changes induce the dikaryotic mycelium to form compact masses that develop into basidiocarps (mushrooms, in this case).
Mating type (–)
Mating type (+)
8 In a suitable environment, the basidiospores germinate and grow into short-lived haploid mycelia.
Haploid mycelia
SEXUAL REPRODUCTION
Gills lined with basidia
Basidiocarp (dikaryotic)
7 When mature, the basidiospores are ejected, fall from the cap, and are dispersed by the wind.
Dispersal and germination
Basidiospores
Basidium
Basidium with four appendages
Basidium containing four haploid nuclei
Basidia (dikaryotic)
4 The basidiocarp gills are lined with terminal dikaryotic cells called basidia.
KARYOGAMY
MEIOSIS
1 µm
Basidiospore
6 Each diploid nucleus yields four haploid nuclei. Each basidium grows four appendages, and one haploid nucleus enters each appendage and develops into a basidiospore (SEM).
Diploid nuclei
5 Karyogamy in the basidia produces diploid nuclei, which then undergo meiosis.
Key
Haploid (n)
Dikaryotic (n + n)
Diploid (2n)

▲ **Figure 31.20 The life cycle of a mushroom-forming basidiomycete.**

organic matter in the soil as it grows. Some giant fairy rings may be centuries old.

The numerous basidia in a basidiocarp are the sources of sexual spores called basidiospores **(Figure 31.20)**. The mushroom's cap supports and protects a large surface area of basidia on gills. A common, store-bought white mushroom has a gill surface area of about 200 cm^2 and may release a billion basidiospores, which drop from the bottom of the cap and are blown away. Asexual reproduction is much less common in basidiomycetes than in ascomycetes.

Concept Check 31.4

1. What feature of chytrids supports the hypothesis that they represent the most primitive fungal lineage?
2. Why are glomeromycetes so ecologically significant?
3. Give different examples of how form fits function in zygomycetes, ascomycetes, and basidiomycetes.

For suggested answers, see Appendix A.

Concept 31.5

Fungi have a powerful impact on ecosystems and human welfare

In our survey of fungal classification, we've touched on some of the ways fungi influence other organisms, including ourselves. We will now look more closely at these impacts.

Decomposers

Fungi are well adapted as decomposers of organic material, including the cellulose and lignin of plant cell walls. However, almost any carbon-containing substrate—including jet fuel and wall paint—can be consumed by at least some fungi. Fungi and bacteria are primarily responsible for keeping ecosystems stocked with the inorganic nutrients essential for plant growth. Without these decomposers, carbon, nitrogen, and other elements would become tied up in organic matter. Plants and the animals that eat them could not exist because elements taken from the soil would not be returned (see Chapter 54).

Symbionts

Fungi form symbiotic relationships with plants, algae, cyanobacteria, and animals. All of these relationships have profound ecological effects.

Mycorrhizae

Mycorrhizae are enormously important in natural ecosystems and agriculture. As you have read, almost all vascular plants have mycorrhizae and rely on their fungal partners for essential nutrients. It is relatively easy to demonstrate the significance of mycorrhizae by comparing the growth of plants with and without mycorrhizae **(Figure 31.21)**. Foresters commonly inoculate pine seedlings with ectomycorrhizal fungi to promote vigorous growth in the trees.

Figure 31.21

Inquiry Does having mycorrhizae benefit a plant?

EXPERIMENT Researchers grew soybean plants in soil treated with fungicide (poison that kills fungi) to prevent the formation of mycorrhizae in the experimental group. A control group was exposed to fungi that formed mycorrhizae in the soybean plants' roots.

RESULTS The soybean plant on the left is typical of the experimental group. Its stunted growth is probably due to a phosphorus deficiency. The taller, healthier plant on the right is typical of the control group and has mycorrhizae.

CONCLUSION These results indicate that the presence of mycorrhizae benefits a soybean plant and support the hypothesis that mycorrhizae enhance the plant's ability to take up phosphate and other needed minerals.

Fungus-Animal Symbiosis

Some fungi share their digestive services with animals, helping break down plant material in the guts of cattle and other grazing mammals. Many species of ants and termites take advantage of the digestive power of fungi by raising them in "farms." The insects scour tropical forests in search of leaves, which they carry back to their nests and feed to the fungi **(Figure 31.22)**. The fungi break down the leaves into a substance the insects can digest. In some tropical forests, the fungi have helped these insects become the major consumers of leaves.

The evolution of such farmer insects and that of their fungal "crops" have been tightly linked for well over 50 million years. The fungi have become so dependent on their caretakers that in many cases they can no longer survive without the insects.

▲ **Figure 31.22 Fungus-gardening insects.** These leaf-cutting ants depend on fungi to convert plant material to a form the insects can digest. The fungi, in turn, depend on the nutrients from the leaves the ants feed them.

(a) A fruticose (shrub-like) lichen

(b) A foliose (leaf-like) lichen

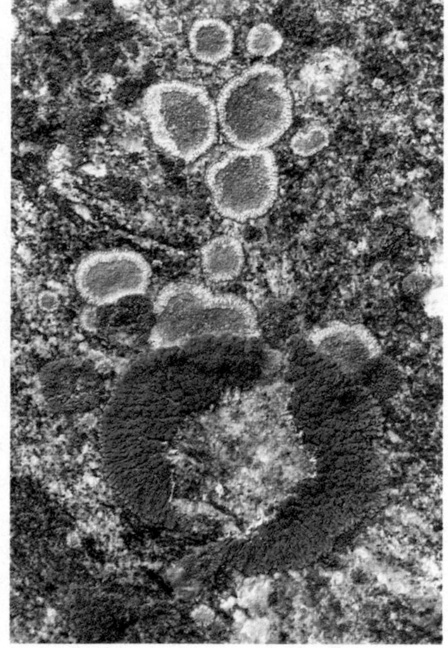
(c) Crustose (crust-like) lichens

▲ **Figure 31.23 Variation in lichen growth forms.**

Lichens

Lichens are a symbiotic association of millions of photosynthetic microorganisms held in a mass of fungal hyphae. They form a surface-hugging carpet that can be found growing on rocks, rotting logs, trees, and roofs in various shrub-like, leaf-like, or encrusting forms **(Figure 31.23)**. The photosynthetic partners are typically unicellular or filamentous green algae or cyanobacteria. The fungal component is most often an ascomycete, but several basidiomycete lichens are known. The fungus usually gives a lichen its overall shape and structure, and tissues formed by hyphae account for most of the lichen's mass. The algae or cyanobacteria generally occupy an inner layer below the lichen surface **(Figure 31.24)**. The merger of fungus and alga or cyanobacterium is so complete that lichens are actually given scientific names as though they were single organisms. More than 13,500 species have been described, accounting for a fifth of all known fungi.

In most lichens that have been studied, each partner provides something the other could not obtain on its own. The algae provide carbon compounds; the cyanobacteria also fix nitrogen (see Chapter 27) and provide organic nitrogen. The fungi provide their photosynthetic partners with a suitable environment for growth. The physical arrangement of hyphae allows for gas exchange, protects the photosynthetic partner, and retains water and minerals, most of which are absorbed either from airborne dust or from rain. The fungi also secrete acids, which aid in the uptake of minerals. Fungal pigments help shade the algae or cyanobacteria from intense sunlight. Some fungal compounds are toxic and prevent lichens from being eaten by animals.

The fungi of many lichens reproduce sexually by forming ascocarps or basidiocarps. Lichen algae reproduce independently of the fungus by asexual cell division. As might be expected of "dual organisms," asexual reproduction as a symbiotic unit also occurs commonly, either by fragmentation of the parental lichen or by the formation of **soredia,** small clusters of hyphae with embedded algae (see Figure 31.24).

Phylogenetic studies of lichen DNA have helped illuminate the evolution of this symbiosis. Molecular studies published in 2001 support the hypothesis that all living lichens can be traced to three original associations involving a fungus and a photosynthetic symbiont. The same studies also indicate that numerous free-living fungi descended from lichen-forming ancestors. *Penicillium,* the

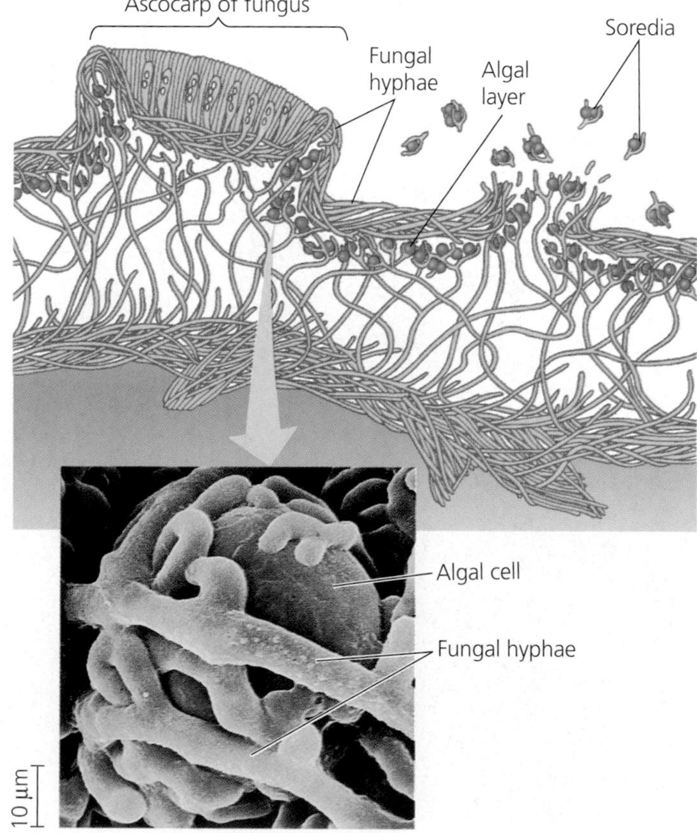

▲ **Figure 31.24 Anatomy of an ascomycete lichen (colorized SEM).**

free-living fungus that provides the antibiotic penicillin, for example, is thought to have descended from a lichen fungus.

Lichens are important pioneers on newly cleared rock and soil surfaces, such as burned forests and volcanic flows. They break down the surface by physically penetrating and chemically attacking it, and they trap windblown soil. Nitrogen-fixing lichens also add organic nitrogen to some ecosystems. These processes make it possible for a succession of plants to grow.

As tough as lichens are, many do not stand up very well to air pollution. Their passive mode of mineral uptake from rain and moist air makes them particularly sensitive to sulfur dioxide and other aerial poisons. The death of sensitive lichens and an increase in the number of hardier species in an area can be an early warning that air quality is deteriorating.

Pathogens

Of the 100,000 known species of fungi, about 30% make their living as parasites, mostly on or in plants **(Figure 31.25)**. For example, *Ophiostoma ulmi,* the ascomycete that causes Dutch elm disease, has drastically changed the landscape of the northeastern United States. Accidentally introduced to the United States on logs that were sent from Europe to help pay World War I debts, the fungus is carried from tree to tree by bark beetles. Another ascomycete, *Cryphonectria parasitica,* has killed 4 billion native American chestnut trees in the eastern United States. Fungi are also serious agricultural pests. Between 10% and 50% of the world's fruit harvest is lost each year to fungal attack. Grain crops suffer major losses each year from fungi such as the basidiomycete *Puccinia graminis,* which causes black stem rust on wheat.

Some of the fungi that attack food crops are toxic to humans. For example, certain species of the ascomycete mold *Aspergillus* contaminate improperly stored grain and peanuts by secreting carcinogenic compounds called aflatoxins. In another example, the ascomycete *Claviceps purpurea* forms purple structures called ergots on rye. If diseased rye is inadvertently milled into flour and then consumed, poisons from the ergots can cause ergotism, a condition characterized by gangrene, nervous spasms, burning sensations, hallucinations, and temporary insanity. An epidemic of ergotism around A.D. 944 killed more than 40,000 people in France. One of the compounds that has been isolated from ergots is lysergic acid, the raw material from which the hallucinogen LSD is made.

Animals are much less susceptible to parasitic fungi than are plants. Only about 50 species of fungi are known to parasitize humans and other animals, but these relatively few species do considerable damage. The general term for such a fungal infection is **mycosis.**

Skin mycoses include the disease ringworm, so named because it appears as circular red areas on the skin. The ascomycetes that cause ringworm can infect almost any skin surface. Most commonly, they grow on the feet, causing the intense itching and blisters known as athlete's foot. Though highly contagious, athlete's foot and other ringworm infections can be treated with fungicidal lotions and powders.

Systemic mycoses, by contrast, spread through the body and usually cause very serious illnesses. They are typically caused

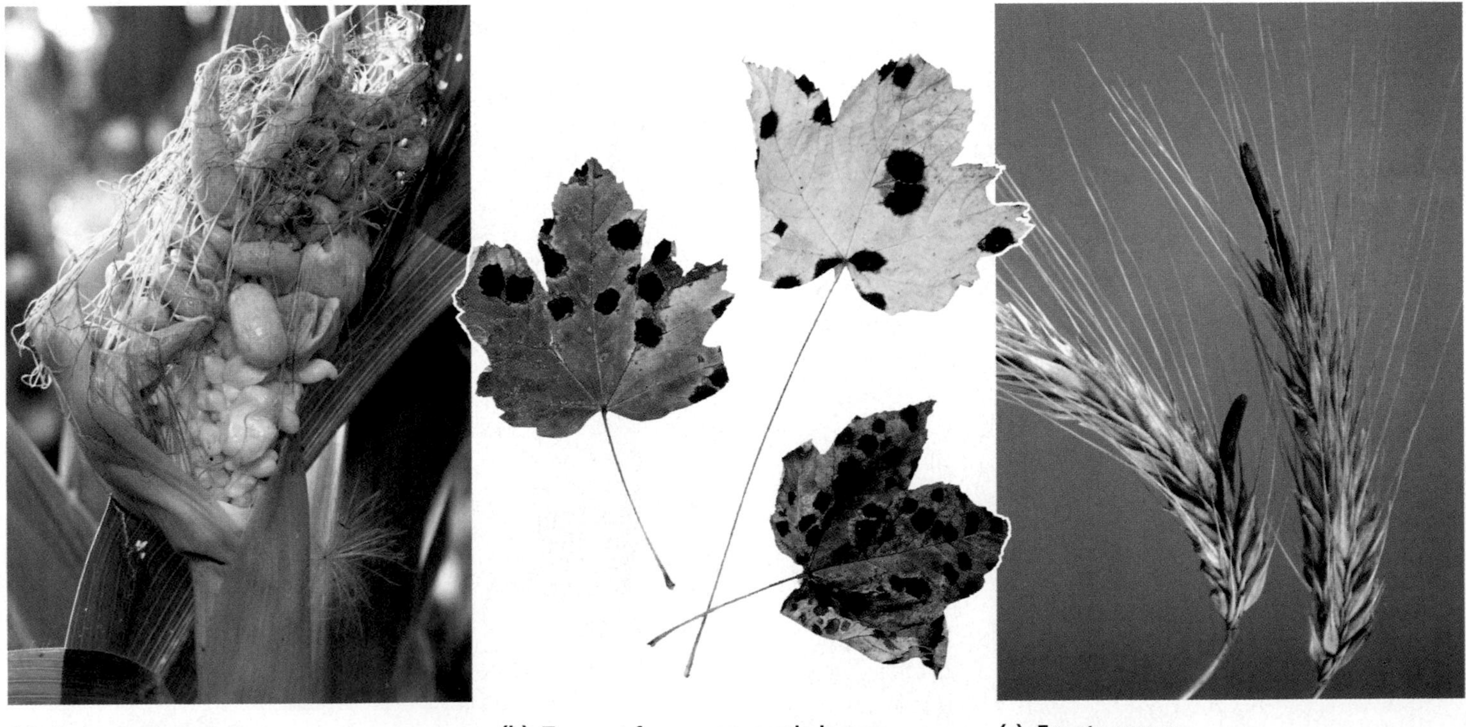

(a) Corn smut on corn **(b) Tar spot fungus on maple leaves** **(c) Ergots on rye**

▲ **Figure 31.25 Examples of fungal diseases of plants.**

by inhaled spores. Coccidioidomycosis is a systemic mycosis that produces tuberculosis-like symptoms in the lungs. It is so deadly that it is now considered a potential biological weapon.

Some mycoses are opportunistic, occurring only when a change in the body's microbiology, chemistry, or immunology allows fungi to grow unchecked. *Candida albicans,* for example, is one of the normal inhabitants of moist epithelia, such as the vaginal lining. Under certain circumstances, *Candida* can grow too rapidly and become pathogenic, leading to so-called "yeast infections." Many other opportunistic mycoses in humans have become more common in recent decades, due in part to AIDS, which compromises the immune system.

Over the past few years, the alleged dangers of black mold (*Stachybotrys chartarum*, an ascomycete) have been the subject of sensationalistic news reports. Black mold thrives in damp buildings and has been implicated in some studies as the cause of a wide variety of diseases. As a result, entire "sick buildings" have been abandoned. Yet there is no well-substantiated evidence linking the presence of *S. chartarum* to these diseases. Many other factors—including bacteria, human-made chemicals, and other fungi—may cause some of the symptoms attributed to black mold.

Practical Uses of Fungi

The dangers posed by fungi should not overshadow the immense benefits we derive from these remarkable eukaryotes. We depend on their ecological services as decomposers and recyclers of organic matter. Without mycorrhizae, our agriculture would be far less productive.

Mushrooms are a popular food, but they are not the only fungi we eat. The distinctive flavors of certain kinds of cheeses, including Roquefort and blue cheese, come from the fungi used to ripen them. The soft drink industry uses a species of *Aspergillus* to produce citric acid for colas. Morels and truffles, the edible fruiting bodies of various ascomycetes, are highly prized for their complex flavors (see Figure 31.16b and c). These fungi can fetch several hundred dollars a pound. Truffles release strong odors that attract mammals and insects, which feed on them and disperse their spores. In some cases, the odors mimic the sex attractants of certain mammals.

Humans have used yeasts to produce alcoholic beverages and raise bread for thousands of years. Under anaerobic conditions, yeasts ferment sugars to alcohol and CO_2, which leavens dough. Only relatively recently have the yeasts involved been separated into pure cultures for more controlled use. The yeast *Saccharomyces cerevisiae* is the most important of all cultured fungi (see Figure 31.7). It is available as many strains of baker's yeast and brewer's yeast.

Many fungi have great medical value as well. For example, a compound extracted from ergots is used to reduce high blood pressure and to stop maternal bleeding after childbirth. Some fungi produce antibiotics that are essential in treating bacterial infections. In fact, the first antibiotic discovered was penicillin, made by the ascomycete mold *Penicillium* **(Figure 31.26)**.

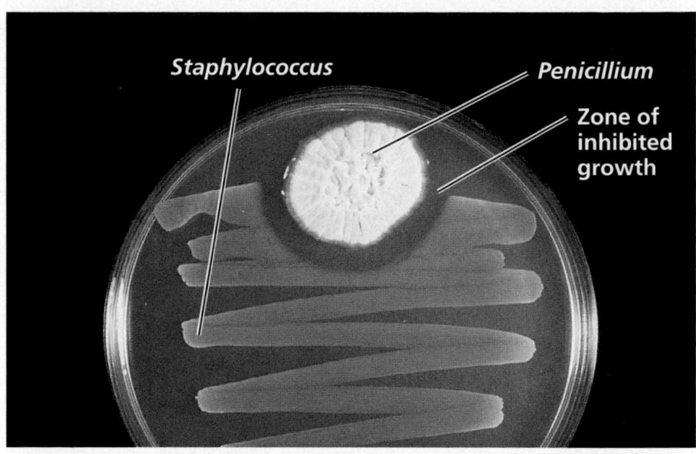

▲ **Figure 31.26 Fungal production of an antibiotic.** The mold *Penicillium* produces an antibiotic that inhibits the growth of *Staphylococcus* bacteria, resulting in the clear area between the mold and the bacteria.

Fungi also figure prominently in research in molecular biology and biotechnology. Researchers use *Saccharomyces* to study the molecular genetics of eukaryotes because its cells are easy to culture and manipulate (see Chapter 20). Scientists are gaining insight into the genes involved in human diseases such as Parkinson's disease and Huntington's disease by examining the interactions of homologous genes in *Saccharomyces*.

Genetically modified fungi hold much promise. Although bacteria such as *Escherichia coli* can produce some useful proteins, they cannot synthesize glycoproteins because they lack enzymes that can attach carbohydrates to proteins. Fungi, on the other hand, contain such enzymes. In 2003, scientists succeeded in engineering a strain of *S. cerevisiae* that produces human glycoproteins, such as insulin. Such fungus-produced glycoproteins could treat people with medical conditions that make them unable to produce these compounds. Meanwhile, other researchers are sequencing the genome of the wood-digesting basidiomycete *Phanerochaete chrysosporium*, also known as white rot. They hope to decipher the metabolic pathways by which white rot breaks down wood, with the goal of harnessing these pathways to produce paper pulp.

Having now completed our survey of the kingdom Fungi, we will turn in the remaining chapters of this unit to its sister kingdom, Animalia, to which we humans belong.

Concept Check 31.5

1. What are some of the benefits that algae in lichens can derive from their relationship with fungi?
2. What characteristics of pathogenic fungi result in their being efficiently transmitted?

For suggested answers, see Appendix A.

Chapter 31 Review

Go to www.campbellbiology.com or the student CD-ROM to explore Activities, Investigations, and other interactive study aids.

SUMMARY OF KEY CONCEPTS

Fungi are heterotrophs that feed by absorption

- **Nutrition and Fungal Lifestyles (pp. 608–609)** All fungi are heterotrophs (including decomposers and symbionts) that acquire nutrients by absorption. They secrete enzymes that break down complex molecules in food to smaller molecules that can be absorbed.
  Activity *Fungal Reproduction and Nutrition*

- **Body Structure (pp. 609–610)** Fungi consist of mycelia, networks of branched hyphae adapted for absorption. Most fungi have cell walls made of chitin. Some fungi have hyphae partitioned into cells by septa, with pores allowing cell-to-cell movement of materials. Coenocytic fungi lack septa. Mycorrhizal fungi have a symbiotic relationship with plants.

Fungi produce spores through sexual or asexual life cycles

- **Sexual Reproduction (pp. 610–611)** The sexual cycle involves cytoplasmic fusion (plasmogamy) and nuclear fusion (karyogamy), with an intervening heterokaryotic stage in which cells have haploid nuclei from two parents. The diploid phase resulting from karyogamy is short-lived and undergoes meiosis, producing haploid spores.

- **Asexual Reproduction (pp. 611–612)** Molds are rapidly growing, asexually reproducing fungi. Yeasts are unicellular fungi adapted to life in liquids such as plant saps. Fungi with no known sexual stage have traditionally been called deuteromycetes, but mycologists are using genetic techniques to assign many of these fungi to particular phyla.

Fungi descended from an aquatic, single-celled, flagellated protist

- **The Origin of Fungi (p. 612)** Molecular evidence supports the hypothesis that fungi and animals diverged from a common ancestor that was unicellular and bore a flagellum.

- **The Move to Land (p. 612)** Fungi were among the earliest colonizers of land, probably as symbionts with early land plants.

Fungi have radiated into a diverse set of lineages

- **Table 31.1**, in the next column, summarizes the fungal phyla and some of their distinguishing characteristics.

- **Chytrids (p. 613)** Chytrids are saprobic or parasitic fungi found in freshwater and terrestrial habitats. They are the only fungi that produce flagellated spores.

- **Zygomycetes (pp. 613–615)** Zygomycetes, such as black bread mold, are named for their sexually produced zygosporangia, which are heterokaryotic structures capable of persisting through unfavorable conditions. Unicellular parasites called microsporidia are now thought to be zygomycetes.

Table 31.1 Review of Fungal Phyla

Phylum	Distinguishing Feature
Chytridiomycota (chytrids)	Motile spores with flagella
Zygomycota	Resistant zygosporangium as sexual stage
Glomeromycota	Arbuscular mycorrhizae
Ascomycota (sac fungi)	Sexual spores borne internally in sacs called asci
Basidiomycota (club fungi)	Elaborate fruiting body called basidiocarp

- **Glomeromycetes (p. 615)** The vast majority of plants have symbiotic relationships with glomeromycetes in the form of arbuscular mycorrhizae.

- **Ascomycetes (pp. 616–617)** Ascomycetes (sac fungi) reproduce asexually by producing vast numbers of asexual spores called conidia. Sexual reproduction involves the formation of spores in sacs, or asci, at the ends of dikaryotic hyphae, which are usually contained in fruiting bodies called ascocarps.

- **Basidiomycetes (pp. 618–619)** Important decomposers of wood, the mycelia of basidomycetes (club fungi) can grow for years in the heterokaryotic stage. Sexual reproduction involves the formation of fruiting bodies called basidiocarps, which produce spores on club-shaped basidia at the ends of dikaryotic hyphae.
  Activity *Fungal Life Cycles*

Concept 31.5

Fungi have a powerful impact on ecosystems and human welfare

- **Decomposers (p. 620)** Fungi perform essential recycling of chemical elements between the living and nonliving world.
  Investigation *How Does the Fungus* Pilobolus *Succeed as a Decomposer?*

- **Symbionts (pp. 620–622)** Mycorrhizae increase plant productivity. Fungi enable animals such as cattle, ants, and termites to digest plant tissue. Lichens are highly integrated symbiotic associations of fungi and algae or cyanobacteria.

- ▶ **Pathogens (pp. 622–623)** About 30% of all known fungal species are parasites, mostly of plants. Some fungi also cause human diseases.
- ▶ **Practical Uses of Fungi (p. 623)** Humans eat many fungi and use others to make cheeses, alcoholic beverages, and bread. Antibiotics produced by fungi treat bacterial infections. Genetic research on fungi is leading to applications in biotechnology.

TESTING YOUR KNOWLEDGE

Self-Quiz

1. *All* fungi share which of the following characteristics?
   a. symbiotic
   b. heterotrophic
   c. flagellated
   d. pathogenic
   e. saprobic

2. Which feature seen in chytrids supports the hypothesis that they diverged earliest in fungal evolution?
   a. the absence of chitin within the cell wall
   b. coenocytic hyphae
   c. flagellated spores
   d. formation of resistant zygosporangia
   e. parasitic lifestyle

3. Which of the following cells or structures are associated with *asexual* reproduction in fungi?
   a. ascospores
   b. basidiospores
   c. conidiophores
   d. zygosporangia
   e. ascocarps

4. Which of the following is an example of an opportunistic pathogen that can cause a mycosis?
   a. *Claviceps purpurea,* which produces ergots on rye that can cause serious symptoms in humans if milled into flour
   b. *Ophiostoma ulmi,* which causes Dutch elm disease
   c. the ascomycetes that cause ringworm
   d. *Candida albicans,* which causes vaginal yeast infections
   e. *Penicillium,* which is grown in culture to produce antibiotics

5. The adaptive advantage associated with the filamentous nature of fungal mycelia is primarily related to
   a. the ability to form haustoria and parasitize other organisms.
   b. avoiding sexual reproduction until the environment changes.
   c. the potential to inhabit almost all terrestrial habitats.
   d. the increased probability of contact between different mating types.
   e. an extensive surface area well suited for absorptive nutrition.

6. Sporangia on upright hyphae that produce asexual spores are characteristic of
   a. ascomycetes.
   b. basidiomycetes.
   c. club fungi.
   d. zygomycetes.
   e. lichens.

7. Basidiomycetes differ from other fungi in that they
   a. have no known sexual stage.
   b. have long-lived dikaryotic mycelia.
   c. produce resistant sporangia that are initially heterokaryotic before karyogamy and meiosis occur.
   d. have members that are symbionts with algae in lichens.
   e. form eight spores that line up in a sac in the order they were formed in meiosis.

8. Which of the following is the best description of a mold?
   a. a deuteromycete, which has no known sexual stage
   b. a coenocytic, rapidly growing mycelium
   c. a mycorrhiza that envelops plant roots and reproduces without forming spores
   d. a unicellular fungus that grows rapidly in moist habitats
   e. the fast-growing mycelia of any asexually reproducing fungus

9. The photosynthetic symbiont of a lichen is often a(n)
   a. moss.
   b. green alga.
   c. brown alga.
   d. ascomycete.
   e. small vascular plant.

10. The closest relatives of fungi are probably
    a. animals.
    b. vascular plants.
    c. mosses.
    d. brown algae.
    e. slime molds.

Go to the website or CD-ROM for more quiz questions.

For Self-Quiz answers, see Appendix A.

Evolution Connection

The fungus-alga symbiosis that makes up a lichen is thought to have evolved several times independently in different fungal groups. However, lichens fall into three well-defined growth forms (see Figure 31.23). What research could you perform to test the following hypotheses? Hypothesis 1: Crustose, foliose, and fruticose lichens each represent a monophyletic group. Hypothesis 2: Each lichen growth form represents convergent evolution by taxonomically diverse fungi.

Scientific Inquiry

Lichens colonize gravestones, such as the one in this photo, almost as soon as the stones are placed and then continue to grow for decades or even centuries. Explain how you could calculate the growth rate of a particular lichen species by collecting data at an old cemetery.

Investigation ***How Does the Fungus*** **Pilobolus** ***Succeed as a Decomposer?***

Science, Technology, and Society

American chestnut trees once made up more than 25% of the hardwood forests of the eastern United States. These trees were wiped out by a fungus accidentally introduced on imported Asian chestnuts, which are not affected. More recently, a fungus has killed many eastern dogwood trees; some experts suspect that the parasite was accidentally introduced from somewhere else. Why are plants particularly vulnerable to fungi imported from other regions? What kinds of human activities might contribute to the spread of plant diseases? Do you think introductions of plant pathogens such as chestnut blight are more or less likely to occur in the future? Why?

32 An Introduction to Animal Diversity

▲ **Figure 32.1 An underwater glimpse of animal diversity on and around a coral reef.**

Key Concepts

32.1 Animals are multicellular, heterotrophic eukaryotes with tissues that develop from embryonic layers

32.2 The history of animals may span more than a billion years

32.3 Animals can be characterized by "body plans"

32.4 Leading hypotheses agree on major features of the animal phylogenetic tree

Overview

Welcome to Your Kingdom

Reading the last few chapters, you may have felt a little like a tourist among some rather unfamiliar organisms, such as slime molds, whisk ferns, and sac fungi. You probably are more at home with the topic introduced in this chapter—the animal kingdom, which of course includes yourself. But as **Figure 32.1** suggests, animal diversity extends far beyond humans or even the dogs, cats, birds, and other animals we humans regularly encounter. Biologists have identified 1.3 million living species of animals, and estimates of the total number of animal species run far higher, from 10 to 20 million to as many as 100 to 200 million. This vast diversity encompasses a spectacular range of morphological variation, from corals to cockroaches to crocodiles.

In this chapter, we embark on a tour of the animal kingdom, which will continue in the next two chapters. We will consider the characteristics that all animals share as well as those that distinguish various taxonomic groups. This information is central to understanding why animal phylogeny is currently one of the liveliest arenas of biological research and debate, as you will read later in the chapter.

Concept 32.1

Animals are multicellular, heterotrophic eukaryotes with tissues that develop from embryonic layers

Constructing a good definition of an animal is not straightforward, as there are exceptions to nearly every criterion for distinguishing animals from other life-forms. However, several characteristics of animals, when taken together, sufficiently define the group for our discussion.

Nutritional Mode

Animals differ from both plants and fungi in their mode of nutrition. Recall that plants are autotrophic eukaryotes capable of generating organic molecules through photosynthesis; fungi are heterotrophs that grow on or near their food, releasing exoenzymes that digest the food outside their bodies. Unlike plants, animals cannot construct all of their own organic molecules and so, in most cases, they ingest them—either by eating other living organisms or by eating nonliving organic material. But unlike fungi, most animals use enzymes to digest their food only after they have ingested it.

Cell Structure and Specialization

Animals are eukaryotes, and like plants and fungi (but unlike most protists), animals are multicellular. In contrast to plants and fungi, however, animals lack the structural support of cell walls. Instead, animal bodies are held together by structural proteins, the most abundant being collagen (see Figure 6.29).

In addition to collagen, which is found mainly in extracellular matrices, animals have three unique types of intercellular junctions—tight junctions, desmosomes, and gap junctions—that consist of other structural proteins (see Figure 6.31).

Among animal cells are two specialized forms not seen in other multicellular organisms: muscle cells and nerve cells. In most animals, these specialized cells are organized into muscle tissue and nervous tissue, respectively, and are responsible for movement and impulse conduction.

Reproduction and Development

Most animals reproduce sexually, and the diploid stage usually dominates the life cycle. In most species, a small, flagellated sperm fertilizes a larger, nonmotile egg, forming a diploid zygote. The zygote then undergoes **cleavage,** a succession of mitotic cell divisions without cell growth between division cycles. During the development of most animals, cleavage leads to the formation of a multicellular stage called a **blastula,** which in many animals takes the form of a hollow ball **(Figure 32.2)**. Following the blastula stage is the process of **gastrulation,** during which layers of embryonic tissues that will develop into adult body parts are produced. The resulting developmental stage is called a **gastrula**.

Some animals develop directly through transient stages of maturation into adults, but the life cycles of many animals also include at least one larval stage. A **larva** is a sexually immature form of an animal that is morphologically distinct from the adult stage, usually eats different food, and may even have a different habitat than the adult, as in the case of the aquatic tadpole (larva) of a terrestrial frog. Animal larvae eventually undergo **metamorphosis,** a resurgence of development that transforms the animal into an adult.

Despite the extraordinary diversity of morphology exhibited by adult animals, the underlying genetic network that controls animal development has been relatively conserved. All eukaryotes have genes that regulate the expression of other genes, and many of these regulatory genes contain common "modules" of DNA sequences called homeoboxes (see Chapter 21). Animals share a unique homeobox-containing family of genes, known as *Hox* genes, suggesting that this gene family evolved in the eukaryote lineage that gave rise to animals. *Hox* genes play important roles in the development of animal embryos, controlling the expression of dozens or even hundreds of other genes. *Hox* genes can thus control cell division and differentiation, producing different morphological features of animals.

The sponges, which represent the lineage of simplest extant (living) animals, have *Hox* genes that regulate the formation of water channels in the body wall, the primary feature of sponge morphology (see Chapter 33). In more complex animals, the *Hox* gene family underwent further duplications, yielding a more versatile "toolkit" for regulating development. In bilaterians (a grouping that includes vertebrates, insects, and most other animals), *Hox* genes regulate patterning of the anterior-posterior axis, as well as other aspects of development. The same conserved genetic network governs the development of both a fly and a human, despite their obvious differences and hundreds of millions of years of divergent evolution.

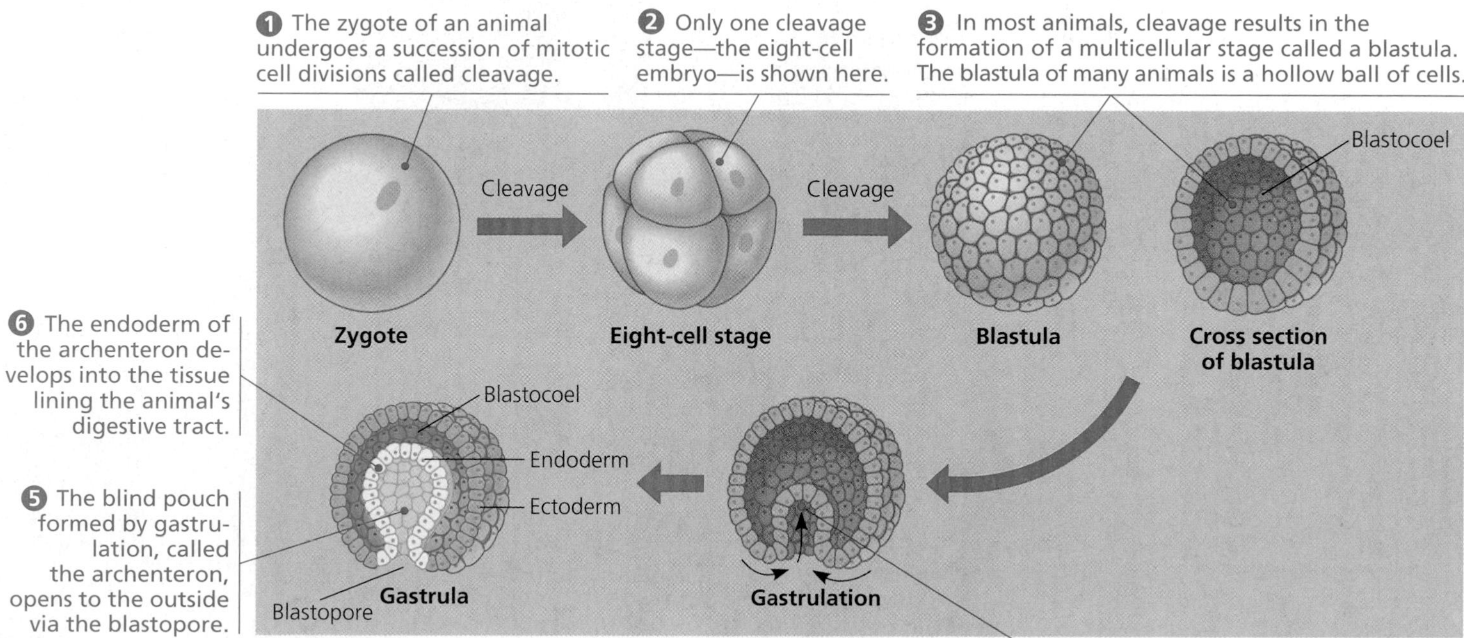

1 The zygote of an animal undergoes a succession of mitotic cell divisions called cleavage.
2 Only one cleavage stage—the eight-cell embryo—is shown here.
3 In most animals, cleavage results in the formation of a multicellular stage called a blastula. The blastula of many animals is a hollow ball of cells.
Cleavage
Cleavage
Blastocoel
Zygote
Eight-cell stage
Blastula
Cross section of blastula
6 The endoderm of the archenteron develops into the tissue lining the animal's digestive tract.
5 The blind pouch formed by gastrulation, called the archenteron, opens to the outside via the blastopore.
Blastocoel
Endoderm
Ectoderm
Blastopore
Gastrula
Gastrulation
4 Most animals also undergo gastrulation, a rearrangement of the embryo in which one end of the embryo folds inward, expands, and eventually fills the blastocoel, producing layers of embryonic tissues: the ectoderm (outer layer) and the endoderm (inner layer).

▲ **Figure 32.2 Early embryonic development in animals.**

Concept Check 32.1

1. Both plants and animals are multicellular eukaryotes. Identify four ways in which plants and animals differ.
2. Complex early developmental patterns such as the formation of a blastula and a gastrula are shared by diverse animals ranging from grasshoppers to clams to humans. What does this observation imply about the timing of the origins of these processes in animal evolution?

For suggested answers, see Appendix A.

Concept 32.2

The history of animals may span more than a billion years

The animal kingdom includes not only the great diversity of living species, but the even greater diversity of extinct ones as well. (Some paleontologists have estimated that 99% of all animal species are extinct.) Various studies suggest that animal diversification began more than a billion years ago. For example, some calculations based on molecular clocks estimate that the ancestors of animals diverged from the ancestors of fungi as far back as 1.5 billion years ago. Similar studies on the common ancestor of living animals suggest that it lived 1.2 billion–800 million years ago. This common ancestor may have resembled modern choanoflagellates **(Figure 32.3)**, protists that are the closest living relatives of animals, and was probably itself a colonial, flagellated protist **(Figure 32.4)**.

In this section, we will survey fossil evidence of animal evolution over four geologic eras.

Neoproterozoic Era (1 Billion–542 Million Years Ago)

Despite the molecular data indicating a much earlier origin of animals, the first generally accepted fossils of animals are only 575 million years old. These fossils are known collectively as the **Ediacaran fauna**, named for the Ediacara Hills of Australia, where they were first discovered **(Figure 32.5)**. Similar fossils have since been found on other continents. Some appear to be related to living cnidarians such as corals (see Chapter 33). Other fossils may represent soft-bodied molluscs, and numerous fossilized tunnels and tracks indicate the presence of several forms of worms.

In addition to these macroscopic fossils, Neoproterozoic rocks have also yielded microscopic signs of early animals. As we discussed in Chapter 26, 570-million-year-old embryos discovered in China clearly exhibit the basic structural organization of present-day animal embryos, though paleontologists are still uncertain which animal clade the embryos represent. Though older fossils of animals will likely be discovered in the future, the fossil record as it is known today strongly suggests

▼ **Figure 32.5 Ediacaran fossils.** Fossils dating from 575 million years ago include animals **(a)** with simple, radial forms and **(b)** with many body segments and legs.

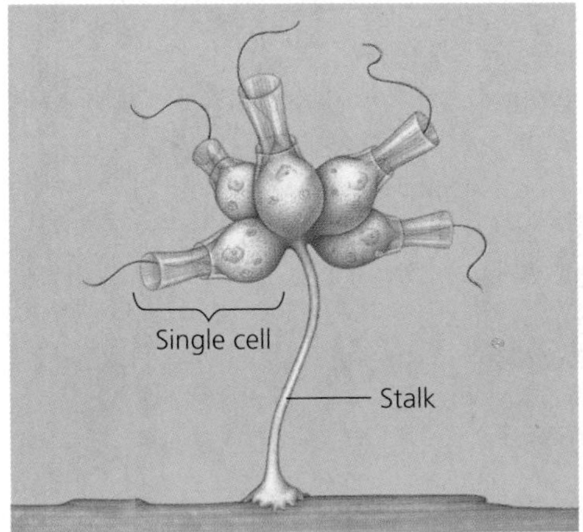

▲ **Figure 32.3 A choanoflagellate colony.** Such a colony is about 0.02 mm high.

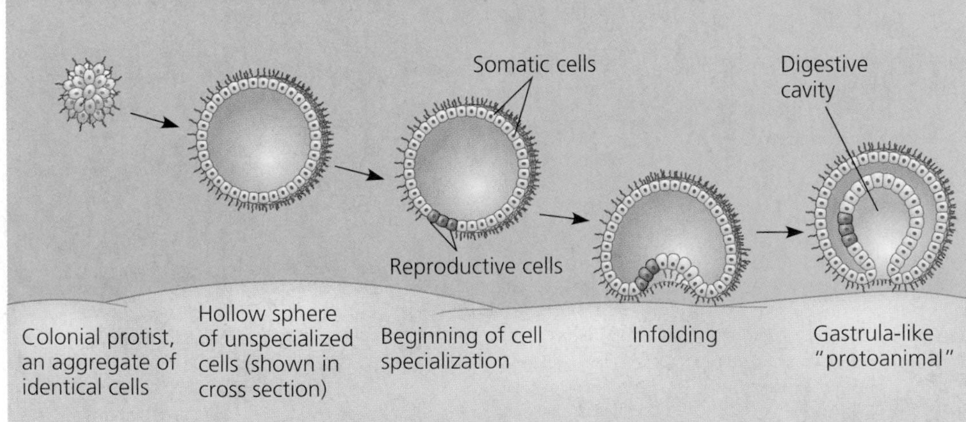

▲ **Figure 32.4 One hypothesis for the origin of animals from a flagellated protist.** (The arrows symbolize evolutionary time.)

that the end of the Neoproterozoic era was a time of rising levels of animal diversity.

Paleozoic Era (542–251 Million Years Ago)

Animal diversification appears to have accelerated dramatically between 542 and 525 million years ago, early in the Cambrian period of the Paleozoic era—a phenomenon often referred to as the **Cambrian explosion.** In strata formed before the Cambrian explosion, only a few animal phyla can be recognized. But in strata that are between 542 and 525 million years old, paleontologists have found the oldest fossils of about half of all extant animal phyla. Many of these distinctive fossils—which include the first animals with hard mineralized skeletons—look quite different from most living animals **(Figure 32.6)**. But for the most part, paleontologists have established that these various Cambrian fossils are members of extant animal phyla—or at least are close relatives.

There are several current hypotheses regarding the cause of the Cambrian explosion. Some evidence suggests that new predator-prey relationships that emerged in the Cambrian period generated diversity through natural selection. Predators acquired adaptations, such as new forms of locomotion that helped them catch prey, while prey species acquired new defenses, such as protective shells. Another hypothesis focuses on the rise in atmospheric oxygen that preceded the Cambrian explosion. With more oxygen available, the opportunity arose for the success of animals with higher metabolic rates and larger body size. A third hypothesis holds that the evolution of the *Hox* gene complex provided the developmental flexibility that resulted in variations in morphology. These hypotheses are not mutually exclusive, however; predator-prey relationships, atmospheric changes, and developmental flexibility may each have played a role.

The Cambrian period was followed by the Ordovician, Silurian, and Devonian periods, when animal diversity continued to increase, although punctuated by episodes of mass extinctions. Vertebrates (fishes) emerged as the top predators of the marine food web. By 460 million years ago, the "innovations" that emerged during the Cambrian period were making an impact on land. Arthropods began to adapt to terrestrial habitats, as indicated by the appearance of millipedes and centipedes. Fern galls—enlarged cavities that resident insects stimulate fern plants to form, providing protection for the insects—date back to at least 302 million years ago, suggesting that insects and plants were influencing each other's evolution by that time.

Vertebrates made the transition to land around 360 million years ago and diversified into numerous terrestrial lineages. Two of these survive today: amphibians (such as frogs and salamanders) and amniotes (such as reptiles and mammals). We will explore these groups, known collectively as tetrapods, in more detail in Chapter 34.

▲ **Figure 32.6 A Cambrian seascape.** This artist's reconstruction depicts a diverse array of organisms represented in fossils from the Burgess Shale site in British Columbia, Canada. The animals include *Pikaia* (swimming eel-like chordate), *Hallucigenia* (animal with toothpick-like spikes on seafloor), *Anomalocaris* (large animal with hooked claws), and *Marella* (arthropod swimming at left).

Mesozoic Era (251–65.5 Million Years Ago)

Few fundamentally new body plans emerged among animals during the Mesozoic era. But the animal phyla that had evolved during the Paleozoic now began to spread into new ecological niches. In the oceans, the first coral reefs formed, providing other animals with new marine habitats. Some reptiles returned to the water and succeeded as large aquatic predators. On land, modification of the tetrapod body plan included wings and other flight equipment in pterosaurs and birds. Large dinosaurs emerged, both as predators and herbivores. At the same time, the first mammals—tiny nocturnal insect-eaters—appeared on the scene.

Cenozoic Era (65.5 Million Years Ago to the Present)

As you read in Chapter 30, insects and flowering plants both underwent a dramatic diversification during the Cenozoic era. The beginning of this era followed mass extinctions of both terrestrial and marine animals. Among the groups of species that disappeared were the large, nonflying dinosaurs and the marine reptiles. The fossil record of the early Cenozoic documents the rise of large mammalian herbivores and carnivores as mammals began to exploit the vacated ecological niches. The global climate gradually cooled throughout the Cenozoic, triggering significant shifts in many animal lineages. Among primates, for example, some species in Africa adapted to the open woodlands and savannas that replaced the former dense forests. The ancestors of our own species were among those grassland apes.

Concept Check 32.2

1. Put the following milestones in animal evolution in chronological order from least recent to most recent: (a) origin of mammals, (b) earliest evidence of terrestrial arthropods, (c) Ediacaran fauna, (d) extinction of large, nonflying dinosaurs.
2. Explain how the relatively rapid Cambrian radiation of animal phyla could have been the product of causes both external and internal to organisms.

For suggested answers, see Appendix A.

Animals can be characterized by "body plans"

One way in which zoologists categorize the diversity of animals is according to general features of morphology and development. A group of animal species that share the same level of organizational complexity is known as a **grade.** Grades, it's important to remember, are not necessarily equivalent to *clades*. Consider the case of slugs: A class of animals called gastropods includes many species that lack shells and are referred to as slugs, along with many shelled species such as snails. You may have encountered terrestrial slugs in a garden; many other slug species live in aquatic habitats. But these species do not all descend from a common ancestor and thus do not form a monophyletic clade (see Figure 25.10). Rather, phylogenetic studies show that several gastropod lineages independently lost their shells and became "slugs." The slug grade, in other words, is polyphyletic.

The set of morphological and developmental traits that define a grade are generally integrated into a functional whole referred to as a **body plan.** Let's now explore some of the major features of animal body plans.

Symmetry

Animals can be categorized according to the symmetry of their bodies (or its absence). Most sponges, for example, lack symmetry altogether. Among the animals that do have symmetrical bodies, symmetry can take different forms. Some animals exhibit **radial symmetry,** the form found in a flowerpot **(Figure 32.7a)**. Sea anemones, for example, have a top (oral, or mouth) side and a bottom (aboral) side. But they have no head and rear end, and no left and right side.

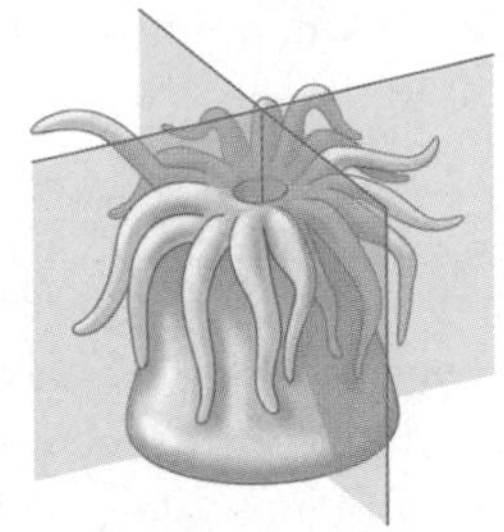
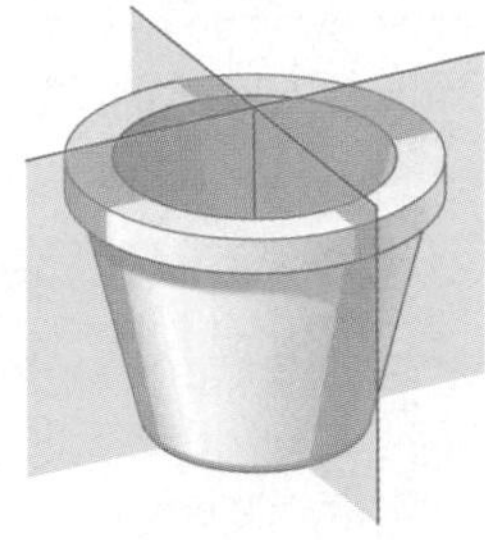

(a) Radial symmetry. The parts of a radial animal, such as a sea anemone (phylum Cnidaria), radiate from the center. Any imaginary slice through the central axis divides the animal into mirror images.

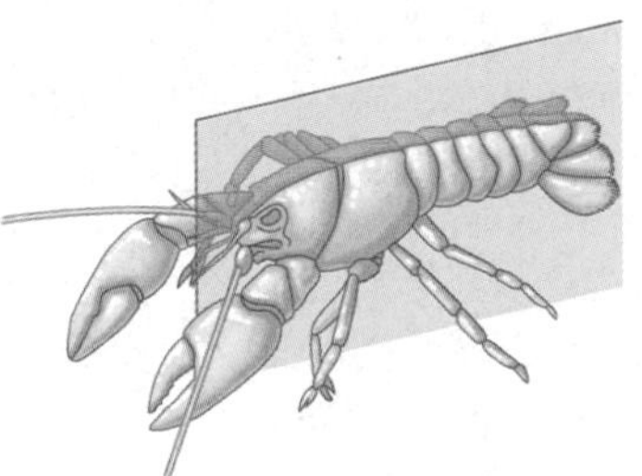
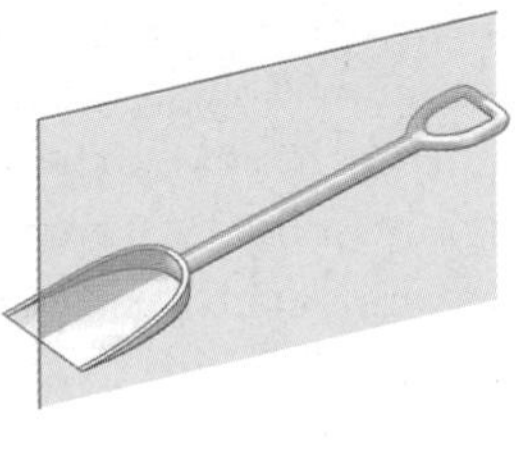

(b) Bilateral symmetry. A bilateral animal, such as a lobster (phylum Arthropoda), has a left side and a right side. Only one imaginary cut divides the animal into mirror-image halves.

▲ **Figure 32.7 Body symmetry.** The flowerpot and shovel are included to help you remember the radial-bilateral distinction.

The two-sided symmetry seen in a shovel is an example of **bilateral symmetry (Figure 32.7b)**. A bilateral animal has a **dorsal** (top) side and a **ventral** (bottom) side, as well as a left and right side and an **anterior** (head) end with a mouth and a **posterior** (tail) end. Many animals with a bilaterally symmetrical body plan (such as arthropods and mammals) have sensory equipment concentrated at the anterior end, along with a central nervous system ("brain") in the head—an evolutionary trend known as **cephalization** (from the Greek *kephale,* head).

The symmetry of an animal generally fits its lifestyle. Many radial animals are sessile (living attached to a substrate) or planktonic (drifting or weakly swimming, such as jellies, or jellyfishes). Their symmetry equips them to meet the environment equally well from all sides. In contrast, bilateral animals generally move actively from place to place. Their central nervous system enables them to coordinate complex movements involved in crawling, burrowing, flying, or swimming. These two fundamentally different kinds of symmetry probably arose very early in the history of animal life (see Figure 32.5).

Tissues

Animal body plans also vary according to the organization of the animal's tissues. True tissues are collections of specialized cells isolated from other tissues by membranous layers. Sponges lack true tissues. In all other animals, the embryo becomes layered through the process of gastrulation, as you read earlier in this chapter (see Figure 32.2). As development progresses, these concentric layers, called **germ layers**, form the various tissues and organs of the body. **Ectoderm**, the germ layer covering the surface of the embryo, gives rise to the outer covering of the animal and, in some phyla, to the central nervous system. **Endoderm**, the innermost germ layer, lines the developing digestive tube, or **archenteron**, and gives rise to the lining of the digestive tract (or cavity) and organs derived from it, such as the liver and lungs of vertebrates.

Animals that have only these two germ layers are said to be **diploblastic.** Diploblasts include the animals called cnidarians (jellies and corals, for example) as well as comb jellies (see Chapter 33). Other animals have a third germ layer, called the **mesoderm**, between the ectoderm and endoderm. These animals are said to be **triploblastic** (having three germ layers). In triploblasts, the mesoderm forms the muscles and most other organs between the digestive tract and the outer covering of the animal. Triploblasts include all bilaterally symmetrical animals, which range from flatworms to arthropods to vertebrates. (Although some diploblasts actually do have a third germ layer, it is not nearly as well developed as the mesoderm of animals considered to be triploblastic.)

Body Cavities

Some triploblastic animals possess a **body cavity**, a fluid-filled space separating the digestive tract from the outer body wall. This body cavity is also known as a **coelom** (from the Greek *koilos*, hollow). A so-called "true" coelom forms from tissue derived from mesoderm. The inner and outer layers of tissue that surround the cavity connect dorsally and ventrally and form structures called mesenteries that suspend the internal organs. Animals that possess a true coelom are known as **coelomates** (**Figure 32.8a**).

Some triploblastic animals have a body cavity formed from the blastocoel, rather than from mesoderm (**Figure 32.8b**). Such a cavity is called a "pseudocoelom" (from the Greek *pseudo*, false), and animals that have one are **pseudocoelomates.** Despite its name, a pseudocoelom is not false; it is a fully functional body cavity.

Finally, some triplobastic animals lack a coelom altogether (**Figure 32.8c**). They are known collectively as **acoelomates** (from the Greek *a*, without).

A body cavity has many functions. Its fluid cushions the suspended organs, helping to prevent internal injury. In soft-bodied coelomates, such as earthworms, the coelom contains noncompressible fluid that acts like a skeleton against which muscles can work. The cavity also enables the internal organs to grow and move independently of the outer body wall. If it were not for your coelom, for example, every beat of your heart or ripple of your intestine could warp your body's surface.

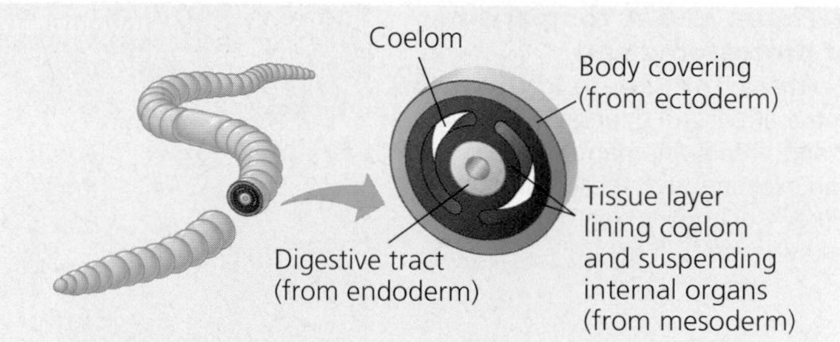

(a) Coelomate. Coelomates such as annelids have a true coelom, a body cavity completely lined by tissue derived from mesoderm.

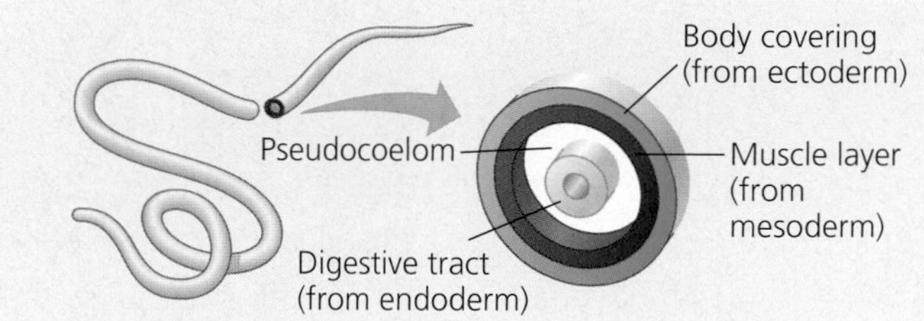

(b) Pseudocoelomate. Pseudocoelomates such as nematodes have a body cavity only partially lined by tissue derived from mesoderm.

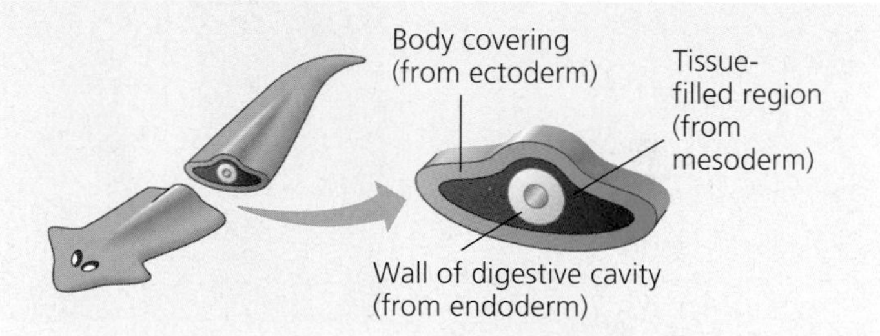

(c) Acoelomate. Acoelomates such as flatworms lack a body cavity between the digestive cavity and outer body wall.

▲ **Figure 32.8 Body plans of triploblastic animals.** The various organ systems of an animal develop from the three germ layers that form in the embryo. Blue represents tissue derived from ectoderm, red from mesoderm, and yellow from endoderm.

Current phylogenetic research suggests that true coeloms and pseudocoeloms have been gained or lost multiple times in the course of animal evolution. Thus, the terms *coelomates* and *pseudocoelomates* refer to grades, not clades.

Protostome and Deuterostome Development

Based on certain features of early development, many animals can be categorized as having one of two developmental modes: **protostome development** or **deuterostome development.** Three features often distinguish these modes.

Cleavage

A pattern in many animals with protostome development is **spiral cleavage**, in which the planes of cell division are diagonal to the vertical axis of the embryo. As seen in the

▶ **Figure 32.9 A comparison of protostome and deuterostome development.** These are useful general distinctions, though there are many variations and exceptions to these patterns. (Blue = ectoderm; red = mesoderm; yellow = endoderm.)

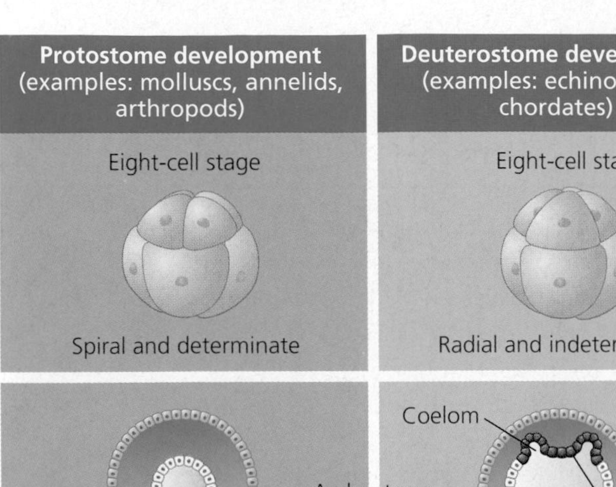

(a) Cleavage. In general, protostome development begins with spiral, determinate cleavage. Deuterostome development is characterized by radial, indeterminate cleavage.

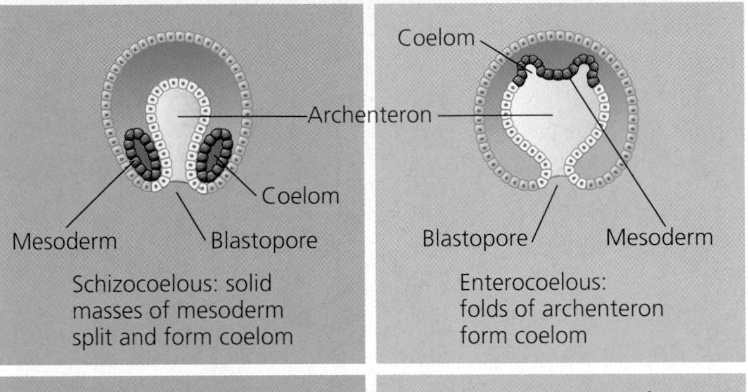

(b) Coelom formation. Coelom formation begins in the gastrula stage. In protostome development, the coelom forms from splits in the mesoderm (schizocoelous development). In deuterostome development, the coelom forms from mesodermal outpocketings of the archenteron (enterocoelous development).

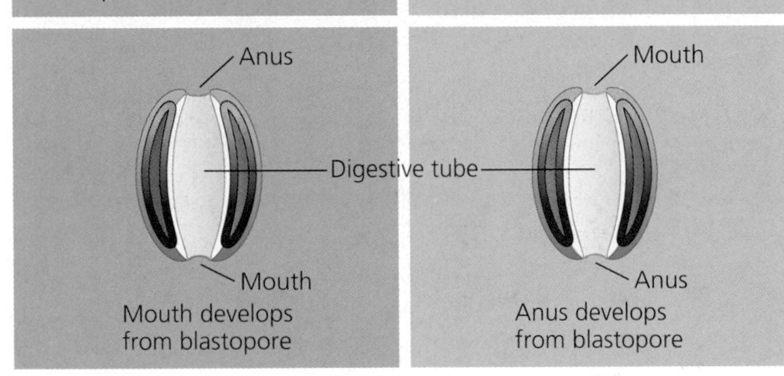

(c) Fate of the blastopore. In protostome development, the mouth forms from the blastopore. In deuterostome development, the mouth forms from a secondary opening.

eight-cell stage resulting from spiral cleavage, smaller cells lie in the grooves between larger, underlying cells **(Figure 32.9a)**. Furthermore, the so-called **determinate cleavage** of some animals with this development pattern rigidly casts ("determines") the developmental fate of each embryonic cell very early. A cell isolated at the four-cell stage from a snail, for example, forms an inviable embryo that lacks many parts.

In contrast to the spiral cleavage pattern, deuterostome development is predominantly characterized by **radial cleavage.** The cleavage planes are either parallel or perpendicular to the vertical axis of the egg; as seen in the eight-cell stage, the tiers of cells are aligned, one directly above the other. Most animals with deuterostome development are further characterized by **indeterminate cleavage,** meaning that each cell produced by early cleavage divisions retains the capacity to develop into a complete embryo. For example, if the cells of a sea star embryo are isolated at the four-cell stage, each will form a larva. It is the indeterminate cleavage of the human zygote that makes identical twins possible. This characteristic also explains the developmental versatility of the embryonic "stem cells" that may provide new ways to overcome a variety of diseases, including Type I diabetes, Parkinson's disease, and Alzheimer's disease (see Chapter 21).

Coelom Formation

Another difference between protostome and deuterostome development is apparent later in the process. In gastrulation, the developing digestive tube of an embryo initially forms as a blind pouch, the archenteron **(Figure 32.9b)**. As the archenteron forms in protostome development, initially solid masses of mesoderm split and form the coelomic cavity; this pattern is called **schizocoelous** development (from the Greek *schizein,* to split). In contrast, formation of the body cavity in deuterostome development is described as **enterocoelous:** The mesoderm buds from the wall of the archenteron, and its cavity becomes the coelom (see Figure 32.9b).

Fate of the Blastopore

The fundamental character that distinguishes the two developmental modes is the fate of the **blastopore,** the indentation that during gastrulation leads to the formation of the archenteron **(Figure 32.9c)**. After the archenteron develops, a second opening forms at the opposite end of the gastrula. Ultimately, the blastopore and this second opening become the two openings of the digestive tube (the mouth and the anus). In protostome development, the mouth generally develops from the first opening, the blastopore, and it is for this characteristic that the term *protostome* derives (from the Greek *protos*, first, and *stoma*, mouth). In deuterostome (from the Greek *deuteros*, second) development, the mouth is derived from the secondary opening, and the blastopore usually forms the anus.

Concept Check 32.3

1. Why is it important to distinguish *grade*-level characteristics from characteristics that unite *clades*?
2. Compare three features of the early development of a snail (a mollusc) and a human (a chordate).

For suggested answers, see Appendix A.

Concept 32.4

Leading hypotheses agree on major features of the animal phylogenetic tree

Zoologists currently recognize about 35 animal phyla. But the relationships between these phyla continue to be debated. Although many biology students must find it frustrating that the phylogenetic trees depicted in textbooks cannot be memorized as set-in-stone truths, the uncertainty inherent in these diagrams is a healthy reminder that science is a process of inquiry and as such is dynamic.

Researchers have long tested their hypotheses about animal phylogeny through morphological studies. In the mid-1990s, zoologists also began to study the molecular systematics of animals. Additional clues to animal phylogeny have come from new studies of lesser-known phyla, along with fossil analyses that can help clarify which traits are primitive in various animal groups and which are derived. Recall that science is partly distinguished from other ways of knowing because its ideas *can* be falsified through testing with experiments, observations, and new analytical methods.

Modern phylogenetic systematics is based on the identification of clades, which are monophyletic sets of taxa as defined by shared derived characters unique to those taxa and their common ancestor. (See Chapter 25 to review cladistic analysis and phylogenetic systematics.) Based on cladistic methods, a phylogenetic tree takes shape as a hierarchy of clades nested within larger clades—the finer branches and thicker branches of the tree, respectively. Defining the shared derived characters is the key to a particular hypothesis. A clade might be defined by key anatomical and embryological similarities that researchers conclude are homologous. More recently, comparisons of monomer sequences in proteins or nucleic acids across species have provided another data source for inferring common ancestry and defining clades. But whether the data are "traditional" morphological characters or "new" molecular sequences or a combination of both, the assumptions and inferences inherent in the resulting trees are the same.

To get a sense of the current debate in animal systematics, let's examine two current phylogenetic hypotheses. These hypotheses are presented on the next two pages as phylogenetic trees. The tree in **Figure 32.10** is based on systematic analyses of morphological characters. **Figure 32.11** is a simplified compilation of results from recent molecular studies.

Points of Agreement

The hypotheses agree on a number of major features of animal phylogeny. After reading each point, see how the statement is reflected by the phylogenetic trees in Figures 32.10 and 32.11.

1. **All animals share a common ancestor.** Both trees indicate that the animal kingdom is monophyletic, representing a clade called Metazoa. In other words, if we could trace all extant and extinct animal lineages back to their origin, the lineages would converge on a common ancestor.
2. **Sponges are basal animals.** Among the extant phyla, sponges (phylum Porifera) branch from the base of both animal trees. They exhibit a **parazoan** (meaning "beside the animals") grade of organization. Tissues evolved only after sponges diverged from other animals. Note that recent molecular analyses suggest that sponges are paraphyletic, as indicated in Figure 32.11.
3. **Eumetazoa is a clade of animals with true tissues.** All animals except for sponges belong to a clade of **eumetazoans** ("true animals"). The common ancestor of living eumetazoans acquired true tissues. Basal members of the eumetazoan clade include the phyla Cnidaria (which includes jellies) and Ctenophora (comb jellies). These basal eumetazoans are diploblastic and generally feature radial symmetry. As a result, they are often placed in the informal grade called Radiata.
4. **Most animal phyla belong to the clade Bilateria.** Bilateral symmetry is a shared derived character that helps to define a clade (*and* grade) containing the majority of animal phyla, called the **bilaterians.** The Cambrian explosion was primarily a rapid diversification of the bilaterians.
5. **Vertebrates and some other phyla belong to the clade Deuterostomia.** The name deuterostome refers not only to an animal development grade, but also to a clade that includes vertebrates. (Note, however, that the two hypotheses depicted disagree as to which other phyla are also deuterostomes.)

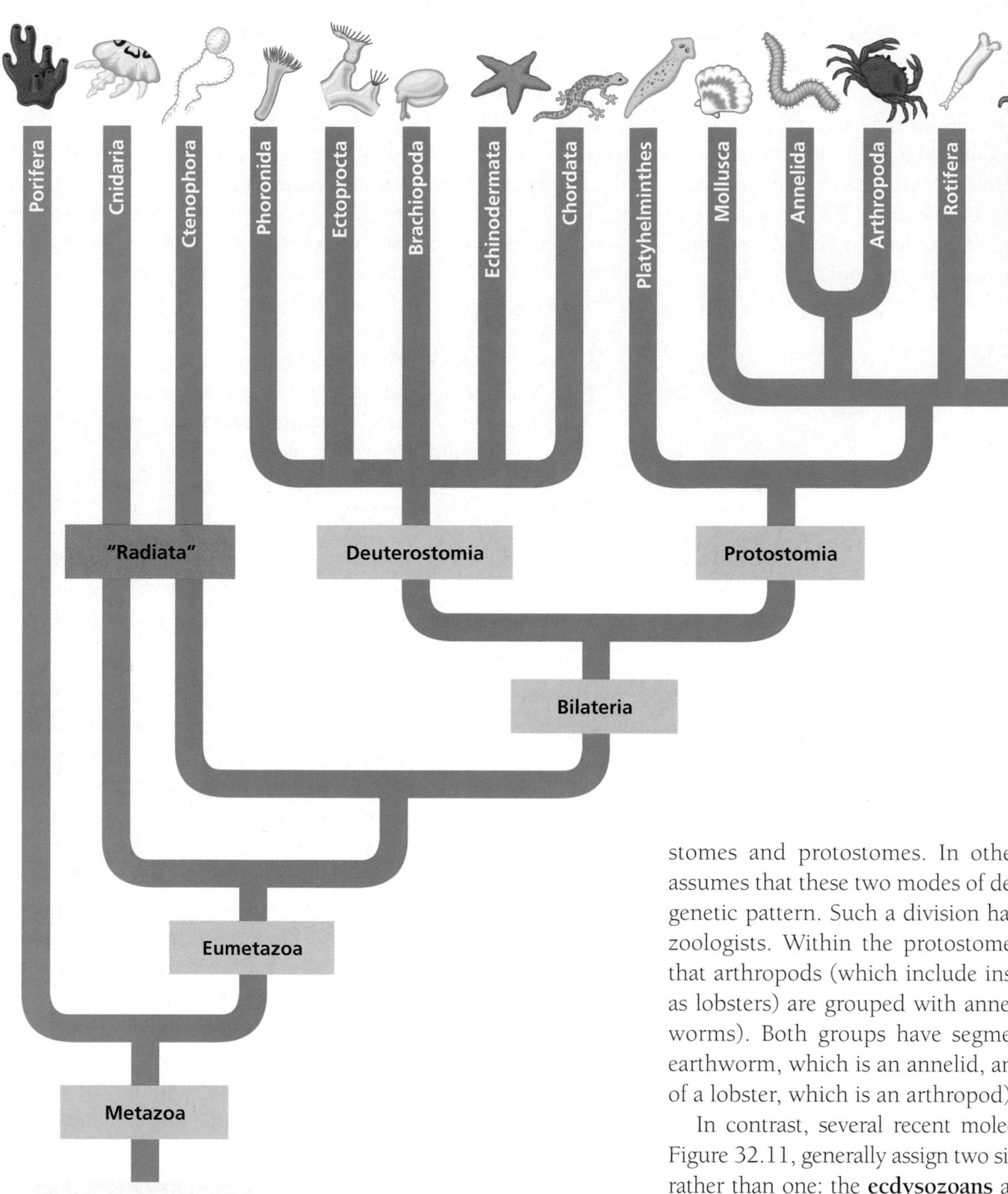

▲ **Figure 32.10 One hypothesis of animal phylogeny based mainly on morphological and developmental comparisons.** The bilaterians are divided into protostomes and deuterostomes.

Disagreement over the Bilaterians

While these two phylogenetic hypotheses agree on the overall structure of the animal tree, they also disagree on some significant points. The most important of these is the relationships among the bilaterians. The morphology-based tree in Figure 32.10 divides the bilaterians into two clades: deuterostomes and protostomes. In other words, this hypothesis assumes that these two modes of development reflect a phylogenetic pattern. Such a division has long been recognized by zoologists. Within the protostomes, Figure 32.10 indicates that arthropods (which include insects and crustaceans such as lobsters) are grouped with annelids (which include earthworms). Both groups have segmented bodies (think of an earthworm, which is an annelid, and the underside of the tail of a lobster, which is an arthropod).

In contrast, several recent molecular studies, as shown in Figure 32.11, generally assign two sister taxa to the protostomes rather than one: the **ecdysozoans** and the **lophotrochozoans.** The name *Ecdysozoa* refers to a characteristic shared by nematodes, arthropods, and some of the other ecdysozoan phyla (which are not included in our survey). These animals secrete external skeletons (exoskeletons); the stiff covering of a cricket is an example. As the animal grows, it molts, squirming out of its old exoskeleton and secreting a new, larger one. The shedding of the old exoskeleton is called *ecdysis,* the process for which the ecdysozoans are named **(Figure 32.12)**. Though named for this characteristic, the clade is actually defined mainly by molecular evidence supporting the common ancestry of its members. Furthermore, some taxa excluded from this clade by their molecular data do in fact molt, such as certain leeches.

The name *Lophotrochozoa* refers to two different structures observed in animals belonging to this clade. Some animals, such

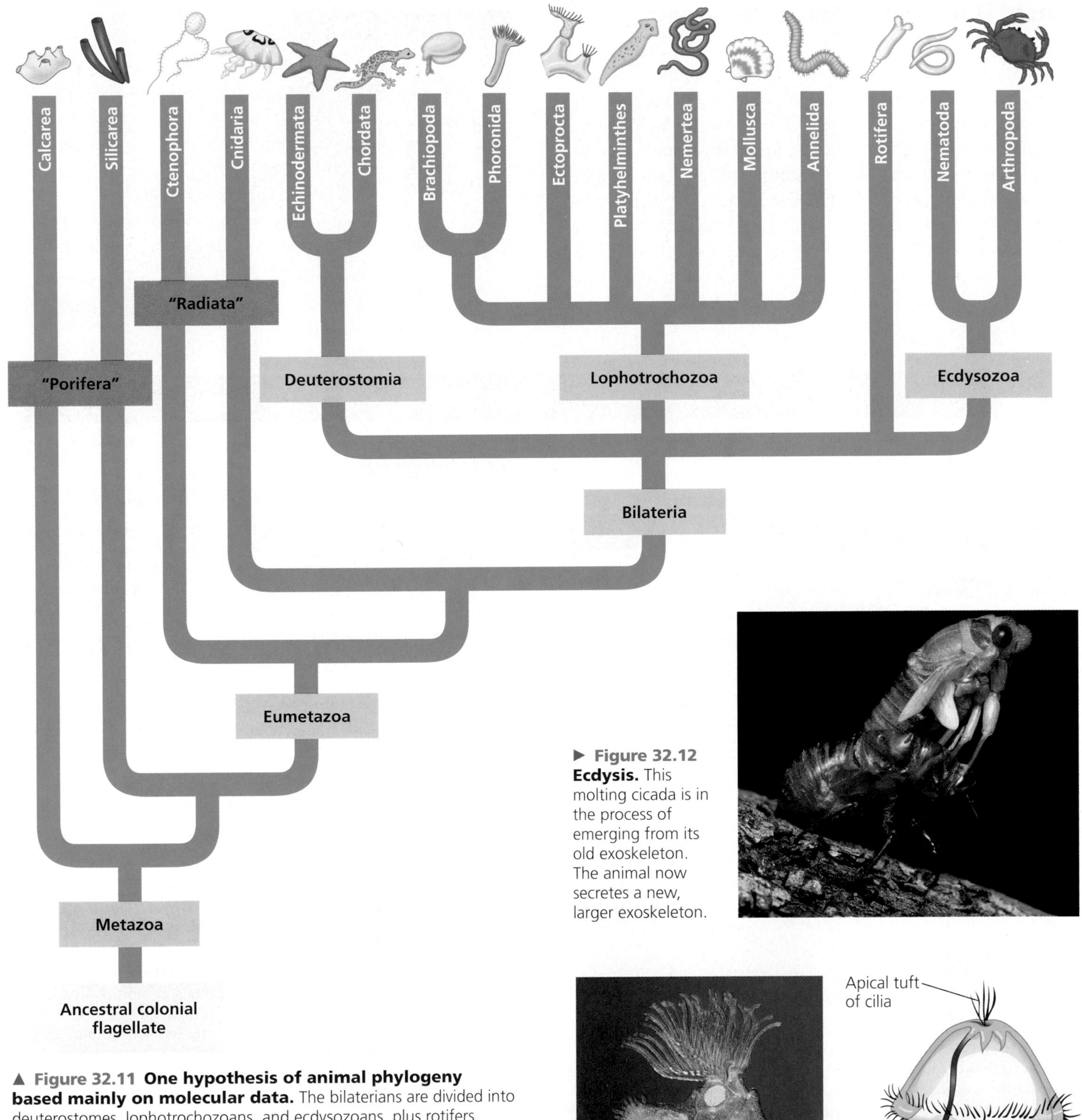

▲ **Figure 32.11 One hypothesis of animal phylogeny based mainly on molecular data.** The bilaterians are divided into deuterostomes, lophotrochozoans, and ecdysozoans, plus rotifers.

► **Figure 32.12 Ecdysis.** This molting cicada is in the process of emerging from its old exoskeleton. The animal now secretes a new, larger exoskeleton.

as ectoprocts, develop a structure called a **lophophore** (from the Greek *lophos*, crest, and *pherein*, to carry), a crown of ciliated tentacles that function in feeding **(Figure 32.13a)**. Other phyla, including annelids and molluscs, go through a distinctive larval stage called the **trochophore larva (Figure 32.13b)**—hence the name lophotrochozoan.

(a) An ectoproct, a lophophorate

(b) Structure of trochophore larva

▲ **Figure 32.13 Characteristics of lophotrochozoans.**

Future Directions in Animal Systematics

Like any area of scientific inquiry, animal systematics is constantly evolving. New sources of information emerge, offering the opportunity to test current hypotheses, which can be modified or replaced by new ones. Systematists are now conducting large-scale analyses of multiple genes across a wide sample of animal phyla. A better understanding of the relationships between these phyla will enable scientists to have a clearer picture of how the diversity of animal body plans arose. In Chapters 33 and 34, we will take a closer look at extant animal phyla and their evolutionary history.

Concept Check 32.4

1. What evidence indicates that cnidarians share a more recent common ancestor with other animals than with sponges?
2. How do the phylogenetic hypotheses presented in Figures 32.10 and 32.11 differ in structuring the major branches within the clade Bilateria?
3. Why is it valuable to collect multiple types of data to evaluate evolutionary relationships among animal phyla?

For suggested answers, see Appendix A.

Chapter 32 Review

Go to www.campbellbiology.com or the student CD-ROM to explore Activities, Investigations, and other interactive study aids.

SUMMARY OF KEY CONCEPTS

Concept 32.1

Animals are multicellular, heterotrophic eukaryotes with tissues that develop from embryonic layers

- **Nutritional Mode** (p. 626) Animals are heterotrophs that ingest their food.
- **Cell Structure and Specialization** (pp. 626–627) Animals are multicellular eukaryotes. Their cells lack cell walls; their bodies are held together instead by structural proteins such as collagen. Nervous tissue and muscle tissue are unique to animals.
- **Reproduction and Development** (pp. 627–628) Gastrulation follows the formation of the blastula, resulting in formation of embryonic tissue layers. All animals, and only animals, have *Hox* genes that regulate the development of body form. Although the *Hox* family of genes has been highly conserved, it can produce a wide diversity of animal morphology.

Concept 32.2

The history of animals may span more than a billion years

- **Neoproterozoic Era (1 Billion–542 Million Years Ago)** (pp. 628–629) Early members of the animal fossil record include the Ediacaran fauna.
- **Paleozoic Era (542–251 Million Years Ago)** (p. 629) The Cambrian explosion marks the earliest fossil appearance of many major groups of living animals.
- **Mesozoic Era (251–65.5 Million Years Ago)** (p. 629) During the Mesozoic era, dinosaurs were dominant terrestrial vertebrates. Coral reefs emerged, providing important marine habitats for other organisms.
- **Cenozoic Era (65.5 Million Years Ago to the Present)** (pp. 629–630) The present-day mammalian orders diversified during the Cenozoic, as did insects.

Concept 32.3

Animals can be characterized by "body plans"

- **Symmetry** (p. 630) Animals may lack any symmetry or have radial or bilateral symmetry. Bilaterally symmetrical animals have dorsal and ventral sides, as well as anterior and posterior ends. Many bilaterally symmetrical animals have undergone cephalization.
- **Tissues** (pp. 630–631) Animal embryos form germ layers and may be diploblastic (having two germ layers) or triploblastic (having three germ layers).
- **Body Cavities** (p. 631) In triploblastic animals, a body cavity may be present or absent. A body cavity can be a pseudocoelom (derived from blastocoel) or a true coelom (derived from mesoderm).
- **Protostome and Deuterostome Development** (pp. 631–633) These two modes of development differ in characteristics of cleavage, coelom formation, and blastopore fate.

Concept 32.4

Leading hypotheses agree on major features of the animal phylogenetic tree

- **Points of Agreement** (p. 633) The animal kingdom is monophyletic. The common ancestor of animals was probably a colonial flagellated protist. Sponges exhibit a parazoan grade of organization. "True animals" possess true tissues and include all animals except sponges. Cnidaria and Ctenophora are sometimes placed in the grade Radiata. Bilateral symmetry is a trait shared by a clade containing the other phyla of animals.
- **Disagreement over the Bilaterians** (pp. 634–635) Different analyses of animal systematics support different bilaterian clades.
- **Future Directions in Animal Systematics** (p. 636) Phylogenetic studies based on larger databases will likely provide further insights into animal evolutionary history.
  Activity *Animal Phylogenetic Tree*
  Investigation *How Do Molecular Data Fit Traditional Phylogenies?*

TESTING YOUR KNOWLEDGE

Self-Quiz

1. Among the characteristics unique to animals is
   a. gastrulation.
   b. multicellularity.
   c. sexual reproduction.
   d. flagellated sperm.
   e. heterotrophic nutrition.

2. Which of the following was the *least* likely factor causing the Cambrian explosion?
   a. the emergence of predator-prey relationships between animals
   b. the accumulation of diverse adaptations, such as shells and different modes of locomotion
   c. the movement of animals onto land
   d. the evolution of *Hox* genes that controlled development
   e. the accumulation of sufficient atmospheric oxygen to support the more active metabolism of mobile animals

3. Bilateral symmetry in the animal kingdom is best correlated with
   a. an ability to sense equally in all directions.
   b. the presence of a skeleton.
   c. motility and active predation and escape.
   d. development of a true coelom.
   e. adaptation to terrestrial environments.

4. Acoelomates are characterized by
   a. the absence of a brain.
   b. the absence of mesoderm.
   c. deuterostome development.
   d. a coelom that is not completely lined with mesoderm.
   e. a solid body without a cavity surrounding internal organs.

5. A direct consequence of indeterminate cleavage is
   a. formation of the archenteron.
   b. the ability of cells isolated from the early embryo to develop into viable individuals.
   c. the arrangement of cleavage planes perpendicular to the egg's vertical axis.
   d. the unpredictable formation of either a schizocoelous or enterocoelous body cavity.
   e. a mouth that forms in association with the blastopore.

6. The distinction between sponges and other animal phyla is based mainly on the absence versus the presence of
   a. a body cavity.
   b. a complete digestive tract.
   c. true tissues.
   d. a circulatory system.
   e. mesoderm.

7. Which of these is a point of conflict between the phylogenetic analyses presented in Figures 32.10 and 32.11?
   a. the monophyly of the animal kingdom
   b. the existence of the bilaterian clades Lophotrochozoa and Ecdysozoa
   c. that sponges are basal animals
   d. that chordates are deuterostomes
   e. the monophyly of the bilaterians

8. What is the main basis for placing the arthropods and nematodes in the Ecdysozoa in one hypothesis of animal phylogeny?
   a. Animals in both groups are segmented.
   b. Animals in both groups undergo ecdysis.
   c. They both have radial, determinate cleavage, and their embryonic development is similar.
   d. The fossil record has revealed a common ancestor to these two phyla.
   e. Analysis of genes shows that their sequences are quite similar, and these sequences differ from those of the lophotrochozoans and deuterostomes.

9. Which of the following combinations of phylum and description is *incorrect*?
   a. Echinodermata—branch of Bilateria, coelom forms from archenteron
   b. Nematoda—roundworms, pseudocoelomates
   c. Cnidaria—radial symmetry, diploblastic
   d. Platyhelminthes—flatworms, acoelomates
   e. Porifera—coelomates, mouth forms from blastopore

10. Which of the following subdivisions of the animal kingdom encompasses all the others in the list?
    a. Protostomia
    b. Bilateria
    c. Radiata
    d. Eumetazoa
    e. Deuterostomia

For Self-Quiz answers, see Appendix A.

Go to the website or CD-ROM for more quiz questions.

Evolution Connection

Some scientists suggest that the phrase "Cambrian fizzle" might be more appropriate than "Cambrian explosion" to describe the diversification of animals during that geologic period. In a similar vein, Lynn Margulis, of the University of Massachusetts, has compared observing an explosion of animal diversity in Cambrian strata to monitoring Earth from a satellite and noticing the emergence of cities only when they are large enough to be visible from that distance. What do these statements imply about the evolutionary history of animals during that time?

Scientific Inquiry

If you were constructing a phylogenetic analysis of the animal kingdom, based on comparing morphological characteristics, why would the presence of flagella be a poor choice of a characteristic to use as a basis for grouping phyla into clades?

Investigation *How Do Molecular Data Fit Traditional Phylogenies?*

Science, Technology, and Society

The study of animal phylogeny is viewed by some people as "science for science's sake," and some organizations that fund scientific research tend to favor projects that have more apparent applications to human needs. On the other hand, general interest in the history of life seems to remain high, with articles on new discoveries frequently featured in popular magazines. Suppose you had the opportunity to join a research team studying animal phylogeny. Write a short letter to a nonbiologist explaining why this research might be worth funding.

33 Invertebrates

▲ **Figure 33.1 A Christmas tree worm, a marine invertebrate.**

Key Concepts

33.1 Sponges are sessile and have a porous body and choanocytes
33.2 Cnidarians have radial symmetry, a gastrovascular cavity, and cnidocytes
33.3 Most animals have bilateral symmetry
33.4 Molluscs have a muscular foot, a visceral mass, and a mantle
33.5 Annelids are segmented worms
33.6 Nematodes are nonsegmented pseudocoelomates covered by a tough cuticle
33.7 Arthropods are segmented coelomates that have an exoskeleton and jointed appendages
33.8 Echinoderms and chordates are deuterostomes

Overview

Life Without a Backbone

At first glance, you might mistake the organism shown in **Figure 33.1** for some type of seaweed. But this colorful inhabitant of coral reefs is actually an animal, not an alga. Specifically, it is a kind of segmented worm commonly called a Christmas tree worm. The two whorls are tentacles, which the worm uses for gas exchange and for filtering small organisms from the water. The tentacles emerge from a tube of calcium carbonate secreted by the worm that protects and supports its soft body. Light-sensitive structures on the tentacles can detect the shadow cast by a predator, triggering muscular contractions that rapidly withdraw the tentacles into the tube.

Christmas tree worms are **invertebrates**—animals that lack a backbone. Invertebrates account for 95% of known animal species and all but one of the roughly 35 animal phyla that have been described. Invertebrates occupy almost every habitat on Earth, from the scalding water released by deep-sea hydrothermal vents to the rocky, frozen ground of Antarctica.

In this chapter, we'll take a brief tour of the invertebrate world, using the phylogenetic tree in **Figure 33.2** as a guide. **Figure 33.3**, on the next three pages, explores two dozen invertebrate phyla. We'll examine 14 of those phyla in more detail in this chapter.

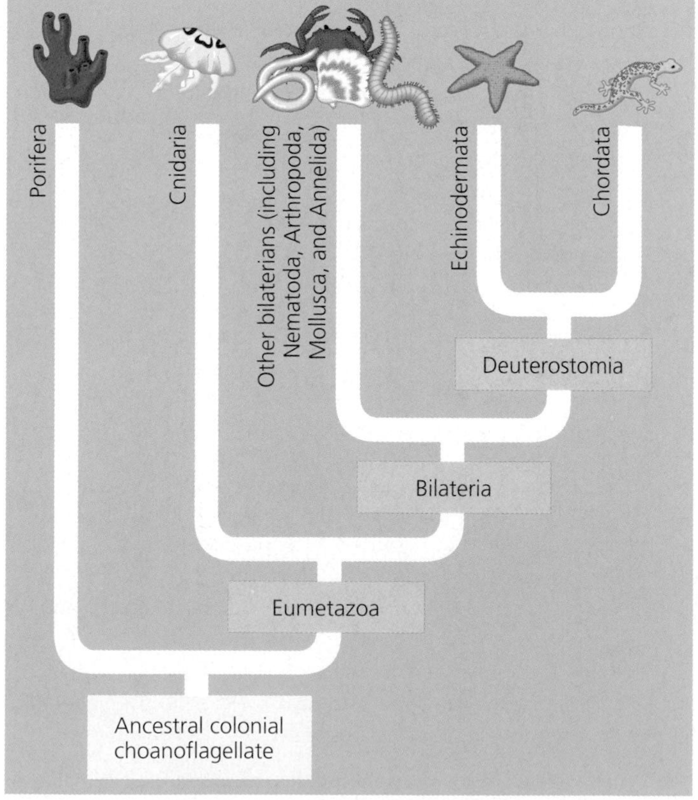

▲ **Figure 33.2 Review of animal phylogeny.**

Figure 33.3

Exploring Invertebrate Diversity

The animal kingdom is divided into about 35 phyla encompassing 1.3 million known species, but estimates of the total number of species range from 10 million to 200 million. Here we explore 24 animal phyla, all of which include invertebrates. Of the phyla surveyed in this figure, those that are illustrated with smaller-sized "preview" photographs are discussed more fully later in this chapter.

PORIFERA (5,500 species)

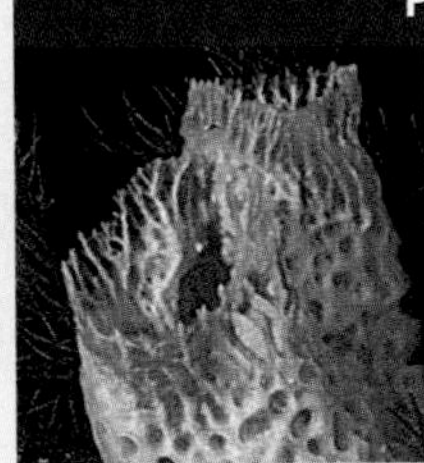

A sponge

Sponges are simple, sessile animals that lack true tissues. They live as suspension feeders, trapping particles that pass through the internal channels of their bodies (see Concept 33.1).

CNIDARIA (10,000 species)

Cnidarians include corals, jellies, and hydras. These animals share a distinctive body plan that includes a gastrovascular cavity with a single opening that serves as both mouth and anus (see Concept 33.2).

A jelly

PLACOZOA (1 species)

On first inspection, the single known species in this phylum, *Trichoplax adhaerens,* does not even look like an animal. It consists of a few thousand cells arranged in a double-layered plate 2 mm across. Feeding on organic detritus, *Trichoplax* can reproduce by dividing into two individuals or by budding off many multicellular individuals.

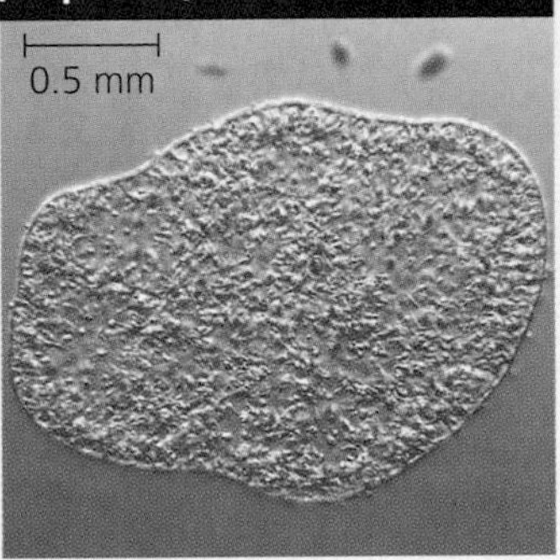

A placozoan (LM)

KINORHYNCHA (150 species)

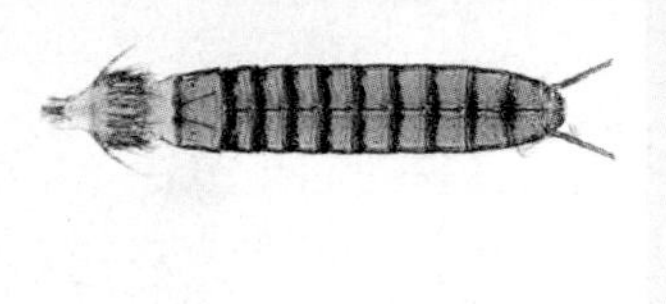

A kinorhynch (LM)

Almost all kinorhynchs are less than 1 mm long. They live in sand and mud in oceans around the world, from the intertidal zone to depths of 8,000 m. A kinorhynch's body consists of 13 segments covered in plates. The mouth is tipped with a ring of spines and can be retracted into the body.

PLATYHELMINTHES (20,000 species)

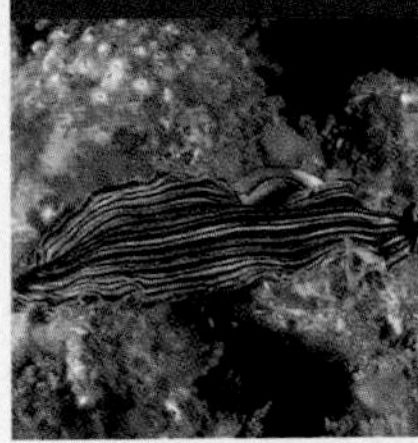

A marine flatworm

Flatworms (including tapeworms, planarians, and flukes) have bilateral symmetry and a central nervous system that processes information from eyes and other sensory structures. They have no body cavity or organs for circulation (see Concept 33.3).

ROTIFERA (1,800 species)

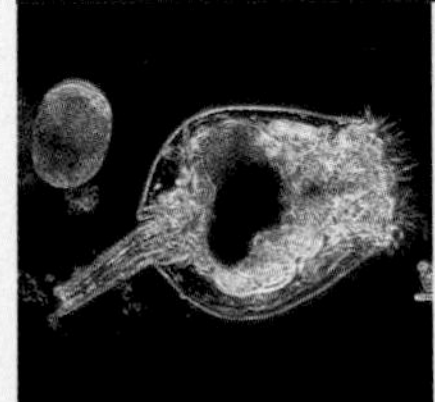

A rotifer (LM)

Despite their microscopic size, rotifers have specialized organ systems, including an alimentary canal (digestive tract). They feed on microorganisms suspended in water (see Concept 33.3).

ECTOPROCTA (4,500 species)

Ectoprocts (also known as bryozoans) live as sessile colonies and are covered by a tough exoskeleton (see Concept 33.3).

Ectoprocts

PHORONIDA (20 species)

Phoronids

Phoronids are marine worms. They live in tunnels in the seafloor, extending tentacles out of the tunnel opening to trap food particles (see Concept 33.3).

continued on the next page

Figure 33.3 (continued)

Exploring Invertebrate Diversity

BRACHIOPODA (335 species)

A brachiopod

Brachiopods, or lamp shells, may be easily mistaken for clams or other molluscs. However, most brachiopods have a unique stalk that anchors them to their substrate (see Concept 33.3).

NEMERTEA (900 species)

Proboscis worms, or ribbon worms, swim through water or burrow in sand, extending a unique proboscis to capture prey. Like flatworms, they lack a true coelom, but they have an alimentary canal (see Concept 33.3).

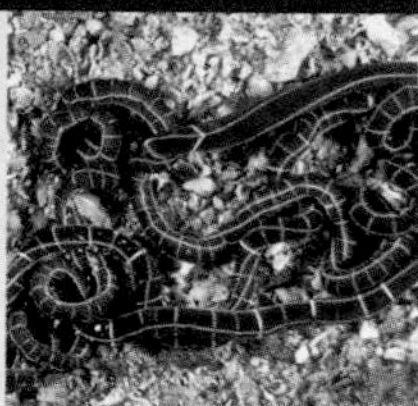

A ribbon worm

ACANTHOCEPHALA (1,100 species)

Acanthocephalans (from the Greek *acanthias,* prickly, and *cephalo,* head) are commonly known as thorny-headed worms because of the curved hooks located on the proboscis at the anterior end of their body. All species are parasites. The larvae develop in arthropods, and the adults live in vertebrates. Some acanthocephalans manipulate their intermediate hosts in ways that increase their chances of reaching their final hosts. For example, acanthocephalans that infect New Zealand mud crabs force their hosts to move to more visible locations on the beach, where the crabs are more likely to be eaten by birds, the worms' final hosts.

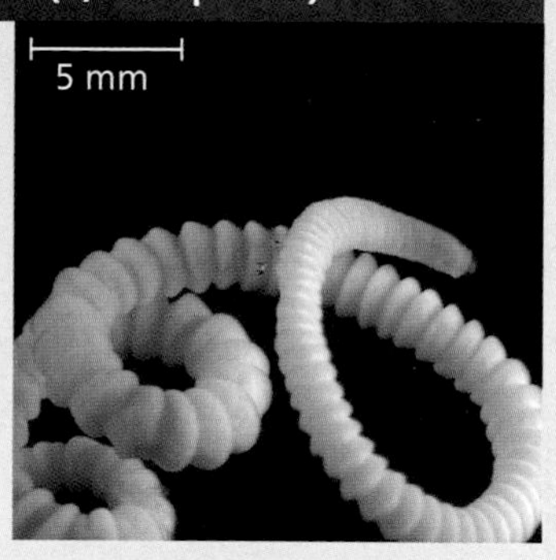

An acanthocephalan

CTENOPHORA (100 species)

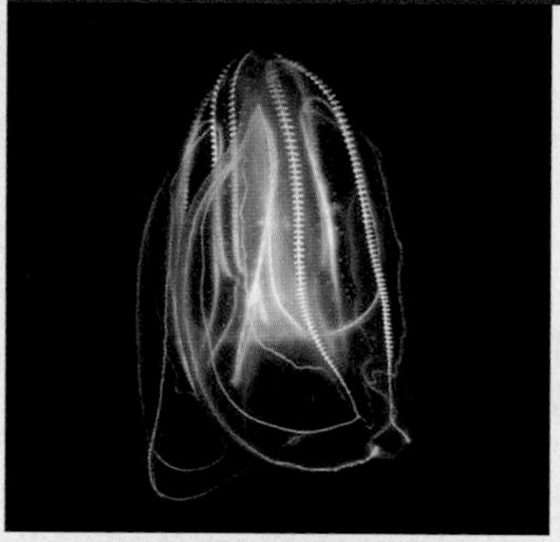

A ctenophore, or comb jelly

Ctenophores (comb jellies) are diploblastic like cnidarians, suggesting that both phyla diverged from other animals very early in evolution. Although they superficially resemble some cnidarians, comb jellies possess a number of distinctive traits, including a set of eight "combs" of cilia that propel the animals through the water. They also have a unique method for catching prey: When a small animal contacts one of their two tentacles, specialized cells burst open, covering the prey with sticky threads. Comb jellies make up a major portion of the ocean's planktonic biomass.

MOLLUSCA (93,000 species)

Molluscs (including snails, clams, squids, and octopuses) have a soft body that in many species is protected by a hard shell (see Concept 33.4).

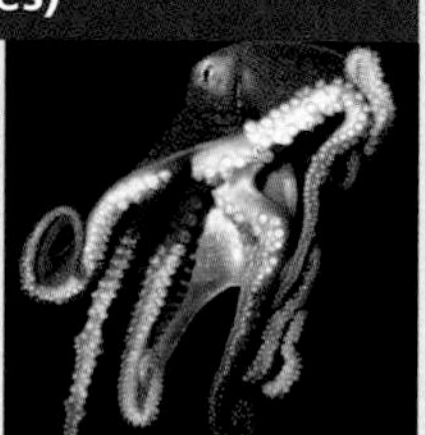

An octopus

ANNELIDA (16,500 species)

Annelids, or segmented worms, are distinguished from other worms by their body segmentation. Earthworms are the most familiar annelids, but the phylum also includes marine and freshwater species (see Concept 33.5).

A marine annelid

LORICIFERA (10 species)

Loriciferans (from the Latin *lorica,* corset, and *ferre,* to bear) are animals measuring only 0.1–0.4 mm in length that inhabit the deep-sea bottom. A loriciferan can telescope its head, neck, and thorax in and out of the lorica, a pocket formed by six plates surrounding the abdomen. Though the natural history of loriciferans is mostly a mystery, at least some species likely eat bacteria.

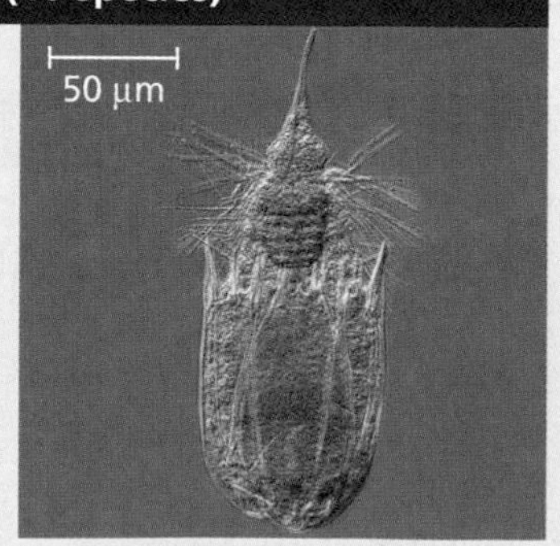

A loriciferan (LM)

PRIAPULA (16 species)

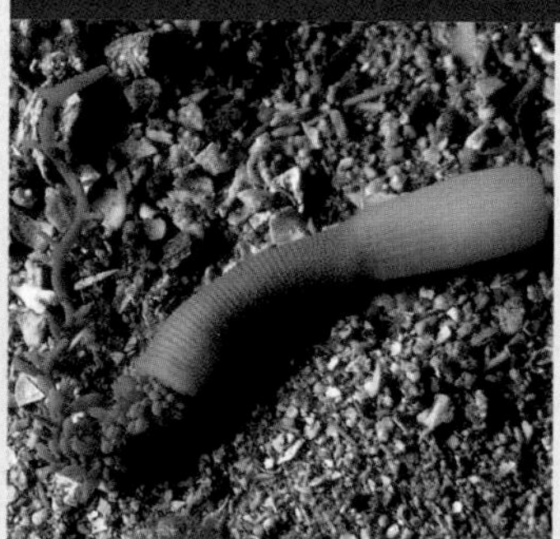

A priapulan

Priapulans are worms with a large, rounded proboscis at the anterior end. (They are named after Priapos, the Greek god of fertility, who was symbolized by a giant penis.) Ranging from 0.5 mm to 20 cm in length, most species burrow through seafloor sediments. Fossil evidence suggests that priapulans were among the major predators during the Cambrian period.

NEMATODA (25,000 species)

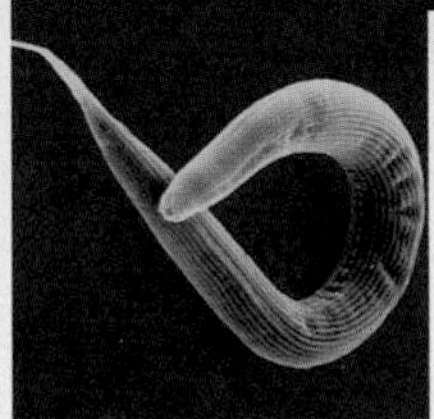

A roundworm

Roundworms are enormously abundant and diverse in the soil and in aquatic habitats; many species parasitize plants and animals. The most distinctive feature of roundworms is a tough cuticle that coats the body (see Concept 33.6).

ARTHROPODA (1,000,000+ species)

The vast majority of known animal species, including insects, crustaceans, and arachnids, are arthropods. All arthropods have a segmented exoskeleton and jointed appendages (see Concept 33.7).

A scorpion (an arachnid)

CYCLIOPHORA (1 species)

The only known species of cycliophoran, *Symbion pandora*, was discovered in 1995 on the mouthparts of a lobster. This tiny, vase-shaped creature has a unique body plan and a particularly bizarre life cycle. Males impregnate females that are still developing in their mothers' bodies. The fertilized females then escape, settle elsewhere on the lobster, and release their offspring. The offspring apparently leave that lobster and search for another one to which they attach.

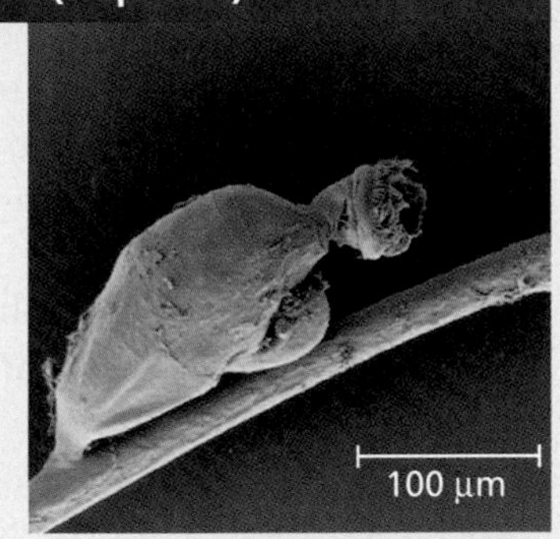

A cycliophoran (colorized SEM)

TARDIGRADA (800 species)

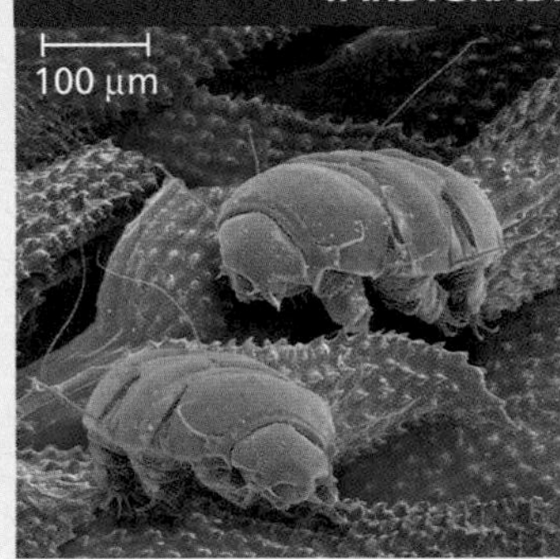

Tardigrades (colorized SEM)

Tardigrades (from the Latin *tardus*, slow, and *gradus*, step) are sometimes called water bears for their rounded shape, stubby appendages, and lumbering, bearlike gait. Most tardigrades are less than 0.5 mm in length. Some live in oceans or fresh water, while others live on plants or animals. As many as 2 million tardigrades can be found on a square meter of moss. Harsh conditions may cause tardigrades to enter a state of dormancy; while dormant, they can survive temperatures as low as −272°C, close to absolute zero!

ONYCHOPHORA (110 species)

Onychophorans, also called velvet worms, originated during the Cambrian explosion (see Chapter 32). Originally, they thrived in the ocean, but at some point they succeeded in colonizing land. Today they live only in humid forests. Onychophorans have fleshy antennae and several dozen pairs of saclike legs.

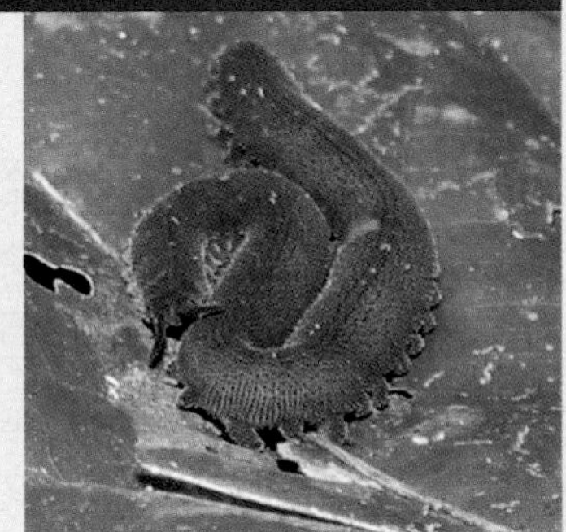

An onychophoran

HEMICHORDATA (85 species)

An acorn worm

Like echinoderms and chordates, hemichordates are deuterostomes (see Chapter 32). Hemichordates also share other traits with chordates, such as gill slits and a dorsal nerve cord. Most hemichordates are known as enteropneusts, or acorn worms. Acorn worms are marine and generally live buried in mud or under rocks; they may reach a length greater than 2 m.

ECHINODERMATA (7,000 species)

A sea urchin

Echinoderms, such as sand dollars, sea stars, and sea urchins, are aquatic animals that display radial symmetry as adults. They move and feed by using a network of internal canals to pump water to different parts of the body (see Concept 33.8).

CHORDATA (52,000 species)

More than 90% of all chordate species are animals with backbones (vertebrates). However, the phylum Chordata also includes three groups of invertebrates: tunicates, lancelets, and hagfishes. See Chapter 34 for a full discussion of this phylum.

A tunicate

Concept 33.1

Sponges are sessile and have a porous body and choanocytes

Sponges (phylum Porifera) are so sedentary that they were mistaken for plants by the ancient Greeks. Living in both fresh and marine waters, sponges are **suspension feeders:** They capture food particles suspended in the water that passes through their body, which typically resembles a sac perforated with pores (the name *Porifera* means "pore bearer"). Water is drawn through the pores into a central cavity, the **spongocoel,** and then flows out of the sponge through a larger opening called the **osculum (Figure 33.4).** More complex sponges have folded body walls, and many contain branched water canals and several oscula. Under certain conditions, the cells around the pores and osculum contract, closing the openings. Sponges range in size from a few millimeters to a few meters.

Unlike eumetazoans, sponges lack true tissues, groups of similar cells that act as a functional unit and are isolated from other tissues by membranous layers. However, the sponge body does contain several different cell types. Lining the interior of the spongocoel or internal water chambers are flagellated **choanocytes,** or collar cells (named for the membranous collar around the base of the flagellum). The flagella generate a water current, and the collars trap food particles that the choanocytes then ingest by phagocytosis. The similarity between choanocytes and the cells of choanoflagellates supports the molecular evidence suggesting that animals evolved from a choanoflagellate-like ancestor (see Chapter 32).

The body of a sponge consists of two layers of cells separated by a gelatinous region, the **mesohyl.** Wandering through the mesohyl are cells called **amoebocytes,** named for their use of pseudopodia. Amoebocytes have many functions. They take up food from the water and from choanocytes, digest it, and carry nutrients to other cells. They also manufacture tough skeletal fibers within the mesohyl. In some groups of sponges, these fibers are sharp spicules made from calcium carbonate or silica; other sponges produce more flexible fibers composed of a collagen protein called spongin. (These pliant skeletons are used as bath sponges.)

Most sponges are **hermaphrodites** (named for the Greek god Hermes and goddess Aphrodite), meaning that each individual functions as both male and female in sexual reproduction by producing sperm *and* eggs. Almost all sponges exhibit sequential hermaphroditism, functioning first as one sex and then as the other.

Gametes arise from choanocytes or amoebocytes. Eggs reside in the mesohyl, but sperm are carried out of the sponge by the water current. Cross-fertilization results from some of the sperm

Azure vase sponge (*Callyspongia plicifera*)

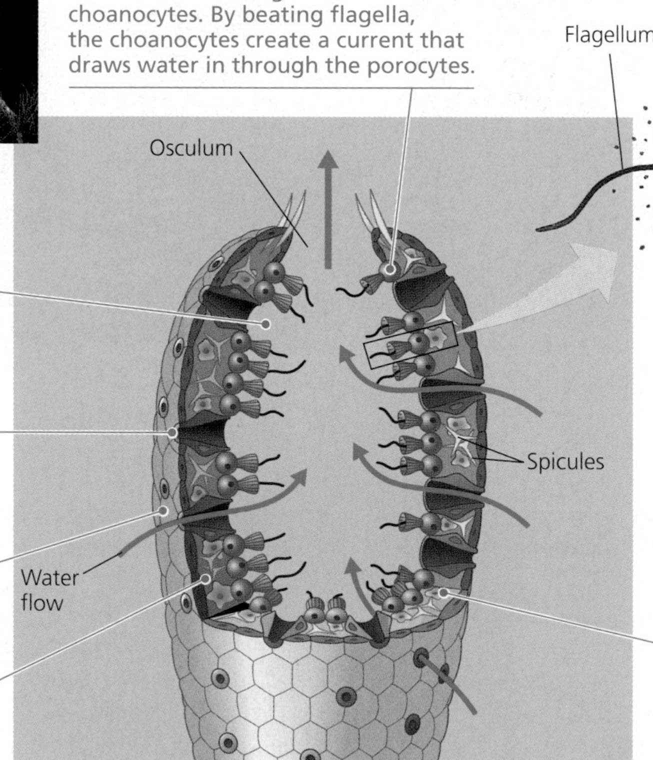

❶ Mesohyl. The wall of this simple sponge consists of two layers of cells separated by a gelatinous matrix, the mesohyl ("middle matter").

❷ Epidermis. The outer layer consists of tightly packed epidermal cells.

❸ Porocytes. Water enters the epidermis through channels formed by porocytes, doughnut-shaped cells that span the body wall.

❹ Spongocoel. Water passing through porocytes enters a cavity called the spongocoel.

❺ Choanocytes. The spongocoel is lined with feeding cells called choanocytes. By beating flagella, the choanocytes create a current that draws water in through the porocytes.

❻ The movement of a choanocyte's flagellum also draws water through its collar of fingerlike projections. Food particles are trapped in the mucus coating the projections, engulfed by phagocytosis, and either digested or transferred to amoebocytes.

❼ Amoebocyte. Amoebocytes transport nutrients to other cells of the sponge body and also produce materials for skeletal fibers (spicules).

▲ **Figure 33.4 Anatomy of a sponge.**

being drawn into neighboring individuals. Fertilization occurs in the mesohyl, where the zygotes develop into flagellated, swimming larvae that disperse from the parent sponge. Upon settling on a suitable substrate, a larva develops into the sessile adult.

Sponges produce a variety of antibiotics and other defensive compounds. Researchers are now isolating these compounds, which hold promise for fighting human diseases. For example, Robert Pettit and his colleagues at Arizona State University have found a compound called cribrostatin in marine sponges that can kill penicillin-resistant strains of the bacterium *Streptococcus*. The team is studying other sponge-derived compounds as anti-cancer agents.

Concept Check 33.1

1. Describe how sponges feed.
2. Explain how changes in water currents can affect sponge reproduction.

For suggested answers, see Appendix A.

Concept 33.2

Cnidarians have radial symmetry, a gastrovascular cavity, and cnidocytes

All animals except sponges belong to the clade Eumetazoa, the animals with true tissues (see Chapter 32). One of the oldest animal groups in this clade is the phylum Cnidaria. Cnidarians have diversified into a wide range of both sessile and floating forms, including jellies (commonly called "jellyfish"), corals, and hydras. Yet cnidarians continue to exhibit a relatively simple, diploblastic, radial body plan that existed some 570 million years ago.

The basic body plan of a cnidarian is a sac with a central digestive compartment, the **gastrovascular cavity.** A single opening to this cavity functions as both mouth and anus. There are two variations on this body plan: the sessile polyp and the floating medusa **(Figure 33.5)**. **Polyps** are cylindrical forms that adhere to the substrate by the aboral end of the body (the end opposite the mouth) and extend their tentacles, waiting for prey. Examples of the polyp form include hydras and sea anemones. A **medusa** is a flattened, mouth-down version of the polyp. It moves freely in the water by a combination of passive drifting and contractions of its bell-shaped body. Medusae include free-swimming jellies. The tentacles of a jelly dangle from the oral surface, which points downward. Some cnidarians exist only as polyps or only as medusae; others have both a medusa stage and a polyp stage in their life cycle.

Cnidarians are carnivores that use tentacles arranged in a ring around their mouth to capture prey and to push the food into their gastrovascular cavity, where digestion begins. The undigested remains are egested through the mouth/anus. The tentacles are armed with batteries of **cnidocytes,** unique cells that function in defense and the capture of prey **(Figure 33.6)**.

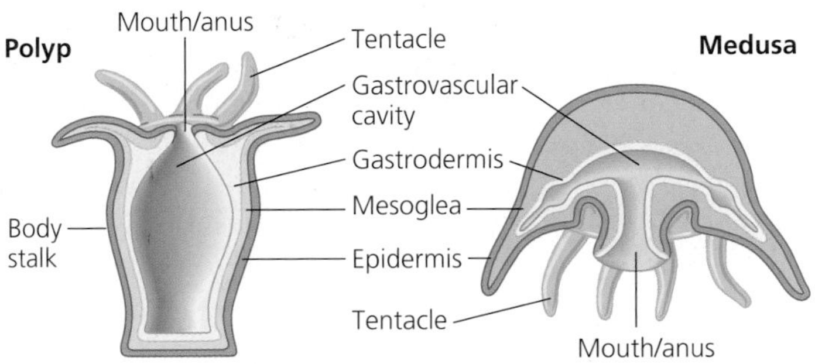

▲ **Figure 33.5 Polyp and medusa forms of cnidarians.** The body wall of a cnidarian has two layers of cells: an outer layer of epidermis (from ectoderm) and an inner layer of gastrodermis (from endoderm). Digestion begins in the gastrovascular cavity and is completed inside food vacuoles in the gastrodermal cells. Flagella on the gastrodermal cells keep the contents of the gastrovascular cavity agitated and help distribute nutrients. Sandwiched between the epidermis and gastrodermis is a gelatinous layer, the mesoglea.

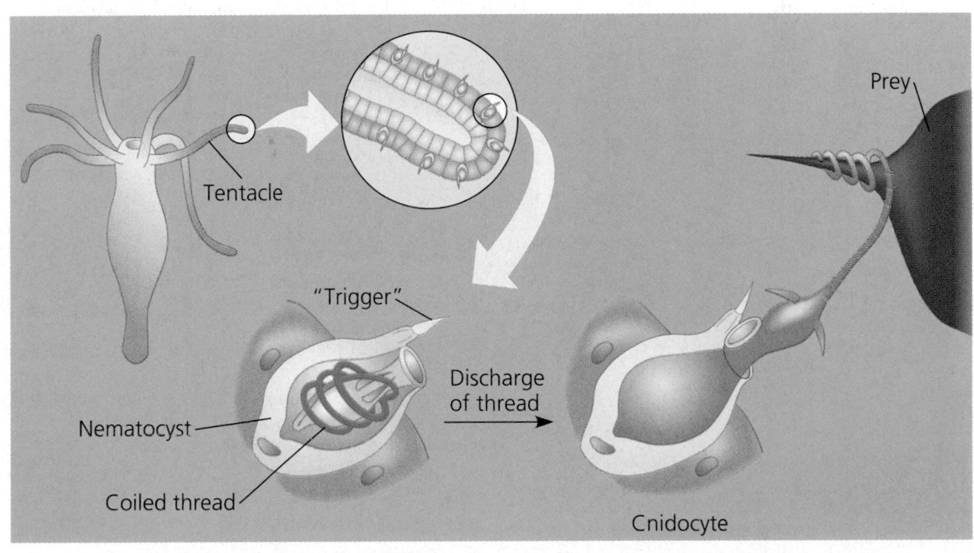

▲ **Figure 33.6 A cnidocyte of a hydra.** This type of cnidocyte contains a stinging capsule, the nematocyst, which itself contains an inverted thread. When a "trigger" is stimulated by touch or by certain chemicals, the thread shoots out, puncturing and injecting poison into prey.

Cnidocytes contain cnidae (from the Greek *cnide,* nettle), capsule-like organelles that are capable of everting and that give phylum Cnidaria its name. Cnidae called **nematocysts** are stinging capsules. Other cnidae have very long threads that stick to or entangle small prey that bump into the tentacles.

Contractile tissues and nerves occur in their simplest forms in cnidarians. Cells of the epidermis (outer layer) and gastrodermis (inner layer) have bundles of microfilaments arranged into contractile fibers (see Chapter 6). The gastrovascular cavity acts as a hydrostatic skeleton against which the contractile cells can work. When a cnidarian closes its mouth, the volume of the cavity is fixed, and contraction of selected cells causes the animal to change shape. Movements are coordinated by a nerve net. Cnidarians have no brain, and the noncentralized nerve net is associated with simple sensory structures that are distributed radially around the body. Thus, the animal can detect and respond to stimuli equally from all directions.

As summarized in **Table 33.1**, the phylum Cnidaria is divided into four major classes: Hydrozoa, Scyphozoa, Cubozoa, and Anthozoa **(Figure 33.7)**.

Hydrozoans

Most hydrozoans alternate between polyp and medusa forms, as in the life cycle of *Obelia* **(Figure 33.8)**. The polyp stage, a colony of interconnected polyps in the case of *Obelia,* is more conspicuous than the medusae. Hydras, among the few cnidarians found in fresh water, are unusual hydrozoans in that they exist only in the polyp form. When environmental conditions are favorable, a hydra reproduces asexually by budding, the formation of outgrowths that pinch off from the parent and live independently (see Figure 13.2). When environmental conditions deteriorate, hydras can reproduce sexually, forming resistant zygotes that remain dormant until the conditions improve.

Scyphozoans

The medusa generally is the predominant stage in the life cycle of the class Scyphozoa. The medusae of most species live among the plankton as jellies. Most coastal scyphozoans go through a small polyp stage during their life cycle, whereas those that live in the open ocean generally lack the polyp stage completely.

Table 33.1 Classes of Phylum Cnidaria

Class and Examples	Main Characteristics
Hydrozoa (Portuguese man-of-war, hydras, *Obelia,* some corals; see Figures 33.7a and 33.8)	Most marine, a few freshwater; both polyp and medusa stages in most species; polyp stage often colonial
Scyphozoa (jellies, sea nettles; see Figure 33.7b)	All marine; polyp stage reduced; free-swimming; medusae up to 2 m in diameter
Cubozoa (box jellies, sea wasps; see Figure 33.7c)	All marine; box-shaped medusae; complex eyes
Anthozoa (sea anemones, most corals, sea fans; see Figure 33.7d)	All marine; medusa stage completely absent; most sessile; many colonial

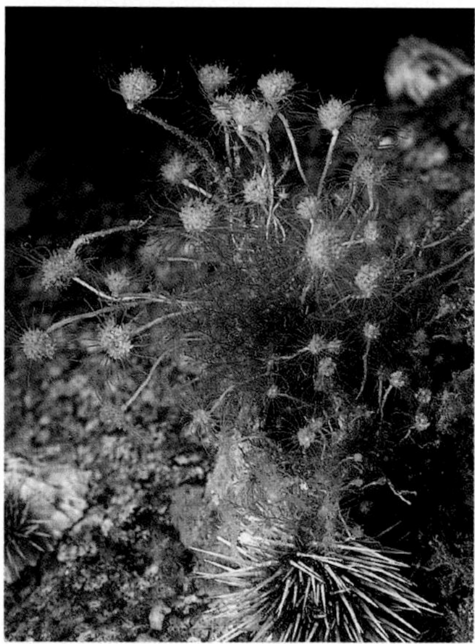

(a) These colonial polyps are members of class Hydrozoa.

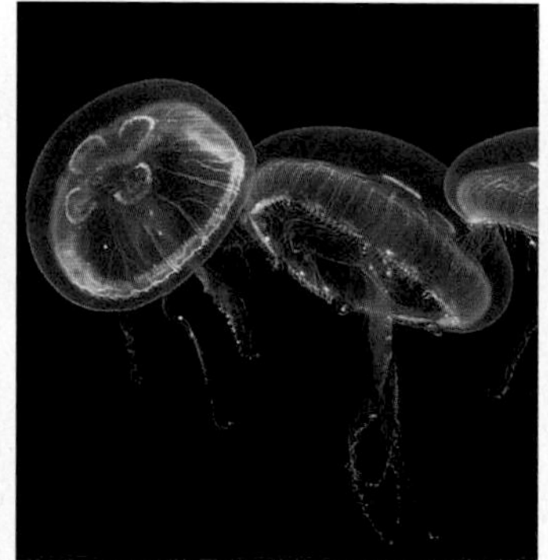

(b) Many species of jellies (class Scyphozoa), including the species pictured here, are bioluminescent. The largest scyphozoans have tentacles more than 100 m long dangling from a bell-shaped body up to 2 m in diameter.

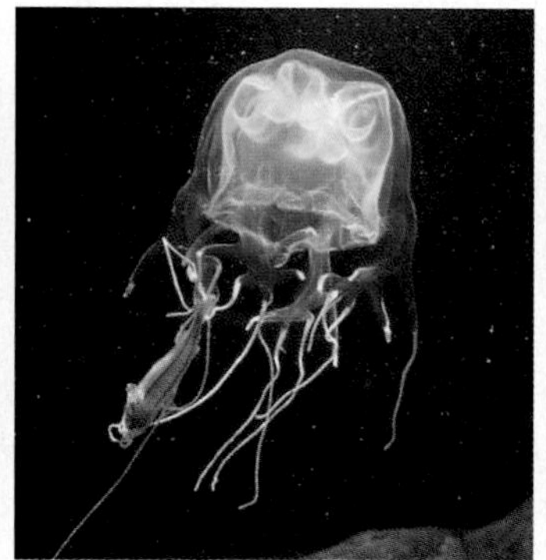

(c) The sea wasp (*Chironex fleckeri*) is a member of class Cubozoa. Its poison, which can subdue fish and other large prey, is more potent than cobra venom.

(d) Sea anemones and other members of class Anthozoa exist only as polyps.

▲ **Figure 33.7 Cnidarians.**

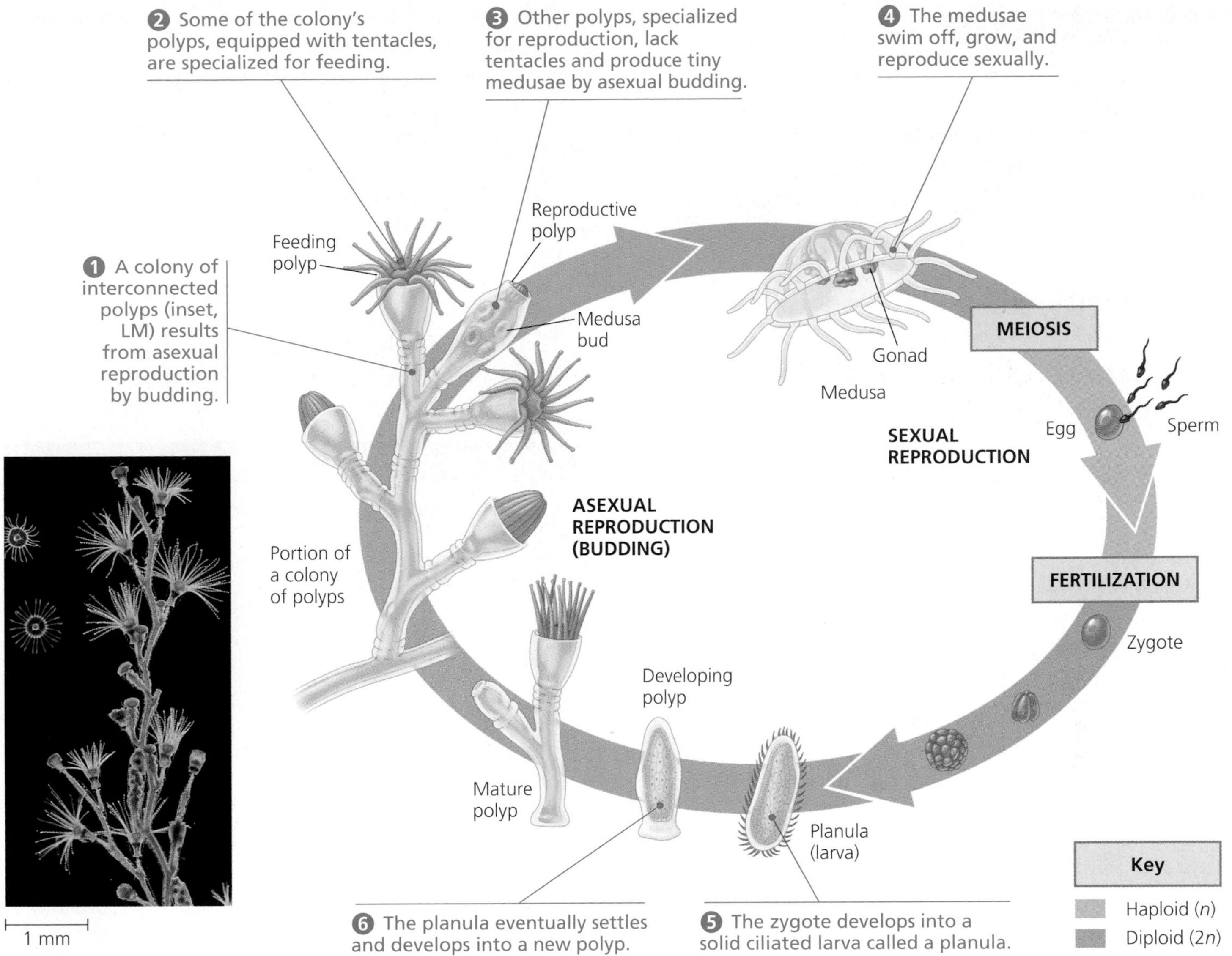

2 Some of the colony's polyps, equipped with tentacles, are specialized for feeding.
3 Other polyps, specialized for reproduction, lack tentacles and produce tiny medusae by asexual budding.
4 The medusae swim off, grow, and reproduce sexually.
Reproductive polyp
Feeding polyp
1 A colony of interconnected polyps (inset, LM) results from asexual reproduction by budding.
Medusa bud
Gonad
Medusa
MEIOSIS
SEXUAL REPRODUCTION
Egg
Sperm
ASEXUAL REPRODUCTION (BUDDING)
Portion of a colony of polyps
FERTILIZATION
Zygote
Developing polyp
Mature polyp
Planula (larva)
Key
Haploid (n)
Diploid (2n)
1 mm
6 The planula eventually settles and develops into a new polyp.
5 The zygote develops into a solid ciliated larva called a planula.

▲ **Figure 33.8 The life cycle of the hydrozoan *Obelia*.** The polyp stage is asexual, and the medusa stage is sexual; these two stages alternate, one producing the other. Do not confuse this with the alternation of generations that occurs in plants and some algae. Both the polyp and the medusa are diploid organisms. (Typical of animals, only the gametes of *Obelia* are haploid.) By contrast, one plant generation is haploid, and the other is diploid.

Cubozoans

As their name (which means "cube animals") suggests, cubozoans have a box-shaped medusa stage. They can be distinguished from scyphozoans in other significant ways, such as having complex eyes embedded in the fringe of their medusae. Cubozoans, which generally live in tropical oceans, are often equipped with highly toxic cnidocytes. The sea wasp (*Chironex fleckeri*), a cubozoan that lives off the coast of northern Australia, is one of the deadliest organisms on Earth: Its sting causes intense pain and can lead to respiratory failure, cardiac arrest, and death within minutes. The amount of poison in one sea wasp is enough to kill 60 people. The poison of sea wasps isn't universally fatal, however; sea turtles have defenses against it, allowing them to eat the cubozoan in great quantities.

Anthozoans

Sea anemones and corals belong to the class Anthozoa ("flower animals"). They occur only as polyps. Corals live as solitary or colonial forms and secrete a hard external skeleton of calcium carbonate. Each polyp generation builds on the skeletal remains of earlier generations, constructing "rocks" with shapes characteristic of their species. It is these skeletons that we call coral.

Coral reefs are to tropical seas what rain forests are to tropical land: They provide habitat for a wealth of other species. Unfortunately, like rain forests, coral reefs are being destroyed at an alarming rate by human activity. Pollution and overfishing are major threats, and global warming (see Chapter 54) may also be contributing to their demise by raising seawater temperatures above the narrow range in which corals thrive.

Concept Check 33.2

1. Compare and contrast the polyp and medusa forms of cnidarians.
2. Describe the structure and function of the stinging cells for which cnidarians are named.

For suggested answers, see Appendix A.

Concept 33.3

Most animals have bilateral symmetry

The vast majority of animal species belong to the clade Bilateria, which consists of animals with bilateral symmetry and triploblastic development (see Chapter 32). Most bilaterians are also coelomates. While the sequence of bilaterian evolution is still a subject of active investigation, researchers generally agree that the most recent common ancestor of living bilaterians probably existed in the late Proterozoic. During the Cambrian explosion, most major groups of bilaterians emerged. This section will focus on just six bilaterian phyla; Concepts 33.4–33.8 will explore six other major bilaterian phyla.

Flatworms

Flatworms (phylum Platyhelminthes) live in marine, freshwater, and damp terrestrial habitats. In addition to many free-living forms, flatworms include many parasitic species, such as flukes and tapeworms. Flatworms are so named because their bodies are thin between the dorsal and ventral surfaces (flattened dorsoventrally; *platyhelminth* means "flat worm"). The smallest are nearly microscopic free-living species, while some tapeworms can be over 20 m long. (Note that *worm* is not a formal taxonomic name but a general term for animals with long, thin bodies.)

Although flatworms undergo triploblastic development, they are acoelomates (animals that lack a body cavity). Their flat shape places all cells close to the surrounding water, enabling gas exchange and the elimination of nitrogenous waste (ammonia) to occur by diffusion across the body surface. Flatworms have no organs specialized for gas exchange or circulation, and their relatively simple excretory apparatus functions mainly to maintain osmotic balance with their surroundings. This apparatus consists of ciliated cells called flame bulbs that waft fluid through branched ducts opening to the outside (see Figure 44.10). Most flatworms have a gastrovascular cavity with only one opening. The fine branches of the gastrovascular cavity distribute food throughout the animal.

Flatworms are divided into four classes **(Table 33.2)**: Turbellaria (mostly free-living flatworms), Monogenea (monogeneans), Trematoda (trematodes, or flukes), and Cestoda (tapeworms).

Turbellarians

Turbellarians are nearly all free-living and mostly marine **(Figure 33.9)**. The best-known turbellarians are members of the genus *Dugesia*, commonly called **planarians.** Abundant in unpolluted ponds and streams, planarians prey on smaller animals or feed on dead animals.

Planarians move by using cilia on their ventral epidermis, gliding along a film of mucus they secrete. Some other turbel-

Table 33.2 Classes of Phylum Platyhelminthes

Class and Examples	Main Characteristics
Turbellaria (mostly free-living flatworms, such as *Dugesia*; see Figures 33.9 and 33.10)	Most marine, some freshwater, a few terrestrial; predators and scavengers; body surface ciliated
Monogenea (monogeneans)	Marine and freshwater parasites; most infect external surfaces of fishes; life history simple; ciliated larva starts infection on host
Trematoda (trematodes, also called flukes; see Figure 33.11)	Parasites, almost always of vertebrates; two suckers attach to host; most life cycles include intermediate hosts
Cestoda (tapeworms; see Figure 33.12)	Parasites of vertebrates; scolex attaches to host; proglottids produce eggs and break off after fertilization; no head or digestive system; life cycle with one or more intermediate hosts

▲ **Figure 33.9 A marine flatworm (class Turbellaria).**

larians also use their muscles to swim through water with an undulating motion.

A planarian's head is equipped with a pair of light-sensitive eyespots and lateral flaps that function mainly to detect specific chemicals. The planarian nervous system is more complex and centralized than the nerve nets of cnidarians **(Figure 33.10)**. Planarians can learn to modify their responses to stimuli.

Planarians can reproduce asexually through regeneration. The parent constricts in the middle, and each half regenerates the missing end. Sexual reproduction also occurs. Although planarians are hermaphrodites, copulating mates cross-fertilize.

Pharynx. The mouth is at the tip of a muscular pharynx that extends from the animal's ventral side. Digestive juices are spilled onto prey, and the pharynx sucks small pieces of food into the gastrovascular cavity, where digestion continues.

Digestion is completed within the cells lining the gastrovascular cavity, which has three branches, each with fine subbranches that provide an extensive surface area.

Undigested wastes are egested through the mouth.

Gastrovascular cavity

Eyespots

Ganglia. Located at the anterior end of the worm, near the main sources of sensory input, is a pair of ganglia, dense clusters of nerve cells.

Ventral nerve cords. From the ganglia, a pair of ventral nerve cords runs the length of the body.

▲ **Figure 33.10 Anatomy of a planarian, a turbellarian.**

Monogeneans and Trematodes

Monogeneans and trematodes live as parasites in or on other animals. Many have suckers for attaching to internal organs or to the outer surfaces of the host. A tough covering helps protect the parasites within their hosts. Reproductive organs occupy nearly the entire interior of these worms.

As a group, trematodes parasitize a wide range of hosts, and most species have complex life cycles with alternating sexual and asexual stages. Many trematodes require an intermediate host in which larvae develop before infecting the final host (usually a vertebrate), where the adult worms live. For example, trematodes that parasitize humans spend part of their lives in snail hosts **(Figure 33.11)**. The 200 million people around the world who are infected with blood flukes (*Schistosoma*) suffer from schistosomiasis, a disease whose symptoms include pain, anemia, and dysentery.

Living within different hosts puts demands on trematodes that free-living animals don't face. A blood fluke, for instance, must evade the immune systems of both snails and humans. By mimicking the surface proteins of its hosts, the blood fluke creates a partial immunological camouflage for itself. It also releases molecules that manipulate the hosts' immune systems into tolerating the parasite's existence. These defenses are so effective that individual flukes can survive in humans for more than 40 years.

Most monogeneans are external parasites of fish. The monogenean life cycle is relatively simple, with a ciliated, free-swimming larva initiating the infection on a host. Although monogeneans have been traditionally aligned with the trematodes, some structural and chemical evidence suggests they are more closely related to tapeworms.

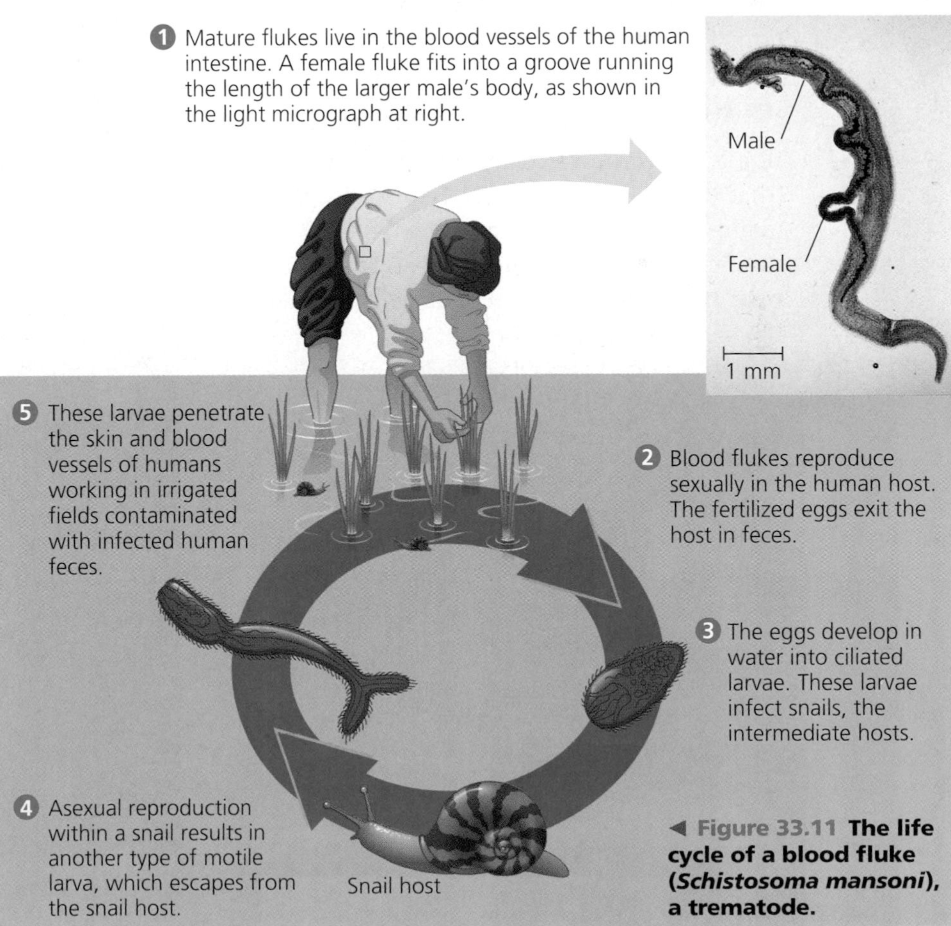

1 Mature flukes live in the blood vessels of the human intestine. A female fluke fits into a groove running the length of the larger male's body, as shown in the light micrograph at right.
Male
Female
1 mm
2 Blood flukes reproduce sexually in the human host. The fertilized eggs exit the host in feces.
3 The eggs develop in water into ciliated larvae. These larvae infect snails, the intermediate hosts.
4 Asexual reproduction within a snail results in another type of motile larva, which escapes from the snail host.
Snail host
5 These larvae penetrate the skin and blood vessels of humans working in irrigated fields contaminated with infected human feces.

◄ **Figure 33.11 The life cycle of a blood fluke (*Schistosoma mansoni*), a trematode.**

Tapeworms

Tapeworms (class Cestoidea) are also parasitic **(Figure 33.12)**. The adults live mostly inside vertebrates, including humans. In many tapeworms, the anterior end, or scolex, is armed with suckers and often hooks that lock the worm to the intestinal lining of the host. Tapeworms lack a gastrovascular cavity; they absorb nutrients released by digestion in the host's intestine. Absorption occurs across the tapeworm's body surface.

Posterior to the scolex is a long ribbon of units called proglottids, which are little more than sacs of sex organs. Mature proglottids, loaded with thousands of eggs, are released from the posterior end of a mature tapeworm and leave the host's body in feces. In one type of life cycle, human feces contaminate the food or water of intermediate hosts, such as pigs or cattle, and the tapeworm eggs develop into larvae that encyst in muscles of these animals. Humans acquire the larvae by eating undercooked meat contaminated with cysts, and the worms develop into mature adults within the human. Large tapeworms can block the intestines and rob enough nutrients from the human host to cause nutritional deficiencies. An orally administered drug named niclosamide kills the adult worms.

Rotifers

Rotifers (phylum Rotifera) are tiny animals that inhabit fresh water, the ocean, and damp soil. Ranging in size from about 50 µm to 2 mm, rotifers are smaller than many protists but nevertheless are truly multicellular and have specialized organ systems **(Figure 33.13)**. In contrast to cnidarians and flatworms, which have a gastrovascular cavity, rotifers have an **alimentary canal,** a digestive tube with a separate mouth and anus. Internal organs lie within the pseudocoelom, a body cavity that is not completely lined by mesoderm (see Figure 32.8b). Fluid in the pseudocoelom serves as a hydrostatic skeleton (see Chapter 49). Movement of a rotifer's body distributes the fluid throughout the body, circulating nutrients and wastes in these tiny animals.

The word *rotifer,* derived from Latin, means "wheel-bearer," a reference to the crown of cilia that draws a vortex of water into the mouth. Posterior to the mouth, a region of the digestive tract called the pharynx bears jaws (trophi) that grind up food, mostly microorganisms suspended in the water.

Rotifers undergo unusual forms of reproduction. Some species consist only of females that produce more females from unfertilized eggs, a type of reproduction called **parthenogenesis.** Other species produce two types of eggs that develop by parthenogenesis, one type forming females and the other type developing into simplified males that cannot even feed themselves. These males survive long enough to produce sperm that fertilize eggs, which form resistant zygotes that can survive when a pond dries up. When conditions are favorable, the zygotes break dormancy and develop into a new female generation that reproduces by parthenogenesis until conditions become unfavorable again.

It is puzzling that so many rotifer species survive without males. The vast majority of animals and plants reproduce sexually at least some of the time, and sexual reproduction has certain advantages over asexual reproduction. For example, species that reproduce asexually tend to accumulate harmful mutations in their genomes faster than sexually reproducing species. As a result, asexual species should experience higher rates of extinction and lower rates of speciation.

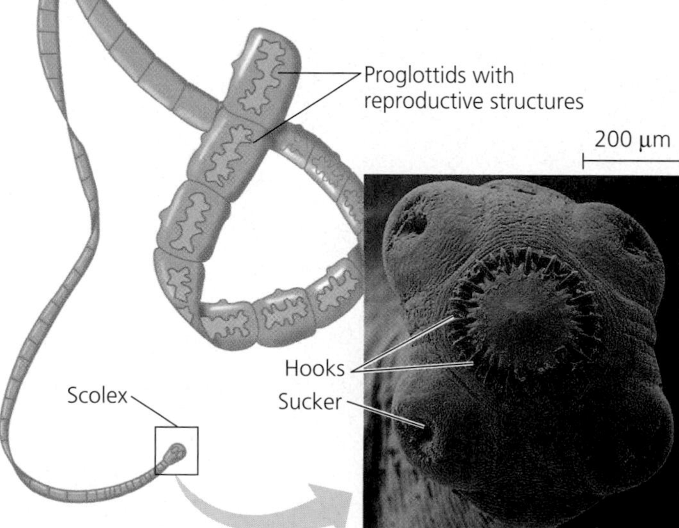

▲ **Figure 33.12 Anatomy of a tapeworm.** The inset shows a closeup of the scolex (colorized SEM).

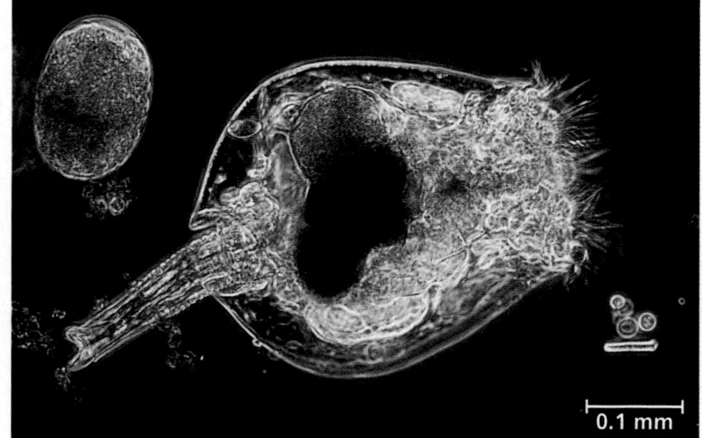

▲ **Figure 33.13 A rotifer.** These pseudocoelomates, smaller than many protists, are generally more anatomically complex than flatworms (LM).

Nobel prize-winning biologist Matthew Meselson, of Harvard University, has been studying a class of asexual rotifers called Bdelloidea. Some 360 species of bdelloid rotifers are known, and all of them reproduce by parthenogenesis without any males. Paleontologists have discovered bdelloid rotifers preserved in 35-million-year-old amber, and the morphology of these fossils resembles only the female form, with no evidence of males. By comparing the DNA of bdelloids with that of their closest sexually reproducing rotifer relatives, Meselson and his colleagues concluded that bdelloids have likely been asexual for much longer than 35 million years. How these animals manage to flout the general rule against long-lived asexuality is a puzzle.

Lophophorates: Ectoprocts, Phoronids, and Brachiopods

Bilaterians in three phyla—Ectoprocta, Phoronida, and Brachiopoda—are traditionally called lophophorates because they all have a lophophore, a horseshoe-shaped or circular crown of ciliated tentacles that surround the mouth (see Figure 32.13a). As the cilia draw water toward the mouth, the tentacles trap suspended food particles. The common occurrence of this complex apparatus in lophophorates suggests that these three phyla are related. Other similarities, such as a U-shaped alimentary canal and the absence of a distinct head, are adaptations to a sessile existence. In contrast to flatworms, which lack a body cavity, and rotifers, which have a pseudocoelom, lophophorates have a true coelom completely lined by mesoderm (see Figure 32.8a).

Ectoprocts (from the Greek *ecto,* outside, and *procta,* anus) are colonial animals that superficially resemble plants. (Their common name, bryozoans, means "moss animals.") In most species, the colony is encased in a hard exoskeleton with pores through which the lophophores extend **(Figure 33.14a)**. Most ectoproct species live in the sea, where they are among the most widespread and numerous sessile animals. Several species are important reef builders. Ectoprocts also live in lakes and rivers. Colonies of the freshwater ectoproct *Pectinatella magnifica* form on submerged sticks or rocks and can grow into a gelatinous, ball-shaped mass more than 10 cm across.

Phoronids are tube-dwelling marine worms ranging from 1 mm to 50 cm in length. Some species live buried in the sand within tubes made of chitin, extending their lophophore from the opening of the tube and withdrawing it into the tube when threatened **(Figure 33.14b)**.

Brachiopods, or lamp shells, superficially resemble clams and other hinge-shelled molluscs, but the two halves of the brachiopod shell are dorsal and ventral rather than lateral, as in clams **(Figure 33.14c)**. All brachiopods are marine. Most live attached to the seafloor by a stalk, opening their shell slightly to allow water to flow over the lophophore. The living brachiopods are remnants of a much richer past that included 30,000 Paleozoic and Mesozoic species. *Lingula,* a living brachiopod genus, is nearly identical to brachiopods that lived 400 million years ago.

(a) Ectoprocts, such as this sea mat (*Membranipora membranacea*), are colonial lophophorates.

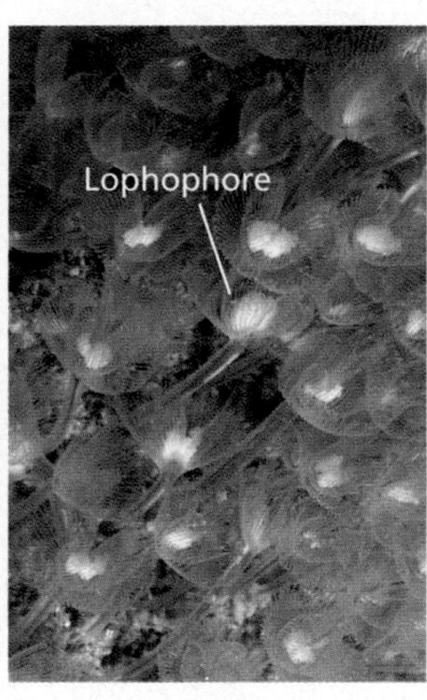

(b) In phoronids such as *Phoronis hippocrepia*, the lophophore and mouth are at one end of an elongated trunk.

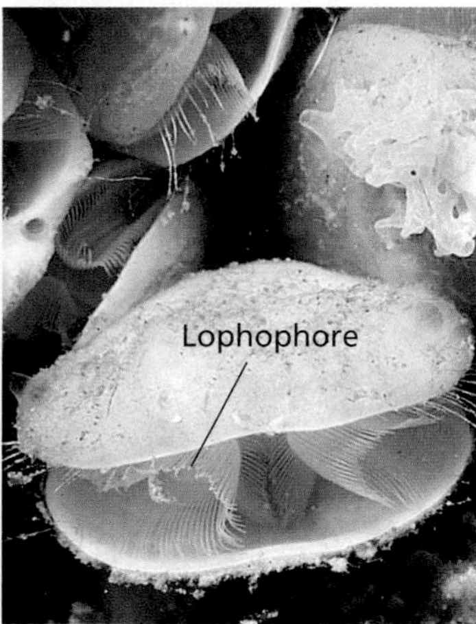

(c) Brachiopods have a hinged shell. The two parts of the shell are dorsal and ventral.

▲ **Figure 33.14 Lophophorates.**

Nemerteans

Members of the phylum Nemertea are commonly called proboscis worms or ribbon worms **(Figure 33.15)**. A nemertean's

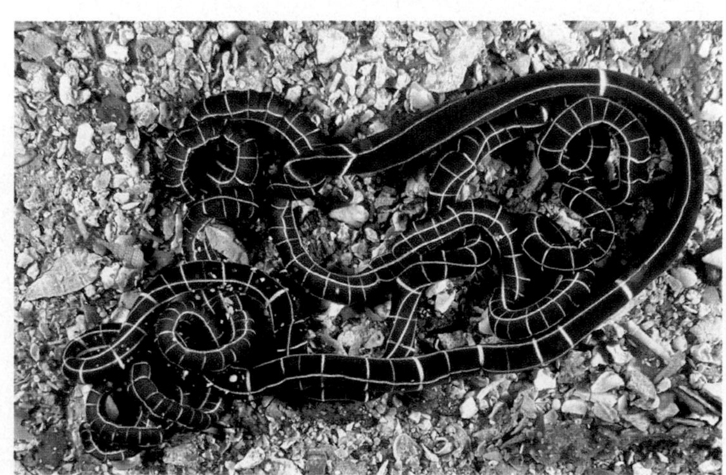

◄ **Figure 33.15 A ribbon worm, phylum Nemertea.**

body is structurally acoelomate, like that of a flatworm, but it contains a small, fluid-filled sac that may be a reduced version of a coelom. The sac and fluid hydraulically operate an extensible proboscis, which rapidly shoots out of the worm's body, in many cases delivering a toxin to its prey.

Nemerteans range in length from less than 1 mm to several meters. Nearly all members of this phylum are marine, but a few species inhabit fresh water or damp soil. Some are active swimmers; others burrow in the sand.

Nemerteans and flatworms have similar excretory, sensory, and nervous systems. But in addition to the unique proboscis apparatus, two anatomical features not found in flatworms have evolved in the phylum Nemertea: an alimentary canal and a **closed circulatory system,** in which the blood is contained in vessels and is therefore distinct from fluid in the body cavity. Nemerteans have no heart; their blood is propelled by muscles squeezing the vessels.

Concept Check 33.3

1. Explain how tapeworms can survive without a coelom, a mouth, a digestive system, or an excretory system.
2. Is the presence or absence of an alimentary canal related to the size of an animal? Support your answer with two examples.
3. Explain how, in terms of function, ectoprocts have more in common with nonbilaterian corals than with their closer bilaterian relatives.

For suggested answers, see Appendix A.

Concept 33.4

Molluscs have a muscular foot, a visceral mass, and a mantle

Snails and slugs, oysters and clams, and octopuses and squids are all molluscs (phylum Mollusca). Most molluscs are marine, though some inhabit fresh water, and there are snails and slugs that live on land. Molluscs are soft-bodied animals (from the Latin *molluscus,* soft), but most are protected by a hard shell made of calcium carbonate. Slugs, squids, and octopuses have a reduced internal shell or have lost their shell completely during their evolution.

Despite their apparent differences, all molluscs have a similar body plan **(Figure 33.16)**. The body has three main parts: a muscular **foot,** usually used for movement; a **visceral mass** containing most of the internal organs; and a **mantle,** a fold of tissue that drapes over the visceral mass and secretes a shell (if one is present). In many molluscs, the mantle extends beyond the visceral mass, producing a water-filled chamber, the **mantle cavity,** which houses the gills, anus, and excretory pores. Many molluscs feed by using a straplike rasping organ called a **radula** to scrape up food.

Most molluscs have separate sexes, with gonads (ovaries or testes) located in the visceral mass. Many snails, however, are hermaphrodites. The life cycle of many marine molluscs includes a ciliated larval stage, the **trochophore,** which is also characteristic of marine annelids (segmented worms) and some other invertebrates (see Figure 32.13b).

The basic body plan of molluscs has evolved in various ways in the eight classes of the phylum. We examine four of

▼ **Figure 33.16 The basic body plan of a mollusc.**

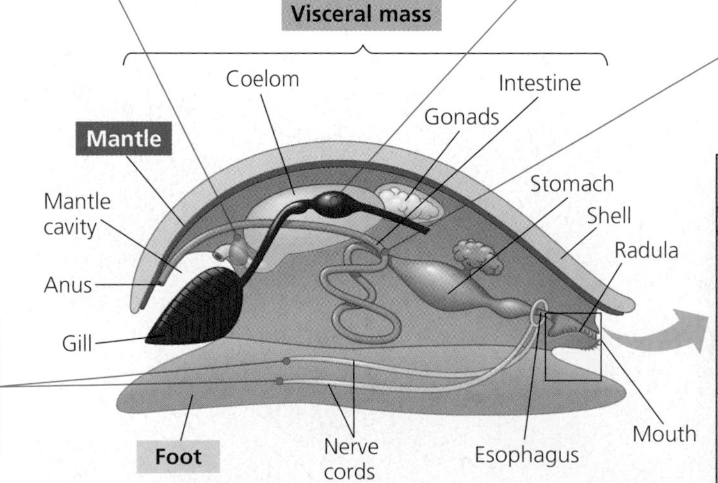

Nephridium. Excretory organs called nephridia remove metabolic wastes from the hemolymph.
Heart. Most molluscs have an open circulatory system. The dorsally located heart pumps circulatory fluid called hemolymph through arteries into sinuses (body spaces). The organs of the mollusc are thus continually bathed in hemolymph.
Visceral mass
The long digestive tract is coiled in the visceral mass.
Coelom
Intestine
Gonads
Mantle
Stomach
Shell
Radula
Mantle cavity
Anus
Gill
The nervous system consists of a nerve ring around the esophagus, from which nerve cords extend.
Foot
Nerve cords
Esophagus
Mouth
Mouth
Radula. The mouth region in many mollusc species contains a rasp-like feeding organ called a radula. This belt of backward-curved teeth slides back and forth, scraping and scooping like a backhoe.

those classes here **(Table 33.3)**: Polyplacophora (chitons), Gastropoda (snails and slugs), Bivalvia (clams, oysters, and other bivalves), and Cephalopoda (squids, octopuses, cuttlefish, and chambered nautiluses).

Chitons

Chitons have an oval-shaped body and a shell divided into eight dorsal plates **(Figure 33.17)**; the body itself, however, is unsegmented. You can find these marine animals clinging to rocks along the shore during low tide. Try to dislodge a chiton by hand, and you will be surprised at how well its foot, acting as a suction cup, grips the rock. A chiton can also use its foot to creep slowly over the rock surface. Chitons use their radula to cut and ingest algae.

Table 33.3 Major Classes of Phylum Mollusca

Class and Examples	Main Characteristics
Polyplacophora (chitons; see Figure 33.17)	Marine; shell with eight plates; foot used for locomotion; radula; no head
Gastropoda (snails, slugs; see Figures 33.18 and 33.19)	Marine, freshwater, or terrestrial; asymmetrical body, usually with a coiled shell; shell reduced or absent in some; foot for locomotion; radula
Bivalvia (clams, mussels, scallops, oysters; see Figures 33.20 and 33.21)	Marine and freshwater; flattened shell with two valves; head reduced; paired gills; no radula; most are suspension feeders; mantle forms siphons
Cephalopoda (squids, octopuses, cuttlefish, chambered nautiluses; see Figure 33.22)	Marine; head surrounded by grasping tentacles, usually with suckers; shell external, internal, or absent; mouth with or without radula; locomotion by jet propulsion using siphon made from foot

▲ **Figure 33.17 A chiton.** Clinging tenaciously to rocks in the intertidal zone, this chiton displays the eight-plate shell characteristic of molluscs in the class Polyplacophora.

Gastropods

About three-quarters of all living species of molluscs are gastropods **(Figure 33.18)**. Most gastropods are marine, but there are also many freshwater species; garden snails and slugs are among the gastropods that have adapted to land.

The most distinctive characteristic of the class Gastropoda is a developmental process known as **torsion.** As a gastropod embryo develops, its visceral mass rotates up to 180°, causing the animal's anus and mantle cavity to wind up above its head **(Figure 33.19)**. After torsion, some of the organs that were bilateral are reduced in size or are lost on one side of the body. Torsion should not be confused with the formation of a coiled shell, which is an independent developmental process.

Most gastropods have a single, spiraled shell into which the animal can retreat when threatened. The shell is often conical but is somewhat flattened in abalones and limpets. Many gastropods have a distinct head with eyes at the tips of tentacles. Gastropods move literally at a snail's pace by a rippling motion of their foot or by means of cilia. Most gastropods use their

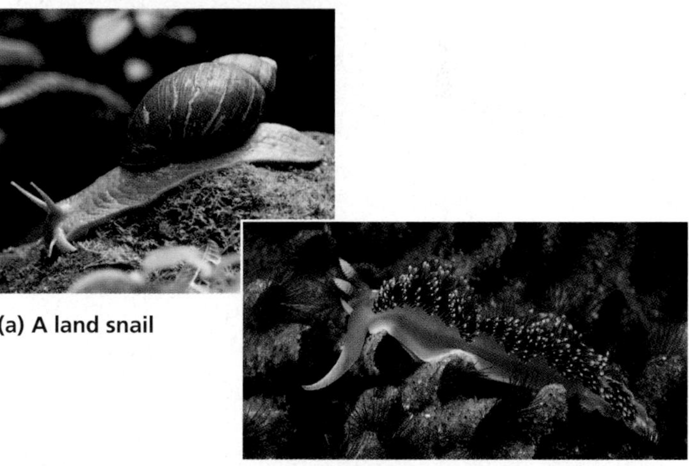

(a) A land snail

(b) A sea slug. Nudibranchs, or sea slugs, lost their shell during their evolution.

▲ **Figure 33.18 Gastropods.**

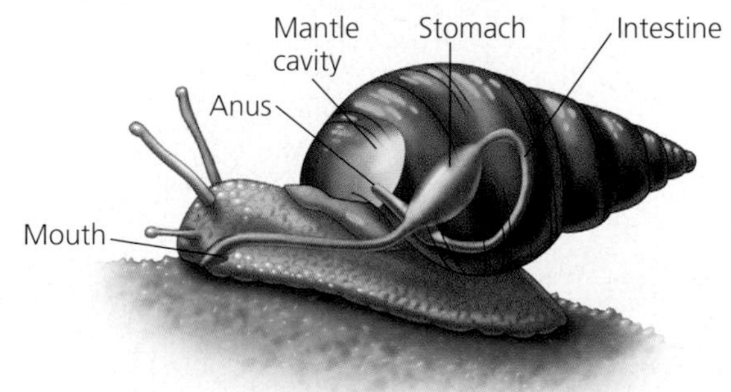

▲ **Figure 33.19 The results of torsion in a gastropod.** Because of torsion (twisting of the visceral mass) during embryonic development, the digestive tract is coiled and the anus is near the anterior end of the animal.

radula to graze on algae or plants. Several groups, however, are predators, and their radula has become modified for boring holes in the shells of other molluscs or for tearing apart prey. In the cone snails, the teeth of the radula form poison darts that are used to subdue prey.

Terrestrial snails lack the gills typical of most aquatic gastropods; instead, the lining of the mantle cavity functions as a lung, exchanging respiratory gases with the air.

Bivalves

The molluscs of class Bivalvia include many species of clams, oysters, mussels, and scallops. Bivalves have a shell divided into two halves **(Figure 33.20)**. The halves are hinged at the mid-dorsal line, and powerful adductor muscles draw them tightly together to protect the soft-bodied animal. Bivalves have no distinct head, and the radula has been lost. Some bivalves have eyes and sensory tentacles along the outer edge of their mantle.

▲ **Figure 33.20 A bivalve.** This scallop has many eyes (dark blue spots) peering out from each half of its hinged shell.

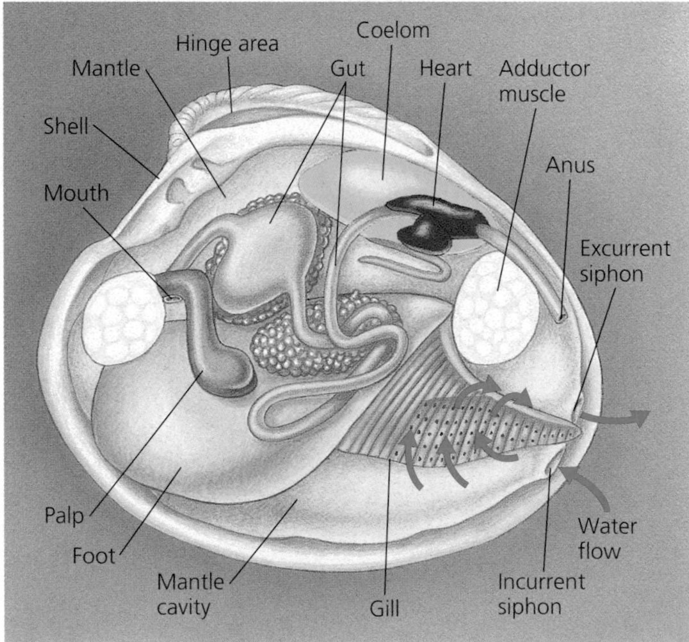

▲ **Figure 33.21 Anatomy of a clam.** The left half of the clam's shell has been removed. Food particles suspended in water that enters through the incurrent siphon are collected by the gills and passed via cilia and elongated flaps called palps to the mouth.

The mantle cavity of a bivalve contains gills that are used for feeding as well as gas exchange **(Figure 33.21)**. Most bivalves are suspension feeders. They trap fine food particles in mucus that coats their gills, and cilia then convey the particles to their mouth. Water enters the mantle cavity through an incurrent siphon, passes over the gills, and then exits the mantle cavity through an excurrent siphon.

Most bivalves lead rather sedentary lives, a characteristic suited to suspension feeding. Sessile mussels secrete strong threads that tether them to rocks, docks, boats, and the shells of other animals. However, clams can pull themselves into the sand or mud, using their muscular foot for an anchor, and scallops can skitter along the seafloor by flapping their shells, rather like the mechanical false teeth sold in novelty shops.

Cephalopods

Cephalopods are active predators. They use their tentacles to grasp prey and their beak-like jaws to inject an immobilizing poison. The foot of a cephalopod has become modified into a muscular excurrent siphon and parts of the tentacles and head. (*Cephalopod* means "head foot.") Most octopuses creep along the seafloor in search of crabs and other food **(Figure 33.22a)**. Squids dart about by drawing water into the mantle cavity and then firing a jet of water through the excurrent siphon **(Figure 33.22b)**. They steer by pointing the siphon in different directions.

A mantle covers the visceral mass of cephalopods, but the shell is reduced and internal (in squids and cuttlefish) or missing altogether (in many octopuses). One small group of shelled cephalopods, the chambered nautiluses, survives today **(Figure 33.22c)**.

Cephalopods are the only molluscs with a closed circulatory system. They also have well-developed sense organs and a complex brain. The ability to learn and behave in a complex manner is probably more critical to fast-moving predators than to sedentary animals such as clams.

The ancestors of octopuses and squids were probably shelled molluscs that took up a predatory lifestyle; the shell was lost in later evolution. Shelled cephalopods called **ammonites,** some of them as large as truck tires, were the dominant invertebrate predators of the seas for hundreds of millions of years until their disappearance during the mass extinctions at the end of the Cretaceous period (see Chapter 26).

Most species of squid are less than 75 cm long, but some are considerably larger. The giant squid (*Architeuthis dux*) was for a long time the largest squid known, with a mantle up to 2.25 m long and a total length of 18 m. In 2003, however, a specimen of the rare species *Mesonychoteuthis hamiltoni* with a mantle length of 2.5 m was caught near Antarctica. Some bi-

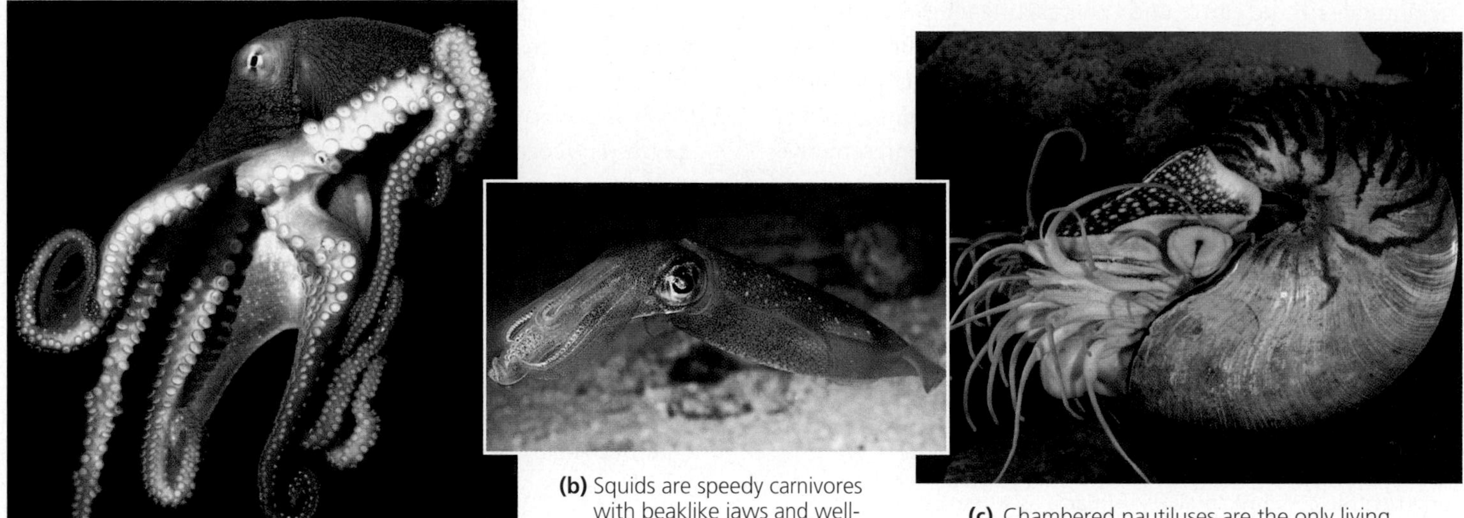

(a) Octopuses are considered among the most intelligent invertebrates.

(b) Squids are speedy carnivores with beaklike jaws and well-developed eyes.

(c) Chambered nautiluses are the only living cephalopods with an external shell.

▲ Figure 33.22 **Cephalopods.**

ologists think this specimen was a juvenile and estimate that adults of its species could be twice as large! Unlike *A. dux,* which has large suckers and small teeth on its tentacles, *M. hamiltoni* has rotating bars at the ends of its tentacles that can deliver deadly lacerations.

It is likely that *A. dux* and *M. hamiltoni* spend most of their time in the deep ocean, where they may feed on large fishes. Remains of both species have been found in the stomachs of sperm whales, which are probably their only natural predator. Scientists have never observed either squid species in its natural habitat. As a result, these marine giants remain among the great mysteries of invertebrate life.

Concept Check 33.4

1. Explain how the modification of the molluscan foot in gastropods and cephalopods relates to their respective lifestyles.
2. How have bivalves diverged from the basic molluscan body plan?

For suggested answers, see Appendix A.

Concept 33.5

Annelids are segmented worms

Annelida means "little rings," referring to the annelid body's resemblance to a series of fused rings. Annelids live in the sea, most freshwater habitats, and damp soil. They range in length from less than 1 mm to 3 m, the length of a giant Australian earthworm. The phylum Annelida is divided into three classes **(Table 33.4)**: Oligochaeta (earthworms and their relatives), Polychaeta (polychaetes), and Hirudinea (leeches).

Oligochaetes

Oligochaetes (from the Greek *oligos,* few, and *chaitē,* long hair) are named for their relatively sparse chaetae, or bristles made of chitin. This class of segmented worms includes the earthworms and a variety of aquatic species. Earthworms eat their way through the soil, extracting nutrients as the soil passes through the alimentary canal. Undigested material, mixed with mucus secreted into the canal, is egested as castings through the anus. Farmers value earthworms because

Table 33.4 Classes of Phylum Annelida

Class and Examples	Main Characteristics
Oligochaeta (freshwater, marine, and terrestrial segmented worms, such as earthworms; see Figure 33.23)	Reduced head; no parapodia, but chaetae present
Polychaeta (mostly marine segmented worms; see Figure 33.24)	Well-developed head; each segment usually has parapodia with chaetae; tube-dwelling and free-living
Hirudinea (leeches; see Figure 33.25)	Body usually flattened, with reduced coelom and segmentation; chaetae absent; suckers at anterior and posterior ends; parasites, predators, and scavengers

the animals till the earth, and their castings improve the texture of the soil. (Darwin estimated that 1 acre of British farmland contained about 50,000 earthworms that produced 18 tons of castings per year.) **Figure 33.23** provides a guided tour of the anatomy of an earthworm, which is representative of annelids.

Earthworms are hermaphrodites, but they cross-fertilize. Two earthworms mate by aligning themselves in such a way that they exchange sperm, and then they separate. The received sperm are stored temporarily while an organ called the clitellum secretes a mucous cocoon. The cocoon slides along the worm, picking up the eggs and then the stored sperm. The cocoon then slips off the worm's head and remains in the soil while the embryos develop. Some earthworms can also reproduce asexually by fragmentation followed by regeneration.

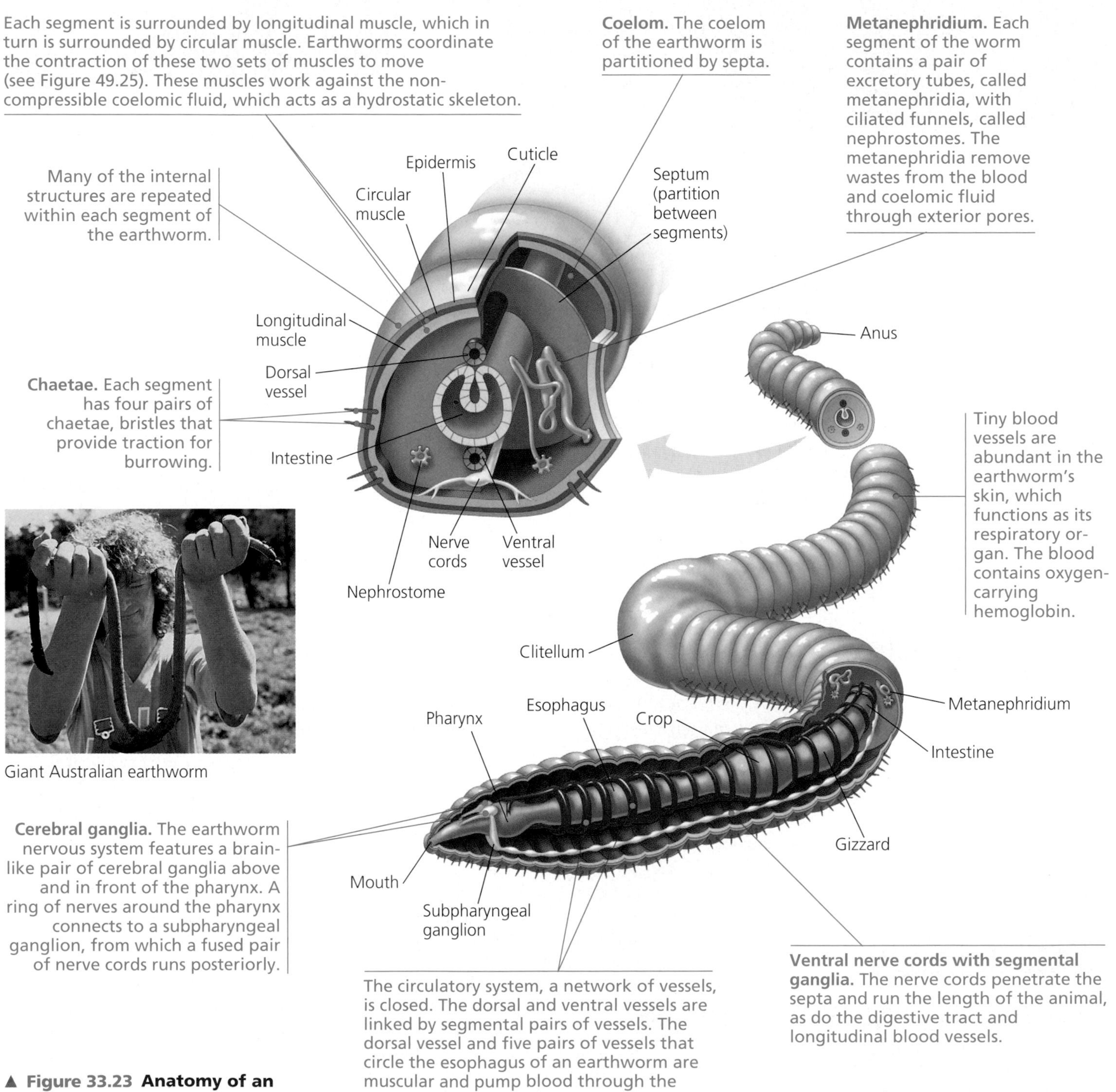

Each segment is surrounded by longitudinal muscle, which in turn is surrounded by circular muscle. Earthworms coordinate the contraction of these two sets of muscles to move (see Figure 49.25). These muscles work against the non-compressible coelomic fluid, which acts as a hydrostatic skeleton.
Coelom. The coelom of the earthworm is partitioned by septa.
Metanephridium. Each segment of the worm contains a pair of excretory tubes, called metanephridia, with ciliated funnels, called nephrostomes. The metanephridia remove wastes from the blood and coelomic fluid through exterior pores.
Many of the internal structures are repeated within each segment of the earthworm.
Epidermis
Cuticle
Circular muscle
Septum (partition between segments)
Longitudinal muscle
Dorsal vessel
Chaetae. Each segment has four pairs of chaetae, bristles that provide traction for burrowing.
Intestine
Nerve cords
Ventral vessel
Nephrostome
Anus
Tiny blood vessels are abundant in the earthworm's skin, which functions as its respiratory organ. The blood contains oxygen-carrying hemoglobin.
Giant Australian earthworm
Clitellum
Esophagus
Pharynx
Crop
Metanephridium
Intestine
Cerebral ganglia. The earthworm nervous system features a brain-like pair of cerebral ganglia above and in front of the pharynx. A ring of nerves around the pharynx connects to a subpharyngeal ganglion, from which a fused pair of nerve cords runs posteriorly.
Gizzard
Mouth
Subpharyngeal ganglion
Ventral nerve cords with segmental ganglia. The nerve cords penetrate the septa and run the length of the animal, as do the digestive tract and longitudinal blood vessels.
The circulatory system, a network of vessels, is closed. The dorsal and ventral vessels are linked by segmental pairs of vessels. The dorsal vessel and five pairs of vessels that circle the esophagus of an earthworm are muscular and pump blood through the circulatory system.

▲ **Figure 33.23 Anatomy of an earthworm.**

Polychaetes

Each segment of a polychaete has a pair of paddle-like or ridge-like structures called parapodia ("almost feet") that function in locomotion **(Figure 33.24)**. Each parapodium has several chaetae, which are more numerous than those in oligochaetes. In many polychaetes, the parapodia are richly supplied with blood vessels and function as gills.

Polychaetes make up a large and diverse class, most of whose members are marine. A few species drift and swim among the plankton, many crawl on or burrow in the seafloor, and many others live in tubes. Some tube-dwellers, such as the fan worms and feather-duster worms, build their tubes by mixing mucus with bits of sand and broken shells. Others, such as Christmas tree worms (see Figure 33.1), construct tubes using only their own secretions.

◀ **Figure 33.24 A polychaete.** *Hesiolyra bergi* lives on the seafloor around deep-sea hydrothermal vents.

Leeches

The majority of leeches inhabit fresh water, but there are also marine species as well as terrestrial leeches found in moist vegetation. Leeches range in length from about 1 to 30 cm. Many are predators that feed on other invertebrates, but some are parasites that suck blood by attaching temporarily to other animals, including humans **(Figure 33.25)**. Some parasitic species use bladelike jaws to slit the skin of the host, whereas others secrete enzymes that digest a hole through the skin. The host is usually oblivious to this attack because the leech secretes an anesthetic. After making the incision, the leech secretes another chemical, hirudin, which keeps the blood of the host from coagulating near the incision. The parasite then sucks as much blood as it can hold, often more than ten times its own weight. After this gorging, a leech can last for months without another meal.

Until this century, leeches were frequently used for bloodletting. Today they are used to drain blood that accumulates in tissues following certain injuries or surgeries. Researchers are also investigating the potential use of hirudin to dissolve unwanted blood clots that form during surgery or as a result of heart disease. A recombinant form of hirudin has been developed and is in clinical trials.

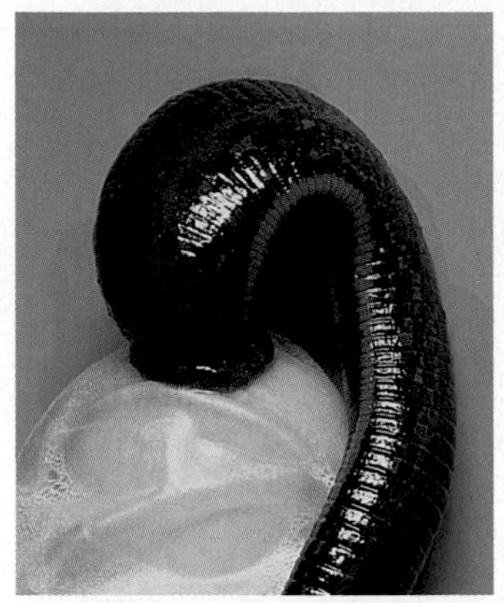

◀ **Figure 33.25 A leech.** A nurse applied this medicinal leech (*Hirudo medicinalis*) to a patient's sore thumb to drain blood from a hematoma (an abnormal accumulation of blood around an internal injury).

Concept Check 33.5

1. Annelid anatomy can be described as "a tube within a tube." Explain.
2. Explain how an earthworm uses its segmental muscles and coelom in movement.

For suggested answers, see Appendix A.

Concept 33.6

Nematodes are nonsegmented pseudocoelomates covered by a tough cuticle

Among the most widespread of all animals, nematodes, or roundworms, are found in most aquatic habitats, in the soil, in the moist tissues of plants, and in the body fluids and tissues of animals. In contrast to annelids, nematodes do not have a segmented body. The cylindrical bodies of nematodes (phylum Nematoda) range from less than 1 mm to more than a meter in length, often tapering to a fine tip at the posterior end and to a more blunt tip at the anterior end **(Figure 33.26)**. The body is covered by a tough coat called a **cuticle**; as the worm grows, it

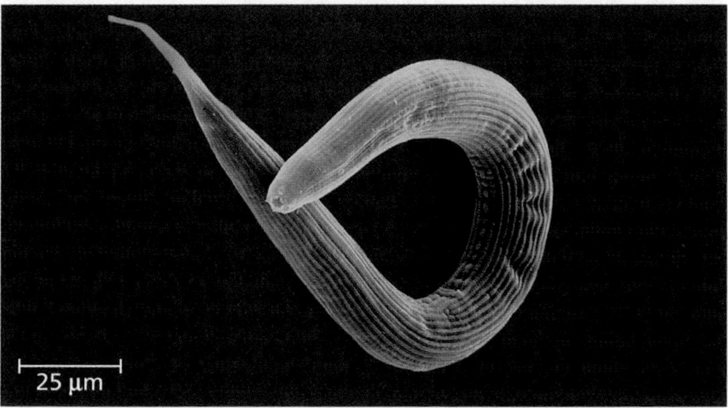

▲ **Figure 33.26 A free-living nematode (colorized SEM).**

periodically sheds its old cuticle and secretes a new, larger one. Nematodes have an alimentary canal, though they lack a circulatory system. Nutrients are transported throughout the body via fluid in the pseudocoelom. The muscles of nematodes are all longitudinal, and their contraction produces a thrashing motion.

Nematode reproduction is usually sexual and involves internal fertilization. In most species, the sexes are separate and females are larger than males. A female may deposit 100,000 or more fertilized eggs per day. The zygotes of most species are resistant cells that can survive harsh conditions.

Great numbers of nematodes live in moist soil and in decomposing organic matter on the bottoms of lakes and oceans. While 25,000 species are known, perhaps 20 times that number actually exist. If nothing but nematodes remained, it has been said, they would still preserve the outline of the planet and many of its features. These extremely numerous free-living worms play an important role in decomposition and nutrient cycling, but little is known about most species. One species of soil nematode, *Caenorhabditis elegans,* is well studied and has become a model research organism in developmental biology (see Chapter 21). Ongoing studies on *C. elegans* are revealing some of the mechanisms involved in aging in humans, among other findings.

Phylum Nematoda includes many important agricultural pests that attack the roots of plants. Other species of nematodes parasitize animals. Humans host at least 50 nematode species, including various pinworms and hookworms. One notorious nematode is *Trichinella spiralis,* the worm that causes trichinosis **(Figure 33.27)**. Humans acquire this nematode by eating undercooked infected pork or other meat with juvenile worms encysted in the muscle tissue. Within the human intestine, the juveniles develop into sexually mature adults. Females burrow into the intestinal muscles and produce more juveniles, which bore through the body or travel in lymphatic vessels to other organs, including skeletal muscles, where they encyst.

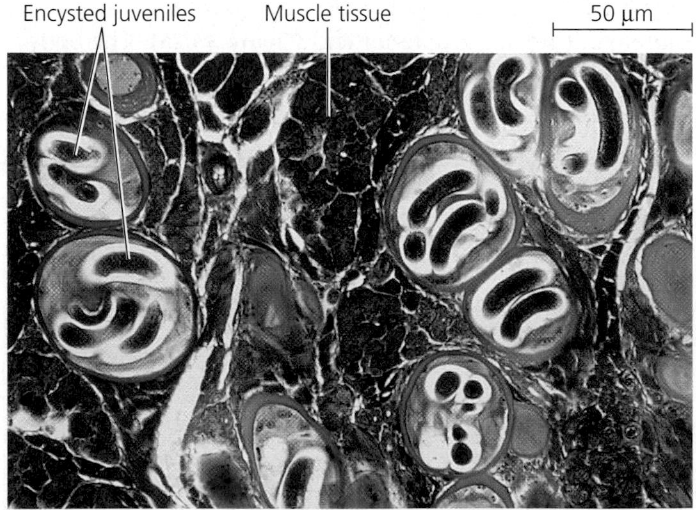

▲ **Figure 33.27 Juveniles of the parasitic nematode *Trichinella spiralis* encysted in human muscle tissue (LM).**

Parasitic nematodes have an extraordinary molecular toolkit that enables them to redirect some of the cellular functions of their hosts. Plant-parasitic nematodes inject molecules that induce the development of root cells, which then supply nutrients to the parasites. *Trichinella* invades individual muscle cells and controls the expression of specific muscle genes, which code for proteins that make the cell elastic enough to house the nematode. Additionally, the cell releases signals that attract blood vessels, which then supply the nematode with nutrients. These extraordinary parasites have been dubbed "animals that act like viruses."

Concept Check 33.6

1. Why would it be risky to order pork chops "rare" in a restaurant?
2. How does the nematode body plan differ from that of annelids?

For suggested answers, see Appendix A.

Concept 33.7

Arthropods are segmented coelomates that have an exoskeleton and jointed appendages

Zoologists estimate that the arthropod population of the world, including crustaceans, spiders, and insects, numbers about a billion billion (10^{18}) individuals. More than 1 million arthropod species have been described, most of which are insects. In fact, two out of every three species known are arthropods, and members of the phylum Arthropoda can be found in nearly all habitats of the biosphere. On the criteria of species diversity, distribution, and sheer numbers, arthropods must be regarded as the most successful of all animal phyla.

General Characteristics of Arthropods

The diversity and success of **arthropods** are largely related to their segmentation, hard exoskeleton, and jointed appendages (*arthropod* means "jointed feet"). Early arthropods, such as the **trilobites,** had pronounced segmentation, but their appendages showed little variation from segment to segment **(Figure 33.28)**. As arthropods continued to evolve, the segments tended to fuse and become fewer in number, and the appendages became specialized for a variety of functions.

These evolutionary changes resulted not only in great diversification but also in an efficient body plan that permits the division of labor among different regions. For example, the various appendages of some living arthropods are modified for walking, feeding, sensory reception, copulation, and defense. **Figure 33.29** illustrates the diverse appendages and other arthropod characteristics of a lobster.

▲ **Figure 33.28 A trilobite fossil.** Trilobites were common denizens of the shallow seas throughout the Paleozoic era but disappeared with the great Permian extinctions about 250 million years ago. Paleontologists have described about 4,000 trilobite species.

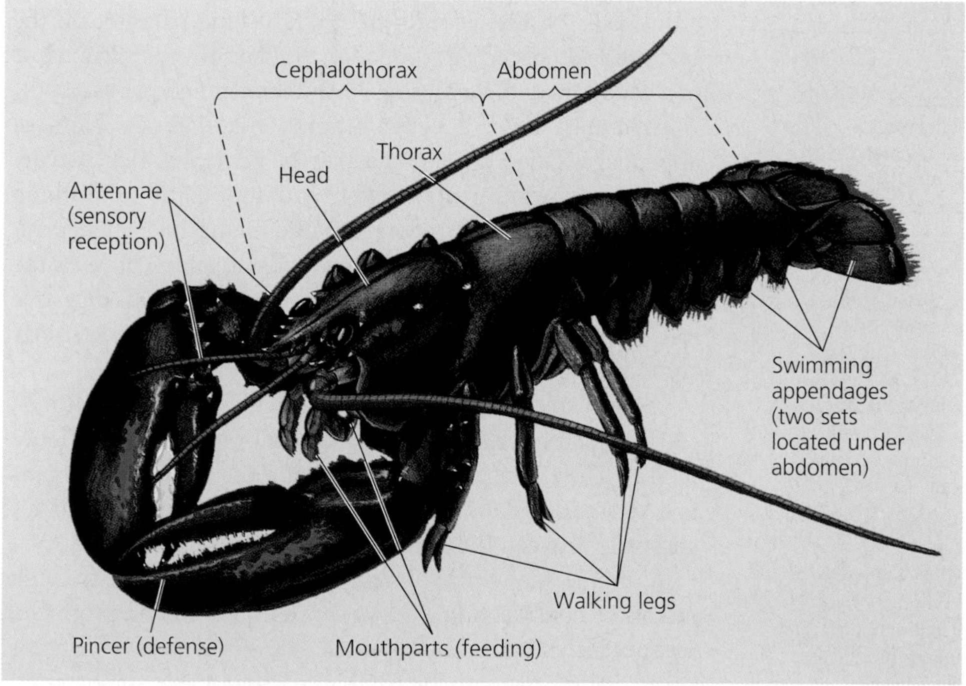

The body of an arthropod is completely covered by the cuticle, an **exoskeleton** (external skeleton) constructed from layers of protein and the polysaccharide chitin. The cuticle can be thick and hard over some parts of the body and paper-thin and flexible over others, such as the joints. The rigid exoskeleton protects the animal and provides points of attachment for the muscles that move the appendages. But it also means that an arthropod cannot grow without occasionally shedding its exoskeleton and producing a larger one. This process, called **molting** or ecdysis, is energetically expensive. A recently molted arthropod is also vulnerable to predation and other dangers until its new, soft exoskeleton hardens.

When the arthropod exoskeleton first evolved in the seas, its main functions were probably protection and anchorage for muscles, but it later additionally enabled certain arthropods to live on land. The exoskeleton's relative impermeability to water helped prevent desiccation, and its strength solved the problem of support when arthropods left the buoyancy of water. Arthropods began to diversify on land following the colonization of land by plants in the early Paleozoic. In 2004, an amateur fossil hunter in Scotland found a 428-million-year-old fossil of a millipede. Fossilized tracks of other terrestrial arthropods date from about 450 million years ago.

Arthropods have well-developed sensory organs, including eyes, olfactory (smell) receptors, and antennae that function in both touch and smell. Most sensory organs are concentrated at the anterior end of the animal.

Like many molluscs, arthropods have an **open circulatory system** in which fluid called hemolymph is propelled by a heart through short arteries and then into spaces called sinuses surrounding the tissues and organs. (The term *blood* is best reserved for fluid in a closed circulatory system.) Hemolymph reenters the arthropod heart through pores that are usually equipped with valves. The body sinuses are collectively called the hemocoel, which is not part of the coelom. In most arthropods, the coelom that forms in the embryo becomes much reduced as development progresses, and the hemocoel becomes the main body cavity in adults. Despite their similarity, the open circulatory systems of molluscs and arthropods probably arose independently.

A variety of organs specialized for gas exchange have evolved in arthropods. These organs allow the diffusion of respiratory gases in spite of the exoskeleton. Most aquatic species have gills with thin, feathery extensions that place an extensive surface area in contact with the

▲ **Figure 33.29 External anatomy of an arthropod.** Many of the distinctive features of arthropods are apparent in this dorsal view of a lobster, along with some uniquely crustacean characteristics. The body is segmented, but this characteristic is obvious only in the abdomen. The appendages (including antennae, pincers, mouthparts, walking legs, and swimming appendages), are jointed. The head bears a pair of compound (multilens) eyes, each situated on a movable stalk. The whole body, including appendages, is covered by an exoskeleton.

surrounding water. Terrestrial arthropods generally have internal surfaces specialized for gas exchange. Most insects, for instance, have tracheal systems, branched air ducts leading into the interior from pores in the cuticle.

Findings from molecular systematics are leading biologists to develop new hypotheses about arthropod evolutionary relationships, just as they are for other branches of the tree of life. Evidence now suggests that living arthropods consist of four major lineages that diverged early in the evolution of the phylum **(Table 33.5)**: **cheliceriforms** (sea spiders, horseshoe crabs, scorpions, ticks, mites, and spiders); **myriapods** (centipedes and millipedes); **hexapods** (insects and their wingless, six-legged relatives); and **crustaceans** (crabs, lobsters, shrimps, barnacles, and many others).

Cheliceriforms

Cheliceriforms (subphylum Cheliceriformes; from the Greek *cheilos*, lips, and *cheir*, arm) are named for clawlike feeding appendages called **chelicerae**, which serve as pincers or fangs. Cheliceriforms have an anterior cephalothorax and a posterior abdomen. They lack antennae, and most have simple eyes (eyes with a single lens).

Table 33.5 Subphyla of Phylum Arthropoda

Subphylum and Examples	Main Characteristics
Cheliceriformes (horseshoe crabs, spiders, scorpions, ticks, mites; see Figures 33.30–33.32)	Body having one or two main parts; six pairs of appendages (chelicerae, pedipalps, and four pairs of walking legs); mostly terrestrial or marine
Myriapoda (millipedes and centipedes; see Figures 33.33 and 33.34)	Distinct head bearing antennae and chewing mouthparts; terrestrial; millipedes are herbivorous and have two pairs of walking legs per trunk segment; centipedes are carnivorous and have one pair of walking legs per trunk segment and poison claws on first body segment
Hexapoda (insects, springtails; see Figures 33.35–33.37)	Body divided into head, thorax, and abdomen; antennae present; mouthparts modified for chewing, sucking, or lapping; three pairs of legs and usually two pairs of wings; mostly terrestrial
Crustacea (crabs, lobsters, crayfish, shrimp; see Figures 33.29 and 33.38)	Body of two or three parts; antennae present; chewing mouthparts; three or more pairs of legs; mostly marine and freshwater

▲ **Figure 33.30 Horseshoe crabs (*Limulus polyphemus*).** Common on the Atlantic and Gulf coasts of the United States, these "living fossils" have changed little in hundreds of millions of years. They have survived from a rich diversity of cheliceriforms that once filled the seas.

The earliest cheliceriforms were **eurypterids**, or water scorpions. These mainly marine and freshwater predators grew up to 3 m long. Most of the marine cheliceriforms, including all of the eurypterids, are extinct; among the marine species that survive today are the sea spiders (pycnogonids) and the horseshoe crabs **(Figure 33.30)**.

The bulk of modern cheliceriforms are arachnids, a group that includes scorpions, spiders, ticks, and mites **(Figure 33.31)**. Ticks and many mites are among a large group of parasitic arthropods. Nearly all ticks are bloodsucking parasites on the body surfaces of reptiles or mammals. Parasitic mites live on or in a wide variety of vertebrates, invertebrates, and plants.

Arachnids have a cephalothorax that has six pairs of appendages: the chelicerae, a pair of pedipalps that usually function in sensing or feeding, and four pairs of walking legs **(Figure 33.32)**. Spiders use their fang-like chelicerae, which are equipped with poison glands, to attack prey. As the chelicerae masticate (chew) the prey, the spider spills digestive juices onto the torn tissues. The food softens, and the spider sucks up the liquid meal.

In most spiders, gas exchange is carried out by **book lungs**, stacked plates contained in an internal chamber (see Figure 33.32). The extensive surface area of these respiratory organs is a structural adaptation that enhances the exchange of O_2 and CO_2 between the hemolymph and air.

A unique adaptation of many spiders is the ability to catch insects by constructing webs of silk, a liquid protein produced by specialized abdominal glands. The silk is spun by organs called spinnerets into fibers that solidify. Each spider engineers a style of web characteristic of its species and builds it perfectly on the first try. This complex behavior is apparently inherited. Various spiders also use silk in other ways: as droplines for rapid escape, as a cover for eggs, and even as "gift wrap" for food that males offer females during courtship.

(a) Scorpions have pedipalps that are pincers specialized for defense and the capture of food. The tip of the tail bears a poisonous stinger.

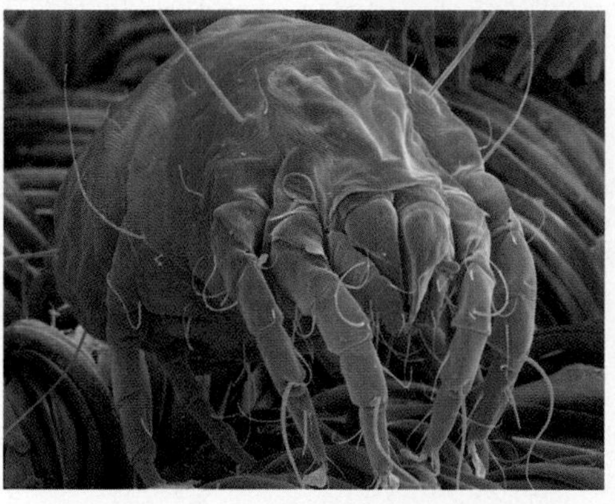

(b) Dust mites are ubiquitous scavengers in human dwellings but are harmless except to those people who are allergic to them (colorized SEM).

(c) Web-building spiders are generally most active during the daytime.

▲ **Figure 33.31 Arachnids.**

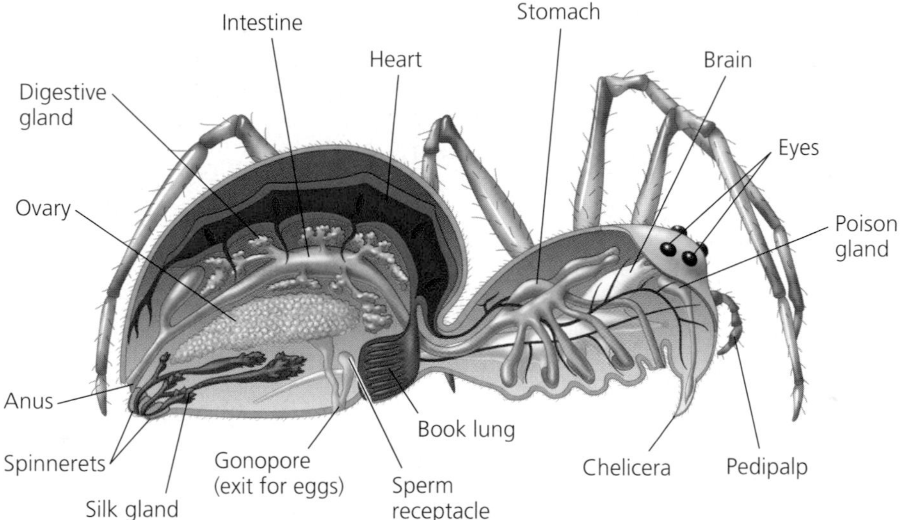

► **Figure 33.32 Anatomy of a spider.**

Myriapods

Millipedes and centipedes belong to the subphylum Myriapoda, the myriapods. All living myriapods are terrestrial. Their head has a pair of antennae and three pairs of appendages modified as mouthparts, including the jaw-like **mandibles.**

Millipedes (class Diplopoda) have a large number of legs, though fewer than the thousand their name implies **(Figure 33.33)**. Each trunk segment is formed from two fused segments and has two pairs of legs. Millipedes eat decaying leaves and other plant matter. They may have been among the earliest animals on land, living on mosses and primitive vascular plants.

Unlike millipedes, centipedes (class Chilopoda) are carnivores. Each segment of a centipede's trunk region has one pair

▲ **Figure 33.33 A millipede.**

▲ **Figure 33.34 A centipede.**

of legs **(Figure 33.34)**. Centipedes have poison claws on their foremost trunk segment that paralyze prey and aid in defense.

Insects

Insects and their relatives (subphylum Hexapoda) are more species-rich than all other forms of life combined. They live in almost every terrestrial habitat and in fresh water, and flying insects fill the air. Insects are rare, though not absent, in the seas, where crustaceans are the dominant arthropods. The internal anatomy of an insect includes several complex organ systems, which are highlighted in **Figure 33.35**.

The oldest insect fossils date from the Devonian period, which began about 416 million years ago. However, when flight evolved during the Carboniferous and Permian periods, it spurred an explosion in insect variety. A fossil record of diverse insect mouthparts indicates that specialized feeding on gymnosperms and other Carboniferous plants also contributed to the adaptive radiation of insects. A widely held hypothesis is that the greatest diversification of insects paralleled the evolutionary radiation of flowering plants during the Cretaceous and early Paleogene periods about 65–60 million years ago. This view is challenged by new research suggesting that insects diversified extensively before the angiosperm radiation. Thus, during the evolution of flowering plants and the herbivorous insects that pollinated them, insect diversity may have been as much a cause of angiosperm radiation as an effect.

Flight is obviously one key to the great success of insects. An animal that can fly can escape many predators, find food and mates, and disperse to new habitats much faster than an animal that must crawl about on the ground. Many insects have one or two pairs of wings that emerge from the dorsal side

▼ **Figure 33.35 Anatomy of a grasshopper, an insect.**

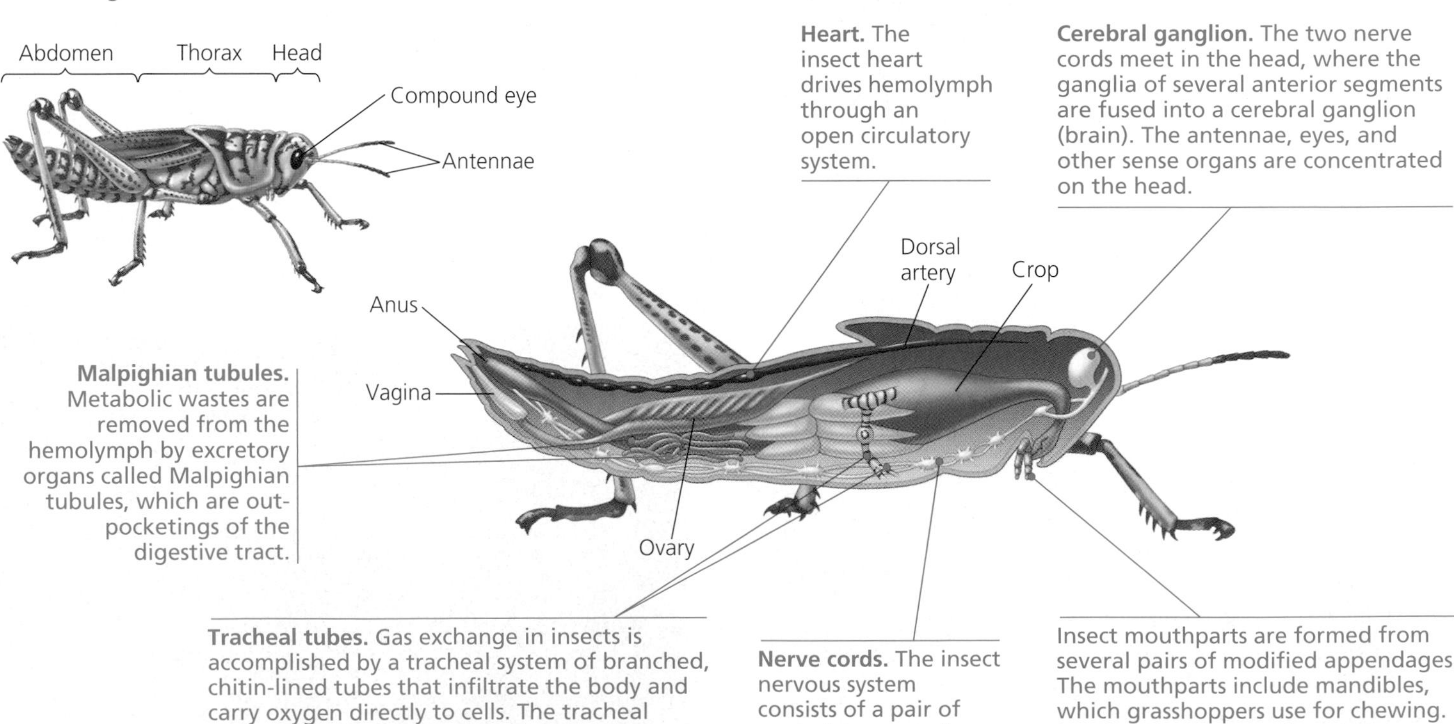

The insect body has three regions: head, thorax, and abdomen. The segmentation of the thorax and abdomen are obvious, but the segments that form the head are fused.
Abdomen
Thorax
Head
Compound eye
Antennae
Heart. The insect heart drives hemolymph through an open circulatory system.
Cerebral ganglion. The two nerve cords meet in the head, where the ganglia of several anterior segments are fused into a cerebral ganglion (brain). The antennae, eyes, and other sense organs are concentrated on the head.
Dorsal artery
Crop
Anus
Vagina
Malpighian tubules. Metabolic wastes are removed from the hemolymph by excretory organs called Malpighian tubules, which are outpocketings of the digestive tract.
Ovary
Tracheal tubes. Gas exchange in insects is accomplished by a tracheal system of branched, chitin-lined tubes that infiltrate the body and carry oxygen directly to cells. The tracheal system opens to the outside of the body through spiracles, pores that can control air flow and water loss by opening or closing.
Nerve cords. The insect nervous system consists of a pair of ventral nerve cords with several segmental ganglia.
Insect mouthparts are formed from several pairs of modified appendages. The mouthparts include mandibles, which grasshoppers use for chewing. In other insects, mouthparts are specialized for lapping, piercing, or sucking.

of the thorax. Because the wings are extensions of the cuticle and not true appendages, insects can fly without sacrificing any walking legs. By contrast, the flying vertebrates—birds and bats—have one of their two pairs of walking legs modified into wings and are generally quite clumsy on the ground.

Insect wings may have first evolved as extensions of the cuticle that helped the insect body absorb heat, only later becoming organs for flight. Other views suggest that wings allowed insects to glide from vegetation to the ground, or even that they served as gills in aquatic insects. Still another hypothesis is that insect wings functioned for swimming before they functioned for flight.

Morphological and molecular data indicate that wings evolved only once in insects. Dragonflies, which have two similar pairs of wings, were among the first insects to fly. Several insect orders that evolved later than dragonflies have modified flight equipment. The wings of bees and wasps, for instance, are hooked together and move as a single pair. Butterfly wings operate in a similar fashion because the anterior pair overlaps the posterior wings. In beetles, the posterior wings function in flight, while the anterior ones are modified as covers that protect the flight wings when the beetle is on the ground or is burrowing.

Many insects undergo metamorphosis during their development. In the **incomplete metamorphosis** of grasshoppers and some other orders, the young (called nymphs) resemble adults but are smaller, have different body proportions, and lack wings. The nymph goes through a series of molts, each time looking more like an adult. With the final molt, the insect reaches full size, acquires wings, and becomes sexually mature. Insects with **complete metamorphosis** have larval stages specialized for eating and growing that are known by such names as maggot, grub, or caterpillar. The larval stage looks entirely different from the adult stage, which is specialized for dispersal and reproduction. Metamorphosis from the larval stage to the adult occurs during a pupal stage **(Figure 33.36)**.

Reproduction in insects is usually sexual, with separate male and female individuals. Adults come together and recognize each other as members of the same species by advertising with bright colors (as in butterflies), sound (as in crickets), or odors (as in moths). Fertilization is generally internal. In most species, sperm are deposited directly into the female's vagina at the time of copulation, though in some species the male deposits a sperm packet outside the female, and the female picks it up. An internal structure in the female called the spermatheca stores the sperm, usually enough to fertilize more than one batch of eggs. Many insects mate only once in a lifetime. After mating, a female often lays her eggs on an appropriate food source where the next generation can begin eating as soon as it hatches.

Insects are classified in about 26 orders, 15 of which are explored in **Figure 33.37**, on the next two pages.

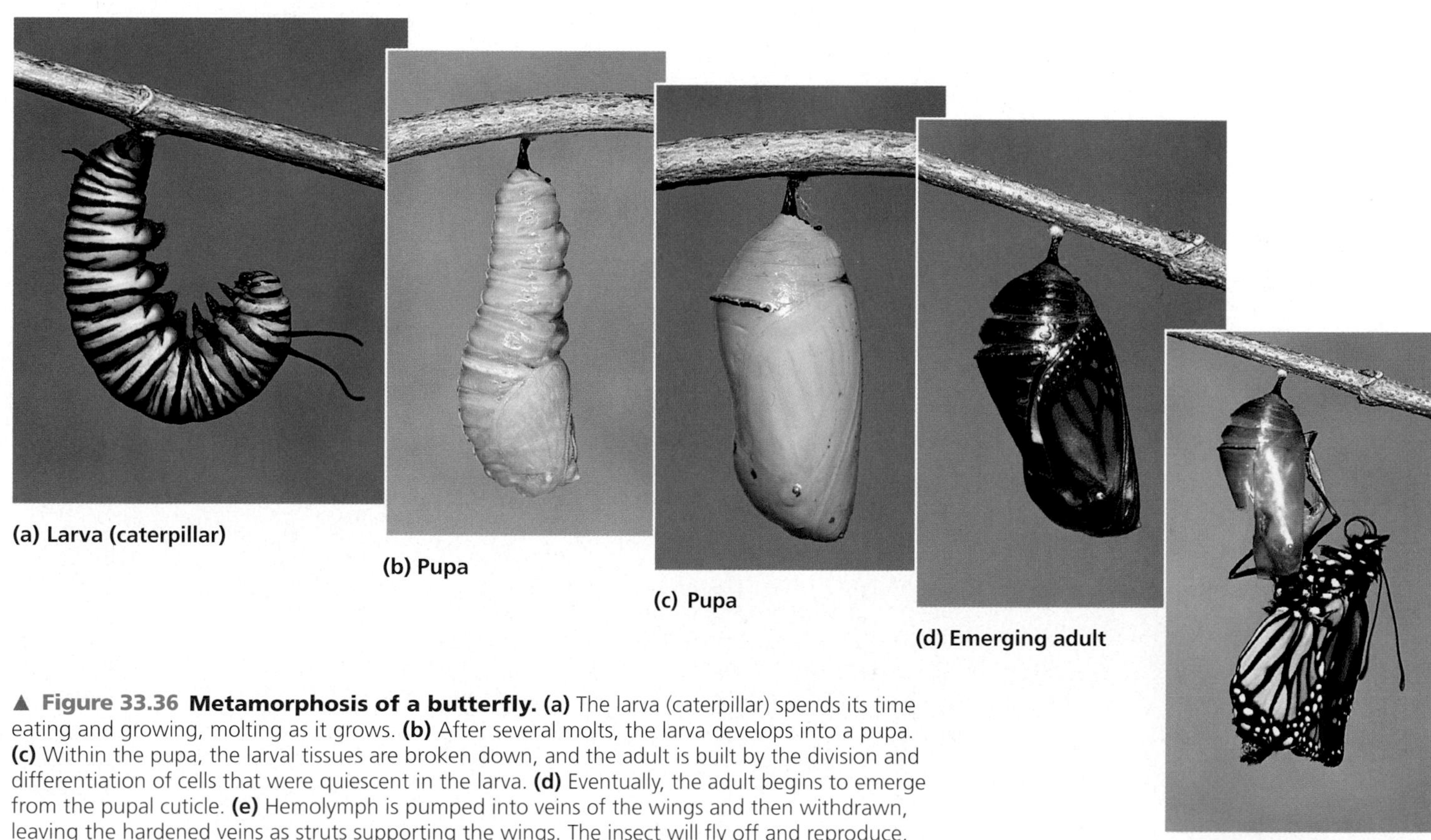

▲ **Figure 33.36 Metamorphosis of a butterfly. (a)** The larva (caterpillar) spends its time eating and growing, molting as it grows. **(b)** After several molts, the larva develops into a pupa. **(c)** Within the pupa, the larval tissues are broken down, and the adult is built by the division and differentiation of cells that were quiescent in the larva. **(d)** Eventually, the adult begins to emerge from the pupal cuticle. **(e)** Hemolymph is pumped into veins of the wings and then withdrawn, leaving the hardened veins as struts supporting the wings. The insect will fly off and reproduce, deriving much of its nourishment from the food reserves stored by the feeding larva.

Figure 33.37

Exploring Insect Diversity

ORDER	APPROXIMATE NUMBER OF SPECIES	MAIN CHARACTERISTICS	EXAMPLES
Blattodea	4,000	Cockroaches have a dorsoventrally flattened body, with legs modified for rapid running. Forewings, when present, are leathery, whereas hind wings are fanlike. Fewer than 40 cockroach species live in houses; the rest exploit habitats ranging from tropical forest floors to caves and deserts.	German cockroach 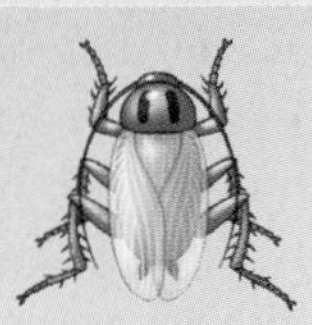
Coleoptera	350,000	Beetles comprise the most species-rich order of insects. They have two pairs of wings, one of which is thick and leathery, the other membranous. They have an armored exoskeleton and mouthparts adapted for biting and chewing. Beetles undergo complete metamorphosis.	Japanese beetle 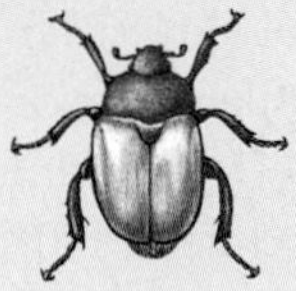
Dermaptera	1,200	Earwigs are generally nocturnal scavengers. While some species are wingless, others have two pairs of wings, one of which is thick and leathery, the other membranous. Earwigs have biting mouthparts and large posterior pincers. They undergo incomplete metamorphosis.	 Earwig
Diptera	151,000	Dipterans have one pair of wings; the second pair has become modified into balancing organs called halteres. Their head is large and mobile; their mouthparts are adapted for sucking, piercing, or lapping. Dipterans undergo complete metamorphosis. Flies and mosquitoes are among the best-known dipterans, which live as scavengers, predators, and parasites.	Horsefly 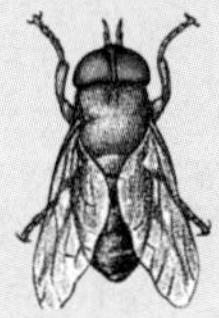
Hemiptera	85,000	Hemipterans are so-called "true bugs," including bed bugs, assassin bugs, and chinch bugs. (Insects in other orders are sometimes erroneously called bugs.) Hemipterans have two pairs of wings, one pair partly leathery, the other membranous. They have piercing or sucking mouthparts and undergo incomplete metamorphosis.	Leaf-footed bug 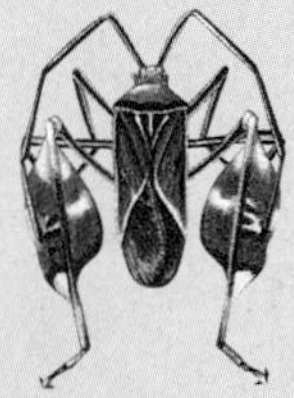
Hymenoptera	125,000	Ants, bees, and wasps are generally highly social insects. They have two pairs of membranous wings, a mobile head, and chewing or sucking mouthparts. The females of many species have a posterior stinging organ. Hymenopterans undergo complete metamorphosis.	 Cicada-killer wasp
Isoptera	2,000	Termites are widespread social insects that produce enormous colonies. It has been estimated that there are 700 kg of termites for every person on Earth! Some termites have two pairs of membranous wings, while others are wingless. They feed on wood with the aid of microbial symbionts carried in specialized chambers in their hindgut.	 Termite

ORDER	APPROXIMATE NUMBER OF SPECIES	MAIN CHARACTERISTICS	EXAMPLES
Lepidoptera	120,000	Butterflies and moths are among the best-known insects. They have two pairs of wings covered with tiny scales. To feed, they uncoil a long proboscis. Most feed on nectar, but some species feed on other substances, including animal blood or tears.	Swallowtail butterfly
Odonata	5,000	Dragonflies and damselflies have two pairs of large, membranous wings. They have an elongated abdomen, large, compound eyes, and chewing mouthparts. They undergo incomplete metamorphosis and are active predators.	Dragonfly
Orthoptera	13,000	Grasshoppers, crickets, and their relatives are mostly herbivorous. They have large hind legs adapted for jumping, two pairs of wings (one leathery, one membranous), and biting or chewing mouthparts. Males commonly make courtship sounds by rubbing together body parts, such as a ridge on their hind leg. Orthopterans undergo incomplete metamorphosis.	Katydid
Phasmida	2,600	Stick insects and leaf insects are exquisite mimics of plants. The eggs of some species even mimic seeds of the plants on which the insects live. Their body is cylindrical or flattened dorsoventrally. They lack forewings but have fanlike hind wings. Their mouthparts are adapted for biting or chewing.	Stick insect
Phthiraptera	2,400	Commonly called sucking lice, these insects spend their entire life as an ectoparasite feeding on the hair or feathers of a single host. Their legs, equipped with clawlike tarsi, are adapted for clinging to their hosts. They lack wings and have reduced eyes. Sucking lice undergo incomplete metamorphosis.	Human body louse
Siphonaptera	2,400	Fleas are bloodsucking ectoparasites on birds and mammals. Their body is wingless and laterally compressed. Their legs are modified for clinging to their hosts and for long-distance jumping. They undergo complete metamorphosis.	Flea
Thysanura	450	Silverfish are small, wingless insects with a flattened body and reduced eyes. They live in leaf litter or under bark. They can also infest buildings, where they can become pests.	Silverfish
Trichoptera	7,100	The larvae of caddisflies live in streams, where they make houses from sand grains, wood fragments, or other material held together by silk. Adults have two pairs of hairy wings and chewing or lapping mouthparts. They undergo complete metamorphosis.	Caddisfly

Animals as numerous, diverse, and widespread as insects are bound to affect the lives of all other terrestrial organisms, including humans. On the one hand, we depend on bees, flies, and many other insects to pollinate our crops and orchards. On the other hand, insects are carriers for many diseases, including African sleeping sickness (spread by tsetse flies that carry *Trypanosoma;* see Figure 28.7) and malaria (spread by mosquitoes that carry *Plasmodium;* see Figure 28.11). Furthermore, insects compete with humans for food. In parts of Africa, for instance, insects claim about 75% of the crops. Trying to minimize their losses, farmers in the United States spend billions of dollars each year on pesticides, spraying crops with massive doses of some of the deadliest poisons ever invented. Try as they may, not even humans have challenged the preeminence of insects and their arthropod kin. As Cornell University entomologist Thomas Eisner puts it: "Bugs are not going to inherit the Earth. They own it now. So we might as well make peace with the landlord."

Crustaceans

While arachnids and insects thrive on land, crustaceans, for the most part, have remained in marine and freshwater environments. Crustaceans (subphylum Crustacea) typically have biramous (branched) appendages that are extensively specialized. Lobsters and crayfish, for instance, have a toolkit of 19 pairs of appendages (see Figure 33.29). The anterior-most appendages are antennae; crustaceans are the only arthropods with two pairs. Three or more pairs of appendages are modified as mouthparts, including the hard mandibles. Walking legs are present on the thorax, and, unlike insects, crustaceans have appendages on the abdomen. A lost appendage can be regenerated.

Small crustaceans exchange gases across thin areas of the cuticle; larger species have gills. Nitrogenous wastes also diffuse through thin areas of the cuticle, but a pair of glands regulates the salt balance of the hemolymph.

Sexes are separate in most crustaceans. In the case of lobsters and crayfish, the male uses a specialized pair of abdominal appendages to transfer sperm to the reproductive pore of the female during copulation. Most aquatic crustaceans go through one or more swimming larval stages.

One of the largest groups of crustaceans (numbering about 10,000 species) is the **isopods,** which include terrestrial, freshwater, and marine species. Some isopod species are abundant in habitats at the bottom of the deep ocean. Among the terrestrial isopods are the pill bugs, or wood lice, common on the undersides of moist logs and leaves.

Lobsters, crayfish, crabs, and shrimp are all relatively large crustaceans called **decapods (Figure 33.38a)**. The cuticle of decapods is hardened by calcium carbonate; the portion that covers the dorsal side of the cephalothorax forms a shield called the carapace. Most decapod species are marine. Crayfish, however, live in fresh water, and some tropical crabs live on land.

Many small crustaceans are important members of marine and freshwater plankton communities. Planktonic crustaceans include many species of **copepods,** which are among the most numerous of all animals, and the shrimplike krill, which grow to about 3 cm long **(Figure 33.38b)**. A major food source for baleen whales (including blue whales and right whales), krill are now being harvested in great numbers by humans for food and agricultural fertilizer. The larvae of many larger-bodied crustaceans are also planktonic.

Barnacles are a group of mostly sessile crustaceans whose cuticle is hardened into a shell containing calcium carbonate **(Figure 33.38c)**. Most barnacles anchor themselves to rocks, boat hulls, pilings, and other submerged surfaces. The adhe-

(a) Ghost crabs (genus *Ocypode*) live on sandy ocean beaches worldwide. Primarily nocturnal, they take shelter in burrows during the day.

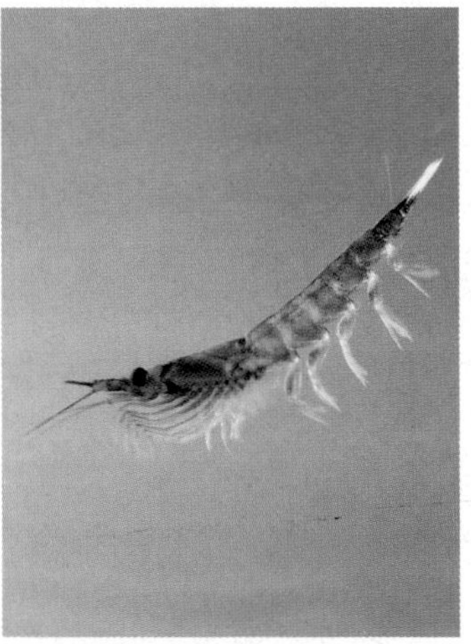

(b) Planktonic crustaceans known as krill are consumed in vast quantities by whales.

(c) The jointed appendages projecting from the shells of these barnacles capture organisms and organic particles suspended in the water.

▲ **Figure 33.38 Crustaceans.**

sive they use is as strong as any synthetic glue. To feed, they extend appendages from their shell to strain food from the water. Other barnacles live as parasites inside hosts such as crabs, where their bodies resemble the roots of a plant. Barnacles were not recognized as crustaceans until the 1800s, when naturalists discovered that barnacle larvae resemble those of other crustaceans. The remarkable mix of unique traits and crustacean homologies found in barnacles was a major inspiration to Charles Darwin as he developed his theory of evolution.

Concept Check 33.7

1. In contrast to our jaws, which move up and down, the mouthparts of arthropods move side to side. Explain this feature of arthropods in terms of the origin of their mouthparts.
2. Would it be reasonable to call phylum Arthropoda the most successful animal phylum? Explain your answer.
3. Describe two adaptations that enabled insects to thrive on land.

For suggested answers, see Appendix A.

Concept 33.8

Echinoderms and chordates are deuterostomes

Sea stars and other echinoderms (phylum Echinodermata) may seem to have little in common with phylum Chordata, which includes the vertebrates—animals that have a backbone. In fact, all these animals share features characteristic of deuterostomes: radial cleavage, development of the coelom from the archenteron, and formation of the mouth at the end of the embryo opposite the blastopore (see Figure 32.9). Molecular systematics has reinforced Deuterostomia as a clade of bilaterian animals.

Echinoderms

Sea stars and most other **echinoderms** (from the Greek *echin*, spiny, and *derma*, skin) are slow-moving or sessile marine animals. A thin skin covers an endoskeleton of hard calcareous plates. Most echinoderms are prickly from skeletal bumps and spines. Unique to echinoderms is the **water vascular system**, a network of hydraulic canals branching into extensions called **tube feet** that function in locomotion, feeding, and gas exchange **(Figure 33.39)**. Sexual reproduction of echinoderms

▼ **Figure 33.39 Anatomy of a sea star, an echinoderm.**

A short digestive tract runs from the mouth on the bottom of the central disk to the anus on top of the disk.

Central disk. The central disk has a nerve ring and nerve cords radiating from the ring into the arms.

Digestive glands secrete digestive juices and aid in the absorption and storage of nutrients.

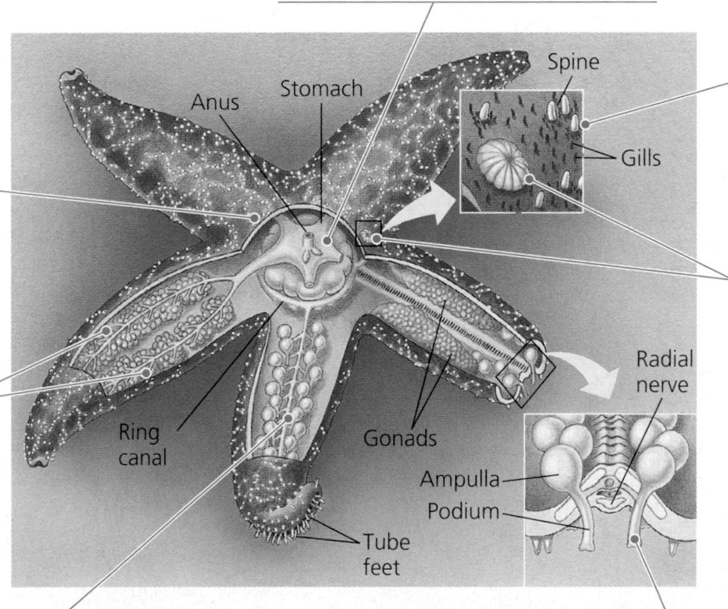

The surface of a sea star is covered by spines that help defend against predators, as well as by small gills that provide gas exchange.

Madreporite. Water can flow in or out of the water vascular system into the surrounding water through the madreporite.

Radial canal. The water vascular system consists of a ring canal in the central disk and five radial canals, each running in a groove down the entire length of an arm.

Branching from each radial canal are hundreds of hollow, muscular tube feet filled with fluid. Each tube foot consists of a bulb-like ampulla and suckered podium (foot portion). When the ampulla squeezes, it forces water into the podium and makes it expand. The podium then contacts the substrate. When the muscles in the wall of the podium contract, they force water back into the ampulla, making the podium shorten and bend.

usually involves separate male and female individuals that release their gametes into the water.

The internal and external parts of most echinoderms radiate from the center, often as five spokes. However, the radial anatomy of adult echinoderms is a secondary adaptation, as echinoderm larvae have bilateral symmetry. Furthermore, the symmetry of adult echinoderms is not perfectly radial. For example, the opening (madreporite) of a sea star's water vascular system is not central but shifted to one side.

Living echinoderms are divided into six classes (**Table 33.6**; **Figure 33.40**): Asteroidea (sea stars), Ophiuroidea (brittle stars), Echinoidea (sea urchins and sand dollars), Crinoidea (sea lilies and feather stars), Holothuroidea (sea cucumbers), and Concentricycloidea (sea daisies).

Sea Stars

Sea stars have multiple arms radiating from a central disk. The undersurfaces of the arms bear tube feet, each of which can act like a suction disk. By a complex set of hydraulic and muscular actions, the suction can be created or released (see Figure 33.39). The sea star adheres firmly to rocks or creeps along slowly as the tube feet extend, grip, contract, release, extend, and grip again. Sea stars also use their tube feet to grasp prey, such as clams and oysters. The arms of the sea star embrace the closed bivalve, hanging on tightly by the tube feet. The sea star then turns its stomach inside out, everting it through its mouth and into the narrow opening between the halves of the bivalve's shell. The digestive system of the sea star secretes juices that begin digesting the soft body of the mollusc within its own shell.

Sea stars and some other echinoderms have considerable powers of regeneration. Sea stars can regrow lost arms, and members of one genus can even regrow an entire body from a single arm.

Table 33.6 Classes of Phylum Echinodermata

Class and Examples	Main Characteristics
Asteroidea (sea stars; see Figures 33.39 and 33.40a)	Star-shaped body with multiple arms; mouth directed to substrate
Ophiuroidea (brittle stars; see Figure 33.40b)	Distinct central disk; long, flexible arms; tube feet lack suckers
Echinoidea (sea urchins, sand dollars; see Figure 33.40c)	Roughly spherical or disk-shaped; no arms; five rows of tube feet enable slow movement; mouth ringed by complex, jaw-like structure
Crinoidea (sea lilies, feather stars; see Figure 33.40d)	Feathered arms surrounding upward-pointing mouth
Holothuroidea (sea cucumbers; see Figure 33.40e)	Cucumber-shaped body; five rows of tube feet; additional tube feet modified as feeding tentacles; reduced skeleton; no spines
Concentricycloidea (sea daisies; see Figure 33.40f)	Disk-shaped body ringed with small spines; incomplete digestive system; live on submerged wood

Brittle Stars

Brittle stars have a distinct central disk and long, flexible arms. They move by serpentine lashing of their arms, as their tube feet lack suckers and thus cannot be used for gripping. Some species are suspension feeders; others are predators or scavengers.

Sea Urchins and Sand Dollars

Sea urchins and sand dollars have no arms, but they do have five rows of tube feet that function in slow movement. Sea urchins also have muscles that pivot their long spines, which aids in locomotion. The mouth of a sea urchin is ringed by complex, jaw-like structures adapted for eating seaweeds and other food. Sea urchins are roughly spherical, whereas sand dollars are flattened and disk-shaped.

Sea Lilies and Feather Stars

Sea lilies live attached to the substrate by a stalk; feather stars crawl about by using their long, flexible arms. Both use their arms in suspension feeding. The arms encircle the mouth, which is directed upward, away from the substrate. Crinoidea is an ancient class whose evolution has been very conservative; fossilized sea lilies some 500 million years old are extremely similar to present-day members of the class.

Sea Cucumbers

On casual inspection, sea cucumbers do not look much like other echinoderms. They lack spines, and their endoskeleton is much reduced. They are also elongated in their oral-aboral axis, giving them the shape for which they are named and further disguising their relationship to sea stars and sea urchins. Closer examination, however, reveals that sea cucumbers have five rows of tube feet. Some of the tube feet around the mouth are developed as feeding tentacles.

Sea Daisies

Sea daisies were discovered in 1986, and only two species are known. Both live on submerged wood off the coast of New Zealand and the Bahamas. Their armless body is typically disk-shaped, has a five-fold symmetry, and measures less than a centimeter in diameter. The edge of the body is ringed with small spines. Sea daisies absorb nutrients through the membrane surrounding their body. The relationship of sea daisies to other echinoderms remains unclear; some taxonomists consider sea daisies to be highly derived sea stars.

(a) A sea star (class Asteroidea)

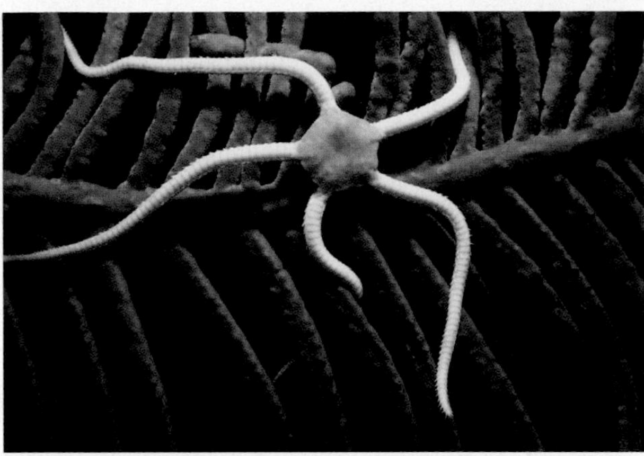

(b) A brittle star (class Ophiuroidea)

(c) A sea urchin (class Echinoidea)

(d) A feather star (class Crinoidea)

(e) A sea cucumber (class Holothuroidea)

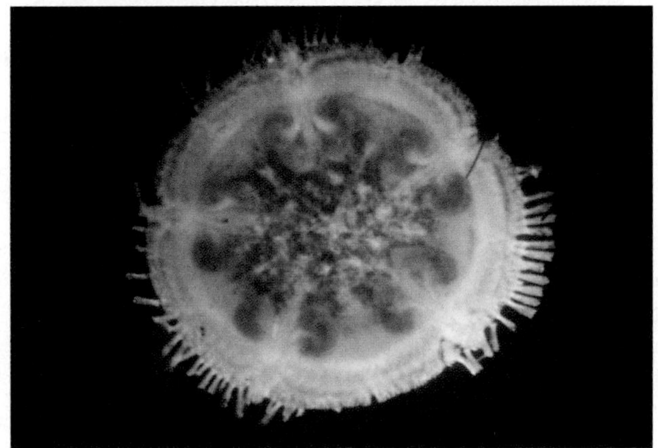

(f) A sea daisy (class Concentricycloidea)

▲ **Figure 33.40 Echinoderms.**

Chordates

Phylum Chordata consists of two subphyla of invertebrates as well as the hagfishes and the vertebrates. The close relationship between echinoderms and chordates does not mean that one phylum evolved from the other. Echinoderms and chordates have existed as distinct phyla for at least half a billion years. We will trace the phylogeny of chordates in Chapter 34, focusing on the history of vertebrates.

Concept Check 33.8

1. Explain how the symmetry of echinoderms and cnidarians exemplifies convergent evolution.
2. Describe the hydraulic and muscular actions by which a sea star moves its tube feet.

For suggested answers, see Appendix A.

Chapter 33 Review

Go to www.campbellbiology.com or the student CD-ROM to explore Activities, Investigations, and other interactive study aids.

SUMMARY OF KEY CONCEPTS

Table 33.7 summarizes the groups of animals surveyed in this chapter.

Concept 33.1

Sponges are sessile and have a porous body and choanocytes

- Sponges lack true tissues and organs. They suspension feed by drawing water through pores; choanocytes (flagellated collar cells) ingest suspended food (pp. 642–643).

Table 33.7 Selected Animal Phyla

Clade	Phylum	Description
Metazoa	Porifera (sponges)	Lack true tissues; have choanocytes (collar cells—unique flagellated cells that ingest bacteria and tiny food particles)
Metazoa; Eumetazoa	Cnidaria (hydras, jellies, sea anemones, corals)	Unique stinging structures (cnidae), each housed in a specialized cell (cnidocyte); gastrovascular cavity (digestive compartment with a single opening)
Metazoa; Eumetazoa; Bilateria	Platyhelminthes (flatworms)	Dorsoventrally flattened, unsegmented acoelomates; gastrovascular cavity or no digestive tract
Metazoa; Eumetazoa; Bilateria	Rotifera (rotifers)	Pseudocoelomates with alimentary canal (digestive tube with mouth and anus); jaws (trophi) in pharynx; head with ciliated crown
Metazoa; Eumetazoa; Bilateria	Lophophorates: Ectoprocta, Phoronida, Brachiopoda	Coelomates with lophophores (feeding structures bearing ciliated tentacles)
Metazoa; Eumetazoa; Bilateria	Nemertea (proboscis worms)	Unique anterior proboscis surrounded by fluid-filled sac; alimentary canal; closed circulatory system
Metazoa; Eumetazoa; Bilateria	Mollusca (clams, snails, squids)	Coelomates with three main body parts (muscular foot, visceral mass, mantle); coelom reduced; most have hard shell made of calcium carbonate
Metazoa; Eumetazoa; Bilateria	Annelida (segmented worms)	Coelomates with segmented body wall and internal organs (except digestive tract, which is unsegmented)
Metazoa; Eumetazoa; Bilateria	Nematoda (roundworms)	Cylindrical, unsegmented pseudocoelomates with tapered ends; no circulatory system
Metazoa; Eumetazoa; Bilateria	Arthropoda (crustaceans, insects, spiders)	Coelomates with segmented body, jointed appendages, and exoskeleton made of protein and chitin
Metazoa; Eumetazoa; Bilateria; Deuterostomia	Echinodermata (sea stars, sea urchins)	Coelomates with secondary radial anatomy (larvae bilateral; adults radial); unique water vascular system; endoskeleton
Metazoa; Eumetazoa; Bilateria; Deuterostomia	Chordata (lancelets, tunicates, vertebrates)	Coelomates with notochord; dorsal, hollow nerve cord; pharyngeal slits; muscular, post-anal tail

Cnidarians have radial symmetry, a gastrovascular cavity, and cnidocytes

▶ Cnidarians are mainly marine carnivores possessing tentacles armed with cnidocytes that aid in defense and the capture of prey. Two body forms are sessile polyps and floating medusae (pp. 643–644).

▶ **Hydrozoans (pp. 644–645)** Class Hydrozoa usually alternates polyp and medusa forms, although the polyp is more conspicuous.

▶ **Scyphozoans (p. 644)** In class Scyphozoa, jellies (medusae) are the prevalent form of the life cycle.

▶ **Cubozoans (p. 645)** In class Cubozoa (box jellies and sea wasps), the medusa is box-shaped and has complex eyes.

▶ **Anthozoans (pp. 645–646)** Class Anthozoa contains the sea anemones and corals, which occur only as polyps.

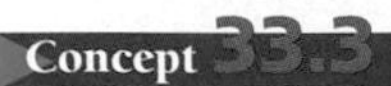

Most animals have bilateral symmetry

▶ **Flatworms (pp. 646–648)** Flatworms are dorsoventrally flattened animals with a gastrovascular cavity. Class Turbellaria is made up of mostly free-living, primarily marine species. Members of the classes Trematoda and Monogenea live as parasites in or on animals. Class Cestoda consists of tapeworms, all of which are parasites and lack a digestive tract.

▶ **Rotifers (pp. 648–649)** Found mainly in fresh water, many rotifer species are parthenogenetic.

▶ **Lophophorates: Ectoprocts, Phoronids, and Brachiopods (p. 649)** Lophophorates are coelomates that have a lophophore, a horseshoe-shaped, suspension-feeding organ bearing ciliated tentacles.

▶ **Nemerteans (pp. 649–650)** Nemerteans have a unique retractable tube (proboscis) used for defense and prey capture. A fluid-filled sac is used to extend the proboscis.

Molluscs have a muscular foot, a visceral mass, and a mantle

▶ **Chitons (p. 651)** Class Polyplacophora is composed of the chitons, oval-shaped marine animals encased in an armor of eight dorsal plates.

▶ **Gastropods (pp. 651–652)** Most members of class Gastropoda, the snails and their relatives, possess a single, spiraled shell. Embryonic torsion of the body is a distinctive characteristic. Many slugs lack a shell or have a reduced shell.

▶ **Bivalves (p. 652)** Class Bivalvia (clams and their relatives) have a hinged shell divided into two halves.

▶ **Cephalopods (pp. 652–653)** Class Cephalopoda includes squids and octopuses, carnivores with beak-like jaws surrounded by tentacles of their modified foot.

Annelids are segmented worms

▶ **Oligochaetes (pp. 653–654)** Class Oligochaeta includes earthworms and various aquatic species.

▶ **Polychaetes (p. 655)** Members of class Polychaeta possess paddle-like parapodia that function as gills and aid in locomotion.

▶ **Leeches (p. 655)** Many members of class Hirudinea are blood-sucking parasites.

Nematodes are nonsegmented pseudocoelomates covered by a tough cuticle

▶ Among the most widespread and numerous animals, nematodes inhabit the soil and most aquatic habitats. Some species are important parasites of animals and plants (pp. 655–656).

Concept 33.7

Arthropods are segmented coelomates that have an exoskeleton and jointed appendages

▶ **General Characteristics of Arthropods (pp. 656–658)** Variation in arthropod morphology consists mainly in specializations of groups of segments and in appendages. The arthropod exoskeleton, made of protein and chitin, undergoes regular ecdysis (molting).

▶ **Cheliceriforms (pp. 658–659)** Cheliceriforms include spiders, ticks, and mites. They have an anterior cephalothorax and a posterior abdomen. The most anterior appendages are modified as chelicerae (either pincers or fangs).

▶ **Myriapods (pp. 659–660)** Millipedes are wormlike, with a large number of walking legs. They were among the first animals to live on land. Centipedes are terrestrial carnivores with poison claws.

▶ **Insects (pp. 660–664)** Insects exceed all other animals combined in species diversity. Flight has been an important factor in the success of insects.
Investigation *How Are Insect Species Identified?*

▶ **Crustaceans (pp. 664–665)** Crustaceans, which include lobsters, crabs, shrimp, and barnacles, are primarily aquatic. They have numerous appendages, many of which are specialized for feeding and locomotion.

Echinoderms and chordates are deuterostomes

▶ **Echinoderms (pp. 665–667)** Echinoderms (sea stars and their relatives) have a water vascular system ending in tube feet used for locomotion and feeding. The radial anatomy of many species evolved secondarily from the bilateral symmetry of ancestors. A thin, bumpy, or spiny skin covers a calcareous endoskeleton.
Activity *Characteristics of Invertebrates*

▶ **Chordates (p. 667)** Chordates include two invertebrate subphyla and all vertebrates. Chordates share many features of embryonic development with echinoderms.

TESTING YOUR KNOWLEDGE

Self-Quiz

1. Which two main clades branch from the earliest Eumetazoan ancestor?
   a. Porifera and Bilateria
   b. Porifera and Cnidaria
   c. Cnidaria and Bilateria
   d. Rotifera and Deuterostomia
   e. Deuterostomia and Bilateria

2. Water movement through a sponge would follow what path?
   a. porocyte → spongocoel → osculum
   b. blastopore → gastrovascular cavity → protostome
   c. choanocyte → mesohyl → spongocoel
   d. porocyte → choanocyte → mesohyl
   e. collar cell → coelom → porocyte

3. Although a diverse group, all cnidarians are characterized by
   a. a gastrovascular cavity.
   b. an alteration between a medusa and a polyp stage.
   c. some degree of cephalization.
   d. muscle tissue of mesodermal origin.
   e. the complete absence of asexual reproduction.

4. A land snail, a clam, and an octopus all share
   a. a mantle.
   b. a radula.
   c. gills.
   d. embryonic torsion.
   e. distinct cephalization.

5. Which of the following is *not* a characteristic of most members of the phylum Annelida?
   a. hydrostatic skeleton
   b. segmentation
   c. metanephridia
   d. pseudocoelom
   e. closed circulatory system

6. Which phylum is characterized by animals that have a segmented body?
   a. Cnidaria
   b. Platyhelminthes
   c. Porifera
   d. Arthropoda
   e. Mollusca

7. Which of the following is *not* true of the cheliceriforms?
   a. They have antennae.
   b. Their body is divided into a cephalothorax and an abdomen.
   c. Horseshoe crabs are surviving marine members.
   d. They include ticks, scorpions, and spiders.
   e. Their anterior appendages are modified as pincers or fangs.

8. Which of the following characteristics is probably *most* responsible for the incredible diversification of insects on land?
   a. segmentation
   b. antennae
   c. tracheal system
   d. bilateral symmetry
   e. flight

9. The water vascular system of echinoderms
   a. functions as a circulatory system that distributes nutrients to body cells.
   b. functions in locomotion, feeding, and gas exchange.
   c. is bilateral in organization, even though the adult animal has radial anatomy.
   d. moves water through the animal's body during suspension feeding.
   e. is analogous to the gastrovascular cavity of flatworms.

10. Which of the following combinations of phylum and description is *incorrect?*
    a. Echinodermata—bilateral and radial symmetry, coelom from archenteron
    b. Nematoda—roundworms, pseudocoelomate
    c. Cnidaria—radial symmetry, polyp and medusa body forms
    d. Platyhelminthes—flatworms, gastrovascular cavity, acoelomate
    e. Porifera—gastrovascular cavity, coelom present

For Self-Quiz answers, see Appendix A.

Go to the website or CD-ROM for more quiz questions.

Evolution Connection

Horseshoe crabs are called "living fossils" because the fossil record shows that they have remained essentially unchanged in morphology for many millions of years. Why might these organisms have retained the same morphology for such a long time? What other aspects of their biology, less obvious than structure, do you think may have evolved over that time?

Scientific Inquiry

A marine biologist has dredged up an unknown animal from the seafloor. Describe some of the characteristics she should look at to determine the phylum to which the animal should be assigned.

Investigation *How Are Insect Species Identified?*

Science, Technology, and Society

Construction of a dam and irrigation canals in an African country has enabled farmers to increase the amount of food they can grow. In the past, crops were planted only after spring floods; the fields were too dry the rest of the year. Now fields can be watered year-round. Improvement in crop yield has had an unexpected cost—a tremendous increase in the incidence of schistosomiasis. Look at the blood fluke life cycle in Figure 33.11 and imagine that your Peace Corps assignment is to help local health officials control the disease. Why do you think the irrigation project increased the incidence of schistosomiasis? It is difficult and expensive to control the disease with drugs. Suggest three other methods that could be tried to prevent people from becoming infected.

34 Vertebrates

▲ **Figure 34.1 The vertebrae and skull of a snake, a terrestrial vertebrate.**

Key Concepts

34.1 Chordates have a notochord and a dorsal, hollow nerve cord
34.2 Craniates are chordates that have a head
34.3 Vertebrates are craniates that have a backbone
34.4 Gnathostomes are vertebrates that have jaws
34.5 Tetrapods are gnathostomes that have limbs and feet
34.6 Amniotes are tetrapods that have a terrestrially adapted egg
34.7 Mammals are amniotes that have hair and produce milk
34.8 Humans are bipedal hominoids with a large brain

Overview

Half a Billion Years of Backbones

By the dawn of the Cambrian period, some 540 million years ago, an astonishing variety of animals inhabited Earth's oceans. Predators used claws and mandibles to skewer their prey. Many animals had protective spikes and armor as well as complex mouthparts that enabled their bearers to filter food from the water. Worms slithered into the muck to feed on organic matter. Amidst all this bustle, it would have been easy to overlook certain slender, 2-cm-long creatures gliding through the water. Although they lacked armor, eyes, and appendages, they would leave behind a remarkable legacy. These animals gave rise to one of the most successful groups of animals ever to swim, walk, or fly: the **vertebrates,** which derive their name from vertebrae, the series of bones that make up the vertebral column, or backbone **(Figure 34.1)**. For nearly 200 million years, vertebrates were restricted to the oceans, but about 360 million years ago the evolution of legs and feet in one lineage of vertebrates accompanied these vertebrates' move to land. There they diversified into amphibians, reptiles (including birds), and mammals.

There are approximately 52,000 species of vertebrates, a relatively small number compared to the 1 million insect species on Earth. But what vertebrates lack in species diversity they have made up for with other statistics. Plant-eating dinosaurs as massive as 40,000 kg were the heaviest animals ever to walk on land. The biggest animal ever to exist on Earth is the blue whale (a mammal), which can exceed a mass of 100,000 kg. Vertebrates are capable of global journeys: For example, birds called Arctic terns, which breed mainly around the shores of the Arctic Ocean, travel to Antarctica to spend the rest of the year before making a return trip. Vertebrates also include the only species capable of full-blown language, complex toolmaking, and symbolic art—humans.

In this chapter, you will learn about current hypotheses regarding the origins of vertebrates from invertebrate ancestors. We will track the stepwise evolution of the vertebrate body plan, from a notochord to a head to a mineralized skeleton, and explore the major groups of vertebrates (both living and extinct) as well as the evolutionary history of our own species.

Concept 34.1

Chordates have a notochord and a dorsal, hollow nerve cord

Vertebrates are a subphylum of the phylum Chordata (the chordates). **Chordates** are bilaterian (bilaterally symmetrical) animals, and within Bilateria, they belong to the clade of

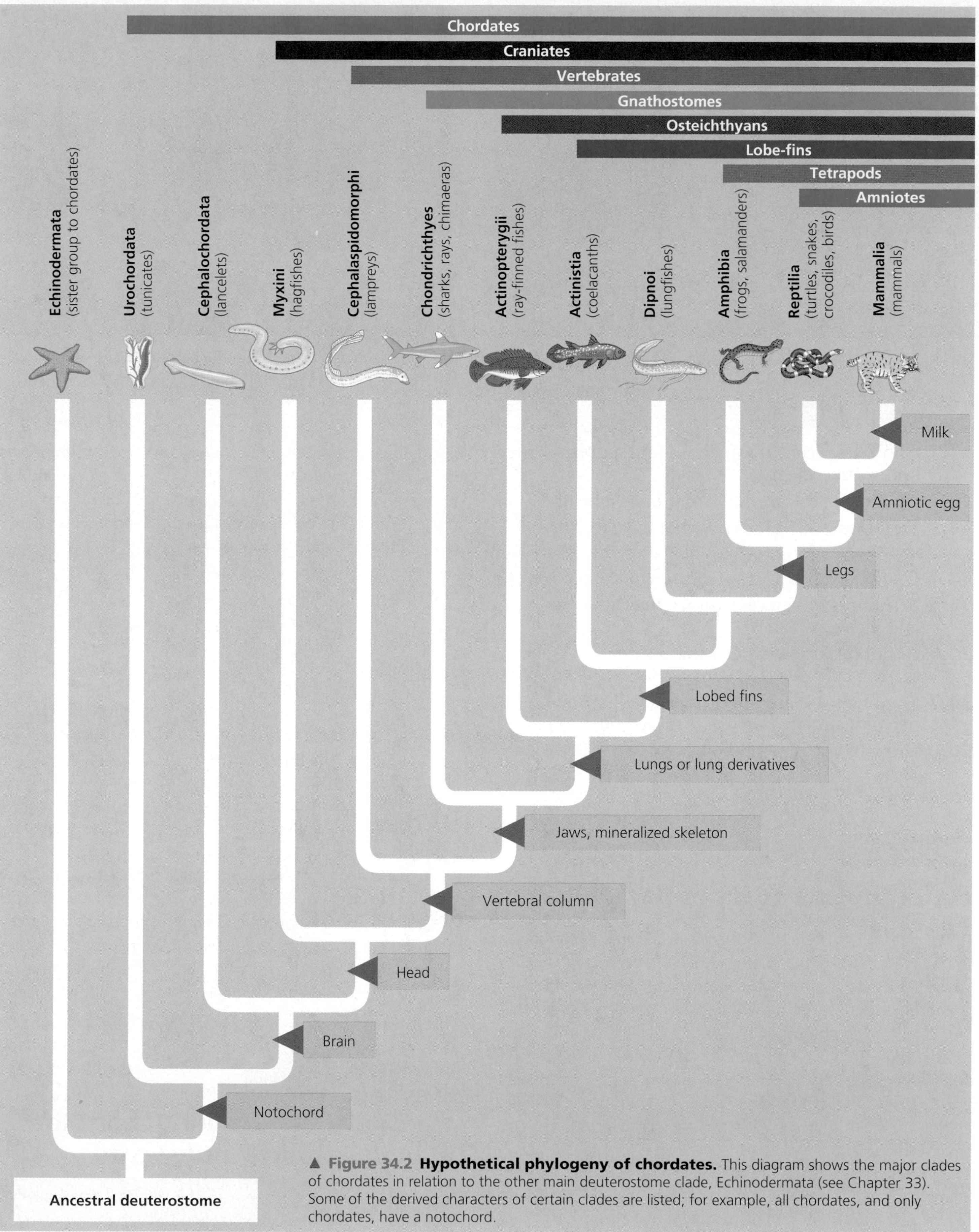

▲ **Figure 34.2 Hypothetical phylogeny of chordates.** This diagram shows the major clades of chordates in relation to the other main deuterostome clade, Echinodermata (see Chapter 33). Some of the derived characters of certain clades are listed; for example, all chordates, and only chordates, have a notochord.

animals known as Deuterostomia (see Chapter 32). The best-known deuterostomes, aside from vertebrates, are the echinoderms, members of the phylum that includes sea stars and sea urchins. However, as shown in **Figure 34.2**, on the facing page, two groups of invertebrate deuterostomes, the urochordates and the cephalochordates, are more closely related to vertebrates than to other invertebrates. Along with the hagfishes and the vertebrates, they make up the chordate phylum.

Derived Characters of Chordates

All chordates share a set of derived characters, though many species possess some of these traits only during embryonic development. **Figure 34.3** illustrates the four key characters of chordates: a notochord; a dorsal, hollow nerve cord; pharyngeal slits or clefts; and a muscular, post-anal tail.

Notochord

Chordates are named for a skeletal structure, the notochord, present in all chordate embryos as well as in some adult chordates. The **notochord** is a longitudinal, flexible rod located between the digestive tube and the nerve cord. It is composed of large, fluid-filled cells encased in fairly stiff, fibrous tissue. The notochord provides skeletal support throughout most of the length of a chordate, and in larvae or adults that retain it, it also provides a firm but flexible structure against which muscles can work during swimming. In most vertebrates, a more complex, jointed skeleton develops, and the adult retains only remnants of the embryonic notochord. In humans, the notochord is reduced to gelatinous disks sandwiched between the vertebrae.

Dorsal, Hollow Nerve Cord

The nerve cord of a chordate embryo develops from a plate of ectoderm that rolls into a tube located dorsal to the notochord. The resulting dorsal, hollow nerve cord is unique to chordates. Other animal phyla have solid nerve cords, and in most cases they are ventrally located. The nerve cord of a chordate embryo develops into the central nervous system: the brain and spinal cord.

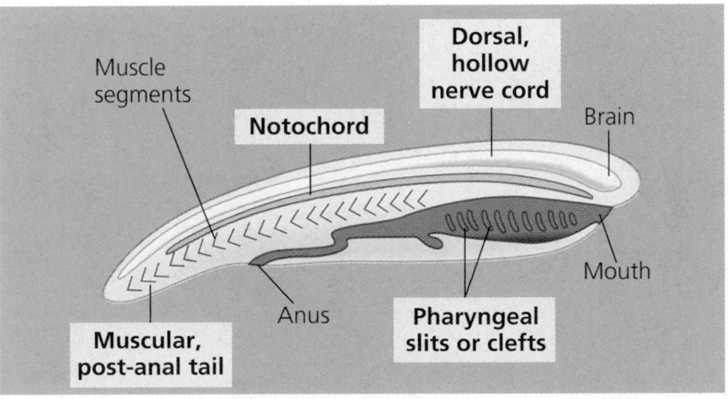

▲ **Figure 34.3 Chordate characteristics.** All chordates possess the four structural trademarks of the phylum at some point during their development.

Pharyngeal Slits or Clefts

The digestive tube of chordates extends from the mouth to the anus. The region just posterior to the mouth is the pharynx. In all chordate embryos, a series of pouches separated by grooves forms along the sides of the pharynx. In most chordates, these grooves (known as **pharyngeal clefts**) develop into slits that open to the outside of the body. These **pharyngeal slits** allow water entering the mouth to exit the body without passing through the entire digestive tract. Pharyngeal slits function as suspension-feeding devices in many invertebrate chordates. In vertebrates (with the exception of terrestrial vertebrates, the tetrapods), these slits and the structures that support them have been modified for gas exchange and are known as gill slits. In tetrapods, the pharyngeal clefts do not develop into slits. Instead, they play an important role in the development of parts of the ear and other structures in the head and neck.

Muscular, Post-Anal Tail

Chordates have a tail extending posterior to the anus, although in many species it is lost during embryonic development. In contrast, most nonchordates have a digestive tract that extends nearly the whole length of the body. The chordate tail contains skeletal elements and muscles, and it provides much of the propelling force in many aquatic species.

Tunicates

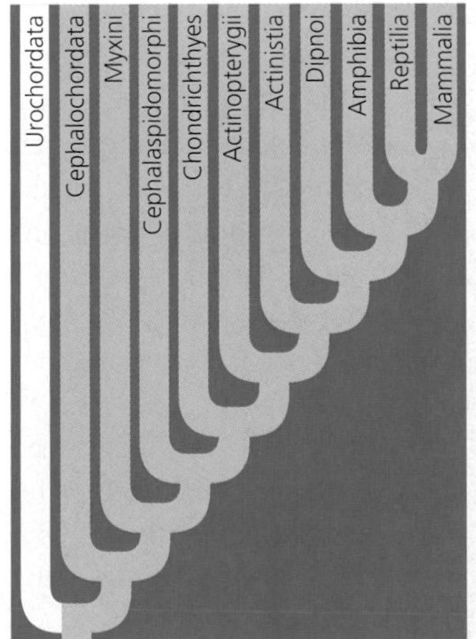

Data from various morphological and molecular studies support the hypothesis that **tunicates** (subphylum Urochordata) belong to the deepest-branching lineage of chordates. The tunicates most resemble other chordates during their larval stage, which may be as brief as a few minutes. In many species, the larva uses its tail muscles and notochord to swim through water in search of a suitable substrate on which it can settle, guided by cues it receives from light- and gravity-sensitive cells.

Once a tunicate has settled on a substrate, it goes through a radical metamorphosis in which many of its chordate

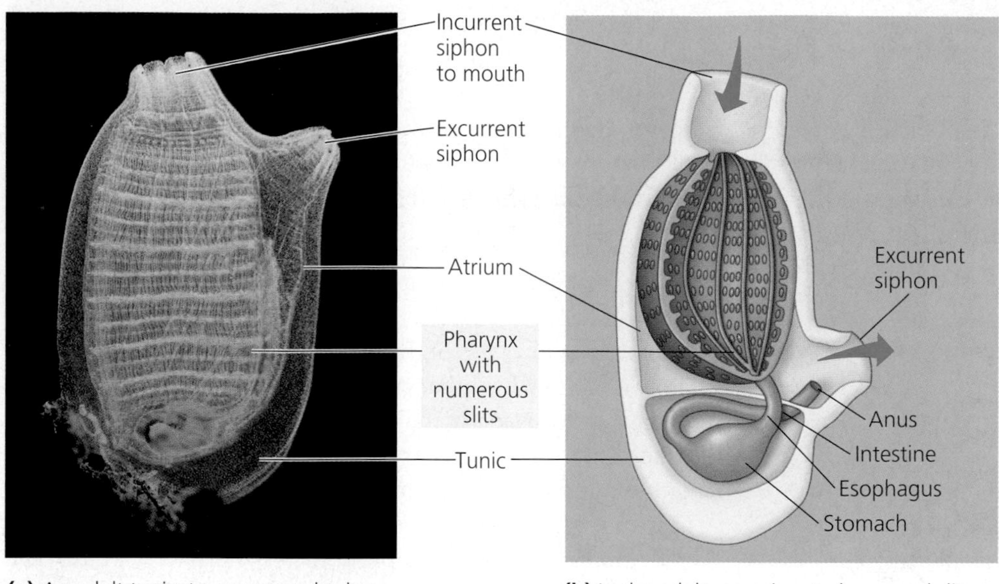

(a) An adult tunicate, or sea squirt, is a sessile animal (photo is approximately life-sized).

(b) In the adult, prominent pharyngeal slits function in suspension feeding, but other chordate characters are not obvious.

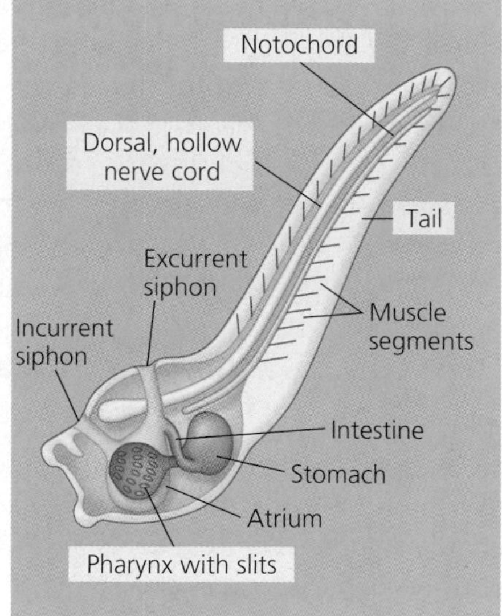

(c) A tunicate larva is a free-swimming but nonfeeding "tadpole" in which all four chief characters of chordates are evident.

▲ **Figure 34.4 A tunicate, a urochordate.**

characters disappear. Its tail and notochord are resorbed; its nervous system degenerates; and its remaining organs rotate 90°. As an adult, a tunicate draws in water through an incurrent siphon; the water then passes through the pharyngeal slits into a chamber called the atrium and exits through an excurrent siphon **(Figure 34.4)**. Food particles are filtered from the water by a mucous net and transported by cilia to the esophagus. The anus empties into the excurrent siphon. Some tunicate species shoot a jet of water through their excurrent siphon when attacked, earning them the informal name of "sea squirts."

Lancelets

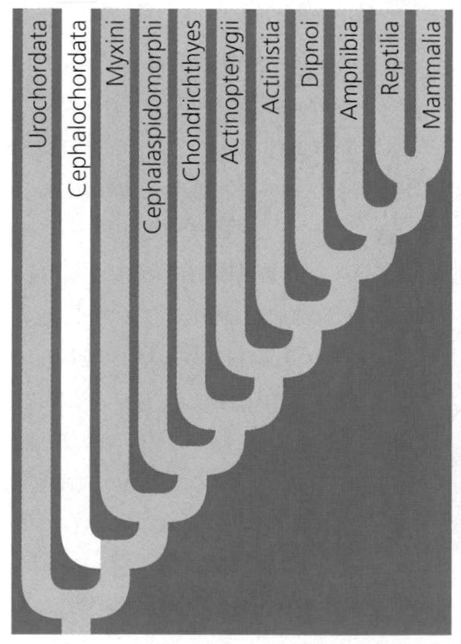

Lancelets (subphylum Cephalochordata) get their name from their bladelike shape **(Figure 34.5)**. As larvae, lancelets develop a notochord, a dorsal, hollow nerve cord, numerous pharyngeal slits, and a post-anal tail. They feed on plankton in the water column, alternating between upward swimming and passive sinking. As they sink, they trap plankton and other suspended matter in their pharynx.

Adult lancelets can be up to 5 cm long. They retain key chordate traits, closely resembling the idealized chordate shown in Figure 34.3. Following metamorphosis, adult lancelets swim down to the seafloor and wriggle backward into the sand, leaving only their anterior end exposed. A mucous net secreted across the pharyngeal slits removes tiny food particles from seawater drawn into the mouth by ciliary pumping. The water passes through the slits, and the trapped food enters the intestine. The pharynx and pharyngeal slits play a minor role in gas exchange, which occurs mainly across the external body surface.

A lancelet frequently leaves its burrow to swim to a new location. Though feeble swimmers, these invertebrate chordates display, in a simple form, the swimming mechanism of fishes. Coordinated contraction of muscles serially arranged like rows of chevrons (<<<<) along the sides of the notochord flexes the notochord, producing side-to-side undulations that thrust the body forward. This serial musculature is evidence of the lancelet's segmentation. The muscle segments develop from blocks of mesoderm called **somites,** which are found along each side of the notochord in all chordate embryos.

Globally, lancelets are rare, but in a few regions (including Tampa Bay, along the Florida coast) they occasionally reach densities in excess of 5,000 individuals per square meter.

Early Chordate Evolution

Although tunicates and lancelets are relatively obscure animals, the rise of evolutionary biology has focused much attention on them. Possessing many—but not all—of the derived characters shared by vertebrates, they can provide clues about the evolutionary origin of the vertebrate body plan.

▶ **Figure 34.5 The lancelet *Branchiostoma*, a cephalochordate.** This small invertebrate displays all four main chordate characters. Water enters the mouth and passes through the pharyngeal slits into the atrium, a chamber that vents to the outside via the atriopore. Food particles trapped by mucus are swept by cilia into the digestive tract. The serially arranged segmental muscles produce the lancelet's undulatory (wavelike) swimming movements.

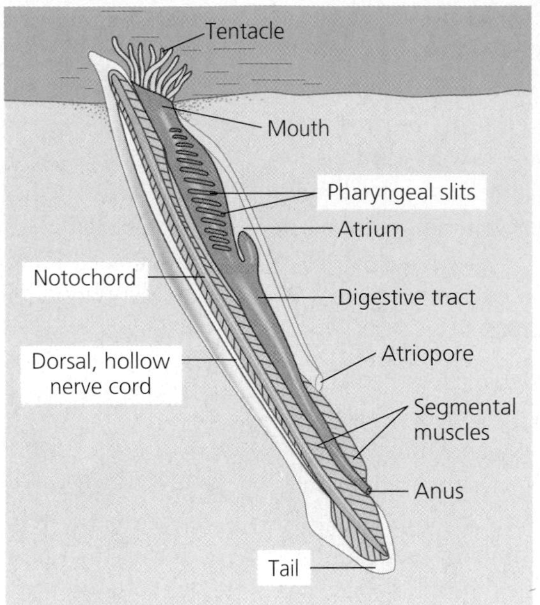

As you have read, tunicates display a number of chordate characters only as larvae, whereas lancelets retain those characters as adults. Thus, an adult lancelet looks much more like a larval tunicate than like an adult tunicate. In the 1920s, based on these observations, biologist William Garstang proposed that tunicates represent an early stage in chordate evolution. He suggested that ancestral tunicate-like chordates accelerated their sexual maturity, becoming mature while still in their larval stage. Thus, they and the chordates that evolved from them retained the notochord and other features as adults. This process, which has been documented in a number of evolutionary transitions, is known as paedomorphosis (see Chapter 24).

While Garstang's idea was popular for several decades, today the weight of evidence is against it in the case of tunicates. The degenerate adult stage of tunicates appears to be a derived trait that evolved only after the tunicate lineage branched off from other chordates. Even the tunicate larva appears to be highly derived, rather than a faithful reproduction of the body plan of early chordates. Studies of *Hox* gene expression (see Chapter 21) suggest that the tunicate larva does not develop the posterior regions of its body axis. Rather, the anterior region is elongated and contains a heart and digestive system.

Research on lancelets has revealed several important clues about the evolution of the chordate brain. Rather than a full-fledged brain, lancelets have only a slightly swollen tip on the anterior end of their dorsal nerve cord. But the same *Hox* genes that organize major regions of the forebrain, midbrain, and hindbrain of vertebrates express themselves in a corresponding pattern in this small cluster of cells in the lancelet's nerve cord **(Figure 34.6)**. This suggests that the vertebrate brain apparently is an elaboration of an ancestral structure similar to the lancelet's simple nerve cord tip.

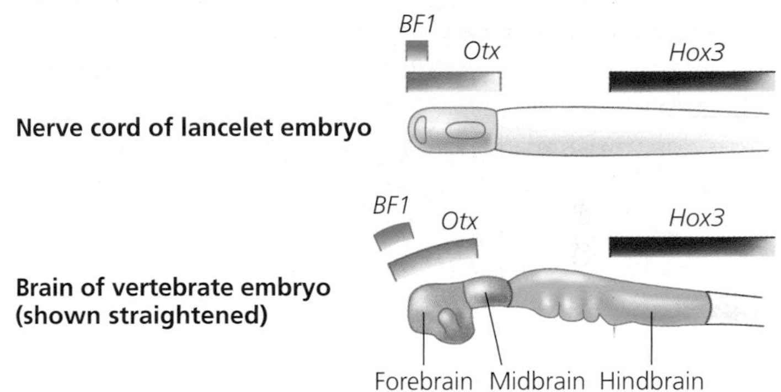

▲ **Figure 34.6 Expression of developmental genes in lancelets and vertebrates.** *Hox* genes (including *BF1*, *Otx*, and *Hox3*) control the development of major regions of the vertebrate brain. These genes are expressed in the same anterior-to-posterior order in lancelets and vertebrates.

Concept Check 34.1

1. Humans are chordates, yet they lack most of the main derived characters of chordates. Explain.
2. How do pharyngeal slits function in feeding in tunicates and lancelets?

For suggested answers, see Appendix A.

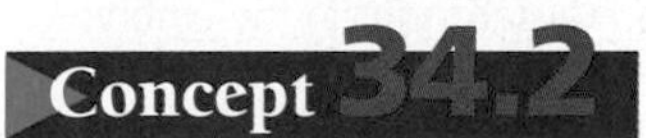

Concept 34.2

Craniates are chordates that have a head

After the evolution of the basic chordate body plan, seen in tunicates and lancelets, the next major transition in chordate evolution was the appearance of a head. Chordates with a head are known as **craniates** (from the word *cranium*, skull). The origin of a head—consisting of a brain at the anterior end of the dorsal nerve cord, eyes and other sensory organs, and a skull—opened up a completely new way of feeding for chordates: active predation. (Note that heads evolved independently in other animal lineages as well, as described in Chapter 33.)

Derived Characters of Craniates

Living craniates share a set of derived characters that distinguish them from other chordates. On a genetic level, they possess two clusters of *Hox* genes (lancelets and tunicates have only one). Other important families of genes that produce signaling molecules and transcription factors are also duplicated in craniates. This additional genetic complexity made it possible for craniates to develop more complex morphologies than those of tunicates and lancelets.

One feature unique to craniates is the **neural crest**, a collection of cells that appears near the dorsal margins of the closing neural tube in an embryo **(Figure 34.7)**. Neural crest cells disperse throughout the body, where they give rise to a variety of structures, including teeth, some of the bones and cartilage of the skull, the inner layer of skin (dermis) of the facial region, several types of neurons, and the sensory capsules in which eyes and other sense organs develop.

In aquatic craniates, the pharyngeal clefts evolved into gill slits. Unlike the pharyngeal slits of lancelets, which are used primarily for suspension feeding, gill slits are associated with muscles and nerves that allow water to be pumped through the slits. This pumping can assist in sucking in food, and it facilitates gas exchange. (In terrestrial craniates, the pharyngeal clefts develop into other structures, as explained later.)

Craniates, which are more active than tunicates and lancelets, also have a higher metabolism and a much more extensive muscular system. Muscles lining their digestive tract aid digestion by moving food through the tract. Craniates also have a heart with at least two chambers, red blood cells, and hemoglobin, as well as kidneys that remove waste products from the blood.

▼ Figure 34.7 The neural crest, embryonic source of many unique vertebrate characters.

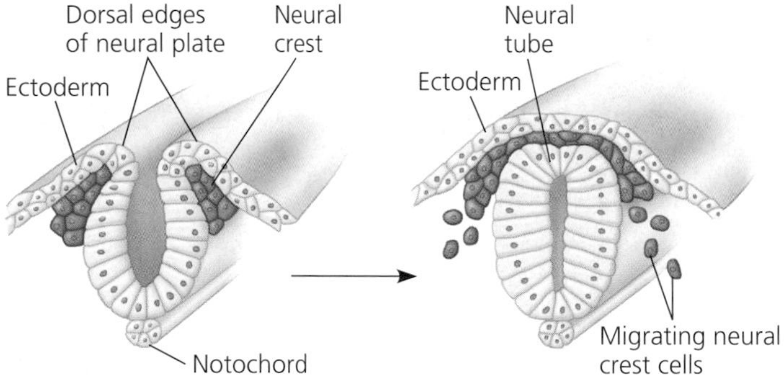

(a) The neural crest consists of bilateral bands of cells near the margins of the embryonic folds that form the neural tube.

(b) Neural crest cells migrate to distant sites in the embryo.

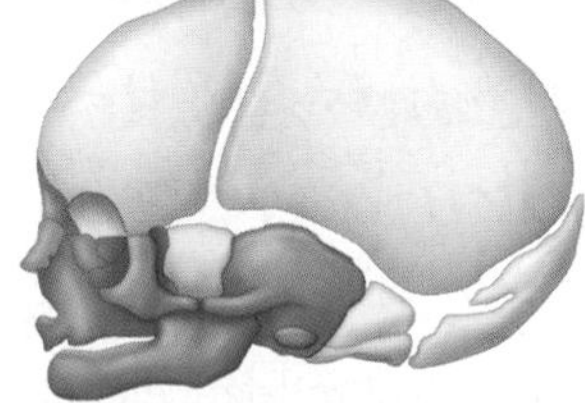

(c) The cells give rise to some of the anatomical structures unique to vertebrates, including some of the bones and cartilage of the skull.

The Origin of Craniates

In the late 1990s, paleontologists working in China discovered a vast supply of fossils of early chordates that appear to straddle the transition to craniates. The fossils were formed during the Cambrian explosion 530 million years ago, when many groups of animals were diversifying (see Chapter 32).

The most primitive of the fossils are those of the 3-cm-long *Haikouella* **(Figure 34.8a)**. In many ways, *Haikouella* resembled a lancelet. Its mouth structure indicates that, like lancelets, it probably was a suspension feeder. However, *Haikouella* also had some of the characters of craniates. For example, it had a small but well-formed brain, eyes, and muscular segments that resemble those of vertebrates. It also had hardened structures in its pharynx that may have been toothlike "denticles." However, *Haikouella* did not have a skull, suggesting that this character emerged with innovations to the chordate nervous system.

In other Cambrian rocks, paleontologists have found fossils of even more advanced chordates, such as *Haikouichthys*. About the same size as *Haikouella, Haikouichthys* had a skull that may have been composed of cartilage **(Figure 34.8b)**. Based on this and other characters, paleontologists have identified *Haikouichthys* as a true craniate.

Hagfishes

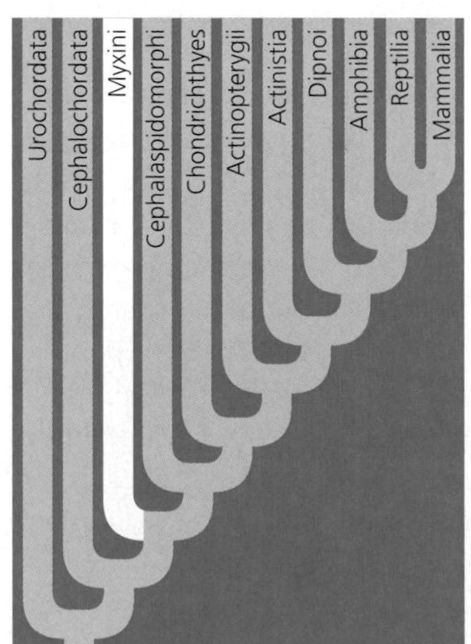

The least derived craniate lineage that still survives is class Myxini, the hagfishes **(Figure 34.9)**. Hagfishes have a skull made of cartilage, but they lack jaws and vertebrae. They swim in a snakelike fashion by using their segmental muscles to exert force against their notochord, which they retain in adulthood as a strong, flexible rod of cartilage. Hagfishes have a small brain, eyes, ears, and a nasal opening

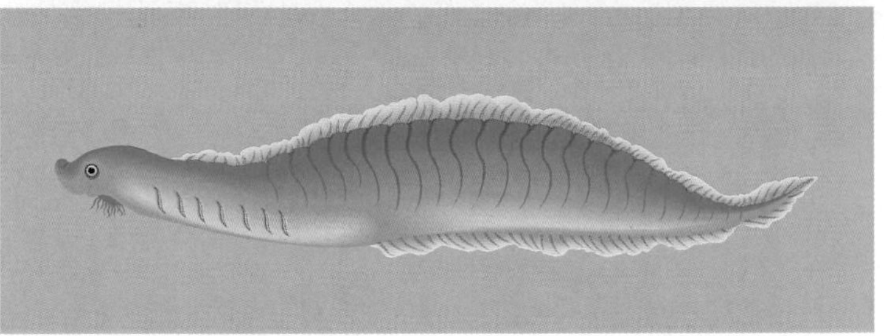

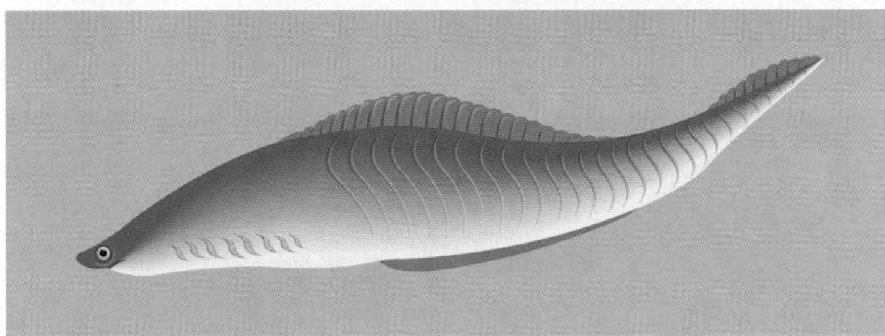

(a) *Haikouella.* Discovered in 1999 in southern China, *Haikouella* had eyes and a brain but lacked a skull, a derived trait of craniates.

(b) *Haikouichthys.* *Haikouichthys* had a skull and thus is considered a true craniate.

▲ **Figure 34.8 Fossils of primitive chordates.** The colors in the illustrations are fanciful.

▲ **Figure 34.9 A hagfish.**

that connects with the pharynx. They also have tooth-like formations made of the protein keratin in their mouth.

All of the 30 living species of hagfishes are marine. Measuring up to 60 cm in length, most are bottom-dwelling scavengers that feed on worms and sick or dead fish. Rows of slime glands on a hagfish's flanks secrete a substance that absorbs water, forming a slime that may repulse other scavengers when a hagfish is feeding (see Figure 34.9). When a hagfish is attacked by a potential predator, it can produce several liters of slime in less than a minute. The slime coats the gills of an attacking fish, sending it into retreat or even suffocating it. Several teams of biologists and engineers are investigating the properties of hagfish slime in hopes of producing an artificial slime that could act as a space-filling gel. Such a gel might be used, for instance, to curtail bleeding during surgery.

Vertebrate systematists do not consider hagfishes to be fishes, despite their common name. Traditionally, the term *fish* has been applied to any craniate except tetrapods. But used in this way, it does not refer to a monophyletic group; therefore, systematists use the term *fish* to refer only to a specific clade of vertebrates, the actinopterygians (see Concept 34.4). We will follow that practice in this chapter.

Concept Check 34.2

1. Which extinct chordate is more closely related to humans, *Haikouichthys* or *Haikouella*? Explain your answer.
2. What characteristics do hagfishes have that tunicates and lancelets lack?

For suggested answers, see Appendix A.

Concept 34.3

Vertebrates are craniates that have a backbone

During the Cambrian period, a lineage of craniates evolved into vertebrates. With a more complex nervous system and a more elaborate skeleton than those of their ancestors, vertebrates became active predators.

Derived Characters of Vertebrates

After vertebrates branched off from other craniates, they underwent another gene duplication, this one involving a group of transcription factor genes called the *Dlx* family. The resulting additional genetic complexity was associated with the development of innovations in the vertebrate nervous system and skeleton, including the presence of a more extensive skull and a backbone composed of vertebrae. In some vertebrates, the vertebrae are little more than small prongs of cartilage arrayed dorsally along the notochord. In the majority of vertebrates, however, the vertebrae enclose the spinal cord and have taken over the mechanical roles of the notochord. Aquatic vertebrates also acquired dorsal, ventral, and anal fins stiffened by fin rays, which provide thrust and steering control during swimming. Faster swimming was supported by other adaptations, including a more efficient gas exchange system in the gills.

Lampreys

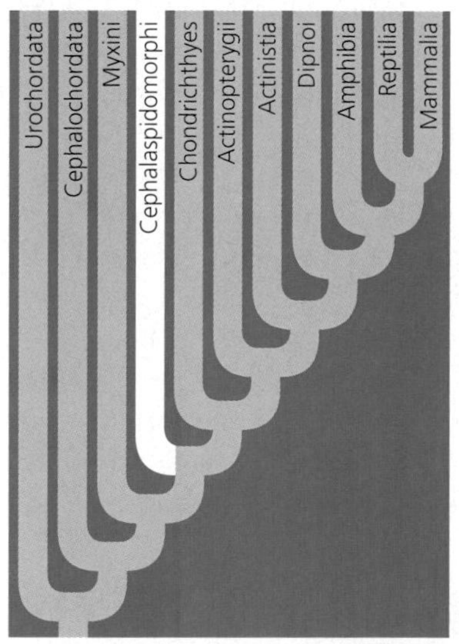

Lampreys (class Cephalaspidomorphi) represent the oldest living lineage of vertebrates. Like hagfishes, lampreys may offer clues to early chordate evolution but have also acquired unique characters.

There are about 35 species of lampreys inhabiting various marine and freshwater environments **(Figure 34.10)**. Most are parasites that feed by clamping their round, jawless mouth onto the flank of a live fish. They then use their rasping tongue to penetrate the skin of the fish and ingest the fish's blood.

As larvae, lampreys live in freshwater streams. The larva is a suspension feeder that resembles a lancelet and spends much of its time partially buried in sediment. Some species of lampreys feed only as larvae; following several years in streams, they attain sexual maturity, reproduce, and die within a few

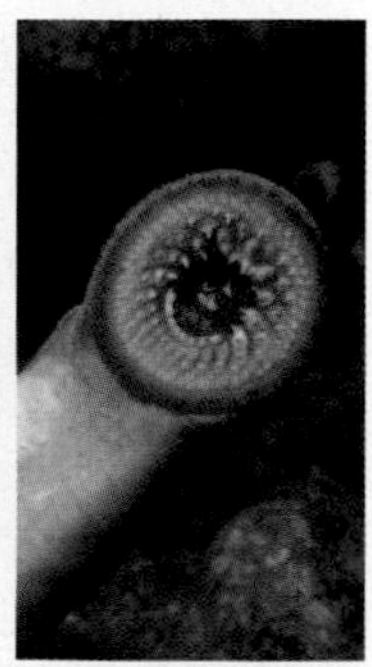

▲ **Figure 34.10 A sea lamprey.** Most lampreys use their mouth (enlarged, right) and tongue to bore a hole in the side of a fish. The lamprey then ingests the blood and other tissues of its host.

days. Most lampreys, however, migrate to the sea or lakes as they mature into adults. The sea lamprey (*Petromyzon marinus*) has invaded the Great Lakes over the past 170 years, where it has devastated a number of fisheries.

The skeleton of lampreys is made of cartilage. Unlike the cartilage found in most vertebrates, lamprey cartilage contains no collagen. Instead, it is a stiff protein matrix. The notochord of lampreys persists as the main axial skeleton in the adult, as it does in hagfishes. However, lampreys also have a cartilaginous pipe around their rodlike notochord. Along the length of this pipe, pairs of cartilaginous projections related to vertebrae extend dorsally, partially enclosing the nerve cord.

Fossils of Early Vertebrates

After the ancestors of lampreys branched off from other vertebrates during the Cambrian period, many other lineages of vertebrates emerged. Like lampreys, the early members of these lineages lacked jaws, but the resemblance stopped there.

Conodonts were slender, soft-bodied vertebrates with prominent eyes controlled by numerous muscles. At the anterior end of their mouth, they had a set of barbed hooks made of mineralized dental tissues **(Figure 34.11)**. Most conodonts were 3–10 cm in length, although some may been as long as 30 cm. They probably hunted with the help of their large eyes, impaling prey on their hooks. The food then passed back to the pharynx, where a different set of dental elements sliced and crushed the food.

Conodonts were extremely abundant for over 300 million years. Their fossilized dental elements are so plentiful that they have been used for decades by petroleum geologists as guides to the age of rock layers in which they search for oil. (These elements also gave conodonts their name, which means "cone teeth.")

Vertebrates with additional innovations emerged during the Ordovician, Silurian, and Devonian periods. These vertebrates had paired fins and an inner ear with two semicircular canals that provided a sense of balance. Although they, too, lacked jaws, they had a muscular pharynx, which they may

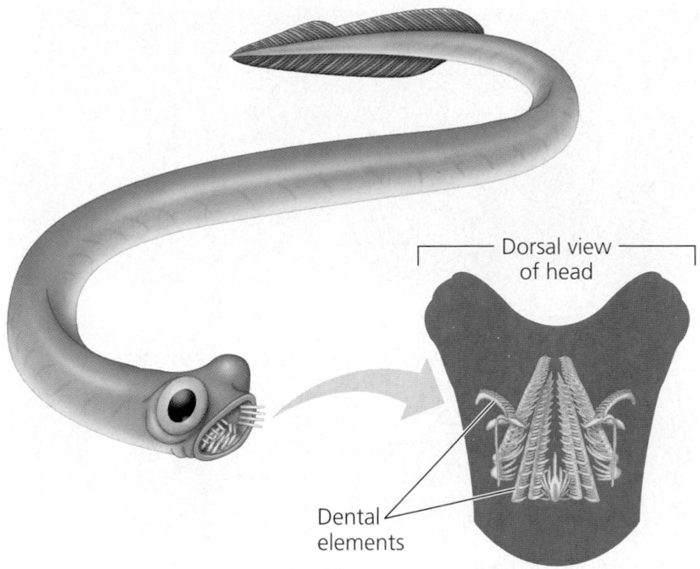

▲ **Figure 34.11 A conodont.** Conodonts were early vertebrates that lived from the late Cambrian until the late Triassic. Unlike lampreys, conodonts had mineralized mouthparts, which they used for either predation or scavenging.

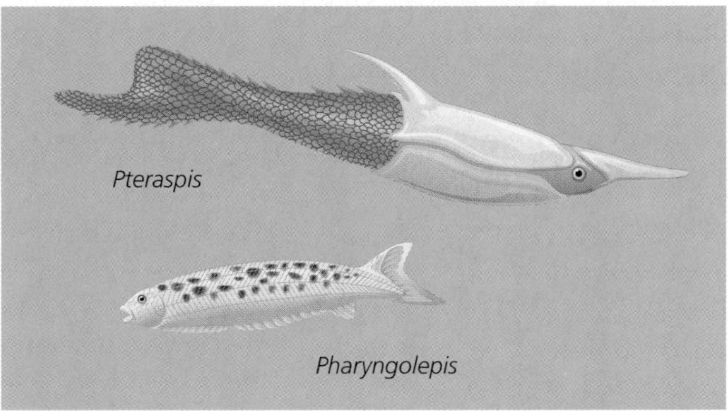

▲ **Figure 34.12 Jawless armored vertebrates.** *Pteraspis* and *Pharyngolepis* were two of many genera of jawless vertebrates that emerged during the Ordovician, Silurian, and Devonian periods.

have used to suck in bottom-dwelling organisms or detritus. They were also armored with mineralized bone, which covered varying amounts of their body **(Figure 34.12)**. The armor, which in some species included spines, may have offered protection from predators. These armored vertebrates were formerly placed in a group called the *ostracoderms* ("shelled skin"). However, more recent research indicates that this group is paraphyletic: Some lineages are more closely related to jawed vertebrates than to other members of the group. These armored jawless vertebrates were exceptionally diverse, but they all became extinct by the end of the Devonian period.

Origins of Bone and Teeth

The human skeleton is heavily mineralized, and cartilage plays a fairly minor role. But this is a relatively recent development in the history of vertebrates. As we've seen, the vertebrate skeleton evolved initially as a structure made of unmineralized cartilage. Its mineralization began only after lampreys diverged from other vertebrates.

What initiated the process of mineralization in vertebrates? Philip Donoghue, of the University of Birmingham, England, hypothesizes that mineralization was associated with a transition in feeding mechanisms. Early chordates probably were suspension feeders, like lancelets, but over time they became larger and were therefore able to ingest larger particles, including some small animals. The earliest known mineralized structures in vertebrates—conodont dental elements—were an adaptation that allowed these animals to become scavengers and predators. The armor seen in later jawless vertebrates was derived from dental mineralization. Thus, mineralization of the vertebrate body began in the mouth, according to Donoghue. Only in more derived vertebrates did the endoskeleton begin to mineralize, starting with the skull. As you'll learn in the next section, younger lineages of vertebrates underwent even further mineralization.

Concept Check 34.3

1. How are differences in lamprey and conodont anatomy reflected in each animal's feeding method?
2. What key roles did mineralized bone play in the first vertebrates?

For suggested answers, see Appendix A.

Gnathostomes are vertebrates that have jaws

Hagfishes and lampreys are survivors from an age when jawless craniates were common. Today they are far outnumbered by jawed vertebrates, known as **gnathostomes.**

Derived Characters of Gnathostomes

Gnathostomes ("jaw mouth") are named for their jaws, hinged structures that, especially with the help of teeth, enable gnathostomes to grip food items firmly and slice them up. According to one hypothesis, gnathostome jaws evolved by modification of the skeletal rods that had previously supported the anterior pharyngeal (gill) slits (**Figure 34.13**, on the next page). The remaining gill slits, no longer required for suspension feeding, remained as the major sites of respiratory gas exchange with the external environment. Other hypotheses posit different structures as the origin of jaws. Researchers comparing genes involved in the development of the mouth in lampreys and gnathostomes are testing these hypotheses.

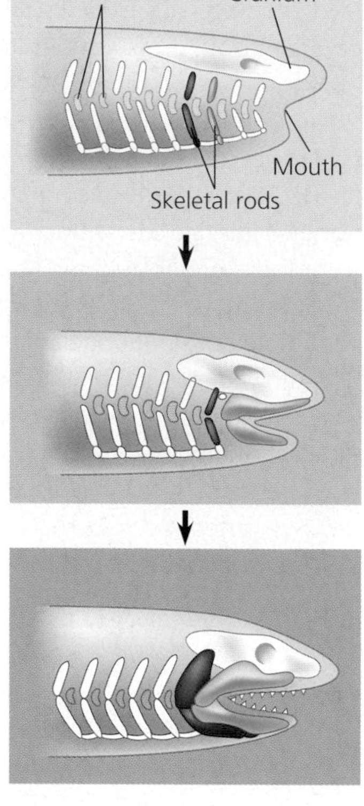

▶ **Figure 34.13 Hypothesis for the evolution of vertebrate jaws.** The skeleton of the jaws and their supports evolved from two pairs of skeletal rods (red and green) located between gill slits near the mouth. Pairs of rods anterior to those that formed the jaws were either lost or incorporated into the cranium or jaws.

(a) ***Coccosteus*, a placoderm**

(b) ***Climatius*, an acanthodian**

▲ **Figure 34.14 Early gnathostomes.**

Gnathostomes share other derived characters besides jaws. The common ancestors of all gnathostomes underwent an additional duplication of *Hox* genes, such that the single cluster present in early chordates became four. Other gene clusters also duplicated, allowing further complexity in the development of gnathostome embryos. The gnathostome forebrain is enlarged compared to that of other craniates, mainly in association with enhanced senses of smell and vision. Running the length of each side of the body in aquatic gnathostomes is the **lateral line system,** a row of microscopic organs sensitive to vibrations in the surrounding water.

As mentioned earlier, the ancestors of gnathostomes began to acquire a mineralized endoskeleton. In the common ancestor of living gnathostomes, the axial skeleton, shoulder girdle, and paired appendages were mineralized.

Fossil Gnathostomes

Gnathostomes appeared in the fossil record in the mid-Ordovician period, about 470 million years ago, and steadily became more diverse. Their success probably lies in two features of their anatomy: Their paired fins and tail allowed them to swim efficiently after prey, and their jaws enabled them to grab prey or simply bite off chunks of flesh.

The earliest gnathostomes in the fossil record are an extinct lineage of armored vertebrates called **placoderms,** which means "plate-skinned" **(Figure 34.14a)**. Most placoderms were less than a meter long, though some giants measured more than 10 m. Another group of jawed vertebrates called **acanthodians** radiated during the Devonian period, and many new forms evolved in fresh and salt water **(Figure 34.14b)**. Acanthodians were closely related to the ancestors of osteichthyans (ray-finned fishes and lobe-fins). Both placoderms and acanthodians had disappeared by the beginning of the Carboniferous period, about 360 million years ago.

Chondrichthyans (Sharks, Rays, and Their Relatives)

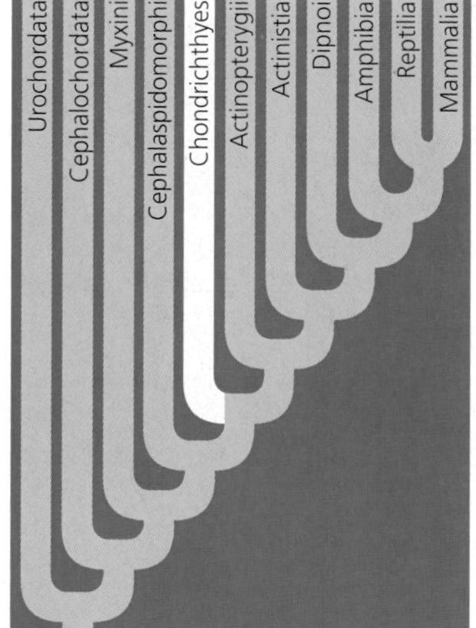

Sharks, rays, and their relatives include some of the biggest and most successful vertebrate predators in the oceans. They belong to the class Chondrichthyes, which means "cartilage fish." As the name indicates, **chondrichthyans** have a skeleton that is composed predominantly of cartilage, often impregnated with calcium.

When the name Chondrichthyes was first coined in the 1800s, scientists thought that chondrichthyans represented a primitive stage in the evolution of the vertebrate skeleton, and that mineralization had evolved only in more derived lineages (such as "bony fishes"). Strengthening this notion was the fact that embryos in these derived lineages first develop a skeleton made largely of cartilage. Later, much of the cartilage is replaced by bone, a hard matrix of calcium phosphate. However, this view has been abandoned. As conodonts and armored jawless vertebrates demonstrate, the mineralization of the vertebrate skeleton had already begun before the chondrichthyan lineage branched off from other vertebrates. Moreover, traces of bone can be found in living chondrichthyans—in their scales, at the base of their teeth, and, in some sharks, in a thin layer on the surface of their vertebrae. The restricted distribution of bone in the chondrichthyan body appears to be a derived condition, which emerged after they diverged from other gnathostomes.

There are about 750 species of living chondrichthyans. The largest and most diverse subclass consists of the sharks and rays (**Figure 34.15a** and **b**). A second subclass is composed of a few dozen species of ratfishes, or chimaeras (**Figure 34.15c**).

Most sharks have a streamlined body and are swift swimmers, but they do not maneuver very well. Powerful movements of the trunk and the caudal (tail) fin propel them forward. The dorsal fins function mainly as stabilizers, and the paired pectoral (fore) and pelvic (hind) fins provide lift when the shark swims. Although a shark gains buoyancy by storing a large amount of oil in its huge liver, the animal is still more dense than water, and if it stops swimming it sinks. Continual swimming also ensures that water flows into the shark's mouth and out through the gills, where gas exchange occurs. However, some sharks and many skates and rays spend a good deal of time resting on the seafloor. When resting, they use muscles of their jaws and pharynx to pump water over the gills.

The largest sharks and rays are suspension feeders that consume plankton. Most sharks, however, are carnivores that swallow their prey whole or use their powerful jaws and sharp teeth to tear flesh from animals too large to swallow in one piece. Sharks have several rows of teeth that gradually move to the front of the mouth as old teeth are lost. The digestive tract of many sharks is proportionately shorter than that of many other vertebrates. Within the shark intestine is a **spiral valve**, a corkscrew-shaped ridge that increases surface area and prolongs the passage of food through the digestive tract.

Acute senses are adaptations that go along with the active, carnivorous lifestyle of sharks. Sharks have sharp vision but cannot distinguish colors. The nostrils of sharks, like those of most aquatic vertebrates, open into dead-end cups. They function only for olfaction (smelling), not for breathing. Sharks also have a pair of regions in the skin of their head that can detect electric fields generated by the muscle contractions of nearby animals. Like other aquatic vertebrates, sharks have no eardrums, structures that terrestrial vertebrates use to transmit sound waves in air to the auditory organs. Sound reaches a shark through water, and the animal's entire body transmits the sound to the hearing organs of the inner ear.

Shark eggs are fertilized internally. The male has a pair of claspers on its pelvic fins that transfer sperm into the reproductive tract of the female. Some species of sharks are **oviparous**; they lay eggs that hatch outside the mother's body. These sharks release their eggs after encasing them in protective coats. Other species are **ovoviviparous**; they retain the fertilized eggs in the oviduct. Nourished by the egg yolk, the embryos develop into young that are born after hatching

▼ **Figure 34.15 Chondrichthyans.**

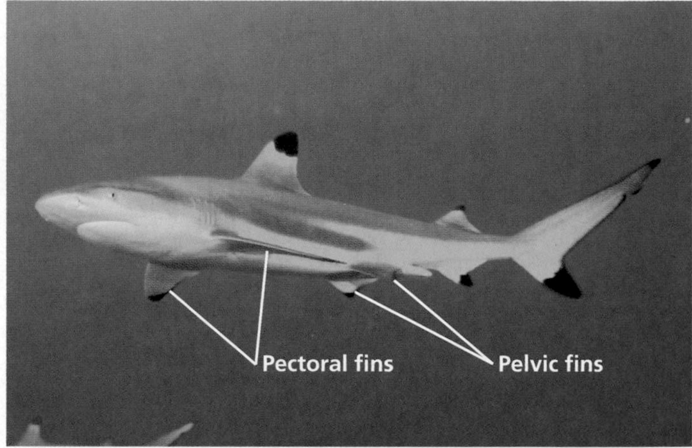

(a) Blacktip reef shark *(Carcharhinus melanopterus).* Fast swimmers with acute senses, sharks have paired pectoral and pelvic fins.

(b) Southern stingray *(Dasyatis americana).* Most rays are flattened bottom-dwellers that crush molluscs and crustaceans for food. Some rays cruise in open water and scoop food into their gaping mouth.

(c) Spotted ratfish *(Hydrolagus colliei).* Ratfishes, or chimaeras, typically live at depths greater than 80 m and feed on shrimps, molluscs, and sea urchins. Some species have a poisonous spine at the front of their dorsal fin.

within the uterus. A few species are **viviparous;** the young develop within the uterus and obtain nourishment prior to birth by receiving nutrients from the mother's blood through a yolk sac placenta, by absorbing a nutritious fluid produced by the uterus, or by eating other eggs. The reproductive tract of the shark empties along with the excretory system and digestive tract into the **cloaca,** a common chamber that has a single opening to the outside.

Although rays are closely related to sharks, they have adopted a very different lifestyle. Most rays are flattened bottom-dwellers that feed by using their jaws to crush molluscs and crustaceans. They use their greatly enlarged pectoral fins like water wings to propel themselves through the water. The tail of many rays is whiplike and, in some species, bears venomous barbs that function in defense.

Chondrichthyans have changed little in over 300 million years. Today, however, they are severely threatened with overfishing. In 2003, researchers reported that shark stocks in the northwest Atlantic declined 75% in 15 years.

Ray-Finned Fishes and Lobe-Fins

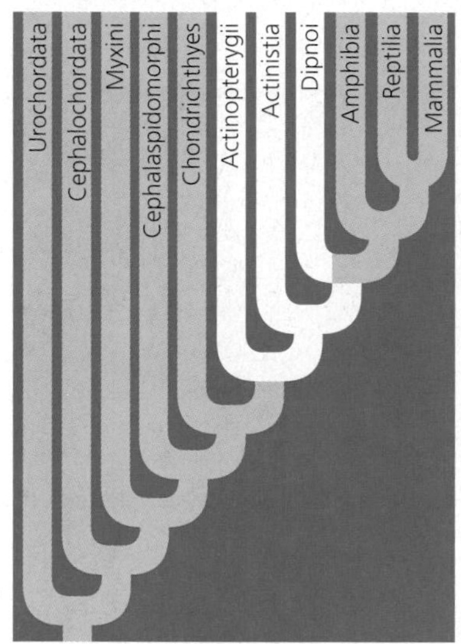

The vast majority of vertebrates belong to a clade of gnathostomes called Osteichthyes. Like many other taxonomic names, the name Osteichthyes ("bony fish") was coined long before the advent of phylogenetic systematics. When it was originally defined, the group excluded tetrapods, but we now know that such a taxon would actually be paraphyletic (see Figure 34.2). Therefore, systematists today include tetrapods along with bony fishes in the clade Osteichthyes. Clearly, the name of the group does not accurately describe all of its members.

Unlike chondrichthyans, nearly all living **osteichthyans** have an ossified (bony) endoskeleton with a hard matrix of calcium phosphate. It is not clear when the shift to a bony skeleton took place during gnathostome evolution. It is possible, for example, that the common ancestor of both chondrichthyans and osteichthyans was already highly ossified and that chondrichthyans subsequently lost much of this bone. Until more fossils of early chondrichthyans and osteichthyans are unearthed, the question will remain open.

Aquatic osteichthyans are the vertebrates we informally call fishes. They breathe by drawing water over four or five pairs of gills located in chambers covered by a protective bony flap called the **operculum (Figure 34.16)**. Water is drawn into the mouth, through the pharynx, and out between the gills by movement of the operculum and contraction of muscles surrounding the gill chambers.

Most aquatic osteichthyans can control their buoyancy with an air sac known as a **swim bladder.** Movement of gases from the blood to the swim bladder increases buoyancy, making the animal rise; a transfer back to the blood causes the animal to sink. Charles Darwin proposed that the lungs of tetrapods evolved from swim bladders, but, strange as it may sound, the opposite seems to be true. Osteichthyans in many early-branching lineages also have lungs, which they use to breathe air as a supplement to gas exchange in their gills. The weight of evidence indicates that lungs arose first and later evolved into swim bladders in some lineages.

The skin of aquatic osteichthyans is often covered by flattened, bony scales that differ in structure from the toothlike scales of sharks. Glands in the skin secrete a slimy mucus over the skin, an adaptation that reduces drag during swimming. Like sharks, aquatic osteichthyans have a lateral line system, which is evident as a row of tiny pits in the skin on either side of the body.

▼ Figure 34.16 Anatomy of a trout, an aquatic osteichthyan.

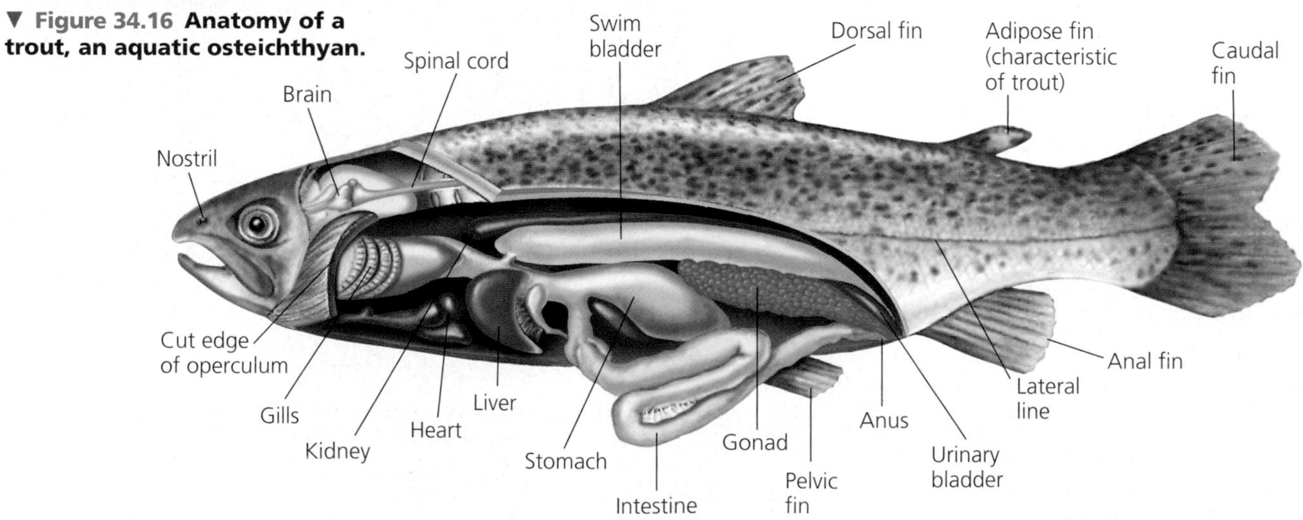

Details about the reproduction of aquatic osteichthyans vary extensively. Most species are oviparous, reproducing by external fertilization after the female sheds large numbers of small eggs. However, internal fertilization and birthing characterize other species.

Ray-Finned Fishes

Nearly all the aquatic osteichthyans familiar to us are among the **ray-finned fishes** (class Actinopterygii, from the Greek *aktin*, ray, and *pteryg*, wing or fin). The various species of bass, trout, perch, tuna, and herring are examples **(Figure 34.17)**. The fins, supported mainly by long, flexible rays, are modified for maneuvering, defense, and other functions.

Ray-finned fishes appear to have originated in fresh water and spread to the seas. (Adaptations that solve the osmotic problems associated with the move to salt water are discussed in Chapter 44.) Numerous species of ray-finned fishes returned to fresh water at some point in their evolution. Some of them, including salmon and sea-run trout, replay their evolutionary round-trip from fresh water to seawater and back to fresh water during their life cycle.

Ray-finned fishes serve as a major source of protein for humans, who have harvested them for tens of thousands of years. However, industrial-scale fishing operations now threaten some of the world's biggest fisheries with collapse. Ray-finned fishes also face other pressures from humans, such as the diversion of rivers by dams.

Lobe-Fins

Ray-finned fishes evolved during the Devonian period, along with another major lineage of osteichthyans, the **lobe-fins** (Sarcopterygii). The key derived character of lobe-fins is the presence of rod-shaped bones surrounded by a thick layer of muscle in their pectoral and pelvic fins. During the Devonian, many lobe-fins lived in brackish waters, such as in coastal wetlands. There they probably used their lobed fins to swim and "walk" underwater across the substrate. Some Devonian lobe-fins were gigantic predators. It is not uncommon to find spike-shaped fossils of Devonian lobe-fin teeth as big as your thumb.

By the end of the Devonian period, lobe-fin diversity was dwindling, and today only three lineages survive. One lineage, the coelacanths (class Actinistia), shifted to the ocean **(Figure 34.18)**. Scientists once thought that coelacanths had become extinct 75 million years ago, but in 1938, fishermen caught a living coelacanth off the Comoros Islands in the western Indian Ocean. Coelacanths were found only in that area until 1999, when a second population was identified in the eastern Indian Ocean, near Indonesia. The Indonesian population may represent a second species.

The second lineage of living lobe-fins is represented today by three genera of lungfishes (class Dipnoi), all of which are

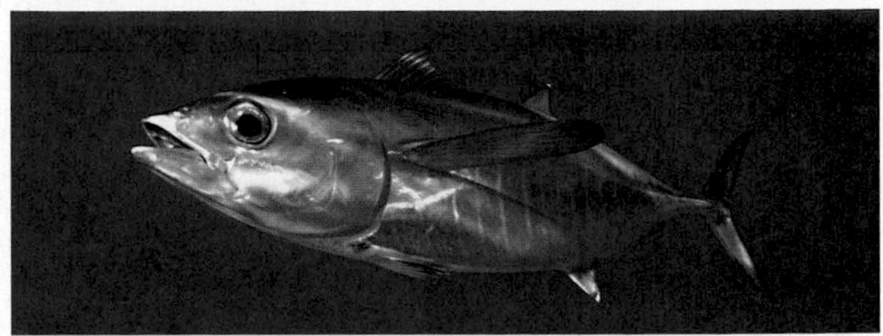

(a) Yellowfin tuna *(Thunnus albacares)*, a fast-swimming, schooling fish that is an important commercial fish worldwide

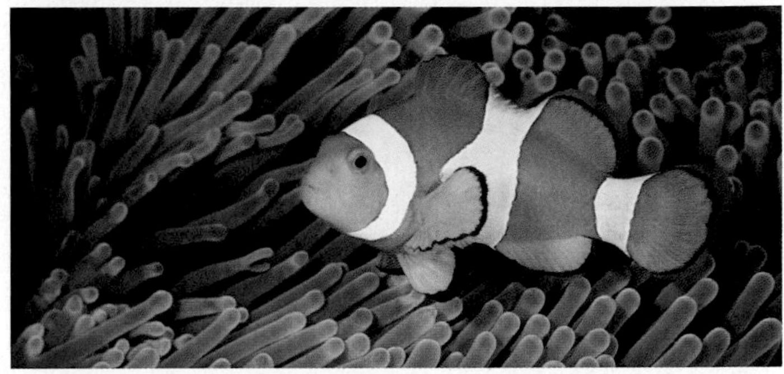

(b) Clownfish *(Amphiprion ocellaris)*, a mutualistic symbiont of sea anemones

(c) Sea horse *(Hippocampus ramulosus)*, unusual in the animal kingdom in that the male carries the young during their embryonic development

(d) Fine-spotted moray eel *(Gymnothorax dovii)*, a predator that ambushes prey from crevices in its coral reef habitat

▲ **Figure 34.17 Ray-finned fishes (class Actinopterygii).**

▲ **Figure 34.18 A coelacanth (*Latimeria*).** These lobe-fins live in deep water off the coasts of southern Africa and Indonesia.

found in the Southern Hemisphere. Lungfishes evolved in the ocean but today are found only in fresh water, generally in stagnant ponds and swamps. They surface to gulp air into lungs connected to their pharynx. Lungfishes also have gills, which are the main organs for gas exchange in Australian lungfishes. When ponds shrink during the dry season, some lungfishes can burrow into the mud and aestivate (wait in a state of torpor; see Chapter 40).

The third lineage of lobe-fins that survives today is far more diverse than the coelacanths or lungfishes. During the mid-Devonian, these organisms gave rise to vertebrates with limbs and feet, called tetrapods—a lineage that includes humans. The tetrapod clade is the topic of the next section.

Concept Check 34.4

1. What derived characters do sharks and tuna share? What are some characteristics that distinguish tuna from sharks?
2. Contrast the habitats of the three surviving lineages of lobe-fins.

For suggested answers, see Appendix A.

Concept 34.5

Tetrapods are gnathostomes that have limbs and feet

One of the most significant events in vertebrate history took place about 360 million years ago, when the fins of some lobe-fins evolved into the limbs and feet of tetrapods. Until then, all vertebrates had shared the same basic fishlike anatomy. Once tetrapods moved onto land, they eventually took on many new forms, from leaping frogs to flying eagles to bipedal humans.

Derived Characters of Tetrapods

The most significant character of **tetrapods** gives the group its name, which means "four feet" in Greek. In place of pectoral and pelvic fins, tetrapods have limbs that can support their weight on land and feet with digits that allow them to transmit muscle-generated forces to the ground when they walk.

Life on land brought numerous other changes to the tetrapod body plan. The bones of the pelvic girdle, to which the hind legs are attached, are fused to the backbone, permitting forces generated by the hind legs against the ground to be transferred to the rest of the body. Living tetrapods do not have gill slits; during embryonic development, the pharyngeal clefts instead give rise to parts of the ears, glands, and other structures. The ears are adapted to the detection of airborne sounds.

As you will see, some of these characters were dramatically altered or lost in various lineages of tetrapods. For example, in birds the pectoral limbs became wings, while in whales the entire body converged toward a fishlike shape.

The Origin of Tetrapods

As you have learned, the Devonian coastal wetlands were home to a wide range of lobe-fins. Those that entered particularly shallow, oxygen-poor water could use their lungs to breathe air. Some species probably used their stout fins to help them move across logs or the muddy bottom. Thus, the tetrapod body plan did not evolve "out of nowhere" but was simply a modification of a preexisting body plan.

In one lineage of lobe-fins, the fins became progressively more limb-like while the rest of the body retained adaptations for aquatic life. For example, *Acanthostega*, a close relative of tetrapods that lived in Greenland 365 million years ago, had fully formed legs, ankles, and digits **(Figure 34.19)**. Yet it also retained adaptations for the water. It had bones that supported gills, and it had rays in its tail that supported a delicate fin that propelled it through water. Its pectoral and pelvic girdles were too weak to carry its body on land. *Acanthostega* may have slithered out of the water from time to time, but for the most part it was aquatic.

Extraordinary fossil discoveries over the past 20 years have allowed paleontologists to reconstruct the origin of tetrapods with confidence **(Figure 34.20)**. A great diversity of tetrapods emerged during the Devonian and Carboniferous periods, and some species reached 2 m in length. Judging from the morphology and locations of their fossils, most of these early tetrapods probably remained tied to the water, a feature they share with some members of a group of living tetrapods called amphibians.

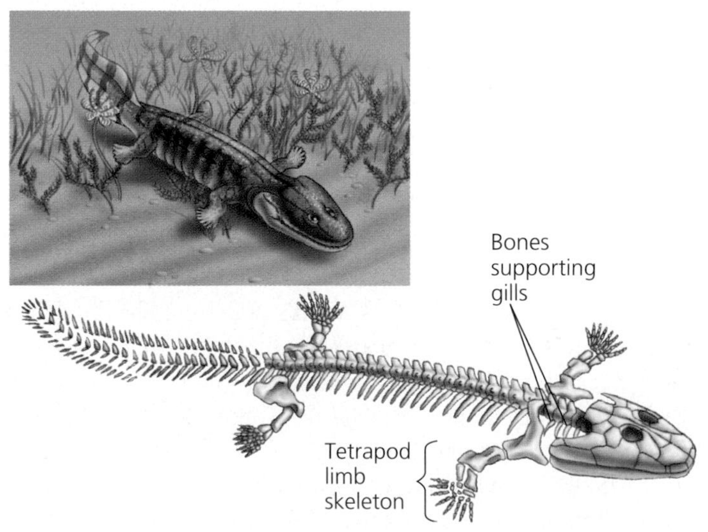

▲ **Figure 34.19 *Acanthostega*, a Devonian relative of tetrapods.** Along with the derived appendages of tetrapods, *Acanthostega* retained primitive aquatic adaptations, such as gills.

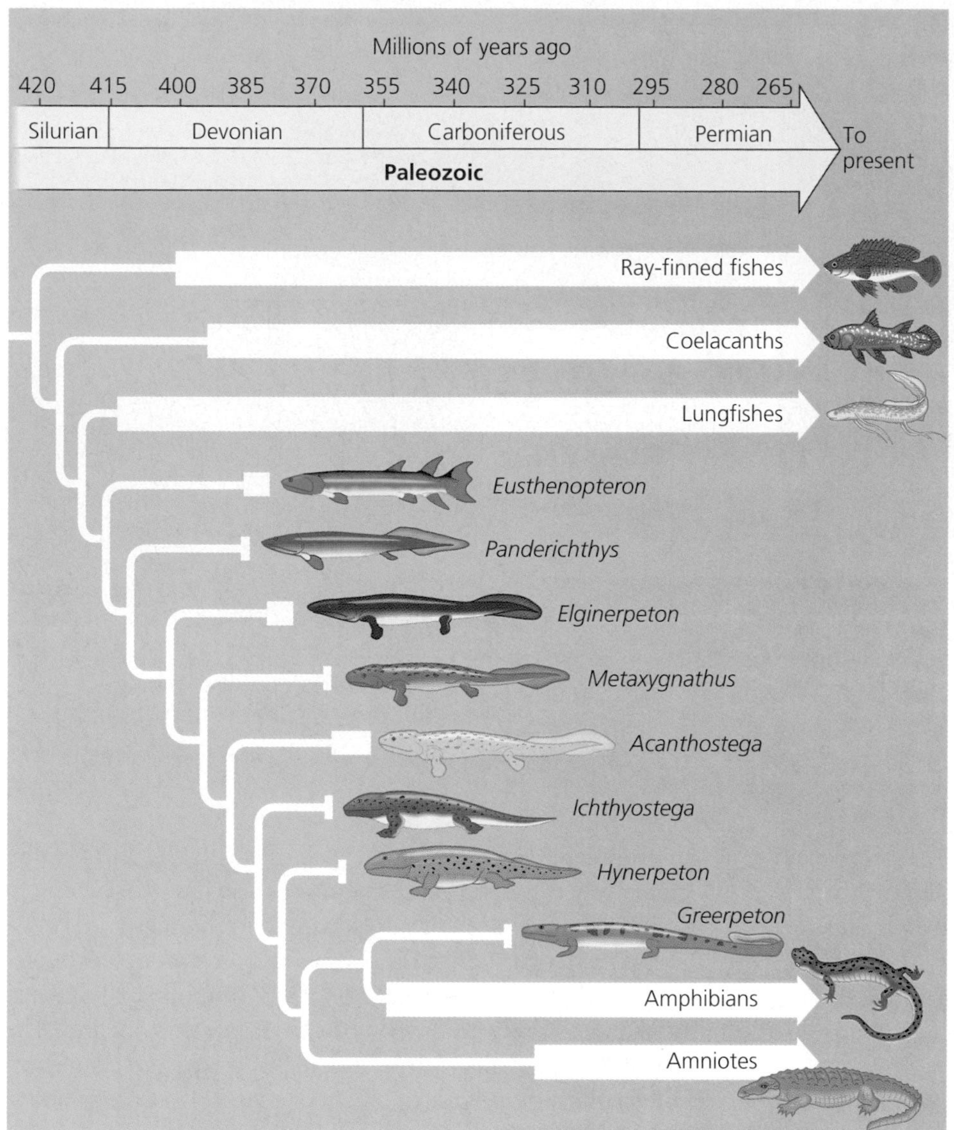

▲ **Figure 34.20 The origin of tetrapods.** The bars on the branches of this diagram place known fossils in time; arrows indicate lineages that extend to today. The drawings of extinct forms are based on fossilized skeletons, but the colors are fanciful.

▼ **Figure 34.21 Amphibians.**

(a) Order Urodela. Urodeles (salamanders) retain their tail as adults.

(b) Order Anura. Anurans, such as this poison arrow frog, lack a tail as adults.

(c) Order Apoda. Apodans, or caecilians, are legless, mainly burrowing amphibians.

Amphibians

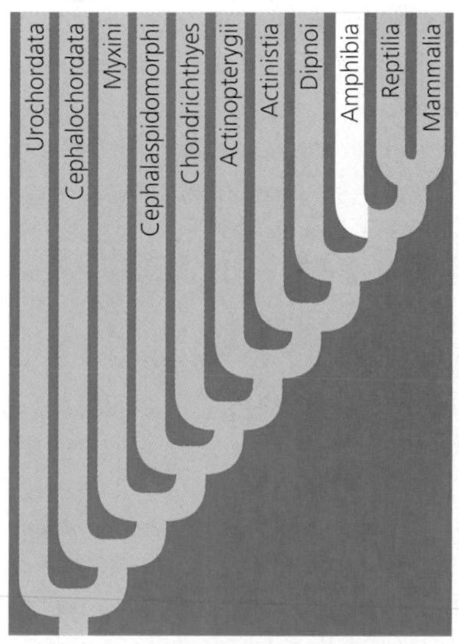

Amphibians (class Amphibia) are represented today by about 4,800 species of salamanders (order Urodela, which means "tailed ones"), frogs (order Anura, meaning "tail-less ones"), and caecilians (order Apoda, meaning "legless ones").

There are only about 500 species of urodeles. Some are entirely aquatic, but others live on land as adults or throughout life **(Figure 34.21a)**. Most salamanders that live on land walk with a side-to-side bending of the body inherited from the early terrestrial tetrapods. Paedomorphosis is common among aquatic salamanders (see Figure 24.17); the mudpuppy (*Necturus*), for instance, retains larval features when sexually mature.

Anurans, numbering nearly 4,200 species, are more specialized than urodeles for moving on land **(Figure 34.21b)**. Adult frogs use their powerful hind legs to hop along the terrain. A frog nabs insects by flicking out its long, sticky tongue, which is attached to the front of the mouth. Frogs display a great variety of adaptations that help them avoid being eaten by larger predators. Their skin glands secrete distasteful or even poisonous mucus. Many poisonous species have bright coloration, which predators apparently associate with danger (see Figure 53.6). Other frogs have color patterns that camouflage (see Figure 53.5).

(a) The male grasps the female, stimulating her to release eggs. The eggs are laid and fertilized in water. They have a jelly coat but lack a shell and would desiccate in air.

(b) The tadpole is an aquatic herbivore with a fishlike tail and internal gills.

(c) During metamorphosis, the gills and tail are resorbed, and walking legs develop.

▲ **Figure 34.22 The "dual life" of a frog (*Rana temporaria*).**

Apodans, the caecilians (about 150 species), are legless and nearly blind, and superficially they resemble earthworms **(Figure 34.21c)**. Their absence of legs is a secondary adaptation, as they evolved from a legged ancestor. Caecilians inhabit tropical areas where most species burrow in moist forest soil. A few South American species live in freshwater ponds and streams.

Amphibian means "two lives," a reference to the metamorphosis of many frog species **(Figure 34.22)**. The larval stage of a frog, called a tadpole, is usually an aquatic herbivore with gills, a lateral line system resembling that of aquatic vertebrates, and a long, finned tail. The tadpole initially lacks legs and swims by undulating its tail. During the metamorphosis that leads to the "second life," the tadpole develops legs, lungs, a pair of external eardrums, and a digestive system adapted to a carnivorous diet. At the same time, the gills disappear; the lateral line system also disappears in most species. The young frog crawls onto shore and becomes a terrestrial hunter. In spite of their name, however, many amphibians do not live a dualistic—aquatic and terrestrial—life. There are some strictly aquatic or strictly terrestrial frogs, salamanders, and caecilians. Moreover, salamander and caecilian larvae look much like adults, and typically both the larvae and the adults are carnivorous.

Most amphibians are found in damp habitats such as swamps and rain forests. Even those that are adapted to drier habitats spend much of their time in burrows or under moist leaves, where the humidity is high. Amphibians generally rely heavily on their moist skin for gas exchange with the environment. Some terrestrial species lack lungs and breathe exclusively through their skin and oral cavity.

Fertilization is external in most amphibians; the male grasps the female and spills his sperm over the eggs as the female sheds them (see Figure 34.22a). Amphibians typically lay their eggs in water or in moist environments on land. The eggs lack a shell and dehydrate quickly in dry air. Some amphibian species lay vast numbers of eggs in temporary pools, and mortality is high. In contrast, other species lay relatively few eggs and display various types of parental care. Depending on the species, either males or females may house eggs on their back, in their mouth, or even in their stomach. Certain tropical tree frogs stir their egg masses into moist, foamy nests that resist drying. There are also some ovoviviparous and viviparous species that retain the eggs in the female reproductive tract, where embryos can develop without drying out.

Many amphibians exhibit complex and diverse social behaviors, especially during their breeding seasons. Frogs are usually quiet, but the males of many species vocalize to defend their breeding territory or to attract females. In some species, migrations to specific breeding sites may involve vocal communication, celestial navigation, or chemical signaling.

Over the past 25 years, zoologists have documented a rapid and alarming decline in amphibian populations throughout the world. There appear to be several causes, including habitat degradation, the spread of a fungal (chytrid) pathogen, and acid precipitation. The last of these is especially damaging to most amphibians because of their dependence on wet places for completing their life cycles.

Concept Check 34.5

1. Was *Acanthostega* a terrestrial tetrapod? Explain your answer.
2. Some amphibians never leave the water, whereas others can survive in relatively dry terrestrial environments. Contrast the adaptations that facilitate these two lifestyles.

For suggested answers, see Appendix A.

Amniotes are tetrapods that have a terrestrially adapted egg

Amniotes are a group of tetrapods whose living members are the reptiles (including birds) and the mammals **(Figure 34.23)**. During their evolution, amniotes acquired a number of new adaptations to life on land.

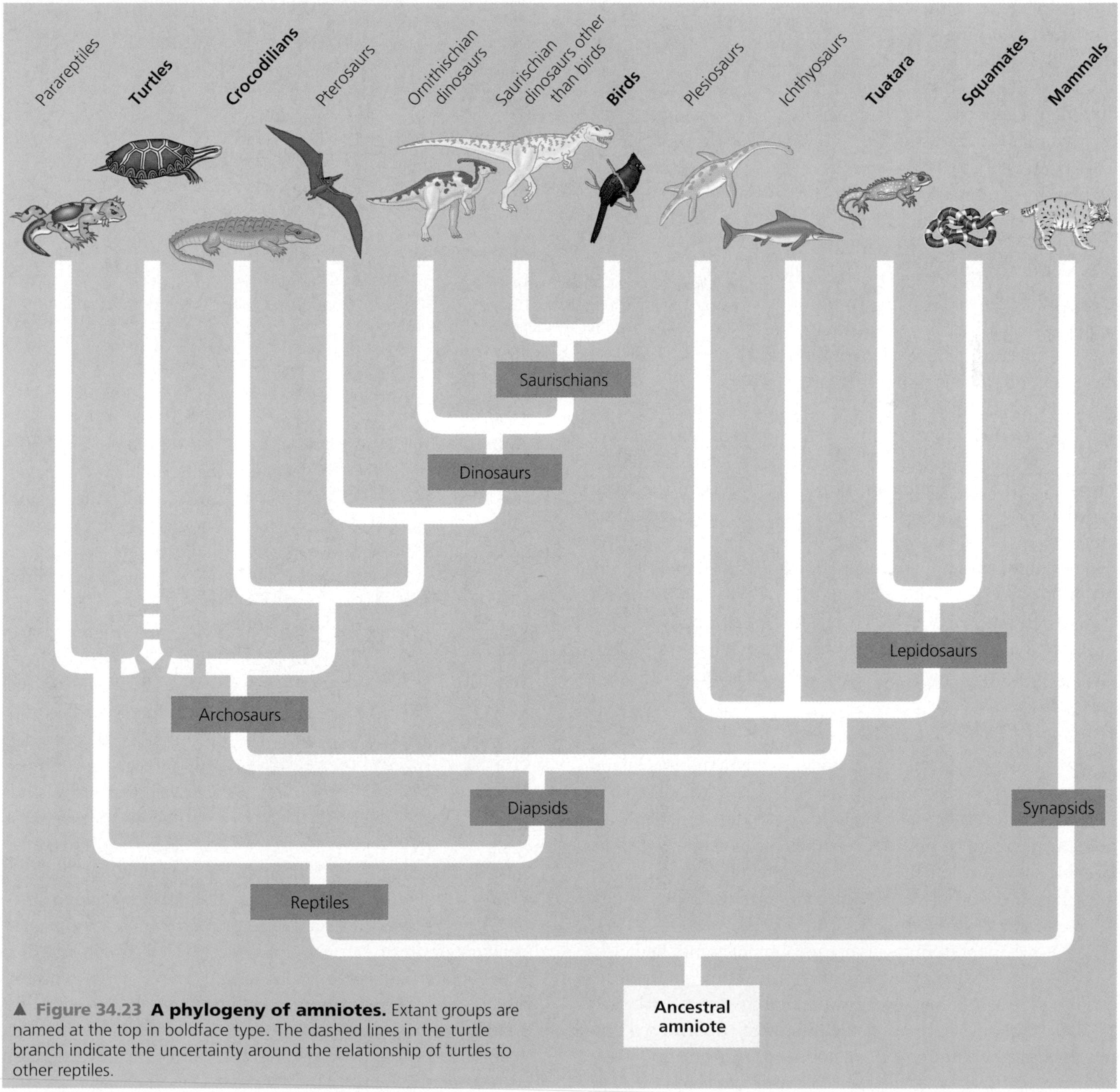

▲ **Figure 34.23 A phylogeny of amniotes.** Extant groups are named at the top in boldface type. The dashed lines in the turtle branch indicate the uncertainty around the relationship of turtles to other reptiles.

Derived Characters of Amniotes

Amniotes are named for the major derived character of the clade, the amniotic egg, which contains specialized membranes that protect the embryo **(Figure 34.24)**. Called **extraembryonic membranes** because they are not part of the body of the embryo itself, these membranes develop from tissue layers that grow out from the embryo. They function in gas exchange, waste storage, and the transfer of stored nutrients to the embryo. The amniotic egg is named for one of these membranes, the amnion, which encloses a compartment of fluid that bathes the embryo and acts as a hydraulic shock absorber.

In contrast to the shell-less eggs of amphibians, the amniotic eggs of most reptiles and some mammals have a shell. The shells of bird eggs are calcareous (made of calcium carbonate) and inflexible, while the shells of many nonbird reptile eggs are leathery and flexible. The shell significantly slows dehydration of the egg in air, an adaptation that allowed amniotes to occupy a wider range of terrestrial habitats than amphibians, their closest living relatives. (Seeds played a similar role in the evolution of land plants, as you learned in Chapter 30.) Most mammals have dispensed with the shell, and the embryo avoids desiccation by developing within the mother.

Amniotes also acquired other adaptations to terrestrial life, including less permeable skin and the ability to use the rib cage to ventilate the lungs. Whereas earlier tetrapods and living amphibians generally sprawl their legs out, living amniotes display, to varying degrees, a more elevated stance.

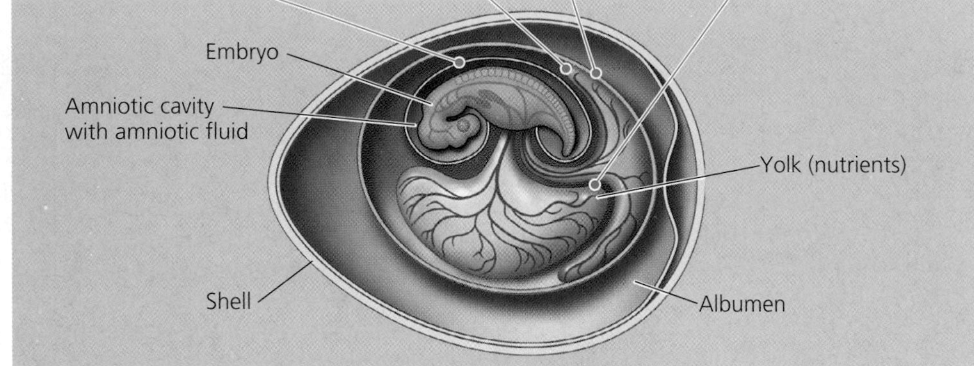

Extraembryonic membranes
Allantois. The allantois is a disposal sac for certain metabolic wastes produced by the embryo. The membrane of the allantois also functions with the chorion as a respiratory organ.
Chorion. The chorion and the membrane of the allantois exchange gases between the embryo and the air. Oxygen and carbon dioxide diffuse freely across the shell.
Amnion. The amnion protects the embryo in a fluid-filled cavity that cushions against mechanical shock.
Yolk sac. The yolk sac contains the yolk, a stockpile of nutrients. Blood vessels in the yolk sac membrane transport nutrients from the yolk into the embryo. Other nutrients are stored in the albumen ("egg white").
Embryo
Amniotic cavity with amniotic fluid
Yolk (nutrients)
Shell
Albumen

▲ **Figure 34.24 The amniotic egg.** The embryos of reptiles and mammals form four extraembryonic membranes: the amnion, allantois, chorion, and yolk sac. This drawing shows these extraembryonic membranes in the shelled egg of a reptile.

Early Amniotes

The most recent common ancestor of living amphibians and amniotes likely lived during the late Devonian period. No fossils of amniotic eggs have been found from that time, which is not surprising given how delicate they are. Thus, it is not yet possible to say when the amniotic egg evolved, although it must have existed in the last common ancestor of living amniotes, which all have amniotic eggs.

What is evident from fossils of early amniotes and their closest relatives is that they lived in drier environments than did earlier tetrapods. Some of them were herbivores, as evidenced by their grinding teeth and other features. Herbivorous amniotes began consuming large amounts of plant matter. They, in turn, became prey for large predatory amniotes.

Reptiles

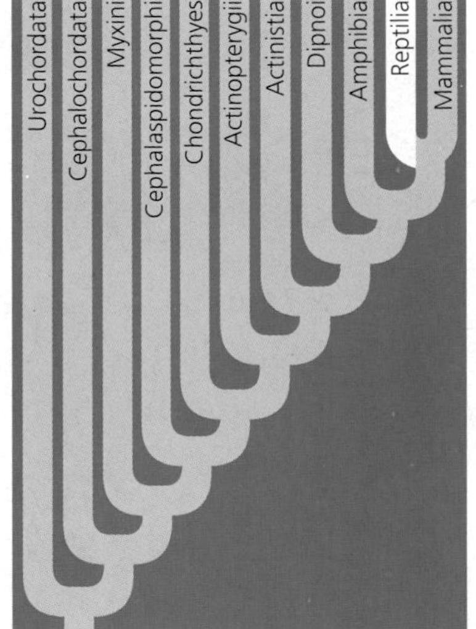

The **reptile** clade includes the tuatara, lizards, snakes, turtles, crocodilians, and birds, along with a number of extinct groups such as the large nonflying dinosaurs. Because all of the living reptile lineages are highly derived, none can serve as a straightforward model for the earliest reptiles that lived some 320 million years ago. Nevertheless, comparative studies allow us to infer some of the derived characters that likely distinguished early reptiles from other tetrapods.

Unlike amphibians, reptiles have scales that contain the protein keratin. Scales create a waterproof barrier that helps prevent dehydration in dry air. (In crocodiles, which have adapted to water, more permeable scales, called scutes, have evolved.) Scales prevent reptiles from breathing through their skin like amphibians; most reptiles rely on their lungs alone for gas exchange. Turtles are the exception to this rule; many turtles also use the moist surfaces of their cloaca for gas exchange.

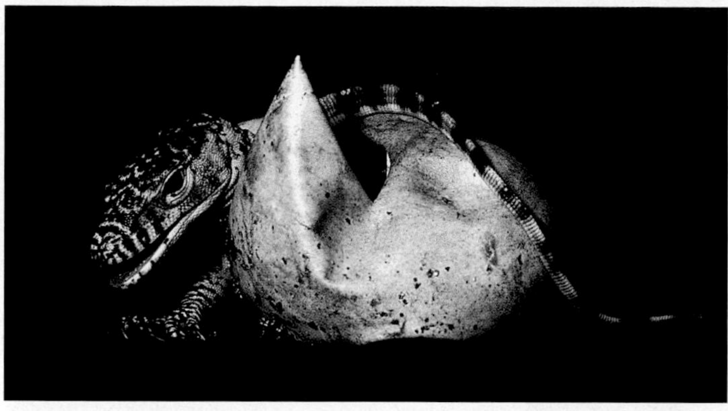

▲ **Figure 34.25 A hatching reptile.** This Komodo dragon is breaking out of a parchment-like shell, a common type of shell among living reptiles other than birds.

Most reptiles lay shelled eggs on land **(Figure 34.25)**. Fertilization must occur internally, before the shell is secreted. Many species of snakes and lizards are viviparous; their extraembryonic membranes form a placenta that enables the embryo to obtain nutrients from its mother.

Many reptiles are sometimes said to be "cold-blooded" because they do not use their metabolism extensively to control their body temperature. However, they do regulate their body temperature by using behavioral adaptations. For example, many lizards bask in the sun when the air is cool and seek shade when the air is too warm. A more accurate description of these reptiles is to say that they are **ectothermic,** which refers to the absorption of external heat as the main source of body heat. (This topic is discussed in more detail in Chapter 40.) By heating directly with solar energy rather than through the metabolic breakdown of food, an ectothermic reptile can survive on less than 10% of the food energy required by a mammal of the same size. The reptile clade is not entirely ectothermic; birds are **endothermic,** capable of keeping the body warm through metabolism.

The Origin and Evolutionary Radiation of Reptiles

The oldest reptilian fossils, found in rocks from Kansas, date from the end of the Carboniferous period, about 300 million years ago. The first major group of reptiles to emerge were the **parareptiles,** which were mostly large, stocky, quadrupedal herbivores. Some parareptiles had dermal plates on their skin that may have provided them with defense against predators. Some systematists have proposed that parareptiles gave rise to turtles, pointing to the possible homology of parareptile dermal plates and turtle shells. But molecular studies point to different reptilian origins for turtles, and the subject remains controversial among experts. The parareptiles died out by about 200 million years ago, at the end of the Triassic period.

As parareptiles were dwindling, an equally ancient clade of reptiles, the **diapsids,** was diversifying. One of the most obvious derived characters of diapsids is a pair of holes on each side of the skull, behind the eye socket. The diapsids are composed of two main lineages. One lineage gave rise to the **lepidosaurs,** which include tuatara, lizards, and snakes. This lineage also produced a number of marine reptiles, including plesiosaurs and ichthyosaurs (see Figure 34.23), some of which rivaled today's whales in length; all of them are extinct. (We'll say more about living lepidosaurs shortly.)

The other diapsid lineage, the **archosaurs,** produced the crocodilians (which we'll discuss later), pterosaurs, and dinosaurs. **Pterosaurs,** which originated in the late Triassic period, were the first tetrapods to take to the air. The pterosaur wing was completely different from the wings of birds and bats. It consisted of a bristle-covered membrane that stretched between the trunk or hind leg and a very long digit on the foreleg. Well-preserved fossils show the presence of muscles, blood vessels, and nerves in the wing membranes, suggesting that pterosaurs could dynamically adjust their membranes to assist their flight.

The smallest pterosaurs were no bigger than a gull, and the largest had a wingspan of nearly 11 m. They appear to have converged on many of the ecological roles later played by birds; some were insect-eaters, others grabbed fish out of the ocean, and still others filtered small animals through a long beak. By the end of the Cretaceous period 65 million years ago, pterosaurs had become extinct.

On land, the **dinosaurs** diversified into a vast range of shapes and sizes, from bipeds the size of a pigeon to 45-m-long quadrupeds with a neck long enough to let them browse the tops of trees. One branch of dinosaurs, the ornithischians, were herbivores; they included many species with elaborate defenses against predators, such as tail clubs and horned crests. The other main branch of dinosaurs, the saurischians, included the long-necked giants and a group called the **theropods,** which were bipedal carnivores. Theropods included the famous *Tyrannosaurus rex* as well as the ancestors of birds.

There is continuing debate about the metabolism of dinosaurs. Some researchers have pointed out that the Mesozoic climate over much of the dinosaurs' range was relatively warm and consistent, and they have suggested that the low surface-to-volume ratios of large dinosaurs combined with behavioral adaptations such as basking may have been sufficient for an ectotherm to maintain a suitable body temperature. However, some anatomical evidence supports the hypothesis that at least some dinosaurs were endotherms. Furthermore, paleontologists have found fossils of dinosaurs in both Antarctica and the Arctic; although the climate in these areas was milder when dinosaurs existed than it is today, it was cool enough that small dinosaurs may have had difficulty maintaining a high body temperature through ectothermy. The dinosaur that gave rise to birds was *certainly* endothermic, as are all birds.

▲ **Figure 34.26 Dinosaur parental care.** A fossil of a nesting dinosaur and her eggs found in Mongolia sheds light on how dinosaurs cared for their eggs and hatchlings. Known as *Oviraptor,* this dinosaur, which lived 80 million years ago, covered her eggs with her limbs, which may have been feathered.

Traditionally, dinosaurs were considered slow, sluggish creatures. Since the early 1970s, however, fossil discoveries and research have led to the conclusion that many dinosaurs were agile, fast moving, and, in some cases, social. Paleontologists have also discovered signs of parental care among dinosaurs **(Figure 34.26)**.

All dinosaurs (except birds) became extinct by the end of the Cretaceous period. However, the fossil record indicates that the number of dinosaur species had begun to decline several million years before the Cretaceous ended. It is unclear whether the extinction of dinosaurs was hastened by the asteroid or comet impact you read about in Chapter 26, as many other terrestrial vertebrates, including most mammals, survived.

Lepidosaurs

One surviving lineage of lepidosaurs is represented by two species of lizard-like reptiles called tuatara **(Figure 34.27a)**. Tuatara relatives lived at least 220 million years ago, and they thrived on many continents well into the Cretaceous period, reaching up to a meter in length. Today, however, tuatara are found only on 30 islands off the coast of New Zealand. When humans arrived in New Zealand 750 years ago, the rats that accompanied them devoured tuatara eggs, eventually eliminating the reptiles on the main islands. The tuatara that remain on the outlying islands are about 50 cm long and feed on insects, small lizards, and bird eggs and chicks. They can live to be over 100 years old. Their survival depends on whether their remaining habitats are kept rat-free.

The other major living lineage of lepidosaurs are the squamates (lizards and snakes). Lizards are the most numerous and diverse reptiles (apart from birds) alive today **(Figure 34.27b)**. Most lizards are small; the Jaragua lizard, discovered in the Dominican Republic in 2001, is only 16 mm long—small enough to fit comfortably on a dime. In contrast, the Komodo dragon of Indonesia can reach a length of 3 m. It hunts deer and other large prey, delivering deadly bacteria with its bite. As its wounded prey dies of the infection, the lizard slowly stalks it.

Snakes are legless lepidosaurs that evolved from lizards closely related to the Komodo dragon **(Figure 34.27c)**. Today, some species of snakes retain vestigial pelvic and limb bones, which provide evidence of their ancestry.

Despite their lack of legs, snakes are quite proficient at moving on land, most often by producing waves of lateral bending that pass from head to tail. Force exerted by the bends against solid objects pushes the snake forward. Snakes can also move by gripping the ground with their belly scales at several points along the body, while the scales at intervening points are lifted slightly off the ground and pulled forward.

Snakes are carnivorous, and a number of adaptations aid them in hunting and eating prey. They have acute chemical sensors, and though they lack eardrums, they are sensitive to ground vibrations, which helps them detect the movements of prey. Heat-detecting organs between the eyes and nostrils of pit vipers, including rattlesnakes, are sensitive to minute temperature changes, enabling these night hunters to locate warm animals. Poisonous snakes inject their toxin through a pair of sharp, hollow or grooved teeth. The flicking tongue is not poisonous but helps fan odors toward olfactory (smell) organs on the roof of the mouth. Loosely articulated jaws and elastic skin enable most snakes to swallow prey larger than the diameter of the snake itself.

Turtles

Turtles are the most distinctive group of reptiles alive today. All turtles have a boxlike shell made of upper and lower shields that are fused to the vertebrae, clavicles (collarbones), and ribs **(Figure 34.27d)**. In most species, the shell is hard, providing excellent defense against predators. The earliest fossils of turtles, dating from about 220 million years ago, have fully developed shells. Without transitional fossils to analyze, the origin of the turtle shell remains a puzzle. As mentioned earlier, some paleontologists have suggested that turtle shells evolved from the dermal plates of parareptiles. But other studies have linked turtles to archosaurs or lepidosaurs.

The earliest turtles could not retract their head into their shell, but mechanisms for doing so evolved independently in two separate branches of turtles. The side-necked turtles

(a) Tuatara *(Sphenodon punctatus)*

(b) Australian thorny devil lizard *(Moloch horridus)*

(c) Wagler's pit viper *(Tropidolaemus wagleri)*, a snake

(d) Eastern box turtle *(Terrapene carolina carolina)*

(e) American alligator *(Alligator mississipiensis)*

▲ Figure 34.27 **Extant reptiles (other than birds).**

(pleurodires) fold their neck horizontally, while the vertical-necked turtles (cryptodires) fold their neck vertically.

Some turtles have adapted to deserts, and others live almost entirely in ponds and rivers. Still others have returned to the sea. Sea turtles have a reduced shell and enlarged forelimbs that function as flippers. This radiation has produced the largest living turtles, the leatherbacks, which can weigh over 1,500 kg. Feeding on jellies, they dive as deep as 60 m. Leatherbacks and other sea turtles are endangered by fishing boats that accidentally catch them in their nets, as well as by the human development of the beaches where the turtles lay their eggs.

Alligators and Crocodiles

Alligators and crocodiles (collectively called crocodilians) belong to an archosaur lineage that reaches back to the late Triassic **(Figure 34.27e)**. The earliest members of this lineage were small terrestrial quadrupeds with long, slender legs. Later species became larger and adapted to aquatic habitats, breathing air through their upturned nostrils. Some Mesozoic crocodilians grew as long as 10 m and may have attacked large dinosaurs.

Living crocodilians are confined to the warm regions of Africa, China, Indonesia, India, Australia, South America, and the southeastern United States. Alligators in the United States have made a strong comeback after spending years on the endangered species list.

Birds

There are 8,600 species of birds in the world. Like crocodilians, birds are archosaurs, but almost every feature of their reptilian anatomy has undergone modification in their adaptation to flight.

Derived Characters of Birds

Many of the characters of birds are adaptations that facilitate flight, including weight-saving modifications that make flying more efficient. For example, birds lack a urinary bladder, and the females of most species have only one ovary. The gonads of both females and males are usually small, except during the breeding season, when they increase in size. Living birds are also toothless, an adaptation that trims the weight of the head. The skull is especially light, although a bird's skeleton as a whole is no lighter in relation to body weight than the skeleton of a mammal of similar size.

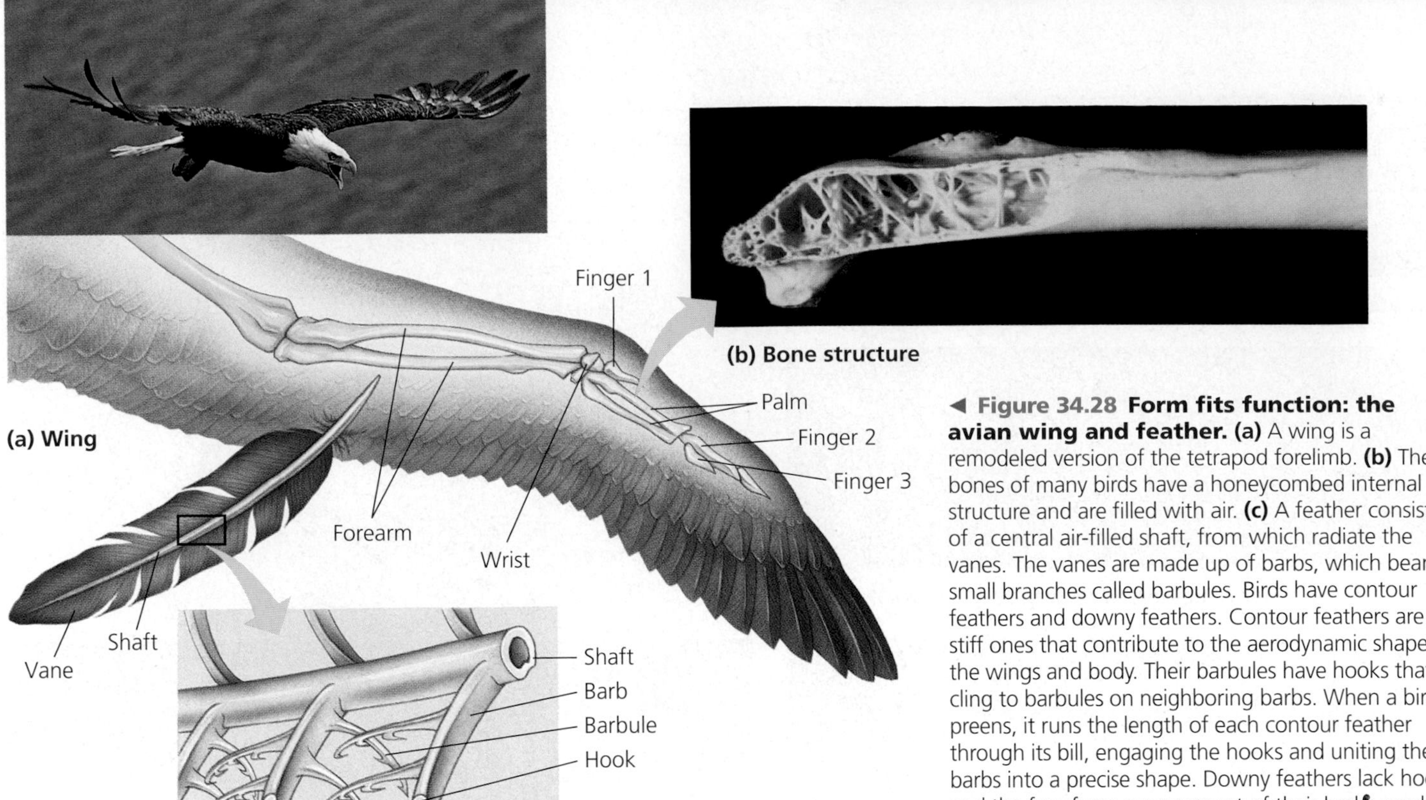

◀ **Figure 34.28 Form fits function: the avian wing and feather. (a)** A wing is a remodeled version of the tetrapod forelimb. **(b)** The bones of many birds have a honeycombed internal structure and are filled with air. **(c)** A feather consists of a central air-filled shaft, from which radiate the vanes. The vanes are made up of barbs, which bear small branches called barbules. Birds have contour feathers and downy feathers. Contour feathers are the stiff ones that contribute to the aerodynamic shapes of the wings and body. Their barbules have hooks that cling to barbules on neighboring barbs. When a bird preens, it runs the length of each contour feather through its bill, engaging the hooks and uniting the barbs into a precise shape. Downy feathers lack hooks, and the free-form arrangement of their barbs produces a fluffiness that provides insulation by trapping air.

A bird's most obvious adaptations for flight are its wings and feathers **(Figure 34.28)**. Feathers are made of a protein called β-keratin that is also found in the scales of other reptiles. The shape and arrangement of the feathers form the wings into airfoils, which illustrate some of the same principles of aerodynamics as the wings of an airplane. Power for flapping the wings comes from contractions of large pectoral (breast) muscles anchored to a keel on the sternum (breastbone). Some birds, such as eagles and hawks, have wings adapted for soaring on air currents and flap their wings only occasionally (see Figure 34.28); other birds, including hummingbirds, must flap continuously to stay aloft. The fastest birds are the appropriately named swifts, which can fly 170 km/hr.

The evolution of flight provides numerous benefits. It enhances hunting and scavenging; many birds consume flying insects, an abundant, highly nutritious food resource. Flight also provides ready escape from earthbound predators and enables some birds to migrate great distances to exploit different food resources and seasonal breeding areas.

Flying requires a great expenditure of energy from an active metabolism. Birds are endothermic; they use their own metabolic heat to maintain a high, constant body temperature. Feathers and in some species a layer of fat provide insulation that enables birds to retain body heat. An efficient respiratory system and a circulatory system with a four-chambered heart keep tissues well supplied with oxygen and nutrients, supporting a high rate of metabolism. The lungs have tiny tubes leading to and from elastic air sacs that help dissipate heat and reduce the density of the body.

Flight also requires both acute vision and fine muscle control. Birds have excellent eyesight, perhaps the best of all the vertebrates. The visual areas of the brain are well developed, as are the motor areas. The avian brain is proportionately larger than those of amphibians and nonbird reptiles.

Birds generally display very complex behaviors, particularly during breeding season, when they engage in elaborate courtship rituals. Because eggs are shelled when laid, fertilization must be internal. Copulation usually involves contact between the mates' vents, the openings to their cloacas. After eggs are laid, the avian embryo must be kept warm through brooding by the mother, father, or both, depending on the species.

The Origin of Birds

Cladistic analyses of birds and of reptilian fossils strongly suggest that birds belong to the group of bipedal saurischians called theropods. Since the late 1990s, Chinese paleontologists have unearthed a spectacular trove of feathered theropod fossils that are shedding light on the first birds. Several species of dinosaurs closely related to birds had feathers with vanes, and a wider range of species had filamentous feathers. Such findings imply that feathers evolved long before powered flight. Among the possible functions of these early feathers were insulation, camouflage, and courtship display.

Scientists have investigated two possible ways in which powered flight may have evolved in theropods. In one scenario, feathers may have enabled small, ground-running dinosaurs chasing prey or escaping predators to gain extra lift as

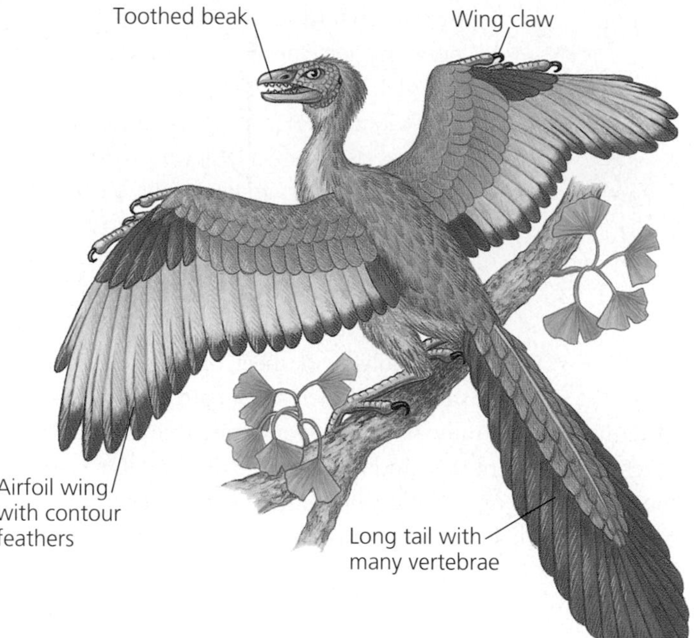

▲ **Figure 34.29 Artist's reconstruction of *Archaeopteryx,* the earliest known bird.** Fossil evidence indicates that *Archaeopteryx* was capable of powered flight but retained many characters of nonbird dinosaurs.

they jumped into the air. In another scenario, some dinosaurs could climb trees and glide, aided by feathers.

By 150 million years ago, feathered theropods had evolved into birds. *Archaeopteryx,* which was discovered in a German limestone quarry in 1861, remains the earliest known bird **(Figure 34.29)**. It had feathered wings but retained primitive characters such as teeth, clawed digits in its wings, and a long tail. Most experts agree that *Archaeopteryx* was a weak flyer at best. Fossils of a number of other birds from the Cretaceous period show a gradual loss of certain features of dinosaur anatomy, such as teeth, as well as the acquisition of innovations that are shared by all birds today, including a short tail covered by a fan of feathers.

Living Birds

Clear evidence of Neornithes, the clade that includes the 28 orders of living birds, can be found after the Cretaceous-Paleogene boundary 65.5 million years ago. Most birds can fly, but there are several orders that include one or more flightless species. The **ratites** (order Struthioniformes), which consist of the ostrich, rhea, kiwi, cassowary, and emu, are all flightless. In ratites (from the Latin *ratitus,* flat-bottomed), the sternal keel is absent, and the pectoral muscles are not greatly enlarged **(Figure 34.30a)**. Penguins make up the flightless order Sphenisciformes, but, like flying birds, they have powerful pectoral muscles, which they use in swimming. Certain species of rails, ducks, and pigeons are also flightless.

The demands of flight have rendered the general body form of many flying birds similar to one another, yet experienced

▼ **Figure 34.30 A small sample of living birds.**

(a) Emu. This ratite lives in Australia.

(b) Mallards. Like many bird species, the mallard exhibits pronounced color differences between the sexes.

(c) Laysan albatrosses. Like most birds, Laysan albatrosses have specific mating behaviors, such as this courtship ritual.

(d) Barn swallows. The barn swallow is a member of the order Passeriformes. Species in this order are called perching birds because the toes of their feet can lock around a branch or wire, enabling the bird to rest in place for long periods.

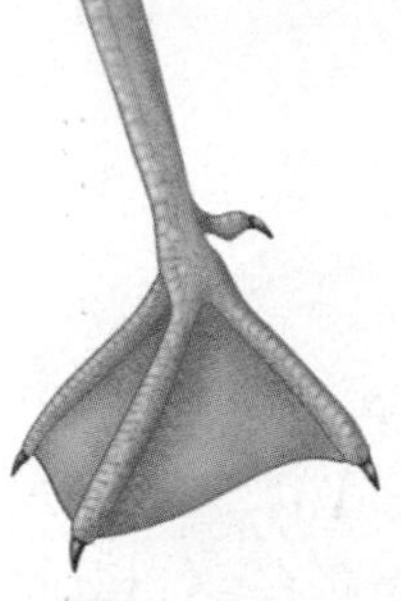

▲ **Figure 34.31 Diversity of form and function in bird feet.**

bird-watchers can distinguish different species by their body profile, flying style, behavior, feather colors, and beak shape **(Figure 34.30b, c,** and **d)**. The beak has proved to be very adaptable during avian evolution, taking on a great variety of shapes suitable for different diets. Foot structure, too, shows considerable variation **(Figure 34.31)**. Various birds use their feet for perching on branches, grasping food, defense, swimming or walking, and even courtship (see Figure 24.4e).

Concept Check 34.6

1. Defend or refute the following statement: The amniotic egg of a reptile is a closed system in which the embryo develops in isolation from the outside environment.
2. Identify four avian adaptations for flight.

For suggested answers, see Appendix A.

Concept 34.7

Mammals are amniotes that have hair and produce milk

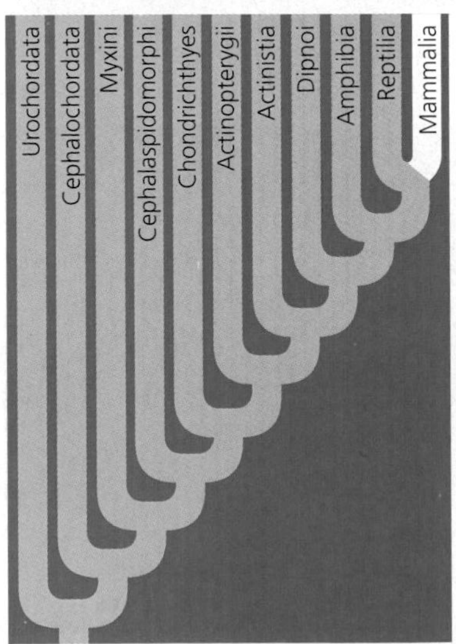

Reptiles (including birds) represent one of two great lineages of amniotes. The other lineage is our own, the **mammals** (class Mammalia). Today, there are more than 5,000 species of mammals on Earth.

Derived Characters of Mammals

Mammary glands, which produce milk for offspring, are a distinctively mammalian character. All mammalian mothers nourish their young with milk, a balanced diet rich in fats, sugars, proteins, minerals, and vitamins. Hair, another mammalian characteristic, and a fat layer under the skin help the body retain heat. Like birds, mammals are endothermic, and most have a high metabolic rate. Efficient respiratory and circulatory systems (including a four-chambered heart) support a mammal's metabolism. A sheet of muscle called the diaphragm helps ventilate the lungs.

Mammals generally have a larger brain than other vertebrates of equivalent size, and many species are capable learners. The relatively long duration of parental care extends the time for offspring to learn important survival skills by observing their parents.

Differentiation of teeth is another important mammalian trait. Whereas the teeth of reptiles are generally conical and uniform in size, the teeth of mammals come in a variety of sizes and shapes adapted for chewing many kinds of foods. Humans, for example, have teeth modified for shearing (incisors and canine teeth) and for crushing and grinding (premolars and molars).

Early Evolution of Mammals

Mammals belong to a group of amniotes known as **synapsids.** Nonmammalian synapsids lacked hair, had a sprawling gait, and laid eggs. A distinctive characteristic of synapsids is the temporal fenestra, a hole behind the eye socket on each side of the skull. Humans retain this feature; your jaw muscles pass through the temporal fenestra and anchor on your temple. The jaw was remodeled during the evolution of mammals from nonmammalian synapsids, and two of the bones that formerly made up the jaw joint were incorporated into the mammalian middle ear **(Figure 34.32)**.

Synapsids evolved into large herbivores and carnivores during the Permian period, and for a time they were the dominant tetrapods. However, the Permian-Triassic extinctions took a heavy toll on them, and their diversity fell during the Triassic. Mammal-like synapsids emerged by the end of the Triassic 200 million years ago. While not true mammals, these synapsids had acquired a number of the derived characters that distinguish mammals from other amniotes. They were small and probably hairy, and they likely fed on insects at night. They probably had a higher metabolism than other synapsids, but they still laid eggs.

During the Jurassic period, the first true mammals arose and diversified into a number of lineages, many of which are extinct. Yet throughout the Mesozoic era, nearly all mammals remained about the size of today's shrews. One possible explanation for

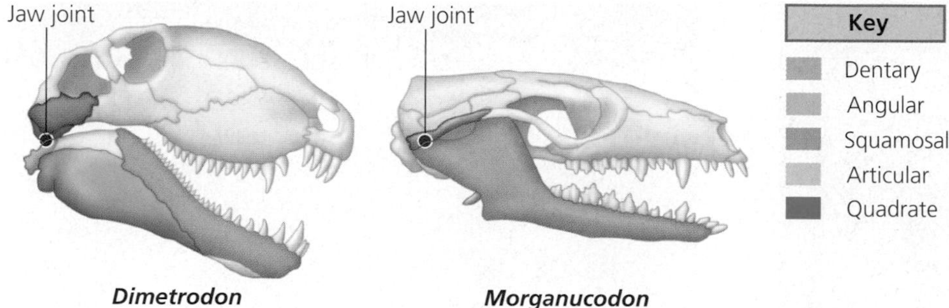

(a) The lower jaw of *Dimetrodon* is composed of several fused bones; two small bones, the quadrate and articular, form part of the jaw joint. In *Morganucodon*, the number of bones in the lower jaw is reduced, and the bones making up the jaw joint have changed.

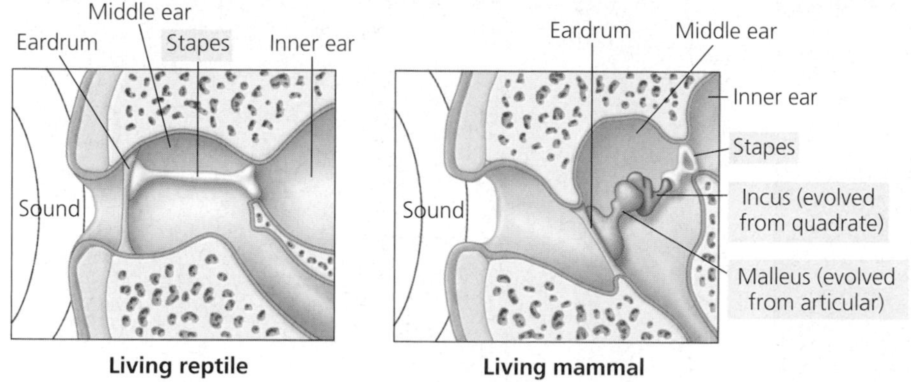

(b) During the evolutionary remodeling of the mammalian skull, the quadrate and articular bones became incorporated into the middle ear as two of the three bones that transmit sound from the eardrum to the inner ear. The steps in this evolutionary remodeling are evident in a succession of fossils.

▲ **Figure 34.32 The evolution of the mammalian jaw and ear bones.** *Dimetrodon* was an early synapsid, a reptile-like lineage that eventually gave rise to the mammals. *Morganucodon* was a mammaliform, a close relative of living mammals that lived 200 million years ago.

their small size is that dinosaurs already occupied ecological niches of large-bodied animals.

During the Mesozoic, the three major lineages of living mammals emerged: monotremes (egg-laying mammals), marsupials (mammals with a pouch), and eutherians (placental mammals). After the extinction of large dinosaurs, pterosaurs, and marine reptiles during the late Cretaceous period, mammals underwent an adaptive radiation, giving rise to large predators and herbivores, as well as flying and aquatic species.

Monotremes

Monotremes are found only in Australia and New Guinea and are represented by one species of platypus and two species of echidnas (spiny anteaters). Monotremes lay eggs, a character that is primitive for amniotes and retained in most reptiles **(Figure 34.33)**. Like all mammals, monotremes have hair and produce milk, but they lack nipples. Milk is secreted by glands on the belly of the mother. After hatching, the baby sucks the milk from the mother's fur.

Marsupials

Opossums, kangaroos, and koalas are examples of **marsupials.** Both marsupials and eutherians share derived characters not found among monotremes. They have a higher metabolic rate and nipples that provide milk, and they give birth to live young. The embryo develops inside the uterus of the female's reproductive tract. The lining of the uterus and the extraembryonic membranes that arise from the embryo form a **placenta,** a structure in which nutrients diffuse into the embryo from the mother's blood.

A marsupial is born very early in its development and completes its embryonic development while nursing. In most species, the nursing young are held within a maternal pouch called a marsupium (**Figure 34.34a**, on the next page). A red kangaroo, for instance, is about the size of a honeybee at its birth, just 33 days after fertilization. Its hind legs are merely buds, but its forelegs are strong enough for it to crawl from the exit of its mother's reproductive tract to a pouch that opens to the front of her body, a journey that lasts a few minutes. In other species, the marsupium opens to the rear of the mother's body; in bandicoots this protects the young as their mother burrows in the dirt (**Figure 34.34b**, on the next page.)

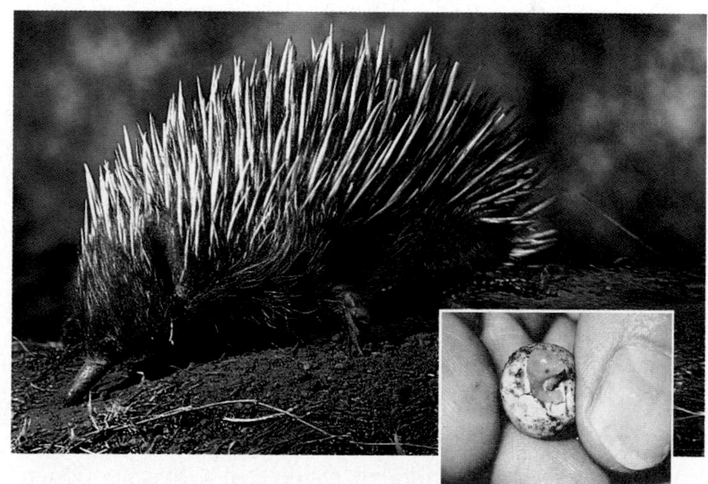

▲ **Figure 34.33 Short-beaked echidna (*Tachyglossus aculeatus*), an Australian monotreme.** Monotremes have hair and produce milk, but they lack nipples. Monotremes are the only mammals that lay eggs (inset).

(a) A young brushtail possum. The young of marsupials are born very early in their development. They finish their growth while nursing from a nipple (in their mother's pouch in most species).

(b) Long-nosed bandicoot. Most bandicoots are diggers and burrowers that eat mainly insects but also some small vertebrates and plant material. Their rear-opening pouch helps protect the young from dirt as the mother digs. Other marsupials, such as kangaroos, have a pouch that opens to the front.

▲ **Figure 34.34 Australian marsupials.**

Marsupials existed worldwide during the Mesozoic era, but today they are found only in the Australian region and in North and South America. The biogeography of marsupials is an example of the interplay between biological and geologic evolution (see Chapter 26). After the breakup of Pangaea, South America and Australia became island continents, and their marsupials diversified in isolation from the eutherians that began an adaptive radiation on the northern continents. Australia has not been in contact with another continent since early in the Cenozoic era, about 65 million years ago. In Australia, convergent evolution has resulted in a diversity of marsupials that resemble eutherians in similar ecological niches in other parts of the world **(Figure 34.35)**. Although South America had a diverse marsupial fauna throughout the Paleogene, it

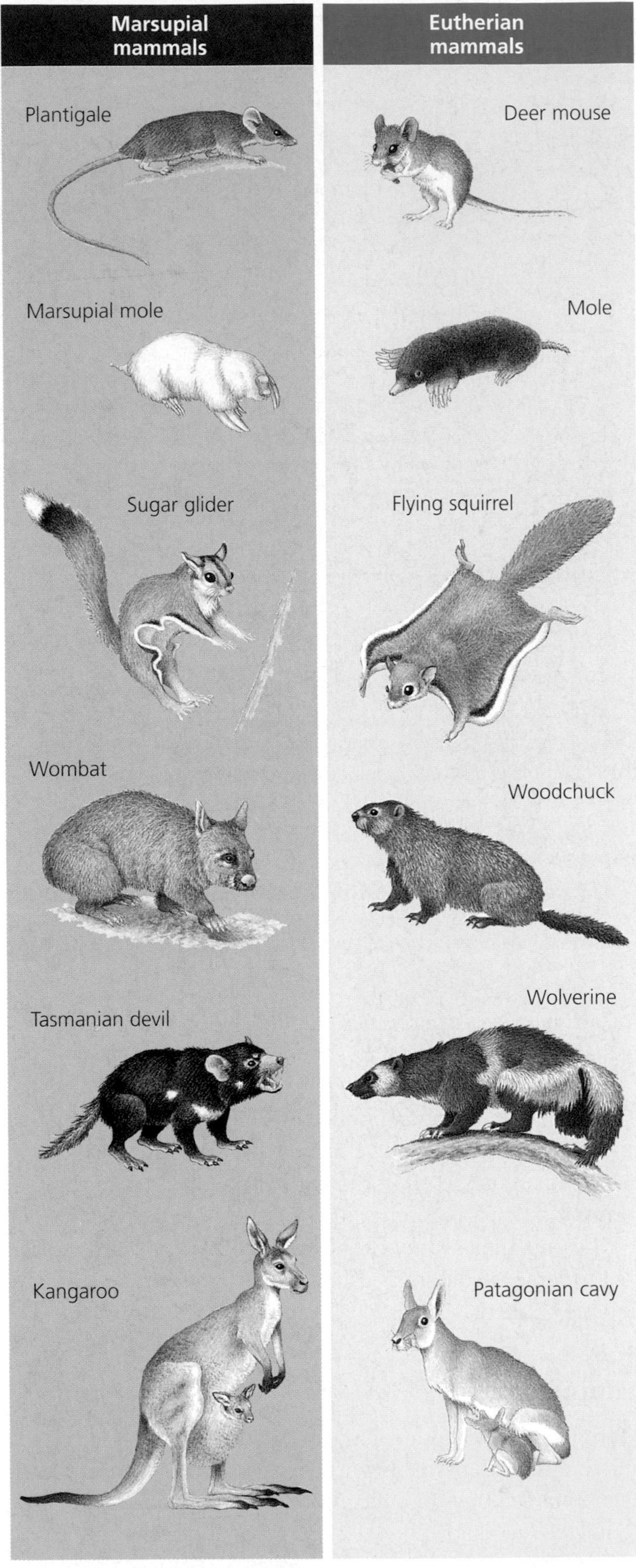

▲ **Figure 34.35 Evolutionary convergence of marsupials and eutherians (placental mammals).** (Drawings are not to scale.)

has experienced several migrations of eutherians. One of the most important migrations occurred about 3 million years ago, when North and South America joined at the Panamanian isthmus and extensive two-way traffic of animals took place over the land bridge. Today, only three families of marsupials live outside the Australian region, and only one species, the opossum, is found in North America.

Eutherians (Placental Mammals)

Eutherians are commonly called placental mammals because their placentas are more complex than those of marsupials. Compared to marsupials, eutherians have a longer period of pregnancy. Young eutherians complete their embryonic development within the uterus, joined to their mother by the placenta. The eutherian placenta provides an intimate and long-lasting association between the mother and her developing young.

The living orders of eutherians originated in the late Cretaceous period. **Figure 34.36**, on the next two pages, explores the major eutherian orders and their possible phylogenetic relationships with each other as well as with the monotremes and marsupials.

Primates

The mammalian order Primates includes the lemurs, the tarsiers, the monkeys, and the apes. Humans are members of the ape group.

Derived Characters of Primates. Most primates have hands and feet adapted for grasping, and their digits have flat nails instead of the narrow claws of other mammals. There are other characteristic features of the hands and feet, such as skin ridges on the fingers (which account for human fingerprints). Relative to other mammals, primates have a large brain and short jaws, giving them a flat face. Their forward-looking eyes are close together on the front of the face. Primates also have relatively well-developed parental care and complex social behavior.

The earliest primates were probably tree-dwellers, and many of the characteristics of primates are adaptations to the demands of living in trees. Grasping hands and feet allow primates to hang onto tree branches. All living primates, except humans, have a big toe that is widely separated from the other toes, enabling them to grasp branches with their feet. All primates have a thumb that is relatively mobile and separate from the fingers, but anthropoids (monkeys and apes) have a fully **opposable thumb**; that is, they can touch the ventral surface (fingerprint side) of the tip of all four fingers with the ventral surface of the thumb of the same hand. In monkeys and apes other than humans, the opposable thumb functions in a grasping "power grip," but in humans, a distinctive bone structure at the base of the thumb allows it to be used for more precise manipulation. The unique dexterity of humans represents descent with modification from our tree-dwelling ancestors. Arboreal maneuvering also requires excellent eye-hand coordination. The overlapping visual fields of the two eyes enhance depth perception, an obvious advantage when brachiating (traveling by swinging from branch to branch in trees).

Living Primates. There are three main groups of living primates: the lemurs of Madagascar **(Figure 34.37)** and the lorises and pottos of tropical Africa and southern Asia; the tarsiers, which live in Southeast Asia; and the **anthropoids,** which include monkeys and apes and are found worldwide. Lemurs, lorises, and pottos probably resemble early arboreal primates. The oldest known anthropoid fossils, discovered in China in mid-Eocene strata about 45 million years old, indicate that tarsiers are more closely related to the anthropoids than to the lemur group (**Figure 34.38**, on page 700). You will see in Figure 34.38 that monkeys do not constitute a monophyletic group.

The first monkeys probably evolved in the Old World (in Africa and Asia). The fossil record indicates that monkeys appeared in the New World (in South America) during the Oligocene. By that time, South America and Africa had drifted apart, and monkeys may have reached South America by rafting on logs or other debris from Africa. What is certain is that New World monkeys and Old World monkeys underwent separate adaptive radiations during their many millions of

◀ **Figure 34.37 Coquerel's sifakas (*Propithecus vereauxi coquereli*), a type of lemur.**

Figure 34.36

Exploring Mammalian Diversity

PHYLOGENETIC RELATIONSHIPS OF MAMMALS

Evidence from numerous fossils and molecular analyses indicates that monotremes diverged from other mammals about 200 million years ago and that marsupials diverged from eutherians (placental mammals) about 180 million years ago. Molecular systematics has helped to clarify the evolutionary relationships between the eutherian orders, though there is still no broad consensus on a phylogenetic tree. One current hypothesis, represented by the tree shown below, clusters the eutherian orders into four main clades.

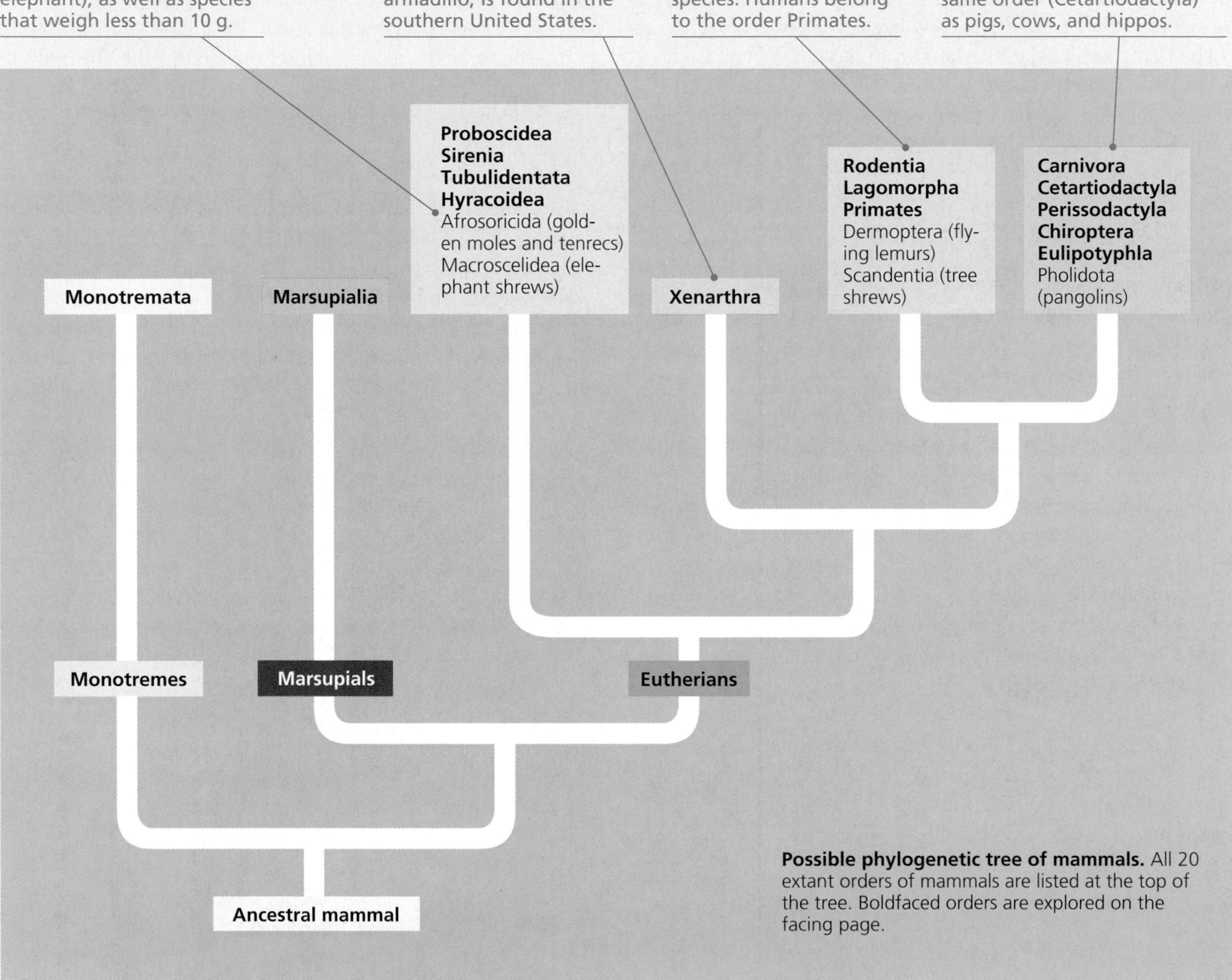
This clade of eutherians evolved in Africa when the continent was isolated from other landmasses. It includes Earth's largest living land animal (the African elephant), as well as species that weigh less than 10 g.
All members of this clade, which underwent an adaptive radiation in South America, belong to the order Xenarthra. One species, the nine-banded armadillo, is found in the southern United States.
This is the largest eutherian clade. It includes the rodents, which make up the largest mammalian order by far, with about 1,770 species. Humans belong to the order Primates.
This diverse clade includes terrestrial and marine mammals as well as bats, the only flying mammals. A growing body of evidence, including Eocene fossils of whales with feet, supports putting whales in the same order (Cetartiodactyla) as pigs, cows, and hippos.
Monotremata
Marsupialia
Proboscidea
Sirenia
Tubulidentata
Hyracoidea
Afrosoricida (golden moles and tenrecs)
Macroscelidea (elephant shrews)
Xenarthra
Rodentia
Lagomorpha
Primates
Dermoptera (flying lemurs)
Scandentia (tree shrews)
Carnivora
Cetartiodactyla
Perissodactyla
Chiroptera
Eulipotyphla
Pholidota (pangolins)
Monotremes
Marsupials
Eutherians
Ancestral mammal

Possible phylogenetic tree of mammals. All 20 extant orders of mammals are listed at the top of the tree. Boldfaced orders are explored on the facing page.

ORDERS AND EXAMPLES	MAIN CHARACTERISTICS
Monotremata Platypuses, echidnas  Echidna	Lay eggs; no nipples; young suck milk from fur of mother
Proboscidea Elephants  African elephant	Long, muscular trunk; thick, loose skin; upper incisors elongated as tusks
Sirenia Manatees, dugongs 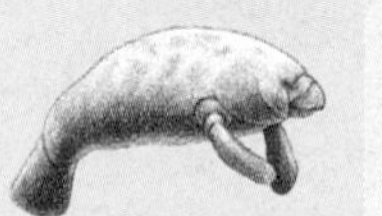  Manatee	Aquatic; finlike forelimbs and no hind limbs; herbivorous
Xenarthra Sloths, anteaters, armadillos  Tamandua	Reduced teeth or no teeth; herbivorous (sloths) or carnivorous (anteaters, armadillos)
Lagomorpha Rabbits, hares, picas  Jackrabbit	Chisel-like incisors; hind legs longer than forelegs and adapted for running and jumping
Carnivora Dogs, wolves, bears, cats, weasels, otters, seals, walruses  Coyote	Sharp, pointed canine teeth and molars for shearing; carnivorous
Cetartiodactyla **Artiodactyls** Sheep, pigs, cattle, deer, giraffes  Bighorn sheep	Hooves with an even number of toes on each foot; herbivorous
Cetaceans Whales, dolphins, porpoises  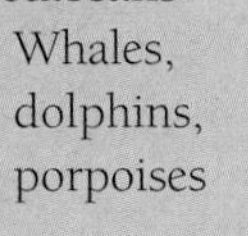  Pacific white-sided porpoise	Aquatic; streamlined body; paddle-like forelimbs and no hind limbs; thick layer of insulating blubber; carnivorous

ORDERS AND EXAMPLES	MAIN CHARACTERISTICS
Marsupialia Kangaroos, opossums, koalas Koala	Embryo completes development in pouch on mother
Tubulidentata Aardvark Aardvark	Teeth consisting of many thin tubes cemented together; eats ants and termites
Hyracoidea Hyraxes  Rock hyrax	Short legs; stumpy tail; herbivorous; complex, multichambered stomach
Rodentia Squirrels, beavers, rats, porcupines, mice 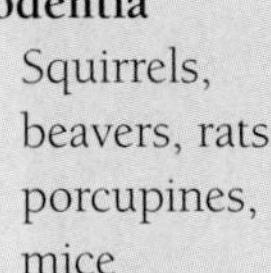  Red squirrel	Chisel-like, continuously growing incisors worn down by gnawing; herbivorous
Primates Lemurs, monkeys, chimpanzees, gorillas, humans  Golden lion tamarin	Opposable thumbs; forward-facing eyes; well-developed cerebral cortex; omnivorous
Perissodactyla Horses, zebras, tapirs, rhinoceroses  Indian rhinoceros	Hooves with an odd number of toes on each foot; herbivorous
Chiroptera Bats Frog-eating bat	Adapted for flight; broad skinfold that extends from elongated fingers to body and legs; carnivorous or herbivorous
Eulipotyphla "Core insectivores": some moles, some shrews  Star-nosed mole	Diet consists mainly of insects and other small invertebrates

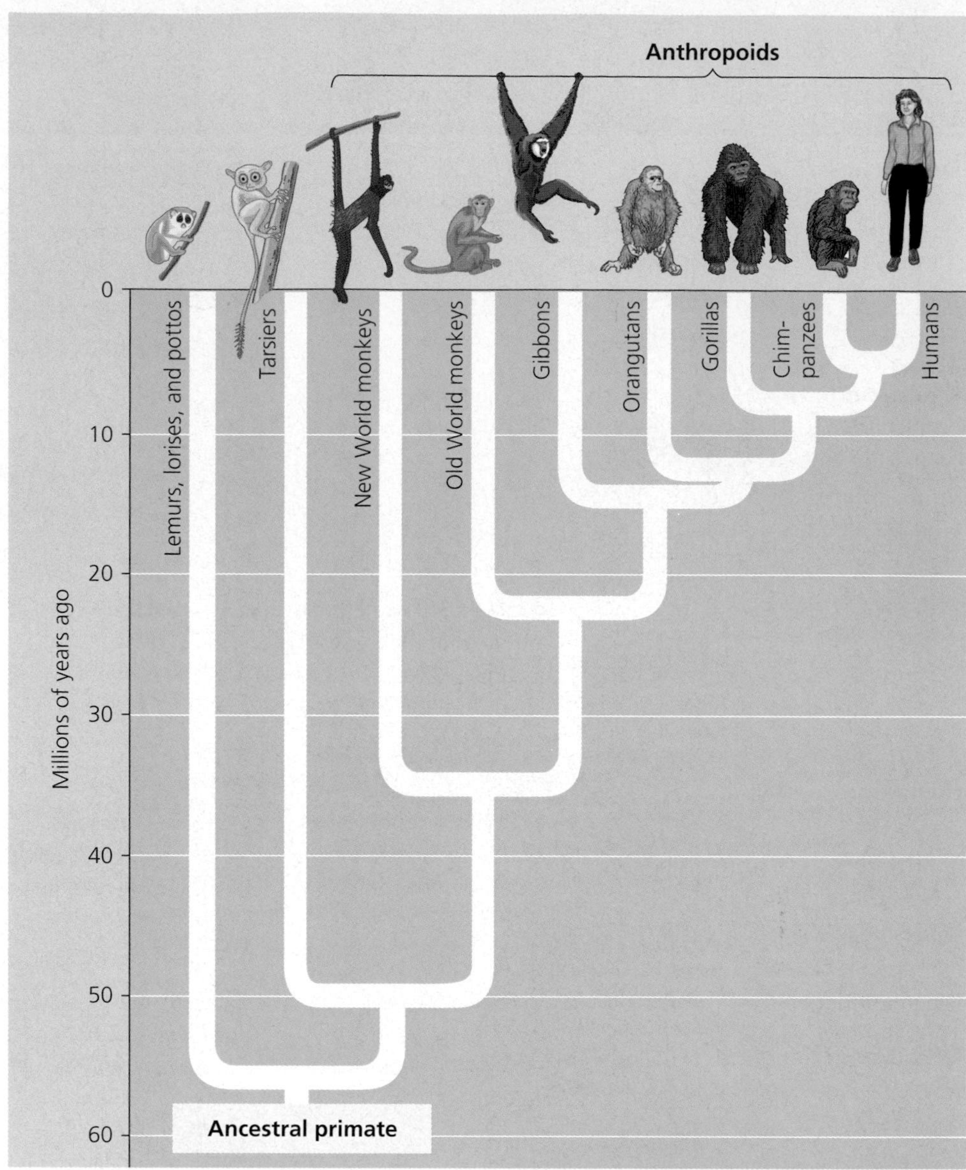

▶ **Figure 34.38 A phylogenetic tree of primates.** The fossil record indicates that anthropoids began diverging from other primates about 50 million years ago. New World monkeys, Old World monkeys, and hominoids (the clade that includes gibbons, orangutans, gorillas, chimpanzees, and humans) have been evolving as separate lineages for over 30 million years. The lineages leading to humans branched off from other hominoids somewhere between 5 and 7 million years ago.

years of separation **(Figure 34.39)**. All species of New World monkeys are arboreal, whereas Old World monkeys include ground-dwelling as well as arboreal species. Most monkeys in both groups are diurnal (active during the day) and usually live in bands held together by social behavior.

The other group of anthropoids, the **hominoids**, consists of primates informally called apes **(Figure 34.40)**. This group includes the genera *Hylobates* (gibbons), *Pongo* (orangutans), *Gorilla* (gorillas), *Pan* (chimpanzees and bonobos), and *Homo* (humans). Hominoids diverged from Old World monkeys about 20–25 million years ago. Today, nonhuman hominoids are found exclusively in tropical regions of the Old World. With the exception of gibbons, living hominoids are larger than monkeys. All living hominoids have relatively long arms, short legs, and no tail. Although all nonhuman hominoids spend time in trees, only gibbons and orangutans are primarily arboreal. Social organization varies among the genera of hominoids; gorillas and chimpanzees are highly social. Hominoids have a larger brain in proportion to their body size than other primates, and hominoid behavior is more flexible.

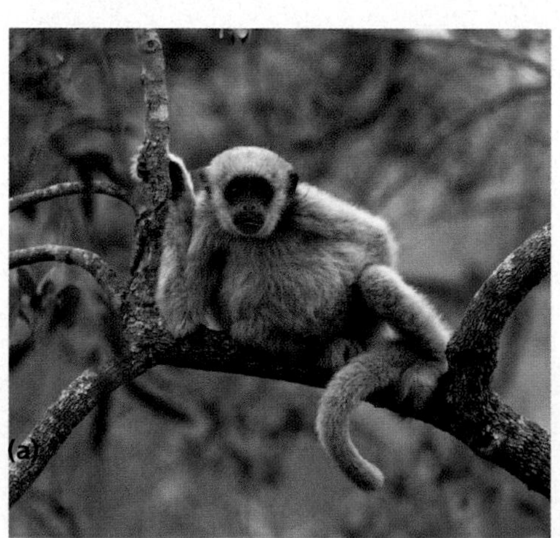

(a) New World monkeys, such as spider monkeys (shown here), squirrel monkeys, and capuchins, have a prehensile tail and nostrils that open to the sides.

(b) Old World monkeys lack a prehensile tail, and their nostrils open downward. This group includes macaques (shown here), mandrills, baboons, and rhesus monkeys.

▶ **Figure 34.39 New World monkeys and Old World monkeys.**

(a) Gibbons, such as this Muller's gibbon, are found only in southeastern Asia. Their very long arms and fingers are adaptations for brachiation.

(b) Orangutans are shy, solitary apes that live in the rain forests of Sumatra and Borneo. They spend most of their time in trees; note the foot adapted for grasping and the opposable thumb.

(c) Gorillas are the largest apes: some males are almost 2 m tall and weigh about 200 kg. Found only in Africa, these herbivores usually live in groups of up to about 20 individuals.

(d) Chimpanzees live in tropical Africa. They feed and sleep in trees but also spend a great deal of time on the ground. Chimpanzees are intelligent, communicative, and social.

(e) Bonobos are closely related to chimpanzees but are smaller. They survive today only in the African nation of Congo.

▲ **Figure 34.40 Hominoids (apes).**

Concept Check 34.7

1. Contrast monotremes, marsupials, and eutherians in terms of how they bear young.
2. Identify at least five derived traits of primates.

For suggested answers, see Appendix A.

Concept 34.8

Humans are bipedal hominoids with a large brain

In our tour of Earth's biodiversity, we come at last to our own species, *Homo sapiens*, which is about 160,000 years old. When you consider that life has existed on Earth for at least 3.5 billion years, we are clearly evolutionary newcomers.

Derived Characters of Humans

A number of characters distinguish humans from other hominoids. Most obviously, humans stand upright and walk on two legs. Humans have a much larger brain than other hominoids and are capable of language, symbolic thought, and the manufacture and use of complex tools. Humans also have reduced jawbones and jaw muscles, along with a shorter digestive tract. The list of derived human characters at the molecular level is growing as scientists compare the genomes of humans and chimpanzees. Although the two genomes are 99% identical, a disparity of 1% can translate into a large number of differences in a genome that contains 3 billion base pairs.

Bear in mind that such genomic differences—and whatever derived phenotypic characters they code for—separate humans from other *living* hominoids. But many of these new characters first emerged in our ancestors, long before our own species appeared. We will consider some of these ancestors to see how these traits originated.

The Earliest Hominids

The study of human origins is known as **paleoanthropology.** Paleoanthropologists have unearthed fossils of approximately 20 species of extinct hominoids that are more closely related to humans than to chimpanzees. These species are known as **hominids** (**Figure 34.41**). Since 1994, four hominid species dating from more than 4 million years ago have been discovered in the fossil record. The oldest of these hominids, *Sahelanthropus tchadensis,* lived about 7–6 million years ago; its fossils were discovered in 2002.

Sahelanthropus and other early hominids shared some of the derived characters of humans. For example, they had reduced canine teeth, and some fossils suggest that they had relatively flat faces. They also show signs of having been more upright and bipedal than other hominoids. One clue to their upright stance can be found in the foramen magnum, the hole at the base of the skull through which the spinal cord exits. In chimpanzees, the foramen magnum is relatively far back on the skull, while in early hominids (and in humans), it is located underneath the skull. This derived position allows us to hold our head directly over our body, and apparently early hominids did as well. Leg bones of *Australopithecus anamensis,* a hominid that lived 4.5–4 million years ago, also suggest that early hominids were increasingly bipedal. (We will return to the subject of bipedalism later.)

Note that the characters that distinguish humans from other living hominoids did not all evolve in tight unison. While early hominids were showing signs of bipedalism, their brains remained small—about 400–450 cm^3 in volume, compared with an average of 1,300 cm^3 for *Homo sapiens*. The earliest hominids also were small (the 4.5-million-year-old

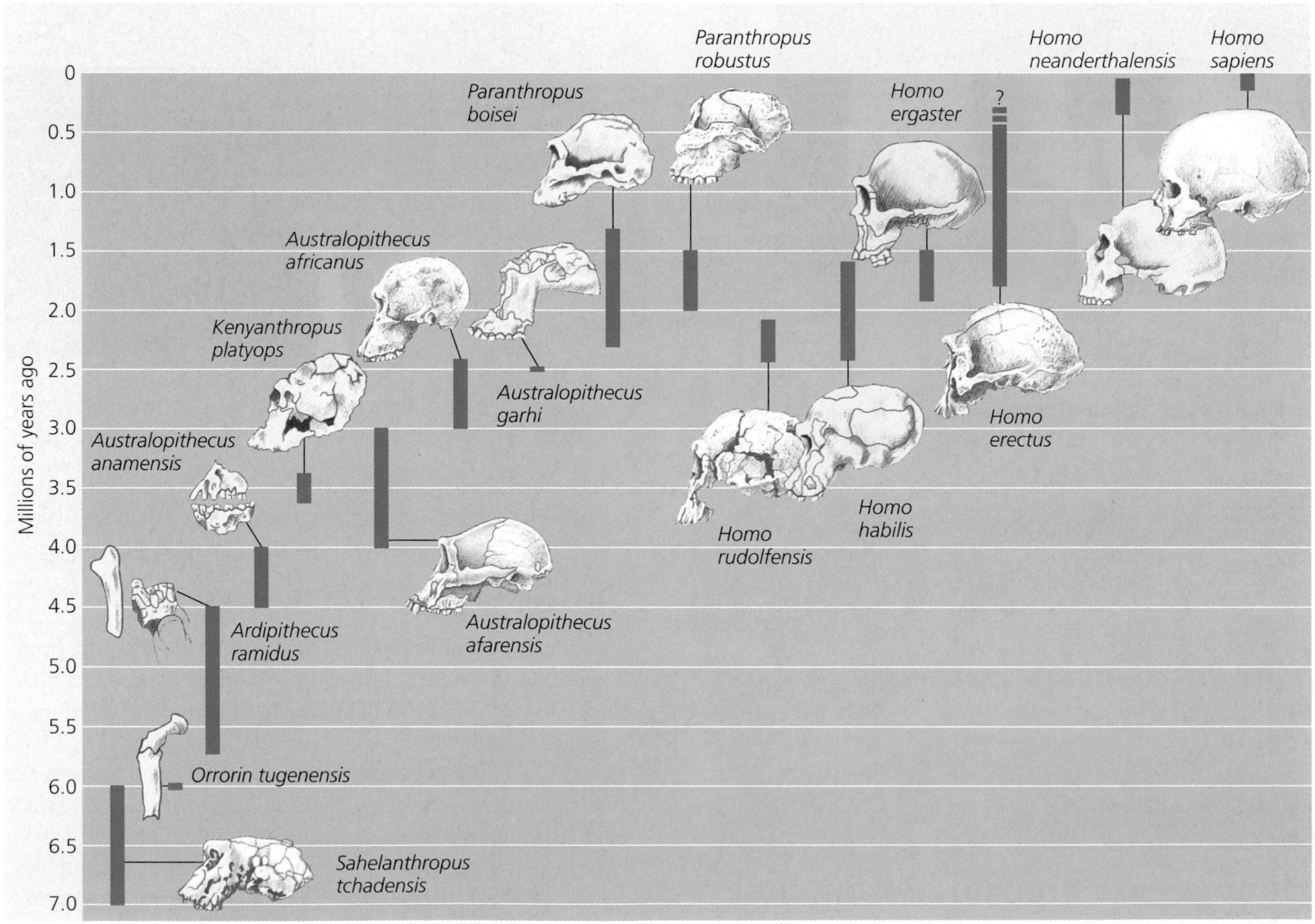

▲ **Figure 34.41 A timeline for some selected hominid species.** Most of these fossils come from sites in eastern and southern Africa. Note that at most times in hominid history, two or more hominid species were contemporaries. The names of some of the species are controversial, reflecting phylogenetic debates about the interpretation of skeletal details and biogeography.

Ardipithecus ramidus is estimated to have weighed only 40 kg) and had relatively large teeth and a lower jaw that projected beyond the upper part of the face. (Humans, in contrast, have a relatively flat face. Compare your own face with that of the chimpanzee in Figure 34.40d.) Differing rates of evolution in different features is known as **mosaic evolution**.

It's important to avoid two common misconceptions when considering these early hominids. One is to think of them as chimpanzees. Chimpanzees represent the tip of a separate branch of hominoid evolution, and they acquired derived characters of their own after they diverged from their common ancestor with humans.

Another misconception is to imagine that human evolution is a ladder leading directly from an ancestral hominoid to *Homo sapiens*. This error is often illustrated as a parade of fossil hominids that become progressively more like ourselves as they march across the page. If human evolution is a parade, it is a very disorderly one, with many groups breaking away from the march to wander down dead-end alleyways. At times, several hominid species coexisted, but all except one lineage—the one that gave rise to *Homo sapiens*—ended in extinction. Some paleoanthropologists suggest that convergence was common in early hominid evolution. Different lineages of hominids acquired different combinations of derived characters and retained different sets of primitive characters. It has even been suggested that some of the fossils claimed to be early hominids, such as *Orrorin tugenensis* (see Figure 34.41) are actually outside the hominid clade altogether. If that's true, they must have independently acquired some hominid-like traits.

Australopiths

The fossil record indicates that hominid diversity increased dramatically between 4 and 2 million years ago. Many of the hominids from this period are collectively called australopiths. Their phylogeny remains unresolved on many points, but as a group, they are almost certainly paraphyletic. *Australopithecus anamensis,* mentioned earlier, links the australopiths to older hominids such as *Ardipithecus ramidus*.

Australopiths got their name from the 1924 discovery in South Africa of *Australopithecus africanus* ("southern ape of Africa"), which lived between 3 and 2.4 million years ago. With the discovery of more fossils, it became clear that *A. africanus* walked fully erect (was bipedal) and had human-like hands and teeth. However, its brain was only about one-third the size of the brain of a present-day human.

In 1974, in the Afar region of Ethiopia, paleoanthropologists discovered a 3.24-million-year-old *Australopithecus* skeleton that was 40% complete. "Lucy," as the fossil was named, was short—only about 1 m tall. Lucy and similar fossils have been considered sufficiently different from *Australopithecus africanus* to be designated as a separate species, *Australopithecus afarensis* (for the Afar region). Fossils discovered in the early 1990s show that *A. afarensis* existed as a species for at least 1 million years.

At the risk of oversimplifying, one could say that *A. afarensis* had fewer derived characters of humans above the neck than below. Lucy's head was the size of a softball, indicating a brain size about the same as that of a chimpanzee of Lucy's body size. *A. afarensis* skulls also have a long lower jaw. Skeletons of *A. afarensis* suggest a capacity for arboreal locomotion, with arms relatively long in proportion to body size (compared to the proportions in humans). However, fragments of the pelvis and skull bones indicate that *A. afarensis* walked on two legs. Fossilized footprints in Laetoli, Tanzania, corroborate the skeletal evidence that hominids living at the time of *A. afarensis* were bipedal **(Figure 34.42)**.

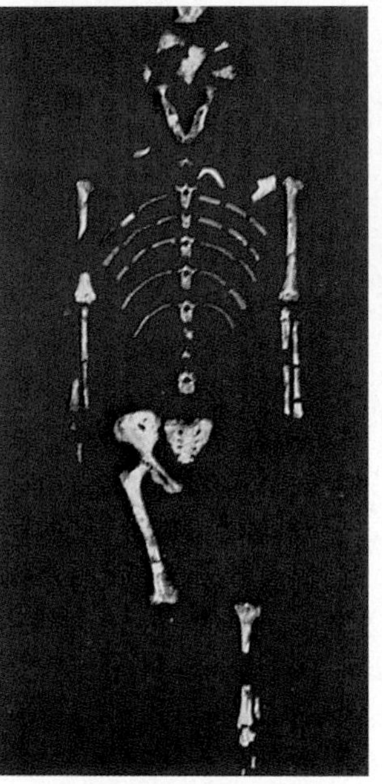

(a) Lucy, a 3.24-million-year-old skeleton, represents the hominid species *Australopithecus afarensis*.

(b) The Laetoli footprints, more than 3.5 million years old, confirm that upright posture evolved quite early in hominid history.

(c) An artist's reconstruction of what *A. afarensis* may have looked like.

▲ **Figure 34.42 Upright posture predates an enlarged brain in human evolution.**

Another lineage of australopiths consisted of the "robust" australopiths. These hominids, which included species such as *Paranthropus boisei,* had sturdy skulls with powerful jaws and large teeth, adapted for grinding and chewing hard, tough foods. They contrast with the "gracile" (slender) australopiths, including *A. afarensis* and *A. africanus,* which had lighter feeding equipment adapted for softer foods.

Combining evidence from the earliest hominids with the much richer fossil record of later australopiths makes it possible to formulate hypotheses about significant trends in hominid evolution. Let's consider two of these trends: the emergence of bipedalism and tool use.

Bipedalism

Our anthropoid ancestors of 30–35 million years ago were still tree-dwellers. But by about 10 million years ago, the Himalayan mountain range had formed, thrust up in the aftermath of the Indian Plate's collision with the Asian Plate (see Figure 26.20). The climate became drier, and the forests of what are now Africa and Asia contracted. The result was an increased area of savanna (grassland) habitat, with fewer trees. For decades, paleoanthropologists have seen a strong connection between the rise of savannas and the rise of bipedal hominids. According to one hypothesis, tree-dwelling hominids could no longer move through the canopy, so natural selection favored adaptations that made moving over open ground more efficient.

Although some elements of this hypothesis survive, the picture now appears somewhat more complex. Although all recently discovered fossils of early hominids show some indications of bipedalism, none of these hominids lived in savannas. Instead, they lived in mixed habitats ranging from forests to open woodlands. Whatever the selective pressure that led to bipedalism, furthermore, hominids did not become more bipedal in a simple, linear fashion. Australopiths seem to have had various locomotor styles, and some spent more time on the ground than others. Only about 1.9 million years ago did hominids begin to walk long distances on two legs. These hominids lived in more arid environments, where bipedal walking requires less energy than walking on all fours.

Tool Use

As you read earlier, the manufacture and use of complex tools is a derived behavioral character of humans. Determining the origin of tool use in hominid evolution is one of paleoanthropology's great challenges today. Other hominoids are capable of surprisingly sophisticated tool use. Orangutans, for example, can fashion probes from sticks for retrieving insects from their nests. Chimpanzees are even more adept, using rocks to smash open food and putting leaves on their feet to walk over thorns. It's likely that early hominids were capable of this sort of simple tool use, but finding fossils of modified sticks and leaves that were used as shoes is practically impossible.

The oldest generally accepted evidence of tool use by hominids is 2.5-million-year-old cut marks on animal bones found in Ethiopia. These marks suggest that hominids cut flesh from the bones of animals using stone tools. Interestingly, the hominids whose fossils were found near the site where the bones were discovered had a relatively small brain. If these hominids, which have been named *Australopithecus garhi,* were indeed the creators of the stone tools used on the bones, that would suggest that stone tool use originated before the evolution of large hominid brains.

Early *Homo*

The earliest fossils that paleoanthropologists place in our genus, *Homo,* are those of the species *Homo habilis.* These fossils, ranging in age from about 2.4 to 1.6 million years, show clear signs of certain derived hominid characters above the neck. Compared to the australopiths, *H. habilis* had a shorter jaw and a larger brain volume, about 600–750 cm^3. Sharp stone tools have also been found with some fossils of *H. habilis* (the name means "handy man").

Fossils from about 1.9 to 1.5 million years ago mark a new stage in hominid evolution. A number of paleoanthropologists recognize these fossils as those of a distinct species, *Homo ergaster. Homo ergaster* had a substantially larger brain than *H. habilis* (over 900 cm^3), as well as long, slender legs with hip joints well adapted for long-distance walking **(Figure 34.43)**. The fingers were relatively short and straight, suggesting that *H. ergaster* did not climb trees like earlier hominids. *Homo ergaster* has been found in far more arid environments than earlier hominids and has been associated with more sophisticated stone tools. Its smaller teeth also suggest that *H. ergaster* either ate different foods than australopiths (more meat and less plant material) or prepared some of its food before chewing, perhaps by cooking or mashing the food.

Homo ergaster marks an important shift in the relative sizes of the sexes. In primates, a size difference between males and females is a major component of sexual dimorphism (see Chapter 23). On average, male gorillas and orangutans weigh about twice as much as females of their species. In chimpanzees and bonobos, males are only about 1.35 times as heavy as females, on average. In *Australopithecus afarensis,* males were 1.5 times as heavy as females. But in early *Homo,* sexual dimorphism was significantly reduced, and this trend continued through our own species: Human males average about 1.2 times the weight of females.

The reduced sexual dimorphism may offer some clues to the social systems of extinct hominids. In living primates, extreme sexual dimorphism is associated with intense male-male competition for multiple females. In species that undergo more pair-bonding (including our own), sexual dimorphism

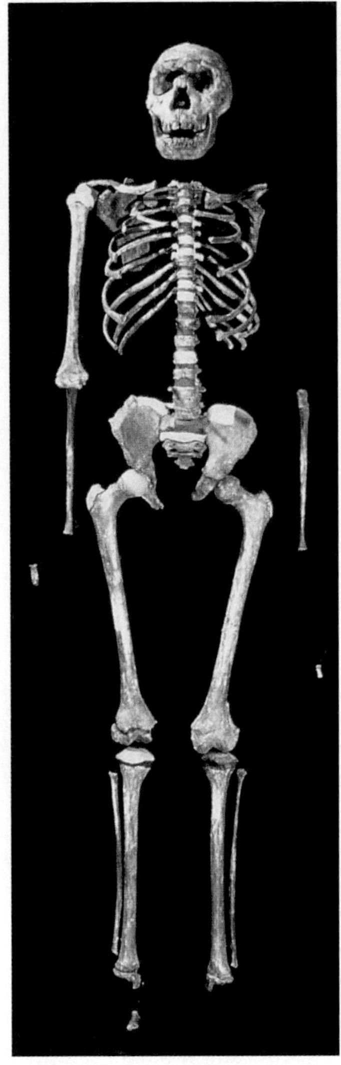

◀ **Figure 34.43 Fossil and artist's reconstruction of *Homo ergaster*.** This 1.7-million-year-old fossil from Kenya belongs to a young male *Homo ergaster*. This individual was tall, slender, and fully bipedal, and he had a relatively large brain.

is less dramatic. Male and female *H. ergaster* may therefore have engaged in more pair-bonding than earlier hominids did. This shift may have been associated with long-term biparental care of babies. Human babies depend on their parents for food and protection much longer than do the young of chimpanzees and other hominoids.

Fossils now recognized as *Homo ergaster* were originally considered early members of another species, *Homo erectus*, and some paleoanthropologists still hold this position. *Homo erectus* originated in Africa and was the first hominid to migrate out of Africa. The oldest fossils of hominids outside Africa, dating back 1.8 million years, were discovered in 2000 in the former Soviet Republic of Georgia. *Homo erectus* eventually migrated as far as the Indonesian archipelago. Comparisons of *H. erectus* fossils with humans and studies of human DNA indicate that *H. erectus* became extinct sometime after 200,000 years ago.

Neanderthals

In 1856, miners discovered some mysterious human fossils in a cave in the Neander Valley in Germany. The 40,000-year-old fossils belonged to a thick-boned, heavy hominid with a prominent brow. The hominid was given the name *Homo neanderthalensis*, commonly called a Neanderthal. At one time, many paleoanthropologists considered Neanderthals to be a stage in the evolution of *Homo erectus* into *Homo sapiens*. Now most have abandoned this view. Neanderthals appear to have descended from an earlier species, *Homo heidelbergensis*, which originated in Africa about 600,000 years ago and then spread to Europe. Neanderthals, which appeared in Europe and the Near East by 200,000 years ago, had a brain as large as that of present-day humans and were capable of making hunting tools from stone and wood. But despite their adaptations, Neanderthals apparently became extinct 30,000 years ago, contributing nothing to the gene pool of living humans.

Evidence of the extinction of Neanderthals can be found in their DNA. Scientists extracted fragments of DNA from fossils of four Neanderthals that lived at different times in different localities in Europe. They then compared the Neanderthal DNA with DNA of living humans from Europe, Africa, and Asia. If Neanderthals had given rise to European humans, then both groups should share a common ancestor, with other humans being more distantly related. Instead, the DNA analysis indicates that all of the Neanderthals form a clade, while living Europeans are more closely related to living Africans and Asians.

Homo sapiens

Evidence from fossils, archaeological relics, and DNA studies is beginning to cohere into a compelling hypothesis about how our own species, *Homo sapiens*, emerged and spread around the world.

It is now clear that the ancestors of humans originated in Africa. Older species (perhaps *H. ergaster* or *H. erectus*) gave rise to newer species, such as *H. heidelbergensis* and ultimately *H. sapiens*. The oldest known fossils of our own species have been found at two different sites in Ethiopia and include specimens that are 195,000 and 160,000 years old **(Figure 34.44)**. These

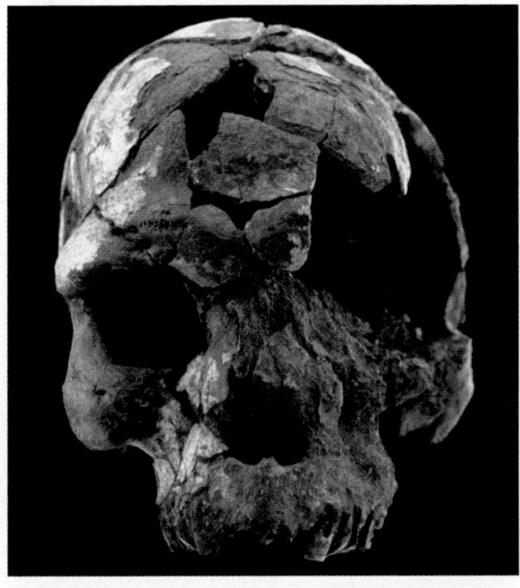

◀ **Figure 34.44 A 160,000-year-old fossil of *Homo sapiens*.** This skull, discovered in Ethiopia in 2003, differs little from the skulls of living humans.

early humans lacked the heavy browridges of *H. erectus* and Neanderthals and were more slender than other hominids.

The Ethiopian fossils support molecular evidence about the origin of humans. As we noted earlier, DNA analysis indicates that all living humans are more closely related to one another than to Neanderthals. Other studies on human DNA show that Europeans and Asians share a relatively recent common ancestor and that many African lineages branched off more ancient positions on the human family tree. These findings strongly suggest that all living humans have ancestors that originated as *H. sapiens* in Africa, which is further supported by analysis of mitochondrial DNA and Y chromosomes from members of various human populations.

The oldest fossils of *H. sapiens* outside Africa date back about 50,000 years. Studies of the human Y chromosome suggest that humans spread beyond Africa in one or more waves, first into Asia and then to Europe and Australia. The date of the first arrival of humans in the New World is still a matter of debate, although the oldest generally accepted evidence puts that date at 15,000 years ago.

New findings continually update our understanding of the context of *H. sapiens*'s evolution. For example, in October 2004, Peter Brown, of the University of New England in New South Wales, Australia, Thomas Sutikna, of the Indonesian Centre for Archaeology, and their colleagues reported an astonishing find unearthed in 2003: the skeleton of an adult hominid dating from just 18,000 years ago and representing a previously unknown species, now named *Homo floresiensis*. Discovered in a limestone cave on the Indonesian island of Flores, the individual was much shorter and had a much smaller brain volume than *H. sapiens*—more similar, in fact, to an australopith. However, the skeleton also displays many derived traits, including skull thickness and proportions and teeth shape, that suggest it is descended from the larger *H. erectus*. One intriguing explanation for this species' apparent "shrinkage" is that isolation on the island may have resulted in selection for greatly reduced size. Such dramatic size reduction is well studied in other dwarf mammalian species endemic to islands; these include primitive pygmy elephants found in the same vicinity as the *H. floresiensis* specimen. Compelling questions that may yet be answered from the cache of anthropological and archeological finds on Flores include how *H. floresiensis* originated, whether its members used tools, and whether they ever encountered *H. sapiens,* which coexisted in Indonesia during the late Pleistocene.

The rapid expansion of our species (and the replacement of Neanderthals) may have been spurred by the evolution of human cognition as *H. sapiens* evolved in Africa. Although Neanderthals and other hominids were able to produce sophisticated tools, they showed little creativity and not much capability for symbolic thought, as far as we can tell. In contrast, researchers are beginning to find evidence of more sophisticated thought as *H. sapiens* evolved. For example, in 2002, researchers reported the discovery in South Africa of 77,000-year-old art—geometric markings made on pieces of ochre **(Figure 34.45)**. And in 2004, archaeologists working in southern and east Africa found 75,000-year-old ostrich eggs and snail shells with holes neatly drilled through them. By 36,000 years ago, humans were producing spectacular cave paintings.

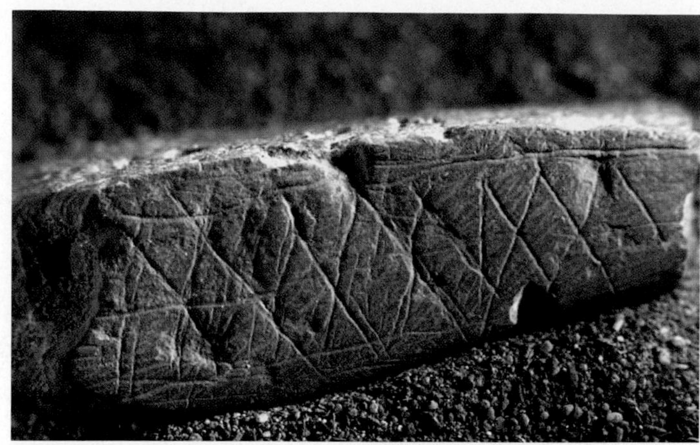

▲ **Figure 34.45 Art, a human hallmark.** The engravings on this 77,000-year-old piece of ochre, discovered in South Africa's Blombos Cave, are among the earliest signs of symbolic thought in humans.

Symbolic thought may have emerged along with full-blown human language. Both may have raised the survival and reproductive fitness of humans by allowing them to construct new tools and teach others how to build them. Long-range trade for scarce resources also became possible. As the population in Africa rose, population pressures may have driven humans to migrate into Asia and then Europe. Neanderthals may have been driven to extinction by the combined stresses of the last ice age and competition from newly arrived humans.

Clues to the cognitive transformation of humans can be found in the human genome as well as in archaeological sites. A gene known as *FOXP2* was identified in 2001 as essential for human language. People who inherit mutated versions of the gene suffer from a range of language impediments and have reduced activity in Broca's area in the brain (see Chapter 48). In 2002, geneticists compared the *FOXP2* gene in humans with the homologous gene in other mammals. They concluded that the gene experienced intense natural selection after the ancestors of humans and chimpanzees diverged. By comparing mutations in flanking regions of the gene, the researchers estimated that this bout of natural selection occurred within the past 200,000 years. Of course, the human capacity for language involves many regions of the brain, and it's almost certain that many other genes are essential for language. But the evolutionary change in *FOXP2* may be the first genetic clue as to how our own species came to play its unique role in the world.

Our discussion of humans brings this unit of chapters on biological diversity to an end. But this organization isn't meant to imply that life consists of a ladder leading from lowly microbes to lofty humanity. Biological diversity is the product of branching phylogeny, not ladderlike "progress," however we choose to measure it. The fact that there are more species of ray-finned fishes alive today than all other vertebrates combined is a clear indication that our finned relatives are not outmoded underachievers that failed to get out of the water. The tetrapods—amphibians, reptiles, and mammals—are derived from one lineage of lobe-finned vertebrates. As tetrapods diversified on land, fishes continued their branching evolution in the greatest portion of the biosphere's volume. Similarly, the ubiquitous presence of diverse prokaryotes throughout the biosphere today is a reminder of the enduring ability of these relatively simple organisms to keep up with the times through adaptive evolution. Biology exalts life's diversity, past and present.

Concept Check 34.8

1. Contrast hominoids with hominids.
2. How do the characteristics of *Homo ergaster* illustrate mosaic evolution?

For suggested answers, see Appendix A.

Chapter 34 Review

Go to www.campbellbiology.com or the student CD-ROM to explore Activities, Investigations, and other interactive study aids.

SUMMARY OF KEY CONCEPTS

Chordates have a notochord and a dorsal, hollow nerve cord

- **Derived Characters of Chordates (p. 673)** Chordate characters include a notochord; a dorsal, hollow nerve cord; pharyngeal slits or clefts; and a muscular, post-anal tail.
- **Tunicates (pp. 673–674)** Tunicates are marine suspension feeders commonly called sea squirts. They lose some of the derived characters of chordates as adults.
- **Lancelets (p. 674)** Lancelets are marine suspension feeders that retain the hallmarks of the chordate body plan as adults.
- **Early Chordate Evolution (pp. 674–675)** The current life history of tunicates probably does not reflect that of the ancestral chordate. Gene expression in lancelets holds clues to the evolution of the vertebrate brain.

Concept 34.2

Craniates are chordates that have a head

- **Derived Characters of Craniates (p. 676)** Craniates have a head, including a skull, a brain, eyes, and other sensory organs. Many structures in craniates develop partly from a novel population of cells, the neural crest.
- **The Origin of Craniates (p. 676)** Craniates evolved at least 530 million years ago, during the Cambrian explosion.
- **Hagfishes (pp. 676–677)** Hagfishes are jawless marine craniates that have a cartilaginous skull and an axial rod of cartilage derived from the notochord. They lack vertebrae.

Concept 34.3

Vertebrates are craniates that have a backbone

- **Derived Characters of Vertebrates (p. 678)** Vertebrates have vertebrae, an elaborate skull, and, in aquatic forms, fin rays.
- **Lampreys (p. 678)** Lampreys are jawless vertebrates that have cartilaginous segments surrounding the notochord and arching partly over the nerve cord.
- **Fossils of Early Vertebrates (pp. 678–679)** Conodonts were the first vertebrates with mineralized skeletal elements in their mouth and pharynx. Armored jawless vertebrates ("ostracoderms") had defensive plates of bone on their skin.
- **Origins of Bone and Teeth (p. 679)** Mineralization appears to have originated with vertebrate mouthparts; the vertebrate endoskeleton became fully mineralized much later.

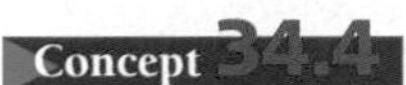

Gnathostomes are vertebrates that have jaws

- **Derived Characters of Gnathostomes (pp. 679–680)** Gnathostomes have jaws, which evolved from skeletal supports of the pharyngeal slits; enhanced sensory systems, including the lateral line system; an extensively mineralized endoskeleton; and paired appendages.
- **Fossil Gnathostomes (p. 680)** Placoderms were close relatives of living gnathostomes. Acanthodians were closely related to osteichthyans.
- **Chondrichthyans (Sharks, Rays, and Their Relatives) (pp. 680–682)** Chondrichthyans, which include sharks and rays, have a cartilaginous skeleton that evolved secondarily from an ancestral mineralized skeleton.
- **Ray-Finned Fishes and Lobe-Fins (pp. 682–684)** Osteichthyans have a skeleton reinforced by calcium phosphate. Aquatic forms have bony gill covers and a swim bladder; some also have lungs. Ray-finned fishes have maneuverable fins supported by long rays. Lobe-fins include coelacanths, lungfishes, and tetrapods. Aquatic lobe-fins have muscular pectoral and pelvic fins.

Concept 34.5

Tetrapods are gnathostomes that have limbs and feet

- **Derived Characters of Tetrapods (p. 684)** Tetrapods have four limbs and feet with digits. They also have other adaptations for life on land, such as ears.
- **The Origin of Tetrapods (p. 684)** Fossil evidence suggests that tetrapod limbs, now mainly used for walking on land, were originally used for paddling in water.
- **Amphibians (pp. 685–687)** Amphibians include salamanders, frogs, and caecilians. Most have moist skin that complements the lungs in gas exchange. Most frogs and some salamanders undergo metamorphosis of an aquatic larva into a terrestrial adult.

Amniotes are tetrapods that have a terrestrially adapted egg

- **Derived Characters of Amniotes (p. 688)** The amniotic egg contains extraembryonic membranes that carry out a variety of functions, including gas exchange and protection. Amniotes also have other terrestrial adaptations, such as relatively impermeable skin.
- **Early Amniotes (p. 688)** Early amniotes appeared in the Carboniferous period. They included large herbivores and predators.
- **Reptiles (pp. 688–691)** Reptiles include tuatara, lizards, snakes, turtles, crocodilians, and birds. Extinct forms include parareptiles, most dinosaurs, pterosaurs, and marine reptiles. Most reptiles are ectothermic, although birds are endothermic (as some other dinosaurs may have been).
- **Birds (pp. 691–694)** Birds probably descended from a group of small, carnivorous dinosaurs known as theropods. They have a variety of adaptations well suited to a lifestyle involving flight.
  Investigation *How Does Bone Structure Shed Light on the Origin of Birds?*

Mammals are amniotes that have hair and produce milk

- **Derived Characters of Mammals (p. 694)** Hair and mammary glands are two derived characteristics of mammals.
- **Early Evolution of Mammals (pp. 694–695)** Mammals evolved from synapsids in the late Triassic period. Living lineages of mammals originated in the Jurassic but did not undergo a significant adaptive radiation until the beginning of the Paleogene.
- **Monotremes (p. 695)** Monotremes are a small group of egg-laying mammals consisting of echidnas and the platypus.
- **Marsupials (pp. 695–697)** Marsupials include opossums, kangaroos, and koalas. Marsupial young begin their embryonic development attached to a placenta in the mother's uterus but complete development inside a maternal pouch.
- **Eutherians (Placental Mammals) (pp. 697–701)** Eutherians have young that complete their embryonic development attached to a placenta. All primates have hands and (with the exception of humans) feet adapted for gripping. Living primates include lemurs and their relatives, tarsiers, and anthropoids. Anthropoids diverged early into New World and Old World monkeys. Hominoids (gibbons, orangutans, gorillas, chimpanzees, bonobos, and humans) evolved from Old World monkeys.
  Activity *Characteristics of Chordates*
  Activity *Primate Diversity*

Concept 34.8

Humans are bipedal hominoids with a large brain

- **Derived Characters of Humans (pp. 701–702)** Humans have bipedal locomotion, a shortened jaw, and an enlarged brain.
- **The Earliest Hominids (pp. 702–703)** Hominids originated in Africa at least 6–7 million years ago. Early hominids had a small brain but probably walked upright.
- **Australopiths (pp. 703–704)** Australopiths are a paraphyletic assemblage of hominids that lived between 4 and 2 million years ago. Some species walked fully erect and had human-like hands and teeth.
- **Bipedalism (p. 704)** Hominids began to walk long distances on two legs about 1.9 million years ago.
- **Tool Use (p. 704)** The oldest evidence of tool use—cut marks on animal bones—is 2.5 million years old.
- **Early *Homo* (pp. 704–705)** *Homo ergaster* was the first fully bipedal, large-brained hominid. *Homo erectus* was the first hominid to leave Africa.
- **Neanderthals (p. 705)** Neanderthals lived in Europe and the Near East from 200,000 to 30,000 years ago. They became extinct a few thousand years after the arrival of *Homo sapiens* in Europe.
- ***Homo Sapiens* (pp. 705–707)** *Homo sapiens* appeared in Africa approximately 195,000 years ago and spread to other continents about 50,000 years ago. This spread may have been preceded by changes to the brain that made symbolic thought and other cognitive innovations possible. Research into the origins and contemporaries of *Homo sapiens* is a lively area of research and debate, including the recent discovery of an apparently new species, *Homo floresiensis,* dating from the late Pleistocene.
  Activity *Human Evolution*

TESTING YOUR KNOWLEDGE

Self-Quiz

1. Vertebrates and tunicates share
   a. jaws adapted for feeding.
   b. a high degree of cephalization.
   c. the formation of structures from the neural crest.
   d. an endoskeleton that includes a skull.
   e. a notochord and a dorsal, hollow nerve cord.

2. Some animals that lived 530 million years ago resembled lancelets but had a brain and a skull. These animals may represent
   a. the first chordates.
   b. a "missing link" between urochordates and cephalochordates.
   c. early craniates.
   d. marsupials.
   e. nontetrapod gnathostomes.

3. Chondrichthyans can be distinguished from osteichthyans by the
   a. presence in osteichthyans of a skull.
   b. presence in osteichthyans of a lateral line system.
   c. presence in chondrichthyans of unpaired fins.
   d. absence in chondrichthyans of a swim bladder and lungs.
   e. absence in chondrichthyans of paired sensory organs.

4. Which of the following could be considered the most recent common ancestor of living tetrapods?
   a. a sturdy-finned, shallow-water lobe-fin whose appendages had skeletal supports similar to those of terrestrial vertebrates
   b. an armored, jawed placoderm that had two sets of paired appendages
   c. an early ray-finned fish that developed bony skeletal supports in its paired fins

d. a salamander that had legs supported by a bony skeleton but moved with the side-to-side bending typical of fishes
e. an early terrestrial caecilian whose legless condition had evolved secondarily

5. Mammals and living birds share all of the following characteristics *except*
   a. endothermy.
   b. descent from a common amniotic ancestor.
   c. a dorsal, hollow nerve cord.
   d. a hole behind the eye socket on each side of the skull.
   e. an amniotic egg.

6. Unlike eutherians, *both* monotremes and marsupials
   a. lack nipples.
   b. have some embryonic development outside the mother's uterus.
   c. lay eggs.
   d. are found in Australia and Africa.
   e. include only insectivores and herbivores.

7. Which of the following characteristics of monkeys is specific to New World monkeys?
   a. well-developed parental care
   b. eyes close together on the front of the skull
   c. use of the tail to hang from a tree limb
   d. occasional bipedal walking
   e. downward orientation of the nostrils

8. Which clade does *not* include humans?
   a. synapsids
   b. lobe-fins
   c. diapsids
   d. craniates
   e. osteichthyans

9. As humans diverged from other primates, which of the following appeared first?
   a. the development of technology
   b. language
   c. a partially erect stance
   d. toolmaking
   e. an enlarged brain

10. Studies on DNA indicate which of the following?
    a. *Homo erectus* had an Asian origin.
    b. *Homo sapiens* originated in Africa.
    c. Neanderthals are the ancestors of present-day humans in Europe.
    d. Australopiths migrated out of Africa.
    e. North America had the first population of modern humans.

For Self-Quiz answers, see Appendix A.

Go to the website or CD-ROM for more quiz questions.

Evolution Connection

Identify one characteristic that qualifies humans for membership in each of the following clades: eukaryotes, animals, deuterostomes, chordates, vertebrates, gnathostomes, amniotes, mammals, primates.

Scientific Inquiry

Scientific inquiry often involves trying to make sense of an interesting observation. One such observation concerns patterns of genetic versus morphological divergence in some vertebrate groups. Amphibians, for example, show great similarity in morphology, yet they are far more genetically diverse than the morphologically diverse birds. A related pattern is evident in chimpanzee-human comparisons: These two species are quite divergent morphologically but are much more similar genetically. Propose one or more hypotheses that might explain these puzzling patterns.

Investigation ***How Does Bone Structure Shed Light on the Origin of Birds?***

Science, Technology, and Society

While human biological evolution is Darwinian, the evolution of human culture could be described as Lamarckian. Explain this distinction after reviewing the ideas of Darwin and Lamarck in Chapter 22.

Plant Form and Function

AN INTERVIEW WITH

Natasha Raikhel

From her early work as an invertebrate zoologist at the University of Leningrad to her research today as one of the most prominent American plant biologists, Natasha Raikhel has traveled very far, both geographically and scientifically. Dr. Raikhel is now the Distinguished Professor of Plant Cell Biology at the University of California, Riverside (UCR), where she is also the director of the Center for Plant Cell Biology. In 2003, the American Society for Cell Biology honored Professor Raikhel's achievements with the Women in Cell Biology Senior Career Recognition Award, and in 2004 she received the Stephen Hales Prize of the American Society of Plant Biologists; the award is given to a scientist who has served the science of plant biology in some noteworthy manner. One of Dr. Raikhel's current interests is making systems biology possible by building multidisciplinary research teams. Dr. Raikhel's service to the biology community and her generosity as a mentor to young researchers exemplify the collaborative spirit of the science culture.

How did you begin your science career in what was then the Soviet Union?

Actually, I didn't go directly into science. I was a musician and was supposed to become a professional pianist. But at some point, I realized that I could not be one of the best pianists, and I didn't like that limitation. I always liked nature, and I was able to study biology at the University of Leningrad. I met my husband when we were both zoology students in Leningrad (now St. Petersburg), where we continued our work at the university through graduate school until we acquired faculty positions. And then we decided to leave the Soviet Union.

What were the circumstances of your emigration?

This was in the late '70s, and as Jews in the Soviet Union, our career opportunities in science were limited, but we were fortunate enough to be allowed to emigrate. After our first son was born in 1975, we wanted him to be exposed to greater opportunities, and that helped us finalize our decision to leave. It was difficult to emigrate from the Soviet Union during that period. But there was a joke at that time that the Carter administration in the United States had a grains-for-Jews exchange policy—the more wheat the U.S. sent to the Soviet Union, the more Jews were allowed out. So we were very lucky that we were able to get out of the Soviet Union at a time when many people would have loved to leave.

How did you find your first jobs here?

It was difficult for scientists emigrating from the Soviet Union to find academic jobs because of a lack of references. But again, we were lucky. The research institute (Institute of Cytology) where I worked had a keen interest in modern biology, and the professor in charge invited visiting professors to spend their sabbatical leaves doing research at the institute. A professor, Jerry Paulin, who came with his family from the University of Georgia, had his desk right next to mine in the lab. Our families became friends. At that time, it was very difficult for American visitors in the Soviet Union to get around because they were told, "You cannot go here, you cannot go there." So we helped the Paulins during their visit to Leningrad. Later, when we planned our immigration to the United States, Jerry Paulin was the one professor in the U.S. who knew us and our work as scientists, and he helped us find jobs at the University of Georgia. And so we started all over again as postdoctoral researchers when we came to this country.

In Leningrad, you worked on protozoans—ciliates. But in Georgia, you began your work with plants. What motivated that change?

My interest is at the basic cellular level, and so the switch of organisms was not too difficult. I realized very soon after arriving in the U.S. that it would be hard to get grants for research on free-living protozoans, which do not cause disease. At the time, plant research was really starting to take off in the U.S., and Georgia was one of the key places where plant biology was very modern. I sensed this and wanted to be part of that environment, and so I learned a lot about plants. I also began to reeducate myself to do molecular biology. Molecular approaches in plant biology were just beginning. I didn't have to catch up because I learned along with many people who were also just beginning in molecular biology. So, once again, I was lucky with good timing.

Speaking of good timing, why is this a good time to be in plant biology?

One reason is that we now know a lot about a research organism, the plant *Arabidopsis*. Its genome has been sequenced, and many of its proteins have been identified. Once you understand the functions of these proteins, how the proteins interact, and how pathways in the cells interact, then you can really start to answer questions about how cells function and how the whole plant works. Systems biology becomes possible, and I think this makes plant biology much more exciting.

What will it take to make plant systems biology work?

The essence of systems biology is to be able to model organisms—to be able to predict how various pathways in the organism interact and

how a change in one pathway will affect the whole organism. This requires our approaches to be more quantitative and much more integrated. We need to infuse plant biology with disciplines such as informatics and engineering, but this is difficult because people are coming from different scientific cultures. Right now, many of us are retraining to understand fields such as bioinformatics, but it will be much better when such multidisciplinary experience becomes the normal environment for training biologists. It is very important that the new generation of plant biologists be very quantitative and comfortable with bioinformatics, and that they are good biologists too. It will take time to create this new generation of scientists. But many people are looking to the community of *Arabidopsis* researchers to make the systems approach work, because this community is exemplary in terms of openness and sharing among colleagues.

The number of protein-coding genes in *Arabidopsis* rivals that of humans and other complex animals. Did that discovery surprise you?

Not really. Because plants are immobile, they have to be very versatile in their ability to respond to environmental stresses. Plants have to be exquisite to survive because they can't run. This requires a lot of different kinds of proteins, including many that are unique to plants.

Are there important agricultural applications from all this work on *Arabidopsis*?

Yes, because *Arabidopsis* is such a convenient model organism for studying processes that are important in all plants, including crop plants. For example, resistance to cold and resistance to pathogens are two traits that are very useful in crop plants. But such traits are much easier to analyze in *Arabidopsis*. So, if we first identify certain genes and regulatory pathways that can help *Arabidopsis* resist environmental stress, we can then transfer some of that knowledge to improve crop plants.

What is the focus of your own lab's research?

For a long time, my lab group has worked on the trafficking of molecules through the cell's vesicles and vacuoles. And in the last few years, we've added another project, the synthesis of the cell wall in plants. Vacuoles and cell walls have unique plant-specific functions and are essential for plant life. I therefore am very interested in fully understanding these mechanisms.

In recent years, your influence has extended beyond your own lab to a leadership role in the international community of plant cell biologists. Why do you think this expanded role is important at this stage in your career?

Well, first of all, I want to make it clear that my research with my own lab group is still my number-one priority. But with time, you start to realize that you have experience that enables you to do more for the science community in your field. I got a lot from the community of plant biologists, and now I want to be constructive in giving back to the community. In fact, at a certain point in your career, you have an obligation to give more than you take. For example, I was not looking to become editor-in-chief of the journal *Plant Physiology*, but colleagues convinced me to take the job. This takes a lot of my time now, but the journal is evolving beautifully, which is very rewarding.

Plants have to be exquisite to survive because they can't run.

As another example of "giving something back," you participate in a program that provides summer research opportunities at UCR for community college students. Is it fun for you to have these students in your lab?

I love to be around these young people because they keep *me* young. At first, when these students join the lab, they are completely naïve about research and just don't know how to get started. But once you put a fire under them and get them excited, it's the most rewarding thing in the world. For example, this past summer, we had a student from San Bernardino Valley College, a community college nearby. He just loved his project, and he will be on a paper that we are writing. He will also be transferring to UCR, and he still works in my lab for many hours each week, even though he's taking lots of courses. This student now wants to become a plant researcher—he's a convert! And that gives me great satisfaction.

With two very busy scientists in the family, how do you and your husband balance the demands of your careers with the responsibilities and enjoyment of family life?

I want young people beginning in science to understand that they will not have to shy away from having a family and having children because of their work. The most wonderful things I have in my life are my two sons. Balancing career and family is not easy, but it's all possible because I have a great husband, and we make decisions and do things together. You can't say, "You stay at home, and that's how we will share." Maybe that does work for some people, but in our case, we both wanted careers in science.

What other advice do you have for students who are interested in becoming scientists?

Because of the way biology is changing, I think students should look for programs that offer interdisciplinary approaches. They should not lock themselves into certain subfields. And they should reach for the stars. You have to be ambitious and work hard, and you have to know what you want, and then don't let anything stop you. You can grow as a scientist here like nowhere else in the world. I think I'm a living example of that. It's a fantastic life!

35 Plant Structure, Growth, and Development

▲ **Figure 35.1 Fanwort *(Cabomba caroliniana).***

Key Concepts

35.1 The plant body has a hierarchy of organs, tissues, and cells
35.2 Meristems generate cells for new organs
35.3 Primary growth lengthens roots and shoots
35.4 Secondary growth adds girth to stems and roots in woody plants
35.5 Growth, morphogenesis, and differentiation produce the plant body

Overview

No Two Plants Are Alike

To some people, the plant in **Figure 35.1** is an intrusive aquatic weed that clogs streams, rivers, and lakes. Others consider it an attractive addition to an aquarium. Whatever else the fanwort (*Cabomba caroliniana*) may be, it is a striking example of **plasticity**—an organism's ability to alter or "mold" itself in response to local environmental conditions. The underwater leaves have a feathery appearance, an adaptation that may provide protection from the stress of moving water. In contrast, the surface leaves are pads that aid in flotation. Both leaf types have genetically identical cells, but the dissimilar environments cause different genes involved in leaf formation to be turned on or off. Such extreme developmental plasticity is much more common in plants than in animals and may help compensate for their lack of mobility. As Natasha Raikhel puts it in the interview preceding this chapter, "Plants have to be exquisite to survive because they can't run." Also, since the form of any plant is controlled by environmental as well as genetic factors, no two plants are exactly alike.

In addition to plastic structural responses by individual plants to specific environments, entire species have by natural selection accumulated characteristics of **morphology**, or external form, that vary little among plants within the species. For example, some species of desert plants, such as cacti, have leaves that are so highly reduced as spines that the stem is actually the primary photosynthetic organ. This reduction in leaf size, and thus in surface area, results in reduced water loss. These leaf adaptations have enhanced survival and reproductive success in the desert environment.

This chapter focuses on how the body of a plant is formed, setting the stage for the rest of this unit on plant biology. Chapters 29 and 30 described the evolution and characteristics of bryophytes, seedless vascular plants, gymnosperms, and angiosperms. This chapter and Unit Six in general focus mainly on vascular plants—especially angiosperms because flowering plants comprise about 90% of plant species and are the base of nearly every terrestrial food web. As the world's population increases, the need for plants to supply food, fuel, fiber, medicine, lumber, and paper has never been greater, heightening the importance of understanding how plants grow and develop.

Concept 35.1

The plant body has a hierarchy of organs, tissues, and cells

Plants, like most animals, have organs composed of different tissues, which in turn are composed of cells. A **tissue** is a group of cells with a common function, structure, or both. An **organ** consists of several types of tissues that together carry out particular functions. In looking at the hierarchy of plant organs, tissues, and cells, we will focus first on organs as the most readily observable features of plant structure.

The Three Basic Plant Organs: Roots, Stems, and Leaves

The basic morphology of vascular plants reflects their evolutionary history as terrestrial organisms that inhabit and draw resources from two very different environments—below-ground and above-ground. Plants must absorb water and minerals from below the ground and CO_2 and light from above the ground. The evolutionary solution to this separation of resources was the development of three basic organs: roots, stems, and leaves. They are organized into a **root system** and a **shoot system,** the latter consisting of stems and leaves **(Figure 35.2)**. With few exceptions, angiosperms and other vascular plants rely completely on both systems for survival. Roots are typically nonphotosynthetic and would starve without the organic nutrients imported from the shoot system. Conversely, the shoot system depends on the water and minerals that roots absorb from the soil.

Later in the chapter, we will discuss the transition from vegetative shoots (shoots that are nonreproductive) to reproductive shoots. In angiosperms, the reproductive shoots are flowers, which are composed of leaves that are highly modified for sexual reproduction.

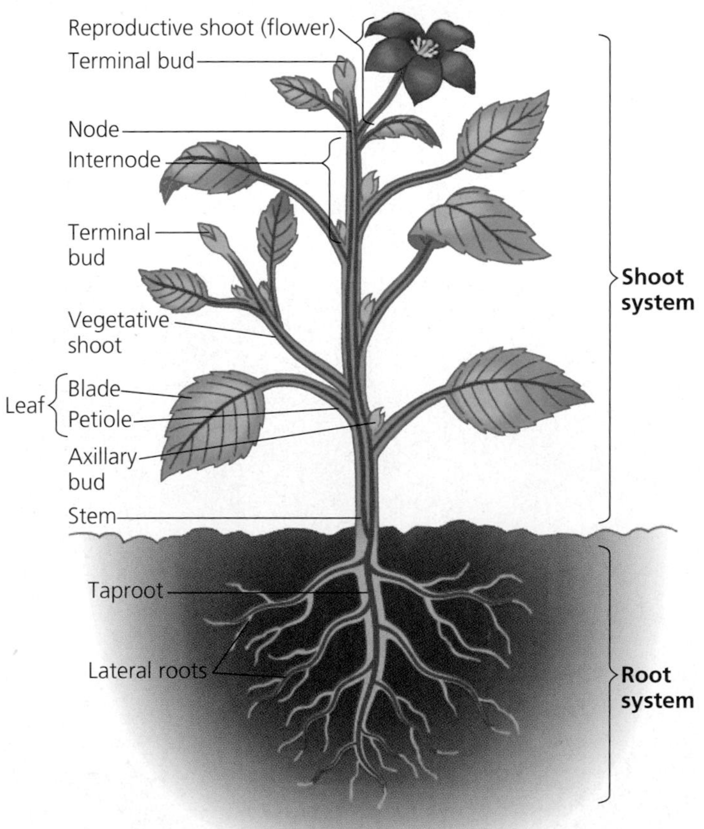

▲ **Figure 35.2 An overview of a flowering plant.** The plant body is divided into a root system and a shoot system, connected by vascular tissue (purple strands in this diagram) that is continuous throughout the plant. The plant shown is an idealized eudicot.

As we take a closer look at roots, stems, and leaves, try to view these organs from the evolutionary perspective of adaptations to living on land. In identifying some variations in these organs, we will focus mainly on the two major groups of angiosperms: monocots and eudicots (see Figure 30.12).

Roots

A **root** is an organ that anchors a vascular plant (usually in the soil), absorbs minerals and water, and often stores organic nutrients. Most eudicots and gymnosperms have a **taproot system,** consisting of one main vertical root (the taproot) that develops from an embryonic root. The taproot gives rise to **lateral roots,** also called branch roots (see Figure 35.2). In angiosperms, the taproot often stores organic nutrients that the plant consumes during flowering and fruit production. For this reason, root crops such as carrots, turnips, and sugar beets are harvested before they flower. Taproot systems generally penetrate deeply into the ground.

In seedless vascular plants and in most monocots, such as grasses, the embryonic root dies and does not give rise to a main root. Instead, many small roots grow from the stem, with each small root forming its own lateral roots. The result is a **fibrous root system**—a mat of generally thin roots spreading out below the soil surface, with no root standing out as the main one (see Figure 30.12). Roots arising from the stem are said to be **adventitious** (from the Latin *adventicius,* extraneous), a term describing any plant part that grows in an unusual location. A fibrous root system is usually shallower than a taproot system. Grass roots are particularly shallow, being concentrated in the upper few centimeters of the soil. Because grass roots hold the topsoil in place, they make excellent ground cover for preventing erosion. Large monocots, such as palms and bamboo, are mainly anchored by sturdy rhizomes, which are horizontal underground stems.

The entire root system helps anchor a plant, but in most plants the absorption of water and minerals occurs primarily near the root tips, where vast numbers of tiny root hairs increase the surface area of the root enormously **(Figure 35.3)**. A

◀ **Figure 35.3 Root hairs and root tip.** Growing by the thousands just behind each root tip, root hairs increase the surface area for the absorption of water and minerals by the roots.

root hair is an extension of a root epidermal cell (protective cell on a plant surface). Root hairs are not to be confused with lateral roots, which are multicellular organs. Absorption is often enhanced by symbiotic relationships between plant roots and fungi and bacteria, as you will see in Chapters 36 and 37.

Many plants have modified roots. Some of these arise from roots, and others are adventitious, developing from stems and, in rare cases, leaves. Some modified roots provide more support and anchorage, while others store water and nutrients or absorb oxygen or water from the air **(Figure 35.4)**.

▼ **Figure 35.4 Modified roots.** Environmental adaptations may result in roots being modified for a variety of functions. Many modified roots are aerial roots that are above the ground during normal development.

(a) Prop roots. The aerial roots shown here in maize are examples of prop roots, so named because they support tall, top-heavy plants. All roots of a mature maize plant are adventitious after the original roots die. The emerging roots shown here will eventually penetrate the soil.

(b) Storage roots. Many plants, such as sweet potatoes, store food and water in their roots.

(c) "Strangling" aerial roots. The seeds of this strangler fig germinate in the branches of tall trees and send numerous aerial roots to the ground. These snake-like roots gradually wrap around the hosts and objects such as this Cambodian temple ruin. Eventually, the host tree dies of strangulation and shading.

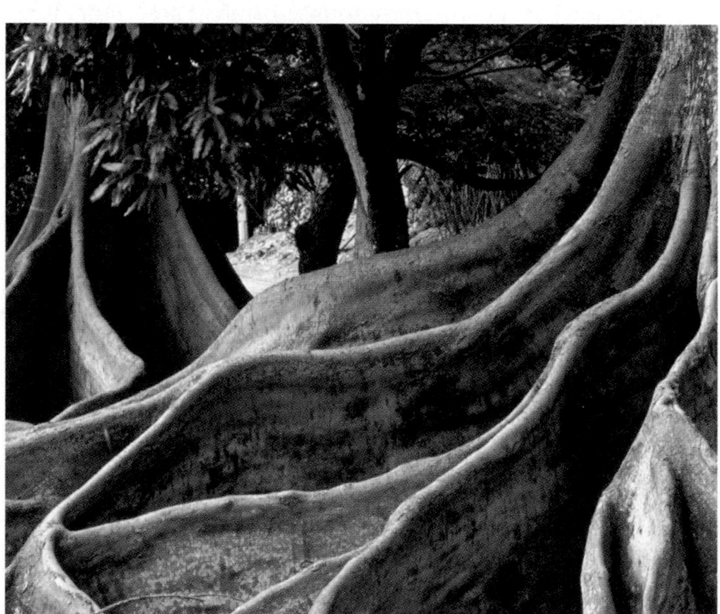

(d) Buttress roots. Aerial roots that look like buttresses support the tall trunks of some tropical trees, such as this ceiba tree in Central America.

(e) Pneumatophores. Also known as air roots, pneumatophores are produced by trees such as mangroves that inhabit tidal swamps. By projecting above the surface, they enable the root system to obtain oxygen, which is lacking in the thick, waterlogged mud.

Stems

A **stem** is an organ consisting of an alternating system of **nodes,** the points at which leaves are attached, and **internodes,** the stem segments between nodes (see Figure 35.2). In the angle (axil) formed by each leaf and the stem is an **axillary bud,** a structure that has the potential to form a lateral shoot, commonly called a branch. Most axillary buds of a young shoot are dormant (not growing). Thus, elongation of a young shoot is usually concentrated near the shoot apex (tip), which consists of a **terminal bud** with developing leaves and a compact series of nodes and internodes.

The proximity of the terminal bud is partly responsible for inhibiting the growth of axillary buds, a phenomenon called **apical dominance.** By concentrating resources toward elongation, apical dominance is an evolutionary adaptation that increases the plant's exposure to light. But what if an animal eats the end of the shoot? Or what if, because of shading objects, light is more intense to the side of a plant than directly above it? Under such conditions, axillary buds break dormancy; that is, they start growing. A growing axillary bud gives rise to a lateral shoot, complete with its own terminal bud, leaves, and axillary buds. Removing the terminal bud usually stimulates the growth of axillary buds, resulting in more lateral shoots. That is why pruning trees and shrubs and "pinching back" houseplants will make them "bushier."

Modified stems with diverse functions have evolved in many plants as environmental adaptations. These modified stems, which include stolons, rhizomes, tubers, and bulbs, are often mistaken for roots **(Figure 35.5)**.

Leaves

The **leaf** is the main photosynthetic organ of most vascular plants, although green stems also perform photosynthesis. Leaves vary extensively in form but generally consist of a flattened **blade** and a stalk, the **petiole,** which joins the leaf to a node of the stem (see Figure 35.2). Among angiosperms, grasses and many other monocots lack petioles; instead, the base of the leaf forms a sheath that envelops the stem. Some monocots, including palms, do have petioles.

Monocots and eudicots differ in the arrangement of **veins,** the vascular tissue of leaves. Most monocots have parallel major veins that run the length of the leaf blade. In contrast,

▶ **Figure 35.5 Modified stems.**

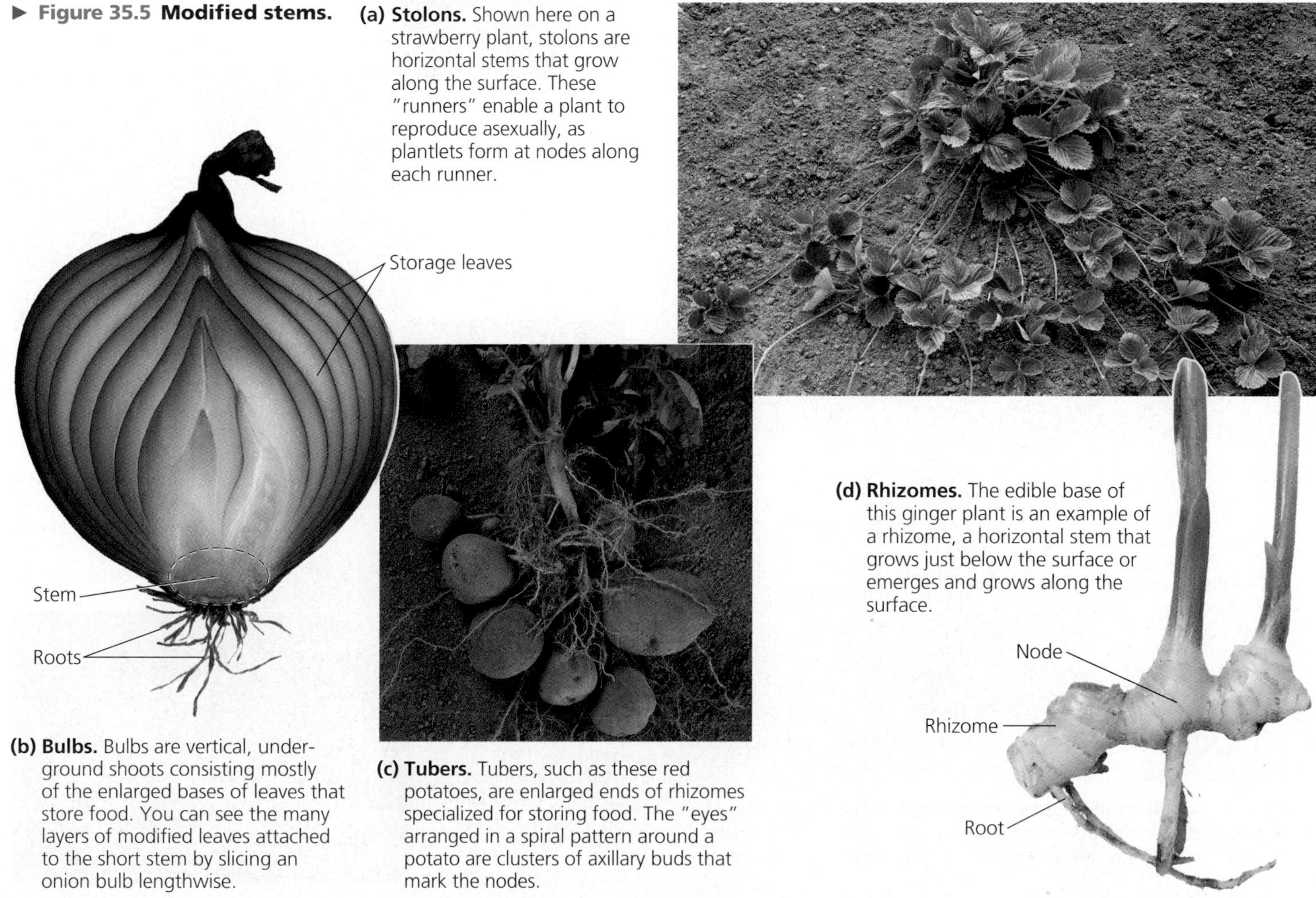

(a) Stolons. Shown here on a strawberry plant, stolons are horizontal stems that grow along the surface. These "runners" enable a plant to reproduce asexually, as plantlets form at nodes along each runner.

(b) Bulbs. Bulbs are vertical, underground shoots consisting mostly of the enlarged bases of leaves that store food. You can see the many layers of modified leaves attached to the short stem by slicing an onion bulb lengthwise.

(c) Tubers. Tubers, such as these red potatoes, are enlarged ends of rhizomes specialized for storing food. The "eyes" arranged in a spiral pattern around a potato are clusters of axillary buds that mark the nodes.

(d) Rhizomes. The edible base of this ginger plant is an example of a rhizome, a horizontal stem that grows just below the surface or emerges and grows along the surface.

eudicot leaves generally have a multibranched network of major veins (see Figure 30.12).

In identifying and classifying angiosperms, taxonomists rely mainly on floral morphology, but they also use variations in leaf morphology, such as leaf shape, spatial arrangement of leaves, and the pattern of a leaf's veins. **Figure 35.6** illustrates a difference in leaf shape: simple versus compound leaves. Most very large leaves are compound or doubly compound. This structural adaptation may enable large leaves to withstand strong wind with less tearing and also confine some pathogens that invade the leaf to a single leaflet, rather than allowing them to spread to the entire leaf.

Most leaves are specialized for photosynthesis. However, some plant species have leaves that have become adapted by evolution for other functions, such as support, protection, storage, or reproduction **(Figure 35.7)**.

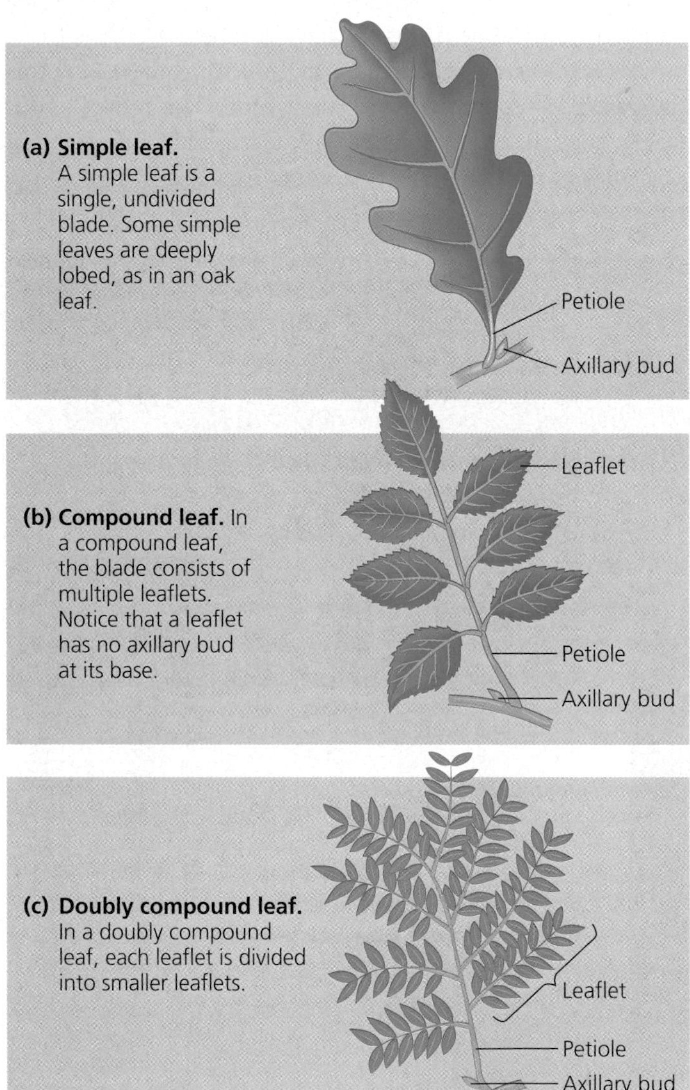

(a) Simple leaf.
A simple leaf is a single, undivided blade. Some simple leaves are deeply lobed, as in an oak leaf.
Petiole
Axillary bud
(b) Compound leaf. In a compound leaf, the blade consists of multiple leaflets. Notice that a leaflet has no axillary bud at its base.
Leaflet
Petiole
Axillary bud
(c) Doubly compound leaf.
In a doubly compound leaf, each leaflet is divided into smaller leaflets.
Leaflet
Petiole
Axillary bud

▲ **Figure 35.6 Simple versus compound leaves.** You can distinguish simple leaves from compound leaves by looking for axillary buds. Each leaf has only one axillary bud, where the petiole attaches to the stem.

▼ **Figure 35.7 Modified leaves.**

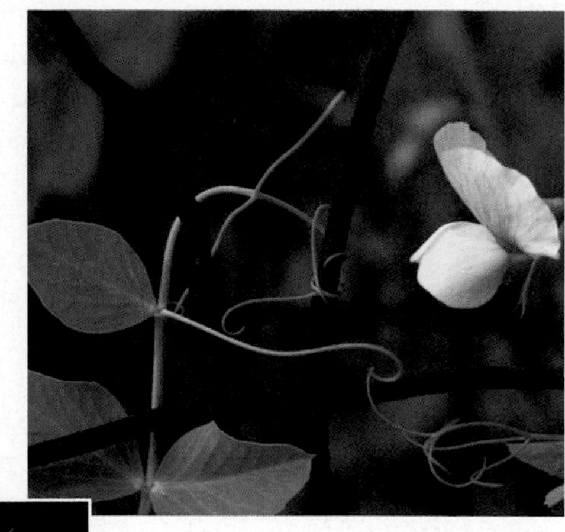

(a) Tendrils. The tendrils by which this pea plant clings to a support are modified leaves. After it has "lassoed" a support, a tendril forms a coil that brings the plant closer to the support. Tendrils are typically modified leaves, but some tendrils are modified stems, as in grapevines.

(b) Spines. The spines of cacti, such as this prickly pear, are actually leaves, and photosynthesis is carried out mainly by the fleshy green stems.

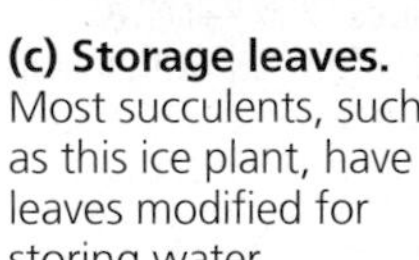

(c) Storage leaves. Most succulents, such as this ice plant, have leaves modified for storing water.

(d) Bracts. Red parts of the poinsettia are often mistaken for petals but are actually modified leaves called bracts that surround a group of flowers. Such brightly colored leaves attract pollinators.

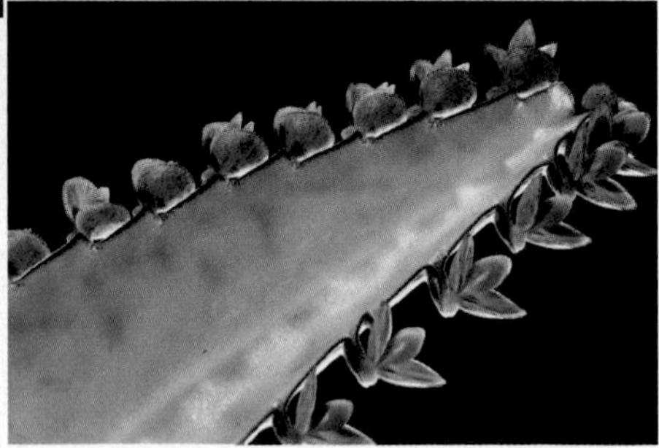

(e) Reproductive leaves. The leaves of some succulents, such as *Kalanchoë daigremontiana*, produce adventitious plantlets, which fall off the leaf and take root in the soil.

The Three Tissue Systems: Dermal, Vascular, and Ground

Each plant organ—root, stem, or leaf—has dermal, vascular, and ground tissues. A **tissue system** consists of one or more tissues organized into a functional unit connecting the organs of a plant. Each tissue system is continuous throughout the entire plant body, but specific characteristics of the tissues and their spatial relationships to one another vary in different organs **(Figure 35.8)**.

The **dermal tissue system** is the outer protective covering. Like our skin, it forms the first line of defense against physical damage and pathogenic organisms. In nonwoody plants, the dermal tissue usually consists of a single layer of tightly packed cells called the **epidermis.** In woody plants, protective tissues known as **periderm** replace the epidermis in older regions of stems and roots by a process discussed later in the chapter. In addition to protecting the plant from water loss and disease, the epidermis has specialized characteristics in each organ. For example, the root hairs so important in the absorption of water and minerals are extensions of epidermal cells near root tips. In the epidermis of leaves and most stems, a waxy coating called the **cuticle** helps prevent water loss—an important adaptation to living on land. Later we will look at specialized leaf cells that regulate CO_2 exchange. Leaf trichomes, which are outgrowths of the epidermis, are yet another example of specialization. For instance, trichomes in aromatic leaves such as mint secrete oils that protect plants from herbivores and disease.

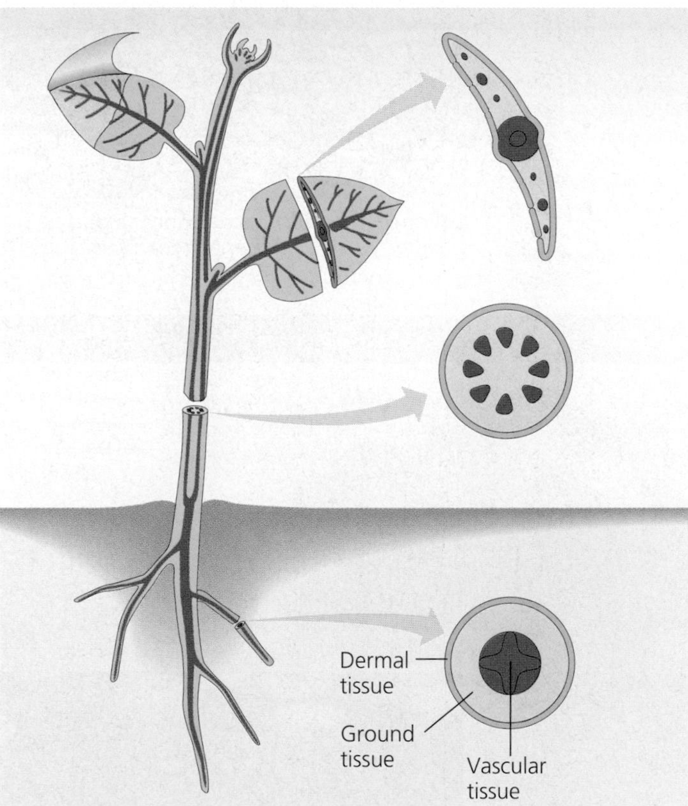

▲ **Figure 35.8 The three tissue systems.** The dermal tissue system (blue) covers the entire body of a plant. The vascular tissue system (purple) is also continuous throughout the plant, but is arranged differently in each organ. The ground tissue system (yellow), responsible for most of the plant's metabolic functions, is located between the dermal tissue and the vascular tissue in each organ.

The **vascular tissue system** carries out long-distance transport of materials between roots and shoots. The two vascular tissues are xylem and phloem. **Xylem** conveys water and dissolved minerals upward from roots into the shoots. **Phloem** transports organic nutrients such as sugars from where they are made (usually the leaves) to where they are needed—usually roots and sites of growth, such as developing leaves and fruits. The vascular tissue of a root or stem is collectively called the **stele** (the Greek word for "pillar"). The arrangement of the stele varies, depending on species and organ. In angiosperms, the stele of the root is in the form of a solid central **vascular cylinder.** In contrast, the stele of stems and leaves is divided into **vascular bundles,** strands consisting of xylem and phloem. Both xylem and phloem are composed of a variety of cell types, including cells highly specialized for transport.

Tissues that are neither dermal nor vascular are part of the **ground tissue system.** Ground tissue that is internal to the vascular tissue is called **pith,** and ground tissue that is external to the vascular tissue is called **cortex.** The ground tissue system is more than just filler. It includes various cells specialized for functions such as storage, photosynthesis, and support.

Common Types of Plant Cells

Like any multicellular organism, a plant is characterized by cellular differentiation, the specialization of cells in structure and function. In plant cells, differentiation is sometimes evident within the **protoplast,** the cell contents exclusive of the cell wall. For example, the protoplasts of some plant cells have chloroplasts, whereas other types of plant cells lack functional chloroplasts. Cell wall modifications also play a role in plant cell differentiation. **Figure 35.9**, on the next two pages, focuses on some major types of plant cells: parenchyma, collenchyma, sclerenchyma, the water-conducting cells of the xylem, and the sugar-conducting cells of the phloem. Notice the structural adaptations that make specific functions possible. You may wish to review Figures 6.9 and 6.28, which show basic plant cell structure.

Concept Check 35.1

1. How does the vascular tissue system enable leaves and roots to combine functions to support growth and development of the whole plant?
2. Describe at least three specializations in plant organs and plant cells that are adaptations to life on land.
3. Describe the role of each tissue system in a leaf.

For suggested answers, see Appendix A.

Figure 35.9

Exploring Examples of Differentiated Plant Cells

PARENCHYMA CELLS

Mature **parenchyma cells** have primary walls that are relatively thin and flexible, and most lack secondary walls. (See Figure 6.28 to review primary and secondary cell walls.) The protoplast generally has a large central vacuole. Parenchyma cells are often depicted as "typical" plant cells because they appear to be the least specialized structurally. Parenchyma cells perform most of the metabolic functions of the plant, synthesizing and storing various organic products. For example, photosynthesis occurs within the chloroplasts of parenchyma cells in the leaf. Some parenchyma cells in stems and roots have colorless plastids that store starch. The fleshy tissue of a typical fruit is composed mainly of parenchyma cells. Most parenchyma cells retain the ability to divide and differentiate into other types of plant cells under special conditions—during the repair and replacement of organs after injury to the plant, for example. It is even possible in the laboratory to regenerate an entire plant from a single parenchyma cell.

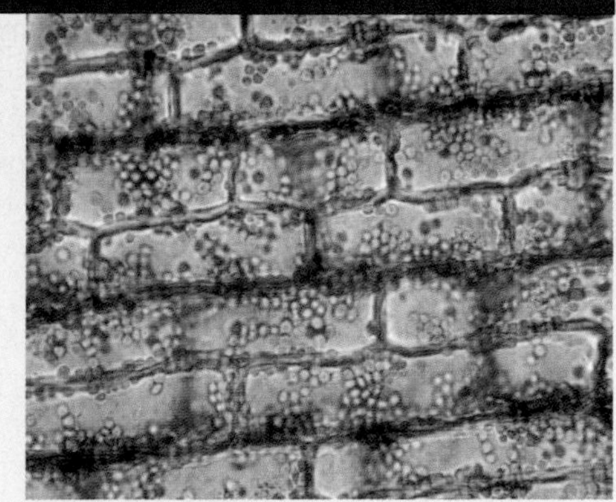

Parenchyma cells in *Elodea* leaf, with chloroplasts (LM)

COLLENCHYMA CELLS

Grouped in strands or cylinders, **collenchyma cells** help support young parts of the plant shoot. Collenchyma cells have thicker primary walls than parenchyma cells, though the walls are unevenly thickened. Young stems and petioles often have strands of collenchyma cells just below their epidermis (the "strings" of a celery stalk, for example). Collenchyma cells lack secondary walls, and the hardening agent lignin is absent in their primary walls. Therefore, they provide flexible support without restraining growth. At functional maturity, collenchyma cells are living and flexible, elongating with the stems and leaves they support—unlike sclerenchyma cells, which we discuss next.

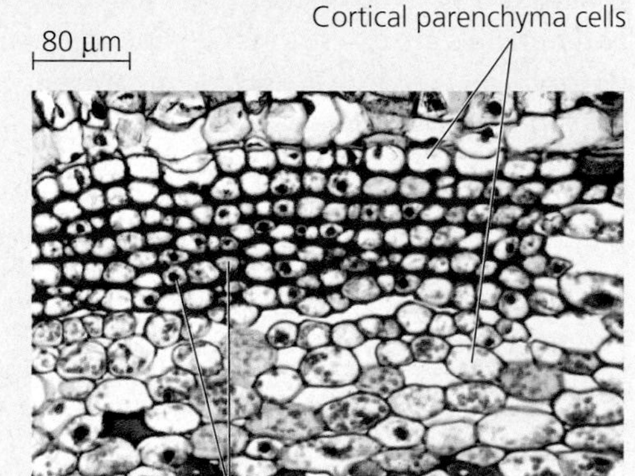

Collenchyma cells (in cortex of *Sambucus*, elderberry; cell walls stained red) (LM)

SCLERENCHYMA CELLS

Also functioning as supporting elements in the plant, but with thick secondary walls usually strengthened by lignin, **sclerenchyma cells** are much more rigid than collenchyma cells. Mature sclerenchyma cells cannot elongate, and they occur in regions of the plant that have stopped growing in length. Sclerenchyma cells are so specialized for support that many are dead at functional maturity, but they produce secondary walls before the protoplast dies. The rigid walls remain as a "skeleton" that supports the plant, in some cases for hundreds of years. In parts of the plant that are still elongating, the secondary walls of immature sclerenchyma are deposited unevenly in spiral or ring patterns. These forms of cell wall thickenings enable the cell wall to stretch like a spring as the cell elongates.

Two types of sclerenchyma cells called **sclereids** and **fibers** are specialized entirely for support and strengthening. Sclereids, which are shorter than fibers and irregular in shape, have very thick, lignified secondary walls. Sclereids impart the hardness to nutshells and seed coats and the gritty texture to pear fruits. Fibers, which are usually arranged in threads, are long, slender, and tapered. Some are used commercially, such as hemp fibers for making rope and flax fibers for weaving into linen.

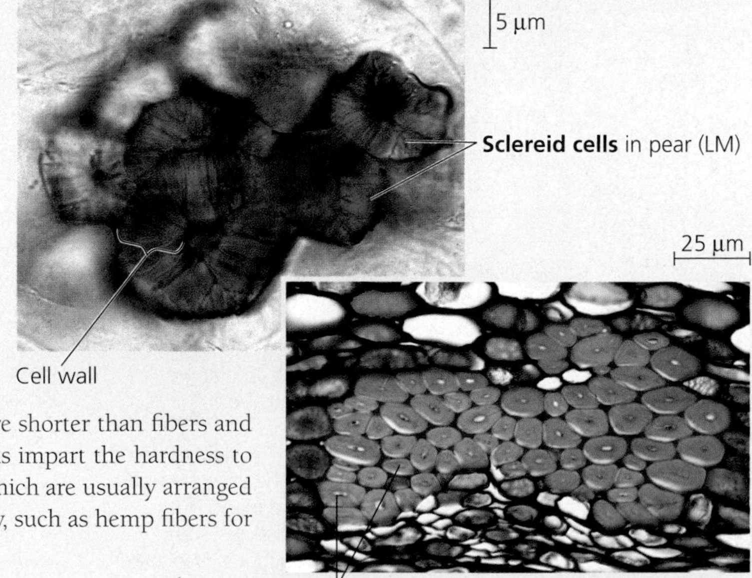

Sclereid cells in pear (LM)

Fiber cells (transverse section from ash tree) (LM)

WATER-CONDUCTING CELLS OF THE XYLEM

The two types of water-conducting cells, **tracheids** and **vessel elements**, are tubular, elongated cells that are dead at functional maturity. Tracheids are found in the xylem of all vascular plants. In addition to tracheids, most angiosperms, as well as a few gymnosperms and a few seedless vascular plants, have vessel elements. When the protoplast of a tracheid or vessel element disintegrates, the cell's thickened walls remain behind, forming a nonliving conduit through which water can flow. The secondary walls of tracheids and vessel elements are often interrupted by pits, thinner regions where only primary walls are present (see Figure 6.28 to review primary and secondary walls). Water can migrate laterally between neighboring cells through pits.

Tracheids are long, thin cells with tapered ends. Water moves from cell to cell mainly through the pits, where it does not have to cross thick secondary walls. The secondary walls of tracheids are hardened with lignin, which prevents collapse under the tensions of water transport and also provides support.

Vessel elements are generally wider, shorter, thinner walled, and less tapered than tracheids. They are aligned end to end, forming long micropipes known as **vessels**. The end walls of vessel elements have perforations that enable water to flow freely through the vessels.

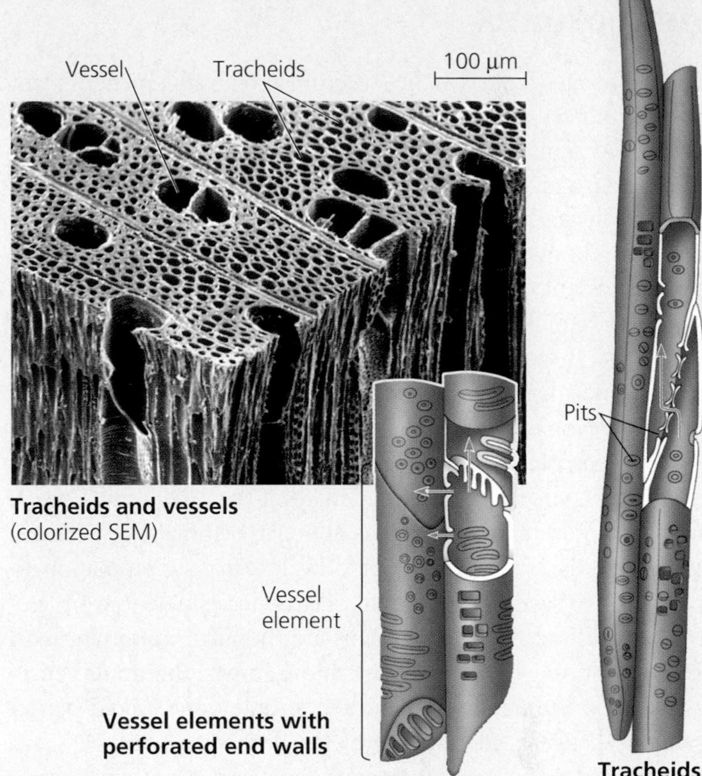

Tracheids and vessels (colorized SEM)

Vessel elements with perforated end walls

Tracheids

SUGAR-CONDUCTING CELLS OF THE PHLOEM

Unlike the water-conducting cells of the xylem, the sugar-conducting cells of the phloem are alive at functional maturity. In seedless vascular plants and gymnosperms, sugars and other organic nutrients are transported through long, narrow cells called sieve cells. In the phloem of angiosperms, these nutrients are transported through sieve tubes, which consist of chains of cells called **sieve-tube members**.

Though alive, sieve-tube members lack such organelles as the nucleus, ribosomes, and a distinct vacuole. This reduction in cell contents enables nutrients to pass more easily through the cell. The end walls between sieve-tube members, called **sieve plates**, have pores that facilitate the flow of fluid from cell to cell along the sieve tube. Alongside each sieve-tube member is a nonconducting cell called a **companion cell**, which is connected to the sieve-tube member by numerous channels, the plasmodesmata (see Figure 6.8). The nucleus and ribosomes of the companion cell may serve not only that cell itself but also the adjacent sieve-tube member. In some plants, companion cells in leaves also help load sugars into the sieve-tube members, which then transport the sugars to other parts of the plant.

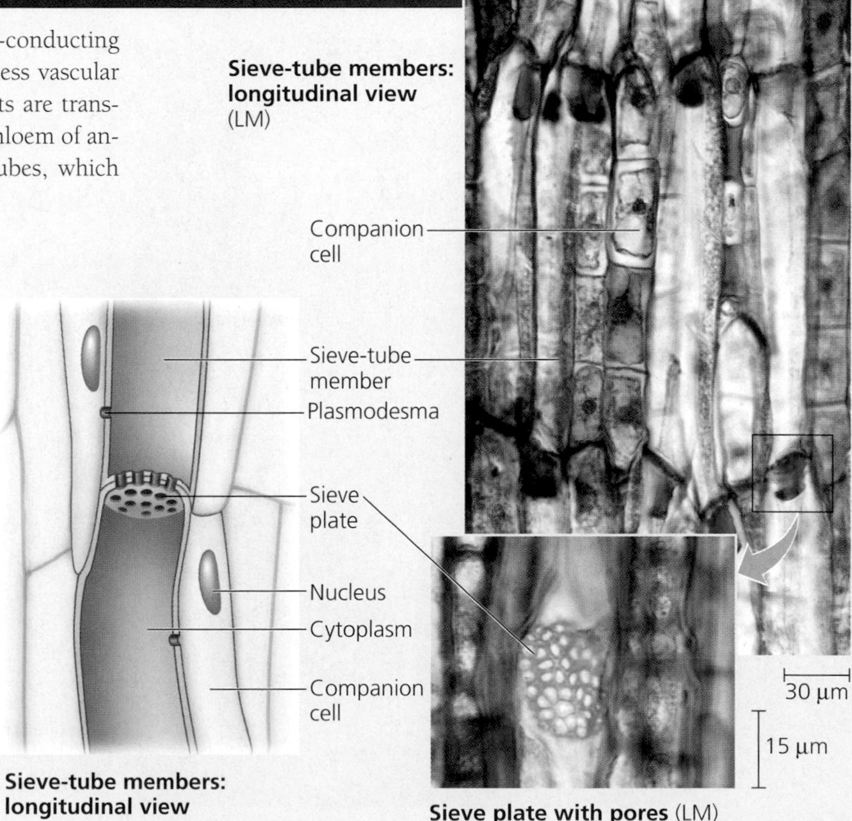

Sieve-tube members: longitudinal view (LM)

Sieve-tube members: longitudinal view

Sieve plate with pores (LM)

Concept 35.2

Meristems generate cells for new organs

So far, we have looked at the structure and arrangement of plant tissues and cells in mature organs. But how does this organization arise? A major difference between plants and most animals is that plant growth is not limited to an embryonic or juvenile period. Instead, growth occurs throughout the plant's life, a condition known as **indeterminate growth.** At any given time, a typical plant consists of embryonic, developing, and mature organs. Except for periods of dormancy, most plants grow continuously. In contrast, most animals and some plant organs, such as most leaves, undergo **determinate growth;** that is, they cease growing after reaching a certain size.

Although plants continue to grow throughout their lives, they do die, of course. Based on the length of their life cycle, flowering plants can be categorized as annuals, biennials, or perennials. **Annuals** complete their life cycle—from germination to flowering to seed production to death—in a single year or less. Many wildflowers are annuals, as are the most important food crops, including the cereal grains and legumes. **Biennials** generally live two years, often including an intervening cold period (winter) between vegetative growth (first spring/summer) and flowering (second spring/summer). Beets and carrots are biennials but are rarely left in the ground long enough to flower. **Perennials** live many years and include trees, shrubs, and some grasses. Some buffalo grass of the North American plains is believed to have been growing for 10,000 years from seeds that sprouted at the close of the last ice age. When a perennial dies, it is not usually from old age, but from an infection or some environmental trauma, such as fire or severe drought.

Plants are capable of indeterminate growth because they have perpetually embryonic tissues called **meristems.** There are two main types: apical meristems and lateral meristems. **Apical meristems,** located at the tips of roots and in the buds of shoots, provide additional cells that enable the plant to grow in length, a process known as **primary growth.** Primary growth allows roots to extend throughout the soil and shoots to increase exposure to light and CO_2. In **herbaceous** (nonwoody) plants, primary growth produces all, or almost all, of the plant body. Woody plants, however, grow in girth in the parts of stems and roots where primary growth has ceased. This growth in thickness, known as **secondary growth,** is caused by the activity of **lateral meristems** called the vascular cambium and cork cambium. These cylinders of dividing cells extend along the length of roots and stems **(Figure 35.10)**. The **vascular cambium** adds layers of vascular tissue called secondary xylem (wood) and secondary phloem. The **cork cambium** replaces the epidermis with periderm, which is thicker and tougher.

The cells within meristems divide relatively frequently, generating additional cells. Some products of this division

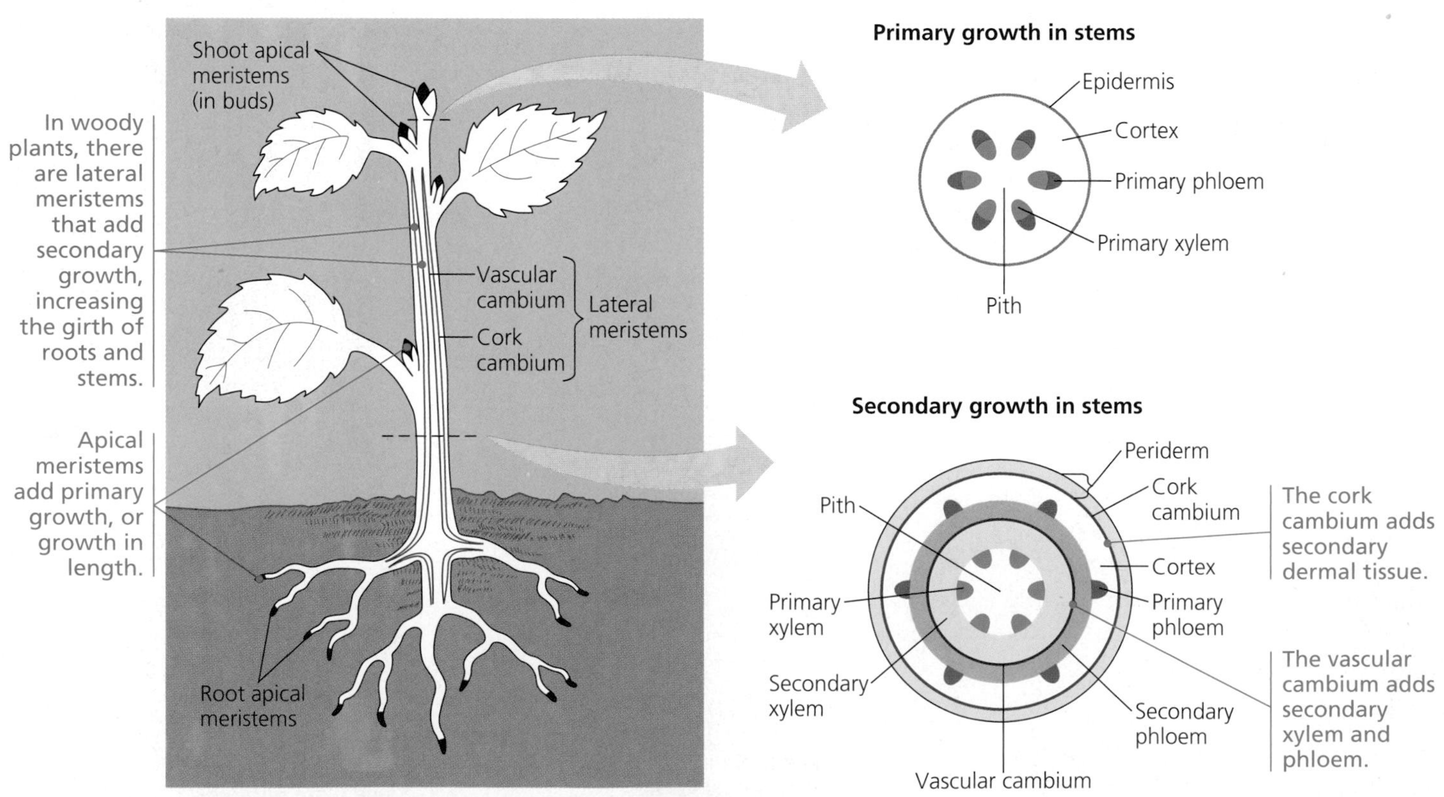

▲ **Figure 35.10 An overview of primary and secondary growth.**

remain in the meristem and produce more cells, while others differentiate and are incorporated into tissues and organs of the growing plant. Cells that remain as sources of new cells are called **initials.** The new cells displaced from the meristem, called **derivatives,** continue to divide until the cells they produce become specialized within developing tissues.

In woody plants, primary and secondary growth occur simultaneously but in different locations. Each growing season, primary growth near the apical meristems produces young extensions of roots and shoots, while lateral meristems produce secondary growth that thickens and strengthens older parts of the plant **(Figure 35.11)**. The oldest parts, such as a tree trunk base, have the most accumulation of tissues produced by secondary growth.

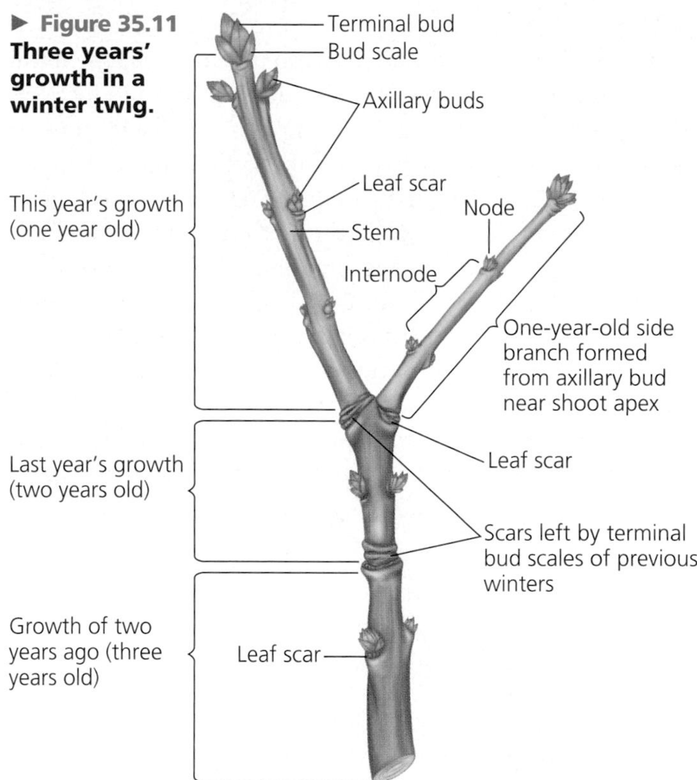

▶ **Figure 35.11 Three years' growth in a winter twig.**

Concept Check 35.2

1. Cells in lower layers of your skin divide and replace dead cells sloughed from the surface. Why is it inaccurate to compare such regions of cell division to a plant meristem?
2. Contrast the types of growth arising from apical and lateral meristems.

For suggested answers, see Appendix A.

Concept 35.3

Primary growth lengthens roots and shoots

Primary growth produces the **primary plant body,** the parts of the root and shoot systems produced by apical meristems. In herbaceous plants, the primary plant body is usually the entire plant. In woody plants, it consists only of the youngest parts, which have not yet become woody. Although apical meristems lengthen both roots and shoots, there are differences in the primary growth of these two systems.

Primary Growth of Roots

The root tip is covered by a thimble-like **root cap,** which protects the delicate apical meristem as the root pushes through the abrasive soil during primary growth. The root cap also secretes a polysaccharide slime that lubricates the soil around the root tip. Growth occurs just behind the root tip, in three zones of cells at successive stages of primary growth. Moving away from the root tip, they are the zones of cell division, elongation, and maturation **(Figure 35.12)**.

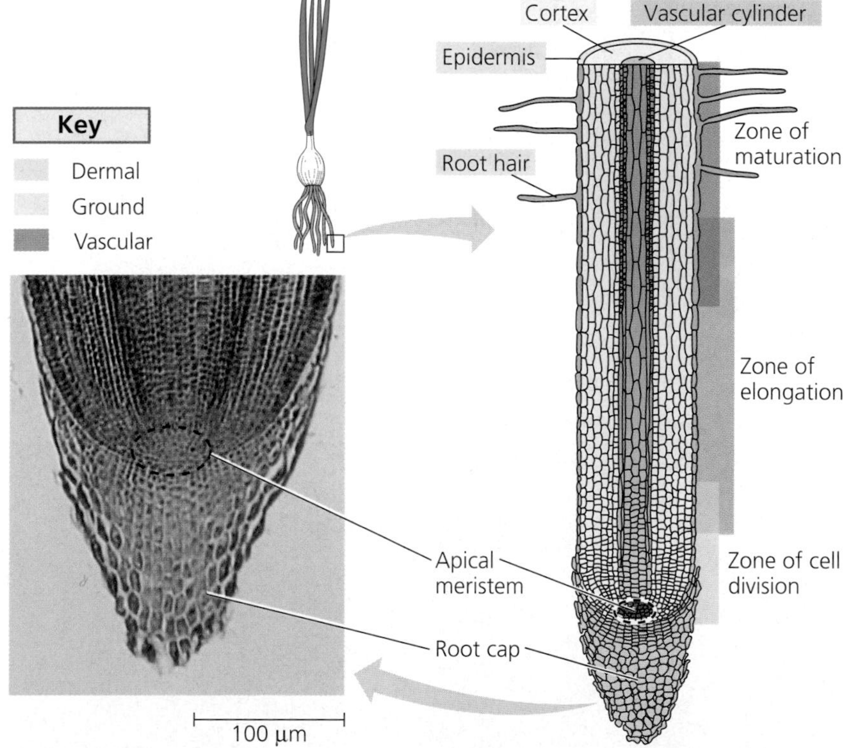

▲ **Figure 35.12 Primary growth of a root.** The diagram and light micrograph take us into the tip of an onion root. Mitosis is concentrated in the zone of cell division, where the apical meristem and its immediate products are located. The apical meristem also maintains the root cap by generating new cells that replace those that are sloughed off. Most lengthening of the root is concentrated in the zone of elongation. Cells become functionally mature in the zone of maturation. The zones grade into one another without sharp boundaries.

The three zones of cells grade together, with no sharp boundaries. The **zone of cell division** includes the root apical meristem and its derivatives. New root cells are produced in this region, including the cells of the root cap. In the **zone of elongation,** root cells elongate, sometimes to more than ten times their original length. Cell elongation is mainly responsible for pushing the root tip farther into the soil. Meanwhile, the root apical meristem keeps adding cells to the younger end of the zone of elongation. Even before the root cells finish lengthening, many begin specializing in structure and function. In the **zone of maturation,** cells complete their differentiation and become functionally mature.

The primary growth of roots produces the epidermis, ground tissue, and vascular tissue. The light micrographs in **Figure 35.13** show in transverse (cross) section the three primary tissue systems in the young roots of a eudicot (*Ranunculus,* buttercup) and a monocot (*Zea,* maize). Water and minerals absorbed from the soil must enter through the epidermis, a single layer of cells covering the root. Root hairs enhance this process by greatly increasing the surface area of epidermal cells.

In most roots, the stele is a vascular cylinder, a solid core of xylem and phloem (see Figure 35.13a). The xylem radiates from the center in two or more spokes, with phloem developing in the wedges between the spokes. In many monocot roots, the vascular tissue consists of a central core of parenchyma cells surrounded by alternating rings of xylem and phloem (see Figure 35.13b). The central region is often called pith but should not be confused with stem pith, which is ground tissue.

The ground tissue of roots, consisting mostly of parenchyma cells, fills the cortex, the region between the vascular cylinder

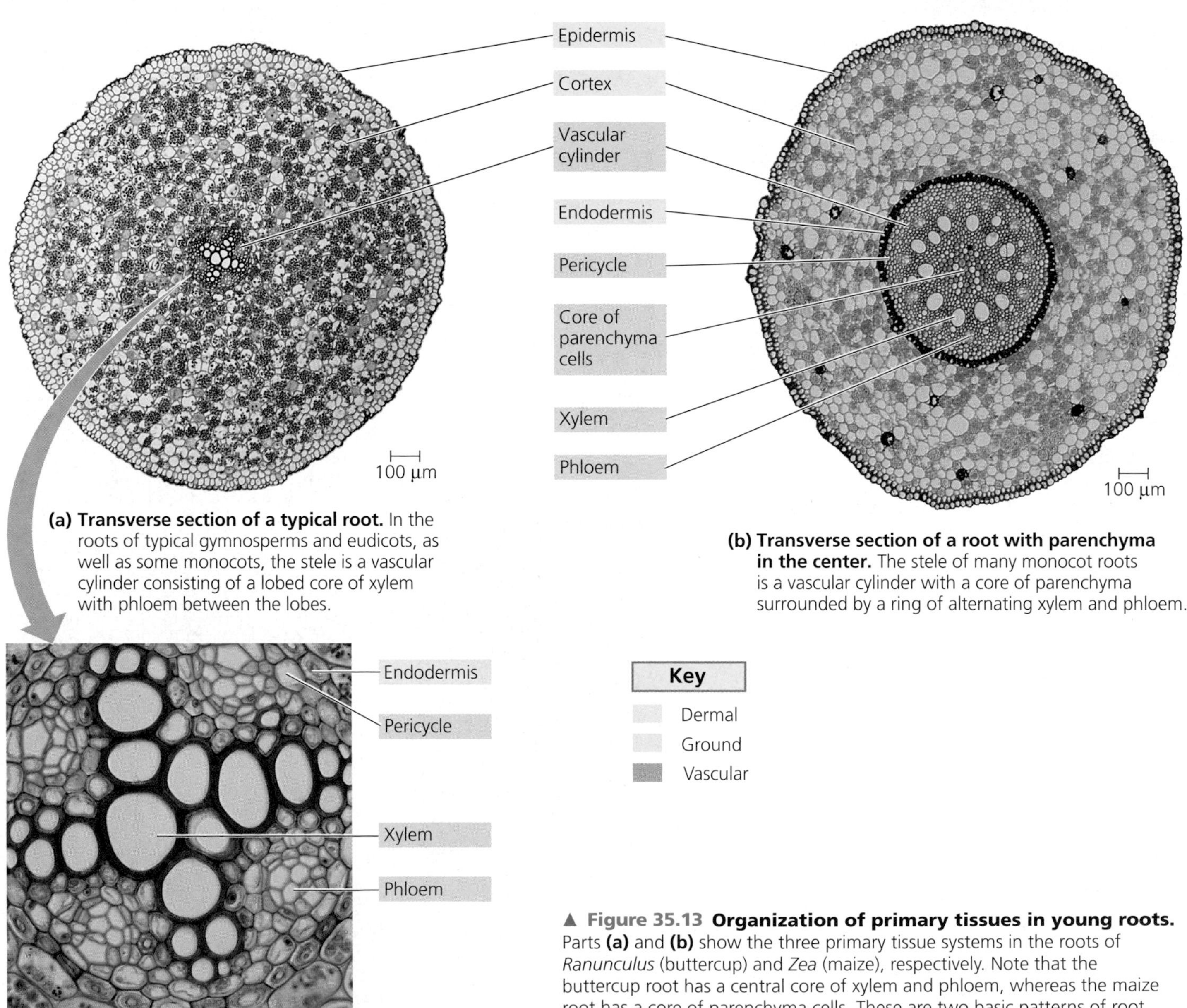

(a) Transverse section of a typical root. In the roots of typical gymnosperms and eudicots, as well as some monocots, the stele is a vascular cylinder consisting of a lobed core of xylem with phloem between the lobes.

(b) Transverse section of a root with parenchyma in the center. The stele of many monocot roots is a vascular cylinder with a core of parenchyma surrounded by a ring of alternating xylem and phloem.

▲ **Figure 35.13 Organization of primary tissues in young roots.** Parts **(a)** and **(b)** show the three primary tissue systems in the roots of *Ranunculus* (buttercup) and *Zea* (maize), respectively. Note that the buttercup root has a central core of xylem and phloem, whereas the maize root has a core of parenchyma cells. These are two basic patterns of root organization, of which there are many variations, depending on the plant species. (All LMs)

▶ **Figure 35.14 The formation of a lateral root.** A lateral root originates in the pericycle, the outermost layer of the vascular cylinder of a root, and grows out through the cortex and epidermis. In this series of micrographs, the view of the original root is a transverse section, while the view of the lateral root is a longitudinal section.

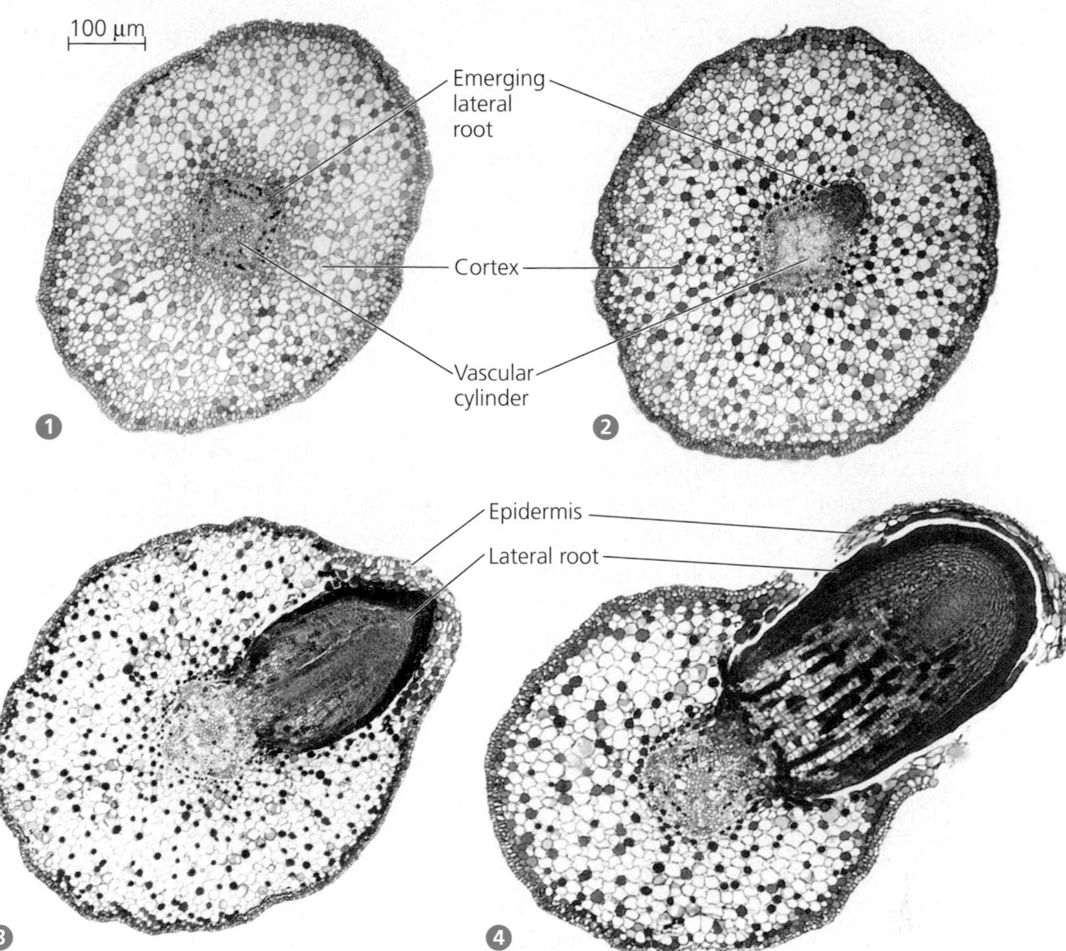

and epidermis. Cells within the ground tissue store organic nutrients, and their plasma membranes absorb minerals from the soil solution. The innermost layer of the cortex is called the **endodermis,** a cylinder one cell thick that forms the boundary with the vascular cylinder. You will learn in Chapter 36 how the endodermis is a selective barrier regulating passage of substances from the soil solution into the vascular cylinder.

Lateral roots arise from within the **pericycle,** the outermost cell layer in the vascular cylinder (see Figure 35.13). A lateral root elongates and pushes through the cortex and epidermis until it emerges from the established root **(Figure 35.14)**. It cannot originate near the root's surface because it must remain connected with the vascular cylinder of the established root as part of the continuous vascular tissue system.

Primary Growth of Shoots

A shoot apical meristem is a dome-shaped mass of dividing cells at the tip of the terminal bud **(Figure 35.15)**. Leaves arise as **leaf primordia** (singular, *primordium*), finger-like projections along the flanks of the apical meristem. Axillary buds develop from islands of meristematic cells left by the apical meristem at the bases of the leaf primordia. Axillary buds can form lateral shoots at some later time (see Figure 35.2).

Within a bud, leaf primordia are crowded close together because internodes are very short. Most of the actual elongation of the shoot occurs by the growth in length of slightly older internodes below the shoot apex. This growth is due to both cell division and cell elongation within the internode. Some plants, including grasses, elongate all along the shoot because there are meristematic regions called intercalary meristems at the base of each leaf. That is why grass continues to grow after being mowed.

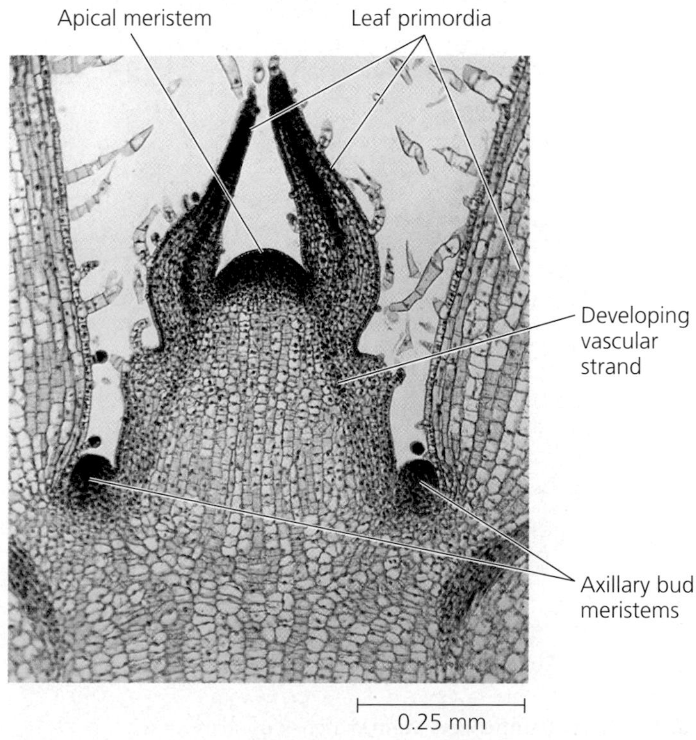

▲ **Figure 35.15 The terminal bud and primary growth of a shoot.** Leaf primordia arise from the flanks of the apical dome. This is a longitudinal section of the shoot tip of *Coleus* (LM).

Tissue Organization of Stems

The epidermis covers stems as part of the continuous dermal tissue system. Vascular tissue runs the length of a stem in vascular bundles. Unlike lateral roots, which arise from vascular tissue deep within a root (see Figure 35.14), lateral shoots arise from preexisting axillary buds on the surface of a stem (see Figure 35.15). The vascular bundles of the stem converge with the root's vascular cylinder in a zone of transition located near the soil surface.

In gymnosperms and most eudicots, the vascular tissue consists of vascular bundles arranged in a ring **(Figure 35.16a)**. The xylem in each vascular bundle faces the pith, and the phloem faces the cortex. In most monocot stems, the bundles are scattered throughout the ground tissue, rather than forming a ring **(Figure 35.16b)**. In both monocot and eudicot stems, ground tissue consists mostly of parenchyma, but collenchyma cells just beneath the epidermis strengthen many stems. Sclerenchyma cells, specifically fiber cells within vascular bundles, also provide support.

Tissue Organization of Leaves

Figure 35.17 provides an overview of leaf structure. The epidermal barrier is interrupted by the **stomata** (singular, *stoma*), which allow CO_2 exchange between the surrounding air and the photosynthetic cells inside the leaf. The term *stoma* can refer to the stomatal pore or to the entire stomatal complex consisting of a pore flanked by two **guard cells**, which regulate the opening and closing of the pore. In addition to regulating CO_2 exchange, stomata are major avenues for the evaporative loss of water, as you will see in Chapter 36.

The ground tissue is sandwiched between the upper and lower epidermis, a region called the **mesophyll** (from the Greek *mesos,* middle, and *phyll,* leaf). Mesophyll consists mainly of parenchyma cells specialized for photosynthesis. The leaves of many eudicots have two distinct areas: palisade mesophyll and spongy mesophyll. The **palisade mesophyll,** or palisade parenchyma, consists of one or more layers of elongated cells on the upper part of the leaf. The **spongy mesophyll,** also called spongy parenchyma, is below the palisade mesophyll. These cells are more loosely arranged, with a labyrinth of air spaces through which CO_2 and oxygen circulate around the cells and up to the palisade region. The air spaces are particularly large in the vicinity of stomata, where gas exchange with the outside air occurs.

The vascular tissue of each leaf is continuous with the vascular tissue of the stem. **Leaf traces,** connections from vascular bundles in the stem, pass through petioles and into leaves. Veins are the leaf's vascular bundles, which subdivide repeatedly and branch throughout the mesophyll. This network brings xylem and phloem into close contact with the photosynthetic tissue, which obtains water and minerals from the xylem and loads its sugars and other organic products into the phloem for shipment to other parts of the plant. The vascular structure also functions as a skeleton that reinforces the shape of the leaf. Each vein is enclosed by a protective **bundle sheath,** consisting of one or more layers of cells, usually parenchyma cells.

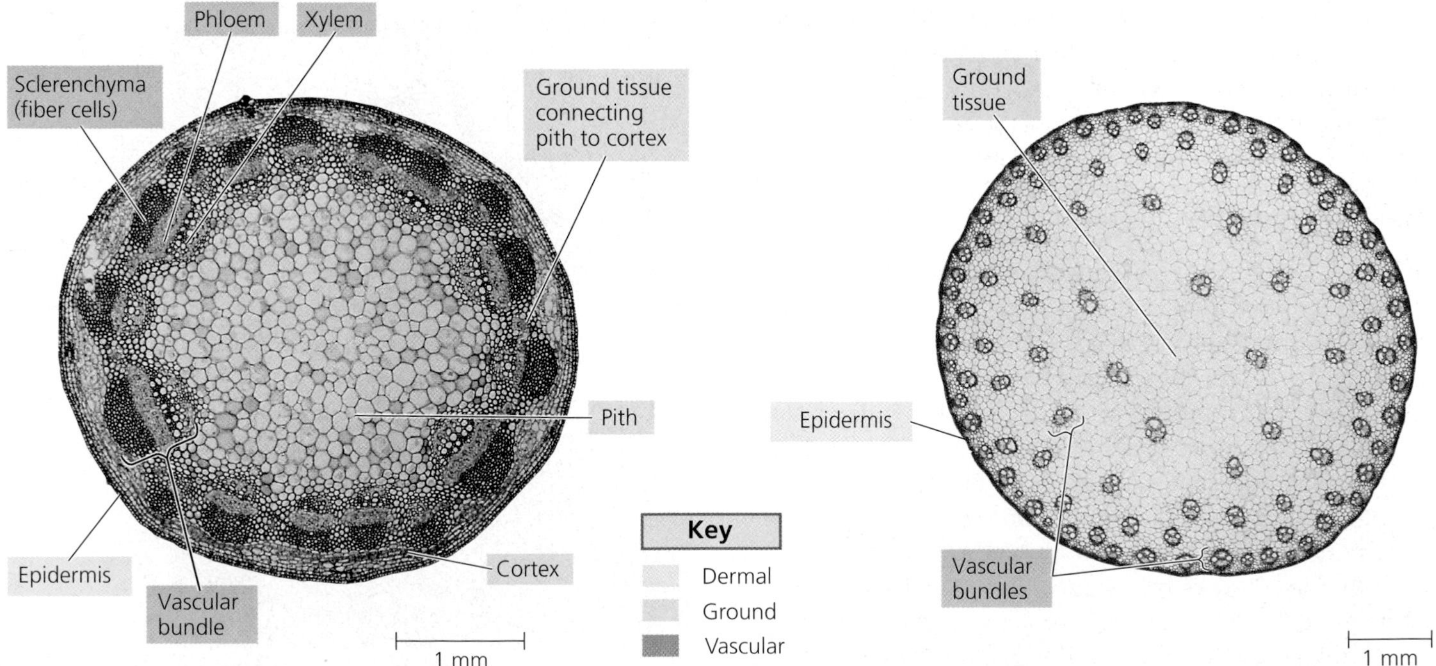

(a) A eudicot (sunflower) stem. Vascular bundles form a ring. Ground tissue toward the inside is called pith, and ground tissue toward the outside is called cortex. (LM of transverse section)

(b) A monocot (maize) stem. Vascular bundles are scattered throughout the ground tissue. In such an arrangement, ground tissue is not partitioned into pith and cortex. (LM of transverse section)

▲ **Figure 35.16 Organization of primary tissues in young stems.**

▼ **Figure 35.17 Leaf anatomy.**

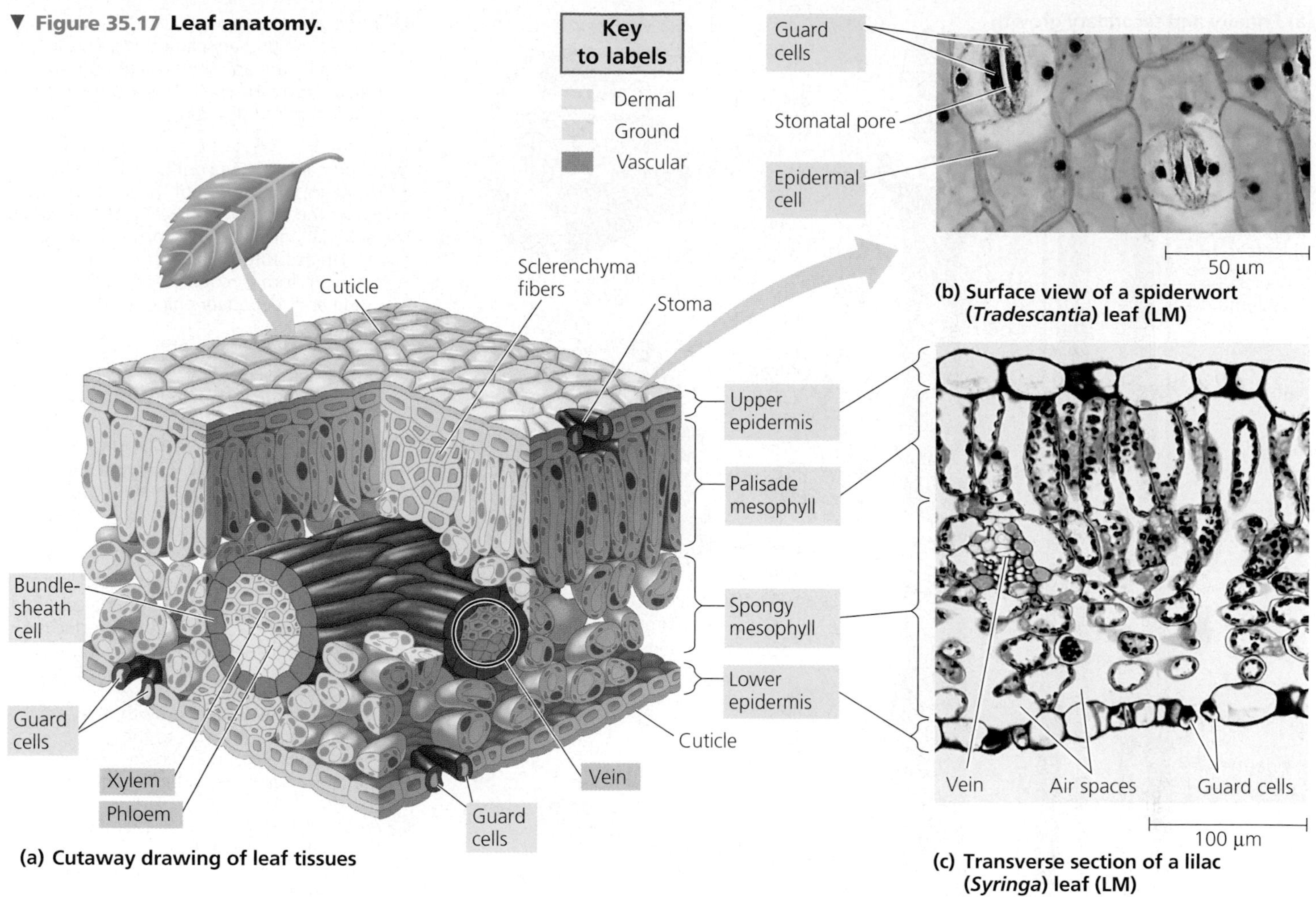

(a) Cutaway drawing of leaf tissues

(b) Surface view of a spiderwort (*Tradescantia*) leaf (LM)

(c) Transverse section of a lilac (*Syringa*) leaf (LM)

Concept Check 35.3

1. Describe how roots and shoots differ in their branching.
2. Contrast primary growth in roots and shoots.
3. Describe the functions of leaf veins.

For suggested answers, see Appendix A.

Concept 35.4

Secondary growth adds girth to stems and roots in woody plants

Secondary growth, the growth in thickness produced by lateral meristems, occurs in stems and roots of woody plants, but rarely in leaves. The **secondary plant body** consists of the tissues produced by the vascular cambium and cork cambium. The vascular cambium adds secondary xylem (wood) and secondary phloem. Cork cambium produces a tough, thick covering consisting mainly of cork cells.

Primary and secondary growth occur simultaneously but in different regions. While an apical meristem elongates a stem or root, secondary growth commences where primary growth has stopped, occurring in older regions of all gymnosperm species and many eudicots, but rarely in monocots. The process is similar in stems and roots, which look much the same after extensive secondary growth. **Figure 35.18**, on the next page, provides an overview of growth in a woody stem.

The Vascular Cambium and Secondary Vascular Tissue

The vascular cambium is a cylinder of meristematic cells one cell thick. It increases in circumference and also lays down successive layers of secondary xylem to its interior and secondary phloem to its exterior, each layer with a larger diameter than the previous layer (see Figure 35.18). In this way, it is primarily responsible for the thickening of a root or stem.

The vascular cambium develops from undifferentiated cells and parenchyma cells that regain the capacity to divide. In a typical gymnosperm or woody eudicot stem, the vascular cambium forms in a layer between the primary xylem and primary phloem of each vascular bundle and in the ground tissue between the bundles. The meristematic bands within and between the

(a) Primary and secondary growth in a two-year-old stem

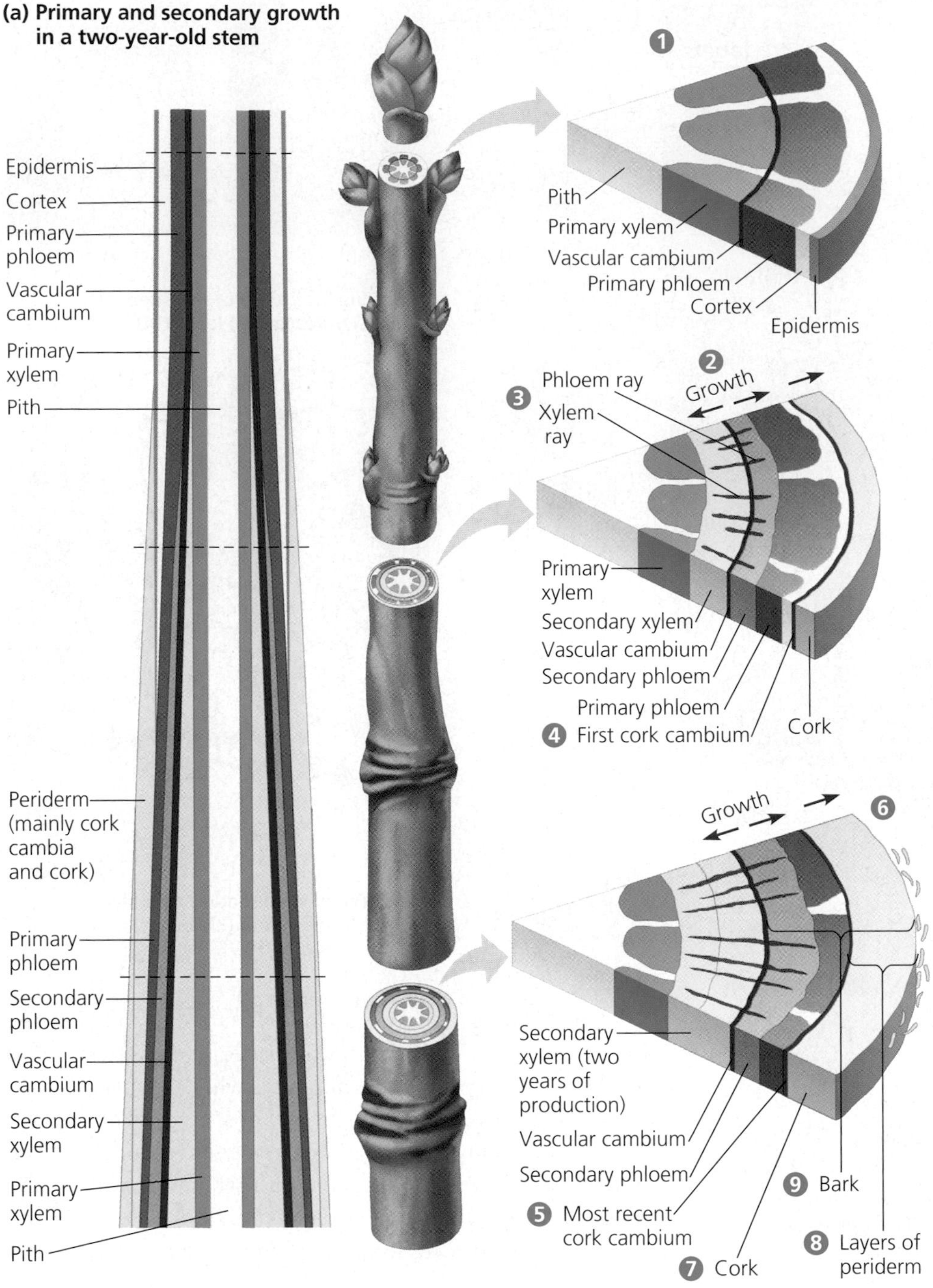

Epidermis
Cortex
Primary phloem
Vascular cambium
Primary xylem
Pith

Pith
Primary xylem
Vascular cambium
Primary phloem
Cortex
Epidermis

Phloem ray
Xylem ray
Growth

Primary xylem
Secondary xylem
Vascular cambium
Secondary phloem
Primary phloem
First cork cambium
Cork

Periderm (mainly cork cambia and cork)
Primary phloem
Secondary phloem
Vascular cambium
Secondary xylem
Primary xylem
Pith

Growth
Secondary xylem (two years of production)
Vascular cambium
Secondary phloem
Most recent cork cambium
Cork
Bark
Layers of periderm

1. In the youngest part of the stem, you can see the primary plant body, as formed by the apical meristem during primary growth. The vascular cambium is beginning to develop.
2. As primary growth continues to elongate the stem, the portion of the stem formed earlier the same year has already started its secondary growth. This portion increases in girth as fusiform initials of the vascular cambium form secondary xylem to the inside and secondary phloem to the outside.
3. The ray initials of the vascular cambium give rise to the xylem and phloem rays.
4. As the diameter of the vascular cambium increases, the secondary phloem and other tissues external to the cambium cannot keep pace with the expansion because the cells no longer divide. As a result, these tissues, including the epidermis, rupture. A second lateral meristem, the cork cambium, develops from parenchyma cells in the cortex. The cork cambium produces cork cells, which replace the epidermis.
5. In year 2 of secondary growth, the vascular cambium adds to the secondary xylem and phloem, and the cork cambium produces cork.
6. As the diameter of the stem continues to increase, the outermost tissues exterior to the cork cambium rupture and slough off from the stem.
7. Cork cambium re-forms in progressively deeper layers of the cortex. When none of the original cortex is left, the cork cambium develops from parenchyma cells in the secondary phloem.
8. Each cork cambium and the tissues it produces form a layer of periderm.
9. Bark consists of all tissues exterior to the vascular cambium.

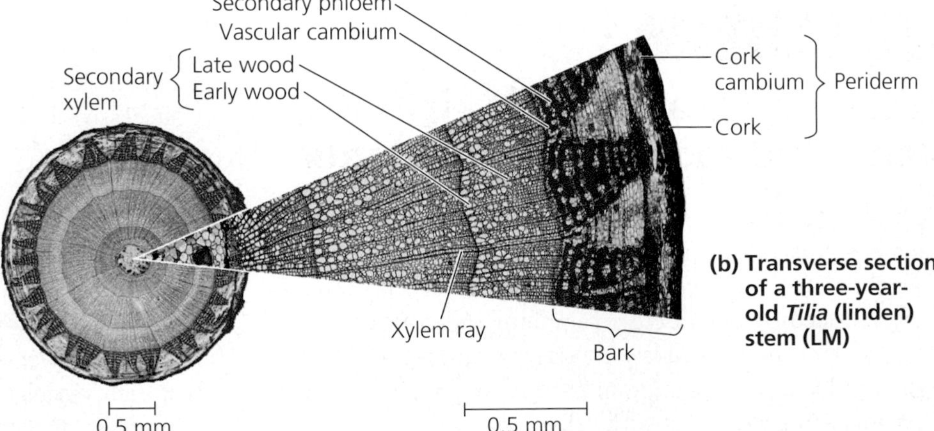

(b) Transverse section of a three-year-old *Tilia* (linden) stem (LM)

▲ **Figure 35.18 Primary and secondary growth of a stem.** You can track the progress of secondary growth by examining the sections through sequentially older parts of the stem. (You would observe the same changes if you could follow the youngest region, near the apex, for the next three years.)

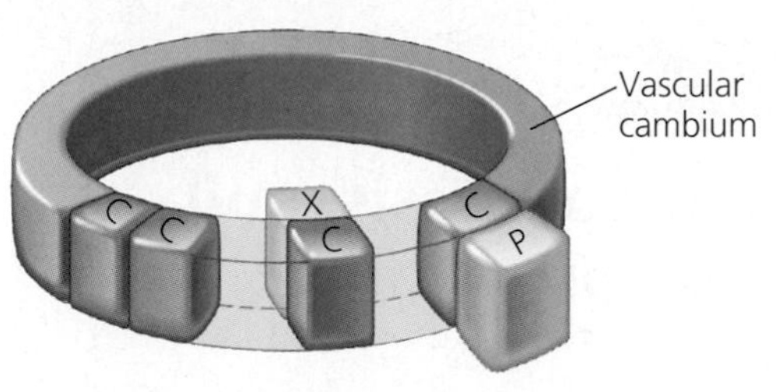

(a) Types of cell division. An initial can divide transversely to form two cambial initials (C) or radially to form an initial and either a xylem (X) or phloem (P) cell.

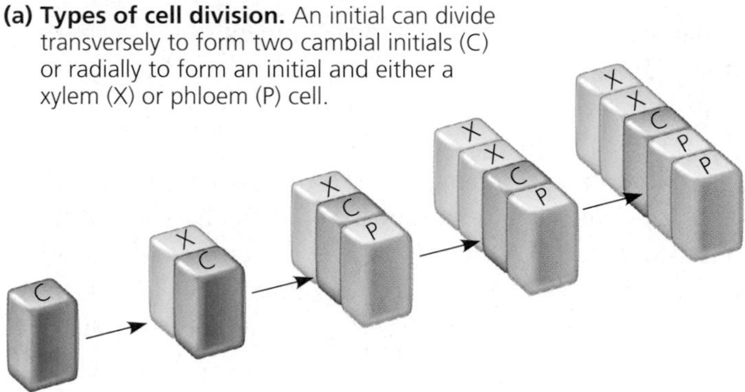

(b) Accumulation of secondary growth. Although shown here as alternately adding xylem and phloem, a cambial initial usually produces much more xylem.

▲ **Figure 35.19 Cell division in the vascular cambium.**

vascular bundles unite to become a continuous cylinder of dividing cells. In a typical gymnosperm or woody eudicot root, the vascular cambium forms in segments between the primary phloem, the lobes of primary xylem, and the pericycle, eventually becoming a cylinder.

Viewed in transverse section, the vascular cambium appears as a ring, with interspersed regions of cells called fusiform initials and ray initials. When these initials divide, they increase the circumference of the cambium itself and add secondary xylem to the inside of the cambium and secondary phloem to the outside **(Figure 35.19)**. **Fusiform initials** produce elongated cells such as the tracheids, vessel elements, and fibers of the xylem, as well as the sieve-tube members, companion cells, parenchyma, and fibers of the phloem. They have tapered (fusiform) ends and are oriented parallel to the axis of a stem or root. **Ray initials,** which are shorter and oriented perpendicular to the stem or root axis, produce vascular rays—radial files consisting mainly of parenchyma cells. Vascular rays are living avenues that move water and nutrients between the secondary xylem and secondary phloem. They also store starch and other organic nutrients. The portion of a vascular ray located in the secondary xylem is known as a xylem ray. The portion located in the secondary phloem is called a phloem ray.

As secondary growth continues over the years, layers of secondary xylem (wood) accumulate, consisting mainly of tracheids, vessel elements, and fibers (see Figure 35.9). Gymnosperms have tracheids, whereas angiosperms have both tracheids and vessel elements. Dead at functional maturity, both types of cells have thick, lignified walls that give wood its hardness and strength. Tracheids and vessel elements that develop early in the growing season, typically in early spring, are known as early wood and usually have relatively large diameters and thin cell walls (see Figure 35.18b). This structure maximizes delivery of water to new, expanding leaves. Tracheids and vessel elements produced later in the growing season, during late summer or early fall, are known as late wood. They are thick-walled cells that do not transport as much water but do add more support than do the thinner-walled cells of early wood.

In temperate regions, secondary growth in perennial plants is interrupted each year when the vascular cambium becomes dormant during winter. When growth resumes the next spring, the boundary between the large cells of the new early wood and the smaller cells of the late wood produced during the previous growing season is usually a distinct ring in the transverse sections of most tree trunks and roots. Therefore, a tree's age can be estimated by counting its annual rings. The rings can have varying thicknesses, reflecting the amount of seasonal growth.

As a tree or woody shrub ages, the older layers of secondary xylem no longer transport water and minerals (xylem sap). These layers are called **heartwood** because they are closer to the center of a stem or root **(Figure 35.20)**. The outer layers still transport xylem sap and are therefore known as **sapwood.** That is why a large tree can still survive even if the center of its trunk is hollow. Because each new layer of secondary xylem has a larger circumference, secondary growth enables the xylem to transport more sap each year, supplying an increasing number of leaves. Heartwood is generally darker than sapwood because of resins and other compounds that clog the cell cavities and help protect the core of the tree from fungi and wood-boring insects.

Only the youngest secondary phloem, closest to the vascular cambium, functions in sugar transport. As a stem or root increases in circumference, the older secondary phloem is sloughed off, which is why secondary phloem does not accumulate as extensively as does secondary xylem.

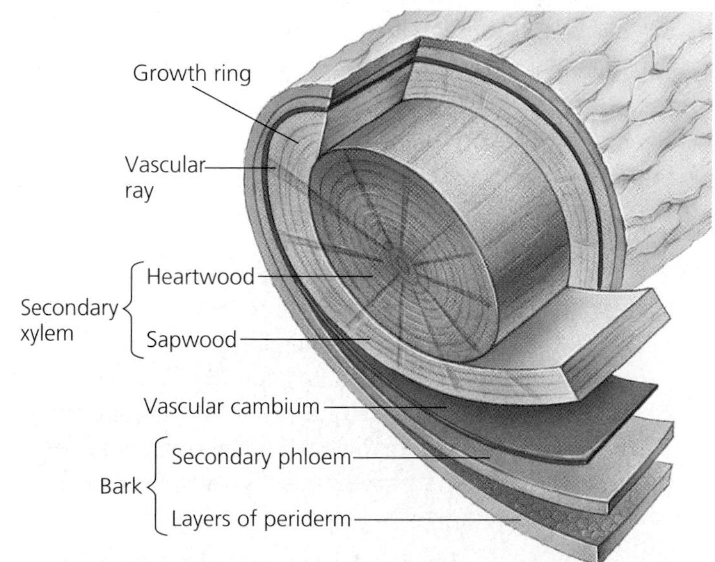

▲ **Figure 35.20 Anatomy of a tree trunk.**

Cork Cambia and the Production of Periderm

During the early stages of secondary growth, the epidermis is pushed outward, causing it to split, dry, and fall off the stem or root. It is replaced by two tissues produced by the first cork cambium, which arises in the outer cortex in stems (see Figure 35.18a) and in the outer layer of the pericycle in roots. One tissue, called *phelloderm,* is a thin layer of parenchyma cells that forms to the interior of the cork cambium. The other tissue consists of cork cells that accumulate to the exterior of the cork cambium. As cork cells mature, they deposit a waxy material called *suberin* in their walls and then die. The cork tissue then functions as a barrier that helps protect the stem or root from water loss, physical damage, and pathogens. A cork cambium and the tissues it produces comprise a layer of periderm.

Because cork cells have suberin and are usually compacted together, most of the periderm is impermeable to water and gases, unlike the epidermis. Therefore, in most plants, absorption of water and minerals takes place primarily in the youngest parts of roots, mainly the root hairs. The older parts anchor the plant and transport water and solutes between roots and shoots. Dotting the periderm are small, raised areas called **lenticels,** in which there is more space between the cork cells, enabling living cells within a woody stem or root to exchange gases with the outside air.

Unlike the vascular cambium, cells of the cork cambium do not continue to divide; thus, there is no increase in its circumference. Instead, the thickening of a stem or root splits the first cork cambium, which loses its meristematic activity and differentiates into cork cells. A new cork cambium forms to the inside, resulting in another layer of periderm. As this process continues, older layers of periderm are sloughed off, as is evident in the cracked, peeling bark of many tree trunks.

Many people think that bark is only the protective outer covering of a woody stem or root. Actually, **bark** includes all tissues external to the vascular cambium. In an outward direction, its main components are the secondary phloem (produced by the vascular cambium), the most recent periderm, and all the older layers of periderm (see Figure 35.20).

Concept Check 35.4

1. A sign is hammered into a tree 2 m from the tree's base. If the tree is 10 m tall and elongates 1 m each year, how high will the sign be after 10 years?
2. A tree can survive even if a tunnel is cut through its center. However, removing a complete ring of bark around the trunk (a process called girdling) will kill the tree. Explain why.

For suggested answers, see Appendix A.

Growth, morphogenesis, and differentiation produce the plant body

So far, we have described the development of the plant body from meristems. Here we will move from a description of plant growth and development to the mechanisms that underlie these processes.

Consider a typical annual weed. It may consist of billions of cells—some large, some small, some highly specialized and others not, but all derived from a single fertilized egg. The increase in mass, or growth, that occurs during the life of the plant results from both cell division and cell expansion, but what controls these processes? Why do leaves stop growing upon reaching a certain size, whereas apical meristems divide perpetually? Also, note that the billions of cells in our hypothetical weed are not an undifferentiated clump of cells. They are organized into recognizable tissues and organs. Leaves arise from nodes; roots (unless adventitious) do not. Epidermis forms on the exterior of the leaf, and vascular tissue in the interior—never vice versa.

The development of body form and organization is called **morphogenesis.** Each cell in the plant body contains the same set of genes, exact copies of the genome present in the fertilized egg. Different patterns of gene expression among cells cause the cellular differentiation that creates a diversity of cell types (see Chapter 21). The three developmental processes of growth, morphogenesis, and cellular differentiation act in concert to transform the fertilized egg into a plant.

Molecular Biology: Revolutionizing the Study of Plants

Modern molecular techniques are helping plant biologists explore how growth, morphogenesis, and cellular differentiation give rise to a plant. In the current renaissance in plant biology, new laboratory methods coupled with clever choices of experimental organisms have catalyzed a research explosion. One research focus is *Arabidopsis thaliana,* a little weed of the mustard family that is small enough to allow researchers to cultivate thousands of plants in a few square meters of lab space **(Figure 35.21)**. Its short generation span, about six weeks from germination to flowering, makes *Arabidopsis* an excellent model for genetic studies. Plant biologists also favor the tiny amount of DNA per cell, among the smallest genomes of all known plants. As a result of these attributes, *Arabidopsis* was the first plant to have its entire genome sequenced—a six-year, multinational effort.

Arabidopsis possesses about 26,000 genes, but many of these are duplicates. There are probably fewer than 15,000 different types of genes, a level of complexity similar to that found in the fly *Drosophila*. Knowing what some of the *Arabidopsis* genes do has already expanded our understanding of plant development (see Figure 35.21). To fill the gaps in our knowledge, plant biologists have launched an ambitious quest to determine the function of every one of the plant's genes by the year 2010. Toward this end, they are attempting to create mutants for every gene in the plant's genome. We will discuss some of these mutants shortly, as we take a closer look at the molecular mechanisms that underlie growth, morphogenesis, and cellular differentiation. By identifying each gene's function and tracking every chemical pathway, researchers aim to establish a blueprint for how plants are built, a major goal of **systems biology**, as discussed by Natasha Raikhel in the interview on pages 710–711. It may one day be possible to create a computer-generated "virtual plant" that enables researchers to visualize which plant genes are activated in different parts of the plant during the entire course of development.

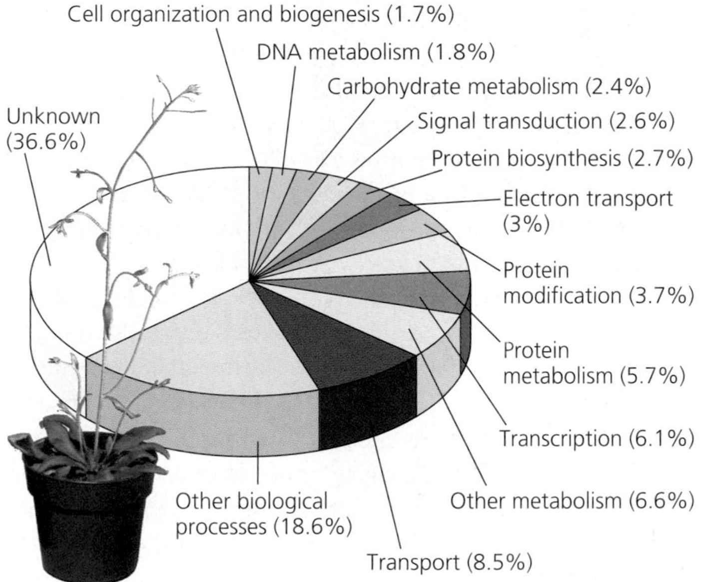

▲ **Figure 35.21** ***Arabidopsis thaliana.*** Owing to its small size, rapid life cycle, and small genome, *Arabidopsis* was the first plant to have its entire genome sequenced (about 26,000 genes). The pie chart represents the proportion of *Arabidopsis* genes in different functional categories. (Data from TAIR, The *Arabidopsis* Information Resource, 2004)

Growth: Cell Division and Cell Expansion

By increasing cell number, cell division in meristems increases the potential for growth. However, it is cell expansion that accounts for the actual increase in plant mass. The process of plant cell division is described more fully in Chapter 12 (see Figure 12.10), and Chapter 39 discusses the process of cell elongation (see Figure 39.8). Here we are more concerned with how these processes contribute to plant form.

The Plane and Symmetry of Cell Division

The plane (direction) and symmetry of cell division are immensely important in determining plant form. Imagine a single cell poised to undergo mitosis. If the planes of division of its descendants are parallel to the plane of the first cell division, a single file of cells will be produced **(Figure 35.22a)**. At the other extreme, if the planes vary randomly, a

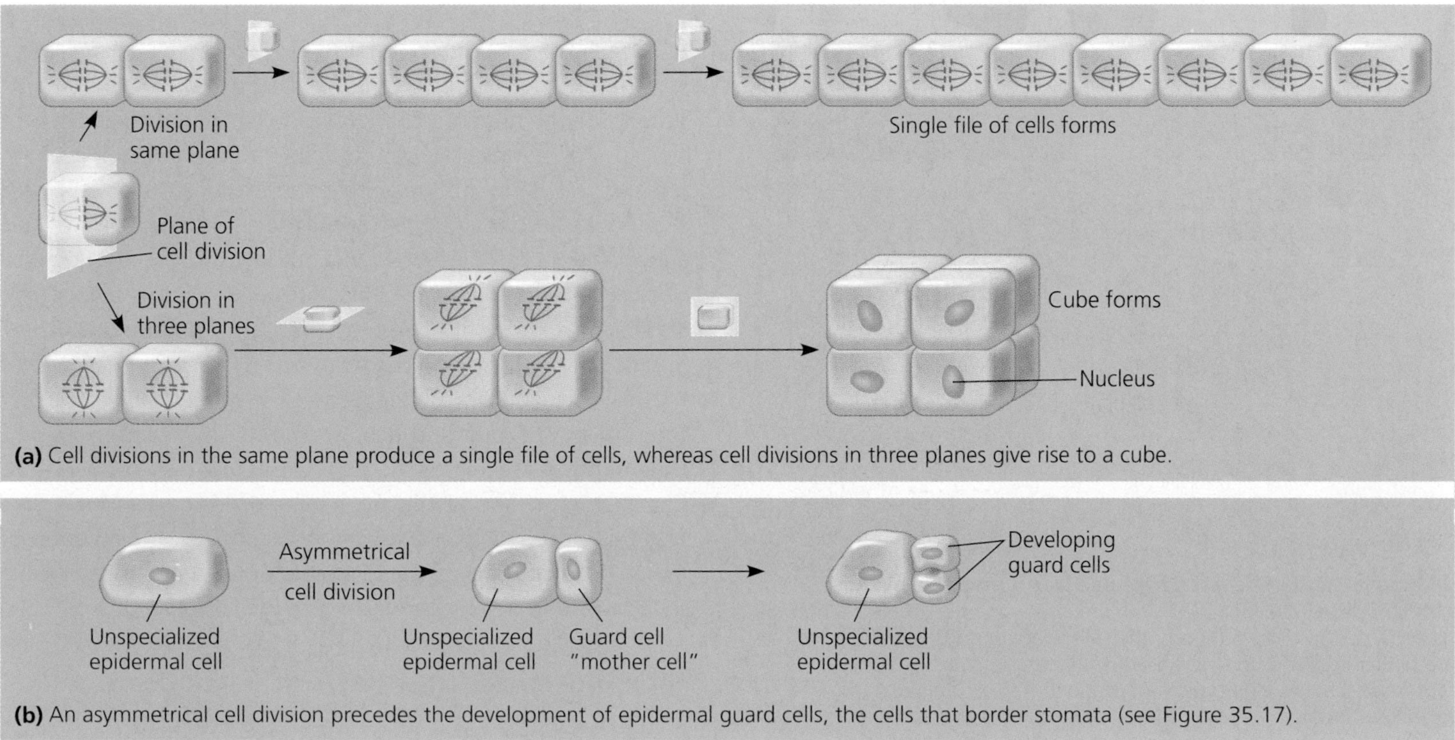

(a) Cell divisions in the same plane produce a single file of cells, whereas cell divisions in three planes give rise to a cube.

(b) An asymmetrical cell division precedes the development of epidermal guard cells, the cells that border stomata (see Figure 35.17).

▲ **Figure 35.22 The plane and symmetry of cell division influence development of form.**

disorganized clump of cells results. Meanwhile, even though chromosomes are allocated to daughter cells equally during mitosis, the cytoplasm may divide asymmetrically. **Asymmetrical cell division,** in which one daughter cell receives more cytoplasm than the other during mitosis, is fairly common in plant cells and usually signals a key event in development. For example, the formation of guard cells typically involves both an asymmetrical cell division and a change in the plane of cell division. An epidermal cell divides asymmetrically to form a large cell that remains an unspecialized epidermal cell and a small cell that becomes the guard cell "mother cell." Guard cells form when this small mother cell divides in a plane perpendicular to the first cell division **(Figure 35.22b)**.

The plane in which a cell divides is determined during late interphase. The first sign of this spatial orientation is a rearrangement of the cytoskeleton. Microtubules in the cytoplasm become concentrated into a ring called the **preprophase band (Figure 35.23)**. The band disappears before metaphase, but it predicts the future plane of cell division. The "imprint" consists of an ordered array of actin microfilaments that remain after the microtubules disperse.

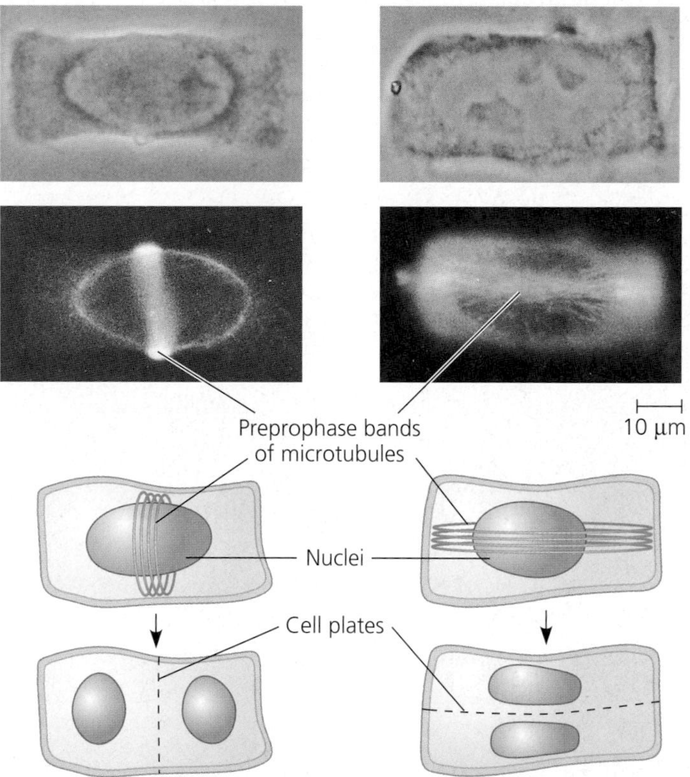

▲ **Figure 35.23 The preprophase band and the plane of cell division.** The location of the preprophase band predicts the plane of cell division. Although the cells shown on the left and right are similar in shape, they will divide in different planes. Each cell is represented by two light micrographs, one (top) unstained and the other (bottom) stained with a fluorescent dye that binds specifically to microtubules. The stained microtubules form a "halo" (preprophase band) around the nucleus in the outer cytoplasm.

Orientation of Cell Expansion

Before discussing how cell expansion contributes to plant form, it is useful to consider the difference in cell expansion between plants and animals. Animal cells grow mainly by synthesizing protein-rich cytoplasm, a metabolically expensive process. Growing plant cells also produce additional protein-rich material in their cytoplasm, but water uptake typically accounts for about 90% of expansion. Most of this water is packaged in the large central vacuole formed by the coalescence of many smaller vacuoles as a cell grows. A plant can grow rapidly and economically because water uptake enables a small amount of cytoplasm to go a long way. Bamboo shoots, for instance, can elongate more than 2 m per week. Rapid extension of shoots and roots increases the exposure to light and soil, an important evolutionary adaptation to the immobile lifestyle of plants.

Plant cells rarely expand equally in all directions. Their greatest expansion is usually oriented along the plant's main axis. For example, cells near the root tip may elongate up to 20 times their original length, with relatively little increase in width. The orientation of cellulose microfibrils in the innermost layers of the cell wall causes this differential growth. The microfibrils cannot stretch much, so the cell expands mainly perpendicular to the "grain" of the microfibrils, as shown in **Figure 35.24**.

As with the plane of cell division, microtubules play a key role in regulating the plane of cell expansion. It is the orientation of microtubules in the cell's outermost cytoplasm that determines the orientation of cellulose microfibrils that are deposited in the cell wall.

Microtubules and Plant Growth

Studies of *fass* mutants of *Arabidopsis* have confirmed the importance of cytoplasmic microtubules in cell division and expansion. The *fass* mutants have unusually squat cells with seemingly random planes of cell division. Their roots and stems have no ordered cell files and layers. Despite these abnormalities, *fass* mutants develop into tiny adult plants, but their organs are compressed longitudinally **(Figure 35.25)**.

The stubby form and disorganized tissue arrangement can be traced to abnormal organization of microtubules. During interphase, the microtubules are randomly positioned, and the preprophase bands do not form prior to mitosis (see Figure 35.23). As a result, there is no orderly "grain" of cellulose microfibrils in the cell wall to determine the direction of elongation (see Figure 35.24). This defect gives rise to cells that expand in all directions and divide haphazardly.

Morphogenesis and Pattern Formation

A plant's body is more than a collection of dividing and expanding cells. Morphogenesis must occur for development to

▶ **Figure 35.24 The orientation of plant cell expansion.** Growing plant cells expand mainly through water uptake. In a growing cell, enzymes weaken cross-links in the cell wall, allowing it to expand as water diffuses into the vacuole by osmosis. As shown in these two examples of different orientations, the orientation of cell growth is mainly in the plane perpendicular to the orientation of cellulose microfibrils in the wall. The microfibrils are embedded in a matrix of other (non-cellulose) polysaccharides, some of which form the cross-links visible in the micrograph (TEM). Loosening of the wall occurs when hydrogen ions secreted by the cell activate cell wall enzymes that break the cross-links between polymers in the wall. This reduces restraint on the turgid cell, which can take up more water and expand. Small vacuoles, which accumulate most of this water, coalesce and form the cell's central vacuole.

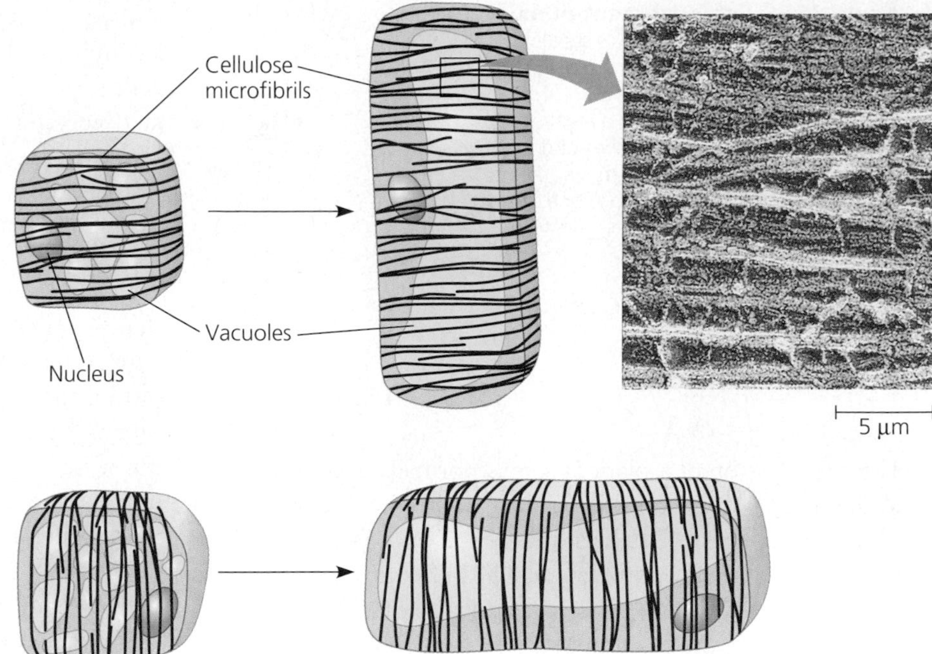

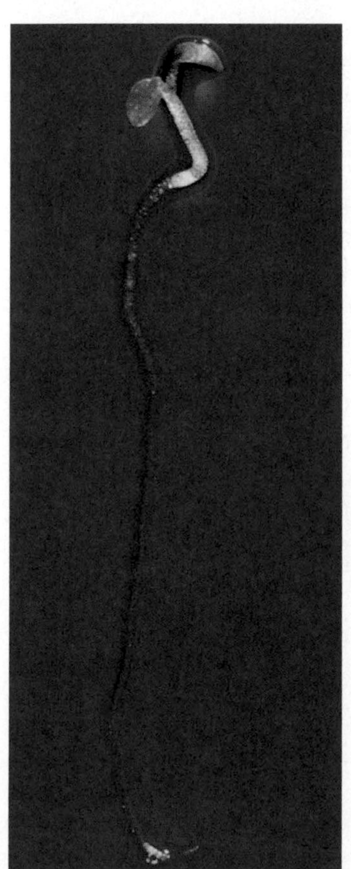

(a) Wild-type seedling

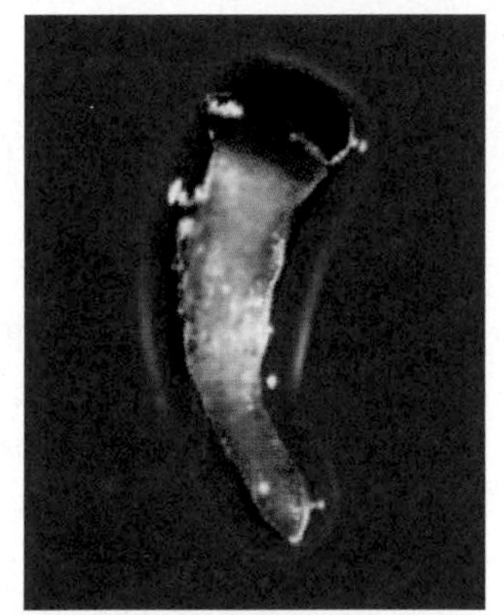

(b) *fass* seedling

(c) Mature *fass* mutant

▲ **Figure 35.25 The *fass* mutant of *Arabidopsis* confirms the importance of cytoplasmic microtubules to plant growth.** The squat body of the *fass* mutant results from cell division and cell elongation being randomly oriented instead of orienting in the direction of the normal plant axis.

proceed properly; that is, cells must be organized into multicellular arrangements such as tissues and organs. The development of specific structures in specific locations is called **pattern formation.**

Many developmental biologists postulate that pattern formation is determined by **positional information** in the form of signals that continuously indicate to each cell its location within a developing structure. According to this idea, each cell within a developing organ responds to positional information from neighboring cells by differentiating into a particular cell type. Developmental biologists are accumulating evidence that gradients of specific molecules, generally proteins or mRNAs, provide positional information. For example, a substance diffusing from a shoot apical meristem may "inform" the cells below of their distance from the shoot apex. Cells possibly gauge their radial positions within the developing organ by detecting a second chemical signal that emanates from the outermost cells. The gradients of these two substances would be sufficient for each cell to "get a fix" on its position relative to the longitudinal and radial axes of the developing organ. This idea of diffusible chemical signals is one of the hypotheses that developmental biologists are testing.

One type of positional information is associated with **polarity,** the condition of having structural differences at opposite ends of an organism. Plants typically have an axis, with a root end and a shoot end. Such polarity is most obvious in morphological differences, but it is also manifest in physiological properties, including the unidirectional movement of the hormone auxin (see Chapter 39) and emergence of adventitious roots and shoots from "cuttings." Adventitious roots form within the root end of a stem cutting, and adventitious shoots arise from the shoot end of a root cutting.

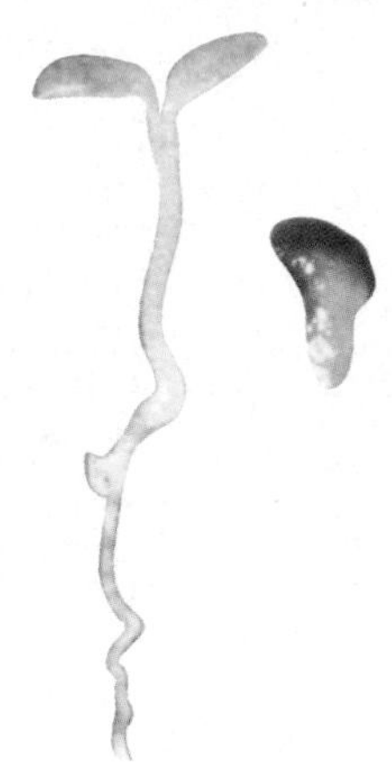

▶ **Figure 35.26 Establishment of axial polarity.** The normal *Arabidopsis* seedling (left) has a shoot end and a root end. In the *gnom* mutant (right), the first division of the zygote was not asymmetrical; as a result, the plant is ball-shaped and lacks leaves and roots. The defect in *gnom* mutants has been traced to an inability to transport the hormone auxin in a polar manner.

The first division of a plant zygote is normally asymmetrical, initiating polarization of the plant body into shoot and root. Once this polarity has been induced, it becomes exceedingly difficult to reverse experimentally. Therefore, the proper establishment of axial polarity is a critical step in a plant's morphogenesis. In the *gnom* mutant of *Arabidopsis*, the establishment of polarity is defective. The first cell division of the zygote is abnormal in being symmetrical, and the resulting ball-shaped plant has neither roots nor leaves **(Figure 35.26)**.

Morphogenesis in plants, as in other multicellular organisms, is often under the control of master regulatory genes called homeotic genes that mediate many of the major events in an individual's development, such as the initiation of an organ (see Chapter 21). For example, the protein product of the *KNOTTED-1* homeotic gene, found in many plant species, is important in the development of leaf morphology, including the production of compound leaves. If the *KNOTTED-1* gene is overexpressed in tomato plants, the normally compound leaves become "super-compound" **(Figure 35.27)**.

Gene Expression and Control of Cellular Differentiation

What makes cellular differentiation so fascinating is that the cells of a developing organism synthesize different proteins and diverge in structure and function even though they share a common genome. The cloning of whole plants from somatic cells supports the conclusion that the genome of a differentiated cell remains intact (see Figure 21.5). If a mature cell removed from a root or leaf can dedifferentiate in tissue culture and give rise to the diverse cell types of a plant, then it must possess all the genes necessary to make any kind of plant cell. This means that cellular differentiation depends, to a large extent, on the control of gene expression—the regulation of transcription and translation leading to specific proteins. Cells selectively express certain genes at specific times during their differentiation. A guard cell has the genes that program the self-destruction of a vessel element protoplast, but it does not express those genes. A xylem vessel element does express those genes, but only does so at a specific time in its differentiation, after the cell has elongated and has produced its secondary wall. Researchers are beginning to unravel the molecular mechanisms that switch specific genes on and off at critical times during a cell's development (see Chapters 19 and 21).

▲ **Figure 35.27 Overexpression of a homeotic gene in leaf formation.** *KNOTTED-1* is a homeotic gene involved in leaf and leaflet formation. Its overexpression in tomato plants results in leaves that are "super-compound" (right) compared with normal leaves (left).

Cellular differentiation to a large extent depends on positional information—where a particular cell is located relative to other cells. For example, two distinct cell types are formed in the root epidermis of *Arabidopsis*: root hair cells and hairless epidermal cells. Cell fate is associated with the position of the epidermal cells. The immature epidermal cells that are in contact with two underlying cells of the root cortex differentiate into root hair cells, whereas the immature epidermal cells in contact with only one cortical cell differentiate into mature hairless cells. Differential expression of a homeotic gene called *GLABRA-2* (from the Latin *glaber,* bald) is required for appropriate root hair distribution. Researchers have demonstrated this by coupling the *GLABRA-2* gene to a "reporter gene" that causes every cell expressing *GLABRA-2* in the root to turn pale blue following a certain treatment **(Figure 35.28)**. The *GLABRA-2* gene is normally expressed only in epidermal cells that will not develop root hairs. If *GLABRA-2* is rendered dysfunctional by mutation, *every* root epidermal cell develops a root hair.

Location and a Cell's Developmental Fate

In the process of shaping a rudimentary organ, patterns of cell division and cell expansion affect the differentiation of cells by placing them in specific locations relative to other cells. Thus, positional information underlies all the processes of development: growth, morphogenesis, and differentiation. One approach to studying the relationships among these processes is clonal analysis, in which the cell lineages (clones) derived from each cell in an apical meristem are mapped as organs develop. Researchers can do this by using radiation or chemicals in order to induce somatic mutations that will alter the chromosome number or otherwise tag a cell in some way that distinguishes it from the neighboring cells in the shoot apex. The lineage of

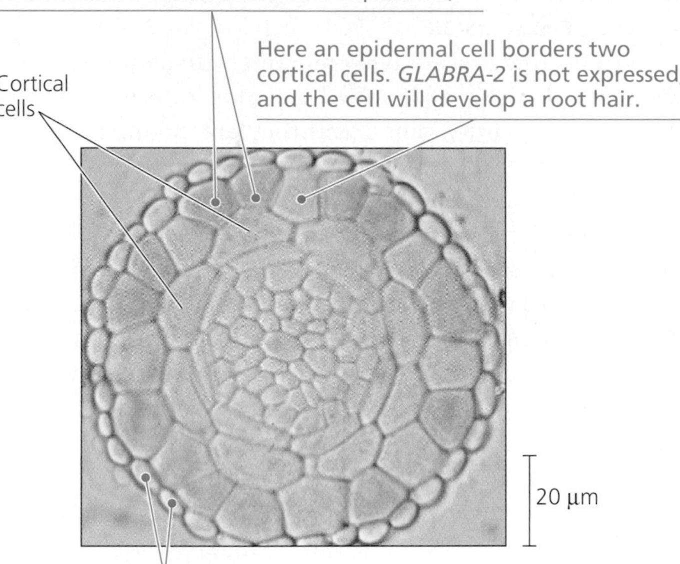

When epidermal cells border a single cortical cell, the homeotic gene *GLABRA-2* is selectively expressed, and these cells will remain hairless. (The blue color in this light micrograph indicates cells in which *GLABRA-2* is expressed.)
Cortical cells
Here an epidermal cell borders two cortical cells. *GLABRA-2* is not expressed, and the cell will develop a root hair.
20 µm
The ring of cells external to the epidermal layer is composed of root cap cells that will be sloughed off as the root hairs start to differentiate.

▲ **Figure 35.28 Control of root hair differentiation by a homeotic gene.** (LM)

cells derived by mitosis and cell division from the mutant meristematic cell will also be "marked." For example, a single cell in the shoot apical meristem may undergo a mutation that prevents chlorophyll from being produced. This cell and all of its descendants will be "albino," and they will appear as a linear file of colorless cells running down the long axis of the otherwise green shoot.

How early is a cell's developmental fate determined by its position in an embryonic structure? To some extent, the developmental fates of cells in the shoot apex are predictable. For example, clonal mapping has shown that almost all the cells derived from the outermost meristematic cells become part of the dermal tissue. However, we cannot pinpoint which meristematic cells will give rise to specific tissues and organs. Apparently random changes in rates and planes of cell division can reorganize the meristem. For example, the outermost cells *usually* divide perpendicular to the surface of the shoot apex, adding cells to the surface layer. Occasionally, however, one of the outermost cells divides parallel to the surface of the shoot apex, placing one daughter cell beneath the surface, among cells derived from different lineages. Such exceptions indicate that meristematic cells are not dedicated early to forming specific tissues and organs. Instead, the cell's *final* position in an emerging organ determines what kind of cell it will become, possibly as a result of positional information.

Shifts in Development: Phase Changes

Multicellular organisms generally pass through developmental phases. In humans, these are infancy, childhood, adolescence, and adulthood, with puberty as the dividing line between the nonreproductive and reproductive phases. Plants also pass through phases, developing from a juvenile phase to an adult vegetative phase to an adult reproductive phase. In animals, these developmental changes take place throughout the entire organism, such as when a larva develops into an adult animal. In plants, in marked contrast, they occur within a single region, the shoot apical meristem. The morphological changes that arise from these transitions in shoot apical meristem activity are called **phase changes.** During the transition from a juvenile phase to an adult phase, the most obvious morphological changes typically occur in leaf size and shape **(Figure 35.29)**. Once the apical meristem has laid down juvenile

▲ **Figure 35.29 Phase change in the shoot system of *Acacia koa*.** This native of Hawaii has compound juvenile leaves, consisting of many small leaflets, and mature, sickle-shaped "leaves" (technically, highly modified petioles). This dual foliage reflects a phase change in the development of the apical meristem of each shoot. In the juvenile vegetative phase of an apical meristem, compound leaves develop at each node. In the adult vegetative phase of an apical meristem, sickle-shaped leaves are produced. Once a node forms, the developmental phase—juvenile or adult—is fixed; that is, compound leaves do not mature into sickle-shaped leaves.

nodes and internodes, they retain that status even after the shoot continues to elongate and the shoot apical meristem has changed to the adult phase. Therefore, any *new* leaves that develop on branches that emerge from axillary buds at juvenile nodes will also be juvenile, even though the apical meristem may have been laying down mature nodes for years.

Phase changes are examples of plasticity in plant development. The transition from juvenile to adult leaves is only one type of phase change. The transition in the fanwort (see Figure 35.1) from feathery underwater leaves to fan-shaped floating leaves is another example. Next, we will examine a common but nevertheless remarkable phase change—the transition of a vegetative shoot apical meristem into a floral meristem.

Genetic Control of Flowering

Flower formation involves a phase change from vegetative growth to reproductive growth. This transition is triggered by a combination of environmental cues, such as day length, and internal signals, such as hormones. (You will learn more about the control of flowering in Chapter 39.) Unlike vegetative growth, which is indeterminate, reproductive growth is determinate: The production of a flower by a shoot apical meristem stops the primary growth of that shoot. The transition from vegetative growth to flowering is associated with the switching-on of floral **meristem identity genes**. The protein products of these genes are transcription factors that regulate the genes required for the conversion of the indeterminate vegetative meristems into determinate floral meristems.

When a shoot apical meristem is induced to flower, each leaf primordium that forms will develop into a specific type of floral organ based on its relative position—a stamen, carpel, sepal, or petal (see Figure 30.7 to review basic flower structure). Viewed from above, the floral organs develop in four concentric circles, or whorls: Sepals form the fourth (outermost) whorl; petals form the third; stamens form the second; and carpels form the first (innermost) whorl. Plant biologists have identified several **organ identity genes** that regulate the development of this characteristic floral pattern. Organ identity genes, also called plant homeotic genes, code for transcription factors. Positional information determines which organ identity genes are expressed in a particular floral organ primordium. The result is development of an emerging leaf into a specific floral organ, such as a petal or a stamen. Just as a mutation in a fruit fly homeotic gene can cause legs to grow in place of antennae, a mutation in a plant organ identity gene can cause abnormal floral development, such as petals growing in place of stamens, as shown in **Figure 35.30**.

By collecting and studying mutants with abnormal flowers, researchers have identified and cloned three classes of floral organ identity genes, and they are beginning to determine how these genes act. The **ABC model** of flower formation, diagrammed in **Figure 35.31a**, identifies how these genes direct the formation of the four types of floral organs. The ABC model proposes that each class of organ identity genes is switched "on" in two specific whorls of the floral meristem. Normally, *A* genes are switched on in the two outer whorls (sepals and petals); *B* genes are switched on in the two middle whorls (petals and stamens); and *C* genes are switched on in the two inner whorls (stamens and carpels). Sepals arise from those parts of the floral meristems in which only *A* genes are active; petals arise where *A* and *B* genes are active; stamens where *B* and *C* genes are active; and carpels where only *C* genes are active. The ABC model can account for the phenotypes of mutants lacking *A*, *B*, or *C* gene activity, with one addition: Where gene *A* activity is present, it inhibits *C*, and vice versa. If either *A* or *C* is missing, the other takes its place. **Figure 35.31b** shows the floral patterns of mutants lacking each of

▶ **Figure 35.30 Organ identity genes and pattern formation in flower development.**

(a) Normal *Arabidopsis* flower. *Arabidopsis* normally has four whorls of flower parts: sepals (Se), petals (Pe), stamens (St), and carpels (Ca).

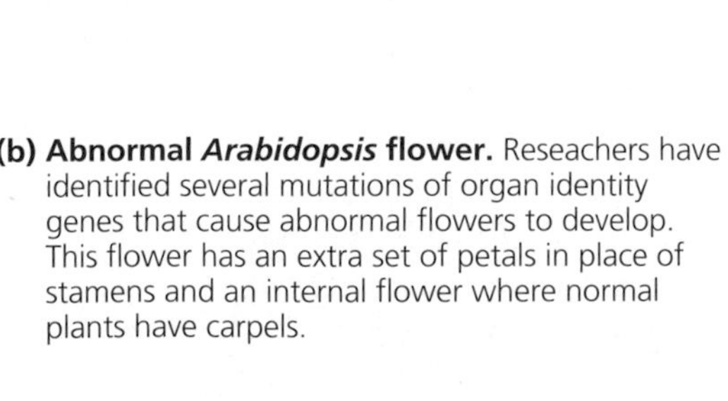

(b) Abnormal *Arabidopsis* flower. Reseachers have identified several mutations of organ identity genes that cause abnormal flowers to develop. This flower has an extra set of petals in place of stamens and an internal flower where normal plants have carpels.

▼ **Figure 35.31 The ABC hypothesis for the functioning of organ identity genes in flower development.**

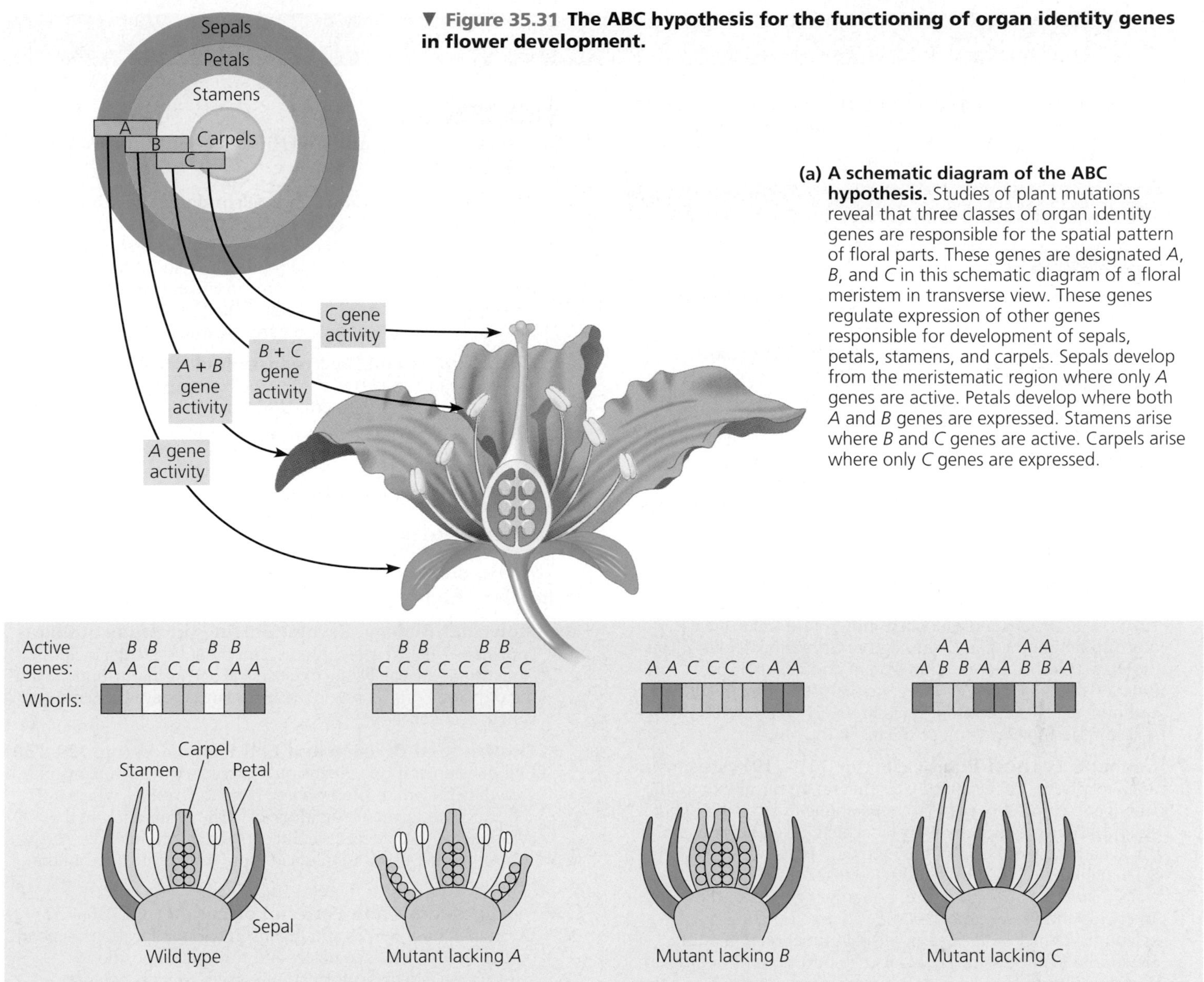

(a) A schematic diagram of the ABC hypothesis. Studies of plant mutations reveal that three classes of organ identity genes are responsible for the spatial pattern of floral parts. These genes are designated *A*, *B*, and *C* in this schematic diagram of a floral meristem in transverse view. These genes regulate expression of other genes responsible for development of sepals, petals, stamens, and carpels. Sepals develop from the meristematic region where only *A* genes are active. Petals develop where both *A* and *B* genes are expressed. Stamens arise where *B* and *C* genes are active. Carpels arise where only *C* genes are expressed.

(b) Side view of organ identity mutant flowers. Combining the model shown in part (a) with the rule that if *A* gene or *C* gene activity is missing, the other activity spreads through all four whorls, we can explain the phenotypes of mutants lacking a functional *A*, *B*, or *C* organ identity gene.

the three classes of organ identity genes and depicts how the model accounts for the floral phenotypes. By constructing such hypotheses and designing experiments to test them, researchers are tracing the genetic basis of plant development.

In dissecting the plant to examine its parts, as we have done in this chapter, we must remember that the whole plant functions as an integrated organism. In the following chapters, you will learn more about how materials are transported within vascular plants (Chapter 36), how plants obtain nutrients (Chapter 37), how plants reproduce (Chapter 38, focusing on flowering plants), and how the various functions of the plant are coordinated (Chapter 39). Remembering that structure fits function and that plant anatomy and physiology reflect evolutionary adaptations to the problems of living on land will enhance your understanding of plants.

Concept Check 35.5

1. How can two cells in a plant have such vastly different structure even though they possess the same genome?
2. Explain how a mutation such as *fass* in *Arabidopsis* results in a stubby plant rather than a normal elongated one.

For suggested answers, see Appendix A.

Chapter 35 Review

Go to www.campbellbiology.com or the student CD-ROM to explore Activities, Investigations, and other interactive study aids.

SUMMARY OF KEY CONCEPTS

Concept 35.1

The plant body has a hierarchy of organs, tissues, and cells

▶ **The Three Basic Plant Organs: Roots, Stems, and Leaves (pp. 713–716)** Roots anchor the plant, absorb and conduct water and minerals, and store food. The shoot system consists of stems, leaves, and (in angiosperms) flowers. Leaves are attached by their petioles to the nodes of the stem, with internodes of the stem separating the nodes. Axillary buds, located in the axils of petioles and stems, have the potential to extend as vegetative or floral shoots. The two main groups of angiosperms, monocots and eudicots, differ in anatomical details.
Activity *Root, Stem, and Leaf Sections*

▶ **The Three Tissue Systems: Dermal, Vascular, and Ground (p. 717)** Dermal tissue (epidermis and periderm), vascular tissue (xylem and phloem), and ground tissue are continuous throughout the plant, although in the various plant organs the three tissues differ in arrangement and in some specialized functions. Vascular tissues integrate the parts of the plant. Water and minerals move up from roots in the xylem. Sugar is exported from leaves or storage organs in the phloem.

▶ **Common Types of Plant Cells (pp. 717–719)** Parenchyma cells, relatively unspecialized cells that retain the ability to divide, perform most of the plant's metabolic functions of synthesis and storage. Collenchyma cells, which have unevenly thickened walls, support young, growing parts of the plant. Sclerenchyma cells—fibers and sclereids—have thick, lignified walls that help support mature, nongrowing parts of the plant. Tracheids and vessel elements, the water-conducting cells of xylem, have thick walls and are dead at functional maturity. Sieve-tube members are the sugar-transporting cells of phloem in angiosperms. Though alive at functional maturity, sieve-tube members depend on the services of neighboring companion cells.

Concept 35.2

Meristems generate cells for new organs

▶ Apical meristems elongate shoots and roots through primary growth. Lateral meristems add girth to woody plants through secondary growth (pp. 720–721).

Concept 35.3

Primary growth lengthens roots and shoots

▶ **Primary Growth of Roots (pp. 721–723)** Apical meristems produce cells that continue to divide as meristematic cells. In roots, the apical meristem is located near the tip, where it regenerates the root cap.

▶ **Primary Growth of Shoots (pp. 723–725)** The apical meristem of a shoot is located in the terminal bud, where it gives rise to a repetition of internodes and leaf-bearing nodes.
Investigation *What Are Functions of Monocot Tissues?*

Concept 35.4

Secondary growth adds girth to stems and roots in woody plants

▶ **The Vascular Cambium and Secondary Vascular Tissue (pp. 725–727)** The vascular cambium develops from parenchyma cells into a meristematic cylinder that produces secondary xylem and secondary phloem. Older layers of secondary xylem (heartwood) become inactive, while younger layers (sapwood) still conduct water. Only the youngest secondary phloem is active in conducting organic nutrients.

▶ **Cork Cambia and the Production of Periderm (p. 728)** The cork cambium gives rise to the secondary plant body's protective covering, or periderm, which consists of the cork cambium plus the layers of cork cells it produces. Bark consists of all the tissues external to the vascular cambium: secondary phloem and periderm.
Activity *Primary and Secondary Growth*

Concept 35.5

Growth, morphogenesis, and differentiation produce the plant body

▶ **Molecular Biology: Revolutionizing the Study of Plants (pp. 728–729)** New techniques and model systems, including *Arabidopsis*, are catalyzing explosive progress in our understanding of plants. *Arabidopsis* is the first plant to have had its entire genome sequenced.

▶ **Growth: Cell Division and Cell Expansion (pp. 729–730)** Cell division and cell expansion are primary determinants of growth and form. A preprophase band determines where a cell plate will form in a dividing cell. The orientation of the cytoskeleton also affects the direction of cell elongation by controlling the orientation of cellulose microfibrils within the cell wall.

▶ **Morphogenesis and Pattern Formation (pp. 730–732)** Development of tissues and organs in specific locations depends on the ability of cells to detect and respond to positional information, such as information associated with polarity. Homeotic genes often control morphogenesis.

▶ **Gene Expression and Control of Cellular Differentiation (p. 732)** The challenge of understanding cellular differentiation is explaining how cells with matching genomes diverge into various cell types.

▶ **Location and a Cell's Developmental Fate (pp. 732–733)** A cell's position in a developing organ determines its pathway of differentiation.

▶ **Shifts in Development: Phase Changes (pp. 733–734)** Internal or environmental cues may cause a plant to switch from one developmental phase to another—for example, from development of juvenile leaves to development of mature leaves. Such morphological changes are called phase changes.

▶ **Genetic Control of Flowering (pp. 734–735)** Research on organ identity genes in developing flowers provides an important model system for the study of pattern formation. The ABC model of flower formation identifies how three classes of organ identity genes control formation of sepals, petals, stamens, and carpels.

36 Transport in Vascular Plants

▲ **Figure 36.1 Coast redwoods (*Sequoia sempervirens*).**

Key Concepts

36.1 Physical forces drive the transport of materials in plants over a range of distances
36.2 Roots absorb water and minerals from the soil
36.3 Water and minerals ascend from roots to shoots through the xylem
36.4 Stomata help regulate the rate of transpiration
36.5 Organic nutrients are translocated through the phloem

Overview

Pathways for Survival

The algal ancestors of plants absorbed water, minerals, and CO_2 directly from the water in which they were immersed; none of their cells were far from these substances. Bryophytes also lack an extensive transport system and are confined to living in very moist environments. For vascular plants, in contrast, the evolutionary journey onto land involved the differentiation of the plant body into roots and shoots. Roots absorb water and minerals from the soil, and shoots absorb light and atmospheric CO_2 for photosynthesis.

Xylem transports water and minerals from roots to shoots. Phloem transports sugars from where they are produced or stored to where they are needed for growth and metabolism. Such transport, which is necessary for a plant to function as a whole, may occur over long distances. For example, the highest leaves of some coast redwoods are more than 100 m (over 300 feet) from the roots (**Figure 36.1**). What enables a vascular plant to conduct water, minerals, and organic nutrients over such long distances? The mechanisms responsible for internal transport are the subject of this chapter.

Concept 36.1

Physical forces drive the transport of materials in plants over a range of distances

Transport in vascular plants occurs on three scales: (1) transport of water and solutes by individual cells, such as root hairs; (2) short-distance transport of substances from cell to cell at the levels of tissues and organs, such as the loading of sugar from photosynthetic leaf cells into the sieve tubes of the phloem; and (3) long-distance transport within xylem and phloem at the level of the whole plant. A variety of physical processes are involved in these different types of transport. **Figure 36.2** provides an overview of long-distance transport in a vascular plant.

Selective Permeability of Membranes: *A Review*

We covered the transport of solutes and water across biological membranes in detail in Chapter 7. Here we reexamine a few of these transport processes in the specific context of plant cells. The selective permeability of a plant cell's plasma membrane controls the movement of solutes into and out of that cell. Recall from Chapter 7 that solutes tend to diffuse down their gradients and that diffusion across a membrane is called **passive transport** because it happens without the cell directly using metabolic energy. **Active transport** is the pumping of solutes across membranes against their electrochemical gradients, the combined effects of the concentration gradient of the solute and the voltage (charge difference) across the membrane. It is called "active" because the cell must expend en-

TESTING YOUR KNOWLEDGE

Self-Quiz

1. Which structure is *incorrectly* paired with its tissue system?
   a. root hair—dermal tissue
   b. palisade parenchyma—ground tissue
   c. guard cell—dermal tissue
   d. companion cell—ground tissue
   e. tracheid—vascular tissue

2. A vessel element would likely lose its protoplast in which area of growth in a root?
   a. zone of cell division
   b. zone of elongation
   c. zone of maturation
   d. root cap
   e. apical meristem

3. Wood consists of
   a. bark.
   b. periderm.
   c. secondary xylem.
   d. secondary phloem
   e. cork

4. Which of the following is *not* part of an older tree's bark?
   a. cork
   b. cork cambium
   c. lenticels
   d. secondary xylem
   e. secondary phloem

5. The phase change of an apical meristem from the juvenile to the mature vegetative phase is often signaled by
   a. a change in the morphology of the leaves that are produced.
   b. the initiation of secondary growth.
   c. the formation of lateral roots.
   d. a change in the orientation of preprophase bands and cytoplasmic microtubules in lateral meristems.
   e. the activation of floral meristem identity genes.

6. Which of the following arise from meristematic activity?
   a. secondary xylem
   b. leaves
   c. trichomes
   d. tubers
   e. all of the above

7. "Pinching off" the tops of snapdragons causes the plants to make many more flowers than they would if left alone. Why does removal of the snapdragon's top cause more flowers to form?
   a. Removal of an apical meristem causes a phase transition from vegetative to floral development.
   b. Removal of an apical meristem causes cell division to become disorganized, much like in the *fass* mutant of *Arabidopsis*.
   c. Removal of an apical meristem allows more nutrients to be delivered to floral meristems.
   d. Removal of an apical meristem causes outgrowth of lateral buds that produce extra branches, which ultimately produce flowers.
   e. Removal of an apical meristem allows the periderm to produce new lateral branches.

8. __________ is to xylem as __________ is to phloem.
   a. Sclerenchyma cell; parenchyma cell
   b. Apical meristem; vascular cambium
   c. Vessel element; sieve-tube member
   d. Cortex; pith
   e. Vascular cambium; cork cambium

9. The type of mature cell that a particular embryonic plant cell will become appears to be determined mainly by
   a. the selective loss of genes.
   b. the cell's final position in a developing organ.
   c. the cell's pattern of migration.
   d. the cell's age.
   e. the cell's particular meristematic lineage.

10. Based on the hypothesis presented in Figure 35.31, predict floral morphology of a mutant lacking activity of *B* genes.
    a. carpel-petal-petal-carpel
    b. petal-petal-petal-petal
    c. sepal-sepal-carpel-carpel
    d. sepal-carpel-carpel-sepal
    e. carpel-carpel-carpel-carpel

For Self-Quiz answers, see Appendix A.

Go to the website or CD-ROM for more quiz questions.

Evolution Connection

The evolution of plant structure can be explored by looking at alternative growth strategies of related plants that occur in different environments. In this respect, Darwin was one of the earliest observers to note that many herbaceous plant species on continental mainlands have woody, tree-like relatives on remote oceanic islands. In the Hawaiian Islands, for example, one can find tree lobelias and tall, woody violets, groups that occur as small herbs in North America. Suggest an evolutionary hypothesis for this trend: Why is it so common for woody, tree-like forms to evolve from herbaceous ancestors that colonize isolated islands?

Scientific Inquiry

Write a paragraph explaining why certain mutants are so useful to researchers who investigate the regulation of plant development. Your paragraph should include at least one specific example.

Investigation ***What Are Functions of Monocot Tissues?***

Science, Technology, and Society

Make a list of the plants and plant products you use in a typical day. How do you use these various plant products? Do you think the number of plants and plant products used in everyday life has increased or decreased in the last century? Do you think the number is likely to increase or decrease in the future? Why?

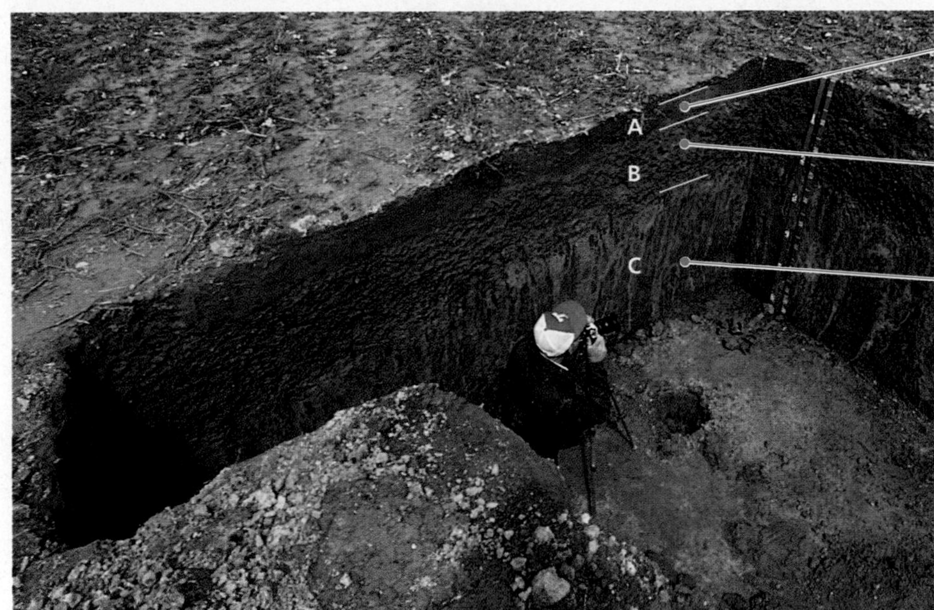

The A horizon is the topsoil, a mixture of broken-down rock of various textures, living organisms, and decaying organic matter.

The B horizon contains much less organic matter than the A horizon and is less weathered.

The C horizon, composed mainly of partially broken-down rock, serves as the "parent" material for the upper layers of soil.

◀ **Figure 37.5 Soil horizons.** This researcher is photographing a vertical profile of three soil layers, or horizons.

bacteria that cohabit with various fungi, algae and other protists, insects, earthworms, nematodes, and roots of plants. The activities of all these organisms affect the soil's physical and chemical properties. Earthworms, for instance, turn over and aerate the soil by their burrowing and add mucus that holds fine soil particles together. Meanwhile, the metabolism of bacteria alters the mineral composition of the soil. Plant roots can also affect soil composition and texture. For instance, they can affect soil pH by releasing organic acids, and they reinforce the soil against erosion.

Humus, an important component of topsoil, consists of decomposing organic material formed by the action of bacteria and fungi on dead organisms, feces, fallen leaves, and other organic refuse. Humus prevents clay from packing together and builds a crumbly soil that retains water but is still porous enough for the adequate aeration of roots. Humus is also a reservoir of mineral nutrients that are returned gradually to the soil as microorganisms decompose the organic matter.

After a heavy rainfall, water drains away from the larger spaces of the soil, but smaller spaces retain water because of its attraction to the negatively charged surfaces of clay and other soil particles. Some water adheres so tightly to soil particles that it cannot be extracted by plants. The film of water bound less tightly to the particles is the water generally available to plants **(Figure 37.6a)**. It is not pure water but a soil solution containing dissolved minerals in the form of ions. Roots absorb this soil solution.

To be available to roots, mineral ions must be released from the soil particles into the soil solution. Negatively charged ions (anions)—such as nitrate (NO_3^-), phosphate ($H_2PO_4^-$), and sulfate (SO_4^{2-})—are not bound tightly to the negatively charged soil particles and are therefore easily released. However, during heavy rain or irrigation, they are leached (drained) quickly into the groundwater, making them less available for uptake by roots. Positively charged ions (cations)—such as potassium (K^+), calcium (Ca^{2+}), and magnesium (Mg^{2+})—are less likely to be leached because they bind closely to soil particle surfaces. Mineral cations become available for absorption when they enter the soil solution after being displaced from soil particles by cations in the form of H^+. This process, called **cation exchange**, is stimulated by the roots, which add H^+ to the soil solution **(Figure 37.6b)**.

Soil Conservation and Sustainable Agriculture

It may take centuries for a soil to become fertile through the breakdown of rock and the accumulation of organic material, but human mismanagement can destroy that fertility within a few years. Soil mismanagement has been a recurring problem in human history. For example, the Dust Bowl was an ecological and human disaster that took place in the southwestern Great Plains region of the United States in the 1930s. Before the arrival of farmers, the region was covered by hardy grasses that held the soil in place in spite of the long recurrent droughts and torrential rains characteristic of the region. However, in the late 1800s and early 1900s, many homesteaders settled in the region, planting wheat and raising cattle. These land uses left the topsoil exposed to erosion by winds that often sweep over the area. Bad luck in the form of a few years of drought made the problem worse. Much of the topsoil was blown away, rendering millions of hectares of

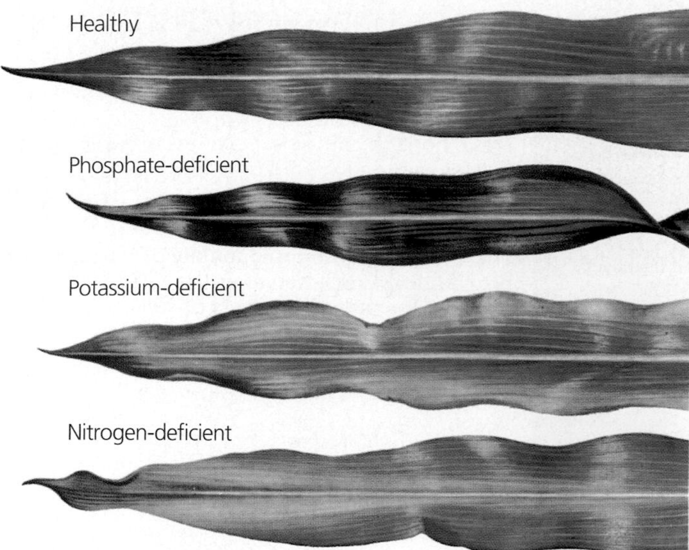

▲ **Figure 37.4 The most common mineral deficiencies, as seen in maize leaves.** Phosphate-deficient plants have reddish purple margins, particularly in young leaves. Potassium-deficient plants exhibit "firing," or drying, along tips and margins of older leaves. Nitrogen deficiency is evident in a yellowing that starts at the tip and moves along the center (midrib) of older leaves.

freely within a plant, and an iron deficiency will cause yellowing of young leaves before any effect on older leaves is visible.

Deficiencies of nitrogen, phosphorus, and potassium are most common. Shortages of micronutrients are less common and tend to occur in certain geographic regions because of differences in soil composition. The symptoms of a mineral deficiency are often distinctive enough for a plant physiologist or farmer to diagnose its cause (**Figure 37.4**). One way to confirm a diagnosis is to analyze the mineral content of the plant and soil. The amount of a micronutrient needed to correct a deficiency is usually quite small. For example, a zinc deficiency in fruit trees can usually be cured by hammering a few zinc nails into each tree trunk. Moderation is important because overdoses of many nutrients can be toxic to plants. Hydroponic culture can ensure optimal mineral nutrition by using nutrient solutions that can be precisely regulated. However, this method is not used widely in agriculture because it is relatively expensive compared with growing crops in soil.

Concept Check 37.1

1. Explain how Table 37.1 can be used to support Hales' hypothesis, yet does not refute van Helmont's hypothesis.
2. Are some essential elements more important than others? Explain.
3. Can a single leaf be used to diagnose all of a plant's mineral deficiencies? Explain.

For suggested answers, see Appendix A.

Concept 37.2

Soil quality is a major determinant of plant distribution and growth

Along with climate, the major factors determining whether particular plants can grow well in a certain location are the texture and composition of the soil. Texture is the soil's general structure, referring to the relative amounts of various sizes of soil particles. Composition refers to the soil's organic and inorganic chemical components. Plants that grow naturally in a certain type of soil are adapted to its texture and composition and can absorb water and extract essential mineral nutrients. In turn, plants affect the soil, as you will soon see. The soil-plant interface is a critical part of the chemical cycles that sustain terrestrial ecosystems.

Texture and Composition of Soils

Soil has its origin in the weathering of solid rock. The freezing of water that has seeped into crevices can mechanically fracture rocks. Acids dissolved in the water can also help break rocks down chemically. When organisms are able to invade the rock, they accelerate breakdown by chemical and mechanical means. Some organisms, for example, secrete acids that dissolve the rock. Roots that grow in fissures lead to mechanical fracturing. The eventual result of all this activity is **topsoil**, a mixture of particles derived from rock, living organisms, and **humus**, the remains of partially decayed organic material. The topsoil and other distinct soil layers, or **horizons**, are often visible in vertical profile where there is a road cut or deep hole (**Figure 37.5**, on the next page). The topsoil, also known as the A horizon, is the richest in organic material and is therefore most important for plant growth.

The texture of topsoil depends on the sizes of its particles, which are classified in a range from coarse sand to microscopic clay particles. The most fertile soils are usually **loams**, made up of roughly equal amounts of sand, silt (particles of intermediate size), and clay. Loamy soils have enough fine particles to provide a large surface area for retaining minerals and water, which adhere to the particles. But loams also have enough coarse particles to provide air spaces containing oxygen that can be used by roots for cellular respiration. If soil does not drain adequately, roots suffocate because the air spaces are replaced by water; the roots may also be attacked by molds that favor soaked soil. These are common hazards for houseplants that are overwatered in pots with poor drainage.

Soil composition includes organic components as well as minerals. Topsoil is home to an astonishing number and variety of organisms. A teaspoon of topsoil has about 5 billion

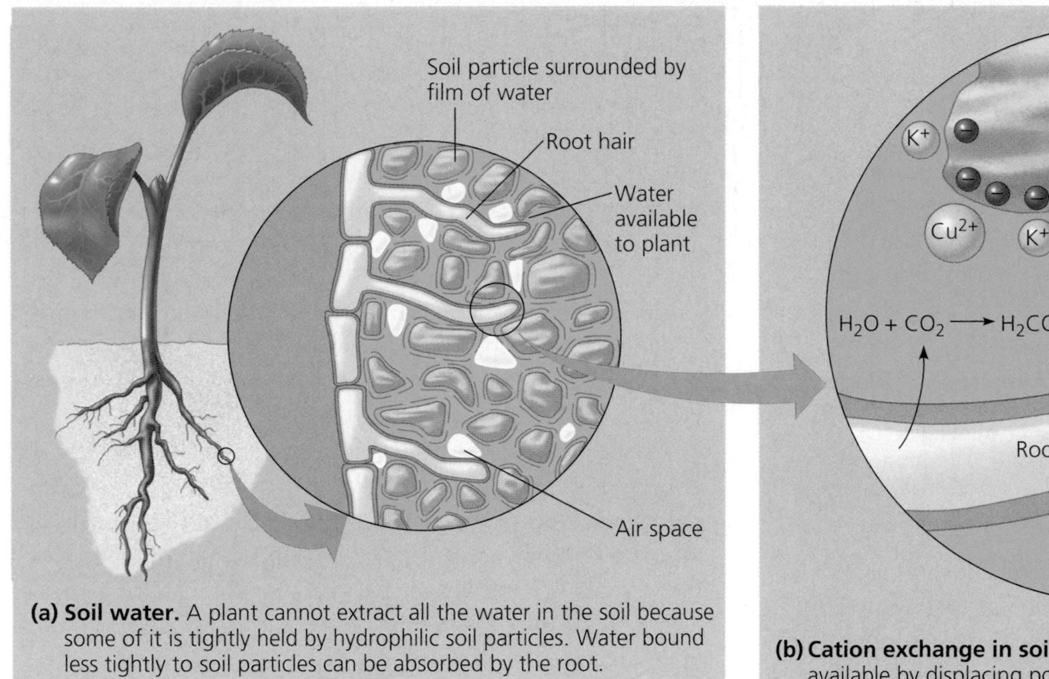

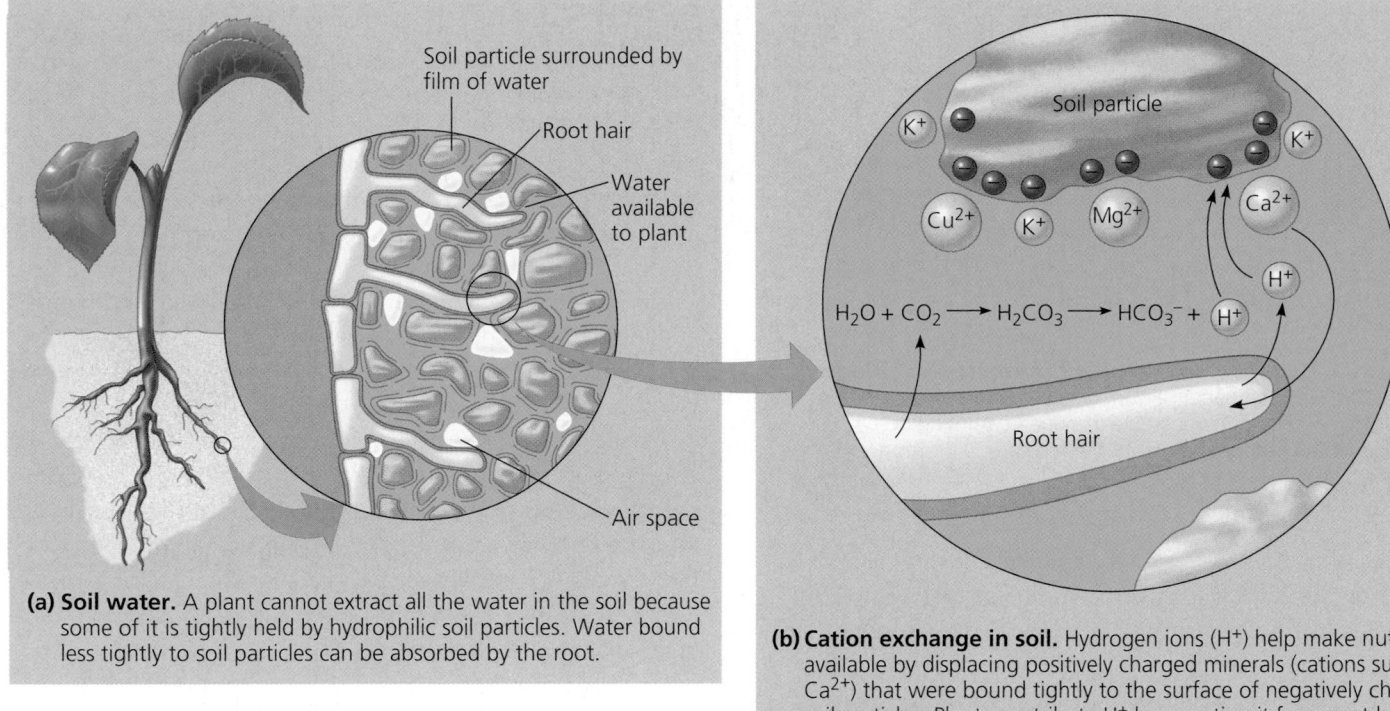

(a) Soil water. A plant cannot extract all the water in the soil because some of it is tightly held by hydrophilic soil particles. Water bound less tightly to soil particles can be absorbed by the root.

(b) Cation exchange in soil. Hydrogen ions (H^+) help make nutrients available by displacing positively charged minerals (cations such as Ca^{2+}) that were bound tightly to the surface of negatively charged soil particles. Plants contribute H^+ by secreting it from root hairs and also by cellular respiration, which releases CO_2 into the soil solution, where it reacts with H_2O to form carbonic acid (H_2CO_3). Dissociation of this acid adds H^+ to the soil solution.

▲ **Figure 37.6 The availability of soil water and minerals.**

farmland useless and forcing hundreds of thousands of people to abandon homes and land, a plight immortalized in John Steinbeck's *The Grapes of Wrath*. Better soil conservation practices could have preserved soil fertility and sustained agricultural productivity.

To understand soil conservation, we must first remember that agriculture can only be sustained by human intervention. In forests, grasslands, and other natural ecosystems, mineral nutrients are usually recycled by the decomposition of dead organic material in the soil. In contrast, when farmers harvest a crop, essential elements are diverted from the chemical cycles going on in that location. In general, agriculture depletes the mineral content of the soil. To grow 1,000 kg of wheat grain, the soil gives up 20 kg of nitrogen, 4 kg of phosphorus, and 4.5 kg of potassium. Each year, the soil fertility diminishes unless fertilizers replace lost minerals such as nitrogen, phosphorus, and potassium. Many crops also use far more water than the vegetation that once grew naturally on that land, forcing farmers to irrigate. Prudent fertilization, thoughtful irrigation, and the prevention of erosion are three of the most important goals of soil conservation. Complementing soil conservation is soil reclamation, the goal of returning agricultural productivity to exhausted or damaged soil. More than 30% of the world's farmland suffers from low productivity stemming from poor soil conditions such as chemical contamination, mineral deficiencies, acidity, salinity, and poor drainage.

Fertilizers

Prehistoric farmers may have started fertilizing their fields after noticing that grass grew faster and greener where animals had defecated. In developed nations today, most farmers use commercially produced fertilizers containing minerals that are either mined or prepared by industrial processes. These fertilizers are usually enriched in nitrogen, phosphorus, and potassium, the macronutrients most commonly deficient in farm and garden soils. Fertilizers are labeled with a three-number code called the N-P-K ratio, indicating the content of these minerals. A fertilizer marked "15-10-5," for instance, is 15% nitrogen (as ammonium or nitrate), 10% phosphorus (as phosphoric acid), and 5% potassium (as the mineral potash).

Manure, fishmeal, and compost are called "organic" fertilizers because they are of biological origin and contain decomposing organic material. Before plants can use organic material, however, it must be decomposed into the inorganic nutrients that roots can absorb. Whether from organic fertilizer or a chemical factory, the minerals a plant extracts are in the same form, but organic fertilizers release minerals gradually, whereas minerals in

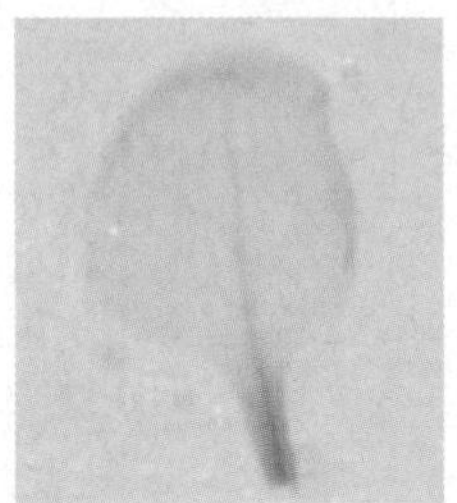
No phosphorus deficiency

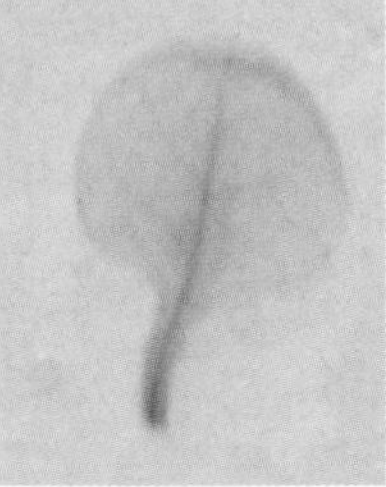
Beginning phosphorus deficiency

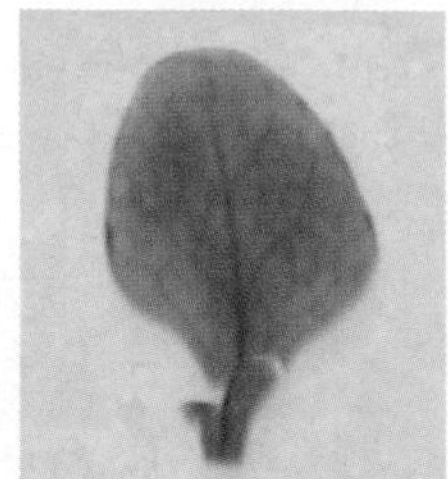
Well-developed phosphorus deficiency

▲ **Figure 37.7 Deficiency warnings from "smart" plants.** Some plants have been genetically modified to signal an impending nutrient deficiency before irreparable damage or stunting occurs. For example, after laboratory treatments, the research plant *Arabidopsis* develops a blue color in response to an imminent phosphate deficiency.

commercial fertilizers are immediately available but may not be retained by the soil for long. Excess minerals not absorbed by roots are usually wasted because they are often leached from the soil by rainwater or irrigation. To make matters worse, mineral runoff may pollute groundwater, streams, and lakes.

Agricultural researchers are developing ways to maintain crop yields while reducing fertilizer use. One approach is to genetically engineer "smart" plants that inform the grower when a nutrient deficiency is imminent—but *before* damage has occurred **(Figure 37.7)**. One type of smart plant takes advantage of a promoter (a DNA sequence that indicates where to begin transcription) that more readily binds RNA polymerase (the enzyme that drives transcription) when the phosphorus content of plant tissues begins to decline. This promoter is linked to a reporter gene that leads to the production of a blue pigment in the leaf cells. When leaves of these smart plants develop a blue tinge, the farmer knows it is time to add phosphate-containing fertilizer.

To fertilize judiciously, a farmer must note the soil pH because it affects cation exchange and the chemical form of minerals. Although an essential element may be abundant, plants may be starving for that element because it is bound too tightly to clay particles or is in a chemical form the plant cannot absorb. Managing soil pH is tricky; a change in H^+ concentration may make one mineral more available but make another less available. At pH 8, for instance, the plant can absorb calcium, but iron is almost completely unavailable. The soil pH should be matched to a crop's mineral needs. If the soil is too alkaline, adding sulfate will lower the pH. Soil that is too acidic can be adjusted by liming (adding calcium carbonate or calcium hydroxide).

A major problem with acidic soils, particularly in tropical areas, is that aluminum dissolves at low pH and becomes toxic to roots. Some plants cope with high aluminum levels by secreting organic anions that bind the aluminum and render it harmless.

Irrigation

Water is most often the limiting factor in plant growth. Irrigation can transform a desert into a garden, but farming in arid regions is a huge drain on water resources. Many rivers in the southwestern United States have been reduced to trickles by diversion of water for irrigation. Another problem is that irrigation in an arid region can gradually make the soil so salty that it may become completely infertile. Salts dissolved in irrigation water accumulate in the soil as the water evaporates, making the water potential of the soil solution more negative, thereby reducing water uptake by lowering the water-potential gradient from soil to roots (see Chapter 36).

As the world population grows, more and more acres of arid land are being cultivated. New irrigation methods may reduce the risks of running out of water or losing farmland to salinization (salt accumulation). Drip irrigation is used for many crops in the western United States, resulting in less evaporation. In another approach, plant breeders are developing plant varieties that require less water.

Erosion

Topsoil from thousands of acres of farmland is lost to water and wind erosion each year in the United States alone. Certain precautions, such as planting rows of trees as windbreaks, terracing hillside crops, and cultivating in a contour pattern, can prevent loss of topsoil **(Figure 37.8)**. Crops such as alfalfa and wheat provide good ground cover and protect the soil better than maize and other crops that are usually planted in more widely spaced rows.

If managed properly, soil is a renewable resource in which farmers can grow food for many generations. The goal of soil management is **sustainable agriculture,** a commitment embracing a variety of farming methods that are conservation-minded, environmentally safe, and profitable.

▲ **Figure 37.8 Contour tillage.** These crops in Wisconsin are planted in rows that go around, rather than up and down, the hills. Contour tillage helps slow the runoff of water and erosion of topsoil after heavy rains.

Soil Reclamation

Some areas are unfit for agriculture because of contamination of soil or groundwater with toxic heavy metals or organic pollutants. Traditionally, soil reclamation has focused on non-biological technologies, such as removal and storage of contaminated soil in landfills, but these techniques are very costly and often disrupt the landscape. A new method known as **phytoremediation** is a biological, nondestructive technology that seeks to reclaim contaminated areas cheaply by using the remarkable ability of some plants to extract soil pollutants and concentrate them in portions of the plant that can be easily removed for safe disposal. For example, alpine pennycress (*Thlaspi caerulescens*) can accumulate zinc in its shoots at concentrations 300 times higher than most plants can tolerate. Such plants show promise for cleaning up areas contaminated by smelters, mining operations, or nuclear testing. Phytoremediation is part of the more general technology of bioremediation, which includes the use of prokaryotes and sometimes protists to detoxify polluted sites (see Chapters 27 and 55).

Concept Check 37.2

1. What are the general characteristics of good soil?
2. Explain how the phrase "too much of a good thing" can apply to watering and fertilizing plants.

For suggested answers, see Appendix A.

Concept 37.3

Nitrogen is often the mineral that has the greatest effect on plant growth

Of all the mineral nutrients, nitrogen contributes the most to plant growth and crop yields. Plants require nitrogen as a component of proteins, nucleic acids, chlorophyll, and other important organic molecules.

Soil Bacteria and Nitrogen Availability

It is ironic that plants can suffer nitrogen deficiencies, for the atmosphere is nearly 80% nitrogen. However, this atmospheric nitrogen is gaseous N_2, a form plants cannot use. For plants to absorb nitrogen, it must first be converted to ammonium (NH_4^+) or nitrate (NO_3^-).

In contrast to other minerals, the NH_4^+ and NO_3^- in soil are not derived from the breakdown of rock. Over the short term, the main source of these minerals is the decomposition of humus by microbes, including ammonifying bacteria **(Figure 37.9)**. Through decomposition, the nitrogen in proteins and other organic compounds is repackaged in inorganic compounds that are recycled when absorbed as minerals by roots. Some nitrogen is lost when soil microbes called denitrifying bacteria convert NO_3^- to N_2, which diffuses from the soil into the atmosphere. However, other bacteria called

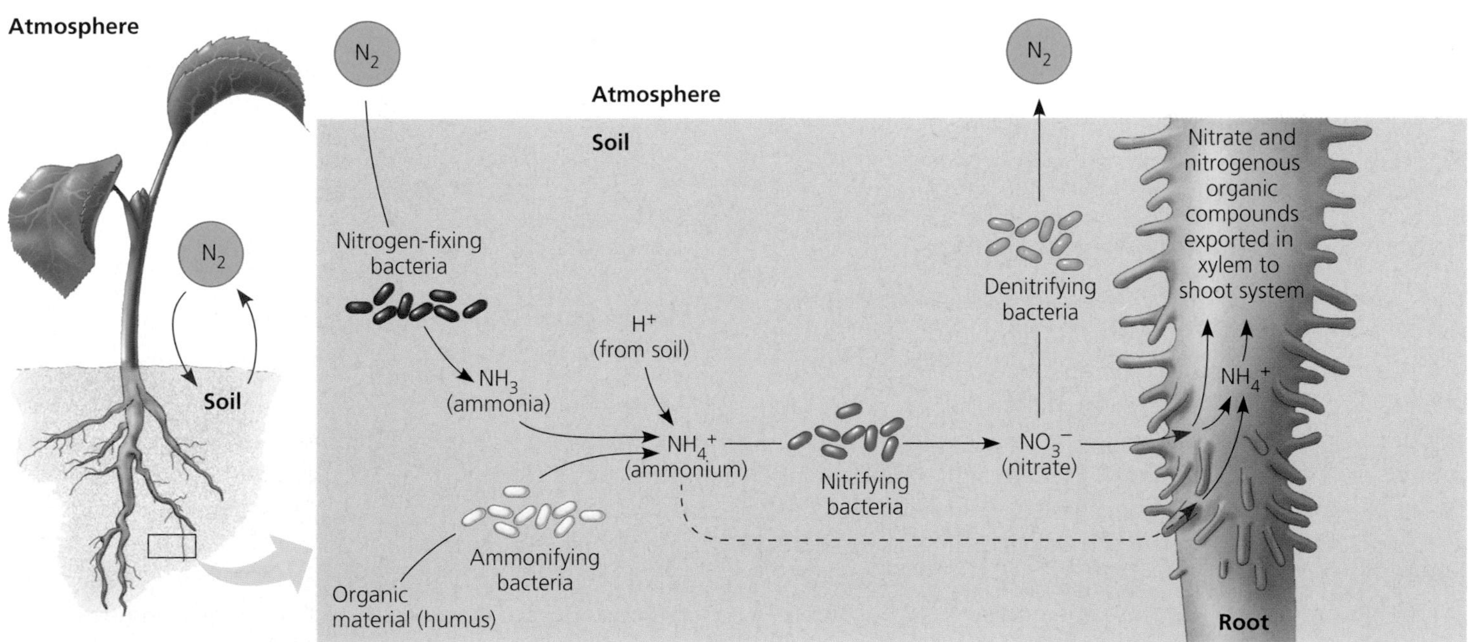

▲ **Figure 37.9 The role of soil bacteria in the nitrogen nutrition of plants.** Ammonium is made available to plants by two types of soil bacteria: those that fix atmospheric N_2 (nitrogen-fixing bacteria) and those that decompose organic material (ammonifying bacteria). Although plants absorb some ammonium from the soil, they absorb mainly nitrate, which is produced from ammonium by nitrifying bacteria. Plants reduce nitrate back to ammonium before incorporating the nitrogen into organic compounds. Xylem transports nitrogen from roots to shoots in the form of nitrate, amino acids, and various other organic compounds, depending on the species.

nitrogen-fixing bacteria restock nitrogenous minerals in the soil by converting N_2 from the atmosphere to NH_3 (ammonia) in a metabolic process called **nitrogen fixation.** The complex cycling of nitrogen in ecosystems is traced in detail in Chapter 54. Here we focus on nitrogen fixation and the other steps that lead directly to nitrogen assimilation by plants.

All life on Earth depends on nitrogen fixation, a process performed only by some bacterial species. Several of these species live freely in the soil, while others live in plant roots in symbiotic relationships. (You will learn more about these relationships in the next section.) The conversion of atmospheric nitrogen (N_2) to ammonia (NH_3) is a complicated, multistep process, but we can simplify nitrogen fixation by just indicating the reactants and products:

$$N_2 + 8\,e^- + 8\,H^+ + 16\,ATP \rightarrow 2\,NH_3 + H_2 + 16\,ADP + 16\,Ⓟ_i$$

The enzyme complex **nitrogenase** catalyzes the entire reaction sequence, which reduces N_2 to NH_3 by adding electrons along with H^+. Notice that nitrogen fixation is very expensive in terms of metabolic energy, costing the bacteria eight ATP molecules for each ammonia molecule synthesized. Nitrogen-fixing bacteria are therefore most abundant in soils rich in organic material, which provides fuel for cellular respiration to produce the ATP.

In the soil solution, ammonia picks up another hydrogen ion to form an ammonium ion (NH_4^+), which plants can absorb. However, plants acquire their nitrogen mainly in the form of nitrate (NO_3^-), which is produced in the soil by nitrifying bacteria that oxidize ammonium (see Figure 37.9). After nitrate is absorbed by the roots, a plant enzyme can reduce the nitrate back to ammonium, which other enzymes can then incorporate into amino acids and other organic compounds. Most plant species export nitrogen from roots to shoots, via the xylem, in the form of nitrate or organic compounds that have been synthesized in the roots.

Improving the Protein Yield of Crops

The incorporation of fixed nitrogen by plants into proteins and other organic substances has a major impact on human welfare because the most common form of malnutrition in humans is protein deficiency. The majority of people in the world, particularly in developing countries, have a predominantly vegetarian diet and thus depend mainly on plants for protein. Unfortunately, many plants have a low protein content, and those proteins may be deficient in one or more amino acids that humans need from their diet (see Figure 41.10). Improving the quality and quantity of proteins in crops is a major goal of agricultural research.

Plant breeding has resulted in new varieties of maize, wheat, and rice that are enriched in protein. However, many of these "super" varieties have an extraordinary demand for nitrogen, usually supplied in commercial fertilizer. Like biological nitrogen fixation, industrial production of ammonia and nitrate from atmospheric nitrogen consumes a lot of energy. A chemical factory making fertilizer consumes large quantities of fossil fuels. The countries that most need high-protein crops are usually the ones least able to pay the fuel bill. In the future, new catalysts based on the mechanism by which nitrogenase fixes nitrogen may make commercial fertilizer production less costly. Several years ago, biochemists determined the structure of nitrogenase in *Rhizobium,* a genus of nitrogen-fixing bacteria, providing a model for chemical engineers to design catalysts by imitating nature. Another strategy that could potentially increase protein yields of crops is to improve the productivity of symbiotic nitrogen fixation, a process examined in the next section.

Concept Check 37.3

1. Explain why nitrogen-fixing bacteria are essential to human welfare.

For suggested answer, see Appendix A.

Concept 37.4

Plant nutritional adaptations often involve relationships with other organisms

Many plants have nutritional adaptations that involve other organisms. Two of these relationships are mutualistic: symbiotic nitrogen fixation, involving roots and bacteria, and mycorrhizae, involving roots and fungi. After examining these mutually beneficial relationships, we will look at adaptations that are one-sided and relatively unusual: epiphytes, parasitic plants, and carnivorous plants.

The Role of Bacteria in Symbiotic Nitrogen Fixation

Symbiotic relationships with nitrogen-fixing bacteria provide some plant species with a built-in source of fixed nitrogen for assimilation into organic compounds. From an agricultural perspective, the most important and efficient symbioses between nitrogen-fixing bacteria and plants occur in the legume family, including peas, beans, soybeans, peanuts, alfalfa, and clover.

Along a legume's roots are swellings called **nodules** composed of plant cells that have been "infected" by nitrogen-fixing *Rhizobium* ("root living") bacteria **(Figure 37.10a)**. Inside the nodule, *Rhizobium* bacteria assume a form called **bacteroids,** which are contained within vesicles formed by the root cell **(Figure 37.10b)**. *Rhizobium* bacteria can fix atmospheric N_2

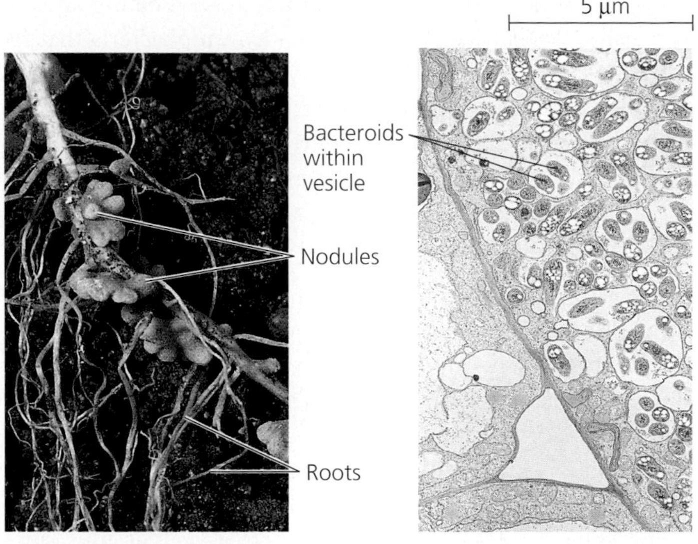

(a) Pea plant root. The bumps on this pea plant root are nodules containing *Rhizobium* bacteria. The bacteria fix nitrogen and obtain photosynthetic products supplied by the plant.

(b) Bacteroids in a soybean root nodule. In this TEM, a cell from a root nodule of soybean is filled with bacteroids in vesicles. The cells on the left are uninfected.

▲ **Figure 37.10 Root nodules on legumes.**

and supply it as ammonium, a form readily used by the plant (see Figure 37.9). Legume-*Rhizobium* symbioses generate more useful nitrogen for plants than all industrial fertilizers, and the symbiosis provides the right amounts of nitrogen at the right time at virtually no cost to the farmer. In addition to supplying the legume with nitrogen, symbiotic nitrogen fixation significantly reduces spending on fertilizers for subsequent crops.

The location of the bacteroids inside living, nonphotosynthetic cells favors nitrogen fixation, which requires an anaerobic environment. Lignified external layers may also limit gas exchange. Some root nodules are reddish, owing to a molecule called leghemoglobin (*leg*- for "legume"), an iron-containing protein that binds reversibly to oxygen (like the hemoglobin of human red blood cells). This protein is an oxygen "buffer," keeping the concentration of free O_2 low and regulating the oxygen supply for the intense respiration that the bacteria require to produce ATP for nitrogen fixation.

Each legume is associated with a particular strain of *Rhizobium*. **Figure 37.11** describes how a root nodule develops after bacteria enter through what is called an infection thread. The symbiotic relationship between a legume and nitrogen-fixing bacteria is mutualistic because the bacteria supply the plant

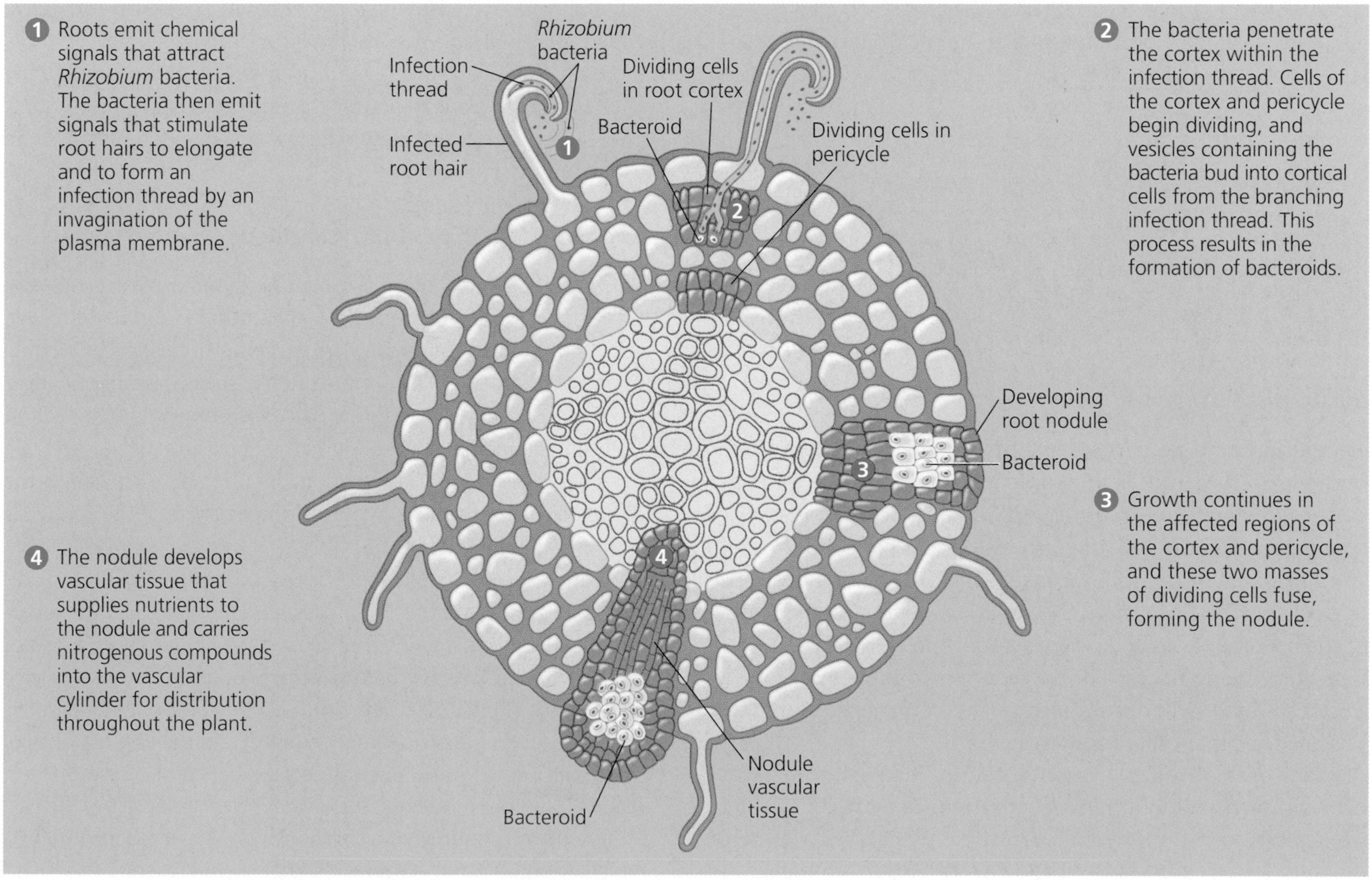

1 Roots emit chemical signals that attract *Rhizobium* bacteria. The bacteria then emit signals that stimulate root hairs to elongate and to form an infection thread by an invagination of the plasma membrane.

Infection thread

Rhizobium bacteria

Infected root hair

Dividing cells in root cortex

Bacteroid

Dividing cells in pericycle

2 The bacteria penetrate the cortex within the infection thread. Cells of the cortex and pericycle begin dividing, and vesicles containing the bacteria bud into cortical cells from the branching infection thread. This process results in the formation of bacteroids.

Developing root nodule

Bacteroid

3 Growth continues in the affected regions of the cortex and pericycle, and these two masses of dividing cells fuse, forming the nodule.

4 The nodule develops vascular tissue that supplies nutrients to the nodule and carries nitrogenous compounds into the vascular cylinder for distribution throughout the plant.

Bacteroid

Nodule vascular tissue

▲ **Figure 37.11 Development of a soybean root nodule.**

with fixed nitrogen while the plant provides the bacteria with carbohydrates and other organic compounds. Most of the ammonium produced by symbiotic nitrogen fixation is used by the nodules to make amino acids, which are then transported to the shoot via the xylem.

The Molecular Biology of Root Nodule Formation

How does a legume species recognize a certain strain of *Rhizobium* among the many bacterial strains in the soil? And how does an encounter with that specific *Rhizobium* lead to development of a nodule? These two questions have led researchers to uncover a chemical dialogue between the bacteria and the root. Each partner responds to the chemical signals from the other by expressing certain genes whose products contribute to nodule formation.

The plant initiates the communication when its roots secrete molecules called flavonoids, which enter *Rhizobium* cells living near the roots. The signal's specificity arises from variations in flavonoid structure, with a particular legume species secreting a flavonoid that only a certain *Rhizobium* strain detects and absorbs. The plant's signal activates a gene-regulating protein in the bacterium, which switches on a cluster of bacterial genes called *nod,* for "nodulation" genes. These *nod* genes produce enzymes that catalyze production of species-specific molecules called Nod factors. Secreted by the bacterial cells, the Nod factors signal the root to initiate the infection process that enables the *Rhizobium* to enter the root and to begin forming the root nodule. The plant's responses require activation of genes called early nodulin genes by a signal transduction pathway involving Ca^{2+} as second messengers (see Chapter 11). By understanding the molecular biology underlying the formation of root nodules, researchers hope to learn how to induce *Rhizobium* uptake and nodule formation in crop plants that do not normally form such nitrogen-fixing symbiotic relationships.

Symbiotic Nitrogen Fixation and Agriculture

The agricultural benefits of symbiotic nitrogen fixation underlie **crop rotation.** In this practice, a non-legume such as maize is planted one year, and the following year alfalfa or some other legume is planted to restore the concentration of fixed nitrogen in the soil. To ensure that the legume encounters its specific *Rhizobium,* the seeds are soaked in a culture of the bacteria or dusted with bacterial spores before sowing. Instead of being harvested, the legume crop is often plowed under so that it will decompose as "green manure," reducing the need for manufactured fertilizers.

Many plant families besides legumes include species that benefit from symbiotic nitrogen fixation. For example, alder trees and certain tropical grasses host gram-positive bacteria of the actinomycetes group (see Figure 27.13). Rice, a crop of great commercial importance, benefits indirectly from symbiotic nitrogen fixation. Rice farmers culture a free-floating aquatic fern called *Azolla,* which has symbiotic cyanobacteria that fix nitrogen and increase the fertility of the rice paddy. The growing rice eventually shades and kills the *Azolla,* and decomposition of this organic material adds more nitrogenous compounds to the paddy.

Mycorrhizae and Plant Nutrition

Mycorrhizae ("fungus roots") are modified roots consisting of mutualistic associations of fungi and roots (see Figures 31.15 and 36.10). The fungus benefits from a steady supply of sugar donated by the host plant. In return, the fungus increases the surface area for water uptake and selectively absorbs phosphate and other minerals from the soil and supplies them to the plant. The fungi of mycorrhizae also secrete growth factors that stimulate roots to grow and branch, as well as antibiotics that may help protect the plant from pathogenic bacteria and fungi in the soil.

Mycorrhizae are not oddities; they are formed by most plant species. In fact, this plant-fungus symbiosis might have been one of the evolutionary adaptations that made it possible for plants to colonize land in the first place. Indeed, fossilized roots from some of the earliest plants include mycorrhizae. When terrestrial ecosystems were young, the soil was probably not very rich in nutrients. The fungi of mycorrhizae, which are more efficient at absorbing minerals than the roots themselves, would have helped nourish the pioneering plants. Even today, the plants that first become established on nutrient-poor soils, such as abandoned farmland or eroded hillsides, are usually mycorrhizae-rich.

The Two Main Types of Mycorrhizae

The modified roots formed from the symbiosis of fungi and plants take two major forms: ectomycorrhizae and endomycorrhizae. In **ectomycorrhizae,** the mycelium (mass of branching hyphae; see Chapter 31) forms a dense sheath, or mantle, over the surface of the root **(Figure 37.12a)**. Fungal hyphae extend from the mantle into the soil, greatly increasing the surface area for water and mineral absorption. Hyphae also grow into the cortex of the root. These hyphae do not penetrate the root cells but form a network in the apoplast, or extracellular space, that facilitates nutrient exchange between the fungus and the plant. Compared with "uninfected" roots, ectomycorrhizae are generally thicker, shorter, and more branched. They typically do not form root hairs, which would be superfluous given the extensive surface area of the fungal mycelium. About 10% of plant families have species that form ectomycorrhizae, and the vast majority of these species are woody, including members of the pine, spruce, oak, walnut, birch, willow, and eucalyptus families.

In contrast, **endomycorrhizae** do not have a dense mantle ensheathing the root **(Figure 37.12b)**. A microscope is needed to see the fine fungal hyphae that extend from the soil into the

(a) Ectomycorrhizae. The mantle of the fungal mycelium ensheathes the root. Fungal hyphae extend from the mantle into the soil, absorbing water and minerals, especially phosphate. Hyphae also extend into the extracellular spaces of the root cortex, providing an extensive surface area for nutrient exchange between the fungus and its host plant (colorized SEMs).

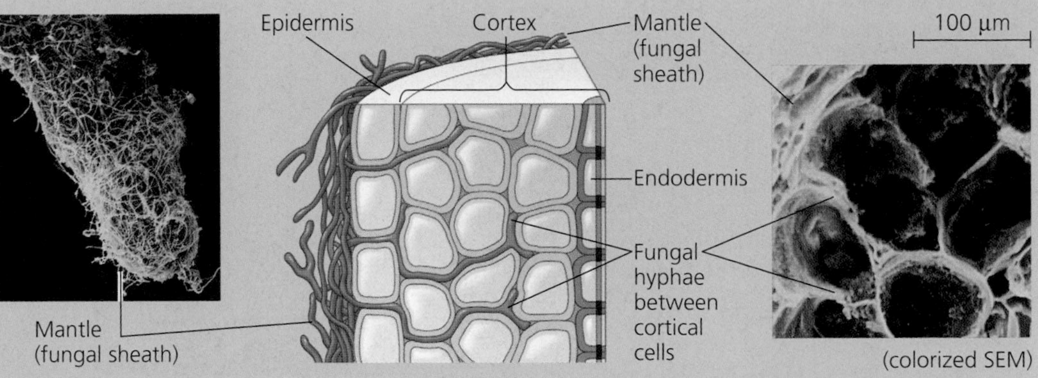

(b) Endomycorrhizae. No mantle forms around the root, but microscopic fungal hyphae extend into the root. Within the root cortex, the fungus makes extensive contact with the plant through branching of hyphae that form arbuscules, providing an enormous surface area for nutrient swapping. The hyphae penetrate the cell walls, but not the plasma membranes, of cells within the cortex.

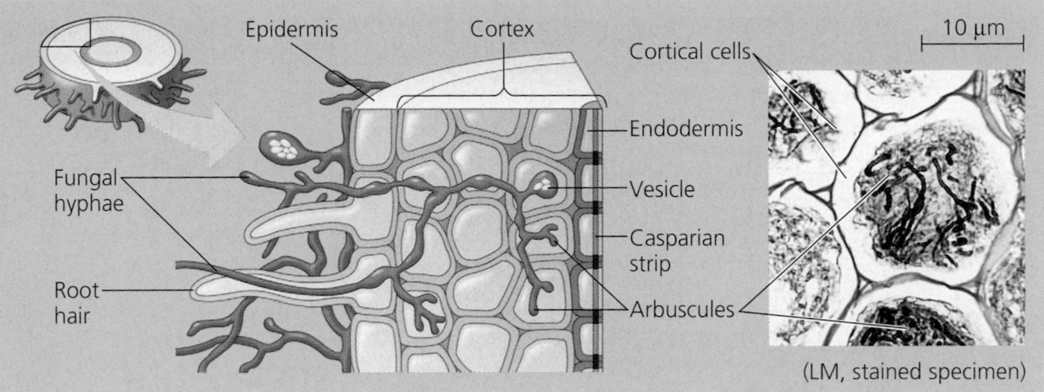

▲ **Figure 37.12 Mycorrhizae.**

root. Hyphae also extend into the root cells (hence the term *endo*mycorrhizae) by digesting small patches of the root cell walls. However, a fungal hypha does not actually pierce the plasma membrane and enter the root cell's cytoplasm. Instead, it grows into a tube formed by invagination of the root cell's membrane. The action is analogous to poking a finger gently into a balloon; your finger is like the fungal hypha, and the balloon skin is like the root cell's membrane. After the fungal hyphae have penetrated in this way, some form densely branched structures called arbuscules ("little trees"), which are important sites of nutrient transfer between the fungus and the plant. Hyphae may also form oval vesicles, which possibly store food for the fungus. To the unaided eye, endomycorrhizae look like "normal" roots with root hairs, but a microscope reveals a symbiotic relationship of enormous importance to plant nutrition. Endomycorrhizae are much more common than ectomycorrhizae and are found in over 85% of plant species, including important crop plants such as maize, wheat, and legumes.

Agricultural Importance of Mycorrhizae

Roots can be transformed into mycorrhizae only if exposed to the appropriate species of fungus. In most natural ecosystems, these fungi are present in the soil, and seedlings develop mycorrhizae. But if seeds are collected in one environment and planted in foreign soil, the plants may show signs of malnutrition (particularly phosphorus deficiency) resulting from absence of the plants' mycorrhizal partners. Researchers observe similar results in experiments in which soil fungi are poisoned. Farmers and foresters are already applying the lessons of such research. For example, inoculating pine seeds with spores of mycorrhizal fungi promotes formation of mycorrhizae by the seedlings. Infected pine seedlings grow more vigorously than trees without the fungal association.

Epiphytes, Parasitic Plants, and Carnivorous Plants

Almost all plant species have mutualistic relationships with fungi, bacteria, or both. Though rarer, there are also plant species with nutritional adaptations that use other organisms in nonmutualistic ways. **Figure 37.13**, on the next page, provides an overview of three unusual adaptations: epiphytes, parasitic plants, and carnivorous plants.

Concept Check 37.4

1. Compare and contrast root nodules and mycorrhizae.
2. Contrast epiphytes with parasitic plants.

For suggested answers, see Appendix A.

Figure 37.13

Exploring Unusual Nutritional Adaptations in Plants

EPIPHYTES

An **epiphyte** (from the Greek *epi*, upon, and *phyton*, plant) nourishes itself but grows on another plant, usually anchored to branches or trunks of living trees. Epiphytes absorb water and minerals from rain, mostly through leaves rather than roots. Some examples are staghorn ferns and many orchids.

Staghorn fern, an epiphyte. This tropical fern (genus *Platycerium*) grows on large rocks, cliffs, and trees. It has two types of fronds: branched fronds resembling antlers and circular fronds that form a collar around the base of the fern.

PARASITIC PLANTS

Unlike epiphytes, parasitic plants absorb sugars and minerals from their living hosts, although some parasitic species are photosynthetic. Many species have roots that function as haustoria, nutrient-absorbing projections that enter the host plant.

Mistletoe, a photosynthetic parasite. Tacked above doorways during the holiday season, mistletoe (genus *Phoradendron*) lives in nature as a parasite on oaks and other trees.

Dodder, a nonphotosynthetic parasite. Dodder (genus *Cuscuta*), the orange "strings" on this pickleweed, draws its nutrients from the host. The transverse section shows haustoria tapping the host's phloem (LM).

Indian pipe, a nonphotosynthetic parasite. Also called the ghost flower, this species (*Monotropa uniflora*) absorbs nutrients from the fungal hyphae of mycorrhizae of green plants.

CARNIVOROUS PLANTS

Carnivorous plants are photosynthetic but obtain some nitrogen and minerals by killing and digesting insects and other small animals. Carnivorous plants live in acid bogs and other habitats where soils are poor in nitrogen and other minerals. Various insect traps consist of modified leaves usually equipped with glands that secrete digestive enzymes. Fortunately for animals, such turnabouts are rare!

Venus' flytrap. Triggered by electrical impulses from sensory hairs, two leaf lobes close in half a second. Despite its common name, *Dionaea muscipula* usually catches ants and grasshoppers.

Pitcher plants. *Nepenthes*, *Sarracenia*, and other genera have water-filled funnels. The insects drown and are digested by enzymes.

Sundews. Sundews (genus *Drosera*) exude a sticky fluid that glitters like dew. Insects get stuck to the leaf hairs, which enfold the prey.

Chapter 37 Review

Go to www.campbellbiology.com or the student CD-ROM to explore Activities, Investigations, and other interactive study aids.

SUMMARY OF KEY CONCEPTS

Concept 37.1

Plants require certain chemical elements to complete their life cycle

- Plants derive most of their organic mass from the CO_2 of air, but they also depend on soil nutrients in the form of water and minerals. The branching of root and shoot systems helps plants come into contact with the resources that they need from the environment (pp. 756–757).
- **Macronutrients and Micronutrients (pp. 757–758)** Macronutrients, elements required in relatively large amounts, include carbon, oxygen, hydrogen, nitrogen, and other major ingredients of organic compounds. Micronutrients, elements required in very small amounts, typically have catalytic functions as cofactors of enzymes.
- **Symptoms of Mineral Deficiency (pp. 758–759)** Deficiency of a mobile nutrient usually affects older organs more than younger ones; the reverse is true for nutrients that are less mobile within a plant. Macronutrient deficiencies are most common, particularly deficiencies of nitrogen, phosphorus, and potassium.

Concept 37.2

Soil quality is a major determinant of plant distribution and growth

- **Texture and Composition of Soils (pp. 759–760)** Various sizes of particles derived from the breakdown of rock are found in soil, along with organic material (humus) in various stages of decomposition. Acids derived from roots contribute to a plant's uptake of minerals when H^+ displaces mineral cations from soil particles.
  Activity *How Plants Obtain Minerals from Soil*
  Investigation *How Does Acid Precipitation Affect Mineral Deficiency?*
- **Soil Conservation and Sustainable Agriculture (pp. 760–763)** In contrast to natural ecosystems, agriculture depletes the mineral content of soil, taxes water reserves, and encourages erosion. The goal of soil conservation strategies is to minimize this damage. A major goal of agricultural researchers is to reduce the amount of fertilizer added to soils without sacrificing high crop yields.
  Graph It *Global Soil Degradation*

Concept 37.3

Nitrogen is often the mineral that has the greatest effect on plant growth

- **Soil Bacteria and Nitrogen Availability (pp. 763–764)** Nitrogen-fixing bacteria convert atmospheric N_2 to nitrogenous minerals that plants can absorb as a nitrogen source for organic synthesis.
  Activity *The Nitrogen Cycle*
- **Improving the Protein Yield of Crops (p. 764)** Research devoted to improving the quality and quantity of crop proteins addresses the most widespread form of human malnutrition: protein deficiency.

Concept 37.4

Plant nutritional adaptations often involve relationships with other organisms

- **The Role of Bacteria in Symbiotic Nitrogen Fixation (pp. 764–766)** The development of nitrogen-fixing root nodules depends on chemical dialogue between *Rhizobium* bacteria and root cells of their specific plant hosts. The bacteria of a nodule obtain sugar from the plant and supply the plant with fixed nitrogen.
- **Mycorrhizae and Plant Nutrition (pp. 766–767)** Mycorrhizae are modified roots consisting of mutualistic associations of fungi and roots. The fungal hyphae of both ectomycorrhizae and endomycorrhizae absorb water and minerals, which they supply to their plant hosts.
- **Epiphytes, Parasitic Plants, and Carnivorous Plants (pp. 767–768)** Epiphytes grow on the surfaces of other plants but acquire water and minerals from rain. Parasitic plants absorb nutrients from host plants. Carnivorous plants supplement their mineral nutrition by digesting animals.

TESTING YOUR KNOWLEDGE

Self-Quiz

1. Most of the mass of organic material of a plant comes from
   a. water.
   b. carbon dioxide.
   c. soil minerals.
   d. atmospheric oxygen.
   e. nitrogen.

2. Micronutrients are needed in very small amounts because
   a. most of them are mobile in the plant.
   b. most function as cofactors of enzymes.
   c. most are supplied in large enough quantities in seeds.
   d. they play only a minor role in the growth and health of the plant.
   e. only the growing regions of the plants require micronutrients.

3. It is valid to consider water a plant nutrient because
   a. plants die without a water source.
   b. cell elongation depends mainly on the osmotic absorption of water by cells.
   c. hydrogen and oxygen atoms from water molecules are incorporated into organic molecules.
   d. transpiration depends on a continuous supply of water to leaves.
   e. most of a plant's mass of organic compounds is derived from water.

4. Based on our retrospective view, the most reasonable conclusion to draw from van Helmont's famous experiment on the growth of a willow is that
   a. the tree increased in mass mainly by photosynthesis.
   b. the increase in the mass of the tree could not be accounted for by the consumption of soil.
   c. most of the increase in the mass of the tree was due to the uptake of O_2.
   d. soil simply provides physical support for the tree without providing nutrients.
   e. trees do not require water to grow.

5. A mineral deficiency is likely to affect older leaves more than younger leaves if
   a. the mineral is a micronutrient.
   b. the mineral is very mobile within the plant.
   c. the mineral is required for chlorophyll synthesis.
   d. the mineral is a macronutrient.
   e. the older leaves are in direct sunlight.

6. Two groups of tomatoes were grown under laboratory conditions, one with humus added to the soil and one a control without the humus. The leaves of the plants grown without humus were yellowish (less green) compared with those of the plants grown in humus-enriched soil. The best explanation for this difference is that
   a. the healthy plants used the food in the decomposing leaves of the humus for energy to make chlorophyll.
   b. the humus made the soil more loosely packed, so water penetrated more easily to the roots.
   c. the humus contained minerals such as magnesium and iron, needed for the synthesis of chlorophyll.
   d. the heat released by the decomposing leaves of the humus caused more rapid growth and chlorophyll synthesis.
   e. the healthy plants absorbed chlorophyll from the humus.

7. The specific relationship between a legume and its symbiotic *Rhizobium* strain probably depends on
   a. each legume having a specific set of early nodulin genes.
   b. each *Rhizobium* strain having a form of nitrogenase that works only in the appropriate legume host.
   c. each legume being found where the soil has only the *Rhizobium* specific to that legume.
   d. specific recognition between the chemical signals and signal receptors of the *Rhizobium* strain and legume species.
   e. destruction of all incompatible *Rhizobium* strains by enzymes secreted from the legume's roots.

8. Mycorrhizae enhance plant nutrition mainly by
   a. absorbing water and minerals through the fungal hyphae.
   b. providing sugar to the root cells, which have no chloroplasts of their own.
   c. converting atmospheric nitrogen to ammonia.
   d. enabling the roots to parasitize neighboring plants.
   e. stimulating the development of root hairs.

9. We would expect the greatest difference in size and general appearance between two groups of plants of the same species, one group with mycorrhizae and one without mycorrhizae, in an environment
   a. where nitrogen-fixing bacteria are abundant.
   b. that has soil with poor drainage.
   c. that has hot summers and cold winters.
   d. in which the soil is relatively deficient in mineral nutrients.
   e. that is near a body of water, such as a pond or river.

10. Carnivorous adaptations of plants mainly compensate for soil that has a relatively low content of
   a. potassium.
   b. nitrogen.
   c. calcium.
   d. water.
   e. phosphate.

For Self-Quiz answers, see Appendix A.

Go to the website or CD-ROM for more quiz questions.

Evolution Connection

Imagine taking the plants out of the picture in Figure 37.9. Write a paragraph explaining how the soil bacteria could sustain the recycling of nitrogen *before* land plants evolved.

Scientific Inquiry

Acid precipitation contains an abnormally high concentration of hydrogen ions (H^+). One effect of acid precipitation is to deplete the soil of plant nutrients such as calcium (Ca^{2+}), potassium (K^+), and magnesium (Mg^{2+}). Suggest a hypothesis to explain how acid precipitation washes these nutrients from the soil. How might you test your hypothesis?

Investigation *How Does Acid Precipitation Affect Mineral Deficiency?*

Science, Technology, and Society

About 10% of U.S. cropland is irrigated. Agriculture is by far the biggest user of water in arid western states, including Colorado, Arizona, and California. The populations of these states are growing, and there is an ongoing conflict between cities and farm regions over water. To ensure adequate water supplies for urban growth, cities are purchasing water rights from farmers. This is often the least expensive way for a city to obtain more water, and it is possible for some farmers to make more money by selling water rights than by growing crops. Discuss the possible consequences of this trend. Is this the best way to allocate water for all concerned? Why or why not?

38 Angiosperm Reproduction and Biotechnology

▲ **Figure 38.1** ***Rafflesia arnoldii,* "monster flower" of Indonesia.**

Key Concepts

38.1 Pollination enables gametes to come together within a flower

38.2 After fertilization, ovules develop into seeds and ovaries into fruits

38.3 Many flowering plants clone themselves by asexual reproduction

38.4 Plant biotechnology is transforming agriculture

Overview

To Seed or Not to Seed

The parasitic plant *Rafflesia arnoldii,* found only in Southeast Asia, spends most of its life invisible to passersby, growing within the woody tissue of a host vine. The plant makes its presence known in a spectacular fashion by producing a cabbage-sized floral bud that eventually develops into a gigantic flower the size of an automobile tire **(Figure 38.1)**. With an odor reminiscent of a decaying corpse, the flower attracts carrion flies that shuttle the pollen from one flower to another. A few days after opening, the flower collapses and shrivels, its function completed. A single flower may produce up to 4 million seeds. However, sexual reproduction, as in *Rafflesia,* is not the only means by which flowering plants (angiosperms) reproduce. Many species also reproduce asexually, creating offspring that are genetically identical to the parent.

The propagation of flowering plants by sexual and asexual reproduction forms the basis of agriculture. Since the dawn of agriculture about 10,000 years ago, plant breeders have genetically manipulated the traits of a few hundred wild angiosperm species by artificial selection, transforming them into the crops we cultivate today. The speed and extent of plant modification have increased dramatically in recent decades with the advent of genetic engineering.

In Chapters 29 and 30, we approached plant reproduction from an evolutionary perspective, tracing the descent of angiosperms and other land plants from their algal ancestors. In this chapter, we will explore the reproductive biology of flowering plants in much greater detail because they are the most important group of plants in most terrestrial ecosystems and in agriculture. After discussing the sexual and asexual reproduction of angiosperms, we will then examine modern plant biotechnology and the role of humans in the genetic alteration of crop species.

Concept 38.1

Pollination enables gametes to come together within a flower

Recall from Chapters 29 and 30 that the life cycles of plants are characterized by an alternation of generations, in which haploid (n) and diploid ($2n$) generations take turns producing each other (see Figures 29.5 and 30.10). The diploid plant, which is the sporophyte, produces haploid spores by meiosis. These spores divide by mitosis, giving rise to the gametophytes, the small male and female haploid plants that produce gametes (sperm and eggs). Fertilization results in diploid zygotes, which divide by mitosis and form new sporophytes.

In angiosperms, the sporophyte is the dominant generation in the sense that it is the largest, most conspicuous, and longest-living plant we see. Over the course of seed plant

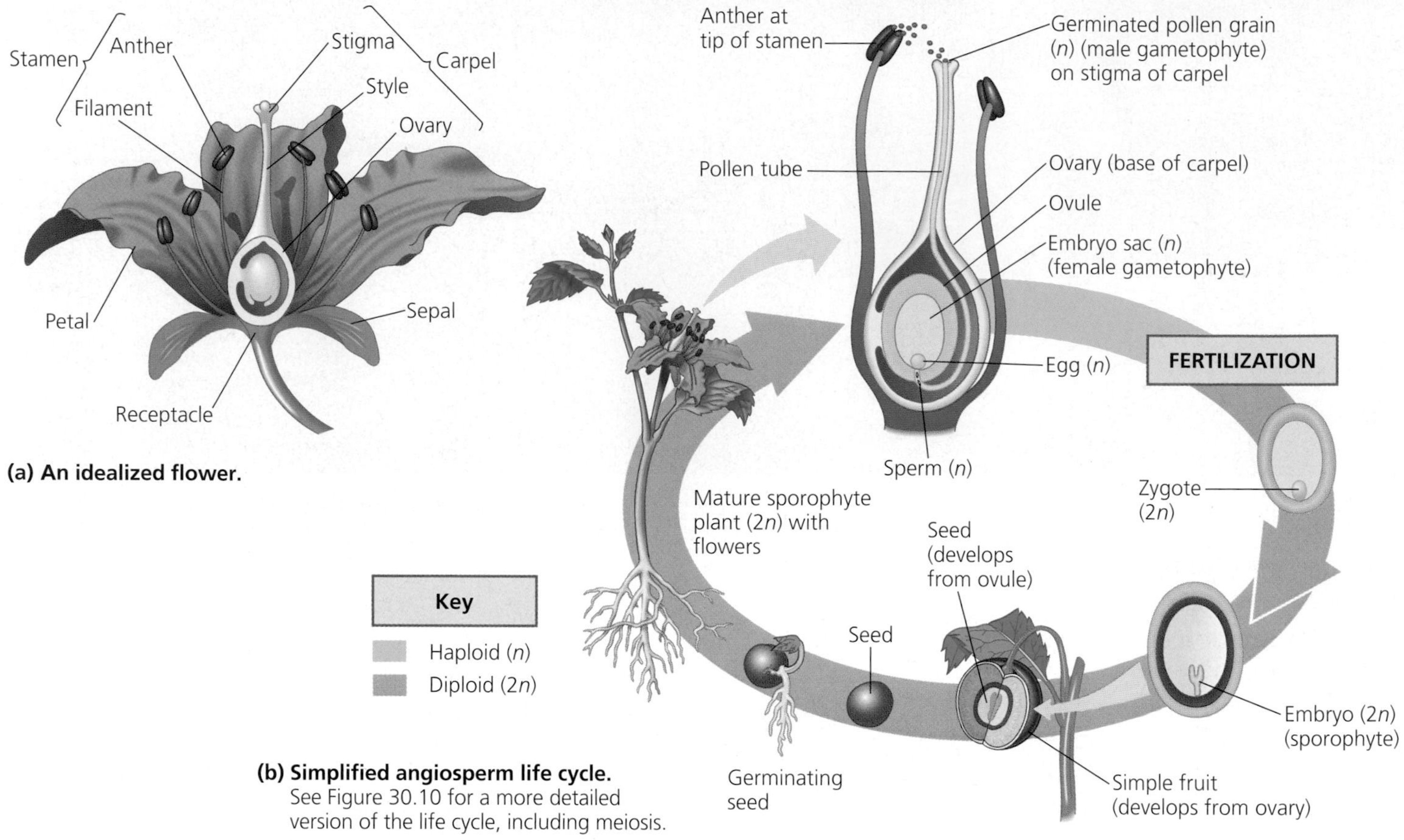

(a) An idealized flower.

(b) Simplified angiosperm life cycle. See Figure 30.10 for a more detailed version of the life cycle, including meiosis.

▲ **Figure 38.2 An overview of angiosperm reproduction.**

evolution, gametophytes became reduced in size and wholly dependent on the sporophyte generation for nutrients. Angiosperm gametophytes are the most reduced of all plants, consisting of only a few cells. Angiosperm sporophytes develop a unique reproductive structure—the flower.

Figure 38.2 reviews the angiosperm life cycle, which is shown in more detail in Figure 30.10. Male and female gametophytes develop within the anthers and ovules, respectively. Pollination by wind, water, or animals brings a pollen grain containing a male gametophyte to the stigma of a flower. Pollen germination brings sperm produced by the male gametophyte to a female gametophyte contained in an ovule embedded in the ovary of a flower. Union of egg and sperm (fertilization) takes place within each ovule in the ovary. Ovules develop into seeds, while the ovary itself becomes a fruit (another unique structure of the angiosperms). In this section, we will focus on the role of the flower in gametophyte development and the process of pollination.

Flower Structure

Flowers, the reproductive shoots of the angiosperm sporophyte, are typically composed of four whorls of highly modified leaves called floral organs, which are separated by very short internodes. Unlike vegetative shoots, which grow indeterminately, flowers are determinate shoots, meaning that they cease growing after the flower and fruit are formed.

Floral organs—**sepals, petals, stamens,** and **carpels**—are attached to a part of the stem called the **receptacle.** Stamens and carpels are reproductive organs, whereas sepals and petals are sterile. Sepals, which enclose and protect the floral bud before it opens, are usually green and more leaflike in appearance than the other floral organs. In many species, petals are more brightly colored than sepals and advertise the flower to insects and other pollinators.

A stamen consists of a stalk called the filament and a terminal structure called the **anther;** within the anther are chambers called pollen sacs, in which pollen is produced. A carpel has an **ovary** at its base and a long, slender neck called the **style.** At the top of the style is a sticky structure called the **stigma** that serves as a landing platform for pollen. Within the ovary are one or more **ovules,** with the number depending on the species. The flower shown in Figure 38.2 has a single carpel, but the flowers of many species have multiple carpels. In most species, two or more carpels are fused into a single structure; the result is an ovary with two or more chambers, each containing one or more ovules. The term **pistil** is sometimes used to refer to a single carpel or to a group of fused carpels. **Figure 38.3** shows examples of variations in floral structure that have evolved during the 140 million years of angiosperm history.

Exploring Floral Variations

Chapter 30 focused on general characteristics of angiosperms (see Figure 30.12). Here we look at some differences in floral traits. Depending on the species, flowers can vary in the presence or absence of sepals, petals, stamens, or carpels. **Complete flowers** have all four basic floral organs (see Figure 38.2a). **Incomplete flowers** lack one or more of these organs. For example, most grass flowers lack petals. Some incomplete flowers are sterile, lacking functional stamens and carpels. Flowers also vary in size, shape, color, odor, and arrangement of floral organs. Much of this diversity represents adaptation to specific groups of pollinators (see Figure 30.13).

SYMMETRY

Flowers can differ in symmetry. In bilateral symmetry, the flower can be divided into two equal parts by a single imaginary line. In radial symmetry, the sepals, petals, stamens, and carpels radiate out from a center. Any imaginary line through the central axis divides the flower into two equal parts. Floral organs can also be either fused or separate. For instance, daffodil petals are fused into a funnel.

OVARY LOCATION

The location of the ovary may vary in relation to the stamens, petals, and sepals. An ovary is called superior if those parts are attached below it, semi-inferior if they are attached alongside it, and inferior if they are attached above it.

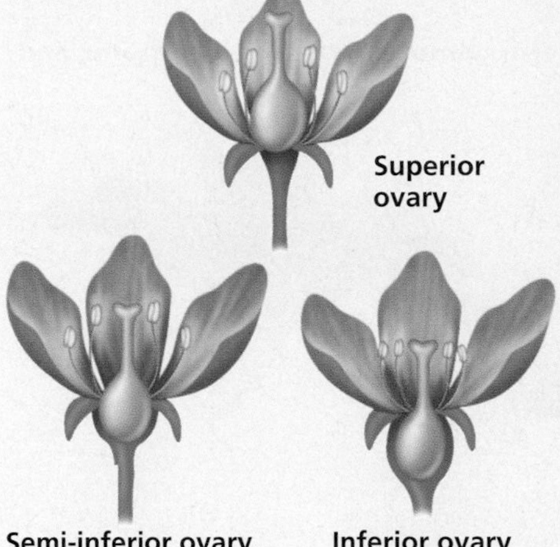

FLORAL DISTRIBUTION

Floral distribution can also differ. Some species have individual flowers, while others have clusters called **inflorescences.**

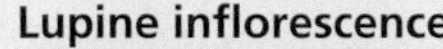

Lupine inflorescence

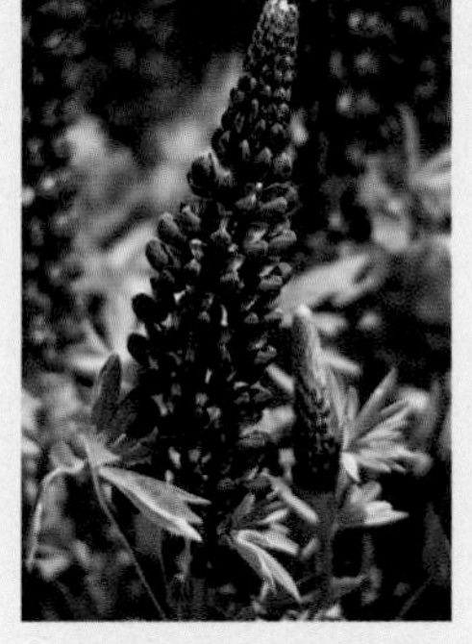

Sunflower inflorescence. A sunflower's central disk actually consists of hundreds of tiny incomplete flowers. What look like petals are actually sterile flowers.

REPRODUCTIVE VARIATIONS

Most species have a single type of flower that has functional stamens and carpels. All complete flowers and some incomplete flowers have functional stamens and carpels. In most incomplete flowers, stamens or carpels are either absent or nonfunctional. Incomplete flowers that have only functional stamens are called staminate, and those with only functional carpels are called carpellate. If staminate and carpellate flowers are on the same plant, the species is said to be **monoecious** (from the Greek word meaning "one house"). A **dioecious** ("two houses") species has staminate flowers and carpellate flowers on separate plants.

Maize, a monoecious species. A maize "ear" (left) consists of kernels (one-seeded fruits) that develop from an inflorescence of fertilized carpellate flowers. Each kernel is derived from a single flower. Each "silk" strand consists of a stigma and long style. The tassels (right) are staminate inflorescences.

Dioecious *Sagittaria latifolia* (common arrowhead). The staminate flower (left) lacks carpels, and the carpellate flower (right) lacks stamens. Having these two types of flowers on separate plants reduces inbreeding.

Gametophyte Development and Pollination

Anthers and ovules bear sporangia, structures where spores are produced by meiosis and gametophytes develop. Pollen grains, each consisting of a mature male gametophyte surrounded by a spore wall, are formed within pollen sacs (microsporangia) of anthers. An egg-producing female gametophyte, or embryo sac, forms within each ovule.

In angiosperms, pollination is the transfer of pollen from an anther to a stigma. If pollination is successful, a pollen grain produces a structure called a pollen tube, which grows and digests its way down into the ovary via the style and discharges sperm in the vicinity of the embryo sac, resulting in fertilization of the egg (see Figure 38.2b). The zygote gives rise to an embryo, and as the embryo grows, the ovule that contains it develops into a seed. The entire ovary, meanwhile, develops into a fruit containing one or more seeds, depending on the species. Fruits, which disperse by dropping to the ground or being carried by wind or animals, help spread seeds some distance from their source plants. When light, soil, and temperature conditions are suitable, seeds germinate and the embryo carried in the seed grows and develops into a seedling.

We will now look more closely at the development of angiosperm gametophytes and the process of pollination. Keep in mind, however, that there are many variations in the details of these processes, depending on the species.

Within the microsporangia (pollen sacs) of an anther are many diploid cells called microsporocytes, also known as microspore mother cells **(Figure 38.4a)**. Each microsporocyte undergoes meiosis, forming four haploid **microspores,** each of which can eventually give rise to a haploid male gametophyte.

▼ Figure 38.4 The development of angiosperm gametophytes (pollen grains and embryo sacs).

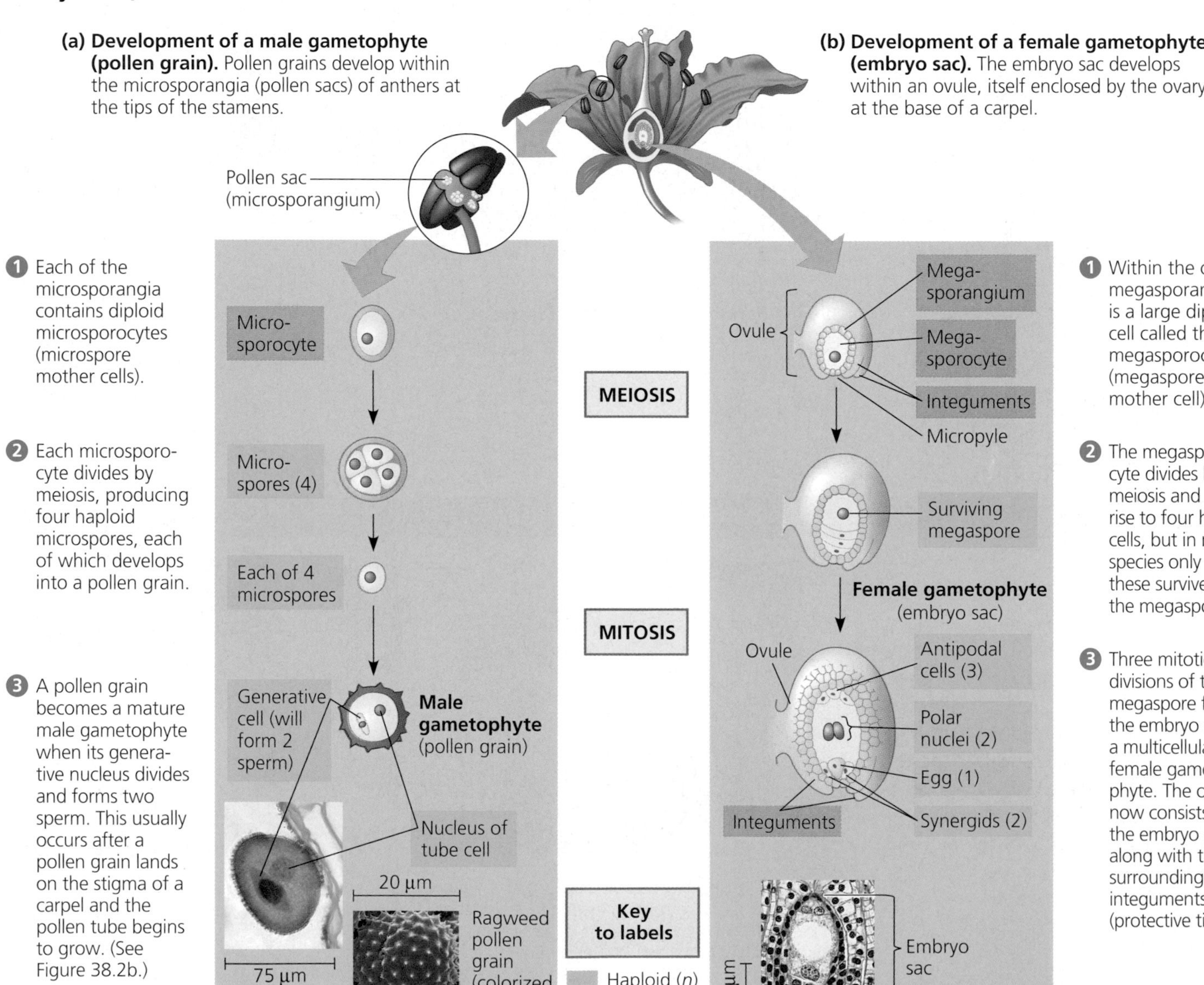

(a) Development of a male gametophyte (pollen grain). Pollen grains develop within the microsporangia (pollen sacs) of anthers at the tips of the stamens.
(b) Development of a female gametophyte (embryo sac). The embryo sac develops within an ovule, itself enclosed by the ovary at the base of a carpel.
Pollen sac (microsporangium)
1 Each of the microsporangia contains diploid microsporocytes (microspore mother cells).
2 Each microsporocyte divides by meiosis, producing four haploid microspores, each of which develops into a pollen grain.
3 A pollen grain becomes a mature male gametophyte when its generative nucleus divides and forms two sperm. This usually occurs after a pollen grain lands on the stigma of a carpel and the pollen tube begins to grow. (See Figure 38.2b.)
Microsporocyte
Microspores (4)
Each of 4 microspores
Generative cell (will form 2 sperm)
Male gametophyte (pollen grain)
Nucleus of tube cell
20 µm
75 µm (LM)
Ragweed pollen grain (colorized SEM)
MEIOSIS
MITOSIS
Key to labels
Haploid (n)
Diploid (2n)
Ovule
Megasporangium
Megasporocyte
Integuments
Micropyle
Surviving megaspore
Female gametophyte (embryo sac)
Ovule
Antipodal cells (3)
Polar nuclei (2)
Egg (1)
Integuments
Synergids (2)
Embryo sac
100 µm
(LM)
1 Within the ovule's megasporangium is a large diploid cell called the megasporocyte (megaspore mother cell).
2 The megasporocyte divides by meiosis and gives rise to four haploid cells, but in most species only one of these survives as the megaspore.
3 Three mitotic divisions of the megaspore form the embryo sac, a multicellular female gametophyte. The ovule now consists of the embryo sac along with the surrounding integuments (protective tissue).

A microspore undergoes mitosis and cytokinesis, producing two separate cells called the generative cell and tube cell. Together, these two cells and the spore wall constitute a pollen grain, which at this stage of its development is an immature male gametophyte. The spore wall usually exhibits an elaborate pattern unique to the particular plant species. During maturation of the male gametophyte, the generative cell passes into the tube cell. The tube cell now has a completely free-standing cell inside it (the generative cell). The tube cell produces the pollen tube, a structure essential for sperm delivery to the egg. During elongation of the pollen tube, the generative cell usually divides and produces two sperm cells, which remain inside the tube cell (see Figure 30.10). The pollen tube grows through the long style of the carpel and into the ovary, where it then releases the sperm cells in the vicinity of an embryo sac.

One or more ovules, each containing a megasporangium, form within the chambers of the ovary **(Figure 38.4b)**. One cell in the megasporangium of each ovule, the megasporocyte (or megaspore mother cell), grows and then goes through meiosis, producing four haploid **megaspores.**

The details of the next steps vary extensively, depending on the species. In most angiosperm species, only one megaspore survives. This megaspore continues to grow, and its nucleus divides by mitosis three times without cytokinesis, resulting in one large cell with eight haploid nuclei. Membranes then partition this mass into a multicellular female gametophyte—the embryo sac. At one end of the embryo sac are three cells: the egg cell and two cells called synergids. The synergids flank the egg cell and function in the attraction and guidance of the pollen tube to the embryo sac. At the opposite end of the embryo sac are three antipodal cells of unknown function. The remaining two nuclei, called polar nuclei, are not partitioned into separate cells but instead share the cytoplasm of the large central cell of the embryo sac. The ovule, which will eventually become a seed, now consists of the embryo sac and two surrounding integuments (layers of protective sporophytic tissue that eventually develop into the seed coat).

Pollination, the transfer of pollen from anther to stigma, is the first step in a chain of events that can lead to fertilization. This step is accomplished in various ways. In some angiosperms, including grasses and many trees, wind is a pollinating agent. In such plants, the release of enormous quantities of pollen compensates for the randomness of this dispersal mechanism. At certain times of the year, the air is loaded with pollen grains, as anyone plagued with pollen allergies can attest. Some aquatic plants rely on water to disperse pollen. Most angiosperms, however, depend on insects, birds, or other animals to transfer pollen directly to other flowers.

Mechanisms That Prevent Self-Fertilization

Generally, one of the great advantages of sexual reproduction is that it increases the genetic diversity of offspring, thereby increasing the likelihood that at least some of the offspring will survive such challenges as environmental change and pathogens (see Chapter 23). Nevertheless, some flowers, such as garden peas, self-fertilize. This process, called "selfing," can be a desirable attribute in some crop plants because it ensures that a seed will develop. Many angiosperm species, however, have mechanisms that make it difficult or impossible for a flower to fertilize itself. The various barriers that prevent self-fertilization contribute to genetic variety by ensuring that sperm and eggs come from different parents. In dioecious species, of course, plants cannot self-fertilize because they have either staminate or carpellate flowers (see Figure 38.3). And in some plants that have flowers with functional stamens and carpels, these floral organs mature at different times or are structurally arranged in such a way that it is unlikely that an animal pollinator could transfer pollen from an anther to a stigma of the same flower **(Figure 38.5)**. However, the most common anti-selfing mechanism in flowering plants is known as **self-incompatibility,** the ability of a plant to reject its own pollen and sometimes the pollen of closely related individuals. If a pollen grain lands on a stigma of a flower on the same plant, a biochemical block prevents the pollen from completing its development and fertilizing an egg.

Researchers are unraveling the molecular mechanisms that are involved in self-incompatibility. This plant response is analogous to the immune response of animals in that both are based on the ability of organisms to distinguish the cells of "self" from those of "nonself." The key difference is that the animal immune system rejects nonself, as when the system mounts a defense against a pathogen or attempts to reject a transplanted organ. Self-incompatibility in plants, in contrast, is a rejection of self.

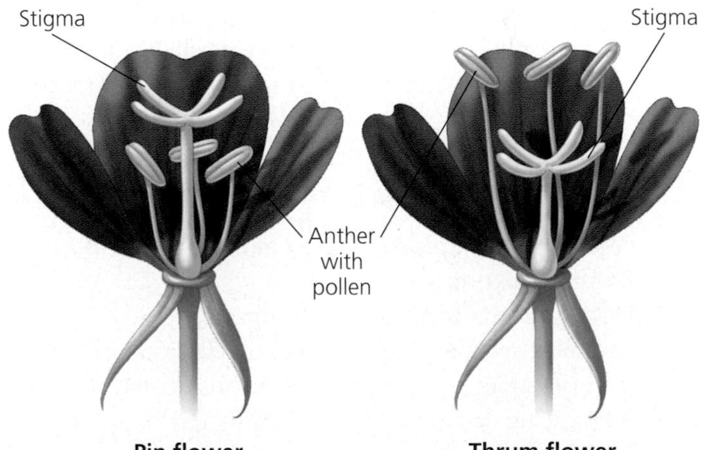

▲ **Figure 38.5 "Pin" and "thrum" flower types reduce self-fertilization.** Some species produce two types of flowers: "pins," which have long styles and short stamens, and "thrums," which have short styles and long stamens. An insect foraging for nectar would collect pollen on different parts of its body; pin pollen would be deposited on thrum stigmas, and vice versa.

Recognition of "self" pollen is based on genes for self-incompatibility, called *S*-genes. In the gene pool of a plant population, there can be dozens of alleles of an *S*-gene. If a pollen grain has an allele that matches an allele of the stigma on which it lands, the pollen tube fails to grow. Depending on the species, self-recognition blocks pollen tube growth by one of two molecular mechanisms: gametophytic self-incompatibility or sporophytic self-incompatibility.

In gametophytic self-incompatibility, the *S*-allele in the pollen genome governs the blocking of fertilization. For example, an S_1 pollen grain from an S_1S_2 parental sporophyte will fail to fertilize eggs of an S_1S_2 flower but will fertilize an S_2S_3 flower. An S_2 pollen grain would not fertilize either flower. Self-recognition of this kind involves the enzymatic destruction of RNA within a pollen tube. RNA-hydrolyzing enzymes are produced by the style and enter the pollen tube. If the pollen tube is a "self" type, these enzymes destroy its RNA.

In sporophytic self-incompatibility, fertilization is blocked by *S*-allele gene products in tissues of the parental sporophyte that adhere to the pollen wall. For example, neither an S_1 nor S_2 pollen grain from an S_1S_2 parental sporophyte will fertilize eggs of an S_1S_2 flower or S_2S_3 flower. Sporophytic incompatibility involves a signal transduction pathway in epidermal cells of the stigma that prevents germination of the pollen grain.

Some crops, such as nonhybrid cultivated varieties of peas, maize, and tomatoes, routinely self-pollinate with satisfactory results. However, plant breeders sometimes hybridize different varieties of a crop plant in order to combine the best traits of the varieties and counter the loss of vigor that can result from excessive inbreeding (see Chapter 14). To obtain hybrid seeds, plant breeders currently must prevent self-fertilization either by laboriously removing the anthers from the parent plants that provide the seeds or by developing male sterile plants. The latter option is increasingly important. Eventually, it may also be possible to impose self-incompatibility on crop species that are now self-compatible. Basic research on mechanisms of self-incompatibility may therefore lead to agricultural applications.

Concept Check 38.1

1. Give some examples of how form fits function in flower structure.
2. Distinguish pollination from fertilization.
3. Given the seeming disadvantages of selfing as a reproductive "strategy" in nature, it is surprising that about 20% of angiosperm species primarily rely on selfing. Although fairly common in nature, self-fertilization has been called an "evolutionary dead end." Suggest a reason why selfing might be selected for in nature and yet still be an "evolutionary dead end."

For suggested answers, see Appendix A.

Concept 38.2

After fertilization, ovules develop into seeds and ovaries into fruits

We have traced the processes of gametophyte development and pollination. Now we will look at fertilization and its products: seeds and fruits.

Double Fertilization

After landing on a receptive stigma, a pollen grain absorbs moisture and germinates; that is, it produces a pollen tube that extends down between the cells of the style toward the ovary **(Figure 38.6)**. The nucleus of the generative cell divides by mitosis and forms two sperm. Directed by a chemical attractant, possibly calcium, the tip of the pollen tube enters the ovary, probes

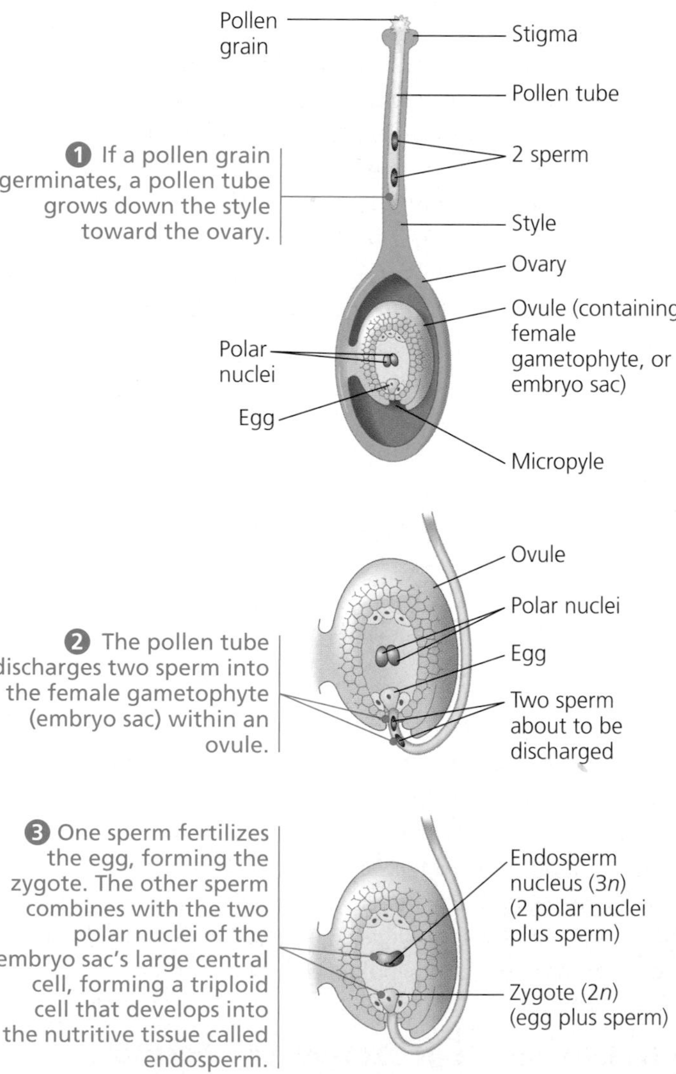

Pollen grain
Stigma
Pollen tube
1 If a pollen grain germinates, a pollen tube grows down the style toward the ovary.
2 sperm
Style
Ovary
Ovule (containing female gametophyte, or embryo sac)
Polar nuclei
Egg
Micropyle
Ovule
Polar nuclei
Egg
2 The pollen tube discharges two sperm into the female gametophyte (embryo sac) within an ovule.
Two sperm about to be discharged
3 One sperm fertilizes the egg, forming the zygote. The other sperm combines with the two polar nuclei of the embryo sac's large central cell, forming a triploid cell that develops into the nutritive tissue called endosperm.
Endosperm nucleus (3n) (2 polar nuclei plus sperm)
Zygote (2n) (egg plus sperm)

▲ **Figure 38.6 Growth of the pollen tube and double fertilization.**

through the micropyle (a gap in the integuments of the ovule), and discharges its two sperm near or within the embryo sac.

The events that follow are a distinctive feature of the angiosperm life cycle. One sperm fertilizes the egg to form the zygote. The other sperm combines with the two polar nuclei to form a triploid ($3n$) nucleus in the center of the large central cell of the embryo sac. This large cell will give rise to the **endosperm,** a food-storing tissue of the seed. The union of two sperm cells with different nuclei of the embryo sac is called **double fertilization.** Double fertilization ensures that the endosperm will develop only in ovules where the egg has been fertilized, thereby preventing angiosperms from squandering nutrients.

The tissues surrounding the embryo sac have prevented researchers from being able to directly observe fertilization in plants grown under normal conditions. Recently, however, scientists have isolated sperm from germinated pollen grains and eggs from embryo sacs and have observed the merging of plant gametes *in vitro* (in an artificial environment). The first cellular event that takes place after gamete fusion is an increase in the cytoplasmic calcium (Ca^{2+}) levels of the egg, as also occurs during animal gamete fusion (see Chapter 47). Another similarity to animals is the establishment of a block to polyspermy, the fertilization of an egg by more than one sperm cell. Thus, maize (*Zea mays*) sperm cannot fuse with zygotes *in vitro.* In maize, this barrier to polyspermy is established as early as 45 seconds after the initial sperm fusion with the egg.

From Ovule to Seed

After double fertilization, each ovule develops into a seed, and the ovary develops into a fruit enclosing the seed(s). As the embryo develops from the zygote, the seed stockpiles proteins, oils, and starch to varying extents, depending on the species. This is why seeds are such major sugar sinks (see Chapter 36). Initially, these nutrients are stored in the endosperm, but later in seed development in many species, the storage function of the endosperm is more or less taken over by the swelling cotyledons of the embryo.

Endosperm Development

Endosperm development usually precedes embryo development. After double fertilization, the triploid nucleus of the ovule's central cell divides, forming a multinucleate "supercell" having a milky consistency. This liquid mass, the endosperm, becomes multicellular when cytokinesis partitions the cytoplasm by forming membranes between the nuclei. Eventually, these "naked" cells produce cell walls, and the endosperm becomes solid. Coconut "milk" is an example of liquid endosperm; coconut "meat" is an example of solid endosperm. The white fluffy part of popcorn is also solid endosperm.

In grains and most other monocots, as well as many eudicots, the endosperm stores nutrients that can be used by the seedling after germination. In other eudicots (including bean seeds), the food reserves of the endosperm are completely exported to the cotyledons before the seed completes its development; consequently, the mature seed lacks endosperm.

Embryo Development

The first mitotic division of the zygote is transverse, splitting the fertilized egg into a basal cell and a terminal cell **(Figure 38.7)**. The terminal cell eventually gives rise to most of the embryo. The basal cell continues to divide transversely, producing a

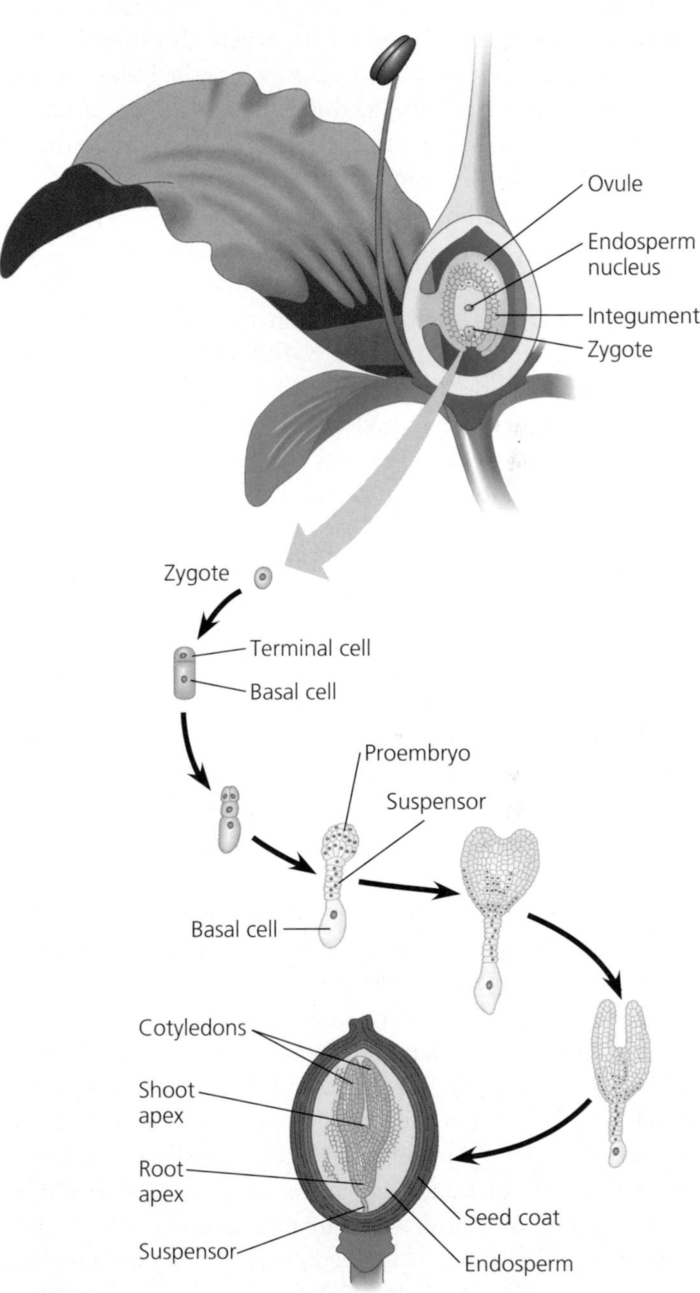

▲ **Figure 38.7 The development of a eudicot plant embryo.** By the time the ovule becomes a mature seed and the integuments harden and thicken to form the seed coat, the zygote has given rise to an embryonic plant with rudimentary organs.

thread of cells called the suspensor, which anchors the embryo to its parent. The suspensor functions in the transfer of nutrients to the embryo from the parent plant and, in some plants, from the endosperm. As the suspensor elongates, it also pushes the embryo deeper into the nutritive and protective tissues. Meanwhile, the terminal cell divides several times and forms a spherical proembryo attached to the suspensor. The cotyledons begin to form as bumps on the proembryo. A eudicot, with its two cotyledons, is heart-shaped at this stage. Only one cotyledon develops in monocots.

Soon after the rudimentary cotyledons appear, the embryo elongates. Cradled between the cotyledons is the embryonic shoot apex, which includes the shoot apical meristem. At the opposite end of the embryo's axis, where the suspensor attaches, is the embryonic root apex, which includes the root apical meristem. After the seed germinates, the apical meristems at the tips of shoots and roots will sustain primary growth as long as the plant lives (see Figure 35.10).

Structure of the Mature Seed

During the last stages of its maturation, the seed dehydrates until its water content is only about 5–15% of its weight. The embryo, surrounded by a food supply (cotyledons, endosperm, or both), becomes dormant; it stops growing and its metabolism nearly ceases. The embryo and its food supply are enclosed by a hard, protective **seed coat** formed from the integuments of the ovule.

We can take a closer look at one type of eudicot seed by splitting open the seed of a common garden bean. At this stage, the embryo consists of an elongate structure, the embryonic axis, attached to fleshy cotyledons **(Figure 38.8a)**. Below the point at which the cotyledons are attached, the embryonic axis is called the **hypocotyl** (from the Greek *hypo,* under). The hypocotyl terminates in the **radicle,** or embryonic root. The portion of the embryonic axis above where the cotyledons are attached is the **epicotyl** (from the Greek *epi,* on, over). It consists of the shoot tip with a pair of miniature leaves.

The cotyledons of the common garden bean are packed with starch before the seed germinates because they absorbed carbohydrates from the endosperm when the seed developed. However, the seeds of some eudicots, such as castor beans (*Ricinus communis*), retain their food supply in the endosperm and have cotyledons that are very thin **(Figure 38.8b)**. The cotyledons absorb nutrients from the endosperm and transfer them to the rest of the embryo when the seed germinates.

The embryo of a monocot has a single cotyledon **(Figure 38.8c)**. Members of the grass family, including maize and wheat, have a specialized type of cotyledon called a **scutellum** (from the Latin *scutella,* small shield, a reference to the scutellum's shape). The scutellum is very thin, with a large surface area pressed against the endosperm, from which the scutellum absorbs nutrients during germination. The embryo of a grass seed

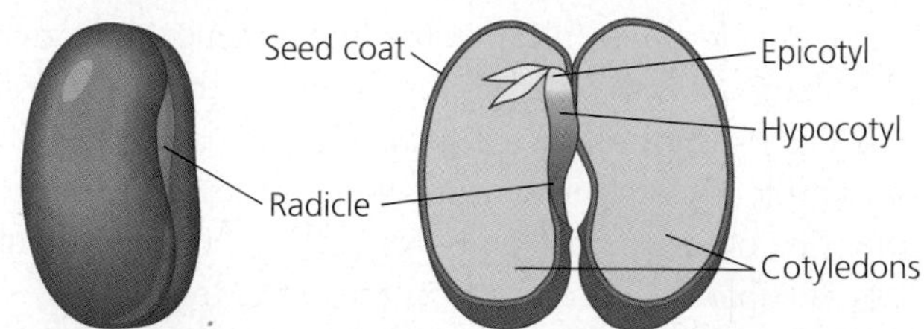

(a) Common garden bean, a eudicot with thick cotyledons. The fleshy cotyledons store food absorbed from the endosperm before the seed germinates.

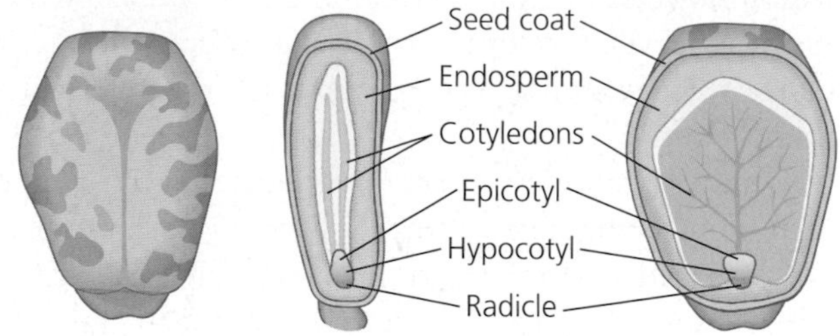

(b) Castor bean, a eudicot with thin cotyledons. The narrow, membranous cotyledons (shown in edge and flat views) absorb food from the endosperm when the seed germinates.

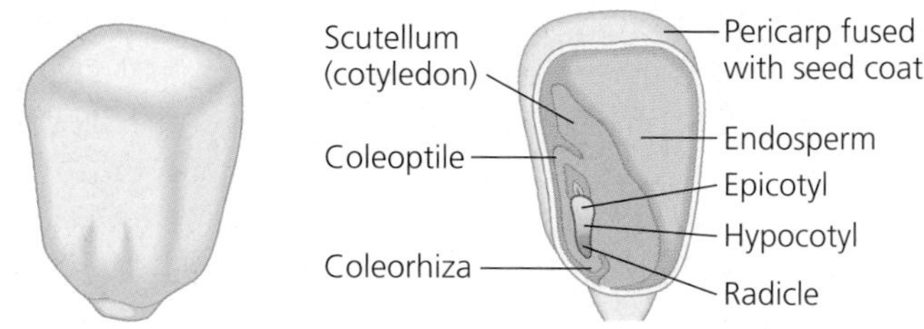

(c) Maize, a monocot. Like all monocots, maize has only one cotyledon. Maize and other grasses have a large cotyledon called a scutellum. The rudimentary shoot is sheathed in a structure called the coleoptile, and the coleorhiza covers the young root.

▲ **Figure 38.8 Seed structure.**

is enclosed by two sheathes: a **coleoptile,** which covers the young shoot, and a **coleorhiza,** which covers the young root.

From Ovary to Fruit

While the seeds are developing from ovules, the ovary of the flower is developing into a **fruit,** which protects the enclosed seeds and, when mature, aids in their dispersal by wind or animals. Fertilization triggers hormonal changes that cause the ovary to begin its transformation into a fruit. If a flower has not been pollinated, fruit usually does not develop, and the entire flower withers and falls away.

During fruit development, the ovary wall becomes the pericarp, the thickened wall of the fruit. As the ovary grows, the other parts of the flower wither and are shed. For example, the pointed tip of a pea pod is the withered remains of the pea flower's stigma.

Fruits are classified into several types, depending on their developmental origin **(Figure 38.9)**. Most fruits are derived from a single carpel or several fused carpels and are called

simple fruits. Some simple fruits are fleshy, such as a peach, whereas others are dry, such as a pea pod or a nut (see Figure 30.8). An **aggregate fruit** results from a single flower that has more than one separate carpel, each forming a small fruit. These "fruitlets" are clustered together on a single receptacle, as in a raspberry. A **multiple fruit** develops from an inflorescence, a group of flowers tightly clustered together. When the walls of the many ovaries start to thicken, they fuse together and become incorporated into one fruit, as in a pineapple.

In some angiosperms, other floral parts in addition to ovaries contribute to what we commonly call the fruit. Such fruits are called accessory fruits. In apple flowers, for example, the ovary is embedded in the receptacle (see Figure 38.2b), and the fleshy part of this simple fruit is derived mainly from the enlarged receptacle; only the apple core develops from the ovary. Another example is the strawberry, an aggregate fruit consisting of an enlarged receptacle embedded with tiny one-seeded fruits.

A fruit usually ripens about the same time that its seeds complete their development. Whereas the ripening of a dry fruit, such as a soybean pod, involves the aging and drying out of fruit tissues, the process in a fleshy fruit is more elaborate. Complex interactions of hormones result in an edible fruit that serves as an enticement to animals that help spread the seeds. The fruit's "pulp" becomes softer as a result of enzymes digesting components of the cell walls. There is usually a color change from green to another color, such as red, orange, or yellow. The fruit becomes sweeter as organic acids or starch molecules are converted to sugar, which may reach a concentration of as much as 20% in a ripe fruit.

Seed Germination

As a seed matures, it dehydrates and enters a phase referred to as **dormancy** (from the Latin word meaning "to sleep"), a condition of extremely low metabolic rate and suspension of growth and development. Conditions required to break dormancy vary between plant species. Some seeds germinate as soon as they are in a suitable environment. Other seeds remain dormant even if sown in a favorable place until a specific environmental cue causes them to break dormancy.

Seed Dormancy: Adaptation for Tough Times

Seed dormancy increases the chances that germination will occur at a time and place most advantageous to the seedling. Breaking dormancy generally requires certain environmental conditions. Seeds of many desert plants, for instance, germinate only after a substantial rainfall. If they were to germinate after a mild drizzle, the soil might become too dry to support the seedlings. Where natural fires are common, many seeds require intense heat to break dormancy; seedlings are therefore

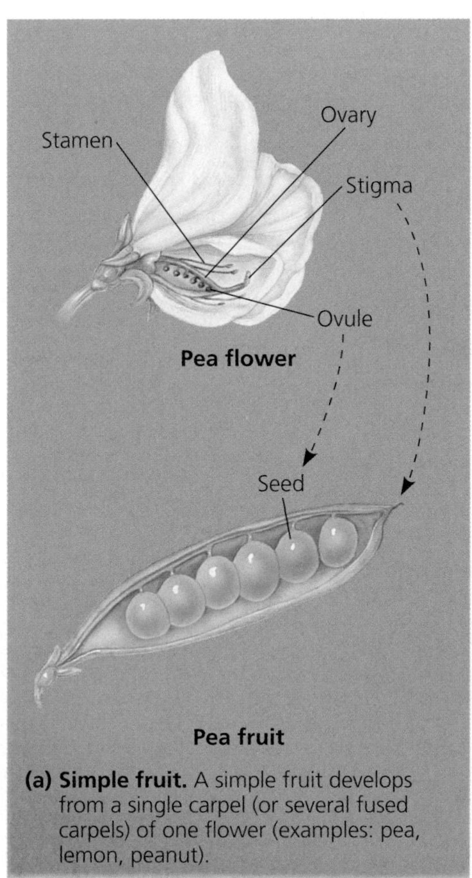

(a) Simple fruit. A simple fruit develops from a single carpel (or several fused carpels) of one flower (examples: pea, lemon, peanut).

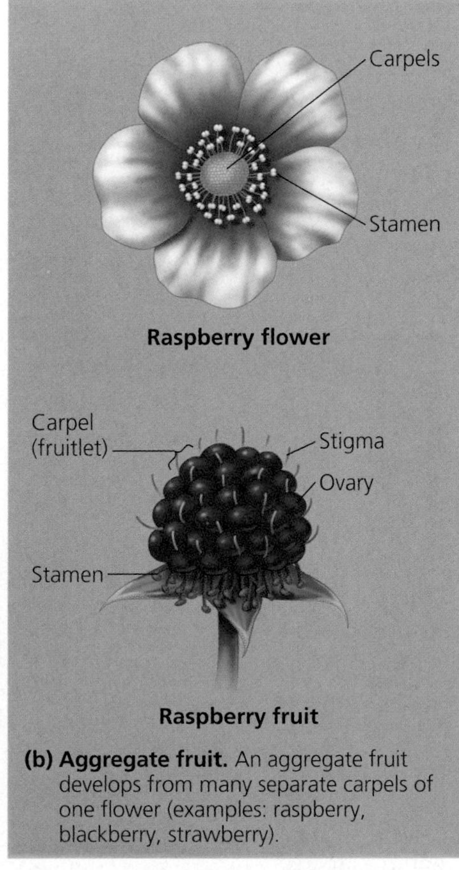

(b) Aggregate fruit. An aggregate fruit develops from many separate carpels of one flower (examples: raspberry, blackberry, strawberry).

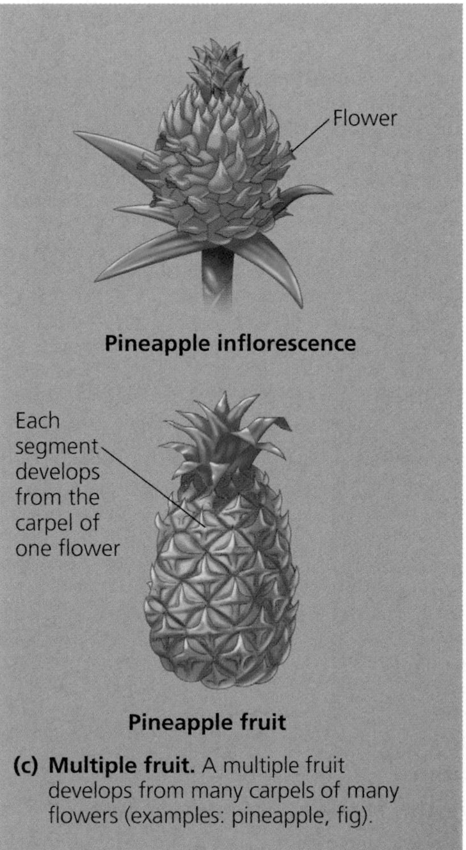

(c) Multiple fruit. A multiple fruit develops from many carpels of many flowers (examples: pineapple, fig).

▲ **Figure 38.9 Developmental origin of fruits.**

most abundant after fire has cleared away competing vegetation. Where winters are harsh, seeds may require extended exposure to cold. Seeds sown during summer or fall do not germinate until the following spring, ensuring a long growth season before the next winter. Some small seeds, such as those of some varieties of lettuce, require light for germination and will break dormancy only if they are buried shallow enough for the seedlings to poke through the soil surface. Some seeds have coats that must be weakened by chemical attack as they pass through an animal's digestive tract, and thus are likely to be carried some distance before germinating.

The length of time a dormant seed remains viable and capable of germinating varies from a few days to decades or even longer, depending on the species and environmental conditions. Most seeds are durable enough to last a year or two until conditions are favorable for germinating. Thus, the soil has a bank of ungerminated seeds that may have accumulated for several years. This is one reason vegetation reappears so rapidly after a fire, drought, flood, or some other environmental disruption.

From Seed to Seedling

Germination of seeds depends on the physical process called **imbibition,** the uptake of water due to the low water potential of the dry seed. Imbibing water causes the seed to expand and rupture its coat and also triggers metabolic changes in the embryo that enable it to resume growth. Following hydration, enzymes begin digesting the storage materials of the endosperm or cotyledons, and the nutrients are transferred to the growing regions of the embryo.

The first organ to emerge from the germinating seed is the radicle, the embryonic root. Next, the shoot tip must break through the soil surface. In garden beans and many other eudicots, a hook forms in the hypocotyl, and growth pushes the hook above ground **(Figure 38.10a)**. Stimulated by light, the hypocotyl straightens, raising the cotyledons and epicotyl. Thus, the delicate shoot apex and bulky cotyledons are pulled upward rather than being pushed tip-first through the abrasive soil. The epicotyl now spreads its first foliage leaves (true leaves, so called to distinguish them from cotyledons, or "seed leaves"). The foliage leaves expand, become green, and begin making food by photosynthesis. The cotyledons shrivel and fall away from the seedling, their food reserves having been exhausted by the germinating embryo.

Maize and other grasses, which are monocots, use a different method for breaking ground when they germinate **(Figure 38.10b)**. The coleoptile, the sheath enclosing and protecting the embryonic shoot, pushes upward through the soil and into the air. The shoot tip then grows straight up through the tunnel provided by the tubular coleoptile and eventually breaks out through the coleoptile's tip.

Seed germination is a precarious stage in a plant's life. The tough seed gives rise to a fragile seedling that will be exposed to predators, parasites, wind, and other hazards. In the wild, only a small fraction of seedlings endure long enough to become parents themselves. Production of enormous numbers of seeds compensates for the odds against individual survival and gives natural selection ample genetic variations to screen. However, this is a very expensive means of reproduction in terms of the resources consumed in flowering and fruiting. Asexual reproduction, generally simpler and less hazardous for offspring, is an alternative means of plant propagation.

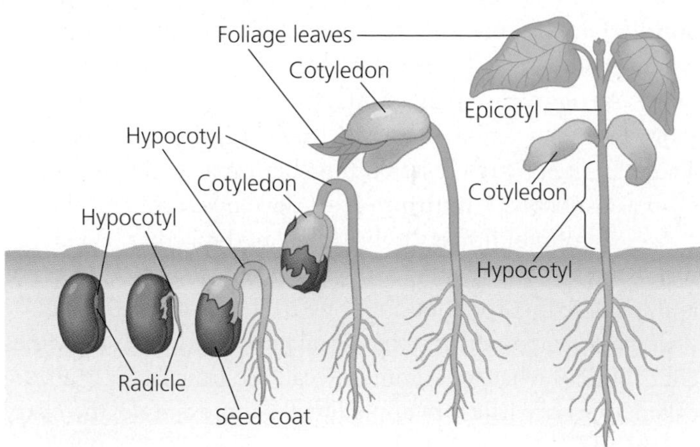

(a) Common garden bean. In common garden beans, straightening of a hook in the hypocotyl pulls the cotyledons from the soil.

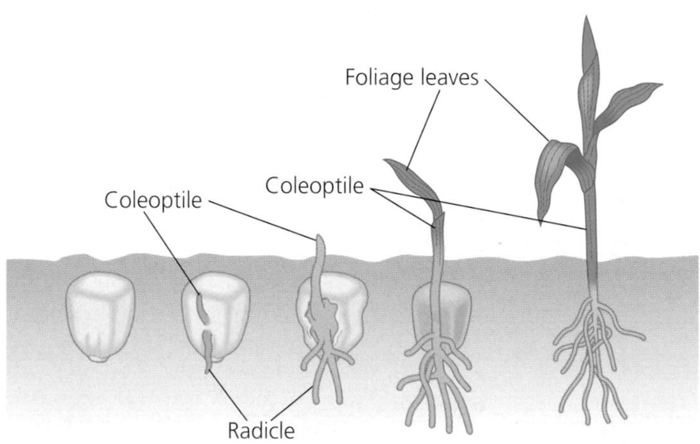

(b) Maize. In maize and other grasses, the shoot grows straight up through the tube of the coleoptile.

▲ **Figure 38.10 Two common types of seed germination.**

Concept Check 38.2

1. Some dioecious species have the XY sexual genotype for male and XX for female. After double fertilization, what would be the genotypes of the endosperm nuclei and the embryos?
2. Explain how the basic structures of seeds and fruits facilitate reproduction.
3. What is the advantage of seed dormancy?

For suggested answers, see Appendix A.

Concept 38.3

Many flowering plants clone themselves by asexual reproduction

Imagine chopping off your finger and watching it develop into an exact copy of you. This would be an example of **asexual reproduction,** in which offspring are derived from a single parent without genetic recombination. (Of course, asexual reproduction does not occur in humans.) The result would be a clone, an asexually produced, genetically identical organism.

Like many seedless plants and some gymnosperms, many angiosperm species reproduce both asexually and sexually. If reproducing asexually, a plant passes on all of its alleles to its offspring. If reproducing sexually, a plant passes on only half of its alleles. Each reproductive mode offers advantages in certain situations.

If a plant is superbly suited to a stable environment, asexual reproduction has advantages. A plant can clone many copies of itself, and if the environmental conditions remain stable, these clones, too, will be well suited for that environment. Also, the offspring are not as frail as seedlings produced by sexual reproduction in seed plants. The offspring are usually mature vegetative fragments from the parent plant, which is why asexual reproduction in plants is also known as **vegetative reproduction.** A sprawling clone of prairie grass may cover an area so thoroughly that seedlings of the same or other species have little chance of competing.

In unstable environments, where evolving pathogens and other variables affect survival and reproductive success, sexual reproduction can be an advantage because it generates variation in offspring and populations. In contrast, the genotypic uniformity of asexually produced plants puts them at great risk of local extinction if there is a catastrophic environmental change, such as a new strain of disease. Moreover, seeds (which are almost always produced sexually) facilitate the dispersal of offspring to more distant locations. And finally, seed dormancy allows growth to be suspended until environmental conditions are more favorable.

Mechanisms of Asexual Reproduction

Asexual reproduction in plants is typically an extension of the capacity for indeterminate growth. Plants, remember, have meristematic tissues of dividing, undifferentiated cells that can sustain or renew growth indefinitely. In addition, parenchyma cells throughout the plant can divide and differentiate into more specialized types of cells, enabling plants to regenerate lost parts. Detached vegetative fragments of some plants can develop into whole offspring; a severed stem, for instance, may develop adventitious roots and become a whole

▲ **Figure 38.11 Asexual reproduction in aspen trees.** Some aspen groves, such as those shown here, actually consist of thousands of trees descended by asexual reproduction. Each grove of trees derives from the root system of one parent. Notice that genetic differences between groves descended from different parents result in different timing for the development of fall color and the loss of leaves.

plant. **Fragmentation,** the separation of a parent plant into parts that develop into whole plants, is one of the most common modes of asexual reproduction. For example, in some species the root system of a single parent gives rise to many adventitious shoots that become separate shoot systems. The result is a clone formed by asexual reproduction from one parent **(Figure 38.11)**. Such asexual propagation has produced the oldest of all known plant clones, a ring of creosote bushes in the Mojave Desert of California, believed to be at least 12,000 years old.

An entirely different mechanism of asexual reproduction has evolved in dandelions and some other plants. These plants can sometimes produce seeds without pollination or fertilization. This asexual production of seeds is called **apomixis** (from the Greek words meaning "away from the act of mixing") because there is no joining of sperm and egg. Instead, a diploid cell in the ovule gives rise to the embryo, and the ovules mature into seeds, which in the dandelion are dispersed by windblown fruits. Thus, these plants clone themselves by an asexual process but have the advantage of seed dispersal, usually associated with sexual reproduction.

Vegetative Propagation and Agriculture

With the objective of improving crops and ornamental plants, humans have devised various methods for asexual propagation of angiosperms. Most of these methods are based on the ability of plants to form adventitious roots or shoots.

Clones from Cuttings

Most houseplants, woody ornamentals, and orchard trees are asexually reproduced from plant fragments called cuttings. In some cases, shoot or stem cuttings are used. At the cut end of the shoot, a mass of dividing, undifferentiated cells called a **callus** forms, and then adventitious roots develop from the callus. If the shoot fragment includes a node, then adventitious roots form without a callus stage. Some plants, including African violets, can be propagated from single leaves rather than stems. For still other plants, cuttings are taken from specialized storage stems. For example, a potato can be cut up into several pieces, each with a vegetative bud, or "eye," that regenerates a whole plant.

Grafting

In a modification of vegetative reproduction from cuttings, a twig or bud from one plant can be grafted onto a plant of a closely related species or a different variety of the same species. Grafting makes it possible to combine the best qualities of different species or varieties into a single plant. The graft is usually done when the plant is young. The plant that provides the root system is called the **stock;** the twig grafted onto the stock is referred to as the **scion.** For example, scions from French varieties of vines that produce superior wine grapes are grafted onto root stocks of American varieties, which are more resistant to certain soil pathogens. The genes of the scion determine the quality of the fruit, so it is not diminished by the genetic makeup of the stock. In some cases of grafting, however, the stock can alter the characteristics of the shoot system that develops from the scion. For example, dwarf fruit trees, which allow for easier harvesting of the fruit, are made by grafting normal twigs onto stocks of dwarf varieties that retard the vegetative growth of the shoot system. Because seeds are produced by the part of the plant derived from the scion, they give rise to plants of the scion species when planted.

Test-Tube Cloning and Related Techniques

Plant biologists have adopted *in vitro* methods to create and clone novel plant varieties. It is possible to grow whole plants by culturing small explants (pieces of tissue cut from the parent) or even single parenchyma cells on an artificial medium containing nutrients and hormones **(Figure 38.12a)**. The cultured cells divide and form an undifferentiated callus. When the hormonal balance is manipulated in the culture medium, the callus can sprout shoots and roots with fully differentiated cells **(Figure 38.12b)**. The test-tube plantlets can then be transferred to soil, where they continue their growth. A single plant can be cloned into thousands of copies by subdividing calluses as they grow. This method is used for propagating orchids and for cloning pine trees that produce wood at unusually high rates.

Plant tissue culture also facilitates genetic engineering. Most techniques for the introduction of foreign genes into plants require small pieces of plant tissue or single plant cells as the starting material. In plant biology, the term **transgenic** is used to describe genetically modified (GM) organisms that have been engineered to express a gene from another species. Test-tube culture makes it possible to regenerate GM plants from a single plant cell into which the foreign DNA has been incorporated. The techniques of genetic engineering are discussed in more detail in Chapter 20.

(a) Just a few parenchyma cells from a carrot gave rise to this callus, a mass of undifferentiated cells.

(b) The callus differentiates into an entire plant, with leaves, stems, and roots.

▲ **Figure 38.12 Test-tube cloning of carrots.** (See also Figure 21.5.)

Some researchers are coupling a technique that is known as **protoplast fusion** with tissue culture methods to invent new plant varieties that can be cloned. Protoplasts are plant cells that have had their cell walls removed by treatment with enzymes (cellulases and pectinases) isolated from fungi **(Figure 38.13)**. Before they are cultured, protoplasts can be screened for mutations that may improve the agricultural value of the plant. In some cases, it is possible to fuse two protoplasts from different plant species that would otherwise be reproductively incompatible, and then culture the hybrid protoplasts. Each protoplast can regenerate a wall and eventually form a hybrid plantlet. One success of this method is a hybrid between a potato and a wild relative called black nightshade. The nightshade is resistant to a herbicide commonly used to kill weeds. The hybrids are also resistant, making it possible to "weed" a field with the herbicide without killing the potato plants.

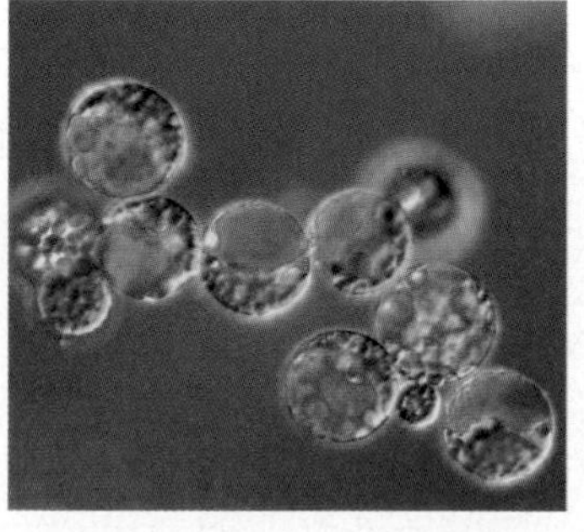

◀ **Figure 38.13 Protoplasts.** These wall-less plant cells are prepared by treating cells or tissues with wall-degrading enzymes isolated from certain types of fungi. Researchers can fuse protoplasts from different species to make hybrids and can also culture the hybrid cells to produce a new plant (LM).

The *in vitro* culturing of plant cells and tissues is fundamental to most types of plant biotechnology. The other basic process is the ability to produce transgenic plants through various methods of genetic engineering. In the next section, we take a closer look at plant biotechnology.

Concept Check 38.3

1. Explain how both asexual and sexual reproduction contribute to the reproductive success of plants.
2. The seedless banana, the world's most popular fruit, is losing the battle against two fungal epidemics. Why do such epidemics generally pose a greater risk to asexually propagated crops?

For suggested answers, see Appendix A.

Concept 38.4

Plant biotechnology is transforming agriculture

Plant biotechnology has two meanings. In the general sense, it refers to innovations in the use of plants (or substances obtained from plants) to make products of use to humans—an endeavor that began in prehistory. In a more specific sense, biotechnology refers to the use of GM organisms in agriculture and industry. Indeed, in the last two decades, genetic engineering has become such a powerful force that the terms *genetic engineering* and *biotechnology* have become synonymous in the media. In this last section, we expand on the discussion in Chapter 20 by examining how humans have altered plants to suit their purposes.

Artificial Selection

Humans have intervened in the reproduction and genetic makeup of plants for thousands of years. It is no exaggeration to say that maize is an unnatural monster created by humans. Left on its own in nature, maize would soon become extinct for the simple reason that it cannot spread its seeds. Maize kernels are not only permanently attached to the central axis (the "cob") but also permanently protected by tough, overlapping leaf sheathes (the "husk"). These attributes arose not by natural selection but by artificial selection by humans **(Figure 38.14)**. (See Chapter 22 to review the basic concept of artificial selection.) Indeed, Neolithic (late Stone Age) humans domesticated most of our crop species over a relatively short period about 10,000 years ago. But genetic modification began long before humans started altering crops by artificial selection. For example, the wheat species we rely on for much of our food are the result of natural hybridization between different species of grasses. Such hybridization is common in plants and has long been exploited by breeders to introduce additional genetic variation for artificial selection and crop improvement. One modern example is the selective breeding of maize.

Maize is a staple in many developing countries, but the most common varieties are poor sources of protein, requiring that diets be supplemented with other sources, such as beans. The proteins in popular maize varieties are very low in lysine and tryptophan, two of the eight essential amino acids that humans cannot synthesize and must obtain from their diet (see Figure 41.10). Forty years ago, researchers discovered a mutant known as *opaque-2* that has much higher levels of lysine and tryptophan. This maize variety is more nutritious, as reflected in the fact that swine fed with *opaque-2* maize gain weight three times faster than those fed with normal maize. However, as is often the case in plant breeding, a highly desirable trait was linked with undesirable ones. The *opaque-2* maize kernels have a soft endosperm that makes them difficult to harvest and more vulnerable to attack by pests. Using the conventional methods of hybridization and artificial selection, plant breeders converted the soft *opaque-2* endosperm into a more desirable hard endosperm type, a transition that took hundreds of scientists nearly 20 years to accomplish. If modern methods of genetic engineering had been available, one laboratory might have inserted genes for the high lysine and tryptophan condition into hard-endosperm maize varieties over a period of only a few years.

Unlike traditional plant breeders, modern plant biotechnologists, using techniques of genetic engineering, are not limited to the transfer of genes between closely related species or varieties of the same species. For example, traditional breeding techniques could not be used to insert a desired gene from daffodil into rice because the many intermediate species between rice and

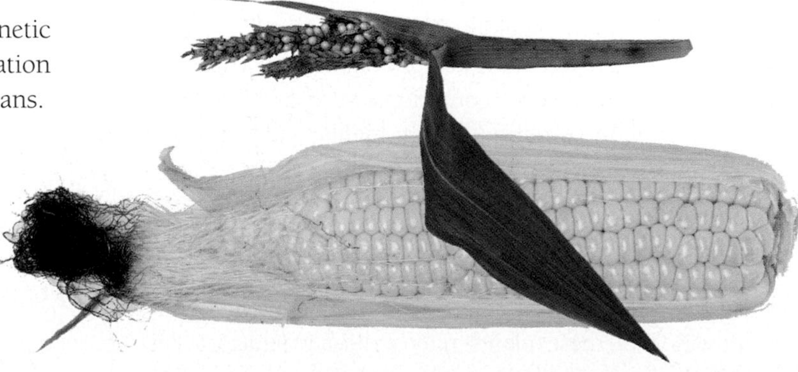

▲ **Figure 38.14 Maize: a product of artificial selection.** Modern maize (bottom) was derived from teosinte (top). Teosinte kernels are tiny, and each row has a husk that must be removed to get at the kernel. The seeds are loose at maturity, allowing dispersal, which probably made harvesting difficult for early farmers. Neolithic farmers selected for larger cob and kernel size as well as the permanent attachment of seeds to the cob and the encasing of the entire cob by a tough husk.

daffodil and their common ancestor are extinct. In theory, if breeders had the intermediate species, over the course of several centuries they could probably introduce a daffodil gene into rice by traditional hybridization and breeding methods. With genetic engineering technology, however, such gene transfers can be done more quickly and without the need for intermediate species.

Reducing World Hunger and Malnutrition

Eight hundred million people on Earth suffer from nutritional deficiencies. Forty thousand people die each day of malnutrition, half of them children. There is much disagreement about the causes of such hunger. Some argue that food shortages arise not from a lack of food production but from inequities in distribution: The dire poor simply cannot afford food. Others regard food shortages as evidence that the world is overpopulated—that the human species has exceeded the carrying capacity of the planet (see Chapter 52). Whatever the social and demographic causes of malnutrition, increasing food production is a humane objective. Because land and water are the most limiting resources for food production, the best option will be to increase yields on the available land. Indeed, there is very little "extra" land that can be put to the plow, especially if the few remaining pockets of wilderness are to be conserved. Based on conservative estimates of population growth, the world's farmers will have to produce 40% more grain per hectare to feed the human population in 2020. Plant biotechnology can help make these crop yields possible.

Already, the commercial adoption by farmers of transgenic crops has been one of the most dramatic examples of rapid technology adoption in the history of agriculture. These crops include transgenic varieties and hybrids of cotton, maize, and potatoes that contain genes from the bacterium *Bacillus thuringiensis*. These "transgenes" encode for a protein (*Bt* toxin) that effectively controls a number of serious insect pests. The use of such plant varieties greatly reduces the need for spraying crops with chemical insecticides. The *Bt* toxin used in crops is produced in the plant as a harmless protoxin that only becomes toxic if activated by alkaline conditions, such as occur in the guts of insects. Since vertebrates have highly acidic stomachs, the protoxin is destroyed without becoming active. Thus, *Bt* toxin is completely harmless to vertebrates, including humans.

Considerable progress also has been made in the development of transgenic plants of cotton, maize, soybeans, sugar beets, and wheat that are tolerant to a number of herbicides. The cultivation of these plants may reduce production costs and enable farmers to "weed" crops with herbicides that do not damage the transgenic crop plants instead of by heavy tillage, which can cause erosion of soil. Researchers are also engineering plants with enhanced resistance to disease. In one case, a transgenic papaya resistant to a ring spot virus was introduced into Hawaii, thereby saving its papaya industry **(Figure 38.15)**.

◀ **Figure 38.15 Genetically engineered papaya.** A ring spot virus has devastated papaya cultivation worldwide. A transgenic papaya variety rescued the industry in Hawaii. The genetically engineered papaya on the left is more resistant to ring spot virus than the nontransgenic papaya on the right.

The nutritional quality of plants is also being improved. "Golden Rice," a transgenic variety with a few daffodil genes that increase quantities of vitamin A, is under development to prevent the blindness that develops in many of the world's poor whose diet is chronically deficient in vitamin A **(Figure 38.16)**.

The Debate over Plant Biotechnology

Much of the debate regarding GM organisms in agriculture is political, social, economic, or ethical and outside the scope of this book. But we *should* consider the biological concerns about GM crops. There are some biologists, particularly ecologists, who are concerned about the unknown risks associated with the release of GM organisms (GMOs) into the environment. The most fundamental debate centers on the extent to which GMOs are an unknown risk that could potentially harm human health or the environment. Those who want to proceed more slowly with agricultural biotechnology (or end it) are concerned about the unstoppable nature of the "experiment." If a drug trial produces unanticipated harmful results, the trial is stopped. But we may not be able to stop the "trial" of novel organisms introduced into the biosphere.

Chapter 20 introduced the key concerns regarding biotechnology in general. Here we take a closer look at issues as they relate to plant biotechnology. Laboratory and field studies continue to examine the possible consequences of using GM crops, including the effects on human health and nontarget organisms and the potential for transgene escape.

Issues of Human Health

One concern is that genetic engineering may transfer allergens, molecules to which some humans are allergic, from a gene source to a plant used for food. However, biotechnologists are engaged in removing genes that encode allergenic proteins from soybeans and other crops. So far, there is no

◀ **Figure 38.16 Grains of "Golden Rice" interspersed with grains of ordinary rice.** The golden color and increased nutritional value of Golden Rice are attributable to its ability to make beta-carotene. The three transgenes that gave Golden Rice the ability to make beta-carotene in its grains were introduced from daffodils (*Narcissus*).

credible evidence that GM plants specifically designed for human consumption have adverse effects on human health. In fact, some GM foods are potentially a healthier alternative to non-GM foods. *Bt* maize, for example, contains 90% less of a cancer-causing mycotoxin called fumonisin that is highly resistant to degradation and has been found in alarmingly high concentrations in some batches of processed maize products ranging from corn flakes to beer. Fumonisin is produced by a fungus (*Fusarium*) that infects insect-damaged maize. Since *Bt* maize generally suffers less insect damage than non-GM maize, it contains much less fumonisin.

Nevertheless, because of health concerns, GMO opponents lobby for the clear labeling of all foods containing products of GMOs. Some also argue for strict regulations against the mixing of GM foods with non-GM foods during food transport, storage, and processing. Some biotechnology advocates, however, note that similar demands were not raised when "transgenic" crops produced by traditional plant-breeding techniques were put on the market. For example, triticale was a completely new crop that was artificially synthesized a few decades ago by hybridizing wheat and rye—two distinct genera that do not interbreed in nature. Triticale is now grown on over 3 million acres worldwide.

Possible Effects on Nontarget Organisms

Many ecologists are concerned that the growing of GM crops might have unforeseen effects on nontarget organisms. One study indicated that the larvae (caterpillars) of monarch butterflies responded adversely and even died following their consumption in the laboratory of milkweed leaves (their preferred food) heavily dusted with pollen from transgenic *Bt* maize. This study has since been discredited and affords a good example of the self-correcting nature of science. As it turns out, when the original researchers shook the male maize inflorescences onto the milkweed leaves in the laboratory, staminal filaments, opened pollen sacs, and other floral parts also rained onto the leaves. Subsequent research found that it was these other floral parts, *not* the pollen, that contained *Bt* toxin in high concentrations. Unlike pollen, these floral parts would not be carried by the wind to neighboring milkweed plants when shed under natural field conditions. Only one *Bt* maize line, accounting for less than 2% of commercial *Bt* maize production (and now discontinued), produced pollen with high *Bt* toxin concentrations.

In considering the negative effects of *Bt* pollen on monarch butterflies, one must also weigh the effects of an alternative to the cultivation of *Bt* maize—the spraying of non-*Bt* maize with chemical pesticides. Recent studies have shown that such spraying was much more harmful to nearby monarch populations than was *Bt* maize production. Although the nontarget effects of *Bt* maize pollen on monarch butterfly larvae appear to be minor, the controversy has emphasized the need for accurate field testing of all GM crops.

Addressing the Problem of Transgene Escape

Perhaps the most serious concern that some scientists raise about GM crops is the possibility of the introduced genes escaping from a transgenic crop into related weeds through crop-to-weed hybridization. The fear is that the spontaneous hybridization between a crop engineered for herbicide resistance and a wild relative, for example, might give rise to a "superweed" that would have a selective advantage over other weeds in the wild and would be much more difficult to control in the field. Some crops *do* hybridize with weedy relatives, and crop-to-weed transgene escape is a possibility. Its likelihood depends on the ability of the crop and weed to hybridize and on how the transgenes affect the overall fitness of the hybrids. A desirable crop trait—a dwarf phenotype, for example—might be disadvantageous to a weed growing in the wild. In other instances, there are no weedy relatives nearby with which to hybridize; soybean, for example, has no wild relatives in the United States. However, canola, sorghum, and many other crops do hybridize readily with weeds.

Because of concerns over transgene escape, efforts are under way to breed male sterility into transgenic crops. These plants will still produce seeds and fruit if pollinated by nearby nontransgenic plants, but they will produce no viable pollen. Another approach is to engineer the transgene into the chloroplast DNA of the crop. Chloroplast DNA in many plant species is inherited strictly from the maternal plant, so transgenes in the chloroplast cannot be transferred by pollen (see Chapter 15 to review maternal inheritance).

"Terminator technology" offers another potential approach toward reducing the problem of transgene escape. Terminator technology employs "suicide" genes that disrupt critical developmental sequences, precluding pollen development or seed maturation. Plants that are genetically modified to undergo the terminator process grow normally until the last stages of seed or pollen maturation. At this point, a gene expressing a "terminator" protein toxic to plants but harmless to animals is activated only in the nearly mature seed or pollen. Although the newly formed seeds or pollen are practically mature, they are inviable. Terminator proteins are produced only if the original seeds are pretreated with a specific chemical. Most research has focused on applying this technology to the killing of seeds, to force farmers to purchase GM seed every growing season. However, applying the technology to kill developing pollen grains could be an effective strategy for reducing transgene escape.

The continuing debate about GMOs in agriculture exemplifies one of this textbook's themes: the relationships of science and technology to society. Technological advances almost always involve the risk of unintended outcomes. In plant biotechnology, zero risk is probably unattainable. On a case-by-case basis, scientists and the public must assess possible benefits of transgenic products versus the risks society is willing to take. The best scenario is for these discussions and decisions to be based on sound scientific information and testing rather than on reflexive fear or blind optimism.

Concept Check 38.4

1. Compare traditional plant breeding with genetic engineering.
2. Explain how GM crops have benefits and risks.

For suggested answers, see Appendix A.

Chapter 38 Review

Go to www.campbellbiology.com or the student CD-ROM to explore Activities, Investigations, and other interactive study aids.

SUMMARY OF KEY CONCEPTS

Concept 38.1

Pollination enables gametes to come together within a flower

- The sporophyte, the dominant generation, produces spores that develop within flowers into male gametophytes (pollen grains) and female gametophytes (embryo sacs) (pp. 771–772).
  Activity *Angiosperm Life Cycle*
- **Flower Structure (pp. 772–773)** The four floral organs are sepals, petals, stamens, and carpels.
- **Gametophyte Development and Pollination (pp. 774–775)** Pollen develops from microspores within the microsporangia of anthers; embryo sacs develop from megaspores within ovules. Pollination, which precedes fertilization, is the placing of pollen on the stigma of a carpel.
- **Mechanisms That Prevent Self-Fertilization (pp. 775–776)** Some plants reject pollen that has an *S*-gene matching an allele in the stigma cells. Recognition of "self" pollen triggers a signal transduction pathway leading to a block in growth of a pollen tube.

Concept 38.2

After fertilization, ovules develop into seeds and ovaries into fruits

- **Double Fertilization (pp. 776–777)** The pollen tube discharges two sperm into the embryo sac; one fertilizes the egg, while the other combines with the polar nuclei, giving rise to the food-storing endosperm.
- **From Ovule to Seed (pp. 777–778)** The seed coat encloses the embryo along with a food supply stocked in either the cotyledons or the endosperm (or both).
  Activity *Seed and Fruit Development*
- **From Ovary to Fruit (pp. 778–779)** The fruit protects the enclosed seeds and aids in wind dispersal or in the attraction of seed-dispersing animals.
- **Seed Germination (pp. 779–780)** Seed dormancy ensures that seeds germinate only when conditions for seedling survival are optimal. The breaking of dormancy often requires environmental cues, such as temperature or lighting changes.
  Investigation *What Tells Desert Seeds When to Germinate?*

Concept 38.3

Many flowering plants clone themselves by asexual reproduction

- Asexual reproduction enables successful clones to spread; sexual reproduction generates the genetic variation that makes evolutionary adaptation possible (p. 781).
- **Mechanisms of Asexual Reproduction (p. 781)** One important mode of asexual reproduction is the fragmentation of a parent plant into parts that become whole plants.
- **Vegetative Propagation and Agriculture (pp. 781–783)** Cloning plants from cuttings is an ancient practice. We can now clone plants from single cells, which can be genetically manipulated before cloning.

Concept 38.4

Plant biotechnology is transforming agriculture

- **Artificial Selection (pp. 783–784)** Interspecific hybridization of plants is common in nature and has been used by breeders, ancient and modern, to introduce new genes.
- **Reducing World Hunger and Malnutrition (p. 784)** Genetically modified plants have the potential of increasing the quality and quantity of food worldwide.
  Activity *Making Decisions About DNA Technology: Golden Rice*
- **The Debate over Plant Biotechnology (pp. 784–786)** There are concerns about the unknown risks of releasing genetically modified organisms into the environment, but the potential benefits of transgenic crops should be considered.

TESTING YOUR KNOWLEDGE

Self-Quiz

1. A plant that has small, green petals is most likely to be
   a. bee-pollinated.
   b. bird-pollinated.
   c. bat-pollinated.
   d. wind-pollinated.

2. Pollen grain is to ______ as ______ is to female gametophyte.
   a. male gametophyte; embryo sac
   b. embryo sac; ovule
   c. ovule; sporophyte
   d. anther; seed
   e. petal; sepal

3. A seed develops from
   a. an ovum.
   b. a pollen grain.
   c. an ovule.
   d. an ovary.
   e. an embryo.

4. A fruit is a(an)
   a. mature ovary.
   b. mature ovule.
   c. seed plus its integuments.
   d. fused carpel.
   e. enlarged embryo sac.

5. Which of the following conditions is needed by almost all seeds to break dormancy?
   a. exposure to light
   b. imbibition
   c. abrasion of the seed coat
   d. exposure to cold
   e. covering of fertile soil

6. Sources of genetic variability in an asexually propagated species may involve all of the following processes *except*
   a. protoplast fusion.
   b. mutation.
   c. hybridization.
   d. genetic engineering.

7. Plant biotechnologists use protoplast fusion mainly to
   a. culture plant cells *in vitro*.
   b. asexually propagate desirable plant varieties.
   c. introduce bacterial genes into a plant genome.
   d. study the early events following fertilization.
   e. produce new hybrid species.

8. The basal cell formed from the first division of a plant zygote will eventually develop into
   a. the suspensor that anchors the embryo and transfers nutrients.
   b. the proembryo.
   c. the endosperm that nourishes the developing embryo.
   d. the root apex of the embryo.
   e. two cotyledons in eudicots, but one in monocots.

9. The development of *Bt* crops continues to raise concerns because
   a. *Bt* crops have been shown to be toxic to humans.
   b. pollen from these crops is harmful to monarch butterfly larvae in the field.
   c. if genes for *Bt* toxin "escape" to related weed species, the hybrid weeds could have harmful ecological effects.
   d. *Bacillus thuringiensis* is a pathogen of humans.
   e. *Bt* toxin reduces the nutritional quality of crops.

10. "Golden Rice" is a transgenic variety that
    a. is resistant to various herbicides, making it practical to weed rice fields with those herbicides.
    b. is resistant to a virus that commonly attacks rice fields.
    c. includes bacterial genes that produce a toxin that reduces damage from insect pests.
    d. produces much larger, golden grains that increase crop yields.
    e. contains daffodil genes that increase the vitamin A content.

For Self-Quiz answers, see Appendix A.

Go to the website or CD-ROM for more quiz questions.

Evolution Connection

With respect to sexual reproduction, some plant species are fully self-fertile, others are fully self-incompatible, and some have adopted a "mixed strategy" with partial self-incompatibility. These reproductive strategies differ in their implications for evolutionary potential. How, for example, might a self-incompatible species fare as small founder populations or remnant populations in a severe population bottleneck, as compared with a self-fertile species?

Scientific Inquiry

Critics of GM foods have argued that because the insertion of foreign genes may cause a disturbance of normal cellular functioning, unexpected and potentially harmful substances may appear. Toxic intermediary substances that normally occur in very small amounts may increase considerably, or new substances may appear. There is also a potential risk that the disruption may lead to the loss of substances that are important for maintaining normal metabolism. If you were your country's chief scientific advisor, how would you respond to these criticisms?

Biology Inquiry: A Workbook of Investigative Cases ***Explore GM crops further with the case "Corn Under Construction."***

Investigation ***What Tells Desert Seeds When to Germinate?***

Science, Technology, and Society

Humans have engaged in genetic manipulation for millennia, producing plant and animal varieties through selective breeding and hybridization processes that modify the genomes of organisms on a significant level. Why do you think modern genetic engineering, which often entails introducing or modifying only one or a few genes, has met with so much public opposition? Should some forms of genetic engineering be of greater concern than others? Explain.

39 Plant Responses to Internal and External Signals

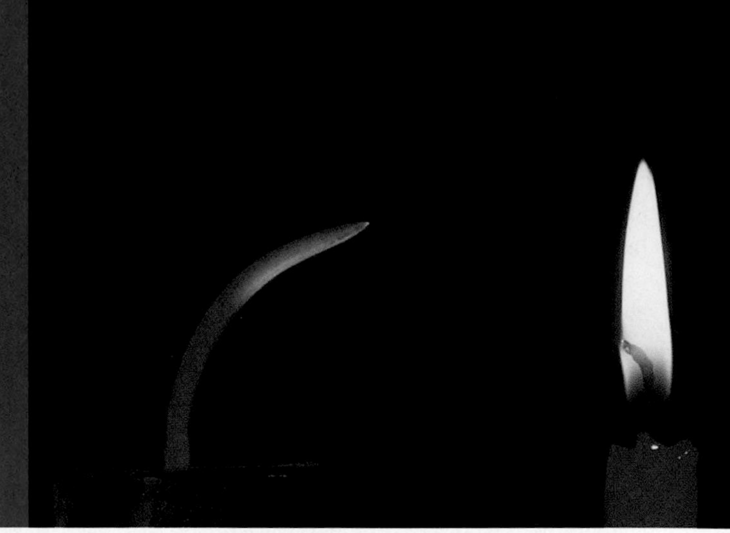

▲ **Figure 39.1 Grass seedling growing toward light.**

Key Concepts

39.1 Signal transduction pathways link signal reception to response
39.2 Plant hormones help coordinate growth, development, and responses to stimuli
39.3 Responses to light are critical for plant success
39.4 Plants respond to a wide variety of stimuli other than light
39.5 Plants defend themselves against herbivores and pathogens

Overview

Stimuli and a Stationary Life

Being rooted to the ground, plants must respond to whatever environmental change comes their way. For example, bending of a grass seedling toward light **(Figure 39.1)** begins with the plant sensing the direction, quantity, and color of the light. After integrating this information, cells in the tip of the seedling undergo a complex series of biochemical changes that influence distribution of chemical signals in the plant. Ultimately, changes in distribution of these growth-regulating chemicals lead to the bending of the seedling toward the light. This is one example of how a plant's morphology and physiology are constantly tuned to its surroundings by complex interactions between environmental stimuli and internal signals.

This chapter focuses on how plants respond to external and internal cues. At the organismal level, plants and animals respond to environmental stimuli by very different means. Animals, being mobile, respond mainly by behavioral mechanisms, moving toward positive stimuli and away from negative stimuli. Being stationary, a plant generally responds to environmental cues by adjusting its pattern of growth and development. For this reason, plants of the same species vary in body form much more than do animals of the same species. All lions have four limbs and approximately the same body proportions, but oak trees vary considerably in their number of limbs and their shape. We begin our study of plant response with the role of signal transduction in plant cells.

Concept 39.1

Signal transduction pathways link signal reception to response

All organisms receive specific environmental signals and respond to them in ways that enhance survival and reproductive success. For example, bees, which possess UV-sensitive photoreceptors in their eyes, can discern nectar-guiding patterns on flower petals—patterns that are completely invisible to humans. Plants, too, have cellular receptors that they use to detect important changes in their environment, whether the change is an increase in the concentration of a growth hormone, an injury from a caterpillar munching on leaves, or a decrease in day length as winter approaches.

For a stimulus to elicit a response, certain cells must have an appropriate receptor, a molecule affected by the stimulus. For example, our blindness to UV-reflecting floral patterns is due to our eyes lacking UV photoreceptors. Upon receiving a stimulus, a receptor initiates a series of biochemical steps, a signal transduction pathway, that couples reception to response. Plants are sensitive to a wide range of stimuli, each initiating a specific signal transduction pathway. Chapter 11 introduced general concepts of signal transduction in cells. Here, we apply those concepts to plants.

Consider a forgotten potato in the back corner of a kitchen cupboard. This modified underground stem, or tuber, has sprouted shoots from its "eyes" (axillary buds). These shoots, however, scarcely resemble those of a typical plant. Instead of broad green leaves and sturdy stems, these plants are ghostly pale, bear leaves that are unexpanded, and lack elongated roots **(Figure 39.2a)**. These morphological adaptations for growing in darkness, collectively referred to as **etiolation**, make sense if we consider that a young potato plant in nature usually encounters continuous darkness when sprouting underground. Under these circumstances, expanded leaves would be a hindrance to soil penetration and would be damaged as the shoots pushed through the soil. Because the leaves are unexpanded and underground, there is little evaporative loss of water and little requirement for an extensive root system to replace the water lost by transpiration. Moreover, the energy expended in producing green chlorophyll would be wasted because there is no light for photosynthesis. Instead, a potato plant growing in the dark allocates as much energy as possible to the elongation of stems. This adaptation enables the shoots to break ground before the nutrient reserves in the tuber are exhausted. The etiolation response is one example of how a plant's morphology and physiology are tuned to its variable surroundings by complex interactions between environmental and internal signals.

When a shoot reaches the sunlight, the plant undergoes profound changes collectively called **de-etiolation** (informally known as greening). The elongation rate of the stems slows; leaves expand; roots elongate; and the shoot produces chlorophyll. In short, it begins to resemble a typical plant **(Figure 39.2b)**. We will use this de-etiolation response as an example of how a plant cell's reception of a signal—in this case, light—is transduced into a response (greening). Along the way, we will explore how studies of mutants provide insights into roles played by molecules in the stages of cell-signal processing: reception, transduction, and response **(Figure 39.3)**.

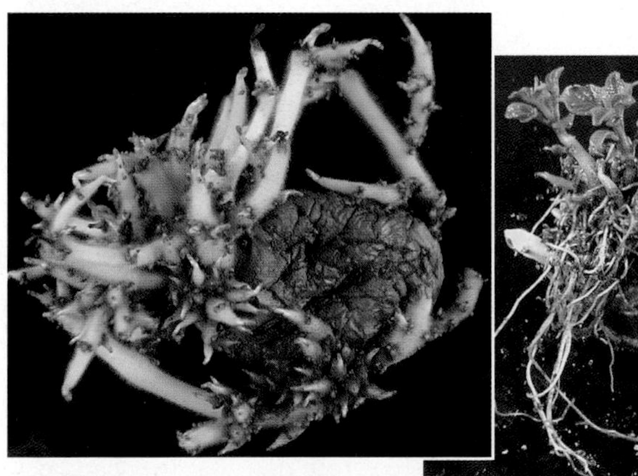

(a) Before exposure to light. A dark-grown potato has tall, spindly stems and nonexpanded leaves—morphological adaptations that enable the shoots to penetrate the soil. The roots are short, but there is little need for water absorption because little water is lost by the shoots.

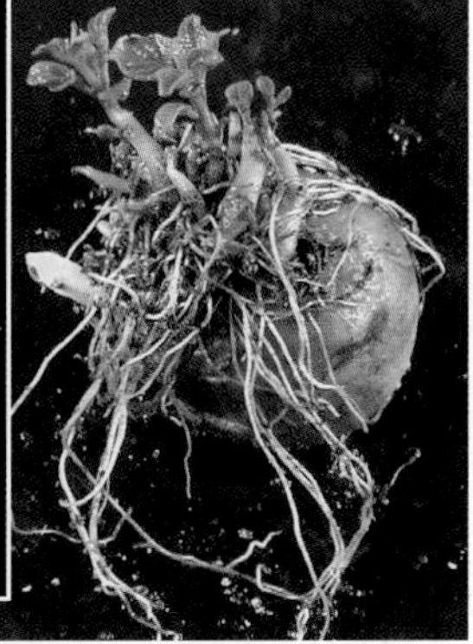

(b) After a week's exposure to natural daylight. The potato plant begins to resemble a typical plant with broad green leaves, short sturdy stems, and long roots. This transformation begins with the reception of light by a specific pigment, phytochrome.

▲ **Figure 39.2 Light-induced de-etiolation (greening) of dark-grown potatoes.**

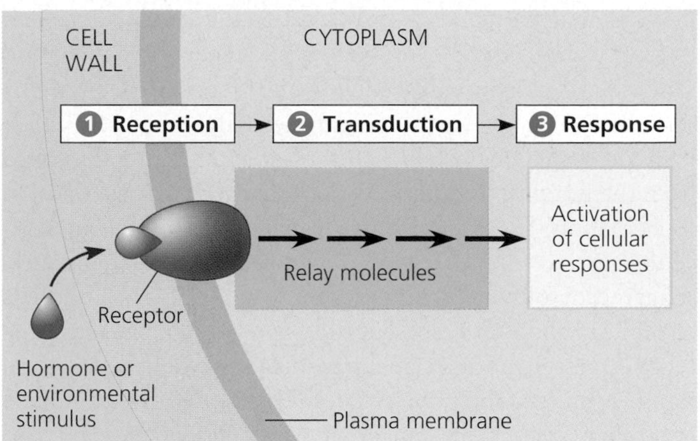

▲ **Figure 39.3 Review of a general model for signal transduction pathways.** As discussed in Chapter 11, a hormone or another signal binding to a specific receptor stimulates the cell to produce relay molecules, such as second messengers. These relay molecules trigger the cell's various responses to the original signal. In this diagram, the receptor is on the surface of the target cell. In other cases, hormones enter cells and bind to specific receptors inside.

Reception

Signals are first detected by receptors, proteins that undergo conformational changes in response to a specific stimulus. The receptor involved in de-etiolation is a type of *phytochrome*, a photoreceptor that we will examine more closely later in the chapter. Unlike many receptors, which are built into the plasma membrane, the phytochrome that functions in de-etiolation occurs in the cytoplasm. Researchers demonstrated the requirement for phytochrome in de-etiolation through studies of the tomato, a close relative of the potato. The *aurea* tomato mutant, which has lower-than-normal levels of phytochrome, greens less than wild-type tomatoes when exposed to light. The name *aurea* comes from the Latin for "gold-colored." In the absence of chlorophyll, the yellow plant pigments called carotenoids are more obvious. Researchers produced a normal de-etiolation response in individual *aurea* leaf cells by injecting phytochrome from other plants and exposing the cells to light. Such experiments indicate that phytochrome functions in light detection during de-etiolation.

Transduction

Receptors are sensitive to very weak environmental and chemical signals. Some de-etiolation responses are triggered by extremely low levels of light. For example, light levels equivalent to a few seconds of moonlight are sufficient to cause a slowing of stem elongation in dark-grown oat seedlings. How is the information from these extremely weak signals amplified, and how is their reception transduced into a specific response by the plant? The answer is **second messengers**—small, internally produced chemicals that transfer and amplify the signal from the receptor to other proteins that cause the response. In de-etiolation, for example, each activated phytochrome molecule may give rise to hundreds of molecules of a second messenger, each of which may lead to the activation of hundreds of molecules of a specific enzyme. By such mechanisms, the second messenger of a signal transduction pathway leads to a rapid amplification of the signal. In Chapter 11, we examined this role of second messengers in general (see Figure 11.10). Here, we consider the production and function of second messengers in de-etiolation. Refer to **Figure 39.4** frequently as you read the following description of this complex process.

Light causes phytochrome to undergo a conformational change that leads to increases in levels of the second messengers cyclic GMP (cGMP) and Ca^{2+}. Changes in cGMP levels can lead to ionic changes within the cell by influencing properties of ion channels. Cyclic GMP also activates protein kinases, a diverse class of enzymes that influence activation of other enzymes by phosphorylation, a modification in which a phosphate group is attached to the protein at a specific site. Injection of cGMP into *aurea* tomato leaf cells induces a partial de-etiolation response, even without the addition of phytochrome. Changes in cytosolic Ca^{2+} levels also play an important role in phytochrome signal transduction. The concentration of Ca^{2+} is generally very low in the cytosol (about 10^{-7} *M*). But phytochrome activation can open Ca^{2+} channels and lead to a transient 100-fold increase in cytosolic Ca^{2+} levels. Like cGMP, cytosolic Ca^{2+} can influence the activity of specific ion channels and protein kinases.

Response

Ultimately, a signal transduction pathway leads to regulation of one or more cellular activities. In most cases, these responses to stimulation involve the increased activity of certain enzymes. There are two main mechanisms by which a signaling pathway can activate an enzyme. One mechanism involves stimulating transcription of mRNA for the enzyme (transcriptional regulation). The other involves activating existing enzyme molecules (post-translational modification).

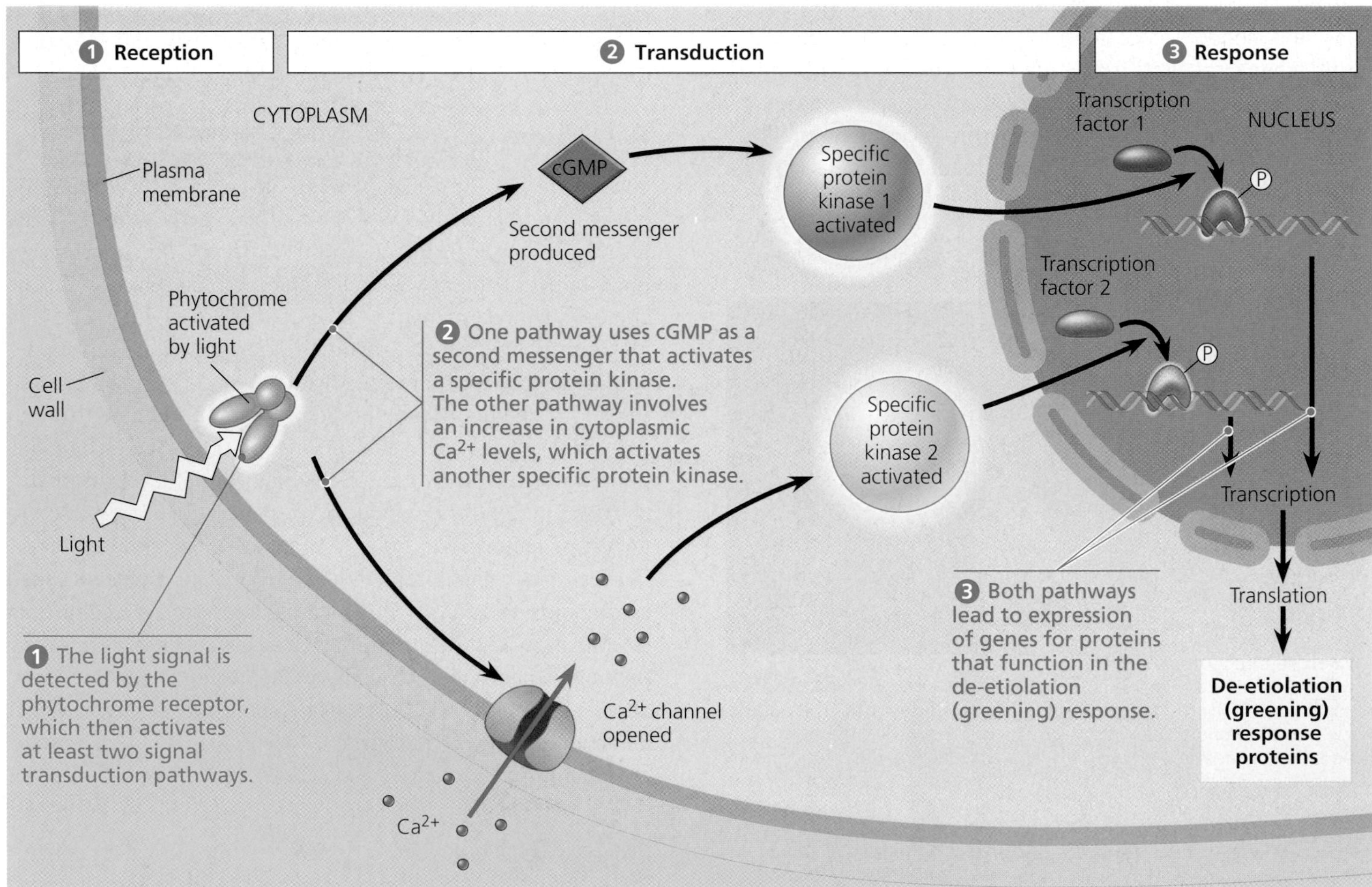

▲ **Figure 39.4 An example of signal transduction in plants: the role of phytochrome in the de-etiolation (greening) response.**

Transcriptional Regulation

Transcription factors bind directly to specific regions of DNA and control the transcription of specific genes (see Figure 19.6). In the case of phytochrome-induced de-etiolation, several transcription factors are activated by phosphorylation in response to the appropriate light conditions. The activation of some of these transcription factors depends on cGMP, whereas the activation of others requires Ca^{2+}.

The mechanism by which a signal promotes a new developmental course may depend on the activation of positive transcription factors (proteins that *increase* transcription of specific genes) or negative transcription factors (proteins that *decrease* transcription) or both. For example, there are *Arabidopsis* mutants that, except for their pale color, have a light-grown morphology (expanded leaves and short, sturdy stems) when grown in the dark (they are not green because the final step in chlorophyll production requires direct light). These mutants have defects in a negative transcription factor that inhibits the expression of other genes normally activated by light. When the negative factor is eliminated by mutation, the pathway that it normally blocks becomes activated. Hence, these mutants, except for their pale color, appear to have been grown in the light.

Post-Translational Modification of Proteins

Although the syntheses of new proteins by transcription and translation are important molecular events associated with de-etiolation, so are post-translational modifications of existing proteins. Most often, these existing proteins are modified by phosphorylation. Many second messengers, including cGMP, and some receptors themselves, including some forms of phytochrome, activate protein kinases directly. Often, one protein kinase will phosphorylate another protein kinase, which then phosphorylates another, and so on. Such kinase cascades may eventually link initial stimuli to responses at the level of gene expression, usually via the phosphorylation of transcription factors. By such mechanisms, many signal pathways ultimately regulate the synthesis of new proteins, usually by turning specific genes on or off.

Signal pathways must also have a means for turning off once the initial signal is no longer present. For example, what if we put the potato back into the cupboard? Protein phosphatases, enzymes that dephosphorylate specific proteins, are involved in these "switch-off" processes. At any moment, the activities of a cell depend on the balance of activity of many types of protein kinases and protein phosphatases.

De-Etiolation ("Greening") Proteins

What sorts of proteins are either newly transcribed or activated by phosphorylation during the de-etiolation process? Many are enzymes that function in photosynthesis directly; others are enzymes involved in supplying the chemical precursors necessary for chlorophyll production; still others affect the levels of plant hormones that regulate growth. For example, the levels of two kinds of hormones, auxin and brassinosteroids, that enhance stem elongation decrease following phytochrome activation. That decrease explains the reduction in stem elongation that accompanies de-etiolation.

We have discussed the signal transduction involved in the de-etiolation response of a potato in some detail to give you a sense of the complexity of biochemical changes that underlie this one process. Every plant hormone and every environmental stimulus triggers one or more signal transduction pathways of comparable complexity. As with the tomato mutant *aurea,* techniques of molecular biology combined with studies of mutants are helping researchers identify these various pathways. But molecular biology builds on a long history of careful physiological and biochemical investigations into how plants work. As you will read in the next section, classical observations and experiments provided biologists with the first clues that chemical signals—hormones—are internal regulators of plant growth and development.

Concept Check 39.1

1. The sexual dysfunction drug Viagra inhibits an enzyme that breaks down cyclic GMP. Assuming that tomato leaf cells have a similar enzyme, would you predict that applying Viagra will cause a normal de-etiolation of *aurea* tomato leaves?
2. Cycloheximide is a drug that inhibits protein synthesis. Predict what effect cycloheximide would have on de-etiolation.

For suggested answers, see Appendix A.

Concept 39.2

Plant hormones help coordinate growth, development, and responses to stimuli

The word *hormone* is derived from a Greek verb meaning "to excite." Found in all multicellular organisms, **hormones** are chemical signals that coordinate the different parts of an organism. By definition, a hormone is a compound that is produced by one part of the body and then transported to other parts of the body, where it binds to a specific receptor and triggers responses in target cells and tissues. Another important characteristic of hormones is that only minute amounts are required to induce substantial changes in an organism. Hormone concentrations or the rates of their transport can change in response to environmental stimuli. Often, the response of a plant is governed by the interaction of two or more hormones.

The Discovery of Plant Hormones

The concept of chemical messengers in plants emerged from a series of classic experiments on how stems respond to light. The shoot of a houseplant on a windowsill grows toward light. If you rotate the plant, it soon reorients its growth until its leaves again face the window. Any growth response that results in curvatures of whole plant organs toward or away from stimuli is called a **tropism** (from the Greek *tropos*, turn). The growth of a shoot toward light is called positive **phototropism**, whereas growth away from light is negative phototropism.

In a forest or other natural ecosystem where plants may be crowded, phototropism directs growing seedlings toward the sunlight that powers photosynthesis. What is the mechanism for this adaptive response? Much of what is known about phototropism has been learned from studies of grass seedlings, particularly oats. The shoot of a sprouting grass seedling is enclosed in a sheath called the coleoptile (see Figure 38.10b), which grows straight upward if the seedling is kept in the dark or if it is illuminated uniformly from all sides. If the growing coleoptile is illuminated from one side, it grows toward the light. This response results from a differential growth of cells on opposite sides of the coleoptile; the cells on the darker side elongate faster than the cells on the brighter side.

Charles Darwin and his son Francis conducted some of the earliest experiments on phototropism in the late 19th century. They observed that a grass seedling could bend toward light only if the tip of the coleoptile was present **(Figure 39.5)**. If the tip was removed, the coleoptile would not curve. The seedling would also fail to grow toward light if the tip was covered with an opaque cap; neither a transparent cap over the tip nor an opaque shield placed below the coleoptile tip prevented the phototropic response. It was the tip of the coleoptile, the Darwins concluded, that was responsible for sensing light. However, the actual growth response, the curvature of the coleoptile, occurred some distance below the tip. The Darwins postulated that some signal was transmitted downward from the tip to the elongating region of the coleoptile. A few decades later, Peter Boysen-Jensen of Denmark demonstrated that the signal was a mobile chemical substance. He separated the tip from the remainder of the coleoptile by a block of gelatin, which would prevent cellular contact but allow chemicals to pass. These seedlings responded normally, bending toward light. However, if the tip was experimentally segregated from the lower coleoptile by an impermeable barrier, such as the mineral mica, no phototropic response occurred.

In 1926, Frits Went, a Dutch graduate student, extracted the chemical messenger for phototropism by modifying the experiments of Boysen-Jensen **(Figure 39.6)**. Went removed the coleoptile tip and placed it on a block of agar, a gelatinous material. The chemical messenger from the tip, Went

Figure 39.5

Inquiry What part of a coleoptile senses light, and how is the signal transmitted?

EXPERIMENT In 1880, Charles Darwin and his son Francis designed an experiment to determine what part of the coleoptile senses light. In 1913, Peter Boysen-Jensen conducted an experiment to determine how the signal for phototropism is transmitted.

RESULTS

Control

Light
Shaded side of coleoptile
Illuminated side of coleoptile

Darwin and Darwin (1880)

Light
Tip removed
Tip covered by opaque cap
Tip covered by transparent cap
Base covered by opaque shield

Boysen-Jensen (1913)

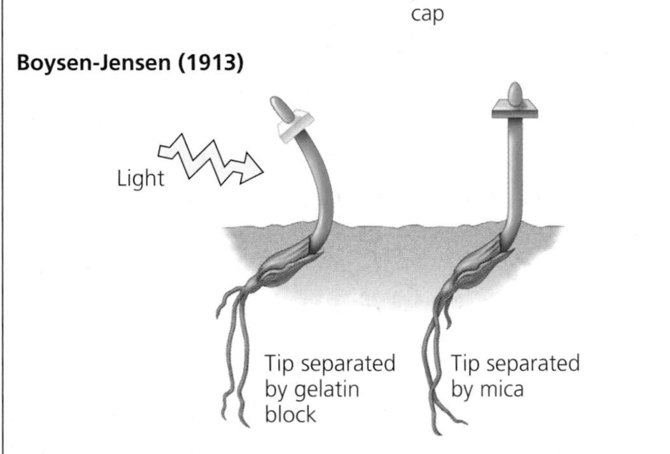

CONCLUSION In the Darwins' experiment, a phototropic response occurred only when light could reach the tip of the coleoptile. Therefore, they concluded that only the tip senses light. Boysen-Jensen observed that a phototropic response occurred if the tip was separated by a permeable barrier (gelatin) but not if separated by an impermeable solid barrier (a mineral called mica). These results suggested that the signal is a light-activated mobile chemical.

Figure 39.6

Inquiry Does asymmetric distribution of a growth-promoting chemical cause a coleoptile to grow toward the light?

EXPERIMENT In 1926, Frits Went's experiment identified how a growth-promoting chemical causes a coleoptile to grow toward light. He placed coleoptiles in the dark and removed their tips, putting some tips on agar blocks that he predicted would absorb the chemical. On a control coleoptile, he placed a block that lacked the chemical. On others, he placed blocks containing the chemical, either centered on top of the coleoptile to distribute the chemical evenly or offset to increase the concentration on one side.

RESULTS The coleoptile grew straight if the chemical was distributed evenly. If the chemical was distributed unevenly, the coleoptile curved away from the side with the block, as if growing toward light, even though it was grown in the dark.

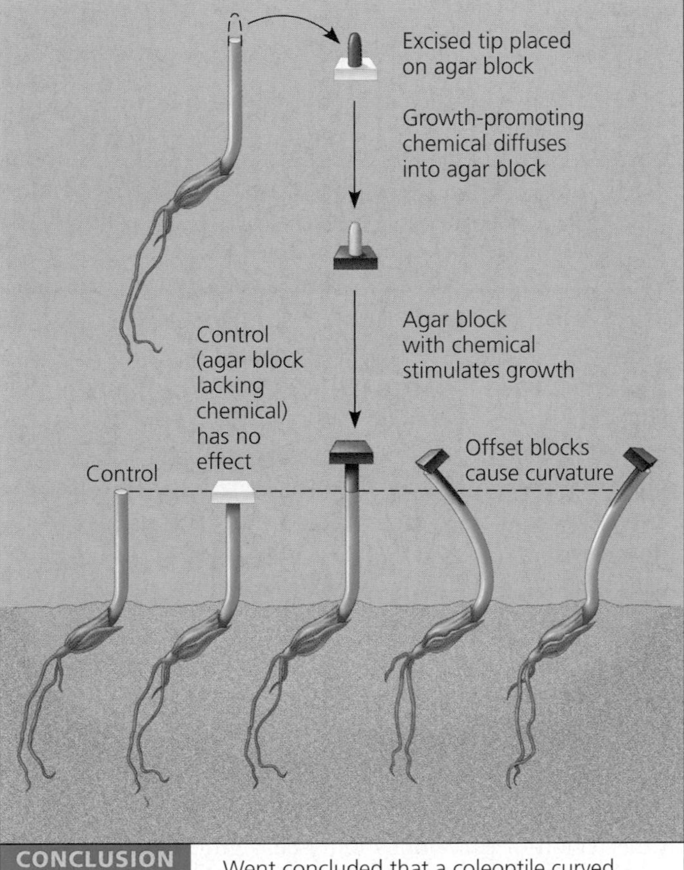
Excised tip placed on agar block
Growth-promoting chemical diffuses into agar block
Agar block with chemical stimulates growth
Control (agar block lacking chemical) has no effect
Control
Offset blocks cause curvature

CONCLUSION Went concluded that a coleoptile curved toward light because its dark side had a higher concentration of the growth-promoting chemical, which he named auxin.

reasoned, should diffuse into the agar, and the agar block should then be able to substitute for the coleoptile tip. Went placed the agar blocks on decapitated coleoptiles that were kept in the dark. A block that was centered on top of the coleoptile caused the stem to grow straight upward. However, if the block was placed off center, then the coleoptile began to bend away from the side with the agar block, as though growing toward light. Went concluded that the agar block contained a chemical produced in the coleoptile tip, that this chemical stimulated growth as it passed down the coleoptile, and that a coleoptile curved toward light because of a higher concentration of the growth-promoting chemical on the darker side of the coleoptile. For this chemical messenger, or hormone, Went chose the name auxin (from the Greek *auxein,* to increase). Auxin was later purified by Kenneth Thimann at the California Institute of Technology, and it was determined to be indoleacetic acid (IAA).

The classical hypothesis for what causes grass coleoptiles to grow toward light, based on the work of the Darwins, Boysen-Jensen, and Went, is that an asymmetrical distribution of auxin moving down from the coleoptile tip causes cells on the darker side to elongate faster than cells on the brighter side. However, studies of phototropism by organs other than grass coleoptiles provide less support for this idea. For example, there is no evidence that illumination from one side causes an asymmetrical distribution of auxin in the stems of sunflowers, radishes, and other eudicots. There *is,* however, an asymmetrical distribution of certain substances that may act as growth *inhibitors,* with these substances more concentrated on the lighted side of a stem.

A Survey of Plant Hormones

Early studies on phototropism provided the basis for subsequent research on plant hormones. **Table 39.1**, on the next page, previews some major classes of plant hormones: auxin, cytokinins, gibberellins, brassinosteroids, abscisic acid, and ethylene. Many molecules that function in defense of the plant against pathogens are probably plant hormones as well. (These molecules will be discussed later in the chapter.) Plant hormones are relatively small molecules. Their transport from cell to cell often involves passage across cell walls, which block movement of large molecules.

In general, hormones control plant growth and development by affecting the division, elongation, and differentiation of cells. Some hormones also mediate shorter-term physiological responses of plants to environmental stimuli. Each hormone has multiple effects, depending on its site of action, its concentration, and the developmental stage of the plant.

Plant hormones are produced in very low concentrations, but a minute amount of hormone can have a profound effect on the growth and development of a plant organ. This implies that the hormonal signal must be amplified in some way. A hormone may act by altering the expression of genes, by affecting the activity of existing enzymes, or by changing properties of membranes. Any of these actions could redirect the metabolism and development of a cell responding to a small number of hormone molecules. Signal transduction pathways amplify the hormonal signal and connect it to a cell's specific responses.

Table 39.1 An Overview of Plant Hormones

Hormone	Where Produced or Found in Plant	Major Functions
Auxin (IAA)	Embryo of seed, meristems of apical buds, young leaves	Stimulates stem elongation (low concentration only), root growth, cell differentiation, and branching; regulates development of fruit; enhances apical dominance; functions in phototropism and gravitropism; promotes xylem differentiation; retards leaf abscission
Cytokinins	Synthesized in roots and transported to other organs	Affect root growth and differentiation; stimulate cell division and growth; stimulate germination; delay senescence
Gibberellins	Meristems of apical buds and roots, young leaves, embryo	Promote seed and bud germination, stem elongation, and leaf growth; stimulate flowering and development of fruit; affect root growth and differentiation
Brassinosteroids	Seeds, fruit, shoots, leaves, and floral buds	Inhibit root growth; retard leaf abscission; promote xylem differentiation
Abscisic acid	Leaves, stems, roots, green fruit	Inhibits growth; closes stomata during water stress; promotes seed dormancy
Ethylene	Tissues of ripening fruit, nodes of stems, aging leaves and flowers	Promotes fruit ripening, opposes some auxin effects; promotes or inhibits growth and development of roots, leaves, and flowers, depending on species

Response to a hormone usually depends not so much on the amount of that hormone as on its relative concentration compared with other hormones. It is hormonal balance, rather than hormones acting in isolation, that may control growth and development. These interactions will become apparent in the following survey of hormone function.

Auxin

The term **auxin** is used for any chemical substance that promotes the elongation of coleoptiles, although auxins actually have multiple functions in flowering plants. The natural auxin occurring in plants is indoleacetic acid, or IAA, but several other compounds, including some synthetic ones, have auxin activity. Throughout this chapter, however, the term *auxin* is used specifically to refer to IAA. Although IAA was the first plant hormone to be discovered, much remains to be learned about auxin signal transduction and the regulation of auxin biosynthesis.

The speed at which auxin is transported down the stem from the shoot apex is about 10 mm/hr, a rate that is much too fast for diffusion, although it is slower than translocation in phloem. Auxin seems to be transported directly through parenchyma tissue, from one cell to the next. It moves only from shoot tip to base, not in the reverse direction. This unidirectional transport of auxin is called polar transport. Polar transport has nothing to do with gravity; experiments have shown that auxin travels upward when a stem or coleoptile segment is placed upside down. Rather, the polarity of auxin movement is attributable to the polar distribution of auxin transport protein in the cells. Concentrated at the basal end of a cell, the auxin transporters move the hormone out of the cell. The auxin can then enter the apical end of the neighboring cell **(Figure 39.7)**.

The Role of Auxin in Cell Elongation. Although auxin affects several aspects of plant development, one of its chief functions is to stimulate elongation of cells within young developing shoots. The apical meristem of a shoot is a major site of auxin synthesis. As auxin from the shoot apex moves down to the region of cell elongation (see Figure 35.15), the hormone stimulates growth of the cells, probably by binding to a receptor in the plasma membrane. Auxin stimulates growth only over a certain concentration range, from about 10^{-8} to 10^{-4} *M*. At higher concentrations, auxin may inhibit cell elongation, probably by inducing production of ethylene, a hormone that generally inhibits elongation. We will return to this hormonal interaction when we discuss ethylene.

According to a model called the acid growth hypothesis, proton pumps play a major role in the growth response of cells to auxin. In a shoot's region of elongation, auxin stimulates the plasma membrane's proton pumps. This pumping of H^+ increases the voltage across the membrane (membrane

Figure 39.7
Inquiry What causes polar movement of auxin from shoot tip to base?

EXPERIMENT To investigate how auxin is transported unidirectionally, researchers designed an experiment to identify the location of the auxin transport protein. They used a greenish-yellow fluorescent molecule to label antibodies that bind to the auxin transport protein. Then they applied the antibodies to longitudinally sectioned *Arabidopsis* stems.

RESULTS The left micrograph shows that the auxin transport protein is not found in all tissues of the stem, but only in the xylem parenchyma. In the right micrograph, a higher magnification reveals that the auxin transport protein is primarily localized at the basal end of the cells. (LMs)

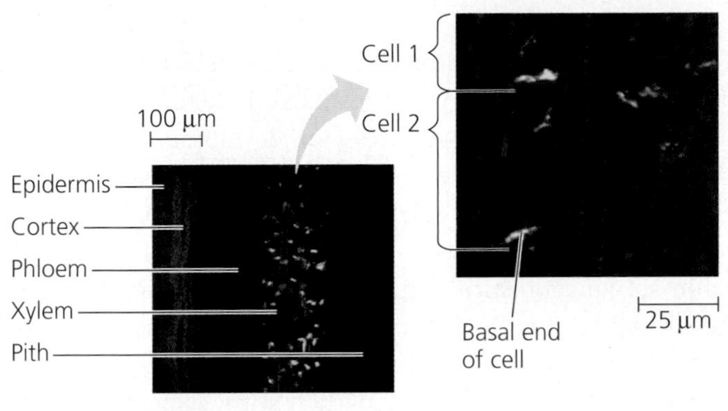

CONCLUSION The results support the hypothesis that concentration of the auxin transport protein at the basal ends of cells is responsible for polar transport of auxin.

potential) and lowers the pH in the cell wall within minutes **(Figure 39.8)**. Acidification of the wall activates enzymes called **expansins** that break the cross-links (hydrogen bonds) between cellulose microfibrils and other cell wall constituents, loosening the wall's fabric. (Expansins can even weaken the integrity of filter paper made of pure cellulose.) Increasing the membrane potential enhances ion uptake into the cell, which causes osmotic uptake of water and increased turgor. Increased turgor and increased cell wall plasticity enable the cell to elongate.

Auxin also alters gene expression rapidly, causing cells in the region of elongation to produce new proteins within minutes. Some of these proteins are short-lived transcription factors that repress or activate the expression of other genes. For sustained growth after this initial spurt, cells must make more cytoplasm and wall material. Auxin also stimulates this sustained growth response.

Lateral and Adventitious Root Formation. Auxins are used commercially in the vegetative propagation of plants by cuttings. Treating a detached leaf or stem with rooting powder containing auxin often causes adventitious roots to form near the cut surface. Auxin is also involved in the branching of roots. Researchers found that an *Arabidopsis* mutant that exhibits extreme proliferation of lateral roots has an auxin concentration 17-fold higher than normal.

Auxins as Herbicides. A number of synthetic auxins, such as 2,4-dichlorophenoxyacteic acid (2,4-D), are widely used as herbicides. Monocots, such as maize and turfgrass, can rapidly

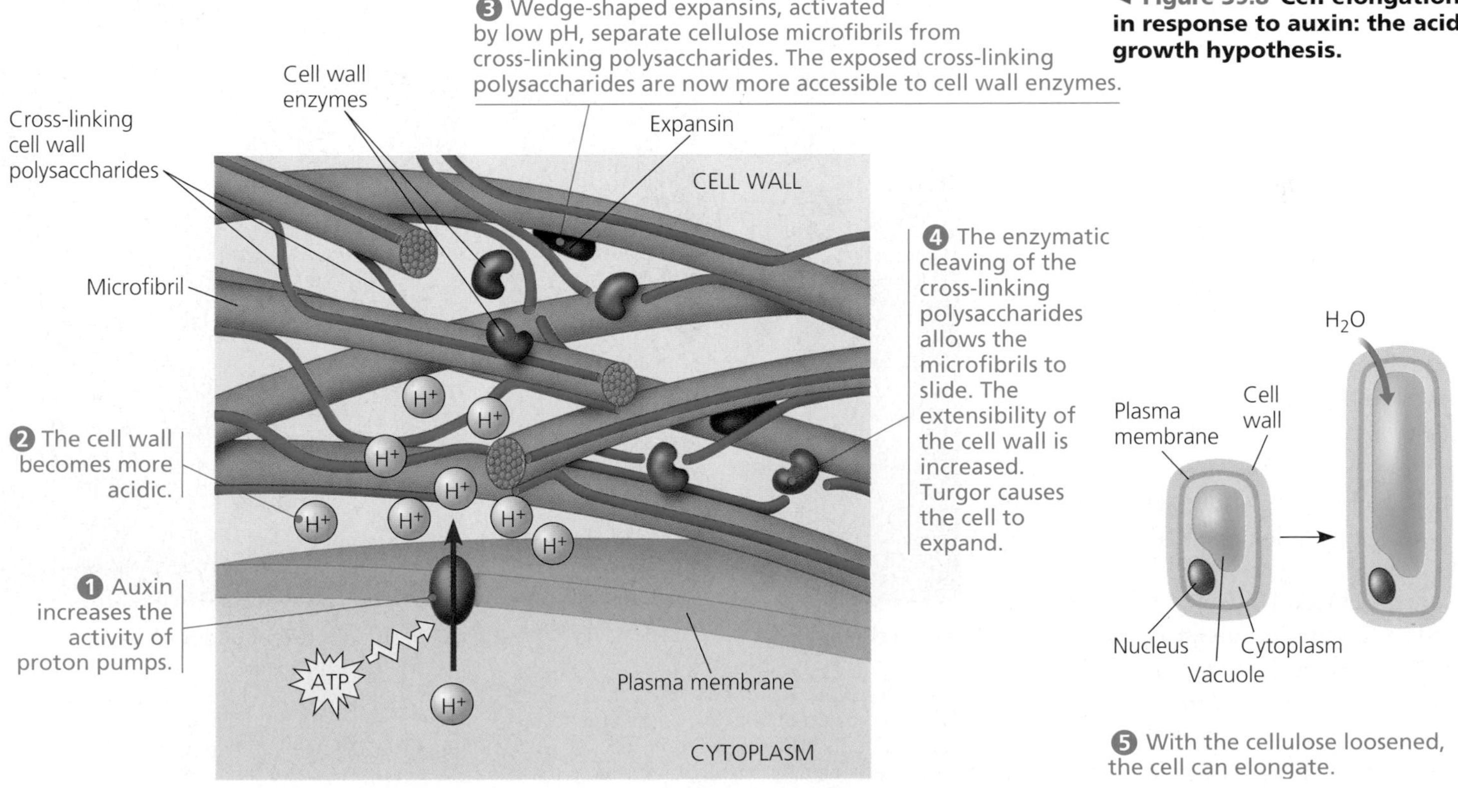

◀ **Figure 39.8 Cell elongation in response to auxin: the acid growth hypothesis.**

inactivate such synthetic auxins. However, eudicots cannot and therefore die from hormonal overdose. Spraying cereal fields or turf with 2,4-D eliminates eudicot (broadleaf) weeds such as dandelions.

Other Effects of Auxin. In addition to stimulating cell elongation for primary growth, auxin affects secondary growth. It induces cell division in the vascular cambium and influences differentiation of secondary xylem (see Figures 35.18 and 35.19).

Developing seeds synthesize auxin, which promotes the growth of fruits in plants. Under greenhouse conditions, seed set is often poor because of the lack of insect pollinators, resulting in poorly developed tomato fruits. Synthetic auxins sprayed on greenhouse-grown tomato vines induce fruit development without a need for pollination. This makes it possible to grow seedless tomatoes by substituting synthetic auxin for the auxin normally synthesized by the developing seeds.

Cytokinins

Trial-and-error attempts to find chemical additives that would enhance the growth and development of plant cells in tissue culture led to the discovery of **cytokinins.** In the 1940s, Johannes van Overbeek, working at the Cold Spring Harbor Laboratory in New York, found he could stimulate the growth of plant embryos by adding coconut milk, the liquid endosperm of a coconut's giant seed, to his culture medium. A decade later, Folke Skoog and Carlos O. Miller, at the University of Wisconsin-Madison, induced cultured tobacco cells to divide by adding degraded samples of DNA. The active ingredients of both experimental additives turned out to be some modified forms of adenine, one of the components of nucleic acids. These growth regulators were named cytokinins because they stimulate cytokinesis, or cell division. Of the variety of cytokinins that occur naturally in plants, the most common is zeatin, so named because it was discovered first in maize (*Zea mays*). Although much remains to be learned about cytokinin synthesis and signal transduction, some of the major effects that cytokinins have on the physiology and development of plants are well documented.

Control of Cell Division and Differentiation. Cytokinins are produced in actively growing tissues, particularly in roots, embryos, and fruits. Cytokinins produced in the root reach their target tissues by moving up the plant in the xylem sap. Acting in concert with auxin, cytokinins stimulate cell division and influence the pathway of differentiation. The effects of cytokinins on cells growing in tissue culture provide clues about how this class of hormones may function in an intact plant. When a piece of parenchyma tissue from a stem is cultured in the absence of cytokinins, the cells grow very large but do not divide. But if cytokinins are added along with auxin, the cells divide. Cytokinins alone have no effect. The ratio of cytokinin to auxin controls cell differentiation. When the concentrations of the two hormones are at the appropriate levels, the mass of cells continues to grow, but it remains a cluster of undifferentiated cells called a callus (see Figure 38.12). If cytokinin levels are raised, shoot buds develop from the callus. If auxin levels are raised, roots form.

Control of Apical Dominance. Cytokinins, auxin, and other factors interact in the control of apical dominance, the ability of the terminal bud to suppress the development of axillary buds **(Figure 39.9a)**. Until recently, the leading hypothesis to explain the hormonal regulation of apical dominance—the direct inhibition hypothesis—proposed that auxin and cytokinin act antagonistically in regulating axillary bud growth. According to this view, auxin transported down the shoot from the terminal bud directly inhibits axillary buds from growing, causing a shoot to lengthen at the expense of lateral branching. Meanwhile, cytokinins entering the shoot system

(a) Intact plant

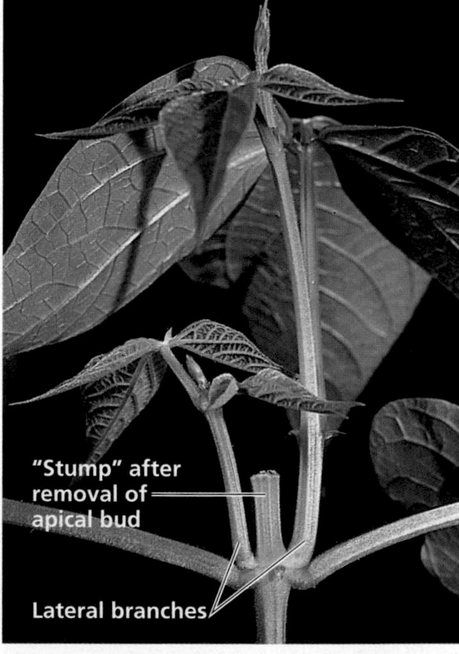

(b) Plant with apical bud removed

▲ **Figure 39.9 Apical dominance. (a)** The inhibition of growth of axillary buds, possibly influenced by auxin from the apical bud, favors elongation of the shoot's main axis. Cytokinins, which are transported upward from roots, may counter auxin, stimulating the growth of axillary buds. **(b)** Removal of the apical bud from the same plant enabled lateral branches to grow.

from roots counter the action of auxin by signaling axillary buds to begin growing. Thus, the ratio of auxin and cytokinin was viewed as the critical factor in controlling axillary bud inhibition. Many observations are consistent with the direct inhibition hypothesis. If the terminal bud, the primary source of auxin, is removed, the inhibition of axillary buds is removed and the plant becomes bushier **(Figure 39.9b)**. Application of auxin to the cut surface of the decapitated shoot resuppresses the growth of the lateral buds. Mutants that overproduce cytokinins or plants treated with cytokinins also tend to be bushier than normal. One prediction of the direct inhibition hypothesis not borne out by experiment is that decapitation, by removing the primary source of auxin, should lead to a decrease in the auxin levels of axillary buds. Biochemical studies, however, have revealed the opposite: Auxin levels actually *increase* in the axillary buds of decapitated plants. The direct inhibition hypothesis, therefore, does not account for all experimental findings. It is likely that plant biologists have not uncovered all the pieces of this puzzle.

Anti-Aging Effects. Cytokinins retard the aging of some plant organs by inhibiting protein breakdown, by stimulating RNA and protein synthesis, and by mobilizing nutrients from surrounding tissues. If leaves removed from a plant are dipped in a cytokinin solution, they stay green much longer than otherwise. Cytokinins also slow the deterioration of leaves on intact plants. Because of this anti-aging effect, florists use cytokinin sprays to keep cut flowers fresh.

Gibberellins

A century ago, farmers in Asia noticed that some rice seedlings in their paddies grew so tall and spindly that they toppled over before they could mature and flower. In 1926, Ewiti Kurosawa, a Japanese plant pathologist, discovered that a fungus of the genus *Gibberella* caused this "foolish seedling disease." By the 1930s, Japanese scientists had determined that the fungus produced hyperelongation of rice stems by secreting a chemical, which was given the name **gibberellin.** In the 1950s, researchers discovered that plants also produce gibberellins. In the past 50 years, scientists have identified more than 100 different gibberellins that occur naturally in plants, although a much smaller number occur in each plant species. "Foolish rice" seedlings, it seems, suffer from an overdose of gibberellins normally found in plants in lower concentrations. Gibberellins have a variety of effects, such as stem elongation, fruit growth, and seed germination.

Stem Elongation. Roots and young leaves are major sites of gibberellin production. Gibberellins stimulate growth of both leaves and stems, but have little effect on root growth. In stems, gibberellins stimulate cell elongation *and* cell division. Like auxin, gibberellins cause cell wall loosening, but not by acidifying the cell wall. One hypothesis proposes that gibberellins stimulate cell-wall-loosening enzymes that facilitate the penetration of expansin proteins into the cell wall. Thus, in a growing stem, auxin, by acidifying the cell wall and activating expansins, and gibberellins, by facilitating the penetration of expansins, act in concert to promote elongation.

The effects of gibberellins in enhancing stem elongation are evident when certain dwarf (mutant) varieties of plants are treated with gibberellins. For instance, some dwarf pea plants (including the variety Mendel studied; see Chapter 14) grow to normal height if treated with gibberellins. If gibberellins are applied to plants of normal size, there is often no response. Apparently, these plants are already producing an optimal dose of the hormone.

The most dramatic example of gibberellin-induced stem elongation is bolting, the rapid growth of the floral stalk. In their vegetative state, some plants, such as cabbage, develop in a rosette form, low to the ground with very short internodes. As the plant switches to reproductive growth, a surge of gibberellins induces internodes to elongate rapidly, elevating floral buds that develop at stem tips.

Fruit Growth. In many plants, both auxin and gibberellins must be present for fruit to set. The most important commercial application of gibberellins is in the spraying of Thompson seedless grapes **(Figure 39.10)**. The hormone makes the individual grapes grow larger, a trait prized by the consumer. The gibberellin sprays also make the internodes of the grape bunch elongate, allowing more space for the individual grapes. This increase in space, by enhancing air circulation between the grapes, also makes it harder for yeast and other microorganisms to infect the fruits.

▲ **Figure 39.10 The effect of gibberellin treatment on Thompson seedless grapes.** The grape bunch on the left is an untreated control. The bunch on the right is growing from a vine that was sprayed with gibberellin during fruit development.

Germination. The embryo of a seed is a rich source of gibberellins. After water is imbibed, the release of gibberellins from the embryo signals the seed to break dormancy and germinate. Some seeds that require special environmental conditions to germinate, such as exposure to light or low temperatures, break dormancy if they are treated with gibberellins. Gibberellins support the growth of cereal seedlings by stimulating the synthesis of digestive enzymes such as α-amylase that mobilize stored nutrients **(Figure 39.11)**.

Brassinosteroids

Brassinosteroids are steroids that are chemically similar to cholesterol and the sex hormones of animals. They induce cell elongation and division in stem segments and seedlings at concentrations as low as 10^{-12} *M*. They also retard leaf abscission (loss of leaves) and promote xylem differentiation. These effects are so qualitatively similar to those of auxin that it took several years for plant physiologists to determine that brassinosteroids were not types of auxins.

Evidence from molecular biology established brassinosteroids as plant hormones. Joanne Chory and colleagues at the Salk Institute in San Diego were interested in an *Arabidopsis* mutant that had morphological features similar to those of light-grown plants even though the mutants were grown in the dark. They discovered that the mutation affects a gene that normally codes for an enzyme similar to one involved in steroid synthesis in mammals. They also showed that mutant plants could be restored to normal phenotype by experimental application of brassinosteroids. The mutant studied by Chory was brassinosteroid-deficient.

Abscisic Acid

In the 1960s, one research group studying the chemical changes that precede bud dormancy and leaf abscission in deciduous trees and another team investigating chemical changes preceding abscission of cotton fruits isolated the same compound, **abscisic acid (ABA).** Ironically, ABA is no longer thought to play a primary role in bud dormancy or leaf abscission, but it is very important in other functions. Unlike the growth-stimulating hormones we have studied so far—auxin, cytokinins, gibberellins, and brassinosteroids—ABA *slows* growth. Often, ABA antagonizes the actions of the growth hormones, and it is the ratio of ABA to one or more growth hormones that determines the final physiological outcome. We will consider two of ABA's many effects: seed dormancy and drought tolerance.

Seed Dormancy. Seed dormancy has great survival value because it ensures that the seed will germinate only when there are sufficient conditions of light, temperature, and moisture (see Chapter 38). What prevents a seed dispersed in autumn from germinating immediately, only to be killed by winter? What mechanisms ensure that such seeds germinate in the spring? For that matter, what prevents seeds from germinating in the dark, moist interior of the fruit? The answer to these questions is ABA. The levels of ABA may increase 100-fold during seed maturation. The high levels of ABA in maturing seeds inhibit germination and induce the production of special proteins that help the seeds withstand the extreme dehydration that accompanies maturation.

Many types of dormant seeds germinate when ABA is removed or inactivated in some way. The seeds of some desert plants break dormancy only when heavy rains wash ABA out of the seed. Other seeds require light or prolonged exposure to cold to trigger the inactivation of ABA. Often, the ratio of ABA to gibberellins determines whether the seed remains dormant or germinates, and the addition of ABA to seeds that are primed to germinate returns them to the dormant condition. A maize mutant that has grains that germinate while still on the cob lacks a functional transcription factor required for ABA to induce expression of certain genes **(Figure 39.12)**.

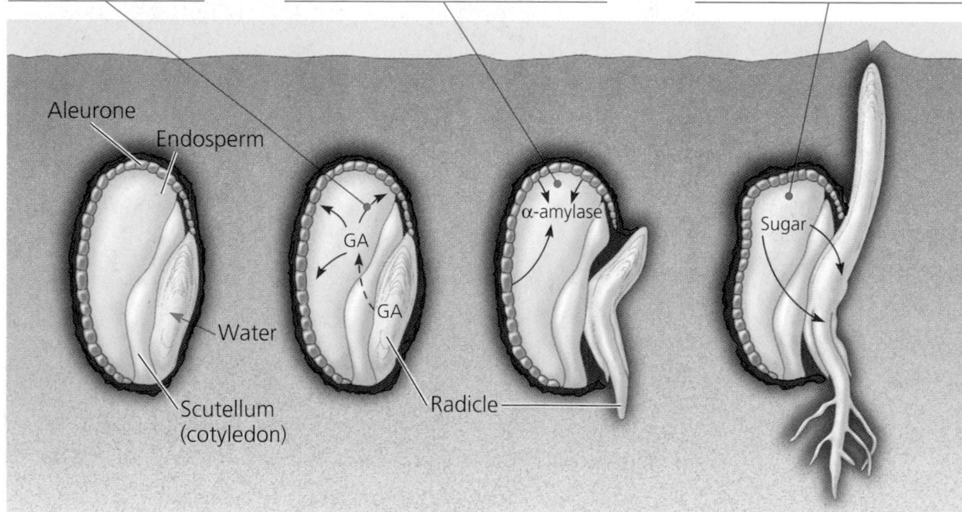

1 After a seed imbibes water, the embryo releases gibberellin (GA), which sends a signal to the aleurone, the thin outer layer of the endosperm.
2 The aleurone responds to GA by synthesizing and secreting digestive enzymes that hydrolyze nutrients stored in the endosperm. One example is α-amylase, which hydrolyzes starch. (A similar enzyme in our saliva helps in digesting bread and other starchy foods.)
3 Sugars and other nutrients absorbed from the endosperm by the scutellum (cotyledon) are consumed during growth of the embryo into a seedling.
Aleurone
Endosperm
Water
Scutellum (cotyledon)
GA
GA
Radicle
α-amylase
Sugar

▲ **Figure 39.11 Gibberellins mobilize nutrients during the germination of grain seeds.**

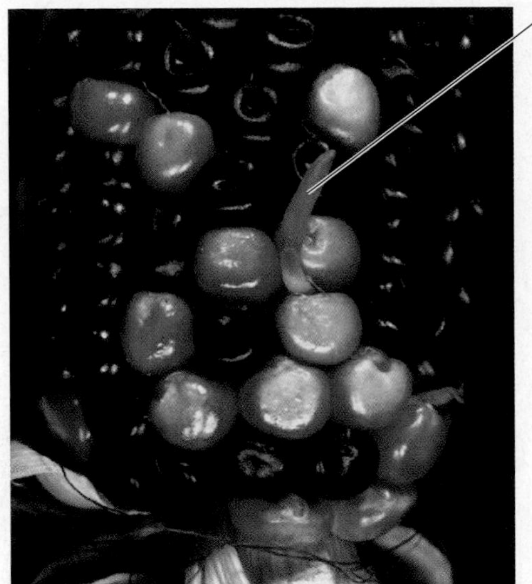

▲ **Figure 39.12 Precocious germination of mutant maize seeds.** Abscisic acid induces dormancy in seeds. When its action is blocked—in this case, by a mutation affecting an abscisic-acid-regulated transcription factor—precocious germination results.

Drought Tolerance. ABA is the primary internal signal that enables plants to withstand drought. When a plant begins to wilt, ABA accumulates in leaves and causes stomata to close rapidly, reducing transpiration and preventing further water loss. ABA, through its effects on second messengers such as calcium, causes the opening of potassium channels in the plasma membrane of guard cells, leading to a massive loss of potassium ions from the cells. The accompanying osmotic loss of water leads to a reduction in guard cell turgor and closing of the stomatal pores (see Figure 36.15). In some cases, water shortage can stress the root system before the shoot system, and ABA transported from roots to leaves may function as an "early warning system." Mutants that are especially prone to wilting are in many cases deficient in the production of ABA.

Ethylene

During the 1800s, when coal gas was used for street illumination, leakage from gas mains caused nearby trees to drop leaves prematurely. In 1901, the Russian scientist Dimitry Neljubow demonstrated that the gas **ethylene** was the active factor in "illuminating gas." The idea that ethylene is a plant hormone was not widely accepted, however, until the advent of gas chromatography simplified measurement.

Plants produce ethylene in response to stresses such as drought, flooding, mechanical pressure, injury, and infection. Production also occurs in fruit ripening and programmed cell death and in response to high concentrations of externally applied auxin. Indeed, many effects previously ascribed to auxin, such as inhibition of root elongation, may be due to auxin-induced ethylene production. We will focus on four of ethylene's many effects: response to mechanical stress, programmed cell death, leaf abscission, and fruit ripening.

The Triple Response to Mechanical Stress. Imagine a pea seedling pushing upward through the soil, only to come up against a stone. As it pushes against the obstacle, the stress in its delicate tip induces the seedling to produce ethylene. The hormone then instigates a growth maneuver known as the **triple response** that enables the shoot to avoid the obstacle. The three parts of this response are a slowing of stem elongation, a thickening of the stem (which makes it stronger), and a curvature that causes the stem to start growing horizontally. As the stem continues to grow, its tip touches upward intermittently. If these movements continue to detect a solid object above, then another pulse of ethylene is generated and the stem continues its horizontal progress. If, however, the upward touch detects no solid object, then ethylene production decreases, and the stem, now clear of the obstacle, resumes its normal upward growth. It is ethylene that induces the stem to grow horizontally rather than the physical obstruction per se; when ethylene is applied to normal seedlings growing free of all physical impediments, they still undergo the triple response **(Figure 39.13)**.

Figure 39.13
Inquiry How does ethylene concentration affect the triple response in seedlings?

EXPERIMENT Germinating pea seedlings were placed in the dark and exposed to varying ethylene concentrations. Their growth was compared with that of an untreated seedling.

RESULTS All the treated seedlings exhibited the triple response. The response was greater with increased concentration.

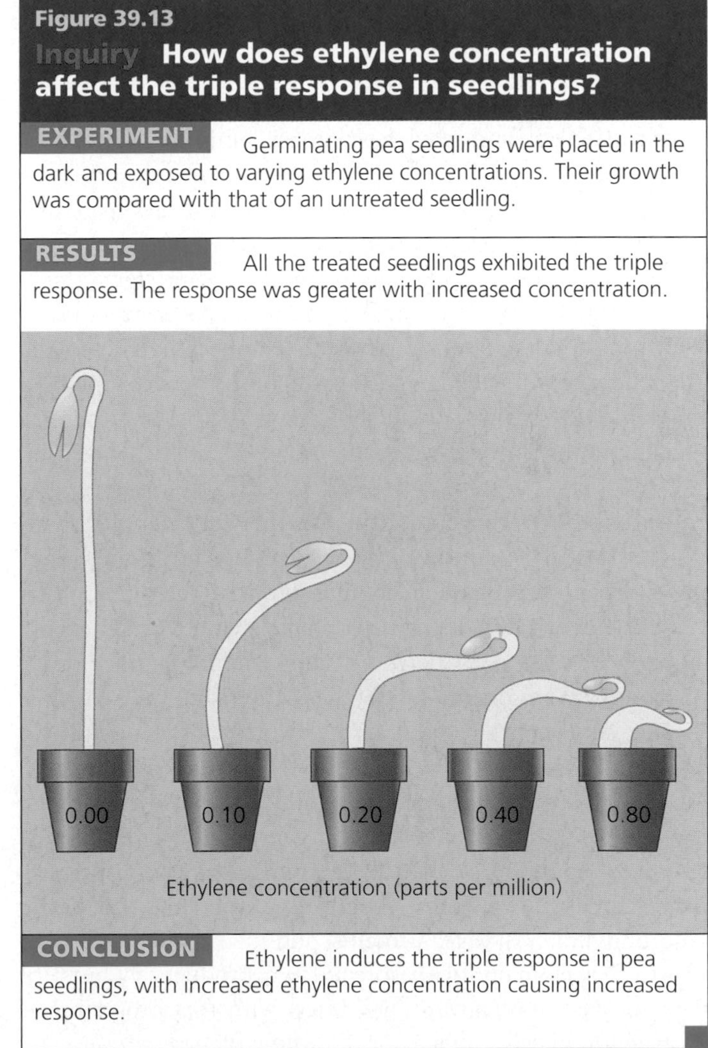

CONCLUSION Ethylene induces the triple response in pea seedlings, with increased ethylene concentration causing increased response.

Studies of *Arabidopsis* mutants with abnormal triple responses are an example of how biologists identify a signal transduction pathway. Scientists isolated ethylene-insensitive (*ein*) mutants that fail to undergo the triple response after exposure to ethylene **(Figure 39.14a)**. Some types of *ein* mutants are insensitive to ethylene because they lack a functional ethylene receptor. Other mutants undergo the triple response even out of soil, in the air, where there are no physical obstacles. Some mutants of this type have a regulatory defect that causes them to produce ethylene at rates 20 times normal. The phenotype of such ethylene-overproducing (*eto*) mutants can be restored to wild-type by treating the seedlings with inhibitors of ethylene synthesis. Still other mutants, called constitutive triple-response (*ctr*) mutants, undergo the triple response in air but do not respond to inhibitors of ethylene synthesis **(Figure 39.14b)**. In this case, ethylene signal transduction is permanently turned on, even though there is no ethylene present. **Figure 39.15** summarizes the responses of *ein, eto,* and *ctr* mutants to ethylene and ethylene synthesis inhibitors.

The affected gene in *ctr* mutants turns out to encode for a protein kinase. The fact that this mutation *activates* the ethylene response suggests that the normal kinase product of the wild-type allele is a *negative* regulator of ethylene signal transduction. Here is one hypothesis for how the pathway works in wild-type plants: Binding of the hormone ethylene to the ethylene receptor leads to inactivation of the kinase; and inactivation of this negative regulator allows synthesis of the proteins required for the triple response.

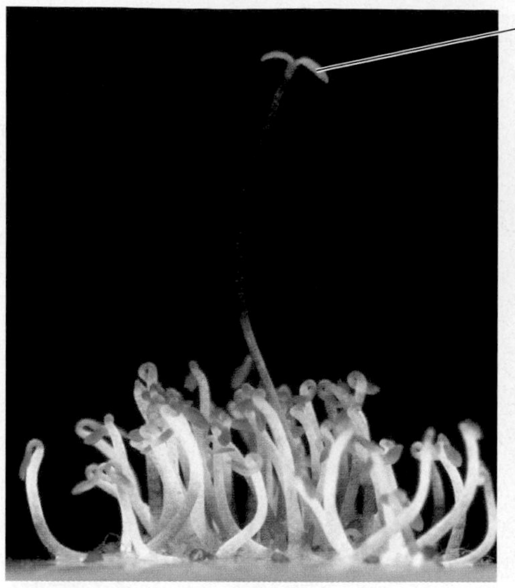

(a) *ein* mutant. An ethylene-insensitive (*ein*) mutant fails to undergo the triple response in the presence of ethylene.

(b) *ctr* mutant. A constitutive triple-response (*ctr*) mutant undergoes the triple response even in the absence of ethylene.

▲ **Figure 39.14 Ethylene triple-response *Arabidopsis* mutants.**

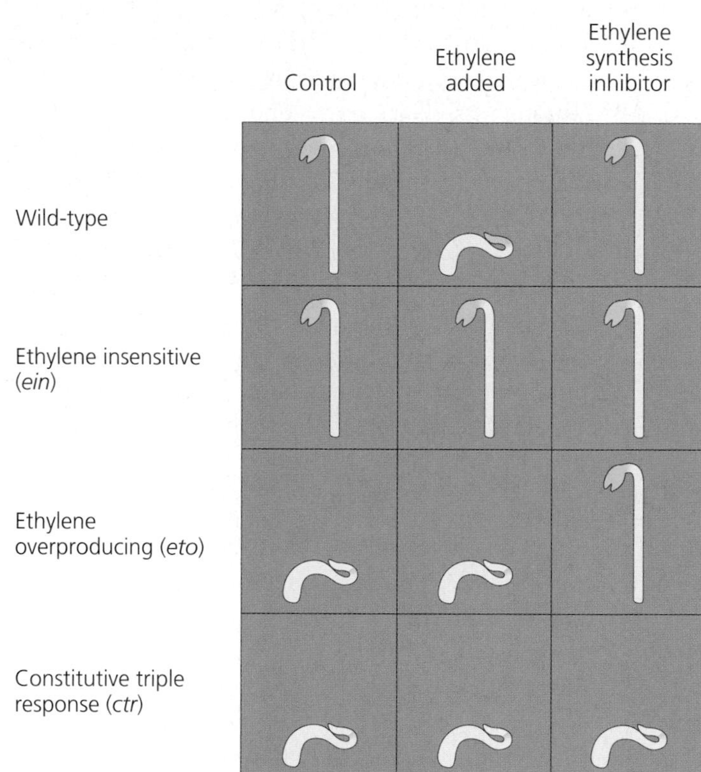

▲ **Figure 39.15 Ethylene signal transduction mutants can be distinguished by their different responses to experimental treatments.**

Apoptosis: Programmed Cell Death. Consider the shedding of a leaf in autumn or the death of an annual after flowering. Or think about the final step in differentiation of a vessel element, when its living contents are destroyed, leaving a hollow tube behind. Such events involve programmed death of certain cells or organs—or of the entire plant. Cells, organs, and plants genetically programmed to die on a schedule do not simply shut down cellular machinery and await death. Rather, the onset of programmed cell death, called **apoptosis**, is a very busy time in a cell's life, requiring new gene expression (see Figure 21.17). During apoptosis, newly formed enzymes break down many chemical components, including chlorophyll, DNA, RNA, proteins, and membrane lipids. The plant salvages many of the breakdown products. A burst of ethylene is almost always associated with this programmed destruction of cells, organs, or the whole plant.

Leaf Abscission. The loss of leaves each autumn is an adaptation that keeps deciduous trees from desiccating during winter when the roots cannot absorb water from the frozen ground. Before leaves abscise, many essential elements are salvaged from the dying leaves and are stored in stem parenchyma cells. These nutrients are recycled back to developing leaves the following spring. Fall color is a combination of new red pigments made during autumn and yellow and orange carotenoids (see Chapter 10) that were already present in the leaf but are rendered visible by the breakdown of the dark green chlorophyll in autumn.

When an autumn leaf falls, the breaking point is an abscission layer that develops near the base of the petiole **(Figure 39.16)**. The small parenchyma cells of this layer have very thin walls, and there are no fiber cells around the vascular tissue. The abscission layer is further weakened when enzymes hydrolyze polysaccharides in the cell walls. Finally, the weight of the leaf, with the help of wind, causes a separation within the abscission layer. Even before the leaf falls, a layer of cork forms a protective scar on the twig side of the abscission layer, preventing pathogens from invading the plant.

A change in the balance of ethylene and auxin controls abscission. An aging leaf produces less and less auxin, and this renders the cells of the abscission layer more sensitive to ethylene. As the influence of ethylene on the abscission layer prevails, the cells produce enzymes that digest the cellulose and other components of cell walls.

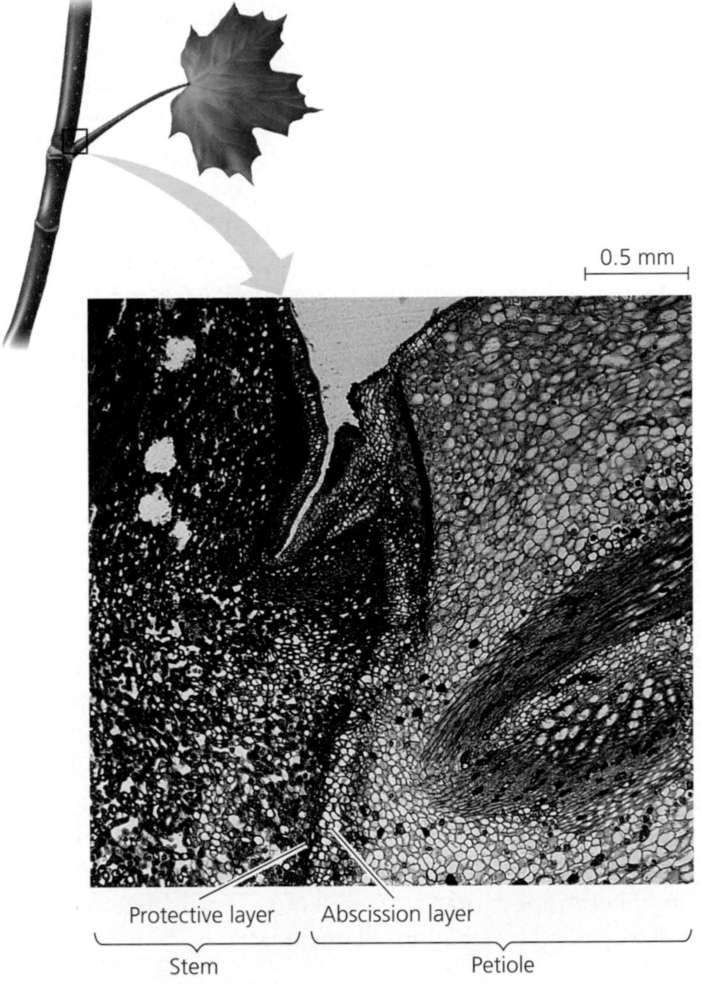

▲ **Figure 39.16 Abscission of a maple leaf.** Abscission is controlled by a change in the balance of ethylene and auxin. The abscission layer can be seen here as a vertical band at the base of the petiole. After the leaf falls, a protective layer of cork becomes the leaf scar that helps prevent pathogens from invading the plant (LM).

Fruit Ripening. Immature fleshy fruits that are tart, hard, and green help protect the developing seeds from herbivores. After ripening, the mature fruits help *attract* animals that disperse the seeds (see Figures 30.8 and 30.9). A burst of ethylene production in the fruit triggers the ripening process. The enzymatic breakdown of cell wall components softens the fruit, and the conversion of starches and acids to sugars makes the fruit sweet. The production of new scents and colors helps advertise ripeness to animals, which eat the fruits and disperse the seeds.

A chain reaction occurs during ripening: Ethylene triggers ripening, and ripening then triggers even more ethylene production—one of the rare examples of positive feedback in physiology (see Figure 1.12). The result is a huge burst in ethylene production. Because ethylene is a gas, the signal to ripen even spreads from fruit to fruit: One bad apple, in fact, *does* spoil the lot. If you pick or buy green fruit, you may be able to speed ripening by storing the fruit in a paper bag, allowing the ethylene gas to accumulate. On a commercial scale, many kinds of fruits are ripened in huge storage containers in which ethylene levels are enhanced. In other cases, measures are taken to retard ripening caused by natural ethylene. Apples, for instance, are stored in bins flushed with carbon dioxide. Circulating the air prevents ethylene from accumulating, and carbon dioxide inhibits synthesis of new ethylene. Stored in this way, apples picked in autumn can still be shipped to grocery stores the following summer.

Given the importance of ethylene in the post-harvest physiology of fruits, the genetic engineering of ethylene signal transduction pathways has potentially important commercial applications. For example, molecular biologists, by engineering a way to block the transcription of one of the genes required for ethylene synthesis, have created tomato fruits that ripen on demand. These fruits are picked while green and will not ripen unless ethylene gas is added. As such methods are refined, they will reduce spoilage of fruits and vegetables, a problem that currently ruins almost half the produce harvested in the United States.

Systems Biology and Hormone Interactions

As we have discussed, plant responses often involve interactions of many hormones and their signal transduction pathways. The study of hormone interactions can be a complex problem. For example, flooding of deepwater rice leads to a 50-fold increase in internal ethylene levels and rapid stem elongation. But ethylene's role in this response is a small part of the story. Flooding also leads to an increase in sensitivity to GA that is mediated by a decrease in ABA levels. Thus, stem elongation is actually the result of an interaction among these three hormones and their signal transduction pathways.

Imagine yourself as a molecular biologist assigned the task of genetically engineering a deepwater rice plant so that it grows even faster when submerged. What would be the best molecular targets for genetic manipulation? An enzyme that inactivates ABA? An enzyme that produces more GA? An ethylene receptor? It is difficult to predict. And this is by no means an isolated problem. Virtually every plant response discussed in this chapter is of comparable complexity. Because of this pervasive and unavoidable problem of complex interactions, many plant biologists, including Natasha Raikhel, who is interviewed on pages 710–711, are promoting a new systems-based approach to plant biology.

Chapter 1 provided a general description of systems biology, which attempts to discover and understand biological properties that emerge from the interactions of many system elements (for example, mRNAs, proteins, hormones, and metabolites). Using genomic techniques, biologists can now identify all the genes in a plant and have already sequenced two plant genomes—the research plant *Arabidopsis* and the crop plant rice (*Oryza sativa*). Moreover, using microarray and proteomic techniques (see Chapter 20), scientists can resolve which genes are activated or inactivated during development or in response to an environmental change. However, identifying all the genes and proteins (system elements) in an organism is comparable to listing all the parts of an airplane. Although such a list provides a catalog of components, it is not sufficient for understanding the complexity underlying the integrated system. What plant biologists really need to know is how all these system elements interact.

A systems-based approach may greatly alter how plant biology is done. The dream is that laboratories will be equipped with fast-moving (high-throughput) robotic scanners that record which genes in a plant's genome are activated in which cells and under what conditions. New hypotheses and avenues of research will emerge from analysis of these comprehensive data sets. Ultimately, one goal of systems biology is to model a living plant predictably. Armed with such detailed knowledge, a molecular biologist attempting to genetically engineer faster stem elongation into rice could proceed much more efficiently. The ability to model a living plant would facilitate predicting the result of a genetic manipulation before even setting foot in the laboratory.

Concept Check 39.2

1. Predict the triple-response phenotype of a plant with the double mutation *ctr* and *ein*. Explain.
2. In a diseased state known as "witch's broom," branches grow and proliferate excessively. Suggest a hypothesis to explain how a pathogen might induce this growth pattern.
3. Fusicoccin is a fungal toxin that stimulates the plasma membrane H^+ pump of plant cells. How might it affect growth of isolated stem sections?

For suggested answers, see Appendix A.

Concept 39.3

Responses to light are critical for plant success

Light is an especially important environmental factor in the lives of plants. In addition to being required for photosynthesis, light cues many key events in plant growth and development. Effects of light on plant morphology are what plant biologists call **photomorphogenesis.** Light reception also allows plants to measure the passage of days and seasons.

Plants can detect not only the presence of light but also its direction, intensity, and wavelength (color). A graph called an **action spectrum** depicts the relative effectiveness of different wavelengths of radiation in driving a particular process. For example, the action spectrum for photosynthesis has two peaks, one in the red and one in the blue (see Figure 10.9). This is because chlorophyll absorbs light primarily in the red and blue portions of the visible spectrum. Action spectra are useful in the study of *any* process that depends on light, such as phototropism **(Figure 39.17)**. By comparing action spectra of different plant responses, researchers determine which responses are mediated by the same photoreceptor (pigment). They also compare action spectra with absorption spectra of pigments. A close correspondence suggests that the pigment may be the photoreceptor mediating the response. Action spectra reveal that red and blue light are the most important colors in regulating a plant's photomorphogenesis. These observations led researchers to two major classes of light receptors: **blue-light photoreceptors** and **phytochromes**, photoreceptors that absorb mostly red light.

Blue-Light Photoreceptors

Blue light initiates diverse responses in plants, including phototropism, the light-induced opening of stomata (see Figure 36.14), and the light-induced slowing of hypocotyl elongation that occurs when a seedling breaks ground. The biochemical identity of the blue-light photoreceptor was so elusive that in the 1970s, plant physiologists began to call this putative receptor cryptochrome (from the Greek *kryptos*, hidden, and *chrom*, pigment). In the 1990s, molecular biologists analyzing *Arabidopsis* mutants found that plants use at least three different types of pigments to detect blue light: *cryptochromes* (for inhibition of hypocotyl elongation), *phototropin* (for phototropism), and a carotenoid-based photoreceptor called *zeaxanthin* (for stomatal opening).

Phytochromes as Photoreceptors

When we introduced signal transduction in plants earlier in the chapter, we discussed the role of a family of plant pigments called phytochromes in the de-etiolation process.

Figure 39.17

Inquiry What wavelengths stimulate phototropic bending toward light?

EXPERIMENT Researchers exposed maize *(Zea mays)* coleoptiles to violet, blue, green, yellow, orange, and red light to test which wavelengths stimulate the phototropic bending toward light.

RESULTS The graph below shows phototropic effectiveness (curvature per photon) relative to effectiveness of light with a wavelength of 436 nm. The photo collages show coleoptiles before and after 90-minute exposure to side lighting of the indicated colors. Pronounced curvature occurred only with wavelengths below 500 nm and was greatest with blue light.

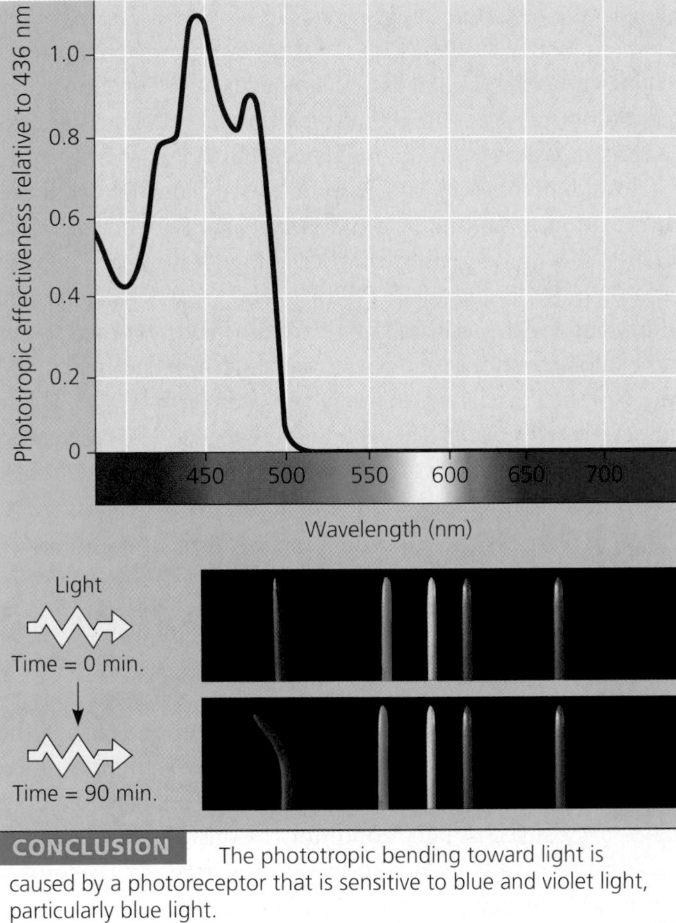

CONCLUSION The phototropic bending toward light is caused by a photoreceptor that is sensitive to blue and violet light, particularly blue light.

Figure 39.18

Inquiry How does the order of red and far-red illumination affect seed germination?

EXPERIMENT During the 1930s, USDA scientists briefly exposed batches of lettuce seeds to red light or far-red light to test the effects on germination. After the light exposure, the seeds were placed in the dark, and the results were compared with control seeds that were not exposed to light.

RESULTS The bar below each photo indicates the sequence of red-light exposure, far-red light exposure, and darkness. The germination rate increased greatly in groups of seeds that were last exposed to red light (left). Germination was inhibited in groups of seeds that were last exposed to far-red light (right).

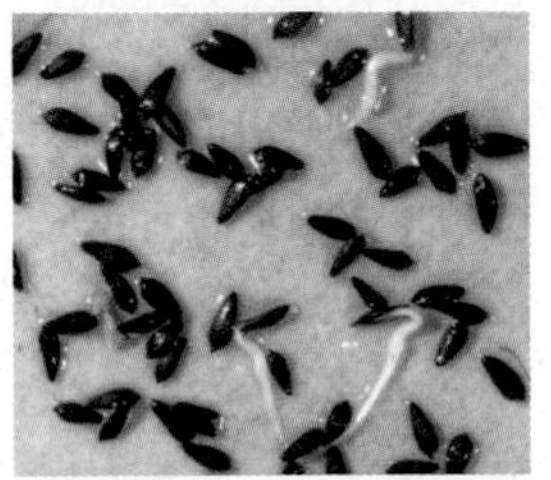

Dark (control)

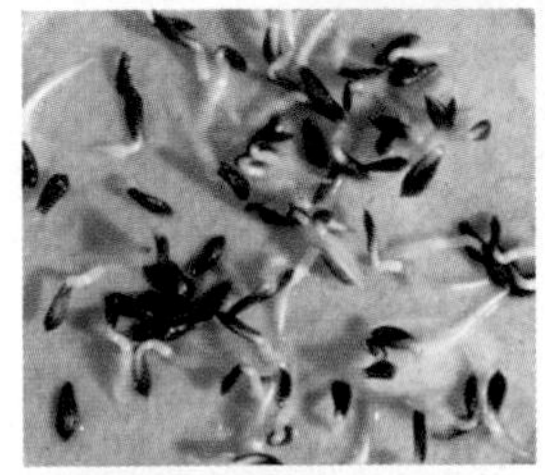

Red | Dark

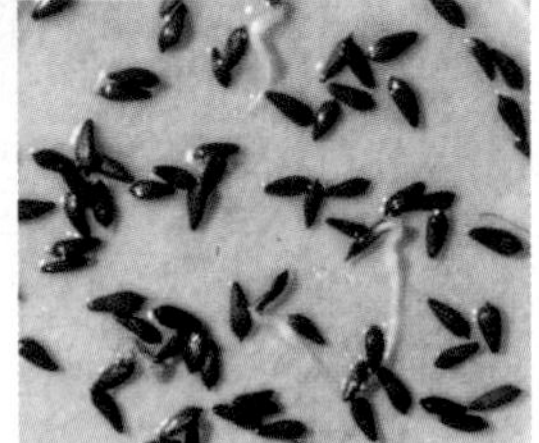

Red | Far-red | Dark

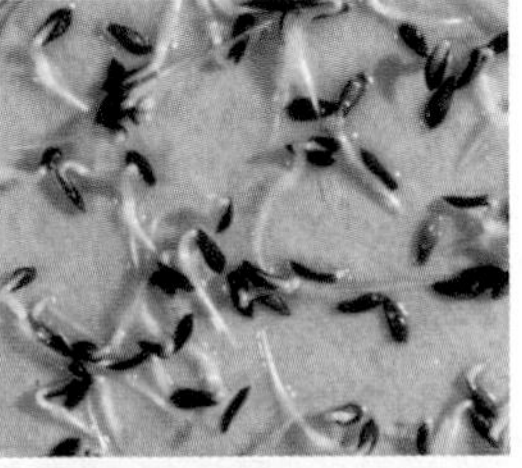

Red | Far-red | Red | Dark

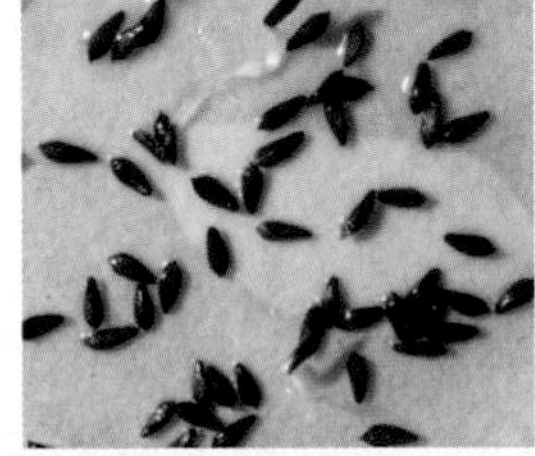

Red | Far-red | Red | Far-red

CONCLUSION Red light stimulated germination, and far-red light inhibited germination. The final exposure was the determining factor. The effects of red and far-red light were reversible.

Phytochromes regulate many of a plant's responses to light throughout its life. Let's look at a couple of examples.

Phytochromes and Seed Germination

Studies of seed germination led to the discovery of phytochromes. Because of their limited nutrient reserves, the successful sprouting of many types of small seeds, such as lettuce, requires germination only when conditions, especially the light environment, are near optimal. Such seeds often remain dormant for many years until a change in light conditions occurs. For example, the death of a shading tree or the plowing of a field may create a favorable light environment.

In the 1930s, scientists at the U.S. Department of Agriculture determined the action spectrum for light-induced germination of lettuce seeds **(Figure 39.18)**. They exposed water-swollen

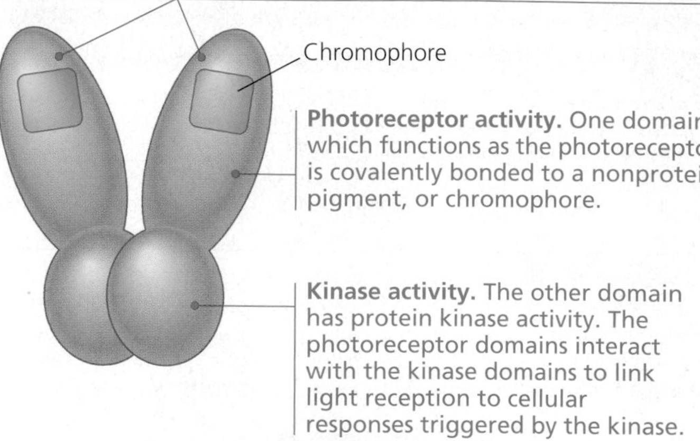

▲ **Figure 39.19 Structure of a phytochrome.**

seeds to a few minutes of monochromatic (single-colored) light of various wavelengths and then stored the seeds in the dark. After two days, the researchers counted the number of seeds that had germinated under each light regimen. They found that red light of wavelength 660 nm increased the germination percentage of lettuce seeds maximally, whereas far-red light—that is, light of wavelengths just near the edge of human visibility (730 nm)—*inhibited* germination compared with dark controls. What happens when the lettuce seeds are subjected to a flash of red (R) light followed by a flash of far-red (FR) light or, conversely, to FR light followed by R light? The *last* flash of light determines the seeds' response. In other words, the effects of red and far-red light are reversible.

A phytochrome consists of two identical proteins joined to form one functional molecule. Each of these proteins has two domains.
Chromophore
Photoreceptor activity. One domain, which functions as the photoreceptor, is covalently bonded to a nonprotein pigment, or chromophore.
Kinase activity. The other domain has protein kinase activity. The photoreceptor domains interact with the kinase domains to link light reception to cellular responses triggered by the kinase.

The photoreceptor responsible for the opposing effects of red and far-red light is a phytochrome. It consists of a protein component covalently bonded to a nonprotein part that functions as a chromophore, the light-absorbing part of the molecule **(Figure 39.19)**. So far, researchers have identified five phytochromes in *Arabidopsis,* each with a slightly different protein component.

The chromophore of a phytochrome is photoreversible, reverting back and forth between two isomeric forms, depending on the color of light provided (see Figure 4.7 to review isomers). In its P_r isomer form, a phytochrome absorbs red light maximally, whereas in its P_{fr} isomer form, it absorbs far-red light:

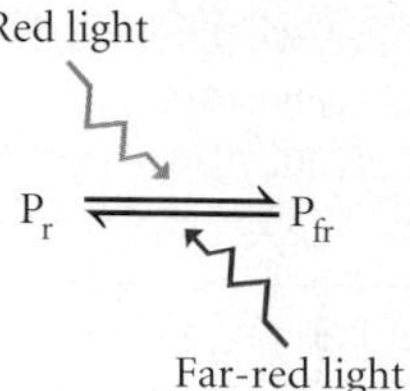

$P_r \rightleftharpoons P_{fr}$ interconversion is a switching mechanism that controls various light-induced events in the life of the plant **(Figure 39.20)**. P_{fr} is the form of phytochrome that triggers many of a plant's developmental responses to light. For example, P_r in lettuce seeds that are exposed to red light is converted to P_{fr}, stimulating the cellular responses that lead to germination. When red-illuminated seeds are then exposed to far-red light, the P_{fr} is converted back to P_r, inhibiting the germination response.

How does phytochrome switching explain light-induced germination in nature? Plants synthesize phytochrome as P_r, and if seeds are kept in the dark, the pigment remains almost entirely in the P_r form (see Figure 39.20). But if the seeds are illuminated with sunlight, the phytochrome is exposed to red light (along with all the other wavelengths in sunlight), and much of the P_r is converted to P_{fr}. The appearance of P_{fr} is one of the ways that plants detect sunlight. When seeds are exposed to adequate sunlight for the first time, it is the appearance of P_{fr} that triggers their germination.

Phytochromes and Shade Avoidance

The phytochrome system also provides the plant with information about the *quality* of light. Sunlight includes both red and far-red radiation. Thus, during the day, the $P_r \rightleftharpoons P_{fr}$ photoreversion reaches a dynamic equilibrium, with the ratio of the two phytochrome forms indicating the relative amounts of red and far-red light. This sensing mechanism enables plants to adapt to changes in light conditions. Consider, for example, the "shade-avoidance" response of a tree that requires relatively high light intensity. If other trees in a forest shade this tree, the phytochrome ratio shifts in favor of P_r because the forest canopy screens out more red light than far-red light.

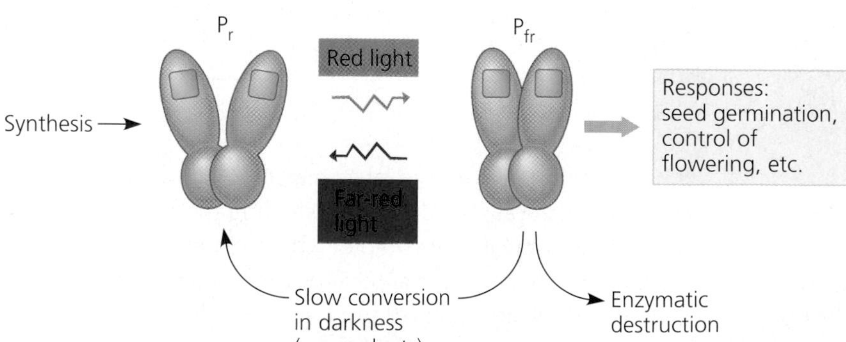

◀ **Figure 39.20 Phytochrome: a molecular switching mechanism.** Absorption of red light causes the bluish P_r to change to the blue-greenish P_{fr}. Far-red light reverses this conversion. In most cases, it is the P_{fr} form of the pigment that switches on physiological and developmental responses in the plant.

This is because the chlorophyll pigments in the leaves of the canopy absorb red light and allow far-red light to pass. This shift in the ratio of red to far-red light induces the tree to allocate more of its resources to growing taller. In contrast, direct sunlight increases the proportion of P_{fr}, which stimulates branching and inhibits vertical growth.

In addition to helping plants detect light, phytochrome helps a plant to keep track of the passage of days and seasons. To understand phytochrome's role in these timekeeping processes, we must first examine the clock itself.

Biological Clocks and Circadian Rhythms

Many plant processes, such as transpiration and synthesis of certain enzymes, oscillate during the course of a day. Some of these cyclic variations are responses to the changes in light levels, temperature, and relative humidity that accompany the 24-hour cycle of day and night. However, one can eliminate these exogenous (external) factors by growing plants in growth chambers under rigidly maintained conditions of light, temperature, and humidity. Even under artificially constant conditions, many physiological processes in plants, such as the opening and closing of stomata and the production of photosynthetic enzymes, continue to oscillate with a frequency of about 24 hours. For example, many legumes lower their leaves in the evening and raise them in the morning **(Figure 39.21)**. A bean plant continues these "sleep movements" even if kept in constant light or constant darkness; the leaves are not simply responding to sunrise and sunset. Such cycles with a frequency that is about 24 hours and not directly paced by any known environmental variable are called **circadian rhythms** (from the Latin *circa,* approximately, and *dies,* day), and they are common to all eukaryotic life. Your pulse, blood pressure, temperature, rate of cell division, blood cell count, alertness, urine composition, metabolic rate, sex drive, and response to medications all fluctuate in a circadian manner.

Current research indicates that the molecular "gears" of the circadian clock are endogenous (internal), rather than a daily response to some subtle but pervasive environmental cycle, such as geomagnetism or cosmic radiation. Organisms, including plants and humans, will continue their rhythms when placed in the deepest mine shafts or when orbited in satellites, conditions that alter these subtle geophysical periodicities. The circadian clock, however, can be entrained (set) to a period of precisely 24 hours by daily signals from the environment.

Noon

Midnight

▲ **Figure 39.21 Sleep movements of a bean plant (*Phaseolus vulgaris*).** The movements are caused by reversible changes in the turgor pressure of cells on opposing sides of the pulvini, motor organs of the leaf.

If an organism is kept in a constant environment, its circadian rhythms deviate from a 24-hour period (a period is the duration of one cycle). These free-running periods, as they are called, vary from about 21 to 27 hours, depending on the particular rhythmic response. The sleep movements of bean plants, for instance, have a period of 26 hours when the plants are kept in free-running conditions of constant darkness. Deviation of the free-running period from exactly 24 hours does not mean that biological clocks drift erratically. Free-running clocks are still keeping perfect time, but they are not synchronized with the outside world.

How do biological clocks work? In attempting to answer this question, we must distinguish between the clock and the rhythmic processes it controls. For example, the leaves of the bean plant in Figure 39.21 are the clock's "hands," but are not the essence of the clock itself. If bean leaves are restrained for several hours and then released, they will rush to the position appropriate for the time of day. We can interfere with a biological rhythm, but the clockwork continues.

Researchers are tracing the clock to a molecular mechanism that may be common to all eukaryotes. A leading hypothesis is that biological timekeeping may depend on the synthesis of a protein that regulates its own production through feedback control. This protein may be a transcription factor that inhibits transcription of the gene that encodes for the transcription factor itself. The concentration of this transcription factor may accumulate during the first half of the circadian cycle and then decline during the second half owing to self-inhibition of its own production.

Researchers have recently used a novel technique to identify clock mutants of *Arabidopsis.* One prominent circadian rhythm in plants is the daily production of certain photosynthesis-related proteins. Molecular biologists traced this rhythm to the promoter that regulates the transcription of the genes for these photosynthesis proteins. To identify clock mutants, scientists spliced the gene for an enzyme called luciferase to the promoter. Luciferase is the enzyme responsible for the bioluminescence of fireflies. When the biological clock turned on the promoter in the *Arabidopsis* genome, it also turned on the production of luciferase. The plants began to glow with a circadian periodicity. Clock mutants were then isolated by selecting specimens that glowed for a longer or shorter time than normal. The altered genes in some of these mutants affect proteins that normally bind photoreceptors. Perhaps these particular mutations disrupt a light-dependent mechanism that sets the biological clock.

The Effect of Light on the Biological Clock

As we have discussed, the free-running period of the circadian rhythm of bean leaf movements is 26 hours. Consider a bean plant placed at dawn in a dark cabinet for 72 hours: Its leaves would not rise again until 2 hours after natural dawn on the second day, and 4 hours after natural dawn on the third day, and so on. Shut off from environmental cues, the plant becomes desynchronized. Desynchronization also happens when we cross several time zones in an airplane; when we reach our destination, the clocks on the wall are not synchronized with our internal clocks. All eukaryotes are probably prone to jet lag.

The factor that entrains the biological clock to precisely 24 hours every day is light. Both phytochrome and blue-light photoreceptors can entrain circadian rhythms in plants, but our understanding of how phytochrome does this is more complete. The mechanism involves turning cellular responses on and off by means of the $P_r \rightleftharpoons P_{fr}$ switch.

Consider again the photoreversible system in Figure 39.20. In darkness, the phytochrome ratio shifts gradually in favor of the P_r form, partly a result of turnover in the overall phytochrome pool. The pigment is synthesized in the P_r form, and enzymes destroy more P_{fr} than P_r. In some plant species, P_{fr} present at sundown slowly converts to P_r. In darkness, there is no means for the P_r to be reconverted to P_{fr}, but upon illumination, the P_{fr} level suddenly increases again by rapid conversion from P_r. This increase in P_{fr} each day at dawn resets the biological clock: Bean leaves reach their most extreme night position 16 hours after dawn.

In nature, interactions between phytochrome and the biological clock enable plants to measure the passage of night and day. The relative lengths of night and day, however, change over the course of the year (except at the equator). Plants use this change to adjust activities in synchrony with the seasons.

Photoperiodism and Responses to Seasons

Imagine the consequences if a plant produced flowers when pollinators were not present or if a deciduous tree produced leaves in the middle of winter. Seasonal events are of critical importance in the life cycles of most plants. Seed germination, flowering, and the onset and breaking of bud dormancy are all stages that usually occur at specific times of the year. The environmental stimulus plants use most often to detect the time of year is the photoperiod, the relative lengths of night and day. A physiological response to photoperiod, such as flowering, is called **photoperiodism.**

Photoperiodism and Control of Flowering

An early clue to how plants detect seasons came from a mutant variety of tobacco, Maryland Mammoth, which grew tall but failed to flower during summer. It finally bloomed in a greenhouse in December. After trying to induce earlier flowering by varying temperature, moisture, and mineral nutrition, researchers learned that the shortening days of winter stimulated this variety to flower. If the plants were kept in light-tight boxes so that lamps could manipulate "day" and "night," flowering occurred only if the day length was 14 hours or shorter. It did not flower during summer because at Maryland's latitude, the days were too long during that season.

The researchers called Maryland Mammoth a **short-day plant** because it apparently required a light period *shorter* than a critical length to flower. Chrysanthemums, poinsettias, and some soybean varieties are some other short-day plants, which generally flower in late summer, fall, or winter. Another group of plants will flower only when the light period is *longer* than a certain number of hours. These **long-day plants** will generally flower in late spring or early summer. Spinach, for example, flowers when days are 14 hours or longer. Radish, lettuce, iris, and many cereal varieties are also long-day plants. **Day-neutral plants** are unaffected by photoperiod and flower when reaching a certain stage of maturity, regardless of day length. Tomatoes, rice, and dandelions are examples.

Critical Night Length. In the 1940s, researchers discovered that flowering and other responses to photoperiod are actually controlled by night length, not day length **(Figure 39.22)**. Many of these scientists worked with cocklebur (*Xanthium strumarium*), a short-day plant that flowers only when days are 16 hours or shorter (and nights are at least 8 hours long). These researchers found that if the daytime portion of the photoperiod is broken by a brief exposure to darkness, there is no effect on flowering. However, if the nighttime part of the photoperiod is interrupted by even a few minutes of dim light, cocklebur will not flower, and this turned out to be true for other short-day plants as well (see Figure 39.22a). Cocklebur is actually unresponsive to *day* length, but it requires at least 8 hours of *continuous darkness* to flower. Short-day plants are really long-night plants, but the older term is embedded firmly in the jargon of plant physiology. Similarly, long-day plants are actually short-night plants. A long-day plant grown on photoperiods of long nights that would not normally induce flowering will flower if the period of continuous darkness is interrupted by a few minutes of light (see Figure 39.22b). Notice that we distinguish long-day from short-day plants *not* by an absolute night length but by whether the critical night length sets a maximum (long-day plants) or minimum (short-day plants) number of hours of darkness required for flowering. In both cases, the actual number of hours in the critical night length is specific to each species of plant.

Red light is the most effective color in interrupting the nighttime portion of the photoperiod. Action spectra and photoreversibility experiments show that phytochrome is the pigment that receives the red light **(Figure 39.23)**. For example, if a flash

Figure 39.22

Inquiry **How does interrupting the dark period with a brief exposure to light affect flowering?**

EXPERIMENT During the 1940s, researchers conducted experiments in which periods of darkness were interrupted with brief exposure to light to test how the light and dark portions of a photoperiod affected flowering in "short-day" and "long-day" plants.

RESULTS

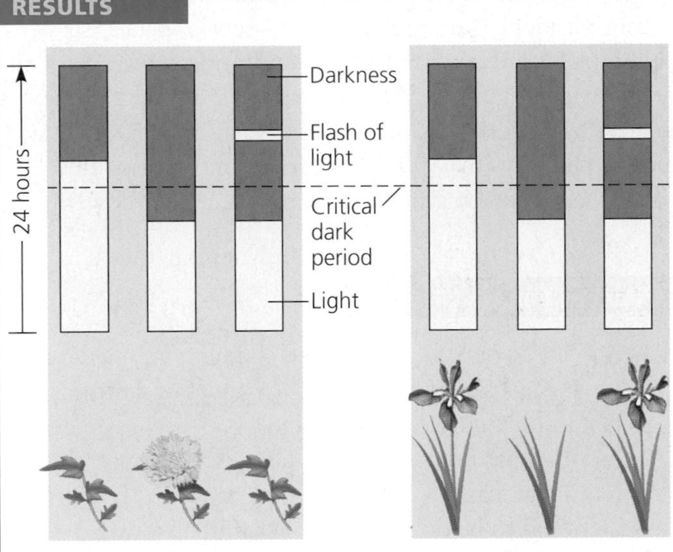

(a) "Short-day" plants flowered only if a period of continuous darkness was *longer* than a critical dark period for that particular species (13 hours in this example). A period of darkness can be ended by a brief exposure to light.

(b) "Long-day" plants flowered only if a period of continuous darkness was *shorter* than a critical dark period for that particular species (13 hours in this example).

CONCLUSION The experiments indicated that flowering of each species was determined by a critical period of *darkness* ("critical night length") for that species, *not* by a specific period of light. Therefore, "short-day" plants are more properly called "long-night" plants, and "long-day" plants are really "short-night" plants.

Figure 39.23

Inquiry **Is phytochrome the pigment that measures the interruption of dark periods in photoperiodic response?**

EXPERIMENT A unique characteristic of phytochrome is reversibility in response to red and far-red light. To test whether phytochrome is the pigment measuring interruption of dark periods, researchers observed how flashes of red light and far-red light affected flowering in "short-day" and "long-day" plants.

RESULTS

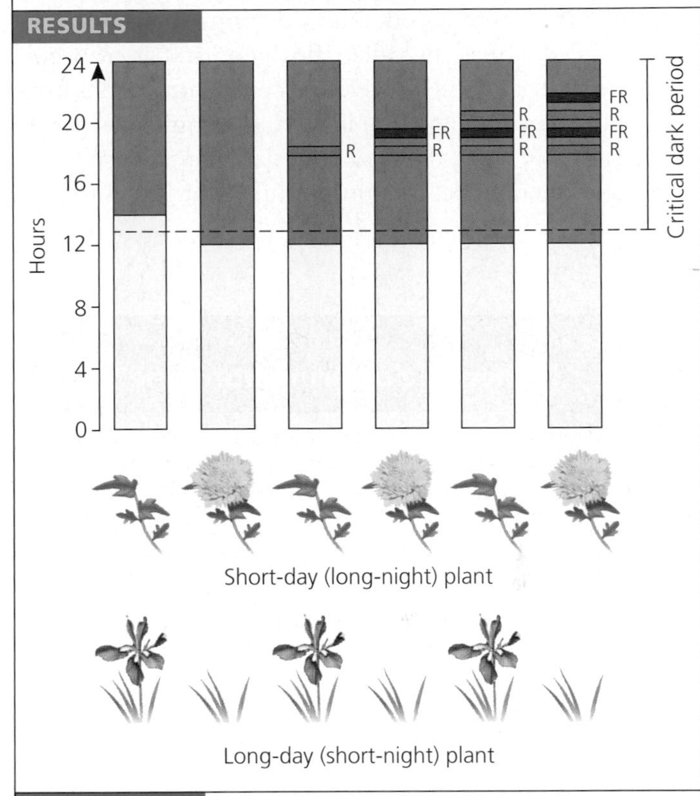

CONCLUSION A flash of red light shortened the dark period. A subsequent flash of far-red light canceled the red light's effect. If a red flash followed a far-red flash, the effect of the far-red light was canceled. This reversibility indicated that it is phytochrome that measures the interruption of dark periods.

of red light (R) during the dark period is followed by a flash of far-red (FR) light, then the plant detects no interruption of night length. As in the case of phytochrome-mediated seed germination, red/far-red photoreversibility occurs.

Plants detect night length very precisely; some short-day plants will not flower if night is even 1 minute shorter than the critical length. Some plant species always flower on the same day each year. It appears that plants use their biological clock, apparently entrained with the help of phytochrome, to tell the season of the year by measuring night length. The floriculture (flower-growing) industry has applied this knowledge to produce flowers out of season. Chrysanthemums, for instance, are short-day plants that normally bloom in fall, but their blooming can be stalled until Mother's Day in May by punctuating each long night with a flash of light, thus turning one long night into two short nights.

Some plants bloom after a single exposure to the photoperiod required for flowering. Other species need several successive days of the appropriate photoperiod. Still other plants will respond to a photoperiod only if they have been previously exposed to some other environmental stimulus, such as a period of cold temperatures. Winter wheat, for example, will not flower unless it has been exposed to several weeks of temperatures below 10°C. This use of pretreatment with cold to induce flowering is called **vernalization** (Latin for "spring"). Several weeks after winter wheat is vernalized, a photoperiod with long days (short nights) induces flowering.

A Flowering Hormone?

Floral buds develop into flowers, but in many species it is leaves that detect photoperiod. When the photoperiodic requirement for flowering is met, signals from leaves cue buds to develop as flowers. To induce a short-day or long-day plant to flower, it is enough in many species to expose one leaf to the appropriate photoperiod. Indeed, if only one leaf is left attached to the plant, photoperiod is detected and floral buds are induced. If all leaves are removed, the plant is insensitive to photoperiod. The flowering signal, not yet chemically identified, is called **florigen**, and it may be a hormone or change in relative concentrations of multiple hormones **(Figure 39.24)**. The flowering stimulus appears to be the same for short-day and long-day plants, despite differing photoperiodic conditions required for leaves to send this signal.

Figure 39.24

Inquiry Is there a flowering hormone?

EXPERIMENT To test whether there is a flowering hormone, researchers conducted an experiment in which a plant that had been induced to flower by photoperiod was grafted to a plant that had not been induced.

RESULTS

Plant subjected to photoperiod that induces flowering

Plant subjected to photoperiod that does not induce flowering

CONCLUSION Both plants flowered, indicating the transmission of a flower-inducing substance. In some cases, the transmission worked even if one was a short-day plant and the other was a long-day plant.

Meristem Transition and Flowering

Whatever combination of environmental cues (such as photoperiod or vernalization) and internal signals (such as hormones) is necessary for flowering to occur, the outcome is the transition of a bud's meristem from a vegetative state to a flowering state. This transition requires changes in the expression of genes that regulate pattern formation. Meristem identity genes that induce the bud to form a flower instead of a vegetative shoot must first be switched on. Then the organ identity genes that specify the spatial organization of the floral organs—sepals, petals, stamens, and carpels—are activated in the appropriate regions of the meristem (see Figure 35.31). Research on flower development is progressing rapidly, and one goal is to identify the signal transduction pathways that link such cues as photoperiod and hormonal changes to the gene expression required for flowering.

Concept Check 39.3

1. A plant flowers in a controlled chamber with a daily cycle of 10 hours of light and 14 hours of darkness. Is it a short-day plant? Explain.
2. Assume that poinsettias grown in a greenhouse require at least 14 hours of darkness to flower. If work has to be done at night, what light source would not disrupt the flowering schedule? Explain.
3. Upon germination, some vine seedlings grow toward darkness until they reach an upright structure. This adaptation helps the vine "find" a shaded object to climb. Suggest an experiment to test whether this negative phototropism is mediated by blue light photoreceptors or by phytochrome.

For suggested answers, see Appendix A.

Concept 39.4

Plants respond to a wide variety of stimuli other than light

Plants can neither migrate to a watering hole when water is scarce nor seek shelter when the weather is too windy. A seed, landing upside down in the soil, cannot maneuver itself into an upright position. Because of their immobility, plants must adjust to a wide range of environmental circumstances through developmental and physiological mechanisms. Natural selection has refined these responses. Light is so important in the life of a plant that we devoted the entire previous section to a plant's reception of and response to this one environmental factor. In this section, we examine responses to some of the other environmental stimuli that a plant commonly faces.

Gravity

Because plants are solar-powered organisms, it is not surprising that mechanisms for growing toward sunlight have evolved. But what environmental cue does the shoot of a young seedling use to grow upward when it is completely underground and there is no light for it to detect? Similarly, what environmental factor prompts the young root to grow downward? The answer to both questions is gravity.

Place a plant on its side, and it adjusts its growth so that the shoot bends upward and the root curves downward. In their responses to gravity, or **gravitropism**, roots display positive gravitropism **(Figure 39.25a)** and shoots exhibit negative gravitropism. Gravitropism functions as soon as a seed germinates, ensuring that the root grows into the soil and the shoot grows toward sunlight, regardless of how the seed is oriented when it lands. Auxin plays a key role in gravitropism.

Plants may detect gravity by the settling of **statoliths**, specialized plastids containing dense starch grains, to the lower portions of cells **(Figure 39.25b)**. In roots, statoliths are located in certain cells of the root cap. According to one hypothesis, the aggregation of statoliths at the low points of these cells triggers redistribution of calcium, which causes lateral transport of auxin within the root. The calcium and auxin accumulate on the lower side of the root's zone of elongation. Because these chemicals are dissolved, they do not respond to gravity but must be actively transported to one side of the root. At high concentration, auxin inhibits cell elongation, an effect that slows growth on the root's lower side. The more rapid elongation of cells on the upper side causes the root to curve as it grows. This tropism continues until the root grows straight down.

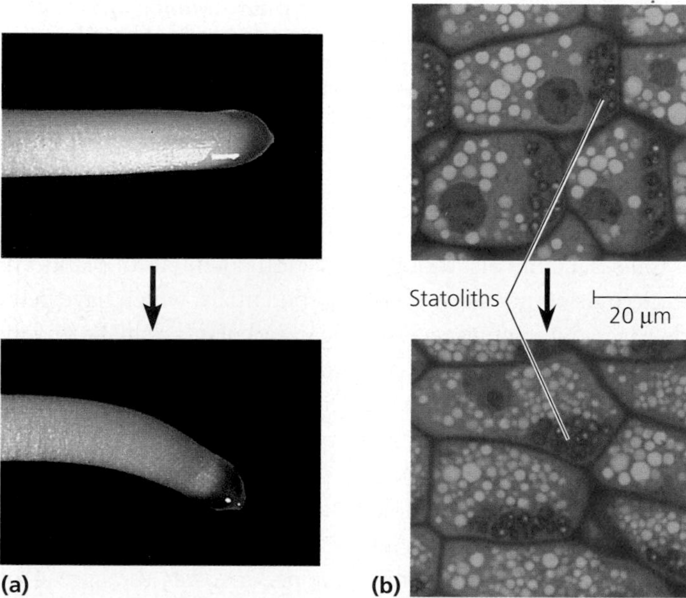

▲ **Figure 39.25 Positive gravitropism in roots: the statolith hypothesis. (a)** Over the course of hours, a horizontally oriented primary root of maize will bend gravitropically until its growing tip becomes vertically oriented. **(b)** Within minutes after the root is placed horizontally, statoliths begin settling to the lowest sides of root cap cells. The statolith hypothesis proposes that this settling may be the gravity-sensing mechanism that leads to redistribution of auxin and differential rates of elongation by cells on opposite sides of the root. (LMs)

Plant physiologists are refining the "falling statolith" hypothesis of root gravitropism as they try new experiments. For example, mutants of *Arabidopsis* and tobacco that lack statoliths are still capable of gravitropism, though the response is slower than in wild-type plants. It could be that the entire cell helps the root sense gravity by mechanically pulling on proteins that tether the protoplast to the cell wall, stretching the proteins on the "up" side and compressing the proteins on the "down" side of the root cells. Dense organelles, in addition to starch granules, may also contribute by distorting the cytoskeleton as they are pulled by gravity. Statoliths, because of their density, may enhance gravitational sensing by a mechanism that works more slowly in their absence.

Mechanical Stimuli

A tree growing on a windy mountain ridge usually has a shorter, stockier trunk than a tree of the same species growing in a more sheltered location. The advantage of this stunted morphology is that it enables the plant to hold its ground against strong gusts of wind. The term **thigmomorphogenesis** (from the Greek *thigma*, touch) refers to the changes in form that result from mechanical perturbation. Plants are very sensitive to mechanical stress: Even the act of measuring the length of a leaf with a ruler alters its subsequent growth.

Rubbing the stems of a young plant a couple of times daily results in plants that are shorter than controls **(Figure 39.26)**.

▲ **Figure 39.26 Altering gene expression by touch in *Arabidopsis*.** The shorter plant on the left was rubbed twice a day. The untouched plant (right) grew much taller.

Mechanical stimulation activates a signal transduction pathway involving an increase in cytosolic Ca^{2+} that mediates the activation of specific genes, some of which encode for proteins that affect cell wall properties.

Some plant species have become, over the course of their evolution, "touch specialists." Acute responsiveness to mechanical stimuli is an integral part of these plants' life "strategies." Most vines and other climbing plants have tendrils that coil rapidly around supports (see Figure 35.7a). These grasping organs usually grow straight until they touch something; the contact stimulates a coiling response caused by differential growth of cells on opposite sides of the tendril. This directional growth in response to touch is called **thigmotropism**, and it allows the vine to take advantage of whatever mechanical supports it comes across as it climbs upward toward a forest canopy.

Other examples of touch specialists are plants that undergo rapid leaf movements in response to mechanical stimulation. For example, when the compound leaf of the sensitive plant *Mimosa pudica* is touched, it collapses and its leaflets fold together **(Figure 39.27)**. This response, which takes only a second or two, results from a rapid loss of turgor by cells within pulvini, specialized motor organs located at the joints of the leaf. The motor cells suddenly become flaccid after stimulation because they lose potassium, which causes water to leave the cells by osmosis. It takes about 10 minutes for the cells to regain their turgor and restore the "unstimulated" form of the leaf. The function of the sensitive plant's behavior invites speculation. Perhaps by folding its leaves and reducing its surface area when jostled by strong winds, the plant conserves water. Or perhaps because the collapse of the leaves exposes thorns on the stem, the rapid response of the sensitive plant discourages herbivores.

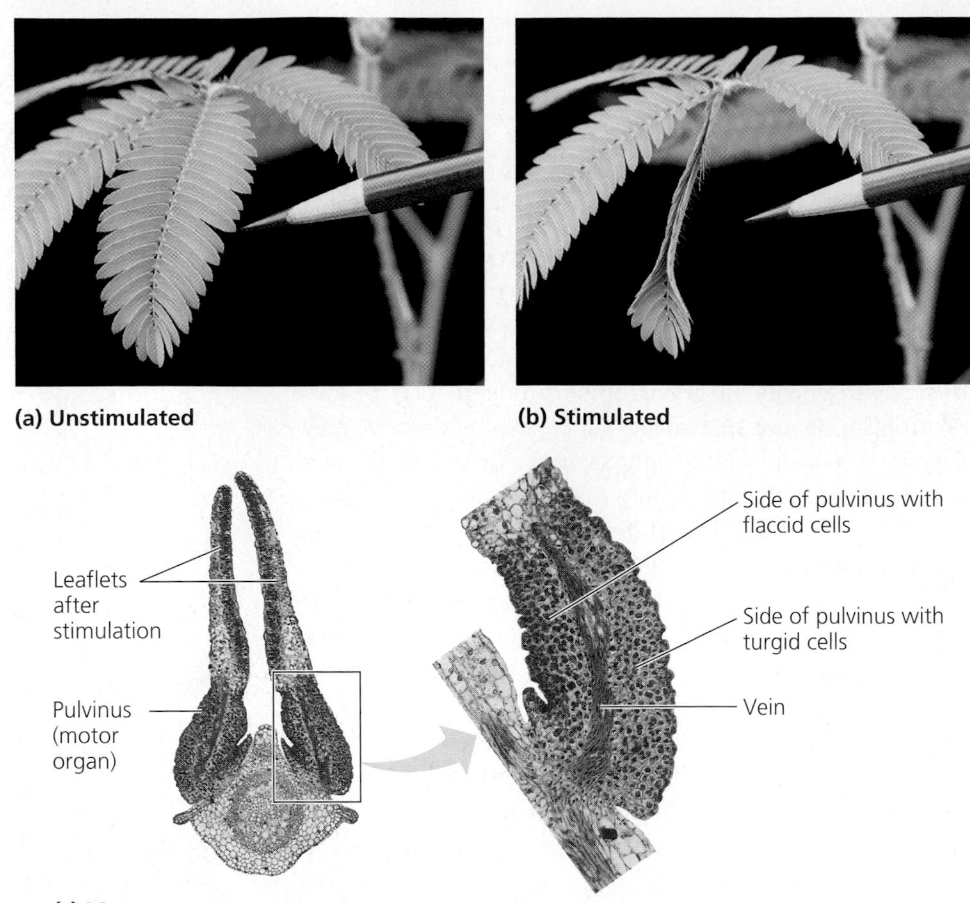

▲ **Figure 39.27 Rapid turgor movements by the sensitive plant (*Mimosa pudica*).** **(a)** In the unstimulated plant, leaflets are spread apart. **(b)** Within a second or two of being touched, the leaflets have folded together. **(c)** In these light micrographs of a leaflet pair in the closed (stimulated) state, you can see the motor cells in the sectioned pulvini (motor organs). The curvature of a pulvinus is caused by motor cells on one side losing water and becoming flaccid while cells on the opposite side retain their turgor.

A remarkable feature of rapid leaf movements is the mode of transmission of the stimulus through the plant. If one leaflet on a sensitive plant is touched, first that leaflet responds, then the adjacent leaflet responds, and so on, until all the leaflet pairs have folded together. From the point of stimulation, the signal that produces this response travels at a speed of about 1 cm/sec. An electrical impulse, traveling at the same rate, can be detected by attaching electrodes to the leaf. These impulses, called **action potentials**, resemble nerve impulses in animals, though the action potentials of plants are thousands of times slower. Action potentials, which have been discovered in many species of algae and plants, may be widely used as a form of internal communication. Another example is the Venus' flytrap (*Dionaea muscipula*), in which action potentials are transmitted from sensory hairs in the trap to the cells that respond by closing the trap (see Figure 37.13). In the case of *Mimosa pudica*, more violent stimuli, such as touching a leaf with a hot needle, causes *all* the leaves and leaflets on a plant to droop, but this systemic response involves the spread of chemical signals released from the injured area to other parts of the shoot.

Environmental Stresses

Occasionally, certain factors in the environment change severely enough to have a potentially adverse effect on a plant's survival, growth, and reproduction. Environmental stresses, such as flooding, drought, or extreme temperatures, can have a devastating impact on crop yields in agriculture. In natural ecosystems, plants that cannot tolerate an environmental stress will either succumb or be outcompeted by other plants, and they will become locally extinct. Thus, environmental stresses are also important in determining the geographic ranges of plants. Here, we will consider some of the more common **abiotic** (nonliving) stresses that plants encounter. In the last section of the chapter, we will examine the defensive responses of plants to common **biotic** (living) stresses, such as pathogens and herbivores.

Drought

On a bright, warm, dry day, a plant may be stressed by a water deficiency because it is losing water by transpiration faster than the water can be restored by uptake from the soil. Prolonged drought can stress crops and the plants of natural ecosystems for weeks or months. Severe water deficit, of course, will kill a plant, as you may know from experience with neglected houseplants. But plants have control systems that enable them to cope with less extreme water deficits.

Many of a plant's responses to water deficit help the plant conserve water by reducing the rate of transpiration. Water deficit in a leaf causes guard cells to lose turgor, a simple control mechanism that slows transpiration by closing stomata (see Figure 36.15). Water deficit also stimulates increased synthesis and release of abscisic acid in the leaf, and this hormone helps keep stomata closed by acting on guard cell membranes. Leaves respond to water deficit in several other ways. Because cell expansion is a turgor-dependent process, a water deficit will inhibit the growth of young leaves. This response minimizes the transpirational loss of water by slowing the increase in leaf surface. When the leaves of many grasses and other plants wilt from a water deficit, they roll into a shape that reduces transpiration by exposing less leaf surface to dry air and wind. While all of these responses of leaves help the plant conserve water, they also reduce photosynthesis. This is one reason a drought diminishes crop yield.

Root growth also responds to water deficit. During a drought, the soil usually dries from the surface down. This inhibits the growth of shallow roots, partly because cells cannot maintain the turgor required for elongation. Deeper roots surrounded by soil that is still moist continue to grow. Thus, the root system proliferates in a way that maximizes exposure to soil water.

Flooding

An overwatered houseplant may suffocate because the soil lacks the air spaces that provide oxygen for cellular respiration in the roots. Some plants are structurally adapted to very wet habitats. For example, the submerged roots of trees called mangroves, which inhabit coastal marshes, are continuous with aerial roots that provide access to oxygen. But how do plants less specialized for aquatic environments cope with oxygen deprivation in waterlogged soils? Oxygen deprivation stimulates the production of the hormone ethylene, which causes some of the cells in the root cortex to undergo apoptosis (programmed cell death). Enzymatic destruction of cells creates air tubes that function as "snorkels," providing oxygen to the submerged roots **(Figure 39.28)**.

Salt Stress

An excess of sodium chloride or other salts in the soil threatens plants for two reasons. First, by lowering the water potential of the soil solution, salt can cause a water deficit in plants even though the soil has plenty of water. As the water potential of the soil solution becomes more negative, the water potential gradient from soil to roots is lowered, thereby reducing water uptake (see Chapter 36). The second problem with saline soil is that sodium and certain other ions are toxic to plants when their concentrations are relatively high. The selectively permeable membranes of root cells impede the uptake of most harmful ions, but this only aggravates the problem of acquiring water from soil that is rich in solutes. Many plants can respond to moderate soil salinity by producing solutes that are well tolerated at high concentrations, mostly organic compounds that keep the water potential of cells more negative than that of the

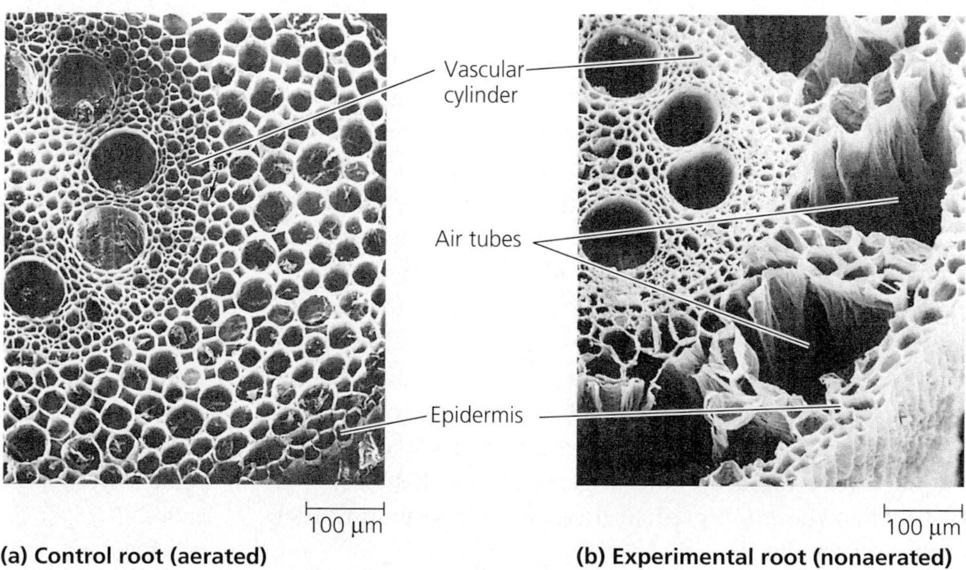

▲ **Figure 39.28 A developmental response of maize roots to flooding and oxygen deprivation. (a)** A transverse section of a control root grown in an aerated hydroponic medium. **(b)** An experimental root grown in a nonaerated hydroponic medium. Ethylene-stimulated apoptosis (programmed cell death) creates the air tubes. (SEMs)

soil solution without admitting toxic quantities of salt. However, most plants cannot survive salt stress for long. The exceptions are halophytes, salt-tolerant plants with adaptations such as salt glands. These glands pump salts out across the leaf epidermis.

Heat Stress

Excessive heat can harm and eventually kill a plant by denaturing its enzymes and damaging its metabolism in other ways. One function of transpiration is evaporative cooling. On a warm day, for example, the temperature of a leaf may be 3–10°C below the ambient air temperature. Of course, hot, dry weather also tends to cause water deficiency in many plants; the closing of stomata in response to this stress conserves water but then sacrifices evaporative cooling. This dilemma is one of the reasons that very hot, dry days take such a toll on most plants.

Most plants have a backup response that enables them to survive heat stress. Above a certain temperature—about 40°C for most plants that inhabit temperate regions—plant cells begin synthesizing relatively large quantities of special proteins called **heat-shock proteins**. Researchers have also discovered this response in heat-stressed animals and microorganisms. Some heat-shock proteins are chaperone proteins, which function in unstressed cells as temporary scaffolds that help other proteins fold into their functional shapes (see Chapter 5). In their roles as heat-shock proteins, perhaps these molecules embrace enzymes and other proteins and help prevent denaturation.

Cold Stress

One problem plants face when the temperature of the environment falls is a change in the fluidity of cell membranes. Recall from Chapter 7 that a biological membrane is a fluid mosaic, with proteins and lipids moving laterally in the plane of the membrane. When a membrane cools below a critical point, it loses its fluidity as the lipids become locked into crystalline structures. This alters solute transport across the membrane and also adversely affects the functions of membrane proteins. Plants respond to cold stress by altering the lipid composition of their membranes. For example, membrane lipids increase in their proportion of unsaturated fatty acids, which have shapes that help keep membranes fluid at lower temperatures by impeding crystal formation (see Figure 7.5b). Such molecular modification of the membrane requires from several hours to days, which is one reason rapid chilling is generally more stressful to plants than the more gradual drop in air temperature that occurs seasonally.

Freezing is a more severe version of cold stress. At subfreezing temperatures, ice forms in the cell walls and intercellular spaces of most plants. The cytosol generally does not freeze at the cooling rates encountered in nature because it contains more solutes than the very dilute solution found in the cell wall, and solutes depress the freezing point of a solution. The reduction in liquid water in the cell wall caused by ice formation lowers the extracellular water potential, causing water to leave the cytoplasm. The resulting increase in the concentration of ionic salts in the cytoplasm is harmful and can lead to cell death. Whether the cell survives depends to a large extent on how resistant it is to dehydration. Plants that are native to regions where winters are cold have special adaptations that enable them to cope with freezing stress. For example, before the onset of winter, the cells of many frost-tolerant species increase their cytoplasmic levels of specific solutes, such as sugars, that are better tolerated at high concentrations and that help reduce the loss of water from the cell during extracellular freezing.

Concept Check 39.4

1. Thermal images are photographs of the heat emitted by an object. Researchers have used thermal imaging of plants to isolate mutants that overproduce abscisic acid. Suggest a reason why these mutants are warmer than wild-type plants under conditions that are normally nonstressful.
2. A worker in a commercial greenhouse finds that the potted chrysanthemums nearest to the aisles are often shorter than those growing in the middle of the bench. Suggest an explanation.

For suggested answers, see Appendix A.

Concept 39.5

Plants defend themselves against herbivores and pathogens

Plants do not exist in isolation but interact with many other species in their communities. Some of these interspecific interactions—for example, the associations of plants with fungi in mycorrhizae (see Figure 37.12) or with pollinators (see Figure 30.13)—are mutually beneficial. Most of the interactions that plants have with other organisms, however, are not beneficial to the plant. As primary producers, plants are at the base of most food webs and are subject to attack by a wide range of plant-eating (herbivorous) animals. A plant is also subject to infection by a diversity of pathogenic viruses, bacteria, and fungi that have the potential to damage tissues or even kill the plant. Plants counter these threats with defense systems that deter herbivory and prevent infection or combat pathogens that do manage to infect the plant.

Defenses Against Herbivores

Herbivory—animals eating plants—is a stress that plants face in any ecosystem. Plants counter excessive herbivory with both physical defenses, such as thorns, and chemical defenses, such as the production of distasteful or toxic compounds. For example, some plants produce an unusual amino acid called *canavanine*, which is named for one of its sources, the jackbean (*Canavalia ensiformis*). Canavanine resembles arginine, one of the 20 amino acids organisms incorporate into their proteins. If an insect eats a plant containing canavanine, the molecule is incorporated into the insect's proteins in place of arginine. Because canavanine is different enough from arginine to adversely affect the conformation and hence the function of the proteins, the insect dies.

Some plants even "recruit" predatory animals that help defend the plant against specific herbivores. For example, insects called parasitoid wasps inject their eggs into their prey, including caterpillars feeding on plants. The eggs hatch within the caterpillars, and the larvae eat through their organic containers from the inside out. The plant, which benefits from the destruction of the herbivorous caterpillars, has an active role in this drama. A leaf damaged by caterpillars releases volatile compounds that attract parasitoid wasps. The stimulus for this response is a combination of physical damage to the leaf caused by the munching caterpillar and a specific compound in the caterpillar's saliva **(Figure 39.29)**.

The volatile molecules a plant releases in response to herbivore damage can also function as an "early warning system" for nearby plants of the same species. Lima bean plants infested with spider mites release volatile chemicals that signal "news" of the attack to neighboring, noninfested lima bean plants. In response to these chemicals, noninfested lima bean leaves activate defense genes. Volatile chemicals released from leaves mechanically wounded in experiments do not have the same effect. Genes activated by infestation-released volatile chemicals overlap considerably with those produced by exposure to **jasmonic acid**, an important molecule in plant defense. As a result of this gene activation, noninfested neighbors become less susceptible to spider mites and more attractive to another species of mite that preys on spider mites.

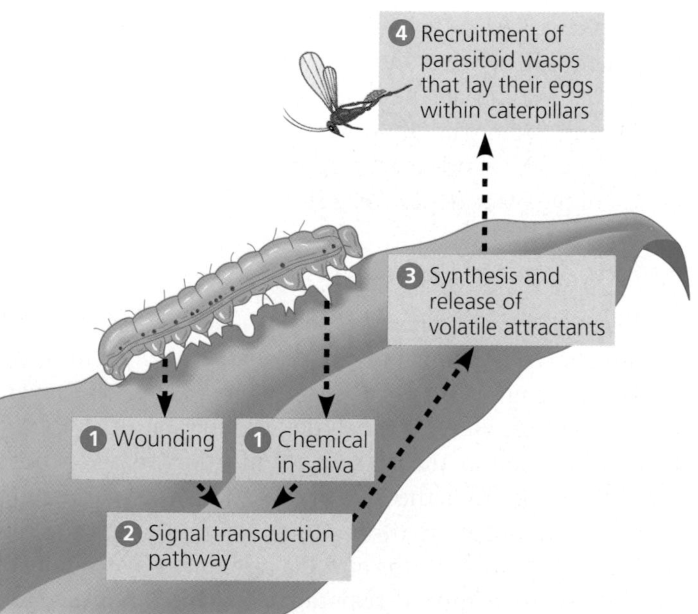

▲ **Figure 39.29 A maize leaf "recruits" a parasitoid wasp as a defensive response to a herbivore, an army-worm caterpillar.**

Defenses Against Pathogens

A plant's first line of defense against infection is the physical barrier of the plant's "skin," the epidermis of the primary plant body and the periderm of the secondary plant body (see Figure 35.18). This first defense system, however, is not impenetrable. Viruses, bacteria, and the spores and hyphae of fungi can still enter the plant through injuries or through the natural openings in the epidermis, such as stomata. Once a pathogen invades, the plant mounts a chemical attack as a second line of defense that destroys the pathogens and prevents their spread from the site of infection. This second defense system is enhanced by the plant's inherited ability to recognize certain pathogens.

Gene-for-Gene Recognition

Plants are generally resistant to most pathogens. This is because plants have the ability to recognize invading pathogens and then to mount successful defenses. In a converse manner, successful pathogens cause disease because they are able to evade recognition or suppress host defense mechanisms. Pathogens against which a plant has little specific defense are said to be **virulent.** They are the exceptions, for if they were not, then hosts and pathogens would soon perish together. A kind of "compromise" has coevolved between plants and most of their pathogens. In such cases, the pathogen gains enough access to its host to enable it to perpetuate itself without severely damaging or killing the plant. Strains of pathogens that only mildly harm but do not kill the host plant are termed **avirulent.**

Gene-for-gene recognition is a widespread form of plant disease resistance that involves recognition of pathogen-derived molecules by the protein products of specific plant disease resistance (*R*) genes. There are many pathogens, and plants have many *R* genes—*Arabidopsis* has at least several hundred. An R protein usually recognizes only a single corresponding pathogen molecule that is encoded by an avirulence (*Avr*) gene. Many Avr proteins play an active role in pathogenesis and are thought to redirect the host metabolism to the advantage of the pathogen. The questions of how R proteins function and how they evolve are currently topics of intense research. In the simplest biochemical model (the receptor-ligand model), an R protein functions as a plant receptor molecule that triggers resistance upon binding to the correct corresponding Avr protein (**Figure 39.30**). If the plant host lacks the *R* gene that counteracts the pathogen's *Avr* gene, then the pathogen can invade and kill the plant. An alternative model of gene-for-gene recognition, dubbed the "guard" hypothesis, proposes that R proteins function as a surveillance system of other plant proteins that undergo activity or conformational changes induced by Avr proteins. Regardless of the exact mechanism involved, recognition of pathogen-derived molecules by R proteins triggers a signal transduction pathway leading to a defense response in the infected plant tissue. This defense includes both an enhancement of the localized response at the site of infection and a more general systemic response of the whole plant.

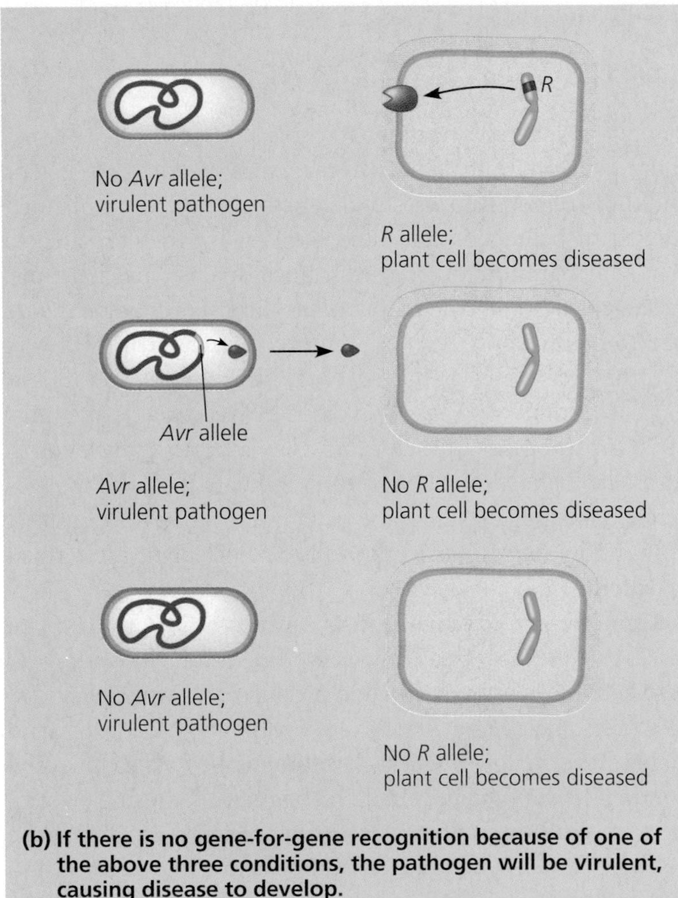

(a) If an *Avr* allele in the pathogen corresponds to an *R* allele in the host plant, the host plant will have resistance, making the pathogen avirulent. *R* alleles probably code for receptors in the plasma membranes of host plant cells. *Avr* alleles produce compounds that can act as ligands, binding to receptors in host plant cells.

(b) If there is no gene-for-gene recognition because of one of the above three conditions, the pathogen will be virulent, causing disease to develop.

▲ **Figure 39.30 Gene-for-gene resistance of plants to pathogens: the receptor-ligand model.**

Plant Responses to Pathogen Invasions

In contrast to the strain-specific *Avr-R* interactions that control plant disease resistance to very narrow groups of pathogens (specifically, strains that contain the appropriate *Avr* allele), molecules called **elicitors** induce a broader type of host defense response. **Oligosaccharins,** derived from cellulose fragments released by cell wall damage, are one of the major classes of elicitors. Elicitors stimulate the production of antimicrobial compounds called **phytoalexins**. Infection also activates genes that produce **PR proteins** (pathogenesis-related proteins). Some of these proteins are antimicrobial, attacking molecules in the cell wall of a bacterium, for example. Others may function as signals that spread "news" of the infection to nearby cells. Infection also stimulates the cross-linking of molecules in the cell wall and the deposition of lignin, responses that set up a local barricade that slows spread of the pathogen to other parts of the plant.

If the pathogen is avirulent based on an *R-Avr* match, then the localized defense response is even more vigorous and is known as a **hypersensitive response** (abbreviated **HR**). There is an enhanced production of phytoalexins and PR proteins, and the "sealing" response that contains the infection is more effective. After the cells at the site of infection mount their chemical defense and seal off the area, they destroy themselves. We can see the result of an HR as lesions on a leaf or other infected organ. As "sick" as such a leaf appears, it will still survive, and its defense response will then help protect the rest of the plant (**Figure 39.31**).

Systemic Acquired Resistance

The hypersensitive response, as you have learned, is localized and specific, a containment response based on gene-for-gene (*R-Avr*) recognition between host and pathogen. However, this defense response also includes production of chemical signals that "sound the alarm" of infection to the whole plant. Released from the site of infection, the alarm hormones are transported throughout the plant, stimulating production of phytoalexins and PR proteins. This response, called **systemic acquired resistance (SAR)**, is nonspecific, providing protection against a diversity of pathogens for days (see Figure 39.31).

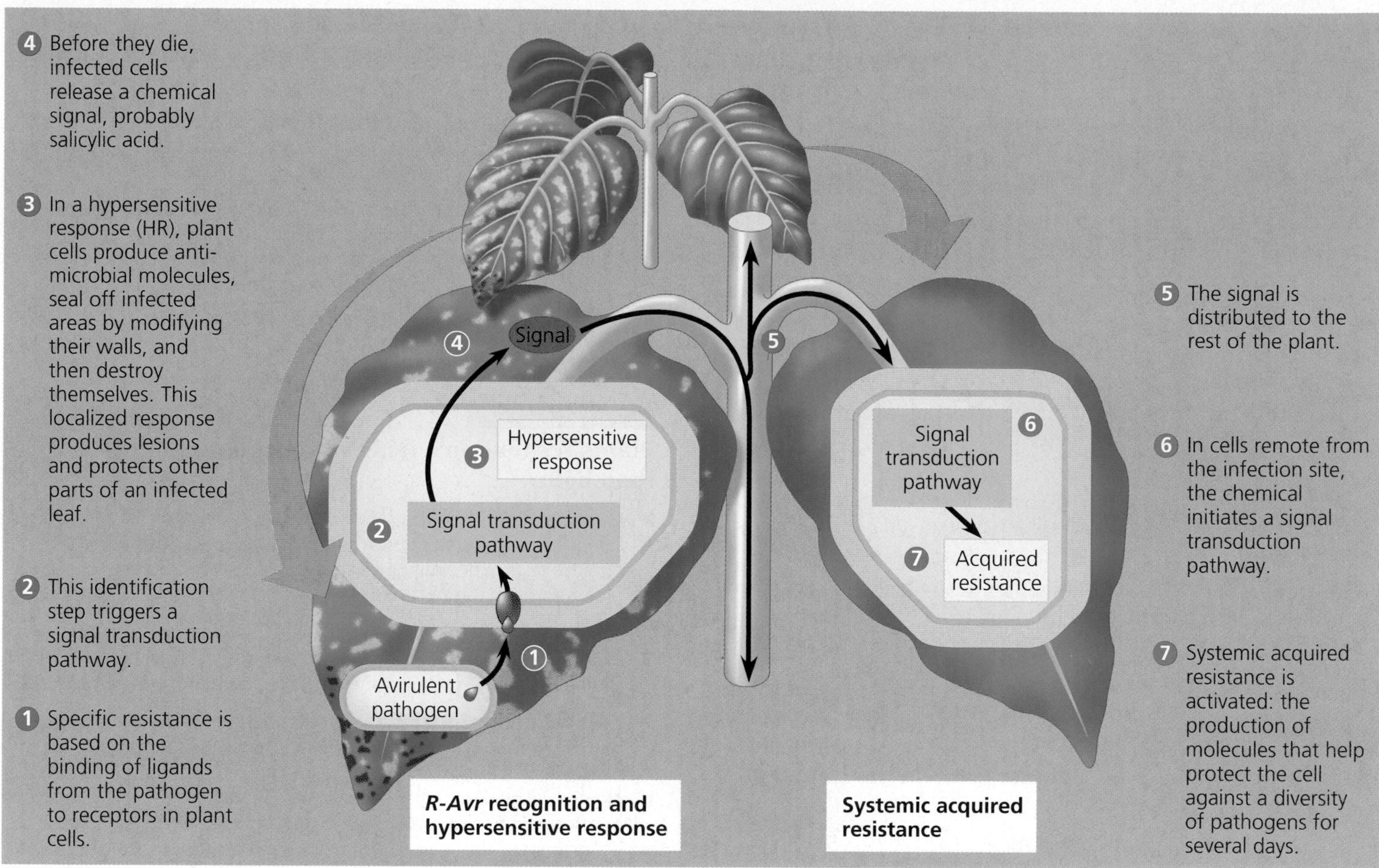

4 Before they die, infected cells release a chemical signal, probably salicylic acid.
3 In a hypersensitive response (HR), plant cells produce anti-microbial molecules, seal off infected areas by modifying their walls, and then destroy themselves. This localized response produces lesions and protects other parts of an infected leaf.
2 This identification step triggers a signal transduction pathway.
1 Specific resistance is based on the binding of ligands from the pathogen to receptors in plant cells.
Signal
Hypersensitive response
Signal transduction pathway
Avirulent pathogen
R-Avr recognition and hypersensitive response
5 The signal is distributed to the rest of the plant.
6 In cells remote from the infection site, the chemical initiates a signal transduction pathway.
7 Systemic acquired resistance is activated: the production of molecules that help protect the cell against a diversity of pathogens for several days.
Signal transduction pathway
Acquired resistance
Systemic acquired resistance

▲ **Figure 39.31 Defense responses against an avirulent pathogen.**

A good candidate for one of the hormones responsible for activating SAR is **salicylic acid.** A modified form of this compound, acetylsalicylic acid, is the active ingredient in aspirin. Centuries before aspirin was sold as a pain reliever, some cultures had learned that chewing the bark of a willow tree (*Salix*) would lessen the pain of a toothache or headache. With the discovery of systemic acquired resistance, biologists have finally learned one function of salicylic acid in plants. Aspirin turns out to be a natural medicine in the plants that produce it, but with effects entirely different from the medicinal action in humans who consume the drug.

Plant biologists investigating disease resistance and other evolutionary adaptations of plants are getting to the heart of how a plant responds to internal and external signals. These scientists, along with thousands of other plant biologists working on other questions and millions of students experimenting with plants in biology courses, are all extending a centuries-old tradition of curiosity about the green organisms that feed the biosphere.

Concept Check 39.5

1. Chewing insects mechanically damage plants and lessen the surface area of leaves for photosynthesis. In addition, these insects make plants more vulnerable to pathogen attack. Suggest a reason why.
2. A scientist finds that a population of plants growing in a breezy location is more prone to defoliation by insects than a population of the same species growing in a sheltered area. Suggest a hypothesis to account for this observation.

For suggested answers, see Appendix A.

Chapter 39 Review

Go to www.campbellbiology.com or the student CD-ROM to explore Activities, Investigations, and other interactive study aids.

SUMMARY OF KEY CONCEPTS

Concept 39.1

Signal transduction pathways link signal reception to response

- **Reception** **(p. 789)** Internal and external signals are detected by receptors, proteins that change in response to specific stimuli.
- **Transduction** **(pp. 789–790)** Second messengers transfer and amplify signals from receptors to proteins that cause specific responses.
- **Response** **(pp. 790–791)** Responses to stimulation typically involve activation of enzymes by stimulating transcription of mRNA for the enzymes (transcriptional regulation) or by activating existing enzyme molecules (post-translational modification of proteins).

Concept 39.2

Plant hormones help coordinate growth, development, and responses to stimuli

- **The Discovery of Plant Hormones** **(pp. 792–793)** Researchers discovered auxin by identifying the compound responsible for transmitting a signal downward through coleoptiles, from the tips to the elongating regions, during phototropism.
- **A Survey of Plant Hormones** **(pp. 793–801)** Auxin, produced primarily in the apical meristem of the shoot, stimulates cell elongation in different target tissues. Cytokinins, produced in actively growing tissues such as roots, embryos, and fruits, stimulate cell division. Gibberellins, produced in roots and young leaves, stimulate growth in leaves and stems. Brassinosteroids, chemically similar to the sex hormones of animals, induce cell elongation and division. Abscisic acid maintains dormancy in seeds. Ethylene helps control fruit ripening.
  Activity *Leaf Abscission*
  Investigation *What Plant Hormones Affect Organ Formation?*
- **Systems Biology and Hormone Interactions** **(pp. 801–802)** Interactions between hormones and their signal transduction pathways makes it difficult to predict what effect a genetic manipulation will have on a plant. Systems biology seeks a comprehensive understanding of plants that will permit successful modeling of plant functions.

Concept 39.3

Responses to light are critical for plant success

- **Blue-Light Photoreceptors** **(p. 802)** Various blue-light photoreceptors control hypocotyl elongation, stomatal opening, and phototropism.
- **Phytochromes as Photoreceptors** **(pp. 802–805)** Phytochromes exist in two photoreversible states, with conversion of P_r to P_{fr} triggering many developmental responses.
- **Biological Clocks and Circadian Rhythms** **(p. 805)** Free-running circadian cycles are approximately 24 hours long but are entrained to exactly 24 hours by the day/night cycle.
- **The Effect of Light on the Biological Clock** **(p. 806)** Phytochrome conversion marks sunrise and sunset, providing the clock with environmental cues.
- **Photoperiodism and Responses to Seasons** **(pp. 806–808)** Some developmental processes, including flowering in many plant species, require a certain photoperiod. For example, a critical night length sets a minimum (in short-day plants) or maximum (in long-day plants) number of hours of darkness required for flowering.
  Activity *Flowering Lab*

Concept 39.4

Plants respond to a wide variety of stimuli other than light

- **Gravity** **(p. 809)** Response to gravity is known as gravitropism. Roots show positive gravitropism, and stems show negative gravitropism. Specialized plastids called statoliths may enable plants to detect gravity.
- **Mechanical Stimuli** **(pp. 809–810)** Growth in response to touch is called thigmotropism. Rapid leaf movements involve transmission of electrical impulses called action potentials.
- **Environmental Stresses** **(pp. 810–812)** During drought, plants respond to water deficit by reducing transpiration. Enzymatic destruction of cells creates air tubes that help plants survive oxygen deprivation during flooding. Plants respond to salt stress by producing solutes tolerated at high concentrations, keeping the water potential of cells more negative than that of the soil solution. Heat-shock proteins help plants survive heat stress. Altering lipid composition of membranes is a response to cold stress.

Concept 39.5

Plants defend themselves against herbivores and pathogens

- **Defenses Against Herbivores** **(p. 813)** Physical defenses include morphological adaptations such as thorns, chemical defenses such as distasteful or toxic compounds, and airborne attractants that bring animals that destroy herbivores.
- **Defenses Against Pathogens** **(pp. 813–815)** A pathogen is avirulent if it has a specific *Avr* gene corresponding to a particular *R* allele in the host plant. A hypersensitive response against an avirulent pathogen seals off the infection and destroys both pathogen and host cells in the region of the infection. Salicylic acid is a signal molecule that triggers systemic acquired resistance (SAR), generalized defense responses in organs distant from the original site of infection.

TESTING YOUR KNOWLEDGE

Self-Quiz

1. Which hormone is *incorrectly* paired with its function?
   a. auxin—promotes stem growth through cell elongation
   b. cytokinins—initiate programmed cell death
   c. gibberellins—stimulate seed germination
   d. abscisic acid—promotes seed dormancy
   e. ethylene—inhibits cell elongation

2. Which of the following is *not* a typical component of a signal transduction pathway?
   a. production of more signal
   b. production of second messengers such as cGMP
   c. expression of specific proteins
   d. activation of protein kinases
   e. phosphorylation of transcription factors

3. Buds and sprouts often form on tree stumps. Which of the following hormones would you expect to stimulate their formation?
   a. auxin
   b. cytokinins
   c. abscisic acid
   d. ethylene
   e. gibberellins

4. Which of the following is *not* part of the acid growth hypothesis?
   a. Auxin stimulates proton pumps in cell membranes.
   b. Lowered pH results in the breakage of cross-links between cellulose microfibrils.
   c. The wall fabric becomes looser (more plastic).
   d. Auxin-activated proton pumps stimulate cell division in meristems.
   e. The turgor pressure of the cell exceeds the restraining pressure of the loosened cell wall, and the cell takes up water and elongates.

5. The signal for flowering could be released earlier than normal in a long-day plant experimentally exposed to flashes of
   a. far-red light during the night.
   b. red light during the night.
   c. red light followed by far-red light during the night.
   d. far-red light during the day.
   e. red light during the day.

6. How might a plant respond to *severe* heat stress?
   a. by orienting leaves toward the sun to increase evaporative cooling
   b. by producing ethylene, which kills some cortex cells and creates air tubes for ventilation
   c. by producing salicylic acid, which initiates a systemic acquired resistance response
   d. by increasing the proportion of unsaturated fatty acids in cell membranes to reduce their fluidity
   e. by producing heat-shock proteins, which may protect the plant's proteins from denaturing

7. If a long-day plant has a critical night length of 9 hours, which of the following 24-hour cycles would prevent flowering?
   a. 16 hours light/8 hours dark
   b. 14 hours light/10 hours dark
   c. 15.5 hours light/8.5 hours dark
   d. 4 hours light/8 hours dark/4 hours light/8 hours dark
   e. 8 hours light/8 hours dark/light flash/8 hours dark

8. The probable role of salicylic acid in systemic acquired resistance of plants is to
   a. destroy pathogens directly.
   b. activate defenses throughout the plant before infection spreads.
   c. close stomata, thus preventing the entry of pathogens.
   d. activate heat-shock proteins.
   e. sacrifice infected tissues by hydrolyzing cells.

9. Auxin triggers the acidification of cell walls that results in rapid growth, but also stimulates sustained, long-term cell elongation. What best explains this dual growth response?
   a. Auxin binds to different receptors in different cells.
   b. Different concentrations of auxin have different effects.
   c. Auxin causes second messengers to activate both proton pumps and certain genes.
   d. The dual effects are due to two different auxins.
   e. Other antagonistic hormones modify auxin's effects.

10. If a scientist discovered an *Arabidopsis* mutant that did not store starch in its plastids but underwent normal gravitropic bending, what aspect of our current understanding of root gravitropism would have to be reevaluated?
   a. the role of auxin in gravitropism
   b. the role of calcium in gravitropism
   c. the role of statoliths in gravitropism
   d. the role of light in gravitropism
   e. the role of differential growth in gravitropic bending

For Self-Quiz answers, see Appendix A.

Go to the website or CD-ROM for more quiz questions.

Evolution Connection

Coevolution is defined as reciprocal adaptations between two species, with each species adapting in how it interacts with the other. In this context of coevolution, write a paragraph explaining the relationship between a plant and an avirulent pathogen.

Scientific Inquiry

A plant biologist observed a peculiar pattern when a tropical shrub was attacked by caterpillars. After a caterpillar ate a leaf, it would skip over nearby leaves and attack a leaf some distance away. The researcher found that when a leaf was eaten, nearby leaves started making a chemical that deterred caterpillars. Simply removing a leaf did not deter them from eating nearby leaves. The biologist suspected that a damaged leaf sent out a chemical that signaled other leaves. How could the researcher test this hypothesis?

Investigation *What Plant Hormones Affect Organ Formation?*

Science, Technology, and Society

Based on your study of this chapter, write a short essay explaining at least three examples of how knowledge about the control systems of plants is applied in agriculture or horticulture.

UNIT 7

Animal Form and Function

AN INTERVIEW WITH Erich Jarvis

Erich Jarvis is an assistant professor at the Duke University Medical Center, where he teaches neuroscience and studies the brain, using vocal communication by zebra finches and other songbirds as his main model system. Growing up in New York City, he attended the High School of the Performing Arts, where he showed great promise as a dancer. However, he chose to go another route, completing a double major in biology and mathematics at Hunter College and then a Ph.D. in molecular neurobiology and animal behavior at Rockefeller University. Dr. Jarvis has won numerous awards, including the National Science Foundation's highest honor for young researchers, the Alan T. Waterman Award. In addition to research and teaching, Dr. Jarvis has long been active in recruiting minority students to science.

What were your educational inspirations and challenges growing up?

Although my parents were separated and we were sometimes poor, I had a lot of support from both sides of my family, and my mother always encouraged me to aspire to something great. Most of my family were in the performing arts, and continuing in that tradition, I had the goal of becoming a dancer. I began with ballet and then went on to jazz and modern dance. But in my senior year of high school, I had a change of heart. I realized I wanted to have a bigger, positive impact in the world, and I thought I could do that best as a scientist. I was an okay student, but I didn't have a strong academic background. So when I got to college, I had to take remedial classes—I had a lot of catching up to do.

Do you see any connection between dance and science?

In terms of what's required for a successful career, I don't really see much of a difference. As a dancer, you need to have discipline. It's a very difficult field, there's a lot of competition, and you have to practice and practice. As a scientist, you also need discipline, and you have to deal with competition. There are actually more opportunities in science, although science is very difficult. Nature doesn't give up its secrets without a struggle!

On top of that, both fields require creativity. As a performing artist, you want to do something new and beautiful. It's the same for science: You're trying to do something new and creative. Also, in both fields, you have to adapt your schedule to what is required. There's so much overlap between being an artist and being a scientist that I'm amazed people don't see it more.

Why did biology appeal to you?

I think it had something to do with my cultural background. Like many of us in this country, I have a mixed ancestry—mine is African, Native American, and European. There was something about the cultural atmosphere in my family that seems especially connected to earthly things, nature, and so forth. Perhaps for that reason, I felt more at home in biology than in chemistry or physics, though I liked those subjects too.

Did you get to do research as an undergraduate?

At Hunter College, I benefited from several programs that promoted minority access to careers in scientific research. For four years, I worked in the laboratory of Professor Rivka Rudner, who was studying bacterial genes for ribosomal RNA. It took me almost a year to start to understand things, but then the excitement of laboratory research overcame me like a passion. Dr. Rudner would have to throw me out of the laboratory sometimes and tell me to go home and sleep. But soon after graduation, I published six papers.

What led you to neuroscience?

I was fascinated by the human brain. I wanted to know how the brain generates behavior, how it perceives it, and how it learns it. Of course, I realized that neurobiologists mostly have to work with nonhuman animals. When I got to graduate school, I started reading lots of papers, and I was disappointed to learn how little was known about the behavior of most common model organisms, such as the rat. At the time, neuroscience consisted of anatomy, neurophysiology (the study of nerve impulses and so forth), and an increasing amount of molecular biology—but very little behavior. But then I learned about Fernando Nottebohm's exciting research with songbirds, and I joined his laboratory.

The wonderful thing about the songbird system was that, when I began, there were already 40 years of research on the ecology and natural behavior of these animals. This research was like a box full of ideas for studying the brain mechanisms of songbird behavior, in particular their imitation of sounds. Very few animals have the ability to imitate what they hear—only songbirds, parrots, hummingbirds, bats, dolphins, and humans. Usually the animal imitates sounds of its own species, though some species can imitate other species and even mechanical sounds like a car. Fernando Nottebohm had identified nearly the complete brain pathway responsible for a songbird's production of a learned sound. I wanted to combine my molecular biology training with Dr. Nottebohm's discoveries and his knowledge of behavior. Already, an assistant professor in the laboratory, David Clayton, was using molecular biology. I was fascinated by the idea of figuring out the molecular genetics of imitative behavior in songbirds.

How might genes be connected to this sort of behavior?

After years of work on this question, I have now come to believe that what makes a vocal learner are genes that control the construction

of brain circuits that connect to vocal muscles. And in addition to studying how genes regulate behavior, we're also studying how behavior regulates genes! The act of producing the learned sounds causes huge, short-term changes in gene expression in certain brain areas. Many of the genes involved encode transcription factors, which regulate the expression of a variety of other genes. When a songbird is singing, we see up to a 60-fold change in the amount of specific mRNAs and proteins in these brain areas.

We think this is happening for most behaviors in most animals, although we don't yet know the full circuits in other cases—in the case of walking, for example. But in songbirds, we can study every single group of neurons involved in the singing pathway and how their genes are regulated.

So now we're trying to find out the relationships between the rapid changes in gene expression in the brain as a result of the singing behavior, the electrical impulses that are going through the neural circuits to regulate both these genes and the behavior, and the behavior itself. To do this, we have to measure the mRNA and protein synthesis of certain brain cells and record electrical impulses from those same cells, both during the behavior.

How do you do these things?

To detect the electrical impulses, we implant hair-thin microelectrodes in different brain areas of a songbird, usually a zebra finch. The wires are connected to what's called a commutator, a mechanism originally used in helicopters. It allows the bird to fly around in the cage without getting tangled in the wires. The bird can sing and communicate with other birds while we record the electrical impulses.

To find out which genes are expressed in the nerve cells that are firing, we use DNA microarrays (see Figure 20.14). For this part of our work, we are sequencing the "transcriptome" of the zebra finch brain—the DNA corresponding to all the different mRNAs made in brain tissue. We want to be able to put all the genes expressed in the songbird brain on microarrays. Then we'll be able to identify which genes are involved in any given behavior.

You have also studied birds in the wild. Why?

I wanted to see if the singing-driven gene expression we find in the lab occurs in the real world. So my colleague Claudio Mello and I visited songbird sites in a station wagon loaded with lab equipment and dry ice. To attract wild birds, we played recorded songs, causing males in the vicinity to sing like crazy. After about half an hour, when the products of gene expression in the brain should have accumulated, we captured the birds with a net. Then we took the brains back to the laboratory, and we did find robust singing-driven gene expression in those brains. But, more exciting, we also found phenomena that we hadn't seen in the lab: Some brain regions showed different patterns of gene expression that, I later discovered, depended on the social context in which the birds had been singing—to another male, to a female, or something else. And there turns out to be evidence for context-dependent patterns of activity in comparable brain areas in other animals, such as monkeys. We think there's a context-dependent brain pathway that is used for learned vocalization, speech, and other learned cognitive behaviors.

You're combining many approaches in your work!

Yes, and in addition to molecular biology, anatomy, electrophysiology, and behavior, we're making a lot of use of bioinformatics, which involves math and computer science. Biologists routinely use the computer to analyze gene expression data from DNA microarrays, but we also include in our computer analyses information about anatomical connectivity, behavioral data from song sonograms, and patterns of neural impulses. Our inference algorithm software helps us identify causal relationships and even proposes models we can test experimentally.

We are trying to look at what we study as an integrated system, not just a pattern of regulated genes or a combination of neural circuits. This is the way I approach science: If I'm going to study how something works, I've got to look at all aspects of it. I really encourage people to look at the big picture. That's probably the only way to reach the ultimate goal of understanding how the brain generates speech and language.

We are trying to look at what we study as an integrated system. . . . This is the way I approach science: If I'm going to study how something works, I've got to look at all aspects of it. I really encourage people to look at the big picture.

Are there potential medical benefits from this research?

One of our projects relates to Parkinson's disease, which is caused by a degeneration of neurons in certain parts of the brain. We're studying birds with similar defects. The songbird brain area that functions differently in different social contexts is the same area affected by Parkinson's disease in humans, and we see some similarities in vocal behavior associated with this area. For example, people with Parkinson's disease often stutter, and when damage occurs to this area of the bird brain, the affected bird sometimes does too. So we think we might be able to use these birds to study speech disorders.

What is your advice to students interested in biology?

It's important for students, especially those coming from a disadvantaged background, to know that biological research can be a very rewarding career, if you have a passion for it. Like an artist, you have to find your work intrinsically interesting. Figuring out how nature works is not an easy job; there are many frustrations. But intellectually, it can be one of the most rewarding and empowering of career choices.

40 Basic Principles of Animal Form and Function

▲ **Figure 40.1 A sphinx moth feeding on orchid nectar.**

Key Concepts

40.1 Physical laws and the environment constrain animal size and shape

40.2 Animal form and function are correlated at all levels of organization

40.3 Animals use the chemical energy in food to sustain form and function

40.4 Many animals regulate their internal environment within relatively narrow limits

40.5 Thermoregulation contributes to homeostasis and involves anatomy, physiology, and behavior

Overview

Diverse Forms, Common Challenges

Animals inhabit almost every part of the biosphere. Despite their amazing diversity of habitat, form, and function, all animals must solve a common set of problems: Animals as different as hydras, halibut, and humans must obtain oxygen, nourish themselves, excrete waste products, and move. How do animals of diverse evolutionary history and varying complexity solve these general challenges of life? As we explore this question throughout Unit 7, note the recurring theme of natural selection and adaptation.

This chapter introduces the unit by presenting some unifying concepts that apply across the animal kingdom. For example, the comparative study of animals reveals that form and function are closely correlated. Consider the long, thin, tonguelike proboscis of the sphinx moth (*Xanthopan morgani*) in **Figure 40.1**. The extended proboscis is a structural adaptation for feeding, serving as a straw through which the moth can suck nectar from deep within tube-shaped flowers. Analyzing a biological structure, such as a sphinx moth's proboscis, gives us clues about what it does and how it functions. **Anatomy** is the study of the *structure* of an organism; **physiology** is the study of the *functions* an organism performs. Natural selection can fit structure to function by selecting, over many generations, what works best among the available variations in a population.

A feeding sphinx moth also illustrates an animal's need for fuel in the form of chemical energy. We will apply concepts of bioenergetics—how organisms obtain, process, and use their energy resources—as another theme throughout our comparative study of animals. One use of an animal's energy resources is to regulate its internal environment. In this chapter, we will begin to explore the concept of homeostasis, using the example of body temperature regulation.

Concept 40.1

Physical laws and the environment constrain animal size and shape

An animal's size and shape, features that biologists often call "body plans" or "designs," are fundamental aspects of form and function that significantly affect the way an animal interacts with its environment. By using the terms *plan* and *design* here, we do not mean to imply that animal body forms are products of conscious invention. The body plan of an animal results from a pattern of development programmed by the genome, itself the product of millions of years of evolution. And the possibilities are not infinite—physical laws and the need to exchange materials with the environment place certain limits on the range of animal forms.

Physical Laws and Animal Form

Imagine seeing a winged serpent several meters long and weighing hundreds of kilograms flying through the air. Fortunately, you won't experience such a horrifying sight outside of a movie. Physical requirements constrain what natural selection can "invent," including the size and shape of flying animals. An animal the size and shape of a mythical dragon could not generate enough lift with its wings to get off the ground. This is just one example of how physical laws—in this case, the physics of flight—limit the evolution of an organism's form.

Consider another example: how the laws of hydrodynamics constrain the shapes that are possible for aquatic animals that swim very fast. Water is about a thousand times denser than air; thus, any bump on the body surface that causes drag would impede a swimmer even more than it would a runner or a flyer.

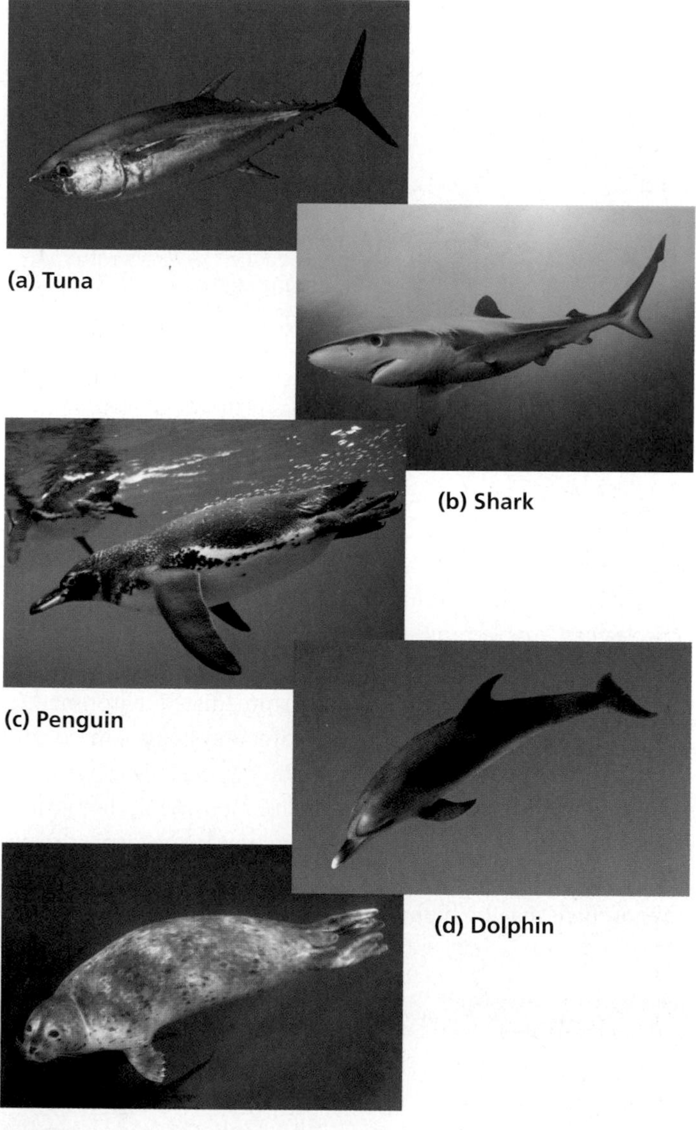

▲ **Figure 40.2 Evolutionary convergence in fast swimmers.**

Tuna and other fast ray-finned fishes can swim at speeds up to 80 km/hr. Sharks, penguins (birds), and aquatic mammals such as dolphins, seals, and whales are also fast swimmers. These animals all have the same streamlined body form: a fusiform shape, which means tapered on both ends **(Figure 40.2)**. The fact that these speedy swimmers have similar shapes is an example of convergent evolution (see Chapter 25). Convergence occurs because natural selection shapes similar adaptations when diverse organisms face the same environmental challenge, such as the resistance of water to fast travel.

Exchange with the Environment

An animal's size and shape have a direct effect on how the animal exchanges energy and materials with its surroundings. An animal's body plan must allow all of its living cells to be bathed in an aqueous medium, a requirement for maintaining the fluid integrity of the plasma membranes. Exchange with the environment occurs as substances dissolved in the aqueous medium diffuse and are transported across the cells' plasma membranes. As shown in **Figure 40.3a**, a single-celled protist living in water has a sufficient surface area of plasma membrane to service its entire volume of cytoplasm. (Thus, surface-to-volume ratio is one of the physical constraints on the size of single-celled protists.)

In contrast, an animal is composed of numerous cells, each with its own plasma membrane that functions as a loading and unloading platform for a modest volume of cytoplasm. But this organization works only if all the cells of the animal have access to a suitable aqueous environment. A hydra, which has a saclike body plan, has a body wall only two cell layers thick **(Figure 40.3b)**. Because its gastrovascular cavity opens to the exterior, both outer and inner layers of cells are

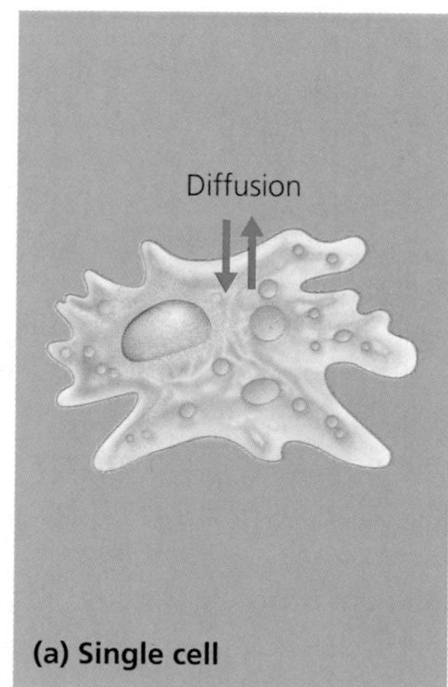

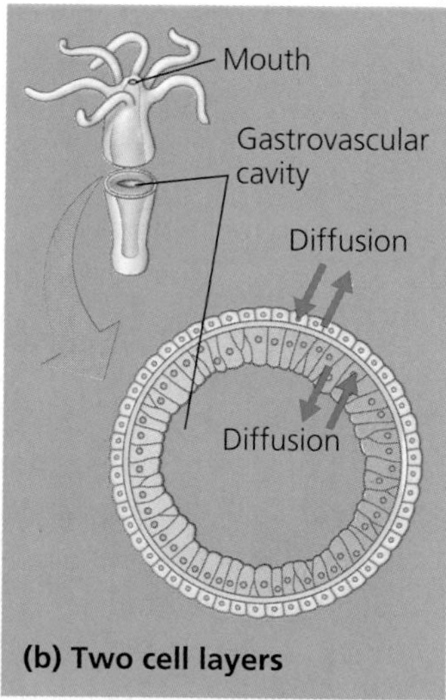

▲ **Figure 40.3 Contact with the environment. (a)** In a unicellular protist, such as this amoeba, the entire surface area contacts the environment. **(b)** A hydra's body consists of two layers of cells. Because the aqueous environment can circulate in and out of the hydra's mouth, virtually every one of its cells directly contacts the environment and exchanges materials with it.

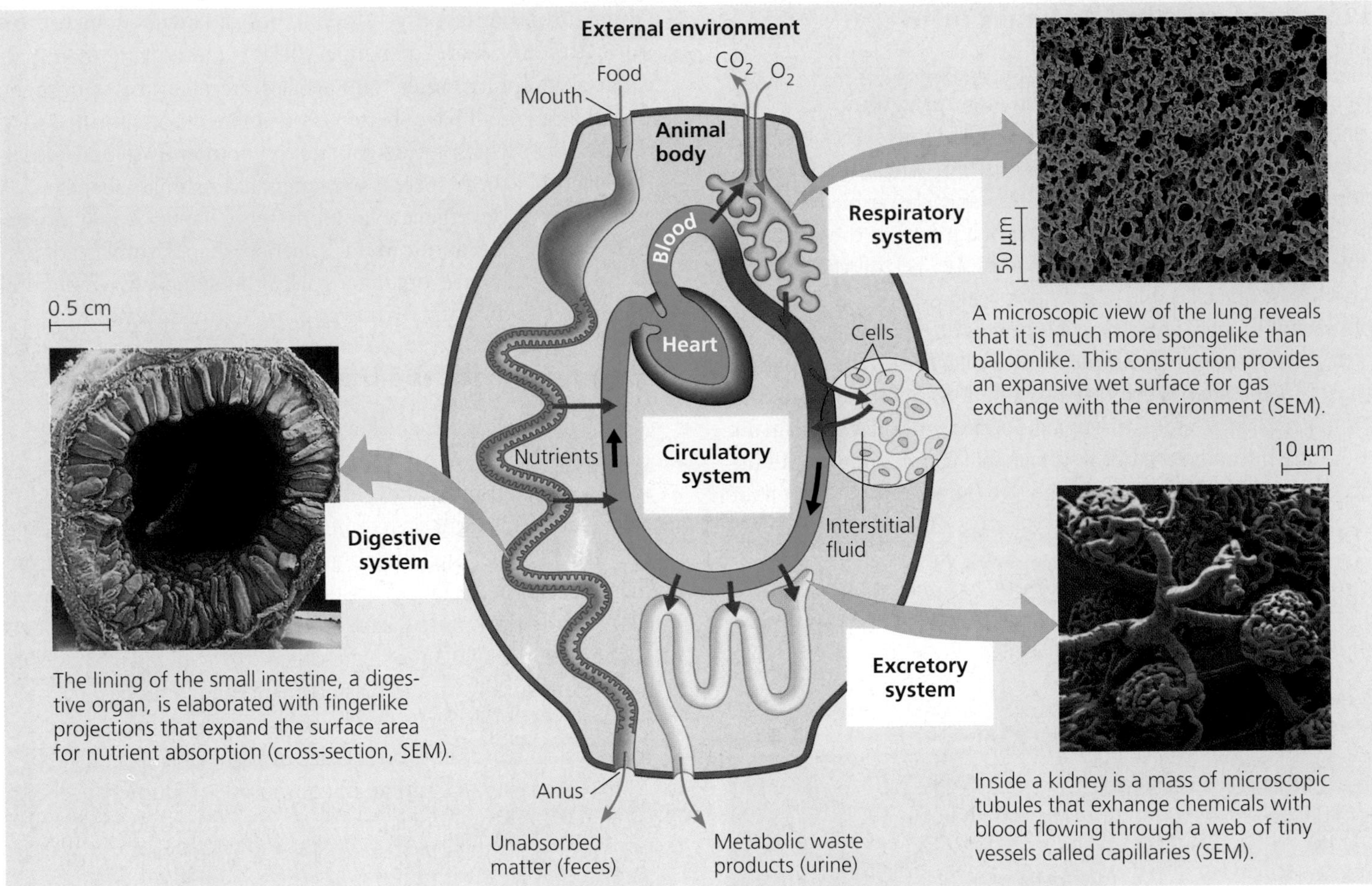
External environment
Food
CO_2 O_2
Mouth
Animal body
Blood
Respiratory system
50 µm
A microscopic view of the lung reveals that it is much more spongelike than balloonlike. This construction provides an expansive wet surface for gas exchange with the environment (SEM).
0.5 cm
Heart
Cells
Nutrients
Circulatory system
10 µm
Interstitial fluid
Digestive system
Excretory system
The lining of the small intestine, a digestive organ, is elaborated with fingerlike projections that expand the surface area for nutrient absorption (cross-section, SEM).
Anus
Inside a kidney is a mass of microscopic tubules that exchange chemicals with blood flowing through a web of tiny vessels called capillaries (SEM).
Unabsorbed matter (feces)
Metabolic waste products (urine)

▲ **Figure 40.4 Internal exchange surfaces of complex animals.** This diagrammatic animal illustrates the logistics of chemical exchange with the environment by a mammal. Most animals have surfaces that are specialized for exchanging certain chemicals with the surroundings. These exchange surfaces are usually internal, but are connected to the environment via openings on the body surface (the mouth, for example). The exchange surfaces are finely branched or folded, giving them a very large area. The digestive, respiratory, and excretory systems all have such exchange surfaces. Chemicals transported across these surfaces are carried throughout the body by the circulatory system.

bathed in water. A flat body shape is another design that maximizes exposure to the surrounding medium. For instance, a parasitic tapeworm may be several meters long, but because it is very thin, most of its cells are bathed in the intestinal fluid of the worm's vertebrate host, the source of nutrients.

Two-layered sacs and flat shapes are designs that put a large surface area in contact with the environment, but these simple forms do not allow much complexity in internal organization. Most animals are more complex and made up of compact masses of cells; their outer surfaces are relatively small compared with their volumes. As an extreme comparison, the surface-to-volume ratio of a whale is hundreds of thousands of times smaller than that of a water flea (*Daphnia*), yet every cell in the whale must be bathed in fluid and have access to oxygen, nutrients, and other resources. Extensively folded or branched internal surfaces facilitate this exchange with the environment in whales and most other animals **(Figure 40.4)**.

Despite the greater challenges of exchange with the environment, complex body forms have distinct benefits. For example, a specialized outer covering can protect against predators; large muscles can enable rapid movement; and internal digestive organs can break down food gradually, controlling the release of stored energy. And because the cells' immediate environment is the internal body fluid, the animal's organ systems can control the composition of this solution, allowing the animal to maintain a relatively stable internal environment while living in a variable external environment. A complex body form is especially well suited for animals living on land, where the external environment may be highly variable.

Concept Check 40.1

1. How does a large surface area contribute to the functions of the small intestine, the lungs, and the kidneys?

For suggested answers, see Appendix A.

Animal form and function are correlated at all levels of organization

As living things, animals exhibit hierarchical levels of organization, each with emergent properties (see Chapter 1). Most animals are composed of specialized cells organized into tissues that have different functions. Tissues are combined into functional units called organs, and groups of organs that work together form organ systems. For example, the digestive system consists of a stomach, small intestine, large intestine, and other organs, each composed of different kinds of tissues.

Tissue Structure and Function

Tissues are groups of cells with a common structure and function. Different types of tissues have different structures that are suited to their functions. For example, a tissue may be held together by a sticky extracellular matrix that coats the cells (see Figure 6.29) or weaves them together in a fabric of fibers. Indeed, the term *tissue* derives from a Latin word meaning "weave."

Tissues are classified into four main categories—epithelial tissue, connective tissue, muscle tissue, and nervous tissue—explored in **Figure 40.5**, on the next three pages.

Epithelial Tissue

Occurring in sheets of tightly packed cells, **epithelial tissue** covers the outside of the body and lines organs and cavities within the body. The cells of an epithelial tissue, or epithelium (plural, *epithelia*), are closely joined, with little material between them. In many epithelia, the cells are riveted together by tight junctions (see Figure 6.31). This tight packing enables the epithelium to function as a barrier against mechanical injury, microbes, and fluid loss. Some epithelia, called **glandular epithelia,** absorb or secrete chemical solutions. For example, the glandular epithelia that line the lumen (cavity) of the digestive and respiratory tracts form a **mucous membrane;** they secrete mucus that lubricates the surface and keeps it moist.

Two criteria for classifying epithelia are the number of cell layers and the shape of the cells on the exposed surface (see Figure 40.5). A **simple epithelium** has a single layer of cells, whereas a **stratified epithelium** has multiple tiers of cells. A "pseudostratified" epithelium is single-layered but appears stratified because the cells vary in length. The shape of the cells at the exposed surface may be **cuboidal** (like dice), **columnar** (like bricks standing on end), or **squamous** (like floor tiles).

Connective Tissue

Connective tissue functions mainly to bind and support other tissues. In contrast to epithelia, with their tightly packed cells, connective tissues have a sparse population of cells scattered through an extracellular matrix. The matrix generally consists of a web of fibers embedded in a uniform foundation that may be liquid, jellylike, or solid. In most cases, the substances that make up the matrix are secreted by the cells of the connective tissue.

Connective tissue fibers, which are made of protein, are of three kinds: collagenous fibers, elastic fibers, and reticular fibers. **Collagenous fibers** are made of collagen, probably the most abundant protein in the animal kingdom. Collagenous fibers are nonelastic and do not tear easily when pulled lengthwise. If you pinch and pull some skin on the back of your hand, it is mainly collagen that keeps the flesh from tearing away from the bone. **Elastic fibers** are long threads made of a protein called elastin. Elastic fibers provide a rubbery quality that complements the nonelastic strength of collagenous fibers. When you release the skin on the back of your hand, elastic fibers quickly restore the skin to its original shape. **Reticular fibers** are very thin and branched. Composed of collagen and continuous with collagenous fibers, they form a tightly woven fabric that joins connective tissue to adjacent tissues.

The major types of connective tissue in vertebrates are loose connective tissue, adipose tissue, fibrous connective tissue, cartilage, bone, and blood (see Figure 40.5). Among the cells scattered in loose connective tissue, two types predominate: fibroblasts and macrophages. **Fibroblasts** secrete the protein ingredients of the extracellular fibers. **Macrophages** are amoeboid cells that roam the maze of fibers, engulfing foreign particles and the debris of dead cells by phagocytosis (see Chapter 6). You will learn more about the specific functions of those cells and various connective tissues later in this unit.

Muscle Tissue

Muscle tissue is composed of long cells called muscle fibers that are capable of contracting, usually when stimulated by nerve signals. Arranged in parallel within the cytoplasm of muscle fibers are large numbers of contracting units called myofibrils made of the proteins actin and myosin. (You will learn more about muscle contraction in Chapter 49.) Muscle is the most abundant tissue in most animals, and muscle contraction accounts for much of the energy-consuming cellular work in an active animal. In the vertebrate body, there are three types of muscle tissue: skeletal muscle, cardiac muscle, and smooth muscle (see Figure 40.5).

Nervous Tissue

Nervous tissue senses stimuli and transmits signals in the form of nerve impulses from one part of the animal to another. The functional unit of nervous tissue is the neuron, or nerve cell, which is uniquely specialized to transmit nerve impulses, as we discuss in detail in Chapter 48. In many animals, nervous tissue is concentrated in the brain, which functions as the control center that coordinates many of the animal's activities.

Figure 40.5

Exploring Structure and Function in Animal Tissues

EPITHELIAL TISSUE

Columnar epithelia, which have cells with relatively large cytoplasmic volumes, are often located where secretion or active absorption of substances is an important function.

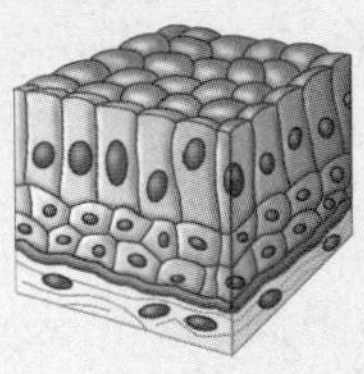

A **stratified columnar epithelium** lines the inner surface of the urethra, the tube through which urine exits the body.

A **simple columnar epithelium** lines the intestines. This epithelium secretes digestive juices and absorbs nutrients.

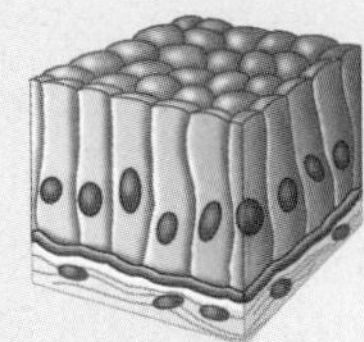

A **pseudostratified ciliated columnar epithelium** forms a mucous membrane that lines the respiratory tract of many vertebrates. The beating cilia move a film of mucus along the surface. The ciliated epithelium of respiratory tubes helps keep lungs clean by trapping dust and other particles and sweeping them back up the trachea (windpipe).

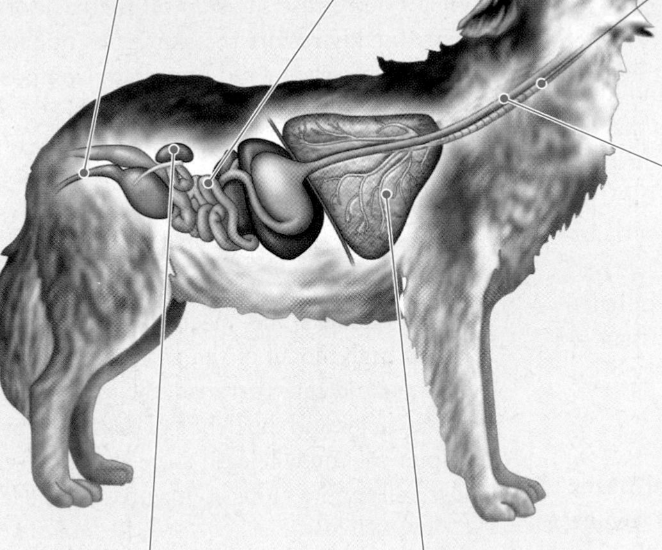

Stratified squamous epithelia regenerate rapidly by cell division near the basement membrane (see below). The new cells are pushed to the free surface as replacements for cells that are continually sloughed off. This type of epithelium is commonly found on surfaces subject to abrasion, such as the outer skin and linings of the esophagus, anus, and vagina. The organization of this type of tissue ensures that abrasion affects the oldest (outermost) cells, while protecting the underlying tissues.

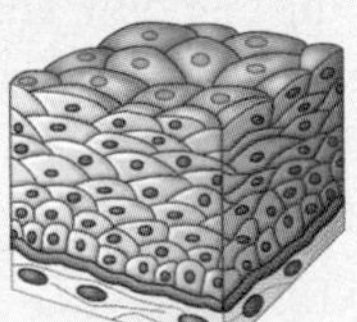

Cuboidal epithelia, with cells specialized for secretion, make up the epithelia of kidney tubules and many glands, including the thyroid gland and salivary glands. Glandular epithelia lining tubules in the thyroid gland secrete a hormone that regulates the body's rate of fuel consumption.

Simple squamous epithelia, which are thin and leaky, function in the exchange of material by diffusion. These epithelia line blood vessels and the air sacs of the lungs, where diffusion of nutrients and gases is critical.

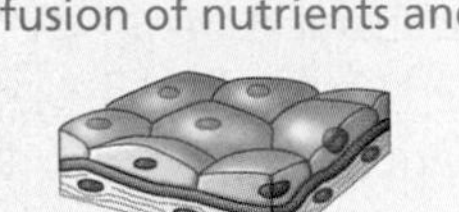

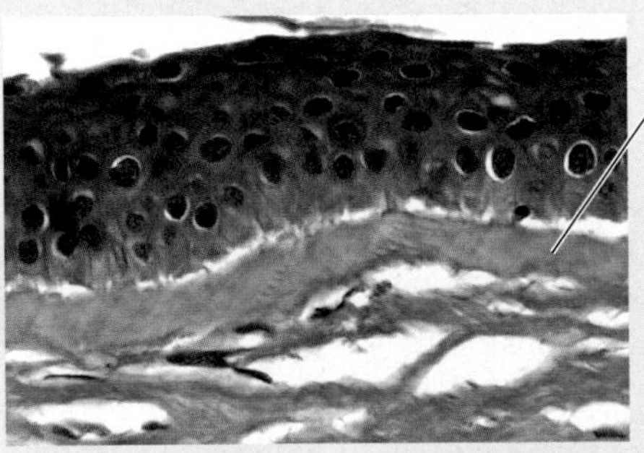

The cells at the base of an epithelial layer are attached to a **basement membrane**, a dense mat of extracellular matrix. The free surface of the epithelium is exposed to air or fluid.

CONNECTIVE TISSUE

The most widespread connective tissue in the vertebrate body is **loose connective tissue**. It binds epithelia to underlying tissues and functions as "packing material," holding organs in place. Loose connective tissue gets its name from the loose weave of its fibers, which include all three fiber types: collagenous, elastic, and reticular. Fibroblasts and macrophages are scattered in the fibrous mesh.

Cartilage has an abundance of collagenous fibers embedded in a rubbery matrix made of a protein-carbohydrate complex called chondroitin sulfate. Chondroitin sulfate and collagen are secreted by cells called **chondrocytes**. The composite of collagenous fibers and chondroitin sulfate makes cartilage a strong yet somewhat flexible support material. Many vertebrates have cartilaginous skeletons during the embryo stage, but most of the cartilage is replaced by bone as the embryo matures. Nevertheless, cartilage is retained in certain locations, such as the discs that act as cushions between vertebrae and the caps on the ends of some bones. These structures' flexibility allows them to absorb considerable physical impact without breaking.

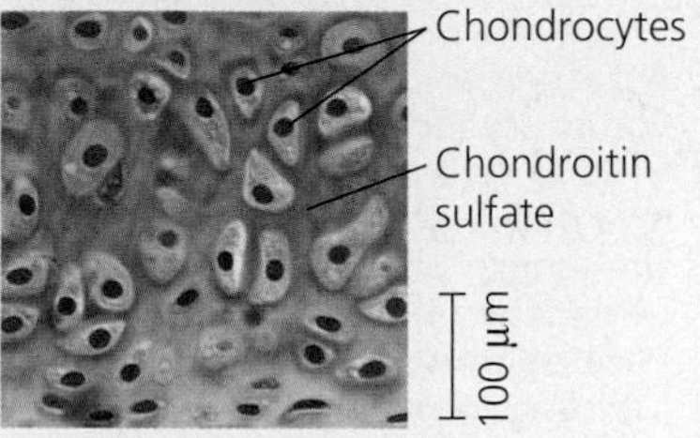

Fibrous connective tissue is dense, owing to its large number of collagenous fibers. The fibers are organized into parallel bundles, an arrangement that maximizes nonelastic strength. Fibrous connective tissue is found in **tendons**, which attach muscles to bones, and in **ligaments**, which join bones together at joints.

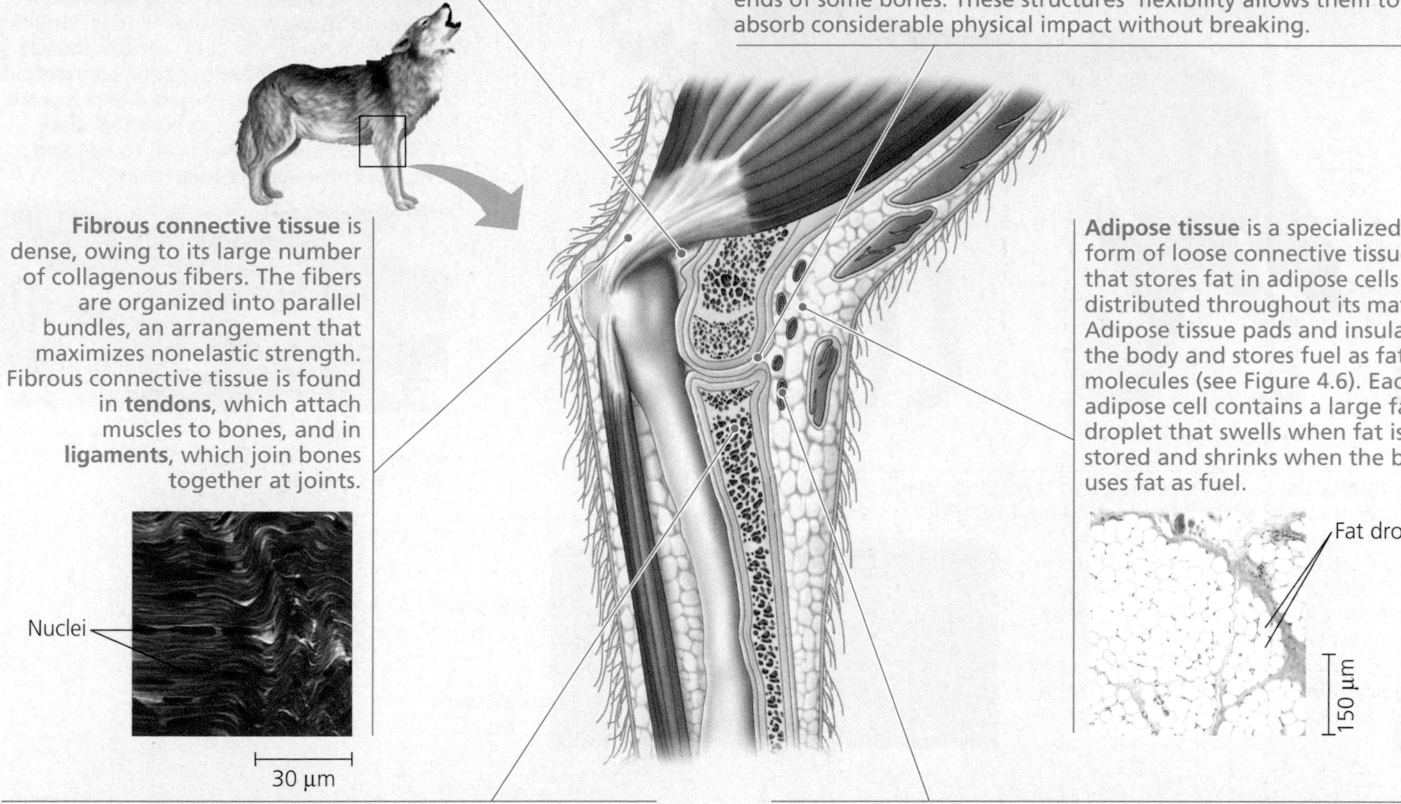

Adipose tissue is a specialized form of loose connective tissue that stores fat in adipose cells distributed throughout its matrix. Adipose tissue pads and insulates the body and stores fuel as fat molecules (see Figure 4.6). Each adipose cell contains a large fat droplet that swells when fat is stored and shrinks when the body uses fat as fuel.

The skeleton supporting the body of most vertebrates is made of **bone**, a mineralized connective tissue. Bone-forming cells called **osteoblasts** deposit a matrix of collagen. Calcium, magnesium, and phosphate ions combine and harden within the matrix into the mineral hydroxyapatite. The combination of hard mineral and flexible collagen makes bone harder than cartilage without being brittle, an important characteristic for supporting the body. The microscopic structure of hard mammalian bone consists of repeating units called **osteons** (or Haversian systems). Each osteon has concentric layers of the mineralized matrix, which are deposited around a central canal containing blood vessels and nerves that service the bone.

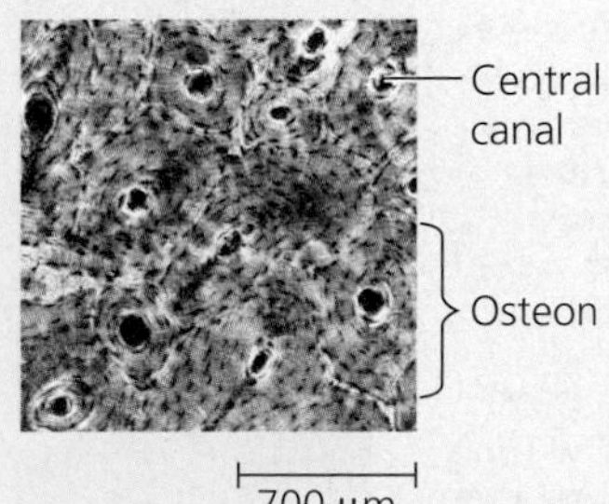

Although **blood** functions differently from other connective tissues, it does meet the criterion of having an extensive extracellular matrix. In this case, the matrix is a liquid called plasma, consisting of water, salts, and a variety of dissolved proteins. Suspended in the plasma are two classes of blood cells, erythrocytes (red blood cells) and leukocytes (white blood cells), and cell fragments called platelets. Red cells carry oxygen; white cells function in defense against viruses, bacteria, and other invaders; and platelets aid in blood clotting. The liquid matrix enables rapid transport of blood cells, nutrients, and wastes through the body.

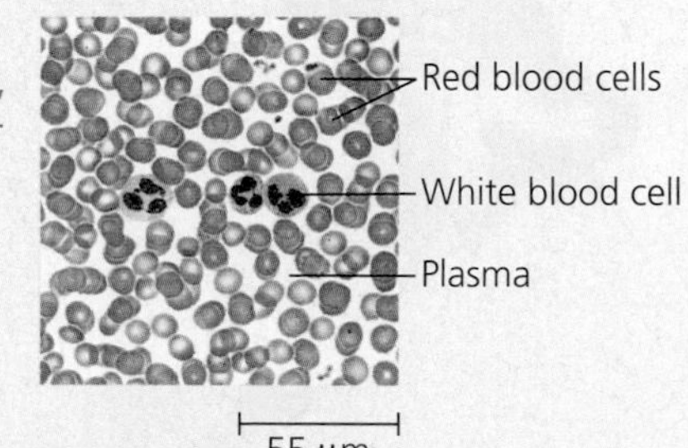

continued on the next page

Exploring Structure and Function in Animal Tissues

MUSCLE TISSUE

Attached to bones by tendons, **skeletal muscle** is responsible for voluntary movements of the body. Skeletal muscle consists of bundles of long cells called fibers; each fiber is a bundle of strands called myofibrils. The arrangement of contractile units, or sarcomeres, along the length of the fibers gives the cells a striped (striated) appearance under the microscope. For this reason, skeletal muscle is also called **striated muscle**. Adult mammals have a fixed number of muscle cells; building muscle does not increase the number of cells but rather enlarges those already present.

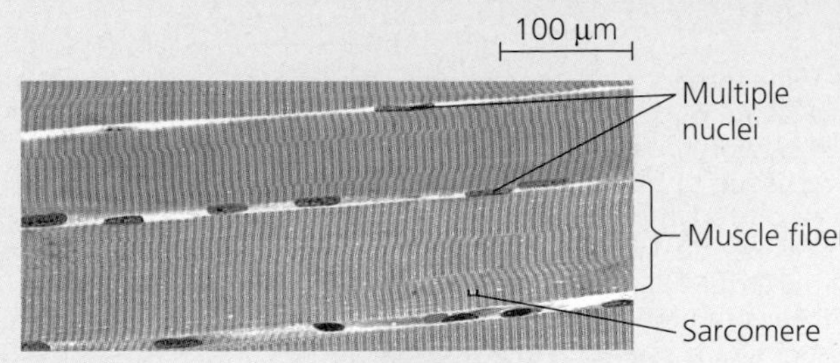

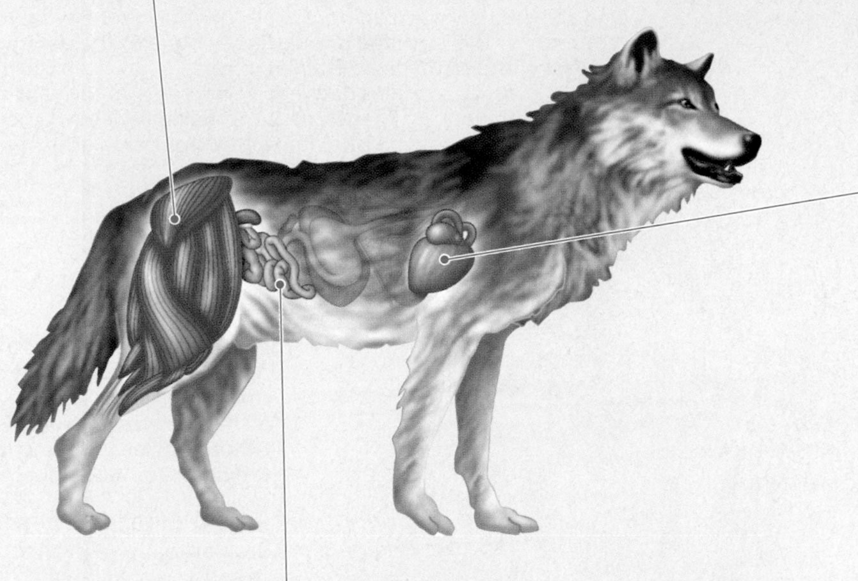

Cardiac muscle forms the contractile wall of the heart. It is striated like skeletal muscle and has contractile properties similar to those of skeletal muscle. Unlike skeletal muscle, however, cardiac muscle carries out an unconscious task: contraction of the heart. Cardiac muscle fibers branch and interconnect via intercalated disks, which relay signals from cell to cell and help synchronize the heartbeat.

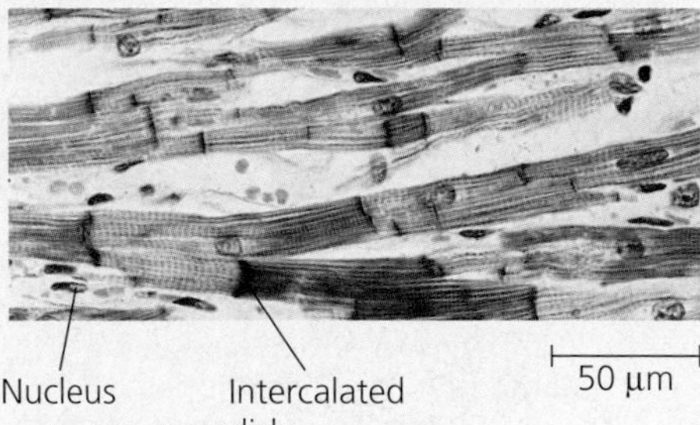

Smooth muscle, so named because it lacks striations, is found in the walls of the digestive tract, urinary bladder, arteries, and other internal organs. The cells are spindle-shaped. They contract more slowly than skeletal muscle cells but can remain contracted longer. Controlled by different kinds of nerves than those controlling skeletal muscles, smooth muscles are responsible for involuntary body activities, such as churning of the stomach or constriction of arteries.

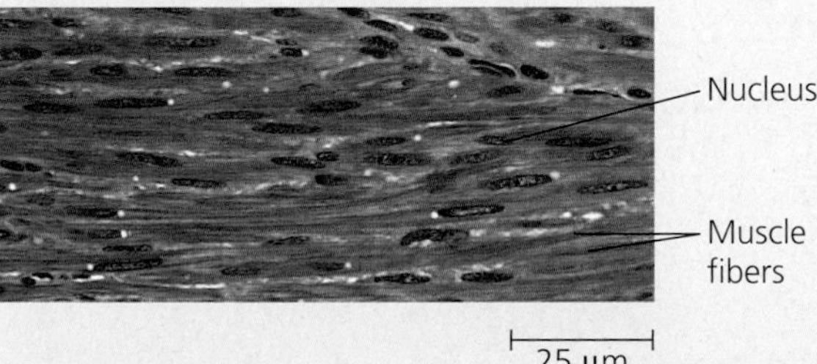

NERVOUS TISSUE

Nerve cells (neurons) are the basic units of the nervous system. A **neuron** consists of a cell body and two or more extensions, or processes, called dendrites and axons, which in certain neurons may be as long as a meter in some animals. Dendrites carry information from their tips toward the rest of the neuron. Axons transmit impulses toward another neuron or toward an effector, a structure such as a muscle cell that carries out a body response. The long axons of some motor neurons enable quick responses by voluntary muscles.

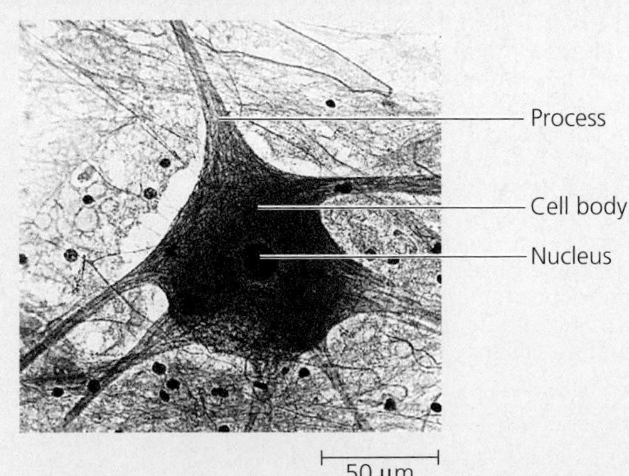

Table 40.1 Organ Systems: Their Main Components and Functions in Mammals

Organ System	Main Components	Main Functions
Digestive	Mouth, pharynx, esophagus, stomach, intestines, liver, pancreas, anus	Food processing (ingestion, digestion, absorption, elimination)
Circulatory	Heart, blood vessels, blood	Internal distribution of materials
Respiratory	Lungs, trachea, other breathing tubes	Gas exchange (uptake of oxygen; disposal of carbon dioxide)
Immune and lymphatic	Bone marrow, lymph nodes, thymus, spleen, lymph vessels, white blood cells	Body defense (fighting infections and cancer)
Excretory	Kidneys, ureters, urinary bladder, urethra	Disposal of metabolic wastes; regulation of osmotic balance of blood
Endocrine	Pituitary, thyroid, pancreas, other hormone-secreting glands	Coordination of body activities (such as digestion, metabolism)
Reproductive	Ovaries, testes, and associated organs	Reproduction
Nervous	Brain, spinal cord, nerves, sensory organs	Coordination of body activities; detection of stimuli and formulation of responses to them
Integumentary	Skin and its derivatives (such as hair, claws, skin glands)	Protection against mechanical injury, infection, drying out; thermoregulation
Skeletal	Skeleton (bones, tendons, ligaments, cartilage)	Body support, protection of internal organs, movement
Muscular	Skeletal muscles	Movement, locomotion

Organs and Organ Systems

In all but the simplest animals (sponges and some cnidarians), different tissues are organized into **organs.** In some organs, the tissues are arranged in layers. For example, the vertebrate stomach has four major tissue layers **(Figure 40.6)**.

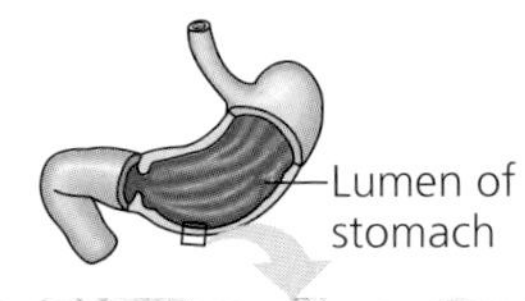

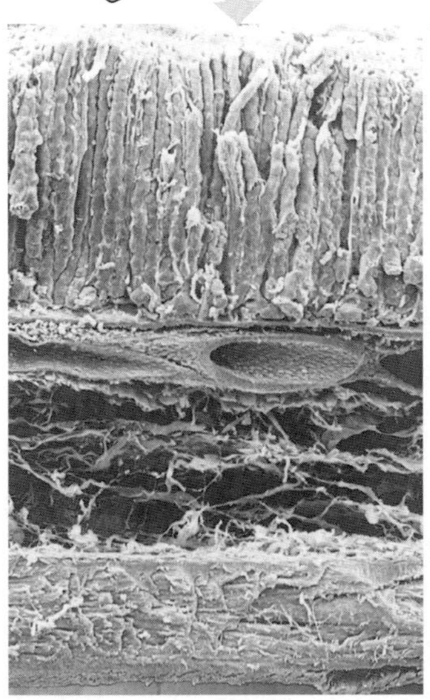

Mucosa. The mucosa is an epithelial layer that lines the lumen.
Submucosa. The submucosa is a matrix of connective tissue that contains blood vessels and nerves.
Muscularis. The muscularis consists mainly of smooth muscle tissue.
Serosa. External to the muscularis is the serosa, a thin layer of connective and epithelial tissue.
0.2 mm

▲ **Figure 40.6 Tissue layers of the stomach, a digestive organ.** The wall of the stomach and other tubular organs of the digestive system has four main tissue layers (SEM).

A thick epithelium lines the lumen and secretes mucus and digestive juices into it. Outside this layer is a zone of connective tissue, surrounded by a thick layer of smooth muscle. Yet another layer of connective tissue encases the entire stomach.

Many of the organs of vertebrates are suspended by sheets of connective tissue called **mesenteries** in moist or fluid-filled body cavities. Mammals have a **thoracic cavity** housing the lungs and heart that is separated from the lower **abdominal cavity** by a sheet of muscle called the diaphragm.

Representing a level of organization higher than organs, **organ systems** carry out the major body functions of most animals **(Table 40.1)**. Each organ system consists of several organs and has specific functions, but the efforts of all systems must be coordinated for the animal to survive. For instance, nutrients absorbed from the digestive tract are distributed throughout the body by the circulatory system. But the heart, which pumps blood through the circulatory system, depends on nutrients absorbed by the digestive tract as well as on oxygen (O_2) obtained by the respiratory system. Any organism, whether single-celled or an assembly of organ systems, is a coordinated living whole greater than the sum of its parts.

Concept Check 40.2

1. Describe how the epithelial tissue that lines the stomach lumen is well suited to its function.
2. Suggest why a disease that damages connective tissue is likely to threaten most of the body's organs.
3. How are muscle and nervous tissue interdependent?

For suggested answers, see Appendix A.

Animals use the chemical energy in food to sustain form and function

All organisms require chemical energy for growth, repair, physiological processes (including movement in the case of animals), regulation, and reproduction. As we have discussed in other chapters, organisms can be classified by how they obtain this energy. Autotrophs such as plants use light energy to build energy-rich organic molecules and then use those organic molecules for fuel. In contrast, heterotrophs such as animals must obtain their chemical energy from food, which contains organic molecules synthesized by other organisms.

Bioenergetics

The flow of energy through an animal—its **bioenergetics**—ultimately limits the animal's behavior, growth, and reproduction and determines how much food it needs. Studying an animal's bioenergetics tells us a great deal about the animal's adaptations.

Energy Sources and Allocation

Animals harvest chemical energy from the food they eat. Food is digested by enzymatic hydrolysis (see Figure 5.2b), and energy-containing molecules are absorbed by body cells. Once absorbed, these energy-containing molecules have several possible fates. Most are used to generate ATP by the catabolic processes of cellular respiration and fermentation (see Chapter 9). The chemical energy of ATP powers cellular work, enabling cells, organs, and organ systems to perform the many functions that keep an animal alive. Because the production and use of ATP generate heat, an animal continuously gives off heat to its surroundings (heat balance is discussed in more detail later in this chapter).

After the energetic needs of staying alive are met, any remaining molecules from food can be used in biosynthesis, including body growth and repair, synthesis of storage material such as fat, and production of gametes **(Figure 40.7)**. Biosynthesis requires both carbon skeletons for new structures and ATP to power their assembly. In some situations, biosynthetic products (such as body fat) can be broken down into fuel molecules for production of additional ATP, depending on the needs of the animal (see Figure 9.19).

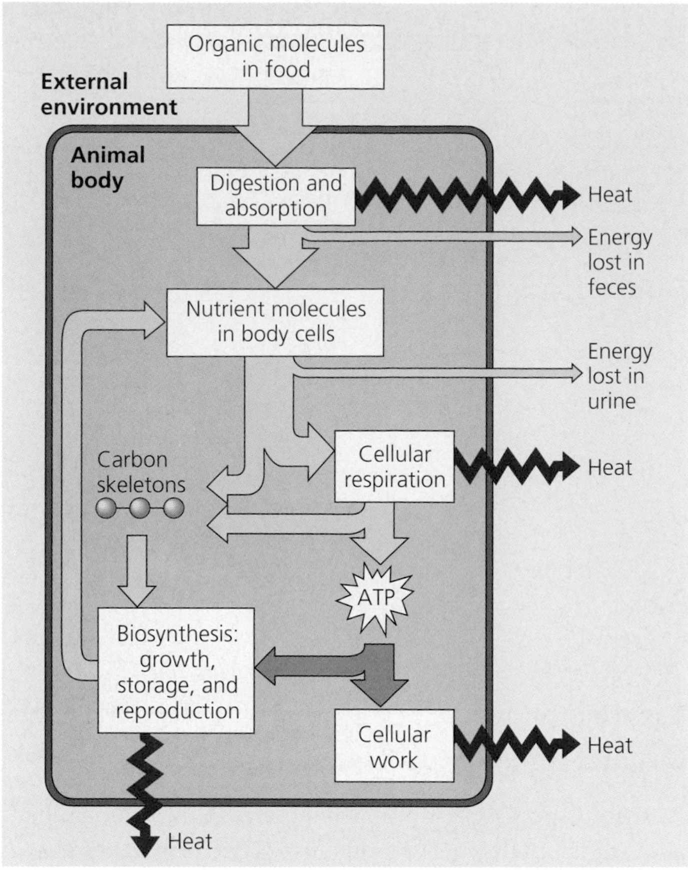

▲ **Figure 40.7 Bioenergetics of an animal: an overview.**

Quantifying Energy Use

Understanding the bioenergetics of animals depends on the ability to quantify their energy needs. How much of the total energy an animal obtains from food does it need just to stay alive? How much energy must be expended to walk, run, swim, or fly from one place to another? What fraction of energy intake can be used for reproduction? Physiologists obtain answers to such questions by measuring the rates at which animals use chemical energy and how these rates change in different circumstances.

The amount of energy an animal uses in a unit of time is called its **metabolic rate**—the sum of all the energy-requiring biochemical reactions occurring over a given time interval. Energy is measured in calories (cal) or kilocalories (kcal). (A kilocalorie is 1,000 calories. The unit *Calorie,* with a capital C, as used by many nutritionists, is actually a kilocalorie.)

Metabolic rate can be determined in several ways. Because nearly all of the chemical energy used in cellular respiration eventually appears as heat, metabolic rate can be measured by monitoring an animal's rate of heat loss. Researchers can use a calorimeter, which is a closed, insulated chamber equipped with a device that records an animal's heat loss, to measure metabolic rate for small animals. A more indirect way to measure metabolic rate is to determine the amount of oxygen consumed or carbon dioxide produced by an animal's cellular respiration **(Figure 40.8)**. Over long periods, the rate of food consumption and the energy content of the food (about 4.5–5 kcal per gram of protein or carbohydrate and about 9 kcal per gram of fat) can be used to estimate metabolic rate. However, this method must account for the energy stored in food that the animal cannot use (the energy lost in feces and urine).

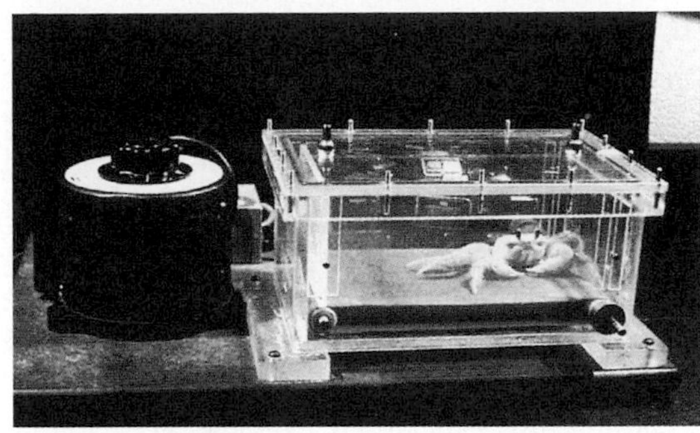

(a) This photograph shows a ghost crab in a respirometer. Temperature is held constant in the chamber, with air of known O_2 concentration flowing through. The crab's metabolic rate is calculated from the difference between the amount of O_2 entering and the amount of O_2 leaving the respirometer. This crab is on a treadmill, running at a constant speed as measurements are made.

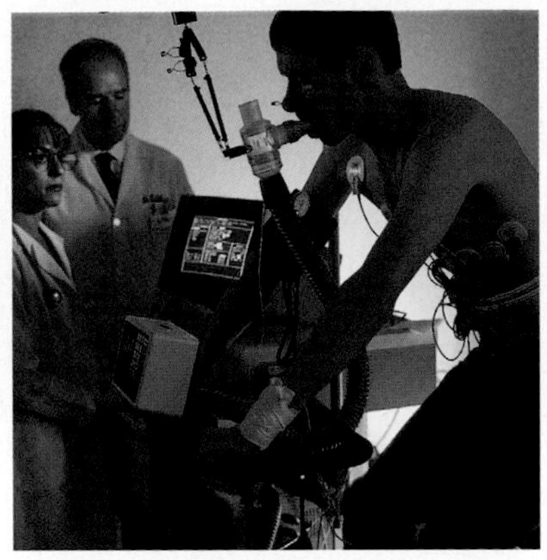

(b) Similarly, the metabolic rate of a man fitted with a breathing apparatus is being monitored while he exercises on a stationary bike.

▲ **Figure 40.8 Measuring metabolic rate.**

course, uses far more calories than the whole mouse). The higher metabolic rate of a smaller animal's tissues demands a proportionately greater rate of oxygen delivery. Correlated with its higher metabolic rate per gram, the smaller animal also has a higher breathing rate, blood volume (relative to its size), and heart rate (pulse), and it must eat much more food per unit of body mass.

The explanation for the inverse relationship between metabolic rate and size is still unclear. One hypothesis is that for endotherms, the smaller the animal, the greater the energy cost of maintaining a stable body temperature. In effect, the smaller an animal is, the greater its surface-to-volume ratio is and thus the faster it loses heat to (or gains heat from) the surroundings. Logical as this hypothesis seems, it cannot be the complete explanation, as it fails to explain the inverse relationship between metabolism and size in *ectotherms*, which do not use metabolic heat to maintain body temperature. Although this relationship has been extensively documented in both endotherms and ectotherms, researchers continue to search for its underlying causes.

Bioenergetic Strategies

An animal's metabolic rate is closely related to its bioenergetic "strategy." There are two basic bioenergetic strategies found in animals. Birds and mammals are mainly **endothermic**, meaning that their bodies are warmed mostly by heat generated by metabolism, and their body temperature is maintained within a relatively narrow range. Endothermy is a high-energy strategy that permits intense, long-duration activity over a wide range of environmental temperatures. In contrast, most fishes, amphibians, reptiles other than birds, and invertebrates are **ectothermic**, meaning that they gain their heat mostly from external sources. The ectothermic strategy requires a much smaller energy expenditure by an animal than the endothermic strategy because of the high energy cost of heating (or cooling) an endothermic body. In general, endotherms have higher metabolic rates than ectotherms. You will read more about endothermic and ectothermic strategies later in this chapter.

Influences on Metabolic Rate

The metabolic rates of animals are affected by many factors besides whether the animal is an endotherm or an ectotherm. One of animal biology's most intriguing (but largely unanswered) questions has to do with the relationship between body size and metabolic rate.

Size and Metabolic Rate

By measuring the metabolic rates of many species of vertebrates and invertebrates, physiologists have shown that the amount of energy it takes to maintain each gram of body weight is inversely related to body size. Each gram of a mouse, for instance, requires about 20 times as many calories as a gram of an elephant (even though the whole elephant, of

Activity and Metabolic Rate

Every animal experiences a range of metabolic rates. A minimum rate powers the basic functions that support life, such as cell maintenance, breathing, and heartbeat. The metabolic rate of a nongrowing endotherm that is at rest, has an empty stomach, and is not experiencing stress is called the **basal metabolic rate (BMR).** The BMR for humans averages about 1,600–1,800 kcal per day for adult males and about 1,300–1,500 kcal per day for adult females. These BMRs are about equivalent to the rate of energy use by a 75-watt light bulb.

In an ectotherm, body temperature changes with the temperature of the animal's surroundings, and so does metabolic rate. Unlike BMRs, which can be determined within a range of environmental temperatures, the minimum metabolic rate of an ectotherm must be determined at a specific temperature. The metabolic rate of a resting, fasting, nonstressed ectotherm at a particular temperature is called its **standard metabolic rate (SMR).**

For both ectotherms and endotherms, activity has a large effect on metabolic rate. Even a person reading quietly at a desk or an insect twitching its wings consumes energy beyond the BMR or SMR. Maximum metabolic rates (the highest rates of ATP utilization) occur during peak activity, such as lifting heavy weights, sprinting, or high-speed swimming.

In general, an animal's maximum possible metabolic rate is inversely related to the duration of activity. **Figure 40.9** contrasts ectothermic and endothermic "strategies" for sustaining activity over different time intervals. Both an alligator (ectotherm) and a human (endotherm) are capable of very intense exercise in short spurts of a minute or less. During these "sprints," the ATP present in muscle cells and ATP generated anaerobically by glycolysis are sufficient to power the activity. Neither the ectotherm nor the endotherm can sustain their maximum metabolic rates and peak levels of activity over longer periods of exercise, though the endotherm has an advantage in such endurance tests. Sustained activity depends on the aerobic process of cellular respiration for ATP supply, and an endotherm's respiration rate is about ten times greater than an ectotherm's. Only a few ectotherms, such as migratory monarch butterflies or salmon, are capable of long-duration activities.

Between the extremes of BMR or SMR and maximum metabolic rate, many factors influence energy requirements, including age, sex, size, body and environmental temperatures, quality and quantity of food, activity level, oxygen availability, and hormonal balance. Time of day also plays a role. Birds, humans, and many insects are usually active (and therefore have their highest metabolic rates) during daylight hours. In contrast, bats, mice, and many other mammals generally are active and have their highest metabolic rates at night or during the hours of dawn and dusk. Metabolic rates measured when animals are performing a variety of activities give a better idea of the energy costs of everyday life. For most terrestrial animals (both endotherms and ectotherms), the average daily rate of energy consumption is 2 to 4 times BMR or SMR. Humans in most developed countries have an unusually low average daily metabolic rate of about 1.5 times BMR—an indication of relatively sedentary lifestyles.

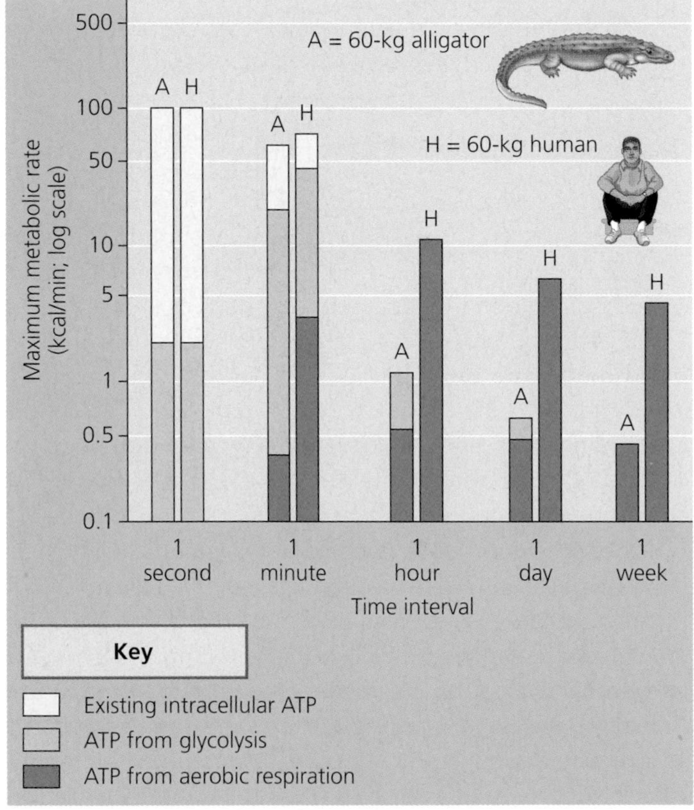

▲ **Figure 40.9 Maximum metabolic rates over different time spans.** The bars compare an ectotherm (alligator) and an endotherm (human) for their *maximum* potential metabolic rates and ATP sources over different durations of time. The human's basal metabolic rate (about 1.2 kcal/min) is much greater than the alligator's standard metabolic rate (about 0.04 kcal/min). The human's higher BMR partly contributes to his ability to sustain a higher maximum metabolic rate over a longer period.

Energy Budgets

Different species of animals use the energy and materials in food in different ways, depending on their environment, behavior, size, and basic strategy of endothermy or ectothermy. For most animals, the majority of food is devoted to the production of ATP, and relatively little goes to growth or reproduction. However, the amount of energy used for BMR (or SMR), activity, and body temperature control varies considerably between species. For example, consider the typical energy "budgets" of four terrestrial vertebrates: a 25-g female deer mouse, a 4-kg female python, a 4-kg male Adélie penguin, and a 60-kg female human **(Figure 40.10)**. We will assume that all of these animals reproduce during the year.

The female human spends a large fraction of her energy budget for BMR and relatively little for activity and body temperature regulation. The small amount of growth, about 1% of her annual energy budget, is equivalent to adding about 1 kg of body fat or 5–6 kg of other tissues. The cost of nine months of pregnancy and several months of breast-feeding amounts to only 5–8% of the mother's annual energy requirements.

A male penguin spends a much larger fraction of his energy expenditures for activity because he swims to catch food. Because the penguin is well insulated and fairly large, he has relatively low costs of temperature regulation in spite of living in the cold antarctic environment. His reproductive costs, about 6% of annual energy expenditures, mainly come from incubating eggs (brooding) and bringing food to his chicks. Penguins, like most birds, do not grow once they are adults.

The female deer mouse spends a large fraction of her energy budget for temperature regulation. Because of the high surface-to-volume ratio that goes with small size, mice lose body heat rapidly and must constantly generate metabolic heat to maintain body temperature. Female deer mice spend about 12% of their energy budget on reproduction.

In contrast to these endothermic animals, the ectothermic python has no temperature regulation costs. Like most reptiles, she grows continuously throughout life. In the example in Figure 40.10, the snake added about 750 g of new body tissue. She also produced about 650 g of eggs. The python's economical ectothermic strategy is revealed by her very low energy expenditure per unit mass, which is only $1/40$ of the

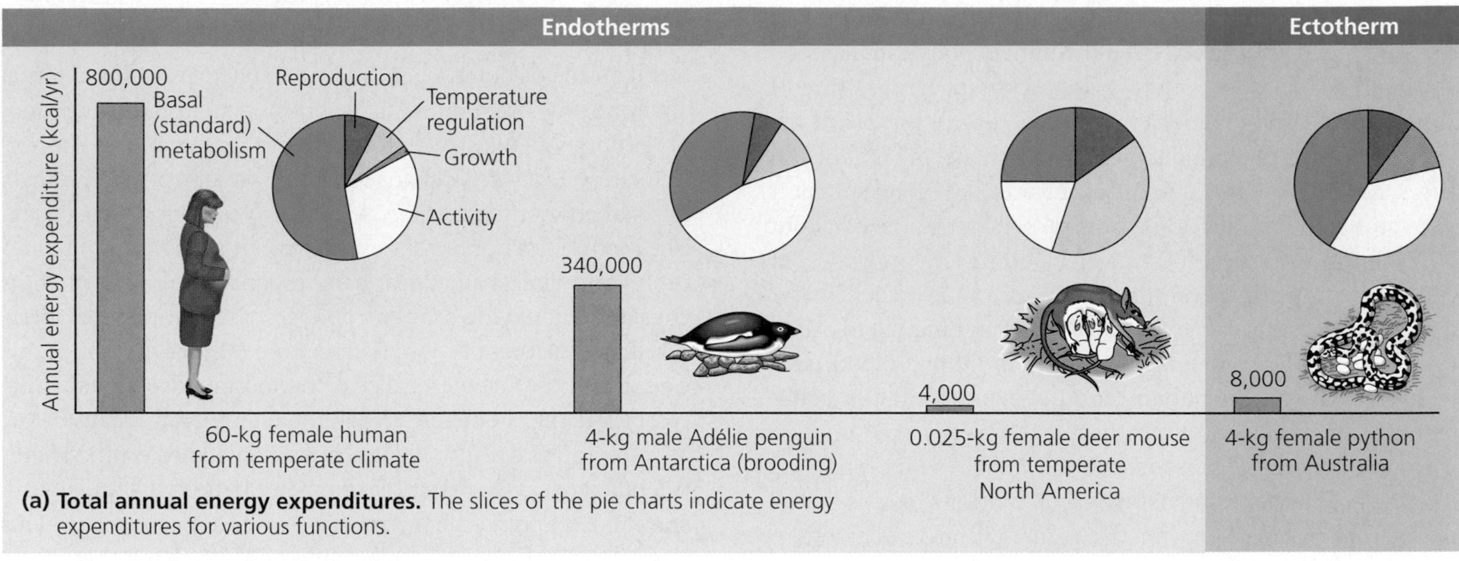

(a) Total annual energy expenditures. The slices of the pie charts indicate energy expenditures for various functions.

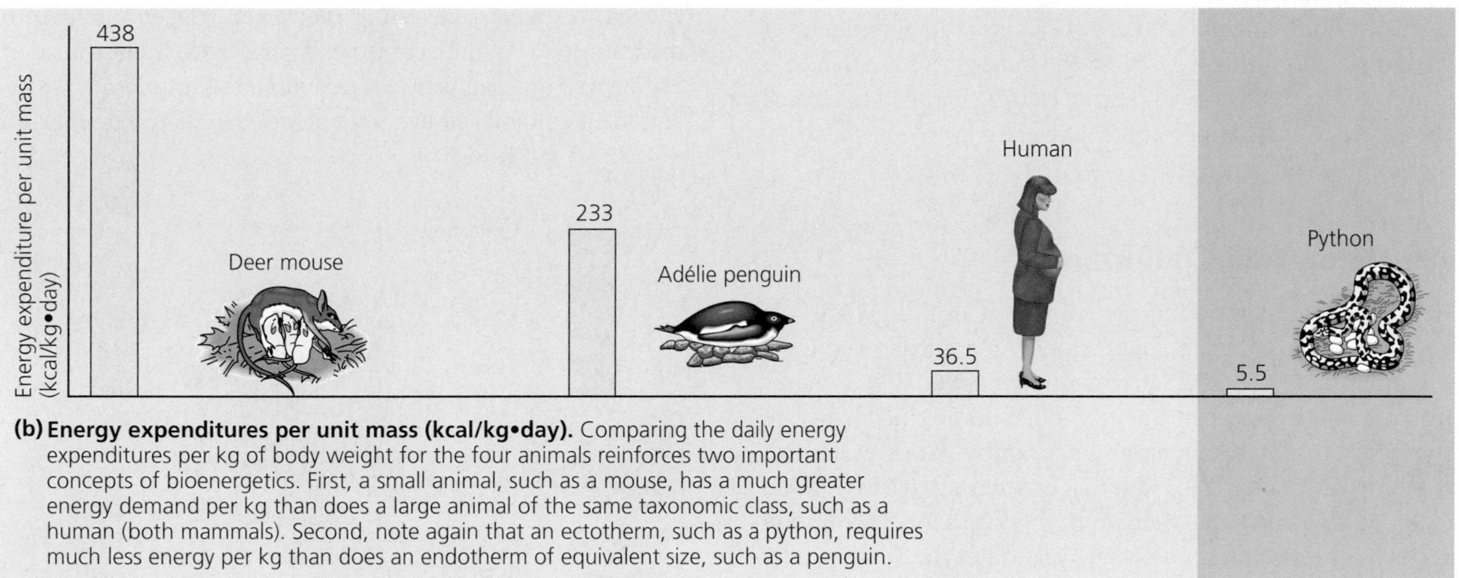

(b) Energy expenditures per unit mass (kcal/kg•day). Comparing the daily energy expenditures per kg of body weight for the four animals reinforces two important concepts of bioenergetics. First, a small animal, such as a mouse, has a much greater energy demand per kg than does a large animal of the same taxonomic class, such as a human (both mammals). Second, note again that an ectotherm, such as a python, requires much less energy per kg than does an endotherm of equivalent size, such as a penguin.

▲ **Figure 40.10 Energy budgets for four animals.**

energy expended by the similarly-sized endothermic penguin (see Figure 40.10b).

Throughout our study of animal biology, we will encounter many other examples of how bioenergetics relates to the form and function of diverse animals.

Concept Check 40.3

1. If a mouse and a small lizard of the same mass (both at rest) were placed in respirometers under identical environmental conditions, which animal would consume oxygen at a higher rate? Explain.
2. Why are alligators not capable of intense activity for periods of more than 1 hour?
3. Which must eat a larger proportion of its weight in food each day: a house cat or an African lion? Explain.

For suggested answers, see Appendix A.

Many animals regulate their internal environment within relatively narrow limits

More than a century ago, French physiologist Claude Bernard made the distinction between the external environment surrounding an animal and the internal environment in which the cells of the animal live. The internal environment of vertebrates is called the **interstitial fluid** (see Figure 40.4). This fluid, which fills the spaces between vertebrate cells, exchanges nutrients and wastes with blood contained in microscopic vessels called capillaries. Bernard also recognized that many animals tend to maintain relatively constant conditions in their internal environment, even when the external environment changes. A pond-dwelling hydra is powerless to affect the temperature of the fluid that bathes its cells, but the human body can maintain its "internal pond" at a more-or-less constant temperature of about 37°C. The human body also can control the pH of the blood and interstitial fluid to within a tenth of a pH unit of 7.4, and it can regulate the amount of sugar in the blood so that it does not fluctuate for long from a concentration of about 90 mg of glucose per 100 mL of blood. There are times, of course, during the development of an animal when major changes in the internal environment are programmed to occur. For example, the balance of hormones in human blood is altered radically during puberty and pregnancy. Still, the stability of the internal environment is remarkable.

Today, Bernard's "constant internal milieu" is incorporated into the concept of **homeostasis,** which means "steady state," or internal balance. One of the main objectives of modern physiology, and a theme of this unit, is to study how animals maintain homeostasis. Actually, the internal environment of an animal always fluctuates slightly. Homeostasis is a dynamic state, an interplay between outside factors that tend to change the internal environment and internal control mechanisms that oppose such changes.

Regulating and Conforming

Regulating and conforming are two extremes in how animals cope with environmental fluctuations. An animal is said to be a **regulator** for a particular environmental variable if it uses internal control mechanisms to moderate internal change in the face of external fluctuation. For example, a freshwater fish is able to maintain a stable internal concentration of solutes in its blood and interstitial fluid, even though that concentration is different from the solute concentration of the water it lives in. The fish's anatomy and physiology enable it to moderate internal changes in solute concentration. (You will learn more about the mechanisms of this regulation in Chapter 44.)

An animal is said to be a **conformer** for a particular environmental variable if it allows its internal condition to vary with certain external changes. For example, many marine invertebrates, such as spider crabs of the genus *Libinia,* live in environments where the solute concentration (salinity) is relatively stable. Unlike freshwater fishes, *Libinia* does not regulate its internal solute concentration but rather conforms to the external environment.

Regulating and conforming represent extremes on a continuum, and no organism is a perfect regulator or conformer. Furthermore, an animal may maintain homeostasis while regulating some internal conditions and allowing others to conform to the environment. For example, even though a freshwater fish regulates its internal solute concentration, it allows its internal temperature to conform to the external water temperature. Next, we will explore in more detail the mechanisms that animals use in regulating certain aspects of their internal environment.

Mechanisms of Homeostasis

Mechanisms of homeostasis moderate changes in the internal environment. Any homeostatic control system has three functional components: a receptor, a control center, and an effector. The *receptor* detects a change in some variable of the animal's internal environment, such as a change in body temperature. The *control center* processes information it receives from the receptor and directs an appropriate response by the *effector.* Let's consider a nonliving example of how these components interact: the regulation of room temperature **(Figure 40.11)**. In this case, the control center, called a thermostat, also contains the receptor (a thermometer). When room temperature falls below a "set point," say 20°C, the thermostat switches on the heater (the effector). When the thermometer detects a temperature above the set point, the thermostat switches the heater off. This type of control circuit is called **negative feedback,** because a change in the variable being monitored triggers the control mechanism to counteract further change in the same direction. Owing to a time lag between reception and response, the variable drifts slightly above and below the set point, but the

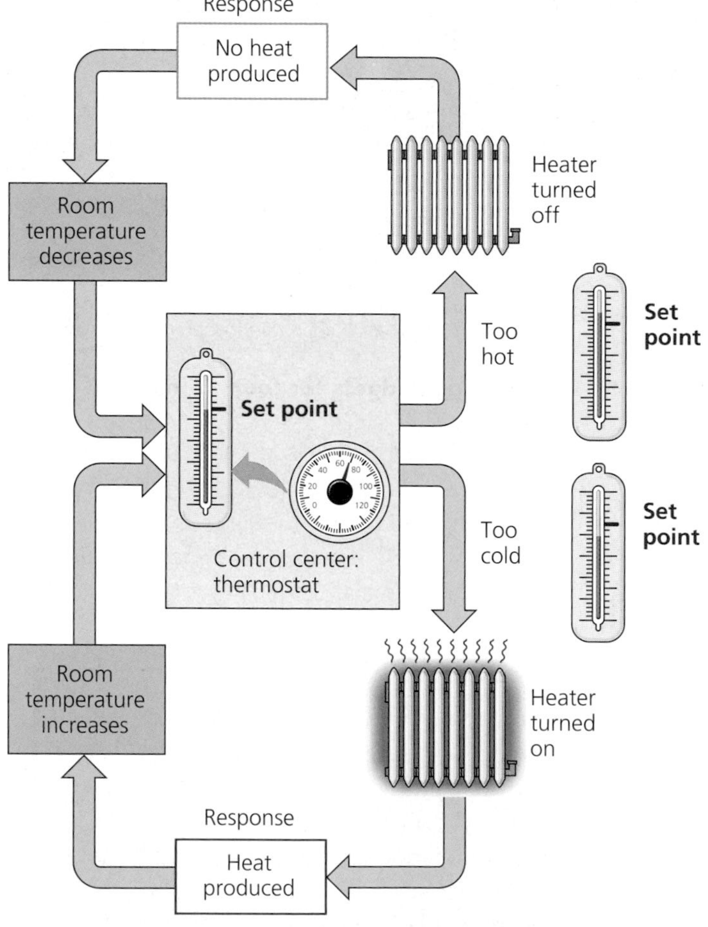

▲ **Figure 40.11 A nonliving example of negative feedback: control of room temperature.** Regulating room temperature depends on a control center that detects temperature change and activates mechanisms that reverse that change.

fluctuations are moderate. Negative-feedback mechanisms prevent small changes from becoming too large. Most homeostatic mechanisms in animals operate on this principle of negative feedback. In fact, your own body temperature is kept close to a set point of 37°C by the cooperation of several negative-feedback circuits, as you will read later.

In contrast to negative feedback, **positive feedback** involves a change in some variable that triggers mechanisms that amplify rather than reverse the change. During childbirth, for instance, the pressure of the baby's head against receptors near the opening of the uterus stimulates uterine contractions, which cause greater pressure against the uterine opening, heightening the contractions, which causes still greater pressure. Positive feedback brings childbirth to completion.

It is important not to overstate the concept of a constant internal environment. In fact, *regulated change* is essential to normal body functions. In some cases, the changes are cyclic, such as the changes in hormone levels responsible for the menstrual cycle in women (see Figure 46.13). In other cases, a regulated change is a reaction to a challenge to the body. For example, the human body reacts to certain infections by raising the set point for temperature to a slightly higher level, and the resulting fever helps fight the infection. Over the short term, homeostatic mechanisms keep body temperature close to a set point, whatever it is at that particular time. But over the longer term, homeostasis allows regulated change in the body's internal environment.

Internal regulation is expensive. Anyone who pays utility bills is aware of the energy costs for heating or cooling a home to maintain a comfortable interior temperature. Similarly, animals use a considerable portion of the energy from the food they eat to maintain favorable internal conditions. In the next section, we will explore in detail how different kinds of animals maintain relatively constant body temperatures.

Concept Check 40.4

1. Does a regulator maintain a constant internal environment? Explain.
2. Describe the difference between negative feedback and positive feedback mechanisms.

For suggested answers, see Appendix A.

Concept 40.5

Thermoregulation contributes to homeostasis and involves anatomy, physiology, and behavior

In this section, we will examine one example of how animal form and function work together in regulating the internal environment—specifically, the regulation of body temperature. Other processes involved in maintaining homeostasis will be discussed in Chapter 44.

Thermoregulation is the process by which animals maintain an internal temperature within a tolerable range. This ability is critical to survival because most biochemical and physiological processes are very sensitive to changes in body temperature. The rates of most enzyme-mediated reactions increase two- to three-fold for every 10°C temperature increase until temperature is high enough to begin to denature proteins. The properties of membranes also change with temperature. These thermal effects dramatically influence animal functioning.

Although different species of animals are adapted to different environmental temperatures, each species has an optimal temperature range. Thermoregulation helps keep body temperature within that optimal range, enabling cells to function most effectively, even as the external temperature fluctuates.

Ectotherms and Endotherms

There are important differences in how various species manage their heat budgets. One way to classify the thermal characteristics of animals is to emphasize the role of metabolic heat in determining body temperature. As you learned earlier, **ectotherms** gain most of their heat from the environment. An ectotherm has such a low metabolic rate that the amount of heat it generates is too small to have much effect on body temperature. In contrast, **endotherms** can use metabolic heat to regulate their body temperature. In a cold environment, an endotherm's high metabolic rate generates enough heat to keep its body substantially warmer than its surroundings. Many endotherms, including humans, maintain high and very stable internal temperatures even as the temperature of their surroundings fluctuates. Many ectotherms can thermoregulate by behavioral means, such as basking in the sun or seeking out shade. But in general, ectotherms tolerate greater variation in internal temperature than do endotherms (**Figure 40.12**, on the next page). Most invertebrates, fishes, amphibians, lizards, snakes, and turtles are ectotherms. Mammals, birds and a few other reptiles, some fishes, and numerous insect species are endotherms.

It is important to note that animals are *not* classified as ectotherms or endotherms based on whether they have variable or constant body temperatures, a common misconception. As mentioned earlier, it is the *source* of heat used to maintain body temperature that distinguishes ectotherms from endotherms. A different set of terms is used to imply variable or constant body temperatures. The term *poikilotherm* refers to animals whose internal temperatures vary widely, and the term *homeotherm* refers to animals that maintain relatively stable internal temperatures. However, as scientists have gained more knowledge of animal thermoregulatory mechanisms, these terms have largely fallen out of use. Many marine fishes and invertebrates, classified as poikilotherms, inhabit water

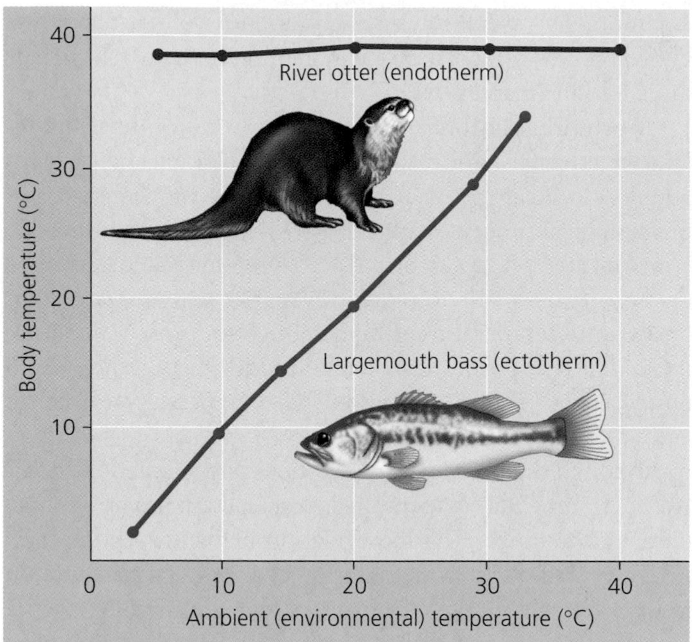

▲ **Figure 40.12 The relationship between body temperature and environmental temperature in an aquatic endotherm and ectotherm.** Using its high metabolic rate to generate heat, the river otter maintains a stable body temperature across a wide range of environmental temperatures. The largemouth bass, meanwhile, generates relatively little metabolic heat and conforms to the water temperature.

with such stable temperatures that their body temperature varies less than that of humans and other mammals. Furthermore, some mammals that were classified as homeotherms experience great variation in internal temperature. For example, a chipmunk sustains a high body temperature while it is active, but its temperature drops as hibernation begins. Because of such exceptions, the terms *ectotherm* and *endotherm* are generally more useful.

Another common misconception is the idea that ectotherms are "cold-blooded" and endotherms are "warm-blooded." Ectotherms do not necessarily have low body temperatures. In fact, when sitting in the sun, many ectothermic lizards have higher body temperatures than mammals. Thus, most biologists avoid the familiar terms *cold-blooded* and *warm-blooded* because they are so often misleading. It is also important to note that ectothermy and endothermy are not mutually exclusive thermoregulatory strategies. For example, a bird is an endotherm, but it may warm itself in the sun on a cold morning, much as an ectothermic lizard does.

Endothermy has several important advantages. Being able to generate a large amount of heat metabolically, along with other biochemical and physiological adaptations associated with endothermy (such as elaborate circulatory and respiratory systems), enables endotherms to perform vigorous activity for much longer than is possible for most ectotherms (see Figure 40.9). Sustained intense activity, such as long-distance running or powered (flapping) flight, is usually feasible only for animals with an endothermic way of life. Endothermy also solves certain thermal problems of living on land, enabling terrestrial animals to maintain stable body temperatures in the face of environmental temperature fluctuations that are generally more severe than in aquatic habitats. For example, few ectotherms are active in the below-freezing weather that prevails during winter over much of Earth's surface, but many endotherms function very well in these conditions. Most of the time, endothermic vertebrates—birds and mammals—are warmer than their surroundings, but these animals also have mechanisms for cooling the body in a hot environment, which enables them to withstand heat loads that are intolerable for most ectotherms. Endotherms are better buffered against external temperature fluctuations compared to ectotherms, but keep in mind that ectotherms can usually tolerate larger fluctuations in their internal temperatures.

Being endothermic is liberating, but it is also energetically expensive. For example, at 20°C, a human at rest has a metabolic rate of 1,300 to 1,800 kcal per day (BMR). In contrast, a resting ectotherm of similar weight, such as an American alligator, has a metabolic rate of only about 60 kcal per day at 20°C (SMR). Thus, endotherms generally need to consume much more food than ectotherms of equivalent size—a serious disadvantage for endotherms if food supplies are limited. For this and other reasons, ectothermy is an extremely effective and successful strategy in most of Earth's environments, as shown by the abundance and diversity of ectothermic animals.

Modes of Heat Exchange

Whether it is an ectotherm or an endotherm, an organism, like any object, exchanges heat by four physical processes: conduction, convection, radiation, and evaporation. **Figure 40.13** distinguishes these processes, which account for the flow of heat within an organism and between an organism and its external environment. Note that heat is always transferred from an object of higher temperature to one of lower temperature.

Balancing Heat Loss and Gain

For endotherms and for those ectotherms that thermoregulate, the essence of thermoregulation is managing the heat budget so that rates of heat gain are equal to rates of heat loss. If the heat budget is unbalanced, the animal becomes either warmer or colder. Five general categories of adaptations help animals thermoregulate.

Insulation

A major thermoregulatory adaptation in mammals and birds is insulation (hair, feathers, or fat layers), which reduces the flow of heat between an animal and its environment and lowers the energy cost of keeping warm. In mammals, the insulating material is associated with the **integumentary system,**

Radiation is the emission of electromagnetic waves by all objects warmer than absolute zero. Radiation can transfer heat between objects that are not in direct contact, as when a lizard absorbs heat radiating from the sun.

Evaporation is the removal of heat from the surface of a liquid that is losing some of its molecules as gas. Evaporation of water from a lizard's moist surfaces that are exposed to the environment has a strong cooling effect.

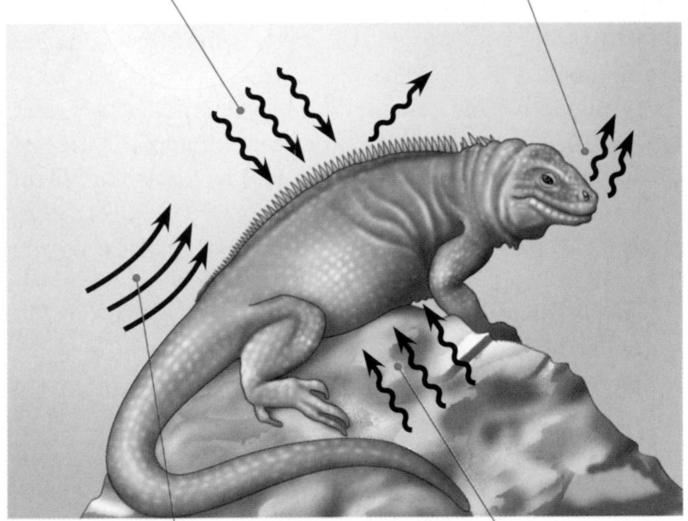

Convection is the transfer of heat by the movement of air or liquid past a surface, as when a breeze contributes to heat loss from a lizard's dry skin, or blood moves heat from the body core to the extremities.

Conduction is the direct transfer of thermal motion (heat) between molecules of objects in direct contact with each other, as when a lizard sits on a hot rock.

▲ **Figure 40.13 Heat exchange between an organism and its environment.**

▼ **Figure 40.14 Mammalian integumentary system.** The skin and its derivatives serve important functions in mammals, including protection and thermoregulation.

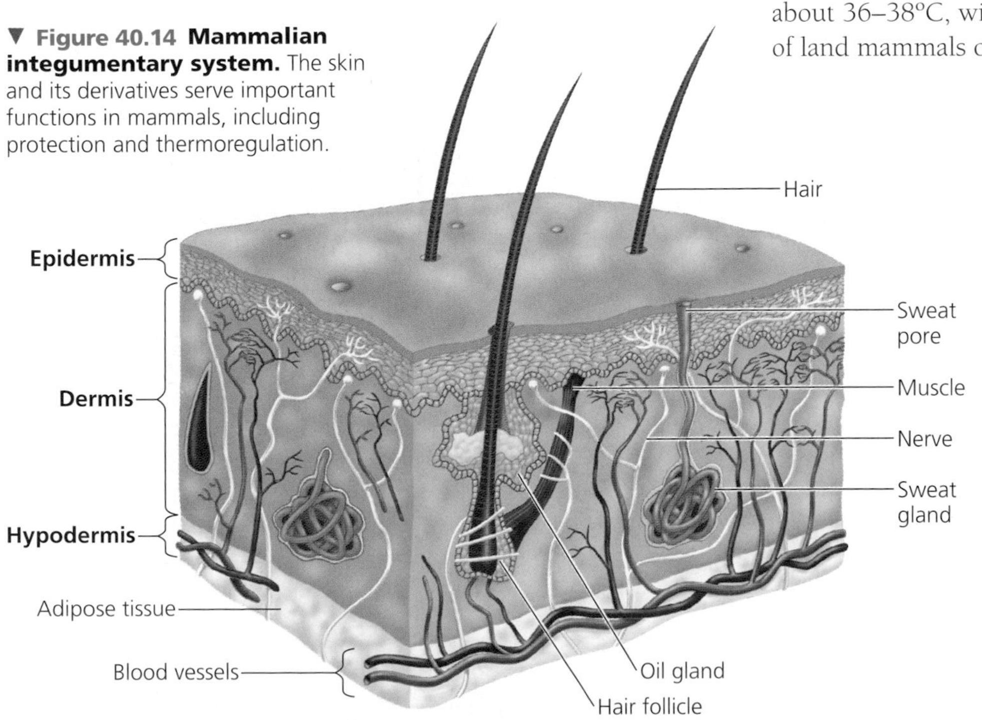

the outer covering of the body, consisting of the skin, hair, and nails (claws or hooves in some species). Skin is a key organ of the integumentary system. In addition to functioning as a thermoregulatory organ by housing nerves, sweat glands, blood vessels, and hair follicles, the skin protects internal body parts from mechanical injury, infection, and drying out. The skin consists of two layers, the epidermis and the dermis, underlain by a tissue layer called the hypodermis **(Figure 40.14)**. The epidermis is the outermost layer of skin and is composed mostly of dead epithelial cells that continually flake and fall off. New cells pushing up from lower layers replace the cells that are lost. The dermis supports the epidermis and contains hair follicles, oil and sweat glands, muscles, nerves, and blood vessels. The hypodermis contains adipose tissue, which includes fat-storing cells and blood vessels. Adipose tissue provides varying degrees of insulation, depending on the species.

The insulating power of a layer of fur or feathers mainly depends on how much still air the layer traps. (Hair loses most of its insulating power when wet.) Most land mammals and birds react to cold by raising their fur or feathers, thereby trapping a thicker layer of air. Humans rely more on a layer of fat just beneath the skin as insulation (see Figure 40.14); goose bumps are a vestige of hair raising inherited from our furry ancestors.

Marine mammals, such as whales and seals, have a very thick layer of insulating fat called blubber just under their skin. Marine mammals swim in water colder than their body core temperature, and many species spend at least part of the year in nearly freezing polar seas. The transfer of heat to water occurs 50 to 100 times more rapidly than heat transfer to air, and the skin temperature of a marine mammal is close to water temperature. Even so, the blubber insulation is so effective that marine mammals maintain body core temperatures of about 36–38°C, with metabolic rates about the same as those of land mammals of similar size.

Circulatory Adaptations

Many endotherms and some ectotherms can alter the amount of blood (and hence heat) flowing between the body core and the skin. Elevated blood flow in the skin normally results from **vasodilation,** an increase in the diameter of superficial blood vessels (those near the body surface) triggered by nerve signals that relax the muscles of the vessel walls. In endotherms, vasodilation usually warms the skin, increasing the transfer of body heat to a cool environment by radiation, conduction, and convection (see Figure 40.13). The reverse process, **vasoconstriction,** reduces blood flow and heat transfer by decreasing the diameter of superficial vessels.

Another circulatory adaptation is an arrangement of blood vessels called a **countercurrent heat exchanger** that is important for reducing heat loss in many endotherms, including marine mammals and birds. **Figure 40.15** explores two examples of countercurrent heat exchangers. In some species, blood can either go through the heat exchanger or bypass it by way of other blood vessels. In this way, the relative amount of blood that flows through the two different paths may vary, adjusting the rate of heat loss as an animal's physiological state or environment changes.

Unlike most fishes, which are thermoconformers with internal body temperatures usually within 1–2°C of the surrounding water temperature, some specialized endothermic bony fishes and sharks have circulatory adaptations that retain metabolic heat in the body. These animals include large, powerful swimmers such as bluefin tuna and swordfish, as well as the great white shark. Large arteries convey most of the cold blood from the gills to tissues just under the skin. Branches deliver blood to the deep muscles, where the small vessels are arranged in a countercurrent heat exchanger **(Figure 40.16)**. Endothermy enables the vigorous, sustained activity that is characteristic of these animals by keeping the main swimming muscles several degrees warmer than the tissues near the animal's surface, which are about the same temperature as the surrounding water.

Some reptiles also have physiological adaptations that regulate heat loss. For example, in the marine iguana, which inhabits the Galápagos Islands, body heat is conserved by vasoconstriction of superficial blood vessels, routing more blood to the central core of the body when the animal is swimming in the cold ocean.

Many endothermic insects (bumblebees, honeybees, and some moths) have a countercurrent heat exchanger that helps maintain a high temperature in the thorax, where the flight muscles are located. For example, the heat exchanger keeps the thorax of certain winter-active moths at about 30°C during flight, even on cold, snowy nights when the external temperature may be subfreezing **(Figure 40.17)**. In contrast, insects flying in hot weather run the risk of overheating because of the large amount of heat produced by working flight muscles. In some species, the countercurrent mechanism can be "shut down," allowing muscle-produced heat to be lost from the thorax to the abdomen, and from there to the environment. A bumblebee queen incubates her eggs this way: She generates

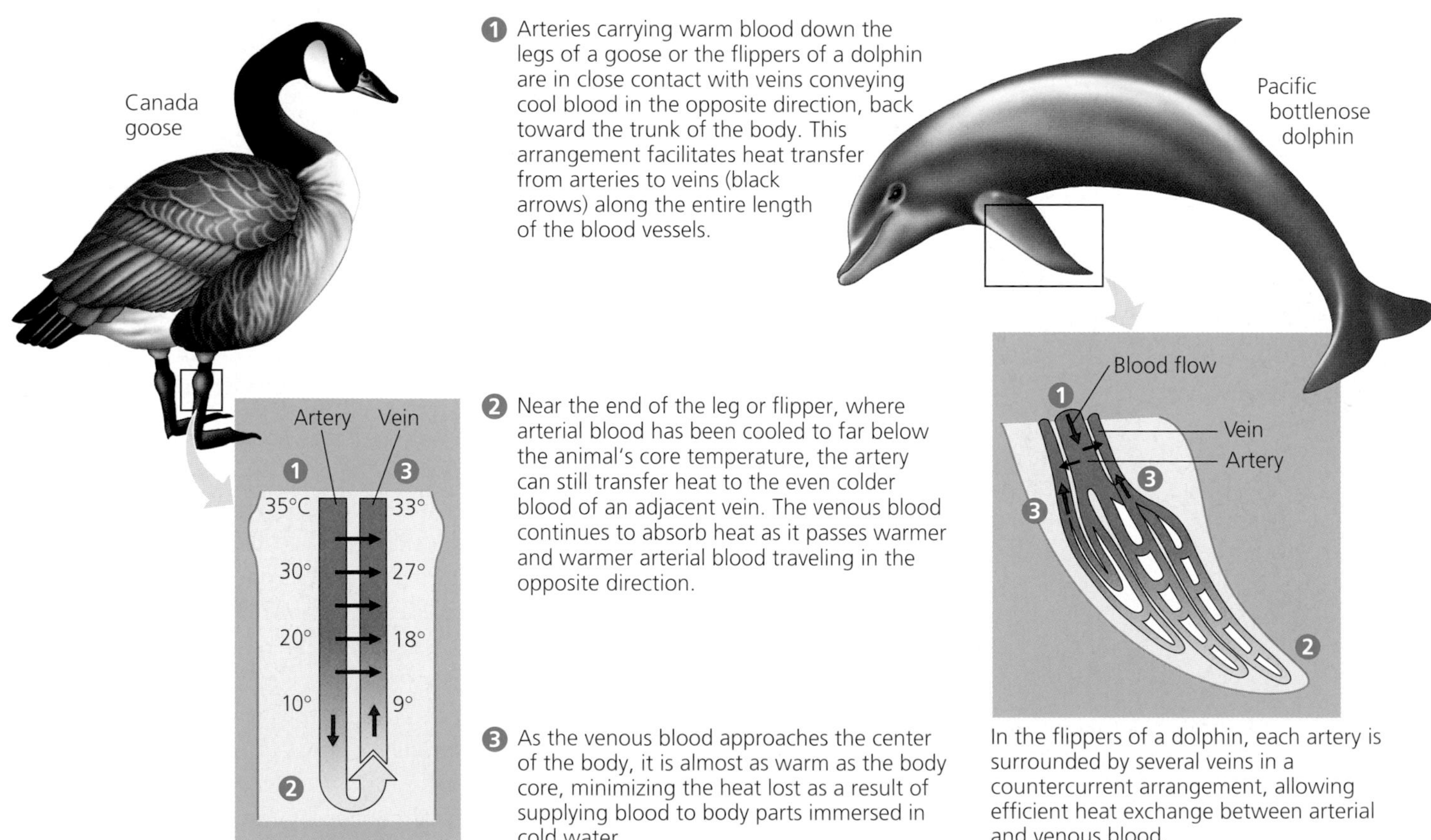

Canada goose
Artery
Vein
1
3
35°C
33°
30°
27°
20°
18°
10°
9°
2
1 Arteries carrying warm blood down the legs of a goose or the flippers of a dolphin are in close contact with veins conveying cool blood in the opposite direction, back toward the trunk of the body. This arrangement facilitates heat transfer from arteries to veins (black arrows) along the entire length of the blood vessels.
2 Near the end of the leg or flipper, where arterial blood has been cooled to far below the animal's core temperature, the artery can still transfer heat to the even colder blood of an adjacent vein. The venous blood continues to absorb heat as it passes warmer and warmer arterial blood traveling in the opposite direction.
3 As the venous blood approaches the center of the body, it is almost as warm as the body core, minimizing the heat lost as a result of supplying blood to body parts immersed in cold water.
Pacific bottlenose dolphin
Blood flow
Vein
Artery
In the flippers of a dolphin, each artery is surrounded by several veins in a countercurrent arrangement, allowing efficient heat exchange between arterial and venous blood.

▲ **Figure 40.15 Countercurrent heat exchangers.** A countercurrent heat exchanger traps heat in the body core, thus reducing heat loss from the extremities, which are often immersed in cold water or in contact with ice or snow. In essence, heat in the arterial blood emerging from the body core is transferred directly to the returning venous blood instead of being lost to the environment.

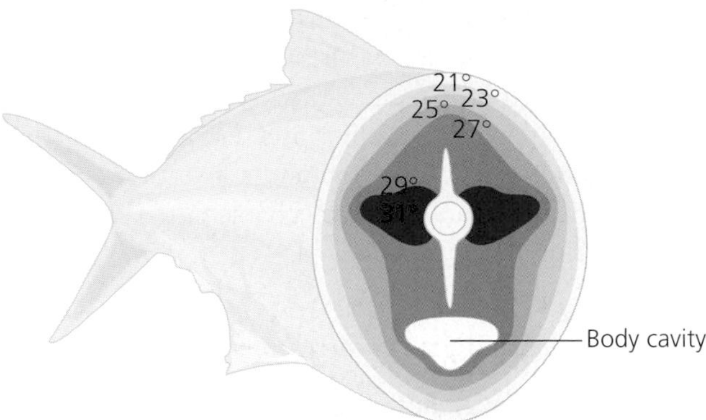

(a) Bluefin tuna. Unlike most fishes, the bluefin tuna maintains temperatures in its main swimming muscles that are much higher than the temperature of the surrounding water (colors indicate swimming muscles cut in transverse section). These temperatures were recorded for a tuna in 19°C water.

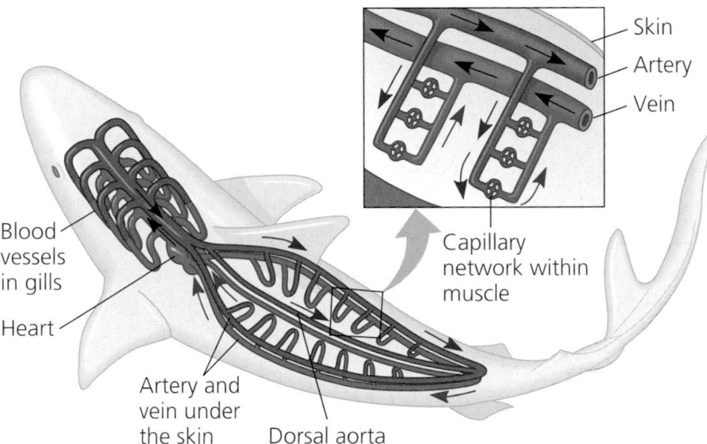

(b) Great white shark. Like the bluefin tuna, the great white shark has a countercurrent heat exchanger in its swimming muscles that reduces the loss of metabolic heat. All bony fishes and sharks lose heat to the surrounding water when their blood passes through the gills. However, endothermic sharks have a small dorsal aorta, and as a result, relatively little cold blood from the gills goes directly to the core of the body. Instead, most of the blood leaving the gills is conveyed via large arteries just under the skin, keeping cool blood away from the body core. As shown in the enlargement, small arteries carrying cool blood inward from the large arteries under the skin are paralleled by small veins carrying warm blood outward from the body core. This countercurrent flow retains heat in the muscles.

▲ **Figure 40.16 Thermoregulation in large, active bony fishes and sharks.**

► **Figure 40.17 Internal temperature in the winter moth.** This infrared map shows the moth's heat distribution immediately after a flight. Red in the thorax region indicates the highest temperature. Moving outward, the variously colored zones correspond to regions of progressively lower body temperatures.

▲ **Figure 40.18 A terrestrial mammal bathing, an adaptation that enhances evaporative cooling.**

heat by shivering her flight muscles and then transfers the heat to her abdomen, which she presses against her eggs.

Cooling by Evaporative Heat Loss

Many mammals and birds live in places where thermoregulation requires cooling as well as warming. If environmental temperature is above body temperature, animals gain heat from the environment as well as from metabolism, and evaporation is the only way to keep body temperature from rising rapidly. Terrestrial animals lose water by evaporation across the skin and when they breathe. Water absorbs considerable heat when it evaporates; this heat is carried away from the body surface with the water vapor.

Some animals have adaptations that can greatly augment this cooling effect. Panting is important in birds and many mammals. Some birds have a pouch richly supplied with blood vessels in the floor of the mouth; fluttering the pouch increases evaporation. Pigeons, for example, can use evaporative cooling to keep body temperature close to 40°C in air temperatures as high as 60°C, as long as they have sufficient water. Sweating or bathing moistens the skin and enhances evaporative cooling **(Figure 40.18)**. Many terrestrial mammals have sweat glands controlled by the nervous system (see Figure 40.14). Other mechanisms that promote evaporative cooling include spreading saliva on body surfaces, an adaptation of some kangaroos and rodents for combating severe heat stress. Some species of amphibians, such as bullfrogs, can vary the amount of mucus they secrete from their surface, a response that regulates evaporative cooling.

Behavioral Responses

Both endotherms and ectotherms use behavioral responses to control body temperature. Many ectotherms maintain a nearly constant body temperature through relatively simple behaviors.

More extreme behavioral adaptations in some animals include hibernation or migration to a more suitable climate.

All amphibians and most reptiles other than birds are ectothermic. Therefore, these organisms control body temperature mainly by behavior. The optimal temperature range for amphibians varies substantially with the species. For example, certain closely related species of salamanders have average body temperatures ranging from 7° to 25°C. When exposed to air, most amphibians lose heat rapidly by evaporation from their moist body surfaces, making it difficult to keep sufficiently warm. However, by moving to a location where solar heat is available, an amphibian can maintain a satisfactory body temperature. And when the surroundings are too warm, amphibians seek cooler microenvironments, such as shaded areas.

Like amphibians, reptiles other than birds thermoregulate mainly by behavior. When cold, they seek warm places, orienting themselves toward heat sources and expanding the body surface exposed to a heat source. When hot, they move to cool areas or turn in another direction. Many reptiles keep their body temperatures very stable over the course of a day by shuttling back and forth between warm and cool spots.

Many terrestrial invertebrates can adjust internal temperature by the same behavioral mechanisms used by vertebrate ectotherms. The desert locust, for example, must reach a certain temperature to become active, and on cold days it orients in a direction that maximizes the absorption of sunlight. Other terrestrial invertebrates have certain postures that enable them to maximize or minimize their absorption of heat from the sun **(Figure 40.19)**.

Honeybees use a thermoregulatory mechanism that depends on social behavior. In cold weather, they increase heat production and huddle together, thereby retaining heat. They maintain a relatively constant temperature by changing the density of the huddling. Individuals move between the cooler outer edges of the cluster and the warmer center, thus circulating and distributing the heat. Even when huddling, honeybees must expend considerable energy to keep warm during long periods of cold weather, and this is the main function of storing large quantities of fuel in the hive in the form of honey. Honeybees also control the temperature of their hive by transporting water to it in hot weather and fanning with their wings, which promotes evaporation and convection. Thus, a honeybee colony uses many of the mechanisms of thermoregulation seen in single organisms.

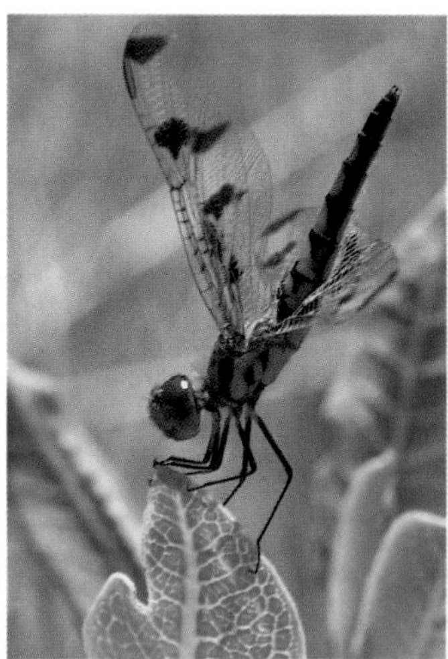

◀ **Figure 40.19 Thermoregulatory behavior in a dragonfly.** This dragonfly's "obelisk" posture is an adaptation that minimizes the amount of body surface exposed to the sun. This posture helps reduce heat gain by radiation.

Adjusting Metabolic Heat Production

Because endotherms generally maintain body temperatures considerably warmer than the environment, they must counteract constant heat loss. Endotherms can vary heat production to match changing rates of heat loss. For example, heat production is increased by such muscle activity as moving or shivering. In some mammals, certain hormones can cause mitochondria to increase their metabolic activity and produce heat instead of ATP. This **nonshivering thermogenesis (NST)** takes place throughout the body, but some mammals also have a tissue called **brown fat** in the neck and between the shoulders that is specialized for rapid heat production. Through shivering and NST, mammals and birds in cold environments can increase their metabolic heat production by as much as five to ten times the minimal levels that occur in warm conditions. For example, small birds called chickadees, which weigh only 20 g, can remain active and hold body temperature nearly constant at 40°C in environmental temperatures as low as −40°C—as long as they have enough food to supply the large amount of energy necessary for heat production.

A few large reptiles become endothermic in particular circumstances. For example, female pythons that are incubating eggs increase their metabolic rate by shivering, generating enough heat to keep their body (and egg) temperatures 5–7°C warmer than the surrounding air for weeks at a time. This temporary endothermy consumes considerable energy. Researchers continue to debate whether certain groups of dinosaurs were endothermic (see Chapter 34).

As mentioned earlier, many species of flying insects, such as bees and moths, are endothermic—the smallest of all endotherms. The capacity of such endothermic insects to elevate body temperature depends on powerful flight muscles, which generate large amounts of heat when operating. Many endothermic insects use shivering to warm up before taking off. They contract their flight muscles in synchrony, so that only slight wing movements occur but considerable heat is produced. Chemical reactions, and hence cellular respiration, speed up in the warmed-up flight "motors," enabling these insects to fly even on cold days or at night **(Figure 40.20)**.

Feedback Mechanisms in Thermoregulation

The regulation of body temperature in humans and other mammals is a complex system facilitated by feedback mecha-

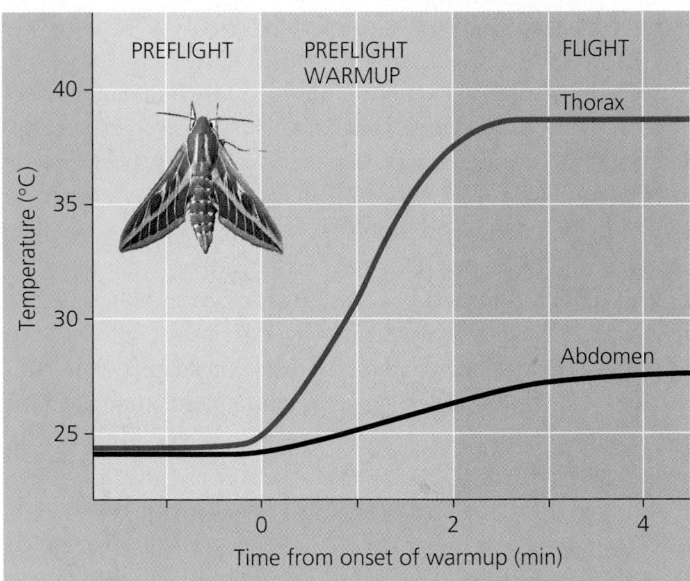

▲ **Figure 40.20 Preflight warmup in the hawkmoth.** The hawkmoth (*Manduca sexta*) is one of many insect species that use a shivering-like mechanism for preflight warmup of thoracic flight muscles. Warming up helps these muscles produce enough power to let the animal take off. Once airborne, flight muscle activity maintains a high thoracic temperature.

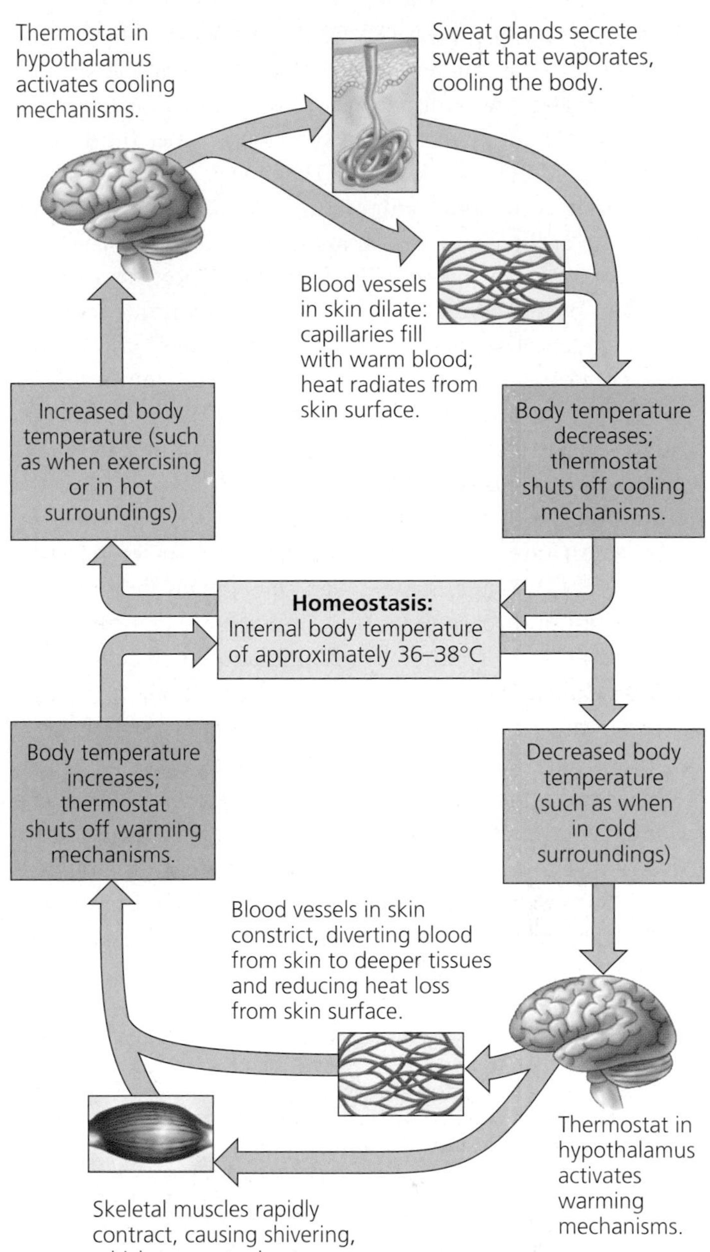

▲ **Figure 40.21 The thermostat function of the hypothalamus in human thermoregulation.**

nisms (see Figure 40.11). Nerve cells that control thermoregulation, as well as those that control many other aspects of homeostasis, are concentrated in a region of the brain called the hypothalamus (discussed in detail in Chapter 48). The hypothalamus contains a group of nerve cells that functions as a thermostat, responding to changes in body temperature above or below a set point (actually above or below a normal range) by activating mechanisms that promote heat loss or gain **(Figure 40.21)**. Nerve cells that sense temperature are in the skin, in the hypothalamus itself, and in several other body regions. Warm receptors signal the hypothalamic thermostat when temperatures increase; cold receptors signal temperature decrease. At body temperatures below the normal range, the thermostat inhibits heat loss mechanisms and activates heat-saving ones such as vasoconstriction of superficial vessels and erection of fur, while stimulating heat-generating mechanisms (shivering and nonshivering thermogenesis). In response to elevated body temperature, the thermostat shuts down heat retention mechanisms and promotes body cooling by vasodilation, sweating, or panting. The thermostat can also respond to external temperature (sensed as skin temperature) even without changes in body core temperature.

Adjustment to Changing Temperatures

Many animals can adjust to a new range of environmental temperatures over a period of days or weeks, a physiological response called **acclimatization.** Both ectotherms and endotherms acclimatize, but in different ways. In birds and mammals, acclimatization often includes adjusting the amount of insulation—by growing a thicker coat of fur in the winter and shedding it in the summer, for example—and sometimes varying the capacity for metabolic heat production in different seasons. These changes help endotherms keep a constant body temperature in both warm and cold seasons. Acclimatization in ectotherms involves compensating for changes in temperature. These adjustments can strongly affect physiology and temperature tolerance. For example, summer-acclimatized bullhead catfish can survive water temperatures up to 36°C but cannot function in cold water; after winter acclimatization, they can easily tolerate cold water, but a temperature above 28°C is lethal.

Acclimatization responses in ectotherms often include adjustments at the cellular level. Cells may increase the production of certain enzymes or produce variants of enzymes that have the same function but different optimal temperatures. Membranes may also change the proportions of saturated and unsaturated lipids they contain, which helps keep membranes fluid at different temperatures (see Figure 7.5). Some ectotherms that experience subzero body temperatures protect themselves by producing "antifreeze" compounds (cryoprotectants) that prevent ice formation in the cells. In arctic regions or on cold mountain peaks, cryoprotectants in the body fluids let overwintering ectotherms, such as some frogs and many arthropods and their eggs, withstand body temperatures considerably below zero. Cryoprotectants are also found in certain species of fishes from arctic and antarctic seas, where water temperatures can be as cold as −1.8°C, well below the freezing point of unprotected body fluids (about −0.7°C).

Cells can often make rapid adjustments to temperature changes. For example, mammalian cells grown in laboratory cultures respond to a marked increase in temperature and to other forms of severe stress by accumulating molecules called **stress-induced proteins,** including **heat-shock proteins.** Within minutes of being "shocked" by a rapid change in temperature from 37°C to about 43°C, cultured mammalian cells begin synthesizing heat-shock proteins. These molecules help maintain the integrity of other proteins that would be denatured by severe heat. Found in bacteria, yeasts, and plant cells as well as in animals, stress-induced proteins help prevent cell death when an organism is challenged by severe changes in the cellular environment.

Torpor and Energy Conservation

Despite their many adaptations for homeostasis, animals may occasionally encounter conditions that severely challenge their abilities to balance heat, energy, and materials budgets. For example, at certain seasons of the year (or certain times of day), temperatures may be extremely hot or cold, or food may be unavailable. An adaptation that enables animals to save energy while avoiding difficult and dangerous conditions is **torpor,** a physiological state in which activity is low and metabolism decreases.

Hibernation is long-term torpor that is an adaptation to winter cold and food scarcity. When vertebrate endotherms (birds and mammals) enter torpor or hibernation, their body temperatures decline—in effect, their body's thermostat is turned down. The temperature reduction may be dramatic: Some hibernating mammals cool to as low as 1–2°C, and a few even drop slightly below 0°C in a supercooled (unfrozen) state. The resulting energy savings due to lower metabolic rate and less heat production are huge; metabolic rates during hibernation can be several hundred times lower than if the animal attempted to maintain normal body temperatures of 36–38°C. This allows hibernators to survive for very long periods on limited supplies of energy stored in the body tissues or as food cached in a burrow.

Certain ground squirrels are favorite research models for biologists studying the physiology of hibernation **(Figure 40.22).** For example, a Belding's ground squirrel (*Spermophilus beldingi*) living in the high mountains of California is active only during spring and summer, when it maintains a body

▶ Figure 40.22 Body temperature and metabolism during hibernation in Belding's ground squirrels.

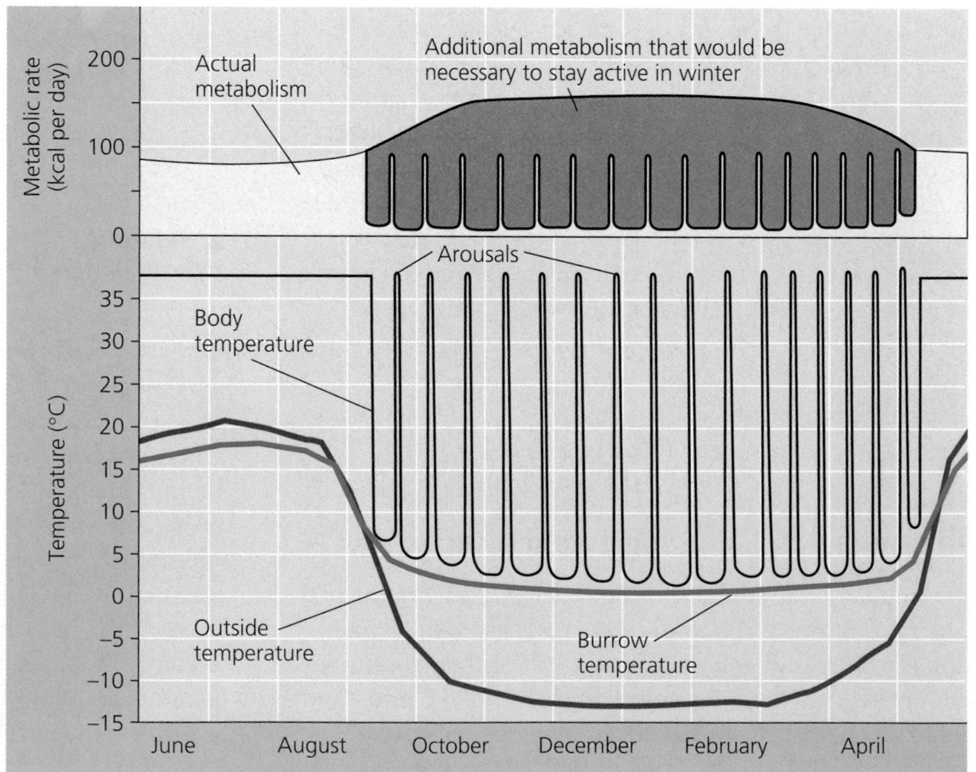

temperature of about 37°C and a metabolic rate of about 85 kcal per day. In September, the squirrel retreats to a safe burrow where it spends the next eight months hibernating. For most of the hibernation season, the squirrel's body temperature is only slightly above burrow temperature (which may be close to freezing), and its metabolic rate is extremely low (see Figure 40.22). Every week or two it arouses for a few hours, using metabolic heat to warm up to about 37°C (these periodic arousals may be needed for maintenance functions that require high body temperature). In late spring, when outside temperature is climbing, the squirrel resumes normal endothermy. By hibernating, Belding's ground squirrels avoid severe cold and greatly reduce the amount of energy they need to survive the winter, when their normal food of grasses and seeds is not available. Instead of having to spend 150 kcal per day to maintain normal body temperatures in winter weather, a squirrel in its burrow spends an average of only 5–8 kcal per day and can live on stored fat—without eating—for the entire hibernation season.

Estivation, or summer torpor, also characterized by slow metabolism and inactivity, enables animals to survive long periods of high temperatures and scarce water supplies. Hibernation and estivation are often triggered by seasonal changes in the length of daylight. As the days shorten, some hibernators prepare for winter by storing food in their burrows; other species eat huge quantities of food. For example, ground squirrels double their weight in a month of gorging.

Many small mammals and birds exhibit a **daily torpor** that seems to be adapted to their feeding patterns. For instance, most bats and shrews feed at night and go into torpor during daylight hours. Chickadees and hummingbirds feed during the day and often go into torpor on cold nights; the body temperature of chickadees drops as much as 10°C at night, and that of hummingbirds can fall 25–30°C. All endotherms that use daily torpor are relatively small; when active, they have high metabolic rates and thus very high rates of energy consumption. During hours when they cannot feed, torpor enables them to survive on stored energy.

An animal's daily cycle of activity and torpor appears to be a built-in rhythm controlled by its biological clock (see Chapter 48). Even if food is made available to a shrew all day, it still goes through its daily torpor. The need for sleep in humans and the slight drop in body temperature that accompanies it may be an evolutionary remnant of a more pronounced daily torpor in our early mammalian ancestors.

Having examined some general principles of animal biology, we are now ready to compare how diverse animals perform such activities as digestion, circulation, gas exchange, excretion of wastes, reproduction, and coordination—the topics of the next several chapters.

Concept Check 40.5

1. Can ectotherms have stable body temperatures? Explain.
2. What mode of heat exchange is involved in "wind chill," when moving air feels colder than still air at the same temperature?
3. Some birds in tropical dry forests periodically go into torpor, especially in the dry season. Explain.

For suggested answers, see Appendix A.

Chapter 40 Review

Go to www.campbellbiology.com or the student CD-ROM to explore Activities, Investigations, and other interactive study aids.

SUMMARY OF KEY CONCEPTS

Concept 40.1

Physical laws and the environment constrain animal size and shape

- **Physical Laws and Animal Form (p. 821)** The ability to perform certain actions, such as flying, depends on an animal's size and shape. Evolutionary convergence reflects different species' independent adaptation to a similar environmental challenge.
- **Exchange with the Environment (pp. 821–822)** Each cell of an animal must have access to an aqueous environment. Simple two-layered sacs and flat shapes maximize exposure to the surrounding medium. More complex body plans have highly folded internal surfaces specialized for exchanging materials.

Concept 40.2

Animal form and function are correlated at all levels of organization

- Animals are composed of cells. Groups of cells with a common structure and function make up tissues. Different tissues make up organs, which together make up organ systems (p. 823).
- **Tissue Structure and Function (pp. 823–826)** Epithelial tissue covers the outside of the body and lines internal organs and cavities. Connective tissues bind and support other tissues. Muscle tissue contracts, usually in response to nerve signals. Nervous tissue transmits nerve signals throughout the animal.
  Activity *Overview of Animal Tissues*
  Activity *Epithelial Tissue*
  Activity *Connective Tissue*
  Activity *Muscle Tissue*
  Activity *Nervous Tissue*
- **Organs and Organ Systems (p. 827)** The body functions as a whole, greater than the sum of its parts, because the activities of all tissues, organs, and organ systems are coordinated.

Concept 40.3

Animals use the chemical energy in food to sustain form and function

- **Bioenergetics (pp. 828–829)** Animals obtain chemical energy from food. Most energy is stored in ATP, which powers cellular work. An animal's metabolic rate is the total amount of energy it uses in a unit of time. Metabolic rates for birds and mammals, which maintain a fairly constant body temperature using metabolic heat (the endothermic strategy), are generally higher than those for most fishes, non-bird reptiles, amphibians, and invertebrates that rely mostly on external sources of heat for maintaining body temperature (the ectothermic strategy).
- **Influences on Metabolic Rate (pp. 829–830)** Metabolic rate per gram is inversely related to body size among similar animals. Activity increases metabolic rate above the BMR (endotherms) or SMR (ectotherms) of an animal. In general, endotherms can sustain high levels of activity longer than ectotherms can.
  Investigation *How Does Temperature Affect Metabolic Rate in* Daphnia?
- **Energy Budgets (pp. 830–831)** Animals use energy for basal (or standard) metabolism, activity, homeostasis (such as temperature regulation), growth, and reproduction.

Many animals regulate their internal environment within relatively narrow limits

- Homeostasis describes an animal's internal steady state. It is a balance between external changes and the animal's internal control mechanisms that oppose the changes (pp. 831–832).
- **Regulating and Conforming (p. 832)** Animals cope with environmental fluctuations by regulating certain internal variables while allowing others to conform to external changes.
- **Mechanisms of Homeostasis (pp. 832–833)** The interstitial fluid surrounding an animal's cells is usually very different from the external environment. Homeostatic mechanisms moderate changes in the internal environment and usually involve negative feedback. These mechanisms enable regulated change.
  Activity *Regulation: Negative and Positive Feedback*

Concept 40.5

Thermoregulation contributes to homeostasis and involves anatomy, physiology, and behavior

- An animal maintains its internal temperature within a tolerable range by the process of thermoregulation (p. 833).
- **Ectotherms and Endotherms (pp. 833–834)** Most invertebrates, fishes, amphibians, and non-bird reptiles are ectotherms. Birds and mammals are endotherms. Endothermy is more energetically expensive than ectothermy. Maintaining a high body temperature enables an animal to maintain a high level of aerobic metabolism.
- **Modes of Heat Exchange (pp. 834–835)** Conduction, convection, radiation, and evaporation account for heat gain or loss.
- **Balancing Heat Loss and Gain (pp. 834–839)** Thermoregulation involves physiological and behavioral adjustments that balance heat gain and loss. Insulation, vasodilation, vasoconstriction, and countercurrent heat exchangers alter the rate of heat exchange. Panting, sweating, and bathing increase evaporation, cooling the body. Both ectotherms and endotherms adjust the rate of heat exchange with their surroundings by behavioral responses. Some animals can even adjust their rate of metabolic heat production.
- **Feedback Mechanisms in Thermoregulation (pp. 838–839)** Mammals regulate their body temperature by complex negative-feedback mechanisms that involve several organ systems, including the nervous, circulatory, and integumentary systems.
- **Adjustment to Changing Temperatures (pp. 839–840)** Acclimatization enables ectotherms and endotherms to adjust to changing environmental temperatures over days or weeks. Acclimatization may involve cellular adjustments or, in the case of birds and mammals, adjustments of insulation and metabolic heat production.
- **Torpor and Energy Conservation (pp. 840–841)** Torpor conserves energy during environmental extremes. Animals may enter torpor in winter (hibernation), summer (estivation), or during sleep periods (daily torpor). Torpor involves a decrease in metabolic rates and enables the animal to temporarily withstand unfavorable temperatures or lack of food and water.

TESTING YOUR KNOWLEDGE

Self-Quiz

1. Consider the energy budgets for a human, an elephant, a penguin, a mouse, and a python. The ______ would have the highest total annual energy expenditure, and the ______ would have the highest energy expenditure per unit mass.
   a. elephant; mouse
   b. elephant; human
   c. human; penguin
   d. mouse; python
   e. penguin; mouse

2. Which of the following structures or substances is *incorrectly* paired with a tissue?
   a. osteon—bone
   b. fibroblasts—skeletal muscle
   c. platelets—blood
   d. chondroitin sulfate—cartilage
   e. basement membrane—epithelium

3. For which of the following animals would the percent of its energy budget spent for homeostatic regulation be the largest?
   a. an amoeba in fresh water
   b. a marine jelly (an invertebrate)
   c. a snake in a temperate forest
   d. a desert insect
   e. a desert bird

4. The involuntary muscles that cause the wavelike contractions pushing food along our intestines are
   a. striated muscles.
   b. cardiac muscles.
   c. skeletal muscles.
   d. smooth muscles.
   e. intercalated muscles.

5. Which of the following statements about bioenergetics is true?
   a. An animal's metabolic rate never changes.
   b. BMR can be determined only at a specific temperature.
   c. Endotherms are warmed by metabolic heat.
   d. SMR is best measured just after an ectotherm eats.
   e. Ectotherms and endotherms use the same basic energy "strategy."

6. Compared to a smaller cell, a larger cell of the same shape has
   a. less surface area.
   b. less surface area per unit of volume.
   c. the same surface-to-volume ratio.
   d. a smaller average distance between its mitochondria and the external source of oxygen.
   e. a smaller cytoplasm-to-nucleus ratio.

7. Which of the following is *not* an adaptation for reducing the rate of heat exchange between an animal and its environment?
   a. feathers or fur
   b. vasoconstriction
   c. nonshivering thermogenesis
   d. countercurrent heat exchanger
   e. blubber or fat layer

8. Which of the following physiological responses is an example of *positive* feedback?
   a. An increase in the concentration of glucose in the blood stimulates the pancreas to secrete insulin, a hormone that lowers blood glucose concentration.
   b. A high concentration of CO_2 in the blood causes deeper, more rapid breathing, which expels CO_2.
   c. Stimulation of a nerve cell causes sodium ions to leak into the cell, and the sodium influx triggers the inward leaking of even more sodium.
   d. The body's production of red blood cells, which transport oxygen from the lungs to other organs, is stimulated by a low concentration of oxygen.
   e. The pituitary gland secretes a hormone called TSH, which stimulates the thyroid gland to secrete another hormone called thyroxine; a high concentration of thyroxine suppresses the pituitary's secretion of TSH.

9. An animal's inputs of energy and materials would exceed its outputs
   a. if the animal is an endotherm, which must always take in more energy because of its high metabolic rate.
   b. if it is actively foraging for food.
   c. if it is hibernating.
   d. if it is growing and increasing its biomass.
   e. never—homeostasis makes these energy and material budgets always balance.

10. You are studying a large tropical reptile that has a high and relatively stable body temperature. How would you determine whether this animal is an endotherm or an ectotherm?
   a. You know from its high and stable body temperature that it must be an endotherm.
   b. You know that it is an ectotherm because it is not a bird or mammal.
   c. You subject this reptile to various temperatures in the lab and find that its body temperature and metabolic rate change with the ambient temperature. You conclude that it is an ectotherm.
   d. You note that its environment has a high and stable temperature. Because its body temperature matches the environmental temperature, you conclude that it is an ectotherm.
   e. You measure the metabolic rate of the reptile, and because it is higher than that of a related species that lives in temperate forests, you conclude that this reptile is an endotherm and its relative is an ectotherm.

For Self-Quiz answers, see Appendix A.

Go to the website or CD-ROM for more quiz questions.

Evolution Connection

The biologist C. Bergmann noted that mammals and birds living at higher latitudes are on average larger and bulkier than related species found at lower latitudes. This observation, sometimes called Bergmann's rule, has exceptions, but appears to hold true in most cases. Suggest an evolutionary hypothesis for this "rule."

Scientific Inquiry

Eastern tent caterpillars (*Malacosoma americanum*) live in sizable groups in silk nests, or tents, which they construct in cherry trees. They are among the first insects to become active in the spring, emerging very early in the season—a time when daily temperature fluctuates from freezing to very hot. Observing a colony over the course of a day, you observe striking differences in group behavior: Early in the morning, the black caterpillars rest in a tightly packed group on the east-facing surface of the tent. In midafternoon, the group is found on the tent undersurface, each caterpillar individually hanging from the tent by just a few of its legs. Propose a hypothesis to explain this behavior. How could you test your hypothesis?

Investigation ***How Does Temperature Affect Metabolic Rate in Daphnia?***

Science, Technology, and Society

Medical researchers are investigating the possibilities of artificial substitutes for various human tissues. Examples are a liquid that could serve as "artificial blood" and a fabric that could temporarily serve as artificial skin for victims of serious burns. In what other situations might artificial blood or skin be useful? What characteristics would these substitutes need in order to function effectively in the body? Why do real tissues work better? Why not use the real things if they work better? Can you think of other artificial tissues that might be useful? What problems do you anticipate in developing and applying them?

41 Animal Nutrition

▲ **Figure 41.1 Modern humans foraging at a farmer's market.**

Key Concepts

41.1 Homeostatic mechanisms manage an animal's energy budget

41.2 An animal's diet must supply carbon skeletons and essential nutrients

41.3 The main stages of food processing are ingestion, digestion, absorption, and elimination

41.4 Each organ of the mammalian digestive system has specialized food-processing functions

41.5 Evolutionary adaptations of vertebrate digestive systems are often associated with diet

Overview

The Need to Feed

Every mealtime is a reminder that we are heterotrophs dependent on a regular supply of food **(Figure 41.1)**. All animals eat other organisms—dead or alive, whole or by the piece. In general, animals fit into one of three dietary categories. **Herbivores,** such as gorillas, cattle, hares, and many snails, eat mainly autotrophs (plants and algae). **Carnivores,** such as sharks, hawks, spiders, and snakes, eat other animals. **Omnivores** regularly consume animals as well as plant or algal matter. Omnivorous animals include cockroaches, crows, bears, raccoons, and humans, who evolved as hunters, scavengers, and gatherers.

The terms *herbivore, carnivore,* and *omnivore* represent the kinds of food an animal *usually* eats and the adaptations enabling it to obtain and process that food. However, keep in mind that most animals are opportunistic feeders, eating foods that are outside their main dietary category when their usual foods aren't available. For example, cattle and deer, which are herbivores, may occasionally eat small animals, such as insects and worms, or bird eggs, along with grass and other plants. Most carnivores obtain some nutrients from plant materials that remain in the digestive tract of the prey they eat. And all animals consume prokaryotes along with their other types of food.

Regardless of what and how an animal eats, an adequate diet must satisfy three nutritional needs: fuel (chemical energy) for all the cellular work of the body; the organic raw materials animals use in biosynthesis (carbon skeletons to make many of their own molecules); and essential nutrients, substances such as vitamins that the animal cannot make for itself from *any* raw material and therefore must obtain in food in prefabricated form. In this chapter, we will examine the nutritional requirements of animals and explore some of the diverse adaptations for obtaining and processing food. You can begin with **Figure 41.2**, which surveys four main mechanisms by which animals feed.

Concept 41.1

Homeostatic mechanisms manage an animal's energy budget

The theme of bioenergetics is integral to our study of nutrition. As we discussed in Chapter 40, the flow of energy into and out of an animal's body can be viewed as a "budget," with the production of ATP accounting for the largest fraction by far of most animals' energy budgets. This ATP powers basal (resting) metabolism, various activities, and in endotherms, thermoregulation.

Figure 41.2

Exploring Four Main Feeding Mechanisms of Animals

SUSPENSION FEEDERS

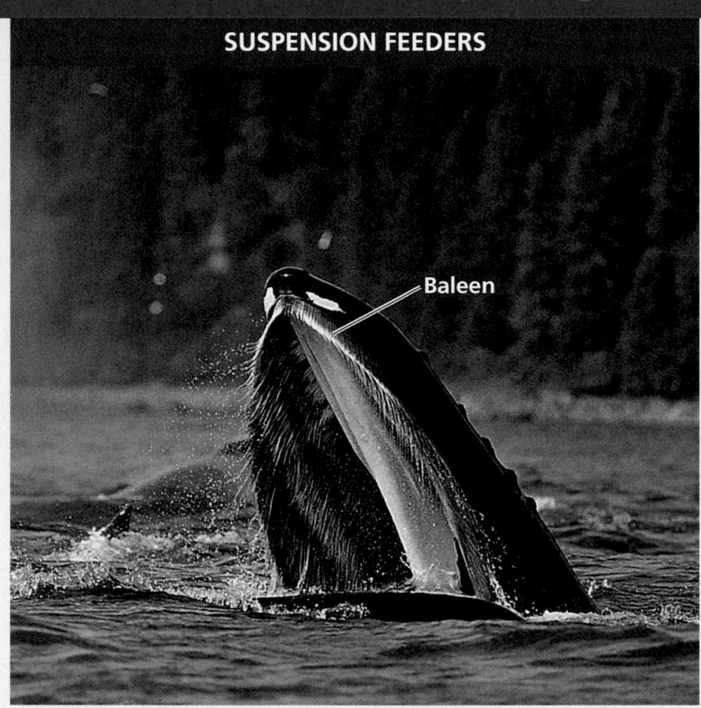

Many species of aquatic animals are **suspension feeders,** which sift small food particles from the water. For example, this humpback whale uses comblike plates called baleen attached to its upper jaw to strain small invertebrates and fish from enormous volumes of water. Among other suspension feeders, clams and oysters use their gills to trap tiny morsels. Cilia sweep the food particles in a film of mucus to the mouth.

SUBSTRATE FEEDERS

Caterpillar

Feces

Substrate feeders are animals that live in or on their food source, eating their way through the food. This leaf miner caterpillar, the larva of a moth, is eating through the soft mesophyll of an oak leaf, leaving a dark trail of feces in its wake. Other substrate feeders include maggots (fly larvae), which burrow into animal carcasses.

FLUID FEEDERS

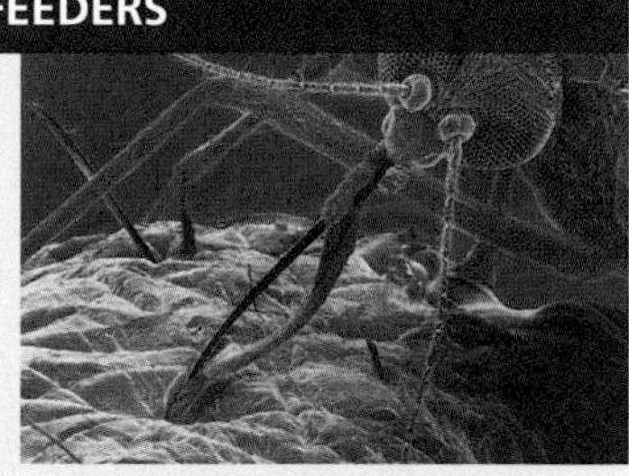

Fluid feeders suck nutrient-rich fluid from a living host. This mosquito has pierced the skin of its human host with hollow, needlelike mouthparts and is filling its digestive tract with a blood meal (colorized SEM). Similarly, aphids are fluid feeders that tap the phloem sap of plants. In contrast to such parasitic fluid feeders, which harm their hosts, some fluid feeders actually benefit their hosts. For example, hummingbirds and bees move pollen between flowers as they fluid-feed on nectar.

BULK FEEDERS

Most animals are **bulk feeders,** which eat relatively large pieces of food. Their adaptations include such diverse utensils as tentacles, pincers, claws, poisonous fangs, jaws, and teeth that kill their prey or tear off pieces of meat or vegetation. In this amazing scene, a rock python is beginning to ingest a gazelle it has captured and killed. Snakes cannot chew their food into pieces and must swallow it whole—even if the prey is much bigger than the diameter of the snake. After swallowing its prey, which may take more than an hour, the python will spend two weeks or more in a quiet place digesting its meal.

Nearly all of an animal's ATP generation is based on the oxidation of energy-rich organic molecules—carbohydrates, proteins, and fats—in cellular respiration. The monomers of any of these substances can be used as fuel, though most animals "burn" proteins only after exhausting their supply of carbohydrates and fats. Fats are especially rich in energy; the oxidation of a gram of fat liberates about twice the energy liberated from a gram of carbohydrate or protein.

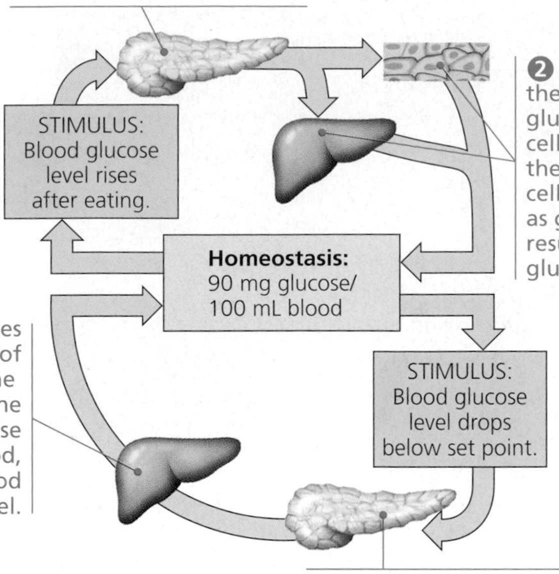

Glucose Regulation as an Example of Homeostasis

When an animal takes in more calories than it needs to produce ATP, the excess can be used for biosynthesis. If the animal isn't growing in size or reproducing, the body tends to store the surplus in energy depots. In humans, the liver and muscle cells store energy in the form of glycogen, a polymer made up of many glucose units (see Figure 5.6b). Glucose is a major fuel for cells, and its metabolism, regulated by hormone action, provides an important example of homeostasis **(Figure 41.3)**. If the body's glycogen depots are full and caloric intake still exceeds caloric expenditure, the excess is usually stored as fat.

❶ When blood glucose level rises, a gland called the pancreas secretes insulin, a hormone, into the blood.

❷ Insulin enhances the transport of glucose into body cells and stimulates the liver and muscle cells to store glucose as glycogen. As a result, blood glucose level drops.

STIMULUS: Blood glucose level rises after eating.

Homeostasis: 90 mg glucose/ 100 mL blood

STIMULUS: Blood glucose level drops below set point.

❸ When blood glucose level drops, the pancreas secretes the hormone glucagon, which opposes the effect of insulin.

❹ Glucagon promotes the breakdown of glycogen in the liver and the release of glucose into the blood, increasing blood glucose level.

▲ **Figure 41.3 Homeostatic regulation of cellular fuel.** After a meal is digested, glucose and other monomers are absorbed into the blood from the digestive tract. The human body regulates the use and storage of glucose, a major cellular fuel. Notice that these regulatory loops are examples of the feedback control we introduced in Chapter 40.

When fewer calories are taken in than are expended—perhaps because of sustained heavy exercise or lack of food—fuel is taken out of storage depots and oxidized. This may cause an animal to lose weight. The human body generally expends liver glycogen first and then draws on muscle glycogen and fat. Most healthy people, even if they are not obese, have enough stored fat to sustain them through several weeks of starvation (an average human's energy needs can be fueled by the oxidation of only 0.3 kg of fat per day).

Caloric Imbalance

Severe problems occur if the energy budget remains out of balance for long periods. If the diet of a human or other animal is chronically deficient in calories, **undernourishment** results. In this condition, the stores of glycogen and fat are used up, the body begins breaking down its own proteins for fuel, muscles begin to decrease in size, and the brain can become protein-deficient. If energy intake remains less than energy expenditures, death will eventually result. Even if a seriously undernourished person survives, some of the damage may be irreversible. Because a diet of a single staple such as rice or corn (maize) can often provide sufficient calories, undernourishment is generally common only where drought, war, or some other crisis has severely disrupted the food supply. Another cause of undernourishment is anorexia nervosa, an eating disorder in which sufferers (who are most often female) compulsively starve themselves to lose weight.

Detrimental effects also result from excessive food intake. In the United States and other affluent nations, **overnourishment**, or obesity, is an increasingly common problem. The human body hoards fat **(Figure 41.4)**; it tends to store any excess fat molecules obtained from food instead of using them for fuel. In contrast, when a person eats an excess of carbohydrates, the body tends to increase its rate of carbohydrate oxidation. Thus, the amount of fat in the diet can have a more direct effect on weight gain than the amount of dietary carbohydrates.

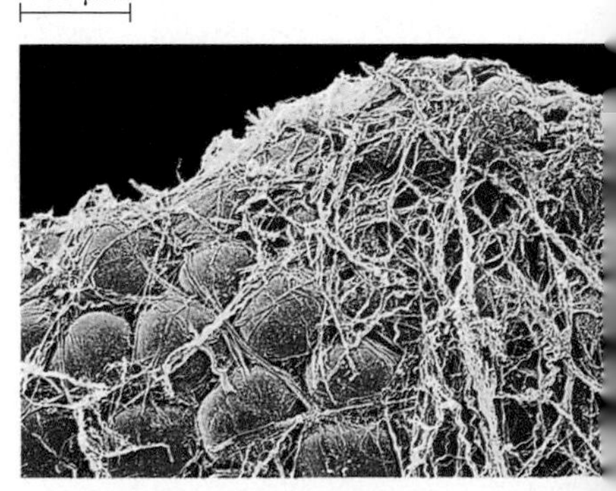

► **Figure 41.4 Fat cells from the abdomen of a human.** Strands of connective tissue (yellow) hold the fat-storing adipose cells in place (colorized SEM).

Obesity as a Human Health Problem

The World Health Organization now recognizes obesity as a major global health problem. The increased availability of fattening foods in many countries combines with more sedentary lifestyles to put excess pounds on bodies. In the United States, the percentage of obese (very overweight) people has doubled to 30% in the past two decades, and another 35% are overweight. Weight problems often begin at a young age; 15% of children and adolescents in the U.S. are overweight.

Obesity contributes to a number of health problems, including the most common type of diabetes, cancer of the colon and breasts, and cardiovascular disease that can lead to heart attacks and strokes. The overall estimate is that obesity is a factor in about 300,000 deaths per year in the U.S. alone.

The obesity epidemic has stimulated an increase in research on the causes and possible treatments for weight-control problems. Researchers have already discovered several of the mechanisms that help regulate body weight. Over the long term, these homeostatic mechanisms are feedback circuits that control the body's storage and metabolism of fat. And several of the chemical signals called hormones regulate both long-term and short-term appetite by affecting a "satiety center" in the brain **(Figure 41.5)**.

Inheritance is a major factor in obesity. Most of the weight-regulating hormones are polypeptides (proteins), and researchers have identified dozens of the genes that code for these hormones. This hereditary connection helps explain why certain people have to struggle so hard to control their weight, while others are seemingly able to eat and eat without gaining a pound.

As researchers continue to study the genes and signaling pathways that regulate weight, there is reason to be somewhat optimistic that obese people who have inherited defects in these weight-controlling mechanisms may someday be treated with a new generation of drugs. But so far, the diversity of defects in the complex systems that regulate weight have made it difficult to develop treatments that are widely useful and free from serious side effects.

The complexity of weight control in humans is evident from studies of the hormone leptin, one of the key long-term appetite regulators in mammals. Leptin is produced by adipose

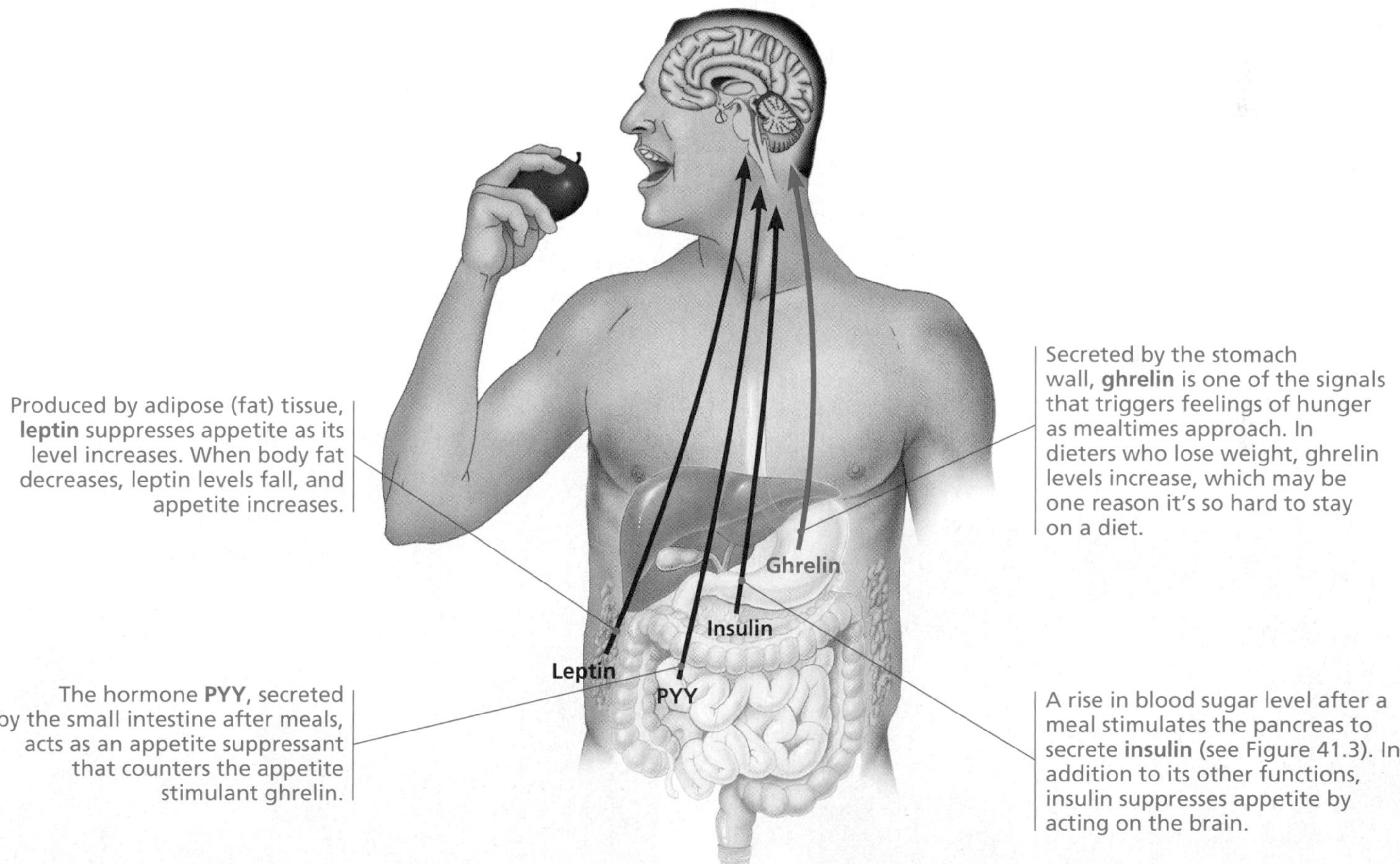

Produced by adipose (fat) tissue, **leptin** suppresses appetite as its level increases. When body fat decreases, leptin levels fall, and appetite increases.
Secreted by the stomach wall, **ghrelin** is one of the signals that triggers feelings of hunger as mealtimes approach. In dieters who lose weight, ghrelin levels increase, which may be one reason it's so hard to stay on a diet.
Ghrelin
Insulin
Leptin
PYY
The hormone **PYY**, secreted by the small intestine after meals, acts as an appetite suppressant that counters the appetite stimulant ghrelin.
A rise in blood sugar level after a meal stimulates the pancreas to secrete **insulin** (see Figure 41.3). In addition to its other functions, insulin suppresses appetite by acting on the brain.

▲ **Figure 41.5 A few of the appetite-regulating hormones.** Secreted by various organs and tissues, the hormones reach the brain via the bloodstream. The hormones act on a region of the brain that in turn controls the "satiety center," which generates the nervous impulses that make us feel either hungry or satiated ("full"). A green arrow indicates an appetite stimulant; red arrows represent appetite suppressants.

(fat) cells. As adipose tissue increases, leptin levels in the blood rise, which normally cues the brain to suppress appetite (see Figure 41.5). This is one of the feedback mechanisms that keeps most people from becoming obese in spite of access to an abundance of food. Conversely, loss of body fat decreases leptin levels, signaling the brain to increase appetite. Mice that inherit a defect in the gene for leptin become very obese **(Figure 41.6)**. Researchers found that they could treat these obese mice by injecting leptin.

The discovery of the leptin-deficiency mutation in mice made front-page news and initially generated much excitement among obesity researchers, as humans also have a leptin gene. And indeed, obese children who have inherited a mutant form of the leptin gene *do* lose weight after leptin treatments. However, relatively few obese people have such defects in leptin production. In fact, most obese humans have an abnormally *high* level of leptin, which, after all, is produced by adipose tissue. For some reason, the brain's satiety center does not respond to the high leptin levels in many obese people. One hypothesis is that in humans, perhaps in contrast to many other mammals, the leptin system's primary function may not be preventing weight gain. Rather, the main function of the leptin system may be to stimulate appetite and prevent weight loss when the hormone level declines due to a loss of body fat. This physiological nuance may be a consequence of our evolutionary history.

Obesity and Evolution

Most of us crave foods that are fatty, such as fries, burgers, cheese, and ice cream. Though fat hoarding can be a health liability today, it may actually have been advantageous in our evolutionary past. It has only been for the past few centuries that large numbers of people have had access to a reliable supply of high-calorie food. Our ancestors on the African savanna were gatherer-hunters who probably survived mainly on seeds and other plant products, a diet only occasionally supplemented by hunting game or scavenging meat from animals killed by other predators. In such a feast-or-famine existence, natural selection may have favored those individuals with a physiology that induced them to gorge on rich, fatty foods on those rare occasions when such treats were abundantly available. Such individuals with genes promoting the storage of high-energy molecules during feasts may have been more likely than their thinner friends to survive famines. So perhaps our present-day taste for fats, which contributes to the obesity epidemic, is partly an evolutionary vestige of less nutritious times. Of course, most of us today only "gather and hunt" in grocery stores and food courts.

What seems to us to be excessive body fat may actually be beneficial in certain animal species. For example, seabirds called petrels must fly long distances to find food. Most of the food petrel parents bring to their chicks is very rich in lipids. This minimizes the weight of the food the parents must carry during their long foraging trips (recall that fat has twice as many calories per gram as other fuels). However, in addition to energy, growing baby petrels need lots of protein for building new tissues. There is relatively little protein in their oily diet, so to get all the protein they need, young petrels have to consume many more calories than they burn in metabolism—and consequently, they become very obese **(Figure 41.7)**. In some petrel species, chicks at the end of the growth period weigh much more than their parents and are far too heavy to fly. These fat youngsters need to fast for several days to lose enough weight to be capable of flight. However, the fat depots in young petrels do serve an important function; the energy reserves help growing chicks survive periods when parents are unable to find enough food. As is often the case, biological oddities seem less bizarre in the context of natural selection.

▲ **Figure 41.6 A ravenous rodent.** The obese mouse on the left has a defect in a gene that normally produces leptin, an appetite-suppressing hormone.

▲ **Figure 41.7 A plump petrel.** Too heavy to fly, this petrel chick (right) will have to lose weight before it takes wing. In the meantime, its stored fat provides energy during times when its parent fails to bring enough food.

Concept Check 41.1

1. In what sense is a stable body weight a matter of caloric bookkeeping?
2. Explain how it is possible for someone to become obese even if his or her intake of dietary fat is relatively low compared to carbohydrate intake.
3. After reviewing Figure 41.5, explain how the hormones PYY and leptin complement each other in regulating body weight.

For suggested answers, see Appendix A.

Concept 41.2

An animal's diet must supply carbon skeletons and essential nutrients

In addition to providing fuel for ATP production, an animal's diet must also supply all the raw materials needed for biosynthesis. To build the complex molecules it needs to grow, maintain itself, and reproduce, an animal must obtain organic precursors (carbon skeletons) from its food. Given a source of organic carbon (such as sugar) and a source of organic nitrogen (usually amino acids from the digestion of protein), animals can fabricate a great variety of organic molecules—carbohydrates, proteins, and lipids.

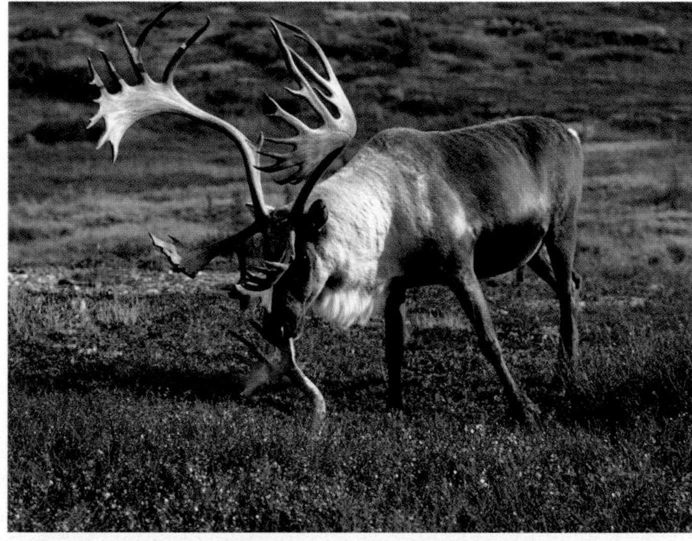

▲ **Figure 41.8 Obtaining essential nutrients.** A caribou, an arctic herbivore, chews on discarded antlers from another animal. Antlers and skeletal bones contain calcium phosphate, and osteophagia ("bone eating") is common among herbivores living where soils and plants are deficient in phosphorus. Animals require phosphorus as a mineral nutrient to make ATP, nucleic acids, phospholipids, and bones.

Besides fuel and carbon skeletons, an animal's diet must also supply **essential nutrients.** These are materials that must be obtained in preassembled form because the animal's cells cannot make them from any raw material. Some of these materials are essential for all animals, but others are needed only by certain species. For instance, ascorbic acid (vitamin C) is an essential nutrient for humans and other primates, guinea pigs, and some birds and snakes, but not for most other animals.

An animal whose diet is missing one or more essential nutrients is said to be **malnourished** (recall that *undernourished* refers to caloric deficiency). For example, cattle and other herbivorous animals may suffer mineral deficiencies if they graze on plants growing in soil lacking key minerals **(Figure 41.8)**. Malnutrition is much more common than undernutrition in human populations, and it is even possible for an overnourished (obese) individual to be malnourished.

There are four classes of essential nutrients: essential amino acids, essential fatty acids, vitamins, and minerals.

Essential Amino Acids

Animals require 20 amino acids to make proteins, and most animal species can synthesize about half of these, as long as their diet includes organic nitrogen. The remaining ones, the **essential amino acids,** must be obtained from food in prefabricated form. Eight amino acids are essential in the adult human diet (a ninth, histidine, is also essential for infants); the same amino acids are essential for most animals.

A diet that provides insufficient amounts of one or more essential amino acids causes a form of malnutrition known as protein deficiency **(Figure 41.9)**. This is the most common type of malnutrition among humans. The victims are usually children, who, if they survive infancy, are likely to be retarded in physical and perhaps mental development. In one variation

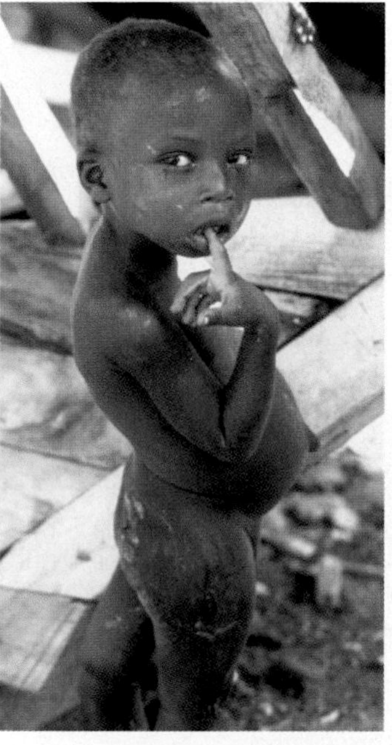

◄ **Figure 41.9 Kwashiorkor (a protein deficiency) in a Haitian boy.** The swelling (edema) of the belly is an osmotic effect: The ability of the blood to take up water from the body cavity by osmosis is reduced because of the deficiency of blood proteins (solutes).

of protein malnutrition, called *kwashiorkor,* the diet provides enough calories but is severely deficient in protein (see Figure 41.9). The syndrome's name is from a Ghanaian word meaning "rejected one," a reference to cases where the malnutrition begins when a child is weaned from the mother's milk.

The most reliable sources of essential amino acids are meat, eggs, cheese, and other animal products. The proteins in animal products are "complete," which means that they provide all the essential amino acids in their proper proportions. Most plant proteins are "incomplete," being deficient in one or more essential amino acids. Corn (maize), for example, is deficient in lysine and tryptophan. People forced by economic necessity or other circumstances to obtain nearly all their calories from corn would show symptoms of protein deficiency, as would those who were to eat only rice, wheat, or potatoes. This problem can be avoided by eating a combination of plant foods that complement one another in supplying all essential amino acids **(Figure 41.10)**. Most cultures have, by trial and error, developed balanced diets that prevent protein deficiency.

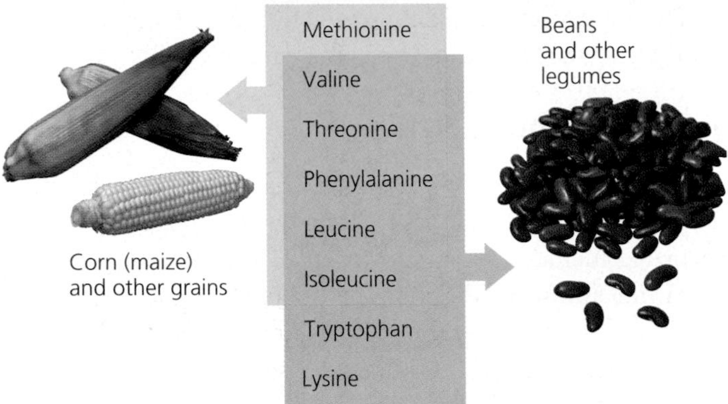

▲ **Figure 41.10 Essential amino acids from a vegetarian diet.** An adult human can obtain all essential amino acids by eating a meal of corn and beans.

Some animals have adaptations that help them through periods when their bodies demand extraordinary amounts of protein. For example, penguins can use their muscle protein as a source of amino acids to make new proteins when they replace their feathers after molting **(Figure 41.11)**.

▲ **Figure 41.11 Storing protein for growth.** Penguins, such as this Adélie from Antarctica, must make an abundance of new protein when they molt (grow new feathers). Because of the temporary loss of their insulating coat of feathers, penguins cannot swim—or feed—when molting. What is the source of amino acids for production of feather protein? Before molting, a penguin greatly increases its muscle mass. The penguin then breaks down the extra muscle protein, which supplies the amino acids for growing new feathers.

Essential Fatty Acids

Animals can synthesize most of the fatty acids they need. The **essential fatty acids**, the ones they cannot make, are certain unsaturated fatty acids (fatty acids having double bonds; see Figure 5.12). In humans, for example, linoleic acid must be present in the diet. This essential fatty acid is required to make some of the phospholipids found in membranes. The diets of humans and other animals generally furnish ample quantities of essential fatty acids, and thus deficiencies are rare.

Vitamins

Vitamins are organic molecules required in the diet in amounts that are quite small compared with the relatively large required quantities of essential amino acids and fatty acids. Tiny amounts of vitamins may suffice—from about 0.01 to 100 mg per day, depending on the vitamin. However, vitamin deficiencies can cause severe problems.

So far, 13 vitamins essential to humans have been identified. They have extremely diverse physiological functions. Vitamins are grouped into two categories: water-soluble vitamins and fat-soluble vitamins **(Table 41.1)**. The water-soluble vitamins include the B complex, which consists of several compounds that generally function as coenzymes in key metabolic processes. Vitamin C, also water-soluble, is required for the production of connective tissue. Excesses of water-soluble vitamins are excreted in urine, and moderate overdoses of these vitamins are probably harmless.

The fat-soluble vitamins are A, D, E, and K. They have a wide variety of functions. Vitamin A is incorporated into visual pigments of the eye. Vitamin D aids in calcium absorption and bone formation. The function of vitamin E is not yet fully understood, but (along with vitamin C) it seems to protect the phospholipids in membranes from oxidation. (You have probably encountered advertisements for dietary supplements containing vitamin E as an "antioxidant.") Vitamin K is required for blood clotting. Excesses of fat-soluble vitamins are not excreted but are deposited in body fat, so overconsumption may result in an accumulation of these compounds to toxic levels.

The subject of vitamin dosage has aroused heated scientific and popular debate. Some think that it is sufficient to meet

Table 41.1 Vitamin Requirements of Humans			
Vitamin	**Major Dietary Sources**	**Some Major Functions in the Body**	**Possible Symptoms of Deficiency or Extreme Excess**
Water-Soluble Vitamins			
Vitamin B_1 (thiamine)	Pork, legumes, peanuts, whole grains	Coenzyme used in removing CO_2 from organic compounds	Beriberi (nerve disorders, emaciation, anemia)
Vitamin B_2 (riboflavin)	Dairy products, meats, enriched grains, vegetables	Component of coenzymes FAD and FMN	Skin lesions such as cracks at corners of mouth
Niacin	Nuts, meats, grains	Component of coenzymes NAD^+ and $NADP^+$	Skin and gastrointestinal lesions, nervous disorders **Liver damage**
Vitamin B_6 (pyridoxine)	Meats, vegetables, whole grains	Coenzyme used in amino acid metabolism	Irritability, convulsions, muscular twitching, anemia **Unstable gait, numb feet, poor coordination**
Pantothenic acid	Most foods: meats, dairy products, whole grains, etc.	Component of coenzyme A	Fatigue, numbness, tingling of hands and feet
Folic acid (folacin)	Green vegetables, oranges, nuts, legumes, whole grains	Coenzyme in nucleic acid and amino acid metabolism	Anemia, gastrointestinal problems **May mask deficiency of vitamin B_{12}**
Vitamin B_{12}	Meats, eggs, dairy products	Coenzyme in nucleic acid metabolism; maturation of red blood cells	Anemia, nervous system disorders
Biotin	Legumes, other vegetables, meats	Coenzyme in synthesis of fat, glycogen, and amino acids	Scaly skin inflammation, neuromuscular disorders
Vitamin C (ascorbic acid)	Fruits and vegetables, especially citrus fruits, broccoli, cabbage, tomatoes, green peppers	Used in collagen synthesis (such as for bone, cartilage, gums); antioxidant; aids in detoxification; improves iron absorption	Scurvy (degeneration of skin, teeth, blood vessels), weakness, delayed wound healing, impaired immunity **Gastrointestinal upset**
Fat-Soluble Vitamins			
Vitamin A (retinol)	Provitamin A (beta-carotene) in deep green and orange vegetables and fruits; retinol in dairy products	Component of visual pigments; maintenance of epithelial tissues; antioxidant; helps prevent damage to cell membranes	Vision problems; dry, scaling skin **Headache, irritability, vomiting, hair loss, blurred vision, liver and bone damage**
Vitamin D	Dairy products, egg yolk (also made in human skin in presence of sunlight)	Aids in absorption and use of calcium and phosphorus; promotes bone growth	Rickets (bone deformities) in children, bone softening in adults **Brain, cardiovascular, and kidney damage**
Vitamin E (tocopherol)	Vegetable oils, nuts, seeds	Antioxidant; helps prevent damage to cell membranes	None well documented in humans; possibly anemia
Vitamin K (phylloquinone)	Green vegetables, tea (also made by colon bacteria)	Important in blood clotting	Defective blood clotting **Liver damage and anemia**

recommended daily allowances (RDAs), the nutrient intakes proposed by nutritionists to maintain health. Others think that RDAs are set too low for some vitamins, and a fraction of those people think, probably mistakenly, that *massive* doses of vitamins confer health benefits. Research is far from complete, and debate continues, especially over optimal doses of vitamins C and E. At this time, all that can be said with any certainty is that people who eat a balanced diet are not likely to develop symptoms of vitamin deficiency.

Minerals

Minerals are simple inorganic nutrients, usually required in small amounts—from less than 1 mg to about 2,500 mg per day (**Table 41.2**, on the next page). As with vitamins, mineral requirements vary in different animal species. Humans and other vertebrates require relatively large quantities of calcium and phosphorus for the construction and maintenance of bone. Calcium is also necessary for the normal functioning of nerves and muscles, and phosphorus is also an ingredient of ATP and nucleic acids. Iron is a component of the cytochromes that function in cellular respiration (see Figure 9.13) and of hemoglobin, the oxygen-binding protein of red blood cells. Magnesium, iron, zinc, copper, manganese, selenium, and molybdenum are cofactors built into the structure of certain enzymes; magnesium, for example, is present in enzymes that split ATP. Vertebrates need iodine to make thyroid hormones, which regulate metabolic rate. Sodium,

Table 41.2 Mineral Requirements of Humans

Mineral	Major Dietary Sources	Some Major Functions in the Body	Possible Symptoms of Deficiency*
Calcium (Ca)	Dairy products, dark green vegetables, legumes	Bone and tooth formation, blood clotting, nerve and muscle function	Retarded growth, possibly loss of bone mass
Phosphorus (P)	Dairy products, meats, grains	Bone and tooth formation, acid-base balance, nucleotide synthesis	Weakness, loss of minerals from bone, calcium loss
Sulfur (S)	Proteins from many sources	Component of certain amino acids	Symptoms of protein deficiency
Potassium (K)	Meats, dairy products, many fruits and vegetables, grains	Acid-base balance, water balance, nerve function	Muscular weakness, paralysis, nausea, heart failure
Chlorine (Cl)	Table salt	Acid-base balance, formation of gastric juice, nerve function, osmotic balance	Muscle cramps, reduced appetite
Sodium (Na)	Table salt	Acid-base balance, water balance, nerve function	Muscle cramps, reduced appetite
Magnesium (Mg)	Whole grains, green leafy vegetables	Cofactor; ATP bioenergetics	Nervous system disturbances
Iron (Fe)	Meats, eggs, legumes, whole grains, green leafy vegetables	Component of hemoglobin and of electron carriers in energy metabolism; enzyme cofactor	Iron-deficiency anemia, weakness, impaired immunity
Fluorine (F)	Drinking water, tea, seafood	Maintenance of tooth (and probably bone) structure	Higher frequency of tooth decay
Zinc (Zn)	Meats, seafood, grains	Component of certain digestive enzymes and other proteins	Growth failure, scaly skin inflammation, reproductive failure, impaired immunity
Copper (Cu)	Seafood, nuts, legumes, organ meats	Enzyme cofactor in iron metabolism, melanin synthesis, electron transport	Anemia, bone and cardiovascular changes
Manganese (Mn)	Nuts, grains, vegetables, fruits, tea	Enzyme cofactor	Abnormal bone and cartilage
Iodine (I)	Seafood, dairy products, iodized salt	Component of thyroid hormones	Goiter (enlarged thyroid)
Cobalt (Co)	Meats and dairy products	Component of vitamin B_{12}	None, except as B_{12} deficiency
Selenium (Se)	Seafood, meats, whole grains	Enzyme cofactor; antioxidant functioning in close association with vitamin E	Muscle pain, possibly heart muscle deterioration
Chromium (Cr)	Brewer's yeast, liver, seafood, meats, some vegetables	Involved in glucose and energy metabolism	Impaired glucose metabolism
Molybdenum (Mo)	Legumes, grains, some vegetables	Enzyme cofactor	Disorder in excretion of nitrogen-containing compounds

*All of these minerals are also harmful when consumed in excess.

potassium, and chlorine are important in nerve functioning and also in maintaining the osmotic balance between cells and the interstitial fluid.

Most people ingest far more salt (sodium chloride) than they need. For example, the average U.S. citizen eats enough salt to provide about 20 times the required amount of sodium. Much of this salt is hidden in packaged, prepared foods, even those that may not taste salty. Ingesting an excess of salt or several other minerals can upset homeostatic balance and cause toxic side effects. For example, too much sodium is associated with high blood pressure, and excess iron can cause liver damage.

In the next section, we will shift our attention from nutritional requirements to how animals process their food.

Concept Check 41.2

1. Contrast undernutrition with malnutrition.
2. Explain how a balanced vegetarian diet can provide all of the essential amino acids.
3. Compare and contrast vitamins with minerals.

For suggested answers, see Appendix A.

Concept 41.3

The main stages of food processing are ingestion, digestion, absorption, and elimination

We began our study of nutrition by surveying the diverse feeding mechanisms that have evolved among animals (see Figure 41.2). But **ingestion,** the act of eating, is only the first stage of food processing. Organic material in food consists largely of proteins, fats, and carbohydrates in the form of starch and other polysaccharides. Animals cannot use these macromolecules directly for two reasons. First, polymers are too large to pass through membranes and enter the cells of the animal. Second, the macromolecules that make up an animal are not identical to those of its food. In building their macromolecules, however, all organisms use common monomers. For example, soybeans, fruit flies, and humans all assemble their proteins from the same 20 amino acids.

Digestion, the second stage of food processing, is the process of breaking food down into molecules small enough for the body to absorb **(Figure 41.12)**. Digestion cleaves macromolecules into their component monomers, which the animal then uses to make its own molecules or as fuel for ATP production. Polysaccharides and disaccharides are split into simple sugars, fats are digested to glycerol and fatty acids, proteins are split into amino acids, and nucleic acids are cleaved into nucleotides.

Recall from Chapter 5 that a cell makes a macromolecule by linking together monomers; it does so by removing a molecule of water for each new covalent bond formed. Digestion reverses this process by breaking bonds with the addition of water (see Figure 5.2). This splitting process is called **enzymatic hydrolysis.** A variety of hydrolytic enzymes catalyze the digestion of each of the classes of macromolecules found in food. This chemical digestion is usually preceded by mechanical fragmentation of the food—by chewing, for instance. Breaking food into smaller pieces increases the surface area exposed to digestive juices containing hydrolytic enzymes.

The last two stages of food processing occur after the food is digested. In the third stage, **absorption,** the animal's cells take up (absorb) small molecules such as amino acids and simple sugars from the digestive compartment. And finally, **elimination** occurs, as undigested material passes out of the digestive compartment (see Figure 41.12).

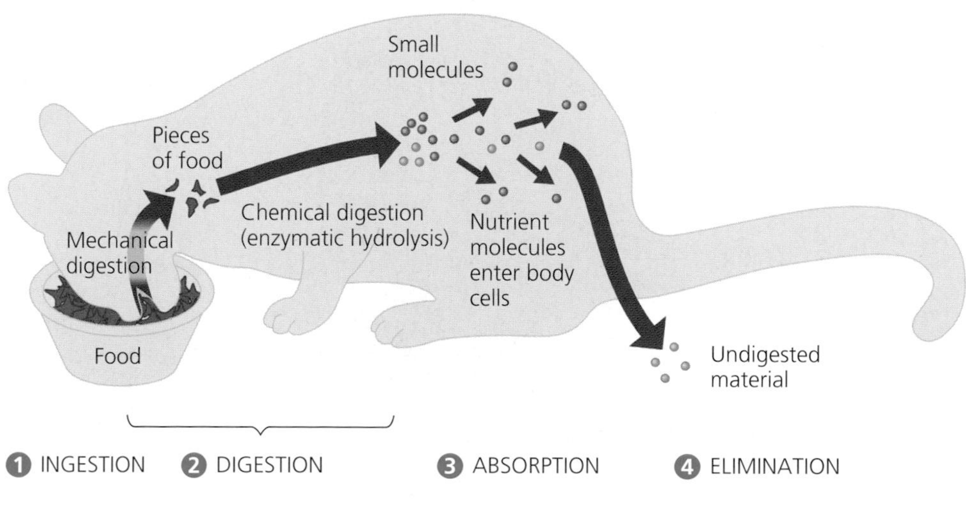

▲ **Figure 41.12 The four stages of food processing.**

Digestive Compartments

How do animals apply their digestive processes to food without digesting their own cells and tissues? After all, digestive enzymes hydrolyze the same biological materials (such as proteins, lipids, and carbohydrates) that animals are made of, and it is obviously important to avoid digesting oneself! Most animals reduce the risk of self-digestion by processing food in specialized compartments.

Intracellular Digestion

Food vacuoles—cellular organelles in which hydrolytic enzymes break down food without digesting the cell's own cytoplasm—are the simplest digestive compartments. This digestion within a cell, called **intracellular digestion,** begins after a cell engulfs food by phagocytosis or pinocytosis (see Figure 7.20). Newly formed food vacuoles fuse with lysosomes, which are organelles containing hydrolytic enzymes. This mixes the food with the enzymes, allowing digestion to occur safely within a compartment that is enclosed by a protective membrane. Sponges are unusual among animals in that they digest their food entirely by the intracellular mechanism (see Figure 33.4).

Extracellular Digestion

In most animals, at least some hydrolysis occurs by **extracellular digestion,** the breakdown of food outside cells. Extracellular digestion occurs within compartments that are continuous with the outside of the animal's body. Having an extracellular cavity for digestion enables an animal to devour much larger prey than can be ingested by phagocytosis and digested intracellularly.

Many animals with relatively simple body plans have a digestive sac with a single opening. This pouch, called a

gastrovascular cavity, functions in both digestion and distribution of nutrients throughout the body (hence the *vascular* part of the term). The cnidarians called hydras provide a good example of how a gastrovascular cavity works. Hydras are carnivores that capture prey with specialized organelles called cnidae and then use tentacles to stuff the food through their mouth into their gastrovascular cavity **(Figure 41.13)**. Specialized gland cells of the gastrodermis, the tissue layer that lines the cavity, then secrete digestive enzymes that break the soft tissues of the prey into tiny pieces. Other gastrodermal cells, called nutritive muscular cells, engulf these food particles, and most of the actual hydrolysis of macromolecules occurs intracellularly, as in sponges. After a hydra has digested its meal, undigested materials remaining in the gastrovascular cavity, such as the exoskeletons of small crustaceans, are eliminated through the single opening, which functions in the

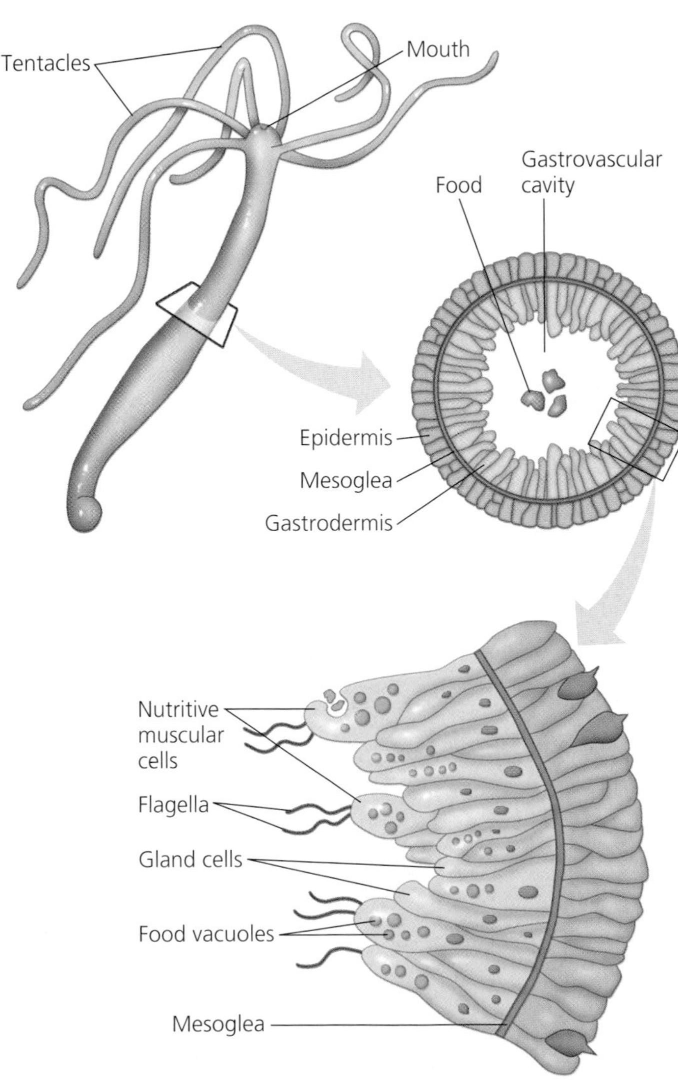

▲ **Figure 41.13 Digestion in a hydra.** The outer epidermis of the hydra has protective and sensory functions, whereas the inner gastrodermis is specialized for digestion. Digestion begins in the gastrovascular cavity and is completed intracellularly after small food particles are engulfed by the gastrodermal cells.

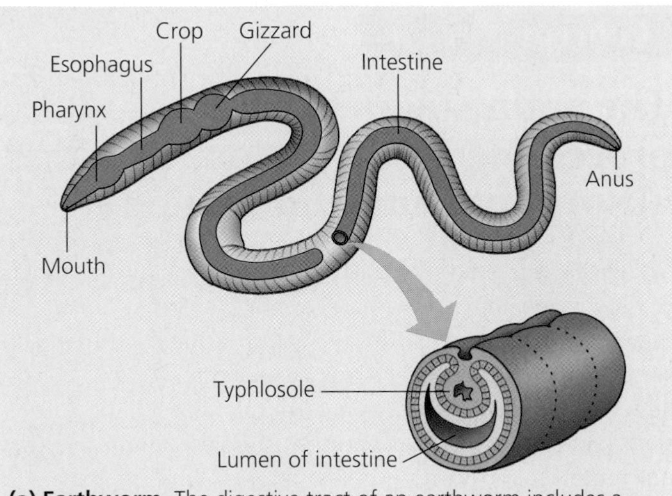

(a) Earthworm. The digestive tract of an earthworm includes a muscular pharynx that sucks food in through the mouth. Food passes through the esophagus and is stored and moistened in the crop. The muscular gizzard, which contains small bits of sand and gravel, pulverizes the food. Further digestion and absorption occur in the intestine, which has a dorsal fold, the typhlosole, that increases the surface area for nutrient absorption.

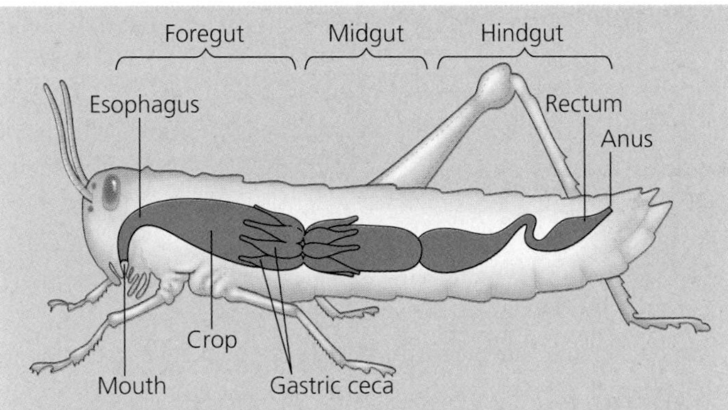

(b) Grasshopper. A grasshopper has several digestive chambers grouped into three main regions: a foregut, with an esophagus and crop; a midgut; and a hindgut. Food is moistened and stored in the crop, but most digestion occurs in the midgut. Gastric ceca, pouches extending from the midgut, absorb nutrients.

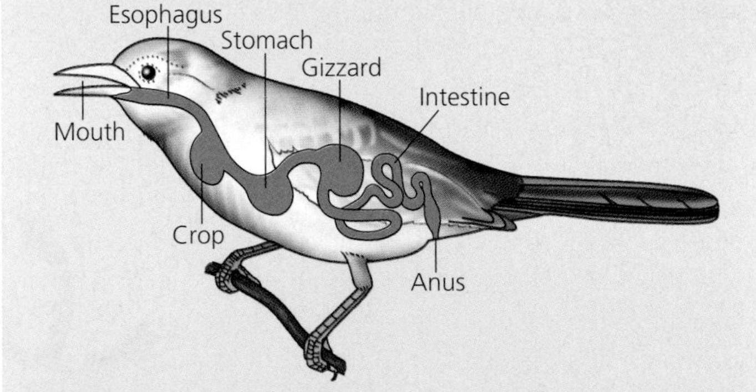

(c) Bird. Many birds have three separate chambers—the crop, stomach, and gizzard—where food is pulverized and churned before passing into the intestine. A bird's crop and gizzard function very much like those of an earthworm. In most birds, chemical digestion and absorption of nutrients occur in the intestine.

▲ **Figure 41.14 Variation in alimentary canals.**

dual role of mouth and anus. Many flatworms also have a gastrovascular cavity with a single opening (see Figure 33.10).

In contrast to cnidarians and flatworms, most animals—including nematodes, annelids, molluscs, arthropods, echinoderms, and chordates—have a digestive tube extending between two openings, a mouth and an anus. Such a tube is called a **complete digestive tract** or an **alimentary canal.** Because food moves along the canal in a single direction, the tube can be organized into specialized regions that carry out digestion and nutrient absorption in a stepwise fashion **(Figure 41.14)**. Another advantage of a complete digestive tract is the ability to ingest additional food before earlier meals are completely digested—which may be difficult or inefficient for animals with gastrovascular cavities.

Concept Check 41.3

1. What is the main anatomical distinction between a gastrovascular cavity and an alimentary canal?
2. Why are nutrients from a recently ingested meal not really "inside" your body prior to the absorption stage of food processing?

For suggested answers, see Appendix A.

Concept 41.4

Each organ of the mammalian digestive system has specialized food-processing functions

The general principles of food processing are similar for a diversity of animals, so we can use the digestive system of mammals as a representative example. The mammalian digestive system consists of the alimentary canal and various accessory glands that secrete digestive juices into the canal through ducts **(Figure 41.15)**. **Peristalsis,** rhythmic waves of contraction by smooth muscles in the wall of the canal, pushes the food along the tract. At some of the junctions between specialized segments of the digestive tube, the muscular layer is modified into ringlike valves called **sphincters,** which close off the tube like drawstrings, regulating the passage of material between chambers of the canal. The accessory glands of the mammalian digestive system are three pairs of **salivary glands,** the **pancreas,** the **liver,** and the **gallbladder,** which stores a digestive juice.

Using the human digestive system as a model, let's now follow a meal through the alimentary canal, examining in more

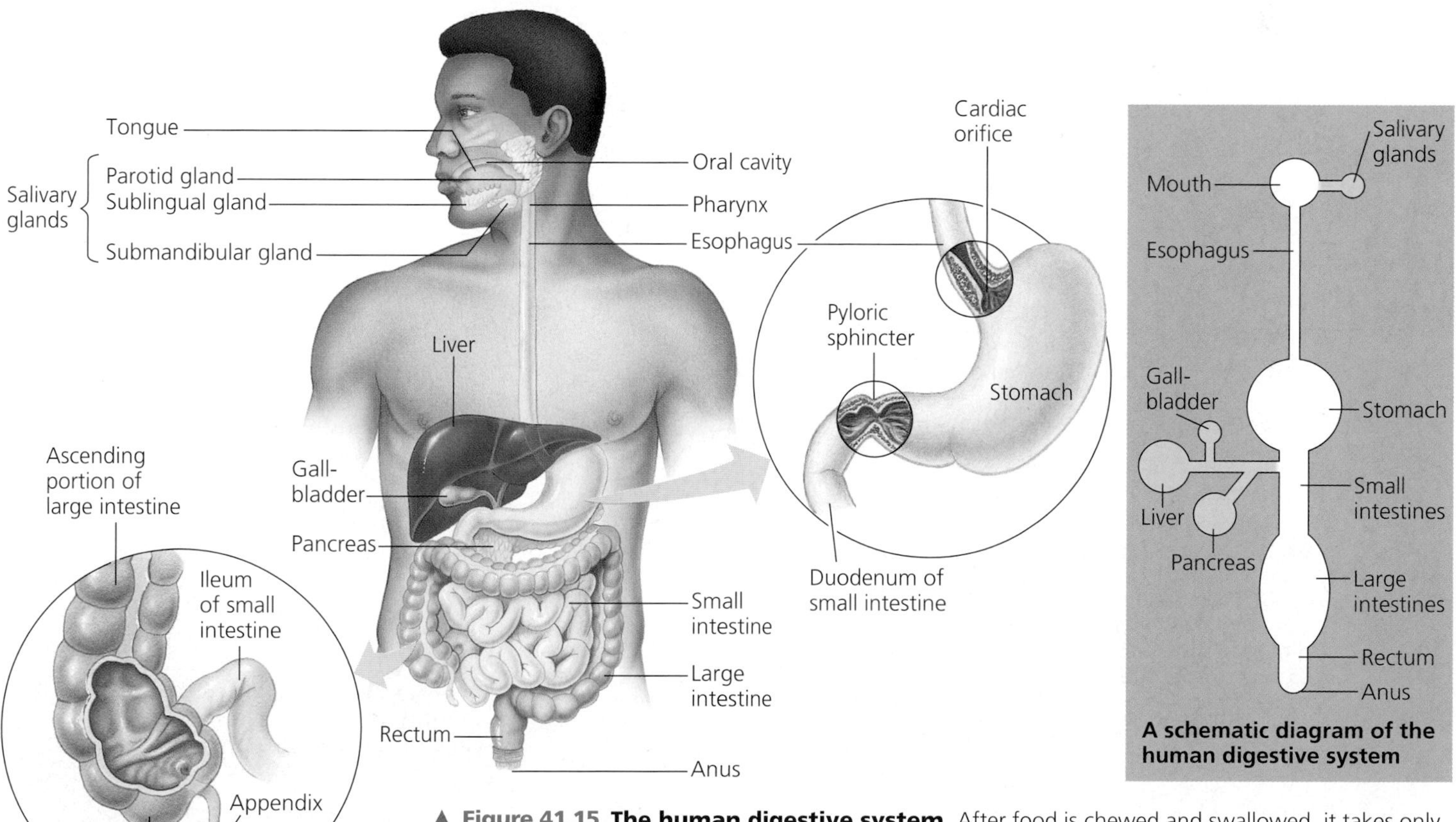

▲ **Figure 41.15 The human digestive system.** After food is chewed and swallowed, it takes only 5–10 seconds for it to pass down the esophagus and into the stomach, where it spends 2–6 hours being partially digested. Final digestion and nutrient absorption occur in the small intestine over a period of 5–6 hours. In 12–24 hours, any undigested material passes through the large intestine, and feces are expelled through the anus.

detail what happens to the food in each of the processing stations along the way (see Figure 41.15).

The Oral Cavity, Pharynx, and Esophagus

Both physical and chemical digestion of food begin in the mouth. During chewing, teeth of various shapes cut, smash, and grind food, making it easier to swallow and increasing its surface area. The presence of food in the **oral cavity** triggers a nervous reflex that causes the salivary glands to deliver saliva through ducts to the oral cavity. Even before food is actually in the mouth, salivation may occur in anticipation because of learned associations between eating and the time of day, cooking odors, or other stimuli. Humans secrete more than a liter of saliva each day.

Saliva contains a slippery glycoprotein (carbohydrate-protein complex) called mucin, which protects the lining of the mouth from abrasion and lubricates food for easier swallowing. Saliva also contains buffers that help prevent tooth decay by neutralizing acid in the mouth. Antibacterial agents in saliva kill many of the bacteria that enter the mouth with food.

Chemical digestion of carbohydrates, a main source of chemical energy, begins in the oral cavity. Saliva contains **salivary amylase**, an enzyme that hydrolyzes starch (a glucose polymer from plants) and glycogen (a glucose polymer from animals). The main products of this enzyme's action are smaller polysaccharides and the disaccharide maltose.

The tongue tastes food, manipulates it during chewing, and helps shape the food into a ball called a **bolus.** During swallowing, the tongue pushes a bolus to the back of the oral cavity and into the pharynx.

The region we call our throat is the **pharynx,** a junction that opens to both the esophagus and the windpipe (trachea). When we swallow, the top of the windpipe moves up so that its opening, the glottis, is blocked by a cartilaginous flap, the **epiglottis.** You can see this motion in the bobbing of the "Adam's apple" during swallowing. This tightly controlled mechanism normally ensures that a bolus is guided into the entrance of the esophagus (**Figure 41.16**, steps 1–4). Food or liquids may go "down the wrong pipe" because the swallowing reflex didn't close the opening of the windpipe in time. The resulting blockage of airflow (choking) stimulates vigorous coughing, which usually expels the material. If it is not expelled quickly, the lack of airflow to the lungs can be fatal.

The **esophagus** conducts food from the pharynx down to the stomach by peristalsis (see Figure 41.16, step 6). The muscles at the very top of the esophagus are striated (voluntary). Thus, the act of swallowing begins voluntarily, but then the involuntary waves of contraction by smooth muscles in the rest of the esophagus take over.

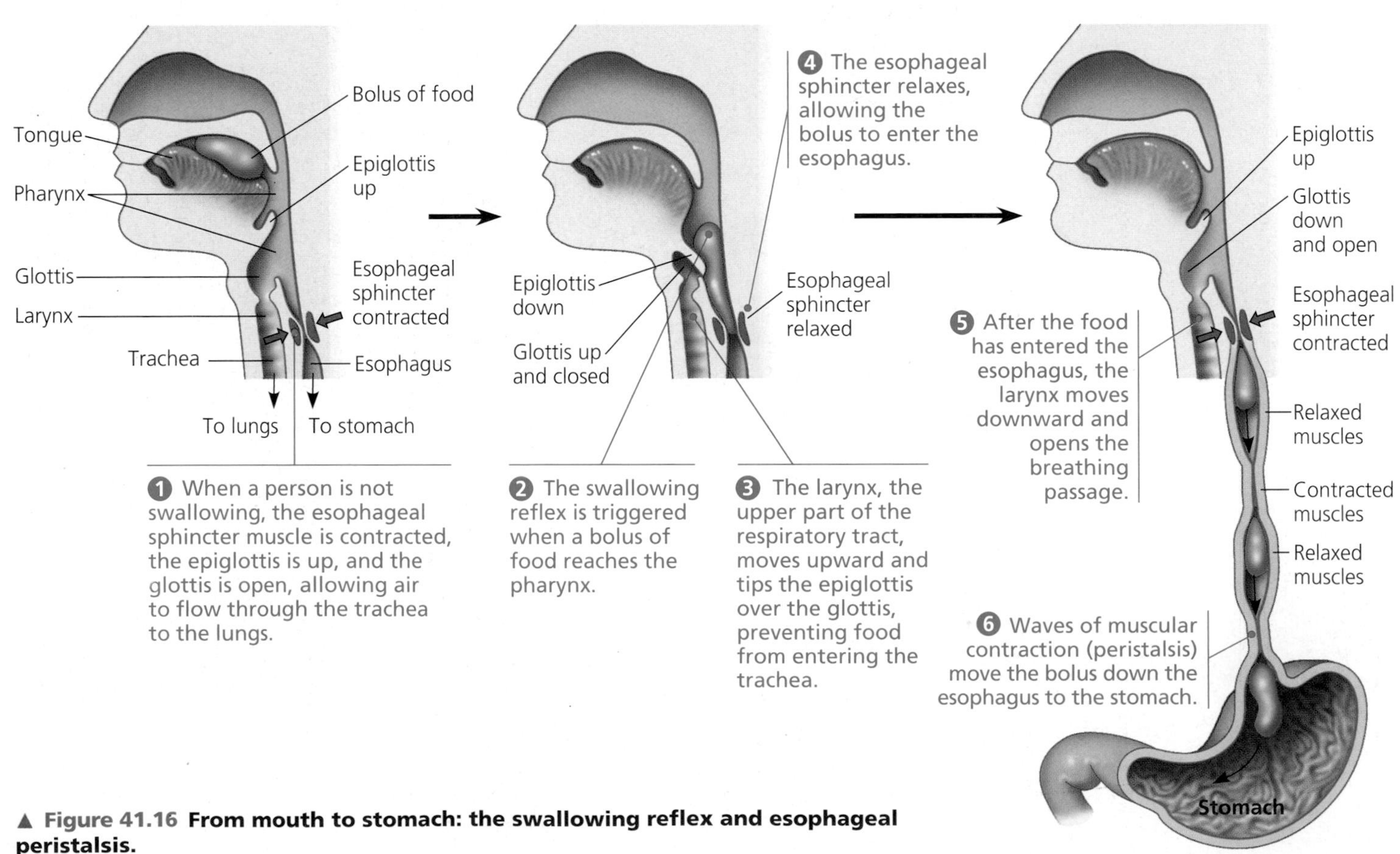

▲ **Figure 41.16 From mouth to stomach: the swallowing reflex and esophageal peristalsis.**

The Stomach

The **stomach** stores food and performs preliminary steps of digestion. This large organ is located in the upper abdominal cavity, just below the diaphragm. With accordionlike folds and a very elastic wall, the stomach can stretch to accommodate about 2 L of food and fluid. It is because the stomach can store an entire meal that we do not need to eat constantly. Besides storing food, the stomach performs important digestive functions: It secretes a digestive fluid called **gastric juice** and mixes this secretion with the food by the churning action of the smooth muscles in the stomach wall.

Gastric juice is secreted by the epithelium lining numerous deep pits in the stomach wall. With a high concentration of hydrochloric acid, gastric juice has a pH of about 2—acidic enough to dissolve iron nails. One function of the acid is to disrupt the extracellular matrix that binds cells together in meat and plant material. The acid also kills most bacteria that are swallowed with food. Also present in gastric juice is **pepsin,** an enzyme that begins the hydrolysis of proteins. Pepsin breaks peptide bonds adjacent to specific amino acids, cleaving proteins into smaller polypeptides, which are later digested completely to amino acids in the small intestine. Pepsin is one of the few enzymes that works best in a strongly acidic environment. The low pH of gastric juice denatures (unfolds) the proteins in food, increasing exposure of their peptide bonds to pepsin.

What prevents pepsin from destroying the cells of the stomach wall? First, pepsin is secreted in an *inactive* form called **pepsinogen** by specialized cells called chief cells located in gastric pits **(Figure 41.17)**. Other cells, called parietal cells, also in the pits, secrete hydrochloric acid. The acid converts pepsinogen to active pepsin by removing a small portion of the molecule and exposing its active site. Because different cells secrete the acid and pepsinogen, the two ingredients do not mix—and pepsinogen is not activated—until they enter the lumen (cavity) of the stomach. Activation of pepsinogen is an example of positive feedback: Once some pepsinogen is activated by acid, activation occurs at an increasingly rapid rate because pepsin itself can activate additional molecules of pepsinogen. Many other digestive enzymes are also secreted in inactive

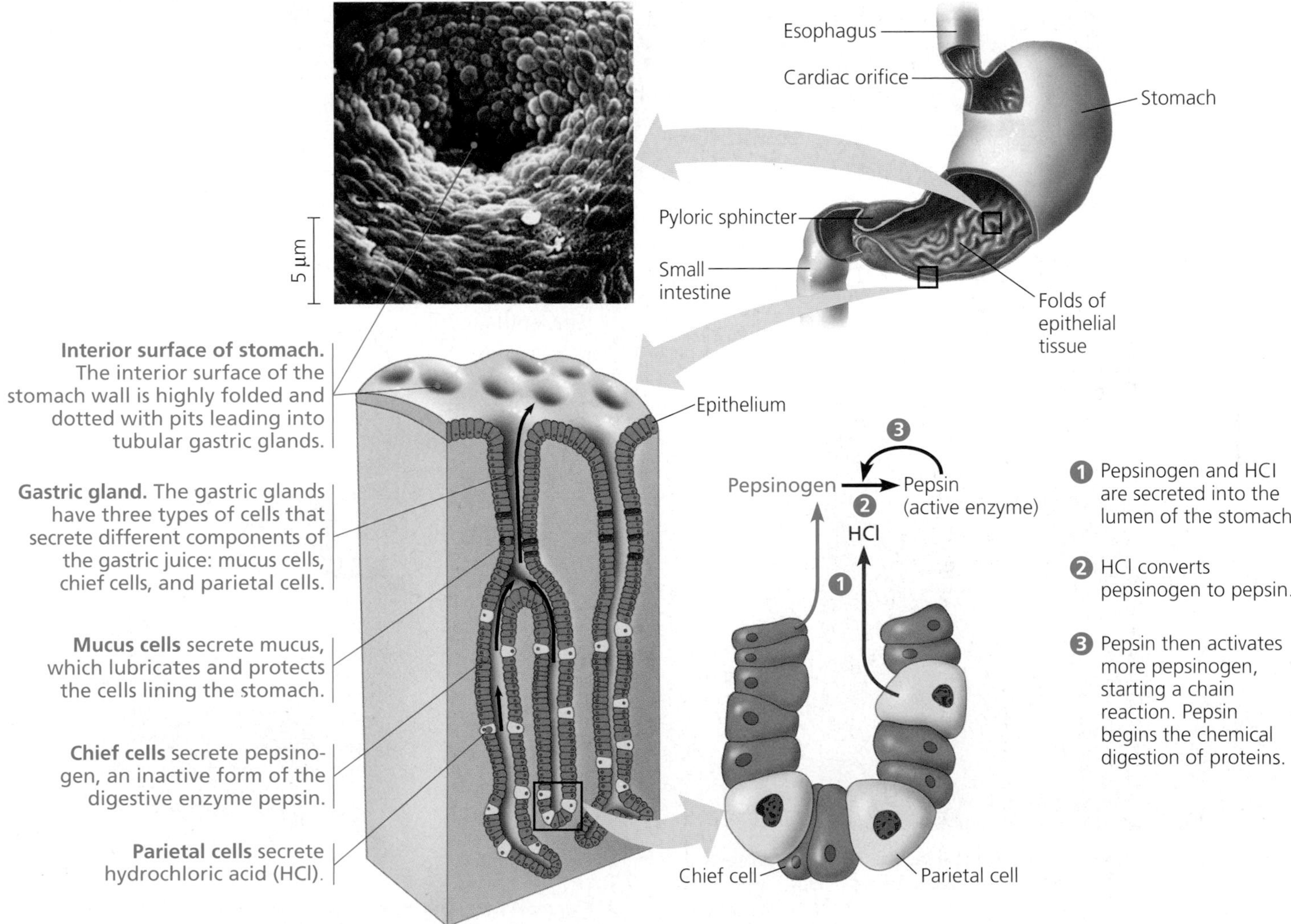
Esophagus
Cardiac orifice
Stomach
Pyloric sphincter
Small intestine
Folds of epithelial tissue
5 µm
Interior surface of stomach. The interior surface of the stomach wall is highly folded and dotted with pits leading into tubular gastric glands.
Epithelium
Gastric gland. The gastric glands have three types of cells that secrete different components of the gastric juice: mucus cells, chief cells, and parietal cells.
Mucus cells secrete mucus, which lubricates and protects the cells lining the stomach.
Chief cells secrete pepsinogen, an inactive form of the digestive enzyme pepsin.
Parietal cells secrete hydrochloric acid (HCl).
3
Pepsinogen
2
Pepsin (active enzyme)
HCl
1
Chief cell
Parietal cell
1 Pepsinogen and HCl are secreted into the lumen of the stomach.
2 HCl converts pepsinogen to pepsin.
3 Pepsin then activates more pepsinogen, starting a chain reaction. Pepsin begins the chemical digestion of proteins.

▲ **Figure 41.17 The stomach and its secretions.** The micrograph (colorized SEM) shows a gastric pit on the interior surface of the stomach, through which digestive juices are secreted.

forms that become active within the lumen of the digestive tract.

The stomach's second defense against self-digestion is a coating of mucus, secreted by the epithelial cells of the stomach lining. Still, the epithelium is constantly eroded, and mitosis generates enough cells to completely replace the stomach lining every three days. Gastric ulcers, lesions in this lining, are caused mainly by the acid-tolerant bacterium *Helicobacter pylori* **(Figure 41.18)**. Though treatable with antibiotics, gastric ulcers may worsen if pepsin and acid destroy the lining faster than it can regenerate.

About every 20 seconds, the stomach contents are mixed by the churning action of smooth muscles. You may feel hunger pangs when your empty stomach churns. (Sensations of hunger are also associated with brain centers that monitor the blood's nutritional status and levels of the appetite-controlling hormones discussed earlier in this chapter.) As a result of mixing and enzyme action, what begins in the stomach as a recently swallowed meal becomes a nutrient-rich broth known as **acid chyme.**

Most of the time, the stomach is closed off at both ends (see Figure 41.15). The opening from the esophagus to the stomach, the cardiac orifice, normally dilates only when a bolus arrives. The occasional backflow of acid chyme from the stomach into the lower end of the esophagus causes heartburn. (If backflow is a persistent problem, an ulcer may develop in the esophagus.) At the opening from the stomach to the small intestine is the **pyloric sphincter,** which helps regulate the passage of chyme into the intestine, one squirt at a time. It takes about 2 to 6 hours after a meal for the stomach to empty in this way.

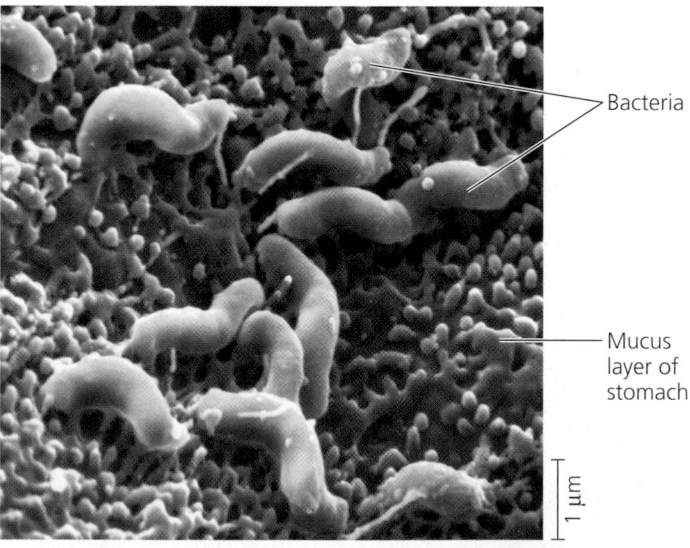

▲ **Figure 41.18 Ulcer-causing bacteria.** The bacteria visible in this colorized SEM, *Helicobacter pylori,* initiate ulcers by destroying protective mucus and causing inflammation of the stomach lining. Then the acidic gastric juice can attack the stomach tissue. In severe ulcers, the erosion can produce a hole in the stomach wall and cause life-threatening internal bleeding and infection.

The Small Intestine

With a length of more than 6 m in humans, the **small intestine** is the longest section of the alimentary canal (its name refers to its small diameter, compared with that of the large intestine). Most of the enzymatic hydrolysis of food macromolecules and most of the absorption of nutrients into the blood occur in the small intestine.

Enzymatic Action in the Small Intestine

The first 25 cm or so of the small intestine is called the **duodenum.** It is here that acid chyme from the stomach mixes with digestive juices from the pancreas, liver, gallbladder, and gland cells of the intestinal wall itself **(Figure 41.19)**.

The pancreas produces several hydrolytic enzymes and an alkaline solution rich in bicarbonate. The bicarbonate acts as a buffer, offsetting the acidity of chyme from the stomach. The pancreatic enzymes include protein-digesting enzymes (proteases) that are secreted into the duodenum in inactive form. In a chain reaction similar to the activation of pepsin in the stomach, the pancreatic proteases are activated once they are safely located in the extracellular space within the duodenum **(Figure 41.20)**.

The liver performs a wide variety of functions in the body, including the production of **bile,** a mixture of substances that is stored in the gallbladder until needed. Bile contains no digestive enzymes, but it does contain bile salts, which act as detergents (emulsifiers) that aid in the digestion and absorption of fats (see Figure 41.24). Bile also contains pigments that are by-products of red blood cell destruction in the liver; these bile pigments are eliminated from the body with the feces.

The epithelial lining of the duodenum, called the brush border, is the source of several digestive enzymes. Some of these enzymes are secreted into the lumen of the duodenum,

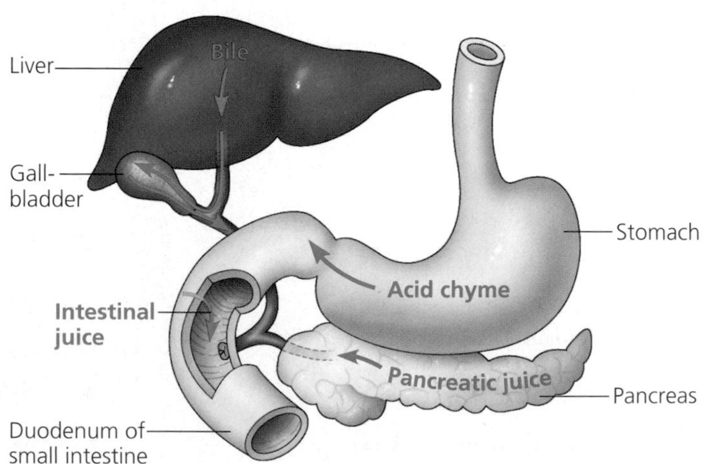

▲ **Figure 41.19 The duodenum.** Hydrolytic enzymes from accessory glands mix with acid chyme in the duodenum, continuing the digestion process. Note that bile is produced in the liver but stored in the gallbladder, which releases bile into the duodenum as needed.

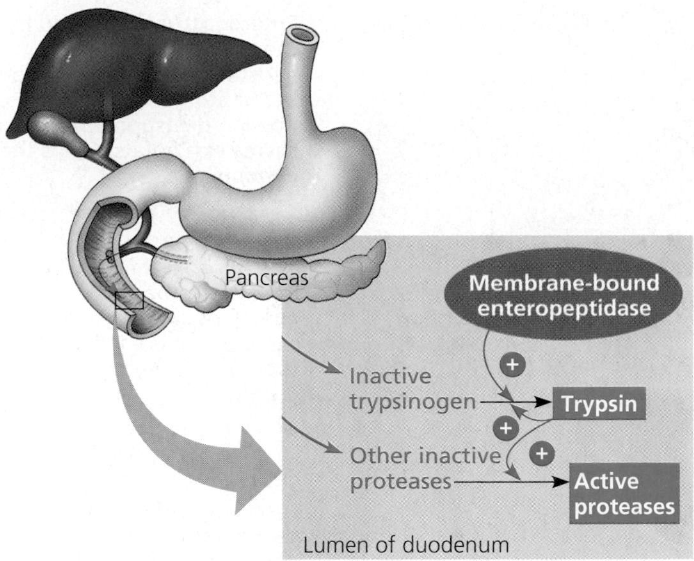

▲ **Figure 41.20 Protease activation.** The pancreas secretes inactive proteases into the duodenum. An enzyme called enteropeptidase, which is bound to the intestinal epithelium, converts trypsinogen to trypsin. Trypsin then activates other proteases. (⊕ indicates activation.)

but other digestive enzymes are actually bound to the surface of epithelial cells.

Enzymatic digestion is completed as peristalsis moves the mixture of chyme and digestive juices along the small intestine **(Figure 41.21)**. Most digestion is completed early in this journey, while the chyme is still in the duodenum. The remaining regions of the small intestine, called the *jejunum* and *ileum*, function mainly in the absorption of nutrients and water. **Figure 41.22**, on the next page, diagrams how hormones help coordinate the secretion of digestive juices into the alimentary canal.

Absorption of Nutrients

To enter the body, nutrients in the lumen must cross the lining of the digestive tract. A few nutrients are absorbed in the stomach and large intestine, but most absorption occurs in the small intestine. This organ has a huge surface area—300 m^2, roughly the size of a tennis court. Large circular folds in the lining bear fingerlike projections called **villi**, and each

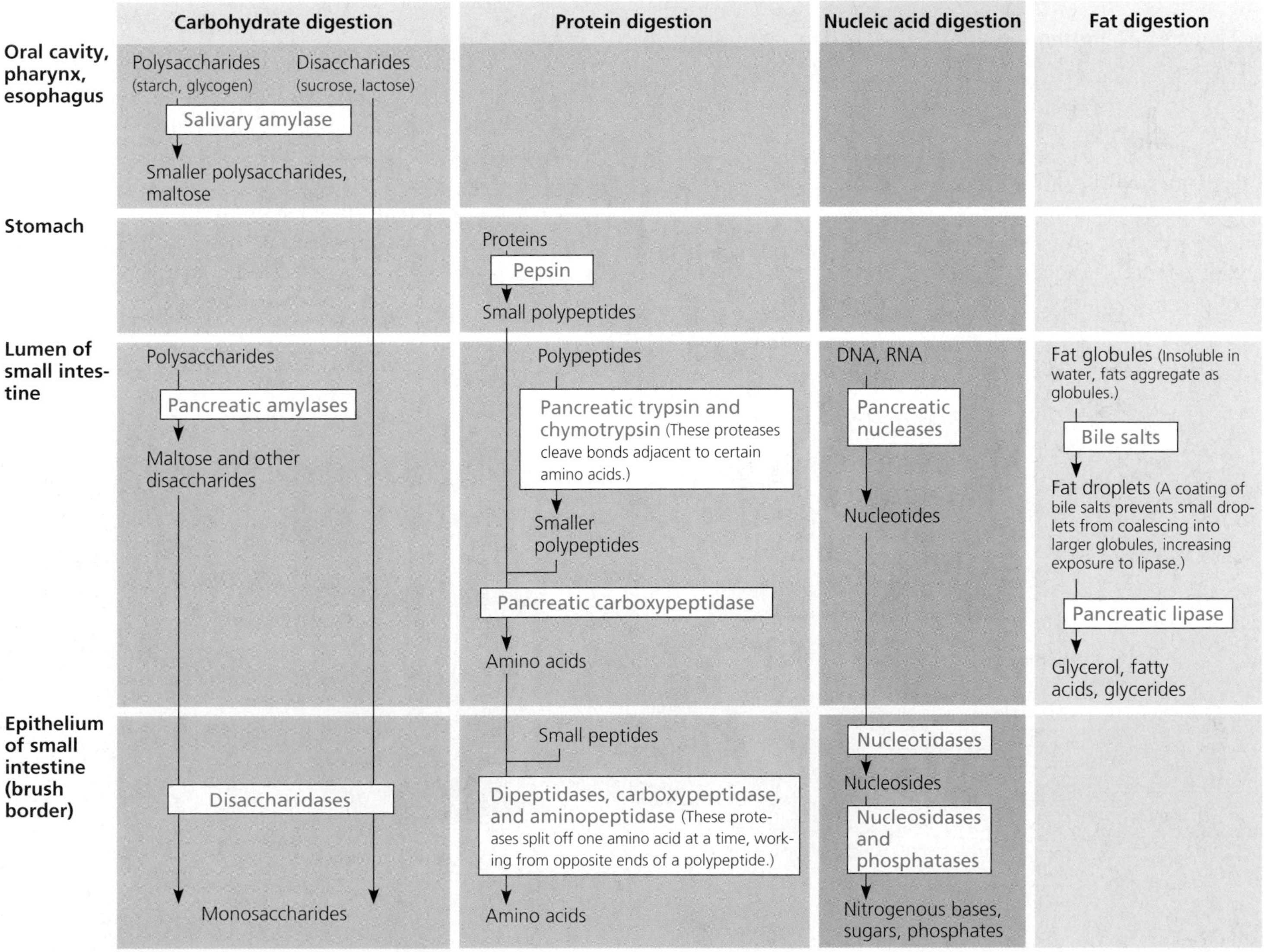

	Carbohydrate digestion	Protein digestion	Nucleic acid digestion	Fat digestion
Oral cavity, pharynx, esophagus	Polysaccharides (starch, glycogen) → Salivary amylase → Smaller polysaccharides, maltose; Disaccharides (sucrose, lactose)			
Stomach		Proteins → Pepsin → Small polypeptides		
Lumen of small intestine	Polysaccharides → Pancreatic amylases → Maltose and other disaccharides	Polypeptides → Pancreatic trypsin and chymotrypsin (These proteases cleave bonds adjacent to certain amino acids.) → Smaller polypeptides → Pancreatic carboxypeptidase → Amino acids	DNA, RNA → Pancreatic nucleases → Nucleotides	Fat globules (Insoluble in water, fats aggregate as globules.) → Bile salts → Fat droplets (A coating of bile salts prevents small droplets from coalescing into larger globules, increasing exposure to lipase.) → Pancreatic lipase → Glycerol, fatty acids, glycerides
Epithelium of small intestine (brush border)	Disaccharidases → Monosaccharides	Small peptides → Dipeptidases, carboxypeptidase, and aminopeptidase (These proteases split off one amino acid at a time, working from opposite ends of a polypeptide.) → Amino acids	Nucleotidases → Nucleosides → Nucleosidases and phosphatases → Nitrogenous bases, sugars, phosphates	

▲ **Figure 41.21 Flowchart of enzymatic digestion in the human digestive system.**

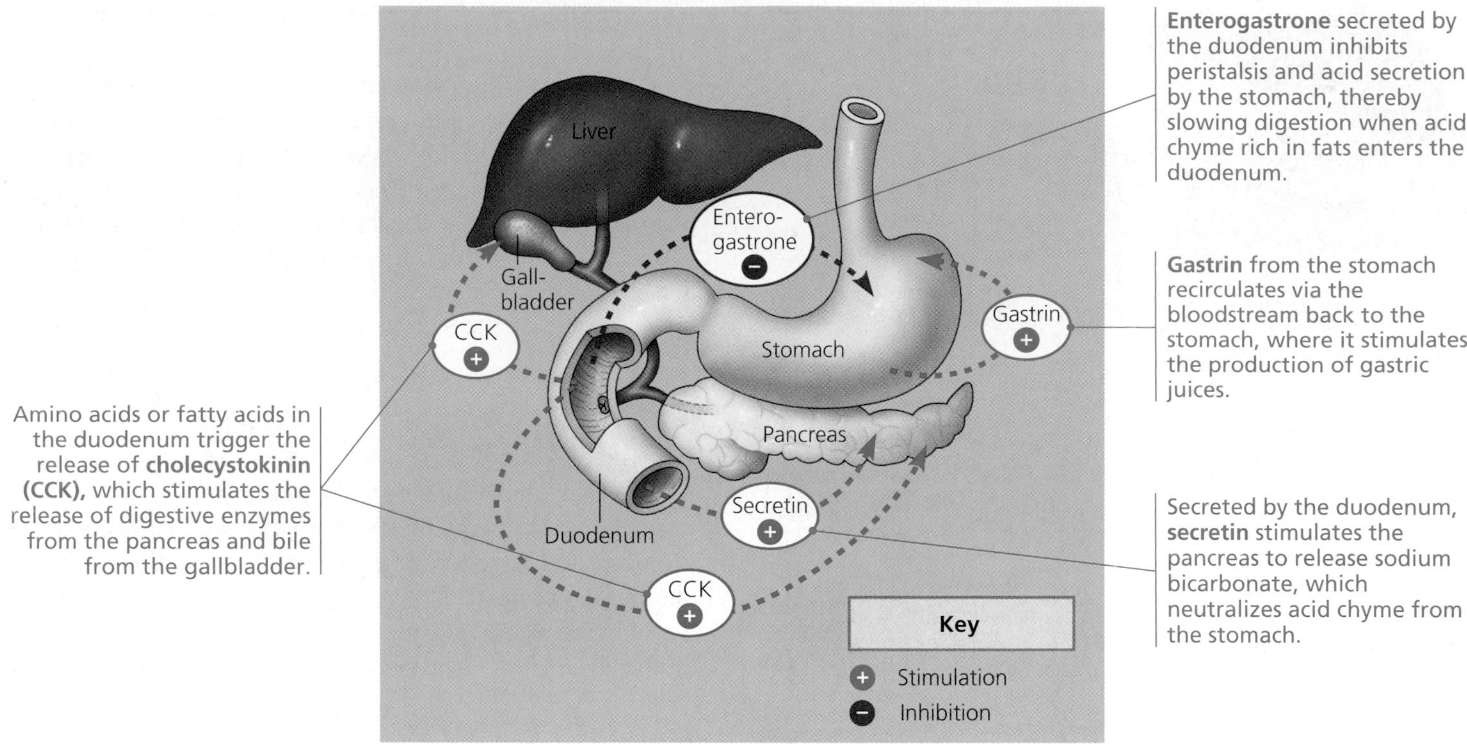

Liver
Gall-
bladder
Entero-
gastrone
Stomach
Gastrin
CCK
Pancreas
Secretin
Duodenum
CCK
Key
Stimulation
Inhibition
Enterogastrone secreted by the duodenum inhibits peristalsis and acid secretion by the stomach, thereby slowing digestion when acid chyme rich in fats enters the duodenum.
Gastrin from the stomach recirculates via the bloodstream back to the stomach, where it stimulates the production of gastric juices.
Amino acids or fatty acids in the duodenum trigger the release of cholecystokinin (CCK), which stimulates the release of digestive enzymes from the pancreas and bile from the gallbladder.
Secreted by the duodenum, secretin stimulates the pancreas to release sodium bicarbonate, which neutralizes acid chyme from the stomach.

▲ **Figure 41.22 Hormonal control of digestion.** Many animals go for long intervals between meals and do not need their digestive systems to run continuously. Hormones released by the stomach and duodenum help ensure that digestive secretions are present only when needed.

epithelial cell of a villus has many microscopic appendages called **microvilli** that are exposed to the intestinal lumen **(Figure 41.23)**. (The microvilli's shape is the basis of the term *brush border* for the intestinal epithelium.) This enormous microvillar surface is an adaptation that greatly increases the rate of nutrient absorption.

Penetrating the core of each villus is a net of microscopic blood vessels (capillaries) and a small vessel of the lymphatic system called a **lacteal.** (In addition to their circulatory system, vertebrates have an associated network of vessels—the lymphatic system—that carries a clear fluid called lymph, discussed in Chapter 43.) Nutrients are absorbed across the

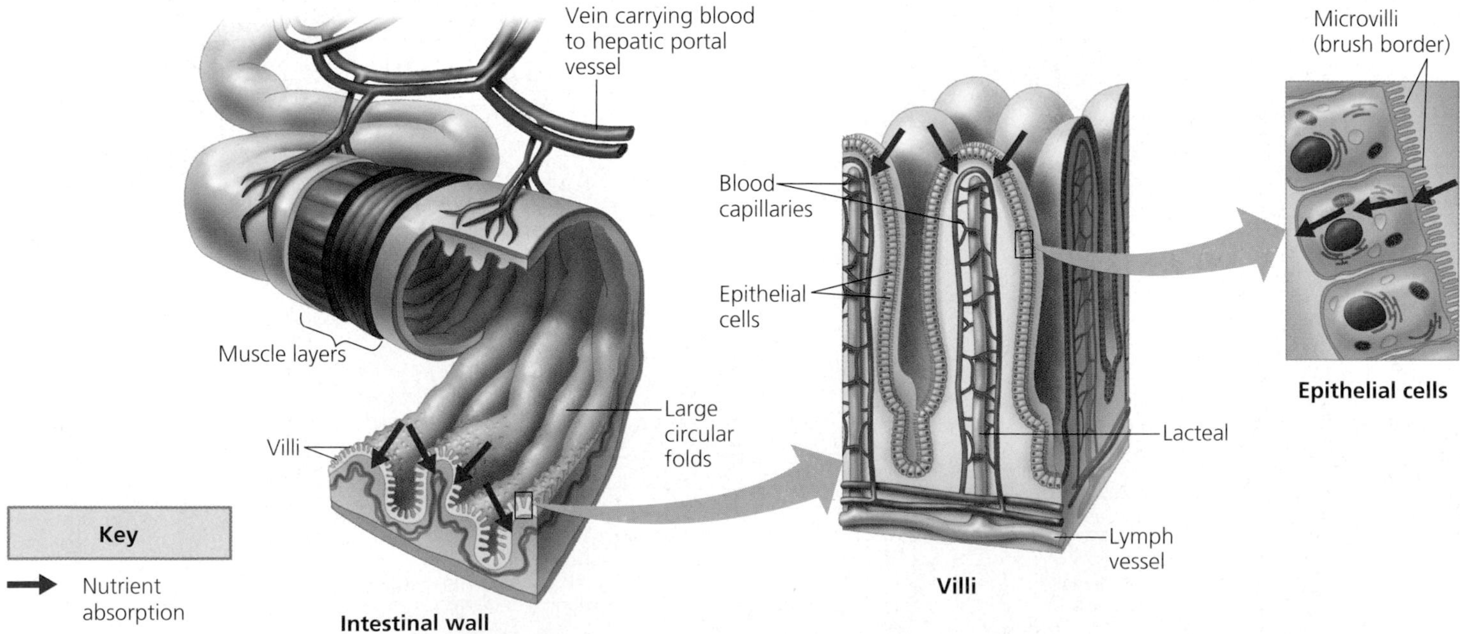

▲ **Figure 41.23 The structure of the small intestine.**

intestinal epithelium and then across the unicellular epithelium of the capillaries or lacteals. Thus, only two layers of epithelial cells separate nutrients in the lumen of the intestine from the bloodstream.

In some cases, transport of nutrients across the epithelial cells is passive. The sugar fructose, for example, apparently moves by facilitated diffusion down its concentration gradient from the lumen of the intestine into the epithelial cells and then into capillaries. Other nutrients, including amino acids, small peptides, vitamins, and glucose and several other simple sugars, are pumped against concentration gradients by the epithelial membranes. This active transport allows the intestine to absorb a much higher proportion of the nutrients in the intestine than would be possible with passive diffusion.

Amino acids and sugars pass through the epithelium, enter capillaries, and are carried away from the intestine by the bloodstream. After glycerol and fatty acids are absorbed by epithelial cells, they are recombined into fats within those cells. The fats are then mixed with cholesterol and coated with proteins, forming small globules called **chylomicrons**, most of which are transported out of the epithelial cells and into lacteals **(Figure 41.24)**. The lacteals converge into the larger vessels of the lymphatic system. Lymph, containing chylomicrons, eventually drains from the lymphatic system into large veins that return blood to the heart.

In contrast to the lacteals, the capillaries and veins that carry nutrient-rich blood away from the villi all converge into the **hepatic portal vein**, a blood vessel that leads directly to the liver. This ensures that the liver—which has the metabolic versatility to interconvert various organic molecules—has first access to amino acids and sugars absorbed after a meal is digested. Therefore, blood that leaves the liver may have a very different balance of these nutrients than the blood that entered via the hepatic portal vein. For example, the liver helps regulate the level of glucose molecules in the blood, and blood exiting the liver usually has a glucose concentration very close to 90 mg/100 mL, regardless of the carbohydrate content of a meal (see Figure 41.3). From the liver, blood travels to the heart, which pumps the blood and the nutrients it contains to all parts of the body.

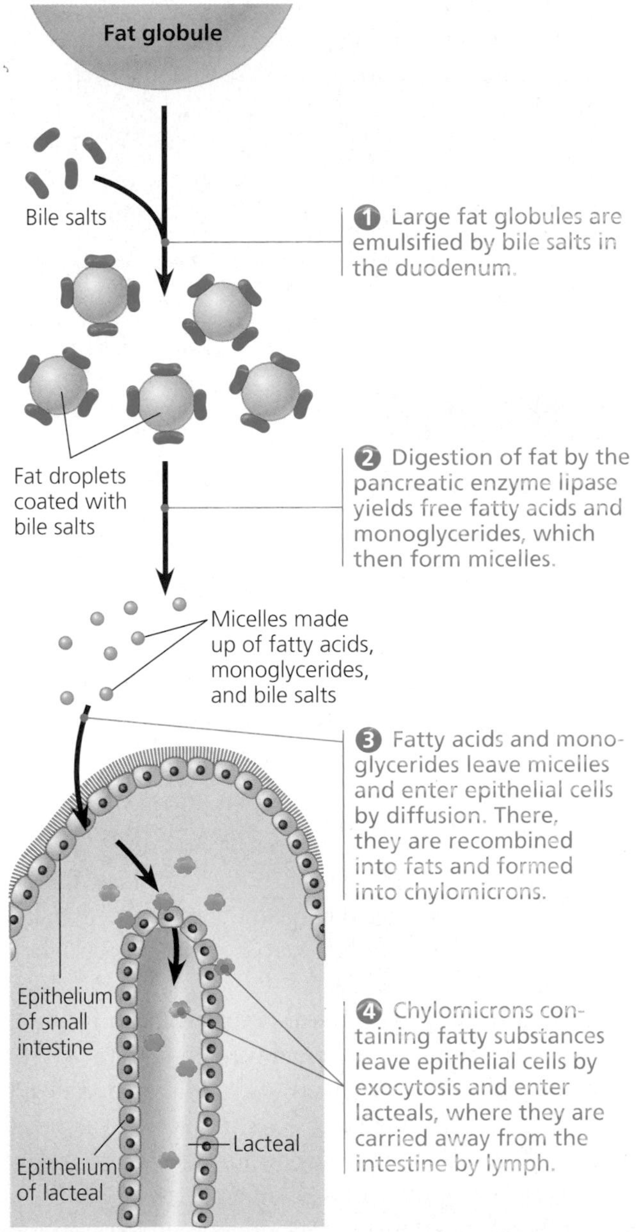

The Large Intestine

The **large intestine**, or **colon (Figure 41.25)**, is connected to the small intestine at a T-shaped junction, where a sphincter

Fat globule

Bile salts

1 Large fat globules are emulsified by bile salts in the duodenum.

Fat droplets coated with bile salts

2 Digestion of fat by the pancreatic enzyme lipase yields free fatty acids and monoglycerides, which then form micelles.

Micelles made up of fatty acids, monoglycerides, and bile salts

3 Fatty acids and monoglycerides leave micelles and enter epithelial cells by diffusion. There, they are recombined into fats and formed into chylomicrons.

Epithelium of small intestine

4 Chylomicrons containing fatty substances leave epithelial cells by exocytosis and enter lacteals, where they are carried away from the intestine by lymph.

Lacteal

Epithelium of lacteal

▲ **Figure 41.24 Digestion and absorption of fats.** Hydrolysis of fats is a digestive challenge because fat molecules are insoluble in water. However, bile salts from the gallbladder secreted into the duodenum coat tiny fat droplets and keep them from coalescing, a process called emulsification. Because the droplets are small, a large surface area of fat is exposed to lipase. Once the fat molecules have been hydrolyzed, they form micelles, which enable the fatty substances to diffuse into the epithelial lining of the small intestine. From the epithelial cells, they can be absorbed into the circulatory system.

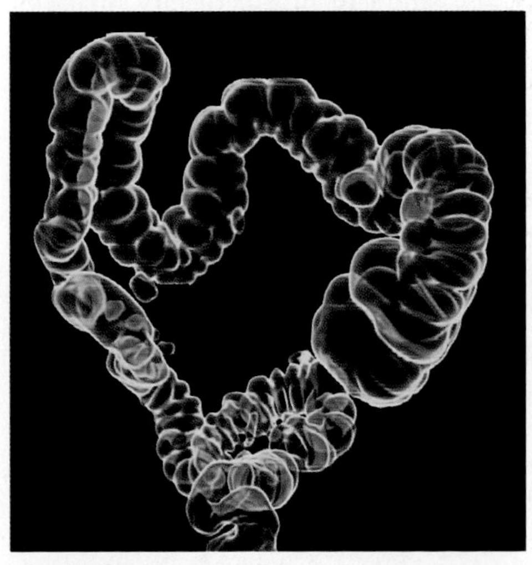

◀ **Figure 41.25 Digital image of a human colon.** This image was produced by integrating two-dimensional sectional views of the large intestine.

(a muscular valve) controls the movement of material. One arm of the T is a pouch called the **cecum** (see Figure 41.15). Compared to many other mammals, humans have a relatively small cecum. The human cecum has a fingerlike extension, the **appendix,** which is dispensable. (Lymphoid tissue in the appendix makes a minor contribution to body defense.) The main branch of the human colon is about 1.5 m long.

A major function of the colon is to recover water that has entered the alimentary canal as the solvent of the various digestive juices. About 7 L of fluid are secreted into the lumen of the digestive tract each day, which is much more liquid than most people drink. Most of this water is reabsorbed when nutrients are absorbed in the small intestine. The colon reclaims much of the remaining water that was not absorbed in the small intestine. Together, the small intestine and colon reabsorb about 90% of the water that enters the alimentary canal.

The wastes of the digestive tract, the **feces,** become more solid as they are moved along the colon by peristalsis. The movement is sluggish, and it generally takes about 12 to 24 hours for material to travel the length of the organ. If the lining of the colon is irritated—by a viral or bacterial infection, for instance—less water than normal may be reabsorbed, resulting in diarrhea. The opposite problem, constipation, occurs when peristalsis moves the feces along the colon too slowly. An excess of water is reabsorbed, and the feces become compacted.

Living in the large intestine is a rich flora of mostly harmless bacteria. One of the common inhabitants of the human colon is *Escherichia coli,* a favorite research organism of molecular biologists (see Chapter 18). The presence of *E. coli* in lakes and streams is an indication of contamination by untreated sewage. Intestinal bacteria live on unabsorbed organic material. As by-products of their metabolism, many colon bacteria generate gases, including methane and hydrogen sulfide. Some of the bacteria produce vitamins, including biotin, folic acid, vitamin K, and several B vitamins. These vitamins, absorbed into the blood, supplement our dietary intake of vitamins.

Feces contain masses of bacteria, as well as cellulose and other undigested materials. Although cellulose fibers have no caloric value to humans, their presence in the diet helps move food along the digestive tract.

The terminal portion of the colon is called the **rectum**, where feces are stored until they can be eliminated. Between the rectum and the anus are two sphincters, one involuntary and the other voluntary. One or more times each day, strong contractions of the colon create an urge to defecate.

We have now followed a meal from one opening (the mouth) of the alimentary canal to the other (the anus). In the last section of this chapter, we'll see how some of the digestive adaptations of animals may have evolved.

Concept Check 41.4

1. In the weightless environment of space, how does food swallowed by an astronaut reach the stomach?
2. Describe two key digestive functions of the hydrochloric acid in gastric juice.
3. What materials are mixed within the duodenum during digestion of a meal?
4. How is the structure of the brush border (epithelium) of the small intestine adapted to its function of nutrient absorption?
5. Explain why treatment of a chronic infection with antibiotics for an extended period of time may cause vitamin K deficiency.
6. After reviewing Figure 41.22, explain how the pancreas times its secretion of digestive juice to mix with a partially digested meal in the duodenum.

For suggested answers, see Appendix A.

Evolutionary adaptations of vertebrate digestive systems are often associated with diet

The digestive systems of mammals and other vertebrates are variations on a common plan, but there are many intriguing adaptations, often associated with the animal's diet. We will examine just a few.

Some Dental Adaptations

Dentition, an animal's assortment of teeth, is one example of structural variation reflecting diet. Particularly in mammals, evolutionary adaptation of teeth for processing different kinds of food is one of the major reasons this vertebrate class has been so successful. Compare the dentition of carnivorous, herbivorous, and omnivorous mammals in **Figure 41.26**. Nonmammalian vertebrates generally have less specialized dentition, but there are interesting exceptions. For example, poisonous snakes, such as rattlesnakes, have fangs, modified teeth that inject venom into prey. Some fangs are hollow, like syringes, while others drip the poison along grooves on the surfaces of the teeth. All snakes have another important anatomical adaptation associated with feeding: They swallow their prey whole, with no chewing, and the lower jaw is loosely hinged to the skull by an elastic ligament that permits the mouth and throat to open very wide for swallowing impressively large prey (once again, witness the astonishing episode recorded in Figure 41.2).

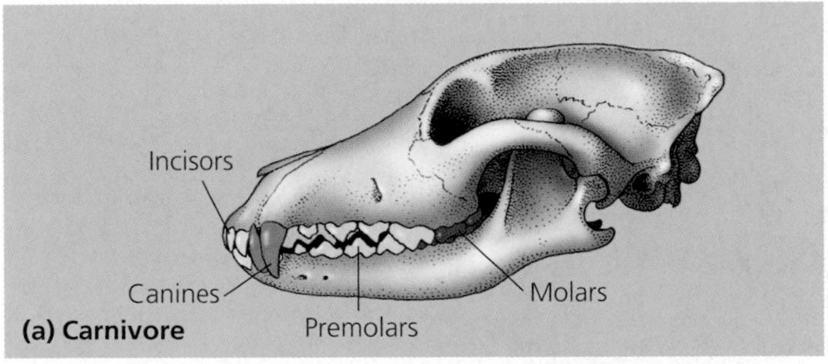

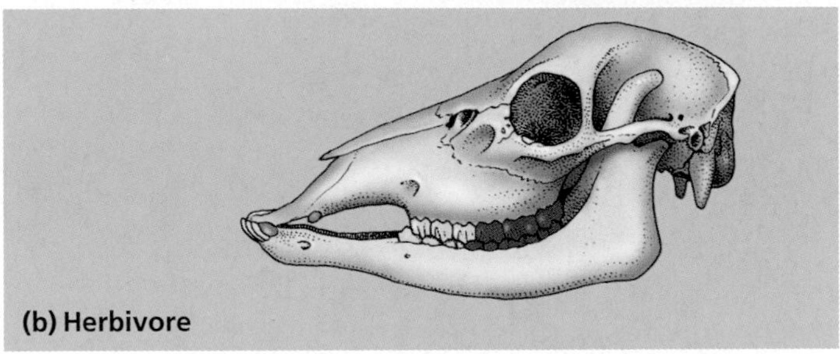

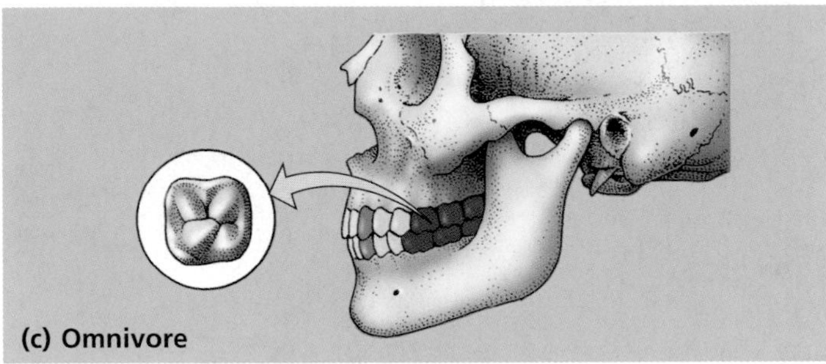

▲ **Figure 41.26 Dentition and diet. (a)** Carnivores, such as members of the dog and cat families, generally have pointed incisors and canines that can be used to kill prey and rip or cut away pieces of flesh. The jagged premolars and molars crush and shred food. **(b)** In contrast, herbivorous mammals, such as horses and deer, usually have teeth with broad, ridged surfaces that grind tough plant material. The incisors and canines are generally modified for biting off pieces of vegetation. In some herbivorous mammals, canines are absent. **(c)** Humans, being omnivores adapted for eating both vegetation and meat, have a relatively unspecialized dentition. The permanent (adult) set of teeth is 32 in number. Beginning at the midline of the upper and lower jaw are two bladelike incisors for biting, a pointed canine for tearing, two premolars for grinding, and three molars for crushing.

Stomach and Intestinal Adaptations

Large, expandable stomachs are common in carnivores, which may go for a long time between meals and therefore must eat as much as they can when they do catch prey. For example, a 200-kg African lion can consume 40 kg of meat in one meal.

The length of the vertebrate digestive system is also correlated with diet. In general, herbivores and omnivores have longer alimentary canals relative to their body size than carnivores **(Figure 41.27)**. Vegetation is more difficult to digest than meat because it contains cell walls. A longer tract furnishes more time for digestion and more surface area for the absorption of nutrients.

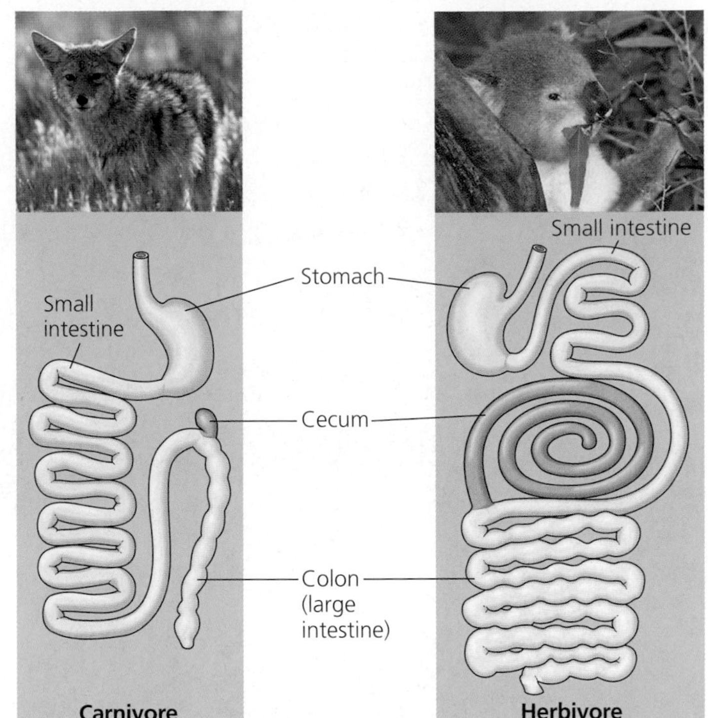

▲ **Figure 41.27 The digestive tracts of a carnivore (coyote) and herbivore (koala) compared.** Although these two mammals are about the same size, the koala's intestines are much longer, an adaptation that enhances processing of fibrous, protein-poor eucalyptus leaves from which it obtains virtually all its food and water. Extensive chewing chops the leaves into very small pieces, increasing exposure of the food to digestive juices. The koala's cecum—at 2 m, the longest of any animal of equivalent size—functions as a fermentation chamber where symbiotic bacteria convert the shredded leaves into a more nutritious diet.

Symbiotic Adaptations

Herbivorous animals face a nutritional challenge: Much of the chemical energy in their diets is contained in the cellulose of plant cell walls, but animals do not produce enzymes that hydrolyze cellulose. Many vertebrates (as well as termites, whose wood diets are largely cellulose) solve this problem by housing large populations of symbiotic bacteria and protists in fermentation chambers in their alimentary canals. These microorganisms *do* have enzymes that can digest cellulose to simple sugars and other compounds that the animal can absorb. In many cases, the microorganisms can also use the sugars from digested cellulose along with minerals to make a variety of nutrients essential to the animal, such as vitamins and amino acids.

The location of symbiotic microbes in the digestive tracts of herbivores varies, depending on the type of animal. The hoatzin, an herbivorous bird that lives in South American rain forests, has a large, muscular crop (an esophageal pouch) that houses symbiotic microorganisms. Hard ridges in the wall of the crop grind plant leaves into small fragments, and the microorganisms break down cellulose. Many herbivorous mammals, including horses, house

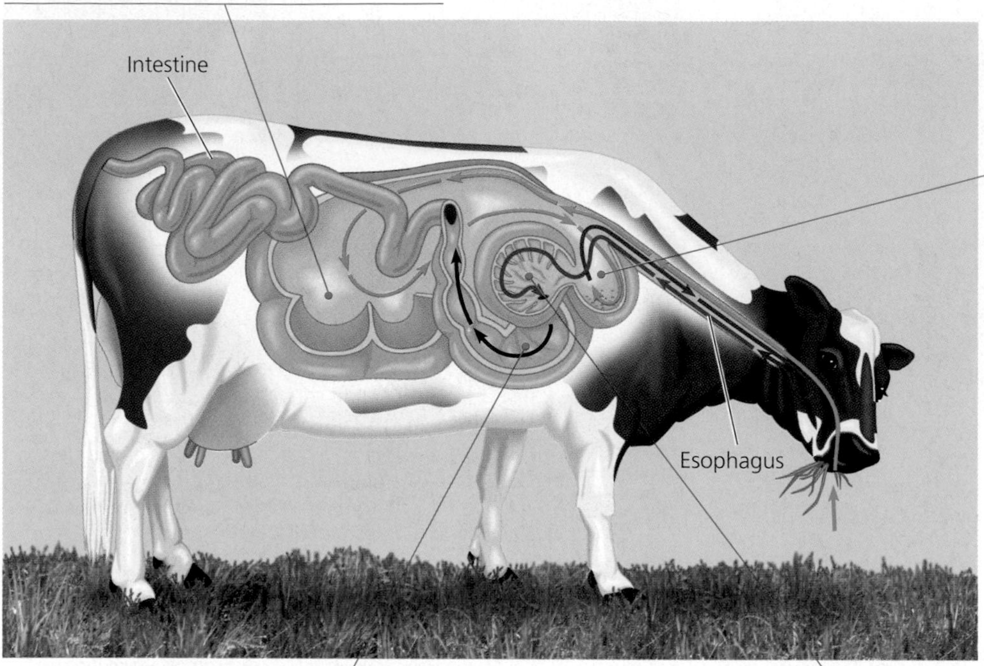

1 Rumen. When the cow first chews and swallows a mouthful of grass, boluses (green arrows) enter the rumen.
Intestine
2 Reticulum. Some boluses also enter the reticulum. In both the rumen and the reticulum, symbiotic prokaryotes and protists (mainly ciliates) go to work on the cellulose-rich meal. As by-products of their metabolism, the microorganisms secrete fatty acids. The cow periodically regurgitates and rechews the cud (red arrows), which further breaks down the fibers, making them more accessible to further microbial action.
Esophagus
4 Abomasum. The cud, containing great numbers of microorganisms, finally passes to the abomasum for digestion by the cow's own enzymes (black arrows).
3 Omasum. The cow then reswallows the cud (blue arrows), which moves to the omasum, where water is removed.

▲ **Figure 41.28 Ruminant digestion.** The stomach of a ruminant has four chambers. Because of the microbial action in the chambers, the diet from which a ruminant actually absorbs its nutrients is much richer than the grass the animal originally ate. In fact, a ruminant eating grass or hay obtains many of its nutrients by digesting the symbiotic microorganisms, which reproduce rapidly enough in the rumen to maintain a stable population.

symbiotic microorganisms in a large cecum, the pouch where the small and large intestines connect. The symbiotic bacteria of rabbits and some rodents live in the large intestine as well as in the cecum. Since most nutrients are absorbed in the small intestine, nourishing by-products of fermentation by bacteria in the large intestine are initially lost with the feces. Rabbits and rodents recover these nutrients by eating some of their feces and passing the food through the alimentary canal a second time. (The familiar rabbit "pellets," which are not reingested, are the feces eliminated after food has passed through the digestive tract twice.) The koala, an Australian marsupial, also has an enlarged cecum, where symbiotic bacteria ferment finely shredded eucalyptus leaves (see Figure 41.27). The most elaborate adaptations for an herbivorous diet have evolved in the animals called **ruminants,** which include deer, cattle, and sheep **(Figure 41.28)**.

In the next chapter, we will see that obtaining food, digesting it, and absorbing nutrients are only parts of a larger story. Provisioning the body also involves distributing nutrients to cells throughout the body (circulation) and exchanging respiratory gases with the environment.

Concept Check 41.5

1. Explain how human dentition is adapted for an omnivorous diet.
2. Compared with an adult frog, a tadpole (frog larva) has a much longer intestine relative to its body size. What does this suggest about the diets of these two stages in a frog's life history?
3. "Chewing its cud" is a common expression about cattle. What is the cud, and what role does it play in bovine nutrition?

For suggested answers, see Appendix A.

Chapter 41 Review

Go to www.campbellbiology.com or the student CD-ROM to explore Activities, Investigations, and other interactive study aids.

SUMMARY OF KEY CONCEPTS

▶ An animal's diet must supply chemical energy, organic raw materials, and essential nutrients. Herbivores mainly eat plants, carnivores mainly eat other animals, and omnivores eat both plant and animal matter. Many aquatic animals are suspension feeders, sifting small particles from the water. Substrate feeders tunnel through their food, eating as they go. Fluid feeders suck nutrient-rich fluids from a living host. Most animals are bulk feeders, eating large pieces of food (pp. 844–845).
Activity *Feeding Mechanisms of Animals*

Homeostatic mechanisms manage an animal's energy budget

▶ **Glucose Regulation as an Example of Homeostasis (p. 846)** Vertebrates store excess calories as glycogen in the liver and muscles and as fat. These energy stores can be tapped when an animal expends more calories than it consumes. Blood glucose level is maintained within a relatively narrow range by a negative feedback mechanism.

▶ **Caloric Imbalance (pp. 846–848)** Undernourished animals have diets deficient in calories. Overnourished (obese) animals consume more calories than they need. Obesity is a serious health problem worldwide and especially in the United States, where lack of exercise and fattening foods make an unhealthy combination. Obesity is also strongly influenced by genes. The problem of maintaining a healthy weight partly stems from our evolutionary past, when fat hoarding may have been important for survival.
Activity *Case Studies of Nutritional Disorders*

Concept 41.2

An animal's diet must supply carbon skeletons and essential nutrients

▶ Carbon skeletons are required in biosynthesis. Essential nutrients must be supplied in preassembled form. Malnourished animals are missing one or more essential nutrients in their diet (p. 849).

▶ **Essential Amino Acids (pp. 849–850)** Animals require 20 amino acids and can synthesize about half of them from the other molecules they obtain from their diet. Essential amino acids are those an animal cannot make. An animal whose diet is lacking one or more essential amino acids will become malnourished and suffer protein deficiency.

▶ **Essential Fatty Acids (p. 850)** Essential fatty acids, which an animal cannot make, are unsaturated, meaning that they have double bonds. Deficiencies in essential fatty acids are rare.

▶ **Vitamins (pp. 850–851)** Vitamins are organic molecules required in small amounts. They are either water-soluble or fat-soluble.

▶ **Minerals (pp. 851–852)** Minerals are inorganic nutrients, usually required in small amounts.
Activity *Analyzing Food Labels*

Concept 41.3

The main stages of food processing are ingestion, digestion, absorption, and elimination

▶ Food processing in animals involves ingestion (the act of eating), digestion (enzymatic breakdown of the macromolecules of food into their monomers), absorption (the uptake of nutrients by body cells), and elimination (the passage of undigested materials out of the body in feces) (p. 853).

▶ **Digestive Compartments (pp. 853–855)** In intracellular digestion, food particles are engulfed by endocytosis and digested within food vacuoles. Most animals use extracellular digestion, with enzymatic hydrolysis occurring outside cells in a gastrovascular cavity or alimentary canal.

Concept 41.4

Each organ of the mammalian digestive system has specialized food-processing functions

▶ The mammalian digestive system is composed of the alimentary canal and accessory glands that secrete digestive juices into the canal (pp. 855–856).

▶ **The Oral Cavity, Pharynx, and Esophagus (p. 856)** Food is lubricated and digestion begins in the oral cavity, where teeth chew food into smaller particles that are exposed to salivary amylase, initiating the breakdown of glucose polymers. The pharynx is the intersection leading to the trachea and esophagus. The esophagus conducts food from the pharynx to the stomach by involuntary peristaltic waves.

▶ **The Stomach (pp. 857–858)** The stomach stores food and secretes gastric juice, which converts a meal to acid chyme. Gastric juice includes hydrochloric acid and the enzyme pepsin.

▶ **The Small Intestine (pp. 858–861)** The small intestine is the major organ of digestion and absorption. Acid chyme from the stomach mixes in the duodenum with intestinal juice, bile, and pancreatic juice. Diverse enzymes complete the hydrolysis of food molecules to monomers, which are absorbed into the blood and lymph across the lining of the small intestine. Hormones help regulate digestive juice secretions.

▶ **The Large Intestine (pp. 861–862)** The large intestine (colon) aids the small intestine in reabsorbing water and houses bacteria, some of which synthesize vitamins. Feces pass through the rectum and out the anus.
Activity *Digestive System Function*
Investigation *What Role Does Amylase Play in Digestion?*
Activity *Hormonal Control of Digestion*

Concept 41.5

Evolutionary adaptations of vertebrate digestive systems are often associated with diet

▶ **Some Dental Adaptations (pp. 862–863)** A mammal's dentition is generally correlated with its diet. In particular, mammals have specialized dentition that best enables them to ingest their usual diet.

▶ **Stomach and Intestinal Adaptations (p. 863)** Herbivores generally have longer alimentary canals than carnivores, reflecting the longer time needed to digest vegetation.

▶ **Symbiotic Adaptations (pp. 863–864)** Many herbivorous animals have fermentation chambers where symbiotic microorganisms digest cellulose.

TESTING YOUR KNOWLEDGE

Self-Quiz

1. Which of the following animals is *incorrectly* paired with its feeding mechanism?
   a. lion—substrate feeder
   b. baleen whale—suspension feeder
   c. aphid—fluid feeder
   d. clam—suspension feeder
   e. snake—bulk feeder

2. If you were to jog a mile a few hours after lunch, which stored fuel would you probably tap?
   a. muscle proteins
   b. muscle and liver glycogen
   c. fat stored in the liver
   d. fat stored in adipose tissue
   e. blood proteins

3. Individuals whose diet consists primarily of corn would likely become
   a. obese.
   b. anorexic.
   c. overnourished.
   d. undernourished.
   e. malnourished.

4. The mammalian trachea and esophagus both open into the
   a. large intestine.
   b. stomach.
   c. pharynx.
   d. rectum.
   e. epiglottis.

5. Which of the following enzymes has the lowest pH optimum?
   a. salivary amylase
   b. trypsin
   c. pepsin
   d. pancreatic amylase
   e. pancreatic lipase

6. Which of the following organs is *incorrectly* paired with its function?
   a. stomach—protein digestion
   b. oral cavity—starch digestion
   c. large intestine—bile production
   d. small intestine—nutrient absorption
   e. pancreas—enzyme production

7. Enteropeptidase, an enzyme bound to the intestinal epithelium, has which of the following actions?
   a. inhibits bile secretion
   b. inhibits duodenal secretion
   c. activates pancreatic enzymes
   d. inhibits peristalsis in the stomach
   e. increases the pH of chyme

8. After surgical removal of an infected gallbladder, a person must be especially careful to restrict his or her dietary intake of
   a. starch.
   b. protein.
   c. sugar.
   d. fat.
   e. water.

9. Our oral cavity, with its dentition, is most functionally analogous to an earthworm's
   a. intestine.
   b. pharynx.
   c. gizzard.
   d. stomach.
   e. anus.

10. The symbiotic microbes that help nourish a ruminant live mainly in specialized regions of the
    a. large intestine.
    b. liver.
    c. small intestine.
    d. pharynx.
    e. stomach.

For Self-Quiz answers, see Appendix A.

Go to the website or CD-ROM for more quiz questions.

Evolution Connection

The human esophagus and trachea share a common passage leading from the mouth and nasal passages, a "design" that occasionally contributes to death by choking. After reviewing vertebrate evolution in Chapter 34, explain the historical (evolutionary) basis for this "imperfect" anatomy.

Scientific Inquiry

Design a controlled experiment to test the hypothesis that human salivary amylase digests starch faster at 37°C (human body temperature) than it does at either 20°C (approximate room temperature) or 43°C (110°F). Your only materials and equipment are a source of human saliva, distilled water, starch, an iodine reagent that stains starch dark purple, beakers, and several water baths that can be maintained at constant temperatures. How would you interpret the results if (a) the rate of enzyme activity were highest at 37°C; or (b) the rate of enzyme activity were highest at 43°C?

Investigation ***What Role Does Amylase Play in Digestion?***
Biological Inquiry: A Workbook of Investigative Cases ***Explore several mammalian mechanisms for starch digestion in the case "Galloper's Gut."***

Science, Technology, and Society

The media report numerous claims and counterclaims about the benefits and dangers of certain foods. Just a few examples are debates about vitamin doses, advocacy of diets enriched in certain food molecules such as carbohydrates or proteins, much discussion about low-carbohydrate diets, and publicity about new products such as cholesterol-lowering margarine. Have you modified your eating habits on the basis of nutritional information disseminated by the media? Why or why not? How should a person evaluate whether such nutritional claims are valid?

42 Circulation and Gas Exchange

▲ **Figure 42.1 The gills of a salmon, which exchange gases between the blood and the environment.**

Key Concepts

42.1 Circulatory systems reflect phylogeny
42.2 Double circulation in mammals depends on the anatomy and pumping cycle of the heart
42.3 Physical principles govern blood circulation
42.4 Blood is a connective tissue with cells suspended in plasma
42.5 Gas exchange occurs across specialized respiratory surfaces
42.6 Breathing ventilates the lungs
42.7 Respiratory pigments bind and transport gases

Overview

Trading with the Environment

Every organism must exchange materials and energy with its environment, and this exchange ultimately occurs at the cellular level. Cells live in aqueous surroundings; the resources they need, such as nutrients and oxygen, move across the plasma membrane into the cytoplasm, and metabolic wastes, such as carbon dioxide, move out of the cell. In unicellular organisms, these exchanges occur directly with the external environment. For most of the cells making up multicellular organisms, however, direct exchange with the environment is not possible. This constraint is associated with the evolution of physiological systems specialized for material transport and exchange.

The feathery gills of a salmon **(Figure 42.1)** present an expansive surface area to the outside environment. Networks of tiny blood vessels (capillaries) lie close to the outside surfaces of the gills. Oxygen dissolved in the surrounding water diffuses across the thin epithelium covering the gills and into the blood, while carbon dioxide diffuses out into the water.

Salmon and most other animals have organ systems specialized for exchanging materials with the environment, and most also have an internal transport system that conveys fluid (blood or interstitial fluid) throughout the body (see Figure 40.4).

In this chapter, we will explore mechanisms of internal transport in animals. We will also examine a key example of chemical transfer between animals and their environment: the exchange of the gases oxygen (O_2) and carbon dioxide (CO_2), a process essential to cellular respiration and bioenergetics.

Concept 42.1

Circulatory systems reflect phylogeny

Diffusion alone is not adequate for transporting substances over long distances in animals—for example, for moving glucose from the digestive tract and oxygen from the lungs to the brain of a mammal. Diffusion is inefficient over distances of more than a few millimeters, because the time it takes for a substance to diffuse from one place to another is proportional to the *square* of the distance. For example, if it takes 1 second for a given quantity of glucose to diffuse 100 µm, it will take 100 seconds for the same quantity to diffuse 1 mm and almost 3 hours to diffuse 1 cm. The circulatory system solves this problem by ensuring that no substance must diffuse very far to enter or leave a cell. By rapidly transporting fluid in bulk throughout the body, the circulatory system functionally connects the aqueous environment of the body cells to the organs that exchange gases, absorb nutrients, and dispose of wastes. In the lungs of a mammal, for example, oxygen from inhaled air diffuses across a thin epithelium and into the blood, while carbon dioxide diffuses in the opposite direction. Bulk fluid movement in the circulatory system, powered by the heart, then quickly carries the oxygen-rich blood to all parts of the body. As the blood streams through the body tissues within capillaries, chemicals are transported between the blood and the interstitial fluid that directly bathes the cells.

Internal transport and gas exchange are functionally related in most animal phyla, and so in this chapter we will focus on both the circulatory and respiratory systems. We will also highlight the role of these two organ systems in maintaining homeostasis (see Chapter 40)—for example, in regulating the interstitial fluid's content of nutrients and wastes. First, let's look at circulation in invertebrate animals.

Invertebrate Circulation

The wide range of invertebrate body size and form is paralleled by diversity in circulatory systems. The different selective pressures of various environments have also led to evolutionary modification of circulatory systems among invertebrates.

Gastrovascular Cavities

Owing to the simplicity of their body plan, hydras and other cnidarians do not require a true circulatory system. In these animals, a body wall only two cells thick encloses a central gastrovascular cavity, which serves both in digestion and in distribution of substances throughout the body (see Figure 41.13). The fluid inside the cavity is continuous with the water outside through a single opening; thus, both the inner and outer tissue layers are bathed by fluid. Thin branches of a hydra's gastrovascular cavity extend into the animal's tentacles, and some cnidarians, such as jellies, have even more elaborate gastrovascular cavities **(Figure 42.2)**. Since digestion begins in the cavity, only the cells of the inner layer have direct access to nutrients, but the nutrients have to diffuse only a short distance to reach the cells of the outer layer.

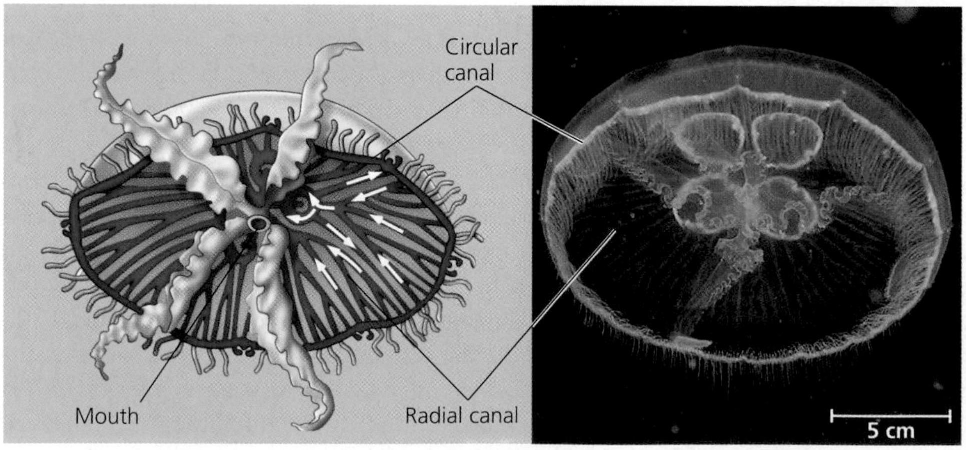

Planarians and most other flatworms also have gastrovascular cavities that exchange materials with the environment through a single opening (see Figure 33.10). The flat shape of the body and the branching of the gastrovascular cavity throughout the animal ensure that all cells are bathed by a suitable medium and that diffusion distances are short.

▲ **Figure 42.2 Internal transport in the cnidarian *Aurelia*.** The animal is viewed here from its underside (oral surface). The mouth leads to an elaborate gastrovascular cavity that has branches radiating to and from a circular canal. Ciliated cells lining the canals circulate fluid in the directions indicated by the arrows.

Open and Closed Circulatory Systems

For animals with many cell layers, gastrovascular cavities are insufficient for internal transport because diffusion distances are too great for adequate exchange of nutrients and wastes. In such animals, two types of circulatory systems that overcome the limitations of diffusion have evolved: open and closed. Both have three basic components: a circulatory fluid (**blood**), a set of tubes (**blood vessels**) through which the blood moves through the body, and a muscular pump (the **heart**). The heart powers circulation by using metabolic energy to elevate the hydrostatic pressure of the blood, which then flows down a pressure gradient through its circuit and back to the heart. This **blood pressure** is the motive force for fluid movement in the circulatory system.

In insects, other arthropods, and most molluscs, blood bathes the organs directly in an **open circulatory system** **(Figure 42.3a)**. There is no distinction between blood and interstitial fluid, and this general body fluid is more correctly termed **hemolymph.** One or more hearts pump the hemolymph into an interconnected system of **sinuses,** which are spaces surrounding the organs. Here, chemical exchange occurs between the hemolymph and body cells. In insects and other arthropods, the heart is an elongated tube located dorsally. When the heart contracts, it pumps hemolymph through vessels out into sinuses. When the heart relaxes, it draws hemolymph into the circulatory system through pores called ostia. Body movements that squeeze the sinuses help circulate the hemolymph.

In a **closed circulatory system,** blood is confined to vessels and is distinct from the interstitial fluid **(Figure 42.3b)**. One or more hearts pump blood into large vessels that branch into smaller ones coursing through the organs. Here, materials are exchanged by diffusion between the blood and the interstitial fluid bathing the cells. Earthworms, squids, octopuses, and all vertebrates have closed circulatory systems.

The fact that open and closed circulatory systems are each widespread among animals suggests that both offer advantages. For example, the lower hydrostatic

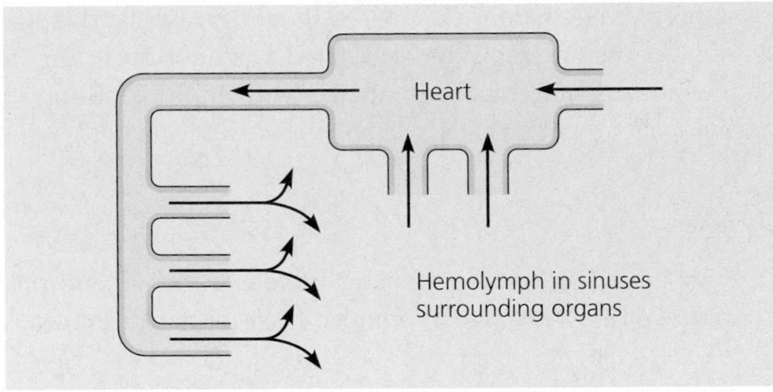

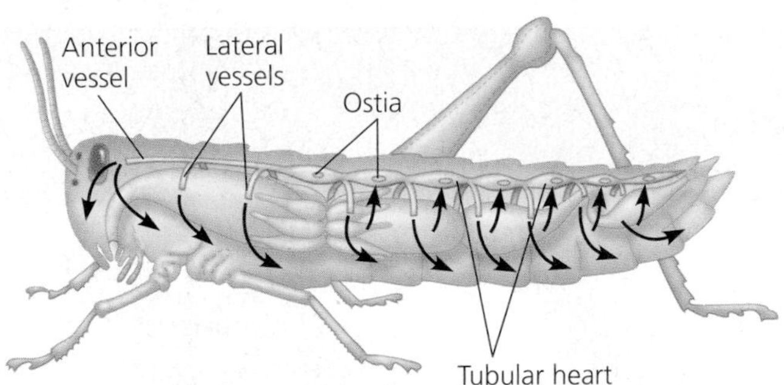

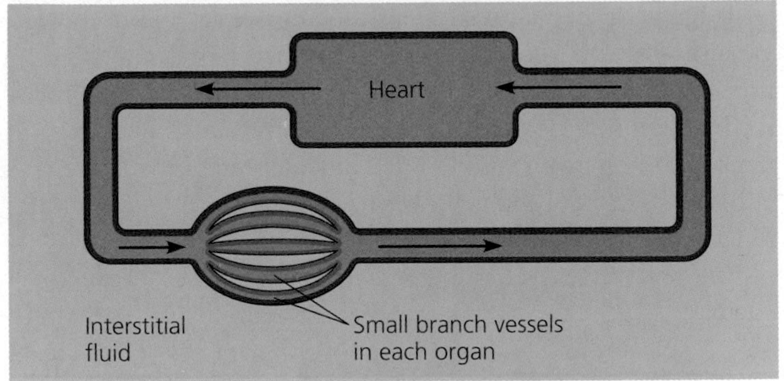

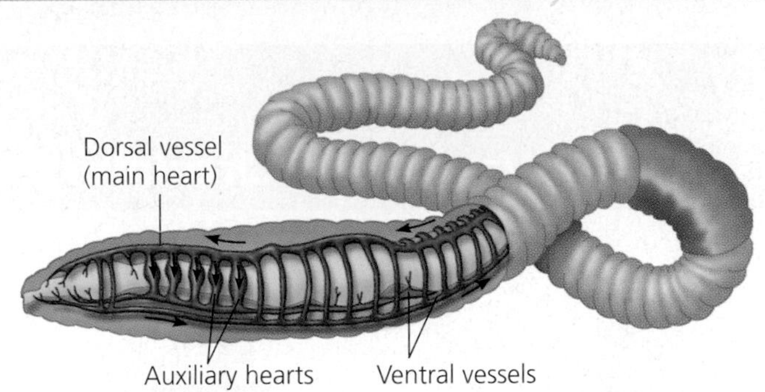

(a) An open circulatory system. In an open circulatory system, such as that of a grasshopper, blood and interstitial fluid are the same, and this fluid is called hemolymph. The heart pumps hemolymph through vessels into sinuses, where materials are exchanged between the hemolymph and cells. Hemolymph returns to the heart through ostia, which are equipped with valves that close when the heart contracts.

(b) A closed circulatory system. Closed circulatory systems circulate blood entirely within vessels, distinct from the interstitial fluid. Chemical exchange occurs between the blood and the interstitial fluid, and between the interstitial fluid and body cells. In an earthworm, three major vessels branch into smaller vessels that supply blood to the various organs. The dorsal vessel functions as the main heart, pumping blood forward by peristalsis. Near the worm's anterior end, five pairs of vessels loop around the digestive tract and function as auxiliary hearts, propelling blood ventrally.

▲ **Figure 42.3 Open and closed circulatory systems.**

pressures associated with open circulatory systems make them less costly than closed systems in terms of energy expenditure. Furthermore, because they lack an extensive system of blood vessels, open systems require less energy to build and maintain. And in some invertebrates, open circulatory systems serve a variety of other functions. For example, in molluscs and freshly molted aquatic arthropods, the open circulatory system functions as a hydrostatic skeleton in supporting the body.

What advantages might be associated with closed circulatory systems? Closed systems, with their higher blood pressure, are more effective at transporting circulatory fluids to meet the high metabolic demands of the tissues and cells of larger and more active animals. For instance, among the molluscs, only the large and active squids and octopuses have closed circulatory systems. And although all arthropods have open circulatory systems, the larger crustaceans, such as the lobsters and crabs, have a more developed system of arteries and veins as well as an accessory pumping organ that helps maintain blood pressure. Closed circulatory systems are most highly developed in the vertebrates.

Survey of Vertebrate Circulation

Humans and other vertebrates have a closed circulatory system, often called the **cardiovascular system.** Generally, the vertebrate heart has one or two **atria** (singular, *atrium*), the chambers that receive blood returning to the heart, and one or two **ventricles,** the chambers that pump blood out of the heart.

Arteries, veins, and **capillaries** are the three main kinds of blood vessels, which in the human body have a total length of about 100,000 km. Arteries carry blood away from the heart to organs throughout the body. Within organs, arteries branch into **arterioles,** small vessels that convey blood to the capillaries.

Capillaries are microscopic vessels with very thin, porous walls. Networks of these vessels, called **capillary beds,** infiltrate each tissue. Across the thin walls of capillaries, chemicals, including dissolved gases, are exchanged by diffusion between blood and the interstitial fluid around the tissue cells.

At their "downstream" end, capillaries converge into **venules,** and venules converge into veins. Generally speaking, veins return blood to the heart. Notice that arteries and veins are distinguished by the *direction* in which they carry blood, not by the characteristics of the blood they contain. All arteries carry blood from the heart *toward* capillaries, and veins return blood to the heart *from* capillaries. A significant exception is the hepatic portal vein that carries blood from capillary beds in the digestive system to capillary beds in the liver. Blood flowing from the liver passes into the hepatic vein, which conducts blood to the heart.

The cardiovascular systems of different vertebrate taxa are variations of this general scheme, modified by natural selection. Metabolic rate (see Chapter 40) is an important factor in the evolution of cardiovascular systems. In general, animals with higher metabolic rates have more complex circulatory systems and more powerful hearts than animals with lower metabolic rates. Similarly, within an animal, the complexity and number of blood vessels in a particular organ are correlated with that organ's metabolic requirements. Perhaps the most fundamental differences in cardiovascular adaptations between animals are associated with gill breathing in most aquatic vertebrates compared with lung breathing in terrestrial vertebrates.

Fishes

A fish heart has two main chambers, one ventricle and one atrium **(Figure 42.4)**. Blood pumped from the ventricle travels

Figure 42.4

Exploring Vertebrate Circulatory Systems

FISHES

Fishes have a two-chambered heart and a single circuit of blood flow.

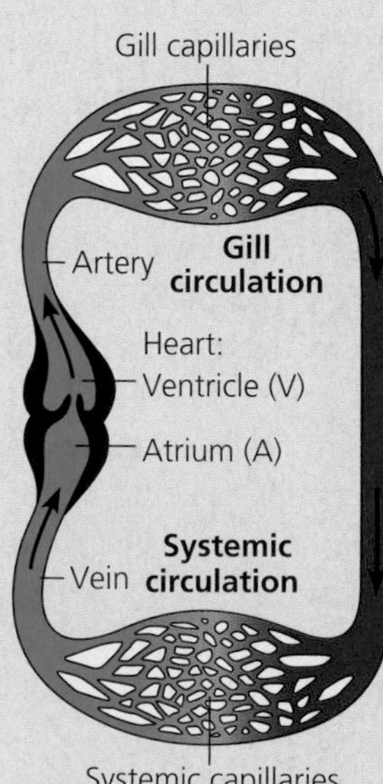

AMPHIBIANS

Amphibians have a three-chambered heart and two circuits of blood flow: pulmocutaneous and systemic. Some mixing of oxygen-rich and oxygen-poor blood occurs in the single ventricle.

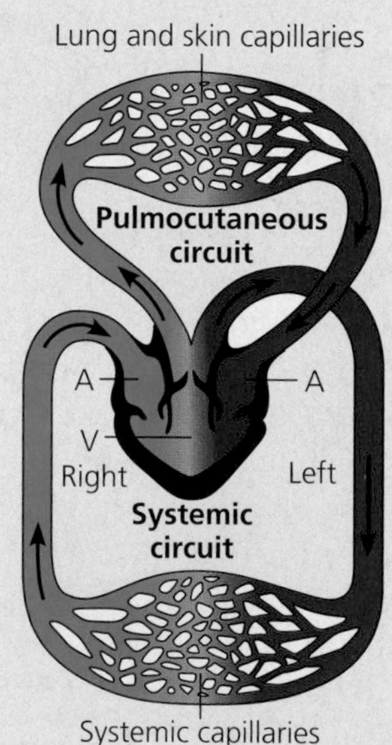

REPTILES (EXCEPT BIRDS)

Lizards, snakes, and turtles have a three-chambered heart. However, a septum partially divides the single ventricle, further reducing mixing of oxygen-rich and oxygen-poor blood. (In crocodilians, the septum completely divides the ventricles.)

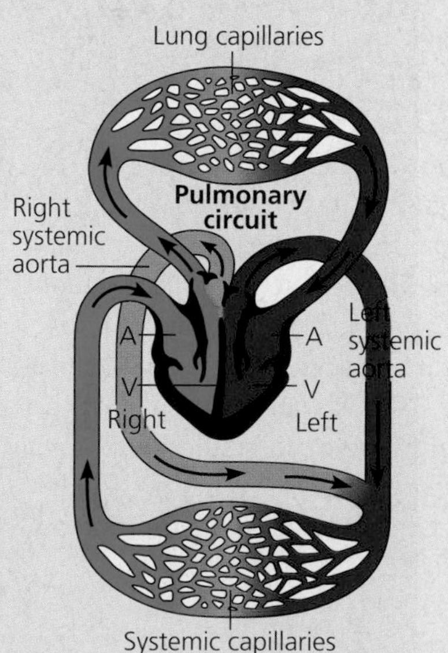

MAMMALS AND BIRDS

Mammals and birds have a four-chambered heart that completely segregates oxygen-rich and oxygen-poor blood. (The major vessels near the heart are slightly different in birds, but the pattern of double circulation is essentially the same as depicted here.)

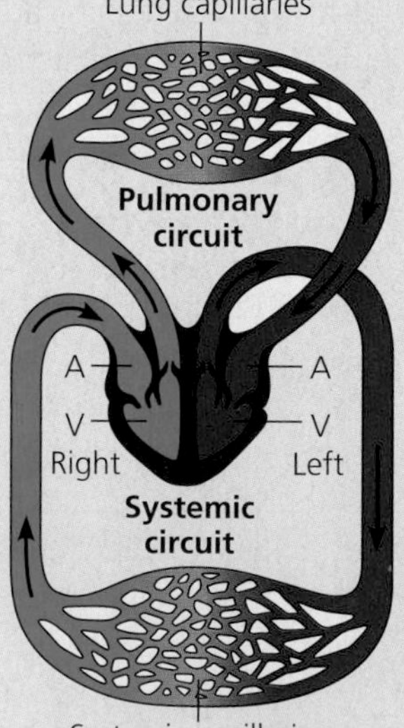

Systemic circuits include all body tissues except lungs. Note that circulatory systems are depicted as if the animal is facing you: with the right side of the heart shown at the left and vice-versa.

first to the gills (the **gill circulation**), where it picks up oxygen (O_2) and disposes of carbon dioxide (CO_2) across capillary walls. The gill capillaries converge into a vessel that carries oxygen-rich blood to capillary beds throughout all other parts of the body (the **systemic circulation**). Blood then returns in veins to the atrium of the heart. Notice that in a fish, blood must pass through *two* capillary beds during each circuit. When blood flows through a capillary bed, blood pressure—the motive force for circulation—drops substantially (for reasons we will explain shortly). Therefore, oxygen-rich blood leaving the gills flows to the systemic circulation quite slowly (although the process is aided by body movements during swimming). This constrains the delivery of O_2 to body tissues and hence the maximum aerobic metabolic rate of fishes.

Amphibians

Frogs and other amphibians have a three-chambered heart, with two atria and one ventricle (see Figure 42.4). The ventricle pumps blood into a forked artery that splits the ventricle's output into the **pulmocutaneous circuit** and the **systemic circuit.** The pulmocutaneous circuit leads to capillaries in the gas exchange organs (the lungs and skin in a frog), where the blood picks up O_2 and releases CO_2 before returning to the heart's left atrium. Most of the returning oxygen-rich blood is pumped into the systemic circuit, which supplies all organs and then returns oxygen-poor blood to the right atrium via the veins. In the ventricle of the frog, there is some mixing of oxygen-rich blood that has returned from the lungs with oxygen-poor blood that has returned from the rest of the body. However, a ridge within the ventricle diverts most of the oxygen-rich blood from the left atrium into the systemic circuit and most of the oxygen-poor blood from the right atrium into the pulmocutaneous circuit.

This organization, called **double circulation,** provides a vigorous flow of blood to the brain, muscles, and other organs because the blood is pumped a second time after it loses pressure in the capillary beds of the lungs or skin. This contrasts sharply with single circulation in fishes, in which blood flows directly from the respiratory organs (gills) to other organs under reduced pressure.

Reptiles (Except Birds)

Reptiles have double circulation with a **pulmonary circuit** (lungs) and a systemic circuit (see Figure 42.4). Turtles, snakes, and lizards have a three-chambered heart, although the ventricle is partially divided by a septum, which results in even less mixing of oxygen-rich and oxygen-poor blood than in amphibians. In the crocodilians, a septum divides the ventricle completely into separate right and left chambers. All reptiles except birds have two arteries leading from the heart to the systemic circuit, and arterial valves allow them to divert most of their blood from the pulmonary circuit to the systemic circuit.

Mammals and Birds

In all mammals and birds, the ventricle is completely divided into separate right and left chambers (see Figure 42.4). The left side of the heart receives and pumps only oxygen-rich blood, while the right side receives and pumps only oxygen-poor blood. Oxygen delivery is enhanced because there is no mixing of oxygen-rich and oxygen-poor blood, and double circulation restores pressure to the systemic circuit after blood has passed through the lung capillaries.

A powerful four-chambered heart was an essential adaptation in support of the endothermic way of life characteristic of mammals and birds. Endotherms use about ten times as much energy as equal-sized ectotherms; therefore, their circulatory systems need to deliver about ten times as much fuel and O_2 to their tissues (and remove ten times as much CO_2 and other wastes). This large traffic of substances is made possible by separate and independent systemic and pulmonary circulations and by large, powerful hearts that pump the necessary volume of blood. As we discussed in Chapter 25, mammals and birds descended from different reptilian ancestors, and their four-chambered hearts evolved independently—an example of convergent evolution.

Concept Check 42.1

1. What fundamental physical constraint necessitates a circulatory system in large organisms?
2. What is one advantage of a closed circulatory system? What is a disadvantage?
3. What are two physiological advantages of separate respiratory (pulmocutaneous or pulmonary) and systemic circuits over a single circuit as seen in fishes, which combines gill and systemic circulation?

For suggested answers, see Appendix A.

Double circulation in mammals depends on the anatomy and pumping cycle of the heart

Because heart disease is such a significant and widespread health problem in human populations, scientists have studied the human circulatory system in much greater detail than any other. Its structure and function can serve as a model for exploring mammalian circulation in general.

Mammalian Circulation: *The Pathway*

As you read this detailed discussion of blood flow through the mammalian cardiovascular system, refer to **Figure 42.5**, which has numbers keyed to the circled numbers in the text. Beginning our tour with the pulmonary (lung) circuit, ❶ the right ventricle pumps blood to the lungs via ❷ the pulmonary arteries. As the blood flows through ❸ capillary beds in the left and right lungs, it loads O_2 and unloads CO_2. Oxygen-rich blood returns from the lungs via the pulmonary veins to ❹ the left atrium of the heart. Next, the oxygen-rich blood flows into ❺ the left ventricle as the ventricle opens and the atrium contracts. The left ventricle pumps the oxygen-rich blood out to body tissues through the systemic circuit. Blood leaves the left ventricle via ❻ the aorta, which conveys blood to arteries leading throughout the body. The first branches from the aorta are the coronary arteries (not shown), which supply blood to the heart muscle itself. Then come branches leading to capillary beds ❼ in the head and arms (forelimbs). The aorta continues in a posterior direction, supplying oxygen-rich blood to arteries leading to ❽ arterioles and capillary beds in the abdominal organs and legs. Within the capillaries, O_2 and CO_2 diffuse along their concentration gradients, with O_2 moving from the blood to the tissues and CO_2 produced by cellular respiration diffusing into the bloodstream. Capillaries rejoin, forming venules, which convey blood to veins. Oxygen-poor blood from the head, neck, and forelimbs is channeled into a large vein called ❾ the anterior (or superior) vena cava. Another large vein called ❿ the posterior (or inferior) vena cava drains blood from the trunk and hind limbs. The two venae cavae empty their blood into ⓫ the right atrium, from which the oxygen-poor blood flows into the right ventricle.

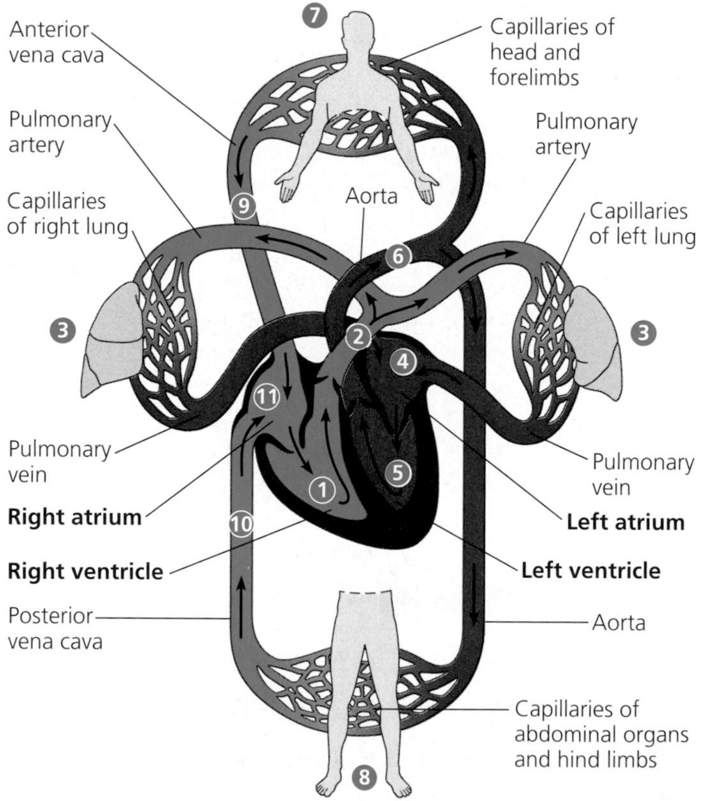

▲ **Figure 42.5 The mammalian cardiovascular system: an overview.** Note that the dual circuits operate simultaneously, not in the serial fashion that the numbering in the diagram suggests. The two ventricles pump almost in unison; while some blood is traveling in the pulmonary circuit, the rest of the blood is flowing in the systemic circuit.

The Mammalian Heart: *A Closer Look*

A closer look at the mammalian heart (using the human heart as an example) provides a better understanding of how double circulation works **(Figure 42.6)**. Located beneath the breastbone (sternum), the human heart is about the size of a clenched fist and consists mostly of cardiac muscle (see Figure 40.5). The two atria have relatively thin walls and serve as collection chambers for blood returning to the heart, most of which flows into the ventricles as they relax. Contraction of the atria completes filling of the ventricles. The ventricles have thicker walls and contract much more strongly than the atria—especially the left ventricle, which must pump blood to all body organs through the systemic circuit.

The heart contracts and relaxes in a rhythmic cycle. When it contracts, it pumps blood; when it relaxes, its chambers fill with blood. The **cardiac cycle** refers to one complete sequence of pumping and filling. The contraction phase of the cycle is called **systole**, and the relaxation phase is **diastole (Figure 42.7)**. The volume of blood per minute that the left

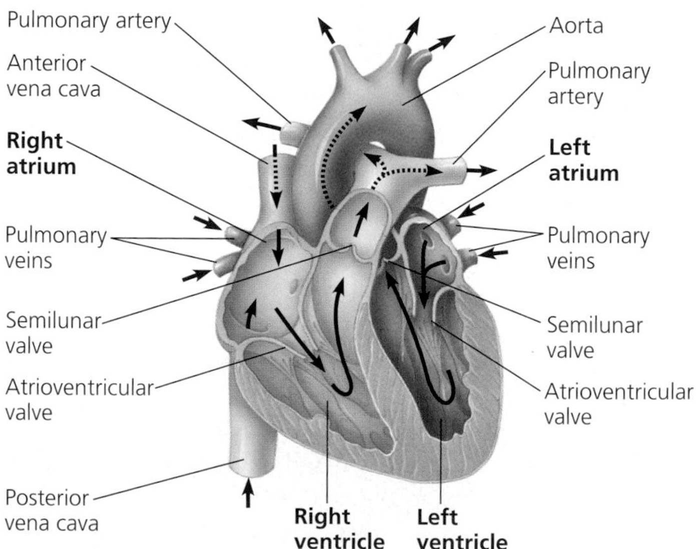

▲ **Figure 42.6 The mammalian heart: a closer look.** Notice the valves, which prevent backflow of blood within the heart, and the relative thickness of the walls of the heart chambers.

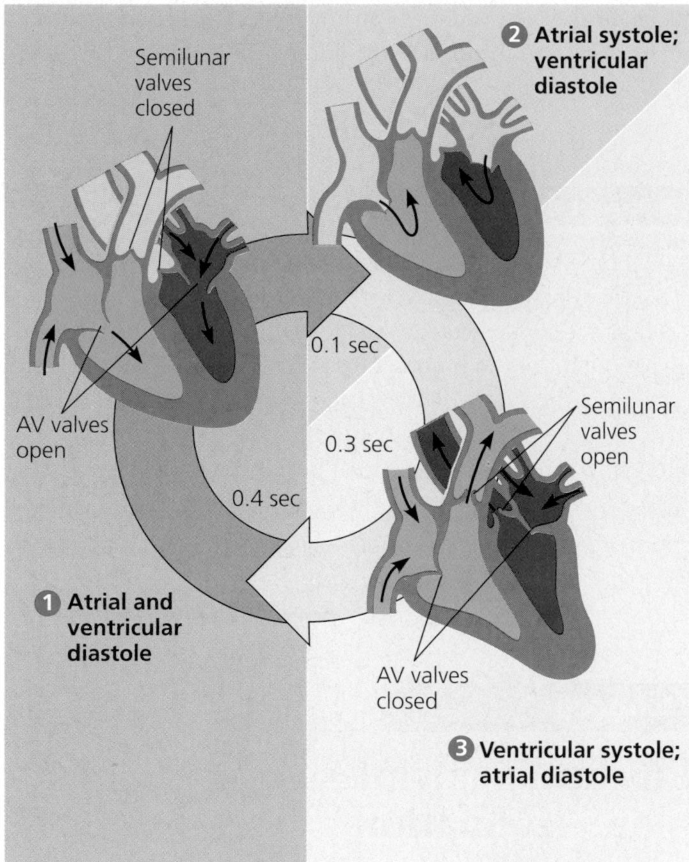

▲ **Figure 42.7 The cardiac cycle.** For an adult human at rest with a pulse of about 75 beats per minute, one complete cardiac cycle takes about 0.8 second. 1 During a relaxation phase (atria and ventricles in diastole), blood returning from the large veins flows into the atria and ventricles. 2 A brief period of atrial systole then forces all remaining blood out of the atria into the ventricles. 3 During the remainder of the cycle, ventricular systole pumps blood into the large arteries. Note that seven-eighths of the time—all but 0.1 second of the cardiac cycle—the atria are relaxed and are filling with blood returning via the veins.

ventricle pumps into the systemic circuit is called **cardiac output.** Cardiac output depends on two factors: the rate of contraction, or **heart rate** (number of beats per minute); and **stroke volume,** the amount of blood pumped by the left ventricle in each contraction. The average stroke volume in humans is about 75 mL. A person with this stroke volume and a heart rate at rest of 70 beats per minute has a cardiac output of 5.25 L/min—about equivalent to the total volume of blood in the human body. Cardiac output can increase as much as about fivefold during heavy exercise. This is equivalent to pumping an amount of blood matching an average person's body mass every 2–3 minutes.

Four valves in the heart, each consisting of flaps made of connective tissue, prevent backflow and keep blood moving in the correct direction (see Figure 42.6). Between each atrium and ventricle is an **atrioventricular (AV) valve.** The AV valves are anchored by strong fibers that prevent them from turning inside out. Pressure generated by the powerful contraction of the ventricles closes the AV valves, keeping blood from flowing back into the atria. **Semilunar valves** are located at the two exits of the heart: where the aorta leaves the left ventricle and the pulmonary artery leaves the right ventricle. These valves are forced open by pressure generated by contraction of the ventricles. When the ventricles relax, blood starts to flow back toward the heart, closing the semilunar valves, which prevents blood from flowing back into the ventricles. The elastic walls of the arteries expand when they receive the blood expelled from the ventricles. By feeling your **pulse**—the rhythmic stretching of arteries caused by the pressure of blood driven by the powerful contractions of the ventricles—you can measure your heart rate.

The heart sounds heard with a stethoscope are caused by the closing of the valves. (Even without a stethoscope, you can hear these sounds by pressing your ear tightly against the chest of a friend—a *close* friend.) The sound pattern is "lub-dup, lub-dup, lub-dup." The first heart sound ("lub") is created by the recoil of blood against the closed AV valves. The second sound ("dup") is the recoil of blood against the semilunar valves.

A defect in one or more valves causes a condition known as a **heart murmur,** which may be detectable as a hissing sound when a stream of blood squirts backward through a valve. Some people are born with heart murmurs; in others, the valves may be damaged by infection (from rheumatic fever, for instance). Most heart murmurs do not reduce the efficiency of blood flow enough to warrant surgery.

Maintaining the Heart's Rhythmic Beat

The timely delivery of oxygen to the body's organs is critical. For example, brain cells die within a few minutes if their oxygen supply is interrupted. Thus, maintaining heart function is crucial for survival. Several mechanisms have evolved that ensure the continuity and control of heartbeat.

Some cardiac muscle cells are self-excitable, meaning they contract without any signal from the nervous system, even if removed from the heart and placed in tissue culture. Each of these cells has its own intrinsic contraction rhythm. How are their contractions coordinated in the intact heart? A region of the heart called the **sinoatrial (SA) node,** or **pacemaker,** sets the rate and timing at which all cardiac muscle cells contract. Composed of specialized muscle tissue, the SA node is located in the wall of the right atrium, near the point where the superior vena cava enters the heart. Because the pacemaker of the human heart (and of other vertebrates) is made up of specialized muscle tissues and located within the heart itself, the vertebrate heart is referred to as a **myogenic heart.** In contrast, the pacemakers of most arthropod hearts originate in motor nerves arising from the outside, an arrangement called a **neurogenic heart.**

The SA node generates electrical impulses much like those produced by nerve cells. Because cardiac muscle cells are electrically coupled (by the intercalated disks between adjacent cells), impulses from the SA node spread rapidly through the walls of the atria, causing both atria to contract in unison **(Figure 42.8)**. The impulses also pass to another region of specialized cardiac muscle tissue, a relay point called the **atrioventricular (AV) node,** located in the wall between the right atrium and right ventricle. Here the impulses are delayed for about 0.1 second before spreading to the walls of the ventricles. The delay ensures that the atria empty completely before the ventricles contract. Specialized muscle fibers called bundle branches and Purkinje fibers then conduct the signals to the apex of the heart and throughout the ventricular walls.

The impulses that travel through cardiac muscle during the heart cycle produce electrical currents that are conducted through body fluids to the skin, where the currents can be detected by electrodes and recorded as an **electrocardiogram** (**ECG** or **EKG**).

The SA node sets the tempo for the entire heart, but is influenced by a variety of physiological cues. Two sets of nerves affect heart rate: one set speeds up the pacemaker, and the other set slows it down. Heart rate is a compromise regulated by the opposing actions of these two sets of nerves. The pacemaker is also influenced by hormones secreted into the blood by glands. For example, epinephrine, the "fight-or-flight" hormone secreted by the adrenal glands, increases heart rate (see Chapter 45). Body temperature is another factor that affects the pacemaker. An increase of only 1°C raises the heart rate by about 10 beats per minute. This is the reason your pulse increases substantially when you have a fever. Heart rate also increases with exercise, an adaptation that enables the circulatory system to provide the additional O_2 needed by muscles hard at work.

Concept Check 42.2

1. Some babies are born with a small hole between their left and right ventricles. Explain how, if not surgically corrected, this hole would affect the O_2 content of the blood entering the systemic circuit from the heart.
2. Why is it important that the AV node of the heart slow or delay the electrical impulse moving from the SA node and the atrial walls to the ventricles?

For suggested answers, see Appendix A.

Concept 42.3

Physical principles govern blood circulation

Blood delivers nutrients and removes wastes throughout an animal's body. These functions are made possible by the circulatory system, a branching network of vessels similar in some ways to the plumbing system that delivers fresh water to a city and removes the city's wastes. The same physical principles that govern the operation of such plumbing systems also influence the functioning of animal circulatory systems.

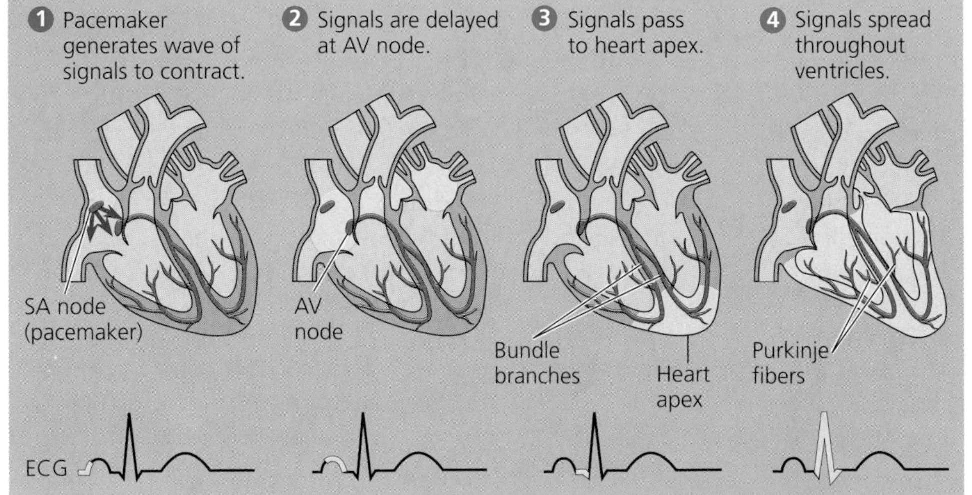
1 Pacemaker generates wave of signals to contract.
2 Signals are delayed at AV node.
3 Signals pass to heart apex.
4 Signals spread throughout ventricles.
SA node (pacemaker)
AV node
Bundle branches
Heart apex
Purkinje fibers
ECG

▲ **Figure 42.8 The control of heart rhythm.** The gold portions of the graphs at the bottom indicate the components of an electrocardiogram (ECG) corresponding to the sequence of electrical events in the heart. In step 4, the black portion of the ECG to the right of the gold "spike" represents electrical activity after the ventricles contract; during this phase, the ventricles become electrically re-primed and thus able to conduct the next round of contraction signals.

Blood Vessel Structure and Function

The "infrastructure" of the circulatory system consists of its network of blood vessels. All blood vessels are built of similar tissues. The walls of both arteries and veins, for instance, have three similar layers **(Figure 42.9)**. On the outside, a layer of connective tissue with elastic fibers allows the vessel to stretch and recoil. A middle layer contains smooth muscle and more elastic fibers. Lining the lumen of all blood vessels, including capillaries, is an **endothelium,** a single layer of flattened cells that provides a smooth surface that minimizes resistance to the flow of blood.

Structural differences correlate with the different functions of arteries, veins, and capillaries. Capillaries lack the two

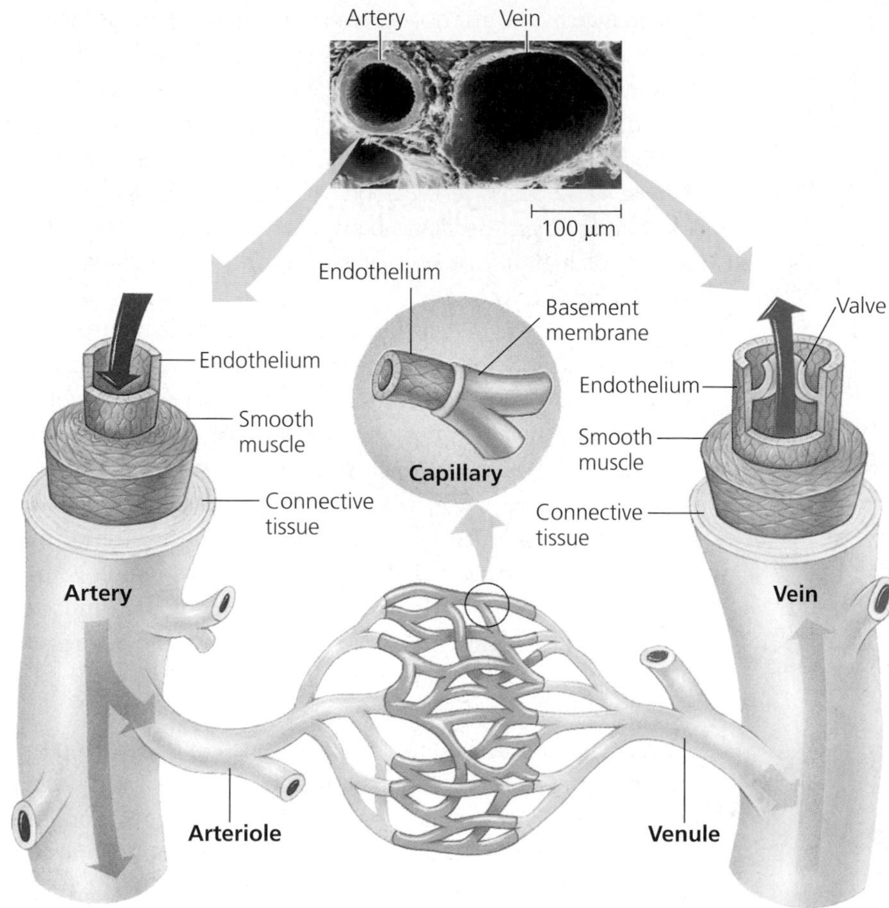

▲ **Figure 42.9 The structure of blood vessels.** This micrograph (SEM) shows an artery next to a thinner-walled vein.

outer layers, and their very thin walls consist only of endothelium and its basement membrane. This facilitates the exchange of substances between the blood and the interstitial fluid that bathes the cells. Arteries have thicker middle and outer layers than veins. Blood flows through the vessels of the circulatory system at uneven speeds and pressures. The thicker walls of arteries provide strength to accommodate blood pumped rapidly and at high pressure by the heart, and their elasticity helps maintain blood pressure even when the heart relaxes between contractions. The thinner-walled veins convey blood back to the heart at low velocity and pressure. Blood flows through the veins mainly as a result of muscle action; whenever you move, your skeletal muscles squeeze your veins and push blood through them. Within large veins, flaps of tissue act as one-way valves that allow blood to flow only toward the heart **(Figure 42.10)**.

Blood Flow Velocity

Blood travels over a thousand times faster in the aorta (about 30 cm/sec on average) than in capillaries (about 0.026 cm/sec). This velocity change follows from the *law of continuity,* which describes fluid movement through pipes. If a pipe's diameter changes over its length, a fluid flows through narrower segments of the pipe faster than it flows through wider segments. Since the *volume* of flow per second must be constant through the entire pipe, the fluid must flow faster as the cross-sectional area of the pipe narrows (think of the velocity of water squirted by a hose with and without a nozzle).

Based on the law of continuity, you might think that blood should travel faster through capillaries than through arteries, because the diameter of capillaries is very small. However, it is the *total* cross-sectional area of capillaries that determines flow rate. Each artery conveys blood to such an enormous number of capillaries that the total cross-sectional area is much greater in capillary beds than in any other part of the circulatory system. For this reason, the blood slows substantially as it enters the arterioles from arteries, and slows further still in the capillary beds. Capillaries are the only vessels with walls thin enough to permit the transfer of substances between the blood and interstitial fluid, and the slower flow of blood through these tiny vessels enhances this exchange. As blood leaves the capillaries and enters the venules and veins, it speeds up again as a result of the reduction in total cross-sectional area (**Figure 42.11**, on the next page).

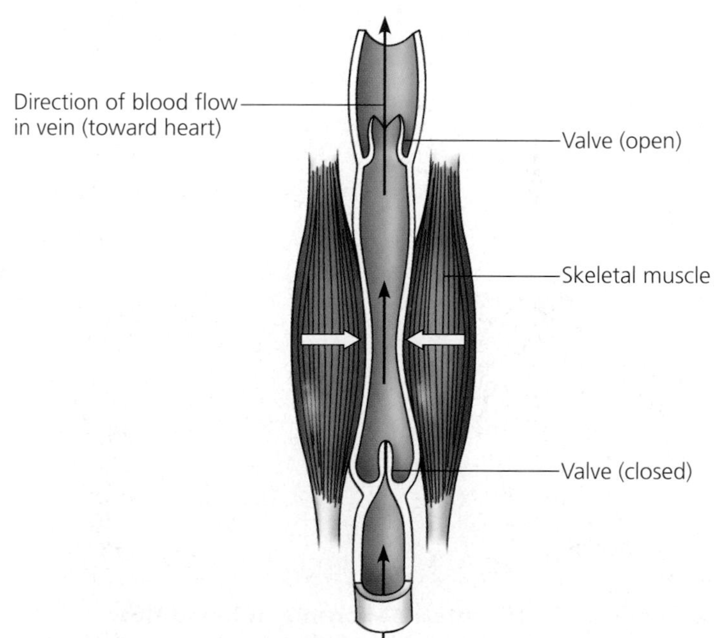

▲ **Figure 42.10 Blood flow in veins.** Contracting skeletal muscles squeeze the veins. Flaps of tissue within the veins act as one-way valves that keep blood moving only toward the heart. If we sit or stand too long, the lack of muscular activity causes our feet to swell with stranded blood unable to return to the heart.

Blood Pressure

Fluids exert a force called hydrostatic pressure against the surfaces they contact, and it is that pressure that drives fluids through pipes. Fluids flow from areas of higher pressure to areas of lower pressure. The hydrostatic pressure that blood exerts against the wall of a vessel and that propels the blood is called blood pressure. Blood pressure is much greater in arteries than in veins and is highest in arteries when the heart contracts during ventricular systole (**systolic pressure**; see Figure 42.11).

When you take your pulse by placing your fingers on your wrist, you can feel an artery bulge with each heartbeat. The surge of pressure is partly due to the narrow openings of arterioles impeding the exit of blood from the arteries. Thus, when the heart contracts, blood enters the arteries faster than it can leave, and the vessels stretch from the pressure. The elastic walls of the arteries snap back during diastole, but the heart contracts again before enough blood has flowed into the arterioles to completely relieve pressure in the arteries. This impedance by the arterioles is called **peripheral resistance**. As a consequence of the elastic arteries working against peripheral resistance, there is a substantial blood pressure even during diastole (**diastolic pressure**), and blood continuously flows into arterioles and capillaries (see Figure 42.11). As shown in **Figure 42.12**, the arterial blood pressure of a healthy 20-year-old human at rest oscillates between about 120 mm Hg (millimeters of mercury; a unit of pressure) at systole and about 70 mm Hg at diastole.

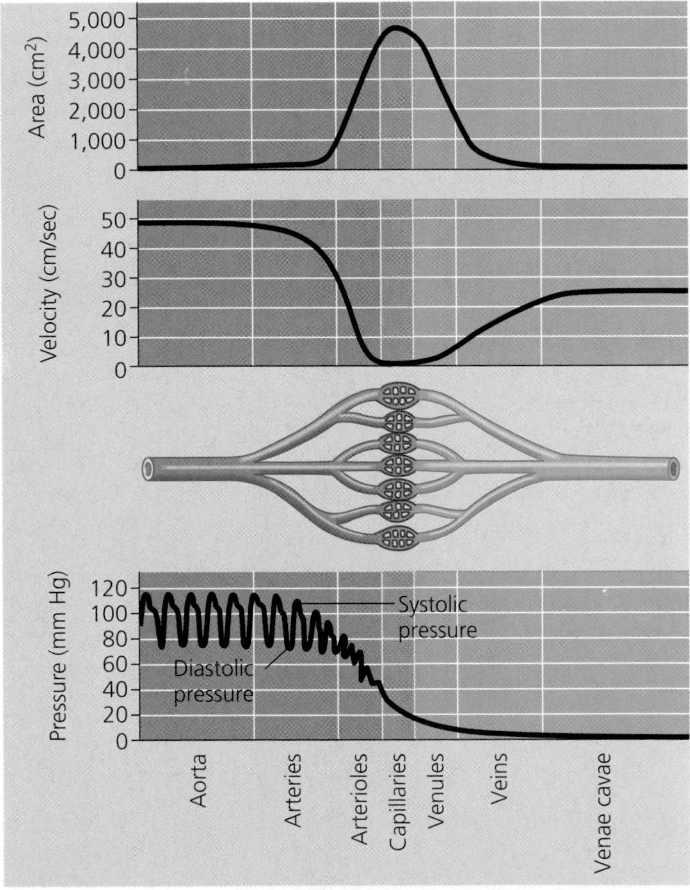

▲ **Figure 42.11 The interrelationship of blood flow velocity, cross-sectional area of blood vessels, and blood pressure.** Blood flow velocity decreases markedly in the arterioles and is lowest in the capillaries, owing to an increase in total cross-sectional area. Blood pressure, the main force driving blood from the heart to the capillaries, is highest in the arteries.

Blood pressure is determined partly by cardiac output and partly by peripheral resistance. Contraction of smooth muscles in the walls of the arterioles reduces the diameter of the tiny vessels, increases peripheral resistance, and therefore increases blood pressure upstream in the arteries. When the smooth muscles relax, the arterioles dilate (increase their diameter). Consequently, blood flow through the arterioles increases and pressure in the arteries falls. Nerve impulses, hormones, and other signals control these arteriole wall muscles. Stress, either physical or emotional, can raise blood pressure by triggering nervous and hormonal responses that constrict blood vessels.

Cardiac output is adjusted in coordination with changes in peripheral resistance. This coordination of regulatory mechanisms maintains adequate blood flow as the body's demands on the circulatory system change. During heavy exercise, for example, the arterioles in working muscles dilate. This response admits a greater flow of oxygen-rich blood to the muscles and also decreases peripheral resistance. By itself, this would cause a drop in blood pressure (and therefore blood flow) in the body as a whole. However, cardiac output increases, maintaining blood pressure and supporting the necessary increase in blood flow.

In large land animals, another factor that affects blood pressure is gravity. Besides the force needed to overcome peripheral resistance, additional pressure is necessary to push blood above the level of the heart. In a standing human, blood must rise about 0.35 m to get from the heart to the brain. This demands an extra 27 mm Hg of pressure, which requires the heart to expend more energy in its contraction cycle. This pumping challenge is significantly greater for animals with long necks. A standing giraffe, for example, needs to pump blood as much as 2.5 m above the heart. That requires about 190 mm Hg of additional blood pressure in the left ventricle, and a giraffe's normal systolic pressure near the heart is over 250 mm Hg. (Systolic pressure that high would be extremely dangerous in a human.) Check valves and sinuses, along with feedback mechanisms that reduce cardiac output, prevent this high pressure from damaging the giraffe's brain when it lowers its head to drink—a body position that causes blood to flow downhill almost 2 m from the heart, adding an extra 150 mm Hg of blood pressure in the arteries leading to the brain. Physiologists speculate about blood pressure and cardiovascular adaptations in dinosaurs—some of which had necks almost

❶ A typical blood pressure reading for a 20-year-old is 120/70. The units for these numbers are mm of mercury (Hg); a blood pressure of 120 is a force that can support a column of mercury 120 mm high.

❹ The cuff is loosened further until the blood flows freely through the artery and the sounds below the cuff disappear. The pressure at this point is the diastolic pressure remaining in the artery when the heart is relaxed.

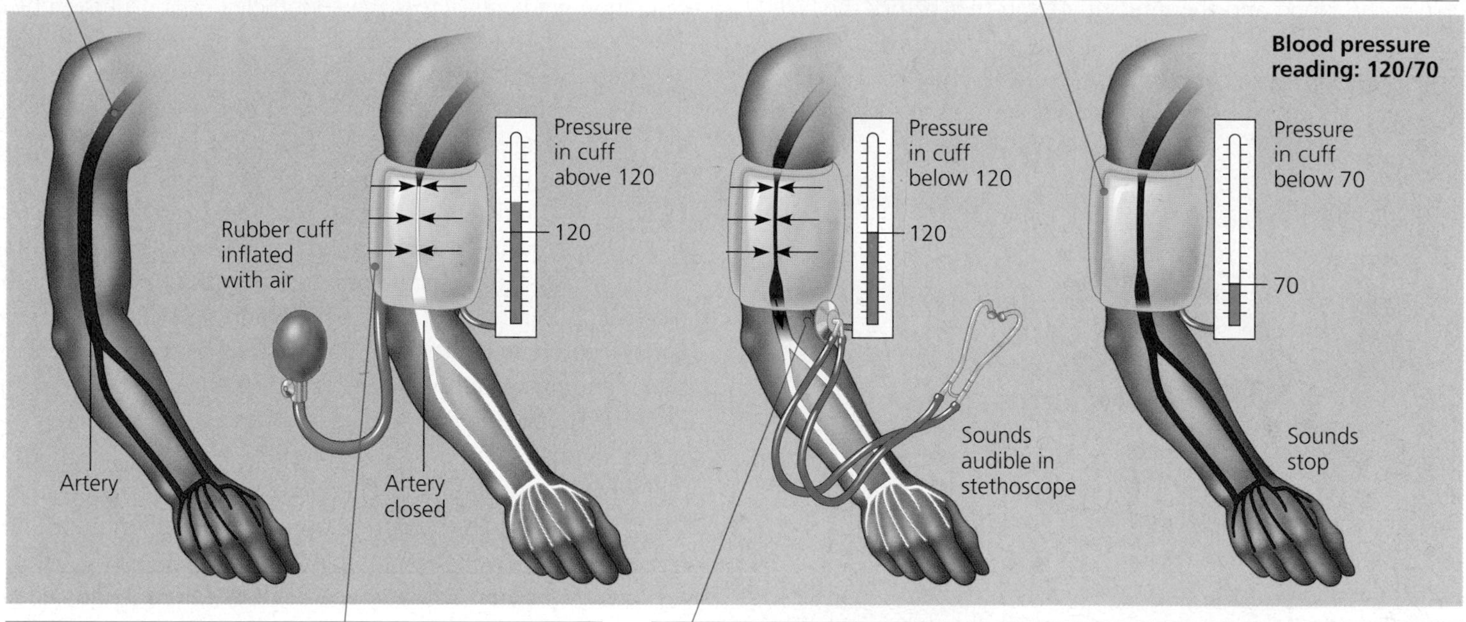

❷ A sphygmomanometer, an inflatable cuff attached to a pressure gauge, measures blood pressure in an artery. The cuff is wrapped around the upper arm and inflated until the pressure closes the artery, so that no blood flows past the cuff. When this occurs, the pressure exerted by the cuff exceeds the pressure in the artery.

❸ A stethoscope is used to listen for sounds of blood flow below the cuff. If the artery is closed, there is no pulse below the cuff. The cuff is gradually deflated until blood begins to flow into the forearm, and sounds from blood pulsing into the artery below the cuff can be heard with the stethoscope. This occurs when the blood pressure is greater than the pressure exerted by the cuff. The pressure at this point is the systolic pressure.

▲ **Figure 42.12 Measurement of blood pressure.** Blood pressure is recorded as two numbers separated by a slash. The first number is the systolic pressure; the second is the diastolic pressure.

10 m long, which would have required a systolic pressure of nearly 760 mm Hg to pump blood to the brain when the head was fully raised. But evidence indicates that dinosaurs probably did not have hearts powerful enough to generate such pressures. Based on this analysis and on studies of neck-bone structure, some biologists have concluded that the long-necked dinosaurs fed close to the ground rather than raising their head to feed on high foliage.

By the time blood reaches the veins, its pressure is not affected much by the action of the heart. This is because the blood encounters so much resistance as it passes through the millions of tiny arterioles and capillaries that the pressure generated by the pumping heart has been dissipated and can no longer propel the blood through the veins. How does blood return to the heart, especially when it must travel from the lower extremities against gravity? Rhythmic contractions of smooth muscles in the walls of venules and veins account for some movement of the blood. More importantly, the activity of skeletal muscles during exercise squeezes blood through the veins (see Figure 42.10). Also, when we inhale, the change in pressure within the thoracic (chest) cavity causes the venae cavae and other large veins near the heart to expand and fill with blood.

Capillary Function

At any given time, only about 5–10% of the body's capillaries have blood flowing through them. However, each tissue has many capillaries, so every part of the body is supplied with blood at all times. Capillaries in the brain, heart, kidneys, and liver are usually filled to capacity, but in many other sites, the blood supply varies over time as blood is diverted from one destination to another. After a meal, for instance, blood supply to the digestive tract increases. During strenuous exercise, blood is diverted from the digestive tract and supplied more generously to skeletal muscles and skin. This is one reason that exercising heavily immediately after eating a big meal may cause indigestion.

Two mechanisms regulate the distribution of blood in capillary beds. Both depend on smooth muscles controlled by nerve signals and hormones. In one mechanism, contraction of the smooth muscle layer in the wall of an arteriole constricts the vessel, reducing its diameter and decreasing blood flow through it to a capillary bed. When the muscle layer relaxes, the arteriole dilates, allowing blood to enter the capillaries. In the other mechanism, rings of smooth muscle—called precapillary sphincters because they are located at the entrance to capillary beds—control the flow of blood between arterioles and venules **(Figure 42.13)**.

As you have read, the critical exchange of substances between the blood and the interstitial fluid that bathes the cells takes place across the thin endothelial walls of the capillaries. Some substances may be carried across an endothelial cell in vesicles that form by endocytosis on one side of the cell and then release their contents by exocytosis on the opposite side. Others simply diffuse between the blood and the interstitial fluid. Small molecules, such as O_2 and CO_2, diffuse down concentration gradients across the endothelial cells. Diffusion can also occur through the clefts between adjoining cells. However, transport through these clefts occurs mainly by bulk flow due to fluid pressure. Blood pressure within the capillary pushes fluid (consisting of water and small solutes such as sugars, salts, O_2, and urea) through the capillary clefts. The outward movement of this fluid causes a net loss of fluid from the upstream end of the capillary near an arteriole. Blood cells suspended in blood and most proteins dissolved in the blood are too large to pass readily through the endothelium and remain in the capillaries. The blood proteins remaining in the capillaries, especially albumin, create approximately constant osmotic pressure from the arteriole to the venule end of a capillary bed. In contrast, blood pressure drops sharply. This difference between blood pressure and osmotic pressure drives fluids out of capillaries at the arteriole end and into capillaries at the venule end **(Figure 42.14)**. About 85% of the fluid that leaves the blood at the arterial end of a capillary bed reenters from the interstitial fluid at the venous end, and the remaining 15% is eventually returned to the blood by the vessels of the lymphatic system.

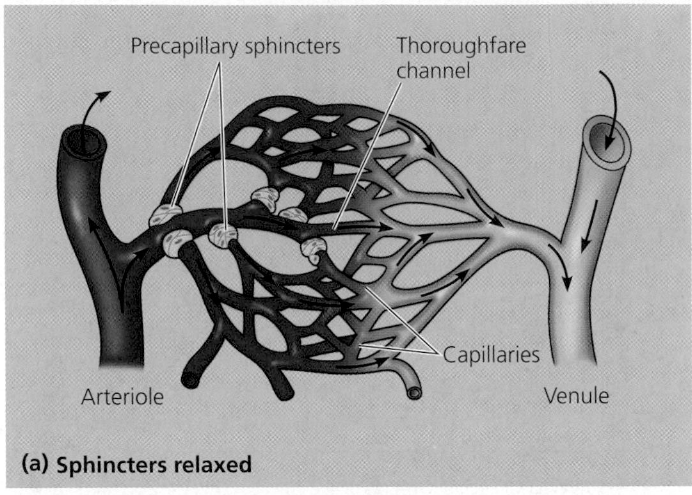

(a) Sphincters relaxed

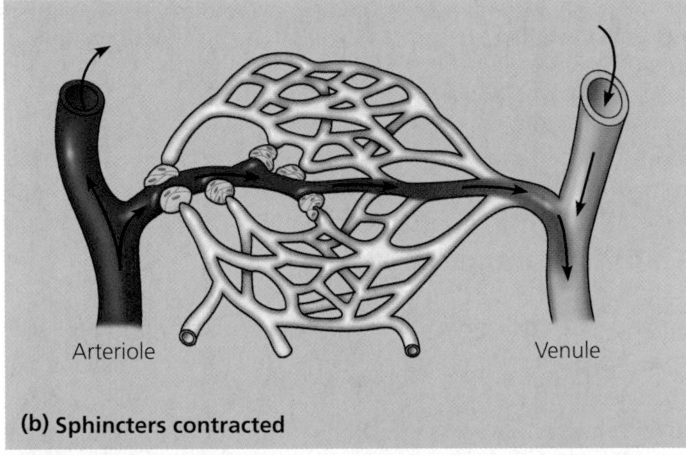

(b) Sphincters contracted

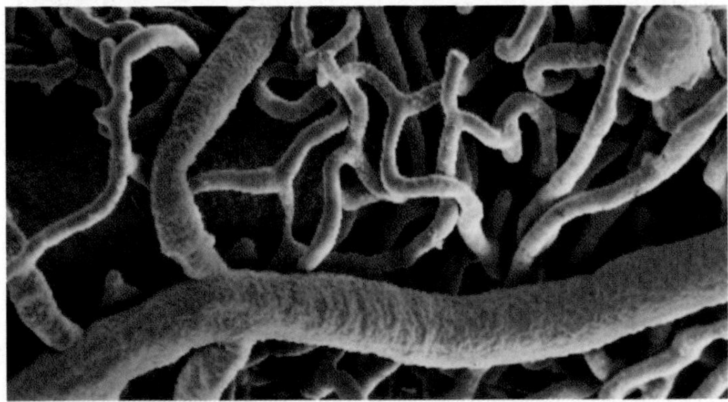

(c) Capillaries and larger vessels (SEM)

▲ **Figure 42.13 Blood flow in capillary beds.** Precapillary sphincters regulate the passage of blood into capillary beds. Some blood flows directly from arterioles to venules through capillaries called thoroughfare channels, which are always open.

Fluid Return by the Lymphatic System

So much blood passes through the capillaries that the cumulative loss of fluid adds up to about 4 L per day. There is also some leakage of blood proteins, even though the capillary wall is not very permeable to large molecules. The lost fluid and proteins return to the blood via the **lymphatic system.** Fluid enters this system by diffusing into tiny lymph capillaries intermingled among capillaries of the cardiovascular system. Once inside the lymphatic system, the fluid is called **lymph;** its composition is about the same as that of interstitial fluid. The lymphatic system drains into the circulatory system near the junction of the venae cavae with the right atrium (see Figure 43.5).

Lymph vessels, like veins, have valves that prevent the backflow of fluid toward the capillaries. Rhythmic contractions of the vessel walls help draw fluid into lymphatic capillaries. Also like veins, lymph vessels depend mainly on the movement of skeletal muscles to squeeze fluid toward the heart.

Along a lymph vessel are organs called **lymph nodes.** By filtering the lymph and attacking viruses and bacteria, lymph nodes play an important role in the body's defense. Inside each lymph node is a honeycomb of connective tissue with spaces filled by white blood cells specialized for defense. When the body is fighting an infection, these cells multiply rapidly, and the lymph nodes become swollen and tender (which is why your doctor checks your neck for swollen lymph nodes when you feel sick).

The lymphatic system helps defend against infection and maintains the volume and protein concentration of the blood.

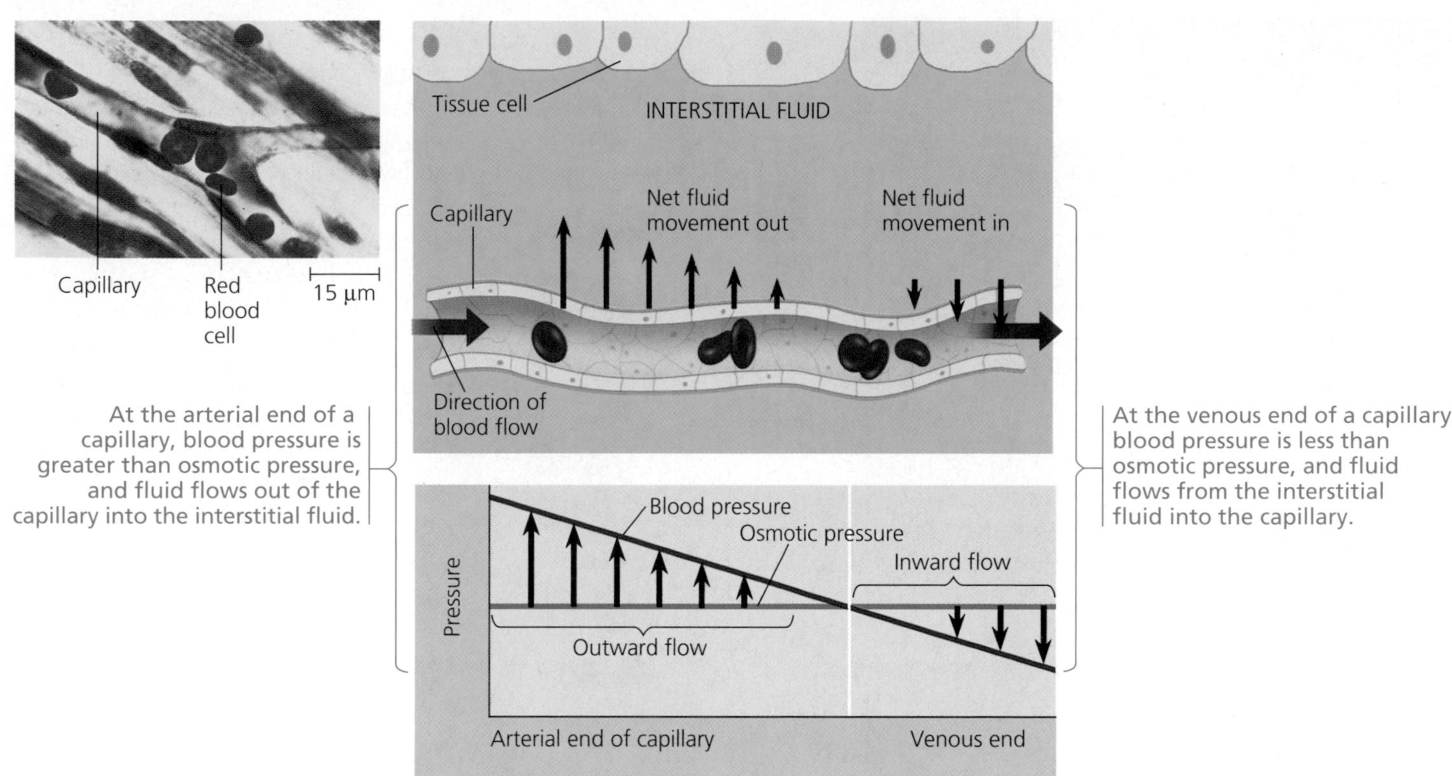
Capillary
Red blood cell
15 µm
Tissue cell
INTERSTITIAL FLUID
Capillary
Net fluid movement out
Net fluid movement in
Direction of blood flow
At the arterial end of a capillary, blood pressure is greater than osmotic pressure, and fluid flows out of the capillary into the interstitial fluid.
At the venous end of a capillary, blood pressure is less than osmotic pressure, and fluid flows from the interstitial fluid into the capillary.
Blood pressure
Osmotic pressure
Inward flow
Pressure
Outward flow
Arterial end of capillary
Venous end

▲ **Figure 42.14 Fluid exchange between capillaries and the interstitial fluid.** The micrograph at top left shows red blood cells traveling through a capillary (LM).

Recall from Chapter 41 that the lymphatic system also transports fats from the digestive tract to the circulatory system.

Concept Check 42.3

1. What is the primary cause of the low velocity of blood flow through capillaries?
2. How does increasing blood pressure by increasing cardiac output combined with diverting most blood flow to the skeletal muscles prepare the body to confront or flee danger?
3. Explain how edema—the accumulation of fluid in body tissues—can result from a decrease in plasma protein due to severe protein deficiency in the diet.

For suggested answers, see Appendix A.

Concept 42.4

Blood is a connective tissue with cells suspended in plasma

We now shift our focus from the tubes and pumps of circulatory systems to the fluids being circulated. As explained earlier, in invertebrates with open circulation, blood (hemolymph) is not different from interstitial fluid. However, blood in the closed circulatory systems of vertebrates is a specialized connective tissue.

Blood Composition and Function

Blood consists of several kinds of cells suspended in a liquid matrix called **plasma**. After a blood sample is collected, the cells can be separated from plasma by spinning the whole blood in a centrifuge. (An anticoagulant must be added to prevent the blood from clotting.) The cellular elements (cells and cell fragments), which occupy about 45% of the volume of blood, settle to the bottom of the centrifuge tube, forming a dense red pellet. Above this cellular pellet is the transparent, straw-colored plasma.

Plasma

Blood plasma is about 90% water. Among its many solutes are inorganic salts in the form of dissolved ions, sometimes referred to as blood electrolytes (**Figure 42.15**, on the next page). The combined concentration of these ions is important in maintaining the osmotic balance of the blood. Some ions also help to buffer the blood, which in humans normally has a pH of 7.4. And the normal functioning of muscles and nerves depends on the concentration of key ions in the interstitial

Plasma 55%	
Constituent	**Major functions**
Water	Solvent for carrying other substances
Ions (blood electrolytes) Sodium Potassium Calcium Magnesium Chloride Bicarbonate	Osmotic balance, pH buffering, and regulation of membrane permeability
Plasma proteins	
Albumin	Osmotic balance, pH buffering
Fibrinogen	Clotting
Immunoglobulins (antibodies)	Defense
Substances transported by blood Nutrients (such as glucose, fatty acids, vitamins) Waste products of metabolism Respiratory gases (O_2 and CO_2) Hormones	

Separated blood elements

Cellular elements 45%		
Cell type	**Number** per µL (mm^3) of blood	**Functions**
Erythrocytes (red blood cells)	5–6 million	Transport oxygen and help transport carbon dioxide
Leukocytes (white blood cells) Basophil, Eosinophil, Lymphocyte, Neutrophil, Monocyte	5,000–10,000	Defense and immunity
Platelets	250,000–400,000	Blood clotting

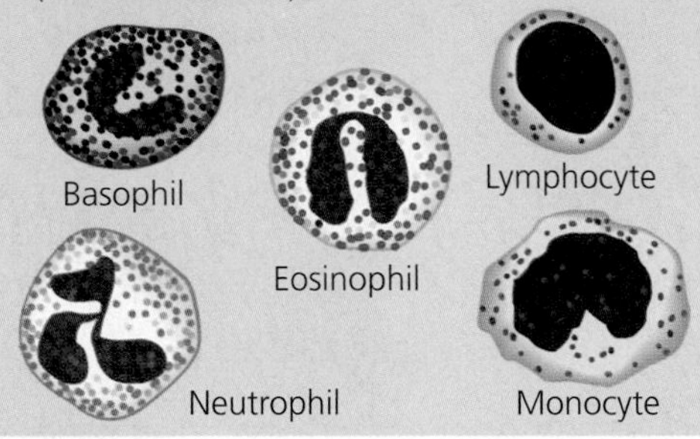

▲ **Figure 42.15 The composition of mammalian blood.**

fluid, which reflects their concentration in plasma. The kidney maintains plasma electrolytes at precise concentrations, an example of homeostasis we will explore in detail in Chapter 44.

Another important class of solutes is the plasma proteins, which have many functions. Collectively, they act as buffers against pH changes, help maintain the osmotic balance between blood and interstitial fluid, and contribute to the blood's viscosity (thickness). The various types of plasma proteins also have specific functions. Some are escorts for lipids, which are insoluble in water and can travel in blood only when bound to proteins. Another class of proteins, the immunoglobulins, or antibodies, help combat viruses and other foreign agents that invade the body (see Chapter 43). And the plasma proteins called fibrinogens are clotting factors that help plug leaks when blood vessels are injured. Blood plasma from which these clotting factors have been removed is called serum.

Plasma also contains a wide variety of substances in transit from one part of the body to another, including nutrients, metabolic wastes, respiratory gases, and hormones. Blood plasma and interstitial fluid are similar in composition, except that plasma has a much higher protein concentration (capillary walls, remember, are not very permeable to proteins).

Cellular Elements

Suspended in blood plasma are two classes of cells (see Figure 42.15): **red blood cells,** which transport oxygen, and **white blood cells,** which function in defense. A third cellular element, **platelets,** are fragments of cells that are involved in the clotting process.

Erythrocytes. Red blood cells, or **erythrocytes,** are by far the most numerous blood cells. Each microliter (µL; or mm^3) of human blood contains 5 to 6 million red cells, and there are about 25 trillion of these cells in the body's 5 L of blood.

Erythrocytes are an excellent example of the close relationship between structure and function. Their main function of O_2 transport depends on rapid diffusion of O_2 across their plasma membranes. Human erythrocytes are small disks (about 7–8.5 µm in diameter) that are biconcave—thinner in the center than at the edges. Their small size and biconcave shape provide a large collective surface area for the total population of erythrocytes, and the greater the total area of erythrocyte membrane in a given volume of blood, the more rapidly O_2 can diffuse. Mammalian erythrocytes lack nuclei, an unusual characteristic that leaves more space in these tiny cells for **hemoglobin,** the iron-containing protein that transports oxygen (see Figure 5.20). Erythrocytes also lack mitochondria and generate their ATP exclusively by anaerobic metabolism. Oxygen transport by erythrocytes would be less efficient if their own metabolism were aerobic and consumed some of the oxygen they carry.

Despite its small size, an erythrocyte contains about 250 million molecules of hemoglobin. Since each hemoglobin binds up to four molecules of O_2, one erythrocyte can transport about a billion O_2 molecules. Researchers have found that hemoglobin also binds the gaseous molecule nitric oxide (NO) as well as O_2. As erythrocytes pass through the capillary beds of lungs, gills, or other respiratory organs, O_2 diffuses into the erythrocytes where hemoglobin binds O_2 and NO. In the systemic capillaries, hemoglobin unloads O_2, which then diffuses into body cells. The NO relaxes the capillary walls, allowing them to expand, which probably helps deliver O_2 to the cells.

Leukocytes. The blood contains five major types of white blood cells, or **leukocytes:** monocytes, neutrophils, basophils, eosinophils, and lymphocytes (see Figure 42.15). Their collective function is to fight infections. For example, monocytes and neutrophils are phagocytes, which engulf and digest bacteria and debris from the body's own dead cells. As we will see in Chapter 43, lymphocytes develop into specialized B cells and T cells, which produce the immune response against foreign substances. White blood cells spend most of their time outside the circulatory system, patrolling through interstitial fluid and the lymphatic system, where most of the battles against pathogens are waged. Normally, a microliter of human blood has about 5,000 to 10,000 leukocytes, but their numbers increase temporarily whenever the body is fighting an infection.

Platelets. The third cellular element of blood, platelets, are fragments of cells about 2–3 μm in diameter. They have no nuclei and originate as pinched-off cytoplasmic fragments of large cells in the bone marrow. Platelets then enter the blood and function in the important process of blood clotting.

Stem Cells and the Replacement of Cellular Elements

The cellular elements of blood (erythrocytes, leukocytes, and platelets) wear out and are replaced constantly throughout a person's life. Erythrocytes, for example, usually circulate for only three to four months and then are destroyed by phagocytic cells in the liver and spleen. Enzymes digest the old cell's macromolecules, and biosynthetic processes construct new macromolecules using many of the monomers, such as amino acids, harvested from the old blood cells, as well as new materials and energy from food. Many of the iron atoms derived from the hemoglobin in old erythrocytes are incorporated into new hemoglobin molecules.

Erythrocytes, leukocytes, and platelets all develop from a common source, a single population of cells called pluripotent **stem cells** in the red marrow of bones, particularly the ribs, vertebrae, breastbone, and pelvis **(Figure 42.16)**. "Pluripotent" means that these cells have the potential to differentiate into any type of blood cell or into cells that produce platelets. Pluripotent stem cells arise in the early embryo, and the population renews itself while replenishing the blood with cellular elements. (See Chapter 21 for more on stem cells.)

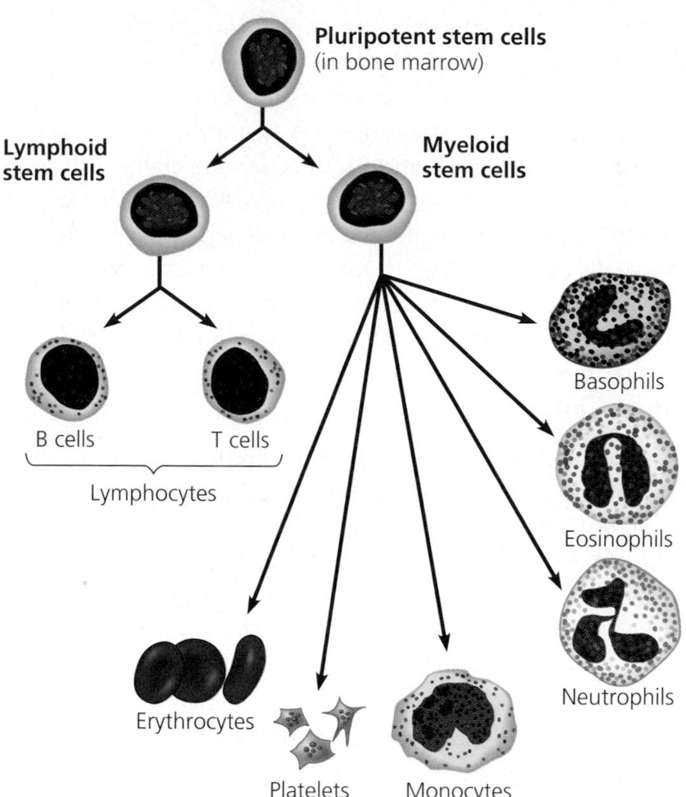

▲ **Figure 42.16 Differentiation of blood cells.** Some of the pluripotent stem cells differentiate into lymphoid stem cells, which then develop into B cells and T cells, two types of lymphocytes that function in the immune response (see Chapter 43). All other blood cells differentiate from myeloid stem cells.

A negative-feedback mechanism, sensitive to the amount of O_2 reaching the body's tissues via the blood, controls erythrocyte production. If the tissues do not receive enough O_2, the kidney synthesizes and secretes a hormone called **erythropoietin (EPO),** which stimulates production of erythrocytes. If blood is delivering more O_2 than the tissues can use, the level of EPO falls and erythrocyte production slows.

Physicians use synthetic erythropoietin to treat people with health problems such as anemia, a condition of lower-than-normal hemoglobin levels. However, some athletes abuse EPO by injecting themselves with the drug to increase their erythrocyte levels. This practice, known as blood doping, is banned by the International Olympic Committee and other sports federations. In 2002, several athletes competing in the winter Olympics in Salt Lake City tested positive for EPO-like chemicals and as a result were stripped of some of their medals.

Recently, researchers succeeded in isolating pluripotent stem cells and growing these cells in laboratory cultures. Purified pluripotent stem cells may soon provide an effective treatment for a number of human diseases, including leukemia. A person with leukemia has a cancerous line of the stem cells that produce leukocytes. The cancerous stem cells crowd out cells that make erythrocytes and produce an unusually high number of leukocytes, many of which are abnormal. One experimental treatment for leukemia involves removing pluripotent stem cells from a patient, destroying the bone marrow, and restocking it with noncancerous stem cells. As few as 30 of these cells can completely repopulate the bone marrow.

Blood Clotting

Most people get cuts and scrapes from time to time, yet we do not bleed to death because blood contains a self-sealing material that plugs leaks. The sealant is always present in an inactive form called **fibrinogen.** A clot forms only when this plasma protein is converted to its active form, **fibrin,** which aggregates into threads that form the framework of the clot. The clotting mechanism usually begins with the release of clotting factors from platelets and involves a complex chain of reactions that ultimately transforms fibrinogen to fibrin **(Figure 42.17)**. More than a dozen clotting factors have been discovered, and the mechanism is still not fully understood. A genetic mutation that affects any step of the clotting process causes **hemophilia,** a disease characterized by excessive bleeding from even minor cuts and bruises.

Anticlotting factors in the blood normally prevent spontaneous clotting in the absence of injury. Sometimes, however, platelets clump and fibrin coagulates within a blood vessel, blocking the flow of blood. Such a clot is called a **thrombus.** These potentially dangerous clots are more likely to form in individuals with cardiovascular disease.

Cardiovascular Disease

More than half the deaths in the United States are caused by **cardiovascular diseases,** disorders of the heart and blood vessels. The tendency to develop cardiovascular disease is inherited to some extent, but lifestyle plays a large role, too. Nongenetic factors that increase the risk of cardiovascular problems include smoking, lack of exercise, a diet rich in animal fat, and high concentrations of cholesterol in the blood.

Cholesterol travels in the blood plasma mainly in the form of particles consisting of thousands of cholesterol molecules

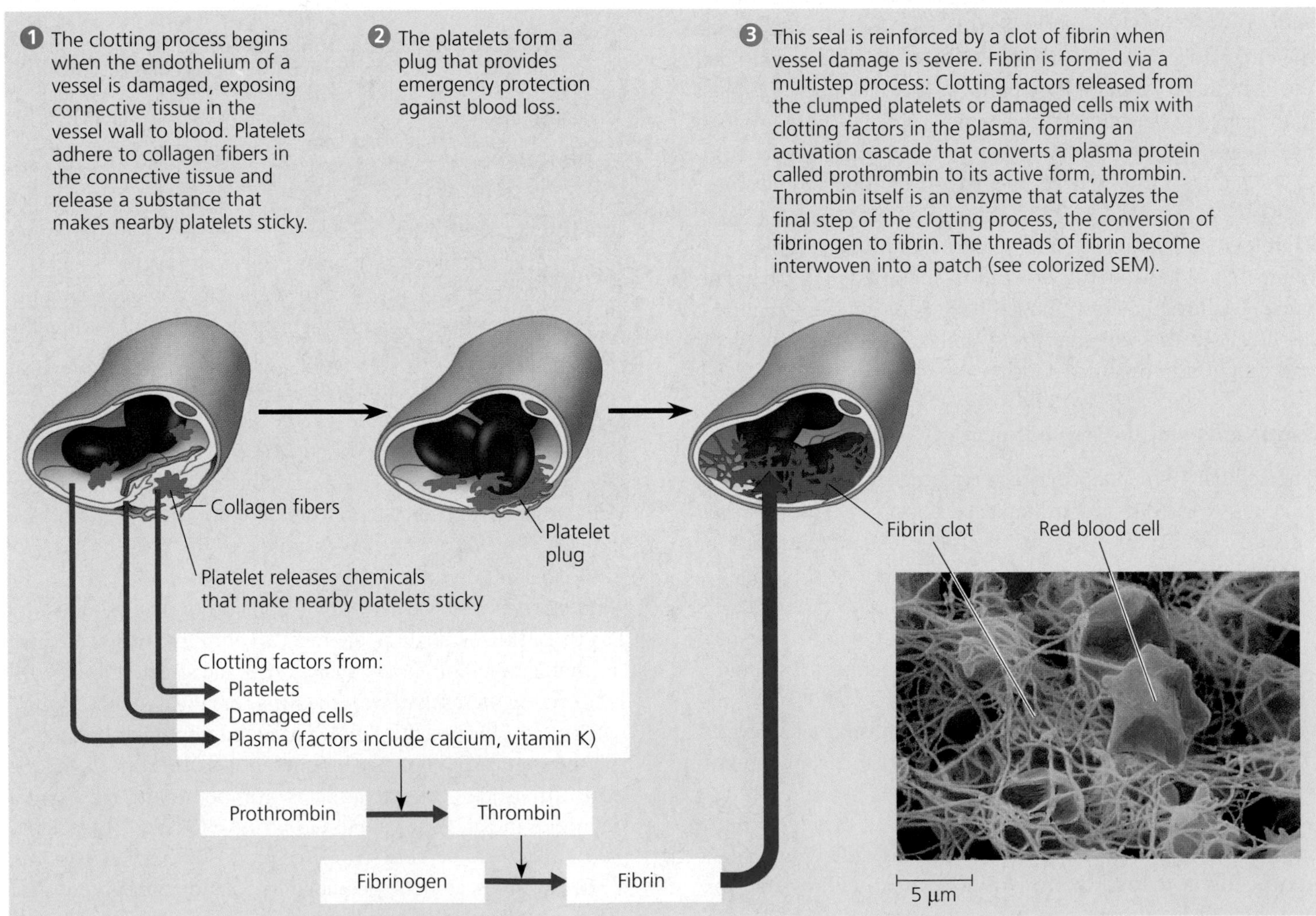

1 The clotting process begins when the endothelium of a vessel is damaged, exposing connective tissue in the vessel wall to blood. Platelets adhere to collagen fibers in the connective tissue and release a substance that makes nearby platelets sticky.
2 The platelets form a plug that provides emergency protection against blood loss.
3 This seal is reinforced by a clot of fibrin when vessel damage is severe. Fibrin is formed via a multistep process: Clotting factors released from the clumped platelets or damaged cells mix with clotting factors in the plasma, forming an activation cascade that converts a plasma protein called prothrombin to its active form, thrombin. Thrombin itself is an enzyme that catalyzes the final step of the clotting process, the conversion of fibrinogen to fibrin. The threads of fibrin become interwoven into a patch (see colorized SEM).
Collagen fibers
Platelet plug
Platelet releases chemicals that make nearby platelets sticky
Fibrin clot
Red blood cell
Clotting factors from:
Platelets
Damaged cells
Plasma (factors include calcium, vitamin K)
Prothrombin
Thrombin
Fibrinogen
Fibrin
5 μm

▲ **Figure 42.17 Blood clotting.**

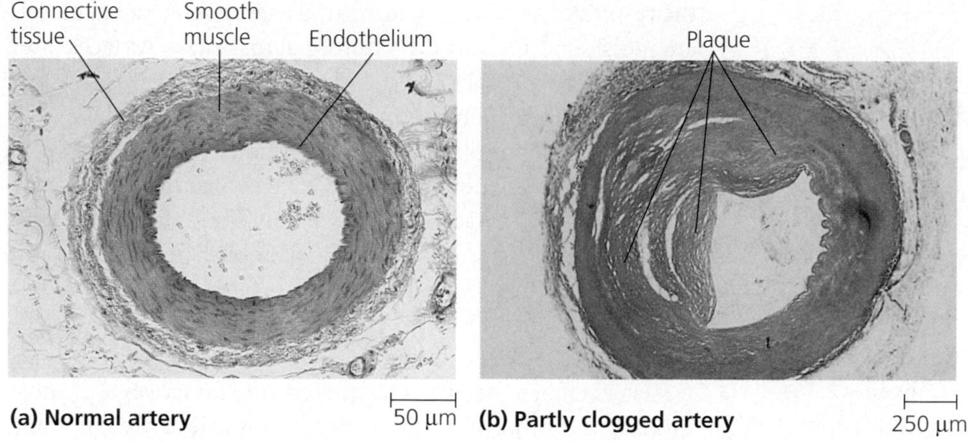

▲ **Figure 42.18 Atherosclerosis.** These light micrographs contrast a cross section of **(a)** a normal (healthy) artery with **(b)** an artery partially blocked by an atherosclerotic plaque. Plaques consist mostly of fibrous connective tissue and smooth muscle cells infiltrated with lipids.

and other lipids bound to a protein. One type of particle—**low-density lipoproteins (LDLs)**, often called the "bad cholesterol"—is associated with the deposition of cholesterol in arterial plaques, growths that develop on the inner walls of arteries. Another type—**high-density lipoproteins (HDLs)**, or "good cholesterol"—appears to reduce the deposition of cholesterol. Exercise increases HDL concentration, whereas smoking has the opposite effect on the LDL/HDL ratio.

Healthy arteries have smooth inner linings that promote unimpeded blood flow. The deposition of cholesterol thickens and roughens this smooth lining. A plaque forms at the site and becomes infiltrated with fibrous connective tissue and still more cholesterol. Such plaques narrow the bore of the artery, leading to a chronic cardiovascular disease known as **atherosclerosis (Figure 42.18)**. The rough lining of an atherosclerotic artery seems to encourage the adhesion of platelets, triggering the clotting process and interfering with circulation.

Hypertension (high blood pressure) promotes atherosclerosis and increases the risk of heart attack and stroke. Atherosclerosis tends to raise blood pressure by narrowing the vessels and reducing their elasticity. According to one hypothesis, chronic high blood pressure damages the endothelium that lines arteries, promoting plaque formation. Fortunately, hypertension is simple to diagnose and can usually be controlled by diet, exercise, medication, or a combination of these. A diastolic pressure above 90 may be cause for concern, and living with extreme hypertension—say, 200/120—is courting disaster.

As atherosclerosis progresses, arteries become narrower, and the threat of heart attack or stroke increases. There may be warning signs. For example, if a coronary artery is only partially blocked, the person may feel occasional chest pain, a condition known as angina pectoris. The pain is most likely to appear when the heart is laboring hard as a result of physical or emotional stress, and it signals that part of the heart is not receiving enough O_2. However, many people with atherosclerosis are completely unaware of their condition until catastrophe strikes.

The final blow is usually a heart attack or a stroke. A **heart attack** is the death of cardiac muscle tissue resulting from prolonged blockage of one or more coronary arteries, the vessels that supply oxygen-rich blood to the heart. Because they are small in diameter to begin with, the coronary arteries are particularly vulnerable. Such blockage can destroy cardiac muscle quickly, since the constantly beating heart muscle cannot survive long without oxygen. A **stroke** is the death of nervous tissue in the brain, usually resulting from rupture or blockage of arteries in the head.

Heart attacks and strokes frequently result from a thrombus, or blood clot, that clogs an artery. A key process leading to the clogging of an artery by a thrombus is an inflammatory response triggered by the accumulation of LDLs in the artery's inner lining. Such an inflammation, which is analogous to the body's response to a cut infected by bacteria (see Figure 43.6), can cause plaques to rupture, releasing fragments that form a thrombus. The thrombus may originate in a coronary artery or an artery in the brain, or it may develop elsewhere in the circulatory system and reach the heart or brain via the bloodstream. The transported clot, called an *embolus*, is swept along until it lodges in an artery too small for the clot to pass. An embolus is more likely to become trapped in a vessel that has been narrowed by plaques. The embolus blocks blood flow, and cardiac or brain tissue downstream from the obstruction may die from O_2 deprivation. If damage in the heart interrupts the conduction of electrical impulses through cardiac muscle, heart rate may change drastically or the heart may stop beating altogether. Still, the victim may survive if a heartbeat is restored by cardiopulmonary resuscitation (CPR) or some other emergency procedure within a few minutes of the attack. The effects of a stroke and the individual's chance of survival depend on the extent and location of the damaged brain tissue.

Concept Check 42.4

1. About how many red blood cells does the bone marrow of a human produce per day, assuming a total red blood cell count of 25 trillion (2.5×10^{13}) and an average longevity of 4 months for the cells?
2. Explain why a physician might order a white-cell count for a patient with symptoms of an infection.
3. How can a few dozen transplanted bone marrow stem cells replace the wide variety of cells that occur in bone marrow?

For suggested answers, see Appendix A.

Concept 42.5

Gas exchange occurs across specialized respiratory surfaces

In the remainder of this chapter, we will focus on the process of **gas exchange.** Although this process is often called respiration, it should not be confused with the energy transformations of cellular respiration. Gas exchange is the uptake of molecular oxygen (O_2) from the environment and the discharge of carbon dioxide (CO_2) to the environment **(Figure 42.19)**. These exchanges are necessary to support the production of ATP in cellular respiration and usually involve both the respiratory system and the circulatory system of an animal.

The source of O_2, called the **respiratory medium,** is air for terrestrial animals and water for most aquatic animals. The atmosphere is Earth's main reservoir of O_2 and is about 21% O_2 (by volume). Oceans, lakes, and other bodies of water contain O_2 in dissolved form. The amount of O_2 dissolved in a given volume of water varies considerably, but is always much less than in an equivalent volume of air.

The part of an animal's body where gases are exchanged with the surrounding environment is called the **respiratory surface.** Animals do not move O_2 and CO_2 across membranes by active transport, so movement of these gases between the respiratory surface and the environment occurs entirely by diffusion. The rate of diffusion is proportional to the surface area across which diffusion occurs and inversely proportional to the *square* of the distance through which molecules must move. As a result, respiratory surfaces tend to be thin and have a large surface area, structural adaptations that maximize the rate of gas exchange. Additionally, all living cells must be bathed in water to maintain their plasma membranes. Thus, the respiratory surfaces of terrestrial as well as aquatic animals are moist, and O_2 and CO_2 diffuse across these surfaces after first dissolving in water.

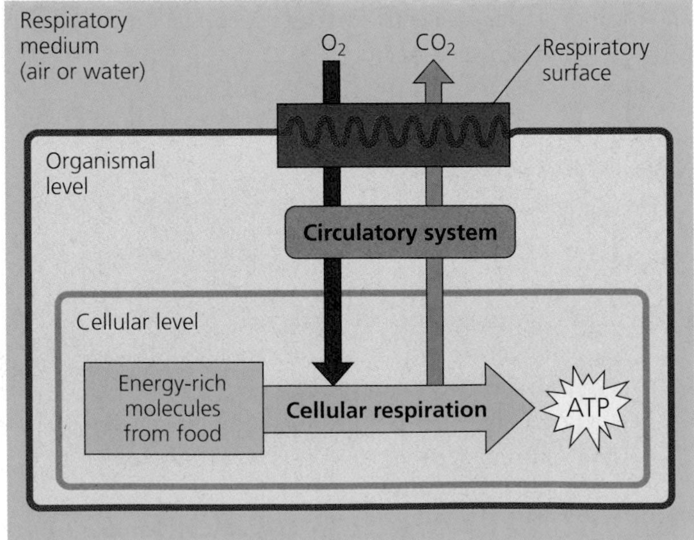

▲ **Figure 42.19 The role of gas exchange in bioenergetics.**

The respiratory surface must supply O_2 and expel CO_2 for the entire body, and a variety of solutions to the problem of providing a large enough surface have evolved. The structure of a respiratory surface depends mainly on the size of the organism and whether it lives in water or on land, but it is also influenced by metabolic demands for gas exchange. Thus, an endotherm generally has a larger area of respiratory surface than a similar-sized ectotherm.

Gas exchange occurs over the entire surface area of most protists and other unicellular organisms. Similarly, for some relatively simple animals, such as sponges, cnidarians, and flatworms, the plasma membrane of every cell in the body is close enough to the external environment for gases to diffuse in and out. In many animals, however, the bulk of the body does not have direct access to the respiratory medium. The respiratory surface in these animals is a thin, moist epithelium separating the respiratory medium from the blood or capillaries, which transport gases to and from the rest of the body (see Figure 42.19).

Some animals use their entire outer skin as a respiratory organ. An earthworm, for example, has moist skin and exchanges gases by diffusion across its general body surface. Just below the earthworm's skin is a dense net of capillaries. Because the respiratory surface must remain moist, earthworms and many other skin-breathers, including some amphibians, must live in water or damp places.

Animals whose only respiratory organ is moist skin are usually small and are either long and thin or flat, with a high ratio of surface area to volume. For most other animals, the general body surface lacks sufficient area to exchange gases for the whole body. The solution is a respiratory organ that is extensively folded or branched, thereby enlarging the available surface area for gas exchange. Gills, tracheae, and lungs are the three most common respiratory organs.

Gills in Aquatic Animals

Gills are outfoldings of the body surface that are suspended in the water. In some invertebrates, such as sea stars, the gills have a simple shape and are distributed over much of the body **(Figure 42.20a)**. Many segmented worms have flaplike gills that extend from each segment of their body **(Figure 42.20b)** or long, feathery gills clustered at the head or tail. The gills of scallops **(Figure 42.20c)**, crayfish **(Figure 42.20d)**, and many other animals are restricted to a local body region. The total surface area of the gills is often much greater than that of the rest of the body.

As a respiratory medium, water has both advantages and disadvantages. There is no problem keeping the plasma membranes of the respiratory surface cells moist, since the

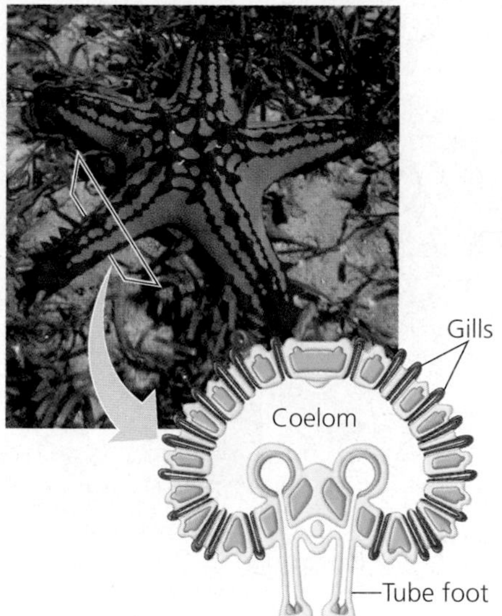

(a) Sea star. The gills of a sea star are simple tubular projections of the skin. The hollow core of each gill is an extension of the coelom (body cavity). Gas exchange occurs by diffusion across the gill surfaces, and fluid in the coelom circulates in and out of the gills, aiding gas transport. The surfaces of a sea star's tube feet also function in gas exchange.

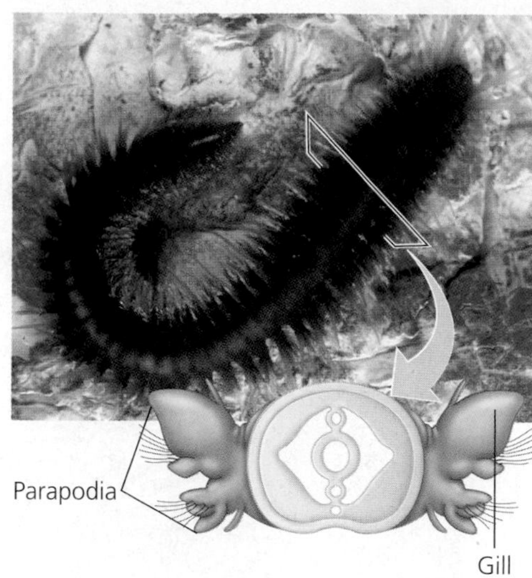

(b) Marine worm. Many polychaetes (marine worms of the phylum Annelida) have a pair of flattened appendages called parapodia on each body segment. The parapodia serve as gills and also function in crawling and swimming.

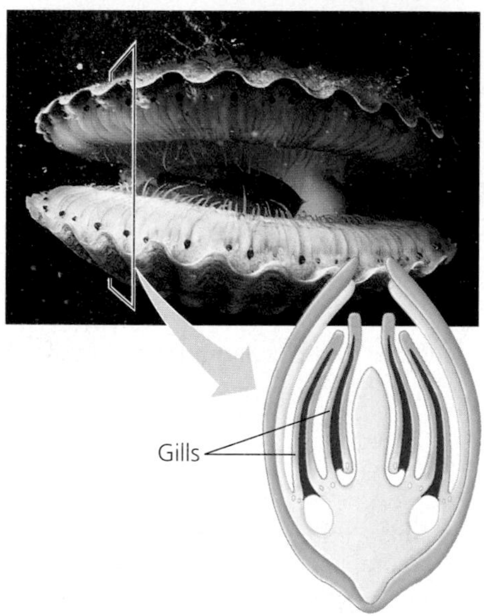

(c) Scallop. The gills of a scallop are long, flattened plates that project from the main body mass inside the hard shell. Cilia on the gills circulate water around the gill surfaces.

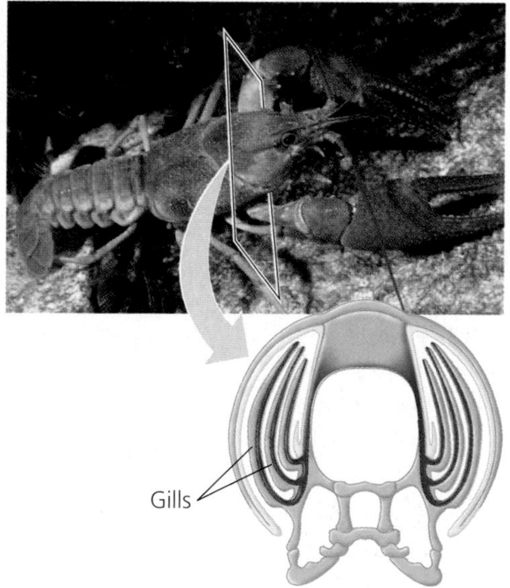

(d) Crayfish. Crayfish and other crustaceans have long, feathery gills covered by the exoskeleton. Specialized body appendages drive water over the gill surfaces.

▲ **Figure 42.20 Diversity in the structure of gills, external body surfaces functioning in gas exchange.**

gills are surrounded by the aqueous environment. However, O_2 concentrations in water are low; and the warmer and saltier the water, the less dissolved O_2 it can hold (water in many marine and freshwater habitats contains only 4–8 mL of dissolved O_2 per liter). Thus, gills must be very effective for the animal to obtain enough O_2. One process that helps is **ventilation,** or increasing the flow of the respiratory medium over the respiratory surface. Without ventilation, a region of low O_2 and high CO_2 concentration can form around the gill as it exchanges gases with the water. Crayfish and lobsters have paddle-like appendages that drive a current of water over the gills. Fish gills are ventilated by a current of water that enters the mouth, passes through slits in the pharynx, flows over the gills, and then exits the body (**Figure 42.21**, on the next page). Because water is dense and contains little O_2 per unit of volume, most fishes must expend considerable energy in ventilating their gills.

The arrangement of capillaries in a fish gill enhances gas exchange and reduces the energy cost of ventilation. Blood flows in the direction opposite to the movement of water past the gills. This makes it possible to transfer O_2 to the blood by a very efficient process called **countercurrent exchange.** As blood moves through a gill capillary, it becomes more and more loaded with O_2, but it simultaneously encounters water with ever higher O_2 concentrations because the water is just beginning its passage over the gills. This means that along the entire length of the capillary, there is a diffusion gradient favoring the transfer of O_2 from water to blood. The countercurrent exchange mechanism is so efficient that the gills can remove more than 80% of the O_2 dissolved in the water passing over the respiratory surface. The basic mechanism of countercurrent exchange is also important in temperature regulation, as we saw in Chapter 40, and in the functioning of the mammalian kidney, as we will see in Chapter 44.

Gills are generally unsuitable for an animal living on land. An expansive surface of wet membrane exposed to air would lose too much water by evaporation, and the gills would collapse as their fine filaments, no longer supported by water, would cling together. Most terrestrial animals house their respiratory surfaces within the body, opening to the atmosphere through narrow tubes.

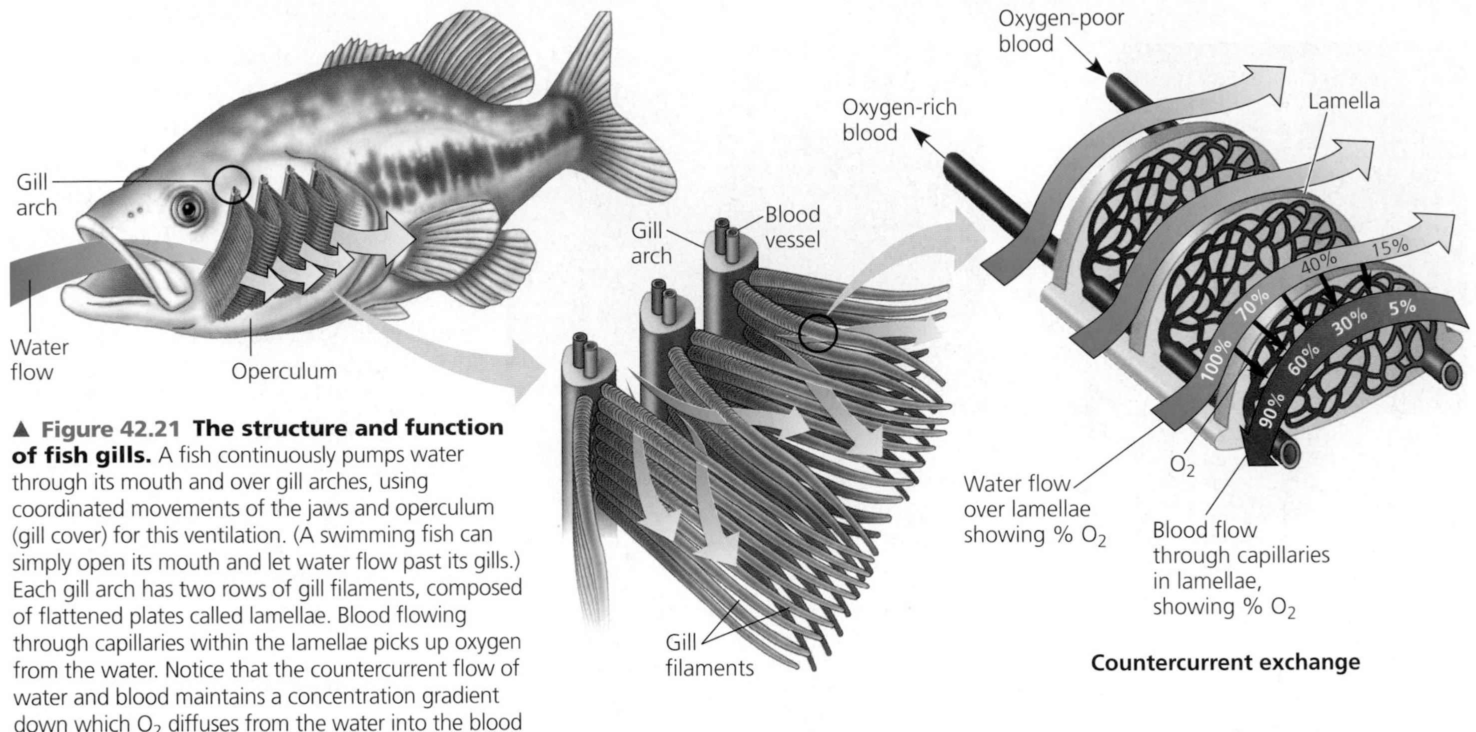

▲ **Figure 42.21 The structure and function of fish gills.** A fish continuously pumps water through its mouth and over gill arches, using coordinated movements of the jaws and operculum (gill cover) for this ventilation. (A swimming fish can simply open its mouth and let water flow past its gills.) Each gill arch has two rows of gill filaments, composed of flattened plates called lamellae. Blood flowing through capillaries within the lamellae picks up oxygen from the water. Notice that the countercurrent flow of water and blood maintains a concentration gradient down which O_2 diffuses from the water into the blood over the entire length of a capillary.

Tracheal Systems in Insects

As a respiratory medium, air has many advantages over water, including a much higher concentration of oxygen (about 210 mL O_2 per liter of air). Also, since O_2 and CO_2 diffuse much faster in air than in water, respiratory surfaces exposed to air do not have to be ventilated as vigorously as gills. As the respiratory surface removes O_2 from the air and expels CO_2, diffusion rapidly brings more O_2 to the respiratory surface and carries the CO_2 away. When a terrestrial animal does ventilate, less energy is needed because air is far lighter and easier to pump than water and because much less volume needs to be breathed to obtain an equal amount of O_2. But along with these advantages comes a problem: The respiratory surface, which must be large and moist, continuously loses water to the air by evaporation. The problem is greatly reduced by a respiratory surface folded into the body. While the most familiar such respiratory structure among terrestrial animals is the lung, the most common is the tracheal system of insects.

The **tracheal system** of insects, made up of air tubes that branch throughout the body, is one variation on the theme of a folded internal respiratory surface. The largest tubes, called tracheae, open to the outside. The finest branches extend to the surface of nearly every cell, where gas is exchanged by diffusion across the moist epithelium that lines the terminal ends of the tracheal system **(Figure 42.22a)**. With virtually all body cells within a very short distance of the respiratory medium, the open circulatory system of insects is not involved in transporting O_2 and CO_2.

For a small insect, diffusion through the trachea brings in enough O_2 and removes enough CO_2 to support cellular respiration. Larger insects with higher energy demands ventilate their tracheal systems with rhythmic body movements that compress and expand the air tubes like bellows. An insect in flight has a very high metabolic rate, consuming 10 to 200 times more O_2 than it does at rest. In many flying insects, alternating contraction and relaxation of the flight muscles compress and expand the body, rapidly pumping air through the tracheal system. The flight muscle cells are packed with mitochondria that support the high metabolic rate, and the tracheal tubes supply each of these ATP-generating organelles with ample O_2 **(Figure 42.22b)**. Thus, adaptations of tracheal systems are directly related to bioenergetics.

Lungs

Unlike tracheal systems that branch throughout the insect body, **lungs** are restricted to one location. Because the respiratory surface of a lung is not in direct contact with all other parts of the body, the gap must be bridged by the circulatory system, which transports gases between the lungs and the rest of the body. Lungs have a dense net of capillaries just under the epithelium that forms the respiratory surface. Lungs have evolved in spiders, terrestrial snails, and vertebrates.

Among the vertebrates, amphibians have relatively small lungs that do not provide a very large surface (many lack lungs altogether), relying heavily on diffusion across other body surfaces for gas exchange. The skin of a frog, for example, supplements gas exchange in the lungs. In contrast, most reptiles (including all birds) and all mammals rely entirely on lungs for gas exchange. Turtles are an exception; they supplement lung breathing with gas exchange across moist epithelial surfaces in

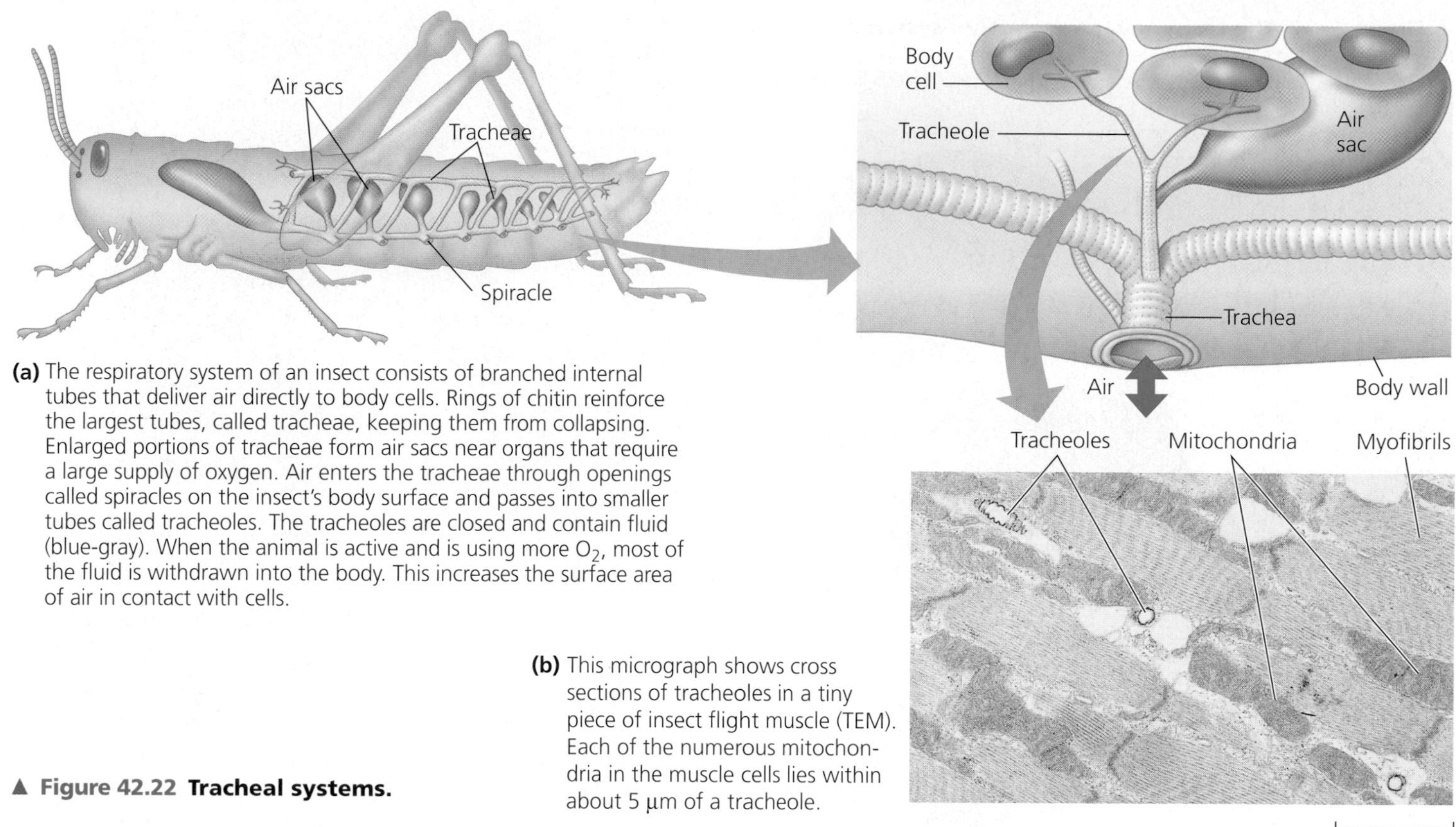

(a) The respiratory system of an insect consists of branched internal tubes that deliver air directly to body cells. Rings of chitin reinforce the largest tubes, called tracheae, keeping them from collapsing. Enlarged portions of tracheae form air sacs near organs that require a large supply of oxygen. Air enters the tracheae through openings called spiracles on the insect's body surface and passes into smaller tubes called tracheoles. The tracheoles are closed and contain fluid (blue-gray). When the animal is active and is using more O_2, most of the fluid is withdrawn into the body. This increases the surface area of air in contact with cells.

(b) This micrograph shows cross sections of tracheoles in a tiny piece of insect flight muscle (TEM). Each of the numerous mitochondria in the muscle cells lies within about 5 μm of a tracheole.

▲ **Figure 42.22 Tracheal systems.**

their mouth and anus. Lungs and air breathing have evolved in a few aquatic vertebrates (lungfishes) as adaptations to living in oxygen-poor water or to spending part of their time exposed to air (for instance, when the water level of a pond recedes).

In general, the size and complexity of lungs are correlated with an animal's metabolic rate (and hence its rate of gas exchange). For example, the lungs of endotherms have a greater area of exchange surface than those of similar-sized ectotherms.

Mammalian Respiratory Systems: A Closer Look

Located in the thoracic (chest) cavity, the lungs of mammals have a spongy texture and are honeycombed with a moist epithelium that functions as the respiratory surface. A system of branching ducts conveys air to the lungs (**Figure 42.23**, on the next page). Air enters through the nostrils and is then filtered by hairs, warmed, humidified, and sampled for odors as it flows through a maze of spaces in the nasal cavity. The nasal cavity leads to the pharynx, an intersection where the paths for air and food cross. When food is swallowed, the **larynx** (the upper part of the respiratory tract) moves upward and tips the epiglottis over the glottis (the opening of the windpipe). This allows food to go down the esophagus to the stomach (see Figure 41.16). The rest of the time, the glottis is open, enabling breathing.

The larynx wall is reinforced with cartilage, and in most mammals, the larynx is adapted as a voice box. Exhaled air rushes by a pair of **vocal cords** in the larynx. Sounds are produced when voluntary muscles in the voice box are tensed, stretching the cords so they vibrate. High-pitched sounds result from tightly stretched cords vibrating rapidly; low-pitched sounds come from less tense cords vibrating slowly.

From the larynx, air passes into the **trachea,** or windpipe. C-shaped rings of cartilage maintain the trachea's shape. The trachea forks into two **bronchi** (singular, *bronchus*), one leading to each lung. Within the lung, the bronchus branches repeatedly into finer and finer tubes called **bronchioles.** The entire system of air ducts has the appearance of an inverted tree, the trunk being the trachea. The epithelium lining the major branches of this respiratory tree is covered by cilia and a thin film of mucus. The mucus traps dust, pollen, and other particulate contaminants, and the beating cilia move the mucus upward to the pharynx, where it can be swallowed into the esophagus. This process helps cleanse the respiratory system.

At their tips, the tiniest bronchioles dead-end as a cluster of air sacs called **alveoli** (singular, *alveolus;* see Figure 42.23). Gas exchange occurs across the thin epithelia of the lung's millions of alveoli, which have a total surface area of about 100 m^2 in humans, sufficient to carry out gas exchange for the entire body. Oxygen in the air entering the alveoli dissolves in the moist film and rapidly diffuses across the epithelium into a web of capillaries that surrounds each alveolus. Carbon dioxide diffuses in the opposite direction, from the capillaries across the epithelium of the alveolus and into the air space.

▼ Figure 42.23 The mammalian respiratory system. From the nasal cavity and pharynx, inhaled air passes through the larynx, trachea, and bronchi to the bronchioles, which end in microscopic alveoli lined by a thin, moist epithelium. Branches of the pulmonary arteries convey oxygen-poor blood to the alveoli; branches of the pulmonary veins transport oxygen-rich blood from the alveoli back to the heart. The left micrograph shows the dense capillary bed that envelops the alveoli. The right micrograph is a cutaway view of alveoli.

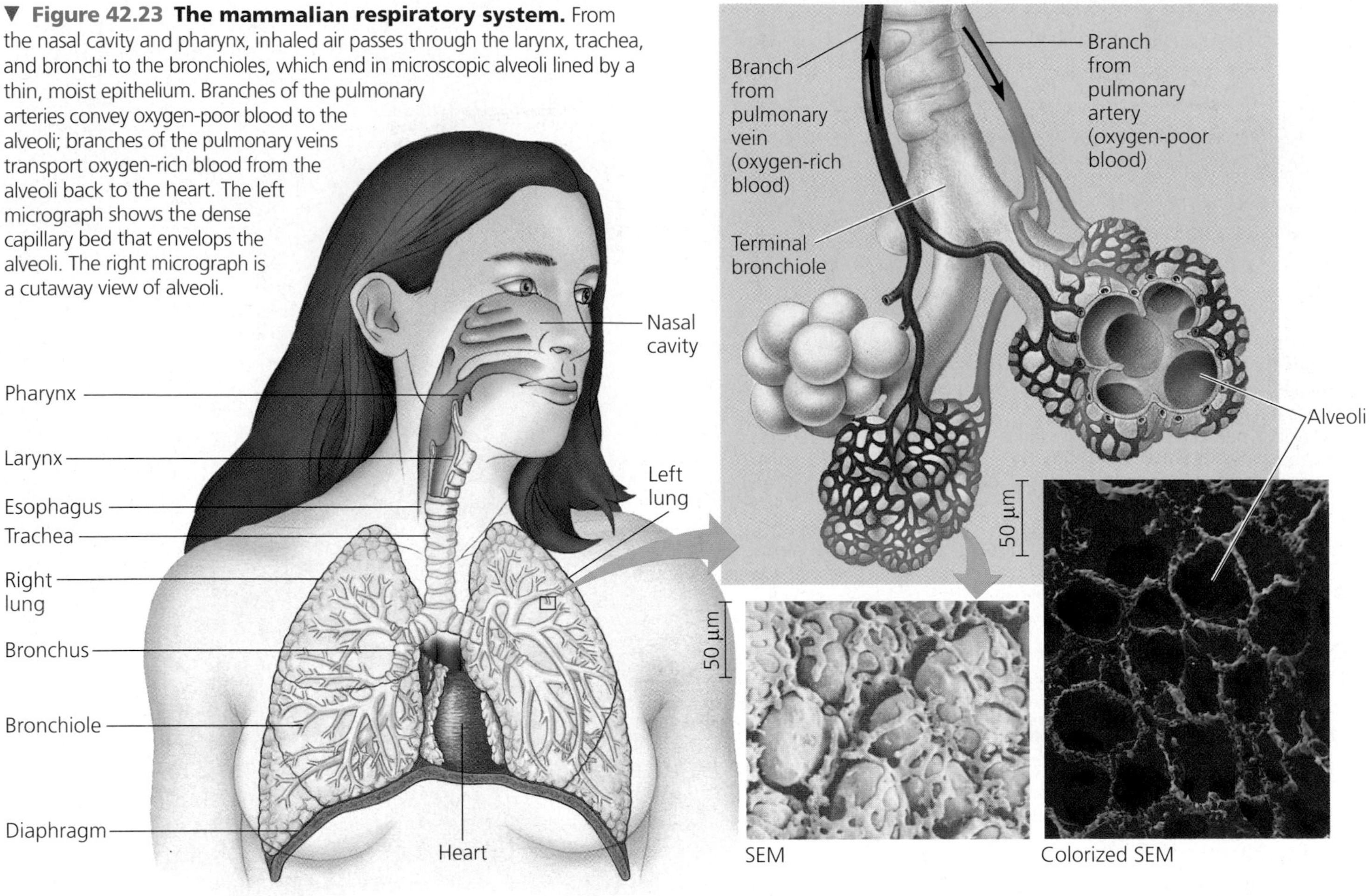

Concept Check 42.5

1. Why is the position of lung tissues *within* the body an advantage for terrestrial animals?
2. Explain how countercurrent exchange maximizes the ability of fish gills to extract dissolved O_2 from water.

For suggested answers, see Appendix A.

Concept 42.6

Breathing ventilates the lungs

Like fishes, terrestrial vertebrates rely on ventilation to maintain high O_2 and low CO_2 concentrations at the gas exchange surface. The process that ventilates lungs is **breathing**, the alternate inhalation and exhalation of air.

How an Amphibian Breathes

An amphibian such as a frog ventilates its lungs by **positive pressure breathing.** During a breathing cycle, muscles lower the floor of the oral cavity, drawing in air through the nostrils. Then, with the nostrils and mouth closed, the floor of the oral cavity rises, forcing air down the trachea. Elastic recoil of the lungs and compression by the muscular body wall force air back out of the lungs during exhalation.

How a Mammal Breathes

Unlike amphibians, mammals ventilate their lungs by **negative pressure breathing,** which works like a suction pump, pulling air instead of pushing it into the lungs **(Figure 42.24)**. Muscle action changes the volume of the rib cage and the chest cavity, and the lungs match these volume changes. The inner layer of the lung's double-walled sac adheres to the outside of the lungs, and the outer layer adheres to the wall of the chest cavity. A thin space filled with fluid separates the two layers. Because of surface tension, the two layers are like two plates of glass stuck together by a film of water: The layers can slide smoothly past each other, but they cannot be pulled apart easily. Surface tension couples movement of the lungs to movement of the rib cage.

Lung volume increases as a result of contraction of the rib muscles and the **diaphragm,** a sheet of skeletal muscle that forms the bottom wall of the chest cavity. Contraction of the rib muscles expands the rib cage by pulling the ribs upward and the breastbone outward. At the same time, the chest cavity expands as the diaphragm contracts and descends like a piston.

All these changes increase the lung volume, and as a result, air pressure within the alveoli becomes lower than atmospheric pressure. Because gas flows from a region of higher pressure to a region of lower pressure, air rushes through the nostrils and mouth and down the breathing tubes to the alveoli. During exhalation, the rib muscles and diaphragm relax, lung volume is reduced, and the increased air pressure within the alveoli forces air up the breathing tubes and out of the body (see Figure 42.24).

The rib muscles and diaphragm change lung volume during shallow breathing, when a mammal is at rest. During exercise, other muscles of the neck, back, and chest further increase ventilation volume by raising the rib cage even more. In some species, running causes visceral organs, including the stomach and liver, to slide forward and backward in the body cavity with each stride. This "visceral pump" further increases ventilation volume.

The volume of air a mammal inhales and exhales with each breath is called **tidal volume.** It averages about 500 mL in resting humans. The maximum tidal volume during forced breathing is the **vital capacity,** which is about 3.4 L and 4.8 L for college-age females and males, respectively. The lungs hold more air than the vital capacity, but since it is impossible to completely collapse the alveoli, a **residual volume** of air remains in the lungs even after you forcefully exhale as much as you can. As lungs lose their resilience owing to aging or disease (such as emphysema), residual volume increases at the expense of vital capacity.

Since the lungs do not completely empty and refill with each breath cycle, newly inhaled air mixes with oxygen-depleted residual air, and the maximum O_2 concentration in alveoli is considerably less than in the atmosphere. Though this limits the effectiveness of gas exchange, the CO_2 in residual air is critical for regulating the pH of blood and breathing rate in mammals.

How a Bird Breathes

Ventilation is much more complex in birds than in mammals. Besides lungs, birds have eight or nine air sacs that penetrate the abdomen, neck, and wings. The air sacs do not function directly in gas exchange, but act as bellows that keep air flowing through the lungs **(Figure 42.25)**. The entire system—lungs and air sacs—is ventilated when the bird breathes. Air flows through the system in a circuit that passes through the lungs in one direction only, regardless of whether the animal is inhaling or exhaling. Instead of alveoli, which are dead ends, the sites of gas exchange in bird lungs are tiny channels called **parabronchi,** through which air flows in one direction.

This system completely renews the air in the lungs with every exhalation, so maximum lung O_2 concentrations are higher in birds than in mammals. Partly because of this advantage, birds perform much better than mammals at high altitude. For example, humans have great difficulty obtaining enough O_2 when climbing the Earth's highest peaks, such as Mount Everest in the Himalayas (8,848 m). But several

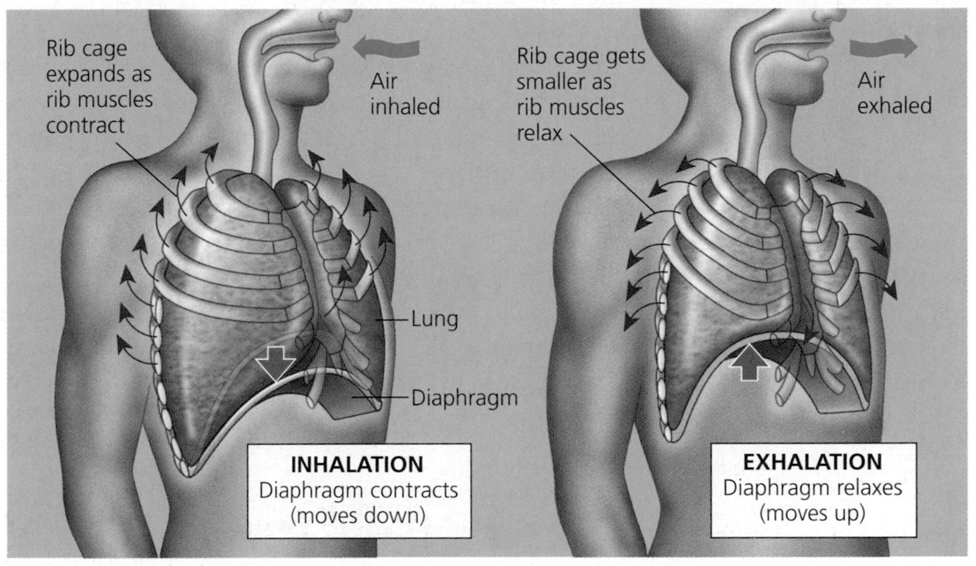

▲ **Figure 42.24 Negative pressure breathing.** A mammal breathes by changing the air pressure within its lungs relative to the pressure of the outside atmosphere.

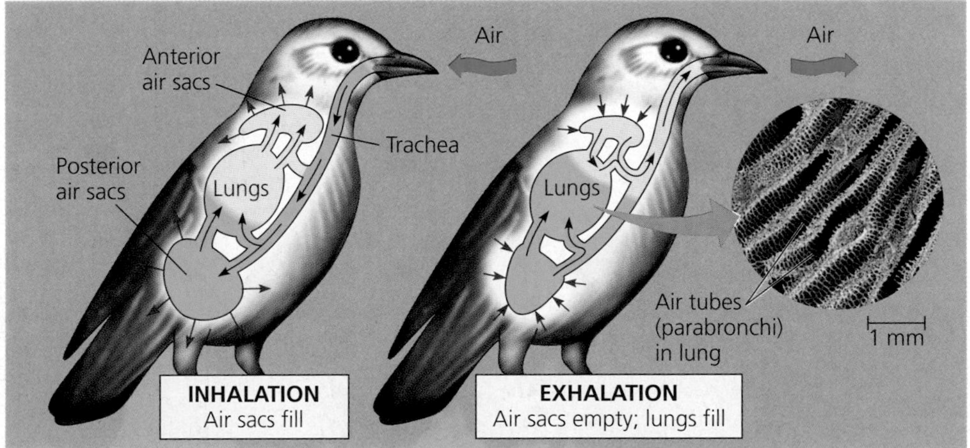

▲ **Figure 42.25 The avian respiratory system.** Contraction and relaxation of the air sacs ventilates the lungs, forcing air in one direction through tiny parallel tubes in the lungs called parabronchi (inset, SEM). Gas exchange occurs across the walls of the parabronchi. During inhalation, both sets of air sacs expand. The posterior sacs fill with fresh air (blue) from the outside, while the anterior sacs fill with stale air (gray) from the lungs. During exhalation, both sets of air sacs deflate, forcing air from the posterior sacs into the lungs, and air from the anterior sacs out of the system via the trachea. Two cycles of inhalation and exhalation are required for the air to pass all the way through the system and out of the bird.

species of birds (notably the bar-headed goose) easily fly over the same mountains during migration.

Control of Breathing in Humans

Humans can voluntarily hold their breath or breathe faster and deeper, but most of the time automatic mechanisms regulate breathing. This ensures that the work of the respiratory system is coordinated with that of the cardiovascular system and with metabolic demands for gas exchange.

The main **breathing control centers** are located in two regions of the brain, the medulla oblongata and the pons **(Figure 42.26)**. Aided by the control center in the pons, the medulla's center sets the basic breathing rhythm. Secondary control over breathing is exerted by sensors in the aorta and carotid arteries that monitor O_2 and CO_2 concentrations in the blood as well as blood pH. During deep breathing, a negative-feedback mechanism prevents the lungs from overexpanding; stretch sensors in the lung tissue send nerve impulses that inhibit the medulla's control center.

The medulla's control center regulates breathing activity in response to changes in pH of the tissue fluid (cerebrospinal fluid) bathing the brain. The pH of the cerebrospinal fluid is determined mainly by the CO_2 concentration in the blood. Carbon dioxide diffuses from the blood to the cerebrospinal fluid, where it reacts with water and forms carbonic acid, which lowers the pH. When the medulla's control center registers a slight drop in pH (increase in CO_2 level) of the cerebrospinal fluid, it increases the depth and rate of breathing, and the excess CO_2 is eliminated in exhaled air. This occurs during exercise, for example.

Oxygen concentrations in the blood usually have little effect on the breathing control centers. However, when the O_2 level is severely depressed (at high altitudes, for instance), O_2 sensors in the aorta and carotid arteries in the neck send alarm signals to the breathing control centers, which respond by increasing the breathing rate. Normally, a rise in CO_2 concentration is a good indication of a fall in O_2 concentration, because CO_2 is produced by the same process that consumes O_2—cellular respiration. However, it is possible to trick the breathing center by hyperventilating. Excessively deep, rapid breathing purges the blood of so much CO_2 that the breathing center temporarily stops sending impulses to the rib muscles and diaphragm. Breathing stops until CO_2 levels increase (or O_2 levels decrease) enough switch the breathing center back on.

The breathing center responds to a variety of nervous and chemical signals and adjusts breathing rate and depth to meet changing demands. However, breathing control is effective only if it is coordinated with control of the cardiovascular system so that there is a good match between lung ventilation and the amount of blood flowing through alveolar capillaries. During exercise, for instance, increased cardiac output is matched to the increased breathing rate, which enhances O_2 uptake and CO_2 removal as blood flows through the lungs.

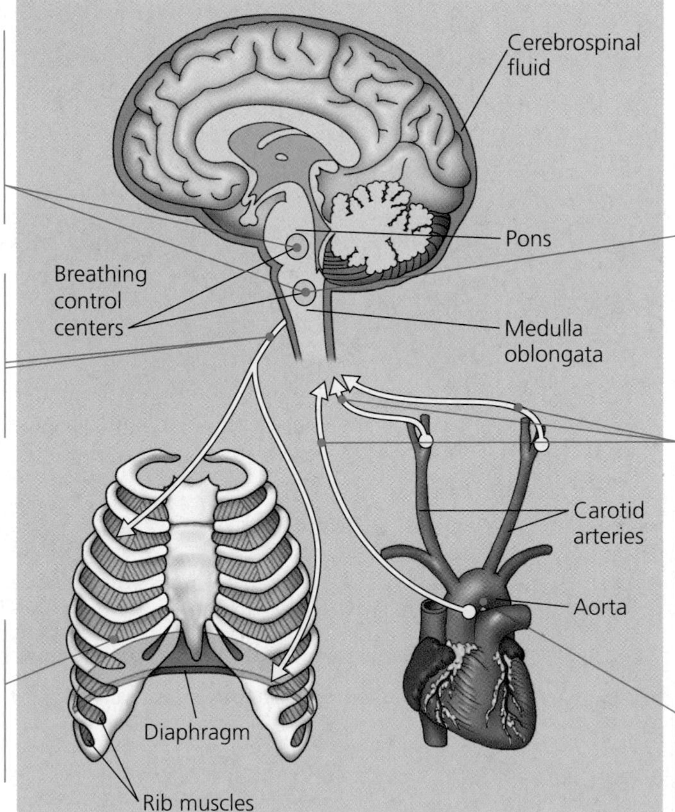

1 A breathing control center in the medulla sets the basic rhythm, and a control center in the pons moderates it, smoothing out the transitions between inhalations and exhalations.
2 Nerves from the medulla's control center send impulses to the diaphragm and rib muscles, stimulating them to contract and causing inhalation.
3 In a person at rest, these nerve impulses result in about 10 to 14 inhalations per minute. Between inhalations, the muscles relax and the person exhales.
Cerebrospinal fluid
Pons
Breathing control centers
Medulla oblongata
Carotid arteries
Aorta
Diaphragm
Rib muscles
4 The medulla's control center also helps regulate blood CO_2 level. Sensors in the medulla detect changes in the pH (reflecting CO_2 concentration) of the blood and cerebrospinal fluid bathing the surface of the brain.
5 Other sensors in the walls of the aorta and carotid arteries in the neck detect changes in blood pH and send nerve impulses to the medulla. In response, the medulla's control center alters the rate and depth of breathing, increasing both if CO_2 levels rise or decreasing both if CO_2 levels are depressed.
6 The sensors in the aorta and carotid arteries also detect changes in O_2 levels in the blood and signal the medulla to increase the breathing rate when levels become very low.

▲ **Figure 42.26 Automatic control of breathing.**

Concept Check 42.6

1. How does an increase in the CO_2 concentration in the blood affect the pH of cerebrospinal fluid?
2. A slight decrease in blood pH causes the heart's pacemaker to speed up. What is the function of this control mechanism?
3. How does breathing differ in mammals and birds?

For suggested answers, see Appendix A.

Concept 42.7

Respiratory pigments bind and transport gases

The high metabolic demands of many organisms require that the blood transport large quantities of O_2 and CO_2. It appears that similar solutions that address the transport requirements set by high metabolic demands have evolved independently in different groups of animals. First, we'll see the role that O_2 and CO_2 gradients play, using the mammalian respiratory system as an example. Then we'll examine blood molecules called respiratory pigments that aid in the process of gas exchange.

The Role of Partial Pressure Gradients

Gases diffuse down pressure gradients in the lungs and other organs. Diffusion of a gas, whether present in air or dissolved in water, depends on differences in a quantity called **partial pressure.** At sea level, the atmosphere exerts a total pressure of 760 mm Hg. This is a downward force equal to that exerted by a column of mercury 760 mm high. Since the atmosphere is 21% O_2 (by volume), the partial pressure of O_2 (abbreviated P_{O_2}) is 0.21×760, or about 160 mm Hg. This is the portion of atmospheric pressure contributed by O_2; hence the term *partial pressure*. The partial pressure of CO_2 (P_{CO_2}) at sea level is only 0.23 mm Hg. When water is exposed to air, the amount of a gas that dissolves in the water is proportional to its partial pressure in the air and its solubility in water. An equilibrium is eventually reached when gas molecules enter and leave the solution at the same rate. At this point, the gas is said to have the same partial pressure in solution as it does in the air. Thus, in a glass of water exposed to air at sea-level air pressure, the P_{O_2} is 160 mm Hg and the P_{CO_2} is 0.23 mm Hg.

A gas always diffuses from a region of higher partial pressure to a region of lower partial pressure. ❶ Blood arriving at the lungs via the pulmonary arteries has a lower P_{O_2} and a higher P_{CO_2} than the air in the alveoli (**Figure 42.27**). Notice that the air in the alveoli has a lower P_{O_2} and a higher P_{CO_2} than air at sea level, since it is not completely replaced with fresh air during breathing. As blood enters the alveolar capillaries, CO_2 diffuses from the blood to the air in the alveoli. Meanwhile, O_2 in the air dissolves in the fluid that coats the epithelium and diffuses into the blood. ❷ By the time the blood leaves the lungs in the pulmonary veins, its P_{O_2} has been raised and its P_{CO_2} has been lowered. After returning to the heart, this blood is pumped through the systemic circuit. ❸ In the tissue capillaries, gradients of partial pressure favor the diffusion of O_2 out of the blood and CO_2 into the blood. This is because cellular respiration removes O_2 from and adds CO_2 to the interstitial fluid (again, by diffusion, from mitochondria in nearby cells). ❹ After the blood unloads O_2 and loads CO_2, it is returned to the heart and pumped to the lungs again, where it exchanges gases with air in the alveoli.

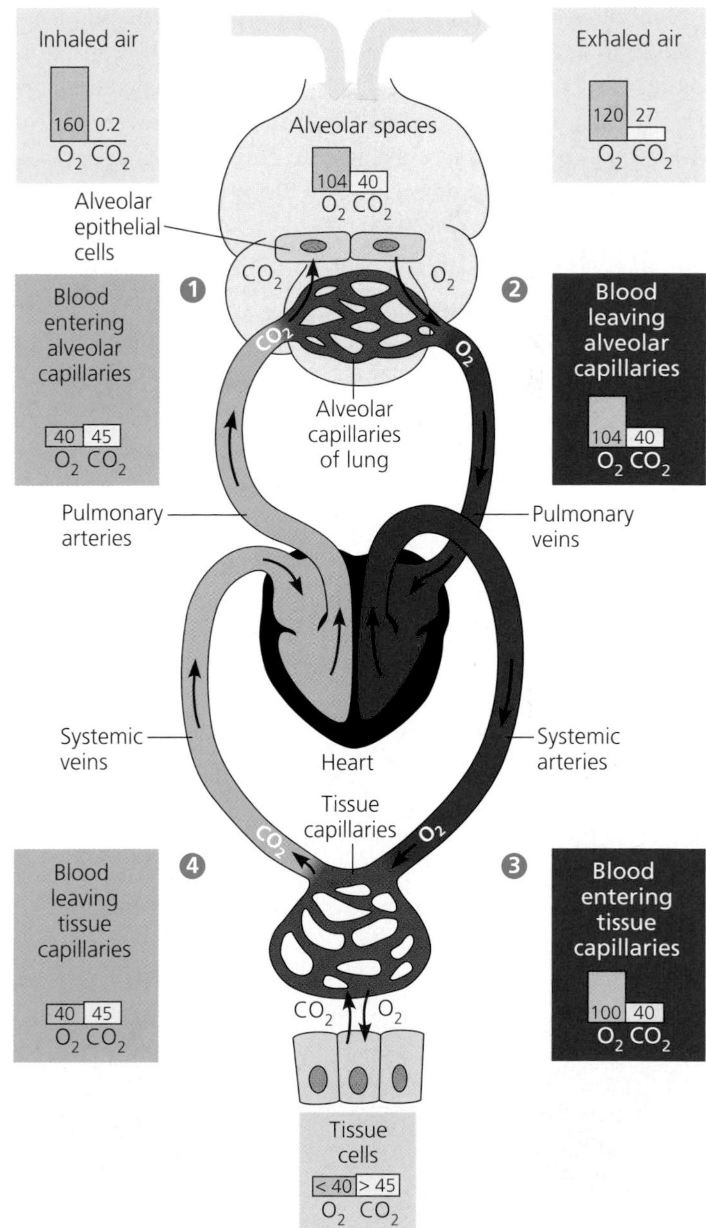

▲ **Figure 42.27 Loading and unloading of respiratory gases.** The colored bars indicate the partial pressures (in mm Hg) of O_2 (P_{O_2}) and CO_2 (P_{CO_2}) in different locations.

Respiratory Pigments

The low solubility of O_2 in water (and thus in blood) is a problem for animals that rely on the circulatory system to deliver O_2. Suppose all our O_2 were delivered in solution in the blood. During intense exercise, a person can consume almost 2 L of O_2 per minute, and all of it must be carried in the blood from the lungs to the active tissues. But at normal body temperature and air pressure, only 4.5 mL of O_2 can dissolve into a liter of blood in the lungs. If 80% of the dissolved O_2 were delivered to the tissues (an unrealistically high percentage), the heart would need to pump 555 L of blood per minute.

In fact, most animals transport most of their O_2 bound to certain proteins called **respiratory pigments** instead of in dissolved form. Respiratory pigments circulate with the blood, often contained within specialized cells. The pigments greatly increase the amount of oxygen that can be carried in blood (to about 200 mL of O_2 per liter in mammalian blood). In our example of an exercising human with a delivery rate of 80%, that greatly reduces the cardiac output necessary for O_2 transport to a manageable 12.5 L of blood per minute.

Oxygen Transport

A diversity of respiratory pigments have evolved in various animal taxa. One example, **hemocyanin,** found in arthropods and many molluscs, has copper as its oxygen-binding component, coloring the blood bluish. The respiratory pigment of almost all vertebrates and a wide variety of invertebrates is the protein hemoglobin, contained in the erythrocytes of vertebrates.

Hemoglobin consists of four subunits, each with a cofactor called a heme group that has an iron atom at its center. The iron binds the O_2; thus, each hemoglobin molecule can carry four molecules of O_2 (see Figure 5.20). Like all respiratory pigments, hemoglobin must bind O_2 reversibly, loading O_2 in the lungs or gills and unloading it in other parts of the body **(Figure 42.28)**. This process depends on cooperativity between the subunits of the hemoglobin molecule (see Chapter 8). Binding of O_2 to one subunit induces the others to change shape slightly, with the result that their affinity for O_2 increases. And when one subunit unloads its O_2, the other three quickly unload too as a shape change lowers their affinity for O_2.

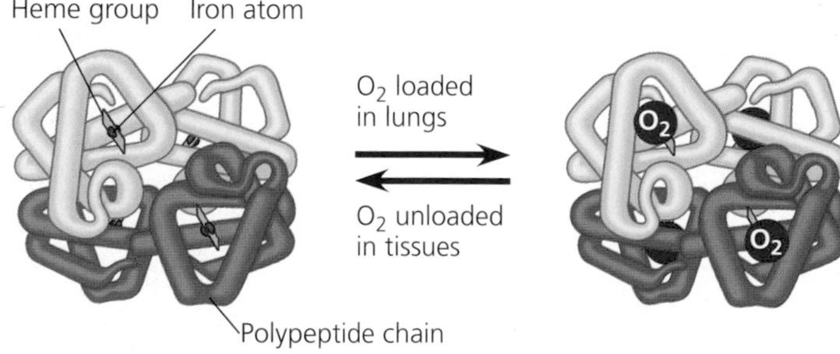

▲ **Figure 42.28 Hemoglobin loading and unloading O_2.**

Cooperative O_2 binding and release is evident in the **dissociation curve** for hemoglobin **(Figure 42.29)**. Over the

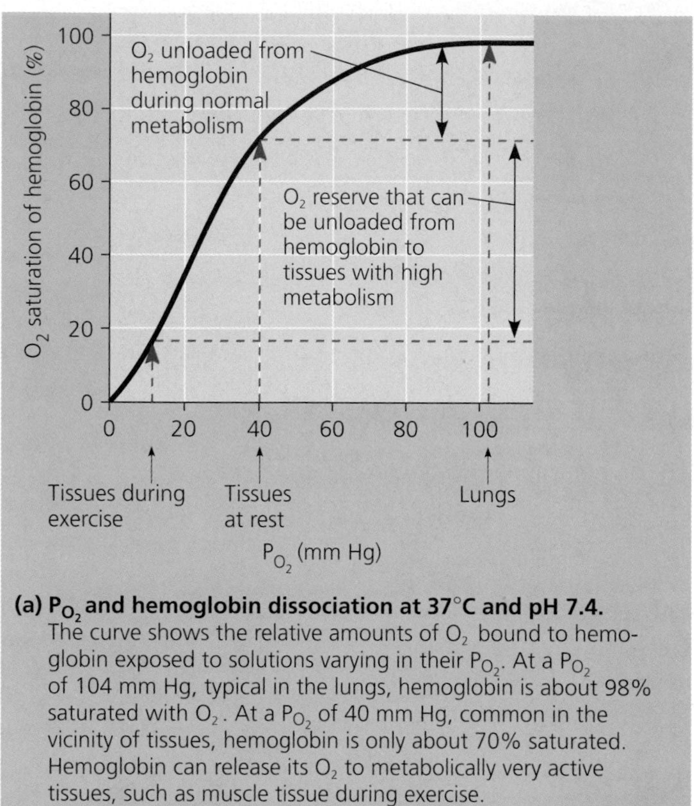

(a) P_{O_2} and hemoglobin dissociation at 37°C and pH 7.4. The curve shows the relative amounts of O_2 bound to hemoglobin exposed to solutions varying in their P_{O_2}. At a P_{O_2} of 104 mm Hg, typical in the lungs, hemoglobin is about 98% saturated with O_2. At a P_{O_2} of 40 mm Hg, common in the vicinity of tissues, hemoglobin is only about 70% saturated. Hemoglobin can release its O_2 to metabolically very active tissues, such as muscle tissue during exercise.

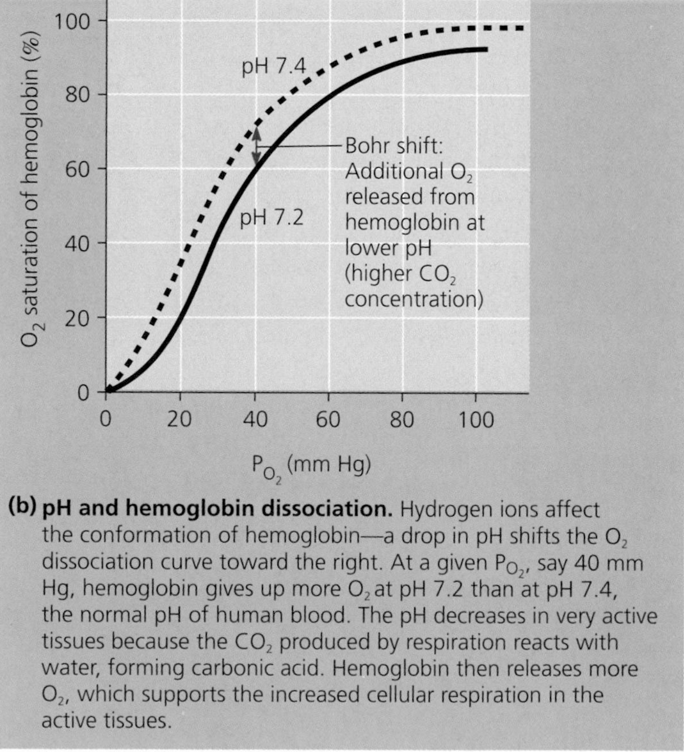

(b) pH and hemoglobin dissociation. Hydrogen ions affect the conformation of hemoglobin—a drop in pH shifts the O_2 dissociation curve toward the right. At a given P_{O_2}, say 40 mm Hg, hemoglobin gives up more O_2 at pH 7.2 than at pH 7.4, the normal pH of human blood. The pH decreases in very active tissues because the CO_2 produced by respiration reacts with water, forming carbonic acid. Hemoglobin then releases more O_2, which supports the increased cellular respiration in the active tissues.

▲ **Figure 42.29 Dissociation curves for hemoglobin.**

range of O_2 partial pressures (P_{O_2}) where the dissociation curve has a steep slope, even a slight change in P_{O_2} causes hemoglobin to load or unload a substantial amount of O_2. Notice that the steep part of the curve corresponds to the range of O_2 partial pressures found in body tissues. When cells in a particular location begin working harder—during exercise, for instance—P_{O_2} dips in their vicinity as the O_2 is consumed in cellular respiration. Because of the effect of subunit cooperativity, a slight drop in P_{O_2} is enough to cause a relatively large increase in the amount of O_2 the blood unloads.

As with all proteins, hemoglobin's conformation is sensitive to a variety of factors. For example, a drop in pH lowers the affinity of hemoglobin for O_2, an effect called the **Bohr shift** (see Figure 42.29b). Because CO_2 reacts with water, forming carbonic acid (H_2CO_3), an active tissue lowers the pH of its surroundings and induces hemoglobin to release more O_2, which can then be used for cellular respiration.

Carbon Dioxide Transport

In addition to its role in oxygen transport, hemoglobin also helps transport CO_2 and assists in buffering—that is, preventing harmful changes in blood pH. Only about 7% of the CO_2 released by respiring cells is transported in solution in blood plasma. Another 23% binds to the multiple amino groups of hemoglobin, and about 70% is transported in the blood in the form of bicarbonate ions (HCO_3^-). Carbon dioxide from respiring cells diffuses into the blood plasma and then into the erythrocytes **(Figure 42.30)**. The CO_2 first reacts with water (assisted by the enzyme carbonic anhydrase) and forms H_2CO_3, which then dissociates into a hydrogen ion (H^+) and HCO_3^-. Most of the H^+ attaches to hemoglobin and other proteins, minimizing the change in blood pH. The HCO_3^- diffuses into the plasma. As blood flows through the lungs, the process is rapidly reversed as diffusion of CO_2 out of the

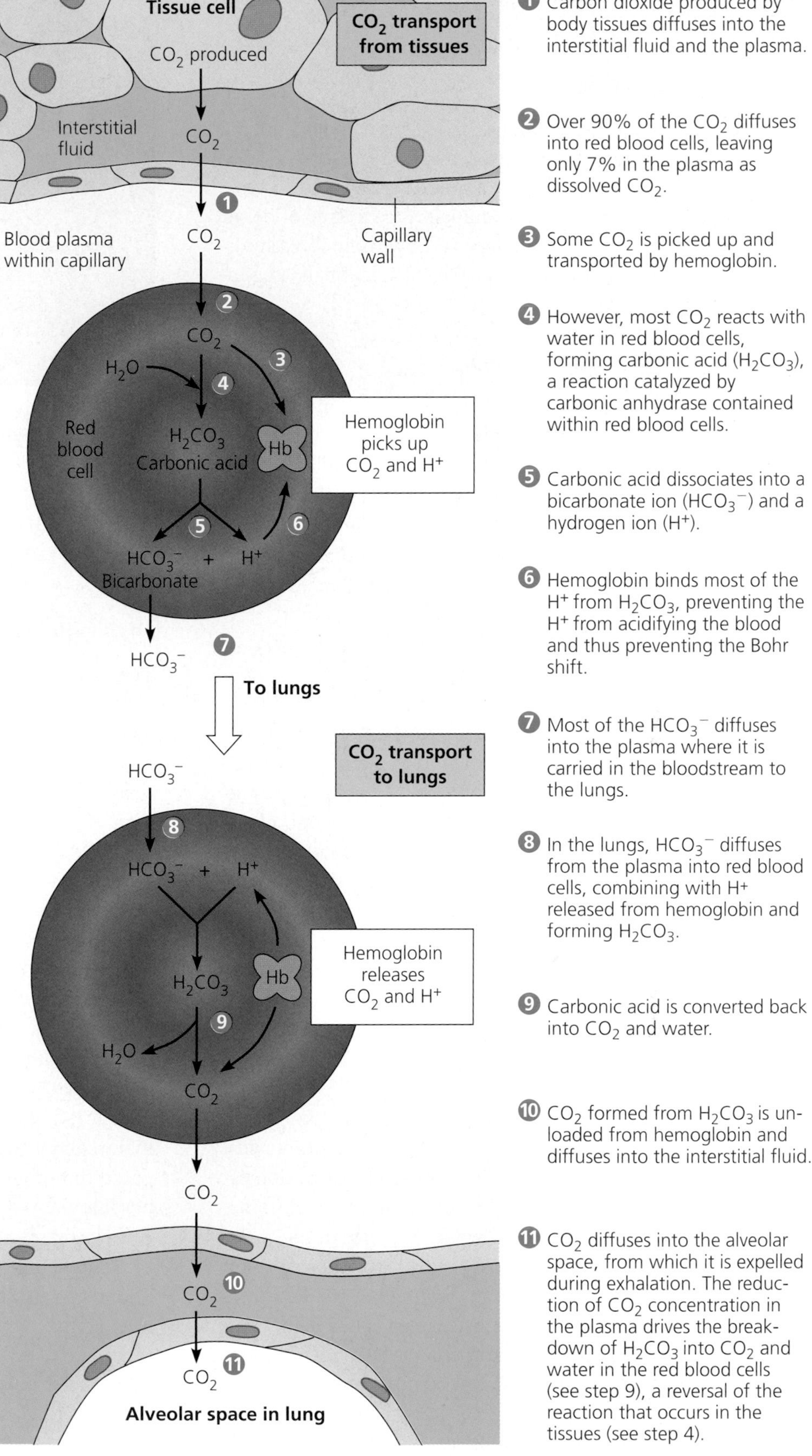

1. Carbon dioxide produced by body tissues diffuses into the interstitial fluid and the plasma.
2. Over 90% of the CO_2 diffuses into red blood cells, leaving only 7% in the plasma as dissolved CO_2.
3. Some CO_2 is picked up and transported by hemoglobin.
4. However, most CO_2 reacts with water in red blood cells, forming carbonic acid (H_2CO_3), a reaction catalyzed by carbonic anhydrase contained within red blood cells.
5. Carbonic acid dissociates into a bicarbonate ion (HCO_3^-) and a hydrogen ion (H^+).
6. Hemoglobin binds most of the H^+ from H_2CO_3, preventing the H^+ from acidifying the blood and thus preventing the Bohr shift.
7. Most of the HCO_3^- diffuses into the plasma where it is carried in the bloodstream to the lungs.
8. In the lungs, HCO_3^- diffuses from the plasma into red blood cells, combining with H^+ released from hemoglobin and forming H_2CO_3.
9. Carbonic acid is converted back into CO_2 and water.
10. CO_2 formed from H_2CO_3 is unloaded from hemoglobin and diffuses into the interstitial fluid.
11. CO_2 diffuses into the alveolar space, from which it is expelled during exhalation. The reduction of CO_2 concentration in the plasma drives the breakdown of H_2CO_3 into CO_2 and water in the red blood cells (see step 9), a reversal of the reaction that occurs in the tissues (see step 4).

▲ **Figure 42.30 Carbon dioxide transport in the blood.**

blood shifts the chemical equilibrium in favor of the conversion of HCO_3^- to CO_2.

Elite Animal Athletes

For some animals, such as long-distance runners and migratory birds and mammals, the O_2 demands of daily activities would overwhelm the capacity of a typical respiratory system. Other animals, such as diving mammals, are capable of being active underwater for extended periods without breathing. What evolutionary adaptations enable these animals to perform such feats?

The Ultimate Endurance Runner

The elite animal marathon runner may be the pronghorn, an antelope-like mammal native to the grasslands of North America, where it has roamed for more than 4 million years. Pronghorns are capable of running as fast as 100 km/hr. Though their top speed does not reach that of the cheetah, pronghorns can sustain high speeds over long distances and have been timed sprinting 11 km in 10 minutes, maintaining an average speed of 65 km/hr.

Stan Lindstedt and his colleagues at the University of Wyoming and University of Bern were curious about how pronghorns sustain their combination of great speed and endurance: through enhancements of normal physiological mechanisms that supply increased O_2 to muscles or through greater energetic efficiency? The researchers exercised pronghorns on a treadmill to estimate their maximum rate of O_2 consumption **(Figure 42.31)** and discovered something surprising: Pronghorns consume O_2 at three times the rate predicted for an animal of their size. Normally, as animals increase in size, their rate of O_2 consumption per gram declines. One gram of shrew tissue, for example, consumes as much O_2 in a day as a gram of elephant tissue consumes in an entire month. But Lindstedt and his colleagues discovered that the rate of O_2 consumption per gram of tissue by a pronghorn was the same as that of a 10-g mouse!

To establish a more appropriate perspective on pronghorn performance, the research team compared various physiological characteristics of pronghorns with those of similar-sized domestic goats, which are adapted to climbing rather than running. They found that the maximum rate of O_2 consumption by pronghorns is five times that of goats. Why? Pronghorns have a larger surface area for O_2 diffusion in the lungs, nearly five times the cardiac output, much higher muscle mass, and a higher volume and density of mitochondria than goats. In addition, pronghorns maintain higher muscle temperatures. The researchers concluded that the pronghorn's extreme O_2 consumption rate, which underlies their ability to run at high speeds over long distances, results from enhancements of the normal physiological mechanisms present in other mammals. We can see in these enhancements the results of natural selection, perhaps exerted by the predators that have chased pronghorns across the open plains of North America for millions of years.

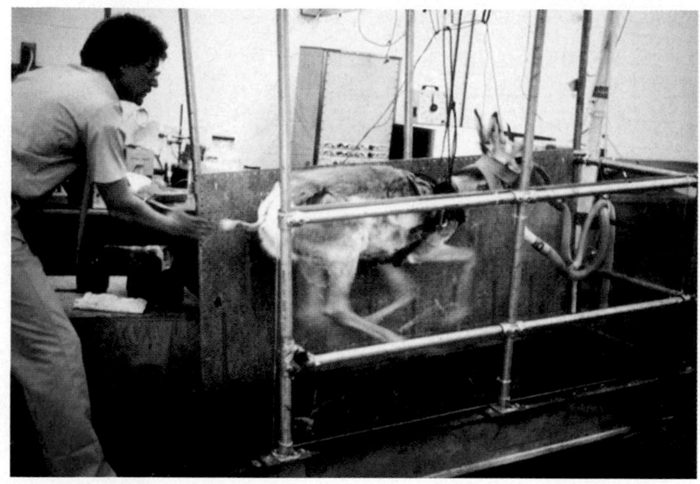

▲ **Figure 42.31 Measuring rate of O_2 consumption in a running pronghorn.** Stan Lindstedt collects respiratory data from a pronghorn running on a treadmill at 40 km/hr.

Diving Mammals

The majority of animals can exchange gases continuously, but sometimes there is no access to the normal respiratory medium—for example, when an air-breather swims underwater. Whereas most humans, even well-trained divers, cannot hold their breath longer than 2 or 3 minutes or swim deeper than 20 m, the Weddell seal of Antarctica routinely plunges to 200–500 m and remains there for about 20 minutes (sometimes for more than an hour). In contrast, humans need to carry extra air—in the form of scuba tanks—to remain submerged for comparable periods. Some species of seals, sea turtles, and whales make even more impressive dives. Elephant seals can reach depths of 1,500 m—almost a mile—and stay submerged for as long as 2 hours! One elephant seal carrying a recording device spent 40 days at sea diving almost continuously, with no surface period longer than 6 minutes.

One adaptation of the Weddell seal (and other diving mammals) is an ability to store large amounts of O_2. Compared to humans, the seal can store about twice as much O_2 per kilogram of body mass, mostly in the blood and muscles. About 36% of our total O_2 is in our lungs, and 51% is in our blood. In contrast, the Weddell seal holds only about 5% of its oxygen in its relatively small lungs (and may exhale before diving, which reduces buoyancy), stockpiling 70% in the blood. The seal has about twice the volume of blood per kilogram of body mass as a human. Another adaptation is the seal's huge spleen, which can store about 24 L of blood. The spleen probably contracts after a dive begins, fortifying the blood with erythrocytes loaded with O_2. Diving mammals also have a high concentration of an oxygen-storing protein called **myoglobin** in their muscles. The Weddell seal can store about 25% of its O_2 in muscle, compared to only 13% in humans.

Diving mammals not only begin an underwater trip with a relatively large O_2 stockpile, but also have adaptations that conserve O_2. They swim with little muscular effort and often use buoyancy changes to glide passively upward or downward. Their heart rate and O_2 consumption rate decrease during a dive, and regulatory mechanisms affecting peripheral resistance route most blood to the brain, spinal cord, eyes, adrenal glands, and placenta (in pregnant seals). Blood supply to the muscles is restricted and is shut off altogether during the longest dives. During dives of more than about 20 minutes, a Weddell seal's muscles deplete the O_2 stored in myoglobin and then derive their ATP from fermentation instead of respiration (see Chapter 9).

The unusual abilities of the Weddell seal and other air-breathing divers to power their bodies during long dives showcase two related themes in our study of organisms—the response to environmental challenges over the short term by physiological adjustments and over the long term as a result of natural selection.

Concept Check 42.7

1. What determines whether O_2 and CO_2 diffuse into or out of the capillaries in the tissues and near the alveolar spaces? Explain.
2. How does the Bohr shift help deliver O_2 to very active tissues?
3. Carbon dioxide within red blood cells in the tissue capillaries combines with water, forming carbonic acid. What causes the reverse of this reaction in red blood cells in capillaries near the alveolar spaces?
4. Describe three adaptations that enable Weddell seals to stay underwater much longer than humans can.

For suggested answers, see Appendix A.

Chapter 42 Review

Go to www.campbellbiology.com or the student CD-ROM to explore Activities, Investigations, and other interactive study aids.

SUMMARY OF KEY CONCEPTS

Concept 42.1

Circulatory systems reflect phylogeny

- Transport systems functionally connect the organs of exchange with the body cells. Most complex animals have internal transport systems that circulate fluid, providing a lifeline between the aqueous environment of living cells and the organs, such as lungs, that exchange chemicals with the outside environment (pp. 867–868).
- **Invertebrate Circulation (pp. 868–869)** Most invertebrates have a gastrovascular cavity or a circulatory system for internal transport.
- **Survey of Vertebrate Circulation (pp. 869–871)** Adaptations of the cardiovascular system reflect vertebrate phylogeny. Blood flows in a closed cardiovascular system consisting of blood vessels and a two- to four-chambered heart. Arteries convey blood to capillaries, the sites of chemical exchange between blood and interstitial fluid. Veins return blood from capillaries to the heart.

Concept 42.2

Double circulation in mammals depends on the anatomy and pumping cycle of the heart

- **Mammalian Circulation: *The Pathway* (p. 872)** Heart valves dictate a one-way flow of blood through the heart, beginning with the right ventricle pumping blood to the lungs, where it loads O_2 and unloads CO_2. Oxygen-rich blood from the lungs enters the heart at the left atrium and is pumped to the body tissues by the left ventricle. Blood returns to the heart through the right atrium.
Activity *Mammalian Cardiovascular System Structure*
- **The Mammalian Heart: *A Closer Look* (pp. 872–873)** The heart rate (pulse) is the number of times the heart beats each minute. The cardiac cycle, one complete sequence of the heart's pumping and filling, consists of a period of contraction, called systole, and a period of relaxation, called diastole. The cardiac output is the volume of blood pumped into the systemic circulation per minute.
- **Maintaining the Heart's Rhythmic Beat (pp. 873–874)** Impulses originating at the sinoatrial (SA) node (pacemaker) of the right atrium pass to the atrioventricular (AV) node. After a delay, they are conducted along the bundle branches and Purkinje fibers. The pacemaker is influenced by nerves, hormones, body temperature, and exercise.

Concept 42.3

Physical principles govern blood circulation

- **Blood Vessel Structure and Function (pp. 874–875)** Structural differences of arteries, veins, and capillaries correlate with their different functions.
- **Blood Flow Velocity (pp. 875–876)** Physical laws governing the movement of fluids through pipes influence blood flow and blood pressure. The velocity of blood flow varies in the circulatory system, being lowest in the capillary beds as a result of the high resistance and large total cross-sectional area of the capillaries.
- **Blood Pressure (pp. 876–877)** Blood pressure, the hydrostatic pressure that blood exerts against the wall of a vessel, is determined by the cardiac output and peripheral resistance due to variable constriction of the arterioles.

- **Capillary Function** **(pp. 877–879)** Transfer of substances between the blood and the interstitial fluid occurs across the thin walls of capillaries.
- **Fluid Return by the Lymphatic System** **(pp. 878–879)** The lymphatic system returns fluid to the blood and aids in body defense. Fluid reenters the circulation directly at the venous end of the capillary and indirectly through the lymphatic system.
  Activity *Path of Blood Flow in Mammals*
  Biology Labs On-Line *CardioLab*
  Activity *Mammalian Cardiovascular System Function*

Blood is a connective tissue with cells suspended in plasma

- **Blood Composition and Function** **(pp. 879–882)** Whole blood consists of cellular elements (cells and fragments of cells called platelets) suspended in a liquid matrix called plasma. Plasma proteins influence blood pH, osmotic pressure, and viscosity and function in lipid transport, immunity (antibodies), and blood clotting (fibrinogens). Red blood cells, or erythrocytes, transport oxygen. Five types of white blood cells, or leukocytes, function in defense by phagocytosing bacteria and debris or by producing antibodies. Platelets function in blood clotting, a cascade of complex reactions that converts plasma fibrinogen to fibrin.
- **Cardiovascular Disease** **(pp. 882–883)** Cardiovascular diseases are the leading cause of death in most developed nations, including the United States.
  Investigation *How Is Cardiovascular Fitness Measured?*

Gas exchange occurs across specialized respiratory surfaces

- Gas exchange supplies oxygen for cellular respiration and disposes of carbon dioxide. Animals require large, moist respiratory surfaces for the adequate diffusion of respiratory gases (O_2 and CO_2) between their cells and the respiratory medium, either air or water (p. 884).
- **Gills in Aquatic Animals** **(pp. 884–886)** Gills, the respiratory adaptations of most aquatic animals, are outfoldings of the body surface specialized for gas exchange. The effectiveness of gas exchange in some gills, including those of fishes, is increased by ventilation and countercurrent flow of blood and water.
- **Tracheal Systems in Insects** **(pp. 886–887)** The tracheae of insects are tiny branching tubes that penetrate the body, bringing O_2 directly to cells.
- **Lungs** **(pp. 886–888)** Spiders, land snails, and most terrestrial vertebrates have internal lungs. In mammals, air inhaled through the nostrils passes through the pharynx into the trachea, bronchi, bronchioles, and dead-end alveoli, where gas exchange occurs.
  Activity *The Human Respiratory System*

Concept 42.6

Breathing ventilates the lungs

- **How an Amphibian Breathes** **(p. 888)** An amphibian ventilates its lungs by positive pressure breathing, which forces air down the trachea.
- **How a Mammal Breathes** **(pp. 888–889)** Mammals ventilate their lungs by negative pressure breathing, which pulls air into the lungs. Lung volume increases as the rib muscles and diaphragm contract.
- **How a Bird Breathes** **(pp. 889–890)** Besides lungs, birds have eight or nine air sacs that act as bellows, keeping air flowing through the lungs. Air passes through the lungs in one direction only. Every exhalation completely renews the air in the lungs.
- **Control of Breathing in Humans** **(p. 890)** Control centers in the medulla oblongata of the brain regulate the rate and depth of breathing. Sensors detect the pH of cerebrospinal fluid (reflecting CO_2 concentration in the blood), and the medulla adjusts breathing rate and depth to match metabolic demands. Secondary control over breathing is exerted by sensors in the aorta and carotid arteries that monitor blood levels of O_2 and CO_2 and blood pH.

Concept 42.7

Respiratory pigments bind and transport gases

- **The Role of Partial Pressure Gradients** **(p. 891)** Gases diffuse down pressure gradients in the lungs and other organs. Oxygen and CO_2 diffuse from where their partial pressures are higher to where they are lower.
- **Respiratory Pigments** **(pp. 892–894)** Respiratory pigments transport gases and help buffer the blood. Respiratory pigments greatly increase the amount of O_2 that blood can carry. Many arthropods and molluscs have copper-containing hemocyanin; vertebrates and a wide variety of invertebrates have hemoglobin.
- **Elite Animal Athletes** **(pp. 894–895)** The pronghorn's extreme O_2 consumption rate underlies its ability to run at high speeds over long distances. Deep-diving air-breathers stockpile O_2 and deplete it slowly.
  Activity *Transport of Respiratory Gases*
  Biology Labs On-Line *HemoglobinLab*

TESTING YOUR KNOWLEDGE

Self-Quiz

1. Which of the following respiratory systems is not closely associated with a blood supply?
   a. vertebrate lungs
   b. fish gills
   c. tracheal systems of insects
   d. the outer skin of an earthworm
   e. the parapodia of a polychaete worm
2. Blood returning to the mammalian heart in a pulmonary vein drains first into the
   a. vena cava.
   b. left atrium.
   c. right atrium.
   d. left ventricle.
   e. right ventricle.
3. Pulse is a direct measure of
   a. blood pressure.
   b. stroke volume.
   c. cardiac output.
   d. heart rate.
   e. breathing rate.

4. The relationship between blood pressure (*bp*), cardiac output (*co*), and peripheral resistance (*pr*) can be expressed as $bp = co \times pr$. All of the following changes would result in an increase in blood pressure *except*
   a. increase in the stroke volume.
   b. increase in the heart rate.
   c. increase in the duration of ventricular diastole.
   d. contraction of the smooth muscle in arteriole walls.
   e. reduction in the diameter of arterioles.

5. The conversion of fibrinogen to fibrin
   a. occurs when fibrinogen is released from broken platelets.
   b. occurs within red blood cells.
   c. is linked to hypertension and may damage artery walls.
   d. is likely to occur too often in an individual with hemophilia.
   e. is the final step of a clotting process that involves multiple clotting factors.

6. In negative pressure breathing, inhalation results from
   a. forcing air from the throat down into the lungs.
   b. contracting the diaphragm.
   c. relaxing the muscles of the rib cage.
   d. using muscles of the lungs to expand the alveoli.
   e. contracting the abdominal muscles.

7. When you hold your breath, which of the following blood gas changes first leads to the urge to breathe?
   a. rising O_2
   b. falling O_2
   c. rising CO_2
   d. falling CO_2
   e. rising CO_2 and falling O_2

8. A decrease in the pH of human blood caused by exercise would
   a. decrease breathing rate.
   b. increase heart rate.
   c. decrease the amount of O_2 unloaded from hemoglobin.
   d. decrease cardiac output.
   e. decrease CO_2 binding to hemoglobin.

9. Compared to the interstitial fluid that bathes active muscle cells, blood reaching these cells in arteries has a
   a. higher P_{O_2}.
   b. higher P_{CO_2}.
   c. greater bicarbonate concentration.
   d. lower pH.
   e. lower osmotic pressure.

10. Which of the following reactions prevails in red blood cells traveling through pulmonary capillaries? (Hb = hemoglobin)
   a. $Hb + 4\,O_2 \rightarrow Hb(O_2)_4$
   b. $Hb(O_2)_4 \rightarrow Hb + 4\,O_2$
   c. $CO_2 + H_2O \rightarrow H_2CO_3$
   d. $H_2CO_3 \rightarrow H^+ + HCO_3^-$
   e. $Hb + 4\,CO_2 \rightarrow Hb(CO_2)_4$

For Self-Quiz answers, see Appendix A.

Go to the website or CD-ROM for more quiz questions.

Evolution Connection

One of the many mutant opponents that the movie monster Godzilla contends with is Mothra, a giant mothlike creature with a wingspan of several dozen feet. Science fiction creatures like these can be critiqued on the grounds of biomechanical and physiological principles. Focusing on respiration and the principles of gas exchange you learned about in this chapter, what physiological problems would Mothra face? The largest insects that have ever lived are Paleozoic dragonflies with half-meter wingspans. Why do you think truly giant insects are improbable?

Scientific Inquiry

The hemoglobin of a human fetus differs from adult hemoglobin. Compare the dissociation curves of the two hemoglobins in the graph below. Propose a hypothesis for the *function* of this difference between these two versions of hemoglobin.

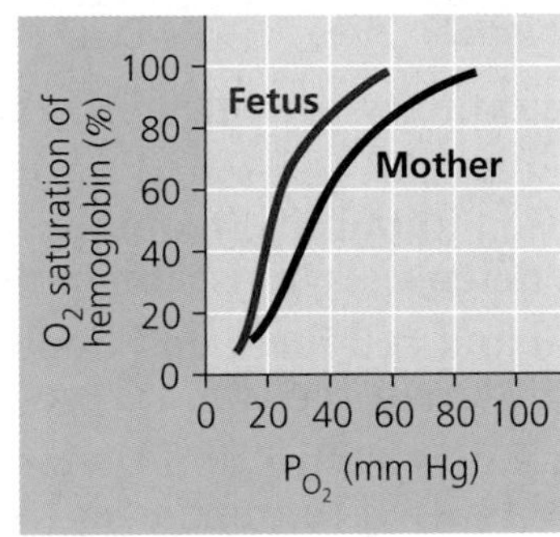

Investigation *How Is Cardiovascular Fitness Measured?*
Biology Labs On-Line *CardioLab*
Biology Labs On-Line *HemoglobinLab*

Science, Technology, and Society

Hundreds of studies have linked smoking with cardiovascular and lung disease. According to most health authorities, smoking is the leading cause of preventable, premature death in the United States. Antismoking and health groups have proposed that cigarette advertising in *all* media be banned *entirely*. What are some arguments in favor of a total ban on cigarette advertising? What are arguments in opposition? Do you favor or oppose such a ban? Defend your position.

43 The Immune System

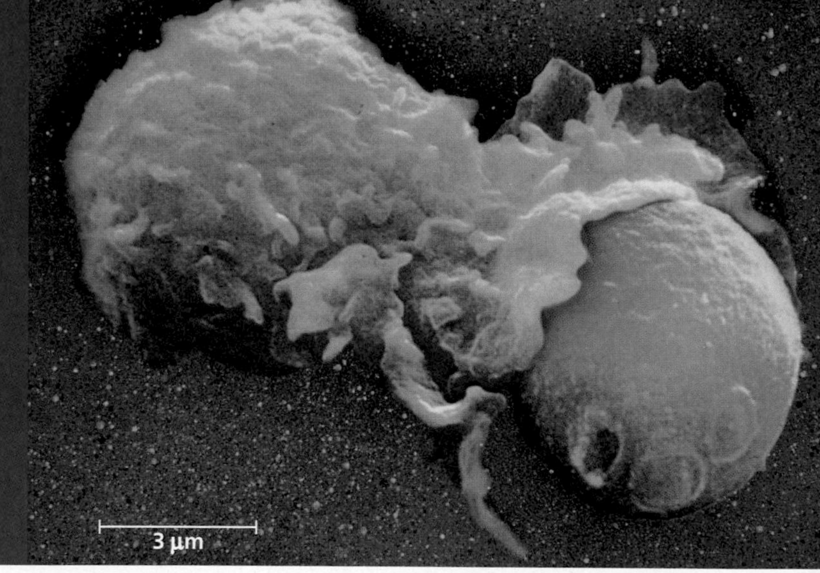

▲ **Figure 43.1 A macrophage (blue) ingesting a yeast cell (green).**

Key Concepts

43.1 Innate immunity provides broad defenses against infection

43.2 In acquired immunity, lymphocytes provide specific defenses against infection

43.3 Humoral and cell-mediated immunity defend against different types of threats

43.4 The immune system's ability to distinguish self from nonself limits tissue transplantation

43.5 Exaggerated, self-directed, or diminished immune responses can cause disease

Overview

Reconnaissance, Recognition, and Response

An animal must defend itself against the many potentially dangerous viruses, bacteria, and other pathogens it encounters in the air, in food, and in water. It must also contend with abnormal body cells that may develop into cancer. Two major kinds of defense have evolved that counter these threats. The first, called **innate immunity**, is present before any exposure to pathogens and is effective from the time of birth. Innate defenses are largely nonspecific, quickly recognizing and responding to a broad range of microbes regardless of their precise identity. Innate immunity consists of external barriers formed by the skin and mucous membranes, plus a set of internal cellular and chemical defenses that combat infectious agents that breach the external barriers. Key players in these internal defenses are macrophages and other phagocytic cells, which ingest and then destroy pathogens. For example, **Figure 43.1** (a colorized SEM) shows a macrophage engulfing a yeast cell.

The second major kind of defense is **acquired immunity**, also called *adaptive immunity*. It develops only after exposure to inducing agents such as microbes, abnormal body cells, toxins, or other foreign substances. Acquired defenses are highly specific—that is, they can distinguish one inducing agent from another, even those that differ only slightly. This recognition is achieved by white blood cells called **lymphocytes**, which produce two general types of immune responses. In the humoral response, cells derived from B lymphocytes secrete defensive proteins called **antibodies** that bind to microbes and mark them for elimination. In the cell-mediated response, cytotoxic lymphocytes directly destroy infected body cells, cancer cells, or foreign tissue.

Figure 43.2 summarizes innate and acquired immunity. In this chapter, you will learn how the various cellular and chemical components of these two kinds of defense together protect vertebrates from various threats. We will also look briefly at immunity in invertebrates, which are protected solely by innate, nonspecific mechanisms.

Concept 43.1

Innate immunity provides broad defenses against infection

An invading microbe must penetrate the external barriers formed by an animal's skin and mucous membranes, which cover the surface and line the openings of the body. A pathogen that successfully breaks through these external defenses soon encounters several innate cellular and chemical mechanisms that impede its attack on the body.

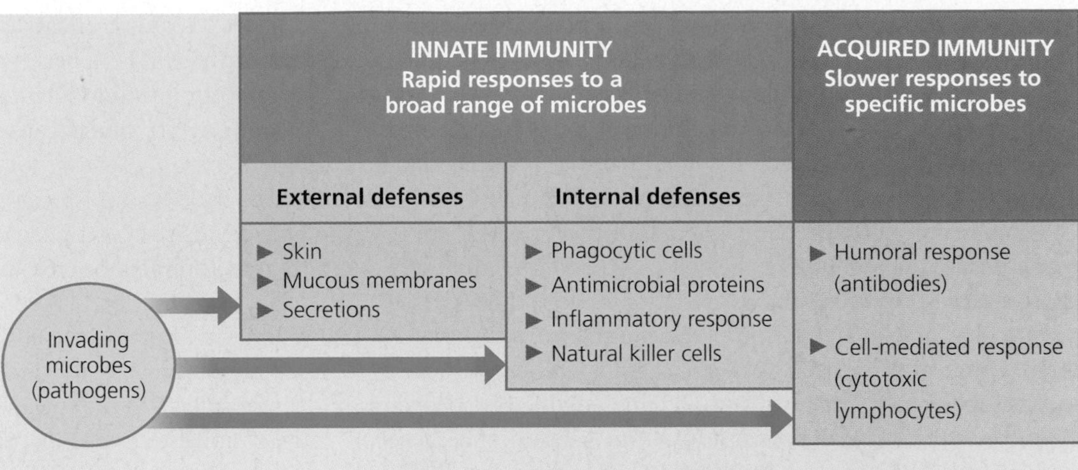

▶ **Figure 43.2 Overview of vertebrate defenses against bacteria, viruses, and other pathogens.** Defenses in vertebrates can be divided into innate and acquired immunity. If an invading pathogen breaches the body's external innate defenses, various internal innate defenses quickly come into play. The defenses provided by acquired immunity against specific pathogens develop more slowly. Some components of innate immunity also function in acquired immunity.

External Defenses

Intact skin is a barrier that cannot normally be penetrated by viruses or bacteria, but even tiny abrasions may allow their passage. Likewise, the mucous membranes lining the digestive, respiratory, and genitourinary tracts bar the entry of potentially harmful microbes. Certain cells of these mucous membranes also produce *mucus*, a viscous fluid that traps microbes and other particles. In the trachea, for example, ciliated epithelial cells sweep mucus and any entrapped microbes upward, preventing the microbes from entering the lungs **(Figure 43.3)**. Microbial colonization of the body is also inhibited by the washing action of the mucous secretions, saliva, and tears that constantly bathe the surfaces of various exposed epithelia.

Beyond their physical role in inhibiting microbe entry, secretions of the skin and mucous membranes provide an environment that is often hostile to microbes. In humans, secretions from sebaceous (oil) glands and sweat glands give the skin a pH ranging from 3 to 5, which is acidic enough to prevent colonization by many microbes. (Bacteria that normally inhabit the skin are adapted to its acidic, relatively dry environment.) Similarly, microbes in food or water and those in swallowed mucus must contend with the acidic environment of the stomach, which destroys most pathogens before they can enter the intestines. But some pathogens, such as the hepatitis A virus, can survive gastric acidity and successfully enter the body via the digestive tract.

Secretions from the skin and mucous membranes also contain antimicrobial proteins. One such protein is **lysozyme**, an enzyme that digests the cell walls of many bacteria. Present in saliva, tears, and mucous secretions, lysozyme can destroy susceptible bacteria as they enter the upper respiratory tract or the openings around the eyes.

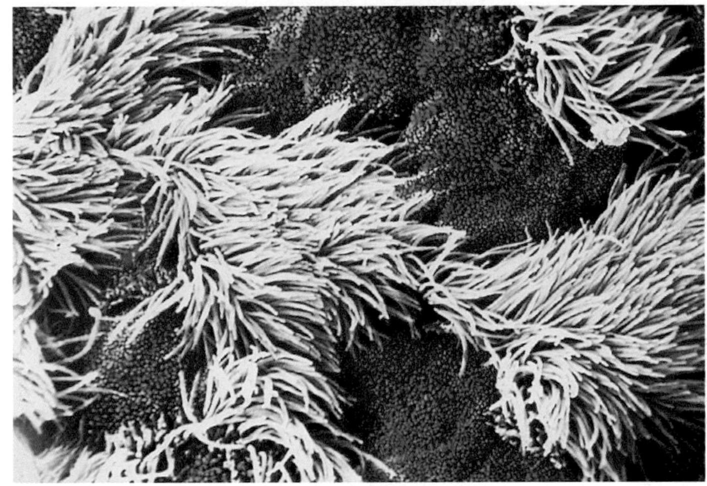

▲ **Figure 43.3 External innate defense by mucous membranes.** The lining of the trachea contains mucus-producing cells (orange) and cells with cilia (yellow). Synchronized beating of the cilia expels mucus and trapped microbes upward into the pharynx (colorized SEM).

Internal Cellular and Chemical Defenses

Microbes that penetrate the body's external defenses, such as those that enter through a break in the skin, must contend with the body's internal mechanisms of innate defense. These defenses depend mainly on **phagocytosis**, the ingestion of invading microorganisms by certain types of white blood cells generically referred to as *phagocytes*. These cells produce certain antimicrobial proteins and help initiate inflammation, which can limit the spread of microbes in the body. Nonphagocytic white blood cells called natural killer cells also play a key role in innate defenses. The various nonspecific mechanisms help limit the spread of microbes before the body can mount acquired, specific immune responses.

Phagocytic Cells

Phagocytes attach to their prey via surface receptors that bind to structures found on many microorganisms but not on normal body cells. Among the structures bound by these receptors are certain polysaccharides on the surface of bacteria.

After attaching to one or more microbes, a phagocyte engulfs the microbes, forming a vacuole that fuses with a lysosome **(Figure 43.4)**. Microbes are destroyed by lysosomes in two ways. First, nitric oxide and other toxic forms of oxygen contained in the lysosomes may poison the engulfed microbes. Second, lysozyme and other enzymes degrade microbial components.

Some microorganisms have adaptations that enable them to evade destruction by phagocytic cells. For example, the outer capsule that surrounds some bacterial cells hides their surface polysaccharides and prevents phagocytes from attaching to them. Other bacteria, such as *Mycobacterium tuberculosis,* which causes tuberculosis, are readily bound and engulfed by phagocytes but are resistant to destruction within lysosomes. Because such microbes can grow and reproduce within their host's cells, they are effectively hidden from the acquired defenses of the body. Evolution of these and other mechanisms that prevent destruction by the immune system has increased the pathogenic threat of many microbes.

Four types of white blood cells (leukocytes) are phagocytic. They differ in their abundance, average life span, and phagocytic ability. By far the most abundant are **neutrophils,** which constitute about 60–70% of all white blood cells. Neutrophils are attracted to and then enter infected tissue, engulfing and destroying the microbes there. However, neutrophils tend to self-destruct in the process of phagocytosis, and their average life span is only a few days.

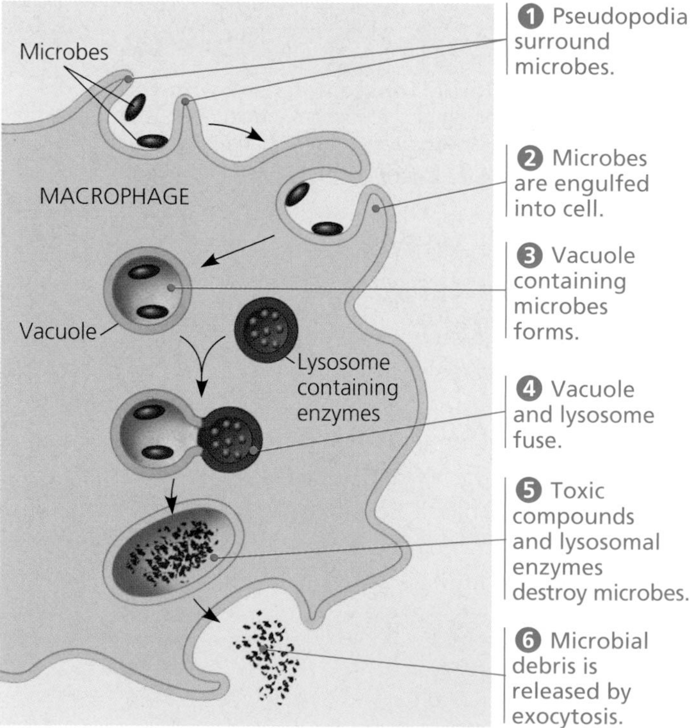

Microbes
MACROPHAGE
Vacuole
Lysosome containing enzymes
❶ Pseudopodia surround microbes.
❷ Microbes are engulfed into cell.
❸ Vacuole containing microbes forms.
❹ Vacuole and lysosome fuse.
❺ Toxic compounds and lysosomal enzymes destroy microbes.
❻ Microbial debris is released by exocytosis.

▲ **Figure 43.4 Phagocytosis.** This schematic depicts events following attachment of one kind of phagocyte, a macrophage, to microbes via its surface receptors (not shown). The process is similar in other types of phagocytic cells.

An even more effective phagocytic defense comes from **macrophages** ("big eaters"). These large, long-lived cells develop from **monocytes,** which constitute about 5% of circulating white blood cells. New monocytes circulate in the blood for only a few hours and then migrate into tissues where they are transformed into macrophages. Carrying out phagocytosis sets off internal signaling pathways that activate the macrophages, increasing their defensive abilities in various ways (described later in this chapter). Some macrophages migrate throughout the body, while others reside permanently in various organs and tissues. Macrophages that are permanent residents in the spleen, lymph nodes, and other tissues of the lymphatic system are particularly well positioned to combat infectious agents. Microbes that enter the blood become trapped in the netlike architecture of the spleen, whereas microbes in interstitial fluid flow into lymph and are trapped in lymph nodes. In either location microbes soon encounter resident macrophages. **Figure 43.5** shows the components of the lymphatic system and summarizes its role in the body's defenses.

The other two types of phagocytes are less abundant and play a more limited role in innate defense than neutrophils and macrophages. **Eosinophils** have low phagocytic activity but are critical to defense against multicellular parasitic invaders, such as the blood fluke *Schistosoma mansoni.* Rather than engulfing such a parasite, eosinophils position themselves against the parasite's body and then discharge destructive enzymes that damage the invader. The fourth type of phagocyte, **dendritic cells,** can ingest microbes like macrophages do. However, as you will learn later in the chapter, their primary role is to stimulate the development of acquired immunity.

Antimicrobial Proteins

Numerous proteins function in innate defense by attacking microbes directly or by impeding their reproduction. You have already learned about the antimicrobial action of lysozyme. Other antimicrobial proteins include about 30 serum proteins that make up the **complement system.** In the absence of infection, these proteins are inactive. Substances on the surface of many microbes, however, can trigger a cascade of steps that activate the complement system, leading to lysis (bursting) of invading cells. Certain complement proteins also help trigger inflammation or play a role in acquired defense.

Two types of **interferon** (α and β) provide innate defense against viral infections. These proteins are secreted by virus-infected body cells and induce neighboring uninfected cells to produce other substances that inhibit viral reproduction. In this way, interferons limit the cell-to-cell spread of viruses in the body, helping control viral infections such as colds and influenza. This innate defense mechanism is not virus-specific; interferons produced in response to one virus may also confer short-term resistance to unrelated viruses. Certain lymphocytes secrete a third type of interferon (γ) that helps activate

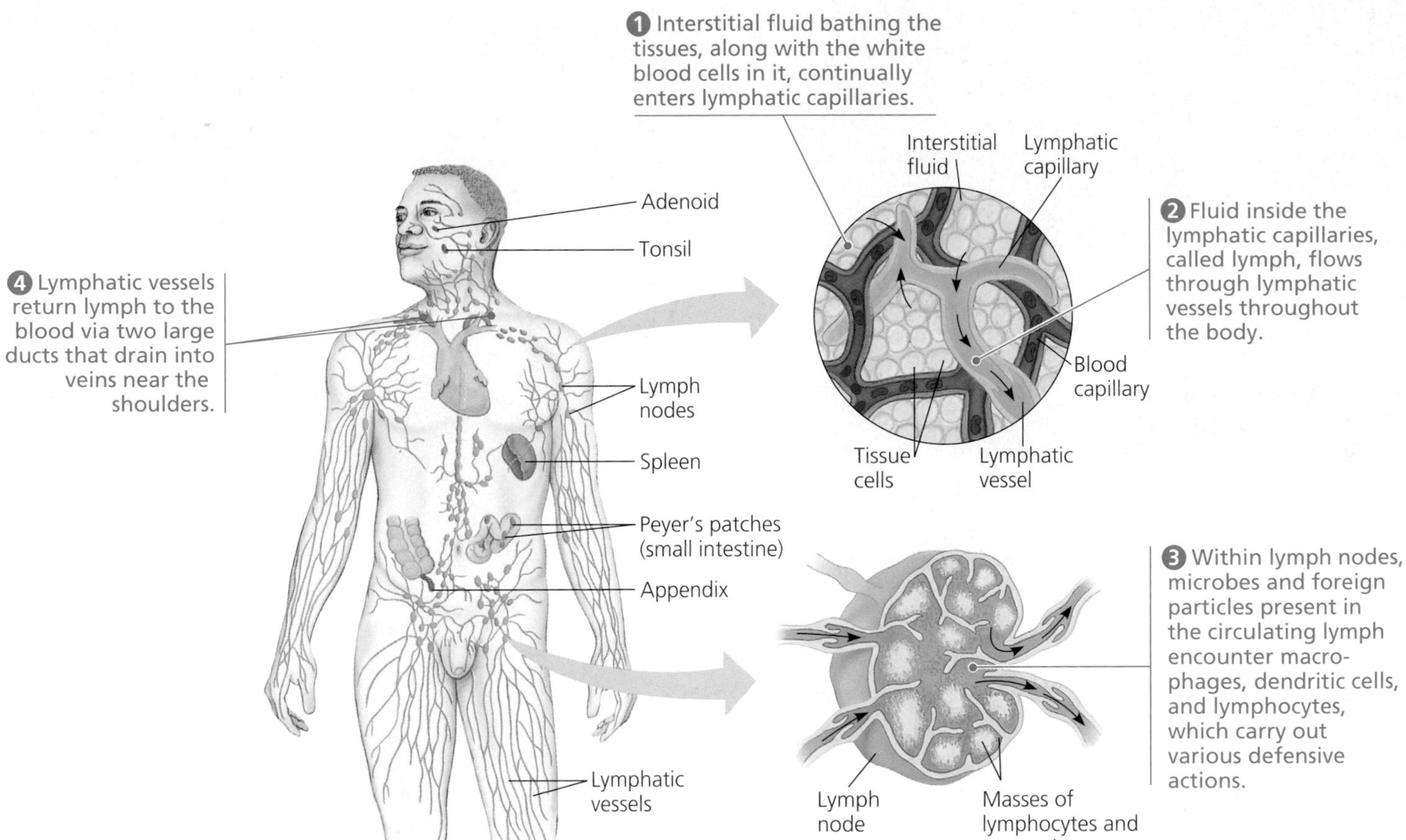

1 Interstitial fluid bathing the tissues, along with the white blood cells in it, continually enters lymphatic capillaries.
Interstitial fluid
Lymphatic capillary
Adenoid
Tonsil
2 Fluid inside the lymphatic capillaries, called lymph, flows through lymphatic vessels throughout the body.
4 Lymphatic vessels return lymph to the blood via two large ducts that drain into veins near the shoulders.
Lymph nodes
Blood capillary
Spleen
Tissue cells
Lymphatic vessel
Peyer's patches (small intestine)
Appendix
3 Within lymph nodes, microbes and foreign particles present in the circulating lymph encounter macrophages, dendritic cells, and lymphocytes, which carry out various defensive actions.
Lymphatic vessels
Lymph node
Masses of lymphocytes and macrophages

▲ **Figure 43.5 The human lymphatic system.** The lymphatic system consists of lymphatic vessels, through which lymph travels, and various structures that trap "foreign" molecules and particles. These structures include the adenoids, tonsils, lymph nodes, spleen, Peyer's patches, and appendix. The flow of lymph is traced in steps 1–4.

macrophages, enhancing their phagocytic ability. Interferons can now be mass-produced by recombinant DNA technology and are being tested for the treatment of viral infections and cancer.

Yet another group of antimicrobial proteins, fittingly named *defensins*, are secreted by activated macrophages. These small proteins damage broad groups of pathogens by various mechanisms without harming body cells.

Inflammatory Response

Damage to tissue by physical injury or the entry of pathogens leads to release of numerous chemical signals that trigger a localized **inflammatory response.** One of the most active chemicals is **histamine,** which is stored in **mast cells** found in connective tissues. When injured, mast cells release their histamine, triggering dilation and increased permeability of nearby capillaries. Activated macrophages and other cells discharge additional signals, such as prostaglandins, that further promote blood flow to the injured site. The resulting increased local blood supply causes the redness and heat typical of inflammation (from the Latin *inflammare,* to set on fire). The blood-engorged capillaries leak fluid into neighboring tissues, causing swelling, another sign of local inflammation.

Although heat and swelling are uncomfortable sensations, the enhanced blood flow and vessel permeability that cause them are critical to innate defense. These vascular changes help deliver antimicrobial proteins and clotting elements to the injured area. Several activated complement proteins, for example, promote the release of histamine or attract phagocytes to the site. Blood clotting begins the repair process and helps block the spread of microbes to other parts of the body. In addition, increased local blood flow and vessel permeability allow more neutrophils and monocyte-macrophages to move from the blood into injured tissues. Small proteins called **chemokines** direct the migration of these phagocytes and signal them to increase production of microbe-killing compounds. Chemokines are secreted by many cell types, including blood vessel endothelial cells, near a site of injury or infection. **Figure 43.6**, on the next page, summarizes the major events in local inflammation resulting from an infected pinprick.

A minor injury causes local inflammation, but the body may also mount a systemic (widespread) response to severe tissue damage or infection. Injured cells often put out a call for reinforcements, secreting chemicals that stimulate the release of additional neutrophils from the bone marrow. In a severe infection, such as meningitis or appendicitis, the

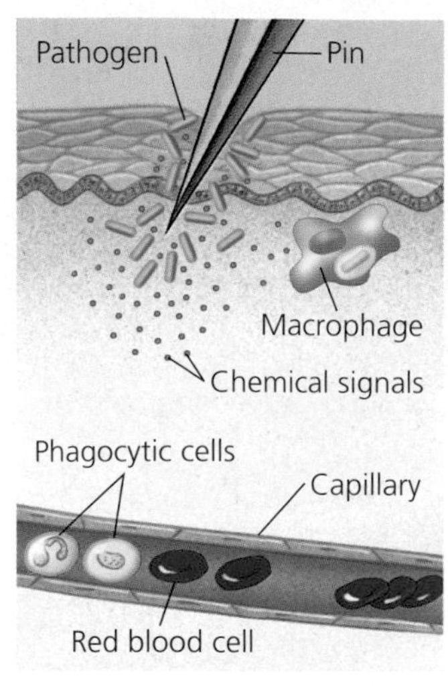

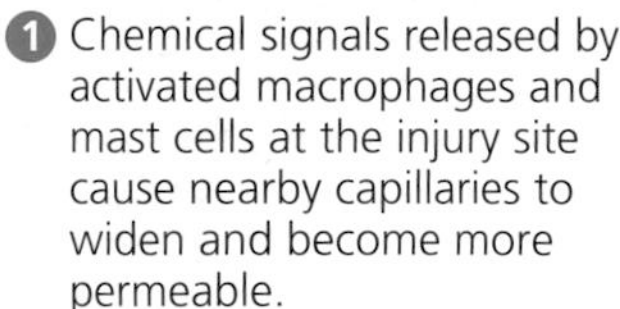

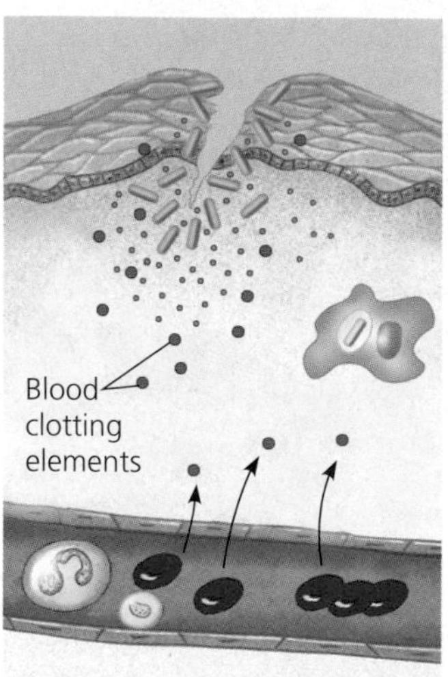

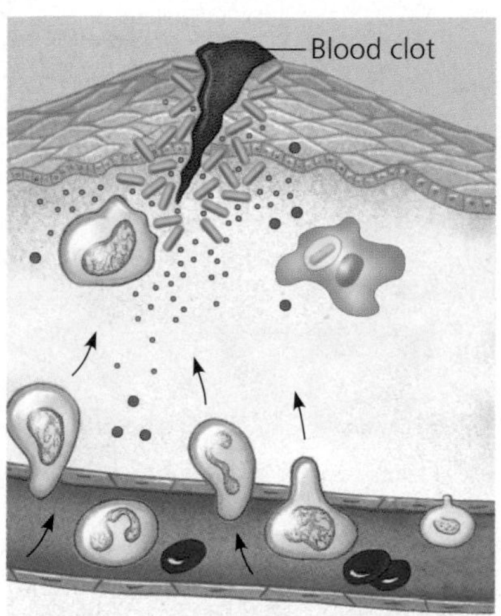

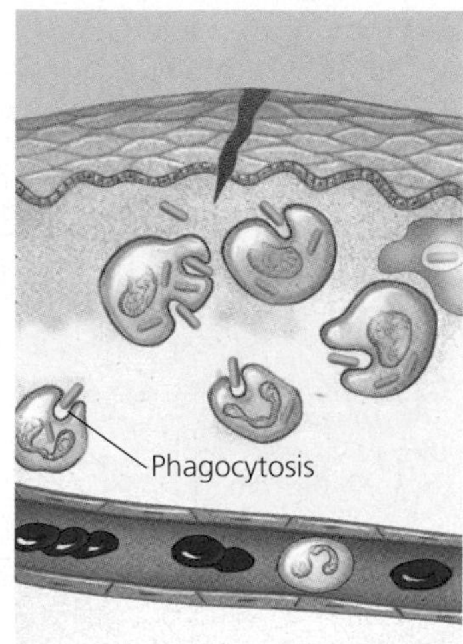

1. Chemical signals released by activated macrophages and mast cells at the injury site cause nearby capillaries to widen and become more permeable.
2. Fluid, antimicrobial proteins, and clotting elements move from the blood to the site. Clotting begins.
3. Chemokines released by various kinds of cells attract more phagocytic cells from the blood to the injury site.
4. Neutrophils and macrophages phagocytose pathogens and cell debris at the site, and the tissue heals.

▲ **Figure 43.6 Major events in the local inflammatory response.**

number of white blood cells in the blood may increase severalfold within a few hours of the initial inflammatory events. Another systemic response to infection is fever. Fever may occur when certain toxins produced by pathogens and substances released by activated macrophages set the body's thermostat at a higher temperature. A very high fever is dangerous, but moderate fever may facilitate phagocytosis and, by speeding up body reactions, hasten the repair of tissues.

Certain bacterial infections can induce an overwhelming systemic inflammatory response, leading to a condition known as *septic shock*. Characterized by very high fever and low blood pressure, septic shock is a common cause of death in hospital critical care units. Clearly, local inflammation is essential for healing, but systemic inflammation can be devastating.

Natural Killer Cells

We wrap up our discussion of vertebrate innate defenses with **natural killer (NK) cells.** NK cells patrol the body and attack virus-infected body cells and cancer cells. Surface receptors on an NK cell recognize general features on the surface of its targets. Once it is attached to a virus-infected cell or cancer cell, the NK cell releases chemicals that lead to the death of the stricken cell by **apoptosis,** or programmed cell death (see Figure 21.18). Athough the defense provided by NK cells is not 100% effective, viral infections and cancer would occur much more often without these innate sentinels of the body.

Invertebrate Immune Mechanisms

Before we move on to examine acquired immunity in vertebrates, we should note that invertebrates also have highly effective innate defenses. For example, sea stars possess amoeboid cells that ingest foreign matter by phagocytosis and secrete molecules that enhance the animal's defensive response. Recent studies of the fruit fly *Drosophila melanogaster* also have revealed striking parallels between insect defenses and vertebrate innate defenses. The exoskeleton of insects, like the skin and mucous membranes of vertebrates, provides an external barrier that can prevent entry of intruders. If an insect's exoskeleton is damaged, pathogens that enter the insect's body must face several internal innate defenses.

The insect equivalent to blood, the hemolymph, contains circulating cells called *hemocytes*. Some hemocytes ingest bacteria and other foreign substances by phagocytosis, while other hemocytes form a cellular capsule around large parasites. The presence of pathogens signals still other hemocytes to make and secrete various antimicrobial peptides that bind to their pathogen targets, leading to death of the pathogens. The internal signaling pathways that trigger hemocytes to produce antimicrobial peptides are comparable to those that activate vertebrate macrophages. In addition, certain hemocytes contain the enzyme phenoloxidase. Once activated, this enzyme converts phenols to reactive compounds that link together into large aggregates. These are deposited around parasites and wounded tissue, helping to prevent the spread of parasites beyond the affected area. Activation of phenoloxidase in insects occurs by a cascade of steps similar to those that activate vertebrate complement proteins.

Recent research indicates that invertebrates lack cells analogous to lymphocytes, the white blood cells responsible for acquired, specific immunity in vertebrates (see Figure 43.2). But even though they depend on innate, nonspecific mechanisms, certain invertebrate defenses do exhibit some features characteristic of acquired immunity. For instance, acquired immunity usually is directed against nonself cells, not normal body (self) cells. The ability to distinguish self from nonself is seen in the least derived animal lineage, the sponges. If cells from two sponges of the same species are mixed, the cells from each sponge sort themselves and reaggregate, excluding cells from the other individual.

Another hallmark of acquired immunity is immunological memory—the ability to respond more quickly to a particular invader or foreign tissue the second time it is encountered. Earthworms exhibit something like this: Phagocytic cells in a worm attack a second graft from the same donor worm much more rapidly than the first graft. Most invertebrates, however, exhibit no such immunological memory.

Following this glimpse at invertebrate host defenses, we begin in the next section to examine the highly developed mechanisms of acquired immunity found in vertebrates.

Concept Check 43.1

1. Innate defenses are nonspecific. How, then, do macrophages recognize an infectious agent, such as a bacterium?
2. What causes the common signs of inflammation—redness, swelling, and heat—and how do these changes help protect the body against infection?
3. State two ways in which the innate defenses of insects (invertebrates) and vertebrates are similar.

For suggested answers, see Appendix A.

In acquired immunity, lymphocytes provide specific defenses against infection

While invading pathogens are under assault by a vertebrate's innate defenses, the intruders inevitably come into contact with lymphocytes, the key cells of acquired immunity—the body's second major kind of defense (see Figure 43.2). Direct contact with microbes and signals from active innate defenses cause lymphocytes to join the fight. For example, as macrophages and dendritic cells phagocytose microbes, the phagocytes begin to secrete **cytokines**, proteins that help activate lymphocytes and other cells of the immune system. This is just one example of how innate and acquired defenses interact.

Any foreign molecule that is specifically recognized by lymphocytes and elicits a response from them is called an **antigen.** Most antigens are large molecules, either proteins or

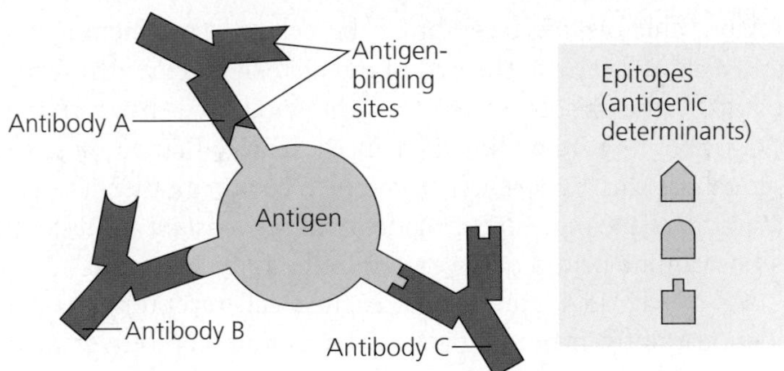

▲ **Figure 43.7 Epitopes (antigenic determinants).** Only small, specific regions on antigens, called epitopes, are bound by the antigen receptors on lymphocytes and by secreted antibodies. In this example, three different secreted antibody molecules react with different epitopes on the same large antigen molecule.

polysaccharides. Some antigens, such as toxins secreted by bacteria, are dissolved in extracellular fluid, but many protrude from the surface of pathogens or transplanted cells. A lymphocyte actually recognizes and binds to just a small, accessible portion of an antigen, called an **epitope** or *antigenic determinant.* A single antigen usually has several different epitopes, each capable of inducing a response from lymphocytes that recognize that epitope. Antibodies, which are secreted by certain lymphocytes in response to antigens, likewise bind to specific epitopes **(Figure 43.7)**.

In this section, we first describe how lymphocytes recognize antigens, or more specifically the epitopes on antigens. We then trace how the vertebrate body becomes populated with a large set of lymphocytes that collectively can recognize and mount a pinpoint attack against any one of a multitude of antigens. Later in the chapter we will examine the various defensive actions of the different types of lymphocytes.

Antigen Recognition by Lymphocytes

The vertebrate body is populated by two main types of lymphocytes: **B lymphocytes (B cells)** and **T lymphocytes (T cells).** Both types of lymphocytes circulate through the blood and lymph and are concentrated in the spleen, lymph nodes, and other lymphoid tissues (see Figure 43.5). B cells and T cells recognize antigens by means of antigen-specific receptors embedded in their plasma membranes. A single B or T cell bears about 100,000 of these **antigen receptors,** and all the receptors on a single cell are identical—that is, they all recognize the same epitope. In other words, each lymphocyte displays *specificity* for a particular epitope on an antigen and defends against that antigen or a small set of closely related antigens.

B Cell Receptors for Antigens

Each **B cell receptor** for an antigen is a Y-shaped molecule consisting of four polypeptide chains: two identical **heavy chains** and two identical **light chains** linked by disulfide bridges. A region in the tail portion of the molecule, the transmembrane region, anchors the receptor in the cell's plasma membrane, and a short region at the end of the tail extends into the cytoplasm. At the tips of the Y are the light- and heavy-chain *variable (V) regions* **(Figure 43.8a)**, so named because their amino acid sequences vary extensively from one B cell to another. The remainder of the molecule is made up of the *constant (C) regions,* whose amino acid sequences vary little from cell to cell.

As shown in Figure 43.8a, each B cell receptor has two identical antigen-binding sites. The unique contour of each binding site is formed from part of a heavy-chain V region and part of a light-chain V region. The interaction between an antigen-binding site and its corresponding antigen is stabilized by multiple noncovalent bonds between chemical groups on the respective molecules. The antigens bound by B cell receptors in this way include molecules that are on the surface of, or are released from, all types of infectious agents. In other words, a B cell recognizes an intact antigen in its native state.

Secreted antibodies, or **immunoglobulins,** are structurally similar to B cell receptors, but they lack the transmembrane regions that anchor receptors in the plasma membrane. Because of this structural similarity, B cell receptors are often called *membrane antibodies* or *membrane immunoglobulins.*

T Cell Receptors for Antigens and the Role of the MHC

Each **T cell receptor** for an antigen consists of two different polypeptide chains, an α *chain* and a β *chain,* linked by a disulfide bridge **(Figure 43.8b)**. Near the base of the molecule is a transmembrane region that anchors the molecule in the cell's plasma membrane. At the outer tip of the molecule, the α and β chain variable (V) regions form a single antigen-binding site. The remainder of the molecule is made up of the constant (C) regions.

T cell receptors recognize and bind with antigens just as specifically as B cell receptors. However, while the receptors on B cells recognize intact antigens, the receptors on T cells recognize small fragments of antigens that are bound to normal cell-surface proteins called MHC molecules. MHC molecules are so named because they are encoded by a family of genes called the **major histocompatibility complex (MHC).** As a newly synthesized MHC molecule is transported toward the plasma membrane, it binds with a fragment of protein (peptide) antigen *within* the cell and brings it to the cell surface, a process called **antigen presentation.** A nearby T cell can detect the antigen fragment thus displayed on the cell's surface **(Figure 43.9)**.

There are two ways in which foreign antigens can end up inside cells of the body. Depending on their source, these peptide antigens are handled by a different class of MHC molecule and recognized by a particular subgroup of T cells:

- **Class I MHC molecules,** found on almost all nucleated cells of the body, bind peptides derived from foreign antigens that have been synthesized within the cell. Any body cell that becomes infected or cancerous can display such peptide antigens by virtue of its class I MHC molecules. Class I MHC molecules displaying bound peptide antigens are recognized by a subgroup of T cells called **cytotoxic T cells** (see Figure 43.9a).

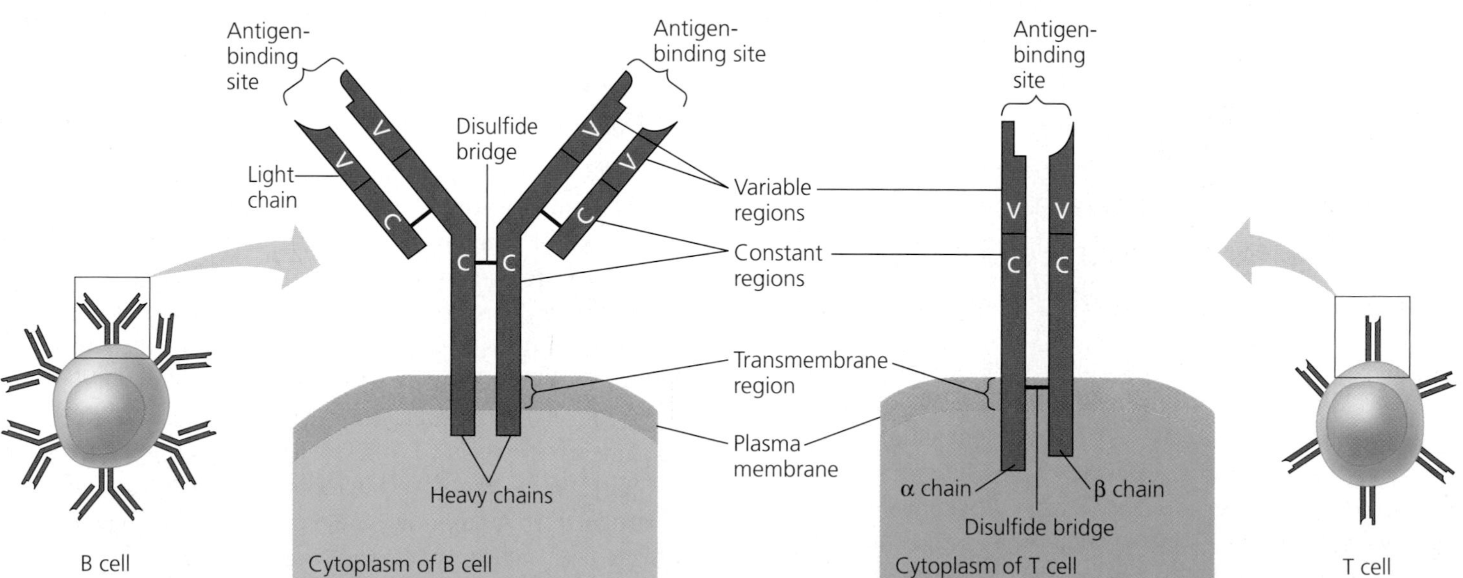

(a) A B cell receptor consists of two identical heavy chains and two identical light chains linked by several disulfide bridges.

(b) A T cell receptor consists of one α chain and one β chain linked by a disulfide bridge.

▲ **Figure 43.8 Antigen receptors on lymphocytes.** All the antigen receptors on a particular B cell or T cell bind the same antigen. The variable (V) regions vary extensively from cell to cell, accounting for the different binding specificities of individual lymphocytes; the constant (C) regions vary little or not at all.

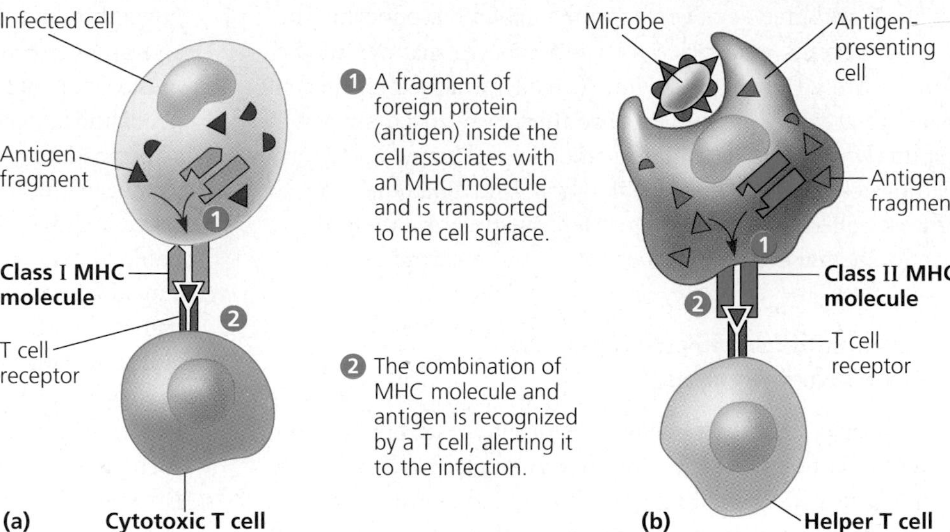

▶ **Figure 43.9 The interaction of T cells with MHC molecules.** Class I and class II MHC molecules display peptide fragments of antigens to **(a)** cytotoxic T cells and **(b)** helper T cells, respectively. In each case, the T cell receptor binds with an MHC molecule–peptide antigen complex. Class I MHC molecules are made by most nucleated cells, whereas class II MHC molecules are made primarily by antigen-presenting cells (macrophages, dendritic cells, and B cells).

- **Class II MHC molecules** are made by just a few cell types, mainly dendritic cells, macrophages, and B cells. In these cells, class II MHC molecules bind peptides derived from foreign materials that have been internalized and fragmented through phagocytosis or endocytosis. Dendritic cells, macrophages, and B cells are known as **antigen-presenting cells** because of their key role in displaying such internalized antigens to another subgroup of T cells called helper T cells (see Figure 43.9b).

Each vertebrate species possesses numerous different alleles for each class I and class II MHC gene; MHC proteins are the most polymorphic known. Because of the large number of different MHC alleles in the human population, most of us are heterozygous for every one of our MHC genes and produce a broad array of MHC molecules. Collectively these molecules are capable of binding to and presenting a large number of peptide antigens. Moreover, any two people, except identical twins, are very unlikely to have exactly the same set of MHC molecules. Thus the MHC provides a biochemical fingerprint, unique to virtually every individual, that marks body cells as "self." In fact, the discovery of the MHC occurred in the process of studying the phenomena of skin graft rejection and acceptance.

Lymphocyte Development

Now that you have read how lymphocytes recognize antigens, let's examine how these cells develop and populate the vertebrate body. Like all blood cells, lymphocytes originate from pluripotent stem cells in the bone marrow (see Chapter 42). Newly formed lymphocytes are all alike, but they later develop into either T cells or B cells, depending on where they continue their maturation **(Figure 43.10)**. Lymphocytes that migrate from the bone marrow to the **thymus**, a gland in the thoracic cavity above the heart, develop into T cells ("T" for thymus).

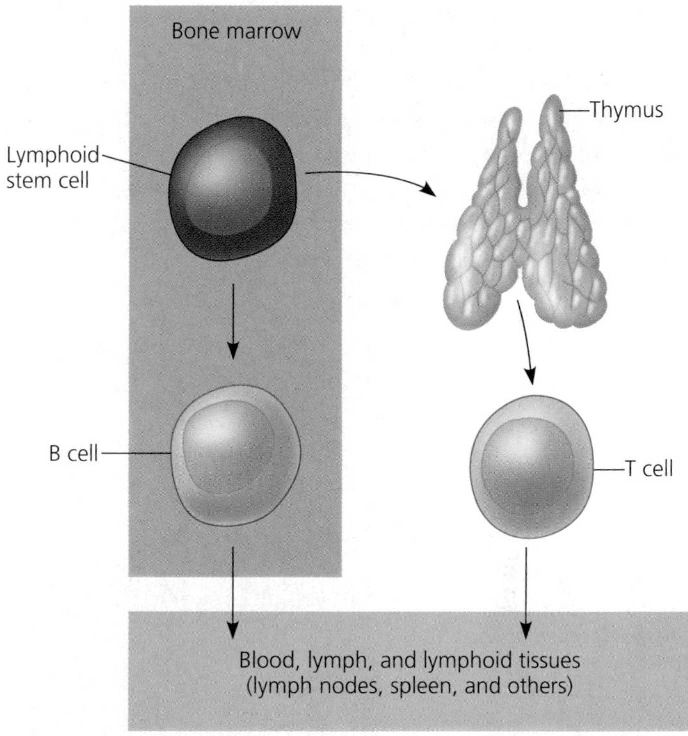

▲ **Figure 43.10 Overview of lymphocyte development.** Lymphocytes arise from stem cells in the bone marrow and differentiate without any contact with antigens. B cells develop entirely in the bone marrow, whereas T cells complete their development in the thymus. Mature lymphocytes, each specific for a particular epitope, circulate in the blood and lymph to various lymphoid tissues, where they encounter antigens.

Lymphocytes that remain in the bone marrow and complete their maturation there become B cells. (The "B" actually stands for the bursa of Fabricius, an organ unique to birds where avian B cells mature and the site where B cells were first discovered. But you can think of "B" as standing for bone marrow, since that is where B cells mature in other vertebrates.)

There are three key events in the life of a lymphocyte. The first two events take place as a lymphocyte matures in the bone marrow (B cell) or thymus (T cell), well before the cell has contact with any antigen. The third event occurs when a mature lymphocyte encounters and binds a specific antigen, leading to its activation, proliferation, and differentiation, a process called clonal selection. Here we describe these three developmental events in the order of their occurrence.

Generation of Lymphocyte Diversity by Gene Rearrangement

Recall that B cells and T cells recognize specific epitopes on antigens via their antigen receptors. A single B cell or T cell bears about 100,000 of these receptors, all identical. However, if we were to compare one B cell to another (or one T cell to another), the chances that both would have the same B cell receptor (or T cell receptor) are very slim. The variable regions at the tip of each antigen receptor chain, which form the antigen-binding site, account for the diversity of lymphocytes (see Figure 43.8). The sequence of amino acids in these regions varies from cell to cell and determines the specificity of an antigen receptor. The variability of these regions, and therefore the possible antigen specificities, is enormous. It is estimated that each person has as many as 1 million different B cells and 10 million different T cells, each with a particular antigen-binding specificity. Thus, our repertoire of lymphocytes can respond to an enormous number of different antigens.

At the core of lymphocyte diversity are the unique genes that encode the antigen receptor chains. These genes consist of numerous coding *gene segments* that undergo random, permanent rearrangement, forming functional genes that can be expressed as receptor chains. In this discussion we will focus on the gene coding for the light chain of the B cell receptor (membrane immunoglobulin), but keep in mind that the genes for the heavy chain and for the α and β chains of the T cell receptor undergo similar rearrangements. The light and heavy chains of the B cell receptor and of secreted antibody are encoded by the same genes, called immunoglobulin (Ig) genes.

The immunoglobulin light-chain gene contains a series of 40 variable (*V*) gene segments separated by a long stretch of DNA from 5 joining (*J*) gene segments **(Figure 43.11)**. Beyond the *J* gene segments is an intron, followed by a single exon designated *C* because it codes for the constant region of the light chain (see Figure 17.10 to review introns and exons). In this state, the light-chain gene is not functional. However, early in B cell development, a set of enzymes collectively called recombinase links one *V* gene segment to one *J* gene segment, eliminating the long stretch of DNA between them and forming a single exon that is part *V* and part *J*. Recombinase acts randomly; that is, it can link any one of the 40 *V* gene segments to any one of the 5 *J* gene segments. So,

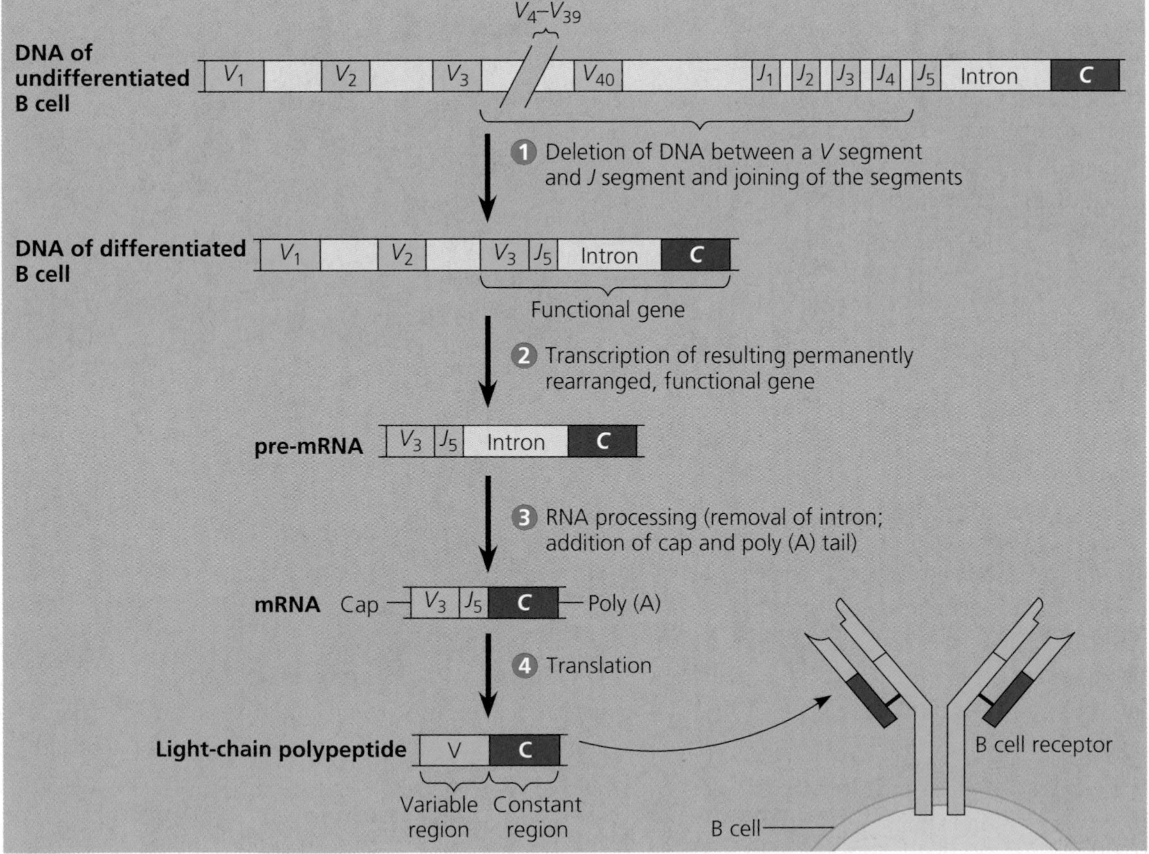

▶ Figure 43.11 Immunoglobulin gene rearrangement. The random joining of *V* and *J* gene segments (V_3 and J_5 in this example) results in a functional gene that encodes the light-chain polypeptide of a B cell receptor. In a cell that produces secreted antibody, the same pre-mRNA is formed, but alternative processing yields an mRNA that lacks coding sequences for the transmembrane region that anchors receptors in the membrane. Gene rearrangement plays a major role in generating a diverse repertoire of lymphocytes and secreted antibodies.

for the light-chain gene, there are 200 possible gene products (40 $V \times 5$ J). In any one cell, however, only one of the 200 possible light chains is made.

Once *V-J* rearrangement has occurred, the gene can be transcribed. The resulting pre-mRNA is processed, forming mRNA that is translated into a light chain containing a variable region and a constant region. The light chains produced in this way combine randomly with the heavy chains that are similarly produced, forming the B cell's antigen receptors (see Figure 43.11).

Testing and Removal of Self-Reactive Lymphocytes

Because the rearrangements of antigen receptor genes are random, a developing lymphocyte may end up with antigen receptors that are specific for some of the body's own molecules. As B cells and T cells are maturing in the bone marrow and thymus, respectively, their antigen receptors are tested for potential self-reactivity. For example, the receptors of maturing T cells are tested against class I and class II MHC molecules, which are both expressed at high levels in the thymus.

For the most part, lymphocytes bearing receptors specific for molecules already present in the body are either destroyed by apoptosis or rendered nonfunctional, leaving only lymphocytes that react to foreign (nonself) molecules. This capacity to distinguish self from nonself continues to develop even as the cells migrate to lymphoid organs. Thus, the body normally has no mature lymphocytes that react against self components: The immune system exhibits the critical feature of *self-tolerance*. Failure of self-tolerance can lead to autoimmune diseases such as multiple sclerosis, as you will read later.

Clonal Selection of Lymphocytes

A soluble antigen or an antigen present on the surface of a microbe, infected body cell, or cancer cell encounters a large repertoire of B cells and T cells in the body. However, a given antigen interacts only with the relatively few lymphocytes bearing receptors specific for epitopes on that antigen. The selection of a B cell or T cell by an antigen activates the lymphocyte, stimulating it to divide many times and to differentiate, forming two clones of daughter cells. One clone consists of a large number of short-lived **effector cells** that combat the same antigen. The nature and function of the effector cells depend on whether the lymphocyte selected is a helper T cell, a cytotoxic T cell, or a B cell. The other clone consists of **memory cells,** long-lived cells bearing receptors specific for the same inducing antigen.

This antigen-driven cloning of lymphocytes is called **clonal selection (Figure 43.12)**. The concept of clonal selection is so fundamental to understanding acquired immunity that it is worth restating: Each antigen, by binding to specific receptors,

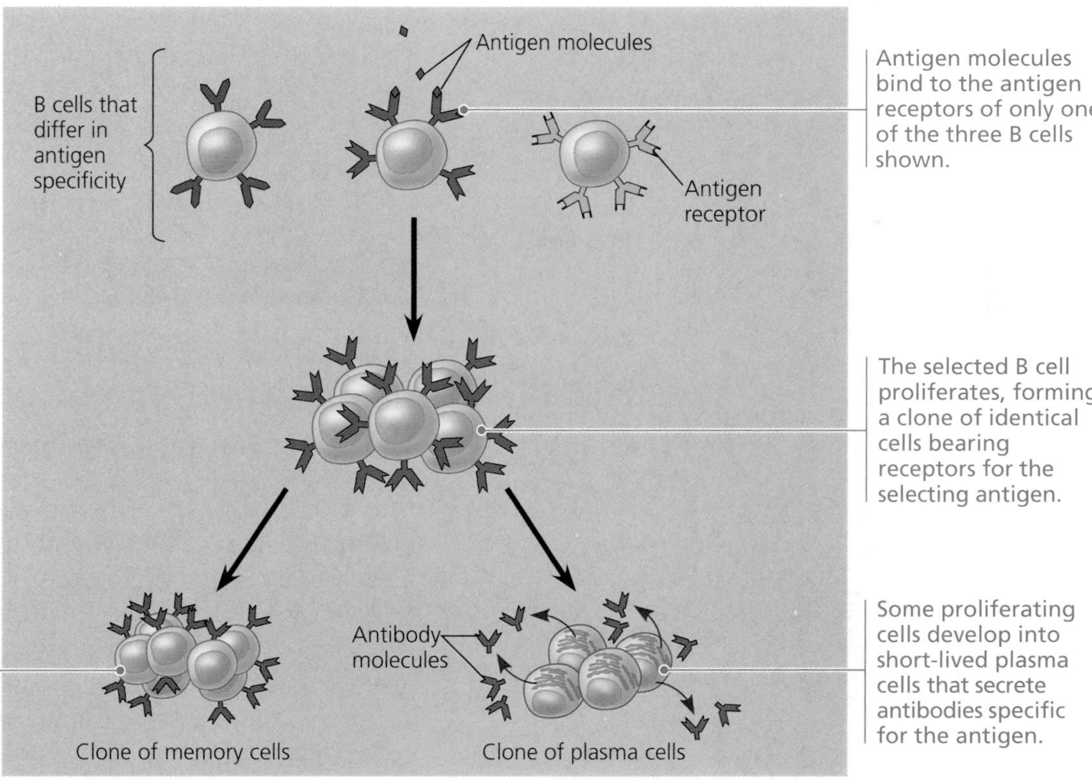

▲ **Figure 43.12 Clonal selection of B cells.** A B cell is selected by an antigen to proliferate and differentiate into memory B cells and antibody-secreting plasma cells. All the B cells whose receptors have a different specificity (indicated by different shapes and colors) do not respond to this antigen. T lymphocytes undergo a similar process, generating memory T cells and effector T cells.

selectively activates a tiny fraction of cells from the body's diverse pool of lymphocytes; this relatively small number of selected cells gives rise to clones of thousands of cells, all specific for and dedicated to eliminating that antigen.

The selective proliferation and differentiation of lymphocytes that occur the first time the body is exposed to a particular antigen represent the **primary immune response.** The primary response does not peak until about 10 to 17 days after the initial exposure to the antigen. During this time, selected B cells generate antibody-secreting effector B cells, called **plasma cells,** and selected T cells are activated to their effector forms, which have distinct functions (discussed in the next section). While these effector cells are developing, a stricken individual may become ill. Eventually, symptoms of illness diminish and disappear as antibodies and effector T cells clear the antigen from the body. If an individual is exposed again to the same antigen, the response is faster (typically only 2 to 7 days), of greater magnitude, and more prolonged. This is the **secondary immune response.** Measures of antibody concentrations in the blood serum over time show clearly the difference between primary and secondary immune responses **(Figure 43.13)**. In addition to being more numerous, antibodies produced in the secondary immune response tend to have greater affinity for the antigen than those secreted in the primary immune response.

The immune system's capacity to generate secondary immune responses, called *immunological memory,* depends on the clones of long-lived T and B memory cells generated following initial exposure to an antigen. These memory cells are poised to proliferate and differentiate rapidly when they later contact the same antigen. The long-term protection developed after exposure to a pathogen was recognized 2,400 years ago by Thucydides of Athens, who described how those sick and dying of plague were cared for by others who had recovered, "for no one was ever attacked a second time."

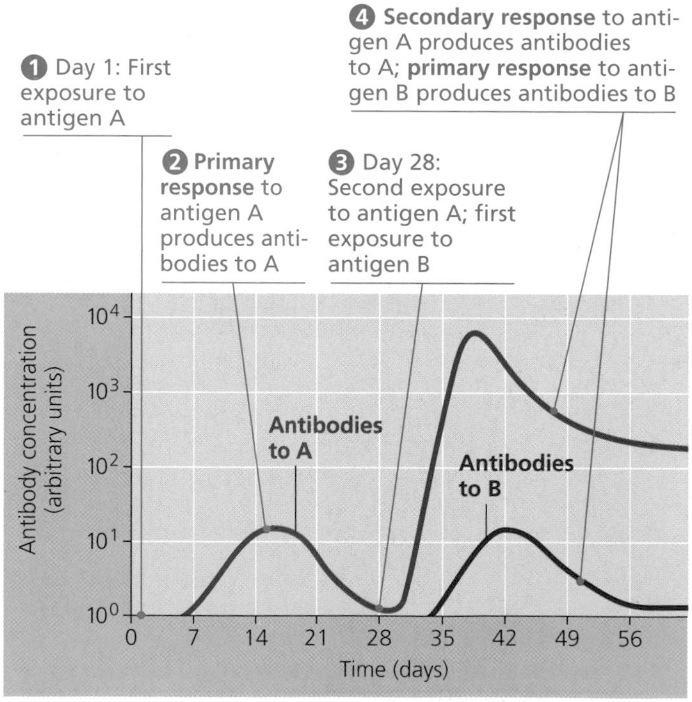

❶ Day 1: First exposure to antigen A
❷ **Primary response** to antigen A produces antibodies to A
❸ Day 28: Second exposure to antigen A; first exposure to antigen B
❹ **Secondary response** to antigen A produces antibodies to A; **primary response** to antigen B produces antibodies to B
Antibodies to A
Antibodies to B
Antibody concentration (arbitrary units)
10^4
10^3
10^2
10^1
10^0
0 7 14 21 28 35 42 49 56
Time (days)

▲ **Figure 43.13 The specificity of immunological memory.** Long-lived memory cells generated in the primary response to antigen A give rise to a heightened secondary response to the same antigen, but do not affect the primary response to a different antigen (B).

Concept Check 43.2

1. Draw a B cell receptor, and label the following: light chains, heavy chains, disulfide bridges, variable (V) regions, constant (C) regions, antigen-binding sites, transmembrane region, and cytoplasmic tails. How does the structure of a secreted antibody differ?
2. What is the major difference in the types of antigens bound by B cell receptors and T cell receptors?
3. Consider the process of clonal selection of B cells shown in Figure 43.12. How does this process demonstrate both the specificity and memory of acquired immunity?
4. A light-chain immunoglobulin gene consists of 40 *V* gene segments and 5 *J* gene segments, and a heavy-chain gene consists of 51 *V* gene segments, 6 *J* gene segments, and another set of gene segments *D*, of which there are 27. How many different antigen-binding specificities can be generated, given random *V-J* and *V-D-J* rearrangements?

For suggested answers, see Appendix A.

Humoral and cell-mediated immunity defend against different types of threats

Early evidence that substances in the blood plasma and lymph play a role in acquired immunity came from experiments performed near the end of the 19th century. Researchers transferred such fluids (long ago called humors) from an animal that had recovered from an infection by a particular microbe to another animal that had not been exposed to that microbe. Later, when the second animal was infected with the same microbe, it did not become ill. The investigators had transferred what we now know were secreted antibodies from one animal to the other. They also found that immunity to some infections could be passed along only if certain cells, later identified as cytotoxic T cells, were transferred.

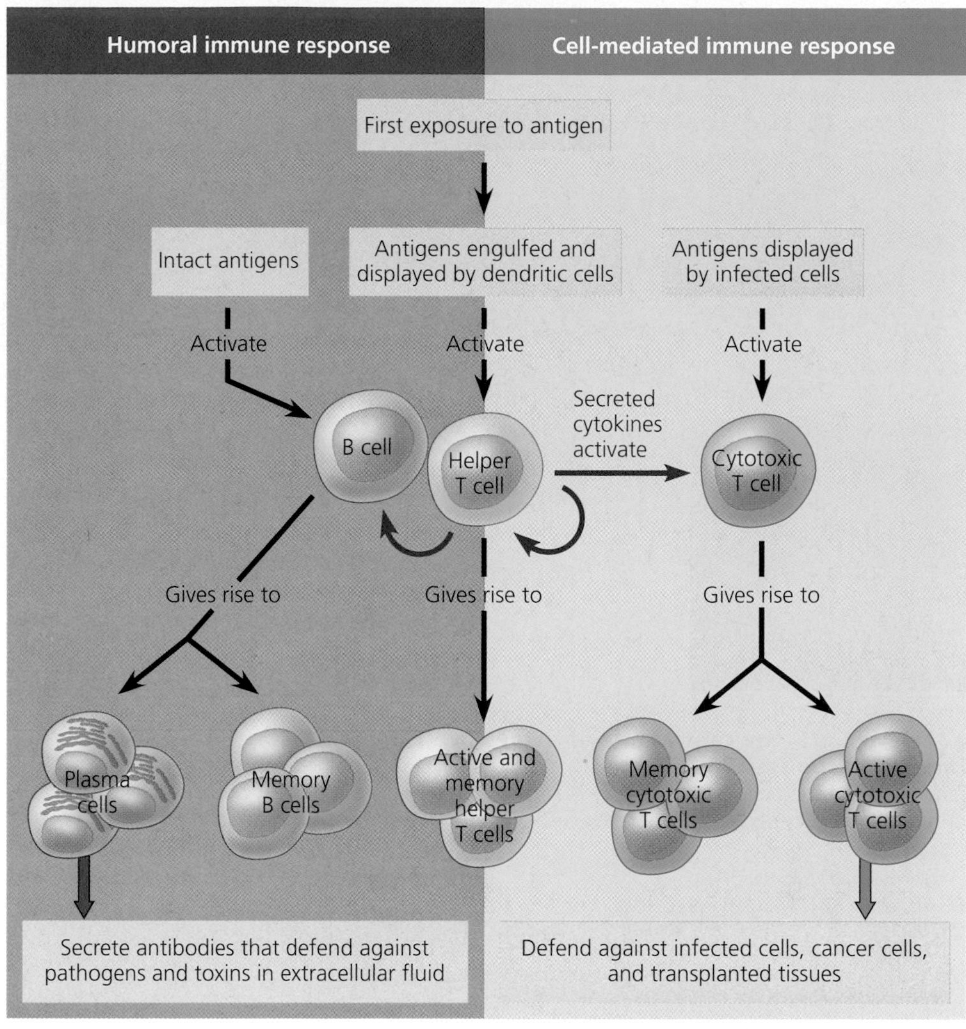

► **Figure 43.14 An overview of the acquired immune response.** Stimulation of helper T cells by an antigen generally requires direct contact between a dendritic cell and helper T cell in a primary response (shown here) or between a macrophage and memory T cell in a secondary response (not shown). Once activated, a helper T cell stimulates the humoral response directly by contacting B cells and indirectly by secreting cytokines. An activated helper T cell stimulates the cell-mediated response indirectly via cytokines.

These and many other studies led to the current understanding that acquired immunity includes two branches. The **humoral immune response** involves the activation and clonal selection of B cells, resulting in production of secreted antibodies that circulate in the blood and lymph. The **cell-mediated immune response** involves the activation and clonal selection of cytotoxic T cells, which directly destroy certain target cells.

Figure 43.14 outlines the roles of the major participants in acquired immune responses. Central to this network of cellular interactions is the **helper T cell,** which responds to peptide antigens displayed on antigen-presenting cells and in turn stimulates the activation of nearby B cells and cytotoxic T cells. Now let's fill in the details of this road map and examine how each branch of the acquired immune response defends against particular types of assault.

Helper T Cells: A Response to Nearly All Antigens

When a helper T cell encounters and recognizes a class II MHC molecule–antigen complex on an antigen-presenting cell, the helper T cell proliferates and differentiates into a clone of activated helper T cells and memory helper T cells. A surface protein called **CD4,** present on most helper T cells, binds the class II MHC molecule. This interaction helps keep the helper T cell and the antigen-presenting cell joined while activation of the helper T cell proceeds.

Activated helper T cells secrete several different cytokines that stimulate other lymphocytes, thereby promoting cell-mediated and humoral responses. The helper T cell itself is also subject to regulation by cytokines. For instance, as a dendritic cell presents an antigen to a helper T cell, the dendritic cell is stimulated to secrete cytokines that, along with an antigen, stimulate the helper T cell to produce its own set of cytokines.

As you've read, the class II MHC molecules recognized by helper T cells are found mainly on dendritic cells, macrophages, and B cells. Dendritic cells are particularly effective in presenting antigens to *naive* helper T cells, so called because they have not previously detected an antigen. In other words, dendritic cells are important in triggering a primary immune response. Dendritic cells are located in the epidermis and in many other tissues, where they efficiently capture antigens. They then migrate from the site of infection to various lymphoid tissues, where they present antigens, via class II MHC molecules, to helper T cells

1 After a dendritic cell engulfs and degrades a bacterium, it displays bacterial antigen fragments (peptides) complexed with a class II MHC molecule on the cell surface. A specific helper T cell binds to the displayed complex via its TCR with the aid of CD4. This interaction promotes secretion of cytokines by the dendritic cell.

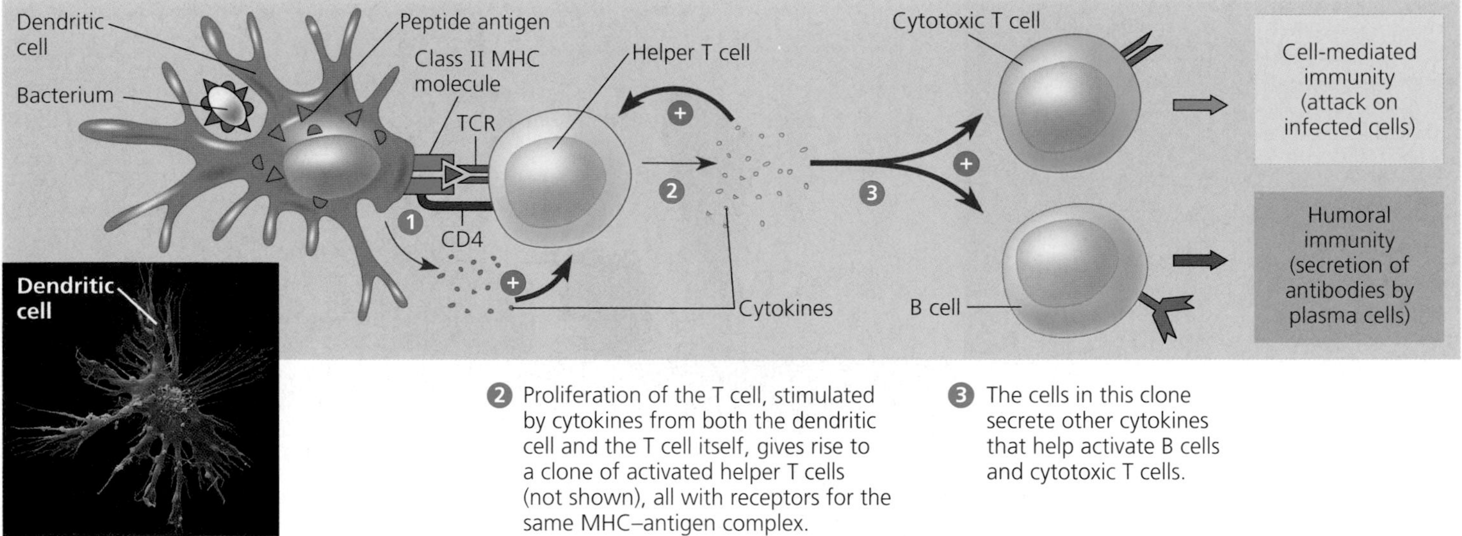

2 Proliferation of the T cell, stimulated by cytokines from both the dendritic cell and the T cell itself, gives rise to a clone of activated helper T cells (not shown), all with receptors for the same MHC–antigen complex.

3 The cells in this clone secrete other cytokines that help activate B cells and cytotoxic T cells.

▲ **Figure 43.15 The central role of helper T cells in humoral and cell-mediated immune responses.** The SEM of a dendritic cell reveals its long branching extensions, reminiscent of the dendrites of a nerve cell. Dendritic cells are the primary antigen-presenting cells during the primary response to an antigen. TCR = T cell receptor. A ⊕ indicates stimulation.

(Figure 43.15). Macrophages play the key role in initiating a secondary immune response by presenting antigens to memory helper T cells, while B cells primarily present antigens to helper T cells in the course of the humoral response.

Cytotoxic T Cells: A Response to Infected Cells and Cancer Cells

Cytotoxic T cells, the effectors of cell-mediated immunity, eliminate body cells infected by viruses or other intracellular pathogens as well as cancer cells and transplanted cells. Fragments of nonself proteins synthesized in such target cells associate with class I MHC molecules and are displayed on the cell surface, where they can be recognized by cytotoxic T cells. A surface protein called **CD8,** present on most cytotoxic T cells, greatly enhances the interaction between a target cell and a cytotoxic T cell. Binding of CD8 to the side of a class I MHC molecule helps keep the two cells in contact during activation of the cytotoxic T cell. Thus, the role of class I MHC molecules and CD8 is similar to that of class II MHC molecules and CD4, except that different cell types are involved.

When a cytotoxic T cell is selected by binding to class I MHC molecule–antigen complexes on an infected body cell, the cytotoxic T cell is activated and differentiates into an active killer. Cytokines secreted from nearby helper T cells promote this activation. The activated cytotoxic T cell then secretes proteins that act on the bound infected cell, leading to its destruction **(Figure 43.16)**. The death of the infected cell not only deprives the pathogen of a place to reproduce but also exposes it to circulating antibodies, which mark it for disposal. After destroying an infected cell, the cytotoxic T cell may move on and kill other cells infected with the same pathogen.

In the same way, cytotoxic T cells defend against malignant tumors. Because tumor cells carry distinctive molecules (tumor antigens) not found on normal body cells, they are identified as foreign by the immune system. Class I MHC molecules on a tumor cell display fragments of tumor antigens to cytotoxic T cells. Interestingly, certain cancers and viruses (such as the Epstein-Barr virus) actively reduce the number of class I MHC molecules on affected cells, helping them escape detection by cytotoxic T cells. But the body has a backup defense: Recall that natural killer (NK) cells, part of the body's nonspecific innate defenses, can induce apoptosis in virus-infected and cancer cells.

B Cells: A Response to Extracellular Pathogens

Antigens that elicit a humoral immune response are typically proteins and polysaccharides present on the surface of bacteria or incompatible transplanted tissue or transfused blood cells. In addition, for some people, proteins of foreign substances such as pollen or bee venom act as antigens that induce an allergic, or hypersensitive, humoral response.

Figure 43.17 depicts the events in the humoral response to a typical protein antigen. The activation of B cells is aided by cytokines secreted from helper T cells activated by the same

1 A specific cytotoxic T cell binds to a class I MHC–antigen complex on a target cell via its TCR with the aid of CD8. This interaction, along with cytokines from helper T cells, leads to the activation of the cytotoxic cell.

2 The activated T cell releases perforin molecules, which form pores in the target cell membrane, and proteolytic enzymes (granzymes), which enter the target cell by endocytosis.

3 The granzymes initiate apoptosis within the target cell, leading to fragmentation of the nucleus, release of small apoptotic bodies, and eventual cell death. The released cytotoxic T cell can attack other target cells.

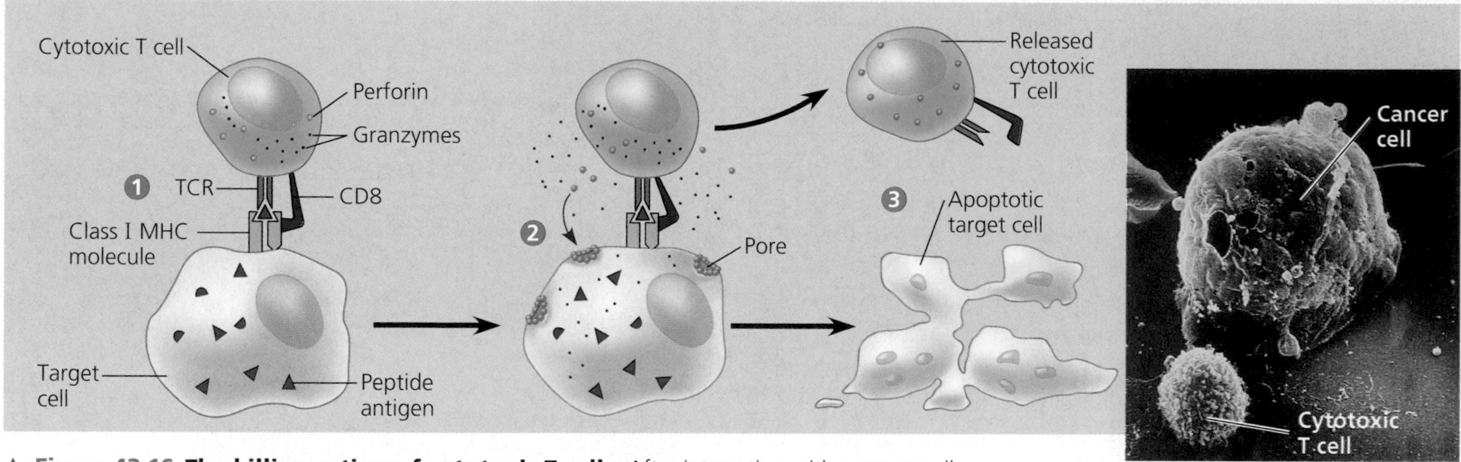

▲ **Figure 43.16 The killing action of cytotoxic T cells.** After interacting with a target cell, such as an infected body cell or cancer cell, an activated cytotoxic T cell releases perforins and proteolytic enzymes (granzymes) that promote death of the target cell. The colorized SEM shows a cancer cell with a perforin-induced pore in the early stages of apoptosis. TCR = T cell receptor.

1 After a macrophage engulfs and degrades a bacterium, it displays a peptide antigen complexed with a class II MHC molecule. A helper T cell that recognizes the displayed complex is activated with the aid of cytokines secreted from the macrophage, forming a clone of activated helper T cells (not shown).

2 A B cell that has taken up and degraded the same bacterium displays class II MHC–peptide antigen complexes. An activated helper T cell bearing receptors specific for the displayed antigen binds to the B cell. This interaction, with the aid of cytokines from the T cell, activates the B cell.

3 The activated B cell proliferates and differentiates into memory B cells and antibody-secreting plasma cells. The secreted antibodies are specific for the same bacterial antigen that initiated the response.

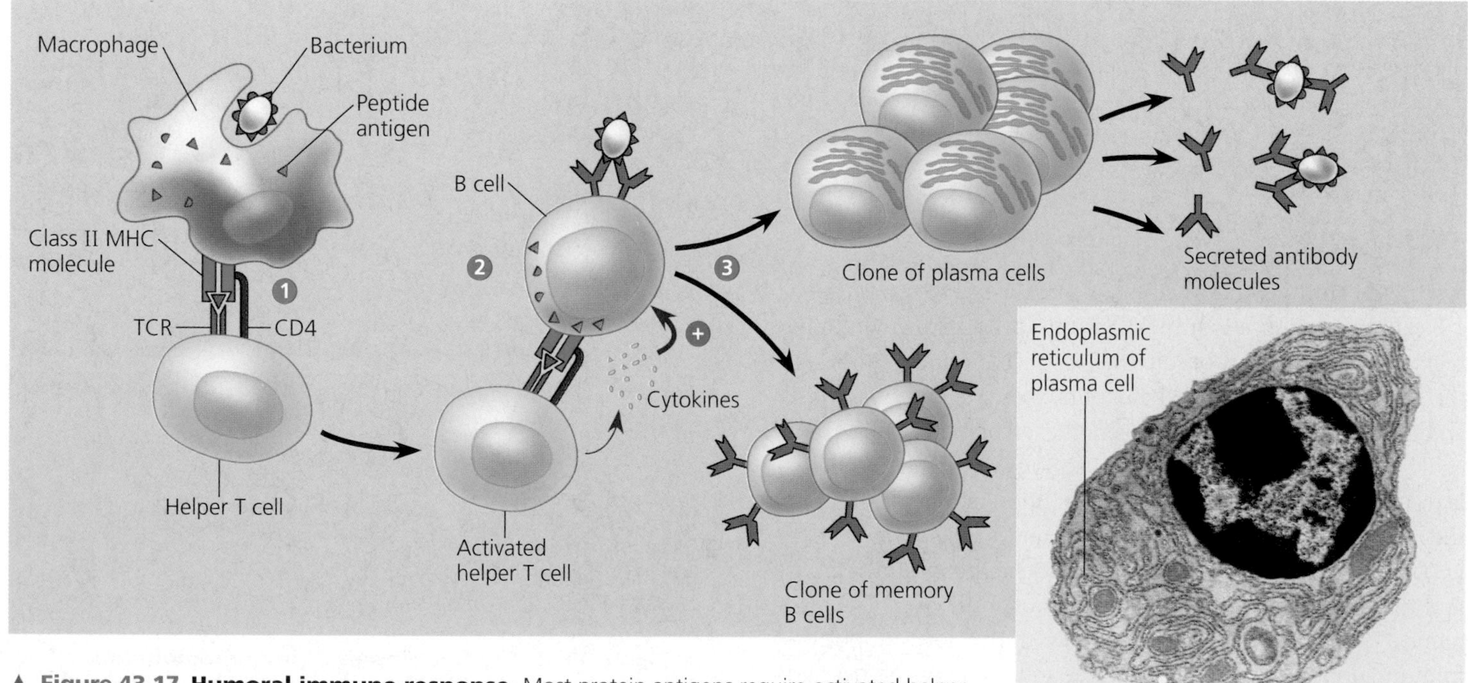

▲ **Figure 43.17 Humoral immune response.** Most protein antigens require activated helper T cells to trigger a humoral response. Either a macrophage (shown here) or a dendritic cell can activate helper T cells. The TEM of a plasma cell reveals abundant endoplasmic reticulum, a common feature of cells dedicated to making proteins for secretion. TCR = T cell receptor. A ⊕ indicates stimulation.

antigen. Stimulated by both an antigen and cytokines, the B cell proliferates and differentiates into a clone of antibody-secreting plasma cells and a clone of memory B cells. When an antigen first binds to receptors on the surface of a B cell, the cell takes in a few of the foreign molecules by receptor-mediated endocytosis. In a process similar to antigen presentation by macrophages and dendritic cells, the B cell then presents antigen fragments to a helper T cell. This achieves the direct cell-cell contact critical to B cell activation (see step 2 in Figure 43.17). Note, however, that a macrophage or dendritic cell can present peptide fragments from a wide variety of antigens, whereas a B cell internalizes and presents only the antigen to which it specifically binds.

Antigens that induce antibody production only with assistance from helper T cells, as shown in Figure 43.17, are known as *T-dependent antigens.* Some antigens, however, can evoke a B cell response without involvement of helper T cells. Such *T-independent antigens* include the polysaccharides of many bacterial capsules and the proteins that make up bacterial flagella. Apparently, the repeated subunits of these molecules bind simultaneously to several antigen receptors on a single B cell, providing enough stimulus to activate the cell without the help of cytokines. The response to T-independent antigens is very important in defending against many bacteria. However, this response is generally weaker than the response to T-dependent antigens and generates no memory B cells.

Most antigens recognized by B cells contain multiple epitopes. For this reason, exposure to a single antigen normally stimulates a variety of different B cells, each giving rise to a clone of thousands of plasma cells. All the plasma cells in one clone secrete antibodies specific for the epitope that provoked their production. Each plasma cell secretes an estimated 2,000 antibody molecules per second over the cell's 4- to 5-day life span. Next we look more closely at antibodies and how they mediate the disposal of antigens.

Antibody Classes

A secreted antibody has the same general Y-shaped structure as a B cell receptor (see Figure 43.8a) but lacks a transmembrane region that would anchor it to the plasma membrane. Although the antigen-binding sites on an antibody are responsible for its ability to identify a specific antigen, the tail of the Y-shaped molecule, formed by the constant (C) regions of the heavy chains, is responsible for the antibody's distribution in the body and for the mechanisms by which it mediates antigen disposal.

There are five major types of heavy-chain constant regions, and these determine five major classes of antibodies. The class names are based on the alternate term for antibody—immunoglobulin (Ig). The structures and functions of these antibody classes are summarized in **Figure 43.18**. Note that two classes exist primarily as polymers of the basic antibody molecule: IgM as a pentamer and IgA as a dimer. The other three classes—IgG, IgE, and IgD—exist exclusively as monomers.

The power of antibody specificity and antigen-antibody binding has been harnessed in laboratory research, clinical diagnosis, and the treatment of diseases. Some of these antibody tools are *polyclonal:* They are the products of many different clones of B cells, each specific for a different epitope. Antibodies

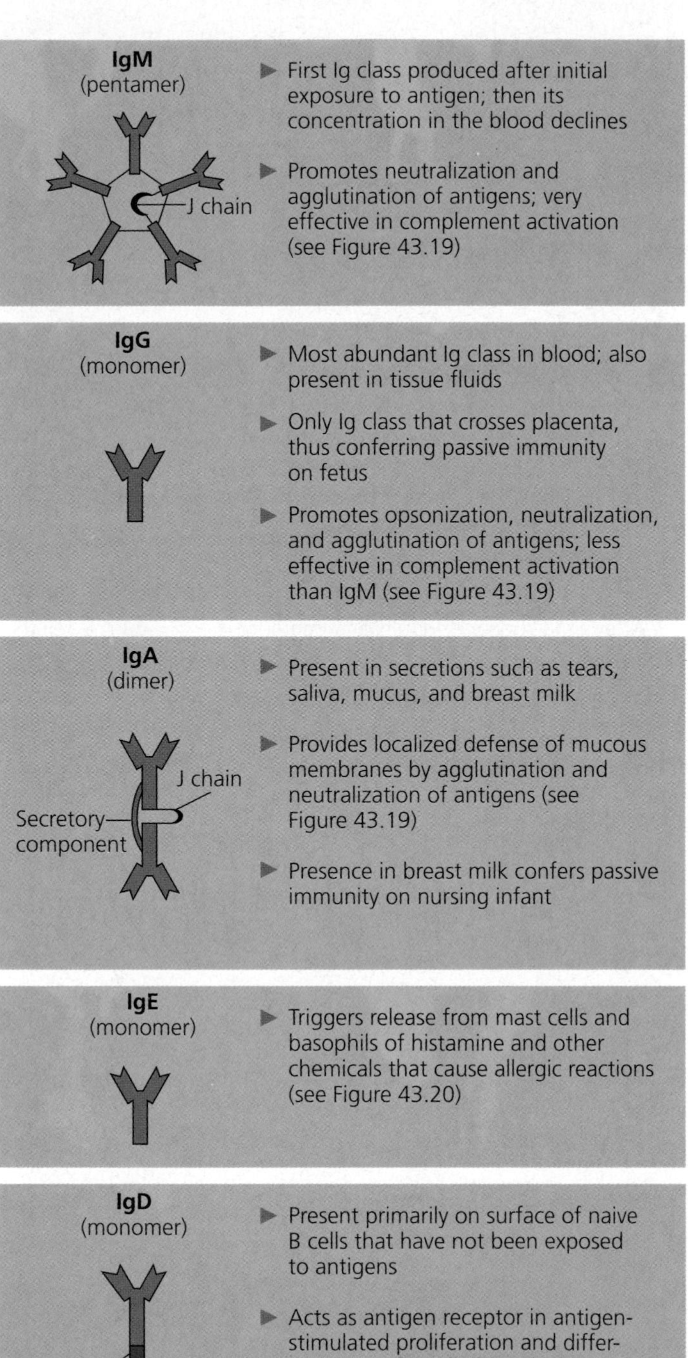

▲ **Figure 43.18 The five classes of immunoglobulins.** All classes consist of similar Y-shaped molecules in which the tail region determines the distribution and functions characteristic of each class. IgM and IgA antibodies contain a J chain (unrelated to the *J* gene segment) that helps hold the monomeric subunits together. As an IgA antibody is secreted across a mucous membrane, it acquires a secretory component that protects it from cleavage by enzymes.

produced in the body following exposure to a microbial antigen are polyclonal. In contrast, other antibody tools are *monoclonal:* They are prepared from a single clone of B cells grown in culture. All the **monoclonal antibodies** produced by such a culture are identical and specific for the same epitope on an antigen. In both basic research and medical applications, monoclonal antibodies are particularly useful for tagging specific molecules. For example, certain types of cancer are treated with tumor-specific monoclonal antibodies bound to toxin molecules. The toxin-linked antibodies carry out a precise search-and-destroy mission, selectively attaching to and killing tumor cells.

Antibody-Mediated Disposal of Antigens

The binding of antibodies to antigens is also the basis of several antigen disposal mechanisms **(Figure 43.19)**. In the simplest of these, *viral neutralization,* antibodies bind to certain proteins on the surface of a virus, thereby blocking the virus's ability to infect a host cell. Similarly, antibodies may bind to a pathogenic bacterium, coating much of the bacterial surface. In a process called *opsonization,* the bound antibodies enhance macrophage attachment to the microbes, and thus increase phagocytosis.

Antibody-mediated *agglutination* (clumping) of bacteria or viruses forms aggregates that can be readily phagocytosed by macrophages. Agglutination is possible because each antibody molecule has at least two antigen-binding sites that can bind to identical epitopes on separate bacterial cells or virus particles, linking them together. Because of their pentameric structure, IgM antibodies can link together five or more viruses or bacteria (as shown in Figure 43.19). In the similar process of *precipitation,* antibodies cross-link soluble antigen molecules dissolved in body fluids, forming immobile aggregates that are disposed of by phagocytes.

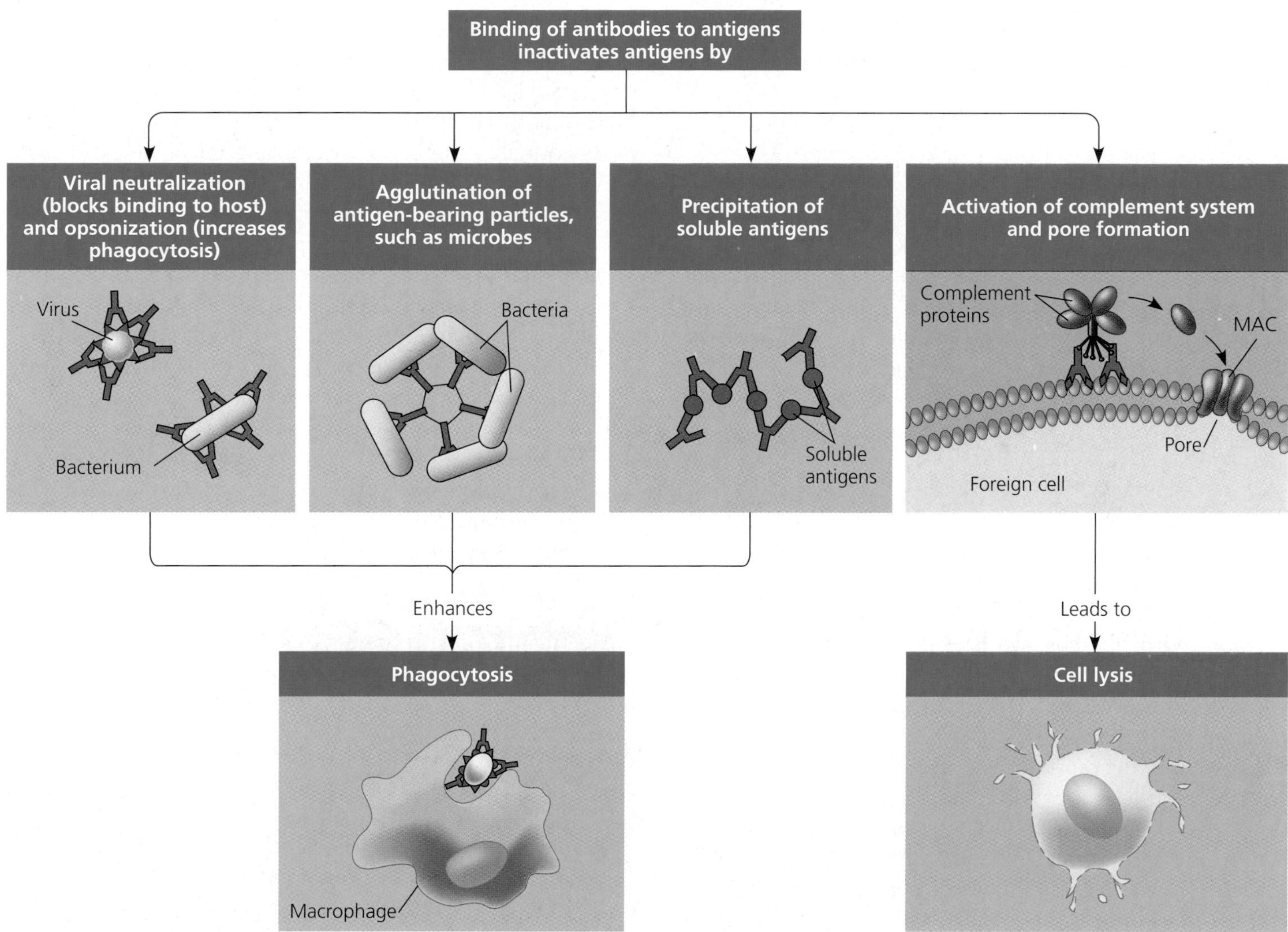

▲ **Figure 43.19 Antibody-mediated mechanisms of antigen disposal.** The binding of antibodies to antigens marks microbes, foreign particles, and soluble antigens for inactivation or destruction. Following activation of the complement system, the membrane attack complex (MAC) forms pores in foreign cells. The pores allow ions and water to rush into the cell, leading to its swelling and eventual lysis.

As you read earlier in the chapter, substances on many microbes activate the complement system as part of the body's innate defenses. The complement system also participates in the antibody-mediated disposal of microbes and transplanted body cells. In this case, binding of antigen–antibody complexes on a microbe or foreign cell to one of the complement proteins triggers a cascade in which each component activates the next. Ultimately, activated complement proteins generate a **membrane attack complex (MAC)** that forms a pore in the membrane. Ions and water rush into the cell, causing it to swell and lyse (see Figure 43.19, right). Whether activated as part of innate or acquired defenses, the complement cascade results in lysis of microbes and produces activated complement proteins that also promote inflammation or stimulate phagocytosis.

As shown in Figure 43.19, antibodies promote phagocytosis in several ways. Recall that phagocytosis enables macrophages and dendritic cells to present antigens to and stimulate helper T cells, which in turn stimulate the very B cells whose antibodies contribute to phagocytosis. This positive feedback between the innate and acquired immune systems contributes to a coordinated, effective response to infection.

Active and Passive Immunization

Immunity conferred by natural exposure to an infectious agent is called **active immunity** because it depends on the action of a person's own lymphocytes and the resulting memory cells specific for the invading pathogen. Active immunity also can develop following **immunization**, often called **vaccination** (from the Latin *vacca,* cow). The first vaccine consisted of the virus causing cowpox, a mild disease usually seen in cows but also occasionally in humans.

In the late 1700s, the English physician Edward Jenner observed that milkmaids who previously had contracted cowpox were resistant to subsequent smallpox infection, a disfiguring and potentially fatal disease. In 1796, in his now-famous experiment, Jenner scratched a farm boy with a needle bearing virus-containing fluid from a sore on a milkmaid who had cowpox. When the boy was later exposed to the smallpox virus, he did not become sick. Cowpox virus protects against smallpox virus because the two viruses are so similar that the immune system cannot distinguish them. Vaccination with the cowpox virus sensitizes the immune system, so that it reacts vigorously if later exposed to cowpox virus or, more importantly, smallpox virus.

Modern vaccines include inactivated bacterial toxins, killed microbes, parts of microbes, viable but weakened microbes that generally do not cause illness, and even genes encoding microbial proteins. All these agents induce an immediate immune response and long-lasting immunological memory (thanks to memory cells). A vaccinated person who encounters the actual pathogen from which the vaccine was derived will have the same quick secondary response as a person previously infected with the pathogen.

A worldwide vaccination campaign led to eradication of smallpox in the late 1970s. Routine active immunization of infants and children has dramatically reduced the incidence of other infectious diseases, such as polio, measles, and whooping cough, in developed countries. Unfortunately, not all infectious agents are easily managed by vaccination. For example, the emergence of new strains of pathogens with slightly altered surface antigens complicates development of vaccines against some microbes, such as the parasite that causes malaria.

Immunity can also be conferred by transferring antibodies from an individual who is immune to a particular infectious agent to someone who is not. This is called **passive immunity** because it does not result from the action of the recipient's own B and T cells. Instead, the transferred antibodies are poised to immediately help destroy any microbes for which they are specific. Passive immunity provides immediate protection, but it persists only as long as the transferred antibodies last (a few weeks to a few months). Passive immunization occurs naturally when IgG antibodies of a pregnant woman cross the placenta to her fetus. In addition, IgA antibodies are passed from a mother to her nursing infant in breast milk. These help protect the infant from infection while the baby's own immune system is maturing.

In artificial passive immunization, antibodies from an immune animal are injected into a nonimmune animal. For example, a person bitten by a rabid animal may be given antibodies isolated from other people who have been vaccinated against rabies. This measure is important because rabies may progress rapidly, and the response to active immunization may be too slow to save the victim. Most people who have been exposed to the rabies virus are both passively and actively immunized. The injected antibodies help keep the virus in check until the victim's own immune response, induced by active immunization and by the infection itself, takes over.

Concept Check 43.3

1. Describe the main role of each of the following cell types, once it is activated by antigens and cytokines: helper T cell, cytotoxic T cell, and B cell.
2. What cells and functions would be deficient in a child born without a thymus?
3. Discuss how antibodies help protect us from infection or the effects of infection.
4. Explain why passive immunization provides short-term protection from an infection, whereas active immunization provides long-term protection.

For suggested answers, see Appendix A.

Concept 43.4

The immune system's ability to distinguish self from nonself limits tissue transplantation

In addition to distinguishing between the body's own cells and invading pathogens, the immune system can wage war against cells from other individuals. For example, skin transplanted from one person to a genetically nonidentical person will look healthy for a week or so, but will then be destroyed (rejected) by the recipient's immune response. (Interestingly, however, a pregnant woman does not reject her fetus as nonself tissue. Apparently, the structure of the placenta—described in Chapter 46—is the key to this tolerance.) Keep in mind that the body's hostile reaction to an incompatible blood transfusion or transplant of other tissues or whole organs is not a disorder of the immune system, but the normal reaction of a healthy immune system exposed to foreign antigens.

Blood Groups and Transfusions

In Chapter 14, we discussed the genetics of ABO blood groups in humans. Recall that type A red blood cells have A antigen molecules on their surface. The A antigen may be recognized as foreign if placed in the body of another person. Similarly, the B antigen is found on type B red blood cells; both A and B antigens are found on type AB red blood cells; and neither antigen is found on type O red blood cells (see Table 14.2).

Individuals with type A blood do not, of course, produce antibodies against the A antigen, because they are self-tolerant. However, these individuals *do* have circulating antibodies against the B antigen, even if they have never been exposed to type B blood! You may find it odd that antibodies to foreign blood group antigens exist in the body even in the absence of exposure to foreign blood cells. The explanation is that these antibodies arise in response to normal bacterial inhabitants of the body that have epitopes very similar to blood group antigens.

For example, a person with type A blood makes antibodies against B-like bacterial epitopes, which the immune system considers foreign, but does not make antibodies against A-like bacterial epitopes, which the immune system recognizes as self. The preexisting anti-B antibodies in a person with type A blood will cause an immediate and devastating transfusion reaction if that person receives a transfusion of type B blood. This reaction involves lysis of the transfused red blood cells, which can lead to chills, fever, shock, and kidney malfunction. By the same token, anti-A antibodies in the donated type B blood can act against the recipient's type A red blood cells. This latter reaction can be minimized by transfusing packed cells instead of whole blood, so that donor antibodies in the fluid portion are not transferred.

Table 43.1 lists the recipient and donor (packed cell) combinations that are safe and those that result in transfusion reactions. Note that the row shaded in blue indicates that a person with type AB blood can safely receive blood of any type (and thus is called a "universal recipient"). The column shaded in green indicates that a person with type O blood can safely donate blood to any recipient (and thus is called a "universal donor").

Blood group antigens and related bacterial epitopes are polysaccharides. Such polysaccharide antigens induce immune responses in which no memory cells are generated. As a result, anti-blood group antibodies are always IgM (generated from primary responses) rather than IgG (generated from secondary responses). This is fortunate in pregnancy because IgM does not cross the placenta, so no harm comes to a fetus with a blood type that is not compatible with its mother's. However, another red blood cell antigen, the **Rh factor,** can cause trouble for a fetus. As a protein antigen, the Rh factor induces immune responses in which memory cells are generated. Later exposure of these memory cells to the Rh factor leads to production of anti-Rh antibodies that are IgG.

Table 43.1 Blood Groups That Can and Cannot Be Safely Combined in Transfusion*

Recipient's Blood Group	Antibodies in Recipient's Blood	Presence (+) or Absence (–) of Tranfusion Reaction: Donated Blood Group (Packed Cells)			
		A	B	AB	O
A	Anti-B	–	+	+	–
B	Anti-A	+	–	+	–
AB	No anti-A or anti-B	–	–	–	–
O	Anti-A and anti-B	+	+	+	–

* Individuals with type AB blood are universal recipients (blue row); those with type O blood are universal donors (green column).

A potentially dangerous situation can arise when a mother who is Rh-negative (lacks the Rh factor) carries a fetus that is Rh-positive, having inherited the factor from its father. If small amounts of fetal blood cross the placenta, which may happen late in pregnancy or during delivery, the mother mounts a humoral response against the Rh factor. The danger occurs in subsequent pregnancies with an Rh-positive fetus, when the mother's Rh-specific memory B cells are exposed to the Rh factor from the fetus. These B cells produce anti-Rh IgG antibodies, which can cross the placenta and destroy the red blood cells of the fetus. To prevent this, the mother is injected with anti-Rh antibodies around the seventh month of pregnancy and again just after delivering an Rh-positive baby. She is, in effect, passively immunized (artificially) to eliminate any fetal Rh-bearing red blood cells that cross the placenta before her own immune system responds to them and generates immunological memory that would endanger future Rh-positive babies.

Tissue and Organ Transplants

Major histocompatibility complex (MHC) molecules are responsible for stimulating the immune response that leads to rejection of tissue and organ transplants, or grafts. As you read earlier, the polymorphism of the MHC almost guarantees that no two people, except identical twins, have exactly the same set of MHC molecules. Thus, a rejection reaction is mounted in the vast majority of graft and transplant recipients because at least some MHC molecules on the donated tissue are foreign to the recipient. There is no danger of rejection if the donor and recipient are identical twins or if tissue is grafted from one part of an individual's body to another part.

To minimize the extent of rejection in nonidentical transplants, attempts are made to use donor tissue bearing MHC molecules that match those of the recipient as closely as possible. In addition, the recipient takes medicines that suppress immune responses. However, these medicines can leave the recipient more susceptible to infections and cancer during the course of treatment.

In a bone marrow transplant, the graft itself, rather than the recipient, is the source of potential immune rejection. Bone marrow transplants are used to treat leukemia and other cancers as well as various hematological (blood cell) diseases. As in any transplant, the MHC of the donor and recipient are matched as closely as possible. Prior to receiving transplanted bone marrow, the recipient is typically treated with irradiation to eliminate his or her own bone marrow cells, including any abnormal cells. This treatment effectively obliterates the recipient's immune system, leaving little chance of graft rejection. However, the great danger in bone marrow transplants is that lymphocytes in the donated marrow will react against the recipient. This **graft versus host reaction** is limited if the MHC molecules of the donor and recipient are well matched. Bone marrow donor programs around the world continually seek volunteers; because of the great variability of the MHC, a diverse pool of potential donors is essential.

Concept Check 43.4

1. Explain why a person who has type AB blood is considered a universal blood recipient.
2. In bone marrow transplantation, there is a danger of a graft versus host reaction. Why is this reaction particular to a bone marrow transplant?
3. Severely burned patients generally must receive numerous skin grafts. What is the advantage of using skin from an unburned part of a patient's own body (an autograft) rather than from another person?

For suggested answers, see Appendix A.

Exaggerated, self-directed, or diminished immune responses can cause disease

The highly regulated interplay of lymphocytes with foreign substances, with each other, and with other body cells provides extraordinary protection against many pathogens. However, if this delicate balance is disrupted by an immune system malfunction, the effects on the individual can range from the minor inconvenience of some allergies to the serious and often fatal consequences of certain autoimmune and immunodeficiency diseases.

Allergies

Allergies are exaggerated (hypersensitive) responses to certain antigens called *allergens*. One hypothesis to explain the origin of allergies is that they are evolutionary remnants of the immune system's response to parasitic worms. The humoral mechanism that combats worms is similar to the allergic response that causes such disorders as hay fever and allergic asthma.

The most common allergies involve antibodies of the IgE class (see Figure 43.18). Hay fever, for instance, occurs when plasma cells secrete IgE antibodies specific for antigens on the surface of pollen grains. Some of these antibodies attach by their tails to mast cells present in connective tissues. Later, when pollen grains again enter the body, they attach to the antigen-binding sites of mast cell-associated IgE, cross-linking adjacent antibody molecules. This induces the mast cell to release histamine and other inflammatory agents from their granules (vesicles), a process called *degranulation* **(Figure 43.20)**. Recall that histamine causes dilation and increased permeability of small blood vessels.

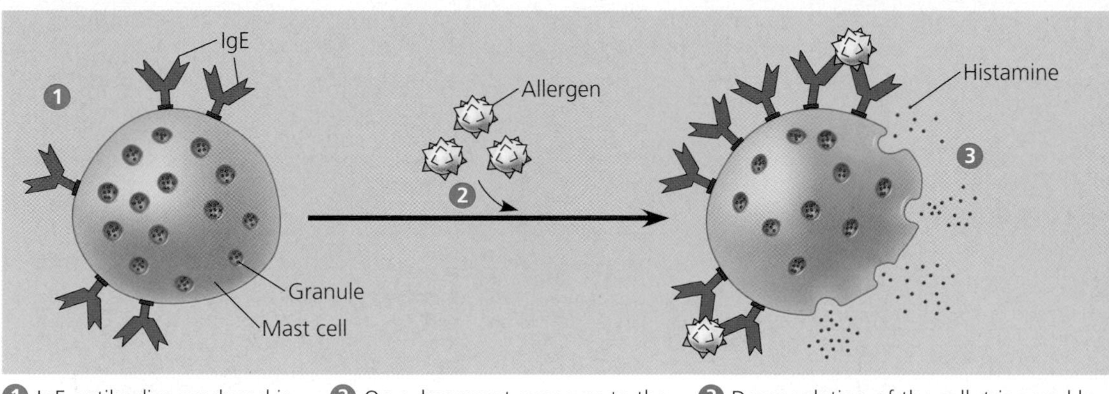

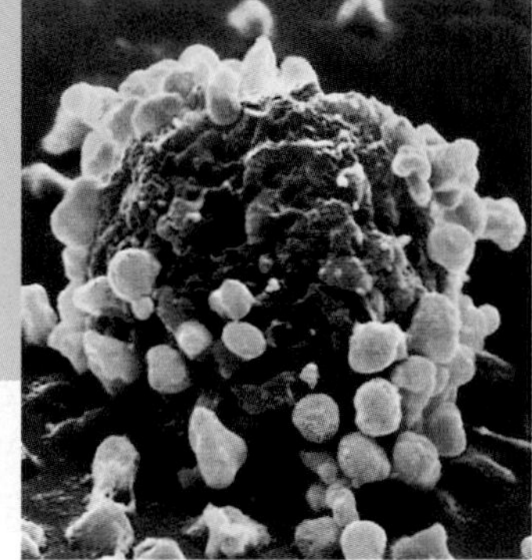

1 IgE antibodies produced in response to initial exposure to an allergen bind to receptors on mast cells.

2 On subsequent exposure to the same allergen, IgE molecules attached to a mast cell recognize and bind the allergen.

3 Degranulation of the cell, triggered by cross-linking of adjacent IgE molecules, releases histamine and other chemicals, leading to allergy symptoms.

▲ **Figure 43.20 Mast cells, IgE, and the allergic response.** The colorized SEM shows a degranulated mast cell that has released granules containing histamine and other inflammatory agents.

Such vascular changes lead to typical allergy symptoms: sneezing, runny nose, tearing eyes, and smooth muscle contractions that can result in breathing difficulty. Antihistamines diminish allergy symptoms by blocking receptors for histamine.

An acute allergic response sometimes leads to **anaphylactic shock,** a whole-body, life-threatening reaction that can occur within seconds of exposure to an allergen. Anaphylactic shock develops when widespread mast cell degranulation triggers abrupt dilation of peripheral blood vessels, causing a precipitous drop in blood pressure. Death may occur within a few minutes. Allergic responses to bee venom or penicillin can lead to anaphylactic shock in people who are extremely allergic to these substances. Likewise, people very allergic to peanuts, fish, or other foods have died from eating only tiny amounts of these allergens. Some individuals with severe hypersensitivities carry syringes containing the hormone epinephrine, which counteracts this allergic response.

Autoimmune Diseases

In some individuals, the immune system loses tolerance for self and turns against certain molecules of the body, causing one of the many **autoimmune diseases.** In *systemic lupus erythematosus (lupus),* the immune system generates antibodies (known as autoantibodies) against a wide range of self molecules, including histones and DNA released by the normal breakdown of body cells. Lupus is characterized by skin rashes, fever, arthritis, and kidney dysfunction. Another antibody-mediated autoimmune disease, *rheumatoid arthritis,* leads to damage and painful inflammation of the cartilage and bone of joints **(Figure 43.21)**. In *insulin-dependent diabetes mellitus,* the insulin-producing beta cells of the pancreas are the targets of autoimmune cytotoxic T cells. Yet another example is *multiple sclerosis,* the most common chronic neurological disease in developed countries. In this disease, T cells infiltrate the central nervous system and destroy the myelin sheath that surrounds some neurons (see Figure 48.8). This results in a number of serious neurological abnormalities.

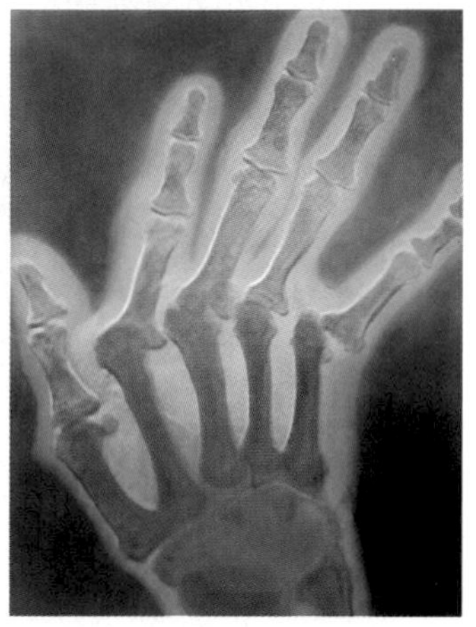

▲ **Figure 43.21 X-ray of a hand deformed by rheumatoid arthritis.**

The mechanisms leading to autoimmunity are not fully understood. For a long time, it was thought that people with autoimmune diseases had self-reactive lymphocytes that had happened to escape elimination during their development. We now know that healthy people also have lymphocytes with the capacity to react against self. However, several regulatory mechanisms make these cells nonfunctional, so they cannot induce autoimmune reactions. Thus, autoimmune disease likely arises from some failure in immune system regulation.

Immunodeficiency Diseases

The inability of the immune system to protect the body from pathogens or cancer cells that it should normally be able to defeat reflects some sort of deficiency in the system. An immunodeficiency disease caused by a genetic or developmental defect in the immune system is classified as an *inborn* or *primary immunodeficiency.* An immunodeficiency disease that develops later in life following exposure to various chemical or biological agents is classified as an *acquired* or *secondary immunodeficiency.* Whatever the cause and nature of the immunodeficiency, a person with this kind of disease is subject to frequent and recurrent infections and also is more susceptible to cancer.

Inborn (Primary) Immunodeficiencies

Inborn immunodeficiencies result from defects in the development of various immune system cells or defects in the production of specific proteins, such as IgA antibodies or complement components. Depending on the specific genetic defect, innate defenses, acquired defenses, or both may be impaired. In *severe combined immunodeficiency (SCID),* both the humoral and cell-mediated branches of acquired immunity fail to function. The long-term survival of people with this genetic disease usually requires a bone marrow transplant that will continue to supply functional lymphocytes.

In one type of SCID, a deficiency of the enzyme adenosine deaminase leads to accumulation of substances that are toxic to both B and T cells. Since the early 1990s, medical researchers have been testing a gene therapy for this disease in which the individual's own bone marrow cells are removed, engineered to contain a functional adenosine deaminase gene, and then returned to the body (see Figure 20.16). Recent successes include a child with SCID who received adenosine deaminase gene therapy when she was 2 years old. About two years after treatment, her T cells and B cells were still functioning normally. In fact, when a family member came down with chicken pox, the treated girl did not develop any sign of the disease, evidence that her immune system was working.

Acquired (Secondary) Immunodeficiencies

Immune dysfunction that develops later in life can be caused by exposure to a number of agents. Drugs used to fight autoimmune diseases or to prevent the rejection of a transplant suppress the immune system, leading to an immunodeficient state. In addition, the immune system is suppressed by certain cancers, especially Hodgkin's disease, which damages the lymphatic system. Acquired immunodeficiencies range from temporary states that may arise from physiological stress to the devastating **acquired immunodeficiency syndrome**, or **AIDS**, which is caused by a virus.

Stress and the Immune System. Healthy immune function appears to depend on both the endocrine system and the nervous systems. Nearly 2,000 years ago, the Greek physician Galen recorded that people suffering from depression were more likely than others to develop cancer. In fact, there is growing evidence that physical and emotional stress can harm immunity. Hormones secreted by the adrenal glands during stress affect the numbers of white blood cells and may suppress the immune system in other ways.

The association between emotional stress and immune function also involves the nervous system. Some neurotransmitters secreted when we are relaxed and happy may enhance immunity. In one study, college students were examined just after a vacation and again during final exams. Their immune systems were impaired in various ways during exam week; for example, interferon levels were lower. These and other observations indicate that general health and state of mind affect immunity. Physiological evidence also points to an immune system–nervous system link: Receptors for neurotransmitters have been discovered on the surface of lymphocytes, and a network of nerve fibers penetrates deep into the thymus.

Acquired Immunodeficiency Syndrome (AIDS). People with AIDS are highly susceptible to opportunistic infections and cancers that take advantage of an immune system in collapse. For example, infection by *Pneumocystis carinii,* a ubiquitous fungus, can cause severe pneumonia in people with AIDS but is successfully rebuffed by individuals with a healthy immune system. Likewise, Kaposi's sarcoma is a rare cancer that occurs most commonly in AIDS patients. Such opportunistic diseases, as well as neurological damage and physiological wasting, can lead to death from AIDS.

Because AIDS arises from the loss of helper T cells, both humoral and cell-mediated immune responses are impaired. This loss of helper T cells results from infection by the **human immunodeficiency virus (HIV),** a retrovirus **(Figure 43.22)**. HIV gains entry into cells by making use of three proteins that participate in normal immune responses. The main receptor for HIV on helper T cells is the cell's CD4 molecule. The virus also infects other cell types, such as macrophages and brain cells, that have low levels of CD4. In addition to CD4, HIV entry requires a second cell-surface protein, a *co-receptor.* One co-receptor, called fusin, is present on all the cell types infected by HIV, while a different co-receptor is present only on macrophages and helper T cells. Both of these HIV co-receptors function as chemokine receptors in uninfected cells. In fact, these proteins were first recognized as HIV co-receptors after it was discovered that chemokines can block HIV entry into cells.

Once inside a cell, the HIV RNA genome is reverse-transcribed, and the product DNA is integrated into the host cell's genome. In this form, the viral genome can direct the production of new virus particles (see Figure 18.10). The infected cell's machinery for transcription and translation thus are hijacked for the benefit of the virus. The death of helper T cells in HIV infection is thought to occur in two ways: Infected cells may succumb to the damaging effects of virus reproduction, and both infected and uninfected cells may undergo inappropriately timed apoptosis triggered by the virus.

At this time, HIV infection cannot be cured, although certain drugs can slow HIV reproduction and the progression to AIDS. However, these drugs are very expensive and not available to all infected people. In addition, the mutational changes that occur in each round of virus reproduction can generate strains of HIV that are drug resistant. The impact of

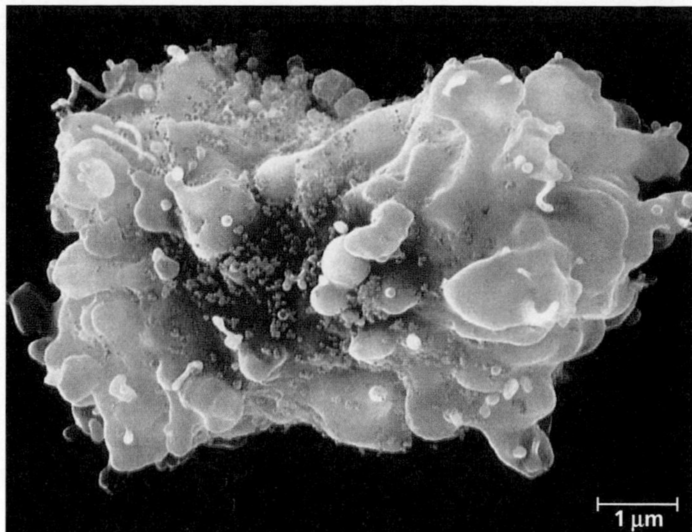

▲ **Figure 43.22 A T cell infected with HIV.** Newly made virus particles (gray) are seen budding from the surface of the T cell in this colorized SEM.

viral drug resistance can be minimized by using a combination of drugs; viruses newly resistant to one drug can be defeated by another. But the appearance of strains resistant to multiple drugs reduces the effectiveness of multidrug "cocktails" in some patients. Frequent mutational changes in HIV surface antigens also have hampered efforts to develop an effective vaccine.

Transmission of HIV requires the transfer of body fluids containing infected cells, such as semen or blood, from person to person. Unprotected sex (that is, without a condom) among male homosexuals and transmission via HIV-contaminated needles (typically among intravenous drug users) account for most of the HIV infections in the United States. However, transmission of HIV among heterosexuals is rapidly increasing as a result of unprotected sex with infected partners. In Africa and Asia, transmission occurs primarily by heterosexual sex.

In a December 2003 report, the Joint United Nations Program on AIDS estimated that 40 million people worldwide are living with HIV/AIDS. The best approach for slowing the spread of HIV is to educate people about the practices that transmit the virus, such as using dirty needles and having sex without a condom. Although condoms do not completely eliminate the risk of transmitting HIV (or other similarly transmitted viruses, such as the hepatitis B virus), they do reduce it. Anyone who has sex—vaginal, oral, or anal—with a potentially infected partner risks exposure to the deadly virus.

Concept Check 43.5

1. How would a macrophage deficiency likely affect a person's innate and acquired defenses?
2. Many anti-allergy medications block responses by mast cells. Explain why these drugs are effective in treating allergies such as hay fever.
3. In myasthenia gravis, antibodies bind to and block acetylcholine receptors at neuromuscular junctions, preventing muscle contraction. Is this disease best classified as an immunodeficiency disease, an autoimmune disease, or an allergic disease? Explain.
4. People who have nonfunctional chemokine receptors because of a genetic mutation are immune to HIV infection. Explain this finding.

For suggested answers, see Appendix A.

Chapter 43 Review

Go to www.campbellbiology.com or the student CD-ROM to explore Activities, Investigations, and other interactive study aids.

SUMMARY OF KEY CONCEPTS

Concept 43.1
Innate immunity provides broad defenses against infection

▶ **External Defenses (p. 899)** Intact skin and mucous membranes form physical barriers that bar the entry of microorganisms and viruses. Mucus produced by cells in these membranes, the low pH of the skin and stomach, and degradation by lysozyme also deter infection by pathogens.

▶ **Internal Cellular and Chemical Defenses (pp. 899–902)** Phagocytic cells ingest microbes that penetrate external innate defenses and help trigger an inflammatory response. Complement proteins, interferons, and other antimicrobial proteins also act against invading microbes. In local inflammation, histamine and other chemicals released from injured cells promote changes in blood vessels that allow fluid, more phagocytes, and antimicrobial proteins to enter the tissues. Natural killer (NK) cells can induce the death of virus-infected or cancer cells via apoptosis.

▶ **Invertebrate Immune Mechanisms (pp. 902–903)** Insects defend themselves by mechanisms similar in many respects to vertebrate innate defenses.

Concept 43.2
In acquired immunity, lymphocytes provide specific defenses against infection

▶ **Antigen Recognition by Lymphocytes (pp. 903–905)** Receptors on lymphocytes bind specifically to small regions of an antigen (epitopes). B cells recognize intact antigens. T cells recognize small antigen fragments (peptide antigens) that are complexed with cell-surface proteins called major histocompatibility (MHC) molecules. Class I MHC molecules, located on all nucleated cells, display peptide antigens to cytotoxic T cells. Class II MHC molecules, located mainly on dendritic cells, macrophages, and B cells (antigen-presenting cells), display peptide antigens to helper T cells.

- **Lymphocyte Development (pp. 905–908)** Lymphocytes arise from stem cells in the bone marrow and complete their maturation in the bone marrow (B cells) or in the thymus (T cells). Early in development, random, permanent gene rearrangement forms functional genes encoding the B or T cell antigen receptor chains. All the antigen receptors produced by a single lymphocyte are specific for the same antigen. Self-reactive lymphocytes, whose receptors bind to normal body components, are destroyed or inactivated. In a primary immune response, binding of antigen to a mature lymphocyte induces the lymphocyte's proliferation and differentiation (clonal selection), generating a clone of short-lived activated effector cells and a clone of long-lived memory cells. These memory cells account for the faster, more efficient secondary immune response.

Concept 43.3

Humoral and cell-mediated immunity defend against different types of threats

- **Helper T Cells: A Response to Nearly All Antigens (pp. 909–910)** Helper T cells make CD4, a surface protein that enhances their binding to class II MHC molecule–antigen complexes on antigen-presenting cells. Activated helper T cells secrete several different cytokines that stimulate other lymphocytes.
- **Cytotoxic T Cells: A Response to Infected Cells and Cancer Cells (pp. 910–911)** Cytotoxic T cells make CD8, a surface protein that enhances their binding to class I MHC molecule–antigen complexes on infected cells, cancer cells, and transplanted tissues. Activated cytotoxic T cells secrete proteins that initiate destruction of their target cells.
- **B Cells: A Response to Extracellular Pathogens (pp. 910–914)** The clonal selection of B cells generates antibody-secreting plasma cells, the effector cells of humoral immunity. The five major antibody classes differ in their distributions and functions within the body. Binding of antibodies to antigens on the surface of pathogens leads to elimination of the microbes by phagocytosis and complement-mediated lysis.
- **Active and Passive Immunization (p. 914)** Active immunity develops naturally in response to an infection; it also develops artificially by immunization (vaccination). In immunization, a nonpathogenic form of a microbe or part of a microbe elicits an immune response to and immunological memory for that microbe. Passive immunity, which provides immediate, short-term protection, is conferred naturally when IgG crosses the placenta from mother to fetus or when IgA passes from mother to infant in breast milk. It also can be conferred artificially by injecting antibodies into a nonimmune person.
  Activity *Immune Responses*

Concept 43.4

The immune system's ability to distinguish self from nonself limits tissue transplantation

- **Blood Groups and Transfusions (pp. 915–916)** Certain antigens on red blood cells determine whether a person has type A, B, AB, or O blood. Because antibodies to nonself blood antigens already exist in the body, transfusion with incompatible blood leads to destruction of the transfused cells. The Rh factor, another red blood cell antigen, creates difficulties when an Rh-negative mother carries successive Rh-positive fetuses.
- **Tissue and Organ Transplants (p. 916)** MHC molecules are responsible for stimulating the rejection of tissue grafts and organ transplants. The chances of successful transplantation are increased if the donor and recipient MHC tissue types are well matched and if immunosuppressive drugs are given to the recipient. Lymphocytes in bone marrow transplants may cause a graft versus host reaction in recipients.

Concept 43.5

Exaggerated, self-directed, or diminished immune responses can cause disease

- **Allergies (pp. 916–917)** In localized allergies such as hay fever, IgE antibodies produced after first exposure to an allergen attach to receptors on mast cells. The next time the same allergen enters the body, it binds to mast cell-associated IgE molecules, inducing the cell to release histamine and other mediators that cause vascular changes and typical symptoms.
- **Autoimmune Diseases (p. 917)** Loss of the immune system's normal self-tolerance can lead to autoimmune diseases such as systemic lupus erythematosus (lupus), multiple sclerosis, rheumatoid arthritis, and insulin-dependent diabetes.
- **Immunodeficiency Diseases (pp. 917–919)** Inborn (primary) immunodeficiencies result from hereditary or congenital defects that prevent proper functioning of innate, humoral, and/or cell-mediated defenses. AIDS is an acquired (secondary) immunodeficiency caused by the human immunodeficiency virus (HIV). HIV infection leads to destruction of helper T cells, leaving the patient prone to opportunistic diseases owing to deficient humoral and cell-mediated immune responses.
  Activity *HIV Reproductive Cycle*
  Investigation *What Causes Infections in AIDS Patients?*
  Investigation *Why Do AIDS Rates Differ Across the U.S.?*

TESTING YOUR KNOWLEDGE

Self-Quiz

1. Which of the following is *not* part of the body's innate, nonspecific defense system?
   a. natural killer (NK) cells
   b. inflammation
   c. phagocytosis by neutrophils
   d. phagocytosis by macrophages
   e. antibodies

2. Which of the following is a characteristic of the early stages of local inflammation?
   a. arteriole constriction
   b. fever
   c. attack by cytotoxic T cells
   d. release of histamine
   e. antibody- and complement-mediated lysis of microbes

3. Which of the following is *not* a component of an insect's defense against infection?
   a. phenoloxidase activation, leading to the formation of large deposits around parasites
   b. activation of natural killer cells
   c. phagocytosis by hemocytes
   d. production of antimicrobial peptides
   e. a protective exoskeleton

4. An epitope associates with which part of an antibody?
   a. the antibody-binding site
   b. the heavy-chain constant regions only
   c. the variable regions of a heavy chain and light chain combined
   d. the light-chain constant regions only
   e. the antibody tail

5. Which of the following is *not* true about helper T cells?
   a. They function in both cell-mediated and humoral immune responses.
   b. They recognize polysaccharide fragments presented by class II MHC molecules.
   c. They bear surface CD4 molecules.
   d. They are subject to infection by HIV.
   e. When activated, they secrete cytokines.

6. Which of the following molecules is *incorrectly* paired with a source?
   a. lysozyme—tears
   b. interferons—virus-infected cells
   c. antibodies—B cells
   d. chemokines—cytotoxic T cells
   e. cytokines—helper T cells

7. Which of the following best describes the difference in the way B cells and cytotoxic T cells respond to invaders?
   a. B cells confer active immunity; cytotoxic T cells confer passive immunity.
   b. B cells kill viruses directly; cytotoxic T cells kill virus-infected cells.
   c. B cells secrete antibodies against a virus; cytotoxic T cells kill virus-infected cells.
   d. B cells accomplish cell-mediated immunity; cytotoxic T cells accomplish humoral immunity.
   e. B cells respond the first time the invader is present; cytotoxic T cells respond subsequent times.

8. Which of the following results in long-term immunity?
   a. the passage of maternal antibodies to a developing fetus
   b. the inflammatory response to a splinter
   c. the administration of serum obtained from people immune to rabies
   d. the administration of the chicken pox vaccine
   e. the passage of maternal antibodies to a nursing infant

9. After an Rh-positive baby is born to an Rh-negative mother, the mother is treated with antibodies specific for the Rh factor. The purpose of this treatment is to
   a. protect her from the baby's red blood cells.
   b. prevent her from generating memory B cells specific for the Rh factor.
   c. protect her future Rh-positive babies.
   d. induce an immune response to Rh antibodies.
   e. both b and c.

10. HIV targets include all of the following *except*
   a. macrophages.
   b. cytotoxic T cells.
   c. helper T cells.
   d. cells bearing CD4 and fusin.
   e. brain cells.

For Self-Quiz answers, see Appendix A.

Go to the website or CD-ROM for more quiz questions.

Evolution Connection

One reason for the success of invertebrates, which make up more than 90% of living animal species, is their effective defense against microbes. Describe one mechanism by which invertebrates combat such invaders, and discuss how this mechanism comprises an evolutionary adaptation that is retained in the vertebrate immune system.

Scientific Inquiry

One effect of interferon-γ is to increase the number of class I MHC molecules on the cell surface. Suppose you want to test its effectiveness in treating viral infections and cancer. What effects would you predict interferon-γ to have on the immune response of laboratory animals against (a) virus-infected cells and (b) cancer cells?

Investigation *What Causes Infections in AIDS Patients?*
Investigation *Why Do AIDS Rates Differ Across the U.S.?*

Science, Technology, and Society

Both an injectable inactivated (killed) vaccine and an oral attenuated (live) vaccine are available for immunization against poliovirus, which can cause paralysis by destroying nerve cells in the brain and spinal cord. The oral vaccine is no longer recommended in western countries, where polio has been eradicated, because the live virus in this vaccine may mutate to a more virulent form, and be reintroduced into the population. However, the oral vaccine continues to be used in countries where polio persists because it is easy to administer (no needle!) and is highly effective. Moreover, the attenuated virus can spread to (and immunize) unvaccinated individuals. Do you feel this risk of mutation to virulence (about 1 in 12 million) is acceptable when compared with the benefits of oral vaccination? How do you think public health decisions of this type should be made?

44 Osmoregulation and Excretion

▲ **Figure 44.1 Salvin's albatrosses (*Diomeda cauta salvini*), birds that can drink seawater with no ill effects.**

Key Concepts

44.1 Osmoregulation balances the uptake and loss of water and solutes

44.2 An animal's nitrogenous wastes reflect its phylogeny and habitat

44.3 Diverse excretory systems are variations on a tubular theme

44.4 Nephrons and associated blood vessels are the functional units of the mammalian kidney

44.5 The mammalian kidney's ability to conserve water is a key terrestrial adaptation

44.6 Diverse adaptations of the vertebrate kidney have evolved in different environments

Overview

A Balancing Act

The physiological systems of animals, from cells and tissues to organs and organ systems, operate within a fluid environment. For such systems to function properly, this environment, particularly the relative concentrations of water and solutes, must be maintained within fairly narrow limits, often in the face of strong challenges from an animal's external environment. For example, freshwater animals, which live in an external environment that threatens to flood and dilute their body fluids, show adaptations that reduce water uptake, conserve solutes, and absorb salts from their surroundings. At the other extreme, desert and marine animals face desiccating environments with the potential to quickly deplete the animals' body water (**Figure 44.1**). Success in such environments depends on conservation of water and, for marine bony fishes, elimination of excess salts.

At the same time, metabolism presents organisms with the problem of waste disposal. The breakdown of proteins and nucleic acids is particularly problematic, since the primary metabolic waste product produced, ammonia, is very toxic. Studies of how animals meet these physiological challenges provide some of the most prominent examples of homeostasis. This chapter focuses on two key homeostatic processes: **osmoregulation**, how animals regulate solute concentrations and balance the gain and loss of water; and **excretion**, how animals get rid of the nitrogen-containing waste products of metabolism.

Concept 44.1

Osmoregulation balances the uptake and loss of water and solutes

Just as thermoregulation depends on balancing heat loss and gain (see Chapter 40), an animal's ability to regulate the chemical composition of its body fluids depends on balancing the uptake and loss of water and solutes. This osmoregulation is based largely on controlled movement of solutes between internal fluids and the external environment. The process also regulates the movement of water, which follows solutes by osmosis. An animal must also remove various metabolic waste products before they accumulate to harmful levels.

Osmosis

All animals—regardless of phylogeny, habitat, or type of waste produced—face the same central problem of osmoregulation: Over time, the rates of water uptake and loss must balance.

Lacking cell walls, animal cells swell and burst if there is a continuous net uptake of water or shrivel and die if there is a substantial net loss of water.

Water enters and leaves cells by osmosis. Recall from Chapter 7 that osmosis, a special case of diffusion, is the movement of water across a selectively permeable membrane. It occurs whenever two solutions separated by the membrane differ in osmotic pressure, or **osmolarity** (total solute concentration expressed as molarity, or moles of solute per liter of solution; see Chapter 3). The unit of measurement for osmolarity used in this chapter is milliosmoles per liter (mosm/L); 1 mosm/L is equivalent to a total solute concentration of 10^{-3} *M*. The osmolarity of human blood is about 300 mosm/L, while seawater has an osmolarity of about 1,000 mosm/L.

If two solutions separated by a selectively permeable membrane have the same osmolarity, they are said to be *isoosmotic*. There is no *net* movement of water by osmosis between isoosmotic solutions; although water molecules are continually crossing the membrane, they do so at equal rates in both directions. When two solutions differ in osmolarity, the one with the greater concentration of solutes is said to be *hyperosmotic* and the more dilute solution is said to be *hypoosmotic*. Water flows by osmosis from a hypoosmotic solution to a hyperosmotic one.*

Osmotic Challenges

There are two basic solutions to the problem of balancing water gain with water loss. One—available only to marine animals—is to be isoosmotic to the surroundings. Such an animal, which does not actively adjust its internal osmolarity, is known as an **osmoconformer.** Because an osmoconformer's internal osmolarity is the same as that of its environment, there is no tendency to gain or lose water. Osmoconformers often live in water that has a very stable composition and hence have a very constant internal osmolarity. In contrast, an **osmoregulator** is an animal that must control its internal osmolarity because its body fluids are not isoosmotic with the outside environment. An osmoregulator must discharge excess water if it lives in a hypoosmotic environment or take in water to offset osmotic loss if it inhabits a hyperosmotic environment. Osmoregulation enables animals to live in environments that are uninhabitable for osmoconformers, such as freshwater and terrestrial habitats; it also allows many marine animals to maintain internal osmolarities different from that of seawater.

Whenever animals maintain an osmolarity difference between the body and the external environment, osmoregulation has an energy cost. Because diffusion tends to equalize concentrations in a system, osmoregulators must expend energy to maintain the osmotic gradients that cause water to move in or out. They do so by using active transport to manipulate solute concentrations in their body fluids.

The energy cost of osmoregulation depends on how different an animal's osmolarity is from its surroundings, how easily water and solutes can move across the animal's surface, and how much work is required to pump solutes across the membrane. Because of the difference in solute concentration between their body fluids (240–450 mosm/L), fresh water (0.5–15 mosm/L), and seawater (approximately 1,000 mosm/L), osmoregulation accounts for nearly 5% of the resting metabolic rate of many marine and freshwater bony fishes. For brine shrimp, small crustaceans that live in Utah's Great Salt Lake and in other extremely salty lakes, the gradient between internal and external osmolarities is very large and the cost of osmoregulation is correspondingly high—as much as 30% of the resting metabolic rate. In contrast, marine osmoconformers, which are isoosmotic with seawater, expend little energy on osmoregulation.

Most animals, whether osmoconformers or osmoregulators, cannot tolerate substantial changes in external osmolarity and are said to be **stenohaline** (from the Greek *stenos*, narrow; *haline* refers to salt). In contrast, **euryhaline** animals (from the Greek *eurys*, broad)—which include both certain osmoconformers and certain osmoregulators—can survive large fluctuations in external osmolarity. Familiar examples of euryhaline osmoregulators are the various species of salmon. A more extreme example is a fish called tilapia (a native of Africa widely grown in fish "farms" for human food), which can adjust to any salt concentration between fresh water and 2,000 mosm/L, twice that of seawater **(Figure 44.2)**.

Next, we'll take a closer look at some of the adaptations for osmoregulation that have evolved in marine, freshwater, and terrestrial animals.

▲ **Figure 44.2 Mozambique tilapia (*Tilapia mossambica*), an extreme euryhaline osmoregulator.**

* In this chapter, we use the terms *isoosmotic, hypoosmotic,* and *hyperosmotic,* which refer specifically to osmolarity, instead of the more familiar terms *isotonic, hypotonic,* and *hypertonic.* The latter set of terms applies to the response of animal cells—whether they swell or shrink—in solutions of known solute concentrations.

Marine Animals

Animals first evolved in the sea, and more animal phyla are found there than in any other environment. Most marine invertebrates are osmoconformers. Their total osmolarity (the sum of the concentrations of all dissolved substances) is the same as that of seawater. However, they differ considerably from seawater in their concentrations of most specific solutes. Thus, even an animal that conforms to the osmolarity of its surroundings regulates its internal composition of solutes.

Marine vertebrates and some marine invertebrates are osmoregulators. For most of these animals, the ocean is a strongly dehydrating environment because it is much saltier than internal fluids, and water tends to be lost from their body by osmosis. Marine bony fishes, such as cod, are hypoosmotic to seawater and constantly lose water by osmosis and gain salt both by diffusion and from the food they eat **(Figure 44.3a)**. The fishes balance the water loss by drinking large amounts of seawater. Their gills dispose of sodium chloride; in the gills, specialized chloride cells actively transport chloride ions (Cl^-) out, and sodium ions (Na^+) follow passively. The kidneys of marine fishes dispose of excess calcium, magnesium, and sulfate ions while excreting only small amounts of water.

Marine sharks and most other chondrichthyans (cartilaginous animals; see Chapter 34) use a different osmoregulatory "strategy." Like bony fishes, their internal salt concentration is much less than that of seawater, so salt tends to diffuse into their body from the water, especially across their gills. The kidneys of sharks remove some of this salt load, and the rest is excreted by an organ called the rectal gland or is lost in feces. Unlike bony fishes, and despite relatively low internal salt concentration, marine sharks do not experience a large and continuous osmotic water loss. The explanation is that sharks maintain high concentrations of the nitrogenous waste urea (a product of protein and nucleic acid metabolism produced by many animals; see Figure 44.8). Another organic solute, trimethylamine oxide (TMAO), protects proteins from damage by urea. (If you have ever prepared shark meat, you know it should be soaked in fresh water to remove urea before cooking.) A shark's total solute concentration of body fluids (salts, urea, TMAO, and other compounds) is somewhat greater than 1,000 mosm/L and therefore slightly hyperosmotic to seawater. Consequently, water slowly *enters* the shark's body by osmosis and in food (sharks do not drink), and this small influx of water is disposed of in urine produced by the kidneys.

Freshwater Animals

The osmoregulatory problems of freshwater animals are opposite those of marine animals. Freshwater animals constantly gain water by osmosis and lose salts by diffusion because the osmolarity of their internal fluids is much higher than that of their surroundings. However, the body fluids of most freshwater animals do have lower solute concentrations compared to their marine relatives, an adaptation to their low-salinity freshwater habitat. For instance, whereas marine molluscs have body fluids with a solute concentration of approximately 1,000 mosm/L, some freshwater mussels maintain the solute concentration of their body fluids at about 40 mosm/L. The reduced osmotic difference between body fluids and the surrounding freshwater environment reduces the energy the animal expends for osmoregulation.

Many freshwater animals, including fishes such as perch, maintain water balance by excreting large amounts of very dilute urine. Salts lost by diffusion and in the urine are replenished by foods and by uptake across the gills; chloride cells in the gills actively transport Cl^-, and Na^+ follows **(Figure 44.3b)**.

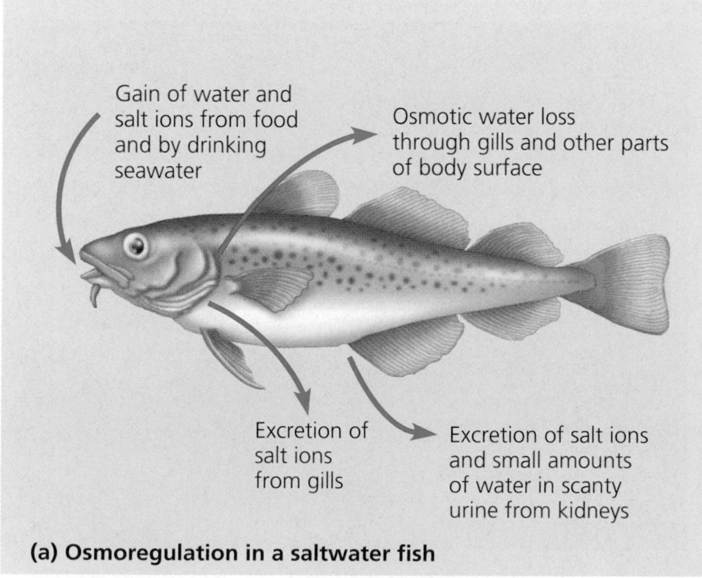

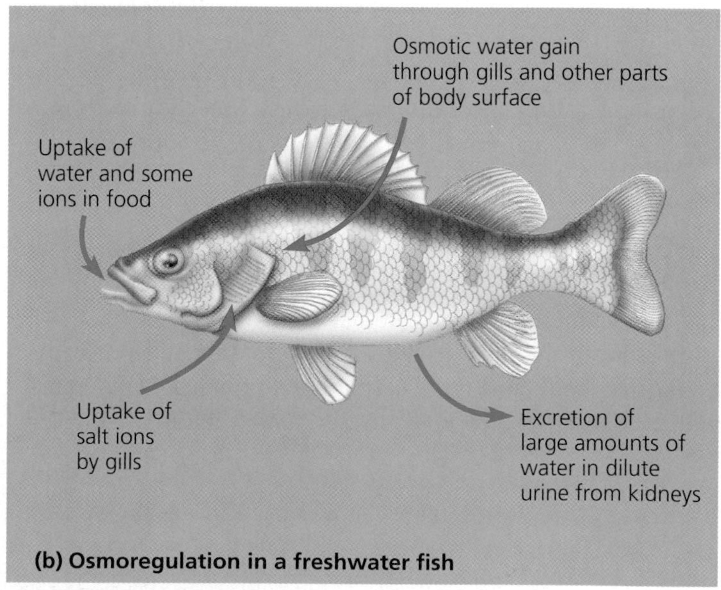

▲ **Figure 44.3 Osmoregulation in marine and freshwater bony fishes: a comparison.**

Salmon and other euryhaline fishes that migrate between seawater and fresh water undergo dramatic and rapid changes in osmoregulatory status. While in the ocean, salmon osmoregulate like other marine fishes by drinking seawater and excreting excess salt from the gills. When they migrate to fresh water, salmon cease drinking and begin to produce large amounts of dilute urine, and their gills start taking up salt from the dilute environment—just like fishes that spend their entire lives in fresh water.

Animals That Live in Temporary Waters

Dehydration is fatal for most animals, but some aquatic invertebrates living in temporary ponds and films of water around soil particles can lose almost all their body water and survive in a dormant state when their habitats dry up. This remarkable adaptation is called **anhydrobiosis** ("life without water"). Among the most striking examples are the tardigrades, or water bears, tiny invertebrates less than 1 mm long **(Figure 44.4)**. In their active, hydrated state (see Figure 44.4a), these animals contain about 85% water by weight, but they can dehydrate to less than 2% water and survive in an inactive state (see Figure 44.4b), dry as dust, for a decade or more. Just add water, and within minutes the rehydrated tardigrades are moving about and feeding.

Anhydrobiotic animals must have adaptations that keep their cells' membranes intact. Researchers are just beginning to learn how tardigrades survive drying out, but studies of anhydrobiotic roundworms (phylum Nematoda) show that dehydrated individuals contain large amounts of sugars. In particular, a disaccharide called trehalose seems to protect the cells by replacing the water that is normally associated with membranes and proteins. Many insects that survive freezing in the winter also utilize trehalose as a membrane protectant.

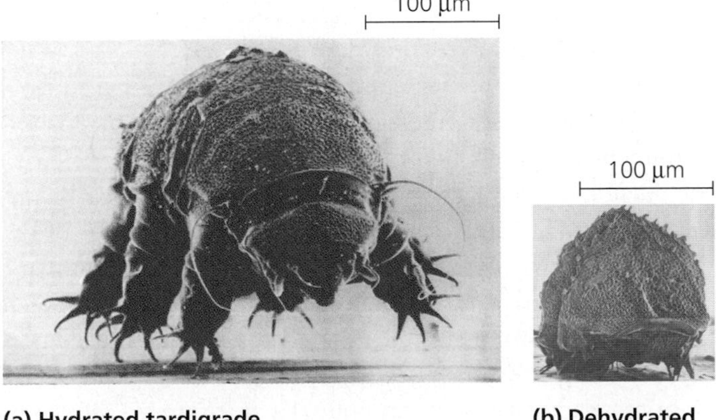

▲ **Figure 44.4 Anhydrobiosis.** Tardigrades (water bears) inhabit temporary ponds and droplets of water in soil and on moist plants (SEMs).

Land Animals

The threat of desiccation is a major regulatory problem for terrestrial plants and animals. Humans die if they lose about 12% of their body water; mammals that evolved in dry enviroments, such as camels, can withstand about twice that level of dehydration. Adaptations that reduce water loss are key to survival on land. Much as a waxy cuticle contributes to the success of land plants, the body coverings of most terrestrial animals help prevent dehydration. Examples are the waxy layers of insect exoskeletons, the shells of land snails, and the layers of dead, keratinized skin cells covering most terrestrial vertebrates. Many terrestrial animals, especially desert-dwellers, are nocturnal; this reduces evaporative water loss by taking advantage of the lower temperature and higher relative humidity of night air.

Despite these adaptations, most terrestrial animals lose considerable water from moist surfaces in their gas exchange organs, in urine and feces, and across their skin. Land animals balance their water budgets by drinking and eating moist foods and by using metabolic water (water produced during cellular respiration). Some animals, such as many insect-eating desert birds and other reptiles, are so well adapted for minimizing water loss that they can survive in deserts without drinking. Kangaroo rats lose so little water that they can recover 90% of the loss by using metabolic water **(Figure 44.5)**, gaining the remaining 10% from the small amount of water in their diet of

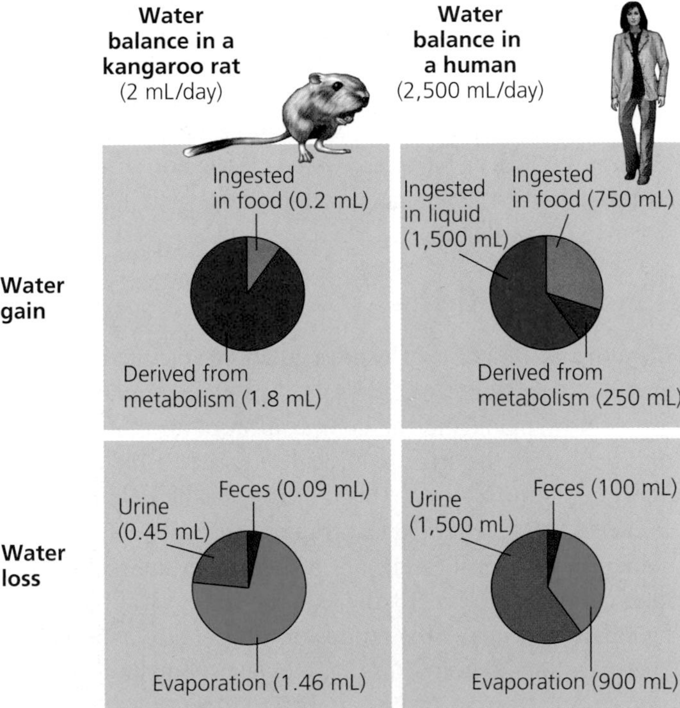

▲ **Figure 44.5 Water balance in two terrestrial mammals.** Kangaroo rats, which live in the American southwest, eat mostly dry seeds and do not drink water. A kangaroo rat loses water mainly by evaporation during gas exchange and gains water mainly from cellular metabolism. In contrast, a human loses a large amount of water in urine and regains it mostly in food and drink.

Figure 44.6

Inquiry What role does fur play in water conservation by camels?

EXPERIMENT Knut and Bodil Schmidt-Nielsen and their colleagues from Duke University observed that the fur of camels exposed to full sun in the Sahara Desert could reach temperatures of over 70°C, while the animals' skin remained more than 30°C cooler. The Schmidt-Nielsens reasoned that insulation of the skin by fur may substantially reduce the need for evaporative cooling by sweating. To test this hypothesis, they compared the water loss rates of unclipped and clipped camels.

RESULTS Removing the fur of a camel increased the rate of water loss through sweating by up to 50%.

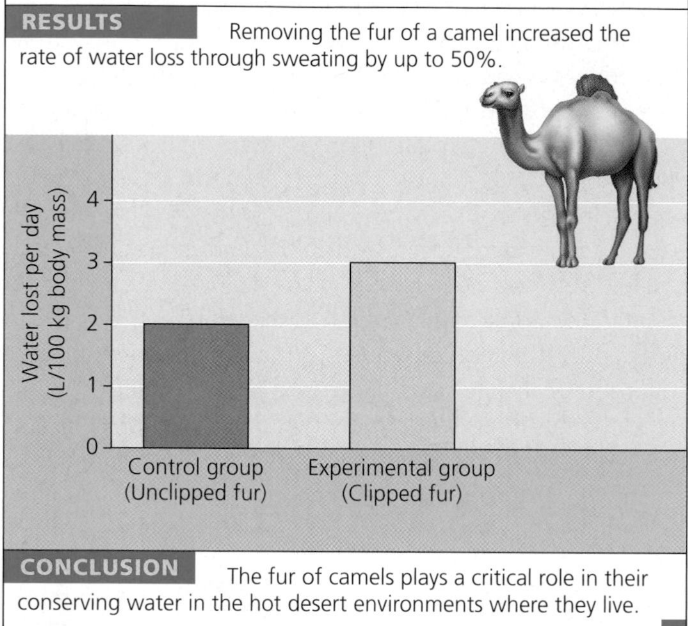

CONCLUSION The fur of camels plays a critical role in their conserving water in the hot desert environments where they live.

seeds. In studying the adaptations of animals to desert environments, physiologists have discovered that major water savings can result from simple anatomical features, such as the fur of the camel **(Figure 44.6)**.

Transport Epithelia

The ultimate function of osmoregulation is to maintain the composition of cellular cytoplasm, but most animals do this indirectly by managing the composition of an internal body fluid that bathes the cells. In insects and other animals with an open circulatory system, this fluid is the hemolymph (see Chapter 42). In vertebrates and other animals with a closed circulatory system, the cells are bathed in an interstitial fluid that is controlled indirectly through the composition of blood. The maintenance of fluid composition depends on specialized structures ranging from cells that regulate solute movement to complex organs such as the vertebrate kidney.

In most animals, one or more different kinds of **transport epithelium**—a layer or layers of specialized epithelial cells that regulate solute movements—are essential components of osmotic regulation and metabolic waste disposal. Transport epithelia move specific solutes in controlled amounts in specific directions. Some transport epithelia face the outside environment directly, while others line channels connected to the outside by an opening on the body surface. Joined by impermeable tight junctions (see Figure 6.31), the cells of the epithelium form a barrier at the tissue-environment boundary. This arrangement ensures that any solutes moving between animal and environment must pass through a selectively permeable membrane.

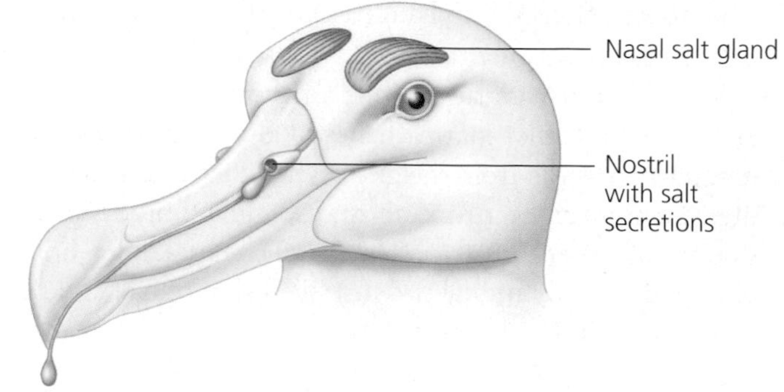

(a) An albatross's salt glands empty via a duct into the nostrils, and the salty solution either drips off the tip of the beak or is exhaled in a fine mist.

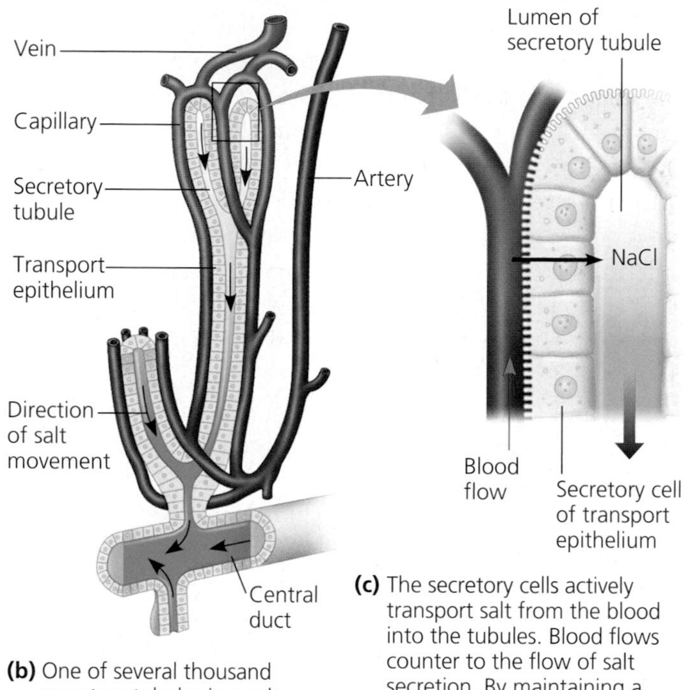

(b) One of several thousand secretory tubules in a salt-excreting gland. Each tubule is lined by a transport epithelium surrounded by capillaries, and drains into a central duct.

(c) The secretory cells actively transport salt from the blood into the tubules. Blood flows counter to the flow of salt secretion. By maintaining a concentration gradient of salt in the tubule (aqua), this countercurrent system enhances salt transfer from the blood to the lumen of the tubule.

▲ **Figure 44.7 Salt-excreting glands in birds.**

In most animals, transport epithelia are arranged into complex tubular networks with extensive surface areas. We find some of the best examples in the salt glands of marine birds, which remove excess sodium chloride from the blood **(Figure 44.7)**. For example, the albatross, which spends months or years at sea and needs to obtain both food and water from the ocean, can drink seawater because its nasal salt glands secrete fluid much saltier than the ocean. Thus, even though drinking seawater brings in a lot of salt, the bird achieves a net gain of water. Humans who drink seawater, by contrast, must use more water to excrete the salt load than was gained by drinking.

The molecular structure of the plasma membrane determines the kinds and directions of solutes that move across a particular type of transport epithelium. In contrast to salt-excreting glands, transport epithelia in the gills of freshwater fishes use active transport to move salts from the dilute surrounding water into the blood. Transport epithelia in excretory organs often have the dual functions of maintaining water balance and disposing of metabolic wastes.

Concept Check 44.1

1. The movement of salt from the surrounding water to the blood of a freshwater fish requires the expenditure of energy in the form of ATP. Why?
2. Why are no freshwater animals osmoconformers?
3. How are sharks able to expend proportionately less energy for osmoregulation compared to marine bony fishes?

For suggested answers, see Appendix A.

An animal's nitrogenous wastes reflect its phylogeny and habitat

Because most metabolic wastes must be dissolved in water when they are removed from the body, the type and quantity of an animal's waste products may have a large impact on its water balance. In terms of their effect on osmoregulation, among the most important waste products are the nitrogenous (nitrogen-containing) breakdown products of proteins and nucleic acids **(Figure 44.8)**. When these macromolecules are broken apart for energy or converted to carbohydrates or fats, enzymes remove nitrogen in the form of **ammonia** (NH_3), a very toxic molecule. Some animals excrete ammonia directly, but many species first convert the ammonia to other compounds that are less toxic but that require energy in the form of ATP to produce.

Forms of Nitrogenous Waste

Different animals excrete nitrogenous wastes in different forms—ammonia, urea, or uric acid—which vary in their toxicity and energy costs.

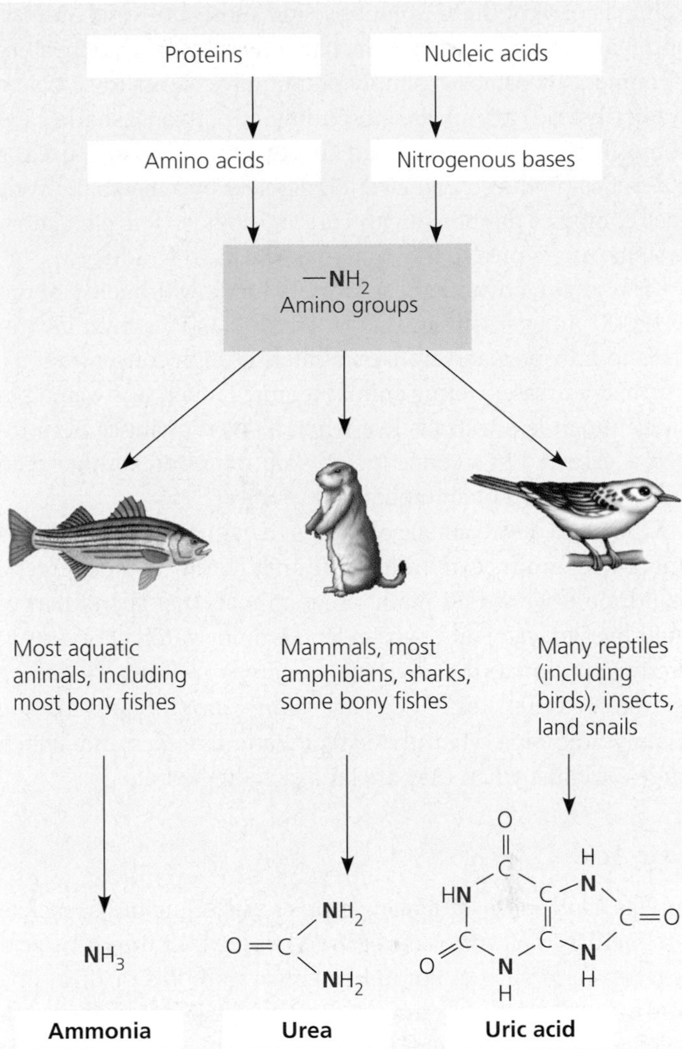

▲ **Figure 44.8 Nitrogenous wastes.**

Ammonia

Because ammonia is very soluble but can only be tolerated at very low concentrations, animals that excrete nitrogenous wastes as ammonia need access to lots of water. Therefore, ammonia excretion is most common in aquatic species. Ammonia molecules easily pass through membranes and are readily lost by diffusion to the surrounding water. In many invertebrates, ammonia release occurs across the whole body surface. In fishes, most of the ammonia is lost as ammonium ions (NH_4^+) across the epithelium of the gills, with kidneys excreting only minor amounts of nitrogenous wastes. In freshwater fishes, the gill epithelium takes up Na^+ from the water in exchange for NH_4^+, which helps to maintain a much higher Na^+ concentration in body fluids than in the surrounding water.

Urea

Although it works well in many aquatic species, ammonia excretion is much less suitable for land animals. Because ammonia is so toxic, it can be transported and excreted only in large volumes of very dilute solutions, and most terrestrial animals and many marine species (which tend to lose water to their environment by osmosis) simply do not have access to sufficient water. Instead, mammals, most adult amphibians, sharks, and some marine bony fishes and turtles excrete mainly **urea**, a substance produced in the vertebrate liver by a metabolic cycle that combines ammonia with carbon dioxide. The circulatory system carries urea to the excretory organs, the kidneys.

The main advantage of urea is its low toxicity, about 100,000 times less than that of ammonia. This permits animals to transport and store urea safely at high concentrations. Further, a urea-excreting animal requires much less water, because much less water is lost when a given quantity of nitrogen is excreted in a concentrated solution of urea rather than a dilute solution of ammonia.

The main disadvantage of urea is that animals must expend energy to produce it from ammonia. From a bioenergetic standpoint, we would predict that animals that spend part of their lives in water and part on land would switch between excreting ammonia (thereby saving energy) and urea (reducing excretory water loss). In fact, many amphibians excrete mainly ammonia when they are aquatic tadpoles and switch largely to urea when they are land-dwelling adults.

Uric Acid

Insects, land snails, and many reptiles, including birds, excrete **uric acid** as their major nitrogenous waste. Like urea, uric acid is relatively nontoxic. But unlike either ammonia or urea, uric acid is largely insoluble in water and can be excreted as a semisolid paste with very little water loss. This is a great advantage for animals with little access to water, but there is a cost: Uric acid is even more energetically expensive to produce than urea, requiring considerable ATP for synthesis from ammonia.

The Influence of Evolution and Environment on Nitrogenous Wastes

In general, the kinds of nitrogenous wastes excreted depend on an animal's evolutionary history and habitat—especially the availability of water (see Figure 44.8). For example, uric acid and urea represent different adaptations for excreting nitrogenous wastes with minimal water loss. One factor that seems to have been important in determining which of these alternatives evolved in a particular group of animals is the mode of reproduction. Soluble wastes can diffuse out of a shell-less amphibian egg or be carried away by the mother's blood in a mammalian embryo. However, the shelled eggs produced by birds and other reptiles are permeable to gases but not to liquids, which means that soluble nitrogenous wastes released by an embryo would be trapped within the egg and could accumulate to dangerous levels (although urea is much less harmful than ammonia, it does become toxic at very high concentrations). The evolution of uric acid as a waste product conveyed a selective advantage because it precipitates out of solution and can be stored within the egg as a harmless solid left behind when the animal hatches.

The type of nitrogenous waste produced by vertebrates depends on habitat as well as on evolutionary lineage. For example, terrestrial turtles (which often live in dry areas) excrete mainly uric acid, whereas aquatic turtles excrete both urea and ammonia. In some species, individuals can change forms of nitrogenous wastes when environmental conditions change. For example, certain tortoises that usually produce urea shift to uric acid when temperature increases and water becomes less available. This is another example of how response to the environment occurs on two levels: Over generations, evolution determines the limits of physiological responses for a species, but during their lives, individual organisms make physiological adjustments within these evolutionary constraints.

The amount of nitrogenous waste produced is coupled to the energy budget, as it strongly depends on how much and what kind of food an animal eats. Because they use energy at high rates, endotherms eat more food—and therefore produce more nitrogenous wastes—per unit volume than ectotherms. Predators, which derive much of their energy from dietary proteins, excrete more nitrogen than animals that rely mainly on lipids or carbohydrates as energy sources.

Concept Check 44.2

1. Dragonfly larvae, which are aquatic, excrete ammonia, whereas adult dragonflies, which are terrestrial, excrete uric acid. Explain.
2. What role does the vertebrate liver play in the body's processing of nitrogenous waste?
3. What advantage does uric acid offer as a nitrogenous waste in arid environments?

For suggested answers, see Appendix A.

Diverse excretory systems are variations on a tubular theme

Although the problems of water balance are very different on land, in salt water, and in fresh water, solving them all depends on the regulation of solute movement between internal fluids and the external environment. Much of this movement is handled by excretory systems, which are central to homeostasis because they dispose of metabolic wastes and control body fluid composition by adjusting rates of solute loss. Before we describe particular excretory systems, let's consider the basic process of excretion.

Excretory Processes

Although excretory systems are diverse, nearly all produce the fluid waste urine in a process that involves several steps (**Figure 44.9**). First, body fluid (blood, coelomic fluid, or hemolymph) is collected. The initial fluid collection usually involves **filtration** through selectively permeable membranes consisting of a single layer of transport epithelium. These membranes retain cells as well as proteins and other large molecules in the body fluid; hydrostatic pressure (blood pressure in many animals) forces water and small solutes, such as salts, sugars, amino acids, and nitrogenous wastes, into the excretory system. This fluid is called the **filtrate.**

Even when filtration occurs, fluid collection is largely nonselective; thus, it is important that essential small molecules are recovered from the filtrate and returned to the body fluids. In the second step of the process, **selective reabsorption,** excretory systems use active transport to reabsorb valuable solutes such as glucose, certain salts, and amino acids from the filtrate. Nonessential solutes and wastes (for example, excess salts and toxins) are left in the filtrate or are added to it by selective **secretion,** which also uses active transport. The pumping of various solutes also adjusts the osmotic movement of water into or out of the filtrate. The processed filtrate is then excreted from the system and from the body as urine.

Capillary

Excretory tubule

Filtrate

Urine

1 Filtration. The excretory tubule collects a filtrate from the blood. Water and solutes are forced by blood pressure across the selectively permeable membranes of a cluster of capillaries and into the excretory tubule.

2 Reabsorption. The transport epithelium reclaims valuable substances from the filtrate and returns them to the body fluids.

3 Secretion. Other substances, such as toxins and excess ions, are extracted from body fluids and added to the contents of the excretory tubule.

4 Excretion. The filtrate leaves the system and the body.

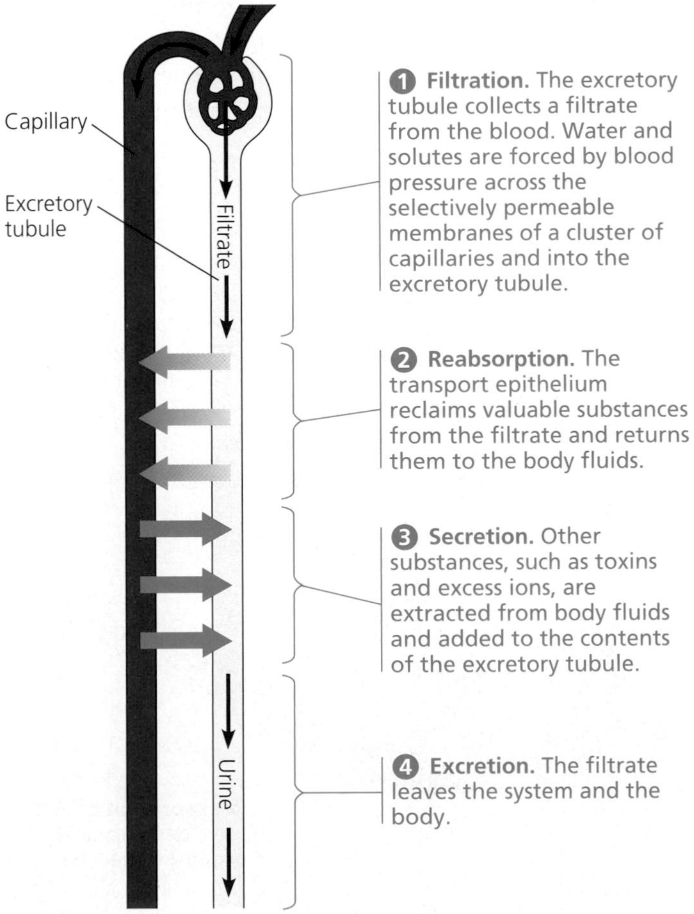

▲ **Figure 44.9 Key functions of excretory systems: an overview.** Most excretory systems produce a filtrate by pressure-filtering body fluids and then modify the filtrate's contents. This diagram is modeled after the vertebrate excretory system.

Survey of Excretory Systems

The systems that perform basic excretory functions vary widely among animal groups. However, they are generally built on a complex network of tubules that provide a large surface area for the exchange of water and solutes, including nitrogenous wastes.

Protonephridia: Flame-Bulb Systems

Flatworms (phylum Platyhelminthes) have excretory systems called protonephridia (singular, *protonephridium*). A **protonephridium** is a network of dead-end tubules lacking internal openings. As shown in **Figure 44.10**, the tubules branch throughout the body, and the smallest branches are capped by a cellular unit called a flame bulb. The flame bulb has a tuft of cilia projecting into the tubule. (The beating cilia resemble a flickering flame, hence, the name flame bulb.) The beating of the cilia draws water and solutes from the interstitial fluid through the flame bulb (filtration) into the tubule system and then moves the urine outward through the tubules until they empty to the external environment through openings called

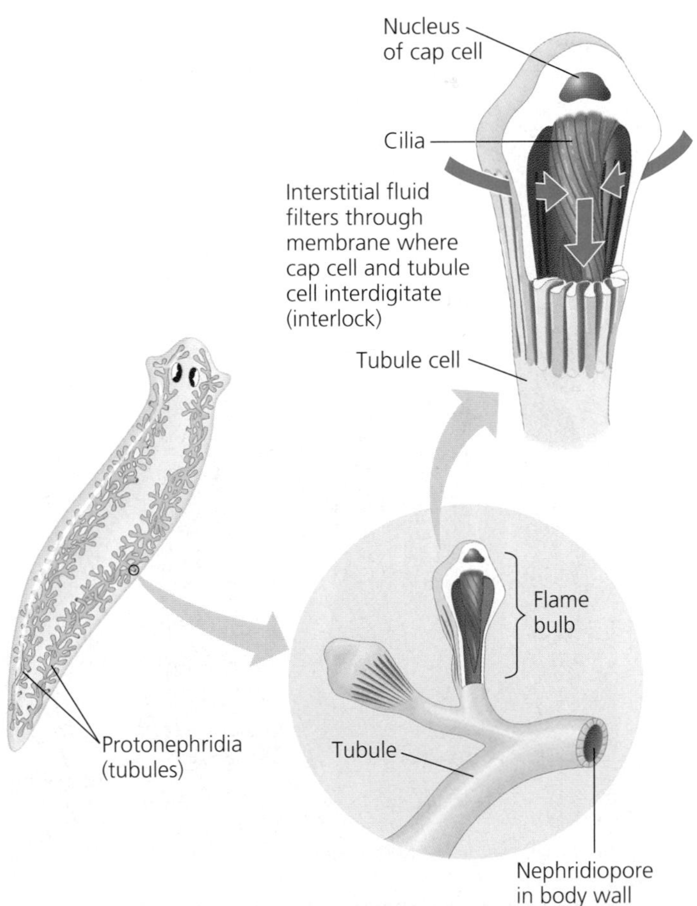

▲ **Figure 44.10 Protonephridia: the flame-bulb system of a planarian.** Protonephridia are branching internal tubules that function mainly in osmoregulation.

nephridiopores. Excreted urine is very dilute in freshwater flatworms, helping balance the osmotic uptake of water from the environment. Apparently, the tubules reabsorb most solutes before the urine exits the body.

The flame-bulb systems of freshwater flatworms seem to function mainly in osmoregulation; most metabolic wastes diffuse out of the animal across the body surface or are excreted into the gastrovascular cavity and eliminated through the mouth (see Figure 33.10). However, in some parasitic flatworms, which are isoosmotic to the surrounding fluids of their host organisms, protonephridia mainly dispose of nitrogenous wastes. This difference in function shows how structures common to a group of organisms can be adapted in diverse ways by evolution in different environments. Protonephridia are also found in rotifers, some annelids, the larvae of molluscs, and lancelets, which are invertebrate chordates. (See Chapters 33 and 34 to review these animal phyla.)

Metanephridia

Another type of tubular excretory system, **metanephridia** (singular, *metanephridium*), has internal openings that collect body fluids **(Figure 44.11)**. Metanephridia are found in most annelids, including earthworms. Each segment of a worm has a pair of metanephridia, which are immersed in coelomic fluid and enveloped by a capillary network. The internal opening of a metanephridium is surrounded by a ciliated funnel, the nephrostome. Fluid enters the nephrostome and passes through a coiled collecting tubule, which includes a storage bladder that opens to the outside through the nephridiopore.

An earthworm's metanephridia have both excretory and osmoregulatory functions. As urine moves along the tubule, the transport epithelium bordering the lumen reabsorbs most solutes and returns them to the blood in the capillaries. Nitrogenous wastes remain in the tubule and are excreted to the outside. Earthworms inhabit damp soil and usually experience a net uptake of water by osmosis through the skin. Their metanephridia balance the water influx by producing dilute urine (hypoosmotic to the body fluids).

Malpighian Tubules

Insects and other terrestrial arthropods have organs called **Malpighian tubules** that remove nitrogenous wastes and also function in osmoregulation **(Figure 44.12)**. The malpighian tubules open into the digestive tract and dead-end at tips that are immersed in hemolymph (circulatory fluid). The transport epithelium lining the tubules secretes certain solutes, including nitrogenous wastes, from the hemolymph into the lumen of the tubule. Water follows the solutes into the tubule by osmosis, and the fluid then passes into the rectum, where most solutes are pumped back into the hemolymph. Water again follows the solutes, and the nitrogenous wastes—mainly insoluble uric acid—are eliminated as nearly dry matter along

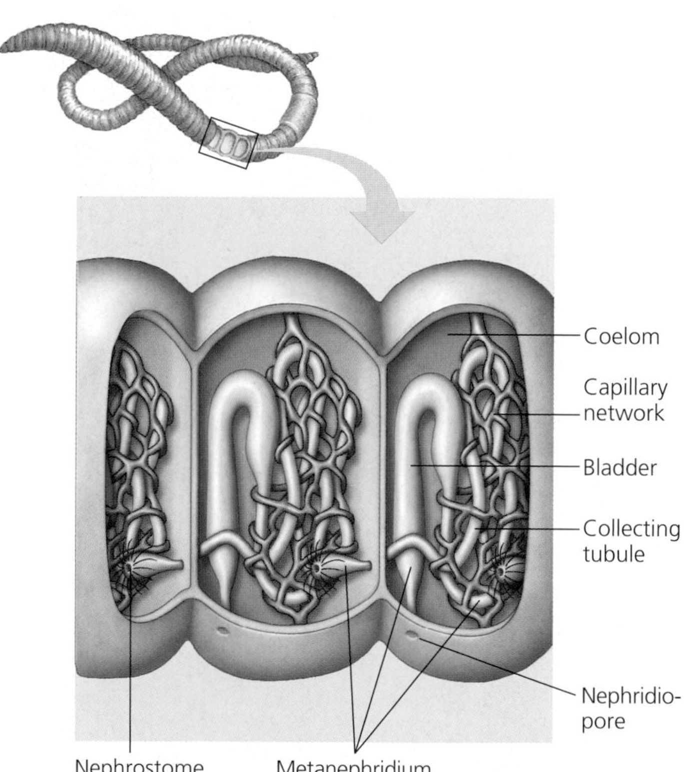

▲ **Figure 44.11 Metanephridia of an earthworm.** Each segment of the worm contains a pair of metanephridia, which collect coelomic fluid from the adjacent anterior segment. (Only one metanephridium of each pair is shown here.)

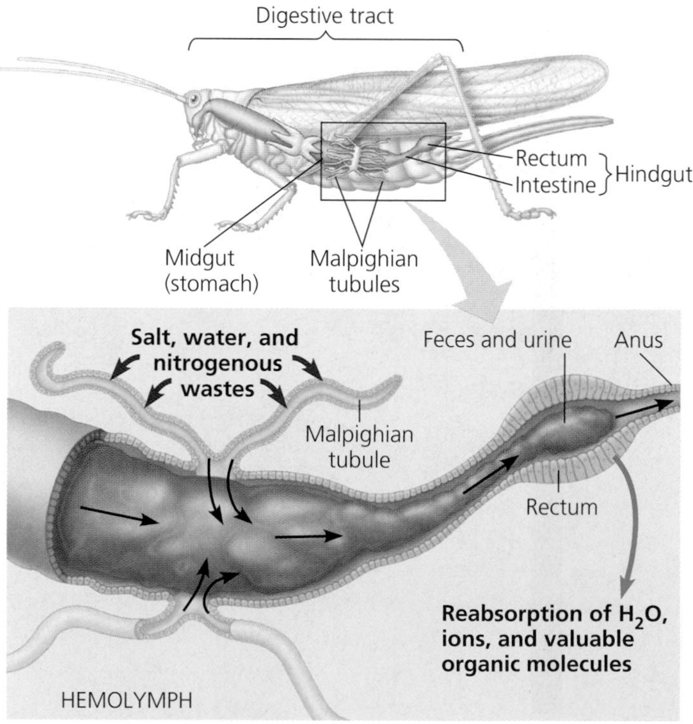

▲ **Figure 44.12 Malpighian tubules of insects.** Malpighian tubules are outpocketings of the digestive tract that remove nitrogenous wastes and function in osmoregulation.

with the feces. Highly effective in conserving water, the insect excretory system is one of several key adaptations contributing to the tremendous success of these animals on land.

Vertebrate Kidneys

The kidneys of vertebrates usually function in both osmoregulation and excretion. Like the excretory organs of most animal phyla, kidneys are built of tubules. Since the osmoconforming hagfishes, which are not vertebrates but are among the most primitive living chordates, have kidneys with segmentally arranged excretory tubules, it is likely that the excretory structures of vertebrate ancestors were segmented. The kidneys of most vertebrates, however, are compact, nonsegmented organs containing numerous tubules arranged in a highly organized manner. A dense network of capillaries intimately associated with the tubules is also an integral part of the vertebrate excretory system, as are the ducts and other structures that carry urine out of the tubules and kidney and eventually out of the body.

In the next two sections, we will focus on the mammalian excretory system, using humans as our primary example. We will then end the chapter by comparing the excretory organs of the various vertebrate classes to see how evolutionary modifications function in different environments.

Concept Check 44.3

1. What are the fundamental processes involved in all excretory systems, regardless of their anatomical differences or evolutionary origins?
2. Describe an advantage of an excretory system built around a network of fine tubules.

For suggested answers, see Appendix A.

Concept 44.4

Nephrons and associated blood vessels are the functional units of the mammalian kidney

The excretory system of mammals centers on the kidneys, which are also the principal site of water balance and salt regulation. Mammals have a pair of kidneys. Each kidney, bean-shaped and about 10 cm long in humans, is supplied with blood by a **renal artery** and drained by a **renal vein** (**Figure 44.13a**, on the following page). Blood flow through the kidneys is voluminous. In humans, the kidneys account for less than 1% of body weight, but they receive about 20% of resting cardiac output. Urine exits each kidney through a duct called the **ureter**, and both ureters drain into a common **urinary bladder**. During urination, urine is expelled from the urinary bladder through a tube called the **urethra**, which empties to the outside near the vagina in females or through the penis in males. Sphincter muscles near the junction of the urethra and the bladder, which are under nervous system control, regulate urination.

Structure and Function of the Nephron and Associated Structures

The mammalian kidney has two distinct regions, an outer **renal cortex** and an inner **renal medulla** (**Figure 44.13b**). Packing both regions are microscopic excretory tubules and their associated blood vessels. The **nephron**—the functional unit of the vertebrate kidney—consists of a single long tubule and a ball of capillaries called the **glomerulus** (**Figure 44.13c** and **d**). The blind end of the tubule forms a cup-shaped swelling, called **Bowman's capsule**, which surrounds the glomerulus. Each human kidney contains about a million nephrons, with a total tubule length of 80 km.

Filtration of the Blood

Filtration occurs as blood pressure forces fluid from the blood in the glomerulus into the lumen of Bowman's capsule (see Figure 44.13d). The porous capillaries, along with specialized cells of the capsule called podocytes, are permeable to water and small solutes but not to blood cells or large molecules such as plasma proteins. Filtration of small molecules is nonselective, and the filtrate in Bowman's capsule contains salts, glucose, amino acids, and vitamins; nitrogenous wastes such as urea; and other small molecules—a mixture that mirrors the concentrations of these substances in blood plasma.

Pathway of the Filtrate

From Bowman's capsule, the filtrate passes through three regions of the nephron: the **proximal tubule;** the **loop of Henle,** a hairpin turn with a descending limb and an ascending limb; and the **distal tubule.** The distal tubule empties into a **collecting duct,** which receives processed filtrate from many nephrons. This filtrate flows from the many collecting ducts of the kidney into the **renal pelvis,** which is drained by the ureter.

In the human kidney, approximately 80% of the nephrons, the **cortical nephrons,** have reduced loops of Henle and are almost entirely confined to the renal cortex. The other 20%, the **juxtamedullary nephrons,** have well-developed loops that extend deeply into the renal medulla. Only mammals and birds have juxtamedullary nephrons; the nephrons of other vertebrates lack loops of Henle. It is the juxtamedullary nephrons that enable mammals to produce urine that is hyperosmotic to body fluids, an adaptation that is extremely important for water conservation.

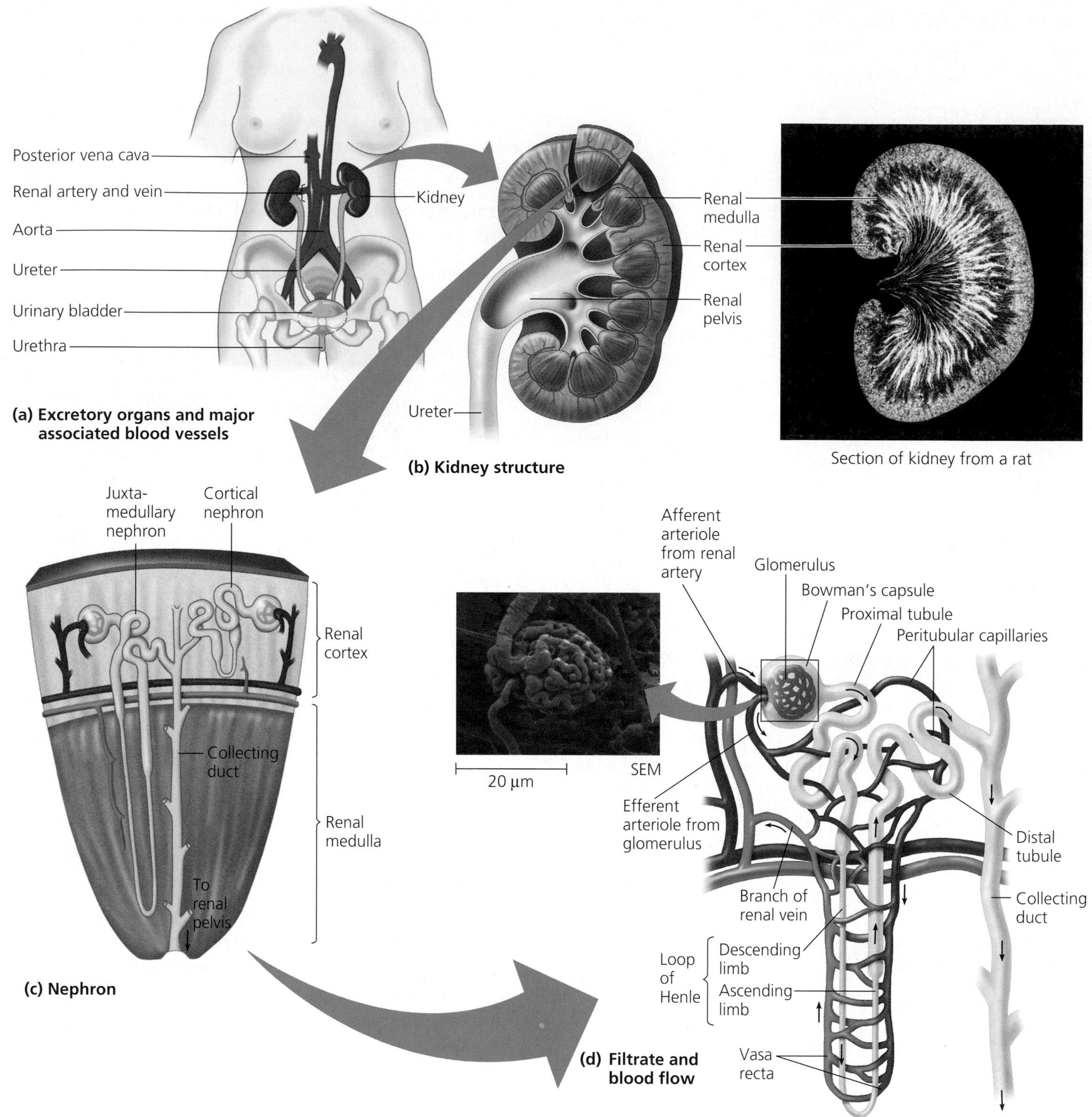

▲ **Figure 44.13 The mammalian excretory system.**

The nephron and the collecting duct are lined by a transport epithelium that processes the filtrate to form the urine. One of this epithelium's most important tasks is reabsorption of solutes and water. Between 1,100 and 2,000 L of blood flows through a pair of human kidneys each day, a volume about 275 times the total volume of blood in the body. From this enormous traffic of blood, the nephrons and collecting ducts process about 180 L of initial filtrate, equivalent to two or three times the body weight of an average person. Of this, nearly all of the sugar, vitamins, and other organic nutrients and about 99% of the water are reabsorbed into the blood, leaving only about 1.5 L of urine to be voided.

Blood Vessels Associated with the Nephrons

Each nephron is supplied with blood by an **afferent arteriole**, a branch of the renal artery that subdivides into the capillaries of the glomerulus (see Figure 44.13d). The capillaries converge as they leave the glomerulus, forming an **efferent arteriole.** This vessel subdivides again, forming the **peritubular capillaries,** which surround the proximal and distal tubules. More capillaries extend downward and form the **vasa recta,** the capillaries that serve the loop of Henle. The vasa recta also form a loop, with descending and ascending vessels conveying blood in opposite directions.

Although the excretory tubules and their surrounding capillaries are closely associated, they do not exchange materials directly. The tubules and capillaries are immersed in interstitial fluid, through which various substances diffuse between the plasma within capillaries and the filtrate within the nephron tubule. This exchange is facilitated by the relative direction of blood flow and filtrate flow in the nephrons.

From Blood Filtrate to Urine: *A Closer Look*

In this section we will concentrate on how the filtrate becomes urine as it flows through the mammalian nephron and collecting duct. The circled numbers correspond to the numbers in **Figure 44.14**.

❶ **Proximal tubule.** Secretion and reabsorption in the proximal tubule substantially alter the volume and composition of filtrate. For example, the cells of the transport epithelium help maintain a relatively constant pH in body fluids by the controlled secretion of H^+. The cells also synthesize and secrete ammonia, which neutralizes the acid and keeps the filtrate from becoming too acidic. The more acidic the filtrate, the more ammonia the cells produce and secrete, and the urine of a mammal usually contains some ammonia from this source (even though most nitrogenous waste is excreted as urea). The proximal tubules also reabsorb about 90% of the important buffer bicarbonate (HCO_3^-). Drugs and other poisons that have been processed in the liver pass from the peritubular capillaries into the interstitial fluid, and then are secreted across the epithelium of the proximal tubule into the nephron's lumen. Conversely, valuable nutrients, including glucose, amino acids, and potassium (K^+), are actively or passively transported from the filtrate to the interstitial fluid and then are moved into the peritubular capillaries.

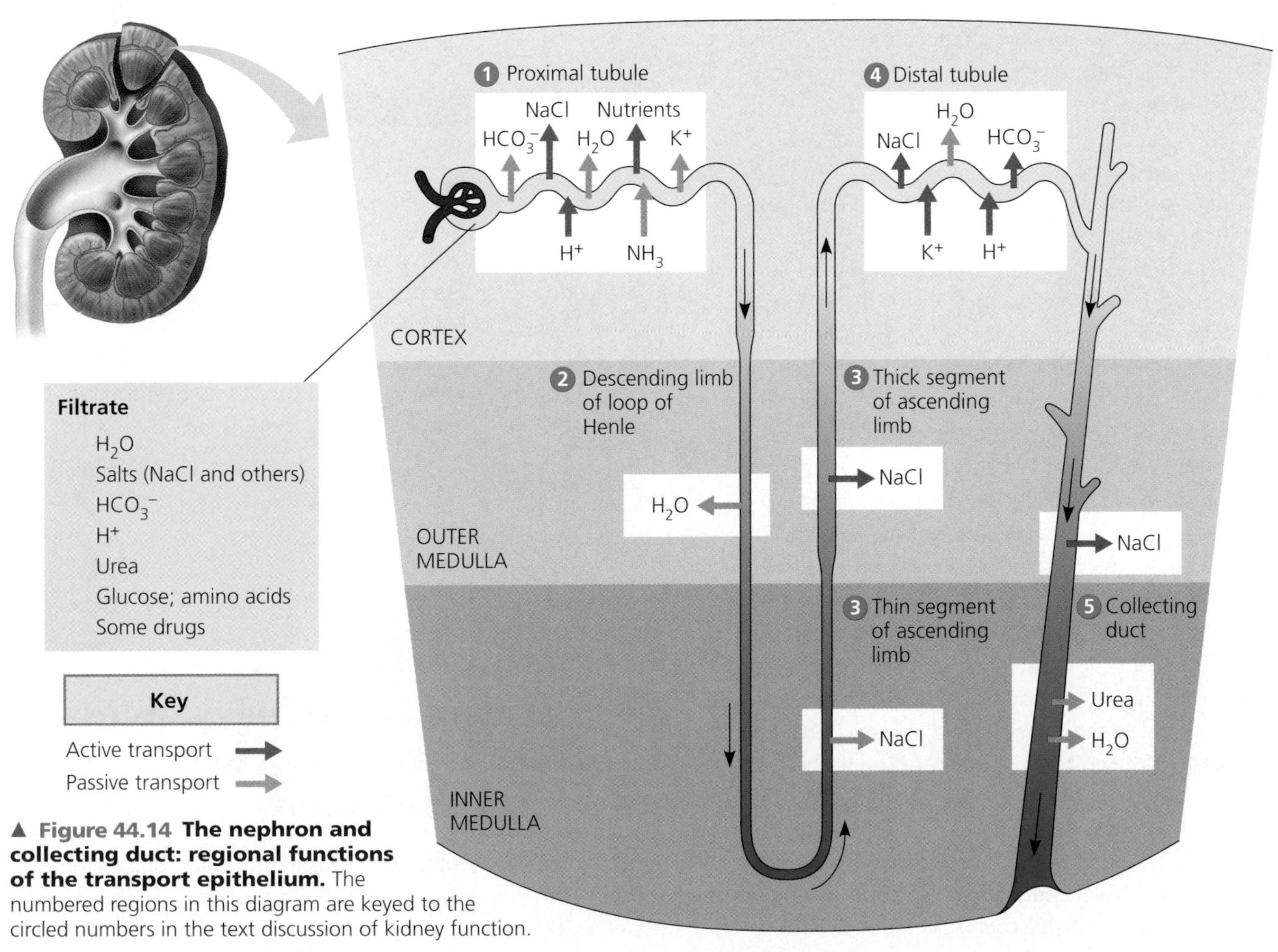

▲ **Figure 44.14 The nephron and collecting duct: regional functions of the transport epithelium.** The numbered regions in this diagram are keyed to the circled numbers in the text discussion of kidney function.

One of the most important functions of the proximal tubule is reabsorption of most of the NaCl (salt) and water from the huge initial filtrate volume. Salt in the filtrate diffuses into the cells of the transport epithelium, and the membranes of the cells actively transport Na^+ into the interstitial fluid. This transfer of positive charge is balanced by the passive transport of Cl^- out of the tubule. As salt moves from the filtrate to the interstitial fluid, water follows by osmosis. The exterior side of the epithelium has a much smaller surface area than the side facing the lumen, minimizing leakage of salt and water back into the tubule. Instead, the salt and water now diffuse from the interstitial fluid into the peritubular capillaries.

2 **Descending limb of the loop of Henle.** Reabsorption of water continues as the filtrate moves into the descending limb of the loop of Henle. Here the transport epithelium is freely permeable to water but not very permeable to salt and other small solutes. For water to move out of the tubule by osmosis, the interstitial fluid bathing the tubule must be hyperosmotic to the filtrate. The osmolarity of the interstitial fluid does in fact become progressively greater from the outer cortex to the inner medulla of the kidney. Thus, filtrate moving downward from the cortex to the medulla within the descending limb of the loop of Henle continues to lose water to interstitial fluid of greater and greater osmolarity, which increases the solute concentration of the filtrate.

3 **Ascending limb of the loop of Henle.** The filtrate reaches the tip of the loop, deep in the renal medulla in the case of juxtamedullary nephrons, then moves back to the cortex within the ascending limb. In contrast to the descending limb, the transport epithelium of the ascending limb is permeable to salt but not to water. The ascending limb has two specialized regions: a thin segment near the loop tip and a thick segment adjacent to the distal tubule. As filtrate ascends in the thin segment, NaCl, which became concentrated in the descending limb, diffuses out of the permeable tubule into the interstitial fluid. This movement increases the osmolarity of the interstitial fluid in the medulla. The exodus of salt from the filtrate continues in the thick segment of the ascending limb, but here the epithelium actively transports NaCl into the interstitial fluid. By losing salt without giving up water, the filtrate is progressively diluted as it moves up to the cortex in the ascending limb of the loop.

4 **Distal tubule.** The distal tubule plays a key role in regulating the K^+ and NaCl concentration of body fluids by varying the amount of the K^+ that is secreted into the filtrate and the amount of NaCl reabsorbed from the filtrate. Like the proximal tubule, the distal tubule also contributes to pH regulation by the controlled secretion of H^+ and reabsorption of bicarbonate (HCO_3^-).

5 **Collecting duct.** The collecting duct carries the filtrate through the medulla to the renal pelvis. By actively reabsorbing NaCl, the transport epithelium of the collecting duct plays a large role in determining how much salt is actually excreted in the urine. Though its degree of permeability is under hormonal control, the epithelium is permeable to water. However, it is not permeable to salt or, in the renal cortex, to urea. Thus, as the collecting duct traverses the gradient of osmolarity in the kidney, the filtrate becomes increasingly concentrated as it loses more and more water by osmosis to the hyperosmotic interstitial fluid. In the inner medulla, the duct becomes permeable to urea. Because of the high urea concentration in the filtrate at this point, some urea diffuses out of the duct and into the interstitial fluid. Along with NaCl, this urea contributes to the high osmolarity of the interstitial fluid in the medulla. This high osmolarity enables the mammalian kidney to conserve water by excreting urine that is hyperosmotic to the general body fluids.

Concept Check 44.4

1. How would a decrease in blood pressure in the afferent arteriole leading to a glomerulus affect the rate of filtration of blood within Bowman's capsule?
2. A variety of drugs make the epithelium of the collecting duct less permeable to water. How would this affect kidney function?
3. List these parts of a nephron in the order in which filtrate moves through them: proximal tubule, Bowman's capsule, distal tubule, loop of Henle

For suggested answers, see Appendix A.

The mammalian kidney's ability to conserve water is a key terrestrial adaptation

The osmolarity of human blood is about 300 mosm/L, but the kidney can excrete urine up to four times as concentrated—about 1,200 mosm/L. Some mammals can do even better. For example, Australian hopping mice, which live in dry desert regions, can produce urine concentrated to 9,300 mosm/L—9 times as concentrated as seawater and 25 times as concentrated as the animal's body fluid.

Solute Gradients and Water Conservation

In a mammalian kidney, the cooperative action and precise arrangement of the loops of Henle and the collecting ducts are largely responsible for the osmotic gradient that concentrates the urine. But even with this highly organized structure, the maintenance of osmotic differences and the production of hyperosmotic urine are possible only because considerable energy is expended for the active transport of solutes against concentration gradients. In essence, the nephrons—especially the loops of Henle—can be thought of as tiny energy-consuming machines whose function is to produce a region of high osmolarity in the kidney, which can then be used to extract water from the filtrate in the collecting duct. The two primary solutes in this osmolarity gradient are NaCl, which is deposited in the renal medulla by the loop of Henle, and urea, which leaks across the epithelium of the collecting duct in the inner medulla (see Figure 44.14).

To better understand the physiology of the mammalian kidney as a water-conserving organ, let's retrace the flow of filtrate through the excretory tubule, this time focusing on how the juxtamedullary nephrons maintain an osmolarity gradient in the kidney and use that gradient to excrete a hyperosmotic urine **(Figure 44.15)**. Filtrate passing from Bowman's capsule to the proximal tubule has an osmolarity of about 300 mosm/L, the same as blood. As the filtrate flows through the proximal tubule in the renal cortex, a large amount of water *and* salt is reabsorbed; thus, the volume of filtrate decreases substantially, but because of the salt loss, its osmolarity remains about the same.

As the filtrate flows from cortex to medulla in the descending limb of the loop of Henle, water leaves the tubule by osmosis. The osmolarity of the filtrate increases as solutes, including NaCl, become more concentrated. The highest osmolarity (about 1,200 mosm/L) occurs at the elbow of the loop of Henle. This maximizes the diffusion of salt out of the tubule as the filtrate rounds the curve and enters the ascending limb, which, remember, is permeable to salt but not to water. Thus, the two limbs of the loop of Henle cooperate in maintaining the gradient of osmolarity in the interstitial fluid of the kidney. The descending limb produces a progressively saltier filtrate, and then NaCl diffuses from the ascending limb to help maintain a high osmolarity in the interstitial fluid of the renal medulla.

Notice that the loop of Henle has several qualities of a countercurrent system, similar in principle to the countercurrent mechanisms that maximize oxygen absorption by fish gills (see Figure 42.21) or reduce heat loss in endotherms (see Figure 40.15). In these cases, the countercurrent mechanisms involve passive movement along either an oxygen concentration gradient or a heat gradient. In contrast, the countercurrent system involving the loop of Henle expends energy to

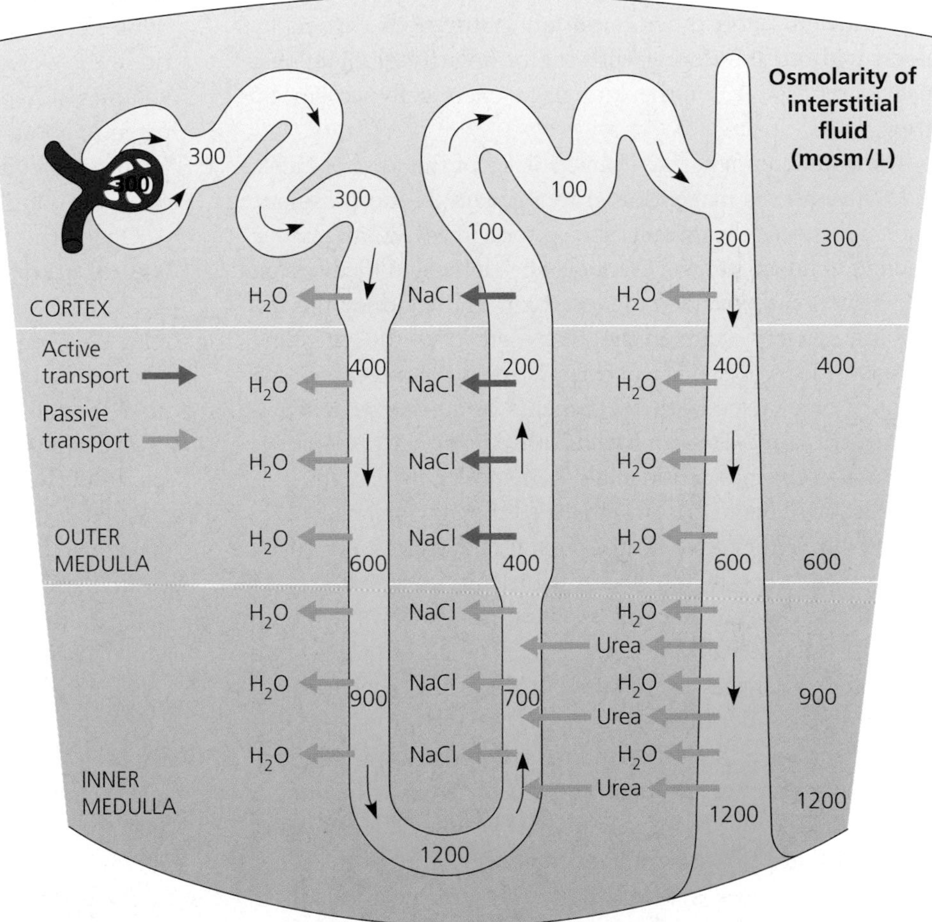

▲ **Figure 44.15 How the human kidney concentrates urine: the two-solute model.** Two solutes contribute to the osmolarity of the interstitial fluid: NaCl and urea. The loop of Henle maintains the interstitial gradient of NaCl, which increases in the descending limb and decreases in the ascending limb. Urea diffuses into the interstitial fluid of the medulla from the collecting duct (most of the urea in the filtrate remains in the collecting duct and is excreted). The filtrate makes three trips between the cortex and medulla: first down, then up, and then down again in the collecting duct. As the filtrate flows in the collecting duct past interstitial fluid of increasing osmolarity, more water moves out of the duct by osmosis, thereby concentrating the solutes, including urea, that are left behind in the filtrate.

actively transport NaCl from the filtrate in the upper part of the ascending limb of the loop. Such countercurrent systems, which expend energy to create concentration gradients, are called **countercurrent multiplier systems.** The countercurrent multiplier system involving the loop of Henle maintains a high salt concentration in the interior of the kidney, enabling the kidney to form concentrated urine.

What prevents the capillaries of the vasa recta from dissipating the gradient by carrying away the high concentration of NaCl in the medulla's interstitial fluid? Notice in Figure 44.13 that the vasa recta is also a countercurrent system, with descending and ascending vessels carrying blood in opposite directions through the kidney's osmolarity gradient. As the descending vessel conveys blood toward the inner medulla, water is lost from the blood and NaCl diffuses into it. These fluxes are reversed as blood flows back toward the cortex in the ascending vessel, with water reentering the blood and salt diffusing out. Thus, the vasa recta can supply the kidney with nutrients and other important substances carried by blood without interfering with the osmolarity gradient that makes it possible for the kidney to excrete hyperosmotic urine.

The countercurrent-like characteristics of the loop of Henle and the vasa recta make it easier to maintain the steep osmotic gradient between the medulla and cortex. However, any osmotic gradient in an animal will eventually be eliminated by diffusion unless energy is expended to preserve it. In the kidney, this expenditure largely occurs in the thick segment of the ascending limb of the loop of Henle, where NaCl is actively transported out of the tubule. Even with the benefits of countercurrent exchange, this process—along with other renal active transport systems—consumes considerable ATP, and for its size, the kidney has one of the highest metabolic rates of any organ.

By the time the filtrate reaches the distal tubule, it is actually hypoosmotic to body fluids because of active transport of NaCl out of the thick segment of the ascending limb. Now the filtrate descends again toward the medulla, this time in the collecting duct, which is permeable to water but not to salt. Therefore, osmosis extracts water from the filtrate as it passes from cortex to medulla and encounters interstitial fluid of increasing osmolarity. This concentrates salt, urea, and other solutes in the filtrate. Some urea leaks out of the lower portion of the collecting duct and contributes to the high interstitial osmolarity of the inner medulla. (This urea is recycled by diffusion into the loop of Henle, but continual leakage from the collecting duct maintains a high interstitial urea concentration.) Before leaving the kidney, the urine may attain the osmolarity of the interstitial fluid in the inner medulla, which can be as high as 1,200 mosm/L. Although *isoosmotic* to the inner medulla's interstitial fluid, the urine is *hyperosmotic* to blood and interstitial fluid elsewhere in the body. This high osmolarity allows the solutes remaining in the urine to be excreted from the body with minimal water loss.

The juxtamedullary nephron, with its urine-concentrating features, is a key adaptation to terrestrial life, enabling mammals to get rid of salts and nitrogenous wastes without squandering water. As we have seen, the remarkable ability of the mammalian kidney to produce hyperosmotic urine is completely dependent on the precise arrangement of the tubules and collecting ducts in the renal cortex and medulla. In this respect, the kidney is one of the clearest examples of how the function of an organ is inseparably linked to its structure.

Regulation of Kidney Function

One of the most important aspects of the mammalian kidney is its ability to adjust both the volume and osmolarity of urine, depending on the animal's water and salt balance and the rate of urea production. In situations of high salt intake and low water availability, a mammal can excrete urea and salt with minimal water loss in small volumes of hyperosmotic urine. But if salt is scarce and fluid intake is high, the kidney can get rid of the excess water with little salt loss by producing large volumes of hypoosmotic urine (as dilute as 70 mosm/L, compared to about 300 mosm/L for human blood). This versatility in osmoregulatory function is managed with a combination of nervous and hormonal controls.

One hormone that is important in regulating water balance is **antidiuretic hormone (ADH)** **(Figure 44.16a)**. ADH is produced in the hypothalamus of the brain and is stored in and released from the posterior pituitary gland, which is positioned just below the hypothalamus. Osmoreceptor cells in the hypothalamus monitor the osmolarity of blood; when it rises above a set point of 300 mosm/L (perhaps due to water loss from sweating or to ingestion of salty food), more ADH is released into the bloodstream and reaches the kidney. The main targets of ADH are the distal tubules and collecting ducts of the kidney, where the hormone increases the permeability of the epithelium to water. This amplifies water reabsorption, which reduces urine volume and helps prevent further increase of blood osmolarity above the set point. By negative feedback, the subsiding osmolarity of the blood reduces the activity of osmoreceptor cells in the hypothalamus, and less ADH is then secreted. But only the gain of additional water in food and drink can bring osmolarity all the way back down to 300 mosm/L.

Conversely, if a large intake of water has reduced blood osmolarity below the set point, very little ADH is released. This decreases the permeability of the distal tubules and collecting ducts, so water reabsorption is reduced, resulting in increased discharge of dilute urine. (Increased urination is called diuresis, and it is because ADH opposes this state that it is called *anti*diuretic hormone.) Alcohol can disturb water balance by inhibiting the release of ADH, causing excessive urinary water loss and dehydration (which may cause some of the symptoms of a hangover). Normally, blood osmolarity, ADH

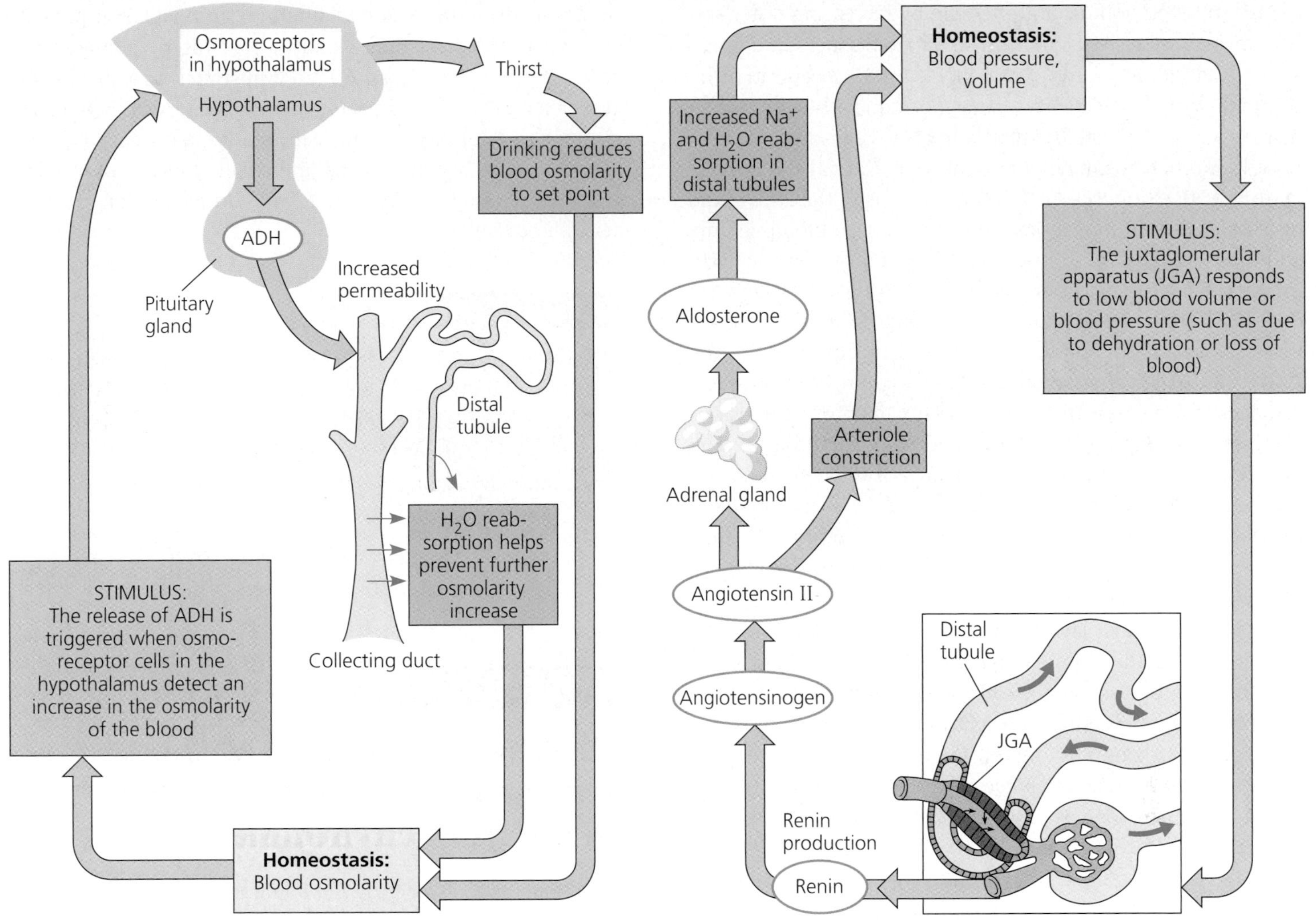

(a) Antidiuretic hormone (ADH) enhances fluid retention by making the kidneys reclaim more water.

(b) The renin-angiotensin-aldosterone system (RAAS) leads to an increase in blood volume and pressure.

▲ **Figure 44.16 Hormonal control of the kidney by negative feedback circuits.**

release, and water reabsorption in the kidney are all linked in a feedback loop that contributes to homeostasis.

A second regulatory mechanism involves a specialized tissue called the **juxtaglomerular apparatus (JGA),** located near the afferent arteriole that supplies blood to the glomerulus **(Figure 44.16b)**. When blood pressure or blood volume in the afferent arteriole drops (for instance, as a result of reduced salt intake or loss of blood), the enzyme renin initiates chemical reactions that convert a plasma protein called angiotensinogen to a peptide called **angiotensin II.** Functioning as a hormone, angiotensin II raises blood pressure by constricting arterioles, decreasing blood flow to many capillaries, including those of the kidney. Angiotensin II also stimulates the proximal tubules of the nephrons to reabsorb more NaCl and water. This reduces the amount of salt and water excreted in the urine and consequently raises blood volume and pressure. Another effect of angiotensin II is stimulation of the adrenal glands to release a hormone called **aldosterone.** This hormone acts on the nephrons' distal tubules, making them reabsorb more sodium (Na^+) and water and increasing blood volume and pressure. In summary, the **renin-angiotensin-aldosterone system (RAAS)** is part of a complex feedback circuit that functions in homeostasis. A drop in blood pressure and blood volume triggers renin release from the JGA. In turn, the rise in blood pressure and volume resulting from the various actions of angiotensin II and aldosterone reduce the release of renin.

The functions of ADH and the RAAS may seem to be redundant, but this is not the case. Both increase water reabsorption, but they counter different osmoregulatory problems. The release of ADH is a response to an increase in the osmolarity of the blood, as when the body is dehydrated from excessive water loss or inadequate intake of water. However, a situation that causes an excessive loss of both salt and body fluids—an injury, for example, or severe diarrhea—will reduce blood volume *without* increasing osmolarity. This will not induce a change in ADH release, but the RAAS will respond to the fall in blood volume and pressure by increasing water and Na^+ reabsorption. ADH and the RAAS are partners in homeostasis; ADH alone would lower blood Na^+ concentration by stimulating water reabsorption in the kidney, but the RAAS helps maintain balance by stimulating Na^+ reabsorption.

Still another hormone, a peptide called **atrial natriuretic factor (ANF),** opposes the RAAS. The walls of the atria of the heart release ANF in response to an increase in blood volume and pressure. ANF inhibits the release of renin from the JGA, inhibits NaCl reabsorption by the collecting ducts, and reduces aldosterone release from the adrenal glands. These actions lower blood volume and pressure. Thus, ADH, the RAAS, and ANF provide an elaborate system of checks and balances that regulate the kidney's ability to control the osmolarity, salt concentration, volume, and pressure of blood. The precise regulatory role of ANF is an area of active research.

The flexibility of the mammalian kidney enables it to adjust rapidly to contrasting osmoregulatory and excretory problems. The South American vampire bat illustrates this versatility **(Figure 44.17)**. Bats of this species feed on the blood of large birds and mammals. The bats use their sharp teeth to make a small incision in the victim's skin and then lap up blood from the wound. Anticoagulants in the bat's saliva prevent the blood from clotting, but the prey animal is usually not seriously harmed. Because vampire bats often search for many hours and fly for long distances to locate a suitable victim, they benefit from consuming as much blood as possible when they do find prey—so much that after feeding, a bat could be too heavy to fly. However, the bat's kidneys offload much of the water absorbed from a blood meal by excreting large volumes of dilute urine as it feeds, up to 24% of body mass per hour. Having lost enough weight to take off, the bat can fly back to its roost in a cave or hollow tree, where it spends the day. In the roost, the bat faces a very different regulatory problem. Its food is mostly protein, which generates large quantities of urea—but roosting bats don't have access to drinking water to dilute it. Instead, their kidneys shift to producing small quantities of highly concentrated urine (up to 4,600 mosm/L), an adjustment that disposes of the urea load while conserving as much water as possible. The vampire bat's ability to alternate rapidly between producing large amounts of dilute urine and small amounts of very hyperosmotic urine is an essential part of its adaptation to an unusual food source.

▲ **Figure 44.17 A vampire bat (*Desmodus rotundas*), a mammal with a unique excretory situation.**

Concept Check 44.5

1. How does alcohol affect regulation of water balance in the body?
2. How does eating salty food affect kidney function?
3. Identify a major functional consequence of the countercurrent-like characteristics of the loop of Henle.

For suggested answers, see Appendix A.

Concept 44.6

Diverse adaptations of the vertebrate kidney have evolved in different environments

Vertebrate animals occupy habitats ranging from rain forests to deserts and from some of the saltiest bodies of water to the dilute waters of high mountain lakes. Variations in nephron structure and function equip the kidneys of different vertebrates for osmoregulation in their various habitats. The adaptations of the vertebrate kidney are most clearly revealed by comparing species that inhabit a wide range of environments or by comparing the responses of different vertebrate groups to similar environmental conditions **(Figure 44.18)**.

In all animals, the intricate physiological machines we call organs work continuously, maintaining solute and water balance and excreting nitrogenous wastes. The details that we have reviewed in this chapter only hint at the great complexity of the neural and hormonal mechanisms involved in regulating these homeostatic processes. The next chapter further explores the hormonal control of homeostasis.

Concept Check 44.6

1. What do the number and length of nephrons indicate about the habitat of fishes? What do they indicate about rates of urine production?

For suggested answers, see Appendix A.

Figure 44.18

Exploring Environmental Adaptations of the Vertebrate Kidney

MAMMALS

Bannertail kangaroo rat (*Dipodomys spectabilis*)

Mammals that excrete the most hyperosmotic urine, such as Australian hopping mice, North American kangaroo rats, and other desert mammals, have exceptionally long loops of Henle. Long loops maintain steep osmotic gradients in the kidney, resulting in urine becoming very concentrated as it passes from cortex to medulla in the collecting ducts.

In contrast, beavers, muskrats, and other aquatic mammals that spend much of their time in fresh water and rarely face problems of dehydration have nephrons with very short loops, resulting in a much lower ability to concentrate urine. Terrestrial mammals living in moist conditions have loops of Henle of intermediate length and the capacity to produce urine intermediate in concentration to that produced by freshwater and desert mammals.

Beaver (*Castor canadensis*)

BIRDS AND OTHER REPTILES

Roadrunner (*Geococcyx californianus*)

Birds, like mammals, have kidneys with juxtamedullary nephrons that specialize in conserving water. However, the nephrons of birds have much shorter loops of Henle; thus, bird kidneys cannot concentrate urine to the high osmolarities achieved by mammalian kidneys. Although they can produce hyperosmotic urine, the main water conservation adaptation of birds is having uric acid as the nitrogen waste molecule: Since uric acid can be excreted as a paste, it reduces urine volume.

The kidneys of other reptiles, having only cortical nephrons, produce urine that is, at most, isoosmotic to body fluids. However, the epithelium of the cloaca (see Chapter 34) helps conserve fluid by reabsorbing some of the water present in urine and feces. Also like birds, most other terrestrial reptiles excrete nitrogenous wastes as uric acid.

Desert iguana (*Dipsosaurus dorsalis*)

FRESHWATER FISHES AND AMPHIBIANS

Rainbow trout (*Oncorrhynchus mykiss*)

Because they are hyperosmotic to their surroundings, freshwater fishes must excrete excess water continuously. In contrast to mammals and birds, freshwater fishes produce large volumes of very dilute urine. Their kidneys, which have a large number of nephrons, produce filtrate at a high rate. Freshwater fishes conserve salts by reabsorbing ions from the filtrate in their distal tubules, leaving water behind.

Amphibian kidneys function much like those of freshwater fishes. When in fresh water, the skin of the frog accumulates certain salts from the water by active transport, and the kidneys excrete dilute urine. On land, where dehydration is the most pressing problem of osmoregulation, frogs conserve body fluid by reabsorbing water across the epithelium of the urinary bladder.

Frog (*Rana temporaria*)

MARINE BONY FISHES

Northern bluefin tuna (*Thunnus thynnus*)

Because they are hypoosmotic to seawater, marine bony fishes lose body water and gain excess salts from their surroundings, environmental challenges opposite those faced by their freshwater relatives. Compared to freshwater fishes, marine fishes have fewer and smaller nephrons, which lack a distal tubule. In addition, the kidneys of most marine fishes have small glomeruli, and some lack glomeruli entirely. Thus, the kidneys of marine fishes have low filtration rates and excrete very little urine. The main function of the kidneys is to get rid of double-charged ions such as calcium (Ca^{2+}), magnesium (Mg^{2+}), and sulfate (SO_4^{2-}), which the fish takes in by its incessant drinking of seawater. Marine fishes rid themselves of these ions by secreting them into the proximal tubules of the nephrons and excreting them with the urine.

Chapter 44 Review

Go to www.campbellbiology.com or the student CD-ROM to explore Activities, Investigations, and other interactive study aids.

SUMMARY OF KEY CONCEPTS

Concept 44.1

Osmoregulation balances the uptake and loss of water and solutes

▶ Osmoregulation is based largely on the controlled movement of solutes between internal fluids and the external environment, and the movement of water, which follows by osmosis (p. 922).

▶ **Osmosis** **(pp. 922–923)** Cells require a balance between osmotic gain and loss of water. Water uptake and loss are balanced by various mechanisms of osmoregulation in different environments.

▶ **Osmotic Challenges** **(pp. 923–926)** Osmoconformers, which are only marine animals, are isoosmotic with their surroundings and do not regulate their osmolarity. Osmoregulators expend energy to control water uptake and loss in a hypoosmotic or hyperosmotic environment, respectively. Among marine animals, most invertebrates are osmoconformers. Sharks have an osmolarity slightly higher than seawater because they retain urea. Marine bony fishes lose water to their hyperosmotic environment and drink seawater. Marine vertebrates excrete excess salt through rectal glands, gills, salt-excreting glands, or kidneys. Freshwater animals, which constantly take in water from their hypoosmotic environment, excrete dilute urine. Lost salt is replaced by eating or by ion uptake by gills. Terrestrial animals combat desiccation through behavioral adaptations, water-conserving excretory organs, and drinking and eating food with high water content. Animals in temporary waters may be anhydrobiotic.

▶ **Transport Epithelia** **(pp. 926–927)** Water balance and waste disposal depend on transport epithelia, layers of specialized epithelial cells that regulate the solute movements required for waste disposal and for tempering changes in body fluids.

Concept 44.2

An animal's nitrogenous wastes reflect its phylogeny and habitat

▶ **Forms of Nitrogenous Waste** **(pp. 927–928)** Protein and nucleic acid metabolism generates ammonia, a toxic waste product. Most aquatic animals excrete ammonia across the body surface or gill epithelia into the surrounding water. The liver of mammals and most adult amphibians converts ammonia to the less toxic urea, which is carried to the kidneys, concentrated, and excreted with a minimal loss of water. Uric acid is a slightly soluble nitrogenous waste excreted in the paste-like urine of land snails, insects, and many reptiles, including birds.

▶ **The Influence of Evolution and Environment on Nitrogenous Wastes** **(p. 928)** The kind of nitrogenous waste excreted depend on an animal's evolutionary history and habitat. The amount of nitrogenous waste produced is coupled to the animal's energy budget and amount of dietary protein.

Diverse excretory systems are variations on a tubular theme

▶ **Excretory Processes** **(p. 929)** Most excretory systems produce urine by refining a filtrate derived from body fluids. Key functions of most excretory systems are filtration (pressure-filtering of body fluids, producing a filtrate) and the production of urine from the filtrate by selective reabsorption, (reclaiming valuable solutes from the filtrate) and secretion (addition of toxins and other solutes from the body fluids to the filtrate).

▶ **Survey of Excretory Systems** **(pp. 929–931)** Extracellular fluid is filtered into the protonephridia of the flame-bulb system in flatworms; these tubules excrete a dilute fluid and also function in osmoregulation. Each segment of an earthworm has a pair of open-ended metanephridia, tubules that collect coelomic fluid and produce dilute urine for excretion. In insects, Malpighian tubules function in osmoregulation and removal of nitrogenous wastes from the hemolymph. Insects produce a relatively dry waste matter, an important adaptation to terrestrial life. Kidneys, the excretory organs of vertebrates, function in both excretion and osmoregulation.

Concept 44.4

Nephrons and associated blood vessels are the functional units of the mammalian kidney

▶ **Structure and Function of the Nephron and Associated Structures** **(pp. 931–933)** Excretory tubules (consisting of nephrons and collecting ducts) and associated blood vessels pack the kidney. Filtration occurs as blood pressure forces fluid from the blood in the glomerulus into the lumen of Bowman's capsule. Filtration of small molecules is nonselective, and the filtrate in Bowman's capsule contains a mixture of small molecules that mirrors the concentrations of these substances in blood plasma. Fluid from several nephrons flows into a collecting duct. A ureter conveys urine from the renal pelvis to the urinary bladder.

▶ **From Blood Filtrate to Urine: *A Closer Look*** **(pp. 933–934)** Nephrons control the composition of the blood by filtration, secretion, and reabsorption. Secretion and reabsorption in the proximal tubule substantially alter the volume and composition of filtrate. The descending limb of the loop of Henle is permeable to water but not to salt; water moves by osmosis into the hyperosmotic interstitial fluid. Salt diffuses out of the concentrated filtrate as it moves through the salt-permeable ascending limb of the loop of Henle. The distal tubule plays a key role in regulating the K^+ and NaCl concentration of body fluids. The collecting duct carries the filtrate through the medulla to the renal pelvis and reabsorbs NaCl.
Activity ***Structure of the Human Excretory System***
Activity ***Nephron Function***

Concept 44.5

The mammalian kidney's ability to conserve water is a key terrestrial adaptation

▶ **Solute Gradients and Water Conservation** **(pp. 935–936)** In a mammalian kidney, the cooperative action and precise arrangement of the loops of Henle and the collecting ducts are largely responsible for the osmotic gradient that concentrates the urine. The countercurrent multiplier system involving the loop of Henle maintains a high salt concentration in the interior of the kidney, which enables the kidney to form concentrated urine. The collecting duct, permeable to water but not to salt, conducts the filtrate through the kidney's osmolarity gradient, and more water exits the filtrate by osmosis. Urea, which diffuses out of the collecting duct as it traverses the inner medulla, forms, along with NaCl, the osmotic gradient that enables the kidney to produce urine that is hyperosmotic to the blood.

▶ **Regulation of Kidney Function (pp. 936–938)** The osmolarity of the urine is regulated by nervous system and hormonal control of water and salt reabsorption in the kidneys. This regulation involves the actions of antidiuretic hormone (ADH), the renin-angiotensin-aldosterone system (RAAS), and atrial natriuretic factor (ANF).
Activity ***Control of Water Reabsorption***
Investigation ***What Affects Urine Production?***

Concept 44.6

Diverse adaptations of the vertebrate kidney have evolved in different environments

▶ The form and function of nephrons in various vertebrates are related primarily to the requirements for osmoregulation in the animal's habitat. Desert mammals, which excrete the most hyperosmotic urine, have exceptionally long loops of Henle, whereas mammals living in moist or aquatic habitats have short loops and excrete less concentrated urine. Although birds can produce a hyperosmotic urine, the main water conservation adaptation of birds is removal of nitrogen as uric acid, which can be excreted as a paste. Most other terrestrial reptiles excrete uric acid. Freshwater fishes and amphibians produce large volumes of very dilute urine. The kidneys of marine bony fishes have low filtration rates and excrete very little urine. (pp. 938–939)

TESTING YOUR KNOWLEDGE

Self-Quiz

1. *Unlike* an earthworm's metanephridia, a mammalian nephron
   a. is intimately associated with a capillary network.
   b. forms urine by changing the composition of fluid inside the tubule.
   c. functions in both osmoregulation and the excretion of nitrogenous wastes.
   d. receives filtrate from blood instead of coelomic fluid.
   e. has a transport epithelium.

2. Which of the following is not a normal response to increased blood osmolarity in humans?
   a. increased permeability of the collecting duct to water
   b. increased thirst
   c. release of ADH by the pituitary gland
   d. production of more dilute urine
   e. reduced urine production

3. The high osmolarity of the renal medulla is maintained by all of the following *except*
   a. diffusion of salt from the ascending limb of the loop of Henle.
   b. active transport of salt from the upper region of the ascending limb.
   c. the spatial arrangement of juxtamedullary nephrons.
   d. diffusion of urea from the collecting duct.
   e. diffusion of salt from the descending limb of the loop of Henle.

4. Select the pair in which the nitrogenous waste is *incorrectly* matched with the benefit of its excretion.
   a. urea—low toxicity relative to ammonia
   b. uric acid—can be stored as a precipitate
   c. ammonia—very soluble in water
   d. uric acid—minimal loss of water when excreted
   e. urea—very insoluble in water

5. The body fluids of freshwater crustaceans generally have a lower osmolarity than the body fluids of their nearest marine crustacean relatives. Which of the following is a benefit of reduced osmolarity of body fluids in freshwater crustaceans?
   a. an increase in the rate of water flow into the body fluids
   b. a decrease in the rate of water loss to the surrounding environment
   c. a reduction in energy expenditures for osmoregulation
   d. an increase in the rate of salt loss to the surrounding environment
   e. a decrease in the rate of salt gain from the environment

6. Which process in the nephron is *least* selective?
   a. secretion
   b. reabsorption
   c. active transport
   d. filtration
   e. salt pumping by the loop of Henle

7. Which of the following animals generally has the lowest volume of urine production?
   a. a marine shark
   b. a salmon in freshwater
   c. a marine bony fish
   d. a freshwater bony fish
   e. a shark inhabiting freshwater Lake Nicaragua

8. African lungfish, which are often found in small stagnant pools of fresh water, produce urea as a nitrogenous waste. What is the advantage of this adaptation?
   a. Urea takes less energy to synthesize than ammonia.
   b. Small stagnant pools do not provide enough water to dilute the toxic ammonia.
   c. The highly toxic urea makes the pool uninhabitable to potential competitors.
   d. Urea forms a precipitate and does not accumulate in the surrounding water.
   e. A buildup of urea in the blood makes a lungfish hypoosmotic to its environment.

9. Natural selection should favor the highest proportion of juxtamedullary nephrons in which of the following species?
   a. a river otter
   b. a mouse species living in a tropical rain forest
   c. a mouse species living in a temperate broadleaf forest
   d. a mouse species living in the desert
   e. a beaver

10. A clinical condition known as diabetes insipidus results in the production of large volumes of dilute urine. Which of the following is consistent with this condition?
    a. high concentration of sodium in the urine
    b. very low production of ADH
    c. overproduction of ADH
    d. high production of aldosterone
    e. high production of angiotensin II

For Self-Quiz answers, see Appendix A.

Go to the website or CD-ROM for more quiz questions.

Evolution Connection

A large part of the evolutionary success of arthropods and vertebrates on land is attributable to their osmoregulatory capabilities. Compare and contrast the Malpighian tubule with the nephron in regard to anatomy, relationship to circulation, and physiological mechanisms for conserving body water.

Scientific Inquiry

Merriam's kangaroo rats (*Dipodomys merriami*) are native to western North America, where they live in a wide range of habitats ranging from moist, cool woodlands to the hottest and driest places on the continent. Assuming that natural selection acting on local populations has resulted in differences in water conservation among Merriam's kangaroo rat populations, propose a hypothesis concerning the relative rates of evaporative water loss by populations that live in dry versus moist environments. Using a humidity sensor to detect evaporative water loss by kangaroo rats, how could you test your hypothesis?

Investigation ***What Affects Urine Production?***

Science, Technology, and Society

Kidneys were the first organs to be successfully transplanted. A donor can live a normal life with a single kidney, making it possible for individuals to donate a kidney to an ailing relative or even an unrelated individual with a similar tissue type. In some countries, poor people *sell* kidneys to transplant recipients through organ brokers. What are some of the ethical issues associated with this organ commerce?

45 Hormones and the Endocrine System

▲ **Figure 45.1 An anise swallowtail butterfly recently emerged from its chrysalis.**

Key Concepts

45.1 **The endocrine system and the nervous system act individually and together in regulating an animal's physiology**

45.2 **Hormones and other chemical signals bind to target cell receptors, initiating pathways that culminate in specific cell responses**

45.3 **The hypothalamus and pituitary integrate many functions of the vertebrate endocrine system**

45.4 **Nonpituitary hormones help regulate metabolism, homeostasis, development, and behavior**

45.5 **Invertebrate regulatory systems also involve endocrine and nervous system interactions**

Overview

The Body's Long-Distance Regulators

People offer hormones as an explanation for the howling of alley cats and the moodiness of teenagers. In the United States more than a million people with diabetes take the hormone insulin, and other hormones are used in cosmetics to keep the skin smooth or in livestock feed to fatten cattle. These powerful substances are also involved in even more dramatic transformations. In becoming an adult, a butterfly like the one in **Figure 45.1** undergoes a complete change of body form, a metamorphosis regulated by hormones. Internal communication involving those hormones enables different parts of the insect's adult body to develop in concert.

An animal **hormone** (from the Greek *horman*, to excite) is a chemical signal that is secreted into the extracellular fluid, is carried by the circulatory system (in blood or hemolymph) and communicates regulatory messages within the body. A hormone may reach all parts of the body, but only certain types of cells, the target cells, are equipped to respond. Thus, a given hormone traveling through the body elicits specific responses—such as a change in metabolism—from its target cells, while other cell types are unaffected by that particular hormone.

In this chapter, we will describe how the basic concepts of biological control systems apply to hormonal pathways and how hormones act on target cells. Here we will focus on hormones that help maintain homeostasis; in Chapters 46 and 47 we will discuss the role of hormones in regulating growth, development, and reproduction. This chapter will examine the major types of hormones in vertebrates, including where they are formed in the body and their major effects. In addition, we will consider comparable regulatory mechanisms in invertebrates.

Concept 45.1

The endocrine system and the nervous system act individually and together in regulating an animal's physiology

Animals have two systems of internal communication and regulation, the nervous system and the endocrine system. The nervous system, which we will discuss in Chapter 48, conveys high-speed electrical signals along specialized cells called neurons. These rapid messages control the movement of body parts in response to sudden environmental changes, such as occur when you jerk your hand away from a hot pan or when your pupils dilate as you enter a dark room.

Collectively, all of an animal's hormone-secreting cells constitute its **endocrine system.** Hormones coordinate slower but longer-acting responses to stimuli such as stress, dehydration, and low blood glucose levels. Hormones also regulate long-term developmental processes by informing different parts of the body how fast to grow or when to develop the characteristics that distinguish male from female or juvenile from adult. Hormone-secreting organs, called **endocrine glands,** are referred to as ductless glands because they secrete their chemical messengers directly into extracellular fluid. From there, the chemicals diffuse into the circulation.

Overlap Between Endocrine and Nervous Regulation

While it is convenient to distinguish between the endocrine and nervous systems, in reality the lines between these two regulatory systems are blurred. In particular, certain specialized nerve cells known as **neurosecretory cells** release hormones into the blood via the extracellular fluid. In animals as distinct as insects and vertebrates, a part of the brain contains neurosecretory cells. The hormones produced by neurosecretory cells are sometimes called *neurohormones* to distinguish them from the "classic" hormones released by endocrine glands.

A few chemicals serve both as hormones in the endocrine system and as chemical signals in the nervous system. Epinephrine, for example, functions in the vertebrate body as the so-called "fight-or-flight" hormone (produced by the adrenal medulla, an endocrine gland) and as a neurotransmitter, a local chemical signal that conveys messages in the nervous system (see Chapter 48). In addition, the nervous system plays a role in certain sustained responses—for example, controlling day/night cycles and reproductive cycles in many animals—often by increasing or decreasing secretion from endocrine glands.

Thus, although the endocrine and nervous systems are anatomically distinct, they interact functionally in regulating a number of physiological processes.

Control Pathways and Feedback Loops

Let's review the fundamental concepts of biological control systems introduced in Chapter 40 and apply them to regulation by hormones. A *receptor,* or *sensor,* detects a stimulus—for example, a change in blood calcium level—and sends the information to a *control center*. After comparing the incoming information to a set point, or "desired" value, the control center sends out a signal that directs an *effector* to respond. In endocrine and neuroendocrine pathways, this outgoing signal, called an *efferent signal,* is a hormone or neurohormone, which acts on particular effector tissues and elicits specific physiological or developmental changes. The three types of simple hormonal pathways depicted in **Figure 45.2** include these basic functional components of a control system. Not depicted in this figure are complex neuroendocrine pathways in which a hormone secreted by one endocrine tissue acts on another endocrine tissue, controlling its release of a different hormone, which then acts on target tissues. Regulation by each of the 20 or so different hormones you will learn about in this and other chapters involves one of these general types of simple or complex pathways.

Another common feature of control pathways is a feedback loop connecting the response to the initial stimulus. In **negative feedback,** the effector response reduces the initial stimulus, and eventually the response ceases. This feedback mechanism prevents overreaction by the system and wild fluctuations in the variable being regulated. Negative feedback operates in many endocrine and nervous pathways, especially those involved in maintaining homeostasis (see Chapter 40). Later in this chapter, we will examine how negative feedback contributes to the hormonal control of blood calcium and glucose levels.

In contrast to negative feedback, which dampens the stimulus, positive feedback reinforces the stimulus and leads to an even greater response. The neurohormone pathway that regulates the release of milk by a nursing mother is an example of positive feedback (see Figure 45.2b). Suckling stimulates sensory nerve cells in the nipples, which send nervous signals that eventually reach the hypothalamus, the control center. An outgoing signal from the hypothalamus triggers the release of the neurohormone oxytocin from the posterior pituitary gland. Oxytocin then causes the mammary glands to secrete milk. The release of milk in turn leads to more suckling and stimulation of the pathway, until the baby is satisfied.

Concept Check 45.1

1. How do neurohormones differ from "classic" hormones? How are they similar?
2. Different biological control systems exhibit common features: a receptor/sensor, a control center, an efferent signal, and an effector. Draw two sketches showing how these components are arranged in a simple endocrine pathway and a simple neurohormone pathway.
3. Explain why, unlike negative feedback, positive feedback is not a common feature of hormonal pathways that help maintain homeostasis.

For suggested answers, see Appendix A.

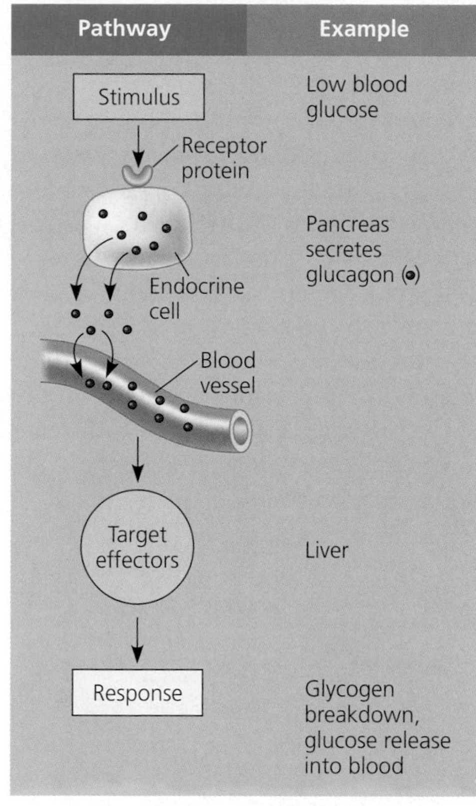

(a) Simple endocrine pathway

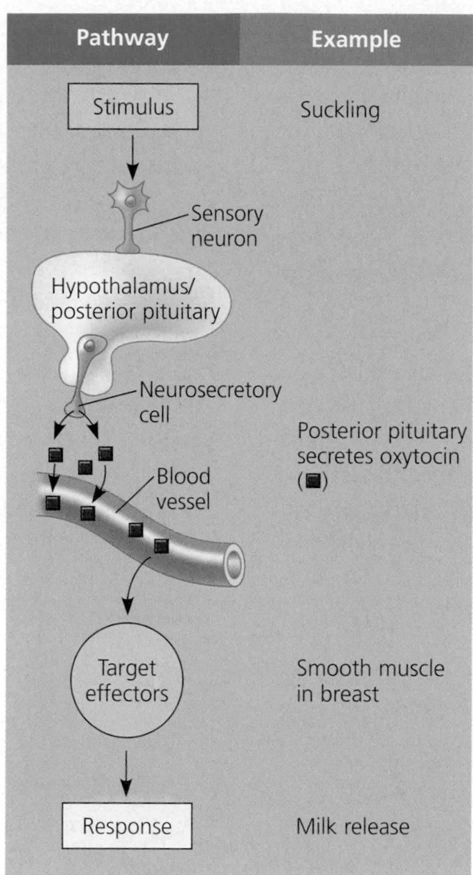

(b) Simple neurohormone pathway

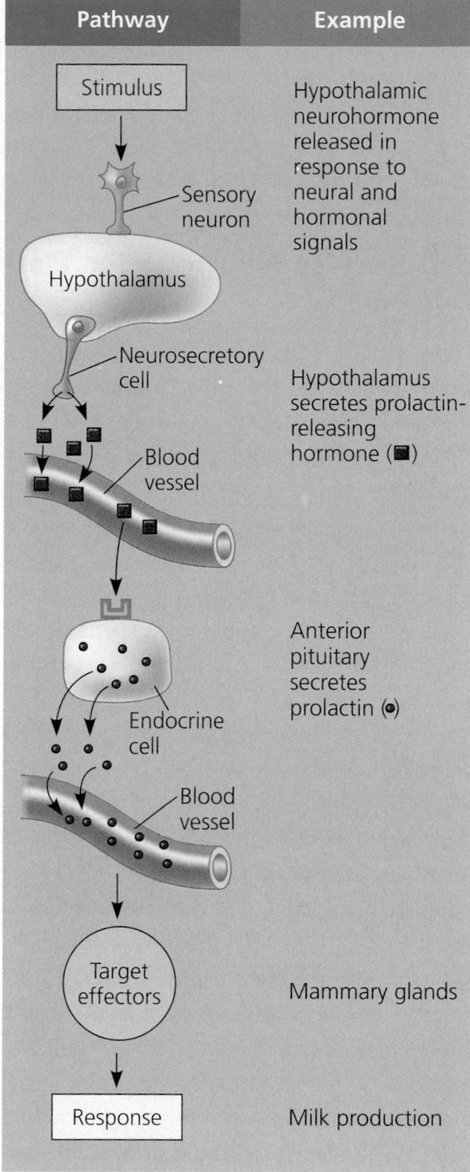

(c) Simple neuroendocrine pathway

▲ **Figure 45.2 Basic patterns of simple hormonal control pathways.** In each pathway, a receptor/sensor (blue) detects a change in some internal or external variable—the stimulus—and informs the control center (gold). The control center sends out an efferent signal, either a hormone (red circles) or neurohormone (red squares). An endocrine cell carries out *both* the receptor and control center functions.

Hormones and other chemical signals bind to target cell receptors, initiating pathways that culminate in specific cell responses

Hormones, the body's long-distance chemical regulators, convey information via the bloodstream to target cells throughout the body. Other chemical signals, called local regulators, transmit information to target cells near the secreting cells. Still other chemical signals, called pheromones, carry messages between different individuals of a species, as in mate attraction. In this chapter, as mentioned earlier, we will concentrate on hormones (and neurohormones) that are not directly involved in reproduction.

Three major classes of molecules function as hormones in vertebrates: proteins and peptides (small polypeptides containing up to 30 amino acids); amines derived from amino acids; and steroids. Most protein/peptide and amine hormones are water-soluble, whereas steroid hormones are not.

Regardless of their chemical nature, however, signaling by any of these molecules involves three key events: reception, signal transduction, and response (see Chapter 11). *Reception* of the signal occurs when the signal molecule binds to a specific receptor protein in or on the target cell. Each signal molecule has a specific shape that can be recognized by that signal's receptors. Receptors may be located in the plasma membrane of a target cell or inside the cell. The binding of a signal molecule to a receptor protein triggers events within the target cell—*signal transduction*—that result in a *response,* a change in the cell's behavior. Cells that lack receptors for a particular chemical signal are unresponsive to that signal.

Let's take a closer look at signal transduction and the kinds of cell responses induced by different types of chemical signals.

Cell-Surface Receptors for Water-Soluble Hormones

The receptors for most water-soluble hormones are embedded in the plasma membrane, projecting outward from the cell surface **(Figure 45.3a)**. Binding of a hormone to its receptor initiates a **signal transduction pathway,** a series of changes in cellular proteins that converts an extracellular chemical signal to a specific intracellular response. Depending on the hormone and target cell, the response may be the activation of an enzyme, a change in the uptake or secretion of specific molecules, or rearrangement of the cytoskeleton. Signal transduction from some cell-surface receptors activates proteins in the cytoplasm that then move into the nucleus and directly or indirectly regulate transcription of specific genes.

Early evidence for the role of cell-surface receptors in triggering signal transduction pathways came from studies on how the hormone epinephrine stimulates breakdown of glycogen to glucose (see Chapter 11). Another demonstration of the role of cell-surface receptors involves changes in a frog's skin color, an adaptation that helps camouflage the frog in changing light. Skin cells called melanocytes contain the dark brown pigment melanin in cytoplasmic organelles called melanosomes. The frog's skin appears light when melanosomes cluster tightly around the cell nuclei and darker when melanosomes spread throughout the cytoplasm. A peptide hormone called melanocyte-stimulating hormone controls the arrangement of melanosomes and thus the frog's skin color. Adding melanocyte-stimulating hormone to the interstitial fluid surrounding the pigment-containing cells causes the melanosomes to disperse. However, direct microinjection of melanocyte-stimulating hormone into individual melanocytes does not induce melanosome dispersion—evidence that interaction between the hormone and a *surface* receptor is required for hormone action.

A particular hormone may cause diverse responses in target cells with different receptors for the hormone, different signal transduction pathways, and/or different proteins for carrying out the response. Consider the multiple effects of epinephrine in mediating the body's response to short-term stress **(Figure 45.4)**. For example, liver cells and the smooth muscle of blood vessels supplying skeletal muscle contain β-type epinephrine receptors, whereas the smooth muscle of intestinal blood vessels have α-type epinephrine receptors. These tissues respond differently to epinephrine, resulting in decreased blood flow to the digestive tract and increased delivery of glucose to major skeletal muscles. These effects help the body react quickly in emergencies.

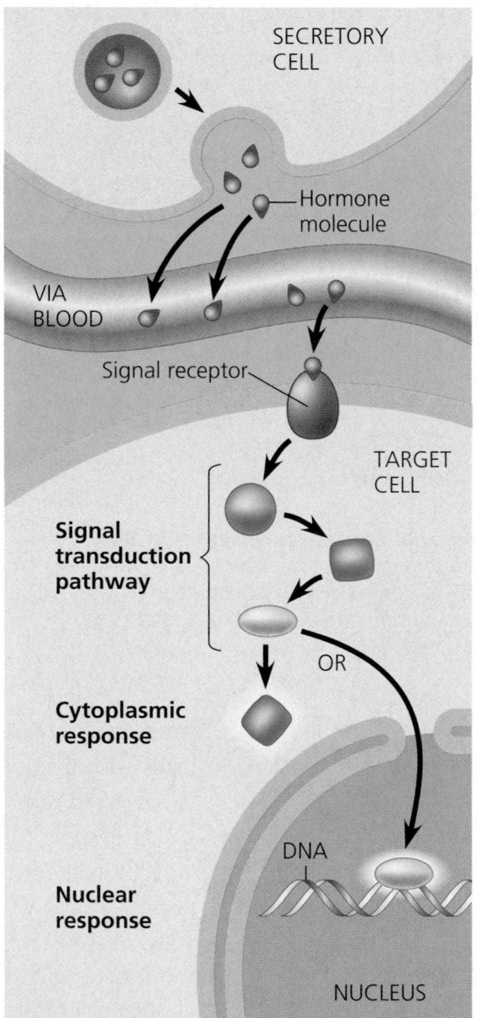

(a) Receptor in plasma membrane

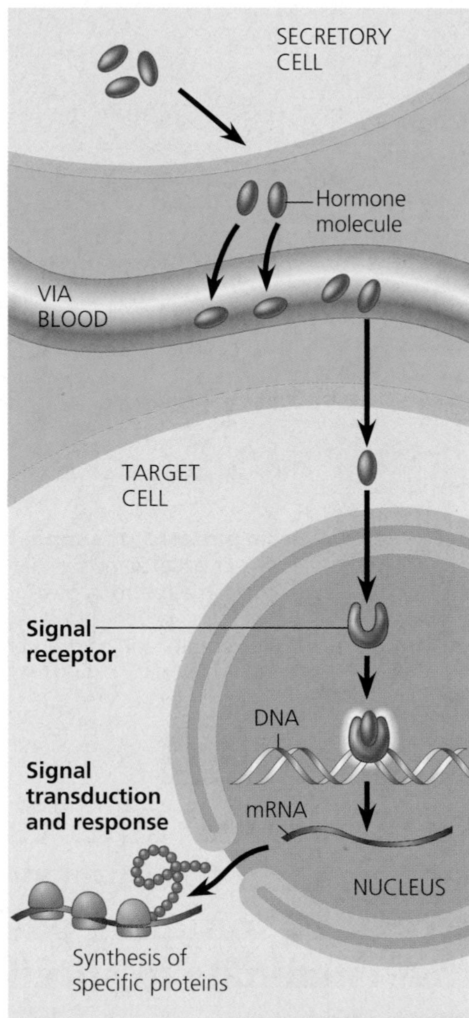

(b) Receptor in cell nucleus

▲ **Figure 45.3 Mechanisms of hormonal signaling: a review. (a)** A water-soluble hormone binds to a receptor protein on the surface of a target cell. This interaction triggers a signal transduction pathway that leads to a change in a cytoplasmic function or a change in gene transcription in the nucleus. **(b)** A lipid-soluble hormone penetrates the target cell's plasma membrane and binds to an intracellular receptor, either in the cytoplasm or in the nucleus (shown here). The signal-receptor complex acts as a transcription factor, typically activating gene expression.

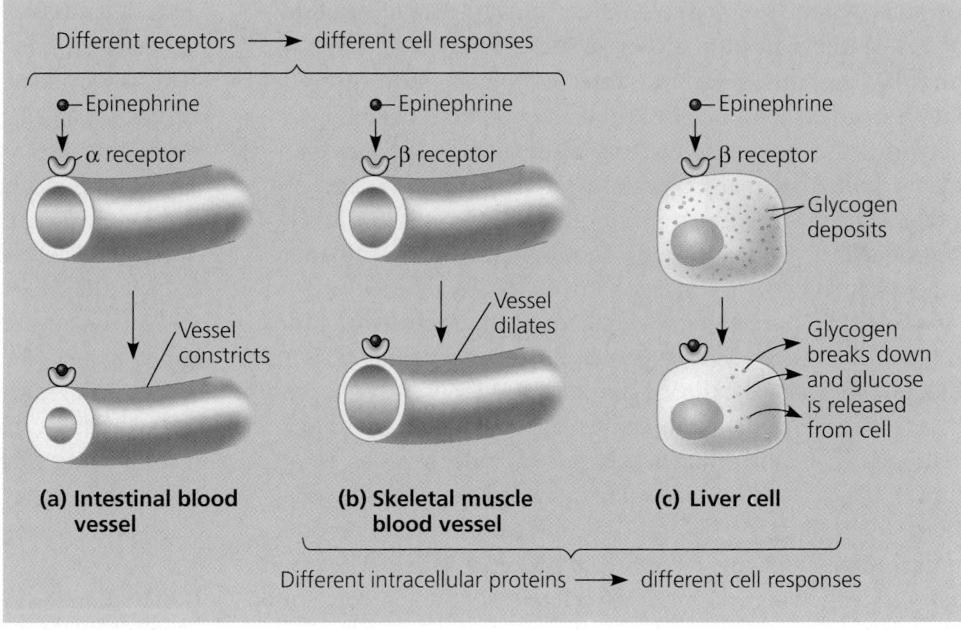

▶ **Figure 45.4 One chemical signal, different effects.** Epinephrine, the primary "fight-or-flight" hormone, produces different responses in different target cells. Responses of target cells may differ if they have different receptors for a hormone [compare (a) with (b)]. Target cells with the same receptor exhibit different responses if they have different signal transduction pathways and/or effector proteins [compare (b) with (c)].

Intracellular Receptors for Lipid-Soluble Hormones

The first indication that the receptors for some hormones are located inside target cells came from studying the vertebrate hormones estrogen and progesterone. For most mammals, including humans, these steroid hormones are necessary for the normal development and function of the female reproductive system. In the early 1960s, researchers demonstrated that estrogen and progesterone accumulate within the nuclei of cells in the reproductive tract of female rats. By contrast, no estrogen accumulated in the cells of tissues that do not respond to estrogen. These observations led to the hypothesis that cells sensitive to a steroid hormone contain internal receptor molecules that bind specifically to that hormone.

Researchers later identified the intracellular proteins that function as receptors for steroid hormones, thyroid hormones, and the hormonal form of vitamin D. All of these hormones are small, mostly nonpolar (hydrophobic) molecules that diffuse easily through the hydrophobic interior of cellular membranes.

Intracellular receptors usually perform the entire task of transducing a signal within a target cell. The chemical signal activates the receptor, which then directly triggers the cell's response. In almost all cases, the intracellular receptor activated by a lipid-soluble hormone is a transcription factor, and the response is a change in gene expression.

Most intracellular receptors are already located in the nucleus (**Figure 45.3b**, on page 946) when they bind hormone molecules, which have diffused in from the bloodstream via the extracellular fluid. The resulting hormone-receptor complexes bind, in turn, to specific sites in the cell's DNA and stimulate the transcription of specific genes. Some steroid hormone receptors, however, are trapped in the cytoplasm when no hormone is present. Binding of a steroid hormone to its cytoplasmic receptor forms a hormone-receptor complex that can move into the nucleus and stimulate transcription of specific genes (see Figure 11.6). In both cases, mRNA produced in response to hormone stimulation is translated into new protein in the cytoplasm. For example, estrogen induces cells in the reproductive system of a female bird to synthesize large amounts of ovalbumin, the main protein of egg white.

As with hormones that bind to cell-surface receptors, hormones that bind to intracellular receptors may exert different effects on different target cells. The estrogen that stimulates a bird's reproductive system to make ovalbumin causes the bird's liver to make other proteins. The same hormone also may have different effects in different *species*. For instance, thyroxine produced by the thyroid gland regulates metabolism in humans and other vertebrates. But in frogs, thyroxine has additional effects: it triggers the metamorphosis of a tadpole into an adult, stimulating resorption of the tadpole's tail and other changes.

Paracrine Signaling by Local Regulators

Before continuing our discussion of the endocrine system and hormonal regulation, let us briefly consider local regulators. In contrast to long-distance endocrine signaling by hormones, local regulators convey messages between neighboring cells—a process referred to as paracrine signaling (see Figure 11.4). Once secreted by the cells that make them, local regulators act on nearby target cells within seconds or even milliseconds, eliciting cell responses more quickly than hormones can. Some local regulators have cell-surface receptors; others have intracellular receptors. Binding of local regulators to their specific receptors triggers events within target cells similar to those elicited by hormones (see Figure 45.3).

Several types of chemical compounds function as local regulators. Many neurotransmitters, the key local regulators in the nervous system, are amino acid derivatives. Among peptide/protein local regulators are **cytokines,** which play a role in immune responses (see Chapter 43), and most **growth factors,** which stimulate cell proliferation and differentiation. Growth factors must be present in the extracellular environment in order for many types of cells to grow, divide, and develop normally. The functions of various growth factors in regulating cell division and tissue development are described in other chapters.

Another important local regulator is the gas **nitric oxide (NO).** When the blood oxygen level falls, endothelial cells in blood vessel walls synthesize and release NO. Nitric oxide activates an enzyme that relaxes the neighboring smooth muscle cells, which in turn dilates the vessels and improves blood flow to tissues. Nitric oxide also plays a role in male sexual function by increasing blood flow into the penis and producing an erection. Highly reactive and potentially toxic, NO usually triggers changes in a target cell within a few seconds of contact and then breaks down. The drug Viagra (sildenafil citrate), a treatment for male erectile dysfunction, sustains an erection by interfering with this breakdown of NO. Nitric oxide has other functions as well: In the nervous system, it can function as a neurotransmitter, and NO secreted by certain white blood cells can kill bacteria and cancer cells in body fluids.

A group of local regulators called **prostaglandins (PGs)** are modified fatty acids, often derived from lipids in the plasma membrane. They are so named because they were first discovered in prostate-gland secretions that contribute to semen. Released from most types of cells into interstitial fluid, prostaglandins regulate nearby cells in various ways, depending on the tissue. In semen that reaches the reproductive tract of a female, prostaglandins stimulate smooth muscles of the female's uterine wall to contract, helping sperm reach an egg. During childbirth, prostaglandins secreted by cells of the female's placenta cause the nearby muscles of the uterus to become more excitable, helping to induce labor (see Figure 46.18).

In the immune system, various prostaglandins help induce fever and inflammation and also intensify the sensation of pain. These responses contribute to the body's defense by sounding an alarm that something harmful is occurring. The anti-inflammatory effects of aspirin and ibuprofen are due to the drugs' inhibition of prostaglandin synthesis.

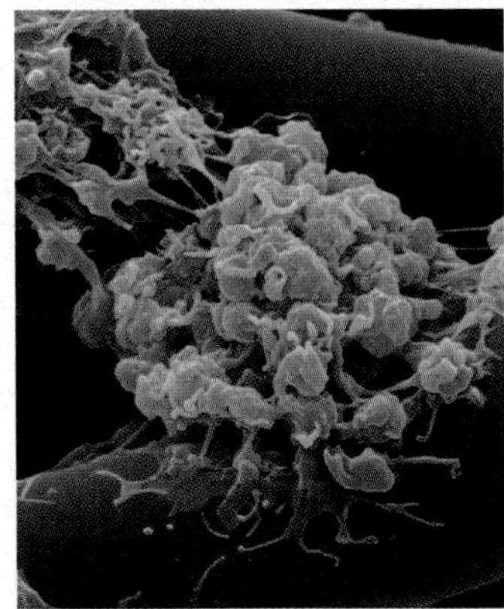

▲ **Figure 45.5 Activated platelets aggregating, a process regulated in part by prostaglandins.** Following injury to a blood vessel wall, platelets (pink and purple) develop a sticky outer surface and adhere to each other, as shown in this colorized SEM.

Prostaglandins also help regulate the aggregation of platelets, an early step in the formation of blood clots **(Figure 45.5)**. This is why some physicians recommend that people who are at risk for a heart attack take aspirin on a regular basis.

In the respiratory system, two prostaglandins with very similar molecular structures have opposite effects on the smooth muscle cells in the walls of blood vessels serving the lungs. Prostaglandin E signals the muscle cells to relax, which dilates the blood vessels and promotes oxygenation of the blood. Prostaglandin F signals the muscle cells to contract, which constricts the vessels and reduces blood flow through the lungs. Shifts in the relative concentrations of these two antagonistic (opposing) signals help maintain homeostasis in changing circumstances. Later in the chapter, we will encounter other antagonistic signals that counterbalance each other.

Concept Check 45.2

1. How do the mechanisms for inducing responses in target cells differ for water-soluble hormones and steroid hormones?
2. Explain how a single hormone, such as epinephrine, can cause different responses in different tissues.
3. Why do local regulators, such as prostaglandins, generally elicit responses in their target cells more rapidly than hormones?

For suggested answers, see Appendix A.

Concept 45.3

The hypothalamus and pituitary integrate many functions of the vertebrate endocrine system

So far, we have looked at the basic components of hormonal regulatory pathways and how a hormonal signal is converted to a cellular response. Now we turn to the physiological effects of the primary vertebrate hormones and the role of the endocrine system in adjusting the body's activities to shifting environmental and developmental conditions. We begin with the hypothalamus and pituitary gland, which control much of the endocrine system.

As you read, you may want to refer to **Table 45.1**, which summarizes the actions of major human hormones, and to **Figure 45.6**, on page 950, which illustrates the major endocrine glands in the human body. Small reference diagrams accompanying each section of the text will help you recall each gland's location.

Table 45.1 Major Human Endocrine Glands and Some of Their Hormones

Gland	Hormone	Chemical Class	Representative Actions	Regulated By
Hypothalamus	Hormones released from the posterior pituitary and hormones that regulate the anterior pituitary (see below)			
Pituitary gland				
Posterior pituitary (releases neurohormones made in hypothalamus)	Oxytocin	Peptide	Stimulates contraction of uterus and mammary gland cells	Nervous system
	Antidiuretic hormone (ADH)	Peptide	Promotes retention of water by kidneys	Water/salt balance
Anterior pituitary	Growth hormone (GH)	Protein	Stimulates growth (especially bones) and metabolic functions	Hypothalamic hormones
	Prolactin (PRL)	Protein	Stimulates milk production and secretion	Hypothalamic hormones
	Follicle-stimulating hormone (FSH)	Glycoprotein	Stimulates production of ova and sperm	Hypothalamic hormones
	Luteinizing hormone (LH)	Glycoprotein	Stimulates ovaries and testes	Hypothalamic hormones
	Thyroid-stimulating hormone (TSH)	Glycoprotein	Stimulates thyroid gland	Thyroxine in blood; hypothalamic hormones
	Adrenocorticotropic hormone (ACTH)	Peptide	Stimulates adrenal cortex to secrete glucocorticoids	Glucocorticoids; hypothalamic hormones
Thyroid gland	Triiodothyronine (T_3) and thyroxine (T_4)	Amine	Stimulate and maintain metabolic processes	TSH
	Calcitonin	Peptide	Lowers blood calcium level	Calcium in blood
Parathyroid glands	Parathyroid hormone (PTH)	Peptide	Raises blood calcium level	Calcium in blood
Pancreas	Insulin	Protein	Lowers blood glucose level	Glucose in blood
	Glucagon	Protein	Raises blood glucose level	Glucose in blood
Adrenal glands				
Adrenal medulla	Epinephrine and norepinephrine	Amine	Raise blood glucose level; increase metabolic activities; constrict certain blood vessels	Nervous system
Adrenal cortex	Glucocorticoids	Steroid	Raise blood glucose level	ACTH
	Mineralocorticoids	Steroid	Promote reabsorption of Na^+ and excretion of K^+ in kidneys	K^+ in blood
Gonads				
Testes	Androgens	Steroid	Support sperm formation; promote development and maintenance of male secondary sex characteristics	FSH and LH
Ovaries	Estrogens	Steroid	Stimulate uterine lining growth; promote development and maintenance of female secondary sex characteristics	FSH and LH
	Progesterone	Steroid	Promotes uterine lining growth	FSH and LH
Pineal gland	Melatonin	Amine	Involved in biological rhythms	Light/dark cycles

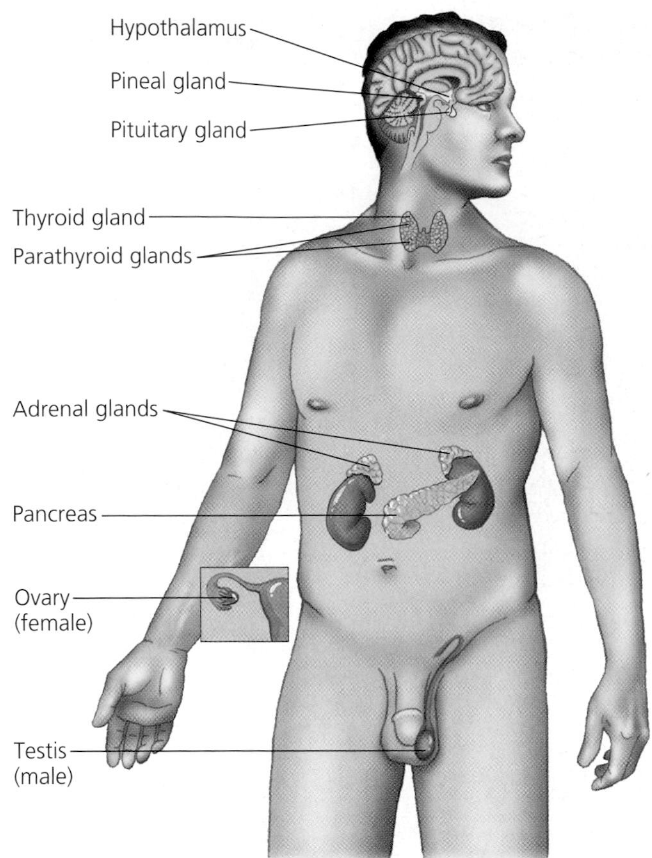

▲ **Figure 45.6 Human endocrine glands surveyed in this chapter.** In addition to the glands shown here, many organs with primarily nonendocrine functions also secrete hormones.

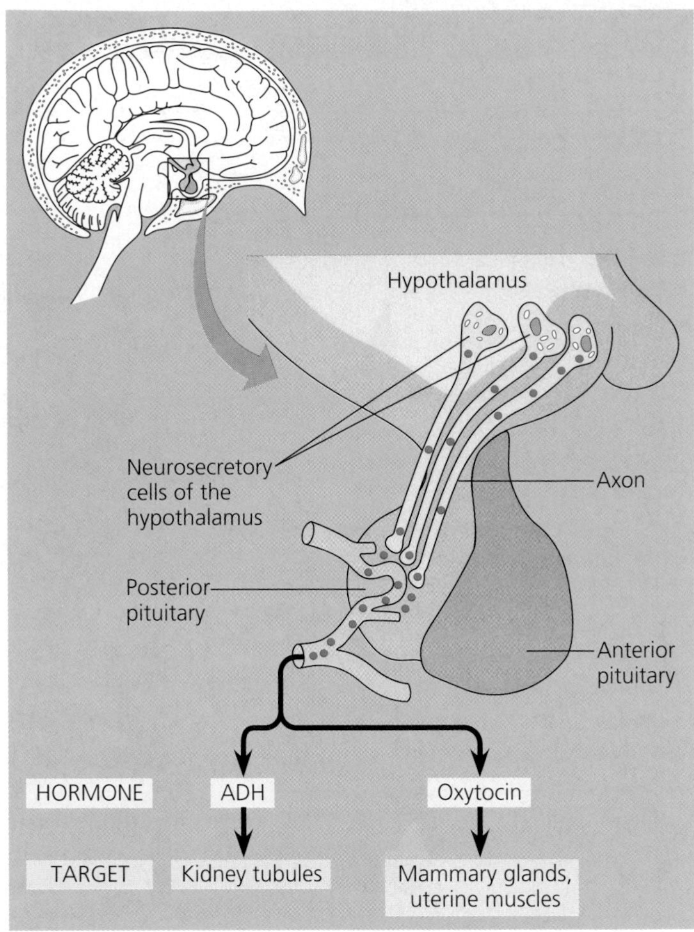

▲ **Figure 45.7 Production and release of posterior pituitary hormones.** The posterior pituitary gland is an extension of the hypothalamus. Certain neurosecretory cells in the hypothalamus make antidiuretic hormone (ADH) and oxytocin, which are transported to the posterior pituitary where they are stored. Nervous signals from the brain trigger release of these neurohormones.

Note that hormone-secreting cells are present in many organs belonging to other systems, including the heart, thymus, liver, stomach, small intestine, kidney, and placenta. Some of the hormones released by these nonendocrine organs are discussed in other chapters.

Relationship Between the Hypothalamus and Pituitary Gland

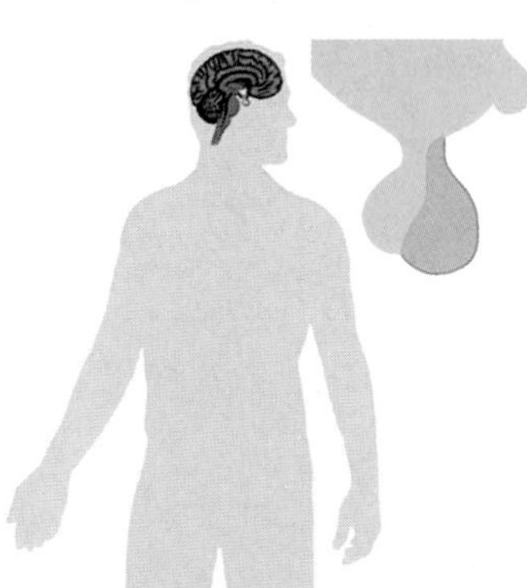

The **hypothalamus** plays an important role in integrating the vertebrate endocrine and nervous systems. This region on the underside of the brain receives information from nerves throughout the body and from other parts of the brain, and it initiates endocrine signals appropriate to environmental conditions. In many vertebrates, for example, the brain passes sensory information about seasonal changes and the availability of a mate to the hypothalamus by means of nerve signals; the hypothalamus then triggers the release of reproductive hormones required for breeding.

The hypothalamus contains two sets of neurosecretory cells whose hormonal secretions are stored in or regulate the activity of the **pituitary gland**, a lima bean-sized organ located at the base of the hypothalamus. No organ illustrates the close structural, functional, and developmental relationships between the endocrine and nervous systems better than the pituitary gland. It has discrete posterior and anterior parts, which are actually two fused glands that develop from separate regions of the embryo and perform very different functions.

The **posterior pituitary**, or **neurohypophysis**, is an extension of the hypothalamus that grows downward toward the mouth during embryonic development. It stores and secretes two hormones that are made by certain neurosecretory cells located in the hypothalamus; the long processes (axons) of these cells carry the hormones to the posterior pituitary **(Figure 45.7)**.

The **anterior pituitary**, or **adenohypophysis**, develops from a fold of tissue at the roof of the embryonic mouth; this tissue grows upward toward the brain and eventually loses its connection to the mouth. The anterior pituitary consists of endocrine cells that synthesize and secrete at least six

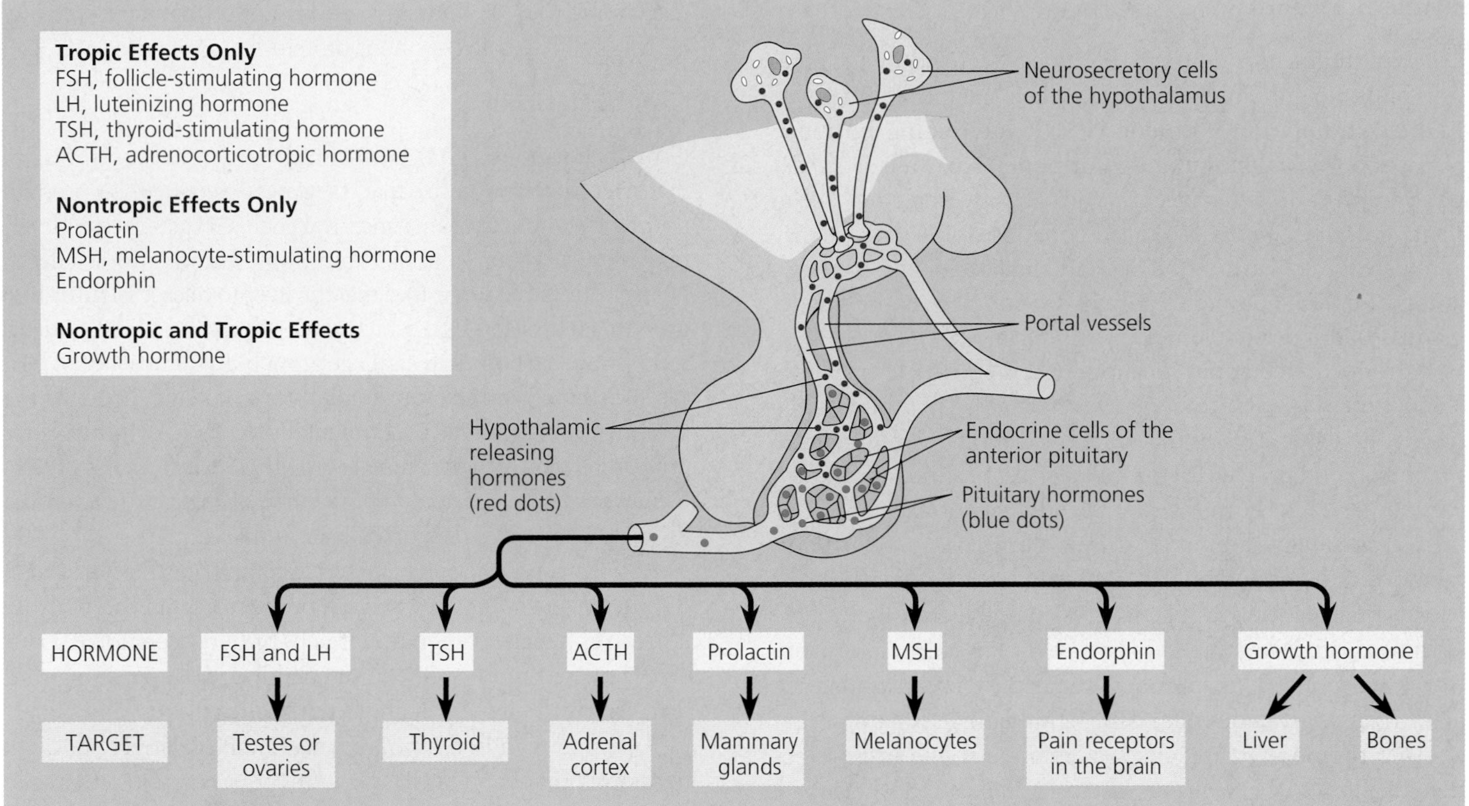

▲ **Figure 45.8 Production and release of anterior pituitary hormones.** The release of hormones synthesized in the anterior pituitary gland is controlled by hypothalamic releasing and inhibiting hormones. The hypothalamic hormones are secreted by neurosecretory cells and enter a capillary network within the hypothalamus. These capillaries drain into portal vessels that connect with a second capillary network in the anterior pituitary. Each hormone made in the anterior pituitary is secreted in response to a specific releasing hormone.

different hormones. Several of these hormones have other endocrine glands as their targets. Hormones that regulate the function of endocrine organs, called **tropic hormones,** are particularly important in coordinating endocrine signaling throughout the body.

The anterior pituitary itself is regulated by tropic hormones produced by a set of neurosecretory cells in the hypothalamus. Some hypothalamic tropic hormones (called releasing hormones) stimulate the anterior pituitary to release its hormones, while others (called inhibiting hormones) inhibit its hormone secretion. The hypothalamic releasing and inhibiting hormones are secreted near capillaries at the base of the hypothalamus **(Figure 45.8)**. The capillaries drain into short blood vessels, called portal vessels, that subdivide into a second capillary bed within the anterior pituitary. In this way, the hypothalamic tropic hormones have direct access to the gland they control. Every anterior pituitary hormone is controlled by at least one releasing hormone, and some have both a releasing hormone and an inhibiting hormone.

Posterior Pituitary Hormones

As shown in Figure 45.7, the posterior pituitary releases two hormones, antidiuretic hormone (ADH) and oxytocin. Both are peptides made by neurosecretory cells in the hypothalamus and thus strictly speaking are neurohormones. They function in simple neurohormone pathways (see Figure 45.2b).

Antidiuretic hormone (ADH) acts on the kidneys, increasing water retention and thus decreasing urine volume. It is part of the elaborate mechanism that helps regulate the osmolarity of the blood. This mechanism illustrates the importance of negative feedback in maintaining homeostasis and the central role of the hypothalamus as a member of both the endocrine system and the nervous system. See Figure 44.16 to review the function of ADH in this complex regulatory circuit.

Oxytocin induces target cells in the uterine muscles to contract during childbirth. As described earlier, it also causes the mammary glands to eject milk during nursing. Oxytocin signaling in both cases exhibits positive feedback.

Anterior Pituitary Hormones

The anterior pituitary produces many different hormones (see Figure 45.8). Four function strictly as tropic hormones, stimulating the synthesis and release of hormones from the thyroid gland, adrenal glands, and gonads. Several others exert only direct, nontropic effects on nonendocrine organs. And one, growth hormone, has both tropic and nontropic actions.

Tropic Hormones

Three of the exclusively tropic hormones secreted by the anterior pituitary are closely related in their chemical structures. **Follicle-stimulating hormone (FSH), luteinizing hormone (LH),** and **thyroid-stimulating hormone (TSH)** are all similar glycoproteins, protein molecules with carbohydrates attached to them. FSH and LH are also called **gonadotropins** because they stimulate the activities of the male and female gonads, the testes and ovaries. TSH promotes normal development of the thyroid gland and the production of thyroid hormones.

Adrenocorticotropic hormone (ACTH), the fourth anterior pituitary tropic hormone, is unrelated structurally to the others. It is a peptide hormone derived by cleavage of a large precursor protein. ACTH stimulates the production and secretion of steroid hormones by the adrenal cortex.

All four anterior pituitary tropic hormones participate in complex neuroendocrine pathways. In each pathway, signals to the brain stimulate release of a hypothalamic neurohormone that in turn stimulates release of an anterior pituitary tropic hormone. The anterior pituitary tropic hormone then acts on its target endocrine tissue, stimulating secretion of yet another hormone that exerts systemic metabolic or developmental effects. We will take a closer look at the hormonal pathways involving TSH and ACTH later in this chapter. In Chapter 46, we will discuss how FSH and LH regulate reproductive functions. As you will see, negative-feedback loops are crucial in regulating these complex pathways.

Nontropic Hormones

Nontropic hormones produced by the anterior pituitary include prolactin, melanocyte-stimulating hormone (MSH), and β-endorphin. These peptide/protein hormones, whose secretion is controlled by hypothalamic hormones, function in simple neuroendocrine pathways (see Figure 45.2c).

The most remarkable characteristic of **prolactin (PRL)** is the great diversity of effects it produces in different vertebrate species. For example, prolactin stimulates mammary gland growth and milk synthesis in mammals; regulates fat metabolism and reproduction in birds; delays metamorphosis in amphibians, where it may also function as a larval growth hormone; and regulates salt and water balance in freshwater fishes. This list suggests that prolactin is an ancient hormone whose functions have diversified during the evolution of the various vertebrate groups.

As described earlier, **melanocyte-stimulating hormone (MSH)** regulates the activity of pigment-containing cells in the skin of some fishes, amphibians, and reptiles. In mammals, MSH appears to act on neurons in the brain, inhibiting hunger.

β-Endorphin belongs to a class of chemical signals called **endorphins.** Certain neurons in the brain also produce endorphins (see Chapter 48). All the endorphins bind to receptors in the brain and dull the perception of pain. Some researchers speculate that the so-called "runner's high" results partly from the release of endorphins when stress and pain in the body reach critical levels. Both MSH and β-endorphin are formed by cleavage of the same precursor protein that gives rise to ACTH.

Growth Hormone

Growth hormone (GH) is so similar structurally to prolactin that scientists hypothesize that the genes directing their production evolved from the same ancestral gene. GH acts on a wide variety of target tissues and has both tropic and nontropic effects. Its major tropic action is to signal the liver to release **insulin-like growth factors (IGFs),** which circulate in the blood and directly stimulate bone and cartilage growth. In the absence of GH, the skeleton of an immature animal stops growing. Injecting the hormone into an animal experimentally deprived of its own supply partially restores the animal's growth. GH also exerts diverse metabolic effects that tend to raise blood glucose, thus opposing the effects of insulin (discussed later in this chapter).

Abnormal production of GH in humans can result in several disorders, depending on when the problem occurs and whether it involves hypersecretion (too much) or hyposecretion (too little). Hypersecretion of GH during childhood can lead to gigantism, in which the person grows unusually tall—as tall as 8 feet (2.4 m)—though body proportions remain relatively normal. Excessive production of GH in adulthood, a condition known as acromegaly, stimulates bony growth in a few tissues that are still responsive to the hormone, such as in the face, hands, and feet.

Hyposecretion of GH in childhood retards long-bone growth and can lead to pituitary dwarfism. Individuals with this disorder generally reach a height of only about 4 feet (1.2 m), though body proportions remain relatively normal. If diagnosed before puberty, pituitary dwarfism can be treated successfully with human GH. For many years, the supply of GH isolated from cadaver pituitaries fell far short of demand. In the mid-1980s, genetic engineers learned to produce human GH by inserting DNA encoding the hormone into bacteria (see Chapter 20). Therapy with genetically engineered GH is now fairly routine for children with pituitary dwarfism.

Some athletes take GH in the belief that it can improve athletic performance. However, research has shown that in healthy adults who are not hormone-deficient, GH has little impact on muscle mass and strength.

Concept Check 45.3

1. How do the two fused glands of the pituitary gland differ in origin and function?
2. Discuss the role of tropic hormones in regulating the endocrine system.
3. Which pathway of hormonal control (see Concept 45.1) is characteristic of (a) prolactin, (b) adrenocorticotropic hormone (ACTH), and (c) oxytocin?

For suggested answers, see Appendix A.

Nonpituitary hormones help regulate metabolism, homeostasis, development, and behavior

Next, let's survey the major functions of several nonpituitary hormones and the endocrine glands that produce them. Tropic hormones from the anterior pituitary control secretion of some, but not all, of these hormones.

Thyroid Hormones

In humans and other mammals, the **thyroid gland** consists of two lobes located on the ventral surface of the trachea (see Figure 42.23). In many other vertebrates, the two halves of the gland are separated on the two sides of the pharynx.

The thyroid gland produces two very similar hormones derived from the amino acid tyrosine: **triiodothyronine (T_3)**, which contains three iodine atoms, and tetraiodothyronine, or **thyroxine (T_4)**, which contains four iodine atoms. In mammals, the thyroid secretes mainly T_4, but target cells convert most of it to T_3 by removing one iodine atom. Although both hormones are bound by the same receptor protein located in the cell nucleus, the receptor has greater affinity for T_3 than for T_4. Thus, it is mostly T_3 that brings about responses in target cells.

The hypothalamus and anterior pituitary control the secretion of thyroid hormones and hence their effects in an animal's body. This process involves a complex neuroendocrine pathway with two negative-feedback loops **(Figure 45.9)**.

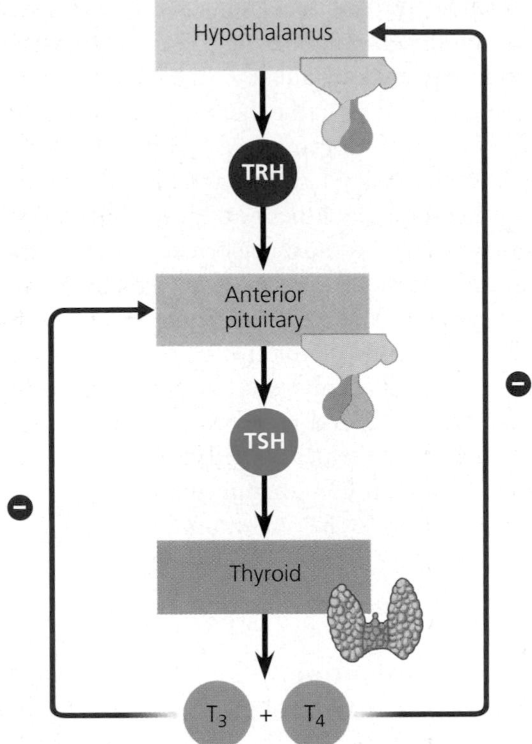

▲ **Figure 45.9 Feedback regulation of T_3 and T_4 secretion from the thyroid gland.** The hypothalamus secretes TSH-releasing hormone (TRH), which stimulates the anterior pituitary to secrete thyroid-stimulating hormone (TSH). TSH then stimulates the thyroid gland to synthesize and release the thyroid hormones T_3 and T_4. These hormones exert negative feedback on the hypothalamus and anterior pituitary by inhibiting release of TRH and TSH.

The thyroid gland plays a crucial role in vertebrate development and maturation. A striking example is the thyroid control of the metamorphosis of a tadpole into a frog, which involves massive reorganization of many different tissues. Studies with other nonhuman animals have shown that thyroid hormones are required for the normal functioning of bone-forming cells and the branching of nerve cells during embryonic development of the brain. The thyroid is equally important in human development. Cretinism, an inherited condition of thyroid deficiency, results in markedly retarded skeletal growth and poor mental development. These defects can often be prevented, at least partially, if treatment with thyroid hormones begins early in life.

The thyroid gland also has important homeostatic functions. In adult mammals, for instance, thyroid hormones help maintain normal blood pressure, heart rate, muscle tone, digestion, and reproductive functions. Throughout the body, T_3 and T_4 are important in bioenergetics, generally increasing the rate of oxygen consumption and cellular metabolism. Too much or too little of these hormones in the blood can result in serious metabolic disorders. In humans, excessive secretion of thyroid hormones, known as hyperthyroidism, can lead to high body temperature, profuse sweating, weight loss, irritability, and high blood pressure. The most common form of hyperthyroidism is Graves' disease; protruding eyes, caused by fluid accumulation behind the eyes, are a typical symptom **(Figure 45.10)**.

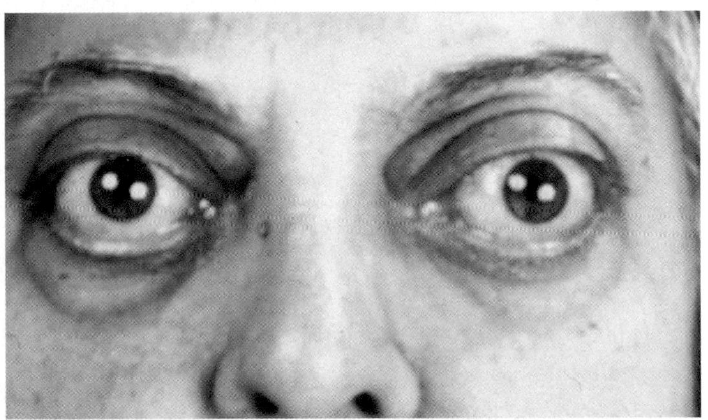

▲ **Figure 45.10 Graves' disease, the most common form of hyperthyroidism in humans.** Tissue behind the eyes can become swollen and fibrous, causing the characteristic symptom of bulging eyes.

Hypothyroidism, the opposite condition, can cause cretinism in infants and produce symptoms such as weight gain, lethargy, and intolerance to cold in adults.

A deficiency of iodine in the diet can result in goiter, an enlargement of the thyroid gland (see Figure 2.3). Without sufficient iodine, the thyroid gland cannot synthesize adequate amounts of T_3 and T_4, and the resulting low blood levels of T_3 and T_4 cannot exert the usual negative feedback on the hypothalamus and anterior pituitary (see Figure 45.9). Consequently, the pituitary continues to secrete TSH, elevating TSH levels and enlarging the thyroid.

In addition to the cells that produce T_3 and T_4, the mammalian thyroid gland contains endocrine cells that produce calcitonin. This hormone acts in conjunction with parathyroid hormone in maintaining calcium homeostasis, as we will describe next.

Parathyroid Hormone and Calcitonin: Control of Blood Calcium

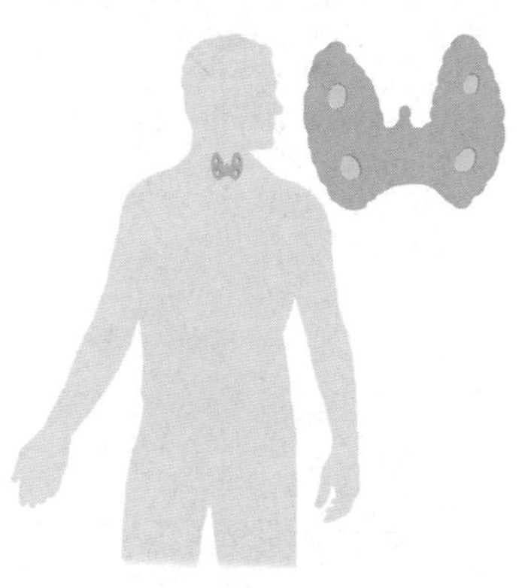

Rigorous homeostatic control of the blood calcium level is critical because calcium ions (Ca^{2+}) are essential to the normal functioning of all cells. If blood Ca^{2+} level falls substantially, skeletal muscles begin to contract convulsively, a condition known as tetany. If not corrected, tetany is fatal. In mammals, two hormones with opposite actions—parathyroid hormone and calcitonin—play a major role in maintaining blood Ca^{2+} level near a set point of about 10 mg/100 mL **(Figure 45.11)**.

When blood Ca^{2+} level falls below this set point, **parathyroid hormone (PTH)** is released. PTH is produced by four small structures, the **parathyroid glands**, that are embedded in the surface of the thyroid.

PTH raises the level of blood Ca^{2+} by direct and indirect effects. In bone, PTH induces specialized cells called osteoclasts to decompose the mineralized matrix of bone and release Ca^{2+} into the blood. In the kidneys, it directly stimulates reabsorption of Ca^{2+} through the renal tubules. PTH also has an indirect effect on the kidneys, promoting the conversion of vitamin D to its active hormonal form. An inactive form of **vitamin D**, a steroid-derived molecule, is obtained from food or synthesized in the skin. Activation of vitamin D begins in the liver and is completed in the kidneys, a process stimulated by PTH. The active form of vitamin D acts directly on the intestines, stimulating the

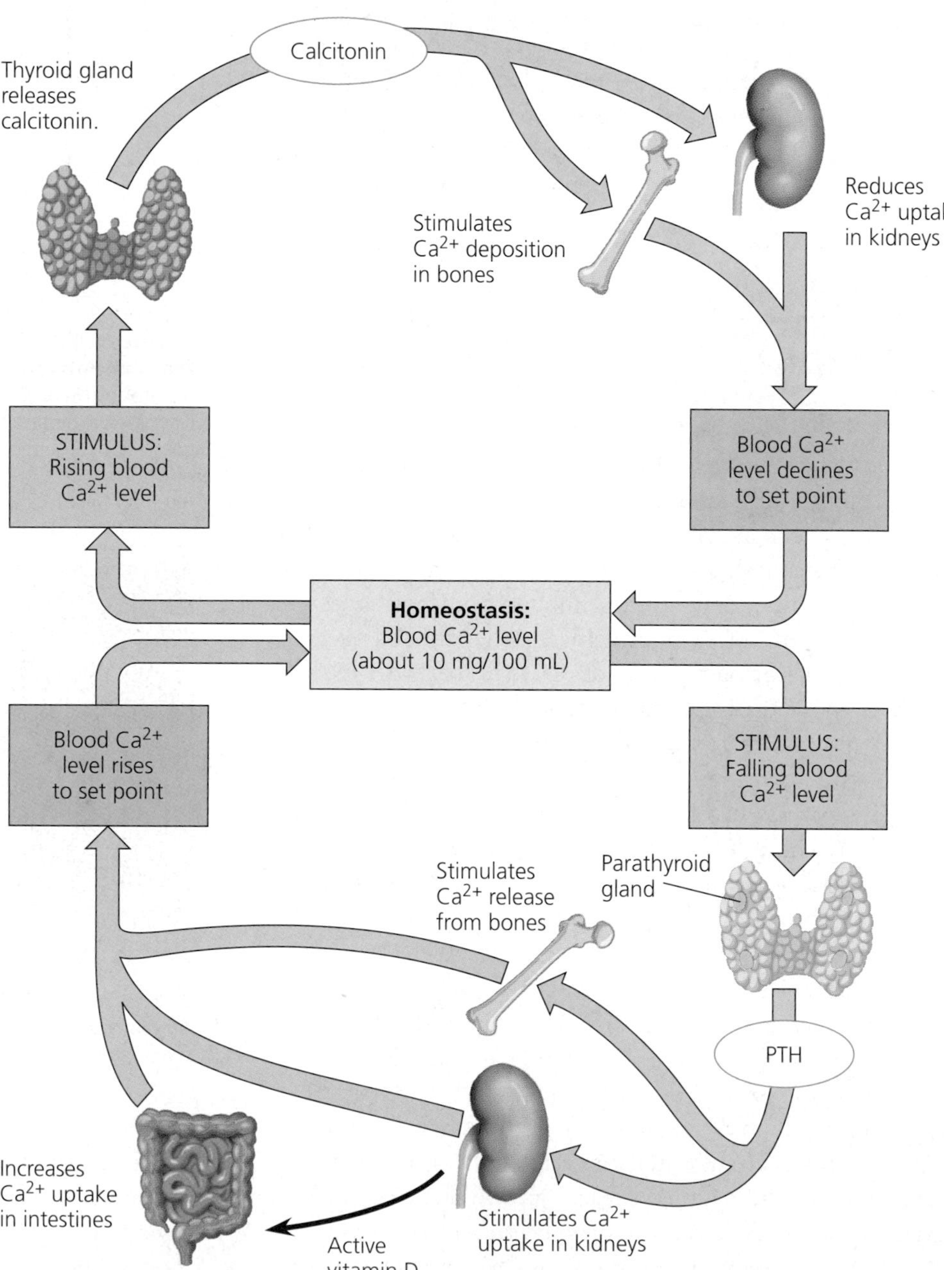

▲ **Figure 45.11 Hormonal control of calcium homeostasis in mammals.** A negative-feedback system involving two antagonistic hormones, calcitonin and parathyroid hormone (PTH), maintains the blood Ca^{2+} concentration near its set point.

uptake of Ca^{2+} from food and thus augmenting the effect of PTH.

A rise in blood Ca^{2+} level above the set point promotes release of **calcitonin** from the thyroid gland. Calcitonin exerts effects on bone and kidneys opposite to those of PTH and thus lowers the blood Ca^{2+} level.

The regulation of blood Ca^{2+} level illustrates how two hormones with opposite effects—in this case, PTH and calcitonin—balance each other, exerting tight regulation and maintaining homeostasis. Each hormone functions in a simple endocrine pathway in which the hormone-secreting cells themselves monitor the variable being regulated (see Figure 45.2a). In classic feedback fashion, the response to either hormone triggers release of the antagonistic hormone, thereby minimizing fluctuations in the concentration of Ca^{2+} in the blood.

Insulin and Glucagon: Control of Blood Glucose

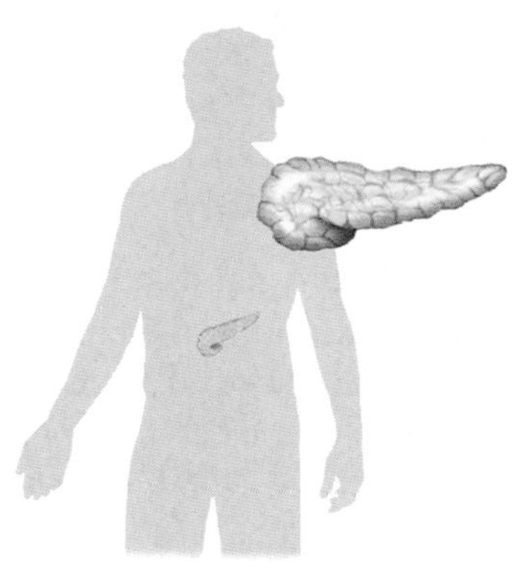

Although the **pancreas** is considered a major endocrine gland, hormone-secreting cells make up only 1–2% of its weight. The rest of the pancreas produces bicarbonate ions and digestive enzymes, which are released into small ducts and carried to the small intestine via the pancreatic duct (see Figure 41.19). Tissues and glands that discharge secretions into ducts are described as *exocrine*. Thus, the pancreas is a dual endocrine and exocrine gland with important functions in both the endocrine and digestive systems.

Clusters of endocrine cells, the **islets of Langerhans**, are scattered throughout the exocrine tissue of the pancreas. Each islet has a population of *alpha cells*, which produce the hormone **glucagon**, and a population of *beta cells*, which produce the hormone **insulin.** Both of these protein hormones, like all endocrine signals, are secreted into the extracellular fluid and enter the circulatory system.

Insulin and glucagon are antagonistic hormones that regulate the concentration of glucose in the blood **(Figure 45.12)**. This is a critical bioenergetic and homeostatic function, because glucose is a major fuel for cellular respiration and a key source of carbon skeletons for the synthesis of other organic compounds. Metabolic balance depends on maintaining blood glucose concentrations near a set point, which is about 90 mg/100 mL in humans. When blood glucose exceeds this level, insulin is released, and its effects lower blood glucose concentration. When blood glucose drops below the set point, glucagon is released, and its effects increase blood

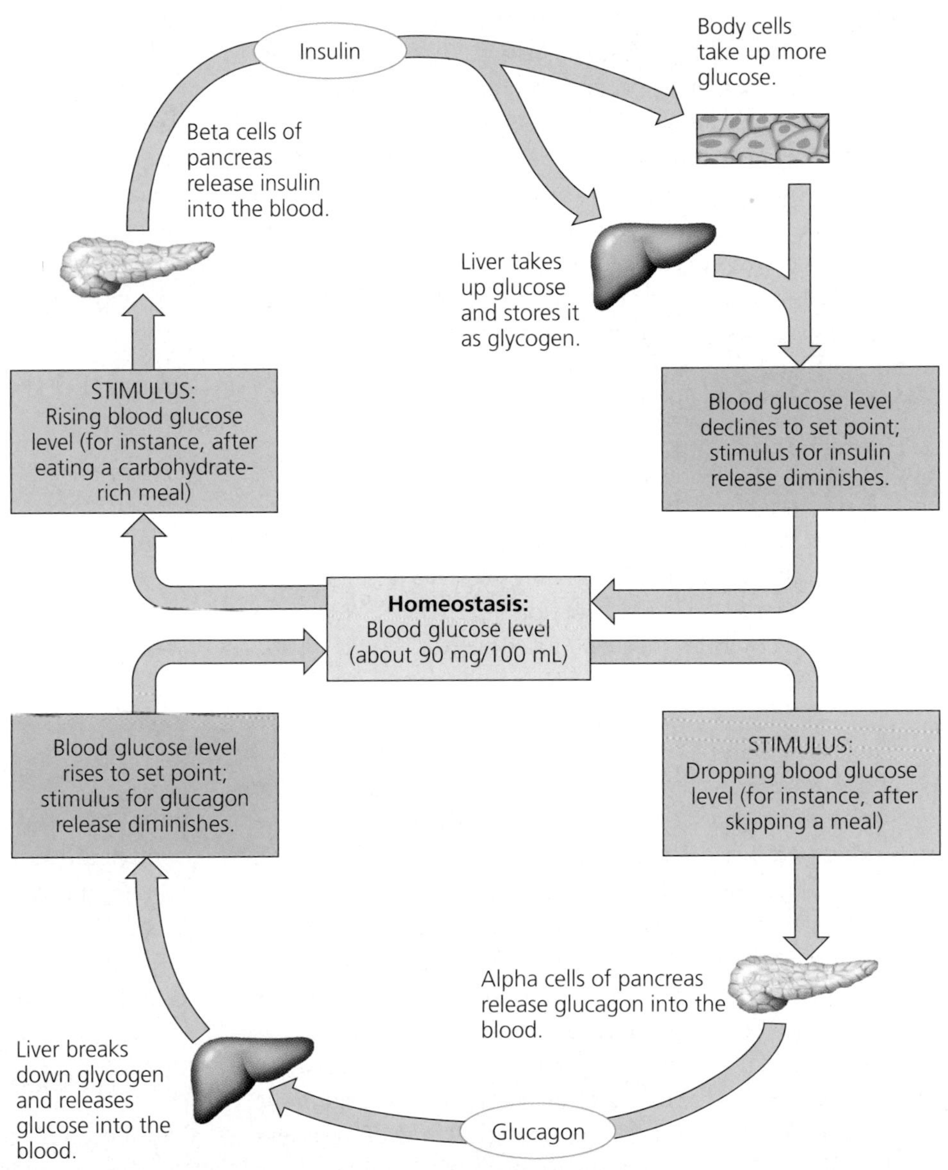

▲ **Figure 45.12 Maintenance of glucose homeostasis by insulin and glucagon.** The antagonistic effects of insulin and glucagon help maintain the blood glucose level near its set point. A rise in blood glucose level above the set point promotes insulin release from the pancreas, leading to removal of excess glucose from the blood and its storage as glycogen. A fall in blood glucose level below the set point stimulates the pancreas to secrete glucagon, which causes the liver to raise the blood glucose level.

glucose concentration. Each hormone operates in a simple endocrine pathway that is regulated by negative feedback. The combination of the two pathways permits precise regulation of blood glucose.

Target Tissues for Insulin and Glucagon

Insulin lowers blood glucose levels by stimulating virtually all body cells except those of the brain to take up glucose from the blood. (Brain cells are unusual in being able to take up glucose without insulin; as a result, the brain has access to circulating fuel almost all the time.) Insulin also decreases blood glucose by slowing glycogen breakdown in the liver and inhibiting the conversion of amino acids and glycerol (from fats) to glucose.

The liver, skeletal muscles, and adipose tissues store large amounts of fuel and are especially important in bioenergetics. The liver and muscles store sugar as glycogen, whereas adipose tissue cells convert sugars to fats. The liver is a key fuel-processing center because only liver cells are sensitive to glucagon. Normally, glucagon starts having an effect before blood glucose levels even drop below the set point. In fact, as soon as excess glucose is cleared from the blood, glucagon signals the liver cells to increase glycogen hydrolysis, convert amino acids and glycerol to glucose, and start slowly releasing glucose back into the circulation.

The antagonistic effects of glucagon and insulin are vital to glucose homeostasis and thus to the precise management of both fuel storage and fuel consumption by body cells. The liver's ability to perform its vital roles in glucose homeostasis results from the metabolic versatility of its cells and its access to absorbed nutrients via the hepatic portal vessel, which carries blood directly from the small intestine to the liver.

Diabetes Mellitus

When the mechanisms of glucose homeostasis go awry, there are serious consequences. **Diabetes mellitus,** perhaps the best-known endocrine disorder, is caused by a deficiency of insulin or a decreased response to insulin in target tissues. There are two major types of diabetes mellitus with very different causes, but each is marked by high blood glucose.

In people with diabetes, elevated blood glucose exceeds the reabsorption capacity of the kidneys, causing them to excrete glucose. This explains why the presence of sugar in urine is one test for diabetes. As glucose is concentrated in the urine, more water is excreted along with it, resulting in excessive volumes of urine and persistent thirst. (*Diabetes,* from the Greek *diabainein,* to pass through, refers to this copious urination; and *mellitus,* from the Greek *meli,* honey, refers to the presence of sugar in urine.) Without sufficient glucose available to meet the needs of most body cells, fat becomes the main substrate for cellular respiration. In severe cases, acidic metabolites formed during fat breakdown accumulate in the blood, threatening life by lowering blood pH.

Type I diabetes mellitus (insulin-dependent diabetes) is an autoimmune disorder in which the immune system destroys the beta cells of the pancreas. Type I diabetes, which usually appears during childhood, destroys the person's ability to produce insulin. Treatment consists of insulin injections, usually several times daily. In the past, insulin for injections was extracted from animal pancreases, but now human insulin can be obtained from genetically engineered bacteria, a relatively inexpensive source (see Figure 20.2).

Type II diabetes mellitus (non-insulin-dependent diabetes) is characterized either by a deficiency of insulin or, more commonly, by reduced responsiveness of target cells due to some change in insulin receptors. Although heredity can play a role in type II diabetes, research indicates that excess body weight and lack of exercise significantly increase the risk. This form of diabetes generally appears after age 40, but young people who are overweight and sedentary can also develop the disease. More than 90% of people with diabetes have type II. Many can manage their blood glucose level with regular exercise and a healthy diet; some require drug therapy.

Adrenal Hormones: Response to Stress

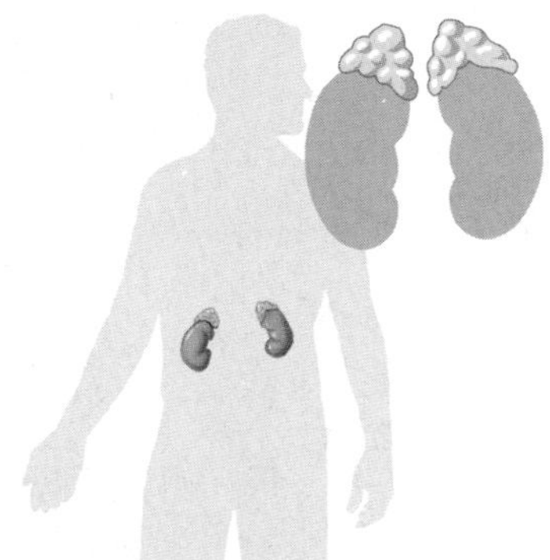

The **adrenal glands** are adjacent to the kidneys. In mammals, each adrenal gland is actually made up of two glands with different cell types, functions, and embryonic origins: the *adrenal cortex,* the outer portion, and the *adrenal medulla,* the central portion. Like the pituitary gland, each adrenal gland is a fused endocrine and neuroendocrine gland. The adrenal cortex consists of true endocrine cells, whereas the secretory cells of the adrenal medulla derive from the neural crest during embryonic development (see Figure 34.7). Nonmammalian vertebrates have quite different arrangements of the same tissues.

Catecholamines from the Adrenal Medulla

Suppose you sense danger or approach a stressful situation, such as speaking in public. Your heart beats faster and your skin develops goose bumps. What causes these reactions? They are part of the so-called fight-or-flight response stimulated by two hormones of the adrenal medulla, **epinephrine** (also known as adrenaline) and **norepinephrine** (noradrenaline). These hormones are members of a class of compounds, the **catecholamines,** that are synthesized from the amino acid tyrosine. Both epinephrine and norepinephrine also function as neurotransmitters in the nervous system.

Either positive or negative stress—ranging from extreme pleasure to a drop in temperature to life-threatening danger—stimulates the adrenal medulla to secrete epinephrine and norepinephrine. These hormones act directly on several target tissues and give the body a rapid bioenergetic boost. They increase the rate of glycogen breakdown in the liver and skeletal muscles, promote glucose release by liver cells, and stimulate the release of fatty acids from fat cells. The released glucose and fatty acids circulate in the blood and can be used by body cells as fuel.

In addition to increasing the availability of energy sources, epinephrine and norepinephrine exert profound effects on the cardiovascular and respiratory systems. For example, they increase both the rate and the stroke volume of the heartbeat and dilate the bronchioles in the lungs, actions that increase the rate of oxygen delivery to body cells. (This is why doctors prescribe epinephrine as a heart stimulant and to open breathing tubes during asthma attacks.) The catecholamines also cause smooth muscles of some blood vessels to contract and muscles of other vessels to relax, with an overall effect of shunting blood away from the skin, digestive organs, and kidneys while increasing the blood supply to the heart, brain, and skeletal muscles (see Figure 45.4). Epinephrine generally has a stronger effect on heart and metabolic rates, while the primary role of norepinephrine is in sustaining blood pressure.

As shown in **Figure 45.13a**, secretion by the adrenal medulla is stimulated by nerve signals carried from the brain via the sympathetic division of the autonomic nervous system (discussed in Chapter 48). In response to a stressful stimulus, nerve impulses from the hypothalamus travel to the adrenal medulla, where they trigger the release of epinephrine. Norepinephrine is released independently of epinephrine. The adrenal medulla hormones provide another example of a simple neurohormone pathway (see Figure 45.2b). In this

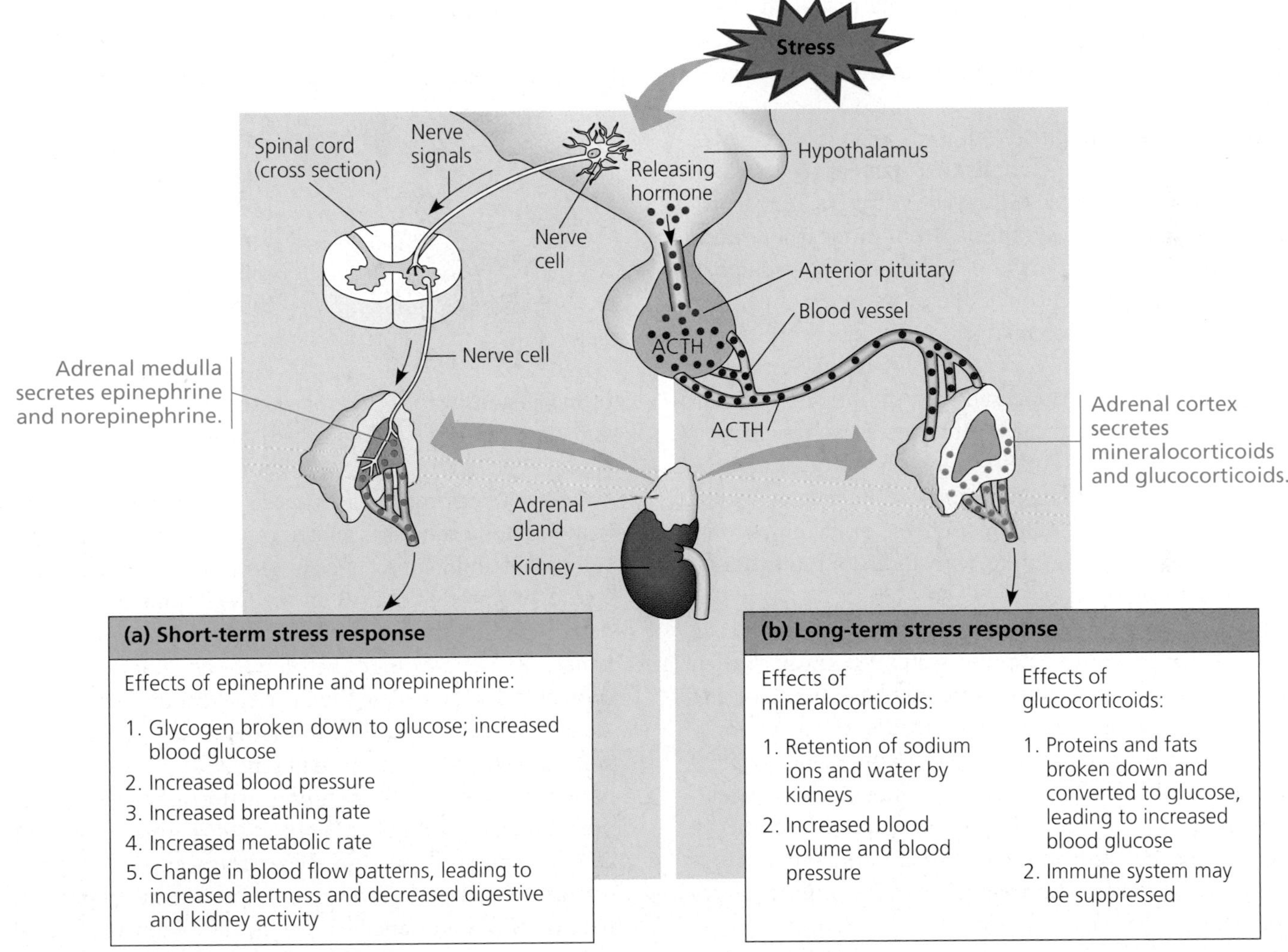

▲ **Figure 45.13 Stress and the adrenal gland.** Stressful stimuli cause the hypothalamus to activate the adrenal medulla via nerve impulses **(a)** and the adrenal cortex via hormonal signals **(b)**. The adrenal medulla mediates short-term responses to stress by secreting the catecholamine hormones epinephrine and norepinephrine. The adrenal cortex controls more prolonged responses by secreting corticosteroids.

case, the neurosecretory cells are modified peripheral nerve cells rather than the hypothalamic neurosecretory cells that release hormones in the posterior pituitary.

Steroid Hormones from the Adrenal Cortex

Hormones from the adrenal cortex also function in the body's response to stress. But in contrast to the adrenal medulla, which reacts to nervous input, the adrenal cortex responds to endocrine signals. Stressful stimuli cause the hypothalamus to secrete a releasing hormone that stimulates the anterior pituitary to release the tropic hormone ACTH. When ACTH reaches the adrenal cortex via the bloodstream, it stimulates the endocrine cells to synthesize and secrete a family of steroids called **corticosteroids.** In another case of negative feedback, elevated levels of corticosteroids in the blood suppress the secretion of ACTH.

The two main types of corticosteroids in humans are the **glucocorticoids,** such as cortisol, and the **mineralocorticoids,** such as aldosterone. Evidence is mounting that both glucocorticoids and mineralocorticoids help maintain homeostasis when the body experiences stress over an extended period of time **(Figure 45.13b)**.

The primary effect of glucocorticoids is on bioenergetics, specifically on glucose metabolism. Augmenting the fuel-mobilizing effects of glucagon from the pancreas, glucocorticoids promote glucose synthesis from noncarbohydrate sources such as proteins, making more glucose available as fuel. Glucocorticoids act on skeletal muscle, causing a breakdown of muscle proteins. The resulting carbon skeletons are transported to the liver and kidneys, where they are converted to glucose and released into the blood. The synthesis of glucose from muscle proteins is a homeostatic mechanism providing circulating fuel when body activities require more than the liver can mobilize from its glycogen stores. It can also be part of a broader role of the glucocorticoids, that of helping the body withstand long-term environmental challenge.

Cortisol and other glucocorticoids also suppress certain components of the body's immune system. Because of their anti-inflammatory effect, glucocorticoids have been used to treat inflammatory diseases such as arthritis. However, long-term glucocorticoid use can have serious side effects owing to their metabolic actions and also can increase susceptibility to infection owing to their immunosuppressive effects. For these reasons, nonsteroidal anti-inflammatory drugs generally are preferred for treating chronic inflammatory conditions.

Mineralocorticoids act principally on salt and water balance. Aldosterone, for example, stimulates cells in the kidneys to reabsorb sodium ions and water from filtrate, raising blood pressure and volume. Aldosterone secretion is stimulated primarily by angiotensin II, as part of the regulatory pathway that controls the kidneys' ability to maintain the ion and water homeostasis of the blood (see Figure 44.16). However, when an individual is under severe stress, the resulting rise in blood ACTH levels can increase the rate at which the adrenal cortex secretes aldosterone as well as glucocorticoids.

The adrenal cortex produces a third group of corticosteroids that function as sex hormones. All the steroid hormones are synthesized from cholesterol (see Figure 5.15), and their structures differ in only minor ways. However, these differences are associated with major differences in their effects. The sex hormones produced by the adrenal cortex are mainly male hormones (androgens), with small amounts of female hormones (estrogens and progestins). There is evidence that adrenal androgens account for the sex drive in adult females, but otherwise the physiological roles of the adrenal sex hormones are not well understood.

Gonadal Sex Hormones

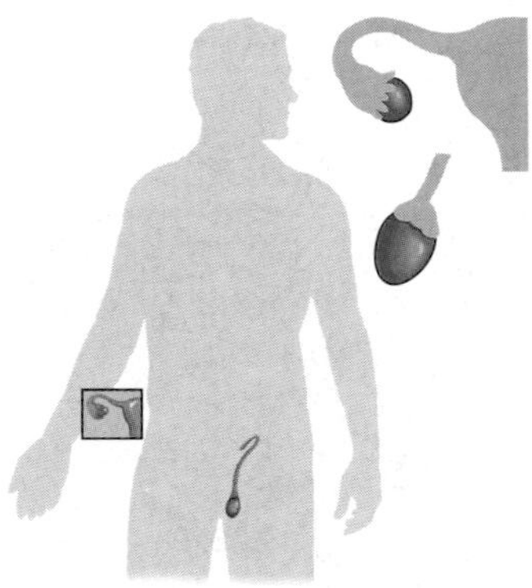

Although the adrenal glands secrete small amounts of sex hormones, the gonads are the primary source of these hormones. The gonads produce and secrete three major categories of steroid hormones: androgens, estrogens, and progestins. All three types are found in both males and females but in different proportions. Produced in the testes of males and ovaries of females, these steroids affect growth and development and also regulate reproductive cycles and sexual behavior.

The testes primarily synthesize **androgens,** the main one being **testosterone.** In general, androgens stimulate the development and maintenance of the male reproductive system. Androgens produced early in the development of an embryo determine that the fetus will develop as a male rather than a female. At puberty, high concentrations of androgens are responsible for the development of human male secondary sex characteristics, such as male patterns of hair growth and a low voice, as well as the increase in muscle and bone mass typical of males. The muscle-building action of testosterone and other anabolic steroids has enticed some athletes to take them as supplements, although these drugs appear to give little competitive advantage in sports requiring fine muscle coordination and endurance. Furthermore, abuse of anabolic steroids carries many health risks **(Figure 45.14)**, and they are banned in most competitive sports.

Estrogens, of which the most important is estradiol, are responsible for the maintenance of the female reproductive system and for the development of female secondary sex

▲ **Figure 45.14 Male breast enlargement due to anabolic steroids.** Steroid abuse disrupts the body's normal production of hormones, which causes both short- and long-term health risks. Men may also experience infertility and testicular atrophy. Women may undergo masculinizing effects such as breast shrinkage, voice deepening, and excessive growth of body hair. Both sexes can develop acne and male-pattern baldness. Potential long-term health risks include heart and liver damage.

characteristics. In mammals, **progestins,** which include progesterone, are primarily involved in preparing and maintaining the uterus, which supports the growth and development of an embryo.

Both estrogens and androgens are components of complex neuroendocrine pathways. Their synthesis is controlled by gonadotropins (FSH and LH) from the anterior pituitary gland (see Figure 45.8). FSH and LH secretion in turn is controlled by a releasing hormone from the hypothalamus, GnRH (gonadotropin-releasing hormone). In Chapter 46 we will examine in detail the feedback relationships that regulate the secretion of gonadal steroids.

Melatonin and Biorhythms

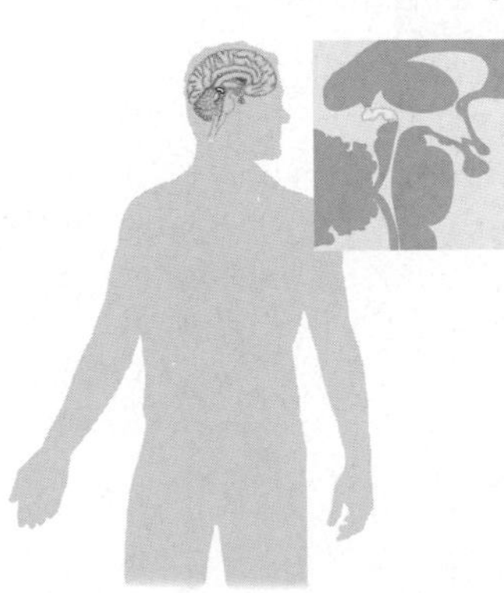

We conclude our discussion of the vertebrate endocrine system with the **pineal gland,** a small mass of tissue near the center of the mammalian brain (it is found closer to the brain surface in some other vertebrates). The pineal gland synthesizes and secretes the hormone **melatonin,** a modified amino acid. Depending on the species, the pineal gland contains light-sensitive cells or has nervous connections from the eyes that control its secretory activity.

Melatonin regulates functions related to light and to seasons marked by changes in day length. Although melatonin affects skin pigmentation in many vertebrates, its primary functions are related to biological rhythms associated with reproduction. Melatonin is secreted at night, and the amount secreted depends on the length of the night. In winter, for example, the days are short and the nights are long, so more melatonin is secreted. Thus, melatonin production is a link between a biological clock and daily or seasonal activities, such as reproduction. Recent evidence suggests that the main target cells of melatonin are in a pair of brain structures called the suprachiasmatic nuclei (SCN), which function as a biological clock. Melatonin seems to decrease the activity of neurons in the SCN, and this may be related to its role in mediating rhythms. However, much remains to be learned about the precise role of melatonin and about biological clocks in general.

Concept Check 45.4

1. How does thyroxine (T_4) control its own production and secretion?
2. How do calcitonin and parathyroid hormone (PTH) maintain blood Ca^{2+} level near the set point?
3. In a glucose tolerance test, a person's blood glucose level is measured periodically following ingestion of a glucose-containing solution. In a healthy individual, the blood glucose level rises moderately at first but then falls to near normal within 2–3 hours. Predict the results of this test in a person with diabetes mellitus. Explain your answer.
4. How would a decrease in the number of corticosteroid receptors in the hypothalamus likely affect levels of corticosteroids in the blood?

For suggested answers, see Appendix A.

Concept 45.5

Invertebrate regulatory systems also involve endocrine and nervous system interactions

Invertebrate animals produce a variety of hormones in typical hormone-secreting endocrine cells and in neurosecretory cells. Some invertebrate hormones have homeostatic functions, such as regulation of water balance. However, we know the most about hormones that function in reproduction and development. In a hydra, for example, one hormone stimulates growth and budding (asexual reproduction) but prevents sexual reproduction. In more complex invertebrates, the endocrine and nervous systems are generally integrated in the control of reproduction and development. In the mollusc *Aplysia*, for instance, specialized nerve cells secrete a neurohormone that stimulates the laying of thousands of eggs and also inhibits feeding and locomotion, activities that interfere with reproduction.

All groups of arthropods possess extensive endocrine systems. Crustaceans have hormones that function in growth and reproduction, water balance, movement of pigments in the integument and eyes, and regulation of metabolism. Having exoskeletons that cannot stretch, crustaceans and insects appear to grow in spurts, shedding the old exoskeleton and secreting a new one with each molt. Further, most insects acquire their adult characteristics in a single terminal molt. In insects and crustaceans (and most likely in all arthropods with exoskeletons), molting is triggered by a hormone.

The hormonal regulation of insect development has been studied extensively. Three hormones play major roles in molting and metamorphosis into the adult form **(Figure 45.15)**. **Brain hormone,** produced by neurosecretory cells in the insect brain, stimulates the release of ecdysone from the prothoracic glands, a pair of endocrine glands just behind the head. **Ecdysone** promotes molting and the development of adult characteristics, as in the change from a caterpillar to a butterfly. Brain hormone and ecdysone are balanced by the third hormone in this system, juvenile hormone. **Juvenile hormone** is secreted by a pair of small endocrine glands just behind the brain, the corpora allata (singular, *corpus allatum*), which are somewhat analogous to the anterior pituitary gland in vertebrates. As its name suggests, juvenile hormone promotes the retention of larval (juvenile) characteristics.

In the presence of a relatively high concentration of juvenile hormone, ecdysone can still stimulate molting, but the product is simply a larger larva. Only when the level of juvenile hor-

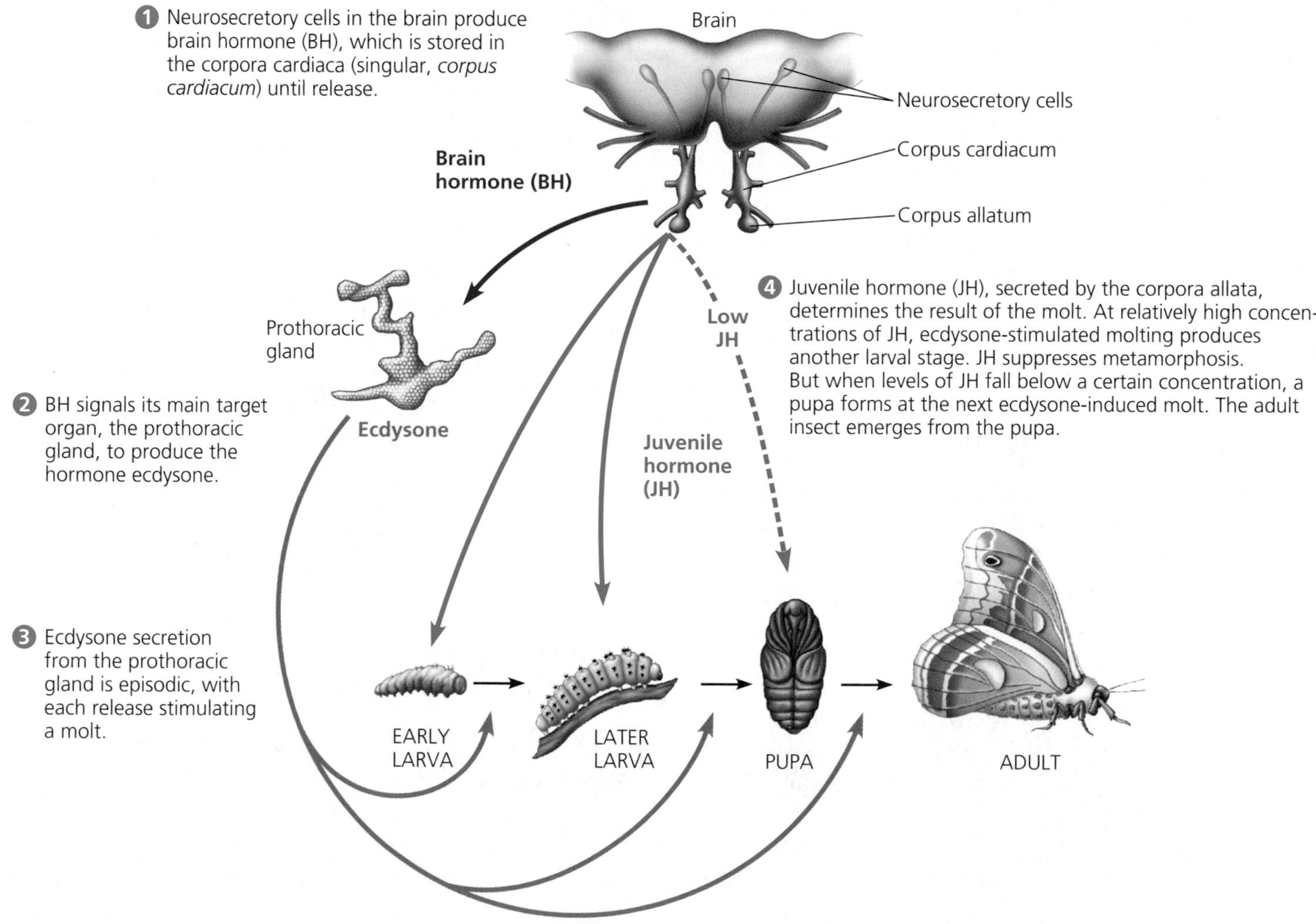

1 Neurosecretory cells in the brain produce brain hormone (BH), which is stored in the corpora cardiaca (singular, *corpus cardiacum*) until release.
Brain
Neurosecretory cells
Corpus cardiacum
Corpus allatum
Brain hormone (BH)
Prothoracic gland
2 BH signals its main target organ, the prothoracic gland, to produce the hormone ecdysone.
Ecdysone
Low JH
Juvenile hormone (JH)
4 Juvenile hormone (JH), secreted by the corpora allata, determines the result of the molt. At relatively high concentrations of JH, ecdysone-stimulated molting produces another larval stage. JH suppresses metamorphosis. But when levels of JH fall below a certain concentration, a pupa forms at the next ecdysone-induced molt. The adult insect emerges from the pupa.
3 Ecdysone secretion from the prothoracic gland is episodic, with each release stimulating a molt.
EARLY LARVA
LATER LARVA
PUPA
ADULT

▲ **Figure 45.15 Hormonal regulation of insect development.** Most insects go through a series of larval stages, with each molt (shedding of the old exoskeleton) leading to a larger larva. Molting of the final larval stage gives rise to a pupa, in which metamorphosis produces the adult form of the insect. Hormones control the progression of stages, as shown here.

mone wanes can ecdysone-induced molting produce a developmental stage called a pupa. Within the pupa, metamorphosis replaces larval anatomy with the insect's adult form. Synthetic versions of juvenile hormone are now being used as insecticides to prevent insects from maturing into reproducing adults.

In the next chapter, we will look at reproduction in both vertebrates and invertebrates. There we will see that the endocrine system is central not only to the survival of the individual but also to the propagation of the species.

Concept Check 45.5

1. How does the nervous system contribute to molting in insects?
2. Juvenile hormone is sometimes used commercially as an insecticide. What effects does it have on insect populations? Explain.

For suggested answers, see Appendix A.

Chapter 45 Review

Go to www.campbellbiology.com or the student CD-ROM to explore Activities, Investigations, and other interactive study aids.

SUMMARY OF KEY CONCEPTS

Concept 45.1

The endocrine system and the nervous system act individually and together in regulating an animal's physiology

▶ **Overlap Between Endocrine and Nervous Regulation (p. 944)** The endocrine and nervous systems often function together in maintaining homeostasis, development, and reproduction. Endocrine glands and various organs with primarily nonendocrine functions secrete hormones, and specialized secretory cells derived from nervous tissue secrete neurohormones. Both classes of hormonal signals circulate through the body to their target tissues, functioning as long-distance regulators.

▶ **Control Pathways and Feedback Loops (pp. 944–945)** There are three major types of hormonal control pathways: endocrine, neurohormone, and neuroendocrine. The basic components of a biological regulatory system are present in each kind of pathway. Negative feedback regulates many hormonal pathways involved in homeostasis.

Concept 45.2

Hormones and other chemical signals bind to target cell receptors, initiating pathways that culminate in specific cell responses

▶ **Cell-Surface Receptors for Water-Soluble Hormones (pp. 946–947)** Peptide/protein hormones and most of those derived from amino acids bind to receptors embedded in the plasma membrane. Hormone binding triggers an intracellular signal transduction pathway leading to specific responses in the cytoplasm or changes in gene expression. The same hormone may have different effects on target cells that have different receptors for the hormone, different signal transduction pathways, or different effector proteins.
Activity *Overview of Cell Signaling*
Activity *Peptide Hormone Action*

▶ **Intracellular Receptors for Lipid-Soluble Hormones (p. 947)** Steroid hormones, thyroid hormones, and the hormonal form of vitamin D enter target cells and bind to specific protein receptors in the cytoplasm or nucleus. The hormone-receptor complexes then act as transcription factors in the nucleus, regulating transcription of specific genes.
Activity *Steroid Hormone Action*

▶ **Paracrine Signaling by Local Regulators (pp. 947–948)** Various types of chemical signals elicit responses in nearby target cells. Local regulators include neurotransmitters, cytokines and growth factors (proteins/peptides), nitric oxide (a gas), and prostaglandins (modified fatty acids).

Concept 45.3

The hypothalamus and pituitary integrate many functions of the vertebrate endocrine system

▶ **Relationship Between the Hypothalamus and Pituitary Gland (pp. 950–951)** The hypothalamus, a region on the underside of the brain, contains different sets of neurosecretory cells. Some produce direct-acting hormones that are stored in and released from the posterior pituitary. Other hypothalamic cells produce tropic hormones that are transported by portal blood vessels to the anterior pituitary, an endocrine gland. These tropic hormones control release of hormones from the anterior pituitary.

▶ **Posterior Pituitary Hormones (p. 951)** The two hormones released from the posterior pituitary act directly on nonendocrine tissues. Oxytocin induces uterine contractions and milk ejection, and antidiuretic hormone (ADH) enhances water reabsorption in the kidneys.

▶ **Anterior Pituitary Hormones (pp. 951–952)** Both tropic and nontropic hormones are produced by the anterior pituitary. The four strictly tropic hormones are thyroid-stimulating hormone (TSH), follicle-stimulating hormone (FSH), luteinizing hormone (LH), and adrenocorticotropic hormone (ACTH). Each acts on its target endocrine tissue to stimulate release of hormone(s) with direct metabolic or developmental effects. Prolactin, melanocyte-stimulating hormone (MSH), and β-endorphin are nontropic anterior pituitary hormones. Prolactin stimulates lactation in mammals but has diverse effects in different vertebrates. MSH influences skin pigmentation in some vertebrates and fat metabolism in mammals. Endorphins inhibit the perception of pain. Growth hormone (GH) promotes growth directly and has diverse metabolic effects; it also stimulates the production of growth factors by other tissues (a tropic effect).

Concept 45.4

Nonpituitary hormones help regulate metabolism, homeostasis, development, and behavior

- **Thyroid Hormones** (pp. 953–954) The thyroid gland produces iodine-containing hormones (T_3 and T_4) that stimulate metabolism and influence development and maturation. Secretion of T_3 and T_4 is controlled by the hypothalamus and pituitary in a complex neuroendocrine pathway involving two negative-feedback loops. The thyroid also secretes calcitonin, which functions in calcium homeostasis.
  Investigation *How Do Thyroxine and TSH Affect Metabolism?*
- **Parathyroid Hormone and Calcitonin: Control of Blood Calcium** (pp. 954–955) Two antagonistic hormones, calcitonin and parathyroid hormone (PTH), play the major role in calcium (Ca^{2+}) homeostasis in mammals. Calcitonin, secreted by the thyroid, stimulates Ca^{2+} deposition in bones and excretion by kidneys, thereby decreasing blood Ca^{2+} levels. PTH, secreted by the parathyroid glands, has the opposite effects on bones and kidneys, thereby increasing blood Ca^{2+} levels. PTH also has an indirect effect, stimulating the kidneys to activate vitamin D, which in turn promotes intestinal uptake of Ca^{2+} from food.
- **Insulin and Glucagon: Control of Blood Glucose** (pp. 955–956) Two types of endocrine cells in the pancreas secrete insulin and glucagon, antagonistic hormones that help maintain glucose homeostasis. Insulin (from beta cells) reduces blood glucose levels by promoting the cellular uptake of glucose, glycogen formation in the liver, protein synthesis, and fat storage. Glucagon (from alpha cells) increases blood glucose levels by stimulating the conversion of glycogen to glucose in the liver and the breakdown of fat and protein to glucose. Diabetes mellitus, which is marked by elevated blood glucose levels, may be caused by inadequate production of insulin (type I) or loss of responsiveness of target cells to insulin (type II).
- **Adrenal Hormones: Response to Stress** (pp. 956–958) Neurosecretory cells in the adrenal medulla release epinephrine and norepinephrine in response to stress-activated impulses from the nervous system. These hormones mediate various fight-or-flight responses. The adrenal cortex releases three functional classes of steroid hormones. Glucocorticoids, such as cortisol, influence glucose metabolism and the immune system; mineralocorticoids, primarily aldosterone, affect salt and water balance. The adrenal cortex also produces small amounts of sex hormones.
- **Gonadal Sex Hormones** (pp. 958–959) The gonads—testes and ovaries—produce most of the body's sex hormones: androgens, estrogens, and progestins. All three types are produced in males and females but in different proportions.
  Activity *Human Endocrine Glands and Hormones*
- **Melatonin and Biorhythms** (p. 959) The pineal gland, located within the brain, secretes melatonin. Release of melatonin is controlled by light/dark cycles. Its primary functions appear to be related to biological rhythms associated with reproduction.

Concept 45.5

Invertebrate regulatory systems also involve endocrine and nervous system interactions

- Diverse hormones regulate different aspects of homeostasis in invertebrates. In insects, molting and development are controlled by three main hormones: brain hormone, a tropic neurohormone; ecdysone, whose release is triggered by brain hormone; and juvenile hormone (pp. 959–961).

TESTING YOUR KNOWLEDGE

Self-Quiz

1. Which of the following is *not* an accurate statement?
   a. Hormones are chemical messengers that travel to target cells through the circulatory system.
   b. Hormones often regulate homeostasis through antagonistic functions.
   c. Hormones of the same chemical class usually have the same function.
   d. Hormones are secreted by specialized cells usually located in endocrine glands.
   e. Hormones are often regulated through feedback loops.

2. A distinctive feature of the mechanism of action of thyroid hormones and steroid hormones is that
   a. these hormones are regulated by feedback loops.
   b. target cells react more rapidly to these hormones than to local regulators.
   c. these hormones bind with specific receptor proteins on the plasma membrane of target cells.
   d. these hormones bind to receptors inside cells.
   e. these hormones affect metabolism.

3. Growth factors are local regulators that
   a. are produced by the anterior pituitary.
   b. are modified fatty acids that stimulate bone and cartilage growth.
   c. are found on the surface of cancer cells and stimulate abnormal cell division.
   d. are proteins that bind to cell-surface receptors and stimulate growth and development of target cells.
   e. convey messages between nerve cells.

4. Which of the following hormones is *incorrectly* paired with its action?
   a. oxytocin—stimulates uterine contractions during childbirth
   b. thyroxine—stimulates metabolic processes
   c. insulin—stimulates glycogen breakdown in the liver
   d. ACTH—stimulates the release of glucocorticoids by the adrenal cortex
   e. melatonin—affects biological rhythms, seasonal reproduction

5. An example of antagonistic hormones controlling homeostasis is
   a. thyroxine and parathyroid hormone in calcium balance.
   b. insulin and glucagon in glucose metabolism.
   c. progestins and estrogens in sexual differentiation.
   d. epinephrine and norepinephrine in fight-or-flight responses.
   e. oxytocin and prolactin in milk production.

6. Which of the following is *not* an example of the structure–function relationship between the nervous and endocrine systems?
   a. the secretion of hormones by neurosecretory cells
   b. the multiple functions of norepinephrine
   c. the stimulation of the adrenal medulla in the short-term response to stress
   d. the embryonic development of the posterior pituitary from the hypothalamus
   e. the alteration of gene expression by steroid hormones

7. A portal vessel carries blood from the hypothalamus directly to the
   a. thyroid.
   b. pineal gland.
   c. anterior pituitary.
   d. posterior pituitary.
   e. liver.

8. Which of the following is the most likely explanation for hypothyroidism in a patient whose iodine level is normal?
   a. a disproportionate production of T_3 to T_4
   b. hyposecretion of TSH
   c. hypersecretion of TSH
   d. hypersecretion of MSH
   e. a decrease in the thyroid secretion of calcitonin

9. The main target organs for tropic hormones are
   a. muscles.
   b. blood vessels.
   c. endocrine glands.
   d. kidneys.
   e. nerves.

10. The relationship between the insect hormones ecdysone and brain hormone
   a. is an example of the interaction between the endocrine and nervous systems.
   b. illustrates homeostasis achieved by positive feedback.
   c. demonstrates that peptide-derived hormones have more widespread effects than steroid hormones.
   d. illustrates homeostasis maintained by antagonistic hormones.
   e. demonstrates competitive inhibition for the hormone receptor.

For Self-Quiz answers, see Appendix A.

Go to the website or CD-ROM for more quiz questions.

Evolution Connection

The intracellular receptors used by all the steroid and thyroid hormones are similar enough in structure that they are all considered members of one "superfamily" of proteins. Propose a hypothesis for how the genes encoding these receptors may have evolved. (*Hint:* See Figure 19.19.)

Scientific Inquiry

In your response to the Evolution Connection question on this page, you came up with a hypothesis. How could you test your hypothesis using DNA sequence data?

Investigation *How Do Thyroxine and TSH Affect Metabolism?*

Science, Technology, and Society

Growth hormone (GH) treatments have enabled hundreds of children with pituitary dwarfism to reach a stature within the normal range. Some parents of children who are extremely short, but who do not suffer from an endocrine disorder, also request GH treatment so that their children will grow faster and taller. There can be potentially harmful effects, such as a reduction in body fat and an increase in muscle mass. And no one yet knows if GH injections could have seriously harmful long-term effects in individuals who do not have a hypopituitary condition. What guidelines would you propose for the use of GH in children? Explain your reasoning.

46 Animal Reproduction

▲ **Figure 46.1 Earthworms mating.**

Key Concepts

46.1 Both asexual and sexual reproduction occur in the animal kingdom
46.2 Fertilization depends on mechanisms that help sperm meet eggs of the same species
46.3 Reproductive organs produce and transport gametes: focus on humans
46.4 In humans and other mammals, a complex interplay of hormones regulates gametogenesis
46.5 In humans and other placental mammals, an embryo grows into a newborn in the mother's uterus

Overview

Doubling Up for Sexual Reproduction

The two earthworms in **Figure 46.1** are mating. Unless disturbed, they will remain above the ground and joined like this for several hours. Each worm produces both sperm and eggs, each donates and receives sperm during mating, and each will produce fertilized eggs. In a few weeks, sexual reproduction will be completed when new worms hatch.

The many aspects of animal form and function we have studied in earlier chapters can be viewed, in the broadest context, as adaptations contributing to reproductive success. Individuals are transient. A population transcends finite life spans only by reproduction, the creation of new individuals from existing ones. Animal reproduction is the subject of this chapter. We will first compare the diverse reproductive mechanisms that have evolved in the animal kingdom and then examine the details of mammalian, particularly human, reproduction. Deferring the cellular and molecular details of embryonic development until the next chapter, we focus here on the physiology of reproduction, mostly from the perspective of the parents.

Concept 46.1

Both asexual and sexual reproduction occur in the animal kingdom

There are two principal modes of animal reproduction. **Asexual** (from the Greek, "without sex") **reproduction** is the creation of new individuals whose genes all come from one parent without the fusion of egg and sperm. In most cases, asexual reproduction relies entirely on mitotic cell division. **Sexual reproduction** is the creation of offspring by the fusion of haploid **gametes** to form a **zygote** (fertilized egg), which is diploid. Animal gametes arise by meiosis (see Figure 13.8). The female gamete, the unfertilized **egg** (also called an **ovum**), is a relatively large cell and not motile. The male gamete, the **sperm**, is generally a much smaller, motile cell. Sexual reproduction increases genetic variability among offspring by generating unique combinations of genes inherited from two parents. By producing offspring having a variety of phenotypes, sexual reproduction may enhance the reproductive success of parents when environmental factors (including pathogens) change relatively rapidly (see Chapter 23, p. 469).

Mechanisms of Asexual Reproduction

Many invertebrates can reproduce asexually by **fission**, the separation of a parent into two or more individuals of approximately

▲ **Figure 46.2 Asexual reproduction of a sea anemone (*Anthopleura elegantissima*).** The individual in the center of this photograph is undergoing fission, a type of asexual reproduction. Two smaller individuals will form as the parent divides approximately in half. The offspring will be genetic copies of the parent.

equal size **(Figure 46.2)**. Also common among invertebrates is **budding**, in which new individuals arise from outgrowths of existing ones. For example, in certain cnidarians and tunicates, new individuals grow out from the body of a parent (see Figure 13.2). The offspring may either detach from the parent or remain joined, eventually forming extensive colonies. Stony corals, which may be more than 1 m across, are cnidarian colonies of several thousand connected individuals. In another form of asexual reproduction, some invertebrates, such as sponges, release specialized groups of cells that can grow into new individuals.

Yet another type of asexual reproduction starts with the process of **fragmentation**, the breaking of the body into several pieces, some or all of which develop into complete adults. For an animal to reproduce this way, fragmentation must be accompanied by **regeneration**, the regrowth of lost body parts. Reproduction by fragmentation and regeneration occurs in many sponges, cnidarians, polychaete annelids, and tunicates. Many animals can also replace lost appendages by regeneration—most sea stars can grow new arms when injured, for example—but this is not reproduction because new individuals are not created. In sea stars of the genus *Linckia*, a whole new individual may develop from an isolated arm. Thus, a single animal with five arms, if broken apart, could asexually give rise to five offspring.

Asexual reproduction has several potential advantages. For instance, it enables animals living in isolation to produce offspring without locating mates. It can also create numerous offspring in a short amount of time, which is ideal for colonizing a habitat rapidly. Theoretically, asexual reproduction is most advantageous in stable, favorable environments because it perpetuates successful genotypes precisely.

Reproductive Cycles and Patterns

Most animals exhibit cycles in reproductive activity, often related to changing seasons. The periodic nature of reproduction allows animals to conserve resources and reproduce when more energy is available than is needed for maintenance and when environmental conditions favor the survival of offspring. Ewes (female sheep), for example, have 15-day reproductive cycles that generally occur during fall and early winter, ensuring that most lambs are born in the late winter or spring. Even animals that live in relatively unvarying habitats, such as the tropics or the ocean, generally reproduce only at certain times of the year. Reproductive cycles are controlled by a combination of hormonal and environmental cues, the latter including seasonal temperature, rainfall, day length, and lunar cycles.

Animals may reproduce exclusively asexually or sexually, or they may alternate between the two modes. In aphids, rotifers, and the freshwater crustacean *Daphnia*, a female can produce eggs of two types, depending on environmental conditions. One type of egg is fertilized, but the other type develops by **parthenogenesis**, a process in which an egg develops without being fertilized. Haploid adults arising from parthenogenesis produce eggs without meiosis. In the case of *Daphnia*, the switch from sexual to asexual reproduction is often related to season. Asexual reproduction occurs under favorable conditions, and sexual reproduction occurs during times of environmental stress.

Parthenogenesis has a role in the social organization of certain species of bees, wasps, and ants. Male honeybees, or drones, are produced parthenogenetically, whereas females, both sterile workers and reproductive females (queens), develop from fertilized eggs.

Among vertebrates, several genera of fishes, amphibians, and lizards reproduce exclusively by a complex form of parthenogenesis that involves the doubling of chromosomes after meiosis, a process that creates diploid "zygotes." For example, about 15 species of whiptail lizards (genus *Cnemidophorus*) reproduce exclusively by parthenogenesis. There are no males in these species, but the lizards imitate courtship and mating behavior typical of sexual species of the same genus. During the breeding season, one female of each mating pair mimics a male **(Figure 46.3a)**. The roles change two or three times during the season, female behavior occurring prior to ovulation (the release of eggs), when the level of the female sex hormone estrogen is high, and male behavior occurring after ovulation, when the level of estrogen drops **(Figure 46.3b)**. Ovulation is more likely to occur if one individual is mounted by another during the critical time of the hormone cycle; isolated lizards lay fewer eggs than those that go through the motions of sex. Apparently, these parthenogenetic lizards, which evolved from species that have two sexes, still require certain sexual stimuli for maximum reproductive success.

(a) Both lizards in this photograph are *C. uniparens* females. The one on top is playing the role of a male. Every two or three weeks during the breeding season, individuals switch sex roles.

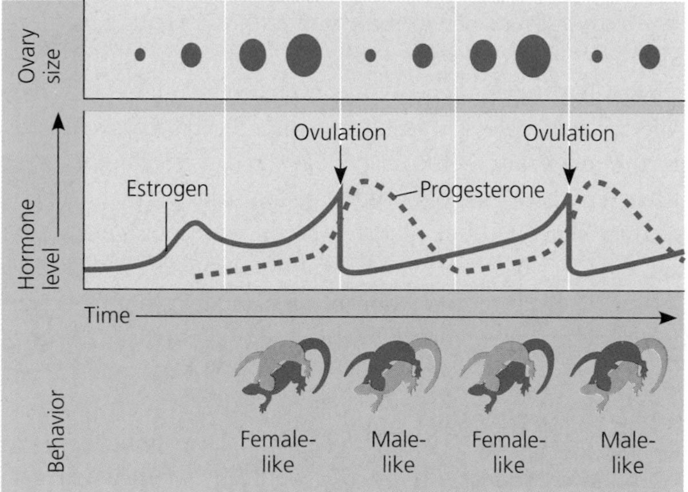

(b) The sexual behavior of *C. uniparens* is correlated with the cycle of ovulation mediated by sex hormones. As the blood level of estrogen rises, the ovaries grow, and the lizard behaves like a female. After ovulation, the estrogen level drops abruptly, and the progesterone level rises; these hormone levels correlate with male behavior.

▲ **Figure 46.3 Sexual behavior in parthenogenetic lizards.** The desert-grassland whiptail lizard (*Cnemidophorus uniparens*) is an all-female species. These reptiles reproduce by parthenogenesis, the development of an unfertilized egg. However, ovulation is stimulated by mating behavior.

Sexual reproduction presents a special problem for sessile or burrowing animals and for parasites, such as tapeworms, which may seldom encounter a member of the opposite sex. One solution to this problem is **hermaphroditism,** in which each individual has both male and female reproductive systems (the term is derived from Hermes and Aphrodite, a Greek god and goddess). Although some hermaphrodites fertilize themselves, most must mate with a member of the same species. In this case, each animal serves as both male and female, donating and receiving sperm, as we saw for earthworms. Every individual is a potential mate, and each hermaphroditic mating can result in twice as many offspring as from a male-female mating, where only one partner's eggs are fertilized.

▲ **Figure 46.4 Sex reversal in a sequential hermaphrodite.** A male Caribbean bluehead wrasse and two smaller females feed on a sea urchin. All wrasses of this species are born female, but the oldest, largest individuals complete their lives as males.

Another remarkable reproductive pattern is **sequential hermaphroditism,** in which an individual reverses its sex during its lifetime. In some species, the sequential hermaphrodite is female first, while in other species it is male first. In various species of reef fishes called wrasses, sex reversal is associated with age and size. For example, the Caribbean bluehead wrasse is a female-first species in which only the largest (usually the oldest) individuals change from female to male **(Figure 46.4)**. These fish live in harems consisting of a single male and several females. If the male dies or is removed in an experiment, the largest female in the harem becomes the new male. Within a week, the transformed individual is producing sperm instead of eggs. In this species, the male defends the harem against intruders, and thus larger size may give a greater reproductive advantage to males than it does to females. In contrast, there are male-first animals that change from male to female when size increases. In such cases, greater size may increase the reproductive success of females more than it does males. For example, the production of huge numbers of gametes is an important asset for sedentary animals, such as oysters, that release their gametes into the surrounding water. Larger females tend to produce more eggs than smaller ones, and species of oysters that are sequential hermaphrodites are generally male first.

The diverse reproductive cycles and patterns we observe in the animal kingdom are adaptations that have evolved by natural selection. We will see many other examples as we survey the various mechanisms of sexual reproduction.

Concept Check 46.1

1. What is the most important difference between the outcomes of asexual and sexual reproduction?
2. How is the term *sequential hermaphroditism* misleading?

For suggested answers, see Appendix A.

Concept 46.2

Fertilization depends on mechanisms that help sperm meet eggs of the same species

The mechanisms of **fertilization**, the union of sperm and egg, play an important part in sexual reproduction. Some species have **external fertilization**: Eggs are released by the female into a wet environment, where they are fertilized by the male **(Figure 46.5)**. Other species have **internal fertilization**: Sperm are deposited in or near the female reproductive tract, and fertilization occurs within the tract. (The cellular and molecular details of fertilization are discussed in Chapter 47.)

A moist habitat is almost always required for external fertilization, both to prevent gametes from drying out and to allow the sperm to swim to the eggs. Many aquatic invertebrates simply shed their eggs and sperm into the surroundings, and fertilization occurs without the parents actually making physical contact. However, timing is crucial to ensure that mature sperm encounter ripe eggs.

For species that use external fertilization, environmental cues such as temperature or day length may cause a whole population to release gametes at once, or chemical signals from one individual releasing gametes may trigger gamete release in others. Alternatively, individuals may exhibit specific mating behaviors leading to the fertilization of the eggs of one female by one male. Such "courtship" behavior has two important benefits: It allows mate selection and, by triggering the release of both sperm and eggs, increases the probability of successful fertilization.

Internal fertilization is an adaptation that, in terrestrial animals, enables sperm to reach an egg when the environment is dry. It requires cooperative behavior that leads to copulation. In some cases, uncharacteristic sexual behavior is eliminated by natural selection in a direct manner; for example, female spiders may eat males that fail to give or respond to specific signals during mating. Internal fertilization requires reproductive systems with structural adaptations, including copulatory organs that deliver sperm and receptacles for its storage and transport to ripe eggs.

However fertilization occurs, the mating animals may make use of **pheromones**, chemical signals released by one organism that influence the physiology and/or behavior of other individuals of the same species. Pheromones are small, volatile or water-soluble molecules that disperse into the environment and, like hormones, are active in minute amounts. Many pheromones function as mate attractants. The mate attractants of some female insects can be detected by males more than a kilometer away. (We will return to mating behavior and pheromones in Chapter 51.)

▲ **Figure 46.5 External fertilization.** Many amphibians shed gametes into the environment, where fertilization occurs. In most species, behavioral adaptations ensure that a male is present when the female releases eggs. Here, a female frog, clasped by a male (on top), has released a mass of eggs. The male released sperm (not visible) at the same time, and external fertilization has already occurred in the water.

Ensuring the Survival of Offspring

All species produce more offspring than can survive long enough to reproduce. Species with external fertilization usually produce enormous numbers of zygotes, but the proportion that survive and develop further is often quite small. Internal fertilization usually produces fewer zygotes, but this may be offset by greater protection of the embryos and parental care of the young. Major types of protection include tough eggshells, development of the embryo within the reproductive tract of the mother, and parental care of the eggs and offspring.

The embryos of many species of terrestrial animals develop in eggs that can withstand harsh environments. Reptiles and monotremes have amniotic eggs with calcium and protein shells that resist water loss and physical damage. (By contrast, the eggs of fishes and amphibians have only a gelatinous coat.)

Rather than secreting a protective shell around the egg, many animals retain the embryo, which develops within the female reproductive tract. Marsupial mammals, such as kangaroos and opossums, retain their embryos for only a short period in the uterus; during that time the embryos are nourished by the mother's blood supply through an organ called the placenta. The embryos then crawl out and complete fetal development attached to a mammary gland in the mother's pouch. However, the embryos of eutherian (placental) mammals, such as humans, develop entirely within the uterus, and the placenta is more complex.

When a kangaroo crawls out of its mother's pouch for the first time, or when a human is born, it is not yet capable of independent existence. We are familiar with adult birds feeding their young and mammals nursing their offspring, but parental care is much more widespread than we might suspect, and it takes a variety of unusual forms. In one species of tropical frog, for instance, the male carries the tadpoles in his stomach until

▲ **Figure 46.6 Parental care in an invertebrate.** Compared with many other insects, giant water bugs of the genus *Belostoma* produce relatively few offspring, but parental protection enhances survival. Fertilization is internal, but the female then glues her fertilized eggs to the back of the male (shown here). Whereas the males of most insect species provide no parental care for their offspring, the male giant water bug carries them for days, frequently fanning water over them. This treatment helps keep the eggs moist, aerated, and free of parasites.

they metamorphose and hop out as young frogs. There are also many cases of parental care among invertebrates **(Figure 46.6)**.

Gamete Production and Delivery

To reproduce sexually, animals must have systems that produce gametes and make them available to the gametes of the opposite sex. These reproductive systems are varied. The least complex systems do not even contain distinct **gonads,** the organs that produce gametes in most animals. Among the simplest systems are those of polychaete worms (phylum Annelida). Most polychaetes have separate sexes but do not have distinct gonads; rather, the eggs and sperm develop from undifferentiated cells lining the coelom. As the gametes mature, they are released from the body wall and fill the coelom. Depending on the species, mature gametes may be shed through the excretory openings, or the swelling mass of eggs may split the body open, killing the parent and spilling the eggs into the environment.

The most complex reproductive systems contain many sets of accessory tubes and glands that carry, nourish, and protect the gametes and the developing embryos. Many animals whose body plans are otherwise relatively simple possess highly complex reproductive systems. The reproductive systems of parasitic flatworms, for example, are among the most complex in the animal kingdom **(Figure 46.7)**. These worms are hermaphrodites.

Most insects have separate sexes with complex reproductive systems **(Figure 46.8)**. In the male, sperm develop in a pair of testes and are conveyed along a coiled duct to two seminal vesicles, where they are stored. During mating, sperm are ejaculated into the female reproductive system. In the female, eggs develop in a pair of ovaries and are conveyed through ducts to the vagina, where fertilization occurs. In many species, the female reproductive system includes a **spermatheca,** a sac in which sperm may be stored for a year or more.

The basic plans of all vertebrate reproductive systems are quite similar, but there are some important variations. In many nonmammalian vertebrates, the digestive, excretory, and reproductive systems have a common opening to the outside, the **cloaca,** which was probably present in the ancestors of all vertebrates. By contrast, most mammals lack a cloaca and have a separate opening for the digestive tract, and most female mammals have separate openings for the excretory and reproductive systems as well. The uterus of most vertebrates is

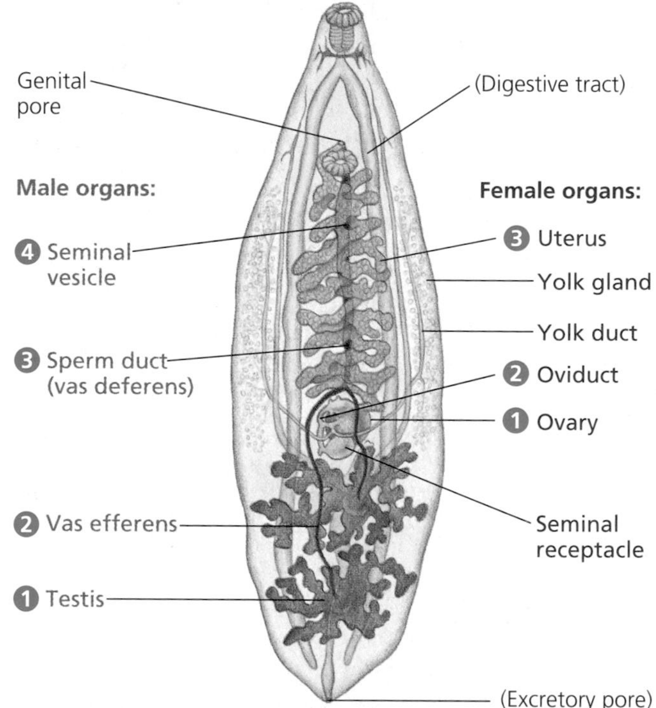

▲ **Figure 46.7 Reproductive anatomy of a parasitic flatworm.** Most flatworms (phylum Platyhelminthes) are hermaphroditic. Both male and female reproductive systems open to the outside via the genital pore. Sperm, made in the testes, travel as shown by the numbered sequence to the seminal vesicle, which stores them. During copulation, sperm are ejaculated into the female system (usually of another flatworm) and then move through the uterus to the seminal receptacle. Eggs from the ovary pass into the oviduct, where they are fertilized by sperm from the seminal receptacle and coated with yolk and shell material secreted by the yolk glands. From the oviduct, the fertilized eggs pass into the uterus and then out of the body.

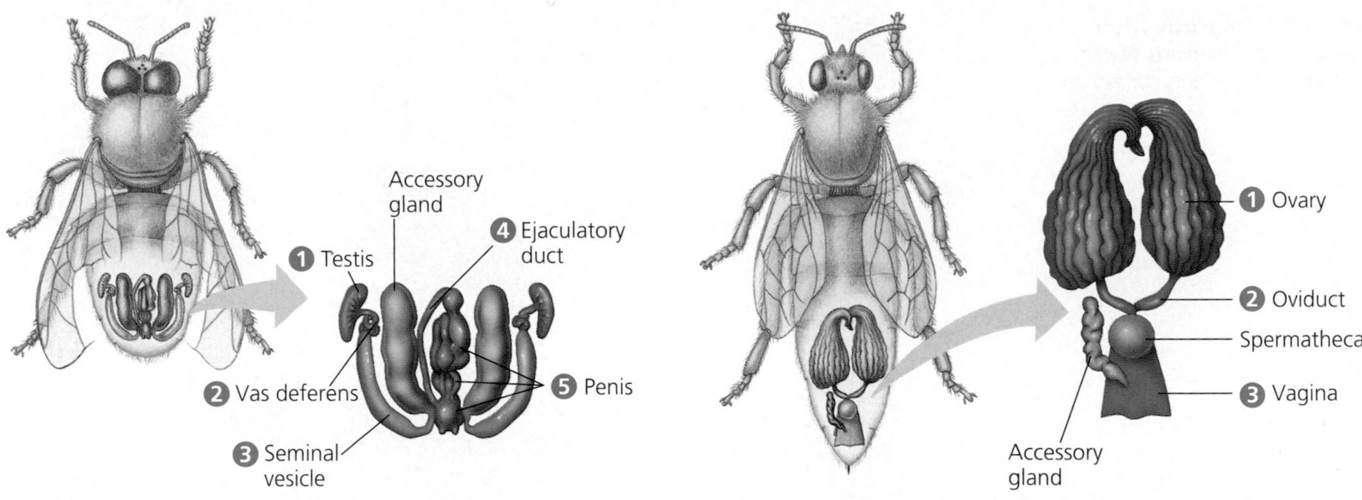

(a) Male honeybee (drone). Sperm form in the testes, pass through the sperm duct (vas deferens), and are stored in the seminal vesicle. The male ejaculates sperm along with fluid from the accessory glands. (Males of some species of insects and other arthropods have appendages called claspers that grasp the female during copulation.)

(b) Female honeybee (queen). Eggs develop in the ovaries and then pass through the oviducts and into the vagina. A pair of accessory glands (only one is shown) add protective secretions to the eggs in the vagina. After mating, sperm are stored in the spermatheca, a sac connected to the vagina by a short duct.

▲ **Figure 46.8 Insect reproductive anatomy.** Circled numbers indicate sequences of sperm and egg movement.

partly or completely divided into two chambers. However, in humans and other mammals that produce only a few young at a time, as well as in birds and many snakes, the uterus is a single structure. Male reproductive systems differ mainly in the copulatory organs. Many nonmammalian vertebrates do not have a well-developed penis and simply turn the cloaca inside out to ejaculate.

Concept Check 46.2

1. In what ways does internal fertilization facilitate terrestrial life?
2. By what "strategies" do animals with (a) external fertilization and (b) internal fertilization ensure that they will have offspring that survive to adulthood?

For suggested answers, see Appendix A.

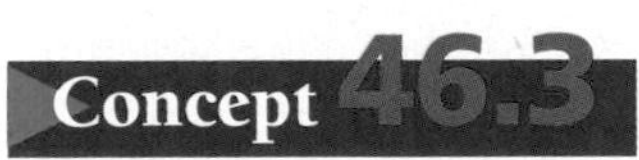

Concept 46.3

Reproductive organs produce and transport gametes: focus on humans

Female Reproductive Anatomy

The female's external reproductive structures are the clitoris and two sets of labia, which surround the clitoris and vaginal opening. The internal organs are a pair of gonads and a system of ducts and chambers that carry gametes and house the embryo and fetus (**Figure 46.9**, on the next page).

Ovaries

The female gonads, the **ovaries,** lie in the abdominal cavity, flanking the uterus and attached to it by a mesentery. Each ovary is enclosed in a tough protective capsule and contains many follicles. A **follicle** consists of one egg cell surrounded by one or more layers of follicle cells, which nourish and protect the developing egg cell. Most or all of the 400,000 follicles a woman will ever have are thought to be formed before her birth (but see the last paragraph in Figure 46.11). Only several hundred follicles will release egg cells during a woman's reproductive years. Starting at puberty and continuing until menopause, usually one follicle matures and releases its egg cell during each menstrual cycle. The cells of the follicle also produce estrogens, the primary female sex hormones. The egg cell is expelled from the follicle in the process of **ovulation**. The remaining follicular tissue then grows within the ovary and forms a solid mass called the **corpus luteum** ("yellow body"). The corpus luteum secretes additional estrogens and progesterone, a hormone that helps maintain the uterine lining during pregnancy. If the egg cell is not fertilized, the corpus luteum disintegrates, and a new follicle matures during the next cycle.

Oviducts and Uterus

The female reproductive system is not completely closed, and the egg cell is released into the abdominal cavity near the opening of the **oviduct,** or fallopian tube. The oviduct has a funnel-like opening, and cilia on the epithelium lining the duct help collect the egg cell by drawing fluid from the body cavity into the duct. The cilia also convey the egg cell down the duct to the **uterus,** also known as the womb. The uterus is a thick, muscular organ that can expand during pregnancy to accommodate

▶ **Figure 46.9 Reproductive anatomy of the human female.** For orientation, some nonreproductive structures are labeled in parentheses.

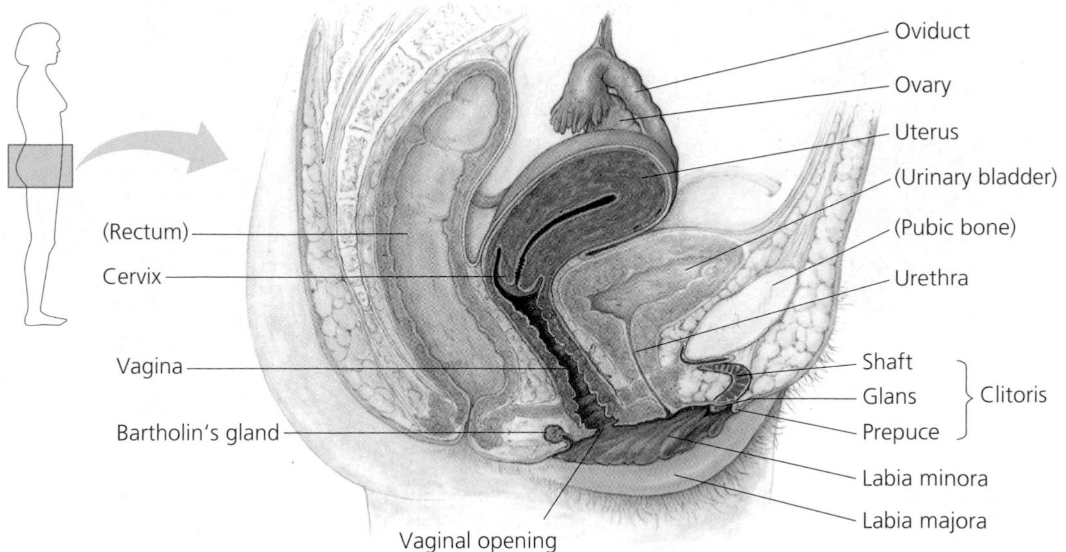

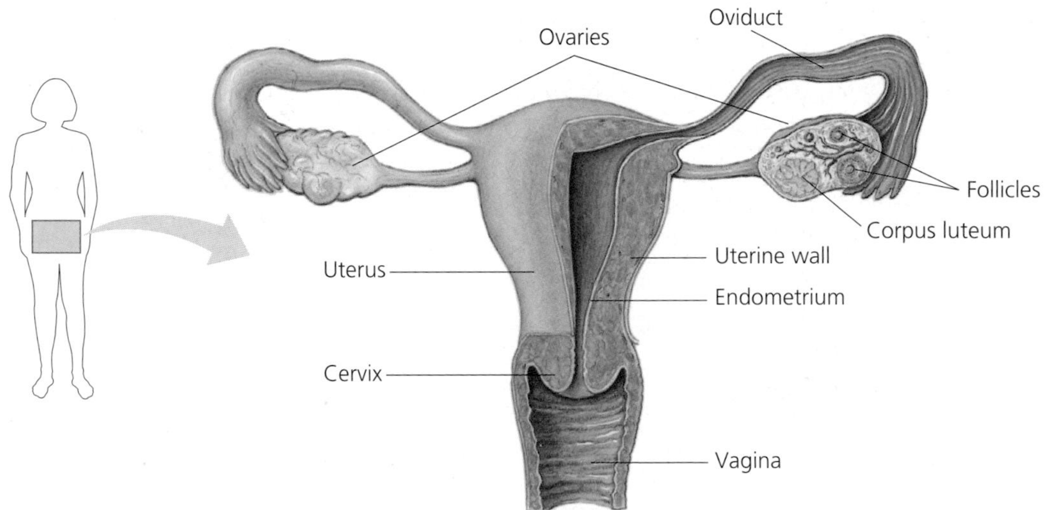

a 4-kg fetus. The inner lining of the uterus, the **endometrium**, is richly supplied with blood vessels. The neck of the uterus is the **cervix**, which opens into the vagina.

Vagina and Vulva

The **vagina** is a thin-walled chamber that is the repository for sperm during copulation and that serves as the birth canal through which a baby is born. It opens to the outside at the **vulva**, the collective term for the external female genitalia.

At birth, and usually until sexual intercourse or vigorous physical activity ruptures it, a thin piece of tissue called the **hymen** partly covers the vaginal opening in humans. The vaginal opening and the separate urethral opening are located within a recess called the **vestibule**, bordered by a pair of slender skin folds, the **labia minora**. A pair of thick, fatty ridges, the **labia majora**, encloses and protects the labia minora and vestibule. Located at the front edge of the vestibule, the **clitoris** consists of a short shaft supporting a rounded glans, or head, covered by a small hood of skin, the prepuce. During sexual arousal, the clitoris, vagina, and labia minora all engorge with blood and enlarge. The clitoris consists largely of erectile tissue. Richly supplied with nerve endings, it is one of the most sensitive points of sexual stimulation. During sexual arousal, **Bartholin's glands**, located near the vaginal opening, secrete mucus into the vestibule, keeping it lubricated and facilitating intercourse.

Mammary Glands

Mammary glands are present in both sexes but normally function only in women. They are not part of the reproductive system but are important to mammalian reproduction. Within the glands, small sacs of epithelial tissue secrete milk, which drains into a series of ducts opening at the nipple. Fatty (adipose) tissue forms the main mass of the mammary gland of a nonlactating mammal. The low level of estrogen in males prevents the development of both the secretory apparatus and the fat deposits, so male breasts remain small, and the nipples are not connected to the ducts.

Male Reproductive Anatomy

In most mammalian species, including humans, the male's external reproductive organs are the scrotum and penis. The internal reproductive organs consist of gonads that produce gametes (sperm cells) and hormones, accessory glands that secrete products essential to sperm movement, and ducts that carry the sperm and glandular secretions **(Figure 46.10)**.

Testes

The male gonads, or **testes** (singular, *testis*), consist of many highly coiled tubes surrounded by several layers of connective tissue. These tubes are the **seminiferous tubules,** where sperm form. The **Leydig cells** that are scattered between the seminiferous tubules produce testosterone and other androgens.

Production of normal sperm cannot occur at the body temperatures of most mammals, and the testes of humans and many other mammals are held outside the abdominal cavity in the **scrotum,** which is a fold of the body wall. The temperature in a scrotum is about 2°C below that in the abdominal cavity. The testes develop high in the abdominal cavity and descend into the scrotum just before birth. In many rodents, the testes are drawn back into the abdominal cavity between breeding seasons, interrupting sperm maturation. Some mammals whose body temperature is low enough to allow sperm maturation, such as monotremes, whales, and elephants, retain the testes within the abdominal cavity permanently.

Ducts

From the seminiferous tubules of a testis, the sperm pass into the coiled tubules of the **epididymis.** It takes about 20 days for sperm to pass through the 6-m-long tubules of each epididymis of a human male. During this passage, the sperm become motile and gain the ability to fertilize an egg. During **ejaculation,** the sperm are propelled from the epididymis through the muscular **vas deferens.** These two ducts (one from each epididymis) run from the scrotum around and behind the

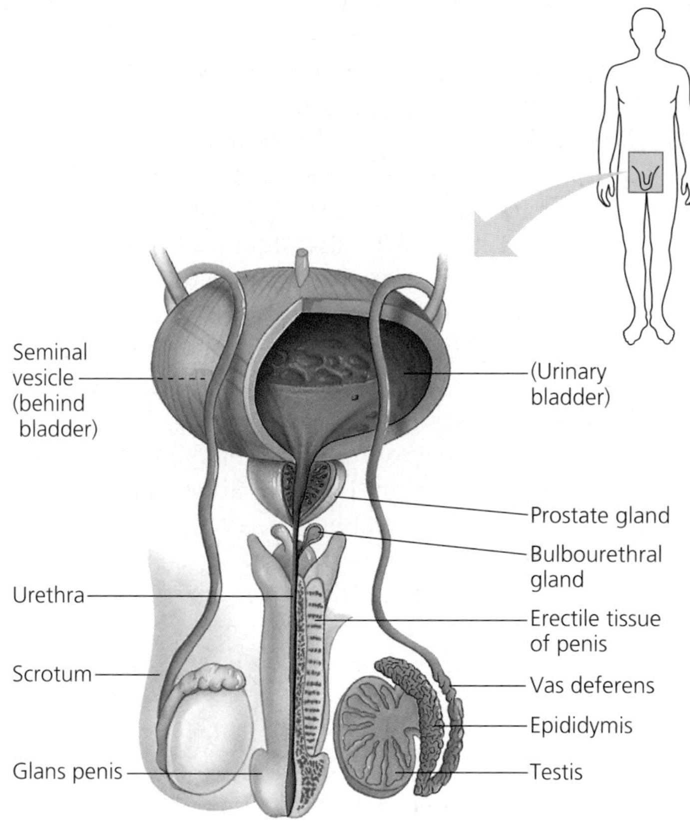

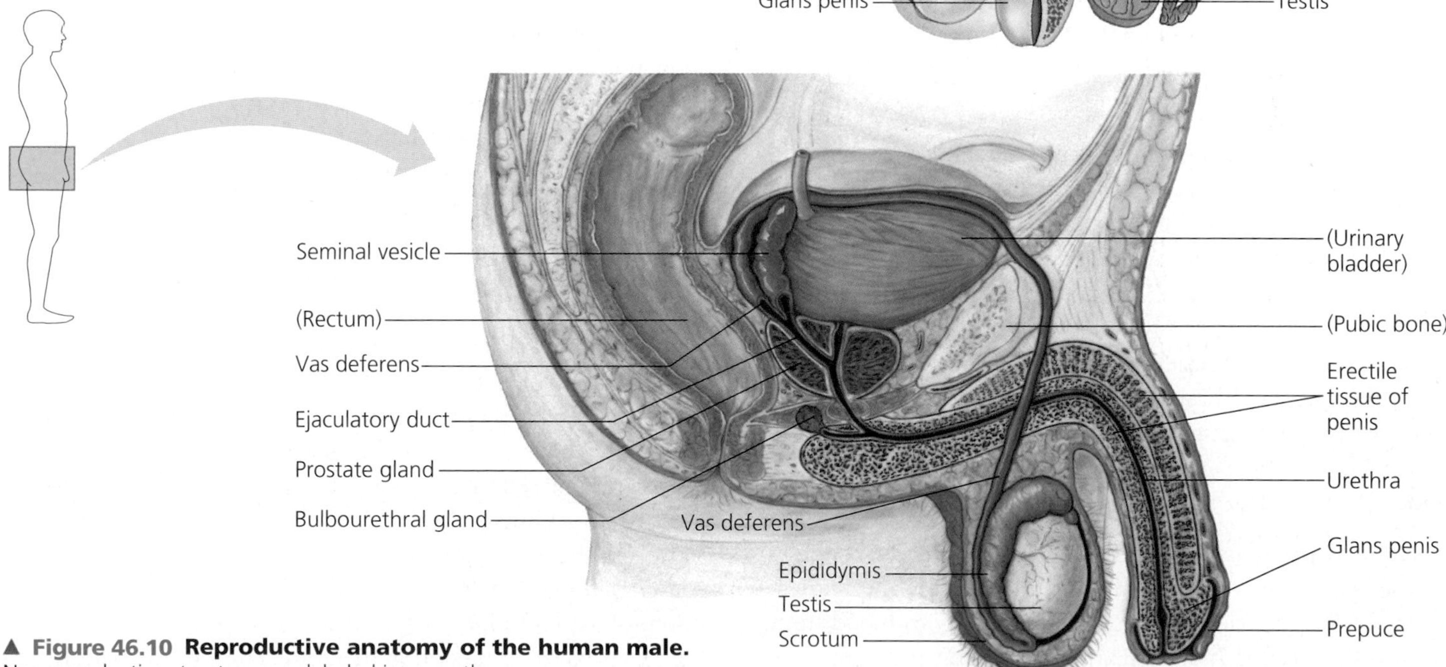

▲ **Figure 46.10 Reproductive anatomy of the human male.** Nonreproductive structures are labeled in parentheses.

urinary bladder, where each joins a duct from the seminal vesicle, forming a short **ejaculatory duct.** The ejaculatory ducts open into the **urethra,** the tube that drains both the excretory system and the reproductive system. The urethra runs through the penis and opens to the outside at the tip of the penis.

Glands

Three sets of accessory glands—the seminal vesicles, prostate gland, and bulbourethral glands—add secretions to the **semen,** the fluid that is ejaculated. A pair of **seminal vesicles** contributes about 60% of the total volume of semen. The fluid from the seminal vesicles is thick, yellowish, and alkaline. It contains mucus, the sugar fructose (which provides most of the energy used by the sperm), a coagulating enzyme, ascorbic acid, and prostaglandins, local regulators discussed in Chapter 45.

The **prostate gland** is the largest of the semen-secreting glands. It secretes its products directly into the urethra through several small ducts. Prostatic fluid is thin and milky; it contains anticoagulant enzymes and citrate (a sperm nutrient). The prostate gland is the source of some of the most common medical problems of men over age 40. Benign (noncancerous) enlargement of the prostate occurs in more than half of all men in this age-group and in virtually all men over 70. Prostate cancer is one of the most common cancers in men. It is treated surgically or with drugs that inhibit gonadotropins, resulting in reduced prostate activity and size.

The **bulbourethral glands** are a pair of small glands along the urethra below the prostate. Before ejaculation, they secrete a clear mucus that neutralizes any acidic urine remaining in the urethra. Bulbourethral fluid also carries some sperm released before ejaculation, which is one reason for the high failure rate of the withdrawal method of birth control.

Semen in the Female Reproductive Tract

A man usually ejaculates 2–5 mL of semen, and each milliliter may contain 50–130 million sperm. Once in the female reproductive tract, prostaglandins in the semen cause thinning of the mucus at the opening of the uterus and stimulate contractions of the uterine muscles, which help move the semen up the uterus. The alkalinity of the semen helps neutralize the acidic environment of the vagina, protecting the sperm and increasing their motility. When first ejaculated, the semen coagulates, making it easier for uterine contractions to move it along; then anticoagulants liquefy the semen, and the sperm begin swimming through the female tract.

Penis

The human **penis** is composed of three cylinders of spongy erectile tissue derived from modified veins and capillaries. During sexual arousal, the erectile tissue fills with blood from the arteries. As this tissue fills, the increasing pressure seals off the veins that drain the penis, causing it to engorge with blood. The resulting erection is essential to insertion of the penis into the vagina. Rodents, raccoons, walruses, whales, and several other mammals also possess a **baculum,** a bone that is contained in, and helps stiffen, the penis. Temporary impotence, a reversible inability to achieve an erection, can result from alcohol consumption, certain drugs, and emotional problems. Several drugs and penile implant devices are available for men with nonreversible impotence due to nervous system or circulatory problems. The oral drug Viagra promotes the action of the local regulator nitric oxide (NO), enhancing relaxation of smooth muscles in the blood vessels of the penis. This allows blood to enter the erectile tissue and sustain an erection.

The main shaft of the penis is covered by relatively thick skin. The head, or **glans penis,** has a much thinner covering and is consequently more sensitive to stimulation. The human glans is covered by a fold of skin called the foreskin, or **prepuce,** which may be removed by circumcision. Circumcision, which arose from religious traditions, has no verified basis in health or hygiene.

Human Sexual Response

As mentioned earlier, many animals have elaborate mating behavior. The arousal of sexual interest in humans is even more complex, involving a variety of psychological as well as physical factors. Nevertheless, human sexual response is characterized by a common physiological pattern.

Two types of physiological reactions predominate in both sexes: **vasocongestion,** the filling of a tissue with blood, caused by increased blood flow through the arteries of that tissue, and **myotonia,** increased muscle tension. Both skeletal and smooth muscle may show sustained or rhythmic contractions, including those associated with orgasm.

The sexual response cycle can be divided into four phases: excitement, plateau, orgasm, and resolution. An important function of the excitement phase is preparation of the vagina and penis for **coitus** (sexual intercourse). During this phase, vasocongestion is particularly evident in erection of the penis and clitoris; enlargement of the testes, labia, and breasts; and vaginal lubrication. Myotonia may occur, resulting in nipple erection or tension of the arms and legs.

In the plateau phase, these responses continue. In females, the outer third of the vagina becomes vasocongested, while the inner two-thirds slightly expands. This change, coupled with the elevation of the uterus, forms a depression that receives sperm at the back of the vagina. Breathing increases and heart rate rises, sometimes to 150 beats per minute—not in response to the physical effort of sexual activity, but as an involuntary response to stimulation of the autonomic nervous system (see Figures 48.21 and 48.22).

Orgasm is characterized by rhythmic, involuntary contractions of the reproductive structures in both sexes. Male orgasm has two stages. Emission is the contraction of the glands and ducts of the reproductive tract, which forces semen into the urethra. Expulsion, or ejaculation, occurs when the urethra contracts and the semen is expelled. During female orgasm, the uterus and outer vagina contract, but the inner two-thirds of the vagina do not. Orgasm is the shortest phase of the sexual response cycle, usually lasting only a few seconds. In both sexes, contractions occur at about 0.8-second intervals and may involve the anal sphincter and several abdominal muscles.

The resolution phase completes the cycle and reverses the responses of the earlier stages. Vasocongested organs return to their normal size and color, and muscles relax. Most of the changes of resolution are completed in 5 minutes. Loss of penile and clitoral erection, however, may take longer. An initial loss of erection is rapid in both sexes, but a return of the organs to their nonaroused size may take as long as an hour.

Concept Check 46.3

1. Arrange the following male ducts in the correct sequence for the travel of sperm: epididymis, seminiferous tubule, urethra, vas deferens.
2. How do human male accessory glands support both sperm motility and fertilization?
3. In human sexual response, what organs undergo vasocongestion?

For suggested answers, see Appendix A.

Concept 46.4

In humans and other mammals, a complex interplay of hormones regulates gametogenesis

How exactly are gametes produced in the mammalian body? The process, **gametogenesis**, is based on meiosis, but details differ in females and males. **Oogenesis**, the development of mature ova (egg cells), is described in **Figure 46.11**, on the next page. **Spermatogenesis**, the production of mature sperm cells, is a continuous and prolific process in the adult male. Each ejaculation of a human male contains 100 to 650 million sperm cells, and males can ejaculate daily with little loss of fertilizing capacity. Spermatogenesis occurs in the seminiferous tubules of the testes. **Figure 46.12** describes the process in some detail.

Oogenesis differs from spermatogenesis in three major ways. First, during the meiotic divisions of oogenesis, cytokinesis is unequal, with almost all the cytoplasm monopolized by a single daughter cell, the secondary oocyte. This large cell can go on to become the ovum; the other products of meiosis, smaller cells called polar bodies, degenerate. By contrast, in spermatogenesis, all four products of meiosis develop into mature sperm (compare Figures 46.11 and 46.12). Second, although the cells from which sperm develop continue to divide by mitosis throughout the male's life, this is thought not to be the case for oogenesis in the human female (but see the text in Figure 46.11). Third, oogenesis has long "resting" periods, in contrast to spermatogenesis, which produces mature sperm from precursor cells in an uninterrupted sequence.

The Reproductive Cycles of Females

In females, the secretion of hormones and the reproductive events they regulate are cyclic. Whereas males produce sperm continuously, females release only one egg or a few eggs at a specific time during each cycle. Hormonal control of the female cycle is complex, as we shall see later.

Menstrual Versus Estrous Cycles

Two different types of cycles occur in female mammals. Humans and certain other primates have **menstrual cycles,** whereas other mammals have **estrous cycles**. In both cases, ovulation occurs at a time in the cycle after the endometrium (lining of the uterus) has started to thicken and develop a rich blood supply, preparing the uterus for the possible implantation of an embryo. One difference between the two types of cycles involves the fate of the uterine lining if pregnancy does not occur. In menstrual cycles, the endometrium is shed from the uterus through the cervix and vagina in a bleeding called **menstruation**. In estrous cycles, the endometrium is reabsorbed by the uterus, and no extensive bleeding occurs.

Other major distinctions include more pronounced behavioral changes during estrous cycles than during menstrual cycles and stronger effects of season and climate on estrous cycles. Whereas human females may be receptive to sexual activity throughout their menstrual cycle, most mammals will copulate only during the period surrounding ovulation. This period of sexual activity, called **estrus** (from the Latin *oestrus*, frenzy, passion), is the only time the condition of the vagina permits mating. Estrus is sometimes called heat, and indeed the female's body temperature increases slightly. The length and frequency of reproductive cycles vary widely among mammals. The human menstrual cycle averages 28 days (although cycles vary, ranging from about 20 to 40 days). In contrast, the estrous cycle of the rat is only 5 days. Bears and dogs have one cycle per year; elephants have several.

The Human Female Reproductive Cycle: A Closer Look

Let's examine the reproductive cycle of the human female in more detail as a case study of how a complex function is coordinated by hormones. The term menstrual cycle refers specifically to the changes that occur in the uterus; therefore, it is also called the **uterine cycle**. It is caused by cyclic events that occur in the ovaries—that is, by the **ovarian cycle**. Thus, the female reproductive cycle is actually one integrated cycle involving two organs, the uterus and the ovaries.

Figure 46.11
Exploring Human Oogenesis

Oogenesis begins in the female embryo with differentiation of primordial germ cells into **oogonia**, ovary-specific stem cells. An oogonium multiplies by mitosis and begins meiosis, but the process stops at prophase I. The cells at this stage, called **primary oocytes**, remain quiescent within small follicles (cavities lined with protective cells) until puberty, when hormones reactivate them. Beginning at puberty, follicle-stimulating hormone (FSH) periodically stimulates a follicle to grow and induces its primary oocyte to complete meiosis I and start meiosis II. Meiosis then stops again. Arrested at metaphase II, the **secondary oocyte** is released at ovulation, when its follicle breaks open. Usually only one oocyte matures and is released each month. Meiosis does not continue until a sperm penetrates the oocyte; only then is oogenesis actually completed, producing an ovum. (In other animal species, the sperm may enter the oocyte at the same stage, earlier, or later.)

The meiotic divisions in oogenesis involve unequal cytokinesis, with the smaller cells becoming polar bodies (the first polar body may or may not divide again). After meiosis is completed, the haploid nuclei of the sperm and the now mature ovum fuse; this event is fertilization.

The ruptured follicle left behind after ovulation develops into the corpus luteum. If the released oocyte is not fertilized, however, the corpus luteum degenerates.

For many years, scientists have believed that women, like most female mammals, are born with all the primary oocytes they will ever have—that no new ones develop after birth. In March of 2004, however, researchers reported that multiplying oogonia exist in the ovaries of adult mice and can develop into oocytes. The researchers are now looking for similar cells in human ovaries. It is possible that the marked decline in fertility that occurs as women age results from the gradual depletion of oogonia rather than solely from the degeneration of aging oocytes.

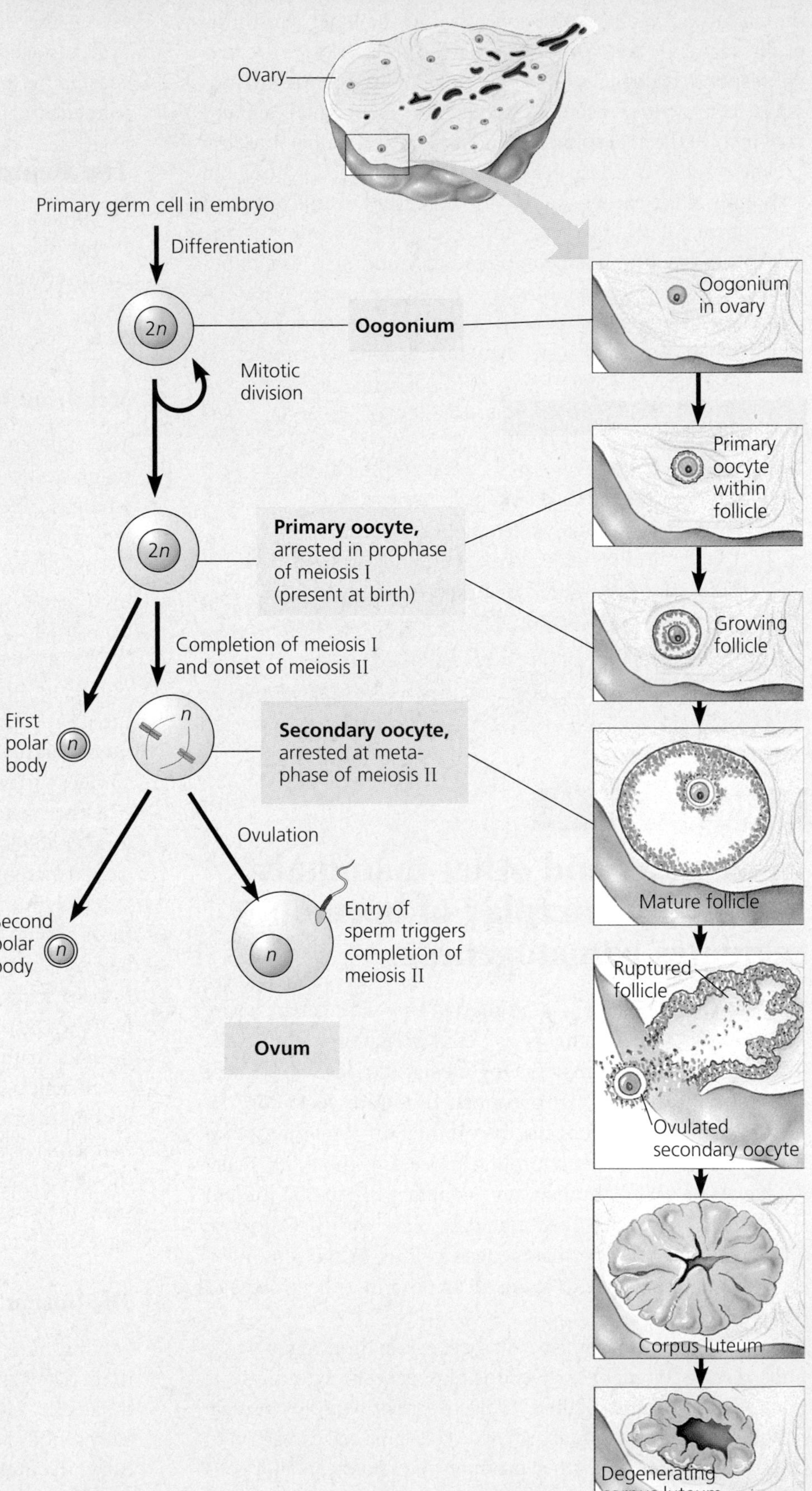

Figure 46.12

Exploring Human Spermatogenesis

These drawings correlate the meiotic stages in sperm development (left) with the microscopic structure of seminiferous tubules. Primordial germ cells of the embryonic testes differentiate into **spermatogonia,** the stem cells that give rise to sperm. As spermatogonia differentiate into spermatocytes and then into spermatids, meiosis reduces the chromosome number from diploid ($2n = 46$ in humans) to haploid ($n = 23$). The developing cells are pushed from a location near the outer wall of the seminiferous tubule toward the lumen (central opening) and then to the epididymis, where they become motile.

The structure of a sperm cell fits its function. In humans, as in most species, a head containing the haploid nucleus is tipped with a special body, the **acrosome,** which contains enzymes that help the sperm penetrate the egg. Behind the head, the sperm cell contains large numbers of mitochondria (or a single large one, in some species) that provide ATP for movement of the tail, which is a flagellum.

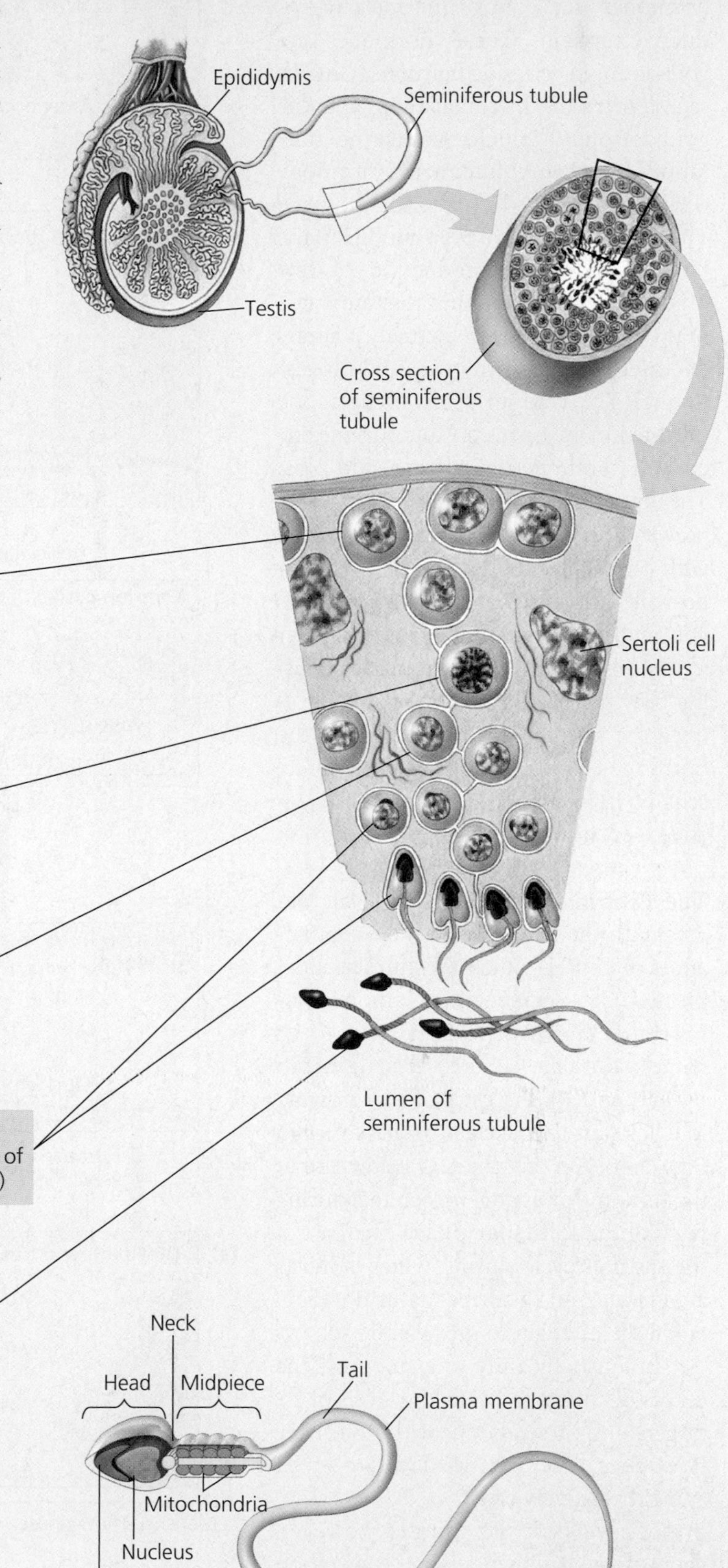

The hormones at the top levels of control of this dual cycle are the same brain hormones that control the male reproductive system. These hormones are gonadotropin-releasing hormone (GnRH), secreted by the hypothalamus, and the gonadotropins follicle-stimulating hormone (FSH) and luteinizing hormone (LH), secreted by the anterior pituitary. The concentrations of FSH and LH in the blood control the production of two kinds of steroid hormones that are made in the ovaries: estrogen (actually a family of closely related hormones) and progesterone. The ovarian cycle of hormone production in turn controls the uterine cycle of endometrial growth and loss. The outcome is that ovarian follicle growth and ovulation are synchronized with preparation of the uterine lining for possible implantation of an embryo. As you read the following discussion, refer to **Figure 46.13** as a guide to understanding how the five kinds of hormones regulate the system. You will learn that the hormones participate in an elaborate scheme involving both positive and negative feedback.

The Ovarian Cycle. ❶ The cycle begins with the release from the hypothalamus of GnRH, which ❷ stimulates the pituitary to secrete small amounts of FSH and LH. ❸ The FSH (true to its name) stimulates follicle growth, aided by LH, and ❹ the cells of the growing follicles start to make estrogen. Notice in Figure 46.13d that there is a slow rise in the amount of estrogen secreted during most of the **follicular phase**, the part of the ovarian cycle during which follicles are growing and oocytes maturing. (Several follicles begin to grow with each cycle, but usually only one matures; the others disintegrate.) The low levels of estrogen inhibit secretion of the pituitary hormones, keeping the levels of FSH and LH relatively low.

The levels of FSH and LH, however, shoot up sharply when ❺ the secretion of estrogen by the growing follicle begins to rise steeply. Whereas a low level of estrogen inhibits the secretion of

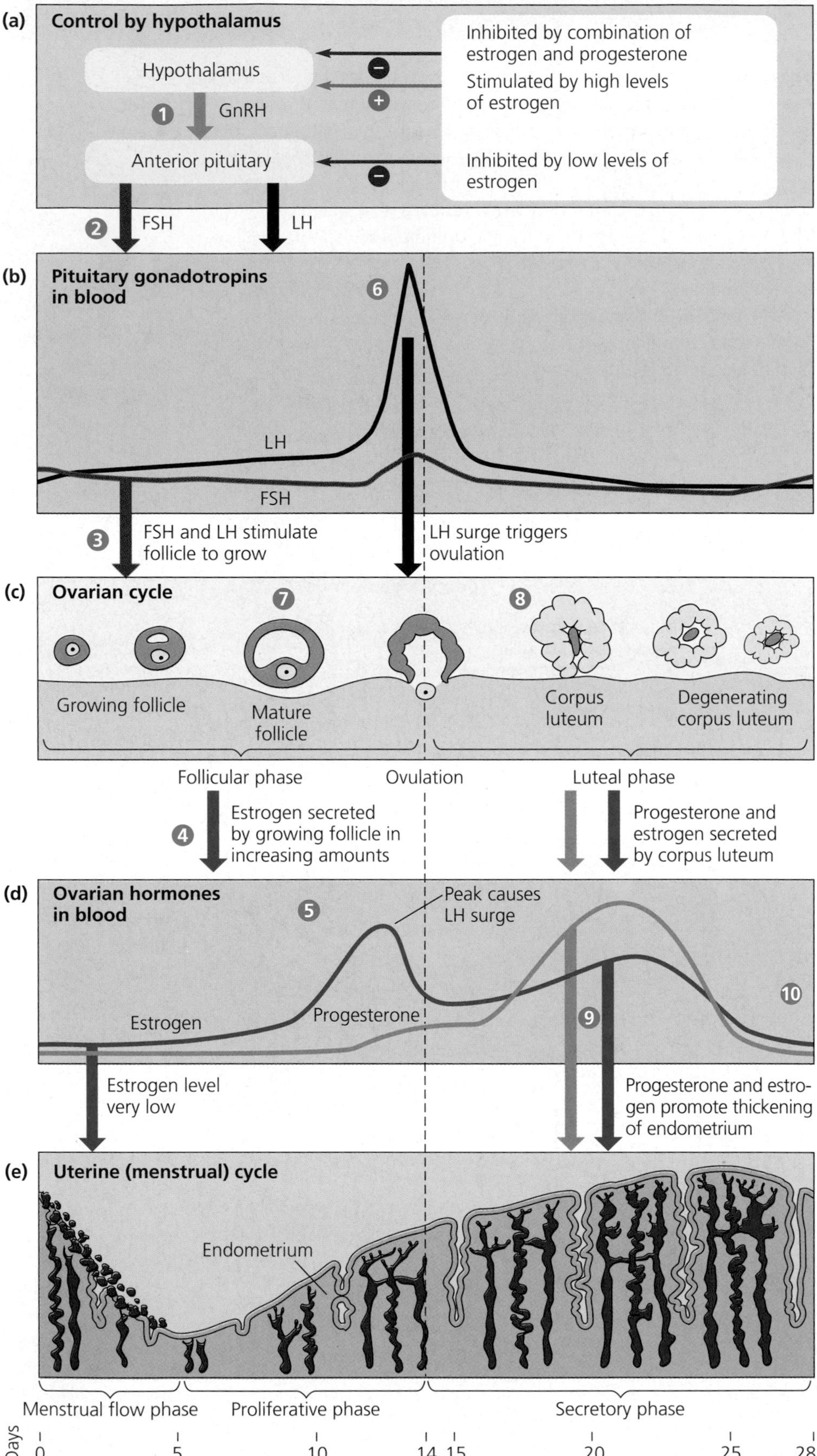

▲ **Figure 46.13 The reproductive cycle of the human female.** This figure shows how (c) the ovarian cycle and (e) the uterine (menstrual) cycle are regulated by changing hormone levels in the blood, depicted in parts (a), (b), and (d). The time scale at the bottom of the figure applies to parts (b)–(e).

pituitary gonadotropins, a high concentration has the opposite effect: It stimulates the secretion of gonadotropins by acting on the hypothalamus to increase its output of GnRH. ❻ You can see this response in Figure 46.13b as steep increases in FSH and LH levels that occur soon after the increase in the concentration of estrogen, indicated in Figure 46.13d. The effect is greater for LH because the high concentration of estrogen also increases the sensitivity of LH-releasing cells in the pituitary to GnRH. By now, the follicles can respond more strongly to LH because more of their cells have receptors for this hormone. The increase in LH concentration caused by increased estrogen secretion from the growing follicle is an example of positive feedback. The LH induces final maturation of the follicle. ❼ The maturing follicle develops an internal fluid-filled cavity and grows very large, forming a bulge near the surface of the ovary. The follicular phase ends, about a day after the LH surge, with **ovulation**: The follicle and adjacent wall of the ovary rupture, releasing the secondary oocyte.

❽ Following ovulation, during the **luteal phase** of the ovarian cycle, LH stimulates the transformation of the follicular tissue left behind in the ovary to form the corpus luteum, a glandular structure (see Figure 46.13c). (LH is named for this "luteinizing" function.) Under continued stimulation by LH during this phase of the ovarian cycle, the corpus luteum secretes progesterone and estrogen (see Figure 46.13d). As the levels of progesterone and estrogen rise, the combination of these hormones exerts negative feedback on the hypothalamus and pituitary, inhibiting the secretion of LH and FSH. Near the end of the luteal phase, the corpus luteum disintegrates, causing concentrations of estrogen and progesterone to decline sharply. The dropping levels of ovarian hormones liberate the hypothalamus and pituitary from the inhibitory effects of these hormones. The pituitary can then begin to secrete enough FSH to stimulate the growth of new follicles in the ovary, initiating the next ovarian cycle.

The Uterine (Menstrual) Cycle. The hormones secreted by the ovaries—estrogen and progesterone—have a major effect on the uterus. Estrogen secreted in increasing amounts by growing follicles signals the endometrium to thicken. In this way, the follicular phase of the ovarian cycle is coordinated with the **proliferative phase** of the uterine cycle (see Figure 46.13e). Before ovulation, the uterus is already being prepared for a possible embryo. After ovulation, ❾ estrogen and progesterone secreted by the corpus luteum stimulate continued development and maintenance of the endometrium, including enlargement of arteries and growth of endometrial glands. These glands secrete a nutrient fluid that can sustain an early embryo even before it actually implants in the uterine lining. Thus, the luteal phase of the ovarian cycle is coordinated with what is called the **secretory phase** of the uterine cycle.

❿ The rapid drop in the level of ovarian hormones when the corpus luteum disintegrates causes spasms of the arteries in the uterine lining that deprive it of blood. The upper two-thirds of the endometrium disintegrates, resulting in menstruation—the **menstrual flow phase** of the uterine cycle—and the beginning of a new cycle. By convention, the first day of menstruation is designated day 1 of the uterine (and ovarian) cycle. Menstrual bleeding usually persists for a few days. During menstruation, a fresh batch of ovarian follicles are just beginning to grow.

Cycle after cycle, the maturation and release of egg cells from the ovary are integrated with changes in the uterus, the organ that must accommodate an embryo if the egg cell is fertilized. If an embryo has not implanted in the endometrium by the end of the secretory phase of the uterine cycle, a new menstrual flow commences, marking day 1 of the next cycle. Later in the chapter, you will learn about override mechanisms that prevent disintegration of the endometrium in pregnancy.

In addition to the roles of estrogen in coordinating the female reproductive cycle, this hormone family is responsible for the secondary sex characteristics of the female. Estrogen induces deposition of fat in the breasts and hips, increases water retention, affects calcium metabolism, stimulates breast development, and influences female sexual behavior.

Menopause. After about 450 cycles, human females undergo **menopause**, the cessation of ovulation and menstruation. Menopause usually occurs between the ages of 46 and 54. Apparently, during these years the ovaries lose their responsiveness to gonadotropins from the pituitary (FSH and LH), and menopause results from a decline in estrogen production by the ovary. Menopause is an unusual phenomenon; in most species, females as well as males retain their reproductive capacity throughout life. Is there an evolutionary explanation for menopause? Why might natural selection have favored females who had stopped reproducing? One intriguing hypothesis proposes that during early human evolution, undergoing menopause after having some children actually increased a woman's fitness; losing the ability to reproduce allowed her to provide better care for her children and grandchildren, thereby increasing the survival of individuals bearing her genes.

Hormonal Control of the Male Reproductive System

In the male, the principal sex hormones are the androgens, of which testosterone is the most important. Androgens are steroid hormones produced mainly by the Leydig cells of the testes, interstitial cells located near the seminiferous tubules.

Testosterone and other androgens are directly responsible for the primary and secondary sex characteristics of the male. Primary sex characteristics are associated with the reproductive system: development of the vasa deferentia and other ducts, development of external reproductive structures, and sperm production. Secondary sex characteristics are features not directly related to the reproductive system, including deepening of the voice, distribution of facial and pubic hair, and muscle growth (androgens stimulate protein synthesis). Androgens are also potent determinants of behavior in mammals and other vertebrates. In addition to specific sexual behaviors and sex drive, androgens increase general aggressiveness and are responsible for vocal behavior such as singing in birds and calling by frogs. Hormones from the anterior pituitary and hypothalamus control androgen secretion and sperm production by the testes. As you study **Figure 46.14**, keep in mind that each hormone acts only on cells that have specific receptors for it (see Chapter 45).

Concept Check 46.4

1. FSH and LH get their names from events of the female reproductive cycle, but they also function in males. How are their functions in females and males similar?
2. How does an estrous cycle differ from a menstrual cycle, and in what types of animals are the two types of cycles found?
3. Why do we regard the ovarian cycle and the uterine (menstrual) cycle as parts of a single cycle?
4. What specific hormonal changes trigger ovulation?

For suggested answers, see Appendix A.

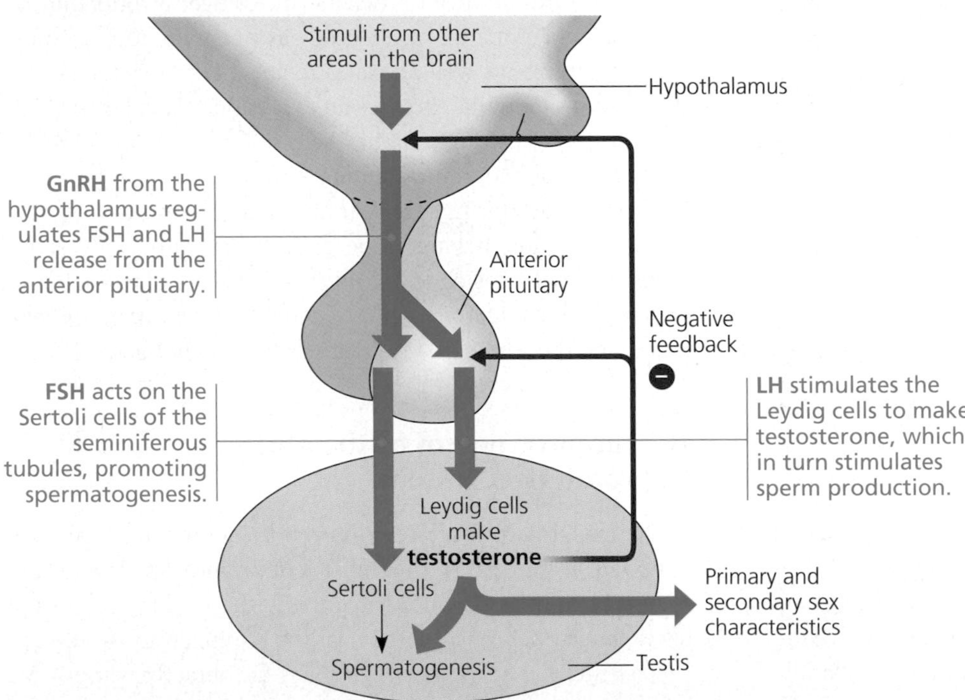
Stimuli from other areas in the brain
Hypothalamus
GnRH from the hypothalamus regulates FSH and LH release from the anterior pituitary.
Anterior pituitary
Negative feedback
FSH acts on the Sertoli cells of the seminiferous tubules, promoting spermatogenesis.
LH stimulates the Leydig cells to make testosterone, which in turn stimulates sperm production.
Leydig cells make testosterone
Sertoli cells
Primary and secondary sex characteristics
Spermatogenesis
Testis

▲ **Figure 46.14 Hormonal control of the testes.** Gonadotropin-releasing hormone (GnRH) from the hypothalamus stimulates the anterior pituitary to secrete two gonadotropic hormones with different effects on the testes, luteinizing hormone (LH) and follicle-stimulating hormone (FSH). FSH acts on Sertoli cells, which nourish developing sperm (see Figure 46.12). LH acts on Leydig cells, which produce androgens, chiefly testosterone. Negative feedback by testosterone on the hypothalamus and anterior pituitary is the main mechanism by which blood levels of LH, FSH, and GnRH are regulated.

Concept 46.5

In humans and other placental mammals, an embryo grows into a newborn in the mother's uterus

In humans and other placental (eutherian) mammals, **pregnancy**, or **gestation**, is the condition of carrying one or more embryos in the uterus. Human pregnancy averages 266 days (38 weeks) from fertilization of the egg, or 40 weeks from the start of the last menstrual cycle. Duration of pregnancy in other species correlates with body size and the maturity of the young at birth. Many rodents have gestation periods of about 21 days, whereas those of dogs are closer to 60 days. In cows, gestation averages 270 days (almost the same as humans) and in elephants, more than 600 days.

Conception, Embryonic Development, and Birth

Fertilization of an egg by a sperm—also called **conception** in humans—occurs in the oviduct **(Figure 46.15)**. About 24 hours later, the resulting zygote begins dividing, a process called **cleavage.** Cleavage continues, with the embryo becoming a ball of cells by the time it reaches the uterus 3 to 4 days after fertilization. By about 1 week after fertilization, cleavage has produced an embryonic stage called the **blastocyst,** a sphere of cells containing a cavity. In a process that takes several more days for completion, the blastocyst implants into the endometrium.

The embryo secretes hormones that signal its presence and control the mother's reproductive system. One embryonic hormone, **human chorionic gonadotropin (HCG),** acts like pituitary LH to maintain secretion of progesterone and estrogens by the corpus luteum through the first few months of pregnancy. In the absence of this hormonal override, the decline in maternal LH due to inhibition of the pituitary would result in menstruation and loss of the embryo. Levels of HCG in the maternal blood are so high that some is excreted in the urine, where it can be detected in pregnancy tests.

First Trimester

Human gestation can be divided for convenience into three **trimesters** of about three months each. The first trimester is the time of most radical change for both

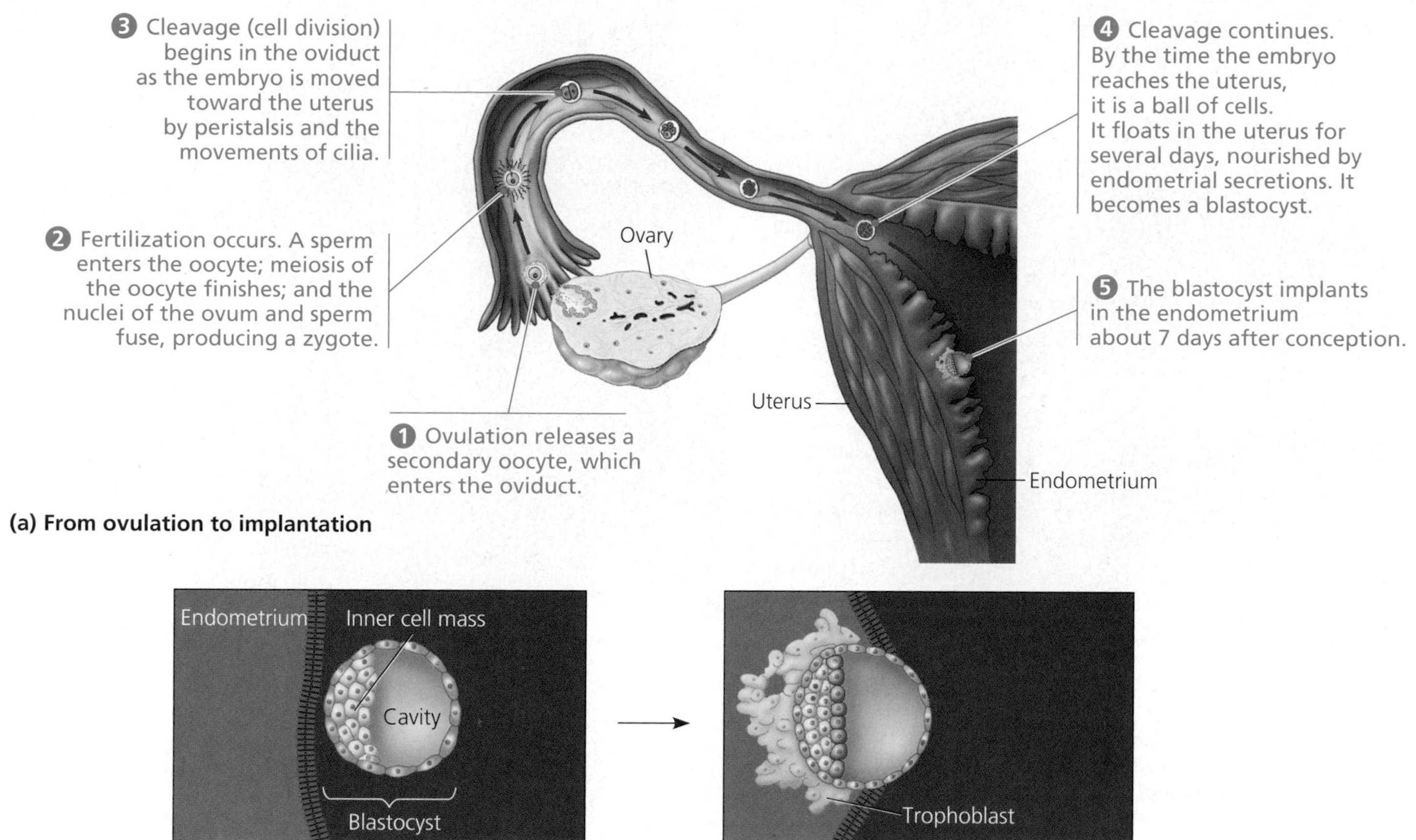

3 Cleavage (cell division) begins in the oviduct as the embryo is moved toward the uterus by peristalsis and the movements of cilia.
2 Fertilization occurs. A sperm enters the oocyte; meiosis of the oocyte finishes; and the nuclei of the ovum and sperm fuse, producing a zygote.
1 Ovulation releases a secondary oocyte, which enters the oviduct.
Ovary
Uterus
Endometrium
4 Cleavage continues. By the time the embryo reaches the uterus, it is a ball of cells. It floats in the uterus for several days, nourished by endometrial secretions. It becomes a blastocyst.
5 The blastocyst implants in the endometrium about 7 days after conception.
(a) From ovulation to implantation
Endometrium
Inner cell mass
Cavity
Blastocyst
Trophoblast
(b) Implantation of blastocyst

▲ **Figure 46.15 Formation of the zygote and early postfertilization events.** You will learn more about fertilization and cleavage in Chapter 47.

the mother and the embryo. Let's take up our story where we left off, at implantation. The endometrium responds to implantation by growing over the blastocyst. Differentiation of the embryo's body structures now begins. (You will learn much more about embryonic development in Chapter 47.)

During its first 2 to 4 weeks of development, the embryo obtains nutrients directly from the endometrium. Meanwhile, the outer layer of the blastocyst, called the **trophoblast,** grows out and mingles with the endometrium, eventually helping to form the **placenta.** This disk-shaped organ, containing both embryonic and maternal blood vessels, grows to about the size of a dinner plate and can weigh close to 1 kg. Diffusion of material between maternal and embryonic circulations provides nutrients, exchanges respiratory gases, and disposes of metabolic wastes for the embryo. Blood from the embryo travels to the placenta through arteries of the umbilical cord and returns via the umbilical vein (**Figure 46.16**, on the next page).

The first trimester is the main period of **organogenesis,** the development of the body organs (**Figure 46.17**). The heart begins beating by the fourth week and can be detected with a stethoscope by the end of the first trimester. By the end of the eighth week, all the major structures of the adult are present in rudimentary form. (It is during organogenesis that the embryo is most sensitive to such threats as radiation and drugs that can cause birth defects.) At 8 weeks, the embryo is called a **fetus.** Although well differentiated, the fetus is only 5 cm long by the end of the first trimester.

Meanwhile, the mother is also undergoing rapid changes. High levels of progesterone initiate changes in her reproductive system. These include increased mucus in the cervix that forms a protective plug, growth of the maternal part of the placenta, enlargement of the uterus, and (by negative feedback on the hypothalamus and pituitary) cessation of ovulation and menstrual cycling. The breasts also enlarge rapidly and are often quite tender.

Second Trimester

During the second trimester, the fetus grows to about 30 cm and is very active. The mother may feel movements during the early part of the second trimester, and fetal activity may be visible through the abdominal wall by the middle of this time period. Hormone levels stabilize as HCG declines, the corpus luteum deteriorates, and the placenta completely takes over the production of progesterone, which maintains the pregnancy. During the second trimester, the uterus grows enough for the pregnancy to become obvious.

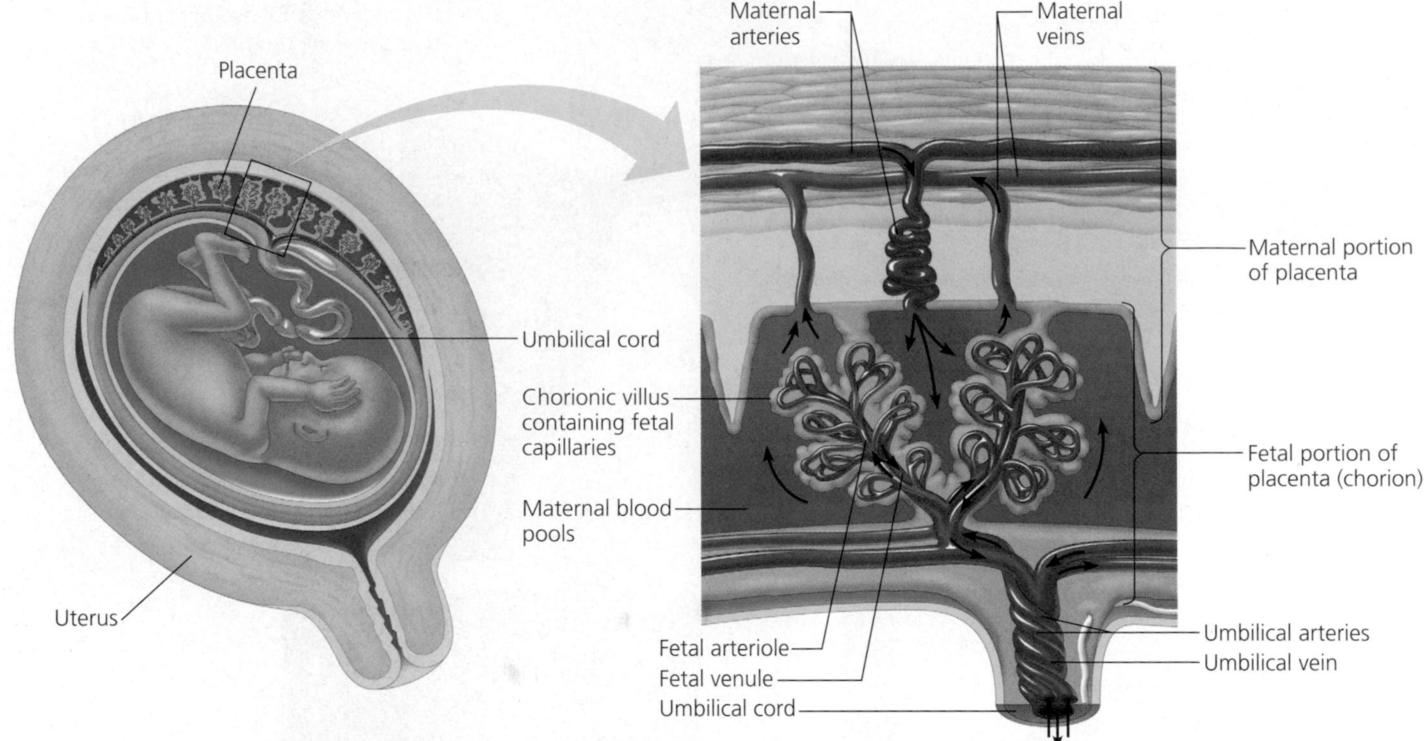

▲ **Figure 46.16 Placental circulation.** From the fourth week of development until birth, the placenta, a combination of maternal and embryonic tissues, transports nutrients, respiratory gases, and wastes between the embryo or fetus and the mother. Maternal blood enters the placenta in arteries, flows through blood pools in the endometrium, and leaves via veins. Embryonic or fetal blood, which remains in vessels, enters the placenta through arteries and passes through capillaries in fingerlike chorionic villi, where oxygen and nutrients are acquired. As indicated in the drawing, the fetal (or embryonic) capillaries and villi project into the maternal portion of the placenta. Fetal blood leaves the placenta through veins leading back to the fetus. Materials are exchanged by diffusion, active transport, and selective absorption between the fetal capillary bed and the maternal blood pools.

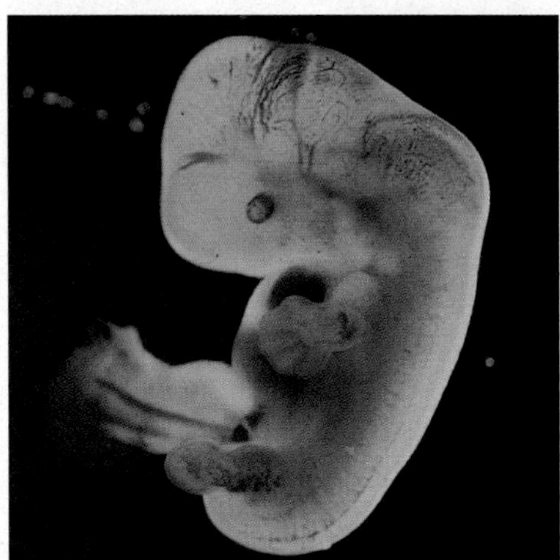

(a) 5 weeks. Limb buds, eyes, the heart, the liver, and rudiments of all other organs have started to develop in the embryo, which is only about 1 cm long.

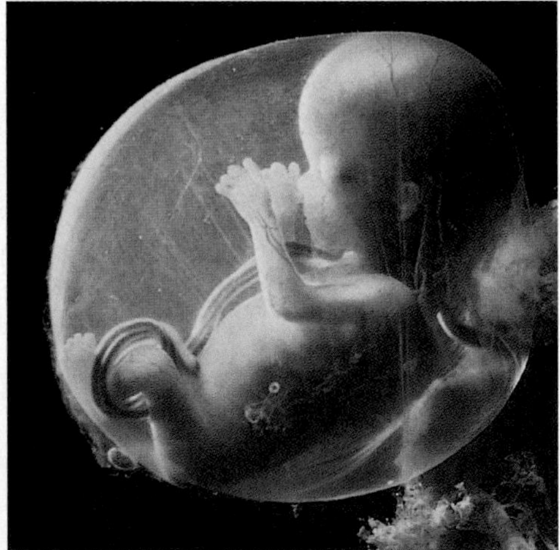

(b) 14 weeks. Growth and development of the offspring, now called a fetus, continue during the second trimester. This fetus is about 6 cm long.

(c) 20 weeks. By the end of the second trimester (at 24 weeks), the fetus grows to about 30 cm in length.

▲ **Figure 46.17 Human fetal development.**

Third Trimester

The final trimester is one of growth of the fetus to about 3–4 kg in weight and 50 cm in length. Fetal activity may decrease as the fetus fills the available space within the embryonic membranes. As the fetus grows and the uterus expands around it, the mother's abdominal organs become compressed and displaced, leading to frequent urination, digestive blockages, and strain in the back muscles. A complex interplay of local regulators (prostaglandins) and hormones (chiefly estrogen and oxytocin) induces and regulates **labor,** the process by which childbirth occurs. The mechanism that triggers labor is not fully understood, but **Figure 46.18** shows one model. Estrogen, which reaches its highest level in the mother's blood during the last weeks of pregnancy, induces the formation of oxytocin receptors on the uterus. Oxytocin, produced by the fetus and the mother's posterior pituitary, stimulates powerful contractions by the smooth muscles of the uterus. Oxytocin also stimulates the placenta to secrete prostaglandins, which enhance the contractions. In turn, the physical and emotional stresses associated with the contractions stimulate the release of more oxytocin and prostaglandins, a positive feedback system that underlies the process of labor.

Birth, or **parturition**, is brought about by a series of strong, rhythmic uterine contractions. The process of labor has three stages **(Figure 46.19)**. The first stage is the opening up and thinning of the cervix, ending with complete dilation. The second stage is expulsion, or delivery, of the baby. Continuous strong contractions force the fetus down and out of the uterus and vagina. The umbilical cord is cut and clamped at this time. The final stage of labor is delivery of the placenta, which normally follows the baby.

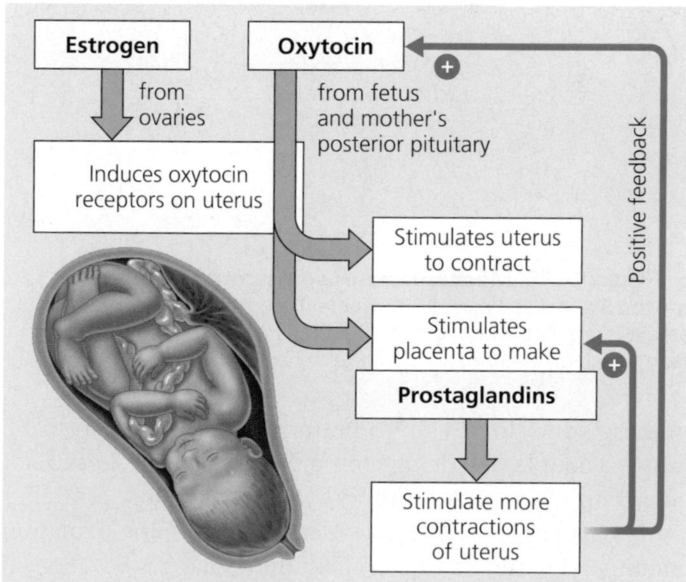

▲ **Figure 46.18 A model for the induction of labor.**

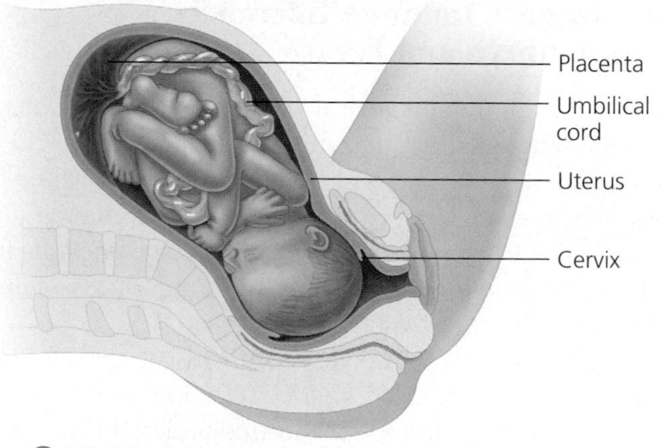

❶ Dilation of the cervix

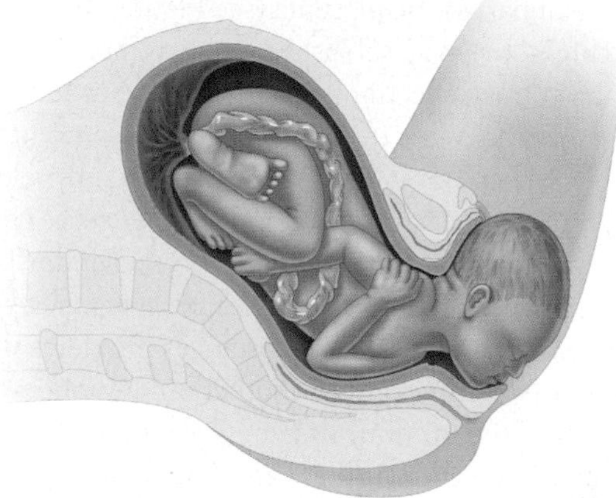

❷ Expulsion: delivery of the infant

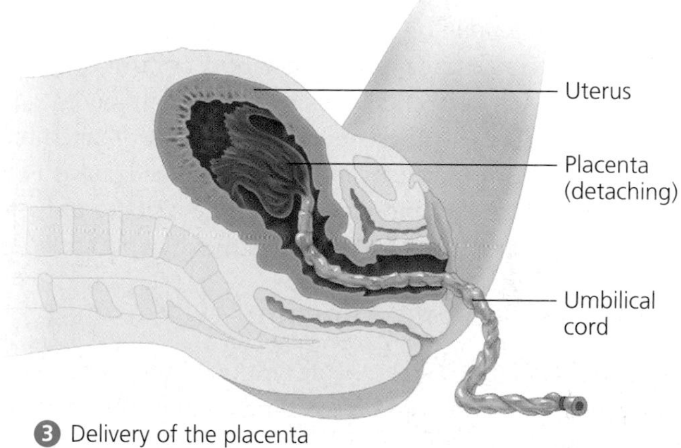

❸ Delivery of the placenta

▲ **Figure 46.19 The three stages of labor.**

Lactation is an aspect of postnatal care unique to mammals. After birth, decreasing levels of progesterone free the anterior pituitary from negative feedback and allow prolactin secretion. Prolactin stimulates milk production after a delay of 2 or 3 days. The release of milk from the mammary glands is controlled by oxytocin (see p. 944 and Figure 45.7).

The Mother's Immune Tolerance of the Embryo and Fetus

Pregnancy is an immunological enigma. Half of the embryo's genes are inherited from the father; thus, many of the chemical markers present on the surface of the embryo will be foreign to the mother. Why, then, does the mother not reject the embryo as a foreign body, as she would a tissue or organ graft bearing antigens from another person? Reproductive immunologists are working to solve this puzzle.

A major key to the puzzle may be the tissue called the trophoblast (see Figures 46.15 and 47.18). Originally the outermost layer of the blastocyst, the trophoblast brings about implantation by growing into the endometrium and later develops into the fetal part of the placenta (see Figures 46.15 and 46.16). How might the trophoblast, and later the placenta, protect the embryo from rejection? Some possibilities follow.

During early pregnancy, the trophoblast seems to prevent the mother's immune system from rejecting the blastocyst by releasing signal molecules with immunosuppressive effects. These include HCG, a variety of protein "factors," a prostaglandin, several interleukins, and an interferon. Several lines of research suggest that some combination of these substances interferes with immune rejection by acting on the mother's T lymphocytes, important players in the immune system (see Chapter 43).

A very different hypothesis is that the trophoblast and later the placenta secrete an enzyme that rapidly breaks down local supplies of tryptophan, an amino acid necessary for T cell survival and function. At least in mice, this enzyme seems to be essential for maintaining pregnancy.

Another possibility is the absence of certain histocompatibility antigens on placental cells and the secretion of a hormone that induces synthesis of a "death activator" membrane protein (FasL) on placental cells. Activated T cells have a complementary "death receptor" (Fas), and the binding of FasL to Fas would cause the T cells to self-destruct by apoptosis.

Contraception and Abortion

Contraception, the deliberate prevention of pregnancy, can be achieved in a number of ways. Some contraceptive methods prevent the release of mature eggs (secondary oocytes) and sperm from gonads, others prevent fertilization by keeping sperm and egg apart, and still others prevent implantation of an embryo **(Figure 46.20)**. The following brief introduction to the biology of the most often used methods makes no pretense of being a contraception manual. For more complete information, you should consult a physician or health center personnel.

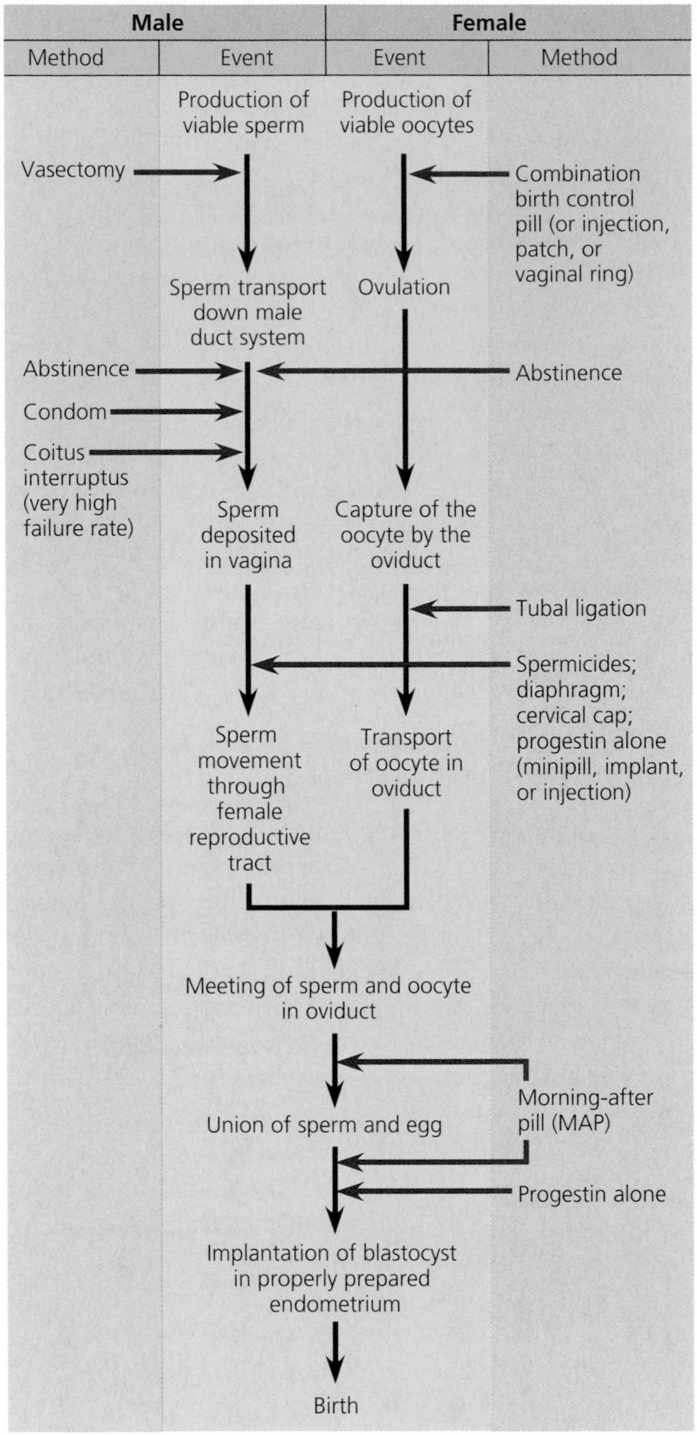

▲ **Figure 46.20 Mechanisms of some contraceptive methods.** Red arrows indicate where these methods, devices, or products interfere with the flow of events from the production of sperm and egg (secondary oocyte) to the birth of a baby.

Fertilization can be prevented by abstinence from sexual intercourse or by any of several barriers that keep live sperm from contacting the egg. Temporary abstinence, often called the **rhythm method** of birth control or **natural family planning**, depends on refraining from intercourse when conception is most likely. Because the egg can survive in the oviduct for 24 to 48 hours and sperm for up to 72 hours, a couple practicing temporary abstinence should not engage in intercourse during several days before and after ovulation. The most effective methods for timing ovulation combine several indicators, including changes in cervical mucus and body temperature during the menstrual cycle. Thus, natural family planning requires that the couple be knowledgeable about these physiological signs. A pregnancy rate of 10–20% is typically reported for couples practicing natural family planning. (Pregnancy rate is the percentage of women using a particular family planning method who become pregnant during one year.) Some couples use the natural family planning method to *increase* the probability of conception.

As a method of preventing fertilization, coitus interruptus, or withdrawal (removal of the penis from the vagina before ejaculation), is unreliable. Sperm may be present in secretions that precede ejaculation, and a lapse in timing or willpower can result in late withdrawal.

The several **barrier methods** of contraception that block the sperm from meeting the egg have pregnancy rates of less than 10%. The **condom,** used by the male, is a thin, latex rubber or natural membrane sheath that fits over the penis to collect the semen. For sexually active individuals, latex condoms are the only contraceptives that offer some protection against sexually transmitted diseases, including AIDS. (This protection is, however, not absolute.) The barrier device most commonly used by females is the **diaphragm,** a dome-shaped rubber cap fitted into the upper portion of the vagina before intercourse. Both of these devices are more effective when used in conjunction with a spermicidal (sperm-killing) foam or jelly. Other barrier devices for women include the cervical cap, which fits tightly around the opening of the cervix and is held in place for a prolonged period by suction, and the vaginal pouch, or "female condom."

Except for complete abstinence from sexual intercourse, the methods that prevent the release of gametes are the most effective means of birth control. Sterilization (discussed later) is almost 100% effective, and chemical contraceptives, most often used in the form of **birth control pills,** have pregnancy rates of less than 1%. The most commonly used birth control pills are a combination of a synthetic estrogen and a synthetic progestin (progesterone-like hormone). This combination acts by negative feedback to stop the release of GnRH by the hypothalamus, and thus of FSH and LH by the pituitary. The prevention of LH release prevents ovulation. As a backup mechanism, the inhibition of FSH secretion by the low dose of estrogen in the pills prevents follicles from developing. A similar combination of hormones is also available as an injection, in a ring inserted into the vagina, and as a patch. Combination birth control pills can be used in high doses as morning-after pills (MAPs). Taken within 3 days of unprotected intercourse, they prevent fertilization or implantation, with an effectiveness of about 75%.

A second type of birth control pill, called the minipill, contains only progestin. It does not effectively block ovulation, and it is not quite as effective a contraceptive as a hormonal combination. The minipill prevents fertilization mainly by causing thickening of a woman's cervical mucus so that it blocks sperm from entering the uterus. The progestin also causes changes in the endometrium that interfere with implantation if fertilization occurs. Progestin can be administered in time-release, match-sized capsules that are implanted under the skin and last for five years or by injections that last for three months, as well as in tablet (minipill) form.

Are there long-term side effects of hormonal contraceptives? For women taking a combination version, cardiovascular problems are the most serious concern. Birth control pills slightly raise a woman's risk of abnormal blood clotting, high blood pressure, heart attack, and stroke. Smoking while using chemical contraception increases the risk of mortality tenfold or more. Although the pill increases the risk for cardiovascular disease, it eliminates the dangers of pregnancy; women on birth control pills have mortality rates about one-half those of pregnant women. Also, the pill decreases the risk of ovarian and endometrial cancers and benign breast disease.

One elusive research goal has been a chemical contraceptive for men. Chemicals that alter testosterone levels have tended to be unsatisfactory because they affect secondary sex characteristics as well as spermatogenesis. Recently, however, researchers have begun to look for drugs that target other kinds of molecules involved in spermatogenesis. One promising drug (actually a sugar) is already approved for another purpose. When administered to mice, it causes nonfunctional sperm to be made, apparently by inhibiting synthesis of certain glycolipids.

Sterilization is the permanent prevention of gamete release. **Tubal ligation** in women usually involves cauterizing or tying off (ligating) a section of the oviducts to prevent eggs from traveling into the uterus. **Vasectomy** in men is the cutting of each vas deferens to prevent sperm from entering the urethra. Both male and female sterilization procedures are relatively safe and free from harmful effects. Both are also difficult to reverse, so the procedures should be considered permanent.

Abortion is the termination of a pregnancy in progress. Spontaneous abortion, or miscarriage, is very common; it occurs in as many as one-third of all pregnancies, often before the woman is even aware she is pregnant. In addition, each year about 1.5 million women in the United States choose abortions performed by physicians. A drug called mifepristone, or RU486, developed in France, enables a woman to terminate pregnancy nonsurgically within the first 7 weeks. An analog of progesterone, RU486 blocks progesterone receptors in the uterus, thus preventing progesterone from maintaining pregnancy. It is taken with a small amount of prostaglandin to induce uterine contractions.

Modern Reproductive Technology

Recent scientific and technological advances have made it possible to deal with many reproductive problems. For example, it is now possible to diagnose many genetic diseases and other congenital disorders (those present at birth) while the

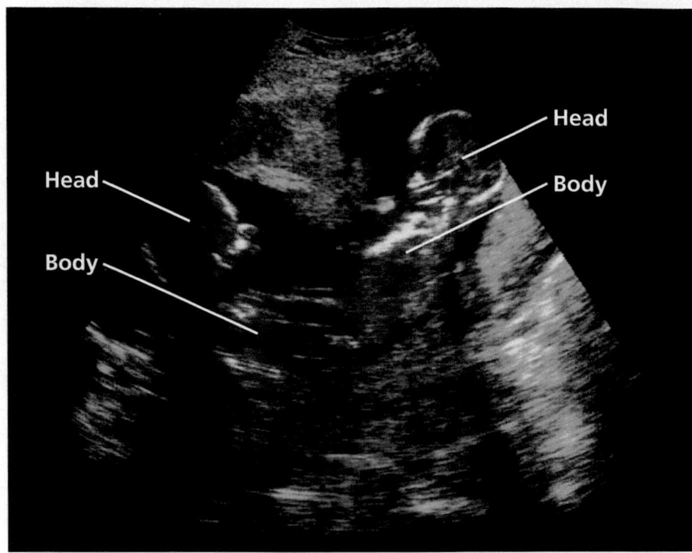

▲ **Figure 46.21 Ultrasound image.** This color-enhanced image shows twins in the uterus. The image was produced on a computer screen when high-frequency sounds from an ultrasound scanner held against the pregnant woman's abdomen bounced off the fetuses.

fetus is in the uterus. Amniocentesis and chorionic villus sampling are invasive techniques in which amniotic fluid or fetal cells are obtained for genetic analysis (see Figure 14.17). Noninvasive procedures usually use ultrasound imaging to detect fetal condition **(Figure 46.21)**. A newer noninvasive technique relies on the fact that a few fetal blood cells leak across the placenta into the mother's bloodstream. A blood sample from the mother yields fetal cells that can be identified with specific antibodies (which bind to proteins on the surface of fetal cells) and then tested for genetic disorders.

Diagnosing genetic diseases in fetuses poses ethical questions. To date, almost all detectable disorders remain untreatable in the uterus, and many cannot be corrected even after birth. Parents may be faced with difficult decisions about whether to terminate a pregnancy or cope with a child who may have profound defects and a short life expectancy. These are complex issues that demand careful, informed thought and competent counseling.

Reproductive technology can help with a number of infertility problems. Hormone therapy will sometimes increase sperm or egg production, and surgery can correct disorders such as blocked oviducts. Many infertile couples turn to fertilization procedures called **assisted reproductive technology (ART)**. These procedures generally involve surgically removing eggs (secondary oocytes) from a woman's ovaries following hormonal stimulation, fertilizing the eggs, and returning them to the woman's body. Unused eggs, sperm, and embryos from such procedures can be frozen for later pregnancy attempts.

With ***in vitro* fertilization (IVF)**, the most common ART procedure, the oocytes are mixed with sperm in culture dishes and incubated for several days to allow the fertilized eggs to start developing. When they have developed into embryos of at least eight cells each, the embryos are carefully inserted into the woman's uterus and allowed to implant. In ZIFT (zygote intrafallopian transfer), eggs are also fertilized *in vitro,* but zygotes are transferred immediately to the woman's oviducts (fallopian tubes). In GIFT (gamete intrafallopian transfer), the eggs are not fertilized *in vitro.* Instead, the eggs and sperm are placed in the woman's oviducts in the hope that fertilization will occur there. Any of these procedures can use sperm or eggs from donors.

These techniques are now performed in major medical centers throughout the world. Though they cost thousands of dollars per attempt, they have resulted in thousands of children. To date, evidence of abnormalities resulting from the procedures has been scanty.

Once conception and implantation have successfully occurred, a developmental program unfolds that transforms the zygote into a baby. The mechanisms of this development in humans and other animals are the subject of Chapter 47.

Concept Check 46.5

1. Describe the state of the embryo just before it implants in the uterine lining.
2. Why does testing for HCG (human chorionic gonadotropin) work as a pregnancy test early in pregnancy but not late in pregnancy? What is the function of HCG in pregnancy?
3. _____ is to males as tubal ligation is to _____.
4. Why is the term *test tube baby* an inaccurate reference to the product of *in vitro* fertilization?

For suggested answers, see Appendix A.

Chapter 46 Review

Go to www.campbellbiology.com or the student CD-ROM to explore Activities, Investigations, and other interactive study aids.

SUMMARY OF KEY CONCEPTS

Concept 46.1

Both asexual and sexual reproduction occur in the animal kingdom

- Asexual reproduction produces offspring whose genes all come from a single parent. Sexual reproduction requires the fusion of male and female gametes, which forms a diploid zygote (p. 964).
- **Mechanisms of Asexual Reproduction (pp. 964–965)** Fission, budding, and fragmentation with regeneration are mechanisms of asexual reproduction in various invertebrates.
- **Reproductive Cycles and Patterns (pp. 965–966)** Animals may reproduce exclusively sexually or asexually, or they may alternate between the two. Variations on these two modes are made possible through parthenogenesis, hermaphroditism, and sequential hermaphroditism. Reproductive cycles are controlled by hormones and environmental cues.

Concept 46.2

Fertilization depends on mechanisms that help sperm meet eggs of the same species

- In external fertilization, eggs shed by the female are fertilized by sperm in the external environment. In internal fertilization, egg and sperm unite within the female's body. In either case, fertilization requires critical timing, often mediated by environmental cues, pheromones, and/or courtship behavior. Internal fertilization requires important behavioral interactions between male and female animals, as well as compatible copulatory organs (p. 967).
- **Ensuring the Survival of Offspring (pp. 967–968)** Greater protection of embryos and parental care usually follow production of relatively few offspring by internal fertilization.
- **Gamete Production and Delivery (pp. 968–969)** Reproductive systems range from the production of gametes by undifferentiated cells in the body cavity to complex assemblages of male and female gonads with accessory tubes and glands that carry and protect gametes and developing embryos.

Concept 46.3

Reproductive organs produce and transport gametes: focus on humans

- **Female Reproductive Anatomy (pp. 969–971)** Externally, the human female has a vulva, consisting of the vestibule (with separate openings of the vagina and urethra), the labia minora, the labia majora, and the clitoris. Internally, the vagina is connected to the uterus, which connects to two oviducts. Two ovaries (female gonads) are stocked with follicles containing developing egg cells (oocytes). After ovulation, the remnant of the follicle forms a corpus luteum, which secretes hormones for a variable duration, depending on whether or not pregnancy occurs. Although separate from the reproductive system, the mammary glands, or breasts, evolved in association with parental care.
  Activity *Reproductive System of the Human Female*
- **Male Reproductive Anatomy (pp. 971–972)** External reproductive structures of the human male are the scrotum and penis. The male gonads, or testes, reside in the cooler environment of the scrotum. They possess hormone-producing cells and sperm-forming seminiferous tubules that successively lead into the epididymis, vas deferens, ejaculatory duct, and urethra, which exits at the tip of the penis.
  Activity *Reproductive System of the Human Male*
  Investigation *What Might Obstruct the Male Urethra?*
- **Human Sexual Response (pp. 972–973)** Both males and females experience the erection of certain body tissues due to vasocongestion and myotonia, culminating in orgasm.

Concept 46.4

In humans and other mammals, a complex interplay of hormones regulates gametogenesis

- Oogenesis is the female form of gametogenesis, the production of gametes, and spermatogenesis is the male form. Sperm develop continuously, whereas the maturation of egg cells is discontinuous and cyclic. Meiosis is central to both processes, but in oogenesis, cytokinesis is unequal, producing only one large ovum. In spermatogenesis, each starting cell becomes four sperm (pp. 973–975).
- **The Reproductive Cycles of Females (pp. 973, 976–977)** Female hormones are secreted in a rhythmic fashion reflected in the menstrual or estrous cycle. In both types of cycles, the endometrium thickens in preparation for possible implantation. The menstrual cycle, however, includes endometrial bleeding, and sexual receptivity is not limited to a heat period, as in the estrous cycle. The female reproductive cycle is orchestrated by cyclic secretion of GnRH from the hypothalamus and of FSH and LH from the anterior pituitary. FSH and LH bring about complex changes in the ovary and, via estrogen and progesterone, in the uterus. The developing follicle produces estrogen, and the corpus luteum secretes progesterone and estrogens. Positive and negative feedback regulate the levels of the five hormones that coordinate the cycle.
- **Hormonal Control of the Male Reproductive System (pp. 977–978)** Androgens (chiefly testosterone) from the testes cause the development of primary and secondary sex characteristics in the male. Androgen secretion and sperm production are both controlled by hypothalamic and pituitary hormones.

Concept 46.5

In humans and other placental mammals, an embryo grows into a newborn in the mother's uterus

- **Conception, Embryonic Development, and Birth (pp. 978–981)** After fertilization of the egg and the completion of meiosis in the oviduct, the zygote undergoes cleavage and develops into a blastocyst before implantation in the endometrium. Human pregnancy can be divided into three trimesters. All major organs have started developing by 8 weeks. Birth, or parturition, results from strong, rhythmic uterine contractions. Positive feedback involving prostaglandins and the hormones estrogen and oxytocin regulates labor.
- **The Mother's Immune Tolerance of the Embryo and Fetus (p. 982)** A pregnant woman's acceptance of her "foreign" offspring is still not fully understood but may be due to the suppression of the immune response in her uterus.

- **Contraception and Abortion (pp. 982–983)** Contraceptive methods may prevent the release of mature gametes from the gonads, fertilization, or implantation of the embryo.
- **Modern Reproductive Technology (pp. 983–984)** In addition to helping detect problems before birth, modern technology can help infertile couples by *in vitro* fertilization.

TESTING YOUR KNOWLEDGE

Self-Quiz

1. Which of the following characterizes parthenogenesis?
   a. An individual may change its sex during its lifetime.
   b. Specialized groups of cells grow into new individuals.
   c. An organism is first a male and then a female.
   d. An egg develops without being fertilized.
   e. Both mates have male and female reproductive organs.

2. Which structure is *incorrectly* paired with its function?
   a. gonads—produce gametes
   b. spermatheca—stores sperm in male honeybees
   c. cloaca—serves as the common opening for the reproductive, excretory, and digestive systems
   d. baculum—stiffens the penis in some mammals
   e. endometrium—forms the maternal part of the placenta

3. Which of the following male and female structures are *least* alike in function?
   a. seminiferous tubules—vagina
   b. Sertoli cells—follicle cells
   c. spermatogonia—oogonia
   d. testes—ovaries
   e. vas deferens—oviduct

4. A difference between estrous and menstrual cycles is that
   a. nonmammalian vertebrates have estrous cycles, whereas mammals have menstrual cycles.
   b. the endometrial lining is shed in menstrual cycles but reabsorbed in estrous cycles.
   c. estrous cycles occur more often than menstrual cycles.
   d. estrous cycles are not controlled by hormones.
   e. ovulation occurs before the endometrium thickens in estrous cycles.

5. Peaks of LH and FSH production occur during
   a. the flow phase of the menstrual (uterine) cycle.
   b. the beginning of the follicular phase of the ovarian cycle.
   c. the period just before ovulation.
   d. the end of the luteal phase of the ovarian cycle.
   e. the secretory phase of the menstrual cycle.

6. In sequential hermaphroditism,
   a. some individuals may change from male to female.
   b. individuals fertilize themselves.
   c. males rather than females release pheromones.
   d. diploid ova are produced.
   e. the adult gonads are undifferentiated.

7. During human gestation, rudiments of all organs develop
   a. in the first trimester.
   b. in the second trimester.
   c. in the third trimester.
   d. while the embryo is in the oviduct.
   e. during the blastocyst stage.

8. Which pharmacological strategy is most likely to result in a successful male contraceptive?
   a. preventing the production of functionally normal sperm
   b. maintaining high circulating concentrations of androgen
   c. blocking testosterone receptors on Leydig cells
   d. blocking androgen receptors within the hypothalamus
   e. maintaining high circulating concentrations of FSH

9. Fertilization of human eggs most often takes place in the
   a. vagina.
   b. ovary.
   c. uterus.
   d. oviduct.
   e. vas deferens.

10. In male mammals, excretory and reproductive systems share
   a. the testes.
   b. the urethra.
   c. the ureter.
   d. the vas deferens.
   e. the prostate.

For Self-Quiz answers, see Appendix A.

Go to the website or CD-ROM for more quiz questions.

Evolution Connection

In animals, hermaphroditism is often found in species that are fixed to a surface. Mobile species are less often hermaphroditic. Why?

Scientific Inquiry

Imagine studying evolution of parental care in a certain animal group. You map the distribution of care behavior on a phylogenetic tree as shown below (see Chapter 25). What is the simplest interpretation of how this behavior evolved? If the outgroup exhibited parental care, how would your interpretation change?

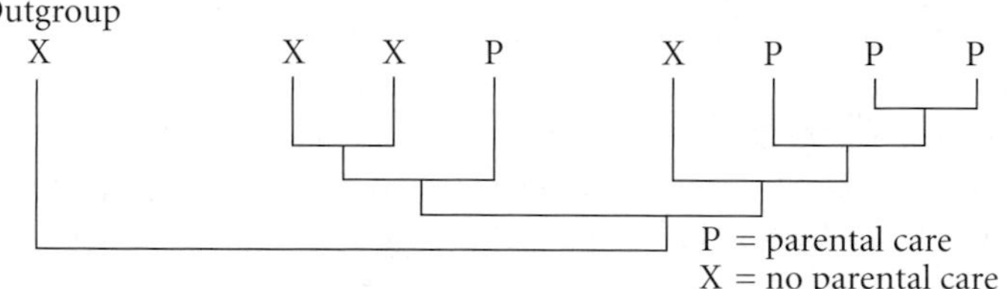

Investigation *What Might Obstruct the Male Urethra?*

Science, Technology, and Society

Techniques for sorting sperm, combined with *in vitro* fertilization, make it possible to choose a baby's sex. What potential problems can you foresee if this procedure becomes widely available?

47 Animal Development

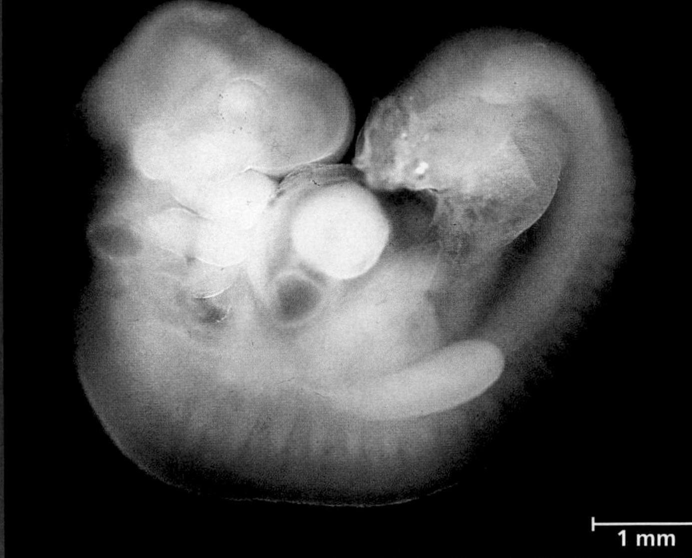

▲ **Figure 47.1 A human embryo about six to eight weeks after conception.**

Key Concepts

47.1 After fertilization, embryonic development proceeds through cleavage, gastrulation, and organogenesis

47.2 Morphogenesis in animals involves specific changes in cell shape, position, and adhesion

47.3 The developmental fate of cells depends on their history and on inductive signals

Overview

A Body-Building Plan for Animals

It is difficult to imagine that each of us began life as a single cell about the size of the period at the end of this sentence. **Figure 47.1** shows a human embryo about six to eight weeks after conception. The brain is forming in the head (upper left), and the developing heart (the red spot in the center) has already begun to pulsate. It takes a total of only about nine months for a single-celled zygote, or fertilized egg, to be transfigured into a newborn human, built of trillions of differentiated cells organized into specialized tissues and organs.

The question of how a zygote becomes an animal has been asked for centuries. As recently as the 18th century, the prevailing notion was *preformation:* the idea that the egg or sperm contains an embryo—a preformed, miniature infant, or "homunculus"—that simply becomes larger during development **(Figure 47.2)**. The competing explanation of embryonic development was *epigenesis,* originally proposed 2,000 years earlier by Aristotle. According to epigenesis, the form of an animal emerges gradually from a relatively formless egg. As microscopy improved during the 19th century, biologists could see that embryos took shape in a series of progressive steps, and epigenesis displaced preformation as the favored explanation among embryologists. The concept of preformation may have some merit, however: Although an embryo's form emerges gradually as it develops, aspects of the developmental plan are already in place in the eggs of many species.

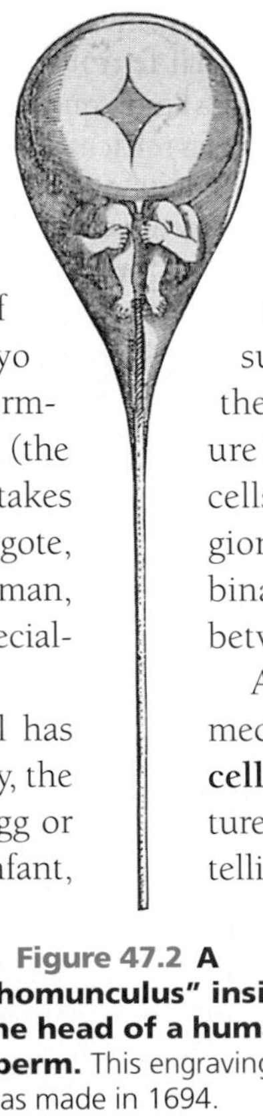

▲ **Figure 47.2 A "homunculus" inside the head of a human sperm.** This engraving was made in 1694.

An organism's development is determined by the genome of the zygote and also by differences that arise between early embryonic cells. These differences set the stage for the expression of different genes in different cells. In some species, early embryonic cells become different because of the uneven distribution within the unfertilized egg of maternal substances called **cytoplasmic determinants.** These substances affect development of the cells that inherit them during early mitotic divisions of the zygote (see Figure 21.11a). In other species, the initial differences between cells are due primarily to their location in embryonic regions with different characteristics. In most species, a combination of these two mechanisms establishes differences between early embryonic cells.

As cell division continues and the embryo develops, mechanisms that selectively control gene expression lead to **cell differentiation,** the specialization of cells in their structure and function. The timely communication of instructions, telling cells precisely what to do and when to do it, occurs by cell signaling among different embryonic cells. Along with cell division and differentiation, development involves **morphogenesis,** the process by which an animal takes shape and the differentiated cells end up in the appropriate locations.

By combining molecular genetics with classical approaches to embryology, developmental biologists are now beginning to answer many questions about how a fertilized egg cell gives rise to a particular animal. In this chapter, we concentrate mainly on organisms, such as the sea urchin, frog, and chick, that have been the subject of classical embryological studies. Although developmental events in these animals are easy to observe in the laboratory, they are more difficult to study genetically than those of the organisms described in Chapter 21. However, molecular techniques are now available for studying the molecular mechanisms of developmental events in these and other species.

In addition to these model organisms, development of our own species has always been of great interest to us and will be covered in this chapter as well. Because ethical concerns preclude experimentation on human embryos, knowledge about human development has been based partly on what we can extrapolate from other mammals, such as the mouse, and partly on observation of very early human development following *in vitro* fertilization.

We will begin with a description of the basic stages of embryonic development common to most animals. Then we will look at the cellular and molecular mechanisms that result in generation of the body form. Finally, we will consider the process by which embryonic cells travel down the appropriate pathways of differentiation, enabling them to play their necessary roles in a functional animal.

Concept 47.1

After fertilization, embryonic development proceeds through cleavage, gastrulation, and organogenesis

Important events regulating development occur during fertilization and each of three successive stages that begin to build the animal's body. During the first stage, called cleavage, cell division creates a hollow ball of cells, the blastula, from the zygote. The second stage, gastrulation, produces a three-layered embryo, the gastrula. The third stage, organogenesis, generates rudimentary organs from which adult structures grow.

In our discussion, we will focus on a few species that have been used as model organisms to investigate each of these processes. As researchers study development in more and more species, they are finding some variations but also many similarities in the processes. For each stage of development, we first consider the species for which the most is known and then compare what is known about the same process in other species. We begin by looking at the fertilization of an egg cell by a sperm.

Fertilization

The sperm and egg (gametes), which unite during fertilization, are both highly specialized cell types produced by a complex series of developmental events in the testes and ovaries of the parents (see Figures 46.11 and 46.12). The main function of fertilization is to combine haploid sets of chromosomes from two individuals into a single diploid cell, the zygote. Another key function is activation of the egg: Contact of the sperm with the egg's surface initiates metabolic reactions within the egg that trigger the onset of embryonic development.

Fertilization has been studied most extensively in sea urchins. Their gametes can simply be combined in seawater in the laboratory, and subsequent events are easily observed. Although sea urchins (phylum Echinodermata) are not vertebrates or even chordates, they share with those two groups the characteristic of being deuterostomes (see Figure 32.9). Despite differences in details, fertilization and early development in sea urchins provide good general models for similar events in vertebrates.

The Acrosomal Reaction

The eggs of sea urchins are fertilized externally after the animals release their gametes into the surrounding seawater. The jelly coat that surrounds the egg provides a source of soluble molecules that attract the sperm, which swim toward the egg.

When the head of a sea urchin sperm cell comes into contact with the jelly coat of a sea urchin egg, molecules present in the egg's coat trigger the **acrosomal reaction** **(Figure 47.3)**. This process begins when a specialized vesicle at the tip of the sperm, called the **acrosome,** discharges hydrolytic enzymes. These enzymes digest the jelly coat, enabling an elongating sperm structure called the *acrosomal process* to penetrate the jelly coat. Molecules of a protein on the tip of the acrosomal process then adhere to molecules of a specific receptor protein on the egg's surface. These receptors extend from the plasma membrane of the egg cell through the *vitelline layer,* a meshwork of extracellular matrix molecules lying under the jelly coat. In sea urchins and many other animals, this "lock-and-key" recognition of molecules ensures that eggs will be fertilized only by sperm of the same species. Such species specificity is especially important when fertilization occurs externally in water, where gametes of other species are likely to be present.

Contact of the tip of the acrosome with the egg membrane leads to the fusion of sperm and egg plasma membranes and the entry of the sperm nucleus into the cytoplasm of the egg. Either contact or fusion of the membranes causes ion channels to open in the egg cell's plasma membrane, allowing sodium ions to flow into the egg cell and change the membrane potential (see Chapter 7). This change in membrane potential, called depolarization, is common among animal species. Occurring within about 1–3 seconds after a sperm binds to an egg,

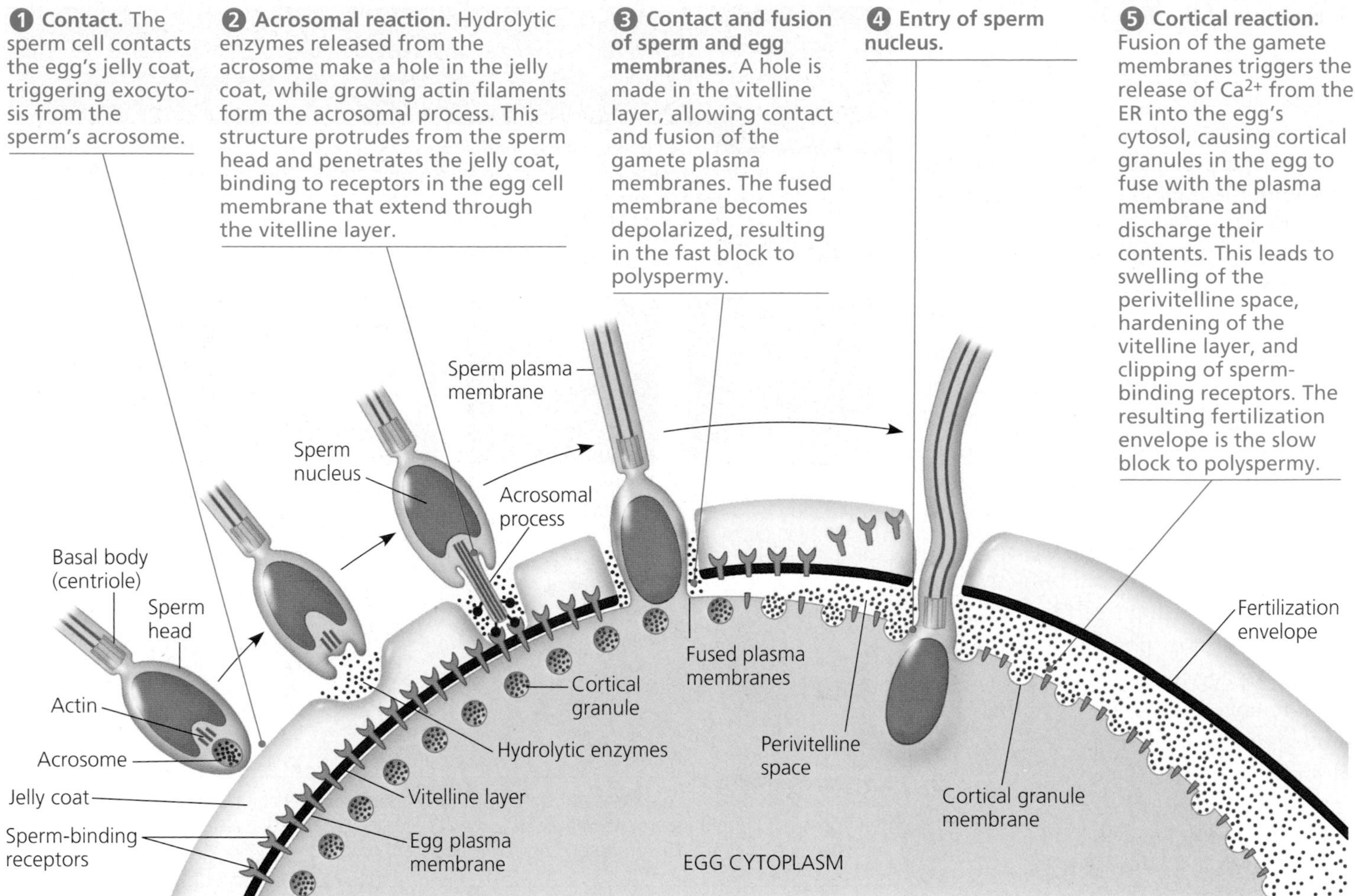

▲ **Figure 47.3 The acrosomal and cortical reactions during sea urchin fertilization.** The events following contact of a single sperm and egg ensure that the nucleus of only one sperm enters the cytoplasm of the egg.

depolarization prevents additional sperm cells from fusing with the egg's plasma membrane. Without this **fast block to polyspermy,** multiple sperm could fertilize the egg, resulting in an aberrant number of chromosomes in the zygote.

1 Contact. The sperm cell contacts the egg's jelly coat, triggering exocytosis from the sperm's acrosome.
2 Acrosomal reaction. Hydrolytic enzymes released from the acrosome make a hole in the jelly coat, while growing actin filaments form the acrosomal process. This structure protrudes from the sperm head and penetrates the jelly coat, binding to receptors in the egg cell membrane that extend through the vitelline layer.
3 Contact and fusion of sperm and egg membranes. A hole is made in the vitelline layer, allowing contact and fusion of the gamete plasma membranes. The fused membrane becomes depolarized, resulting in the fast block to polyspermy.
4 Entry of sperm nucleus.
5 Cortical reaction. Fusion of the gamete membranes triggers the release of Ca2+ from the ER into the egg's cytosol, causing cortical granules in the egg to fuse with the plasma membrane and discharge their contents. This leads to swelling of the perivitelline space, hardening of the vitelline layer, and clipping of sperm-binding receptors. The resulting fertilization envelope is the slow block to polyspermy.
Sperm plasma membrane
Sperm nucleus
Acrosomal process
Basal body (centriole)
Sperm head
Actin
Acrosome
Jelly coat
Sperm-binding receptors
Cortical granule
Hydrolytic enzymes
Vitelline layer
Egg plasma membrane
Fused plasma membranes
Perivitelline space
Cortical granule membrane
Fertilization envelope
EGG CYTOPLASM

The Cortical Reaction

The membrane depolarization lasts for only a minute or so, thus functioning only as a short-term block to polyspermy. However, fusion of egg and sperm plasma membranes also triggers a series of changes in the egg that cause a longer-lasting block. Sperm binding appears to activate a signal transduction pathway involving two second messengers, IP_3 and DAG, that cause calcium (Ca^{2+}) to be released from the egg's endoplasmic reticulum (ER) into the cytosol (see Figure 11.12). The Ca^{2+} release from the ER begins at the site of sperm entry and then propagates in a wave across the fertilized egg (**Figure 47.4**, on the next page).

Within seconds, the high concentration of Ca^{2+} brings about the **cortical reaction,** fusion with the egg's plasma membrane of numerous vesicles lying in the egg's cortex, the area just beneath the membrane. These **cortical granules**, formed during oogenesis, now release their contents into the *perivitelline space,* the space between the plasma membrane and the vitelline layer (see Figure 47.3, step 5). Enzymes released from the cortical granules degrade proteins holding the vitelline layer to the plasma membrane, while mucopolysaccharides produce an osmotic gradient, drawing water into the perivitelline space and swelling it. The swelling pushes the vitelline layer away from the plasma membrane, and other enzymes harden the vitelline layer. Different enzymes clip off and release the external portions of the remaining receptor proteins from the vitelline layer. These changes transform the vitelline layer into the **fertilization envelope,** which resists the entry of additional sperm. The fertilization envelope and other changes in the egg's surface function together as a longer-term **slow block to polyspermy.** Although studied in greatest detail in sea urchins, the cortical reaction also occurs in vertebrates such as fishes and mammals.

Activation of the Egg

Another outcome of the sharp rise in Ca^{2+} concentration in the egg's cytosol is a substantial increase in the rates of cellular respiration and protein synthesis by the egg cell. With these rapid changes in metabolism, the egg is said to be activated.

Although the binding and fusion of sperm are triggers for egg activation, sperm cells do not contribute any materials required for activation. Indeed, the unfertilized eggs of many species can be artificially activated by the injection of Ca^{2+} or by a variety of mildly injurious treatments, such as temperature shock. This artificial activation switches on the metabolic responses of the egg and causes it to begin developing by parthenogenesis (without fertilization by a sperm). It is even possible to artificially activate an egg that has had its own nucleus removed. This finding shows that proteins and mRNAs present in the cytoplasm of the unfertilized egg are sufficient for egg activation.

While the metabolism of the activated sea urchin egg increases, the sperm nucleus within the egg starts to swell. After about 20 minutes, the sperm nucleus merges with the egg nucleus, creating the diploid nucleus of the zygote. DNA synthesis begins, and the first cell division occurs after about 90 minutes in the case of sea urchins and some frogs. The events and timeline of fertilization in sea urchins are summarized in **Figure 47.5**.

Fertilization in other species shares many features with the process in sea urchins. However, the timing differs with the species, and the stage of meiosis the egg has reached at fertilization also varies among species. At the time of release from the female, sea urchin eggs have completed meiosis. In other species, the eggs are arrested at a specific stage of meiosis; upon fertilization, meiosis is quickly completed along with many of the events already described. Human eggs, for example, are arrested at metaphase of meiosis II (see Figure 46.11) until they are fertilized in the female reproductive tract.

Figure 47.4

Inquiry What is the effect of sperm binding on Ca^{2+} distribution in the egg?

EXPERIMENT A fluorescent dye that glows when it binds free Ca^{2+} was injected into unfertilized sea urchin eggs. After sea urchin sperm were added, researchers observed the eggs with a fluorescence microscope.

RESULTS

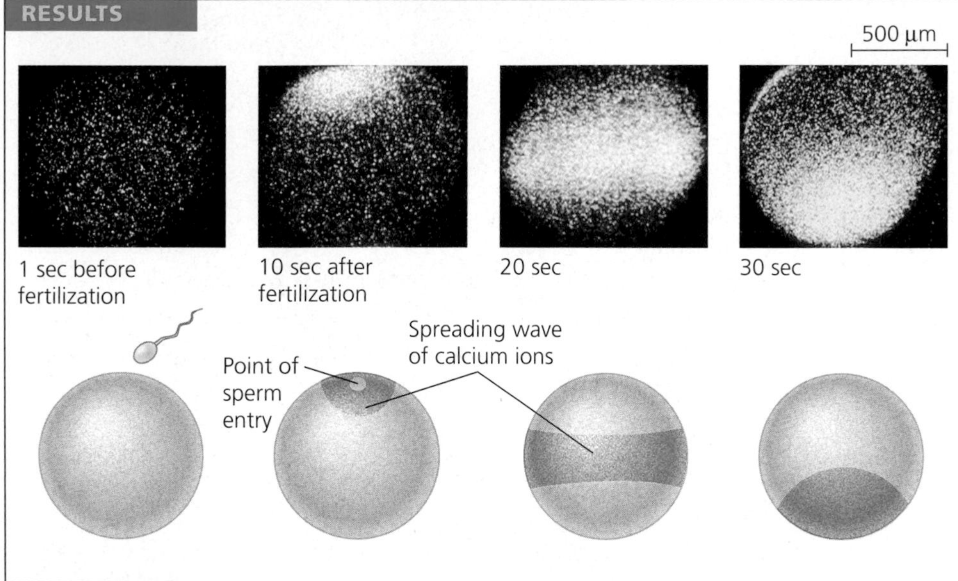

CONCLUSION The release of Ca^{2+} from the endoplasmic reticulum into the cytosol at the site of sperm entry triggers the release of more and more Ca^{2+} in a wave that spreads to the other side of the cell. The entire process takes about 30 seconds.

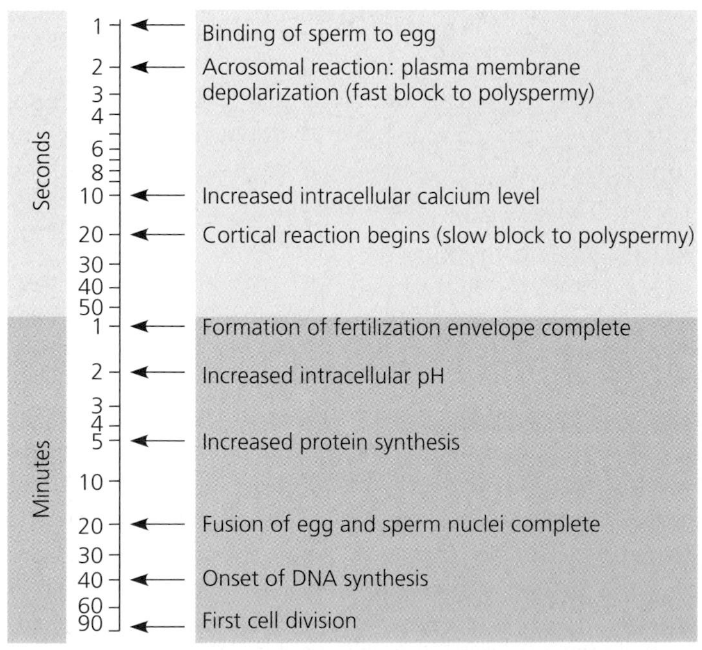

▲ **Figure 47.5 Timeline for the fertilization of sea urchin eggs.** The process begins when a sperm cell contacts the jelly coat of an egg (top of chart). Notice that the scale is logarithmic.

Fertilization in Mammals

In contrast to the external fertilization of sea urchins and most other marine invertebrates, fertilization in terrestrial animals, including mammals, is generally internal. Secretions in the mammalian female reproductive tract alter certain molecules on the surface of sperm cells and also increase sperm motility.

In humans, this enhancement of sperm function requires about 6 hours of exposure to the female reproductive tract.

The mammalian egg is cloaked by follicle cells released along with the egg during ovulation. A sperm cell must migrate through this layer of follicle cells before it reaches the **zona pellucida,** the extracellular matrix of the egg. One component of the zona pellucida functions as a sperm receptor, binding to a complementary molecule on the surface of the sperm head. Binding of the sperm head to receptor molecules induces an acrosomal reaction similar to that of sea urchin sperm **(Figure 47.6)**. Hydrolytic enzymes spilled from the acrosome enable the sperm cell to penetrate the zona pellucida and reach the plasma membrane of the egg. The acrosomal reaction also exposes a protein in the sperm membrane that binds with the egg's plasma membrane.

As in sea urchin fertilization, the binding of a sperm cell to the egg triggers changes within the egg that lead to a cortical reaction, the release of enzymes from cortical granules to the outside of the cell via exocytosis. The released enzymes catalyze alterations of the zona pellucida, which then functions as the slow block to polyspermy. (There is no known fast block to polyspermy in mammals.)

After the egg and sperm membranes fuse, the whole sperm, tail and all, is taken into the egg, which lacks a centrosome. A centrosome forms around the centriole that acted as the basal body of the sperm's flagellum. This centrosome, which now includes a second centriole, duplicates to form two centrosomes in the zygote. These will generate the mitotic spindle for the first cell division. The haploid nuclei of mammalian sperm and egg do not fuse immediately as in sea urchin fertilization. Instead, the envelopes of both nuclei disperse, and the two sets of chromosomes (one set from each gamete) share a common spindle apparatus during the first mitotic division of the zygote. Thus, only after this first division, as diploid nuclei form in the two daughter cells, do the chromosomes from the two parents come together in a common nucleus. Fertilization is much slower in mammals than in the sea urchin; the first cell division occurs 12–36 hours after sperm binding in mammals, compared with about 90 minutes in sea urchins.

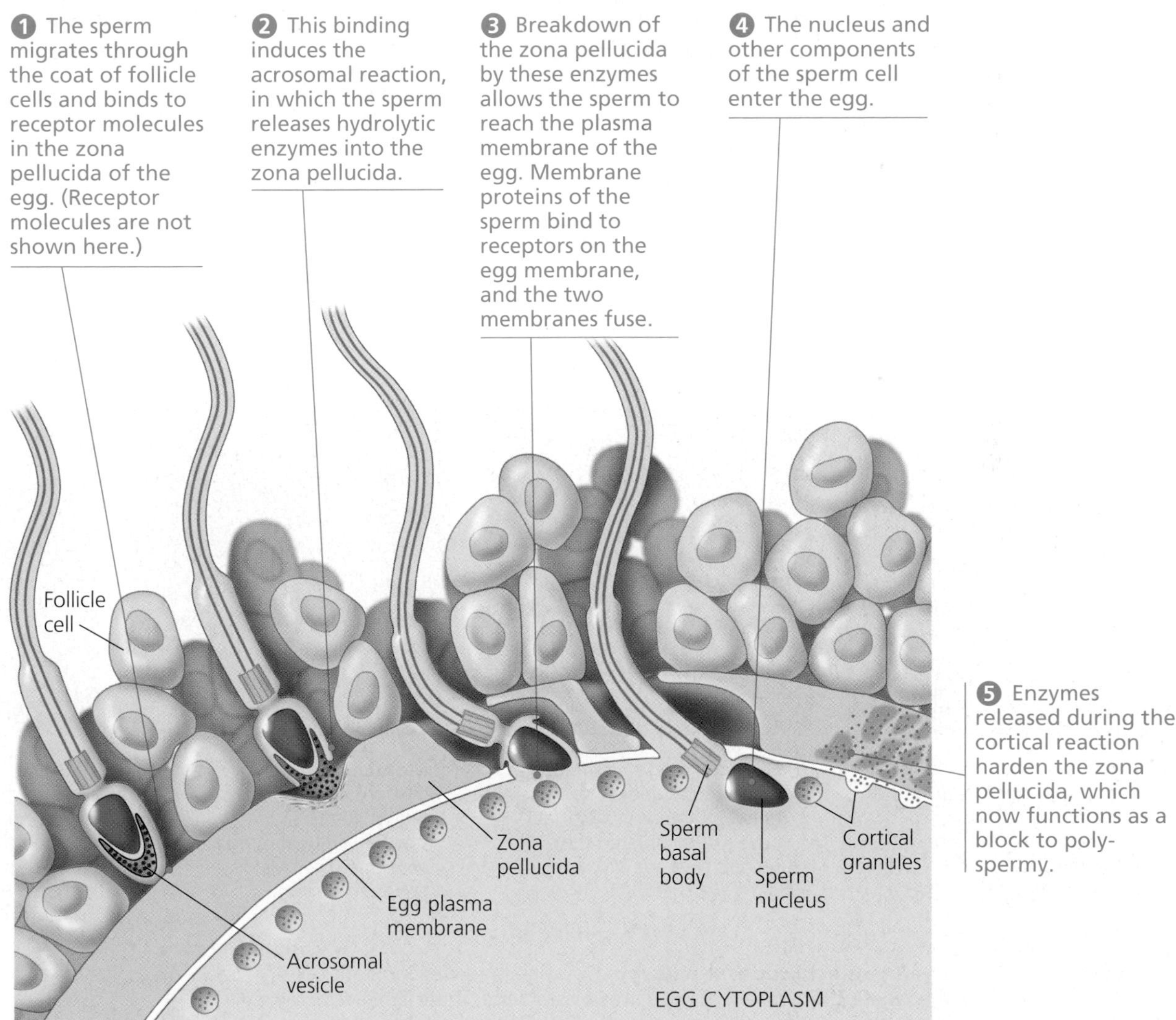

1 The sperm migrates through the coat of follicle cells and binds to receptor molecules in the zona pellucida of the egg. (Receptor molecules are not shown here.)
2 This binding induces the acrosomal reaction, in which the sperm releases hydrolytic enzymes into the zona pellucida.
3 Breakdown of the zona pellucida by these enzymes allows the sperm to reach the plasma membrane of the egg. Membrane proteins of the sperm bind to receptors on the egg membrane, and the two membranes fuse.
4 The nucleus and other components of the sperm cell enter the egg.
5 Enzymes released during the cortical reaction harden the zona pellucida, which now functions as a block to polyspermy.
Follicle cell
Zona pellucida
Egg plasma membrane
Acrosomal vesicle
Sperm basal body
Sperm nucleus
Cortical granules
EGG CYTOPLASM

▶ **Figure 47.6 Early events of fertilization in mammals.** As in sea urchins, events that occur during fertilization ensure that only one sperm enters the cytoplasm of the egg.

Cleavage

Once fertilization is completed, a succession of rapid cell divisions ensues. During this period, called **cleavage,** the cells undergo the S (DNA synthesis) and M (mitosis) phases of the cell cycle, but often virtually skip the G_1 and G_2 phases, so little or no protein synthesis occurs (see Figure 12.5). The embryo does not enlarge during this period of development. Cleavage simply partitions the cytoplasm of one large cell, the zygote, into many smaller cells called **blastomeres,** each with its own nucleus **(Figure 47.7)**.

The first five to seven divisions form a cluster of cells known as the **morula** (Latin for "mulberry," in reference to the lobed surface of the embryo at this stage). A fluid-filled cavity called the **blastocoel** begins to form within the morula and is fully formed in the **blastula**, which is a hollow ball of cells. During cleavage, different regions of cytoplasm present in the original undivided egg cell end up in separate blastomeres. Because the regions may contain different cytoplasmic determinants, in many species this partitioning sets the stage for subsequent developmental events.

The eggs and zygotes of sea urchins and other animals, with the possible exception of mammals, have a definite polarity. During cleavage in such organisms, the planes of division follow a specific pattern relative to the poles of the zygote. The polarity is defined by the uneven distribution of substances in the cytoplasm, including specific mRNAs, proteins, and **yolk** (stored nutrients). In many frogs and other animals, the distribution of yolk is a key factor influencing the pattern of cleavage. Yolk is most concentrated toward one pole of the egg, called the **vegetal pole**; the yolk concentration decreases significantly toward the opposite pole, the **animal pole.** The animal pole is also the site where the polar bodies of oogenesis are budded from the cell (see Figure 46.11).

The three body axes shown in **Figure 47.8a** are established early in development. This process has been well studied in particular frog species where different regions of the egg and early embryo can be distinguished by color and followed easily. The animal and vegetal hemispheres of the zygote, named for their respective poles, differ in color. The animal hemisphere is a deep gray, because dark-colored melanin granules are embedded in the outer layer (cortex) of the cell in this region. The lack of melanin granules in the vegetal hemisphere allows the yellow color of the yolk to be visible.

Following fusion of the egg and the sperm, rearrangement of the amphibian egg cytoplasm establishes one of the body axes **(Figure 47.8b)**. The plasma membrane and associated cortex rotate with respect to the inner cytoplasm. The animal pole cortex moves toward the point of sperm entry, and the vegetal hemisphere cortex across from the point of sperm entry moves toward the animal hemisphere. Molecules in the vegetal cortex are now able to interact with inner cytoplasmic molecules in the animal hemisphere, leading to the formation of cytoplasmic determinants that will later initiate development of dorsal structures. Thus, *cortical rotation* establishes the dorsal-ventral (back-belly) axis of the zygote. In some species, this event also exposes a light gray region of cytoplasm, the **gray crescent,** that is covered by the pigmented animal cortex near the equator of the egg prior to rotation. Located on the side opposite sperm entry, the gray crescent can be used as a marker for the future dorsal side of the embryo. The lighter pigment of the gray crescent can persist through many rounds of cell division.

Figure 47.9 shows the cleavage planes during the initial cell divisions in frogs. The first two divisions in frogs are meridional

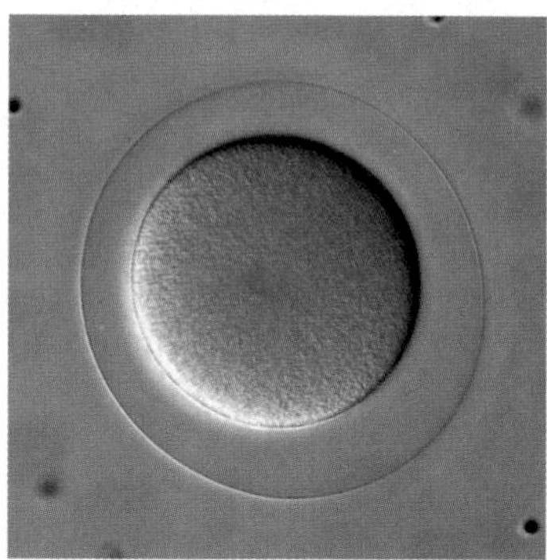

(a) Fertilized egg. Shown here is the zygote shortly before the first cleavage division, surrounded by the fertilization envelope. The nucleus is visible in the center.

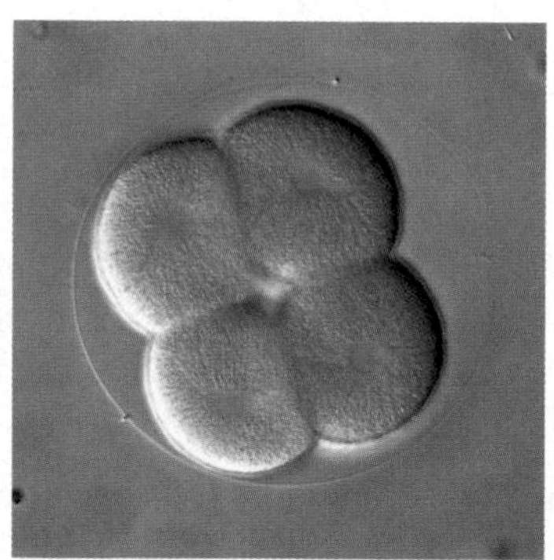

(b) Four-cell stage. Remnants of the mitotic spindle can be seen between the two cells that have just completed the second cleavage division.

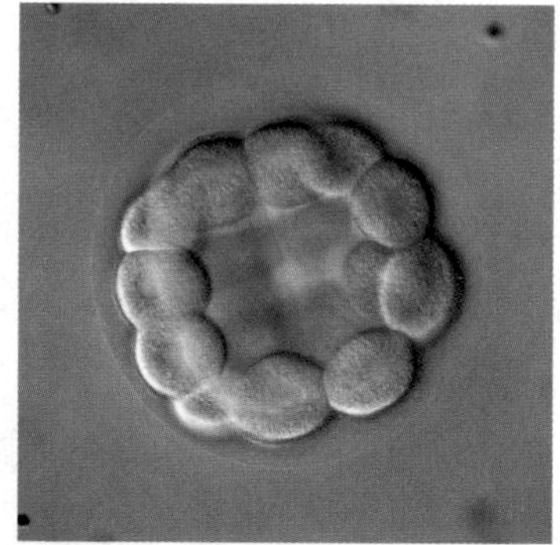

(c) Morula. After further cleavage divisions, the embryo is a multicellular ball that is still surrounded by the fertilization envelope. The blastocoel has begun to form.

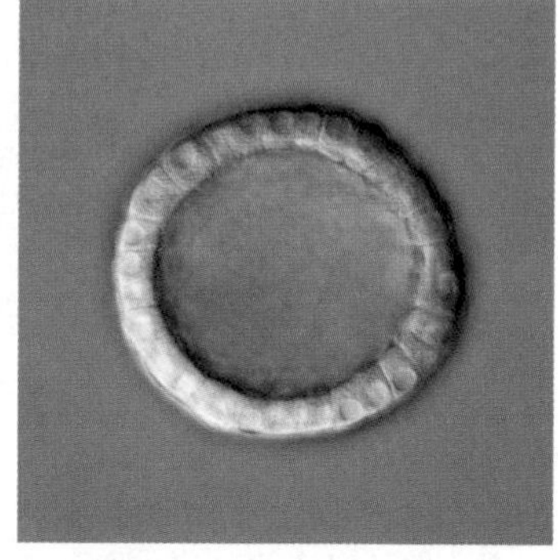

(d) Blastula. A single layer of cells surrounds a large blastocoel. Although not visible here, the fertilization envelope is still present; the embryo will soon hatch from it and begin swimming.

▲ **Figure 47.7 Cleavage in an echinoderm embryo.** Cleavage is a series of mitotic cell divisions that transform the zygote into a sphere of cells called blastomeres. These light micrographs show the embryonic stages of a sand dollar, which are virtually identical to those of a sea urchin.

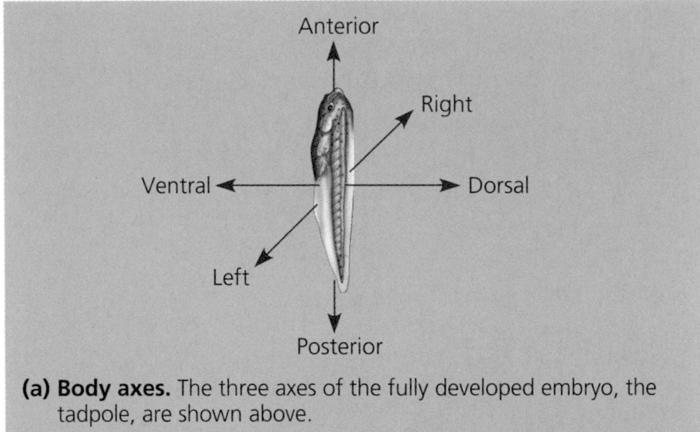

(a) Body axes. The three axes of the fully developed embryo, the tadpole, are shown above.

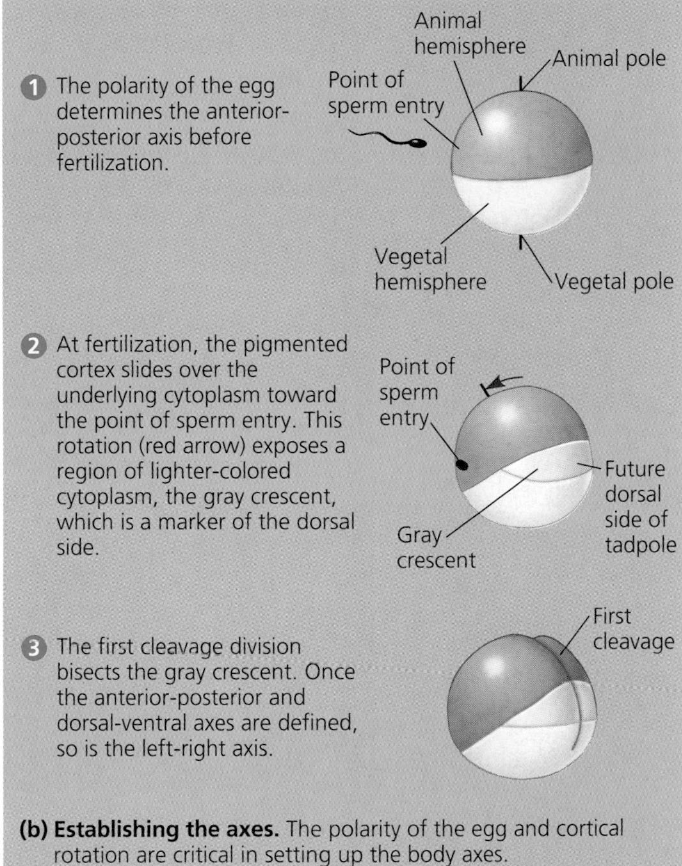

1 The polarity of the egg determines the anterior-posterior axis before fertilization.
Animal hemisphere
Animal pole
Point of sperm entry
Vegetal hemisphere
Vegetal pole
2 At fertilization, the pigmented cortex slides over the underlying cytoplasm toward the point of sperm entry. This rotation (red arrow) exposes a region of lighter-colored cytoplasm, the gray crescent, which is a marker of the dorsal side.
Point of sperm entry
Future dorsal side of tadpole
Gray crescent
3 The first cleavage division bisects the gray crescent. Once the anterior-posterior and dorsal-ventral axes are defined, so is the left-right axis.
First cleavage

(b) Establishing the axes. The polarity of the egg and cortical rotation are critical in setting up the body axes.

▲ **Figure 47.8 The body axes and their establishment in an amphibian.** All three axes are established before the zygote begins cleavage.

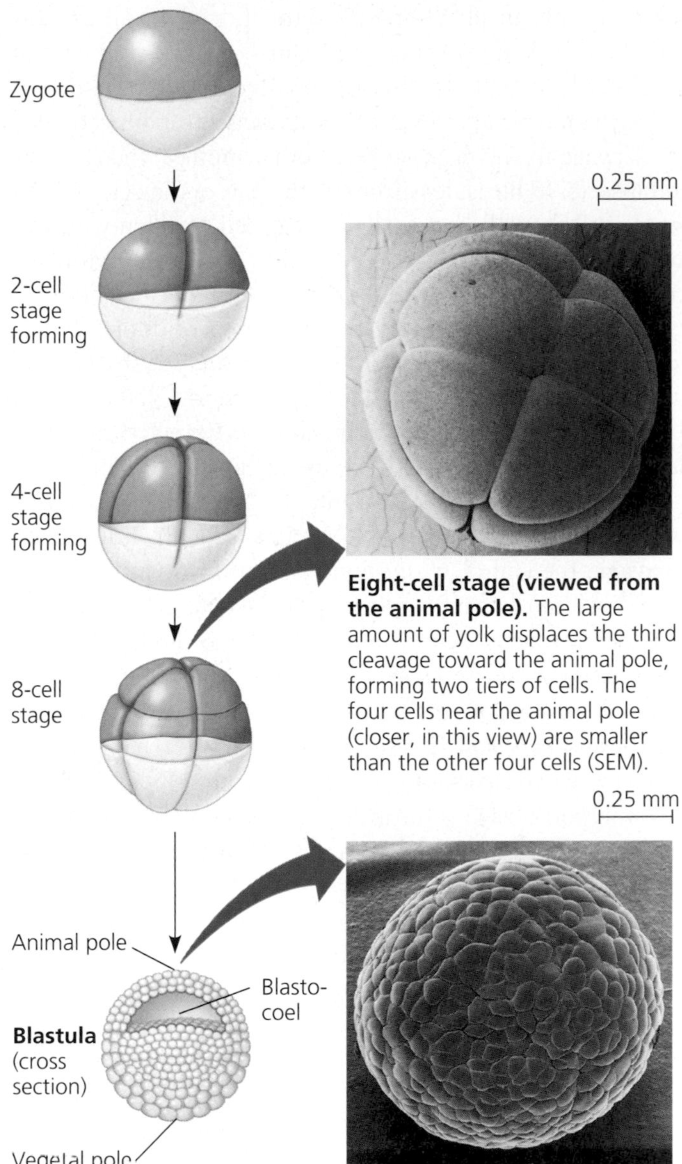

Eight-cell stage (viewed from the animal pole). The large amount of yolk displaces the third cleavage toward the animal pole, forming two tiers of cells. The four cells near the animal pole (closer, in this view) are smaller than the other four cells (SEM).

Blastula (at least 128 cells). As cleavage continues, a fluid-filled cavity, the blastocoel, forms within the embryo. Because of unequal cell division due to the large amount of yolk in the vegetal hemisphere, the blastocoel is located in the animal hemisphere, as shown in the cross section. The SEM shows the outside of a blastula with about 4,000 cells, viewed from the animal pole.

▲ **Figure 47.9 Cleavage in a frog embryo.** The cleavage planes in the first and second divisions extend from the animal to the vegetal pole, but the third cleavage is perpendicular to the polar axis.

(vertical), resulting in four blastomeres of equal size, each extending from the animal pole to the vegetal pole. The third division is equatorial (horizontal), producing an eight-celled embryo. However, the highly uneven distribution of yolk in the frog zygote displaces the mitotic apparatus and eventual cytokinesis toward the animal end of the dividing cells in equatorial divisions. Thus, the four blastomeres in the animal hemisphere are smaller than those in the vegetal hemisphere at the eight-cell stage. The displacing effect of the yolk persists in subsequent divisions. Continued divisions produce a morula and then a blastula. In frogs, because of unequal cell division, the blastocoel is located in the animal hemisphere (see Figure 47.9).

Although the eggs of sea urchins and some other animals have less yolk than frog eggs, they still have an animal-vegetal axis, owing to uneven distribution of other substances. Without the restraint imposed by yolk, however, the blastomeres formed during cleavage are more likely to be of similar size, particularly during the first few divisions (see Figure 47.7). Nonetheless, the general cleavage pattern seen in frogs is seen in sea urchins and other echinoderms, in all chordates, and indeed in all animals classified as deuterostomes. In animals whose eggs contain relatively little yolk, the blastocoel is centrally located.

Yolk is most plentiful and has its most pronounced effect on cleavage in the eggs of birds, other reptiles, many fishes, and insects. In birds, for example, the part of the egg we commonly call the yolk is actually the egg cell, swollen with yolk nutrients. There is a very small disk of actual cytoplasm located at the animal pole (**Figure 47.10**, step ❶). This enormous cell is surrounded by a protein-rich solution (the egg white) that will provide additional nutrients for the growing embryo. Cleavage of the fertilized egg is restricted to the small disk of yolk-free cytoplasm and cannot penetrate through the dense yolk; thus, the yolk remains uncleaved (Figure 47.10, step ❷). This incomplete division of a yolk-rich egg is known as **meroblastic cleavage.** It contrasts with **holoblastic cleavage,** the complete division of eggs having little yolk (as in sea urchins) or a moderate amount of yolk (as in frogs).

Early cleavage divisions in a bird embryo produce a cap of cells called the **blastoderm,** which rests on the undivided egg yolk. The blastomeres then sort into upper and lower layers, the *epiblast* and *hypoblast* (Figure 47.10, step ❸). The cavity between these two layers is the avian version of the blastocoel, and this embryonic stage is the avian equivalent of the blastula, although its form is different from the hollow ball of an early sea urchin or frog embryo.

In insects such as *Drosophila,* the zygote's nucleus is situated *within* a mass of yolk. Cleavage begins with the nucleus undergoing mitotic divisions that are not accompanied by cytokinesis (see Figure 21.12). These mitotic divisions produce several hundred nuclei, which at first spread throughout the yolk and later migrate to the outer edge of the embryo. After several more rounds of mitosis, a plasma membrane forms around each nucleus, and the embryo, now the equivalent of a blastula, consists of a single layer of about 6,000 cells surrounding a mass of yolk.

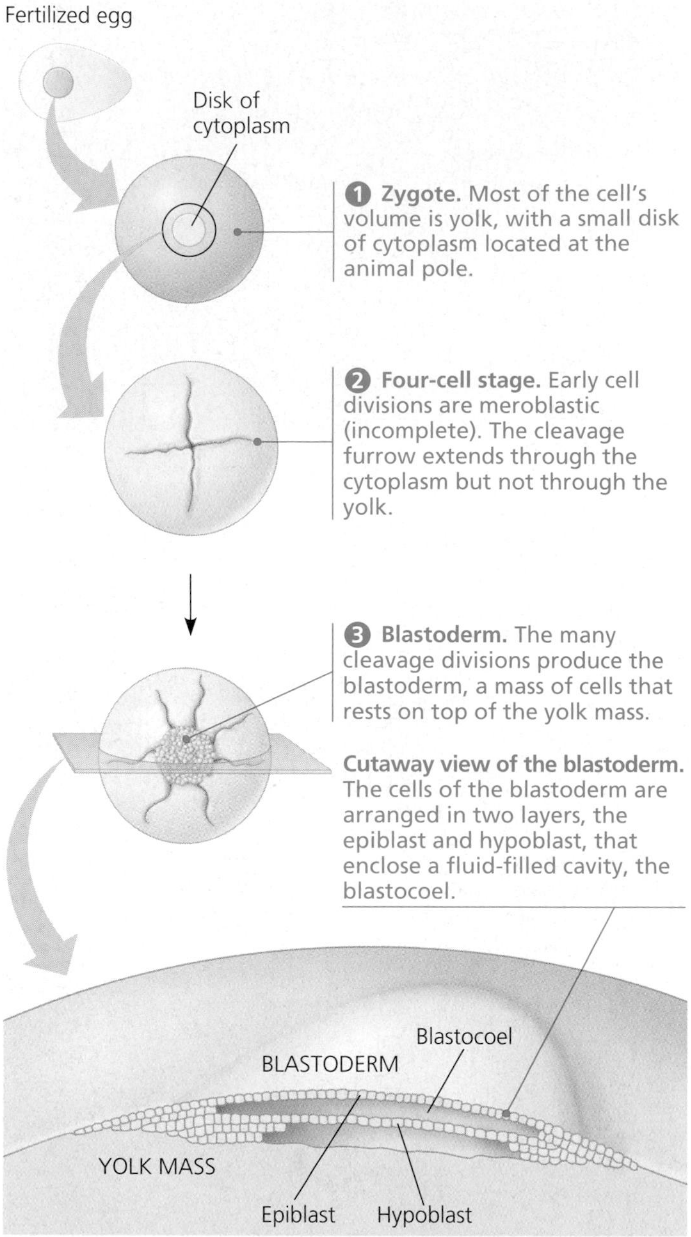

Fertilized egg
Disk of cytoplasm
❶ **Zygote.** Most of the cell's volume is yolk, with a small disk of cytoplasm located at the animal pole.
❷ **Four-cell stage.** Early cell divisions are meroblastic (incomplete). The cleavage furrow extends through the cytoplasm but not through the yolk.
❸ **Blastoderm.** The many cleavage divisions produce the blastoderm, a mass of cells that rests on top of the yolk mass.
Cutaway view of the blastoderm. The cells of the blastoderm are arranged in two layers, the epiblast and hypoblast, that enclose a fluid-filled cavity, the blastocoel.
Blastocoel
BLASTODERM
YOLK MASS
Epiblast
Hypoblast

▲ **Figure 47.10 Cleavage in a chick embryo.** The view in the top three drawings is from above, looking toward the animal pole. The bottom drawing is a cutaway view of the embryo with just a portion of the yolk mass showing. The blastoderm that forms during cleavage is the avian equivalent of the blastula in frogs (see Figure 47.9).

Gastrulation

The morphogenetic process called **gastrulation** is a dramatic rearrangement of the cells of the blastula to form a three-layered embryo with a primitive gut. Although gastrulation differs in detail from one animal group to another, the process is driven by the same general mechanisms in all species: changes in cell motility, changes in cell shape, and changes in cellular adhesion to other cells and to molecules of the extracellular matrix. The result of gastrulation is that some of the cells at or near the surface of the blastula move to an interior location, and three cell layers are established. The three-layered embryo is called the **gastrula** (plural, *gastrulae*). The positioning of the cell layers in the gastrula allows cells to interact with each other in new ways.

The three layers produced by gastrulation are embryonic tissues collectively termed the embryonic **germ layers.** The **ectoderm** forms the outer layer of the gastrula; the **endoderm** lines the embryonic digestive tract; and the **mesoderm** partly fills the space between the ectoderm and the endoderm. Eventually, these three cell layers develop into all the tissues and organs of the adult animal. Here we examine the events that occur during gastrulation in the sea urchin, frog, and chick.

Figure 47.11 outlines gastrulation in a sea urchin embryo. The sea urchin blastula consists of a single layer of cells enclosing the centrally located blastocoel. Gastrulation begins at the vegetal pole, where individual cells detach from the blastula wall and enter the blastocoel as migratory cells called *mesenchyme* *cells.* The remaining cells near the vegetal pole flatten slightly and form a *vegetal plate* that buckles inward, a process called **invagination.** The buckled vegetal plate then undergoes extensive rearrangement of its cells, transforming the shallow invagination into a deeper, narrower, blind-ended tube called the **archenteron,** or primitive gut. The open end of the archenteron, which will become the anus, is called the **blastopore.** A second opening forms when the other end of the archenteron touches the inside of the ectoderm and the two layers fuse, forming the mouth end of what is now a rudimentary digestive tube.

Gastrulation produces a sea urchin embryo with a primitive gut and three germ layers, which are commonly color-coded blue for ectoderm, red for mesoderm, and yellow for endoderm (see Figure 47.11, step 5). This three-layered body plan is characteristic of most animal phyla and is established very early in development. In the sea urchin, the gastrula eventually develops into a ciliated larva that drifts in ocean surface waters as plankton, feeding on bacteria and unicellular algae. After some time, the larva metamorphoses into the adult form of the sea urchin, which takes up existence on the bottom of the ocean.

In the frog, gastrulation also produces a three-layered embryo with an archenteron. The mechanics of gastrulation are more complicated in a frog, however, because of the large, yolk-laden cells of the vegetal hemisphere and because the wall of the blastula is more than one cell thick in most species. Gastrulation begins on the dorsal side of the blastula, when a group of cells begin to invaginate in a line along the region where the gray crescent formed in the zygote (see Figure 47.8). This invagination becomes the dorsal side of the blastopore, called the **dorsal lip** (**Figure 47.12**, on the next page). The blastopore lip extends as invagination continues to be initiated at each end, until the two ends of the blastopore meet on the ventral side. The blastopore is now a complete circle.

All along the blastopore, future endoderm and mesoderm cells on the surface of the embryo roll over the edge of the lip into the interior of the embryo, a process called **involution.** Once inside the embryo, these cells move away from the blastopore and become organized into layers of endoderm and mesoderm, with the endoderm on the inside. The blastocoel collapses during this process, displaced by the archenteron cavity that is formed by the tube of endoderm. As gastrulation is completed, the circular lip of the blastopore encircles a **yolk plug** consisting of the outer food-laden cells; these protruding cells will move inward as expansion of the ectoderm causes the blastopore to shrink further. At this point, the cells remaining on the surface make up the ectoderm, surrounding the layers of mesoderm and endoderm. As in the sea urchin, the amphibian anus develops from the blastopore, and the mouth eventually forms at the opposite end of the archenteron after it extends to the ventral side near the animal pole.

Gastrulation in the chick is similar to frog gastrulation in that it involves cells moving from the surface of the embryo to an interior location. In birds, however,

Key
Future ectoderm
Future mesoderm
Future endoderm
Animal pole
Blastocoel
Mesenchyme cells
Vegetal plate
Vegetal pole
1 The blastula consists of a single layer of ciliated cells surrounding the blastocoel. Gastrulation begins with the migration of mesenchyme cells from the vegetal pole into the blastocoel.
2 The vegetal plate invaginates (buckles inward). Mesenchyme cells migrate throughout the blastocoel.
Blastocoel
Filopodia pulling archenteron tip
Archenteron
Mesenchyme cells
Blastopore
50 µm
3 Endoderm cells form the archenteron (future digestive tube). New mesenchyme cells at the tip of the tube begin to send out thin extensions (filopodia) toward the ectoderm cells of the blastocoel wall (inset, LM).
Blastocoel
Archenteron
Blastopore
4 Contraction of these filopodia then drags the archenteron across the blastocoel.
Ectoderm
Mouth
Mesenchyme (mesoderm forms future skeleton)
Digestive tube (endoderm)
Anus (from blastopore)
5 Fusion of the archenteron with the blastocoel wall completes formation of the digestive tube with a mouth and an anus. The gastrula has three germ layers and is covered with cilia, which function in swimming and feeding.

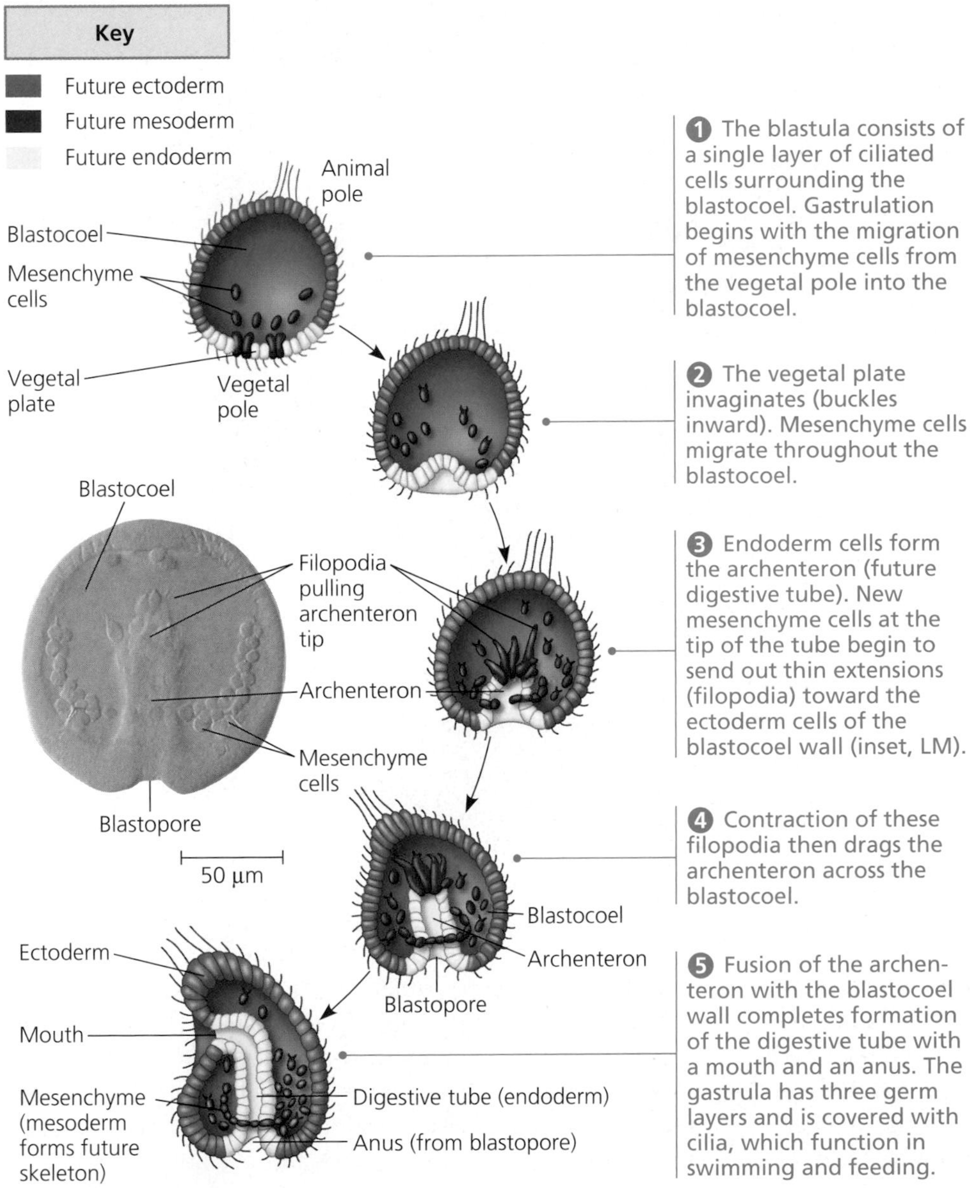

▲ **Figure 47.11 Gastrulation in a sea urchin embryo.** The movement of cells during gastrulation forms an embryo with a primitive gut and three germ layers. Some of the mesenchyme (mesoderm) cells that migrate inward (step 1) will eventually secrete calcium carbonate and form a simple internal skeleton. Embryos in steps 1–3 are viewed from the front, those in 4 and 5 from the side.

1 Gastrulation begins when a small indented crease, the dorsal lip of the blastopore, appears on one side of the blastula. The crease is formed by cells changing shape and pushing inward from the surface (invagination). Additional cells then roll inward over the dorsal lip (involution) and move into the interior, where they will form endoderm and mesoderm. Meanwhile, cells of the animal pole, the future ectoderm, change shape and begin spreading over the outer surface.

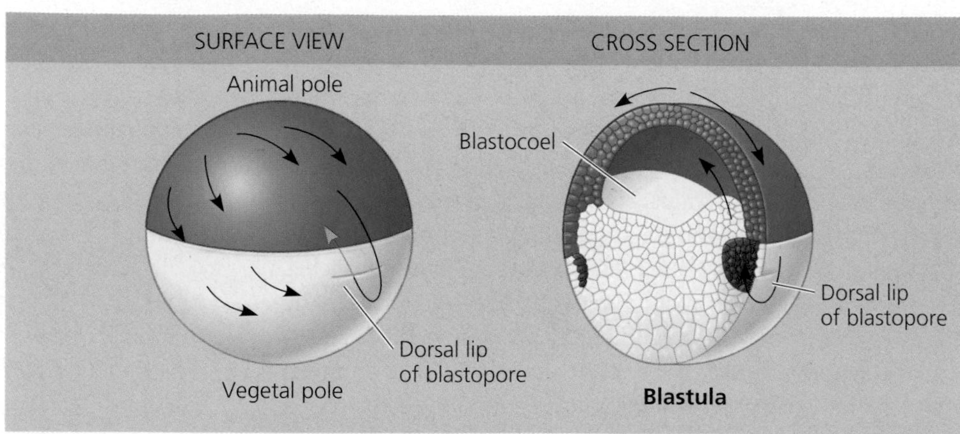

2 The blastopore lip grows on both sides of the embryo, as more cells invaginate. When the sides of the lip meet, the blastopore forms a circle that becomes smaller as ectoderm spreads downward over the surface. Internally, continued involution expands the endoderm and mesoderm, and the archenteron begins to form; as a result, the blastocoel becomes smaller.

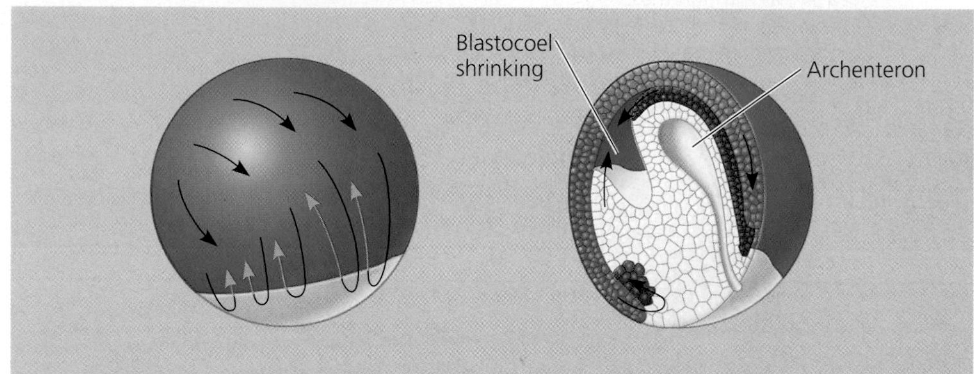

3 Late in gastrulation, the endoderm-lined archenteron has completely replaced the blastocoel and the three germ layers are in place. The circular blastopore surrounds a plug of yolk-filled cells.

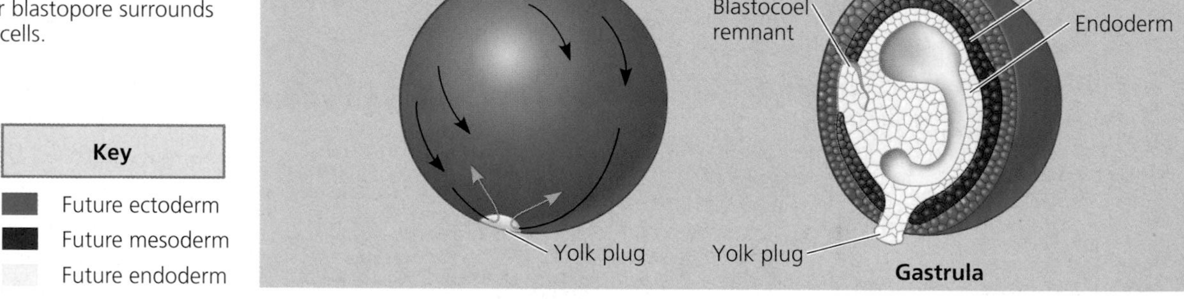

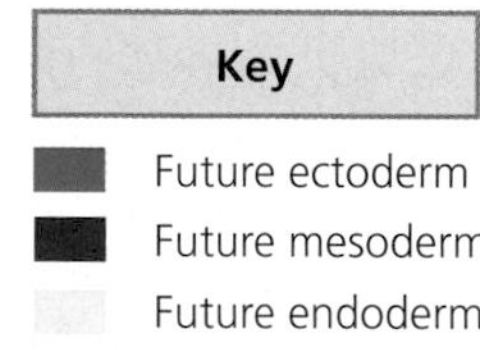

▲ **Figure 47.12 Gastrulation in a frog embryo.** In the frog blastula, the blastocoel is displaced toward the animal pole and is surrounded by a wall several cells thick. The cell movements that begin gastrulation occur on the dorsal side of the blastula, where the gray crescent was located in the zygote (see Figure 47.8b). Although still visible as gastrulation begins, the gray crescent is not shown here.

the inward movement of cells during gastrulation is affected by the large mass of yolk pressing against the bottom of the embryo. Recall that cleavage in the chick results in a blastoderm consisting of upper and lower layers—the epiblast and hypoblast—lying atop the yolk mass (see Figure 47.10). All the cells that will form the embryo come from the epiblast. During gastrulation, some epiblast cells move toward the midline of the blastoderm, then detach and move inward toward the yolk **(Figure 47.13)**. The pileup of cells moving inward at the blastoderm's midline produces a thickening called the **primitive streak**, which runs along what will become the bird's anterior-posterior axis. The primitive streak is functionally equivalent to the lip of the blastopore in the frog, but the two structures are oriented differently in the two embryos. Some of the inward-moving epiblast cells displace hypoplast cells and form the endoderm; other epiblast cells move laterally into the blastocoel, forming the mesoderm. The epiblast cells that remain on the surface give rise to the ectoderm. Although the hypoblast contributes no cells to the embryo, it seems to help direct the formation of the primitive streak before the onset of gastrulation and is required for normal development. The hypoblast cells later segregate from the endoderm and eventually form portions of a sac that surrounds the yolk and a stalk that connects the yolk mass to the embryo.

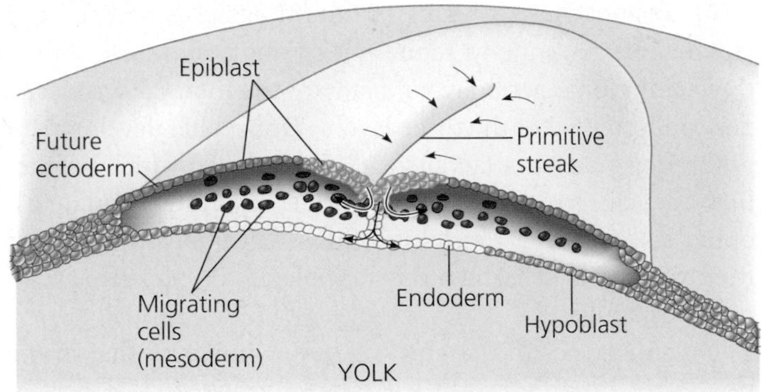

▲ **Figure 47.13 Gastrulation in a chick embryo.** During gastrulation, some cells of the epiblast migrate (arrows) into the interior of the embryo through the primitive streak, shown here in transverse section. Some of these cells move downward and form endoderm, while others migrate laterally and form mesoderm. The cells left behind on the surface of the embryo at the end of gastrulation will become ectoderm.

Despite variations in how the three germ layers form in different species, once they are in place, gastrulation is complete. Now is the time when the embryo's organs begin to form.

Organogenesis

Various regions of the three embryonic germ layers develop into the rudiments of organs during the process of **organogenesis.** While gastrulation involves mass cell movements, organogenesis involves more localized morphogenetic changes in tissue and cell shape. The first evidence of organ building is the appearance of folds, splits, and dense clustering (condensation) of cells.

Figure 47.14 shows some events during early organogenesis in a frog. The organs that begin to take shape first in the embryos of frogs and other chordates are the neural tube and the notochord, the skeletal rod characteristic of all chordate embryos. The **notochord** is formed from dorsal mesoderm

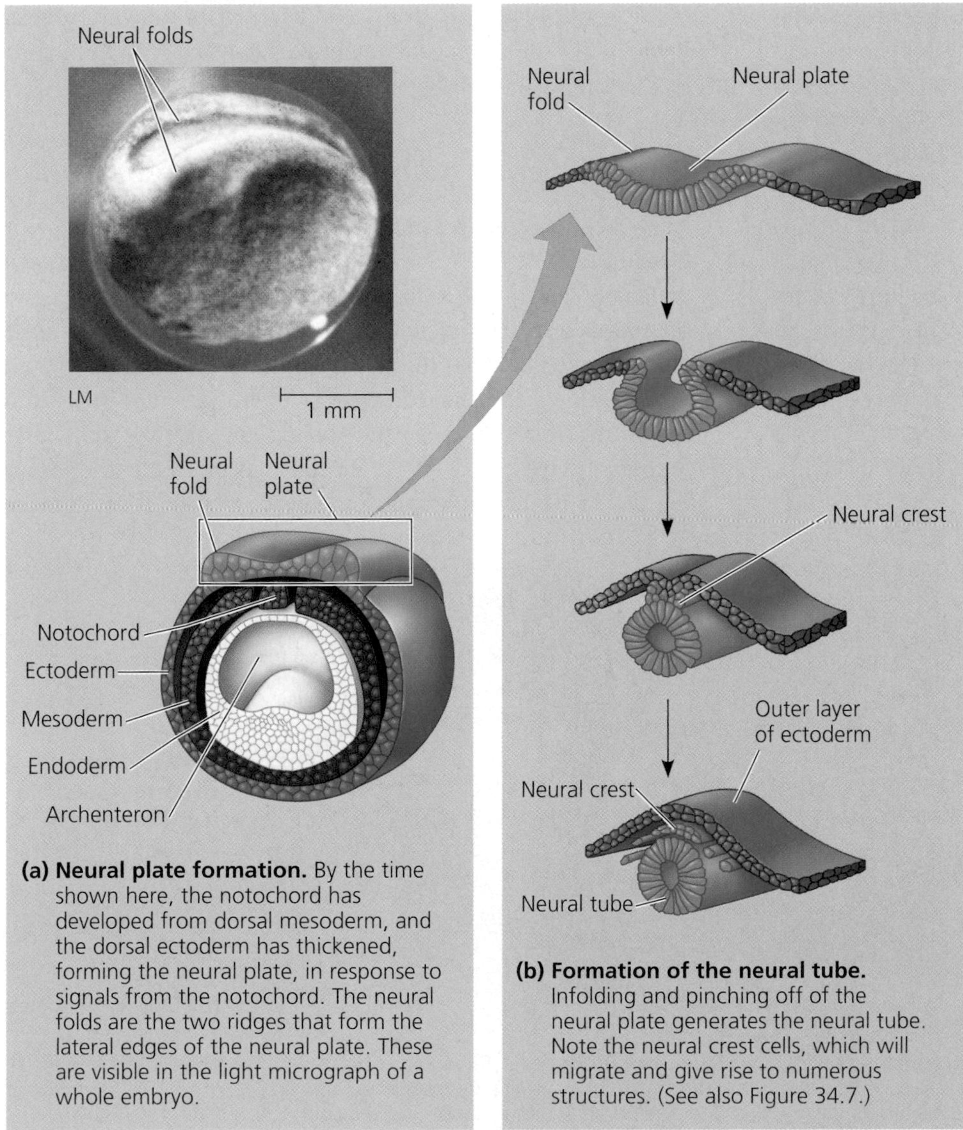

(a) Neural plate formation. By the time shown here, the notochord has developed from dorsal mesoderm, and the dorsal ectoderm has thickened, forming the neural plate, in response to signals from the notochord. The neural folds are the two ridges that form the lateral edges of the neural plate. These are visible in the light micrograph of a whole embryo.

(b) Formation of the neural tube. Infolding and pinching off of the neural plate generates the neural tube. Note the neural crest cells, which will migrate and give rise to numerous structures. (See also Figure 34.7.)

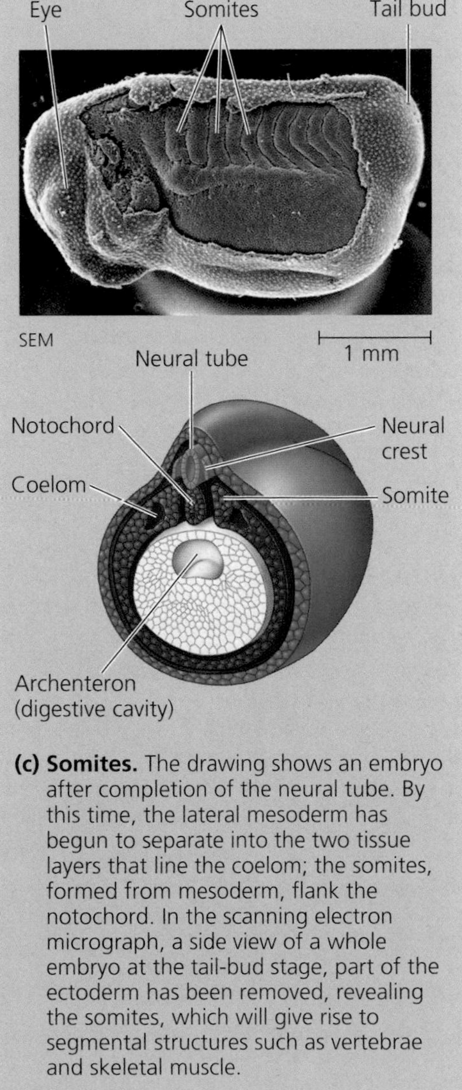

(c) Somites. The drawing shows an embryo after completion of the neural tube. By this time, the lateral mesoderm has begun to separate into the two tissue layers that line the coelom; the somites, formed from mesoderm, flank the notochord. In the scanning electron micrograph, a side view of a whole embryo at the tail-bud stage, part of the ectoderm has been removed, revealing the somites, which will give rise to segmental structures such as vertebrae and skeletal muscle.

▲ **Figure 47.14 Early organogenesis in a frog embryo.**

that condenses just above the archenteron **(Figure 47.14a)**. Signals sent from the notochord to the ectoderm above it cause that region of ectoderm to become the neural plate. The neural plate soon curves inward, rolling itself into the **neural tube,** which runs along the anterior-posterior axis of the embryo **(Figure 47.14b)**. The neural tube will become the central nervous system—the brain in the head and the spinal cord down the rest of the body. The signaling by the mesodermally derived notochord to the ectoderm is a good example of a process seen often during organogenesis: One germ layer signals another, thereby determining the fate of the second.

Unique to vertebrate embryos, a band of cells called the **neural crest** develops along the border where the neural tube pinches off from the ectoderm. Cells of the neural crest subsequently migrate to various parts of the embryo, forming peripheral nerves, teeth, skull bones, and so many other different cell types that some developmental biologists have proposed considering the neural crest as a "fourth germ layer."

Other condensations occur in strips of mesoderm lateral to the notochord, which separate into blocks called **somites (Figure 47.14c)**. The somites are arranged serially on both sides along the length of the notochord. Parts of the somites dissociate into individual mesenchymal (wandering) cells, which migrate to new locations. The notochord functions as a core around which these mesodermal cells gather and form the vertebrae. Parts of the notochord between the vertebrae persist as the inner portions of the vertebral disks in adults. (These are the disks that can "slip," causing back pain.) Somite cells also form the muscles associated with the axial skeleton. This serial origin of the axial skeleton and muscles reinforces the idea that chordates are basically segmented animals, although the segmentation becomes less obvious later in development. Lateral to the somites, the mesoderm splits into two layers that form the lining of the body cavity, or coelom.

As organogenesis progresses, morphogenesis and cell differentiation continue to refine the organs that arise from the three embryonic germ layers; many of the internal organs are derived from two of the three layers. Embryonic development of the frog leads to a larval stage, the tadpole, which hatches from the jelly coat that originally protected the egg and developing embryo. Later, metamorphosis will transform the frog from the aquatic, herbivorous tadpole to the terrestrial, carnivorous adult.

Organogenesis in the chick is quite similar to that in the frog. After the three germ layers are formed, the borders of the blastoderm fold downward and come together, pinching the embryo into a three-layered tube joined underneath the middle of the body to the yolk **(Figure 47.15a)**. Neural tube formation, development of the notochord and somites, and other events in organogenesis occur much as in the frog embryo. The rudiments of the major organs are evident in a 2- to 3-day-old chick embryo **(Figure 47.15b)**.

Figure 47.16 lists the embryonic sources of the major organs and tissues in frogs, chicks, and other vertebrates. Review this figure to become familiar with the embryonic origins of the various body structures.

Developmental Adaptations of Amniotes

All vertebrate embryos require an aqueous environment for development. In the case of fishes and amphibians, the egg is usually laid in the surrounding sea or pond and needs no special water-filled enclosure. The movement of vertebrate animals onto land required the evolution of structures that would allow reproduction in dry environments. Two such structures exist today: (1) the shelled egg of birds and other reptiles, as well as a few mammals (monotremes), and (2) the uterus of marsupial and eutherian (placental) mammals.

Neural tube
Notochord
Somite
Coelom
Archenteron
Endoderm
Lateral fold
Mesoderm
Ectoderm
Yolk stalk
YOLK
Yolk sac
Form extraembryonic membranes

(a) Early organogenesis. The archenteron forms when lateral folds pinch the embryo away from the yolk. The embryo remains open to the yolk, attached by the yolk stalk, about midway along its length, as shown in this cross section. The notochord, neural tube, and somites subsequently form much as they do in the frog.

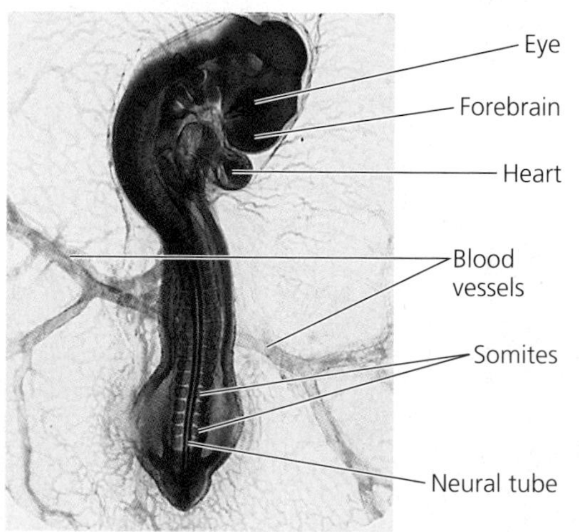

(b) Late organogenesis. Rudiments of most major organs have already formed in this chick embryo, which is about 56 hours old and about 2–3 mm long (LM).

◄ **Figure 47.15 Organogenesis in a chick embryo.** **(a)** The germ layers lateral to the embryo itself give rise to extraembryonic membranes (discussed later). **(b)** These membranes eventually are supplied by blood vessels extending from the embryo; several major blood vessels are seen here.

▶ Figure 47.16 Adult derivatives of the three embryonic germ layers in vertebrates.

ECTODERM	MESODERM	ENDODERM
• Epidermis of skin and its derivatives (including sweat glands, hair follicles) • Epithelial lining of mouth and rectum • Sensory receptors in epidermis • Cornea and lens of eye • Nervous system • Adrenal medulla • Tooth enamel • Epithelium of pineal and pituitary glands	• Notochord • Skeletal system • Muscular system • Muscular layer of stomach, intestine, etc. • Excretory system • Circulatory and lymphatic systems • Reproductive system (except germ cells) • Dermis of skin • Lining of body cavity • Adrenal cortex	• Epithelial lining of digestive tract • Epithelial lining of respiratory system • Lining of urethra, urinary bladder, and reproductive system • Liver • Pancreas • Thymus • Thyroid and parathyroid glands

Within the shell or uterus, the embryos of these animals are surrounded by fluid within a sac formed by a membrane called the amnion. Reptiles (including birds) and mammals are therefore called **amniotes** (see Chapter 34).

We have already seen that embryonic development of the chick, an amniote, is very similar to that of the frog, a vertebrate that lacks an amnion. However, in the chick, development also includes the formation of **extraembryonic membranes,** membranes located outside the embryo. Notice in Figure 47.15a that only part of each germ layer contributes to the embryo itself. The germ layers located outside the embryo proper develop into the amnion and three other membranes—the yolk sac, the chorion, and the allantois. These four extraembryonic membranes provide a "life-support system" for further embryonic development within the shelled egg or the uterus of an amniote. Each membrane is a sheet of cells derived from two germ layers. **Figure 47.17** reviews the locations and functions of these membranes in birds and other reptiles.

Mammalian extraembryonic membranes will be discussed next, as we describe early development in mammalian embryos. Formation of the placenta, a structure unique to marsupial and eutherian mammals, is an important part of this process.

Mammalian Development

In most mammalian species, fertilization takes place in the oviduct, and the earliest stages of development occur while the embryo completes its journey down the oviduct to the uterus (see Figure 46.15). In contrast to the large, yolky eggs of birds, other reptiles, and monotremes, the eggs of most mammals are quite small, storing little in the way of food reserves. As already mentioned, the mammalian egg cell and zygote have not been shown to exhibit obvious polarity with respect to the contents of the cytoplasm, and cleavage of the zygote, which lacks yolk, is holoblastic. However, mammalian gastrulation and early organogenesis follow a pattern similar to that of birds and other reptiles. (Recall from Chapter 34 that the mammalian and reptilian lineages diverged during the Mesozoic era.)

Amnion. The amnion protects the embryo in a fluid-filled cavity that prevents dehydration and cushions mechanical shock.

Allantois. The allantois functions as a disposal sac for certain metabolic wastes produced by the embryo. The membrane of the allantois also functions with the chorion as a respiratory organ.

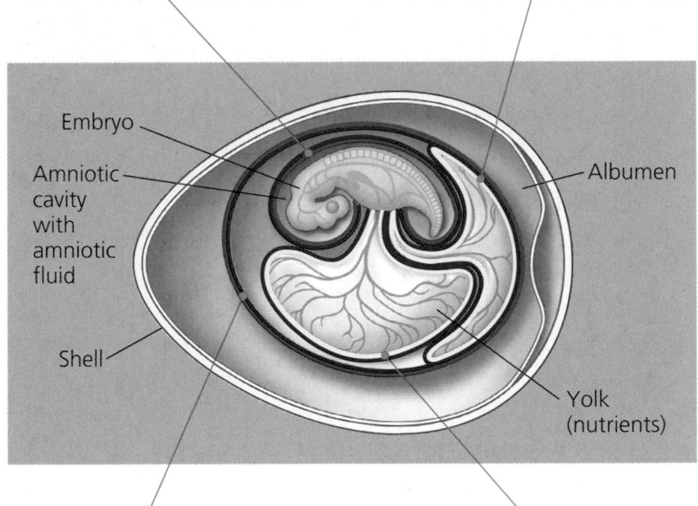

Chorion. The chorion and the membrane of the allantois exchange gases between the embryo and the surrounding air. Oxygen and carbon dioxide diffuse freely across the egg's shell.

Yolk sac. The yolk sac expands over the yolk, a stockpile of nutrients stored in the egg. Blood vessels in the yolk sac membrane transport nutrients from the yolk into the embryo. Other nutrients are stored in the albumen (the "egg white").

▲ Figure 47.17 Extraembryonic membranes in birds and other reptiles. There are four extraembryonic membranes: the amnion, the allantois, the chorion, and the yolk sac. Each membrane is a sheet of cells that develops from epithelial sheets of two of the three germ layers external to the embryo proper (see Figure 47.15a).

Early cleavage is relatively slow in mammals. In the case of humans, the first division is complete about 36 hours after fertilization, the second division at about 60 hours, and the third division at about 72 hours. The blastomeres are equal in size. At the eight-cell stage, the blastomeres become tightly adhered to one another, causing the outer surface of the embryo to appear smooth. **Figure 47.18** depicts development of the human embryo starting about 7 days after fertilization (see key numbers in the figure):

1. At the completion of cleavage, the embryo has over 100 cells arranged around a central cavity and has traveled down the oviduct to the uterus. This embryonic stage, the **blastocyst,** is the mammalian version of a blastula. Clustered at one end of the blastocyst cavity is a group of cells called the **inner cell mass,** which will subsequently develop into the embryo proper and form or contribute to all the extraembryonic membranes.
2. The **trophoblast,** the outer epithelium of the blastocyst, initiates implantation by secreting enzymes that break down molecules of the endometrium, the lining of the uterus. This allows the blastocyst to invade the endometrium. As the trophoblast thickens through cell division, it extends fingerlike projections into the surrounding maternal tissue, which is rich in blood vessels. Invasion by the trophoblast leads to erosion of capillaries in the endometrium, causing blood to spill out and bathe trophoblast tissues. Around the time of implantation, the inner cell mass of the blastocyst forms a flat disk with an upper layer of cells, the *epiblast,* and a lower layer, the *hypoblast,* which are homologous to the epiblast and hypoblast of birds (see Figure 47.10, step 3). As in birds, the human embryo develops almost entirely from epiblast cells.
3. As implantation is completed, gastrulation begins. Cells move inward from the epiblast through a primitive streak to form mesoderm and endoderm, just as in the chick. At the same time, extraembryonic membranes begin to form. The trophoblast continues to expand into the endometrium. The invading trophoblast, mesodermal cells derived from the epiblast, and adjacent endometrial tissue all contribute to formation of the placenta (see Figure 46.16).
4. By the end of gastrulation, the embryonic germ layers have formed. The three-layered embryo is now surrounded by proliferating extraembryonic mesoderm and the four extraembryonic membranes.

The extraembryonic membranes in mammals are homologous to those of birds and other reptiles and develop in a similar way. The **chorion,** which completely surrounds the embryo and the other extraembryonic membranes, functions in gas exchange. The **amnion** eventually encloses the embryo in a protective fluid-filled amniotic cavity. (The fluid from this cavity is the "water" expelled from the vagina of the mother when the amnion breaks just prior to childbirth.) Below the developing embryo proper, the **yolk sac** encloses another fluid-filled cavity. Although this cavity contains no yolk, the membrane that surrounds it is given the same name as the homologous membrane in birds and other reptiles. The yolk sac membrane

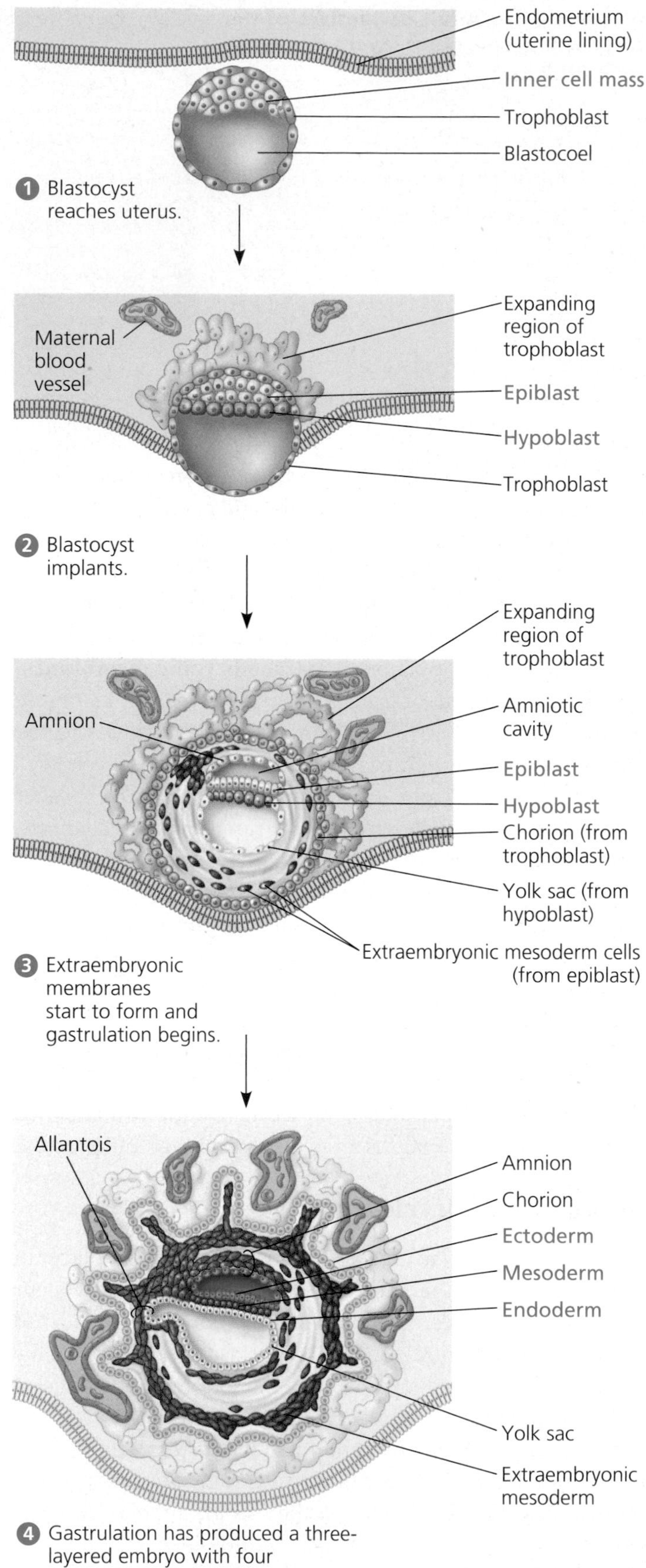

▲ **Figure 47.18 Four stages in early embryonic development of a human.** The epiblast gives rise to the three germ layers, which form the embryo proper. See the text for a description of each stage.

of mammals is a site of early formation of blood cells, which later migrate into the embryo proper. The fourth extraembryonic membrane, the **allantois**, is incorporated into the umbilical cord. Here it forms blood vessels that transport oxygen and nutrients from the placenta to the embryo and rid the embryo of carbon dioxide and nitrogenous wastes. Thus, the extraembryonic membranes of shelled eggs, where embryos are nourished with yolk, were conserved in mammals as they diverged from reptiles in the course of evolution, but with modifications adapted to development within the reproductive tract of the mother.

The completion of gastrulation is followed by formation of the notochord, neural tube, and somites, the first events in organogenesis. By the end of the first trimester of human development, rudiments of all the major organs have developed from the three germ layers (see Figure 47.16).

Concept Check 47.1

1. Predict what would happen if you injected Ca^{2+} into an unfertilized sea urchin egg.
2. A frog zygote and frog blastula are nearly the same size. Explain this observation.
3. Although gastrulation looks different in sea urchin, frog, and chick embryos, what is its common result in these animals?
4. Compare cell movements during gastrulation and organogenesis in animals.

For suggested answers, see Appendix A.

Concept 47.2

Morphogenesis in animals involves specific changes in cell shape, position, and adhesion

Now that you have learned about the main events of embryonic development in animals, we will address the cellular and molecular mechanisms by which they occur. Although biologists are far from fully understanding these mechanisms, several key principles have emerged as fundamental to the development of all animals.

Morphogenesis is a major aspect of development in both animals and plants, but only in animals does it involve the *movement* of cells. Movement of parts of a cell can bring about changes in cell shape or enable a cell to migrate from one place to another within the embryo. Changes in both cell shape and cell position are involved in cleavage, gastrulation, and organogenesis. Here we consider some of the cellular components that contribute to these events.

The Cytoskeleton, Cell Motility, and Convergent Extension

Changes in the shape of a cell usually involve reorganization of the cytoskeleton (see Table 6.1). Consider, for example, how the cells of the neural plate form the neural tube **(Figure 47.19)**. First, microtubules oriented parallel to the dorsal-ventral axis of the embryo apparently help lengthen the cells in that direction. At the dorsal end of each cell is a parallel array of actin filaments oriented crosswise. These contract, giving the cells a wedge shape that forces the ectoderm layer to bend inward. Similar changes in cell shape occur during other invaginations (inpocketings) and evaginations (outpocketings) of tissue layers throughout development.

The cytoskeleton also drives cell migration, the active movement of cells from one place to another in developing animals. Cells "crawl" within the embryo by using cytoskeletal fibers to extend and retract cellular protrusions. This type of motility is akin to the amoeboid movement described in Figure 6.27b, but in contrast to the thick pseudopodia of some amoeboid cells, the cellular protrusions of migrating embryonic cells are usually flat sheets (lamellipodia) or spikes (filopodia).

During gastrulation in some organisms, invagination is initiated by the wedging of cells on the surface of the blastula, but the movement of cells deeper into the embryo involves the extension of filopodia by cells at the leading edge of the migrating tissue. The cells that first move through the blastopore

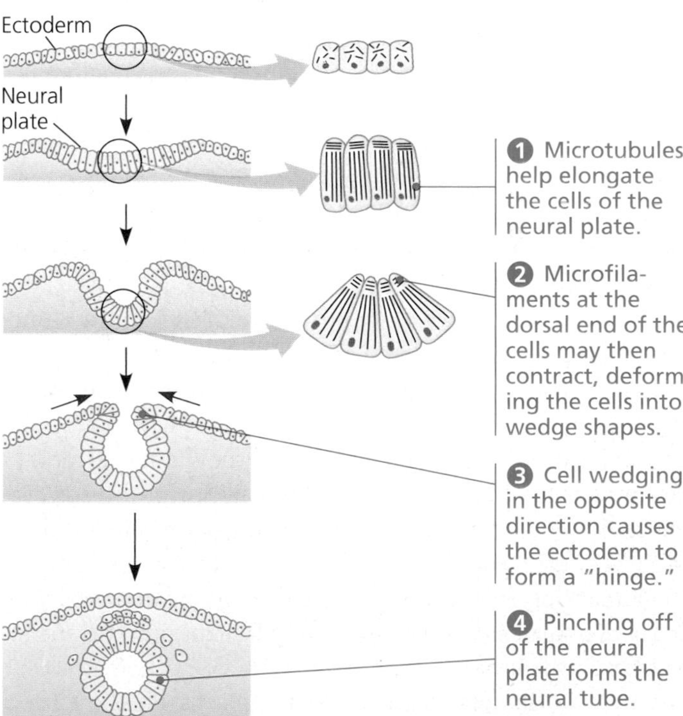
Ectoderm
Neural plate
1 Microtubules help elongate the cells of the neural plate.
2 Microfilaments at the dorsal end of the cells may then contract, deforming the cells into wedge shapes.
3 Cell wedging in the opposite direction causes the ectoderm to form a "hinge."
4 Pinching off of the neural plate forms the neural tube.

▲ **Figure 47.19 Change in cellular shape during morphogenesis.** Reorganization of the cytoskeleton is associated with morphogenetic changes in embryonic tissues, as shown here for the formation of the neural tube in vertebrates.

and up along the inside of the blastocoel wall drag others behind them, thus helping direct movement of the entire sheet of cells from the embryo's surface into appropriate locations in the blastocoel. The involuted sheet then forms the endoderm and mesoderm of the embryo (see Figure 47.12). There are also many situations in which cells migrate individually, as when the cells of a somite or the neural crest disperse to various parts of the embryo.

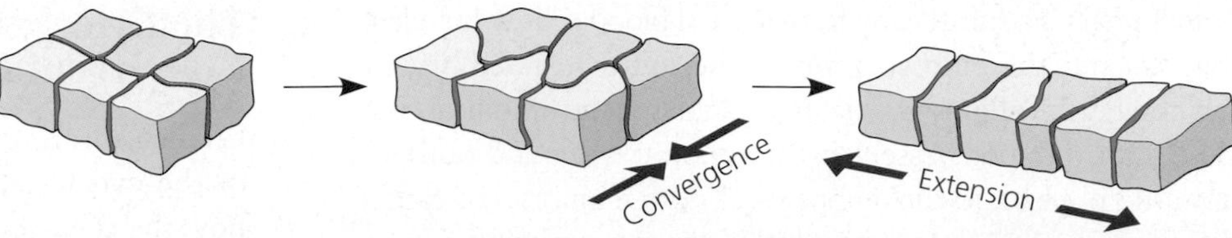

▲ **Figure 47.20 Convergent extension of a sheet of cells.** In this simplified diagram, changes in cell shape and position cause the layer of cells to become narrower and longer. Molecular signals are believed to cause the cells to elongate in a particular direction and to crawl between each other (*convergence*), leading to *extension* of the cell sheet in a perpendicular direction.

Cell crawling is also involved in **convergent extension,** a type of morphogenetic movement in which the cells of a tissue layer rearrange themselves so that the sheet becomes narrower (converges) while it becomes longer (extends) **(Figure 47.20)**. When many cells wedge between one another, the tissue can extend dramatically. Convergent extension is important in early embryonic development. It occurs, for example, as the archenteron elongates in sea urchin embryos and during involution in the frog gastrula. In the latter case, convergent extension is responsible for the change in shape from the spherical gastrula to the submarine-shaped frog embryo seen in Figure 47.14c.

Roles of the Extracellular Matrix and Cell Adhesion Molecules

Scientists are coming closer to understanding the signaling pathways that trigger and guide cell movement. The process appears to involve the extracellular matrix (ECM), the mixture of secreted glycoproteins lying outside the plasma membranes of cells (see Figure 6.29). The ECM is known to help guide cells in many types of morphogenetic movements. ECM fibers may function as tracks, directing migrating cells along particular routes. Several kinds of extracellular glycoproteins, including fibronectin, promote cell migration by providing specific molecular anchorage for moving cells. The experiment depicted in **Figure 47.21** provides evidence for this function of fibronectin in guiding cell migration.

Evidence also exists that fibronectin plays a similar role at the leading edge of the involuting tissue during frog gastrulation. Fibronectin fibers line the roof of the blastocoel, and as the future mesoderm moves into the interior of the embryo, the cells at the free edge of the involuting sheet migrate along the fibers (see Figure 47.12). Researchers can prevent the attachment of cells to fibronectin by injecting embryos with antibodies against fibronectin. This treatment prevents the inward movement of the mesoderm.

Other substances in the ECM keep cells on the correct paths by *inhibiting* migration in certain directions. Thus, depending on the substances they secrete, nonmigratory cells situated along migration pathways may promote or inhibit movement of other cells.

Figure 47.21

Inquiry Does fibronectin promote cell migration?

EXPERIMENT Researchers placed a strip of fibronectin on an artificial underlayer. After positioning migratory neural crest cells at one end of the strip, the researchers observed the movement of the cells in a light microscope.

RESULTS In this micrograph, the dashed lines indicate the edges of the fibronectin layer. Note that cells are migrating along the strip, not off of it.

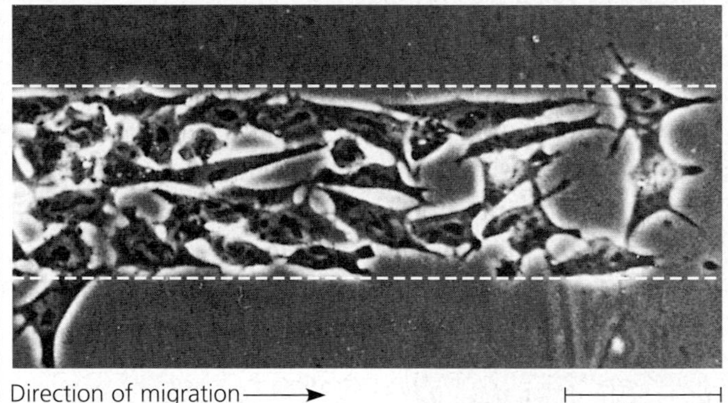

CONCLUSION Fibronectin helps promote cell migration, possibly by providing anchorage for the migrating cells.

As migrating cells move along specific paths through the embryo, receptor proteins on their surfaces pick up directional cues from the immediate environment. Such signals from the ECM environment can direct the orientation of cytoskeletal elements in a way that propels the cell in the proper direction.

Also contributing to cell migration and stable tissue structure are glycoproteins called **cell adhesion molecules (CAMs),** which are located on the surfaces of cells and bind to CAMs on other cells. CAMs vary in amount, chemical identity, or both from one type of cell to another, and these differences help regulate morphogenetic movements and tissue building.

One important class of cell-to-cell adhesion molecule is the **cadherins,** which require calcium ions for proper function.

There are many different cadherins, and the gene for each cadherin is expressed in specific locations at specific times during embryonic development. Researchers have vividly demonstrated the importance of one particular cadherin in the formation of the frog blastula **(Figure 47.22)**. Cadherins are also involved in the tight adhesion of cells in the mammalian embryo that first occurs at the eight-cell stage, when cadherin production begins.

As we have seen, cell behavior and molecular mechanisms contribute to the morphogenesis of the embryo. The same basic cellular and genetic processes ensure that the various types of cells end up in the right places in each embryo.

Figure 47.22

Inquiry Is cadherin required for development of the blastula?

EXPERIMENT Researchers injected frog eggs with nucleic acid complementary to the mRNA encoding a cadherin known as EP cadherin. This "antisense" nucleic acid leads to destruction of the mRNA for normal EP cadherin, so no EP cadherin protein is produced. Frog sperm were then added to control (noninjected) eggs and to experimental (injected) eggs. The control and experimental embryos that developed were observed with a scanning electron microscope.

RESULTS As shown in these micrographs, fertilized control eggs developed into normal blastulas, but fertilized experimental eggs did not. In the absence of EP cadherin, the blastocoel did not form properly, and the cells were arranged in a disorganized fashion.

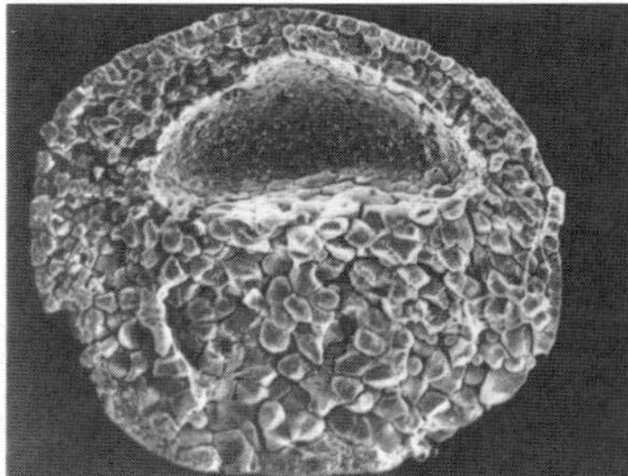

Control embryo

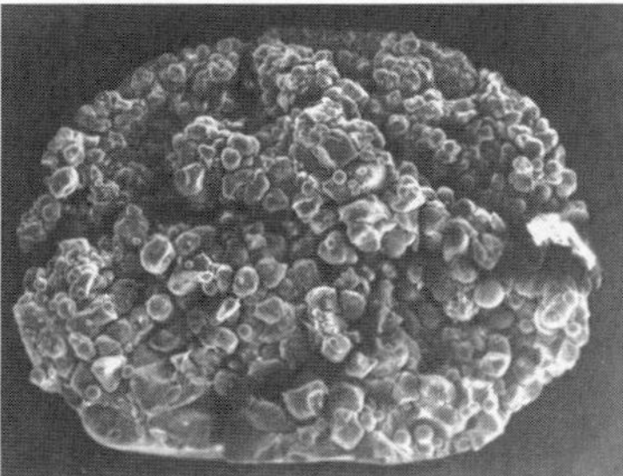

Experimental embryo

CONCLUSION Proper blastula formation in the frog requires EP cadherin.

Concept Check 47.2

1. During formation of the neural tube, cube-shaped cells change to wedge-shaped cells. Describe the roles of microtubules and microfilaments in this process.
2. In the frog embryo, convergent extension is thought to elongate the notochord along the anterior-posterior axis. How does this morphogenetic process work?

For suggested answers, see Appendix A.

Concept 47.3

The developmental fate of cells depends on their history and on inductive signals

Coupled with the morphogenetic changes that give an animal and its parts their characteristic shapes, development also requires the timely differentiation of many kinds of cells in specific locations. Two general principles integrate our current knowledge of the genetic and cellular mechanisms that underlie differentiation during embryonic development.

First, *during early cleavage divisions, embryonic cells must somehow become different from one another.* In many animal species, initial differences between cells result from the uneven distribution of cytoplasmic determinants in the unfertilized egg. By partitioning the heterogeneous cytoplasm of a polarized egg, cleavage parcels out different mRNAs, proteins, and other molecules to blastomeres in a type of asymmetrical cell division (see Figure 21.11a). The resulting differences in the cells' cytoplasmic composition help specify the body axes and influence the expression of genes that affect the developmental fate of the cells. In amniotes, local environmental differences play the major role in establishing early differences between embryonic cells. For example, cells of the inner cell mass are located internally in the early human embryo, whereas trophoblast cells are located on the outside surface of the blastocyst. The different environments of these two groups of cells appear to determine their very different fates.

Second, *once initial cell asymmetries are set up, subsequent interactions among the embryonic cells influence their fate, usually by causing changes in gene expression.* This mechanism, termed **induction**, eventually brings about the differentiation of the many specialized cell types making up an animal. Induction may be mediated by diffusible chemical signals or, if the cells are in contact, by cell-surface interactions.

It will help to keep these two principles in mind as we delve into the molecular and cellular mechanisms of differentiation and pattern formation during embryonic development of the species we're focusing on in this chapter. To ask questions about *how* the fate of an early embryonic cell is determined, we need to know what that fate is. So let's first look at some historic experiments that provided early researchers with information about cell fates.

Fate Mapping

As discussed in Chapter 21, researchers have worked out the developmental history of every cell in *Caenorhabditis elegans*, beginning with the first cleavage division of the zygote (see Figure 21.15). Although such a complete cell lineage has not yet been determined for any other animal, biologists have been making more general territorial diagrams of embryonic development, called **fate maps**, for many years.

In classic studies performed in the 1920s, German embryologist W. Vogt charted fate maps for different regions of early amphibian embryos. Vogt's results were among the earliest indications that the lineage of cells making up the three germ layers created by gastrulation is traceable to cells in the blastula, before gastrulation has begun **(Figure 47.23a)**. Later researchers developed more sophisticated techniques that allowed them to mark an individual blastomere during cleavage and then follow the marker as it was distributed to all the mitotic descendants of that cell **(Figure 47.23b)**.

Developmental biologists have combined fate-mapping studies with experimental manipulation of parts of embryos in which they study whether a cell's fate can be changed by moving it elsewhere in the embryo. Two important conclusions have emerged. First, in most animals, specific tissues of the older embryo can be attributed to certain early "founder cells." Second, as development proceeds, a cell's *developmental potential*—the range of structures that it can give rise to—becomes restricted. (For review, see the discussion of cell fate determination in Chapter 21, pp. 418–420.) Starting with the normal embryo's fate map, researchers can examine how the differentiation of cells is altered in experimental situations or in mutant embryos.

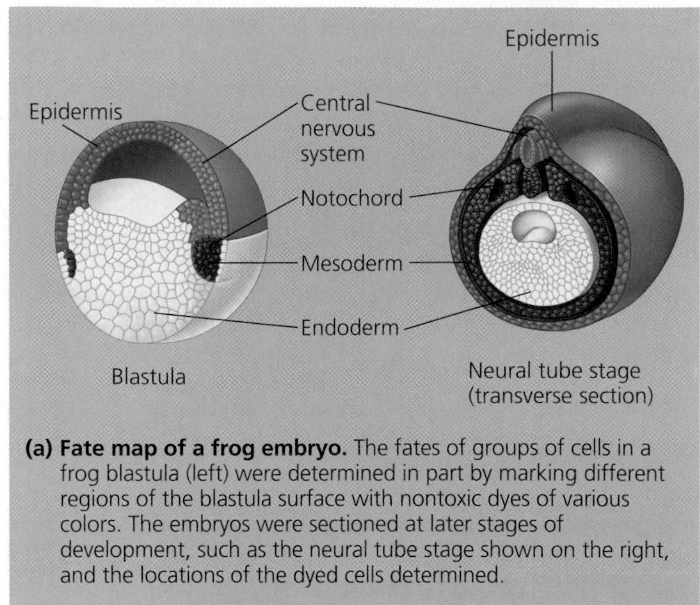

(a) Fate map of a frog embryo. The fates of groups of cells in a frog blastula (left) were determined in part by marking different regions of the blastula surface with nontoxic dyes of various colors. The embryos were sectioned at later stages of development, such as the neural tube stage shown on the right, and the locations of the dyed cells determined.

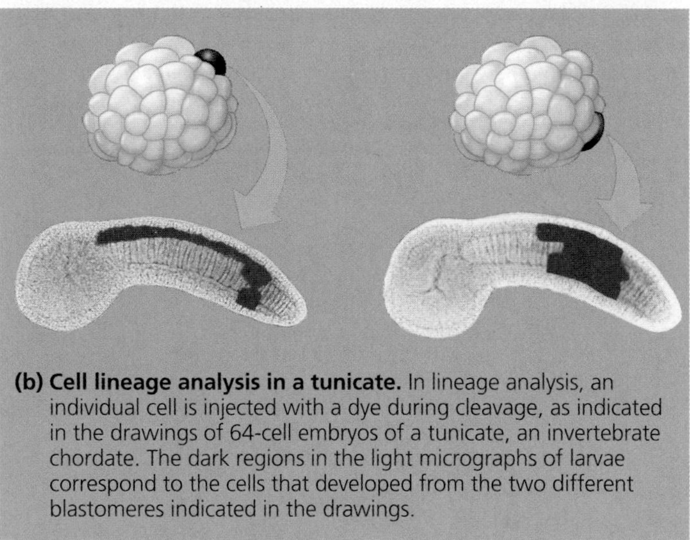
(b) Cell lineage analysis in a tunicate. In lineage analysis, an individual cell is injected with a dye during cleavage, as indicated in the drawings of 64-cell embryos of a tunicate, an invertebrate chordate. The dark regions in the light micrographs of larvae correspond to the cells that developed from the two different blastomeres indicated in the drawings.

▲ **Figure 47.23 Fate mapping for two chordates.**

Establishing Cellular Asymmetries

To understand at the molecular level how embryonic cells acquire their fates, it is helpful to think first about how the basic axes of the embryo are established. This can often be traced to a specific event in early development that sets up a particular cellular asymmetry, thus beginning to lay out the body plan.

The Axes of the Basic Body Plan

As you have learned, a bilaterally symmetrical animal has an anterior-posterior axis, a dorsal-ventral axis, and left and right sides (see Figure 47.8a). Establishing this basic body plan is a first step in morphogenesis and is a prerequisite for the development of tissues and organs.

In nonamniotic vertebrates, basic instructions for establishing the body axes are set down early, during oogenesis or fertilization. For example, in many frogs, the locations of melanin and yolk in the unfertilized egg define the animal and vegetal hemispheres, respectively. The animal-vegetal axis indirectly determines the anterior-posterior body axis. Fertilization then triggers cortical rotation, which establishes the dorsal-ventral axis and at the same time leads to the appearance of the gray crescent, whose position marks the dorsal side (see Figure 47.8b). Once any two axes are established, the third (in this case, the left-right) axis is specified by default. (Of course, specific molecular mechanisms must then carry out the pattern associated with that axis.)

In amniotes, the body axes are not fully established until later. In chicks, gravity is apparently involved in establishing the anterior-posterior axis as the egg travels down the oviduct in the hen before being laid. Later, pH differences between the two sides of the blastoderm cells establish the dorsal-ventral axis (see Figure 47.10). If the pH is artificially reversed above and below the blastoderm, the part facing the egg white will turn into the belly (ventral side) and the side facing the yolk will turn into the back (dorsal side), opposite to their normal fates. In mammals, no polarity is obvious until after cleavage, although recent research suggests that the orientation of the egg and sperm nuclei before their fusion may play a role in determining the axes.

Restriction of Cellular Potency

In many species that have cytoplasmic determinants, only the zygote is **totipotent**—that is, capable of developing into all the cells types found in the adult. In these organisms, the first cleavage is asymmetrical, with the two blastomeres receiving different cytoplasmic determinants. However, even in species that have cytoplasmic determinants, the first cleavage may occur along an axis that produces two identical blastomeres, which will have equal developmental potential. This is what happens in amphibians, for instance, as demonstrated in 1938 in an experiment by German zoologist Hans Spemann **(Figure 47.24)**. Thus, the fates of embryonic cells can be affected not only by the distribution of cytoplasmic determinants but also by how their distribution is affected by the zygote's characteristic pattern of cleavage.

Figure 47.24

Inquiry How does distribution of the gray crescent at the first cleavage affect the potency of the two daughter cells?

EXPERIMENT

1
Left (control): Fertilized salamander eggs were allowed to divide normally, resulting in the gray crescent being evenly divided between the two blastomeres.
Right (experimental): Fertilized eggs were constricted by a thread so that the first cleavage plane restricted the gray crescent to one blastomere.
Gray crescent
Gray crescent
2 The two blastomeres were then separated and alllowed to develop.
Normal
Belly piece
Normal

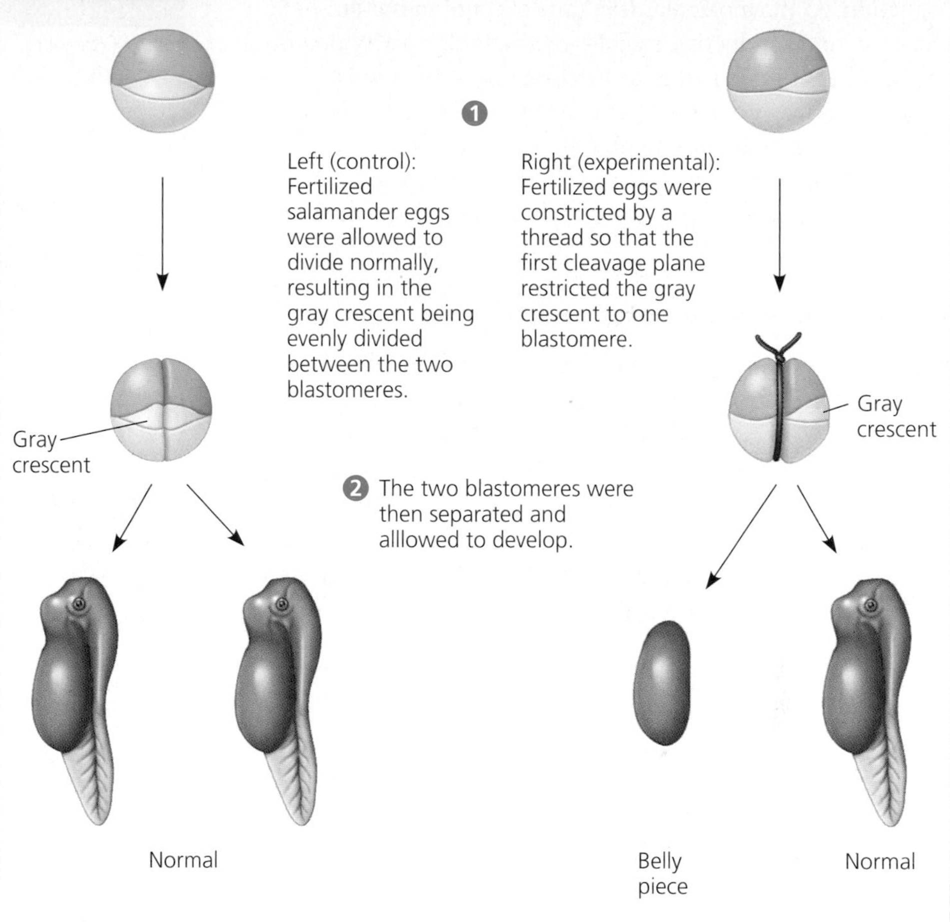

RESULTS Blastomeres that receive half or all of the gray crescent develop into normal embryos, but a blastomere that receives none of the gray crescent gives rise to an abnormal embryo without dorsal structures. Spemann called it a "belly piece."

CONCLUSION The totipotency of the two blastomeres normally formed during the first cleavage division depends on cytoplasmic determinants localized in the gray crescent.

In contrast to many other animals, the cells of mammalian embryos remain totipotent until the 16-cell stage, when they become arranged into the precursors of the trophoblast and inner cell mass of the blastocyst. At this time, their different locations determine their fates. The early blastomeres of mammals seem to receive equivalent amounts of cytoplasmic components from the egg. Indeed, up to the 8-cell stage, the blastomeres of a mammalian embryo all look alike, and each can form a complete embryo if isolated.

Regardless of how similar or different early embryonic cells are in a particular species, the progressive restriction of potency is a general feature of development in all animals. In some species, the cells of the early gastrula retain the capacity to give rise to more than one kind of cell, though they have lost their totipotency. If left alone, the dorsal ectoderm of an early amphibian gastrula will develop into a neural plate above the notochord. If the dorsal ectoderm is experimentally replaced with ectoderm from some other location in the same gastrula, the transplanted tissue will form a neural plate. If the same experiment is performed on a late-stage gastrula, however, the transplanted ectoderm will not respond to its new location and will not form a neural plate. In general, the tissue-specific fates of cells in a late gastrula are fixed. Even when they are manipulated experimentally, these cells usually give rise to the same types of cell as in the normal embryo, indicating that their fate is already sealed.

Cell Fate Determination and Pattern Formation by Inductive Signals

Once embryonic cell division creates cells that differ from each other, the cells begin to influence each other's fates by induction. At the molecular level, the effect of induction—the response to an inductive signal—is usually the switching on of a set of genes that make the receiving cells differentiate into a specific tissue. You have already learned about the role of induction in vulval development in the nematode *C. elegans* (see Figure 21.16b). Here we examine two other examples of induction, an essential process in the development of many tissues in most animals.

The "Organizer" of Spemann and Mangold

The importance of induction during development of amphibians was dramatically demonstrated in transplantation experiments performed by Hans Spemann and his student Hilde Mangold in the 1920s. Based on the results of their most famous experiment, summarized in **Figure 47.25**, they concluded that the dorsal lip of the blastopore in the early gastrula functions as an organizer of the embryo by initiating a chain of inductions that results in the formation of the notochord, the neural tube, and other organs.

Developmental biologists are working intensively to identify the molecular basis of induction by *Spemann's organizer* (also called the *gastrula organizer*, or simply the *organizer*). An important clue has come from studies of a growth factor called bone morphogenetic protein 4 (BMP-4). (Bone morphogenetic proteins, a family of related proteins with a variety of developmental roles, derive their name from members of the family that are important in bone formation.) Amphibian BMP-4 is active exclusively in cells on the ventral side of the gastrula, inducing these cells to travel down the pathway toward formation of ventral structures. One major function of the cells of the organizer seems to be to inactivate BMP-4 on the dorsal side of the embryo by producing proteins that bind to BMP-4, rendering it unable to signal. This, in turn, allows formation of dorsal structures such as the notochord and neural tube. Proteins related to BMP-4 and its inhibitors are also found in other animals, including invertebrates such as the fruit fly. The ubiquity of these molecules suggests that they evolved long ago and may participate in the development of many different organisms.

The induction that causes dorsal ectoderm to develop into the neural tube is only one of many cell-cell interactions that transform the three germ layers into organ systems. Many inductions seem to involve a sequence of inductive steps that progressively determine the fate of cells. For example, in the late gastrula of the frog, ectoderm cells destined to become the lenses of the eyes receive inductive signals from the ectodermal cells that will become the neural plate. Additional inductive signals probably come from endodermal cells and mesodermal cells. Finally, inductive signals from the optic cup, an outgrowth of the developing brain, complete the determination of the lens-forming cells.

Figure 47.25

Inquiry Can the dorsal lip of the blastopore induce cells in another part of the amphibian embryo to change their developmental fate?

EXPERIMENT Spemann and Mangold transplanted a piece of the dorsal lip of a pigmented newt gastrula to the ventral side of the early gastrula of a nonpigmented newt.

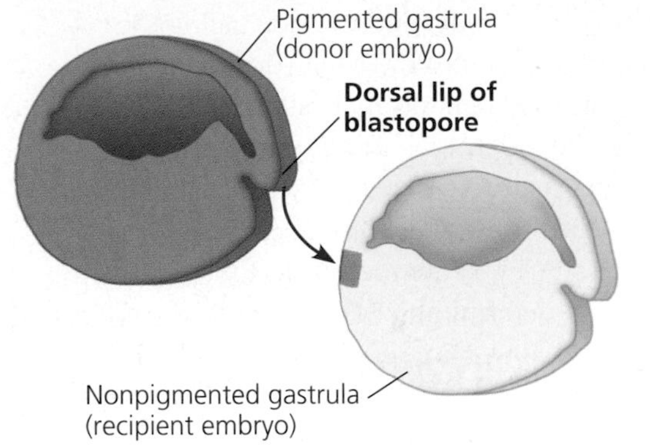

Pigmented gastrula (donor embryo)
Dorsal lip of blastopore
Nonpigmented gastrula (recipient embryo)

RESULTS During subsequent development, the recipient embryo formed a second notochord and neural tube in the region of the transplant, and eventually most of a second embryo. Examination of the interior of the double embryo revealed that the secondary structures were formed in part from host tissue.

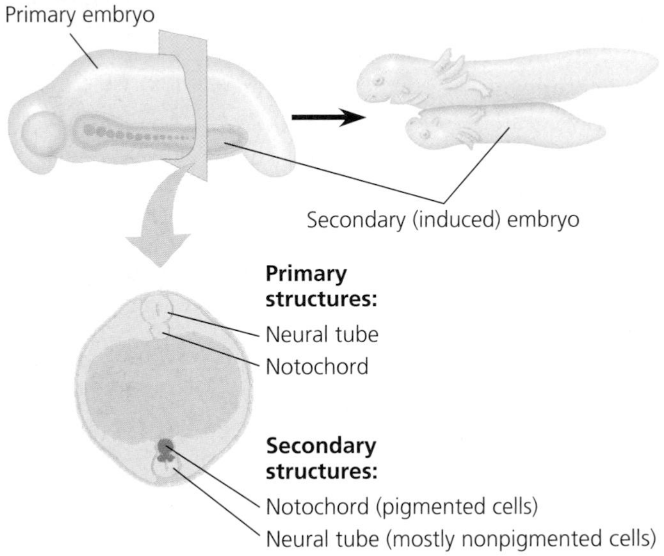

Primary embryo
Secondary (induced) embryo
Primary structures:
Neural tube
Notochord
Secondary structures:
Notochord (pigmented cells)
Neural tube (mostly nonpigmented cells)

CONCLUSION The transplanted dorsal lip was able to induce cells in a different region of the recipient to form structures different from their normal fate. In effect, the dorsal lip "organized" the later development of an entire embryo.

Formation of the Vertebrate Limb

The action of the gastrula organizer is a classic example of induction, and we can see that the organizer induces cells to take on their fates in appropriate locations relative to each other. Thus, inductive signals play a major role in **pattern formation**—the development of an animal's spatial organization, the arrangement of organs and tissues in their characteristic places in three-dimensional space. The molecular cues that control pattern formation, called **positional information,** tell a cell where it is with respect to the animal's body axes and help to determine how the cell and its descendants respond to future molecular signals.

In Chapter 21, we discussed pattern formation in the development of the body segments of *Drosophila.* For understanding pattern formation in vertebrates, a classic model system has been limb development in the chick. The wings and legs of chicks, like all vertebrate limbs, begin as bumps of tissue called limb buds **(Figure 47.26a)**. Each component of a chick limb, such as a specific bone or muscle, develops with a precise location and orientation relative to three axes: the proximal-distal axis (the "shoulder-to-fingertip" axis), the anterior-posterior axis (the "thumb-to-little finger" axis), and the dorsal-ventral axis (the "knuckle-to-palm" axis). The embryonic cells within a limb bud respond to positional information indicating location along these three axes **(Figure 47.26b)**.

A limb bud consists of a core of mesodermal tissue covered by a layer of ectoderm. Two critical organizer regions in the limb bud have profound effects on the limb's development. These two organizer regions are present in all vertebrate limb buds, including those that will develop into forelimbs (such as wings or arms) and those destined to become hind limbs. The cells of these regions secrete proteins that provide key positional information to the other cells of the bud.

One limb-bud organizer region is the **apical ectodermal ridge (AER),** a thickened area of ectoderm at the tip of the bud (see **Figure 47.26a**). The AER is required for the outgrowth of the limb along the proximal-distal axis and for patterning along this axis. The cells of the AER produce several secreted protein signals, belonging to the fibroblast growth factor (FGF) family, that promote limb-bud outgrowth. If the AER is surgically removed and beads soaked with FGF are put in its place, a nearly normal limb will develop. The AER and other ectoderm of the limb bud also appear to guide pattern formation along the limb's dorsal-ventral axis. In experiments where the ectoderm of a limb bud, including the AER, is detached from the mesoderm and then replaced with its orientation rotated 180° back-to-front, the limb elements that form have reversed dorsal-ventral orientation. (This is equivalent to reversing the palm and back of your hand.)

The second major limb-bud organizer region is the **zone of polarizing activity (ZPA),** a block of mesodermal tissue located underneath the ectoderm where the posterior side of the bud is attached to the body (see Figure 47.26a). The ZPA is necessary for proper pattern formation along the anterior-posterior axis of the limb. Cells nearest the ZPA give rise to the posterior structures, such as the most posterior of the chick's

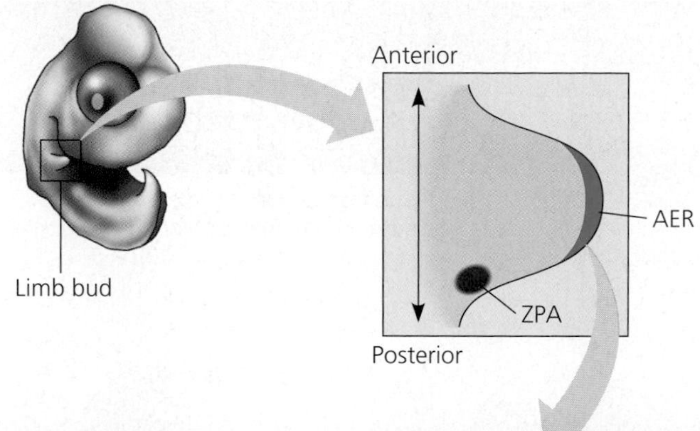

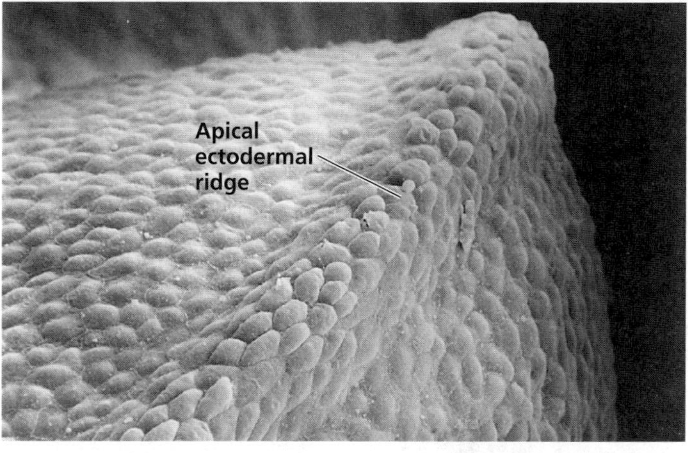

(a) Organizer regions. Vertebrate limbs develop from protrusions called limb buds, each consisting of mesoderm cells covered by a layer of ectoderm. Two regions, termed the apical ectodermal ridge (AER, shown in this SEM) and the zone of polarizing activity (ZPA), play key organizer roles in limb pattern formation.

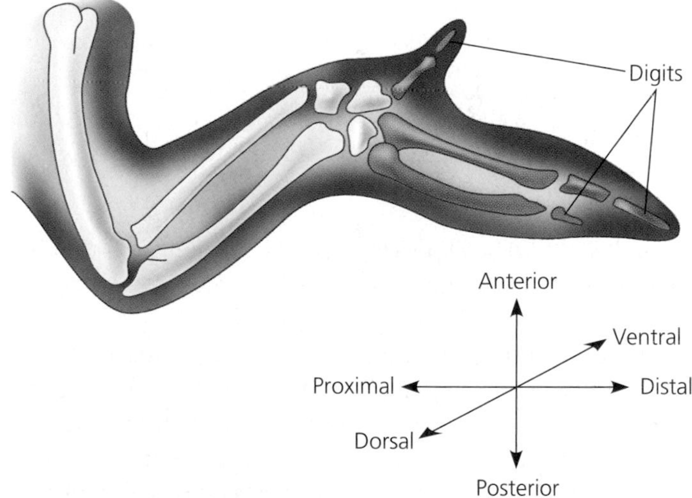

(b) Wing of chick embryo. As the bud develops into a limb, a specific pattern of tissues emerges. In the chick wing, for example, the three digits are always present in the arrangement shown here. Pattern formation requires each embryonic cell to receive some kind of positional information indicating location along the three axes of the limb. The AER and ZPA secrete molecules that help provide this information.

▲ **Figure 47.26 Vertebrate limb development.**

Figure 47.27

Inquiry What role does the zone of polarizing activity (ZPA) play in limb pattern formation in vertebrates?

EXPERIMENT ZPA tissue from a donor chick embryo was transplanted under the ectoderm in the anterior margin of a recipient chick limb bud.

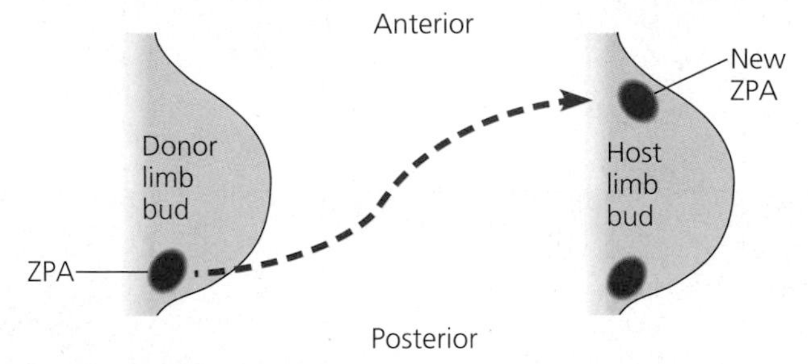

RESULTS In the grafted host limb bud, extra digits developed from host tissue in a mirror-image arrangement to the normal digits, which also formed (see Figure 47.26b for a diagram of a normal chick wing).

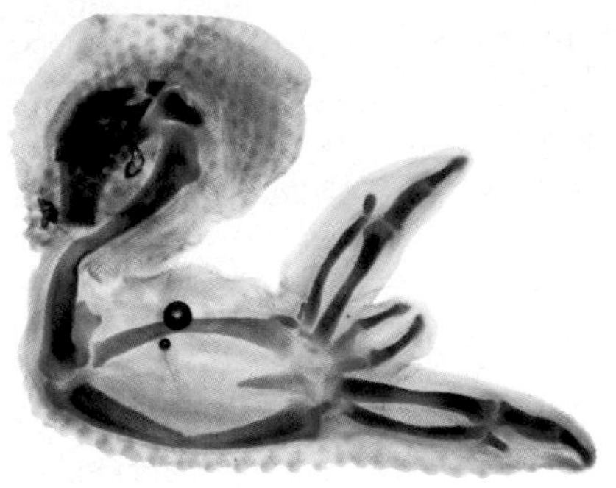

CONCLUSION The mirror-image duplication observed in this experiment suggests that ZPA cells secrete a signal that diffuses from its source and conveys positional information indicating "posterior." As the distance from the ZPA increases, the signal concentration decreases and hence more anterior digits develop.

three digits (like our little finger); cells farthest from the ZPA form anterior structures, including the most anterior digit (like our thumb).

The tissue transplantation experiment outlined in **Figure 47.27** supports the hypothesis that the ZPA produces some sort of inductive signal that conveys positional information indicating "posterior." Indeed, researchers have discovered that the cells of the ZPA secrete an important protein growth factor called Sonic hedgehog.* If cells genetically engineered to produce large amounts of Sonic hedgehog are implanted in the anterior region of a normal limb bud, a mirror-image limb results—just as if a ZPA had been grafted there. Evidence from studying the mouse version of Sonic hedgehog suggests that extra toes in mice—and perhaps also in humans—can result from the production of this protein in the wrong part of the limb bud.

We can conclude from experiments like these that pattern formation requires cells to receive and interpret environmental cues that vary from one location to another. These cues, acting together along three axes, tell cells where they are in the three-dimensional realm of a developing organ. In vertebrate limb development, for instance, we now know that specific proteins serve as some of these cues. In other words, organizer regions such as the AER and the ZPA function as signaling centers. Researchers have recently established that these two regions also interact with each other by way of signaling molecules and signaling pathways to influence each other's developmental fates. Such mutual signaling interactions between mesoderm and ectoderm also occur during formation of the neural tube and during development of many other tissues and organs.

What determines whether a limb bud develops into a forelimb or a hind limb? The cells receiving the signals from the AER and ZPA respond according to their developmental histories. Before the AER or ZPA issues its signals, earlier developmental signals have set up patterns of gene expression distinguishing the future forelimbs from the future hind limbs. These differences cause the limb-bud cells of the forelimb and hind limb to react differently to the same positional cues.

Thus, construction of the fully formed animal involves a sequence of events that include many steps of signaling and differentiation. Initial cell asymmetries allow different types of cells to influence each other to express specific sets of genes. The products of these genes then direct cells to differentiate into specific types. Coordinated with morphogenesis, various pathways of pattern formation occur in all the different parts of the developing embryo. These processes ultimately produce a complex arrangement of multiple tissues and organs, each functioning in the appropriate location to form a single coordinated organism.

Concept Check 47.3

1. Although there are three body axes, only two must be determined during early development. Why?
2. Predict what would happen if you transplanted tissue from the organizer region (above the dorsal lip) of a late-stage frog gastrula to its ventral side.
3. If the ventral cells of an early frog gastrula are experimentally induced to express a protein that inhibits BMP-4, could a second embryo develop? Explain.

For suggested answers, see Appendix A.

* Sonic hedgehog gets its name from two sources, its similarity to a *Drosophila* protein called Hedgehog, which is involved in segmentation of the fly embryo, and a video game character.

Chapter 47 Review

Go to www.campbellbiology.com or the student CD-ROM to explore Activities, Investigations, and other interactive study aids.

SUMMARY OF KEY CONCEPTS

Concept 47.1

After fertilization, embryonic development proceeds through cleavage, gastrulation, and organogenesis

- **Fertilization** (pp. 988–991) Fertilization brings together the nuclei of sperm and egg, forming a diploid zygote, and activates the egg, initiating embryonic development. The acrosomal reaction, which is triggered when the sperm meets the egg, releases hydrolytic enzymes that digest material surrounding the egg. Gamete contact and/or fusion depolarizes the egg cell membrane and sets up a fast block to polyspermy in many animals. Sperm-egg fusion also initiates the cortical reaction, in which a rise in Ca^{2+} concentration stimulates cortical granules to release their contents outside the egg. This forms a fertilization envelope that functions as a slow block to polyspermy. In mammalian fertilization, the cortical reaction modifies the zona pellucida as a slow block to polyspermy.
- **Cleavage** (pp. 992–994) Fertilization is followed by cleavage, a period of rapid cell division without growth, which results in the production of a large number of cells called blastomeres. Holoblastic cleavage, or division of the entire egg, occurs in species whose eggs have little or moderate amounts of yolk, such as sea urchins, frogs, and mammals. Meroblastic cleavage, incomplete division of the egg, occurs in species with yolk-rich eggs, such as birds and other reptiles. Cleavage planes usually follow a specific pattern relative to the animal and vegetal poles of the zygote. In many species, cleavage creates a multicellular ball called the blastula, which contains a fluid-filled cavity, the blastocoel.
Activity *Sea Urchin Development*
- **Gastrulation** (pp. 994–997) Gastrulation transforms the blastula into a gastrula, which has a primitive digestive cavity (the archenteron) and three embryonic germ layers: the ectoderm, mesoderm, and endoderm.
Investigation *What Determines Cell Differentiation in the Sea Urchin?*
- **Organogenesis** (pp. 997–999) The organs of the animal body develop from specific portions of the three embryonic germ layers. Early events in organogenesis in vertebrates include formation of the notochord by condensation of dorsal mesoderm, development of the neural tube from infolding of the ectodermal neural plate, and formation of the coelom from splitting of lateral mesoderm.
Activity *Frog Development*
- **Developmental Adaptations of Amniotes** (pp. 998–999) The embryos of birds, other reptiles, and mammals develop within a fluid-filled sac that is contained within a shell or the uterus. In these organisms, the three germ layers give rise not only to embryonic tissue but also to the four extraembryonic membranes: the amnion, chorion, yolk sac, and allantois.
- **Mammalian Development** (pp. 999–1001) The eggs of marsupial and eutherian mammals are small and store few nutrients. They exhibit holoblastic cleavage and show no obvious polarity. Gastrulation and organogenesis, however, resemble the processes in birds and other reptiles. After fertilization and early cleavage in the oviduct, the blastocyst implants in the uterus. The trophoblast initiates formation of the fetal portion of the placenta, and the embryo proper develops from a single layer of cells, the epiblast, within the blastocyst. Extraembryonic membranes homologous to those of birds and other reptiles function in intrauterine development.

Concept 47.2

Morphogenesis in animals involves specific changes in cell shape, position, and adhesion

- **The Cytoskeleton, Cell Motility, and Convergent Extension** (pp. 1001–1002) Cytoskeletal rearrangements are responsible for changes in both the shape and position of cells. Both kinds of changes are involved in tissue invaginations, as occurs in gastrulation, for example. In convergent extension, cell movements cause a sheet of cells to become narrower and longer.
- **Roles of the Extracellular Matrix and Cell Adhesion Molecules** (pp. 1002–1003) Fibers of the extracellular matrix provide anchorage for cells and also help guide migrating cells toward their destinations. Fibronectin and other glycoproteins located on cell surfaces are important for cell migration and for holding cells together in tissues.

Concept 47.3

The developmental fate of cells depends on their history and on inductive signals

- **Fate Mapping** (p. 1004) Experimentally derived fate maps of embryos have shown that specific regions of the zygote or blastula develop into specific parts of older embryos.
- **Establishing Cellular Asymmetries** (pp. 1004–1005) In nonamniotic species, unevenly distributed cytoplasmic determinants in the egg cell are important in establishing the body axes and in setting up differences between the blastomeres resulting from cleavage of the zygote. Cells that receive different cytoplasmic determinants undergo different fates. In amniotes, local environmental differences play the major role in establishing initial differences between cells and later the body axes. As embryonic development proceeds, the potency of cells becomes progressively more limited in all species.
- **Cell Fate Determination and Pattern Formation by Inductive Signals** (pp. 1006–1008) Cells in a developing embryo receive and interpret positional information that varies with location. This information is often in the form of signal molecules secreted by cells in special "organizer" regions of the embryo, such as the dorsal lip of the blastopore in the amphibian gastrula and the AER and ZPA of the vertebrate limb bud. The signal molecules influence gene expression in the cells that receive them, leading to differentiation and the development of particular structures.

TESTING YOUR KNOWLEDGE

Self-Quiz

1. The cortical reaction of sea urchin eggs functions directly in the
   a. formation of a fertilization envelope.
   b. production of a fast block to polyspermy.
   c. release of hydrolytic enzymes from the sperm cell.
   d. generation of an electrical impulse by the egg cell.
   e. fusion of egg and sperm nuclei.

2. Which of the following is common to both avian and mammalian development?
   a. holoblastic cleavage
   b. epiblast and hypoblast
   c. trophoblast
   d. yolk plug
   e. gray crescent

3. The archenteron develops into
   a. the mesoderm.
   b. the blastocoel.
   c. the endoderm.
   d. the placenta.
   e. the lumen of the digestive tract.

4. In a frog embryo, the blastocoel is
   a. completely obliterated by yolk platelets.
   b. lined with endoderm during gastrulation.
   c. located primarily in the animal hemisphere.
   d. the cavity that becomes the coelom.
   e. the cavity that later forms the archenteron.

5. What structural adaptation in chickens allows them to lay their eggs in arid environments, rather than in water?
   a. extraembryonic membranes.
   b. yolk.
   c. cleavage.
   d. gastrulation.
   e. development of the brain from ectoderm.

6. In an amphibian embryo, a band of cells called the neural crest
   a. rolls up to form the neural tube.
   b. develops into the main sections of the brain.
   c. produces cells that migrate to form teeth, skull bones, and other structures in the embryo.
   d. has been shown by experiments to be the organizer region of the developing embryo.
   e. induces the formation of the notochord.

7. Differences in the development of different cells in the early frog embryo (zygote to blastula) are due to
   a. the differences between meroblastic and holoblastic cleavage.
   b. the heterogeneous distribution of cytoplasmic determinants, such as proteins and mRNA.
   c. inductive interactions occurring between the developing cells.
   d. concentration gradients of regulatory molecules such as BMP-4.
   e. the position of the cells relative to the zone of polarizing activity (ZPA).

8. During convergent extension,
   a. cells on the opposite side of the embryo follow converging developmental pathways leading to bilateral symmetry.
   b. the cells of the neural folds adhere to one another to complete the neural tube.
   c. the cells of a tissue layer reorganize, forming a narrowed, elongated sheet.
   d. the dorsal-ventral axis is established.
   e. cell adhesion molecules are expressed, causing the eight blastomeres to adhere tightly to one another.

9. In the early development of an amphibian embryo, an important "organizer" is located in the
   a. neural tube.
   b. notochord.
   c. archenteron roof.
   d. dorsal ectoderm.
   e. dorsal lip of the blastopore.

10. Any blastomere removed from an 8-cell mammalian embryo can develop into a normal late-stage embryo. This finding supports the idea that
   a. only the zygote is totipotent.
   b. the progressive restriction of potency hypothesis applies.
   c. the first cleavage event must be transverse to the animal-vegetal axis of the zygote.
   d. cell divisions producing the earliest blastomeres do not result in an asymmetrical distribution of cytoplasmic determinants.
   e. there is no organizer in mammals.

For Self-Quiz answers, see Appendix A.

Go to the website or CD-ROM for more quiz questions.

Evolution Connection

Evolution in insects and vertebrates has involved the repeated duplication of body segments, followed by fusion of some segments and specialization of their structure and function. What parts of vertebrate anatomy reflect the vertebrate segmentation pattern? Can you guess what vertebrate body part is a product of segment fusion and specialization?

Scientific Inquiry

The "snout" of a frog tadpole bears a sucker. A salamander tadpole has a mustache-shaped structure called a balancer in the same area. You perform an experiment in which you transplant ectoderm from the side of a young salamander embryo to the snout of a frog embryo. You find that the tadpole that develops has a balancer. When you transplant ectoderm from the side of a slightly older salamander embryo to the snout of a frog embryo, the frog tadpole ends up with a patch of salamander skin on its snout. Suggest a hypothesis to explain the results of this experiment in terms of the mechanisms of development. How might you test your hypothesis?

Investigation *What Determines Cell Differentiation in the Sea Urchin?*

Science, Technology, and Society

Many scientists believe that fetal tissue transplants offer great potential for treating Parkinson's disease, epilepsy, diabetes, Alzheimer's disease, and spinal cord injuries. Why might tissues from a fetus be particularly useful for replacing diseased or damaged cells in such conditions? Some people would allow only tissues from miscarriages to be used in fetal transplant research. But most researchers prefer to use tissues from surgically aborted fetuses. Why? What is your position on this controversial issue, and why?

48 Nervous Systems

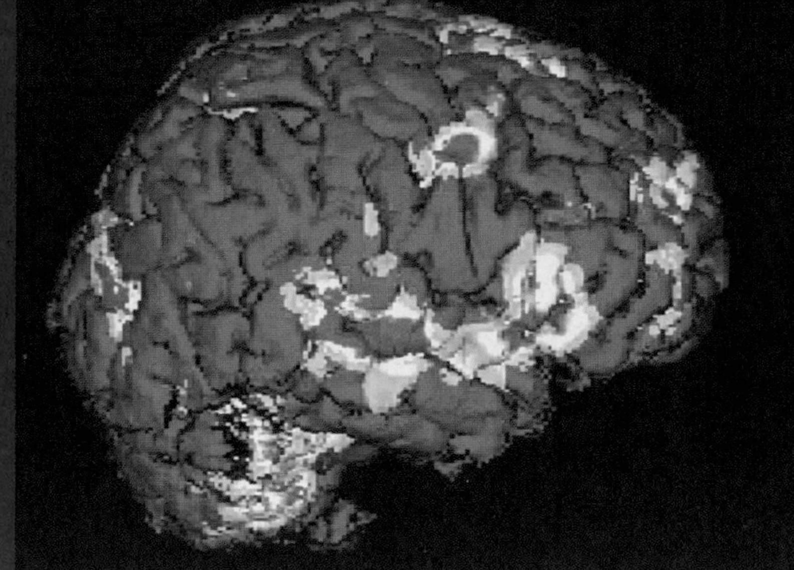

▲ **Figure 48.1 A functional magnetic resonance image of brain areas activated during language processing.**

Key Concepts

48.1 Nervous systems consist of circuits of neurons and supporting cells

48.2 Ion pumps and ion channels maintain the resting potential of a neuron

48.3 Action potentials are the signals conducted by axons

48.4 Neurons communicate with other cells at synapses

48.5 The vertebrate nervous system is regionally specialized

48.6 The cerebral cortex controls voluntary movement and cognitive functions

48.7 CNS injuries and diseases are the focus of much research

Overview

Command and Control Center

What happens in your brain when you picture something with your "mind's eye"? Until recently, scientists had little hope of answering that question. The human brain contains an estimated 10^{11} (100 billion) nerve cells, or **neurons.** Each neuron may communicate with thousands of other neurons in complex information-processing circuits that make the most powerful electronic computers look primitive. An engineer who wants to learn how a computer functions can just open the computer's housing and trace the circuits. But except for rare glimpses (such as during brain surgery), the circuitry of the living human brain has been hidden from view.

That's no longer the case, however, thanks in part to recent technologies that can record brain activity from outside a person's skull. One technique is functional magnetic resonance imaging (fMRI). During fMRI, a subject lies with his or her head in the hole of a large, doughnut-shaped magnet that records increased blood flow in brain areas with active neurons. A computer then uses the data to construct a three-dimensional map of the subject's brain activity, like the one shown in **Figure 48.1**. These recordings can be made while the subject is doing various tasks, such as speaking, moving a hand, looking at pictures, or forming a mental image of an object or a person's face.

The results of brain imaging and other research methods, such as those described by Erich Jarvis in the interview on pages 818–819, reveal that groups of neurons function in specialized circuits dedicated to different tasks. These circuits are responsible for the feats of sensing and moving (see Chapter 49) and for the many types of animal behavior (see Chapter 51).

The ability to sense and react originated billions of years ago with prokaryotes that could detect changes in their environment and respond in ways that enhanced their survival and reproductive success—for example, locating food sources by chemotaxis (see Chapter 27). Later, modification of this simple process provided multicellular organisms with a mechanism for communication between cells of the body. By the time of the Cambrian explosion more than 500 million years ago (see Chapter 32), systems of neurons that allowed animals to sense and move rapidly had evolved in essentially their present-day forms.

In this chapter, we will discuss the organization of animal nervous systems and the mechanisms by which neurons transmit information. We'll also investigate some of the functions performed by specific parts of the vertebrate brain. Finally, we'll examine several mental illnesses and neurological disorders that are the subject of intense research today.

Concept 48.1

Nervous systems consist of circuits of neurons and supporting cells

All animals except the sponges have some type of nervous system. What distinguishes the nervous systems of different animal groups is not so much their basic building blocks—the neurons themselves—but how the neurons are organized into circuits.

Organization of Nervous Systems

The simplest animals with nervous systems, the cnidarians, have radially symmetrical bodies organized around a gastrovascular cavity (see Figure 33.5). In some cnidarians, such as the hydra shown in **Figure 48.2a**, the neurons controlling the contraction and expansion of the gastrovascular cavity are arranged in diffuse **nerve nets.** The nervous systems of more complex animals contain nerve nets as well as **nerves,** which are bundles of fiberlike extensions of neurons. For example, sea stars have a nerve net in each arm, connected by radial nerves to a central nerve ring **(Figure 48.2b)**; this organization is better suited than a diffuse nerve net for controlling more complex movements.

Greater complexity of nervous systems and more complex behavior evolved with cephalization, which included the clustering of neurons in a brain near the anterior (front) end in animals with elongated, bilaterally symmetrical bodies. In flatworms, such as the planarian shown in **Figure 48.2c**, a small brain and longitudinal nerve cords constitute the simplest clearly defined **central nervous system (CNS).** In more complex invertebrates, such as annelids **(Figure 48.2d)** and arthropods **(Figure 48.2e)**, behavior is regulated by more complicated brains and ventral nerve cords containing segmentally arranged clusters of neurons called **ganglia** (singular, *ganglion*). Nerves that connect the CNS with the rest of an animal's body make up the **peripheral nervous system (PNS).**

Molluscs are good examples of how nervous system organization correlates with an animal's lifestyle. Sessile or slow-moving molluscs such as clams and chitons have little or no cephalization and relatively simple sense organs **(Figure 48.2f)**. In contrast, cephalopod molluscs (squids and octopuses) have the most sophisticated nervous systems of any invertebrates, rivaling even those of some vertebrates. The large brain of cephalopods, accompanied by large, image-forming eyes and rapid signaling along nerves, supports the active, predatory life of these animals **(Figure 48.2g)**. Researchers have demonstrated that octopuses can learn to discriminate among visual patterns and can perform complex tasks. In vertebrates, the CNS consists of the brain and the spinal cord, which runs along the dorsal (back) side of the body; nerves and ganglia

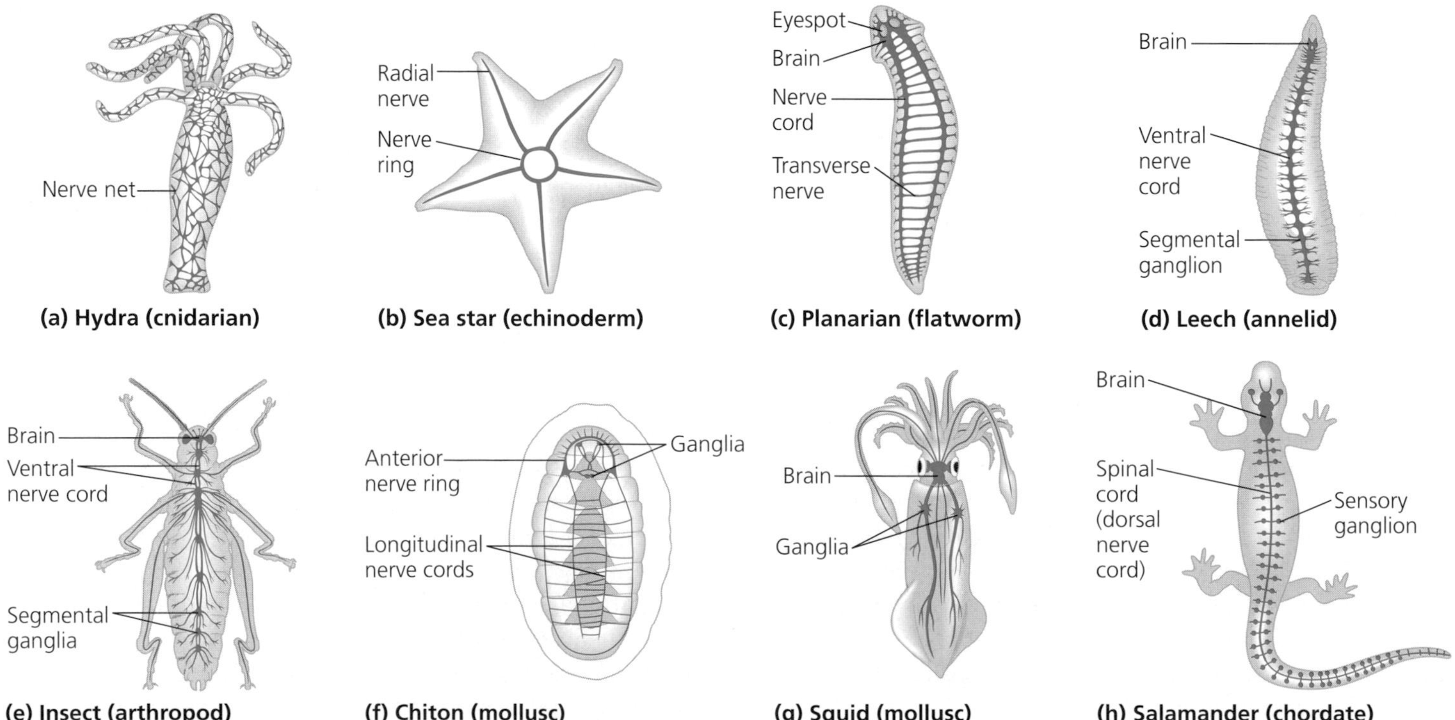

▲ **Figure 48.2 Organization of some nervous systems.**

comprise the PNS **(Figure 48.2h)**. We will examine the vertebrate nervous system in more detail later in this chapter.

Information Processing

In general, there are three stages in the processing of information by nervous systems: sensory input, integration, and motor output **(Figure 48.3)**. These three stages are handled by specialized populations of neurons. **Sensory neurons** transmit information from sensors that detect external stimuli (light, sound, touch, heat, smell, and taste) and internal conditions (such as blood pressure, blood CO_2 level, and muscle tension). This information is sent to the CNS, where **interneurons** integrate (analyze and interpret) the sensory input, taking into account the immediate context as well as what has happened in the past. The greatest complexity in neural circuits exists in the connections between interneurons. Motor output leaves the CNS via **motor neurons,** which communicate with **effector cells** (muscle cells or endocrine cells). The stages of sensory input, integration, and motor output are easiest to study in the relatively simple nerve circuits that produce **reflexes,** the body's automatic responses to stimuli. **Figure 48.4** diagrams the nerve circuit that underlies the knee-jerk reflex in humans.

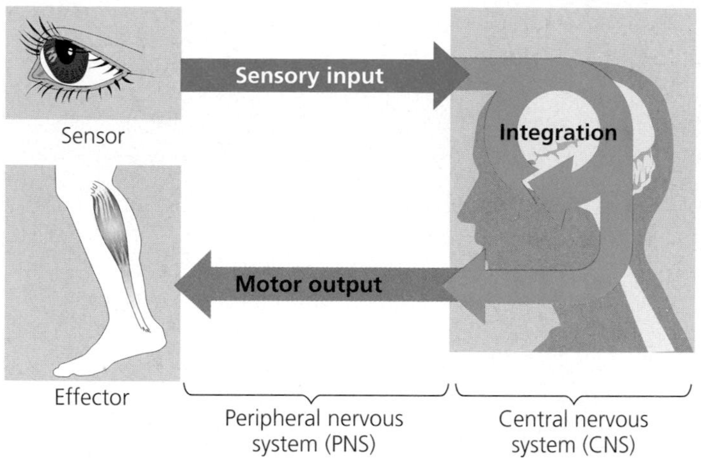

▲ **Figure 48.3 Overview of information processing by nervous systems.**

Neuron Structure

As you read in Chapter 1, one of the major themes of biology is that form fits function. This theme, which applies at all levels from molecules to organisms, is clearly illustrated at the cellular level by neurons. The ability of neurons to receive and transmit information depends on their elaborate structure. Most of a neuron's organelles, including its nucleus,

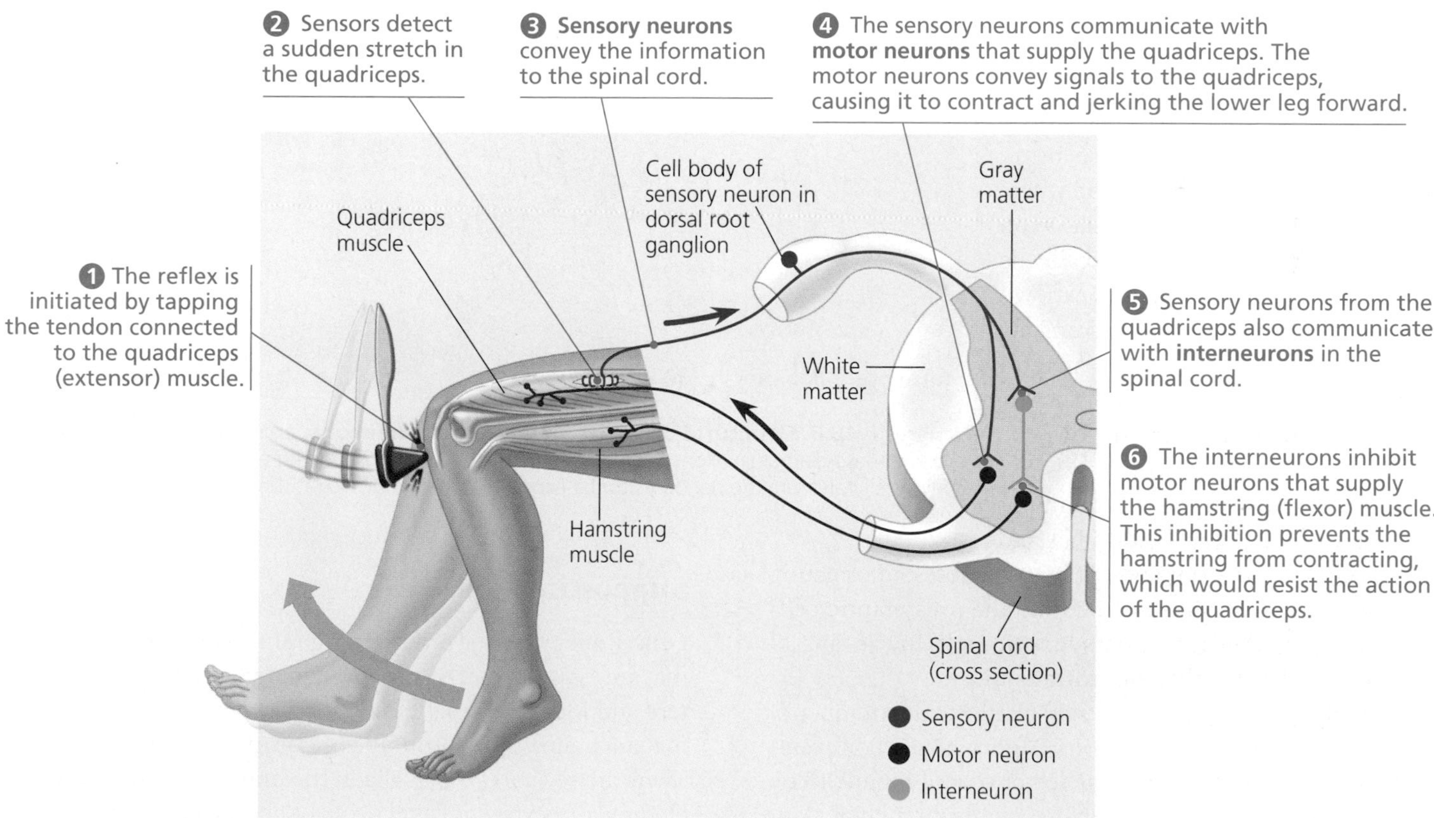

▲ **Figure 48.4 The knee-jerk reflex.** Many neurons are involved in the reflex, but for simplicity, only one neuron of each type is shown.

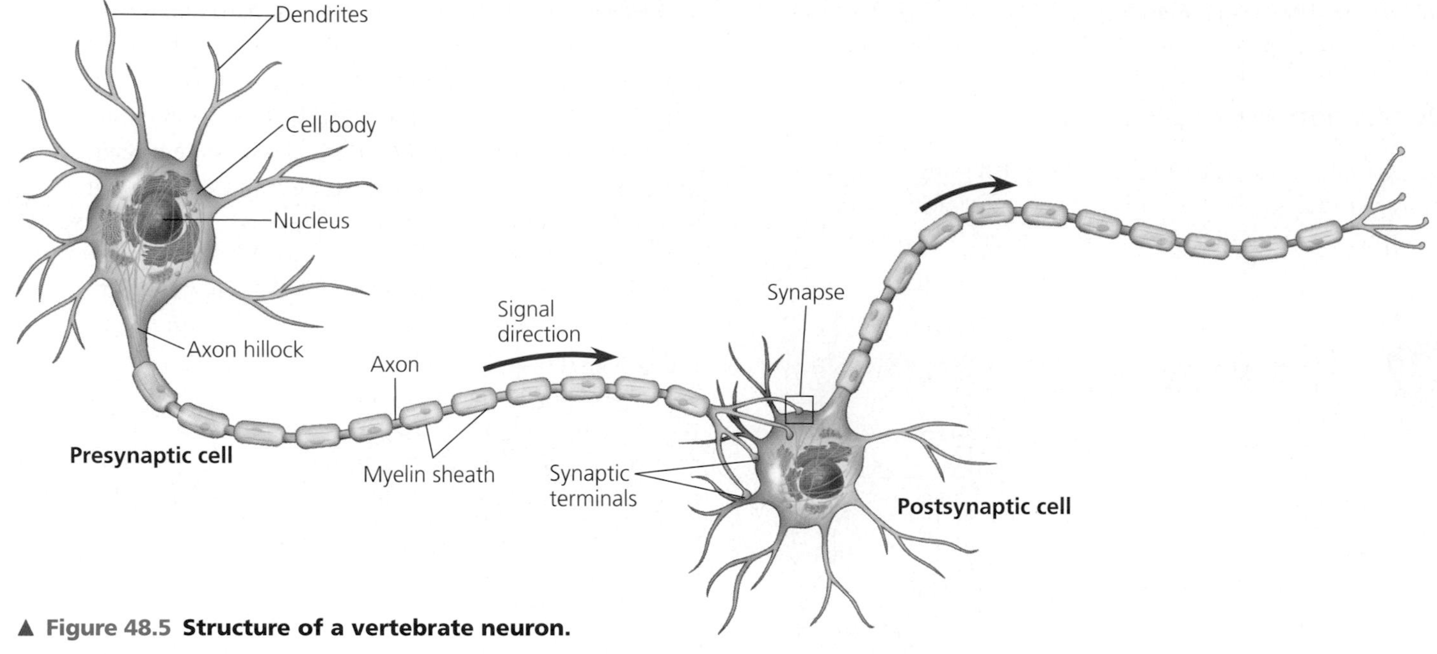

▲ **Figure 48.5 Structure of a vertebrate neuron.**

are located in the **cell body (Figure 48.5)**. Arising from the cell body are two types of extensions: numerous dendrites and a single axon. **Dendrites** (from the Greek *dendron*, tree) are highly branched extensions that *receive* signals from other neurons. The **axon** is typically a much longer extension that *transmits* signals to other cells, which may be neurons or effector cells. Some axons, such as the ones that reach from your spinal cord to muscle cells in your feet, may be over a meter long. The conical region of an axon where it joins the cell body is called the **axon hillock;** as we will see, this is typically the region where the signals that travel down the axon are generated. Many axons are enclosed by a layer called the **myelin sheath,** described later in this section. Near its end, an axon usually divides into several branches, each of which ends in a **synaptic terminal.** The site of communication between a synaptic terminal and another cell is called a **synapse.** At most synapses, information is passed from the transmitting neuron (the **presynaptic cell**) to the receiving cell (the **postsynaptic cell**) by means of chemical messengers called **neurotransmitters.**

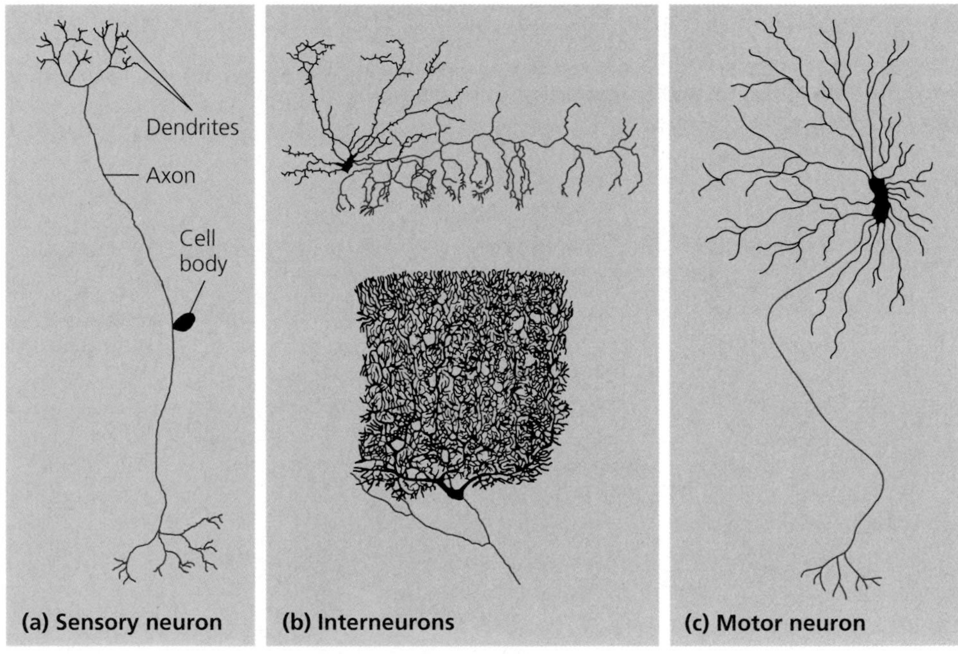

▲ **Figure 48.6 Structural diversity of vertebrate neurons.** Cell bodies and dendrites are black in these diagrams; axons are red. In (a), the cell body is connected only to the axon, which conveys signals from the dendrites to the axon's terminal branches (at the bottom).

The complexity of a neuron's shape reflects the number of synapses it has with other neurons **(Figure 48.6)**. For example, the interneuron shown in the lower part of Figure 48.6b has about 100,000 synapses on its highly branched dendrites, whereas neurons with simpler dendrites have far fewer synapses.

Supporting Cells (Glia)

Glia (from a Greek word meaning "glue") are supporting cells that are essential for the structural integrity of the nervous system and for the normal functioning of neurons. In the mammalian brain, glia outnumber neurons 10-fold to 50-fold. There are several types of glia in the brain and spinal cord, including astrocytes, radial glia, oligodendrocytes, and Schwann cells. As a group, these cells do much more than just hold neurons together.

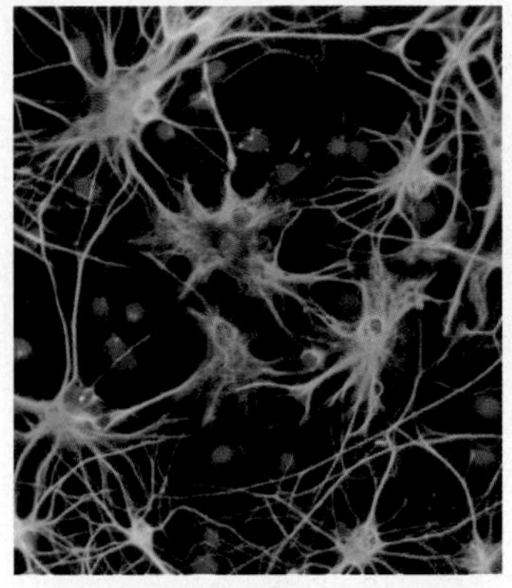

◀ **Figure 48.7 Astrocytes.** In this section through a mammalian cerebral cortex, astrocytes appear green after being labeled with a fluorescent antibody (LM). The blue dots are cell nuclei, labeled with a different antibody. The term "astrocyte" refers to the star-like shape of the cells.

In the CNS, **astrocytes** provide structural support for neurons and regulate the extracellular concentrations of ions and neurotransmitters **(Figure 48.7)**. Some astrocytes respond to activity in neighboring neurons by facilitating information transfer at those neurons' synapses. Scientists hypothesize that this facilitation may be part of the cellular mechanism of learning and memory. Astrocytes adjacent to active neurons also cause nearby blood vessels to dilate, which increases blood flow to the area, enabling the neurons to obtain oxygen and glucose more quickly. During development, astrocytes induce the formation of tight junctions (see Figure 6.31) between cells that line the capillaries in the brain and spinal cord. The result is the **blood-brain barrier,** which restricts the passage of most substances into the CNS, allowing the extracellular chemical environment of the CNS to be tightly controlled.

In an embryo, **radial glia** form tracks along which newly formed neurons migrate from the neural tube, the structure that gives rise to the CNS (see Figures 47.14 and 47.15). Both radial glia and astrocytes can also act as stem cells, generating neurons and other glia. Researchers view these multipotent precursors as a potential way to replace neurons and glia that are lost to injury or disease, a topic explored in Concept 48.7.

Oligodendrocytes (in the CNS) and **Schwann cells** (in the PNS) are glia that form the myelin sheaths around the axons of many vertebrate neurons. The structure of a myelinated PNS axon is shown in **Figure 48.8**. Neurons become myelinated during development when Schwann cells or oligodendrocytes grow around axons, wrapping them in many layers of membrane, somewhat like a jelly roll. These membranes are mostly lipid, which is a poor conductor of electrical currents. Thus, the myelin sheath provides electrical insulation of the axon, analogous to plastic insulation that covers many electrical wires. In the disease multiple sclerosis, myelin sheaths gradually deteriorate, resulting in a progressive loss of body function due to the disruption of nerve signal transmission.

Concept Check 48.1

1. a. Arrange the following neurons in the correct sequence for information flow during the knee-jerk reflex: interneuron, sensory neuron, motor neuron.
b. Which of the neuron types is located entirely within the CNS?
2. Would severing a neuron's axon stop the neuron from receiving or from transmitting information? Explain.
3. What would be the most obvious structural abnormality in the nervous system of a mouse lacking oligodendrocytes?

For suggested answers, see Appendix A.

Concept 48.2

Ion pumps and ion channels maintain the resting potential of a neuron

We noted in Chapter 7 that all cells have an electrical potential difference (voltage) across their plasma membrane. This voltage is called the **membrane potential.** In neurons, the membrane potential is typically between −60 and −80 mV (millivolts) when the cell is not transmitting signals. The minus

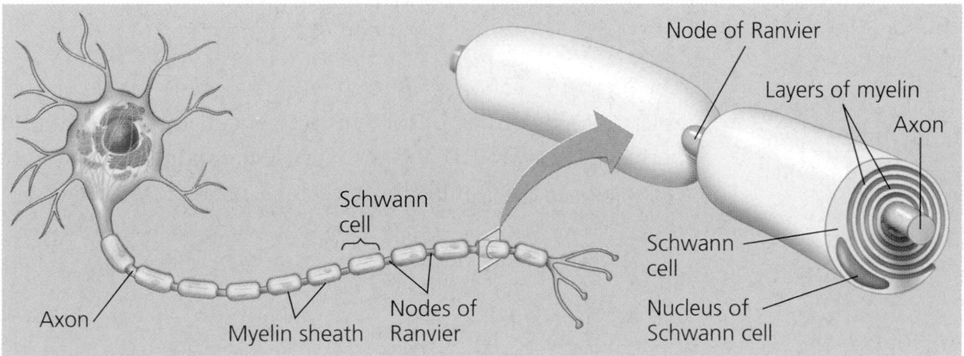

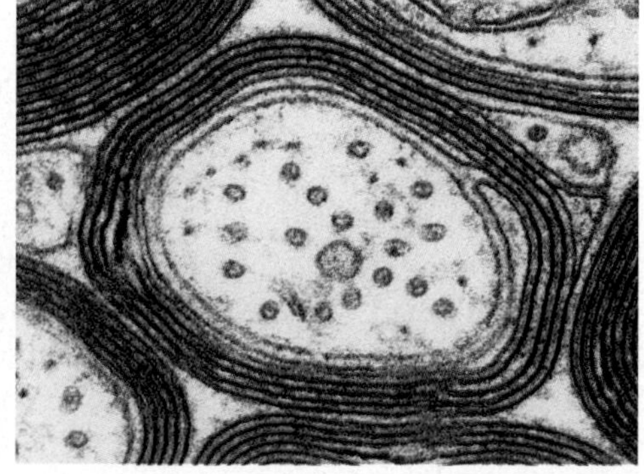

▲ **Figure 48.8 Schwann cells and the myelin sheath.** In the PNS, glia called Schwann cells wrap themselves around axons, forming layers of myelin. Gaps between adjacent Schwann cells are called nodes of Ranvier. The TEM shows a cross section through a myelinated axon.

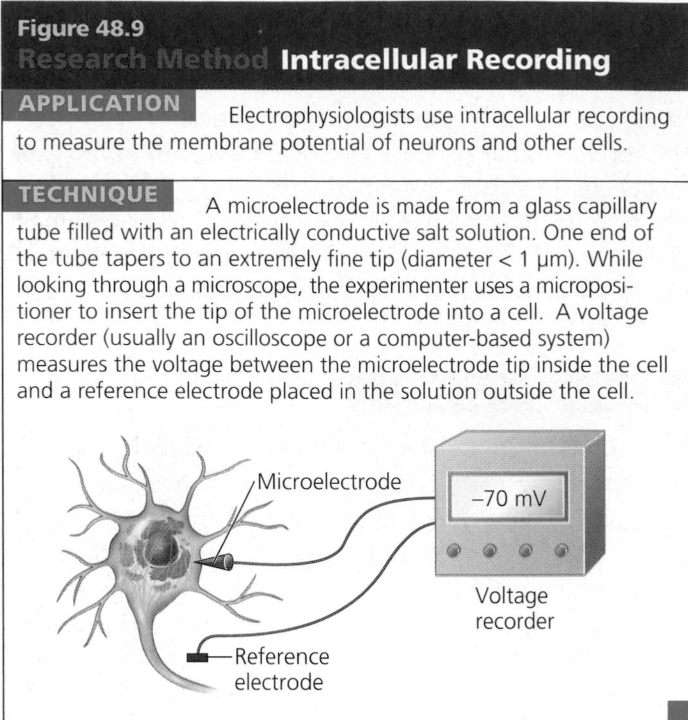

Figure 48.9
Research Method Intracellular Recording
APPLICATION
Electrophysiologists use intracellular recording to measure the membrane potential of neurons and other cells.
TECHNIQUE
A microelectrode is made from a glass capillary tube filled with an electrically conductive salt solution. One end of the tube tapers to an extremely fine tip (diameter < 1 μm). While looking through a microscope, the experimenter uses a micropositioner to insert the tip of the microelectrode into a cell. A voltage recorder (usually an oscilloscope or a computer-based system) measures the voltage between the microelectrode tip inside the cell and a reference electrode placed in the solution outside the cell.
Microelectrode
–70 mV
Voltage recorder
Reference electrode

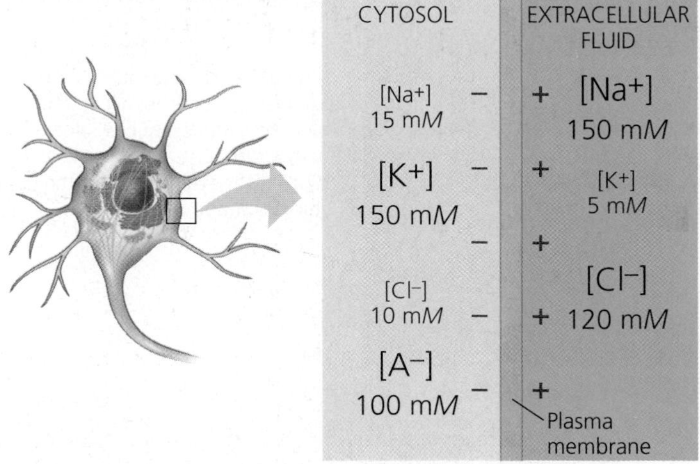

▲ **Figure 48.10 Ionic gradients across the plasma membrane of a mammalian neuron.** The concentrations of Na^+ and Cl^- are higher in the extracellular fluid than in the cytosol. The reverse is true for K^+. The cytosol also contains a variety of organic anions [A^-], including charged amino acids.

sign indicates that the inside of the cell is negative relative to the outside. **Figure 48.9** explains how electrophysiologists measure a cell's membrane potential.

The Resting Potential

The membrane potential of a neuron that is not transmitting signals is called the **resting potential.** In all neurons, the resting potential depends on the ionic gradients that exist across the plasma membrane **(Figure 48.10)**. In mammals, for example, the extracellular fluid has a sodium ion (Na^+) concentration of 150 millimolar (m*M*) and a potassium ion (K^+) concentration of 5 m*M* (1 millimolar = 1 millimole/liter = 10^{-3} mol/L). In the cytosol, the Na^+ concentration is 15 m*M* and the K^+ concentration is 150 m*M*. Thus, the Na^+ concentration gradient, expressed as the ratio of outside concentration/inside concentration, is 150/15, or 10. The K^+ concentration gradient is 5/150, or 1/30. (There are also anion gradients, but we will ignore them for the moment.) The Na^+ and K^+ gradients are maintained by the sodium-potassium pump (see Figure 7.16). The fact that the gradients are responsible for the resting potential is shown by a simple experiment: If the pump is disabled by the addition of a specific poison, the gradients gradually disappear, and so does the resting potential.

It is particularly challenging to understand how the resting potential arises from ionic gradients, but it is an essential step in learning how neurons, including our own, work. We start with a model of a mammalian neuron consisting of two chambers separated by an artificial membrane **(Figure 48.11a)**. The membrane contains many ion channels (see Chapter 7) that allow only K^+ to diffuse across the membrane. To produce a concentration gradient for K^+ like that of a mammalian neuron, we add 150 m*M* potassium chloride (KCl) to the inner chamber and 5 m*M* KCl to the outer chamber. Like any solute, K^+ tends to diffuse down its concentration gradient, from an area of higher concentration (inner chamber) to an area of lower concentration (outer chamber). But because the channels are selective for K^+, chloride ions (Cl^-) cannot cross the membrane. As a result, a separation of charge (voltage) develops across the membrane, with an excess of negative charge on the side of the membrane facing the inner chamber. The developing membrane voltage opposes the efflux of K^+ because the excess negative charges attract the positively charged K^+. Thus, an electrical gradient builds up whose direction is opposite that of the concentration gradient. When the electrical gradient exactly balances the concentration gradient, an equilibrium is established. At equilibrium, there is no net diffusion of K^+ across the membrane.*

The magnitude of the membrane voltage at equilibrium, called the **equilibrium potential (E_{ion})**, is given by a formula called the Nernst equation. For an ion with a net charge of +1, such as K^+, at 37°C, the Nernst equation is

$$E_{ion} = 62 \text{ mV}\left(\log \frac{[\text{ion}]_{\text{outside}}}{[\text{ion}]_{\text{inside}}}\right)$$

The Nernst equation applies to any membrane that is permeable to a single type of ion. In our model, the membrane is permeable only to K^+, and the Nernst equation can be used to calculate E_K, the equilibrium potential for K^+:

$$E_K = 62 \text{ mV}\left(\log \frac{5 \text{ mM}}{150 \text{ mM}}\right) = -92 \text{ mV (at 37°C)}$$

* The charge separation needed to generate the resting potential is extremely small: about 10^{-12} mol/cm^2 of membrane. Thus, the ionic diffusion that leads to the resting potential does not appreciably change the ion concentrations on either side of the membrane.

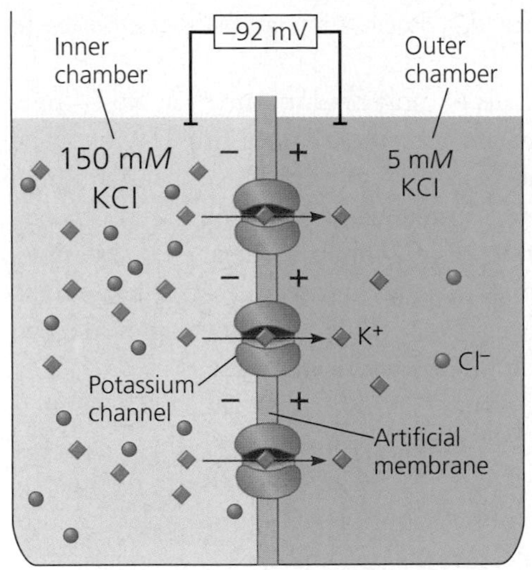

(a) Membrane selectively permeable to K+

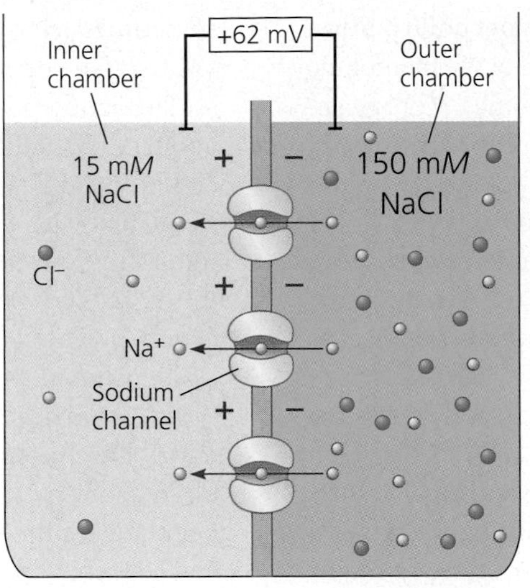

(b) Membrane selectively permeable to Na+

◀ **Figure 48.11 Modeling a mammalian neuron.** Each beaker is divided into two chambers by an artificial membrane. **(a)** The membrane is selectively permeable to K^+, and the inner chamber contains a 30-fold higher concentration of K^+ than the outer chamber; at equilibrium, the inside of the membrane is −92 mV relative to the outside. **(b)** The membrane is selectively permeable to Na^+, and the inner chamber contains a ten-fold lower concentration of Na^+ than the outer chamber; at equilibrium, the inside of the membrane is +62 mV relative to the outside.

The minus sign indicates that with this K^+ concentration gradient, K^+ is at equilibrium when the inside of the membrane is 92 mV more negative than the outside.

Now imagine that we change our model neuron by using a membrane that contains ion channels selective for Na^+ **(Figure 48.11b)**. We also replace the contents of the chambers to produce a concentration gradient for Na^+ like that of a mammalian neuron: 15 m*M* sodium chloride (NaCl) in the inner chamber and 150 m*M* NaCl in the outer chamber. Under these conditions, the Nernst equation can be used to calculate E_{Na}, the equilibrium potential for Na^+:

$$E_{Na} = 62 \text{ mV} \left(\log \frac{150 \text{ m}M}{15 \text{ m}M}\right) = +62 \text{ mV (at 37°C)}$$

The plus sign indicates that with this Na^+ concentration gradient, Na^+ is at equilibrium when the inside of the membrane is 62 mV more positive than the outside.

How does a real mammalian neuron differ from these model neurons? The plasma membrane of a real neuron at rest has many potassium channels that are open, but it also has a relatively small number of open sodium channels. Consequently, the resting potential is around −60 to −80 mV, between E_K and E_{Na}. Because neither K^+ nor Na^+ is at equilibrium, there is a net flow of each ion (a current) across the membrane at rest. The resting potential remains steady, which means that the K^+ and Na^+ currents are equal and opposite. The reason the resting potential is closer to E_K than to E_{Na} is that the membrane is more permeable to K^+ than to Na^+. If something causes the membrane's permeability to Na^+ to increase, the membrane potential will move toward E_{Na} and away from E_K. *This is the basis of nearly all electrical signals in the nervous system: The membrane potential can change from its resting value when the membrane's permeability to particular ions changes.* Sodium and potassium ions play major roles, but there are also important roles for chloride (Cl^-) and calcium (Ca^{2+}) ions; they follow the same rules (as described by the Nernst equation).

Gated Ion Channels

The resting potential results from the diffusion of K^+ and Na^+ through ion channels that are always open; these channels are said to be *ungated.* Neurons also have **gated ion channels,** which open or close in response to one of three kinds of stimuli. **Stretch-gated ion channels** are found in cells that sense stretch (see Figure 48.4) and open when the membrane is mechanically deformed. **Ligand-gated ion channels** are found at synapses and open or close when a specific chemical, such as a neurotransmitter, binds to the channel. **Voltage-gated ion channels** are found in axons (and in the dendrites and cell body of some neurons, as well as in some other types of cells) and open or close when the membrane potential changes. As we'll explain in the next section, gated ion channels are responsible for generating the signals of the nervous system.

Concept Check 48.2

1. What is the equilibrium potential (E_X) at 37°C for an ion X^+ if $[X^+]_{outside}$ = 10 m*M* and $[X^+]_{inside}$ = 100 m*M*?
2. Suppose a cell's membrane potential shifts from −70 mV to −50 mV. What changes in the cell's permeability to K^+ or Na^+ could cause such a shift?
3. Contrast ligand-gated and voltage-gated ion channels in terms of the stimuli that open them.

For suggested answers, see Appendix A.

Action potentials are the signals conducted by axons

If a cell has gated ion channels, its membrane potential may change in response to stimuli that open or close those channels. Some such stimuli trigger a **hyperpolarization** **(Figure 48.12a)**, an increase in the magnitude of the membrane potential (the inside of the membrane becomes more negative). Hyperpolarizations may be caused by the opening of gated K^+ channels, which increases the membrane's permeability to K^+ and causes the potential to approach E_K (−92 mV at 37°C). Other stimuli trigger a **depolarization** **(Figure 48.12b)**, a reduction in the magnitude of the membrane potential (the inside of the membrane becomes less negative). Depolarizations may be due to the opening of gated Na^+ channels, which increases the membrane's permeability to Na^+ and causes the potential to approach E_{Na} (+62 mV at 37°C). These changes in membrane potential are called **graded potentials** because the magnitude of the hyperpolarization or depolarization varies with the strength of the stimulus: A larger stimulus causes a larger change in permeability and thus a larger change in the membrane potential.

Production of Action Potentials

In most neurons, depolarizations are graded only up to a certain membrane voltage, called the **threshold.** A stimulus strong enough to produce a depolarization that reaches the threshold triggers a different type of response, called an **action potential** **(Figure 48.12c)**. An action potential is an all-or-none phenomenon: Once triggered, it has a magnitude that is independent of the strength of the triggering stimulus. Action potentials are the signals that carry information along axons, sometimes over great distances, such as from your toes to your spinal cord.

The action potentials of most neurons are very brief—only about 1–2 milliseconds (msec) in duration. Having brief action potentials enables a neuron to produce them at a high frequency. This feature is significant because neurons encode information in their action potential frequency. For example, in the sensory neurons that function in the knee-jerk reflex, action potential frequency is related to the magnitude and suddenness of stretch in the quadriceps muscle.

As **Figure 48.13** illustrates, both voltage-gated Na^+ channels and voltage-gated K^+ channels are involved in the production of an action potential. Both types of channels are opened by depolarizing the membrane, but they respond independently and sequentially: Na^+ channels open before K^+ channels.

Each voltage-gated Na^+ channel has two gates, an activation gate and an inactivation gate, and both must be open for Na^+ to diffuse through the channel. ❶ At the resting potential, the activation gate is closed and the inactivation gate is open on most Na^+ channels. Depolarization of the membrane *rapidly opens* the activation gate and *slowly closes* the inactivation gate.

Each voltage-gated K^+ channel has just one gate, an activation gate. At the resting potential, the activation gate on most K^+ channels is closed. Depolarization of the membrane *slowly opens* the K^+ channel's activation gate.

How do these channel properties contribute to the production of an action potential? ❷ When a stimulus depolarizes the

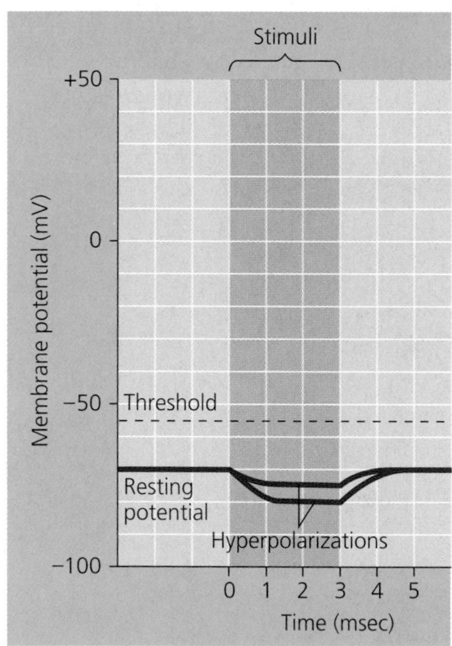

(a) Graded hyperpolarizations produced by two stimuli that increase membrane permeability to K^+. The larger stimulus produces a larger hyperpolarization.

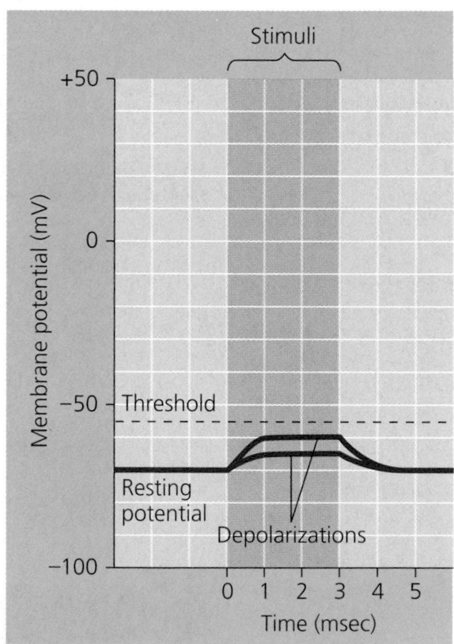

(b) Graded depolarizations produced by two stimuli that increase membrane permeability to Na^+. The larger stimulus produces a larger depolarization.

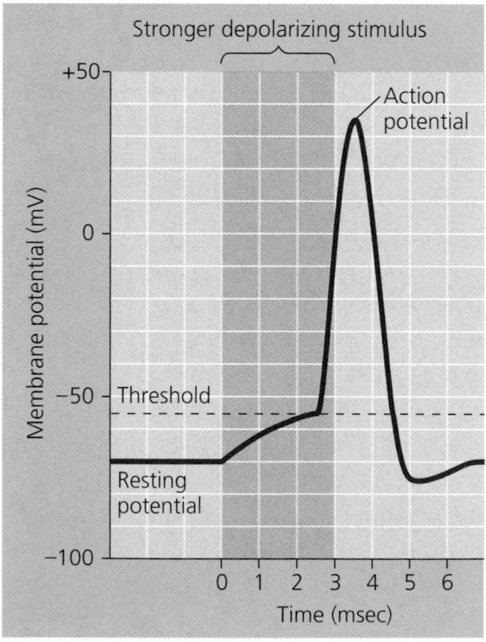

(c) Action potential triggered by a depolarization that reaches the threshold.

▲ **Figure 48.12 Graded potentials and an action potential in a neuron.**

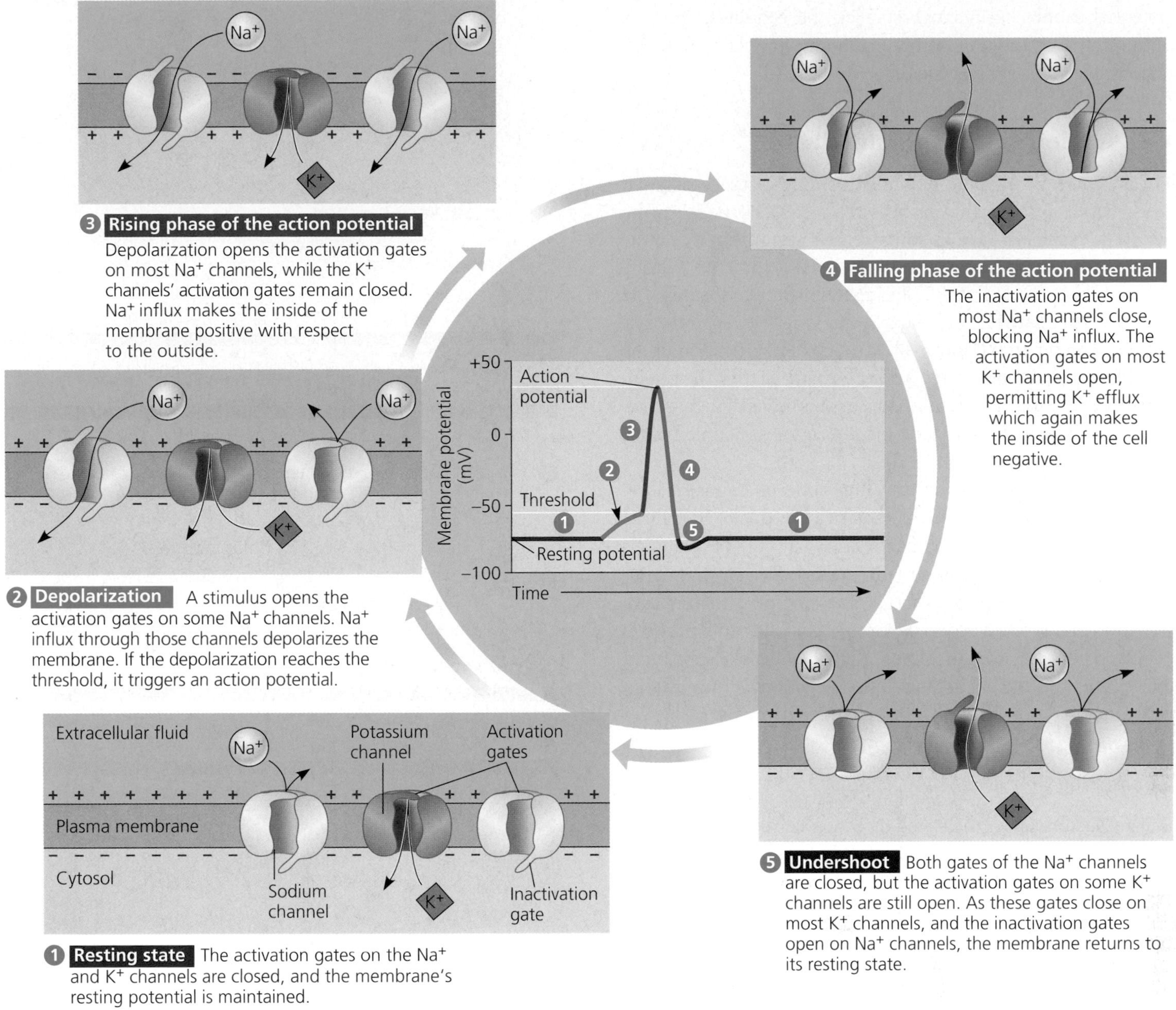

Na+
K+
3 Rising phase of the action potential
Depolarization opens the activation gates on most Na+ channels, while the K+ channels' activation gates remain closed. Na+ influx makes the inside of the membrane positive with respect to the outside.
4 Falling phase of the action potential
The inactivation gates on most Na+ channels close, blocking Na+ influx. The activation gates on most K+ channels open, permitting K+ efflux which again makes the inside of the cell negative.
Membrane potential (mV)
+50
0
−50
−100
Action potential
Threshold
Resting potential
Time
1
2
3
4
5
2 Depolarization A stimulus opens the activation gates on some Na+ channels. Na+ influx through those channels depolarizes the membrane. If the depolarization reaches the threshold, it triggers an action potential.
Extracellular fluid
Potassium channel
Activation gates
Plasma membrane
Cytosol
Sodium channel
Inactivation gate
1 Resting state The activation gates on the Na+ and K+ channels are closed, and the membrane's resting potential is maintained.
5 Undershoot Both gates of the Na+ channels are closed, but the activation gates on some K+ channels are still open. As these gates close on most K+ channels, and the inactivation gates open on Na+ channels, the membrane returns to its resting state.

▲ **Figure 48.13 The role of voltage-gated ion channels in the generation of an action potential.** The circled numbers on the graph in the center and the colors of the action potential phases correspond to the five diagrams showing voltage-gated Na^+ and K^+ channels in a neuron's plasma membrane. (Ungated ion channels are not illustrated.)

membrane, the activation gates on some Na^+ channels open, allowing more Na^+ to diffuse into the cell. The Na^+ influx causes further depolarization, which opens the activation gates on still more Na^+ channels, allowing even more Na^+ to diffuse into the cell, and so on. ❸ Once the threshold is crossed, this positive-feedback cycle rapidly brings the membrane potential close to E_{Na} during the *rising phase*. ❹ However, two events prevent the membrane potential from actually reaching E_{Na}: (a) The inactivation gates on most Na^+ channels close, halting Na^+ influx; and (b) the activation gates on most K^+ channels open, causing a rapid efflux of K^+. Both events quickly bring the membrane potential back toward E_K during the *falling phase*. ❺ In fact, in the final phase of an action potential, called the *undershoot*, the membrane's permeability to K^+ is higher than at rest, so the membrane potential is closer to E_K than it is at the resting potential. The K^+ channels' activation gates eventually close, and the membrane potential returns to the resting potential.

The Na^+ channels' inactivation gates remain closed during the falling phase and the early part of the undershoot. As a result, if a second depolarizing stimulus occurs during this period, it will be unable to trigger an action potential. The "downtime" following an action potential when a second action potential cannot be initiated is called the **refractory period.** This interval sets a limit on the maximum frequency at which action potentials can be generated.

Conduction of Action Potentials

For an action potential to function as a long-distance signal, it must travel without diminishing from the cell body to the synaptic terminals. It does so by regenerating itself along the axon. At the site where an action potential is initiated (usually the axon hillock), Na^+ influx during the rising phase creates an electrical current that depolarizes the neighboring region of the axon membrane **(Figure 48.14)**. The depolarization in the neighboring region is large enough to reach the threshold, causing the action potential to be re-initiated there. This process is repeated many times as the action potential travels the length of the axon.

Immediately behind the traveling zone of depolarization due to Na^+ influx is a zone of repolarization due to K^+ efflux. In the repolarized zone, the inactivation gates of Na^+ channels are still closed. Consequently, the inward current that depolarizes the axon membrane *ahead* of the action potential cannot produce another action potential *behind* it. This prevents action potentials from traveling back toward the cell body. Thus, once an action potential starts, it normally moves in only one direction—toward the synaptic terminals.

Conduction Speed

Several factors affect the speed at which action potentials are conducted. One factor is the diameter of the axon: The larger (wider) the axon's diameter, the faster the conduction. This is because resistance to the flow of electrical current is inversely proportional to the cross-sectional area of a conductor (such as a wire or an axon). As an analogy, think about how a wide hose offers less resistance to the flow of water than a narrow hose does. Similarly, a wide axon provides less resistance to the depolarizing current associated with an action potential than a narrow axon does. Hence, the resulting depolarization can spread farther along the interior of a wide axon, bringing more distant regions of the membrane to the threshold sooner. In invertebrates, conduction speed varies from several centimeters per second in very narrow axons to about 100 m/sec in the giant axons of squids and some arthropods. These giant axons function in behavioral responses requiring great speed, such as the backward tail-flip that enables a lobster or crayfish to escape from a predator.

A different means of increasing the conduction speed of action potentials has evolved in vertebrates. Recall that many vertebrate axons are surrounded by a myelin sheath (see Figure 48.8). Myelin increases the conduction speed of action potentials by insulating the axon membrane. Insulation has

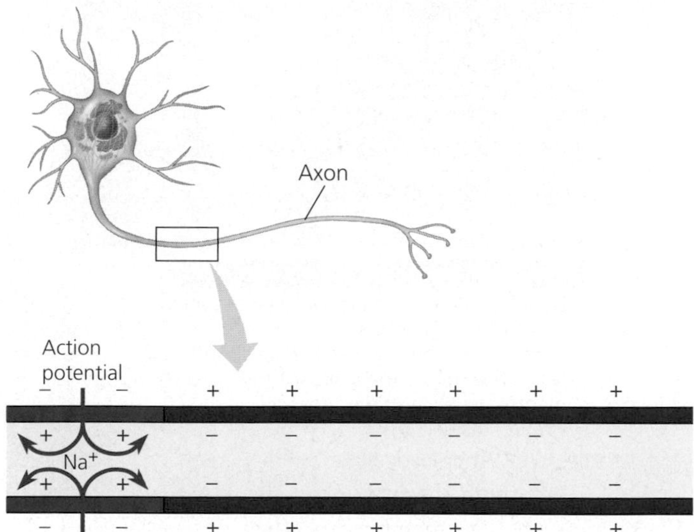

❶ An action potential is generated as Na^+ flows inward across the membrane at one location.

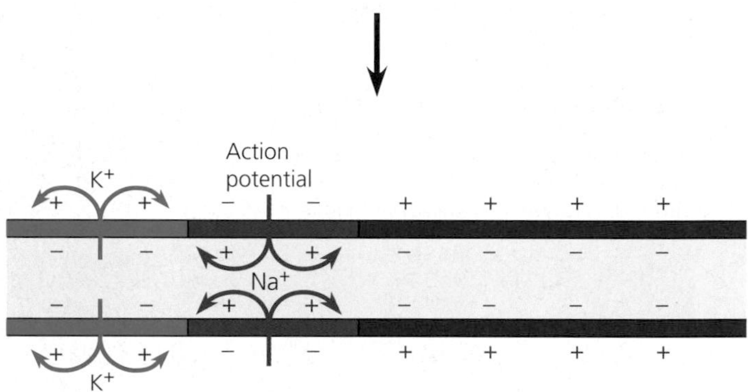

❷ The depolarization of the action potential spreads to the neighboring region of the membrane, re-initiating the action potential there. To the left of this region, the membrane is repolarizing as K^+ flows outward.

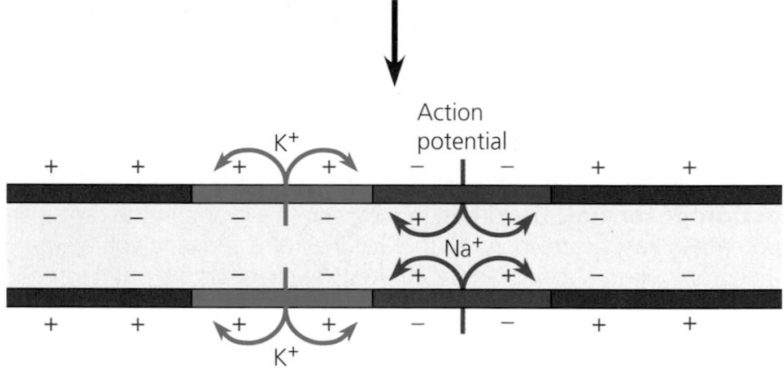

❸ The depolarization-repolarization process is repeated in the next region of the membrane. In this way, local currents of ions *across* the plasma membrane cause the action potential to be propagated *along* the length of the axon.

▲ **Figure 48.14 Conduction of an action potential.** The three parts of this figure show events that occur in an axon at three successive times as an action potential passes from left to right. At each point along the axon, voltage-gated ion channels go through the sequence of changes described in Figure 48.13. The colors of membrane regions shown here correspond to the action potential phases in Figure 48.13.

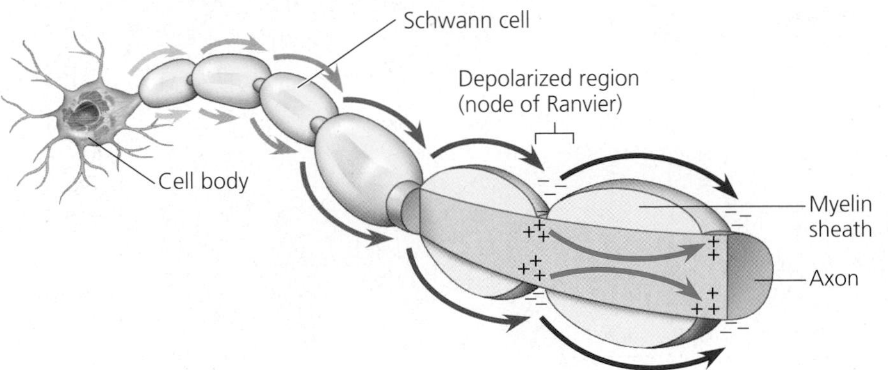

◄ **Figure 48.15 Saltatory conduction.** In a myelinated axon, the depolarizing current during an action potential at one node of Ranvier spreads along the interior of the axon to the next node (blue arrows), where it re-initiates itself. Thus, the action potential jumps from node to node as it travels along the axon (red arrows).

the same effect as increasing the axon's diameter: It causes the depolarizing current associated with an action potential to spread farther along the interior of the axon, bringing more distant regions of the membrane to the threshold sooner. The great advantage of myelination is its space efficiency. A myelinated axon 20 μm in diameter has about the same conduction speed as a squid giant axon (diameter 1 mm), but more than 2,000 of those myelinated axons could be packed into the space occupied by just one giant axon.

In a myelinated axon, voltage-gated Na^+ and K^+ channels are concentrated at gaps in the myelin sheath called nodes of Ranvier (see Figure 48.8). The extracellular fluid is in contact with the axon membrane only at the nodes. As a result, action potentials are not generated in the regions between the nodes. Rather, the inward current produced during the rising phase of the action potential at a node travels all the way to the next node, where it depolarizes the membrane and generates a new action potential **(Figure 48.15)**. This mechanism is called **saltatory conduction** (from the Latin *saltare*, to leap) because the action potential appears to jump along the axon from node to node. Saltatory conduction can transmit action potentials at speeds up to 120 m/sec in myelinated axons.

Concept Check 48.3

1. How does an action potential differ from a graded potential?
2. Suppose a mutation caused the inactivation gates on Na^+ channels to remain closed for a longer time following an action potential. How would that affect the maximum frequency at which action potentials could be generated?
3. Arrange the following from lowest to highest conduction speed: (a) myelinated, small-diameter axon; (b) myelinated, large-diameter axon; (c) unmyelinated, small-diameter axon.

For suggested answers, see Appendix A.

Concept 48.4

Neurons communicate with other cells at synapses

When an action potential reaches the terminals of an axon, it generally stops there. In most cases, action potentials are not transmitted from neurons to other cells. However, information is transmitted, and this transmission occurs at the synapses. Some synapses, called *electrical synapses*, contain gap junctions (see Figure 6.31), which *do* allow electrical current to flow directly from cell to cell. In both vertebrates and invertebrates, electrical synapses synchronize the activity of neurons responsible for certain rapid, stereotypical behaviors. For example, electrical synapses associated with the giant axons of lobsters and other crustaceans facilitate the swift execution of their escape responses.

The vast majority of synapses are *chemical synapses*, which involve the release of a chemical neurotransmitter by the presynaptic neuron. The presynaptic neuron synthesizes the neurotransmitter and packages it in **synaptic vesicles**, which are stored in the neuron's synaptic terminals. Hundreds of synaptic terminals may interact with the cell body and dendrites of a postsynaptic neuron **(Figure 48.16)**. When an action potential

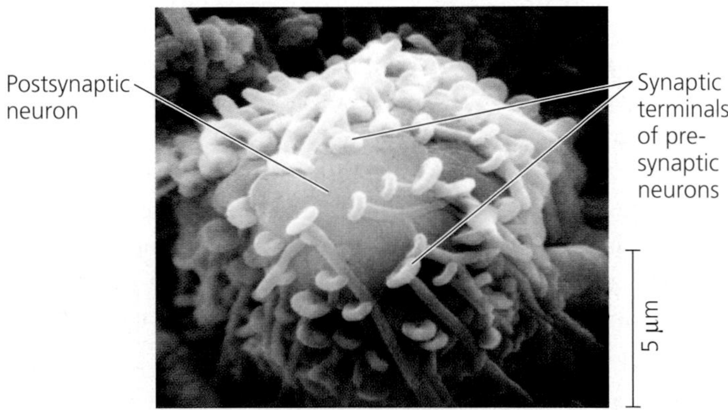

▲ **Figure 48.16 Synaptic terminals on the cell body of a postsynaptic neuron (colorized SEM).**

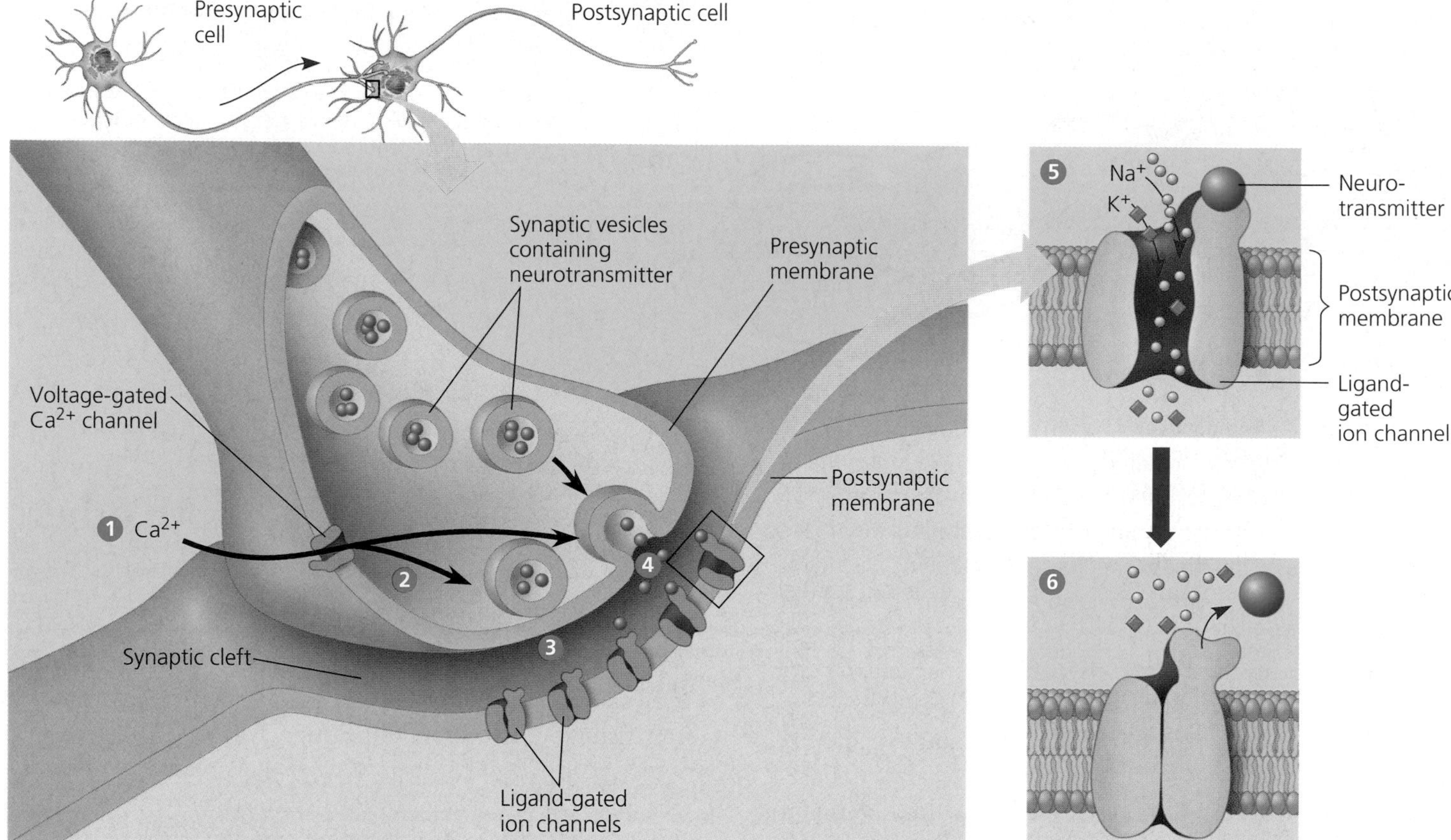

▲ **Figure 48.17 A chemical synapse.** ❶ When an action potential depolarizes the plasma membrane of the synaptic terminal, it ❷ opens voltage-gated Ca^{2+} channels in the membrane, triggering an influx of Ca^{2+}. ❸ The elevated Ca^{2+} concentration in the terminal causes synaptic vesicles to fuse with the presynaptic membrane. ❹ The vesicles release neurotransmitter into the synaptic cleft. ❺ The neurotransmitter binds to the receptor portion of ligand-gated ion channels in the postsynaptic membrane, opening the channels. In the synapse illustrated here, both Na^+ and K^+ can diffuse through the channels. ❻ The neurotransmitter releases from the receptors, and the channels close. Synaptic transmission ends when the neurotransmitter diffuses out of the synaptic cleft, is taken up by the synaptic terminal or another cell, or is degraded by an enzyme.

reaches a synaptic terminal, it depolarizes the terminal membrane, opening voltage-gated calcium channels in the membrane **(Figure 48.17)**. Calcium ions (Ca^{2+}) then diffuse into the terminal, and the rise in Ca^{2+} concentration in the terminal causes some of the synaptic vesicles to fuse with the terminal membrane, releasing the neurotransmitter by exocytosis (see Chapter 7). The neurotransmitter diffuses across the **synaptic cleft,** a narrow gap that separates the presynaptic neuron from the postsynaptic cell. The effect of the neurotransmitter on the postsynaptic cell may be either direct or indirect (described next).

Information transfer is much more modifiable at chemical synapses than at electrical synapses. A variety of factors can affect the amount of neurotransmitter that is released or the responsiveness of the postsynaptic cell. Such modifications underlie an animal's ability to alter its behavior in response to change and form the basis for learning and memory.

Direct Synaptic Transmission

At many chemical synapses, ligand-gated ion channels capable of binding to the neurotransmitter are clustered in the membrane of the postsynaptic cell, directly opposite the synaptic terminal (see Figure 48.17). Binding of the neurotransmitter to a particular part of the channel, the receptor, opens the channel and allows specific ions to diffuse across the postsynaptic membrane. This mechanism of information transfer is called *direct synaptic transmission.* The result is generally a *postsynaptic potential,* a change in the membrane potential of the postsynaptic cell. At some synapses, for example, the neurotransmitter binds to a type of channel through which both Na^+ and K^+ can diffuse. When those channels open, the postsynaptic membrane depolarizes as the membrane potential approaches a value roughly midway between E_K and E_{Na}. Since these depolarizations bring the membrane potential toward the threshold, they are called **excitatory postsynaptic potentials (EPSPs).** At other synapses, a different neurotransmitter binds to channels that are selective for K^+ only. When those channels open, the postsynaptic membrane hyperpolarizes. Hyperpolarizations produced in this manner are called **inhibitory postsynaptic potentials (IPSPs)** because they move the membrane potential farther from the threshold.

Various mechanisms terminate the effect of neurotransmitters on postsynaptic cells. At many synapses, the neurotransmitter simply diffuses out of the synaptic cleft. At others, the neurotransmitter is taken up by the presynaptic neuron through active transport and repackaged into synaptic vesicles. Glia actively take up the neurotransmitter at some synapses and metabolize it as fuel. The neurotransmitter acetylcholine (discussed shortly) is degraded by an enzyme, acetylcholinesterase, which resides in the synaptic cleft.

Summation of Postsynaptic Potentials

Unlike action potentials, which are all-or-none, postsynaptic potentials are graded; their magnitude varies with a number of factors, including the amount of neurotransmitter released by the presynaptic neuron. Another difference is that postsynaptic potentials usually do *not* regenerate themselves as they spread along the membrane of a cell; they become smaller with distance from the synapse. Recall that most synapses on a neuron are located on its dendrites or cell body, whereas action potentials are generally initiated at the axon hillock. Therefore, a single EPSP is usually too small to trigger an action potential in a postsynaptic neuron **(Figure 48.18a)**.

However, if two EPSPs occur in rapid succession at a single synapse, the second EPSP may begin before the postsynaptic neuron's membrane potential has returned to the resting potential after the first EPSP. When that happens, the EPSPs add together, an effect called **temporal summation (Figure 48.18b)**. Moreover, EPSPs produced nearly simultaneously by *different* synapses on the same postsynaptic neuron can also add together, an effect called **spatial summation (Figure 48.18c)**. Through spatial and temporal summation, several EPSPs can depolarize the membrane at the axon hillock to the threshold, causing the postsynaptic neuron to produce an action potential. Summation also applies to IPSPs: Two or more IPSPs occurring nearly simultaneously or in rapid succession have a larger hyperpolarizing effect than a single IPSP. Through summation, an IPSP can also counter the effect of an EPSP **(Figure 48.18d)**.

This interplay between multiple excitatory and inhibitory inputs is the essence of integration in the nervous system. The axon hillock is the neuron's integrating center, the region where the membrane potential at any instant represents the summed effect of all EPSPs and IPSPs. Whenever the membrane potential at the axon hillock reaches the threshold, an action potential is generated and travels along the axon to its synaptic terminals. After the refractory period, the neuron may produce another action potential if the threshold is reached again at the axon hillock. On the other hand, the summed effect of EPSPs and IPSPs may hold the membrane potential below the threshold, preventing production of action potentials.

Indirect Synaptic Transmission

So far, we have focused on direct synaptic transmission, in which a neurotransmitter binds directly to an ion channel, causing the channel to open. In *indirect synaptic transmission,* a neurotransmitter binds to a receptor that is not part of an ion channel. This activates a signal transduction pathway involving

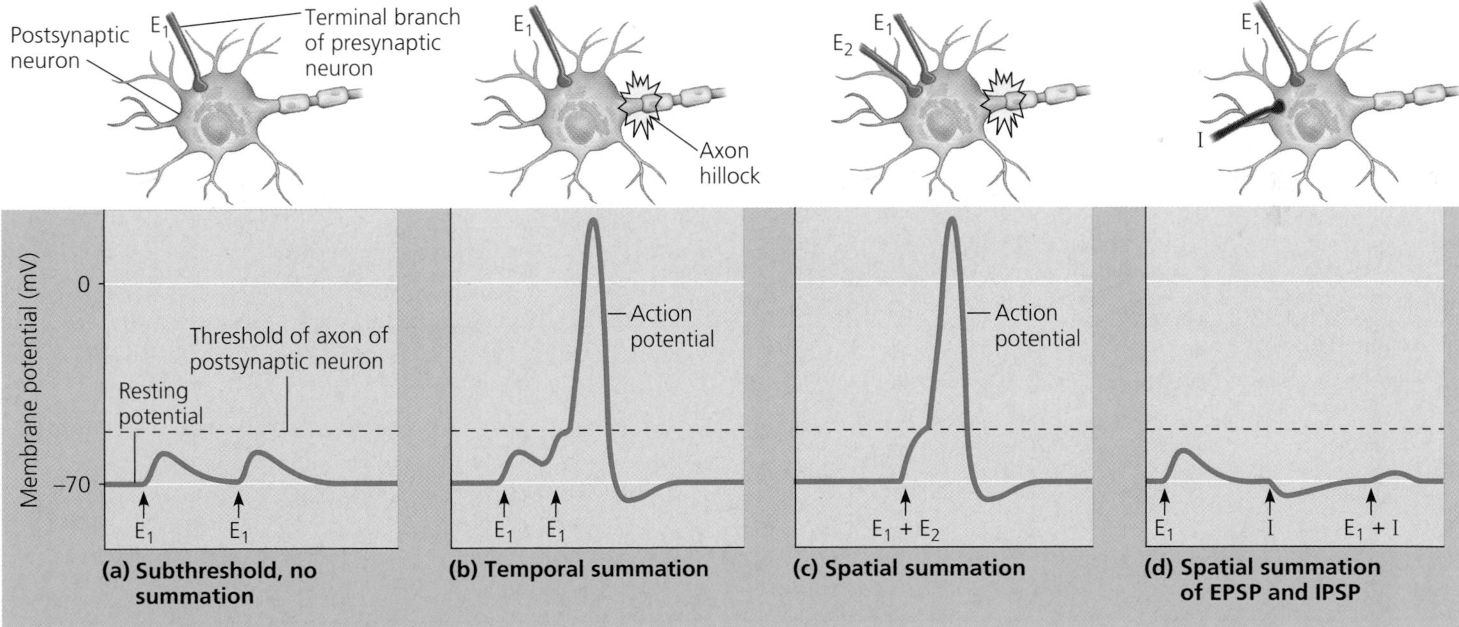

▲ **Figure 48.18 Summation of postsynaptic potentials.** These graphs trace changes in the membrane potential at a postsynaptic neuron's axon hillock. The arrows indicate times when postsynaptic potentials occur at two excitatory synapses (E_1 and E_2, green in the diagrams above the graphs) and at one inhibitory synapse (I, red). Like most EPSPs, those produced at E_1 or E_2 do not reach the threshold at the axon hillock without summation.

a second messenger in the postsynaptic cell (see Chapter 11). Compared to the postsynaptic potentials produced by direct synaptic transmission, the effects of indirect synaptic transmission have a slower onset but last longer (up to several minutes).

A variety of signal transduction pathways play a role in indirect synaptic transmission. One of the best-studied pathways involves cyclic AMP (cAMP) as a second messenger. For example, when the neurotransmitter norepinephrine binds to its receptor, the neurotransmitter-receptor complex activates a G protein, which in turn activates adenylyl cyclase, the enzyme that converts ATP to cAMP (see Chapter 11). Cyclic AMP activates protein kinase A, which phosphorylates specific channel proteins in the postsynaptic membrane, causing them to open or, in some cases, to close. Because of the amplifying effect of the signal transduction pathway, the binding of a neurotransmitter molecule to a single receptor can open or close many channels.

Neurotransmitters

Table 48.1 lists some of the major known neurotransmitters. Each neurotransmitter binds to its own group of receptors; some neurotransmitters have a dozen or more different receptors, which can produce very different effects in postsynaptic cells. Drugs that target specific receptors are powerful tools in the treatment of nervous system diseases.

Acetylcholine

Acetylcholine is one of the most common neurotransmitters in both invertebrates and vertebrates. In the vertebrate CNS, it can be either inhibitory or excitatory, depending on the type of receptor. At the vertebrate neuromuscular junction, the synapse between a motor neuron and a skeletal muscle cell, acetylcholine released by the motor neuron binds to receptors on ligand-gated channels in the muscle cell, producing an EPSP via direct synaptic transmission. The drug nicotine binds to the same receptors, which are also found elsewhere in the PNS and at various sites in the CNS. Nicotine's physiological and psychological effects result from its affinity for this type of acetylcholine receptor. In vertebrate cardiac (heart) muscle, acetylcholine released by parasympathetic neurons (discussed later) activates a signal transduction pathway whose G proteins have two effects: inhibition of adenylyl cyclase and opening of K^+ channels in the muscle cell membrane. Both effects reduce the strength and rate of contraction of cardiac muscle cells.

Table 48.1 Major Neurotransmitters

Neurotransmitter	Structure	Functional Class	Secretion Sites
Acetylcholine	$H_3C-C(=O)-O-CH_2-CH_2-N^+-[CH_3]_3$	Excitatory to vertebrate skeletal muscles; excitatory or inhibitory at other sites	CNS; PNS; vertebrate neuromuscular junction
Biogenic Amines			
Norepinephrine	$(HO)_2C_6H_3-CH(OH)-CH_2-NH_2$	Excitatory or inhibitory	CNS; PNS
Dopamine	$(HO)_2C_6H_3-CH_2-CH_2-NH_2$	Generally excitatory; may be inhibitory at some sites	CNS; PNS
Serotonin	HO-indole (ring C=CH–NH)$-C-CH_2-CH_2-NH_2$	Generally inhibitory	CNS
Amino Acids			
GABA (gamma aminobutyric acid)	$H_2N-CH_2-CH_2-CH_2-COOH$	Inhibitory	CNS; invertebrate neuromuscular junction
Glycine	H_2N-CH_2-COOH	Inhibitory	CNS
Glutamate	$H_2N-CH(COOH)-CH_2-CH_2-COOH$	Excitatory	CNS; invertebrate neuromuscular junction
Aspartate	$H_2N-CH(COOH)-CH_2-COOH$	Excitatory	CNS
Neuropeptides (a very diverse group, only two of which are shown)			
Substance P	Arg—Pro—Lys—Pro—Gln—Gln—Phe—Phe—Gly—Leu—Met	Excitatory	CNS; PNS
Met-enkephalin (an endorphin)	Tyr—Gly—Gly—Phe—Met	Generally inhibitory	CNS

Biogenic Amines

Biogenic amines are neurotransmitters derived from amino acids. One group, known as catecholamines, consists of neurotransmitters produced from the amino acid tyrosine. This group includes **epinephrine** and **norepinephrine,** which also function as hormones (see Chapter 45), and a closely related compound called **dopamine.** Another biogenic amine, **serotonin,** is synthesized from the amino acid tryptophan. The biogenic amines are often involved in indirect synaptic transmission, most commonly in the CNS. However, norepinephrine also functions in a branch of the PNS called the autonomic nervous system, discussed in Concept 48.5.

Dopamine and serotonin are released at many sites in the brain and affect sleep, mood, attention, and learning. Imbalances of these neurotransmitters are associated with several disorders. For example, the degenerative illness Parkinson's disease is associated with a lack of dopamine in the brain (see Concept 48.7). Some psychoactive drugs, including LSD and mescaline, apparently produce their hallucinatory effects by binding to brain receptors for serotonin and dopamine. Depression is often treated with drugs that increase the brain concentrations of biogenic amines such as norepinephrine or serotonin. Prozac, for instance, enhances the effect of serotonin by inhibiting its uptake after release.

Amino Acids and Peptides

Four amino acids are known to function as neurotransmitters in the CNS: **gamma aminobutyric acid (GABA), glycine, glutamate,** and **aspartate.** GABA, believed to be the neurotransmitter at most inhibitory synapses in the brain, produces IPSPs by increasing the permeability of the postsynaptic membrane to Cl^-.

Several **neuropeptides,** relatively short chains of amino acids, serve as neurotransmitters. Most neurons release one or more neuropeptides as well as a nonpeptide neurotransmitter. Many neuropeptides are produced by post-translational modification of much larger protein precursors. For example, cleavage of the 267-amino-acid precursor proenkephalin yields four copies of the pentapeptide met-enkephalin as well as other peptides. In common with the biogenic amines, neuropeptides often operate via signal transduction pathways.

The neuropeptide **substance P** is a key excitatory neurotransmitter that mediates our perception of pain, while other neuropeptides, called **endorphins,** function as natural analgesics, decreasing pain perception. Neurochemists Candace Pert and Solomon Snyder, of Johns Hopkins University, discovered endorphins in the 1970s when they found specific receptors for the opiates morphine and heroin on brain neurons. Further research showed that opiates bind to these receptors by mimicking endorphins (see Figure 2.17), which are produced in the brain during times of physical or emotional stress, such as childbirth. In addition to relieving pain, endorphins also decrease urine output by stimulating ADH secretion (see Chapter 44), depress respiration, produce euphoria, and have other emotional effects. An endorphin released from the anterior pituitary gland as a hormone also affects specific regions of the brain, providing an example of the overlap between endocrine and nervous system control.

Gases

In common with many other types of cells, some neurons of the vertebrate PNS and CNS release dissolved gases, notably nitric oxide (NO; see Chapter 45) and carbon monoxide (CO), as local regulators. For example, during sexual arousal of human males, certain neurons release NO into the erectile tissue of the penis. In response, smooth muscle cells in the blood vessel walls of the erectile tissue relax, which causes the blood vessels to dilate and fill the spongy erectile tissue with blood, producing an erection. As you read in Chapter 45, the male impotence drug Viagra increases the ability to achieve and maintain an erection by inhibiting an enzyme that slows the muscle-relaxing effects of NO.

Carbon monoxide is synthesized by the enzyme heme oxygenase, one form of which is localized to certain populations of neurons in the brain and the PNS. In the brain, CO regulates the release of hypothalamic hormones. In the PNS, it acts as an inhibitory neurotransmitter that hyperpolarizes intestinal smooth muscle cells.

Unlike typical neurotransmitters, NO and CO are not stored in cytoplasmic vesicles; cells synthesize them on demand. These substances diffuse into neighboring target cells, produce a change, and are broken down—all within a few seconds. In many of its targets, including smooth muscle cells, NO works like many hormones, stimulating a membrane-bound enzyme to synthesize a second messenger that directly affects cellular metabolism.

In the rest of this chapter, we will consider how the cellular and biochemical mechanisms we have discussed so far contribute to neural function on the system level.

Concept Check 48.4

1. Cone snails produce a toxin that blocks voltage-gated calcium channels. Which of the two main types of synapses would be most affected by the toxin? Why?
2. Organophosphate pesticides work by inhibiting acetylcholinesterase, the enzyme that breaks down the neurotransmitter acetylcholine. Explain how these toxins would affect EPSPs produced by acetylcholine.
3. How is it possible for a given neurotransmitter to produce opposite effects in different tissues?

For suggested answers, see Appendix A.

Concept 48.5

The vertebrate nervous system is regionally specialized

In all vertebrates, the nervous system shows cephalization and distinct CNS and PNS components **(Figure 48.19)**. The brain provides the integrative power that underlies the complex behavior of vertebrates. The spinal cord, which runs lengthwise inside the vertebral column (spine), integrates simple responses to certain kinds of stimuli (such as the knee-jerk reflex) and conveys information to and from the brain. Unlike the ventral nerve cord of many invertebrates, the vertebrate spinal cord runs along the dorsal side of the body and does not contain segmental ganglia. However, there are segmental ganglia just outside the spinal cord (see Figures 48.2h and 48.19), and the arrangement of neurons within the spinal cord clearly shows an underlying segmental organization.

The vertebrate CNS is derived from the dorsal embryonic nerve cord, which is hollow—one of the phylogenetic hallmarks of chordates (see Chapter 34). In the adult, this feature persists as the narrow **central canal** of the spinal cord and the

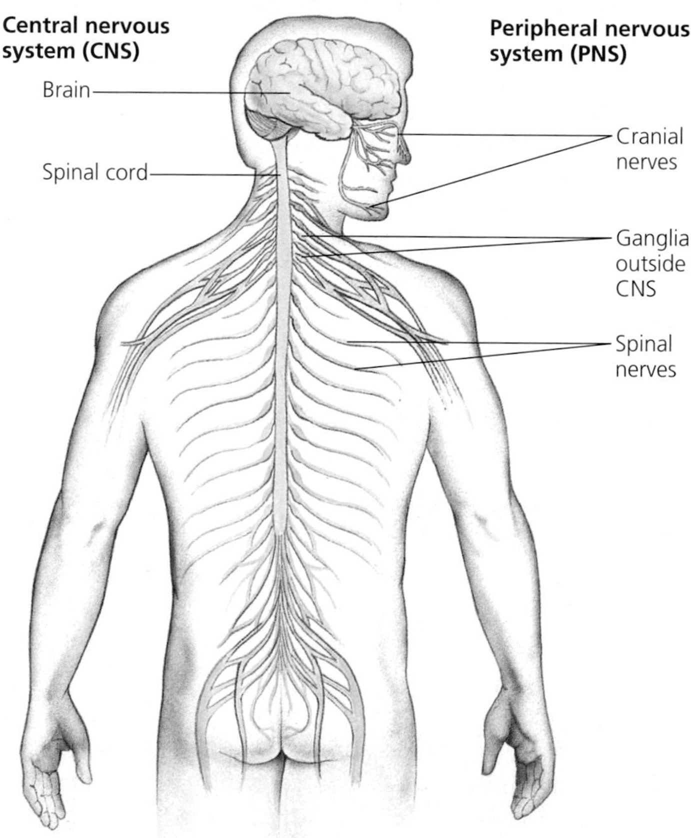

▲ **Figure 48.19 The vertebrate nervous system.** The central nervous system consists of the brain and spinal cord. Cranial nerves, spinal nerves, and ganglia outside the central nervous system make up the peripheral nervous system.

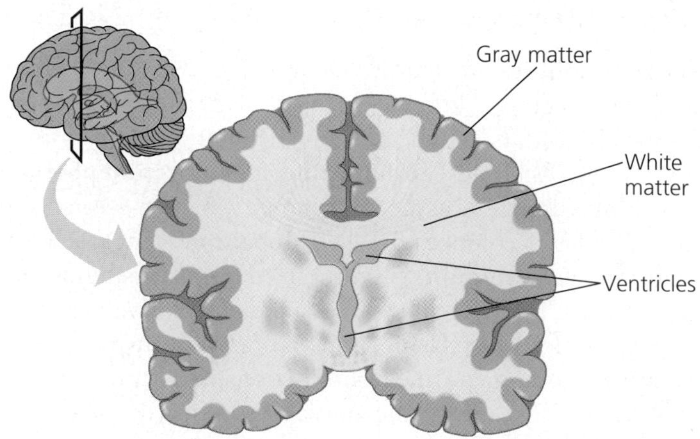

▲ **Figure 48.20 Ventricles, gray matter, and white matter.** Ventricles deep in the interior of the brain contain cerebrospinal fluid. Most of the gray matter is on the surface of the brain, surrounding the white matter.

four **ventricles** of the brain **(Figure 48.20)**. Both the central canal and the ventricles are filled with **cerebrospinal fluid**, which is formed in the brain by filtration of the blood. The cerebrospinal fluid circulates slowly through the central canal and ventricles and then drains into the veins, assisting in the supply of nutrients and hormones to different parts of the brain and in the removal of wastes. In mammals, the cerebrospinal fluid also cushions the brain and spinal cord by circulating between two of the meninges, layers of connective tissue that surround the CNS.

Axons within the CNS are often found in well-defined bundles, or tracts, whose myelin sheaths give them a whitish appearance. In cross sections of the brain and spinal cord, this **white matter** is clearly distinguishable from **gray matter**, which consists mainly of dendrites, unmyelinated axons, and neuron cell bodies (see Figures 48.4 and 48.20).

The Peripheral Nervous System

The PNS transmits information to and from the CNS and plays a large role in regulating a vertebrate's movement and internal environment. Structurally, the vertebrate PNS consists of left-right pairs of cranial and spinal nerves and their associated ganglia (see Figure 48.19). The **cranial nerves** originate in the brain and terminate mostly in organs of the head and upper body. The **spinal nerves** originate in the spinal cord and extend to parts of the body below the head. Mammals have 12 pairs of cranial nerves and 31 pairs of spinal nerves. Most of the cranial nerves and all of the spinal nerves contain axons of both sensory and motor neurons; a few of the cranial nerves (the olfactory and optic nerves, for example) are sensory only.

The PNS can be divided into two functional components: the somatic nervous system and the autonomic nervous system **(Figure 48.21)**. The **somatic nervous system** carries signals to and from skeletal muscles, mainly in response to *external* stimuli. It is often considered voluntary because it is subject to conscious control, but much skeletal muscle activity is actually controlled by reflexes mediated by the spinal cord or a part of the brain called the brainstem. The **autonomic nervous system** regulates the *internal* environment by controlling smooth and cardiac muscles and the organs of the digestive, cardiovascular, excretory, and endocrine systems. This control is generally involuntary. Three divisions—sympathetic, parasympathetic, and enteric—together make up the autonomic nervous system.

Figure 48.22 compares the organization and actions of the sympathetic and parasympathetic divisions. Activation of the **sympathetic division** corresponds to arousal and energy generation (the "fight-or-flight" response). For example, the heart beats faster, the liver converts

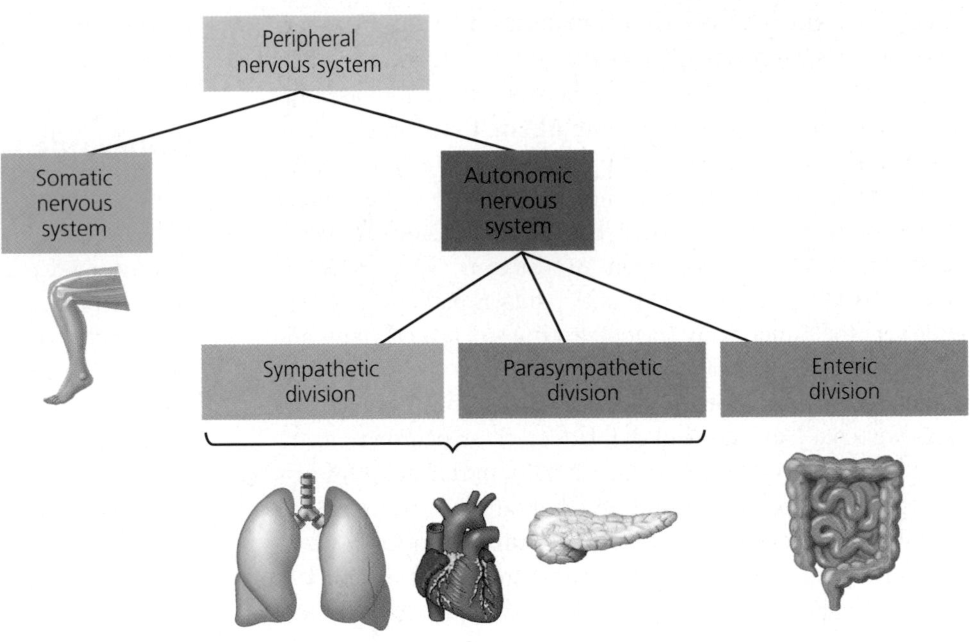

▲ **Figure 48.21 Functional hierarchy of the vertebrate peripheral nervous system.**

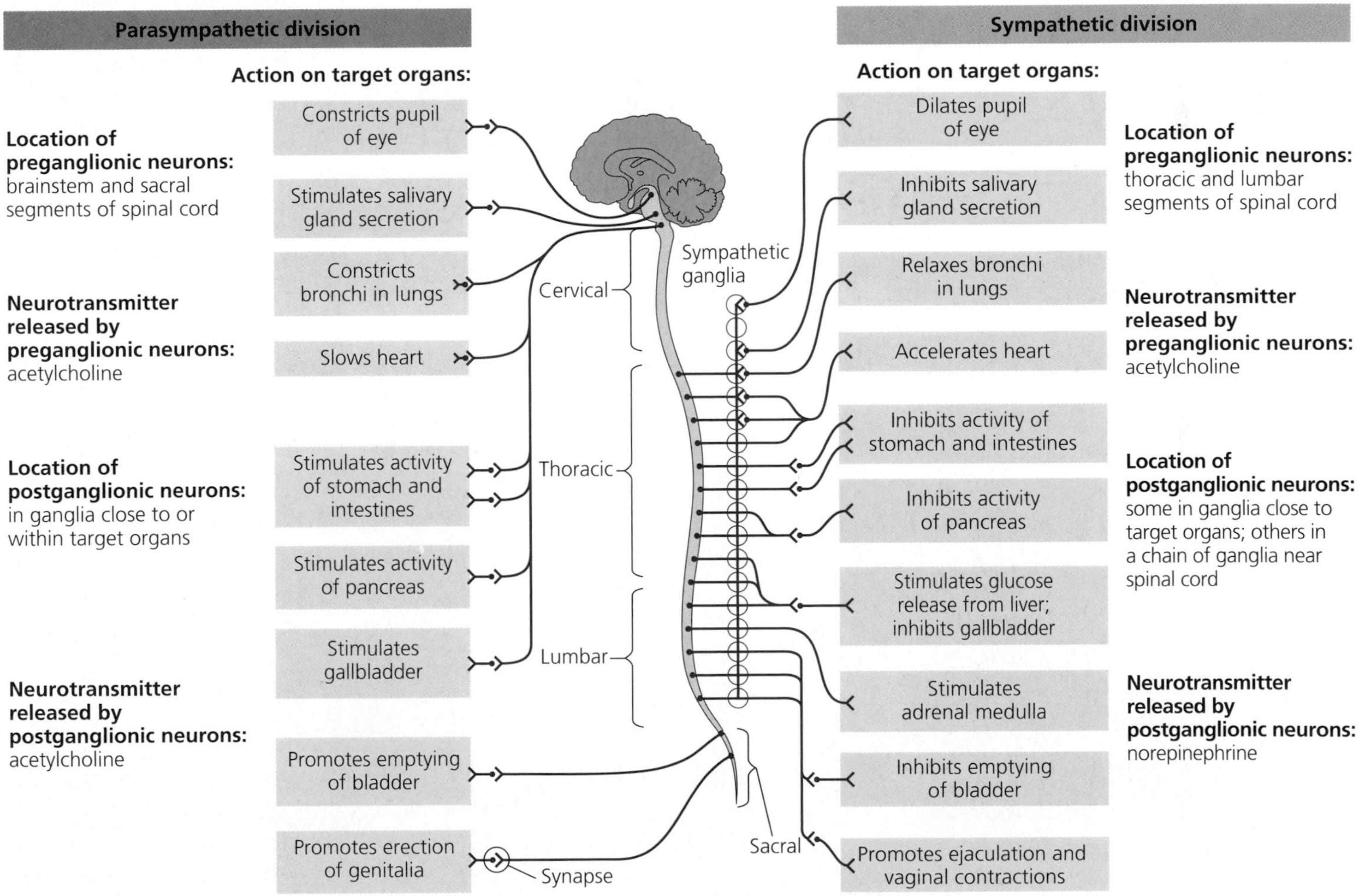

Parasympathetic division

Action on target organs:

Constricts pupil of eye

Stimulates salivary gland secretion

Constricts bronchi in lungs

Slows heart

Stimulates activity of stomach and intestines

Stimulates activity of pancreas

Stimulates gallbladder

Promotes emptying of bladder

Promotes erection of genitalia

Location of preganglionic neurons: brainstem and sacral segments of spinal cord

Neurotransmitter released by preganglionic neurons: acetylcholine

Location of postganglionic neurons: in ganglia close to or within target organs

Neurotransmitter released by postganglionic neurons: acetylcholine

Cervical

Thoracic

Lumbar

Sacral

Synapse

Sympathetic ganglia

Sympathetic division

Action on target organs:

Dilates pupil of eye

Inhibits salivary gland secretion

Relaxes bronchi in lungs

Accelerates heart

Inhibits activity of stomach and intestines

Inhibits activity of pancreas

Stimulates glucose release from liver; inhibits gallbladder

Stimulates adrenal medulla

Inhibits emptying of bladder

Promotes ejaculation and vaginal contractions

Location of preganglionic neurons: thoracic and lumbar segments of spinal cord

Neurotransmitter released by preganglionic neurons: acetylcholine

Location of postganglionic neurons: some in ganglia close to target organs; others in a chain of ganglia near spinal cord

Neurotransmitter released by postganglionic neurons: norepinephrine

▲ **Figure 48.22 The parasympathetic and sympathetic divisions of the autonomic nervous system.** Most pathways in each division consist of preganglionic neurons (with cell bodies in the CNS) and postganglionic neurons (with cell bodies in ganglia in the PNS).

glycogen to glucose, bronchi of the lungs dilate and support increased gas exchange, digestion is inhibited, and secretion of epinephrine (adrenaline) from the adrenal medulla is stimulated. Activation of the **parasympathetic division** generally causes opposite responses that promote calming and a return to self-maintenance functions ("rest and digest"). For example, increased activity in the parasympathetic division decreases heart rate, increases glycogen production, and enhances digestion. When sympathetic and parasympathetic neurons innervate the same organ, they often (but not always) have antagonistic (opposite) effects.

The **enteric division** consists of networks of neurons in the digestive tract, pancreas, and gallbladder; these neurons control these organs' secretions as well as activity in the smooth muscles that produce peristalsis (see Chapter 41). Although the enteric division can function independently, it is normally regulated by the sympathetic and parasympathetic divisions.

The somatic and autonomic nervous systems often cooperate in maintaining homeostasis. In response to a drop in body temperature, for example, the hypothalamus signals the autonomic nervous system to constrict surface blood vessels, which reduces heat loss. At the same time, the hypothalamus signals the somatic nervous system to cause shivering, which increases heat production.

Embryonic Development of the Brain

In all vertebrates, three bilaterally symmetrical, anterior bulges of the neural tube—the **forebrain, midbrain,** and **hindbrain**—become evident as the embryo develops **(Figure 48.23a)**. During vertebrate evolution, the brain further divided structurally and functionally. This regionalization increased capacity for complex integration, with the forebrain becoming much larger in birds and mammals than in other vertebrates.

By the fifth week of human embryonic development, five brain regions have formed from the three primary bulges **(Figure 48.23b)**: The *telencephalon* and *diencephalon* develop from the forebrain; the *mesencephalon* develops from the midbrain (and is also called the midbrain in the adult); and the *metencephalon* and *myelencephalon* develop from the hindbrain.

As a human brain develops further, the most profound changes occur in the telencephalon, the region of the forebrain that gives rise to the **cerebrum (Figure 48.23c)**. Rapid, expansive growth of the telencephalon during the second

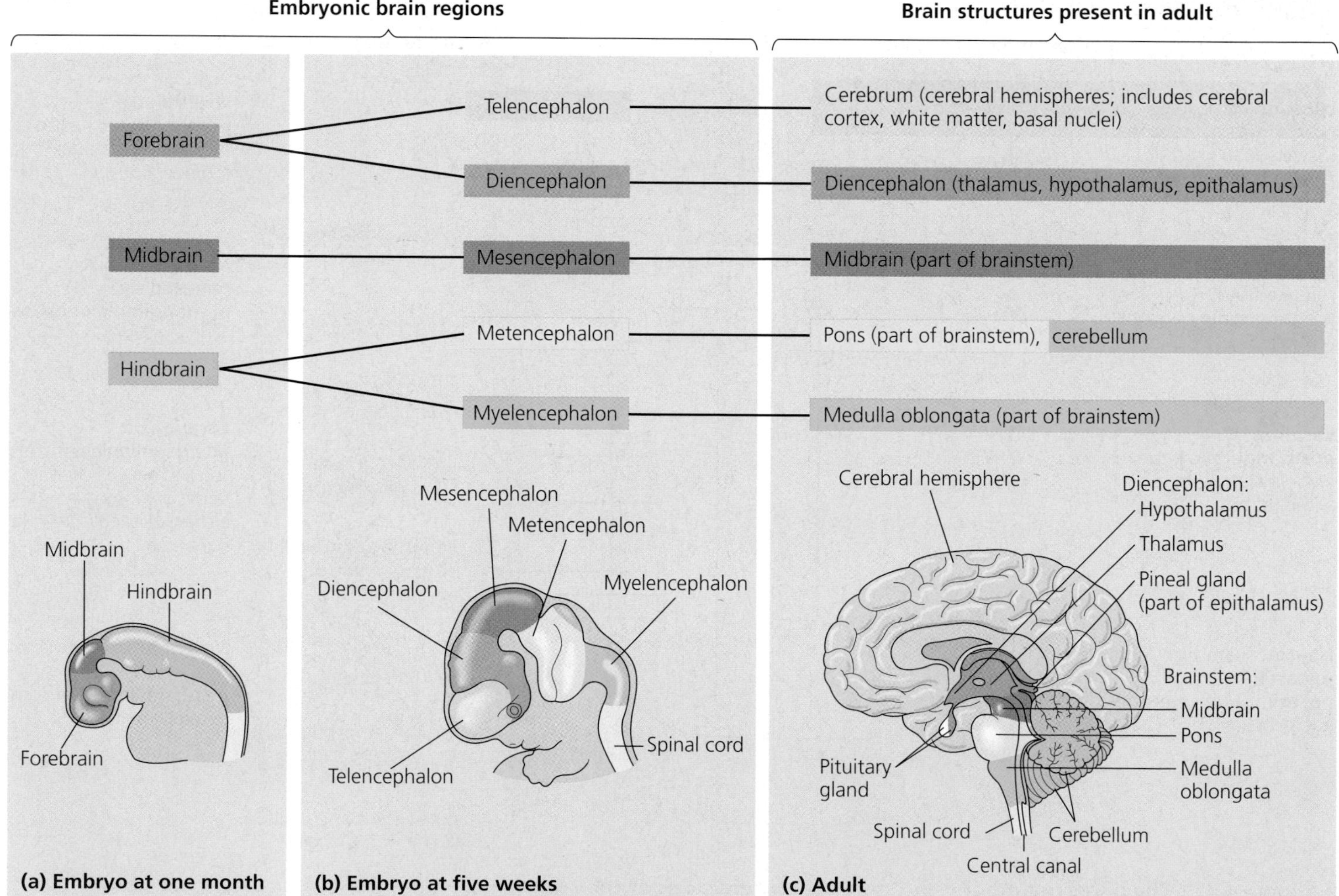

▲ **Figure 48.23 Development of the human brain.**

and third months causes the outer portion of the cerebrum, called the **cerebral cortex,** to extend over and around much of the rest of the brain. Major centers that develop from the diencephalon—the forebrain division that evolved earliest in vertebrate history—are the thalamus, hypothalamus, and epithalamus.

The three regions derived from the midbrain and hindbrain give rise to the brainstem, a set of structures deep within the brain. The adult **brainstem** consists of the midbrain (derived from the mesencephalon), the pons (derived from the metencephalon), and the medulla oblongata (derived from the myelencephalon). The metencephalon also gives rise to another major brain center, the cerebellum, which is not part of the brainstem.

The Brainstem

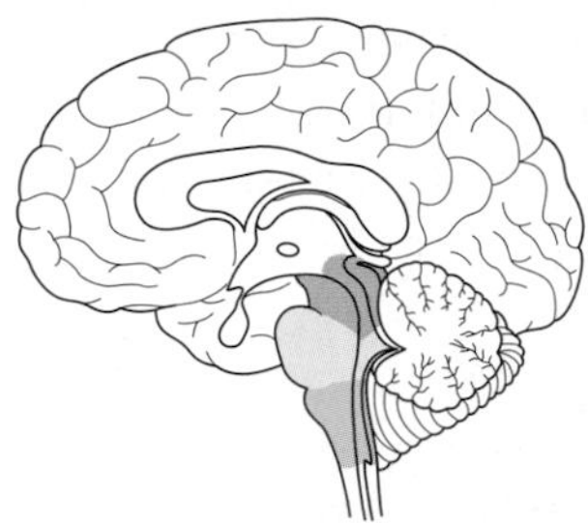

The brainstem is one of the evolutionarily older parts of the vertebrate brain. Sometimes called the "lower brain," it consists of a stalk with caplike swellings at the anterior end of the spinal cord. The three parts of the brainstem—the medulla oblongata, the pons, and the midbrain—function in homeostasis, coordination of movement, and conduction of information to higher brain centers.

Several centers in the brainstem contain neuron cell bodies that send axons to many areas of the cerebral cortex and cerebellum, releasing neurotransmitters such as norepinephrine, dopamine, serotonin, and acetylcholine. Signals in these pathways cause changes in attention, alertness, appetite, and motivation. The **medulla oblongata** (commonly called the **medulla**) contains centers that control several visceral (automatic, homeostatic) functions, including breathing, heart and blood vessel activity, swallowing, vomiting, and digestion. The **pons** also participates in some of these activities; for example, it regulates the breathing centers in the medulla.

All axons carrying sensory information to and motor instructions from higher brain regions pass through the brainstem, making information transmission one of the most important functions of the medulla and pons. These two regions also help coordinate large-scale body movements, such as walking. Most of the axons carrying instructions about movement from the midbrain and forebrain to the spinal cord cross from one side of the CNS to the other in the medulla. As a result, the right side of the brain controls much of the movement of the left side of the body, and vice versa.

The midbrain contains centers for the receipt and integration of several types of sensory information. It also sends coded sensory information along neurons to specific regions of the forebrain. Prominent centers of the midbrain are the inferior and superior colliculi, which are part of the auditory and visual systems, respectively. All sensory axons involved in hearing either terminate in or pass through the inferior colliculi on their way to the cerebrum. In nonmammalian vertebrates, the superior colliculi take the form of prominent optic lobes and may be the only visual centers. In mammals, vision is integrated in the cerebrum, leaving the superior colliculi to coordinate visual reflexes, such as automatically turning your head when your peripheral vision picks up something moving toward you from the side.

Arousal and Sleep

As anyone who has sat through a lecture on a warm spring afternoon knows, attentiveness and mental alertness vary from moment to moment. Arousal is a state of awareness of the external world. The counterpart of arousal is sleep, when an individual continues to receive external stimuli but is not conscious of them. Arousal and sleep are controlled by several centers in the brainstem and cerebrum.

A diffuse network of neurons called the **reticular formation,** containing over 90 separate clusters of cell bodies, is present in the core of the brainstem **(Figure 48.24)**. A part of the reticular formation, the reticular activating system (RAS), regulates sleep and arousal. Acting as a sensory filter, the RAS selects which information reaches the cerebral cortex, and the more information the cortex receives, the more alert and aware a person is. But arousal is not just a generalized phenomenon; certain stimuli can be ignored while the brain is actively processing other input. Also, sleep and wakefulness are regulated by specific parts of the brainstem: The pons and medulla contain centers that cause sleep when stimulated, and the midbrain has a center that causes arousal. Serotonin may be the neurotransmitter of the sleep-producing centers. Drinking

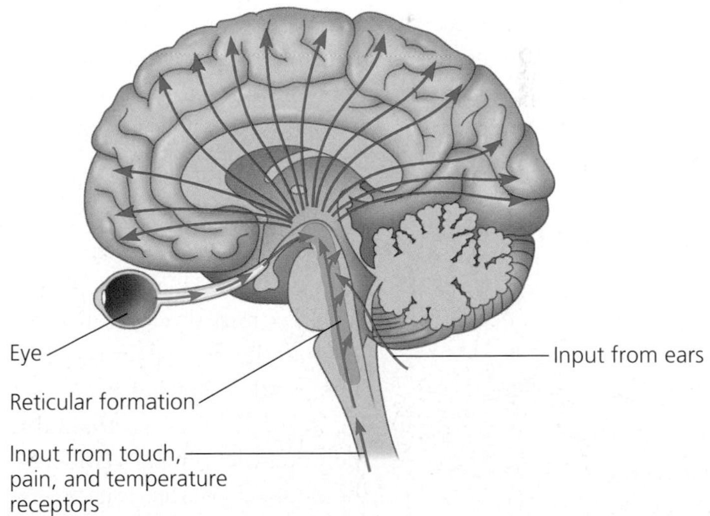

▲ **Figure 48.24 The reticular formation.** This system of neurons distributed throughout the brainstem filters sensory input (blue arrows), blocking familiar and repetitive information that constantly enters the nervous system. It sends the filtered input to the cerebral cortex (green arrows).

milk before bedtime may induce sleep because milk contains large amounts of tryptophan, the amino acid from which serotonin is synthesized.

All birds and mammals show characteristic sleep/wake cycles, and melatonin, a hormone produced by the pineal gland, also appears to play an important role in these cycles. As explained in Chapter 45, peak melatonin secretion occurs at night. Melatonin has been promoted as a dietary supplement to treat sleep disturbances, such as those associated with jet lag, insomnia, seasonal affective disorder, and depression.

Understanding the function of sleep remains a compelling research topic. One hypothesis is that sleep is involved in the consolidation of learning and memory, and experiments show that regions of the brain activated during a learning task can become active again during sleep.

The Cerebellum

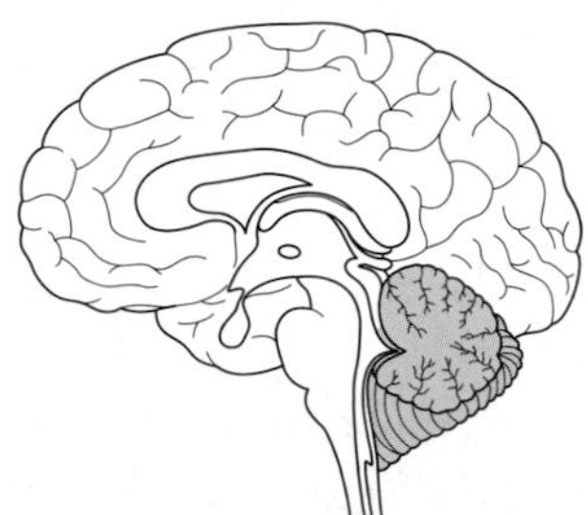

The **cerebellum** develops from part of the metencephalon (see Figure 48.23). It is important for coordination and error checking during motor, perceptual, and cognitive functions. (Cognitive functions include learning, decision making, consciousness, and an integrated sensory awareness of the surroundings.) The cerebellum is likely involved in learning and remembering motor skills, such as those involved in riding a bicycle, because such learning can be blocked by damage to one of its major subdivisions. The cerebellum receives sensory information about the position of the joints and the length of the muscles, as well as information from the auditory and visual systems. It also receives input concerning motor commands issued by the cerebrum. The cerebellum integrates this sensory and motor information as it coordinates movements and balance. Hand-eye coordination is an example of cerebellar control; if the cerebellum is damaged, the eyes can follow a moving object, but they will not stop at the same place as the object.

The Diencephalon

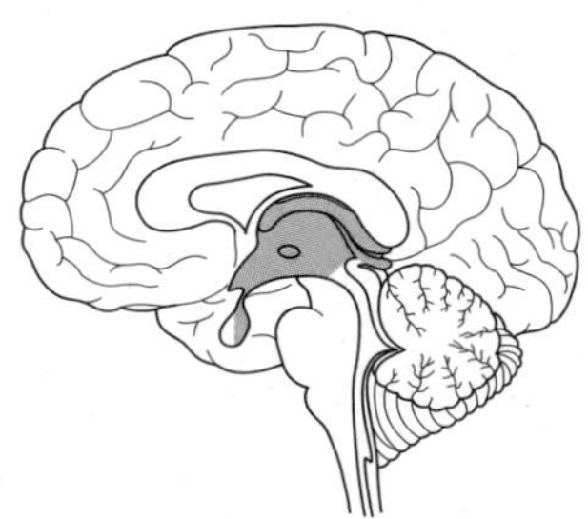

The embryonic diencephalon develops into three adult brain regions: the epithalamus, thalamus, and hypothalamus (see Figure 48.23). The **epithalamus** includes the pineal gland and the choroid plexus, one of several clusters of capillaries that produce cerebrospinal fluid from blood. The thalamus and hypothalamus are major integrating centers.

The **thalamus** is the main input center for sensory information going to the cerebrum and the main output center for motor information leaving the cerebrum. Incoming information from all the senses is sorted in the thalamus and sent to the appropriate cerebral centers for further processing. The thalamus also receives input from the cerebrum and other parts of the brain that regulate emotion and arousal.

Although it weighs only a few grams, the **hypothalamus** is one of the most important brain regions for homeostatic regulation. We saw in Chapter 45 that the hypothalamus is the source of two sets of hormones, posterior pituitary hormones and releasing hormones that act on the anterior pituitary (see Figure 45.8). The hypothalamus also contains the body's thermostat, as well as centers for regulating hunger, thirst, and many other basic survival mechanisms. Hypothalamic centers also play a role in sexual and mating behaviors, the fight-or-flight response, and pleasure. Stimulation of specific centers can cause what are known as "pure" behaviors. For example, rats placed in an experimental situation where they can press a lever to stimulate a "pleasure" center will do so to the exclusion of eating and drinking. Stimulation of another hypothalamic area can produce rage.

Circadian Rhythms

We have already discussed circadian (daily) rhythms in plants (see Chapter 39). Animals also exhibit circadian rhythms, one example being the sleep/wake cycle. Numerous studies have shown that animals usually have an internal timekeeper, known as a **biological clock**, that is involved in maintaining circadian rhythms. Biological clocks regulate a variety of physiological phenomena, including hormone release, hunger, and heightened sensitivity to external stimuli. In mammals, the biological clock is a pair of hypothalamic structures called the **suprachiasmatic nuclei (SCN).** (Some clusters of neurons in the CNS are referred to as "nuclei.") In contrast, fruit flies (*Drosophila*) appear to have many biological clocks in various parts of their body, such as the outer edges of their wings.

Biological clocks usually require external cues to remain synchronized with environmental cycles. For example, visual information about light intensity transmitted to the SCN synchronizes the mammalian clock with the natural cycles of day length and darkness (**Figure 48.25**). Experiments with rodents have revealed that cells in the SCN produce specific proteins in response to changing light/dark cycles.

Human circadian rhythms have been the subject of particularly intense study because upsetting them can cause sleep disorders. In a famous series of experiments in the 1970s, researchers housed subjects in comfortable rooms deep underground, where the subjects could set their own schedules without external cues. Those experiments suggested that the human biological clock has a cycle length of about 25 hours, with much individual variation. In the late 1990s, however, a research team at Harvard University challenged those findings, pointing out that even indoor lighting can influence circadian rhythms. Using more rigorous experimental conditions, the Harvard scientists found that the human biological clock has a cycle length of 24 hours and 11 minutes, with very little variation between individuals.

Figure 48.25

Inquiry Are mammalian biological clocks influenced by external cues?

EXPERIMENT In the northern flying squirrel (*Glaucomys sabrinus*), activity normally begins with the onset of darkness and ends at dawn, which suggests that light is an important external cue for the squirrel. To test this idea, researchers monitored the activity of captive squirrels for 23 days under two sets of conditions: (a) a regular cycle of 12 hours of light and 12 hours of darkness and (b) constant darkness. The squirrels were given free access to an exercise wheel and a rest cage. A recorder automatically noted when the wheel was rotating and when it was still.

RESULTS When the squirrels were exposed to a regular light/dark cycle, their wheel-turning activity (indicated by the dark bars) occurred at roughly the same time every day. However, when they were kept in constant darkness, their activity phase began about 21 minutes later each day.

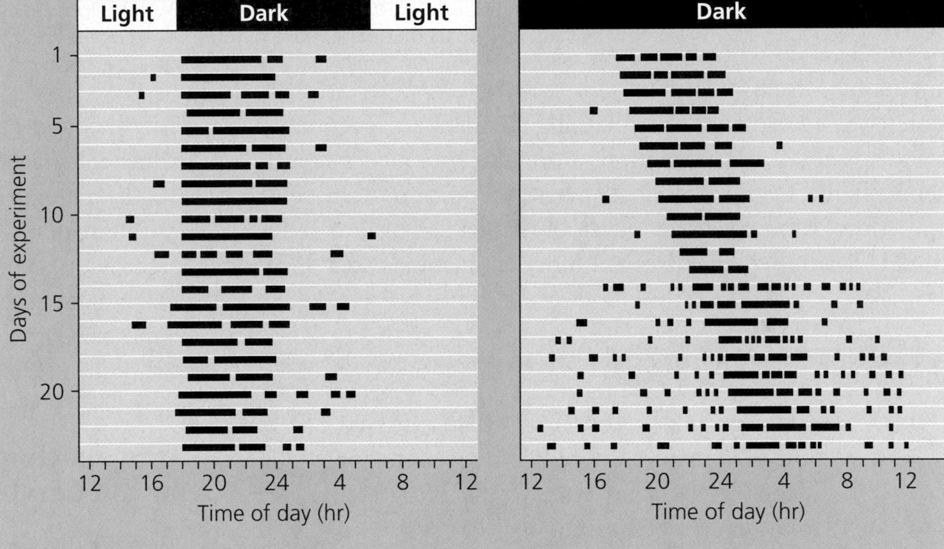

CONCLUSION The northern flying squirrel's internal clock can run in constant darkness, but it does so on its own cycle, which lasts about 24 hours and 21 minutes. External (light) cues keep the clock running on a 24-hour cycle.

The Cerebrum

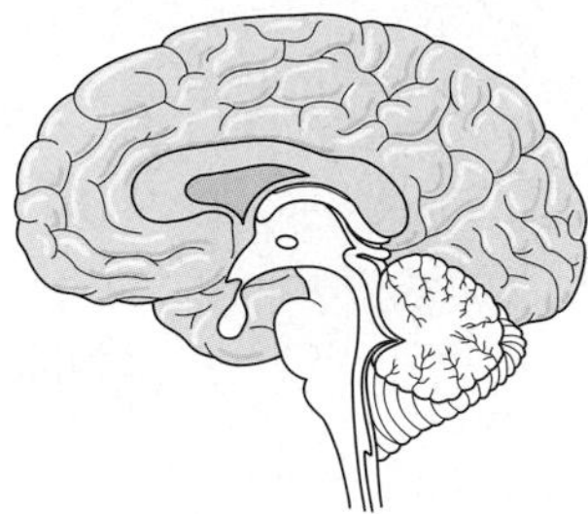

The cerebrum develops from the embryonic telencephalon, an outgrowth of the forebrain that arose early in vertebrate evolution as a region supporting olfactory reception as well as auditory and visual processing. The cerebrum is divided into right and left **cerebral hemispheres.** Each hemisphere consists of an outer covering of gray matter, the cerebral cortex; internal white matter; and groups of neurons collectively called **basal nuclei** located deep within the white matter **(Figure 48.26)**. The basal nuclei are important centers for planning and learning movement sequences. Damage in this region can prevent motor commands from being sent to the muscles, rendering a person passive and immobile.

In humans, the largest, most complex part of the brain is the cerebral cortex. Here, sensory information is analyzed, motor commands are issued, and language is generated. The cerebral cortex underwent a dramatic expansion when the ancestors of mammals diverged from those of reptiles. Most significantly, mammals have a region of the cerebral cortex known as the **neocortex.** The neocortex forms the outermost part of the mammalian cerebrum, consisting of six parallel layers of neurons arranged tangential to the brain surface. Whereas the neocortex of a rat is relatively smooth, the human neocortex is highly convoluted (see Figure 48.26). The convolutions allow the neocortex to have a large surface area and still fit inside the skull: Although less than 5 mm thick, the human neocortex has a surface area of about 0.5 m^2 and accounts for about 80% of total brain mass. Nonhuman primates and cetaceans (whales and porpoises, for example) also have exceptionally large, convoluted neocortices. In fact, the surface area (relative to body size)

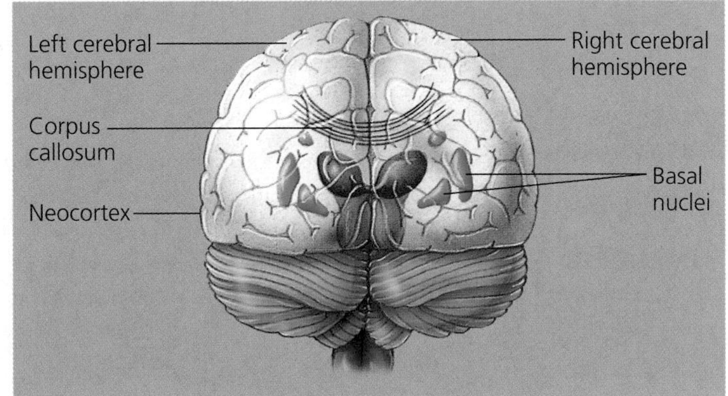

▲ **Figure 48.26 The human cerebrum viewed from the rear.** The corpus callosum and basal nuclei are not visible from the surface because they are completely covered by the left and right cerebral hemispheres.

of a porpoise's neocortex is second only to that of a human.

Like the rest of the cerebrum, the cerebral cortex is divided into right and left sides, each of which is responsible for the opposite half of the body. The left side of the cortex receives information from, and controls the movement of, the right side of the body, and vice versa. A thick band of axons known as the **corpus callosum** enables communication between the right and left cerebral cortices (see Figure 48.26).

Damage to one area of the cerebrum early in development can frequently cause redirection of its normal functions to other areas. Perhaps the most dramatic example of this phenomenon occurs after an entire cerebral hemisphere is removed in infants as a treatment for severe epilepsy. Amazingly, the remaining hemisphere eventually assumes most of the functions normally provided by both hemispheres. Even in adults, damage to a portion of the cerebral cortex can trigger the development or use of new brain circuits, leading to a recovery of function in some cases.

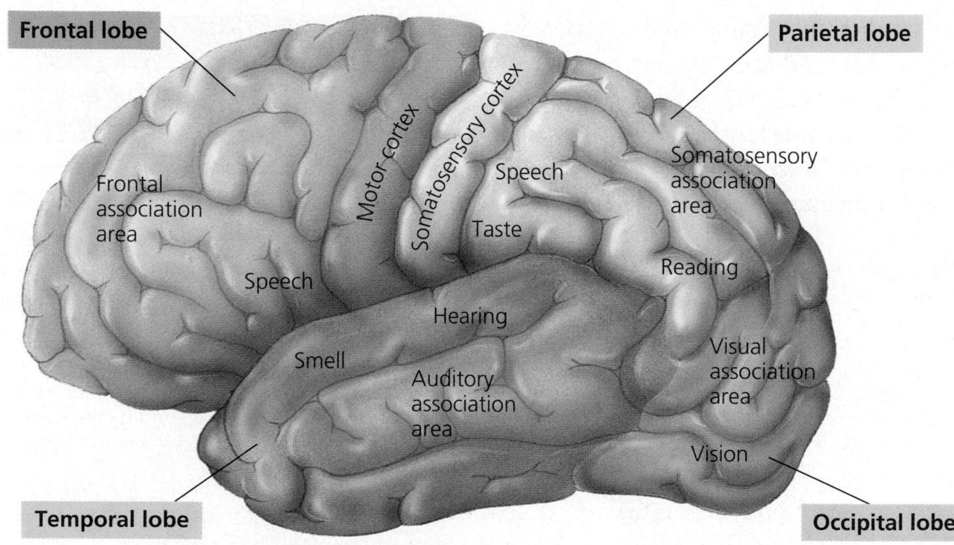

▲ **Figure 48.27 The human cerebral cortex.** Each side of the cerebral cortex is divided into four lobes, and each lobe has specialized functions. Some of the association areas on the left side (shown here) have different functions than those on the right side.

Concept Check 48.5

1. Which division of your autonomic nervous system would most likely be activated if you were to come to class and learn that an exam you had forgotten about were scheduled for that day? Explain.
2. List at least three functions of the medulla oblongata.
3. Compare the roles of the reticular formation and thalamus in the transmission of sensory information to the cerebrum.

For suggested answers, see Appendix A.

Concept 48.6

The cerebral cortex controls voluntary movement and cognitive functions

Each side of the cerebral cortex is customarily described as having four lobes, called the frontal, temporal, occipital, and parietal lobes. Researchers have identified a number of functional areas within each lobe **(Figure 48.27)**. These areas include *primary sensory areas*, each of which receives and processes a specific type of sensory information, and *association areas*, which integrate the information from various parts of the brain.

The major increase in the size of the neocortex that occurred during mammalian evolution was mostly an expansion of the association areas that integrate higher cognitive functions and make more complex behavior and learning possible. Whereas a rat's neocortex is occupied mainly by primary sensory areas, the human neocortex consists largely of association areas.

Information Processing in the Cerebral Cortex

Most sensory information coming into the cortex is directed via the thalamus to primary sensory areas within the lobes: visual information to the occipital lobe; auditory input to the temporal lobe; and somatosensory information about touch, pain, pressure, temperature, and the position of muscles and limbs to the parietal lobe (see Figure 48.27). Information about taste goes to a separate sensory region of the parietal lobe. Olfactory information is sent first to "primitive" regions of the cortex (that is, regions that are similar in mammals and reptiles) and then via the thalamus to an interior part of the frontal lobe.

The primary sensory areas send information to nearby association areas that can process particular features in the sensory input. In the primary visual cortex, for example, some neurons are sensitive to bars of light that have a certain width and orientation. Information related to such features is combined in association areas dedicated to recognizing complex images, such as faces.

Based on the integrated sensory information, the cerebral cortex may generate motor commands that cause specific behaviors—moving a limb or saying hello, for example. These commands consist of action potentials produced by neurons in the primary motor cortex, which lies at the rear of the frontal lobe, adjacent to the primary somatosensory cortex (see Figure 48.27). The action potentials travel along axons to the brainstem and spinal cord, where they excite motor neurons, which in turn excite skeletal muscle cells.

In both the somatosensory cortex and the motor cortex, neurons are distributed in an orderly fashion according to the part of the body that generates the sensory input or receives the motor commands **(Figure 48.28)**. For example, neurons that process sensory information from the legs and feet are located in the region of the somatosensory cortex that lies closest to the midline. Neurons that control muscles in the legs and feet are located in the corresponding region of the motor cortex. Notice in Figure 48.28 that the cortical surface area devoted to each body part is not related to the size of that part. Instead, it is related to the number of sensory neurons that innervate that part (for the somatosensory cortex) or to the amount of skill needed to control muscles in that part (for the motor cortex). Thus, the surface area devoted to the face is much larger than that devoted to the trunk.

Lateralization of Cortical Function

During brain development after birth, competing functions segregate and displace each other in the cortex of the left and right cerebral hemispheres, resulting in **lateralization** of functions. The left hemisphere becomes more adept at language, math, logical operations, and the serial processing of sequences of information. It has a bias for the detailed, speed-optimized activities required for skeletal muscle control and the processing of fine visual and auditory details. The right hemisphere is stronger at pattern recognition, face recognition, spatial relations, nonverbal thinking, emotional processing in general, and the simultaneous processing of many kinds of information. Understanding and generating the stress and intonation patterns of speech that convey its emotional content emphasize right-hemisphere function, as does music. The right hemisphere appears to specialize in perceiving the relationship between images and the whole context in which they occur, whereas the left hemisphere is better at focused perception. Most right-handed people use their left hand

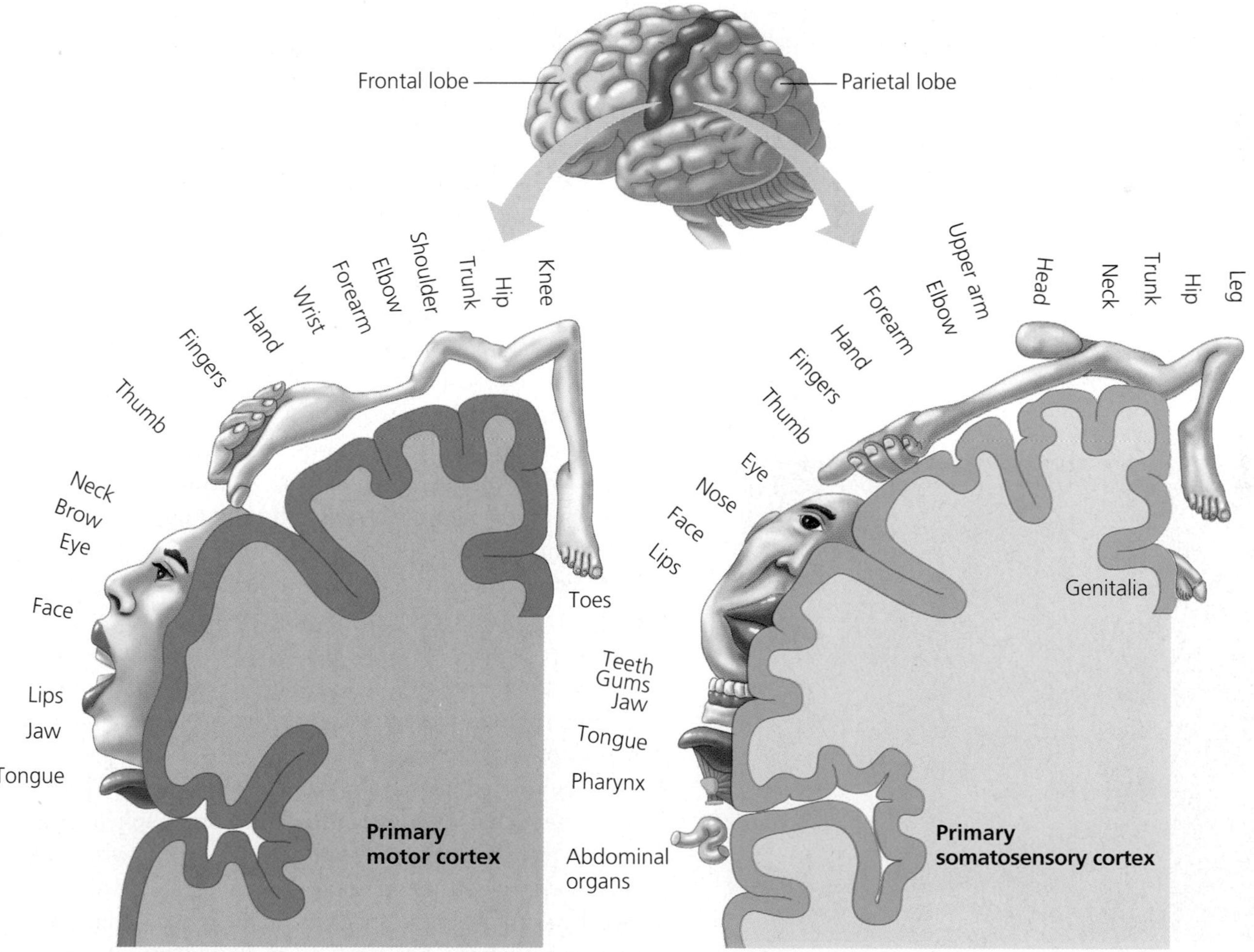

▲ **Figure 48.28 Body representations in the primary motor and primary somatosensory cortices.** In these cross-sectional maps of the cortices, the cortical surface area devoted to each body part is represented by the relative size of that part in the cartoons.

(controlled by the right hemisphere) for context or holding and use their right hand (controlled by the left hemisphere) for fine, detailed movement.

The two hemispheres normally work together harmoniously, trading information back and forth through the fibers of the corpus callosum. The importance of this exchange is revealed in patients who have had their corpus callosum severed to control their epileptic seizures. When such "split-brain" patients see a familiar word in their left field of vision, they cannot read the word because the sensory information that travels from the left field of vision to the right hemisphere cannot reach the language centers in the left hemisphere. Each hemisphere in these patients functions independently of the other.

Language and Speech

The systematic mapping of higher cognitive functions to specific brain areas began in the 19th century when physicians learned that damage to particular regions of the cortex by injuries, strokes, or tumors can produce distinctive changes in a person's behavior. The French physician Pierre Broca conducted postmortem examinations of patients who could understand language but could not speak. He discovered that many of these patients had defects in a small region of the left frontal lobe. That region, now known as *Broca's area,* is located in front of the part of the primary motor cortex that controls muscles in the face. The German physician Karl Wernicke also conducted examinations and found that damage to a posterior portion of the temporal lobe, now called *Wernicke's area,* abolished the ability to comprehend speech but left speech generation intact. Over a century later, studies of brain activity using fMRI and positron-emission tomography (PET; see Chapter 2) have confirmed that Broca's area is active during the generation of speech (**Figure 48.29**, lower left image) and that Wernicke's area is active when speech is heard (Figure 48.29, upper left image).

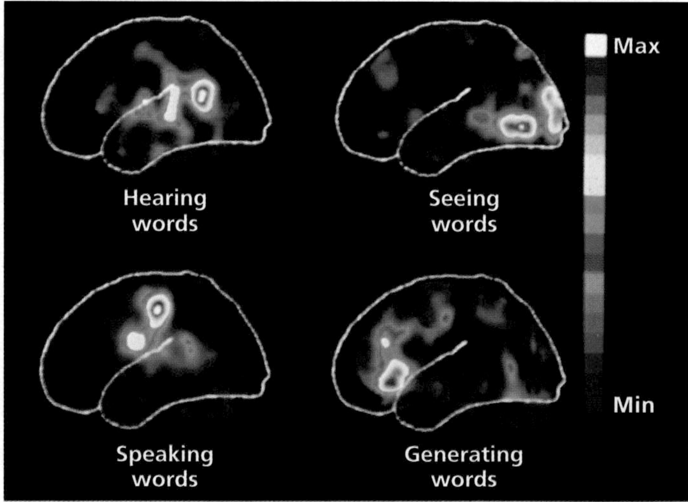

▲ **Figure 48.29 Mapping language areas in the cerebral cortex.** These PET images show regions with different activity levels in one person's brain during four activities, all related to speech.

Broca's area and Wernicke's area are part of a much larger network of brain regions involved in language. Reading a printed word without speaking activates the visual cortex (Figure 48.29, upper right image), whereas reading a printed word out loud activates both the visual cortex and Broca's area. Frontal and temporal areas become active when meaning must be attached to words, such as when a person generates verbs to go with nouns or groups related words or concepts (Figure 48.29, lower right image).

Emotions

Emotions are the result of a complex interplay of many regions of the brain. Prominent among these regions is the **limbic system,** a ring of structures around the brainstem (**Figure 48.30**). The limbic system includes three parts of the cerebral cortex—the amygdala, hippocampus, and olfactory bulb—along with some inner portions of the cortex's lobes and sections of the thalamus and hypothalamus. These structures interact with sensory areas of the neocortex and other higher brain centers, mediating primary emotions that manifest themselves in behaviors such as laughing and crying. It also attaches emotional "feelings" to basic, survival-related functions controlled by the brainstem, including aggression, feeding, and sexuality. Moreover, the limbic system is central to some of the behaviors—such as extended nurturing of infants and emotional bonding to other individuals—that distinguish mammals from most reptiles and amphibians.

Structures of the limbic system form early in development and provide a foundation for the higher cognitive functions that appear later, during the development of neocortical areas. We are born with brain circuits prepared to recognize the crude elements of faces, to support bonding to a caregiver, to have visual and vocal interactions with that caregiver, and to express fear, distress, and anger. Learning and memory processes then build a history of the particular sensing and motor actions that are successful in obtaining warmth and food. We also begin very early to distinguish right from wrong—for example, by perceiving happy or angry facial or vocal expressions from a caregiver. The amygdala, a structure in the temporal lobe (see Figure 48.30), is central in recognizing the emotional content of facial expressions and laying down emotional memories.

This emotional memory system seems to appear earlier in development than the system that supports explicit recall of events, which requires the hippocampus. Adults who learn to avoid an aversive situation, such as an image that is always followed by a mild electric shock, remember the image and experience autonomic arousal—as measured by an increased heart rate or sweating—if the image is presented again. Some adults with damage to the hippocampus report that they do

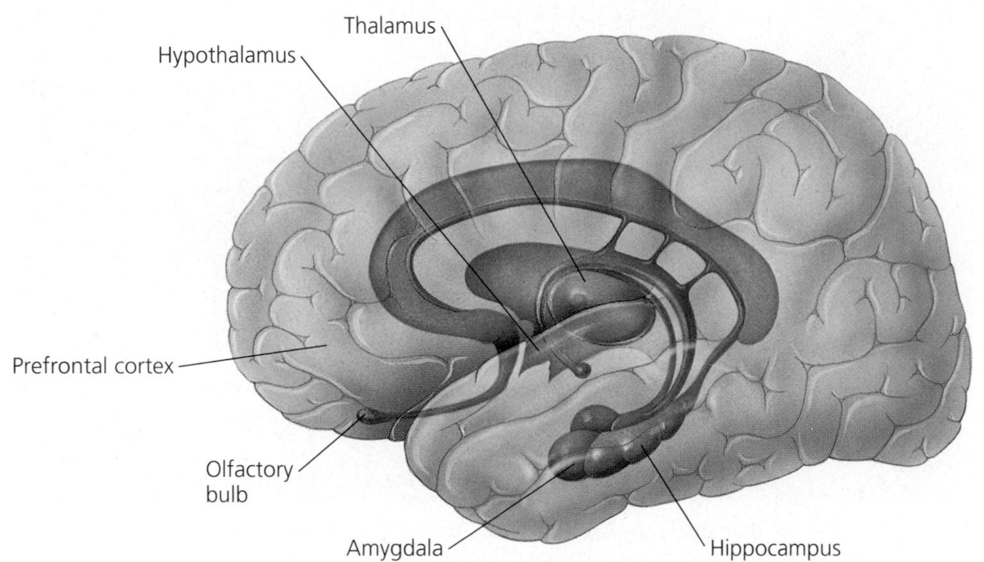

▲ **Figure 48.30 The limbic system.**

not recognize the image, but the autonomic arousal still occurs because the emotional memory mediated by the amygdala is still intact. Conversely, other patients with damage confined to the amygdala do not exhibit autonomic arousal, but they recall the image because explicit memory mediated by the hippocampus is intact.

As children develop, primary emotions such as pleasure and fear are associated with different situations in a process that requires portions of the neocortex, especially the forwardmost part of the frontal lobes, called the prefrontal cortex. The remarkable medical case of Phineas Gage reveals just how important this emotional learning can be. In 1848, Gage was working on a railroad construction site when a dynamite explosion drove a meter-long iron rod through his head. The rod entered his skull just below his left eye and exited through the top of his head. Astonishingly, Gage recovered, but his personality was drastically changed. Once an efficient, capable foreman, he had become impatient, profane, and capricious. In the late 1990s, neuroscientists studied Gage's skull and determined that the rod had destroyed portions of his frontal lobe known to mediate emotions. Today, patients who are diagnosed with tumors or lesions in these regions sometimes suffer the same combination of symptoms. Their intellect and memory seem intact, but their motivation, foresight, goal formation, and decision making are flawed. Their emotions and feelings also are diminished.

Frontal lobotomy, a surgical procedure in which the connection between the prefrontal cortex and the limbic system is disrupted, was once widely performed to treat severe emotional disorders. However, the resulting docility in the patient was usually accompanied by a loss of the ability to concentrate, plan, and work toward goals. Therefore, drug therapy has replaced frontal lobotomy for treating such severely ill individuals.

Memory and Learning

Though we may not be aware of it, we are constantly checking what is happening right now against what just happened a few moments ago. We hold information, anticipation, or goals for a time in **short-term memory** locations in the frontal lobes and then release them if they become irrelevant. Should we wish to retain knowledge of a face or a phone number, the mechanisms of **long-term memory** are activated in a process that requires the hippocampus. If we later need to recall a name or number, we can fetch it from long-term memory and return it to short-term memory. The transfer of information from short-term to long-term memory is enhanced by rehearsal ("practice makes perfect"), positive or negative emotional states mediated by the amygdala, and the association of new data with data previously learned and stored in long-term memory. For example, it's easier to learn a new card game if you already have "card sense" from playing other games.

Many sensory and motor association areas of the cerebral cortex outside Broca's area and Wernicke's area are involved in storing and retrieving words and images. Studies of patients with brain lesions and imaging studies on unimpaired subjects suggest, for example, that recognition of people is associated with the anterior part of the left temporal lobe, recognition of animals with the lower middle part of this lobe, and recognition of tools with the lower posterior part.

The memorization of phone numbers, facts, and places—which can be very rapid and may require only one exposure to the relevant item—may rely mainly on rapid changes in the strength of existing neural connections. In contrast, the slow learning and remembering of skills and procedures, such as those required to improve one's tennis game, appear to involve cellular mechanisms very similar to ones responsible for brain growth and development. In such cases, neurons actually make new connections.

Motor skills, such as walking, tying your shoes, riding a bicycle, or writing, are usually learned by repetition. You can perform such skills without consciously recalling the individual steps required to do these tasks correctly. Once a skill memory is learned, it is difficult to unlearn. For example, a person who has played tennis for years with a self-taught, awkward backhand has a much tougher time learning the correct form than a beginner just learning the game. Habits, whether good or bad, are hard to break. Our gait, gestures, and accent are individual, and we are very aware of these traits in other people when we interact with them.

Cellular Mechanisms of Learning

Nobel laureate Eric Kandel and his colleagues at Columbia University have studied the cellular basis of learning using an animal with a small, experimentally accessible nervous system, the sea hare (*Aplysia californica*). The researchers were able to explain the mechanism of simple forms of learning in this mollusc in terms of changes in the strength of synaptic transmission between specific sensory and motor neurons. **Figure 48.31** describes one of their experiments.

In the vertebrate brain, a form of learning called **long-term potentiation (LTP)** involves an increase in the strength of synaptic transmission that occurs when presynaptic neurons produce a brief, high-frequency series of action potentials. Because LTP can last for days or weeks, it may be a fundamental process by which memories are stored or learning takes place. The cellular mechanism of LTP has been studied most thoroughly at synapses in the hippocampus, where presynaptic neurons release the excitatory neurotransmitter glutamate **(Figure 48.32)**. The postsynaptic neurons possess two types of glutamate receptors: AMPA receptors and NMDA receptors. AMPA receptors are part of ligand-gated ion channels; when glutamate binds to them, Na^+ and K^+ diffuse through the channels, and the postsynaptic membrane depolarizes. NMDA receptors are part of channels that are both ligand gated and voltage gated: The channels open only if glutamate is bound *and* the membrane is depolarized. As detailed in Figure 48.32, the binding of glutamate to these two types of receptors can lead to LTP through changes in both the presynaptic and the postsynaptic neurons.

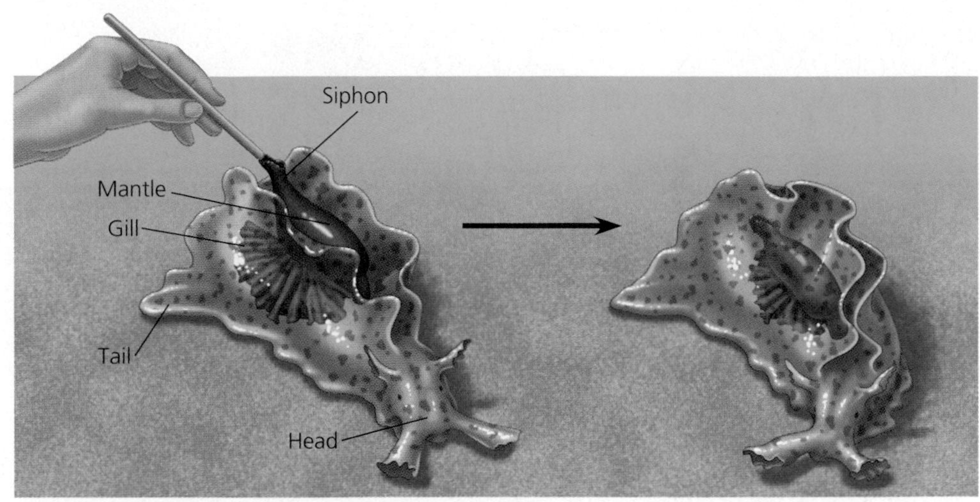

(a) Touching the siphon triggers a reflex that causes the gill to withdraw. If the tail is shocked just before the siphon is touched, the withdrawal reflex is stronger. This strengthening of the reflex is a simple form of learning called sensitization.

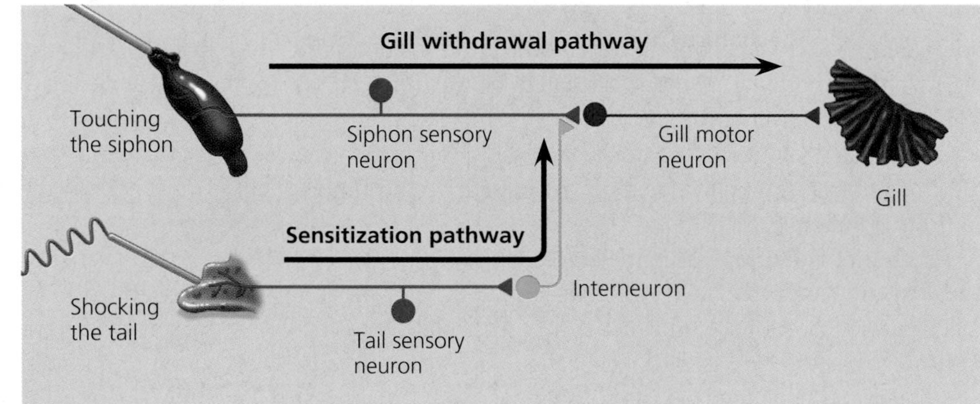

(b) Sensitization involves interneurons that make synapses on the *synaptic terminals* of the siphon sensory neurons. When the tail is shocked, the interneurons release serotonin, which activates a signal transduction pathway that closes K^+ channels in the synaptic terminals of the siphon sensory neurons. As a result, action potentials in the siphon sensory neurons produce a prolonged depolarization of the terminals. That allows more Ca^{2+} to diffuse into the terminals, which causes the terminals to release more of their excitatory neurotransmitter onto the gill motor neurons. In response, the motor neurons generate action potentials at a higher frequency, producing a more forceful gill withdrawal.

▲ **Figure 48.31 Sensitization in the sea hare (*Aplysia californica*).**

Consciousness

The study of human consciousness was long considered outside the province of hard science, more appropriate as a subject for philosophy or religion. One reason for this view is that consciousness is both broad—encompassing our awareness of ourselves and of our own experiences—and subjective.

Over the past few decades, however, neuroscientists have begun studying consciousness using brain-imaging techniques such as fMRI (see Figure 48.1). It is now possible to compare activity in the human brain during different states of consciousness—for example, before and after a person is aware of seeing an object. These imaging techniques can also be used to compare the conscious and unconscious processing of sensory information. Such studies do not pinpoint a "consciousness center" in the brain; rather, they offer an increasingly detailed picture of how neural activity correlates with conscious experiences. There is a growing consensus among neuroscientists that consciousness is an emergent property of the brain (see Chapter 1), one that recruits activities in

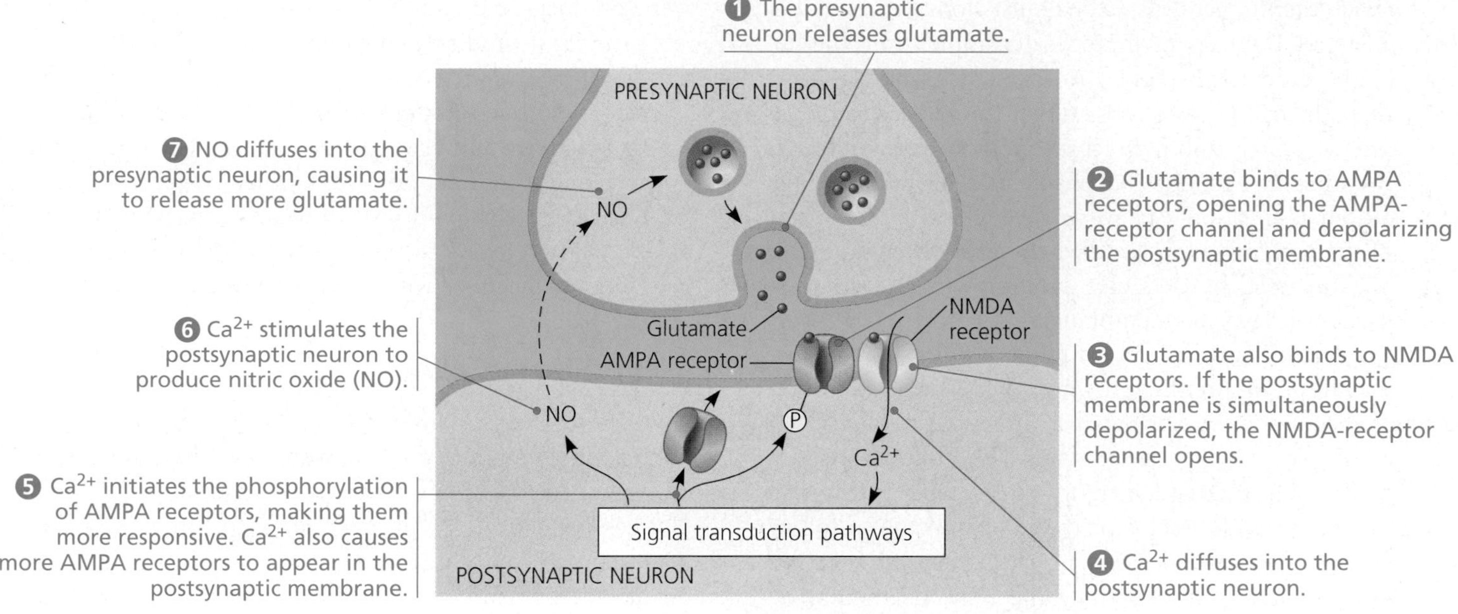
1 The presynaptic neuron releases glutamate.
PRESYNAPTIC NEURON
7 NO diffuses into the presynaptic neuron, causing it to release more glutamate.
NO
2 Glutamate binds to AMPA receptors, opening the AMPA-receptor channel and depolarizing the postsynaptic membrane.
NMDA receptor
Glutamate
AMPA receptor
6 Ca^{2+} stimulates the postsynaptic neuron to produce nitric oxide (NO).
3 Glutamate also binds to NMDA receptors. If the postsynaptic membrane is simultaneously depolarized, the NMDA-receptor channel opens.
NO
P
Ca^{2+}
5 Ca^{2+} initiates the phosphorylation of AMPA receptors, making them more responsive. Ca^{2+} also causes more AMPA receptors to appear in the postsynaptic membrane.
Signal transduction pathways
POSTSYNAPTIC NEURON
4 Ca^{2+} diffuses into the postsynaptic neuron.

▲ **Figure 48.32 Mechanism of long-term potentiation in the vertebrate brain.**

many areas of the cerebral cortex. Several models postulate the existence of a sort of "scanning mechanism" that repetitively sweeps across the brain, integrating widespread activity into a unified, conscious moment.

Still, a well-supported theory of consciousness may have to wait until brain-imaging technology becomes more sophisticated. Analyzing dynamic patterns of activity that span the entire brain may reveal that these patterns bear no more direct relationship to individual nerve cells than do hurricanes to their constituent water molecules.

Concept Check 48.6

1. Based on Figure 48.28, what can you infer about the relative number of sensory neurons that innervate the hand and the neck? Explain.
2. If a man with a severed corpus callosum were asked to view a photograph of a familiar face, first in his left field of vision and then in his right field, why would it be difficult for him to put a name to the face in either field?
3. Two brain areas important in speech are Broca's area and Wernicke's area. How is the function of each area related to the area's location in the cerebral cortex?
4. Explain how the binding of glutamate to NMDA receptors in the hippocampus exhibits properties of both direct and indirect synaptic transmission.

For suggested answers, see Appendix A.

Concept 48.7

CNS injuries and diseases are the focus of much research

Unlike the PNS, the mammalian CNS cannot fully repair itself when damaged or assaulted by disease. Surviving neurons in the brain can make new connections and thus sometimes compensate for damage, as in the remarkable recoveries of some stroke victims. Generally speaking, however, brain and spinal cord injuries, strokes, and diseases that destroy CNS neurons, such as Alzheimer's disease and Parkinson's disease, have devastating effects. Current research on nerve cell development and the discovery of neural stem cells enhance our fundamental knowledge of the nervous system and may one day make it possible for physicians to repair or replace damaged neurons.

Nerve Cell Development

Among the key questions in neurobiology are how certain cells in a developing animal differentiate into neurons and how neurons migrate to their proper locations, grow axons to specific places, and make synapses with the correct postsynaptic cells (target cells). The labs of Corey Goodman (University of California, Berkeley) and Marc Tessier-Lavigne (University of California, San Francisco) have been studying how neurons "find their way" during development of the CNS. Their work combines elements of cell-to-cell communication (see Chapter 11), control of gene expression (see Chapter 19), and the genetic basis of development (see Chapter 21).

To reach their target cells, axons must elongate from a few micrometers to a meter or more (for example, from the human spinal cord to the foot). An axon does not follow a straight path to its target cells; rather, molecular signposts along the way direct and redirect the growing axon in a series of mid-course corrections that result in a meandering, but not random, elongation. The responsive region at the leading edge of the growing axon is called the **growth cone.** Signal molecules released by cells along the growth route bind to receptors on the plasma membrane of the growth cone, triggering a signal transduction pathway **(Figure 48.33)**. The axon may respond by either growing toward the source of the signal molecules (attraction) or away from it (repulsion). Cell adhesion molecules on the axon's growth cone also play a role; they attach to complementary molecules on surrounding cells that provide tracks for the growing axon to follow. Nerve growth factor released by astrocytes and growth-promoting proteins produced by the neurons themselves contribute to the process by stimulating axonal elongation. The growing axon expresses different genes at different times during its development, and it is influenced by surrounding cells that it moves away from. This complex process has been conserved during millions of years of evolution, for the genes, gene products, and mechanism of axon guidance are remarkably similar in studied nematode worms (*C. elegans*), insects (*Drosophila*), and vertebrates.

Work continues on deciphering the mechanisms of axonal growth, with the ultimate goal of repairing CNS damage. Using the right combination of attractants, repellants, growth-associated proteins, and growth factors, researchers hope to coax damaged axons to regrow, follow the correct pathway, and form connections with the correct target cells.

Neural Stem Cells

Until 1998, it was "common knowledge" that you were born with all the brain neurons you would ever have. In that year, however, Fred Gage from the Salk Institute for Biological Studies in California and Peter Ericksson from Sahlgrenska University Hospital in Sweden made a startling announcement: The human brain *does* produce new neurons in adulthood. Ericksson was working in Gage's lab, where researchers injected mice with the marker bromodeoxyuridine (BrdU) to label the DNA of dividing cells. After Ericksson returned to Sweden, he learned that a group of terminally ill cancer patients were receiving BrdU as part of a study to monitor tumor

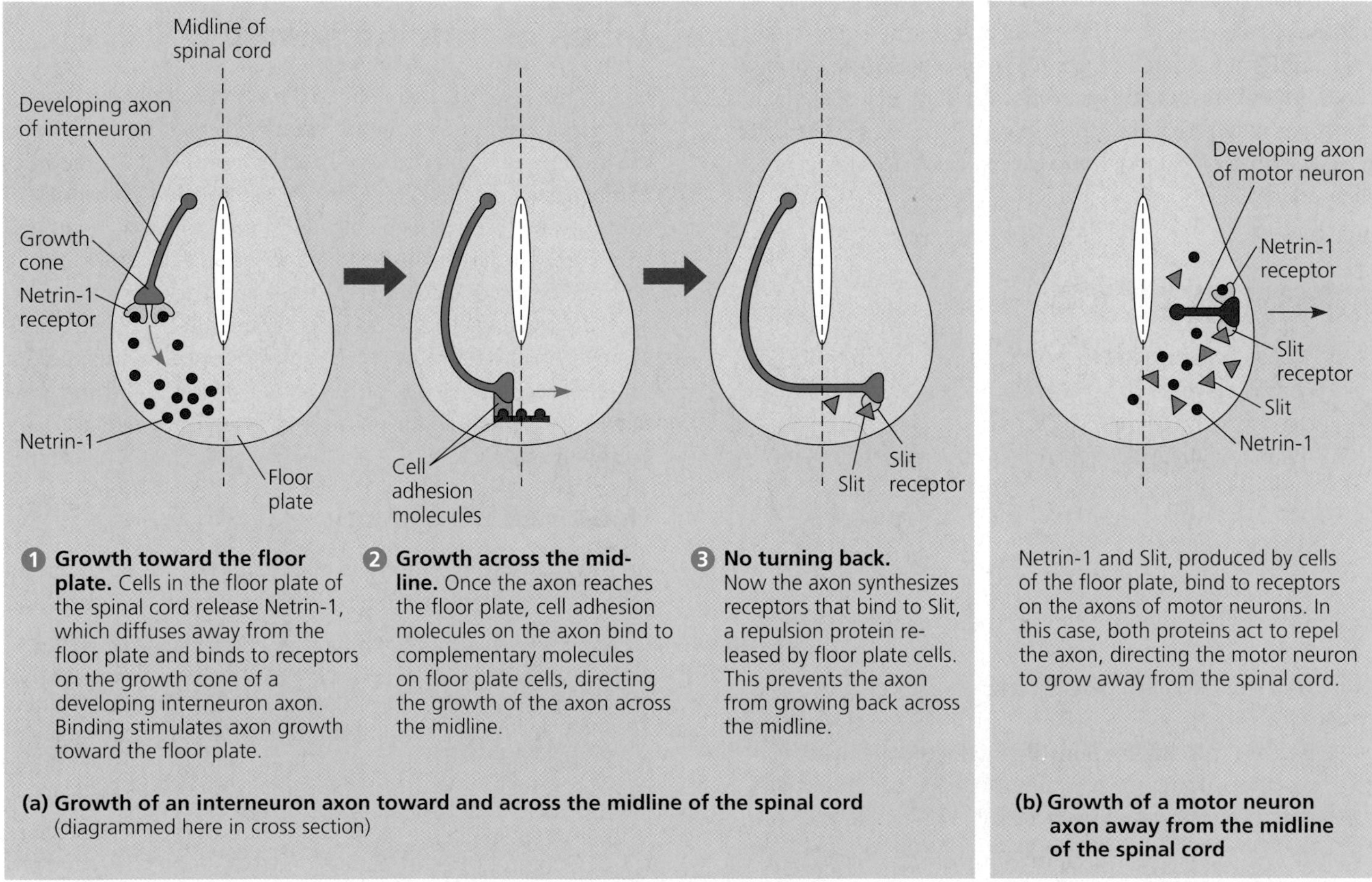

Midline of spinal cord
Developing axon of interneuron
Growth cone
Netrin-1 receptor
Netrin-1
Floor plate
Cell adhesion molecules
Slit
Slit receptor
Developing axon of motor neuron
Netrin-1 receptor
Slit receptor
Slit
Netrin-1
1 Growth toward the floor plate. Cells in the floor plate of the spinal cord release Netrin-1, which diffuses away from the floor plate and binds to receptors on the growth cone of a developing interneuron axon. Binding stimulates axon growth toward the floor plate.
2 Growth across the midline. Once the axon reaches the floor plate, cell adhesion molecules on the axon bind to complementary molecules on floor plate cells, directing the growth of the axon across the midline.
3 No turning back. Now the axon synthesizes receptors that bind to Slit, a repulsion protein released by floor plate cells. This prevents the axon from growing back across the midline.
Netrin-1 and Slit, produced by cells of the floor plate, bind to receptors on the axons of motor neurons. In this case, both proteins act to repel the axon, directing the motor neuron to grow away from the spinal cord.
(a) Growth of an interneuron axon toward and across the midline of the spinal cord (diagrammed here in cross section)
(b) Growth of a motor neuron axon away from the midline of the spinal cord

▲ **Figure 48.33 Molecular signals direct the growth of developing axons.**

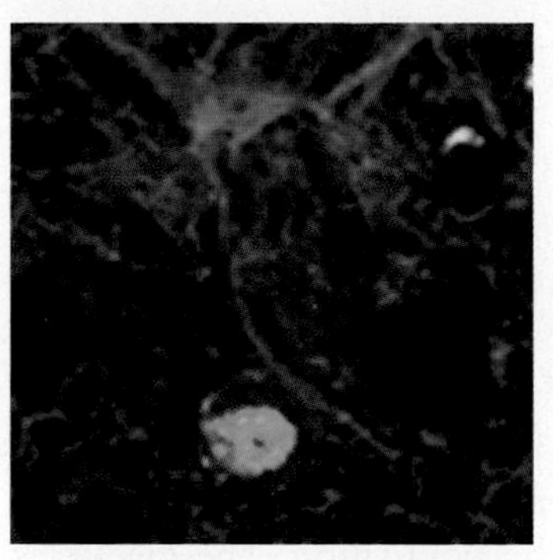

◀ **Figure 48.34 A newly born neuron in the hippocampus of a human adult.** All the red cells in this LM are neurons. The cell that is both red and green is a neuron that has incorporated BrdU, indicating that it resulted from a recent cell division.

growth. The patients agreed to donate their brains for research upon their death. In a postmortem study, Ericksson found newly divided neurons in the hippocampus of all the patients **(Figure 48.34)**.

It is not clear what function these new cells play in the human brain. However, mice that live in stimulating environments and run on exercise wheels have more new neurons in their hippocampus and perform better on learning tasks than genetically identical caged mice that receive little stimulation.

Mature neurons, with their extensive processes and intricate connections with other cells, clearly are not able to undergo cell division. Therefore, the new brain neurons must have come from stem cells. Recall from Chapter 21 that stem cells are relatively unspecialized cells that continually divide. While some of their progeny remain undifferentiated, others may differentiate into specialized cells under the right conditions.

One of the difficulties of conducting research on stem cells is finding a source of human stem cells. Embryonic stem cells can be obtained from embryos produced through *in vitro* fertilization (see Chapter 46), but because harvesting these cells requires destroying the embryo, various ethical and political issues surround their use. Certain adult tissues, such as bone marrow (see Figure 42.16), also have stem cells, but those cells may be less developmentally versatile than embryonic stem cells.

In 2001, Gage and his colleagues announced that they had cultured neural progenitor cells from the brains of recently deceased individuals and surgical tissue samples. The term *progenitor* refers to the fact that these stem cells are committed to becoming either neurons or glia. In culture, the cells divided 30 to 70 times and differentiated into neurons and astrocytes. One of the goals of further research is to find a way to induce the body's own neural progenitor cells to differentiate into specific types of neurons or glia when and where they are needed. Another goal is to transplant cultured neural progenitor cells into the damaged CNS.

Diseases and Disorders of the Nervous System

Mental illnesses, including schizophrenia and depression, and neurological disorders such as Alzheimer's disease and Parkinson's disease take an enormous toll on society, in both the patient's potential loss of a productive life and the high cost of long-term medical care. They also severely disrupt the lives of patients' families. Finding proper treatments and cures is a research priority.

For many years, the only treatment for patients with mental illness was to sequester them in institutions, where most of them spent the rest of their lives. Today, with the advent of drugs that can treat (though not yet cure) many of these conditions, the average stay in a mental hospital is about two or three weeks. Attitudes toward the mentally ill are slowly changing as people become more aware that mental illnesses are caused by chemical or anatomical changes in the brain.

Schizophrenia

About 1% of the world's population suffers from **schizophrenia**, a severe mental disturbance characterized by psychotic episodes in which patients lose the ability to distinguish reality. The symptoms of schizophrenia typically include hallucinations (most often "voices" telling the patient that they are worthless and evil) and delusions (generally paranoid), as well as blunted emotions, distractibility, lack of initiative, and poverty of speech. Despite the commonly held belief, schizophrenics do not necessarily exhibit a split personality. There seem to be several different forms of schizophrenia, and it is unclear whether they represent different disorders or variations of the same underlying disease.

The cause of schizophrenia is unknown, although the disease has a strong genetic component. Studies of identical twins show that if one twin has schizophrenia, there is a 50% chance that the other twin has it, too. Since identical twins have identical genes, the fact that this likelihood is not 100% implies that schizophrenia must have an equally strong environmental component, which is unidentified. Now that the human genome has been sequenced, there is a vigorous effort under way to find the mutant genes that predispose a person to the disease. This effort includes DNA sequencing of individuals in families with a high incidence of schizophrenia. Multiple genes must be involved because the inheritance does not follow the Mendelian pattern expected for a single-gene mutation.

Treatments for schizophrenia have focused mostly on brain pathways that use dopamine as a neurotransmitter. Two lines of evidence support this protocol. First, amphetamine ("speed"), which stimulates dopamine release, can produce symptoms indistinguishable from those of schizophrenia. Second, many of the drugs that alleviate the symptoms block dopamine receptors, especially a subform called the D_2 receptor. However, additional neurotransmitters may also be involved because other drugs successful in treating schizophrenia, called atypical antipsychotics, block the D_2 receptor only very weakly but have stronger effects on serotonin and/or norepinephrine receptors. Furthermore, there are indications that glutamate receptors may play a role in schizophrenia: The street drug "angel dust," or PCP, blocks the NMDA type of glutamate receptor (but not dopamine receptors) and induces strong schizophrenia-like symptoms.

Despite their ability to alleviate the major symptoms, many of the existing drugs have such negative side effects that patients frequently stop taking them. For example, the drugs often produce motor deficits that resemble those of Parkinson's disease. In about 25% of cases, chronic drug therapy produces a new and frequently irreversible condition called tardive dyskinesia, in which the patient has uncontrolled facial writhing movements. The atypical antipsychotics mostly avoid such side effects, but some of these drugs produce a blood disorder in certain individuals. Identification of the genetic mutations responsible for schizophrenia may yield new insights about the causes of the disease, which may in turn lead to new therapies.

Depression

Two broad forms of depressive illness are known: bipolar disorder and major depression. **Bipolar disorder,** or manic-depressive disorder, involves swings of mood from high to low and affects about 1% of the world's population. In contrast, people with **major depression** have a low mood most of the time; they constitute roughly 5% of the population.

In bipolar disorder, the manic phase is characterized by high self-esteem, increased energy, a flow of ideas, and overtalkativeness, as well as behaviors that often bring disaster, such as increased risk taking, promiscuity, and reckless spending. In its milder forms, this phase is sometimes associated with great creativity, and some well-known artists, musicians, and literary figures (including Keats, Tolstoy, Hemingway, and Schumann, to name a few) have had periods of intense output during manic phases. The depressive phase comes with lowered ability to feel pleasure, loss of interest, sleep disturbances, and feelings of worthlessness. Some people with bipolar disorder even attempt to commit suicide during the depressive phase. Nevertheless, some patients prefer to endure the pain of the depressive phase rather than take medication and risk losing the enhanced creative output of their manic phase.

Both bipolar disorder and major depression have a genetic component, as identical twins have about a 50% chance of sharing this mental illness. As with schizophrenia, this percentage means that there is also a strong environmental component, and there are indications that stress, especially severe stress in childhood, may be an important factor.

Several treatments for depression are available, some of which were identified serendipitously from the mood changes produced as side effects of drugs or procedures aimed at other conditions, including high blood pressure and tuberculosis. Many drugs effective against depression, including Prozac, increase the activity of biogenic amines in the brain. The mechanisms of various other treatments, such as electroconvulsive therapy, lithium administration, and talk therapy, are not well understood.

Alzheimer's Disease

Alzheimer's disease (AD) is a mental deterioration, or dementia, characterized by confusion, memory loss, and a variety of other symptoms. Its incidence is age related, rising from about 10% at age 65 to about 35% at age 85. Thus, by helping humans live longer, modern medicine is increasing the proportion of AD patients in the population. The disease is progressive, with patients gradually becoming less able to function and eventually needing to be dressed, bathed, and fed by others. There are also personality changes, almost always for the worse. Patients often lose their ability to recognize people, including their immediate family, and may act toward them with suspicion and hostility.

At present, a firm diagnosis of AD is difficult to make while the patient is alive because AD is one of several forms of dementia. However, AD results in a characteristic pathology of the brain: Neurons die in huge areas of the brain, often leading to massive shrinkage of brain tissue. This shrinkage can be seen with brain imaging but is not enough to positively identify AD. What is diagnostic is the postmortem finding of two features—neurofibrillary tangles and senile plaques—in remaining brain tissue **(Figure 48.35)**. Neurofibrillary tangles are bundles of degenerated neuronal and glial processes. Senile plaques are aggregates of β-amyloid, an insoluble peptide that is cleaved from a membrane protein normally present in neurons. The role of the uncleaved protein is not known. Membrane enzymes, called secretases, catalyze the cleavage, causing β-amyloid to accumulate outside the neurons and to aggregate in the form of plaques. The plaques appear to trigger the death of surrounding neurons.

There has been an enormous effort devoted to developing a treatment for AD. In 2004, a team of researchers at Northwestern University used genetic engineering to eliminate one

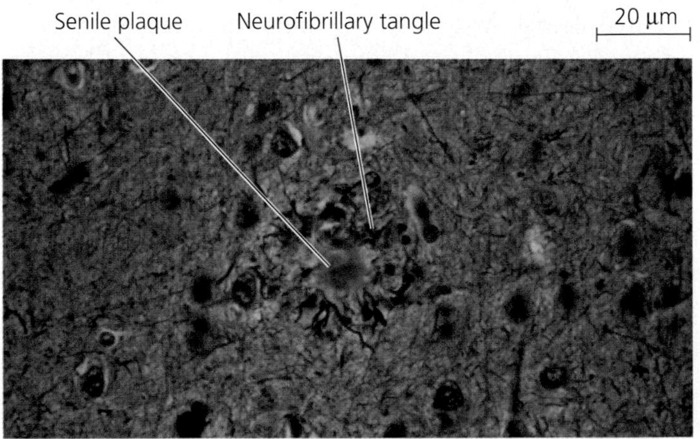

▲ **Figure 48.35 Microscopic signs of Alzheimer's disease.** A hallmark of AD is the presence in brain tissue of neurofibrillary tangles surrounding senile plaques made of β-amyloid (LM).

of the secretases in a strain of mice prone to AD. The genetically engineered mice accumulated much less β-amyloid and did not experience the age-related memory deficits typical of mice in that strain. A successful treatment for AD in humans may hinge on the early detection of senile plaques, which commonly form before overt symptoms of the disease appear. One step in this direction may be a chemical nicknamed Pittsburgh Compound-B (PIB), which was synthesized by scientists at the University of Pittsburgh in 2004. The scientists linked PIB to a short-lived radioactive isotope and used a PET scanner to track the distribution of PIB in the brains of AD patients. They found that PIB accumulated selectively in brain regions that contain large amounts of β-amyloid deposits. PIB may be useful in testing the effectiveness of drugs aimed at reducing the rate of production of β-amyloid or of dispersing already-formed plaques. Several such drugs are currently in clinical trials.

Parkinson's Disease

Approximately 1 million people in the United States suffer from **Parkinson's disease,** a motor disorder characterized by difficulty in initiating movements, slowness of movement, and rigidity. Patients often have a masked facial expression, muscle tremors, poor balance, a flexed posture, and a shuffling gait. Like Alzheimer's disease, Parkinson's disease is a progressive brain illness whose risk increases with advancing age. The incidence of Parkinson's disease is about 1% at age 65 and about 5% at age 85.

The symptoms of Parkinson's disease result from the death of neurons in a midbrain nucleus called the substantia nigra. These neurons normally release dopamine from their synaptic terminals in the basal nuclei. The degeneration of dopamine neurons is associated with the accumulation of protein aggregates containing α-synuclein, a protein typically found in presynaptic nerve terminals.

Most cases of Parkinson's disease lack a clearly identifiable cause. However, the consensus among scientists who study the disease is that it results from a combination of environmental and genetic factors. For example, some families with an increased incidence of Parkinson's disease have a mutated form of the gene for α-synuclein. One hypothesis is that changes in the structure of α-synuclein cause dopamine neurons in the substantia nigra to become susceptible to oxidative cell damage.

At present there is no cure for Parkinson's disease, although various approaches are used to manage the symptoms, including brain surgery, deep-brain stimulation, and drugs such as L-dopa, a dopamine precursor that can cross the blood-brain barrier. One potential cure is to implant dopamine-secreting neurons, either in the substantia nigra or in the basal ganglia. Embryonic stem cells can be stimulated or genetically engineered to develop into dopamine-secreting neurons, and transplantation of these cells into rats with an experimentally induced condition that mimics Parkinson's disease has led to a recovery of motor control. Whether this kind of regenerative medicine will also work in humans is one of many important questions on the frontier of modern brain research.

Concept Check 48.7

1. Based on Figure 48.33, what would be the likely effect of a mutation preventing spinal cord interneurons from having Netrin-1 receptors? What might result from interneurons lacking Slit receptors?
2. What evidence indicates that schizophrenia, bipolar disorder, and major depression have both genetic and environmental components?
3. What are some similarities between Alzheimer's disease and Parkinson's disease?

For suggested answers, see Appendix A.

Chapter 48 Review

Go to www.campbellbiology.com or the student CD-ROM to explore Activities, Investigations, and other interactive study aids.

SUMMARY OF KEY CONCEPTS

Nervous systems consist of circuits of neurons and supporting cells

▶ **Organization of Nervous Systems (pp. 1012–1013)** Invertebrate nervous systems range in complexity from simple nerve nets to highly centralized nervous systems having complicated brains and ventral nerve cords. In vertebrates, the central nervous system (CNS) consists of the brain and the spinal cord, which is located dorsally.

▶ **Information Processing (p. 1013)** Nervous systems process information in three stages: sensory input, integration, and motor output to effector cells. The CNS integrates information, while the nerves of the peripheral nervous system (PNS) transmit sensory and motor signals between the CNS and the rest of the body. The three stages are illustrated in the knee-jerk reflex.

▶ **Neuron Structure (pp. 1013–1014)** Most neurons have highly branched dendrites that receive signals from other neurons. They also typically have a single axon that transmits signals to other

cells at synapses. Neurons have a wide variety of shapes that reflect their input and output interactions.
Activity *Neuron Structure*

▶ **Supporting Cells (Glia) (pp. 1014–1015)** Glia perform a number of functions, including providing structural support for neurons, regulating the extracellular concentrations of certain substances, guiding the migration of developing neurons, and forming myelin, which electrically insulates axons.

Ion pumps and ion channels maintain the resting potential of a neuron

▶ Every cell has a voltage across its plasma membrane called a membrane potential. The inside of the cell is negative relative to the outside (pp. 1015–1016).

▶ **The Resting Potential (pp. 1016–1017)** The membrane potential depends on ionic gradients across the plasma membrane: The concentration of Na^+ is higher in the extracellular fluid than in the cytosol, while the reverse is true for K^+. A neuron that is not transmitting signals contains many open K^+ channels and fewer open Na^+ channels in its plasma membrane. The diffusion of K^+ and Na^+ through these channels leads to the separation of charges across the membrane, producing the resting potential.

▶ **Gated Ion Channels (p. 1017)** Gated ion channels open or close in response to membrane stretch, the binding of a specific ligand, or a change in the membrane potential.

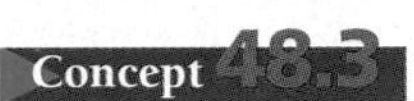

Action potentials are the signals conducted by axons

▶ An increase in the magnitude of the membrane potential is called a hyperpolarization; a decrease in magnitude is called a depolarization. Changes in membrane potential that vary with the strength of a stimulus are known as graded potentials (pp. 1017–1018).

▶ **Production of Action Potentials (pp. 1018–1020)** An action potential is a brief, all-or-none depolarization of a neuron's plasma membrane. When a graded depolarization brings the membrane potential to the threshold, many voltage-gated Na^+ channels open, triggering an influx of Na^+ that rapidly brings the membrane potential to a positive value. The membrane potential is restored to its normal resting value by the inactivation of Na^+ channels and by the opening of many voltage-gated K^+ channels, which increases K^+ efflux. A refractory period follows the action potential, corresponding to the interval when the Na^+ channels are inactivated.
Activity *Nerve Signals: Action Potentials*
Investigation *What Triggers Nerve Impulses?*

▶ **Conduction of Action Potentials (pp. 1020–1021)** An action potential travels from the axon hillock to the synaptic terminals by regenerating itself along the axon. The speed of conduction of an action potential increases with the diameter of the axon and, in many vertebrate axons, with myelination. Action potentials in myelinated axons jump between the nodes of Ranvier, a process called saltatory conduction.

Neurons communicate with other cells at synapses

▶ In an electrical synapse, electrical current flows directly from one cell to another via a gap junction. In a chemical synapse, depolarization of the synaptic terminal causes synaptic vesicles to fuse with the terminal membrane and to release neurotransmitter into the synaptic cleft (pp. 1021–1022).

▶ **Direct Synaptic Transmission (pp. 1022–1023)** The neurotransmitter binds to ligand-gated ion channels in the postsynaptic membrane, producing an excitatory or inhibitory postsynaptic potential (EPSP or IPSP). After release, the neurotransmitter diffuses out of the synaptic cleft, is taken up by surrounding cells, or is degraded by enzymes. A single neuron has many synapses on its dendrites and cell body. Whether it generates an action potential depends on the temporal and spatial summation of EPSPs and IPSPs at the axon hillock.
Activity *Signal Transmission at a Chemical Synapse*

▶ **Indirect Synaptic Transmission (pp. 1023–1024)** The binding of neurotransmitter to some receptors activates signal transduction pathways, which produce slowly developing but long-lasting effects in the postsynaptic cell.

▶ **Neurotransmitters (pp. 1024–1025)** The same neurotransmitter can produce different effects on different types of cells. Major known neurotransmitters include acetylcholine, biogenic amines (epinephrine, norepinephrine, dopamine, and serotonin), various amino acids and peptides, and the gases nitric oxide and carbon monoxide.

The vertebrate nervous system is regionally specialized

▶ **The Peripheral Nervous System (pp. 1026–1028)** The PNS consists of paired cranial and spinal nerves and associated ganglia. Functionally, the PNS is divided into the somatic nervous system, which carries signals to skeletal muscles, and the autonomic nervous system, which regulates the primarily automatic, visceral functions of smooth and cardiac muscles. The autonomic nervous system has three divisions: the sympathetic and parasympathetic divisions, which usually have antagonistic effects on target organs, and the enteric division, which controls the activity of the digestive tract, pancreas, and gallbladder.

▶ **Embryonic Development of the Brain (pp. 1028–1029)** The vertebrate brain develops from three embryonic regions: the forebrain, the midbrain, and the hindbrain. In humans, the most expansive growth occurs in the part of the forebrain that gives rise to the cerebrum.

▶ **The Brainstem (pp. 1029–1030)** The medulla oblongata, pons, and midbrain make up the brainstem, which controls homeostatic functions such as breathing rate, conducts sensory and motor signals between the spinal cord and higher brain centers, and regulates arousal and sleep.

▶ **The Cerebellum (p. 1030)** The cerebellum helps coordinate motor, perceptual, and cognitive functions. It also is involved in learning and remembering motor skills.

▶ **The Diencephalon (pp. 1030–1031)** The thalamus is the main center through which sensory and motor information passes to and from the cerebrum. The hypothalamus regulates homeostasis; basic survival behaviors such as feeding, fighting, fleeing, and reproducing; and circadian rhythms.

▶ **The Cerebrum (pp. 1031–1032)** The cerebrum has two hemispheres, each of which consists of cerebral cortex overlying white matter and basal nuclei, which are important in planning and learning movements. In mammals, the cerebral cortex has a convoluted surface called the neocortex. A thick band of axons, the corpus callosum, provides communication between the right and left cerebral cortices.

Concept 48.6

The cerebral cortex controls voluntary movement and cognitive functions

- Each side of the cerebral cortex has four lobes—frontal, temporal, occipital, and parietal—which contain primary sensory areas and association areas (p. 1032).
- **Information Processing in the Cerebral Cortex (pp. 1032–1033)** Specific types of sensory input enter the primary sensory areas. Adjacent association areas process particular features in the sensory input and integrate information from different sensory areas. In the somatosensory cortex and the motor cortex, neurons are distributed according to the part of the body that generates sensory input or receives motor commands.
- **Lateralization of Cortical Function (pp. 1033–1034)** The left hemisphere is normally specialized for high-speed serial information processing essential to language and logic operations. The right hemisphere is stronger at pattern recognition, nonverbal ideation, and emotional processing.
- **Language and Speech (p. 1034)** Portions of the frontal and temporal lobes, including Broca's area and Wernicke's area, are essential for generating and understanding language.
- **Emotions (pp. 1034–1035)** The limbic system, a ring of cortical and noncortical centers around the brainstem, mediates primary emotions and attaches emotional "feelings" to survival-related functions. The association of primary emotions with different situations during human development requires parts of the neocortex, especially the prefrontal cortex.
- **Memory and Learning (pp. 1035–1036)** The frontal lobes are a site of short-term memory and can interact with the hippocampus and amygdala in consolidating long-term memory. Experiments on invertebrates and vertebrates have revealed the cellular basis of some simple forms of learning, including sensitization and long-term potentiation.
- **Consciousness (pp. 1036–1037)** Modern brain-imaging techniques suggest that consciousness may be an emergent property of the brain based on activity in many areas of the cortex.

Concept 48.7

CNS injuries and diseases are the focus of much research

- **Nerve Cell Development (pp. 1037–1038)** Signal molecules direct an axon's growth by binding to receptors on the plasma membrane of the growth cone. The genes and basic events involved in axon guidance are similar in invertebrates and vertebrates. Knowledge of these events may be applied one day to stimulate axonal regrowth following CNS damage.
- **Neural Stem Cells (pp. 1038–1039)** The adult human brain contains stem cells that can differentiate into mature neurons. The induction of stem cell differentiation and transplantation of cultured stem cells are potential methods for replacing neurons lost to trauma or disease.
- **Diseases and Disorders of the Nervous System (pp. 1039–1041)** Schizophrenia is characterized by hallucinations, delusions, blunted emotions, and other symptoms. Depression includes bipolar disorder, characterized by manic (high-mood) and depressive (low-mood) phases, and major depression, in which patients have a persistent low mood. Alzheimer's disease is an age-related dementia in which neurofibrillary tangles and senile plaques form in the brain. Parkinson's disease is a motor disorder caused by the death of dopamine-secreting neurons in the substantia nigra.

TESTING YOUR KNOWLEDGE

Self-Quiz

1. What happens when a neuron's membrane depolarizes?
   a. There is a net diffusion of Na^+ out of the cell.
   b. The equilibrium potential for K^+ (E_K) becomes more positive.
   c. The magnitude of the neuron's membrane voltage is reduced.
   d. The neuron becomes less likely to generate an action potential.
   e. The inside of the cell becomes more negative relative to the outside.

2. Why are action potentials usually conducted in only one direction along an axon?
   a. The nodes of Ranvier can conduct potentials in only one direction.
   b. The brief refractory period prevents reopening of voltage-gated Na^+ channels.
   c. The axon hillock has a higher membrane potential than the terminals of the axon.
   d. Ions can flow along the axon in only one direction.
   e. Voltage-gated channels for both Na^+ and K^+ open in only one direction.

3. A common feature of action potentials is that they
   a. cause the membrane to hyperpolarize and then depolarize.
   b. can undergo temporal and spatial summation.
   c. are triggered by a depolarization that reaches the threshold.
   d. move at the same speed along all axons.
   e. result from the diffusion of Na^+ and K^+ through ligand-gated channels.

4. Which of the following is a *direct* result of depolarizing the presynaptic membrane of an axon terminal?
   a. Voltage-gated Ca^{2+} channels in the membrane open.
   b. Synaptic vesicles fuse with the membrane.
   c. The postsynaptic cell produces an action potential.
   d. Ligand-gated channels open, allowing neurotransmitters to enter the synaptic cleft.
   e. An EPSP or IPSP is generated in the postsynaptic cell.

5. Where are neurotransmitter receptors located?
   a. on the nuclear membrane
   b. at nodes of Ranvier
   c. on the postsynaptic membrane
   d. on the membranes of synaptic vesicles
   e. in the myelin sheath

6. Which of the following structures or regions is *incorrectly* paired with its function?
   a. limbic system—motor control of speech
   b. medulla oblongata—homeostatic control
   c. cerebellum—coordination of movement and balance
   d. corpus callosum—communication between the left and right cerebral cortices
   e. hypothalamus—regulation of temperature, hunger, and thirst

7. What is the neocortex?
   a. a primitive brain region that is common to reptiles and mammals
   b. a region deep in the cortex that is associated with the formation of emotional memories
   c. a central part of the cortex that receives olfactory information
   d. an additional outer layer of neurons in the cerebral cortex that is unique to mammals
   e. an association area of the frontal lobe that is involved in higher cognitive functions
8. Which of the following provides evidence that brain circuits involved in emotion form early during human development?
   a. Humans are more likely to be able to recall emotional memories from childhood than factual memories.
   b. Infants can understand language before they themselves can speak.
   c. Such circuits involve parts of the brain that evolved before the neocortex evolved.
   d. Young infants can bond to a caregiver and express fear, distress, and anger.
   e. Individuals with damage to the amygdala no longer have autonomic responses to stressful stimuli.
9. Which of the following best describes how an axon grows toward its target cells?
   a. The axon grows in a direct path, attracted by signal molecules released by the target cells.
   b. Cells along the growth path release signal molecules that either attract or repel the axon, and the interaction of CAMs on the growth cone and neighboring cells may provide tracks that guide axon growth.
   c. Nerve growth factor released by astrocytes stimulates a neural progenitor cell to differentiate into a neuron, whose axon then grows toward an increasing concentration of signal molecules.
   d. The axon produces growth-promoting proteins only in its growth cone, causing the axon to grow in an outward direction toward its target cells.
   e. Glia first migrate to the target cells, leaving a trail of CAMs along the path that the growth cone of the axon follows.
10. Which disease or disorder is caused by the death of brain neurons that release dopamine?
   a. schizophrenia
   b. bipolar disorder
   c. major depression
   d. Alzheimer's disease
   e. Parkinson's disease

For Self-Quiz answers, see Appendix A.

Go to the website or CD-ROM for more quiz questions.

Evolution Connection

An action potential is an all-or-none event. This on/off signaling is an evolutionary adaptation of animals that must sense and act in a complex environment. It is possible to imagine a nervous system in which the action potentials were graded, with the amplitude depending on the size of the stimulus. What advantage might on/off signaling have over a graded (continuously variable) kind of signaling?

Scientific Inquiry

From what you know about action potentials and synapses, propose two or three hypotheses for how various anesthetics might prevent pain.

Investigation *What Triggers Nerve Impulses?*

Science, Technology, and Society

Alcohol's depressant effects on the nervous system cloud judgment and slow reflexes. Alcohol consumption is a factor in most fatal traffic accidents in the United States. What are some other impacts of alcohol abuse on society? What are some of the responses of people and society to alcohol abuse? Do you think that alcohol abuse is primarily an individual problem or a societal problem? Do you think that society's responses to alcohol abuse are appropriate and proportional to the seriousness of the problem? Defend your position.

49 Sensory and Motor Mechanisms

▲ **Figure 49.1 A bat using sonar to locate its prey.**

Key Concepts

49.1 Sensory receptors transduce stimulus energy and transmit signals to the central nervous system

49.2 The mechanoreceptors involved with hearing and equilibrium detect settling particles or moving fluid

49.3 The senses of taste and smell are closely related in most animals

49.4 Similar mechanisms underlie vision throughout the animal kingdom

49.5 Animal skeletons function in support, protection, and movement

49.6 Muscles move skeletal parts by contracting

49.7 Locomotion requires energy to overcome friction and gravity

Overview

Sensing and Acting

In the gathering dusk, a male moth's antennae detect the chemical attractant of a female moth somewhere upwind. The male takes to the air, following the scent trail toward the female. Suddenly, vibration sensors in the male moth's abdomen detect the ultrasonic chirps of a rapidly approaching bat. The bat's sonar enables it to locate moths and other flying insect prey. Reflexively, the moth's nervous system alters the motor output to its wing muscles, sending the insect into an evasive spiral toward the ground. Although it is probably too late for the moth in **Figure 49.1**, many moths can escape in such situations because they can detect the sonar of a bat as much as 30 m away. The bat must be within 3 m to sense the moth, but since the bat flies faster, it may still have time to detect, home in on, and catch its prey.

The outcome of this interaction depends on the abilities of both predator and prey to sense important environmental stimuli and to produce appropriate coordinated movement. We can trace the origins of sensing and acting back to the appearance, in prokaryotes, of cellular structures that sense pressure or chemicals in the environment and then direct movement in an appropriate direction. These structures have been transformed during evolution into diverse mechanisms that sense various types of energy and generate many different levels of physical movement in response. The detection and processing of sensory information and the generation of motor output provide the physiological basis for all animal behavior.

It is customary to think of animal behavior as a linear sequence of sensing, brain analysis, and action—similar to a computer passively waiting for instructions before it acts. This is not the case. When animals are in motion, they are probing the environment through that motion, sensing changes, and using the information to generate the next action. This is a continuous cycle rather than a linear sequence, as the brain carries on background activity that is constantly updated as sensing and acting proceed.

In Chapter 48, we examined how nervous systems transmit and integrate sensory and motor information. Here we will explore, in several invertebrate and vertebrate groups, the processes of sensing and acting. We will start with sensory processes that convey information about the external and internal environment to the brain. We will then consider the structure and function of skeletons and muscles that carry out movements as instructed by the brain. Finally, we will investigate various mechanisms of animal movement.

Concept 49.1

Sensory receptors transduce stimulus energy and transmit signals to the central nervous system

Information is transmitted through the nervous system in the form of nerve impulses, or action potentials, which are all-or-none events (see Figure 48.12c). An action potential triggered by light striking the eye is the same as an action potential triggered by air vibrating in the ear. The ability to distinguish any type of stimulus, such as sight or sound, depends on the part of the brain that receives the action potentials. What matters is where action potentials go, not what triggers them.

Sensations are action potentials that reach the brain via sensory neurons. Once the brain is aware of sensations, it interprets them, giving the **perception** of stimuli. Perceptions—such as colors, smells, sounds, and tastes—are constructions formed in the brain and do not exist outside it. If a tree falls and no animal is present to hear it, is there a sound? The fall certainly produces pressure waves in the air, but if sound is defined as a perception, then there is none unless an animal senses the waves and its brain perceives them.

Sensations and perceptions begin with **sensory reception,** the detection of a stimulus by sensory cells, which are called **sensory receptors.** Most sensory receptors are specialized neurons or epithelial cells that exist singly or in groups with other cell types in sensory organs, such as the eyes and ears. **Exteroreceptors** are sensory receptors that detect stimuli coming from outside the body, including heat, light, pressure, and chemicals. **Interoreceptors** detect stimuli coming from within the body, such as blood pressure and body position.

Functions Performed by Sensory Receptors

All stimuli represent forms of energy. Sensation involves converting this energy into a change in the membrane potential of sensory receptors, which leads to a change in frequency of action potentials transmitted to the central nervous system (CNS). Sensory receptors perform four functions in this process: sensory transduction, amplification, transmission, and integration. These functions are shown in **Figure 49.2** for two types of sensory receptors: a stretch receptor in a crayfish and a **hair cell,** which detects motion in the vertebrate ear and in lateral line systems of fishes and amphibians.

Sensory Transduction

The conversion of stimulus energy into a change in the membrane potential of a sensory receptor is called **sensory transduction,** and the change in membrane potential itself is known as a **receptor potential.** Note in Figure 49.2a that receptor potentials are graded potentials; their magnitude varies with the strength of the stimulus.

All receptor potentials result from the opening or closing of ion channels in the sensory receptor's plasma membrane, which changes the ionic permeability of the membrane. In the examples in Figure 49.2, stretching or bending of the membrane is the stimulus that causes ion channels to open or close. In other sensory receptors, channels open or close when substances outside the cell bind to proteins on the membrane or when pigments in the sensory receptor absorb light. We will examine these mechanisms later in the chapter.

One amazing feature of many sensory receptors is extreme sensitivity: They can detect the smallest physical unit of stimulus possible. Most light receptors can detect a single quantum (photon) of light; chemical receptors can detect a single molecule; and hair cells of the inner ear can detect a motion of only a fraction of a nanometer. Receptor sensitivity also changes with conditions. For example, the sensitivity of glucose receptors in the human mouth can vary over several orders of magnitude of sugar concentration as both the general state of nutrition and amount of sugar in the diet change.

Amplification

The strengthening of stimulus energy by cells in sensory pathways is called **amplification.** For example, an action potential conducted from the eye to the brain has about 100,000 times as much energy as the few photons of light that triggered it. Some amplification occurs in sensory receptors, and signal transduction pathways involving second messengers often contribute to it. Amplification may also take place in accessory structures of a complex sense organ, as when sound waves are enhanced by a factor of more than 20 before reaching receptors of the inner ear.

Transmission

After energy in a stimulus has been transduced into a receptor potential, action potentials are transmitted to the CNS. Some sensory receptors, such as the crayfish stretch receptor, are sensory neurons that can produce action potentials and have an axon that extends into the CNS (see Figure 49.2a). Others, such as hair cells, lack an axon and cannot generate action potentials themselves; these receptors release neurotransmitters at synapses with sensory neurons (see Figure 49.2b). In almost all pathways where receptors synapse with sensory neurons, the receptor releases an excitatory neurotransmitter, causing the sensory neuron to transmit action potentials to the CNS. (An exception is in the vertebrate visual system, discussed in Concept 49.4.)

The magnitude of a receptor potential affects the frequency of action potentials that travel as sensations to the CNS. If a receptor is also a sensory neuron, a larger receptor potential

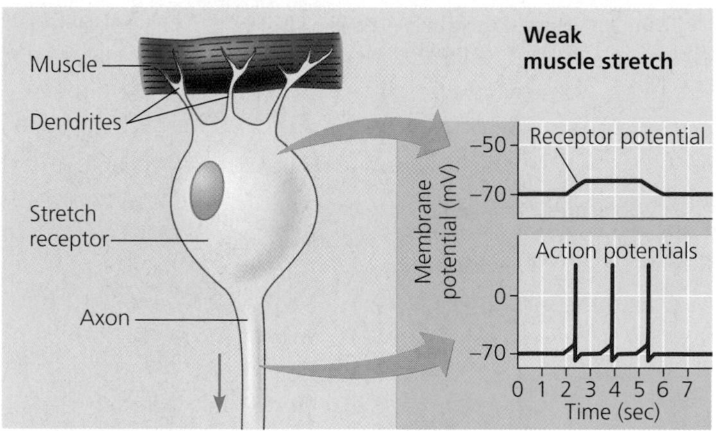

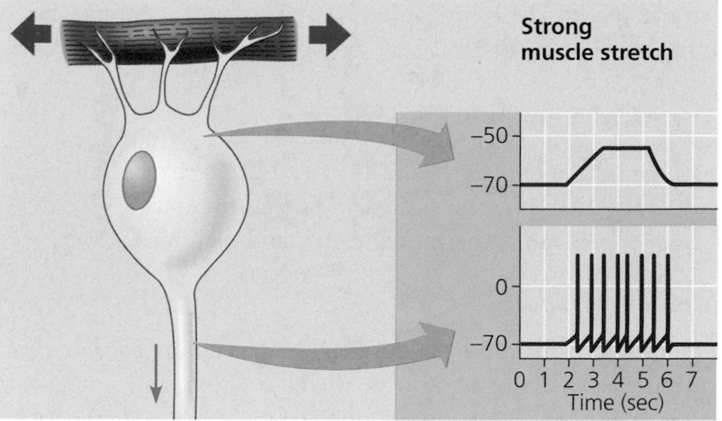

(a) Crayfish stretch receptors have dendrites embedded in abdominal muscles. When the abdomen bends, muscles and dendrites stretch, producing a receptor potential in the stretch receptor. The receptor potential triggers action potentials in the axon of the stretch receptor. A stronger stretch produces a larger receptor potential and higher frequency of action potentials.

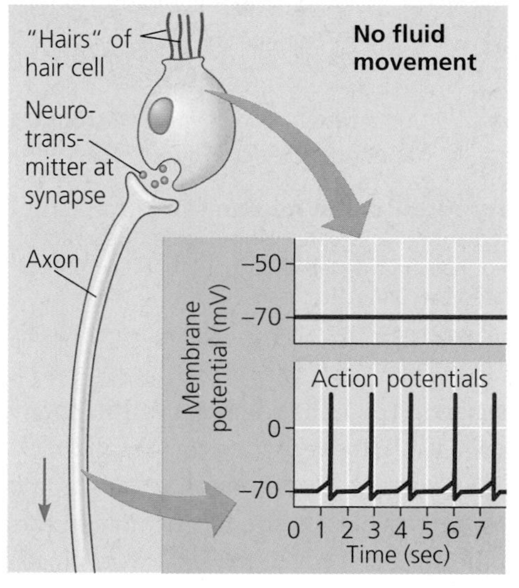

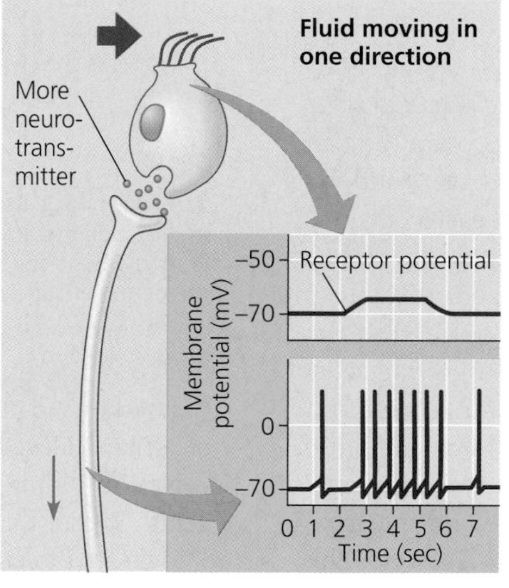

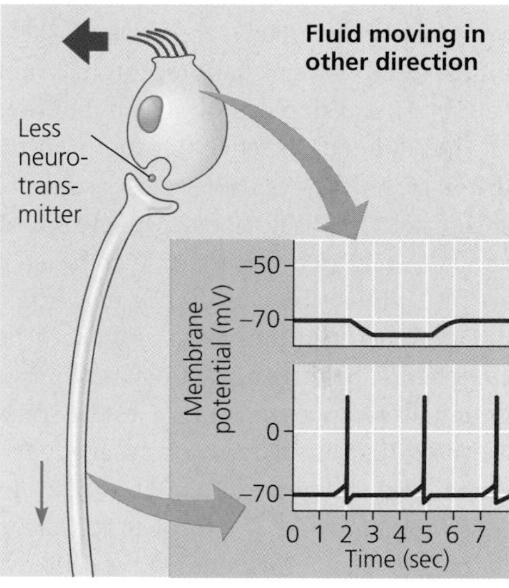

(b) Vertebrate hair cells have specialized cilia or microvilli ("hairs") that bend when surrounding fluid moves. Each hair cell releases an excitatory neurotransmitter at a synapse with a sensory neuron, which conducts action potentials to the CNS. Bending in one direction depolarizes the hair cell, causing it to release more neurotransmitter and increasing frequency of action potentials in the sensory neuron. Bending in the other direction has the opposite effects. Thus, hair cells respond to the direction of motion as well as to its strength and speed.

▲ **Figure 49.2 Sensory reception: two mechanisms.**

reaches the threshold sooner, resulting in more frequent action potentials (see Figure 49.2a). If a receptor synapses with a sensory neuron, a larger receptor potential causes more neurotransmitter to be released at the synapse, usually with the same result. Many sensory neurons spontaneously generate action potentials at a low rate. Therefore, a stimulus does not switch the production of action potentials on or off in these neurons; rather, it modulates action potential frequency (see Figure 49.2b). Such neurons alert the CNS not only to the presence or absence of a stimulus but also to changes in intensity or direction.

Integration

The processing, or integration, of sensory information begins as soon as the information is received. Receptor potentials produced by stimuli delivered to different parts of a sensory receptor are integrated through summation, as are postsynaptic potentials in sensory neurons that synapse with multiple receptors. Another type of integration by receptors is **sensory adaptation**, a decrease in responsiveness during continued stimulation (not to be confused with the evolutionary sense of the term *adaptation*). Some receptors adapt to stimuli more quickly than others. Without sensory adaptation, you would feel every beat of your heart and every bit of clothing on your body. Receptors are selective in the information they send to the CNS, and adaptation reduces the likelihood that a maintained stimulus will result in a continuous sensation.

The integration of sensory information occurs at all levels within the nervous system, and the cellular actions just described are only the first steps. Complex sensory structures such as the eyes have higher levels of integration, and the CNS further processes all incoming signals.

Types of Sensory Receptors

Based on the energy they transduce, sensory receptors fall into five categories: mechanoreceptors, chemoreceptors, electromagnetic receptors, thermoreceptors, and pain receptors.

Mechanoreceptors

Mechanoreceptors sense physical deformation caused by stimuli such as pressure, touch, stretch, motion, and sound—all forms of mechanical energy. Bending or stretching of a mechanoreceptor's plasma membrane increases the membrane's permeability to sodium and potassium ions. When this happens, the membrane potential moves to a value between E_K and E_{Na}, resulting in depolarization (see Chapter 48).

The crayfish stretch receptor and the vertebrate hair cell in Figure 49.2 are mechanoreceptors. Another example is the vertebrate stretch receptor, which monitors the length of skeletal muscles. The mechanoreceptors in this case are the dendrites of sensory neurons that spiral around the middle of small skeletal muscle fibers. **Muscle spindles,** each containing about 2 to 12 of these fibers surrounded by connective tissue, are distributed throughout the muscle, parallel to other muscle fibers. When the muscle is stretched, the spindle fibers are stretched, depolarizing sensory neurons and triggering action potentials that are transmitted to the spinal cord. Muscle spindles and the sensory neurons that innervate them are part of nerve circuits that underlie reflexes (see Figure 48.4).

The mammalian sense of touch also relies on mechanoreceptors that are the dendrites of sensory neurons. Often they are embedded in layers of connective tissue **(Figure 49.3)**. The structure of the connective tissue and the location of the receptors dramatically affect the type of mechanical energy (light touch, vibration, or strong pressure) that best stimulates them. Receptors that detect light touch are close to the surface of the skin; they transduce very slight inputs of mechanical energy into receptor potentials. Receptors responding to strong pressure and vibrations are in deep skin layers. Other receptors sense movement of hairs (not to be confused with the hair cells in Figure 49.2); those at the base of the stout whiskers of mammals such as cats and many rodents are extremely sensitive and enable the animal to detect nearby objects in the dark.

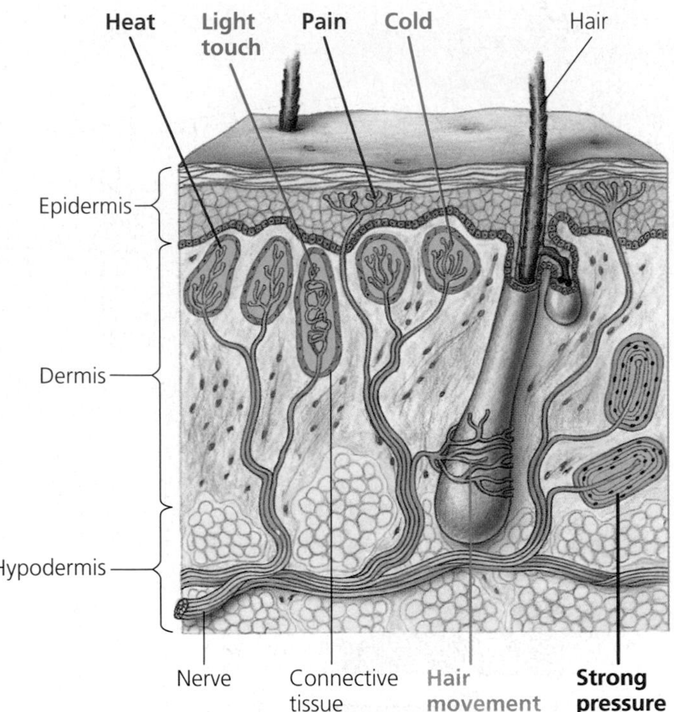

▲ **Figure 49.3 Sensory receptors in human skin.** Receptors in the epidermis are naked dendrites, as are hair movement receptors wound around the base of hairs in the dermis. Most other receptors in the dermis are encapsulated by connective tissue.

Chemoreceptors

Chemoreceptors include both general receptors that transmit information about the total solute concentration of a solution and specific receptors that respond to individual kinds of molecules. Osmoreceptors in the mammalian brain, for example, are general receptors that detect changes in the total solute concentration of the blood and stimulate thirst when osmolarity increases (see Chapter 44). Water receptors in the feet of houseflies respond to pure water or to a dilute solution of virtually any substance. Most animals also have receptors for specific molecules, including glucose, oxygen, carbon dioxide, and amino acids. Two of the most sensitive and specific chemoreceptors known are present in the antennae of the male silkworm moth **(Figure 49.4)**. They detect the two chemical components of the female moth sex pheromone. In all these examples, the stimulus molecule binds to a specific site on the membrane of the receptor cell and initiates changes in membrane permeability.

Electromagnetic Receptors

Electromagnetic receptors detect various forms of electromagnetic energy, such as visible light, electricity, and magnetism. **Photoreceptors,** electromagnetic receptors that detect the radiation we know as visible light, are often organized into eyes. Some snakes have very sensitive infrared receptors that detect body heat of prey against a colder background **(Figure 49.5a)**. Some fishes generate electric currents and use electroreceptors to locate objects, such as prey, that disturb those currents. The platypus, a monotreme mammal, has electroreceptors on its bill that probably detect electric fields generated by the muscles of crustaceans, frogs, small fish, and other prey. Many animals appear to use Earth's magnetic field lines to orient themselves as they migrate **(Figure 49.5b)**. The iron-containing mineral magnetite is found in the skulls of many vertebrates (including salmon, pigeons, sea turtles, and humans), in the abdomen of

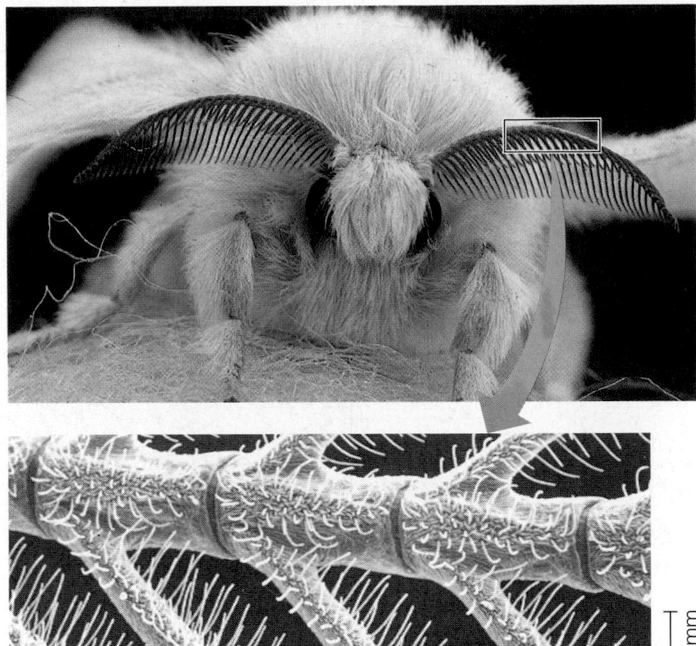

▲ **Figure 49.4 Chemoreceptors in an insect.** The antennae of the male silkworm moth *Bombyx mori* are covered with sensory hairs, visible in the SEM enlargement. The hairs have chemoreceptors that are highly sensitive to the sex pheromone released by the female.

(a) This rattlesnake and other pit vipers have a pair of infrared receptors, one between each eye and nostril. The organs are sensitive enough to detect the infrared radiation emitted by a warm mouse a meter away. The snake moves its head from side to side until the radiation is detected equally by the two receptors, indicating that the mouse is straight ahead.

(b) Some migrating animals, such as these beluga whales, apparently sense Earth's magnetic field and use the information, along with other cues, for orientation.

▲ **Figure 49.5 Specialized electromagnetic receptors.**

bees, in the teeth of some molluscs, and in certain protists and prokaryotes that orient to Earth's magnetic field. Once used by sailors as a compass, magnetite may be part of an orienting mechanism in many animals.

Thermoreceptors

Thermoreceptors, which respond to heat or cold, help regulate body temperature by signaling both surface and body core temperature. Debate continues about the identity of thermoreceptors in mammalian skin. Possible candidates are two receptors consisting of encapsulated, branched dendrites (see Figure 49.3). Many researchers, however, think that these structures are modified pressure receptors and that thermoreceptors consist of naked dendrites of certain sensory neurons. There is general agreement that thermoreceptors in the skin and in the anterior hypothalamus send information to the body's thermostat, located in the posterior hypothalamus.

Pain Receptors

In humans, **pain receptors,** also called **nociceptors** (from the Latin *nocere,* to hurt) are a class of naked dendrites in the epidermis (see Figure 49.3). Most animals probably experience pain, although we cannot say what perceptions other animals associate with stimulation of pain receptors. Pain is an important sensation because the stimulus usually leads to a defensive reaction, such as withdrawal from danger. Rare individuals who are born without any pain sensation may die from conditions such as a ruptured appendix because they cannot feel the associated pain and are unaware of the danger.

Different groups of pain receptors respond to excess heat, pressure, or specific classes of chemicals released from damaged or inflamed tissues. Some chemicals that trigger pain include histamine and acids. Prostaglandins increase pain by sensitizing receptors—that is, lowering their threshold (see Chapter 45); aspirin and ibuprofen reduce pain by inhibiting prostaglandin synthesis. While density of nociceptors is highest in the skin, some are associated with other organs.

Concept Check 49.1

1. Why do drugs that interfere with synaptic transmission block some but not all sensations?
2. If you were to apply an anesthetic to your skin, which senses would be affected first? Which would be affected last or not at all? Explain.

For suggested answers, see Appendix A.

The mechanoreceptors involved with hearing and equilibrium detect settling particles or moving fluid

Hearing and the perception of body equilibrium, or balance, are related in most animals. Both involve mechanoreceptors that produce receptor potentials when some part of the membrane is bent by settling particles or moving fluid.

Sensing Gravity and Sound in Invertebrates

Most invertebrates have sensory organs called **statocysts** that contain mechanoreceptors and function in their sense of equilibrium **(Figure 49.6)**. A common type of statocyst consists of a layer of ciliated receptor cells surrounding a chamber that contains one or more **statoliths**, which are grains of sand or other dense granules. Gravity causes the statoliths to settle to the low point in the chamber, stimulating receptors in that location. Statocysts are located in different parts of invertebrates' bodies. For example, many jellies have statocysts at the fringe of their "bell," giving the animals an indication of body position. Lobsters and crayfish have statocysts near the base of their antennules. In experiments, crayfish have been "tricked" into swimming upside down when their statoliths were replaced with metal shavings that were pulled to the upper end of the statocysts with magnets.

Many invertebrates demonstrate a general sensitivity to sound, although structures specialized for hearing seem to be less widespread than gravity sensors. Hearing structures have been studied most extensively in terrestrial insects.

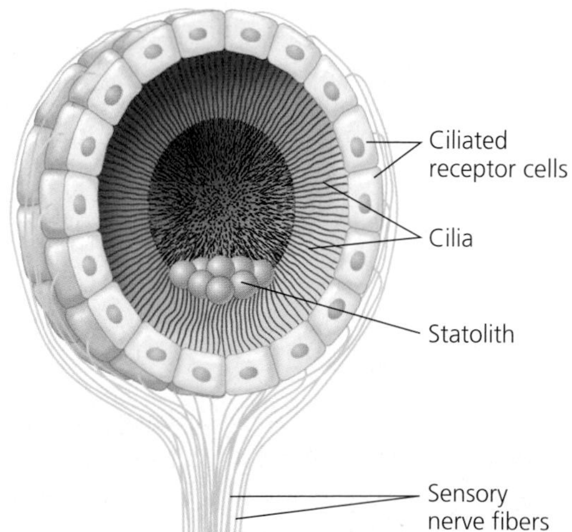

▲ **Figure 49.6 The statocyst of an invertebrate.** The settling of statoliths to the low point in the chamber bends cilia on receptor cells in that location, providing the brain with information about the orientation of the body with respect to gravity.

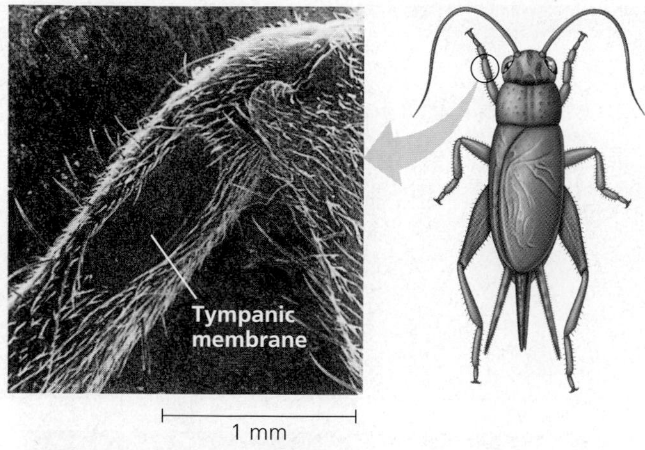

▲ **Figure 49.7 An insect ear.** The tympanic membrane, visible in this SEM of a cricket's front leg, vibrates in response to sound waves. The vibrations stimulate mechanoreceptors attached to the inside of the tympanic membrane.

Many (perhaps most) insects have body hairs that vibrate in response to sound waves. Hairs of different stiffnesses and lengths vibrate at different frequencies. The hairs are commonly tuned to frequencies of sounds produced by other organisms. For example, a male mosquito locates a mate by means of fine hairs on his antennae. The hairs vibrate in a specific way in response to the hum produced by the beating wings of flying females. A tuning fork that vibrates at the same frequency as a female mosquito's wings will also attract males. Some caterpillars (larval moths and butterflies) have vibrating body hairs that can detect the buzzing wings of predatory wasps, warning the caterpillars of the danger.

Many insects also have localized "ears," in which a tympanic membrane (eardrum) is stretched over an internal air chamber **(Figure 49.7)**. Sound waves vibrate the tympanic membrane, stimulating receptor cells attached to the inside of the membrane and resulting in nerve impulses that are transmitted to the brain. Some moths can hear the high-pitched sounds that bats produce for sonar, and perceiving these sounds triggers the moths' escape maneuver, as mentioned at the beginning of this chapter. This also explains why it is so difficult to step on a cockroach; the insect senses your descending foot and moves very quickly to avoid it.

Hearing and Equilibrium in Mammals

In mammals, as in most other terrestrial vertebrates, the sensory organs for hearing and equilibrium are closely associated in the ear. **Figure 49.8** explores the structure of these organs in the human ear.

Hearing

How does the ear convert the energy of pressure waves traveling through air into nerve impulses that the brain perceives as sound? Vibrating objects, such as a plucked guitar string or the

Figure 49.8

Exploring The Structure of the Human Ear

1 Overview of ear structure. The mammalian ear can be divided into three regions: the outer ear, middle ear, and inner ear. The **outer ear** consists of the external pinna and the auditory canal, which collect sound waves and channel them to the **tympanic membrane** (eardrum) separating the outer ear from the middle ear.

2 The middle ear and inner ear. In the **middle ear,** three small bones—the **malleus** (hammer), **incus** (anvil), and **stapes** (stirrup)—transmit vibrations to the **oval window,** which is a membrane beneath the stapes. The middle ear also opens into the **Eustachian tube,** which connects with the pharynx and equalizes pressure between the middle ear and the atmosphere, enabling you to "pop" your ears when you change altitude, for example.

The **inner ear** consists of a labyrinth of fluid-filled chambers within the temporal bone of the skull. These chambers include the **semicircular canals,** which function in equilibrium, and the coiled **cochlea** (Latin, "snail"), which is involved in hearing.

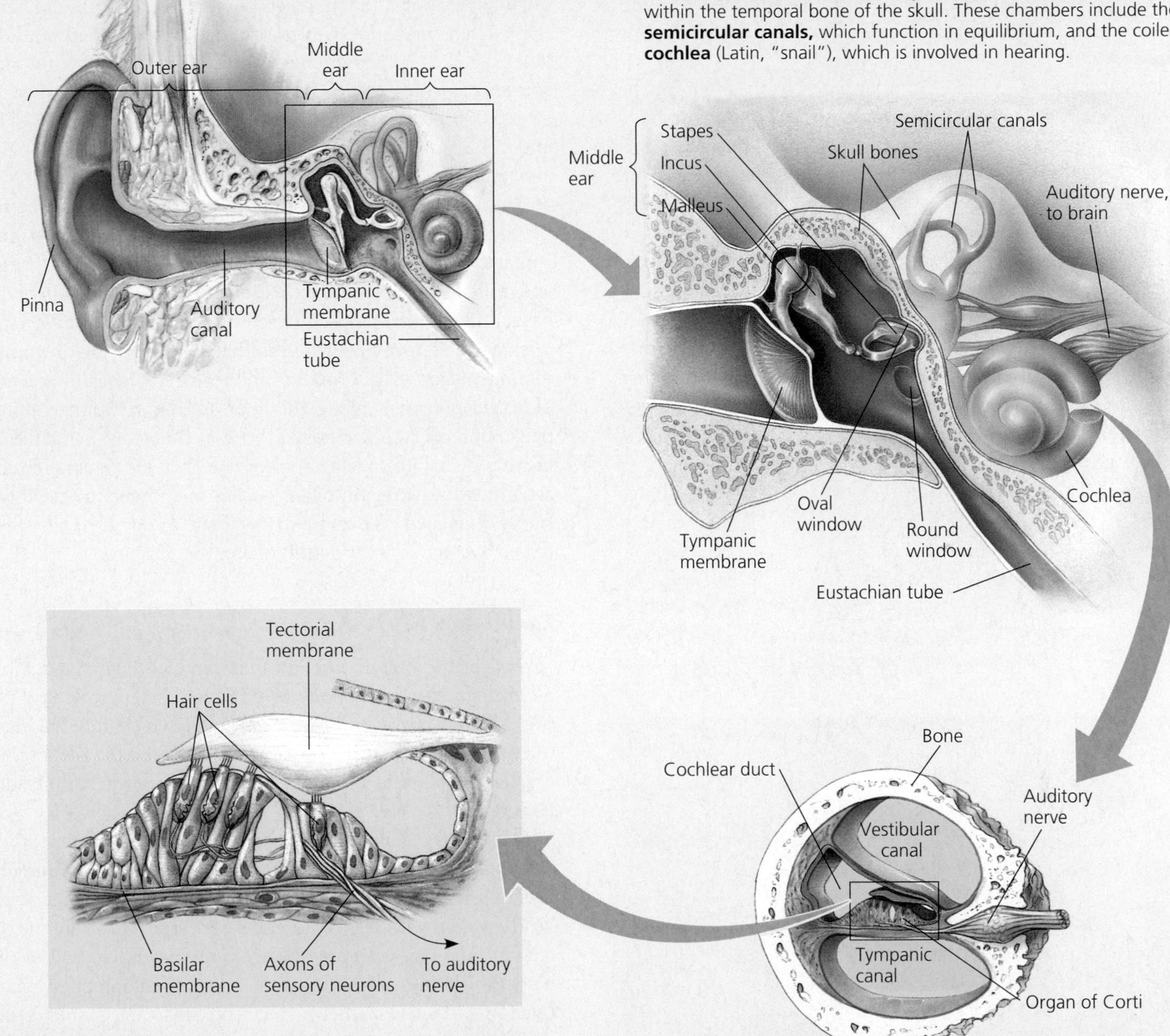

3 The cochlea. The cochlea has two large canals—an upper vestibular canal and a lower tympanic canal—separated by a smaller cochlear duct. The vestibular and tympanic canals contain a fluid called perilymph, and the cochlear duct is filled with a fluid called endolymph.

4 The organ of Corti. The floor of the cochlear duct, the basilar membrane, bears the **organ of Corti,** which contains the mechanoreceptors of the ear, hair cells with hairs projecting into the cochlear duct. Many of the hairs are attached to the tectorial membrane, which hangs over the organ of Corti like a shelf. Sound waves make the basilar membrane vibrate, which results in bending of the hairs and depolarization of the hair cells.

vocal cords of a speaking person, create percussion waves in the surrounding air. These waves cause the tympanic membrane to vibrate with the same frequency as the sound. The three bones of the middle ear transmit the vibrations to the oval window, a membrane on the cochlea's surface. When one of those bones, the stapes, vibrates against the oval window, it creates pressure waves in the fluid of the cochlea. The waves travel through the vestibular canal, pass around the apex (tip) of the cochlea, and continue through the tympanic canal, dissipating as they strike the **round window** **(Figure 49.9)**.

Pressure waves in the vestibular canal push down on the cochlear duct and basilar membrane. In response, the basilar membrane vibrates up and down, and its hair cells alternately brush against and are withdrawn from the tectorial membrane. That causes the hairs to bend first in one direction and then the other with each vibration. As shown in Figure 49.2b, bending in one direction depolarizes the hair cells, increasing neurotransmitter release and the frequency of action potentials in the sensory neurons with which they synapse. These neurons carry sensations to the brain through the auditory nerve. Bending of the hairs in the other direction hyperpolarizes the hair cells, reducing neurotransmitter release and the frequency of sensations in the auditory nerve.

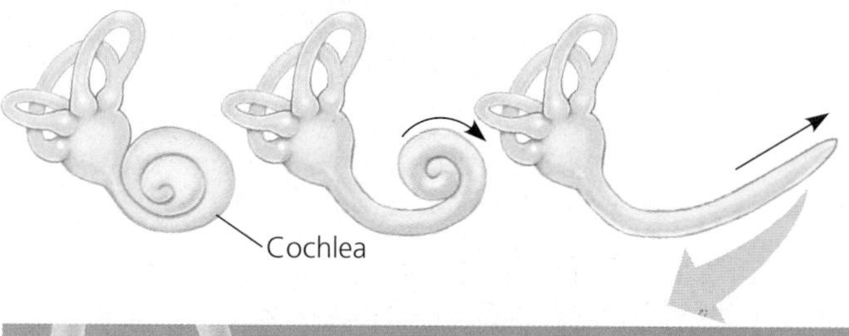

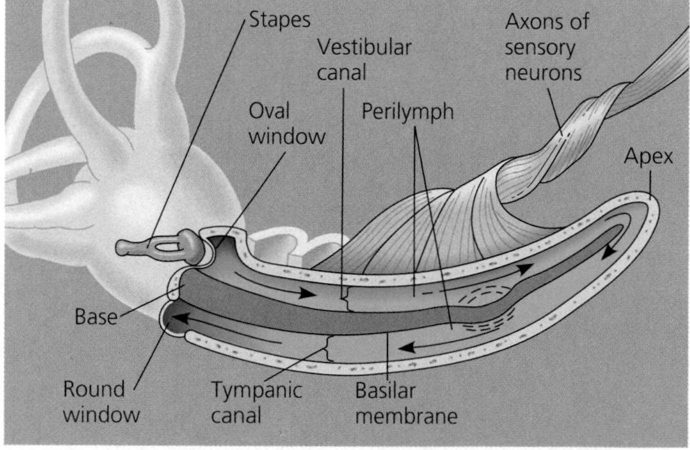

▲ **Figure 49.9 Transduction in the cochlea.** For simplicity, the cochlea is shown uncoiled here and in Figure 49.10, even though it does not physically uncoil. Vibrations of the stapes against the oval window produce pressure waves in the perilymph of the cochlea. The waves (black arrows) travel to the apex of the cochlea through the vestibular canal and back toward the base through the tympanic canal. The energy in the waves causes the basilar membrane to vibrate, stimulating hair cells.

If sound is represented by changes in the frequency of sensations in the auditory nerve, how is the quality of sound conveyed? Two important sound variables are volume and pitch. Volume (loudness) is determined by the amplitude, or height, of the sound wave. A large-amplitude sound wave causes a more vigorous vibration of the basilar membrane, a greater bending of the hairs on hair cells, and more action potentials in the sensory neurons. **Pitch** is a function of a sound wave's frequency, or number of vibrations per second, expressed in hertz (Hz). High-frequency waves produce high-pitched sounds, whereas the low-frequency waves produce low-pitched sounds. Healthy young humans can hear in the range of 20–20,000 Hz; dogs can hear sounds as high as 40,000 Hz; and bats can emit and hear clicking sounds at frequencies above 100,000 Hz, using this ability to locate objects.

The cochlea can distinguish pitch because the basilar membrane is not uniform along its length: It is relatively narrow and stiff at the base of the cochlea near the oval window and wider and more flexible at the apex **(Figure 49.10)**. Each region of the basilar membrane is most affected by a particular vibration frequency. The sensory neurons associated with the region vibrating most vigorously at any instant send the highest frequency of action potentials along the auditory nerve. But the actual perception of pitch occurs in the brain. Axons in the auditory nerve project into specific auditory areas of the cerebral cortex according to the region of the basilar membrane in which the signal originated. When a particular site in our cortex is stimulated, we perceive the sound of a particular pitch.

Equilibrium

Several of the organs in the inner ear of humans and most other mammals detect body position and balance. Behind the oval window is a vestibule that contains two chambers, called the **utricle** and **saccule.** The utricle opens into the three semicircular canals that complete the apparatus for equilibrium **(Figure 49.11)**.

Sensations related to body position are generated much like the sensations of sound in humans and most other mammals. Hair cells in the utricle and the saccule respond to changes in head position with respect to gravity and movement in one direction. The hair cells are arranged in clusters, and all of the hairs project into a gelatinous material containing many small calcium carbonate particles called otoliths ("ear stones"). Because this material is heavier than the endolymph within the utricle and the saccule, gravity is always pulling downward on the hairs of the receptor cells, sending a constant series of action potentials along the sensory neurons of the vestibular branch of the auditory nerve. This mechanism is similar to how statocysts function in invertebrates, and indeed the utricle and saccule are considered to be specialized types of statocysts.

Different body angles cause different hair cells and their sensory neurons to be stimulated. When the position of the

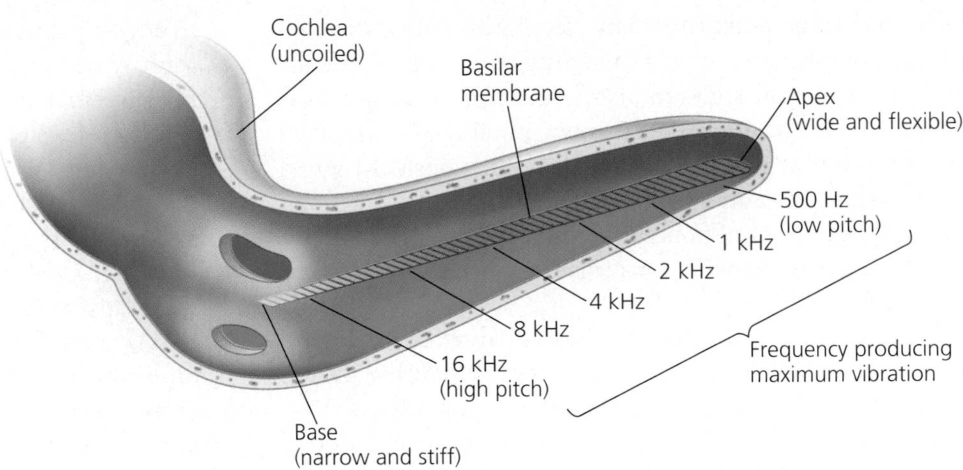

▶ **Figure 49.10 How the cochlea distinguishes pitch.** Variation in the width and stiffness of the basilar membrane (pink) along its length "tunes" specific regions of the basilar membrane to specific frequencies. As a result, different frequencies of pressure waves in the cochlea cause different places along the basilar membrane to vibrate, stimulating particular hair cells and sensory neurons. The selective stimulation of hair cells is perceived in the brain as sound of a certain pitch.

▼ **Figure 49.11 Organs of equilibrium in the inner ear.**

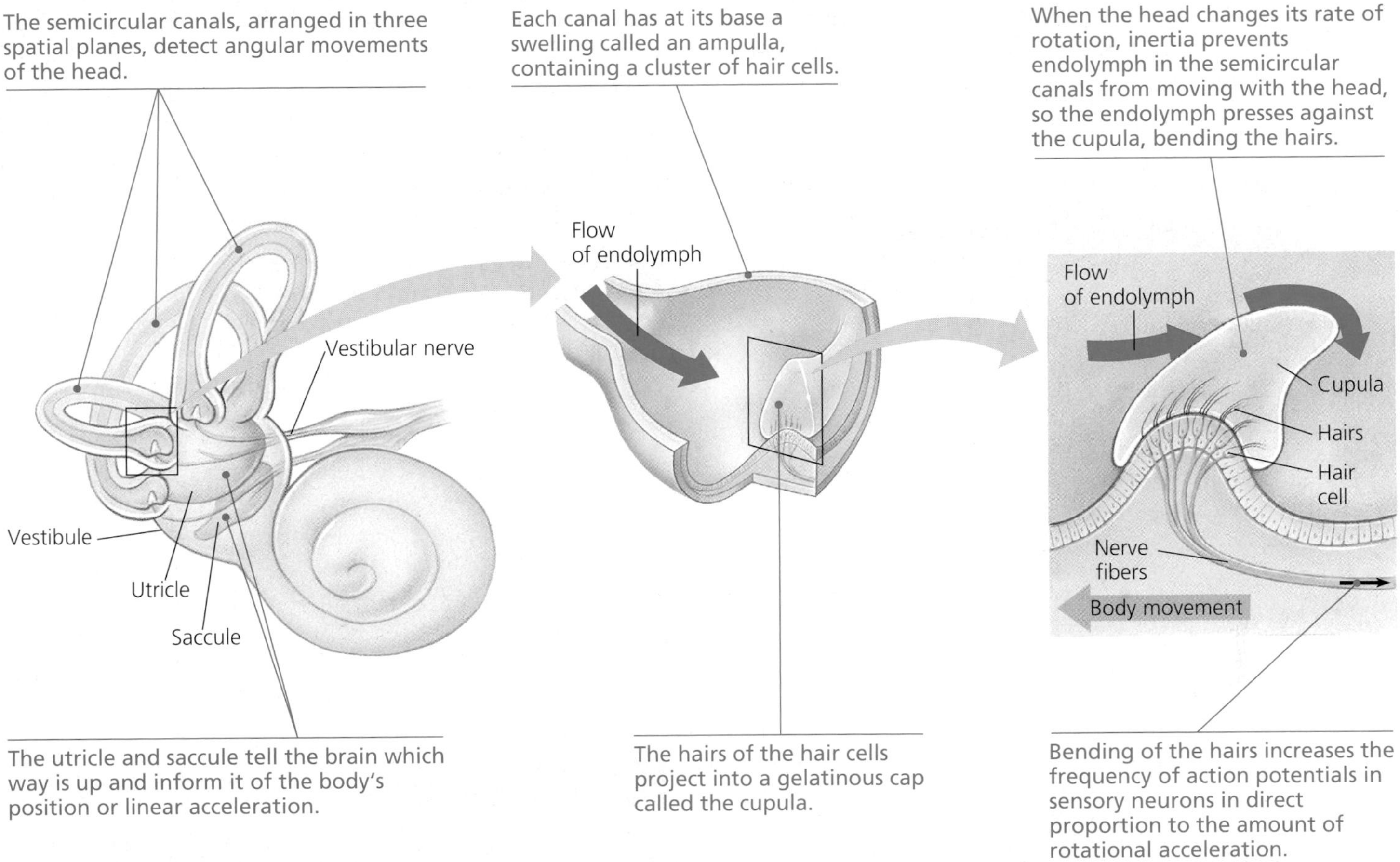

head changes with respect to gravity (as when the head bends forward), the force on a hair cell changes, and the cell increases or decreases its release of neurotransmitter. The brain interprets the resulting changes in impulse production by the sensory neurons as a change in the position of the head. By a similar mechanism, the semicircular canals, arranged in the three spatial planes, detect changes in the rate of rotation or angular movements of the head (see Figure 49.11). When you spin around, the equilibrium in the semicircular canals is disrupted, which is why you become dizzy.

The semicircular canals, arranged in three spatial planes, detect angular movements of the head.
Each canal has at its base a swelling called an ampulla, containing a cluster of hair cells.
When the head changes its rate of rotation, inertia prevents endolymph in the semicircular canals from moving with the head, so the endolymph presses against the cupula, bending the hairs.
Vestibular nerve
Flow
of endolymph
Flow
of endolymph
Cupula
Hairs
Hair
cell
Nerve
fibers
Body movement
Vestibule
Utricle
Saccule
The utricle and saccule tell the brain which way is up and inform it of the body's position or linear acceleration.
The hairs of the hair cells project into a gelatinous cap called the cupula.
Bending of the hairs increases the frequency of action potentials in sensory neurons in direct proportion to the amount of rotational acceleration.

Hearing and Equilibrium in Other Vertebrates

Like other vertebrates, fishes and aquatic amphibians also have inner ears located near the brain. There is no cochlea, but there are a saccule, a utricle, and semicircular canals, structures homologous to the equilibrium sensors of human ears. Within these chambers in the inner ear of a fish, the movement of otoliths stimulates sensory hairs. Unlike the mammalian hearing apparatus, the ear of a fish has no eardrum and does not open to the outside of the body. The vibrations of the water caused by sound waves are conducted through the skeleton of the head to the inner ears, setting the otoliths in motion and stimulating the hair cells. The fish's air-filled swim bladder (see Chapter 34) also vibrates in response to sound and may contribute to the transfer of sound to the inner ear. Some fishes, including catfishes and minnows, have a series of bones, called the Weberian apparatus, which conducts the vibrations from the swim bladder to the inner ear.

Most fishes and aquatic amphibians also have a **lateral line system** along both sides of their body **(Figure 49.12)**. The system contains mechanoreceptors that detect low-frequency waves by a mechanism similar to the function of the inner ear. Water from the animal's surroundings enters the lateral line system through numerous pores and flows along a tube past the mechanoreceptors. The receptor units, called neuromasts, resemble the ampullae in our semicircular canals. Each neuromast has a cluster of hair cells whose hairs are embedded in a gelatinous cap, the cupula. Water movement bends the cupula, depolarizing the hair cells and leading to action potentials that are transmitted along the axons of sensory neurons to the brain. This information helps the fish perceive its movement through water or the direction and velocity of water currents flowing over its body. The lateral line system also detects water movements or vibrations that are generated by other moving objects, including prey and predators.

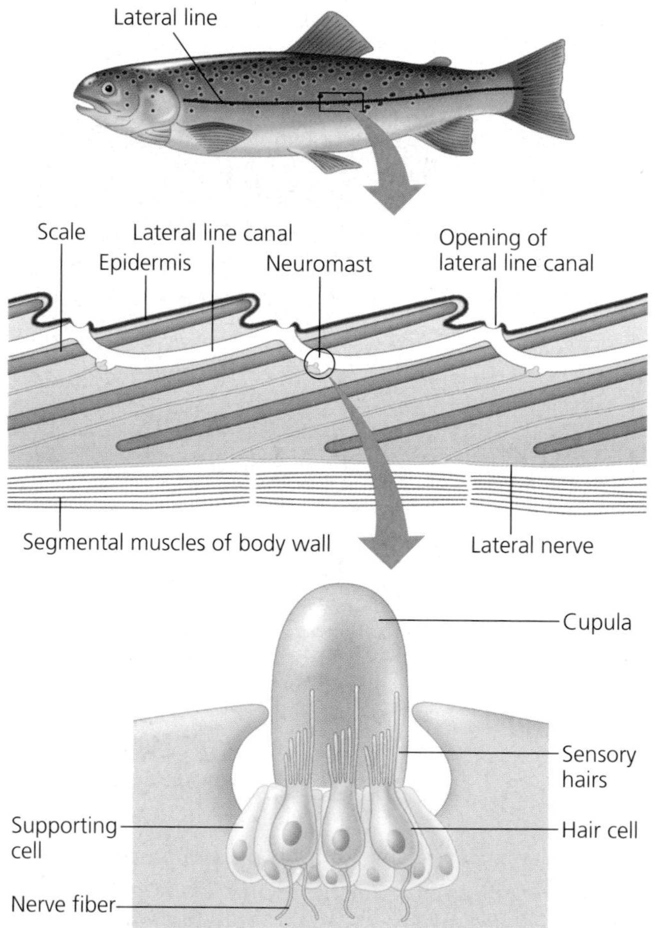

▲ **Figure 49.12 The lateral line system in a fish.** Water flowing through the system bends hair cells. The hair cells transduce the energy into receptor potentials, triggering action potentials that are conveyed to the brain. The lateral line system enables a fish to monitor water currents, pressure waves produced by moving objects, and low-frequency sounds conducted through the water.

In terrestrial vertebrates, the inner ear has evolved as the main organ of hearing and equilibrium. Some amphibians have a lateral line system as tadpoles, but not as adults living on land. In the ear of a terrestrial frog or toad, sound vibrations in the air are conducted to the inner ear by a tympanic membrane on the body surface and a single middle ear bone. There also is evidence that the lungs of a frog vibrate in response to sound and transmit their vibrations to the eardrum via the auditory tube. A small side pocket of the saccule functions as the main hearing organ of the frog, and it is this outgrowth of the saccule that gave rise to the more elaborate cochlea during the evolution of mammals. Birds also have a cochlea. However, as in amphibians, sound is conducted from the tympanic membrane to the inner ear by a single bone, the stapes.

Concept Check 49.2

1. How are statocysts adaptive for animals that burrow underground or live deep in the ocean?
2. In otosclerosis, the stapes becomes fused to the other middle ear bones or to the oval window. Explain how this condition affects hearing.
3. Suppose a series of pressure waves in your cochlea causes a vibration of the basilar membrane that moves gradually from the apex toward the base. How would your brain interpret this stimulus?

For suggested answers, see Appendix A.

Concept 49.3

The senses of taste and smell are closely related in most animals

Many animals use their chemical senses to find mates (as when male silk moths respond to pheromones emitted by females), to recognize territory that has been marked by some chemical substance (as when dogs and cats sniff boundaries that have been staked out by their spraying neighbors), and to help navigate during migration (as when salmon use the unique scent of their streams of origin to return for breeding). Chemical "conversation" is especially important for animals, such as ants and bees, that live in large social groups. In all animals, chemical senses are important in feeding behavior. For example, a hydra begins to make ingestive movements when it detects the compound glutathione, which is released from prey captured by the hydra's tentacles.

The perceptions of **gustation** (taste) and **olfaction** (smell) are both dependent on chemoreceptors that detect specific chemicals in the environment. In the case of terrestrial animals, taste is the detection of chemicals that are present in a solution, and smell is the detection of chemicals that are carried through the air. There is no distinction between taste and smell in aquatic animals.

The taste receptors of insects are located within sensory hairs called sensilla, which are located on the feet and in mouthparts. These animals use their sense of taste to select food. A tasting hair contains several chemoreceptors, each especially responsive to a particular class of chemical stimuli, such as sugar or salt **(Figure 49.13)**. Insects are also capable of smelling airborne chemicals using olfactory hairs, usually located on the antennae (see Figure 49.4).

Figure 49.13
Inquiry How do insects detect different tastes?

EXPERIMENT Insects taste using gustatory sensilla (hairs) on their feet and mouthparts. Each sensillum contains four chemoreceptors with dendrites that extend to a pore at the tip of the sensillum. To study the sensitivity of each chemoreceptor, researchers immobilized a blowfly (*Phormia regina*) by attaching it to a rod with wax. They then inserted the tip of a microelectrode into one sensillum to record action potentials in the chemoreceptors, while they used a pipette to touch the pore with various test substances.

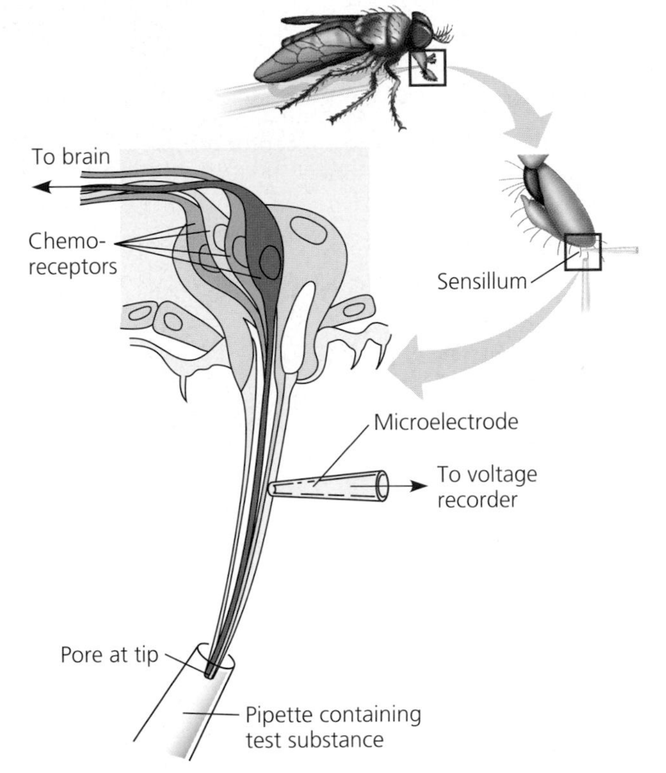

RESULTS Each chemoreceptor is especially sensitive to a particular class of substance, but this specificity is relative; each cell can respond to some extent to a broad range of different chemical stimuli.

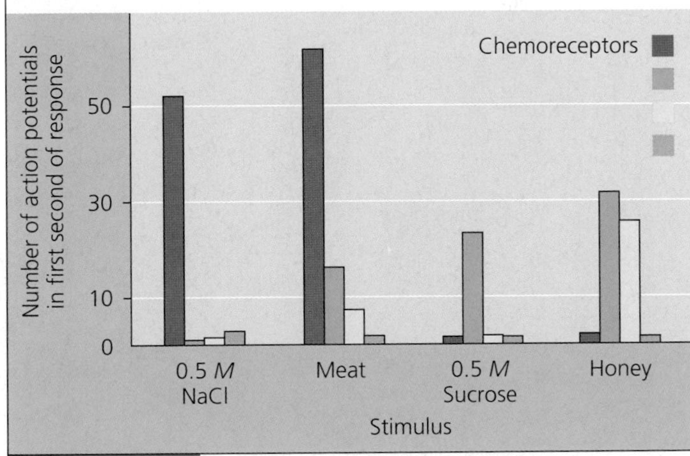

CONCLUSION Any natural food probably stimulates multiple chemoreceptors. By integrating sensations, the insect's brain can apparently distinguish a very large number of tastes.

Taste in Humans

The receptor cells for taste in humans are modified epithelial cells organized into **taste buds,** which are scattered in several areas of the tongue and mouth, as shown in **Figure 49.14** on the next page. Most taste buds on the tongue are associated with nipple-shaped projections called papillae. In addition to the four familiar taste perceptions—sweet, sour, salty, and bitter—a fifth, called umami (Japanese for "delicious"), is elicited by the amino acid glutamate and perhaps others. Glutamate is the key component of the flavor enhancer monosodium glutamate (MSG) and is abundant in foods such as meat and aged cheese. Each type of taste receptor can be stimulated by a broad range of chemicals but is most responsive to a particular type of substance. Complex flavors are perceived as the brain integrates the differential input from many taste receptors.

Transduction in taste receptors occurs by several mechanisms. Chemoreceptors that detect saltiness—mainly the presence of sodium ions (Na^+)—and sourness—hydrogen ions (H^+) generated by acids—have channels in their plasma membrane through which these ions can diffuse. The influx of Na^+ or H^+ depolarizes the cell. The mechanism for umami chemoreceptors may involve the binding of glutamate to Na^+ channels; when glutamate is bound, the channels open, Na^+ diffuses into the cell, and the cell depolarizes. In some chemoreceptors that detect bitter substances, such as quinine, the substance binds to potassium ion (K^+) channels and *closes* them; the resulting decrease in the membrane's permeability to K^+ depolarizes the cell. Finally, sweetness is detected by chemoreceptors that have receptor proteins for sugars; binding of a sugar molecule to a receptor protein triggers a signal transduction pathway that results in depolarization (see Figure 49.14). In all taste receptors, the depolarization causes

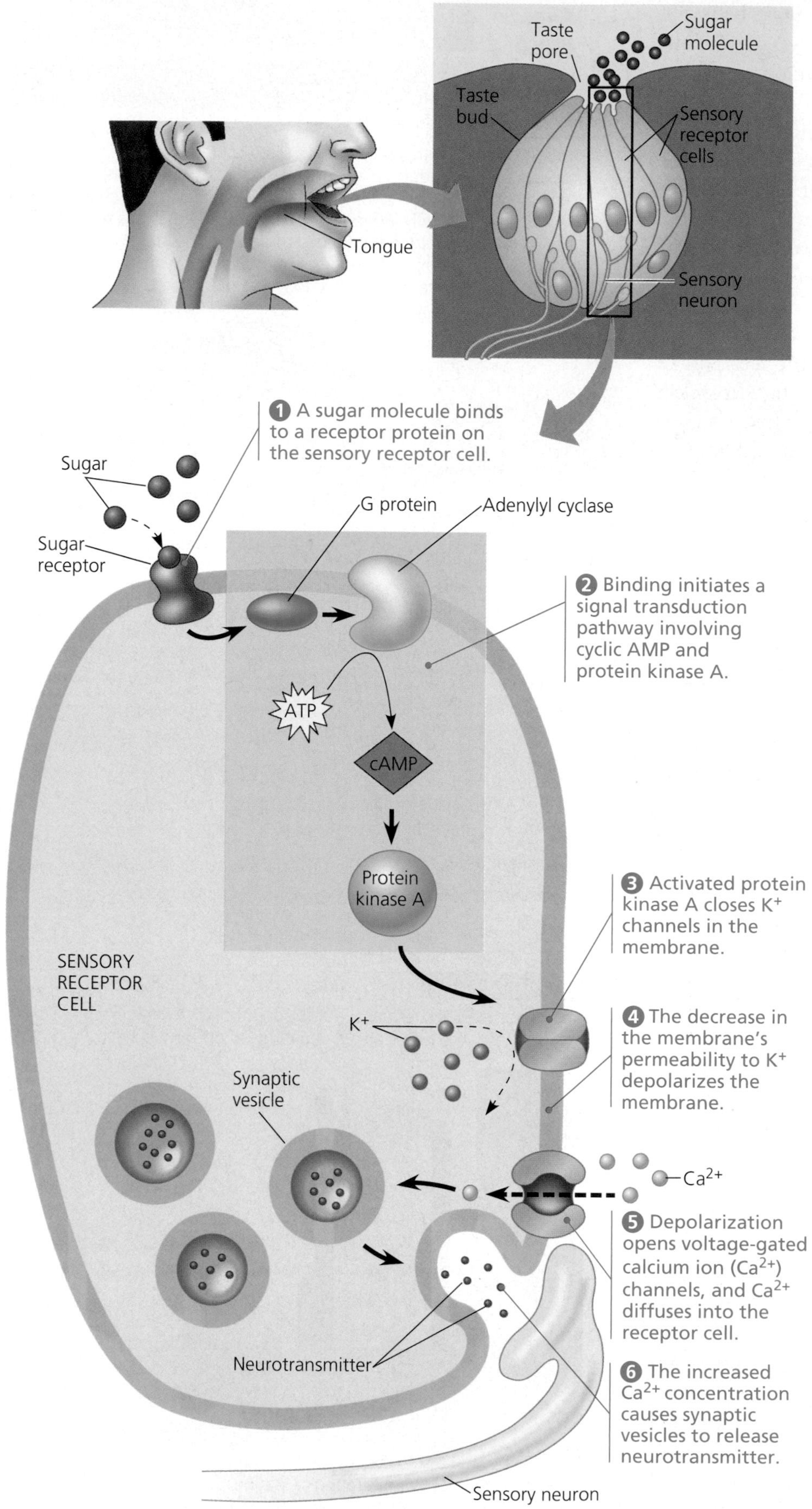

▲ **Figure 49.14 Sensory transduction by a sweetness receptor.**

the cell to release neurotransmitter onto a sensory neuron, which transmits action potentials to the brain.

Smell in Humans

Olfactory receptor cells are neurons that line the upper portion of the nasal cavity and send impulses along their axons directly to the olfactory bulb of the brain **(Figure 49.15)**. The receptive ends of the cells contain cilia that extend into the layer of mucus coating the nasal cavity. When an odorous substance, or odorant, diffuses into this region, it binds to specific proteins called odorant receptors (ORs) on the plasma membrane of the olfactory cilia. The binding triggers a signal transduction pathway involving a G protein, the enzyme adenylyl cyclase, and the second messenger cyclic AMP. The second messenger opens channels in the plasma membrane that are permeable to both Na^+ and Ca^{2+} ions. The influx of these ions depolarizes the membrane, causing the receptor cell to generate action potentials.

Humans can distinguish thousands of different odors, each caused by a structurally distinct odorant. This level of sensory discrimination requires many different ORs. There are more than 1,000 OR genes—about 3% of all human genes. Each olfactory receptor cell expresses only one or at most a few OR genes. Cells with different odorant selectivities are interspersed in the nasal cavity, but their axons sort themselves out in the olfactory bulb. Cells that express the same OR gene transmit action potentials to the same small region of the olfactory bulb. In 2004, Richard Axel, of Columbia University, and Linda Buck, of the Fred Hutchinson Cancer Research Center in Seattle, shared a Nobel Prize for their studies of the gene family and receptors that function in olfaction.

Although the receptors and brain pathways for taste and smell are independent, the two senses do interact. Indeed, much of what we call taste is really smell. If the olfactory system is blocked, as by a head cold, the perception of taste is sharply reduced.

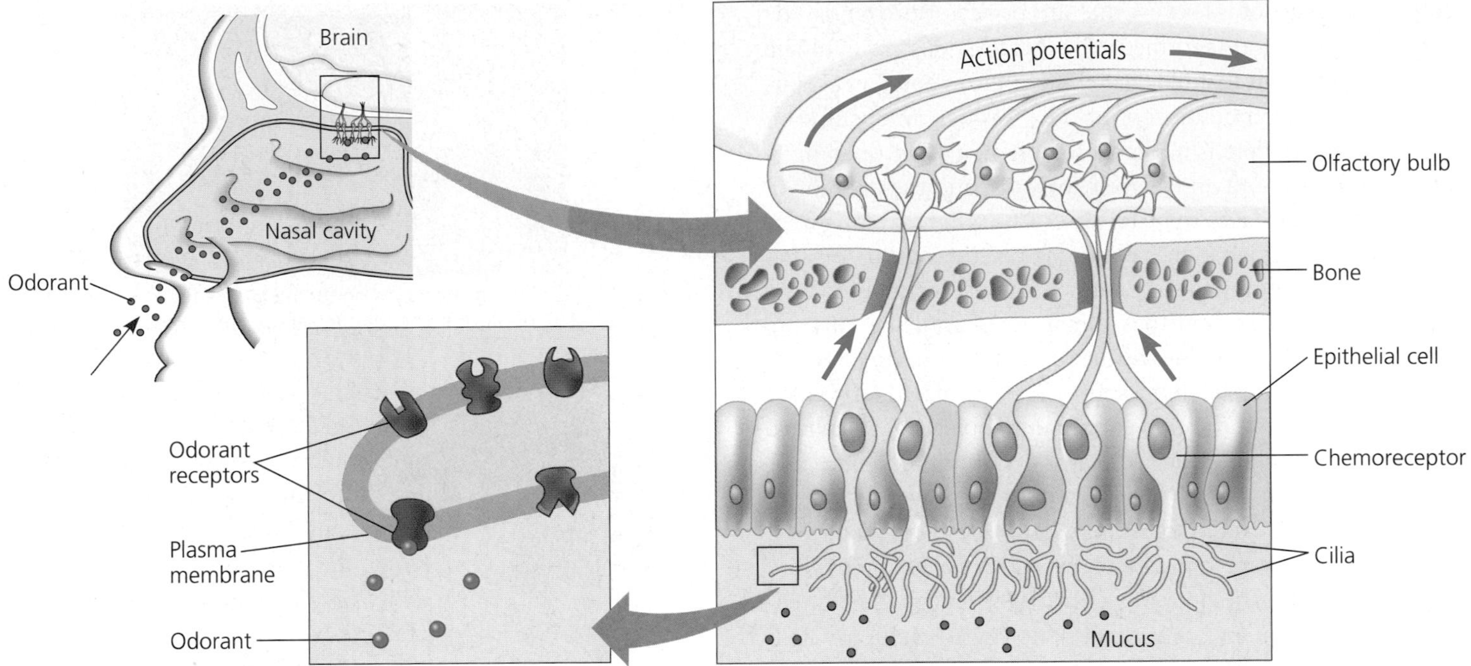

▲ **Figure 49.15 Smell in humans.** Odorant molecules bind to specific receptor proteins in the plasma membrane of chemoreceptors, triggering action potentials.

Concept Check 49.3

1. How is the location of gustatory sensilla on a fly an adaptation to the fly's feeding behavior?
2. Compare sensory transduction in gustatory chemoreceptors for sweetness and olfactory chemoreceptors.

For suggested answers, see Appendix A.

Concept 49.4

Similar mechanisms underlie vision throughout the animal kingdom

Many types of light detectors have evolved in the animal kingdom, from simple clusters of cells that detect only the direction and intensity of light to complex organs that form images. Despite their diversity, all photoreceptors contain similar pigment molecules that absorb light, and most, if not all, photoreceptors in the animal kingdom may be homologous. Animals as diverse as flatworms, annelids, arthropods, and vertebrates share genes associated with the embryonic development of photoreceptors. Thus, the genetic underpinnings of all photoreceptors may have evolved in the earliest bilateral animals. The specific types of eyes that form in an animal depend on developmental patterns regulated by genetic mechanisms that evolved later, and effects of those mechanisms appear to be superimposed on the common ancestral mechanism.

Vision in Invertebrates

Most invertebrates have some kind of light-detecting organ. One of the simplest is the ocellus (plural, *ocelli*) of planarians, which provides information about light intensity and direction but does not form images **(Figure 49.16)**. Sometimes

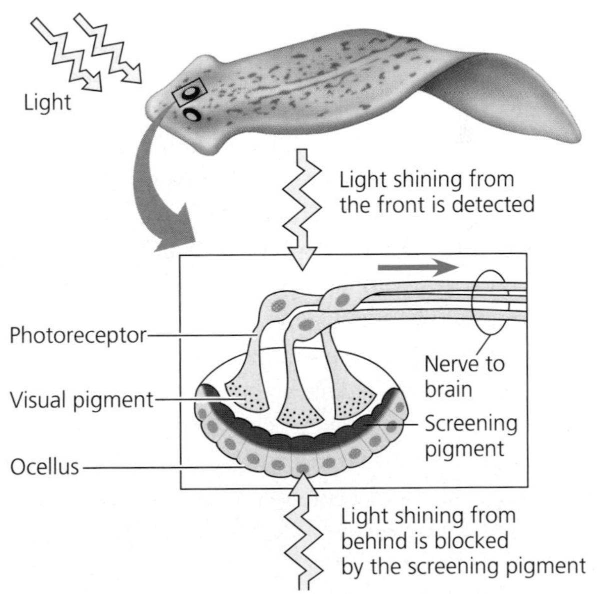

▲ **Figure 49.16 Ocelli and orientation behavior of a planarian.** The brain directs the body to turn until the sensations from the two ocelli are equal and minimal, causing the animal to move away from light.

called an eyespot or eye cup, an ocellus surrounds the photoreceptors and is formed by a layer of cells containing a screening pigment that blocks light. Light can enter an ocellus and stimulate the photoreceptors only through an opening on one side where there is no screening pigment. The opening of one ocellus faces left and slightly forward, and the opening of the other ocellus faces right-forward. Thus, light shining from one side of the planarian can enter only the ocellus on that side. The brain compares the rate of action potentials coming from the two ocelli, and the animal turns until the rates are equal and minimal. The result is that the animal moves directly away from the light source until it reaches a shaded location beneath a rock or some other object, a behavioral adaptation that helps hide the planarian from predators.

Two major types of image-forming eyes have evolved in invertebrates: the compound eye and the single-lens eye. **Compound eyes** are found in insects and crustaceans (phylum Arthropoda) and in some polychaete worms (phylum Annelida). A compound eye consists of up to several thousand light detectors called **ommatidia** (the "facets" of the eye), each with its own light-focusing lens **(Figure 49.17)**. Each ommatidium detects light from a tiny portion of the visual field. Differences in the intensity of light entering the many ommatidia result in a mosaic image. The animal's brain may sharpen the image when it integrates the visual information. The compound eye is extremely capable of detecting movement, an important adaptation for flying insects and small animals constantly threatened with predation. For comparison, note that the human eye can distinguish light flashes up to about 50 flashes per second, whereas the compound eyes of some insects can detect the flickering of a light flashing 330 times per second. Such an insect viewing a movie could easily resolve each frame of the film as a separate still image. Insects also have excellent color vision, and some (including bees) can see into the ultraviolet range of the electromagnetic spectrum, which is invisible to us. In studying animal behavior, we cannot extrapolate our sensory world to other species; different animals have different sensitivities and different brain organizations.

Among invertebrates, **single-lens eyes** are found in some jellies, polychaetes, spiders, and many molluscs. A single-lens eye works on a camera-like principle. The eye of an octopus or squid, for example, has a small opening, the pupil, through which light enters. Analogous to a camera's adjustable aperture (f-stop), the iris changes the diameter of the pupil; behind the pupil, a single lens focuses light on a layer of photoreceptors. Also similar to a camera's action, muscles in an invertebrate's single-lens eye move the lens forward or backward, focusing on objects at different distances.

The Vertebrate Visual System

Like the single-lens eyes of many invertebrates, the eyes of vertebrates are also camera-like, but they evolved independently in the vertebrate lineage and differ from the single-lens eyes of invertebrates in several details. Although the eye is the first stage in vision, remember that it is actually the brain that "sees." Thus, to understand vision, we must examine how the vertebrate eye generates sensations (action potentials) and then follow these signals to the visual centers of the brain, where images are perceived.

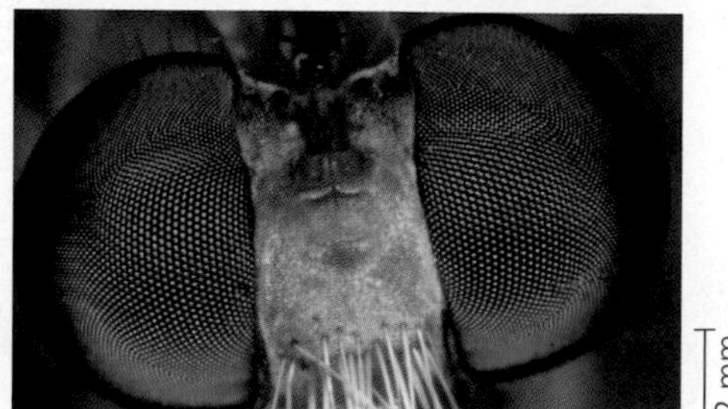

(a) The faceted eyes on the head of a fly, photographed with a stereomicroscope.

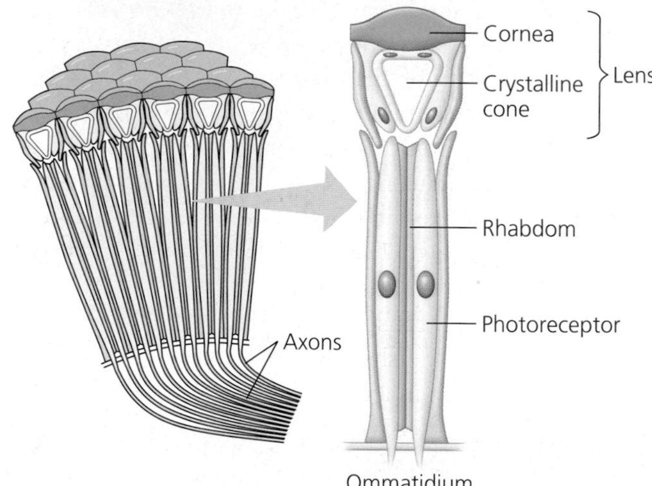

(b) The cornea and crystalline cone of each ommatidium function as a lens that focuses light on the rhabdom, a stack of pigmented plates inside a circle of photoreceptors. The rhabdom traps light and guides it to photoreceptors. The image formed by a compound eye is a mosaic of dots produced by different intensities of light entering the many ommatidia from different angles.

▲ **Figure 49.17 Compound eyes.**

Structure of the Eye

The globe of the vertebrate eye, or eyeball, consists of a tough, white outer layer of connective tissue called the **sclera** and a thin, pigmented inner layer that is called the **choroid** **(Figure 49.18)**. A delicate layer of epithelial cells forms a mucous membrane, the **conjunctiva**, that covers the outer surface of the sclera and helps keep the eye moist. At the front of the eye, the sclera becomes the transparent **cornea**, which lets light into the eye and acts as a fixed lens. The conjunctiva does not cover the cornea. The anterior choroid forms the doughnut-shaped **iris**, which gives the eye its color. By changing size, the iris regulates the amount of light entering the **pupil**, the hole in the center of the iris. Just inside the choroid, the **retina** forms the innermost layer of the eyeball and contains the photoreceptors. Information from the photoreceptors leaves the eye at the optic disk, where the optic nerve attaches to the eye. Because there are no photoreceptors in the optic disk, this spot on the lower outside of the retina is a "blind spot": Light focused onto that part of the retina is not detected.

The **lens** and **ciliary body** divide the eye into two cavities, an anterior cavity between the lens and the cornea and a much larger posterior cavity behind the lens within the eyeball itself. The ciliary body constantly produces the clear, watery **aqueous humor** that fills the anterior cavity. Blockage of the ducts that drain the aqueous humor can produce glaucoma, a condition in which increased pressure compresses the retina, causing blindness. The posterior cavity, filled with the jellylike **vitreous humor**, constitutes most of the volume of the eye. The aqueous and vitreous humors function as liquid lenses that help focus light onto the retina. The lens itself is a transparent protein disk. Like squids and octopuses, many fishes focus by moving the lens forward or backward, as in a camera. Humans and other mammals, however, focus by changing the shape of the lens **(Figure 49.19)**. When focusing on a close object, the lens becomes almost spherical, a change called **accommodation.** When viewing a distant object, the lens is flattened.

The human retina contains about 125 million **rods** and about 6 million **cones,** two types of photoreceptors that are named for their shapes. These cells account for about 70% of all the sensory receptors in the body, a fact that underscores the importance of the eyes and visual information in how humans perceive their environment.

Rods and cones have different functions in vision, and the relative numbers of these two photoreceptors in the retina are partly correlated with whether an animal is most active during the day or at night. Rods are more sensitive to light but do not distinguish colors; they enable us to see at night, but only in black and white. Because cones are less sensitive, they contribute very little to night vision, but they can distinguish colors in daylight. Color vision is found in all vertebrate classes, though not in all species. Most fishes, amphibians, and reptiles, including birds, have strong color

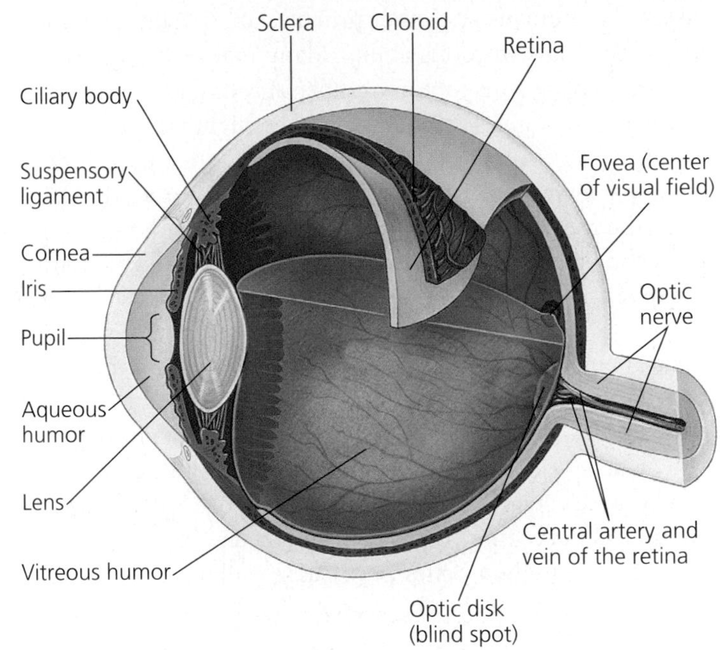

▲ **Figure 49.18 Structure of the vertebrate eye.** In this longitudinal section of the eye, the jellylike vitreous humor is illustrated only in the lower half of the eyeball. The conjunctiva, a mucous membrane that surrounds the sclera, is not shown.

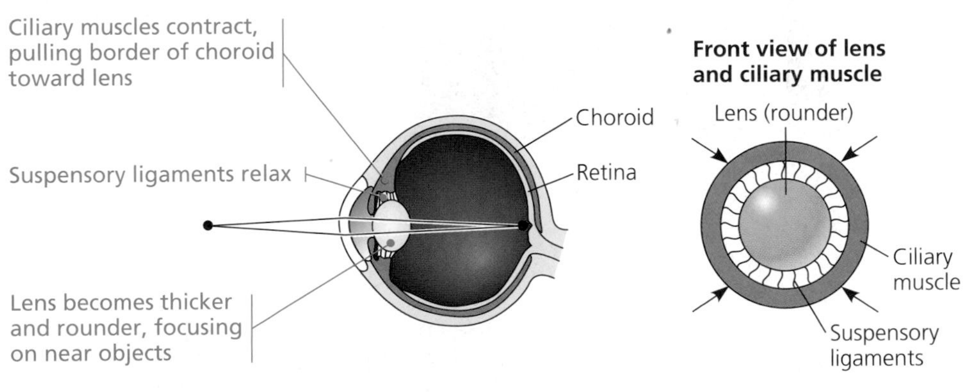

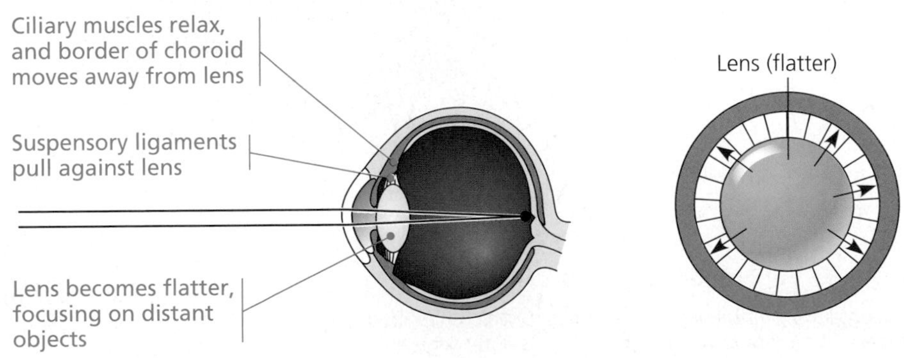

▲ **Figure 49.19 Focusing in the mammalian eye.** Ciliary muscles control the shape of the lens, which bends light and focuses it on the retina. The thicker the lens, the more sharply the light is bent.

vision, but humans and other primates are among the minority of mammals with this ability. Many mammals are nocturnal, and having a maximum number of rods in the retina is an adaptation that gives these animals keen night vision. Cats, for instance, are usually most active at night; they have limited color vision and probably see a pastel world during the day.

In the human eye, rods are found in greatest density at the peripheral regions of the retina and are completely absent from the **fovea,** the center of the visual field (see Figure 49.18). You cannot see a dim star at night by looking at it directly because the star's image is focused on the fovea; if you view the star at an angle, however, focusing the image on a region of the retina that contains rods, you will be able to see the star. In daylight, you achieve your sharpest vision by looking directly at an object because cones are most dense at the fovea, where there are about 150,000 cones per square millimeter. Some birds have more than a million cones per square millimeter, which enables species such as hawks to spot mice and other small prey from high in the sky. In the retina, as in all biological structures, variations represent evolutionary adaptations.

Sensory Transduction in the Eye

Each rod or cone in the vertebrate retina contains visual pigments that consist of a light-absorbing molecule called **retinal** (a derivative of vitamin A) bonded to a membrane protein called an **opsin.** Opsins vary in structure from one type of photoreceptor to another, and the light-absorbing ability of retinal is affected by the specific identity of its opsin partner. Rods have one type of opsin, which, combined with retinal, makes up the visual pigment **rhodopsin.** To understand the relationship between the structure of rods and light absorption, examine **Figure 49.20**.

As Figure 49.20 explains, rhodopsin changes shape when it absorbs light, a process referred to as "bleaching." Bright light keeps rhodopsin bleached, and the rods become unresponsive. When you move from a very bright environment into a dark one, such as when you enter a movie theater on a sunny afternoon, you are initially almost blind to the faint light. There is too little light to stimulate your cones, and it takes at least a few minutes for your bleached rods to become fully responsive again.

▼ **Figure 49.20 Rod structure and light absorption.**

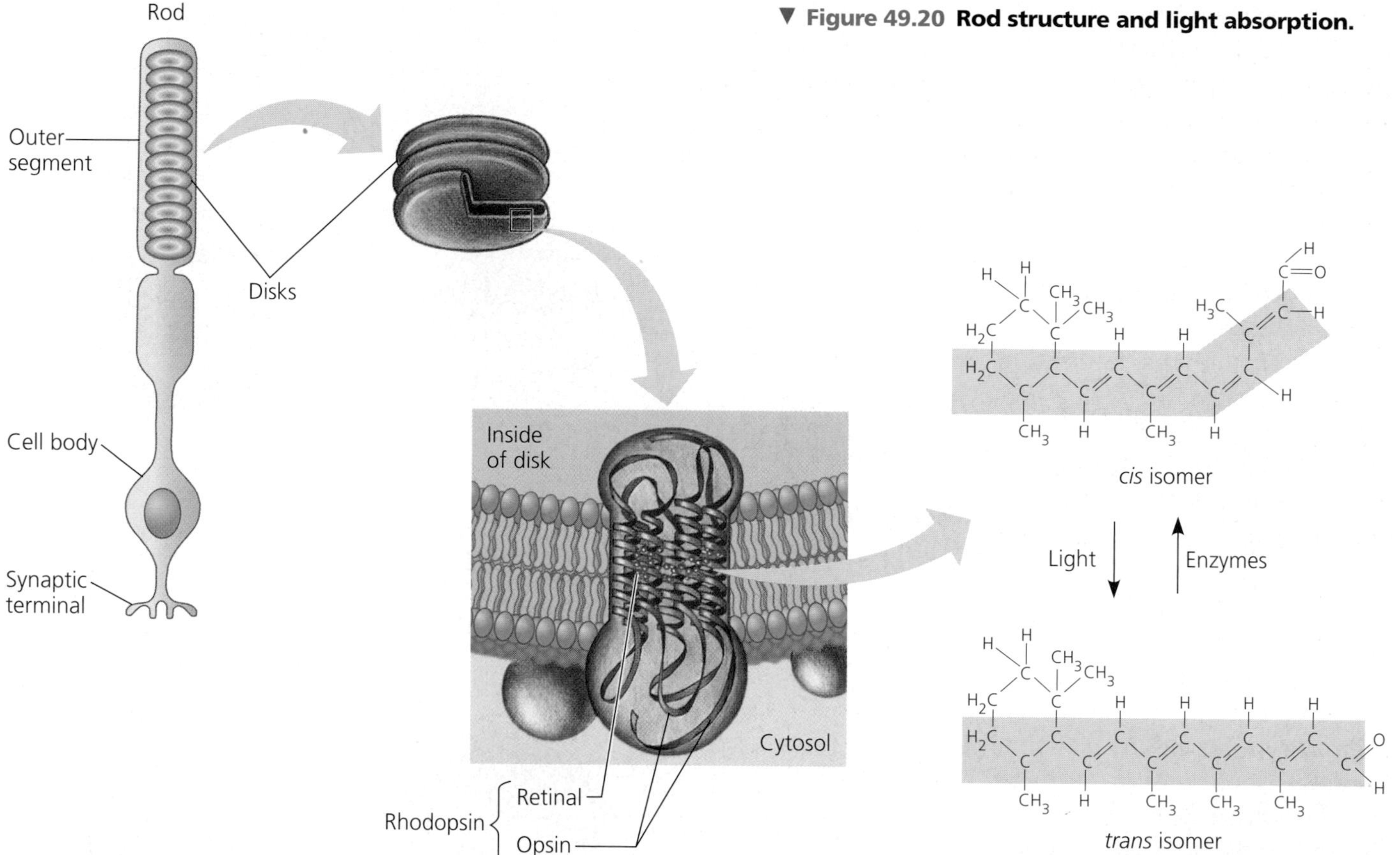

(a) Rods contain the visual pigment rhodopsin, which is embedded in a stack of membranous disks in the rod's outer segment. Rhodopsin consists of the light-absorbing molecule retinal bonded to opsin, a protein. Opsin has seven α helices that span the disk membrane.

(b) Retinal exists as two isomers. Absorption of light converts the *cis* isomer to the *trans* isomer, which causes opsin to change its conformation (shape). After a few minutes, retinal detaches from opsin. In the dark, enzymes convert retinal back to its *cis* form, which recombines with opsin, forming rhodopsin.

Color vision involves even more complex signal processing than the rhodopsin mechanism in rods. There are three classes of cones in the retina—red, green, and blue—each with its own type of opsin associated with retinal in a visual pigment. The three visual pigments of cones are collectively called **photopsins.** The name of each cone class refers to the color that its type of photopsin absorbs best. These visual pigments have overlapping absorption spectra, and the brain's perception of intermediate hues depends on the differential stimulation of two or more classes of cones. For example, when both red and green cones are stimulated, we may see yellow or orange, depending on which class is more strongly stimulated. Color blindness, which is more common in males than in females because it is generally inherited as a sex-linked trait (see Figure 15.10), is the result of a deficiency or absence of one or more types of photopsin.

Processing Visual Information

The processing of visual information begins in the retina itself, where both rods and cones make synapses with neurons called **bipolar cells.** In the dark, rods and cones are depolarized, and they continually release the neurotransmitter glutamate (see Table 48.1) at these synapses. This steady glutamate release in the dark depolarizes some bipolar cells and hyperpolarizes others, depending on the type of postsynaptic receptor molecules they contain. In the light, rods and cones hyperpolarize, shutting off their release of glutamate **(Figure 49.21)**. In response, the bipolar cells that are depolarized by glutamate hyperpolarize, and those that are hyperpolarized by glutamate depolarize **(Figure 49.22)**. Three other types of neurons contribute to information processing in the retina: ganglion cells, horizontal cells, and amacrine cells

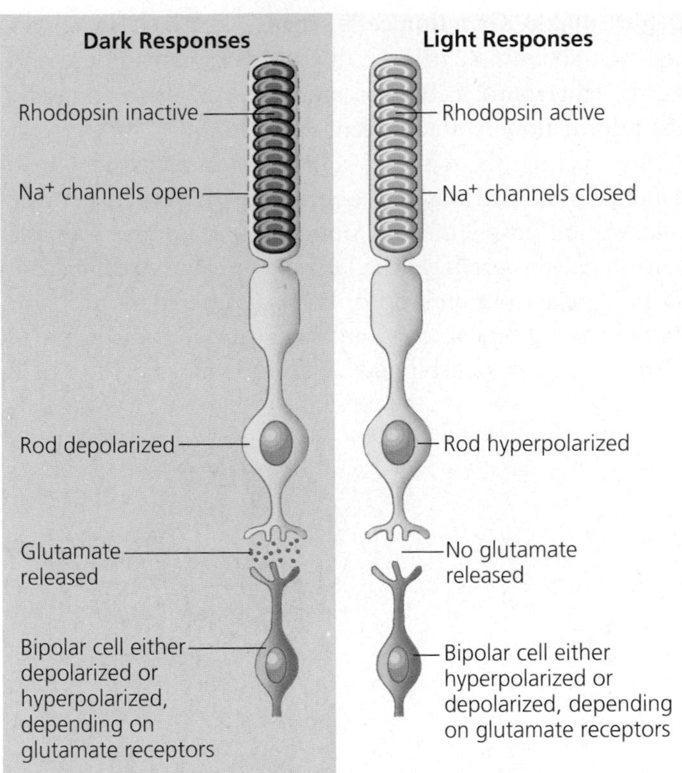

▲ **Figure 49.22 The effect of light on synapses between rod cells and bipolar cells.**

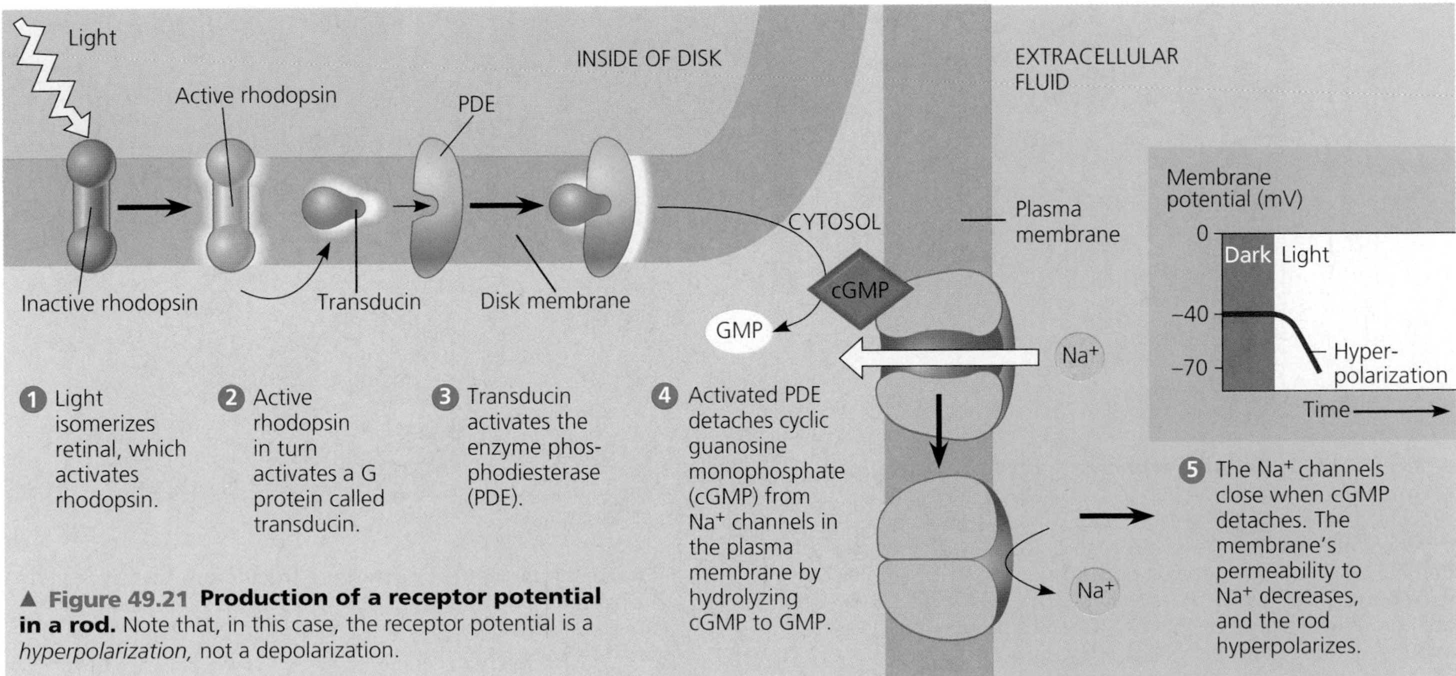

▲ **Figure 49.21 Production of a receptor potential in a rod.** Note that, in this case, the receptor potential is a *hyperpolarization,* not a depolarization.

(Figure 49.23). **Ganglion cells** synapse with bipolar cells and transmit action potentials to the brain via axons in the optic nerve. **Horizontal cells** and **amacrine cells** help integrate the information before it is sent to the brain.

Signals from the rods and cones may follow either a vertical or a lateral pathway in the retina. In the vertical pathway, information passes directly from photoreceptors to bipolar cells to ganglion cells. In the lateral pathway, horizontal cells carry signals from one rod or cone to other photoreceptors and to several bipolar cells, and amacrine cells distribute information from one bipolar cell to several ganglion cells.

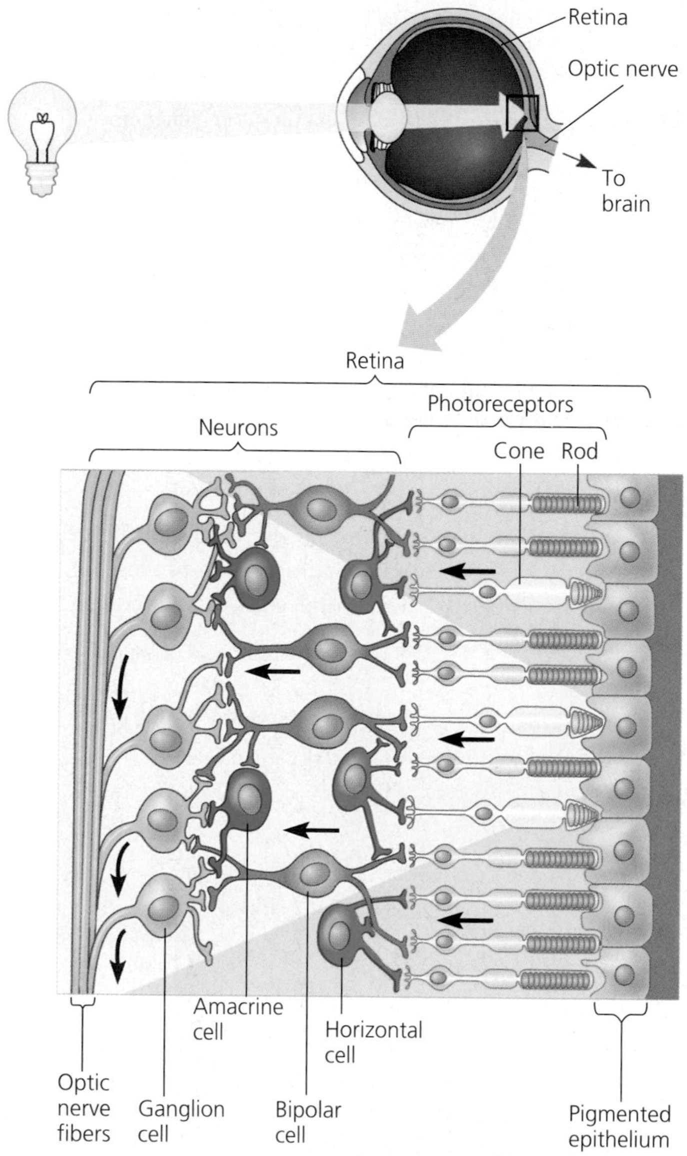

▲ **Figure 49.23 Cellular organization of the vertebrate retina.** Light must pass through several relatively transparent layers of cells before reaching the rods and cones. These photoreceptors communicate via bipolar cells with ganglion cells, which have axons that transmit visual sensations (action potentials) to the brain. Each bipolar cell receives information from several rods or cones, and each ganglion cell from several bipolar cells. Horizontal and amacrine cells integrate information across the retina. Black arrows indicate the pathway of visual information from the photoreceptors to the optic nerve.

When an illuminated rod or cone stimulates a horizontal cell, the horizontal cell inhibits more distant photoreceptors and bipolar cells that are not illuminated, making the light spot appear lighter and the dark surroundings even darker. This form of integration, called **lateral inhibition**, sharpens edges and enhances contrast in the image. Lateral inhibition is repeated by the interactions of the amacrine cells with the ganglion cells and occurs at all levels of visual processing in the brain.

All the rods or cones that feed information to one ganglion cell form the *receptive field* for that cell. A larger receptive field (in which more rods or cones supply a ganglion cell) results in a less sharp image than a smaller receptive field because the larger field provides less information about exactly where light struck the retina. The ganglion cells of the fovea have very small receptive fields, so visual acuity is high in the fovea.

Axons of ganglion cells form the optic nerves that transmit sensations from the eyes to the brain **(Figure 49.24)**. The two

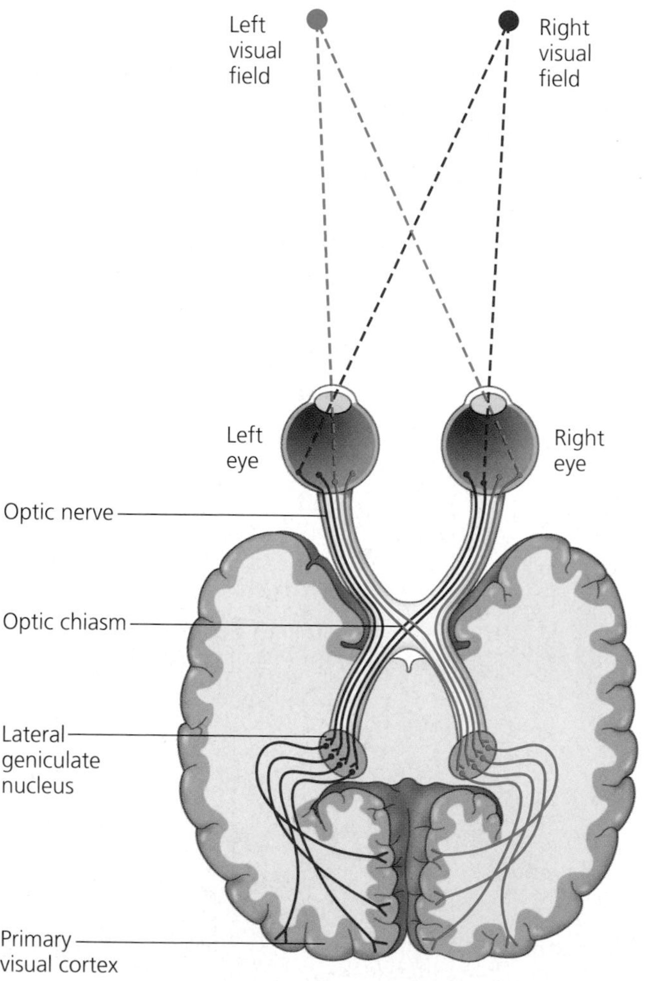

▲ **Figure 49.24 Neural pathways for vision.** Each optic nerve contains about a million axons that synapse with interneurons in the lateral geniculate nuclei. The nuclei relay sensations to the primary visual cortex, one of many brain centers that cooperate in constructing our visual perceptions.

optic nerves meet at the **optic chiasm** near the center of the base of the cerebral cortex. Axons in the optic nerves are routed at the optic chiasm such that sensations from the left visual field of both eyes are transmitted to the right side of the brain, and sensations from the right visual field are transmitted to the left side of the brain. Most ganglion cell axons lead to the **lateral geniculate nuclei** of the thalamus. Neurons in the lateral geniculate nuclei have axons that extend to the **primary visual cortex** in the occipital lobe of the cerebrum. Additional interneurons carry the information to higher-order visual processing and integrating centers elsewhere in the cortex.

Point-by-point information in the visual field is projected along neurons onto the visual cortex. How does the cortex convert a complex set of action potentials representing two-dimensional images focused on our retinas into three-dimensional perceptions of our surroundings? Researchers estimate that at least 30% of the cerebral cortex—hundreds of millions of interneurons in perhaps dozens of integrating centers—take part in formulating what we actually "see." Determining how these centers integrate such components of our vision as color, motion, depth, shape, and detail is the goal of an exciting, fast-moving research effort.

Concept Check 49.4

1. Contrast the light-detecting organs of planarians and flies. How is each organ adaptive for the lifestyle of the animal?
2. In a condition called presbyopia, the eye's lenses lose their elasticity and maintain a constant, flat shape. Explain how this condition would affect a person's vision.
3. Concentrating on the vertical pathway through the retina, explain how illuminating a photoreceptor can lead to an increased frequency of action potentials in ganglion cells.

For suggested answers, see Appendix A.

Concept 49.5

Animal skeletons function in support, protection, and movement

Throughout our discussions of sensory mechanisms, we have seen several examples of how sensory inputs to the nervous system result in specific behaviors: the escape maneuver of a moth that hears a bat's sonar, the upside-down swimming of a crayfish with manipulated statocysts, the feeding movements of a hydra when it tastes glutathione, and the movement of planarians away from light. Animal behavior flows in a seamless cycle involving continuous brain operations that generate actions, note the consequences of those actions via input from sensory receptors, and use that input to determine the next action. Underlying the diverse forms of behavior in animals are common fundamental mechanisms. Swimming, crawling, running, hopping, and flying all result from muscles working against some type of skeleton.

Types of Skeletons

The three main functions of a skeleton are support, protection, and movement. Most land animals would sag from their own weight if they had no skeleton to support them. Even an animal living in water would be a formless mass with no framework to maintain its shape. In many animals, a hard skeleton provides protection for soft tissues. For example, the vertebrate skull protects the brain, and the ribs of terrestrial vertebrates form a cage around the heart, lungs, and other internal organs. Skeletons also aid in movement by giving muscles something firm to work against. There are three main types of skeletons: hydrostatic skeletons, exoskeletons, and endoskeletons.

Hydrostatic Skeletons

A **hydrostatic skeleton** consists of fluid held under pressure in a closed body compartment. This is the main type of skeleton in most cnidarians, flatworms, nematodes, and annelids (see Chapter 33). These animals control their form and movement by using muscles to change the shape of fluid-filled compartments.

Among the cnidarians, for example, a hydra can elongate by closing its mouth and using contractile cells in the body wall to constrict the central gastrovascular cavity. Because water cannot be compressed very much, decreasing the diameter of the cavity forces the cavity to increase in length.

In planarians and other flatworms, the interstitial fluid is kept under pressure and functions as the main hydrostatic skeleton. Planarian movement results mainly from muscles in the body wall exerting localized forces against the hydrostatic skeleton.

Nematodes (roundworms) hold fluid in their body cavity, which is a pseudocoelom (see Figure 32.8b). The fluid is under a high pressure, and contractions of longitudinal muscles result in thrashing movements.

In earthworms and other annelids, the coelomic fluid functions as a hydrostatic skeleton. The coelomic cavity is divided by septa between the segments in many annelids, allowing the animal to change the shape of each segment individually, using both circular and longitudinal muscles. Such annelids use

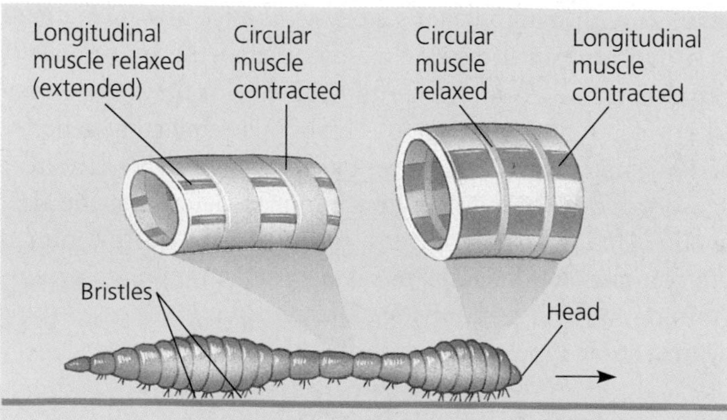

(a) Body segments at the head and just in front of the rear are short and thick (longitudinal muscles contracted; circular muscles relaxed) and anchored to the ground by bristles. The other segments are thin and elongated (circular muscles contracted; longitudinal muscles relaxed.)

(b) The head has moved forward because circular muscles in the head segments have contracted. Segments behind the head and at the rear are now thick and anchored, thus preventing the worm from slipping backward.

(c) The head segments are thick again and anchored in their new positions. The rear segments have released their hold on the ground and have been pulled forward.

▲ **Figure 49.25 Peristaltic locomotion in an earthworm.** Contraction of the longitudinal muscles thickens and shortens the worm; contraction of the circular muscles constricts and elongates it.

their hydrostatic skeleton for **peristalsis,** a type of movement on land produced by rhythmic waves of muscle contractions passing from front to back **(Figure 49.25)**.

Hydrostatic skeletons are well suited for life in aquatic environments. They may also cushion internal organs from shocks and provide support for crawling and burrowing in terrestrial animals. However, a hydrostatic skeleton cannot support terrestrial activities in which an animal's body is held off the ground, such as walking or running.

Exoskeletons

An **exoskeleton** is a hard encasement deposited on the surface of an animal. For example, most molluscs are enclosed in a calcareous (calcium carbonate) shell secreted by the mantle, a sheetlike extension of the body wall (see Figure 33.16). As the animal grows, it enlarges the shell by adding to its outer edge. Clams and other bivalves close their hinged shell using muscles attached to the inside of this exoskeleton.

The jointed exoskeleton of arthropods is a cuticle, a nonliving coat secreted by the epidermis. Muscles are attached to knobs and plates of the cuticle that extend into the interior of the body. About 30–50% of the cuticle consists of **chitin,** a polysaccharide similar to cellulose. Fibrils of chitin are embedded in a protein matrix, forming a composite material that combines strength and flexibility. Where protection is most important, the cuticle is hardened with organic compounds that cross-link the proteins of the exoskeleton. Some crustaceans, such as lobsters, harden portions of their exoskeleton even more by adding calcium salts. In contrast, there is little cross-linking of proteins or inorganic salt deposition in places where the cuticle must be thin and flexible, such as leg joints. An arthropod must shed its exoskeleton (molt) and produce a larger one with each growth spurt (see Figure 5.10).

Endoskeletons

An **endoskeleton** consists of hard supporting elements, such as bones, buried within the soft tissues of an animal. Sponges are reinforced by hard spicules of inorganic material or by softer fibers made of protein (see Figure 33.4). Echinoderms have an endoskeleton of hard plates called ossicles beneath the skin. The ossicles are composed of magnesium carbonate and calcium carbonate crystals and are usually bound together by protein fibers. Whereas the ossicles of sea urchins are tightly bound, the ossicles of sea stars are more loosely linked, allowing a sea star to change the shape of its arms.

Chordates have an endoskeleton consisting of cartilage, bone, or some combination of these materials (see Figure 40.5). The mammalian skeleton is built from more than 200 bones, some fused together and others connected at joints by ligaments that allow freedom of movement **(Figure 49.26)**. Anatomists divide the vertebrate skeleton into two main parts: the axial skeleton, consisting of the skull, vertebral column (backbone), and rib cage; and the appendicular skeleton, made up of limb bones and the pectoral and pelvic girdles that anchor the appendages to the axial skeleton. In each appendage, several types of joints provide flexibility for body movements.

Physical Support on Land

An engineer designing a bridge or tall building must take into account the effects of changes in size, or scale. An increase in size from a small-scale model to the real thing has a significant impact on building design. For example, the strength of a building support depends on its cross-sectional area, which increases with the square of its diameter. In

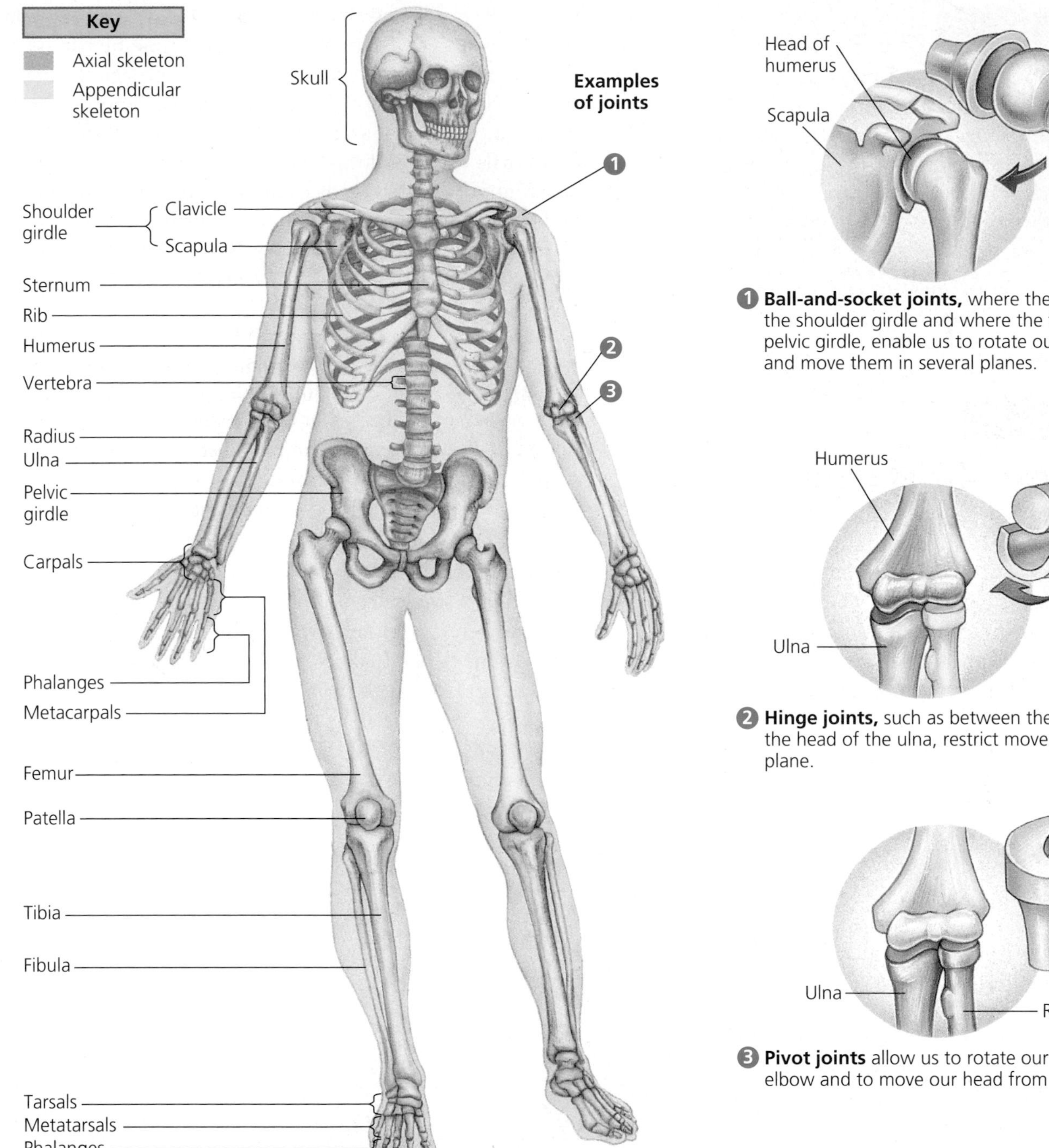

Key
Axial skeleton
Appendicular skeleton
Skull
Examples of joints
Shoulder girdle
Clavicle
Scapula
Sternum
Rib
Humerus
Vertebra
Radius
Ulna
Pelvic girdle
Carpals
Phalanges
Metacarpals
Femur
Patella
Tibia
Fibula
Tarsals
Metatarsals
Phalanges
Head of humerus
Scapula
1 Ball-and-socket joints, where the humerus contacts the shoulder girdle and where the femur contacts the pelvic girdle, enable us to rotate our arms and legs and move them in several planes.
Humerus
Ulna
2 Hinge joints, such as between the humerus and the head of the ulna, restrict movement to a single plane.
Ulna
Radius
3 Pivot joints allow us to rotate our forearm at the elbow and to move our head from side to side.

▲ **Figure 49.26 Bones and joints of the human skeleton.**

sharp contrast, the strain on that support depends on the building's weight, which increases with the cube of its height or other linear dimension. In common with the structure of a bridge or building, an animal's body structure must support its size. Consequently, a large animal has very different body proportions than a small animal. If a mouse were scaled up to an elephant's size, its slender legs would buckle under its weight.

In simply applying the building analogy, we might predict the size of an animal's leg bones to be directly proportional to the strain imposed by its body weight. However, our prediction would be inaccurate; animal bodies are complex and nonrigid, and the building analogy only partly explains the relationship between body structure and support. An animal's leg size relative to its body size is only part of the story. It turns out that body posture—the position of the legs relative to the main body—is a more important structural feature in supporting body weight, at least in mammals and birds. Muscles and tendons (connective tissue that joins muscle to bone), which hold the legs of large mammals relatively straight and positioned under the body, bear most of the load.

Concept Check 49.5

1. Explain how an earthworm would use its muscles to extend its body to its full length.
2. Contrast the properties of a lobster's exoskeleton on the gripping surfaces of its claws (pincers) and on the hinge joints of the claws.
3. Contrast the motions allowed by the two types of joints found in the elbow.

For suggested answers, see Appendix A.

Concept 49.6

Muscles move skeletal parts by contracting

At the cellular level, all animal movement is based on one of two basic contractile systems, both of which consume energy in moving protein strands against one another. These two systems of cell motility—one involving microtubules and the other microfilaments—were discussed in Chapter 6. Microtubules are responsible for the beating of cilia and the undulations of flagella. Microfilaments play a major role in amoeboid movement, and they are also the contractile elements of muscle cells.

The action of a muscle is always to contract; muscles can extend only passively. Therefore, the ability to move parts of the body in opposite directions requires that muscles be attached to the skeleton in antagonistic pairs, each member of the pair working against the other **(Figure 49.27)**. We flex an arm, for instance, by contracting the biceps, with the hinge joint of the elbow acting as the fulcrum of a lever. To extend the arm, we relax the biceps while the triceps on the opposite side contracts.

To understand how a muscle contracts, we must analyze its structure. We will begin by examining the structure and mechanism of contraction of vertebrate skeletal muscle and then turn our attention to other types of muscle.

Vertebrate Skeletal Muscle

Vertebrate **skeletal muscle**, which is attached to the bones and is responsible for their movement, is characterized by a hierarchy of smaller and smaller units **(Figure 49.28)**. A skeletal muscle consists of a bundle of long fibers running parallel to the length of the muscle. Each fiber is a single cell with multiple nuclei, reflecting its formation by the fusion of many embryonic cells. A muscle fiber is itself a bundle of smaller **myofibrils** arranged longitudinally. The myofibrils, in turn, are composed of two kinds of **myofilaments:** thin filaments and thick filaments. **Thin filaments** consist of two strands of actin and one strand of regulatory protein coiled around one another. **Thick filaments** are staggered arrays of myosin molecules.

Skeletal muscle is also called striated muscle because the regular arrangement of the myofilaments creates a pattern of light and dark bands. Each repeating unit is a **sarcomere**, the basic contractile unit of the muscle. The borders of the sarcomere, the **Z lines**, are lined up in adjacent myofibrils and contribute to the striations visible with a light microscope. Thin filaments are attached to the Z lines and project toward the center of the sarcomere, while thick filaments are centered in the sarcomere. In a muscle fiber at rest, thick and thin filaments do not overlap completely, and the area near the edge of the sarcomere where there are only thin filaments is called the **I band.** The **A band** is the broad region that corresponds to the length of the thick filaments. The thin filaments do not

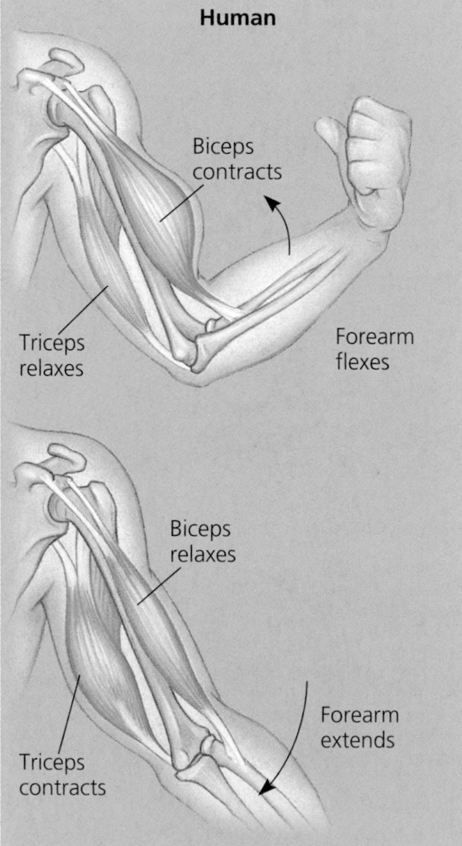

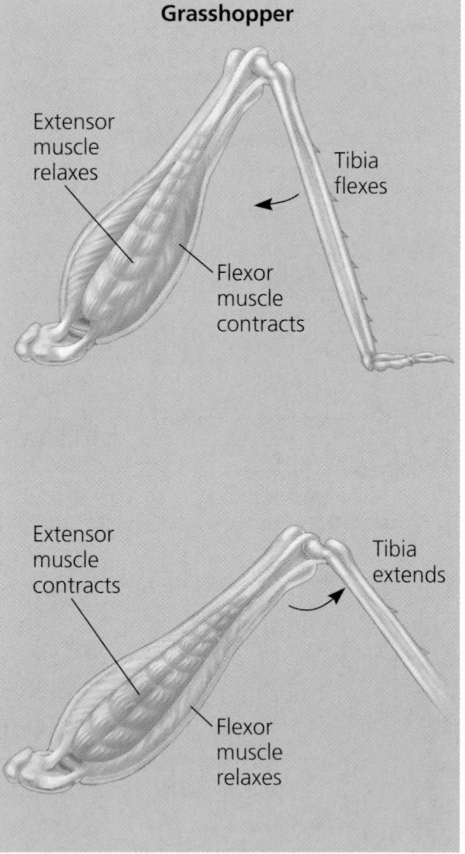

▲ **Figure 49.27 The interaction of muscles and skeletons in movement.** Back-and-forth movement of a body part is generally accomplished by antagonistic muscles. This arrangement works with either an endoskeleton or an exoskeleton.

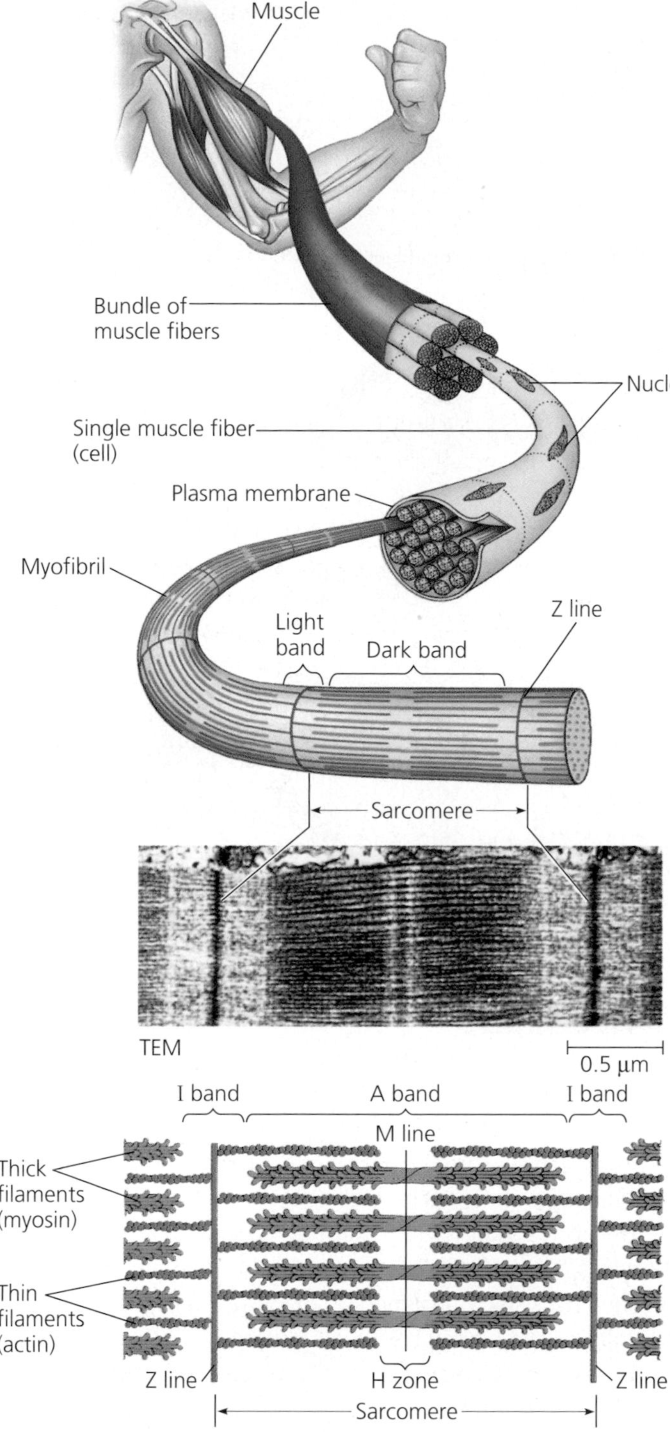

▲ **Figure 49.28 The structure of skeletal muscle.**

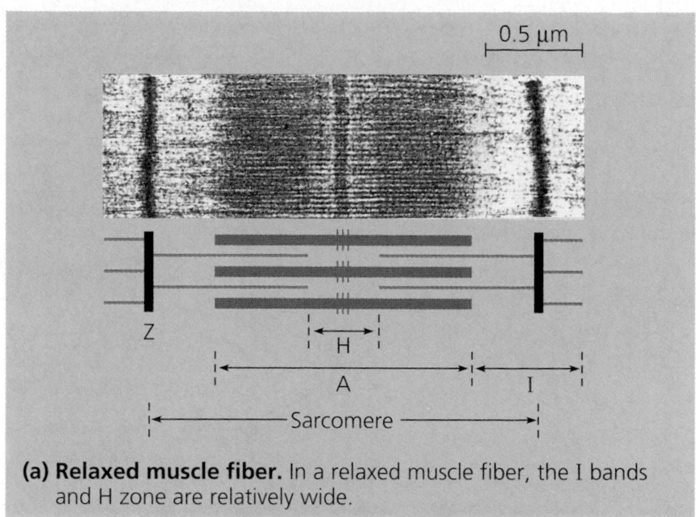

(a) Relaxed muscle fiber. In a relaxed muscle fiber, the I bands and H zone are relatively wide.

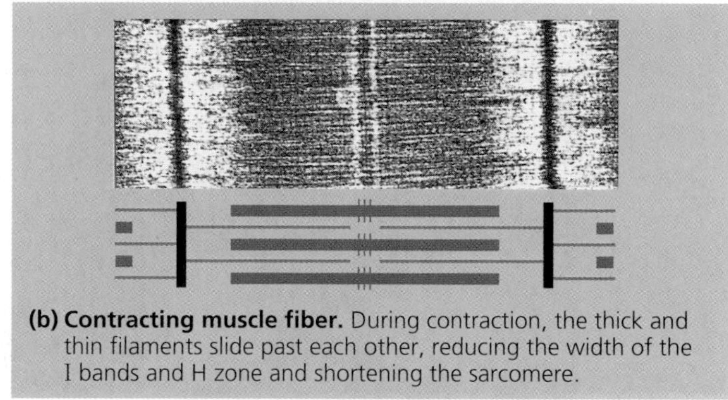

(b) Contracting muscle fiber. During contraction, the thick and thin filaments slide past each other, reducing the width of the I bands and H zone and shortening the sarcomere.

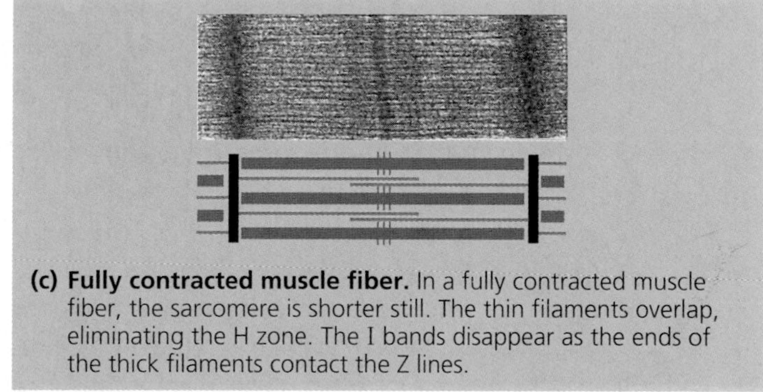

(c) Fully contracted muscle fiber. In a fully contracted muscle fiber, the sarcomere is shorter still. The thin filaments overlap, eliminating the H zone. The I bands disappear as the ends of the thick filaments contact the Z lines.

▲ **Figure 49.29 The sliding-filament model of muscle contraction.** These transmission electron micrographs show that the lengths of the thick (myosin) filaments (purple) and thin (actin) filaments (orange) remain the same as a muscle fiber contracts.

extend completely across the sarcomere, so the H zone in the center of the A band contains only thick filaments. This arrangement of thick and thin filaments is the key to how the sarcomere, and hence the whole muscle, contracts.

The Sliding-Filament Model of Muscle Contraction

We can explain much of what happens during contraction of a whole muscle by focusing on a single muscle fiber **(Figure 49.29)**. According to the **sliding-filament model** of muscle contraction, neither the thin filaments nor the thick filaments change in length when the sarcomere shortens; rather, the filaments slide past each other longitudinally, producing more overlap between the thin and thick filaments. As a result, both the region occupied only by thin filaments (the I band) and the region occupied only by thick filaments (the H zone) will shrink.

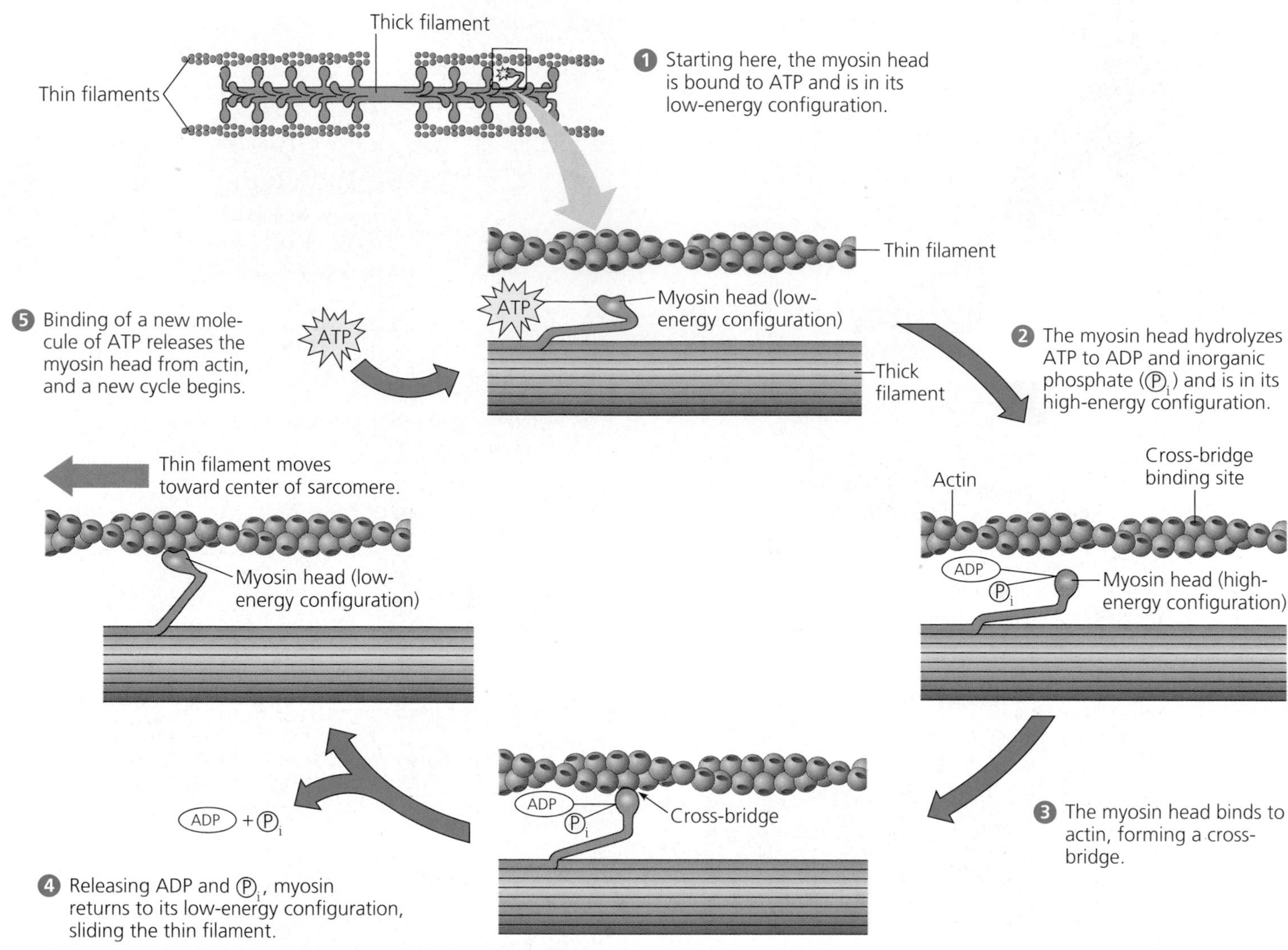
Thick filament
Thin filaments
1 Starting here, the myosin head is bound to ATP and is in its low-energy configuration.
Thin filament
ATP
Myosin head (low-energy configuration)
5 Binding of a new molecule of ATP releases the myosin head from actin, and a new cycle begins.
ATP
Thick filament
2 The myosin head hydrolyzes ATP to ADP and inorganic phosphate (Pi) and is in its high-energy configuration.
Thin filament moves toward center of sarcomere.
Myosin head (low-energy configuration)
Actin
Cross-bridge binding site
ADP
Pi
Myosin head (high-energy configuration)
ADP + Pi
ADP
Pi
Cross-bridge
3 The myosin head binds to actin, forming a cross-bridge.
4 Releasing ADP and Pi, myosin returns to its low-energy configuration, sliding the thin filament.

▲ **Figure 49.30 Myosin-actin interactions underlying muscle fiber contraction.**

The sliding of the filaments is based on the interaction between the actin and myosin molecules that make up the thick and thin filaments. Each myosin molecule consists of a long "tail" region and a globular "head" region extending to the side. The tail adheres to the tails of other myosin molecules that form the thick filament. The head is the center of bioenergetic reactions that power muscle contractions. It can bind ATP and hydrolyze it into ADP and inorganic phosphate. As shown in **Figure 49.30**, hydrolysis of ATP triggers steps in which myosin binds to actin, forming a cross-bridge and pulling the thin filament toward the center of the sarcomere. The cross-bridge is broken when a new molecule of ATP binds to the myosin head. In a repeating cycle, the free head cleaves the new ATP and attaches to a new binding site on another actin molecule farther along the thin filament. Each of the approximately 350 heads of a thick filament forms and re-forms about five cross-bridges per second, driving filaments past each other.

A typical muscle fiber at rest contains only enough ATP for a few contractions. The energy needed for repetitive contractions is stored in two other compounds: creatine phosphate and glycogen. Creatine phosphate can quickly make ATP by transferring a phosphate group to ADP. The resting supply of creatine phosphate is sufficient to sustain contractions for about 15 seconds. Glycogen is broken down to glucose, which can be used to generate ATP via glycolysis or aerobic respiration (see Chapter 9). Using the glucose from a typical muscle fiber's glycogen store, glycolysis can support about 1 minute of sustained contraction, while aerobic respiration can power contractions for nearly an hour.

The Role of Calcium and Regulatory Proteins

A skeletal muscle fiber contracts only when stimulated by a motor neuron. When the muscle fiber is at rest, the myosin-binding sites on the thin filament are blocked by the regulatory protein **tropomyosin (Figure 49.31a)**. For the muscle fiber to contract, those binding sites must be uncovered. This occurs when calcium ions (Ca^{2+}) bind to another set of regulatory proteins, the **troponin complex,** which controls the position of tropomyosin on the thin filament. Calcium binding rearranges

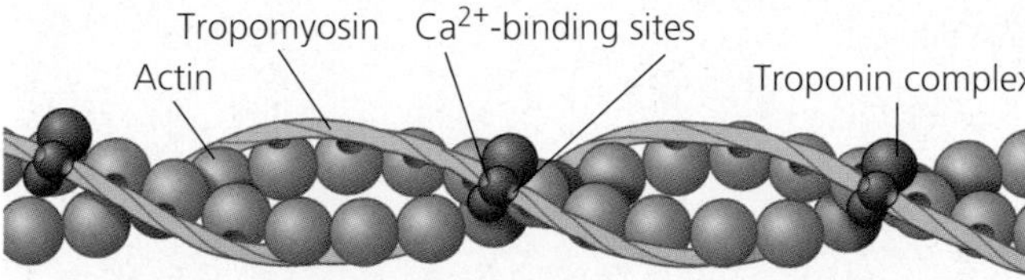

(a) Myosin-binding sites blocked.

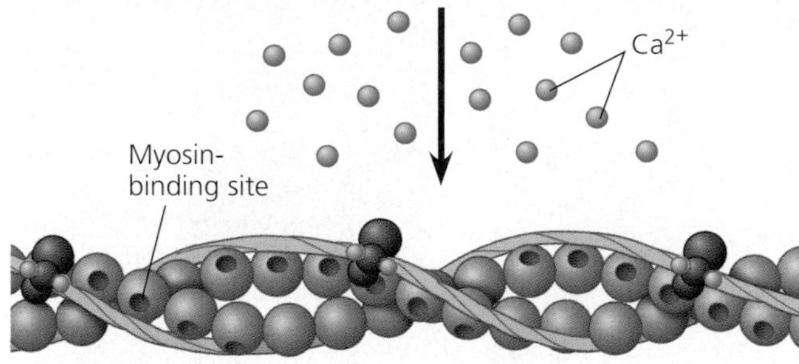

(b) Myosin-binding sites exposed.

▲ **Figure 49.31 The role of regulatory proteins and calcium in muscle fiber contraction.** Each thin filament consists of two strands of actin, tropomyosin, and the troponin complex.

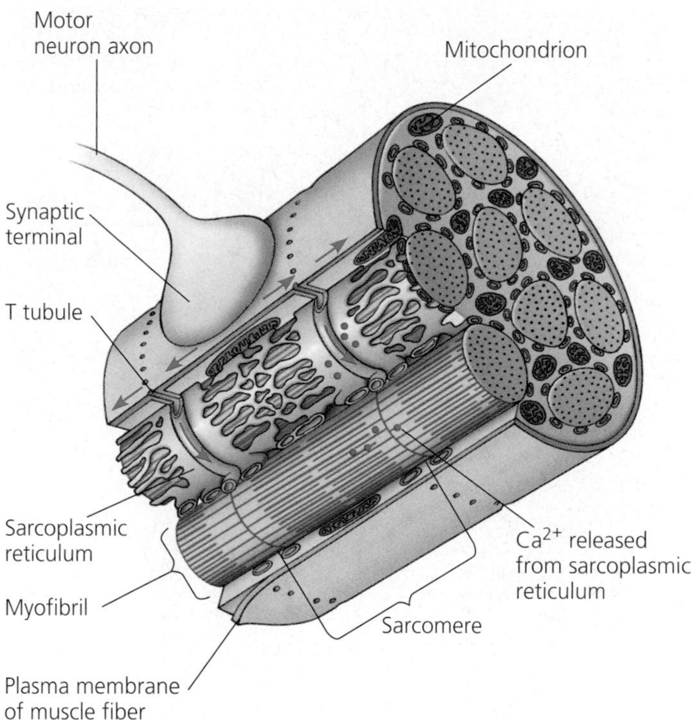

▲ **Figure 49.32 The roles of the sarcoplasmic reticulum and T tubules in muscle fiber contraction.** The synaptic terminal of a motor neuron releases acetylcholine, which depolarizes the plasma membrane of the muscle fiber. The depolarization causes action potentials (blue arrows) to sweep across the muscle fiber and deep into it along the transverse (T) tubules. The action potentials trigger the release of calcium (green dots) from the sarcoplasmic reticulum into the cytosol. Calcium initiates the sliding of filaments by allowing myosin to bind to actin.

the tropomyosin-troponin complex, exposing the myosin-binding sites on the thin filament (**Figure 49.31b**). When Ca^{2+} is present in the cytosol, the thin and thick filaments slide past each other, and the muscle fiber contracts. When the Ca^{2+} concentration falls, the binding sites are covered, and contraction stops.

The stimulus leading to the contraction of a skeletal muscle fiber is an action potential in a motor neuron that makes a synapse with the muscle fiber (**Figure 49.32**). The synaptic terminal of the motor neuron releases the neurotransmitter acetylcholine, depolarizing the muscle fiber and causing it to produce an action potential. That action potential spreads deep into the interior of the muscle fiber along infoldings of the plasma membrane called **transverse (T) tubules.** The T tubules make close contact with the **sarcoplasmic reticulum (SR),** a specialized endoplasmic reticulum. When the muscle fiber is at rest, the membrane of the SR pumps Ca^{2+} from the cytosol into the interior of the SR, which is thus an intracellular storehouse for Ca^{2+}. When the muscle fiber produces an action potential, however, the action potential opens Ca^{2+} channels in the SR, allowing Ca^{2+} to enter the cytosol. Calcium ions bind to the troponin complex, triggering contraction of the muscle fiber. Contraction stops when the SR pumps the Ca^{2+} back out of the cytosol, and tropomyosin again blocks the myosin-binding sites on the thin filaments. **Figure 49.33**, on the next page, reviews the steps in the contraction of a skeletal muscle fiber.

Several diseases cause paralysis by interfering with the excitation of skeletal muscle fibers by motor neurons. In amyotrophic lateral sclerosis (ALS), formerly called Lou Gehrig's disease, motor neurons in the spinal cord and brainstem degenerate, and the muscle fibers with which they synapse atrophy. ALS is progressive and usually fatal within five years after symptoms appear; there is no cure or treatment at this time. Botulism results from the consumption of an exotoxin secreted by the bacterium *Clostridium botulinum* in improperly preserved foods (see Chapter 27). The toxin, which paralyzes muscles by blocking the release of acetylcholine from motor neurons, is also injected into certain facial muscles to eliminate frown lines in "Botox" treatments. Myasthenia gravis is an autoimmune disease in which a person produces antibodies to the acetylcholine receptors on skeletal muscle fibers. The number of these receptors decreases, and the synaptic transmission between motor neurons and the muscle fibers becomes less effective.

Neural Control of Muscle Tension

When an action potential in a motor neuron releases acetylcholine on a skeletal muscle fiber, the muscle fiber responds by producing a brief, all-or-none contraction, called a twitch. However, our everyday experience shows that the contraction of a *whole muscle,* such as the biceps, is graded; we can voluntarily alter the extent and strength of its contraction. Experimental studies confirm this observation. There are two basic mechanisms by which the nervous system produces graded contractions of whole muscles: (1) by varying the number of muscle

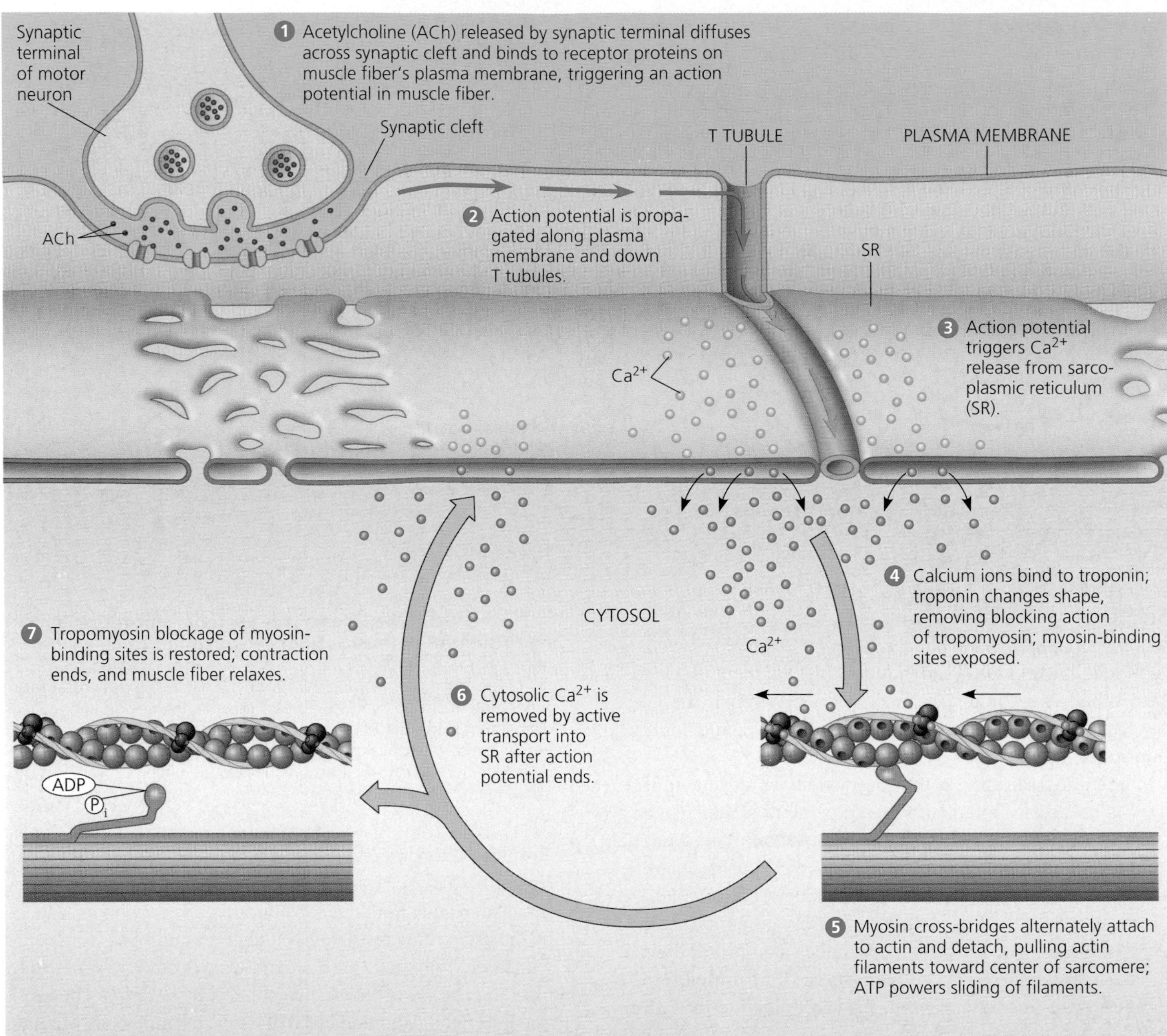

Synaptic terminal of motor neuron
1 Acetylcholine (ACh) released by synaptic terminal diffuses across synaptic cleft and binds to receptor proteins on muscle fiber's plasma membrane, triggering an action potential in muscle fiber.
Synaptic cleft
T TUBULE
PLASMA MEMBRANE
ACh
2 Action potential is propagated along plasma membrane and down T tubules.
SR
3 Action potential triggers Ca2+ release from sarcoplasmic reticulum (SR).
Ca2+
4 Calcium ions bind to troponin; troponin changes shape, removing blocking action of tropomyosin; myosin-binding sites exposed.
CYTOSOL
Ca2+
7 Tropomyosin blockage of myosin-binding sites is restored; contraction ends, and muscle fiber relaxes.
6 Cytosolic Ca2+ is removed by active transport into SR after action potential ends.
ADP
Pi
5 Myosin cross-bridges alternately attach to actin and detach, pulling actin filaments toward center of sarcomere; ATP powers sliding of filaments.

▲ **Figure 49.33 Review of contraction in a skeletal muscle fiber.**

fibers that contract and (2) by varying the rate at which muscle fibers are stimulated. Let's consider each mechanism in turn.

In a vertebrate skeletal muscle, each muscle fiber is innervated by only one motor neuron, but each branched motor neuron may synapse with many muscle fibers **(Figure 49.34)**. There may be hundreds of motor neurons controlling a muscle, each with its own pool of muscle fibers scattered throughout the muscle. A **motor unit** consists of a single motor neuron and all the muscle fibers it controls. When a motor neuron produces an action potential, all the muscle fibers in its motor unit contract as a group. The strength of the resulting contraction therefore depends on how many muscle fibers the motor neuron controls. In most muscles, the number of muscle fibers in different motor units ranges from a few to hundreds. The nervous system can thus regulate the strength of contraction in a whole muscle by determining how many motor units are activated at a given instant and by selecting large or small motor units to activate. The force (tension) developed by a muscle progressively increases as more and more of the motor neurons controlling the muscle are activated, a process called **recruitment** of motor neurons. Depending on the number of motor neurons your brain recruits and the size of their motor units, you can lift a fork or something much heavier, like your biology textbook.

Some muscles, especially those that hold the body up and maintain posture, are almost always partially contracted. Prolonged contraction can result in muscle fatigue caused by the depletion of ATP, dissipation of ion gradients required for

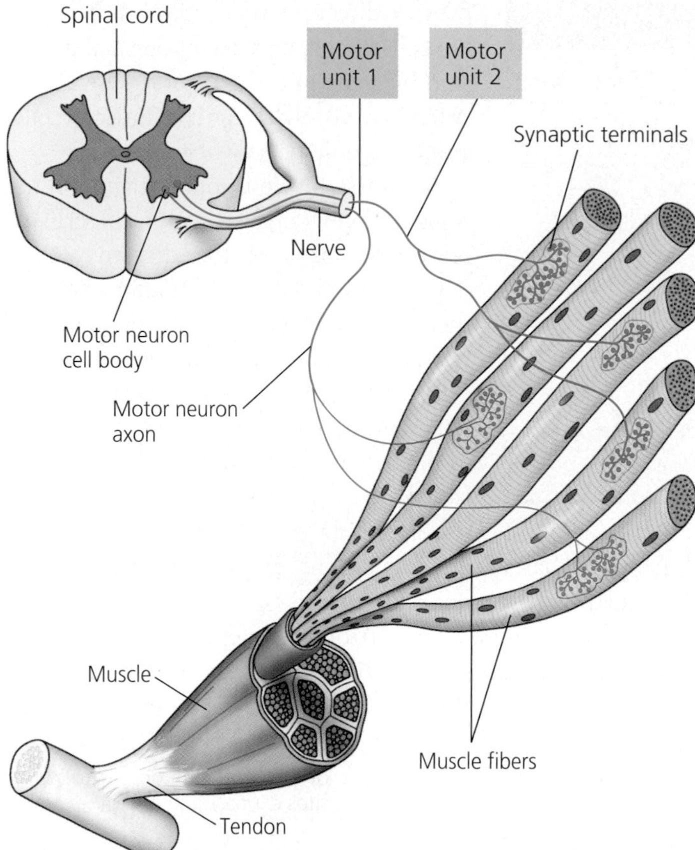

▲ **Figure 49.34 Motor units in a vertebrate skeletal muscle.** Each muscle fiber (cell) has a single synapse with one motor neuron, but each motor neuron typically synapses with several or many muscle fibers. A motor neuron and all the muscle fibers it controls constitute a motor unit.

normal electrical signaling, and accumulation of lactate (see Figure 9.17). Recent research suggests that although lactate may contribute to muscle fatigue, it may have a beneficial effect on muscle function. In a mechanism that reduces fatigue, the nervous system alternates activation among the motor units in a muscle, allowing different motor units to take turns maintaining the prolonged contraction.

The second mechanism by which the nervous system produces graded whole-muscle contractions is by varying the rate of muscle fiber stimulation. A single action potential will produce a twitch lasting about 100 milliseconds (msec) or less. If a second action potential arrives before the muscle fiber has completely relaxed, the two twitches will sum, resulting in greater tension **(Figure 49.35)**. Further summation occurs as the rate of stimulation increases. When the rate is high enough that the muscle fiber cannot relax at all between stimuli, the twitches fuse into one smooth, sustained contraction called **tetanus** (not to be confused with the disease of the same name). Motor neurons usually deliver their action potentials in rapid-fire volleys, and the resulting summation of tension results in the smooth contraction typical of tetanus rather than the jerky actions of individual twitches.

The increase in tension during summation and tetanus occurs because muscle fibers are not directly attached to bones

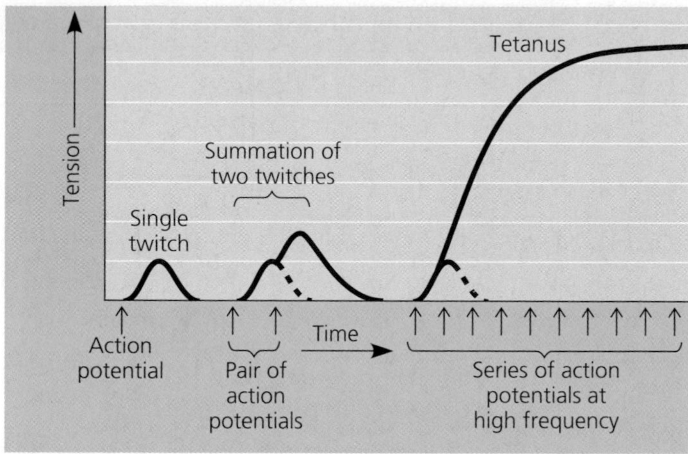

▲ **Figure 49.35 Summation of twitches.** This graph compares the tension developed in a muscle fiber in response to a single action potential in a motor neuron, a pair of action potentials, and a series of action potentials. The dashed lines show the tension that would have developed if only the first action potential had occurred.

but instead are connected via tendons and connective tissues. When a muscle fiber contracts, it stretches these elastic structures, which then transmit tension to the bones. In a single twitch, the muscle fiber begins to relax before the elastic structures are fully stretched. During summation, however, the high-frequency action potentials maintain an elevated concentration of calcium in the muscle fiber's cytosol, prolonging cross-bridge cycling and causing greater stretching of the elastic structures. During tetanus, the elastic structures are fully stretched, and all of the tension generated by the muscle fiber is transmitted to the bones.

Types of Muscle Fibers

All skeletal muscle fibers contract when stimulated by an action potential in a motor neuron, but the speed at which they contract differs among muscle fibers. This difference is mainly due to the rate at which their myosin heads hydrolyze ATP. Based on their speed of contraction, we can classify muscle fibers as either fast or slow. Fast fibers are used for brief, rapid, powerful contractions. In contrast, slow fibers, often found in muscles that maintain posture, can sustain long contractions. A slow fiber has less sarcoplasmic reticulum and slower calcium pumps than a fast fiber, so calcium remains in the cytosol longer. This causes a twitch in a slow fiber to last about five times as long as one in a fast fiber.

Another criterion for classifying muscle fibers is the major metabolic pathway they use for producing ATP. Fibers that rely mostly on aerobic respiration are called oxidative fibers, while those that primarily use glycolysis are called glycolytic fibers. Oxidative fibers are specialized to make use of a steady supply of energy: They have many mitochondria, a rich blood supply, and a large amount of an oxygen-storing protein called myoglobin. **Myoglobin**, the brownish-red pigment in the dark meat of poultry and fish, binds oxygen more tightly than does

Table 49.1 Types of Skeletal Muscle Fibers

	Slow Oxidative	Fast Oxidative	Fast Glycolytic
Contraction speed	Slow	Fast	Fast
Myosin ATPase activity	Slow	Fast	Fast
Major pathway for ATP synthesis	Aerobic respiration	Aerobic respiration	Glycolysis
Rate of fatigue	Slow	Intermediate	Fast
Fiber diameter	Small	Intermediate	Large
Mitochondria	Many	Many	Few
Capillaries	Many	Many	Few
Myoglobin content	High	High	Low
Color	Red	Red to pink	White

hemoglobin, so it can effectively extract oxygen from the blood. All glycolytic fibers are fast, but oxidative fibers can be either fast or slow. Therefore, considering both contraction speed and ATP synthesis, we can identify three main types of skeletal muscle fibers: slow oxidative, fast oxidative, and fast glycolytic. **Table 49.1** compares some of their characteristics.

Most human skeletal muscles contain all three fiber types, although the muscles of the eye and hand lack slow oxidative fibers. In muscles that have a mixture of fast and slow fibers, the relative proportions of each are genetically determined. However, if such a muscle is used repeatedly for activities requiring high endurance, some fast glycolytic fibers can develop into fast oxidative fibers. Since fast oxidative fibers fatigue more slowly than fast glycolytic fibers, the muscle as a whole will become more resistant to fatigue.

Other Types of Muscle

There are many types of muscles in the animal kingdom. As noted previously, however, they share the same fundamental mechanism of contraction: the sliding of actin and myosin filaments past each other. In addition to skeletal muscle, vertebrates have cardiac muscle and smooth muscle (see Figure 40.5).

Vertebrate **cardiac muscle** is found in only one place—the heart. Like skeletal muscle, cardiac muscle is striated. However, structural differences between skeletal and cardiac muscle fibers result in differences in their electrical and membrane properties. Whereas skeletal muscle fibers will not produce action potentials unless stimulated by a motor neuron, cardiac muscle cells have ion channels in their plasma membrane that cause rhythmic depolarizations, triggering action potentials without input from the nervous system. Action potentials of cardiac muscle cells last up to 20 times longer than those of the skeletal muscle fibers and play a key role in controlling the duration of contraction. Plasma membranes of adjacent cardiac muscle cells interlock at specialized regions called **intercalated disks,** where gap junctions (see Figure 6.31) provide direct electrical coupling between the cells. Thus, an action potential generated by a cell in one part of the heart will spread to all other cardiac muscle cells, and the whole heart will contract.

Smooth muscle is found mainly in the walls of hollow organs, such as blood vessels and organs of the digestive tract. Smooth muscle cells lack striations because their actin and myosin filaments are not regularly arrayed along the length of the cell. Instead, the thick filaments are scattered throughout the cytoplasm, and the thin filaments are attached to structures called dense bodies, some of which are tethered to the plasma membrane. There is less myosin than in striated muscle fibers, and the myosin is not associated with specific actin strands. Further, smooth muscle cells have no troponin complex or T tubules, and their SR is not well developed. During an action potential, calcium ions enter the cytosol mainly through the plasma membrane, and the amount reaching the filaments is rather small. Calcium ions cause contraction by binding to a protein called calmodulin, which activates an enzyme that phosphorylates the myosin head. Smooth muscles contract relatively slowly but over a much greater range of length than striated muscle. Some smooth muscle cells contract only when stimulated by neurons of the autonomic nervous system. Others can generate action potentials without neural input and are electrically coupled to one another.

Invertebrates have muscle cells similar to vertebrate skeletal and smooth muscle cells. Arthropod skeletal muscles are nearly identical to vertebrate skeletal muscles. However, the flight muscles of insects are capable of independent, rhythmic contraction, so the wings of some insects can actually beat faster than action potentials can arrive from the central nervous system. Another interesting evolutionary adaptation has been discovered in the muscles that hold clam shells closed. The thick filaments in these muscles contain a protein called paramyosin that enables the muscles to remain contracted with a low rate of energy consumption for as long as a month.

Concept Check 49.6

1. Summarize the microscopic evidence indicating that thick and thin filaments slide past each other when a skeletal muscle fiber contracts.
2. How can the nervous system cause a skeletal muscle to produce the most forceful contraction it is capable of?
3. Contrast the role of calcium ions in the contraction of a skeletal muscle fiber and a smooth muscle cell.

For suggested answers, see Appendix A.

Concept 49.7

Locomotion requires energy to overcome friction and gravity

Movement is a hallmark of animals. To catch food, an animal must either move through its environment or move the surrounding water or air past itself. Although all sponges are sessile, they use beating flagella to generate water currents that draw and trap small food particles, and sessile cnidarians wave tentacles that capture prey (see Chapter 33). Most animals, however, are mobile and spend a considerable portion of their time and energy actively searching for food, as well as escaping from danger and looking for mates. Our focus here is **locomotion,** or active travel from place to place.

The modes of animal locomotion are diverse. Most animal phyla include species that swim. On land and in the sediments on the floor of the sea and lakes, animals crawl, walk, run, or hop. Active flight (in contrast to gliding downward from a tree or elevated ground) has evolved in only a few animal groups: insects, reptiles (including birds), and, among the mammals, bats. A group of large flying reptiles died out millions of years ago, leaving birds and bats as the only flying vertebrates.

In all its modes, locomotion requires that an animal expend energy to overcome two forces that tend to keep it stationary: friction and gravity. Exerting force requires energy-consuming cellular work.

Swimming

Because most animals are reasonably buoyant in water, overcoming gravity is less of a problem for swimming animals than for species that move on land or through the air. On the other hand, water is a much denser and more viscous medium than air, and thus drag (friction) is a major problem for aquatic animals. A sleek, fusiform (torpedo-like) shape is a common adaptation of fast swimmers (see Figure 40.2).

Animals swim in diverse ways. For instance, many insects and four-legged vertebrates use their legs as oars to push against the water. Squids, scallops, and some cnidarians are jet-propelled, taking in water and squirting it out in bursts. Sharks and bony fishes swim by moving their body and tail from side to side, while whales and other aquatic mammals move by undulating their body and tail up and down.

Locomotion on Land

In general, the problems of locomotion on land are the opposite of those in water. On land, a walking, running, hopping, or crawling animal must be able to support itself and move against gravity, but air poses relatively little resistance, at least at moderate speeds. When a land animal walks, runs, or hops, its leg muscles expend energy both to propel it and to keep it from falling down. With each step, the animal's leg muscles must overcome inertia by accelerating a leg from a standing start. For moving on land, powerful muscles and strong skeletal support are more important than a streamlined shape.

Diverse adaptations for traveling on land have evolved in various vertebrates. For example, kangaroos have large, powerful muscles in their hind legs, suitable for locomotion by hopping **(Figure 49.36)**. When a kangaroo lands, tendons in its hind legs momentarily store energy. The farther the animal hops, the more energy the tendons store. Analogous to the energy in a compressed spring on a pogo stick, the energy stored in the tendons is available for the next jump and reduces the total amount of energy the animal must expend to travel. The pogo stick analogy applies to many land animals; for instance, the legs of an insect, a dog, or a human retain some energy during walking or running, although considerably less than those of a kangaroo.

Maintaining balance is another prerequisite for walking, running, or hopping. A kangaroo's large tail helps balance its body during leaps and also forms a stable tripod with its hind legs when the animal sits or moves slowly. Illustrating the same principle, a walking cat, dog, or horse keeps three feet on the ground. Bipedal animals, such as humans and birds, keep part of at least one foot on the ground when walking. When an animal runs, all four feet (or both feet for bipeds) may be off the ground briefly, but at running speeds it is momentum more than foot contact that keeps the body upright.

Crawling poses a very different situation. Because much of its body is in contact with the ground, a crawling animal must exert considerable effort to overcome friction. You have read how earthworms crawl by peristalsis. Many snakes crawl by undulating their entire body from side to side. Assisted by

▲ **Figure 49.36 Energy-efficient locomotion on land.** Members of the kangaroo family travel from place to place mainly by leaping on their large hind legs. Kinetic energy momentarily stored in tendons after each leap provides a boost for the next leap. In fact, a large kangaroo hopping at 30 km/hr uses no more energy per minute than it does at 6 km/hr. The large tail helps balance the kangaroo when it leaps as well as when it sits.

large, movable scales on the underside, a snake's body pushes against the ground, driving the animal forward. Boa constrictors and pythons creep straight forward, driven by muscles that lift belly scales off the ground, tilt the scales forward, and then push them backward against the ground.

Flying

Gravity poses a major problem for flight because wings must develop enough lift to overcome gravity's downward force. The key to flight is wing shape. All types of wings, including those of airplanes, are airfoils—structures whose shape alters air currents in a way that helps them stay aloft.

Flying animals are relatively light, with body masses ranging from less than a gram for some insects to about 20 kg for the largest flying birds. Many flying animals have structural adaptations that reduce body mass. Birds, for example, have no urinary bladder or teeth (see Chapter 34). A fusiform shape helps reduce drag in air as it does in water.

Comparing Costs of Locomotion

The study of locomotion returns us to the theme of animal bioenergetics, discussed in Chapter 40. The energy cost of locomotion depends on the mode of locomotion and the environment **(Figure 49.37)**. Running animals generally expend more energy per meter traveled than equivalently sized animals specialized for swimming, partly because running or walking requires energy to overcome gravity. Swimming is the most energy-efficient mode of locomotion (assuming that an animal is specialized for swimming). And if we were to compare the energy consumption per minute rather than per meter, we would find that flying animals use more energy than swimming or running animals with the same body mass.

The downward slope of each line on the graph in Figure 49.37 also shows that a larger animal travels more efficiently than a smaller animal specialized for the same mode of transport. For example, a 450-kg horse expends less energy *per kilogram of body mass* than a 4-kg cat running the same distance. Of course, the total amount of energy expended in locomotion is greater for the larger animal.

An animal's use of energy to move determines how much energy in the food it consumes is available for other activities, such as growth and reproduction. Therefore, structural and behavioral adaptations that maximize the efficiency of locomotion increase an animal's evolutionary fitness.

Although we have discussed sensory receptors and muscles separately in this chapter, they are part of a single integrated system linking together brain, body, and the external world. An animal's behavior is the product of this system. Unit Eight discusses behavior within the broader context of ecology, the study of interactions between organisms and their environment.

Figure 49.37

Inquiry What are the energy costs of locomotion?

EXPERIMENT Physiologists typically determine an animal's rate of energy use during locomotion by measuring its oxygen consumption or carbon dioxide production while it swims in a water flume, runs on a treadmill, or flies in a wind tunnel. For example, the trained parakeet shown below is wearing a plastic face mask connected to a tube that collects the air the bird exhales as it flies.

RESULTS This graph compares the energy cost, in joules per kilogram of body mass per meter traveled, for animals specialized for running, flying, and swimming (1 J = 0.24 cal). Notice that both axes are plotted on logarithmic scales.

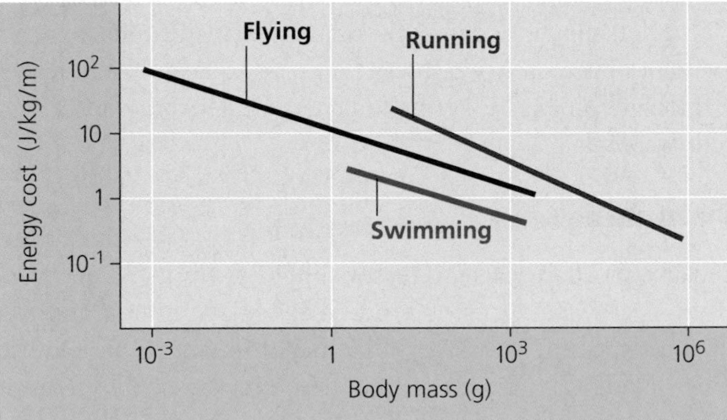

CONCLUSION For animals of a given body mass, swimming is the most energy-efficient and running the least energy-efficient mode of locomotion. In any mode, a small animal expends more energy per kilogram of body mass than a large animal.

Concept Check 49.7

1. Contrast swimming and flying in terms of the main problems they pose and the adaptations that allow animals to overcome those problems.
2. Based on Figure 49.37, which animal uses more energy per kilogram of body mass to move 1 m—a 1-g flyer or a 1-kg runner?

For suggested answers, see Appendix A.

Chapter 49 Review

Go to www.campbellbiology.com or the student CD-ROM to explore Activities, Investigations, and other interactive study aids.

SUMMARY OF KEY CONCEPTS

Concept 49.1

Sensory receptors transduce stimulus energy and transmit signals to the central nervous system

▶ **Functions Performed by Sensory Receptors (pp. 1046–1048)** Sensory receptors are usually specialized neurons or epithelial cells that detect environmental stimuli. Exteroreceptors detect external stimuli; interoreceptors detect internal stimuli. Sensory transduction is the conversion of stimulus energy into a change in membrane potential called a receptor potential. Signal transduction pathways in receptor cells often amplify the signal, which causes the receptor cell either to produce action potentials or to release neurotransmitter at a synapse with a sensory neuron.

▶ **Types of Sensory Receptors (pp. 1048–1049)** Mechanoreceptors respond to stimuli such as pressure, touch, stretch, motion, and sound. Chemoreceptors detect either total solute concentrations or specific molecules. Electromagnetic receptors detect different forms of electromagnetic radiation. Various types of thermoreceptors signal surface and core temperatures of the body. Pain is detected by a group of diverse receptors that respond to excess heat, pressure, or specific classes of chemicals.

Concept 49.2

The mechanoreceptors involved with hearing and equilibrium detect settling particles or moving fluid

▶ **Sensing Gravity and Sound in Invertebrates (p. 1050)** Most invertebrates sense their orientation with respect to gravity by means of statocysts. Many arthropods sense sounds with body hairs that vibrate and with localized "ears," consisting of a tympanic membrane and receptor cells.

▶ **Hearing and Equilibrium in Mammals (pp. 1050–1053)** The tympanic membrane (eardrum) transmits sound waves to three small bones of the middle ear, which transmit the waves through the oval window to the fluid in the coiled cochlea of the inner ear. Pressure waves in the fluid vibrate the basilar membrane, depolarizing hair cells in the organ of Corti and triggering action potentials that travel via the auditory nerve to the brain. Each region of the basilar membrane vibrates most vigorously at a particular frequency and leads to excitation of a specific auditory area of the cerebral cortex. The utricle, saccule, and three semicircular canals in the inner ear function in balance and equilibrium.

▶ **Hearing and Equilibrium in Other Vertebrates (pp. 1053–1054)** The detection of water movement in fishes and aquatic amphibians is accomplished by a lateral line system containing clustered hair cells.

Concept 49.3

The senses of taste and smell are closely related in most animals

▶ **Taste in Humans (pp. 1055–1056)** Taste and smell both depend on the stimulation of chemoreceptors by small, dissolved molecules that bind to proteins on the plasma membrane. In humans, taste receptors are organized into taste buds on the tongue and in the mouth. Five taste perceptions—sweet, sour, salty, bitter, and umami (elicited by glutamate)—involve several different transduction mechanisms.

▶ **Smell in Humans (pp. 1056–1057)** Olfactory receptor cells line the upper part of the nasal cavity and send their axons to the olfactory bulb of the brain. More than 1,000 genes code for membrane proteins that bind to specific classes of odorants, and each receptor cell expresses only one or at most a few of those genes.

Concept 49.4

Similar mechanisms underlie vision throughout the animal kingdom

▶ **Vision in Invertebrates (pp. 1057–1058)** The light detectors of invertebrates include the simple, light-sensitive eyespots of planarians; the image-forming compound eyes of insects, crustaceans, and some polychaetes; and the single-lens eyes of some jellies, polychaetes, spiders, and many molluscs.

▶ **The Vertebrate Visual System (pp. 1058–1063)** The main parts of the vertebrate eye are the sclera, which includes the cornea; the conjunctiva; the choroid, which includes the iris; the retina, which contains the photoreceptors; and the lens, which focuses light on the retina. Photoreceptors (rods and cones) contain a pigment, retinal, bonded to a protein (opsin). Absorption of light by retinal triggers a signal transduction pathway that hyperpolarizes the photoreceptors, causing them to release less neurotransmitter. Synapses transmit information from photoreceptors to bipolar cells and then to ganglion cells, whose axons in the optic nerve convey action potentials to the brain. Other neurons in the retina integrate information before it is sent to the brain. Most axons in the optic nerves go to the lateral geniculate nuclei of the thalamus, which relay information to the primary visual cortex. Several integrating centers in the cerebral cortex are active in creating visual perceptions.
Activity *Structure and Function of the Eye*

Concept 49.5

Animal skeletons function in support, protection, and movement

▶ **Types of Skeletons (pp. 1063–1064)** A hydrostatic skeleton, found in most cnidarians, flatworms, nematodes, and annelids, consists of fluid under pressure in a closed body compartment. Exoskeletons, found in most molluscs and arthropods, are hard coverings deposited on the surface of an animal. Endoskeletons, found in sponges, echinoderms, and chordates, are hard supporting elements embedded within an animal's body.
Activity *Human Skeleton*

▶ **Physical Support on Land (pp. 1064–1066)** In addition to the skeleton, muscles and tendons help support large land vertebrates.

Concept 49.6

Muscles move skeletal parts by contracting

▶ **Vertebrate Skeletal Muscle (pp. 1066–1072)** Skeletal muscles, often present in antagonistic pairs, provide movement by contracting and pulling against the skeleton. Vertebrate skeletal muscle consists of a bundle of muscle cells (fibers), each of which contains myofibrils composed of thin filaments of actin and thick filaments of myosin. Myosin heads, energized by the hydrolysis of ATP, bind to the thin filaments, forming cross-bridges. Bending of the myosin heads exerts force on the thin filaments. When ATP binds to the myosin heads, they release, ready to start a new cycle. Repeated cycles cause the thick and thin filaments to slide past each other, shortening the sarcomere and contracting the muscle fiber.

A motor neuron initiates contraction by releasing acetylcholine, which depolarizes the muscle fiber. Action potentials travel to the interior of the muscle fiber along the T tubules, stimulating the release of Ca^{2+} from the sarcoplasmic reticulum. The calcium ions reposition the tropomyosin-troponin complex on the thin filaments, exposing the myosin-binding sites on actin and allowing the cross-bridge cycle to proceed. A motor unit consists of a motor neuron and the muscle fibers it innervates. Recruitment of multiple motor units results in stronger contractions. A twitch results from a single action potential in a motor neuron. More rapidly delivered action potentials produce a graded contraction by summation. Tetanus is a state of smooth and sustained contraction produced when motor neurons deliver a volley of action potentials. Skeletal muscle fibers are classified as slow oxidative, fast oxidative, or fast glycolytic based on their contraction speed and major pathway for producing ATP.

Activity ***Skeletal Muscle Structure***
Activity ***Muscle Contraction***
Investigation ***How Do Electrical Stimuli Affect Muscle Contraction?***

- **Other Types of Muscle (p. 1072)** Cardiac muscle, found only in the heart, consists of striated cells that are electrically connected by intercalated disks and can generate action potentials without neural input. In smooth muscle, contractions are slow and may be initiated by the muscles themselves or by stimulation from neurons in the autonomic nervous system.

Locomotion requires energy to overcome friction and gravity

- **Swimming (p. 1073)** Overcoming friction is a major problem for swimmers. Gravity is less of a problem for swimming animals than for those that move on land or fly.
- **Locomotion on Land (pp. 1073–1074)** Walking, running, hopping, or crawling on land requires an animal to support itself and to move against gravity.
- **Flying (p. 1074)** Flight requires that wings develop enough lift to overcome the downward force of gravity.
- **Comparing Costs of Locomotion (p. 1074)** Animals that are specialized for swimming expend less energy per meter traveled than equivalently sized animals specialized for flying or running.

TESTING YOUR KNOWLEDGE

Self-Quiz

1. Which of the following sensory receptors is *incorrectly* paired with its category?
   a. hair cell—mechanoreceptor
   b. muscle spindle—mechanoreceptor
   c. taste receptor—chemoreceptor
   d. rod—electromagnetic receptor
   e. olfactory receptor—electromagnetic receptor

2. Some sharks close their eyes just before they bite. Although they cannot see their prey, their bites are on target. Researchers have noted that sharks often misdirect their bites at metal objects, and that sharks can find batteries buried under the sand of an aquarium. This evidence suggests that sharks keep track of their prey during the split second before they bite in the same way that a
   a. rattlesnake finds a mouse in its burrow.
   b. male silkworm moth locates a mate.
   c. bat finds moths in the dark.
   d. platypus locates its prey in a muddy river.
   e. flatworm avoids light places.

3. The transduction of sound waves into action potentials takes place
   a. within the tectorial membrane as it is stimulated by the hair cells.
   b. when hair cells are bent against the tectorial membrane, causing them to depolarize and release neurotransmitter that stimulates sensory neurons.
   c. as the basilar membrane becomes more permeable to sodium ions and depolarizes, initiating an action potential in a sensory neuron.
   d. as the basilar membrane vibrates at different frequencies in response to the varying volume of sounds.
   e. within the middle ear as the vibrations are amplified by the malleus, incus, and stapes.

4. Which of the following is an *incorrect* statement about the vertebrate eye?
   a. The vitreous humor regulates the amount of light entering the pupil.
   b. The transparent cornea is an extension of the sclera.
   c. The fovea is the center of the visual field and contains only cones.
   d. The ciliary muscle functions in accommodation.
   e. The retina lies just inside the choroid and contains the photoreceptor cells.

5. When light strikes the rhodopsin in a rod, retinal isomerizes, initiating a signal transduction pathway that
   a. depolarizes the neighboring bipolar cells and initiates an action potential in a ganglion cell.
   b. depolarizes the rod, causing it to release the neurotransmitter glutamate, which excites bipolar cells.
   c. hyperpolarizes the rod, reducing its release of glutamate, which excites some bipolar cells and inhibits others.
   d. hyperpolarizes the rod, increasing its release of glutamate, which excites amacrine cells but inhibits horizontal cells.
   e. converts cGMP to GMP, opening sodium channels and hyperpolarizing the membrane, causing the rhodopsin to become bleached.

6. Clams and lobsters both have an exoskeleton, but lobsters have much greater mobility. Why?
   a. Clams have only adductor muscles that hold the shell closed, whereas lobsters have both abductor and adductor muscles.
   b. The paramyosin of clam muscles holds the muscles in a low-energy state of contraction, whereas lobster muscles are very similar to vertebrate striated muscles.

c. Clams can grow only by adding to the outer edge of the shell, whereas lobsters molt and repeatedly replace their exoskeleton with a larger, more flexible one.
d. The lobster skeleton can actively contract, while the clam skeleton lacks its own contractile mechanism.
e. Lobsters have a jointed exoskeleton, allowing for the flexible movement of appendages and body parts at the joints.

7. During the contraction of a vertebrate skeletal muscle fiber, calcium ions
   a. break cross-bridges by acting as a cofactor in the hydrolysis of ATP.
   b. bind with troponin, changing its shape so that the myosin-binding sites on actin are exposed.
   c. transmit action potentials from the motor neuron to the muscle fiber.
   d. spread action potentials through the T tubules.
   e. reestablish the polarization of the plasma membrane following an action potential.

8. Tetanus refers to
   a. the partial sustained contraction of major supporting muscles.
   b. the all-or-none contraction of a single muscle fiber.
   c. a stronger contraction resulting from the activation of multiple motor units.
   d. a smooth and sustained contraction resulting from summation.
   e. a state of muscle fatigue caused by the depletion of ATP and the accumulation of lactate.

9. Which of the following is a true statement about cardiac muscle cells?
   a. They lack an orderly arrangement of actin and myosin filaments.
   b. They have a less extensive sarcoplasmic reticulum and thus contract more slowly than smooth muscle cells.
   c. They are connected by intercalated disks, through which action potentials spread to all cells in the heart.
   d. They have a resting potential more positive than an action potential threshold.
   e. They contract only when stimulated by neurons.

10. Which of the following changes occurs when a skeletal muscle fiber contracts?
   a. The A bands shorten.
   b. The I bands shrink.
   c. The Z lines slide farther apart.
   d. The thin filaments contract.
   e. The thick filaments contract.

For Self-Quiz answers, see Appendix A.

Go to the website or CD-ROM for more quiz questions.

Evolution Connection

In general, locomotion on land requires more energy than locomotion in water. By integrating what you have learned throughout these chapters on animal functions, discuss some of the evolutionary adaptations of mammals that support the high energy requirements for moving on land.

Scientific Inquiry

Although skeletal muscles generally fatigue fairly rapidly, clam shell muscles have a protein called paramyosin that allows them to sustain contraction for up to a month. From your knowledge of the cellular mechanism of contraction, propose a hypothesis to explain how paramyosin might work. How would you test your hypothesis experimentally?

Investigation *How Do Electrical Stimuli Affect Muscle Contraction?*

Science, Technology, and Society

You may know an older person who has broken a bone (often a hip) partly because of osteoporosis, a loss of bone density that affects many women after menopause. As a means of avoiding osteoporosis, researchers recommend exercise and maximum calcium intake when a person is between 10 and 30 years old. Is it realistic to expect young people to view themselves as future senior citizens? How would you suggest that they be encouraged to develop good health habits that might not pay off for 40 or 50 years?

AN INTERVIEW WITH

Gene E. Likens

If scientists wore medals on their chests like military generals, ecologist Gene Likens would be one of biology's most decorated heroes. His recent awards include the 2001 National Medal of Science and the 2003 Blue Planet Prize, which honors "outstanding scientific research that helps solve environmental problems." Dr. Likens shared the latter award with his longtime collaborator, Herbert Bormann, with whom Likens pioneered the field of "long-term ecological research" at the Hubbard Brook Experimental Forest in the White Mountains of New Hampshire. As President and Director of the Institute of Ecosystem Studies in upstate New York, Dr. Likens manages a large team of scientists and educators, but he returns to Hubbard Brook each summer to continue his inquiry about how forest and aquatic ecosystems work. We conducted this interview at one of Hubbard Brook's historic research sites.

How did the Hubbard Brook Experimental Forest become the focus of your ecosystem research?

When I joined the Dartmouth faculty in 1961, I met Herb Bormann, who had this idea for a small-watershed approach for studying forest ecosystems. The Hubbard Brook Experimental Forest was ideal for such studies and was close by. And so it was one of those chance situations where it all just fell together. Much of my scientific career has been the result of such serendipity—that is, keeping your eyes, your ears, and your mind open, and when there's an opportunity, jump in and try to make something productive from it.

Why is Hubbard Brook so attractive as a field station for ecosystem research?

Our challenge from the beginning was to confront the complexity of this ecosystem and make sense of it. So, we started with the small-watershed approach. The idea was very simple: We used a medical analogy. We figured we could measure the chemistry of stream water, much like a physician would measure the chemistry of blood, to analyze the health of the watershed, the patient. We wanted to measure all the inputs to the system, such as in precipitation, and all the outputs such as in the stream water draining the system, and from that, try to understand the health of this patient. In each watershed, we find a place where the bedrock is close to the surface, and then build a gauging station, also called a weir. All the water coming down the hill is forced to pass through the weir, so we can measure the amount and chemical composition of the stream water very accurately year round.

This is your forty-second straight summer of such studies here at Hubbard Brook. It must be tough to get funding to support research over such a long term.

Our proposals for research grants mostly emphasize short-term questions that require long-term data. For example, long-term data from Hubbard Brook were critical in establishing that acid rain exists, and also provide the basis for shorter-term measurements of whether the problem of acid rain is getting worse or better. Long-term ecological research has become a hallmark of Hubbard Brook studies.

What was the public impact of evidence for acid rain in the Hubbard Brook Forest?

Although Swedish scientists studied acid rain earlier, our research was the first documentation of acid rain in North America. When we published our first paper on acid rain in 1972, the press really picked up on it. The evidence that there was something wrong with rain was a powerful issue that the public could relate to. The long-term data from Hubbard Brook made it very difficult to argue that there was no such thing as acid rain, and I think it's fair to say that the quality of those long-term data contributed to the 1990 amendments to the Clean Air Act that addressed the link between acid rain and combustion of fossil fuels.

Hubbard Brook is also famous for experiments on the effects of deforestation. How do you do such large-scale experiments, and what were some key results?

An example of these experimental manipulations is when we cut down all the trees in one of the watersheds and left them where they fell. We then used herbicides to prevent regrowth for three years. We compared the inputs and outputs for this experimental watershed with a reference watershed that was not altered. We expected more water to run off the experimental watershed, and that turned out to be the case. What we *didn't* expect was that the chemistry of the runoff would change so dramatically. For example, nitrate levels in the water shot up to a level that exceeded public health standards for drinking. Cutting the trees totally disrupted the recycling of nitrogen in the soil, and the nitrate that accumulated moved downstream. The nitrogen loss slowed after the second year, but some soil nutrients, such as calcium, still haven't come back to the levels and patterns that existed before the trees were cut in 1965.

How do these tree-cutting experiments relate to issues of forest management?

The research is fundamentally important to our basic understanding of how an ecosystem like this works, but the popular press picked up the story in terms of the clear-cutting controversy. Based on our long-term data, we proposed in a 1978 *Science* article that you can do clear-cutting in these kinds of forests if you follow specific guidelines. Clear cutting shouldn't be done on really steep slopes or in very large blocks, and certainly the stream channel should be protected. But most importantly, a site should not be clear cut very often. If a forest is harvested too frequently, then the system will become depleted of critical nutrients. It's just like a garden; if you continue to take away and never put anything back, you're going to deplete the soil. The Forest Service has now adopted a 100-year rotation policy (a particular section of forest can be logged only once every 100 years), and we're very proud of that. That forest management protocol is based on science.

Which current research projects at Hubbard Brook do you think are most likely to capture the attention of both the ecology community and the general public?

Richard Holmes at Dartmouth College and his colleagues are conducting really elegant long-term studies of bird populations. I think one of their newsworthy findings is that the total number of birds of all species present in the forest is currently about a third of what it was in 1969. Some species have actually increased in abundance, while others have almost disappeared from here. The researchers are now focusing on how predation, food availability, and other factors may be affecting the bird populations. Another project that I think is quite newsworthy involves a beautiful little lake at the bottom of this valley called Mirror Lake. One of the things I started at the very beginning of my work here, about 1965, was to record the period of ice cover on the lake. The ice cover is related to the total heat budget for the lake, so it's an indication of temperature and climate change. Well, as you might expect, the period of ice cover is highly variable over the short term because the weather varies from year to year. But our long-term data show that there is a highly statistically significant decline in the period of ice cover—about 19 fewer days of ice cover on the lake now than in the 1960s. This is a clear indication of global warming.

Along with your work at Hubbard Brook, you are President and Director of the Institute of Ecosystem Studies (IES). What motivated you 20 years ago to trade a prestigious endowed chair at Cornell for the challenge of leading a not-for-profit research and education institute?

I was asked that question many times by my colleagues, and some thought I was crazy. The job description for the IES position basically said that I could do anything I wanted as long as I did it well. And I liked that a lot. I was the founding director of IES, and I'm extremely proud of how the Institute has built such an outstanding international reputation. We have a superb scientific staff, often working together on projects ranging from a study of invasive zebra mussels in the Hudson River to research on the ecology of infectious diseases, including Lyme disease.

What do you find so attractive about the multidisciplinary, team-oriented research you direct at IES?

Many brains can often make faster and better advances than a single brain. The only way you can do big projects in science today is with teams, particularly in fields that emphasize biocomplexity. But such collaborative science requires trust in your colleagues. I think we should offer more training to graduate students to prepare them for working with research teams.

Speaking of training, IES complements its scientific mission with education programs. Are there any opportunities there for undergrads?

We have an NSF Research Experiences for Undergraduates (REU) program. Each summer, about 10 to 12 REU students are selected. In addition to participating in research, the REUs receive instruction on writing scientific papers, finding jobs, and scientific ethics. Two of the students this year are doing research here at Hubbard Brook, and the rest of the group will come for a visit next week.

What was it about your own education, formal or otherwise, that made you so interested in nature?

I grew up on a small farm in Indiana, and my early childhood was spent barefoot, out in the forests and particularly around lakes. When I went to college, I was planning to become an elementary school teacher and a baseball coach. But I had a professor who just kept insisting that I go to graduate school. So to get him off my back, I applied to graduate schools and went to the University of Wisconsin, where I studied lake systems. It was another wonderful serendipitous event for me.

How have your 42 years of experience here at Hubbard Brook affected your view of nature?

Hubbard Brook has provided me with just wonderful questions to explore. I've gone from being a naive farm kid to using very sophisticated techniques to answer questions about this complex system and trying to help people understand why it's important to protect this kind of complexity in natural ecosystems. You must respect nature because it provides the basic life support system: clean air, clean water, clean soil, and nourishing food. And if you give these ecosystems a chance, they're resilient and will continue to provide the life support we need. Learning that here at Hubbard Brook changed my life. And can you believe someone pays me to work in such a beautiful place?

Much of my scientific career has been the result of such serendipity—that is, keeping your eyes, your ears, and your mind open, and when there's an opportunity, jump in and try to make something productive from it.

50 An Introduction to Ecology and the Biosphere

▲ **Figure 50.1 The richness of the biosphere evident in one area of a Panamanian forest.**

Key Concepts

50.1 Ecology is the study of interactions between organisms and the environment
50.2 Interactions between organisms and the environment limit the distribution of species
50.3 Abiotic and biotic factors influence the structure and dynamics of aquatic biomes
50.4 Climate largely determines the distribution and structure of terrestrial biomes

Overview

The Scope of Ecology

Organisms are open systems that interact continuously with their environment—a theme that has already surfaced many times in this book. The scientific study of the interactions between organisms and the environment is called **ecology** (from the Greek *oikos*, home, and *logos*, to study). These interactions determine both the distribution of organisms and their abundance, leading to three questions that ecologists often ask about organisms: Where do they live? Why do they live where they do? And how many are there? Ecologists also study how interactions between organisms and the environment affect phenomena such as the number of species living in a particular area, the cycling of nutrients in a forest or lake, and the growth of populations.

Because of its great scope, ecology is an enormously complex and exciting area of biology, as well as one of critical importance. Ecology reveals the richness of the biosphere—the entire portion of Earth inhabited by life—and can provide the basic understanding that will help us to conserve and sustain that richness, now threatened more than ever by human activity. The richness of the biosphere is particularly apparent in tropical forests, such as the Panamanian forest where the Hercules scarab beetle (*Dynastes hercules*) pictured in **Figure 50.1** lives. Earth's tropical forests are home to millions of species, including an estimated 5–30 million still undescribed species of insects, spiders, and other arthropods. In fact, every part of the biosphere is inhabited by diverse life-forms, most of which, especially the microbial species, are unknown to science. This chapter introduces the science of ecology and describes some of the factors, both living and nonliving, that affect the distribution of organisms. It also surveys the major types of aquatic and terrestrial habitats where organisms live and where ecologists go to study them.

Concept 50.1

Ecology is the study of interactions between organisms and the environment

Humans have always had an interest in the distribution and abundance of other organisms. As hunters and gatherers, prehistoric people had to learn where game and edible plants could be found in abundance. With the development of agriculture and the domestication of animals, people learned more about how the environment affects the growth, survival, and reproduction of plants and animals. Later, naturalists from Aristotle to Darwin and beyond observed and described organisms in their natural habitats and systematically recorded their observations. Because extraordinary insight can still be gained through this descriptive approach to discovery science (see Chapter 1), natural history remains a fundamental part of the science of ecology.

In addition to its long history as a descriptive science, ecology is also a rigorous experimental science. Despite the difficulties of conducting experiments in natural environments, ecologists often test their hypotheses through field experiments. The long-term research by Gene Likens and his colleagues in the Hubbard Brook Experimental Forest is one example (see the interview on pages 1078–1079). Other examples of field experiments include studies in which researchers measure the impact of herbivory (plant-eating) on plant species diversity by comparing open control plots to experimental "exclosures" designed to keep out certain herbivore species. Perhaps because of ecology's focus on complex systems that challenge the capacity of scientists to produce consistent results, ecologists have been innovators in the areas of experimental design and the application of statistical inference. You will encounter many examples of ecological experiments throughout this unit.

Ecology and Evolutionary Biology

Ecology and evolutionary biology are closely related sciences. Darwin's extensive observations of the distribution of organisms and their adaptation to specific environments led him to propose that environmental factors interacting with variation within populations could cause evolutionary change (see Chapter 22). Today, we have ample evidence that events that occur in the framework of ecological time (minutes, months, and years) translate into effects over the longer scale of evolutionary time (decades, centuries, millennia, and longer). For instance, hawks feeding on field mice have an immediate impact on the prey population by killing certain individuals, thereby reducing population size (an ecological effect) and altering the gene pool (an evolutionary effect). One long-term evolutionary effect of this predator-prey interaction may be selection for mice with fur coloration that camouflages the animals.

Organisms and the Environment

The environment of any organism includes **abiotic**, or nonliving, components—chemical and physical factors such as temperature, light, water, and nutrients—and **biotic**, or living, components—all the organisms, or the **biota**, that are part of the individual's environment. Other organisms may compete with an individual for food and other resources, prey upon it, parasitize it, provide its food, or change its physical and chemical environment.

Questions about the relative influence of various environmental components, both abiotic and biotic, are frequently at the heart of ecological studies. For example, **Figure 50.2** shows the geographic range of the red kangaroo in Australia and illustrates the basis for two common ecological questions: What environmental factors limit the geographic range, or *distribution*, of a species such as the red kangaroo? And what factors determine its *abundance*? Some of the climatic factors that ecologists might consider in the case of the red kangaroo are indicated in Figure 50.2. Notice that red kangaroos do not live in regions around the periphery of Australia, where the climate varies from moist to wet. Notice also that red kangaroos are most abundant in a few areas, shaded dark brown to orange, where the climate is drier and more variable from year to year. Near the edges of their range, where the climate is wetter than in the interior and where the map is shaded pale tan, red kangaroos are scarce.

These patterns suggest that an abiotic factor, the amount and variability of precipitation, influences the distribution of red kangaroos in Australia. However, are the controls on red kangaroo distribution only abiotic? Perhaps climate influences red kangaroo populations indirectly through biotic factors, such as pathogens, parasites, competitors, predators, and food availability. For instance, do wetter regions harbor pathogens and parasites deadly to red kangaroos? Ecologists generally need to consider multiple factors and alternative hypotheses when attempting to explain patterns of distribution and abundance.

Kangaroos/km^2
> 20
10–20
5–10
1–5
0.1–1
< 0.1
Limits of distribution

Climate in northern Australia is hot and wet, with seasonal drought.

Red kangaroos occur in most semiarid and arid regions of the interior, where precipitation is relatively low and variable from year to year.

Southeastern Australia has a wet, cool climate.

Southern Australia has cool, moist winters and warm, dry summers.

Tasmania

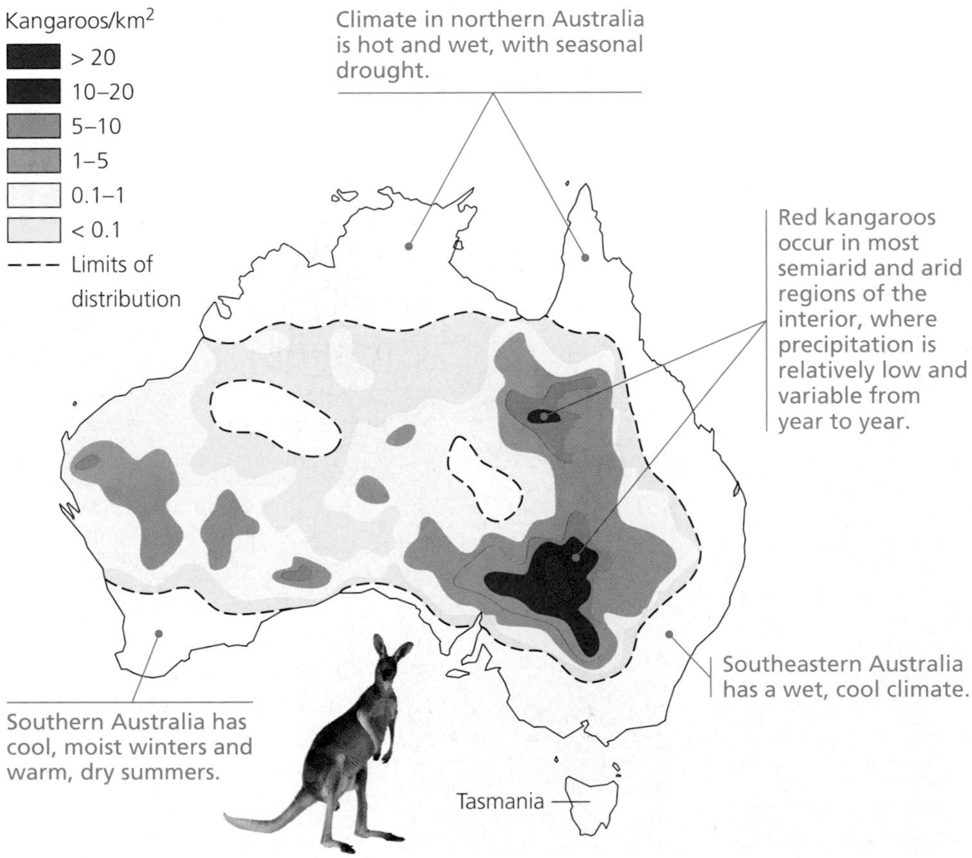

▲ **Figure 50.2 Distribution and abundance of the red kangaroo in Australia, based on aerial surveys.**

Subfields of Ecology

Ecology can be divided into areas of study ranging from the ecology of individual organisms to the dynamics of ecosystems and landscapes **(Figure 50.3)**. While each of these subfields has some unique terminology and scientific concerns, modern ecological studies increasingly cross the boundaries between traditionally separate areas.

Organismal ecology, which may be subdivided into the disciplines of physiological ecology, evolutionary ecology, and behavioral ecology, concerns how an organism's structure, physiology, and (for animals) behavior meet the challenges posed by the environment.

A **population** is a group of individuals of the same species living in a particular geographic area. **Population ecology** concentrates mainly on factors that affect how many individuals of a particular species live in an area.

A **community** consists of all the organisms of all the species that inhabit a particular area; it is an assemblage of populations of many different species. Thus, **community ecology** deals with the whole array of interacting species in a community. This area of research focuses on how interactions such as predation, competition, and disease, as well as abiotic factors such as disturbance, affect community structure and organization.

An **ecosystem** consists of all the abiotic factors in addition to the entire community of species that exist in a certain area. An ecosystem—a lake, for example—may contain many different communities. In **ecosystem ecology,** the emphasis is on energy flow and chemical cycling among the various biotic and abiotic components.

Landscape ecology deals with arrays of ecosystems—just two examples are a corridor of trees lining a river flowing through a sparsely vegetated plain, or patches of coral reef surrounded by turtle grass—and how they are arranged in a geographic region. Every landscape or seascape consists of a mosaic of different types of "patches," an environmental characteristic ecologists refer to as **patchiness.** Landscape ecological research focuses on the factors controlling exchanges of energy, materials, and organisms among the ecosystem patches making up a landscape or seascape.

(a) Organismal ecology. How do humpback whales select their calving areas?

(b) Population ecology. What enviromental factors affect the reproductive rate of deer mice?

(c) Community ecology. What factors influence the diversity of species that make up a particular forest?

(d) Ecosystem ecology. What factors control photosynthetic productivity in a temperate grassland ecosystem?

(e) Landscape ecology. To what extent do the trees lining the drainage channels in this landscape serve as corridors of dispersal for forest animals?

▲ **Figure 50.3 Examples of questions in different subfields of ecology.**

The **biosphere** is the global ecosystem—the sum of all the planet's ecosystems. This broadest area of ecology includes the entire portion of Earth inhabited by life: the atmosphere to an altitude of several kilometers, the land down to and including water-bearing rocks at least 3 kilometers below-ground, lakes and streams, caves, and the oceans to a depth of several kilometers. An example of research at the biosphere level is the analysis of how changes in atmospheric CO_2 concentration may affect Earth's climate and, in turn, all life.

Ecology and Environmental Issues

Because the term *ecology* is so often misused in popular writing to refer to environmental concerns, it is important to clarify the difference between ecology and environmentalism (advocating for the protection or preservation of the natural environment). To address environmental problems, we need to understand the often complicated and delicate relationships between organisms and the environment. The science of ecology provides that understanding. If we know that phosphate promotes the growth of algae in lakes, for instance, we may decide to limit the use of phosphate-rich fertilizers in surrounding areas to protect lakes from becoming clogged with algae.

Much of society's current awareness of environmental issues had its beginnings in 1962 with Rachel Carson's *Silent Spring* **(Figure 50.4)**. In that now-classic book, Carson warned that the widespread use of pesticides such as DDT was causing population declines in many nontarget organisms. Today, acid precipitation, localized famine aggravated by land misuse and population growth, the poisoning of soil and streams with toxic wastes, and the growing list of species extinct or endangered because of habitat destruction are just a few of the problems that threaten the home we share with millions of other forms of life.

Many influential ecologists, including Gene Likens, recognize a responsibility to educate legislators and the general public about decisions that affect the environment. An important part of this responsibility is communicating the scientific complexity of environmental issues. Politicians and lawyers often want definitive answers to such environmental questions as how much old-growth forest is needed to "save the spotted owls." While ecological studies can certainly provide essential information for making policy decisions on habitat preservation, responses to such questions often include further questions: How many owls must be saved? With what certainty must they be saved? How long can they survive in this amount of forest? Ecologists can help answer these questions so that the public and policymakers can make informed decisions about environmental issues.

◀ **Figure 50.4 Rachel Carson.** In *Silent Spring*, a book seminal to the modern environmental movement, Carson had a broad message: "The 'control of nature' is a phrase conceived in arrogance, born of the Neanderthal age of biology and philosophy, when it was supposed that nature exists for the convenience of man."

Although our ecological information is always incomplete, we cannot abstain from making decisions about environmental issues until all the answers are known. But given what we do know about the interconnectedness of the biosphere, it is probably wise to follow the **precautionary principle**, which can be expressed simply as "An ounce of prevention is worth a pound of cure." Aldo Leopold, the famous wildlife conservationist, expressed the precautionary principle well when he wrote, "To keep every cog and wheel is the first precaution to intelligent tinkering."

Concept Check 50.1

1. Contrast the terms *ecology* and *environmentalism*. How does ecology relate to environmentalism?
2. How can an event that occurs on the ecological time scale affect events that occur on an evolutionary time scale?
3. Within which area of ecology is each of the following directly working: (a) an ecologist studying the distribution and abundance of red kangaroos in Australia; (b) an ecologist studying changes in which plant species are most abundant in a forest following a wildfire?

For suggested answers, see Appendix A.

Concept 50.2

Interactions between organisms and the environment limit the distribution of species

In Chapter 22, we discussed *biogeography*, the study of the past and present distribution of individual species, in the context of evolutionary theory. Ecologists have long recognized global and regional patterns in the distribution of organisms within the biosphere. Kangaroos occur in Australia but not in North America, whereas pronghorn antelope occur in the western United States but not in Europe or Africa. More than a century ago, Darwin, Wallace, and other naturalists began to identify broad patterns of distribution by naming biogeographic

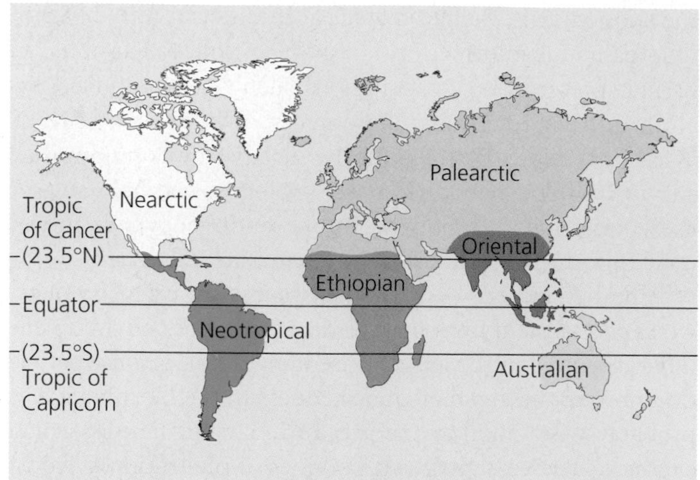

▲ **Figure 50.5 Biogeographic realms.** Continental drift and barriers such as deserts and mountain ranges all contribute to the distinctive flora and fauna found in Earth's major biogeographic realms. The realms are not sharply delineated but grade together in zones where species from adjacent realms coexist.

realms **(Figure 50.5)**. We now associate these biogeographic realms with patterns of continental drift that followed the breakup of the supercontinent Pangaea (see Chapter 26).

Biogeography provides a good starting point for understanding what limits the geographic distribution of a species. To see how ecologists might arrive at such an understanding, let's work our way through the series of questions in the flowchart in **Figure 50.6**.

Dispersal and Distribution

The movement of individuals away from centers of high population density or from their area of origin, called **dispersal**, contributes to the global distribution of organisms. For example, why are there no kangaroos in North America? A biogeographer might propose this simple hypothesis: They could not get there because a barrier to their dispersal existed; the area has been inaccessible to kangaroos since they first appeared. However, while land-bound kangaroos have not reached North America under their own power, some other organisms that are adapted for long-distance dispersal, such as some birds, have. The dispersal of organisms is critical to understanding both geographic isolation in evolution (see Chapter 24) and the broad patterns of current geographic distributions of species.

Natural Range Expansions

The role of dispersal is demonstrated when organisms expand their range by moving into areas where they did not exist previously. For instance, the cattle egret (see Figure 53.10), a species widely distributed in Africa, Eurasia, and Australia, was not found in the Americas 200 years ago. But in the late 1800s, some of these strong-flying birds managed to fly across the Atlantic Ocean and colonize northeastern South America. From there, cattle egrets gradually spread northward through Central America and into North America, reaching Florida in the 1950s. They have since spread widely and today have breeding populations as far west as the Pacific Coast and as far north as southern Canada. The great-tailed grackle, a large, conspicuous bird related to blackbirds and orioles, is another species that has expanded its range over time, moving northward from the coast of the Gulf of Mexico and the Rio Grande Valley **(Figure 50.7)**.

Natural range expansions clearly show the influence of dispersal on distribution, but opportunities to observe such dispersal are rare. As a consequence, ecologists have often turned to experimental methods to better understand the role of dispersal in limiting the distribution of species.

Species Transplants

One direct way to determine if dispersal is a key factor limiting distribution is to observe the results of intentional or accidental transplants of a species to areas where it was previously

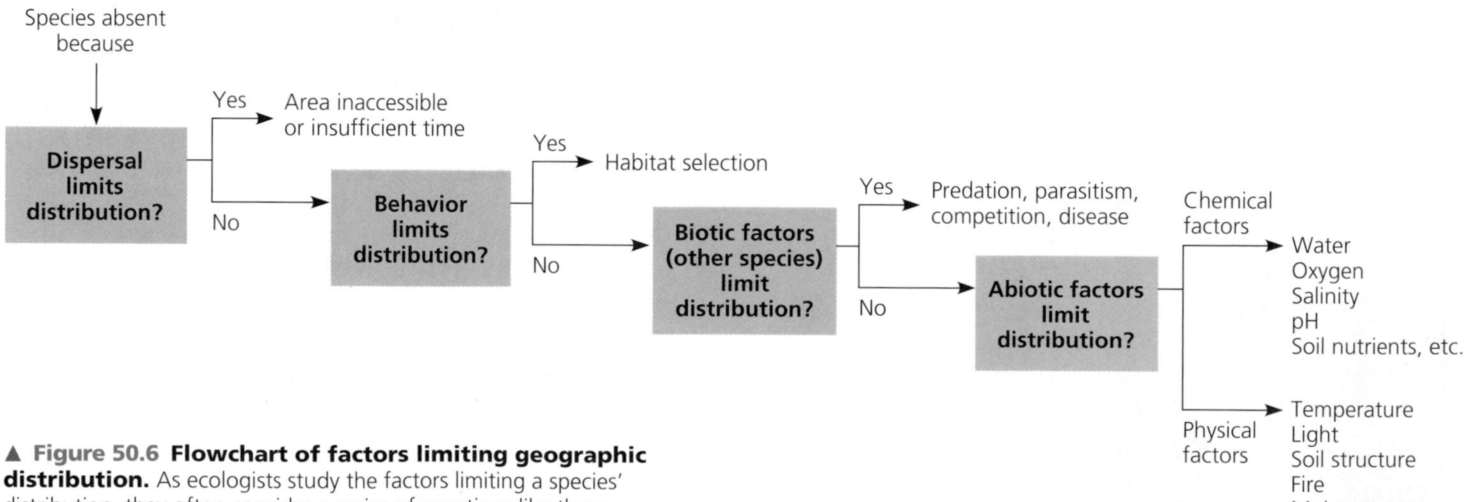

▲ **Figure 50.6 Flowchart of factors limiting geographic distribution.** As ecologists study the factors limiting a species' distribution, they often consider a series of questions like these.

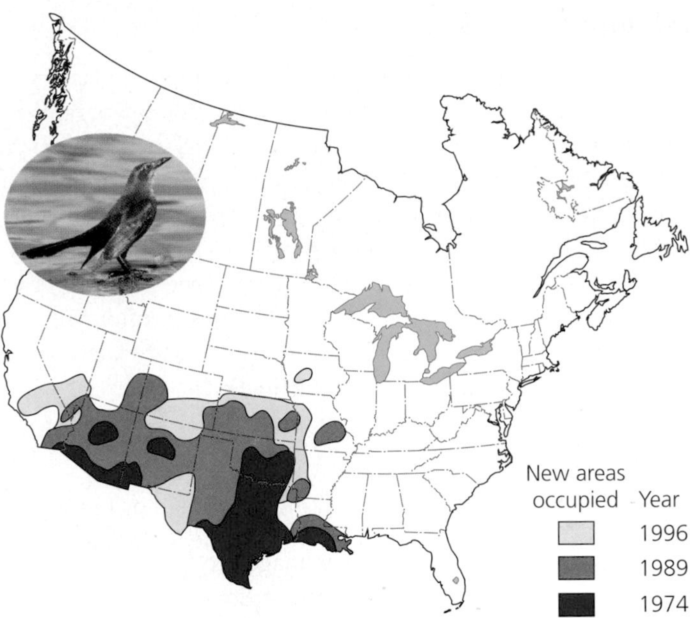

▲ **Figure 50.7 Spread of breeding populations of the great-tailed grackle in the United States from 1974 to 1996.** The grackle expanded its breeding range substantially in just 22 years.

absent. For a transplant to be considered successful, some of the organisms must not only survive in the new area but also reproduce there. Thus, a transplant's success may not be determined until at least one life cycle is complete. If a transplant is successful, then we can conclude that the *potential* range of the species is larger than its *actual* range; in other words, the species *could* live in certain areas where it currently does not.

Species introduced to new geographic locations often disrupt the communities and ecosystems to which they have been introduced and spread far beyond the area of intended introduction (see Chapter 55). Consequently, ecologists rarely conduct transplant experiments today. Instead, they document the outcome when a species has been transplanted for other purposes, such as to introduce game animals or sport fish, or when a species has been accidentally transplanted.

Behavior and Habitat Selection

As transplant experiments show, some organisms do not occupy all of their potential range, even though they may be physically able to disperse into the unoccupied areas. To follow our line of questioning from Figure 50.6, does behavior play a role in limiting distribution in such cases? When individuals seem to avoid certain habitats, even when the habitats are suitable, their distribution may be limited by habitat selection behavior. While habitat selection is typically applied to animals, plants may also select their habitats—for instance, by producing seeds that germinate only under a restricted set of environmental conditions—even though individual plants cannot move once they are rooted.

Although habitat selection is one of the least understood ecological processes, some instances have been closely studied, including habitat selection by several insect species. Female insects often oviposit (deposit eggs) only in response to a very narrow set of stimuli, which may restrict distribution of the insects to certain host plants. Larvae of the European corn borer, for example, can feed on a wide variety of plants but occur almost exclusively on corn because the ovipositing females are attracted by odors produced by the corn plant. And consider the ecology of anopheline mosquitoes. Important carriers of disease, these insects have been extensively studied as part of efforts to eradicate malaria in tropical areas. Each anopheline species is usually associated with a particular type of habitat. Surprisingly, large areas of apparently suitable tropical habitat are completely free of dangerous mosquitoes. Habitat selection for oviposition sites by female mosquitoes appears to restrict their distribution.

It might seem that a tropical mosquito is deficient because she does not lay eggs in apparently suitable rice field habitats. But this behavior may simply reflect the fact that rice fields are an evolutionarily recent habitat, and the mosquito is not adapted to it. Adaptation is not instantaneous, nor does evolution produce organisms perfectly adapted to every environment (see Chapter 23).

Biotic Factors

If behavior does not limit the distribution of a species, our next question is whether biotic factors—that is, other species—are responsible. In many cases, a species cannot complete its full life cycle if transplanted to a new area. This inability to survive and reproduce may be due to negative interactions with other organisms in the form of predation, parasitism, disease, or competition. Or survival and reproduction may be limited by the absence of other species on which the transplanted species depends. For instance, the absence of particular pollinator species could prevent reproduction by a transplanted plant species, though such specific dependence is rare. More common examples of biotic limitation on geographic distribution involve predators (organisms that kill their prey) and herbivores (organisms that eat plants or algae). Simply put, organisms that eat can limit the distribution of organisms that get eaten.

Let's examine one specific case of an herbivore limiting the distribution of a food species. In certain marine ecosystems, there is often an inverse relationship between the abundance of sea urchins and the abundance of seaweeds (large marine algae, such as kelp). Where sea urchins that graze on seaweeds and other algae are common, large stands of seaweeds do not become established. Thus, the local distribution of seaweeds appears to be limited by sea urchins. This kind of interaction can be tested by "removal and addition" experiments. In studies near Sydney, Australia, W. J. Fletcher, of the University of

Sydney, tested the hypothesis that sea urchins are a biotic factor limiting seaweed distribution. Fletcher reasoned that if this hypothesis is correct, then more seaweeds should invade an area from which sea urchins have been removed. Conversely, if urchins are added to an area rich in seaweeds, the seaweeds should be diminished. One complication is that there are often several other herbivores in addition to sea urchins in the habitats where seaweeds may grow; thus, Fletcher needed to perform a series of manipulative field experiments to isolate the influence of sea urchins on seaweeds in his study area **(Figure 50.8)**.

In addition to predation and herbivory, the presence or absence of food resources, parasites, diseases, and competing organisms can act as biotic limitations on species distribution. Unfortunately, some of the most dramatic cases of limitation occur when humans introduce (either accidentally or intentionally) exotic predators or diseases into new areas and wipe out native species. You will encounter examples of these impacts in Chapter 55, which discusses conservation ecology.

Figure 50.8

Inquiry Does feeding by sea urchins and limpets affect seaweed distribution?

EXPERIMENT W. J. Fletcher tested the effects of two algae-eating animals, sea urchins and limpets, on seaweed abundance near Sydney, Australia. In areas adjacent to a control site, either the urchins, the limpets, or both were removed.

RESULTS Fletcher observed a large difference in seaweed growth between areas with and without sea urchins.

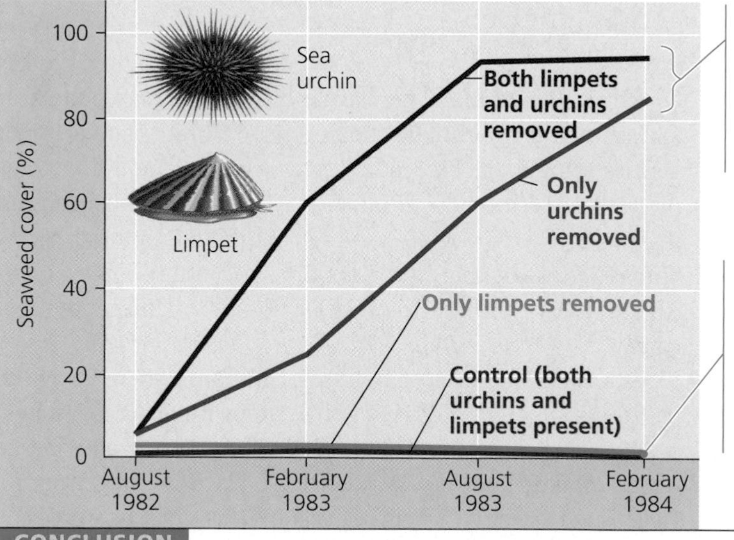

Removing both limpets and urchins or removing only urchins increased seaweed cover dramatically.

Almost no seaweed grew in areas where both urchins and limpets were present, or where only limpets were removed.

CONCLUSION Removing both limpets and urchins resulted in the greatest increase of seaweed cover, indicating that both species have some influence on seaweed distribution. But since removing only urchins greatly increased seaweed growth while removing only limpets had little effect, Fletcher concluded that sea urchins have a much greater effect than limpets in limiting seaweed distribution.

Abiotic Factors

The last question in our flowchart considers abiotic factors as limits to distribution. The global distribution of organisms broadly reflects the influence of abiotic factors, such as regional differences in temperature, water, and sunlight. Throughout this discussion, keep in mind that the environment is characterized by both *spatial heterogeneity* and *temporal heterogeneity*. In other words, it varies in both space and time. Although two regions of Earth may experience different conditions at any given time, daily and annual fluctuations of abiotic factors may either blur or accentuate regional distinctions.

Temperature

Environmental temperature is an important factor in the distribution of organisms because of its effect on biological processes. Cells may rupture if the water they contain freezes (at temperatures below 0°C), and the proteins of most organisms denature at temperatures above 45°C. In addition, few organisms can maintain an appropriately active metabolism at very low or very high temperatures. Extraordinary adaptations enable some organisms, such as thermophilic prokaryotes (see Chapter 27), to live outside the temperature range habitable by other life.

An organism's internal temperature is affected by heat exchange with its environment, and most organisms cannot maintain tissue temperatures more than a few degrees above or below the ambient temperature (see Chapter 40). As endotherms, mammals and birds are the major exceptions, but even endothermic species function best within various, quite narrow environmental temperature ranges.

Water

The dramatic variation in water availability among habitats is another important factor in species distribution. Freshwater and marine organisms live submerged in aquatic environments, but most are restricted to either freshwater or saltwater habitats by their limited ability for osmoregulation. Terrestrial organisms face a nearly constant threat of desiccation, and their global distribution reflects their ability to obtain and conserve water. Desert organisms exhibit a variety of adaptations for water acquisition and conservation in a desiccating environment, as described in Chapter 44.

Sunlight

Sunlight provides the energy that drives nearly all ecosystems, although only plants and other photosynthetic organisms use this energy source directly. Light intensity is not the most important factor limiting plant growth in many terrestrial environments, although shading by a forest canopy makes competition for light in the understory intense. In aquatic environments, however, the intensity and quality of light limit the distribution of photosynthetic organisms. Every meter of water depth selectively absorbs about 45% of the red light and about 2% of the blue light passing through it. As a result, most photosynthesis in aquatic environments occurs relatively near the surface. The photosynthetic organisms themselves absorb some of the light that penetrates, further reducing light levels in the waters below.

Light is also important to the development and behavior of the many organisms that are sensitive to photoperiod, the relative lengths of daytime and nighttime. Photoperiod is a more reliable indicator than temperature for cuing seasonal events, such as flowering by plants (see Chapter 39) or migration by animals.

Wind

Wind amplifies the effects of environmental temperature on organisms by increasing heat loss due to evaporation and convection (see Chapter 40). It also contributes to water loss in organisms by increasing the rate of evaporative cooling in animals and transpiration in plants. In addition, wind can have a substantial effect on the morphology of plants by inhibiting the growth of limbs on the windward side of trees, resulting in a "flagged" appearance **(Figure 50.9)**.

Rocks and Soil

The physical structure, pH, and mineral composition of rocks and soil limit the distribution of plants and thus of the animals that feed upon them, contributing to the patchiness of terrestrial ecosystems. In streams and rivers, the composition of the substrate (bottom surface) can affect water chemistry, which in turn influences the resident organisms. In freshwater and marine environments, the structure of the substrate determines the organisms that can attach to or burrow in it.

▲ **Figure 50.9 "Flagging" of tree limbs due to wind.**

Now that we have surveyed the abiotic factors that affect the distribution of organisms, let's focus on how those factors vary with climate as we consider the major role that climate plays in determining species distribution.

Climate

Four abiotic factors—temperature, water, sunlight, and wind—are the major components of **climate,** the prevailing weather conditions in a particular area. Climatic factors, particularly temperature and water, have a major influence on the distribution of organisms. We can describe climate patterns on two scales: **macroclimate,** patterns on the global, regional, and local level; and **microclimate,** very fine patterns, such as those encountered by the community of organisms underneath a fallen log. First, let's consider Earth's macroclimate.

Global Climate Patterns

Earth's global climate patterns are determined largely by the input of solar energy and the planet's movement in space. The sun's warming effect on the atmosphere, land, and water establishes the temperature variations, cycles of air movement, and evaporation of water that are responsible for dramatic latitudinal variations in climate. **Figure 50.10**, on the next two pages, summarizes Earth's climate patterns and how they are formed.

Regional, Local, and Seasonal Effects on Climate

Proximity to bodies of water and topographic features such as mountain ranges create regional climatic variations, and smaller features of the landscape contribute to local climatic variation. These regional and local variations in climate contribute to the patchiness of the biosphere. Seasonal variation is another influence on climate.

Bodies of Water. Ocean currents influence climate along the coasts of continents by heating or cooling overlying air masses, which may then pass across the land. Coastal regions are generally moister than inland areas at the same latitude. The cool, misty climate produced by the cold California current that flows southward along the western United States supports a temperate rain forest ecosystem dominated by large coniferous trees in the Pacific Northwest and large redwood groves farther south. Similarly, the warm Gulf Stream flowing northward from the Caribbean Sea and across the North Atlantic tempers the climate on the west coast of northern Europe. As a result, northwest Europe is warmer during winter than New England, which is actually farther south but is cooled by the Labrador Current flowing south from the coast of Greenland.

Figure 50.10

Exploring Global Climate Patterns

LATITUDINAL VARIATION IN SUNLIGHT INTENSITY

Earth's curved shape causes latitudinal variation in the intensity of sunlight. Because sunlight strikes the **tropics** (those regions that lie between 23.5° north latitude and 23.5° south latitude) most directly, the most heat and light per unit of surface area are delivered there. At higher latitudes, sunlight strikes Earth at an oblique angle, and thus the light energy is more diffuse on Earth's surface.

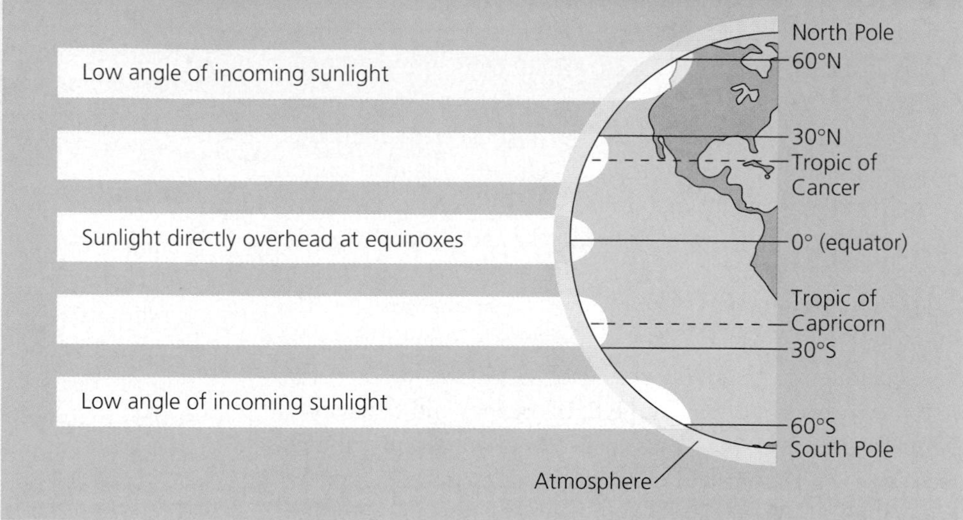

Low angle of incoming sunlight
Sunlight directly overhead at equinoxes
Low angle of incoming sunlight
North Pole
60°N
30°N
Tropic of Cancer
0° (equator)
Tropic of Capricorn
30°S
60°S
South Pole
Atmosphere

SEASONAL VARIATION IN SUNLIGHT INTENSITY

Earth's tilt causes seasonal variation in the intensity of solar radiation. Because the planet is tilted on its axis by 23.5° relative to its plane of orbit around the sun, the tropics experience the greatest annual input of solar radiation and the least seasonal variation. The seasonal variations of light and temperature increase steadily toward the poles.

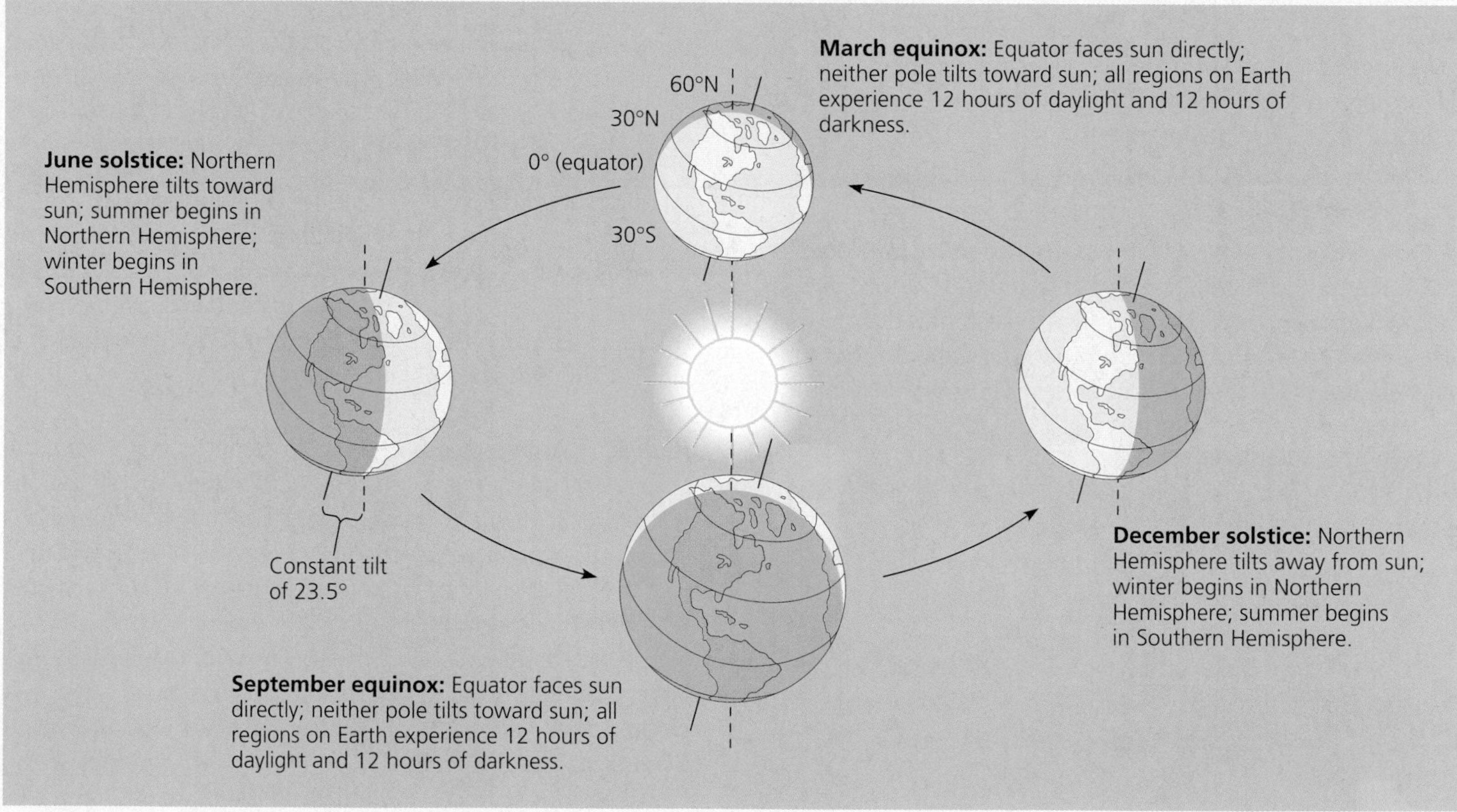

March equinox: Equator faces sun directly; neither pole tilts toward sun; all regions on Earth experience 12 hours of daylight and 12 hours of darkness.
60°N
30°N
0° (equator)
30°S
June solstice: Northern Hemisphere tilts toward sun; summer begins in Northern Hemisphere; winter begins in Southern Hemisphere.
Constant tilt of 23.5°
December solstice: Northern Hemisphere tilts away from sun; winter begins in Northern Hemisphere; summer begins in Southern Hemisphere.
September equinox: Equator faces sun directly; neither pole tilts toward sun; all regions on Earth experience 12 hours of daylight and 12 hours of darkness.

GLOBAL AIR CIRCULATION AND PRECIPITATION PATTERNS

Intense solar radiation near the equator initiates a global pattern of air circulation and precipitation. High temperatures in the tropics evaporate water from Earth's surface and cause warm, wet air masses to rise (blue arrows) and flow toward the poles. The rising air masses release much of their water content, creating abundant precipitation in tropical regions. The high-altitude air masses, now dry, descend (brown arrows) toward Earth, absorbing moisture from the land and creating an arid climate conducive to the development of the deserts that are common at latitudes around 30° north and south. Some of the descending air then flows toward the poles. At latitudes around 60° north and south, the air masses again rise and release abundant precipitation (though less than in the tropics). Some of the cold, dry rising air then flows to the poles, where it descends and flows back toward the equator, absorbing moisture and creating the comparatively rainless and bitterly cold climates of the polar regions.

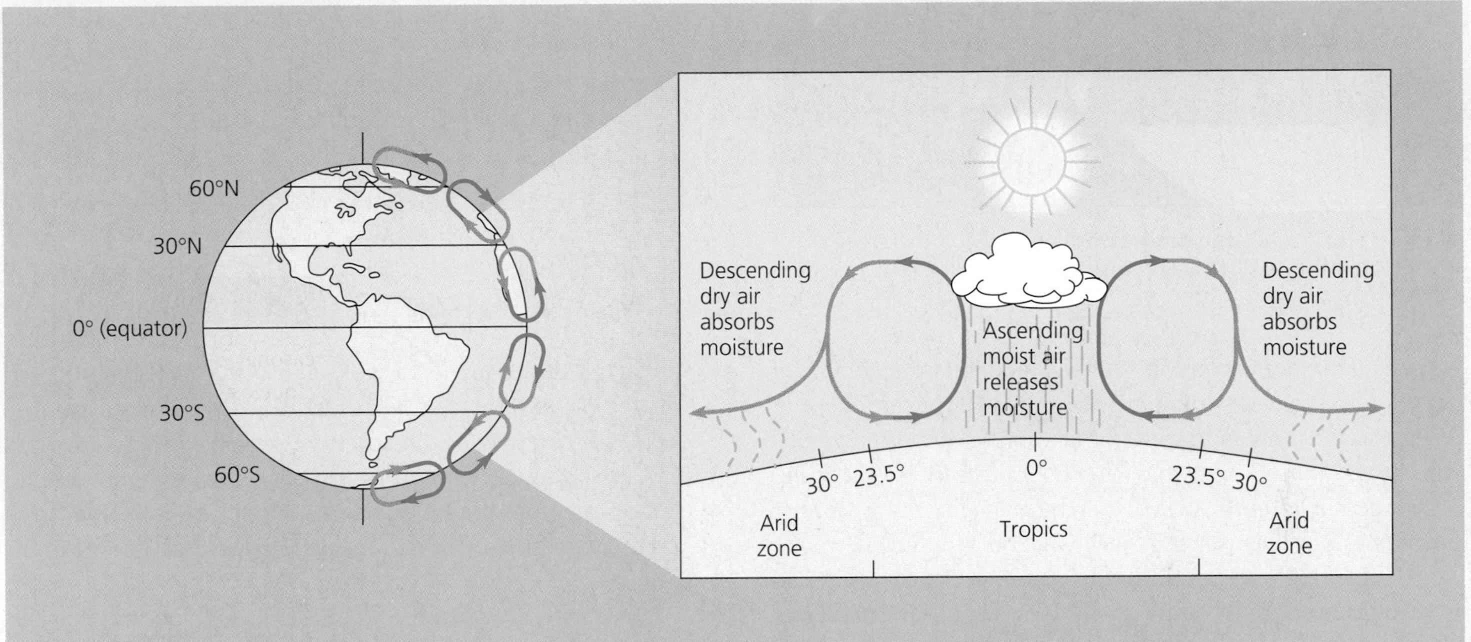

GLOBAL WIND PATTERNS

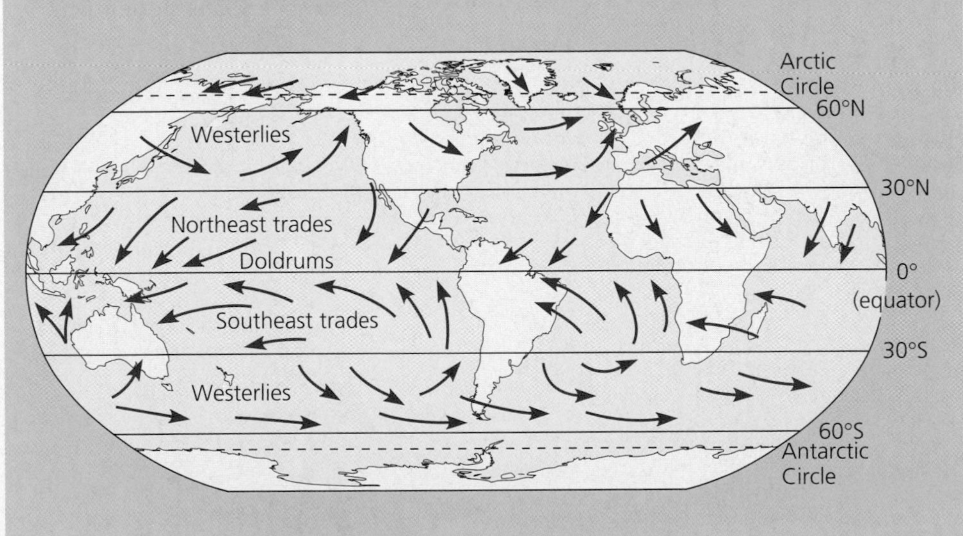

Air flowing close to Earth's surface creates predictable global wind patterns. As Earth rotates on its axis, land near the equator moves faster than that at the poles, deflecting the winds from the vertical paths shown above and creating more easterly and westerly flows. Cooling trade winds blow from east to west in the tropics; prevailing westerlies blow from west to east in the temperate zones, the regions between the tropics and the Arctic Circle or the Antarctic Circle.

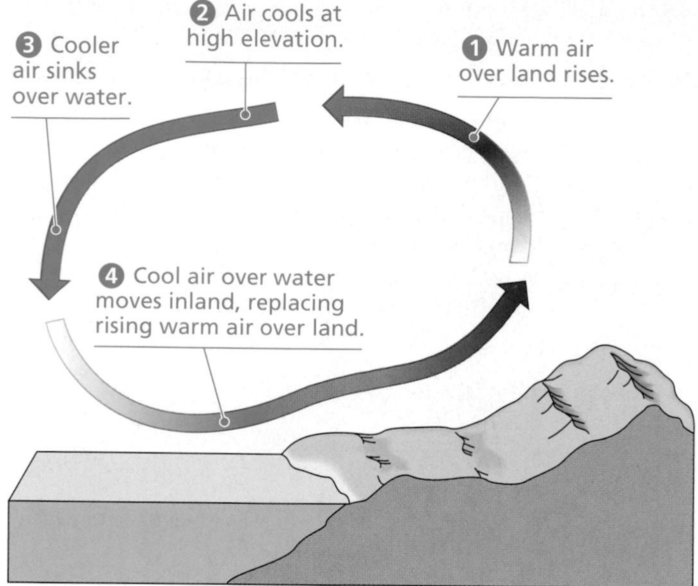

❸ Cooler air sinks over water.
❶ Warm air over land rises.
❹ Cool air over water moves inland, replacing rising warm air over land.

▲ **Figure 50.11 Moderating effects of large bodies of water on climate.** This figure illustrates what happens on a warm summer day.

In general, oceans and large lakes moderate the climate of nearby terrestrial environments. During a warm summer day, when the land is hotter than a large lake or the ocean, air over the land heats up and rises, drawing a cool breeze from the water across the land **(Figure 50.11)**. At night, air over the now warmer ocean or lake rises, reversing the circulation and drawing cooler air from the land out over the water, replacing it with warmer air from offshore. Proximity to water does not always moderate climate, however. In summer in certain regions, including the coast of central and southern California, cool, dry ocean breezes are warmed when they contact the land, absorbing moisture and creating a hot, rainless climate just a few miles inland. This climate pattern also occurs in the area around the Mediterranean Sea, which gives it the name *Mediterranean climate*.

Mountains. Mountains have a significant effect on the amount of sunlight reaching an area, as well as on local temperature and rainfall. South-facing slopes in the Northern Hemisphere receive more sunlight than nearby north-facing slopes and are therefore warmer and drier. These abiotic differences influence species distribution; for example, in many mountains of western North America, spruce and other conifers occupy the north-facing slopes, whereas shrubby, drought-resistant plants inhabit the south-facing slopes. In addition, at any particular latitude, air temperature declines approximately 6°C with every 1,000-m increase in elevation, paralleling the decline of temperature with increasing latitude. In the north temperate zone, for example, a 1,000-m increase in elevation produces a temperature change equivalent to that produced by an 880-km increase in latitude. This is one reason the biological communities of mountains are similar to those at lower elevations but farther from the equator.

When warm, moist air approaches a mountain, the air rises and cools, releasing moisture on the windward side of the peak **(Figure 50.12)**. On the leeward side of the mountain, cooler, dry air descends, absorbing moisture and producing a "rain shadow." Deserts commonly occur on the leeward side of mountain ranges, a phenomenon evident in the Great Basin and the Mojave Desert of western North America, the Gobi Desert of Asia, and the small deserts that characterize the southwest corners of some Caribbean islands.

Seasonality. In addition to the global changes in day length, solar radiation, and temperature described earlier, the changing angle of the sun over the course of the year affects local environments. For example, the belts of wet and dry air on either side of the equator move slightly northward and southward with the changing angle of the sun, producing marked wet and dry seasons around 20° latitude, where many tropical deciduous forests grow. In addition, seasonal changes in wind patterns produce variations in ocean currents, sometimes causing the upwelling of nutrient-rich, cold water from deep ocean layers, thus nourishing organisms that live near the surface.

Lakes are also sensitive to seasonal temperature changes. During the summer and winter, many lakes in temperate regions are thermally stratified—that is, layered vertically according to temperature. Such lakes undergo a semiannual mixing of their waters as a result of changing temperature

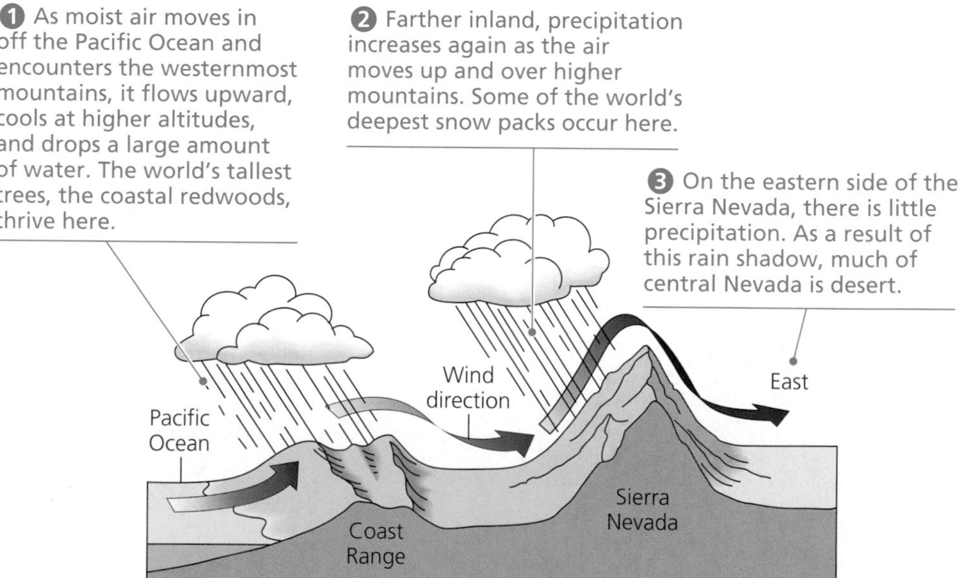

❶ As moist air moves in off the Pacific Ocean and encounters the westernmost mountains, it flows upward, cools at higher altitudes, and drops a large amount of water. The world's tallest trees, the coastal redwoods, thrive here.
❷ Farther inland, precipitation increases again as the air moves up and over higher mountains. Some of the world's deepest snow packs occur here.
❸ On the eastern side of the Sierra Nevada, there is little precipitation. As a result of this rain shadow, much of central Nevada is desert.
Pacific Ocean
Wind direction
East
Coast Range
Sierra Nevada

▲ **Figure 50.12 How mountains affect rainfall.**

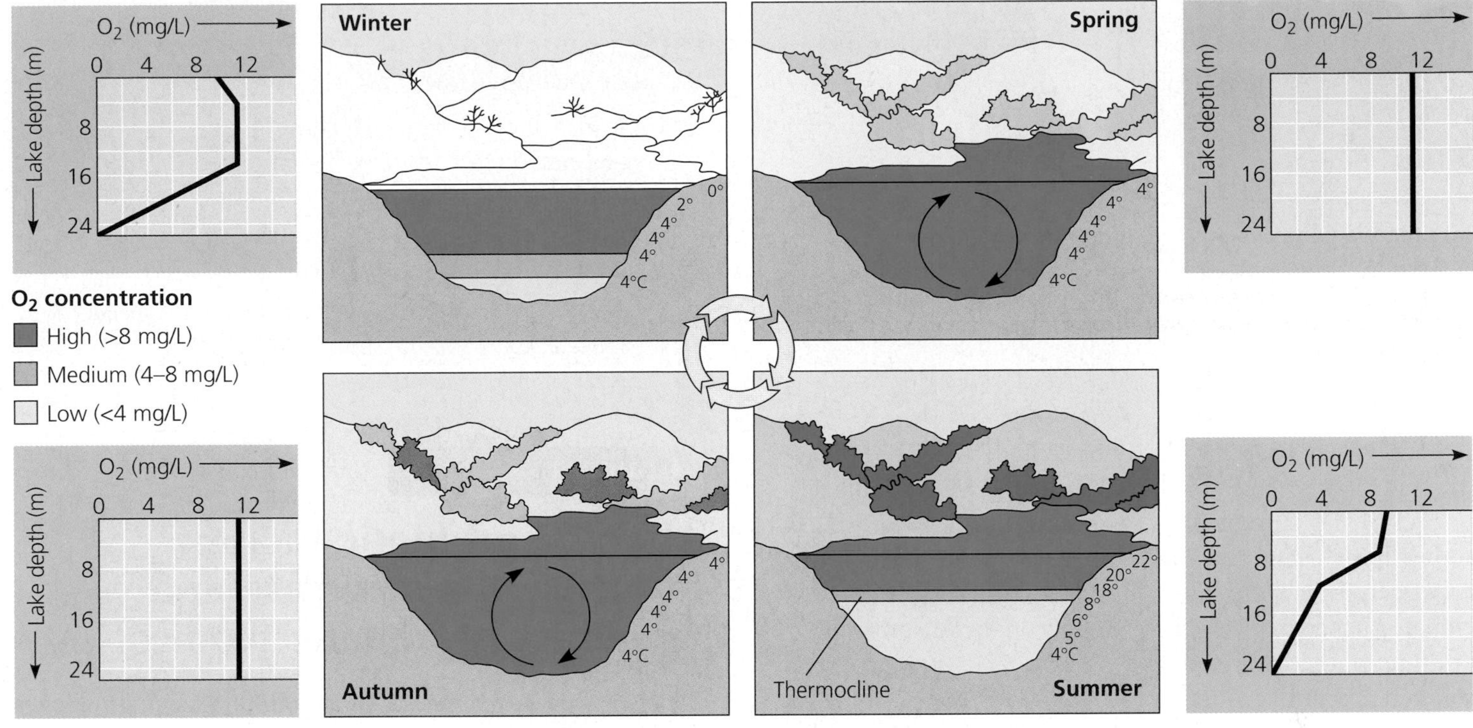

1 In winter, the coldest water in the lake (0°C) lies just below the surface ice; water is progressively warmer at deeper levels of the lake, typically 4°C at the bottom.
2 In spring, as the sun melts the ice, the surface water warms to 4°C and sinks below the cooler layers immediately below, eliminating the thermal stratification. Spring winds mix the water to great depth, bringing oxygen (O_2) to the bottom waters (see graphs) and nutrients to the surface.
O_2 (mg/L)
Lake depth (m)
Winter
Spring
Autumn
Summer
Thermocline
O_2 concentration
High (>8 mg/L)
Medium (4–8 mg/L)
Low (<4 mg/L)
4 In autumn, as surface water cools rapidly, it sinks below the underlying layers, remixing the water until the surface begins to freeze and the winter temperature profile is reestablished.
3 In summer, the lake regains a distinctive thermal profile, with warm surface water separated from cold bottom water by a narrow vertical zone of rapid temperature change, called a thermocline.

▲ **Figure 50.13 Seasonal turnover in lakes with winter ice cover.**

profiles **(Figure 50.13)**. This **turnover**, as it is called, brings oxygenated water from a lake's surface to the bottom and nutrient-rich water from the bottom to the surface in both spring and autumn. These cyclic changes in the abiotic properties of lakes are essential for the survival and growth of organisms at all levels within this ecosystem.

Microclimate

Many features in the environment influence microclimates by casting shade, affecting evaporation from soil, and changing wind patterns. For example, forest trees frequently moderate the microclimate below them. Consequently, cleared areas generally experience greater temperature extremes than the forest interior because of greater solar radiation and wind currents that are established by the rapid heating and cooling of open land; evaporation is generally greater in clearings as well. Within a forest, low-lying ground is usually wetter than high ground and tends to be occupied by different species of trees. A log or large stone shelters organisms such as salamanders, worms, and insects, buffering them from the extremes of temperature and moisture. Every environment on Earth is similarly characterized by a mosaic of small-scale differences in the abiotic factors that influence the local distributions of organisms.

Long-Term Climate Change

If temperature and moisture are the master factors limiting the geographic ranges of plants and animals, the global climate change currently under way will profoundly affect the biosphere (see Chapter 54). One way to predict the possible effects is to look back at the changes that have occurred in temperate regions since the last ice age ended.

Until about 16,000 years ago, continental glaciers covered much of North America and Eurasia. As the climate warmed and the glaciers retreated, tree distribution expanded northward. A detailed record of these migrations is captured in fossil pollen deposited in lakes and ponds. (It may seem odd to think of trees "migrating," but recall from Chapter 38 that wind and animals can disperse seeds, sometimes over great distances.) If researchers can determine the climatic limits of current geographic distributions for organisms, they can make predictions about how distributions will change with climatic warming. A major question when applying this approach to plants is whether seed dispersal is rapid enough to sustain the migration of each species as climate changes. For example, fossils suggest that the eastern hemlock was delayed nearly 2,500 years in its movement north at the end of the last ice age partly because of relatively slow seed dispersal.

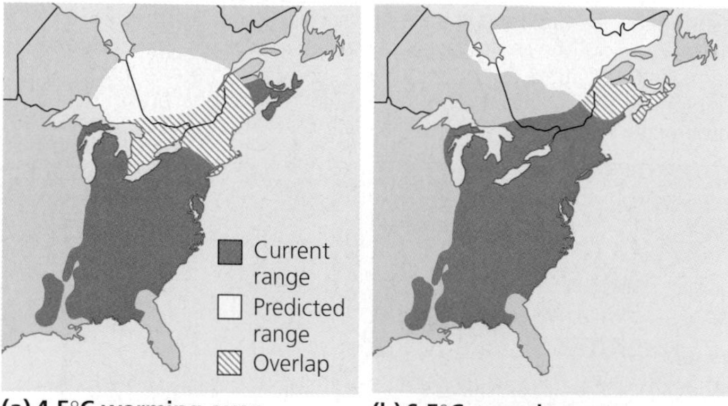

▲ **Figure 50.14 Current range and predicted range for the American beech (*Fagus grandifolia*) under two scenarios of climate change.**

Let's look at a specific case of how the fossil record of past tree migrations can inform predictions about the biological impact of the current global warming trend. **Figure 50.14** shows the current and predicted geographic ranges of the American beech (*Fagus grandifolia*) under two different climate-change models. These models predict that the northern limit of the beech's range will move 700–900 km northward in the next century, and its southern range limit will move northward an even greater distance. If these predictions are even approximately correct, the beech must move 7–9 km per year northward to keep pace with the warming climate. However, since the end of the last ice age, the beech has migrated into its present range at a rate of only 0.2 km per year. The unhappy conclusion is clear: Without human assistance in moving into new ranges where they can survive as the climate warms, species such as the American beech will have much smaller ranges and may become extinct.

Concept Check 50.2

1. Give examples of human actions that could expand a species' distribution by changing its (a) dispersal or (b) biotic interactions.
2. Explain how the sun's unequal heating of Earth's surface influences global climate patterns.

For suggested answers, see Appendix A.

Concept 50.3

Abiotic and biotic factors influence the structure and dynamics of aquatic biomes

We have seen how both biotic and abiotic factors influence the distribution of organisms on Earth. Varying combinations of these factors determine the nature of Earth's many **biomes**, major types of ecological associations that occupy broad geographic regions of land or water. We'll begin by examining Earth's aquatic biomes **(Figure 50.15)**.

▼ **Figure 50.15 The distribution of major aquatic biomes.**

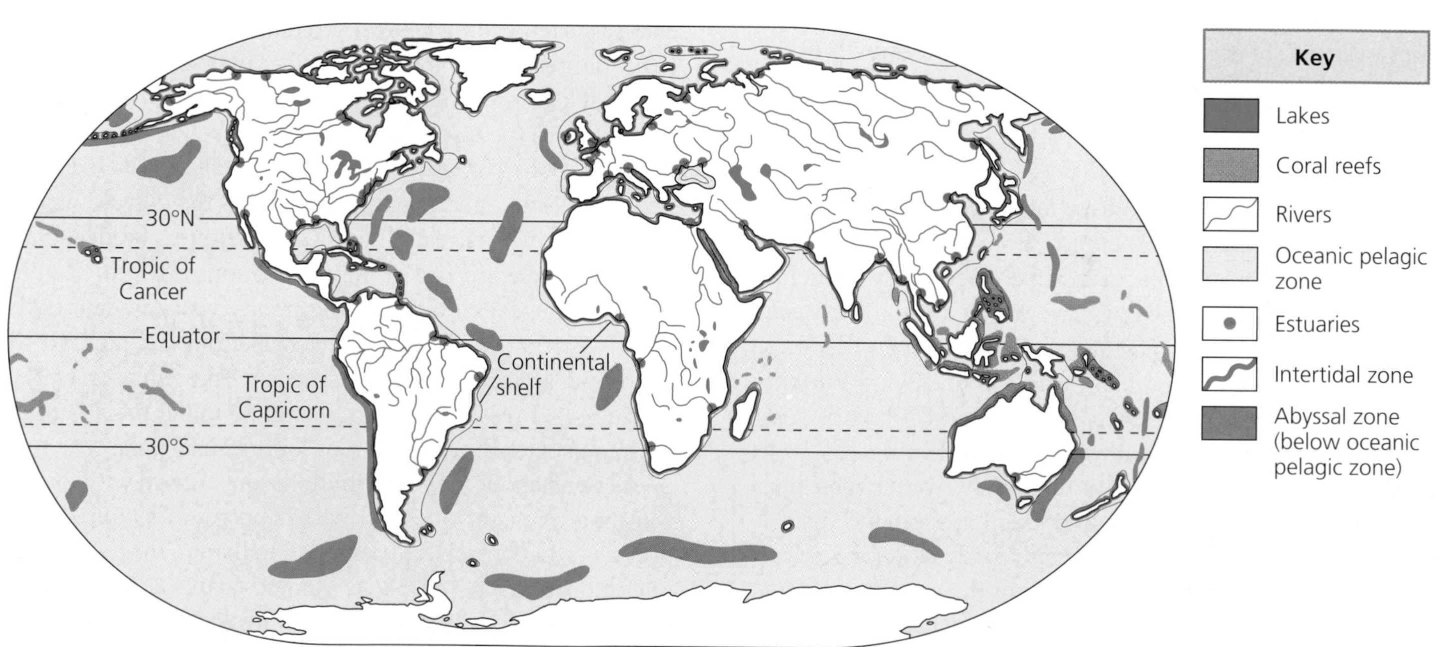

Aquatic biomes account for the largest part of the biosphere in terms of area, and all types are found around the globe. Ecologists distinguish between freshwater biomes and marine biomes on the basis of physical and chemical differences. For example, marine biomes generally have salt concentrations that average 3%, whereas freshwater biomes are usually characterized by a salt concentration of less than 1%.

The largest marine biomes, the oceans, cover about 75% of Earth's surface and thus have an enormous impact on the biosphere. The evaporation of water from the oceans provides most of the planet's rainfall, and ocean temperatures have a major effect on world climate and wind patterns. In addition, marine algae and photosynthetic bacteria supply a substantial portion of the world's oxygen and consume huge amounts of atmospheric carbon dioxide.

Freshwater biomes are closely linked to the soils and biotic components of the terrestrial biomes through which they pass or in which they are situated. The particular characteristics of a freshwater biome are also influenced by the patterns and speed of water flow and the climate to which the biome is exposed.

Many aquatic biomes are physically and chemically stratified, as illustrated in **Figure 50.16** for both a lake and a marine environment. Light is absorbed by both the water itself and the photosynthetic organisms in it, so its intensity decreases rapidly with depth, as mentioned earlier. Ecologists distinguish between the upper **photic zone**, where there is sufficient light for photosynthesis, and the lower **aphotic zone**, where little light penetrates. At the bottom of all aquatic biomes, the substrate is called the **benthic zone.** Made up of sand and organic and inorganic sediments ("ooze"), the benthic zone is occupied by communities of organisms collectively called **benthos.** A major source of food for the benthos is dead organic matter called **detritus,** which "rains" down from the productive surface waters of the photic zone. The deepest regions of the ocean floor are known as the abyssal zone.

Thermal energy from sunlight warms surface waters to whatever depth the sunlight penetrates, but the deeper waters remain quite cold. As a result, water temperature in lakes tends to be stratified, especially during summer and winter (see Figure 50.13). In the ocean and in most lakes, a narrow stratum of rapid temperature change called a **thermocline** separates the more uniformly warm upper layer from more uniformly cold deeper waters.

In both freshwater and marine environments, communities are distributed according to depth of the water, degree of light penetration, distance from shore, and open water versus bottom. Marine communities illustrate the limitations on species distribution that result from these abiotic factors. Phytoplankton, zooplankton, and many fish species occur in the relatively shallow photic zone (see Figure 50.16b). Because water absorbs light so well and the ocean is so deep, most of the ocean volume is virtually devoid of light (the aphotic zone) and harbors relatively little life, except for microorganisms and relatively sparse populations of luminescent fishes and invertebrates.

Figure 50.17, on the next four pages, surveys the major aquatic biomes.

(a) Zonation in a lake. The lake environment is generally classified on the basis of three physical criteria: light penetration (photic and aphotic zones), distance from shore and water depth (littoral and limnetic zones), and whether it is open water (pelagic zone) or bottom (benthic zone).

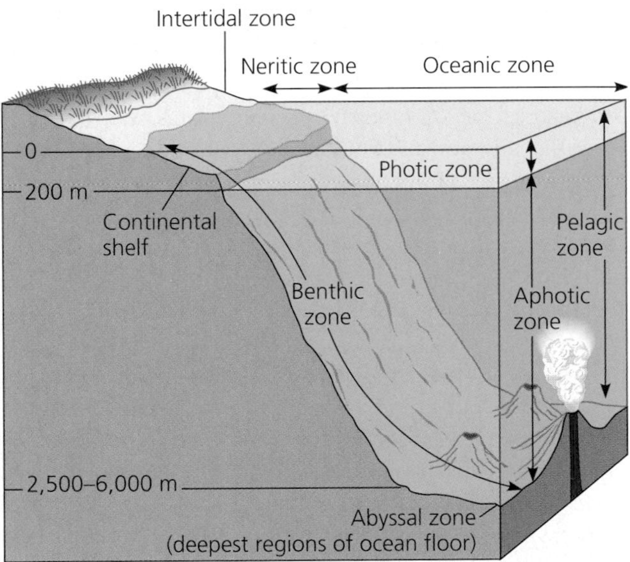

(b) Marine zonation. Like lakes, the marine environment is generally classified on the basis of light penetration (photic and aphotic zones), distance from shore and water depth (intertidal, neritic, and oceanic zones), and whether it is open water (pelagic zone) or bottom (benthic and abyssal zones).

▲ **Figure 50.16 Zonation in aquatic environments.**

Figure 50.17

Exploring Aquatic Biomes

LAKES

Physical Environment Standing bodies of water range from ponds a few square meters in area to lakes covering thousands of square kilometers. Light decreases with depth, creating stratification (see Figure 50.16a). Temperate lakes may have a seasonal thermocline (see Figure 50.13); tropical lowland lakes have a thermocline year-round.

An oligotrophic lake in Grand Teton, Wyoming

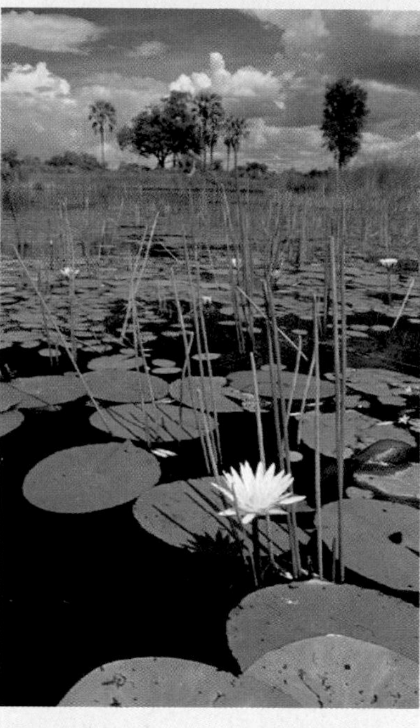
A eutrophic lake in Okavango delta, Botswana

Chemical Environment The salinity (salt content), oxygen concentration, and nutrient content differ greatly among lakes and can vary substantially with season. **Oligotrophic lakes** are nutrient-poor and generally oxygen-rich; **eutrophic lakes** are nutrient-rich and often depleted of oxygen if ice-covered in winter and in the deepest zone during summer. The amount of decomposable organic matter in bottom sediments is low in oligotrophic lakes and high in eutrophic lakes.

Geologic Features Oligotrophic lakes tend to have less surface area relative to their depth than eutrophic lakes. Over long periods of time, an oligotrophic lake may become more eutrophic as runoff adds sediments and nutrients to the lake.

Photosynthetic Organisms Rates of photosynthesis are higher in eutrophic lakes than in oligotrophic lakes. Rooted and floating aquatic plants live in the **littoral zone**, the shallow, well-lighted waters close to shore. Further away from shore, the **limnetic zone**, where water is too deep to support rooted aquatic plants, is inhabited by a variety of phytoplankton and cyanobacteria.

Animals In the limnetic zone, small drifting animals, or zooplankton, graze on the phytoplankton. The benthic zone is inhabited by a variety of invertebrate animals, with the species composition depending partly on oxygen levels. Fishes live in all zones in lakes with sufficient oxygen.

Human Impact Pollution by runoff from fertilized land and dumping of municipal wastes leads to nutrient enrichment, which can produce algal blooms, oxygen depletion, and fish kills.

WETLANDS

Physical Environment A **wetland** is an area covered with water for a long enough period to support aquatic plants. Wetlands range from those that are permanently inundated to those that flood infrequently.

Chemical Environment Because of the high organic production and decomposition in wetlands, both the water and the soils are periodically low in dissolved oxygen. Wetlands have a high capacity to filter dissolved nutrients and chemical pollutants.

Geologic Features *Basin wetlands* develop in shallow basins, ranging from upland depressions to filled-in lakes and ponds. *Riverine wetlands* develop along shallow and periodically flooded banks of rivers and streams. *Fringe wetlands* occur along the coasts of large lakes and seas, where water flows back and forth because of rising lake levels or tidal action. Thus, fringe wetlands include both freshwater and marine biomes.

Photosynthetic Organisms Wetlands are among the most productive biomes on Earth. Their water-saturated soils favor the growth of plants, such as floating pond lilies and emergent cattails, many sedges, tamarack, and black spruce, that have adaptations enabling them to grow in water or in soil that is periodically anaerobic owing to the presence of unaerated water. Woody plants dominate the vegetation of swamps, while bogs are dominated by sphagnum mosses.

Animals Wetlands are home to a diverse community of invertebrates, which in turn support a wide variety of birds. Herbivores, from crustaceans and aquatic insect larvae to muskrats, consume algae, detritus, and plants. Carnivores are also varied and may include dragonflies, otters, alligators, and owls.

Okefenokee National Wetland Reserve in Georgia

Human Impact Draining and filling have destroyed up to 90% of wetlands in some regions.

STREAMS AND RIVERS

Physical Environment The most prominent physical characteristic of **streams** and **rivers** is current. Headwater streams are generally cold, clear, turbulent, and swift. Farther downstream, where numerous tributaries may have joined, forming a river, the water is generally warmer and more turbid, since rivers generally carry more sediment than their headwaters. Streams and rivers are stratified into vertical zones, extending from surface water through groundwater.

Chemical Environment The salt and nutrient content of streams and rivers increases from the headwaters to the mouth. In streams, headwaters are generally rich in oxygen. River water may also contain substantial oxygen, except where there has been organic enrichment from either natural or human sources.

Geologic Features Headwater stream channels are often narrow, with a rocky bottom alternating between riffles and pools. The downstream reaches of rivers are generally wide and meandering. River bottoms are often silty from sediments deposited over long periods of time.

Photosynthetic Organisms Headwater streams that flow through grasslands or deserts may be rich in algae or rooted aquatic plants; but in streams flowing through temperate or tropical forests, leaves and other organic matter produced by terrestrial vegetation are the primary source of food for aquatic consumers. In rivers, a large fraction of the organic matter consists of dissolved and highly fragmented material that is carried by the current from forested headwater streams.

Animals A great diversity of fishes and invertebrates inhabit unpolluted rivers and streams, distributed according to, and throughout, the vertical zones.

Human Impact Municipal, agricultural, and industrial pollution degrade water quality and kill aquatic organisms. Damming and flood control impair natural functioning of stream and river ecosystems and threaten migratory species such as salmon.

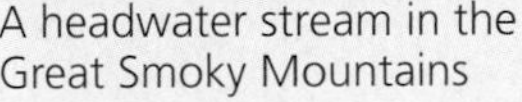

A headwater stream in the Great Smoky Mountains

The Mississippi River far from its headwaters

ESTUARIES

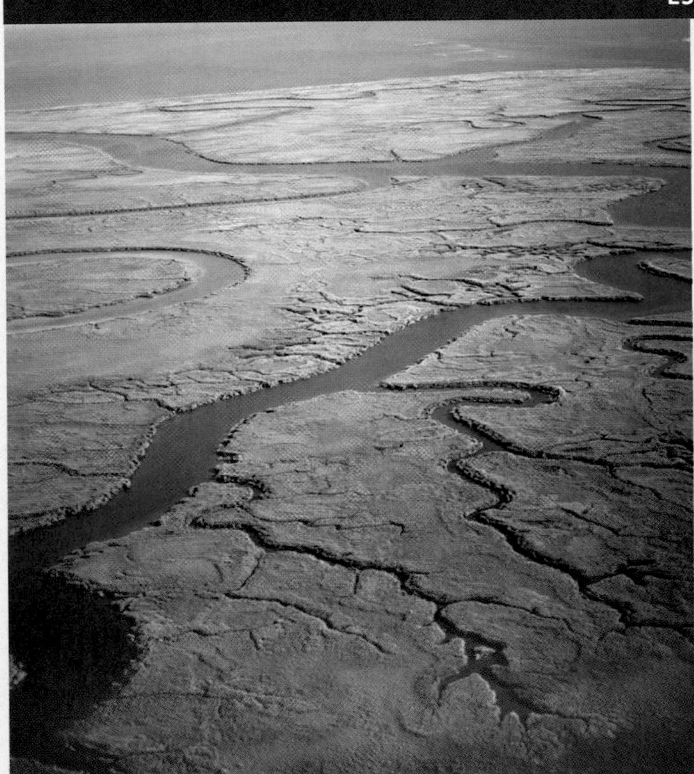

An estuary in a low coastal plain of Georgia

Physical Environment An **estuary** is a transition area between river and sea. Estuaries have very complex flow patterns. During a rising tide, seawater flows up the estuary channel, flowing back down again during the falling tide. Often higher-density seawater occupies the bottom of an estuary channel, while lower-density river water forms a surface layer that mixes little with the salty bottom layer.

Chemical Environment Salinity varies spatially within estuaries, from nearly that of fresh water to that of seawater. Salinity also varies with the rise and fall of the tides. Nutrients from the river make estuaries, like wetlands, among the most productive biomes.

Geologic Features Estuarine flow patterns combined with the sediments carried by river and tidal water create a complex network of tidal channels, islands, natural levees, and mudflats.

Photosynthetic Organisms Saltmarsh grasses and algae, including phytoplankton, are the major producers in estuaries.

Animals Estuaries support an abundance of worms, oysters, crabs, and many of the fish species that humans consume. Because of the abundant food in estuaries, many marine invertebrates and fishes use them as a breeding ground; others migrate through estuaries to freshwater habitats upstream. Estuaries are also crucial feeding areas for many semi-aquatic vertebrates, particularly waterfowl.

Human Impact Pollution from upstream, and also filling and dredging, have disrupted estuaries worldwide.

continued on next page

Figure 50.17 (continued)

Exploring Aquatic Biomes

INTERTIDAL ZONES

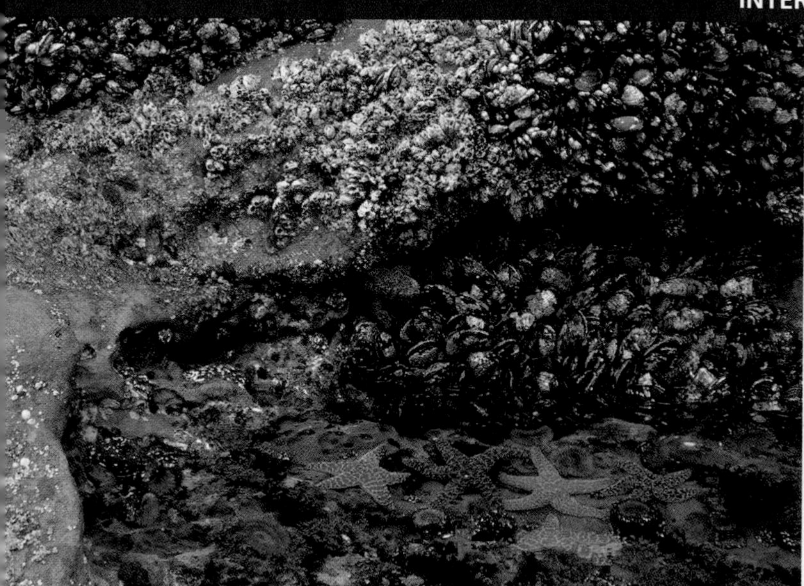

Rocky intertidal zone on the Oregon coast

Physical Environment An **intertidal zone** is periodically submerged and exposed by the tides, twice daily on most marine shores. Upper zones experience longer exposures to air and greater variations in physical environment. Among the physical challenges faced by intertidal organisms are variations in temperature and salinity and the mechanical forces of wave action. Changes in physical conditions from the upper intertidal zone to the lower intertidal zone limit the distributions of many organisms to particular strata, as shown in the photograph.

Chemical Environment Oxygen and nutrient levels are generally high and are renewed with each turn of the tides.

Geologic Features The substrates of intertidal zones, which are generally either rocky or sandy, select for particular behavior and anatomy among intertidal organisms. The configuration of bays or coastlines influences the magnitude of tides and the relative exposure of intertidal organisms to wave action.

Photosynthetic Organisms A high diversity and biomass of attached marine algae inhabit rocky intertidal zones, especially in the lower zone. Because of the instability of the substrate, sandy intertidal zones exposed to vigorous wave action generally lack attached plants or algae, while sandy intertidal zones in protected bays or lagoons often support rich beds of sea grass and algae.

Animals Many of the animals of rocky intertidal environments have structural adaptations that enable them to attach to the hard substrate. The composition, density, and diversity of intertidal animals change markedly from the upper to lower intertidal zones. Many of the animals in sandy or muddy intertidal environments, such as suspension-feeding worms and clams and predatory crustaceans, bury themselves in sand or mud, feeding as the tides bring sources of food. Other common animals are sponges, sea anemones, molluscs, echinoderms, and small fishes.

Human Impact Oil pollution has disrupted many intertidal areas. Recreational use has caused a severe decline in the numbers of beach-nesting birds and sea turtles.

OCEANIC PELAGIC BIOME

Physical Environment The **oceanic pelagic biome** is a vast realm of open blue water, constantly mixed by wind-driven oceanic currents. The surface waters of temperate ocean areas turn over during fall through spring. Because of higher water clarity, the photic zone extends to greater depths than in coastal marine waters.

Chemical Environment Oxygen levels are generally high. Nutrient concentrations are generally lower than in coastal waters. Because they are thermally stratified year-round, some tropical areas of the oceanic pelagic biome have lower nutrient concentrations than temperate oceans. Turnover during fall through spring renews nutrients in the photic zones of temperate and high-latitude ocean areas.

Geologic Features The most prominent geologic characteristic of the oceanic pelagic biome is its vastness and the great depth of the ocean basins. This biome covers approximately 70% of Earth's surface and has an average depth of nearly 4,000 m. The deepest point in the ocean is more than 10,000 m beneath the surface.

Photosynthetic Organisms The dominant photosynthetic organisms are phytoplankton, including photosynthetic bacteria, that drift with the oceanic currents. Spring turnover and renewal of nutrients in temperate oceans produces a surge of phytoplankton growth. Despite the large extent of their biome, photosynthetic plankton account for less than half of the photosynthetic activity on Earth.

Animals The most abundant animals and other heterotrophs in this biome are zooplankton. These protozoans, worms, copepods, shrimp-like krill, jellies, and the small larvae of invertebrates and fishes graze on photosynthetic plankton. The oceanic pelagic biome also includes free-swimming animals, such as large squids, fishes, sea turtles, and marine mammals.

Human Impact Overfishing has depleted fish stocks in all Earth's oceans, which have also been polluted by waste dumping and oil spills.

Open ocean off the island of Hawaii

CORAL REEFS

Physical Environment Reef-building corals are limited to the photic zone of relatively stable tropical marine environments with high water clarity. They are sensitive to temperatures below about 18°–20°C and above 30°C.

Chemical Environment Corals require high oxygen levels and are excluded by high inputs of fresh water and nutrients.

Geologic Features Corals require a solid substrate for attachment. A **coral reef,** which is formed largely from the calcium carbonate skeletons of corals, develops over a long time on oceanic islands. It begins as a fringing reef on a young, high island, forming an offshore barrier reef later in the history of the island, and becoming a coral atoll as the older island submerges.

Photosynthetic Organisms Unicellular algae live within the tissues of the corals, forming a mutualistic symbiotic relationship that provides the corals with organic molecules. Diverse multicellular red and green algae also contribute substantial amounts of photosynthesis on coral reefs.

Animals Corals, a diverse group of cnidarians (see Chapter 33), are themselves the predominant animals on coral reefs. However, fish and invertebrate diversity is exceptionally high. Overall animal diversity on coral reefs rivals that of tropical forests.

Human Impact Collecting of coral skeletons, as well as overfishing for food and for the aquarium trade, often using poisons and explosives, have reduced populations of corals and reef fishes. Global warming and pollution may be contributing to large-scale coral mortality.

A coral reef in the Red Sea

MARINE BENTHIC ZONE

Physical Environment The **marine benthic zone** consists of the seafloor below the surface waters of the coastal, or **neritic,** zone and the offshore, pelagic zone (see Figure 50.16b). Although in shallow, near-coastal waters the benthic zone receives enough sunlight to support photosynthetic organisms, most of the ocean's benthic zone receives no sunlight. Water temperatures decline with depth, while pressure increases. As a result, organisms in the very deep benthic, or **abyssal,** zone are adapted to continuous cold (about 3°C) and extremely high water pressure.

Chemical Environment Except in some areas of organic enrichment, oxygen is present at sufficient concentrations to support a diversity of animals.

Geologic Features Soft sediments cover most of the benthic zone. However, there are areas of rocky substrate on reefs, submarine mountains, and new oceanic crust created by seafloor volcanoes.

Food-Producing Organisms Photosynthetic organisms, mainly seaweeds and filamentous algae, are limited to shallow benthic areas with sufficient light to support them. Unique assemblages of organisms, such as those shown in the photo, are associated with **deep-sea hydrothermal vents** of volcanic origin on mid-ocean ridges. In these dark, hot, oxygen-deficient environments, the food producers are chemoautotrophic prokaryotes (see Chapter 27) that obtain energy by oxidizing H_2S formed by a reaction of the hot water with dissolved sulfate (SO_4^{2-}).

Animals Neritic benthic communities include numerous invertebrates and fishes. Beyond the photic zone, most consumers depend entirely on organic matter raining down from above. Among the animals of the deep-sea hydrothermal vent communities are giant tube-dwelling worms (pictured at left), some more than 1 m long. They are apparently nourished by chemosynthetic prokaryotes that live as symbionts within the worms. Many other invertebrates, including arthropods and echinoderms, are also abundant around the hydrothermal vents.

Human Impact Overfishing has decimated important benthic fish populations such as the cod of the Grand Banks off Newfoundland. Dumping of organic wastes has created oxygen-deprived benthic areas.

A deep-sea hydrothermal vent community

Concept Check 50.3

The following questions refer to Figure 50.17.

1. Are stoneflies, benthic aquatic insects that require relatively high concentrations of oxygen, more likely to live in oligotrophic lakes or eutrophic lakes? Why?
2. Why are phytoplankton, and not benthic algae or rooted aquatic plants, the dominant photosynthetic organisms of the oceanic pelagic biome?

For suggested answers, see Appendix A.

Concept 50.4

Climate largely determines the distribution and structure of terrestrial biomes

All the abiotic factors discussed in this chapter, but especially climate, are important in determining why a particular terrestrial biome is found in a certain area. Because there are latitudinal patterns of climate over Earth's surface (see Figure 50.10), there are also latitudinal patterns of biome distribution.

Climate and Terrestrial Biomes

We can see the great impact of climate on the distribution of organisms by constructing a **climograph**, a plot of the temperature and precipitation in a particular region. For example, **Figure 50.18** is a climograph denoting annual mean temperature and precipitation for some of the biomes found in North America. Notice that the range of precipitation in northern coniferous forests is similar to that in temperate forests, but the temperature ranges are different. Grasslands are generally drier than either kind of forest, and deserts are drier still.

Annual means for temperature and rainfall are reasonably well correlated with the biomes that exist in different regions. However, it is important to distinguish *correlation* from *causation*. Although the climograph provides circumstantial evidence that temperature and rainfall are important to the distribution of biomes, it does not confirm that these variables govern the biomes' location. Only a detailed analysis of the water and temperature tolerances of individual species could establish the controlling effects of these variables.

As Figure 50.18 shows, there are regions where biomes overlap. Thus factors other than mean temperature and precipitation must play a role in determining where biomes exist. For example, certain areas in North America with a particular temperature and precipitation combination support a temperate broadleaf forest, but other areas with similar values for these variables support a coniferous forest. How do we explain this variation? First, remember that the climograph is based on annual *averages*. Often it is not only the mean or average climate

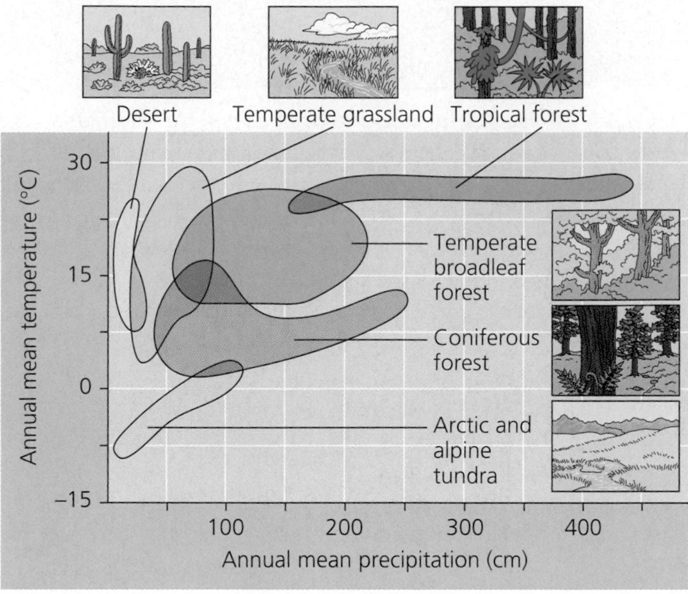

▲ **Figure 50.18 A climograph for some major types of biomes in North America.** The areas plotted here encompass the range of annual mean temperature and precipitation in the biomes.

that is important but also the *pattern* of climatic variation. For example, some areas may receive regular precipitation throughout the year, whereas other areas with the same amount of annual precipitation have distinct wet and dry seasons. A similar phenomenon may occur with respect to temperature. Other factors, such as the bedrock in an area, may greatly affect mineral nutrient availability and soil structure, which in turn affect the kind of vegetation that will develop.

The general distribution of the major terrestrial biomes is shown in **Figure 50.19**.

General Features of Terrestrial Biomes

Most terrestrial biomes are named for major physical or climatic features and for their predominant vegetation. Temperate grasslands, for instance, are generally found in middle latitudes, where the climate is more moderate than in the tropics or polar regions, and are dominated by various grass species. Each biome is also characterized by microorganisms, fungi, and animals adapted to that particular environment. For example, temperate grasslands are more likely than forests to be populated by large grazing mammals.

Vertical stratification is an important feature of terrestrial biomes, and the shapes and sizes of plants largely define the layering. For example, in many forests, the layers consist of the upper **canopy**, then the low-tree stratum, the shrub understory, the ground layer of herbaceous plants, the forest floor (litter layer), and finally the root layer. Nonforest biomes have similar, though usually less pronounced, strata. Grasslands have an herbaceous layer of grasses and forbs (small broadleaf plants), a litter layer, and a root layer. Stratification of vegetation provides many different habitats for animals, which often occupy well-defined feeding groups, from the

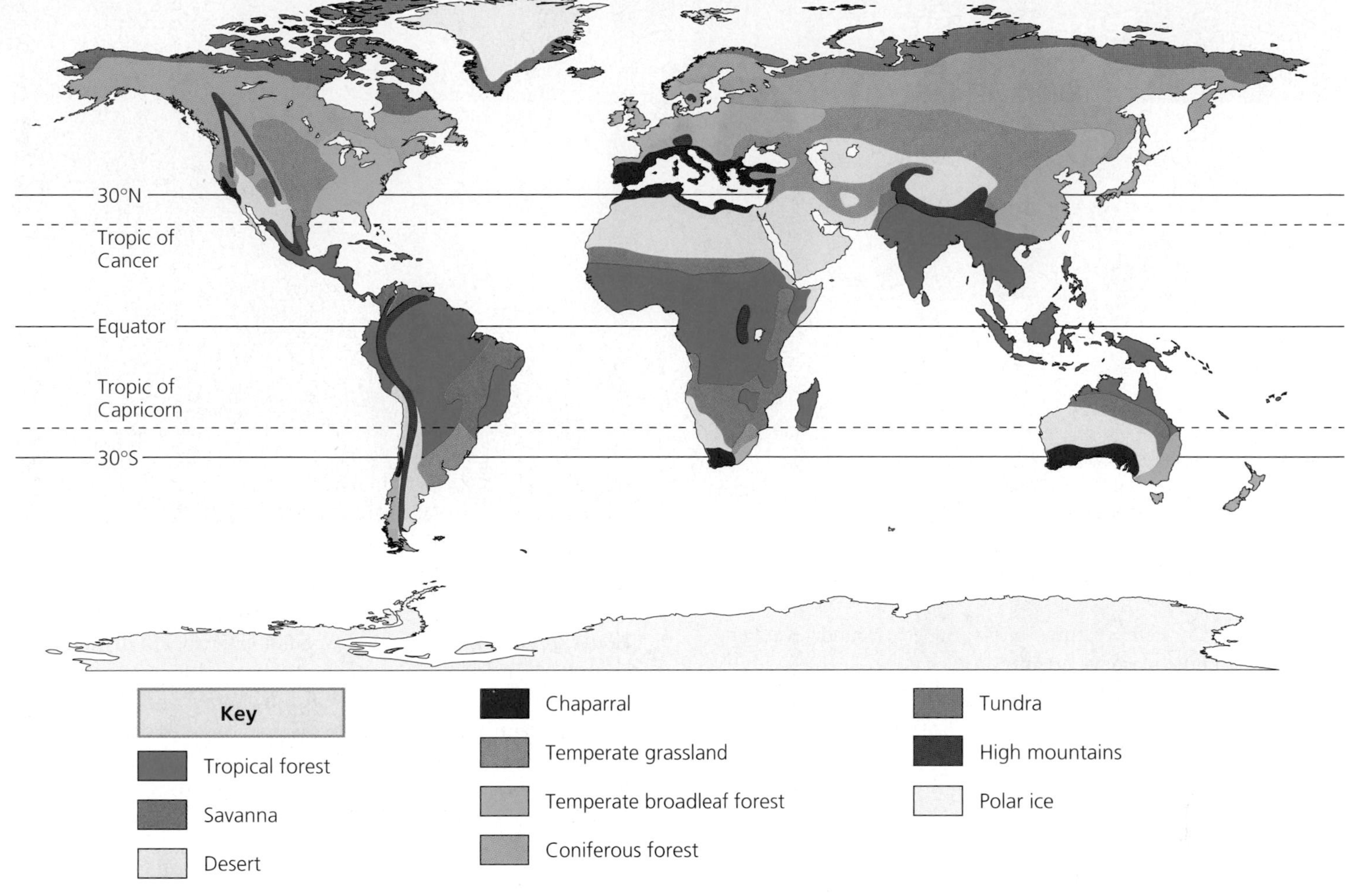

▲ **Figure 50.19 The distribution of major terrestrial biomes.** Although terrestrial biomes are mapped here with sharp boundaries, biomes actually grade into one another, sometimes over relatively large areas.

insectivorous and carnivorous birds and bats that feed above canopies to the small mammals, numerous worms, and arthropods that forage in the litter and root layers for food.

Although Figure 50.19 shows distinct boundaries between the biomes, in actuality, terrestrial biomes usually grade into each other, without sharp boundaries. The area of intergradation, called an **ecotone,** may be wide or narrow.

The actual species composition of any one kind of biome varies from one location to another. For instance, in the northern coniferous forest (taiga) of North America, red spruce is common in the east but does not occur in most other areas, where black spruce and white spruce are abundant. Although the vegetation of African deserts superficially resembles that of North American deserts, many similar-appearing plants are actually in different families. Such "ecological equivalents" can arise because of convergent evolution (see Figure 25.5).

Biomes are dynamic, and disturbance rather than stability tends to be the rule. For example, hurricanes create openings for new species in tropical and temperate forests. In northern coniferous forests, old trees die and fall over, or snowfall may break branches, producing gaps that allow deciduous species, such as aspen and birch, to grow. As a result, biomes usually exhibit extensive patchiness, with several different communities represented in any particular area.

In many biomes, the dominant plants depend on periodic disturbance. For example, natural wildfires are an integral component of grasslands, savannas, chaparral, and many coniferous forests. Before agricultural and urban development, much of the southeastern United States was dominated by a single conifer species, the longleaf pine. Without periodic burning, broadleaf trees tended to replace the pines. Forest managers now use fire as a tool to help maintain many coniferous forests.

In many biomes today, extensive human activity has radically altered the natural patterns of periodic physical disturbance. Fires, which used to be part of life on the Great Plains, are now controlled for the sake of agricultural land use. Humans have altered much of Earth's surface, replacing original biomes with urban and agricultural ones. Most of the eastern United States, for example, is classified as temperate broadleaf forest, but human activity has eliminated all but a tiny percentage of the original forest.

Figure 50.20, on the next four pages, surveys the major terrestrial biomes.

Figure 50.20

Exploring Terrestrial Biomes

TROPICAL FOREST

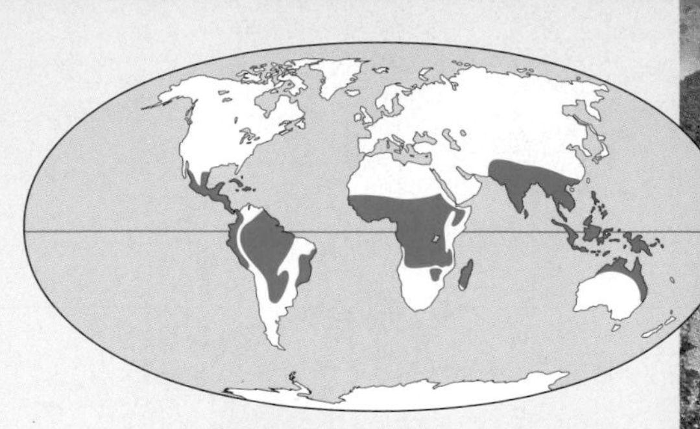

A tropical rain forest in Borneo

Distribution Equatorial and subequatorial regions.

Precipitation In **tropical rain forests**, rainfall is relatively constant, about 200–400 cm annually. In tropical dry forests, precipitation is highly seasonal, about 150–200 cm annually, with a six- to seven-month dry season.

Temperature Air temperatures are warm year-round, averaging 25–29°C with little seasonal variation.

Plants Tropical forests are stratified, and competition for light is intense. Strata in rain forests include a layer of emergent trees that grow above a closed canopy, the canopy trees, one or two layers of subcanopy trees, and shrub and herb layers. There are generally fewer strata in tropical dry forests. Broadleaf evergreen trees are dominant in tropical rain forests, whereas tropical dry forest trees drop their leaves during the dry season. Epiphytes such as bromeliads and orchids generally cover tropical forest trees, but are less abundant in dry forests. Thorny shrubs and succulents are common in tropical dry forests.

Animals Animal diversity is higher in tropical forests than in any other terrestrial biome. The animals, including amphibians, birds and other reptiles, mammals, and arthropods, are adapted to the three-dimensional environment and are often inconspicuous.

Human Impact Humans long ago established thriving communities in tropical forests. Rapid population growth leading to agriculture and development is now destroying tropical forests.

DESERT

Distribution **Deserts** occur in a band near 30° north and south latitude or at other latitudes in the interior of continents (for instance, the Gobi Desert of north central Asia).

Precipitation Precipitation is low and highly variable, generally less than 30 cm per year.

Temperature Temperature is variable seasonally and daily. Maximum air temperature in hot deserts may exceed 50°C; in cold deserts air temperature may fall below −30°C.

Plants Desert landscapes are dominated by low, widely scattered vegetation; the proportion of bare ground is high compared with other terrestrial biomes. The plants include succulents such as cacti, deeply rooted shrubs, and herbs that grow during the infrequent moist periods. Desert plant adaptations include heat and desiccation tolerance, water storage, and reduced leaf surface area. Physical defenses, such as spines, and chemical defenses, such as toxins in the leaves of shrubs, are common. Many of the plants exhibit C_4 or CAM photosynthesis (see Chapter 10).

Animals Common desert animals include many kinds of snakes and lizards, scorpions, ants, beetles, migratory and resident birds, and seed-eating rodents. Many species are nocturnal. Water conservation is a common adaptation, with some species surviving on water from metabolic breakdown of carbohydrates in seeds.

Human Impact Long-distance transport of water and deep groundwater wells have allowed humans to maintain substantial populations in deserts. Conversion to irrigated agriculture and urbanization have reduced the natural biodiversity of deserts.

The Sonoran Desert in southern Arizona

SAVANNA

A typical savanna in Kenya

Distribution Equatorial and subequatorial regions.

Precipitation Rainfall, which is seasonal, averages 30–50 cm per year. The dry season can last up to eight or nine months.

Temperature The **savanna** is warm year-round, averaging 24–29°C, but with somewhat more seasonal variation than in tropical forests.

Plants The scattered trees found in the savanna are often thorny with reduced leaf surface area, an apparent adaptation to the relatively dry conditions. Fires are common in the dry season, and the dominant plant species are fire-adapted and tolerant of seasonal drought. Grasses and forbs, which make up most of the ground cover, grow rapidly in response to seasonal rains and are tolerant of grazing by large mammals and other herbivores.

Animals Large herbivorous mammals, such as wildebeests and zebras, and their predators, including lions and hyenas, are common inhabitants. However, the dominant herbivores are actually insects, especially termites. During seasonal droughts, grazing mammals must migrate to other parts of the savanna with more forage and scattered watering holes.

Human Impact The earliest humans appear to have lived on savannas. Fires set by humans may help maintain this biome. Cattle ranching and overhunting have led to declines in large-mammal populations.

CHAPARRAL

Distribution This biome occurs in midlatitude coastal regions on several continents, and its many names reflect its far-flung distribution: **chaparral** in North America, matorral in Spain and Chile, garigue and maquis in southern France, and fynbos in South Africa.

Precipitation Precipitation is highly seasonal, with rainy winters and long, dry summers. Annual precipitation generally falls within the range of 30–50 cm.

Temperature Fall, winter, and spring are cool, with average temperatures ranging from 10–12°C. Average summer temperature can reach 30°C, and daytime maximum temperature can exceed 40°C.

Plants The chaparral is dominated by shrubs and small trees, along with a high diversity of grasses and herbs. Plant diversity is high, with many species confined to a specific, relatively small geographic area. Adaptations to drought include the tough evergreen leaves of woody plants, which reduce water loss. Adaptations to fire are also prominent. Some of the shrubs produce seeds that will germinate only after a hot fire; food reserves stored in their fire-resistant roots enable them to resprout quickly and use nutrients released by the fire.

Animals Native mammals include browsers, such as deer and goats, that feed on twigs and buds of woody vegetation and a high diversity of small mammals. Chaparral areas also support a high diversity of amphibians, birds and other reptiles, and insects.

Human Impact Chaparral areas have been heavily settled and reduced through conversion to agriculture and urbanization. Humans contribute to the fires that sweep across the chaparral.

An area of chaparral in California

continued on next page

TEMPERATE GRASSLAND

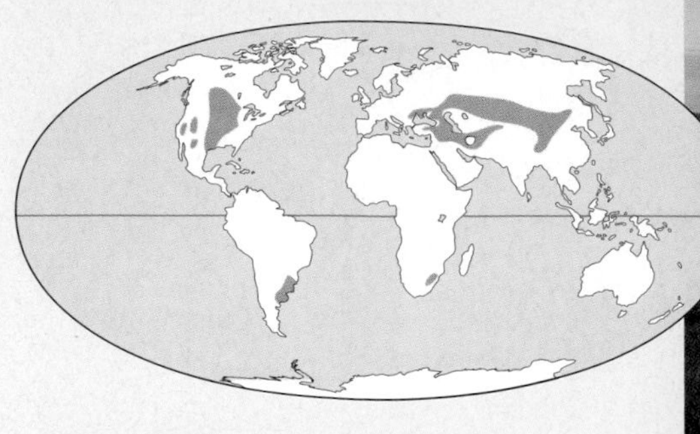

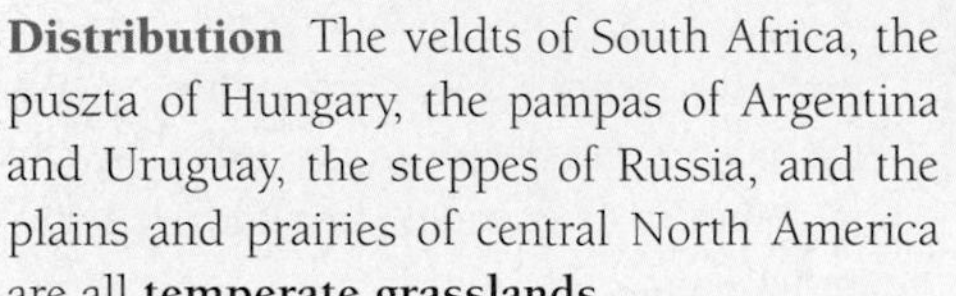

Sheyenne National Grassland in North Dakota

Distribution The veldts of South Africa, the puszta of Hungary, the pampas of Argentina and Uruguay, the steppes of Russia, and the plains and prairies of central North America are all **temperate grasslands**.

Precipitation Precipitation is highly seasonal, with relatively dry winters and wet summers. Annual precipitation generally averages between 30 and 100 cm. Periodic drought is common.

Temperature Winters are generally cold, with average temperatures frequently falling well below −10°C. Summers, with average temperatures frequently approaching 30°C, are hot.

Plants The dominant plants are grasses and forbs, which vary in height from a few centimeters to 2 meters in tallgrass prairie. Some of the main adaptations of plants are to periodic, protracted droughts and to fire. The grasses of temperate grassland sprout quickly following fire. Grazing by large mammals helps prevent establishment of woody shrubs and trees.

Animals Native mammals include large grazers such as bison and wild horses. Temperate grasslands are also inhabited by a wide variety of burrowing mammals, such as prairie dogs in North America.

Human Impact Deep, fertile soils make temperate grasslands ideal places for agriculture, especially for growing grains. As a consequence, most grassland in North America and much of Eurasia has been converted to farmland.

CONIFEROUS FOREST

Distribution Extending in a broad band across northern North America and Eurasia to the edge of the arctic tundra, the northern **coniferous forest**, or *taiga,* is the largest terrestrial biome on Earth.

Precipitation Precipitation generally ranges from 30 to 70 cm, and periodic droughts are common. However, some coastal coniferous forests of the United States Pacific Northwest are temperate rain forests that may receive over 300 cm of annual precipitation.

Temperature Winters are usually cold and long; summers may be hot. Some areas of coniferous forest in Siberia range in temperature from −70°C in the winter to over 30°C in summer.

Plants Cone-bearing trees, such as pine, spruce, fir, and hemlock, dominate coniferous forests. The conical shape of many conifers prevents too much snow from accumulating and breaking their branches. The diversity of plants in the shrub and herb layers of coniferous forests is lower than in temperate broadleaf forests.

Animals While many migratory birds nest in coniferous forests, many species reside there year-round. The mammals of coniferous forests, which include moose, brown bears, and Siberian tigers, are diverse. Periodic outbreaks of insects that feed on the dominant trees can kill vast tracts of trees.

Human Impact Although they have not been heavily settled by human populations, coniferous forests are being logged at an alarming rate, and the old-growth stands of these trees may soon disappear.

Rocky Mountain National Park in Colorado

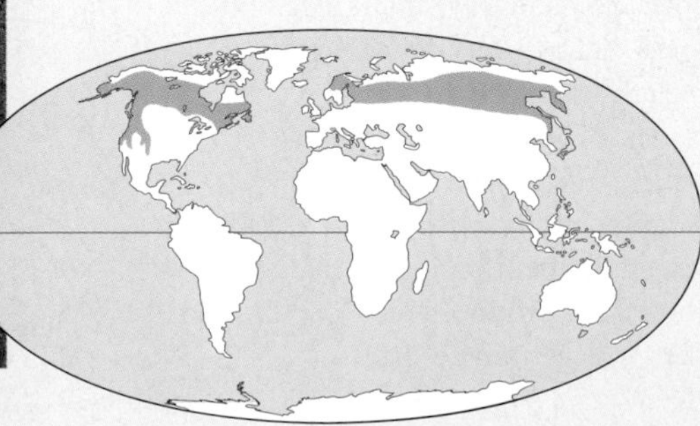

TEMPERATE BROADLEAF FOREST

Great Smoky Mountains National Park in North Carolina

Distribution Found mainly at midlatitudes in the Northern Hemisphere, with smaller areas in New Zealand and Australia.

Precipitation Precipitation can average from about 70 to over 200 cm annually. Significant amounts fall during all seasons, including summer rain and winter snow.

Temperature Winter temperatures average around 0°C. Summers, with maximum temperatures near 30°C, are hot and humid.

Plants A mature **temperate broadleaf forest** has distinct, highly diverse, vertical layers, including a closed canopy, one or two strata of understory trees, a shrub layer, and an herbaceous stratum. There are few epiphytes. The dominant plants in the Northern Hemisphere are deciduous trees, which drop their leaves before winter, when low temperatures would reduce photosynthesis and make water uptake from frozen soil difficult. In Australia, evergreen eucalyptus dominate these forests.

Animals In the Northern Hemisphere, many mammals hibernate in winter, while many bird species migrate to warmer climates. The mammals, birds, and insects make use of all vertical layers of the forest.

Human Impact Temperate broadleaf forest has been heavily settled on all continents. Logging and land clearing for agriculture and urban development destroyed virtually all the original deciduous forests in North America. However, owing to their capacity for recovery, these forests are returning over much of their former range.

TUNDRA

Distribution **Tundra** covers expansive areas of the Arctic, amounting to 20% of Earth's land surface. High winds and cold temperatures create similar plant communities, called *alpine tundra,* on very high mountaintops at all latitudes, including the tropics.

Precipitation Precipitation averages from 20 to 60 cm annually in arctic tundra but may exceed 100 cm in alpine tundra.

Temperature Winters are long and cold, with averages in some areas below −30°C. Summers are short with cool temperatures, generally averaging less than 10°C.

Plants The vegetation of tundra is mostly herbaceous, consisting of a mixture of lichens, mosses, grasses, and forbs, along with some dwarf shrubs and trees. A permanently frozen layer of soil called **permafrost** generally prevents water infiltration.

Animals Large grazing musk ox are resident, while caribou and reindeer are migratory. Predators include bears, wolves, and foxes. Migratory birds use tundra intensely during the summer as nesting grounds.

Human Impact Tundra is sparsely settled but has become the focus of significant mineral and oil extraction in recent years.

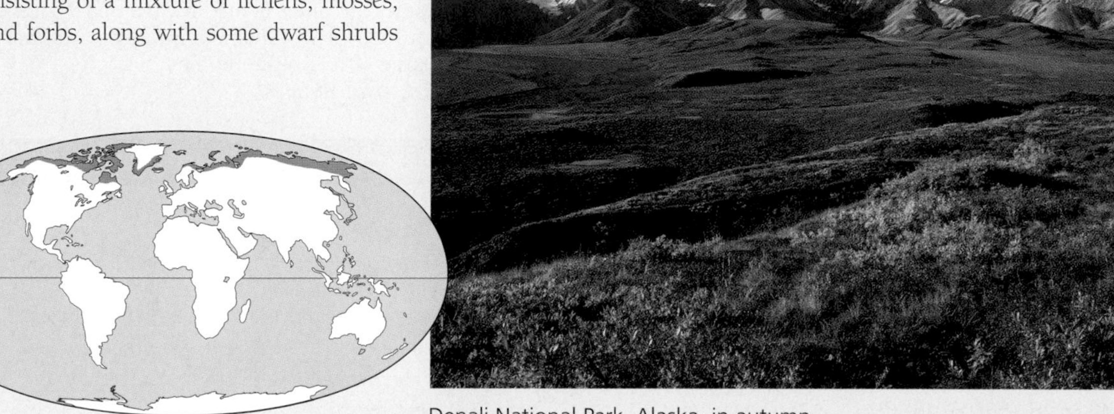

Denali National Park, Alaska, in autumn

Throughout your exploration in this chapter of Earth's varied aquatic and terrestrial biomes, you have seen many examples of the considerable impact of the abiotic factors in an organism's environment. In the next chapter, we will look more closely at organisms themselves, examining how behavioral mechanisms and adaptations play key roles in the interactions between an organism and both the nonliving and living parts of its environment.

Concept Check 50.4

1. Judging from the climograph in Figure 50.18, what mainly differentiates dry tundra and deserts?
2. Identify the natural biome in which you live and summarize its abiotic and biotic characteristics. Do these reflect your actual surroundings? Explain.

For suggested answers, see Appendix A.

Chapter 50 Review

Go to www.campbellbiology.com or the student CD-ROM to explore Activities, Investigations, and other interactive study aids.

SUMMARY OF KEY CONCEPTS

Ecology is the study of interactions between organisms and the environment

- **Ecology and Evolutionary Biology (p. 1081)** Events that occur in ecological time affect life on the scale of evolutionary time.
- **Organisms and the Environment (p. 1081)** Examples of questions that ecologists ask are, Who lives where? Why do they live there? and How many are there? Ecologists use observations and experiments to test explanations for the distribution and abundance of species and other ecological phenomena. The environment of any organism includes both abiotic and biotic components.
- **Subfields of Ecology (pp. 1082–1083)** Ecology can be divided into several subfields of study, ranging from the ecology of organisms to the dynamics of ecosystems, landscapes, and the biosphere. Modern ecological studies cross boundaries between traditionally separate areas.
- **Ecology and Environmental Issues (p. 1083)** Ecology provides the scientific understanding underlying environmental issues. Many ecologists favor the precautionary principle of "An ounce of prevention is worth a pound of cure."
  Activity *Science, Technology, and Society: DDT*

Concept 50.2

Interactions between organisms and the environment limit the distribution of species

- **Dispersal and Distribution (pp. 1084–1085)** The dispersal of organisms results in broad patterns of geographic distribution. Natural range expansions and species transplants suggest hypotheses to explain why species are found where they are. Transplanted species may disrupt the ecosystem at the new site.
- **Behavior and Habitat Selection (p. 1085)** Some organisms do not occupy all of their potential range. Species distribution may be limited by habitat selection behavior.
- **Biotic Factors (pp. 1085–1086)** Biotic factors that affect the distribution of organisms include interactions with other species, such as predation and competition.
- **Abiotic Factors (pp. 1086–1087)** Among important abiotic factors affecting species distribution are temperature, water, sunlight, wind, rocks, and soil.
  Activity *Adaptations to Biotic and Abiotic Factors*
  Investigation *How Do Abiotic Factors Affect Distribution of Organisms?*
- **Climate (pp. 1087–1092)** Global climate patterns are largely determined by the input of solar energy and Earth's rotation around the sun. Regional, local, and seasonal effects on climate are influenced by bodies of water, mountains, and the changing angle of the sun over the year. Fine-scale differences in abiotic factors determine microclimates.

Abiotic and biotic factors influence the structure and dynamics of aquatic biomes

- Aquatic biomes account for the largest part of the biosphere in terms of area and are generally stratified with regard to light penetration, temperature, and community structure. Marine biomes have a higher salt concentration than freshwater biomes (pp. 1092–1098).
  Activity *Aquatic Biomes*

Concept 50.4

Climate largely determines the distribution and structure of terrestrial biomes

- **Climate and Terrestrial Biomes (p. 1098)** Climographs show that temperature and precipitation are correlated with biomes, but because biomes overlap, other abiotic factors must play a role in biome location.
- **General Features of Terrestrial Biomes (pp. 1098–1104)** Terrestrial biomes are often named for major physical or climatic factors and for their predominant vegetation. Stratification is an important feature of terrestrial biomes.
  Activity *Terrestrial Biomes*

TESTING YOUR KNOWLEDGE

Self-Quiz

1. Which of the following areas of study focuses on the exchange of energy, organisms, and materials between ecosystems?
   a. population ecology
   b. organismal ecology
   c. landscape ecology
   d. ecosystem ecology
   e. community ecology

2. Which statement about dispersal is *false*?
   a. Dispersal is a common component of the life cycles of plants and animals.
   b. Colonization of devastated areas after floods or volcanic eruptions depends on dispersal.
   c. Dispersal occurs only on an evolutionary time scale.
   d. Seeds are important dispersal stages in the life cycles of most flowering plants.
   e. The ability to disperse can expand the geographic distribution of a species.

3. Imagine some cosmic catastrophe that jolts Earth so that it is no longer tilted. Instead, its axis is perpendicular to the plane of its orbit around the sun. The most predictable effect of this change would be
   a. no more night and day.
   b. a big change in the length of the year.
   c. a cooling of the equator.
   d. a loss of seasonal variation at northern and southern latitudes.
   e. the elimination of ocean currents.

4. When climbing a mountain, we can observe transitions in biological communities that are analogous to the changes
   a. in biomes at different latitudes.
   b. at different depths in the ocean.
   c. in a community through different seasons.
   d. in an ecosystem as it evolves over time.
   e. across the United States from east to west.

5. The oceans affect the biosphere in all of the following ways *except*
   a. producing a substantial amount of the biosphere's oxygen.
   b. removing carbon dioxide from the atmosphere.
   c. moderating the climate of terrestrial biomes.
   d. regulating the pH of freshwater biomes and terrestrial groundwater.
   e. being the source of most of Earth's rainfall.

6. Which lake zone would be absent in a very shallow lake?
   a. benthic zone
   b. aphotic zone
   c. pelagic zone
   d. littoral zone
   e. limnetic zone

7. Which of the following is *true* with respect to oligotrophic lakes and eutrophic lakes?
   a. Oligotrophic lakes are more subject to oxygen depletion.
   b. Rates of photosynthesis are lower in eutrophic lakes.
   c. Eutrophic lake water contains lower concentrations of nutrients.
   d. Eutrophic lakes are richer in nutrients.
   e. Sediments in oligotrophic lakes contain larger amounts of decomposable organic matter.

8. Which of the following is characteristic of most terrestrial biomes?
   a. annual average rainfall in excess of 250 cm
   b. a distribution predicted almost entirely by rock and soil patterns
   c. clear boundaries between adjacent biomes
   d. vegetation demonstrating stratification
   e. cold winter months

9. Which of the following biomes is *correctly* paired with the description of its climate?
   a. savanna—cool temperature, precipitation uniform during the year
   b. tundra—long summers, mild winters
   c. temperate broadleaf forest—relatively short growing season, mild winters
   d. temperate grasslands—relatively warm winters, most rainfall in summer
   e. tropical forests—nearly constant day length and temperature

10. Suppose that the number of bird species is determined mainly by the number of vertical strata found in the environment. If so, in which of the following biomes would you find the greatest number of bird species?
    a. tropical rain forest
    b. savanna
    c. taiga
    d. temperate broadleaf forest
    e. temperate grassland

For Self-Quiz answers, see Appendix A.

Go to the website or CD-ROM for more quiz questions.

Evolution Connection

Discuss how the concept of time applies to ecological situations and evolutionary changes. Do ecological time and evolutionary time ever correspond? If so, what are some examples?

Scientific Inquiry

Hiking up a mountain, you notice a plant species that has one growth form at low elevations and a very different growth form at high elevations. You wonder if these represent two genetically distinct populations of this species, each adapted to the prevailing conditions, or if this species has developmental flexibility and can assume either growth form, depending on local conditions. What experiments could you design to test these two hypotheses?

Investigation ***How Do Abiotic Factors Affect Distribution of Organisms?***

Science, Technology, and Society

In pet shops throughout North America, you can purchase a variety of nonnative fishes, birds, and other animals. Describe some scenarios in which such pet trade could endanger native plants and animals. Should governments regulate the pet trade? Are there currently any restrictions on what species a pet shop can sell in your city? How would you balance such regulation against a person's individual rights?

51 Behavioral Ecology

▲ **Figure 51.1 A courting pair of East Asian red-crowned cranes (*Grus japonicus*).**

Key Concepts

51.1 Behavioral ecologists distinguish between proximate and ultimate causes of behavior
51.2 Many behaviors have a strong genetic component
51.3 Environment, interacting with an animal's genetic makeup, influences the development of behaviors
51.4 Behavioral traits can evolve by natural selection
51.5 Natural selection favors behaviors that increase survival and reproductive success
51.6 The concept of inclusive fitness can account for most altruistic social behavior

Overview

Studying Behavior

We humans have probably studied animal behavior for as long as we have lived on Earth. As hunters—and sometimes the hunted—knowledge of animal behavior was essential to human survival. But other animals are also a source of fascination beyond the need for practical information. For instance, cranes are animals that have long captivated people's interest, perhaps because cranes are large and their behavior is easily observed **(Figure 51.1)**. Male and female cranes engage in elaborate courtship rituals involving graceful dance-like movements and synchronized vocalizations. Observing these rituals, many people have viewed the birds as symbols of fidelity and devotion. Some of the most conspicuous crane behaviors are associated with their annual migrations. Each spring, thousands of cranes fly from wintering grounds in southern Eurasia, North Africa, and North America to northern nesting grounds. Various species of cranes fly hundreds or thousands of kilometers, stopping periodically to rest and feed. Because migrating cranes fly at high altitudes and call as they fly, some cultures have traditionally viewed them as messengers between Earth and the heavens.

The modern scientific discipline of **behavioral ecology** extends such observations of animal behavior by studying how such behavior is controlled and how it develops, evolves, and contributes to survival and reproductive success. For example, a behavioral ecologist might ask how similarities or differences in courtship displays may be related to genetic similarities or differences among crane species and how learning contributes to the development of courtship displays. Questions about migration might include why migrating cranes call, what environmental cues trigger migration, or how migration contributes to the reproductive success of cranes. Behavioral ecology is essential to solving critically important problems ranging from the conservation of endangered species to the control of emerging infectious diseases. This chapter focuses on such questions and others in a quest to understand how behavior is related to genetics, environment, and evolution.

Concept 51.1

Behavioral ecologists distinguish between proximate and ultimate causes of behavior

The questions that can be posed about any behavior can be divided generally into two classes: those that focus on the immediate stimulus and mechanism for the behavior and those that explore how the behavior contributes to survival and reproduction. First, however, let's consider an even more fundamental question: what the term *behavior* encompasses.

What Is Behavior?

Behavioral traits are as much a part of an animal's phenotype as the length of its appendages or the color of its fur. Most of what we call behavior is the visible result of an animal's muscular activity, as when a predator chases its prey or a fish raises its fins as part of a territorial display **(Figure 51.2)**. In some behaviors, muscular activity is involved but less obvious, as when a bird uses muscles to force air from its lungs and shape the sounds in its throat, producing a song. Some nonmuscular activities are also considered behaviors, as when an animal secretes a chemical that attracts members of the opposite sex. Furthermore, we can consider learning to be a behavioral process. For example, a juvenile bird may learn to reproduce a song that it hears an adult of its species singing, even though the muscular activity based on this memory may not occur until months later, when the young bird itself begins to sing the song. Thus, in addition to studying observable behaviors, mainly in the form of muscle-powered activities, behavioral ecologists also study the mechanisms underlying those behaviors, which may not involve muscles at all. Put more simply, we can think of **behavior** as everything an animal does and how it does it.

Proximate and Ultimate Questions

When we observe a certain behavior, we may ask both *proximate* and *ultimate* questions. **Proximate questions** about behavior focus on the environmental stimuli, if any, that trigger a behavior, as well as the genetic, physiological, and anatomical mechanisms underlying a behavioral act. Proximate questions are often referred to as "how" questions. For example, the red-crowned cranes in Figure 51.1, like many animals, breed in spring and early summer. A proximate question about the timing of breeding by this species might be, How does day length influence breeding by red-crowned cranes? A reasonable hypothesis for the proximate cause of this behavior is that breeding is triggered by the effect of increased day length on an animal's production of and responses to particular hormones. Indeed, experiments with various animals demonstrate that lengthening daily exposure to light produces neural and hormonal changes that induce behavior associated with reproduction, such as singing and nest building in birds.

In contrast to proximate questions, **ultimate questions** address the evolutionary significance of a behavior. Ultimate questions take such forms as, Why did natural selection favor this behavior and not a different one? Hypotheses addressing "why" questions propose that the behavior increases fitness in some particular way. A reasonable hypothesis for why the red-crowned crane reproduces in spring and early summer is that breeding is most productive at that time of year. For instance, at that time, parent birds can find ample food for rapidly growing offspring, providing an advantage in reproductive success compared to birds that breed in other seasons.

Although proximate causation is distinct from ultimate causation, the two concepts are nevertheless connected. Proximate mechanisms produce behaviors that have evolved because they reflect fitness in some particular way. For instance, increased day length itself has little adaptive significance for red-crowned cranes, but since it corresponds to seasonal conditions that increase reproductive success, such as the availability of food for feeding young birds, breeding when days are longer is a proximate mechanism that has evolved in cranes.

▲ **Figure 51.2 A male African cichlid (*Neolamprologus tetracephalus*) with erect fins.** Muscular contraction that raises the fins is a behavioral response to a threat to the fish's territory.

Ethology

In the mid-20th century, a number of pioneering behavioral biologists developed the discipline of **ethology,** the scientific study of how animals behave, particularly in their natural environments. Ethologists such as Niko Tinbergen, of the Netherlands, and Karl von Frisch and Konrad Lorenz, of Austria, who shared a Nobel Prize in 1973, established the conceptual foundations on which modern behavioral ecology is based. In a 1963 paper, Tinbergen suggested four questions that must be answered to fully understand any behavior. His questions, which remain at the core of behavioral ecology today, can be summarized as follows:

1. What is the mechanistic basis of the behavior, including chemical, anatomical, and physiological mechanisms?
2. How does development of the animal, from zygote to mature individual, influence the behavior?
3. What is the evolutionary history of the behavior?
4. How does the behavior contribute to survival and reproduction (fitness)?

Tinbergen's list includes both proximate and ultimate questions. The first two, which concern mechanism and development of the behavior, are proximate questions, while the second two are ultimate, or evolutionary, questions. The complementary nature of proximate and ultimate perspectives can be demonstrated with behaviors frequently studied by the classical ethologists, such as fixed action patterns and imprinting.

Fixed Action Patterns

A type of behavior studied extensively by the ethologists is the **fixed action pattern (FAP),** a sequence of unlearned behavioral acts that is essentially unchangeable and, once initiated, is usually carried to completion. A FAP is triggered by an external sensory stimulus known as a **sign stimulus.** Tinbergen studied what has become a classic example of sign stimuli and FAPs in the male three-spined stickleback fish, which attacks other males that invade its nesting territory. The stimulus for the attack behavior is the red underside of the intruder; the stickleback will not attack an intruding fish lacking a red belly (note that female sticklebacks never have red bellies), but will readily attack unrealistic models as long as some red is present **(Figure 51.3).** In fact, Tinbergen was inspired to look into the matter by his casual observation that his fish responded aggressively when a red truck passed their tank. As a consequence of his research, which was first reported in 1937, Tinbergen discovered that the color red is a key component of the sign stimulus releasing aggression in male sticklebacks.

▼ **Figure 51.3 Sign stimuli in a classic fixed action pattern.**

(a) A male three-spined stickleback fish shows its red underside.

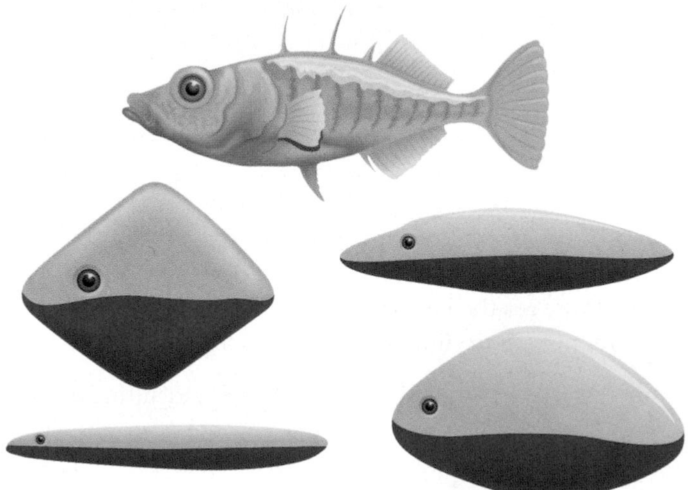

(b) The realistic model at the top, without a red underside, produces no aggressive response in a male three-spined stickleback fish. The other models, with red undersides, produce strong responses.

BEHAVIOR: A male stickleback fish attacks other male sticklebacks that invade its nesting territory.

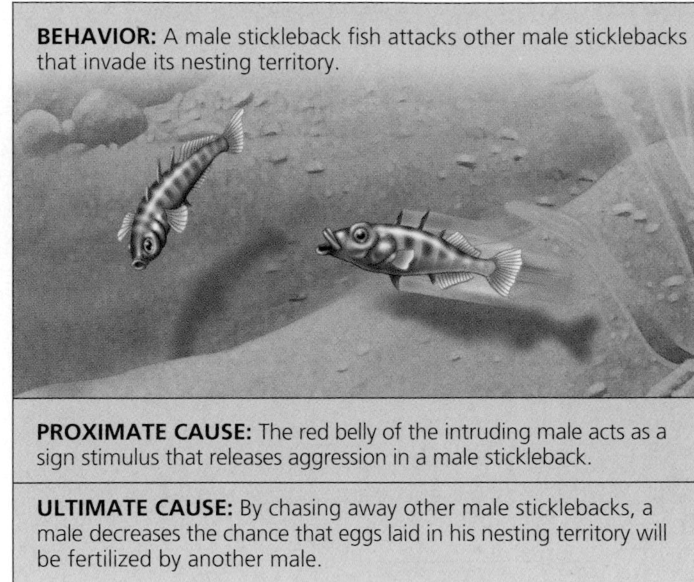

PROXIMATE CAUSE: The red belly of the intruding male acts as a sign stimulus that releases aggression in a male stickleback.

ULTIMATE CAUSE: By chasing away other male sticklebacks, a male decreases the chance that eggs laid in his nesting territory will be fertilized by another male.

▲ **Figure 51.4 Proximate and ultimate perspectives on aggressive behavior by male sticklebacks.**

Figure 51.4 offers both proximate and ultimate explanations for this particular FAP among male stickleback fish.

Imprinting

Another phenomenon studied by the classical ethologists is **imprinting,** a type of behavior that includes both learning and innate components and is generally irreversible. Imprinting is distinguished from other types of learning by having a **sensitive period,** a limited phase in an animal's development that is the only time when certain behaviors can be learned. An example of imprinting is young geese following their mother. In species that provide parental care, parent-offspring bonding is a critical part of the life cycle. During the period of bonding, the young imprint on their parent and learn the basic behaviors of their species, while the parent learns to recognize its offspring. Among gulls, for instance, the sensitive period for parental bonding on young lasts one to two days. If bonding does not occur, the parent will not initiate care of the infant, leading to certain death for the offspring and a decrease in reproductive success for the parent.

But how do the young know on whom—or what—to imprint? How do young geese know that they should follow the mother goose? The tendency to respond is innate in the birds; the outside world provides the *imprinting stimulus,* something to which the response will be directed. Experiments with many species of waterfowl indicate that they have no innate recognition of "mother." They respond to and identify with the first object they encounter that has certain key characteristics. In classic experiments done in the 1930s, Konrad Lorenz showed that the most important imprinting stimulus in graylag geese is movement of an object away from the young. When incubator-hatched goslings spent their first few hours with Lorenz rather than with a goose, they imprinted on him, and from then on, they steadfastly followed him and showed no recognition of their biological mother or other adults of their own species. Again, there are both proximate and ultimate explanations, as outlined in **Figure 51.5**.

Cranes also imprint as hatchlings, creating both problems and opportunities in captive rearing programs designed to save endangered crane species. For instance, a group of 77 endangered whooping cranes hatched and raised by sandhill cranes imprinted on the sandhill foster parents; none of these whooping cranes ever formed a mating pair-bond with another whooping crane. As a consequence, captive breeding programs now isolate young cranes and expose them to the sights and sounds of members of their own species. But imprinting can also be used to aid crane conservation **(Figure 51.6)**. Young whooping cranes imprinted on humans in "crane suits" have been taught to follow these "parents" flying ultralight aircraft along new migration routes. And importantly, such cranes have formed mating pair-bonds with other whooping cranes.

BEHAVIOR: Young geese follow and imprint on their mother.

PROXIMATE CAUSE: During an early, critical developmental stage, the young geese observe their mother moving away from them and calling.

ULTIMATE CAUSE: On average, geese that follow and imprint on their mother receive more care and learn necessary skills, and thus have a greater chance of surviving than those that do not follow their mother.

▲ **Figure 51.5 Proximate and ultimate perspectives on imprinting in graylag geese.**

▲ **Figure 51.6 Imprinting for conservation.** Conservation biologists have taken advantage of imprinting by young whooping cranes as a means to teach the birds a migration route. A pilot wearing a crane suit in an ultralight plane acts as a surrogate parent.

While research on imprinting and fixed action patterns is much less active than it once was, the early study of these behaviors helped to make the distinction between proximate and ultimate causes of behavior. These studies also helped establish a strong tradition of experimental approaches in behavioral ecology.

Concept Check 51.1

1. A ground squirrel that sees a predator may utter a brief, loud call. List four questions about this behavior, one for each of Tinbergen's four questions. State whether each is a proximate or an ultimate question.
2. If an egg rolls out of the nest, a mother graylag goose will retrieve it by nudging with her beak and head. If researchers remove the egg or substitute a ball during this process, the goose will not alter her response. What type of behavior is this? Suggest a proximate and an ultimate explanation.

For suggested answers, see Appendix A.

Concept 51.2

Many behaviors have a strong genetic component

Extensive research shows that behavioral traits, like anatomical and physiological aspects of a phenotype, are the result of complex interactions between genetic and environmental factors. This conclusion contrasts sharply with the popular conception that behavior is due *either* to genes (nature) *or* to environment (nurture). In biology, nature versus nurture is not a debate. Rather, biologists study how *both* genes *and* the environment influence the development of phenotypes, including behavioral phenotypes. Though we first discuss genetic influences on behavior and defer the discussion of environmental influences, you should bear in mind that all behaviors are affected by both genes and environment.

One approach to studying the influence of different factors on a particular behavior is to view the behavior in terms of the norm of reaction (see Figure 14.13). For instance, we might measure the behavioral phenotypes for a particular genotype that develop in a range of environments. In some cases, the behavior is variable, depending on environmental experience. In other cases, nearly all individuals in a population exhibit virtually the same behavior, despite internal and external environmental differences during development and throughout life. Behavior that is *developmentally fixed* in this way is called **innate behavior.** Innate behaviors are under strong genetic influence, as we will discuss in the following examples.

Directed Movements

Many animal movements, ranging from simple ones that take place within a few millimeters to complex movements that span hundreds or thousands of kilometers, are under substantial genetic influence. Because of the clear role of genes in the control of these animal movements, we may refer to them as *directed movements.*

Kinesis

A **kinesis** is a simple change in activity or turning rate in response to a stimulus. For example, sow bugs (also called wood lice), terrestrial crustaceans that survive best in moist environments, exhibit a kinesis in response to variation in humidity **(Figure 51.7a)**. The sow bugs become more active in dry areas and less active in humid areas. Though sow bugs do not move toward or away from specific conditions, their increased movement under dry conditions increases the chance that they will leave a dry area and encounter a moist area. And since they slow down in a moist area, they tend to stay there once they encounter it.

Taxis

In contrast to a kinesis, a **taxis** is a more or less automatic, oriented movement toward (a positive taxis) or away from (a negative taxis) some stimulus. For example, many stream fish, such as trout, exhibit positive rheotaxis (from the Greek *rheos*, current); they automatically swim or orient themselves in an upstream direction (toward the current). This taxis keeps the fish from being swept away and keeps them facing the direction from which food will come **(Figure 51.7b)**.

▼ **Figure 51.7 A kinesis and a taxis.**

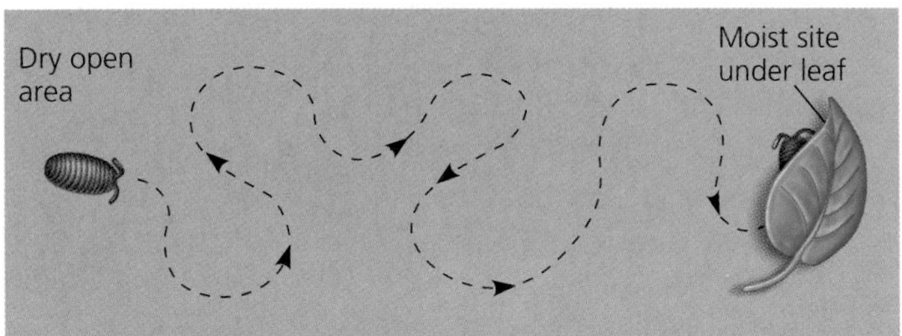

(a) Kinesis increases the chance that a sow bug will encounter and stay in a moist environment.

(b) Positive rheotaxis keeps trout facing into the current, the direction from which most food comes.

Migration

It is easy to assume that a simple behavior, such as kinesis shown by sow bugs or positive rheotaxis by trout, is under strong genetic control. However, genetic influence can be substantial even for more complex behaviors. For example, ornithologists have found that many features of migratory behavior in birds are genetically programmed **(Figure 51.8)**.

One of the most thoroughly studied migratory birds is the blackcap (*Sylvia atricapilla*), a small warbler that ranges from the Cape Verde Islands off the coast of West Africa to northern Europe. The migratory behavior of blackcaps differs greatly among populations; for instance, while all blackcaps in the northern portion of the range migrate, usually at night, those in the Cape Verde Islands do not migrate at all. During the normal season for migration, captive migratory blackcaps spend their nights hopping restlessly about their cages or rapidly flapping their wings while sitting on a perch.

Peter Berthold and his colleagues at the Max Planck Research Center for Ornithology in Radolfzell, Germany, studied the genetic basis of this behavior, known as "migratory restlessness," in several populations of blackcaps. In one study, the research team crossed (mated) migratory blackcaps from southern Germany with nonmigratory blackcaps from Cape Verde and subjected their offspring to environments simulating one location or the other. Approximately 40% of the offspring raised in both conditions showed migratory restlessness,

▲ **Figure 51.8 Bird migration, a behavior that is largely under genetic control.** Each spring, migrating western sandpipers (*Calidris mauri*), such as those shown here, migrate from their wintering grounds, which may be as far south as Peru, to their breeding grounds in Alaska. In the autumn, they return to the wintering grounds.

leading Berthold to conclude that migratory restlessness is under genetic control and follows a polygenic inheritance pattern (see Chapter 14). Other breeding experiments in Berthold's laboratory have demonstrated genetic influences on many components of blackcap migration, as we will see later in this chapter.

Animal Signals and Communication

Much of the social interaction between animals involves transmitting information through specialized behaviors called signals (also known as displays). In behavioral ecology, a **signal** is a behavior that causes a change in another animal's behavior. The transmission of, reception of, and response to signals constitute animal **communication,** an essential element of interactions between individuals. Though the environment makes significant contributions to all communication systems, some of their features are under strong genetic control.

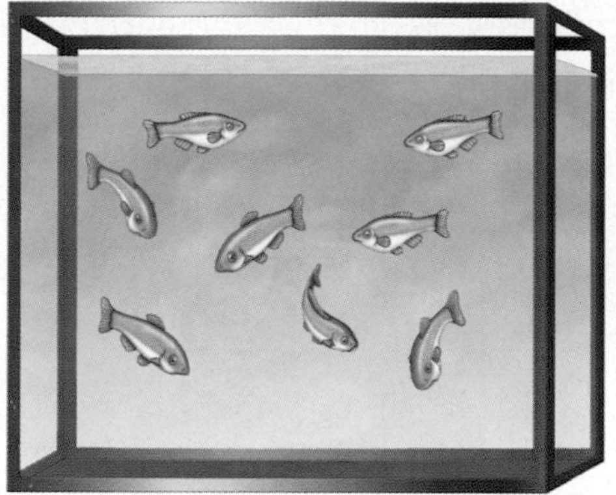

(a) Minnows are widely dispersed in an aquarium before an alarm substance is introduced.

(b) Within seconds of the alarm substance being introduced, minnows aggregate near the bottom of the aquarium and reduce their movement.

▲ **Figure 51.9 Minnows responding to the presence of an alarm substance.**

Many signals are very efficient in energy costs. For instance, a common signal between territorial fish is to erect their fins, which gives them a larger profile and is generally enough to drive off an intruding individual (see Figure 51.2). It takes less energy to erect fins than it does to push an intruding male from a territory.

Animals communicate using visual, auditory, chemical (olfactory), tactile, and electrical signals. The type of signal used to transmit information is closely related to an animal's lifestyle and environment. For example, most terrestrial mammals are nocturnal, which makes visual displays relatively ineffective. But olfactory and auditory signals work as well in the dark as in the light, and most mammalian species signal by these means. Birds, by contrast, are mostly diurnal and have a relatively poor olfactory sense; they communicate primarily by visual and auditory signals. Unlike most mammals, humans are diurnal and, like birds, use mainly visual and auditory communication. Therefore, we can detect the songs and bright colors that birds use to communicate with each other. If humans had the well-developed olfactory abilities of most mammals and could detect the rich world of chemical cues, we would have a very different perspective of nature.

Chemical Communication

Many animals that communicate through odors emit chemical substances called **pheromones.** In most cases, both the production of pheromones and animal responses to them are controlled genetically. Pheromones are especially common among mammals and insects and often relate to reproductive behavior. Many moths, for example, emit pheromones that can attract mates from several kilometers away. Once the moths are together, pheromones also trigger specific courtship behaviors.

The context of a chemical signal can be as important as the chemical itself. In a honeybee colony, pheromones produced by the queen and her daughters, the workers, maintain the hive's very complex social order. When male honeybees (drones) are outside the hive, where they can mate with a queen, they are attracted to her pheromone; when drones are inside the hive, they are unaffected by the queen's pheromone.

Pheromones also function in nonreproductive behavior. For example, when a minnow or catfish is injured, an alarm substance that is stored in glands in the fish's skin disperses in the water, inducing a fright response among other fish in the area. These nearby fish become more vigilant and group together in tightly packed schools, often near the bottom, where they are safer from attack **(Figure 51.9)**. Pheromones can be very effective at low concentrations. For instance, just 1 cm^2 of skin from a fathead minnow contains sufficient alarm substance to induce an alarm reaction even when diluted by 58,000 L of water.

Auditory Communication

The songs of most bird species are at least partly learned. By contrast, in many species of insects, mating rituals include characteristic songs that are generally under direct genetic control. In *Drosophila* species, males produce a song by vibrating their wings, and females are able to recognize the songs of males of their species through such details as the intervals between pulses of wing vibrations, the rhythm of the song, and the length of song pulses. A variety of evidence indicates that song structure in *Drosophila* males is controlled genetically and is under strong selective pressure. For instance, males reared in isolation produce a characteristic song for their species despite having no exposure to other singing males. In addition, the male song shows very little variation among individuals within a *Drosophila* species.

While *Drosophila* species differ morphologically as well as in their courtship songs, some other insect species can be identified only through their courtship songs or behavior. For example, morphologically identical green lacewings found throughout central to northern Eurasia and North America were once thought to belong to a single species, *Chrysoperla carnea.* However, studies of their courtship songs by Charles Henry and his colleagues and students at the University of Connecticut revealed the presence of at least 15 different species, each singing a different courtship song **(Figure 51.10)**. Several of these morphologically identical species may occur together in the same habitat (an example of behavioral isolation of species; see Figure 24.4).

Figure 51.10

Inquiry Are the different songs of closely related green lacewing species under genetic control?

EXPERIMENT Charles Henry, Lucía Martínez, and Kent Holsinger crossed males and females of *Chrysoperla plorabunda* and *Chrysoperla johnsoni,* two morphologically identical species of lacewings that sing different courtship songs.

SONOGRAMS

Chrysoperla plorabunda parent

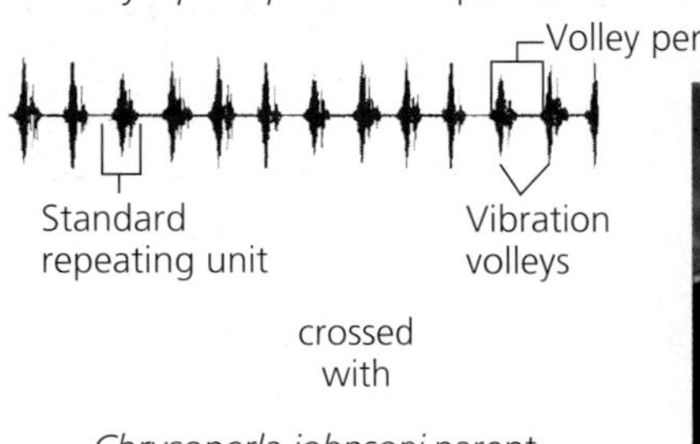

crossed with

Chrysoperla johnsoni parent

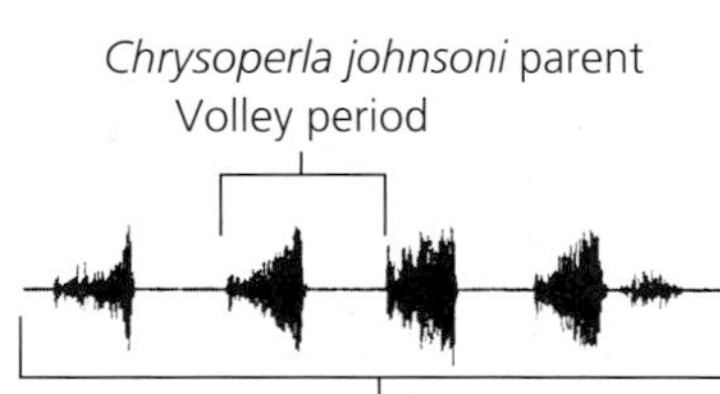

Standard repeating unit

The researchers recorded and compared the songs of the male and female parents with those of the hybrid offspring that had been raised in isolation from other lacewings.

RESULTS The F_1 hybrid offspring sing a song in which the length of the standard repeating unit is similar to that sung by the *Chrysoperla plorabunda* parent, but the volley period, that is, the interval between vibration volleys, is more similar to that of the *Chrysoperla johnsoni* parent.

F_1 hybrids, typical phenotype

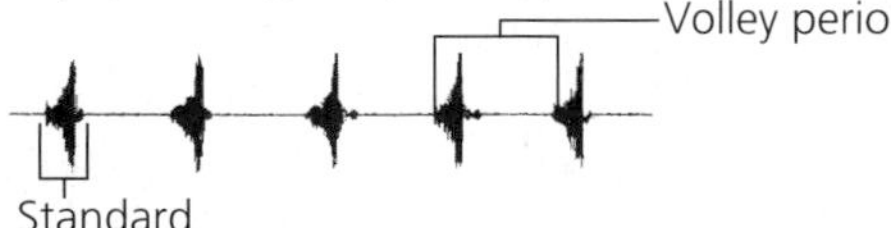

Standard repeating unit

CONCLUSION The results of this experiment indicate that the songs sung by *Chrysoperla plorabunda* and *Chrysoperla johnsoni* are under genetic control.

Though they never observed hybrids of the different green lacewing species in the wild during more than 20 years of fieldwork, Henry and his colleagues have been able to produce hybrids between different species under laboratory conditions. While all individuals in each species sing the same song, laboratory-raised hybrid offspring sing songs that contain elements of the songs of both parental species. These data led the researchers to conclude that the various songs sung by the different green lacewing species are genetically controlled.

Genetic Influences on Mating and Parental Behavior

Behavioral research has uncovered a variety of mammalian behaviors that are under relatively strong genetic control, as well as the physiological mechanisms for these behaviors. One of the most striking lines of research concerns mating and parental behavior by male prairie voles (*Microtus ochrogaster*).

Prairie voles and a few other vole species are monogamous, a social trait found among only about 3% of mammalian species. Male prairie voles also help their mates care for young, another relatively uncommon trait among male mammals. In contrast with most other vole species, a male prairie vole forms a strong pair-bond with a single female after they mate, associating closely with his mate and engaging in grooming and huddling behaviors **(Figure 51.11)**. Further, though an unmated male prairie vole shows little aggression toward other prairie voles, whether male or female, a mated male becomes intensely aggressive toward any strange male or female prairie voles, while remaining nonaggressive toward his mate. A few days following the birth of pups, a male prairie vole will spend a great deal of time hovering over them, licking them, and carrying them around, while remaining vigilant for intruders.

The genetic and physiological controls on the complex social and parental behaviors of prairie voles have been revealed over the past decade by the research of Thomas R. Insel and his colleagues at Emory University. Early research suggested

◄ **Figure 51.11 A pair of prairie voles (*Microtus ochrogaster*) huddling.** North American prairie voles are monogamous, with males associating closely with their mates, as shown here, and contributing substantially to the care of young.

that arginine-vasopressin (AVP), a nine-amino-acid neurotransmitter released during mating, might mediate both pair-bond formation and aggression by male prairie voles. In the central nervous system, AVP binds with a receptor called the V_{1a} receptor. The Emory researchers found significant differences between the distribution of V_{1a} receptors in the brains of monogamous prairie voles and their distribution in promiscuous montane voles (which live in the mountainous regions of western North America).

To test whether the distribution of V_{1a} receptors is a key factor controlling the mating and parental behavior of prairie voles, Insel and his colleagues inserted the prairie vole V_{1a} receptor gene into laboratory mice. Not only did the transgenic mice develop brains in which the distribution of V_{1a} receptors was remarkably similar to the distribution found in prairie voles, but they also showed many of the same mating behaviors as monogamous male prairie voles. In contrast, male wild-type mice did not have the same V_{1a} receptor distribution and did not show such behavior. Thus, though many genes influence the mating behavior of these voles, it appears that a single gene may mediate a considerable amount of the complex mating behavior (and perhaps parental behavior) that occurs in prairie voles. It remains to be seen whether other complex behaviors are under similarly simple genetic influences. In addition, although genes influence behaviors in a multitude of ways, environment also has major effects on behaviors, as we shall see in the next section.

Concept Check 51.2

1. Which has greater influence on the development of behaviors, "nature" or "nurture"? Explain.
2. Use an example to explain how researchers study whether a particular behavior has a strong genetic component.

For suggested answers, see Appendix A.

Concept 51.3

Environment, interacting with an animal's genetic makeup, influences the development of behaviors

While experimental demonstrations of genetic influences on behaviors accumulate, research also reveals that environmental conditions modify many of the same behaviors. Environmental factors, such as the quality of the diet, the nature of social interactions, and opportunities for learning can influence the development of behaviors in every group of animals.

Dietary Influence on Mate Choice Behavior

One example of environmental influence on behavior is the role of diet in mate selection by *Drosophila mojavensis*, which mates and lays its eggs on the rotting tissues of cactus. *D. mojavensis* populations in Baja California, Mexico, breed almost entirely on agria cactus, whereas most populations in Sonora, Mexico, and in Arizona use organ pipe cactus. These cacti serve as food for the developing larvae.

When *D. mojavensis* from Baja and Sonora were raised on an artificial banana-based medium in the laboratory, researchers noticed that females from the Sonoran population tended to avoid mating with Baja males. Following up on these initial observations, William Etges and Mitchell Ahrens, of the University of Arkansas, showed that the food eaten by *D. mojavensis* larvae strongly influences later mate selection by females, especially those from Sonoran populations (**Figure 51.12**).

Figure 51.12
Inquiry How does dietary environment affect mate choice by female *Drosophila mojavensis*?

EXPERIMENT William Etges raised a *D. mojavensis* population from Baja California and a *D. mojavensis* population from Sonora on three different culture media: artificial medium, agria cactus (the Baja host plant), and organ pipe cactus (the Sonoran host plant). From each culture medium, Etges collected 15 male and female Baja *D. mojavensis* pairs and 15 Sonoran pairs and observed the numbers of matings between males and females from the two populations.

RESULTS When *D. mojavensis* had been raised on artificial medium, females from the Sonoran population showed a strong preference for Sonoran males (a). When *D. mojavensis* had been raised on cactus medium, the Sonoran females mated with Baja and Sonoran males in approximately equal frequency (b).

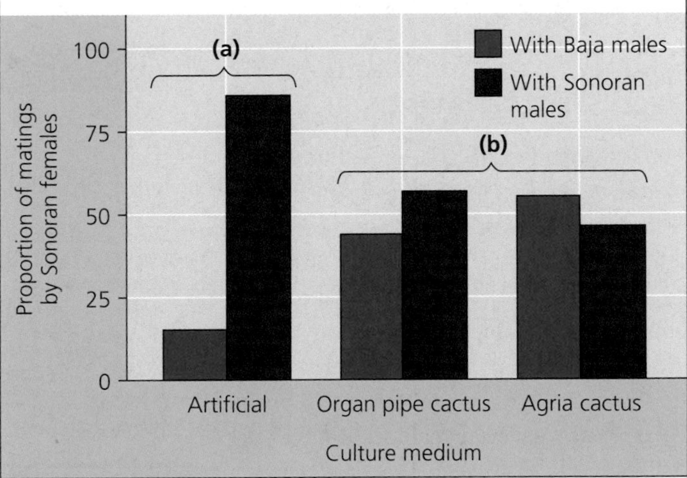

CONCLUSION The difference in mate selection shown by females that developed on different diets indicates that mate choice by females of Sonoran populations of *D. mojavensis* is strongly influenced by the dietary environment in which larvae develop.

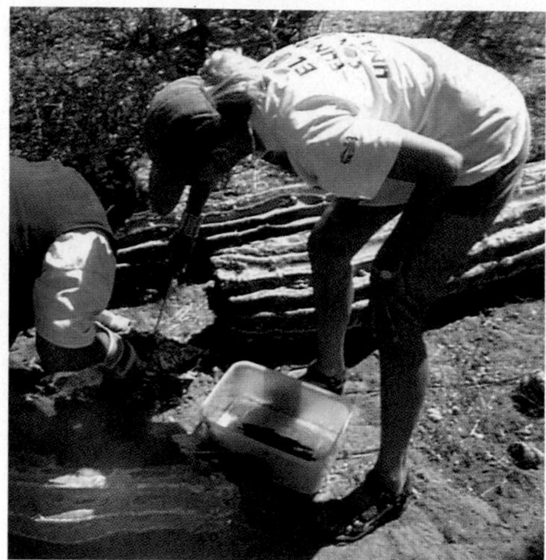

◀ **Figure 51.13 Therese Markow (right) and a colleague collecting *Drosophila mojavensis* from a cactus.**

But why do Sonoran females avoid Baja males that develop on some diets but not those that develop on other diets? In other words, what is the proximate cause of the behavior? Therese Markow, of Arizona State University, and Eric Toolson, of the University of New Mexico, proposed that the physiological basis for the observed mate preferences was differences in hydrocarbons in the exoskeletons of the flies **(Figure 51.13)**. This was a reasonable hypothesis, since in addition to using courtship songs and other sensory information, *Drosophila* use their sense of taste to assess the hydrocarbons in the exoskeleton of possible mates. To test the influence of exoskeleton hydrocarbons, Etges and Ahrens raised Baja males on artificial medium and then perfumed them with hydrocarbons extracted from the exoskeletons of males from the Sonoran population. Instead of rejecting these Baja males, Sonoran females accepted them as frequently as they accepted Sonoran males.

Etges and Ahrens's study showed how the effects of the nutritional environment can modify a critical behavior. Next we will see how social environment has been found to modify the behavior of at least one species.

Social Environment and Aggressive Behavior

The discoveries of genetic influence on mating behaviors in male prairie voles are complemented by parallel studies of another monogamous rodent, the California mouse (*Peromyscus californicus*). Like prairie voles, male California mice are highly aggressive toward other mice and provide extensive parental care. In contrast to prairie voles, however, even male California mice that have not mated are already aggressive.

Janet Bester-Meredith and Catherine Marler, of the University of Wisconsin at Madison, studied what influence the social environment might have on the behavior of male California mice. To manipulate the early social environment, they placed newborn California mice in the nests of white-footed mice (*Peromyscus leucopus*), a species in which males are not monogamous and engage in little parental care. They also placed newborn white-footed mice in the nests of California mice. This "cross-fostering," in which the young of each species were placed in the nests of the other species, altered the behavior of both species **(Table 51.1)**. For instance, when male California mice raised by white-footed mice crawled away from the nest, they were not retrieved as frequently as those raised by their own species. When these cross-fostered California mice became parents, they, too, spent less time retrieving their pups. The cross-fostered California mice were also less aggressive toward intruders. In contrast, male white-footed mice raised by California mice were more aggressive than those raised by white-footed mice. In one study by Bester-Meredith and Marler, the brains of California mice raised by white-footed mice were found to contain reduced levels of the same neurotransmitter (AVP) associated with elevated aggression and parental care in male prairie voles.

The fact that cross-fostered California and white-footed mice adopted some of the behaviors of their foster parents suggests that experience during development can lead to changes in parental and aggressive behaviors in these rodents that can be passed from one generation to the next.

Learning

One of the most powerful ways that environmental conditions can influence behavior is through **learning**, the modification of behavior based on specific experiences. Learned behaviors range from the very simple, such as imprinting, which results in a young bird following a particular individual it has learned to recognize as its parent, to the highly complex. Before we consider some complex forms of learning, let's look briefly at another very simple one: habituation.

Table 51.1 Influence of Cross-Fostering on Male Mice*

Species	Aggression Toward an Intruder	Aggression in Neutral Situation	Paternal Behavior	Arginine-Vasopressin (AVP) Content in Brain
California mice fostered by white-footed mice	Reduced	No difference	Reduced	Reduced
White-footed mice fostered by California mice	No difference	Increased	No difference	No difference

*Comparisons are with mice raised by parents of their own species.

Data from J. K. Bester-Meredith and C. A. Marler, Vasopressin and the transmission of paternal behavior across generations in mated, cross-fostered *Peromyscus* mice. *Behavioral Neuroscience* 117(2003):455–463.

Habituation

Habituation is a loss of responsiveness to stimuli that convey little or no information. Examples are widespread. A hydra contracts when disturbed by a slight touch; if it is repeatedly disturbed by such a stimulus without any further consequences, it stops responding. Many mammals and birds recognize alarm calls of members of their species, but they eventually stop responding if these calls are not followed by an actual attack (the "cry-wolf" effect). In terms of ultimate causation, habituation may increase fitness by allowing an animal's nervous system to focus on stimuli that signal the presence of food, mates, or real danger instead of wasting time or energy on a vast number of other stimuli that are irrelevant to the animal's survival and reproduction. Next, let's shift our attention to some more complex forms of learning.

Spatial Learning

Every natural environment shows some degree of spatial variation. For instance, sites suitable for nesting may be more common in some places than in others. As a consequence, the fitness of an organism may be enhanced by the capacity for **spatial learning,** which is the modification of behavior based on experience with the spatial structure of the environment, including the locations of nest sites, hazards, food, and prospective mates.

In a classic experiment done in 1932, Niko Tinbergen studied how digger wasps find their nest entrances. Tinbergen moved a circle of pinecones that had previously surrounded a nest entrance and observed that the wasp landed in the center of the pinecones, even though the nest entrance was no longer there **(Figure 51.14)**. The wasp was using the pinecones as a **landmark,** or location indicator. The use of landmarks is a more complex cognitive mechanism than a taxis or kinesis, since it involves learning. The wasp flies toward a stimulus (in this case, the center of the pinecones), as in a taxis, but the stimulus is an arbitrary landmark the animal must learn rather than a constant stimulus such as light. One nest entrance may have pinecones around it, while another may be next to a pile of stones. Each wasp has to learn the unique landmarks of each individual nest site.

Tinbergen's experiment reveals that for spatial learning to be a reliable way to navigate through the environment, the landmarks used must be stable (within the time frame of a particular activity). For example, the pinecones that indicate the presence of the wasp's nest should have a low probability of moving. Using unreliable information for learning may result in a significant cost to the animal. If a nesting wasp learns to locate its nest using objects that might blow away, for instance, the cost may be an inability to locate and provision the nest and, consequently, reduced reproductive success for the wasp.

Figure 51.14

Inquiry Does a digger wasp use landmarks to find her nest?

EXPERIMENT A female digger wasp excavates and cares for four or five separate underground nests, flying to each nest daily with food for the single larva in the nest. To test his hypothesis that the wasp uses visual landmarks to locate the nests, Niko Tinbergen marked one nest with a ring of pinecones.

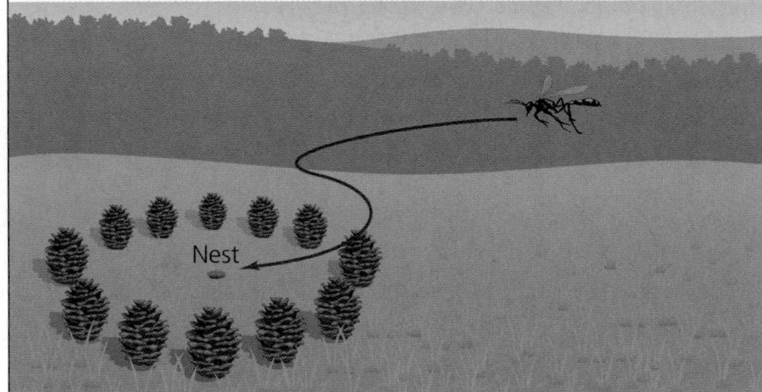

After the mother visited the nest and flew away, Tinbergen moved the pinecones a few feet to one side of the nest.

RESULTS When the wasp returned, she flew to the center of the pinecone circle instead of to the nearby nest. Repeating the experiment with many wasps, Tinbergen obtained the same results.

CONCLUSION The experiment supported the hypothesis that digger wasps use landmarks to keep track of their nests.

Because some environments are more stable than others, animals may use different kinds of information for spatial learning in different environments. Lucy Odling-Smee and Victoria Braithwaite, of the University of Edinburgh, hypothesized that three-spined stickleback fish from stable pond environments would rely more on landmarks than would sticklebacks from more changeable river environments. Odling-Smee and Braithwaite trained 20 sticklebacks from rivers and 20 from ponds to navigate a T-shaped maze to reach a reward (a combination of food and other fish the sticklebacks could school with). In the first phase of the experiment, the researchers placed the reward at one end of the maze and marked the correct direction to turn for the reward with landmarks in the form of two plastic plants.

Fish that learned to find the reward were given a second set of trials, in which Odling-Smee and Braithwaite put rewards at both ends of the maze and switched the location of the landmarks. The results showed that the sticklebacks from stable pond environments located the rewards using a combination of learning which direction to turn and landmarks, while most river fish learned simply to turn in a particular direction. These results suggest that the degree of environmental variability influences the spatial learning strategies of animals.

Cognitive Maps

An animal can move around its environment in a flexible and efficient manner using landmark orientation alone. Honeybees, for instance, might learn ten or so landmarks and locate their hive and flowers in relation to those landmarks. A more powerful mechanism is a **cognitive map**, an internal representation, or code, of the spatial relationships between objects in an animal's surroundings.

It is very difficult to distinguish experimentally between an animal that is using landmarks and one that is using a true cognitive map. Researchers have collected evidence that specifically supports the use of cognitive maps by corvids, a family of birds that includes ravens, crows, jays, and nutcrackers. Many corvids store food in caches, from which the birds can retrieve the food later. For instance, pinyon jays (*Gymnorhinus cyanocephalus*) and Clark's nutcrackers (*Nucifraga columbiana*) store nuts in as many as thousands of caches that may be widely dispersed. The birds not only relocate each cache, but also keep track of food quality, bypassing caches in which the food was relatively perishable and would have decayed since it was stored. Research by Alan Kamil, of the University of Nebraska, suggests that pinyon jays and Clark's nutcrackers use cognitive maps to memorize the locations of their food caches. In one experiment with Clark's nutcrackers, Kamil varied the distance between landmarks and discovered that the birds could learn to find the halfway point between them. Such behavior suggests that the birds are capable of using a general abstract geometric rule for locating a seed cache. In Kamil's experiment, the rule was something like, "Seed caches are found halfway between particular landmarks." Such rules form the basis of cognitive maps, which benefit the organism by reducing the amount of detail that must be remembered to relocate an object.

Associative Learning

As we have seen, in learning, an animal modifies its behavior based on information from the environment. For instance, an inexperienced white-footed mouse may readily attack a brightly colored, slow-moving caterpillar, such as the larva of a monarch butterfly, only to find its mouth full of a distasteful poisonous fluid. Following this experience, the white-footed mouse may avoid attacking insects with similar coloration and behavior. The ability of many animals to associate one feature of the environment (a stimulus, such as color) with another (bad taste) is called **associative learning.**

Associative learning and its genetic and neurological bases have been extensively studied in the fruit fly *Drosophila melanogaster.* William Quinn, William Harris, and Seymour Benzer, of the California Institute of Technology, published the first demonstration of associative learning by *Drosophila* over 30 years ago. The research team trained *Drosophila* flies to avoid air carrying a particular scent by coupling exposure to the odor with an electric shock. This is an example of a type of associative learning called **classical conditioning,** in which an arbitrary stimulus, in this case an odor, is associated with a reward or punishment, the electric shock. *Drosophila* individuals trained in this way would avoid the particular odor for as long as 24 hours. Further research over the past three decades has revealed a surprising capacity for learning by *Drosophila,* which has become a model for the study of this component of behavior.

The laboratory study of associative learning has been extended to natural environments. The role that associative learning may play in helping animals avoid predators has been particularly well studied in fishes and aquatic insects. Recall our earlier discussion about the alarm substance in the skins of minnows and related fishes that induces an automatic fright response in nearby fish. Nichole Korpi and Brian Wisenden, of Minnesota State University at Moorhead, conducted a study to determine whether this alarm substance might be involved in a process of associative learning through which fish learn to avoid predators.

Korpi and Wisenden studied the zebrafish (*Danio rerio*), a common aquarium fish. For the predator they chose a pike. Zebrafish, a minnow from southern Asia, and pike, a fish of northern lakes, do not occur together in nature. Would the zebrafish learn to associate the scent of this unfamiliar predator with their alarm substance if there were a delay between exposure to the alarm substance and exposure to the odor of the predator? Korpi and Wisenden proposed that a delay of 5 minutes would realistically model a possible delay between the occurrence of an alarm and encountering an actual predator.

The researchers exposed zebrafish in an experimental group to an influx of 20 mL of water containing an alarm substance and then, 5 minutes later, to 20 mL of water with pike odor. Zebrafish in a control group were exposed to an influx of water alone (without alarm substance) and then to water containing pike odor. On day 1 of the experiment, the zebrafish in the control group did not change their activity in response to the influx of either water alone or pike odor **(Figure 51.15)**. This lack of response by the control group is critical to the interpretation of the results, because it indicates that the zebrafish had no innate negative reaction to pike odor. As expected, the zebrafish in the experimental group reduced their activity markedly in response to the alarm substance. When pike odor was added 5 minutes later, their activity actually increased

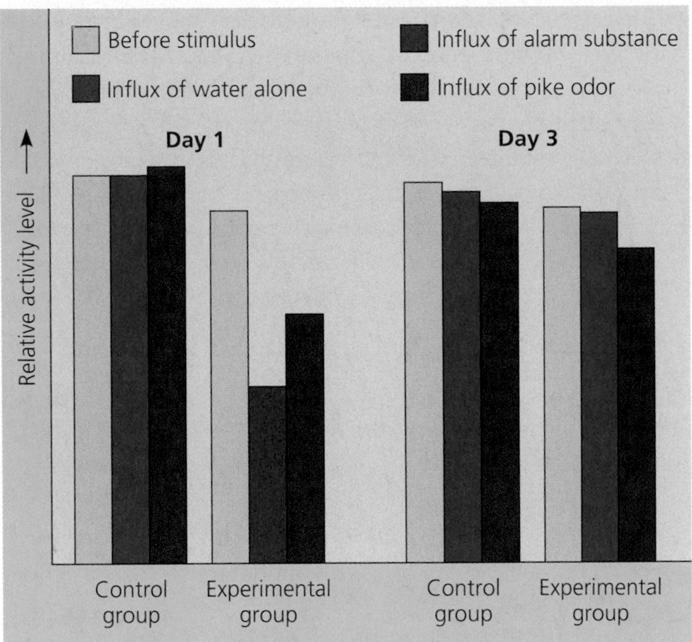

▲ **Figure 51.15 Associative learning in zebrafish.** Changes in activity indicated that the experimental group of zebrafish had learned to associate the odor of pike with the presence of alarm substance.

▲ **Figure 51.16 Operant conditioning.** Having received a face full of quills, a young coyote has probably learned to avoid porcupines.

somewhat, a difference from the typical alarm response. Three days later, the zebrafish in the control group still showed no response to the influx of either water alone or pike odor (see Figure 51.15). The experimental fish also showed no significant response to an influx of water alone. However, they did reduce their activity somewhat in response to pike odor. Korpi and Wisenden concluded that the zebrafish in the experimental group had learned to associate pike odor with the alarm substance, even though their exposure to pike odor on the first day of the study was delayed for 5 minutes. In terms of classical conditioning, the zebrafish had been *conditioned* to associate pike odor with the alarm substance. The experimental results show that this association was retained for three days.

Another type of associative learning is **operant conditioning,** also called trial-and-error learning. Here, an animal learns to associate one of its own behaviors with a reward or punishment and then tends to repeat or avoid that behavior **(Figure 51.16)**. As in the example of a mouse eating a distasteful caterpillar, for instance, predators quickly learn to associate certain kinds of potential prey with painful experiences and modify their behavior accordingly.

Cognition and Problem Solving

The study of cognition connects behavior with nervous system function. The term *cognition* is variously defined. In a narrow sense, it is synonymous with consciousness, or awareness. In a broad sense—the way we use the term in this book—**cognition** is the ability of an animal's nervous system to perceive, store, process, and use information gathered by sensory receptors. The study of animal cognition, called **cognitive ethology,** examines the connection between an animal's nervous system and its behavior.

One area of research in cognitive ethology investigates how an animal's brain represents objects in the environment. For example, researchers have discovered that many animals, including insects, are capable of categorizing objects according to concepts such as "same" and "different." Martin Giurfa, of the Centre of Animal Cognition Research in Toulouse, France, and his team trained honeybees to match a sample color to the same color marking one of two arms in a Y-maze. Later, the trained bees were able to match black-and-white patterns in the same orientation on the maze, indicating that they recognized the "sameness" of the problem.

Watching an animal solve a problem makes us aware that its nervous system has a substantial ability to process information. For example, if a chimpanzee is placed in a room with a banana hung high out of reach and several boxes on the floor, the chimp can "size up" the situation and stack the boxes, enabling it to reach the food. Such novel problem-solving behavior is highly developed in some mammals, especially primates and dolphins, and notable examples have also been observed in some bird species, especially crows, ravens, and jays. Behavioral ecologist Bernd Heinrich, of the University of Vermont, performed an experiment in which ravens had to obtain food hanging from a string. Several ravens solved the problem by using one foot to pull up the string in increments and the other foot to secure the string so the food did not drop down again; other ravens showed great individual variation in their problem-solving attempts.

One powerful source of information used by many animals to solve problems is the behavior of other individuals. Long-term research by Tetsuro Matsuzawa, of the Primate Research Institute of Kyoto University, Japan, has revealed that chimpanzees learn to solve problems by copying the behavior of other chimpanzees. Matsuzawa and his research team have shown that

▲ **Figure 51.17 Chimpanzees learning to crack oil palm nuts by observing an experienced chimpanzee.**

young wild chimpanzees at Bossou, Guinea, learn how to use two stones as a hammer and anvil to crack oil palm nuts by observing and copying experienced chimpanzees **(Figure 51.17)**.

Genetic and Environmental Interaction in Learning

Considerable research on the development of bird songs has revealed varying degrees of genetic and environmental influence on the learning of complex behavior. In some species, learning appears to play only a small part in song development. For instance, New World flycatchers that are reared away from adults of their species will develop the song characteristic of their own species without ever having heard it: In other words, their songs are innate. In contrast, among the songbirds—a group that includes the sparrows, robins, and canaries—learning plays a key role in song development.

Techniques for learning songs vary from species to species. Some have a sensitive period for developing their songs. For example, if a white-crowned sparrow is isolated for the first 50 days of life and unable to hear either real sparrows or recordings of sparrow songs, it fails to develop the adult song of its species. Although the young bird does not sing during the sensitive period, it memorizes the song of its species by listening to other white-crowned sparrows sing. During the sensitive period, fledglings seem to be stimulated more by the songs of their own species than by songs of other species, chirping more in response. Thus, young white-crowned sparrows *learn* the songs they will sing as adults, but the learning appears to be bounded by genetically controlled preferences.

The sensitive period when a white-crowned sparrow memorizes its species' song is followed by a second learning phase, when the juvenile bird sings some tentative notes that researchers call a subsong. The juvenile bird hears its own singing and compares it with the song that it memorized during the sensitive period. Once a sparrow's own song matches the one it memorized, the song "crystallizes" as the final song, and the bird sings only this adult song for the rest of its life.

There are important exceptions to the song-learning scenario seen in white-crowned sparrows. Canaries, for example, do not have a single sensitive period for song learning. A young canary begins with a subsong, but the full song it develops is not crystallized in the same way as it is in white-crowned sparrows. Between breeding seasons, the song becomes flexible again, and an adult male may learn new song "syllables" each year, adding to the song it already sings.

Concept Check 51.3

1. How can cognitive maps increase an animal's capacity to learn spatial relationships?
2. Describe three ways in which an animal's environment can influence the development of its behavior.
3. How might associative learning explain why unrelated distasteful or stinging insects have similar colors?

For suggested answers, see Appendix A.

Concept 51.4

Behavioral traits can evolve by natural selection

Because of the influence of genes on behavior, natural selection can result in the evolution of behavioral traits in populations. One of the primary sources of evidence for this evolution is behavioral variation between and within species.

Behavioral Variation in Natural Populations

Behavioral differences between closely related species are common. As you have already seen, the males of different species of *Drosophila* sing different courtship songs. Another example is found in voles: In some species, both males and females care for young, while in other species, the young receive only maternal care. Though often less obvious, significant differences in behavior can also be found *within* animal species. When behavioral variation within a species corresponds to variation in environmental conditions, it may be evidence of past evolution.

Variation in Prey Selection

One of the best-known examples of genetically based variation in behavior within a species is prey selection by the garter snake *Thamnophis elegans* **(Figure 51.18a)**. Stevan Arnold, of the University of Chicago, discovered that the natural diet of this species differs widely across its range in California. Coastal populations feed on salamanders, frogs, and toads, but predominantly on slugs. Inland populations feed on frogs,

(a) A garter snake (*Thamnophis elegans*)

(b) A banana slug (*Ariolimus californicus*); not to scale

▲ **Figure 51.18 Predator and potential prey.**

leeches, and fish, but not on slugs. In fact, slugs are rare or absent in the inland habitats of *T. elegans*.

The contrast in prey availability led Arnold to compare the responses of coastal and inland populations of *T. elegans* when offered bits of banana slug **(Figure 51.18b)**, which is found throughout the coastal habitats of *T. elegans* but is absent from the inland habitats Arnold studied. In a first test, Arnold offered banana slugs to *T. elegans* from both wild populations. Whereas most of the coastal snakes readily ate the slugs, the inland snakes tended to refuse them.

Because the contrast in feeding behavior may have been conditioned by experience in their natural habitat, Arnold tested the responses of young snakes born in the laboratory. He found that 73% of the young snakes from coastal mothers attacked the slugs they were offered, while only 35% of the young snakes from inland mothers attacked the slugs.

Arnold proposed that when inland snakes colonized coastal habitats more than 10,000 years ago, a fraction of the population had the ability to recognize slugs by chemoreception. Because these snakes took advantage of the abundant food source, they had higher fitness than snakes in the population that ignored slugs; thus, the capacity to recognize slugs as prey increased in frequency in the population over hundreds or thousands of generations. The variation in behavior today between the two populations is evidence of this evolution.

Variation in Aggressive Behavior

Although the differences in feeding behavior documented by Arnold occur over hundreds of kilometers, such distances are not necessary for intraspecific variation. In arid regions, the presence of water can produce striking environmental contrast over small distances, particularly where a desert meets a riparian (riverside) woodland. The obvious contrast in vegetation is only one of many differences, which extend to the behavioral ecology of species inhabiting riparian zones and the surrounding arid environments.

Agelenopsis aperta is a species of funnel web spider that lives in both riparian zones and the surrounding arid environments in the western United States. The web of *A. aperta* consists of a silken sheet ending in a funnel hidden in some sheltering feature of the habitat. While **foraging** (behavior associated with recognizing, searching for, capturing, and consuming food), the spider sits in the mouth of the funnel. When prey strikes the web, the spider runs out to make its capture.

Susan Riechert, of the University of Tennessee at Knoxville, and several colleagues found a striking contrast in the behavior of *A. aperta* spiders inhabiting riparian forests and those in arid habitats in Arizona and New Mexico. In arid, food-poor habitats, *A. aperta* is more aggressive toward potential prey and toward other spiders, and it returns to foraging more quickly following disturbance.

Ann Hedrick and Riechert tested whether the higher aggressiveness of arid-habitat *A. aperta* spiders reflects a genetic difference between the two populations or a learned response to living in a food-poor environment **(Figure 51.19)**. The researchers compared each spider's time to attack, the time that elapsed between a prey's first contact with the web and the point at which the spider first touched it. They found that in their natural environments, the arid-habitat spiders attacked all 15 prey types more quickly than did the riparian spiders. Hedrick and Riechert then brought female spiders from the two habitats into the laboratory, where the spiders laid eggs. Again using attack time as a measure, the researchers compared the aggressiveness of laboratory-raised spiders from

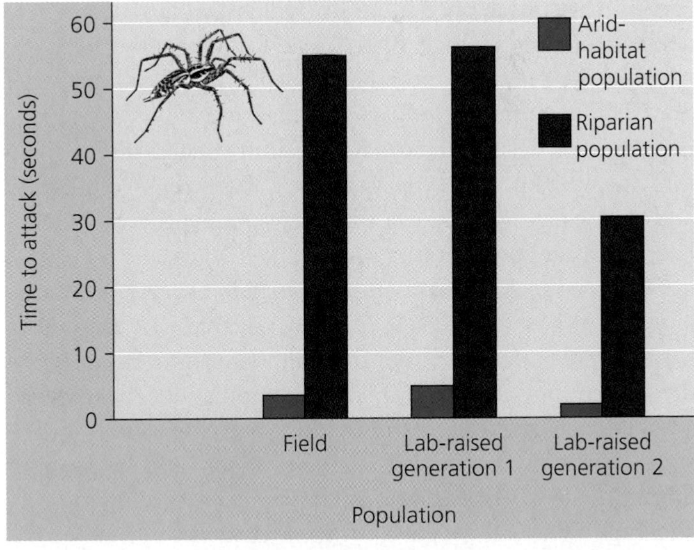

▲ **Figure 51.19 Aggressiveness of funnel web spiders (*Agelenopsis aperta*) from different environments.** Field-collected and lab-raised funnel web spiders from arid habitats delay less before attacking potential prey compared to *A. aperta* from riparian environments.

arid and riparian habitats and repeated the experiment with the laboratory-raised offspring of each population.

The experiments revealed a strikingly consistent difference between arid-habitat and riparian *A. aperta* in their time to attack prey. The highly productive riparian sites are rich in prey for the spiders, but the density of potential predators, particularly birds, is high. Riechert had proposed that this higher risk of predation was the cause of the more timid behavior of *A. aperta* in riparian habitats. Based on their results, Hedrick and Riechert concluded that the difference in aggressiveness between arid-habitat and riparian *A. aperta* spiders is genetically based and is the product of natural selection for differences in foraging and territorial behavior—a genetically determined rather than a learned response. Their findings were reinforced by later experiments with *A. aperta* transplanted from one environment to the other.

Experimental Evidence for Behavioral Evolution

Looking for more direct ways to demonstrate the evolution of behavior, researchers are increasingly turning to laboratory and field experiments using organisms with short lifespans, enabling the observation of change over many generations.

Laboratory Studies of Drosophila *Foraging Behavior*

In a few cases, researchers have been able to link behaviors to specific genes. For instance, Marla Sokolowski, of the University of Toronto, studied a polymorphism in a gene for foraging in *Drosophila melanogaster* called *for.* One allele, for^s, results in a behavioral phenotype called "sitter," in which the fly larva moves less than average. A different allele, for^R, results in a "rover" behavioral phenotype, in which the fly larva moves about more than average. Sokolowski found that the allele frequencies in a natural population of *Drosophila melanogaster* are 70% for^R and 30% for^s. She noted, however, that while the mechanistic basis for the differences in rover and sitter behaviors are well known, little is known about the evolutionary and ecological significance of different frequencies of these alleles in natural populations.

Sokolowski teamed with Sofia Pereira and Kimberly Hughes, of York University, Ontario, to conduct a laboratory study of how population density might affect the frequency of the for^R and for^s alleles in *Drosophila* populations. Sokolowski and her colleagues raised *Drosophila* at both high and low densities, starting with equal frequencies of for^R and for^s in both lineages. After 74 generations, larvae from the low-density and high-density lineages showed a clear divergence in behavior, as measured by differences in average foraging path length **(Figure 51.20)**. The researchers concluded that the for^s allele had increased in frequency in the low-density populations, in which short-distance foraging would yield sufficient food, while long-distance foraging would result in unnecessary energy expenditure. Meanwhile, the for^R allele had increased in frequency in the high-density lineage, where long-distance foraging could enable larvae to move beyond areas of food depletion. In other words, they had observed an evolutionary change in behavior in their laboratory populations.

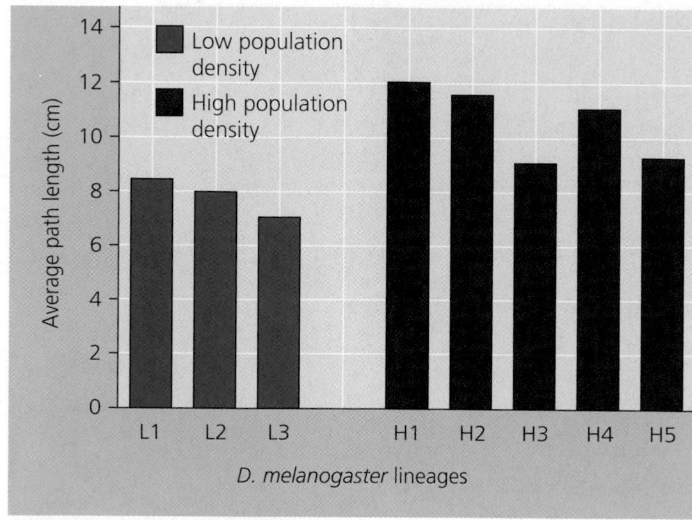

▲ **Figure 51.20 Evolution of foraging behavior by laboratory populations of *Drosophila melanogaster.*** After 74 generations of living at low population density, *D. melanogaster* followed a foraging path significantly shorter than that of *D. melanogaster* populations that had lived at high density.

Migratory Patterns in Blackcaps

An example of evolutionary change in the behavior of a wild population comes from studies on migratory behavior in a population of blackcaps, which we introduced in our earlier discussion of migration. The migratory patterns of blackcaps are well known in western Europe because of a great interest in bird study and a long history of bird banding. Blackcaps that breed in northwestern Europe generally migrate to the western Mediterranean region for the winter. As a consequence, blackcaps in Germany have historically migrated in a southwest direction, while those in Britain have migrated south. In the 1950s, a few blackcaps began to spend their winters in Britain, and over time, the population of blackcaps wintering in Britain has grown to many thousands. Surprisingly, these wintering birds are not birds that bred in Britain and then stayed for the winter. Leg bands on some of the birds show that they had come from continental Europe and migrated westward to Britain instead of south to the Mediterranean.

Peter Berthold and his colleagues captured 20 male and 20 female blackcaps wintering in Britain and transported them back to their laboratory in southwestern Germany. They compared the migratory orientation of these birds and their F_1 offspring to that of young birds originally from southwestern Germany. Behavioral ecologists have long known that the migratory restlessness that Berthold had studied previously is not random activity but rather is directed. That is, caged birds showing migratory restlessness orient their activity in the direction that they would migrate if they were not caged. In the autumn when the birds were showing migratory restlessness, Berthold's team placed the blackcaps from their three study groups in large glass-covered funnel cages lined with carbon-coated paper for 1.5–2 hours. As the birds moved around the funnels, the marks they made on the paper indicated the direction in which they were trying to "migrate" **(Figure 51.21a)**. The results showed that the migratory orientation of wintering adult birds captured in Britain was very similar to the orientation of their laboratory-raised offspring **(Figure 51.21b)**. That orientation contrasted sharply with the migratory orientation of the young birds originally from Germany. This study indicates a genetic basis for migratory orientation, since the young of the British blackcaps and the young birds from Germany were raised under similar conditions but showed very different migratory orientations.

▼ Figure 51.21 Evidence of a genetic basis for migratory orientation.

(a) Blackcaps placed in a funnel cage left marks indicating the direction in which they were trying to migrate.

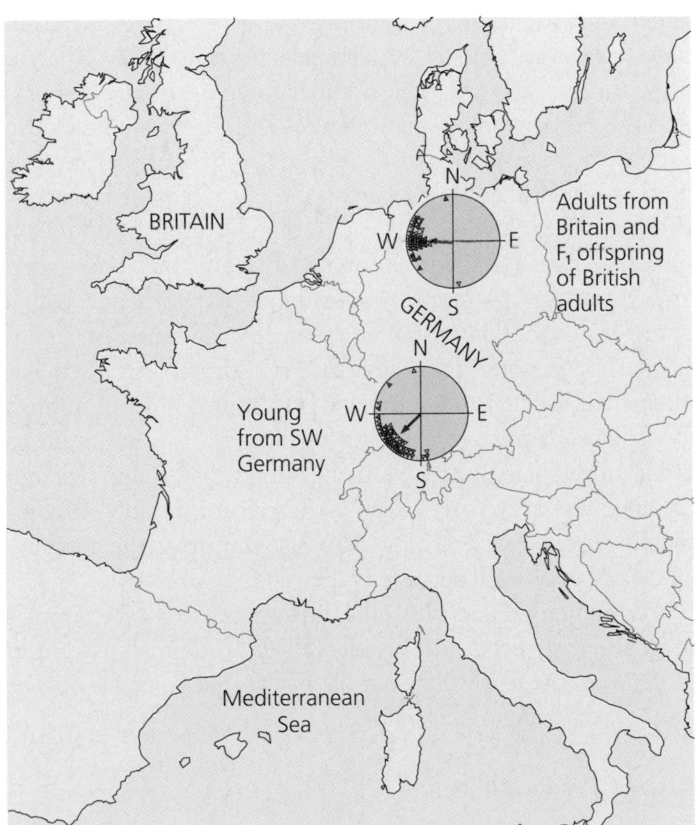

(b) Wintering blackcaps captured in Britain and their laboratory-raised offspring had a migratory orientation toward the west, while young birds from Germany were oriented toward the southwest.

But has the behavior actually evolved over time? Berthold's study also indicates that the change in migratory behavior in western European blackcaps is both recent and rapid, having taken place over the past few decades. Before 1960, there were no known westward-migrating blackcaps in Germany, but by the 1990s, westward migrants made up 7–11% of the blackcap populations of Germany and parts of Austria. Berthold suggests that once westward migration began, it persisted and increased in frequency as a result of several factors, including milder winter climate in Britain and improved winter food partly due to the widespread use of winter bird feeders in Britain. These hypotheses touch on the subject of the next section, which concerns how behavior can influence survival and reproduction and thus lead to evolution.

Concept Check 51.4

1. Explain why geographic variation in garter snake foraging behavior might demonstrate that the behavior evolved by natural selection.
2. Why did Hedrick and Riechert examine behavioral variation between populations of both laboratory-raised spiders and wild ones?
3. What conclusions did Berthold and his colleagues draw from their studies of blackcap migratory patterns?

For suggested answers, see Appendix A.

Concept 51.5

Natural selection favors behaviors that increase survival and reproductive success

The genetic components of behavior, like all aspects of phenotype, evolve through natural selection for traits that enhance survival and reproductive success in a population. Two of the most direct ways a behavior can affect fitness are through its influences on foraging and mate choice behavior.

Foraging Behavior

Because adequate nutrition is essential to an animal's survival and reproductive success, we should expect natural selection to refine behaviors that enhance the efficiency of feeding. Food-obtaining behavior, or foraging, includes not only eating, but also any mechanisms an animal uses to recognize, search for, and capture food items. **Optimal foraging theory** views foraging behavior as a compromise between the benefits of nutrition and the costs of obtaining food, such as the energy expenditure or the risk of being eaten by a predator while foraging. According to this theory, natural selection should favor foraging behavior that minimizes the costs of foraging and maximizes the benefits. Some behavioral ecologists are applying cost-benefit analysis to study the proximate and ultimate causes of diverse foraging strategies.

Energy Costs and Benefits

Reto Zach, of the University of British Columbia, conducted a cost-benefit analysis of feeding behavior in crows of the Pacific Northwest. Northwestern crows (*Corvus caurinus*), like other crow species, are opportunistic feeders that eat a variety of food items. On Mandarte Island, off British Columbia, crows search the rocky tide pools for gastropod molluscs called whelks. A bird grasps a whelk in its beak, then flies upward and drops it onto the rocks to break the shell. If the drop is successful, the crow can dine on the mollusc's soft parts. If the shell doesn't break, the crow flies up and drops the whelk again and again until the shell breaks. The higher the bird flies before dropping a whelk, the fewer drops are required to break the shell. But there is an energy cost correlated with the height of the crow's ascent. Zach predicted that crows would, on average, fly to a height that would provide the most food relative to the amount of energy required to break the whelk shells.

To determine the optimal height, Zach erected a 15-m pole and then dropped shells of relatively uniform size onto rocks from different heights along the pole. He recorded the number of drops required to break a shell from each height and multiplied it by the drop height to arrive at the *total flight height* **(Figure 51.22)**. He then calculated the *average total height* required to break shells as the average number of drops from a particular height times the drop height. Zach's experiments indicated that a drop height of about 5 m is optimal, breaking the shells with the lowest total flight height—in other words, with the least work. When Zach measured the actual average flight height for crows in their whelk-eating behavior, it was 5.23 m, very close to the prediction based on an optimal trade-off between energy gained (food) versus energy expended.

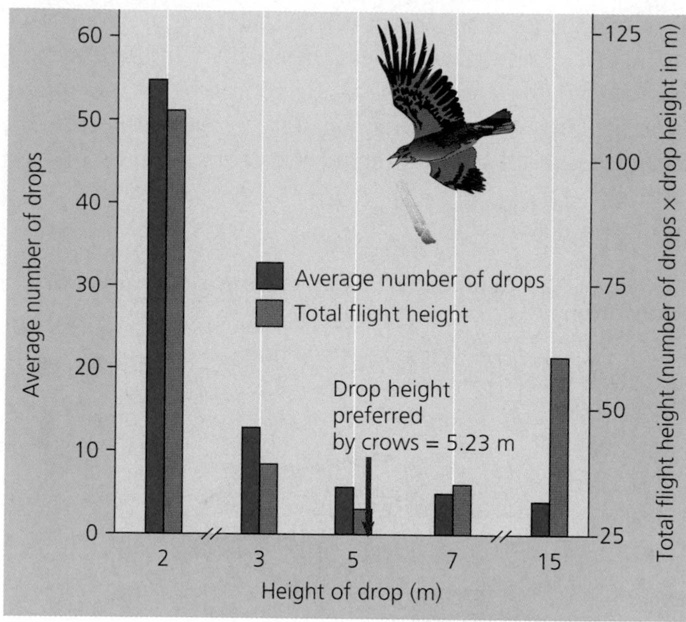

▲ **Figure 51.22 Energy costs and benefits in foraging behavior.** Experimental results indicate that dropping shells from a height of 5 m results in breakage with the least amount of work. The actual drop height preferred by crows corresponds almost exactly to the height that minimizes total flight height.

The feeding behavior of the bluegill sunfish provides further support for the optimal foraging theory. These animals feed on small crustaceans called *Daphnia,* generally selecting larger individuals, which supply the most energy per unit time. However, bluegill sunfish select smaller prey if larger prey are too far away. The optimal foraging theory predicts that the proportion of small prey to large prey eaten will also vary with the overall density of prey. When prey density is high, it is efficient to concentrate on larger crustaceans, whereas at very low prey densities, bluegill sunfish should exhibit little size selectivity, because all available prey are needed to meet energy requirements. In actual experiments, young bluegill sunfish did become more selective at higher prey densities, though not to the extent that would theoretically maximize their efficiency **(Figure 51.23)**. Young bluegill sunfish forage fairly efficiently but not as close to the optimum as older individuals. It may be that younger fish judge size and distance less accurately because their vision is not yet completely developed; learning may also improve the foraging efficiency of bluegill sunfish as they age.

Though crows and bluegill sunfish provide good examples of how energy costs and benefits affect an animal's choice of food items, the following example shows that energy is not the only cost of foraging behavior.

Risk of Predation

One of the most significant potential costs to a forager is risk of predation. Clearly, an animal that feeds in a way that maximizes energy benefits and minimizes energy costs without regard to its

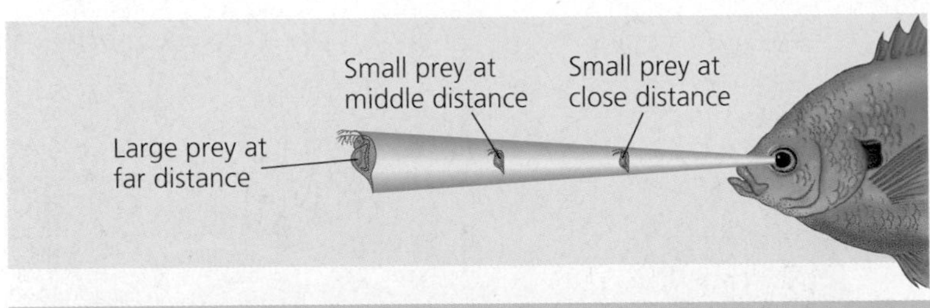

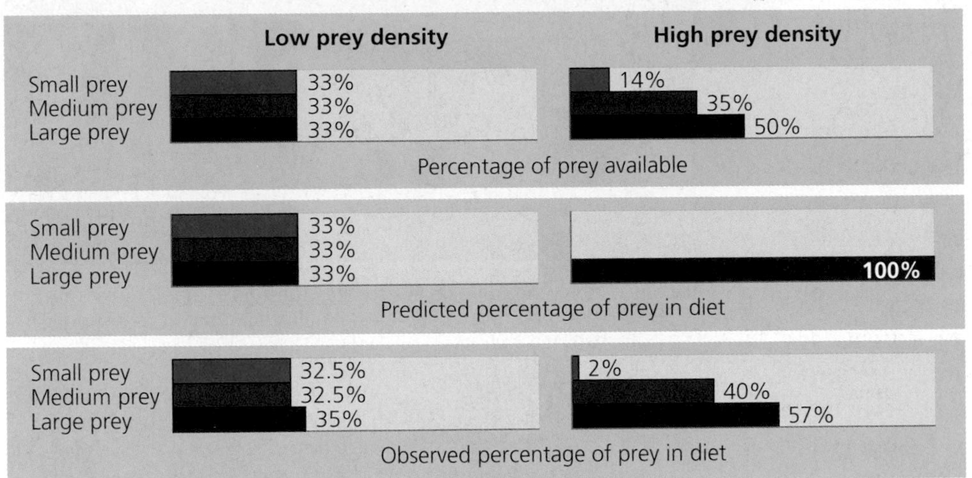

▲ **Figure 51.23 Feeding by bluegill sunfish.** In feeding on *Daphnia* (water fleas), the fish do not feed randomly but select prey based on both size and distance, tending to pursue prey that looks largest. Small prey (low energy yield) at the middle distance may be ignored. But small prey at close distances may be taken with a relatively small energy expenditure. In the experiments described here, when prey was at low density, young bluegill sunfish foraged nonselectively, as predicted by optimal foraging theory.

risk of becoming a meal for some predator has not optimized its foraging behavior. Researchers have investigated the influence of predation risk on foraging behavior for a number of species, including the mule deer (*Odocoileus hemionus*). A major predator of mule deer throughout their range in the mountainous regions of western North America is the mountain lion (*Puma concolor*). A team of researchers from several institutions studied populations of mule deer in Idaho to determine if the mule deer forage in a way that reduces their risk of falling prey to mountain lions. The team investigated how food availability and risk of predation varied across the study landscape, which consisted of both patches of forest and open, nonforested areas. The researchers found that the food available for mule deer was fairly uniform across the potential foraging areas, though somewhat lower in open areas. In contrast, risk of predation differed greatly; mountain lions killed most mule deer at forest edges and only a small number in open areas and forest interiors **(Figure 51.24)**.

How does mule deer feeding behavior respond to the dramatic differences in predation risk in different foraging areas? The researchers found that the mule deer in their study area feed predominantly in open areas, avoiding both forest edges and forest interiors (see Figure 51.24). In addition, when deer are at forest edges, they spend significantly more time scanning their surroundings than when they are in either open areas or forest interiors. The evidence indicates that mule deer foraging behavior reflects variation in predation risk more than variation in food availability. This result, like the results of the bluegill sunfish study, underscores the fact that behavior often reflects a compromise among competing selective pressures.

Mating Behavior and Mate Choice

Mating behavior, which includes seeking or attracting mates, choosing among potential mates, and competing for mates, is the product of a form of natural selection called sexual selection (see Chapter 23). How mating behavior enhances reproductive success varies, depending on the species' mating system.

Mating Systems and Parental Care

The mating relationship between males and females varies a great deal from species to species. In many species, mating is **promiscuous**, with no strong pair-bonds or lasting relationships. In species in which the mates remain together for a longer period, the relationship may be **monogamous** (one male mating with one female) or **polygamous** (an individual of one sex mating with several of the other). Polygamous relationships most often involve a single male and many females,

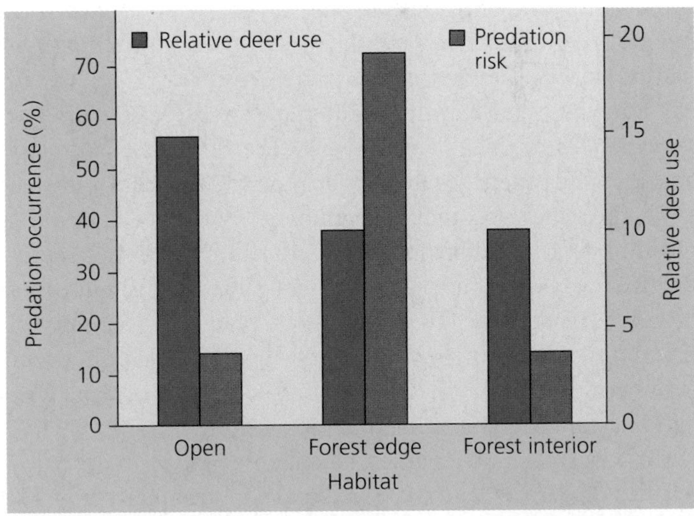

▲ **Figure 51.24 Risk of predation and use of foraging areas by mule deer.** Optimal foraging theory predicts that prey will forage in a way that minimizes the risk of predation. The risk of predation by mountain lions is lowest in open areas and forest interiors and highest at forest edges. Mule deer feed preferentially in open areas, avoiding both forest edges and interiors.

a system called **polygyny**, though some species exhibit **polyandry**, in which a single female mates with several males. Among monogamous species, males and females are often so much alike morphologically that they may be difficult or impossible to distinguish based on external characteristics **(Figure 51.25a)**. Polygynous species are generally dimorphic, with males being more showy and often larger than females **(Figure 51.25b)**. Polyandrous species are also dimorphic, but in this case, females are generally ornamented and larger than males **(Figure 51.25c)**.

The needs of the young are an important factor constraining the evolution of mating systems. Most newly hatched birds, for instance, cannot care for themselves and require a large, continuous food supply that a single parent may not be able to provide. In such cases, a male may ultimately leave more viable offspring by helping a single mate than by going off to seek more mates. This may explain why most birds are monogamous. In contrast, birds with young that can feed and care for themselves almost immediately after hatching, such as pheasants and quail, have less need for their parents to stay together. Males of such species can maximize their reproductive success by seeking other mates, and polygyny is relatively common in such birds. In the case of mammals, the lactating female is often the only food source for the young; males usually play no role in raising the young. In mammalian species where males protect the females and young, such as lions, a male or small group of males typically takes care of many females at once in a harem.

Another factor influencing mating behavior and parental care is certainty of paternity. Young born or eggs laid definitely contain the mother's genes. But even within a normally monogamous relationship, these young may have been fathered by a male other than the female's usual mate. The certainty of paternity is relatively low in most species with internal fertilization because the acts of mating and birth (or mating and egg laying) are separated over time. This could explain why exclusively male parental care occurs in very few species of birds and mammals. However, the males of many species with internal fertilization engage in behaviors that appear to increase their certainty of paternity, including guarding females, removing any sperm from the female reproductive tract before copulation, and introducing large quantities of sperm to displace the sperm of other males. Certainty of paternity is much higher when egg laying and mating occur together, as in external fertilization. This may explain why parental care in aquatic invertebrates, fishes, and amphibians, when it occurs at all, is at least as likely to be by males as by females **(Figure 51.26)**. Male parental care occurs in only 7% of fish and amphibian families with internal fertilization, but in 69% of families with external fertilization. In fishes, even when parental care is provided exclusively by males, the mating system is often polygynous, with several females laying eggs in a nest tended by one male.

(a) Since monogamous species, such as these trumpeter swans, are often monomorphic, males and females are difficult to distinguish using external characteristics only.

(b) Among polygynous species, such as elk, the male (left) is often highly ornamented.

(c) In polyandrous species, such as these Wilson's phalaropes, females (top) are generally more ornamented than males.

▲ **Figure 51.25 Relationship between mating system and male and female forms.**

▲ **Figure 51.26 Paternal care by a male jawfish.** Male jawfish, which live in tropical marine environments, hold the eggs they have fertilized in their mouths, keeping them aerated and protecting them from egg predators until the young hatch.

▲ **Figure 51.27 Zebra finches, native to Australia.** The male zebra finch (left) is more strikingly patterned and more colorful than the female zebra finch.

It is important to point out that when behavioral ecologists use the expression *certainty of paternity,* they do not mean that animals are aware of those factors when they behave a certain way. Parental behavior correlated with certainty of paternity exists because it has been reinforced over generations by natural selection. The relationship between certainty of paternity and male parental care remains an area of active research, enlivened by controversy.

Sexual Selection and Mate Choice

As you read in Chapter 23, the degree of sexual dimorphism within a species results from sexual selection, a form of natural selection in which differences in reproductive success among individuals are a consequence of differences in mating success. Recall from that chapter that sexual selection can take the form of *intersexual selection,* in which members of one sex choose mates on the basis of particular characteristics of the other sex, such as courtship songs, or *intrasexual selection,* which involves competition among members of one sex for mates. Let's look next at some experimental evidence for sexual selection.

Mate Choice by Females. Mate preferences by females may play a central role in the evolution of male behavior and anatomy through intersexual selection. Klaudia Witte and Nadia Sawka, of the University of Bielefeld, Germany, used experiments to test whether imprinting by young zebra finches on their parents may influence their choice of mates when they mature. Male zebra finches are more ornate than females, though neither have crests on their heads **(Figure 51.27)**. Witte and Sawka taped a 2.5-cm-long red feather to the forehead feathers of both male and female parents, to the male parent only, or to the female parent only when the chicks were 8 days old, approximately 2 days before they opened their eyes. (The artificial ornaments represent novel anatomical features that might appear naturally in populations as a consequence of genetic mutations.) A control group of zebra finches were raised by unadorned parents.

When the chicks from the experiments matured, Witte and Sawka tested their mate preferences by giving them a choice between prospective mates that were either ornamented with a red feather or nonornamented. Males showed no preference for either ornamented or nonornamented females. Females raised by nonornamented parents or by parents in which only the female was ornamented also showed no preference for either ornamented or nonornamented males (**Figure 51.28**, on the next page). However, females raised by parents that were both ornamented or by a pair in which the male was ornamented preferred ornamented males as their own mates. These experiments suggest that the females imprint on their fathers rather than on their mothers and that mate choice by female zebra finches has played a key role in the evolution of ornamentation in male zebra finches.

As another example of how female choice affects the evolution of males, consider the courtship behavior of stalk-eyed flies. The eyes of these insects are at the tips of stalks, which are longer in males than in females (**Figure 51.29**, on the next page). During courtship, a male presents himself to a female, front end on. Researchers have documented that females are more likely to mate with those males that have relatively long eyestalks; thus, female choice has been a strong selection factor in the evolution of long eyestalks in males. But why would the females favor this seemingly arbitrary trait? Behavioral ecologists have correlated certain genetic disorders in the male flies with an inability to develop long eyestalks. Such studies support the hypothesis that females are basing their mate choices on characteristics that correlate with male quality. As

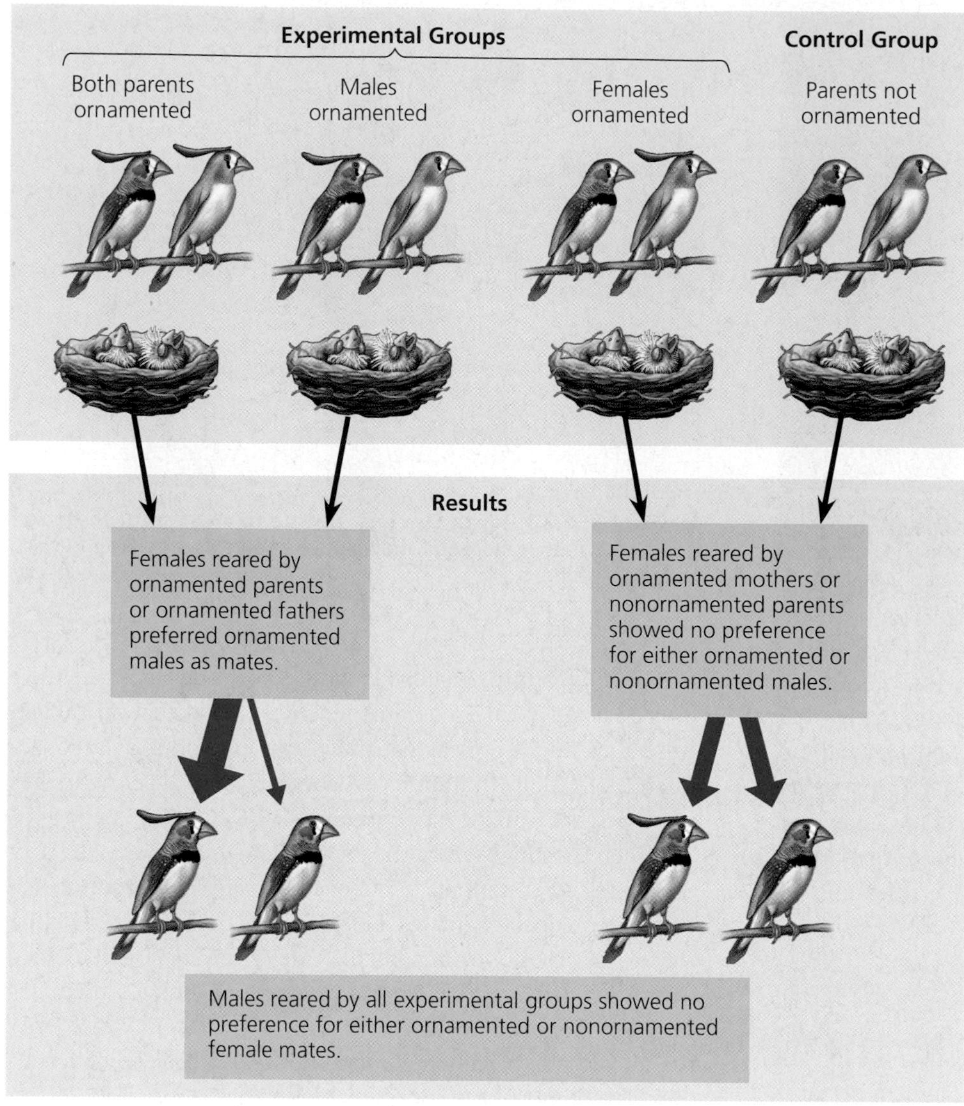

Experimental Groups
Control Group
Both parents ornamented
Males ornamented
Females ornamented
Parents not ornamented
Results
Females reared by ornamented parents or ornamented fathers preferred ornamented males as mates.
Females reared by ornamented mothers or nonornamented parents showed no preference for either ornamented or nonornamented males.
Males reared by all experimental groups showed no preference for either ornamented or nonornamented female mates.

▲ **Figure 51.28 Sexual selection influenced by imprinting.** Experiments demonstrated that female zebra finch chicks that had imprinted on artificially ornamented fathers preferred ornamented males as adults.

► **Figure 51.29 Male stalk-eyed flies.**

we discussed in Chapter 23, in general, ornaments such as long eye stalks in these flies or brightly colored feathers in male birds correlate with the males' health and vitality. A female that chooses a healthy male increases the probability of producing healthy offspring.

Male Competition for Mates. The previous examples show how female choice can select for one best type of male in a given situation, resulting in low variation among males. Male competition for mates is a source of intrasexual selection that also can reduce variation among males. Such competition may involve **agonistic behavior,** an often ritualized contest that determines which competitor gains access to a resource, such as food or mates **(Figure 51.30)**. The outcome of such contests may be determined by strength, size, or the form of horns, teeth, and so forth, but the victories may be psychological rather than physical. Despite the potential for such competition to select for reduced variation, male behavioral and morphological variation is extremely high in some vertebrate species ranging from fish to deer and in a wide variety of invertebrates. In some species, more than one mating behavior can result in successful reproduction; in these cases, intrasexual selection has led to the evolution of alternative male mating behavior and morphology.

One invertebrate in which alternative male mating behaviors and morphology have been well documented is the marine intertidal isopod *Paracerceis sculpta,* which lives within sponges in the Gulf of California. This species includes three genetically distinct male types **(Figure 51.31)**. Stephen M. Shuster, of the University of New Mexico, discovered that large α males defend harems of females within intertidal sponges, mainly against other α males. Meanwhile, β males mimic female morphology and behavior and do not elicit a defensive response in α males; thus, they are able to gain access to guarded harems. Tiny γ males invade and live within large harems, which can be inhabited by all three types of males at the same time.

Shuster reported that the mating success of each type of *Paracerceis* male depends on the relative densities of males and females within sponges. The α males father the majority of young when defending a single female. If more than one female is present, β males sire approximately 60% of the offspring. Finally, the reproductive rate of γ males increases linearly with harem size. Using this information and the patterns of distribution of *Paracerceis* females and males in their natural environment, Shuster and Michael Wade, of the University of Chicago, concluded that α, β, and γ males have approximately equal mating success, indicating that a diversity of mating behavior and anatomy can result from intrasexual selection due to male competition for mates. Variation among males in this species is sustained, since none of the three male types is most successful in all circumstances; each type of male has high mating success at some female densities but reduced mating success at other female densities (an example of balanced polymorphism, which we discussed in Chapter 23).

▲ **Figure 51.30 Agonistic behavior.** These two male polar bears (*Ursus maritimus*) are engaging in "play-fighting," ritualized contests that usually do not injure the bears. By contrast, during the spring breeding season, male polar bears fight fiercely as they compete for access to females in estrus.

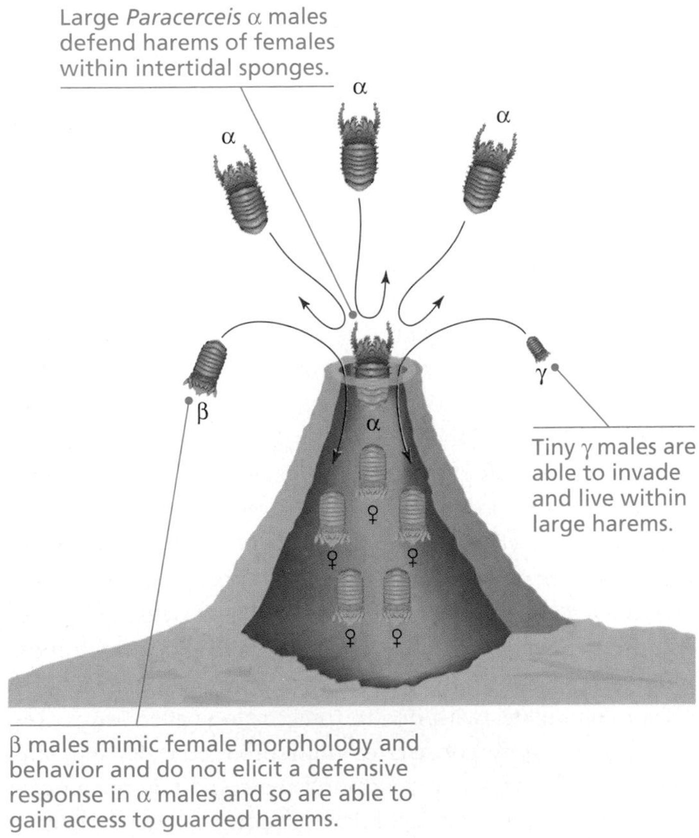

Large *Paracerceis* α males defend harems of females within intertidal sponges.

α

α

α

β

γ

α

♀

♀

♀

♀

♀

Tiny γ males are able to invade and live within large harems.

β males mimic female morphology and behavior and do not elicit a defensive response in α males and so are able to gain access to guarded harems.

▲ **Figure 51.31 Male polymorphism in the marine intertidal isopod *Paracerceis sculpta*.** Different morphology affects the mating behavior of three genetically different types.

Applying Game Theory

The three types of *Paracerceis* males discovered by Shuster can coexist because they have equal mating success, which likely gives them equal evolutionary fitness. However, in some other species, the fitness of different types of males appears to be unequal. How can alternative reproductive strategies exist in such a situation? Why doesn't natural selection favor one strategy to the exclusion of the others? In studying this question, behavioral ecologists use a range of tools, including game theory. Originally developed by John Nash and other mathematicians to model human economic behavior, **game theory** evaluates alternative strategies in situations where the outcome depends not only on each individual's strategy but also on the strategies of other individuals. Currently applied to a wide variety of problems in behavioral ecology, game theory is a way of thinking about evolution in situations where the fitness of a particular behavioral phenotype is influenced by other behavioral phenotypes in the population.

Barry Sinervo and Curt Lively, of Indiana University, used game theory to account for the coexistence of three different male phenotypes in populations of the side-blotched lizard (*Uta stansburiana*) in the inner Coast Ranges of California **(Figure 51.32)**. Sinervo and Lively identified males with three different types of genetically controlled coloration: orange throats, blue throats, and yellow throats. Orange-throat males are the most aggressive and defend large territories that contain many females. Blue-throat males are also territorial but defend smaller territories and fewer females. Yellow-throats

▲ **Figure 51.32 Male polymorphism in the side-blotched lizard (*Uta stansburiana*).** An orange-throat male, left; a blue-throat male, center; a yellow-throat male, right.

are nonterritorial males that mimic females and use "sneaky" tactics to obtain matings. Sinervo and Lively found that the three types go through cycles of relative abundance; over a period of several years, their study population went from high frequency of blue-throats to high frequency of orange-throats to high frequency of yellow-throats and finally back to high frequency of blue-throats.

Sinervo and Lively connected their study population to game theory when they realized that the relative mating success of each male type is not fixed but changes with the relative abundance of the other male types in the populations. Sinervo and Lively suggested that the cycling of male abundance in side-blotched lizards is like the children's game of paper-rock-scissors, in which paper defeats rock, rock defeats scissors, and scissors defeats paper. When blue-throats are abundant, they can successfully defend the few females in their territories from the advances of the sneaky yellow-throat males; in terms of the game, blue-throats "defeat" yellow-throats. However, blue-throat males cannot defend their territories against the hyperaggressive orange-throat males. As a consequence, orange-throat males take over blue-throat territories and increase in abundance; orange-throats "defeat" blue throats. But the numerical dominance by orange-throat males is temporary, since they cannot defend all the females in their large territories from the sneaky yellow-throat males. So eventually, orange-throat males, the most aggressive males, are replaced as the most numerous by yellow-throat males, the least aggressive males; yellow-throats "defeat" orange-throats. And then the cycle begins again: Since the blue-throat tactics of guarding small territories and few females can restrict yellow-throat access to females, blue-throat males soon replace yellow-throats as the most abundant males. Game theory gives behavioral ecologists a way to think about complex evolutionary problems in which relative performance, not absolute performance, is the key to understanding the evolution of behavior. This makes game theory an important tool, since the relative performance of one phenotype compared to others is how evolutionary biologists assess Darwinian fitness.

Concept Check 51.5

1. Why is male parental care more likely to evolve among species with external fertilization than among species with internal fertilization?
2. How does optimal foraging theory explain why mule deer spend more time foraging in open areas than near or in forests?
3. How is a female bird's fitness associated with her ability to choose a mate by discerning among displays and adornments that "advertise" the health of the male?

For suggested answers, see Appendix A.

Concept 51.6

The concept of inclusive fitness can account for most altruistic social behavior

Many social behaviors are selfish, meaning that they benefit the individual at the expense of others, especially competitors. Even in species in which individuals do not engage in agonistic behavior, most adaptations that benefit one individual will indirectly harm others. For example, superior foraging ability by one individual may leave less food for others. It is easy to understand the pervasive nature of selfishness if natural selection shapes behavior. Behavior that maximizes an individual's survival and reproductive success is favored by selection, regardless of how much damage such behavior does to another individual, a local population, or even an entire species. How, then, can we explain observed examples of what appears to be "unselfish" behavior?

Altruism

On occasion, some animals do behave in ways that reduce their individual fitness but increase the fitness of other individuals in the population; this is our functional definition of **altruism**, or selflessness. Consider the Belding's ground squirrel, which lives in some mountainous regions of the western United States and is vulnerable to predators such as coyotes and hawks. If a squirrel sees a predator approach, the squirrel often gives a high-pitched alarm call, which alerts unaware individuals, who then retreat to their burrows. Field observations have confirmed that the conspicuous alarm behavior increases the risk of being killed, because it brings attention to the caller's location.

Another example of altruistic behavior occurs in bee societies, in which the workers are sterile. The workers themselves never reproduce, but they labor on behalf of a single fertile queen. Furthermore, the workers sting intruders, a behavior that helps defend the hive but results in the death of those workers.

Still another example of altruism is seen in naked mole rats (*Heterocephalus glaber*), highly social rodents that live in underground chambers and tunnels in southern and northeastern Africa **(Figure 51.33)**. The naked mole rat, which is almost hairless and nearly blind, lives in colonies of 75 to 250 or more individuals. Each colony has only one reproducing female, the queen, who mates with one to three males, called kings. The rest of the colony consists of nonreproductive females and males who forage for underground roots and tubers and care for the queen, the kings, and new offspring still dependent on the queen. The nonreproductive individuals may sacrifice their own lives in trying to protect the queen or kings from snakes or other predators that invade the colony.

▲ **Figure 51.33 Naked mole rats, a species of colonial mammal that exhibits altruistic behavior.** Pictured here is a queen nursing offspring while surrounded by other members of the colony.

Inclusive Fitness

How can a naked mole rat, a worker bee, or a Belding's ground squirrel enhance its fitness by aiding other members of the population, which may be its closest competitors? How can altruistic behavior be maintained by evolution if it does not enhance—and in fact may even reduce—the survival and reproductive success of the self-sacrificing individuals? It is easiest to see how such behavior might be selected for when it applies to parents sacrificing for offspring. When parents sacrifice their own personal well-being to produce and aid offspring, this actually increases the fitness of the parents, because it maximizes their genetic representation in the population.

But individuals sometimes help others who are not their offspring. Like parents and offspring, siblings have half their genes in common. Therefore, selection might also favor helping siblings or helping one's parents produce more siblings. Evolutionary biologist William Hamilton was the first to realize that selection could result in an animal's increasing its genetic representation in the next generation by "altruistically" helping close relatives other than its own offspring. This realization led to the concept of **inclusive fitness,** the total effect an individual has on proliferating its genes by producing its own offspring *and* by providing aid that enables other close relatives, who share many of those genes, to produce offspring.

Hamilton's Rule and Kin Selection

Hamilton proposed a quantitative measure for predicting when natural selection would favor altruistic acts among related individuals. The three key variables in an act of altruism are the benefit to the recipient (B), the cost to the altruist (C), and the coefficient of relatedness (r). The benefit and cost measure the change in the average number of offspring produced by the recipient and altruist, respectively, resulting from the altruistic act. Thus, B, the benefit, is the average number of *extra* offspring that the beneficiary of an altruistic act produces; and C, the cost, is how many *fewer* offspring the altruist produces. Suppose, for example, that members of a human population average two children each. Now consider two brothers who are close in age, reproductively mature, equally fertile, but not yet fathers. One of these young men is close to drowning in heavy surf, and his brother risks his own life to swim out and pull his sibling to safety. The benefit to the almost-drowned brother, the recipient of this altruistic act, is two offspring. Had he drowned, his reproductive output would have been zero; but now, if we use the average, he can father two children: $B = 2.0$. The cost to the heroic brother depends on the risk to his own life in saving his sibling. Let's say that in this kind of surf, an average swimmer has a 5% chance of drowning. Thus, the cost of the altruistic act is 5% of the number of offspring we would expect if the altruist had not taken the risky plunge. The cost is $0.05 \times 2 = 0.1$.

We now know that $B = 2.0$ and $C = 0.1$ for this hypothetical act of altruism, but what about r? The **coefficient of relatedness** equals the probability that if two individuals share a common parent or ancestor, a particular gene present in one individual will also be present in the second individual. With two siblings who are not identical twins, there is a 50% chance that a gene in one sibling will also be present in the other: $r = 0.5$. One way to see this is in terms of the segregation of homologous chromosomes that occurs during meiosis of gametes (**Figure 51.34**; also see Chapter 13).

We can now use values of B, C, and r to evaluate whether natural selection would favor the altruistic act in our imaginary scenario. Natural selection favors altruism when the

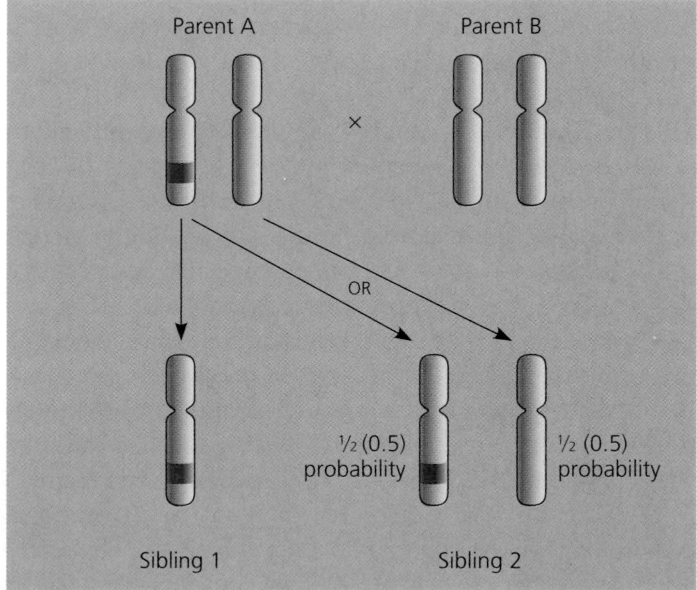

▲ **Figure 51.34 The coefficient of relatedness between siblings.** The red band indicates a particular gene on a chromosome of a homologous chromosome pair in one parent. Sibling 1 has inherited the gene from parent A. There is a probability of ½ that sibling 2 will also inherit this gene from parent A. The coefficient of relatedness between the two siblings is thus ½, or 0.5.

benefit to the recipient multiplied by the coefficient of relatedness exceeds the cost to the altruist—in other words, when $rB > C$. This inequality is called **Hamilton's rule.** For our swimming brothers, $rB = 0.5 \times 2 = 1$ and $C = 0.1$. This satisfies Hamilton's rule; thus, natural selection would favor this altruistic act of one brother saving another. Any particular gene in the altruist will, on average, be passed on to more offspring if that brother risks the rescue than if he does not. (And among those genes may be some that actually contribute to altruistic behavior, so these genes, too, are propagated.) The natural selection that favors this kind of altruistic behavior by enhancing reproductive success of relatives is called **kin selection.**

Kin selection weakens with hereditary distance. Siblings have an r of 0.5, but between an aunt and her niece, $r = 0.25$ ($\frac{1}{4}$), and between first cousins, $r = 0.125$ ($\frac{1}{8}$). Notice that as the degree of relatedness decreases, the rB term in the Hamilton inequality also decreases. Would natural selection favor our strong swimmer rescuing his cousin? For this altruistic act, $rB = 0.125 \times 2 = 0.25$, which, luckily for the drowning cousin, is still much greater than $C = 0.1$, the cost to the altruist. Of course, the degree of risk the altruist takes comes into play, too. If the potential rescuer is a poor swimmer, he may have a 50% chance of drowning instead of the 5% chance for a strong swimmer. In this case, the cost to the altruist would be $0.5 \times 2 = 1$. That's greater than the rB of 0.25 we calculated for the drowning cousin, who'd better hope a lifeguard is near.

The British geneticist J. B. S. Haldane anticipated the concepts of inclusive fitness and kin selection by jokingly saying that he would lay down his life for two brothers or eight cousins. In today's terms, we would say that he would do this because either two brothers or eight cousins would result in as much representation of Haldane's genes as would two of his own offspring.

If kin selection explains altruism, then the examples of unselfish behavior we observe among diverse animal species should involve close relatives. This is in fact the case, but often in complex ways. Like most mammals, female Belding's ground squirrels settle close to their site of birth, whereas males settle at distant sites. Since nearly all alarm calls are given by females **(Figure 51.35)**, they are most likely aiding close relatives. If all of a female's close relatives are dead, she rarely gives alarm calls. In the case of worker bees, who are all sterile, anything they do to help the entire hive benefits the only permanent member who is reproductively active—the queen, who is their mother.

In the case of naked mole rats, DNA analyses have shown that all the individuals in a colony are closely related. Genetically, the queen appears to be a sibling, daughter, or mother of the kings, and the nonreproductive rats are the queen's direct descendants or her siblings. Hence, when a nonreproductive individual enhances a queen's or king's chances of reproducing, it increases the chances that some genes identical to its own will be passed to the next generation.

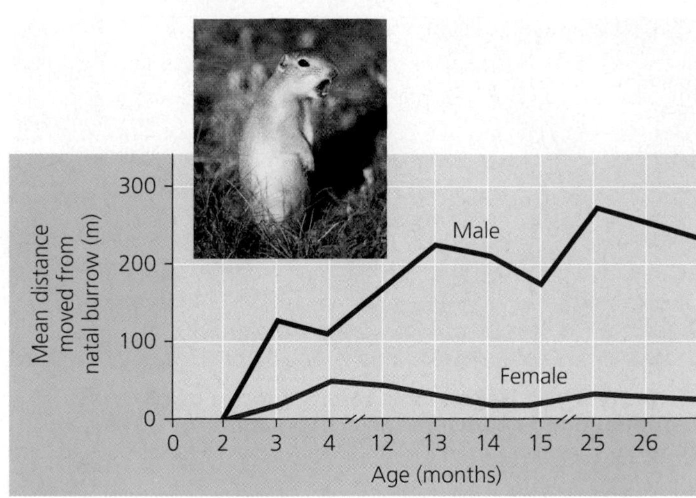

▲ **Figure 51.35 Kin selection and altruism in Belding's ground squirrels.** This graph helps explain the male-female difference in altruistic behavior of ground squirrels. After they are weaned, females are more likely than males to live near close relations. Alarm calls that warn these relatives increase the inclusive fitness of the female altruist.

Reciprocal Altruism

Some animals occasionally behave altruistically toward others who are not relatives. A baboon may help an unrelated companion in a fight, or a wolf may offer food to another wolf even though they share no kinship. Such behavior can be adaptive if the aided individual returns the favor in the future. This sort of exchange of aid, called **reciprocal altruism,** is commonly invoked to explain altruism between unrelated humans. Reciprocal altruism is rare in other animals; it is limited largely to species (such as chimpanzees) with social groups stable enough that individuals have many chances to exchange aid. It is generally thought to be most likely where individuals are likely to meet again and where there would be negative consequences associated with not returning favors to individuals who had been helpful in the past, a pattern of behavior that behavioral ecologists refer to as "cheating." It is likely that all behavior that seems altruistic actually has at least the potential to increase fitness in some way.

However, because cheating may provide a large benefit to the cheater, behavioral ecologists have questioned how reciprocal altruism could evolve. To find answers to this question, many behavioral ecologists have turned to game theory. In 1981, Robert Axelrod and William Hamilton, then at the University of Michigan, proposed that reciprocal altruism can evolve and persist in a population where individuals adopt a behavioral strategy they called *tit for tat.* In the tit-for-tat strategy, an individual treats another in the same way it was treated the last time they met. Individuals adopting this behavior are always altruistic, or cooperative, on the first encounter with another individual and will remain so as long as their altruism is reciprocated. When their cooperation is not reciprocated, however, individuals employing tit for tat will retaliate immediately but return to cooperative behavior as soon as the other individual becomes cooperative. The tit-for-tat strategy has been used to explain the few apparently reciprocal altruistic interactions observed in animals—ranging from blood sharing between nonrelated vampire bats to social grooming in primates.

Social Learning

When we discussed learning earlier in this chapter, we focused largely on the genetic and environmental influences that lead animals to acquire new behaviors. Learning can also have a significant social component, as seen in **social learning**, which is learning through observing others (see Figure 51.17). Social learning forms the roots of **culture,** which can be defined as a system of information transfer through social learning or teaching that influences the behavior of individuals in a population. Cultural transfer of information has the potential to alter behavioral phenotypes and, in turn, to influence the fitness of individuals. Culturally based changes in the phenotype occur on a much shorter time scale than changes resulting from natural selection. Because we recognize social learning most easily in humans, we may take the process for granted or assume that social learning occurs only in humans. However, social learning can be seen among lineages of animals that diverged from ours very long ago, some of which we describe next.

Mate Choice Copying

We have seen how female mate choice can lead to intersexual selection for elaborate male ornamentation (see Figure 51.28). In many species, mate choice is strongly influenced by social learning.

Mate choice copying, a behavior in which individuals in a population copy the mate choice of others, has been extensively studied in the guppy *Poecilia reticulata.* Female wild guppies generally prefer to mate with males showing a high percentage of orange coloration. However, they are also known to copy the mate choices of other females. In other words, female guppies appear to mate with males that have been successful in attracting other females. With this background in mind, Lee Dugatkin, of the University of Louisville, designed an ingenious experiment to compare the influences of male phenotype and social learning on mate choice by female guppies. He gave female guppies a choice of mating with males with varying degrees of orange coloration. In the control samples, a female chose between males with no other females present. In the experimental treatments, males with the same range of orange coloration as the control males were present, but the experimental female also observed a model of a female engaging in courtship with a male with relatively little orange. In the control samples, females overwhelmingly chose males with higher percentages of orange

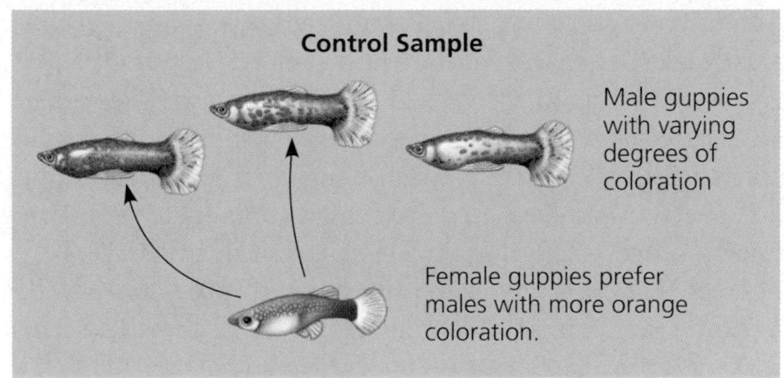

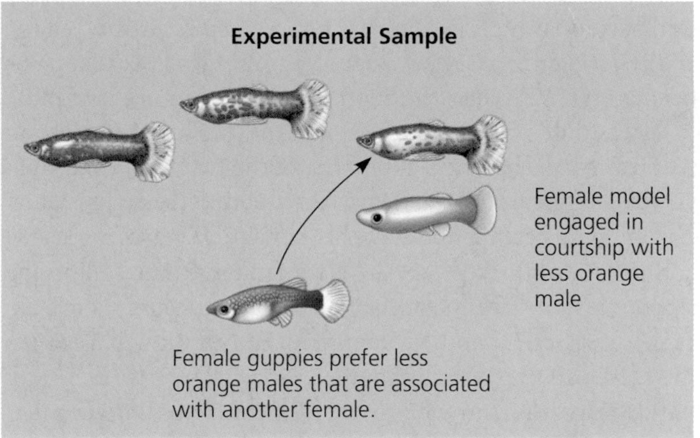

▲ **Figure 51.36 Mate choice copying by female guppies (*Poecilia reticulata*).** Female guppies generally choose the males with more orange coloration. However, when males were matched for orange or differed in amount of orange by 12% or 24%, experimental females chose the *less* orange male that had been presented with a model female. Females ignored the apparent choice of the model female only where the alternative male had 40% more orange coloration.

coloration **(Figure 51.36)**. However, experimental females in most cases chose the less orange male that had been presented in association with a model female, choosing a male not associated with a model female only when he had a *much* higher percentage of orange coloration. Dugatkin concluded that below a certain threshold of difference in male color, mate choice copying by female guppies can mask genetically controlled female preference for orange males. A female that mates with males that are attractive to other females may increase the probability that her male offspring will also be attractive and have high reproductive success. Mate choice copying, one form of social learning, has also been observed in several other fish and bird species.

Social Learning of Alarm Calls

In their studies of vervet monkeys (*Cercopithecus aethiops*) in Amboseli National Park, Kenya, Dorothy Cheney and Richard Seyfarth, of the University of Pennsylvania, determined that performance of a behavior by the vervets could improve through learning. Vervet monkeys, which are about the size of a domestic cat, produce a complex set of alarm calls. The

Amboseli vervets give distinct alarm calls when they see leopards, eagles, or snakes, all of which prey on vervets. When a vervet sees a leopard, it gives a loud barking sound; when it sees an eagle, it gives a short double-syllabled cough; and the snake alarm call is a "chutter." Upon hearing a particular alarm call, other vervets in the group behave in an appropriate way: They run up a tree on hearing the alarm for a leopard (vervets are nimbler than leopards in the trees); look up on hearing the alarm for an eagle; and look down on hearing the alarm for a snake **(Figure 51.37)**.

Infant vervet monkeys give alarm calls, but in a relatively undiscriminating way. For example, they give the "eagle" alarm on seeing any bird, including harmless birds such as bee-eaters. With age, the monkeys improve their accuracy. In fact, adult vervet monkeys give the eagle alarm only on seeing an eagle belonging to either of the two species that eat vervets. Infants probably learn how to give the right call by observing other members of the group and receiving social confirmation. For instance, if the infant gives the call on the right occasion—for instance, an eagle alarm when there is an eagle overhead—another member of the group will also give the eagle call. But if the infant gives the call when a bee-eater flies by, the adults in the group are silent. Thus, vervet monkeys have an initial, unlearned tendency to give calls on seeing potentially threatening objects in the environment. Learning fine-tunes the call so that by adulthood, vervets give calls only in response to genuine danger and are prepared to fine-tune the alarm calls of the next generation. However, neither vervets nor any other species comes close to matching the social learning and cultural transmission that occurs among humans **(Figure 51.38)**.

▲ **Figure 51.37 Vervet monkeys learning correct use of alarm calls.** On seeing a python (foreground), vervet monkeys give a distinct "snake" alarm call (inset), and the members of the group stand upright and look down.

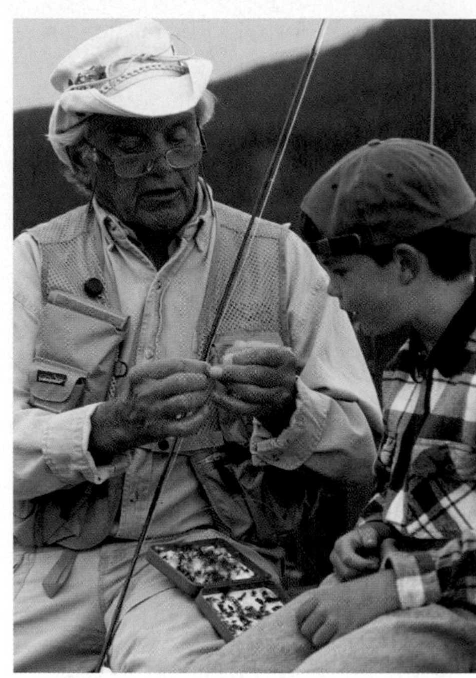

◀ **Figure 51.38 Both genes and culture build human nature.** Teaching of a younger generation by an older generation is one of the basic ways in which all cultures are transmitted.

Evolution and Human Culture

Human culture is related to evolutionary theory in the discipline of **sociobiology**, whose main premise is that certain behavioral characteristics exist because they are expressions of genes that have been perpetuated by natural selection. In his seminal 1975 book *Sociobiology: The New Synthesis,* E. O. Wilson speculated about the evolutionary basis of certain kinds of social behavior mainly in nonhuman animals, but also including human culture, sparking a debate that remains heated today.

The spectrum of possible human social behaviors may be influenced by our genetic makeup, but this is very different from saying that genes are rigid determinants of behavior. This distinction is at the core of the debate about evolutionary perspectives on human behavior. Skeptics fear that evolutionary interpretations of human behavior could be used to justify the status quo in human society, thus rationalizing current social injustices. Evolutionary biologists argue that this is a gross oversimplification and misunderstanding of what the data tell us about human biology. Evolutionary explanations of human behavior do not reduce us to robots stamped out of rigid genetic molds. Just as individuals vary extensively in anatomical features, we should expect inherent variations in behavior as well. Environment intervenes in the pathway from genotype to phenotype for physical traits and even more so for behavioral traits. And because of our capacity for learning and versatility, human behavior is probably more plastic than that of any other animal. Over our recent evolutionary history, we have built up a diversity of structured societies with governments, laws, cultural values, and religions that define what is acceptable behavior and what is not, even when unacceptable behavior might enhance an individual's Darwinian fitness.

And yet we read in the media about newly discovered genes for complex human behavioral traits, such as depression, violence, or alcoholism. Doesn't this reinforce the idea that our behavior is, in fact, "hardwired"? According to Robert Plomin, director of the Center for Developmental and Health Genetics at Pennsylvania State University, research into the heritability of behavior is the best demonstration of the importance of environment. As Plomin puts it, genes and nongenetic, environmental factors "build on each other." For instance, it might seem that the human ability to speak is completely genetic. However, the ability to learn a specific language, such as English or Spanish, is a function of a complex brain that develops in a particular environmental context under the guidance of a human genome and with the aid of social learning. If the behavior of humans, like that of other species, is the result of interactions between genes and environment, what is unique about our species? Perhaps it is our social and cultural institutions that make us truly unique and that provide the only feature in which there is no continuum between humans and other animals.

Concept Check 51.6

1. What is the ultimate cause for altruistic behavior among kin?
2. What hypothesis could explain cooperative behavior among nonrelated animals? Explain.
3. Could the changes in behavior produced by cross-fostering of white-footed mice and California mice (described in Concept 51.3) extend beyond one generation? Explain.

For suggested answers, see Appendix A.

Chapter 51 Review

Go to www.campbellbiology.com or the student CD-ROM to explore Activities, Investigations, and other interactive study aids.

SUMMARY OF KEY CONCEPTS

Concept 51.1

Behavioral ecologists distinguish between proximate and ultimate causes of behavior

- **What Is Behavior? (p. 1107)** Behavior, which includes muscular as well as nonmuscular activity, is everything that an animal does and how it does it. Learning is also generally considered a behavioral process.
- **Proximate and Ultimate Questions (p. 1107)** Proximate, or "how," questions focus on the environmental stimuli, if any, that trigger a behavior, as well as the genetic, physiological, and anatomical mechanisms underlying a behavioral act. Ultimate, or "why," questions address the evolutionary significance of a behavior.
- **Ethology (pp. 1107–1109)** Ethology is the scientific study of animal behavior, particularly in natural environments. The mid-20th-century ethologists developed a conceptual framework defined by a set of questions that highlight the complementary nature of proximate and ultimate perspectives.
  Investigation *How Can Pillbug Responses to Environments Be Tested?*

Concept 51.2

Many behaviors have a strong genetic component

- Biologists study the ways both genes and the environment influence the development of behavioral phenotypes (pp. 1109–1110).
- **Directed Movements (pp. 1110–1111)** Kinesis, a behavior involving a change in activity or turning rate, and taxis, an oriented movement toward or away from some stimulus, are examples of innate behaviors that are under strong genetic influence. Experiments have demonstrated that bird migration, a very complex behavior, is at least partly under genetic control.
- **Animal Signals and Communication (pp. 1111–1112)** The transmission of, reception of, and response to signals constitute animal communication. Animals communicate using visual, auditory, chemical (olfactory), tactile, and electrical signals.
  Activity *Honeybee Waggle Dance Video*
- **Genetic Influences on Mating and Parental Behavior (pp. 1112–1113)** A variety of mammalian behaviors are under relatively strong genetic control. For instance, research has revealed the genetic and neural basis for the mating and parental behavior of male prairie voles.

Concept 51.3

Environment, interacting with an animal's genetic makeup, influences the development of behaviors

- **Dietary Influence on Mate Choice Behavior (pp. 1113–1114)** Laboratory experiments have demonstrated that the type of food eaten during larval development strongly influences later mate selection by *Drosophila mojavensis* females.
- **Social Environment and Aggressive Behavior (p. 1114)** Cross-fostering studies of California mice and white-footed mice have uncovered an influence of social environment on the aggressive and parental behaviors of these mice.
- **Learning (pp. 1114–1118)** Learning is the modification of behavior based on specific experiences. Types of learning include habituation, spatial learning, use of cognitive maps, and associative learning. Studies of cognition and problem solving are revealing levels of mental sophistication in animals ranging from insects and birds to primates. Environment and genetics can interact in influencing the learning process.

Concept 51.4

Behavioral traits can evolve by natural selection

- **Behavioral Variation in Natural Populations (pp. 1118–1120)** When behavioral variation within a species corresponds to variation in environmental conditions, it may be evidence of past evolution.
- **Experimental Evidence for Behavioral Evolution (pp. 1120–1121)** Laboratory studies of *Drosophila* populations raised in high- and low-density conditions show a clear divergence in behavior linked to specific genes. Field and laboratory studies have documented a change in migratory behavior in western European blackcaps over a period of a few decades.

Concept 51.5

Natural selection favors behaviors that increase survival and reproductive success

- **Foraging Behavior (pp. 1122–1123)** Several studies provide support for optimal foraging theory, which posits that natural selection should favor foraging behavior that minimizes the costs of foraging and maximizes the benefits.
- **Mating Behavior and Mate Choice (pp. 1123–1127)** How mate choice enhances reproductive success varies, depending on the species' mating system. The mating relationship between males and females, which includes monogamous, polygynous, and polyandrous mating systems, varies a great deal from species to species. Certainty of paternity has a significant influence on mating behavior and parental care. Mate preferences by females may play a central role in the evolution of male behavior and anatomy through intersexual selection. Male competition for mates is a source of intrasexual selection that also can reduce variation among males. Intrasexual selection can lead to the evolution of alternative male mating behavior and morphology.
- **Applying Game Theory (pp. 1127–1128)** Applied to problems in behavioral ecology, game theory provides a way of thinking about evolution in situations where the fitness of a particular behavioral phenotype is influenced by other behavioral phenotypes in the population.

Concept 51.6

The concept of inclusive fitness can account for most altruistic social behavior

- **Altruism (pp. 1128–1129)** On occasion, animals behave in altruistic ways that reduce their individual fitness but increase the fitness of the recipient of the behavior.
- **Inclusive Fitness (pp. 1129–1131)** Altruistic behavior can be explained by the concept of inclusive fitness, the total effect an individual has on proliferating its genes by producing its own offspring *and* by providing aid that enables close relatives to produce offspring. Kin selection favors altruistic behavior by enhancing the reproductive success of relatives. Altruistic behavior toward unrelated individuals can be adaptive if the aided individual returns the favor in the future, an exchange of aid called reciprocal altruism.
- **Social Learning (pp. 1131–1132)** Social learning forms the roots of culture, which can be defined as a system of information transfer through observation or teaching that influences the behavior of individuals in a population.
- **Evolution and Human Culture (pp. 1132–1133)** Human behavior, like that of other species, is the result of interactions between genes and environment. However, our social and cultural institutions may provide the only feature in which there is no continuum between humans and other animals.

TESTING YOUR KNOWLEDGE

Self-Quiz

1. Which of the following is true of innate behaviors?
   a. Genes have very little influence on the expression of innate behaviors.
   b. Innate behaviors tend to vary considerably among members of a population.
   c. Innate behaviors are limited to invertebrate animals.
   d. Innate behaviors are expressed in most individuals in a population across a wide range of environmental conditions.
   e. Innate behaviors occur in invertebrates and some vertebrates but not in mammals.

2. Which of the following is an example of a taxis?
   a. navigation of a sparrow during seasonal migration
   b. pulling your hand away from a hot stove
   c. a wasp locating its nest with landmarks
   d. a fish orienting itself into the river current
   e. a crow dropping a whelk from a specific height

3. Which of the following statements provides an ultimate explanation for the observation that adult salmon return from the ocean to spawn in the stream in which they hatched?
   a. Young salmon imprint on the chemical scent of their home stream.
   b. Adult salmon use stellar navigation to relocate their home stream.
   c. Salmon navigate to their home stream using their ability to detect Earth's magnetic field.
   d. Spawning in the home stream results in higher survival of young salmon.
   e. Oceanic currents aid salmon in their search for their home stream.

4. Researchers have found that a region of the canary forebrain shrinks during the nonbreeding season and then enlarges when breeding season begins. This annual enlargement of brain tissue is probably associated with
   a. the annual addition of new syllables to a canary's song repertoire.
   b. the annual crystallization of subsong into adult songs.
   c. the annual sensitive period in which canary parents imprint on new offspring.
   d. the annual renewal of mating and nest-building behaviors.
   e. the annual elimination of the memorized template for songs sung the previous year.

5. Although many chimpanzee populations live in environments containing oil-palm nuts, members of only a few populations use stones to crack open the nuts. The most likely explanation for this behavioral difference between populations is that
   a. the behavioral difference is caused by genetic differences between populations.
   b. members of different populations have different nutritional requirements.
   c. the cultural tradition of using stones to crack nuts has arisen in only some populations.

d. members of different populations differ in learning ability.
e. members of different populations differ in manual dexterity.

6. Which of the following is *not* required for a behavioral trait to evolve by natural selection?
   a. In each individual, the form of the behavior is determined entirely by genes.
   b. The behavior varies among individuals.
   c. An individual's reproductive success depends in part on how the behavior is performed.
   d. Some component of the behavior is genetically inherited.
   e. An individual's genotype influences its behavioral phenotype.

7. Which of the following is *not* true of agonistic behavior?
   a. It is most common among members of the same species.
   b. It may be used to establish and defend territories.
   c. It may be a basis for competition between individuals for mates.
   d. It usually results in death or serious injury to one or both of the competitors.
   e. It may be used to establish dominance over other individuals.

8. Female spotted sandpipers aggressively court males and then, after mating, leave the clutch of young for the male to incubate. This sequence may be repeated several times with different males until no available males remain, forcing the female to incubate her last clutch. Which of the following terms best describes this behavior?
   a. monogamy
   b. polygyny
   c. polyandry
   d. promiscuity
   e. certainty of paternity

9. According to the inequality known as Hamilton's rule ($rB > C$),
   a. natural selection does not favor altruistic behavior that causes the death of the altruist.
   b. natural selection favors altruistic acts when the resulting benefit to the beneficiary, multiplied by the coefficient of relatedness, exceeds the cost to the altruist.
   c. natural selection is more likely to favor altruistic behavior that benefits an offspring than altruistic behavior that benefits a sibling.
   d. the effects of kin selection are larger than the effects of direct natural selection on individuals.
   e. altruism is always reciprocal.

10. The core idea of sociobiology is that
   a. human behavior is rigidly determined by inheritance.
   b. humans cannot choose to change their social behavior.
   c. much human behavior has evolved by natural selection.
   d. the social behavior of humans has many similarities to that of social insects such as honeybees.
   e. the environment plays a larger role than genes in shaping human behavior.

For Self-Quiz answers, see Appendix A.

Go to the website or CD-ROM for more quiz questions.

Evolution Connection

In human affairs, we often explain our behavior in terms of subjective feelings, motives, or reasons; but evolutionary explanations for behavior are based on reproductive fitness. What is the relationship between the two kinds of explanation? For instance, is a human explanation for behavior, such as "falling in love," incompatible with an evolutionary explanation? Does falling in love become more meaningful or less meaningful (or neither) if it has an evolutionary basis?

Scientific Inquiry

Scientists studying scrub jays found that it is common for "helpers" to assist mated pairs of birds in raising their young. The helpers lack territories and mates of their own. Instead, they help the territory owner gather food for their offspring. Propose a hypothesis to explain what advantage there might be for the helpers to engage in this behavior instead of seeking their own territories and mates. How would you test your hypothesis? If your hypothesis is correct, what kind of results would you expect your tests to yield?

Investigation *How Can Pillbug Responses to Environments Be Tested?*

Biological Inquiry: A Workbook of Investigative Cases *Explore the behavior of a large gull population in a marina, and human attempts to control the population, in the case "Back to the Bay."*

Science, Technology, and Society

Researchers are very interested in studying identical twins who were separated at birth and raised apart. So far, the data suggest that such twins are much more alike than researchers would have predicted; they frequently have similar personalities, mannerisms, habits, and interests. What general question do you think researchers hope to answer by studying twins that have been raised apart? Why do identical twins make good subjects for this kind of research? What are the potential pitfalls of this research? What abuses might occur if the studies are not evaluated critically and if the results are carelessly cited in support of a particular social agenda?

52 Population Ecology

▲ **Figure 52.1 Population of northern fur seals *(Callorhinus ursinus)* on St. Paul Island, off the coast of Alaska.**

Key Concepts

52.1 Dynamic biological processes influence population density, dispersion, and demography
52.2 Life history traits are products of natural selection
52.3 The exponential model describes population growth in an idealized, unlimited environment
52.4 The logistic growth model includes the concept of carrying capacity
52.5 Populations are regulated by a complex interaction of biotic and abiotic influences
52.6 Human population growth has slowed after centuries of exponential increase

Overview

Earth's Fluctuating Populations

The size of the human population and its impact are now among Earth's most significant problems. With a population of more than 6 billion individuals, our species requires vast amounts of materials and space, including places to live, land to grow our food, and places to dump our waste. By rapidly expanding our presence on Earth, we have devastated the environment for many other species and now threaten to make it unfit for ourselves.

To understand human population growth, we must consider the general principles of population ecology. No population, including the human population, can continue to grow indefinitely. Though many species exhibit population explosions, their numbers inevitably decline. In contrast to such radical booms and busts, other populations are relatively stable over time, with only minor changes in size.

Our earlier study of populations in Chapter 23 emphasized the relationship between population genetics—the structure and dynamics of gene pools—and evolution. Evolution remains our central theme as we now view populations in the context of ecology. **Population ecology,** the subject of this chapter, is the study of populations in relation to the environment, including environmental influences on population density and distribution, age structure, and variations in population size. The fur seal population of St. Paul Island **(Figure 52.1)**, off the coast of Alaska, is one population that has experienced dramatic fluctuations in size. Reduced to low numbers by hunting early in the 20th century, the seal population grew rapidly once it was protected. In this case, the population had been reduced by human predation, but there are many reasons why populations grow or decline.

Later in this chapter, we will apply the basic population concepts we develop to the human population. Let's begin by exploring some of the structural and dynamic aspects of populations that apply to all species.

Concept 52.1

Dynamic biological processes influence population density, dispersion, and demography

A **population** is a group of individuals of a single species living in the same general area. Members of a population rely on the same resources, are influenced by similar environmental factors, and have a high likelihood of interacting with and breeding with one another. Populations can evolve through natural selection acting on heritable variations among individuals and changing the frequencies of various traits over time (see Chapter 23).

Density and Dispersion

At any given moment, every population has specific boundaries and a specific size (the number of individuals living within those boundaries). Ecologists usually begin investigating a population by defining boundaries appropriate to the organisms under study and to the questions being asked. A population's boundaries may be natural ones, as in the case of terns nesting on a particular island in Lake Superior, or they may be arbitrarily defined by an investigator, as in the case of oak trees within a specific county in Minnesota. Once defined, the population can be described in terms of its density and its dispersion. **Density** is the number of individuals per unit area or volume—the number of oak trees per square kilometer in the Minnesota county, for example, or the number of *Escherichia coli* bacteria per milliliter in a test tube. **Dispersion** is the pattern of spacing among individuals within the boundaries of the population.

Density: A Dynamic Perspective

In rare cases, it is possible to determine population size and density by actually counting all individuals within the boundaries of the population. We could count all the sea stars in a tide pool, for example. Herds of large mammals, such as buffalo or elephants, can sometimes be counted accurately from airplanes. In most cases, however, it is impractical or impossible to count all individuals in a population. Instead, ecologists use a variety of sampling techniques to estimate densities and total population sizes. For example, they might count the number of oak trees in several randomly located 10×100 m plots (samples), calculate the average density in the samples, and then extrapolate to estimate the population size in the entire area. Such estimates are most accurate when there are many sample plots and when the habitat is homogeneous. In some cases, instead of counting individual organisms, population ecologists estimate density from some index of population size, such as the number of nests, burrows, tracks, or fecal droppings.

Another sampling technique commonly used to estimate wildlife populations is the **mark-recapture method.** The researchers place traps within the boundaries of the population under study. Captured animals are marked with tags, collars, bands, or spots of dye, and then released. After a few days or weeks—enough time for the marked individuals to mix with unmarked members of the population—traps are set again. This second capture yields both marked and unmarked individuals. From these data, researchers can estimate the total number of individuals in the population. Note that the mark-recapture method assumes that each marked individual has the same probability of being trapped as each unmarked individual. This is not always a safe assumption, because an animal that has been trapped once may be wary of traps in the future.

Density is not a static property of a population, but is rather the result of a dynamic interplay between processes that add individuals to a population and those that remove individuals from it **(Figure 52.2)**. Additions to populations occur through birth (which we will define here to include all forms of reproduction) and **immigration**, the influx of new individuals from other areas. The factors that remove individuals from a population are death (mortality) and **emigration**, the movement of individuals out of a population.

While birth and death rates have obvious influences on the density of all populations, immigration and emigration can

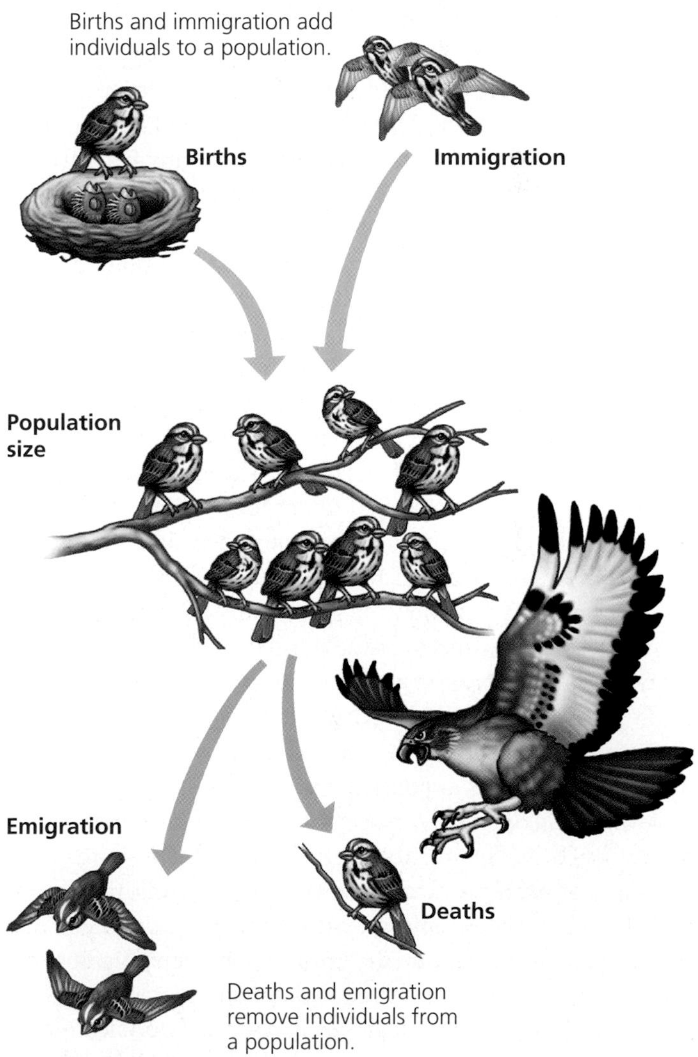
Births and immigration add individuals to a population.
Births
Immigration
Population size
Emigration
Deaths
Deaths and emigration remove individuals from a population.

▲ **Figure 52.2 Population dynamics.**

also make important contributions to the density of local populations. For instance, long-term studies of Belding's ground squirrels *(Spermophilus beldingi)* in the vicinity of Tioga Pass, in the Sierra Nevada mountains of California, show that some of the squirrels move nearly 2 km from where they are born, making them immigrants to other populations. Paul Sherman and Martin Morton, at that time at Cornell University and Occidental College, respectively, estimated that immigrants made up 1–8% of the males and 0.7–6% of the females in the study population. While these may seem like small percentages, over time such rates represent biologically significant exchanges between populations.

Patterns of Dispersion

Within a population's geographic range, local densities may vary substantially. Variations in local density are among the most important characteristics that a population ecologist might study, since they provide insight into the environmental associations and social interactions of individuals in the population. Environmental differences—even at a local level—contribute to variation in population density; some habitat patches are simply more suitable for a species than are others. Social interactions between members of the population, which may maintain patterns of spacing between individuals, can also contribute to variation in population density.

The most common pattern of dispersion is *clumped,* with the individuals aggregated in patches. Plants or fungi are often clumped where soil conditions and other environmental factors favor germination and growth. For example, mushrooms may be clumped on a rotting log. Many animals spend much of their time in a particular microenvironment that satisfies their requirements. Forest insects and salamanders, for instance, are frequently clumped under logs, where the humidity tends to be higher than in more exposed areas. Clumping of animals may also be associated with mating behavior. For example, mayflies often swarm in great numbers, a behavior that increases mating chances for these insects, which survive only a day or two as reproductive adults. Group living may also increase the effectiveness of certain predators; for example, a wolf pack is more likely than a single wolf to subdue a large prey animal, such as a moose **(Figure 52.3a)**.

A *uniform,* or evenly spaced, pattern of dispersion may result from direct interactions between individuals in the population. For example, some plants secrete chemicals that inhibit the germination and growth of nearby individuals that could compete for resources. Animals often exhibit uniform dispersion as a result of antagonistic social interactions, such as **territoriality**—the defense of a bounded physical space against encroachment by other individuals **(Figure 52.3b)**. Uniform patterns are not as common in populations as clumped patterns.

Random dispersion (unpredictable spacing) occurs in the absence of strong attractions or repulsions among individuals of a population or where key physical or chemical factors are relatively homogeneous across the study area; the position of each individual is independent of other individuals. For example,

(a) Clumped. For many animals, such as these wolves, living in groups increases the effectiveness of hunting, spreads the work of protecting and caring for young, and helps exclude other individuals from their territory.

(b) Uniform. Birds nesting on small islands, such as these king penguins on South Georgia Island in the South Atlantic Ocean, often exhibit uniform spacing, maintained by aggressive interactions between neighbors.

(c) Random. Dandelions grow from windblown seeds that land at random and later germinate.

▲ **Figure 52.3 Patterns of dispersion within a population's geographic range.**

plants established by windblown seeds, such as dandelions, are sometimes randomly distributed in a fairly consistent habitat **(Figure 52.3c)**. Random patterns are not as common in nature as one might expect; most populations show at least a tendency toward a clumped distribution.

Demography

The factors that influence population density and dispersion patterns—ecological needs of a species, structure of the environment, and interactions between individuals within the population—also influence other characteristics of populations. **Demography** is the study of the vital statistics of populations and how they change over time. Of particular interest to demographers are birth rates and how they vary among individuals (specifically among females) and death rates. A useful way of summarizing some of the vital statistics of a population is with a life table.

Life Tables

About a century ago, when life insurance first became available, insurance companies needed to estimate how long, on average, individuals of a given age could be expected to live. To do this, demographers developed **life tables**, age-specific summaries of the survival pattern of a population. Population ecologists adapted this approach to the study of nonhuman populations and developed quantitative demography as a branch of population ecology.

The best way to construct a life table is to follow the fate of a **cohort**, a group of individuals of the same age, from birth until all are dead. To build the life table, we need to determine the number of individuals that die in each age group and calculate the proportion of the cohort surviving from one age to the next. Cohort life tables are difficult to construct for wild animals and plants and are available for only a limited number of species. Sherman and Morton's studies of the Tioga Pass Belding's ground squirrels resulted in the life table in **Table 52.1**. The table reveals many things about the population. For instance, the third and eighth columns list, respectively, the proportions of females and males in the cohort that are still alive at each age. A comparison of the fifth and tenth columns reveals that males have higher death rates than females.

Survivorship Curves

A graphic way of representing the data in a life table is a **survivorship curve**, a plot of the proportion or numbers in a cohort still alive at each age. Let's use the data for Belding's ground squirrels in Table 52.1 to construct a survivorship curve for this population. Generally, a survivorship curve is constructed beginning with a cohort of 1,000 individuals from a population. We can do this for the Belding's ground squirrel population by multiplying the proportion alive at the start of each year (third and eighth columns of Table 52.1) by 1,000 (the hypothetical beginning cohort). The result is the

Table 52.1 Life Table for Belding's Ground Squirrels *(Spermophilus beldingi)* at Tioga Pass, in the Sierra Nevada Mountains of California*

	FEMALES					MALES				
Age (years)	Number Alive at Start of Year	Proportion Alive at Start of Year	Number of Deaths During Year	Death Rate†	Average Additional Life Expectancy (years)	Number Alive at Start of Year	Proportion Alive at Start of Year	Number of Deaths During Year	Death Rate†	Average Additional Life Expectancy (years)
0–1	337	1.000	207	0.61	1.33	349	1.000	227	0.65	1.07
1–2	252††	0.386	125	0.50	1.56	248††	0.350	140	0.56	1.12
2–3	127	0.197	60	0.47	1.60	108	0.152	74	0.69	0.93
3–4	67	0.106	32	0.48	1.59	34	0.048	23	0.68	0.89
4–5	35	0.054	16	0.46	1.59	11	0.015	9	0.82	0.68
5–6	19	0.029	10	0.53	1.50	2	0.003	0	1.00	0.50
6–7	9	0.014	4	0.44	1.61	0				
7–8	5	0.008	1	0.20	1.50					
8–9	4	0.006	3	0.75	0.75					
9–10	1	0.002	1	1.00	0.50					

*Males and females have different mortality schedules, so they are tallied separately.
†The death rate is the proportion of individuals dying in the specific time interval.
††Includes 122 females and 126 males first captured as one-year-olds and therefore not included in the count of squirrels age 0–1.

Source: Data from P. W. Sherman and M. L. Morton, "Demography of Belding's Ground Squirrel," *Ecology* 65(1984): 1617–1628.

number alive at the start of each year. Plotting these numbers versus age for female and male Belding's ground squirrels yields **Figure 52.4**. The relatively straight lines of the plots indicate relatively constant rates of death; however, males have a lower survival rate overall than females.

Figure 52.4 represents just one of many patterns of survivorship exhibited by natural populations. Though diverse, survivorship curves can be classified into three general types **(Figure 52.5)**. A Type I curve is flat at the start, reflecting low death rates during early and middle life, then drops steeply as death rates increase among older age groups. Humans and many other large mammals that produce few offspring but provide them with good care often exhibit this kind of curve. In contrast, a Type III curve drops sharply at the start, reflecting very high death rates for the young, but then flattens out as death rates decline for those few individuals that have survived to a certain critical age. This type of curve is usually associated with organisms that produce very large numbers of offspring but provide little or no care, such as long-lived plants, many fishes, and marine invertebrates. An oyster, for example, may release millions of eggs, but most offspring die as larvae from predation or other causes. Those few that survive long enough to attach to a suitable substrate and begin growing a hard shell will probably survive for a relatively long time. Type II curves are intermediate, with a constant death rate over the organism's life span. This kind of survivorship occurs in Belding's ground squirrels (see Figure 52.4) and some other rodents, various invertebrates, some lizards, and some annual plants.

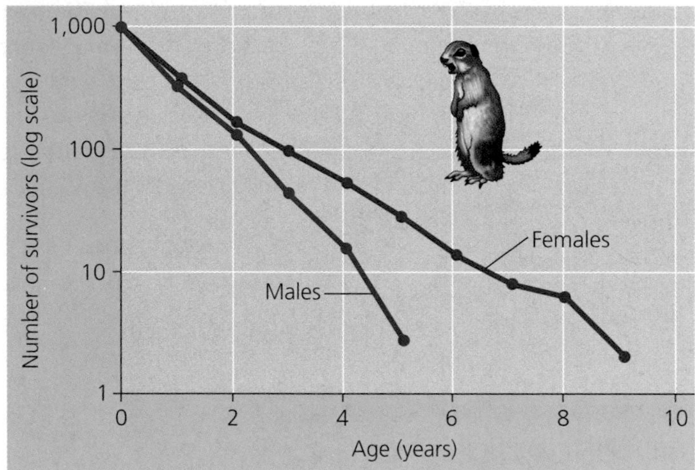

▲ **Figure 52.4 Survivorship curves for male and female Belding's ground squirrels.** The logarithmic scale on the *y*-axis makes changes in number of survivors visible across the entire range (2 to 1,000 individuals) on the graph.

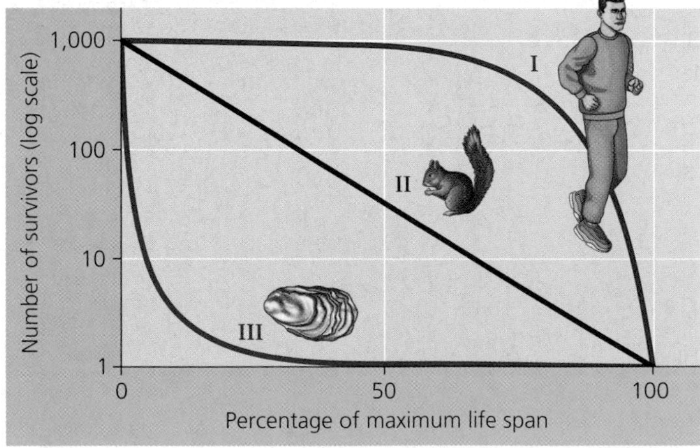

▲ **Figure 52.5 Idealized survivorship curves: Types I, II, and III.** The *y*-axis is logarithmic and the *x*-axis is on a relative scale, so species with widely varying life spans can be compared on the same graph.

Many species fall somewhere between these basic types of survivorship or show more complex patterns. In birds, for example, mortality is often high among the youngest individuals (as in a Type III curve) but fairly constant among adults (as in a Type II curve). Some invertebrates, such as crabs, may show a "stair-stepped" curve, with brief periods of increased mortality during molts, followed by periods of lower mortality when the exoskeleton is hard.

In populations without immigration or emigration, survivorship is one of the two key factors determining changes in population size. In such populations, the other key factor determining population trends is reproductive rate.

Reproductive Rates

Demographers who study sexually reproducing species generally ignore males and concentrate on females in a population because only females produce offspring. Demographers view populations in terms of females giving rise to new females; males are important only as distributors of genes. The simplest way to describe the reproductive pattern of a population is to ask how reproductive output varies with the ages of females.

A **reproductive table,** or fertility schedule, is an age-specific summary of the reproductive rates in a population. The best way to construct a fertility schedule is to measure the reproductive output of a cohort from birth until death. For sexual species, the reproductive table tallies the number of female offspring produced by each age group. **Table 52.2** illustrates a reproductive table for Belding's ground squirrels. Reproductive output for sexual species such as birds and mammals is the product of the proportion of females of a given age that are breeding and the number of female offspring of those breeding females. Multiplying these numbers gives the average number of female offspring for each female in a given age class (the last column in Table 52.2). For Belding's ground squirrels, which begin to reproduce at age 1 year, reproductive output rises to a peak at 4 years and then falls off in older females.

Reproductive tables vary greatly, depending on the species. Squirrels have a litter of two to six young once a year for less than a decade, whereas oak trees drop thousands of acorns each year for tens or hundreds of years. Mussels and other invertebrates may release hundreds of thousands of eggs in a

Table 52.2 Reproductive Table for Belding's Ground Squirrels at Tioga Pass

Age (years)	Proportion of Females Weaning a Litter	Mean Size of Litters (Males + Females)	Mean Number of Females in a Litter	Average Number of Female Offspring*
0–1	0.00	0.00	0.00	0.00
1–2	0.65	3.30	1.65	1.07
2–3	0.92	4.05	2.03	1.87
3–4	0.90	4.90	2.45	2.21
4–5	0.95	5.45	2.73	2.59
5–6	1.00	4.15	2.08	2.08
6–7	1.00	3.40	1.70	1.70
7–8	1.00	3.85	1.93	1.93
8–9	1.00	3.85	1.93	1.93
9–10	1.00	3.15	1.58	1.58

*The average number of female offspring is the proportion weaning a litter multiplied by the mean number of females in a litter.

Data from P. W. Sherman and M. L. Morton, "Demography of Belding's Ground Squirrel," *Ecology* 65 (1984): 1617–1628.

spawning cycle. Why a particular type of reproductive pattern evolves in a particular population is one of the many questions at the interface of population ecology and evolutionary biology. This is the subject of life history studies.

Concept Check 52.1

1. One species of forest bird is highly territorial, while a second lives in flocks. What is each species' likely pattern of dispersion? Explain.
2. Each female of a particular fish species produces millions of eggs per year. What is its likely survivorship pattern? Explain.
3. What is the average proportion of females born into the population of Belding's ground squirrels described by Table 52.2?

For suggested answers, see Appendix A.

Concept 52.2

Life history traits are products of natural selection

Natural selection favors traits that improve an organism's chances of survival and reproductive success. In every species, there are trade-offs between survival and traits such as frequency of reproduction, the number of offspring produced (the number of seeds produced by plants and litter or clutch size for animals), and investment in parental care. The traits that affect an organism's schedule of reproduction and survival (from birth through reproduction to death) make up its **life history.** Life histories entail three basic variables: when reproduction begins (the age at first reproduction or age at maturity), how often the organism reproduces, and how many offspring are produced during each reproductive episode.

Keep in mind that, with the important exception of humans (which we will consider later in the chapter), organisms do not choose consciously when to reproduce or how many offspring to have. Life history traits are evolutionary outcomes reflected in the development, physiology, and behavior of an organism.

Life History Diversity

Life histories are very diverse. Pacific salmon, for example, hatch in the headwaters of a stream, then migrate to open ocean, where they require one to four years to mature. The salmon eventually return to freshwater streams to spawn, producing thousands of small eggs in a single reproductive opportunity, and then they die. This "one-shot" pattern of **big-bang reproduction,** or **semelparity** (from the Latin *semel,* once, and *parito,* to beget), also occurs in some plants, such as the agave **(Figure 52.6)**. Agaves, or century plants, generally grow in arid climates with sparse and unpredictable rainfall. Agaves grow for several years, then send up a large flowering stalk, produce seeds, and die. The shallow roots of agaves catch water after rain showers but are dry during droughts. This irregular water supply may prevent seed production or seedling establishment for several years at a time. By growing and storing nutrients until an unusually wet year and then putting all its resources into reproduction, the agave's life history is an adaptation to erratic climate.

▲ **Figure 52.6 An agave, or century plant, an example of big-bang reproduction.**

In contrast to semelparity is **iteroparity** (from the Latin *itero,* to repeat), or **repeated reproduction.** For example, some lizards produce a few large eggs during their second year of life, then repeat the reproductive act annually for several years.

What factors contribute to the evolution of semelparity versus iteroparity? In other words, how much does an individual gain in reproductive success through one pattern versus the other? The critical factor is survival rate of the offspring. Where the survival rate of offspring is low, as in highly variable or unpredictable environments, big-bang reproduction (semelparity) is favored. Production of large numbers of offspring in such environments increases the probability that at least some will survive. Repeated reproduction (iteroparity) will be favored in more dependable environments where competition for resources may be intense. In such environments, a few relatively large, well-provisioned offspring will have a better chance of surviving to reproductive age.

"Trade-offs" and Life Histories

We might imagine an organism that could produce as many offspring as a semelparous species, provision them well like an iteroparous species, and do so repeatedly, but such organisms do not exist. Time, energy, and nutrients that are used for one thing cannot be used for something else. As a consequence, natural selection cannot maximize all these reproductive variables simultaneously. Organisms have finite resources, and limited resources mean trade-offs. In the broadest sense, there is a trade-off between reproduction and survival, as demonstrated by studies of a number of organisms. For example, a study of red deer in Scotland showed that females that reproduced in one summer suffered higher mortality over the next winter than females that did not reproduce. And a study of European kestrels demonstrated the survival cost to parents of caring for young **(Figure 52.7)**.

Figure 52.7

Inquiry How does caring for offspring affect parental survival in kestrels?

EXPERIMENT Researchers in the Netherlands studied the effects of parental caregiving in European kestrels over 5 years. The researchers transferred chicks among nests to produce reduced broods (three or four chicks), normal broods (five or six), and enlarged broods (seven or eight). They then measured the percentage of male and female parent birds that survived the following winter. (Both males and females provide care for chicks.)

RESULTS

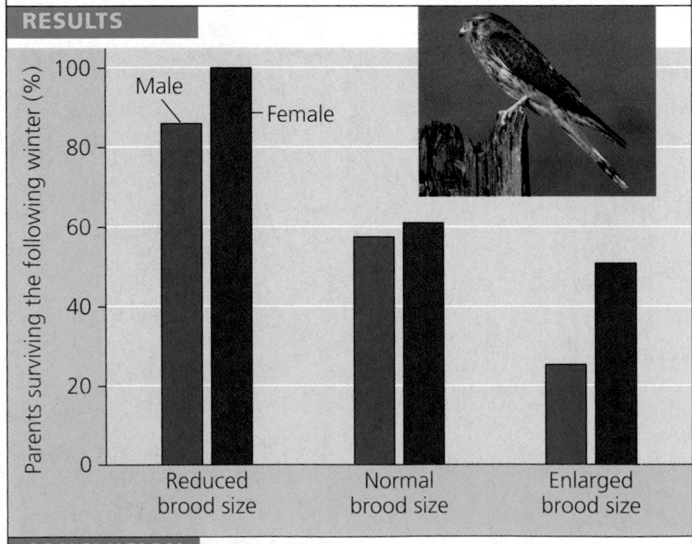

CONCLUSION The lower survival rates of kestrels with larger broods indicate that caring for more offspring negatively affects survival of the parents.

Selective pressures also influence the trade-off between the number and size of offspring. Plants and animals whose young are subject to high mortality rates often produce large numbers of relatively small offspring. Plants that colonize disturbed environments, for example, usually produce many small seeds, only a few of which may reach a suitable habitat. Small size may also increase the chance of seedling establishment by enabling the seeds to be carried longer distances to a broader range of habitats **(Figure 52.8a)**. Animals that suffer high predation rates, such as quail, sardines, and mice, also tend to produce large numbers of offspring.

In other organisms, extra investment on the part of the parent greatly increases the offspring's chances of survival. Oak,

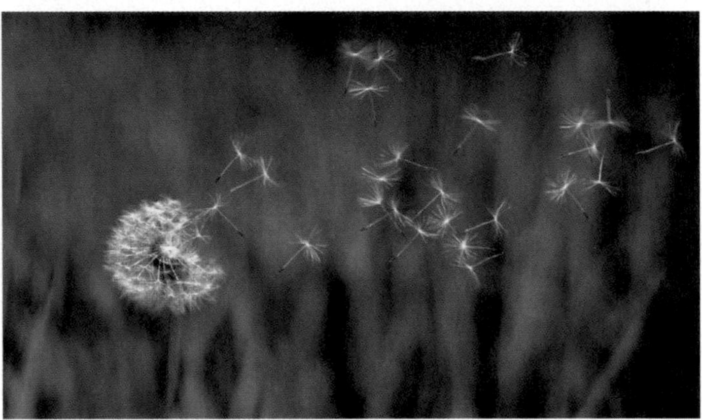

(a) Most weedy plants, such as this dandelion, grow quickly and produce a large number of seeds, ensuring that at least some will grow into plants and eventually produce seeds themselves.

(b) Some plants, such as this coconut palm, produce a moderate number of very large seeds. The large endosperm provides nutrients for the embryo, an adaptation that helps ensure the success of a relatively large fraction of offspring.

▲ **Figure 52.8 Variation in seed crop size in plants.**

walnut, and coconut trees all have large seeds with a large store of energy and nutrients that help the seedlings become established **(Figure 52.8b)**. In animals, parental investment in offspring does not always end after incubation or gestation. For instance, primates generally bear only one or two offspring at a time. Parental care and an extended period of learning in the first several years of life are very important to offspring fitness in these species.

Concept Check 52.2

1. Consider two rivers: One is spring fed and is constant in water volume and temperature year-round; the other drains a desert landscape and floods and dries out at unpredictable intervals. Which is more likely to support many species of iteroparous animals? Why?

For suggested answers, see Appendix A.

The exponential model describes population growth in an idealized, unlimited environment

The concepts of life history provide a biological foundation for a quantitative understanding of population growth. To appreciate the potential for population increase, consider a single bacterium that can reproduce by fission every 20 minutes under ideal laboratory conditions. There would be 2 bacteria after 20 minutes, 4 after 40 minutes, 8 after 60 minutes, and so on. If reproduction continued at this rate, with no mortality, for only a day and a half—a mere 36 hours—there would be enough bacteria to form a layer a foot deep over the entire globe. At the other life history extreme, an elephant may produce only 6 offspring in a 100-year life span. Still, Charles Darwin once estimated that the descendants of a single pair of mating elephants would number 19 million within only 750 years. Though Darwin's estimate may not have been precisely correct, such analyses led him to recognize the tremendous capacity for growth in all populations. However, unlimited population increase does not occur indefinitely for any species, either in the laboratory or in nature. A population that begins at a small size in a favorable environment may increase rapidly for a while, but eventually, as a result of limited resources and other factors, its numbers must stop growing. Nevertheless, it is useful to study population growth in an idealized, unlimited environment because such studies reveal the capacity of species for increase and the conditions in which that capacity may be expressed.

Per Capita Rate of Increase

Imagine a hypothetical population consisting of a few individuals living in an ideal, unlimited environment. Under these conditions, there are no restrictions on the abilities of individuals to harvest energy, grow, and reproduce, aside from the inherent biological limitations of their life history traits. The population will increase in size with every birth and with the immigration of individuals from other populations, and it will decrease in size with every death and with the emigration of individuals out of the population. For simplicity here, we will ignore the effects of immigration and emigration (a more complex formulation would certainly include these factors). We can thus define a change in population size during a fixed time interval with the following verbal equation:

$$\frac{\text{Change in population}}{\text{size during time interval}} = \frac{\text{Births during}}{\text{time interval}} - \frac{\text{Deaths during}}{\text{time interval}}$$

We can use mathematical notation to express this relationship more concisely. If N represents population size and t represents time, then ΔN is the change in population size and Δt is the time interval (appropriate to the life span or generation time of the species) over which we are evaluating population growth. (The Greek letter delta, Δ, indicates change, such as change in time.) We can now rewrite the verbal equation as

$$\frac{\Delta N}{\Delta t} = B - D$$

where B is the number of births in the population during the time interval and D is the number of deaths.

Next, we can convert this simple model into one in which births and deaths are expressed as the average number of births and deaths per individual during the specified time interval. The *per capita birth rate* is the number of offspring produced per unit time by an average member of the population. If, for example, there are 34 births per year in a population of 1,000 individuals, the annual per capita birth rate is $^{34}/_{1000}$, or 0.034. If we know the annual per capita birth rate (expressed as b), we can use the formula $B = bN$ to calculate the expected number of births per year in a population of any size. For example, if the annual per capita birth rate is 0.034 and the population size is 500,

$$B = bN$$
$$B = 0.034 \times 500$$
$$B = 17 \text{ per year}$$

Similarly, the *per capita death rate* (symbolized as m, for mortality) allows us to calculate the expected number of deaths per unit time in a population of any size. If $m = 0.016$ per year, we would expect 16 deaths per year in a population of 1,000 individuals. For natural populations or those in the laboratory, the per capita birth and death rates can be calculated from estimates of population size and data in life tables and reproductive tables (for example, Tables 52.1 and 52.2).

Now we can revise the population growth equation again, this time using per capita birth and death rates rather than the numbers of births and deaths:

$$\frac{\Delta N}{\Delta t} = bN - mN$$

One final simplification is in order. Population ecologists are most interested in the difference between the per capita birth rate and per capita death rate. This difference is the *per capita rate of increase,* or r:

$$r = b - m$$

The value of r indicates whether a given population is growing ($r > 0$) or declining ($r < 0$). **Zero population growth (ZPG)** occurs when the per capita birth and death rates are equal ($r = 0$). Births and deaths still occur in such a population, of course, but they balance each other exactly.

Using the per capita rate of increase, we now rewrite the equation for change in population size as

$$\frac{\Delta N}{\Delta t} = rN$$

Most ecologists use differential calculus to express population growth as growth rate at a particular instant in time:

$$\frac{dN}{dt} = rN$$

If you have not yet studied calculus, don't be intimidated by the form of the last equation; it is essentially the same as the previous one, except that the time intervals Δt are very short and are expressed in the equation as dt.

Exponential Growth

Earlier we described a population whose members all have access to abundant food and are free to reproduce at their physiological capacity. Population increase under these ideal conditions is called **exponential population growth,** also known as geometric population growth. Under these conditions, the per capita rate of increase may assume the maximum rate for the species, called the *intrinsic rate of increase* and denoted as r_{max}. The equation for exponential population growth is

$$\frac{dN}{dt} = r_{max}N$$

The size of a population that is growing exponentially increases at a constant rate, resulting eventually in a J-shaped growth curve when population size is plotted over time (**Figure 52.9**). Although the intrinsic *rate* of increase is constant, the population accumulates more new individuals per unit of time when it is large than when it is small; thus, the curves in Figure 52.9 get progressively steeper over time. This occurs because population growth depends on N as well as r_{max}, and larger populations experience more births (and deaths) than small ones growing at the same per capita rate. It is also clear from Figure 52.9 that a population with a higher intrinsic rate of increase ($dN/dt = 1.0N$) will grow faster than one with a lower rate of increase ($dN/dt = 0.5N$).

The J-shaped curve of exponential growth is characteristic of some populations that are introduced into a new or unfilled environment or whose numbers have been drastically reduced by a catastrophic event and are rebounding. For example, **Figure 52.10** illustrates the exponential population growth that occurred in the population of elephants in Kruger National Park, South Africa, after they were protected from hunting. After approximately 60 years of exponential growth, the large number of elephants had caused enough damage to the park vegetation

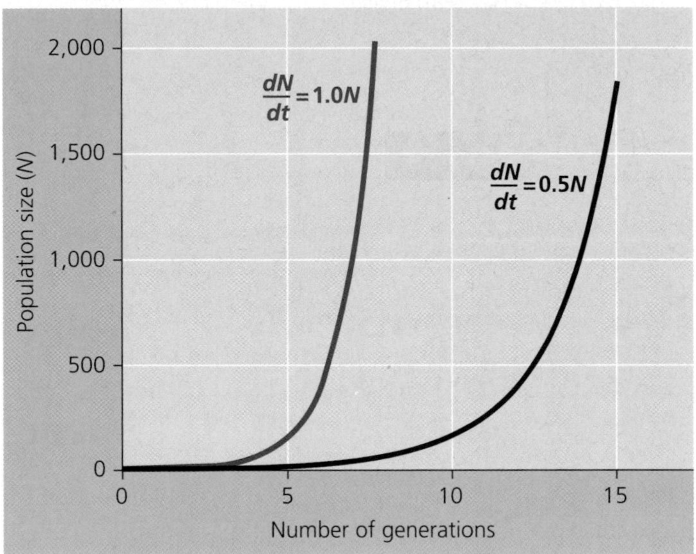

▲ **Figure 52.9 Population growth predicted by the exponential model.** This graph compares growth in two populations with different values of r_{max}. Increasing the value from 0.5 to 1.0 increases the rate of rise in population size with time, as reflected by the relative slopes of the curves.

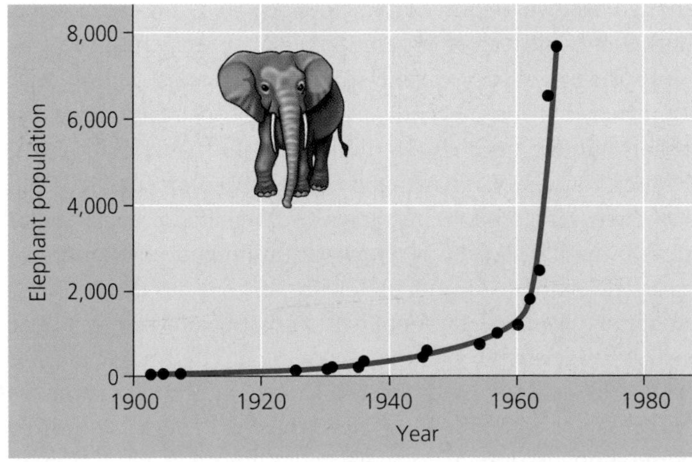

▲ **Figure 52.10 Exponential growth in the African elephant population of Kruger National Park, South Africa.**

that a collapse in the elephant food supply was likely, leading to an end to population growth through starvation. To protect other species and the park ecosystem before that happened, park managers began limiting the elephant population by using birth control and exporting elephants to other countries.

Concept Check 52.3

1. Explain why a constant rate of increase (r_{max}) for a population produces a growth graph that is J-shaped rather than a straight line.
2. Where is exponential growth by a plant population more likely—on a newly formed volcanic island or in a mature, undisturbed rain forest? Why?

For suggested answers, see Appendix A.

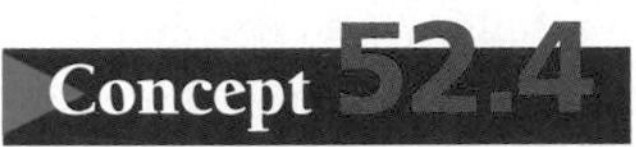

The logistic growth model includes the concept of carrying capacity

The exponential growth model assumes resources are unlimited, never the case in the real world. As population density increases, each individual has access to fewer resources. Ultimately, there is a limit to the number of individuals that can occupy a habitat. Ecologists define **carrying capacity**, symbolized as K, as the maximum population size that a particular environment can support. Carrying capacity is not fixed, but varies over space and time with the abundance of limiting resources. Energy, shelters, refuges from predators, soil nutrients, water, and suitable nesting sites can all be limiting factors. For example, the carrying capacity for bats may be high in a habitat with abundant flying insects and roosting sites but lower where there is abundant food but fewer suitable shelters.

Crowding and resource limitation can have a profound effect on population growth rate. If individuals cannot obtain sufficient resources to reproduce, the per capita birth rate (b) will decline. If they cannot find and consume enough energy to maintain themselves, the per capita death rate (m) may increase. A decrease in b or an increase in m results in a lower per capita rate of increase, r.

The Logistic Growth Model

We can modify our mathematical model to incorporate changes in growth rate as the population size nears the carrying capacity. In the **logistic population growth** model, the per capita rate of increase declines as carrying capacity is reached.

To construct the logistic model, we start with the exponential population growth model and add an expression that reduces the per capita rate of increase as N increases **(Figure 52.11)**. If the maximum sustainable population size (carrying capacity) is K, then $K - N$ is the number of additional individuals the environment can accommodate, and $(K-N)/K$ is the fraction of K that is still available for population growth. By multiplying the exponential rate of increase $r_{max}N$ by $(K - N)/K$, we modify the growth rate of the population as N increases:

$$\frac{dN}{dt} = r_{max}N\frac{(K-N)}{K}$$

Table 52.3 shows calculations of population growth rate at various population sizes for a hypothetical population growing

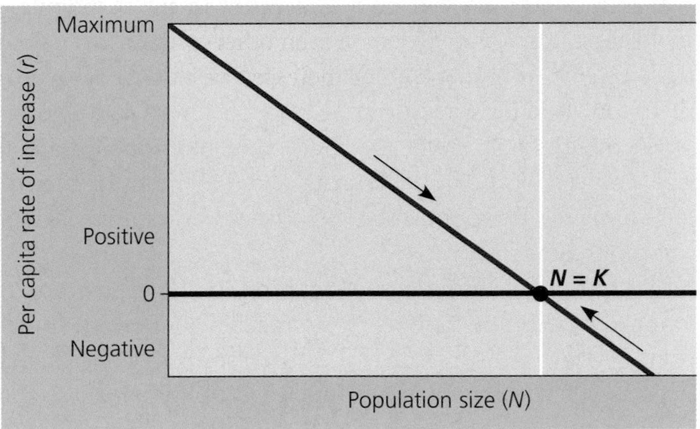

▲ **Figure 52.11 Influence of population size (*N*) on per capita rate of increase (*r*).** The logistic model assumes that the per capita rate of increase decreases as N increases. If N is greater than K, then the population growth rate is negative, and population size decreases. An equilibrium is reached at the white line when $N = K$.

Table 52.3 A Hypothetical Example of Logistic Population Growth, Where K = 1,000 and r_{max} = 0.05 per Individual per Year

Population Size: N	Intrinsic Rate of Increase: r_{max}	$\left(\frac{K-N}{K}\right)$	Per Capita Growth Rate: $r_{max}\left(\frac{K-N}{K}\right)$	Population Growth Rate:* $r_{max}N\left(\frac{K-N}{K}\right)$
20	0.05	0.98	0.049	+1
100	0.05	0.90	0.045	+5
250	0.05	0.75	0.038	+9
500	0.05	0.50	0.025	+13
750	0.05	0.25	0.013	+9
1,000	0.05	0.00	0.000	0

*Rounded to the nearest whole number.

according to the logistic model. When N is small compared to K, the term $(K - N)/K$ is large, and the per capita rate of increase $r_{max}(K - N)/K$, is close to the intrinsic (maximum) rate of increase. But when N is large and resources are limiting, then $(K - N)/K$ is small, and so is the per capita rate of increase. When N equals K, the population stops growing. Notice in Table 52.3 that the overall population growth rate is highest, +13, over a year when the population size is 500, or half the carrying capacity. Why is population growth rate greatest at 500 and not at smaller population sizes? This is due to the balance between per capita rate of increase and population size. At a population size of 500, the per capita rate of increase remains relatively high (one-half the maximum rate), and there are many more reproducing individuals in the population than at lower population sizes.

The logistic model of population growth produces a sigmoid (S-shaped) growth curve when N is plotted over time (the red line in **Figure 52.12**). New individuals are added to the population most rapidly at intermediate population sizes, when there is not only a breeding population of substantial size, but also lots of available space and other resources in the environment. The population growth rate slows dramatically as N approaches K.

Notice that we haven't said anything about *why* the population growth rate slows as N approaches K. For a population's growth rate to decrease, either the birth rate b must decrease, the death rate m must increase, or both. Later in the chapter, we will consider some of the factors affecting these rates.

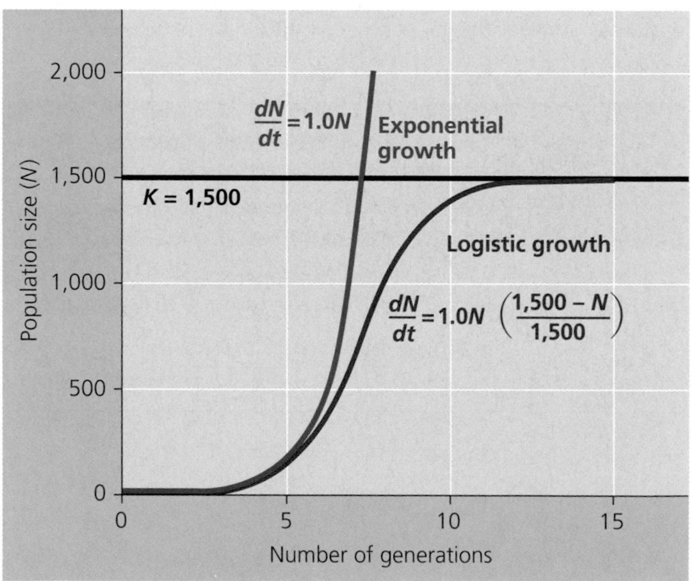

▲ **Figure 52.12 Population growth predicted by the logistic model.** The rate of population growth slows as population size (N) approaches the carrying capacity (K) of the environment. The red line shows logistic growth in a population where $r_{max} = 1.0$ and $K = 1,500$ individuals. For comparison, the blue line illustrates a population continuing to grow exponentially with the same r_{max}.

The Logistic Model and Real Populations

The growth of laboratory populations of some small animals, such as beetles and crustaceans, and of some microorganisms, such as paramecia, yeasts, and bacteria, fits an S-shaped curve fairly well **(Figure 52.13a)**. These populations are grown in a constant environment lacking predators and other species that may compete for resources, conditions that rarely occur in nature.

Some of the basic assumptions built into the logistic model clearly do not apply to all populations. For example, the logistic model assumes that populations adjust instantaneously to growth and approach carrying capacity smoothly. In most natural populations, however, there is a lag time before the negative effects of an increasing population are realized. If food becomes limiting for a population, for instance, reproduction will be reduced. However, the birth rate may not decline immediately, since females may use their energy reserves to continue reproducing for a short time. This may cause the population to overshoot its carrying capacity before settling down to a relatively stable density. **Figure 52.13b** illustrates this overshoot for a laboratory population of water fleas *(Daphnia)*. If the population then drops below carrying capacity, there will be a delay in population growth until the increasing number of offspring are actually born. And still other populations fluctuate greatly, making it difficult even to define carrying capacity. For example, **Figure 52.13c** shows marked population changes in the song sparrow on a small island in southern British Columbia. The population increases rapidly but suffers periodic catastrophes in winter; consequently, there is no stable population size. We will examine some possible reasons for these fluctuations later in the chapter.

The logistic model also incorporates the idea that regardless of population density, each individual added to a population has the same negative effect on population growth rate. However, some populations show an *Allee effect* (named after W. C. Allee, of the University of Chicago, who first described it), in which individuals may have a more difficult time surviving or reproducing if the population size is too small. For example, a single plant may be damaged by excessive wind if it is standing alone, but it would be protected in a clump of individuals. Conservation biologists fear that if populations of certain solitary animals, such as rhinoceroses, drop below a critical size, individuals will not be able to locate mates in the breeding season.

Although the logistic model fits few, if any, real populations closely, it is a useful starting point for thinking about how populations grow and for constructing more complex models. The model is also useful in conservation biology for estimating how rapidly a particular population might increase in numbers after it has been reduced to a small size, or for estimating sustainable harvest rates for fish and wildlife populations. And like any good starting hypothesis, the logistic model has stimulated research leading to a better understanding of the factors affecting population growth.

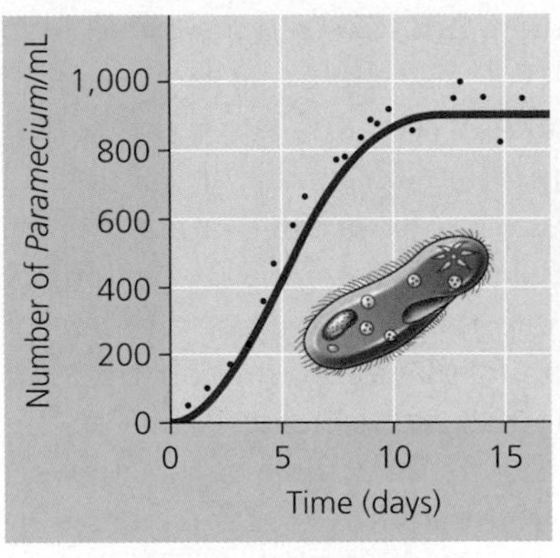

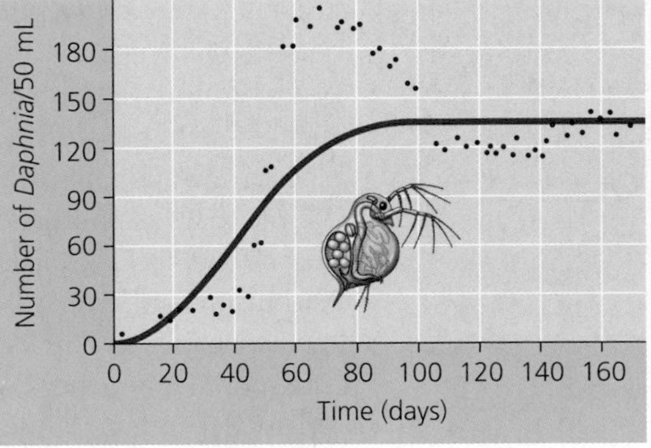

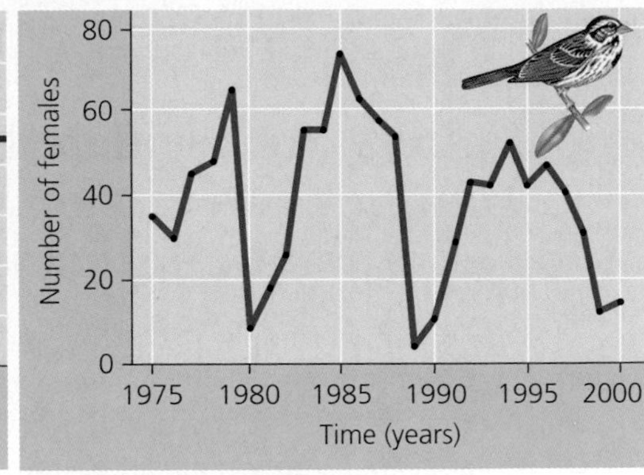

(a) A *Paramecium* population in the lab. The growth of *Paramecium aurelia* in small cultures (black dots) closely approximates logistic growth (red curve) if the experimenter maintains a constant environment.

(b) A *Daphnia* population in the lab. The growth of a population of *Daphnia* in a small laboratory culture (black dots) does not correspond well to the logistic model (red curve). This population overshoots the carrying capacity of its artificial environment and then settles down to an approximately stable population size.

(c) A song sparrow population in its natural habitat. The population of female song sparrows nesting on Mandarte Island, British Columbia, is periodically reduced by severe winter weather, and population growth is not well described by the logistic model.

▲ **Figure 52.13 How well do these populations fit the logistic growth model?**

The Logistic Model and Life Histories

The logistic model predicts different per capita growth rates for populations of low or high density relative to the carrying capacity of the environment. At high densities, each individual has few resources available, and the population grows slowly, if at all. At low densities, the opposite is true: Per capita resources are relatively abundant, and the population can grow rapidly. Different life history features are favored under each condition. At high population density, selection favors adaptations that enable organisms to survive and reproduce with few resources. Thus, competitive ability and efficient use of resources should be favored in populations that are at or near their carrying capacity. These are the traits we associated earlier with iteroparity. At low population density, on the other hand, even in the same species, adaptations that promote rapid reproduction, such as the production of numerous, small offspring, should be favored.

Ecologists have attempted to connect these differences in favored traits at different population densities with the logistic growth model. Selection for life history traits that are sensitive to population density is known as ***K*-selection,** or density-dependent selection. In contrast, selection for life history traits that maximize reproductive success in uncrowded environments (low densities) is called ***r*-selection,** or density-independent selection. These names follow from the variables of the logistic equation. *K*-selection tends to maximize population size and operates in populations living at a density near the limit imposed by their resources (the carrying capacity, *K*). By contrast, *r*-selection tends to maximize *r*, the rate of increase, and occurs in environments in which population densities fluctuate well below carrying capacity or individuals are likely to face little competition.

Laboratory experiments have shown that different populations of the same species may show a different balance of *K*-selected and *r*-selected traits, depending on conditions. For example, cultures of the fruit fly *Drosophila melanogaster* raised for 200 generations under crowded conditions with minimal food are more productive at high density than other populations raised for many generations in uncrowded conditions with abundant food. Apparently, when given equal access to food, larvae from cultures selected for living in crowded conditions feed faster than larvae selected for living in uncrowded cultures. The fruit fly genotypes that are most fit at low density do not have high fitness at high density, as predicted by *r*- and *K*-selection theory.

The concepts of *r*- and *K*-selection have been criticized as an oversimplification of the variation seen in the natural histories of species. The characteristics of most species place them somewhere in between the extremes represented by *r*- and *K*-selection. The critical assessment of *r*- and *K*-selection has led ecologists to propose alternative theories of life history evolution. These alternative theories, in turn, have stimulated more thorough study of how factors such as disturbance, stress, and the frequency of opportunities for successful reproduction affect the evolution of life histories.

Concept Check 52.4

1. Explain why a population that fits the logistic growth model increases more rapidly at intermediate size than at relatively small or large sizes.

For suggested answers, see Appendix A.

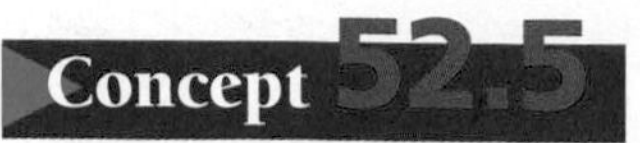

Populations are regulated by a complex interaction of biotic and abiotic influences

In this section, we will apply the theme of *regulation* (see Chapter 1) to populations. There are two general questions we can ask about regulation of population growth. First, what environmental factors stop a population from growing? Second, why do some populations show radical fluctuations in size over time, while others remain more stable?

These questions have many practical applications. In agriculture, the objective may be to reduce the size of a pest population. If an introduced weed is spreading rapidly, what can be done to stop its growth? Why do agricultural pests have severe effects in some areas and negligible effects in others? Conversely, what environmental factors would create a favorable feeding or breeding habitat for an endangered species, such as the humpback whale or whooping crane? Such questions, which involve population-regulating factors, are at the core of management programs that have helped prevent the extinction of certain endangered species.

Population Change and Population Density

The first step in understanding why a population stops growing is to study how the rates of birth, death, immigration, and emigration change as population density rises. If immigration and emigration offset each other, then a population grows when the birth rate exceeds the death rate and declines when the death rate exceeds the birth rate.

A birth rate or death rate that does *not* change with population density is said to be **density independent.** In a classic study of population regulation, Andrew Watkinson and John Harper, plant population ecologists from the University of Wales, found that the mortality of dune fescue grass is mainly due to physical factors that kill similar proportions of a local population, regardless of its density. In contrast, a death rate that rises as population density rises is said to be **density dependent,** as is a birth rate that falls with rising density. Watkinson and Harper found that reproduction by dune fescue declines as population density increases. Thus, in this grass population, the key factors regulating birth rate were density dependent, while death rate was largely regulated by density-independent factors. **Figure 52.14** models how a population may stop increasing and reach equilibrium due to various combinations of density-dependent and density-independent regulation.

Density-Dependent Population Regulation

Density-dependent birth and death rates are an example of negative feedback, a type of regulation described in Chapter 1. Without some type of negative feedback between population density and the vital rates of birth and death, a population would not stop growing. However, at increased densities, birth rates decline and/or death rates increase, providing the negative feedback that halts continued population growth. Once we know how birth and death rates change with population density, we need to determine the mechanisms causing these changes, which may involve many factors.

Competition for Resources

In crowded populations, increasing population density intensifies intraspecific competition for declining nutrients and other resources, resulting in a lower birth rate. For example, crowding

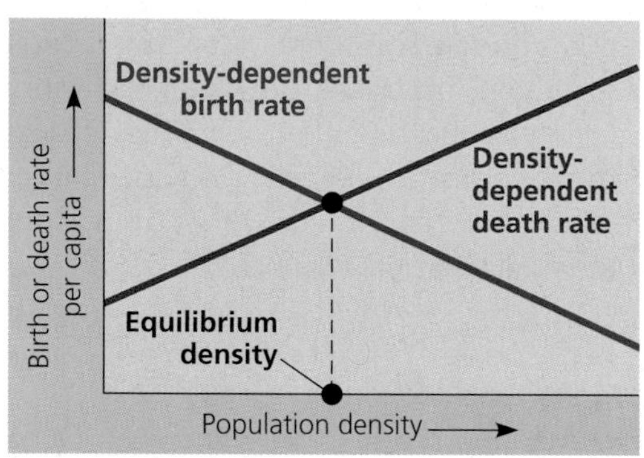

(a) Both birth rate and death rate change with population density.

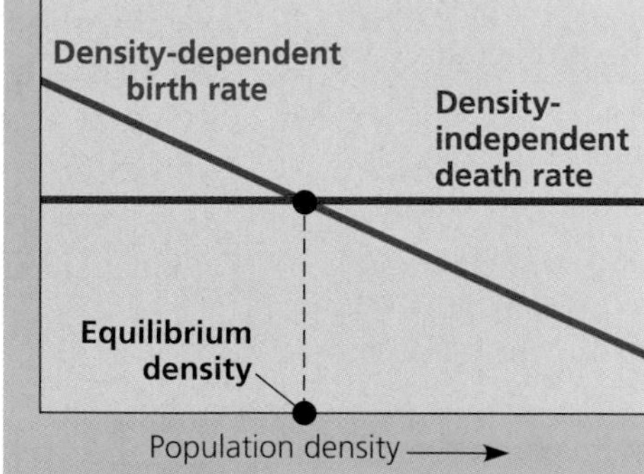

(b) Birth rate changes with population density while death rate is constant.

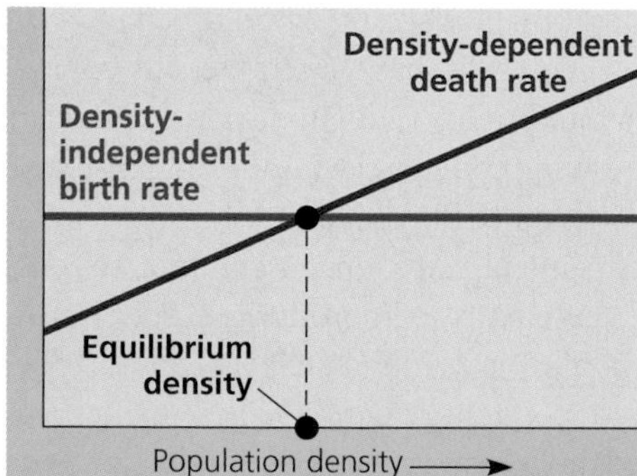

(c) Death rate changes with population density while birth rate is constant.

▲ **Figure 52.14 Determining equilibrium for population density.** This simple model considers only birth and death rates (immigration and emigration rates are assumed to be either zero or equal).

can reduce seed production by plants **(Figure 52.15a)**. And available food supplies often limit the reproductive output of songbirds; as bird population density increases in a particular habitat, each female lays fewer eggs, a density-dependent response **(Figure 52.15b)**. In an experiment in which female song sparrows living in high-density conditions were given extra food, they did not suffer a reduction in clutch size.

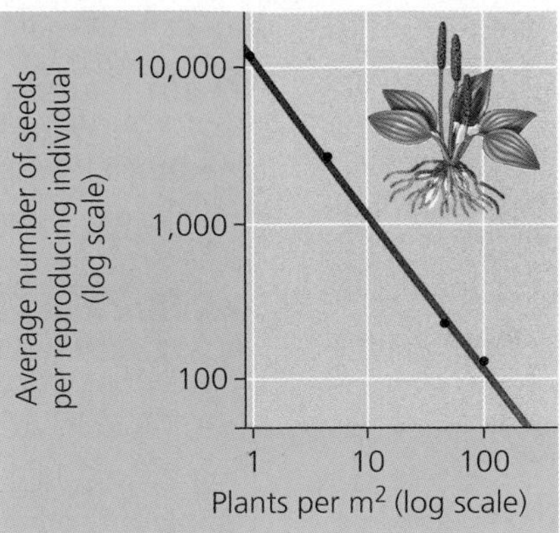

(a) Plantain. The number of seeds produced by plantain (*Plantago major*) decreases as density increases.

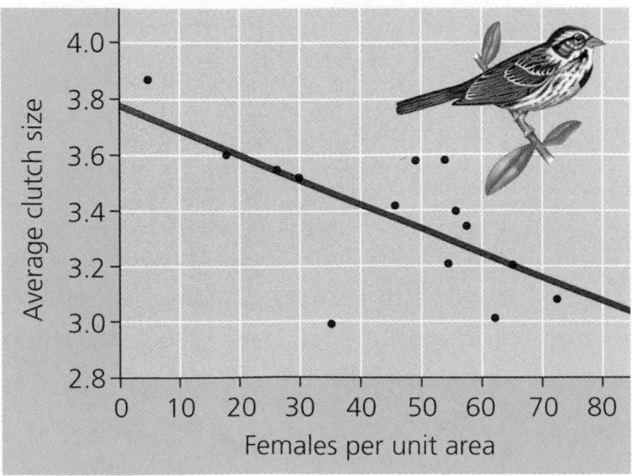

(b) Song sparrow. Clutch size in the song sparrow on Mandarte Island, British Columbia, decreases as density increases and food is in short supply.

▲ **Figure 52.15 Decreased reproduction at high population densities.**

Territoriality

In many vertebrates and some invertebrates, territoriality may limit density. In this case, territory space becomes the resource for which individuals compete. Cheetahs, for example, are highly territorial, using chemical communication to warn other cheetahs of their territorial boundaries **(Figure 52.16)**. Oceanic birds, such as gannets, often nest on rocky islands, where they are relatively safe from predators **(Figure 52.17)**. Up to a certain population density, most birds can find a suitable nest site, but beyond that threshold, few birds breed successfully. Thus, the limiting resource that determines breeding population density for gannets is safe nesting territory. Birds that cannot obtain a nesting spot do not reproduce. The presence of surplus, or nonbreeding, individuals is a good indication that territoriality is restricting population growth, as it does in many bird populations.

▲ **Figure 52.16 Cheetah staking out a territory with a chemical marker.**

Health

Population density can also influence the health and thus the survival of organisms. If the transmission rate of a disease depends on a certain level of crowding in a population, the disease's impact may be density dependent. Field experiments conducted by Charles Mitchell, David Tilman, and James Groth, of the University of Minnesota, demonstrated that the severity of infection of plants by fungal pathogens is greater where the density of the host plant population is higher. Animals, too, can experience increased infection by pathogens at high population densities. Steven Kohler and Wade Hoiland of the Illinois Natural History Survey showed that in case-building caddis flies (stream-dwelling insects) peaks in disease-related mortality followed years of high insect abundance. Their study indicated that such disease-related mortality was largely responsible for cyclic fluctuations in the density of the insect population. Human pathogens can also show density-dependent infection rates. For example, tuberculosis, which is caused by bacteria that spread through the air when an infected person sneezes or coughs, strikes a greater percentage of people living in high-density cities than those in rural areas.

▲ **Figure 52.17 Territories.** Gannets nest virtually a peck apart and defend their territories by calling and pecking at one another.

Predation

Predation may be an important cause of density-dependent mortality for some prey populations if a predator encounters and captures more food as the population density of the prey increases. As a prey population builds up, predators may feed preferentially on that species, consuming a higher percentage of individuals. For example, trout may concentrate for a few days on a particular species of insect that is emerging from its aquatic larval stage, then switch prey as another insect species becomes more abundant.

Toxic Wastes

The accumulation of toxic wastes can contribute to density-dependent regulation of population size. In laboratory cultures of small microorganisms, for example, metabolic by-products accumulate as the populations grow, poisoning the organisms within this limited, artificial environment. For example, ethanol accumulates as a by-product of yeast fermentation. The alcohol content of wine is usually less than 13% because that is the maximum concentration of ethanol that most wine-producing yeast cells can tolerate.

Intrinsic Factors

For some animal species, intrinsic (physiological) factors, rather than the extrinsic (environmental) factors we've just discussed, appear to regulate population size. White-footed mice in a small field enclosure will multiply from a few to a colony of 30 to 40 individuals, but eventually reproduction will decline until the population ceases to grow. This drop in reproduction is associated with aggressive interactions that increase with population density, and it occurs even when food and shelter are provided in abundance. Although the exact mechanisms by which aggressive behavior affects reproductive rate are not yet understood, researchers have learned that high population densities in mice induce a stress syndrome in which hormonal changes can delay sexual maturation, cause reproductive organs to shrink, and depress the immune system. In this case, high densities cause an increase in mortality and a decrease in birth rates. Similar effects of crowding occur in some other wild rodent populations.

These various examples of population regulation by negative feedback show how increased densities cause population growth rates to decline by affecting reproduction, growth, and survivorship. This helps answer our first question about populations: What causes a population to stop increasing? Now let's turn to our second question: Why do some populations fluctuate dramatically over time while others remain more stable?

Population Dynamics

While some populations appear to be more stable in size than others, over the long term, all populations for which we have data show some fluctuation in numbers. Although ecologists can determine an average population size for many species, the average is often of less interest than the variation in numbers from year to year or place to place. The study of **population dynamics** focuses on the complex interactions between biotic and abiotic factors that cause variation in population size.

Stability and Fluctuation

Populations of large mammals, such as deer and moose, were once thought to remain relatively stable over time. But some long-term studies have challenged that hypothesis. The moose population on Isle Royale, an island in Lake Superior, is a striking example. Moose from the mainland colonized the island around 1900 by walking across the frozen lake. However, the lake has not frozen over in recent years, so the moose population has been isolated from immigration and emigration. Nevertheless, as **Figure 52.18** shows, the population has been anything but stable, with two major increases and collapses over the last 40 years. In general, the severity of winter loss in large grazers and browsers inhabiting temperate and polar regions is proportional to the harshness of the winter. Colder temperatures increase energy requirements (and therefore the need for food), while deeper snow makes it harder to find food.

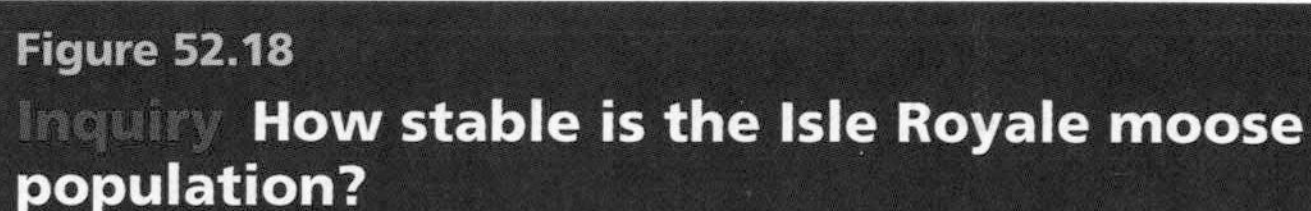

Figure 52.18
Inquiry **How stable is the Isle Royale moose population?**

FIELD STUDY Researchers regularly surveyed the population of moose on Isle Royale, Michigan, from 1960 to 2003. During that time, the lake never froze over, and so the moose population was isolated from the effects of immigration and emigration.

RESULTS Over 43 years, this population experienced two significant increases and collapses, as well as several less severe fluctuations in size.

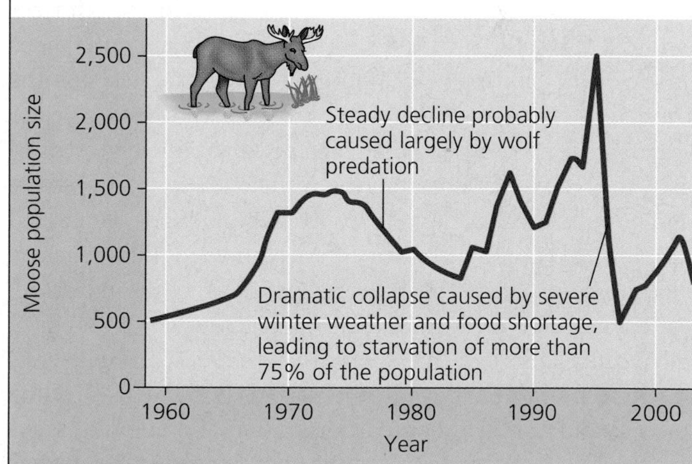
Moose population size
2,500
2,000
1,500
1,000
500
0
1960
1970
1980
1990
2000
Year
Steady decline probably caused largely by wolf predation
Dramatic collapse caused by severe winter weather and food shortage, leading to starvation of more than 75% of the population

CONCLUSION The pattern of population dynamics observed in this isolated population indicates that various biotic and abiotic factors can result in dramatic fluctuations over time in a moose population.

Although large mammal populations are more dynamic than once thought, they show much more stability than other populations. The Dungeness crab is a classic example of an erratically fluctuating population. As shown in **Figure 52.19**, the crab population at Fort Bragg on the northern California coast varied between 10,000 and hundreds of thousands over a 40-year period. By contrast, over this same period, the moose population of Isle Royale varied only between 500 and 2,500 (see Figure 52.18).

One key factor involved in the large fluctuations in the Dungeness crab population is cannibalism. Females release up to 2 million eggs each fall, and large numbers of juvenile crabs are eaten by older juveniles and by adult crabs. In addition, successful settlement of larval crabs occurs only in shallow waters and depends on ocean currents and water temperature. If winds and currents move larval crabs too far offshore, they cannot reach the ocean bottom and settle successfully. Small changes in environmental variables seem to be magnified by density-dependent cannibalism, and together these factors explain the marked fluctuations in Dungeness crab populations. These results support the hypothesis that the dynamics of many populations result from a complex interaction of biotic and abiotic factors.

Metapopulations and Immigration

To this point in our discussion of population dynamics, we have focused mainly on the contributions of births and deaths. However, as you have read, immigration and emigration can also influence populations. This is particularly true when a group of populations is linked, forming a **metapopulation.** For example, immigration and emigration link the Belding's ground squirrel population we discussed earlier to other populations of the species, forming a metapopulation.

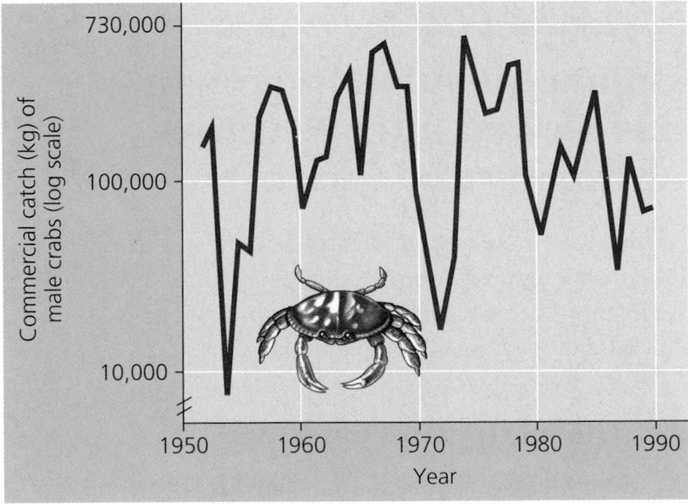

▲ **Figure 52.19 Extreme population fluctuations.** This graph shows the commercial catch of male Dungeness crabs *(Cancer magister)* at Fort Bragg, California. Note the log scale of the *y*-axis.

How might immigration influence population dynamics? A metapopulation of song sparrows on Mandarte Island and a cluster of smaller islands off the coast of British Columbia provides an example. On relatively isolated Mandarte Island, differences between births and deaths are the principal influences maintaining the highly dynamic population of song sparrows (see Figure 52.13c). In 1988–1991, only 11% of the breeding song sparrows were immigrants. By contrast, during the same period, immigrants made up 57% of the breeding song sparrow population on a cluster of small islands near Vancouver Island. This higher level of song sparrow immigration combined with a higher survival rate on small islands resulted in greater stability in song sparrow populations on the small islands than on Mandarte Island **(Figure 52.20)**. The metapopulation concept underscores the significance of immigration and emigration in the contrasting song sparrow populations, and is important for understanding populations in patchy habitats.

Population Cycles

While many populations fluctuate at unpredictable intervals, others undergo regular boom-and-bust cycles, fluctuating in density with a remarkable regularity. For example, some small herbivorous mammals, such as voles and lemmings, tend to have 3- to 4-year cycles, and some birds, such as ruffed grouse and ptarmigans, have 9- to 11-year cycles.

One striking example of population cycles is the 10-year cycles of snowshoe hares and lynx in the far northern forests

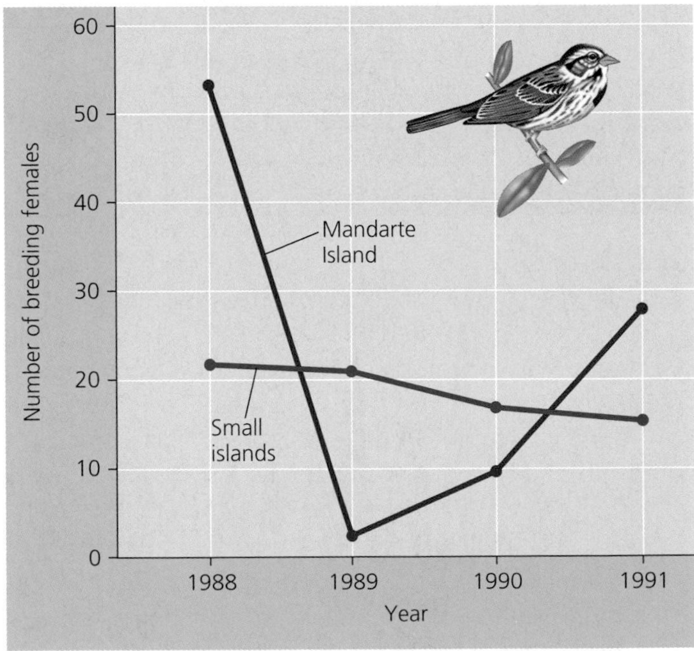

▲ **Figure 52.20 Song sparrow populations and immigration.** Song sparrow populations on a cluster of small islands make up a metapopulation. Immigration keeps the linked populations more stable than the isolated population on larger Mandarte Island.

of Canada and Alaska. Lynx are specialist predators of snowshoe hares, so it is not surprising that lynx numbers rise and fall with the numbers of hares **(Figure 52.21)**. But why do hare numbers rise and fall in 10-year cycles? Three main hypotheses have been proposed. First, the cycles may be caused by food shortage during winter. Hares eat the terminal twigs of small shrubs such as willow and birch in winter and may suffer from malnutrition due to overgrazing. Second, the cycles may be due to predator-prey interactions. Many predators other than lynx eat hares, and they may overexploit their prey. Third, the cycles may be affected by a combination of food resource limitation and excessive predation.

If hare cycles are due to winter food shortage, then they should stop if extra food is added to a field population. Researchers have conducted such experiments in the Yukon for 20 years—over two hare cycles—and report two results. First, hare populations in the areas with extra food increased about threefold in density. The carrying capacity of a habitat for hares can clearly be increased by adding food. Second, the hares with extra food continued to cycle in the same way as the unfed control populations. In particular, cyclic collapses occurred in both experimental and control areas, and the decline in numbers could not be stopped by adding food. Thus, food supplies alone are not the cause of the hare cycle shown in Figure 52.21, so we can discard the first hypothesis.

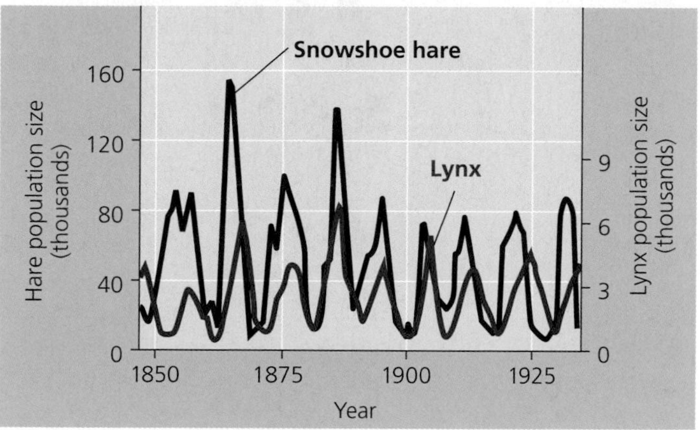

▲ **Figure 52.21 Population cycles in the snowshoe hare and lynx.** Population counts are based on the number of pelts sold by trappers to the Hudson Bay Company.

By putting radio collars on hares, field ecologists can find individual hares as soon as they die and determine the immediate cause of death. Almost 90% of the hares that died were killed by predators; none appeared to have died of starvation. These data support either the second or third hypothesis. Ecologists tested these hypotheses by excluding predators from one area with electric fences and by both excluding predators and adding food to another area. The results supported the hypothesis that the hare cycle is driven largely by excessive predation but that food availability also plays an important role, particularly in the winter. Perhaps better-fed hares are more likely to escape from predators. Many different predators contribute to these losses; the cycle is not simply a hare-lynx cycle.

For the lynx, great-horned owls, weasels, and other predators that depend heavily on a single prey species, the availability of prey is the major factor influencing their population changes. When prey become scarce, predators often turn on one another. Coyotes kill foxes and lynx, and great-horned owls kill smaller birds of prey as well as weasels, accelerating the collapse of the predator populations. Long-term experimental studies will help to unravel the complex causes of such population cycles.

Concept Check 52.5

1. Identify three density-dependent factors that limit population size, and explain how each exerts negative feedback.

For suggested answers, see Appendix A.

Concept 52.6

Human population growth has slowed after centuries of exponential increase

As you have read, no population can grow indefinitely, and humans are no exception. In this last section of the chapter, we'll apply the concepts of population dynamics to the specific case of the human population.

The Global Human Population

The exponential growth model in Figure 52.9 approximates the population explosion of humans since 1650. Ours is a singular case; it is unlikely that any other population of large animals has ever sustained so much growth for so long. The human

population increased relatively slowly until about 1650, at which time approximately 500 million people inhabited Earth **(Figure 52.22)**. The population doubled to 1 billion within the next two centuries, doubled again to 2 billion between 1850 and 1930, and doubled still again by 1975 to more than 4 billion. The global population now numbers over 6 billion people and is increasing by about 73 million each year. The population grows by approximately 201,000 people each day, the equivalent of adding a city the size of Amarillo, Texas, or Madison, Wisconsin. Every week the population increases by the size of San Antonio, Milwaukee, or Indianapolis. It takes only four years for world population growth to add the equivalent of another United States. Population ecologists predict a population of 7.3–8.4 billion people on Earth by the year 2025.

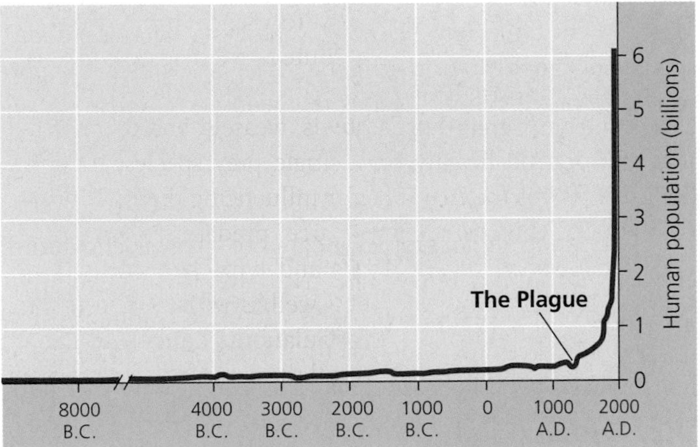

▲ **Figure 52.22 Human population growth (data as of 2003).** The global human population has grown almost continuously throughout history, but it skyrocketed after the Industrial Revolution. Though it is not apparent at this scale, the rate of population growth has slowed in recent decades, mainly as a result of decreased birth rates throughout the world.

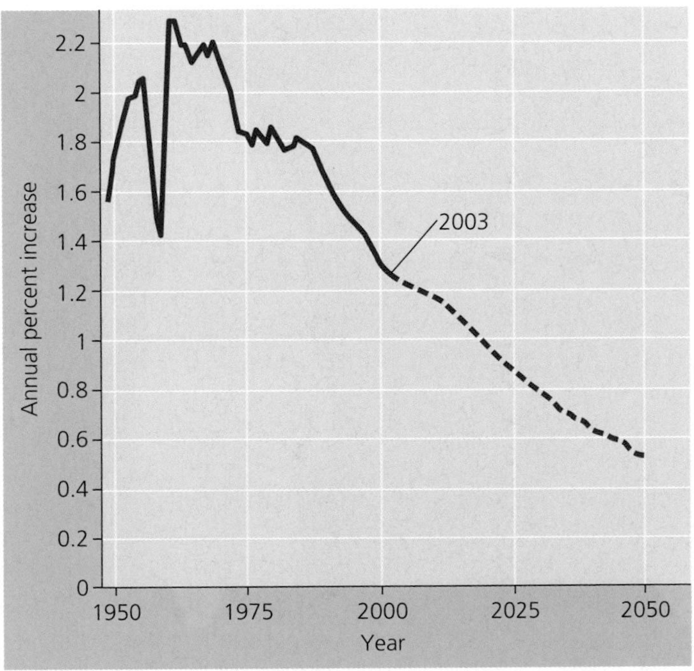

▲ **Figure 52.23 Annual percent increase in the global human population (data as of 2003).** The dashed portion of the curve indicates projected data. The sharp dip in the 1960s is due mainly to a famine in China in which about 60 million people died.

Though the global population is still growing, the *rate* of growth began to slow during the 1960s. **Figure 52.23** presents the percentage increase in the global population from 1950 to 2003 and projected to the year 2050. The annual rate of increase in the global population peaked at 2.19% in 1962; by 2003 it had declined to 1.16%. Current models project a decline in the annual growth rate to just over 0.4% by 2050. This reduction in growth rate shows that the human population has departed from true exponential growth, which assumes a constant rate. These declines are the result of fundamental changes in population dynamics due to diseases, such as AIDS, and voluntary population control.

Regional Patterns of Population Change

So far we have described changes in the global population. But population dynamics vary widely from region to region. In a stable regional human population, birth rate equals death rate (disregarding the effects of immigration and emigration). Two possible configurations for a stable population are

Zero population growth = High birth rate − High death rate

or

Zero population growth = Low birth rate − Low death rate

The movement from the first toward the second state is called the **demographic transition. Figure 52.24** compares the demographic transition in one of the most economically developed countries, Sweden, and in a developing country, Mexico. The demographic transition in Sweden took about 150 years,

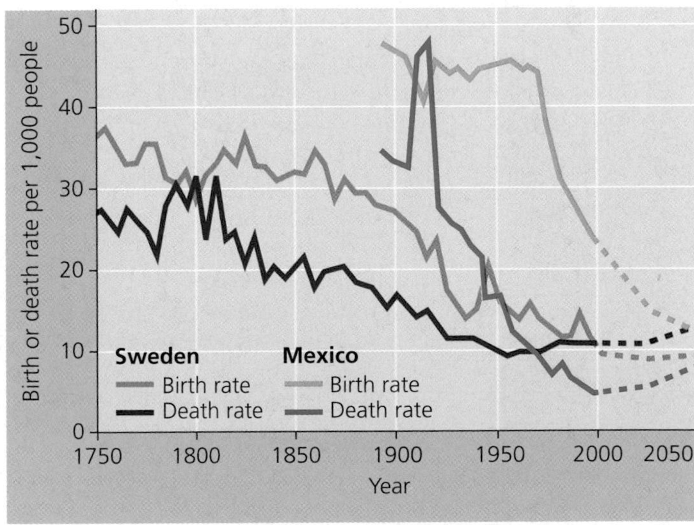

▲ **Figure 52.24 Demographic transition in Sweden and Mexico, 1750–2050 (data as of 2003).**

from 1810 to 1960; in Mexico, the changes are projected to continue until sometime after 2050, almost the same length of time. Demographic transition is associated with an increase in the quality of health care and sanitation as well as improved access to education, especially for women.

After 1950, mortality rates declined rapidly in most developing countries, but birth rates have declined in a more variable manner. Birth rate decline has been most dramatic in China. In 1970, the Chinese birth rate predicted an average of 5.9 children per woman per lifetime (total fertility rate); by 2004, largely because of the government's strict one-child policy, the expected total fertility rate was 1.7 children. In India, birth rates have fallen more slowly. In some countries of Africa, the transition to lower birth rates has been dramatic, though birth rates remain high in most of sub-Saharan Africa.

How do such variable birth rates affect the growth of the world's population? In the developed nations, populations are near equilibrium (growth rate about 0.1% per year), with reproductive rates near the replacement level (total fertility rate = 2.1 children per female). In many developed countries, including Canada, Germany, Japan, Italy, and the United Kingdom, total reproductive rates are in fact *below* replacement. These populations will eventually decline if there is no immigration and if the birth rate does not change. In fact, the population is already declining in many eastern and central European countries. Most of the current global population growth (1.4% per year) is concentrated in developing countries, where about 80% of the world's people now live.

A unique feature of human population growth is the ability to control it with family planning and voluntary contraception. Reduced family size is the key to the demographic transition. Social change and the rising educational and career aspirations of women in many cultures encourage women to delay marriage and postpone reproduction. Delayed reproduction helps to decrease population growth rates and to move a society toward zero population growth under conditions of low birth rates and low death rates. However, there is a great deal of disagreement among world leaders as to how much support should be provided for global family planning efforts.

Age Structure

One important demographic variable in present and future growth trends is a country's **age structure,** the relative number of individuals of each age. Age structure is commonly represented in "pyramids," such as those shown in **Figure 52.25**. For Italy, the pyramid has a small base, indicating that individuals younger than reproductive age are relatively underrepresented in the population. This situation contributes to the projection of continuing population decrease in Italy. In contrast, Afghanistan has an age structure that is bottom-

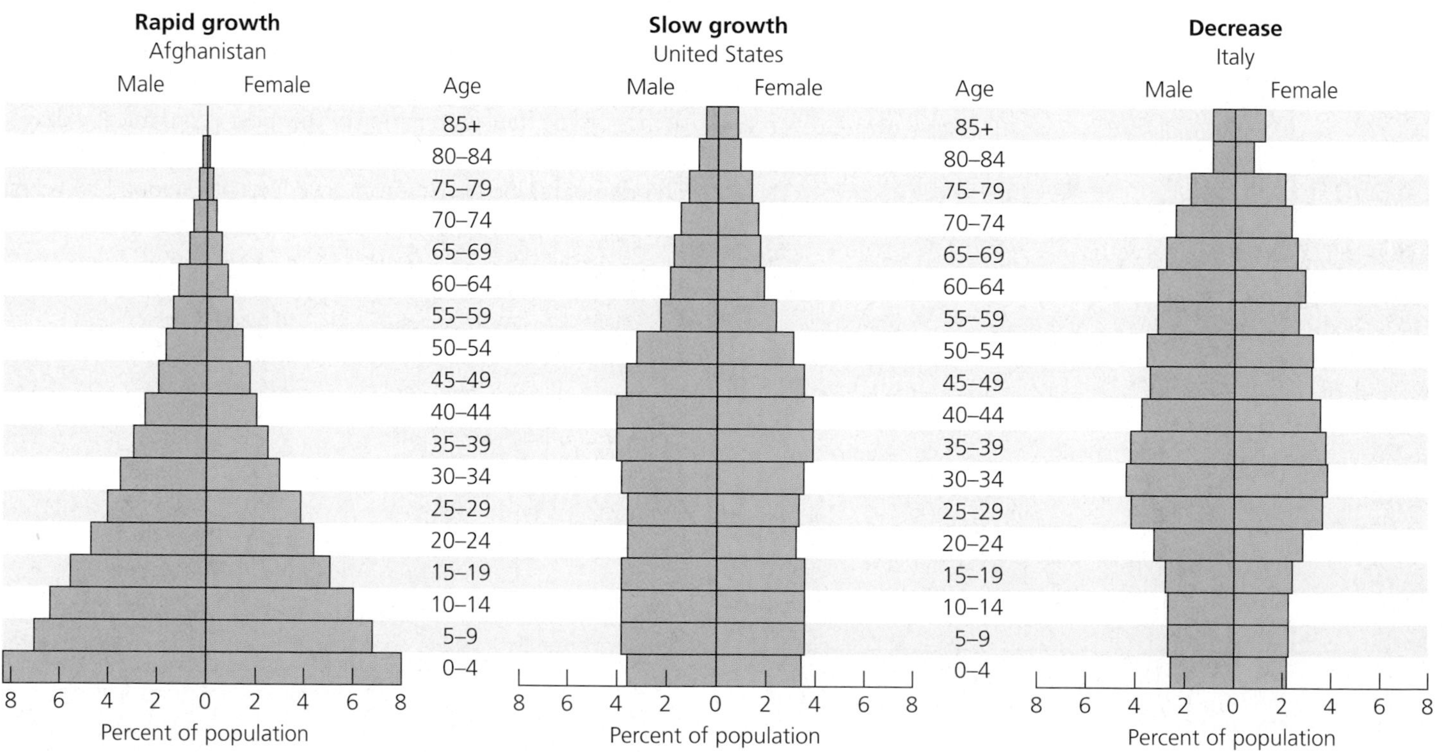

▲ **Figure 52.25 Age-structure pyramids for the human population of three countries (data from 2000).** As of 2004, Afghanistan was growing at 2.6% per year, the United States was growing at 0.6% per year, and Italy was declining at −0.1% per year.

heavy, skewed toward young individuals who will grow up and may sustain the explosive growth with their own reproduction. The age structure for the United States is relatively even (until the older, postreproductive ages) except for a bulge that corresponds to the "baby boom" that lasted for about two decades after the end of World War II. Even though couples born during those years have had an average of fewer than two children, the nation's overall birth rate still exceeds the death rate because there are still so many "boomers" and their offspring of reproductive age. Moreover, although the current total reproductive rate in the United States is 2.1 children per woman—approximately replacement rate—the population is projected to grow slowly through 2050 as a result of immigration.

Age-structure diagrams not only predict a population's growth trends, but they can also illuminate social conditions. Based on the diagrams in Figure 52.25, we can predict, for instance, that employment and education opportunities will continue to be a significant problem for Afghanistan in the foreseeable future. The large number of young entering the Afghan population could also be a source of continuing social and political unrest, particularly if their needs and aspirations are not met. In Italy and the United States, a decreasing proportion of younger working-age people will soon be supporting an increasing population of retired "boomers." In the United States, this demographic feature has made the future of Social Security and Medicare a major political issue. Understanding age structures can help us plan for the future.

Infant Mortality and Life Expectancy

Infant mortality, the number of infant deaths per 1,000 live births, and **life expectancy at birth**, the predicted average length of life at birth, vary widely among different human populations. These differences reflect the quality of life faced by children at birth. **Figure 52.26** contrasts average infant mortality and average life expectancy in the developed and developing countries of the world in 2000. While these averages are markedly different, they do not capture the broad range of the human condition. In 2003, for example, Afghanistan had an infant mortality rate of 143 (14.3%). By contrast, in Japan only 3 children out of every 1,000 born died in infancy. Moreover, life expectancy at birth in Afghanistan was 47 years, compared to 81 years in Japan. While global life expectancy has been increasing since about 1950, more recently it has dropped in a number of regions, including countries of the former Soviet Union and in sub-Saharan Africa. In these regions, the combination of social upheaval, decaying infrastructure, and infectious diseases such as AIDS and tuberculosis is reducing life expectancy. In the East African country of Rwanda, for instance, life expectancy in 2003 was approximately 39 years, about half that in Japan, Sweden, Italy, and Spain.

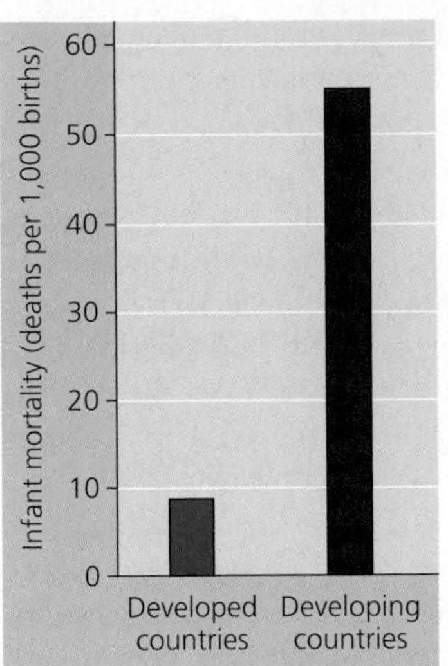

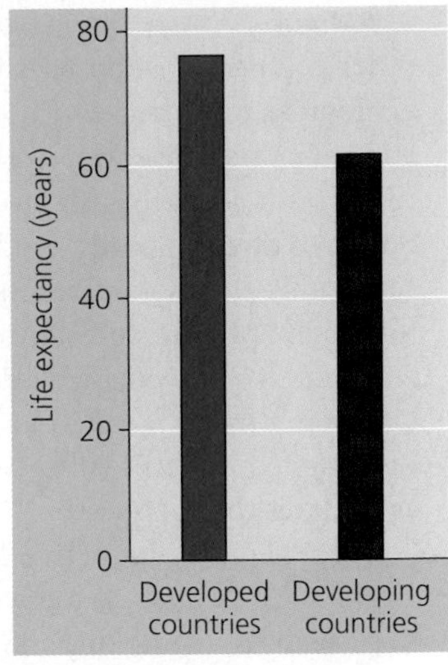

▲ **Figure 52.26 Infant mortality and life expectancy at birth in developed and developing countries.** (Data as of 2003.)

Global Carrying Capacity

No ecological question is more important than the future size of the human population. The projected worldwide population size depends on assumptions about future changes in birth and death rates. For 2050, the United Nations projects a global population of approximately 7.5 to 10.3 billion people. In other words, without some catastrophe, an estimated 1.2–4.0 billion people will be added to the population in the next four decades because of the momentum of population growth. But just how many humans can the biosphere support? Will the world be overpopulated in 2050? Is it *already* overpopulated?

Estimates of Carrying Capacity

For more than three centuries, scientists have asked, What is the carrying capacity of Earth for humans? The first known estimate, 13.4 billion, was made by Anton van Leeuwenhoek in 1679. Since then, estimates have varied from less than 1 billion to over 1,000 billion (1 trillion) people, with an average of 10–15 billion.

Carrying capacity is difficult to estimate, and the scientists who provide these estimates use different methods to get their answers. Some researchers use curves like that produced by the logistic equation (see Figure 52.12) to predict the future maximum of the human population. Others generalize from existing "maximum" population density and multiply this by the area of habitable land. Still other estimates are based on a single limiting factor, such as food, and require many assumptions (such as the amount of available farmland, the average yield of crops, the prevalent diet—vegetarian or meat-based—and the number of calories needed per person per day).

Ecological Footprint

A more comprehensive approach to estimating the carrying capacity of Earth is to recognize that humans have multiple constraints: We need food, water, fuel, building materials, and other requisites, such as clothing and transportation. The **ecological footprint** concept summarizes the aggregate land and water area appropriated by each nation to produce all the resources it consumes and to absorb all the waste it generates. Six types of ecologically productive areas are distinguished in calculating the ecological footprint: arable land (land suitable for crops), pasture, forest, ocean, built-up land, and fossil energy land. (Fossil energy land is calculated on the basis of the land required for vegetation to absorb the CO_2 produced by burning fossil fuels.) All measures are converted to land area as hectares (ha) per person (1 ha = 2.47 acres). Adding up all the ecologically productive land on the planet yields about 2 ha per person. Reserving some land for parks and conservation means reducing this allotment to 1.7 ha per person—the benchmark for comparing actual ecological footprints.

Figure 52.27 graphs the ecological footprints for 13 countries and for the whole world as of 1997. We can draw two key conclusions from the graph. First, countries vary greatly in their individual footprint size and in their available **ecological capacity** (the actual resource base of each country). The United States has an ecological footprint of 8.4 ha per person but has only 6.2 ha per person of available ecological capacity. In other words, the U.S. population is already above carrying capacity. By contrast, New Zealand has a larger ecological footprint of 9.8 ha per person but an available capacity of 14.3 ha per person, so it is below its carrying capacity.

The second conclusion is that, in general, the world was *already* in ecological deficit when the study was conducted. The overall analysis suggests that the world is now at or slightly above its carrying capacity.

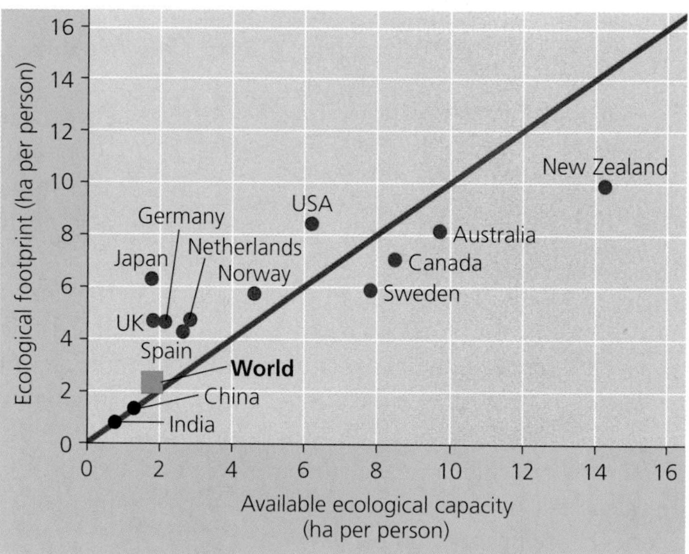

▲ **Figure 52.27 Ecological footprint in relation to available ecological capacity.** Countries indicated by red dots were in an ecological deficit in 1997 when the study was conducted. Countries indicated by blue dots still had resource surpluses relative to the demands of their populations.

We can only speculate about Earth's ultimate carrying capacity for the human population or about what factors will eventually limit our growth. Perhaps food will be the main factor. Malnutrition and famines are common in some regions, but they result mainly from unequal distribution, rather than inadequate production, of food. So far, technological improvements in agriculture have allowed food supplies to keep up with global population growth. However, the principles of energy flow through ecosystems (explained in Chapter 54) tell us that environments can support a larger number of herbivores than carnivores. If everyone ate as much meat as the wealthiest people in the world, less than half of the present world population could be fed on current food harvests.

Perhaps we will eventually be limited by suitable space, like the gannets on ocean islands. Certainly, as our population grows, the conflict over how space will be utilized will intensify, and agricultural land will be developed for housing. There seem to be few limits, however, on how closely humans can be crowded together.

Humans could also run out of nonrenewable resources, such as certain metals and fossil fuels. The demands of many populations have already far exceeded the local and even regional supplies of one renewable resource—water. More than 1 billion people do not have access to sufficient water to meet their basic sanitation needs. It is also possible that the human population will eventually be limited by the capacity of the environment to absorb its wastes. In such cases, Earth's current human occupants could lower the planet's long-term carrying capacity for future generations.

Some optimists have suggested that because of our ability to develop technology, human population growth has no practical limits. Technology has undoubtedly increased Earth's carrying capacity for humans, but as we have emphasized, no population can continue to grow indefinitely. Exactly what the world's human carrying capacity is and under what circumstances we will approach it are topics of great concern and debate. Unlike other organisms, we can decide whether zero population growth will be attained through social changes based on human choices or through increased mortality due to resource limitation, plagues, war, and environmental degradation.

Concept Check 52.6

1. How does a population's age structure affect its growth rate?
2. What are the relationships among carrying capacity, ecological capacity, and ecological footprint for a country's population? Define each term in your answer.

For suggested answers, see Appendix A.

Chapter 52 Review

Go to www.campbellbiology.com or the student CD-ROM to explore Activities, Investigations, and other interactive study aids.

SUMMARY OF KEY CONCEPTS

Dynamic biological processes influence population density, dispersion, and demography

- **Density and Dispersion (pp. 1137–1139)** Population density—the number of individuals per unit area or volume—results from the interplay of births, deaths, immigration, and emigration. Environmental and social factors influence the spacing of individuals into clumped, uniform, or random dispersion patterns.
  Activity *Techniques for Estimating Population Density and Size*
- **Demography (pp. 1139–1141)** Populations increase from births and immigration and decrease from deaths and emigration. Life tables, survivorship curves, and reproductive tables summarize specific demographic trends.
  Activity *Investigating Survivorship Curves*

Life history traits are products of natural selection

- Life history traits are evolutionary outcomes reflected in the development, physiology, and behavior of an organism (p. 1141).
- **Life History Diversity (pp. 1141–1142)** Big-bang, or semelparous, organisms reproduce a single time and die. In contrast, iteroparous organisms produce offspring repeatedly.
- **"Trade-offs" and Life Histories (pp. 1142–1143)** Life history traits such as brood size, age at maturity, and parental caregiving represent trade-offs between conflicting demands for limited time, energy, and nutrients.

Concept 52.3

The exponential model describes population growth in an idealized, unlimited environment

- **Per Capita Rate of Increase (pp. 1143–1144)** If immigration and emigration are ignored, a population's growth rate (the per capita rate of increase) equals birth rate minus death rate.
- **Exponential Growth (pp. 1144–1145)** The exponential growth equation $dN/dt = r_{max}N$ represents a population's potential growth in an unlimited environment, where r_{max} is the maximum per capita, or intrinsic, rate of increase and N is the number of individuals in the population. This model predicts that the larger a population becomes, the faster it grows. Plotting population size over time in such a population yields a J-shaped graph.

The logistic growth model includes the concept of carrying capacity

- Exponential growth cannot be sustained for long in any population. A more realistic population model limits growth by incorporating carrying capacity (K), the maximum population size the environment can support (p. 1145).
- **The Logistic Growth Model (pp. 1145–1146)** According to the logistic equation $dN/dt = r_{max}N(K - N)/K$, growth levels off as population size approaches the carrying capacity.
- **The Logistic Model and Real Populations (p. 1146)** The logistic model fits few real populations, but it is useful for estimating possible growth.
- **The Logistic Model and Life Histories (p. 1147)** Two hypothetical, but controversial, life history patterns are *K*-selection, or density-dependent selection, and *r*-selection, or density-independent selection.

Concept 52.5

Populations are regulated by a complex interaction of biotic and abiotic influences

- **Population Change and Population Density (p. 1148)** In density-dependent population regulation, death rates rise and birth rates fall with increasing density. In density-independent population regulation, birth and death rates do not change with increasing density.
- **Density-Dependent Population Regulation (pp. 1148–1150)** Density-dependent changes in birth and death rates curb population increase through negative feedback and can eventually stabilize a population near its carrying capacity. Density-dependent limiting factors include intraspecific competition for limited food or space, increased predation, disease, stress due to crowding, and buildup of toxins.
- **Population Dynamics (pp. 1150–1151)** Because changing environmental conditions periodically disrupt them, all populations exhibit some size fluctuations. Metapopulations are groups of populations linked by immigration and emigration.
- **Population Cycles (pp. 1151–1152)** Many populations undergo regular boom-and-bust cycles that are influenced by complex interactions between biotic and abiotic factors.
  Biology Labs On-Line *Population EcologyLab*

Concept 52.6

Human population growth has slowed after centuries of exponential increase

- **The Global Human Population (pp. 1152–1155)** Since about 1650, the global human population has grown exponentially, but within the last 40 years, the rate of growth has fallen by nearly 50%. Differences in age structure show that while some nations are growing rapidly, others are stable or declining in size. Infant mortality rates and life expectancy at birth differ markedly between developed and developing countries.
  Activity *Human Population Growth*
  Activity *Analyzing Age-Structure Pyramids*
  Graph It *Age Pyramids and Population Growth*
  Biology Labs On-Line *DemographyLab*
- **Global Carrying Capacity (pp. 1155–1156)** The carrying capacity of Earth for humans is uncertain. Ecological footprint, the aggregate land and water area needed to sustain the people of a nation, is one measure of how close we are to the carrying capacity of Earth. At more than 6 billion people, the world is already in ecological deficit.

TESTING YOUR KNOWLEDGE

Self-Quiz

1. The observation that members of a population are uniformly distributed suggests that
   a. the size of the area occupied by the population is increasing.
   b. resources are distributed unevenly.
   c. the members of the population are competing for access to a resource.
   d. the members of the population are neither attracted to nor repelled by one another.
   e. the density of the population is low.

2. Population ecologists follow the fate of same-age cohorts to
   a. determine a population's carrying capacity.
   b. determine if a population is regulated by density-dependent processes.
   c. determine the birth rate and death rate of each group in a population
   d. determine the factors that regulate the size of a population.
   e. determine if a population's growth is cyclical.

3. In a population that is growing as described by the logistic growth model,
   a. the number of individuals added per unit time is greatest when N is close to zero. (The key equation is $dn/dt = r_{max}$ N($K - N/K$).)
   b. the per capital growth rate (r) increases as N approaches K.
   c. population growth is zero when N equals K.
   d. the population grows exponentially when K is small.
   e. the birth rate (b) approaches zero as N approaches K.

4. A population's carrying capacity
   a. can be accurately calculated using the logistic growth model.
   b. generally remains constant over time.
   c. increases as the per capita growth rate (r) decreases.
   d. may change as environmental conditions change.
   e. can never be exceeded.

5. Which pair of terms most accurately describes life-history traits for a stable population of wolves?
   a. semelparous; r-selected
   b. semelparous; K-selected
   c. iteroparous; r-selected
   d. iteroparous; K-selected
   e. iteroparous; N-selected

6. The infant mortality rate is ________ in developing countries than in developed countries (see Figure 52.26).
   a. no different
   b. about two times higher
   c. slightly lower
   d. about three times higher
   e. more than six times higher

7. Scientific study of the population cycles of the snowshoe hare and its predator, the lynx, has revealed that
   a. the prey population is controlled by predators alone.
   b. hares and lynx are so mutually dependent that each species cannot survive without the other.
   c. the most obvious, plausible hypothesis about the cause of population cycles is not necessarily the correct one.
   d. both hare and lynx populations are regulated mainly by abiotic factors.
   e. the hare population is r-selected; the lynx population is K-selected.

8. The current size of the human population is closest to
   a. 2 million.
   b. 3 billion.
   c. 4 billion.
   d. 6 billion.
   e. 10 billion.

9. Which of the following statements about human population in developed countries is INCORRECT?
   a. Average family size is relatively small.
   b. The population has undergone the demographic transition.
   c. Life history is r-selected.
   d. The survivorship curve is Type I.
   e. Age distribution is relatively uniform.

10. A recent study of ecological footprints (described in the text) concluded that
    a. Earth's carrying capacity for humans is about 10 billion.
    b. Earth's carrying capacity would increase if per capita meat consumption increased.
    c. current demand by industrialized countries for resources is much smaller than the ecological footprint of those countries.
    d. the ecological footprint of the United States is larger than the ecological capacity of its land.
    e. it is not possible for technological improvements to increase Earth's carrying capacity for humans.

For Self-Quiz answers, see Appendix A.

Go to the website or CD-ROM for more quiz questions.

Evolution Connection

Write a paragraph contrasting the conditions that favor the evolution of semelparous (one-time) versus iteroparous (repeated) reproduction.

Scientific Inquiry

You are testing the hypothesis that population density of a particular plant species influences the rate at which a pathogenic fungus infects the plant. Because the fungus causes visible scars on the leaves, you can easily determine whether a plant is infected. Design an experiment to test your hypothesis. Include your experimental treatments and control, the data you will collect, and the results expected if your hypothesis is correct.

Biology Labs On-Line *Population EcologyLab*
Biology Labs On-Line *DemographyLab*

Science, Technology, and Society

Many people regard the rapid population growth of developing countries as our most serious environmental problem. Others think that the population growth in developed countries, though smaller, is actually a greater environmental threat. What problems result from population growth in (a) developing countries and (b) the industrialized world? Which do you think is a greater threat, and why?

53 Community Ecology

▲ **Figure 53.1 A savanna community in Chobe National Park, Botswana.**

Key Concepts

53.1 A community's interactions include competition, predation, herbivory, symbiosis, and disease

53.2 Dominant and keystone species exert strong controls on community structure

53.3 Disturbance influences species diversity and composition

53.4 Biogeographic factors affect community biodiversity

53.5 Contrasting views of community structure are the subject of continuing debate

Overview

What Is a Community?

On your next walk through a field or a woodland or even across campus or a park, try to observe some of the interactions between the species present. You may see birds using trees as nesting sites, bees pollinating flowers, spiders trapping insects in their webs, ferns growing in shade provided by trees—a tiny sample of the many interactions between species that exist in any ecological theater. In addition to the physical and chemical factors discussed in Chapter 50, an organism's environment includes biotic factors: other individuals of the same species as well as individuals of other species. Such an assemblage of populations of various species living close enough for potential interaction is called a biological **community**.

Ecologists define the boundaries of a particular community to fit their research questions. They might study, for example, the community of decomposers and other organisms living on a rotting log, the benthic community in Lake Superior, or the community of trees and shrubs in Shenandoah National Park. The various animals as well as the grass and trees surrounding the watering hole in **Figure 53.1** are all members of a savanna community in southern Africa.

In this chapter, we will examine the factors that are most significant in structuring a community—in determining how many species there are overall, which particular species are present, and the relative abundance of these species. We begin with a fundamental factor influencing community structure: the interactions between the organisms in a community.

Concept 53.1

A community's interactions include competition, predation, herbivory, symbiosis, and disease

Some key relationships in the life of an organism are its interactions with other species in the community. Ecologists refer to these relationships as **interspecific interactions.** We will start with the simplest situation: interactions between populations of just two species.

The possible interactions that can link species include competition, predation, herbivory, symbiosis (parasitism, mutualism, and commensalism), and disease. Here we will use the symbols + and − to indicate how each interspecific interaction affects survival and reproduction of the two species engaged in the interaction. For example, in mutualism, the survival and reproduction of each species is increased in the presence of the other; therefore, this is a +/+ interaction. Predation is an example of a +/− interaction, with a positive effect on the survival and reproduction of the predator population and a negative effect on that of the prey population. A 0 indicates that a population is not affected by the interaction in any known way.

Historically, most ecological research has focused on interactions with a negative effect on at least one species, such as competition and predation. However, positive interactions are ubiquitous, and their contributions to community structure are the subject of much study.

Competition

Interspecific competition occurs when species compete for a particular resource that is in short supply. For instance, the weeds growing in a garden compete with garden plants for soil nutrients and water. Grasshoppers and bison in the Great Plains compete for the grass they both eat. Lynx and foxes in the northern forests of Alaska and Canada compete for prey such as snowshoe hares. In contrast, some resources, such as oxygen, are rarely in short supply; thus, although almost all species use this resource, they do not compete for it. When two species *do* compete for a resource, the result is detrimental to both species (−/−). Strong competition can lead to the local elimination of one of the two competing species, a process called **competitive exclusion.**

The Competitive Exclusion Principle

In 1934, the Russian ecologist G. F. Gause studied the effects of interspecific competition in laboratory experiments with two closely related species of protists, *Paramecium aurelia* and *Paramecium caudatum.* He cultured the protists under stable conditions with a constant amount of food added every day. When Gause grew the two species in separate cultures, each population grew rapidly and then leveled off at what was apparently the carrying capacity of the culture. But when Gause cultured the two species together, *P. caudatum* was driven to extinction in the culture. Gause inferred that *P. aurelia* had a competitive edge in obtaining food, and he concluded that two species competing for the same limiting resources cannot coexist in the same place. One species will use the resources more efficiently and thus reproduce more rapidly than the other. Even a slight reproductive advantage will eventually lead to local elimination of the inferior competitor. Ecologists call Gause's concept the *competitive exclusion principle.*

Ecological Niches

The sum total of a species' use of the biotic and abiotic resources in its environment is called the species' **ecological niche.** One way to grasp the concept is through an analogy made by ecologist Eugene Odum: If an organism's habitat is its "address," the niche is the organism's "profession." Put another way, an organism's niche is its ecological role—how it "fits into" an ecosystem. For example, the niche of a tropical tree lizard consists of, among many components, the temperature range it tolerates, the size of branches on which it perches, the time of day when it is active, and the sizes and kinds of insects it eats.

We can use the niche concept to restate the competitive exclusion principle: Two species cannot coexist in a community if their niches are identical. However, ecologically similar species *can* coexist in a community if there are one or more significant differences in their niches **(Figure 53.2)**. As a result of competition, a species' *fundamental niche,* which is the niche potentially

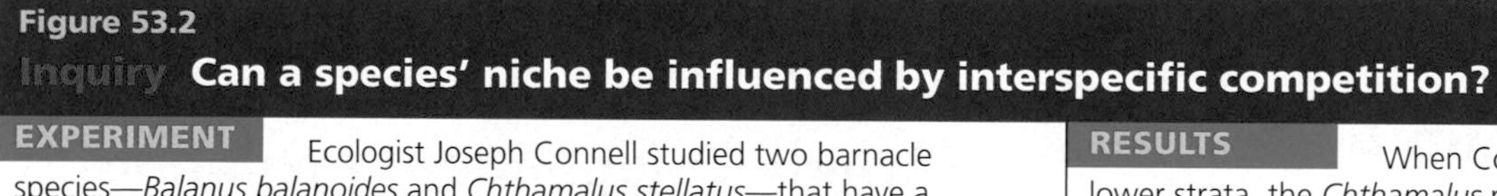

Figure 53.2
Inquiry Can a species' niche be influenced by interspecific competition?

EXPERIMENT Ecologist Joseph Connell studied two barnacle species—*Balanus balanoides* and *Chthamalus stellatus*—that have a stratified distribution on rocks along the coast of Scotland.

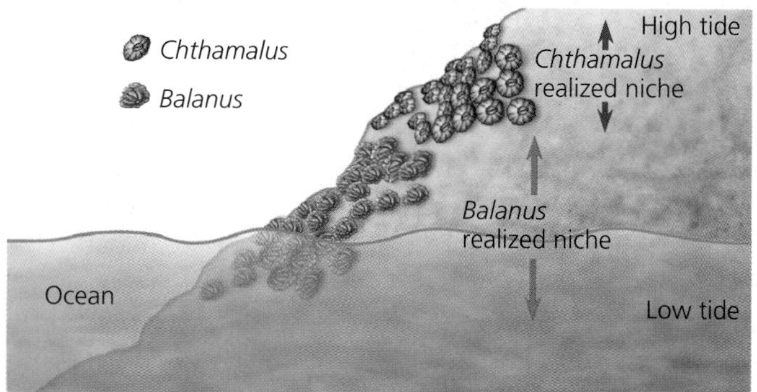

In nature, *Balanus* fails to survive high on the rocks because it is unable to resist desiccation (drying out) during low tides. Its realized niche is therefore similar to its fundamental niche. In contrast, *Chthamalus* is usually concentrated on the upper strata of rocks. To determine the fundamental niche of *Chthamalus*, Connell removed *Balanus* from the lower strata.

RESULTS When Connell removed *Balanus* from the lower strata, the *Chthamalus* population spread into that area.

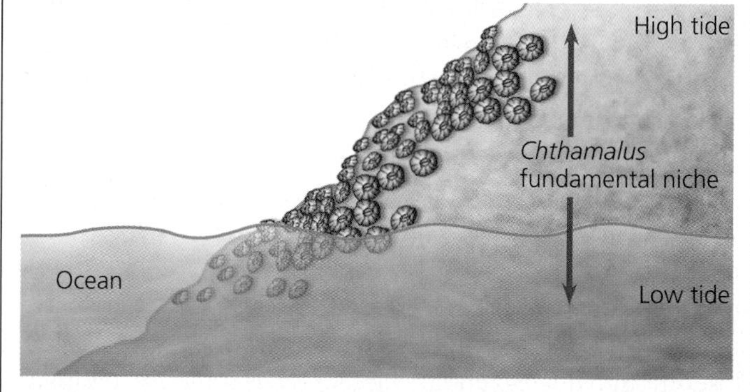

CONCLUSION The spread of *Chthamalus* when *Balanus* was removed indicates that competitive exclusion makes the realized niche of *Chthamalus* much smaller than its fundamental niche.

occupied by that species, may be different from its *realized niche,* the niche it actually occupies in a particular environment.

Resource Partitioning

When competition between species having identical niches does not lead to local extinction of either species, it is generally because one species' niche becomes modified. In other words, evolution by natural selection can result in one of the species using a different set of resources. The differentiation of niches that enables similar species to coexist in a community is called **resource partitioning (Figure 53.3)**. You can think of resource partitioning in a community as "the ghost of competition past"—indirect evidence of earlier interspecific competition resolved by the evolution of niche differentiation.

Character Displacement

A related line of indirect evidence of the effects of competition comes from comparisons of closely related species whose populations are sometimes allopatric (geographically separate; see Chapter 24) and sometimes sympatric (geographically overlapping). In some cases, the allopatric populations of such species are morphologically similar and use similar resources. By contrast, sympatric populations, which would potentially compete for resources, show differences in body structures and in the resources they use. This tendency for characteristics to be more divergent in sympatric populations of two species than in allopatric populations of the same two species is called **character displacement.** An example of character displacement is the variation in beak size between different populations of the Galápagos finches *Geospiza fuliginosa* and *Geospiza fortis* **(Figure 53.4)**.

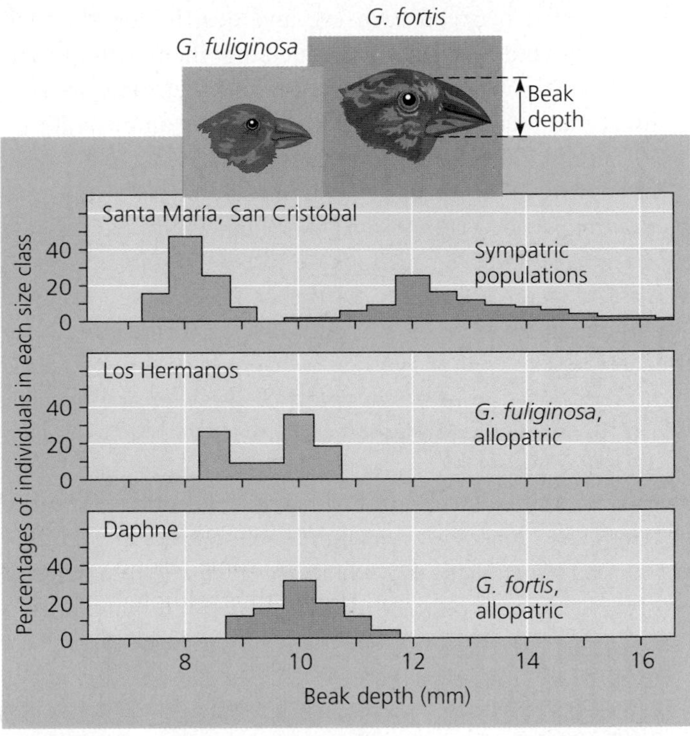

▲ **Figure 53.4 Character displacement: indirect evidence of past competition.** Allopatric populations of *Geospiza fuliginosa* and *Geospiza fortis* on Los Hermanos and Daphne Islands have similar beak morphologies (bottom two graphs) and presumably eat similarly sized seeds. However, where the two species are sympatric on Santa María and San Cristóbal, *G. fuliginosa* has a shallower, smaller beak and *G. fortis* a deeper, larger one (top graph), adaptations that favor eating different sizes of seeds.

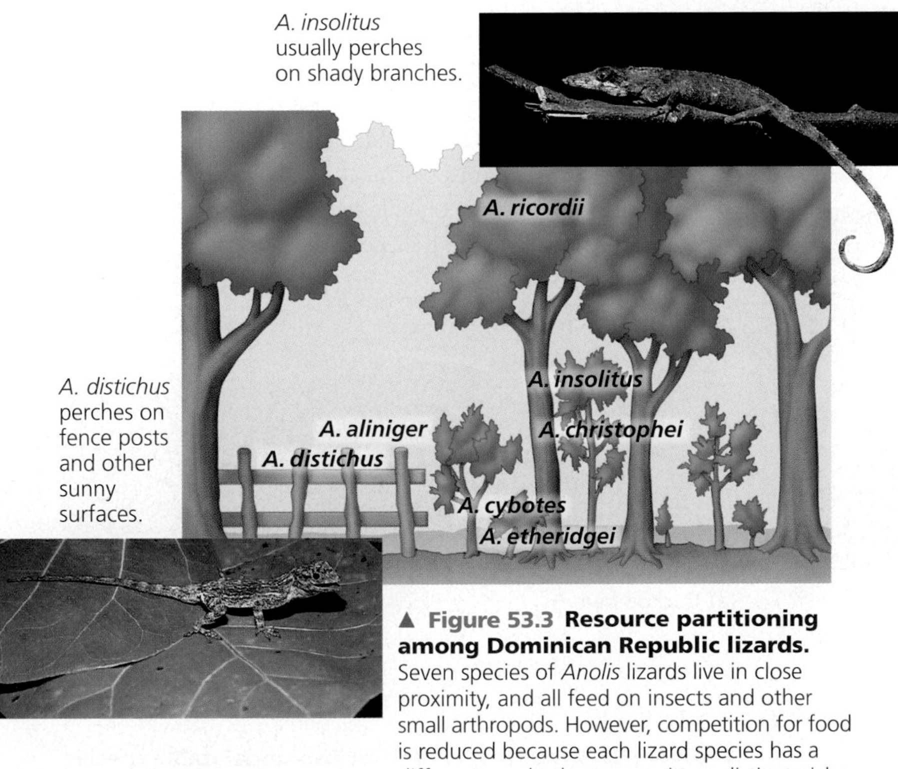

▲ **Figure 53.3 Resource partitioning among Dominican Republic lizards.** Seven species of *Anolis* lizards live in close proximity, and all feed on insects and other small arthropods. However, competition for food is reduced because each lizard species has a different perch, thus occupying a distinct niche.

Predation

Predation refers to a +/− interaction between species in which one species, the predator, kills and eats the other, the prey. Though the term *predation* generally elicits such images as a lion attacking and eating an antelope, it applies to a wide range of interactions defined by the killing of prey. For instance, animals such as certain weevils that chew up or digest plant seeds, thereby killing them, are called *seed predators*. Because eating and avoiding being eaten are prerequisite to reproductive success, the adaptations of both predators and prey tend to be refined through natural selection.

Many important feeding adaptations of predators are both obvious and familiar. Most predators have acute senses that enable them to locate and identify potential prey. In addition, many predators have adaptations such as claws, teeth, fangs, stingers, or poison that help catch and subdue the organisms on which they feed. Rattlesnakes and other pit vipers, for example, locate their prey with heat-sensing organs located between each eye and nostril, and they kill small birds and mammals by injecting them with toxins through their fangs. Predators that pursue their prey are generally fast and agile, whereas those that lie in ambush are often disguised in their environments.

Just as predators possess adaptations for securing prey, prey animals have adaptations that help them avoid being eaten. Behavioral defenses include hiding, fleeing, and self-defense. Active self-defense is not common, though some large grazing mammals vigorously defend their young from predators such as lions. Other behavioral defenses include alarm calls that summon many individuals of the prey species, which then mob the predator.

Animals also display a variety of morphological and physiological defensive adaptations. For example, **cryptic coloration**, or camouflage, makes prey difficult to spot **(Figure 53.5)**. Other animals have mechanical or chemical defenses. For example, most predators are strongly discouraged by the familiar defenses of porcupines and skunks. Some animals, such as the poison arrow frog of Costa Rica, can synthesize toxins, while others passively acquire a chemical defense by accumulating toxins from the plants they eat. Animals with effective chemical defenses often exhibit bright warning coloration, or **aposematic coloration**, like that of the poison arrow frog **(Figure 53.6)**. Aposematic coloration seems to be adaptive; there is evidence that predators are particularly cautious in dealing with potential prey having bright color patterns (see Chapter 1).

Sometimes, one prey species may gain significant protection by mimicking the appearance of another. In **Batesian mimicry**, a palatable or harmless species mimics an unpalatable or harmful model. For example, the larva of the hawkmoth puffs up its head and thorax when disturbed, looking like the head of a small poisonous snake **(Figure 53.7)**. In this case, the mimicry even involves behavior; the larva weaves its head back and forth and hisses like a snake. In **Müllerian mimicry**, two or more unpalatable species, such as the cuckoo bee and yellow jacket, resemble each other **(Figure 53.8)**. Presumably, each species gains an additional advantage because the greater the number of unpalatable prey, the more quickly and intensely predators adapt, avoiding any prey with that particular appearance. The shared appearance thus becomes a kind of aposematic coloration. In an example of convergent evolution, unpalatable animals in several different taxa have similar patterns of coloration: Black and yellow or red stripes characterize unpalatable animals as diverse as yellow jackets and coral snakes (see Figure 1.27).

Predators also use mimicry in a variety of ways. For example, some snapping turtles have tongues that resemble a wriggling worm, thus luring small fish; any fish that tries to eat the "bait" is itself quickly consumed as the turtle's strong jaws snap closed.

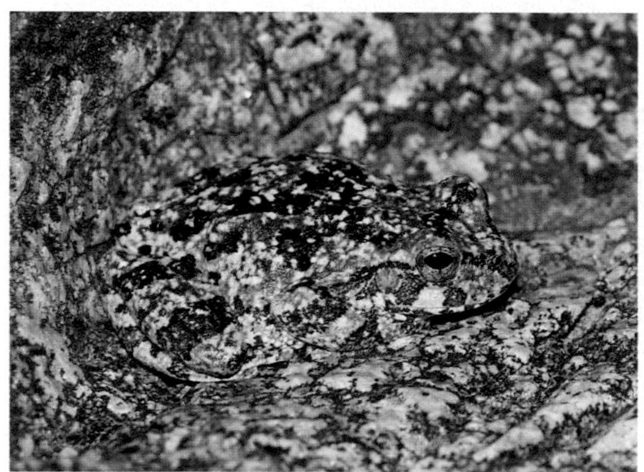

▲ **Figure 53.5 Cryptic coloration: canyon tree frog.**

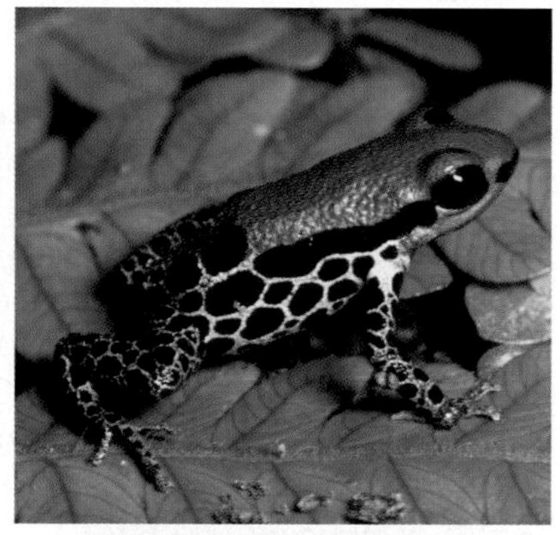

▲ **Figure 53.6 Aposematic coloration: poison arrow frog.**

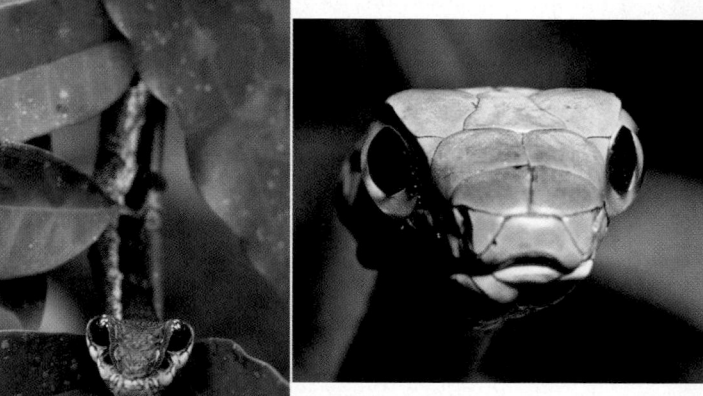

(b) Green parrot snake

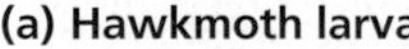

(a) Hawkmoth larva

▲ **Figure 53.7 Batesian mimicry: A harmless species mimics a harmful one.**

(a) Cuckoo bee

(b) Yellow jacket

▲ **Figure 53.8 Müllerian mimicry: Two unpalatable species mimic each other.**

Herbivory

Ecologists use the term **herbivory** to refer to a +/− interaction in which an herbivore eats parts of a plant or alga. While the large mammalian herbivores such as cattle, sheep, and water buffalo may be most familiar, most herbivores are small invertebrates, such as grasshoppers and beetles. In the ocean, herbivores include snails, sea urchins, and some tropical fishes, such as the surgeon fish. Like predators, herbivores have specialized adaptations.

Because plants are so chemically diverse, it is advantageous for herbivores to be able to distinguish between toxic and nontoxic plants as well as between more nutritious and less nutritious plants. Many herbivorous insects have chemical sensors on their feet that recognize appropriate food plants. Some mammalian herbivores, such as goats, use their sense of smell to examine plants, rejecting some and eating others. In some cases, they eat just a specific part of a plant, such as the flowers. Many herbivores also have specialized teeth or digestive systems adapted for processing vegetation (see Chapter 41).

Unlike prey animals, of course, plants cannot run away to avoid being eaten. Instead, a plant's main arsenal against herbivores includes chemical toxins, often in combination with spines and thorns. Among the plant compounds that serve as chemical weapons are the poison strychnine, produced by the tropical vine *Strychnos toxifera;* nicotine, from the tobacco plant; and tannins, from a variety of plant species. Compounds that are not toxic to humans but may be distasteful to many herbivores are responsible for the familiar flavors of cinnamon, cloves, and peppermint. Certain plants produce chemicals that cause abnormal development in some insects that eat them.

Parasitism

Parasitism is a +/− symbiotic interaction in which one organism, the **parasite,** derives its nourishment from another organism, its **host,** which is harmed in the process. (This text adopts the most general definition of *symbiosis* as an interaction in which two organisms of different species live together in direct contact, including parasitism, mutualism, and commensalism. However, some biologists use symbiosis more specifically as a synonym for mutualism.) Parasites that live within the body of their host, such as tapeworms and malarial parasites, are called **endoparasites;** parasites that feed on the external surface of a host, such as ticks and lice, are called **ectoparasites.** In a particular type of parasitism called **parasitoidism,** insects—usually small wasps—lay eggs on or in living hosts. The larvae then feed on the body of the host, eventually killing it.

Many parasites have a complex life cycle involving a number of hosts. For instance, the life cycle of the blood fluke, which currently infects approximately 200 million people around the world, involves two hosts: humans and freshwater snails (see Figure 33.11). Some parasites change the behavior of their hosts in a way that increases the probability of the parasite being transferred from one host to another. For instance, the presence of parasitic acanthocephalan (spiny-headed) worms leads their crustacean hosts to engage in a variety of atypical behaviors, including leaving the protection of cover and moving into the open. As a result of their modified behavior, the crustaceans have a greater chance of being eaten by the birds that are the second host in the parasitic worm's life cycle.

Parasites can have a significant effect on the survival, reproduction, and density of their host population, either directly or indirectly. For example, ticks that live as ectoparasites on moose weaken their hosts by withdrawing blood and causing hair breakage and loss, increasing the chance that the moose will die from cold stress or predation by wolves. Some of the declines of the moose population on Isle Royale, Michigan, have been attributed to tick outbreaks (see Figure 52.18).

Disease

In terms of their effect on host organisms, **pathogens,** or disease-causing agents, are similar to parasites (+/−). Pathogens are typically bacteria, viruses, or protists, but fungi and prions (protein bodies; see Chapter 18) may also be pathogenic. In contrast to many parasites, which are relatively large, multicellular organisms, most pathogens are microscopic. Further, while most parasites cause nonlethal damage to their hosts—by pilfering nutrients, for example—many pathogens inflict lethal harm.

Despite the potential of pathogenic organisms to limit populations, pathogens have been the subject of relatively few ecological studies. This imbalance is now being addressed as dramatic events highlight the ecological importance of disease. One such event was the appearance of sudden oak death, a tree disease caused by a fungus-like protistan pathogen called *Phytophthora ramorum.* By 2004, sudden oak death had spread 650 km from where it was first discovered in Marin County, California, in 1994. In one decade, the disease killed tens of thousands of oaks of several species from the central California coast to southern Oregon and in the process changed the structure of the affected forest communities.

Animal populations also tend to decline in the face of disease epidemics. From 1999 to 2004, the mosquito-transmitted West Nile virus spread from New York to 46 other states, killing hundreds of thousands of birds as it moved across the United States. Crows appear to be particularly susceptible to this disease. Meanwhile, the number of human cases in the United States rose from 62 in 1999 to 9,862 in 2003, resulting in 7 and 264 deaths, respectively. These statistics indicate a fraction of the total exposure, since most people exposed to the virus show no symptoms.

▲ **Figure 53.9 Mutualism between acacia trees and ants.** Certain species of acacia trees have hollow thorns that house stinging ants of the genus *Pseudomyrmex*. The ants feed on sugar produced by nectaries on the tree and on protein-rich swellings (orange in the photograph) at the tips of leaflets. The acacia benefits because the pugnacious ants, which attack anything that touches the tree, remove fungal spores and other debris and clip vegetation that grows close to the acacia.

▲ **Figure 53.10 A possible example of commensalism between cattle egrets and water buffalo.**

Mutualism

Mutualistic symbiosis, or **mutualism**, is an interspecific interaction that benefits both species (+/+). **Figure 53.9** illustrates mutualism between ants and acacia trees in Central and South America. We have described many other examples of mutualism in previous chapters: nitrogen fixation by bacteria in the root nodules of legumes; the digestion of cellulose by microorganisms in the digestive systems of termites and ruminant mammals; photosynthesis by unicellular algae in the tissues of corals; and the exchange of nutrients in mycorrhizae, the association of fungi and the roots of plants.

Mutualistic relationships sometimes involve the evolution of related adaptations in both species, with changes in either species likely to affect the survival and reproduction of the other. For example, most flowering plants have adaptations such as fruit or nectar that attract animals that function in pollination or seed dispersal. The hypothesis is that natural selection might favor evolution of such adaptations because by sacrificing organic materials such as nectar rather than pollen or seeds to a consumer, the plant loses fewer gametes and thereby increases its relative reproductive success. In turn, many animals have adaptations that help them find and consume nectar.

Commensalism

Commensalism is defined as an interaction between species that benefits one of the species but neither harms nor helps the other (+/0). Commensal interactions have been difficult to document in nature because any close association between species likely affects both species, if only slightly. For instance, "hitchhiking" species, such as algae that grow on the shells of aquatic turtles or barnacles that attach to whales, are sometimes considered commensal. The hitchhikers gain access to a substrate while seeming to have little effect on their ride. However, the hitchhikers may in fact slightly decrease the reproductive success of their hosts by reducing the hosts' efficiency of movement in searching for food or escaping from predators. Conversely, the hitchhikers may provide a benefit in the form of camouflage.

Some putative commensal associations involve one species obtaining food that is inadvertently exposed by another. For instance, cowbirds and cattle egrets feed on insects flushed out of the grass by grazing bison, cattle, horses, and other herbivores **(Figure 53.10)**. Because the birds increase their feeding rates when following the herbivores, they clearly benefit from the association. Much of the time, the herbivores may be unaffected by the relationship. However, they, too, may sometimes derive some benefit; the birds tend to be opportunistic feeders that occasionally remove and eat ticks and other ectoparasites from the herbivores. They may also give warning to the herbivores of a predator's approach.

Interspecific Interactions and Adaptation

Earlier in this book, we discussed the concept of **coevolution**, reciprocal evolutionary adaptations of two interacting species. A change in one species acts as a selective force on another species, whose adaptation in turn acts as a selective force on the first species. This linkage of adaptations requires that genetic change in one of the interacting populations of the two species be tied to genetic change in the other population. An example of such dual adaptation that probably qualifies as coevolution is the gene-for-gene recognition between a plant species and a species of avirulent pathogen (see Figure 39.31). In contrast, the aposematic coloration of various tree frogs and the corresponding aversion reactions of various predators do *not* qualify as coevolution because these are adaptations to multiple species in the community rather than coupled genetic changes in just two interacting species.

In fact, the term *coevolution* may often be used too loosely in describing the adaptations of certain organisms to the presence of other organisms in a community. There is little evidence for true coevolution in most cases of interspecific interactions. Nevertheless, the more generalized adaptation of organisms to other organisms in their environment is a fundamental characteristic of life. At present, much evidence seems to indicate that competition and predation are the key processes driving community dynamics. But this conclusion is based mainly on research in temperate communities. Far fewer data exist for interspecific interactions in tropical communities. In addition, the hypothesis that competition and predation control community structure is being challenged by ecologists exploring the influences of parasitism, disease, mutualism, and commensalism on communities.

Concept Check 53.1

1. Explain how interspecific competition, predation, and mutualism differ in their effects on the interacting populations of two species.
2. According to the competitive exclusion principle, what outcome is expected when two species with identical niches compete for a resource? Why?
3. Is the evolution of Batesian mimicry an example of coevolution? Explain your answer.

For suggested answers, see Appendix A.

Concept 53.2

Dominant and keystone species exert strong controls on community structure

In general, a small number of the species in a community exert strong control on that community's structure, particularly on the composition, relative abundance, and diversity of its species. Before examining the effects of these particularly influential species, we first need to consider two fundamental features of community structure: species diversity and feeding relationships.

Species Diversity

The **species diversity** of a community—the variety of different kinds of organisms that make up the community—has two components. One is **species richness**, the total number of different species in the community. The other is the **relative abundance** of the different species, the proportion each species represents of the total individuals in the community. For example, imagine two small forest communities, each with 100 individuals distributed among four different tree species (A, B, C, and D) as follows:

Community 1: 25A, 25B, 25C, 25D
Community 2: 80A, 5B, 5C, 10D

The species richness is the same for both communities because they both contain four species, but the relative abundance is very different **(Figure 53.11)**. You would easily notice the four different types of trees in community 1, but without looking carefully, you might see only the abundant species A in the second forest. Most observers would intuitively describe community 1 as the more diverse of the two communities. Indeed, for an ecologist, the species diversity of the community is dependent on *both* species richness and relative abundance.

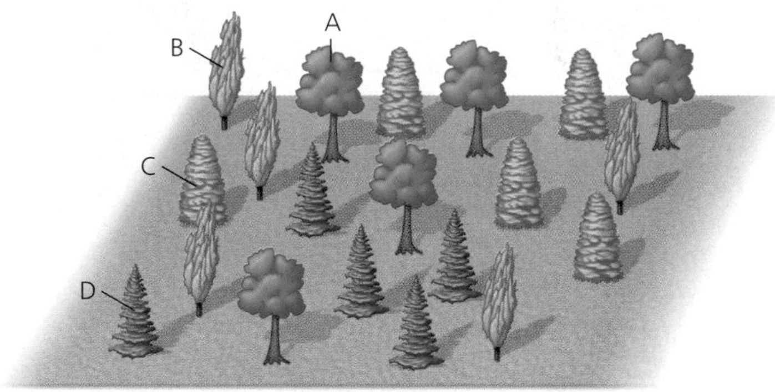

▲ **Figure 53.11 Which forest is more diverse?** Ecologists would say that community 1 has greater species diversity, a measure that includes both species richness and relative abundance.

Determining the number of species in a community is easier said than done. Various sampling techniques can be employed, but since most species in a community are relatively rare, it may be hard to come up with a sample size large enough to be representative. It is particularly difficult to accurately census the highly motile or less visible members of communities, such as mites and nematodes. Nevertheless, measuring species diversity is essential not only for understanding community structure, but for conserving biodiversity, as you will read in Chapter 55.

Trophic Structure

The structure and dynamics of a community depend to a large extent on the feeding relationships between organisms—the **trophic structure** of the community. The transfer of food energy up the trophic levels from its source in plants and other photosynthetic organisms (primary producers) through herbivores (primary consumers) to carnivores (secondary and tertiary consumers) and eventually to decomposers is referred to as a **food chain** (**Figure 53.12**).

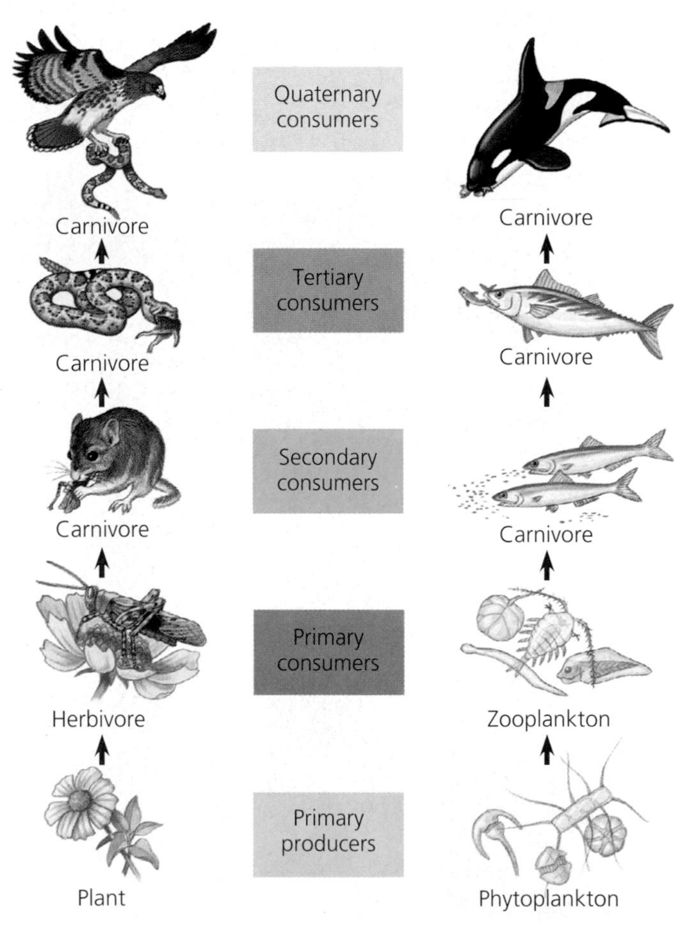

▲ **Figure 53.12 Examples of terrestrial and marine food chains.** The arrows trace energy and nutrients that pass through the trophic levels of a community when organisms feed on one another. Decomposers, which "feed" on organisms from all trophic levels, are not shown here.

Food Webs

In the 1920s, Oxford University biologist Charles Elton recognized that food chains are not isolated units but are linked together in **food webs.** An ecologist can summarize the trophic relationships of a community by diagramming a food web with arrows linking species according to who eats whom. For example, **Figure 53.13** is a simplified food web for an antarctic pelagic community. The primary producers in the community are phytoplankton, which serve as food for the dominant grazing zooplankton, especially euphausids (krill) and copepods, both of which are crustaceans. These zooplankton species are in turn eaten by various carnivores, including other plankton, penguins, seals, fishes, and baleen whales. Squids, which are carnivores that feed on fishes as well as zooplankton, are another important link in these food webs, as they are in turn eaten by seals and toothed whales. During the time

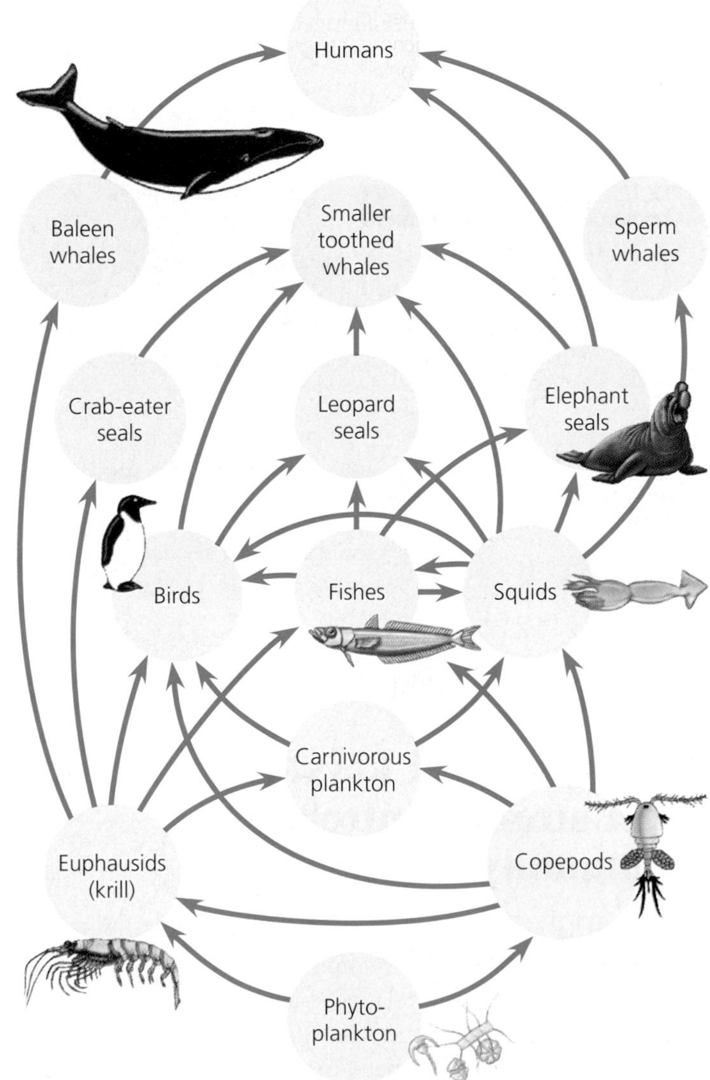

▲ **Figure 53.13 An antarctic marine food web.** Arrows follow the transfer of food from the producers (phytoplankton) up through the trophic levels. For simplicity, this diagram omits decomposers.

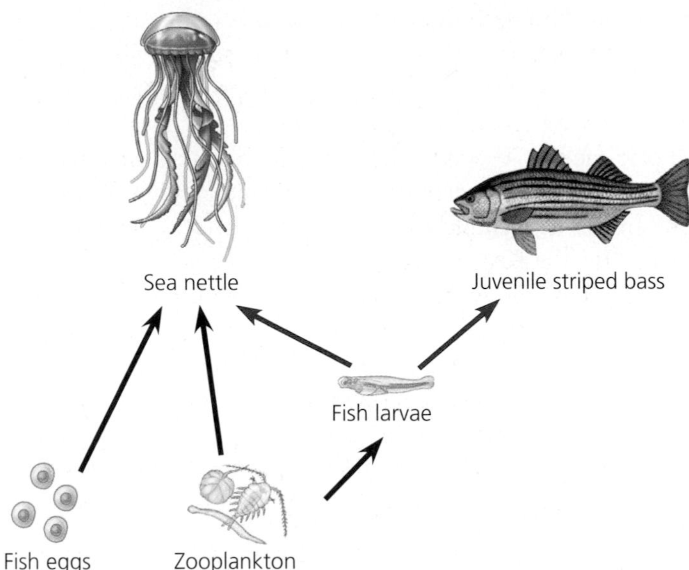

▲ **Figure 53.14 Partial food web for the Chesapeake Bay estuary on the U.S. Atlantic coast.** The sea nettle (*Chrysaora quinquecirrha*) and juvenile striped bass (*Morone saxatilis*) are the main predators of fish larvae (bay anchovy and several other species). Note that sea nettles are secondary consumers (black arrows) when they eat zooplankton and tertiary consumers (red arrows) when they eat fish larvae, which are themselves secondary consumers of zooplankton.

when whales were commonly hunted for food, humans became the top predator in this food web. Having hunted many whale species to low numbers, humans are now harvesting at lower trophic levels, catching krill as well as fishes.

How are food chains linked into food webs? First, a given species may weave into the web at more than one trophic level. For example, in the antarctic food web, euphausids feed on phytoplankton as well as on other grazing zooplankton, such as copepods. Such "nonexclusive" consumers are also found in terrestrial communities. For example, foxes are omnivores with diets that include berries and other plant materials, herbivores such as mice, and other predators, such as weasels. Humans are among the most versatile of omnivores.

Food webs can be very complicated, but we can simplify them for easier study in two ways. First, we can group species with similar trophic relationships in a given community into broad functional groups. For example, in Figure 53.13, more than 100 phytoplankton species are grouped as the primary producers in the food web. A second way to simplify a food web for closer study is to isolate a portion of the web that interacts very little with the rest of the community. **Figure 53.14** illustrates a partial food web for sea nettles (jellies) and juvenile striped bass in Chesapeake Bay.

Limits on Food Chain Length

Each food chain within a food web is usually only a few links long. In the antarctic web of Figure 53.13, there are rarely more than seven links from the producers to any top-level predator, and there are even fewer links in most chains. In fact, most food webs studied to date consist of five or fewer links, as Charles Elton first pointed out in the 1920s.

Why are food chains relatively short? There are two main hypotheses. One, the **energetic hypothesis,** suggests that the length of a food chain is limited by the inefficiency of energy transfer along the chain. As you will read in Chapter 54, only about 10% of the energy stored in the organic matter of each trophic level is converted to organic matter at the next trophic level. Thus, a producer level consisting of 100 kg of plant material can support about 10 kg of herbivore biomass and 1 kg of carnivore biomass. The energetic hypothesis predicts that food chains should be relatively longer in habitats of higher photosynthetic productivity, since the starting amount of energy is greater.

A second hypothesis, the **dynamic stability hypothesis,** proposes that long food chains are less stable than short chains. Population fluctuations at lower trophic levels are magnified at higher levels, potentially causing the local extinction of top predators. In a variable environment, top predators must be able to recover from environmental shocks (such as extreme winters) that can reduce the food supply all the way up the food chain. The longer the food chain, the slower the recovery from environmental setbacks for top predators. This hypothesis predicts that food chains should be shorter in unpredictable environments.

Most of the data available support the energetic hypothesis. For example, ecologists have used tree-hole communities in tropical forests as experimental models to test the energetic hypothesis. Many trees have small branch scars that rot, forming small holes in the tree trunk. The tree holes hold water and provide a habitat for tiny communities consisting of decomposer microorganisms and insects that feed on leaf litter, as well as predatory insects. **Figure 53.15** shows the results of

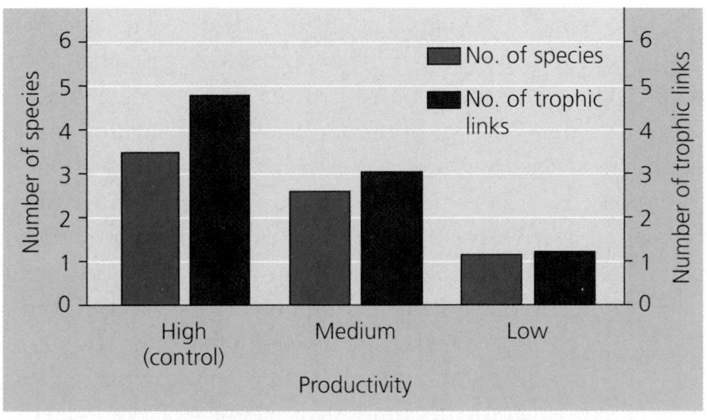

▲ **Figure 53.15 Test of the energetic hypothesis for the restriction of food chain length.** Researchers manipulated the productivity of tree-hole communities in Queensland, Australia, by providing leaf litter input at three levels: high litter input = natural (control) rate of litter fall; medium = $1/10$ natural rate; and low = $1/100$ natural rate. Reducing energy input reduced food chain length, a result consistent with the energetic hypothesis.

a set of experiments in which researchers manipulated productivity (leaf litter falling into the tree holes). As predicted by the energetic hypothesis, holes with the most leaf litter, and hence the greatest total food supply at the producer level, supported the longest food chains.

Another factor that may limit food chain length is that animals in a food chain tend to be larger at successive trophic levels (except for parasites). The size of an animal and its feeding mechanism put some upper limit on the size of food it can take into its mouth. And except in a few cases, large carnivores cannot live on very small food items because they cannot procure enough food in a given time to meet their metabolic needs. Among the exceptions are baleen whales, huge suspension feeders with adaptations that enable them to consume enormous quantities of krill and other small organisms (see Figure 41.2).

Species with a Large Impact

Certain species have an especially large impact on the structure of entire communities either because they are highly abundant or because they play a pivotal role in community dynamics. The impact of these species can occur either through their trophic interactions or through their influences on the physical environment.

Dominant Species

Dominant species are those species in a community that are the most abundant or that collectively have the highest **biomass** (the total mass of all individuals in a population). As a result, dominant species exert a powerful control over the occurrence and distribution of other species. For example, the abundance of sugar maples, the dominant plant species in many eastern North American forest communities, has a major impact on abiotic factors such as shading and soil, which in turn affect which other species live there.

There is no single explanation for why a species becomes dominant in a community. One hypothesis suggests that dominant species are most competitive in exploiting limited resources such as water or nutrients. Another explanation is that dominant species are most successful at avoiding predation or the impact of disease. This latter idea could explain the high biomass that **invasive species** (species, generally introduced by humans, that take hold outside their native range) can attain in environments lacking their natural predators and pathogens.

One way to discover the impact of a dominant species is to remove it from the community. Humans have carried out this type of experiment many times by accident. For example, the American chestnut was a dominant tree in eastern North American deciduous forests before 1910, making up more than 40% of the canopy (top-story) trees. Then humans accidentally introduced the fungal disease called chestnut blight to New York City via nursery stock imported from Asia. Between 1910 and 1950, this fungus, which attacks only chestnuts, killed all the chestnut trees in eastern North America. The once-dominant tree of many of these forests was eliminated.

In this case, removing the dominant species apparently had a relatively small impact on most of the other species. Species of oak, hickory, beech, and red maple that were already present in the forest increased in abundance and replaced the chestnut. No mammals or birds seemed to be seriously affected by the loss of this dominant species. Some insect species, however, were significantly affected. Fifty-six species of moths and butterflies fed on the American chestnut. Of these, 7 species became extinct, though the other 49 species, which did not feed exclusively on the chestnut, still survive. This is only one example of a community response to the loss of a dominant species. More research is needed before we can generalize about the overall effects of such losses.

Keystone Species

In contrast to dominant species, **keystone species** are not necessarily abundant in a community **(Figure 53.16a)**. They

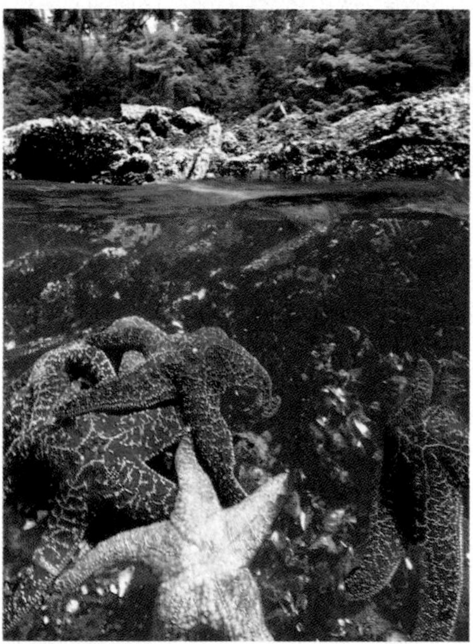

(a) The sea star *Pisaster ochraceous* feeds preferentially on mussels but will consume other invertebrates.

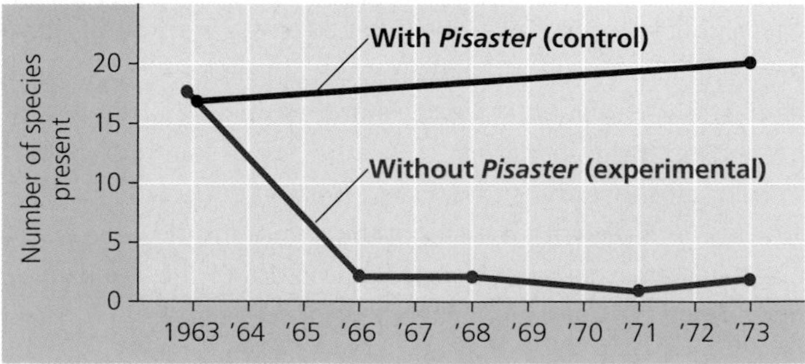

(b) When *Pisaster* was removed from an intertidal zone, mussels eventually took over the rock face and eliminated most other invertebrates and algae. In a control area from which *Pisaster* was not removed, there was little change in species diversity.

▲ **Figure 53.16 Testing a keystone predator hypothesis.**

exert strong control on community structure not by numerical might but by their pivotal ecological roles, or niches. A good way to identify keystone species is by removal experiments, which is how ecologist Robert Paine of the University of Washington first developed the concept of keystone species. The sea star *Pisaster ochraceous* is a predator of mussels such as *Mytilus californianus* in rocky intertidal communities of western North America. *M. californianus* is a dominant species, a superior competitor for space. Predation by *Pisaster* offsets this competitive edge and allows other species to use the space vacated by *Mytilus*. After Paine removed *Pisaster* from rocky intertidal areas, the abundant mussels were able to monopolize space and exclude other invertebrates and algae from attachment sites **(Figure 53.16b)**. When sea stars were present, 15 to 20 species of invertebrates and algae occurred in the intertidal zone. But after experimental removal of *Pisaster,* species diversity quickly declined to fewer than 5 species. *Pisaster* thus acts as a keystone species, exerting an influence on the whole community that is not reflected in its abundance.

The sea otter, *Enhydra lutris,* a keystone predator in the North Pacific, offers another example. Sea otters feed on sea urchins, and sea urchins feed mainly on kelp. In areas where sea otters are abundant, sea urchins are rare and kelp forests are well developed. Where sea otters are rare, sea urchins are common and kelp is almost absent. Over the last 20 years, killer whales have been preying on sea otters as the whales' usual prey has declined. As a result, sea otter populations have declined precipitously in large areas off the coast of western Alaska **(Figure 53.17)**, sometimes at rates as high as 25% per year. The loss of this keystone species has allowed sea urchin populations to increase, resulting in the destruction of kelp forests.

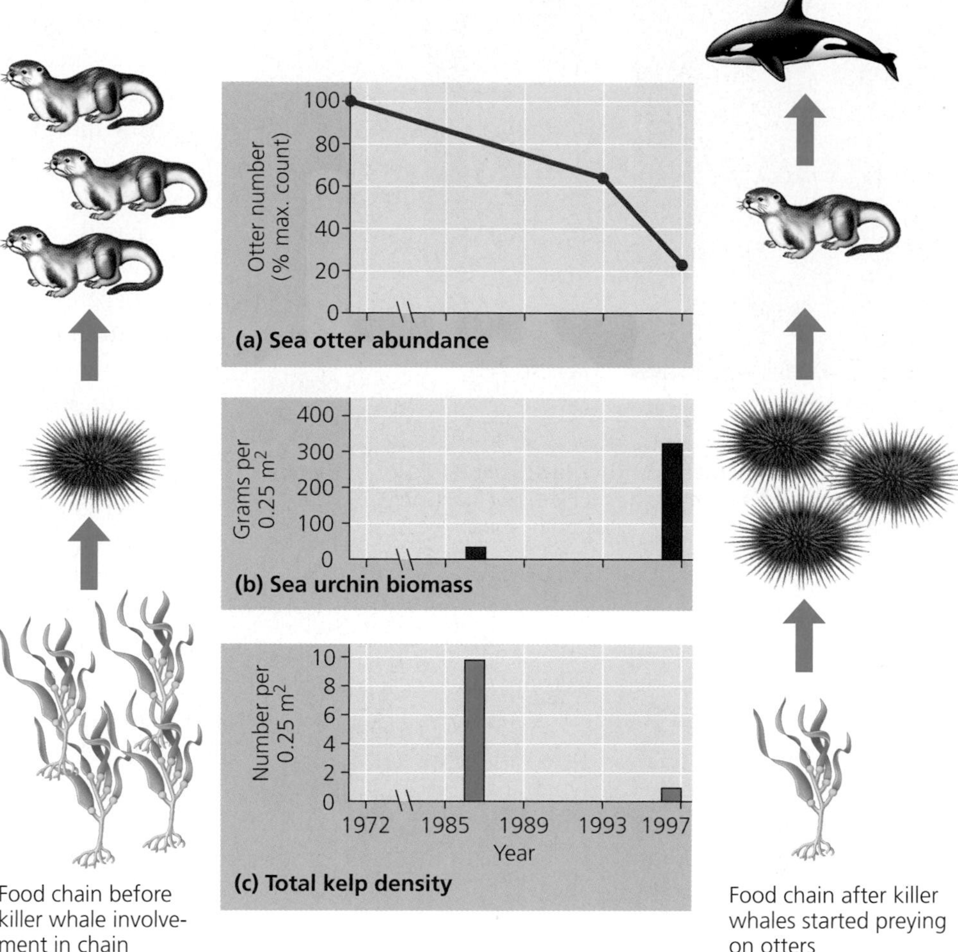

▲ **Figure 53.17 Sea otters as keystone predators in the North Pacific.** The graphs correlate changes over time in sea otter abundance **(a)** with changes in sea urchin biomass **(b)** and changes in kelp density **(c)** in kelp forests at Adak Island (part of the Aleutian Island chain). The food chains alongside the graphs symbolically indicate the relative abundance of species before and after killer whales entered the chain.

Ecosystem "Engineers" (Foundation Species)

Some organisms exert their influence not through their trophic interactions but by causing physical changes in the environment that affect the structure of the community. Such organisms may alter the environment through their behavior or by virtue of their large collective biomass.

An example of a species that dramatically alters its physical environment is the beaver **(Figure 53.18)**, which, through tree

▲ **Figure 53.18 Beavers as ecosystem "engineers" in temperate and boreal forests.** By felling trees, building dams, and creating ponds, beavers can transform large areas of forest into flooded wetlands.

Salt marsh with *Juncus* (foreground)

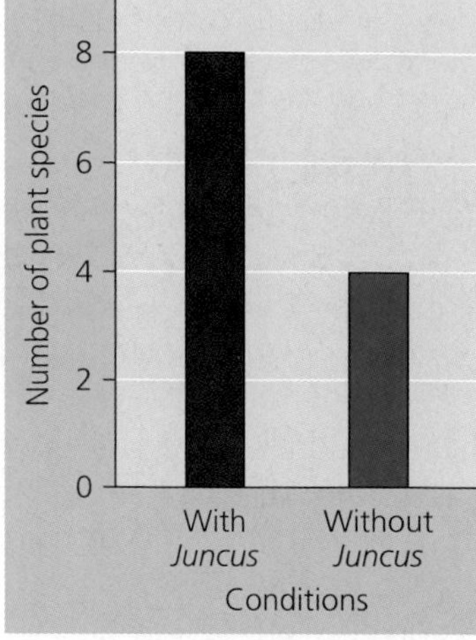

▲ **Figure 53.19 Facilitation by black rush (*Juncus gerardi*) in New England salt marshes.** Black rush facilitates the occupation of the middle upper zone of the marsh, which increases local plant species richness.

felling and dam building, can transform landscapes on a very large scale. Species with such effects are called ecosystem "engineers" or, to avoid implications of conscious intent, "foundation species."

By altering the structure or dynamics of the environment, foundation species act as **facilitators** that have positive effects on the survival and reproduction of some of the other species in the community. For example, by modifying soils, the black rush *Juncus gerardi* increases the species richness in some zones of New England salt marshes. *Juncus* helps prevent salt buildup in the soil by shading the soil surface, which reduces evaporation. *Juncus* also prevents the salt marsh soils from becoming anoxic as it transports oxygen to its below-ground tissues. Sally Hacker and Mark Bertness, of Brown University, uncovered some of *Juncus*'s facilitation effects through removal experiments **(Figure 53.19)**. When they removed *Juncus* from study plots in a New England salt marsh, three other plant species died out. Hacker and Bertness suggest that without *Juncus*, the upper middle intertidal zone would support approximately 50% fewer plant species.

Bottom-Up and Top-Down Controls

Simplified models based on relationships between adjacent trophic levels are useful for discussing how biological communities might be organized. For example, let's consider the three possible relationships between plants (V for vegetation) and herbivores (H):

$$V \rightarrow H \qquad V \leftarrow H \qquad V \leftrightarrow H$$

The arrows indicate that a change in the biomass (total mass) of one trophic level causes a change in the other trophic level. $V \rightarrow H$ means that an increase in vegetation will impact (increase) the numbers or biomass of herbivores, but not vice versa. In this situation, herbivores are limited by vegetation, but vegetation is not limited by herbivory. In contrast, $V \leftarrow H$ means that an increase in herbivore biomass will have an impact on vegetation (decreasing it), but not vice versa. A double-headed arrow indicates that feedback flows in both directions, with each trophic level sensitive to changes in the biomass of the other.

Based on the possible interactions in our example, we can define two models of community organization: the bottom-up model and the top-down model. The $V \rightarrow H$ linkage suggests a **bottom-up model,** which postulates a unidirectional influence from lower to higher trophic levels. In this case, the presence or absence of mineral nutrients (N) controls plant (V) numbers, which control herbivore (H) numbers, which in turn control predator (P) numbers. The simplified bottom-up model is thus $N \rightarrow V \rightarrow H \rightarrow P$. To change the community structure of a bottom-up community, you need to alter biomass at the lower trophic levels. For example, if you add mineral nutrients to stimulate growth of vegetation, then the higher trophic levels should also increase in biomass. If you add predators to or remove predators from a bottom-up community, however, the effect will not extend down to the lower trophic levels.

In contrast, the **top-down model** postulates that the influence moves in the opposite direction: It is mainly predation that controls community organization because predators limit herbivores, which in turn limit plants, which in turn limit nutrient levels through their uptake of nutrients during growth and reproduction. The simplified top-down model is thus $N \leftarrow V \leftarrow H \leftarrow P$ and is also called the *trophic cascade model.* For example, in a lake community with four trophic levels, the top-down model predicts that removing the top carnivores will increase the abundance of primary carnivores, in turn decreasing herbivores, increasing phytoplankton, and eventually decreasing concentrations of mineral nutrients. If there were only three trophic levels in a lake, removing the primary carnivores would increase the herbivores and decrease the phytoplankton, causing nutrient levels to rise. The effects of any manipulation thus moves down the trophic structure as a series of +/− effects.

Many intermediate models between the bottom-up and top-down extremes are possible. For example, all the interactions between trophic levels may be reciprocal (↔). Or, the direction of interaction may alternate over time. For instance, one long-term experimental study of a desert shrub community in Chile showed that controls on primary producer biomass shift periodically from bottom-up to top-down, depending on the amount of rainfall **(Figure 53.20)**. During wet years triggered by the El Niño Southern Oscillation, top-down controls by predators and herbivores prevail. In contrast, during dry years, bottom-up control by moisture limitation of plant growth is most important. Despite the discovery of complex scenarios such as this shifting direction of

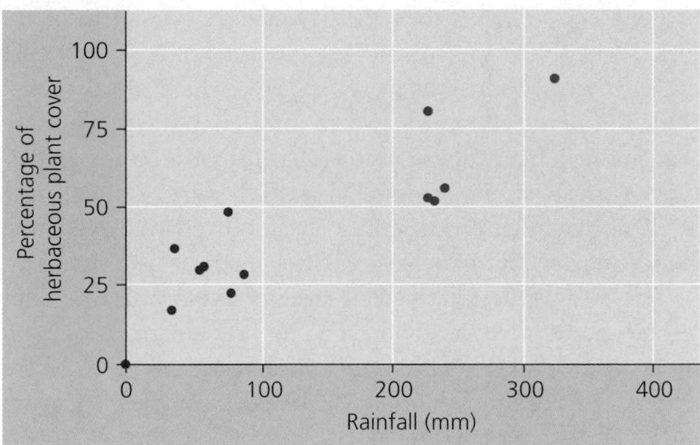

▲ **Figure 53.20 Relationship between rainfall and herbaceous plant cover in a desert shrub community in Chile.** Moisture limitation on plant growth during dry non-El Niño years (red points) creates strong bottom-up control on this community, whereas abundant moisture during El Niño years (blue points) stimulates increased plant and animal abundance, inducing strong top-down control.

control, the simplified models discussed in this section remain valuable as a starting point for the analysis of communities. Let's look at one example where ecologists have applied these simple models to counter complex pollution problems in lakes.

Pollution has degraded the freshwater lakes in many countries. Because many freshwater lake communities seem to be structured according to the top-down model, ecologists have a potential means of improving water quality. We can summarize this strategy, called **biomanipulation,** with a diagram:

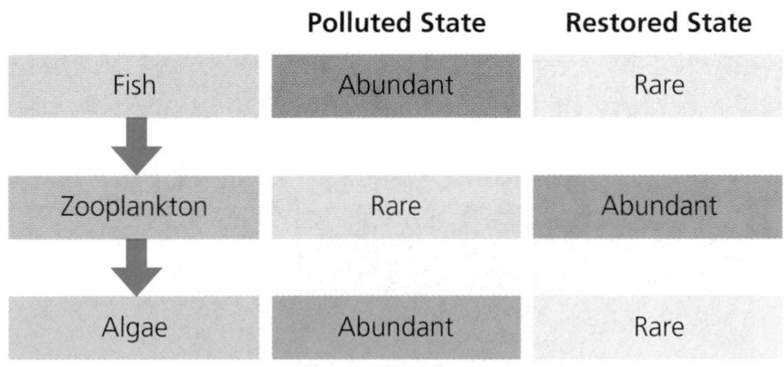

In lakes with three trophic levels, for example, removing fish should improve water quality by increasing zooplankton and thereby decreasing algal populations. In lakes with four trophic levels, adding top predators should have the same effect.

Ecologists applied biomanipulation on a large scale in Lake Vesijärvi in southern Finland. Lake Vesijärvi is a large (110 km^2), shallow lake that was heavily polluted with city sewage and industrial wastewater until 1976. Pollution controls then stopped these inputs, and the water quality of the lake began to recover. But by 1986, massive blooms of cyanobacteria began to occur in the lake. These blooms coincided with a very dense population of roach, a fish that had accumulated during the years of mineral nutrient (pollution) input. Roach eat zooplankton, and by reducing zooplankton, they also reduced feeding on cyanobacteria and algae, which then increased in abundance. To reverse these changes, ecologists removed 1,018 tons of fish from Lake Vesijärvi between 1989 and 1993, reducing roach to about 20% of their former abundance. At the same time, the ecologists stocked the lake with pike perch, a predatory fish that eats roach. This added a fourth trophic level to the lake, which kept down the population of roach (the main carnivore in the lake). Biomanipulation was a success in Lake Vesijärvi. The water became clear, and cyanobacterial blooms ended in 1989. The lake remains clear even though the roach removal ended in 1993.

As these examples show, communities vary in their relative degree of bottom-up and top-down control. To manage agricultural landscapes, national parks, reservoirs, and marine fisheries, scientists need to understand the dynamics of each particular community.

Concept Check 53.2

1. Describe the two components of species diversity. Explain how two communities that contain the same number of species can differ in species diversity.
2. Describe two hypotheses that explain why food chains are usually short, and a key prediction of each hypothesis.
3. How does a dominant species' effect on community structure differ from a keystone species' effect?
4. How do top-down and bottom-up controls on community organization differ?

For suggested answers, see Appendix A.

Concept 53.3

Disturbance influences species diversity and composition

Decades ago, most ecologists favored the traditional view that biological communities are in a state of equilibrium, a more or less stable balance, unless seriously disturbed by human activities. This "balance of nature" view focused on interspecific competition as a key factor determining community composition and maintaining stability in communities. *Stability* in this context refers to the tendency of a community to reach and maintain a relatively constant composition of species in the face of disturbances. But in many communities, at least on a local scale, change seems to be more common than stability. This recent emphasis on change has led to a **nonequilibrium model,** which describes communities as constantly changing after being buffeted by disturbances. We will now take a look at the influence of disturbances on community structure and composition.

What Is Disturbance?

A **disturbance** is an event, such as a storm, fire, flood, drought, overgrazing, or human activity, that changes a community, removes organisms from it, and alters resource availability. The types of disturbances and their frequency and severity vary from community to community. A high level of disturbance is generally the result of a high intensity *and* high frequency of disturbance, while low disturbance levels can result from either a low intensity or low frequency of disturbance. Storms disturb almost all communities, even those in deep oceans. Fire is a significant disturbance in most terrestrial communities; in fact, grasslands and chaparral biomes are dependent on regular burning **(Figure 53.21)**. Freezing is a frequent occurrence in many rivers, lakes, and ponds, and many streams and ponds are disturbed by spring flooding and seasonal drying. By gathering data from specific communities over many years, ecologists are beginning to appreciate the impact of disturbances.

Although the term *disturbance* implies a negative impact on communities, this is not always the case. For example, disturbances often create opportunities for species that have not previously occupied habitats in the community. Small-scale disturbances sometimes enhance environmental patchiness, which can be a key to maintaining species diversity in a community. The **intermediate disturbance hypothesis,** which is supported by a broad range of studies of terrestrial and aquatic communities, suggests that moderate levels of disturbance can create conditions that foster greater species diversity than low or high levels of disturbance. High levels of disturbance reduce species diversity by creating environmental stresses that exceed the tolerances of many species, which are then excluded from the community, or by subjecting the community to such a high frequency of disturbance that slowly colonizing or slowly growing species are excluded. At the other extreme, low levels of disturbance can reduce species diversity by allowing competitively dominant species to exclude less competitive species. Meanwhile, intermediate levels of disturbance can foster greater species diversity by opening up habitats for occupation by less competitive species yet not creating conditions so severe that they exceed the environmental tolerances or rate of recovery by potential community members. Frequent small-scale disturbances can also prevent large-scale disturbances in some communities. However, large-scale disturbances are a natural part of others, such as the forests of Yellowstone National Park. Much of this park is dominated by lodgepole pine, a tree that requires the rejuvenating influence of periodic fires. Lodgepole cones remain closed until exposed to intense heat. When a forest fire burns the trees, the cones open and the seeds are released. The new generation of lodgepole pines can then thrive on nutrients released from the burned trees and in the sunlight which is no longer blocked by taller trees.

In the summer of 1988, major fires occurred in Yellowstone, resulting largely from a severe drought accompanied by high winds. By 1989, burned areas in Yellowstone were largely covered with new vegetation, suggesting that the species in this community are adapted to rapid recovery following massive disturbance by fire **(Figure 53.22)**. And in fact, large-scale fires have swept through the lodgepole pine forests of Yellowstone and other northern areas for thousands of years. In contrast, more southerly pine forests were historically affected by frequent but low-intensity fires. In these forests, a century of human intervention to suppress small

(a) Before a controlled burn. A prairie that has not burned for several years has a high proportion of detritus (dead grass).

(b) During the burn. The detritus serves as fuel for fires.

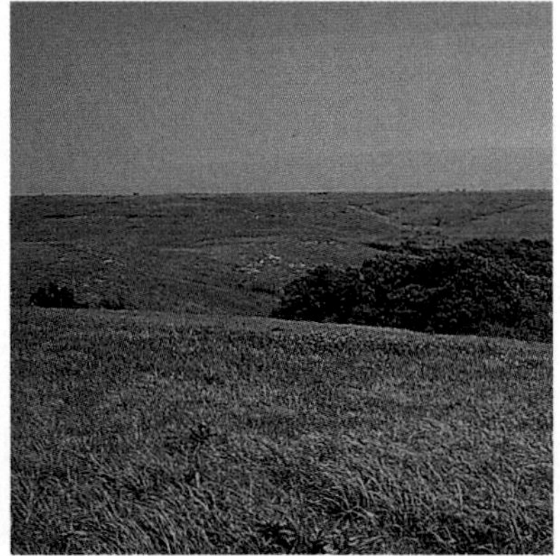

(c) After the burn. Approximately one month after the controlled burn, virtually all of the biomass in this prairie is living.

▲ **Figure 53.21 Long-term effects of fire on a tallgrass prairie community in Kansas.** The trees in the photographs were growing along a stream and were not burned during the study.

(a) Soon after fire. As this photo taken soon after the fire shows, the burn left a patchy landscape. Note the unburned trees in the distance.

(b) One year after fire. This photo of the same general area taken the following year indicates how rapidly the community began to recover. A variety of herbaceous plants, different from those in the former forest, cover the ground.

▲ **Figure 53.22 Patchiness and recovery following a large-scale disturbance.** The 1988 Yellowstone National Park fire destroyed large areas of forests dominated by lodgepole pines.

fires has allowed an unnatural buildup of fuels and elevated the risk of large, severe fires to which the biota is not adapted.

Studies of the Yellowstone forest community and many others indicate that they are nonequilibrium communities, changing continually because of natural disturbances and the internal processes of growth and reproduction. And there is mounting evidence that some amount of nonequilibrium resulting from disturbance is the norm for *most* communities.

Human Disturbance

Of all animals, humans have the greatest impact on biological communities worldwide. Agricultural development has disrupted what were once the vast grasslands of the North American prairie. Logging and clearing for urban development, mining, and farming have reduced large tracts of forests to small patches of disconnected woodlots in many parts of the United States and throughout Europe. After forests are clear-cut and left alone, weedy and shrubby vegetation often colonizes the area and dominates it for many years. This type of vegetation is also found in agricultural fields that are no longer under cultivation and in vacant lots and construction sites that are periodically cleared.

Human disturbance of communities is by no means limited to the United States and Europe; nor is it a recent problem. Tropical rain forests are quickly disappearing as a result of clear-cutting for lumber and pastureland. And centuries of overgrazing and agricultural disturbance have contributed to the current famine in parts of Africa by turning seasonal grasslands into great barren areas.

Because human disturbance is often severe, it usually reduces species diversity in communities. In Chapter 55, we will take a closer look at how community disturbance by human activities is affecting the diversity of life.

Ecological Succession

Changes in community composition and structure are most apparent after some disturbance, such as a glacier or a volcanic eruption, strips away all the existing vegetation. The disturbed area may be colonized by a variety of species, which are gradually replaced by other species, which are in turn replaced by still other species—a process called **ecological succession.**

When this process begins in a virtually lifeless area where soil has not yet formed, such as on a new volcanic island or on the rubble (moraine) left behind by a retreating glacier, it is called **primary succession.** Often the only life-forms initially present are autotrophic prokaryotes. Lichens and mosses, which grow from windblown spores, are commonly the first macroscopic photosynthesizers to colonize such areas. Soil develops gradually, as rocks weather and organic matter accumulates from the decomposed remains of the early colonizers. Once soil is present, the lichens and mosses are usually overgrown by grasses, shrubs, and trees that sprout from seeds blown in from nearby areas or carried in by animals. Eventually, an area is colonized by plants that become the community's prevalent form of vegetation. Producing such a community through primary succession may take hundreds or thousands of years.

Secondary succession occurs when an existing community has been cleared by some disturbance that leaves the soil intact, as in Yellowstone following the 1988 fires (see Figure 53.22). Often the area begins to return to something like its original state. For instance, in a forested area that has been cleared for farming and later abandoned, the earliest plants to recolonize are often herbaceous species that grow from windblown or animal-borne seeds. If the area has not been burned or heavily grazed, woody shrubs may in time replace most of the herbaceous species, and forest trees may eventually replace most of the shrubs.

Early arrivals and later-arriving species may be linked in one of three key processes. The early arrivals may *facilitate* the appearance of the later species by making the environment more favorable—for example, by increasing the fertility of the soil. Alternatively, the early species may *inhibit* establishment of the later species, so that successful colonization by later species occurs in spite of, rather than because of, the activities of the early species. Finally, the early species may be completely independent of the later species, which *tolerate* conditions created early in succession but are neither helped nor hindered by early species. Let's look at how these various processes contribute to primary succession in one specific example.

During the past 300 years, there has been a gradual retreat of glaciers in the Northern Hemisphere. As the glaciers retreat, they leave moraines. Researchers can determine the age of these post-glaciation areas of moraine from the age of the new trees growing on the moraine or, in the last 80 years, by direct observation. Ecologists have conducted the most extensive research on moraine succession at Glacier Bay in southeastern Alaska. Since about 1760, the glaciers there have retreated about 98 km, an extraordinary retreat rate of almost 400 m per year **(Figure 53.23)**. Rocky moraines form a nearly pristine but harsh environment that plants can colonize.

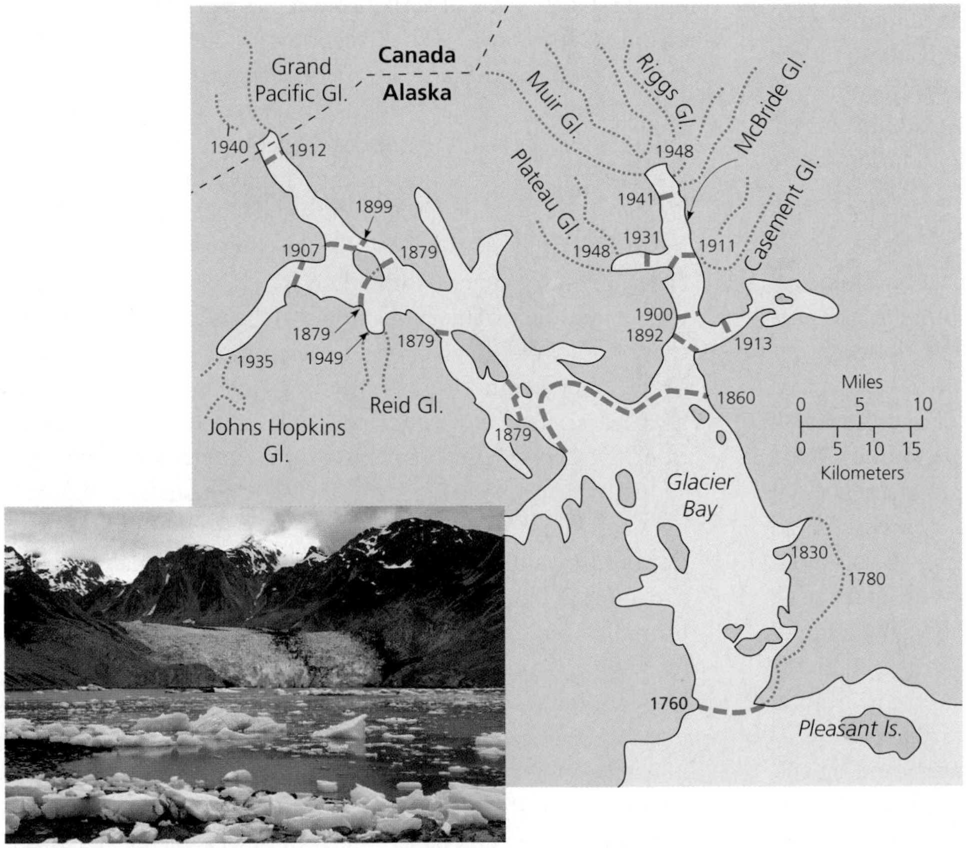

Succession on the moraines in Glacier Bay follows a predictable pattern of change in vegetation and soil characteristics. The exposed moraine is colonized first by pioneering plant species, including liverworts, mosses, fireweed, and scattered *Dryas* (a mat-forming shrub), willows, and cottonwood **(Figure 53.24a)**. After about three decades, *Dryas* dominates the plant community **(Figure 53.24b)**. A few decades later, the area is invaded by alder (*Alnus*), which eventually forms dense, pure thickets up to 9 m tall. These alder stands are overgrown by Sitka spruce, which, after another century, forms a dense forest **(Figure 53.24c)**. Western hemlock and mountain hemlock invade the spruce stands, and after another century, the community is a spruce-hemlock forest. This forest, however, endures only on well-drained slopes. In areas of poor drainage, the forest floor of this spruce-hemlock forest is invaded by sphagnum mosses, which hold large amounts of water and acidify the soil. The trees in such an area die because the soil is waterlogged and too oxygen-deficient to sustain their roots, and the area becomes a sphagnum bog. Thus, by about 300 years after glacial retreat, the vegetation consists of sphagnum bogs on the poorly drained flat areas and spruce-hemlock forest on the well-drained slopes.

How is succession on glacial moraines related to the environmental changes caused by transitions in the vegetation? The bare soil exposed as the glacier retreats is quite basic, with a pH of 8.0–8.4 due to the carbonate compounds in the parent rocks. The soil pH falls rapidly as vegetation arrives, with the rate of change depending on the vegetation type. The most striking change is caused by spruce: Decomposition of acidic spruce needles reduces the pH of the soil from 7.0 to approximately 4.0. The soil concentrations of mineral nutrients also show marked changes with time. One of the characteristic features of the bare soil after glacial retreat is its low nitrogen content. Almost all the pioneer species begin the succession with very poor growth and yellow leaves due to inadequate nitrogen supply. The exceptions are *Dryas* and, particularly, alder; these species have symbiotic bacteria that fix atmospheric nitrogen (see Chapter 37). Soil nitrogen increases rapidly during the alder stage of succession **(Figure 53.24d)**. The spruce forest develops by spending the capital of soil nitrogen accumulated

McBride glacier retreating

▲ **Figure 53.23 A glacial retreat in southeastern Alaska.** The dated locations chronicle recession of the glacier since 1760, based on historical description. As the ice retreats (inset), it leaves moraines along the edge of the bay, on which primary succession occurs.

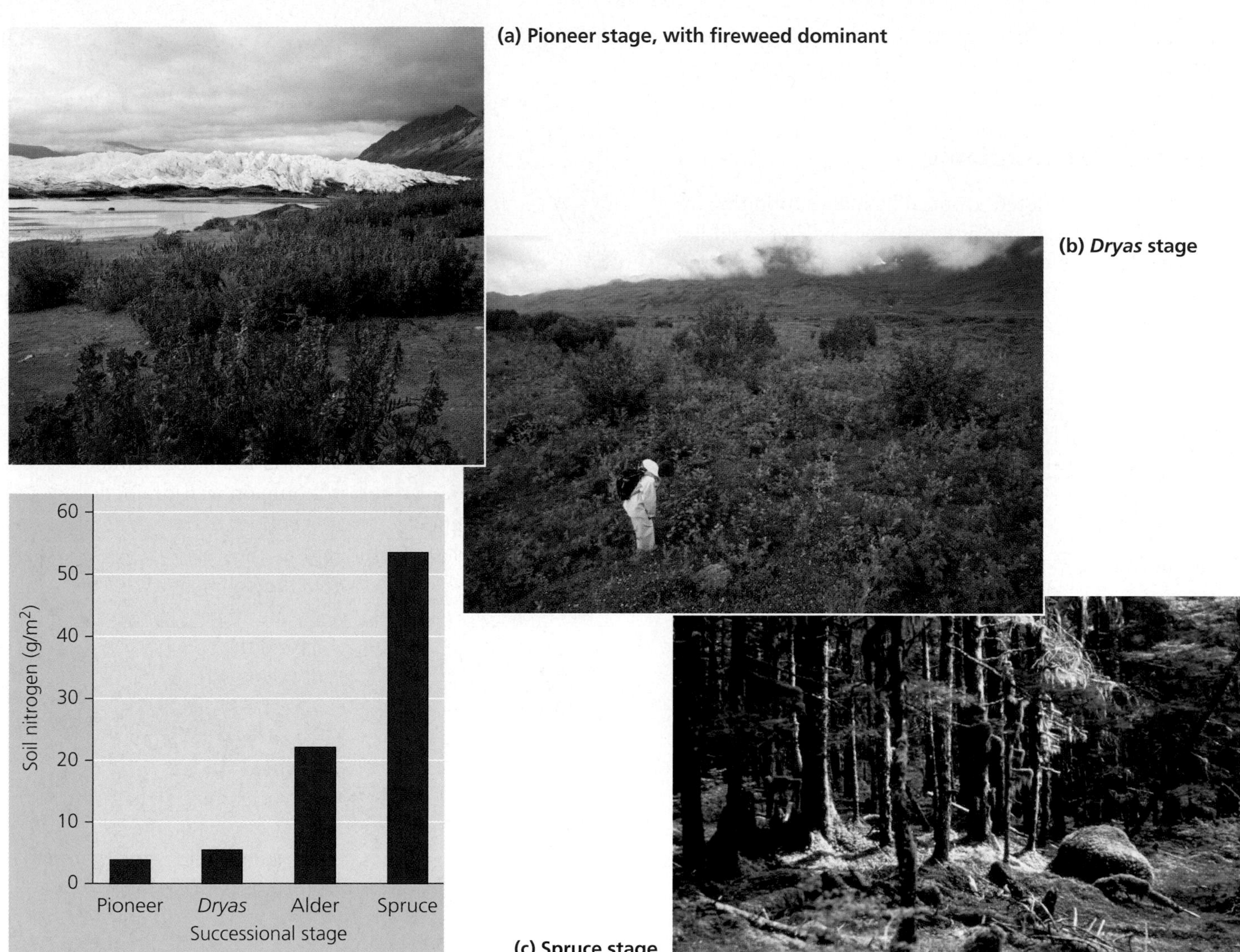

(d) Nitrogen fixation by *Dryas* and alder increases the soil nitrogen content.

▲ Figure 53.24 **Changes in plant community structure and soil nitrogen during succession at Glacier Bay, Alaska.** Plant species composition changes dramatically during succession at Glacier Bay.

by the alder. By altering soil properties, pioneer plant species permit new plant species to grow, and the new plants in turn alter the environment in different ways, contributing to succession.

Concept Check 53.3

1. Why do high and low levels of disturbance usually reduce species diversity? Why does an intermediate level of disturbance promote species diversity?
2. How do primary and secondary succession differ?
3. During succession, how might the early species facilitate the arrival of other species?

For suggested answers, see Appendix A.

Concept 53.4

Biogeographic factors affect community biodiversity

Communities vary greatly in the number of species they contain. Two key factors correlated with a community's species diversity are its geographic location and its size. In the 1850s, both Charles Darwin and Alfred Wallace pointed out that plant and animal life was generally more abundant and varied in the tropics than in other parts of the globe. Darwin and Wallace also noted that small or remote islands have fewer species than large islands or those nearer continents. Such observations imply that biogeographic patterns in biodiversity conform to a set of basic principles rather than being historical coincidences. One way to understand such large-scale patterns in the diversity of life is to examine variation along environmental gradients.

Equatorial-Polar Gradients

As Darwin and Wallace noted, tropical habitats support many more species than do temperate and polar regions. For example, one study found that a 6.6-hectare [1 hectare (ha) = 10,000 m^2] area in Sarawak (part of Malaysia) contained 711 tree species, while a 2-ha plot of deciduous forest in Michigan contained just 10 to 15 tree species. And the whole of Western Europe north of the Alps, an area of more than 2 million km^2, has only 50 tree species. Similarly, there are more than 200 species of ants in Brazil, 73 in Iowa, and only 7 in Alaska.

The two key factors in equatorial-polar gradients of species richness are probably evolutionary history and climate. Over the course of evolutionary time, species diversity may increase in a community as more speciation events occur. And tropical communities are generally older than temperate or polar communities. This age difference stems partly from the fact that the growing season in tropical forests is about five times longer than the growing season in the tundra communities of high latitudes. In effect, biological time, and hence time for speciation, runs about five times faster in the tropics than near the poles. And many polar and temperate communities have repeatedly "started over" as a result of major disturbances in the form of glaciations.

Climate is likely the primary cause of the latitudinal gradient in biodiversity. The two main climatic factors correlated with biodiversity are solar energy input and water availability. These factors can be considered together by measuring a community's rate of **evapotranspiration,** the evaporation of water from soil plus the transpiration of water from plants. *Actual evapotranspiration,* which is determined by the amount of solar radiation, temperature, and water availability, is much higher in hot areas with abundant rainfall than in areas with low temperatures or low precipitation. *Potential evapotranspiration,* a measure of energy availability but not water availability, is determined by the amount of solar radiation and temperature and is highest in regions of high solar radiation and temperature. The species richness of plants and animals correlates with measures of evapotranspiration **(Figure 53.25)**.

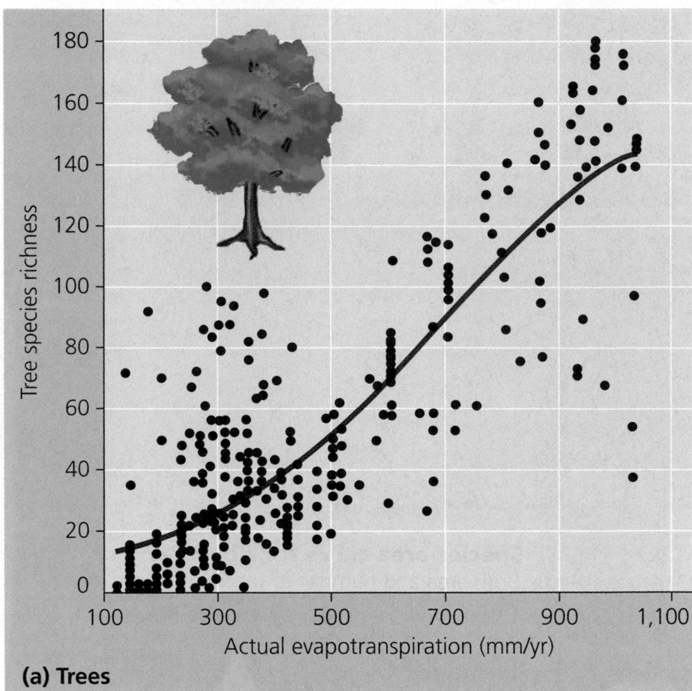

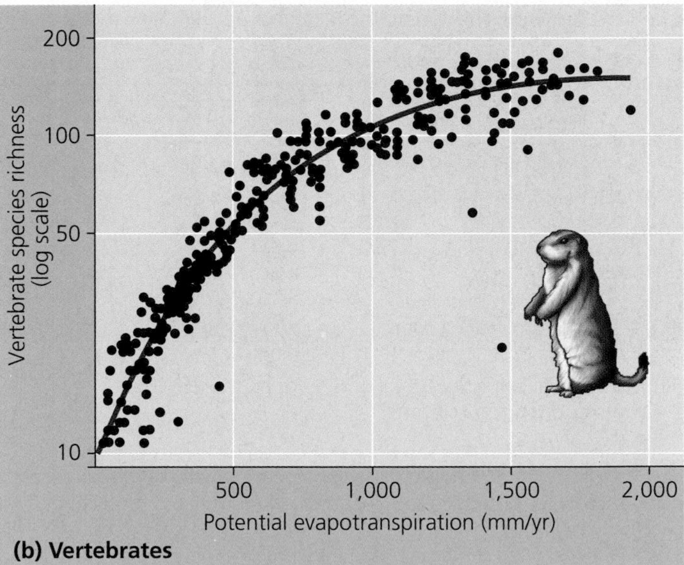

▲ **Figure 53.25 Energy, water, and species richness.** **(a)** Species richness of North American trees increases most predictably with actual evapotranspiration, while **(b)** vertebrate species richness in North America increases most predictably with potential evapotranspiration. Evapotranspiration values are expressed as rainfall equivalents in millimeters per year.

Area Effects

In 1807, Alexander von Humboldt described one of the first patterns of biodiversity to be recognized, the **species-area curve**, which quantifies what probably seems obvious: All other factors being equal, the larger the geographic area of a community, the greater the number of species. The likely explanation for this pattern is that larger areas offer a greater diversity of habitats and microhabitats than smaller areas. In conservation biology, developing species-area curves for the key taxa in a community helps ecologists to predict how the potential loss of a certain area of habitat is likely to affect the community's biodiversity.

Figure 53.26 is a species-area curve for North American breeding birds (birds with breeding populations in the mapped area, as opposed to migrant populations). The slope indicates the extent to which species richness increases with community area. The slopes of different species-area curves vary, depending on the taxon being sampled and the type of

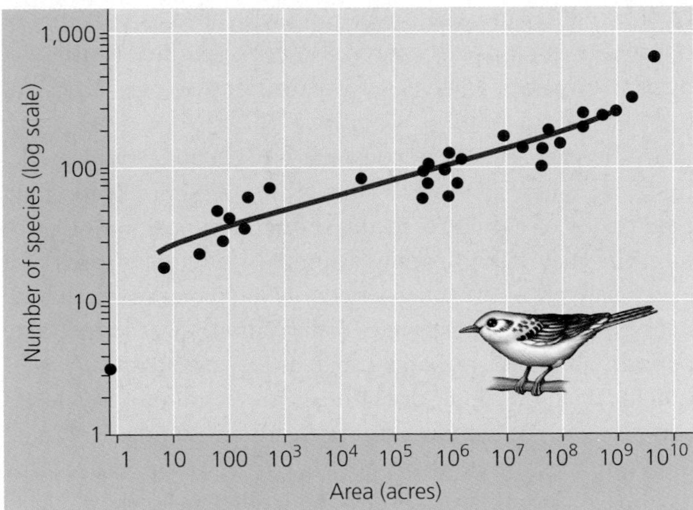

1,000
100
10
1
Number of species (log scale)
1 10 100 10³ 10⁴ 10⁵ 10⁶ 10⁷ 10⁸ 10⁹ 10¹⁰
Area (acres)

▲ **Figure 53.26 Species-area curve for North American breeding birds.** Both area and number of species are plotted on a logarithmic scale. The data points range from a 0.5-acre plot with 3 species in Pennsylvania to the whole United States and Canada (4.6 billion acres) with 625 species.

community. But the basic concept of diversity increasing with increasing area applies in a variety of situations, from surveys of ant diversity in New Guinea to the number of plant species on islands of different sizes. In fact, island biogeography provides some of the best examples of species-area curves, as we see next.

Island Equilibrium Model

Because of their isolation and limited size, islands provide excellent opportunities for studying the biogeographic factors that affect the species diversity of communities. By "islands," we mean not only oceanic islands, but also habitat islands on land, such as lakes, mountain peaks separated by lowlands, or natural woodland fragments surrounded by areas disturbed by humans—in other words, any patch surrounded by an environment not suitable for the "island" species. In the 1960s, American ecologists Robert MacArthur and E. O. Wilson developed a general model of island biogeography identifying the key determinants of species diversity on an island with a given set of physical characteristics **(Figure 53.27)**.

Consider a newly formed oceanic island that receives colonizing species from a distant mainland. Two factors determine the number of species that will eventually inhabit the island: the rate at which new species immigrate to the island and the rate at which species become extinct on the island. And two physical features of the island affect immigration and extinction rates: its size and its distance from the mainland. Small islands generally have lower immigration rates because potential colonizers are less likely to reach a small island. For instance, birds blown out to sea by a storm are more likely to land by chance on a larger island than on a small one. Small islands also have higher extinction rates, as they generally contain fewer resources and less diverse habitats for colonizing species to partition. Distance from the mainland is also important; for two islands of equal size, a closer island generally has a higher immigration rate than one farther away. And because of their higher immigration rates, closer islands also have lower extinction rates, as arriving colonists help sustain the presence of a species on a near island and prevent its extinction.

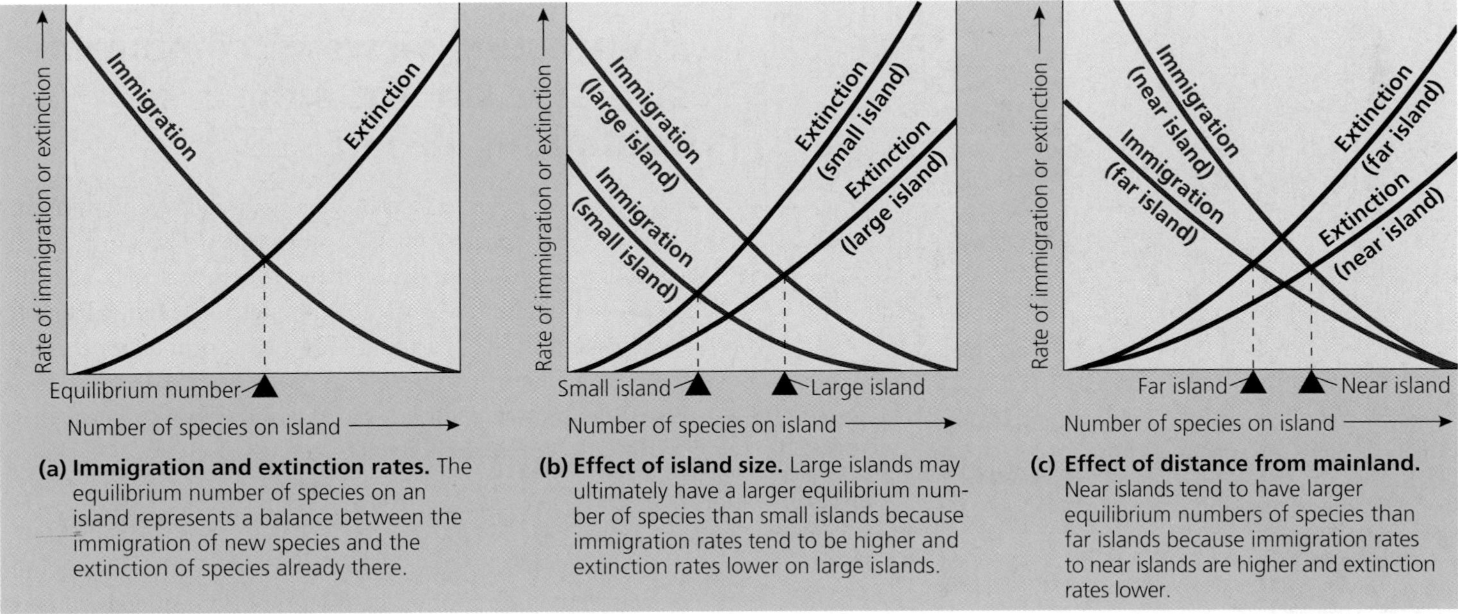

Rate of immigration or extinction →
Immigration
Extinction
Equilibrium number
Number of species on island →
(a) Immigration and extinction rates. The equilibrium number of species on an island represents a balance between the immigration of new species and the extinction of species already there.
Rate of immigration or extinction →
Immigration (large island)
Immigration (small island)
Extinction (small island)
Extinction (large island)
Small island
Large island
Number of species on island →
(b) Effect of island size. Large islands may ultimately have a larger equilibrium number of species than small islands because immigration rates tend to be higher and extinction rates lower on large islands.
Rate of immigration or extinction →
Immigration (near island)
Immigration (far island)
Extinction (far island)
Extinction (near island)
Far island
Near island
Number of species on island →
(c) Effect of distance from mainland. Near islands tend to have larger equilibrium numbers of species than far islands because immigration rates to near islands are higher and extinction rates lower.

▲ **Figure 53.27 The equilibrium model of island biogeography.** Black triangles represent equilibrium numbers of species.

At any given time, an island's immigration and extinction rates are also affected by the number of species already present. As the number of species on the island increases, the immigration rate of new species decreases, because any individual reaching the island is less likely to represent a species that is not already present. At the same time, as more species inhabit an island, extinction rates on the island increase because of the greater likelihood of competitive exclusion.

These relationships make up MacArthur and Wilson's model of island biogeography (see Figure 53.27). Immigration and extinction rates are plotted as a function of the number of species present on the island. This mathematical model is called the island equilibrium model because an equilibrium will eventually be reached where the rate of species immigration equals the rate of species extinction. The number of species at this equilibrium point is correlated with the island's size and distance from the mainland. Like any ecological equilibrium, this species equilibrium is dynamic; immigration and extinction continue, and the exact species composition may change over time.

MacArthur and Wilson's studies of the diversity of plants and animals on many island chains, including the Galápagos Islands, support the prediction that species richness increases with island size, in keeping with species-area theory **(Figure 53.28)**. Species counts also fit the prediction that the number of species decreases with increasing remoteness of the island.

Figure 53.28
Inquiry How does species richness relate to area?

FIELD STUDY Ecologists Robert MacArthur and E. O. Wilson studied the number of plant species on the Galápagos Islands, which vary greatly in size, in relation to the area of each island.

RESULTS

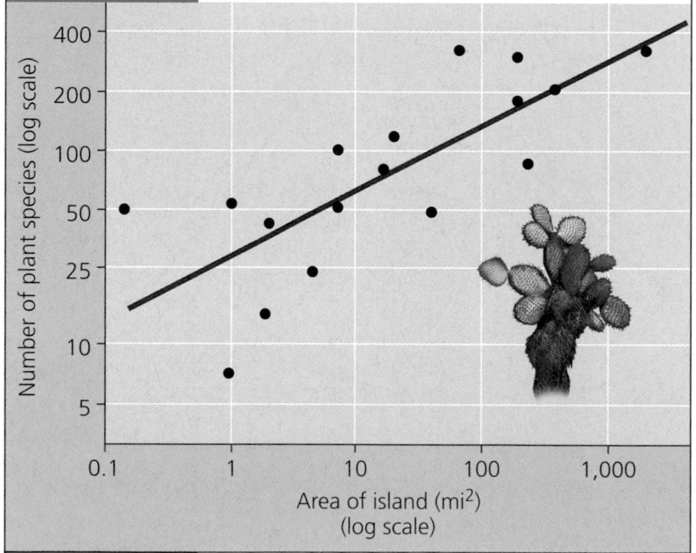

Number of plant species (log scale): 400, 200, 100, 50, 25, 10, 5

Area of island (mi^2) (log scale): 0.1, 1, 10, 100, 1,000

CONCLUSION The results of the study showed that plant species richness increased with island size, supporting the species-area theory.

In the past several decades, the island equilibrium model has come under considerable fire as an oversimplification. Its predictions of equilibria in the species composition of communities may apply in only a limited number of cases and over relatively short time periods, where colonization is the main process affecting species composition. Over longer time periods, abiotic disturbances such as storms, adaptive evolutionary changes, and speciation generally alter the species composition and community structure on islands. However, more important than the model's widespread applicability is its impact in stimulating discussion and research on the effects of habitat size on species diversity, a topic of vital importance for conservation biology.

Concept Check 53.4

1. Summarize the general relationship between area and species richness.
2. Describe two hypotheses that explain why species diversity is greater in tropical regions than in temperate and polar regions.
3. Describe how an island's size and distance from the mainland affect the island's species richness.

For suggested answers, see Appendix A.

Concept 53.5

Contrasting views of community structure are the subject of continuing debate

Now that we have looked at various aspects of community structure and species diversity, how can we account for the particular species found together as members of a community? Two different views on this question emerged among ecologists in the 1920s and 1930s, based primarily on observations of plant distribution. We will briefly describe these historical arguments because they are the forerunners of two community models still debated today.

Integrated and Individualistic Hypotheses

In the early 1900s, F. E. Clements, of the Carnegie Institute of Washington, proposed the **integrated hypothesis** of community structure. This hypothesis describes a community as an assemblage of closely linked species, locked into association by mandatory biotic interactions that cause the community to function as an integrated unit—in effect, as a superorganism. Evidence for the integrated view includes the observation that certain species of plants are consistently found together as a group. For example, deciduous forests in the northeastern United States usually include particular species of oak, maple, birch, and beech, along with a specific assemblage of shrubs and vines.

Shortly after the integrated hypothesis was proposed, H. A. Gleason, of the University of Chicago, enunciated a different view of community structure. His **individualistic hypothesis** depicts a plant community as a chance assemblage of species found in the same area simply because they happen to have similar abiotic requirements—for example, for temperature, rainfall, and soil type.

These two very different ways of thinking about community structure—integrated versus individualistic—suggest different priorities in studying biological communities. The integrated hypothesis emphasizes entire assemblages of species as the essential units for understanding the interrelationships and distributions of organisms, whereas the individualistic hypothesis emphasizes studying single species.

Both the integrated and the individualistic hypotheses see communities as involving interspecific interactions. However, they offer different perspectives on how species are assembled into a community. The two hypotheses also make contrasting predictions about how plant species should be distributed along an environmental gradient, such as a gradual change in moisture or temperature over some distance. The integrated hypothesis predicts that species should be clustered into discrete communities with noticeable boundaries, because the presence or absence of a particular species is largely governed by the presence or absence of other species with which it interacts in its group **(Figure 53.29a)**. In contrast, the individualistic hypothesis predicts that communities should generally lack discrete geographic boundaries because each species has an independent (that is, individualistic) distribution along the environmental gradient. In other words, each species is distributed according to its tolerance ranges for abiotic factors that vary along a gradient, and communities should change continuously along the gradient, with the addition or loss of particular species **(Figure 53.29b)**.

In most actual cases, especially where there are broad regions characterized by gradients of environmental variation, the composition of plant communities does seem to change continuously, with each species more or less independently distributed **(Figure 53.29c)**. Such distributions generally support the view of plant communities as relatively loose associations without discrete boundaries. However, where some key factor in the physical environment, such as soil type, changes abruptly, adjacent communities are delineated by correspondingly sharp boundaries. Although such sharp boundaries between plant communities occur rarely in nature, humans have produced many artificial sharp boundaries by altering the landscape through agriculture and forestry, as you will read in Chapter 55.

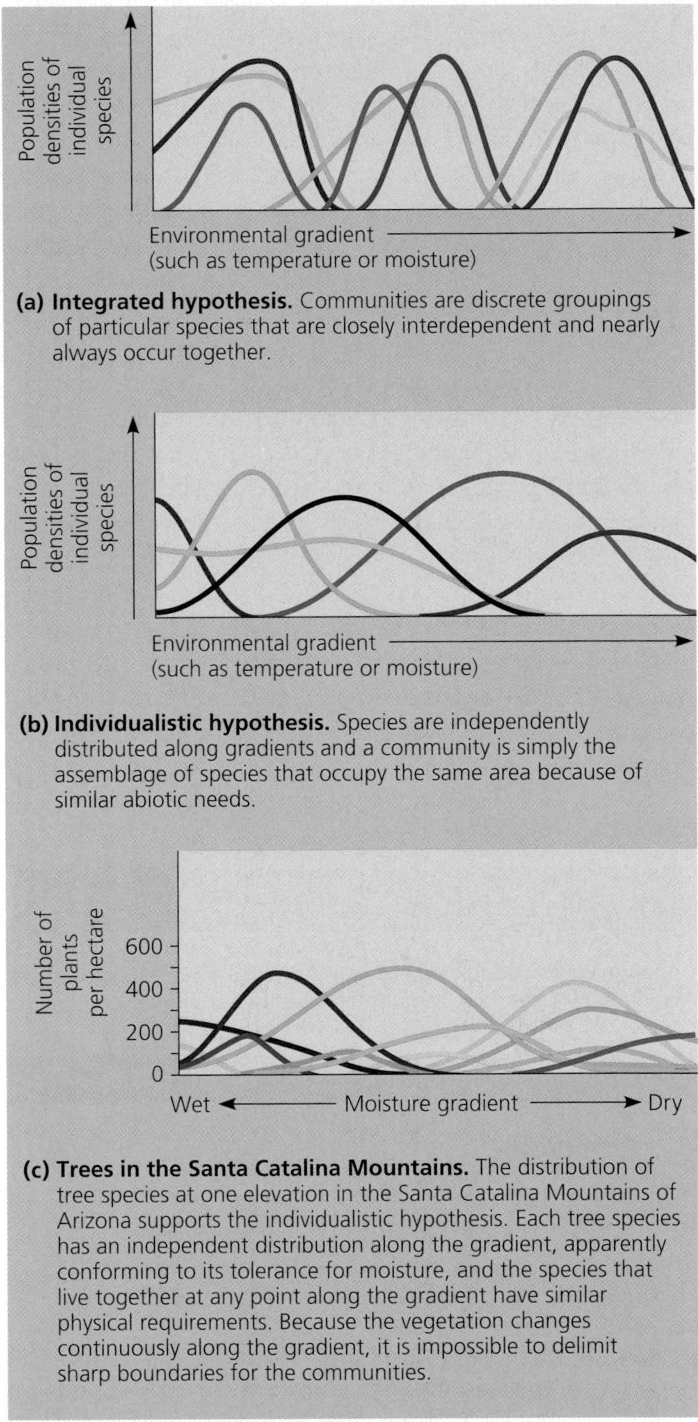

(a) Integrated hypothesis. Communities are discrete groupings of particular species that are closely interdependent and nearly always occur together.

(b) Individualistic hypothesis. Species are independently distributed along gradients and a community is simply the assemblage of species that occupy the same area because of similar abiotic needs.

(c) Trees in the Santa Catalina Mountains. The distribution of tree species at one elevation in the Santa Catalina Mountains of Arizona supports the individualistic hypothesis. Each tree species has an independent distribution along the gradient, apparently conforming to its tolerance for moisture, and the species that live together at any point along the gradient have similar physical requirements. Because the vegetation changes continuously along the gradient, it is impossible to delimit sharp boundaries for the communities.

▲ **Figure 53.29 Testing the integrated and individualistic hypotheses of communities.** Ecologist Robert Whittaker tested these two hypotheses by graphing the abundance of different plant species (*y*-axis) along environmental gradients of abiotic factors such as temperature or moisture (*x*-axis). Each colored curve represents the abundance of one species.

Rivet and Redundancy Models

The individualistic hypothesis is generally accepted by plant ecologists today, but further debate arises in the application of these ideas to the animals in a community. In 1981, American ecologists Paul and Anne Ehrlich proposed that species in a community are like the rivets in the wings of airplanes: Not all the rivets are required to hold the wing together, but if someone started taking out the rivets one by one, we would become concerned about flying in that airplane. The **rivet model** of communities reincarnates the integrated model that Clements suggested for plant communities. It suggests that most of the species in a community are associated tightly with particular other species in a web of life. Thus, reducing or increasing the abundance of one species in a community affects many other species.

In 1992, Australian ecologist Brian Walker suggested an opposing view of communities, the **redundancy model.** According to this model, most of the species in a community are not tightly associated with one another, and the web of life is very loose. An increase or decrease in one species in a community has little effect on other species, which operate independently, just as Gleason had suggested 80 years before in his individualistic model of the plant community. Species within a community are redundant. For example, if one predator disappears, another predatory species in the community will usually take its place as a consumer of specific prey. If one pollinator ceases to visit a particular species of flowering plant because the pollinator has disappeared from the area, another pollinator species will do the job.

It is important to keep in mind that the rivet and redundancy models, like the integrated and individualistic models, represent extremes; most communities probably lie somewhere in the middle. So why is the debate that began early in the 20th century continuing? The main reason is that we still do not have enough information to answer the fundamental questions raised by Clements and Gleason: Are communities loose associations of species or highly integrated units? To fully assess the contrasting models, we need to study how species interact in communities and how tight these associations are. For example, how would the impacts on a community differ if the one species lost from it were a keystone species compared to a rare species with relatively little influence on community structure?

Such questions about community structure are important because they underlie many current environmental problems, such as how the extinction of particular species may affect the functioning of ecosystems. We will return to these models and the questions they stimulate in Chapter 55. But first, we will take a look at ecosystem ecology in Chapter 54.

Concept Check 53.5

1. Describe the individualistic and integrated hypotheses of community structure and how these hypotheses are related to the rivet and redundancy models.

For suggested answers, see Appendix A.

Chapter 53 Review

Go to www.campbellbiology.com or the student CD-ROM to explore Activities, Investigations, and other interactive study aids.

SUMMARY OF KEY CONCEPTS

A community's interactions include competition, predation, herbivory, symbiosis, and disease

- Populations are linked by interspecific interactions that affect the survival and reproduction of the species engaged in the interaction (pp. 1159–1160).
- **Competition (pp. 1160–1161)** Interspecific competition occurs when species compete for a particular resource that is in short supply. The ecological niche is the total of the organism's use of the biotic and abiotic resources in its environment. The competitive exclusion principle states that two species cannot coexist in the same community if their niches are identical.
- **Predation (pp. 1161–1162)** Predation refers to an interaction in which one species, the predator, kills and eats the other, the prey. Predation has led to diverse adaptations, including mimicry.

Table 53.1 Interspecific Interactions

Interaction	Effects on Interacting Species
Competition (−/−)	The interaction can be detrimental to both species.
Predation (+/−) Herbivory (+/−) Parasitism (+/−) Disease (+/−)	The interaction is beneficial to one species and detrimental to the other.
Mutualism (+/+)	The interaction is beneficial to both species.
Commensalism (+/0)	One species benefits from the interaction, and the other species is unaffected by it.

- **Herbivory (p. 1163)** Herbivory, an interaction in which an herbivore eats parts of a plant or an alga, has led to the evolution of various chemical and mechanical defenses in plant species as well as consequent adaptations by herbivorous species.

▶ **Parasitism** (p. 1163) In parasitism, one organism, the parasite, derives its nourishment from another organism, its host, which is harmed in the process. Parasitism exerts substantial influence on populations and the structure of communities.

▶ **Disease** (p. 1163) The effects of disease on populations and communities is similar to that of parasites. Most pathogens are microscopic.

▶ **Mutualism** (p. 1164) Both species benefit in a mutualistic interaction.

▶ **Commensalism** (p. 1164) In commensalism, one species benefits and the other is not affected. There are few if any cases of pure commensalism.

▶ **Interspecific Interactions and Adaptation** (pp. 1164–1165) Evidence for coevolution, involving reciprocal genetic change by interacting populations, is scarce. However, generalized adaptation of organisms to other organisms in their environment is a fundamental feature of life.
Activity *Interspecific Interactions*
Biology Labs On-Line *PopulationEcologyLab*

Concept 53.2

Dominant and keystone species exert strong controls on community structure

▶ **Species Diversity** (pp. 1165–1166) Species diversity measures the number of species in a community—its species richness—and their relative abundance. A community with an even species abundance is more diverse than one in which one or two species are abundant and the remainder rare.
Investigation *How Are Impacts on Community Diversity Measured?*

▶ **Trophic Structure** (pp. 1166–1168) Trophic structure is a key factor in community dynamics. Food chains link the trophic levels from producers to top carnivores. Branching food chains and complex trophic interactions form food webs. The energetic hypothesis suggests that the length of a food chain is limited by the inefficiency of energy transfer along the chain. The dynamic stability hypothesis proposes that long food chains are less stable than short chains.
Activity *Food Webs*

▶ **Species with a Large Impact** (pp. 1168–1170) Dominant species and keystone species exert strong controls on community structure. Dominant species are the most abundant species in a community, and their dominance is achieved by having high competitive ability. Keystone species are usually less abundant species that exert a disproportionate influence on community structure because of their ecological niche. Ecosystem "engineers," also called foundation species, exert influence on community structure through their effects on the physical environment.

▶ **Bottom-Up and Top-Down Controls** (pp. 1170–1171) The bottom-up model proposes a unidirectional influence from lower to higher trophic levels, in which nutrients and other abiotic factors are the main determinants of community structure, including the abundance of primary producers. The top-down model proposes that control of each trophic level comes from the trophic level above, with the result that predators control herbivores, which in turn control primary producers.

Disturbance influences species diversity and composition

▶ **What Is Disturbance?** (pp. 1172–1173) Increasingly, evidence suggests that disturbance and nonequilibrium rather than stability and equilibrium are the norm for most communities. According to the intermediate disturbance hypothesis, moderate levels of disturbance can foster higher species diversity than can low or high levels of disturbance.

▶ **Human Disturbance** (p. 1173) Humans are the most widespread agents of disturbance, and their disturbance to communities usually reduces species diversity. Humans also prevent some naturally occurring disturbances, such as fire, which can be important to community structure.

▶ **Ecological Succession** (pp. 1173–1175) Ecological succession is the sequence of community and ecosystem changes after a disturbance. Primary succession occurs where no soil exists when succession begins; secondary succession begins in an area where soil remains after a disturbance. Mechanisms producing community change during succession include facilitation and inhibition.
Activity *Primary Succession*

Concept 53.4

Biogeographic factors affect community biodiversity

▶ **Equatorial-Polar Gradients** (p. 1176) Species richness generally declines along an equatorial-polar gradient and is especially great in the tropics. The greater age of tropical environments may account for the greater species richness of the tropics. Climate also influences the biodiversity gradient through energy (heat and light) and water.

▶ **Area Effects** (pp. 1176–1177) Species richness is directly related to a community's geographic size, a principle formalized in the species-area curve.

▶ **Island Equilibrium Model** (pp. 1177–1178) Species richness on islands depends on island size and distance from the mainland. The equilibrium model of island biogeography maintains that species richness on an ecological island levels off at some dynamic equilibrium point, where new immigrations are balanced by extinctions. The island equilibrium model has been questioned in recent years.
Activity *Exploring Island Biogeography*
Graph It *Species-Area Effect and Island Biogeography*

Concept 53.5

Contrasting views of community structure are the subject of continuing debate

▶ **Integrated and Individualistic Hypotheses** (pp. 1178–1179) The integrated hypothesis states that the species within a community are locked into particular biotic interactions. The individualistic hypothesis proposes that communities are loosely organized associations of independently distributed species with the same abiotic requirements.

▶ **Rivet and Redundancy Models** (p. 1180) The rivet model suggests that all the species in a community are linked together in a tight web of interactions, so that the loss of even a single species has strong repercussions for the community. The redundancy model proposes that if a species is lost from a community, other species will fill the gap.

TESTING YOUR KNOWLEDGE

Self-Quiz

1. The feeding relationships among the species in a community determine the community's
   a. secondary succession.
   b. ecological niche.
   c. trophic structure.
   d. species-area curve.
   e. species richness.

2. The competitive exclusion principle states that
   a. two species cannot coexist in the same habitat.
   b. competition between two species always causes extinction or emigration of one species.
   c. competition in a population promotes survival of the best-adapted individuals.
   d. two species with the exact same niche cannot coexist in a community.
   e. species that compete usually coevolve.

3. Keystone predators maintain species diversity in a community by
   a. competitively excluding other predators.
   b. preying on the community's dominant species.
   c. allowing immigration of other predators.
   d. reducing the number of disruptions in the community.
   e. coevolving with their prey.

4. Food chains are usually short mainly because
   a. only a single species of herbivore feeds on each plant species.
   b. local extinction of a species causes extinction of the other species in its food chain.
   c. most of the energy in a trophic level is lost as it passes to the next higher level.
   d. predator species tend to be less diverse and less abundant than prey species.
   e. most producers are inedible.

5. According to the rivet model of community structure,
   a. species share a community because of similar abiotic requirements.
   b. communities generally lack sharp geographic boundaries.
   c. if a species disappears from the community, its role will be assumed by another species.
   d. all species in a natural community contribute to the community's integrity.
   e. dependence on a particular type of soil rivets plant species into a tightly networked community.

6. A community's species diversity is
   a. increased by frequent massive disturbance.
   b. increased by stable conditions with no disturbance.
   c. increased by moderate levels of disturbance.
   d. increased when humans intervene to eliminate disturbance.
   e. increased by intensive disturbance by humans.

7. Which of the following is an example of Müllerian mimicry?
   a. a butterfly that resembles a leaf
   b. two poisonous frogs with similar color patterns
   c. a minnow with spots that look like large eyes
   d. a beetle that resembles a scorpion
   e. a carnivorous fish with a wormlike tongue that lures prey

8. Which of the following could qualify as a top-down control on a grassland community?
   a. limitation of plant biomass by rainfall amount
   b. influence of temperature on competition among plants
   c. influence of soil nutrients on the abundance of grasses versus wildflowers
   d. effect of grazing intensity by bison on plant species diversity
   e. effect of humidity on plant growth rates

9. The most plausible hypothesis to explain why species richness is higher in tropical than in temperate regions is that
   a. tropical communities are younger.
   b. tropical regions have more available water and higher levels of solar radiation.
   c. warmer temperatures cause more rapid speciation.
   d. biodiversity increases as evapotranspiration decreases.
   e. tropical regions have very high rates of immigration and very low rates of extinction.

10. According to the equilibrium model of island biogeography, species richness would be greatest on an island that is
   a. small and remote.
   b. large and remote.
   c. large and close to a mainland.
   d. small and close to a mainland.
   e. environmentally homogeneous.

For Self-Quiz answers, see Appendix A.

Go to the website or CD-ROM for more quiz questions.

Evolution Connection

Explain why adaptations of particular organisms to interspecific competition may not necessarily represent instances of character displacement. What would a researcher have to demonstrate about two competing species to make a convincing case for character displacement?

Scientific Inquiry

An ecologist studying plants in the desert performed the following experiment. She staked out two identical plots, each of which included a few sagebrush plants and numerous small annual wildflowers. She found the same five wildflower species in roughly equal numbers on both plots. She then enclosed one of the plots with a fence to keep out kangaroo rats, the most common grain-eaters of the area. After two years, four of the wildflower species were no longer present in the fenced plot, but one species had increased drastically. The control plot had not changed in species diversity. Using the principles of community ecology, propose a hypothesis to explain her results. What additional evidence would support your hypothesis?

Biology Labs On-Line *PopulationEcologyLab*
Investigation *How Are Impacts on Community Diversity Measured?*

Science, Technology, and Society

By 1935, hunting and trapping had eliminated wolves from the United States except for Alaska. Because wolves have since been protected as an endangered species, they have moved south from Canada and have become reestablished in the Rocky Mountains and northern Great Lakes region. Conservationists who would like to speed up wolf recovery have reintroduced wolves into Yellowstone National Park. Local ranchers are opposed to bringing back the wolves because they fear predation on their cattle and sheep. What are some reasons for reestablishing wolves in Yellowstone National Park? What effects might the reintroduction of wolves have on the ecological communities in the region? What might be done to mitigate the conflicts between ranchers and wolves?

54 Ecosystems

▲ **Figure 54.1 An aquarium, an ecosystem bounded by glass.**

Key Concepts

54.1 Ecosystem ecology emphasizes energy flow and chemical cycling

54.2 Physical and chemical factors limit primary production in ecosystems

54.3 Energy transfer between trophic levels is usually less than 20% efficient

54.4 Biological and geochemical processes move nutrients between organic and inorganic parts of the ecosystem

54.5 The human population is disrupting chemical cycles throughout the biosphere

Overview

Ecosystems, Energy, and Matter

An **ecosystem** consists of all the organisms living in a community as well as all the abiotic factors with which they interact. Ecosystems can range from a microcosm, such as the aquarium in **Figure 54.1**, to a large area such as a lake or forest. As with populations and communities, the boundaries of ecosystems are usually not discrete. Cities and farms are examples of human-dominated ecosystems. Many ecologists regard the entire biosphere as a global ecosystem, a composite of all the local ecosystems on Earth.

Regardless of an ecosystem's size, its dynamics involve two processes that cannot be fully described by population or community processes and phenomena: energy flow and chemical cycling. Energy enters most ecosystems in the form of sunlight. It is then converted to chemical energy by autotrophic organisms, passed to heterotrophs in the organic compounds of food, and dissipated in the form of heat. Chemical elements, such as carbon and nitrogen, are cycled among abiotic and biotic components of the ecosystem. Photosynthetic organisms assimilate these elements in inorganic form from the air, soil, and water and incorporate them into organic molecules, some of which are consumed by animals. The elements are returned in inorganic form to the air, soil, and water by the metabolism of plants and animals and by other organisms, such as bacteria and fungi, that break down organic wastes and dead organisms.

Both energy and matter move through ecosystems via the transfer of substances during photosynthesis and feeding relationships. However, because energy, unlike matter, cannot be recycled, an ecosystem must be powered by a continuous influx of energy from an external source—in most cases, the sun. Energy flows through ecosystems, while matter cycles within them.

Resources critical to human survival and welfare, ranging from the food we eat to the oxygen we breathe, are products of ecosystem processes. In this chapter, we will explore the dynamics of energy flow and chemical cycling in ecosystems and consider some of the impacts of human activities on these processes.

Concept 54.1

Ecosystem ecology emphasizes energy flow and chemical cycling

Ecosystem ecologists view ecosystems as transformers of energy and processors of matter. By grouping the species in a community into trophic levels of feeding relationships (see Chapter 53), we can follow the transformation of energy in the ecosystem and map the movements of chemical elements through the biotic community.

Ecosystems and Physical Laws

Because ecosystem ecologists work at the interface between organisms and the physical environment, much of their analysis of ecosystem dynamics derives from well-established laws of physics and chemistry. The principle of conservation of energy, which you studied in Chapter 8, states that energy cannot be created or destroyed but only transformed. Thus, we can potentially account for the transfer of energy through an ecosystem from its input as solar radiation to its release as heat from organisms. Plants and other photosynthetic organisms convert solar energy to chemical energy, but the total amount of energy does not change. The total amount of energy stored in organic molecules plus the amounts reflected and dissipated as heat must equal the total solar energy intercepted by the plant. One area of ecosystem ecology involves computing such energy budgets and tracing energy flow through ecosystems in order to understand the factors controlling these energy transfers.

One implication of the second law of thermodynamics is that energy conversions cannot be completely efficient; some energy is always lost as heat in any conversion process (see Chapter 8). This idea suggests that we can measure the efficiency of ecological energy conversions in the same way we measure the efficiency of lightbulbs and car engines.

Energy flowing through ecosystems is ultimately dissipated into space as heat, so if the sun were not continuously providing energy to Earth, ecosystems would vanish. In contrast, chemical elements are continually recycled. A carbon or nitrogen atom moves from trophic level to trophic level and eventually to the decomposers and then back again in an endless cycle. Elements are not lost on a global scale, although they may move from one ecosystem to another. The measurement and analysis of chemical cycling in ecosystems and in the entire biosphere represent another area of study in ecosystem ecology.

Trophic Relationships

As you read in Chapter 53, ecologists assign species to trophic levels on the basis of their main source of nutrition and energy. **Figure 54.2** summarizes the trophic relationships in an ecosystem.The trophic level that ultimately supports all others consists of autotrophs, also called the **primary producers** of the ecosystem. Most autotrophs are photosynthetic organisms that use light energy to synthesize sugars and other organic compounds, which they then use as fuel for cellular respiration and as building material for growth. Plants, algae, and photosynthetic prokaryotes are the biosphere's main autotrophs, although chemosynthetic prokaryotes are the primary producers in certain ecosystems, such as deep-sea hydrothermal vents (see Figure 50.17).

Organisms in trophic levels above the primary producers are heterotrophs, which directly or indirectly depend on the photosynthetic output of primary producers. Herbivores, which eat plants and other primary producers, are **primary consumers.** Carnivores that eat herbivores are **secondary consumers,** and carnivores that eat other carnivores are **tertiary consumers.** Another important group of heterotrophs consists of the detritivores. **Detritivores,** or **decomposers,** are consumers that get their energy from **detritus,** which is nonliving organic material, such as the remains of dead organisms, feces, fallen leaves, and wood.

Decomposition

The prokaryotes, fungi, and animals that feed as detritivores form a major link between the primary producers and the

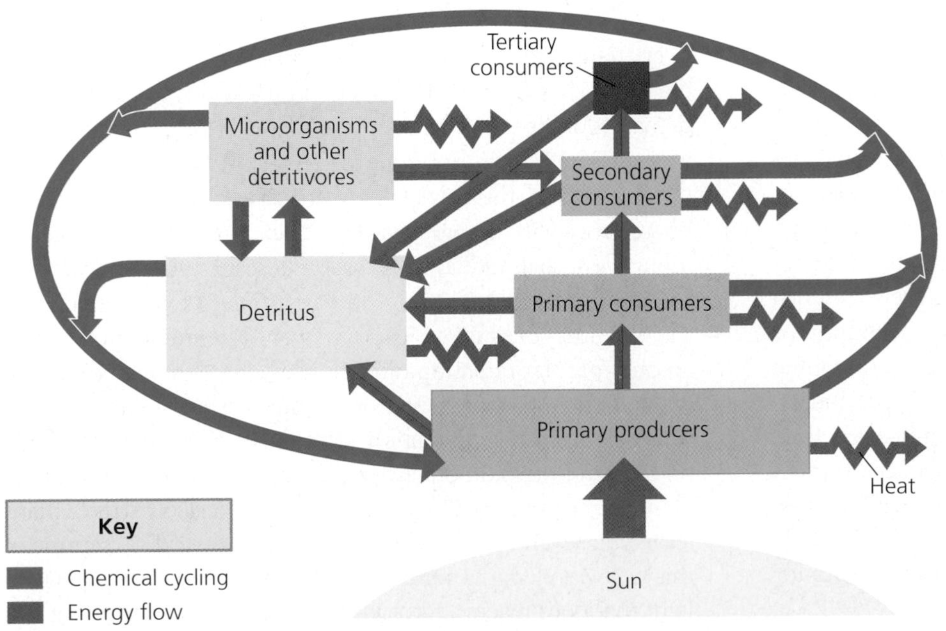

◀ **Figure 54.2 An overview of energy and nutrient dynamics in an ecosystem.** Energy enters, flows through, and exits an ecosystem, whereas chemical nutrients cycle within it. In this generalized scheme, energy (orange arrows) enters from the sun as radiation, moves as chemical energy transfers through the food web, and exits as heat radiated into space. Most transfers of nutrients (blue arrows) through the trophic levels lead eventually to detritus; the nutrients then cycle back to the primary producers.

▲ **Figure 54.3 Fungi decomposing a dead tree.**

consumers in an ecosystem. In a forest, for example, birds might feed on earthworms that have been feeding on leaf litter and its associated prokaryotes and fungi. But even more important than this channeling of resources from producers to consumers is the role that detritivores play in making vital chemical elements available to producers.

Detritivores decompose the organic material in an ecosystem and transfer the chemical elements in inorganic forms to abiotic reservoirs such as soil, water, and air. Producers can then recycle these elements into organic compounds. All organisms perform some decomposition, breaking down organic molecules during cellular respiration, for example. But an ecosystem's main decomposers are prokaryotes and fungi, which secrete enzymes that digest organic material; these decomposers then absorb the breakdown products **(Figure 54.3)**. Decomposition by prokaryotes and fungi accounts for most of the conversion of organic materials from all trophic levels to inorganic compounds usable by primary producers, thereby closing the loop of an ecosystem's chemical cycling. Because most detritivores are not readily visible to the human eye, decomposition is an underappreciated ecological process. Yet if decomposition stopped, all life on Earth would cease as detritus piled up and the supply of chemical ingredients for new organic matter was exhausted.

Concept Check 54.1

1. Why is the transfer of energy in an ecosystem referred to as energy flow, not energy cycling?
2. How does the second law of thermodynamics explain why an ecosystem's energy supply must be continually replenished?
3. How are detritivores essential to sustaining ecosystems?

For suggested answers, see Appendix A.

Concept 54.2

Physical and chemical factors limit primary production in ecosystems

The amount of light energy converted to chemical energy (organic compounds) by autotrophs during a given time period is an ecosystem's **primary production.** This photosynthetic product is the starting point for studies of ecosystem metabolism and energy flow.

Ecosystem Energy Budgets

Most primary producers use light energy to synthesize energy-rich organic molecules, which can subsequently be broken down to generate ATP (see Chapter 10). Consumers acquire their organic fuels secondhand (or even third- or fourthhand) through food webs such as that in Figure 53.13. Therefore, the extent of photosynthetic production sets the spending limit for the energy budget of the entire ecosystem.

The Global Energy Budget

Every day, Earth is bombarded by about 10^{22} joules of solar radiation (1 J = 0.239 cal). This is enough energy to supply the demands of the entire human population for approximately 25 years at 2004 consumption levels. As described in Chapter 50, the intensity of the solar energy striking Earth and its atmosphere varies with latitude, with the tropics receiving the greatest input. Most solar radiation is absorbed, scattered, or reflected by the atmosphere, depending on variations in cloud cover and the quantity of dust in the air over different regions. The amount of solar radiation ultimately reaching the surface of the globe limits the photosynthetic output of ecosystems.

Much of the solar radiation that reaches Earth's surface lands on bare ground and bodies of water that either absorb or reflect the incoming energy. Only a small fraction actually strikes plant leaves, algae, and photosynthetic prokaryotes, and only some of this is of wavelengths suitable for photosynthesis. Of the visible light that does reach photosynthetic organisms, only about 1%

is converted to chemical energy by photosynthesis, though this yield varies with the type of organism, light level, and other factors. Although the fraction of the total incoming solar radiation that is ultimately trapped by photosynthesis is very small, primary producers on Earth collectively create about 170 billion tons of organic material per year.

Gross and Net Primary Production

Total primary production in an ecosystem is known as that ecosystem's **gross primary production (GPP)**—the amount of light energy that is converted to chemical energy by photosynthesis per unit time. Not all of this production is stored as organic material in the growing plants, because the plants use some of the molecules as fuel in their own cellular respiration. **Net primary production (NPP)** is equal to gross primary production minus the energy used by the primary producers for respiration (R):

$$NPP = GPP - R$$

To ecologists, net primary production is the key measurement because it represents the storage of chemical energy that will be available to consumers in the ecosystem. For example, in forests, net primary production may be as little as one-fourth that of gross primary production. Trees must support large masses of trunks, branches, and roots through respiration; thus, more energy is spent in respiration in forests than in herbaceous and crop communities.

Net primary production can be expressed as energy per unit area per unit time ($J/m^2/yr$) or as biomass (weight) of vegetation added to the ecosystem per unit area per unit time ($g/m^2/yr$). (Note that biomass is usually expressed in terms of the dry weight of organic material because water molecules contain no usable energy and because the water content of plants varies over short periods of time.) An ecosystem's net primary production should not be confused with the *total* biomass of photosynthetic autotrophs present at a given time, a measure called the *standing crop*. Net primary production is the amount of *new* biomass added in a given period of time. Although a forest has a very large standing crop biomass, its net primary production may actually be less than that of some grasslands, which do not accumulate vegetation because animals consume the plants rapidly and because some of the plants are annuals.

Different ecosystems vary considerably in their net primary production as well as in their contribution to the total net primary production on Earth **(Figure 54.4)**. Tropical rain forests are among the most productive terrestrial ecosystems and contribute a large portion of the planet's overall net primary

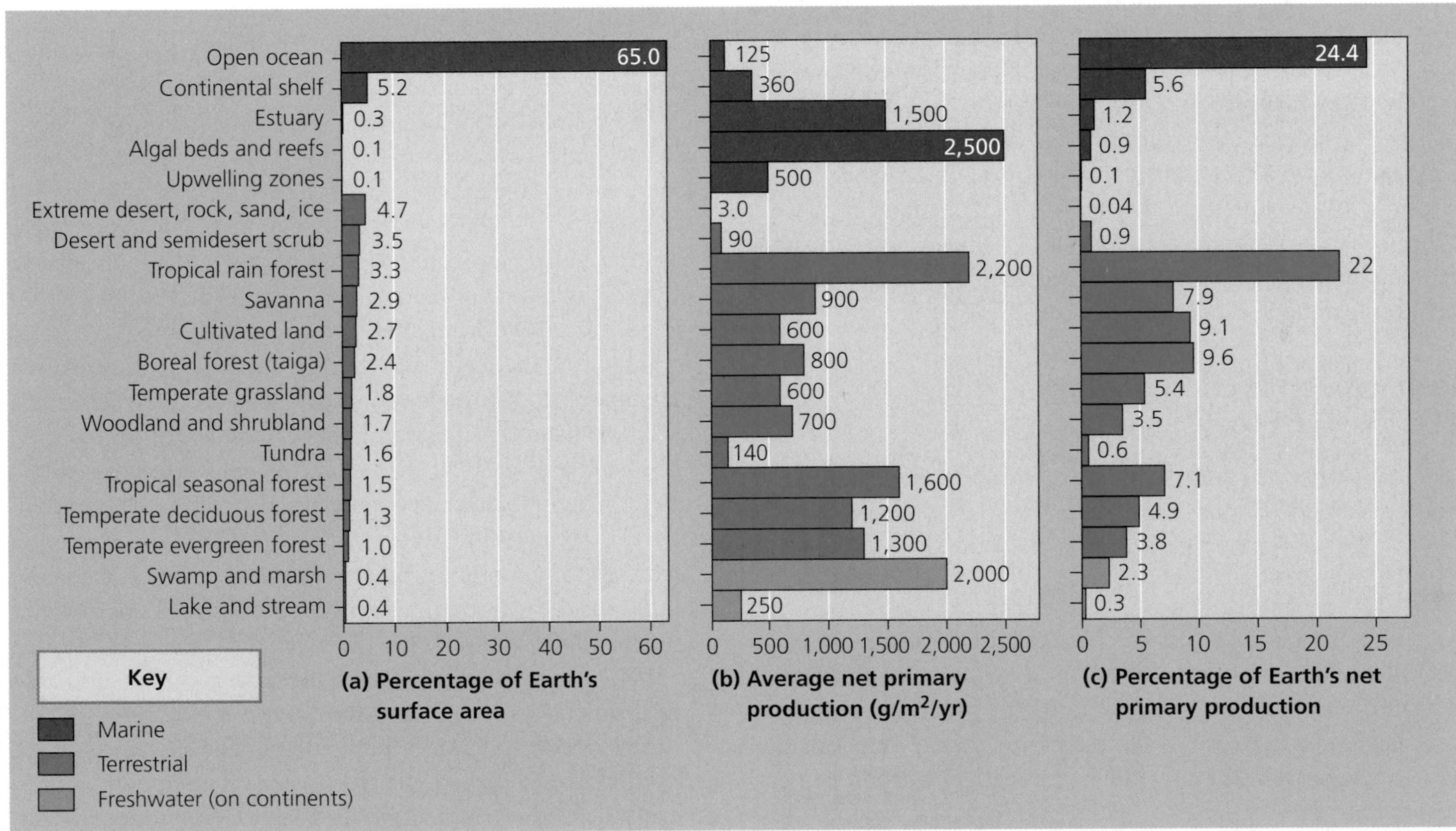

▲ **Figure 54.4 Net primary production of different ecosystems.** The geographic extent **(a)** and the net primary production per unit area **(b)** of different ecosystems determine their total contribution to worldwide net primary production **(c)**.

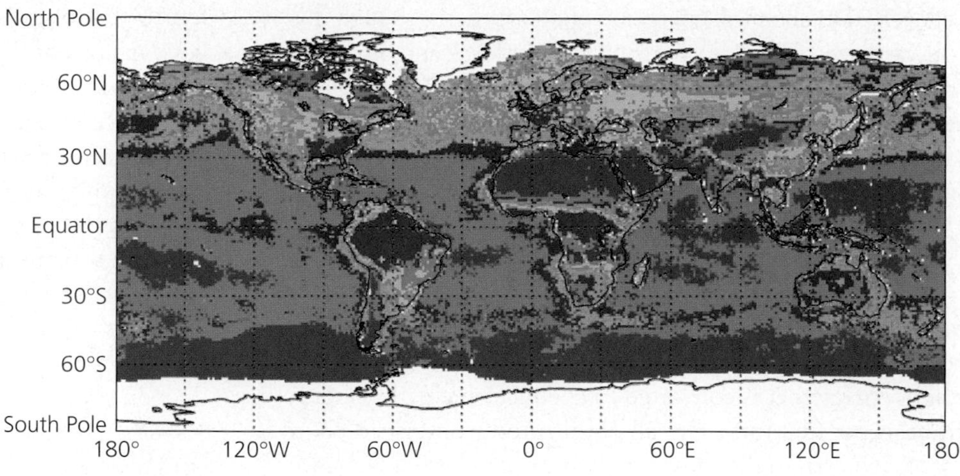

▶ **Figure 54.5 Regional annual net primary production for Earth.** The image is based on data, such as chlorophyll density, collected by satellites. Lighter violet indicates regions of lowest net primary production, with increasing net primary production indicated by darker violet, light blue, dark green, light green, yellow, orange, and red in that order. Ocean data are averages from 1978 to 1983. Land data are averages from 1982 to 1990.

production. Estuaries and coral reefs also have very high net primary production, but their total contribution to global net primary production is relatively small because these ecosystems cover only about one-tenth the area covered by tropical rain forests. The open ocean contributes more net primary production than any other single ecosystem, but this is because of its very large size; net primary production per unit area is relatively low.

Satellite images now provide a means of studying global patterns of primary production. The most striking impression from such maps is how unproductive most of the oceans are, per unit area, in contrast to the high production of tropical forests. Overall, terrestrial ecosystems contribute approximately two-thirds of global net primary production and marine ecosystems approximately one-third **(Figure 54.5)**.

What controls, or limits, primary production in ecosystems? To ask this question another way, what factors could we change to increase or decrease primary production for a given ecosystem? We look next at the factors that limit primary production in aquatic ecosystems.

Primary Production in Marine and Freshwater Ecosystems

In aquatic (marine and freshwater) ecosystems, both light and nutrients are important in controlling primary production.

Light Limitation

Because solar radiation drives photosynthesis, you might expect that light is a key variable in controlling primary production in oceans. Indeed, the depth of light penetration affects primary production throughout the photic zone of an ocean or lake (see Figure 50.16). More than half of the solar radiation is absorbed in the first meter of water. Even in "clear" water, only 5–10% of the radiation may reach a depth of 20 m.

If light is the main variable limiting primary production in the ocean, we would expect production to increase along a gradient from the poles toward the equator, which receives the greatest intensity of light. However, you can see in Figure 54.5 that there is no such gradient. Some parts of the tropics and subtropics are very unproductive, while some high-latitude ocean regions are relatively productive. Another factor must influence primary production in the ocean.

Nutrient Limitation

More than light, nutrients limit primary production both in different geographic regions of the ocean and in lakes. A **limiting nutrient** is the element that must be added in order for production to increase in a particular area. The nutrient most often limiting marine production is either nitrogen or phosphorus. Concentrations of these nutrients are very low in the photic zone, where phytoplankton live; ironically, they are more abundant in deeper water, where it is too dark for photosynthesis.

As detailed in **Figure 54.6**, nutrient enrichment experiments confirmed that nitrogen was limiting phytoplankton growth off the south shore of Long Island, New York. Practical applications of this work include preventing algal blooms caused by pollution that fertilizes the phytoplankton: Eliminating phosphates from sewage will not help solve the problem of algal blooms unless nitrogen pollution is also controlled.

Several large areas of the ocean, however, have low phytoplankton densities in spite of relatively high nitrogen concentrations. For example, the waters of the Sargasso Sea, a subtropical region of the Atlantic Ocean, are among the most transparent in the world because of their very low density of phytoplankton. A series of nutrient enrichment experiments revealed that in this case, it is availability of the micronutrient iron that limits primary production **(Table 54.1)**.

Such evidence that iron limits production in some oceanic ecosystems encouraged marine ecologists to mount several large-scale field experiments in the Pacific Ocean during the last two decades. For example, in one study, researchers spread low concentrations of dissolved iron over 72 km^2 of ocean and then measured the change in phytoplankton density over a seven-day

Figure 54.6

Inquiry Which nutrient limits phytoplankton production along the coast of Long Island?

EXPERIMENT Pollution from duck farms concentrated near Moriches Bay adds both nitrogen and phosphorus to the coastal water off Long Island. Researchers cultured the phytoplankton *Nannochloris atomus* with water collected from several bays.

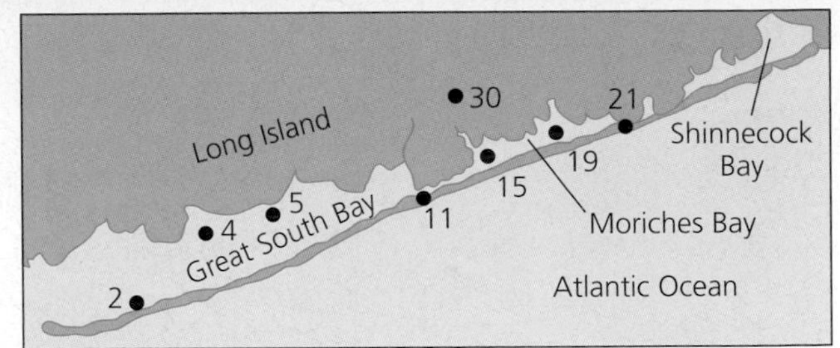

Coast of Long Island, New York. The numbers on the map indicate the data collection stations.

RESULTS Phytoplankton abundance parallels the abundance of phosphorus in the water (a). Nitrogen, however, is immediately taken up by algae, and no free nitrogen is measured in the coastal waters. The addition of ammonium (NH_4^+) caused heavy phytoplankton growth in bay water, but the addition of phosphate (PO_4^{3-}) did not induce algal growth (b).

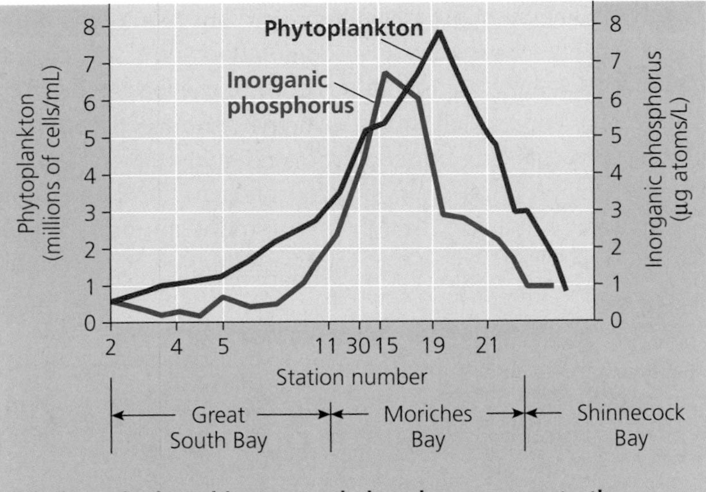

(a) Phytoplankton biomass and phosphorus concentration

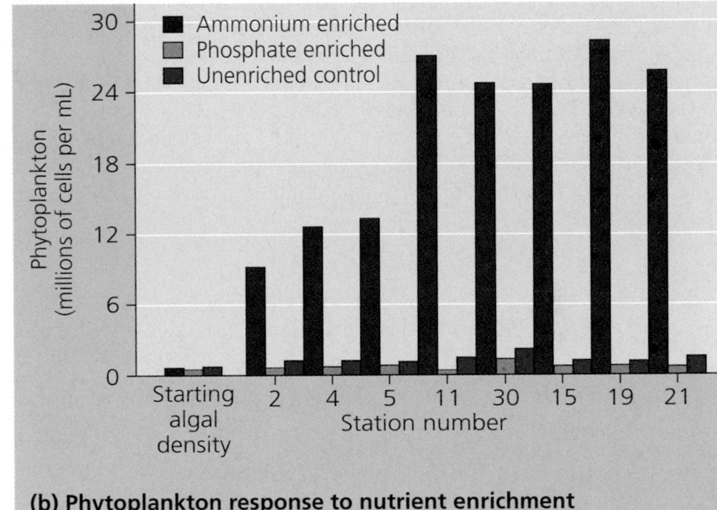

(b) Phytoplankton response to nutrient enrichment

CONCLUSION Since adding phosphorus, which was already in rich supply, had no effect on *Nannochloris* growth, whereas adding nitrogen increased algal density dramatically, researchers concluded that nitrogen was the nutrient limiting phytoplankton growth in this ecosystem.

Table 54.1 Nutrient Enrichment Experiment for Sargasso Sea Samples

Nutrients Added to Experimental Culture	Relative Uptake of ^{14}C by Cultures*
None (controls)	1.00
Nitrogen (N) + phosphorus (P) only	1.10
N + P + metals (excluding iron)	1.08
N + P + metals (including iron)	12.90
N + P + iron	12.00

*^{14}C uptake by cultures measures primary production.

Data from Menzel and Ryther, *Deep Sea Research* 7(1961): 276–281.

period. A massive phytoplankton bloom occurred, as indicated by a 27-fold increase in chlorophyll concentration in water samples from the test sites.

Why are iron concentrations naturally low in certain oceanic regions? Windblown dust from the land is the main process delivering iron to the ocean, and relatively little dust reaches the central Pacific and central Atlantic Oceans.

The iron factor in marine ecosystems is actually related to the nitrogen factor. Where iron is limiting, adding iron stimulates growth of cyanobacteria that fix nitrogen, converting atmospheric N_2 to nitrogenous compounds (see Chapter 27). These nitrogenous nutrients in turn stimulate proliferation of eukaryotic phytoplankton. This relationship can be summarized as:

$$\text{Iron} \xrightarrow{+} \text{Cyanobacteria} \xrightarrow{+} \text{Nitrogen fixation} \xrightarrow{+} \text{Phytoplankton production}$$

▲ **Figure 54.7 The experimental eutrophication of a lake.** In 1974, the far basin of this lake was separated from the near basin by a plastic curtain and fertilized with inorganic sources of carbon, nitrogen, and phosphorus. Within two months, the fertilized basin was covered with a cyanobacterial bloom, which appears white in the photograph. The near basin, which was treated with only carbon and nitrogen, remained unchanged. In this case, phosphorus was the key limiting nutrient, and its addition stimulated the explosive growth of cyanobacteria.

Areas of upwelling, where nutrient-rich deep waters circulate to the ocean surface, have exceptionally high primary production, which supports the hypothesis that nutrient availability determines marine primary production. Because the steady supply of nutrients stimulates growth of the phytoplankton populations that form the base of marine food webs, upwelling areas are prime fishing locations. The largest areas of upwelling occur in the Southern Ocean (Antarctic Ocean) and in the coastal waters off Peru, California, and parts of western Africa.

Nutrient limitation is also common in freshwater lakes. During the 1970s, scientists, including David Schindler from the University of Alberta, Canada, noted that sewage and fertilizer runoff from farms and yards added large amounts of nutrients to lakes. In many lakes, phytoplankton communities that had been dominated by diatoms or green algae became dominated by cyanobacteria. This process, known as **eutrophication** (from the Greek *eutrophos,* well nourished), has a wide range of ecological impacts, including the eventual loss of all but the most tolerant fish species from the lakes (see Figure 50.17). Controlling eutrophication requires knowing which polluting nutrient enables the cyanobacteria to bloom; nitrogen is rarely the limiting factor for primary production in lakes. Schindler conducted a series of whole-lake experiments that pointed to phosphorus as the nutrient that limited cyanobacterial growth **(Figure 54.7)**. His research led to the use of phosphate-free detergents and other water quality reforms.

Primary Production in Terrestrial and Wetland Ecosystems

On a large geographic scale, temperature and moisture are the key factors controlling primary production in terrestrial and wetland ecosystems. Note again in Figure 54.4b that tropical rain forests, with their warm, wet conditions that promote plant growth, are the most productive of all terrestrial ecosystems. In contrast, low-productivity terrestrial ecosystems are generally dry—for example, deserts, which receive little precipitation—or cold and dry—for instance, arctic tundra. Between these extremes lie the temperate forest and grassland ecosystems, which have moderate climates and intermediate productivity levels. These contrasts in climate can be represented by a measure called **actual evapotranspiration,** which is the annual amount of water transpired by plants and evaporated from a landscape, usually measured in millimeters. Actual evapotranspiration increases with the amount of precipitation in a region and the amount of solar energy available to drive evaporation and transpiration. **Figure 54.8** shows the significant positive relationship between net primary production and actual evapotranspiration in selected ecosystems, ranging from desert shrubland to tropical forest.

On a more local scale, mineral nutrients in the soil can play a key role in limiting primary production in terrestrial and wetland ecosystems. Primary production removes soil nutrients, sometimes faster than they are replaced. At some point, a single nutrient deficiency may cause plant growth to slow or cease. It is unlikely that all nutrients will be exhausted simultaneously. If a certain nutrient limits production, adding a nonlimiting nutrient, even if that nutrient is in short supply, will not stimulate production. For example, if nitrogen is limiting, adding phosphorus will not stimulate production, whereas additional nitrogen will act as a stimulant until some other nutrient—say, phosphorus—becomes

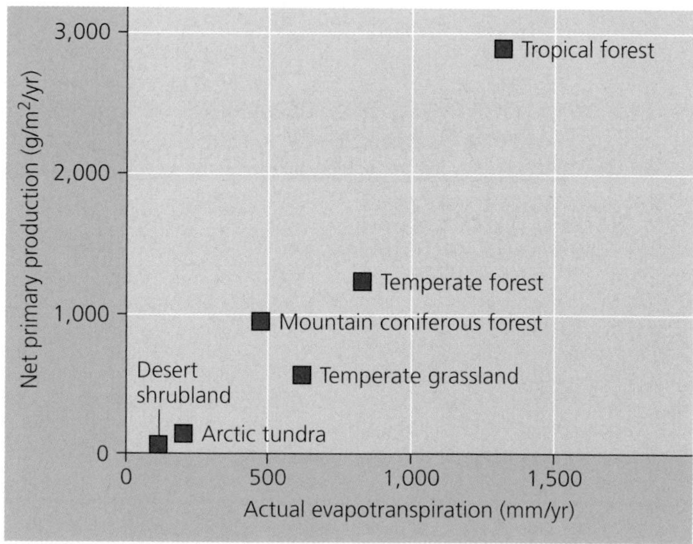

▲ **Figure 54.8 Actual evapotranspiration (temperature and moisture) related to terrestrial net primary production in selected ecosystems.**

Figure 54.9

Inquiry Is phosphorus or nitrogen the limiting nutrient in a Hudson Bay salt marsh?

EXPERIMENT Over the summer of 1980, researchers added phosphorus to some experimental plots in the salt marsh, nitrogen to other plots, and both phosphorus and nitrogen to others. Some plots were left unfertilized as controls.

RESULTS

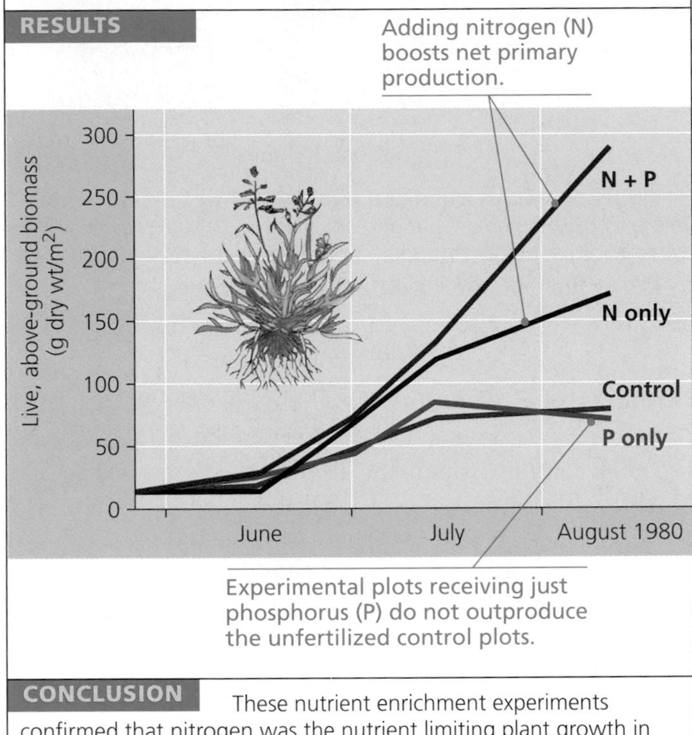

CONCLUSION These nutrient enrichment experiments confirmed that nitrogen was the nutrient limiting plant growth in this salt marsh.

limiting **(Figure 54.9)**. In fact, nitrogen and phosphorus are most often the nutrients limiting terrestrial and wetland production.

Studies relating nutrients to terrestrial primary production have practical applications in agriculture. Farmers maximize their crop yields by using fertilizers with the right balance of nutrients for the local soil and the type of crop.

Concept Check 54.2

1. Why is only a small portion of the solar energy that strikes Earth's atmosphere stored by primary producers?
2. How can ecologists experimentally determine the factor limiting primary production in an ecosystem?
3. Why does the open ocean account for almost 25% of Earth's primary production despite its relatively low rate of primary production?
4. Why is an ecosystem's net primary production lower than its gross primary production?

For suggested answers, see Appendix A.

Energy transfer between trophic levels is usually less than 20% efficient

The amount of chemical energy in consumers' food that is converted to their own new biomass during a given time period is called the **secondary production** of the ecosystem. Consider the transfer of organic matter from primary producers to herbivores, the primary consumers. In most ecosystems, herbivores eat only a small fraction of plant material produced. And they cannot digest all the plant material that they *do* eat, as anyone who has walked through a dairy farm will attest. Thus, much of primary production is not used by consumers. Let's analyze this process of energy transfer more closely.

Production Efficiency

First let's examine secondary production in an individual organism—a caterpillar. When a caterpillar feeds on a plant leaf, only about 33 J out of 200 J (48 cal), or one-sixth of the energy in the leaf, is used for secondary production, or growth **(Figure 54.10)**. The caterpillar uses some of the remaining energy for cellular respiration and passes the rest in its feces. The energy contained in the feces remains in the ecosystem, as it will be consumed by detritivores. The energy used for the caterpillar's respiration, however, is lost

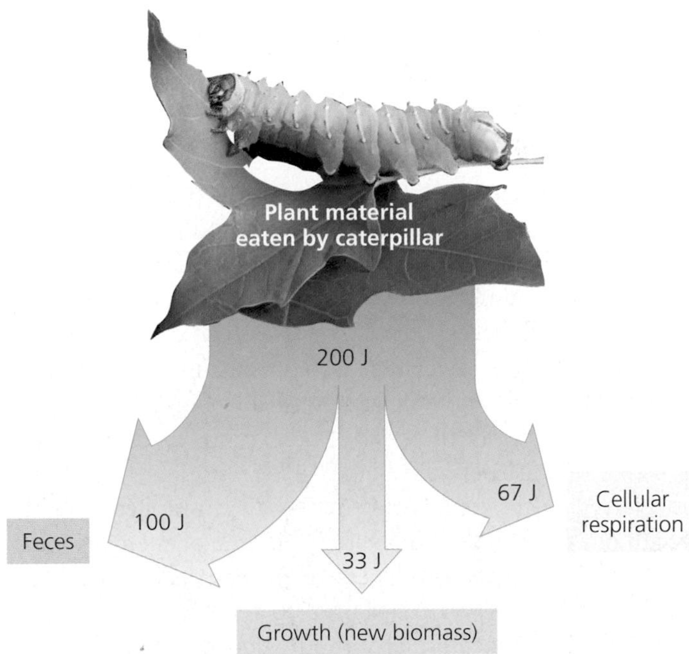

▲ **Figure 54.10 Energy partitioning within a link of the food chain.** Less than 17% of the caterpillar's food is actually used for secondary production (growth).

from the ecosystem as heat. This is why energy is said to flow through, not cycle within, ecosystems. Only the chemical energy stored by herbivores as biomass (through growth or the production of offspring) is available as food to secondary consumers.

Our caterpillar example is based on the leaf tissue that is actually consumed but says nothing about the available plant material not eaten. The greenness of most terrestrial landscapes is visible evidence that most net primary production is in fact *not* consumed by herbivores and converted to secondary production.

We can measure the efficiency of animals as energy transformers using the following equation:

$$\text{Production efficiency} = \frac{\text{Net secondary production}}{\text{Assimilation of primary production}}$$

Net secondary production is the energy stored in biomass represented by growth and reproduction. Assimilation consists of the total energy taken in and used for growth, reproduction, and respiration. **Production efficiency,** therefore, is the fraction of energy stored in food that is *not* used for respiration. For the caterpillar in Figure 54.10, production efficiency is 33%; 67 J of the 100 J of assimilated energy is used for respiration. (Note that the energy lost as undigestible material in feces does not count toward assimilation.) Birds and mammals typically have low production efficiencies, in the range of 1–3%, because they use so much energy in maintaining a constant, warm body temperature. Fishes, which are ectotherms (see Chapter 40), have production efficiencies around 10%. Insects are even more efficient, with production efficiencies averaging 40%.

Trophic Efficiency and Ecological Pyramids

Let's scale up now from the production efficiencies of individual consumers to the flow of energy through whole trophic levels.

Trophic efficiency is the percentage of production transferred from one trophic level to the next. Trophic efficiencies must always be less than production efficiencies because they take into account not only the energy lost through respiration and contained in feces, but also the energy in organic material in a lower trophic level that is not consumed by the next trophic level. Trophic efficiencies usually range from 5% to 20%, depending on the type of ecosystem. In other words, 80–95% of the energy available at one trophic level is not transferred to the next. And this loss is multiplied over the length of a food chain. For example, if 10% of energy is transferred from primary producers to primary consumers, and 10% of that energy is transferred to secondary consumers, then only 1% of net primary production is available to secondary consumers (10% of 10%).

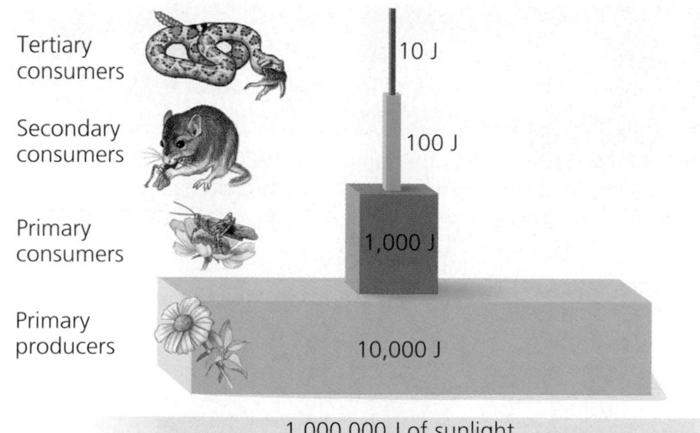

▲ **Figure 54.11 An idealized pyramid of net production.** This example is based on a trophic efficiency of 10% for each link in the food chain. Notice that primary producers convert only about 1% of the energy in the sunlight available to them into net primary production.

Pyramids of Production. This loss of energy with each transfer in a food chain can be represented by a *pyramid of net production,* in which the trophic levels are stacked in blocks, with primary producers forming the foundation of the pyramid **(Figure 54.11)**. The size of each block is proportional to the net production, expressed in energy units, of each trophic level.

Pyramids of Biomass. One important ecological consequence of low trophic efficiencies can be represented in a *biomass pyramid,* in which each tier represents the standing crop (the total dry weight of all organisms) in one trophic level. Most biomass pyramids narrow sharply from primary producers at the base to top-level carnivores at the apex because energy transfers between trophic levels are so inefficient **(Figure 54.12a)**. Certain aquatic ecosystems, however, have inverted biomass pyramids: Primary consumers outweigh the producers **(Figure 54.12b)**. Such inverted biomass pyramids occur because the producers—phytoplankton—so quickly grow, reproduce, and are consumed by the zooplankton that they never develop a large population size, or standing crop. In other words, phytoplankton have a short **turnover time,** which means they have a small standing crop biomass compared to their production:

$$\text{Turnover time} = \frac{\text{Standing crop biomass (mg/m}^2)}{\text{Production (mg/m}^2\text{/day)}}$$

Because the phytoplankton continue to replace their biomass at such a rapid rate, they can support a biomass of zooplankton bigger than their own biomass. Nevertheless, because phytoplankton have much higher production than zooplankton, the pyramid of *production* for this ecosystem is bottom-heavy, like the one in Figure 54.11.

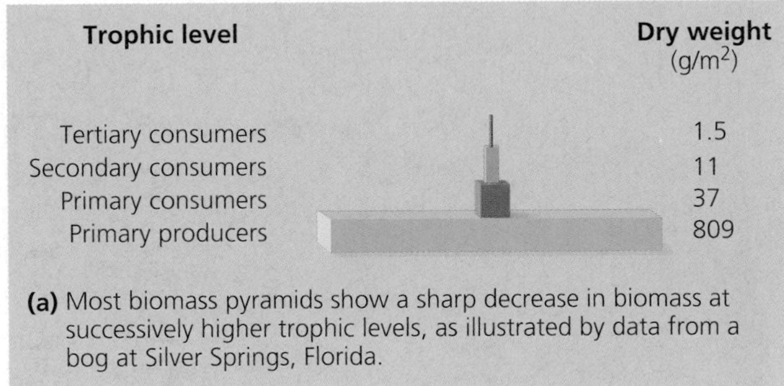

(a) Most biomass pyramids show a sharp decrease in biomass at successively higher trophic levels, as illustrated by data from a bog at Silver Springs, Florida.

Trophic level	Dry weight (g/m^2)
Primary consumers (zooplankton)	21
Primary producers (phytoplankton)	4

(b) In some aquatic ecosystems, such as the English Channel, a small standing crop of primary producers (phytoplankton) supports a larger standing crop of primary consumers (zooplankton).

▲ **Figure 54.12 Pyramids of biomass (standing crop).** Numbers denote the dry weight (g/m^2) for all organisms at a trophic level.

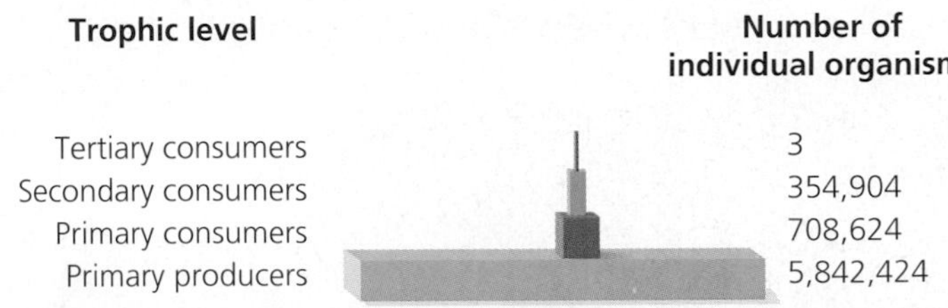

▲ **Figure 54.13 A pyramid of numbers.** In a bluegrass field in Michigan, only three top carnivores are supported in an ecosystem based on production by nearly 6 million plants.

Pyramids of Numbers. The progressive loss of energy along a food chain severely limits the overall biomass of top-level carnivores that any ecosystem can support. Only about 0.1% of the chemical energy fixed by photosynthesis can flow all the way through a food web to a tertiary consumer, such as a snake or a shark (see Figure 54.11). This explains why most food webs include only about four or five trophic levels (see Chapter 53).

Although there are situations in which many small consumers feed on a few larger primary producers, such as insects feeding on trees, predators are usually larger than the prey they eat. Thus, top-level predators tend to be fairly large animals. As a result, the limited biomass at the top of an ecological pyramid is concentrated in a relatively small number of large individuals. This phenomenon is reflected in a *pyramid of numbers*, in which the size of each block is proportional to the number of individual organisms present in that trophic level **(Figure 54.13)**. Because populations of top predators are typically small, and the animals may be widely spaced within their habitats, many predators are highly susceptible to extinction (as well as to the evolutionary consequences of small population size discussed in Chapter 23).

The dynamics of energy flow through ecosystems have important implications for the human population. Eating meat is a relatively inefficient way of tapping photosynthetic production. A human obtains far more calories by eating a certain amount of grain directly as a primary consumer than by processing that same amount of grain through another trophic level and eating meat from grain-fed animals (beef, poultry, pork, lamb, farmed fish). Worldwide agriculture could, in fact, successfully feed many more people than it does today and require less cultivated land to support those people if humans all fed more efficiently—as primary consumers, eating only plant material **(Figure 54.14)**.

The Green World Hypothesis

With so many primary consumers (herbivores) feeding on plants, how can we explain why most terrestrial ecosystems are actually quite green, boasting large standing crops of vegetation? According to the **green world hypothesis**, terrestrial herbivores consume relatively little plant biomass because they are held in check by a variety of factors, including predators, parasites, and disease.

◀ **Figure 54.14 Relative food energy available to the human population at different trophic levels.** Most humans have a diet between these two extremes.

▲ **Figure 54.15 A green ecosystem.** Most terrestrial ecosystems have large standing crops of vegetation despite the large number of resident herbivores. The green world hypothesis offers possible explanations for this observation.

Just how green is our green world? A total of approximately 83×10^{10} metric tons of carbon are stored in the plant biomass of terrestrial ecosystems. And the global rate of terrestrial primary production is about 5×10^{10} metric tons of plant biomass per year. On a global scale, herbivores annually consume less than 17% of the total net primary production by plants; (most of the rest is eventually consumed by detritivores). Thus, overall, herbivores represent only a minor nuisance to plants. Of course, some herbivores have the potential to completely strip local vegetation over the short term; an example is the ability of exploding gypsy moth populations to occasionally defoliate areas of forest in the northeastern United States. Such exceptions only heighten our curiosity about why Earth is so green **(Figure 54.15)**.

The green world hypothesis proposes several factors that keep herbivores in check:

- **Plants have defenses against herbivores.** These defenses include spines and noxious chemicals, such as those described in Chapter 39.
- **Nutrients, not energy supply, usually limit herbivores.** Animals need certain nutrients, such as organic nitrogen (protein), that plants tend to supply in relatively small amounts. Even in a world of plentiful green energy, the growth and reproduction of many herbivores are limited by availability of essential nutrients.
- **Abiotic factors limit herbivores.** Unfavorable seasonal changes in temperature and moisture are examples of abiotic factors that can set a carrying capacity for herbivores in a particular environment far below the number that would strip vegetation there.
- **Intraspecific competition can limit herbivore numbers.** Territorial behavior and other consequences of competition may maintain herbivores' population densities below what the vegetation could feed.
- **Interspecific interactions keep herbivore densities in check.** The green world hypothesis postulates that predators, parasites, and disease are the most important factors limiting the growth of herbivore populations. This applies the top-down model of community structure, which we discussed in Chapter 53.

In the next section, we will look at how the transfer of chemical nutrients along with energy through food webs is part of a larger picture of nutrient recycling in ecosystems.

Concept Check 54.3

1. If an insect that eats plant seeds containing 100 J of energy uses 30 J of that energy for respiration and excretes 50 J in its feces, what is the insect's net secondary production? What is its production efficiency?
2. On a global scale, herbivores consume only about 17% of net primary production by terrestrial plants, yet most plant biomass is eventually consumed. Explain.
3. Why does the production pyramid have the same general shape as the biomass pyramid in most ecosystems? Under what circumstances might the shapes of the two pyramids differ?

For suggested answers, see Appendix A.

Biological and geochemical processes move nutrients between organic and inorganic parts of the ecosystem

Although most ecosystems receive an essentially inexhaustible influx of solar energy, chemical elements are available only in limited amounts. (The meteorites that occasionally strike Earth are the only extraterrestrial source of new matter.) Life on Earth therefore depends on the recycling of essential chemical elements. While an individual organism is alive, much of its chemical stock is rotated continuously as nutrients are assimilated and waste products released. Atoms present in the complex molecules of an organism when it dies are returned as simpler compounds to the atmosphere, water, or soil by the action of decomposers. This decomposition replenishes the pools of inorganic nutrients that plants and other autotrophs use to build new organic matter. Because nutrient circuits involve both biotic and abiotic components, they are also called **biogeochemical cycles.**

A General Model of Chemical Cycling

A chemical's specific route through a biogeochemical cycle varies with the particular element and the trophic structure of the ecosystem. We can, however, recognize two general categories of biogeochemical cycles: global and local. Gaseous forms of carbon, oxygen, sulfur, and nitrogen occur in the atmosphere, and cycles of these elements are essentially global. For example, some of the carbon and oxygen atoms a plant acquires from the air as CO_2 may have been released into the atmosphere by the respiration of an organism in a distant locale. Other, less mobile elements, including phosphorus, potassium, and calcium, generally cycle on a more localized scale, at least over the short term. Soil is the main abiotic reservoir of these elements, which are absorbed by plant roots and eventually returned to the soil by decomposers, usually in the same vicinity.

Before examining the details of individual cycles, let's look at a general model of nutrient cycling that includes the main reservoirs of elements and the processes that transfer elements between reservoirs **(Figure 54.16)**. Each reservoir is defined by two characteristics: whether it contains organic or inorganic materials and whether or not the materials are directly available for use by organisms.

The nutrients in living organisms themselves and in detritus (reservoir a in Figure 54.16) are available to other organisms when consumers feed and when detritivores consume nonliving organic matter. Some material moved from the living organic reservoir to the fossilized organic reservoir (reservoir b) long ago, when dead organisms were buried by sedimentation

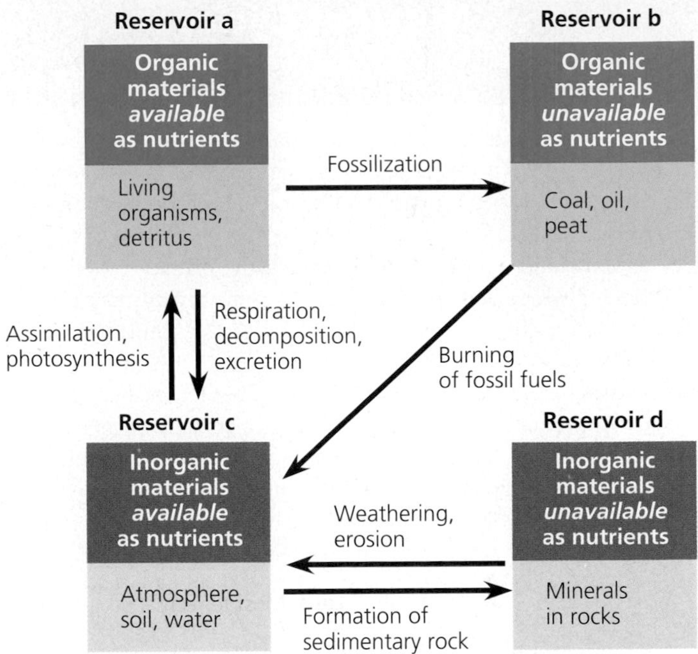

▲ **Figure 54.16 A general model of nutrient cycling.** Arrows indicate the processes that move nutrients between reservoirs.

over millions of years, becoming coal, oil, or peat (fossil fuels). Nutrients in these deposits cannot be assimilated directly.

Inorganic materials (elements and compounds) that are dissolved in water or present in soil or air (reservoir c) are available for use. Organisms assimilate materials from this reservoir directly and return chemicals to it through the relatively rapid processes of cellular respiration, excretion, and decomposition. Although organisms cannot directly tap into the inorganic elements tied up in rocks (reservoir d), these nutrients may slowly become available through weathering and erosion. Similarly, unavailable organic materials move into the available reservoir of inorganic nutrients when fossil fuels are burned, releasing exhaust into the atmosphere.

Tracing elements through particular biogeochemical cycles is much more complex, especially since ecosystems exchange elements with other ecosystems. Even in a pond, which has discrete boundaries, dust or leaves blowing into the pond and the emergence of aquatic insects can add or remove key nutrients. Keeping track of inflow and outflow is even more challenging in less clearly delineated terrestrial ecosystems. Nevertheless, ecologists have worked out the schemes for chemical cycling in several ecosystems. Ecologists study chemical cycling by adding tiny amounts of radioactive isotopes of the elements they want to track or by following the movement of naturally occurring stable, nonradioactive isotopes through the various biotic and abiotic components of an ecosystem.

Biogeochemical Cycles

Figure 54.17, on the next two pages, provides a detailed look at the cycling of water, carbon, nitrogen, and phosphorus. The

Figure 54.17
Exploring Nutrient Cycles

THE WATER CYCLE

Biological importance: Water is essential to all organisms (see Chapter 3), and its availability influences rates of ecosystem processes, particularly primary production and decomposition in terrestrial ecosystems.
Forms available to life: Liquid water is the primary physical phase in which water is used, though some organisms can harvest water vapor. Freezing of soil water can limit water availability to terrestrial plants.
Reservoirs: The oceans contain 97% of the water in the biosphere. Approximately 2% is bound in glaciers and polar ice caps, and the remaining 1% is in lakes, rivers, and groundwater, with a negligible amount in the atmosphere.
Key processes: The main processes driving the water cycle are evaporation of liquid water by solar energy, condensation of water vapor into clouds, and precipitation. Transpiration by terrestrial plants also moves significant volumes of water. Surface and groundwater flow can return water to the oceans, completing the water cycle. The widths of the arrows in the diagram reflect the relative contribution of each process to the movement of water in the biosphere.

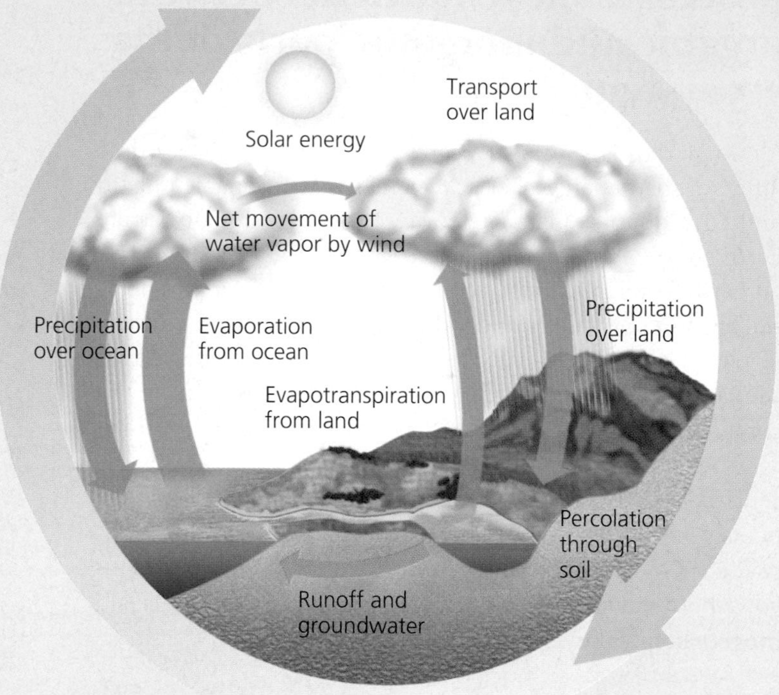

THE CARBON CYCLE

CO$_2$ in atmosphere
Photosynthesis
Cellular respiration
Burning of fossil fuels and wood
Higher-level consumers
Primary consumers
Carbon compounds in water
Detritus
Decomposition

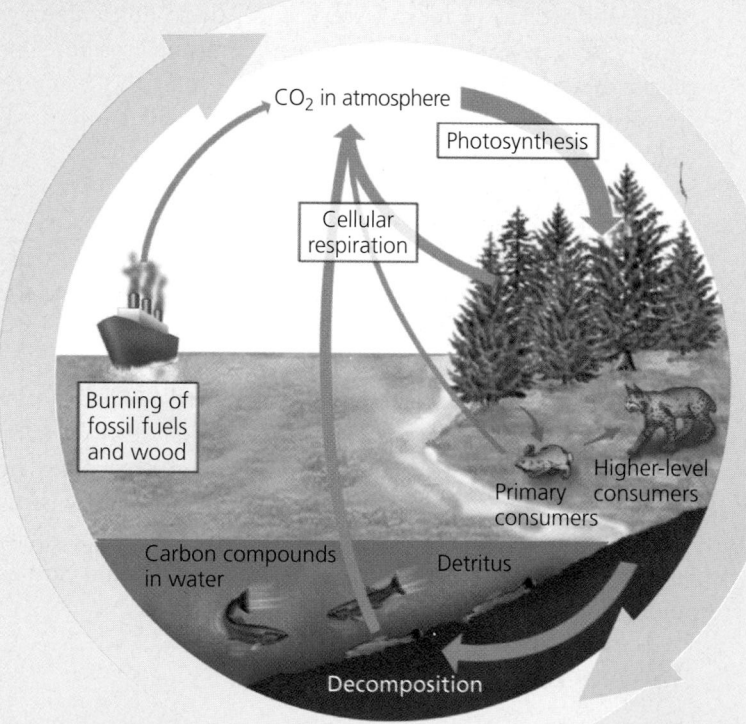

Biological importance: Carbon forms the framework for the organic molecules essential to all organisms.
Forms available to life: Photosynthetic organisms utilize CO_2 during photosynthesis and convert the carbon to organic forms that are used by consumers, including heterotrophic prokaryotes (see Chapter 27).
Reservoirs: The major reservoirs of carbon include fossil fuels, soils, the sediments of aquatic ecosystems, the oceans (dissolved carbon compounds), plant and animal biomass, and the atmosphere (CO_2). The largest reservoir is sedimentary rocks such as limestone; however, this pool turns over very slowly.
Key processes: Photosynthesis by plants and phytoplankton removes substantial amounts of atmospheric CO_2 each year. This quantity is approximately equaled by CO_2 added to the atmosphere through cellular respiration by producers and consumers. Over geologic time, volcanoes are a substantial source of CO_2. The burning of fossil fuels is adding significant amounts of additional CO_2 to the atmosphere. The widths of the arrows reflect the relative contribution of each process.

THE NITROGEN CYCLE

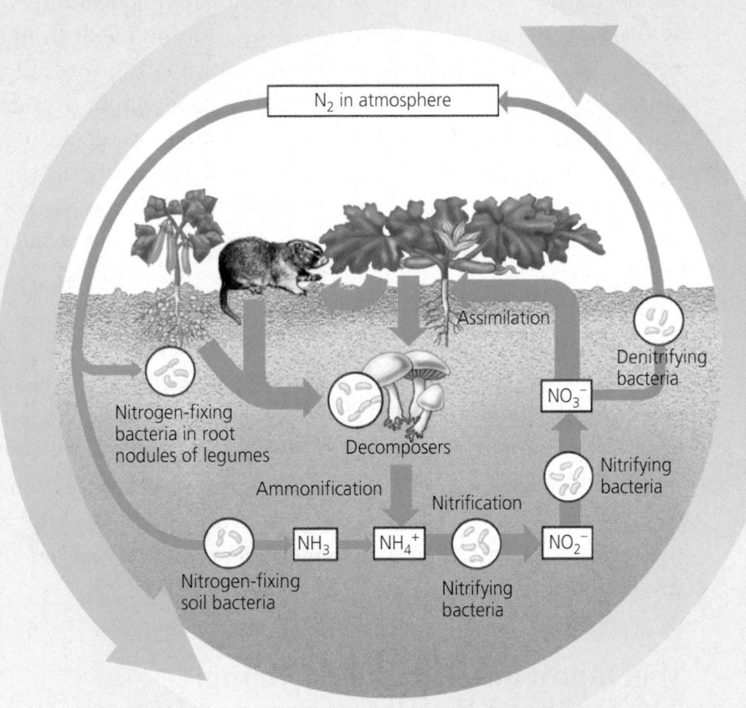

Biological importance: Nitrogen is a component of amino acids, proteins, and nucleic acids and is a crucial and often limiting plant nutrient.

Forms available to life: Plants and algae can use two inorganic forms of nitrogen: ammonium (NH_4^+) or nitrate (NO_3^-). Various bacteria can also use NH_4^+ and NO_3^-, as well as nitrite (NO_2^-). Animals can utilize only organic forms of nitrogen (such as amino acids or proteins).

Reservoirs: The main reservoir of nitrogen is the atmosphere, which is 80% nitrogen gas (N_2). The other reservoirs are bound nitrogen in soils; in the sediments of lakes, rivers, and oceans; dissolved in surface water and groundwater systems; and in the biomass of living organisms.

Key processes: The major pathway for nitrogen to enter an ecosystem is via *nitrogen fixation,* the conversion of N_2 by bacteria to forms that can be used to synthesize nitrogenous organic compounds (see Chapter 37). Some nitrogen is also fixed by lightning. Nitrogen fertilizer, precipitation, and blowing dust can also provide substantial inputs of NH_4^+ and NO_3^- to ecosystems. ***Ammonification*** decomposes organic nitrogen to NH_4^+. In ***nitrification***, NH_4^+ is converted to NO_3^- by nitrifying bacteria. Under anaerobic conditions, denitrifying bacteria use NO_3^- in their metabolism instead of O_2, releasing N_2 in a process known as ***denitrification***. The widths of the arrows reflect the relative contribution of each process.

THE PHOSPHORUS CYCLE

Biological importance: Organisms require phosphorus as a major constituent of nucleic acids, phospholipids, and ATP and other energy-storing molecules, and as a mineral constituent of bones and teeth.

Forms available to life: The only biologically important inorganic form of phosphorus is phosphate (PO_4^{3-}), which plants absorb and use to synthesize organic compounds.

Reservoirs: The largest accumulations of phosphorus are in sedimentary rocks of marine origin. There are also large quantities of phosphorus in soils, in the oceans (in dissolved form), and in organisms. Because humus and soil particles bind phosphate, the recycling of phosphorus tends to be quite localized in ecosystems.

Key processes: Weathering of rocks gradually adds phosphate to soil; some leaches into groundwater and surface water and may eventually find its way into the sea. Phosphate taken up by producers and incorporated into biological molecules may be eaten by consumers and distributed through the food web. Phosphate is returned to soil or water through either decomposition of biomass or excretion by consumers. Because there are no significant phosphorus-containing gases, only relatively small amounts of phosphorus move through the atmosphere, usually in the form of dust and sea spray. The widths of the arrows reflect the relative contribution of each process.

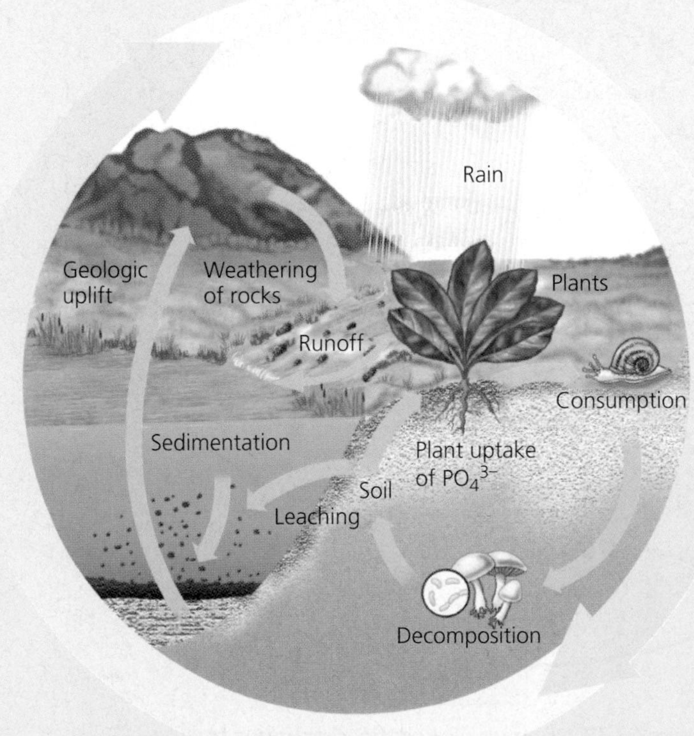

diagrams focus on four important factors that ecologists consider as they research biogeochemical cycles:

1. The biological importance of each chemical
2. The forms in which each chemical is available or used by organisms
3. The major repositories or reservoirs for each chemical
4. The key processes that drive the movement of each chemical through its biogeochemical cycle

Decomposition and Nutrient Cycling Rates

Now that we have examined several individual biogeochemical cycles, we can review the general pattern of chemical cycling as illustrated in **Figure 54.18**. Note again the key role of decomposers (detritivores). The rates at which nutrients cycle in different ecosystems are extremely variable, mostly as a result of differences in rates of decomposition. In tropical rain forests, for instance, most organic material decomposes in a few months to a few years, while in temperate forests, decomposition takes four to six years, on average. The difference is largely the result of the warmer temperatures and more abundant precipitation in the tropical forests. Temperature and the availability of water affect rates of decomposition and thus nutrient cycling times. Like net primary production, the rate of decomposition in terrestrial ecosystems increases with actual evapotranspiration. (Other factors that can also influence nutrient cycling are the local soil chemistry and the frequency of fires.)

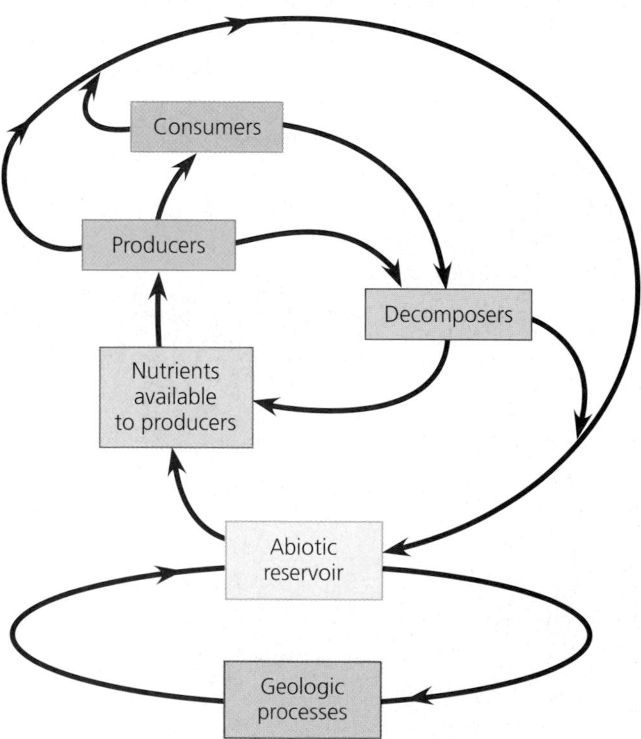

▲ **Figure 54.18 Review: Generalized scheme for biogeochemical cycles.**

When decomposition occurs in a tropical rain forest, relatively little organic material accumulates as leaf litter on the forest floor; about 75% of the nutrients in the ecosystem are present in the woody trunks of trees, and about 10% are contained in the soil. Thus, the relatively low concentrations of some nutrients in the soil of tropical rain forests result from a fast cycling time, not from an overall scantiness of these elements in the ecosystem. In temperate forests, where decomposition is much slower, the soil may contain as much as 50% of all the organic material in the ecosystem. The nutrients that are present in temperate forest detritus and soil may remain there for fairly long periods of time before being assimilated by plants.

In aquatic ecosystems decomposition in anaerobic muds can take 50 years or more. Bottom sediments are comparable to the detritus layer in terrestrial ecosystems; however, algae and aquatic plants usually assimilate nutrients directly from the water. Thus, the sediments often constitute a nutrient sink, and aquatic ecosystems are very productive only when there is interchange between the bottom layers of water and the surface (as in the upwelling regions described earlier in this chapter).

Vegetation and Nutrient Cycling: The Hubbard Brook Experimental Forest

In one of the longest-running examples of long-term ecological research (LTER) in North America, a team of scientists led by Herbert Bormann and Gene Likens (see interview on pages 1078–1079) has been studying nutrient cycling in a forest ecosystem since 1963. The study site, the Hubbard Brook Experimental Forest in the White Mountains of New Hampshire, is a deciduous forest with several valleys, each drained by a small creek that is a tributary of Hubbard Brook. Bedrock impenetrable to water is close to the surface of the soil, and each valley constitutes a watershed that can drain only through its creek.

The research team first determined the mineral budget for each of six valleys by measuring the input and outflow of several key nutrients. They collected rainfall at several sites to measure the amount of water and dissolved minerals added to the ecosystem. To monitor the loss of water and minerals, they constructed a small concrete dam with a V-shaped spillway across the creek at the bottom of each valley **(Figure 54.19a)**. About 60% of the water added to the ecosystem as rainfall and snow exits through the stream, and thus the remaining 40% is lost by transpiration from plants and evaporation from other organisms and the soil.

Preliminary studies confirmed that internal cycling within a terrestrial ecosystem conserves most of the mineral nutrients. For example, only about 0.3% more calcium (Ca^{2+}) left a valley via its creek than was added by rainwater, and this small net loss was probably replaced by chemical

(a) Concrete dams and weirs built across streams at the bottom of watersheds enabled researchers to monitor the outflow of water and nutrients from the ecosystem.

(b) One watershed was clear cut to study the effects of the loss of vegetation on drainage and nutrient cycling.

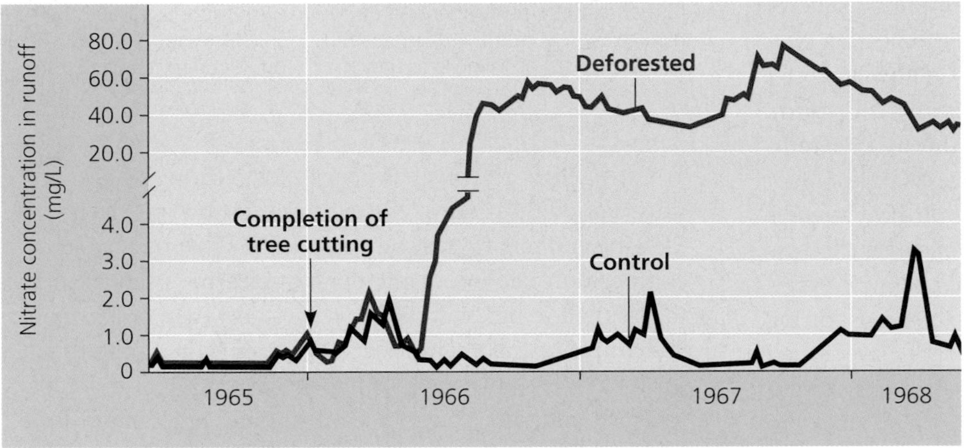

(c) The concentration of nitrate in runoff from the deforested watershed was 60 times greater than in a control (unlogged) watershed.

▲ **Figure 54.19 Nutrient cycling in the Hubbard Brook Experimental Forest: an example of long-term ecological research.**

decomposition of the bedrock. During most years, the forest actually registered small net gains of a few mineral nutrients, including nitrogen.

In one experiment, the trees in one valley were cut down and then the valley was sprayed with herbicides for three years to prevent regrowth of plants **(Figure 54.19b)**. All the original plant material was left in place to decompose. The inflow and outflow of water and minerals in this experimentally altered watershed were compared with the inflow and outflow in a control watershed. Over the three years, water runoff from the altered watershed increased by 30–40%, apparently because there were no plants to absorb and transpire water from the soil. Net losses of minerals from the altered watershed were huge. The concentration of Ca^{2+} in the creek increased fourfold, for example, and the concentration of K^{+} increased by a factor of 15. Most remarkable was the loss of nitrate, which increased in concentration in the creek 60-fold **(Figure 54.19c)**. Not only was this vital mineral nutrient drained from the ecosystem, but nitrate in the creek reached a level considered unsafe for drinking water.

This study demonstrated that the amount of nutrients leaving an intact forest ecosystem is controlled mainly by the plants. These effects of removing the trees are almost immediate, occurring within a few months, and continue as long as living plants are absent.

The 40 years of data from Hubbard Brook reveal some other trends. For instance, since the 1950s, acid rain and snow have dissolved most of the Ca^{2+} from the forest soil, and the streams have carried it away. By the 1990s, the forest biomass at Hubbard Brook had stopped growing, apparently because of a lack of Ca^{2+}. To test the possibility of growth limitation by inadequate supplies of calcium, ecologists at Hubbard Brook began a massive experiment in 1998. They first established a control and an experimental watershed, which they monitored over two years before using a helicopter to add Ca^{2+} to the experimental watershed. As Gene Likens, interviewed on pages 1078–1079, points out, the trees growing in Ca^{2+}-enriched soil are already showing signs of increased growth.

Results of the Hubbard Brook studies, as well as 25 other LTER projects funded by the National Science Foundation, not only assess natural ecosystem dynamics but also provide important insight into the mechanisms by which human activities affect these processes.

Concept Check 54.4

1. For each of the four biogeochemical cycles detailed in Figure 54.17, draw a simple diagram that shows one possible path for an atom or molecule of that chemical from abiotic to biotic reservoirs and back.
2. Why does deforestation of a watershed increase the concentration of nitrates in streams draining the watershed?

For suggested answers, see Appendix A.

Concept 54.5

The human population is disrupting chemical cycles throughout the biosphere

As the human population has grown in size, our activities and technological capabilities have disrupted the trophic structure, energy flow, and chemical cycling of ecosystems in most areas of the world. As you will read in this section, the effects are sometimes local or regional, but the ecological impact of humans can even be global.

Nutrient Enrichment

Human activity often intrudes in nutrient cycles by removing nutrients from one part of the biosphere and adding them to another. On the simplest level, someone eating a piece of broccoli in Washington, DC, is consuming nutrients that only days before might have been in the soil in California; and a short time later, some of these nutrients will be in the Potomac River on their way to the sea, having passed through the individual's digestive system and the local sewage facilities. On a larger scale, nutrients in farm soil may run off into streams and lakes, resulting in nutrient depletion in one area, excesses in another, and the disruption of the natural chemical cycles in both locations. Furthermore, in addition to transporting nutrients from one location to another, humans have added entirely new materials—many of them toxic—to ecosystems.

Humans have intruded on nutrient cycles to such an extent that it is no longer possible to understand any cycle without taking these effects into account. Let's examine a few specific examples of how humans are impacting the biosphere's chemical dynamics.

Agriculture and Nitrogen Cycling

After natural vegetation is cleared from an area, the existing reserve of nutrients in the soil is sufficient to grow crops for some time without nutrient supplementation. However, in agricultural ecosystems, a substantial fraction of these nutrients is not recycled but is exported from the area in the form of crop biomass **(Figure 54.20)**. The "free" period for crop production—when there is no need to add nutrients to the soil—varies greatly. When some of the early North American prairie lands were first tilled, for example, good crops could be produced for many years because the large store of organic materials in the soil continued to decompose and provide nutrients. By contrast, some cleared land in the tropics can be farmed for only one or two years because so little of the ecosystems' nutrient load is contained in the soil. Despite such variations, in any area under intensive agriculture, the natural store of nutrients eventually becomes exhausted.

▲ **Figure 54.20 Agricultural impact on soil nutrients.** Removal of harvested plant biomass for market removes mineral nutrients that would otherwise be cycled back to the local soil. To replace the lost nutrients, farmers must apply fertilizers—either organic fertilizers, such as manure or mulch, or manufactured fertilizers.

Nitrogen is the main nutrient lost through agriculture; thus, agriculture has a great impact on the nitrogen cycle. Plowing and mixing the soil are processes that increase the decomposition rate of organic matter, releasing usable nitrogen that is then removed from the ecosystem when crops are harvested. Industrially synthesized fertilizer is then used to make up for the loss of usable nitrogen from agricultural ecosystems. In addition, as we saw in the case of Hubbard Brook, without plants to take up nitrates from the soil, the nitrates are likely to be leached from the ecosystem.

Recent studies indicate that human activities have approximately doubled Earth's supply of fixed nitrogen available to primary producers. The main cause is industrial nitrogen fixation for fertilizers, but increased cultivation of legumes, with their nitrogen-fixing symbionts, is also important. In addition, the deliberate burning of fields after harvest, as well as burning to clear tropical forest for agriculture, releases nitrogen compounds stored in soil and vegetation, thereby enhancing nitrogen cycling. Excessive supplements of fixed nitrogen are also associated with a greater release of N_2 and nitrogen oxides into the air by denitrifying bacteria (see Figure 54.17). (Nitrogen oxides can contribute to atmospheric warming, to the depletion of atmospheric ozone, and in some ecosystems to acid precipitation.)

Contamination of Aquatic Ecosystems

The key problem with excess nitrogen seems to be **critical load**, the amount of added nutrient, usually nitrogen or phosphorus, that can be absorbed by plants without damaging ecosystem integrity. Nitrogenous minerals in the soil that exceed the critical load eventually leach into groundwater or run off directly into freshwater and marine ecosystems, contaminating water supplies, choking waterways, and killing fish. Many rivers contaminated with nitrates and ammonium from agricultural runoff and sewage drain into the North Atlantic Ocean, with the highest nitrogen inputs coming from northern Europe. Groundwater concentrations of nitrate are also increasing in agricultural areas, sometimes exceeding the maximum level considered safe for drinking water (10 mg of nitrate per liter).

As we discussed in Chapter 50, lakes are classified by nutrient availability as oligotrophic or eutrophic (see Figure 50.17). In an oligotrophic lake, primary productivity is relatively low because the mineral nutrients required by phytoplankton are scarce. In other lakes, basin and watershed characteristics result in the addition of more nutrients. These nutrients are captured by the primary producers and then continuously recycled through the lake's food webs. Thus, the overall productivity is higher in lakes that are eutrophic.

Human intrusion has disrupted freshwater ecosystems by what is termed *cultural eutrophication*. Sewage and factory wastes; runoff of animal waste from pastures and stockyards; and the leaching of fertilizer from agricultural, recreational, and urban areas have all overloaded many streams, rivers, and lakes with inorganic nutrients. This enrichment often results in an explosive increase in the density of photosynthetic organisms, whose population growth, as described earlier in this chapter, is generally limited by either nitrogen or phosphorus. As a consequence, shallower areas become choked with weeds, making boating and fishing impossible. Large "blooms" of algae and cyanobacteria become common (see Figure 54.7). A eutrophic lake may be supersaturated with oxygen produced during the day while photosynthesis is taking place, but anoxic (oxygen-depleted) at night when respiration occurs in the absence of photosynthesis. As the photosynthetic organisms die and organic material accumulates at the lake bottom, detritivores use all the oxygen in the deeper waters.

All these conditions threaten the survival of some organisms. For example, cultural eutrophication of Lake Erie wiped out commercially important fishes such as blue pike, whitefish, and lake trout by the 1960s. Since then, tighter regulations on waste dumping into the lake have enabled some fish populations to rebound, but many of the native species of fishes and invertebrates have not recovered.

Acid Precipitation

The burning of wood and of coal and other fossil fuels releases oxides of sulfur and nitrogen that react with water in the atmosphere, forming sulfuric and nitric acid, respectively. The acids eventually fall to Earth's surface as acid precipitation—rain, snow, sleet, or fog that has a pH less than 5.6. Acid precipitation lowers the pH of aquatic ecosystems and affects the soil chemistry of terrestrial ecosystems. Although acid precipitation due to fuel combustion has been occurring ever since the Industrial Revolution, emissions have increased during the past hundred years, mainly as a result of ore smelters and electrical generating plants.

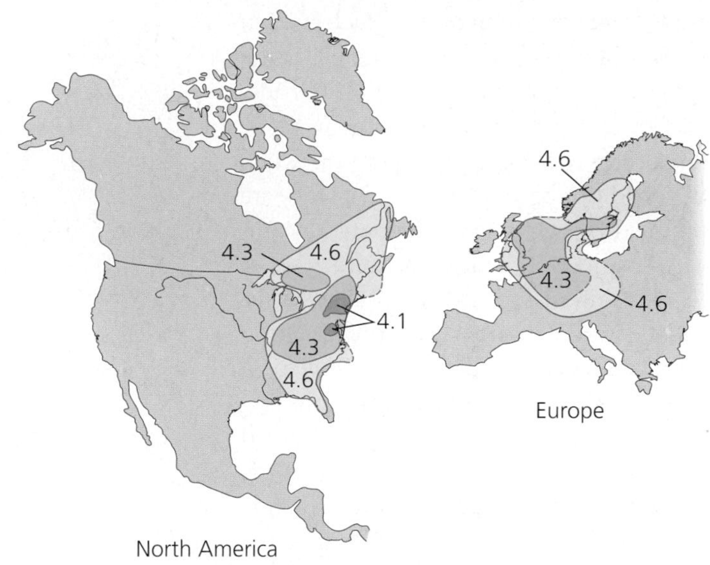

▲ **Figure 54.21 Distribution of acid precipitation in North America and Europe, 1980.** The numbers identify the average pH of precipitation in the shaded areas.

Acid precipitation is a regional and even global problem rather than a local one. To avoid local pollution problems, smelters and generating plants are built with very tall exhaust stacks (more than 300 m high). This reduces pollution at ground level, but exports the problem far downwind. The sulfurous and nitrogenous pollutants from fuel combustion may drift hundreds of kilometers before falling as acid precipitation.

Gene Likens and other ecologists first documented damage to forests and lakes in eastern North America and Europe in the 1960s. Lake-dwelling organisms in eastern Canada were dying because of air pollution from factories in the midwestern United States. Lakes and streams in southern Norway and Sweden were losing fish because of acid rain from pollutants generated in Great Britain and central and eastern Europe. By 1980, precipitation in large areas of North America and Europe averaged pH 4.0–4.5, with "record" storms occasionally dropping rain as acidic as pH 3.0 **(Figure 54.21)**.

In terrestrial ecosystems, such as the deciduous forests of New England, the change in soil pH due to acid precipitation causes calcium and other nutrients to leach from the soil (as we saw earlier in the chapter from the Hubbard Brook studies). The nutrient deficiencies affect the health of plants and limit their growth. Acid precipitation can also damage plants directly, mainly by leaching nutrients from leaves.

Freshwater ecosystems are particularly sensitive to acid precipitation. The lakes in North America and northern Europe that are most readily damaged by acid precipitation are those underlain by granite bedrock. Such lakes generally have relatively poor buffering capacity because the concentration of bicarbonate, an important buffer, is low. Fish populations have declined in thousands of such lakes in Norway and Sweden, where the pH of the water has dropped below 5.0. In Canada, lake trout are keystone predators in many lakes. Newly hatched lake trout die when the pH drops below 5.4, and when they are replaced by acid-tolerant fish, the dynamics of food webs change dramatically.

The entire contiguous United States has been affected by acid precipitation **(Figure 54.22)**. But there is also some positive news. Environmental regulations and new industrial technologies have enabled many developed countries, including the United States, to reduce sulfur dioxide emissions during the past 30 years. In the United States, for example, sulfur dioxide emissions were reduced 31% between 1993 and 2002. The water chemistry in the streams and freshwater lakes of New England is slowly improving after decades of severe acid precipitation. However, ecologists estimate that it will take another 10 to 20 years for these ecosystems to recover, even if sulfur dioxide emissions continue to decrease. Meanwhile, massive emissions of sulfur dioxide and acid precipitation continue in parts of central and eastern Europe, contributing to forest destruction over vast areas.

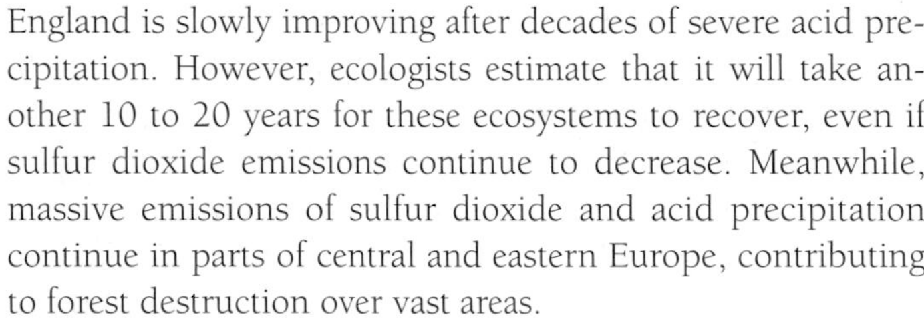

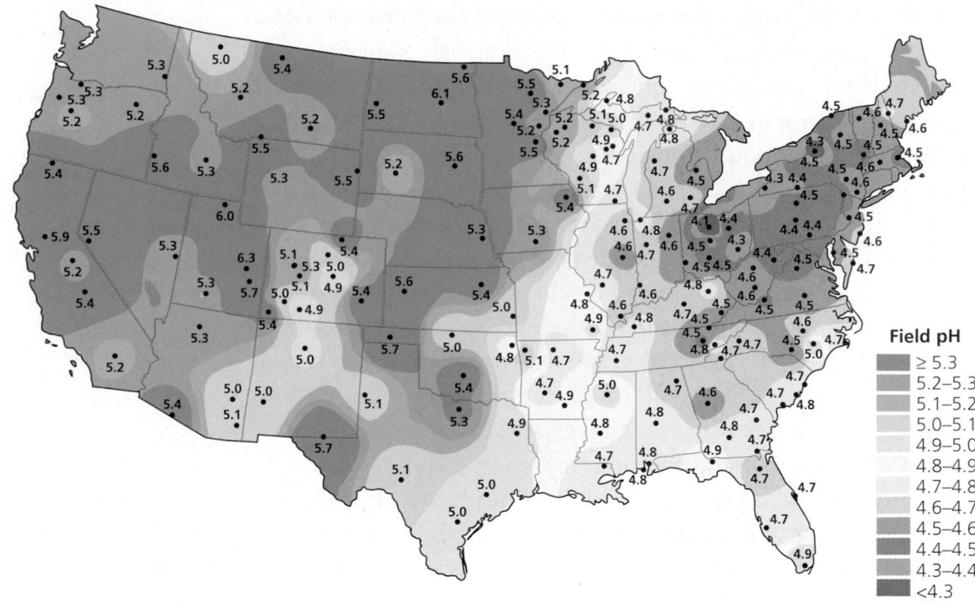

▲ **Figure 54.22 Average pH for precipitation in the contiguous United States in 2002.**

Toxins in the Environment

Humans release an immense variety of toxic chemicals, including thousands of synthetics previously unknown in nature, with little regard for the ecological consequences. Organisms acquire toxic substances from the environment along with nutrients and water. Some of the poisons are metabolized and excreted, but others accumulate in specific tissues, especially fat. One of the reasons such toxins are particularly harmful is that they become more concentrated in successive trophic levels of a food web, a process called **biological magnification.** Magnification occurs because the biomass at any given trophic level is produced from a much larger biomass ingested from the level below. Thus, top-level carnivores tend to be the organisms most severely affected by toxic compounds in the environment.

One class of industrially synthesized compounds that have demonstrated biological magnification are the chlorinated hydrocarbons, which include many pesticides, such as DDT, and the industrial chemicals called PCBs (polychlorinated biphenyls). Current research implicates many of these compounds in endocrine system disruption in a large number of animal species, including humans. Biological magnification of PCBs has been found in the food web of the Great Lakes, where the concentration of PCBs in herring gull eggs at the top of the food web is nearly 5,000 times greater than in phytoplankton at the base of the food web **(Figure 54.23)**.

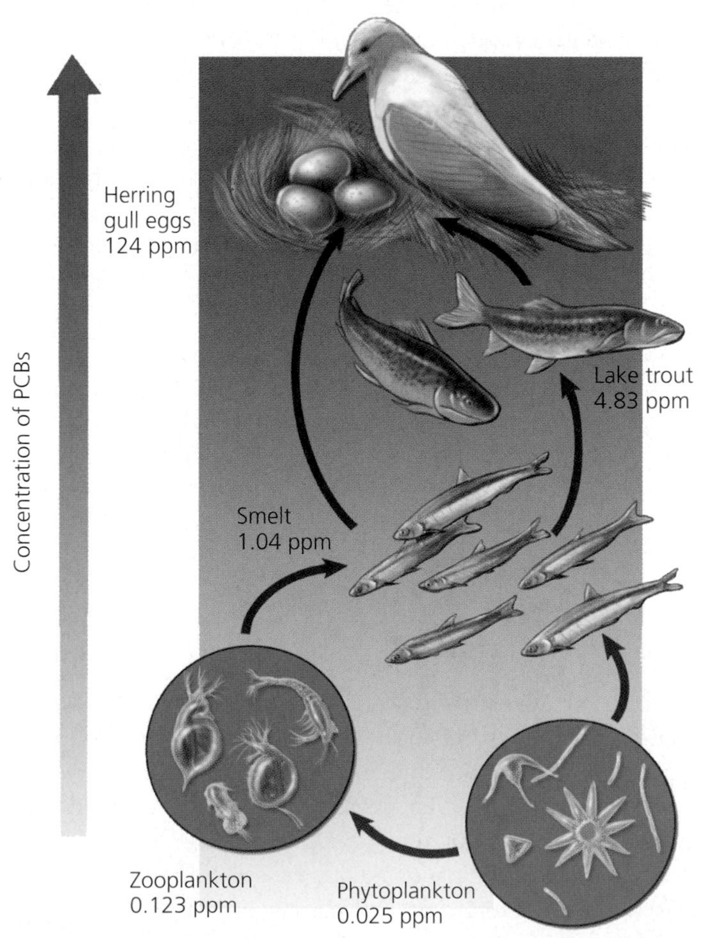

▲ **Figure 54.23 Biological magnification of PCBs in a Great Lakes food web.**

An infamous case of biological magnification that harmed top-level carnivores involved DDT, a chemical used to control insects such as mosquitoes and agricultural pests. In the decade after World War II, the pesticide industry promoted the benefits of DDT before anybody really understood the ecological consequences. By the 1950s, scientists were learning that DDT persists in the environment and is transported by water to areas far from where it is applied. But by then, the poison had already become a global problem. One of the first signs that DDT was a serious environmental problem was a decline in the populations of pelicans, ospreys, and eagles, birds that feed at the top of food webs. The accumulation of DDT (and DDE, a product of its partial breakdown) in the tissues of these birds interfered with the deposition of calcium in their eggshells, a trend that may have already begun because of other environmental contaminants. When these birds tried to incubate their eggs, the weight of the parents broke the shells of affected eggs, resulting in catastrophic declines in their reproduction rates. Rachel Carson's *Silent Spring* helped bring the problem to public attention in the 1960s (see Chapter 50), and DDT was banned in the United States in 1971. A dramatic recovery in populations of the affected bird species followed. The pesticide is still used in many other parts of the world, however.

Many toxins cannot be degraded by microorganisms and consequently persist in the environment for years or even decades. In other cases, chemicals released into the environment may be relatively harmless but are converted to more toxic products by reaction with other substances or by the metabolism of microorganisms. For example, mercury, a by-product of plastic production and coal-fired power generation, has been routinely expelled into rivers and the sea in an insoluble form. Bacteria in the bottom mud convert the waste to methyl mercury, an extremely toxic soluble compound that accumulates in the tissues of organisms, including humans who consume fish from the contaminated waters.

Atmospheric Carbon Dioxide

Many human activities release a variety of gaseous waste products. People once thought that the vastness of the atmosphere could absorb these materials without significant consequences, but we now know that such additions can cause fundamental changes in the composition of the atmosphere and in its interactions with the rest of the biosphere. One pressing problem is the rising level of atmospheric carbon dioxide.

Rising Atmospheric CO_2

Since the Industrial Revolution, the concentration of CO_2 in the atmosphere has been increasing as a result of the combustion of fossil fuels and the burning of enormous quantities of wood removed by deforestation. Scientists estimate that the average CO_2 concentration in the atmosphere before 1850 was about 274 ppm. In 1958, a monitoring station began taking very accurate measurements on Hawaii's Mauna Loa peak, where the air is free from the variable short-term effects that occur near large urban areas. At that time, the CO_2 concentration was 316 ppm **(Figure 54.24)**. Today, the concentration of CO_2 in the atmosphere exceeds 370 ppm, an increase of about 17%. If CO_2 emissions continue to increase at the present rate, by the year 2075, the atmospheric concentration of this gas will be double what it was at the start of the Industrial Revolution.

Increased productivity by vegetation is one predictable consequence of increasing CO_2 levels. In fact, when CO_2 concentrations are raised in experimental chambers such as greenhouses, most plants respond with increased growth. However, because C_3 plants are more limited than C_4 plants by CO_2 availability (see Chapter 10), one effect of increasing global CO_2 concentration may be the spread of C_3 species into terrestrial habitats previously favoring C_4 plants. This may

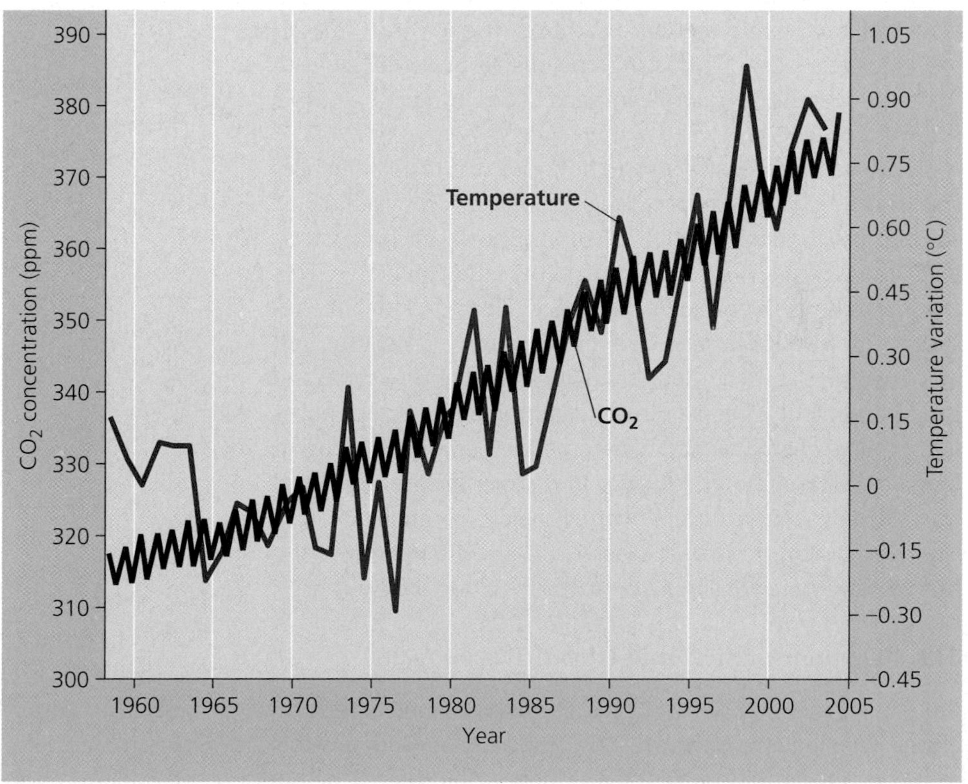

▲ **Figure 54.24 The increase in atmospheric carbon dioxide at Mauna Loa, Hawaii, and average global temperatures over land from 1958 to 2004.** Aside from normal seasonal fluctuations, the total amount of CO_2 has increased steadily (black). Though average global land temperatures over the same period fluctuate a great deal (red), there is a warming trend.

have important agricultural implications. For example, corn (maize), a C_4 plant and the most important grain crop in the United States, may be replaced by wheat and soybeans, C_3 crops that outproduce corn in a CO_2-enriched environment. To predict the gradual and complex effects of rising CO_2 levels on species composition in nonagricultural communities, scientists are turning to long-term field experiments.

How Elevated CO_2 Affects Forest Ecology: The FACTS-I Experiment

To assess how increasing atmospheric concentration of CO_2 might affect temperate forests, scientists at Duke University began the Forest-Atmosphere Carbon Transfer and Storage (FACTS-I) experiment in 1995. The researchers are manipulating a single environmental factor, the concentration of CO_2, to which tracts of forest are exposed. All other factors, such as temperature, precipitation, and wind speed and direction vary normally for both experimental plots and adjacent control plots exposed to only atmospheric CO_2.

The FACTS-I experiment includes six plots in a 200-acre (80-hectare) tract of loblolly pine within Duke University's experimental forest. Each plot consists of a circular area, approximately 30 m in diameter, ringed by 16 towers **(Figure 54.25)**. Within three of the plots (the experimental plots), the towers produce air containing about 1½ times present-day CO_2 concentrations. Instruments on a tall tower in the center of each experimental plot measure the direction and speed of the wind, adjusting the distribution of CO_2 to maintain a stable CO_2 concentration. The other three plots serve as controls.

The FACTS-I study is testing how elevated CO_2 influences tree growth, carbon concentration in soils, insect populations, soil moisture, the growth of plants in the forest understory, and other factors over a ten-year period. Researchers expect that the study will provide a sound basis for predicting the response of whole forests to elevated future CO_2 conditions. Some of the data recorded so far include higher weight and lipid content of loblolly pine seeds, higher soil respiration, higher rates of soil weathering, and higher rates of photosynthesis by sweetgum trees *(Liquidamber styraciflua)* in the experimental plots. As in the Hubbard Brook experimental forest, the responses of this forest ecosystem will take many years to unfold, and there are sure to be surprises along the way.

The Greenhouse Effect and Global Warming

One factor that complicates predictions about the long-term effects of rising atmospheric CO_2 concentration is its possible influence on Earth's heat budget. Much of the solar radiation that strikes the planet is reflected back into space. Although CO_2 and water vapor in the atmosphere are transparent to visible light, they intercept and absorb much of the reflected infrared radiation, re-reflecting some of it back toward Earth. This process retains some of the solar heat. If it were not for this **greenhouse effect**, the average air temperature at Earth's surface would be −18°C (−2.4°F), and most life as we know it could not exist. The marked increase in atmospheric CO_2 concentrations over the last 150 years concerns many scientists because of its potential to increase global temperature. (See Figure 54.24 for circumstantial evidence of a link between increasing CO_2 and global warming.)

▲ **Figure 54.25 Large-scale experiment on the effects of elevated CO_2.** Rings of towers in the Duke University Experimental Forest emit enough carbon dioxide to raise and maintain CO_2 levels 200 ppm above present-day concentrations.

To date, no models are sophisticated enough to include all the biotic and abiotic factors that can influence atmospheric gas concentrations and temperature (for example, cloud cover, CO_2 uptake by photosynthetic organisms, and the effects of particles in the air). However, a number of studies predict that by the end of the 21st century, atmospheric CO_2 concentration will double, and average global temperature will increase by about 2°C. Supporting these models is a correlation between CO_2 levels and temperatures in prehistoric times. Climatologists can actually measure CO_2 levels in bubbles trapped in glacial ice at different times in Earth's history. Prehistoric temperatures are inferred by several methods, including analysis of past vegetation based on fossils.

An increase of only 1.3°C would make the world warmer than at any time in the past 100,000 years. A worst-case scenario suggests that the warming would be greatest near the poles. The resultant melting of polar ice might raise sea level by an estimated 100 m, gradually flooding areas 150 km (or more) inland from the current coastline; New York, Miami, Los Angeles, and many other cities would then be under water. Recent collapses in the Antarctic ice shelf, which in 2002 released an iceberg four times the area of Manhattan, may be an early sign of global warming.

A warming trend would also alter the geographic distribution of precipitation, making major agricultural areas of the central United States much drier, for example. However, the various mathematical models disagree about the details of how climate in each region will be affected. By studying how *past* periods of global warming and cooling affected plant communities, ecologists are trying to predict the consequences of *future* temperature changes. Analysis of fossilized pollen provides evidence that plant communities change dramatically with changes in temperature. However, past climate changes occurred gradually, and plant and animal populations could migrate into areas where abiotic conditions allowed them to survive. Many organisms, especially plants that cannot disperse rapidly over long distances, will probably not be able to survive the high rates of climate change projected to result from global warming.

The global warming that apparently is now under way as a result of the addition of CO_2 to the atmosphere is a problem of uncertain consequences and no simple solutions. Coal, natural gas, gasoline, wood, and other organic fuels that cannot be burned without releasing CO_2 are central to our increasingly industrialized societies. Stabilizing CO_2 emissions will require concerted international effort and the acceptance of dramatic changes in both personal lifestyles and industrial processes. Many ecologists think that this effort suffered a major setback in 2001, when the United States pulled out of the Kyoto Protocol, a 1997 pledge by the industrialized nations to reduce their CO_2 output by about 5% over a ten-year period.

Depletion of Atmospheric Ozone

Life on Earth is protected from the damaging effects of ultraviolet (UV) radiation by a protective layer of ozone molecules (O_3) that is present in the lower stratosphere between 17 and 25 km above Earth's surface. Ozone absorbs UV radiation, preventing much of it from reaching organisms in the biosphere.

Satellite studies of the atmosphere suggest that the ozone layer has been gradually "thinning" since 1975 **(Figure 54.26)**. The destruction of atmospheric ozone probably results mainly from the accumulation of chlorofluorocarbons (CFCs), chemicals used for refrigeration, as propellants in aerosol cans, and in certain manufacturing processes. When the breakdown products from these chemicals rise to the stratosphere, the chlorine they contain reacts with ozone, reducing it to molecular O_2 **(Figure 54.27)**. Subsequent chemical reactions liberate the chlorine, allowing it to react with other ozone molecules in a catalytic chain reaction. The effect is most apparent over Antarctica, where cold winter temperatures facilitate these atmospheric reactions. Scientists first described the "ozone hole" over Antarctica in 1985 and have since documented that it is a seasonal phenomenon that grows and shrinks in an annual cycle. However, the magnitude of ozone depletion and the size of the ozone hole have generally increased in recent years, and the hole sometimes extends as far as the southernmost portions of Australia, New Zealand, and South

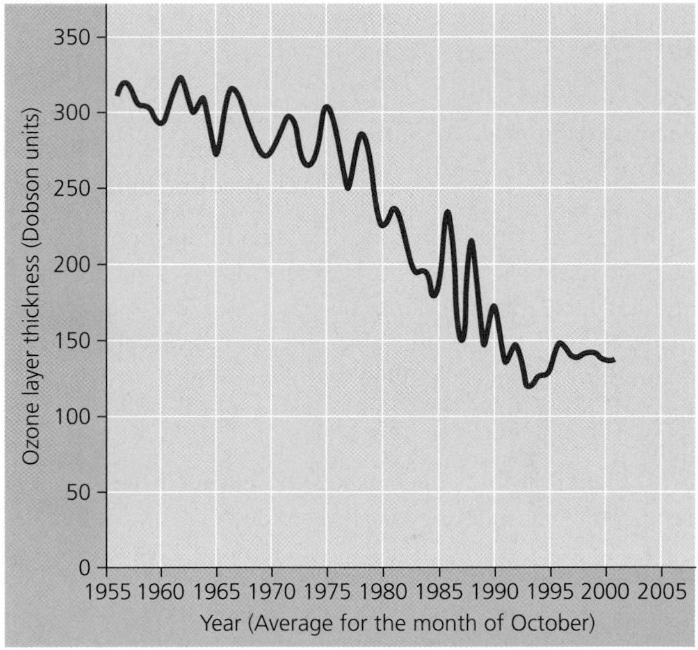

▲ **Figure 54.26 Thickness of the ozone layer over Antarctica in units called Dobsons.**

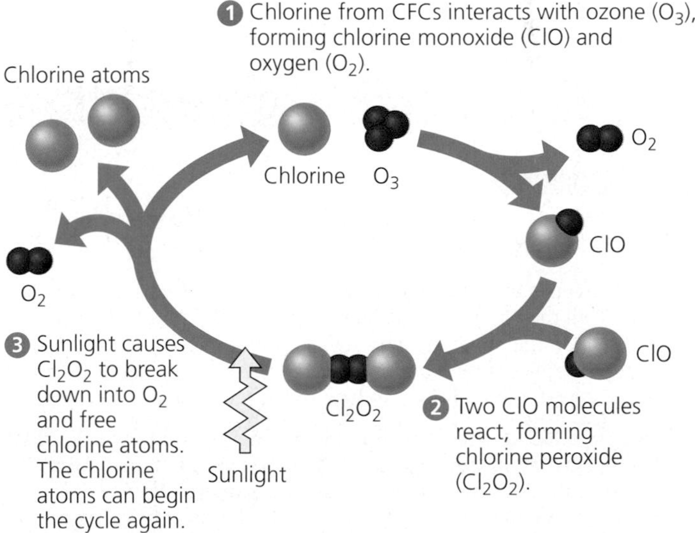

▲ **Figure 54.27 How free chlorine in the atmosphere destroys ozone.**

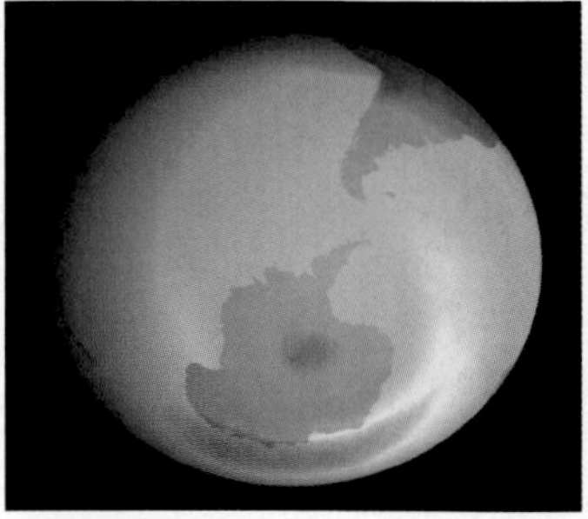

(a) October 1979

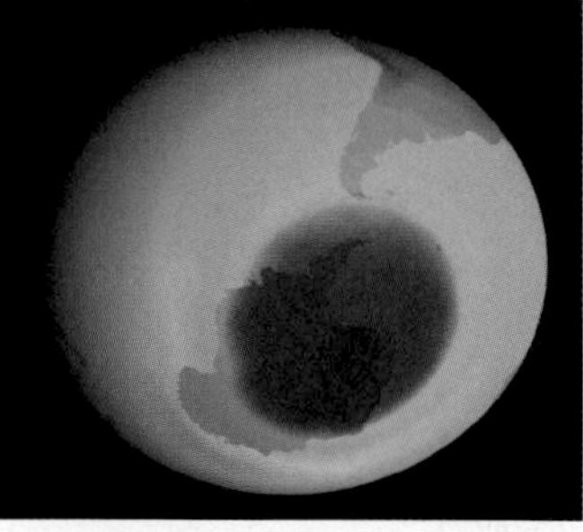

(b) October 2000

▲ **Figure 54.28 Erosion of Earth's ozone shield.** The ozone hole over Antarctica is visible as the blue patch in these images based on atmospheric data.

America **(Figure 54.28)**. At the more heavily populated middle latitudes, ozone levels have decreased 2–10% during the past 20 years.

The consequences of ozone depletion for life on Earth may be quite severe. Some scientists expect increases in both lethal and nonlethal forms of skin cancer and in cataracts among humans, as well as unpredictable effects on crops and natural communities, especially the phytoplankton that are responsible for a large proportion of Earth's primary production. The danger posed by ozone depletion is so great that 180 countries, including the United States, have signed the Montreal Protocol, which calls for the elimination of chemicals that deplete ozone. Many nations, again including the United States, have ended the production of chlorofluorocarbons. As a consequence of these actions, there are signs that the rate of ozone depletion is slowing. Unfortunately, even if all chlorofluorocarbons were globally banned today, chlorine molecules that are already in the atmosphere will continue to influence stratospheric ozone levels for at least a century.

Destruction of Earth's ozone shield is one more example of how much humans have been able to disrupt the dynamics of ecosystems and the entire biosphere. In this book's final chapter, we'll explore how scientists in fields of conservation biology and restoration ecology are studying and responding to human impact on Earth's biodiversity.

Concept Check 54.5

1. How can the addition of excess nutrients to a lake threaten its fish population?
2. How can clear cutting a forest (removing all trees) damage the water quality of nearby lakes?
3. In the face of biological magnification of toxins, is it healthier to feed at a lower or higher trophic level? Explain.
4. There are vast stores of organic matter in the frozen soils of the Arctic. Why might this be a cause for concern by scientists studying global warming?

For suggested answers, see Appendix A.

Chapter 54 Review

Go to www.campbellbiology.com or the student CD-ROM to explore Activities, Investigations, and other interactive study aids.

SUMMARY OF KEY CONCEPTS

Concept 54.1
Ecosystem ecology emphasizes energy flow and chemical cycling

▶ **Ecosystems and Physical Laws (p. 1185)** An ecosystem consists of all the organisms in a community and all the abiotic factors with which they interact. The laws of physics and chemistry apply to ecosystems, particularly in regard to the flow of energy. Energy is conserved, but degraded to heat during ecosystem processes.

▶ **Trophic Relationships (p. 1185)** Energy and nutrients pass from primary producers (autotrophs) to primary consumers (herbivores) and then to secondary consumers (carnivores). Energy *flows through* an ecosystem, entering as light and exiting as heat. Nutrients *cycle within* an ecosystem.

▶ **Decomposition (pp. 1185–1186)** Decomposition connects all trophic levels. Detritivores, mainly bacteria and fungi, recycle essential chemical elements by decomposing organic material and returning elements to inorganic reservoirs.

Concept 54.2
Physical and chemical factors limit primary production in ecosystems

▶ **Ecosystem Energy Budgets (pp. 1186–1188)** The energy assimilated during photosynthesis is a tiny fraction of the solar radiation reaching Earth, but primary production sets the spending limit for the global energy budget. Gross primary production is the total energy assimilated by an ecosystem in a given time period. Net primary production, the energy accumulated in autotroph biomass, equals gross primary production minus the energy used by the primary producers for respiration. Only net primary production is available to consumers.
Investigation ***How Do Temperature and Light Affect Primary Production?***

▶ **Primary Production in Marine and Freshwater Ecosystems (pp. 1188–1190)** In marine and freshwater ecosystems, light and nutrients limit primary production. Within the photic zone, the factor that most often limits primary production is a nutrient such as nitrogen or iron.

▶ **Primary Production in Terrestrial and Wetland Ecosystems (pp. 1190–1191)** In terrestrial and wetland ecosystems, climatic factors such as temperature and moisture affect primary production on a large geographic scale. More locally, a soil nutrient is often the limiting factor in primary production.

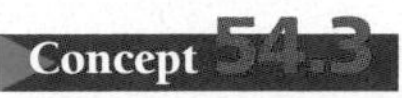

Energy transfer between trophic levels is usually less than 20% efficient

▶ **Production Efficiency (pp. 1191–1193)** The amount of energy available to each trophic level is determined by the net primary production and the efficiencies with which food energy is converted to biomass at each link of the food chain. The percentage of energy transferred from one trophic level to the next, called trophic efficiency, is generally 5–20%. Pyramids of production, biomass, and numbers reflect low trophic efficiency.
Activity *Pyramids of Production*
Graph It *Animal Food Production Efficiency and Food Policy*

▶ **The Green World Hypothesis (pp. 1193–1194)** According to the green world hypothesis, herbivores consume a small percentage of vegetation because predators, disease, competition, nutrient limitations, and other factors keep their populations in check.

Biological and geochemical processes move nutrients between organic and inorganic parts of the ecosystem

▶ **A General Model of Chemical Cycling (p. 1195)** Gaseous forms of carbon, oxygen, sulfur, and nitrogen occur in the atmosphere and cycle globally. Other less mobile elements, including phosphorus, potassium, and calcium, cycle on a more localized scale, at least over the short term. All elements cycle between organic and inorganic reservoirs.

▶ **Biogeochemical Cycles (pp. 1195–1198)** Water moves in a global cycle driven by solar energy. The carbon cycle primarily reflects the reciprocal processes of photosynthesis and cellular respiration. Nitrogen enters ecosystems through atmospheric deposition and nitrogen fixation by prokaryotes, but most of the nitrogen cycling in natural ecosystems involves local cycles between organisms and soil or water. The phosphorus cycle is relatively localized.
Activity *Energy Flow and Chemical Cycling*
Activity *The Carbon Cycle*
Activity *The Nitrogen Cycle*

▶ **Decomposition and Nutrient Cycling Rates (p. 1198)** The proportion of a nutrient in a particular form and its cycling time in that form vary among ecosystems, largely because of differences in the rate of decomposition.

▶ **Vegetation and Nutrient Cycling: The Hubbard Brook Experimental Forest (pp. 1198–1199)** Nutrient cycling is strongly regulated by vegetation. Long-term ecological research projects monitor ecosystem dynamics over relatively long periods of time. The Hubbard Brook study has shown that logging increases water runoff and can cause huge losses of minerals.

Concept 54.5

The human population is disrupting chemical cycles throughout the biosphere

▶ **Nutrient Enrichment (pp. 1200–1201)** Agriculture constantly removes nutrients from ecosystems, so large supplements are continually required. Considerable amounts of the nutrients in fertilizer pollute groundwater and surface-water aquatic ecosystems, where they can stimulate excess algal growth (cultural eutrophication).
Activity *Water Pollution from Nitrates*

▶ **Acid Precipitation (pp. 1201–1202)** Combustion of fossil fuels is the main cause of acid precipitation. North American and European ecosystems downwind from industrial regions have been damaged by rain and snow containing nitric acid and sulfuric acid.

▶ **Toxins in the Environment (pp. 1202–1203)** Toxins can become concentrated in successive trophic levels of food webs. The release of toxic wastes has polluted the environment with harmful substances that often persist for long periods of time and become concentrated along the food chain by biological magnification.

▶ **Atmospheric Carbon Dioxide (pp. 1203–1205)** Because of the burning of wood and fossil fuels and other human activities, atmospheric concentration of CO_2 has been steadily increasing. The ultimate effects may include significant warming and other climate change.
Activity *The Greenhouse Effect*
Graph It *Atmospheric CO_2 and Temperature Changes*

▶ **Depletion of Atmospheric Ozone (pp. 1205–1206)** The ozone layer reduces the penetration of UV radiation through the atmosphere. Human activities, including release of chlorine-containing pollutants, are eroding the ozone layer, with dangerous results.

TESTING YOUR KNOWLEDGE

Self-Quiz

1. Which of the following organisms is *incorrectly* paired with its trophic level?
   a. cyanobacteria—primary producer
   b. grasshopper—primary consumer
   c. zooplankton—primary producer
   d. eagle—tertiary consumer
   e. fungi—detritivore

2. A production pyramid such as the one shown in Figure 54.11 implies that
   a. only half the energy in a trophic level is transferred to the next higher level.
   b. most of the energy in a trophic level is incorporated into the next higher level.
   c. as energy is transferred from one trophic level to another, about 10% of the energy is lost.
   d. the most efficient transfer of energy is from producers to primary consumers.
   e. eating grain-fed beef is an inefficient way to obtain the energy captured by photosynthesis.

3. Nitrifying bacteria participate in the nitrogen cycle mainly by
   a. converting nitrogen gas to ammonia.
   b. releasing ammonia from organic compounds, thus returning it to the soil.
   c. converting ammonia to nitrogen gas, which returns to the atmosphere.
   d. converting ammonia to nitrate, which is absorbed by plants.
   e. incorporating nitrogen into amino acids and organic compounds.

4. The Hubbard Brook watershed deforestation experiment supported all of the following conclusions *except* that
   a. most minerals were recycled within a forest ecosystem.
   b. the flow of minerals out of a natural watershed was offset by minerals flowing in.
   c. deforestation increased water runoff.
   d. the nitrate concentration in waters draining the deforested area became dangerously high.
   e. calcium levels remained high in the soil of deforested areas.

5. The main cause of the recent increase in the amount of CO_2 in Earth's atmosphere is
   a. increased worldwide primary production.
   b. increased worldwide standing crop biomass.
   c. an increase in the amount of infrared radiation absorbed by the atmosphere.
   d. the burning of larger amounts of wood and fossil fuels.
   e. additional respiration by the rapidly growing human population.

6. Which of the following is a consequence of biological magnification?
   a. Toxic chemicals in the environment pose greater risk to top-level predators than to primary consumers.
   b. Populations of top-level predators are generally smaller than populations of primary consumers.
   c. The biomass of producers in an ecosystem is generally higher than the biomass of primary consumers.
   d. Only a small portion of the energy captured by producers is transferred to consumers.
   e. The amount of biomass in the producer level of an ecosystem decreases if the producer turnover time increases.

7. Which of these ecosystems has the *lowest* net primary production per square meter?
   a. a salt marsh
   b. an open ocean
   c. a coral reef
   d. a grassland
   e. a tropical rain forest

8. Tropical forest soils contain comparatively low levels of mineral nutrients because
   a. the standing crop biomass of tropical forests is comparatively small.
   b. tropical soil microorganisms do not break down organic matter as efficiently as temperate zone soil microorganisms do.
   c. organic matter decomposes more rapidly and plants assimilate soil nutrients more rapidly in the topics.
   d. nutrients cycle more slowly in the tropics.
   e. many nutrients are destroyed by high temperatures in the tropics.

9. Imagine that you test samples of coastal water polluted with runoff from farms and find detectable levels of phosphates, but not of nitrogen. In a follow-up experiment, you find that if you enrich some of your samples with nitrogen, algal growth is much greater than in unenriched, control samples. Enriching samples with phosphate, however, does not increase algal growth. From your results, you can conclude that
   a. algae populations in these waters could be reduced by decreasing phosphate runoff.
   b. eutrophication of these waters could be reduced by adding nitrogen.
   c. algae populations in these waters are kept low by high levels of phosphorus.
   d. nitrogen is the limiting nutrient in these waters.
   e. phosphate is the limiting nutrient in these waters.

10. Which of the following has the greatest effect on the rate of chemical cycling in an ecosystem?
    a. the ecosystem's rate of primary production
    b. the production efficiency of the ecosystem's consumers
    c. the rate of decomposition in the ecosystem
    d. the trophic efficiency of the ecosystem
    e. the location of the nutrient reservoirs in the ecosystem

For Self-Quiz answers, see Appendix A.

Go to the website or CD-ROM for more quiz questions.

Evolution Connection

Some biologists, struck by the complex interdependence of biotic and abiotic factors that make up ecosystems, have suggested that ecosystems themselves are emergent, "living" systems capable of evolving. One manifestation of this is James Lovelock's Gaia hypothesis, which views Earth itself as a living, homeostatic entity—a kind of superorganism. Critique the idea that ecosystems and the biosphere can evolve by applying the principles of evolution you have learned in this book. If ecosystems are capable of evolving, is this a form of Darwinian evolution? Why or why not?

Scientific Inquiry

With two nearby ponds in a forest as your study site, how would you design a controlled experiment to measure the effect of falling leaves on net primary production in a pond?

Investigation *How Do Temperature and Light Affect Primary Production?*

Science, Technology, and Society

The amount of CO_2 in the atmosphere is increasing, and global temperature has increased over the past century. Most scientists agree that the two phenomena are related and say that greenhouse warming is under way. These scientists stress that we need to take action now to avoid drastic environmental change. However, some scientists say that it is still too soon to tell and we should gather more data before we act. What are the advantages and disadvantages of doing something now to slow global warming? What are the advantages and disadvantages of waiting until more data are available?

55 Conservation Biology and Restoration Ecology

▲ **Figure 55.1 Tropical deforestation in West Kalimantan, Borneo.**

Key Concepts

55.1 Human activities threaten Earth's biodiversity

55.2 Population conservation focuses on population size, genetic diversity, and critical habitat

55.3 Landscape and regional conservation aim to sustain entire biotas

55.4 Restoration ecology attempts to restore degraded ecosystems to a more natural state

55.5 Sustainable development seeks to improve the human condition while conserving biodiversity

Overview

The Biodiversity Crisis

Biology is the science of life. Thus, it is fitting that our final chapter focuses on two disciplines that seek to preserve life. **Conservation biology** integrates ecology (including behavioral ecology), physiology, molecular biology, genetics, and evolutionary biology to conserve biological diversity at all levels. Efforts to sustain ecosystem processes and stem the loss of biodiversity also connect the life sciences with the social sciences, economics, and humanities. **Restoration ecology** applies ecological principles in an effort to return degraded ecosystems to conditions as similar as possible to their natural, predegraded state.

To date, scientists have described and formally named about 1.8 million species of organisms. Some biologists think that about 10 million more species currently exist; others estimate the number to be as high as 200 million. These are not evenly distributed around the globe. Some of the greatest concentrations of species are found in the tropics. Unfortunately, in tropical areas, scenes such as the one in **Figure 55.1** are commonplace: Tropical forests are being destroyed at an alarming rate to make room for and support a burgeoning human population.

Throughout the biosphere, human activities are altering trophic structures, energy flow, chemical cycling, and natural disturbance—ecosystem processes on which we and other species depend (see Chapter 54). The amount of human-altered land surface is approaching 50%, and we use over half of all accessible surface fresh water. In the oceans, stocks of many fishes are being depleted by overharvesting, and some of the most productive and diverse aquatic areas, such as coral reefs and estuaries, are being severely stressed. By some estimates, we are in the process of doing more damage to the biosphere and pushing more species toward extinction than the large asteroid that may have triggered the mass extinctions at the close of the Cretaceous period 65.5 million years ago (see Figure 26.9). Globally, the rate of species loss may be as much as 1,000 times higher than at any time in the past 100,000 years.

In this chapter, we will take a closer look at the biodiversity crisis and examine some of the conservation and restoration strategies that biologists are using in attempting to slow the rate of species loss.

Concept 55.1

Human activities threaten Earth's biodiversity

Extinction is a natural phenomenon that has been occurring almost since life first evolved; it is the current *rate* of extinction that underlies the biodiversity crisis. Because we can only estimate the number of species currently existing, we cannot determine the actual rate of species loss or the real magnitude of the biodiversity crisis. We do know for certain that the rate of species extinction is high and that it largely results from an escalating rate of ecosystem degradation by a single species—*Homo sapiens*. Simply put, humans are threatening Earth's biodiversity.

The Three Levels of Biodiversity

Biodiversity—short for biological diversity—has three main components, or levels: genetic diversity, species diversity, and ecosystem diversity **(Figure 55.2)**. All three are being decreased by human activity.

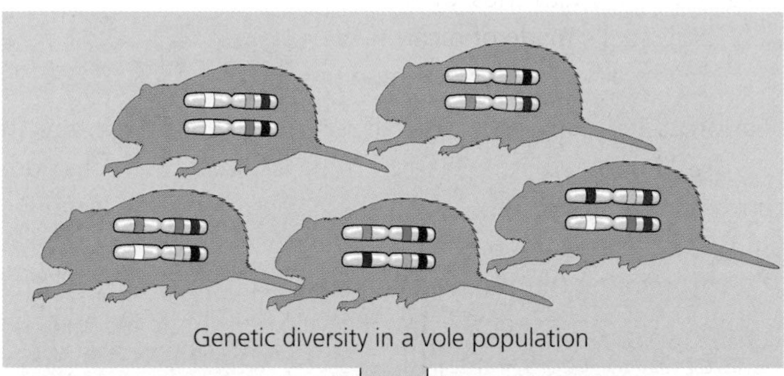

Community and ecosystem diversity
across the landscape of an entire region

▲ **Figure 55.2 Three levels of biodiversity.** The oversized chromosomes symbolize the genetic variation within the population.

Genetic Diversity

Genetic diversity comprises not only the individual genetic variation *within* a population, but also the genetic variation *between* populations that is often associated with adaptations to local conditions (see Chapter 23). If one population becomes extinct, then a species may have lost some of the genetic diversity that makes microevolution possible. This erosion of genetic diversity is, of course, detrimental to the overall adaptive prospects of the species. But the loss of genetic diversity throughout the biosphere also has implications for human welfare. For example, if we lose wild populations of plants closely related to agricultural species, we lose genetic resources that could potentially be used to improve certain crop qualities, such as disease resistance, through plant breeding.

Species Diversity

Much of the public discussion of the biodiversity crisis centers on species diversity—the variety of species in an ecosystem or throughout the entire biosphere, or what we called *species richness* in Chapter 53. The U.S. Endangered Species Act (ESA) defines an **endangered species** as one that is "in danger of extinction throughout all or a significant portion of its range." Also defined for protection by the ESA, **threatened species** are those that are considered likely to become endangered in the foreseeable future. Here are just a few statistics that illustrate the problem of species loss:

- According to the International Union for Conservation of Nature and Natural Resources (IUCN), 12% of the nearly 10,000 known species of birds and 24% of the nearly 5,000 known species of mammals in the world are threatened with extinction.
- A recent survey conducted by the Center for Plant Conservation showed that of the approximately 20,000 known plant species in the United States, 200 species have become extinct since such records have been kept. Another 730 plant species in the United States are endangered or threatened.
- About 20% of the known freshwater fishes in the world have either become extinct during historical times or are seriously threatened. One of the largest rapid extinction events yet recorded is the ongoing loss of freshwater fishes in East Africa's Lake Victoria. About 200 of the more than 500 species of cichlids in the lake have been lost, mainly as a result of the introduction of a nonnative predator species, the Nile perch, in the 1960s.
- Since 1900, 123 freshwater vertebrate and invertebrate species have become extinct in North America, and hundreds more species are threatened. Extinction rates for North American freshwater fauna are about five times higher than those for terrestrial animals.

- According to a 2004 report in the journal *Science* that was based on a global assessment of amphibians by more than 500 scientists, 32% of all known amphibian species are now either very near extinction or endangered.
- Several researchers estimate that at current rates of extinction, more than half of all currently living plant and animal species will have disappeared by the end of the 21st century **(Figure 55.3)**.

Extinction of species may be local; for example, a species may be lost in one river system but survive in an adjacent one. Global extinction of a species means that it is lost from *all* its locales. However, extinction is often an unseen process. To know for certain that a given species is extinct, we must know its exact distribution. Without a more complete catalog of species diversity and knowledge of the geographic distribution and ecological roles of Earth's species, our efforts to understand the structure and function of ecosystems on which our survival depends will remain incomplete.

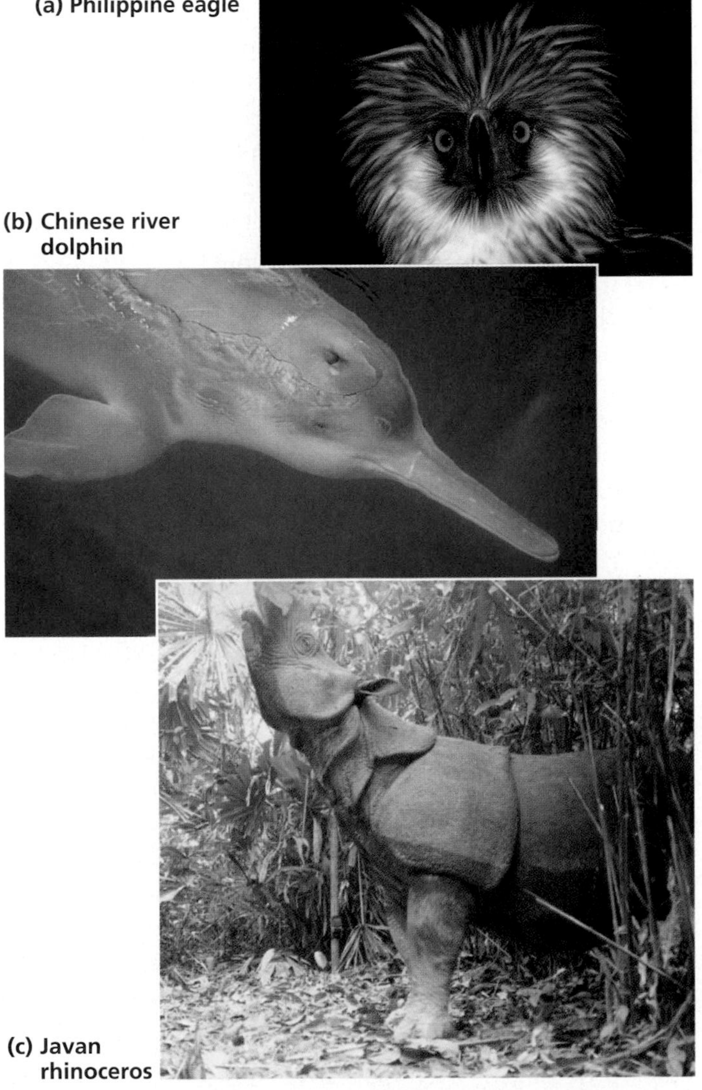

▲ **Figure 55.3 A hundred heartbeats from extinction.** These are just three of the members of what Harvard biologist E. O. Wilson grimly calls the Hundred Heartbeat Club, species with fewer than 100 individuals remaining on Earth.

Ecosystem Diversity

The variety of the biosphere's ecosystems is the third level of biological diversity. Because of the network of community interactions among populations of different species within an ecosystem, the local extinction of one species—say, a keystone predator—can have a negative impact on the overall species richness of the community (see Figure 53.16). More broadly, each ecosystem has characteristic patterns of energy flow and chemical cycling that can affect the whole biosphere. For example, the productive "pastures" of phytoplankton in the oceans may help moderate the greenhouse effect by consuming massive quantities of CO_2 for photosynthesis and for building shells made of bicarbonate.

Some ecosystems have already been heavily impacted by humans, and others are being destroyed at an astonishing pace. For example, within the contiguous United States, wetland and riparian (riverbank) ecosystems have been altered dramatically within just a few centuries. Since European colonization, more than 50% of wetlands have been drained and converted to other ecosystems, primarily agricultural ones. Meanwhile, in California, Arizona, and New Mexico, approximately 90% of native riparian communities have been destroyed by overgrazing, flood control, water diversions, lowering of water tables, and invasive (nonnative) plants.

Biodiversity and Human Welfare

Why should we care about the loss of biodiversity? Perhaps the purest reason is what E. O. Wilson calls *biophilia,* our sense of connection to nature and other forms of life. The belief that other species are entitled to life is a pervasive theme of many religions and the basis of a moral argument that we should protect biodiversity. There is also a concern for future human generations: Is it fair to deprive them of Earth's species richness? Paraphrasing an old proverb, G. H. Brundtland, a former prime minister of Norway, put it this way: "We must consider our planet to be on loan from our children, rather than being a gift from our ancestors." But species diversity brings us many practical benefits as well.

Benefits of Species and Genetic Diversity

Many species that are threatened could potentially provide crops, fibers, and medicines for human use, making biodiversity a crucial natural resource. In the United States, 25% of all prescriptions dispensed from pharmacies contain substances originally derived from plants. For example, in the 1970s, researchers discovered that the rosy periwinkle, which grows on Madagascar, an island off the coast of Africa, contains alkaloids

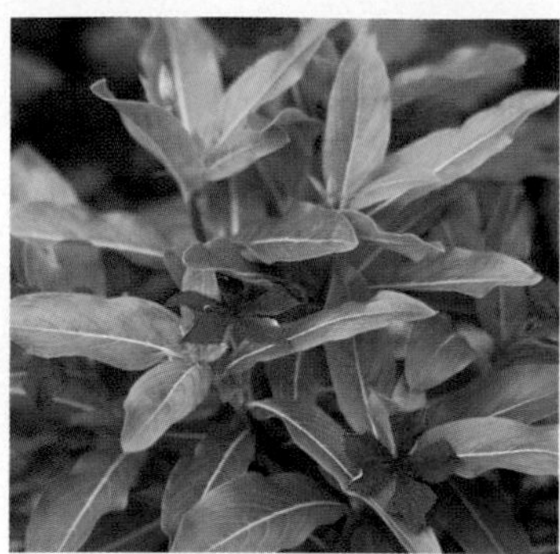

◄ **Figure 55.4 The rosy periwinkle (*Catharanthus roseus*), a plant that saves lives.**

that inhibit cancer cell growth **(Figure 55.4)**. The result of this discovery for most people with either of two potentially deadly forms of cancer—Hodgkin's disease and a childhood leukemia—is remission. Madagascar is also home to five other species of periwinkles, one of which is approaching extinction. The loss of these species would mean the loss of any possible medicinal benefits they might offer.

The loss of species also means the loss of genes. Each species has certain unique genes, and biodiversity represents the sum of all the genomes of all organisms on Earth. Such enormous genetic diversity has the potential for great human benefit. Consider the historical example of the polymerase chain reaction (PCR), the gene-cloning technology based on an enzyme extracted from thermophilic prokaryotes from hot springs (see Figure 20.7). Currently, corporations anticipate using DNA extracted from the prokaryotes in the numerous hot springs in Yellowstone National Park to mass-produce useful enzymes. Many researchers and industry officials are enthusiastic about the potential that such "bioprospecting" holds for the development of new medicines, foods, petroleum substitutes, industrial chemicals, and other products. However, because many millions of species may become extinct before we even know about them, we stand to lose irretrievably the valuable genetic potential held in their unique libraries of genes.

Ecosystem Services

The benefits that individual species provide to humans are often substantial, but saving individual species is only part of the rationale for saving ecosystems. Humans evolved in Earth's ecosystems, and we are finely adjusted to these systems and the other species within them.

In the urban and suburban settings in which most of us live today, it is easy to lose sight of the ways our ecosystems support us. **Ecosystem services** encompass all the processes through which natural ecosystems and the species they contain help sustain human life on Earth. The following are just a few of these ecosystem services:

- Purification of air and water
- Reduction of the severity of droughts and floods
- Generation and preservation of fertile soils
- Detoxification and decomposition of wastes
- Pollination of crops and natural vegetation
- Dispersal of seeds
- Cycling of nutrients
- Control of many agricultural pests by natural enemies
- Protection of shorelines from erosion
- Protection from ultraviolet rays
- Moderation of weather extremes
- Provision of beauty and recreational opportunities

A growing body of evidence indicates that the functioning of ecosystems, and hence their capacity to perform services, is linked to biodiversity. As human activities reduce biodiversity, we may be reducing the capacity of Earth's ecosystems to perform processes critical to our own survival.

Perhaps it is because we don't attach a monetary value to the services of natural ecosystems that we generally undervalue them. In a 1997 article, ecologist Robert Costanza and his colleagues estimated the dollar value of Earth's ecosystem services at $33 trillion per year, nearly twice the gross national product of all the countries an Earth ($18 trillion). Perhaps it is more realistic, and more meaningful, to do the accounting on a small scale. What, for example, is the true price of building a dam or clear-cutting a patch of forest if we include the dollar loss of ecosystem services in the cost column?

Four Major Threats to Biodiversity

Many different human activities threaten biodiversity on a local, regional, and global scale. Of this long list, most species loss can be traced to four major threats: habitat destruction, introduced species, overexploitation, and disruption of "interaction networks" such as food webs.

Habitat Destruction

Human alteration of habitat is the single greatest threat to biodiversity throughout the biosphere. Massive destruction of habitats has been brought about by agriculture, urban development, forestry, mining, and pollution. When no alternative habitat is available or a species is unable to move, habitat loss may mean extinction. The International Union for Conservation of Nature and Natural Resources implicates destruction of physical habitat for 73% of the species that have become extinct, endangered, vulnerable, or rare during modern history.

Habitat destruction may occur over immense regions. For instance, approximately 98% of the tropical dry forests of Central America and Mexico have been cleared (cut down). In addition, many natural landscapes have been broken up, fragmenting habitat into small patches **(Figure 55.5)**. Forest fragmentation is occurring at a rapid rate in tropical forests. For example, clearing of tropical rain forest in the state of Veracruz, Mexico, mostly for cattle ranching, has resulted in the loss of approximately 91% of the original forest, leaving a fragmented archipelago of small forest islands.

▲ **Figure 55.5 Habitat fragmentation in the Mount Hood National Forest, western United States.**

In almost all cases, habitat fragmentation leads to species loss, since the smaller populations in habitat fragments have a higher probability of local extinction. The prairies of North America are an example: Prairie covered about 800,000 hectares of southern Wisconsin when Europeans first arrived, but now occupies less than 0.1% of its original area. Plant diversity surveys of 54 Wisconsin prairie remnants were conducted in 1948–1954 and then repeated in 1987–1988. During the few decades between surveys, the various prairie fragments lost between 8% and 60% of their plant species.

Though most studies have focused on terrestrial ecosystems, habitat loss is also a major threat to marine biodiversity, especially along continental coasts and around coral reefs. About 93% of coral reefs, among Earth's most species-rich aquatic communities, have been damaged by human activities. At the current rate of destruction, 40–50% of the reefs, home to one-third of marine fish species, could be lost in the next 30 to 40 years. Aquatic habitat destruction and species loss also result from the dams, reservoirs, channel modification, and flow regulation now affecting most of the world's rivers. Fishes that migrate up the heavily dammed Columbia River system in the northwestern United States, for example, have shown sharp declines in abundance.

Introduced Species

Introduced species, also called invasive, nonnative, or exotic species, are those that humans move, either intentionally or accidentally, from the species' native locations to new geographic regions. The modern ease of travel by ship and airplane has accelerated the transplant of species. Free from the predators, parasites, and pathogens that limit their populations in their native habitats, such transplanted species may spread through a new region at exponential rates.

Introduced species that gain a foothold usually disrupt their adopted community, often by preying on native organisms or outcompeting them for resources. For instance, the brown tree snake was accidentally introduced to the island of Guam as a "stowaway" in military cargo after World War II **(Figure 55.6a)**. Since then, 12 species of birds and 6 species of lizards on which the snakes prey have become extinct on Guam. Another particularly devastating accidental introduction is the fingernail-sized zebra mussel (*Dreissena polymorpha*), which was introduced into the Great Lakes of North America in 1988, most likely in the ballast water of ships arriving from Europe. Efficient suspension-feeding molluscs that attain high population densities, zebra mussels have extensively disrupted freshwater ecosystems, threatening native aquatic species. Zebra mussels have also clogged water intake structures, disrupting domestic and industrial water supplies and causing billions of dollars in damages.

Humans have deliberately introduced many species with good intentions but disastrous effects. For example, a Japanese plant called kudzu, which the U.S. Department of Agriculture introduced in the South to help control erosion, has taken over vast expanses of the southern landscape **(Figure 55.6b)**. The

(a) Brown tree snake, introduced to Guam in cargo

(b) Introduced kudzu thriving in South Carolina

▲ **Figure 55.6 Two introduced species.**

European starling, brought intentionally into New York's Central Park in 1890 by a citizens' group intent on introducing all the plants and animals mentioned in Shakespeare's plays, quickly spread across North America, increasing to a population of more than 100 million and displacing many native songbirds.

Introduced species are a worldwide problem, contributing to approximately 40% of the extinctions recorded since 1750 and costing billions of dollars annually in damage and control efforts. There are more than 50,000 introduced species in the United States alone.

Overexploitation

Overexploitation refers generally to the human harvesting of wild plants or animals at rates exceeding the ability of populations of those species to rebound. It is possible for overexploitation to endanger certain plant species, such as rare trees that produce valuable wood or some other commercial product. However, the term more often refers to commercial fishing, hunting, collecting, and trading of animals.

Especially susceptible to overexploitation are large organisms with low intrinsic reproductive rates, such as elephants, whales, and rhinoceroses. The decline of Earth's largest extant terrestrial animals, the African elephants, is a classic example of the impact of overhunting. Largely because of the trade in ivory, elephant populations have been declining in most of Africa during the last 50 years. Unfortunately, an international ban on the sale of new ivory resulted in increased poaching (illegal hunting), so the ban had little effect in central and east Africa. Only in South Africa, where once-decimated herds have been well protected for nearly a century, have elephant populations been stable or increasing (see Chapter 52).

Species with restricted habitats, such as small islands, are also very vulnerable to overexploitation. For example, by the 1840s, humans had hunted the great auk to extinction on islands in the Atlantic Ocean. This large, flightless seabird had been in great demand for its feathers, eggs, and meat. And many populations of commercially important marine fishes, once thought to be inexhaustible, have been dramatically reduced by overfishing. The exploding human population's increasing demand for protein, coupled with new harvesting technologies, such as long-line fishing and modern trawlers, have reduced these fish populations to levels that cannot sustain further exploitation. The fate of the North Atlantic bluefin tuna is just one example. Until the past few decades, this big tuna was considered a sport fish of little commercial value—just a few cents per pound for cat food. Then, in the 1980s, wholesalers began airfreighting fresh, iced bluefin to Japan for sushi and sashimi. In that market, the fish now brings up to $100 per pound **(Figure 55.7)**. With that kind of demand, it took just ten years to reduce the North American bluefin population to less than 20% of its 1980 size. The collapse of the northern cod fishery off Newfoundland in the 1990s is a more recent example of how it is possible to overharvest what was formerly a very common species.

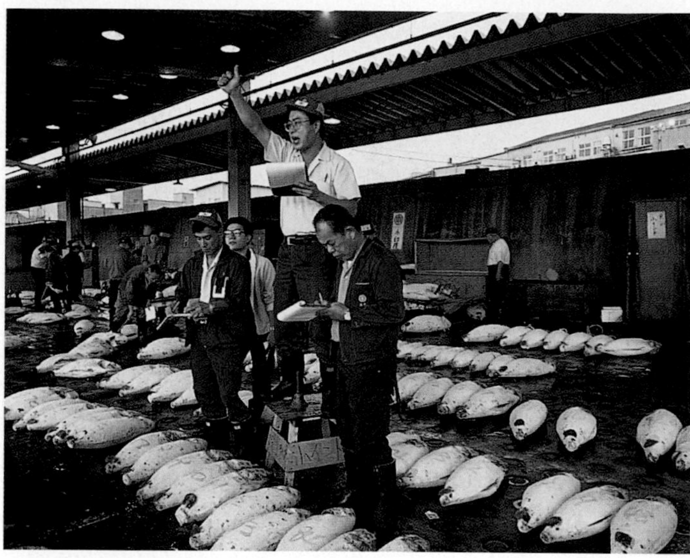

▲ **Figure 55.7 Overexploitation.** North Atlantic bluefin tuna are auctioned in a Japanese fish market.

Disruption of Interaction Networks

Ecosystem dynamics depend on the networks of interspecific interactions within biological communities. Like falling dominoes, the extinction of one species can doom others, particularly when the extinction involves a keystone species, an ecosystem engineer, or a species with a highly specialized relationship to others (see Chapter 53). Though this threat to biodiversity has been less explored than the other three we have discussed, there are many examples of how disruption of interaction networks can threaten other species.

Sea otters are a keystone species whose elimination over most of their historic range has led to major changes in the structure of shallow-water benthic communities along the west coast of North America (see Figure 53.17). And the extermination of beavers, one of the best-known ecosystem engineers, resulted in a large reduction in wetland and pond habitats across much of North America. In yet another example, global declines in native pollinators due to habitat destruction and excessive use of pesiticides have the potential to disrupt reproduction in both wild and domesticated plants. Bats called "flying foxes" are important plant pollinators in the Pacific Islands, where they have been subject to increasing pressure from hunters who sell them as luxury food items **(Figure 55.8)**. Because of this hunting pressure, conservation biologists are concerned about the fate of the native plants of the Samoan islands, where more than 79% of the trees depend on flying foxes for pollination or seed dispersal.

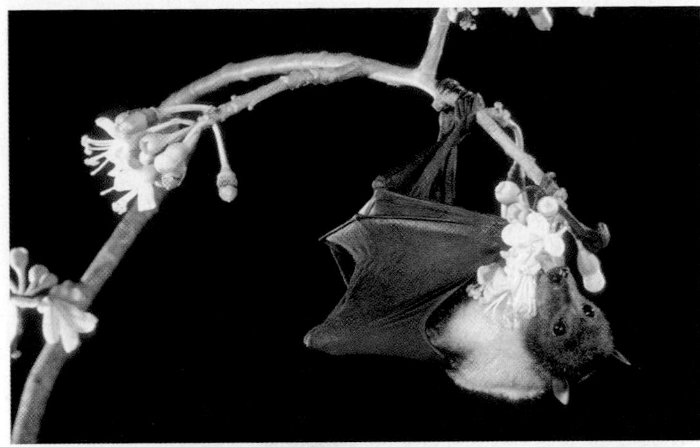

▲ **Figure 55.8 Disruption of interaction networks.** The endangered Marianas "flying fox" bat (*Pteropus mariannus*) is an important pollinator.

Concept Check 55.1

1. Explain why it is too narrow to define the biodiversity crisis as simply a loss of species.
2. Describe the four main threats to biodiversity and how each one damages diversity.
3. In what ways would humans benefit by preserving biodiversity?

For suggested answers, see Appendix A.

Concept 55.2

Population conservation focuses on population size, genetic diversity, and critical habitat

Biologists focusing on conservation at the population and species levels follow two main approaches—the small-population approach and the declining-population approach.

Small-Population Approach

A species is designated as endangered when its populations are very small. After such factors as habitat loss have taken their toll on population size, a population's smallness itself can drive it to extinction. Conservation biologists who adopt the small-population approach study the processes that can cause very small populations finally to become extinct.

The Extinction Vortex

A small population is prone to positive-feedback loops of inbreeding and genetic drift that draw the population down an **extinction vortex** toward smaller and smaller population size until no individuals exist **(Figure 55.9)**. The key factor driving the extinction vortex is the loss of the genetic variation necessary to enable evolutionary responses to environmental change, such as new strains of pathogens. Both inbreeding and genetic drift can cause a loss of genetic variation (see Chapter 23), and the effects of both processes become more significant as a population shrinks.

Not all populations are doomed by low genetic diversity. Low genetic variability does not necessarily lead to permanently small populations. For example, as we discussed in Chapter 23, overhunting of northern elephant seals in the 1890s reduced the species to only 20 individuals—clearly a bottleneck with reduced genetic variation. Since that time, however, the northern elephant seal populations have rebounded to about 150,000 individuals today, though their genetic variation remains relatively low. Furthermore, a number of plant species, such as the lousewort *Pedicularis* and several grasses, seem to have inherently low genetic variability. Many populations of cord grass (*Spartina anglica*), which thrives in salt marshes, are genetically uniform at many loci. *S. anglica* arose from a few parent plants only about a century ago by hybridization and allopolyploidy (see Figure 24.9). Having spread by cloning, this species now dominates large areas of tidal mudflats in Europe and Asia. Thus, in some cases, low genetic diversity is associated with population expansion rather than decline, but these cases may stand out precisely because they are so unusual.

How small does a population have to be before it starts down an extinction vortex? The answer depends on the type

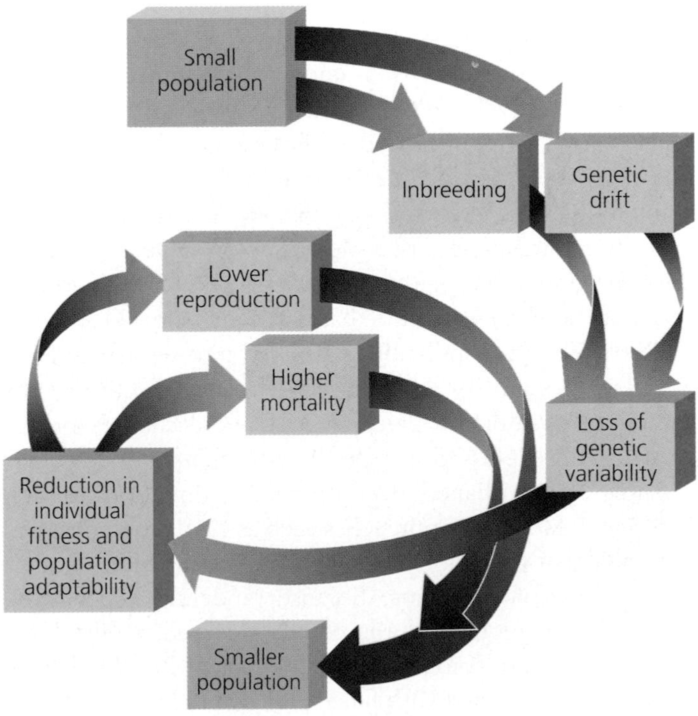

▲ **Figure 55.9 Processes culminating in an extinction vortex.**

of organism and several other factors, and must be determined case by case. For example, large predators that feed high on the food chain usually require very large individual ranges, resulting in very low population densities. Thus, not all rare species concern conservation biologists. Whatever the number, however, most populations presumably require some minimum size to remain viable.

Case Study: The Greater Prairie Chicken and the Extinction Vortex

When Europeans arrived in North America, the greater prairie chicken (*Tympanuchus cupido*) was common from New England to Virginia and all across the western prairies of the United States and Canada. Agriculture later fragmented the populations of the greater prairie chicken in the central and western states and provinces. For example, in Illinois alone, greater prairie chickens numbered in the millions in the 19th century but declined to 25,000 birds by 1933. By 1993, the Illinois population of prairie chickens was down to only 50, though large populations remained in Kansas, Minnesota, and Nebraska. Researchers found that the decline in the Illinois population was associated with a decrease in fertility. As a test of the extinction vortex hypothesis, the scientists imported genetic variation by transplanting more than 270 birds from larger populations in Kansas, Minnesota, and Nebraska (**Figure 55.10**). The Illinois population rebounded, confirming that it had been on its way down the extinction vortex until rescued by a transfusion of genetic variation.

Minimum Viable Population Size

The minimal population size at which a species is able to sustain its numbers and survive is known as the **minimum viable population (MVP).** MVP is usually estimated for a given species using computer models that integrate many factors. The calculation may include, for example, an estimate of how many individuals in a small population are likely to be killed by some natural catastrophe such as a storm. Once in the extinction vortex, two or three years in a row of bad weather could finish off a population that is already below MVP.

The MVP of a population is factored into what is called a **population viability analysis (PVA).** The objective of the analysis is to reasonably predict a population's chances for survival, usually expressed as a specific probability of survival (for example, a 99% chance) over a particular time (for instance, 100 years). Modeling approaches such as PVA allow conservation biologists to explore the potential consequences of alternative management plans. Because modeling depends on reliable information about the populations under study, however, conservation biology is most robust when theoretical modeling is combined with field studies of the managed populations. The case study on grizzly bears presented later in this section is an example of such an approach.

Figure 55.10

Inquiry What caused the drastic decline of the Illinois greater prairie chicken population?

EXPERIMENT Researchers observed that the population collapse of the greater prairie chicken was mirrored in a reduction in fertility, as measured by the hatching rate of eggs. Comparison of DNA samples from the Jasper County, Illinois, population with DNA from feathers in museum specimens showed that genetic variation had declined in the study population. In 1992, researchers began experimental translocations of prairie chickens from Minnesota, Kansas, and Nebraska in an attempt to increase genetic variation.

RESULTS After translocation (blue arrow), the viability of eggs rapidly improved, and the population rebounded.

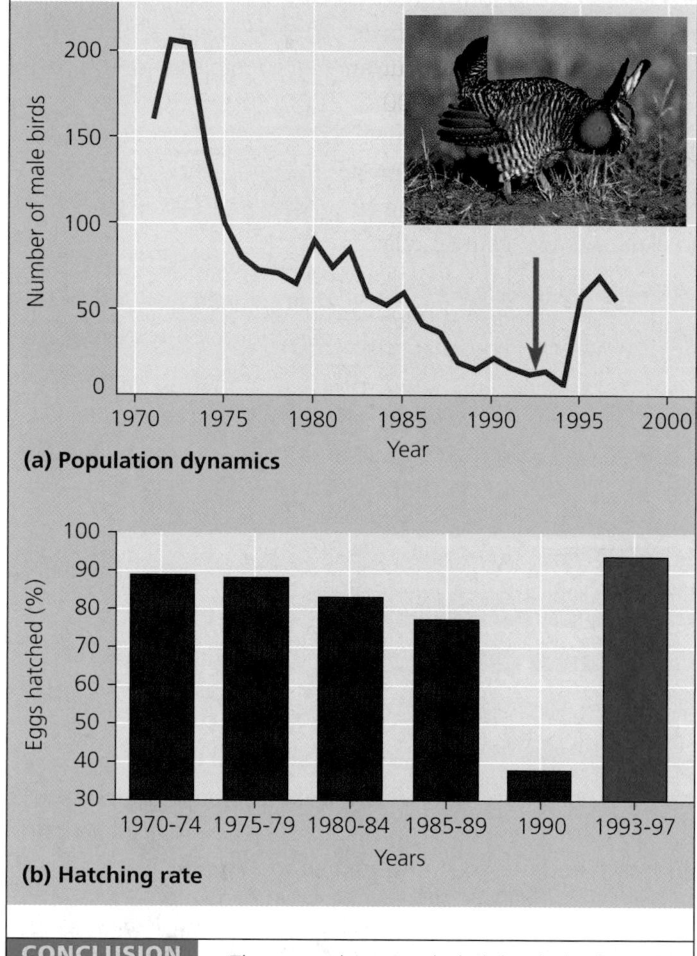

CONCLUSION The researchers concluded that lack of genetic variation had started the Jasper County population of prairie chickens down the extinction vortex.

Effective Population Size

Genetic variation is the key issue in the small-population approach. The *total* size of a population may be misleading because only certain members of the population breed successfully and pass their alleles on to offspring. Therefore, a meaningful estimate of MVP requires the researcher to determine the **effective population size,** which is based on the breeding potential of the population. The following formula incorporates the sex ratio of breeding individuals into the estimate of effective population size, abbreviated N_e:

$$N_e = \frac{4N_f N_m}{N_f + N_m}$$

where N_f and N_m are, respectively, the number of females and the number of males that successfully breed. Applying this formula to an idealized population whose total size is 1,000 individuals, N_e will also be 1,000 if every individual breeds and the sex ratio is 500 females to 500 males. In this case, $N_e = (4 \times 500 \times 500)/(500 + 500) = 1{,}000$. Deviation from these conditions (not all individuals breed and/or there is not a 1:1 sex ratio) reduces N_e. For instance, if the total population size is 1,000 but only 400 females and 400 males breed, then $N_e = (4 \times 400 \times 400)/(400 + 400) = 800$, or 80% of the total population size.

Numerous life history traits can influence N_e, and alternative formulas for estimating N_e take into account family size, maturation age, genetic relatedness among population members, the effects of gene flow between geographically separated populations, and population fluctuations.

In actual study populations, N_e is always some fraction of the total population. Thus, simply determining the total number of individuals in a small population does not provide a good measure of whether the population is large enough to avoid extinction. Whenever possible, conservation programs are geared to sustain total population sizes that include at least the minimum viable number of *reproductively active* individuals. The conservation goal of sustaining effective population size (N_e) above MVP stems from the concern that populations retain enough genetic diversity to be evolutionarily adaptable.

Case Study: Analysis of Grizzly Bear Populations

One of the first population viability analyses was conducted in 1978 by Mark Shaffer, of Duke University, as part of a long-term study of grizzly bears in Yellowstone National Park and its surrounding areas **(Figure 55.11)**. A threatened species in the United States, the grizzly bear (*Ursus arctos horribilis*) is currently found in only 4 of the 48 contiguous states. Its populations in those states have been drastically reduced and fragmented: In 1800, an estimated 100,000 grizzlies ranged over about 500 million hectares of more or less continuous habitat, while today there are six relatively isolated populations totaling about 1,000 individuals with a total range of less than 5 million hectares.

Shaffer attempted to determine viable sizes for the U.S. grizzly populations. Using life history data obtained for individual Yellowstone bears over a 12-year period, he simulated the effects of environmental factors on survival and reproduction. His models predicted that, given a suitable habitat, a total grizzly bear population of 70 to 90 individuals would have about a 95% chance of surviving for 100 years, whereas a population of 100 bears would have a 95% chance of surviving for 200 years.

▲ **Figure 55.11 Long-term monitoring of a grizzly bear population.** Studies by John and Frank Craighead, shown here placing a radio collar on a tranquilized grizzly bear, resulted in population estimates that were essential to Mark Shaffer's population viability analyses.

How does the actual size of the Yellowstone grizzly population compare with Shaffer's estimates of a minimum viable population size? Several sources of information indicate that the grizzly population of Yellowstone is growing. One of the best indicators of that growth is the number of females observed with cubs each year. Increases in this number from 1973 to 2002 support the conclusion that the Yellowstone grizzly population has grown substantially **(Figure 55.12)**. Moreover, since not all females with cubs are detected, the numbers shown in Figure 55.12 are minimal estimates. And because female grizzly bears have young approximately once every three years, the number of females with cubs each year is only about one-third

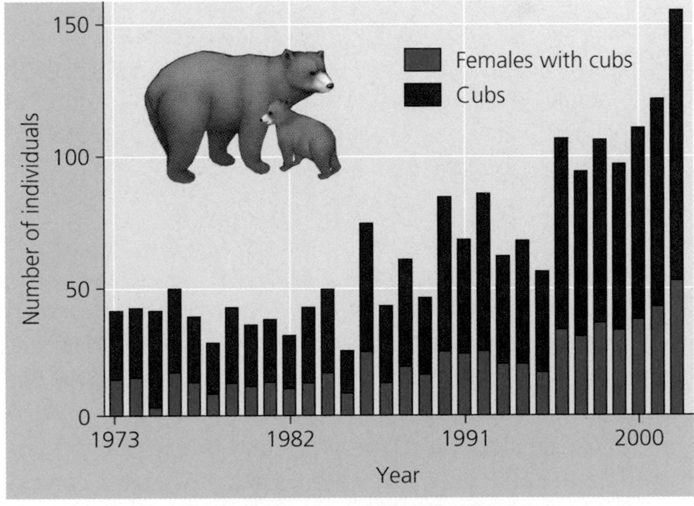

▲ **Figure 55.12 Growth of the Yellowstone grizzly bear population, as indicated by the number of females observed with cubs and the number of cubs.**

of the total number of mature females in the population. Adding immature males and females and mature males to estimates of mature females and cubs yields a total grizzly population size for the greater Yellowstone ecosystem exceeding 400.

The relationship of this estimate of the total grizzly population to the effective population size, N_e is dependent on several factors. Usually, only a few dominant males breed, and it may be difficult for them to locate females, since individuals inhabit such extensive areas. Moreover, females may reproduce only when there is abundant food. As a result, N_e is only about 25% of the total population size, which for the Yellowstone population of about 400 would be only about 100 bears.

Because small populations tend to lose genetic variation over time, a number of research teams have analyzed proteins, mitochondrial DNA, and nuclear microsatellite DNA to assess the genetic variability in the Yellowstone grizzly bear population. All results to date indicate that the Yellowstone population has less genetic variability than other grizzly bear populations in North America. However, the isolation and decline in genetic variability in the Yellowstone grizzly bear population were gradual during the 20th century and not as severe as feared. Museum specimens collected in the early 1900s demonstrate that genetic variability among the Yellowstone grizzly bears has historically been low. These studies also indicate that the effective population size of the Yellowstone grizzly population is larger than formerly thought—approximately 80 individuals through most of the 20th century and now slightly over 100.

How might conservation biologists increase the effective size and genetic variation of the Yellowstone grizzly bear population? Migration between isolated populations of grizzlies could increase both effective and total population sizes. Computer models predict that introducing only two unrelated bears each decade into a population of 100 individuals would reduce the loss of genetic variation by about half. For the grizzly bear, and probably for many other species whose populations are very small, finding ways to promote dispersal among populations may be one of the most urgent conservation needs.

This case study, as well as that of the greater prairie chicken, bridge small-population theory to practical applications in conservation. Next, we look at an alternative approach to understanding the biology of extinction.

Declining-Population Approach

The declining-population approach focuses on threatened and endangered populations that show a downward trend, even if the population is far above minimum viable size. The distinction between a declining population (which may be small) and a small population (which may be declining) is less important than the different priorities of the two basic conservation approaches. The small-population approach emphasizes smallness itself as an ultimate cause of a population's extinction, especially through loss of genetic diversity. In contrast, the declining-population approach emphasizes the environmental factors that caused a population decline in the first place. If, for example, an area is deforested, then species that depend on trees will decline and become locally extinct, whether or not they retain genetic variation.

Steps for Analysis and Intervention

The declining-population approach requires that population declines be evaluated on a case-by-case basis, with researchers carefully dissecting the causes of a decline before recommending or trying corrective measures. If, for example, the biological magnification of a particular toxic pollutant is causing a decline in some top-level consumer such as a predatory bird (see Chapter 54), then only reduction or elimination of the poison in the environment can save that particular species. Rarely is the situation so straightforward. However, the following steps for analyzing declining populations and determining interventions are useful even in complex cases.

1. Assess population trends and distribution to confirm that the species is presently in decline or that it was formerly more widely distributed or more abundant.
2. Study the natural history of this and related species, including reviewing the research literature, to determine the species' environmental requirements.
3. Develop hypotheses for all possible causes of the decline, including human activities and natural events, and list the predictions of each hypothesis.
4. Because many factors may be correlated with the decline, test the most likely hypothesis first. For example, remove the suspected agent of decline to see if the experimental population rebounds relative to a control population.
5. Apply the results of the diagnosis to management of the threatened species and monitor recovery.

The following case study is an example of how the declining-population approach was applied to one endangered species.

Case Study: Decline of the Red-Cockaded Woodpecker

The red-cockaded woodpecker (*Picoides borealis*) is an endangered species, endemic (found nowhere else) to the southeastern United States and once living throughout that region. This species requires mature pine forests, preferably ones dominated by the longleaf pine, for its habitat. Most woodpeckers nest in dead trees, but the red-cockaded woodpecker drills its nest holes in mature, living pine trees **(Figure 55.13a)**. Red-cockaded woodpeckers also drill small holes around the entrance to their nest cavity, which causes resin from the tree to ooze down the trunk. The resin seems to repel certain predators, such as corn snakes, that eat bird eggs and nestlings.

Another critical habitat factor for this woodpecker species is that the understory of plants around the pine trunks must be low **(Figure 55.13b)**. Breeding birds tend to abandon nests

(a) A red-cockaded woodpecker perches at the entrance to its nest site in a longleaf pine.

(b) Forest that can sustain red-cockaded woodpeckers has low undergrowth.

(c) Forest that cannot sustain red-cockaded woodpeckers has high, dense undergrowth that impacts the woodpeckers' access to feeding grounds.

▲ **Figure 55.13 Habitat requirements of the red-cockaded woodpecker.**

when vegetation among the pines is thick and higher than about 15 feet **(Figure 55.13c)**. Apparently, the birds require a clear flight path between their home trees and the neighboring feeding grounds. Historically, periodic fires swept through longleaf pine forests, keeping the undergrowth low.

One factor leading to decline of the red-cockaded woodpecker is the destruction or fragmentation of suitable habitats by logging and agriculture. Recognition of the key habitat factors, protection of some longleaf pine forests, and the use of controlled fires to reduce forest undergrowth have helped restore habitat that can support viable populations. However, designing a recovery program was complicated by the birds' social organization. Red-cockaded woodpeckers live in groups of one breeding pair and up to four "helpers," mostly males. Helpers are offspring who do not disperse and breed but remain behind and assist in incubating eggs and feeding nestlings. They may eventually attain breeding status within the flock when older birds die, but the wait may take years, and even then, helpers must compete to breed. Young birds that do disperse as members of new groups also have a tough path to reproductive success. New groups usually occupy abandoned territories or start at a new site and excavate nesting cavities, which can take several years. Individuals generally have a better chance of reproducing by remaining behind than by dispersing and excavating homes in new territories.

To test the hypothesis that this social behavior contributes to the decline of the red-cockaded woodpecker, Carole Copeyon, Jeffrey Walters, and Jay Carter, of North Carolina State University, constructed cavities in pine trees at 20 sites. The results were dramatic: cavities in 18 of the 20 sites were colonized by red-cockaded woodpeckers, and new breeding groups formed only in these sites. The experiment supported the hypothesis that this woodpecker species had been leaving much suitable habitat unoccupied because of an absence of breeding cavities. On the basis of this experiment, a combination of habitat maintenance and excavation of new breeding cavities has enabled a once-endangered species to rebound.

Weighing Conflicting Demands

Determining population numbers and habitat needs is only part of the effort to save species. It is also necessary to weigh a species' biological and ecological needs against other conflicting demands. Conservation biology often highlights the relationship between science, technology, and society—one of the themes of this book. For example, an ongoing, sometimes bitter debate in the U.S. Pacific Northwest pits habitat preservation for northern spotted owl, timber wolf, grizzly bear, and bull trout populations against job opportunities in the timber, mining, and other resource extraction industries. Programs to restock wolves in Yellowstone Park are opposed by some recreationists concerned for human safety and by many ranchers concerned with potential loss of livestock.

Large, high-profile vertebrates are not always the focal point in such conflicts, but habitat use is almost always at issue. Should work proceed on a new highway bridge if it destroys the only remaining habitat of a species of freshwater mussel? If you were the owner of a coffee plantation growing varieties that thrive in bright sunlight, would you be willing to change to shade-tolerant coffee varieties that are less productive and less profitable but that support large numbers of songbirds?

Another important consideration is the ecological roles of species. Because we will not be able to save every endangered species, we must determine which are most important for conserving biodiversity as a whole. Species do not exert equal influence on community and ecosystem processes. Identifying keystone species and finding ways to sustain their populations can be central to the survival of whole communities.

Management aimed at conserving a single species carries with it the possibility of negatively affecting populations of other species. For example, management of pine forests for the red-cockaded woodpecker might impact migratory birds associated with broadleaf temperate forests. To test for such impacts, ecologists compared bird communities near clusters of nest cavities in managed pine forests with communities in forests not managed for the woodpeckers. Contrary to expectations, the managed sites supported higher numbers and a higher diversity of other birds than the control forests. In this case, managing for one bird species enhanced the diversity of an entire community of birds. In many situations, conservation must look beyond single species and consider the whole community and ecosystem as an important unit of biodiversity.

Concept Check 55.2

1. Why does the reduced genetic diversity of small populations make them more vulnerable to extinction?
2. Contrast the small-population and declining-population approaches in terms of the remedies they recommend for preventing extinction of a species.
3. Why is a population's effective size (N_e) almost always smaller than its total size (N)?

For suggested answers, see Appendix A.

Concept 55.3

Landscape and regional conservation aim to sustain entire biotas

Historically, most preservation efforts have focused on saving endangered species, but increasingly, conservation biology aims to sustain the biodiversity of entire communities, ecosystems, and landscapes. Such a broad view requires understanding and applying the principles of community, ecosystem, and landscape ecology as well as human population dynamics and economics. One goal of **landscape ecology** (see Chapter 50), of which ecosystem management is part, is to understand past, present, and future patterns of landscape use and to make biodiversity conservation part of land-use planning.

Landscape Structure and Biodiversity

The biodiversity of a given landscape is in large part a function of the structure of the landscape. Understanding landscape dynamics is critically important in conservation because many species use more than one kind of ecosystem, and many live on the borders between ecosystems.

Fragmentation and Edges

The boundaries, or *edges*, between ecosystems (between a lake and the surrounding forest, for example, or between cropland and suburban housing tracts) and within ecosystems (roadsides and rock outcroppings, for instance) are defining features of landscapes **(Figure 55.14)**. An edge has its own set of physical conditions, which differ from those on either side of it. For instance, the soil surface of an edge between a forest patch and a burned area receives more sunlight and is usually

(a) Natural edges. Grasslands give way to forest ecosystems in Yellowstone National Park.

(b) Edges created by human activity. Pronounced edges (roads) surround clear-cuts in this photograph of a heavily logged rain forest in Malaysia.

▲ **Figure 55.14 Edges between ecosystems.**

hotter and drier than the forest interior, but it is cooler and wetter than the soil surface in the burned area.

Some organisms thrive in edge communities because they have access to the resources of both adjacent areas. For instance, the ruffed grouse (*Bonasa umbellatus*) is a bird that needs forest habitat for nesting, winter food, and shelter, but it also needs forest openings with dense shrubs and herbs for summer food. White-tailed deer also thrive in edge habitats, where they can browse on woody shrubs; deer populations often expand when forests are logged and more edges are exposed.

The proliferation of edge species can have positive or negative effects on a community's biodiversity. For example, a 1997 study in Cameroon comparing edge and interior populations of the little greenbul (a tropical rain forest bird) suggested that forest edges may be important sites of speciation. On the other hand, communities in which edges have resulted from human alterations often have reduced biodiversity because the relatively large percentage of edge habitat leads to a preponderance of edge-adapted species. For example, the brown-headed cowbird (*Molothrus ater*) is an edge-adapted species that lays its eggs in the nests of other birds, particularly migratory songbirds. Cowbirds need forests, where they can parasitize the nests of other birds, and also open fields, where they forage on insects. Thus, their populations are burgeoning where forests are being heavily cut and fragmented, creating more edge habitat and open land. Increasing cowbird parasitism and loss of habitat are correlated with declining populations of several of the cowbird's host species.

The influence of fragmentation on the structure of communities has been explored for two decades in the long-term Biological Dynamics of Forest Fragments Project. Located in the heart of the Amazon River basin, approximately 80 km north of the city of Manaus, the study area consists of a series of forest fragments **(Figure 55.15)**. The fragments are isolated patches of forest separated from surrounding continuous tropical rain forest by distances of 80 to 1,000 m. Researchers from all over the world have clearly documented the physical and biological effects of this fragmentation in taxa ranging from bryophytes and beetles to birds. Among their findings is the repeated discovery of two groups of species—those that live in forest edge habitats and those that live in the forest interior. Species adapted to forest interiors show the greatest declines in the smallest fragments, suggesting that landscapes dominated by small fragments will support fewer species, mainly due to a loss of interior-adapted species.

▲ **Figure 55.15 Amazon rain forest fragments, isolated sections of forest ranging in area from 1 to 100 ha.**

Corridors That Connect Habitat Fragments

Where habitats have been severely fragmented, the presence of a **movement corridor**, a narrow strip or series of small clumps of quality habitat connecting otherwise isolated patches, can be a deciding factor in conserving biodiversity. Streamside habitats often serve as corridors, and in some nations, government policy prohibits destruction of these riparian areas. In areas of heavy human use, artificial corridors are sometimes constructed. For example, bridges or tunnels can reduce the number of animals killed trying to cross highways **(Figure 55.16)**.

Movement corridors also can promote dispersal and reduce inbreeding in declining populations. Corridors have been shown to increase the exchange of individuals among subpopulations of a metapopulation in organisms including butterflies, voles, and various aquatic plants. Corridors are especially important to species that migrate between different habitats seasonally. However, a corridor can also be harmful—as, for example, in the spread of disease, especially among small populations in closely situated habitat patches. In a 2003 study, Agustin Estrada-Pena, of the University of Zaragoza, Spain, showed that habitat corridors facilitate the movement of disease-carrying ticks among forest patches in northern Spain. All the effects of

▲ **Figure 55.16 An artificial corridor.** This bridge in Banff National Park, Canada, helps animals cross a human-created barrier.

corridors are not yet understood, and their impact is an area of active research in conservation biology and restoration ecology.

Establishing Protected Areas

Conservation biologists are applying their understanding of community, ecosystem, and landscape dynamics in establishing protected areas to slow the loss of biodiversity. Currently, governments have set aside about 7% of the world's land in various forms of reserves. Choosing locations for protection and designing nature reserves pose many challenges. If a community is subject to fire, grazing, and predation, for example, should the reserve be managed to minimize the risks of these processes to endangered or threatened species? Or should the reserve be left as natural as possible, with such processes as fires ignited by lightning allowed to play out without any human intervention? This is just one of the debates that arise among people who share an interest in the health of national parks and other protected areas. Much of the focus has been on hot spots of biological diversity.

Finding Biodiversity Hot Spots

A **biodiversity hot spot** is a relatively small area with an exceptional concentration of endemic species and a large number of endangered and threatened species **(Figure 55.17)**. For example, nearly 30% of all bird species are confined to only about 2% of Earth's land area. And about 50,000 plant species, or about 17% of all known plant species, inhabit just 18 hot spots that comprise a total of only 0.5% of the global land surface. Together, the "hottest" of the biodiversity hot spots total less than 1.5% of Earth's land but are home to a third of all species of plants and vertebrates. Hot spots also include aquatic ecosystems, such as coral reefs and certain river systems.

Biodiversity hot spots are obviously good choices for nature reserves, but identifying them is not always straightforward. And even if all hot spots could be protected, that effort would fall woefully short of conserving the planet's biodiversity. One problem is that a hot spot for one taxonomic group, such as butterflies, may not be a hot spot for some other taxonomic group, such as birds. Designating an area as a biodiversity hot spot is often biased toward saving vertebrates and plants, with less attention paid to invertebrates and microorganisms. Some biologists are also concerned that the hot-spot strategy focuses too much of the conservation effort on such a small fraction of Earth's land.

Philosophy of Nature Reserves

Nature reserves are biodiversity islands in a sea of habitat degraded to varying degrees by human activity. It is important to realize, however, that protected "islands" are not isolated from their surroundings and that the nonequilibrium model we described in Chapter 53 applies to nature reserves as well as to the larger landscapes in which they are embedded.

An earlier policy—that protected areas should be set aside to remain unchanged forever—was based on the concept that ecosystems are balanced, self-regulating units. However, as we saw in Chapter 53, disturbance is a functional component of all ecosystems, and management policies that ignore natural disturbances or attempt to prevent them have generally proved to be self-defeating. For instance, setting aside an area of a fire-dependent community, such as a portion of a tallgrass prairie, chaparral, or dry pine forest, with the intention of saving it is unrealistic if periodic burning is excluded. Without the dominant disturbance, the fire-adapted species are usually outcompeted by other species, and biodiversity is reduced.

Because human disturbance and fragmentation are increasingly common landscape features, the dynamics of disturbance, population dynamics, edges, and corridors are important in the design and management of protected areas. A major conservation question is whether it is better to create one large reserve or a group of smaller reserves. One argument for extensive preserves is that large, far-ranging animals with low-density populations, such as the grizzly bear, require extensive habitats. In addition, more extensive areas have proportionately smaller perimeters than smaller areas and are therefore less affected by edges. As conservation

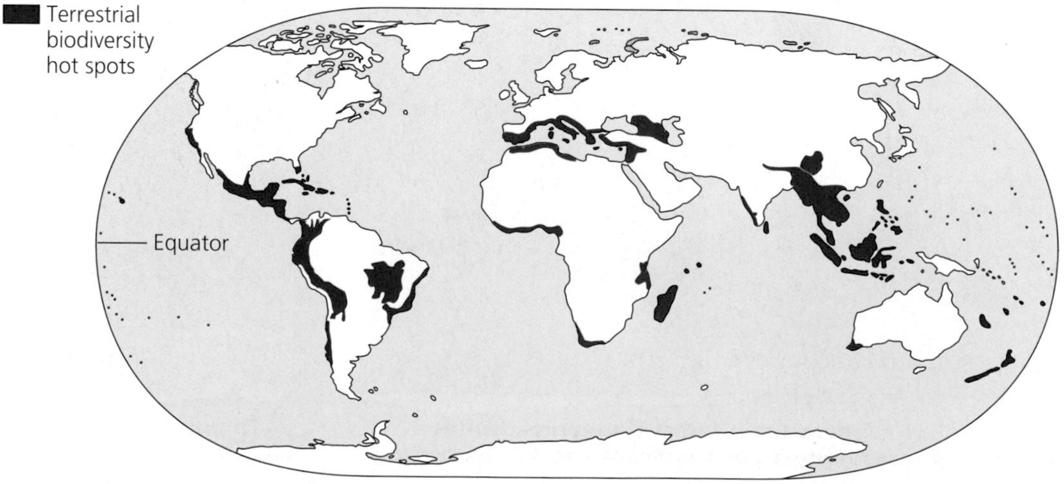

▶ **Figure 55.17 Earth's terrestrial biodiversity hot spots.**

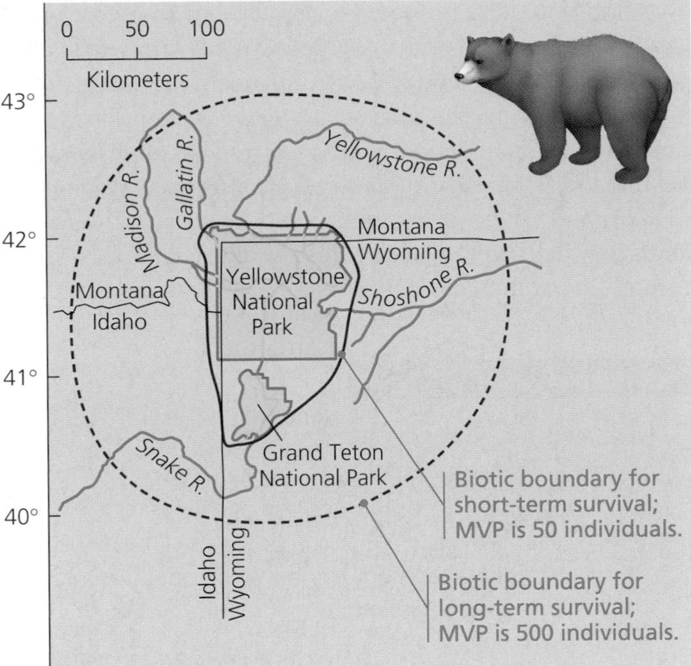

▲ **Figure 55.18 The legal (green border) and biotic (red borders) boundaries for grizzly bears in Yellowstone and Grand Teton National Parks.** The biotic boundaries are defined by the entire watershed for this region and the area necessary to support a minimum viable population (MVP) of grizzly bears.

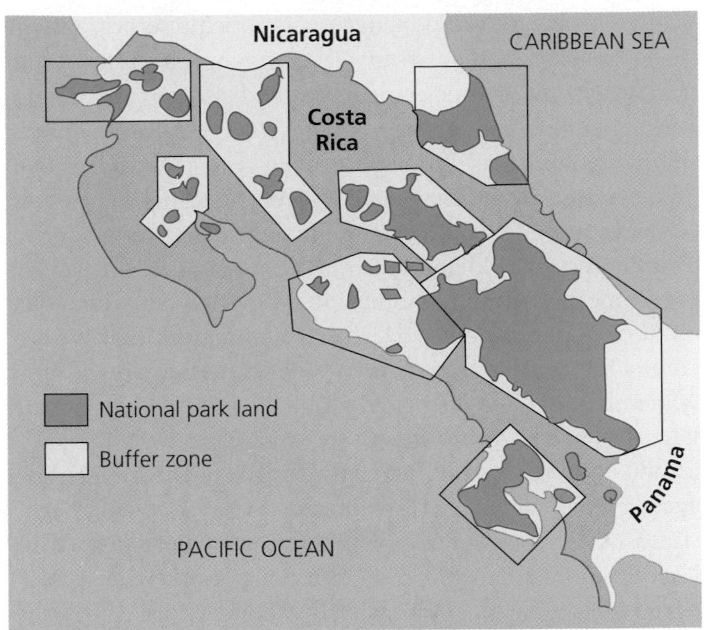

(a) Boundaries of the zoned reserves are indicated by black outlines.

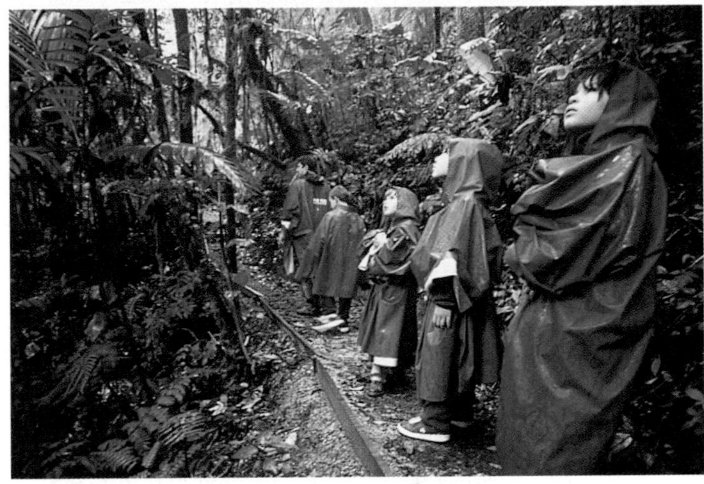

(b) Local schoolchildren marvel at the diversity of life in one of Costa Rica's reserves.

▲ **Figure 55.19 Zoned reserves in Costa Rica.**

biologists learn more about the requirements for achieving minimum viable population sizes for endangered species, it is becoming clear that most national parks and other reserves are far too small. For example, **Figure 55.18** compares the boundaries of Yellowstone and Grand Teton National Parks with the actual area required to prevent extinction of grizzly bears. The *biotic boundary,* the area needed to sustain the grizzly, is more than ten times as large as the *legal boundary,* the actual area of the parks. Given political and economic realities, it is unlikely that many existing parks will be enlarged, and most newly created reserves will also be too small. Areas of private and public land surrounding reserves will likely have to contribute to biodiversity conservation. On the other side of the argument, smaller, unconnected preserves may slow the spread of disease throughout a population.

In practical terms, land use by humans may outweigh all other considerations and ultimately dictate the size and shape of protected areas. Much of the land left for conservation efforts is useless for exploitation by agriculture or forestry. But in some cases, as when reserve land is surrounded by commercially viable property, the use of land for agriculture or forestry must be integrated into conservation strategies.

Zoned Reserves

Several nations have adopted a zoned reserve approach to landscape management. A **zoned reserve** is an extensive region of land that includes one or more areas undisturbed by humans surrounded by lands that have been changed by human activity and are used for economic gain. The key challenge of the zoned reserve approach is to develop a social and economic climate in the surrounding lands that is compatible with the long-term viability of the protected core area. These surrounding areas continue to be used to support the human population, but with regulations that prevent the types of extensive alterations likely to impact the protected area. As a result, the surrounding tracts of land serve as buffer zones against further intrusion into the undisturbed area.

The small Central American nation of Costa Rica has become a world leader in establishing zoned reserves **(Figure 55.19)**. In an international agreement that reduced its financial debt to other countries, the Costa Rican government established eight zoned reserves, called "conservation areas," which contain designated national park land. Costa Rica is making progress toward managing its zoned reserves, and the buffer zones provide a steady, lasting supply of forest products, water, and hydroelectric power and also support sustainable agriculture and tourism. An important goal is providing a stable economic base for people living there. As University of Pennsylvania ecologist Daniel Janzen, a leader in tropical conservation, has said, "The likelihood of long-term survival of a conserved wildland area is directly proportional to the economic health and stability of the society in which that wildland is embedded." Destructive practices that are not compatible with long-term ecosystem conservation and from which there is often little local profit, such as massive logging, large-scale single-crop agriculture, and extensive mining, are ideally confined to the outermost fringes of the buffer zones and are gradually being discouraged.

Costa Rica looks to its zoned reserve system to maintain at least 80% of its native species, but the system is not without negative aspects. A 2003 analysis of land cover change between 1960 and 1997 showed negligible deforestation within Costa Rica's national parks and a gain in forest cover in the 1-km buffer around the parks. However, significant losses in forest cover were discovered in the 10-km buffer zones around all national parks, which threatens to turn the parks into isolated habitat islands.

The continued high rate of human exploitation of ecosystems leads some analysts to predict that considerably less than 10% of the biosphere will ever be protected as nature reserves. Sustaining biodiversity often involves working in landscapes that are almost entirely human dominated. Even vast marine ecosystems have become heavily impacted by human exploitation. For example, commercially important fish populations around the world have collapsed in the face of mounting fishing pressure from increasingly sophisticated equipment that puts nearly all potential fishing grounds within reach. In response, Fiona Gell and Callum Roberts of the University of York, England, have proposed the establishment of marine reserves around the world that would be off limits to fishing **(Figure 55.20)**. Gell and Roberts present strong evidence that a patchwork of marine reserves can serve as a means of both increasing fish populations within the reserves and improving fishing success in nearby areas. Their proposed system is a modern application of a centuries-old practice in the Fiji Islands in which some areas have historically remained closed to fishing—a traditional example of the zoned reserve concept.

▲ **Figure 55.20 A monitoring team preparing to count and measure clams inside an intertidal marine reserve in Fiji.**

Concept Check 55.3

1. What is a biodiversity hot spot?
2. How do zoned reserves provide economic incentives for long-term protection of protected areas?
3. How can corridors connecting habitat fragments help protect endangered populations? How might such corridors harm populations?

For suggested answers, see Appendix A.

Concept 55.4

Restoration ecology attempts to restore degraded ecosystems to a more natural state

Eventually, some areas that are altered by human activity are abandoned. For instance, the soils of many tropical areas become unproductive and are abandoned fewer than five years after being cleared for farming. Mining activities may last for several decades, but the lands are then abandoned in a degraded state. Many ecosystems are also damaged inadvertently by the dumping of toxic chemicals or such mishaps as oil spills. Such degraded habitats and ecosystems are increasing in area because the natural rate of recovery by successional processes is slower than the rate of degradation by human activities.

Given enough time, biological communities can recover naturally from many types of disturbances through the various stages of ecological succession that we discussed in Chapter 53. The amount of time required for such natural recovery is more closely related to the spatial scale of the disturbance than the type of disturbance: The larger the area disturbed, the longer the time required for recovery. Whether the disturbance is natural or caused by humans seems to make little difference in this size-time relationship **(Figure 55.21)**.

One of the basic assumptions of restoration ecology is that most environmental damage is reversible, but this optimism must be balanced by a second assumption—that communities are not infinitely resilient. Restoration ecologists

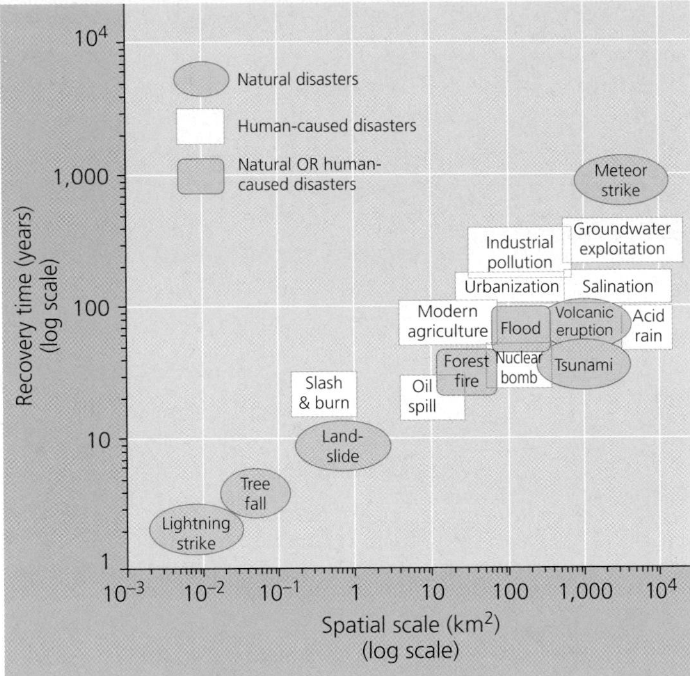

▲ **Figure 55.21 The size-time relationship for community recovery from natural and human-caused disasters.** Note that the scales are logarithmic.

work to identify and manipulate the processes that most limit the speed of recovery in order to reduce the time it takes for a community to bounce back from the impact of disturbances. It is important to remember, however, that natural disturbance such as periodic fire or flooding is part of the dynamics of many ecosystems and needs to be considered in restoration strategies.

Two key strategies in restoration ecology are bioremediation and augmentation of ecosystem processes.

Bioremediation

Bioremediation is the use of living organisms, usually prokaryotes, fungi, or plants, to detoxify polluted ecosystems (see Chapter 27). For example, some plants adapted to soils containing heavy metals are capable of accumulating high concentrations of potentially toxic metals such as zinc, nickel, lead, and cadmium. Restoration ecologists can use these plants to revegetate sites polluted by mining and other human activities and then harvest the plants to remove the metals from the ecosystem. A number of researchers are also focusing on the ability of certain prokaryotes and lichens to concentrate metals. Researchers in the United Kingdom discovered a lichen species that grows on soil polluted with uranium dust left over from mining. The lichen concentrates uranium in a dark pigment, making it useful as a biological monitor and potentially as a remediator. Restoration ecologists have also achieved some success in using the bacterium *Pseudomonas* to clean up oil spills on beaches. More common still is the use of certain prokaryotes to metabolize toxins in dump sites. In the future, genetic engineering may become increasingly important as a tool for improving the performance of certain species as bioremediators.

Biological Augmentation

In contrast to bioremediation, which is a strategy for *removing* harmful substances, **biological augmentation** uses organisms to *add* essential materials to a degraded ecosystem. Augmenting ecosystem processes requires determining what factors, such as chemical nutrients, have been removed from an area and thus are limiting its rate of recovery. Encouraging the growth of plants that thrive in nutrient-poor soils often speeds up the rate of successional changes that can lead to recovery of damaged sites. An example is the rapid regrowth of indigenous plant communities along roadsides in Puerto Rico, which was overseen by Ariel Lugo, director of the U.S. Forest Service's Institute of Tropical Forestry in Puerto Rico. Lugo used *Albizzia procera,* a nonnative plant that thrives on nitrogen-poor soils, to colonize roadside areas after the original forest was removed and soils were depleted of nutrients. Apparently, the rapid buildup of organic material from dense stands of *Albizzia* enabled indigenous plants to recolonize the area and overgrow the introduced plant in a relatively brief time.

Exploring Restoration

Because of the newness of restoration ecology as a discipline, the complexity of ecosystems, and the unique features of each situation, restoration ecologists usually must learn as they go. Many restoration ecologists advocate adaptive management: experimenting with several promising types of management to learn what works best. The key to adaptive management, and the key to restoration ecology, is to consider alternative ways of accomplishing goals and to learn from mistakes as well as successes. The long-term objective of restoration is to speed the reestablishment of an ecosystem as similar as possible to the predisturbance ecosystem. **Figure 55.22**, on the next two pages, identifies several ambitious and successful restoration projects around the world. The great number of such projects, the dedication of the people engaged in them, and the successes that have been achieved suggest that this is just the beginning.

Concept Check 55.4

1. What are the goals of restoration ecology?
2. How do bioremediation and biological augmentation differ?

For suggested answers, see Appendix A.

Figure 55.22
Exploring Restoration Ecology Worldwide

The examples highlighted on these pages are just a few of the many restoration ecology projects taking place around the world. The color-coded dots on the map indicate the locations of the projects.

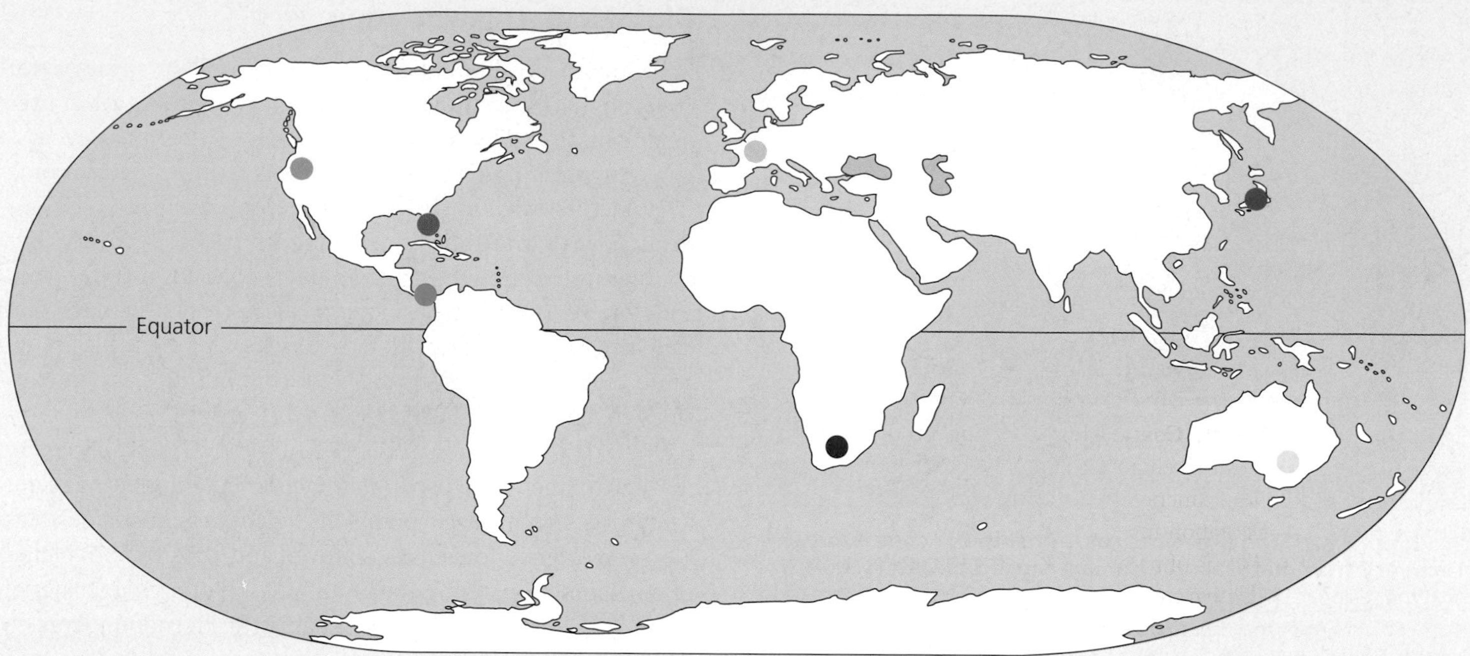

Truckee River, Nevada. Damming and water diversions during the 20th century reduced flow in the Truckee River, Nevada, leading to declines in riparian forests. Restoration ecologists worked with water managers to ensure that sufficient water would be released during the short season of seed release by the native cottonwood and willow trees for seedlings to become established. Nine years of controlled-flow release led to the result shown here: a dramatic recovery of cottonwood–willow riparian forest.

Australia. One of the most difficult aspects of restoration ecology is determining how human activity has altered ecosystems and what the goals of restoration should be. In Australia, environmental narratives from people who have lived on the land are an invaluable source of scarce baseline information to help create restoration goals.

Kissimmee River, Florida. The Kissimmee River in south-central Florida was converted from a meandering river into a 90-km canal, with significant negative impacts on fish and wetland bird populations. Kissimmee River restoration has filled 12 km of drainage canal and reestablished 24 km of the original 167 km of natural river channel. Pictured here is a section of the Kissimmee canal that has been plugged (wide, light strip on right side of photo), diverting flow into remnant river channels in the center of the photo. The project will also restore the natural flow regime, which will foster self-sustaining populations of wetland birds and fishes.

Tropical dry forest, Costa Rica. Clearing for agriculture, mainly for livestock grazing, eliminated approximately 98% of tropical dry forest in Central America and Mexico. In a reversal of roles, tropical dry forest restoration in Costa Rica has used domestic livestock to disperse the seeds of native trees into open grasslands. The photo shows one of the first trees (right center), dispersed as seed by livestock, to colonize former pastureland. This project is a model for joining restoration ecology with the local economy and educational institutions.

Rhine River, Europe. Centuries of dredging and channeling for navigation (see the barges in the wide, main channel on the right side of the photo) have straightened the once-meandering Rhine River and disconnected it from its floodplain and associated wetlands. The countries along the Rhine, particularly France, Germany, Luxembourg, the Netherlands, and Switzerland, are cooperating to reconnect the river to side channels, such as the one shown on the left side of the photo. Such side channels increase the diversity of habitats available to aquatic biota, improve water quality, and provide flood protection.

Succulent Karoo, South Africa. In this desert region of southern Africa, as in many arid regions, overgrazing by livestock has damaged vast areas. Reversing this trend, private landowners and government agencies in South Africa are restoring large areas of this unique region. The photo shows a small sample of the exceptional plant diversity of the Succulent Karoo; its 5,000 plant species include the highest diversity of succulent plants in the world.

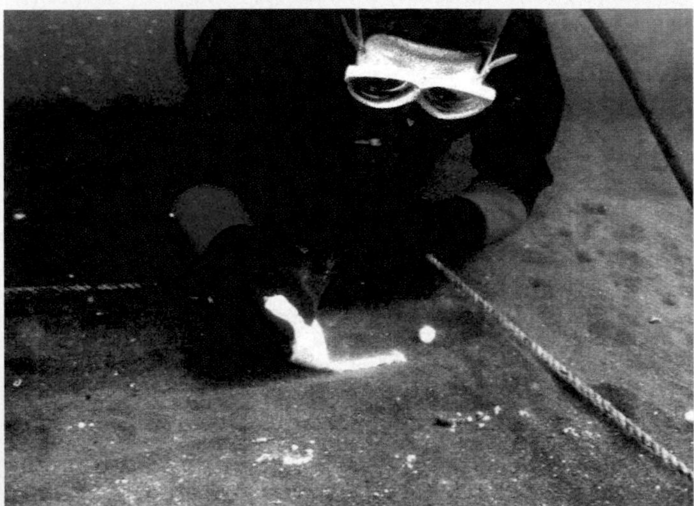

Coastal Japan. Seaweed and seagrass beds are important nursery grounds for a wide variety of fishes and shellfish. Once extensive but now reduced by development, these beds are being restored in the coastal areas of Japan. Techniques include construction of suitable seafloor habitat, transplantion from natural beds using artificial substrates, and hand seeding (shown in this photograph).

Concept 55.5

Sustainable development seeks to improve the human condition while conserving biodiversity

Facing increasing loss and fragmentation of habitats, how can we best manage Earth's resources? If we are to conserve most of a nation's species, which habitat patches are most crucial? Among the choices, which areas are most practical to protect and manage if we are to save endangered species or the greatest number of species?

Sustainable Biosphere Initiative

We must understand the complex interconnections of the biosphere in order to make sensible decisions about how to conserve these networks. To this end, many nations, scientific societies, and private foundations have embraced the concept of **sustainable development,** the long-term prosperity of human societies and the ecosystems that support them. The forward-looking Ecological Society of America, the world's largest organization of professional ecologists, endorses a research agenda called the Sustainable Biosphere Initiative. The goal of this initiative is to define and acquire the basic ecological information necessary for the intelligent and responsible development, management, and conservation of Earth's resources. The research agenda includes studies of global change, including interactions between climate and ecological processes; biological diversity and its role in maintaining ecological processes; and the ways in which the productivity of natural and artificial ecosystems can be sustained. This initiative requires a strong commitment of human and economic resources.

Sustainable development is not *just* about science. To sustain ecosystem processes and stem the loss of biodiversity, we must connect life science with the social sciences, economics, and humanities. And equally important, we must reassess our values. Those of us living in affluent developed nations are responsible for the greatest amount of environmental degradation. Reality demands that we learn to revere the natural processes that sustain us, and reduce our orientation toward short-term personal gain. The following case study illustrates how the combination of scientific and personal efforts can make a significant difference in creating a truly sustainable world.

Case Study: Sustainable Development in Costa Rica

The success of conservation in Costa Rica that we discussed in Concept 55.3 has involved leadership by the national government as well as an essential partnership between the national government, nongovernment organizations (NGOs), and private citizens. For instance, many nature reserves established by individuals have been recognized by the government as national wildlife reserves and given significant tax benefits. However, conservation and restoration of biodiversity is only one side of sustainable development; the other key facet is improving the human condition.

How have the living conditions of the Costa Rican people fared as the country pursued its conservation goals? We can gain insights into conditions through population statistics. As we discussed in Chapter 52, two of the most fundamental indicators of living conditions are infant mortality rate and life expectancy. **Figure 55.23** shows that the infant mortality rate in Costa Rica declined sharply during the 20th century, and life expectancy at birth increased. In 1930, out of each 1,000 live births, over 170 infants died. By 2003, infant mortality had fallen to 10 per 1,000 live births. Meanwhile, life expectancy climbed from just over 35 years in 1900 to over 78 years in 2003, which is approximately one year longer than in the United States, where life expectancy in 2003 was just over 77 years. Another indicator of living conditions is literacy rate. The 2003 literacy rate in Costa Rica was 96%, compared to 97% in the United States.

Such statistics show that living conditions in Costa Rica have improved greatly over the period in which the country dedicated itself to conservation and restoration. While a correlation does not confirm a cause-effect relationship, we can certainly infer that Costa Rica's conservation initiatives have not compromised human welfare. And we can say that development in Costa Rica has attended to both nature and people. Nevertheless, many problems remain. One of the challenges that the country will have to face is maintaining its commitment to conservation in the face of a growing population.

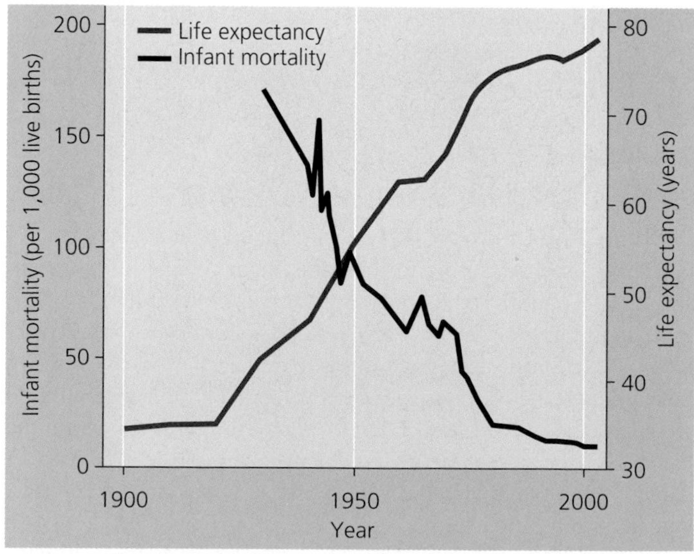

▲ **Figure 55.23 Infant mortality and life expectancy at birth in Costa Rica.**

Costa Rica is in the middle of a rapid demographic transition (see Chapter 52), and even though birth rates are dropping rapidly, its population continues to grow at about 1.5% annually (compared to 0.9% growth in the United States). Costa Rica's population, which is currently about 4 million, is predicted to continue to grow until the middle of this century, when it will level off at approximately 6 million. Given their recent history, it seems probable that the Costa Rican people will confront the remaining challenges of sustainable development with the same energy and optimism that they have shown in the past.

Biophilia and the Future of the Biosphere

Despite the uncertainties about the future of the biosphere, now is not the time for gloom and doom but the time to reconnect with the rest of nature. Not many people today live in truly wild environments or even visit such places often. Our modern lives are very different from those of early humans, who hunted and gathered and painted wildlife murals on cave walls **(Figure 55.24a)**. But our behavior reflects remnants of our ancestral attachment to nature and the diversity of life—the concept of *biophilia* that we introduced early in this chapter. We evolved in natural environments rich in biodiversity, and we still have an affinity for such settings **(Figure 55.24b)**. E. O. Wilson makes the case that our biophilia is innate, an evolutionary product of natural selection acting on a brainy species whose survival depended on a close connection to the environment and a practical appreciation of plants and animals.

It will come as no surprise that most biologists have embraced the concept of biophilia. After all, these are people who have turned their passion for nature into careers. But biophilia strikes a harmonic chord with biologists for another reason. If biophilia is evolutionarily embedded in our genome, then there is hope that we can become better custodians of the biosphere. If we all pay more attention to our biophilia, a new environmental ethic could catch on among individuals and societies. That ethic is a resolve never to allow a species to become extinct or any ecosystem to be destroyed through human activities as long as there are reasonable ways to prevent such ecological violence. It is an environmental ethic that balances out another human trait—our tendency to "subdue" Earth. Yes, we should be motivated to preserve biodiversity because we depend on it for food, medicine, building materials, fertile soil, flood control, habitable climate, drinkable water, and breathable air. But maybe we can also work harder to prevent the extinction of other forms of life just because it is the ethical thing for us to do as the most thoughtful species in the biosphere. Again, Wilson sounds the call: "Right now, we're pushing the species of the world through a bottleneck. We've got to make it a major moral principle to get as many of them through this as possible. It's the challenge now and for the next century. And there's one good thing about our species: We like a challenge!"

It is appropriate that we end this textbook with a discussion of biophilia, for biology is a scientific expression of our desire to know nature. We are most likely to protect what we appreciate, and we are most likely to appreciate what we understand. By learning about the processes and diversity of life, we also become more aware of ourselves and our place in the biosphere. We hope this book serves you well in this lifelong adventure.

(a) Detail of animals in a 36,000-year-old cave painting, Lascaux, France

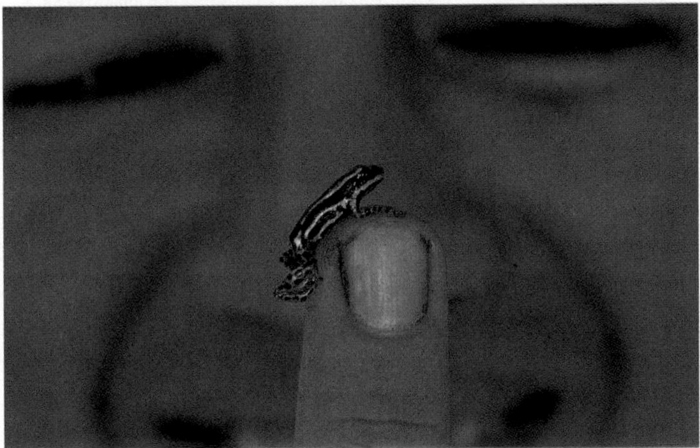

(b) Biologist Carlos Rivera Gonzales examining a tiny tree frog in Peru

▲ **Figure 55.24 Biophilia, past and present.**

Concept Check 55.5

1. What is meant by the term *sustainable development*?
2. How might biophilia influence environmental ethics?

For suggested answers, see Appendix A.

Chapter 55 Review

Go to www.campbellbiology.com or the student CD-ROM to explore Activities, Investigations, and other interactive study aids.

SUMMARY OF KEY CONCEPTS

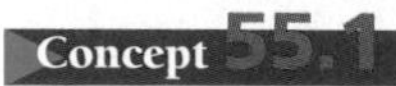

Human activities threaten Earth's biodiversity

- **The Three Levels of Biodiversity (pp. 1210–1211)** Biodiversity consists of the varied ecosystems in the biosphere, the species richness within those ecosystems, and the genetic variation within and among populations of each species.
- **Biodiversity and Human Welfare (pp. 1211–1212)** Our biophilia enables us to recognize the value of biodiversity for its own sake. Other species also provide humans with food, fiber, medicines, and ecosystem services.
- **Four Major Threats to Biodiversity (pp. 1212–1215)** The four major threats to biodiversity are habitat destruction, introduced species, overexploitation, and disruption of interaction networks.
  Activity *Madagascar and the Biodiversity Crisis*
  Activity *Introduced Species: Fire Ants*
  Graph It *Forestation Changes*
  Graph It *Global Fisheries and Overfishing*
  Graph It *Municipal Solid Waste Trends in the U.S.*

Population conservation focuses on population size, genetic diversity, and critical habitat

- **Small-Population Approach (pp. 1215–1218)** When a population drops below a minimum viable population (MVP) size, its loss of genetic variation due to nonrandom mating and genetic drift can trap it in an extinction vortex.
- **Declining-Population Approach (pp. 1218–1219)** The declining-population approach focuses on the environmental factors that cause decline, regardless of absolute population size. It follows a step-by-step proactive conservation strategy.
- **Weighing Conflicting Demands (pp. 1219–1220)** Conserving species often requires resolving conflicts between the habitat needs of endangered species and human demands.

Landscape and regional conservation aim to sustain entire biotas

- **Landscape Structure and Biodiversity (pp. 1220–1222)** The structure of a landscape can strongly influence biodiversity. As habitat fragmentation increases and edges become more extensive, biodiversity tends to decrease. Movement corridors can promote dispersal and help sustain populations.
- **Establishing Protected Areas (pp. 1222–1224)** Biodiversity hot spots are also hot spots of extinction and thus prime candidates for protection. Sustaining biodiversity in parks and reserves requires management to ensure that human activities in the surrounding landscape do not harm the protected habitats. The zoned reserve model recognizes that conservation efforts often involve working in landscapes that are largely human dominated.

Concept 55.4

Restoration ecology attempts to restore degraded ecosystems to a more natural state

- **Bioremediation (p. 1225)** Restoration ecologists are harnessing living organisms to detoxify polluted ecosystems.
- **Biological Augmentation (p. 1225)** Ecologists also use organisms to add essential materials to ecosystems.
- **Exploring Restoration (pp. 1225–1227)** The newness and complexity of restoration ecology require scientists to consider alternative solutions and adjust approaches based on experience.
  Investigation *How Are Potential Prairie Restoration Sites Analyzed?*

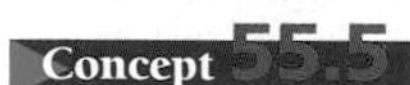

Sustainable development seeks to improve the human condition while conserving biodiversity

- **Sustainable Biosphere Initiative (p. 1228)** The goal of the Sustainable Biosphere Initiative is to acquire the ecological information needed for the development, management, and conservation of Earth's resources.
- **Case Study: Sustainable Development in Costa Rica (pp. 1228–1229)** Costa Rica's success in conserving tropical biodiversity has involved partnerships between the government, other organizations, and private citizens. Human living conditions in Costa Rica have improved along with ecological conservation.
  Activity *Conservation Biology Review*
  Graph It *Global Freshwater Resources*
  Graph It *Prospects for Renewable Energy*
- **Biophilia and the Future of the Biosphere (p. 1229)** Our innate sense of connection to nature may eventually motivate a realignment of our environmental priorities.

TESTING YOUR KNOWLEDGE

Self-Quiz

1. Ecologists conclude there is a biodiversity crisis because
   a. biophilia causes humans to feel ethically responsible for protecting other species.
   b. scientists have at last discovered and counted most of Earth's species and can now accurately calculate the current extinction rate.
   c. the current extinction rate is far higher than the rate at any time in the past 100,000 years.
   d. many potential life-saving medicines are being lost as species become extinct.
   e. there are too few biodiversity hot spots.

2. Which of the following directly addresses the growing concern over the loss of ecosystem diversity?
   a. the small-population approach
   b. restoration ecology
   c. the declining-population approach
   d. increasing effective population sizes of endangered species
   e. managing populations to increase genetic diversity

3. What is the effective population size (N_e) of a population of 50 strictly monogamous swans that includes 40 males and 10 females?
a. 50 b. 40 c. 30 d. 20 e. 10

4. One characteristic that distinguishes a population in an extinction vortex from most other populations is that
a. its habitat is fragmented.
b. it is a rare, top-level predator.
c. its effective population size is much lower than its total population size
d. its genetic diversity is very low.
e. it is not well adapted to edge conditions.

5. The discipline that applies ecological principles to returning degraded ecosystems to more natural states is known as
a. population viability analysis.
b. landscape ecology.
c. conservation ecology.
d. restoration ecology.
e. resource conservation.

6. What is the single greatest threat to biodiversity?
a. overexploitation of commercially important species
b. introduced species that compete with or prey on native species
c. pollution of Earth's air, water, and soil
d. disruption of trophic relationships as more and more prey species become extinct
e. habitat alteration, fragmentation, and destruction

7. Which of the following is *not* a step in the declining-population approach to conservation biology?
a. Gather data to determine whether a population is in decline.
b. Implement a conservation plan at the outset of a study, as it is too risky to wait until data are gathered and analyzed.
c. Develop multiple alternative hypotheses for the cause of population decline.
d. Include human activities and natural events as possible causes of a population decline.
e. Test the hypotheses for the cause of the decline, beginning with the hypothesis most likely to be correct.

8. Which of the following strategies would most rapidly increase the genetic diversity of a population in an extinction vortex?
a. Capture all remaining individuals in the population for captive breeding followed by reintroduction to the wild.
b. Establish a reserve that protects the population's habitat.
c. Introduce new individuals transported from other populations of the same species.
d. Sterilize the least fit individuals in the population.
e. Control populations of the endangered population's predators and competitors.

9. Of the following statements about protected areas that have been established to preserve biodiversity, which one is *not* correct?
a. About 25% of Earth's land area is now protected.
b. National parks are one of many types of protected area.
c. Most protected areas are too small to protect species.
d. Management of a protected area should be coordinated with management of the land surrounding the area.
e. It is especially important to protect biodiversity hot spots.

10. What is the Sustainable Biosphere Initiative?
a. a plan to convert all natural ecosystems in the biosphere to carefully engineered ones
b. a research agenda to study biodiversity and support sustainable development
c. a conservation practice that sets up zoned reserves surrounded by buffer zones
d. the declining-population approach to conservation that seeks to identify and remedy causes of species' declines
e. a program that uses adaptive management to experiment and learn while working with disturbed ecosystems

For Self-Quiz answers, see Appendix A.

Go to the website or CD-ROM for more quiz questions.

Evolution Connection

One factor favoring rapid population growth by an introduced species is the absence of the predators, parasites, and pathogens that controlled its population in the region where it evolved. Over the long term, how should evolution by natural selection influence the rate at which the native predators, parasites, and pathogens in a region of introduction attack an introduced species?

Scientific Inquiry

Suppose that you are in charge of planning a forest reserve, and one of your main goals is to help sustain local populations of woodland birds suffering from parasitism by the brown-headed cowbird. Reading research reports, you note that female cowbirds are usually reluctant to penetrate more than about 100 m into a forest and that nest parasitism is reduced when some woodland birds nest only in denser, more central forest regions. The forested area you have to work with is about 1,000 m by 6,000 m. A recent logging operation removed about half of the trees on one of the 6,000-m sides; the other three sides are adjacent to deforested pastureland. Your plan must include space for a small maintenance building, which you estimate to take up about 100 m^2. It will also be necessary to build a road, 10 m by 1,000 m, across the reserve. Where would you construct the road and the building, and why?

Investigation *How Are Potential Prairie Restoration Sites Analyzed?*

Science, Technology, and Society

Some organizations, such as the Ecological Society of America, are starting to envision a sustainable society—one in which each generation inherits sufficient natural and economic resources and a relatively stable environment. The Worldwatch Institute estimates that to reach sustainability by 2030, we must begin shaping a sustainable society during the next ten years or so. In what ways is our current system not sustainable? What might we do to work toward sustainability, and what are the major roadblocks to achieving it? How would your life be different in a sustainable society?

APPENDIX A Answers

CHAPTER 1

Concept Check 1.1

1. Examples: A molecule consists of *atoms* bonded together. Each organelle has an orderly arrangement of *molecules*. Photosynthetic plant cells contain the *organelles* called chloroplasts. An animal tissue consists of a group of similar *cells*. Organs such as the heart are constructed from several *tissues*. A complex organism, such as a plant, has several types of *organs*, including leaves and roots, in the case of the plant. A population is a set of *organisms* of the same species. A community consists of *populations* of the various species inhabiting a specific area. An ecosystem consists of a biological *community* along with the nonliving factors important to life, such as air, soil, and water. The biosphere is made up of all of Earth's *ecosystems*.
2. DNA is the chemical substance of genes. Genes are the hereditary units arranged along DNA molecules. The DNA molecules are built into the cellular structures called chromosomes.
3. Both plants and animals consist of eukaryotic cells, while the cells of bacteria are prokaryotic.

Concept Check 1.2

1. The meaning of a sentence is a property that emerges from the specific sequence of letters and spaces.
2. High-throughput data collection is the source of the enormous and expanding databases of biological information that make bioinformatics a necessary and productive field.
3. Negative feedback

Concept Check 1.3

1. An address pinpoints a location by tracking from broader to narrower categories—a state, city, zip code, street, and building number. This is analogous to the groups-subordinate-to-groups structure of biological taxonomy.
2. Organisms of domain Eukarya are made of eukaryotic cells, in contrast to the prokaryotic cells of domains Bacteria and Archaea.

Concept Check 1.4

1. Natural selection does not "create" the variation that makes adaptation possible, but "edits" by selecting in favor of certain heritable traits in a naturally varying population.
2.

Concept Check 1.5

1. Inductive reasoning derives generalizations from specific cases; deductive reasoning predicts specific outcomes from general premises.
2. It's usually impossible to exclude all unwanted variables; instead, a controlled experiment cancels out those variables by comparing an experimental group with a control group that differs only in the variable of interest.
3. To test the prediction of the mimicry hypothesis that king snakes will benefit from their coral snake coloration only in environments where poisonous coral snakes also live
4. Compared to a hypothesis, a scientific theory is usually more general and more substantiated by an accumulation of evidence.

Concept Check 1.6

1. Example: Scientific inquiry and the technology it informs have an enormous impact on society.

Self-Quiz

1. b	2. d	3. a	4. c	5. c
6. c	7. c	8. d	9. b	10. c

CHAPTER 2

Concept Check 2.1

1. Table salt consists of two elements, whereas oxygen consists of only one.
2. Carbon, oxygen, hydrogen, and nitrogen

Concept Check 2.2

1. 7
2. Atomic number = 7; mass number = 15; ${}^{15}_{7}N$
3. Atomic number = 12; 12 protons, 12 electrons; three electron shells; 2 electrons in the valence shell
4. The electrons in the shell farthest from the nucleus have the most potential energy, and the electrons in the shell closest to the nucleus have the least.
5. 9 electrons; two electron shells; 1*s*, 2*s*, 2*p* (three of them); 1 unpaired electron

Concept Check 2.3

1. Each carbon atom has only three covalent bonds instead of the required four.
2. The attractions between oppositely charged ions form ionic bonds.

Concept Check 2.4

1.

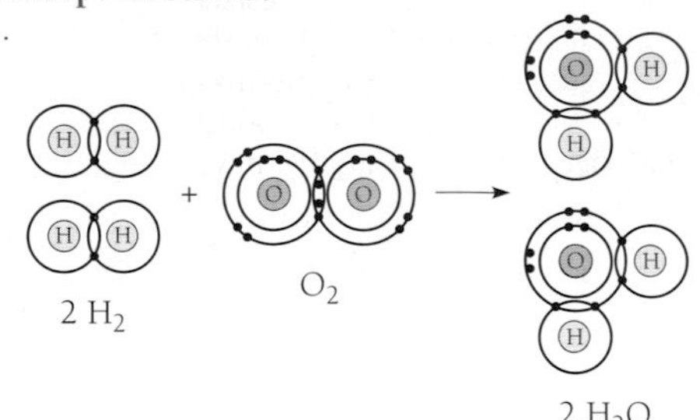

2. At equilibrium, the forward and reverse reactions occur at the same rate.

Self-Quiz

1. b	2. a	3. b	4. b	5. c
6. b	7. b	8. a	9. b	10. b

CHAPTER 3

Concept Check 3.1

1. Electronegativity is the attraction of an atom for the electrons of a covalent bond. Since oxygen is more electronegative than hydrogen, the oxygen atom in H_2O pulls electrons toward itself, resulting in a partial negative charge on the oxygen atom and partial positive charges on the hydrogen atoms. Oppositely charged ends of water molecules are attracted to each other, forming a hydrogen bond.
2. The hydrogen atoms of one molecule, with their partial positive charges, would repel the hydrogen atoms of the adjacent molecule.

Concept Check 3.2

1. Hydrogen bonds hold neighboring water molecules together; this cohesion helps the molecules resist the downward pull of gravity. Adhesion between water molecules and the walls of water-conducting cells also counters the downward pull of gravity. As water evaporates from the leaves, the chain of water molecules in water-conducting cells moves upward.
2. High humidity hampers cooling by suppressing the evaporation of sweat.
3. Water expands as it freezes, because the water molecules move farther apart in forming ice crystals. When there is water in a crevice of a boulder, expansion of the water due to freezing may crack the rock.
4. The sum of the atomic masses of Na and Cl equals 58.5 daltons, so a mole of NaCl would have a mass of 58.5 g. You would measure out 0.5 mole, or 29.3 g of NaCl, and gradually add water, stirring until it is dissolved. You would add water to bring the final volume to 1 L.

Concept Check 3.3

1. 10^5, or 100,000
2. $[H^+] = 0.01\ M = 10^{-2}\ M$, so pH = 2

Self-Quiz

1. d	2. d	3. b	4. c	5. b
6. c	7. c	8. d	9. c	10. c

CHAPTER 4

Concept Check 4.1

1. Urea is a molecule synthesized by organisms and found in urine. Its synthesis from gases thought to have been present in the primitive atmosphere on Earth demonstrated that life's molecules may initially have been synthesized abiotically.

Concept Check 4.2

1.
```
H\     /H
   C=C
H/     \H
```
2. The butanes in (b) are structural isomers, as are the butenes in (c).
3. Both consist largely of hydrocarbon chains.

Concept Check 4.3

1. It has both an amino group ($—NH_2$) and a carboxyl group (—COOH), which makes it a carboxylic acid.
2. The ATP molecule loses a phosphate, becoming ADP.

Self-Quiz

1. b	2. c	3. d	4. d	5. a
6. b	7. b	8. a	9. d	

10. b, because there are not only the two electronegative oxygens of the carboxyl group, but also an oxygen on the next (carbonyl) carbon. All of these oxygens help make the bond between the O and H of the —OH group more polar, thus making the dissociation of H^+ more likely.

CHAPTER 5

Concept Check 5.1

1. Proteins, carbohydrates, lipids, and nucleic acids
2. Nine, with one water required to hydrolyze each connected pair of monomers
3. The amino acids in the apple protein are released in hydrolysis reactions and incorporated into your proteins in dehydration reactions.

Concept Check 5.2

1. $C_3H_6O_3$ or $C_3(H_2O)_3$
2. $C_{12}H_{22}O_{11}$
3. Both molecules are polymers of glucose made by plants, but the glucose monomers are arranged differently. Starch functions mainly for sugar storage. Cellulose is a structural polysaccharide that is the main material of plant cell walls.

Concept Check 5.3

1. Both have a glycerol molecule attached to fatty acids. The glycerol of a fat has three fatty acids attached, whereas the glycerol of a phospholipid is attached to two fatty acids and one phosphate group.
2. The fatty acids on a saturated fat have no double bonds in their hydrocarbon chains, whereas at least one fatty acid on an unsaturated fat has a double bond. Saturated fats tend to be solid at room temperature, while unsaturated fats are liquid.
3. Human sex hormones are steroids, a type of hydrophobic compound.

Concept Check 5.4

1. The function of each protein is a consequence of its specific shape, which is lost when a protein becomes denatured.
2. Secondary structure involves hydrogen bonds between atoms of the polypeptide backbone. Tertiary structure involves bonding between atoms of the R groups of the amino acid subunits.
3. Primary structure, the amino acid sequence, affects the secondary structure, which affects the tertiary structure, which affects the quaternary structure (if any). In short, the amino acid sequence affects the shape of the protein, and the function of a protein depends on its shape.

Concept Check 5.5

1.

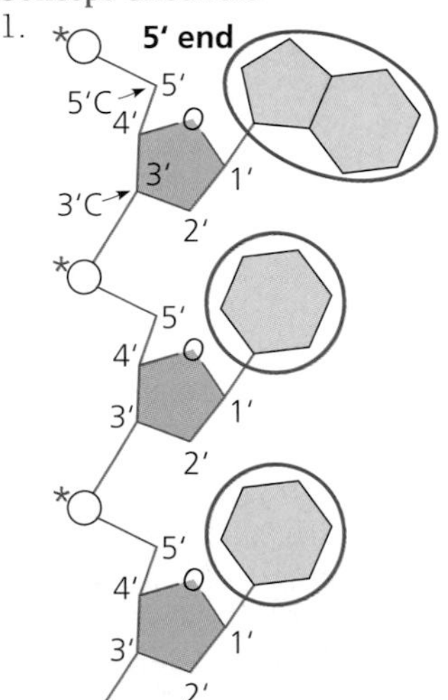

2. 3′-ATCCGGA-5′

Self-Quiz Answers

1. d	2. c	3. c, d, and e	4. a	5. b
6. a	7. d	8. b	9. b	10. b

CHAPTER 6

Concept Check 6.1

1. (a) Light microscope, (b) scanning electron microscope, (c) transmission electron microscope

Concept Check 6.2

1. See Figure 6.9.

Concept Check 6.3

1. Ribosomes in the cytoplasm translate the genetic message, carried from the DNA in the nucleus by mRNA, into a polypeptide chain.
2. Chromatin is composed of DNA and associated proteins; it carries the cell's genetic information. Nucleoli consist of RNA and proteins and are associated with particular regions of the DNA. In the nucleoli, rRNA is synthesized and ribosomal subunits are assembled.

Concept Check 6.4

1. The primary distinction between rough and smooth ER is the presence of bound ribosomes on the rough ER. While both types of ER synthesize phospholipids, membrane proteins are all produced on the ribosomes of the rough ER. The smooth ER also functions in detoxification, carbohydrate metabolism, and storage of calcium ions.
2. The mRNA is synthesized in the nucleus, and then passes out through a nuclear pore to be translated on a bound ribosome, attached to the rough ER. The protein is synthesized into the lumen of the ER and perhaps modified there. A transport vesicle carries the protein to the Golgi apparatus. After further modification in the Golgi, another transport vesicle carries it back to the ER, where it will perform its cellular function.
3. Transport vesicles move membranes and substances they enclose between other components of the endomembrane system.

Concept Check 6.5

1. Both organelles are involved in energy transformation, mitochondria in cellular respiration and chloroplasts in photosynthesis. They are both enclosed by two or more separate membranes.
2. Mitochondria and chloroplasts contain DNA, which encodes some of their proteins. They are not connected physically or via transport vesicles to organelles of the endomembrane system.

Concept Check 6.6

1. Microtubules are hollow tubes that resist bending; microfilaments are more like strong cables that resist stretching. These opposing properties work together to define and maintain cell shape. Intermediate filaments also resist stretching, and they are more permanent tension-bearing elements that reinforce the shape of the cell.
2. Dynein arms, powered by ATP, move neighboring doublets of microtubules relative to one another. Because they are anchored within the organelle and with respect to each other, the doublets bend instead of sliding past one another.

Concept Check 6.7

1. The most obvious difference is the presence of direct cytoplasmic connections between cells of plants (plasmodesmata) and between cells of animals (gap junctions). These connections result in the cytoplasm being continuous between adjacent cells.

2. The cell wall or ECM must be permeable to the materials that enter and leave the cell, and to molecules that provide information about the cell's environment.

Self-Quiz

1. c	2. c	3. b	4. d	5. d
6. b	7. c	8. b	9. e	10. a

CHAPTER 7

Concept Check 7.1

1. Plants adapted to cold environments would be expected to have more unsaturated fatty acids in their membranes, since those remain fluid at lower temperatures. Plants adapted to hot environments would be expected to have more saturated fatty acids, which would allow the fatty acids to "stack" more closely, making the membranes less fluid and therefore helping them to stay intact at higher temperatures.
2. They are on the inner side of the transport vesicle membrane.

Concept Check 7.2

1. O_2 and CO_2 are both small, uncharged molecules that can easily pass through the hydrophobic core of a membrane.
2. Water is a very polar molecule, so it cannot pass very rapidly through the hydrophobic region in the middle of a phospholipid bilayer.

Concept Check 7.3

1. The activity of the *Paramecium*'s contractile vacuole would decrease. The vacuole pumps out excess water that flows into the cell; this flow occurs only in a hypotonic environment.

Concept Check 7.4

1. The pump would use ATP. To establish a voltage, ions would have to be pumped against their gradient, which requires energy.
2. Each ion is being transported against its electrochemical gradient. If either ion were flowing down its electrochemical gradient, this *would* be considered cotransport.

Concept Check 7.5

1. Exocytosis. When a transport vesicle fuses with the plasma membrane, the vesicle membrane becomes part of the plasma membrane.
2. Receptor-mediated endocytosis, because in this case one specific molecule needs to be taken up at a particular time; pinocytosis takes up substances in a nonspecific manner.

Self-Quiz Answers

1. b 2. c 3. a 4. d 5. b
6. Fructose
7. Glucose
8. Hypotonic
9. Into the cell
10. b, c, and e. Regarding e: Even though sucrose can't cross the membrane, water flow (osmosis) will equalize the solute concentrations on each side of the membrane.

CHAPTER 8

Concept Check 8.1

1. The second law is the trend toward randomness. Equal concentrations of a substance on both sides of a membrane is a more random distribution than unequal concentrations. Diffusion of a substance to a region where it is initially less concentrated increases entropy, as mandated by the second law.
2. Energy is the capacity to cause change, and some forms of energy can do work.
3. The apple has potential energy in its position hanging on the tree, and the sugars and other nutrients it contains have chemical energy. The apple has kinetic energy as it falls from the tree to the ground. Finally, thermal energy is lost as the apple is digested and its molecules broken down.

Concept Check 8.2

1. Cellular respiration is a spontaneous and exergonic process. The energy released from glucose is used to do work in the cell, or is lost as heat.
2. H^+ ions can perform work only if their concentrations on each side of a membrane differ. When the H^+ concentrations are the same, the system is at equilibrium and can do no work.

Concept Check 8.3

1. ATP transfers energy to endergonic processes by phosphorylating other molecules.
2. A set of coupled reactions can transform the first group into the second group. Since, overall, this is an exergonic process, ΔG is negative and the first group must have more free energy.

Concept Check 8.4

1. A spontaneous reaction is a reaction that is exergonic. However, if it has a high activation energy that is rarely attained, the rate of the reaction may be low.
2. Only the specific substrate(s) will fit into the active site of an enzyme, the part of the enzyme that carries out catalysis.
3. As a competitive inhibitor, malonate binds to the active site of succinate dehydrogenase and so prevents the normal substrate, succinate, from binding.

Concept Check 8.5

1. The activator binds in such a way that it stabilizes the active conformation of an enzyme, whereas the inhibitor stabilizes the inactive conformation.

Self-Quiz

1. b	2. c	3. b	4. b	5. e
6. a	7. e	8. c	9. a	10. c

CHAPTER 9

Concept Check 9.1

1. $C_4H_6O_5$ is oxidized and NAD^+ is reduced.

Concept Check 9.2

1. NAD^+ acts as the oxidizing agent in step 6, accepting electrons from glyceraldehyde-3-phosphate, which thus acts as the reducing agent.

Concept Check 9.3

1. NADH and $FADH_2$; they will donate electrons to the electron transport chain.
2. CO_2 is removed from pyruvate, which is produced by glycolysis, and CO_2 is produced by the citric acid cycle.

Concept Check 9.4

1. Oxidative phosphorylation would stop entirely, resulting in no ATP production. Without oxygen to "pull" electrons down the electron transport chain, H^+ would not be pumped into the mitochondrion's intermembrane space and chemiosmosis would not occur.
2. Because addition of H^+ (decreasing the pH) would establish a proton gradient even without the function of the electron transport chain, we would expect ATP synthase to function and synthesize ATP. (In fact, it was experiments like this one that provided support for chemiosmosis as an energy-coupling mechanism.)

Concept Check 9.5

1. A derivative of pyruvate—either acetaldehyde during alcohol fermentation or pyruvate itself during lactic acid fermentation; oxygen
2. The cell would need to consume glucose at a rate about 19 times the consumption rate in the aerobic environment (2 ATP are generated by fermentation per molecule of glucose versus up to 38 ATP by cellular respiration).

Concept Check 9.6

1. The fat is much more reduced; it has many $—CH_2—$ units. The electrons present in a carbohydrate molecule are already somewhat oxidized, as some of them are bound to oxygen.
2. When we consume more food than necessary for metabolic processes, our bodies synthesize fat as a way of storing energy for later.
3. AMP will accumulate, stimulating phosphofructokinase, which increases the rate of glycolysis. Since oxygen is not present, the cell will convert pyruvate to lactate in lactic acid fermentation, providing a supply of ATP.

Self-Quiz

1. b	2. d	3. c	4. c	5. a
6. a	7. b	8. d	9. b	10. b

CHAPTER 10

Concept Check 10.1

1. CO_2 enters leaves via stomata, and water enters via roots and is carried to leaves through veins.
2. Using a heavy isotope of oxygen as a label, ^{18}O, van Niel was able to show that the oxygen produced during photosynthesis originates in water, not in carbon dioxide.

3. The Calvin cycle depends on the NADPH and ATP that the light reactions generate, and the light reactions depend on the $NADP^+$ and ADP and Ⓟ$_i$ that the Calvin cycle generates.

Concept Check 10.2

1. Green, because green light is mostly transmitted and reflected—not absorbed—by photosynthetic pigments
2. In chloroplasts, light-excited electrons are trapped by a primary electron acceptor, which prevents them from dropping back to the ground state. In isolated chlorophyll, there is no electron acceptor, so the photoexcited electrons immediately drop back down to the ground state, with the emission of light and heat.
3. Water (H_2O) is the electron donor; $NADP^+$ accepts electrons at the end of the electron transport chain, becoming reduced to NADPH.

Concept Check 10.3

1. 6, 18, 12
2. The more potential energy that a molecule stores, the more energy and reducing power required for the formation of that molecule. Glucose is a valuable energy source because it is highly reduced, storing lots of potential energy in its electrons. To reduce CO_2 to glucose, much energy and reducing power are required in the form of a high number of ATP and NADPH molecules, respectively.
3. The light reactions require ADP and $NADP^+$, which would not be formed from ATP and NADPH if the Calvin cycle stopped.

Concept Check 10.4

1. Photorespiration decreases photosynthetic output by adding oxygen, instead of carbon dioxide, to the Calvin cycle. As a result, no sugar is generated (no carbon is fixed), and O_2 is used rather than generated.
2. C_4 and CAM species would replace many of the C_3 species.

Self-Quiz

1. d	2. b	3. d	4. b	5. b
6. c	7. d	8. e	9. d	10. c

CHAPTER 11

Concept Check 11.1

1. The secretion of neurotransmitter molecules at a synapse is an example of local signaling. The electrical signal that travels along a very long nerve cell and is passed to the next nerve cell can be considered an example of long-distance signaling. (Note, however, that local signaling at the synapse between two cells is necessary for the signal to pass from one cell to the next.)
2. No glucose-1-phosphate is generated, because the activation of the enzyme requires an intact cell membrane with an intact receptor in the membrane. The enzyme cannot be activated directly by interaction with the signal molecule in the test tube.

Concept Check 11.2

1. The NGF receptor is in the plasma membrane. The water-soluble NGF molecule cannot pass through the lipid membrane and reach intracellular receptors, as steroid hormones can.

Concept Check 11.3

1. A protein kinase is an enzyme that transfers a phosphate group from ATP to a protein, usually activating that protein (often a second type of protein kinase). Many signal transduction pathways include a series of such interactions, in which each phosphorylated protein kinase in turn phosphorylates the next protein kinase in the series. Such phosphorylation cascades carry a signal from outside the cell to the cellular protein(s) that will carry out the response.
2. Protein phosphatases reverse the effects of the kinases.
3. The IP_3-gated channel opens, allowing calcium ions to flow out of the ER, which raises the cytosolic Ca^{2+} concentration.

Concept Check 11.4

1. By a cascade of sequential activations, at each step of which one molecule may activate numerous molecules functioning in the next step
2. Scaffolding proteins hold molecular components of signaling pathways in a complex with each other. Different scaffolding proteins would assemble different collections of proteins, leading to different responses in the two cells.

Self-Quiz

1. c	2. d	3. a	4. c	5. c
6. c	7. a	8. b	9. c	10. a

CHAPTER 12

Concept Check 12.1

1. 32 cells
2. 2
3. 39; 39; 78; 39

Concept Check 12.2

1. From the end of S phase of interphase through the end of metaphase of mitosis
2. 4; 8
3. Cytokinesis results in two genetically identical daughter cells in both plant cells and animal cells, but the mechanism of dividing the cytoplasm is different in animals and plants. In an animal cell, cytokinesis occurs by cleavage, which divides the parent cell in two with a contractile ring of actin filaments. In a plant cell, a cell plate forms in the middle of the cell and grows until its membrane fuses with the plasma membrane of the parent cell. A new cell wall is also produced from the cell plate, resulting in two daughter cells.
4. They elongate the cell during anaphase.
5. Sample answer: Each type of chromosome consists of a single molecule of DNA with attached proteins. If stretched out, the molecules of DNA would be many times longer than the cells in which they reside. During cell division, the two copies of each type of chromosome actively move apart, and one copy ends up in each of the two daughter cells.

Concept Check 12.3

1. G_1
2. The nucleus on the right was originally in the G_1 phase; therefore, it had not yet duplicated its chromosome. The nucleus on the left was in the M phase, so it had already duplicated its chromosome.
3. A sufficient amount of MPF has to build up for a cell to pass the G_2 checkpoint.
4. The cells might divide even in the absence of PDGF, in which case they would not stop when the surface was covered; they would continue to divide, piling on top of one another.
5. Most body cells are not in the cell cycle, but rather are in a nondividing state called G_0.
6. Both types of tumors consist of abnormal cells. A benign tumor stays at the original site and can usually be surgically removed. Cancer cells from a malignant tumor spread from the original site by metastasis and may impair the functions of one or more organs.

Self-Quiz

1. b	2. b	3. a	4. a	5. c
6. c	7. c	8. e	9. a	10. b

11. See Figure 12.6.

CHAPTER 13

Concept Check 13.1

1. Parents pass genes to their offspring that program cells to make specific enzymes and other proteins whose cumulative action produces an individual's inherited traits.
2. Such organisms reproduce by mitosis, which generates offspring whose genomes are virtually exact copies of the parent's genome.
3. Offspring resemble their parents but are not genetically identical to them or their siblings because sexual reproduction generates different combinations of genetic information.

Concept Check 13.2

1. A female has two X chromosomes; a male has an X and Y.
2. In meiosis, the chromosome count is reduced from diploid to haploid; the union of two haploid gametes in fertilization restores the diploid chromosome count.
3. Haploid number (n) is 39; diploid number ($2n$) is 78.
4. Meiosis is involved in the production of gametes in animals. Mitosis is involved in the production of gametes in plants and most fungi (see Figure 13.6).

Concept Check 13.3

1. In mitosis, a single replication of the chromosomes is followed by one division of the cell, so the number of chromosome sets in daughter cells is the same as in the parent cell. In meiosis, a single replication of the chromosomes is followed by two cell divisions that

Answers

reduce the number of chromosome sets from two (diploid) to one (haploid).

2. The chromosomes are similar in that each is composed of two sister chromatids, and the individual chromosomes are positioned similarly on the metaphase plate. The chromosomes differ in that in a mitotically dividing cell, sister chromatids of each chromosome are genetically identical, but in a meiotically dividing cell, sister chromatids are genetically distinct because of crossing over in meiosis I. Moreover, the chromosomes in metaphase of mitosis can be composed of a diploid set or a haploid set, but the chromosomes in metaphase of meiosis II always consist of a haploid set.

Concept Check 13.4

1. Even in the absence of crossing over, independent assortment of chromosomes during meiosis I theoretically can generate 2^n possible haploid gametes, and random fertilization can produce $2^n \times 2^n$ possible diploid zygotes. Since the haploid number (n) of honeybees is 16 and that of fruit flies is 4, two honeybees would be expected to produce a greater variety of zygotes than would two fruit flies.
2. If the segments of the maternal and paternal chromatids that undergo crossing over are genetically identical, then the recombinant chromosomes will be genetically equivalent to the parental chromosomes. Crossing over contributes to genetic variation only when it involves rearrangement of different versions of genes.

Self-Quiz

1. d	2. b	3. d	4. d	5. c
6. d	7. a	8. c	9. c	10. d

CHAPTER 14

Concept Check 14.1

1. First, all the F_1 plants had flowers with the same color (purple) as the flowers of one of the parental varieties, rather than an intermediate color as predicted by the "blending" hypothesis. Second, the reappearance of white flowers in the F_2 generation indicates that the allele controlling the white-flower trait was not lost in the F_1 generation; rather its phenotypic effect was masked by the effect of the dominant purple-flower allele.
2. According to the law of independent assortment, 25 plants ($1/16$ of the offspring) are predicted to be *aatt,* or recessive for both characters. The actual result is likely to differ slightly from this value.

Concept Check 14.2

1. $1/2$ dominant homozygous (*CC*), 0 recessive homozygous (*cc*), and $1/2$ heterozygous (*Cc*)
2. $1/4$ *BBDD*; $1/4$ *BbDD*; $1/4$ *BBDd*; $1/4$ *BbDd*
3. 0; since only one of the parents has a recessive allele for each character, there is no chance of producing homozygous recessive offspring that would display the recessive traits.

Concept Check 14.3

1. The black and white alleles are incompletely dominant, with heterozygotes being gray in color. A cross between a gray rooster and a black hen should yield approximately equal numbers of gray and black offspring.
2. Height is at least partially hereditary and appears to exhibit polygenic inheritance with a wide norm of reaction, indicating that environmental factors have a strong influence on phenotype.

Concept Check 14.4

1. $1/9$ (Since cystic fibrosis is caused by a recessive allele, Beth and Tom's siblings who have CF must be homozygous recessive. Therefore, each parent must be a carrier of the recessive allele. Since neither Beth nor Tom has CF, this means they each have a $2/3$ chance of being a carrier. If they are both carriers, there is a $1/4$ chance that they will have a child with CF. $2/3 \times 2/3 \times 1/4 = 1/9$); 0 (Both Beth and Tom would have to be carriers to produce a child with the disease.)
2. Joan's genotype is *Dd*. Because the allele for polydactyly (*D*) is dominant to the allele for five digits per appendage (*d*), the trait is expressed in people with either the *DD* or *Dd* genotype. If Joan's mother were homozygous dominant (*DD*), then all of her children would have polydactyly. But since some of Joan's siblings do not have this condition, her mother must be heterozygous (*Dd*). All the children born to her mother (*Dd*) and father (*dd*) have either the *dd* genotype (normal phenotype) or the *Dd* genotype (polydactyly phenotype).

Genetics Problems

1. Parental cross is $AAC^RC^R \times aaC^WC^W$. Genotype of F_1 is AaC^RC^W, phenotype is all axial-pink. Genotypes of F_2 are 4 AaC^RC^W : 2 AaC^RC^R : 2 AAC^RC^W : 2 aaC^RC^W : 2 AaC^WC^W : 1 AAC^RC^R : 1 aaC^RC^R : 1 AAC^WC^W : 1 aaC^WC^W. Phenotypes of F_2 are 6 axial-pink : 3 axial-red : 3 axial-white : 2 terminal-pink : 1 terminal-white : 1 terminal-red.
2. a. $1/64$ b. $1/64$ c. $1/8$ d. $1/32$
3. Albino (*b*) is a recessive trait; black (*B*) is dominant. First cross: parents *BB* × *bb;* gametes *B* and *b;* offspring all *Bb* (black coat). Second cross: parents *Bb* × *bb;* gametes $1/2$ *B* and $1/2$ *b* (heterozyous parent) and *b;* offspring $1/2$ *Bb* and $1/2$ *bb.*
4. a. *PPLl* × *PPLl*, *PPLl* × *PpLl*, or *PPLl* × *ppLl.* b. *ppLl* × *ppLl.* c. *PPLL* × any of the 9 possible genotypes or *PPll* × *ppLL.* d. *PpLl* × *Ppll.* e. *PpLl* × *PpLl.*
5. Man I^Ai; woman I^Bi; child *ii.* Other genotypes for children are $1/4$ I^AI^B, $1/4$ I^Ai, $1/4$ I^Bi.
6. a. $3/4 \times 3/4 \times 3/4 = 27/64$
   b. $1 - 27/64 = 37/64$
   c. $1/4 \times 1/4 \times 1/4 = 1/64$
   d. $1 - 1/64 = 63/64$
7. a. $1/256$ b. $1/16$ c. $1/256$ d. $1/64$ e. $1/128$
8. a. 1 b. $1/32$ c. $1/8$ d. $1/2$
9. $1/9$
10. Matings of the original mutant cat with true-breeding noncurl cats will produce both curl and noncurl F_1 offspring if the curl allele is dominant, but only noncurl offspring if the curl allele is recessive. You would obtain some true-breeding offspring homozygous for the curl allele from matings between the F_1 cats resulting from the original curl × noncurl crosses whether the curl trait is dominant or recessive. You know that cats are true-breeding when curl × curl matings produce only curl offspring. As it turns out, the allele that causes curled ears is dominant.
11. $1/16$
12. 25% will be cross-eyed; all of the cross-eyed offspring will also be white.
13. The dominant allele *I* is epistatic to the *P/p* locus, and thus the genotypic ratio for the F_1 generation will be 9 *I_P_* (colorless) : 3 *I_pp* (colorless) : 3 *iiP_* (purple) : 1 *iipp* (red). Overall, the phenotypic ratio is 12 colorless : 3 purple : 1 red.
14. Recessive. All affected individuals (Arlene, Tom, Wilma, and Carla) are homozygous recessive *aa.* George is *Aa,* since some of his children with Arlene are affected. Sam, Ann, Daniel, and Alan are each *Aa,* since they are all unaffected children with one affected parent. Michael also is *Aa,* since he has an affected child (Carla) with his heterozygous wife Ann. Sandra, Tina, and Christopher can each have the *AA* or *Aa* genotype.
15. $1/2$
16. $1/6$
17. 9 *B_A_* (agouti) : 3 *B_aa* (black) : 3 *bbA_* (white) : 1 *bbaa* (white). Overall, 9 agouti : 3 black : 4 white.

CHAPTER 15

Concept Check 15.1

1. The law of segregation relates to the inheritance of alleles for a single character; the physical basis is the separation of homologues in anaphase I. The law of independent assortment of alleles relates to the inheritance of alleles for two characters; the physical basis is the alternative arrangements of homologous chromosome pairs in metaphase I.
2. About $3/4$ of the F_2 offspring would have red eyes, and about $1/4$ would have white eyes. About half of the white-eyed flies would be female and half would be male; about half of the red-eyed flies would be female and half would be male.

Concept Check 15.2

1. Crossing over during meiosis I in the heterozygous parent produces some gametes with recombinant genotypes for the two genes. Offspring with a recombinant phenotype arise from fertilization of the recombinant gametes by homozygous recessive gametes from the double-mutant parent.
2. In each case, the alleles contributed by the female parent determine the phenotype of the offspring because the male only contributes recessive alleles in this cross.

3. No, the order could be *A-C-B* or *C-A-B*. To determine which possibility is correct, you need to know the recombination frequency between *B* and *C*.

Concept Check 15.3

1. Because the gene for this eye-color character is located on the X chromosome, all female offspring will be red-eyed and heterozygous ($X^{w^+}X^{w}$); all male offspring will be white-eyed ($X^{w}Y$).
2. 1/4; 1/2 chance that the child will inherit a Y chromosome from the father and be male × 1/2 chance that he will inherit the X carrying the disease allele from his mother.

Concept Check 15.4

1. At some point during development, one of the embryo's cells may have failed to carry out mitosis after duplicating its chromosomes. Subsequent normal cell cycles would produce genetic copies of this tetraploid cell.
2. In meiosis, a combined 14-21 chromosome will behave as one chromosome. If a gamete receives the combined 14-21 chromosome and a normal copy of chromosome 21, trisomy 21 will result when this gamete combines with a normal gamete during fertilization.
3. An aneuploid male cat with more than one X chromosome could have a tortoiseshell phenotype if its X chromosomes have different alleles of the fur-color gene.

Concept Check 15.5

1. Inactivation of an X chromosome in females and genomic imprinting. Because of X inactivation, the effective dose of genes on the X chromosome is the same in males and females. As a result of genomic imprinting, only one allele of certain genes is phenotypically expressed.
2. The genes for leaf coloration are located in plastids within the cytoplasm. Normally, only the maternal parent transmits plastid genes to offspring. Since variegated offspring are produced only when the female parent is of the B variety, we can conclude that variety B contains both the wild-type and mutant alleles of pigment genes, producing variegated leaves.
3. Each cell contains numerous mitochondria, and in affected individuals, most cells contain a variable mixture of normal and abnormal mitochondria.

Genetics Problems

1. 0; 1/2, 1/16
2. Recessive; if the disorder were dominant, it would affect at least one parent of a child born with the disorder. The disorder's inheritance is sex-linked because it is seen only in boys. For a girl to have the disorder, she would have to inherit recessive alleles from *both* parents. This would be very rare, since males with the recessive allele on their X chromosome die in their early teens.
3. 1/4 for each daughter (1/2 chance that child will be female × 1/2 chance of a homozygous recessive genotype); 1/2 for first son
4. 17%
5. 6%. Wild type (heterozygous for normal wings and red eyes) × recessive homozygote with vestigial wings and purple eyes
6. The disorder would always be inherited from the mother.
7. The inactivation of two X chromosomes in XXX women would leave them with one genetically active X, as in women with the normal number of chromosomes. Microscopy should reveal two Barr bodies in XXX women.
8. *D–A–B–C*
9. Fifty percent of the offspring would show phenotypes that resulted from crossovers. These results would be the same as those from a cross where *A* and *B* were not linked. Further crosses involving other genes on the same chromosome would reveal the linkage and map distances.
10. Between *T* and *A*, 12%; between *A* and *S*, 5%
11. Between *T* and *S*, 18%; sequence of genes is *T-A-S*
12. No. The child can be either I^AI^Ai or I^Aii. An ovum with the genotype *ii* could result from nondisjunction in the mother, while a sperm of genotype I^AI^A could result from nondisjunction in the father.
13. 450 each of blue-oval and white-round (parentals) and 50 each of blue-round and white-oval (recombinants)
14. About one-third of the distance from the vestigial-wing locus to the brown-eye locus

CHAPTER 16

Concept Check 16.1

1. DNA from the dead pathogenic S cells was somehow taken up by the living, nonpathogenic R cells. The DNA from the S cells enabled the R cells to make a capsule, which protected them from the mouse's defenses. In this way, the R cells were transformed into pathogenic S cells.
2. When the proteins were radioactively labeled (batch 1), the radioactivity would have been found in the pellet of bacterial cells.
3. Chargaff's rules state that in DNA, the percentages of A and T and of G and C are essentially the same, and the fly data are consistent with those rules. (Slight variations are most likely due to limitations of analytical technique.)
4. Each A hydrogen-bonds to a T, so in a DNA double helix, their numbers are equal; the same is true for G and C.

Concept Check 16.2

1. Complementary base pairing ensures that the two daughter molecules are exact copies of the parent molecule. When the two strands of the parent molecule separate, each serves as a template on which nucleotides are arranged, by the base-pairing rules, into new complementary strands.
2. DNA pol III covalently adds nucleotides to new DNA strands and proofreads each added nucleotide for correct base pairing.
3. The leading strand is initiated by an RNA primer, which must be removed and replaced with DNA, a task performed by DNA pol I. In Figure 16.16 just to the left of the origin of replication, DNA pol I would replace the primer of the leading strand with DNA nucleotides.
4. The ends of eukaryotic chromosomes become shorter with each round of DNA replication, and telomeres at the ends of DNA molecules ensure that genes are not lost after numerous rounds of replication.

Self-Quiz

1. c 2. d 3. b 4. c 5. b 6. a 7. c 8. d 9. a 10. b

CHAPTER 17

Concept Check 17.1

1. The nontemplate strand would read 5′-TGGTTTGGCTCA-3′. The 5′ → 3′ direction is the same as that for the mRNA; the base sequence is the same except for the presence of U in the mRNA where there is T in the nontemplate strand of DNA.
2. A polypeptide made up of 10 Gly (glycine) amino acids

Concept Check 17.2

1. Both assemble nucleic acid chains from monomer nucleotides using complementary base pairing to a template strand. Both synthesize in the 5′ → 3′ direction, antiparallel to the template. DNA polymerase requires a primer, but RNA polymerase can start a nucleotide chain from scratch. DNA polymerase uses nucleotides with the sugar deoxyribose and the base T, whereas RNA polymerase uses nucleotides with the sugar ribose and the base U.
2. Upstream end
3. In a prokaryote, RNA polymerase recognizes the gene's promoter and binds to it. In a eukaryote, transcription factors mediate the binding of RNA polymerase to the promoter.
4. A prokaryotic primary transcript is immediately usable as mRNA, but a eukaryotic primary transcript must be modified before it can be used as mRNA.

Concept Check 17.3

1. The 5′ cap and poly-A tail facilitate mRNA export from the nucleus, prevent the mRNA from being degraded by hydrolytic enzymes, and facilitate ribosome attachment.
2. snRNPs join with other proteins and form spliceosomes that cut introns out of a pre-mRNA molecule and join its exons together.
3. Alternative RNA splicing produces different mRNA molecules from a pre-mRNA molecule depending on which exons are included in the mRNA and which are not. By yielding more than one version of mRNA, a single gene can code for more than one polypeptide.

Concept Check 17.4

1. First, each aminoacyl-tRNA synthetase specifically recognizes a single amino acid and will attach it only to an appropriate tRNA. Second, a tRNA charged with its specific amino acid has an anticodon that will bind only to an mRNA codon for that amino acid.
2. Polyribosomes enable the cell to produce multiple copies of a polypeptide in a short time.
3. A signal peptide on the leading end of the polypeptide being synthesized is recognized by a signal-recognition particle that brings the ribosome to the ER membrane. There the ribosome attaches and continues to synthesize the polypeptide, depositing it in the ER lumen.

Concept Check 17.5

1. RNA can form hydrogen bonds with either DNA or RNA, take on a specific three-dimensional shape, and catalyze chemical reactions. These abilities enable RNA to interact functionally with all the major types of molecules in a cell.

Concept Check 17.6

1. The RNA polymerase farthest to the right is first, since it has traveled the farthest along the DNA (and its mRNA is longest). The first ribosome is at the top of each mRNA, because it has traveled the farthest along the mRNA starting from the 5′ end; the second is immediately below it, and so on.
2. No, the processes of transcription and translation are separated in space and time in a eukaryotic cell, a result of the eukaryotic cell's compartmental organization.

Concept Check 17.7

1. In the mRNA, the reading frame downstream from the deletion is shifted, leading to a long string of incorrect amino acids in the polypeptide and, in most cases, premature termination. The polypeptide will most likely be nonfunctional.
2. The amino acid sequence of the wild-type protein is Met-Asn-Arg-Leu. The amino acid sequence of the mutant protein would be the same, because the mRNA codons 5′-CUA-3′ and 5′-UUA-3′ both code for Leu.

Self-Quiz

1. c	2. b	3. d	4. d	5. a
6. a	7. c	8. d	9. e	10. b

CHAPTER 18

Concept Check 18.1

1. Lytic phages can only carry out lysis of the host cell, whereas lysogenic phages may either lyse the host cell or integrate into the host chromosome. In the latter case, the viral DNA (prophage) is simply replicated along with the host chromosome. Under certain conditions, a prophage may exit the host chromosome and initiate a lytic cycle.
2. The genetic material of these viruses is RNA, which is replicated inside the infected cell by special enzymes encoded by the virus. The viral genome (or a complementary copy of it) serves as mRNA for the synthesis of viral proteins.
3. Because it synthesizes DNA from its RNA genome. This is the reverse ("retro") of the usual DNA → RNA information flow.

Concept Check 18.2

1. Mutations can lead to a new strain of a virus that can no longer be recognized by the immune system, even if an animal has been exposed to the original strain; a virus can jump from one species to a new host; and a rare virus can spread if a population becomes less isolated.
2. In horizontal transmission, a plant is infected from an external source of virus, which could enter through a break in the plant's epidermis due to damage by insects or other animals. In vertical transmission, a plant inherits viruses from its parent either via infected seeds (sexual reproduction) or via an infected cutting (asexual reproduction).
3. A source of infection, such as prion-infected cattle, may show no symptoms for many years. Beef prepared from such animals before symptoms appear would not be recognized as hazardous and could transmit infection to people who eat the meat.

Concept Check 18.3

1. In transformation, naked, foreign DNA from the environment is taken up by a bacterial cell. In transduction, phages carry bacterial genes from one bacterial cell to another. In conjugation, a bacterial cell directly transfers plasmid or chromosomal DNA to another cell via a mating bridge that temporarily connects the two cells.
2. Both are episomes—that is, they can exist as part of the bacterial chromosome or independently. However, phage DNA can leave the cell in a protein coat (as a complete phage), whereas a plasmid cannot. Also, plasmids are generally beneficial to the cell, while phage DNA can direct the production of complete phages that may harm or kill the cell.
3. In an $F^+ \times F^-$ mating, only plasmid genes are transferred, but in an Hfr $\times$ F^- mating, bacterial genes may be transferred, because the F factor is integrated into the donor cell's chromosome. In the latter case, the transferred genes may then recombine with the recipient F^- cell's chromosome.

Concept Check 18.4

1. The cell would continuously produce β-galactosidase and the two other enzymes for lactose utilization, even in the absence of lactose, thus wasting cell resources.
2. Binding by the *trp* corepressor (tryptophan) activates the *trp* repressor, shutting off transcription of the *trp* operon; binding by the *lac* inducer (allolactose) inactivates the *lac* repressor, leading to transcription of the *lac* operon.

Self-Quiz

1. d	2. b	3. b	4. e	5. d
6. a	7. d	8. e	9. c	10. d

CHAPTER 19

Concept Check 19.1

1. A nucleosome is made up of eight histone proteins, two each of four different types, around which DNA is wound. Linker DNA runs from one nucleosome to the next one.
2. Histones contain many basic (positively charged) amino acids, such as lysine and arginine, which can form weak bonds with the negatively charged phosphate groups on the sugar-phosphate backbone of the DNA molecule.
3. RNA polymerase and other proteins required for transcription do not have access to the DNA in tightly packed regions of a chromosome.

Concept Check 19.2

1. Histone acetylation is generally associated with gene expression, while DNA methylation is generally associated with lack of expression.
2. General transcription factors function in assembling the transcription initiation complex at the promoters for all genes. Specific transcription factors bind to control elements associated with a particular gene and, once bound, either increase (activators) or decrease (repressors) transcription of that gene.
3. The three genes should have some similar or identical sequences in the control elements of their enhancers. Because of this similarity, the same specific transcription factors could bind to the enhancers of all three genes and stimulate their expression coordinately.
4. Degradation of the mRNA, regulation of translation, activation of the protein (by chemical modification, for example), and protein degradation

Concept Check 19.3

1. The protein product of a proto-oncogene is usually involved in a pathway that stimulates cell division. The protein product of a tumor-suppressor gene is usually involved in a pathway that inhibits cell division.
2. A cancer-causing mutation in a proto-oncogene usually makes the gene product overactive, whereas a cancer-causing mutation in a tumor-suppressor gene usually makes the gene product nonfunctional.
3. When an individual has inherited an oncogene or a mutant allele of a tumor-suppressor gene

Concept Check 19.4

1. The number of genes is 5–15 times higher in mammals, and the amount of noncoding DNA is about 10,000 times greater. The presence of introns in mammalian genes makes them about 27 times longer, on average, than prokaryotic genes.

2. Introns are interspersed within the coding sequences of genes. Many copies of each transposable element are scattered throughout the genome. Simple sequence DNA is concentrated at the centromeres and telomeres.
3. In the rRNA gene family, identical transcription units encoding three different RNA products are present in long, tandemly repeated arrays. The large number of copies of the rRNA genes enable organisms to produce the rRNA for enough ribosomes to carry out active protein synthesis. Each globin gene family consists of a relatively small number of nonidentical genes clustered near each other. The differences in the globin proteins encoded by these genes result in production of hemoglobin molecules adapted to particular developmental stages of the organism.

Concept Check 19.5

1. If cytokinesis is faulty, two copies of the entire genome can end up in a single cell. Errors in crossing over during meiosis can lead to one segment being duplicated while another is deleted. During DNA replication, slippage backward along the template strand can result in a duplication.
2. Gene duplication and divergence by mutation. Movement of genes to different chromosomes also occurred.
3. For either gene, a mistake in crossing over during meiosis could have occurred between the two copies of that gene, such that one ended up with a duplicated exon. This could have happened several times, resulting in the multiple copies of a particular exon in each gene.
4. Homologous transposable elements scattered throughout the genome provide sites where recombination can occur between different chromosomes. Movement of these elements into coding or regulatory sequences may change expression of genes. Transposable elements also can carry genes with them, leading to dispersion of genes and in some cases different patterns of expression. Or transport of an exon during transposition and its insertion into a gene may add a new functional domain to the originally encoded protein, a type of exon shuffling.

Self-Quiz

1. c	2. a	3. a	4. a	5. e
6. a	7. c	8. c	9. b	10. b

CHAPTER 20

Concept Check 20.1

1. White (no functional *lacZ* gene is present)
2. A cDNA library, made using mRNA from developing red blood cells, which would be expected to contain many copies of β-globin mRNAs
3. Some human genes are too large to be incorporated into bacterial plasmids. Bacterial cells lack the means to process RNA transcripts, and even if the need for RNA processing is avoided by using cDNA, bacteria lack enzymes to catalyze the post-translational processing that many human proteins undergo.

Concept Check 20.2

1. Any restriction enzyme will cut genomic DNA in many places, generating such a large number of fragments that they would appear as a smear rather than distinct bands when the gel is stained after electrophoresis.
2. RFLPs are inherited in a Mendelian fashion, and variations in RFLPs among individuals can be detected by Southern blotting.

Concept Check 20.3

1. In a genetic linkage map, genes and other markers are ordered with respect to each other, but only the relative distances between them are known. In a physical map, the actual distances between markers, expressed in base pairs, are known.
2. The three-stage approach employed in the Human Genome Project involves genetic mapping, physical mapping, and then sequencing of short, overlapping fragments that previously have been ordered relative to each other (see Figure 20.11). The shotgun approach eliminates the genetic mapping and physical mapping stages; instead, short fragments generated by multiple restriction enzymes are sequenced and then subsequently ordered by computer programs that identify overlapping regions (see Figure 20.13).

Concept Check 20.4

1. Alternative splicing of RNA transcripts from a gene and post-translational processing of polypeptides
2. It allows expression of thousands of genes to be examined simultaneously, thus providing a genome-wide view of which genes are expressed in different tissues, under particular conditions, or at different stages of development.
3. Because the human species arose more recently than many other species, there has been less time for genetic variations in coding and noncoding DNA to accumulate.

Concept Check 20.5

1. Stem cells continue to reproduce themselves.
2. Herbicide resistance, pest resistance, disease resistance, delayed ripening, and improved nutritional value

Self-Quiz

1. b	2. b	3. c	4. b	5. a
6. c	7. e	8. c	9. d	10. c

CHAPTER 21

Concept Check 21.1

1. Cells undergo differentiation during embryonic development, becoming different from each other; in the adult organism, there are many highly specialized cell types.
2. During animal development, movement of cells and tissues is a major mechanism, which is not the case in plants. In plants, growth and morphogenesis continue throughout the life of the plant. This is true of only a few types of animal cells.

Concept Check 21.2

1. Information deposited by the mother in the egg (cytoplasmic determinants) is required for embryonic development.
2. No, primarily because of subtle (and perhaps not so subtle) differences in their environments
3. By binding to a receptor on the receiving cell's surface and triggering a signal transduction pathway that affects gene expression

Concept Check 21.3

1. Because their products, made by the mother, determine the head and tail ends, as well as the back and belly, of the egg (and eventually the adult fly)
2. The prospective vulval cells require an inductive signal from the anchor cell before they can differentiate into vulval cells.
3. A shoot is a differentiated structure, yet some of the cells that make it up are able to dedifferentiate and redifferentiate, forming all of the organs of an entire new plant.

Concept Check 21.4

1. Homeotic genes differ in their *non*homeobox sequences, which determine their interactions with other transcription factors and hence which genes are regulated by the homeotic genes. These interactions differ in the two organisms, as do the expression patterns of the homeobox genes.

Self-Quiz

1. c	2. e	3. b	4. a	5. a
6. d	7. c	8. a	9. c	10. e

CHAPTER 22

Concept Check 22.1

1. Aristotle, Linnaeus, and Cuvier viewed species as fixed (though Cuvier noted that the species present in a particular location could change over time). Lamarck, Erasmus Darwin, and Charles Darwin thought species could change.
2. Lamarck observed evidence of changes in species over time and noted that evolution could result in organisms' adaptations to their environments, though his theory was based on an incorrect mechanism for evolution: that modifications an organism acquires during its lifetime can be passed to its offspring.

Concept Check 22.2

1. Species have the potential to produce more offspring than survive (overreproduction), leading to a struggle for resources, which are limited. Populations exhibit a range of heritable variations, some of which confer advantages to their bearers that make them more likely to leave more offspring than less well-suited individuals. Over time this natural selection can result in a greater proportion of favorable traits in a population (evolutionary adaptation).

2. Though an individual may become modified during its lifetime through interactions with its environment, this does not represent evolution. Evolution can be measured only as a change in proportions of heritable variations from generation to generation.

Concept Check 22.3

1. An environmental factor such as a drug does not create new traits such as drug resistance, but rather selects for traits among those that are already present in the population.
2. Despite their different functions, the forelimbs of different mammals are structurally similar because they all represent modifications of a structure found in the common ancestor. The similarities between the sugar glider and flying squirrel indicate that similar environments selected for similar adaptations despite different ancestry.
3. If molecular biology or biogeography indicates a particular branching pattern of descent from a single group of ancestral organisms, representatives of the ancestral group should appear earlier in the fossil record than representatives of the later organisms. Likewise, the many transitional forms that link ancient organisms to present-day species are evidence of descent with modification.

Self-Quiz

1. c	2. b	3. c	4. d	5. c
6. b	7. d	8. c	9. c	10. d

CHAPTER 23

Concept Check 23.1

1. Mendel showed that inheritance is particulate, and subsequently it was shown that this type of inheritance can preserve the variation on which natural selection acts.
2. 750. Half the loci (250) are fixed, meaning only one allele exists for each locus: $250 \times 1 = 250$. There are two alleles each for the other loci: $250 \times 2 = 500$. $250 + 500 = 750$.
3. $2pq + q^2$; $2pq$ represents heterozygotes with one PKU allele and q^2 represents homozygotes with two PKU alleles.

Concept Check 23.2

1. Most mutations occur in somatic cells that do not produce gametes and so are lost when the organism dies. Of mutations that do occur in cell lines that produce gametes, many do not have a phenotypic effect on which natural selection can act. Others have a harmful effect and are thus unlikely to spread in a population from generation to generation because they decrease the reproductive success of their bearers.
2. A population contains a vast number of possible mating combinations, and fertilization brings together the gametes of individuals with different genetic backgrounds. Sexual reproduction reshuffles alleles into fresh combinations every generation.

Concept Check 23.3

1. Natural selection is more "predictable" in that it tends to increase or decrease the frequency of alleles that correspond to variations that increase or decrease an organism's reproductive success in its environment. Alleles subject to genetic drift all have the same likelihood of increasing or decreasing in frequency.
2. Genetic drift results from chance fluctuations of allele frequencies from generation to generation; it tends to decrease variation over time. Gene flow is the exchange of alleles between populations; it tends to increase variation within a population but decrease allele frequency differences between populations.

Concept Check 23.4

1. No; many nucleotides are in noncoding portions of DNA or in pseudogenes that have been inactivated by mutations. A change in a nucleotide may not even change the amino acid encoded because of the redundancy of the genetic code.
2. Zero, because fitness includes reproductive contribution to the next generation, and a sterile mule cannot produce offspring.
3. In sexual selection, organisms may compete for mates through behaviors or displays of secondary sexual characteristics; only the competing sex is selected for these characteristics.
4. Only *half* of the members (the females) of a sexual population actually produce offspring, while *all* the members of an asexual population can produce offspring.

Self-Quiz

1. c	2. d	3. c	4. a	5. c
6. b	7. c	8. b	9. b	10. b

CHAPTER 24

Concept Check 24.1

1. Since the birds are known to breed successfully in captivity, the reproductive barrier in nature must be prezygotic. Given the species differences in habitat preference, the reproductive barrier is most likely to be habitat isolation.
2. a. All species concepts except the biological species concept can be applied to both asexual and sexual species because they define species on the basis of characteristics other than ability to reproduce. b. The biological species concept can be applied only to extant sexual species. c. The easiest species concept to apply in the field would be the morphological species concept because it is based only on the appearance of the organism. Additional information about its ecological habits, evolutionary history, and reproduction are not required.

Concept Check 24.2

1. Continued gene flow between mainland populations and those on a nearby island reduces the chance that enough genetic divergence will take place for allopatric speciation to occur.
2. The diploid and tetraploid watermelons are separate species. Their hybrids are triploid and as a result are sterile because of problems carrying out meiosis.
3. According to the model of punctuated equilibrium, in most cases the time during which speciation (that is, the distinguishing evolutionary changes) occurs is relatively short compared with the overall duration of the species' existence. Thus, on the vast geologic time scale of the fossil record, the transition of one species to another seems abrupt, and instances of gradual change in the fossil record are rare. Furthermore, some of the changes that transitional species underwent may not be apparent in fossils.

Concept Check 24.3

1. Such complex structures do not evolve all at once, but in increments, with natural selection selecting for adaptive variants of the earlier versions.
2. Although an exaptation is co-opted for new or additional functions in a new environment, it existed in the first place because it worked as an adaptation to the original environment.
3. The timing of different developmental pathways in organisms can change in different ways (heterochrony). This can result in differential growth patterns, such as those producing different patterns of webbing in salamander feet.

Self-Quiz

1. b	2. b	3. a	4. c	5. e
6. d	7. b	8. a	9. c	10. c.

CHAPTER 25

Concept Check 25.1

1. (a) Analogy, since porcupines and cacti are not closely related and since most other animals and plants do not have similar structures; (b) homology, since cats and humans are both mammals and have homologous forelimbs, of which the hand and paw are the lower part; (c) analogy, since owls and hornets are not closely related, and since the structure of their wings is very different.
2. The latter two are more likely to be closely related, since small genetic changes can produce divergent physical appearances, but if genes have diverged greatly, that implies the lineages have been separate for some time.

Concept Check 25.2

1. We are classified the same down to the class level; both the leopard and human are mammals. Leopards belong to order Carnivora, whereas humans do not.
2. The branching pattern of the tree indicates that the skunk and the wolf share a common ancestor that is more recent than the ancestor these two animals share with the leopard.

Concept Check 25.3

1. No; hair is a shared primitive character common to all mammals and thus cannot be helpful in distinguishing different mammalian subgroups.

2. The principle of maximum parsimony states that the hypothesis about nature we investigate first should be the simplest explanation found to be consistent with the facts. But nature does not always take the simplest course; thus, the most parsimonious tree (reflecting the fewest evolutionary changes) may not reflect reality.

Concept Check 25.4

1. Proteins are gene products. Their amino acid sequences are determined by the nucleotide sequences of the DNA that codes for them. Thus, differences between comparable proteins in two species reflect underlying genetic differences.
2. Orthologous genes are homologous genes that have ended up in different gene pools, whereas paralogous genes are found in multiple copies in a single genome because they are the result of gene duplication.

Concept Check 25.5

1. A molecular clock is a method of estimating the actual time of evolutionary events based on numbers of base changes in orthologous genes. It is based on the assumption that the regions of genomes being compared evolve at constant rates.
2. There are many portions of the genome that do not code for genes, in which many base changes could accumulate through drift without affecting an organism's fitness. Even in coding regions of the genome, some mutations may not have a critical effect on genes or proteins.

Self-Quiz

1. b	2. c	3. a	4. d	5. e
6. c	7. d	8. d	9. a	10. b

CHAPTER 26

Concept Check 26.1

1. The hypothesis that conditions on the early Earth could have permitted the synthesis of organic molecules from inorganic ingredients
2. In contrast to random mingling of molecules in an open solution, segregation of molecular systems by membranes could concentrate organic molecules, and electrical charge gradients across the membrane could assist biochemical reactions.
3. An RNA molecule that functions as a catalyst

Concept Check 26.2

1. 22,920 years (four half-life reductions)
2. About 1,300 million years, or 1.3 billion years

Concept Check 26.3

1. Prokaryotes must have existed at least 3.5 billion years ago, when the oldest fossilized stromatolites were formed.
2. Free oxygen attacks chemical bonds and can inhibit enzymes and damage cells. Some organisms were able to survive in anaerobic habitats, however.

Concept Check 26.4

1. All eukaryotes have mitochondria or genetic remnants of these organelles, but not all eukaryotes have plastids.
2. The chimera of Greek mythology contained parts from different animals. Similarly, a eukaryotic cell contains parts from various prokaryotes: mitochondria from one type of bacterium, plastids from another type, and a nuclear genome from parts of the genomes of these endosymbionts and at least one other cell.

Concept Check 26.5

1. A single-celled organism must carry out all of the functions required to stay alive. Most multicellular organisms have many types of specialized cells, and life functions are divided among specific cell types.
2. Fossils of most major animal phyla appear suddenly in the first 20 million years of the Cambrian period. Molecular clocks suggest that many animal phyla originated much earlier.

Concept Check 26.6

1. Protista, Plantae, Fungi, and Animalia
2. Monera included both bacteria and archaea, but archaea are more closely related to eukaryotes than to bacteria.

Self-Quiz

1. b	2. d	3. b	4. e	5. d
6. c	7. d	8. a	9. e	10. c

CHAPTER 27

Concept Check 27.1

1. Adaptations include the capsule (shields prokaryotes from host's immune system), plasmids (confer "contingency" functions such as antibiotic resistance), and the formation of endospores (enable cells to survive harsh conditions and to revive when the environment becomes favorable).
2. Prokaryotic cells generally lack the internal compartmentalization of eukaryotic cells. Prokaryotic genomes have much less DNA than eukaryotic genomes, and most of this DNA is contained in a single ring-shaped chromosome located in a nucleoid region rather than within a true membrane-bounded nucleus. In addition, many prokaryotes also have plasmids, small ring-shaped DNA molecules containing a few genes.
3. Rapid reproduction enables a favorable mutation to spread quickly through a prokaryotic population by natural selection.

Concept Check 27.2

1. Chemoheterotrophy; the bacterium must rely on chemical sources of energy, since it is not exposed to light, and it must be a heterotroph if it requires an organic source of carbon rather than CO_2.
2. *Anabaena* is a photoautotroph that obtains its carbon from CO_2. As a nitrogen-fixing prokaryote, *Anabaena* obtains its nitrogen from N_2.

Concept Check 27.3

1. Before molecular systematics, taxonomists classified prokaryotes according to phenotypic characters that did not clarify evolutionary relationships. Molecular comparisons indicate key divergences in prokaryotic lineages.
2. Both diseases are caused by spirochetes.
3. The ability of various archaea to use hydrogen, sulfur, and other chemicals as energy sources and to survive or even thrive without oxygen enables them to live in environments where more commonly needed resources are not present.

Concept Check 27.4

1. Although prokaryotes are small, mostly unicellular organisms, they play key roles in ecosystems by decomposing wastes, recycling chemicals, and providing nutrients to other organisms.
2. *Bacteroides thetaiotaomicron,* which lives inside the human intestine, benefits by obtaining nutrients from the digestive system and by receiving protection from competing bacteria from host-produced antimicrobial compounds to which it is not sensitive. The human host benefits because the bacterium manufactures carbohydrates, vitamins, and other nutrients.

Concept Check 27.5

1. Exotoxins are proteins secreted by prokaryotes; endotoxins are lipopolysaccharides released from the outer membrane of gram-negative bacteria that have died.
2. Their quick reproduction can make it difficult to combat them with antibiotics, particularly as they may evolve resistance to the drugs. Some also have the ability to form endospores and withstand harsh environments, surviving until conditions become more favorable.
3. Sample answers: eating fermented foods such as yogurt or cheese; receiving clean water from sewage treatment; taking prokaryote-produced medicines

Self-Quiz

1. d	2. a	3. d	4. d	5. c
6. d	7. c	8. d	9. a	10. a

CHAPTER 28

Concept Check 28.1

1. Sample response: Protists include unicellular, colonial, and multicellular organisms; photoautotrophs, heterotrophs, and mixotrophs; species that reproduce asexually, sexually, or both ways; and species that live in marine, freshwater, and moist terrestrial habitats.

2. Four: the inner and outer membranes of the bacterium, and the food vacuole membrane and plasma membrane of the eukaryotic cell

Concept Check 28.2

1. Their mitochondria do not have DNA, an electron transport chain, or citric-acid cycle enzymes.
2. Its flagella and undulating membrane enable it to move along the mucus-coated lining of these tracts inside its host.

Concept Check 28.3

1. The proteins have slightly different structures, but only one protein at a time is expressed. Frequent changes in expression prevent the host from developing immunity.
2. *Euglena* could be considered an alga because it is a photosynthetic autotroph; however, it could also be considered a fungus-like protist because it can absorb organic nutrients from its environment.

Concept Check 28.4

1. Membrane-bounded sacs under the plasma membrane
2. A red tide is a bloom of dinoflagellates, some of which produce deadly toxins that accumulate in molluscs and can affect people who eat the molluscs.
3. During conjugation, two ciliates exchange micronuclei, but no new individuals are produced.

Concept Check 28.5

1. A pair of flagella, one hairy and one smooth
2. Oomycetes acquire nutrition mainly as decomposers or parasites; golden algae are photosynthetic, but some also absorb dissolved organic compounds or ingest food particles and prokaryotes by phagocytosis.
3. The holdfast anchors the alga to the rocks, while the wide, flat blades provide photosynthetic surfaces. The cellulose and algin in the alga's cell walls cushion the thallus from waves and protect it from drying out.

Concept Check 28.6

1. Because foram tests are hardened with calcium carbonate, they form long-lasting fossils in marine sediments and sedimentary rocks.
2. Forams feed by extending their pseudopodia through pores in their tests. Radiolarians ingest smaller microorganisms by phagocytosis using their pseudopodia; cytoplasmic streaming carries the engulfed prey to the main part of the cell.

Concept Check 28.7

1. Amoebozoans have lobe-shaped pseudopodia, whereas forams have threadlike pseudopodia.
2. Slime molds are fungus-like in that they produce fruiting bodies that aid in the dispersal of spores, and they are animal-like in that they are motile and ingest food. However, slime molds are more closely related to gymnamoebas and entamoebas than to fungi or animals.
3. Yes. In the life cycle of a cellular slime mold, individual amoebas may congregate in response to a chemical signal, forming a sluglike aggregate form that can move. Then some of the cells form a stalk that supports an asexual fruiting body.

Concept Check 28.8

1. Many red algae contain an accessory pigment called phycoerythrin, which gives them a reddish color and allows them to carry out photosynthesis in relatively deep coastal water. Also unlike brown algae, red algae have no flagellated stages in their life cycle and must depend on water currents to bring gametes together for fertilization.
2. *Ulva*'s thallus contains many cells and is differentiated into leaflike blades and a rootlike holdfast. *Caulerpa*'s thallus is composed of multinucleate filaments without cross-walls, so it is essentially one large cell.

Self-Quiz

1. d	2. b	3. b	4. c	5. e
6. d	7. c	8. a	9. d	10. b

CHAPTER 29

Concept Check 29.1

1. Land plants share some key traits only with charophyceans: rosette cellulose-producing complexes, presence of peroxisome enzymes, similarity in sperm structure, and similarity in cell division (the formation of a phragmoplast). Comparisons of nuclear and chloroplast genes also point to a common ancestry.

Concept Check 29.2

1. Spore walls toughened by sporopollenin; multicellular, dependent embryos; cuticle. Such traits help prevent drying out.
2. a. diploid, b. haploid; c. haploid; d. diploid; e. haploid; f. haploid

Concept Check 29.3

1. Bryophytes are described as nonvascular plants because they do not have an extensive transport system. Another difference is that their life cycle is dominated by gametophytes rather than sporophytes.
2. Answers may include the following: large surface area of protonema enhances absorption of water and minerals; the vase-shaped archegonia protect eggs during fertilization and transport nutrients to the embryos via placental transfer cells; the stalklike seta conducts nutrients from the gametophyte to the capsule where spores are produced; the peristome enables gradual spore discharge; stomata enable CO_2/O_2 exchange while minimizing water loss; lightweight spores are wind-dispersed; mosses can lose water without dying and rehydrate when moisture is available.

Concept Check 29.4

1. Some characteristics that distinguish seedless vascular plants from bryophytes are a sporophyte-dominant life cycle, the presence of xylem and phloem, and the evolution of true roots and leaves.
2. Most lycophytes have microphylls, whereas ferns and most fern relatives have megaphylls.

Self-Quiz

1. b	2. c	3. d	4. a	5. a
6. b	7. c	8. c	9. a	10. a

CHAPTER 30

Concept Check 30.1

1. To have any chance of reaching the eggs, the flagellated sperm of seedless vascular plants must rely on swimming through a film of water, usually limited to a range of less than a few centimeters. In contrast, the sperm of seed plants are produced within durable pollen grains that can be carried long distances by wind or by animal pollinators. Although flagellated in some species, the sperm of most seed plants do not require water because pollen tubes convey them directly to the eggs.
2. The reduced gametophytes of seed plants are nurtured by sporophytes and protected from stress, such as drought conditions and UV radiation. Pollen grains have tough protective coats and can be carried long distances, facilitating widespread sperm transfer without reliance on water. Seeds are more resilient than spores, enabling better resistance to environmental stresses and wider distribution.

Concept Check 30.2

1. Although gymnosperms are similar in not having their seeds enclosed in ovaries and fruits, their seed-bearing structures vary greatly. For instance, cycads have large cones, whereas some gymnosperms, such as *Ginkgo* and *Gnetum*, have small cones that look somewhat like berries, even though they are not fruits. Leaf shape also varies greatly, from the needles of many conifers to the palmlike leaves of cycads to *Gnetum* leaves that look like those of flowering plants.
2. The life cycle illustrates heterospory, as ovulate cones produce megaspores and pollen cones produce microspores. The reduced gametophytes are evident in the form of the microscopic pollen grains and the microscopic female gametophyte within the megaspore. The egg is shown developing within an ovule, and the pollen tube is shown conveying the sperm. The figure also shows the protective and nutritive features of a seed.

Concept Check 30.3

1. In the oak's life cycle, the tree (the sporophyte) produces flowers, which contain gametophytes in pollen grains and ovules; the eggs in ovules are fertilized; the mature ovaries develop into dry fruits called acorns; and the acorn seeds germinate, resulting in embryos giving rise to seedlings and finally to mature trees, which produce flowers and then acorns.

2. Pine cones and flowers both have sporophylls, modified leaves that produce spores. Pine trees have separate pollen cones (with pollen grains) and ovulate cones (with ovules inside cone scales). In flowers, pollen grains are produced by the anthers of stamens, and ovules are within the ovaries of carpels. Unlike pine cones, many flowers produce both pollen and ovules.
3. Traditionally, angiosperms have been classified as either monocots or dicots, based on certain traits, such as the number of cotyledons. However, recent molecular evidence reveals that while monocots are a clade, dicots are not. Based on phylogenetic relationships, *most* dicots form a clade, now known as eudicots.

Concept Check 30.4

1. Because extinction is irreversible, it decreases the total diversity of plants, many of which may have brought important benefits to humans.

Self-Quiz

1. d	2. a	3. b	4. a	5. e
6. d	7. b	8. d	9. c	10. a

CHAPTER 31

Concept Check 31.1

1. Both a fungus and a human are heterotrophs. A fungus digests its food externally by secreting enzymes into the food and then absorbing the small molecules that result from digestion. In contrast, humans (and other animals) ingest relatively large pieces of food and digest the food within their bodies.
2. The extensive network of hyphae puts a large surface area in contact with the food source, and rapid growth of the mycelium extends hyphae into new territory.

Concept Check 31.2

1. The majority of the fungal life cycle consists of haploid stages, whereas the majority of the human life cycle consists of diploid stages.
2. The two mushrooms might be reproductive structures of the same mycelium (the same organism). Or they might be parts of two separate organisms that have arisen from a single parent organism through asexual reproduction and thus carry the same genetic information.

Concept Check 31.3

1. The fungal lineage thought to be the most primitive, the chytrids, have posterior flagella, as do most other opisthokonts. This suggests that other fungal lineages lost their flagella after diverging from the chytrid lineage.
2. This indicates that fungi had already established symbiotic relationships with plants by the time the first vascular plants evolved.

Concept Check 31.4

1. Flagellated spores
2. Most plants form arbuscular mycorrhizae with glomeromycetes; without the fungi, the plants would be poorly nourished.
3. Possible answers include the following: In zygomycetes, the sturdy, thick-walled zygosporangium can withstand harsh conditions and then undergo karyogamy and meiosis when the environment is favorable for reproduction. In ascomycetes, the asexual spores (conidia) are produced in chains or clusters at the tips of conidiophores, where they are easily dispersed by wind. The often cup-shaped ascocarps house the sexual spore-forming asci. In basidiomycetes, the basidiocarp supports and protects a large surface area of basidia, from which spores are dispersed.

Concept Check 31.5

1. A suitable environment for growth, retention of water and minerals, protection from sunlight, and protection from being eaten
2. A hardy spore stage enables dispersal to host organisms through a variety of mechanisms; their ability to grow rapidly in a favorable new environment enables them to capitalize on the host's resources.

Self-Quiz

1. b	2. c	3. c	4. d	5. e
6. d	7. b	8. e	9. b	10. a

CHAPTER 32

Concept Check 32.1

1. Plants are autotrophs; animals are heterotrophs. Plants have cell walls that provide structural support; animals lack strong cell walls (their bodies are held together by structural proteins, including collagen). Animals have unique cell types and tissues (muscle and nerve) and unique patterns of development, including the multicellular blastula stage.
2. These basic patterns of early development arose early in animal evolution and have been conserved across the diverse phyla within this clade today.

Concept Check 32.2

1. c, b, a, d
2. This diversification may have resulted from such external factors as changing ecological relationships (for instance, predator-prey interactions) and environmental conditions (for instance, increased oxygen levels) and from such internal factors as the evolution of the *Hox* complex.

Concept Check 32.3

1. Grade-level characteristics are those that multiple lineages share regardless of evolutionary history. Some grade-level characteristics may have evolved multiple times independently. Features that unite clades are those that were possessed by a common ancestor and were passed on to the various descendants.
2. Snail has spiral and determinate cleavage pattern; human has radial, indeterminate cleavage. Snail has schizocoelous development (characterized by a coelomic cavity formed by splitting of mesoderm masses); human has enterocoelous development (coelom forms from folds of archenteron). In a snail, the mouth forms from the blastopore; in a human, the anus develops from the blastopore.

Concept Check 32.4

1. Cnidarians possess true tissues, while sponges do not. Also unlike sponges, cnidarians exhibit body symmetry, though it is radial and not bilateral as in most other animal phyla.
2. The morphology-based tree divides Bilateria into two clades: Deuterostomia and Protostomia. The molecule-based tree recognizes three clades: Deuterostomia, Ecdysozoa, and Lophotrochozoa, as well as the phylum Rotifera.
3. Each type of data contributes to scientists' ability to test hypotheses about relationships; the more lines of data that support a particular hypothesis, the more likely it is to be valid.

Self-Quiz

1. a	2. c	3. c	4. e	5. b
6. c	7. b	8. e	9. e	10. d

CHAPTER 33

Concept Check 33.1

1. The flagella of choanocytes draw water through their collars, which trap food particles. The particles are engulfed by phagocytosis and digested, either by choanocytes or by amoebocytes.
2. Sponges release their sperm into the surrounding water; changes in current direction will affect the odds that the sperm will be drawn into neighboring individuals.

Concept Check 33.2

1. Both the polyp and the medusa are composed of an outer epidermis and an inner gastrodermis separated by a gelatinous layer, the mesoglea. The polyp is a cylindrical form that adheres to the substrate by its aboral end; the medusa is a flattened, mouth-down form that moves freely in the water.
2. Cnidarian stinging cells (cnidocytes) function in defense and prey capture. They contain capsule-like organelles (cnidae), which in turn contain inverted threads. The threads either inject poison or stick to and entangle small prey.

Concept Check 33.3

1. Tapeworms can absorb food from their environment and release ammonia into their environment through their body surface because their body is very flat.
2. No. Rotifers are microscopic but have an alimentary canal, whereas tapeworms can be very large but lack any digestive system.
3. Ectoprocts and corals are both sessile animals that collect suspended food with their tentacles and build reefs with their exoskeletons.

Concept Check 33.4

1. The function of the foot reflects the method of locomotion in each class. Gastropods use their foot as a holdfast or to move slowly on the substrate. In cephalopods, the foot functions as a siphon and tentacles.
2. The shell has become divided in two halves connected by a hinge, and the mantle cavity has gills that function in feeding as well as gas exchange. The radula has been lost, as bivalves have become adapted to suspension feeding.

Concept Check 33.5

1. The inner tube is the alimentary canal, which runs the length of the body. The outer tube is the body wall. The two tubes are separated by the coelom.
2. Each segment is surrounded by longitudinal and circular muscles. These muscles work against the fluid-filled coelom, which acts as a hydrostatic skeleton. Coordinated contraction of the muscles produces movement.

Concept Check 33.6

1. Incomplete cooking doesn't kill nematodes and other parasites that might be present in the meat.
2. Nematodes lack body segments and a true coelom; annelids have both.

Concept Check 33.7

1. Arthropod mouthparts are modified appendages, which are bilaterally paired.
2. Yes. Two-thirds of all known animal species are arthropods, which are found in nearly all habitats of the biosphere.
3. The arthropod exoskeleton, which had already evolved in the ocean, allowed terrestrial species to retain water and support their bodies on land. Wings allowed insects to disperse quickly to new habitats and to find food and mates.

Concept Check 33.8

1. Both echinoderms and cnidarians have radial symmetry. However, the ancestors of echinoderms had bilateral symmetry, and adult echinoderms develop from bilaterally symmetrical larvae. Therefore, the radial symmetry of echinoderms and cnidarians is analogous (resulting from convergent evolution), not homologous.
2. Each tube foot consists of an ampulla and a podium. When the ampulla squeezes, it forces water into the podium and makes the podium expand. When the muscles in the wall of the podium contract, they force water back into the ampulla, making the podium shorten and bend.

Self-Quiz

1. c	2. a	3. a	4. a	5. d
6. d	7. a	8. e	9. b	10. e

CHAPTER 34

Concept Check 34.1

1. In humans, these characters are present only in the embryo. The notochord becomes disks between the vertebrae, the tail is almost completely lost, and the pharyngeal clefts develop into various adult structures.
2. As water passes through the slits, food particles are filtered from the water and transported to the digestive system.

Concept Check 34.2

1. *Haikouichthys;* it had a skull and thus was a craniate, as are humans. *Haikouella* did not have a skull.
2. Hagfishes have a head and a skull made of cartilage, plus a small brain, sensory organs, and tooth-like structures. They have a neural crest, gill slits, and more extensive organ systems. In addition, hagfishes have slime glands that ward off predators and may repel competing scavengers.

Concept Check 34.3

1. Lampreys have a round, rasping mouth, which they use to attach to fish. Conodonts had two sets of mineralized dental elements, which may have been used to impale prey and cut it into smaller pieces.
2. Mineralized dental elements allowed vertebrates to become scavengers and predators. In armored jawless vertebrates, bone served as external defensive armor.

Concept Check 34.4

1. Both are gnathostomes and have jaws, four clusters of *Hox* genes, enlarged forebrains, and lateral line systems. Sharks secondarily lost much mineralization in their skeletons, which consist mainly of cartilage, whereas tuna have bony skeletons. Sharks also have a spiral valve. Tuna have an operculum and a swim bladder, as well as flexible rays supporting their fins.
2. Coelacanths live in deep marine waters, lungfishes live in ponds and swamps, and terrestrial vertebrates live on land.

Concept Check 34.5

1. No. Though it had four limb-like appendages with fully formed legs, ankles, and digits, its pectoral and pelvic girdles could not carry its body on land. It had gills and a tail fin that propelled it in water.
2. Some fully aquatic species are paedomorphic, retaining larval features as adults. Species that live in dry environments may avoid dehydration by burrowing or living under moist leaves, and they protect their eggs with foam nests, viviparity, or other adaptations.

Concept Check 34.6

1. The amniotic egg is not an entirely closed system. Nutrients used by the embryo are stored within the egg (in the yolk sac and albumen) as are some metabolic wastes produced by the embryo (in the allantois). However, the embryo exchanges oxygen and carbon dioxide with the outside environment via the chorion, allantois, and egg shell.
2. Birds have weight-saving modifications, including having no teeth or urinary bladder and only one ovary in females. The wings and feathers are adaptations that facilitate flight, as are efficient respiratory and circulatory systems, which support a high metabolic rate.

Concept Check 34.7

1. Monotremes lay eggs. Marsupials give birth to very small live young that remain attached to the mother in a pouch. Eutherians give birth to more developed live young.
2. Hands and feet adapted for grasping, flat nails, large brain, forward-looking eyes on a flat face, parental care, mobile big toe and thumb

Concept Check 34.8

1. Hominoids are a clade including gibbons, orangutans, gorillas, chimpanzees, bonobos, and humans, along with extinct species that descended from the same ancestor. Hominids are a clade including humans and all species more closely related to humans than to other living hominoids.
2. *Homo ergaster* was fully upright, bipedal, and as tall as modern humans, but its brain was significantly smaller than that of modern humans. This difference in the evolutionary change of different body parts is known as mosaic evolution.

Self-Quiz

1. e	2. c	3. d	4. a	5. d
6. b	7. c	8. c	9. c	10. b

CHAPTER 35

Concept Check 35.1

1. The vascular tissue system connects leaves and roots, allowing sugars to move from leaves to roots in the phloem, and allowing water and minerals to move to the leaves in the xylem.
2. Here are a few examples: The tubular, hollow structures of the tracheids and vessel elements of the xylem and the sieve plates in the sieve-tube members of the phloem facilitate transport. Root hairs aid in absorption of water and nutrients. The cuticle in leaves and stems protects from desiccation and pathogens. Leaf trichomes protect from herbivores and pathogens. Collenchyma and sclerenchyma cells have thick walls that provide support for plants.
3. The dermal tissue system is the leaf's protective covering. The vascular tissue system consists of the transport tissues xylem and phloem. The ground tissue system performs metabolic functions such as photosynthesis.

Concept Check 35.2

1. Your dividing cells are normally limited in the types of cells they can form. In contrast, the products of cell division in a plant meristem differentiate into all the diverse types of plant cells.

Answers

2. Primary growth arises from apical meristems and involves the production and elongation of organs. Secondary growth arises from lateral meristems and adds to the girth of roots and stems.

Concept Check 35.3

1. Lateral roots emerge from the root's interior (from the pericycle), pushing through cortical and epidermal cells. In contrast, shoot branches arise on the exterior of a shoot (from axillary buds).
2. In roots, primary growth occurs in three successive stages, moving away from the root tip: the zones of cell division, elongation, and maturation. In shoots, it occurs at the tip of terminal buds, with leaf primordia arising along the sides of apical meristems. Most growth in length occurs in older internodes below the shoot apex.
3. Veins are a network of vascular tissue that provides water and minerals to leaf cells and carries organic products of photosynthesis to other parts of the plant.

Concept Check 35.4

1. The sign will still be 2 m above the ground because only secondary growth occurs in this part of the tree.
2. A hollow tree can survive because water, minerals, and organic nutrients are conducted by the younger secondary vascular tissues, some of which remain intact: the outer secondary xylem (sapwood) and youngest secondary phloem. However, girdling removes an entire ring of secondary phloem (part of the bark), completely preventing transport of organic nutrients from the shoots to the roots.

Concept Check 35.5

1. Differences in structure result from differential gene expression.
2. In *fass* mutants, the arrangement of microtubules is disrupted so that the preprophase band does not form. This results in random planes of cell division, rather than the ordered planes of division that normally occur. Disruption of microtubule organization also prevents the alignment of cellulose microfibrils that sets the direction of cell elongation. Because of this randomness, directional growth is disrupted, and the plant becomes stubby instead of elongated.

Self-Quiz

1. d	2. c	3. c	4. d	5. a
6. e	7. d	8. c	9. b	10. c

CHAPTER 36

Concept Check 36.1

1. The relatively high concentration of salts might cause the soil's water potential to be more negative, thereby reducing water uptake by lowering the water potential gradient from the soil to the roots.
2. The cell's ψ_P is 0.7 MPa. In a solution with a ψ of -0.4 MPa, the cell's ψ_P at equilibrium would be 0.3 MPa.

Concept Check 36.2

1. The fungicide may kill the mycorrhizal fungi that function in phosphate uptake.
2. The endodermis regulates the passage of water-soluble solutes by requiring all such molecules to cross a selectively permeable membrane.

Concept Check 36.3

1. By lowering the solute potential (and water potential) of the soil, the fertilizer would make it harder for the plant to absorb water.
2. The humid air has higher water potential than the leaves.
3. After the flowers are cut, transpiration from any leaves and from the petals (which are modified leaves) will continue to draw water up the xylem. If cut flowers are transferred directly to a vase, air pockets in xylem vessels prevent delivery of water from the vase to the flowers. Cutting stems again underwater, a few centimeters from the original cut, will sever the xylem above the air pockets. The water droplets prevent other air pockets from forming while the flowers are transferred to a vase.

Concept Check 36.4

1. Accumulation of potassium by guard cells results in osmotic water uptake, and the turgid condition of the cells keeps the stomata open. This enables the mold to grow into the leaf interior via the stomata.
2. A sunny, warm, but not hot, day; high humidity; low wind speed

Concept Check 36.5

1. In both cases, the long-distance transport is a bulk flow driven by a pressure difference at opposite ends of tubes. Pressure is generated at the source end of a sieve tube by the loading of sugar and resulting osmotic flow of water into the phloem, and this pressure *pushes* sap from the source end to the sink end of the tube. In contrast, transpiration generates a negative pressure (tension) as a force that *pulls* the ascent of xylem sap.
2. At low temperature, the higher sugar content of a growing potato tuber would lower the solute potential (and water potential) of the tuber and reduce the bulk flow of sugar into it.

Self-Quiz

1. e	2. c	3. d	4. c	5. b
6. c	7. a	8. b	9. c	10. c

CHAPTER 37

Concept Check 37.1

1. Table 37.1 shows that CO_2 is the source of 90% of a plant's dry weight, supporting Hales' view that plants are nourished mostly by air. However, van Helmont's hypothesis is correct with respect to a plant's overall increase in size, which is based mainly on accumulation of water in cell vacuoles.
2. No, because even though macronutrients are required in greater amounts, all essential elements are necessary for the plant to complete its life cycle.
3. No, because deficiencies of nutrients that are more mobile show up first in older leaves, whereas deficiencies in nutrients that are less mobile show up first in younger leaves.

Concept Check 37.2

1. The topsoil has a mixture of larger particles (which provide aeration) and smaller particles (which facilitate water and mineral retention), as well as an adequate amount of humus (which supplies mineral nutrients) and a suitable pH.
2. Overwatering deprives roots of oxygen and can lead to mold, and overfertilizing can result in waste and pollution of groundwater.

Concept Check 37.3

1. Nitrogen-fixing bacteria provide the long-term supply of nitrogenous minerals essential for the survival of plants, which are directly or indirectly the source of food for humans.

Concept Check 37.4

1. Both involve mutualistic symbiotic relationships in which other organisms interact with plant roots. Root nodules involve nitrogen-fixing bacteria, whereas mycorrhizae involve fungi that facilitate absorption of both water and minerals. In both relationships, the plant provides organic compounds. Unlike root nodules, mycorrhizae occur in most plant species, but both relationships are important agriculturally.
2. Epiphytes use another plant as a substrate without obtaining nutrients from the other plant. In contrast, parasitic plants extract nutrients from their host plants.

Self-Quiz

1. b	2. b	3. c	4. b	5. b
6. c	7. d	8. a	9. d	10. b

CHAPTER 38

Concept Check 38.1

1. Sepals usually protect the unopened floral bud, and petals often help attract animal pollinators to a flower. Stamens, the pollen-producing parts, are long structures that facilitate distribution of pollen. Carpels, parts that produce female gametophytes, have stigmas, platforms that facilitate receipt of pollen. The ovary of a carpel provides protection for developing eggs in ovules. Variations in arrangement of floral organs reflect adaptations to animal pollinators and may also reduce self-fertilization, as in the case of "pin" and "thrum" flower types.
2. In angiosperms, pollination is the transfer of pollen from an anther to a stigma. It is the subsequent development of the pollen tube that eventually enables fertilization, the fusion of egg and sperm to form the zygote.

3. In the short term, selfing may be advantageous in a population that is so dispersed and sparse that pollen delivery is unreliable. In the long term, however, selfing is an evolutionary dead end because it leads to a loss of genetic diversity that may preclude adaptive evolution, including reversion from selfing to outcrossing.

Concept Check 38.2

1. Fifty percent of the ovules would have XXX endosperm and XX embryos, and 50% would have XXY endosperm and XY embryos.
2. Seeds contain endosperm or cotyledons (or both), which nourish a developing embryo, and they have a seed coat, which protects the embryo until conditions are suited for germination. Fruits, regardless of whether they are dry or fleshy, enhance seed dispersal by being eaten by animals or carried by wind.
3. Seed dormancy prevents the premature germination of seeds. A seed will germinate only when the environmental conditions are optimal for the survival of its embryo as a young seedling.

Concept Check 38.3

1. Sexual reproduction produces genetic variety, which may be advantageous in an unstable environment. The likelihood is better that at least one offspring of sexual reproduction will survive in a changed environment. Asexual reproduction can be advantageous in a stable environment because individual plants that are well-suited to that environment pass on all their genes to offspring without mates. Asexual reproduction also generally results in offspring that are less fragile than the seedlings produced by sexual reproduction. However, sexual reproduction offers the advantage of dispersal of tough seeds.
2. Asexually propagated crops lack genetic diversity. Genetically diverse populations are less likely to become extinct in the face of an epidemic because there is a greater likelihood that a few individuals in the population may be resistant.

Concept Check 38.4

1. Traditional breeding and genetic engineering both involve artificial selection for desired traits. However, genetic engineering techniques facilitate faster gene transfer and are not limited to transferring genes between closely related varieties or species.
2. GM crops may be more nutritious and less susceptible to insect damage or pathogens that invade insect-damaged plants. They also do not require as much chemical spraying. However, continued field testing of GM crops remains important to avoid adverse effects on human health and nontarget organisms and the possibility of transgene escape.

Self-Quiz

1. d 2. a 3. c 4. a 5. b
6. c 7. e 8. a 9. c 10. e

CHAPTER 39

Concept Check 39.1

1. No. Viagra, like injection of cyclic GMP, should cause only a partial de-etiolation response.
2. Cycloheximide should inhibit de-etiolation by preventing the synthesis of new proteins necessary for de-etiolation.

Concept Check 39.2

1. The plant will exhibit a constitutive triple response. Because the kinase that normally prevents the triple response is dysfunctional, the plant will undergo the triple response regardless of whether ethylene is present or the ethylene receptor is functional.
2. The pathogen might induce an increase in cytokinin concentration or a decrease in auxin concentration in the infected host plant.
3. Fusicoccin, like auxin, causes an increase in plasma H^+ pump activity and, like auxin, promotes elongation of cells in stems.

Concept Check 39.3

1. It is impossible to say. To identify this species as a short-day plant, it would be necessary to determine the critical night length for flowering and to establish that this species flowers only when the night is longer than the critical night length.
2. The use of far-red light would maintain phytochrome in its P_r form, allowing flowering to occur.
3. An experiment could involve use of an action spectrum to determine which wavelengths of light are most effective. If the action spectrum indicates phytochrome, further testing could involve red/far-red experiments to check for photoreversibility.

Concept Check 39.4

1. A plant that overproduces ABA would undergo less evaporative cooling because its stomata would be less open.
2. Plants growing close to the aisles may be more subject to mechanical stresses caused by passing workers and air currents within the greenhouse. The plants nearer to the middle of the bench may also be taller as a result of shading.

Concept Check 39.5

1. Mechanical damage breaches a plant's first line of defense against infection, its protective dermal tissue.
2. Perhaps the breeze dilutes the local concentration of a volatile defensive compound that the plants produce.

Self-Quiz

1. b 2. a 3. b 4. d 5. b
6. e 7. b 8. b 9. c 10. c

CHAPTER 40

Concept Check 40.1

1. The small intestine, the lungs, and the kidneys contain internal exchange surfaces through which nutrients, gases, and chemicals, respectively, flow. A large surface area facilitates this exchange and enables the body to carry out the exchange more efficiently than if there were less surface area available.

Concept Check 40.2

1. The epithelium lining the inner surface of the stomach secretes mucus, which lubricates and protects the surface, and digestive juices. Also, the tight packing of the epithelial tissue serves as a protective barrier.
2. Connective tissue is an important part of most organs, and sheets of connective tissue support many of the body's organs.
3. Both nervous tissue and muscle tissue are required to carry out responses to many stimuli. Skeletal muscle tissue contracts in response to nerve impulses transmitted by nerve cells.

Concept Check 40.3

1. The mouse, because it is an endotherm and therefore its basal metabolic rate is higher than the ectothermic lizard's standard metabolic rate
2. Intense activity quickly depletes existing ATP. Because alligators are ectotherms, they are relatively slow to generate more ATP by aerobic respiration.
3. The house cat; the smaller an animal, the higher its metabolic rate and its demand for food per unit of body mass.

Concept Check 40.4

1. No; even though an animal regulates some aspects of its internal environment, often by negative feedback mechanisms, the internal environment fluctuates slightly around a set point. Homeostasis is a dynamic state. And some changes, such as radical increases in hormones at particular times in development, are programmed to occur.
2. In negative feedback, a change triggers control mechanisms that counteract further change in that direction. In positive feedback, a change triggers mechanisms that amplify the change.

Concept Check 40.5

1. Yes, ectotherms in the deep sea and in constant-temperature freshwater springs have constant body temperatures. And terrestrial ectotherms can maintain relatively constant body temperatures by behavioral means.
2. Heat loss through convection
3. Food and water supplies may be short during the dry season, and torpor enables animals to survive at a much lower metabolic rate.

Self-Quiz

1. a 2. b 3. e 4. d 5. c
6. b 7. c 8. c 9. d 10. c

CHAPTER 41

Concept Check 41.1

1. Body weight is stable when caloric intake (food) is balanced by caloric expenditure (metabolic rate).

2. Over the long term, the body converts excess calories to fat whether those calories are consumed as fat, carbohydrate, or protein.
3. Both hormones have appetite-suppressing effects on the brain's satiety center. During the course of a day, PYY, secreted by the small intestine, suppresses appetite after meals. Over the longer term, leptin, produced by adipose tissue, normally reduces appetite as fat storage increases.

Concept Check 41.2

1. Undernutrition is a deficiency of calories in the diet. In contrast, malnutrition results from a deficiency of one or more essential nutrients, even though total caloric intake may be adequate.
2. A balanced vegetarian meal combines vegetables and fruits that complement one another, each providing certain essential amino acids that may be deficient in the other food items.
3. Vitamins and minerals are essential nutrients required in relatively small daily amounts. Vitamins are organic nutrients, whereas minerals are inorganic nutrients.

Concept Check 41.3

1. A gastrovascular cavity is a digestive sac with a single opening that functions in both ingestion and elimination; an alimentary canal is a digestive tube with a separate mouth and anus on opposite ends.
2. As long as nutrients are within the cavity of the alimentary canal, they are in a compartment that is continuous with the outside environment via the mouth and anus and have not yet entered the body by crossing a membrane.

Concept Check 41.4

1. Peristalsis can squeeze food through the esophagus even without the help of gravity.
2. The acid breaks down the tissues in plant and animal materials by its harsh, nonenzymatic, chemical attack. The acid also activates the protein-digesting enzyme pepsin and destroys any bacteria that may have been ingested with food.
3. The partially digested meal entering from the stomach as acid chyme + pancreatic juice containing hydrolytic enzymes + intestinal juice with enzymes + bile, which includes bile salts that aid in digestion by emulsifying fats
4. The villi and microvilli of the intestinal epithelium provide an enormous surface area for absorption, the transport of nutrients from the lumen of the small intestine into blood capillaries and lacteals.
5. Long-term use of antibiotics can kill colon bacteria that augment nutrition by producing vitamin K.
6. As it receives acid chyme from the stomach, the duodenum secretes the hormone cholecystokinin (CCK), which reaches the pancreas via the bloodstream and stimulates release of pancreatic juice.

Concept Check 41.5

1. The sharp incisors are adapted for cutting pieces of meat and plants. The broad, ridged surfaces of molars are adapted for grinding tough foods.
2. The tadpole is herbivorous (eats mostly algae), while the adult frog is carnivorous (eats insects, for example).
3. The cud is a regurgitated brew of material from the rumen—once-chewed vegetation with fatty acids and other metabolic byproducts of the rumen bacteria. After a second chewing, which increases the surface area of the plant material, the cow swallows the cud, and microbial action in the stomach chambers continues to convert cellulose to a diversity of nutrients.

Self-Quiz

1. a	2. b	3. e	4. c	5. c
6. c	7. c	8. d	9. c	10. e

CHAPTER 42

Concept Check 42.1

1. The inability of diffusion to deliver nutrients and remove wastes at high enough rates to sustain a large organism
2. Advantage: high rate of delivery of nutrients and removal of wastes. Disadvantage: requires more energy to build, operate, and maintain.
3. Two main advantages of separate respiratory and systemic circuits are higher blood pressure in the systemic circuit and a higher rate of blood circulation.

Concept Check 42.2

1. This condition would reduce the oxygen content by mixing O_2-depleted blood returned to the right ventricle from the systemic circuit with the O_2-rich blood of the left ventricle.
2. The delay ensures that the atria empty completely before the ventricles contract.

Concept Check 42.3

1. The large total cross-sectional area of the capillaries
2. These changes increase the capacity for action by increasing the rate of blood circulation and delivery of oxygen and nutrients to the skeletal muscles.
3. Plasma proteins remaining in the blood in a capillary maintain fairly constant osmotic pressure, whereas blood pressure drops from the arteriole end to the venule end. This difference allows fluid to reenter the capillary at the venule end; if plasma proteins are deficient, fluid will remain in the tissues and cause swelling.

Concept Check 42.4

1. About 200 billion, or 2.08×10^{11}, calculated by dividing the total number of cells, 2.5×10^{13}, by 120 days
2. An increase in the number of leukocytes may indicate that the person is combating an infection.
3. Bone marrow stem cells divide repeatedly and are pluripotent.

Concept Check 42.5

1. If the respiratory surfaces of lungs extended out into the terrestrial environment, they would quickly dry out, and diffusion of O_2 and CO_2 across the membrane would stop.
2. The countercurrent results in a diffusion gradient for O_2 over the entire length of capillaries in the gill lamellae. As water flows over the gill lamellae, the opposite direction of blood flow in the capillaries enables the blood to continue loading O_2 because the O_2-rich blood "downstream" encounters the even more O_2-rich water just beginning to flow over the lamellae.

Concept Check 42.6

1. An increase in CO_2 in the blood increases the rate of CO_2 diffusion into the cerebrospinal fluid, where the CO_2 combines with water, forming carbonic acid. Dissociation of carbonic acid releases hydrogen ions, decreasing the pH of cerebrospinal fluid.
2. Increased heart rate increases the rate at which CO_2-rich blood is delivered to the lungs, where CO_2 is removed.
3. Air passes through the lungs of birds in one direction only. In mammals, the direction of air flow reverses between inhalation and exhalation.

Concept Check 42.7

1. Differences in partial pressure; gases diffuse from areas of higher partial pressure to areas of lower partial pressure.
2. The Bohr shift causes hemoglobin to release more oxygen at a lower pH, which occurs in the vicinity of tissues with high rates of respiration and carbon dioxide release.
3. The decrease in concentration of CO_2 in the plasma as it diffuses into the alveolar spaces causes the carbonic acid within the red blood cells to break down, yielding CO_2, which diffuses into the plasma.
4. Examples: more blood volume relative to body mass; much larger spleen; more oxygen-storing myoglobin in muscles; heart rate and metabolic rate decrease during dive

Self-Quiz

1. c	2. b	3. d	4. c	5. e
6. b	7. c	8. b	9. a	10. a

CHAPTER 43

Concept Check 43.1

1. Macrophages have receptors that bind to polysaccharides present on the surface of bacterial cells but not on body cells.
2. Vessel dilation, which allows enhanced blood flow, and increased vessel permeability result in the common signs of inflammation. These vascular changes aid in delivering clotting factors, antimicrobial proteins, and phagocytic cells to the tissue of the affected region; all of these help in repairing tissue damage and stopping the spread of infection.
3. The exoskeleton of insects provides an external barrier similar to the skin and mucous membranes of vertebrates. Phagocytic cells and antimicrobial proteins also contribute to innate defenses in both insects and vertebrates.

Answers

Concept Check 43.2

1. See Figure 43.8a; a secreted antibody lacks a transmembrane region and cytoplasmic tail.
2. B cell receptors bind intact extracellular antigens present on the surface of microbes or free in body fluids. T cell receptors bind small fragments of antigens that are complexed with class I or class II MHC molecules on the surface of infected body cells or antigen-presenting cells, respectively.
3. Specificity: Only B cells with receptors that bind to the antigen are selected to proliferate and differentiate into plasma cells secreting antibodies specific for the antigen and memory B cells bearing receptors specific for the same antigen. Memory: The large number of memory B cells generated respond more rapidly to the same antigen the next time it enters the body.
4. 40 $V \times 5 J = 200$ possible light chains; 51 $V \times 6 J \times 27 D = 8{,}262$ possible heavy chains. Each antigen-binding site is formed from a region on a light chain and heavy chain. The number of possible random combinations is 200 light chains $\times$ 8,262 heavy chains $= 1.65 \times 10^6$ possible antigen-binding specificities.

Concept Check 43.3

1. An activated helper T cell secretes cytokines that promote activation of both cytotoxic T cells and B cells. An activated cytotoxic T cell kills infected cells and tumor cells by apoptosis. An activated B cell differentiates into plasma cells that secrete antibodies.
2. A child lacking a thymus would have no functional T cells. Without helper T cells to help activate B cells, the child would be unable to produce antibodies against extracellular bacteria. Without cytotoxic T cells or helper T cells to help activate them, the child's immune system would be unable to kill virus-infected cells.
3. Antibodies bound to viruses can block their attachment to potential host cells (viral neutralization). Coating of bacteria or other particles by antibodies bound to surface antigens increases their phagocytosis by macrophages (opsonization). Antibodies bound to antigens on bacterial cells also can activate a cascade of complement proteins leading to lysis of the bacteria (complement activation). Cross-linking of antigens on many bacterial cells or viruses by binding of multiple antibody molecules can lead to formation of large clumps (agglutination), which are then phagocytosed.
4. Passive immunization, the transfer of antibodies from one individual to another, is protective only as long as the antibody molecules last. Active immunization, the introduction of antigen, induces an immune response in the recipient that can lead to the generation of long-lived memory cells. Someone who is actively immunized may be immune to that antigen for life.

Concept Check 43.4

1. Because individuals with type AB blood do not produce antibodies against either the A or the B antigens, they can safely receive type A blood, type B blood, type AB blood, or type O blood—that is, they are universal recipients. In the case of donated type O blood, packed cells should be used, since the donor serum (fluid part of blood) would contain antibodies to A and to B, which could react with the recipient's red blood cells.
2. The danger of the graft rejecting the host arises because transplanted bone marrow contains lymphocytes that could react against components of the recipient's body.
3. An autograft will not trigger a rejection reaction.

Concept Check 43.5

1. A person with a macrophage deficiency would have frequent infections. This would be due to poor innate responses, particularly diminished phagocytosis and inflammation, and poor acquired responses because of the role of macrophages in presenting antigens to helper T cells.
2. Binding of antigens by IgE molecules attached to mast cells induces degranulation of these cells, releasing histamine and other inflammatory agents, which cause typical allergy symptoms. Drugs that block the degranulation response prevent the release of inflammatory agents and hence the symptoms they cause.
3. Myasthenia gravis is considered an autoimmune disease because the immune system produces antibodies against self molecules (acetylcholine receptors).
4. To enter a host cell, HIV requires CD4 and a co-receptor. The co-receptor for HIV normally functions as a chemokine receptor. If a person's chemokine receptors are faulty, HIV cannot use them for entry into cells.

Self-Quiz

1. e	2. d	3. b	4. c	5. b
6. d	7. c	8. d	9. e	10. b

CHAPTER 44

Concept Check 44.1

1. Because the salt is moved against the concentration gradient, from a hypoosmotic to a hyperosmotic environment
2. A freshwater osmoconformer would have body fluids too dilute to carry out life's processes.
3. By maintaining high concentrations of urea and TMAO in their blood, sharks reduce the osmotic gradient between their blood and seawater.

Concept Check 44.2

1. The aquatic larvae can dispose of the very toxic ammonia continuously by secreting it across epithelia into the surrounding water. The adults conserve water by excreting the nontoxic uric acid.
2. The liver is the site of urea synthesis.
3. Because uric acid is largely insoluble in water, it can be excreted as a semisolid paste, thereby reducing an animal's water loss.

Concept Check 44.3

1. Filtration (of blood, hemolymph, or coelomic fluid) and selective reabsorption or secretion of solutes
2. A large surface area for exchange of water and solutes

Concept Check 44.4

1. A decline in blood pressure in the afferent arteriole would reduce the rate of filtration.
2. The kidney medulla would absorb less water and thus the drug would increase water loss in the urine.
3. Bowman's capsule, proximal tubule, loop of Henle, distal tubule

Concept Check 44.5

1. Alcohol inhibits the release of ADH, causing an increase in urinary water loss and increasing the chance of dehydration.
2. Consuming salty food increases the osmolarity of the blood, which triggers osmoreceptor cells in the hypothalamus to stimulate drinking and to release ADH, which increases the rate of water reabsorption by the distal tubules and collecting ducts.
3. The capacity to conserve water by producing hyperosmotic urine

Concept Check 44.6

1. Numerous nephrons and well-developed glomeruli are characteristic of the kidneys of freshwater fishes, while reduced numbers of nephrons and smaller glomeruli indicate marine environments. The numerous nephrons and well-developed glomeruli of freshwater fishes produce urine at a high rate, while small numbers of nephrons and smaller glomeruli produce urine at a low rate.

Self-Quiz

1. d	2. d	3. e	4. e	5. c
6. d	7. c	8. b	9. d	10. b

CHAPTER 45

Concept Check 45.1

1. Hormones are produced by endocrine cells, whereas neurohormones are produced by specialized nerve cells called neurosecretory cells. Both hormones and neurohormones are secreted into the extracellular fluid and act on target tissues.
2. See Figure 45.2a and b.
3. In negative feedback, the effector response reduces the initial stimulus, so eventually the response ceases as the variable being controlled reaches the set point. In positive feedback, the effector response causes an increase in the stimulus, leading to an even greater response.

Concept Check 45.2

1. Water-soluble hormones, which cannot penetrate the plasma membrane, bind to cell-surface receptors. This interaction triggers an intracellular multicomponent signal transduction pathway that ultimately alters the activity of a preexisting cytoplasmic protein and/or changes transcription of specific genes

in the nucleus. Steroid hormones are lipid-soluble and can cross the plasma membrane into the cell interior, where they bind to receptors located in the cytoplasm or nucleus. In both cases, the hormone-receptor complex functions directly as a transcription factor that binds to the cell's DNA and activates or inhibits transcription of specific genes.

2. A particular hormone may cause diverse responses in target cells having different receptors for the hormone, different signal transduction pathways, and/or different proteins for carrying out the response.
3. Once secreted, local regulators can diffuse rapidly to their nearby target cells. Hormones circulate in the bloodstream or hemolymph from their sites of release to their target tissues, a slower process.

Concept Check 45.3

1. The posterior pituitary, an extension of the hypothalamus that contains the axons of neurosecretory cells, is the storage and release site for two neurohormones, oxytocin and antidiuretic hormone (ADH). The anterior pituitary, derived from tissue of the embryonic mouth, contains endocrine cells that make at least six different hormones. Secretion of anterior pituitary hormones is controlled by hypothalamic hormones that travel via portal vessels to the anterior pituitary.
2. Tropic hormones control the synthesis and/or secretion of hormones from other endocrine tissues. Releasing and inhibiting hormones produced in the hypothalamus control hormone secretion by the anterior pituitary. The anterior pituitary produces several tropic hormones that control the hormonal function of the thyroid gland, adrenal cortex, and gonads.
3. (a) Prolactin functions in a simple neuroendocrine pathway. (b) ACTH functions in a complex neuroendocrine pathway. (c) Oxytocin functions in a neurohormone pathway.

Concept Check 45.4

1. By negative feedback on the hypothalamus and anterior pituitary (see Figure 45.9)
2. The hormone-secreting cells themselves monitor the blood Ca^{2+} levels. An increase in blood Ca^{2+} above the set point stimulates the thyroid gland to release calcitonin. By promoting deposition of Ca^{2+} in the bones and excretion of Ca^{2+} by the kidneys, calcitonin decreases the blood Ca^{2+} level. PTH, secreted by the parathyroid glands in response to low blood Ca^{2+}, has the opposite effects on bones and kidneys, thereby increasing blood Ca^{2+}. The response to one hormone triggers release of the antagonistic hormone, a feedback mechanism that minimizes extreme deviations of blood Ca^{2+} from the set point.
3. In a person with diabetes mellitus, the initial increase in blood glucose is greater than in a healthy person, and it remains high for a prolonged period because inadequate production of insulin or nonresponsiveness of target cells decreases the body's ability to clear excess glucose from the blood.
4. The levels of these hormones in the blood would become very high owing to the absence of negative feedback on the hypothalamic neurons that secrete the releasing hormone that stimulates the secretion of ACTH by the anterior pituitary.

Concept Check 45.5

1. During larval stages, neurosecretory cells produce a tropic hormone (brain hormone) that stimulates production of ecdysone, the molting hormone, by endocrine cells in the prothoracic glands.
2. Juvenile hormone promotes the retention of larval characteristics. In insecticides, it prevents larvae from maturing into adults that can reproduce.

Self-Quiz

1. c	2. d	3. d	4. c	5. b
6. e	7. c	8. b	9. c	10. a

CHAPTER 46

Concept Check 46.1

1. The offspring of sexual reproduction are genetically diverse.
2. The term could be considered misleading in the sense that a sequentially hermaphroditic organism is never of two sexes (hermaphroditic) at the same time; first it is of one sex and later of the other sex.

Concept Check 46.2

1. Internal fertilization allows the sperm to reach the egg without either gamete drying out.
2. (a) Animals with external fertilization tend to release many gametes at once, resulting in the production of enormous numbers of zygotes. This increases the chances that some will survive to adulthood. (b) Animals with internal fertilization produce fewer offspring but generally provide greater care of the embryos and the young.

Concept Check 46.3

1. Seminiferous tubule, epididymis, vas deferens, urethra
2. The seminal vesicles provide most of the fluid in which the sperm swim, as well as a sugar that is a source of energy for the sperm and prostaglandins that cause changes in the uterus that help move the sperm toward the egg after coitus; also, the fluid's alkalinity helps neutralize the acidic vaginal environment, which could harm the sperm. The fluid from the prostate contains another sperm nutrient and anticoagulants that help the sperm swim by keeping the semen liquid. The fluid from the bulbourethral glands, secreted just before ejaculation, neutralizes any acidic urine remaining in the urethra.
3. Primarily the penis and clitoris, but also the testes, labia, breasts, and outer third of the vagina

Concept Check 46.4

1. In the testis, FSH stimulates the Sertoli cells, which nourish developing sperm. LH stimulates the production of androgens (mainly testosterone), which in turn stimulate sperm production. In both females and males, FSH encourages the growth of cells that support and nourish developing gametes (follicle cells in females and Sertoli cells in males), and LH stimulates the production of sex hormones that promote gametogenesis (estrogen in females and androgens, especially testosterone, in males).
2. In estrous cycles, which occur in most female mammals, the endometrium is reabsorbed (rather than shed) if fertilization does not occur. Estrous cycles often occur just one or a few times a year, and the female is usually receptive to copulation only during the period around ovulation. Menstrual cycles are found only in humans and some other primates.
3. Hormones produced in the ovarian cycle control the uterine cycle (see Figure 46.13). Also, the occurrence of pregnancy (implantation in the uterus) turns off the ovarian cycle.
4. Ovulation is triggered by a surge in LH. The secretion of LH has been stimulated by the influence of a rising estrogen level on GnRH.

Concept Check 46.5

1. The embryo is a blastocyst, a ball of cells containing a cavity.
2. HCG secreted by the early embryo stimulates the corpus luteum to make progesterone, which helps maintain the pregnancy. During the second trimester, however, HCG production drops, the corpus luteum disintegrates, and the placenta completely takes over progesterone production.
3. Vasectomy, females
4. Because, in all cases where IVF is used, the embryo completes most of its growth and development in a woman's uterus

Self-Quiz

1. d	2. b	3. a	4. b	5. c
6. a	7. a	8. a	9. d	10. b

CHAPTER 47

Concept Check 47.1

1. The increased Ca^{2+} concentration would cause the cortical granules to fuse with the plasma membrane, releasing their contents and causing a fertilization envelope to form, even though no sperm had entered. This would prevent fertilization.
2. During the cleavage stage in frogs and many other animals, the cell cycle is modified so that it lacks G_1 and G_2, the growth phases. As a result, the early cleavage divisions divide the zygote's cytoplasm into many smaller cells; the embryo's size thus remains the same.
3. Gastrulation organizes the embryo's cells into three tissue layers: ectoderm on the outside, endoderm on the inside, and mesoderm between them.

4. Gastrulation involves global rearrangement of cells in the embryo, generating three tissue layers. Organogenesis involves local changes in cell position and cell shape.

Concept Check 47.2

1. Microtubules elongate, lengthening the cell along one axis, while microfilaments oriented crosswise at one end of the cell contract, making that end smaller and the whole cell wedge-shaped.
2. The cells of the notochord migrate toward the midline of the embryo, rearranging themselves so there are fewer cells across the notochord, which thus becomes longer overall (see Figure 47.20).

Concept Check 47.3

1. Once the first two axes are specified, the third one is automatically determined. (Think of your own body: If you know where your head and feet are and where your left and right sides are, you automatically know where your front and back are.)
2. A second embryo probably would not form, since the cells of a late gastrula, including the ventral cells, are already determined and cannot change their fate, even if an organizer is present.
3. A second embryo could develop because inhibiting BMP-4 activity would have the same effect as transplanting an organizer.

Self-Quiz

1. a	2. b	3. e	4. c	5. a
6. c	7. b	8. c	9. e	10. d

CHAPTER 48

Concept Check 48.1

1. (a) Sensory neuron → interneuron → motor neuron (b) Interneuron
2. Transmitting information, as the axon transmits away from the cell body
3. The axons in the CNS would not have myelin sheaths.

Concept Check 48.2

1. $E_X = 62 \text{ mV} \log(10/100) = -62 \text{ mV}$
2. A decrease in permeability to K^+, an increase in permeability to Na^+, or both
3. Ligand-gated ion channels open when a specific chemical binds to the channel, whereas voltage-gated ion channels open when the membrane potential changes.

Concept Check 48.3

1. A graded potential has a magnitude that varies with stimulus strength, whereas an action potential has an all-or-none magnitude that is independent of stimulus strength.
2. The maximum frequency would decrease.
3. c, a, b

Concept Check 48.4

1. Chemical synapses would be most affected because they require the influx of Ca^{2+} into the synaptic terminal for neurotransmitter release.
2. These poisons would prolong the EPSPs that acetylcholine produces.
3. It can bind to different types of receptors, each triggering a specific response in postsynaptic cells.

Concept Check 48.5

1. The sympathetic division, which mediates the "fight-or-flight" response in stressful situations
2. Functions include controlling breathing, heart and blood vessel activity, swallowing, vomiting, and digestion and coordinating large-scale body movements such as walking.
3. A part of the reticular formation, the reticular activating system, acts as a sensory filter, selecting which information reaches the cerebral cortex. The thalamus sorts information from all the senses and sends it to the appropriate cerebral centers for further processing.

Concept Check 48.6

1. More sensory neurons innervate the hand than the neck. This conclusion is supported by the fact that the cortical surface area devoted to the hand is larger than that devoted to the neck.
2. Each cerebral hemisphere is specialized for different parts of this task—the right for face recognition and the left for language. Without an intact corpus callosum, neither hemisphere can take advantage of the other's processing abilities.
3. Broca's area, which is active during the generation of speech, is located near the part of the primary motor cortex that controls muscles in the face. Wernicke's area, which is active when speech is heard, is located near the part of the temporal lobe that is involved in hearing.
4. Direct: The receptor is part of an ion channel; when glutamate binds, Ca^{2+} diffuses through the channel. Indirect: Ca^{2+} influx through the channel activates signal transduction pathways that produce long-lasting changes.

Concept Check 48.7

1. Without Netrin-1 receptors, the axons of interneurons would not be attracted toward the floor plate and might grow randomly through the spinal cord. Without Slit receptors, the axons would not be repelled by the floor plate and might grow back across the midline.
2. In all three diseases, identical twins have about a 50% chance of sharing the disease, which implies that genetic and environmental components are of nearly equal importance. Also, there is evidence that stress may be an environmental factor in bipolar disorder and major depression.
3. Both are progressive brain diseases whose risk increases with advancing age. Both result from the death of brain neurons and are associated with the accumulation of peptide or protein aggregates.

Self-Quiz

1. c	2. b	3. c	4. a	5. c
6. a	7. d	8. d	9. b	10. e

CHAPTER 49

Concept Check 49.1

1. Such drugs block sensations transmitted by sensory neurons that synapse with receptors. Sensory neurons that *are* receptors (such as stretch receptors) transmit sensations without a synapse, so they are unaffected by such drugs.
2. Pain and light touch should be affected first because they involve receptors in the epidermis. Senses related to hair movement, strong pressure, and vibrations are affected last or not at all because they involve receptors deep in the dermis.

Concept Check 49.2

1. Statocysts detect the animal's orientation with respect to gravity, providing information that is essential in environments such as these, where light cues are absent.
2. The stapes and the other middle ear bones transmit vibrations from the tympanic membrane to the oval window. Fusion of these bones blocks this transmission, resulting in hearing loss.
3. As a sound that changes gradually from a very low to a very high pitch

Concept Check 49.3

1. As a fly walks on an object, gustatory sensilla on the feet determine whether the object contains food-related molecules, such as sugars. Sensilla on the mouthparts also detect these molecules when the fly begins to feed.
2. Both chemoreceptors have receptor proteins in their plasma membrane that bind certain substances, leading to membrane depolarization through a signal transduction pathway involving cAMP. In gustatory chemoreceptors for sweetness, depolarization triggers neurotransmitter release; in olfactory chemoreceptors, it triggers action potential production.

Concept Check 49.4

1. Planarians have ocelli that cannot form images but can sense the intensity and direction of light, providing enough information to enable the animals to find protection in shaded places. Flies have compound eyes that form images and excel at detecting movement; both are adaptations for more active behaviors, including flight.
2. The person would be able to focus on distant but not close objects (without glasses) because close focusing requires the lens to become almost spherical.
3. When illuminated, the photoreceptor hyperpolarizes, stopping its release of glutamate. If bipolar cells that it synapses with are hyperpolarized by glutamate, they will depolarize in the light and release more neurotransmitter. If that neurotransmitter is excitatory, the ganglion cells they synapse with will produce action potentials at a higher frequency.

Concept Check 49.5

1. The earthworm would elongate each body segment by contracting all of its circular muscles and relaxing all of its longitudinal muscles.

Answers

2. The exoskeleton on its gripping surfaces is hardened with calcium salts and with organic compounds that cross-link proteins. The exoskeleton on its joints lacks these compounds and salts and therefore is flexible.
3. The hinge joint between the humerus and ulna allows the ulna and radius to move toward or away from the humerus in a single plane. The pivot joint between the ulna and radius allows the radius to rotate.

Concept Check 49.6

1. During contraction, the Z lines, which mark the ends of a sarcomere, move closer together; the A band, which marks the length of the thick filaments, stays the same length; and both the I band, which contains only thin filaments, and the H zone, which contains only thick filaments, shrink.
2. By causing all of the motor neurons that innervate the muscle to generate action potentials at a rate high enough to produce tetanus in all of the muscle fibers
3. In a skeletal muscle fiber, calcium ions bind to the troponin complex, which moves tropomyosin away from the myosin-binding sites on actin and allows cross-bridges to form. In a smooth muscle cell, calcium ions bind to calmodulin, which activates an enzyme that phosphorylates the myosin head.

Concept Check 49.7

1. The main problem in swimming is drag; a fusiform body minimizes drag. The main problem in flying is overcoming gravity; wings shaped like airfoils provide lift, and structural adaptations reduce body mass.
2. A 1-g flyer

Self-Quiz

1. e	2. d	3. b	4. a	5. c
6. e	7. b	8. d	9. c	10. b

CHAPTER 50

Concept Check 50.1

1. *Ecology* is the scientific study of the interactions between organisms and their environment; *environmentalism* is advocacy for the environment. Ecology provides scientific understanding that can inform decision making about environmental issues.
2. Interactions in ecological time that affect the survival or reproduction of organisms could result in changes to the population's gene pool, and ultimately result in change in the population on an evolutionary time scale.
3. a. Population ecology; b. Community ecology

Concept Check 50.2

1. a. Humans could transplant a species to a new area that it could not previously reach because of a geographic barrier (dispersal change). b. Humans could change a species' biotic interactions by eliminating a predator species, such as sea urchins, from an area.
2. The sun's unequal heating of Earth's surface produces temperature variations between the warmer tropics and colder polar regions, and influences the movement of air masses and thus the distribution of moisture at different latitudes.

Concept Check 50.3

1. Oligotrophic lakes, because they tend to be nutrient-poor and oxygen-rich
2. The benthic zone lies below the photic zone, so the water is too deep for enough light to support photosynthetic organisms on the bottom.

Concept Check 50.4

1. Higher average temperature in deserts
2. Answers will vary by location, but should be based on the information and maps in Figure 50.20. How much your local area has been altered from its natural state will influence how much it reflects the expected characteristics of your biome, particularly the expected plants and animals.

Self-Quiz

1. c	2. c	3. d	4. a	5. d
6. b	7. d	8. d	9. e	10. a

CHAPTER 51

Concept Check 51.1

1. Sample response: How does the squirrel produce the sound (proximate)? Does the behavior change over the course of the squirrel's development (proximate)? Do other closely related squirrel species utter similar calls (ultimate)? How does the call affect the squirrel's reproductive fitness (ultimate)?
2. It is an example of a fixed action pattern. The proximate explanation might be that nudging and rolling are released by the sign stimulus of an object located outside the nest, and the behavior is carried to completion once initiated. The ultimate explanation might be that ensuring that eggs remain in the nest increases the chance of producing healthy offspring.

Concept Check 51.2

1. All behaviors are influenced by both "nature" (genes) and "nurture" (environment), though certain behaviors have especially strong genetic components or environmental components.
2. Sample response: In the green lacewing example, researchers isolated hybrid offspring of two lacewing species with distinct songs to eliminate the environmental (learning) component and demonstrate the genetic basis of the hybrids' song.

Concept Check 51.3

1. Using general rules to learn spatial relationships reduces the details that must be memorized to relocate objects of interest.
2. Diet (as in the *D. mojavensis* example, in which mate choice by females is influenced by the medium on which the larvae feed); social environment (as in the study of aggressive behavior in which cross-fostering mice caused them to adopt some behaviors of their foster parents' species); and learning (in which animals modify their behavior based on specific experiences).
3. Natural selection would tend to favor convergence in color pattern, since a predator learning to associate a particular pattern with a sting or bad taste would avoid all other individuals with that same color pattern, regardless of species.

Concept Check 51.4

1. Because this geographic variation corresponds to differences in prey availability between two garter snake habitats, it seems likely that snakes with characteristics enabling them to feed on the abundant prey in their particular locale would have had increased survival and reproductive success, and thus natural selection would have resulted in the divergent foraging behaviors.
2. To show that their differences in aggressiveness are due to genetic differences between the populations
3. Berthold and his colleagues concluded that migratory orientation in blackcaps has a genetic basis, since laboratory-raised offspring of both British blackcaps and German blackcaps showed different migratory orientations. Furthermore, the westward-migrating blackcaps appear to have evolved over about the last 50 years.

Concept Check 51.5

1. Certainty of paternity is higher with external fertilization.
2. Optimal foraging theory predicts that mule deer will forage in a way that incorporates the risk of predation as well as the benefit of food availability. This explains why, despite the fact that food is slightly less abundant in open areas than in forest edges or interiors, mule deer feed most frequently in open areas, where risk of predation by mountain lions is significantly lower than in forest edges.
3. If the female chooses a mate that is in fact healthy, there is an increased chance that their offspring will inherit some of these healthy genes, making them both inherently more likely to survive and reproduce as well as to be favored themselves by choosy females in the next generation.

Concept Check 51.6

1. Enhanced reproductive success of closely related individuals that have many genes in common with the altruist, including genes for altruism
2. Reciprocal altruism, the exchange of helpful behaviors for future similar behaviors, can explain cooperative behaviors between unrelated animals, though often the behavior has some potential benefit to the benefactor as well.
3. Yes, since parental behavior in these mice changes the future parental behavior of offspring, the changes in behavior produced by

cross-fostering could be passed on through cultural inheritance.

Self-Quiz

1. d	2. d	3. d	4. a	5. c
6. a	7. d	8. c	9. b	10. c

CHAPTER 52

Concept Check 52.1

1. The territorial species likely has a uniform pattern of dispersion, since the interactions between individuals will maintain constant space between them. The flocking species is probably clumped, since most individuals probably live in one of the clumps (flocks).
2. A Type III survivorship curve, since very few of the young probably survive
3. 0.5; divide the mean number of females in a litter by the mean size of litters for each age when reproduction occurs and take the average.

Concept Check 52.2

1. The constant, spring-fed stream. In more constant physical conditions, where populations are more stable and competition for resources more likely, larger, well-provisioned young have a better chance of surviving.

Concept Check 52.3

1. Though r_{max} is constant, *N*, the population size, is increasing. As r_{max} is applied to an increasingly large *N*, population growth ($r_{max}N$) steepens to produce the J-shaped curve.
2. On the new island. The first plants that found suitable habitat on the island would encounter an abundance of space, nutrients, and light. In the rain forest, competition among plants for these resources is intense.

Concept Check 52.4

1. When *N* (population size) is small, there are relatively few individuals producing offspring. When *N* is large, near the carrying capacity, per capita growth is relatively small because it is limited by available resources. The steepest part of the logistic growth curve corresponds to a population with a number of reproducing individuals that is substantial but not yet near carrying capacity.

Concept Check 52.5

1. Competition for resources and space can negatively impact population growth by limiting reproductive output. Diseases that are transmitted more easily in crowded populations can exert negative feedback on increasing population size. Some predators feed preferentially on species at higher population densities, since they are easier to find than are the members of less dense prey populations. In crowded populations, toxic metabolic wastes can build up and poison the organisms.

Concept Check 52.6

1. A bottom-heavy age structure, with a disproportionate number of young people, portends continuing growth of the population as these young people begin reproducing. In contrast, a more evenly distributed age structure predicts a more stable population size.
2. Carrying capacity is the sustainable number of people that can be supported by available resources in a country. Ecological capacity is the actual resource base of the country, as measured by hectares per person. Ecological footprint is a country's actual resource consumption, again converted to hectares per person. A country exceeds its carrying capacity if its footprint is greater than its ecological capacity.

Self-Quiz

1. c	2. c	3. c	4. d	5. d
6. e	7. c	8. d	9. c	10. d

CHAPTER 53

Concept Check 53.1

1. Interspecific competition has negative effects on both species (−/−). In predation, the predator population benefits at the expense of the prey population (+/−). Mutualism is a symbiosis in which both species benefit (+/+).
2. One of the competing species will become locally extinct because of the greater reproductive success of the more efficient competitor.
3. No. Coevolution involves adaptive responses of two species to each other. In the case of Batesian mimicry, the model species is not generally adapting to changes in the mimic.

Concept Check 53.2

1. Species richness is the number of species in the community. Relative abundance is the proportions of the community represented by the various species. Compared to a community with a very high proportion of one species, one with more even proportions of species is considered to be more diverse. Higher species richness and a more even distribution of abundance among species both contribute to higher species diversity.
2. The energetic hypothesis suggests that the length of a food chain is limited by the inefficiency of energy transfer along the chain, while the dynamic stability hypothesis proposes that long food chains are less stable than short chains. The energetic hypothesis predicts that food chains will be longer in habitats with higher primary productivity. The dynamic stability hypothesis predicts that food chains will be longer in more predictable environments.
3. Dominant species have major effects on communities owing to their high abundance or high biomass. Keystone species exert strong control on community structure by virtue of their ecological roles.
4. In top-down control, consumer organisms control community structure, particularly the abundance of primary producers. In bottom-up control, environmental factors, such as nutrients and moisture, control primary producers.

Concept Check 53.3

1. High levels of disturbance are generally so disruptive that they eliminate many species from communities, leaving the community dominated by a few tolerant species. Low levels of disturbance permit competitively dominant species to exclude other species from the community. In contrast, moderate levels of disturbance can facilitate coexistence of more species in a community by preventing competitively dominant species from becoming abundant enough to eliminate other species from the community.
2. The initial absence of soil in primary succession and its presence at the beginning of secondary succession
3. Early successional species can facilitate the arrival of other species in many ways, including increasing the fertility or water-holding capacity of soils or providing shelter to seedlings from wind and intense sunlight.

Concept Check 53.4

1. Larger areas are thought to support more species mainly because larger areas include a greater diversity of habitats.
2. Ecologists propose that the greater species richness of tropical regions is the result of their longer evolutionary history and the greater solar energy input and water availability in tropical regions.
3. Immigration of species to islands declines with distance from the mainland and increases with island area. Extinction of species is lower on larger islands and on less isolated islands. Since the number of species on islands is largely determined by the difference between rates of immigration and extinction, the number of species will be highest on large islands near the mainland and lowest on small islands far from the mainland.

Concept Check 53.5

1. The individualistic hypothesis of communities proposes that communities are assemblages of species distributed independently of other species along environmental gradients. This hypothesis is closely related to the more recent redundancy model, which proposes that most species in a community are not tightly associated with one another. The integrated hypothesis proposes that communities are highly integrated assemblages of interdependent species. This idea has been reincarnated as the rivet model.

Self-Quiz

1. c	2. d	3. b	4. c	5. d
6. c	7. b	8. d	9. b	10. c

CHAPTER 54

Concept Check 54.1

1. Energy passes (flows) through an ecosystem, entering as sunlight, moving as transfers of chemical energy in the food web, and leaving the ecosystem as heat. It is not recycled within the ecosystem.

2. The second law states that in any energy transfer or transformation, some of the energy is dissipated to the surroundings as heat. This "escape" of energy from an ecosystem is offset by the continuous influx of solar radiation; otherwise, the ecosystem would eventually run out of energy.
3. By decomposing organic material and returning vital chemical elements to the environment in inorganic form, detritivores are central to nutrient recycling. They also provide key linkages between primary producers and consumers.

Concept Check 54.2

1. Only a small fraction of solar radiation strikes photosynthetic organisms, only a portion of that fraction is of wavelengths suitable for photosynthesis, and only a portion of that radiation is converted to chemical energy by the photosynthetic organisms.
2. By manipulating the factors of interest, such as light availability, nutrient availability or soil moisture, and recording the primary producers' responses
3. Because the open ocean covers such a large proportion of Earth's surface
4. Net primary production reflects the loss of organic material due to the cellular respiration of the primary producers.

Concept Check 54.3

1. 20 J; 40%
2. Decomposer organisms (detritivores) consume most of what herbivores leave.
3. The biggest producers in most ecosystems comprise the most biomass, like plants in terrestrial ecosystems. However, in some ecosystems, such as aquatic systems supported by phytoplankton that reproduce rapidly, a relatively small standing crop of primary producers can sustain a much larger biomass of primary consumers. In these cases, the biomass pyramid is top-heavy, whereas the production pyramid is bottom-heavy.

Concept Check 54.4

1. For example, for the carbon cycle:

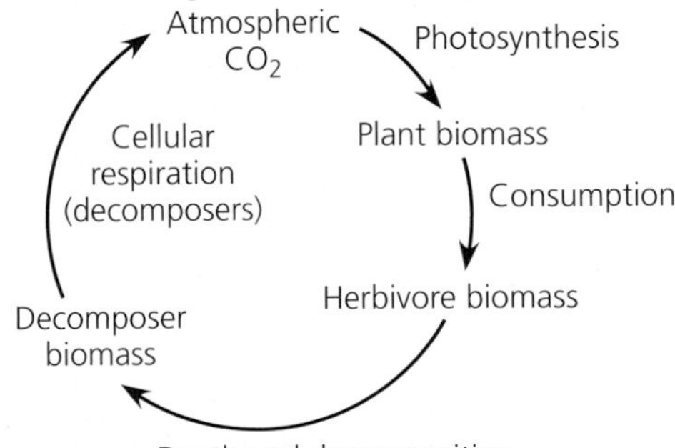

Cycling of a carbon atom

2. Removal of the trees disrupts nitrogen cycling in the forest, and the nitrate that accumulates in the soil runs off into the streams.

Concept Check 54.5

1. Adding nutrients causes population explosions of algae and the organisms that feed on them. Increased respiration by algae and consumers, including detritivores, depletes the lake's oxygen, which the fish need to survive.
2. Without the growing trees to assimilate minerals from the soil, more of the minerals run off and end up polluting nearby water resources.
3. At a lower trophic level, because biological magnification increases the concentration of toxins up the food chain
4. Thawing of these frozen soils could lead to decomposition of the organic matter, further increasing the carbon dioxide in the atmosphere.

Self-Quiz

1. c	2. e	3. d	4. e	5. d
6. a	7. b	8. c	9. d	10. c

CHAPTER 55

Concept Check 55.1

1. In addition to species loss, the biodiversity crisis includes the loss of genetic diversity within populations and species, and the degradation of entire ecosystems.
2. Habitat destruction, such as deforestation, channelizing of rivers, or conversion of natural ecosystems to agriculture or cities, deprives species of places to live. Introduced species, which are transported by humans to regions outside of their native range where they are not controlled by their natural pathogens or predators, often reduce the population sizes of native species through competition or predation. Overexploitation has reduced populations of plants and animals or driven them to extinction. Disruption of interaction networks, such as food webs or networks of mutualism, threatens species that depend on interactions, such as pollination, for their survival.
3. The benefits of biodiversity to humans include the potential use of genetic diversity to improve crop qualities; the capacity of threatened species to provide food, fibers, or medicines for human use; and ecosystem services.

Concept Check 55.2

1. Reduced genetic variation decreases the capacity of a population to evolve in the face of change.
2. The small-population approach focuses on management that increases the genetic diversity within small populations and on increasing effective population size. The declining-population approach emphasizes improvement of environmental conditions, such as habitat destruction, pollution, or overexploitation, that may be the source of a population's decline.
3. The effective population size is almost always smaller than the total population size, because effective population size is based on only the numbers of males and females that actually breed.

Concept Check 55.3

1. A small area supporting an exceptionally large number of endemic species, as well as a disproportionate number of endangered and threatened species.
2. Zoned reserves may provide sustained supplies of forest products, water, hydroelectric power, educational opportunities, and income from ecotourism.
3. Habitat corridors could increase the rate of movement between habitat patches and thus the rate of gene flow between subpopulations. However, corridors may also speed the transmission of disease in some endangered species.

Concept Check 55.4

1. The main goal is to restore degraded ecosystems to a more natural state. Specific examples include removing toxic metals from soils polluted by mining, cleaning up oil spills, and adding nutrients to soils depleted of nutrients after deforestation.
2. Bioremediation uses organisms, generally prokaryotes, fungi, or plants, to detoxify or remove pollutants from ecosystems. Biological augmentation uses organisms, such as nitrogen-fixing plants, to add essential materials to degraded ecosystems.

Concept Check 55.5

1. Sustainable development is an approach to development that works toward the long-term prosperity of human societies and the ecosystems that support them, which requires linking the biological sciences with the social sciences, economics, and humanities.
2. Biophilia, our sense of connection to nature and other forms of life, may act as a significant motivation for the development of an environmental ethic that resolves not to allow species to become extinct or ecosystems to be destroyed. Such an ethic is necessary if we are to become more attentive and effective custodians of the environment.

Self-Quiz

1. c	2. b	3. d	4. d	5. d
6. e	7. b	8. c	9. a	10. b

APPENDIX B The Metric System

Measurement	Unit and Abbreviation	Metric Equivalent	Metric-to-English Conversion Factor	English-to-Metric Conversion Factor
Length	1 kilometer (km)	= 1000 (10^3) meters	1 km = 0.62 mile	1 mile = 1.61 km
	1 meter (m)	= 100 (10^2) centimeters = 1000 millimeters	1 m = 1.09 yards 1 m = 3.28 feet 1 m = 39.37 inches	1 yard = 0.914 m 1 foot = 0.305 m
	1 centimeter (cm)	= 0.01 (10^{-2}) meter	1 cm = 0.394 inch	1 foot = 30.5 cm 1 inch = 2.54 cm
	1 millimeter (mm)	= 0.001 (10^{-3}) meter	1 mm = 0.039 inch	
	1 micrometer (µm) (formerly micron, µ)	= 10^{-6} meter (10^{-3}mm)		
	1 nanometer (nm) (formerly millimicron, µm)	= 10^{-9} meter (10^{-3}µm)		
	1 angstrom (Å)	= 10^{-10} meter (10^{-4}µm)		
Area	1 hectare (ha)	= 10,000 square meters	1 ha = 2.47 acres	1 acre = 0.405 ha
	1 square meter (m^2)	= 10,000 square centimeters	1 m^2 = 1.196 square yards 1 m^2 = 10.764 square feet	1 square yard = 0.8361 m^2 1 square foot = 0.0929 m^2
	1 square centimeter (cm^2)	= 100 square millimeters	1 cm^2 = 0.155 square inch	1 square inch = 6.4516 cm^2
Mass	1 metric ton (t)	= 1000 kilograms	1 t = 1.103 tons	1 ton = 0.907 t
	1 kilogram (kg)	= 1000 grams	1 kg = 2.205 pounds	1 pound = 0.4536 kg
	1 gram (g)	= 1000 milligrams	1 g = 0.0353 ounce 1 g = 15.432 grains	1 ounce = 28.35 g
	1 milligram (mg)	= 10^{-3} gram	1 mg =approx. 0.015 grain	
	1 microgram (µg)	= 10^{-6} gram		
Volume (solids)	1 cubic meter (m^3)	= 1,000,000 cubic centimeters	1m^3 = 1.308 cubic yards 1 m^3 = 35.315 cubic feet	1 cubic yard = 0.7646 m^3 1 cubic foot = 0.0283 m^3
	1 cubic centimeter (cm^3 or cc)	= 10^{-6} cubic meter	1 cm^3 = 0.061 cubic inch	1 cubic inch = 16.387 cm^3
	1 cubic millimeter (mm^3)	= 10^{-9} cubic meter (10^{-3} cubic centimeter)		
Volume (liquids and gases)	1 kiloliter (kl or kL)	= 1000 liters	1 kL = 264.17 gallons	1 gallon = 3.785 L
	1 liter (L)	= 1000 milliliters	1 L = 0.264 gallons 1 L = 1.057 quarts	1 quart = 0.946 L
	1 milliliter (mL)	= 10^{-3} liter = 1 cubic centimeter	1 mL = 0.034 fluid ounce 1 mL = approx. ¼ teaspoon 1 mL = approx. 15 – 16 drops (gtt.)	1 quart = 946 mL 1 pint = 473 mL 1 fluid ounce = 29.57 mL 1 teaspoon = approx. 5 mL
	1 microliter (µl or µL)	= 10^{-6} liter (10^{-3} milliliters)		
Time	1 second (s)	= 1/60 minute		
	1 millisecond (ms)	= 10^{-3} second		
Temperature	Degrees Celsius (°C) (Absolute zero, when all molecular motion ceases, is −273°C. The Kelvin [K] scale, which has the same size degrees as Celsius, has its zero point at absolute zero. Thus, 0°K = −273°C.)		°F = ⁹⁄₅°C + 32	°C = ⁵⁄₉(°F − 32)

APPENDIX C A Comparison of the Light Microscope and the Electron Microscope

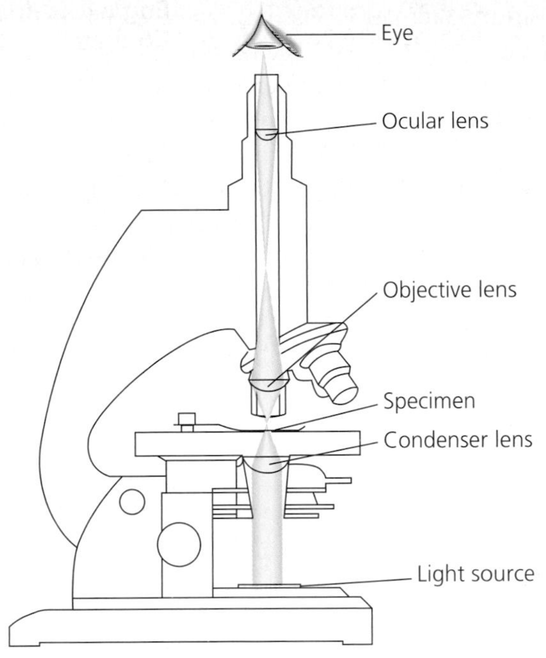

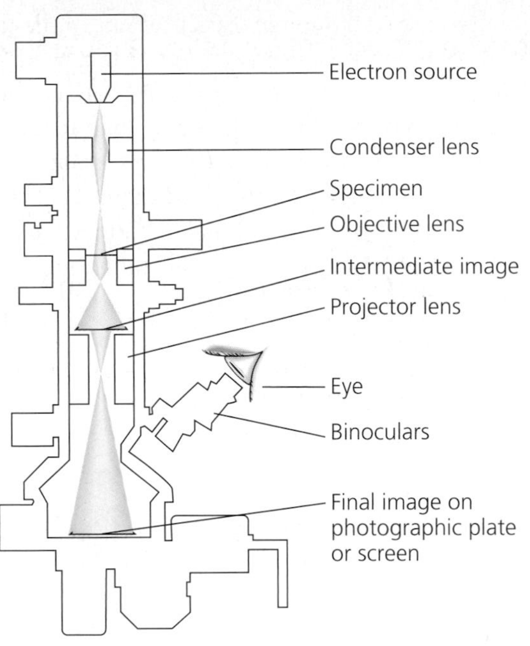

Light Microscope

In light microscopy, light is focused on a specimen by a glass condenser lens; the image is then magnified by an objective lens and an ocular lens, for projection on the eye or on photographic film.

Electron Microscope

In electron microscopy, a beam of electrons (top of the microscope) is used instead of light, and electromagnets are used instead of glass lenses. The electron beam is focused on the specimen by a condenser lens; the image is magnified by an objective lens and a projector lens for projection on a screen or on photographic film.

APPENDIX D Classification of Life

This appendix presents a taxonomic classification for the major extant groups of organisms discussed in this text; not all phyla are included. The classification presented here is based on the three-domain system, which assigns the two major groups of prokaryotes, archaea and bacteria, to separate domains (with eukaryotes making up the third domain). This classification contrasts with the traditional five-kingdom system, which groups all prokaryotes in a single kingdom, Monera (see Chapter 26).

Various alternative classification schemes are discussed in Unit Five of the text. The taxonomic turmoil includes debates about the number and boundaries of kingdoms. In this review, asterisks (*) indicate "candidate kingdoms," major clades of prokaryotes and protists that many systematists have elevated to the kingdom level. Other debates concern the alignment of the Linnaean classification hierarchy with the findings of modern cladistic analysis. Daggers (†) indicate currently recognized phyla thought by some systematists to be paraphyletic.

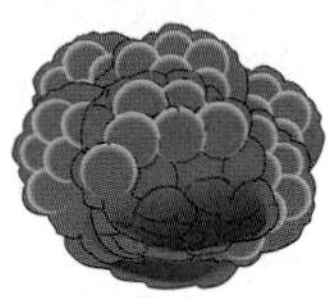

DOMAIN ARCHAEA

***Kingdom Euryarchaeota**

***Kingdom Crenarchaeota**

***Kingdom Korarchaeota**

***Kingdom Nanoarchaeota**

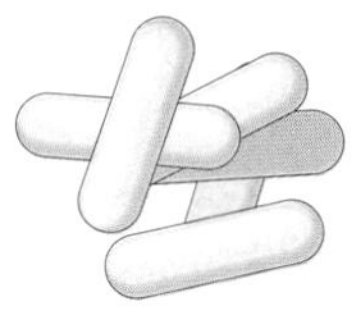

DOMAIN BACTERIA

***Kingdom Proteobacteria**

***Kingdom Gram-Positive Bacteria**

***Kingdom Cyanobacteria**

***Kingdom Spirochetes**

***Kingdom Chlamydia**

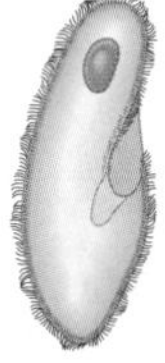

DOMAIN EUKARYA

The five-kingdom classification scheme unites all the eukaryotes generally called protists in a single kingdom, Protista. This book adopts the cladistic argument for dividing the protists into several "candidate kingdoms" (*). This review also includes some protistan groups of less certain phylogeny (see Chapter 28).

***Kingdom Parabasala** (parabasalids)

***Kingdom Diplomonadida** (diplomonads)

***Kingdom Euglenozoa**
- Phylum Euglenophyta (euglenids)
- Phylum Kinetoplastida (kinetoplastids)

***Kingdom Alveolata**
- Phylum Dinoflagellata (dinoflagellates)
- Phylum Apicomplexa (apicomplexans)
- Phylum Ciliophora (ciliates)

***Kingdom Stramenopila**
- Phylum Phaeophyta (brown algae)
- Phylum Oomycota (water molds)
- Phylum Chrysophyta (golden algae)
- Phylum Bacillariophyta (diatoms)

***Kingdom Rhodophyta** (red algae)

***Kingdom Chlorophyta** (green algae: chlorophytes and charophyceans, which some biologists now place with plants in the kingdom Viridiplantae)

***Kingdom Amoebozoa**
- Phylum Myxogastrida (plasmodial slime molds)
- Phylum Dictyostelida (cellular slime molds)
- Phylum Gymnamoeba (gymnamoebas)
- Phylum Entamoeba (entamoebas)

Protists of less certain phylogenetic affinities:
- Phylum Radiolaria (radiolarians)
- Phylum Foraminifera (forams)

Kingdom Plantae

- Phylum Hepatophyta (liverworts) } Bryophytes (nonvascular plants)
- Phylum Anthocerophyta (hornworts) } Bryophytes (nonvascular plants)
- Phylum Bryophyta (mosses) } Bryophytes (nonvascular plants)
- Phylum Lycophyta (lycophytes) } Seedless vascular plants
- Phylum Pterophyta (ferns, horsetails, whisk ferns) } Seedless vascular plants
- Phylum Ginkgophyta (ginkgo) } Gymnosperms } Seed plants
- Phylum Cycadophyta (cycads) } Gymnosperms } Seed plants
- Phylum Gnetophyta (gnetophytes) } Gymnosperms } Seed plants
- Phylum Coniferophyta (conifers) } Gymnosperms } Seed plants
- Phylum Anthophyta (flowering plants) } Angiosperms } Seed plants

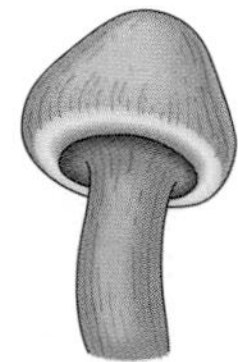

Kingdom Fungi

- †Phylum Chytridiomycota (chytrids)
- †Phylum Zygomycota (zygomycetes)
- Phylum Glomeromycota (glomeromycetes)
- Phylum Ascomycota (sac fungi)
- Phylum Basidiomycota (club fungi)

Kingdom Animalia

- †Phylum Porifera (sponges)
- Phylum Cnidaria (cnidarians)
  - Class Hydrozoa (hydrozoans)
  - Class Scyphozoa (jellies)
  - Class Cubozoa (box jellies and sea wasps)
  - Class Anthozoa (sea anemones and most corals)
- Phylum Placozoa (placozoans)
- Phylum Kinorhyncha (kinorhynchs)
- Phylum Platyhelminthes (flatworms)
  - Class Turbellaria (free-living flatworms)
  - Class Trematoda (flukes)
  - Class Monogenea (monogeneans)
  - Class Cestoda (tapeworms)
- Phylum Nemertea (proboscis worms)
- Phylum Ectoprocta (ectoprocts) } Lophophorates
- Phylum Phoronida (phoronids) } Lophophorates
- Phylum Brachiopoda (brachiopods) } Lophophorates
- Phylum Rotifera (rotifers)
- Phylum Mollusca (molluscs)
  - Class Polyplacophora (chitons)
  - Class Gastropoda (gastropods)
  - Class Bivalvia (bivalves)
  - Class Cephalopoda (cephalopods)
- Phylum Acanthocephala (spiny-headed worms)
- Phylum Ctenophora (comb jellies)
- Phylum Loricifera (loriciferans)
- Phylum Priapula (priapulans)
- Phylum Annelida (segmented worms)
  - Class Oligochaeta (oligochaetes)
  - Class Polychaeta (polychaetes)
  - Class Hirudinea (leeches)
- Phylum Nematoda (roundworms)
- Phylum Arthropoda (This survey groups arthropods into a single phylum, but some zoologists now split the arthropods into multiple phyla.)
  - Subphylum Cheliceriformes (horseshoe crabs, arachnids)
  - Subphylum Myriapoda (millipedes, centipedes)
  - Subphylum Hexapoda (insects, springtails)
  - Subphylum Crustacea (crustaceans)
- Phylum Cycliophora (cycliophorans)
- Phylum Tardigrada (tardigrades)
- Phylum Onychophora (velvet worms)
- Phylum Hemichordata (hemichordates)
- Phylum Echinodermata (echinoderms)
  - Class Asteroidea (sea stars)
  - Class Ophiuroidea (brittle stars)
  - Class Echinoidea (sea urchins and sand dollars)
  - Class Crinoidea (sea lilies)
  - Class Concentricycloidea (sea daisies)
  - Class Holothuroidea (sea cucumbers)
- Phylum Chordata (chordates)
  - Subphylum Urochordata (urochordates: tunicates)
  - Subphylum Cephalochordata (cephalochordates: lancelets)
  - Subphylum Craniata (craniates)
    - Class Myxini (hagfishes)
    - Class Cephalaspidomorphi (lampreys) } Vertebrates
    - Class Chondrichthyes (sharks, rays, chimaeras) } Vertebrates
    - Class Actinopterygii (ray-finned fishes) } Vertebrates
    - Class Actinistia (coelacanths) } Vertebrates
    - Class Dipnoi (lungfishes) } Vertebrates
    - Class Amphibia (amphibians) } Vertebrates
    - Class Reptilia (tuatara, lizards, snakes, turtles, crocodilians, birds) } Vertebrates
    - Class Mammalia (mammals) } Vertebrates

Credits

This section constitutes a continuation of the copyright page.

PHOTO CREDITS
Cover Image Linda Broadfoot

Unit Opening Interviews
Unit I Paul Ossa, Benjamin Cummings Publishing; **Unit II** Sam Kittner; **Unit III** Justin Allardyce Knight; **Unit IV** Angela Elbern, Benjamin Cummings Publishing; **Unit V** Lee W. Wilcox, Benjamin Cummings Publishing; **Unit VI** Steve Walag, Benjamin Cummings Publishing; **Unit VII** Les Todd, Duke University Photography; **Unit VIII** John Sherman, Benjamin Cummings Publishing

Detailed Contents
p. xxv Wolfgang Kaehler/Liaison; **p. xxvi** Lennart Nilsson/Albert Bonniers Forlag **p. xxviii** Andrew Syred/Photo Researchers; **p. xxxi** Steve P. Hopkin/Taxi; **p. xxxii** Garry McCarthy; **p. xxxv** Robert Holmes/CORBIS; **p. xxxvi** Dave Watts/NHPA/Photo Researchers; **p. xxxix** Gerry Ellis/Minden Pictures

Chapter 1
1.1 Linda Broadfoot; **1.2a** John Foxx/Image State; **1.2b** Fred Bavendam/Minden Pictures; **1.2c** Dorling Kindersley; **1.2d** Joe McDonald/CORBIS; **1.2e** Michael & Patricia Fogden/CORBIS; **1.2f** Frans Lanting/Minden Pictures; **1.2g** Frans Lanting/Minden Pictures; **1.3.1** WorldSat International/Photo Researchers; **1.3.2** Yann Arthus-Bertrand/CORBIS; **1.3.3** Gary Carter/Visuals Unlimited; **1.3.4** Michael Orton/Photographer's Choice; **1.3.5** Carol Fuegi/CORBIS; **1.3.6** Photodisc; **1.3.7** Jeremy Burgess/SPL/Photo Researchers; **1.3.8** Manfred Kage/Peter Arnold; **1.3.9** E. H. Newcomb & W.P.Wergin/ Biological Photo Service; **1.3.10** Benjamin Cummings; **1.4** Photodisc; **1.5 both** Conly Rieder; **1.6** Camille Tokerud/Stone; **1.7** Photodisc; **1.8 left** Dr. D. W. Fawcett/Visuals Unlimited; **1.8 right** S. C. Holt, University of Texas Health Center/Biological Photo Service; **1.9** David Parker/SPL/Photo Researchers; **1.13** Charles H. Phillips; **1.15 top left** Oliver Meckes/Nicole Ottawa/Photo Researchers; **1.15 bottom left** Ralph Robinson/Visuals Unlimited; **1.15 top middle** D. P. Wilson/Photo Researchers; **1.15 top right** Konrad Wothe/Minden Pictures; **1.15 bottom middle** Peter Lilja/Taxi; **1.15 bottom right** Anup Shah/Nature Picture Library; **1.16 top** VVG/SPL/Photo Researchers; **1.16 bottom** OMIKRON/Photo Researchers; **1.16 middle** W. L. Dentler, University of Kansas/Biological Photo Sevices; **1.17** Mike Hettwer; **1.18** American Museum of Natural History; **1.19 top** Hal Horwitz, CORBIS; **1.19 middle and bottom** Dorling Kindersley; **1.22** Dorling Kindersley; **1.24 bottom** Karl Ammann/CORBIS; **1.24 top** Dorling Kindersley; **1.26 bottom** Hans Pfletschinger/Peter Arnold; **1.26 top** John Alcock/Visuals Unlimited; **1.27 top and bottom** Breck P. Kent; **1.27 middle** E. R. Degginger/Photo Resarchers; **1.28 both** David Pfennig; **1.31** Don Hammerman/NYU; **1.32** Stone; **Table 1.1 left** Dr. D. W. Fawcett/Visuals Unlimited; **Table 1.1 right** S. C. Holt, University of Texas Health Center/Biological Photo Service; **Table 1.2** Photodisc; **Table 1.5** Photodisc; **Table 1.6** Michael & Patricia Fogden/CORBIS; **Table 1.7 left** Hal Horwitz, CORBIS; **Table 1.7 right** Dorling Kindersley; **Table 1.9** Dorling Kindersley; **Table 1.10** Karl Ammann/CORBIS; **Table 1.11** Stone

Chapter 2
2.1 Thomas Eisner; **2.2 left** Chip Clark; **2.2 middle & right** Benjamin Cummings; **2.3a** Grant Heilman Photography; **2.3b** Ivan Polunin/Bruce Coleman; **2.5** Terraphotographics/Biological Photo Service; **2.6** CTI; **2.14** Benjamin Cummings; **p42** Dorling Kindersley; **2.18** Runk/Schoenberger/Grant Heilman Photography; **p46** E. R. Degginger/Color-Pic

Chapter 3
3.1 NASA; **3.3 left** PhotoDisc; **3.3 right** Richard Kessel & Gene Shih/Visuals Unlimited; **3.4** George Bernard/Animals Animals; **3.5** Flip Nicklin/Minden Pictures; **3.9** Oliver Strewe/Stone

Chapter 4
4.1 Gerry Ellis/Minden Pictures; **4.2** Roger Ressmeyer/CORBIS; **4.6** Manfred Kage/Peter Arnold; **4.9 both** Digital Vision

Chapter 5
5.1 Lester Lefkowitz/CORBIS; **5.6a** John N. A. Lott, McMaster University/Biological Photo Service; **5.6b** H. Shio and P.B. Lazarow; **5.8** J. Litray/Visuals Unlimited; **5.9 left** Jeremy Woodhouse/PhotoDisc; **5.9 right** T.J. Beveridge/Visuals Unlimited; **5.10b** F. Collet/Photo Researchers; **5.10c** CORBIS; **5.12a** Dorling Kindersley; **5.12b** Photodisc Green; **5.20** Wolfgang Kaehler/Liaison; **5.21 both** Eye of Science/Photo Researchers; **5.23** P. B. Sigler from Z. Xu, A. L. Horwich, and P. B. Sigler, Nature (1997) 388:741-750. ©1997 Macmillan Magazines, Ltd.; **5.24a** Marie Green, University of California, Riverside

Chapter 6
6.1 Molecular Probes; **6.3a** Biophoto Associates/Photo Researchers; **6.3b** Ed Reschke; **6.3c-d** David M. Phillips/Visuals Unlimited; **6.3e** Molecular Probes; **6.3f both** Karl Garsha; **6.4a-b** William L. Dentler, University of Kansas/Biological Photo Service; **6.6** S. C. Holt, University of Texas Health Center/Biological Photo Service; **6.8** Daniel Friend; **6.10 top left** From L. Orci and A. Perelet, Freeze-Etch Histology. (Heidelberg: Springer-Verlag, 1975) © 1975 Springer-Verlag; **6.10 bottom left** From A. C. Faberge, Cell Tiss. Res. 151 (1974):403. © 1974 Springer-Verlag; **6.10 right** U. Aebi et al. Nature 323 (1996):560-564, figure 1a. Used with permission; **6.11** D. W. Fawcett/Photo Researchers; **6.12** R. Bolender, D. Fawcett/Photo Researchers; **6.13** Don Fawcett/Visuals Unlimited; **6.14a** R. Rodewald, University of Virginia/Biological Photo Service; **6.14b** Daniel S. Friend, Harvard Medical School; **6.15** E. H. Newcomb; **6.17** Daniel S. Friend, Harvard Medical School; **6.18** W. P. Wergin and E. H. Newcomb, University of Wisconsin, Madison/Biological Photo Service; **6.19** From S. E. Fredrick and E. H. Newcomb, The Journal of Cell Biology 43 (1969):343. Provided by E. H. Newcomb; **6.20** John E. Heuser, Washington University School of Medicine, St. Louis, MO; **6.21** B. J. Schapp et al., 1985, Cell 40:455; **Table 6.1 left** Mary Osborn, Max Planck Institute; **middle** Frank Solomon and J. Dinsmore, Massachusetts Institute of Technology; **right** Mark S. Ladinsky and J. Richard McIntosh, University of Colorado; **6.22** Kent McDonald; **6.23a** Biophoto Associates/Photo Researchers; **6.23b** Oliver Meckes & Nicole Ottawa/Photo Researchers; **6.24a** OMIKRON/Science Source/Photo Researchers; **6.24b-c** W. L. Dentler, University of Kansas/Biological Photo Service; **6.26** From Hirokawa Nobutaka, The Journal of Cell Biology 94 (1982):425 by copyright permission of The Rockefeller University Press; **6.28** G. F. Leedale/Photo Researchers; **6.30** W. P. Wergin, provided by E. H. Newcomb; **6.31 top** From Douglas J. Kelly, The Journal of Cell Biology 28 (1966):51 by copyright permission of The Rockefeller University Press; **6.31 middle** From L. Orci and A. Perrelet, Freeze-Etch Histology. (Heidelberg: Springer-Verlag, 1975) copyright 1975 Springer-Verlag; **6.31 bottom** From C. Peracchia and A. F. Dulhunty, The Journal of Cell Biology 70 (1976):419 by copyright permission of The Rockefeller University Press; **6.32** Lennart Nilsson/Albert Bonniers Forlag AB.

Chapter 7
7.4 both D. W. Fawcett/Photo Researchers; **7.14a-b** Cabisco/Visuals Unlimited; **7.20 top** R. N. Band and H. S. Pankratz, Michigan State University/Biological Photo Service; **7.20 middle** D. W. Fawcett/Photo Researchers; **7.20 bottom, both** M. M. Perry and A. B. Gilbert, J. Cell Science 39 (1979) 257. Copyright 1979 by The Company of Biologists Ltd.

Chapter 8
8.1 Jean-Marie Bassot/Photo Researchers; **8.2** David W. Hamilton/Image Bank; **8.3a** Joe McDonald/CORBIS; **8.3b** Manoj Shah/Stone; **8.4** Brian Capon; **8.16a-b** Thomas Steitz, Yale University; **8.22** R. Rodewald, University of Virginia/Biological Photo Service

Chapter 9
9.1 Frans Lanting/Minden Pictures

Chapter 10
10.1 Bob Rowan, Progressive Image/CORBIS; **10.2a** Jim Brandenburg/Minden Pictures; **10.2b** Bob Evans/Peter Arnold; **10.2c** Michael Abbey/Visuals Unlimited; **10.2d** Susan Barns; **10.3 middle** M. Eichelberger/Visuals Unlimited; **10.3 bottom** W. P. Wergin and E. H. Newcomb, University of Wisconsin/Biological Photo Service; **10.11b** Christine L. Case, Skyline College; **10.20a** David Muench/CORBIS; **10.20b** Dave Bartruff/CORBIS

Chapter 11
11.1 CrystalGenomics

Chapter 12
12.1 J. M. Peters; **12.2a** Biophoto Associates/Photo Researchers; **12.2b** C. R. Wyttenback, University of Kansas/Biological Photo Service; **12.2c** Biophoto/Science Source/Photo Researchers; **12.3** J. M. Murray, University of Pennsylvania; **12.4** Biophoto/Photo Researchers; **12.6 all** Conly Rieder; **12.7 top** Matthew Schibler, from Protoplasma 137 (1987):29-44; **12.7 bottom** Richard Mcintosh; **12.9a** David M. Phillips/Visuals Unlimited; **12.9b** Micrograph by B. A. Palevitz. Courtesy of E. H. Newcomb, University of Wisconsin; **12.10 all** Carolina Biological Supply/Phototake; **12.17** Gunter Albrecht-Buehler, Northwestern University; **12.18a-b** Lan Bo Chen; **p 235** Carolina Biological Supply/Phototake

Chapter 13
13.1 Getty Images; **13.2** Roland Birke/OKAPIA/Photo Researchers; **13.3 top** Veronique Burger/Photo Researchers; **13.3 bottom** CNRI/SPL/Photo Researchers; **13.11** Carolina Biological/Visuals Unlimited

Chapter 14
14.1 Bettmann/CORBIS; **14.13 both** Photodisc; **14.14a, both** Photodisc; **14.14b, both** Benjamin Cummings; **14.15** Dick Zimmerman/Shooting Star International Photo Agency; **14.16** Nancy Wexler, Columbia University; **p 272** Breeder/owner: Patricia Speciale, photographer: Norma JubinVille

Chapter 15
15.1 Peter Lichter and David Ward, Science 247 (1990). Copyright 1990 American Association for the Advancement of Science; **15.3** Carolina Biological/Visuals Unlimited; **p. 282** Andrew Syred/Photo Researchers; **15.11** Dorling Kindersley; **15.13** Martin Gallardo, Universidad Austral de Chile; **15.15 left** Greenlar/The Image Works; **15.15 right** CNRI/Science Photo Library/Photo Researchers; **15.18** Ken Wagner/Phototake

Chapter 16
16.1 National Cancer Institute; **16.3** Oliver Meckes/Photo Researchers; **16.6a** Elliott & Fry, National Portrait Gallery; **16.6b** From J.D. Watson, The Double Helix, NY: Atheneum Press, 1968, p. 215. ©1968 by J.D. Watson. Courtesy of Cold Spring Harbor Laboratory Archives; **16.7c** Richard Wagner, UCSF Graphics; **16.12b** From D. J. Burks and P. J. Stambrook, The Journal of Cell Biology 77 (1978): 762 by copyright permission of The Rockefeller University Press. Photo provided by P. J. Stambrook; **16.19** Peter Lansdorp

Chapter 17
17.1 Harry Noller, UC Santa Cruz, from *Science* Vol. 291, p. 2526; **17.6** Keith V. Wood; **17.16a** Joachim Frank, Howard Hughes Medical Institute; **17.20** B. Hamkalo and O.L. Miller, Jr.; **17.22** Reprinted with permission from O. L. Miller, Jr., B. A. Hamkalo, and C. A Thomas, Jr., Science 169 (1970):392. Copyright ©1970 American Association for the Advancement of Science.

Chapter 18
18.1 Science Photo Library/Photo Researchers; **18.3** Eric Lam, Naohiro Kato & Michael Lawton. Programmed cell death, mitochondria and the plant hypersensitive response. NATURE, Vol 411, 14 June 2001, fig. 1, p. 849; **18.4a-b, d** Robley C. Williams, University of California Berkeley/Biological Photo Service; **18.4c** G. Murti/Visuals Unlimited; **18.10 all** C. Dauguet/Institute Pasteur/Photo Researchers; **18.11a** AP World Wide Photos; **18.11b** Dr. Linda Stannard, UCT/Science Photo Library/Photo Researchers; **18.12 left** Arden Sherf, Department of Plant Pathology, Cornell University; **18.12 top right** Wayside/Visuals Unlimited; **18.12 bottom right** Dennis Mayhew, California Department of Food and Agriculture; **18.17** Dennis Kunkel/Phototake

Chapter 19
19.1 Mark B. Roth and Joseph G. Gall, Department of Embryology, Carnegie Institution; **19.2a top** S. C. Holt, University of Texas, Health Science Center, San Antonio/BPS; **19.2a bottom** Courtesy of Victoria Foe; **19.2b** Barbara Hamkalo; **19.2c** From J. R. Paulsen and U. K. Laemmli, Cell 12 (1977):817-828; **19.2d both** G. F. Bahr/AFIP; **19.15 left** Associated Press/Wide World Photos; **19.15 right** Virginia Walbot, Stanford University; **19.17** O. L. Miller Jr., Department of Biology, University of Virginia

Chapter 20
20.1 Incyte Pharmaceuticals, Inc., Palo Alto, CA, from R. F. Service, Science (1998) 282:396-399, with permission from Science; **20.8** Repligen Corporation; **20.14** Incyte Pharmaceuticals, Inc., Palo Alto, CA, from R. F. Service, Science (1998) 282:396-399, with permission from Science; **20.17** Cellmark Diagnostics, Inc., Germantown, MD; **20.18** PPL Therapeutics

Chapter 21
21.1 Walter Gehring; **21.2 p412 left** N. A. Callow/NHPA/Photo Researchers; **right** Brad Mogen/Visuals Unlimited; **21.2 p413 left** Stanton Short/Jackson Laboratory; **middle** Dorling Kindersley; **right** Wally Eberhart/Visuals Unlimited; **21.3a** Carolina Biological/Visuals Unlimited; **21.3b** Hans Pfletschinger/Peter Arnold; **21.8** Richard Olsenius/National Geographic Image Collection; **21.13 both** F. R. Turner, Indiana University; **21.14a both** Wolfgang Driever; **21.14b both** Dr. Ruth Lehmann, The Whitehead Institution; **21.15** J.E. Sulston and H.R. Horvitz, Dev. Biol. 56 (1977):110-156; **21.17** Dr. Gopal Murti/Visuals Unlimited; **21.19 all** Development 127, 5245-5252 (2000). Mesenchymal cells engulf and clear apoptotic footplate cells in macrophageless PU.1 null mouse embryos. William Wood, Mark Turmaine, Roberta Weber, Victoria Camp, Richard A. Maki, Scott R. McKercher and Paul Martin; **21.20** Dwight Kuhn; **21.22 both** Elliot M. Meyerowitz, Plants Compared to Animals: The Broadest Comparative Study of Development, Science, vol. 295, Feb. 22, 2002, pp.1482-1485, fig. 2

Chapter 22
22.1 Craig Lovell/CORBIS; **22.3** CORBIS; **22.3 insets** Dorling Kindersley; **22.4** Michael S. Yamashita/CORBIS; **22.5 left** ARCHIV/Photo Researchers; **22.5 right** National Maritime Museum, London; **22.6 all** Tui De Roy/Minden Pictures; **22.8** Dorling Kindersley; **22.9** Steve P. Hopkin/Taxi; **22.10** Jack Wilburn/Earth Scenes/Animals Animals; **22.11a** Edward. S. Ross, California Academy of Sciences; **22.11b** Michael & Patricia Fogden/Minden Pictures; **22.15 left** Phototake; **22.15 right** Lennart Nilsson/Albert Bonniers Forlag AB; **22.17** Tom Van Sant/Geosphere Project, Santa Monica/Science Photo Library/Photo Researchers; **22.18** Philip Gingerich, *Discover* Magazine

Chapter 23
23.1 Chip Clark; **23.2** J. Antonovics/Visuals Unlimited; **23.3 top** Michio Hosino/Minden Pictures; **23.3 bottom** James L. Davis/ProWildlife; **23.6** Eastcott Momatiuk/Stone; **23.8** Kennan Ward/CORBIS ; **23.9a-b** H. F. Nijhout; **23.10 map** NASA; **23.10 karyotype** Janice Britton-Davidian, ISEM, UMR 5554 CNRS, Universite Montpellier II. Reprinted by permission from Nature, Vol. 403, 13 January 2000, p. 158. © 2000 Macmillan Magazines Ltd.; **23.10 mice** Dorling Kindersley; **23.14 all** Alan C. Kamil, George Holmes University; **23.15** Frans Lanting/Minden Pictures

Chapter 24
24.1 George Harrison/Grant Heilman Photography; **24.3a left** John Shaw/Tom Stack & Associates; **24.3a right** Don & Pat Valenti/Tom Stack & Associates; **24.3b top left** zefa/Masterfile; **24.3b all others** Photodisc; **24.4a** Joe McDonald/Bruce Coleman; **24.4b** Joe McDonald/CORBIS; **24.4c** Roger Barbour; **24.4d** Stephen Krasemann/Photo Researchers; **24.4e** Barbara Gerlach/Tom Stack & Associates; **24.4f** Mike Zens/CORBIS; **24.4g** Dennis Johnson, Papilio/CORBIS; **24.4h** William E. Ferguson; **24.4i** Charles W. Brown; **24.4j** Photodisc; **24.4k** Ralph A. Reinhold/Animals Animals/Earth Scenes; **24.4l** Grant Heilman/Grant Heilman Photography; **24.4m** Kazutoshi Okuno, National Institute of Agrobiological Sciences, Tsukuba; **24.6** CORBIS; **24.6 left inset** John Shaw/Bruce Coleman; **24.6 right inset** Michael Fogden/Bruce Coleman; **24.10 all** Ole Seehausen, University of Leiden; **24.11** Kevin Schafer; **24.12 all** Gerald D. Carr; **24.16a** Gary Meszaros/Visuals Unlimited; **24.16b** Tom McHugh/Photo Researchers; **24.17** Jane Burton/Bruce Coleman

Chapter 25
25.1 John Cancalosi/Peter Arnold; **25.2 left** Photodisc Green; **25.2 middle, right** Dorling Kindersley; **25.4a** Georg Gerster/Photo Researchers; **25.4b** Yva Momatiuk/John Eastcott/Minden Pictures; **25.4c** Manfred Kage/Peter Arnold; **25.4d** Chip Clark; **25.4e** Dr. Martin Lockley, University of Colorado; **25.4f** Jeff Daly/Visuals Unlimited; **25.4g** F. Latreille/Cerpolex/Cercles Polaires Expeditions

Chapter 26
26.1 Chip Clark NMNH, artist Peter Sawyer; **26.3** George Luther, University of Delaware Graduate College of Marine Studies; **26.4** F. M. Menger and Kurt Gabrielson, Emory University; **26.6** L. K. Broman/Photo Researchers; **26.11a top** John Stolz; **26.11a bottom** S. M. Awramik, University of California/BPS; **26.11b top** Mitsuaki Iwago/Minden Pictures; **26.11b bottom** S. H. Awramik, University of California/BPS, **26.12** Theodore J. Bornhorst, Michigan Technological University; **26.14** Dean Soulia, University of Massachusetts; **26.15 both** Shuhai Xiao; **26.16** Sinclair Stammers/SPL/Photo Researchers

Chapter 27
27.1 Jack Dykinga/Stone; **27.2a-b** Dr. Dennis Kunkel/Visuals Unlimited; **27.2c** Stem Jems/Photo Researchers; **27.3** Jack Bostrack/Visuals Unlimited; **27.4** Dr. Immo Rantala/Photo Researchers; **27.5** David Hasty, Fran Heyl Associates; **27.6** J.

Adler; **27.7a** S. W. Watson. © Journal of Bacteriology, American Society of Microbiology; **27.7b** N. J. Lang/University of California/Biological Photo Service; **27.8** Gopal Murti/Science Photo Library/Photo Researchers; **27.9** H. S. Pankratz, T. C. Beaman/Biological Photo Service; **27.10** Sue Barns; **27.11** Dr. Tony Brain/Science Photo Library/Photo Researchers; **27.13.1** L. Evans Roth/ Biological Photo Service; **27.13.2** Yuichi Suwa; **27.13.4 left** Phototake; **27.13.4 right** Alfred Pasieka/Peter Arnold; **27.13.5** Photo Researchers; **27.13.6** Moredon Animal Health/SPL/Photo Researchers; **27.13.7** CNRI/SPL/Photo Researchers; **27.13.8 top** Frederick P. Mertz/Visuals Unlimited; **27.13.8 bottom** David M. Phillips, Visuals Unlimited; **27.13.9** T. E. Adams/Visuals Unlimited; **27.14** Helen E. Carr/ Biological Photo Service; **27.15** Ken Lucas/Biological Photo Service; **27.16 left** Scott Camazine/Photo Researchers; **27.16 middle** David M. Phillips/Photo Researchers, Inc.; **27.16 right** Centers for Disease Control and Prevention; **27.17** Exxon Corporation

Chapter 28

28.1 M. I. Walker/Photo Researchers; **28.2a** Eric Grave/Photo Researchers; **28.2b** Eric Grave/Phototake; **28.2c** M.D. Guiry; **28.2d right** Jeremy Burgess/Photo Researchers; **28.2d left** Ann Clemens/UTEX Culture Collection of Algae, University of Texas at Austin; **28.5a** Jerome Paulin, Visuals Unlimited; **28.5b** David M. Phillips/Visuals Unlimited; **28.6** David J. Patterson/micro*scope; **28.7** Oliver Meckes/Science Source/Photo Researchers; **28.8** Michael Abbey/Visuals Unlimited; **28.9** Guy Brugerolle, Universitad Clearmont, Ferrand; **28.10** Virginia Institute of Marine Science; **28.11** Masamichi Aikawa; **28.12** M. Abbey/Visuals Unlimited; **28.13** R. R. Powers; **28.14** Fred Rhoades/Mycena Consulting; **28.15** Eric Condliffe/Visuals Unlimited; **28.16** Kent Wood/Science Source/Photo Researchers; **28.17** Biological Photo Service; **28.18** Luis A. Solorzano and Warren E. Savary; **28.19** Jeffrey Rotman/Photo Researchers; **28.20a** Biophoto Associates/Photo Researchers; **28.20b** Michael Yamashita/IPN/Aurora & Quanta Productions; **28.20c** Dorling Kindersley; **28.21** J. R. Waaland, University of Washington/Biological Photo Service; **28.22** Manfred Kage/Peter Arnold; **28.22 inset** Richard Kessel & Gene Shih/Visuals Unlimited; **28.23** Robert Brons/Biological Photo Service; **28.24 all** Akiro Kihara, Hosei University; **28.25** George Barron; **28.26** R. Calentine/Visuals Unlimited; **28.27 both** Robert Kay, MRC Cambridge; **28.28a** D. P. Wilson, Eric & David Hosking/Photo Researchers; **28.28b** M.D. Guiry; **28.28c** Gary Robinson/Visuals Unlimited; **28.29** Gerald & Buff Corsi/Visuals Unlimited; **28.30a** Manfred Kage/Peter Arnold; **28.30a inset** David J. Patterson/micro*scope; **28.30b** David L. Ballantine, Department of Marine Sciences, University of Puerto Rico; **28.30c** Laurie Campbell/NHPA; **28.31** W. L. Dentler, University of Kansas

Chapter 29

29.1 Martin Rugner/AGE fotostock; **29.2** S. C. Mueller and R. M. Brown, Jr.; **29.3a** Heather Angel/Natural Visions; **29.3b** Linda Graham; **29.5 p576 left** Ed Reschke; **right** F. A. L. Clowes; **29.5 p577 top left** Alan S. Heilman; **top right** Michael Clayton; **middle** Barry Runk/Stan/Grant Heilman Photography; **bottom left** Karen Renzaglia, Southern Illinois University; **bottom right** Linda Graham; **29.6** Charles H. Wellman; **29.8** Richard Kessel & Gene Shih/Visuals Unlimited; **29.9 top left** Runk/Schoenberger/Grant Heilman Photography; **29.9 top middle** Linda Graham; **29.9 top right** The Hidden Forest, hiddenforest.co.nz; **29.9 bottom left** The Hidden Forest, hiddenforest.co.nz; **29.9 bottom right** Tony Wharton; Frank Lane Picture Agency/CORBIS; **29.10a** Brian Lightfoot/AGE fotostock; **29.10b-c** Linda Graham; **29.10d** Chris Lisle/CORBIS; **29.14 top left** Jane Grushow/Grant Heilman Photography; **29.14 top middle** Murray Fagg, Australian National Botanic Gardens; **29.14 top right** Helga and Kurt Rasbach; **29.14 bottom left** Michael Viard/Peter Arnold; **29.14 bottom middle** Milton Rand/Tom Stack & Associates; **29.14 bottom right** Barry Runk/Stan/Grant Heilman Photography; **29.15** The Field Museum of Natural History

Chapter 30

30.1 National Museum of Natural History, Smithsonian Institution; **30.4 p594 top left** George Loun/Visuals Unlimited; **top middle** Travis Amos, Benjamin Cummings; **top right** Grant Heilman Photography; **center left** Michael and Patricia Fogden/Minden Pictures; **center right** Michael Clayton; **bottom left** Thomas Schoepke; **bottom right** Doug Sokell/Visuals Unlimited; **30.4 p595 top left** Raymond Gehman/CORBIS; **center left** William Mullins/Photo Researchers; **bottom left** David Muench/CORBIS; **bottom right** Breck P. Kent/Animals Animals/Earth Scenes; **top right** Gunter Marx Photography/CORBIS; **center right** Jaime Plaza, Wildlight; **inset** Royal Botanic Gardens Sydney; **30.8a-c, e** Dorling Kindersley; **30.8d** Bill Steele/Stone; **30.9a** C. P. George/Visuals Unlimited; **30.9b** Hans Dieter Brandl, Frank Lane Picture Agency/CORBIS; **30.9c left** Scott Camazine/Photo Researchers; **30.9c right** Dorling Kindersley; **30.11a** David Dilcher and Ge Sun; **30.11b** K. Simons and David Dilcher; **30.12 p602 top** Stephen McCabe; **bottom left** Dorling Kindersley; **bottom middle** Bob & Ann Simpson/Visuals Unlimited; **bottom right** Andrew Butler/Dorling Kindersley; **30.12 p603 left top to bottom** Eric Crichton/Dorling Kindersley; J.Dransfield; Dorling Kindersley; Terry W. Eggers/CORBIS; **30.12 right top to bottom** Ed Reschke/PeterArnold; Dorling Kindersley; Tony Wharton, Frank Lane Picture Agency/CORBIS; Dorling Kindersley; Gerald D. Carr; **30.13a** D. Wilder; **30.13b** Connie Toops; **30.13c** Merlin D. Tuttle, Bat Conservation International; **Table 30.1** Dorling Kindersley

Chapter 31

31.1 Barry Runk/Grant Heilman Photography; **31.2 top** Hans Reinhard/Taxi; **31.2 bottom** Fred Rhoades/Mycena Consulting **31.2 inset** Elmer Koneman/Visuals Unlimited; **31.4** George L. Barron; **31.6 left** Jack M. Bostrack/Visuals Unlimited; **31.6 right** David Scharf/Peter Arnold; **31.7** Stephen J. Kron, University of Chicago; **31.8** Dirk Redecker, Robin Kodner, and Linda E. Graham. Glomalean Fungi from the Ordovician. Science 15 September 2000; 289: 1920-1921; **31.10 left** John W. Taylor; **31.10 right** William Barstow, Department of Botany, University of Georgia, Athens; **31.12 bottom left** George Barron; **31.12 right** Ed Reschke/Peter Arnold; **31.12 top left, middle** Barry Runk/Stan/Grant Heilman Photography; **31.13** G. L. Barron, University of Guelph/Biological Photo Service; **31.14** Centers for Disease Control; **31.15** M. F. Brown/Biological Photo Service; **31.16a** Frank Young/Papilio/CORBIS; **31.16b** David M. Dennis/Animals Animals/Earth Scenes; **31.16c** Viard/Jacana/Photo Researchers; **31.16d** Matt Springer; **31.17** Fred Spiegel, University of Arkansas; **31.18a** Phil Dotson/Photo Researchers; **31.18b** Fletcher and Baylis/Photo Researchers; **31.18c** Konrad Wothe/Minden Pictures; **31.18d** Michael Fogden/DRK Photo; **31.19** Rob Simpson/Visuals Unlimited; **31.20** Biophoto Associates/Photo Researchers; **31.21** R. Ronacordi/Visuals Unlimited; **31.22** Mark Moffett/Minden Pictures; **31.23a** Gerald & Buff Corsi/Visuals Unlimited; **31.23b** David Sieren/Visuals Unlimited; **31.23c** Fritz Polking/Visuals Unlimited; **31.24** V. Ahmadijian/Visuals Unlimited; **31.25a** Brad Mogen/Visuals Unlimited; **31.25b** Dorling Kindersley; **31.25c** Robert Calentine/Visuals Unlimited; **31.26** Christine Case; **p625** Dr. Jeremy Burgess/The National Audubon Society Collection/Photo Researchers

Chapter 32

32.1 Jeff Hunter/The Image Bank; **32.5a** The Natural History Museum, London; **32.5b** The Natural History Museum/Dorling Kindersley; **32.6** J. Sibbick/The Natural History Museum, London; **32.12** Kent Wood/Photo Researchers; **32.13a** Carolina Biological/Visuals Unlimited

Chapter 33

33.1 C. Wolcott Henry III/National Geographic; **33.3 p639 left top to bottom:** Andrew J. Martinez/Photo Researchers; Garry McCarthy; Colin Milkins, OSF/Animals Animals; Stephen L. Dellaporta; **right top to bottom:** Robert Brons/Biological Photo Service; Reinhart Møberg Kristensen; W. I. Walker/Photo Researchers; Ron Offermans; **33.3 p640 left top to bottom:** Fred Bavendam/Peter Arnold; Darlynne A. Murawski/Peter Arnold; Robert Pickett/Alamy Images; Reinhart Møberg Kristensen; **right top to bottom:** Erling Svensen/UW Photo; Gregory G. Dimijian/Photo Researchers; Peter Batson/Image Quest; Erling Svensen/UW Photo; **33.3 p641 left top to bottom:** Reprinted with permission from A. Eizinger and R. Sommer, Max Planck Institut fur entwicklungsbiologie, Tubingen. ©2000, American Association for the Advancement of Science; Peter Funch, University of Copenhagen; Peter Funch; Thomas Stromberg; Robert Harding World Imagery/Alamy; **right top to bottom:** Tim Flach/Stone; Andrew Syred/SPL/Photo Researchers; Heather Angel/Natural Visions; Robert Brons/Biological Photo Service; **33.4** Andrew J. Martinez/Photo Researchers; **33.7a** Andrew J. Martinez/Photo Researchers; **33.7b** Robert Brons/Biological Photo Service; **33.7c** Great Barrier Reef Marine Park Authority; **33.7d** Neil G. McDaniel/Photo Researchers; **33.8** Robert Brons/Biological Photo Services; **33.9** Garry McCarthy; **33.11** Center for Disease Control; **33.12** Stanley Fleger/Visuals Unlimited; **33.13** W. I. Walker/Photo Researchers; **33.14a** Colin Milkins, Oxford Scientific Films/Animals Animals; **33.14b** Ron Offermans; **33.14c** Fred Bavendam/Peter Arnold; **33.15** Erling Svensen/UW Photo; **33.17** Jeff Foott/Tom Stack and Associates; **33.18a** Gerry Ellis/Minden Pictures; **33.18b** CORBIS; **33.20** H. W. Pratt/Biological Photo Service; **33.22a** Robert Pickett/Papilio/Alamy Images; **33.22b** Mike Severns/Tom Stack & Associates; **33.22c** Jonathan Blair/CORBIS; **33.23** A.N.T./NHPA; **33.24** Peter Batson/Image Quest Marine; **33.25** Astrid & Hanns-Frieder Michler/SPL/Photo Researchers; **33.26** Reprinted with permission from A. Eizinger and R. Sommer, Max Planck Institut fur entwicklungsbiologie, Tubingen. ©2000, American Association for the Advancement of Science; **33.27** L. S. Stepanowicz/Photo Researchers; **33.28** Chip Clark; **33.30** MIlton Tierney, Jr./Visuals Unlimited; **33.31a** Tim Flach/Stone; **33.31b** Andrew Syred/Photo Researchers; **33.31c** Gunter Ziesler/Peter Arnold; **33.33** John R. Macgregor/Peter Arnold; **33.34** Tom McHugh/Photo Researchers; **33.36 all** John Shaw/Tom Stack and Associates; **33.38a** Maximilian Weinzierl/Alamy Images; **33.38b** Peter Herring/Image Quest Marine; **33.38c** Peter Parks/Image Quest Marine; **33.40a** Fred Bavendam/Minden Pictures; **33.40b** Jeff Rotman/Photo Researchers; **33.40c** Robert Harding World Imagery/Alamy Images; **33.40d** Jurgen Freund/Nature Picture Library; **33.40e** Hal Beral/CORBIS; **33.40f** Daniel Janies

Chapter 34

34.1 Biophoto Associates/Photo Researchers; **34.4a** Robert Brons/Biological Photo Service; **34.5** Runk Schoenenberg/Grant Heilman Photography; **34.8a** Nanjing Institute of Palaeontology and Geology; **34.8b** D. G. Shu et al, Nature, vol. 421, 30 Jan. 2003; **34.9** Tom McHugh/Photo Researchers; **34.10 both** Breck P. Kent/Animals Animals/Earth Scenes; **34.15a** Carlos Villoch/Image Quest; **34.15b** Masa Ushioda/Image Quest; **34.15c** Jeff Mondragon/Mondragon Photography; **34.17a** James D. Watt/Image Quest; **34.17b** Fred Bavendam/Minden Pictures; **34.17c** Marevision/AGE Fotostock; **34.17d** Fred McConnaughey/Photo Researchers; **34.18** Tom McHugh/Photo Researchers; **34.21a** Alberto Fernandez/AGE Fotostock; **34.21b-c** Michael Fogden/Bruce Coleman; **34.22 all** Hans Pfletschinger/Peter Arnold; **34.25** Jessie Cohen, National Zoological Park, © Smithsonian Institution; **34.26 left** Department of Library Services, American Museum of Natural History; **34.26 right** John Sibbick/National Geographic Magazine; **34.27a** Doug Wechsler; **34.27b** Matt Lee; **34.27c** Michael & Patricia Fogden/Minden Pictures; **34.27d** Medford Taylor/National Geographic Image Collection; **34.27e** Carl & Ann Purcell/CORBIS; **34.28a** Stephen J. Kraseman/ DRK Photo; **34.28b** Janice Sheldon; **34.30a** Russell Mountford/Alamy Images; **34.30b** CORBIS; **34.30c** Frans Lanting/Minden Pictures; **34.30d** franzfoto.com/ Alamy Images; **34.33** Mervyn Griffiths/CSIRO; **34.33 inset** D. Parer and E. Parer Cook/Auscape; **34.34a** Dan Hadden/Ardea Ltd; **34.34b** Fritz Prenzel/Animals Animals; **34.37** Frans Lanting/Minden Pictures; **34.39a** Kevin Schafer/AGE Fotostock; **34.39b** Frans Lanting/Minden Pictures; **34.40a** Morales/AGE Fotostock; **34.40b** Anup Shah/ImageState/Alamy Images; **34.40c** T. J. Rich/Nature Photo Library; **34.40d** E. A. Janes/AGE Fotostock; **34.40e** Frans Lanting/Minden Pictures; **34.42a** Cleveland Museum of Natural History; **34.42b** John Reader/SPL/Photo Researchers; **34.42c** John Gurche; **34.43 left** Alan Walker, National Museum of Kenya; **34.43 right** Jay Matternes; **34.44** David L. Brill; **34.45** C. Henshilwood & F. d'Errico

Chapter 35

35.3 Robert & Linda Mitchell/Robert & Linda Mitchell Photography; **35.4a** James Strawser/Grant Heilman Photography; **35.4b** Dorling Kindersley; **35.4c** Drew Weiner; **35.4d** Robert Holmes/CORBIS; **35.4e** Geoff Tompkinson/Science Photo Library/Photo Researchers; **35.5a** Dorling Kindersley; **35.5b** Gusto Productions/Science Photo Library/Photo Researchers; **35.5c** Barry Runk/Stan/Grant Heilman Photography; **35.5d** Dorling Kindersley; **35.7a** Scott Camazine/Photo Researchers; **35.7b** Fritz Polking/Visuals Unlimited; **35.7c** Mike Zens/CORBIS; **35.7d** James Strawser/Grant Heilman Photography; **35.7e** Jerome Wexler/Photo Researchers; **35.9 p718 top** Brian Capon; **middle** Graham Kent/Benjamin Cummings; **bottom both** Graham Kent/Benjamin Cummings; **35.9 p719 top** Richard Kessel and Gene Shih/Visuals Unlimited; **35.9 bottom both** Graham Kent/Benjamin Cummings; **35.12** Carolina Biological Supply/Phototake; **35.13a-b top** Ed Reschke; **35.13 bottom** Carolina Biological Supply/Phototake; **35.14** Michael Clayton; **35.15** Ed Reschke; **35.16 both** Ed Reschke; **35.17b-c** Ed Reschke; **35.18 left** Michael Clayton; **35.18 right** Alison W. Roberts **35.21** Janet Braam, from Cell 60 (9 February 1990) copyright 1990 by Cell Press; **35.23 all** Susan Wick, University of Minnesota; **35.24** B. Wells and Kay Roberts; **35.25a-b** From figure 1 in B. Scheres et al, Development 121:53-62. © 1995 The Company of Biologists Ltd.; **35.25c** From figure 6c in R. Torres Ruiz and G. Jurgens, Development 120:2967-2978. © 1994 The Company of Biologists Ltd.; **35.26** From figure 1a in U. Mayer et al, Development 117 (1): 149-162. © 1993 The Company of Biologists Ltd.; **35.27 both** From figure 1 in D. Hareven et al, Cell 84 (5): 735-744. 1996 reprinted with permission of Elsevier Science; **35.28** From figure 2g in Hung et al, Plant Physiology 117:73-84. © 1998 Plant Physiology. Photo courtesy John Schiefelbein/University of Michigan; **35.29** Gerald D. Carr; **35.30 both** Elliot M Meyerowitz and John Bowman, Development 112 1991:1-2 31.2

Chapter 36

36.1 Gary Randall/Taxi; **36.7** Nigel Cattlin/Holt Studios International/Photo Researchers; **36.10** Dana Richter/Visuals Unlimited; **36.11** Scott Camazine/Photo Researchers; **36.14** Jeremy Burgess/Science Photo Library/Photo Researchers; **36.16** John D. Cunningham/Visuals Unlimited; **36.16 inset** Dorling Kindersley; **36.19 all** M. H. Zimmerman, courtesy of Professor P. B. Tomlinson, Harvard University

Chapter 37

37.1 Adam Hart-Davis/SPL/Photo Researchers; **37.4** Maurice Reece. From the Country Gentleman, courtesy of the Curtis Publishing Co.; **37.5** U. S. Department of Agriculture; **37.7** White et al, Plant Physiology, June 2003; **37.8** Kevin Horan/Stone; **37.10a** Breck P. Kent/Earth Scenes; **37.10b** E. H. Newcomb/Biological Photo Service; **37.12a both** Gerald Van Dyke/Visuals Unlimited; **37.12b** Carolina Biological Supply/Phototake NYC; **37.13 top** Wolfgang Kaehler/CORBIS; **37.13 middle left to right** James Strawser/Grant Heilman Photography; Kevin Schafer/CORBIS; Andrew Syred/Science Photo Library/Photo Researchers; Gary W. Carter/CORBIS; **37.13 bottom left to right** Dorling Kindersley; Biophoto Associates/Photo Researchers; Dorling Kindersley; Paul A. Zahl/Photo Researchers; Fritz Polking/Frank Lane Picture Agency/CORBIS

Chapter 38

38.1 COMPOST/VISAGE/Peter Arnold; **38.3 top left to right** Karen Tweedy-Holmes/CORBIS; Dorling Kindersley; Craig Lovell/CORBIS; John Cancalosi/Nature Photo Library; **38.3 bottom left to right** D. Cavagnaro/Visuals Unlimited; David Sieren/Visuals Unlimited; Marcel E. Dorken (2 photos); **38.4a left** Ed Reschke; **38.4a right** David Scharf/Peter Arnold; **38.4b** Ed Reschke; **38.11** David Cavagnaro/DRK; **38.12a, b** Bruce Iverson; **38.13** Sinclair Stammers/Photo Researchers; **38.14 both** Dorling Kindersley; **38.15** Steve Ferreira, University of Hawaii and Dennis Gonsalves, Cornell University, courtesy American Phytopathological Society; **38.16** Peter Berger, Institut für Biologie, Freiburg

Chapter 39

39.1 Malcolm Wilkins, University of Glasgow; **39.2 both** Natalie Bronstein; **39.7 both** Regulation of Polar Auxin Transport ATPIN1 in Arabidopsis Vascular Tissue," by Leo Galweiler, et al) Science 18 Dec 1998 , vol. 282, pp. 2226-2229; **39.9 both** Malcolm Wilkins, University of Glasgow; **39.10** Fred Jensen/Kearney Agricultural Center; **39.12** Karen E. Koch; **39.14a** Kurt Stepnitz, DOE Plant Research Laboratory, Michigan State University; **39.14b** Joe Kieber, University of North Carolina; **39.16** Ed Reschke; **39.17 all** Malcolm Wilkins, University of Glasgow; **39.18 all** Malcolm Wilkins, University of Glasgow; **39.21 both** Malcolm Wilkins, University of Glasgow; **39.25 all** Michael Evans, Ohio State University; **39.26** Janet Braam, from Cell 60 (9 February 1990): cover. © 1990 by Cell Press; **39.27a-b** David Sieren/Visuals Unlimited; **39.27c** From K. Esau, anatomy of Seed Plants, 2nd ed. (New York: John Wiley and sons, 1977), fig. 19.4, p. 358; **39.28** J. L. Basq and M. C. Drew

Chapter 40

40.1 Mitsuhiko Imamori/Minden Pictures; **40.2a** Flip Nicklin/Minden Pictures; **40.2b** Andrew Sallmon/Mondragon Photography; **40.2c** Tui De Roy/Minden Pictures; **40.2d** Bill Varie/CORBIS; **40.2e** Norbert Wu/Minden Pictures; **40.4 top right** D. M. Phillips/Visuals Unlimited; **40.4 left** G. Shih-R. Kessel/Visuals Unlimited; **40.4 bottom right** Dr. Richard Kessel & Dr. Randy Kardon/Tissues & Organs/Visuals Unlimited; **40.5 p824** CNRI/SPL/Photo Researchers; **40.5 p825 top left** Nina Zanetti; **top right** Chuck Brown/Photo Researchers; **middle left** Science VU/Visuals Unlimited; **middle right** Nina Zanetti; **bottom left** Nina Zanetti; **bottom right** Dr. Gopal Murti/SPL/Photo Researchers; **40.5 p826 top to bottom** Nina Zanetti; Manfred Kage/Peter Arnold; Gladden Willis, M. D./Visuals Unlimited; Ed Reschke; **40.6** Dr. Richard Kessel & Dr. Randy Kardon/Tissues & Organs/Visuals Unlimited; **40.8a** Robert Full, University of California; **40.8b** Yoav Levy/Phototake NY; **40.18** Daniel Lyons/Bruce Coleman; **40.19** Robert Ganz; **40.22** John Gerlach/Visuals Unlimited

Chapter 41

41.1 AP World Wide Photos; **41.2 top left** Brandon Cole; **top right** Tom Eisner, Cornell University; **41.2 middle right** Lennart Nilsson; **41.2 bottom** Gunter Ziesler/Peter Arnold, Inc.; **41.4** Susumu Nishinaga/SPL/Photo Researchers; **41.6** The Jackson Laboratory; **41.7** Wolfgang Kaehler/CORBIS; **41.8** Thomas Mangelsen/Minden Pictures; **41.9** Dagmar Fabricius/Stock Boston; **41.10 both** Digital Vision/CORBIS; **41.11** Roland Seitre/Peter Arnold; **41.17** Fred E. Hossler/Visuals Unlimited; **41.18** Eye of Science/Photo Researchers; **41.25** Kelley Wise/Vital Imaging; **41.27 both** PhotoDisc

Chapter 42

42.1 George Bernard/Animals Animals; **42.2** Norbert Wu/Mo Young Productions; **42.9** Dr. Richard Kessel & Dr. Randy Kardon/Tissues & Organs/Visuals Unlimited; **42.13** CNRI/Phototake; **42.14** Lennart Nilsson, The Body Victorious, Dell Publishing Company; **42.17** Science Source/Photo Researchers; **42.18a** Ed Reschke; **42.18b** W. Ober/Visuals Unlimited; **42.20a** Frans Lanting/Minden Pictures; **42.20b** Peter Batson/Image Quest Marine; **42.20c** H. W. Pratt/Biological Photo Service; **42.20d** Dave Haas/The Image Finders; **42.22b** Thomas Eisner; **42.23 left** Dr. Richard Kessel & Dr. Randy Kardon/Tissues & Organs/Visuals Unlimited; **42.23 right** CNRI/ Photo Researchers; **42.25** Hans Rainer Dunker, Justus Leibig University, Giessen; **42.31** Stan Lindstedt

Chapter 43

43.1 Biology Media/Photo Researchers; **43.3** Science Photo Library/Photo Researchers; **43.15** David Scharf/Peter Arnold; **43.16** Lennart Nilsson/Boehringer Ingelheim International GmbH; **43.17** Gopal Murti/Phototake; **43.20** Lennart Nilsson/Boehringer Ingelheim International GmbH; **43.21** CNRI/SPL/Photo Researchers; **43.22** Lennart Nilsson/Albert Bonniers Forlag AB

Chapter 44

44.1 Peter Reese/Nature Photo Library; **44.2** Nigel J. Dennis/NHPA/Photo Researchers; **44.4 both** John Crowe; **44.13b** Lise Bankir. From Urinary concentrating ability: insights from comparative anatomy, Bankir and de Rouffignac, Am J. Physiol Regul Integr Comp Physiol, 1985; 249: 643-666; **44.13d** Dr. Richard Kessel & Dr. Randy Kardon/Tissues & Organs/Visuals Unlimited; **44.17** Michael

& Patricia Fogden/Minden Pictures; **44.18 kangaroo rat** Mary McDonald/Nature Photo Library; **44.18 beaver** Michael Quinton/Minden Pictures; **44.18 road runner** Gerry Ellis/Minden Pictures; **44.18 iguana** Daniel Heuclin/NHPA/Photo Researchers; **44.18 trout** Tim Martin/Nature Photo Library; **44.18 frog** Laurie Campbell/NHPA/Photo Researchers; **44.18 tuna** Juan Carlos Calvin/AGE fotostock

Chapter 45
45.1 Ralph A. Clevenger/CORBIS; **45.5** VVG/SPL/Photo Researchers; **45.10** Custom Medical Stock Photo; **45.14** Photodisc

Chapter 46
46.1 Robin Chittenden/CORBIS; **46.2** David Wrobel, Monterey Bay Aquarium; **46.3** David Crews, photo by P. de Vries; **46.4** Stephan Myers; **46.5** Dwight Kuhn; **46.6** William Ferguson; **46.17 all** Lennart Nilsson/Albert Bonniers Vorlag AB; **46.21** UHB Trust/Stone

Chapter 47
47.1 Lennart Nilsson/Albert Bonniers Vorlag AB; **47.2** Historical Collections, College of Physicians, Philadelphia; **47.4 all** Jerry Schatten et al; **47.7a-d** George von Dassow; **47.9 both** Dr. Richard Kessel & Dr. Gene Shih/Visuals Unlimited; **47.11** Charles A Ettensohn, Carnegie Mellon University; **47.14a** CABISCO/Visuals Unlimited; **47.14c** Thomas Poole, SUNY Health Science Center; **47.15** Carolina Biological Supply/Phototake; **47.21** Reproduced form Dr. Jean Paul Thiery Journal of Cell Biology 96 (1983): 462-473; **47.22 both** Janet Heasman, University of Minnesota; **47.23 both** Hiroki Nishida, Developmental Biology 121 (1987): 526. Reprinted by permission of Academic Press; **47.26a** Kathryn Tosney, University of Michigan; **47.27** Courtesy of Dennis Summerbell

Chapter 48
48.1 G. Fesl MD, Dept. of Neuroradiology, LMU Munich, Germany; **48.7** N. Kedersha/Photo Researchers; **48.8** Alan Peters, from Bear, Connors, and Paradiso, Neuroscience: Exploring the Brain © 1996, p. 43; **48.16** E. R Lewis, University of California; **48.29** Marcus Raichle, MD, Washington University School of Medicine; **48.34** Fred H. Gage, The Salk Institute, Laboratory of Genetics; **48.35** Martin M. Rotker/Photo Researchers

Chapter 49
49.1 Stephen Dalton/NHPA; **49.4 top** OSF/Animals Animals; **49.4 bottom** R. A. Steinbrecht, Max Planck Institute; **49.5a** Joe McDonald/Animals Animals; **49.5b** Flip Nicklin/Minden Pictures; **49.7** John L. Pontier/Animals Animals; **49.16** From Richard Elzinga, Fundamentals of Entomology 3 ed. ©1987, p. 185. Reprinted by permission of Prentice-Hall, Upper Saddle River, NJ; **49.28** Clara Franzini-Armstrong, University of Pennsylvania; **49.29 all** Dr. H. E Huxley; **49.36** Dave Watts/NHPA/Photo Researchers; **49.37** Vance A. Tucker

Chapter 50
50.1 Mark Moffett/Minden Pictures; **50.2** PhotoDisc; **50.3a** Amos Nachoum/The Image Bank; **50.3b** Joe McDonald/CORBIS; **50.3c** Tom Bean/CORBIS; **50.3d** B. Tharp/Photo Researchers; **50.3e** Yann Arthus-Bertrand/CORBIS; **50.4** Erich Hartmann/Magnum Photos; **50.7** Richard Ditch; **50.9** Hubert Stadler/CORBIS; **50.17 p1094 top left** Allen Russell/Index Stock; **top right** Gerry Ellis/Minden Pictures; **bottom** David Muench/CORBIS **50.17 p1095 top left** Ron Watts/CORBIS; **top right** Charles McDowell/Grant Heilman Photography; **bottom** James Randklev/The Image Bank; **50.17 p1096 both** Stuart Westmorland/CORBIS; **50.17 p1097 top** Digital Vision; **bottom** William Lange/Woods Hole Oceanographic Institution; **50.20 p1100 top** Frans Lanting/Minden Pictures; **bottom** Joe McDonald/CORBIS; **50.20 p1101 top** Wolfgang Kaehler/CORBIS; **bottom** John D. Cunningham/Visuals Unlimited; **50.20 p1102 top** Tom Bean/The Image Bank; **bottom** Bill Ross/CORBIS; **50.20 p1103** Kennan Ward/CORBIS; **bottom** Darrell Gulin/CORBIS

Chapter 51
51.1 Steve Kaufman/CORBIS; **51.2** Anthony Calfo; **51.3** Kim Taylor/Bruce Coleman; **51.6** Operation Migration; **51.8** Michael Quinton/Minden Pictures; **51.10** Robert Pickett/CORBIS; **51.11** Lowell L. Getz and Lisa Davis; **51.13** Therese Markow and Tom Watts; **51.16** Harry Engels/Animals Animals; **51.17** Clive Bromhall/OSF/Animals Animals; **51.18a** Gary A. Nafis; **51.18b** Breck P. Kent/Animals Animals/Earth Scenes; **51.25a** Thomas Mangelsen/Minden Pictures; **51.25b** James H. Robinson/Animals Animals/Earth Scenes; **51.25c** Bill Schomoker; **51.26** Fred Bavendam/Minden Pictures; **51.27** Dorling Kindersley; **51.29** G. S. Wilkinson; **51.30** Michio Hoshino/Minden Pictures; **51.32** Erik Svensson; **51.33** Jennifer Jarvis, University of Cape Town; **51.35** Stephen Kraseman/Peter Arnold; **51.37 main** Richard Wrangham, Harvard University; **51.37 inset** Alissa Crandall/CORBIS; **51.38** Bob Winsett/Index Stock Imagery

Chapter 52
52.1 Yva Momatiuk, John Eastcott/Minden Pictures; **52.3a** Stone; **52.3b** Frans Lanting/Minden Pictures; **52.3c** Niall Benvie/CORBIS; **52.6** Tom Bean/CORBIS; **52.7** H. Willcox/Wildlife Pictures/Peter Arnold; **52.8a** Jean Louis Batt/Taxi; **52.8b** Christine Osborne/CORBIS; **52.16** Adrian Bailey/Aurora Photos; **52.17** Wolfgang Kaehler/CORBIS; **52.21** Joe McDonald/CORBIS

Chapter 53
53.1 Martin Harvey/Peter Arnold; **53.3 left** Joseph T. Collins/Photo Researchers; **53.3 right** Kevin de Queiroz, National Museum of Natural History; **53.5** C. Allan Morgan/Peter Arnold; **53.6** Michael & Patricia Fogden/CORBIS; **53.7a** Stephen J. Krasemann/Photo Researchers; **53.7b** Michael & Patricia Fogden/Minden Pictures; **53.8a** Edward S. Ross; **53.8b** Runk/Schoenberger/Grant Heilman Photography; **53.9** Michael & Patricia Fogden/CORBIS; **53.10** Peter Johnson/CORBIS; **53.16a** Bill Curtsinger/National Geographic Image Collection; **53.18** Darrell Gulin/CORBIS; **53.19** Sally D. Hacker, Oregon State University; **53.21 all** Frank Gilliam, Marshall University; **53.22a** Michael Quinton/Minden Pictures; **53.22b** Scott T. Smith/CORBIS; **53.23** Tom Bean/CORBIS; **53.24a** Charles Mauzy/CORBIS; **53.24b** Tom Bean/DRK; **53.24c** Glacier Bay National Park and Preserve

Chapter 54
54.1 Alex Wetmore; **54.3** Fritz Poelking/AGE fotostock; **54.5** C. B. Field, M. J. Behrenfeld, J. T. Randerson, and P. Falkowski. 1998. "Primary production of the bio-sphere: Integrating terrestrial and oceanic components." Science 281:237–240; **54.7** Reprinted with permission from D.W. Schindler, Science 184 (1974): 897, Figure 1.49. ©1974 American Association for the Advancement of Science; **54.15** Thomas Del Brase/Photographer's Choice; **54.19a-b** Hubbard Brook Research Foundation; **54.20** Javier Larrea/AGE Fotostock; **54.25** Will Owens Photography; **54.28a-b** NASA/Goddard Space Flight Center Scientific Visualization Studio

Chapter 55
55.1 Wayne Lawler/CORBIS; **55.3a** Neil Lucas/Nature Picture Library; **55.3b** Mark Carwardine/Still Pictures/Peter Arnold; **55.3c** Foead, Yahya & Sumiadi/WWF & BTNUK; **55.4** Scott Camazine/Photo Researchers; **55.5** Gary Braasch/Woodfin Camp & Associates; **55.6a** Michael Fogden/Animals Animals; **55.6b** Mark E. Gibson/Visuals Unlimited; **55.7** Richard Vogel/Gamma Liaison/Getty Images; **55.8** Merlin D. Tuttle, Bat Conservation International; **55.10** William Ervin/SPL/Photo Researchers; **55.11** Craighead Environmental Research Institute; **55.13a** Tim Thompson/CORBIS; **55.13b** David Sieren/Visuals Unlimited; **55.13c** Blanche Haning/The Lamplighter; **55.14a** Yann Arthus-Bertrand/CORBIS; **55.14b** James P. Blair/National Geographic Image Collection; **55.15** Richard O. Bierregaard, Jr.; **55.16** Alan Sirulnikoff/SPL/Photo Researchers; **55.19** Frans Lanting/Minden Pictures; **55.20** Photo by John Parks/WWF with approval from Verata District, Fiji; **55.22 p1226 left** Stewart Rood, University of Lethbridge; **right** South Florida Water Management District (WPB); **55.22 p1227 top left** Daniel H. Janzen; **bottom left** Bert Boekhoven; **top right** Jean Hall/Holt Studios/Photo Researchers; **bottom right** Kenji Morita/Environment Division, Tokyo Kyuei Co., Ltd.; **55.24a** Serge de Sazo/Photo Researchers; **55.24b** Frans Lanting/Minden Pictures

ILLUSTRATION CREDITS

The following figures are adapted from Christopher K. Matthews and K. E. van Holde, *Biochemistry,* 2nd ed., Menlo Park, CA: Benjamin Cummings. © 1996 The Benjamin Cummings Publishing Company, Inc.: **4.6**, **9.9**, and **17.16b** and **c**.

The following figures are adapted from Wayne M. Becker, Jane B. Reece, and Martin F. Poenie, *The World of the Cell,* 3rd ed., Menlo Park, CA: Benjamin Cummings. © 1996 The Benjamin Cummings Publishing Company, Inc.: **4.7**, **6.7**, **7.8**, **11.7a**, **11.10**, **17.10**, **19.13**, **19.16**, and **20.7**

Figures 6.9 and **6.23** and cell organelle drawings in Figures **6.12**, **6.13**, **6.14**, and **6.20** are adapted from illustrations by Tomo Narashima in Elaine N. Marieb, *Human Anatomy and Physiology,* 5th ed., San Francisco, CA: Benjamin Cummings. © 2001 Benjamin Cummings, an imprint of Addison Wesley Longman, Inc. Figures **6.12**, **49.10**, and **49.11** are also from *Human Anatomy and Physiology,* 5th ed.

Figures **46.16**, **48.22**, **48.24**, **49.26**, **49.29**, and **49.33** are adapted from Elaine N. Marieb, *Human Anatomy and Physiology,* 4th ed., Menlo Park, CA: Benjamin Cummings. © 1998 Benjamin Cummings, an imprint of Addison Wesley Longman, Inc.

The following figures are adapted from Murray W. Nabors, *Introduction to Botany,* San Francisco, CA: Benjamin Cummings. © 2004 Pearson Education, Inc., Upper Saddle River, New Jersey: **30.12j**, **38.3c**, **39.13**, and **41.10**.

Some illustrations used in *BIOLOGY*, Seventh Edition, are adapted from Neil Campbell, Brad Williamson, and Robin Heyden, *Biology: Exploring Life*, Needham, MA, Prentice Hall School Division. © 2004 by Pearson Education, Inc., Upper Saddle River, NJ. Artists: Jennifer Fairman; Mark Foerster; Carlyn Iverson; Phillip Guzy; Steve McEntee; Stephen McMath; Karen Minot; Quade and Emi Paul, Fivth Media; and Nadine Sokol.

1.10 and **graphic in Table 1.1** Adapted from Figure 4 from L. Giot et al., "A Protein Interaction Map of *Drosophila melanogaster*," *Science*, Dec. 5, 2003, p. 1733 Copyright © 2003 AAAS. Reprinted with permission from the American Association for the Advancement of Science; **1.27** Map provided courtesy of David W. Pfennig, University of North Carolina at Chapel Hill; **1.29** Map provided courtesy of David W. Pfennig, University of North Carolina at Chapel Hill. Data in pie charts based on D. W. Pfennig et al. 2001. Frequency-dependent Batesian mimicry. *Nature* 410: 323.

3.7a Adapted from *Scientific American*, Nov. 1998, p. 102.

4.8 Adapted from an illustration by Clark Still, Columbia University.

5.13 From *Biology: The Science of Life*, 3/e by Robert Wallace et al. Copyright © 1991. Reprinted by permission of Pearson Education, Inc.; **5.20a** and **b** Adapted from D. W. Heinz, W. A. Baase, F. W. Dahlquist, B. W. Matthews. 1993. How amino-acid insertions are allowed in an alpha-helix of T4 lysozyme. *Nature* 361:561; **5.20e** and **f** © Illustration, Irving Geis. Rights owned by Howard Hughes Medical Institute. Not to be reproduced without permission; **Table 6.1** Adapted from W. M. Becker, L. J. Kleinsmith, and J. Hardin, *The World of the Cell*, 4th ed. (San Francisco, CA: Benjamin Cummings, 2000), p. 753.

9.5a and **b** Copyright © 2002 from *Molecular Biology of the Cell*, 4th ed. by Bruce Alberts et al., fig. 2.69, p. 92. Garland Science/Taylor & Francis Books, Inc.

10.14 Adapted from Richard and David Walker. *Energy, Plants and Man*, fig. 4.1, p. 69. Sheffield: University of Sheffield. © Richard Walker. Used courtesy of Oxygraphics.

12.12 Copyright © 2002 from *Molecular Biology of the Cell*, 4th ed., by Bruce Alberts et al., fig. 18.41, p. 1059. Garland Science/Taylor & Francis Books, Inc.

17.12 Adapted from L. J. Kleinsmith and V. M. Kish. 1995. *Principles of Cell and Molecular Biology*, 2nd ed. New York, NY: HarperCollins. Reprinted by permission of Addison Wesley Educational Publishers.

19.17b © Illustration, Irving Geis. Rights owned by Howard Hughes Medical Institute. Not to be reproduced without permission; **Table 19.1** From A. Griffiths et al. 2000. *An Introduction to Genetic Analysis*, 7/e, Table 26-4, p. 787. New York: W. H. Freeman and Company. Copyright © 2000 W. H. Freeman and Company.

20.9 Adapted from Peter Russell, *Genetics*, 5th ed., fig. 15.24, p. 481, San Francisco, CA: Benjamin Cummings. © 1998 Pearson Education, Inc., Upper Saddle River, New Jersey; **20.11** Adapted from a figure by Chris A. Kaiser and Erica Beade.

21.15 Copyright © 2002 from *Molecular Biology of the Cell*, 4th ed., by Bruce Alberts et al., fig. 21.17, p. 1172. Garland Science/Taylor & Francis Books, Inc.; **21.23** Adapted from an illustration by William McGinnis; **21.24** Brine shrimp adapted from M. Akam, "*Hox* genes and the evolution of diverse body plans," *Philosophical Transactions B*, 1995, 349:313-319. © Royal Society of London.

22.13 Adapted from R. Shurman et al. 1995. *Journal of Infectious Diseases* 171:1411.

23.13 Adapted from A. C. Allison. 1961. Abnormal hemoglobin and erythrocyte enzyme-deficiency traits. In *Genetic Variation in Human Populations*, ed. G. A. Harrison. Oxford: Elsevier Science.

24.7 Adapted from D. M. B. Dodd, *Evolution* 11: 1308-1311; **24.14** Adapted from M. Strickberger. 1990. *Evolution*. Boston: Jones & Bartlett; **24.16** Adapted from L. Wolpert. 1998. *Principles of Development*. Oxford University Press; **24.18** Adapted from M. I. Coates. 1995. *Current Biology* 5:844-848.

25.18 Adapted from S. Blair Hedges. The origin and evolution of model organisms. fig. 1, p. 840. *Nature Reviews Genetics* 3: 838-849.

26.5 Figure 4c from "The Antiquity of RNA-based Evolution" by G.F. Joyce et al., *Nature*, Vol. 418, p. 217. Copyright © 2002 Nature Publishing Co.; **26.7** Adapted from D. Futuyma. 1998. *Evolutionary Biology*, 3rd ed., p. 128. Sunderland, MA: Sinauer Associates; **26.8** Data from M. J. Benton. 1995. Diversification and extinction in the history of life. *Science* 268:55; **26.10** Adapted from David J. Des Marais. September 8, 2000. When did photosynthesis emerge on Earth? *Science* 289:1703-1705; **26.17** Data from A. H. Knoll and S. B. Carroll, June 25, 199. *Science* 284:2129-2137; **26.18** Map adapted from http://geology.er.usgs.gov/eastern/plates.html.

27.6 Adapted from Gerard J. Tortora, Berdell R. Funke, and Christine L. Case. 1998. *Microbiology: An Introduction*, 6th ed. Menlo Park, CA: Benjamin Cummings. © 1998 Benjamin Cummings, an imprint of Addison Wesley Longman, Inc.

28.3 Figure 3 from Archibald and Keeling, "Recycle Plastics," *Trends in Genetics*, Vol. 18, No. 1, 2, 2002, p. 352. Copyright © 2002, with permission from Elsevier; **28.12** Adapted from R. W. Bauman. 2004. *Microbiology*, fig. 12.7, p. 350. San Francisco, CA: Benjamin Cummings. © 2004 Pearson Education, Inc., Upper Saddle River, New Jersey.

29.13 Adapted from Raven et al. *Biology of Plants*, 6th ed., fig. 19.7.

Table 30.1 Adapted from Randy Moore et al., *Botany*, 2nd ed. Dubuque, IA: Brown, 1998, Table 2.2, p. 37.

34.8a Adapted From J. Mallatt and J. Chen, "Fossil sister group of craniates: predicted and found," *Journal of Morphology*, Vol. 251, no. 1, fig.1, 5/15/03,© 2003 Wiley-Liss, Inc., a Wiley Co.; **34.8b** Adapted from D.-G. Shu et al. 2003. Head and backbone of the early Cambrian vertebrate *Haikouichthys*. *Nature* 421:528, fig. 1, part l. © 2003 Nature Publishing Group; **34.12** Adapted from K. Kardong, *Vertebrates: Comparative Anatomy, Function and Evolution*, 3/e, © 2001McGraw-Hill Science/Engineering/Mathematics; **34.19** From C. Zimmer, *At the Water's Edge*. Copyright © 1999 by Carl Zimmer. Reprinted by permission of Janklow & Nesbit; **34.20** From C. Zimmer, *At the Water's Edge*. Copyright © 1999 by Carl Zimmer. Reprinted by permission of Janklow & Nesbit; **34.32** Adapted from Stephen J. Gould, *The Book of Life*, 2/e. Copyright © 2001 by Stephen J. Gould. Reprinted by permission of the Estate of Stephen J. Gould; **34.41** Drawn from photos of fossils: *O. tugenensis* photo in Michael Balter, Early hominid sows division, *ScienceNow*, Feb. 22, 2001, © 2001 American Association for the Advancement of Science. *A. ramidus kadabba* photo by Timothy White, 1999/Brill Atlanta. *A. anamensis*, *A. garhi*, and *H. neanderthalensis* adapted from *The Human Evolution Coloring Book*. *K platyops* drawn from photo in Meave Leakey et al., New hominid genus from eastern Africa shows diverse middle Pliocene lineages, *Nature*, March 22, 2001, 410:433. *P. boisei* drawn from a photo by David Bill. *H. ergaster* drawn from a photo at www.inhandmuseum.com. *S. tchadensis* drawn from a photo in Michel Brunet et al., A new hominid from the Upper Miocene of Chad, Central Africa, *Nature*, July 11, 2002, 418:147, fig. 1b.

35.21 Pie chart adapted from *Nature*, Dec. 14, 2000, 408:799.

39.17(graph), 38.18 Adapted from M. Wilkins. 1988. *Plant Watching*. Facts of File Publ.; **39.29** Reprinted with permission from Edward Framer, 1997, *Science* 276:912. Copyright © 1997 American Association for the Advancement of Science.

40.17 Adapted from an illustration by Enid Kotschnig in B. Heinrich, 1987. Thermoregulation in a winter moth. *Scientific American* 105; **40.20** Adapted with permission from B. Heinrich, 1974, *Science* 185:747-756. © 1974 American Association for the Advancement of Science.

41.5 Adapted from J. Marx, "Cellular Warriors at the Battle of the Bulge," *Science*, Vol. 299, p. 846. Copyright © 2003 American Association for the Advancement of Science. Illustration: Katharine Sutliff; **41.13** Adapted from Lawrence G. Mitchell, John A. Mutchmor, and Warren D. Dolphin. 1988. *Zoology*. Menlo Park, CA: Benjamin Cummings. © 1988 The Benjamin Cummings Publishing Company; **41.15** Adapted from R. A. Rhoades and R. G. Pflanzer. 1996. *Human Physiology*, 3/e., fig. 22-1, p. 666. Copyright © 1996 Saunders.

43.7 Adapted from Gerard J. Tortora, Berdell R. Funke, and Christine L. Case. 1998. *Microbiology: An Introduction*, 6th ed. Menlo Park, CA: Benjamin/Cummings. © 1998 Benjamin Cummings, an imprint of Addison Wesley Longman, Inc.

44.5 Kangaroo rat data adapted from Schmidt-Nielsen. 1990. *Animal Physiology: Adaptation and Environment*, 4th ed., p. 339. Cambridge: Cambridge University Press; **44.6** Adapted from K.B. Schmidt-Nielsen et al., "Body temperature of the camel and its relation to water economy," *American Journal of Physiology*, Vol. 10, No. 188, (Dec.), 1956, figure 7. Copyright © 1956 American Physiological Society. Used with permission; **44.8** Adapted from Lawrence G. Mitchell, John A. Mutchmor, and Warren D. Dolphin. 1988. *Zoology*. Menlo Park, CA: Benjamin Cummings. © 1988 The Benjamin Cummings Publishing Company.

47.20 From Wolpert et al. 1998. *Principles of Development,* fig. 8.25, p. 251 (right). Oxford: Oxford University Press. By permission of Oxford University Press; **47.23b** From Hiroki Nishida, *Developmental Biology* Vol. 121, p. 526, 1987. Copyright © 1987, with permission from Elsevier; **47.25 Experiment** and **left side of "Results"**: From Wolpert et al. 1998. *Principles of Development,* fig. 1.10, Oxford: Oxford University Press. By permission of Oxford University Press; **Right side of "Results"**: Figure 15.12, p. 604 from *Developmental Biology,* 5th ed. by Gilbert et al. Copyright © 1997 Sinauer Associates. Used with permission.

48.13 From G. Matthews, *Cellular Physiology of Nerve and Muscle.* Copyright © 1986 Blackwell Science. Used with permission; **48.33** Adapted from John G. Nicholls et al. 2001. *From Neuron to Brain,* 4th ed., fig. 23.24. Sunderland, MA: Sinauer Associates Inc. © 2001 Sinauer Associates.

49.19 Adapted from Bear et al. 2001. *Neuroscience: Exploring the Brain,* 2nd ed., figs. 11.8 and 11.9, pp. 281 and 283. Hagerstown, MD: Lippincott Williams & Wilkins © 2001 Lippincott Williams & Wilkins; **49.22** Adapted from Shepherd. 1988. *Neurobiology,* 2nd ed., fig. 11.4, p. 227. Oxford University Press. (From V. G. Dethier. 1976. *The Hungry Fly.* Cambridge, MA: Harvard University Press.); **49.23 (Lower)** Adapted from Bear et al. 2001. *Neuroscience: Exploring the Brain,* 2nd ed., fig. 8.7, p. 196. Hagerstown, MD: Lippincott Williams & Wilkins. © 2001 Lippincott Williams & Wilkins; **49.27b** Grasshopper adapted from Hickman et al. 1993. *Integrated Principles of Zoology,* 9th ed., fig. 22.6, p. 518. New York: McGraw-Hill Higher Education. © 1995 The McGraw-Hill Companies.

50.2 Adapted from G. Caughly, N. Shepherd, and J. Short. 1987. *Kangaroos: Their Ecology and Management in the Sheep Rangelands of Australia,* fig. 1.2, p. 12, Cambridge: Cambridge University Press. Copyright © 1987 Cambridge University Press; **50.7a** Data from U. S. Geological Survey; **50.8** Data from W. J. Fletcher. 1987. Interactions among subtidal Australian sea urchins, gastropods and algae: effects of experimental removals. *Ecological Monographs* 57:89-109; **50.14** Adapted from L. Roberts. 1989. How fast can trees migrate? *Science* 243:736, fig. 2. © 1989 by the American Association for the Advancement of Science; **50.19** Adapted from Heinrich Walter and Siegmar-Walter Breckle. 2003. *Walter's Vegetation of the Earth,* fig. 16, p. 36. Springer-Verlag, © 2003.

51.3b Adapted from N. Tinbergen. 1951. *The Study of Instinct.* Oxford: Oxford University Press. By permission of Oxford University Press; **51.10** Adapted from C. S. Henry et al. 2002. The inheritance of mating songs in two cryptic, sibling lacewings species (Neuroptera: Chrysopidae: *Chrysoperla*). *Genetica* 116: 269-289, fig. 2; **51.14** Adapted from Lawrence G. Mitchell, John A. Mutchmor, and Warren D. Dolphin. 1988. *Zoology.* Menlo Park, CA: Benjamin/Cummings. © 1988 The Benjamin/Cummings Publishing Company; **51.15** Adapted from N. L. Korpi and B. D. Wisenden. 2001. Learned recognition of novel predator odour by zebra danios, *Danio rerio,* following time-shifted presentation of alarm cue and predator odour. *Environmental Biology of Fishes* 61: 205-211, fig. 1; **51.19** Adapted from M. B. Sokolowski et al. 1997. Evolution of foraging behavior in *Drosophila* by density-dependent selection. *Proceedings of the National Academy of Sciences of the United States of America.* 94: 7373-7377, fig. 2b; **51.21a** Adapted from a photograph by Jonathan Blair in Alcock. 2002. *Animal Behavior,* 7th ed. Sinauer Associates, Inc., Publishers; **51.21b** From P. Berthold et al., "Rapid microevolution of migratory behaviour in a wild bird species," *Nature,* Vol. 360, 12/17/92, p. 668. Copyright © 1992 Nature Publishing, Inc. Used with permission; **51.28** K. Witte and N. Sawka. 2003. Sexual imprinting on a novel trait in the dimorphic zebra finch: sexes differ. *Animal Behaviour* 65: 195-203. Art adapted from http://www.uni-bielefeld.de/biologie/vhf/KW/Forschungsprojekte2.html; **Table 51.1** Source: J. K. Bester-Meredith and C. A. Marler. 2003. Vasopressin and the transmission of paternal behavior across generations in mated, cross-fostered *Peromyscus* mice. *Behavioral Neuroscience* 117:455-463.

52.4 Adapted from P. W. Sherman and M. L. Morton, "Demography of Belding's ground squirrels," *Ecology,* Vol. 65, No. 5, p. 1622, 1984. Copyright © 1984 Ecological Society of America. Used by permission; **52.13c** Data courtesy of P. Arcese and J. N. M. Smith, 2001; **52.14** Adapted from J. T. Enright. 1976. Climate and population regulation: the biogeographer's dilemma. *Oecologia* 24:295-310; **52.15b** Data from J. N. M. Smith and P. Arcese; **52.18** Data courtesy of Rolf O. Peterson, Michigan Technological University, 2004; **52.19** Data from Higgins et al. May 30, 1997. Stochastic dynamics and deterministic skeletons: population behavior of Dungeness crab. *Science;* **52.20** Adapted from J.N.M. Smith et al., "A metapopulation approach to the population biology of the song sparrow *Melospiza melodia,*" *IBIS,* Vol. 138, fig. 3, 1996. Copyright © 1996; **52.23** Data from U. S. Census Bureau International Data Base; **52.24** Data from Population Reference Bureau 2000 and U. S. Census Bureau International Data Base, 2003; **52.25** Data from U. S. Census Bureau International Data Base; **52.26** Data from U. S. Census Bureau International Data Base 2003; **52.27** Data from J. Wackernagel et al. 1999. National natural capital accounting with the ecological footprint concept. *Ecological Economics* 29: 375-390. **Tables 52.1** and **52.2** Data from P. W. Sherman and M. L. Morton. 1984. Demography of Belding's Ground Squirrels. *Ecology* 65:1617-1628. © 1984 by the Ecological Society of America.

53.3 A. S. Rand and E. E. Williams. 1969. The anoles of La Palma: aspects of their ecological relationships. *Breviora* 327. Museum of Comparative Zoology, Harvard University. © Presidents and Fellows of Harvard College; **53.13** Adapted from E. A. Knox. 1970. Antarctic marine ecosystems. In *Antarctic Ecology,* ed. M. W. Holdgate, 69-96. London: Academic Press; **53.14** Adapted from D. L. Breitburg et al. 1997. Varying effects of low dissolved oxygen on trophic interactions in an estuarine food web. *Ecological Monographs* 67: 490. Copyright © 1997 Ecological Society of America; **53.15** Adapted from B. Jenkins. 1992. Productivity, disturbance and food web structure at a local spatial scale in experimental container habitats. *Oikos* 65: 252. Copyright © 1992 Oikos, Sweden; **53.17** Adapted from J. A. Estes et al. 1998. Killer whale predation on sea otters linking oceanic and nearshore ecosystems. *Science* 282:474. Copyright © 1998 by the American Association for the Advancement of Science; **53.19** Data from S. D. Hacker and M. D. Bertness. 1999. Experimental evidence for factors maintaining plant species diversity in a New England salt marsh. *Ecology* 80: 2064-2073; **53.23** Adapted from R. L. Crocker and J. Major. 1955. Soil Development in relation to vegetation and surface age at Glacier Bay, Alaska. *Journal of Ecology* 43: 427-448; **53.24d** Data from F. S. Chapin, III, et al. 1994. Mechanisms of primary succession following deglaciation at Glacier Bay, Alaska. *Ecological Monographs* 64: 149-175. **53.25** Adapted from D. J. Currie. 1991. Energy and large-scale patterns of animal- and plant-species richness. *American Naturalist* 137: 27-49; **53.26** Adapted from F. W. Preston. 1960. Time and space and the variation of species. *Ecology* 41: 611-627; **53.28** Adapted from F. W. Preston. 1962. The canonical distribution of commonness and rarity. *Ecology* 43: 185-215, 410-432.

54.2 Adapted from D. L. DeAngelis. 1992. *Dynamics of Nutrient Cycling and Food Webs.* New York: Chapman & Hall; **54.6** Adapted from J. H. Ryther and W. M. Dunstan. 1971. Nitrogen, phosphorus, and eutrophication in the coastal marine environment. *Science* 171:1008-1013; **54.8** Data from M. L. Rosenzweig. 1968. New primary productivity of terrestrial environments: Predictions from climatologic data, *American Naturalist* 102:67-74. **54.9** Adapted from S. M. Cargill and R. L. Jefferies. 1984. Nutrient limitation of primary production in a sub-arctic salt marsh. *Journal of Applied Ecology* 21:657-668; **54.17a** Adapted from R. E. Ricklefs. 1997. *The Economy of Nature,* 4th ed. © 1997 by W. H. Freeman and Company. Used with permission; **54.21** Adapted from G. E. Likens et al. 1981. Interactions between major biogeochemical cycles in terrestrial ecosystems. In *Some Perspectives of the Major Biogeochemical Cycles,* ed. G. E. Likens, 93-123. New York: Wiley; **54.22** Adapted from National Atmospheric Deposition Program (NRSP-3) National Trends Network. (2004). NADP Program Office, Illinois State Water Survey, 2204 Griffith Dr., Champaign, IL 61820. http://nadp.sws.uiuc.edu; **54.24** Temperature data from U. S. National Climate Data Center, NOAA. CO_2 data from C. D. Keeling and T. P. Whorf, Scripps Institution of Oceanography; **54.26** Data from British Antarctic Survey; **Table 54.1** Data from Menzel and Ryther, *Deep Sea Ranch* 7(1961):276-281.

55.9 Adapted from Charles J. Krebs. 2001. *Ecology,* 5th ed., fig. 19.1. San Francisco, CA: Benjamin Cummings. © 2001 Benjamin Cummings, an imprint of Addison Wesley Longman, Inc.; **55.10** Adapted from R. L. Westemeier et al. 1998. Tracking the long-term decline and recovery of an isolated population. *Science* 282:1696. © 1998 by the American Association for the Advancement of Science; **55.12** Data from K. A. Keating et al. 2003. Estimating numbers of females with cubs-of-the-year in the Yellowstone grizzly bear population. *Ursus* 13:161-174 and from M. A. Haroldson. 2003. Unduplicated females. Pages 11-17 in C.C. Schwartz and M. A. Haroldson, eds. Yellowstone grizzly bear investigations. Annual Report of the Interagency Grizzly Bear Study Team, 2002. U.S. Geological Survey, Bozeman, Montana; **55.17** From N. Myers et al., "Biodiversity hotspots for conservation priorities," *Nature,* Vol. 403, p. 853, 2/24/2000. Copyright © 2000 Nature Publishing, Inc. Used with permission; **55.18** Adapted from W. D. Newmark. 1985. Legal and biotic boundaries of western North American national parks: a problem of congruence. *Biological Conservation* 33:199. © 1985 Elsevier Applied Science Publishers Ltd., Barking, England; **55.21** Adapted from A. P. Dobson et al. 1997. Hopes for the future: restoration ecology and conservation biology. *Science* 277:515. © 1997 by the American Association for the Advancement of Science; **55.23** Data from Instituto Nacional de Estadistica y Censos de Costa Rica and Centro Centroamericano de Poblacion, Universidad de Costa Rica.

Glossary

5′ cap The 5′ end of a pre-mRNA molecule modified by the addition of a cap of guanine nucleotide.

A band The broad region that corresponds to the length of the thick filaments of myofibrils.

A site One of a ribosome's three binding sites for tRNA during translation. The A site holds the tRNA carrying the next amino acid to be added to the polypeptide chain. (A stands for aminoacyl tRNA.)

ABC model A model of flower formation identifying three classes of organ identity genes that direct formation of the four types of floral organs.

abdominal cavity The body cavity in mammals that primarily houses parts of the digestive, excretory, and reproductive systems. It is separated from the thoracic cavity by the diaphragm.

abiotic (ā´-bī-ot´-ik) Nonliving.

ABO blood groups Genetically determined classes of human blood that are based on the presence or absence of carbohydrates A and B on the surface of red blood cells. The ABO blood group phenotypes, also called blood types, are A, B, AB, and O.

abortion The termination of a pregnancy in progress.

abscisic acid (ABA) (ab-sis´-ik) A plant hormone that slows down growth, often antagonizing actions of growth hormones. Two of its many effects are to promote seed dormancy and facilitate drought tolerance.

absorption The uptake of small nutrient molecules by an organism's own body; the third main stage of food processing, following digestion.

absorption spectrum The range of a pigment's ability to absorb various wavelengths of light.

abyssal Referring to the very deep benthic zone of the ocean.

acanthodian (ak´-an-thō´-dē-un) Any of a group of ancient jawed fishes from the Devonian period.

acclimatization (uh-klī´-muh-tī-zā´-shun) Physiological adjustment to a change in an environmental factor.

accommodation The automatic adjustment of an eye to focus on near objects.

acetyl CoA Acetyl coenzyme A; the entry compound for the citric acid cycle in cellular respiration, formed from a fragment of pyruvate attached to a coenzyme.

acetylcholine (as´-uh-til-kō´-lēn) One of the most common neurotransmitters; functions by binding to receptors and altering the permeability of the postsynaptic membrane to specific ions, either depolarizing or hyperpolarizing the membrane.

acid A substance that increases the hydrogen ion concentration of a solution.

acid chyme (kīm) A mixture of recently swallowed food and gastric juice.

acid precipitation Rain, snow, or fog that is more acidic than pH 5.6.

acoelomate (uh-sē´-lō-māt) A solid-bodied animal lacking a cavity between the gut and outer body wall.

acquired immunity The kind of defense that is mediated by B lymphocytes (B cells) and T lymphocytes (T cells). It exhibits specificity, memory, and self-nonself recognition. Also called adaptive immunity.

acrosomal reaction The discharge of a sperm's acrosome when the sperm approaches an egg.

acrosome (ak´-ruh-sōm) A vesicle at the tip of a sperm cell that helps the sperm penetrate the egg.

actin (ak´-tin) A globular protein that links into chains, two of which twist helically about each other, forming microfilaments in muscle and other contractile elements in cells.

action potential A rapid change in the membrane potential of an excitable cell, caused by stimulus-triggered, selective opening and closing of voltage-sensitive gates in sodium and potassium ion channels.

action spectrum A graph that depicts the relative effectiveness of different wavelengths of radiation in driving a particular process.

activation energy *See* free energy of activation.

activator A protein that binds to DNA and stimulates transcription of a specific gene.

active immunity Long-lasting immunity conferred by the action of a person's B cells and T cells and the resulting B and T memory cells specific for a pathogen. Active immunity can develop as a result of natural infection or immunization.

active site The specific portion of an enzyme that attaches to the substrate by means of weak chemical bonds.

active transport The movement of a substance across a biological membrane against its concentration or electrochemical gradient with the help of energy input and specific transport proteins.

actual evapotranspiration The amount of water annually transpired by plants and evaporated from a landscape, usually measured in millimeters.

adaptive radiation The emergence of numerous species from a common ancestor introduced into an environment that presents a diversity of new opportunities and challenges.

adenohypophysis (ad´-uh-nō-hī-pof´-uh-sis) *See* anterior pituitary.

adenylyl cyclase (ad´-en-uh-lil) An enzyme that converts ATP to cyclic AMP in response to a chemical signal.

adhesion The attraction between different kinds of molecules.

adipose tissue A connective tissue that insulates the body and serves as a fuel reserve; contains fat-storing cells called adipose cells.

adrenal gland (uh-drē´-nul) One of two endocrine glands located adjacent to the kidneys in mammals. Endocrine cells in the outer portion (cortex) respond to ACTH by secreting steroid hormones that help maintain homeostasis during long-term stress. Neurosecretory cells in the central portion (medulla) secrete epinephrine and norepinephrine in response to nervous inputs triggered by short-term stress.

adrenocorticotropic hormone (ACTH) A tropic hormone produced and secreted by the anterior pituitary that stimulates the production and secretion of steroid hormones by the adrenal cortex.

adventitious A term describing any plant organ that grows in an atypical location, such as roots growing from stems.

aerobic (ār-ō´-bik) Containing oxygen; referring to an organism, environment, or cellular process that requires oxygen.

afferent arteriole (af´-er-ent) The blood vessel supplying a nephron.

age structure The relative number of individuals of each age in a population.

aggregate fruit A fruit derived from a single flower that has more than one carpel.

agonistic behavior (a´-gō-nis´-tik) A type of behavior involving a contest of some kind that determines which competitor gains access to some resource, such as food or mates.

AIDS (acquired immunodeficiency syndrome) The name of the late stages of HIV infection, defined by a specified reduction of T cells and the appearance of characteristic secondary infections.

alcohol fermentation The conversion of pyruvate to carbon dioxide and ethyl alcohol.

aldosterone (al-dos´-tuh-rōn) An adrenal hormone that acts on the distal tubules of the kidney to stimulate the reabsorption of sodium (Na^+) and the passive flow of water from the filtrate.

alga (plural, **algae**) (al´-guh, al´-jē) A photosynthetic, plant-like protist.

alimentary canal (al´-uh-men´-tuh-rē) A digestive tract consisting of a tube running between a mouth and an anus.

allantois (al´-an-tō´-is) One of four extraembryonic membranes; serves as a repository for the embryo's nitrogenous waste.

alleles (uh-lē´-ulz) Alternative versions of a gene that produce distinguishable phenotypic effects.

allometric growth (al´-ō-met´-rik) The variation in the relative rates of growth of various parts of the body, which helps shape the organism.

allopatric speciation (al´-ō-pat´-rik) A mode of speciation induced when an ancestral population becomes segregated by a geographic barrier or is itself divided into two or more geographically isolated subpopulations.

allopolyploid (al´-ō-pol´-ē-ployd) A common type of polyploid species resulting from two different species interbreeding and combining their chromosomes.

allosteric regulation The binding of a molecule to a protein that affects the function of the protein at a different site.

alpha (α) helix (al´-fuh hē´-liks) A spiral shape constituting one form of the secondary structure of proteins, arising from a specific hydrogen-bonding structure.

alternation of generations A life cycle in which there is both a multicellular diploid form, the sporophyte, and a multicellular haploid form, the gametophyte; characteristic of plants and some algae.

alternative RNA splicing A type of regulation at the RNA-processing level in which different mRNA molecules are produced from the same primary transcript, depending on which RNA segments are treated as exons and which as introns.

altruism (al´-trū-iz-um) Behavior that reduces an individual's fitness while increasing the fitness of another individual.

alveolus (al-vē´-uh-lus) (plural, **alveoli**) One of the dead-end, multilobed air sacs that constitute the gas exchange surface of the lungs.

Alzheimer's disease An age-related dementia (mental deterioration) characterized by confusion, memory loss, and other symptoms.

amacrine cell (am´-uh-krin) A neuron of the retina that helps integrate information before it is sent to the brain.

amino acid (uh-mēn´-ō) An organic molecule possessing both carboxyl and amino groups. Amino acids serve as the monomers of proteins.

amino group A functional group that consists of a nitrogen atom bonded to two hydrogen atoms; can act as a base in solution, accepting a hydrogen ion and acquiring a charge of 1+.

aminoacyl-tRNA synthetase An enzyme that joins each amino acid to the correct tRNA.

ammonia A small, very toxic molecule made up of three hydrogen atoms and one nitrogen atom; produced by nitrogen fixation and as a metabolic waste product of protein and nucleic acid metabolism.

ammonite A shelled cephalopod that was the dominant invertebrate predator for millions of years until the end of the Cretaceous period.

amniocentesis (am´-nē-ō-sen-tē´-sis) A technique of prenatal diagnosis in which amniotic fluid, obtained by aspiration from a needle inserted into the uterus, is analyzed to detect certain genetic and congenital defects in the fetus.

amnion (am´-nē-on) The innermost of four extraembryonic membranes; encloses a fluid-filled sac in which the embryo is suspended.

amniote Member of a clade of tetrapods that have an amniotic egg containing specialized membranes that protect the embryo, including mammals and birds and other reptiles.

amoeba (uh-mē´-buh) A protist grade characterized by the presence of pseudopodia.

amoebocyte (uh-mē´-buh-sīt) An amoeba-like cell that moves by pseudopodia, found in most animals; depending on the species, may digest and distribute food, dispose of wastes, form skeletal fibers, fight infections, and change into other cell types.

amphibian Member of the tetrapod class Amphibia, including salamanders, frogs, and caecilians.

amphipathic molecule (am´-fē-path´-ik) A mol-ecule that has both a hydrophilic region and a hydrophobic region.

amplification The strengthening of stimulus energy that is otherwise too weak to be carried into the nervous system.

anabolic pathway (an´-uh-bol´-ik) A metabolic pathway that synthesizes a complex molecule from simpler compounds.

anaerobic (an´-ār-ō´-bik) Lacking oxygen; referring to an organism, environment, or cellular process that lacks oxygen and may be poisoned by it.

anaerobic respiration The use of inorganic molecules other than oxygen to accept electrons at the "downhill" end of electron transport chains.

analogy Similarity between two species that is due to convergent evolution rather than to descent from a common ancestor with the same trait.

anaphase The fourth stage of mitosis, in which the chromatids of each chromosome have separated and the daughter chromosomes are moving to the poles of the cell.

anaphylactic shock (an´-uh-fi-lak-tic) An acute, whole-body, life-threatening, allergic response.

anatomy The study of the structure of an organism.

anchorage dependence The requirement that to divide, a cell must be attached to the substratum.

androgen (an´-drō-jen) Any steroid hormone, such as testosterone, that stimulates the development and maintenance of the male reproductive system and secondary sex characteristics.

aneuploidy (an´-yū-ploy-dē) A chromosomal aberration in which one or more chromosomes are present in extra copies or are deficient in number.

angiosperm (an´-jē-ō-sperm) A flowering plant, which forms seeds inside a protective chamber called an ovary.

angiotensin II A hormone that stimulates constriction of precapillary arterioles and increases reabsorption of NaCl and water by the proximal tubules of the kidney, increasing blood pressure and volume.

anhydrobiosis (an´-hī´-drō-bī-ō´-sis) The ability to survive in a dormant state when an organism's habitat dries up.

animal pole The portion of the egg where the least yolk is concentrated; opposite of vegetal pole.

Animalia The kingdom that consists of multicellular eukaryotes that ingest their food.

anion (an´-ī-on) A negatively charged ion.

annual A flowering plant that completes its entire life cycle in a single year or growing season.

anterior Referring to the head end of a bilaterally symmetrical animal.

anterior pituitary Also called the adenohypophysis; portion of the pituitary that develops from nonneural tissue; consists of endocrine cells that synthesize and secrete several tropic and nontropic hormones.

anther (an´-ther) In an angiosperm, the terminal pollen sac of a stamen, where pollen grains with male gametes form.

antheridium (an´-thuh-rid´-ē-um) (plural, **antheridia**) In plants, the male gametangium, a moist chamber in which gametes develop.

anthropoid (an´-thruh-poyd) A member of a primate group made up of the apes (gibbon, orangutan, gorilla, chimpanzee, and bonobo), monkeys, and humans.

antibody A protein secreted by plasma cells (differentiated B cells) that binds to a particular antigen and marks it for elimination; also called immunoglobulin. All antibody molecules have the same Y-shaped structure and in their monomer form consist of two identical heavy chains and two identical light chains joined by disulfide bridges.

anticodon (an´-tī-kō´-don) A specialized base triplet at one end of a tRNA molecule that recognizes a particular complementary codon on an mRNA molecule.

antidiuretic hormone (ADH) A hormone produced in the hypothalamus and released from the posterior pituitary. It promotes water retention by the kidneys as part of an elaborate feedback scheme that helps regulate the osmolarity of the blood.

antigen (an´-tuh-jen) A macromolecule that elicits an immune response by lymphocytes.

antigen presentation The process by which an MHC molecule binds to a fragment of an intracellular protein antigen and carries it to the cell surface, where it is displayed and can be recognized by a T cell.

antigen receptor The general term for a surface protein, located on B cells and T cells, that binds to antigens, initiating acquired immune responses. The antigen receptors on B cells are called B cell receptors (or membrane immunoglobulins), and the antigen receptors on T cells are called T cell receptors.

antigen-presenting cell A cell that ingests bacteria and viruses and destroys them, generating peptide fragments that are bound by class II MHC molecules and subsequently displayed on the cell surface to helper T cells. Macrophages, dendritic cells, and B cells are the primary antigen-presenting cells.

antiparallel The opposite arrangement of the sugar-phosphate backbones in a DNA double helix.

aphotic zone (ā´-fō´-tik) The part of the ocean beneath the photic zone, where light does not penetrate sufficiently for photosynthesis to occur.

apical dominance (ā´-pik-ul) Concentration of growth at the tip of a plant shoot, where a terminal bud partially inhibits axillary bud growth.

apical ectodermal ridge A limb-bud organizer region consisting of a thickened area of ectoderm at the tip of a limb bud.

apical meristem (ā´-pik-ul mār´-uh-stem) Embryonic plant tissue in the tips of roots and in the buds of shoots that supplies cells for the plant to grow in length.

apicomplexan (ap´-ē-kom-pleks´-un) A parasitic protozoan. Some apicomplexans cause human diseases.

apomixis (ap´-uh-mik´-sis) The asexual production of seeds.

apoplast (ap´-ō-plast) In plants, the continuum of cell walls plus the extracellular spaces.

apoptosis The changes that occur within a cell as it undergoes programmed cell death, which is brought about by signals that trigger the activation of a cascade of suicide proteins in the cell destined to die.

aposematic coloration (ap´-ō-sō-mat´-ik) The bright coloration of animals with effective physical or chemical defenses that acts as a warning to predators.

appendix A small, fingerlike extension of the vertebrate cecum; contains a mass of white blood cells that contribute to immunity.

aquaporin A transport protein in the plasma membrane of a plant or animal cell that specifically facilitates the diffusion of water across the membrane (osmosis).

aqueous humor (ā´-kwē-us hyū´-mer) Plasma-like liquid in the space between the lens and the cornea in the vertebrate eye; helps maintain the shape of the eye, supplies nutrients and oxygen to its tissues, and disposes of its wastes.

aqueous solution (ā´-kwē-us) A solution in which water is the solvent.

arbuscular mycorrhiza A distinct type of endomycorrhiza formed by glomeromycete fungi, in which the tips of the fungal hyphae that invade the plant roots branch into tiny treelike structures called arbuscules.

Archaea (ar´-kē-uh) One of two prokaryotic domains, the other being Bacteria.

archegonium (ar-ki-gō´-nē-um) (plural, **archegonia**) In plants, the female gametangium, a moist chamber in which gametes develop.

archenteron (ar-ken´-tuh-ron) The endoderm-lined cavity, formed during the gastrulation process, that develops into the digestive tract of an animal.

archosaur Member of the reptilian group that includes crocodiles, alligators, dinosaurs, and birds.

arteriole (ar-ter´-ē-ōl) A vessel that conveys blood between an artery and a capillary bed.

artery A vessel that carries blood away from the heart to organs throughout the body.

arthropod A segmented coelomate with a chitinous exoskeleton, jointed appendages, and a body formed of distinct groups of segments.

artificial selection The selective breeding of domesticated plants and animals to encourage the occurrence of desirable traits.

ascocarp The fruiting body of a sac fungus (ascomycete).

ascomycete *See* sac fungus.

ascus (plural, **asci**) A saclike spore capsule located at the tip of a dikaryotic hypha of a sac fungus.

asexual reproduction A type of reproduction involving only one parent that produces genetically identical offspring by budding or by the division of a single cell or the entire organism into two or more parts.

aspartate An amino acid that functions as a CNS neurotransmitter.

assisted reproductive technology (ART) Fertilization procedures that generally involve the surgical removal of eggs (secondary oocytes) from a woman's ovaries after hormonal stimulation, fertilizing the eggs, and returning them to the woman's body.

associative learning The acquired ability to associate one stimulus with another; also called classical conditioning.

aster A radial array of short microtubules that extends from each centrosome toward the plasma membrane in a cell undergoing mitosis.

astrocyte A glial cell that provides structural and metabolic support for neurons.

atherosclerosis A cardiovascular disease in which growths called plaques develop in the inner walls of the arteries, narrowing their inner diameters.

atom The smallest unit of matter that retains the properties of an element.

atomic mass The total mass of an atom, which is the mass in grams of one mole of the atom.

atomic nucleus An atom's central core, containing protons and neutrons.

atomic number The number of protons in the nucleus of an atom, unique for each element and designated by a subscript to the left of the elemental symbol.

ATP (adenosine triphosphate) (a-den´-ō-sēn trī-fos´-fāt) An adenine-containing nucleoside triphosphate that releases free energy when its phosphate bonds are hydrolyzed. This energy is used to drive endergonic reactions in cells.

ATP synthase A cluster of several membrane proteins found in the mitochondrial crista (and bacterial plasma membrane) that function in chemiosmosis with adjacent electron transport chains, using the energy of a hydrogen ion concentration gradient to make ATP. ATP synthases provide a port through which hydrogen ions diffuse into the matrix of a mitochondrion.

atrial natriuretic factor (ANF) (ā´-trē-al na´-trē-ū-ret´-ik) A peptide hormone that opposes the renin-angiotensin-aldosterone system (RAAS).

atrioventricular (AV) node A region of specialized muscle tissue between the right atrium and right ventricle where electrical impulses are delayed for about 0.1 second before spreading to the ventricles and causing them to contract.

atrioventricular (AV) valve A valve in the heart between each atrium and ventricle that prevents a backflow of blood when the ventricles contract.

atrium (ā´-trē-um) (plural, **atria**) A chamber that receives blood returning to the vertebrate heart.

autoimmune disease An immunological disorder in which the immune system turns against self.

autonomic nervous system (ot´-ō-nom´-ik) A subdivision of the motor nervous system of vertebrates that regulates the internal environment; consists of the sympathetic, parasympathetic, and enteric divisions.

autopolyploid (ot´-ō-pol´-ē-ploid) An individual that has more than two chromosome sets, all derived from a single species.

autosome (ot´-ō-sōm) A chromosome that is not directly involved in determining sex, as opposed to a sex chromosome.

autotroph (ot´-ō-trōf) An organism that obtains organic food molecules without eating other organisms or substances derived from other organisms. Autotrophs use energy from the sun or from the oxidation of inorganic substances to make organic molecules from inorganic ones.

auxin (ok´-sin) A term that primarily refers to indoleacetic acid (IAA), a natural plant hormone that has a variety of effects, including cell elongation, root formation, secondary growth, and fruit growth.

average heterozygosity The percent, on average, of a population's loci that are heterozygous in members of the population.

avirulent A term describing a pathogen that can only mildly harm, but not kill, the host plant.

axillary bud (ak´-sil-ār-ē) A structure that has the potential to form a lateral shoot, or branch. The bud appears in the angle formed between a leaf and a stem.

axon (ak´-son) A typically long extension, or process, from a neuron that carries nerve impulses away from the cell body toward target cells.

axon hillock The conical region of a neuron's axon where it joins the cell body; typically the region where nerve signals are generated.

B cell receptor The antigen receptor on B cells: a Y-shaped, membrane-bound molecule consisting of two identical heavy chains and two identical light chains linked by disulfide bridges and containing two antigen-binding sites; also called a membrane immunoglobulin or membrane antibody.

B lymphocyte (B cell) A type of lymphocyte that develops to maturity in the bone marrow. After encountering antigen, B cells differentiate into antibody-secreting plasma cells, the effector cells of humoral immunity.

Bacteria One of two prokaryotic domains, the other being Archaea.

bacterial artificial chromosome (BAC) An artificial version of a bacterial chromosome that can carry inserts of 100,000 to 500,000 base pairs.

bacteriophage (bak-tēr´-ē-ō-fāj) A virus that infects bacteria; also called a phage.

bacteroids A form of *Rhizobium* contained within the vesicles formed by the root cells of a root nodule.

baculum (bak´-ū-lum) A bone that is contained in, and helps stiffen, the penis of rodents, raccoons, walruses, whales, and several other mammals.

balanced polymorphism The ability of natural selection to maintain diversity in a population.

balancing selection Natural selection that maintains stable frequencies of two or more phenotypic forms in a population (balanced polymorphism).

bark All tissues external to the vascular cambium, consisting mainly of the secondary phloem and layers of periderm.

Barr body A dense object lying along the inside of the nuclear envelope in female mammalian cells, representing an inactivated X chromosome.

barrier method Contraception that relies on a physical barrier to block the passage of sperm. Examples include condoms and diaphragms.

Bartholin's glands (bar´-tō-linz) Glands near the vaginal opening in a human female that secrete lubricating fluid during sexual arousal.

basal angiosperms The most primitive lineages of flowering plants, including *Amborella,* water lilies, and star anise and relatives.

basal body (bā´-sul) A eukaryotic cell organelle consisting of a 9 + 0 arrangement of microtubule triplets; may organize the microtubule assembly of a cilium or flagellum; structurally identical to a centriole.

basal metabolic rate (BMR) The metabolic rate of a resting, fasting, and nonstressed endotherm.

basal nuclei A cluster of nuclei deep within the white matter of the cerebrum.

base A substance that reduces the hydrogen ion concentration of a solution.

basement membrane The floor of an epithelial membrane on which the basal cells rest.

base-pair substitution A type of point mutation; the replacement of one nucleotide and its partner in the complementary DNA strand by another pair of nucleotides.

basidiocarp Elaborate fruiting body of a dikaryotic mycelium of a club fungus.

basidiomycete *See* club fungus.

basidium (plural, **basidia**) A reproductive appendage that produces sexual spores on the gills of mushrooms (club fungi).

Batesian mimicry (bāt´-zē-un mim´-uh-krē) A type of mimicry in which a harmless species looks like a species that is poisonous or otherwise harmful to predators.

behavior Everything an animal does and how it does it, including muscular activities such as chasing prey, certain nonmuscular processes such as secreting a hormone that attracts a mate, and learning.

behavioral ecology The scientific study of animal behavior, including how it is controlled and how it develops, evolves, and contributes to survival and reproductive success.

benign tumor A mass of abnormal cells that remains at the site of origin.

benthic zone The bottom surface of an aquatic environment.

benthos (ben´-thōz) The communities of organisms living in the benthic zone of an aquatic biome.

beta oxidation A metabolic sequence that breaks fatty acids down to two-carbon fragments that enter the citric acid cycle as acetyl CoA.

beta (β) pleated sheet One form of the secondary structure of proteins in which the polypeptide chain folds back and forth. Two regions of the chain lie parallel to each other and are held together by hydrogen bonds.

biennial (bī-en´-ē-ul) A flowering plant that requires two years to complete its life cycle.

big-bang reproduction A life history in which adults have but a single reproductive opportunity to produce large numbers of offspring, such as the life history of the Pacific salmon; also known as semelparity.

bilateral symmetry Characterizing a body form with a central longitudinal plane that divides the body into two equal but opposite halves.

bilaterian (bī´-luh-tēr´-ē-uhn) Member of the clade Bilateria, animals with bilateral symmetry.

bile A mixture of substances that is produced in the liver, stored in the gallbladder, and acts as a detergent to aid in the digestion and absorption of fats.

binary fission The type of cell division by which prokaryotes reproduce. Each dividing daughter cell receives a copy of the single parental chromosome.

binomial The two-part latinized name of a species, consisting of genus and specific epithet.

biodiversity hot spot A relatively small area with an exceptional concentration of endemic species and a large number of endangered and threatened species.

bioenergetics The flow of energy through an animal, taking into account the energy stored in the food it consumes, the energy used for basic functions, activity, growth, reproduction, and regulation, and the energy lost to the environment as heat or in waste.

biofilm A surface-coating colony of prokaryotes that engage in metabolic cooperation.

biogenic amine A neurotransmitter derived from an amino acid.

biogeochemical cycle Any of the various nutrient circuits, which involve both biotic and abiotic components of ecosystems.

biogeography The study of the past and present distribution of species.

bioinformatics Using computing power, software, and mathematical models to process and integrate biological information from large data sets.

biological augmentation An approach to restoration ecology that uses organisms to add essential materials to a degraded ecosystem.

biological clock An internal timekeeper that controls an organism's biological rhythms. The biological clock marks time with or without environmental cues but often requires signals from the environment to remain tuned to an appropriate period. *See also* circadian rhythm.

biological magnification A trophic process in which retained substances become more concentrated with each link in the food chain.

biological species concept Definition of a species as a population or group of populations whose members have the potential to interbreed in nature and produce viable, fertile offspring, but are not able to produce viable, fertile offspring with members of other populations.

biology The scientific study of life.

biomanipulation A technique for restoring eutrophic lakes that reduces populations of algae by manipulating the higher-level consumers in the community rather than by changing nutrient levels or adding chemical treatments.

biomass The dry weight of organic matter comprising a group of organisms in a particular habitat.

biome (bī´-ōm) Any of the world's major ecosystems, classified according to the predominant vegetation and characterized by adaptations of organisms to that particular environment.

bioremediation The use of living organisms to detoxify and restore polluted and degraded ecosystems.

biosphere (bī´-ō-sfēr) The entire portion of Earth inhabited by life; the sum of all the planet's ecosystems.

biota All the organisms that are part of an ecosystem.

biotechnology The manipulation of living organisms or their components to produce useful products.

biotic (bī-ot´-tik) Pertaining to the living organisms in the environment.

bipolar cell A neuron that synapses with the axon of a rod or cone in the retina of the eye.

bipolar disorder Depressive mental illness characterized by swings of mood from high to low; also called manic-depressive disorder.

birth control pills Chemical contraceptives that inhibit ovulation, retard follicular development, or alter a woman's cervical mucus to prevent sperm from entering the uterus.

blade (1) A leaflike structure of a seaweed that provides most of the surface area for photosynthesis. (2) The flattened portion of a typical leaf.

blastocoel (blas´-tuh-sēl) The fluid-filled cavity that forms in the center of the blastula embryo.

blastocyst An embryonic stage in mammals; a hollow ball of cells produced one week after fertilization in humans.

blastoderm An embryonic cap of dividing cells resting on a large undivided yolk.

blastomere A small cell of an early embryo.

blastopore (blas´-tō-pōr) The opening of the archenteron in the gastrula that develops into the mouth in protostomes and the anus in deuterostomes.

blastula (blas´-tyū-luh) The hollow ball of cells marking the end stage of cleavage during early embryonic development.

blood A type of connective tissue with a fluid matrix called plasma in which blood cells are suspended.

blood pressure The hydrostatic force that blood exerts against the wall of a vessel.

blood vessels A set of tubes through which the blood moves through the body.

blood-brain barrier A specialized capillary arrangement in the brain that restricts the passage of most substances into the brain, thereby preventing dramatic fluctuations in the brain's environment.

blue-light photoreceptors A class of light receptors in plants. Blue light initiates a variety of responses, such as phototropism and slowing of hypocotyl elongation.

body cavity A fluid-containing space between the digestive tract and the body wall.

body plan In animals, the set of morphological and developmental traits that define a grade (level of organizational complexity).

Bohr shift A lowering of the affinity of hemoglobin for oxygen, caused by a drop in pH; facilitates the release of oxygen from hemoglobin in the vicinity of active tissues.

bolus A lubricated ball of chewed food.

bone A type of connective tissue, consisting of living cells held in a rigid matrix of collagen fibers embedded in calcium salts.

book lung An organ of gas exchange in spiders, consisting of stacked plates contained in an internal chamber.

bottleneck effect Genetic drift resulting from the reduction of a population, typically by a natural disaster, such that the surviving population is no longer genetically representative of the original population.

bottom-up model A model of community organization in which mineral nutrients control community organization because nutrients control plant numbers, which in turn control herbivore numbers, which in turn control predator numbers.

Bowman's capsule (bō´-munz) A cup-shaped receptacle in the vertebrate kidney that is the initial, expanded segment of the nephron where filtrate enters from the blood.

brachiopod A marine lophophorate with a shell divided into dorsal and ventral halves. Brachiopods are also called lamp shells.

brain hormone A hormone, produced by neurosecretory cells in the insect brain, that promotes development by stimulating the prothoracic glands to secrete ecdysone.

brainstem Collection of structures in the adult brain, including the midbrain, the pons, and the medulla oblongata; functions in homeostasis, coordination of movement, and conduction of information to higher brain centers.

brassinosteroids Steroid hormones in plants that have a variety of effects, including cell elongation, retarding leaf abscission, and promoting xylem differentiation.

breathing The process involving alternate inhalation and exhalation of air that ventilates the lungs.

breathing control center A brain center that directs the activity of organs involved in breathing.

bronchiole One of the fine branches of the bronchus that transport air to alveoli.

bronchus (bron´-kus) (plural, **bronchi**) One of a pair of breathing tubes that branch from the trachea into the lungs.

brown alga A phaeophyte; a marine, multicellular, autotrophic protist that is the most common type of seaweed. Brown algae include the kelps.

brown fat A tissue in some mammals, located in the neck and between the shoulders, that is specialized for rapid heat production.

bryophyte (brī´-uh-fīt) A moss, liverwort, or hornwort; a nonvascular plant that inhabits the land but lacks many of the terrestrial adaptations of vascular plants.

budding An asexual means of propagation in which outgrowths from the parent form and pinch off to live independently or else remain attached to eventually form extensive colonies.

buffer A substance that consists of acid and base forms in a solution and that minimizes changes in pH when extraneous acids or bases are added to the solution.

bulbourethral gland (bul´-bō-yū-rē´-thrul) One of a pair of glands near the base of the penis in the human male that secretes fluid that lubricates and neutralizes acids in the urethra during sexual arousal.

bulk feeder An animal that eats relatively large pieces of food.

bulk flow The movement of water due to a difference in pressure between two locations.

bundle sheath A protective covering around a leaf vein, consisting of one or more cell layers, usually parenchyma.

bundle-sheath cell A type of photosynthetic cell arranged into tightly packed sheaths around the veins of a leaf.

C_3 plant A plant that uses the Calvin cycle for the initial steps that incorporate CO_2 into organic material, forming a three-carbon compound as the first stable intermediate.

C_4 plant A plant that prefaces the Calvin cycle with reactions that incorporate CO_2 into a four-carbon compound, the end product of which supplies CO_2 for the Calvin cycle.

cadherins An important class of cell-to-cell adhesion molecules.

calcitonin (kal´-si-tō´-nin) A hormone secreted by the thyroid gland that lowers blood calcium levels by promoting calcium deposition in bone and calcium excretion from the kidneys.

callus A mass of dividing, undifferentiated cells at the cut end of a shoot.

calorie (cal) The amount of heat energy required to raise the temperature of 1 g of water by 1°C; also the amount of heat energy that 1 g of water releases when it cools by 1°C. The Calorie (with a capital C), usually used to indicate the energy content of food, is a kilocalorie.

Calvin cycle The second of two major stages in photosynthesis (following the light reactions), involving atmospheric CO_2 fixation and reduction of the fixed carbon into carbohydrate.

calyptra A protective cap of gametophyte tissue that wholly or partially covers an immature capsule in many mosses.

CAM plant A plant that uses crassulacean acid metabolism, an adaptation for photosynthesis in arid conditions, first discovered in the family Crassulaceae. Carbon dioxide entering open stomata during the night is converted into organic acids, which release CO_2 for the Calvin cycle during the day, when stomata are closed.

Cambrian explosion A burst of evolutionary origins when most of the major body plans of animals appeared in a relatively brief time in geologic history; recorded in the fossil record about 542–525 million years ago.

canopy The uppermost layer of vegetation in a terrestrial biome.

capillary (kap´-il-ār-ē) A microscopic blood vessel that penetrates the tissues and consists of a single layer of endothelial cells that allows exchange between the blood and interstitial fluid.

capillary bed A network of capillaries that infiltrate every organ and tissue in the body.

capsid The protein shell that encloses a viral genome. It may be rod-shaped, polyhedral, or more complex in shape.

capsule (1) A sticky layer that surrounds the cell walls of some prokaryotes, protecting the cell surface and sometimes helping to glue the cell to surfaces. (2) The sporangium of a bryophyte (moss, liverwort, or hornwort).

carbohydrate (kar´-bō-hī´-drāt) A sugar (monosaccharide) or one of its dimers (disaccharides) or polymers (polysaccharides).

carbon fixation The incorporation of carbon from CO_2 into an organic compound by an autotrophic organism (a plant, another photosynthetic organism, or a chemoautotrophic prokaryote).

carbonyl group (kar´-buh-nēl´) A functional group present in aldehydes and ketones and consisting of a carbon atom double-bonded to an oxygen atom.

carboxyl group (kar-bok´-sil) A functional group present in organic acids and consisting of a single carbon atom double-bonded to an oxygen atom and also bonded to a hydroxyl group.

cardiac cycle (kar´-dē-ak) The alternating contractions and relaxations of the heart.

cardiac muscle A type of muscle that forms the contractile wall of the heart. Its cells are joined by intercalated disks that relay each heartbeat.

cardiac output The volume of blood pumped per minute by the left ventricle of the heart.

cardiovascular disease (kar´-dē-ō-vas´-kyū-ler) Diseases of the heart and blood vessels.

cardiovascular system A closed circulatory system with a heart and branching network of arteries, capillaries, and veins. The system is characteristic of vertebrates.

carnivore An animal, such as a shark, hawk, or spider, that eats other animals.

carotenoid (kuh-rot´-uh-noyd) An accessory pigment, either yellow or orange, in the chloroplasts of plants. By absorbing wavelengths of light that chlorophyll cannot, carotenoids broaden the spectrum of colors that can drive photosynthesis.

carpel (kar´-pul) The ovule-producing reproductive organ of a flower, consisting of the stigma, style, and ovary.

carrier In genetics, an individual who is heterozygous at a given genetic locus, with one normal allele and one potentially harmful recessive allele. The heterozygote is phenotypically normal for the character determined by the gene but can pass on the harmful allele to offspring.

carrying capacity The maximum population size that can be supported by the available resources, symbolized as *K*.

cartilage (kar´-til-ij) A type of flexible connective tissue with an abundance of collagenous fibers embedded in chondroitin sulfate.

Casparian strip (kas-par´-ē-un) A water-impermeable ring of wax in the endodermal cells of plants that blocks the passive flow of water and solutes into the stele by way of cell walls.

catabolic pathway (kat´-uh-bol´-ik) A metabolic pathway that releases energy by breaking down complex molecules to simpler compounds.

catalyst A chemical agent that changes the rate of a reaction without being consumed by the reaction.

catastrophism The hypothesis by Georges Cuvier that each boundary between strata corresponded in time to a catastrophe, such as a flood or drought, that had destroyed many of the species living there at that time.

catecholamine Any of a class of compounds, including the hormones epinephrine and norepinephrine, that are synthesized from the amino acid tyrosine.

cation (kat´-ī-on) An ion with a positive charge, produced by the loss of one or more electrons.

cation exchange A process in which positively charged minerals are made available to a plant when hydrogen ions in the soil displace mineral ions from the clay particles.

CD4 A surface protein, present on most helper T cells, that binds to class II MHC molecules on antigen-presenting cells, enhancing the interaction between the T cell and the antigen-presenting cell.

CD8 A surface protein, present on most cytotoxic cells, that binds to class I MHC molecules on target cells, enhancing the interaction between the T cell and the target cell.

cDNA library A limited gene library using complementary DNA. The library includes only the genes that were transcribed in the cells examined.

cecum (sē´-kum) (plural, **ceca**) A blind outpocket of a hollow organ such as an intestine.

cell Life's fundamental unit of structure and function.

cell adhesion molecules (CAMs) Glycoproteins that contribute to cell migration and stable tissue structure.

cell body The part of a neuron that houses the nucleus and other organelles.

cell cycle An ordered sequence of events in the life of a eukaryotic cell, from its origin in the division of a parent cell until its own division into two; composed of the M, G_1, S, and G_2 phases.

cell cycle control system A cyclically operating set of molecules in the cell that triggers and coordinates key events in the cell cycle.

cell differentiation The structural and functional divergence of cells as they become specialized during a multicellular organism's development; dependent on the control of gene expression.

cell division The reproduction of cells.

cell fractionation The disruption of a cell and separation of its organelles by centrifugation.

cell lineage The ancestry of a cell.

cell-mediated immune response The branch of acquired immunity that involves the activation of cytotoxic T cells, which defend against infected cells, cancer cells, and transplanted cells.

cell plate A double membrane across the midline of a dividing plant cell, between which the new cell wall forms during cytokinesis.

cellular respiration The most prevalent and efficient catabolic pathway for the production of ATP, in which oxygen is consumed as a reactant along with the organic fuel.

cellular slime mold A type of protist that has unicellular amoeboid cells and aggregated reproductive bodies in its life cycle.

cellulose (sel´-yū-lōs) A structural polysaccharide of cell walls, consisting of glucose monomers joined by β-1,4-glycosidic linkages.

cell wall A protective layer external to the plasma membrane in plant cells, prokaryotes, fungi, and some protists. In plant cells, the wall is formed of cellulose fibers embedded in a polysaccharide-protein matrix. The primary cell wall is thin and flexible, whereas the secondary cell wall is stronger and more rigid and is the primary constituent of wood.

Celsius scale (sel´-sē-us) A temperature scale (°C) equal to 5/9 (°F − 32) that measures the freezing point of water at 0°C and the boiling point of water at 100°C.

central canal The narrow cavity in the center of the spinal cord that is continuous with the fluid-filled ventricles of the brain.

central nervous system (CNS) In vertebrate animals, the brain and spinal cord.

central vacuole A membranous sac in a mature plant cell with diverse roles in reproduction, growth, and development.

centriole (sen´-trē-ōl) A structure in an animal cell composed of cylinders of microtubule triplets arranged in a 9 + 0 pattern. An animal cell usually has a pair of centrioles involved in cell division.

centromere (sen´-trō-mēr) The centralized region joining two sister chromatids.

centrosome (sen´-trō-sōm) Material present in the cytoplasm of all eukaryotic cells, important during cell division; the microtubule-organizing center.

cephalization (sef´-uh-luh-zā´-shun) An evolutionary trend toward the concentration of sensory equipment on the anterior end of the body.

cerebellum (sār´-ruh-bel´-um) Part of the vertebrate hindbrain located dorsally; functions in unconscious coordination of movement and balance.

cerebral cortex (suh-rē´-brul) The surface of the cerebrum; the largest and most complex part of the mammalian brain, containing sensory and motor nerve cell bodies of the cerebrum; the part of the vertebrate brain most changed through evolution.

cerebral hemisphere The right or left side of the vertebrate brain.

cerebrospinal fluid (suh-rē´-brō-spī´-nul) Blood-derived fluid that surrounds, protects against infection, nourishes, and cushions the brain and spinal cord.

cerebrum (suh-rē´-brum) The dorsal portion of the vertebrate forebrain, composed of right and left hemispheres; the integrating center for memory, learning, emotions, and other highly complex functions of the central nervous system.

cervix (ser´-viks) The neck of the uterus, which opens into the vagina.

chaparral (shap´-uh-ral´) A scrubland biome of dense, spiny evergreen shrubs found at midlatitudes along coasts where cold ocean currents circulate offshore; characterized by mild, rainy winters and long, hot, dry summers.

chaperonin A protein molecule that assists the proper folding of other proteins.

character An observable heritable feature.

character displacement The tendency for characteristics to be more divergent in sympatric populations of two species than in allopatric populations of the same two species.

checkpoint A critical control point in the cell cycle where stop and go-ahead signals can regulate the cycle.

chelicera One of a pair of clawlike feeding appendages characteristic of cheliceriforms.

cheliceriform An arthropod that has chelicerae and a body divided into a cephalothorax and an abdomen. Living cheliceriforms include sea spiders, horseshoe crabs, scorpions, ticks, and spiders.

chemical bond An attraction between two atoms, resulting from a sharing of outer-shell electrons or the presence of opposite charges on the atoms. The bonded atoms gain complete outer electron shells.

chemical energy Energy stored in the chemical bonds of molecules; a form of potential energy.

chemical equilibrium In a reversible chemical reaction, the point at which the rate of the forward reaction equals the rate of the reverse reaction.

chemical reaction A process leading to chemical changes in matter; involves the making and/or breaking of chemical bonds.

chemiosmosis (kem´-ē-oz-mō´-sis) An energy-coupling mechanism that uses energy stored in the form of a hydrogen ion gradient across a membrane to drive cellular work, such as the synthesis of ATP. Most ATP synthesis in cells occurs by chemiosmosis.

chemoautotroph (ke′-mō-ot′-ō-trōf) An organism that needs only carbon dioxide as a carbon source but that obtains energy by oxidizing inorganic substances.

chemoheterotroph (kē´-mō-het´-er-ō-trōf) An organism that must obtain both energy and carbon by consuming organic molecules.

chemokine Any of about 50 different proteins, secreted by many cell types near a site of injury or infection, that help direct migration of white blood cells to an injury site and induces other changes central to inflammation.

chemoreceptor A receptor that transmits information about the total solute concentration in a solution or about individual kinds of molecules.

chiasma (plural, **chiasmata**) (kī-az´-muh, kī-az´-muh-tuh) The X-shaped, microscopically visible region representing homologous chromatids that have exchanged genetic material through crossing over during meiosis.

chimera An organism with a mixture of genetically different cells.

chitin (kī-tin) A structural polysaccharide of an amino sugar found in many fungi and in the exoskeletons of all arthropods.

chlorophyll (klōr´-ō-fil) A green pigment located within the chloroplasts of plants. Chlorophyll *a* can participate directly in the light reactions, which convert solar energy to chemical energy.

chlorophyll *a* A type of blue-green photosynthetic pigment that participates directly in the light reactions.

chlorophyll *b* A type of yellow-green accessory photosynthetic pigment that transfers energy to chlorophyll *a*.

chloroplast (klōr´-ō-plast) An organelle found only in plants and photosynthetic protists that absorbs sunlight and uses it to drive the synthesis of organic compounds from carbon dioxide and water.

choanocyte (kō-an´-uh-sīt) A flagellated feeding cell found in sponges. Also called a collar cell, it has a collar-like ring that traps food particles around the base of its flagellum.

cholesterol (kō-les´-tuh-rol) A steroid that forms an essential component of animal cell membranes and acts as a precursor molecule for the synthesis of other biologically important steroids.

chondrichthyan Member of the class Chondrichthyes, vertebrates with skeletons made mostly of cartilage, such as sharks and rays.

chondrocyte Cartilage cell that secretes collagen and chondroitin sulfate.

chordate (kōr´-dāt) Member of the phylum Chordata, animals that at some point during their development have a notochord; a dorsal, hollow nerve cord; pharyngeal slits or clefts; and a muscular, post-anal tail.

chorion (kōr´-ē-on) The outermost of four extraembryonic membranes; contributes to the formation of the mammalian placenta.

chorionic villus sampling (CVS) (kōr´-ē-on´-ik vil´-us) A technique of prenatal diagnosis in which a small sample of the fetal portion of the placenta is removed and analyzed to detect certain genetic and congenital defects in the fetus.

choroid A thin, pigmented inner layer of the vertebrate eye.

chromatin (krō´-muh-tin) The complex of DNA and proteins that makes up a eukaryotic chromosome. When the cell is not dividing, chromatin exists as a mass of very long, thin fibers that are not visible with a light microscope.

chromosome (krō´-muh-sōm) A threadlike, gene-carrying structure found in the nucleus. Each chromosome consists of one very long DNA molecule and associated proteins. *See* chromatin.

chromosome theory of inheritance A basic principle in biology stating that genes are located on chromosomes and that the behavior of chromosomes during meiosis accounts for inheritance patterns.

chylomicron (kī´-lō-mī´-kron) One of the small intracellular globules composed of fats that are mixed with cholesterol and coated with special proteins.

chytrid (kī´-trid) Member of the fungal phylum Chytridiomycota, mostly aquatic fungi with flagellated zoospores that probably represent the most primitive fungal lineage.

ciliary body A portion of the vertebrate eye associated with the lens. It produces the clear, watery aqueous humor that fills the anterior cavity of the eye.

ciliate (sil´-ē-it) A type of protozoan that moves by means of cilia.

cilium (sil´-ē-um) (plural, **cilia**) A short cellular appendage specialized for locomotion, formed from a core of nine outer doublet microtubules and two inner single microtubules ensheathed in an extension of plasma membrane.

circadian rhythm (ser-kā´-dē-un) A physiological cycle of about 24 hours that is present in all eukaryotic organisms and that persists even in the absence of external cues.

cis Arrangement of two noncarbon atoms, each bound to one of the carbons in a carbon-carbon double bond, where the two noncarbon atoms are on the same side relative to the double bond.

citric acid cycle A chemical cycle involving eight steps that completes the metabolic breakdown of glucose molecules to carbon dioxide; occurs within the mitochondrion; the second major stage in cellular respiration.

clade A group of species that includes an ancestral species and all its descendants.

cladistics (kluh-dis´-tiks) The analysis of how species may be grouped into clades.

cladogram A diagram depicting patterns of shared characteristics among species.

class In classification, the taxonomic category above order.

class I MHC molecules A collection of cell surface proteins encoded by a family of genes called the major histocompatibility complex. Class I MHC molecules are found on nearly all nucleated cells.

class II MHC molecules A collection of cell surface proteins encoded by a family of genes called the major histocompatibility complex. Class II MHC molecules are restricted to a few specialized cell types, commonly called antigen-presenting cells (dendritic cells, macrophages, and B cells).

classical conditioning A type of associative learning; the association of a normally irrelevant stimulus with a fixed behavioral response.

cleavage The process of cytokinesis in animal cells, characterized by pinching of the plasma membrane. Also, the succession of rapid cell divisions without growth during early embryonic development that converts the zygote into a ball of cells.

cleavage furrow The first sign of cleavage in an animal cell; a shallow groove in the cell surface near the old metaphase plate.

climate The prevailing weather conditions at a locality.

climograph A plot of the temperature and precipitation in a particular region.

cline A graded variation in a trait that parallels a gradient in the environment.

clitoris (klit´-uh-ris) An organ in the female that engorges with blood and becomes erect during sexual arousal.

cloaca (klō-ā´-kuh) A common opening for the digestive, urinary, and reproductive tracts found in many nonmammalian vertebrates but in few mammals.

clonal selection The process by which an antigen selectively binds to and activates only those lymphocytes bearing receptors specific for the antigen. The selected lymphocytes proliferate and differentiate into a clone of effector cells and a clone of memory cells specific for the stimulating antigen. Clonal selection accounts for the specificity and memory of acquired immune responses.

clone (1) A lineage of genetically identical individuals or cells. (2) In popular usage, a single individual organism that is genetically identical to another individual. (3) As a verb, to make one or more genetic replicas of an individual or cell. *See also* gene cloning.

cloning Using a somatic cell from a multicellular organism to make one or more genetically identical individuals.

cloning vector An agent used to transfer DNA in genetic engineering. A plasmid that moves recombinant DNA from a test tube back into a cell is an example of a cloning vector, as is a virus that transfers recombinant DNA by infection.

closed circulatory system A circulatory system in which blood is confined to vessels and is kept separate from the interstitial fluid.

club fungus The common name for members of the phylum Basidiomycota. The name comes from the club-like shape of the basidium.

cnidocyte (nī´-duh-sīt) A specialized cell for which the phylum Cnidaria is named; contains a capsule containing a fine coiled thread, which, when discharged, functions in defense and prey capture.

cochlea (kok´-lē-uh) The complex, coiled organ of hearing that contains the organ of Corti.

codominance The situation in which the phenotypes of both alleles are exhibited in the heterozygote.

codon (kō´-don) A three-nucleotide sequence of DNA or mRNA that specifies a particular amino acid or termination signal; the basic unit of the genetic code.

coefficient of relatedness The probability that a particular gene present in one individual will also be inherited from a common parent or ancestor in a second individual.

coelom (sē´-lōm) A body cavity completely lined with mesoderm.

coelomate (sē´-lō-māt) Animal that possesses a true coelom (fluid-filled body cavity lined by tissue completely derived from mesoderm).

coenocytic (sē´-nō-sit´-ik) Referring to a multinucleated condition resulting from the repeated division of nuclei without cytoplasmic division.

coenzyme (kō-en´-zīm) An organic molecule serving as a cofactor. Most vitamins function as coenzymes in important metabolic reactions.

coevolution The mutual evolutionary influence between two different species interacting with each other and reciprocally influencing each other's adaptations.

cofactor Any nonprotein molecule or ion that is required for the proper functioning of an enzyme. Cofactors can be permanently bound to the active site or may bind loosely with the substrate during catalysis.

cognition The ability of an animal's nervous system to perceive, store, process, and use information obtained by its sensory receptors.

cognitive ethology The scientific study of cognition; the study of the connection between data processing by nervous systems and animal behavior.

cognitive map A representation within the nervous system of spatial relations between objects in an animal's environment.

cohesion The binding together of like molecules, often by hydrogen bonds.

cohort A group of individuals of the same age, from birth until all are dead.

coitus (kō´-uh-tus) The insertion of a penis into a vagina, also called sexual intercourse.

coleoptile (kō´-lē-op´-tul) The covering of the young shoot of the embryo of a grass seed.

coleorhiza (kō´-lē-uh-rī´-zuh) The covering of the young root of the embryo of a grass seed.

collagen A glycoprotein in the extracellular matrix of animal cells that forms strong fibers, found extensively in connective tissue and bone; the most abundant protein in the animal kingdom.

collagenous fiber A tough fiber of the extracellular matrix. Collagenous fibers are made of collagen, are nonelastic, and do not tear easily when pulled lengthwise.

collecting duct The location in the kidney where filtrate from renal tubules is collected; the filtrate is now called urine.

collenchyma cell (kō-len´-kim-uh) A flexible plant cell type that occurs in strands or cylinders that support young parts of the plant without restraining growth.

colloid A mixture made up of a liquid and particles that (because of their large size) remain suspended in that liquid.

colon *See* large intestine.

colony A collection of autonomously replicating cells of the same species.

columnar The column shape of a type of epithelial cell.

commensalism (kuh-men´-suh-lizm) A symbiotic relationship in which one organism benefits but the other is neither helped nor harmed.

communication Animal behavior involving transmission of, reception of, and response to signals.

community All the organisms that inhabit a particular area; an assemblage of populations of different species living close enough together for potential interaction.

community ecology The study of how interactions between species affect community structure and organization.

companion cell A type of plant cell that is connected to a sieve-tube member by many plasmodesmata and whose nucleus and ribosomes may serve one or more adjacent sieve-tube members.

competitive exclusion The concept that when populations of two similar species compete for the same limited resources, one population will use the resources more efficiently and have a reproductive advantage that will eventually lead to the elimination of the other population.

competitive inhibitor A substance that reduces the activity of an enzyme by entering the active site in place of the substrate whose structure it mimics.

complement system A group of about 30 blood proteins that may amplify the inflammatory response, enhance phagocytosis, or directly lyse pathogens. The complement system is activated in a cascade initiated by surface antigens on microorganisms or by antigen-antibody complexes.

complementary DNA (cDNA) A DNA molecule made *in vitro* using mRNA as a template and the enzyme reverse transcriptase. A cDNA molecule therefore corresponds to a gene, but lacks the introns present in the DNA of the genome.

complete digestive tract A digestive tube that runs between a mouth and an anus; also called an alimentary canal. An incomplete digestive tract has only one opening.

complete dominance The situation in which the phenotypes of the heterozygote and dominant homozygote are indistinguishable.

complete flower A flower that has all four basic floral organs: sepals, petals, stamens, and carpels.

complete metamorphosis The transformation of a larva into an adult that looks very different, and often functions very differently in its environment, than the larva.

compound A substance consisting of two or more elements in a fixed ratio.

compound eye A type of multifaceted eye in insects and crustaceans consisting of up to several thousand light-detecting, focusing ommatidia; especially good at detecting movement.

concentration gradient An increase or decrease in the density of a chemical substance in an area. Cells often maintain concentration gradients of ions across their membranes. When a gradient exists, the ions or other chemical substances involved tend to move from where they are more concentrated to where they are less concentrated.

conception The fertilization of the egg by a sperm cell in humans.

condensation reaction A reaction in which two molecules become covalently bonded to each other through the loss of a small molecule, usually water; also called a dehydration reaction.

condom A thin, latex rubber or natural membrane sheath that fits over the penis to collect semen.

condont Ancient lineage of jawless vertebrates that arose during the Cambrian period.

conduction The direct transfer of thermal motion (heat) between molecules of objects in direct contact with each other.

cone cell One of two types of photoreceptors in the vertebrate eye; detects color during the day.

conformer A characterization of an animal in regard to environmental variables. A conformer allows some conditions within its body to vary with certain external changes.

conidia Naked, asexual spores produced at the ends of specialized hyphae in ascomycetes.

conifer A member of the largest gymnosperm phylum. Most conifers are cone-bearing trees, such as pines and firs.

coniferous forest A terrestrial biome characterized by long, cold winters and dominated by cone-bearing trees.

conjugation (kon´-jū-gā-shun) In prokaryotes, the direct transfer of DNA between two cells that are temporarily joined. In ciliates, a sexual process in which two cells exchange haploid micronuclei.

conjunctiva (kon´-junk-tī´-vuh) A mucous membrane that helps keep the eye moist; lines the inner surface of the eyelid and covers the front of the eyeball, except the cornea.

connective tissue Animal tissue that functions mainly to bind and support other tissues, having a sparse population of cells scattered through an extracellular matrix.

conodont An early, soft-bodied vertebrate with prominent eyes and dental elements.

conservation biology The integrated study of ecology, evolutionary biology, physiology, molecular biology, genetics, and behavioral biology in an effort to sustain biological diversity at all levels.

contraception The prevention of pregnancy.

contractile vacuole A membranous sac that helps move excess water out of certain cells.

control element A segment of noncoding DNA that helps regulate transcription of a gene by binding proteins called transcription factors.

controlled experiment An experiment in which an experimental group is compared to a control group that varies only in the factor being tested.

convection The mass movement of warmed air or liquid to or from the surface of a body or object.

convergent extension A mechanism of cell crawling in which the cells of a tissue layer rearrange themselves in such a way that the sheet of cells becomes narrower while it becomes longer.

cooperativity An interaction of the constituent subunits of a protein whereby a conformational change in one subunit is transmitted to all the others.

copepod (kō´-puh-pod) Any of a group of small crustaceans that are important members of marine and freshwater plankton communities.

coral reef A warm-water, tropical ecosystem dominated by the hard skeletal structures secreted primarily by the resident cnidarians.

corepressor A small molecule that cooperates with a repressor protein to switch an operon off.

cork cambium (kam´-bē-um) A cylinder of meristematic tissue in woody plants that replaces the epidermis with thicker, tougher cork cells.

cornea (kor´-nē-uh) The transparent frontal portion of the sclera, which admits light into the vertebrate eye.

corpus callosum (kor´-pus kuh-lō-sum) The thick band of nerve fibers that connect the right and left cerebral hemispheres in placental mammals, enabling the hemispheres to process information together.

corpus luteum (kor´-pus lū´-tē-um) A secreting tissue in the ovary that forms from the collapsed follicle after ovulation and produces progesterone.

cortex Ground tissue that is between the vascular tissue and dermal tissue in a root or dicot stem.

cortical granules Vesicles located just under the plasma membrane of an egg cell that undergo exocytosis during the cortical reaction.

cortical nephrons Nephrons located almost entirely in the renal cortex. These nephrons have a reduced loop of Henle.

cortical reaction Exocytosis of enzymes from cortical granules in the egg cytoplasm during fertilization.

corticosteroid Any steroid hormone produced and secreted by the adrenal cortex.

cotransport The coupling of the "downhill" diffusion of one substance to the "uphill" transport of another against its own concentration gradient.

cotyledon (kot´-uh-lē´-don) A seed leaf of an angiosperm embryo. Some species have one cotyledon, others two.

countercurrent exchange The opposite flow of adjacent fluids that maximizes transfer rates; for example, blood in the gills flows in the opposite direction in which water passes over the gills, maximizing oxygen uptake and carbon dioxide loss.

countercurrent heat exchanger An arrangement of blood vessels that helps trap heat in the body core and is important in reducing heat loss in many endotherms.

countercurrent multiplier system A countercurrent system in which energy is expended in active transport to facilitate exchange of materials and create concentration gradients. For example, the loop of Henle actively transports NaCl from the filtrate in the upper part of the ascending limb of the loop, making the urine-concentrating function of the kidney more effective.

covalent bond (kō-vā´-lent) A type of strong chemical bond in which two atoms share one or more pairs of valence electrons.

cranial nerve A nerve that leaves the brain and innervates an organ of the head or upper body.

craniate A chordate with a head.

crassulacean acid metabolism (CAM) A type of metabolism in which carbon dioxide is taken in at night and incorporated into a variety of organic acids.

crista (plural, **cristae**) (kris´-tuh, kris´-tē) An infolding of the inner membrane of a mitochondrion that houses the electron transport chain and the enzyme catalyzing the synthesis of ATP.

critical load The amount of added nutrient, usually nitrogen or phosphorus, that can be absorbed by plants without damaging ecosystem integrity.

crop rotation The alternation of planting a nonlegume one year and a legume the next year to restore concentration of fixed nitrogen in the soil.

crossing over The reciprocal exchange of genetic material between nonsister chromatids during prophase I of meiosis.

cross-pollination In angiosperms, the transfer of pollen from an anther of a flower on one plant to the stigma of a flower on another plant of the same species.

crustacean A member of a subphylum of arthropods that includes lobsters, crayfish, crabs, shrimps, and barnacles.

cryptic coloration Camouflage, making potential prey difficult to spot against its background.

cuboidal The cubic shape of a type of epithelial cell.

culture The ideas, customs, skills, rituals, and similar activities of a people or group that are passed along to succeeding generations.

cuticle (kyū´-tuh-kul) (1) A waxy covering on the surface of stems and leaves that acts as an adaptation to prevent desiccation in terrestrial plants. (2) The exoskeleton of an arthropod, consisting of layers of protein and chitin that are variously modified for different functions. (3) A tough coat that covers the body of a nematode.

cyclic AMP (cAMP) Cyclic adenosine monophosphate, a ring-shaped molecule made from ATP that is a common intracellular signaling molecule (second messenger) in eukaryotic cells (for example, in vertebrate endocrine cells). It is also a regulator of some bacterial operons.

cyclic electron flow A route of electron flow during the light reactions of photosynthesis that involves only photosystem I and that produces ATP but not NADPH or oxygen.

cyclin (sī´-klin) A regulatory protein whose concentration fluctuates cyclically.

cyclin-dependent kinase (Cdk) A protein kinase that is active only when attached to a particular cyclin.

cystic fibrosis A human genetic disorder caused by a recessive allele for a chloride channel protein; characterized by an excessive secretion of mucus and consequent vulnerability to infection; fatal if untreated.

cytochrome (sī´-tō-krōm) An iron-containing protein, a component of electron transport chains in mitochondria and chloroplasts.

cytogenetic map Chart of a chromosome that locates genes with respect to chromosomal features.

cytokine Any of a group of proteins secreted by a number of cell types, including macrophages and helper T cells, that regulate the function of lymphocytes and other cells of the immune system.

cytokinesis (sī´-tō-kuh-nē´-sis) The division of the cytoplasm to form two separate daughter cells immediately after mitosis.

cytokinins (sī´-tō-kī´-nins) A class of related plant hormones that retard aging and act in concert with auxin to stimulate cell division, influence the pathway of differentiation, and control apical dominance.

cytoplasm (sī´-tō-plaz´-um) The entire contents of the cell, exclusive of the nucleus, and bounded by the plasma membrane.

cytoplasmic determinants The maternal substances in the egg that influence the course of early development by regulating the expression of genes that affect the developmental fate of cells.

cytoplasmic streaming A circular flow of cytoplasm, involving myosin and actin filaments, that speeds the distribution of materials within cells.

cytoskeleton A network of microtubules, microfilaments, and intermediate filaments that branch throughout the cytoplasm and serve a variety of mechanical and transport functions.

cytosol (sī´-tō-sol) The semifluid portion of the cytoplasm.

cytotoxic T cell A type of lymphocyte that, when activated, kills infected cells, cancer cells, and transplanted cells.

daily torpor A daily decrease in metabolic activity and body temperature during times of inactivity for some small mammals and birds.

dalton A measure of mass for atoms and subatomic particles.

data Recorded observations.

day-neutral plant A plant whose flowering is not affected by photoperiod.

decapod A member of the group of crustaceans that includes lobsters, crayfish, crabs, and shrimps.

decomposer Any of the saprobic fungi and prokaryotes that absorb nutrients from nonliving organic material such as corpses, fallen plant material, and the wastes of living organisms, and convert them into inorganic forms.

deductive reasoning A type of logic in which specific results are predicted from a general premise.

deep-sea hydrothermal vent A dark, hot, oxygen-deficient environment associated with volcanic activity. The food producers are chemoautotrophic prokaryotes.

de-etiolation The changes a plant shoot undergoes in response to sunlight; also known informally as greening.

dehydration reaction A chemical reaction in which two molecules covalently bond to each other with the removal of a water molecule.

deletion (1) A deficiency in a chromosome resulting from the loss of a fragment through breakage. (2) A mutational loss of one or more nucleotide pairs from a gene.

demographic transition A shift from zero population growth in which birth rates and death rates are high to zero population growth characterized instead by low birth and death rates.

demography The study of statistics relating to births and deaths in populations.

denaturation (dē-nā´-chur-ā´-shun) In proteins, a process in which a protein unravels and loses its native conformation, thereby becoming biologically inactive. In DNA, the separation of the two strands of the double helix. Denaturation occurs under extreme conditions of pH, salt concentration, and temperature.

dendrite (den´-drīt) One of usually numerous, short, highly branched processes of a neuron that convey nerve impulses toward the cell body.

dendritic cell An antigen-presenting cell, located mainly in lymphatic tissues and skin, that is particularly efficient in presenting antigens to naive helper T cells, thereby initiating a primary immune response.

density The number of individuals per unit area or volume.

density dependent Referring to any characteristic that varies according to an increase in population density.

density-dependent inhibition The phenomenon observed in normal animal cells that causes them to stop dividing when they come into contact with one another.

density independent Referring to any characteristic that is not affected by population density.

deoxyribonucleic acid (DNA) (dē-ok´-sē-rī´-bō-nū-klā´-ik) A double-stranded, helical nucleic acid molecule capable of replicating and determining the inherited structure of a cell's proteins.

deoxyribose The sugar component of DNA, having one less hydroxyl group than ribose, the sugar component of RNA.

depolarization An electrical state in an excitable cell whereby the inside of the cell is made less negative relative to the outside than at the resting membrane potential. A neuron membrane is depolarized if a stimulus decreases its voltage from the resting potential of −70 mV in the direction of zero voltage.

derivatives New cells that are displaced from an apical meristem and continue to divide until the cells they produce become specialized.

dermal tissue system The outer protective covering of plants.

descent with modification Darwin's initial phrase for the general process of evolution.

desert A terrestrial biome characterized by very low precipitation.

desmosome (dez´-muh-sōm) A type of intercellular junction in animal cells that functions as an anchor.

determinate cleavage A type of embryonic development in protostomes that rigidly casts the developmental fate of each embryonic cell very early.

determinate growth A type of growth characteristic of most animals and some plant organs, in which growth stops after a certain size is reached.

determination The progressive restriction of developmental potential, causing the possible fate of each cell to become more limited as the embryo develops.

detritivore A consumer that derives its energy from nonliving organic material; a decomposer.

detritus (di-trī´-tus) Dead organic matter.

deuteromycete Traditional classification for a fungus with no known sexual stage. When a sexual stage for a so-called deuteromycete is discovered, the species is assigned to a phylum. Also called an imperfect fungus.

deuterostome development In animals, a developmental mode distinguished by the development of the anus from the blastopore; often also characterized by enterocoelous development of the body cavity and by radial cleavage.

diabetes mellitus An endocrine disorder marked by inability to maintain glucose homeostasis. The type I form results from autoimmune destruction of insulin-secreting cells; treatment usually requires insulin injections several times a day. The type II form most commonly results from reduced responsiveness of target cells to insulin; obesity and lack of exercise are risk factors.

diacylglycerol (DAG) A second messenger produced by the cleavage of a certain kind of phospholipid in the plasma membrane.

diaphragm (1) A sheet of muscle that forms the bottom wall of the thoracic cavity in mammals; active in ventilating the lungs. (2) A dome-shaped rubber cup fitted into the upper portion of the vagina before sexual intercourse. It serves as a physical barrier to block the passage of sperm.

diapsid Member of an amniote clade distinguished by a pair of holes on each side of the skull, including the lepidosaurs and archosaurs.

diastole (dī-as´-tō-lē) The stage of the heart cycle in which the heart muscle is relaxed, allowing the chambers to fill with blood.

diastolic pressure Blood pressure that remains between heart contractions.

diatom (dī´-uh-tom) A unicellular photosynthetic alga with a unique, glassy cell wall containing silica.

dicots A term traditionally used to refer to flowering plants that have two embryonic seed leaves, or cotyledons. Recent molecular evidence indicates that dicots do not form a clade (*see* eudicots).

differential gene expression The expression of different sets of genes by cells with the same genome.

diffusion The spontaneous tendency of a substance to move down its concentration gradient from a more concentrated to a less concentrated area.

digestion The process of breaking down food into molecules small enough for the body to absorb.

dihybrid (dī´-hī´-brid) An organism that is heterozygous with respect to two genes of interest. All the offspring from a cross between parents doubly homozygous for different alleles are dihybrids. For example, parents of genotypes *AABB* and *aabb* produce a dihybrid of genotype *AaBb*.

dikaryotic Referring to a fungal mycelium with two haploid nuclei per cell, one from each parent.

dinoflagellate (dī´-nō-flaj´-uh-let) Member of a group of mostly unicellular photosynthetic algae with two flagella situated in perpendicular grooves in cellulose plates covering the cell.

dinosaur Member of an extremely diverse group of ancient reptiles varying in body shape, size, and habitat.

dioecious (dī-ē´-shus) A term typically used to describe an angiosperm species in which carpellate and staminate flowers are on separate plants.

diploblastic Having two germ layers.

diploid cell (dip´-loid) A cell containing two sets of chromosomes ($2n$), one set inherited from each parent.

diplomonad A protist that has modified mitochondria, two equal-sized nuclei, and multiple flagella.

directional selection Natural selection that favors individuals at one end of the phenotypic range.

disaccharide (dī-sak´-uh-rīd) A double sugar, consisting of two monosaccharides joined by dehydration synthesis.

discovery science The process of scientific inquiry that focuses on describing nature.

dispersal The distribution of individuals within geographic population boundaries.

dispersion The pattern of spacing among individuals within geographic population boundaries.

disruptive selection Natural selection that favors individuals on both extremes of a phenotypic range over intermediate phenotypes.

dissociation curve A chart showing the relative amounts of oxygen bound to hemoglobin when the pigment is exposed to solutions varying in their partial pressure of dissolved oxygen, pH, or other characteristics.

distal tubule In the vertebrate kidney, the portion of a nephron that helps refine filtrate and empties it into a collecting duct.

disturbance A force that changes a biological community and usually removes organisms from it. Disturbances, such as fire and storms, play pivotal roles in structuring many biological communities.

disulfide bridge A strong covalent bond formed when the sulfur of one cysteine monomer bonds to the sulfur of another cysteine monomer.

DNA fingerprint An individual's unique collection of DNA restriction fragments, detected by electrophoresis and nucleic acid probes.

DNA ligase (lī´-gās) A linking enzyme essential for DNA replication; catalyzes the covalent bonding of the 3′ end of a new DNA fragment to the 5′ end of a growing chain.

DNA microarray assay A method to detect and measure the expression of thousands of genes at one time. Tiny amounts of a large number of single-stranded DNA fragments representing different genes are fixed to a glass slide. These fragments, ideally representing all the genes of an organism, are tested for hybridization with various samples of cDNA molecules.

DNA polymerase (puh-lim´-er-ās) An enzyme that catalyzes the elongation of new DNA at a replication fork by the addition of nucleotides to the existing chain.

domain (1) A taxonomic category above the kingdom level. The three domains are Archaea, Bacteria, and Eukarya. (2) An independently folding part of a protein.

dominant allele An allele that is fully expressed in the phenotype of a heterozygote.

dominant species Those species in a community that have the highest abundance or highest biomass. These species exert a powerful control over the occurrence and distribution of other species.

dopamine A biogenic amine closely related to epinephrine and norepinephrine.

dormancy A condition typified by extremely low metabolic rate and a suspension of growth and development.

dorsal Pertaining to the back (top) of a bilaterally symmetrical animal.

dorsal lip The dorsal side of the blastopore.

double circulation A circulation scheme with separate pulmonary and systemic circuits, which ensures vigorous blood flow to all organs.

double fertilization A mechanism of fertilization in angiosperms, in which two sperm cells unite with two cells in the embryo sac to form the zygote and endosperm.

double helix The form of native DNA, referring to its two adjacent polynucleotide strands wound into a spiral shape.

Down syndrome A human genetic disease caused by presence of an extra chromosome 21; characterized by mental retardation and heart and respiratory defects.

Duchenne muscular dystrophy (duh-shen´) A human genetic disease caused by a sex-linked recessive allele; characterized by progressive weakening and a loss of muscle tissue.

duodenum (dū-ō-dē´-num) The first section of the small intestine, where acid chyme from the stomach mixes with digestive juices from the pancreas, liver, gallbladder, and gland cells of the intestinal wall.

duplication An aberration in chromosome structure due to fusion with a fragment from a homologous chromosome, such that a portion of a chromosome is duplicated.

dynamic stability hypothesis The idea that long food chains are less stable than short chains.

dynein (dī´-nin) A large contractile protein forming the side-arms of microtubule doublets in cilia and flagella.

E site One of a ribosome's three binding sites for tRNA during translation. The E site is the place where discharged tRNAs leave the ribosome. (E stands for exit.)

ecdysone (ek´-duh-sōn) A steroid hormone, secreted by the prothoracic glands, that triggers molting in arthropods.

ecdysozoan Member of a group of animal phyla with protostome development that some systematists hypothesize form a clade, including many molting animals.

echinoderm (uh-kī´-nō-derm) A slow-moving or sessile marine deuterostome with a water vascular system and, in adults, radial anatomy. Echinoderms include sea stars, brittle stars, sea urchins, feather stars, and sea cucumbers.

ecological capacity The actual resource base of a country.

ecological footprint A method of using multiple constraints to estimate the human carrying capacity of Earth by calculating the aggregate land and water area in various ecosystem categories appropriated by a nation to produce all the resources it consumes and to absorb all the waste it generates.

ecological niche (nich) The sum total of a species' use of the biotic and abiotic resources in its environment.

ecological species concept Defining species in terms of ecological roles (niches).

ecological succession Transition in the species composition of a biological community, often following ecological disturbance of the community; the establishment of a biological community in an area virtually barren of life.

ecology The study of how organisms interact with their environment.

ecosystem All the organisms in a given area as well as the abiotic factors with which they interact; a community and its physical environment.

ecosystem ecology The study of energy flow and the cycling of chemicals among the various biotic and abiotic factors in an ecosystem.

ecosystem services Functions performed by natural ecosystems that directly or indirectly benefit humans.

ecotone The transition from one type of habitat or ecosystem to another, such as the transition from a forest to a grassland.

ectoderm (ek´-tō-derm) The outermost of the three primary germ layers in animal embryos; gives rise to the outer covering and, in some phyla, the nervous system, inner ear, and lens of the eye.

ectomycorrhiza (ek´-tō-mī´-kō-rī´-zuh) A type of mycorrhiza in which the mycelium forms a dense sheath, or mantle, over the surface of the root. Hyphae extend from the mantle into the soil, greatly increasing the surface area for water and mineral absorption.

ectomycorrhizal fungus A fungus that forms ectomycorrhizae with plant roots.

ectoparasite A parasite that feeds on the external surface of a host.

ectoproct A sessile, colonial lophophorate commonly called a bryozoan.

ectotherm (ek´-tō-therm) An animal, such as a reptile (other than birds), fish, or amphibian, that must use environmental energy and behavioral adaptations to regulate its body temperature.

ectothermic Referring to organisms that do not produce enough metabolic heat to have much effect on body temperature.

Ediacaran fauna Earliest generally accepted animal fossils, dating from about 575 million years ago.

effective population size An estimate of the size of a population based on the numbers of females and males that successfully breed; generally smaller than the total population.

effector cell A muscle cell or gland cell that performs the body's responses to stimuli; responds to signals from the brain or other processing center of the nervous system.

efferent arteriole The blood vessel draining a nephron.

egg-polarity gene Another name for a maternal effect gene, a gene that helps control the orientation (polarity) of the egg.

ejaculation The propulsion of sperm from the epididymis through the muscular vas deferens, ejaculatory duct, and urethra.

ejaculatory duct The short section of the ejaculatory route in mammals formed by the convergence of the vas deferens and a duct from the seminal vesicle. The ejaculatory duct transports sperm from the vas deferens to the urethra.

elastic fiber A long thread made of the protein elastin. Elastic fibers provide a rubbery quality to the extracellular matrix that complements the nonelastic strength of collagenous fibers.

electrocardiogram (ECG or EKG) A record of the electrical impulses that travel through cardiac muscle during the heart cycle.

electrochemical gradient The diffusion gradient of an ion, representing a type of potential energy that accounts for both the concentration difference of the ion across a membrane and its tendency to move relative to the membrane potential.

electrogenic pump An ion transport protein that generates voltage across a membrane.

electromagnetic receptor A receptor of electromagnetic energy, such as visible light, electricity, and magnetism.

electromagnetic spectrum The entire spectrum of radiation ranging in wavelength from less than a nanometer to more than a kilometer.

electron A subatomic particle with a single negative charge. One or more electrons move around the nucleus of an atom.

electron microscope (EM) A microscope that focuses an electron beam through a specimen, resulting in resolving power a thousandfold greater than that of a light microscope. A transmission electron microscope (TEM) is used to study the internal structure of thin sections of cells. A scanning electron microscope (SEM) is used to study the fine details of cell surfaces.

electron shell An energy level represented as the distance of an electron from the nucleus of an atom.

electron transport chain A sequence of electron carrier molecules (membrane proteins) that shuttle electrons during the redox reactions that release energy used to make ATP.

electronegativity The attraction of an atom for the electrons of a covalent bond.

electroporation A technique to introduce recombinant DNA into cells by applying a brief electrical pulse to a solution containing cells. The electricity creates temporary holes in the cells' plasma membranes, through which DNA can enter.

element Any substance that cannot be broken down to any other substance.

elicitor A molecule that induces a broad type of host defense response.

elimination The passing of undigested material out of the digestive compartment.

embryo New developing individual.

embryo sac (em´-brē-ō) The female gametophyte of angiosperms, formed from the growth and division of the megaspore into a multicellular structure with eight haploid nuclei.

embryonic lethal A mutation with a phenotype leading to death at the embryo or larval stage.

embryophyte Another name for land plants, recognizing that land plants share the common derived trait of multicellular, dependent embryos.

emergent properties New properties that emerge with each step upward in the hierarchy of life, owing to the arrangement and interactions of parts as complexity increases.

emigration The movement of individuals out of a population.

enantiomer (en-an´-tē-ō-mer) One of two molecules that are mirror images of each other.

endangered species A species that is in danger of extinction throughout all or a significant portion of its range.

endemic (en-dem´-ik) Referring to a species that is confined to a specific, relatively small geographic area.

endergonic reaction (en´-der-gon´-ik) A nonspontaneous chemical reaction, in which free energy is absorbed from the surroundings.

endocrine gland (en´-dō-krin) A ductless gland that secretes hormones directly into the interstitial fluid, from which they diffuse into the bloodstream.

endocrine system The internal system of chemical communication involving hormones, the ductless glands that secrete hormones, and the molecular receptors on or in target cells that respond to hormones; functions in concert with the nervous system to effect internal regulation and maintain homeostasis.

endocytosis (en´-dō-sī-tō´-sis) The cellular uptake of macromolecules and particulate substances by localized regions of the plasma membrane that surround the substance and pinch off to form an intracellular vesicle.

endoderm (en´-dō-derm) The innermost of the three primary germ layers in animal embryos; lines the archenteron and gives rise to the liver, pancreas, lungs, and the lining of the digestive tract.

endodermis The innermost layer of the cortex in plant roots; a cylinder one cell thick that forms the boundary between the cortex and the vascular cylinder.

endomembrane system The collection of membranes inside and around a eukaryotic cell, related either through direct physical contact or by the transfer of membranous vesicles.

endometrium (en´-dō-mē´-trē-um) The inner lining of the uterus, which is richly supplied with blood vessels.

endomycorrhiza (en´-dō-mī´-kō-rī´-zuh) A type of mycorrhiza that, unlike ectomycorrhizae, does not have a dense mantle ensheathing the root. Instead, microscopic fungal hyphae extend from the root into the soil.

endomycorrhizal fungus A fungus that forms endomycorrhizae with plant roots.

endoparasite A parasite that lives within a host.

endoplasmic reticulum (ER) (en´-dō-plaz´-mik ruh-tik´-yū-lum) An extensive membranous network in eukaryotic cells, continuous with the outer nuclear membrane and composed of ribosome-studded (rough) and ribosome-free (smooth) regions.

endorphin (en-dōr´-fin) Any of several hormones produced in the brain and anterior pituitary that inhibits pain perception.

endoskeleton A hard skeleton buried within the soft tissues of an animal, such as the spicules of sponges, the plates of echinoderms, and the bony skeletons of vertebrates.

endosperm A nutrient-rich tissue formed by the union of a sperm cell with two polar nuclei during double fertilization, which provides nourishment to the developing embryo in angiosperm seeds.

endospore A thick-coated, resistant cell produced within a bacterial cell exposed to harsh conditions.

endothelium (en´-dō-thē´-lē-um) The innermost, simple squamous layer of cells lining the blood vessels; the only constituent structure of capillaries.

endotherm An animal, such as a bird or mammal, that uses metabolic heat to regulate body temperature.

endothermic Referring to organisms with bodies that are warmed by heat generated by metabolism. This heat is usually used to maintain a relatively stable body temperature higher than that of the external environment.

endotoxin A toxic component of the outer membrane of certain gram-negative bacteria that is released only when the bacteria die.

energetic hypothesis The concept that the length of a food chain is limited by the inefficiency of energy transfer along the chain.

energy The capacity to do work (to move matter against an opposing force).

energy coupling In cellular metabolism, the use of energy released from an exergonic reaction to drive an endergonic reaction.

energy level Any of several different states of potential energy for electrons in an atom.

enhancer A DNA segment containing multiple control elements that may be located far away from the gene it regulates.

enteric division Complex networks of neurons in the digestive tract, pancreas, and gallbladder; normally regulated by the sympathetic and parasympathetic divisions of the autonomic nervous system.

enterocoelous (en´-ter-ō-sē´-lus) Pattern of formation of the body cavity common in deuterostome development, in which the mesoderm buds from the wall of the archenteron and hollows, forming the body cavity.

entropy (en´-truh-pē) A quantitative measure of disorder or randomness, symbolized by *S*.

enzymatic hydrolysis The process in digestion that splits macromolecules from food by the enzymatic addition of water.

enzyme (en´-zīm) A protein serving as a catalyst, a chemical agent that changes the rate of a reaction without being consumed by the reaction.

enzyme-substrate complex A temporary complex formed when an enzyme binds to its substrate molecule(s).

eosinophil A type of white blood cell with low phagocytic activity that is thought to play a role in defense against parasitic worms by releasing enzymes toxic to these invaders.

epicotyl (ep´-uh-cot´-ul) In an angiosperm embryo, the embryonic axis above the point of attachment of the cotyledon(s).

epidermis (1) The dermal tissue system of nonwoody plants, usually consisting of a single layer of tightly packed cells. (2) The outer covering of animals.

epididymis (ep´-uh-did´-uh-mus) A coiled tubule located adjacent to the testes where sperm are stored.

epigenetic inheritance Inheritance of traits transmitted by mechanisms not directly involving the nucleotide sequence.

epiglottis A cartilaginous flap that blocks the top of the windpipe, the glottis, during swallowing, which prevents the entry of food or fluid into the respiratory system.

epinephrine A catecholamine hormone secreted from the adrenal medulla that mediates "fight-or-flight" responses to short-term stress; also functions as a neurotransmitter.

epiphyte (ep´-uh-fīt) A plant that nourishes itself but grows on the surface of another plant for support, usually on the branches or trunks of tropical trees.

episome (ep´-uh-sōm) A genetic element that can exist either as a plasmid or as part of the bacterial chromosome.

epistasis A type of gene interaction in which one gene alters the phenotypic effects of another gene that is independently inherited.

epithalamus A brain region, derived from the diencephalon, that contains several clusters of capillaries that produce cerebrospinal fluid.

epithelial tissue (ep´-uh-thē´-lē-ul) Sheets of tightly packed cells that line organs and body cavities.

epitope A small, accessible region of an antigen to which an antigen receptor or antibody binds; also called an antigenic determinant.

equilibrium potential (E_{ion}) The magnitude of a cell's membrane voltage at equilibrium; calculated using the Nernst equation.

erythrocyte (eh-rith´-rō-sīt) A red blood cell; contains hemoglobin, which functions in transporting oxygen in the circulatory system.

erythropoietin (EPO) (eh-rith´-rō-poy´-uh-tin) A hormone produced in the kidney when tissues of the body do not receive enough oxygen. This hormone stimulates the production of erythrocytes.

esophagus (eh-sof´-uh-gus) A channel that conducts food, by peristalsis, from the pharynx to the stomach.

essential amino acid An amino acid that an animal cannot synthesize itself and must be obtained from food. Eight amino acids are essential in the human adult.

essential element In plants, a chemical element that is required for the plant to grow from a seed and complete the life cycle, producing another generation of seeds.

essential fatty acids Certain unsaturated fatty acids that animals cannot make.

essential nutrient A substance that an organism must absorb in preassembled form because it cannot be synthesized from any other material. In humans, there are essential vitamins, minerals, amino acids, and fatty acids.

estivation (es´-tuh-vā´-shun) Summer torpor; a physiological state that is characterized by slow metabolism and inactivity and that permits survival during long periods of elevated temperature and diminished water supplies.

estrogen (es´-trō-jen) Any steroid hormone, such as estradiol, that stimulates the development and maintenance of the female reproductive system and secondary sex characteristics.

estrous cycle (es´-trus) A type of reproductive cycle in all female mammals except higher primates, in which the nonpregnant endometrium is reabsorbed rather than shed, and sexual response occurs only during mid-cycle at estrus.

estrus A period of sexual activity associated with ovulation.

estuary The area where a freshwater stream or river merges with the ocean.

ethology The study of animal behavior in natural conditions.

ethylene (eth´-uh-lēn) The only gaseous plant hormone. Among its many effects are response to mechanical stress, programmed cell death, leaf abscission, and fruit ripening.

etiolation Plant morphological adaptations for growing in darkness.

euchromatin (ū-krō´-muh-tin) The more open, unraveled form of eukaryotic chromatin that is available for transcription.

eudicots (ū´-di-kots) A clade consisting of the vast majority of flowering plants that have two embryonic seed leaves, or cotyledons.

euglenid A protist, such as *Euglena* or its relatives, characterized by an anterior pocket, or chamber, from which one or two flagella emerge.

Eukarya The domain that includes all eukaryotic organisms.

eukaryotic cell (ū´-kār-ē-ot´-ik) A type of cell with a membrane-enclosed nucleus and membrane-enclosed organelles. Organisms with eukaryotic cells (protists, plants, fungi, and animals) are called eukaryotes.

eumetazoan (ū´-met-uh-zō´-uhn) Member of the clade Eumetazoa, animals with true tissues (all animals except sponges).

euryhaline Referring to organisms that can tolerate substantial changes in external osmolarity.

eurypterid An extinct carnivorous cheliceriform also called a water scorpion.

Eustachian tube The tube that connects the middle ear to the pharynx.

eutherian (ū-thēr´-ē-un) Placental mammal; mammal whose young complete their embryonic development within the uterus, joined to the mother by the placenta.

eutrophic lake A nutrient-rich and oxygen-poor lake, having a high rate of biological productivity.

eutrophication A process by which nutrients, particularly phosphorus and nitrogen, become highly concentrated in a body of water, leading to increased growth of organisms such as algae. Cultural eutrophication refers to situations where the nutrients added to the water body originate mainly from human sources, such as agricultural drainage or sewage.

evaporation The removal of heat energy from the surface of a liquid that is losing some of its molecules.

evaporative cooling The property of a liquid whereby the surface becomes cooler during evaporation, owing to a loss of highly kinetic molecules to the gaseous state.

evapotranspiration The evaporation of water from soil plus the transpiration of water from plants.

evolution All the changes that have transformed life on Earth from its earliest beginnings to the diversity that characterizes it today.

evolutionary adapation An accumulation of inherited characteristics that enhance organisms' ability to survive and reproduce in specific environments.

excitatory postsynaptic potential (EPSP) An electrical change (depolarization) in the membrane of a postsynaptic neuron caused by the binding of an excitatory neurotransmitter from a presynaptic cell to a postsynaptic receptor; makes it more likely for a postsynaptic neuron to generate an action potential.

excretion The disposal of nitrogen-containing waste products of metabolism.

exergonic reaction (ek´-ser-gon´-ik) A spontaneous chemical reaction, in which there is a net release of free energy.

exocytosis (ek´-sō-sī-tō´-sis) The cellular secretion of macromolecules by the fusion of vesicles with the plasma membrane.

exoenzyme A powerful hydrolytic enzyme secreted by a fungus outside its body to digest food.

exon A coding region of a eukaryotic gene. Exons, which are expressed, are separated from each other by introns.

exoskeleton A hard encasement on the surface of an animal, such as the shell of a mollusc or the cuticle of an arthropod, that provides protection and points of attachment for muscles.

exotoxin (ek´-sō-tok´-sin) A toxic protein that is secreted by a prokaryote and that produces specific symptoms even in the absence of the prokaryote.

expansins Plant enzymes that break the cross-links (hydrogen bonds) between cellulose microfibrils and other cell wall constituents, loosening the wall's fabric.

exponential population growth The geometric increase of a population as it grows in an ideal, unlimited environment.

expression vector A cloning vector that contains the requisite prokaryotic promoter just upstream of a restriction site where a eukaryotic gene can be inserted.

external fertilization The fusion of gametes that parents have discharged into the environment.

exteroreceptor A sensory receptor that detects stimuli outside the body, such as heat, light, pressure, and chemicals.

extinction vortex A downward population spiral in which positive-feedback loops of inbreeding and genetic drift cause a small population to shrink and, unless reversed, become extinct.

extracellular digestion The breakdown of food outside cells.

extracellular matrix (ECM) The substance in which animal tissue cells are embedded, consisting of protein and polysaccharides.

extraembryonic membranes Four membranes (yolk sac, amnion, chorion, allantois) that support the developing embryo in mammals and birds and other reptiles.

extreme halophile A prokaryote that lives in a highly saline environment, such as the Great Salt Lake or the Dead Sea.

extreme thermophile A prokaryote that thrives in hot environments (often 60–80°C or hotter).

extremophile A prokaryote that lives in an extreme environment. Extremophiles include methanogens, extreme halophiles, and extreme thermophiles.

F factor A fertility factor in bacteria; a DNA segment that confers the ability to form pili for conjugation and associated functions required for the transfer of DNA from donor to recipient. It may exist as a plasmid or be integrated into the bacterial chromosome.

F plasmid The plasmid form of the F factor.

F_1 generation The first filial, or hybrid, offspring in a series of genetic crosses.

F_2 generation Offspring resulting from interbreeding of the hybrid F_1 generation.

facilitator A species that has a positive effect on the survival and reproduction of other species in a community and that contributes to community structure.

facilitated diffusion The spontaneous passage of molecules and ions, bound to specific carrier proteins, across a biological membrane down their concentration gradients.

facultative anaerobe (fak´-ul-tā´-tiv an´-uh-rōb) An organism that makes ATP by aerobic respiration if oxygen is present but that switches to fermentation under anaerobic conditions.

family In classification, the taxonomic category above genus.

fast block to polyspermy The depolarization of the egg membrane within 1–3 seconds after sperm binding to the vitelline layer. The reaction prevents additional sperm from fusing with the egg's plasma membrane.

fast muscle fibers Muscle cells used for rapid, powerful contractions.

fat (triacylglycerol) (trī-as´-ul-glis´-uh-rol) A biological compound consisting of three fatty acids linked to one glycerol molecule.

fate map Territorial diagram of embryonic development that reveals the future development of individual cells and tissues.

fatty acid A long carbon chain carboxylic acid. Fatty acids vary in length and in the number and location of double bonds; three fatty acids linked to a glycerol molecule form fat.

feces The wastes of the digestive tract.

feedback inhibition A method of metabolic control in which the end product of a metabolic pathway acts as an inhibitor of an enzyme within that pathway.

fermentation A catabolic process that makes a limited amount of ATP from glucose without an electron transport chain and that produces a characteristic end product, such as ethyl alcohol or lactic acid.

fertilization The union of haploid gametes to produce a diploid zygote.

fertilization envelope The swelling of the vitelline layer away from the plasma membrane.

fetus (fē´-tus) A developing human from the ninth week of gestation until birth; has all the major structures of an adult.

fiber A lignified cell type that reinforces the xylem of angiosperms and functions in mechanical support; a slender, tapered sclerenchyma cell that usually occurs in bundles.

fibrin (fī´-brin) The activated form of the blood-clotting protein fibrinogen, which aggregates into threads that form the fabric of the clot.

fibrinogen The inactive form of the plasma protein that is converted to the active form fibrin, which aggregates into threads that form the framework of a blood clot.

fibroblast (fī´-brō-blast) A type of cell in loose connective tissue that secretes the protein ingredients of the extracellular fibers.

fibronectin A glycoprotein that helps cells attach to the extracellular matrix.

fibrous connective tissue A dense tissue with large numbers of collagenous fibers organized into parallel bundles. This is the dominant tissue in tendons and ligaments.

fibrous root system A root system common to monocots consisting of a mat of thin roots spreading out below the soil surface.

filament The stalk of a stamen.

filtrate Fluid extracted by the excretory system from the body fluid. The excretory system produces urine from the filtrate after extracting valuable solutes from it and concentrating it.

filtration The extraction of water and small solutes, including metabolic wastes, from the body fluid into the excretory system.

fimbria (plural, **fimbriae**) A short, hairlike prokaryotic appendage that functions in adherence to the substrate or to other cells.

first law of thermodynamics The principle of conservation of energy. Energy can be transferred and transformed, but it cannot be created or destroyed.

fission The separation of a parent into two or more individuals of approximately equal size.

fitness The contribution an individual makes to the gene pool of the next generation, relative to the contributions of other individuals.

fixed action pattern (FAP) A sequence of behavioral acts that is essentially unchangeable and usually carried to completion once initiated.

flaccid (flas´-id) Limp. A walled cell is flaccid in surroundings where there is no tendency for water to enter.

flagellum (fluh-jel´-um) (plural, **flagella**) A long cellular appendage specialized for locomotion. The flagella of prokaryotes and eukaryotes differ in both structure and function.

florigen A flowering signal, not yet chemically identified, that may be a hormone or may be a change in relative concentrations of multiple hormones.

flower In an angiosperm, a short stem with up to four sets of modified leaves, bearing structures that function in sexual reproduction.

fluid feeder An animal that lives by sucking nutrient-rich fluids from another living organism.

fluid mosaic model The currently accepted model of cell membrane structure, which envisions the membrane as a mosaic of individual protein molecules drifting laterally in a fluid bilayer of phospholipids.

follicle (fol´-uh-kul) A microscopic structure in the ovary that contains the developing ovum and secretes estrogens.

follicle-stimulating hormone (FSH) A tropic hormone produced and secreted by the anterior pituitary that stimulates the production of eggs by the ovaries and sperm by the testes.

follicular phase That part of the ovarian cycle during which follicles are growing and oocytes maturing.

food chain The pathway along which food is transferred from trophic level to trophic level, beginning with producers.

food vacuole A membranous sac formed by phagocytosis.

food web The elaborate, interconnected feeding relationships in an ecosystem.

foot (1) The portion of a bryophyte sporophyte that gathers sugars, amino acids, water, and minerals from the parent gametophyte via transfer cells. (2) One of the three main parts of a mollusc; a muscular structure usually used for movement.

foraging Behavior necessary to recognize, search for, capture, and consume food.

foraminiferan (foram) An aquatic protist that secretes a hardened shell containing calcium carbonate and extends pseudopodia through pores in the shell.

forebrain One of three ancestral and embryonic regions of the vertebrate brain; develops into the thalamus, hypothalamus, and cerebrum.

fossil A preserved remnant or impression of an organism that lived in the past.

fossil record The chronicle of evolution over millions of years of geologic time engraved in the order in which fossils appear in rock strata.

founder effect Genetic drift that occurs when a few individuals become isolated from a larger population, with the result that the new population's gene pool is not reflective of the original population.

fovea (fō´-vē-uh) An eye's center of focus and the place on the retina where photoreceptors are highly concentrated.

fragmentation A means of asexual reproduction whereby a single parent breaks into parts that regenerate into whole new individuals.

frameshift mutation A mutation occurring when the number of nucleotides inserted or deleted is not a multiple of three, resulting in the improper grouping of the following nucleotides into codons.

free energy The portion of a system's energy that can perform work when temperature and pressure are uniform throughout the system. The change in free energy of a system is calculated by the equation $\Delta G = \Delta H - T\Delta S$, where T is absolute temperature.

free energy of activation The amount of energy that reactants must absorb before a chemical reaction will start; also called activation energy.

frequency-dependent selection A decline in the reproductive success of a morph resulting from the morph's phenotype becoming too common in a population; a cause of balanced polymorphism in populations.

fruit A mature ovary of a flower that protects dormant seeds and aids in their dispersal.

functional group A specific configuration of atoms commonly attached to the carbon skeletons of organic molecules and usually involved in chemical reactions.

Fungi (fun´-jē) The eukaryotic kingdom that includes organisms that absorb nutrients after decomposing organic material.

fusiform initials Cells within the vascular cambrium that produce elongated cells such as tracheids, vessel elements, fibers, and sieve-tube members.

G protein A GTP-binding protein that relays signals from a plasma membrane signal receptor, known as a G-protein-linked receptor, to other signal transduction proteins inside the cell. When such a receptor is activated, it in turn activates the G protein, causing it to bind a molecule of GTP in place of GDP. Hydrolysis of the bound GTP to GDP inactivates the G protein.

G_0 phase A nondividing state in which a cell has left the cell cycle.

G_1 phase The first growth phase of the cell cycle, consisting of the portion of interphase before DNA synthesis begins.

G_2 phase The second growth phase of the cell cycle, consisting of the portion of interphase after DNA synthesis occurs.

gallbladder An organ that stores bile and releases it as needed into the small intestine.

gametangium (gam´-uh-tan´-jē-um) (plural, **gametangia**) Multicellular plant structures in which gametes are formed. Female gametangia are called archegonia, and male gametangia are called antheridia.

gamete (gam´-ēt) A haploid cell, such as an egg or sperm. Gametes unite during sexual reproduction to produce a diploid zygote.

game theory An approach to evaluating alternative strategies in situations where the outcome depends not only on each individual's strategy but also on the strategies of other individuals; a way of thinking about behavioral evolution in situations where the fitness of a particular behavioral phenotype is influenced by other behavioral phenotypes in the population.

gametogenesis The process by which gametes are produced in the mammalian body.

gametophore The mature gamete-producing structure of a gametophyte body of a moss.

gametophyte (guh-mē´-tō-fīt) In organisms undergoing alternation of generations, the multicellular haploid form that mitotically produces haploid gametes that unite and grow into the sporophyte generation.

gamma aminobutyric acid (GABA) An amino acid that functions as a CNS neurotransmitter.

ganglion (gang´-glē-un) (plural, **ganglia**) A cluster (functional group) of nerve cell bodies in a centralized nervous system.

ganglion cell A type of neuron in the retina that synapses with bipolar cells and transmits action potentials to the brain via axons in the optic nerve.

gap junction A type of intercellular junction in animal cells that allows the passage of material or current between cells.

gas exchange The uptake of molecular oxygen from the environment and the discharge of carbon dioxide to the environment.

gastric juice A digestive fluid secreted by the stomach.

gastrovascular cavity An extensive pouch that serves as the site of extracellular digestion and a passageway to disperse materials throughout most of an animal's body.

gastrula (gas´-trū-luh) The three-layered, cup-shaped embryonic stage.

gastrulation (gas´-trū-lā´-shun) The formation of a gastrula from a blastula.

gated channel A protein channel in a cell membrane that opens or closes in response to a particular stimulus.

gated ion channel A gated channel for a specific ion. When ion channels are opened or closed, the membrane potential of the cell is altered.

gel electrophoresis (ē-lek´-trō-fōr-ē´-sis) The separation of nucleic acids or proteins, on the basis of their size and electrical charge, by measuring their rate of movement through an electrical field in a gel.

gene A discrete unit of hereditary information consisting of a specific nucleotide sequence in DNA (or RNA, in some viruses).

gene cloning The production of multiple copies of a gene.

gene flow Genetic additions to or substractions from a population resulting from the movement of fertile individuals or gametes.

gene-for-gene recognition A widespread form of plant disease resistance involving recognition of pathogen-derived molecules by the protein products of specific plant disease resistance genes.

gene pool The total aggregate of genes in a population at any one time.

gene therapy The alteration of the genes of a person afflicted with a genetic disease.

genetic annealing The production of a new genome through the transfer of part of the genome of one organism to another organism.

genetic drift Unpredictable fluctuations in allele frequencies from one generation to the next because of a population's finite size.

genetic engineering The direct manipulation of genes for practical purposes.

genetic map An ordered list of genetic loci (genes or other genetic markers) along a chromosome.

genetic polymorphism The existence of two or more distinct alleles at a given locus in a population's gene pool.

genetic recombination General term for the production of offspring that combine traits of the two parents.

genetically modified (GM) organism An organism that has acquired one or more genes by artificial means; also known as a transgenic organism.

genetics The scientific study of heredity and hereditary variation.

genome (jē´-nōm) The complete complement of an organism's genes; an organism's genetic material.

genomic imprinting Phenomenon in which expression of an allele in offspring depends on whether the allele is inherited from the male or female parent.

genomic library A set of thousands of DNA segments from a genome, each carried by a plasmid, phage, or other cloning vector.

genomics (juh-nō´-miks) The study of whole sets of genes and their interactions.

genotype (jē´-nō-tīp) The genetic makeup, or set of alleles, of an organism.

genus (jē´-nus) (plural, **genera**) A taxonomic category above the species level, designated by the first word of a species' two-part scientific name.

geographic variation Differences between the gene pools of separate populations or population subgroups.

geologic record The division of Earth's history into time periods, grouped into three eras: Archaean, Proterozoic, and Phanerozoic, and further subdivided into eras and epochs.

geometric isomer One of several organic compounds that have the same molecular formula but differ in the spatial arrangements of their atoms.

germ layers Three main layers that form the various tissues and organs of an animal body.

gestation (jes-tā´-shun) Pregnancy; the state of carrying developing young within the female reproductive tract.

gibberellins (jib´-uh-rel´-inz) A class of related plant hormones that stimulate growth in the stem and leaves, trigger the germination of seeds and breaking of bud dormancy, and stimulate fruit development with auxin.

gill A localized extension of the body surface of many aquatic animals, specialized for gas exchange.

gill circulation The flow of blood through gills.

glandular epithelium An epithelium that absorbs or secretes chemical solutions.

glans penis The head end of the penis.

glia Supporting cells that are essential for the structural integrity of the nervous system and for the normal functioning of neurons.

glomeromycete Member of the fungal phylum Glomeromycota, characterized by forming a distinct branching form of endomycorrhizae (symbiotic relationships with plant roots) called arbuscular mycorrhizae.

glomerulus (glō-mār´-ū-lus) A ball of capillaries surrounded by Bowman's capsule in the nephron and serving as the site of filtration in the vertebrate kidney.

glucagon (glū´-kuh-gon) A hormone secreted by pancreatic alpha cells that raises blood glucose levels. It promotes glycogen breakdown and release of glucose by the liver.

glucocorticoid A steroid hormone secreted by the adrenal cortex that influences glucose metabolism and immune function.

glutamate An amino acid that functions as a CNS neurotransmitter.

glyceraldehyde-3-phosphate (G3P) (glis´-er-al´-de-hīd) The carbohydrate produced directly from the Calvin cycle.

glycine (glī´-sēn) An amino acid that functions as a CNS neurotransmitter.

glycogen (glī´-kō-jen) An extensively branched glucose storage polysaccharide found in the liver and muscle of animals; the animal equivalent of starch.

glycolipid A lipid covalently attached to a carbohydrate.

glycolysis (glī-kol´-uh-sis) The splitting of glucose into pyruvate. Glycolysis is the one metabolic pathway that occurs in all living cells, serving as the starting point for fermentation or aerobic respiration.

glycoprotein A protein covalently attached to a carbohydrate.

glycosidic linkage A covalent bond formed between two monosaccharides by a dehydration reaction.

gnathostome (nā´-thuh-stōm) Member of the vertebrate subgroup possessing jaws.

golden alga A chrysophyte; a typically unicellular, biflagellated alga with yellow and brown carotenoid pigments.

Golgi apparatus (gol´-jē) An organelle in eukaryotic cells consisting of stacks of flat membranous sacs that modify, store, and route products of the endoplasmic reticulum.

gonadotropin (gō-nah´-dō-trō´-pin) A hormone that stimulates the activities of the testes and ovaries. Follicle-stimulating hormone and luteinizing hormone are gonadotropins.

gonads (gō´-nadz) The male and female sex organs; the gamete-producing organs in most animals.

G-protein-linked receptor A signal receptor protein in the plasma membrane that responds to the binding signal molecule by activating a G protein.

grade Group of animal species that share the same level of organizational complexity.

graded potential A local voltage change in a neuron membrane induced by stimulation of a neuron, with strength proportional to the strength of the stimulus and lasting about a millisecond.

gradualism A view of Earth's history that attributes profound change to the cumulative product of slow but continuous processes.

graft versus host reaction An attack against a patient's body cells by lymphocytes received in a bone marrow transplant.

gram-negative Describing the group of bacteria with a cell wall that is structurally more complex and contains less peptidoglycan than that of gram-positive bacteria. Gram-negative bacteria are often more toxic than gram-positive bacteria.

gram-positive Describing the group of bacteria with a cell wall that is structurally less complex and contains more peptidoglycan than that of gram-negative bacteria. Gram-positive bacteria are usually less toxic than gram-negative bacteria.

Gram stain A staining method that distinguishes between two different kinds of bacterial cell walls.

granum (gran´-um) (plural, **grana**) A stacked portion of the thylakoid membrane in the chloroplast. Grana function in the light reactions of photosynthesis.

gravitropism (grav´-uh-trō´-pizm) A response of a plant or animal to gravity.

gray crescent A light-gray region of cytoplasm located near the equator of the egg on the side opposite the sperm entry.

gray matter Regions of dendrites and clusters of neuron cell bodies within the CNS.

green alga A unicellular, colonial, or multicellular photosynthetic protist that has grass-green chloroplasts. Green algae are closely related to true plants.

greenhouse effect The warming of planet Earth due to the atmospheric accumulation of carbon dioxide, which absorbs reflected infrared radiation and re-reflects some of it back toward Earth.

green world hypothesis The conjecture that terrestrial herbivores consume relatively little plant biomass because they are held in check by a variety of factors, including predators, parasites, and disease.

gross primary production (GPP) The total primary production of an ecosystem.

ground tissue system Plant tissues that are neither vascular nor dermal, fulfilling a variety of functions, such as storage, photosynthesis, and support.

growth cone Responsive region at the leading edge of a growing axon.

growth factor A protein that must be present in the extracellular environment (culture medium or animal body) for the growth and normal development of certain types of cells; a local regulator that acts on nearby cells to stimulate cell proliferation and differentiation.

growth hormone (GH) A hormone produced and secreted by the anterior pituitary that has both direct (nontropic) effects and tropic effects on a wide variety of tissues.

guard cells The two cells that flank the stomatal pore and regulate the opening and closing of the pore.

gustatory receptor Taste receptor.

guttation The exudation of water droplets, caused by root pressure in certain plants.

gymnosperm (jim´-nō-sperm) A vascular plant that bears naked seeds—seeds not enclosed in specialized chambers.

habituation A very simple type of learning that involves a loss of responsiveness to stimuli that convey little or no information.

hair cell A type of mechanoreceptor that detects sound waves and other forms of movement in air or water.

half-life The number of years it takes for 50% of a sample of an isotope to decay.

Hamilton's rule The principle that for natural selection to favor an altruistic act, the benefit to the recipient, devalued by the coefficient of relatedness, must exceed the cost to the altruist.

haploid cell (hap´-loyd) A cell containing only one set of chromosomes (n).

Hardy-Weinberg equilibrium The condition describing a non-evolving population (one that is in genetic equilibrium).

Hardy-Weinberg theorem The principle that frequencies of alleles and genotypes in a population remain constant from generation to generation, provided that only Mendelian segregation and recombination of alleles are at work.

haustorium (plural, **haustoria**) In certain symbiotic fungi, specialized hyphae that can penetrate the tissues of host organisms.

heart A muscular pump that uses metabolic energy to elevate hydrostatic pressure of the blood. Blood then flows down a pressure gradient through blood vessels that eventually return blood to the heart.

heart attack The death of cardiac muscle tissue resulting from prolonged blockage of one or more coronary arteries.

heart murmur A hissing sound that occurs when blood squirts backward through a leaky valve in the heart.

heart rate The rate of heart contraction.

heartwood Older layers of secondary xylem, closer to the center of a stem or root, that no longer transport xylem sap.

heat The total amount of kinetic energy due to molecular motion in a body of matter. Heat is energy in its most random form.

heat of vaporization The quantity of heat a liquid must absorb for 1 g of it to be converted from the liquid to the gaseous state.

heat-shock protein A protein that helps protect other proteins during heat stress. Heat-shock proteins are found in plants, animals, and microorganisms.

heavy chain One of the two types of polypeptide chains that make up an antibody molecule and B cell receptor; consists of a variable region, which contributes to the antigen-binding site, and a constant region.

helicase An enzyme that untwists the double helix of DNA at the replication forks.

helper T cell A type of T cell that, when activated, secretes cytokines that promote the response of B cells (humoral response) and cytotonic T cells (cell-mediated response) to antigens.

hemocyanin (hē´-muh-sī´-uh-nin) A type of respiratory pigment that uses copper as its oxygen-binding component. Hemocyanin is found in the hemolymph of arthropods and many molluscs.

hemoglobin (hē´-mō-glō-bin) An iron-containing protein in red blood cells that reversibly binds oxygen.

hemolymph In invertebrates with an open circulatory system, the body fluid that bathes tissues.

hemophilia A human genetic disease caused by a sex-linked recessive allele; characterized by excessive bleeding following injury.

hepatic portal vein A large circulatory channel that conveys nutrient-laden blood from the small intestine to the liver, which regulates the blood's nutrient content.

herbaceous Referring to nonwoody plants.

herbivore A heterotrophic animal that eats plants.

herbivory An interaction in which an herbivore eats parts of a plant or alga.

heredity The transmission of traits from one generation to the next.

hermaphrodite (her-maf´-rō-dīt) An individual that functions as both male and female in sexual reproduction by producing both sperm and eggs.

hermaphroditism (her-maf´-rō-dī-tizm) A condition in which an individual has both female and male gonads and functions as both a male and female in sexual reproduction by producing both sperm and eggs.

heterochromatin (het´-er-ō-krō´-muh-tin) Nontranscribed eukaryotic chromatin that is so highly compacted that it is visible with a light microscope during interphase.

heterochrony (het´-uh-rok´-ruh-nē) Evolutionary change in the timing or rate of an organism's development.

heterocyte (het´-er-ō-sīt) A specialized cell that engages in nitrogen fixation in some filamentous cyanobacteria (formerly called heterocyst).

heterokaryon A fungal mycelium formed by the fusion of two hyphae that have genetically different nuclei.

heteromorphic (het´-er-ō-mōr´-fik) Referring to a condition in the life cycle of all living plants and certain algae in which the sporophyte and gametophyte differ in morphology.

heterosporous (het´-er-os´-pōr-us) A term referring to a plant species that has two kinds of spores: microspores that develop into male gametophytes and megaspores that develop into female gametophytes.

heterotroph (het´-er-ō-trōf) An organism that obtains organic food molecules by eating other organisms or their by-products.

heterozygote advantage Greater reproductive success of heterozygous individuals compared to homozygotes; tends to preserve variation in gene pools.

heterozygous (het´-er-ō-zī´-gus) Having two different alleles for a given gene.

hexapod An insect or closely related wingless, six-legged arthropod.

hibernation A physiological state that allows survival during long periods of cold temperatures and reduced food supplies, in which metabolism decreases, the heart and respiratory system slow down, and body temperature is maintained at a lower level than normal.

high-density lipoprotein (HDL) A cholesterol-carrying particle in the blood, made up of cholesterol and other lipids surrounded by a single layer of phospholipids in which proteins are embedded. An HDL particle carries less cholesterol than a related lipoprotein, LDL, and may be correlated with a decreased risk of blood vessel blockage.

hindbrain One of three ancestral and embryonic regions of the vertebrate brain; develops into the medulla oblongata, pons, and cerebellum.

histamine (his´-tuh-mēn) A substance released by mast cells that causes blood vessels to dilate and become more permeable during an inflammatory response.

histone (his´-tōn) A small protein with a high proportion of positively charged amino acids that binds to the negatively charged DNA and plays a key role in its chromatin structure.

histone acetylation The attachment of acetyl groups to certain amino acids of histone proteins.

HIV (human immunodeficiency virus) The infectious agent that causes AIDS. HIV is a retrovirus.

holdfast A rootlike structure that anchors a seaweed.

holoblastic cleavage (hō´-lō-blas´-tik) A type of cleavage in which there is complete division of the egg, as in eggs having little yolk (sea urchin) or a moderate amount of yolk (frog).

homeobox (hō´-mē-ō-boks´) A 180-nucleotide sequence within homeotic genes and some other developmental genes that is widely conserved in animals. Related sequences occur in plants and prokaryotes.

homeostasis (hō´-mē-ō-stā´-sis) The steady-state physiological condition of the body.

homeotic gene (hō´-mē-ot´-ik) Any of the genes that control the overall body plan of animals and plants by controlling the developmental fate of groups of cells.

hominid (hah´-mi-nid) A species on the human branch of the evolutionary tree; a member of the family Hominidae, including *Homo sapiens* and our ancestors.

hominoid A term that refers to great apes and humans.

homologous chromosomes (hō-mol´-uh-gus) Chromosome pairs of the same length, centromere position, and staining pattern that possess alleles of the same genes at corresponding loci. One homologous chromosome is inherited from the organism's father, the other from the mother.

homologous structures Structures in different species that are similar because of common ancestry.

homology (hō-mol´-uh-jē) Similarity in characteristics resulting from a shared ancestry.

homoplasy Similar (analogous) structure or molecular sequence that has evolved independently in two species.

homosporous (hō-mos´-pōr-us) A term referring to a plant species that has a single kind of spore, which typically develops into a bisexual gametophyte.

homozygous (hō´-mō-zī´-gus) Having two identical alleles for a given gene.

horizon A distinct layer of soil, such as topsoil.

horizontal cell A neuron of the retina that helps integrate information before it is sent to the brain.

hormone In multicellular organisms, one of many types of circulating chemical signals that are formed in specialized cells, travel in body fluids, and act on specific target cells to change their functioning.

hornwort A small, herbaceous nonvascular plant that is a member of the phylum Anthocerophyta.

host The larger participant in a symbiotic relationship, serving as home and feeding ground to the symbiont.

host range The limited range of host cells that each type of virus can infect and parasitize.

human chorionic gonadotropin (HCG) (kōr´-ē-on´-ik gō-nah´-dō-trō´-pin) A hormone secreted by the chorion that maintains the corpus luteum of the ovary during the first three months of pregnancy.

Human Genome Project An international collaborative effort to map and sequence the DNA of the entire human genome.

humoral immune response (hyū´-mer-al) The branch of acquired immunity that involves the activation of B cells and that leads to the production of antibodies, which defend against bacteria and viruses in body fluids.

humus (hyū´-mus) Decomposing organic material found in topsoil.

Huntington's disease A human genetic disease caused by a dominant allele; characterized by uncontrollable body movements and degeneration of the nervous system; usually fatal 10 to 20 years after the onset of symptoms.

hybridization In genetics, the mating, or crossing, of two true-breeding varieties.
hydration shell The sphere of water molecules around each dissolved ion.
hydrocarbon An organic molecule consisting only of carbon and hydrogen.
hydrogen bond A type of weak chemical bond formed when the slightly positive hydrogen atom of a polar covalent bond in one molecule is attracted to the slightly negative atom of a polar covalent bond in another molecule.
hydrogen ion A single proton with a charge of 1+. The dissociation of a water molecule (H_2O) leads to the generation of a hydroxide ion (OH^-) and a hydrogen ion (H^+).
hydrolysis (hī-drol´-uh-sis) A chemical process that lyses, or splits, molecules by the addition of water.
hydrophilic (hī´-drō-fil´-ik) Having an affinity for water.
hydrophobic (hī´-drō-fō´-bik) Having an aversion to water; tending to coalesce and form droplets in water.
hydrophobic interaction A type of weak chemical bond formed when molecules that do not mix with water coalesce to exclude the water.
hydroponic culture A method in which plants are grown without soil by using mineral solutions.
hydrostatic skeleton (hī´-drō-stat´-ik) A skeletal system composed of fluid held under pressure in a closed body compartment; the main skeleton of most cnidarians, flatworms, nematodes, and annelids.
hydroxide ion A water molecule that has lost a proton; OH^-.
hydroxyl group (hī-drok´-sil) A functional group consisting of a hydrogen atom joined to an oxygen atom by a polar covalent bond. Molecules possessing this group are soluble in water and are called alcohols.
hymen A thin membrane that partly covers the vaginal opening in the human female. The hymen is ruptured by sexual intercourse or other vigorous activity.
hyperpolarization An electrical state in which the inside of the cell is more negative relative to the outside than at the resting membrane potential. A neuron membrane is hyperpolarized if a stimulus increases its voltage from the resting potential of −70 mV, reducing the chance that the neuron will transmit a nerve impulse.
hypersensitive response (HR) A plant's localized defense response to a pathogen.
hypertension Chronically high blood pressure within the arteries.
hypertonic In comparing two solutions, referring to the one with a greater solute concentration.
hypha (plural, **hyphae**) (hī´-fuh, hī´-fē) A filament that collectively makes up the body of a fungus.
hypocotyl (hī´-puh-cot´-ul) In an angiosperm embryo, the embryonic axis below the point of attachment of the cotyledon(s) and above the radicle.
hypothalamus (hī´-pō-thal´-uh-mus) The ventral part of the vertebrate forebrain; functions in maintaining homeostasis, especially in coordinating the endocrine and nervous systems; secretes hormones of the posterior pituitary and releasing factors that regulate the anterior pituitary.
hypothesis A tentative answer to a well-framed question.
hypotonic In comparing two solutions, referring to the one with a lower solute concentration.
I band The area near the edge of the sarcomere where there are only thin filaments.
immigration The influx of new individuals from other areas.
immunization The process of generating a state of immunity by artifical means. In active immunization, a nonpathogenic version of a normally pathogenic microbe is administered, inducing B and T cell responses and immunological memory. In passive immunization, antibodies specific for a particular microbe are administered, conferring immediate but temporary protection. Also called vaccination.
immunoglobulin (Ig) (im´-ū-nō-glob´-ū-lin) Any of the class of proteins that function as antibodies. Immunoglobulins are divided into five major classes that differ in their distribution in the body and antigen disposal activities.
imperfect fungus *See* deuteromycete.
imprinting A type of learned behavior with a significant innate component, acquired during a limited critical period.
***in vitro* fertilization** (vē´-trō) Fertilization of ova in laboratory containers followed by artificial implantation of the early embryo in the mother's uterus.
***in vitro* mutagenesis** A technique to discover the function of a gene by introducing specific changes into the sequence of a cloned gene, reinserting the mutated gene into a cell, and studying the phenotype of the mutant.
inclusive fitness The total effect an individual has on proliferating its genes by producing its own offspring and by providing aid that enables other close relatives to increase the production of their offspring.
incomplete dominance The situation in which the phenotype of heterozygotes is intermediate between the phenotypes of individuals homozygous for either allele.
incomplete flower A flower in which one or more of the four basic floral organs (sepals, petals, stamens, or carpels) are either absent or nonfunctional.
incomplete metamorphosis A type of development in certain insects, such as grasshoppers, in which the young (called nymphs) resemble adults but are smaller and have different body proportions. The nymph goes through a series of molts, each time looking more like an adult, until it reaches full size.
incus The second of the three middle ear bones.
indeterminate cleavage A type of embryonic development in deuterostomes, in which each cell produced by early cleavage divisions retains the capacity to develop into a complete embryo.
indeterminate growth A type of growth characteristic of plants, in which the organism continues to grow as long as it lives.
individualistic hypothesis The concept, put forth by H. A. Gleason, that a plant community is a chance assemblage of species found in the same area simply because they happen to have similar abiotic requirements.
induced fit The change in shape of the active site of an enzyme so that it binds more snugly to the substrate, induced by entry of the substrate.
inducer A specific small molecule that inactivates the repressor in an operon.
induction The ability of one group of embryonic cells to influence the development of another.
inductive reasoning A type of logic in which generalizations are based on a large number of specific observations.
infant mortality The number of infant deaths per 1,000 live births.
inflammatory response A localized innate immune defense triggered by physical injury or infection of tissue in which changes to nearby small blood vessels enhance the infiltration of white blood cells, antimicrobial proteins, and clotting elements that aid in tissue repair and destruction of invading pathogens; may also involve systemic effects such as fever and increased production of white blood cells.
inflorescence A group of flowers tightly clustered together.
ingestion A heterotrophic mode of nutrition in which other organisms or detritus are eaten whole or in pieces.
ingroup In a cladistic study of evolutionary relationships among taxa of organisms, the group of taxa that is actually being analyzed.
inhibitory postsynaptic potential (IPSP) (pōst´-sin-ap´-tik) An electrical charge (hyperpolarization) in the membrane of a postsynaptic neuron caused by the binding of an inhibitory neurotransmitter from a presynaptic cell to a postsynaptic receptor; makes it more difficult for a postsynaptic neuron to generate an action potential.
initials Cells that remain within an apical meristem as sources of new cells.
innate behavior Behavior that is developmentally fixed and under strong genetic control. Innate behavior is exhibited in virtually the

same form by all individuals in a population despite internal and external environmental differences during development and throughout their lifetimes.

innate immunity The kind of defense that is mediated by phagocytic cells, antimicrobial proteins, the inflammatory response, and natural killer (NK) cells. It is present before exposure to pathogens and is effective from the time of birth.

inner cell mass A cluster of cells in a mammalian blastocyst that protrudes into one end of the cavity and subsequently develops into the embryo proper and some of the extraembryonic membranes.

inner ear One of three main regions of the vertebrate ear; includes the cochlea, organ of Corti, and semicircular canals.

inositol trisphosphate (IP_3) (in-ō´-suh-tol) A second messenger that functions as an intermediate between certain nonsteroid hormones and a third messenger, a rise in cytoplasmic Ca^{2+} concentration.

inquiry The search for information and explanation, often focused by specific questions.

insertion A mutation involving the addition of one or more nucleotide pairs to a gene.

insertion sequence The simplest kind of transposable element, consisting of inverted repeats of DNA flanking a gene for transposase, the enzyme that catalyzes transposition.

insulin (in´-sū-lin) A hormone secreted by pancreatic beta cells that lowers blood glucose levels. It promotes the uptake of glucose by most body cells and the synthesis and storage of glycogen in the liver and also stimulates protein and fat synthesis.

insulin-like growth factor A hormone produced by the liver whose secretion is stimulated by growth hormone. It directly stimulates bone and cartilage growth.

integral protein Typically a transmembrane protein with hydrophobic regions that completely spans the hydrophobic interior of the membrane.

integrated hypothesis The concept, put forth by F. E. Clements, that a community is an assemblage of closely linked species, locked into association by mandatory biotic interactions that cause the community to function as an integrated unit, a sort of superorganism.

integration The interpretation of sensory signals within neural processing centers of the central nervous system.

integrin A receptor protein built into the plasma membrane that interconnects the extracellular matrix and the cytoskeleton.

integument (in-teg´-ū-ment) Layer of sporophyte tissue that contributes to the structure of an ovule of a seed plant.

integumentary system The outer covering of a mammal's body, including skin, hair, and nails.

intercalated disk A specialized junction between cardiac muscle cells that provides direct electrical coupling between cells.

interferon (in´-ter-fēr´-on) A protein that has antiviral or immune regulatory functions. Interferon-α and interferon-β, secreted by virus-infected cells, help nearby cells resist viral infection; interferon-γ, secreted by T cells, helps activate macrophages.

intermediate disturbance hypothesis The concept that moderate levels of disturbance can foster greater species diversity than low or high levels of disturbance.

intermediate filament A component of the cytoskeleton that includes all filaments intermediate in size between microtubules and microfilaments.

internal fertilization Reproduction in which sperm are typically deposited in or near the female reproductive tract and fertilization occurs within the tract.

interneuron (in´-ter-nūr´-on) An association neuron; a nerve cell within the central nervous system that forms synapses with sensory and motor neurons and integrates sensory input and motor output.

internode A segment of a plant stem between the points where leaves are attached.

interoreceptor A sensory receptor that detects stimuli within the body, such as blood pressure and body position.

interphase The period in the cell cycle when the cell is not dividing. During interphase, cellular metabolic activity is high, chromosomes and organelles are duplicated, and cell size may increase. Interphase accounts for 90% of the cell cycle.

intersexual selection Selection whereby individuals of one sex (usually females) are choosy in selecting their mates from individuals of the other sex; also called mate choice.

interspecific competition Competition for resources between plants, between animals, or between decomposers when resources are in short supply.

interspecific interaction Relationships between species of a community.

interstitial fluid The internal environment of vertebrates, consisting of the fluid filling the spaces between cells.

intertidal zone The shallow zone of the ocean where land meets water.

intracellular digestion The joining of food vacuoles and lysosomes to allow chemical digestion to occur within the cytoplasm of a cell.

intrasexual selection A direct competition among individuals of one sex (usually the males in vertebrates) for mates of the opposite sex.

introduced species A species moved by humans, either intentionally or accidentally, from its native location to a new geographic region; also called an *exotic species*.

intron (in´-tron) A noncoding, intervening sequence within a eukaryotic gene.

invagination The infolding of cells.

invasive species A species that takes hold outside of its native range; usually introduced by humans.

inversion An aberration in chromosome structure resulting from reattachment in a reverse orientation of a chromosomal fragment to the chromosome from which the fragment originated.

invertebrate An animal without a backbone. Invertebrates make up 95% of animal species.

involution Cells rolling over the edge of the lip of the blastopore into the interior of the embryo during gastrulation.

ion (ī´-on) An atom that has gained or lost electrons, thus acquiring a charge.

ion channel Protein channel in a cell membrane that allows passage of a specific ion down its concentration gradient.

ionic bond (ī-on´-ik) A chemical bond resulting from the attraction between oppositely charged ions.

ionic compound A compound resulting from the formation of an ionic bond; also called a salt.

islets of Langerhans Clusters of endocrine cells within the pancreas that produce and secrete the hormones glucagon (alpha cells) and insulin (beta cells).

isomer (ī´-sō-mer) One of several organic compounds with the same molecular formula but different structures and therefore different properties. The three types of isomers are structural isomers, geometric isomers, and enantiomers.

isomorphic Referring to alternating generations in plants and certain algae in which the sporophytes and gametophytes look alike, although they differ in chromosome number.

isopod A member of one of the largest groups of crustaceans, which includes terrestrial, freshwater, and marine species. Among the terrestrial isopods are the pill bugs, or wood lice.

isotonic (ī´-sō-ton´-ik) Having the same solute concentration as another solution.

isotope (ī´-sō-tōp) One of several atomic forms of an element, each containing a different number of neutrons and thus differing in atomic mass.

iteroparity A life history in which adults produce large numbers of offspring over many years; also known as repeated reproduction.

jasmonic acid An important molecule in plant defense against herbivores.

joule (J) A unit of energy: $1\ J = 0.239\ cal$; $1\ cal = 4.184\ J$.

juvenile hormone A hormone in arthropods, secreted by the corpora allata glands, that promotes the retention of larval characteristics.

juxtaglomerular apparatus (JGA) (juks´-tuh-gluh-mār´-ū-ler) A specialized tissue that releases the enzyme renin when blood pressure or blood volume drops in the afferent arteriole that supplies blood to the glomerulus.

juxtamedullary nephrons Nephrons with well-developed loops of Henle that extend deeply into the renal medulla.

karyogamy (kār´-ē-og´-uh-mē) The fusion of nuclei of two cells, as part of syngamy.

karyotype (kār´-ē-ō-tīp) A display of the chromosome pairs of a cell arranged by size and shape.

keystone species A species that is not necessarily abundant in a community yet exerts strong control on community structure by the nature of its ecological role or niche.

kilocalorie (kcal) A thousand calories; the amount of heat energy required to raise the temperature of 1 kg of water by 1°C.

kin selection A phenomenon of inclusive fitness, used to explain altruistic behavior between related individuals.

kinesis (kuh-nē´-sis) A change in activity or turning rate in response to a stimulus.

kinetic energy (kuh-net´-ik) The energy of motion, which is directly related to the speed of that motion. Moving matter does work by imparting motion to other matter.

kinetochore (kuh-net´-uh-kōr) A specialized region on the centromere that links each sister chromatid to the mitotic spindle.

kinetoplastid A protist, such as *Trypanosoma*, which has a single large mitochondrion that houses extranuclear DNA.

kingdom A taxonomic category, the second broadest after domain.

***K*-selection** The concept that in certain (*K*-selected) populations, life history is centered around producing relatively few offspring that have a good chance of survival.

labia majora A pair of thick, fatty ridges that enclose and protect the labia minora and vestibule.

labia minora A pair of slender skin folds that enclose and protect the vestibule.

labor A series of strong, rhythmic contractions of the uterus that expel a baby out of the uterus and vagina during childbirth.

lactation The continued production of milk.

lacteal (lak´-tē-al) A tiny lymph vessel extending into the core of an intestinal villus and serving as the destination for absorbed chylomicrons.

lactic acid fermentation The conversion of pyruvate to lactate with no release of carbon dioxide.

lagging strand A discontinuously synthesized DNA strand that elongates in a direction away from the replication fork.

lancelet Member of the subphylum Cephalochordata, small blade-shaped marine chordates that lack a backbone.

landmark A point of reference for orientation during navigation.

landscape Several different, primarily terrestrial ecosystems linked by exchanges of energy, materials, and organisms.

landscape ecology The study of past, present, and future patterns of landscape use, as well as ecosystem management and the biodiversity of interacting ecosystems.

large intestine (colon) (kō´-len) The tubular portion of the vertebrate alimentary tract between the small intestine and the anus; functions mainly in water absorption and the formation of feces.

larva (lar´-vuh) (plural, **larvae**) A free-living, sexually immature form in some animal life cycles that may differ from the adult in morphology, nutrition, and habitat.

larynx (lār´-inks) The voice box, containing the vocal cords.

lateral geniculate nuclei The destination in the thalamus for most of the ganglion cell axons that form the optic nerves.

lateral inhibition A process that sharpens the edges and enhances the contrast of a perceived image by inhibiting receptors lateral to those that have responded to light.

lateralization Segregation of functions in the cortex of the left and right hemispheres of the brain.

lateral line system A mechanoreceptor system consisting of a series of pores and receptor units (neuromasts) along the sides of the body in fishes and aquatic amphibians; detects water movements made by the animal itself and by other moving objects.

lateral meristem (mār´-uh-stem) A meristem that thickens the roots and shoots of woody plants. The vascular cambium and cork cambium are lateral meristems.

lateral root A root that arises from the outermost layer of the pericycle of an established root.

law of independent assortment Mendel's second law, stating that each pair of alleles segregates independently during gamete formation; applies when genes for two characters are located on different pairs of homologous chromosomes.

law of segregation Mendel's first law, stating that each allele in a pair separates into a different gamete during gamete formation.

leading strand The new continuous complementary DNA strand synthesized along the template strand in the mandatory $5' \rightarrow 3'$ direction.

leaf The main photosynthetic organ of vascular plants.

leaf primordia Fingerlike projections along the flanks of a shoot apical meristem, from which leaves arise.

leaf trace A small vascular bundle that extends from the vascular tissue of the stem through the petiole and into a leaf.

learning A behavioral change resulting from experience.

lens The structure in an eye that focuses light rays onto the retina.

lenticels Small raised areas in the bark of stems and roots that enable gas exchange between living cells and the outside air.

lepidosaur Member of the reptilian group that includes lizards, snakes, and two species of New Zealand animals called tuataras.

leukocyte (lū´-kō-sīt) A white blood cell; typically functions in immunity, such as phagocytosis or antibody production.

Leydig cell A cell that produces testosterone and other androgens and is located between the seminiferous tubules of the testes.

lichen (lī´-ken) The symbiotic collective formed by the mutualistic association between a fungus and a photosynthetic alga or cyanobacterium.

life cycle The generation-to-generation sequence of stages in the reproductive history of an organism.

life expectancy at birth The predicted average length of life at birth.

life history The series of events from birth through reproduction and death.

life table A table of data summarizing mortality in a population.

ligament A type of fibrous connective tissue that joins bones together at joints.

ligand (lig´-und) A molecule that binds specifically to a receptor site of another molecule.

ligand-gated ion channel A protein pore in the plasma membrane that opens or closes in response to a chemical signal, allowing or blocking the flow of specific ions.

light chain One of the two types of polypeptide chains that make up an antibody molecule and B cell receptor; consists of a variable region, which contributes to the antigen-binding site, and a constant region.

light-harvesting complex Complex of proteins associated with pigment molecules (including chlorophyll *a*, chlorophyll *b*, and carotenoids) that captures light energy and transfers it to reaction-center pigments in a photosystem.

light microscope (LM) An optical instrument with lenses that refract (bend) visible light to magnify images of specimens.

light reactions The steps in photosynthesis that occur on the thylakoid membranes of the chloroplast and that convert solar energy to the chemical energy of ATP and NADPH, evolving oxygen in the process.

lignin (lig´-nin) A hard material embedded in the cellulose matrix of vascular plant cell walls that functions as an important adaptation for support in terrestrial species.

limbic system (lim´-bik) A group of nuclei (clusters of nerve cell bodies) in the lower part of the mammalian forebrain that interact with the cerebral cortex in determining emotions; includes the hippocampus and the amygdala.

limiting nutrient An element that must be added for production to increase in a particular area.

limnetic zone In a lake, the well-lit, open surface waters farther from shore.

linkage map A genetic map based on the frequencies of recombination between markers during crossing over of homologous chromosomes.

linked genes Genes located close enough together on a chromosome to be usually inherited together.

lipid (lip´-id) One of a family of compounds, including fats, phospholipids, and steroids, that are insoluble in water.

littoral zone In a lake, the shallow, well-lit waters close to shore.

liver The largest organ in the vertebrate body. The liver performs diverse functions, such as producing bile, preparing nitrogenous wastes for disposal, and detoxifying poisonous chemicals in the blood.

liverwort A small, herbaceous nonvascular plant that is a member of the phylum Hepatophyta.

loam The most fertile of all soils, made up of roughly equal amounts of sand, silt, and clay.

lobe-fin Member of the vertebrate subgroup Sarcopterygii, osteichthyans with rod-shaped muscular fins, including coelacanths and lungfishes, as well as the lineage that gave rise to tetrapods.

local regulator A chemical messenger that influences cells in the vicinity.

locomotion Active movement from place to place.

locus (lō´-kus) (plural, **loci**) A specific place along the length of a chromosome where a given gene is located.

logistic population growth A model describing population growth that levels off as population size approaches carrying capacity.

long-day plant A plant that flowers (usually in late spring or early summer) only when the light period is longer than a critical length.

long-term memory The ability to hold, associate, and recall information over one's life.

long-term potentiation (LTP) An enhanced responsiveness to an action potential (nerve signal) by a receiving neuron.

loop of Henle The long hairpin turn, with a descending and ascending limb, of the renal tubule in the vertebrate kidney; functions in water and salt reabsorption.

loose connective tissue The most widespread connective tissue in the vertebrate body. It binds epithelia to underlying tissues and functions as packing material, holding organs in place.

lophophore (lof´-uh-fōr) A horseshoe-shaped or circular fold of the body wall bearing ciliated tentacles that surround the mouth.

lophotrochozoan Member of a group of animal phyla with protostome development that some systematists hypothesize form a clade, characterized by lophophores or trochophore larvae.

low-density lipoprotein (LDL) A cholesterol-carrying particle in the blood, made up of cholesterol and other lipids surrounded by a single layer of phospholipids in which proteins are embedded. An LDL particle carries more cholesterol than a related lipoprotein, HDL, and high LDL levels in the blood correlate with a tendency to develop blocked blood vessels and heart disease.

lung An invaginated respiratory surface of terrestrial vertebrates, land snails, and spiders that connects to the atmosphere by narrow tubes.

luteal phase That portion of the ovarian cycle during which endocrine cells of the corpus luteum secrete female hormones.

luteinizing hormone (LH) (lū´-tē-uh-nī´-zing) A tropic hormone produced and secreted by the anterior pituitary that stimulates ovulation in females and androgen production in males.

lycophyte An informal name for any member of the phylum Lycophyta, which includes club mosses, spike mosses, and quillworts.

lymph The colorless fluid, derived from interstitial fluid, in the lymphatic system of vertebrate animals.

lymph node Organ located along a lymph vessel. Lymph nodes filter lymph and help attack viruses and bacteria.

lymphatic system A system of vessels and lymph nodes, separate from the circulatory system, that returns fluid, proteins, and cells to the blood.

lymphocyte A type of white blood cell that mediates acquired immunity. Lymphocytes that complete their development in the bone marrow are called B cells, and those that mature in the thymus are called T cells.

lysogenic cycle (lī´-sō-jen´-ik) A phage replication cycle in which the viral genome becomes incorporated into the bacterial host chromosome as a prophage and does not kill the host.

lysosome (lī´-so-sōm) A membrane-enclosed sac of hydrolytic enzymes found in the cytoplasm of eukaryotic cells.

lysozyme (lī´-sō-zīm) An enzyme in sweat, tears, and saliva that attacks bacterial cell walls.

lytic cycle (lit´-ik) A type of viral (phage) replication cycle resulting in the release of new phages by lysis (and death) of the host cell.

M phase *See* mitotic (M) phase.

macroclimate Large-scale variations in climate; the climate of an entire region.

macroevolution Evolutionary change above the species level, including the appearance of major evolutionary developments, such as flight, that we use to define higher taxa.

macromolecule A giant molecule formed by the joining of smaller molecules, usually by a condensation reaction. Polysaccharides, proteins, and nucleic acids are macromolecules.

macronutrient A chemical substance that an organism must obtain in relatively large amounts. *See also* micronutrient.

macrophage (mak´-rō-fāj) A phagocytic cell present in many tissues that functions in innate immunity by destroying microbes and in acquired immunity as an antigen-presenting cell.

magnetic reversal A reversal of the polarity of Earth's magnetic field.

magnoliids A flowering plant clade that evolved later than basal angiosperms but before monocots and eudicots. Extant examples are magnolias, laurels, and black pepper plants.

major depression Depressive mental illness characterized by experiencing a low mood most of the time.

major histocompatibility complex (MHC) A family of genes that encode a large set of cell surface proteins called MHC molecules. Class I and class II MHC molecules function in antigen presentation to T cells. Foreign MHC molecules on transplanted tissue can trigger T cell responses that may lead to rejection of the transplant.

malignant tumor A cancerous tumor that is invasive enough to impair the functions of one or more organs.

malleus The first of the three middle ear bones.

malnourished Referring to an animal whose diet is missing one or more essential nutrients.

Malpighian tubule (mal-pig´-ē-un) A unique excretory organ of insects that empties into the digestive tract, removes nitrogenous wastes from the hemolymph, and functions in osmoregulation.

mammal Member of the class Mammalia, amniotes with mammary glands that produce milk.

mammary glands Exocrine glands that secrete milk to nourish the young. These glands are characteristic of mammals.

mandible One of a pair of jaw-like feeding appendages found in myriapods, hexapods, and crustaceans.

mantle A fold of tissue in molluscs that drapes over the visceral mass and may secrete a shell.

mantle cavity A water-filled chamber that houses the gills, anus, and excretory pores of a mollusc.

map unit A unit of measurement of the distance between genes. One map unit is equivalent to a 1% recombination frequency.

marine benthic zone The ocean floor.

mark-recapture method A sampling technique used to estimate wildlife populations.

marsupial (mar-sū´-pē-ul) A mammal, such as a koala, kangaroo, or opossum, whose young complete their embryonic development inside a maternal pouch called the marsupium.

mass number The sum of the number of protons and neutrons in an atom's nucleus.

mast cell A vertebrate body cell that produces histamine and other molecules that trigger the inflammatory response.

mate choice copying Behavior in which individuals in a population copy the mate choice of others, apparently as a result of social learning.

maternal effect gene A gene that, when mutant in the mother, results in a mutant phenotype in the offspring, regardless of the genotype.

matter Anything that takes up space and has mass.

maximum likelihood A principle that states that when considering multiple phylogenetic hypotheses, one should take into account the one that reflects the most likely sequence of evolutionary events, given certain rules about how DNA changes over time.

maximum parsimony A principle that states that when considering multiple explanations for an observation, one should first investigate the simplest explanation that is consistent with the facts.

mechanoreceptor A sensory receptor that detects physical deformations in the body's environment associated with pressure, touch, stretch, motion, and sound.

medulla oblongata The lowest part of the vertebrate brain, commonly called the medulla; a swelling of the hindbrain dorsal to the anterior spinal cord that controls autonomic, homeostatic functions, including breathing, heart and blood vessel activity, swallowing, digestion, and vomiting.

medusa (muh-dū´-suh) The floating, flattened, mouth-down version of the cnidarian body plan. The alternate form is the polyp.

megapascal (MPa) (meg´-uh-pas-kal´) A unit of pressure equivalent to 10 atmospheres of pressure.

megaphyll A leaf with a highly branched vascular system, characteristic of the vast majority of vascular plants.

megaspore A spore from a heterosporous plant species that develops into a female gametophyte.

meiosis (mī-ō´-sis) A two-stage type of cell division in sexually reproducing organisms that results in cells with half the chromosome number of the original cell.

meiosis I The first division of a two-stage process of cell division in sexually reproducing organisms that results in cells with half the chromosome number of the original cell.

meiosis II The second division of a two-stage process of cell division in sexually reproducing organisms that results in cells with half the chromosome number of the original cell.

melanocyte-stimulating hormone (MSH) A hormone produced and secreted by the anterior pituitary that regulates the activity of pigment-containing cells in the skin of some vertebrates.

melatonin A hormone secreted by the pineal gland that regulates body functions related to seasonal day length.

membrane attack complex (MAC) A molecular complex consisting of a set of complement proteins that forms a pore in the membrane of bacterial and transplanted cells, causing the cells to die by lysis.

membrane potential The charge difference between a cell's cytoplasm and the extracellular fluid, due to the differential distribution of ions. Membrane potential affects the activity of excitable cells and the transmembrane movement of all charged substances.

memory cell One of a clone of long-lived lymphocytes, formed during the primary immune response, that remains in a lymphoid organ until activated by exposure to the same antigen that triggered its formation. Activated memory cells mount the secondary immune response.

menopause The cessation of ovulation and menstruation.

menstrual cycle (men´-strū-ul) A type of reproductive cycle in higher female primates, in which the nonpregnant endometrium is shed as a bloody discharge through the cervix into the vagina.

menstrual flow phase That portion of the uterine (menstrual) cycle when menstrual bleeding occurs.

menstruation The shedding of portions of the endometrium during a uterine (menstrual) cycle.

meristem (mār´-uh-stem) Plant tissue that remains embryonic as long as the plant lives, allowing for indeterminate growth.

meristem identity gene A plant gene that promotes the switch from vegetative growth to flowering.

meroblastic cleavage (mār´-ō-blas´-tik) A type of cleavage in which there is incomplete division of yolk-rich egg, characteristic of avian development.

mesentery (mez´-en-tār-ē) A membrane that suspends many of the organs of vertebrates inside fluid-filled body cavities.

mesoderm (mez´-ō-derm) The middle primary germ layer of an early embryo that develops into the notochord, the lining of the coelom, muscles, skeleton, gonads, kidneys, and most of the circulatory system.

mesohyl (mes´-uh-hil) A gelatinous region between the two layers of cells of a sponge.

mesophyll (mez´-ō-fil) The ground tissue of a leaf, sandwiched between the upper and lower epidermis and specialized for photosynthesis.

mesophyll cell A loosely arranged photosynthetic cell located between the bundle sheath and the leaf surface.

messenger RNA (mRNA) A type of RNA, synthesized from DNA, that attaches to ribosomes in the cytoplasm and specifies the primary structure of a protein.

metabolic pathway A series of chemical reactions that either builds a complex molecule (anabolic pathway) or breaks down a complex molecule into simpler compounds (catabolic pathway).

metabolic rate The total amount of energy an animal uses in a unit of time.

metabolism (muh-tab´-uh-lizm) The totality of an organism's chemical reactions, consisting of catabolic and anabolic pathways.

metamorphosis (met´-uh-mōr´-fuh-sis) The resurgence of development in an animal larva that transforms it into a sexually mature adult.

metanephridium (met´-uh-nuh-frid´-ē-um) (plural, **metanephridia**) In annelid worms, a type of excretory tubule with internal openings called nephrostomes that collect body fluids and external openings called nephridiopores.

metaphase The third stage of mitosis, in which the spindle is complete and the chromosomes, attached to microtubules at their kinetochores, are all aligned at the metaphase plate.

metaphase plate An imaginary plane during metaphase in which the centromeres of all the duplicated chromosomes are located midway between the two poles.

metapopulation A subdivided population of a single species.

metastasis (muh-tas´-tuh-sis) The spread of cancer cells to locations distant from their original site.

methanogen A microorganism that obtains energy by using carbon dioxide to oxidize hydrogen, producing methane as a waste product.

microclimate Very fine scale variations of climate, such as the specific climatic conditions underneath a log.

microevolution Evolutionary change below the species level; change in the genetic makeup of a population from generation to generation.

microfilament A solid rod of actin protein in the cytoplasm of almost all eukaryotic cells, making up part of the cytoskeleton and acting alone or with myosin to cause cell contraction.

micronutrient An element that an organism needs in very small amounts and that functions as a component or cofactor of enzymes. *See also* macronutrient.

microphyll In lycophytes, a small leaf with a single unbranched vein.

micropyle A pore in the integument(s) of an ovule.

micro-RNA (miRNA) A small, single-stranded RNA molecule that binds to a complementary sequence in mRNA molecules and directs associated proteins to degrade or prevent translation of the target mRNA.

microspore A spore from a heterosporous plant species that develops into a male gametophyte.

microsporidia Unicellular parasites of animals and protists that molecular comparisons suggest may be most closely related to zygomycete fungi.

microtubule A hollow rod of tubulin protein in the cytoplasm of all eukaryotic cells and in cilia, flagella, and the cytoskeleton.

microvillus (plural, **microvilli**) One of many fine, fingerlike projections of the epithelial cells in the lumen of the small intestine that increase its surface area.

midbrain One of three ancestral and embryonic regions of the vertebrate brain; develops into sensory integrating and relay centers that send sensory information to the cerebrum.

middle ear One of three main regions of the vertebrate ear; a chamber containing three small bones (the hammer, anvil, and stirrup) that convey vibrations from the eardrum to the oval window.

middle lamella (luh-mel´-uh) A thin layer of adhesive extracellular material, primarily pectins, found between the primary walls of adjacent young plant cells.

mineral In nutrition, a chemical element other than hydrogen, oxygen, or nitrogen that an organism requires for proper body functioning.

mineral nutrient An essential chemical element absorbed from the soil in the form of inorganic ions.

mineralocorticoid A steroid hormone secreted by the adrenal cortex that regulates salt and water homeostasis.

minimum viable population (MVP) The smallest population size at which a species is able to sustain its numbers and survive.

mismatch repair The cellular process that uses special enzymes to fix incorrectly paired nucleotides.

missense mutation The most common type of mutation, a base-pair substitution in which the new codon makes sense in that it still codes for an amino acid.

mitochondrial matrix The compartment of the mitochondrion enclosed by the inner membrane and containing enzymes and substrates for the Krebs cycle.

mitochondrion (mī´-tō-kon´-drē-on) (plural, **mitochondria**) An organelle in eukaryotic cells that serves as the site of cellular respiration.

mitosis (mī-tō´-sis) A process of nuclear division in eukaryotic cells conventionally divided into five stages: prophase, prometaphase, metaphase, anaphase, and telophase. Mitosis conserves chromosome number by equally allocating replicated chromosomes to each of the daughter nuclei.

mitotic (M) phase The phase of the cell cycle that includes mitosis and cytokinesis.

mitotic spindle An assemblage of microtubules and associated proteins that is involved in the movements of chromosomes during mitosis.

mixotroph An organism that is capable of both photosynthesis and heterotrophy.

model A representation of a theory or process.

model organism An organism chosen to study broad biological principles.

modern synthesis A comprehensive theory of evolution emphasizing populations as units of evolution and integrating ideas from many fields, including genetics, statistics, paleontology, taxonomy, and biogeography.

molarity A common measure of solute concentration, referring to the number of moles of solute per liter of solution.

mold A rapidly growing fungus that reproduces asexually by producing spores.

mole (mol) The number of grams of a substance that equals its molecular weight in daltons and contains Avogadro's number of molecules.

molecular clock An evolutionary timing method based on the observation that at least some regions of genomes evolve at constant rates.

molecular formula A type of molecular notation indicating only the quantity of the constituent atoms.

molecular mass The sum of the masses of all the atoms in a molecule; sometimes called molecular weight.

molecular systematics The comparison of nucleic acids or other molecules in different species to infer relatedness.

molecule Two or more atoms held together by covalent bonds.

molting A process in arthropods in which the exoskeleton is shed at intervals, allowing growth by the production of a larger exoskeleton.

monoclonal antibody (mon´-ō-klōn´-ul) Any of a preparation of antibodies that have been produced by a single clone of cultured cells and thus are all specific for the same epitope.

monocots A clade consisting of flowering plants that have one embryonic seed leaf, or cotyledon.

monocyte A type of white blood cell that migrates into tissues and develops into a macrophage.

monoecious (muh-nē´-shus) A term typically used to describe an angiosperm species in which carpellate and staminate flowers are on the same plant.

monogamous A type of relationship in which one male mates with just one female.

monohybrid An organism that is heterozygous with respect to a single gene of interest. All the offspring from a cross between parents homozygous for different alleles are monohybrids. For example, parents of genotypes *AA* and *aa* produce a monohybrid of genotype *Aa*.

monomer (mon´-uh-mer) The subunit that serves as the building block of a polymer.

monophyletic (mon´-ō-fī-let´-ik) Pertaining to a grouping of species consisting of an ancestral species and all its descendants; a clade.

monosaccharide (mon´-ō-sak´-uh-rīd) The simplest carbohydrate, active alone or serving as a monomer for disaccharides and polysaccharides. Also known as simple sugars, the molecular formulas of monosaccharides are generally some multiple of CH_2O.

monosomic Referring to a cell that has only one copy of a particular chromosome, instead of the normal two.

monotreme (mon´-uh-trēm) An egg-laying mammal, represented by the platypus and echidna.

morphogen A substance, such as Bicoid protein, that provides positional information in the form of a concentration gradient along an embryonic axis.

morphogenesis (mōr´-fō-jen´-uh-sis) The development of body shape and organization.

morphological species concept Defining species by measurable anatomical criteria.

morula (mōr´-yuh-luh) A solid ball of blastomeres formed by early cleavage.

mosaic evolution The evolution of different features of organisms at different rates.

moss A small, herbaceous nonvascular plant that is a member of the phylum Bryophyta.

motor neuron A nerve cell that transmits signals from the brain or spinal cord to muscles or glands.

motor unit A single motor neuron and all the muscle fibers it controls.

movement corridor A series of small clumps or a narrow strip of quality habitat (usable by organisms) that connects otherwise isolated patches of quality habitat.

MPF Maturation-promoting factor (M-phase-promoting factor); a protein complex required for a cell to progress from late interphase to mitosis. The active form consists of cyclin and a protein kinase.

mucous membrane (myū´-kus) Smooth moist epithelium that lines the digestive tract and air tubes leading to the lungs.

Müllerian mimicry (myū-lār´-ē-un) A mutual mimicry by two unpalatable species.

multifactorial Referring to a phenotypic character that is influenced by multiple genes and environmental factors.

multigene family A collection of genes with similar or identical sequences, presumably of common origin.

multiple fruit A fruit derived from an inflorescence, a group of flowers tightly clustered together.

muscle spindle A mechanoreceptor stimulated by mechanical distortion.

muscle tissue Tissue consisting of long muscle cells that are capable of contracting when stimulated by nerve impulses.

mutagen (myū´-tuh-jen) A chemical or physical agent that interacts with DNA and causes a mutation.

mutation (myū-tā´-shun) A change in the DNA of a gene, ultimately creating genetic diversity.

mutualism (myū´-chū-ul-izm) A symbiotic relationship in which both participants benefit.

mycelium (mī-sē´-lē-um) The densely branched network of hyphae in a fungus.

mycorrhizae (mī´-kō-rī´-zē) Mutualistic associations of plant roots and fungi.

mycosis The general term for a fungal infection.

myelin sheath (mī´-uh-lin) In a neuron, an insulating coat of cell membrane from Schwann cells that is interrupted by nodes of Ranvier, where saltatory conduction occurs.

myofibril (mī´-ō-fī´-bril) A fibril collectively arranged in longitudinal bundles in muscle cells (fibers); composed of thin filaments of actin and a regulatory protein and thick filaments of myosin.

myofilaments The thick and thin filaments that form the myofibrils.

myogenic heart A type of heart, such as in vertebrate animals, in which the pacemaker is made up of specialized muscle tissues and located within the heart itself.

myoglobin (mī´-uh-glō´-bin) An oxygen-storing, pigmented protein in muscle cells.

myosin (mī´-uh-sin) A type of protein filament that interacts with actin filaments to cause cell contraction.

myotonia Increased muscle tension.

myriapod A terrestrial arthropod with many body segments and one or two pairs of legs per segment. Millipedes and centipedes comprise the two classes of living myriapods.

NAD$^+$ Nicotinamide adenine dinucleotide, a coenzyme present in all cells that helps enzymes transfer electrons during the redox reactions of metabolism.

NADP$^+$ Nicotinamide adenine dinucleotide phosphate, an acceptor that temporarily stores energized electrons produced during the light reactions.

natural family planning A form of contraception that relies on refraining from sexual intercourse when conception is most likely to occur; also called the rhythm method.

natural killer (NK) cell A type of white blood cell that can kill tumor cells and virus-infected cells; an important component of innate immunity.

natural selection Differential success in the reproduction of different phenotypes resulting from the interaction of organisms with their environment. Evolution occurs when natural selection causes changes in relative frequencies of alleles in the gene pool.

negative feedback A primary mechanism of homeostasis, whereby a change in a physiological variable that is being monitored triggers a response that counteracts the initial fluctuation.

negative pressure breathing A breathing system in which air is pulled into the lungs.

nematocyst (nem´-uh-tuh-sist) A stinging, capsule-like organelle in a cnidocyte.

neocortex In the mammalian brain, the outermost region of the cerebral cortex.

nephron (nef´-ron) The tubular excretory unit of the vertebrate kidney.

neritic zone (nuh-rit´-ik) The shallow region of the ocean overlying the continental shelf.

nerve A ropelike bundle of neuron fibers (axons and dendrites) tightly wrapped in connective tissue.

nerve cord A ropelike arrangement of neurons characteristic of animals with bilateral symmetry and cephalization.

nerve net A weblike system of neurons, characteristic of radially symmetrical animals, such as hydra.

nervous tissue Tissue made up of neurons and supportive cells.

net primary production (NPP) The gross primary production of an ecosystem minus the energy used by the producers for respiration.

neural crest A band of cells along the border where the neural tube pinches off from the ectoderm. The cells migrate to various parts of the embryo and form the pigment cells in the skin, bones of the skull, the teeth, the adrenal glands, and parts of the peripheral nervous system.

neural tube A tube of cells running along the dorsal axis of the body, just dorsal to the notochord. It will give rise to the central nervous system.

neurogenic heart A type of heart, such as in insects, in which the pacemakers originate in motor nerves arising from outside the heart.

neurohypophysis (ner´-ō-hī-pof´-uh-sis) *See* posterior pituitary.

neuron (ner´-on) A nerve cell; the fundamental unit of the nervous system, having structure and properties that allow it to conduct signals by taking advantage of the electrical charge across its cell membrane.

neuropeptide A relatively short chain of amino acids that serves as a neurotransmitter.

neurosecretory cell A specialized nerve cell that releases a hormone into the bloodstream in response to signals from other nerve cells; located in the hypothalamus and adrenal medulla.

neurotransmitter A chemical messenger released from the synaptic terminal of a neuron at a chemical synapse that diffuses across the synaptic cleft and binds to and stimulates the postsynaptic cell.

neutral theory The hypothesis that much evolutionary change in genes and proteins has no effect on fitness and therefore is not influenced by Darwinian natural selection.

neutral variation Genetic diversity that confers no apparent selective advantage.

neutron An electrically neutral particle (a particle having no electrical charge), found in the nucleus of an atom.

neutrophil The most abundant type of white blood cell. Neutrophils are phagocytic and tend to self-destruct as they destroy foreign invaders, limiting their life span to a few days.

niche *See* ecological niche.

nitric oxide (NO) A gas produced by many types of cells that functions as a local regulator, a neurotransmitter, and an antibacterial agent.

nitrogen fixation The assimilation of atmospheric nitrogen by certain prokaryotes into nitrogenous compounds that can be directly used by plants.

nitrogenase (nī-troj´-uh-nāz) An enzyme complex, unique to certain prokaryotes, that reduces N_2 to NH_3.

nitrogen-fixing bacteria Microorganisms that restock nitrogenous minerals in the soil by converting nitrogen to ammonia.

nociceptor A class of naked dendrites in the epidermis of the skin.

node A point along the stem of a plant at which leaves are attached.

nodule A swelling on the root of a legume. Nodules are composed of plant cells that contain nitrogen-fixing bacteria of the genus *Rhizobium*.

noncompetitive inhibitor A substance that reduces the activity of an enzyme by binding to a location remote from the active site, changing its conformation so that it no longer binds to the substrate.

noncyclic electron flow A route of electron flow during the light reactions of photosynthesis that involves both photosystems and produces ATP, NADPH, and oxygen. The net electron flow is from water to NADP$^+$.

nondisjunction An error in meiosis or mitosis, in which both members of a pair of homologous chromosomes or both sister chromatids fail to move apart properly.

nonequilibrium model The model of communities that emphasizes that they are not stable in time but constantly changing after being buffeted by disturbances.

nonpolar covalent bond A type of covalent bond in which electrons are shared equally between two atoms of similar electronegativity.

nonsense mutation A mutation that changes an amino acid codon to one of the three stop codons, resulting in a shorter and usually nonfunctional protein.

nonshivering thermogenesis (NST) The increased production of heat in some mammals by the action of certain hormones that cause mitochondria to increase their metabolic activity and produce heat instead of ATP.

norepinephrine A hormone that is chemically and functionally similar to epinephrine.

norm of reaction The range of phenotypes produced by a single genotype, due to environmental influences.

notochord A longitudinal, flexible rod that runs along the dorsal axis of an animal's body in the future position of the vertebral column.

nuclear envelope The membrane in eukaryotes that encloses the nucleus, separating it from the cytoplasm.

nuclear lamina A netlike array of protein filaments that maintains the shape of the nucleus.

nuclease An enzyme that hydrolyzes DNA and RNA into their component nucleotides.

nucleic acid A polymer (polynucleotide) consisting of many nucleotide monomers; serves as a blueprint for proteins and, through the actions of proteins, for all cellular activities. The two types are DNA and RNA.

nucleic acid hybridization Base pairing between a gene and a complementary sequence on another nucleic acid molecule.

nucleic acid probe (nū-klā´-ik) In DNA technology, a labeled single-stranded nucleic acid molecule used to tag a specific nucleotide sequence in a nucleic acid sample. Molecules of the probe hydrogen-bond to the complementary sequence wherever it occurs; radioactive or other labeling of the probe allows its location to be detected.

nucleoid (nū´-klē-oid) A dense region of DNA in a prokaryotic cell.

nucleoid region The region in a prokaryotic cell consisting of a concentrated mass of DNA.

nucleolus (nū-klē´-ō-lus) (plural, **nucleoli**) A specialized structure in the nucleus, formed from various chromosomes and active in the synthesis of ribosomes.

nucleosome (nū´-klē-ō-sōm) The basic, beadlike unit of DNA packaging in eukaryotes, consisting of a segment of DNA wound around a protein core composed of two copies of each of four types of histone.

nucleotide (nū´-klē-ō-tīd) The building block of a nucleic acid, consisting of a five-carbon sugar covalently bonded to a nitrogenous base and a phosphate group.

nucleotide excision repair The process of removing and then correctly replacing a damaged segment of DNA using the undamaged strand as a guide.

nucleus (1) An atom's central core, containing protons and neutrons. (2) The chromosome-containing organelle of a eukaryotic cell. (3) A cluster of neurons.

obligate aerobe (ob´-lig-et âr´-ōb) An organism that requires oxygen for cellular respiration and cannot live without it.

obligate anaerobe (ob´-lig-et an´-uh-rōb) An organism that cannot use oxygen and is poisoned by it.

oceanic pelagic biome Most of the ocean's waters far from shore, constantly mixed by ocean currents.

oceanic zone The region of water lying over deep areas beyond the continental shelf.

Okazaki fragment A short segment of DNA synthesized on a template strand during DNA replication. Many Okazaki fragments make up the lagging strand of newly synthesized DNA.

olfactory receptor Smell receptor.

oligodendrocyte (ol´-ig-ō-den´-druh-sīt) A type of glial cell that forms insulating myelin sheaths around the axons of neurons in the central nervous system.

oligosaccharin A type of elicitor (molecule that induces a broad defense response in plants) that is derived from cellulose fragments released by cell wall damage.

oligotrophic lake A nutrient-poor, oxygen-rich clear, deep lake with few phytoplankton.

ommatidium (plural, **ommatidia**) (ōm´-uh-tid´-ē-um) One of the facets of the compound eye of arthropods and some polychaete worms.

omnivore A heterotrophic animal that consumes both meat and plant material.

oncogene (on´-kō-jēn) A gene found in viruses or as part of the normal genome that is involved in triggering cancerous characteristics.

one gene–one polypeptide hypothesis The premise that a gene is a segment of DNA that codes for one polypeptide.

oogenesis (ō´-uh-jen´-uh-sis) The process in the ovary that results in the production of female gametes.

oogonia Ovary-specific stem cells.

oomycete A protist with flagellated cells, such as a water mold, white rust, or downy mildew, that acquires nutrition mainly as a decomposer or plant parasite.

open circulatory system A circulatory system in which fluid called hemolymph bathes the tissues and organs directly and there is no distinction between the circulating fluid and the interstitial fluid.

operant conditioning (op´-er-ent) A type of associative learning in which an animal learns to associate one of its own behaviors with a reward or punishment and then tends to repeat or avoid that behavior; also called trial-and-error learning.

operator In prokaryotic DNA, a sequence of nucleotides near the start of an operon to which an active repressor can attach. The binding of the repressor prevents RNA polymerase from attaching to the promoter and transcribing the genes of the operon.

operculum In aquatic osteichthyans, a protective bony flap that covers and protects the gills.

operon (op´-er-on) A unit of genetic function common in bacteria and phages, consisting of coordinately regulated clusters of genes with related functions.

opisthokont Member of the clade Opisthokonta, organisms that descended from an ancestor with a posterior flagellum, including fungi, animals, and certain protists.

opposable thumb An arrangement of the fingers such that the thumb can touch the ventral surface of the fingertips of all four fingers.

opsin A membrane protein bonded to a light-absorbing pigment molecule.

optic chiasm The arrangement of the nerve tracts of the eye such that the visual sensations from the left visual field of both eyes are transmitted to the right side of the brain and the sensations from the right visual field of both eyes are transmitted to the left side of the brain.

optimal foraging theory The basis for analyzing behavior as a compromise of feeding costs versus feeding benefits.

oral cavity The mouth of an animal.

orbital The three-dimensional space where an electron is found 90% of the time.

order In classification, the taxonomic category above family.

organ A specialized center of body function composed of several different types of tissues.

organ identity genes Plant homeotic genes that use positional information to determine which emerging leaves develop into which types of floral organs.

organ of Corti The actual hearing organ of the vertebrate ear, located in the floor of the cochlear canal in the inner ear; contains the receptor cells (hair cells) of the ear.

organ system A group of organs that work together in performing vital body functions.

organelle (ōr-guh-nel´) One of several formed bodies with specialized functions, suspended in the cytoplasm of eukaryotic cells.

organic chemistry The study of carbon compounds (organic compounds).

organism An individual living thing.

organismal ecology The branch of ecology concerned with the morphological, physiological, and behavioral ways in which individual organisms meet the challenges posed by their biotic and abiotic environments.

organogenesis (ōr-gan´-ō-jen´-uh-sis) The development of organ rudiments from the three germ layers.

orgasm Rhythmic, involuntary contractions of certain reproductive structures in both sexes during the human sexual response cycle.

origin of replication Site where the replication of a DNA molecule begins.

orthologous genes Homologous genes that are passed in a straight line from one generation to the next, but have ended up in different gene pools because of speciation.

osculum A large opening in a sponge that connects the spongocoel to the environment.

osmoconformer An animal that does not actively adjust its internal osmolarity because it is isoosmotic with its environment.

osmolarity (oz´-mō-lār´-uh-tē) Solute concentration expressed as molarity.

osmoregulation How organisms regulate solute concentrations and balance the gain and loss of water.

osmoregulator An animal whose body fluids have a different osmolarity than the environment and that must either discharge excess water if it lives in a hypoosmotic environment or take in water if it inhabits a hyperosmotic environment.

osmosis (oz-mō´-sis) The diffusion of water across a selectively permeable membrane.

osmotic potential A component of water potential that is proportional to the number of dissolved solute molecules in a solution and measures the effect of solutes on the direction of water movement; also called solute potential, it can be either zero or negative.

osteichthyan Member of a vertebrate subgroup with jaws and mostly bony skeletons.

osteoblast A bone-forming cell that deposits collagen.

osteon The repeating organizational unit forming the microscopic structure of hard mammalian bone.

outer ear One of three main regions of the ear in reptiles, birds, and mammals; made up of the auditory canal and, in many birds and mammals, the pinna.

outgroup A species or group of species that is closely related to the group of species being studied, but clearly not as closely related as any study-group members are to each other.

oval window In the vertebrate ear, a membrane-covered gap in the skull bone, through which sound waves pass from the middle ear to the inner ear.

ovarian cycle (ō-vār´-ē-un) The cyclic recurrence of the follicular phase, ovulation, and the luteal phase in the mammalian ovary, regulated by hormones.

ovary (ō´-vuh-rē) (1) In flowers, the portion of a carpel in which the egg-containing ovules develop. (2) In animals, the structure that produces female gametes and reproductive hormones.

overexploitation Harvesting by humans of wild plants or animals at rates exceeding the ability of populations of those species to rebound.

overnourishment A diet that is chronically excessive in calories.

oviduct (ō´-vuh-duct) A tube passing from the ovary to the vagina in invertebrates or to the uterus in vertebrates.

oviparous (ō-vip´-uh-rus) Referring to a type of development in which young hatch from eggs laid outside the mother's body.

ovoviviparous (ō´-vō-vī-vip´-uh-rus) Referring to a type of development in which young hatch from eggs that are retained in the mother's uterus.

ovulation The release of an egg from ovaries. In humans, an ovarian follicle releases an egg during each uterine (menstrual) cycle.

ovule (ō´-vyūl) A structure that develops within the ovary of a seed plant and contains the female gametophyte.

ovum (ō´-vum) The female gamete; the haploid, unfertilized egg, which is usually a relatively large, nonmotile cell.

oxidation The loss of electrons from a substance involved in a redox reaction.

oxidative phosphorylation (fos´-for-uh-lā-shun) The production of ATP using energy derived from the redox reactions of an electron transport chain.

oxidizing agent The electron acceptor in a redox reaction.

oxytocin A hormone produced by the hypothalamus and released from the posterior pituitary. It induces contractions of the uterine muscles and causes the mammary glands to eject milk during nursing.

P generation The parent individuals from which offspring are derived in studies of inheritance; P stands for "parental."

P site One of a ribosome's three binding sites for tRNA during translation. The P site holds the tRNA carrying the growing polypeptide chain. (P stands for peptidyl tRNA.)

***p53* gene** The "guardian angel of the genome," a gene that is expressed when a cell's DNA is damaged. Its product, p53 protein, functions as a transcription factor for several genes.

pacemaker A specialized region of the right atrium of the mammalian heart that sets the rate of contraction; also called the sinoatrial (SA) node.

paedomorphosis (pē´-duh-mōr´-fuh-sis) The retention in an adult organism of the juvenile features of its evolutionary ancestors.

pain receptor A kind of interoreceptor that detects pain; also called a nociceptor.

paleoanthropology The study of human origins and evolution.

paleontological species concept Definition of species based on morphological differences known only from the fossil record.

paleontology (pā´-lē-un-tol´-ō-jē) The scientific study of fossils.

palisade mesophyll One or more layers of elongated photosynthetic cells on the upper part of a leaf; also called palisade parenchyma.

pancreas (pan´-krē-us) A gland with dual functions: The nonendocrine portion secretes digestive enzymes and an alkaline solution into the small intestine via a duct; the endocrine portion secretes the hormones insulin and glucagon into the blood.

Pangaea (pan-jē´-uh) The supercontinent formed near the end of the Paleozoic era when plate movements brought all the landmasses of Earth together.

parabasalid A protist such as a trichomonad, with modified mitochondria.

parabronchus (plural, **parabronchi)** A site of gas exchange in bird lungs. Parabronchi allow air to flow past the respiratory surface in just one direction.

paralogous genes Homologous genes that are found in the same genome due to gene duplication.

paraphyletic (pār´-uh-fī-let´-ik) Pertaining to a grouping of species that consists of an ancestral species and some, but not all, of its descendants.

parareptile First major group of reptiles to emerge, mostly large, stocky quadrupedal herbivores; died out in the late Triassic period.

parasite (pār´-uh-sīt) An organism that benefits by living in or on another organism at the expense of the host.

parasitism (pār´-uh-sit-izm) A symbiotic relationship in which the symbiont (parasite) benefits at the expense of the host by living either within the host (as an endoparasite) or outside the host (as an ectoparasite).

parasitoidism A type of parasitism in which an insect lays eggs on or in a living host; the larvae then feed on the body of the host, eventually killing it.

parasympathetic division One of three divisions of the autonomic nervous system; generally enhances body activities that gain and conserve energy, such as digestion and reduced heart rate.

parathyroid gland Any of four small endocrine glands, embedded in the surface of the thyroid gland, that secrete parathyroid hormone.

parathyroid hormone (PTH) A hormone secreted by the parathyroid glands that raises blood calcium level by promoting calcium release from bone and calcium retention by the kidneys.

parazoan Animal belonging to a grade of organization lacking true tissues (collections of specialized cells isolated from other tissues by membranes); a sponge (phylum Porifera).

parenchyma cell (puh-ren´-kim-uh) A relatively unspecialized plant cell type that carries out most of the metabolism, synthesizes and stores organic products, and develops into a more differentiated cell type.

parental type An offspring with a phenotype that matches one of the parental phenotypes.

Parkinson's disease A motor disorder caused by a progressive brain disease and characterized by difficulty in initiating movements, slowness of movement, and rigidity.

parthenogenesis (par´-thuh-no´-jen´-uh-sis) A type of reproduction in which females produce offspring from unfertilized eggs.

partial pressure A measure of the concentration of one gas in a mixture of gases; the pressure exerted by a particular gas in a mixture of gases (for instance, the pressure exerted by oxygen in air).

parturition The expulsion of a baby from the mother; also called birth.

passive immunity Short-term immunity conferred by the administration of ready-made antibodies or the transfer of maternal antibodies to a fetus or nursing infant; lasts only a few weeks or months because the immune system has not been stimulated by antigens.

passive transport The diffusion of a substance across a biological membrane.

patchiness Localized variation in environmental conditions within an ecosystem, arranged spatially into a complex of discrete areas that may be characterized by distinctive groups of species or ecosystem processes.

pathogen A disease-causing agent.

pattern formation The ordering of cells into specific three-dimensional structures, an essential part of shaping an organism and its individual parts during development.

peat Extensive deposits of undecayed organic material formed primarily from the wetland moss *Sphagnum*.

pedigree A diagram of a family tree showing the occurrence of heritable characters in parents and offspring over multiple generations.

pelagic zone (puh-laj´-ik) The area of the ocean past the continental shelf, with areas of open water often reaching to very great depths.

penis The copulatory structure of male mammals.

PEP carboxylase An enzyme that adds carbon dioxide to phosphoenolpyruvate (PEP) to form oxaloacetate.

pepsin An enzyme present in gastric juice that begins the hydrolysis of proteins.

pepsinogen The inactive form of pepsin that is first secreted by specialized (chief) cells located in gastric pits of the stomach.

peptide bond The covalent bond between two amino acid units, formed by a dehydration reaction.

peptidoglycan (pep´-tid-ō-glī´-kun) A type of polymer in bacterial cell walls consisting of modified sugars cross-linked by short polypeptides.

perception The interpretation of sensations by the brain.

perennial (puh-ren´-ē-ul) A flowering plant that lives for many years.

pericarp The thickened wall of a fruit.

pericycle (pār´-uh-sī-kul) The outermost layer of the vascular cylinder of a root, where lateral roots originate.

periderm (pār-uh-derm) The protective coat that replaces the epidermis in plants during secondary growth, formed of the cork and cork cambium.

periodic table of the elements A chart of the chemical elements, arranged in three rows, corresponding to the number of electron shells in their atoms.

peripheral nervous system (PNS) The sensory and motor neurons that connect to the central nervous system.

peripheral protein A protein appendage loosely bound to the surface of a membrane and not embedded in the lipid bilayer.

peripheral resistance The impedance of blood flow by the arterioles.

peristalsis (pār´-uh-stal´-sis) (1) Rhythmic waves of contraction of smooth muscle that push food along the digestive tract. (2) A type of movement on land produced by rhythmic waves of muscle contractions passing from front to back, as in many annelids.

peristome The upper part of the moss capsule (sporangium) often specialized for gradual spore discharge.

peritubular capillaries The network of tiny blood vessels that surrounds the proximal and distal tubules in the kidney.

permafrost A permanently frozen stratum below the arctic tundra.

peroxisome (puh-rok´-suh-sōm) A microbody containing enzymes that transfer hydrogen from various substrates to oxygen, producing and then degrading hydrogen peroxide.

petal A modified leaf of a flowering plant. Petals are the often colorful parts of a flower that advertise it to insects and other pollinators.

petiole (pet´-ē-ōl) The stalk of a leaf, which joins the leaf to a node of the stem.

pH A measure of hydrogen ion concentration equal to $-\log [H^+]$ and ranging in value from 0 to 14.

phage (fāj) A virus that infects bacteria; also called a bacteriophage.

phagocytosis (fag´-ō-sī-tō´-sis) A type of endocytosis involving large, particulate substances, accomplished mainly by macrophages, neutrophils, and dendritic cells.

pharyngeal clefts In chordate embryos, grooves that separate a series of pouches along the sides of the pharynx and may develop into pharyngeal slits.

pharyngeal slits In chordate embryos, slits that form from the pharyngeal clefts and communicate to the outside, later developing into gill slits in many vertebrates.

pharynx (fār´-inks) An area in the vertebrate throat where air and food passages cross; in flatworms, the muscular tube that protrudes from the ventral side of the worm and ends in the mouth.

phase change A shift from one developmental phase to another.

phenotype (fē´-nō-tīp) The physical and physiological traits of an organism, which are determined by its genetic makeup.

phenotypic polymorphism The existence of two or more distinct morphs (discrete forms), each represented in a population in high enough frequencies to be readily noticeable.

pheromone (fār´-uh-mōn) In animals and fungi, a small, volatile chemical that functions in communication and that in animals acts much like a hormone in influencing physiology and behavior.

phloem (flō´-um) Vascular plant tissue consisting of living cells arranged into elongated tubes that transport sugar and other organic nutrients throughout the plant.

phoronids A tube-dwelling marine lophophorate.

phosphate group (fos´-fāt) A functional group important in energy transfer.

phospholipid (fos´-fō-lip´-id) A molecule that is a constituent of the inner bilayer of biological membranes, having a polar, hydrophilic head and a nonpolar, hydrophobic tail.

phosphorylated Referring to a molecule that has been the recipient of a phosphate group.

photic zone (fō´-tic) The narrow top slice of the ocean, where light permeates sufficiently for photosynthesis to occur.

photoautotroph (fō´-tō-ot´-ō-trōf) An organism that harnesses light energy to drive the synthesis of organic compounds from carbon dioxide.

photoheterotroph (fō´-tō-het´-uh-rō-trōf) An organism that uses light to generate ATP but that must obtain carbon in organic form.

photomorphogenesis Effects of light on plant morphology.

photon (fō´-ton) A quantum, or discrete amount, of light energy.

photoperiodism (fō´-tō-pēr´-ē-ō-dizm) A physiological response to photoperiod, the relative lengths of night and day. An example of photoperiodism is flowering.

photophosphorylation (fō´-tō-fos´-fōr-uh-lā´-shun) The process of generating ATP from ADP and phosphate by means of a proton-motive force generated by the thylakoid membrane of the chloroplast during the light reactions of photosynthesis.

photopsin (fō-top´-sin) One of a family of visual pigments in the cones of the vertebrate eye that absorb bright, colored light.

photoreceptor An electromagnetic receptor that detects the radiation known as visible light.

photorespiration A metabolic pathway that consumes oxygen, releases carbon dioxide, generates no ATP, and decreases photosynthetic output; generally occurs on hot, dry, bright days, when stomata close and the oxygen concentration in the leaf exceeds that of carbon dioxide.

photosynthesis The conversion of light energy to chemical energy that is stored in glucose or other organic compounds; occurs in plants, algae, and certain prokaryotes.

photosystem Light-capturing unit located in the thylakoid membrane of the chloroplast, consisting of a reaction center surrounded by numerous light-harvesting complexes. There are two types of photosystems, I and II; they absorb light best at different wavelengths.

photosystem I One of two light-capturing units in a chloroplast's thylakoid membrane; it has two molecules of P700 chlorophyll *a* at its reaction center.

photosystem II One of two light-capturing units in a chloroplast's thylakoid membrane; it has two molecules of P680 chlorophyll *a* at its reaction center.

phototropism (fō´-tō-trō´-pizm) Growth of a plant shoot toward or away from light.

phragmoplast An alignment of cytoskeletal elements and Golgi-derived vesicles across the midline of a dividing plant cell.

phylogenetic species concept Defining a species as a set of organisms with a unique genetic history.

phylogenetic tree A branching diagram that represents a hypothesis about evolutionary relationships.

phylogeny (fī-loj´-uh-nē) The evolutionary history of a species or group of related species.

phylogram A phylogenetic tree in which the lengths of the branches reflect the number of genetic changes that have taken place in a particular DNA or RNA sequence in the various lineages.

phylum (fī´-lum) In classification, the taxonomic category above class.

physical map A genetic map in which the actual physical distances between genes or other genetic markers are expressed, usually as the number of base pairs along the DNA.

physiology The study of the functions of an organism.

phytoalexin (fī´-tō-uh-lek´-sin) An antibiotic, produced by plants, that destroys microorganisms or inhibits their growth.

phytochromes (fī´-tuh-krōmz) A class of light receptors in plants. Mostly absorbing red light, these photoreceptors regulate many plant responses, including seed germination and shade avoidance.

phytoplankton (fī´-tō-plank´-ton) Algae and photosynthetic bacteria that drift passively in the pelagic zone of an aquatic environment.

phytoremediation An emerging nondestructive technology that seeks to cheaply reclaim contaminated areas by taking advantage of the remarkable ability of some plant species to extract heavy metals and other pollutants from the soil and to concentrate them in easily harvested portions of the plant.

pilus (plural, **pili**) (pī´-lus, pī´-lī) A long, hairlike prokaryotic appendage that functions in adherence or in the transfer of DNA during conjugation.

pineal gland (pin´-ē-ul) A small gland on the dorsal surface of the vertebrate forebrain that secretes the hormone melatonin.

pinocytosis (pī´-nō-sī-tō´-sis) A type of endocytosis in which the cell ingests extracellular fluid and its dissolved solutes.

pistil A single carpel or a group of fused carpels.

pit A thinner region in the walls of tracheids and vessels where only primary wall is present.

pitch A function of a sound wave's frequency, or number of vibrations per second, expressed in hertz.

pith Ground tissue that is internal to the vascular tissue in a stem; in many monocot roots, parenchyma cells that form the central core of the vascular cylinder.

pituitary gland (puh-tū´-uh-tār-ē) An endocrine gland at the base of the hypothalamus; consists of a posterior lobe (neurohypophysis), which stores and releases two hormones produced by the hypothalamus, and an anterior lobe (adenohypophysis), which produces and secretes many hormones that regulate diverse body functions.

placenta (pluh-sen´-tuh) A structure in the pregnant uterus for nourishing a viviparous fetus with the mother's blood supply; formed from the uterine lining and embryonic membranes.

placental transfer cell A plant cell that enhances the transfer of nutrients from parent to embryo.

placoderm (plak´-ō-derm) A member of an extinct class of fishlike vertebrates that had jaws and were enclosed in a tough, outer armor.

planarian A free-living flatworm found in unpolluted ponds and streams.

plankton Mostly microscopic organisms that drift passively or swim weakly near the surface of oceans, ponds, and lakes.

Plantae (plan´-tā) The kingdom that consists of multicellular eukaryotes that carry out photosynthesis.

plasma (plaz´-muh) The liquid matrix of blood in which the cells are suspended.

plasma cell The antibody-secreting effector cell of humoral immunity; arises from antigen-stimulated B cells.

plasma membrane The membrane at the boundary of every cell that acts as a selective barrier, thereby regulating the cell's chemical composition.

plasmid (plaz´-mid) A small ring of DNA that carries accessory genes separate from those of a bacterial chromosome; also found in some eukaryotes, such as yeast.

plasmodesma (plaz´-mō-dez´-muh) (plural, **plasmodesmata**) An open channel in the cell wall of a plant through which strands of cytosol connect from an adjacent cell.

plasmodial slime mold (plaz-mō´-dē-ul) A type of protist that has amoeboid cells, flagellated cells, and a plasmodial feeding stage in its life cycle.

plasmodium A single mass of cytoplasm containing many diploid nuclei that forms during the life cycle of some slime molds.

plasmogamy The fusion of the cytoplasm of cells from two individuals; occurs as one stage of syngamy.

plasmolysis (plaz-mol´-uh-sis) A phenomenon in walled cells in which the cytoplasm shrivels and the plasma membrane pulls away from the cell wall when the cell loses water to a hypertonic environment.

plasmolyze To shrink and pull away from a cell wall, or when a plant cell protoplast pulls away from the cell wall as a result of water loss.

plasticity An organism's ability to alter or "mold" itself in response to local environmental conditions.

plastid One of a family of closely related plant organelles that includes chloroplasts, chromoplasts, and amyloplasts (leucoplasts).

platelet A small enucleated blood cell important in blood clotting; derived from large cells in the bone marrow.

pleiotropy (plī´-uh-trō-pē) The ability of a single gene to have multiple effects.

pluripotent Describing a stem cell, from an embryo or adult organism, that can give rise to multiple but not all differentiated cell types.

point mutation A change in a gene at a single nucleotide pair.

polar covalent bond A covalent bond between atoms that differ in electronegativity. The shared electrons are pulled closer to the more electronegative atom, making it slightly negative and the other atom slightly positive.

polar molecule A molecule (such as water) with opposite charges on opposite sides.

polarity A lack of symmetry. Structural differences in opposite ends of an organism or structure, such as the root end and shoot end of a plant.

pollen grains The structures that contain the male gametophyte of seed plants.

pollination (pol´-uh-nā´-shun) The transfer of pollen to the part of a seed plant containing the ovules, a process that is a prerequisite for fertilization.

poly-A tail The modified end of the 3′ end of an mRNA molecule consisting of the addition of some 50 to 250 adenine nucleotides.

polyandry (pol´-ē-an´-drē) A polygamous mating system involving one female and many males.

polygamous A type of relationship in which an individual of one sex mates with several of the other.

polygenic inheritance (pol´-ē-jen´-ik) An additive effect of two or more gene loci on a single phenotypic character.

polygyny (puh-lij´-en-ē) A polygamous mating system involving one male and many females.

polymer (pol´-uh-mer) A long molecule consisting of many similar or identical monomers linked together.

polymerase chain reaction (PCR) (puh-lim′-uh-rās) A technique for amplifying DNA *in vitro* by incubating with special primers, DNA polymerase molecules, and nucleotides.

polymorphism (pol´-ē-mōr´-fizm) The coexistence of two or more distinct forms in the same population.

polynucleotide (pol´-ē-nū´-klē-ō-tīd) A polymer consisting of many nucleotide monomers; serves as a blueprint for proteins and, through the actions of proteins, for all cellular activities. The two types are DNA and RNA.

polyp (pol´-ip) The sessile variant of the cnidarian body plan. The alternate form is the medusa.

polypeptide (pol´-ē-pep´-tīd) A polymer (chain) of many amino acids linked together by peptide bonds.

polyphyletic Pertaining to a grouping of species derived from two or more different ancestral forms.

polyploidy (pol´-ē-ploy´-dē) A chromosomal alteration in which the organism possesses more than two complete chromosome sets.

polyribosome (polysome) (pol´-ē-rī´-bō-sōm) An aggregation of several ribosomes attached to one messenger RNA molecule.

polysaccharide (pol´-ē-sak´-uh-rīd) A polymer of up to over a thousand monosaccharides, formed by dehydration reactions.

pons Portion of the brain that participates in certain automatic, homeostatic functions, such as regulating the breathing centers in the medulla.

population A localized group of individuals that belong to the same biological species (that are capable of interbreeding and producing fertile offspring).

population dynamics The study of how complex interactions between biotic and abiotic factors influence variations in population size.

population ecology The study of populations in relation to the environment, including environmental influences on population density and distribution, age structure, and variations in population size.

population genetics The study of how populations change genetically over time.

population viability analysis (PVA) A method of predicting whether or not a population will persist.

positional information Signals to which genes regulating development respond, indicating a cell's location relative to other cells in an embryonic structure.

positive feedback A physiological control mechanism in which a change in some variable triggers mechanisms that amplify the change.

positive pressure breathing A breathing system in which air is forced into the lungs.

posterior Pertaining to the rear, or tail end, of a bilaterally symmetrical animal.

posterior pituitary Also called the neurohypophysis; an extension of the hypothalamus composed of nervous tissue that secretes oxytocin and antidiuretic hormone made in the hypothalamus; a temporary storage site for these hormones.

postsynaptic cell The target cell at a synapse.

postzygotic barrier (pōst´-zī-got´-ik) Any of several species-isolating mechanisms that prevent hybrids produced by two different species from developing into viable, fertile adults.

potential energy The energy stored by matter as a result of its location or spatial arrangement.

precautionary principle A guiding principle in making decisions about the environment, cautioning to consider carefully the potential consequences of actions.

predation An interaction between species in which one species, the predator, eats the other, the prey.

pregnancy The condition of carrying one or more embryos in the uterus.

preprophase band Microtubules in the cortex (outer cytoplasm) of a cell that are concentrated into a ring.

prepuce (prē´-pyūs) A fold of skin covering the head of the clitoris and penis.

pressure potential (Ψ_P) A component of water potential that consists of the physical pressure on a solution, which can be positive, zero, or negative.

presynaptic cell The transmitting cell at a synapse.

prezygotic barrier (prē´-zī-got´-ik) A reproductive barrier that impedes mating between species or hinders fertilization of ova if interspecific mating is attempted.

primary cell wall A relatively thin and flexible layer first secreted by a young plant cell.

primary consumer An herbivore; an organism in the trophic level of an ecosystem that eats plants or algae.

primary electron acceptor A specialized molecule sharing the reaction center with the pair of reaction-center chlorophyll *a* molecules; it accepts an electron from one of these two chlorophylls.

primary growth Growth produced by apical meristems, lengthening stems and roots.

primary immune response The initial acquired immune response to an antigen, which appears after a lag of about 10 to 17 days.

primary oocyte (ō´-uh-sīt) A diploid cell, in prophase I of meiosis, that can be hormonally triggered to develop into an ovum.

primary plant body The tissues produced by apical meristems, which lengthen stems and roots.

primary producer An autotroph, usually a photosynthetic organism. Collectively, autotrophs make up the trophic level of an ecosystem that ultimately supports all other levels.

primary production The amount of light energy converted to chemical energy (organic compounds) by autotrophs in an ecosystem during a given time period.

primary structure The level of protein structure referring to the specific sequence of amino acids.

primary succession A type of ecological succession that occurs in a virtually lifeless area, where there were originally no organisms and where soil has not yet formed.

primary transcript An initial RNA transcript; also called pre-mRNA when transcribed from a protein-coding gene.

primary visual cortex The destination in the occipital lobe of the cerebrum for most of the axons from the lateral geniculate nuclei.

primase An enzyme that joins RNA nucleotides to make the primer.

primer A polynucleotide with a free 3′ end, bound by complementary base pairing to the template strand, that is elongated during DNA replication.

primitive streak A groove on the surface of an early avian embryo along the future long axis of the body.

prion An infectious form of protein that may increase in number by converting related proteins to more prions.

product An ending material in a chemical reaction.

production efficiency The fraction of energy stored in food that is not used for respiration.

progestin (prō-jes´-tin) One of a family of steroid hormones, including progesterone, that prepare the uterus for pregnancy.

progymnosperms Extinct seedless vascular plants that may be ancestral to seed plants.

prokaryotic cell (prō´-kār´-ē-ot´-ik) A type of cell lacking a membrane-enclosed nucleus and membrane-enclosed organelles. Organisms with prokaryotic cells (bacteria and archaea) are called prokaryotes.

prolactin (PRL) A hormone produced and secreted by the anterior pituitary with a great diversity of effects in different vertebrate species. In mammals, it stimulates growth of and milk production by the mammary glands.

proliferative phase That portion of the uterine (menstrual) cycle when the endometrium regenerates and thickens.

prometaphase The second stage of mitosis, in which discrete chromosomes consisting of identical sister chromatids appear, the nuclear envelope fragments, and the spindle microtubules attach to the kinetochores of the chromosomes.

promiscuous A type of relationship in which mating occurs with no strong pair-bonds or lasting relationships.

promoter A specific nucleotide sequence in DNA that binds RNA polymerase and indicates where to start transcribing RNA.

prophage (prō´-faj) A phage genome that has been inserted into a specific site on the bacterial chromosome.

prophase The first stage of mitosis, in which the chromatin is condensing and the mitotic spindle begins to form, but the nucleolus and nucleus are still intact.

prostaglandin (PG) (pros′-tuh-glan′-din) One of a group of modified fatty acids secreted by virtually all tissues and functioning as local regulators.

prostate gland (pros´-tāt) A gland in human males that secretes an acid-neutralizing component of semen.

proteasome A giant protein complex that recognizes and destroys proteins tagged for elimination by the small protein ubiquitin.

protein (prō´-tēn) A three-dimensional biological polymer constructed from a set of 20 different monomers called amino acids.

protein kinase An enzyme that transfers phosphate groups from ATP to a protein.

protein phosphatase An enzyme that removes phosphate groups from proteins, often functioning to reverse the effect of a protein kinase.

proteoglycan (prō´-tē-ō-glī´-kun) A glycoprotein in the extracellular matrix of animal cells, rich in carbohydrate.

proteomics The systematic study of the full protein sets (proteomes) encoded by genomes.

protist An informal term applied to any eukaryote that is not a plant, animal, or fungus. Most protists are unicellular, though some are colonial or multicellular.

protobiont An aggregate of abiotically produced molecules surrounded by a membrane or membrane-like structure.

proton (prō´-ton) A subatomic particle with a single positive electrical charge, found in the nucleus of an atom.

proton pump An active transport mechanism in cell membranes that uses ATP to force hydrogen ions out of a cell, generating a membrane potential in the process.

protonema A mass of green, branched, one-cell-thick filaments produced by germinating moss spores.

protonephridium (prō´-tō-nuh-frid´-ē-um) An excretory system, such as the flame-bulb system of flatworms, consisting of a network of closed tubules having external openings called nephridiopores and lacking internal openings.

proton-motive force The potential energy stored in the form of an electrochemical gradient, generated by the pumping of hydrogen ions across biological membranes during chemiosmosis.

proto-oncogene (prō´-tō-on´-kō-jēn) A normal cellular gene corresponding to an oncogene; a gene with a potential to cause cancer but that requires some alteration to become an oncogene.

protoplast The contents of a plant cell exclusive of the cell wall.

protoplast fusion The fusing of two protoplasts from different plant species that would otherwise be reproductively incompatible.

protostome development In animals, a developmental mode distinguished by the development of the mouth from the blastopore; often also characterized by schizocoelous development of the body cavity and by spiral cleavage.

protozoan (prō´-tō-zō´-un) A protist that lives primarily by ingesting food, an animal-like mode of nutrition.

provirus Viral DNA that inserts into a host genome.

proximal tubule In the vertebrate kidney, the portion of a nephron immediately downstream from Bowman's capsule that conveys and helps refine filtrate.

proximate question In animal behavior, an inquiry that focuses on the environmental stimuli, if any, that trigger a particular behavioral act, as well as the genetic, physiological, and anatomical mechanisms underlying it.

PR protein A protein involved in plant responses to pathogens (PR = pathogenesis-related).

pseudocoelomate (sū´-dō-sē´-lō-māt) An animal whose body cavity is not completely lined by mesoderm.

pseudogene A DNA segment very similar to a real gene but which does not yield a functional product; a gene that has become inactivated in a particular species because of mutation.

pseudopodium (sū´-dō-pō´-dē-um) (plural, **pseudopodia**) A cellular extension of amoeboid cells used in moving and feeding.

pterophyte An informal name for any member of the phylum Pterophyta, which includes ferns, horsetails, whisk ferns, and the genus *Tmesipteris*.

pterosaur Winged reptile that lived during the time of dinosaurs.

pulmocutaneous circuit The route of circulation that directs blood to the skin and lungs.

pulmonary circuit The branch of the circulatory system that supplies the lungs.

pulse The rhythmic stretching of the arteries caused by the pressure of blood forced through the arteries by contractions of the ventricles during systole.

punctuated equilibrium In evolutionary theory, long periods of apparent stasis (no change) interrupted by relatively brief periods of sudden change.

Punnett square A diagram used in the study of inheritance to show the results of random fertilization in genetic crosses.

pupil The opening in the iris, which admits light into the interior of the vertebrate eye. Muscles in the iris regulate its size.

purine (pyū´-rēn) One of two types of nitrogenous bases found in nucleotides. Adenine (A) and guanine (G) are purines.

pyloric sphincter (pī-lōr´-ik sfink´-ter) In the vertebrate digestive tract, a muscular ring that regulates the passage of food out of the stomach and into the small intestine.

pyrimidine (puh-rim´-uh-dēn) One of two types of nitrogenous bases found in nucleotides. Cytosine (C), thymine (T), and uracil (U) are pyrimidines.

quantitative character A heritable feature that varies continuously over a range rather than in an either-or fashion.

quaternary structure (kwot´-er-nār-ē) The particular shape of a complex, aggregate protein, defined by the characteristic three-dimensional arrangement of its constituent subunits, each a polypeptide.

R plasmid A bacterial plasmid carrying genes that confer resistance to certain antibiotics.

radial cleavage A type of embryonic development in deuterostomes in which the planes of cell division that transform the zygote into a ball of cells are either parallel or perpendicular to the polar axis, thereby aligning tiers of cells one above the other.

radial glia In an embryo, supporting cells that form tracks along which newly formed neurons migrate from the neural tube; can also act as stem cells that give rise to neurons and other glia.

radial symmetry Characterizing a body shaped like a pie or barrel, with many equal parts radiating outward like the spokes of a wheel; present in cnidarians and echinoderms; also can refer to flower structure.

radiation The emission of electromagnetic waves by all objects warmer than absolute zero.

radicle An embryonic root of a plant.

radioactive isotope An isotope (an atomic form of a chemical element) that is unstable; the nucleus decays spontaneously, giving off detectable particles and energy.

radiolarian A protist, usually marine, with a shell generally made of silica and pseudopodia that radiate from the central body.

radiometric dating A method paleontologists use for determining the ages of rocks and fossils on a scale of absolute time, based on the half-life of radioactive isotopes.

radula A straplike rasping organ used by many molluscs during feeding.

***ras* gene** A gene that codes for Ras protein, a G protein that relays a growth signal from a growth factor receptor on the plasma membrane to a cascade of protein kinases that ultimately results in the stimulation of the cell cycle. Many *ras* oncogenes have a point mutation that leads to a hyperactive version of the Ras protein that can lead to excessive cell division.

ratite (rat´-īt) Member of the group of flightless birds.

ray initials Cells within the vascular cambrium that produce xylem and phloem rays, radial files that consist mostly of parenchyma cells.

ray-finned fish Member of the class Actinopterygii, aquatic osteichthyans with fins supported by long, flexible rays, including tuna, bass, and herring.

reactant A starting material in a chemical reaction.

reaction center Complex of proteins associated with two special chlorophyll *a* molecules and a primary electron acceptor. Located centrally in a photosystem, this complex triggers the light reactions of photosynthesis. Excited by light energy, one of the chlorophylls donates an electron to the primary electron acceptor, which passes an electron to an electron transport chain.

reading frame The way a cell's mRNA-translating machinery groups the mRNA nucleotides into codons.

receptacle The base of a flower; the part of the stem that is the site of attachment of the floral organs.

reception In cellular communication, the target cell's detection (by binding to a receptor protein) of a signal molecule from outside the cell.

receptor-mediated endocytosis (en´-dō-sī-tō´-sis) The movement of specific molecules into a cell by the inward budding of membranous vesicles containing proteins with receptor sites specific to the molecules being taken in; enables a cell to acquire bulk quantities of specific substances.

receptor potential An initial response of a receptor cell to a stimulus, consisting of a change in voltage across the receptor membrane proportional to the stimulus strength. The intensity of the receptor potential determines the frequency of action potentials traveling to the nervous system.

receptor tyrosine kinase A receptor protein in the plasma membrane that responds to the binding of a signal molecule by catalyzing the transfer of phosphate groups from ATP to tyrosines on the cytoplasmic side of the receptor. The phosphorylated tyrosines activate other signal transduction proteins within the cell.

recessive allele An allele whose phenotypic effect is not observed in a heterozygote.

reciprocal altruism Altruistic behavior between unrelated individuals, whereby the current altruistic individual benefits in the future when the current beneficiary reciprocates.

recombinant An offspring whose phenotype differs from that of the parents; also called recombinant type.

recombinant chromosome A chromosome created when crossing over combines the DNA from two parents into a single chromosome.

recombinant DNA A DNA molecule made *in vitro* with segments from different sources.

recruitment The process of progressively increasing the tension of a muscle by activating more and more of the motor neurons controlling the muscle.

rectum The terminal portion of the large intestine where the feces are stored until they are eliminated.

red alga A photosynthetic marine protist that contains the accessory pigment phycoerythrin. Most are multicellular.

red blood cell A blood cell containing hemoglobin, which transports O_2; also called an erythrocyte.

redox reaction (rē´-doks) A chemical reaction involving the transfer of one or more electrons from one reactant to another; also called oxidation-reduction reaction.

reducing agent The electron donor in a redox reaction.

reduction The addition of electrons to a substance involved in a redox reaction.

reductionism Reducing complex systems to simpler components that are more manageable to study.

redundancy model The concept, put forth by H. A. Gleason and Brian Walker, that most of the species in a community are not tightly coupled with one another (that is, the web of life is very loose). According to this model, an increase or decrease in one species in a community has little effect on other species, which operate independently.

reflex An automatic reaction to a stimulus, mediated by the spinal cord or lower brain.

refractory period (rē-frak´-tōr-ē) The short time immediately after an action potential in which the neuron cannot respond to another stimulus, owing to an increase in potassium permeability.

regeneration The regrowth of body parts from pieces of an organism.

regulator A characterization of an animal in regard to environmental variables. A regulator uses mechanisms of homeostasis to moderate internal changes in the face of external fluctuations.

regulatory gene A gene that codes for a protein, such as a repressor, that controls the transcription of another gene or group of genes.

relative abundance Differences in the abundance of different species within a community.

relative fitness The contribution of one genotype to the next generation compared to that of alternative genotypes for the same locus.

renal artery The blood vessel bringing blood to the kidney.

renal cortex The outer portion of the vertebrate kidney.

renal medulla The inner portion of the vertebrate kidney, beneath the renal cortex.

renal pelvis Funnel-shaped chamber that receives processed filtrate from the vertebrate kidney's collecting ducts and is drained by the ureter.

renal vein The blood vessel draining the kidney.

renin-angiotensin-aldosterone system (RAAS) A part of a complex feedback circuit that helps regulate blood pressure and blood volume.

repeated reproduction A life history in which adults produce large numbers of offspring over many years; also known as interoparity.

repetitive DNA Nucleotide sequences, usually noncoding, that are present in many copies in a eukaryotic genome. The repeated units may be short and arranged tandemly (in series) or long and dispersed in the genome.

replication fork A Y-shaped region on a replicating DNA molecule where new strands are growing.

repressor A protein that suppresses the transcription of a gene.

reproductive isolation The existence of biological factors (barriers) that impede members of two species from producing viable, fertile hybrids.

reproductive table An age-specific summary of the reproductive rates in a population.

reptile Member of the clade of amniotes that includes tuatara, lizards, snakes, turtles, crocodilians, and birds.

residual volume The amount of air that remains in the lungs after forcefully exhaling.

resource partitioning The division of environmental resources by coexisting species such that the niche of each species differs by one or more significant factors from the niches of all coexisting species.

respiratory medium The source of oxygen. It is typically air for terrestrial animals and water for aquatic organisms.

respiratory pigment A protein that transports most of the oxygen in blood.

respiratory surface The part of an animal where gases are exchanged with the environment.

response In cellular communication, the change in a specific cellular activity brought about by a transduced signal from outside the cell.

resting potential The membrane potential characteristic of a nonconducting, excitable cell, with the inside of the cell more negative than the outside.

restoration ecology A goal-directed science that applies ecological principles in an effort to return degraded ecosystems to conditions as similar as possible to their natural, predegraded state.

restriction enzyme A degradative enzyme that recognizes and cuts up DNA (including that of certain phages) that is foreign to a bacterium.

restriction fragment DNA segment resulting from cutting of DNA by a restriction enzyme.

restriction fragment length polymorphisms (RFLPs) Differences in DNA sequence on homologous chromosomes that can result in different patterns of restriction fragment lengths (DNA segments resulting from treatment with restriction enzymes); useful as genetic markers for making linkage maps.

restriction site A specific sequence on a DNA strand that is recognized as a "cut site" by a restriction enzyme.

reticular fiber A very thin and branched fiber made of collagen. Reticular fibers form a tightly woven fabric that is continuous with the collagenous fibers of the extracellular matrix.

reticular formation A system of neurons, containing over 90 separate nuclei, that passes through the core of the brainstem.

retina (ret´-uh-nuh) The innermost layer of the vertebrate eye, containing photoreceptor cells (rods and cones) and neurons; transmits images formed by the lens to the brain via the optic nerve.

retinal The light-absorbing pigment in rods and cones of the vertebrate eye.

retrotransposon A transposable element that moves within a genome by means of an RNA intermediate, a transcript of the retrotransposon DNA.

retrovirus (ret´-trō-vī´-rus) An RNA virus that reproduces by transcribing its RNA into DNA and then inserting the DNA into a cellular chromosome; an important class of cancer-causing viruses.

reverse transcriptase (tran-skrip´-tās) An enzyme encoded by some certain viruses (retroviruses) that uses RNA as a template for DNA synthesis.

Rh factor A protein antigen on the surface of red blood cells designated Rh-positive. If an Rh-negative mother is exposed to blood from an Rh-positive fetus, she produces anti-Rh antibodies of the IgG class.

rhizoid Long tubular single cell or filament of cells that anchors bryophytes to the ground. Rhizoids are not composed of tissues, lack specialized conducting cells, and do not play a primary role in water and mineral absorption.

rhodopsin A visual pigment consisting of retinal and opsin. When rhodopsin absorbs light, the retinal changes shape and dissociates from the opsin, after which it is converted back to its original form.

rhythm method A form of contraception that relies on refraining from sexual intercourse when conception is most likely to occur; also called natural family planning.

ribonucleic acid (RNA) (rī´-bō-nū-klā´-ik) A type of nucleic acid consisting of nucleotide monomers with a ribose sugar and the nitrogenous bases adenine (A), cytosine (C), guanine (G), and uracil (U); usually single-stranded; functions in protein synthesis and as the genome of some viruses.

ribose The sugar component of RNA.

ribosomal RNA (rRNA) (rī´-buh-sō´-mul) The most abundant type of RNA, which together with proteins forms the structure of ribosomes. Ribosomes coordinate the sequential coupling of tRNA molecules to mRNA codons.

ribosome A cell organelle constructed in the nucleolus and functioning as the site of protein synthesis in the cytoplasm; consists of rRNA and protein molecules, which make up two subunits.

ribozyme (rī´-bō-zīm) An enzyme-like RNA molecule that catalyzes reactions during RNA splicing.

river A flowing body of water.

rivet model The concept, put forth by Paul and Anne Ehrlich, that many or most of the species in a community are associated tightly with other species in a web of life. According to this model, an increase or decrease in one species in a community affects many other species.

RNA interference (RNAi) A technique to silence the expression of selected genes in nonmammalian organisms. The method uses synthetic double-stranded RNA molecules matching the sequence of a particular gene to trigger the breakdown of the gene's messenger RNA.

RNA polymerase An enzyme that links together the growing chain of ribonucleotides during transcription.

RNA processing Modification of RNA before it leaves the nucleus, a process unique to eukaryotes.

RNA splicing The removal of noncoding portions (introns) of the RNA molecule after initial synthesis.

rod cell One of two kinds of photoreceptors in the vertebrate retina; sensitive to black and white and enables night vision.

root An organ in vascular plants that anchors the plant and enables it to absorb water and nutrients from the soil.

root cap A cone of cells at the tip of a plant root that protects the apical meristem.

root hair A tiny extension of a root epidermal cell, growing just behind the root tip and increasing surface area for absorption of water and minerals.

root pressure The upward push of xylem sap in the vascular tissue of roots.

root system All of a plant's roots that anchor it in the soil, absorb and transport minerals and water, and store food.

rosette cellulose-synthesizing complex Rose-shaped array of proteins that synthesize the cellulose microfibrils of the cell walls of charophyceans and land plants.

rough ER That portion of the endoplasmic reticulum studded with ribosomes.

round window The point of contact between the stapes and the cochlea. It is where the vibrations of the stapes create a traveling series of pressure waves in the fluid of the cochlea.

***r*-selection** The concept that in certain (*r*-selected) populations, a high reproductive rate is the chief determinant of life history.

rubisco Ribulose carboxylase, the enzyme that catalyzes the first step of the Calvin cycle (the addition of CO_2 to RuBP, or ribulose bisphosphate).

ruminant An animal, such as a cow or a sheep, with an elaborate, multicompartmentalized stomach specialized for an herbivorous diet.

S phase The synthesis phase of the cell cycle; the portion of interphase during which DNA is replicated.

sac fungus Member of the phylum Ascomycota. Sac fungi range in size and complexity from unicellular yeasts to minute leafspot fungi to elaborate cup fungi and morels. About half of the sac fungi live with algae or cyanobacteria in the mutualistic associations called lichens.

saccule A chamber in the vestibule behind the oval window that participates in the sense of balance.

salicylic acid A plant hormone that may be partially responsible for activating systemic acquired resistance to pathogens.

salivary amylase A salivary gland enzyme that hydrolyzes starch and glycogen.

salivary glands Exocrine glands associated with the oral cavity. The secretions of salivary glands contain substances to lubricate food, adhere together chewed pieces into a bolus, and begin the process of chemical digestion.

salt A compound resulting from the formation of an ionic bond; also called an ionic compound.

saltatory conduction (sol´-tuh-tōr´-ē) Rapid transmission of a nerve impulse along an axon, resulting from the action potential jumping from one node of Ranvier to another, skipping the myelin-sheathed regions of membrane.

sapwood Outer layers of secondary xylem that still transport xylem sap.

sarcomere (sar´-kō-mēr) The fundamental, repeating unit of striated muscle, delimited by the Z lines.

sarcoplasmic reticulum (SR) A specialized endoplasmic reticulum that regulates the calcium concentration in the cytosol.

saturated fatty acid A fatty acid in which all carbons in the hydrocarbon tail are connected by single bonds, thus maximizing the number of hydrogen atoms that can attach to the carbon skeleton.

savanna (suh-van´-uh) A tropical grassland biome with scattered individual trees, large herbivores, and three distinct seasons based primarily on rainfall, maintained by occasional fires and drought.

scaffolding protein A type of large relay protein to which several other relay proteins are simultaneously attached to increase the efficiency of signal transduction.

scanning electron microscope (SEM) A microscope that uses an electron beam to scan the surface of a sample to study details of its topography.

schizocoelous Pattern of formation of the body cavity common in protostome development, in which initially solid masses of mesoderm split, forming the body cavity.

schizophrenia Severe mental disturbance characterized by psychotic episodes in which patients lose the ability to distinguish reality from hallucination.

Schwann cell A type of glial cell that forms insulating myelin sheaths around the axons of neurons in the peripheral nervous system.

scion (sī´-un) The twig grafted onto the stock when making a graft.

sclera (sklār´-uh) A tough, white outer layer of connective tissue that forms the globe of the vertebrate eye.

sclereid (sklār´-ē-id) A short, irregular sclerenchyma cell in nutshells and seed coats and scattered through the parenchyma of some plants.

sclerenchyma cell (skluh-ren´-kē-muh) A rigid, supportive plant cell type usually lacking protoplasts and possessing thick secondary walls strengthened by lignin at maturity.

scrotum A pouch of skin outside the abdomen that houses a testis; functions in cooling sperm, thereby keeping them viable.

scutellum (skū-tel´-um) A specialized type of cotyledon found in the grass family.

seascape Several different, primarily aquatic ecosystems linked by exchanges of energy, materials, and organisms.

second law of thermodynamics The principle whereby every energy transfer or transformation increases the entropy of the universe. Ordered forms of energy are at least partly converted to heat, and in spontaneous reactions, the free energy of the system also decreases.

second messenger A small, nonprotein, water-soluble molecule or ion, such as calcium ion or cyclic AMP, that relays a signal to a cell's interior in response to a signal received by a signal receptor protein.

secondary cell wall A strong and durable matrix often deposited in several laminated layers for plant cell protection and support.

secondary consumer A member of the trophic level of an ecosystem consisting of carnivores that eat herbivores.

secondary endosymbiosis A process in eukaryotic evolution in which a heterotrophic eukaryotic cell engulfed a photosynthetic eukaryotic cell, which survived in a symbiotic relationship inside the heterotrophic cell.

secondary growth Growth produced by lateral meristems, thickening the roots and shoots of woody plants.

secondary immune response The acquired immune response elicited on second or subsequent exposures to a particular antigen. The secondary immune response is more rapid, of greater magnitude, and of longer duration than the primary immune response.

secondary oocyte A haploid cell resulting from meiosis I in oogenesis, which will become an ovum after meiosis II.

secondary plant body The tissues produced by the vascular cambium and cork cambium, which thicken the stems and roots of woody plants.

secondary production The amount of chemical energy in consumers' food that is converted to their own new biomass during a given time period.

secondary structure The localized, repetitive coiling or folding of the polypeptide backbone of a protein due to hydrogen bond formation between peptide linkages.

secondary succession A type of succession that occurs where an existing community has been cleared by some disturbance that leaves the soil intact.

secretion (1) The discharge of molecules synthesized by a cell. (2) The discharge of wastes from the body fluid into the filtrate.

secretory phase That portion of the uterine (menstrual) cycle when the endometrium continues to thicken, becomes more vascularized, and develops glands that secrete a fluid rich in glycogen.

sedimentary rock (sed´-uh-men´-tuh-rē) Rock formed from sand and mud that once settled in layers on the bottom of seas, lakes, and marshes. Sedimentary rocks are often rich in fossils.

seed An adaptation for terrestrial plants consisting of an embryo packaged along with a store of food within a resistant coat.

seed coat A tough outer covering of a seed, formed from the outer coat of an ovule. In a flowering plant, the seed coat encloses and protects the embryo and endosperm.

seedless vascular plants The informal collective name for the phyla Lycophyta (club mosses and their relatives) and Pteridophyta (ferns and their relatives).

segmentation gene A gene of the embryo that directs the actual formation of segments after the embryo's axes are defined.

selective permeability A property of biological membranes that allows some substances to cross more easily than others.

selective reabsorption The selective uptake of solutes from a filtrate of blood, coelomic fluid, or hemolymph in the excretory organs of animals.

self-incompatibility The ability of a seed plant to reject its own pollen and sometimes the pollen of closely related individuals.

semelparity A life history in which adults have but a single reproductive opportunity to produce large numbers of offspring, such as the life history of the Pacific salmon; also known as big-bang reproduction.

semen (sē´-mun) The fluid that is ejaculated by the male during orgasm; contains sperm and secretions from several glands of the male reproductive tract.

semicircular canals A three-part chamber of the inner ear that functions in maintaining equilibrium.

semiconservative model Type of DNA replication in which the replicated double helix consists of one old strand, derived from the old molecule, and one newly made strand.

semilunar valve A valve located at the two exits of the heart, where the aorta leaves the left ventricle and the pulmonary artery leaves the right ventricle.

seminal vesicle (sem´-uh-nul ves´-uh-kul) A gland in males that secretes a fluid component of semen that lubricates and nourishes sperm.

seminiferous tubule (sem´-uh-nif´-uh-rus) A highly coiled tube in the testis in which sperm are produced.

sensation An impulse sent to the brain from activated receptors and sensory neurons.

sensitive period A limited phase in an individual animal's development when learning of particular behaviors can take place.

sensory adaptation The tendency of sensory neurons to become less sensitive when they are stimulated repeatedly.

sensory neuron A nerve cell that receives information from the internal and external environments and transmits the signals to the central nervous system.

sensory reception The detection of the energy of a stimulus by sensory cells.

sensory receptor A cellular system that collects information about the physical world outside the body and inside the organism.

sensory transduction The conversion of stimulus energy to a change in the membrane potential of a sensory receptor.

sepal (sē´-pul) A modified leaf in angiosperms that helps enclose and protect a flower bud before it opens.

septum (plural, **septa**) One of the cross-walls that divide a fungal hypha into cells. Septa generally have pores large enough to allow ribosomes, mitochondria, and even nuclei to flow from cell to cell.

sequential hermaphroditism A reproductive pattern in which an individual reverses its sex during its lifetime.

serial endosymbiosis A model of the origin of eukaryotes consisting of a sequence of endosymbiotic events in which mitochondria, chloroplasts, and perhaps other cellular structures were derived from small prokaryotes that had been engulfed by larger cells.

serotonin A biogenic amine synthesized from the amino acid tryptophan.

seta (sē´-tuh) The elongated stalk of a bryophyte sporophyte, such as in a moss.

sex chromosome One of the pair of chromosomes responsible for determining the sex of an individual.

sex-linked gene A gene located on a sex chromosome.

sexual dimorphism (dī-mōr´-fizm) A special case of polymorphism based on the distinction between the secondary sex characteristics of males and females.

sexual reproduction A type of reproduction in which two parents give rise to offspring that have unique combinations of genes inherited from the gametes of the two parents.

sexual selection Natural selection for mating success.

shared derived character An evolutionary novelty that evolved within a particular clade.

shared primitive character A character displayed in species outside a particular taxon.

shoot system The aerial portion of a plant body, consisting of stems, leaves, and (in angiosperms) flowers.

short-day plant A plant that flowers (usually in late summer, fall, or winter) only when the light period is shorter than a critical length.

short-term memory The ability to hold information, anticipations, or goals for a time and then release them if they become irrelevant.

sickle-cell disease A human genetic disease caused by a recessive allele that results in the substitution of a single amino acid in the hemoglobin protein; characterized by deformed red blood cells that can lead to numerous symptoms.

sieve plate An end wall in a sieve-tube member, which facilitates the flow of phloem sap in angiosperm sieve tubes.

sieve-tube member A living cell that conducts sugars and other organic nutrients in the phloem of angiosperms. They form chains called sieve tubes.

sign stimulus An external sensory stimulus that triggers a fixed action pattern.

signal A behavior that causes a change in behavior in another animal.

signal peptide A stretch of amino acids on a polypeptide that targets the protein to a specific destination in a eukaryotic cell.

signal-recognition particle (SRP) A protein-RNA complex that recognizes a signal peptide as it emerges from the ribosome.

signal transduction pathway A mechanism linking a mechanical or chemical stimulus to a specific cellular response.

simple fruit A fruit derived from a single carpel or several fused carpels.

simple epithelium An epithelium consisting of a single layer of cells that all touch the basal lamina.

single nucleotide polymorphisms (SNPs) One base-pair variation in the genome sequence.

single-lens eye The camera-like eye found in some jellies, polychaetes, spiders, and many molluscs.

single-strand binding protein During DNA replication, molecules that line up along the unpaired DNA strands, holding them apart while the DNA strands serve as templates for the synthesis of complementary strands of DNA.

sinoatrial (SA) node A region of the heart composed of specialized muscle tissue that sets the rate and timing at which all cardiac muscle cells contract; the pacemaker.

sinus Any of the spaces surrounding the organs of the body in animals with open circulatory systems.

sister chromatids Replicated forms of a chromosome joined together by the centromere and eventually separated during mitosis or meiosis II.

skeletal muscle (striated muscle) Muscle generally responsible for the voluntary movements of the body.

sliding-filament model The theory explaining how muscle contracts, based on change within a sarcomere, the basic unit of muscle organization, stating that thin (actin) filaments slide across thick (myosin) filaments, shortening the sarcomere. The shortening of all sarcomeres in a myofibril shortens the entire myofibril.

slow block to polyspermy The formation of the fertilization envelope and other changes in the egg's surface that prevent fusion of the egg with more than one sperm.

slow muscle fibers Muscle cells that can sustain long contractions.

small intestine The longest section of the alimentary canal; the principal site of the enzymatic hydrolysis of food macromolecules and the absorption of nutrients.

smooth ER That portion of the endoplasmic reticulum that is free of ribosomes.

smooth muscle A type of muscle lacking the striations of skeletal and cardiac muscle because of the uniform distribution of myosin filaments in the cell; responsible for involuntary body activities.

snowball Earth hypothesis The hypothesis that glaciers covered the planet's landmasses from pole to pole 750–570 million years ago, confining life to very limited areas.

social learning Modification of behavior through the observation of other individuals.

sociobiology The study of social behavior based on evolutionary theory.

sodium-potassium pump A special transport protein in the plasma membrane of animal cells that transports sodium out of the cell and potassium into the cell against their concentration gradients.

solute (sol´-ūt) A substance that is dissolved in a solution.

solute potential (Ψ_S) A component of water potential that is proportional to the number of dissolved solute molecules in a solution and measures the effect of solutes on the direction of water movement; also called osmotic potential, it can be either zero or negative.

solution A liquid that is a homogeneous mixture of two or more substances.

solvent The dissolving agent of a solution. Water is the most versatile solvent known.

somatic cell (sō-mat´-ik) Any cell in a multicellular organism except a sperm or egg cell.

somatic nervous system The branch of the motor division of the vertebrate peripheral nervous system composed of motor neurons that carry signals to skeletal muscles in response to external stimuli.

somites Paired blocks of mesoderm just lateral to the notochord of a vertebrate embryo.

soredia In lichens, small clusters of fungal hyphae with embedded algae.

sorus (plural, **sori**) A cluster of sporangia on a fern sporophyll. Sori may be arranged in various patterns, such as parallel lines or dots, that are useful in fern identification.

Southern blotting A hybridization technique that enables researchers to determine the presence of certain nucleotide sequences in a sample of DNA.

spatial learning Modification of behavior based on experience of the spatial structure of the environment.

spatial summation A phenomenon of neural integration in which the membrane potential of the postsynaptic cell is determined by the combined effect of EPSPs or IPSPs produced nearly simultaneously by different synapses.

speciation (spē´-sē-ā´-shun) The origin of new species in evolution.

species A group whose members possess similar anatomical characteristics and have the ability to interbreed.

species-area curve The biodiversity pattern, first noted by Alexander von Humboldt, that illustrates that the larger the geographic area of a community, the greater the number of species.

species diversity The number and relative abundance of species in a biological community.

species richness The number of species in a biological community.

species selection A theory maintaining that species living the longest and generating the greatest number of species determine the direction of major evolutionary trends.

specific epithet The second part of a binomial, which is unique for each species in a genus.

specific heat The amount of heat that must be absorbed or lost for 1 g of a substance to change its temperature by 1°C.

spectrophotometer An instrument that measures the proportions of light of different wavelengths absorbed and transmitted by a pigment solution.

sperm The male gamete.

spermatheca (sper´-muh-thē´-kuh) A sac in the female reproductive system where sperm are stored.

spermatogenesis The continuous and prolific production of mature sperm cells in the testis.

spermatogonia Stem cells that give rise to sperm.

sphincter (sfink´-ter) A ringlike valve consisting of modified muscles in a muscular tube, such as a digestive tract; closes off the tube like a drawstring.

spinal nerve In the vertebrate peripheral nervous system, a nerve that carries signals to or from the spinal cord.

spiral cleavage A type of embryonic development in protostomes, in which the planes of cell division that transform the zygote into a ball of cells occur obliquely to the polar axis, resulting in cells of each tier sitting in the grooves between cells of adjacent tiers.

spiral valve A corkscrew-shaped ridge that increases surface area and prolongs the passage of food along the short digestive tract.

spliceosome (splī´-sē-ō-sōm) A complex assembly that interacts with the ends of an RNA intron in splicing RNA, releasing the intron and joining the two adjacent exons.

spongocoel (spon´-jō-sēl) The central cavity of a sponge.

spongy mesophyll Loosely arranged photosynthetic cells located below the palisade mesophyll cells in a leaf.

sporangium (plural, **sporangia**) A capsule in fungi and plants in which meiosis occurs and haploid spores develop.

spore In the life cycle of a plant or alga undergoing alternation of generations, a meiotically produced haploid cell that divides mitotically, generating a multicellular individual, the gametophyte, without fusing with another cell.

sporocyte A diploid cell, also known as a spore mother cell, that undergoes meiosis and generates haploid spores.

sporophyll A leaf specialized for reproduction.

sporophyte (spōr´-ō-fīt) In organisms undergoing alternation of generations, the multicellular diploid form that results from a union of gametes and that meiotically produces haploid spores that grow into the gametophyte generation.

sporopollenin (spōr´-uh-pol´-uh-nin) A duarable polymer that covers exposed zygotes of charophycean algae and forms walls of plant spores, preventing them from drying out.

sporozoite (spōr´-uh-zō´-īt) A tiny infectious cell that represents a stage in the apicomplexan life cycle.

squamous The flat, tile-like shape of a type of epithelial cell.

stabilizing selection Natural selection that favors intermediate variants by acting against extreme phenotypes.

stamen (stā´-men) The pollen-producing reproductive organ of a flower, consisting of an anther and filament.

standard metabolic rate (SMR) The metabolic rate of a resting, fasting, and nonstressed ectotherm.

stapes The third of the three middle ear bones.

starch A storage polysaccharide in plants consisting entirely of glucose.

statocyst (stat´-uh-sist´) A type of mechanoreceptor that functions in equilibrium in invertebrates through the use of statoliths, which stimulate hair cells in relation to gravity.

statolith (1) In plants, a specialized plastid that contains dense starch grains and may play a role in detecting gravity. (2) In invertebrates, a grain or other dense granule that settles in response to gravity and is found in sensory organs that function in equilibrium.

stele The vascular tissue of a stem or root.

stem A vascular plant organ consisting of an alternating system of nodes and internodes that support the leaves and reproductive structures.

stem cell Any relatively unspecialized cell that can divide during a single division into one identical daughter cell and one more specialized daughter cell, which can undergo further differentiation.

stenohaline Referring to organisms that cannot tolerate substantial changes in external osmolarity.

steroid A type of lipid characterized by a carbon skeleton consisting of four rings with various functional groups attached.

sticky end A single-stranded end of a double-stranded DNA restriction fragment.

stigma (plural, **stigmata**) The sticky part of a flower's carpel, which traps pollen grains.

stipe A stemlike structure of a seaweed.

stock The plant that provides the root system when making a graft.

stoma (stō´-muh) (plural, **stomata**) A microscopic pore surrounded by guard cells in the epidermis of leaves and stems that allows gas exchange between the environment and the interior of the plant.

stomach An organ of the digestive system that stores food and performs preliminary steps of digestion.

stratified epithelium An epithelium consisting of more than one layer of cells in which some but not all cells touch the basal lamina.

stream A flowing body of water that is generally small, cold, and clear.

stress-induced proteins Molecules, including heat-shock proteins, that are produced within cells in response to exposure to marked increases in temperature and to other forms of severe stress, such as toxins, rapid pH changes, and viral infections.

stretch-gated ion channel Protein pore in a cell's plasma membrane that opens when the membrane is mechanically deformed, allowing the passage of certain ions.

striated muscle *See* skeletal muscle.

strobili The technical term for clusters of sporophylls known commonly as cones, found in most gymnosperms and some seedless vascular plants.

stroke The death of nervous tissue in the brain, usually resulting from rupture or blockage of arteries in the head.

stroke volume The amount of blood pumped by the left ventricle in each contraction.

stroma (strō´-muh) The fluid of the chloroplast surrounding the thylakoid membrane; involved in the synthesis of organic molecules from carbon dioxide and water.

stromatolite Rocklike structure composed of layers of prokaryotes and sediment.

structural formula A type of molecular notation in which the constituent atoms are joined by lines representing covalent bonds.

structural isomer One of several organic compounds that have the same molecular formula but differ in the covalent arrangements of their atoms.

style The stalk of a flower's carpel, with the ovary at the base and the stigma at the top.

substance P A neuropeptide that is a key excitatory signal that mediates our perception of pain.

substrate The reactant on which an enzyme works.

substrate feeder An organism that lives in or on its food source, eating its way through the food.

substrate-level phosphorylation The formation of ATP by directly transferring a phosphate group to ADP from an intermediate substrate in catabolism.

sugar sink A plant organ that is a net consumer or storer of sugar. Growing roots, shoot tips, stems, and fruits are sugar sinks supplied by phloem.

sugar source A plant organ in which sugar is being produced by either photosynthesis or the breakdown of starch. Mature leaves are the primary sugar sources of plants.

sulfhydryl group A functional group consisting of a sulfur atom bonded to a hydrogen atom (—SH).

suprachiasmatic nuclei (SCN) A pair of structures in the hypothalamus of mammals that functions as a biological clock.

surface tension A measure of how difficult it is to stretch or break the surface of a liquid. Water has a high surface tension because of the hydrogen bonding of surface molecules.

survivorship curve A plot of the number of members of a cohort that are still alive at each age; one way to represent age-specific mortality.

suspension feeder An aquatic animal, such as a clam or a baleen whale, that sifts small food particles from the water.

sustainable agriculture Long-term productive farming methods that are environmentally safe.

sustainable development The long-term prosperity of human societies and the ecosystems that support them.

swim bladder In aquatic osteichthyans, an air sac that enables the animal to control its buoyancy in the water.

symbiont (sim´-bē-unt) The smaller participant in a symbiotic relationship, living in or on the host.

symbiosis An ecological relationship between organisms of two different species that live together in direct contact.

sympathetic division One of three divisions of the autonomic nervous system of vertebrates; generally increases energy expenditure and prepares the body for action.

sympatric speciation (sim-pat´-rik) A mode of speciation occurring as a result of a radical change in the genome of a subpopulation, reproductively isolating the subpopulation from the parent population.

symplast In plants, the continuum of cytoplasm connected by plasmodesmata between cells.

synapse (sin´-aps) The locus where one neuron communicates with another neuron in a neural pathway; a narrow gap between a synaptic terminal of an axon and a signal-receiving portion (dendrite or cell body) of another neuron or effector cell. Neurotransmitter molecules released by synaptic terminals diffuse across the synapse, relaying messages to the dendrite or effector.

synapsid Member of an amniote clade distinguished by a single hole on each side of the skull, including the mammals.

synapsis The pairing of replicated homologous chromosomes during prophase I of meiosis.

synaptic cleft (sin-ap´-tik) A narrow gap separating the synaptic knob of a transmitting neuron from a receiving neuron or an effector cell.

synaptic terminal A bulb at the end of an axon in which neurotransmitter molecules are stored and released.

synaptic vesicle Membranous sac containing neurotransmitter molecules at the tip of the presynaptic axon.

system A more complex organization formed from a combination of components.

systematics The analytical study of the diversity and relationships of organisms, both present-day and extinct.

systemic acquired resistance (SAR) A defensive response in infected plants that helps protect healthy tissue from pathogenic invasion.

systemic circuit The branch of the circulatory system that supplies all body organs and then returns oxygen-poor blood to the right atrium via the veins.

systemic circulation Movement of blood through the systemic circuit.

systems biology An approach to studying biology that aims to model the dynamic behavior of whole biological systems.

systole (sis´-tō-lē) The stage of the heart cycle in which the heart muscle contracts and the chambers pump blood.

systolic pressure Blood pressure in the arteries during contraction of the ventricles.

T cell receptor The antigen receptor on T cells; a membrane-bound molecule consisting of one α chain and one β chain linked by a disulfide bridge and containing one antigen-binding site.

T lymphocyte (T cell) A type of lymphocyte, including the helper T cells and cytotoxic T cells, that develops to maturity in the thymus. After encountering antigen, T cells are responsible for cell-mediated immunity.

taproot system A root system common to eudicots, consisting of one large, vertical root (the taproot) that produces many smaller lateral, or branch, roots.

taste buds Collections of modified epithelial cells that are scattered in several areas of the tongue and mouth and are receptors for taste in humans.

TATA box A promoter DNA sequence crucial in forming the transcription initiation complex.

taxis (tak´-sis) Movement toward or away from a stimulus.

taxon (plural, **taxa**) The named taxonomic unit at any given level of classification.

taxonomy (tak-son´-uh-mē) Ordered division of organisms into categories based on a set of characteristics used to assess similarities and differences, leading to a classification scheme; the branch of biology concerned with naming and classifying the diverse forms of life.

Tay-Sachs disease A human genetic disease caused by a recessive allele for a dysfunctional enzyme, leading to accumulation of certain lipids in the brain. Seizures, blindness, and degeneration of motor and mental performance usually become manifest a few months after birth.

technology The application of scientific knowledge for a specific purpose.

telomerase An enzyme that catalyzes the lengthening of telomeres. The enzyme includes a molecule of RNA that serves as a template for new telomere segments.

telomere (tel´-uh-mēr) The protective structure at each end of a eukaryotic chromosome. Specifically, the tandemly repetitive DNA at the end of the chromosome's DNA molecule. *See also* repetitive DNA.

telophase The fifth and final stage of mitosis, in which daughter nuclei are forming and cytokinesis has typically begun.

temperate broadleaf forest A biome located throughout midlatitude regions where there is sufficient moisture to support the growth of large, broadleaf deciduous trees.

temperate grassland A terrestrial biome dominated by grasses and forbs.

temperate phage A phage that is capable of reproducing by either the lytic or lysogenic cycle.

temperature A measure of the intensity of heat in degrees, reflecting the average kinetic energy of the molecules.

template strand The DNA strand that provides the template for ordering the sequence of nucleotides in an RNA transcript.

temporal summation A phenomenon of neural integration in which the membrane potential of the postsynaptic cell in a chemical synapse is determined by the combined effect of EPSPs or IPSPs produced in rapid succession.

tendon A type of fibrous connective tissue that attaches muscle to bone.

terminal bud Embryonic tissue at the tip of a shoot, made up of developing leaves and a compact series of nodes and internodes.

terminator In prokaryotes, a special sequence of nucleotides in DNA that marks the end of a gene. It signals RNA polymerase to release the newly made RNA molecule, which then departs from the gene.

territoriality A behavior in which an animal defends a bounded physical space against encroachment by other individuals, usually of its own species. Territory defense may involve direct aggression or indirect mechanisms such as scent marking or singing.

tertiary consumer A member of the trophic level of an ecosystem consisting of carnivores that eat mainly other carnivores.

tertiary structure (ter´-shē-ār-ē) Irregular contortions of a protein molecule due to interactions of side chains involved in hydrophobic interactions, ionic bonds, hydrogen bonds, and disulfide bridges.

test The hardened shell of some protists, including forams and rediolarians, or the rigid endoskeleton of a sea urchin or sand dollar.

testcross Breeding of an organism of unknown genotype with a homozygous recessive individual to determine the unknown genotype. The ratio of phenotypes in the offspring determines the unknown genotype.

testis (plural, **testes**) The male reproductive organ, or gonad, in which sperm and reproductive hormones are produced.

testosterone The most abundant androgen hormone in the male body.

tetanus (tet´-uh-nus) The maximal, sustained contraction of a skeletal muscle, caused by a very fast frequency of action potentials elicited by continual stimulation.

tetrad A paired set of homologous chromosomes, each composed of two sister chromatids. Tetrads form during prophase I of meiosis.

tetrapod A vertebrate with two pairs of limbs, including mammals, amphibians, and birds and other reptiles.

thalamus (thal´-uh-mus) One of two integrating centers of the vertebrate forebrain. Neurons with cell bodies in the thalamus relay neural input to specific areas in the cerebral cortex and regulate what information goes to the cerebral cortex.

thallus (plural, **thalli**) A seaweed body that is plantlike but lacks true roots, stems, and leaves.

theory An explanation that is broad in scope, generates new hypotheses, and is supported by a large body of evidence.

thermal energy *See* heat.

thermocline A narrow stratum of rapid temperature change in the ocean and in many temperate-zone lakes.

thermodynamics (ther´-mō-dī-nam´-iks) (1) The study of energy transformations that occur in a collection of matter. *See* first law of thermodynamics and second law of thermodynamics. (2) A phenomenon in which external DNA is taken up by a cell and functions there.

thermoreceptor An interoreceptor stimulated by either heat or cold.

thermoregulation The maintenance of internal body temperature within a tolerable range.

theropod A member of an ancient group of dinosaurs that were bipedal carnivores.

thick filament A filament composed of staggered arrays of myosin molecules; a component of myofibrils in muscle fibers.

thigmomorphogenesis A response in plants to chronic mechanical stimulation, resulting from increased ethylene production. An example is thickening stems in response to strong winds.

thigmotropism (thig´-mō-trō´-pizm) A directional growth of a plant in response to touch.

thin filament The smaller of the two myofilaments consisting of two strands of actin and two strands of regulatory protein coiled around one another.

thoracic cavity The body cavity in mammals that houses the lungs and heart. It is surrounded in part by ribs and separated from the lower abdominal cavity by the diaphragm.

threatened species A species that is considered likely to become endangered in the foreseeable future.

three-domain system A system of taxonomic classification based on three "superkingdoms": Bacteria, Archaea, and Eukarya.

threshold The potential an excitable cell membrane must reach for an action potential to be initiated.

thrombus A clump of platelets and fibrin that blocks the flow of blood through a blood vessel.

thylakoid (thī´-luh-koyd) A flattened membrane sac inside the chloroplast, used to convert light energy to chemical energy.

thymus (thī´-mus) A small organ in the thoracic cavity of vertebrates where maturation of T cells is completed.

thyroid gland An endocrine gland, located on the ventral surface of the trachea, that secretes two iodine-containing hormones, triiodothyronine (T_3) and thyroxine (T_4), and calcitonin.

thyroid-stimulating hormone (TSH) A tropic hormone produced and secreted by the anterior pituitary that regulates the release of thyroid hormones.

thyroxine (T_4) One of two iodine-containing hormones that are secreted by the thyroid gland and help regulate metabolism, development, and maturation in vertebrates.

Ti plasmid A plasmid of a tumor-inducing bacterium that integrates a segment of its DNA into the host chromosome of a plant; frequently used as a carrier for genetic engineering in plants.

tidal volume The volume of air an animal inhales and exhales with each breath.

tight junction A type of intercellular junction in animal cells that prevents the leakage of material between cells.

tissue An integrated group of cells with a common function, structure, or both.

tissue system One or more tissues organized into a functional unit connecting the organs of a plant.

tonicity The ability of a solution to cause a cell within it to gain or lose water.

tonoplast A membrane that encloses the central vacuole in a plant cell, separating the cytosol from the vacuolar contents, called cell sap; also known as the vacuolar membrane.

top-down model A model of community organization in which predation controls community organization because predators control herbivores, which in turn control plants, which in turn control nutrient levels; also called the trophic cascade model.

topoisomerase A protein that functions in DNA replication, helping to relieve strain in the double helix ahead of the replication fork.

topsoil A mixture of particles derived from rock, living organisms, and humus.

torpor In animals, a physiological state that conserves energy by slowing down metabolism.

torsion A characteristic of gastropods in which the visceral mass rotates during development.

totipotent Describing a cell that can give rise to all parts of an organism.

trace element An element indispensable for life but required in extremely minute amounts.

trachea (trā´-kē-uh) The windpipe; that portion of the respiratory tube that has C-shaped cartilagenous rings and passes from the larynx to two bronchi.

tracheal system A gas exchange system of branched, chitin-lined tubes that infiltrate the body and carry oxygen directly to cells in insects.

tracheid (trā´-kē-id) A long, tapered water-conducting cell that is dead at maturity and is found in the xylem of all vascular plants.

trait Any detectable variation in a genetic character.

trans Arrangement of two noncarbon atoms, each bound to one of the carbons in a carbon-carbon double bond, where the two noncarbon atoms are on opposite sides relative to the double bond.

transcription The synthesis of RNA on a DNA template.

transcription factor A regulatory protein that binds to DNA and stimulates transcription of specific genes.

transcription initiation complex The completed assembly of transcription factors and RNA polymerase bound to the promoter.

transcription unit A region of a DNA molecule that is transcribed into an RNA molecule.

transduction (1) A DNA transfer process in which phages carry bacterial genes from one host cell to another. (2) In cellular communication, the conversion of a signal from outside the cell to a form that can bring about a specific cellular response.

transfer cell A companion cell with numerous ingrowths of its wall, increasing the cell's surface area and enhancing the transfer of solutes between apoplast and symplast.

transfer RNA (tRNA) An RNA molecule that functions as an interpreter between nucleic acid and protein language by picking up specific amino acids and recognizing the appropriate codons in the mRNA.

transformation (1) The conversion of a normal animal cell to a cancerous cell. (2) A change in genotype and phenotype due to the assimilation of external DNA by a cell.

transgenic Pertaining to an individual plant or animal whose genome contains a gene introduced from another organism, either from the same or a different species.

translation The synthesis of a polypeptide using the genetic information encoded in an mRNA molecule. There is a change of "language" from nucleotides to amino acids.

translocation (1) An aberration in chromosome structure resulting from attachment of a chromosomal fragment to a nonhomologous chromosome. (2) During protein synthesis, the third stage in the elongation cycle when the RNA carrying the growing polypeptide moves from the A site to the P site on the ribosome. (3) The transport of organic nutrients in the phloem of vascular plants.

transmission The conduction of impulses to the central nervous system.

transmission electron microscope (TEM) A microscope that passes an electron beam through very thin sections; primarily used to study the internal ultrastructure of cells.

transpiration The evaporative loss of water from a plant.

transport epithelium One or more layers of specialized epithelial cells that regulate solute movements.

transport protein A transmembrane protein that helps a certain substance or class of closely related substances to cross the membrane.

transport vesicle A tiny membranous sac in a cell's cytoplasm carrying molecules produced by the cell.

transposable element A segment of DNA that can move within the genome of a cell by means of a DNA or RNA intermediate; also called a transposable element.

transposon A transposable element that moves within a genome by means of a DNA intermediate.

transverse (T) tubules Infoldings of the plasma membrane of skeletal muscle cells.

triacylglycerol Three fatty acids linked to one glycerol molecule.

triiodothyronine (T_3) (trī´-ī-ō´-dō-thī´-rō-nēn) One of two iodine-containing hormones that are secreted by the thyroid gland and help regulate metabolism, development, and maturation in vertebrates.

trilobite An extinct arthropod with pronounced segmentation and appendages that varied little from segment to segment.

trimester In human development, one of three 3-month-long periods of pregnancy.

triple response A plant growth maneuver in response to mechanical stress, involving slowing of stem elongation, a thickening of the stem, and a curvature that causes the stem to start growing horizontally.

triplet code A set of three-nucleotide-long words that specify the amino acids for polypeptide chains.

triploblastic Possessing three germ layers: the endoderm, mesoderm, and ectoderm. Most eumetazoans are triploblastic.

trisomic Referring to a cell that has three copies of a particular chromosome, instead of the normal two.

trochophore larva Distinctive larval stage observed in certain invertebrates, including some annelids and molluscs.

trophic efficiency The percentage of production transferred from one trophic level to the next.

trophic structure The different feeding relationships in an ecosystem, which determine the route of energy flow and the pattern of chemical cycling.

trophoblast The outer epithelium of the blastocyst, which forms the fetal part of the placenta.

tropical rain forest A terrestrial biome characterized by high levels of precipitation and warm temperatures year-round.

tropic hormone A hormone that has another endocrine gland as a target.

tropics Latitudes between 23.5° north and south.

tropism A growth response that results in the curvature of whole plant organs toward or away from stimuli owing to differential rates of cell elongation.

tropomyosin The regulatory protein that blocks the myosin-binding sites on the actin molecules.

troponin complex The regulatory proteins that control the position of tropomyosin on the thin filament.

true-breeding Referring to plants that produce offspring of the same variety when they self-pollinate.

tubal ligation A means of sterilization in which a woman's two oviducts (Fallopian tubes) are tied closed to prevent eggs from reaching the uterus. A segment of each oviduct is removed.

tube foot One of numerous extensions of an echinoderm's water vascular system. Tube feet function in locomotion, feeding, and gas exchange.

tumor-suppressor gene A gene whose protein products inhibit cell division, thereby preventing uncontrolled cell growth (cancer).

tundra A biome at the extreme limits of plant growth. At the northernmost limits, it is called arctic tundra, and at high altitudes, where plant forms are limited to low shrubby or matlike vegetation, it is called alpine tundra.

tunicate Member of the subphylum Urochordata, sessile marine chordates that lack a backbone.

turgid (ter´-jid) Very firm. A walled cell becomes turgid if it has a greater solute concentration than its surroundings, resulting in entry of water.

turgor pressure The force directed against a cell wall after the influx of water and the swelling of a walled cell due to osmosis.

turnover The mixing of waters as a result of changing water-temperature profiles in a lake.

turnover time The time required to replace the standing crop of a population or group of populations (for example, of phytoplankton), calculated as the ratio of standing crop biomass to production.

tympanic membrane Another name for the eardrum.

tyrosine kinase An enzyme that catalyzes the transfer of phosphate groups from ATP to the amino acid tyrosine on a substrate protein.

ultimate question In animal behavior, an inquiry that focuses on the evolutionary significance of a behavioral act.

ultrametric tree A phylogenetic tree in which the lengths of the branches reflect measurements of geologic time.

undernourishment A diet that is chronically deficient in calories.

uniformitarianism Charles Lyell's idea that geologic processes have not changed throughout Earth's history.

unsaturated fatty acid A fatty acid possessing one or more double bonds between the carbons in the hydrocarbon tail. Such bonding reduces the number of hydrogen atoms attached to the carbon skeleton.

urea A soluble nitrogenous waste excreted by mammals, most adult amphibians, sharks, and some marine bony fishes and turtles; produced in the liver by a metabolic cycle that combines ammonia with carbon dioxide.

ureter A duct leading from the kidney to the urinary bladder.

urethra A tube that releases urine from the body near the vagina in females and through the penis in males; also serves in males as the exit tube for the reproductive system.

uric acid An insoluble precipitate of nitrogenous waste excreted by land snails, insects, and many reptiles, including birds.

urinary bladder The pouch where urine is stored prior to elimination.

uterine cycle The changes that occur in the uterus during the reproductive cycle of the human female; also called the menstrual cycle.

uterus A female organ where eggs are fertilized and/or development of the young occurs.

utricle A chamber behind the oval window that opens into the three semicircular canals.

vaccination *See* immunization.

vaccine A harmless variant or derivative of a pathogen that stimulates a host's immune system to mount defenses against the pathogen.

vacuolar membrane A membrane that encloses the central vacuole in a plant cell, separating the cytosol from the vacuolar contents, called cell sap; also known as the tonoplast.

vagina Part of the female reproductive system between the uterus and the outside opening; the birth canal in mammals; also accommodates the male's penis and receives sperm during copulation.

valence The bonding capacity of an atom, generally equal to the number of unpaired electrons in the atom's outermost shell.

valence electron An electron in the outermost electron shell.

valence shell The outermost energy shell of an atom, containing the valence electrons involved in the chemical reactions of that atom.

van der Waals interactions Weak attractions between molecules or parts of molecules that are brought about by localized charge fluctuations.

variation Differences between members of the same species.

vas deferens The tube in the male reproductive system in which sperm travel from the epididymis to the urethra.

vasa recta The capillary system that serves the loop of Henle.

vascular bundle (vas´-kyū-ler) A strand of vascular tissues (both xylem and phloem) in a stem or leaf.

vascular cambium A cylinder of meristematic tissue in woody plants that adds layers of secondary vascular tissue called secondary xylem (wood) and secondary phloem.

vascular cylinder The central cylinder of vascular tissue in a root.

vascular plant A plant with vascular tissue. Vascular plants include all living species except mosses, liverworts, and hornworts.

vascular tissue Plant tissue consisting of cells joined into tubes that transport water and nutrients throughout the plant body.

vascular tissue system A system formed by xylem and phloem throughout a vascular plant, serving as a transport system for water and nutrients, respectively.

vasectomy The cutting of each vas deferens to prevent sperm from entering the urethra.

vasocongestion The filling of a tissue with blood, caused by increased blood flow through the arteries of that tissue.

vasoconstriction A decrease in the diameter of superficial blood vessels triggered by nerve signals that contract the muscles of the vessel walls.

vasodilation An increase in the diameter of superficial blood vessels triggered by nerve signals that relax the muscles of the vessel walls.

vegetal pole The portion of the egg where most yolk is concentrated; opposite of animal pole.

vegetative reproduction Cloning of plants by asexual means.

vein (1) In animals, a vessel that returns blood to the heart. (2) In plants, a vascular bundle in a leaf.

ventilation Any method of increasing contact between the respiratory medium and the respiratory surface.

ventral Pertaining to the underside, or bottom, of a bilaterally symmetrical animal.

ventricle (ven´-truh-kul) (1) A heart chamber that pumps blood out of a heart. (2) A space in the vertebrate brain, filled with cerebrospinal fluid.

venule (ven´-ūl) A vessel that conveys blood between a capillary bed and a vein.

vernalization The use of cold treatment to induce a plant to flower.

vertebrate A chordate animal with a backbone: the mammals, reptiles (including birds), amphibians, sharks and rays, ray-finned fishes, and lobe-fins.

vesicle A sac made of membrane inside of cells.

vessel element A short, wide, water-conducting cell found in the xylem of most angiosperms and a few nonflowering vascular plants. Dead at maturity, vessel elements are aligned end to end to form micropipes called vessels.

vessels Continuous water-conducting micropipes found in most angiosperms and a few nonflowering vascular plants.

vestibule The cavity enclosed by the labia minora; the space into which the vagina and urethral opening empty.

vestigial organ A structure of marginal, if any, importance to an organism. Vestigial organs are historical remnants of structures that had important functions in ancestors.

villus (plural, **villi**) (1) A fingerlike projection of the inner surface of the small intestine. (2) A fingerlike projection of the chorion of the mammalian placenta. Large numbers of villi increase the surface areas of these organs.

viral envelope A membrane that cloaks the capsid that in turn encloses a viral genome.

viroid (vī´-royd) A plant pathogen composed of molecules of naked circular RNA only several hundred nucleotides long.

virulent A term describing a pathogen against which a plant has little specific defense.

virulent phage A phage that reproduces only by a lytic cycle.

visceral mass One of the three main parts of a mollusc, containing most of the internal organs.

visible light That portion of the electromagnetic spectrum detected as various colors by the human eye, ranging in wavelength from about 380 nm to about 750 nm.

vital capacity The maximum volume of air that a respiratory system can inhale and exhale.

vitamin An organic molecule required in the diet in very small amounts. Vitamins serve primarily as coenzymes or parts of coenzymes.

vitamin D One of the fat-soluble vitamins. The active form functions as a hormone, acting in concert with parathyroid hormone in bone and promoting the uptake of calcium from food within the intestines.

vitreous humor The jellylike material that fills the posterior cavity of the vertebrate eye.

viviparous (vī-vip´-uh-rus) Referring to a type of development in which the young are born alive after having been nourished in the uterus by blood from the placenta.

vocal cord One of a pair of stringlike tissues in the larynx. Air rushing past the tensed vocal cords makes them vibrate, producing sounds.

voltage-gated ion channel A specialized ion channel that opens or closes in response to changes in membrane potential.

vulva Collective term for the female external genitalia.

water potential The physical property predicting the direction in which water will flow, governed by solute concentration and applied pressure.

water vascular system A network of hydraulic canals unique to echinoderms that branches into extensions called tube feet, which function in locomotion, feeding, and gas exchange.

wavelength The distance between crests of waves, such as those of the electromagnetic spectrum.

wetland An ecosystem intermediate between an aquatic one and a terrestrial one. Wetland soil is saturated with water permanently or periodically.

white blood cell A blood cell that functions in defending the body against infections and cancer cells; also called a leukocyte.

white matter Tracts of axons within the CNS.

wild type An individual with the normal (most common) phenotype.

wilting The drooping of leaves and stems as a result of plant cells becoming flaccid.

wobble A violation of the base-pairing rules in that the third nucleotide (5′ end) of a tRNA anticodon can form hydrogen bonds with more than one kind of base in the third position (3′ end) of a codon.

xerophyte A plant adapted to an arid climate.

X-ray crystallography A technique that depends on the diffraction of an X-ray beam by the individual atoms of a molecule to study the three-dimensional structure of the molecule.

xylem (zī´-lum) Vascular plant tissue consisting mainly of tubular dead cells that conduct most of the water and minerals upward from roots to the rest of the plant.

yeast Single-celled fungi that inhabit liquid or moist habitats and reproduce asexually by simple cell division or by the pinching of small buds off a parent cell.

yeast artificial chromosome (YAC) A vector that combines the essentials of a eukaryotic chromosome—an origin for DNA replication, a centromere, and two telomeres—with foreign DNA.

yolk Nutrients stored in an egg.

yolk plug Large food-laden endodermal cells surrounded by the blastopore of an amphibian gastrula.

yolk sac One of four extraembryonic membranes that support embryonic development; the first site of blood cells and circulatory system function.

Z lines The borders of a sarcomere.

zero population growth (ZPG) A period of stability in population size, when the per capita birth rate and death rate are equal.

zona pellucida The extracellular matrix of a mammalian egg.

zone of cell division The zone of primary growth in roots consisting of the root apical meristem and its derivatives. New root cells are produced in this region.

zone of elongation The zone of primary growth in roots where new cells elongate, sometimes up to ten times their original length.

zone of maturation The zone of primary growth in roots where cells complete their differentiation and become functionally mature.

zone of polarizing activity (ZPA) A limb-bud organizer region consisting of a block of mesoderm located where the posterior side of the bud is attached to the body.

zoned reserve An extensive region of land that includes one or more areas undisturbed by humans surrounded by lands that have been changed by human activity and are used for economic gain.

zoospore Flagellated spore occurring in chytrid fungi.

zygomycete Member of the fungal phylum Zygomycota, characterized by forming a sturdy structure called a zygosporangium during sexual reproduction.

zygosporangium In zygomycete fungi, a sturdy multinucleate structure in which karyogamy and meiosis occur.

zygote The diploid product of the union of haploid gametes in conception; a fertilized egg.

Index

NOTE: A **bold** page number indicates where a term is bold and defined; an *f* following a page number indicates a figure; a *t* following a page number indicates a table.

3' end, DNA, 88, 89, 302*f*, 323
5' cap, DNA, **317**
5' end, DNA, 88, 89, 302*f*, 317, 323
30-nm fiber, 360, 361*f*
300-nm fiber, 360, 361*f*

A band, muscle, **1066**, 1067*f*
ABC model of flower formation, **734**, 735*f*
Abdominal cavity, **827**
Abiotic components of environment, **1081**, 1086–87
  climate as, 1081
  distribution of organisms affected by, 1081*f*
  disturbances in communities as, 1171–75
  influence of, on aquatic biomes, 1092–93
  influence of, on populations, 1148–52
Abiotic stresses affecting plants, **811**
Abiotic synthesis
  of organic compounds, 59*f*, 513–14
  of polymers, 514–15
ABO blood groups, 462
  carbohydrates on red blood cell surface and, 129
  immune response to transfusions from, 915*t*
  multiple alleles as determiners of, 262*t*
Abortion, **983**
Abscisic acid (ABA), **798**–99
  drought tolerance in plants and, 799
  seed dormancy and, 798, 799*f*
Absorption, **853**
  in animal food processing, 853, 859–61
  as fungal nutritional mode, 608–10
  in small intestine, 859–61
  of water and minerals by plant roots, 744, 745f, 746
Absorption spectrum, **187**
  determining, with spectrophotometers, 187*f*
  photosynthetically important, 187*f*
Abyssal zone, 1093*f*, **1097**
Acacia koa, phase change in shoot system of, 733*f*
Acanthocephalans (thorny-headed worms), 640*f*
Acanthodians, **680**
Acanthostega, 684*f*
Acclimatization, **839**
Accomodation in vision, **1059***f*
Acetic acid, **64***f*
Acetone, **64***f*
Acetylcholine as neurotransmitter, **1024***t*
Acetyl CoA, conversion of pyruvate to, in cellular respiration, **168***f*
Achillea (yarrow), 464*f*
Achondroplasia (dwarfism), 267
Acid, **53**–54
Acid chyme, **858**
Acid growth hypothesis in plants, 794, 795*f*
Acid precipitation, **55**–56
  deforestation and, 1199
  distribution of, in North America and Europe, 1201*f*
  effects on wildlife, 686
  human activities as cause of, 1201–2
  G. Likens on, 1078
  in United States, 1202*f*
Acid rain. *See* Acid precipitation
Acoelomates, **631***f*
Acquired characteristics, Lamarck's theory of, 441
Acquired immunity, **898**, 903–8
  active and passive immunization and, 914
  antigen recognition by lymphocytes in, 903–5
  cell-mediated immune response and humoral immune response as branches of, 908–14
  innate immunity versus, 899*t*
  lymphocyte development in, 905–8
  overview of, 909*f*
Acquired immunodeficiency syndrome, **918**. *See also* AIDS (acquired immunodeficiency syndrome)
Acrosomal process, 988
Acrosomal reaction, **988**, 989*f*, 991*f*
Acrosome, **975**, **988**, 989*f*
Actin, **116**
  cell cleavage and, 226
  cytokinesis and, 226
  microfilaments, motility, and, 116, 117*f*, 118
  muscle contraction and, 117*f*, 1068*f*
Actin filaments. *See* Microfilaments
Actinopterygii (ray-finned fishes), 683
Action potential, **1018**
  changes in membrane potential and triggering of, 1017–18
  conduction of, 1020*f*, 1021
  graded potentials and, 1018*f*
  in plants, **810**
  production of, 1018–20
  refractory period following, 1020
  role of voltage-gated ion channels in, 1018, 1019*f*
  saltatory conduction and, 1021*f*
  sensory perception and transmission of, 1046–47
Action spectrum, **188, 802**
  for photosynthesis, 187*f*, 188
Activation energy (E_A), 150, **151**, 152
  for exergonic reaction, 151*f*
  lowering barrier of, by enzymes, 152f
Activator, **356**, **365**
  eukaryotic gene expression and role of, 365, 366*f*
  prokaryotic gene expression and role of, 356
Active immunity, **914**
Active site, enzyme, **152**, 153*f*
  catalytic cycle and, 153*f*
  induced fit between substrate and, 152, 153*f*
Active transport, **134**–36, **738**
  cotransport, 136
  energy required for, 134
  maintenance of membrane potential by ion pumps and, 134–36
  passive transport compared with, 135*f*
Actual evapotranspiration, 1176, **1190**
  terrestrial net primary production and, 1190*f*
Adaptation. *See* Evolutionary adaptation
Adaptive immunity. *See* Acquired immunity
Adaptive radiation, **480**, 481*f*
  of Galápagos finches, 18*f*
  on Hawaiian Islands, 436, 480, 481*f*
Addition rule and Mendelian genetics, 258–59
Adenine (A), 88
  DNA structure and, 296*f*, 297*f*, 298*f*
Adenohypophysis (anterior pituitary), 949t, **950**, 951*f*, 952
Adenomatous polyposis coli (APC) tumor-suppressor gene, 373*f*, 374
Adenosine triphosphate. *See* ATP (adenosine triphosphate)
Adenoviruses, structure, 336*f*
Adenylyl cyclase, **210**, 1056
ADH. *See* Antidiuretic hormone (ADH)
Adhesion, **49**
  in ascent of xylem sap, 748*f*
  water transport in plants and, 48*f*
Adipose tissue, 61*f*, 76, **825***f*
  obesity and, 847*f*
  thermoregulation and, 835
Adrenal cortex, 958
Adrenal glands, **956**
  catecholamine hormones produced by, 956–58
  hormones secreted by, 949*t*, 956–58
  nervous system regulation of, 957
  steroid hormones produced by, 958
  stress and response of, 957*f*
Adrenaline. *See* Epinephrine
Adrenal medulla, 956–58
Adrenocorticotropic hormone (ACTH), **952**
Adult stem cells, 418*f*
Adventitious roots, **713**
Aerobic conditions, **174**
Afferent arteriole, **933**
Afghanistan, population age structure in, 1154*f*
Aflatoxins, 622
Africa
  cichlid fish speciation in Lake Victoria, 479, 480*f*
  human origins in, 705
  malaria and sickle-cell allele in, 466*f*
African-Americans, sickle-cell trait in, 267
African cichlid (*Neolamprologus tetracephalus*), 1107*f*
Agave, reproductive life history traits in, 1141*f*
Age structure of populations, **1154**–55
  in Afghanistan, Italy, and United States, 1154*f*
Agglutination of pathogens, 913
Aggregate fruit, **779**
Aging, cytokinins and retardation of plant, 797
Aglaophyton major, 584*f*
Agonistic behavior, **1126**, 1127*f*
Agre, Peter, 130
  on aquaporins and cell biology, 92–93
Agriculture. *See also* Crop plants
  angiosperms and, 605, 771
  Arabidopsis studies applied to, 711
  DNA technology applied in, 406–7
  hydroponic, 757*f*
  impact of, on freshwater ecosystems, 1200–1201
  impact of, on soil and nutrient cycling, 1200
  mycorrhizae and, 20, 767
  soil conservation and sustainable, 761–63
  symbiotic nitrogen fixation and, 766
Agrobacterium, 390, 542*f*
Agrobacterium tumefaciens, Ti plasmid from, 406, 407*f*
Agrostis tenuis (bent grass), 454*f*
Ahrens, Mitchell, 1113
AIDS (acquired immunodeficiency syndrome), **341**, **918**, 1155
  drug therapy for, 918–19
  HIV infection and, 447–48, 918–19
Air circulation, global patterns of, 1089*f*
Alarm calls
  Belding's ground squirrel and, 1128
  social learning of, in vervet monkeys, 1131–32
Alarm signals, minnow response to, 1111*f*, 1116
Alaska, glacial retreat and ecological succession in, 1174*f*, 1175*f*
Albatross, 926*f*
  Laysan, 693*f*
  Salvin's (*Diomeda canta salvini*), 922*f*
Albizzia procera, biological augmentation project using, 1226
Albumin, 419
  regulation of gene expression for synthesis of, 367*f*
Alcohol(s), **64***f*
Alcohol fermentation, **175**

Aldehydes, **64**f
Aldoses, 70f
Aldosterone, **937**
Alevria avrantia, 616f
Algae
alternation of generations in, 242
blue-green (*see* Cyanobacteria)
brown (phaeophytes), 560–62
diatoms (bacillariophytes), 559–60
fungi and (lichens), 621–22
golden (chrysophytes), 560
green (chlorophytes), 567–69
life cycles, 242f, 562f, 569f
photosynthesis in, 560
red algae (rhodophytes), 567
secondary endosymbiosis and diversity of, 551
Alimentary canal, **648**, 854f, **855**
evolutionary adaptations in, 863f
peristalsis in, 855
Alkaptonuria, 310
Allantois, 688, **1001**
Allee, W. C., 1146
Allee effect, 1146
Alleles, **254**. *See also* Gene(s)
as alternate form of gene, 253–54, 255f
for blood groups, 260
codominant, 260
complete dominance of, 260
correlating behavior of chromosome pair with, 276, 277f
dominant, 254, 260, 261–62, 267–68
elimination of, by bottleneck effect, 461
fertilization and segregation of, as chance events, 259f
frequencies of, in populations, 455–62
frequency of dominant, 261–62
genetic disorders and (*see* Genetic disorders)
incomplete dominance of, 260, 261f
law of independent assortment of, 256–58, 274, 275f
law of segregation of, 253–55, 275f
microevolution and change in frequencies of, 454, 455–56, 459–70
multiple, 262
mutations in (*see* Mutations)
natural selection and changes in frequencies of, 460, 462–70
recessive, 254, 283–84
relationship between phenotype and dominance of, 261
spectrum of dominance of, 260–62
Allergens, 916
Allergies
anaphylactic shock and, 917
histamine release and, 901, 916–17
mast cells, IgE, and, 916, 917f
Alligators, 691f
metabolic rate, 834
Allolactose, 355, 356
Allometric growth, **484**f
Allopatric speciation, 476f, **477**, 480
Allopolyploidy, **478**, 479f, 480
Allosteric regulation of enzymes, **156**f–57
Almonds, 605
Alpha (α) helix of protein, **82**f, 85
Alpine pennycress (*Thlaspi caerulescens*), 763
Alternation of generations, **242**, **461**, **561**, **576**
heteromorphic and isomorphic, 562
in plants, 242f, 576f
in protists (algae), 242f, 561–62
Alternative RNA splicing, **319**, **368**, 399
Altman, Sidney, 515
Altruism, **1128**–33
Hamilton's rule and kin selection, 1130
inclusive fitness and, 1129–31
reciprocal, 1130–31
Alu elements, 376, 381
Alvarez, Walter and Luis, 520
Alveolates (Alveolata), 555–58
apicomplexans, 555–56
ciliates, 556–58
dinoflagellates, 555
Alveoli, 555f, **887**
Alvin research submarine, 514f
Alzheimer's disease, 290, **1040**–41
Amacrine cells, 1061f, **1062**
Amborella trichopoda, 602f
Amebic dysentery, 564
American alligator (*Alligator mississipiensis*), 691f
American beech tree (*Fagus grandifolia*), climate change and range of, 1092f
American chestnut tree, diseases of, 1168
Amines, **65**f
Amino acid(s), **78**. *See also* Protein(s)
binding of tRNA to specific, 321f
as essential nutrients, 849–50
monomers, 78–80
as neurotransmitters, 1024t, 1025
polymers, 80
of proteins, 79f
sequence of, in polypeptides, 80–81
triplet code for, 312–14
twenty, of proteins, 79f
Aminoacyl-tRNA synthetase, **321**f
Amino functional group, **65**f
Ammonia, **927**
hydrogen ions and, 54
as nitrogenous waste, 927
soil bacteria and production of nitrates from, 763f, 764
Ammonites, **652**
Ammospermophilus harrisi (Harris's antelope squirrel), 477f
Ammospermophilus leucurus (white-tailed antelope squirrel), 477f
Amniocentesis, **269**, 270f
Amnion, **1000**
Amniotes, **687**–94, **999**
birds as, 691–94
derived characters of, 688
developmental adaptations of, 998–99
early, 688
mammals as, 694–701
phylogeny of, 687f
reptiles as, 688–94
Amniotic egg, 688f
Amoebas, **563**
exchange with environment and body structure of, 821f
movement of, 117f, 118
phagocytosis feeding, 107, 564f
Amoeba sp., 564f
Amoebocytes, **642**
Amoebozoans, 564–66
entamoebas, 564
gymnamoebas, 564
slime molds, 564–66
AMPA receptors, 1036, 1037f
Amphibians (Amphibia), **685**–86
body axes in, 992, 993f, 1004
breathing in, 888
cadherins and formation of blastula in, 1003f
circulatory system, 870f, 871
cleavage in embryo of, 992, 993f
decline in populations of, 686
fate map of embryo of, 1004f
frog skin color, 946
gastrulation in, 995, 996f
hearing and equilibrium in, 1053–54
kidney adaptations in, 939f
lateral line system in, 680, 1054
metamorphosis in, 686f
nitrogenous wastes of, 927f, 928
orders of, 685f
organogenesis in, 997f
reproduction in, 967f
thermoregulation in, 833, 837, 838
Amphipathic molecule, **124**
Amplification
DNA, 391–92
of cell signals, 214
of genes, 371
of sensory stimuli energy, **1046**
amp^R gene, 387f
Amygdala, 1034, 1035f
β-Amyloid, 1040, 1041
Amylopectin, 71, 72f
Amyloplasts, 110
Amylose, 71, 72f
Amylotrophic lateral sclerosis (ALS), 1069
Anabaena, 539
Anabolic steroids, 959f
Anabolism and anabolic pathways, **142**, 177
products of glycolysis and citric acid cycle diverted to, 177
protein synthesis as, 309–33
Anaerobic conditions, **174**
Anaerobic respiration, **539**
Anagenesis as evolutionary pattern, 472f
Analogy, homology versus, **493**–94
parsimony principle and, 504f
Anaphase
of meiosis, 244f, 245f
of mitosis, **221**, 223f, 225f, 226f
Anaphylactic shock, **917**
Anatomy. *See* Animal anatomy and physiology; Plant structure
Anchorage dependence, **231**, 232f
Anchor cells, 426
Androgens, **958**
Anemia, 881. *See also* Sickle-cell disease
Aneuploidy, **285**
of sex chromosomes, 287
Angina pectoris, 883
Angiosperms, **579**, 598–604
agriculture and, 605, 771
basal, 602
body structure of, 713f (*see also* Plant structure)
characteristics of, 598–601
dicots and monocots, 602, 780f
diversity of, 602–3f
evolutionary links between animals and, 604
evolution of, 601–2
flower of, 598f (*see also* Flower)
fossil, 601
fruit of, 598, 599f, 778, 779f
leaves, and classification of, 716f
life cycle of, 599, 600f, 601
phylogeny, 602f
pressure flow as translocation mechanism in, 753f
seed of, 593 (*see also* Seed)
sexual reproduction in, 600f, 772f
translocation in, 753f
vascular tissue of (*see* Vascular tissue system)
Angiotensin II, kidney function, osmoregulation, and, **937**
Anhydrobiosis, **925**
Animal(s), 626–36
anatomy and physiology of (*see* Animal anatomy and physiology)
aquatic (*see* Aquatic animals)
behavior of (*see* Behavior)
bioenergetics of, 828–31
body of, 630–33
Cambrian explosion and diversity of, 526, 527f, 629
chemical signals in (*see* Chemical signals in animals)
circulation (*see* Circulatory system)
cloning of, 416–17
cognition in, 1117–18
defenses of (*see* Animal defense systems)
definitions and characteristics of, 626–28
development (*see* Animal development; Development)
diseases in (*see* Diseases and disorders; Genetic disorders: Pathogens)

  evolutionary links between angiosperms and, 604
  evolution of, 628–30
  excretion in, 927–39
  gas exchange (*see* Gas exchange in animals)
  homeostasis in (*see* Homeostasis)
  human (*see* Human(s))
  invertebrate (*see* Invertebrates)
  land colonization by, 527, 573, 574, 575–79
  movement in (*see* Motor mechanisms)
  nervous system (*see* Nervous system)
  nutrition (*see* Animal nutrition)
  phylogeny and diversity of, 633–36
  polyploid speciation in, 479*f*
  polyploidy in, 286*f*
  regulatory systems (*see* Animal regulatory systems)
  reproduction in (*see* Animal reproduction)
  sensory mechanisms (*see* Sensory mechanisms)
  sexual life cycle in, 242*f*
  stenohaline versus euryhaline, 923
  territoriality in (*see* Territoriality)
  transgenic, 406
  vertebrate (*see* Vertebrates)
  viruses of (*see* Animal viruses)
Animal anatomy and physiology, 820–43
  bioenergetics and, 828–31
  body plans and, 630–33
  development and (*see* Animal development)
  evidence for evolution in homologies of, 448–51
  form and function of, 17*f*, 820
  as intrinsic factor limiting population density, 1150
  physical and environmental constraints on, 820–22
  regulation of internal environment and, 831–33
  tissue organization, 630–31
  tissue structure and function, 823–27
Animal cell, 626–27
  calcium ion maintenance in, 211*f*
  cholesterol within membranes of, 126*f*
  cytokinesis and cleavage in, 225*f*
  dehydration in, 136
  expansion of, 730
  extracellular matrix in, 119*f*–20
  fate of, during development process, 1003–8
  glycogen storage in, 72*f*
  intercellular junctions and, 120, 121*f*
  meiotic division of, 244–45*f*
  microvilli, 116, 117*f*
  plasma membrane of, 127*f*
  size of, compared with virus and bacterium, 335*f*
  structure of generalized, 100*f*
  water balance in, 132, 133*f*
Animal control systems. *See* Animal regulatory systems
Animal defense systems
  acquired immune system as, 898, 903–14
  chemical, 32*f*
  innate immune system as, 898–903
  lymphatic system and, 878–79
  against predators, 32*f*
Animal development, 412–14, 627*f*, 987–1010. *See also* Development
  apoptosis in, 427–29
  brain, 1028*f*, 1029
  cell differentiation and, 420, 421*f*, 987
  cell fate and fate mapping in, 1003–8
  cell lineage in, 426*f*
  cleavage of zygote in, 627*f*, 978, 979*f*, 992–94
  conservation of developmental genes, 431*f*, 432
  cytoplasmic determinants of, 420, 421*f*, 423–25, 987
  of embryo, 412–13, 414*f*, 485–86, 627*f*, 988–1001
  epigenesis and animal form, 987
  evolutionary novelty, 482–83
  evolution of genes controlling, 484–86
  fertilization, 978, 979*f*, 988
  genomic imprinting and, 289, 364
  hormonal regulation of insect, 960*f*
  Hox genes and, 432*f*, 485–86, 627
  induction and cell signaling in, 420, 421*f*, 1006–8
  model organisms for study of, 412–13*f*
  morphogenesis and, 987, 1001–3
  nuclear transplantation and, 415–17
  organogenesis in, 997*f*, 998*f*
  overview, 987–88
  pattern formation in, 421–29, 1006–8
  plant development compared with, 414*f*, 433
  transcriptional gene regulation during, 418–20
Animal growth. *See* Growth
Animalia (kingdom), 626. *See also* Animal(s)
Animal nutrition, 626, 855–66
  carbon skeletons and essential nutrients provided by diet, 849–52
  evolutionary adaptations of vertebrate digestive systems, 862–64
  feeding categories, overview, 844
  feeding mechanisms of animals, four types, 845*f*
  food processing stages and, 853–55
  homeostatic mechanisms for management of animal energy budget, 844–48
  mammalian digestive system and organs and, 855–62
  nutritional requirements, 33*f*, 34, 849–52
Animal pole, **992**
Animal regulatory systems, 831–33
  endocrine and nervous system overlap in, 943–44 (*see also* Endocrine system; Nervous system)
  energy budget, nutrition, and, 844–48
  excretory systems as, 927–39
  homeostatic mechanisms, overview of, 832–33
  hormonal controls, 860*f*, 943–63
  interstitial fluid and, 831–32, 879*f*
  in invertebrates, 959–61
  osmoregulation, 132, 922–27
  regulation of body temperature (thermoregulation), 833–41
  regulation versus conforming and, 832
  thermoregulation, 833–41
  water balance (osmoregulation) and waste disposal, 132
Animal reproduction, 627, 964–86
  asexual, 648–49, 964–65
  cartilaginous fishes, 681
  embryonic and fetal development in, 978–82
  hormonal regulation of, 973–78
  mammalian (human), 969–84
  in placental mammals, 978–84
  sexual, mechanisms of, 967–69
  sexual, overview of, 964
  sexual reproductive organs, 969–73
  variations in cycles and patterns of, 965–66
Animal viruses
  classes of, 340*t*
  diseases caused by, 340*t*
  reproductive cycles of, 339–42
  RNA as genetic material of, 340–42
  viral envelopes of, 340, 341*f*
Anion, **41**
  cotransport of, 740*f*
Annelids (Annelida), 640*f*, 653–55
  classes of phylum, 653*t*
  coelomate of, 631*f*
  Hirudinea (leeches), 655*f*
  hydrostatic skeleton in, 1063, 1064*f*
  Oligochaeta (earthworms), 654*f*
  Polychaeta (polychaetes), 655
Annuals, plant, **720**
Anopheles mosquito, 555, 556*f*
Antelope squirrels (*Ammospermophilus* sp.), allopatric speciation of, 477*f*
Anterior pituitary gland, **950**
  hormones secreted by, 949t, 951*f*, 952, 976*f*
Anterior side of body, **630**
Anther, flower, **598**, **772**, 774
Antheridium (antheridia), **577*f***
Anthocerophyta (hornworts), 580, 582*f*
Anthozoans (sea anemones, corals), 644*f*, 645
Anthrax (*Bacillus anthracis*), 538*f*, 543*f*, 546
Anthroceros hornwort, 582*f*
Anthropoids (Anthropoidea), **697**. *See also* Human evolution
Antibiotics, 343, **898**
  action of, on prokaryotic cell walls, 535–36
  action of, on prokaryotic ribosomes, 322, 537
  bacterial resistance to, and R plasmids, 351
  as enzyme inhibitors, 155
  fungal production of, 623*f*
  sponge production of, 643
Antibodies
  immunoglobulins as classes of, 880, 912*f*
  monoclonal, 913
  polyclonal, 912
Anticodon, **320**
  protein synthesis, binding of RNA codons and, 320f
Antidiuretic hormone (ADH), **936**, **951**
  kidney function and water balance regulated by, 936, 937*f*
  secretion of, by posterior pituitary, 950*f*, 951
Antigen, **903**
  antibody-mediated disposal of, 913*f*, 914
  B cell receptors for, 903–4*f*
  B cell response to, 910–14
  binding site for, 903*f*
  clonal selection of lymphocytes in response to, 907–8
  cytotoxic T cell response to, 910, 911*f*
  epitope of, 903
  helper T cell response to, 909–10
  lymphocyte development and, 905–8
  lymphocyte recognition of, 903–5
  T cell receptors for, and role of MHC, 904f, 905
  T-dependent, and T-independent, 912
Antigen presentation, **904**, 905*f*
Antigen-presenting cells (APCs), **905**
Antigen receptors, **903**
  B cell, 903, 904*f*
  T cell, 904, 905*f*
Antimicrobial proteins, 900–901
Antioxidants, 188
Antiparallel, **89**
Antiviral drugs, 343–44
Ants
  as invasive species, 437
  mutualism between acacia trees and, 1164*f*
  symbiosis between fungus and leaf-cutting, 620*f*
Anurans (frogs, toads), 685. *See also* Frogs (Anura)
APC (adenomatous polyposis coli) tumor-suppressor gene, 373*f*, 374
Aphotic zone, **1093**
Apical dominance, **715**
  cytokinins and control of, 796*f*, 797
Apical ectodermal ridge (AER), **1007*f***
Apical meristems, **414**, **576**, **720**
  primary growth and, 576*f*, 720*f*–21, 722–25
Apicomplexans, **555**, 556*f*
Apodans (caecilians), 685*f*, 686
Apomixis, **781**
Apoplast, **743**
Apoptosis (programmed cell death), 108, **427–29**, **800**, **902**
  cancer and failed, 373
  cytotoxic T cells and, 911*f*
  in human white blood cells, 428*f*
  molecular basis of, in *C. elegans*, 428*f*
  mouse paw development and effects of, 429*f*
  natural killer cells and, 902
  in plants, 800
Aposematic coloration, **1162*f***
Appendix, **862**
Apple maggot fly, North American (*Rhagoletis pomonella*), 479
Aquaporins, **130**, 131, 133, **742**
  P. Agre on discovery and function of, 92–93
  water transport and, 742
Aquarium as ecosystem, 1184*f*
Aquatic animals. *See also* Fish(es); Marine animals
  body plan of, and swimming, 821*f*
  gas exchange and gills in, 884, 885*f*, 886*f*
  lateral line system in select, 680, 1054*f*
  nitrogenous waste of, 927*f*

osmoregulation in, 923f, 924–25
thermoregulation in, 833, 834*f*, 835, 836*f*, 837*f*
Aquatic biomes, 1092–97
coral reefs, 1097*f*
distribution of, 1092*f*
estuaries, 1095*f*
freshwater, 1093*f*, 1094*f*, 1095*f*
human impact on, 1200–1201
intertidal zone, 1096*f*
lakes, 1094*f*
marine, 1093*f*, 1096*f*, 1097*f*
marine benthic zone, 1097*f*
ocean pelagic, 1096*f*
streams and rivers as, 1095*f*
wetlands, 1094*f*
zonation in, 1093*f*
Aquatic ecosystems
freshwater, 1187f, 1188–91
marine, 1166f, 1187f, 1188–90
nutrient enrichment of, 1200–1201
Aqueous humor, **1059**
Aqueous solution, **51**
concentration of solute in, 52–53
pH of select, 54*f*
Arabidopsis thaliana, 728
auxin transport in, 795*f*
brassinosteroid deficiency in, 798
clock mutants, 805
ethylene triple response in, 800*f*
fass mutants, and growth of, 730, 731*f*
genome of, 400, 729*f*, 802
gnom mutant,and axial polarity of, 732*f*
homeobox genes in, 432
as model organism, 413*f*
normal and abnormal flowering in, 734*f*
N. Raikhel on systems biology and study of, 710–11
responses to light and photoreceptors, 791, 802, 804
root gravitropism in, 809
thigmomorphogenesis in, 809*f*
Arachnids (scorpions, spider, mites), 641*f*, 658, 659*f*
Araschnia levana (map butterflies), 462*f*
Arbuscular mycorrhizae, **615***f*
Archaea (domain), **13**, 14*f*, 98, 507, 521, 534. *See also* Prokaryotes
compared with Bacteria and Eukarya, 541*t*
Archaefructus liaoningensis, 601*f*
Archaefructus sinensis, 601*f*
Archaeopteris, 596*f*
Archaeopteryx fossil, 693*f*
Archegonium (archegonia), **577**
Archenteron, **631**, 632, **995**
Archosaurs, **689**
Ardipithecus ramidus, 703
Arginine
biosynthesis, 310, 311*f*
genetic recombination of bacteria synthesizing, 347*f*
Arginine-vasopressin (AVP), 1113
Argyroxiphium sandwicense, 481*f*
Aristotle, 439, 756
Arnold, Stevan, 1118–19
Arousal, human brain and state of, 1029–30
Art
as human characteristic, 706*f*
Paleolithic, in Lascaux, France, 1230*f*
Arteries, **869**
blood flow velocity through, 875f
countercurrent heat exchanger in, 836*f*
diseases and blockages in, 883*f*
structure and function of, 874, 875*f*
Arterioles, **869**, 875*f*
Arthrobotrys fungus, 610*f*
Arthropods (Arthropoda), 640*f*, **656**–65
characteristics of, 656–58
cheliceriforms (spiders, mites), 658, 659*f*
circulatory system of, 657
crustaceans (lobsters, crabs), 657*f*, 658*t*, 664–65
exoskeleton of, 74*f*, 657*f*, 1064
hexapoda (insects), 658*t*, 660–64
myriapods (millipedes, centipedes), 658*t*, 659–60
nervous sytem, 1012*f*
subphyla of, 658*t*
Artificial chromosomes, bacterial, 396
Artificial selection, **445**
plant breeding and, 764, 783–84
of vegetables from wild mustard, 445*f*
Asci (ascus), **616**
Ascocarps, **616**
Ascomycetes (Ascomycota), **616**–17
examples of, 616*f*
life cycle of, 617*f*
as pathogen, 622
Asexual reproduction, **239**, **781**, **964**. *See also* Clone; Cloning
in animals, 649
in bacteria, 226, 227*f*, 346, 537
in fungi, 611–12
in hydra, 239*f*
mechanisms of, 964–65
in seed plants, 781–83
sexual reproduction compared with, 239
A site, tRNA binding, **322***f*
Aspen trees, asexual reproduction in, 781*f*
Aspergillus mold, 622, 623
Assisted reproductive technology (ART), **984**
Association areas, 1032
Associative learning, **1116**–17
Aster (microtubule), **221**, 222*f*
Asteroid, mass extinctions and indications of, 520*f*
Asteroidea (sea stars), 666*t*, 667*f*
Astrocytes, **1015***f*
Asymmetrical cell division, **730**
Atherosclerosis, 76, **883**
Athyrium filix-femina lady fern, 587*f*
Atmosphere
carbon dioxide concentrations in, 1203–5
depletion of ozone in upper, 1205–6
effect of human activities on (*see* Human environmental impact)
global air circulation and winds in, 1089*f*
greenhouse effect in, and global warming, 1204–5
organic compounds and Earth's earliest, 59*f*, 513–14
origins of oxygen in, 522–23
Atom(s), **34**–39
chemical bonding of, 39–42
electron configuration and chemical properties, 37–38
electron orbitals of, 37*f*, 38–39
electron transfer between,and ion formation, 41*f*
energy levels of electrons in, 36–37
isotopes, 35–36
number and mass of, 34–35
subatomic particles, 34
tracking, through photosynthesis, 183, 184*f*
valence, 40
Atomic mass, **35**
Atomic mass unit (amu), 34
Atomic nucleus, **34**
Atomic number, **34**–35
ATP (adenosine triphosphate), **66**
active transport and, 134
allosteric regulation of enzymes and role of, 156
animal nutrition and, 844, 846
bioenergetics and, 828, 844, 846
Calvin cycle and conversion of carbon dioxide to glucose using, 193–95
catabolic pathways and, 161
cellular respiration and synthesis of, 170–72, 173*f*, 174
cellular work powered by, 149*f*–150
citric acid cycle and synthesis of, 168–70
dynein "walking" powered by, 116*f*
energy coupling by using hydrolysis of, 149*f*
fermentation and production of, 174–76
glycolysis and synthesis of, 165–67
hydrolysis of, 148*f*, 149
metabolic rate and, 830
oxidative phosphorylation and synthesis of 164–65, 170–74
photosynthetic light reactions and synthesis of, 184, 185*f*, 190–91, 193*f*
regeneration of, 150
structure of, 148*f*, 149
ATP cycle, 150*f*
ATP synthase, **171**
ATP synthesis in cellular respiration and, 171–73, 192
Atria (atrium), heart, **869**
Atrial natriuretic factor (ANF), **938**
Atrioventricular (AV) node, **874**
Atrioventricular (AV) valve, **873**
Auditory communication, 1111–12
Aurelia sp.,internal transport in, 868*f*
Australia
distribution and abundance of red kangaroos in, 1081*f*
marsupials and monotremes of, 695, 696*f*
Australian thorny devil lizard (*Moloch horridus*), 691*f*
Australopithecus afarensis, 703, 704*f*
Australopithecus africanus, 703
Australopithecus anamensis, 702–3
Autoimmune diseases, **917**
Type I diabetes mellitus as, 956
Autonomic nervous system, **1027**
neurotransmitters in, 1025
sympathetic division of, 957
Autophagy, 107–8
lysosomes and, 107*f*
Autopolyploidy, **478***f*
Autosomes, **240**
Autotrophs, **181**, 538, 539*t*
Auxin, **794**–97
cell elongation, plant growth, and, 791, 792, 793*f*, 794, 795*f*
fruit growth and, 797
as herbicides, 795–96
lateral and adventitious root formation and, 795
polar transport of, 794, 795*f*
secondary growth in plants and, 796
Average heterozygosity, **463**
Avery, Oswald, DNA studies by, 294
Avirulence (*avr*) gene, 813, 814*f*
Avirulent pathgens, **813**
plant response to, 814*f*, 815*f*
Avogadro's number, 52
Axelrod, Robert, 1130
Axillary bud, **715**
Axis establishment in early development, 423–25
Axolotl, 485*f*
Axon(s), **1014**
action potentials conducted along, 1020*f*, 1021
migration of vesicles to tips of neuron, 112*f*
molecular signals directing growth of developing, 1038*f*
myelinated, 1015*f*
Axon hillock, **1014**

Bacillariophytes (diatoms), 559–60
Bacillus coagulans, 98*f*
Bacillus thuringiensis, 784
Backbone, as homologous structure, 498
Bacteria. *See also* Prokaryotes; Prokaryotic cell
alcohol fermentation by, 175
antibiotic action on, 535–36
antibiotic resistance in, 535
binary fission in, 226, 227*f*, 346, 537
cell walls of, 534–36
compared with Archaea and Eukarya, 541*t*
as decomposers, 544, 1186
disease-causing(*see* Pathogens)
DNA replication in, 300–305
DNA transformation of, in F. Griffith experiments, 294
expression systems, 390
gene expression in, regulation, 352–56
genetic recombination in, 346–51

genome, 346–56, 374, 537
major groups of, 542–43*f*
metabolism and nutrition in, 538–40
motility of, 536*f*, 537
nitrogen fixation by, 763*f*, 764–66
photosynthesis in, 184
plasmids of (*see* Plasmids)
regulation of metabolic pathways in, 352–56
reproduction and adaptation in, 537–38
size of, 98, 335*f*
structure, 98*f*
translation and transcription coupled in, 328*f*
transposable genetic elements in, 351–52
ulcer-causing, 858*f*
viral infection of (*see* Bacteriophage (phage))
Bacteria (domain), **13**, 14*f*, 98, 507*f*, 521, 534, 541*t*
Bacterial artificial chromosome (BAC), **396**
Bacterial chromosome, replication of, 346, 347*f*
Bacterial mats, 522*f*
Bacterial plasmid. *See* Plasmids
Bacteriophage (phage), **294**–96, **336**
bacteria infected by virus, 295*f*
as cloning vector, 388
genetic recombination by bacterial transduction due to, 348–49
genomic libraries of, 389*f*
Hershey and Chase experiment on bacterial infection by, 294, 295*f*, 296
phage λ, 338, 339*f*
prophage, 339, 349
reproductive (lytic and lysogenic cycles) of, 337, 338*f*, 339*f*
T2, infection of *E. coli* by, 294, 295f
T4, 334*f*, 336*f*, 338*f*
temperate, 338, 349
Bacteriorhodopsin, transmembrane protein structure in, 128*f*
Bacteroid, **764**
in soybean roots, 765*f*
Bacteroides thetaiotaomicron, 545
Baculum, **972**
Balanced polymorphism, **466**
Balancing selection, **466**–68
Baleen, 845*f*
Ball-and-socket joint, 1065*f*
Banana slug (*Ariolimus californicus*), 1119*f*
Bandicoot, 696
Bannertail kangaroo rat (*Dipodomys spectabilis*), 939*f*
Bark, 727*f*, **728**
Barnacles, 664*f*–65
competition between species of, 1160*f*
Barn swallows, 693*f*
Barr body, **284**
Barrier methods of contraception, **983**
Bartholin's glands, **970**
Basal angiosperms, **602**
Basal body, **114**, 115*f*
Basal metabolic rate (BMR), **829**
Basal nuclei, **1031**
Base(s), **54**
Base(s), nitrogenous. *See* Nitrogenous bases
Basement membrane, **824**
Basidiocarps, **618**
Basidiomycetes (Basidiomycota), **618***f*–19
life cycle, 619*f*
Basidium, **618**
Basilar membrane, 1052*f*, 1053*f*
Basophils, 880*f*, 881
Bat(s)
disruption of interaction networks involving, 1214, 1215*f*
excretory system of vampire, 938*f*
form and function in anatomy of, 17*f*
as pollinators, 604*f*
predation using sonar by, 1045*f*
Bates, Henry, 22
Batesian mimicry, **1162***f*
B cell receptors, **903**
for antigens, 903, 904*f*
B cells. *See* B lymphocytes (B cells)
Bdellovibrio bacteria, 542*f*
Beadle, George, one gene-one enzyme hypothesis of, 310, 311*f*
Beagle, H.M.S., Charles Darwin's voyage on, 441, 442*f*
Bean (*Phaseolus vulgaris*)
germination of, 780*f*
seed of, 778*f*
sleep movements in, 805*f*
Bears, 465
classification of American black, 13*f*
extinction vortex and case study of grizzly, 1217–18
Beaver (*Castor canadensis*)
as foundation species, 1169*f*, 1214
kidney, 939*f*
Bees, 22*f*, 661
chemical signals among, 1111
cognitive maps of, 1116
cuckoo, 1162*f*
learning in, 1117
as pollinator, 604*f*
reproductive anatomy of, 969*f*
thermoregulation in, 836–37
Beetle
bombardier, 32*f*
Hercules scarab, 1080*f*
Behavior, **1107**
agonistic, 1126, 1127*f*
altruistic, 1128–33
bird songs and, 1111, 1118
cognition in animals and, 1117–18
communication methods and, 1111–12
competitive, 160–61
courtship and mating, 1106
defined, 1107
ethology as study of, 1107–9
foraging, 1119, 1120, 1122–23
genetic component of, 1109–13
genetic and environmental interaction resulting in, 1113–18
imprinting, 1108, 1109*f*
mating behaviors in, 1123–27
natural selection and, 1118–28
parental (*see* Parental behavior)
proximate and ultimate causes of, 1106, 1107, 1108*f*, 1109*f*
as reproductive barrier, 474*f*
sensory input and, 1063–64
social, 1128–33
species distribution, habitat, and, 1085
thermoregulation and responses of, 837–38
Behavioral ecology, **1106**–35
altruistic behavior and, 1128–33
genetics and, 1109–13
genetics and environment interactions resulting in behaviors, 1113–18
natural selection and, 1118–28
overview, 1106
proximate and ultimate causes of behavior and, 1106–9
Behavioral isolation as prezygotic reproductive barrier, 474*f*
Behavioral variation in natural populations, 1118–20
in aggressive behavior, 1119–20
in prey selection, 1118–19
Beijerinck, Martinus, 335
Belding's ground squirrel (*Spermophilus beldingi*)
altruistic behavior in, 1128, 1130*f*
hibernation and thermoregulation, 840*f*
immigration of, 1138, 1151
life table for, 1139*t*
reproductive table for, 1141*t*
survivorship curves for, 1140*f*
Benign tumor, **232**
Bent grass (*Agrostis tenuis*), 454*f*
Benthic zone, **1093***f*
in lakes, **1094***f*
in oceans, 1097*f*
Benthos, **1093**
Benzer, Seymour, 1116
Bernard, Claude, 831
Berthold, Peter, 1110–11, 1121
Bertness, Mark, 1170
Berzelius,Jōns Jakob, 59
Bester-Meredith, Janet, 1114
β-Amyloid, 1040, 1041
β-Galactosidase, 335*f*, 387
β globin gene
evolution of, 379*f*
gene families for, 377*f*, 378
RNA processing in transcription, 318*f*
restriction fragment analysis of, 392, 393*f*, 394
beta oxidation, **177**
β pleated sheet, protein, **82***f*, 85
Bicarbonate
ion (HCO_3^-), 55
produced by pancreas, 858
Bicoid gene and protein, effects of, 424*f*
Biennials, **720**
Big-bang reproduction, **1141**
Bilateral symmetry, **630**
in animal body plan, 630*f*, 646
in flower structure, 773*f*
Bilaterians (Bilateria), **633**, 646–50
Bile, **858**
Binary fission, **226**, 227*f*, 346, 347*f*, 537, 558
Binomial name, **496**
Biochemistry, 97
Biodiversity, 12–15. *See also* Tree of life
as biological theme, 27*f*
in communities, 1165–66, 1175–78
conservation of (*see* Conservation biology)
hot spots for, 437, 1222
human activities and threats to, 1211–14
human welfare dependent upon, 1211–12
mass extinctions of, 518*f*, 520–21
nature reserves and parks for protection of, 1222–24
threats to, 606, 1211–14
three levels of, 1210*f*–11
unity in, 14, 15*f*
Biodiversity hotspot, 437, **1222**
Earth's terrestrial, 1222*f*
Bioenergetics, **142**, **828**
in animal body, 828–31, 844–48
ATP and cellular work, 148–50, 828 (*see also* ATP (adenosine triphosphate))
energy budgets and, 830–31
energy forms and, 142–43
enzyme activity and, 150–57
free energy, 145–48
gas exchange and, 884*f*
glucose and, 958
homeostasis and (*see* Homeostasis)
laws of energy transformation (thermodynamics), 143–44
metabolic pathways and, 141–42 (*see also* Metabolism)
metabolic rate and, 828, 829–30
thyroid hormones and, 953
Biofilms, **539***f*–40
Biogenic amines as neurotransmitters, 1024*t*, **1025**
Biogeochemical cycles, **1195**–98
carbon, 1196*f*
generalized scheme for, 1198*f*
human disruption of, 1200–1206
nitrogen, 1197*f*
phosphorus, 1197*f*
water, 1196*f*
Biogeographic realms, 1083, 1084*f*
Biogeography, **450***f*, 1083
distribution of organisms and, 1083–92
as evidence for natural selection in, 450
of islands, 1177–78
Bioinformatics, **11**, 819
Biological augmentation, **1226**
Biological clocks, **1030**
circadian rhythms and (*see* Circadian rhythms)
light and, in animals, 1030, 1031*f*

light and, in plants, 805, 806
pineal gland, melatonin, and, 959
Biological diversity. *See* Biodiversity; Tree of life
Biological Dynamics of Forest Fragments Project, Brazil, 1221
Biological magnification, **1202**
Biological order and disorder, 144
Biological organization, 3–6
cell as basic unit of, 5f, 6–8
ecosystems as level of, 4f, 6
emergent properties in levels of, 9
levels of, 4–5*f*
Biological species concept, **473**–76
alternatives to, 476
limitations of, 476
reproductive isolation of gene pools and, 473, 474–75*f*
Biology, **2**
conservation, 1209–32
evolution as core theme of, 15–19
forms of inquiry in, 19–26
grouping of diverse species and life domains by, 12–15
organizational hierarchy of, 3, 4–5f, 6–8
themes unifying, 26, 27*f*
systems concept applied to, 9–12
Bioluminescence
in bacteria, 545*f*
in fungi, 141*f*
in protists, 555
Biomass, **1168**
standing crop of, 1187, 1194*f*
Biomass pyramid, 1192
Biomanipulation, **1171**
Biomes, **1092**. *See also* Ecosystem(s)
abiotic factors affecting, 1086–87
aquatic, 1092–97
climate and, 1098
terrestrial, 1098–1104
Biophilia, 1230
Bioremediation, **546**–47, **1226**
Biorhythms, 959. *See also* Biological clocks
Biosafety Protocol, 408
Biosphere, 4*f*, **1083**
abiotic factors in, 1081, 1086–87
autotrophs and heterotrophs in, 181
biomes of, 1092–1104
biophilia and future of, 1230
human impact on, 1200–1206
organisms in (*see* Organism(s); Species)
Biosynthesis, 177. *See also* Anabolism and anabolic pathways
Biotechnology, **384**, 783–86. *See also* DNA technology
Bioterrorism, 546
Biota, **1081**
Biotic boundary versus legal boundary of reserves, 1222, 1223*f*
Biotic components of environment, **1081**. *See also* Organism(s); Species
distribution of species affected by, 1085–86
influence of, on aquatic biomes, 1092–93
influence of, on populations, 1148–52
Biotic stresses affecting plants, **811**
Bipedal posture, human evolution and, 703–4
Bipolar cells in retina, **1061***f*
Bipolar disorder, **1040**
Birds, 691–94
bones of, 483
circulatory system, 870*f*, 871
cleavage in embryo of, 992, 993*f*
countercurrent heat exchanger in, 836*f*
courtship behaviors in, 1106*f*
derived characters of, 691–92
digestion in, 854*f*, 863
edge-adapted species, 1221
feathers and wings of, 692*f*
feet, 694*f*
finches (*see* Finches)
flight in, 483, 692
fossil, 693*f*
gastrulation in embryo of, 995–96, 997*f*
imprinting of graylag geese, 1108, 1109*f*
kidney adaptations in, 939*f*
learning and problem solving, 1117
limb development in, 1007*f*
living, 693–94
migration of, 1106, 1109*f*, 1110–11
nitrogenous wastes of, 927*f*, 928
organogenesis in, 998*f*
origin of, 692–93
parental caregiving in, 1142*f*
as pollinators, 604*f*
respiratory system and breathing by, 889*f*
salt excretion in marine, 926*f*
songs of (*see* Bird songs)
species-area curves for North American breeding, 1177*f*
thermoregulation in, 833, 836*f*, 838
vision in, 1060
Bird songs, 1111
E. Jarvis on genetics of, 818–19
sensitive period for learning, 1118
Birth control pills, **983**
Birth rate in populations, per capita, 1143–44
Bivalves (Bivalvia), 652*f*
Black bread mold (*Rhizopus stolonifer*), life cycle of, 614*f*
Blackcap (*Sylvia atricapilla*), 1110–11
migratory patterns in, 1120, 1121*f*
Black mold (*Stachybotrys chartarum*), 623
Black rush (*Juncos gerardi*) as facilitator species, 1170*f*
Blacktip reef shark (*Carcharhinus melanopterus*), 681*f*
Blade, algae, **560**, 561*f*
Blade, leaf, **715**
Blastocoel, **992**, 993*f*
Blastocyst, **978**, 979*f*, **1000**
Blastoderm, **994**
Blastomeres, **992**
Blastopore, **632**, **995**
developmental modes and fate of, 632–33
dorsal lip of, as induction organizer, 1006*f*
Blastula, **627**, **992**
cadherins and formation of frog, 1003*f*
cleavage and creation of, 992*f*, 993*f*
gastrulation and rearrangement of, 994
Blending hypothesis of heredity, 251
Blood, **825**, **868**, 879–83
capillary exchange between interstitial fluid and, 875, 879*f*
cellular elements in, 880*f*, 881
clotting of, 880*f*, 882*f*
composition of, 879, 880*t*, 879–82
as connective tissue, 825*f*
filtration of, by kidneys, 931, 932*f*, 933*f*, 934
groups of (*see* ABO blood groups; Blood groups, human)
osmolarity of, 936, 937*f*, 938
pH of, 55
plasma of, 879–80
Rh factor in, 92, 915–16
stem cells and, 881–82
transfusions of, immune system and, 262, 915–16
volume of, and cardiac output, 872–73
Blood-brain barrier, **1015**
Blood doping, 881
Blood electrolytes, 880*t*
Blood flow velocity, 875
blood pressure and, 876*f*
in veins, 875*f*
Blood fluke (*Schistosoma manson*), 30, 1163
life cycle, 647*f*
Blood groups, human
ABO, 129, 262*t*, 915*t* (*see also* ABO blood groups)
M,N,and MN, 269
multiple alleles and, 262*f*
Rh factor and, 92, 915–16
self-nonself recognition of, and blood transfusions, 915–16
Blood pressure, **868**, 876–77
blood flow velocity and, 876*f*
high (hypertension), 883
measurement of, 877*f*
norepinephrine and, 957
Blood transfusions, 262, 915–16
Blood vascular system in animals. *See* Circulatory system
Blood vessels, **868**, 869–70
associated with nephrons, 933
blood flow velocity in, 875*f*
blood pressure in, 876, 877*f*
structure and function of, 874, 875*f*
Blowfly (*Phormia regina*), taste in, 1055*f*
Bluegill sunfish, energy costs and benefits of feeding behavior in, 1122, 1123*f*
Bluehead wrasse fish, hermaphroditism in, 966*f*
Blue jays, studying effects of selection in populations of North American, 467*f*
Blue-light photoreceptors, 750, **802**
B lymphocytes (B cells), 881, **903**, 910–14
antibody classes and, 912–13
antibody-mediated disposal of antigens and, 913*f*, 914
clonal selection of, 907*f*, 908
development of, 905*f*
gene rearrangement in, 906*f*, 907
humoral immune response and, 911*f*
polyclonal and monoclonal antibodies and, 912, 913
as receptors for antigens, 903, 904*f*
response of, to extracellular pathogens, 910–14
Body cavity, **631***f*
Body plan, 421, 630–33. *See also* Pattern formation
acoelomate, coelomate, and pseudocoelomate, 631*f*
animal development and, 987–88, 992, 993*f*
axial polarity in plant, 731, 732*f*
bilateral symmetry, 630*f*, 646
body axes, 993*f*, 1004–5
body cavities, 631*f*
cnidarian, 643
cytoplasmic determinants affecting animal, 420, 421*f*, 423–25, 987
exchanges with environment and, 821s, 822*f*
homeotic genes and, 485–86
molluscs, 650*f*
physical laws constraining, 820, 821
protostome versus deuterostome, 631–33
segmentation, 425
symmetry in, 630
tissues and, 630–31
Body size
metabolic rate and, 829
skeleton and, 1064–65
Body temperature
heart rate and, 874
regulation of, 833–41
Bohr shift, **893**
Bolus, **856**
Bombardier beetle, 32*f*
Bond, Alan, 467*f*
Bone(s), **825***f*
of birds, 483, 692*f*
as connective tissue, 825*f*
mammalian jaw and ear, 695*f*
origins of, 679, 680
Bone morphogenic protein (BMP-4), 1006
Bonnemaisonia hamifera, 567*f*
Bony fishes (Osteichthyes), 682–84
Boobies, blue-footed, 474*f*
Book lungs, **658**
Bormann, Herbert, 1078
Borneo, deforestation in, 1209*f*
Botswana, savanna community in, 1159*f*
Bottleneck effect, **461***f*
founder effect and isolation, 462
Bottom-up model of community organization, **1170**
Botulism (*Clostridium botulinum*), 206*f*, 339, 543*f*, 1069
Bound ribosomes, 102, 103*f*, 325
Boveri, Theodor, 274
Bowman's capsule, **931**, 932*f*

Index

Boysen-Jensen, Peter, 792
Brachiopods (lamp shells), 640*f*, **649**
Bracts, 716*f*
Brain
brainstem, 1028*f*, 1029–30
cerebellum, 1030
cerebral cortex, 1028f, 1029, 1031, 1032–37
cerebrum, 1028*f*, 1031–32
control of breathing by, 890
development of embryonic, 1028*f*–29
diencephalon, 1028, 1030–31
evolution of chordate, 675
human (*see* Human brain)
injuries to, 1035, 1037–41
mammalian, 694
neural processing of vision in, 1062*f*, 1063
primary motor and somatosensory cortices of, 1033*f*
thalamus and epithalamus of, 1030
ventricles of, 1026*f*
Brain hormone, **960**
Brainstem, 1028*f*, **1029**–30
Braithwaite, Victoria, 1113–14
Branchiostoma, 675*f*
Brassinosteroids, 791, **798**
Brazil, Biological Dynamics of Forest Fragments Project, 1221
BRCA1 and BRCA2 genes and breast cancer, 374
Bread mold (*Neurospora crassa*), 310, 311*f*, 616*f*
life cycle, 617*f*
Breast cancer, 400
genetic basis of some, 374
growth and metastasis of malignant, 233*f*
Breathing, **888**–90
in amphibians, 888
in birds, 889*f*
centers in human brain for automatic control of, 890
in mammals, 888, 889*f*
negative pressure, 888, 889*f*
positive pressure, 888
Breathing control centers, **890**
Brenner, Sidney, 429
Briggs, Robert, 416
Brightfield microscopes, 96*f*
Bristlecone pine (*Pinus longaeva*), 595*f*
Brittle stars (Ophiuroidea), 666, 667*f*
Broad Institute, Eric Lander at, 236–37
Broca, Pierre, 1034
Broca's area, brain, 1034, 1035
Bronchi, **887**
Bronchioles, **887**
Brown algae (phaeophytes), **560**–62
alternation of generations in, 561–62*f*
human uses of, 561
structure, 560, 561*f*
Brown fat, **838**
Brush border, 858, 860
Bryophyta (division), 580
Bryophytes, **578***t*, 580–83
diversity of, 582*f*
ecological and economic importance of, 583
gametophyte and sporophyte of, 580–81, 583
L. Graham on evolution of, 510
life cycle, 581*f*
Bryozoans (ectoprocts), 640*f*
B*t* toxin, 784
Budding, asexual reproduction by, **965**
in yeast, 611*f*
Buffers, **55**
Bulbourethral glands, **972**
Bulbs, 715*f*
Bulk feeders, **845***f*
Bulk flow, **743**
in circulatory systems, 868, 878
long distance transport in plants and, 743–44
transport of xylem sap as, 748*f*, 749
Bulk transport across plasma membrane, 137, 138*f*
Bundle-sheath cells, **196***f*, **724**
Burgess Shale, 629*f*
Burkitt's lymphoma, 370
Butterflies
metamorphosis in, 661*f*, 943*f*
nonheritable variation with populations of map, 462*f*
Buttress roots, 714*f*

C_3 plants, **195**
C_4 plants, **196**
anatomy and pathway of, 196*f*
CAM plants compared to, 197*f*
Cacti, 749
Cactus ground finch (*Geospiza scandens*), 443*f*
Cadherins, **1002**
role in development of blastula, 1003*f*
Caecilians (Apoda), 685*f*, 686
Caenorhabditis elegans (nematode)
cell lineage of, 426*f*
cell signaling and induction in, 425–26, 427*f*
genome, 394, 399*t*, 400
hydrostatic skeleton in, 1063
as model organism, 412*f*
programmed cell death in, 427–29
as pseudocoelomate, 631*f*
Calcitonin, 954, **955**
calcium homeostasis and, 954*f*, 955
Calcium
cortical reaction in egg fertilization and role of, 989, 990*f*
deforestation and loss of soil, 1199
hormonal control of homeostasis in, 954*f*, 955
maintenance of calcium ions in animal cells, 211*f*
muscle contraction and role of, 1068, 1069*f*
neural chemical synapses and role of, 1022*f*
regulation of blood concentrations of, in kidney, 934
as second messenger, 211, 212*f*, 790
storage of, in smooth ER tissue, 104–5
Callus, **782**
Calmodulin, 1072
Caloric imbalance, 846–48
Calorie, **49**, 828
Calvin, Melvin, 185
Calvin cycle, photosynthesis, **184**–85, 198*f*
conversion of carbon dioxide to glucose in, 193, 194f, 195
cooperation between light reactions and, 185*f*
role of G3P in, 194
Calyptra, **581***f*
Cambrian explosion, 526–27, **629**
Camel, role of fur in water conservation by, 926*f*
Camouflage, 446*f*
CAM (crassulacean acid metabolism) plants, 196, **197**
C_4 plants compared to, 197*f*
cAMP. *See* Cyclic AMP
Campylobacter, 542*f*
Canavanine, 813
Cancer, 370–74
abnormal cell cycle and, 370, 371–73
breast, 232*f*
cell-signaling pathways, interference with, 371, 372*f*, 373
colon/colorectal, 305, 373*f*, 374
cytotoxic T cell response to, 910, 911*f*
environmental factors and, 370
faulty gene expression as cause of, 362
inherited predisposition to, 374
leukemia, 288*f*, 404, 882
loss of cell cycle controls and, 232–33, 371–73
multistep model of development of, 373*f*, 374
mutations as cause of, 370, 330
obesity and, 847
oncogenes, proto-oncogenes, and, 371
skin, 305*f*, 306
tumor-suppressor genes, faulty, 371
types of genes associated with, 370–71
viruses as cause of, 370, 374
Candida albicans, 623
Cannibalism, population fluctuations and, 1151
Canopy, forest, **1098**
Canyon tree frog, cryptic coloration in, 1162*f*
CAP (catabolite activator protein), 356
Capases, 428
Capillaries, **869**
as base of hypothalamus, 951
blood flow through, 877, 878*f*
blood flow velocity through, 875
exchange between interstitial fluid and blood at, 875, 879*f*
structure of, 874, 875*f*
Capillary bed, **869**
blood flow through, 877, 878*f*
Capsid, viral, **335**, 336*f*
Capsomeres, 335, 336*f*
Capsule
bryophyte sporangium, **580**, 581*f*
prokaryotic, **536***f*
Carbohydrates, **69**–74
catabolism of, 161, 176, 177*f*
cell-cell recognition and membrane, 129
digestion of, 859*f*
disaccharides, 70, 71*f*
monosaccharides (sugars), 70
polysaccharides (sugar polymers), 71–74
role of membrane, in cell-cell recognition, 129
α Carbon, 78
Carbon, 58–67
as backbone of biological molecules, 58
formation of chemical bonds with, 59–61
functional groups attached to compounds containing, 63–66
organic chemistry and study of, 58–59
valences of, 60*f*
variations in skeletons of, 61*f*, 62, 63
versatility of, in molecular formation, 59–63
Carbon cycle in ecosystems, 1196*f*
Carbon dioxide (CO_2)
atmospheric, and climate change, 1196*f*, 1203–5
atmospheric, in Carboniferous period, 588
Calvin cycle and conversion of, to sugar, 193–95
covalent bonding of, 60
gas exchange in animals and disposal of, 750
peat bogs and absorption of, 511
plant guard-cell opening/closing and levels of, 751
respiratory pigments and transport of respiratory, 893*f*, 894
Carbon fixation, **185**
alternative mechanisms of, 195–98
Calvin cycle and, 185, 193–95
Carbonic acid, 54, 55
Carboniferous period, 596
swamp forests of, 588*f*
Carbon monoxide (CO) as neurotransmitter, 1025
Carbon skeletons, 61–63
animal diet and provision of, 844, 849–52
Carbonyl functional group, **64***f*
Carboxyl end, protein, 82*f*, 323
Carboxyl functional group, **64***f*
Carboxylic acids, **64***f*
Cardiac cycle, **872**, 873*f*
Cardiac muscle, **826***f*, **1072**
acetylcholine, G proteins and synaptic transmissions in, 1024
Cardiac output, **873**
Cardiovascular disease, 76, 847, **882**–83
Cardiovascular system, **869**–71
blood composition and function, 879–82
blood flow velocity in, 875*f*
blood pressure in, 876–77
blood vessels of, 874, 875*f*
diseases of, in humans, 76
heart of, 868, 871–74
lymphatic system and, 878, 879*f*
metabolic rate and, 870
Caribou, 849*f*
populations of, 455*f*
Carnivores, **844**
digestive tract and dentition in, 862, 863*f*
plants as, 767, **768**f
Carotenoids, **188**, 189

Carpels, flower, 429f, 430, **598**, **772**
Carrier proteins, 130
- facilitated diffusion and, 133, 134f

Carriers of genetic disorders, **266**
- testing to identify, 269

Carrots, test-tube cloning of, 782f
Carrying capacity, **1145**
- estimates of, 1155
- human ecological footprint and, 1155–56

Carson, Rachel, 1083f
Cartilage, **825f**
- vertebrate skeletons of, 678

Casparian strip, **745**
Castor bean (*Rincinus communis*), 778f
Cat
- cloning of, 417f
- four stages of food processing in, 853f
- X chromosome inactivation and, fur color in, 284f

Catabolism and catabolic pathways, **142**
- cellular respiration as, 142, 161–74
- connection of, to other metabolic pathways, 176–78
- fermentation as, 174–76

Catalyst, **77**, **150**. *See also* Enzyme(s); Ribozymes
Catalytic cycle, 78f
Catastrophism, **440**
Catecholamines, **957**
Cation, **41**
- membrane potential and uptake of, 740f

Cation exchange in soil, **760**, 761f
Catolite activator protein (CAP), 356f
Cattle egret
- distribution of, 1084
- mutualism between water buffalo and, 1164f

Caulerpa sp., 568f, 569
CD4 cell surface protein, **909**
CD8 cell surface protein, **910**
Cdks (cyclin-dependent kinases), **229**–30
cDNA (complementary DNA), **390**, 400
cDNA library, **390**
Cech, Thomas, 515
Cecum, **862**
ced-3 gene, 427–28
ced-4 gene, 427–28
ced-9 gene, 428
Celera Genomics, 398
Cell(s), 5f, 6–8, 94–123
- animal (*see* Animal cell)
- ATP and work performed by, 148–50, 828
- as biological theme, 27f
- cell fractionation and study of, 97
- cell-to-cell communication (*see* Cell signaling)
- differentiation of (*see* Cell differentiation)
- division of (*see* Cell division)
- eukaryotic, 8, 9f (*see also* Eukaryotic cells)
- heritable information in, 7f
- importance of, 94
- information flow in (DNA → RNA → protein), 86f, 87
- membranes of (*see* Plasma membrane)
- microscopes and study of, 95–97
- morphogenesis and changes in shape of, 1001f
- motility of (*see* Cell motililty)
- organization of genetic material in, 219
- plant (*see* Plant cell)
- programmed death of (*see* Apoptosis)
- prokaryotic, 8, 9f (*see also* Prokaryotic cell)
- protobionts as precursors to, 513, 515
- size of, compared with virus and bacterium, 335f
- size range of, 95f, 98
- surface-to-volume ratio of, 99f
- three types of work performed by, 148
- ultrastructure, 95
- water channel proteins of, 93

Cell adhesion molecules (CAMs), **1002**
Cell body (neuron), **1014**
Cell-cell communication. *See* Cell signaling
Cell-cell recognition
- membrane carbohydrates, and, 129
- membrane proteins and, 128f

Cell crawling during morphogenesis, 1001–2
Cell cycle, 218–35
- alternation of mitotic phase and interphase in, 221–28
- cancer and abnormal, 232–33, 370, 371–73
- cell division, roles of, in, 218–20
- cell division and formation of daughter cells in, 219–20
- cell division by binary fission and, 226–27
- cytokinesis in, 224–26
- defined, **219**
- phases of, 221f
- regulation of, by cytoplasmic signals, 228–33

Cell-cycle control system, **229**–31
- cancer as failure of, 232–33
- clock for, 229–30
- G_1 checkpoint in, 229f
- G_2 checkpoint in, 230f
- mechanical analogy for, 229f
- stop and go (internal and external signals) in, 230–31

Cell death, programmed. *See* Apoptosis
Cell differentiation, **362**, **413**, **987**
- cell-cell signals and, 420, 421f, 429–30
- cell determination and, 418–20, 1003–8
- cytoplasmic determinants affecting, 420, 421f, 987
- determination of muscle cells and, 419f, 420
- embryonic development and (*See* Embryonic development)
- in plants, 732, 733f

Cell division, 218–28
- asymmetrical, 730
- binary fission as bacterial, 226, 227f
- cancer as uncontrolled, 232–33
- cellular organization of genetic material and, 219
- chromosome distribution during, 219–20
- cytokinesis and, 224–26
- daughter cells produced by, 219–20
- density-dependent inhibition of, 231, 232f
- embryonic development and role of, 218f
- functions of, 218f
- key roles of, 218–19
- mitosis and, 221, 222–23f, 224, 225f, 226f, 227–28
- plane and symmetry of, and plant morphogenesis, 729f
- platelet-derived growth factor (PDGF) and, 231f
- preprophase band and plane of, 730f
- reproduction and role of, 218f
- tissue renewal and role of, 218f

Cell expansion, 730
- in plants, 731f

Cell fate in animal development, 1003–8
- blastopore dorsal lip, effect on, 1006f
- determination of, and pattern formation by inductive signals, 1006–8
- determination of, and transcriptional regulation of gene expression, 418–20f
- establishing cellular asymmetries, 1004–5
- fate mapping, 1004f

Cell fate in plant development, 732–33
Cell fractionation, **97f**
Cell lineage, **426**
- of nematodes, 426f

Cell-mediated immune response, 899f, 908, **909**
- cytotoxic T cell function against pathogens in, 910, 911f
- helper T lymphocyte function in, 909, 910f
- overview of, 909f

Cell migration in animal morphogenesis, 1001–2
- role of fibronectin in, 1002f

Cell motility, 112, 114, 115f, 116f, 117f, 118f
- in animal morphogenesis, 1001f, 1002–3

Cell plate, 225f, **226**
Cell signaling, 201–17
- in animal development, 425–29, 1006–8
- cell differentiation and, 420, 421f, 429–30
- control of coordinate gene expression by, 367
- conversion of external signals into cell responses, 201–4
- efficiency of, 215
- in embryonic development, 420, 421f
- endocrine system and, 945–48
- evolution of, 201–2
- interference with normal, and development of cancer, 371, 372f, 373
- local and distant, 202f, 203f
- nerve cell development and, 1037, 1038f
- nervous system, neurotransmitters, and, 1021–25
- overview of, 201, 204f
- pattern formation in animal development, and inductive, 1006–8
- in plant development, 429–30
- responses to, through regulation or transcription, 204, 212–15
- root nodule formation in legumes and role of, 766
- reception of signals, 204–8, 945–46
- signal amplification, 214
- signal transduction pathways in, 202, 208, 945–48
- specificity of, 214f
- termination of, 215
- three stages of, 203–4
- transduction and relay of signals to target cells, 204, 208–12

Cell size, 335f
Cell-surface receptors for water-soluble hormones, 946–47
Cellular energetics. *See* Bioenergetics
Cellular respiration, 142, 160, **161**–80
- anabolic pathways and, 177
- ATP synthesis in, 170–74
- catabolic pathways of, 161–65
- citric acid cycle of, 168, 169f, 170
- disequilibrium in, 147f
- electron transport chain in, 163f, 164
- as exergonic reaction, 146, 163f
- feedback mechanisms for regulation of, 177, 178f
- fermentation compared to, 175–76
- formula, 146
- glycolysis in, 165, 166f, 167
- mitochondria as site of, 109, 110f, 164–65, 168–74
- redox reactions in, 161–64
- stages of, overview, 164f, 65
- versatility of catabolism and, 176, 177f

Cellular slime molds, **565**
- life cycle of *Dictyostelium*, 566f

Cellulose, 52, **72**, 106, 558
- arrangement of,in plant cell walls, 73f
- bacterial digestion of, 74f
- cell expansion and orientation of microfibrils of, 72, 73f
- structure of, compared with starch, 73f

Cell wall (plant), **118**, 119f
- transport in plants and, 743
- water balance and, 132, 133f

Cell wall (prokaryote), 534–36
- water balance and, 132

Celsius scale, **49**
Cenozoic era, 519t, 521f, 629
Center for Conservation Research and Training, Hawaii, 436–37
Centipedes (Chilopoda), 659, 660f
Central canal of central nervous system, **1026**
Central nervous system (CNS), **1012**, 1026f
- brain, 1028–37
- central canal of, 1026
- injuries and diseases of, 1035, 1037–41
- oligodendrocytes of, 1015
- organization of vertebrate, 1012–13

Central vacuole, **108f**
Centrioles, **114f**
Centromere, **220**, 222f, 991
Centrosome, **114f**, **221**, 222f, 223f
Cephalization, **630**
Cephalochordates (Cephalochordata), 674, 675f
Cephalopods (Cephalopoda), 652, 653f
Ceratium tripos, 550f
Cercozoans (Cercozoa), 563
Cerebellum, **1030**

Cerebral cortex, 1028*f*, **1029**, 1031, 1032–37
consciousness and, 1036–37
emotions and, 1034–35
information processing in, 1032–33
language, speech, and, 706, 1034*f*
lateralization of function in, 1033–34
lobes of, 1032*f*, 1035
memory, learning, and, 1035–36
primary motor and somatosensory cortices of, 1033*f*
Cerebral hemispheres, **1031**
lateralization of cortical function and, 1033–34
Cerebrospinal fluid, **1026**
Cerebrum, **1028***f*, 1031–32. *See also* Cerebral cortex
hemispheres of, 1031*f*
Certainty of paternity, 1124–25
Certhidea olivacea(green warbler finch), 443*f*
Cervix, **970**
Cestoda (tapeworms), 646*t*, 648*f*
cGMP (cyclic GMP), 211, 790
Chagas' disease, 554
Chambered nautilus, 653*f*
Channel proteins, 130
facilitated diffusion and, 133, 134*f*
Chaparral, **1101***f*
Chaperonin, **85***f*, 325
Chaperon protein, 85, 325
Character, genetic, **252**. *See also* Trait, genetic
discrete, 463
law of independent assortment and, 256–58
law of segregation and, 253–56
multifactorial, 264
quantitative, 263, 463
Character displacement, **1161**
Characters
shared derived, and shared primitive, 498
table of, 499*f*
Chara *sp*., 574*f*
Chargaff, Erwin, experiments on DNA nucleotides by, 296
Chargaff's rules, 296, 298, 313*f*
wobble and relaxation of, 322
Charophyceans, 567, 569
evolution of land plants from, 510–11, 573–74
examples of, 574*f*
Chase, Martha, experiments on genetic material by, 294, 295*f*, 296
Cheating and noncheating cells on cellular slime mold, 566
Checkpoints in cell-cycle control system, **229***f*, 230*f*
Cheetah
energy transformations and metabolism in, 143*f*
territoriality in, 1149*f*
Chelicerae, **658**
Cheliceriforms, **658***f*, 659*f*
Chemical(s)
as mutagens, 330
protein denaturation caused by, 85
Chemical bonds, **39**–43
of carbon, 59–63
chemical reactions and, 44–45
covalent, 39*f*, 40, 60
hydrogen, 42
ionic, 41f, 42
van der Waals interactions and, 42
weak, 42
Chemical cycling in ecosystems, 160*f*, 1184–86, 1195–98
carbon cycle, 1196*f*
decomposition and, 1185–86
disruption of, by human population, 1200–1206
general model of, 1195
nitrogen cycle, 1197*f*
overview, 1185*f*
phosphorus cycle, 1197*f*
regulation of, by vegetation, 1198–99
role of prokaryotes in, 544
trophic relationships as determinants of, 1185
water cycle, 1196*f*
Chemical elements. *See* Element(s)
Chemical energy, **142**. *See also* Chemical reactions
conversion of light energy into, in photosynthesis, 182–85
Chemical equilibrium, **44**
Chemical markers of territory, 1149*f*
Chemical messengers, cell signaling using, 205. *See also* Signal transduction pathways
Chemical properties, electron configuration and, 37–38
Chemical reactions
effects of enzymes on (*see* Enzyme(s))
energy levels of electrons and, 36
exergonic and endergonic, 146, 147*f*, 148, 149*f*
free energy and, 146
making and breaking chemical bonds in, 44–45
reactants and products of, 44
Chemical signals in animals. *See also* Cell signaling
animal communication behavior and, 1149*f*
cell-surface receptors for, 946, 947*f*
endocrine system and, 205, 943–44, 945*f* (*see also* Endocrine system)
hormones as, 945–48, 949t, 951–59
intracellular receptors for, 205*f*, 947
local, 202–3
nervous system and, 943–44, 945*f*, 1021–25 (*see also* Nervous system)
neurosecretory cells as, 944
neurotransmitters as, 947–48, 1024*f*, 1025
paracrine signaling, 947–48
pheromones as, 437, 610, 945
signal transduction pathways initiated by, 1024
Chemical signals in plants. *See* Cell signaling; Plant hormones
Chemical synapses, 1021–25
diagram of, 1022*f*
direct synaptic transmission and, 1022–23
indirect synaptic transmission and, 1023–24
neurotransmitters in, 1024–25
synaptic vesicles and, 1021
Chemical work, cellular, 148
driven by ATP, 149*f*, 828
Chemiosmosis, **171**, **740**
in chloroplasts, during photosynthesis, 110, 111*f*, 173
in chloroplasts versus in mitochondria, 109–11, 173, 192–93
in mitochondria, during cellular respiration, 109, 110*f*, 164–65, 171–73
in prokaryotes, 521–22
transport in plants and, 740
Chemistry, 32–46
atomic structure and properties of elements, 34–39
chemical bonding of atoms and molecule formation and function, 39–43
chemical reactions, 44–45
foundations of biology in, 32
L. Makhubu on medicine and, 30–31
matter, elements, and compounds, 32–34
of metabolism (*see* Metabolism)
molecular shape and biological function, 42–43
Chemoautotrophs, **538**, 539*t*
Chemoheterotrophs, **539***t*
Chemokines, **901**
Chemoreceptors, **1048**
in insects, 1048, 1049*f*, 1055*f*
for taste and olfaction, 1054–56
Chemotaxis, 537
Chemotrophs, 538–39
Cheny, Dorothy, 1131–32
Chesapeake Bay, partial food web for, 1167*f*
Chiasma, **247**
Chick embryo, homologous structures in human and, 449*f*
Chicxulub asteroid and crater, Caribbean Sea, 520*f*
Chief cells, 857*f*
Childbirth, 981*f*
positive feedback mechanisms in, 833*f*
Chile, relationship between rainfall and herbaceous plant cover in desert community in, 1171*f*
Chilopoda (centipedes), 659, 660*f*
Chimeras, **430**
Chimpanzees, 701*f*
cognition and problem solving in, 1117, 1118*f*
skull of, 484*f*, 494
China
birth rate and population growth in, 1154
fossils found in, 676, 677*f*
Chinese river dolphin, 1211*f*
Chitin, **74**, **609**, **1064**
in exoskeletons, 74*f*, 1064
in fungal walls, 609
medical uses of, 74*f*
structure of, 74*f*
Chitons (Polyplacophora), 651*f*
nervous system, 1012*f*
Chlamydia trachomatis, 543*f*
Chlamydomonas sp., 567, 568*f*, 569*f*
Chlorofluorocarbons (CFCs), atmospheric ozone depletion and, 1205–6
Chlorophyll(s), **182**
chlorophyll *a*, **187**, 188, 189, 190
chlorophyll b, **188**, 189, 190
light reception by, 187f, 189
photoexcitation of, 188, 189*f*
structure of, 188*f*
Chlorophytes (green algae), 567–69, 575*f*
colonial and multicellular, 568*f*
life cycle of *Chlamydomonas*, 569*f*
Chloroplasts, **109**, 110, 111*f*
capture of light energy by, 110, 111*f*
chemiosmosis in mitochondria versus in, 109–11, 173, 192–93
chlorophylls in, 182 (*see also* Chlorophyll(s))
in green algae, 567
green leaves as interaction of light with, 186*f*
as site of photosynthesis, 109, 110, 111*f*, 182, 183*f*
starch storage in, 71, 72*f*
structure of chlorophyll in, 188*f*
Choanocytes, **642**
Choanoflagellates, 628*f*
Cholecystokinin (CCK), 860*f*
Cholera (*Vibrio cholerae*), 206*f*, 211, 542*f*, 546
Cholesterol, **77**
diseases associated with, 882–83
low-density and high-density lipoproteins of, 137, 883
membrane fluidity and role of, 126f
structure of, 77*f*
transport of, 137, 882–83
Chondrichthyans, **680**–82
Chondrocytes, **825***f*
Chordates (Chordata), 641*f*, **671**
craniates as, 675–77
derived characters of, 673
early evolution of, 674–75
fate mapping for two, 1004*f*
invertebrate, 667
lancelets as, 674, 675*f*
nervous system, 1012*f* (*see also* Nervous system)
phylogeny, 671, 672*f*, 673
skeleton of, 1064, 1065*f*
tunicates as, 673–74
Chorion, **1000**
Chorionic villus sampling (CVS), **269**, 270*f*
Choroid, **1058**, 1059*f*
Chory, Joanne, 798
Christmas tree worm, 638*f*
Chromatids
chiasmata of nonsister, 247
crossing over of nonsister, 247
sister, 219, 220*f*
tetrad of, 247
Chromatin, **102**, **219**, **359**
DNA packing and structure of eukaryotic, 350–60, 361*f*
euchromatin and heterochromatin in, 360
nuclear transplantation and changes in, 416
nucleosomes of, 360, 361*f*

structural modifications of, as regulation of eukaryotic gene expression, 363–64
Chromatium, 542*f*
Chromoplasts, 110
Chromosomal basis of inheritance, 274–92
alterations of chromosomes, genetic disorders, and 285–88
behavior of chromosomes, and Medelian inheritance, 274–77
linked genes and, 277–82
locating genes on chromosomes and, 274
select inheritance patterns as exceptions to theory of, 288–90
sex-linked genes and, 282–84
Chromosome(s), 98, **102**, **219**. *See also* DNA (deoxyribonucleic acid); Gene(s)
alterations in structure of, 286*f*, 287, 288
bacterial, 226, 227*f*, 346, 347*f*
bacterial artificial, 396
during cell division, 218*f*, 219, 220*f*
chromatin in (*see* Chromatin)
description and terms related to, 241*f*
distribution of, to daughter cells during mitosis, 219, 220*f*, 222–23*f*, 225*f*
eukaryotic, 219*f*, 359*f*
extra set of (polyploidy), 285, 378
gene locus on, 239
genetic disorders due to structural alterations of, 288
genetic maps of, 279–81
genetic recombination and, 278–79, 280*f*
independent assortment of, 247, 248*f*, 274, 275*f*, 278
homologous, 240 (*see also* Homologous chromosomes)
karyotype of, 240*f*
locating genes on, 274
Mendelian inheritance based in behavior of, 274–77
mutations in, leading to genetic variation, 463–64
nondisjunction of, 285*f*–86
number of human (*see* Chromosome number in humans)
Philadelphia, 288
prokaryotic, 537*f*
reduction of, from diploid to haploid during meiosis, 242, 243–47
sex, 282–83 (*see also* X chromosome; Y chromosome)
translocations of, 286*f*, 287, 288*f*, 459
"walking" of, during mitosis, 224, 225*f*
Chromosome number in humans, 102, 240–41
abnormal, 285*f*–86
aneuploidy and polyploidy, 285, 378
genetic disorders caused by alteration of, 285–88
reduction of, by meiosis, 243–47
in somatic cells, 219, 240
Chromosome theory of inheritance, **274**–77
exceptions to, 288–90
Mendelian inheritance related to, 274, 275*f*, 276–77 (*see also* Mendelian inheritance)
T.H. Morgan's experimental evidence supporting, 276–77
sex chromosomes and, 285–88
Chronic myelogenous leukemia (CML), 288*f*
Chrysophytes (golden algae), 560
Chylomicrons, **861***f*
Chytridium, 613*f*
Chytrids (Chytridiomycota), **613**
Cichlid fishes
speciation in, 479, 480*f*
territoriality in, 1107*f*
Cilia, 15*f*, **114**–16
dynein "walking" and movement of, 116*f*
motion of, 115*f*
ultrastructure of, 115*f*
Ciliary body, **1059**
Ciliates, 550*f*, **556**, 557*f*, 558
Circadian rhythms, **751**, **805**
in animals, 1030–31
in plants, 751, 805
Circular DNA, 289
Circulatory system, 867–83
adaptations of, for thermoregulation, 835–37
amphibian, 870*f*, 871
animal phylogeny reflected in, 867–71
arthropod, 657
blood composition and function, 879–82
blood flow velocity, 875*f*
blood pressure, 876–77
blood vessels, 874, 875*f*
capillary function, 877–78
closed, 868, 869*f*
diseases of, in humans, 882–83
double, in mammals, 871–74
in fish, 870*f*, 871
heart, 871–74
human, 871–83
invertebrate, 868–69
gas exchange and, 885, 891–95
lymphatic system and, 878–79
mammalian, 870*f*, 871–83
open, 657, 868, 869*f*
physical principles governing, 874–79
placenta, and fetal, 980*f*
reptile, 870*f*, 871
vertebrate, 869–71
Cisternae of golgi apparatus, 105, 106*f*
Cisternal maturation model, 106
Citric acid cycle, **164**, 168–70
connection of, to other metabolic pathways, 176–78
junction between glycolysis and, 168*f*
overview, 168*f*
steps of, 169*f*
Clade(s), **497**
grades versus, 630
lizard-bird, 504*f*
mammal-bird, 504*f*
plant kingdom, 575*f*
Cladistics, **497**, 498
Cladogenesis as evolutionary pattern, 472*f*
Cladogram, **497**
construction of, 499*f*
Clams, 639*f*, 652
anatomy of, 652*f*
Clark's nutcrackers (*Nucifraga columbiana*), 1116
Class (taxonomy), **496**
Classical conditioning, **1116**–17
Classification of life, 12, 13*f*, 14–15. *See also* Taxonomy
Claviceps purpurea, 622
Clayton, David, 818
Cleavage, **224**–**26**, **627**, **978**, **992**
in amphibian, 992, 993*f*
in animal cells, 225*f*, 627*f*, 978, 979*f*, 992–94
in bird embryo, 994*f*
cleavage of zygote, 627*f*, 978, 979*f*, 992–94
cytokinesis and, 224, 225*f*, 226
in frog embryo, 992, 993*f*
gray crescent, distribution at first, 1005*f*
in human zygote, 978, 979*f*, 999, 1000*f*
in mammals, 999, 1000*f*, 1001
meroblastic, and holoblastic, 994
in plant cells, 225*f*
in protostomes versus deuterostomes, 631–32
in sand dollar embryo, 992*f*
Cleavage furrow, **224**
Clements, F. E., 1178
Climate, **1087**–92
as abiotic factor, 1087–92
air circulation, wind and, 1089*f*
alternative carbon fixation in plants living in arid, 195–98
changes in (global warming), 1204–5
Cretaceous mass extinctions and, 518–20
Earth's latitude and, 1088*f*
effect of bodies of water on, 1987, 1090*f*
effect of mountains on, 1090*f*
effect of seasonality on, 1088*f*, 1090, 1091*f*
global patterns of, 1087, 1088–89*f*
latitudinal gradient in biodiversity and, 1176
long-term change/warming in, 1091, 1092*f*
microclimate, 1087, 1091
terrestrial biome distribution and structure determined by, 1098–1104
Climate change, 1091–92
effect of, on forests and trees, 1092*f*
global warming, greenhouse effect, and, 1204–5
Climatius acanthodian, 680*f*
Climograph, **1098***f*
Cline, **464**
Clitoris, **970**
Cloaca, **682**, **968**
Clock mutants, 805
Clonal selection, **907**
Clone, **239**, **385**, **415**. *See also* Gene clones
animal, 415–17
asexual reproduction and production of, 239
plant, 415, 781–83
Cloning, **415**
of animals, 415–17
of DNA and genes, 385–92
of plants, 415, 782–83
Cloning vectors, **386**
expression vector as, 390
phages as, 388, 389*f*
plasmids as, 386–88
Closed circulatory system, **650**, **868**
open circulatory system versus, 869*f*
in vertebrates (see Circulatory system)
Closed system, 143
equilibrium and work in, 147*f*
Clostridium botulinum (botulism), 206*f*, 339, 543*f*, 1069
Clownfish (*Amphiprion ocellaris*), 683*f*
Club fungus (Basidiomycota), **618**–19
Club moss, 586, 587*f*
Clumped pattern of population dispersion, 1138*f*
Cnidarians (Cnidaria), 633, 639*f*, 643–46
anthozoans, 645
cubozoans, 645
classes of, 644*t*
digestion in, 854*f*
hydra (*see Hydra*)
hydrozoans, 644, 645*f*
hydrostatic skeleton in, 1063
internal transport in *Aurelia*, 868
life cycle of *Obelia*, 645*f*
nervous system, 1012*f*
origins of, 526
polyp and medusa forms of, 643*f*
scyphozoans, 644
Cnidocytes (cnidarians), **643**
Coal formation, 588
Cocci (spherical-shaped) prokaryotes, 534, 535*f*
Coccidioidomycosis, 623
Coccosteus placoderm, 680*f*
Cochlea, hearing and, **1051***f*
pitch and, 1052, 1053*f*
transduction in, 1052*f*
Coconut, seed crop size in, 1142*f*
Codominance of allele, **260**
Codons, 312, **313**
start and stop, 313, 314*f*, 324
Coefficient of relatedness, **1129**–30
Coelocanth (*Latimeria*), 683*f*, 684
Coelom, **631**
Coelomates, **631***f*
deuterostomes (*see* Deuterostomes (Deuterostomia))
protostomes, 369, 370
Coenocytic fungi, **609***f*
Coenzyme, **155**
Coevolution, **1164**–65
relationship between angiosperms and animals as, 604
Cofactors, enzyme, **155**
Cognition, **1117**–18
Cognitive ethology, **1117**
Cognitive maps, **1116**
Cohesion, **48**
in ascent of xylem sap, 748*f*

Cohort, population, **1139**
Coitus, **972**
Cold, plant response to, 812
Cold virus, 337, 343
Coleochaete sp., 574*f*
Coleoptile, **778**
  response of, to light, 792*f*
Coleorhiza, **778**
Collagen, **119**, 380
  animal cells, 626–27
  in extracellular matrix of animal cell, 119
  quaternary structure of protein in, 83*f*
Collagenous fibers, **823**
Collecting duct, **931**, 932*f*, 933*f*, 934
Collenchyma cells, **718**
Colloid, **52**
Colon, **861**–62
  cancer of, 305
Colonies, **526**
  of eukaryotes, 526
  of prokaryotes, 539*f*
  of protists, 568*f*
Coloration
  aposematic, 1162*f*
  cryptic, 1162
  in plant leaves, 789*f*, 790*f*, 791
  in skin, 685, 946
  warning, in snakes, 21–24
Color blindness, 283, 1061
Colorectal cancer, model for development of, 373*f*, 374
Color vision, 1059–60, 1061
Colpidium, cilia motion in, 115*f*
Columnar epithelia, **823**, **824***f*
Comb jellies (Ctenophora), 633, 640*f*
Commensalism, **545**, **1164**
Commercial applications/uses
  of angiosperms, 605
  of fungi, 623
  pharmaceuticals (*see* Pharmaceutical products)
Common juniper (*Juniperus communis*), 595*f*
Communication, animal, **1111**–12
Community(communities), 4*f*, **1082**, **1159**
  adaptation and interspecific interactions in, 1164–65
  biogeographical factors affecting biodiversity in, 1175–78
  bottom-up versus top-down models of structure in, 1170–71
  commensalism in, 545, 1164
  competition in, 1160–61
  conservation of (*see* Conservation biology)
  contrasting views on structure of, 1178–80
  controls in, 1170–71
  disease in, 1163
  disturbances in, 1171–75
  dominant species in, 1168
  ecological succession in, 1173–75
  foundation species in, 1169–70
  herbivory in, 1163
  humans as agents of disturbance in, 1173
  individualistic versus interactive hypotheses on structure of, 1178, 1179*f*
  interspecific interactions in, overview, 1159–60
  keystone species in, 1168*f*, 1169*f*, 1214
  mutualism in, 1164
  parasitism in, 1163
  predation in, 1161–62
  recovery time for disasters in, 1226*f*
  rivet and redundancy models on structure of, 1180
  species diversity and richness in, 1165–66
  stability and change in, 1171–72
  trophic structure of, 1166–68
Community ecology, **1082**, 1159–82
  biogeographic factors affecting biodiversity, 1175–78
  debate on structure of communities, 1178–80
  defining "community," 1159
  disturbances affecting species diversity and composition, 1171–75
  dominant and keystone species and, 1165–71
  interspecific interactions, 1159–65
Companion cells, **719**
Competition, **16**, 1160–61
  character displacement and, 1161*f*
  competitive exclusion principle and, 1160
  ecological niches and, 1160–61
  resource partitioning and, 1161*f*
Competitive enzyme inhibitors, **155***f*
Competitive exclusion, **1160**
Complementary DNA (cDNA), **390**, 400
Complement system, **900**, 914
Complete digestive tract, **855**. *See also* Alimentary canal
Complete dominance of allele, **260**
Complete flowers, **773**
Complete metamorphosis, **661**
Compound(s), **33**
  elements and 33–34
  emergent properties of, 33*f*
  organic, 59–66, 513–14
Compound eyes, **1058***f*
Compound leaf, 716*f*
Concentration gradient, **131**, 135
  cotransport as active transport driven by, 136
Concentricycloidea (sea daisies), 666*t*, 667*f*
Conception, **978**, 979*f*
Condensation reactions, **68**, 69*f*
Condom, contraceptive, **983**
Conduction, **835**
Cone cells, eye, **1059**, 1061
Confocal microscopy, 96*f*
Conformers versus regulators, **832**
Conidia, **616**
Coniferous forest, **1102***f*
Conifers (Coniferophyta), **593**
  diversity of, 595*f*
  forest biome of, 1102
  life cycle of pine, 596, 597*f*
Conjugation, **349**, **558**
  in bacteria, 349*f*, 536, 538
  in ciliate *Paramecium*, 557*f*
  plasmids and bacterial, 349–51
Conjunctiva, **1058**, 1059*f*
Connective tissue, **823**, **825**
  types of, 825*f*
Connell, Joseph, studies of competition and niches of barnacles, 1160*f*
Conodonts, **678**, 679*f*
Consciousness, human, 1036–37
Conservation biology, **1209**–32
  biodiversity and human welfare in, 1211–12
  human threats to biodiversity and, 1212–14
  K. Kaneshiro on, 437
  at landscape and regional level, 1220–24
  levels of biodiversity addressed by, 1210–11
  at population and species levels, 1215–20
  restoration ecology and, 1224–28
  species area curves used in, 1176
  sustainable development and, 1229–30
  threat of human activities and need for, 1209–14
Conservation of energy, 143
Conservative model of DNA replication, 300*f*
Constant (C) regions, lymphocytes, 904*f*
Consumers, **6**
Continental drift, 527, 528*f*
  history of, during Phanerozoic, 529*f*
  plate movements and, 528*f*
  species distribution and, 528
Contour tillage, 762*f*
Contraception, **982**–83
  mechanisms of, 982*f*
  reproductive technology and, 983–84
Contractile vacuoles, **108**
  in *Paramecium*, 132, 133*f*
Control center, animal's internal environment and, 832, 944
Control elements, eukaryotic gene, **364**–66, 367*f*
  proximal, and distal, 364–65
Controlled experiment, **23**–24
Convection, **835**
Convergent evolution
  of aquatic animals, 821*f*
  of marsupials and eutherian mammals, 450, 494*f*, 495, 696*f*
  of oomycetes and fungi, 558
  in terrestrial biomes, 1099
Convergent extension, **1002**
  animal morphogenesis and cell sheet, 1002*f*
Cooperativity, enzyme activity and, 156*f*, **157**
Copepods, **664**
Coppola, Francis Ford, and family, 238*f*
Coquerel's sifakas (*Propitheaus vereauxi coquereli*), 697*f*
Coral reef, **1097***f*
  algae and, 567*f*
Corals, 639*f*, 645
Coral snake, eastern, case study on mimicry of warning coloration in, 21–24
Corepressor, **354**
Cork cambium, **720**
  production of periderm and, 728
Corn. *See* Maize (*Zea mays*)
Corn borer, European, 1085
Cornea, **1058**, 1059*f*
Coronavirus, 344
Corpus callosum, 1031*f*, **1032**
Corpus luteum, **969**
Correns, Karl, 289
Corridors between ecosystems, 1221–22
Cortex, plant tissue, **717**
  in roots, 722–23
Cortical granules, **989**
Cortical nephrons, **931**, 932*f*
Cortical reaction, **989***f*, 990*f*
Corticosteroids, **958**
Coruzzi, Gloria, 25*f*
Costa Rica
  infant mortality and life expectancy in, 1229*f*
  parks and zoned reserves in, 1223*f*, 1224
  restoration project in tropical dry forest of, 1228*f*
  sustainable development in, 1229–30
Cotransport, **136**, **740**
  solute transport in plants and, 740*f*
Cotyledons, **600**, 780*f*
Countercurrent exchange, **885**
  in blood vessels of fish gills, 885, 886*f*
  thermoregulation and, 836*f*
Countercurrent heat exchanger, **836**
Countercurrent multiplier systems, **936**
Courtship behaviors, 1106. *See also* Mating
  in *Drosophila*, 436–37
Covalent bonds, **39**–40
  carbon and, 60
  double, 40
  formation of, 39*f*
  in four molecules, 40*f*
  nonpolar and polar, 40, 41*f*
Cowbird, brown-headed (*Molothrus ater*), 1221
Cowpox, 914
Cows
  cellulose-digesting bacteria in, 74*f*
  ruminant digestion in, 864*f*
Coyote, digestive tract of, 863*f*
Crabs, 664
  ghost crabs (*Ocypode*), 664*f*, 829*f*
  population fluctuations in Dungeness, 1151*f*
Cranial nerves, **1026**
Craniates, 675, **676**
  derived characters of, 676
  hagfishes as, 676, 677*f*
  neural crest of, 676*f*
  origin of craniates, 676, 677*f*
Crassulacean acid metabolism (CAM), **197**
Crawling, 1073–74
Crayfish, 664
  gills, 885*f*
  stretch receptors in, 1047*f*
Crenarchaeota, 544
Cretaceous period mass extinctions, 518–21
Cretinism, 953, 954
Creutzfeldt-Jacob disease, 345

Cribrostatin, 643
Crick, Francis, 9
on DNA replication, 299
modeling of DNA structure by J. Watson and, 88–89, 293*f*
Cri du chat syndrome, 288
Crinoidea (sea lilies, feather star), 666*t*, 667*f*
Cristae, **110***f*
Critical load of nutrients in ecosystems, **1200**
Crocodiles, 691*f*
Crop plants. *See also* Agriculture
angiosperms, 605
genetic engineered, 407
improving protein yield in, 764
polyploidy in, 478
Crop rotation, **766**
Cross-fostering, influence of, on male mice, 1114*t*
Crossing over, **247**, **278**, 348
genetic recombination of linked genes by, 278, 279*f*, 280*f*
introns and, 319
production of recombinant chromosomes by, 248, 249*f*
unequal, and gene duplication, 378*f*, 379
Cross-pollination, **599**
Crows (*Corvus caurinus*), energy costs/benefits of feeding behavior in, 1122
Crustaceans (lobsters, shrimp, crabs), 641*f*, **658***t*, 664*f*–65
exoskeleton in, 657*f*
Hox gene expression in, 432*f*
Cryoprotectants, 840
Cryphonectria parasitica, 622
Cryptic coloration, **1162**
Cryptochromes, 802
Crystallin, 419
regulation of gene expression for synthesis of, 367*f*
Ctenophora (comb jellies), 633, 640*f*
ctr (constitutive triple-response) mutant, 800*f*
Cuban tree snails, 454*f*
Cuboidal epithelia, **823**, **824***f*
Cubozoans (box jellies, sea wasps), 644*f*, 645
Cultural eutrophication of lakes, 1201
Culture, **1131**
Cuticle, **575**, **655**
plant, 575, **717**
Cuttings, plant, 781–82
Cuvier, Georges, 440
Cyanobacteria, 537*f*, 539*f*, 543*f*
blooms of, 1171, 1190*f*
lichens and, 621–22
oxygen production by, 522
plastids and, 551*f*
Cycads (Cycadophyta), 593, 594*f*
Cycas revolute, 594*f*
Cyclic AMP (cAMP), **210**, **356**
as second messenger, 210*f*, 211*f*, 1024, 1056
Cyclic electron flow, photosynthesis, **191***f*, 192
Cyclin, **229**–30
Cyclin-dependent kinases (Cdks), **229**–30
Cycliophoran *Symbion Pandora*, 641*f*
Cystic fibrosis, **266**, 403
as recessively inherited disorder, 266
Cystinuria, 134
Cytochromes (cyt), **170**, 851
cytochrome *c*, 428
Cytogenetic maps, **281**, 396*f*
Cytokines, **903**, **948**
Cytokinesis, **220**, 223*f*, 245*f*
in animal and plant cells, 225*f*
cleavage and, 224, 225*f*, 226
Cytokinins, **796**–97
anti-aging effects of, 797
control of apical dominance in plants by, 796*f*, 797
control of plant cell division and differentiation by, 796
Cytology, microscopes as tool in, 97
Cytoplasm, **98**
cell cycle regulated by signals in, 228–33
cytokinesis and division of, 220, 224–26
response of, to cell signaling, 212, 213*f*
Cytoplasmic determinants, **420**, **987**
cell differentiation and, 420, 421*f*
pattern formation and, 423–25
Cytoplasmic streaming, 117*f*, **118**
Cytosine (C), 88
DNA structure and, 296*f*, 297*f*, 298*f*, 364
Cytoskeleton, **112**–18
animal morphogenesis and changes in, 1001*f*
components of, 113–18
intermediate filaments of, 113*t*, 118
membrane proteins and, 128*f*
microfilaments of, 113*t*, 116–18
microtubules of, 113*t*, 114–16
motor molecules in, 112*f*
role of, in cell support, motility and regulation, 112*f*–13
structure and function of, 113*t*
Cytosol, **98**
ionic gradients across neuron membrane and, 1016*f*
transport in plant cells and, 743
Cytotoxic T cells, **904**, 905*f*
function of, against intracellular pathogens, 910, 911*f*
interaction of, with MHC molecules, 905*f*

Daffodil, radial symmetry in, 773*f*
DAG (diacylglycerol), **212**
Daily torpor, **841**
Dalton, **34**, 52
Dalton, John, 34
Dandelion, seed crop size, 1142*f*
Danielli, James, 125
Darwin, Charles, 15*f*–19, 438–51, 482, 497, 601, 682, 1081, 1143, 1175
adaptation in research and theory of, 442, 443*f*, 444–46
Beagle voyage and field research of, 441–43
descent with modification and theory of, 443–44
evidence supporting theory of, 446–51
genetic variation and theory of, 248–49
historical context of life and ideas of, 439*f*
modern synthesis of Mendelian inheritance and evolutionary theory of, 455
natural selection and evolutionary adaptation as theory of, 16, 17*f*, 438, 441–46
On the Origin of Species by, 15, 438, 439, 443–46, 455
on resistance to idea of evolution and, 439–40
studies on phototropism by, 792
Darwin, Francis, 792
Data, scientific, **19**–20
Davson, Hugh, 125
Davson-Danielli membrane model, 125
Day-neutral plants, **806**
DCC (deleted in colorectal cancer) tumor-suppressor gene, 373*f*
DdeI restriction site, 393*f*
D-dopa, 62, 63*f*
DDT pesticide, 155
biological magnification of, in food webs, 1202–3
Death rate in populations, per capita, 1143–44
Decapods, **664**
Declining-population approach to species conservation, 1218–19
case study of red-cockaded woodpecker, 1218, 1219*f*
steps for analysis and intervention, 1218
Decomposers, **544**, **1185**
fungi as, 620, 1186*f*
prokaryotes as, 544, 1186
Decomposition in ecosystems, 1185–86
rates of, and nutrient cycling, 1198
Deductive reasoning, **20**–21
Deep-diving mammals, respiratory adaptations of, 894–95
Deep-sea hydrothermal vents, 514*f*, 541, **1097***f*
De-etiolation in plants, **789***f*
as response to cell signaling, 789–91
Defensins, 901
Deforestation, 606
in Hubbard Brook Experimental Forest, 1078–79, 1198, 1199*f*
of tropical forests, 1209*f*, 1212, 1221*f*
Dehydration in animal cells, 136
Dehydration reactions, **68**, 69*f*
synthesis of fats, 75*f*
synthesis of maltose and sucrose, 71*f*
Delesseria sanguinea, 550*f*
Deletions
chromosomal, **286***f*
of nucleotides, as point mutation, **329**, 330*f*
Demographic transition, **1153**–54, 1230
Demography, **1139**
age structure and sex ratio as factors in, 1154–55
life tables and survivorship curves in, 1139–40
reproductive rates and, 1140–41
Denaturation
of DNA, **388**
of proteins, **84**, 85*f*
Dendrites, **1014**
stretch receptors and, 1047*f*
thermoreceptors and, 1048*f*, 1049
Dendritic cells, **900**, 909*f*, 910*f*
Dense bodies, 1072
Density-dependent birth/death rates, **1148**–50
Density-dependent inhibition of cell division, **231**, 232*f*
Density-independent birth/death rates, **1148**
Density of populations, **1137**–38
density dependent birth/death rates, 1148–50
density independent birth/death rates, 1148
determining equilibrium for, 1148*f*
dispersion patterns and, 1137–39
population change and, 1148
Dental plaque, 539*f*
Dentition, evolutionary adaptions of, 862, 863*f*
Deoxyribonucleic acid. *See* DNA (deoxyribonucleic acid)
Deoxyribose, **88**
Depolarization, membrane potential and, **1018***f*
Depression, mental, 1025, 1040
Derivatives, **721**
Derived traits of plants, 575, 576–77*f*
Dermal tissue system, **717**
Dermis, animal, 835*f*
Descent with modification, 16, 18*f*, **443**, 444*f*. *See also* Natural selection
Desert biomes, 1100*f*
Desmosomes, 120, **121***f*, 627
Determinate cleavage, **632**
Determinate growth, **720**
Determination, **419**. *See also* Cell differentiation
Detritus, **1093**, **1185**
Detrivores, **1185**, 1186. *See also* Decomposers
Deuteromycetes (imperfect fungi), **612**
Deuterostome, **633**–34
Chordata, 641*f* (*see also* Chordates)
cleavage and, 631–32
coelom formation in, 632
Echinodermata, 641*f*
fate of blastopore, 632–33
protostome development versus, 632*f*
Deuterostome development mode, **631–33**
Development, 411–33. *See also* Animal development; Embryonic development; Plant development
cell division, cell differentiation, and morphogenesis during, 412–14
cell fate during, 1003–8
differential gene expression in, 415–20
evolutionary novelty and, 482–83
evolution of, and morphological diversity, 431–33
evolution of genes controlling, 484–86
hormonal regulation of insect, 960*f*
model organisms for study of, 412–13*f*
pattern formation mechanisms in process of, 421–31
as property of life, 3

Developmental potential of cells, 418, 1004
Devonian period, 629
Diabetes mellitus, **956**
two forms of, 956
Diacylglycerol (DAG), **212**
Diaphragm, contraceptive, **983**
Diaphragm, respiratory, **888**
Diapsids, **689**
Diastole, cardiac cycle, **872**
Diastolic pressure, **876**
Diatoms (bacillariophyta), **559**, 560*f*
Dicots (Dicotyledon), **602**
germination and cotyledons of, 780*f*
Dictyostelium discoideum, life cycle, 566*f*
Dideoxy chain-termination method, sequencing DNA using, 397*f*, 398
Diencephalon, 1028
Diet. *See also* Food; Nutrition
dentition and, 862, 863*f*
disease and, 847, 849*f*, 851*t*, 852*t*
evolutionary adaptations of vertebrate digestive systems and, 862–64
mate selection and, in *Drosophila*, 1113*f*
nutritional requirements in animal, 844, 849–52
vegetarian, 850*f*
Differential gene expression, **362**–63
Differential-interference-contrast (Nomarski) microscopy, 96*f*
Diffusion, **130**
facilitated, 133–34
free energy and, 146*f*
of solutes, 131*f*
Digestion, **853**–55
in alimentary canals, 854*f*, 855
enzyme function in, 857, 858, 859*f*
extracellular, 853–55
in gastrovascular cavities, 854
hormonal control of, 860*f*
intracellular, 853
in small intestine, 858–61
in stomach, 857–58
Digestive system, 855–62
bacteria in, 862
evolutionary adaptations of, 862–64
hormones and, 860*f*
human, 855*f*–62
large intestine, 861–62
oral cavity, pharynx, and esophagus, 856
small intestine, 858–61
stomach, 857–58
Digestive tract
bird, 854*f*, 863
diet and evolutionary adaptions in, 862–64
earthworm, 854*f*
insect, 660*f*, 854*f*
mammalian, 855–62
ruminant, 864*f*
Digger wasp nest-locating behavior, 1115*f*
Dihybrid(s), **257**
Dihybrid cross, 257
Dikaryon, **611**
Dimetrodon, jaw and ear bones of, 695*f*
Dinobryon, 560*f*
Dinoflagellates (Dinoflagellata), **555***f*
mitosis in, 228
Dinosaurs, 629, **689**
blood pressure and feeding in, 877
parental care among, 690*f*
P. Sereno on searching for fossils of, 15*f*
Dioecious plants, **773**
Diphasiastrum tristachyum club moss, 587*f*
Diphtheria, 339
Diploblastic animals, **631**
Diploid cells, **241**
genetic variation preserved in, 466
reduction of chromosomes from, to haploid cells, 242, 243–47
Diplomonads, **552**, 553f
Diplopodes (millipedes), 659*f*
Directional selection, **465***f*
Direct synaptic transmission, 1022–23
Disaccharides, **70**, 71*f*. *See also* Sucrose
Discovery science, **19**–20
Discrete characters, 463
Diseases and disorders, 1163. *See also* Genetic disorders; *names of specific diseases*
allergies, 916–17
autoimmune, 917, 956
of brain and nervous system, 1039–41
cancer (*see* Cancer)
cardiovascular, 76, 882–83
DNA technology applied to diagnosis of, 403
erectile disfunction, 211, 948
fungal, 622–23
gene therapy for, 403–4
G-proteins and, 206*f*, 211
growth disorders, 952, 953
hormonal, 952, 953*f*, 954, 956
hypercholesterolemia, 137
immunodeficiency, 917–19
of muscle fiber excitation by motor neurons, 1069
obesity, 847–48
parasites as cause of (*see* Parasite(s))
pathogens as cause of (*see* Pathogens)
in plants (*see* Plant disease)
sexual reproduction and resistance to, 469
vaccination against, 914
viral, in animals, 343–44
viroids and prions as cause of, 345, 346*f*
Disease vectors, insects as, 554*f*, 555, 556*f*, 664. *See also* Pathogens
Dispersal of organisms, **1084**–85
Dispersion patterns, population, **1137**, 1138–39
Dispersive model of DNA replication, 299, 300*f*
Disruptive selection, 465*f*, **466**
Dissociation curve of oxygen for hemoglobin, **892***f*
Distal control elements, 365
Distal tubule, **931**, 932*f*
regulation of calcium and potassium concentration by, 934
Distribution of species, 1083–92
abiotic factors affecting, 1086–87
behavior, habitat selection, and, 1085
biotic factors affecting, 1085–86
climate and, 1087–92
dispersal and, 1084–85
species transplants and, 1084–85
Disturbances in communities, **1172**–73
community recovery from, 1226*f*
ecological succession following, 1173–75
fire as, 1172*f*, 1173*f*
humans as agents of, 1173
immediate, 1172
Disulfide bridges, **83**
Diuresis, 936
DNA (deoxyribonucleic acid), **7***f*, **86**
amplification of, by polymerase chain reaction, 391*f*
chromatin structure and packing of, 359–60, 361*f*
chromosomes formed from (*see* Chromosome(s))
circular, 289 (*see also* Plasmids)
complementary (cDNA), 390, 400
denaturation of, 388
discovery of, as key genetic material, 293–98
DNA-protein complex (*see* Chromatin)
double helix structure of, 7*f*, 88*f*, 89, 293*f*, 296–98
duplication and divergence in segments of, 378–80
genes as segments of (see Gene(s))
intron and exon segments of, 318 (*see also* Exons; Introns)
methylation of, 284, 289, 364, 417
nitrogenous bases of, 87*f*, 88, (*see* also Nitrogenous bases)
physical mapping of fragments of, 396*f*
plasmids, 537
proteins specified by, 86*f* (*see also* Protein synthesis)
radioactive isotopes and research on, 35*f*
recombinant, 385*f*, 386*f*
repetitive, 375, 376–77
replication of (*see* DNA replication)
restriction fragment analysis of differences in, 392–94
satellite, 377
sequences of (*see* DNA sequences)
sticky ends, 386
strands of (*see* DNA strand(s))
telomeres ending molecules of, 306*f*, 307
transcription of RNA under direction of (*see* Transcription)
DNA cloning, 385–92
of eukaryotic gene in bacterial plasmid, 386, 387*f*, 388, 389*f*
gene cloning and applications, 385*f*, 386
storing cloned genes in DNA libraries, 388, 389*f*, 390
using restriction enzymes to make recombinant DNA, 386*f*
DNA fingerprinting, forensic investigations and, **405***f*
DNA ligase, **303***f*, 304*t*, **386**
creating recombinant DNA with, 386*f*
DNA methylation, 284, 417
control of eukaryotic gene expression and effects of, 364
genomic imprinting and, 289
DNA microarray, 384*f*, 400
DNA microarray assays, **400**, 401f
DNA nucleotide sequence. *See* DNA sequences
DNA polymerases, **301**, 302, 303, 304*t*, 305, 306
DNA replication, 88*f*, 89
antiparallel arrangement of DNA strands and, 89, 297*f*
in bacteria, 346, 347*f*, 350*f*
base pairing to template strands in, 89, 299*f*, 300*f*
models of, 299*f*, 300*f*
origins of, in eukaryotes, 302*f*
proofreading and repair of, 305–6
priming, 303
in prokarytes and eukaryotes, 300–305, 537
proteins facilitating in, 301–3, 304*t*
radioactive tracers used to study, 35*f*
rolling, 350*f*
summary of bacterial, 304*f*
DNA sequences, 88
analysis of, 392–94
homeobox, 431–32, 627
identifying protein-coding genes in, 399
insertion, 351, 352*f*
introns and exons, 318 (*see also* Exons; Introns)
mapping, 396, 397*f*, 398
molecular homologies in, 494, 495*f*
mutations in (*see* Mutations)
noncoding, 374–78
phylogeny inferred from, 492, 494–95
principle of maximum likelihood applied to, 501*f*
promoter, 315*f*, 316*f*
repetitive, 375
TATA box , 315, 316*f*, 364
terminator, 315*f*, 317
types of, in human genome, 375*f*
DNA strand(s), 7*f*
antiparallel arrangement and elongation of, 89, 297*f*, 302–3
complementary, 89
directionality (5' end and 3' end), 88, 89, 302*f*
elongation of, in replication process, 301–2
incorporation of nucleotide into, 302*f*
nontemplate, 313
structure of, 88*f*, 296*f*
synthesis of leading and lagging strands of, 302*f*, 303*f*
telomeres at end of, 306*f*, 307
template, 313
DNA technology, 26*f*, 384–409
DNA cloning and, 385–92
genomics and, 398–402
Human Genome Project (*see* Human Genome Project)
manipulation of genomes as, 384

mapping genomes at DNA level, 394–98
practical applications of, 402–7
restriction fragment analysis, 392–94
safety and ethical questions related to, 26, 404, 407–8
Dobzhansky, Theodosius, 455
Dodd, Diane, on reproductive isolation in divergent *Drosophila* populations, 477*f*
Dodder (*Cuscuta*), 768*f*
Dolphin, 821*f*
Chinese river, 1211*f*
countercurrent heat exchanger in, 836*f*
Domains, protein, **319**
correspondence between exons and, 319*f*
Domains of life, **13**–14, **496**. *See also* Archaea (domain); Bacteria (domain); Eukarya (domain)
phylogeny of (*see* Phylogeny)
three-domain system, 14*f*, 530–31*f*, 532
Dominant allele, **254**
complete dominance of, 260
frequency of, 261–62
genetic disorders due to, 267–68
incomplete dominance, 260, 261*f*
relationship between phenotype and, 261
Dominant species, **1168**
Dominant trait, 253
Donoghue, Philip, 679
Doolittle, Ford, 525
Dopamine, **1025**, 1039, 1041
Doppler, Christian, 252
Dormancy, seed, **779**–80
abscisic acid and, 798, 799*f*
Dorsal lip, blastopore, **995**
as induction organizer, 1006*f*
Dorsal side, body, **630**
Double bond, **40**
Double circulation, **871**
in mammals, 871–74
Double fertilization, **599**, 600*f*, 776, **777**
Double helix, 7*f*, **88*f***, 89, 296, **297*f***, 298. *See also* DNA (deoxyribonucleic acid)
antiparallel structure of, 89, 297*f*
base pairing in, 89, 298*f*
complementary strands of, 89
key features of, 297*f*
Douglas fir (*Pseudotsuga menziesii*), 595*f*
Down syndrome, **287*f***
Downy mildews (Oomycota), 558
Dragonfly, 661
fossil, 491*f*
Drosophila. *See* Fruit fly (*Drosophila* sp.)
Drought
abscisic acid and plant tolerance of, 799
plant response to, 811
Drug(s). *See also* Pharmaceutical products
antibiotics (*see* Antibiotics)
antiviral, 343–44
DNA technology and development of new, 404
enantiomers as, 62, 63*f*
for HIV and AIDS, 918–19
opiates, as mimics of endorphins, 43, 81, 1025
penicillin, 623
Prozac, 1025, 1040
psychoactive, 1025
smooth endoplasmic reticulum and detoxification of, 104
Viagra, 201, 211
Drug resistance
in bacteria, 535
in HIV, 447, 448*f*
Dubautia laxa, 481*f*
Dubautia linearis, 481*f*
Dubautia scabra, 481*f*
Dubautia waialealae, 481*f*
Duchenne muscular dystrophy, **283**–84, 403
Dugatkin, Lee, 1131
Dulse (*Palmaria palmate*), 567*f*
Dungeness crab, population fluctuations in, 1151*f*
Duodenum, **858*f***–59f
Duplication
chromosomal, **286*f***
gene, **459**
Dust mites, 659*f*
Dutch elm disease (*Ophiostoma ulmi*), 622
Dwarfism, 952
Dynamic stability hypothesis on food chain length, **1167**
Dynein, **114**–16
"walking" of, and cilia and flagella movement, 116*f*
Dyson, Freeman, 516

Ear
balance, equilibrium,and, 1052–53
evolution of mammalian jaw and bones of, 695*f*
hearing in mammalian (human), 1050, 1051*f*, 1052
Earlobes, attached, as recessive trait, 265*f*
Earth. *See also* Environment
atmosphere of (*see* Atmosphere)
biodiversity hotspots on, 1222*f*
carrying capacity of, 1145, 1155
chemical conditions of early, and origins of life, 513–16
classifying life on (*see* Systematics; Taxonomy)
clock analogy for key events in history of, 521*f*
continental drift and plate tectonics of, 527–29
continuum of life on, 512
global climate patterns, 1088–89*f*
global energy budget, 1186–87
moderation of temperatures on, by water, 50
prokaryote evolution and changes to, 521–23
regional annual net primary production for, 1188*f*
Earthworm (Oligochaeta), 640*f*, 653–54
anatomy, 654*f*
closed circulatory system, 869*f*
defense systems, 903
digestion in, 854*f*
excretory system (metanephridia) of, 930*f*
gas exchange in, 884
peristalsis and locomotion in, 1064*f*
reproduction in, 964*f*
Eastern box turtle (*Terrapene carolina carolina*), 691*f*
Ebola virus, 344
Ecdysis, 634, 635*f*, 657
Ecdysone, **960**
Ecdysozoans (Ecdysozoa), **634**
Echinoidea (sea urchins), 666*t*, 667*f*, 988–990
Echinoderms (Echinodermata), 641*f*, 665–67
anatomy, 665*f*
brittle stars, 666, 667*f*
classes of, 666*t*
nervous system, 1012*f*
sea daisies, 666, 667*f*
sea lilies, feather stars, 666, 667*f*
sea stars, 665*f*, 666, 667*f*
sea urchins, sand dollars, 666, 667*f* (*see also* Sea urchin)
sea cucumbers, 666, 667*f*
E. coli. *See Escherichia coli* (*E. coli*)
Ecological capacity, **1156**
Ecological footprint, **1156**. *See also* Carrying capacity
in relation to available ecological capacity, 1156*f*
Ecological niche, **1160**
Ecological pyramids, 1192–93
Ecological species concept, **476**
Ecological succession, **1173**
Ecology, **1080**–1105
abiotic and biotic factors and, 1081
of aquatic biomes, 1092–93
of behavior (*see* Behavioral ecology)
biosphere and, 1083 (*see also* Biosphere)
of communities (*see* Community ecology)
conservation and (*see* Conservation biology)
distribution of species and, 1083–92
ecosystem approach to, 1082 (*see also* Ecosystem(s))
environmental issues and, 1083
evolutionary biology related to, 1081
landscape, 1082, 1220–24
of organisms, 1082
organisms and environment in study of, 1081
of populations (*see* Population ecology)
prokaryotes and, 544–45
restoration of degraded, 1224–28
scope of, 1080
subfields of, 1082–83
of terrestrial biomes, 1098–1104
Ecosystem(s), 4f, **1082**, **1184**–1208. *See also* Biomes
aquatic, 1188–90
as biodiversity level, 1211
chemical cycling in, 160*f*, 1184–86, 1195–99
communities of species in (*see* Community(ies))
disruption of chemical cycles by humans, 1200–1206
edges between, 1220*f*, 1221
effects of acid precipitation on, 55–56, 1201–2
energy and matter in, 1184
energy budgets of, 1186–88
energy conversion in, 6
energy flow in, 160*f*, 1184–86
energy transfer between trophic levels in, 1191–94
"engineers" (foundation species) in, 1169–70
fragmentation of, 1220–21
human impact on, 1200–1206
limitations on primary production in marine and freshwater, 1188–91
producers and consumers in, 6
role of fungi in, 620
role of prokaryotes in, 544–45
services of, promoting human welfare, 1212
terrestrial and wetland, 1190–91
trophic structure of, 1166–68
Ecosystem ecology, **1082**. *See also* Ecosystem(s)
Ecosystem services, **1212**
Ecotone, **1099**
Ectoderm, **631**, **994**
gastrulation and formation of, 994
organs and tissues formed from, 999*t*
Ectomycorrhizae, **610**, **766**, 767*f*
Ectoparasites, **1163**
Ectoprocts (bryozoans), 640*f*, **649*f***
Ectotherms (ectothermic animals), **689**, **829**, **833**
energy budgets in, 830, 831*f*
thermoregulation in, 833–34
reptiles as, 689
Edges (boundaries) between ecosystems, 1220*f*, 1221
Ediacaran fauna, **628**
Effective population size, **1216**–17
Effector cells, **907**, 944, **1013**
Efferent arteriole, **933**
Efferent signal, 944
Egg, **964**. *See also* Ovum
activation of, 990
amniotic, 688*f*
cytoplasmic determinants in, 987
development of human, 973, 974*f*
fertilization of human, 978, 979*f*, 991
fertilization of mammalian, 990–91*f*
fertilization of sea urchin, 988, 989f, 990*f*
Egg-polarity genes, **424**
Ehrlich, Anne, 1180
Ehrlich, Paul, 1180
ein (ethylene-insensitive) mutants, 800*f*
Ejaculation, **971**
Ejaculatory duct, **972**
Elastic fibers, **823**
Eldredge, Niles, 481
Electrical membrane potential. *See* Membrane potentials
Electrical synapses, 1021
Electrocardiogram (EKG/ECG), **874**
Electrochemical gradient, **135**
Electrogenic pump, **136*f***
Electromagnetic receptors, **1048**, 1049*f*
Electromagnetic spectrum, **186*f***
Electron(s), **34**
configuration of,and chemical properties, 37–38
cyclic and noncyclic flow of, in photosynthetic light reactions, 190*f*, 191*f*, 192

  energy levels of, 36–37f
  ionic bonding and transfer of, 41f
  orbitals of, 38f, 39
  transport of, in cellular respiration, 170–3
  valence, 38
Electronegativity, **40**
Electron microscope (EM), **95–97**
  freeze-fracture preparation of cells for, 125, 126f
  scanning (SEM), and transmission (TEM), **96f**
Electron shells, **36**
  of eighteen elements, 37f
  valences for elements of major organic molecules, 60f
Electron transport chain, **163–64**
  in cellular respiration, versus in uncontrolled reaction, 163f
  energy coupling to chemiosmosis, and ATP synthesis, 171, 172f, 173
  evolution of, 521–22
  free energy change during, 171f
  pathway of, during oxidative phosphorylation, 170–71
  in photosynthetic light-harvesting, 190f
Electroporation, **390–91**
Element(s), **33**
  atomic structure and properties of, 34
  cycling of, in ecosystems, 160f, 1184–86, 1195–98
  electron shells of eighteen, 37f
  essential, required by life, 33–34
  in human body, 33t
  periodic table of, 37
  plant nutritional requirements for, 756–59
  trace, 34
Elephants
  evolutionary tree of, 443, 444f
  exponential growth in population of African, 1144f
  overexploitation of African, 1214
  thermoregulation in, 837f
Elephant seals, bottleneck effect in populations of northern, 461f
Elicitors, **814**
Elimination as stage in food processing, **853**
Elk, mating behaviors in, 1124f
El Niño Southern Oscillation, 1170, 1171f
Elongation factors, polypeptide synthesis, 323, 324f
Elton, Charles, 1166, 1167
Embolus, 883
Embryo
  cleavage of zygote and formation of human, 978, 979f
  development (Embryonic development)
  ensuring survival of animal, 967–68
  epigenesis and, 987
  fertilization and formation of animal, 967, 978, 979f, 988–91
  gastrulation and tissue formation in, 994–97
  homologous structures in, 448, 449f
  maternal effect (egg-polarity) genes, effects of, 424f
  organogenesis in, 997f, 998f
  pattern formation in (*see* Pattern formation)
  plant, 577f, 777f, 778
  restriction of cellular potency in, 1005
  stem cells of, 418
Embryology, 411. *See also* Development; Embryonic development
Embryonic development, 412–14
  of animal, 412–13, 414f, 485–86, 627f, 988–1001
  cell division, cell differentiation and, 413
  cell fate during, 1003–8
  development of brain in, 1028f, 1029
  differential gene expression and, 415–20
  effects of DNA methylation on, 364
  Hox genes and, 485–86, 627
  human, 978, 979f, 980f, 987f, 1028–29
  morphogenesis and, 413, 431–33, 987
  pattern formation and, 421–31
  plant, 414f, 777f–78
  restriction of cellular potency in, 1005
Embryonic lethals, **423**
Embryonic stem cells, 418f, 1039
Embryophytes, plant, 575f, **577f**
Embryo sac, plant, **599**
Encephalitis virus, 337
Emergent properties, **9**
  of biological systems, 9
  as biological theme, 27f
  cellular functions, 120, 121f
  chemistry of life, 89
  of compound, 33f
  of water, 48–53
Emigration, population density and influence of, **1137**
Emotions, human brain and, 1034–35
Emu, 693f
Enantiomers, **62**, 63f
  amino acids in proteins, 79f
  pharmaceuticals and, 62, 63f
Endangered species, **1210–11**
Endangered Species Act, U.S., 1210
Endemic species, **450**
Endergonic reactions, **146**
  coupling of, to exergonic reactions by ATP, 148, 149f
  free energy changes in, 146, 147f
Endler, John, research on guppy evolution by, 446, 447f
Endocrine glands, **944**
  adrenal, 956–58
  gonads, 958–59
  hypothalamus, 948, 950–51
  major glands and hormones, 949t, 950t
  pancreas, 955–56
  parathyroid, 954–55
  pineal, 959
  pituitary, 948–52
  thyroid, 953–54
Endocrine pathway, 945f
Endocrine signaling, 203
Endocrine system, 943–63
  control pathways and feedback loops in, 944, 945f
  defined, **944**
  hormones as chemical signals of, 945–48
  invertebrate, 959–61
  major glands and hormones of, 949t, 950f
  nervous system overlap with, 943, 944
  nonpituitary glands of, 953–59
  pituitary gland and hypothalamus of, 948–52
Endocytosis, transport by, **137**, 138f
Endoderm, **631**, **994**
  gastrulation and formation of, 994
  organs and tissues formed from, 999t
Endodermis, plant root, **723**, **744**
  water/mineral absorption and, 744, 745f, 746
Endolymph in inner ear, 1053f
Endomembrane system, **104–9**
  endoplasmic reticulum, 104, 105f
  Golgi apparatus, 105, 106f, 107
  lysosomes, 107f, 108
  relationships of organelles within, 108, 109f
  review of, 108
  vacuoles, 108f
Endometrium, **970**
Endomycorrhizae, **610**, **766**, 767f
Endoparasites, **1163**
Endoplasmic reticulum (ER), **104**, 105f
  origins of, 525
  rough, functions of, 105
  smooth, 104, 105f
  transitional, 105
Endorphin(s), **952**, **1025**
  opiate drugs as mimics of, 43, 81, 1025
  pain reduced by, 43, 81, 952
  structure, 43f
Endorphin receptors, 43
Endoskeleton, **1064**
  human, 1065f
  vertebrate, 682
Endosperm, **600**, **777**
  development of, 777
Endospores, **537**, 538f
Endosymbiont, 523
Endosymbiosis, 550–51
  origins of mitochondria and plastids in, 523–25
  secondary, 551f
  serial, 524
Endothelium, **874**
  in blood vessels, 874
Endotherms (endothermic animals), **689**, **829**, 833
  birds as, 689
  energy budgets in, 830, 831f, *871*
  thermoregulation in, 833–34
Endotoxins, **546**
Energetic hypothesis on food chain length, **1167**
Energy, **36**, **142**. *See also* Bioenergetics
  animal communication and cost of, 1111
  chemical, 142, 828
  free, 145–48
  kinetic, 49, 142
  forms of, 142–43
  laws of thermodynamics applied to, 143–44
  life and, as biological theme, 27f
  locomotion and requirements for, 1074f
  potential, 36, 142
  species richness, water, and, 1176f
  storage of, in fats, 76
  thermal (heat), 142 (*see also* Heat)
  transfer of, between ecosystem trophic levels, 1191–94
  transformations of, in metabolic processes, 141–44
Energy budget
  of animals, 830, 831f, 844–48, 928
  of ecosystems, 1186–88
Energy costs
  of animal locomotion, 1074
  of animal signals, 1111
  of foraging behavior, 1122
Energy coupling, **148**
Energy flow in ecosystems, 6f, 160f, 1184–86
  decomposition and, 1185–86
  overview of, 1185f
  physical laws and, 1185
  primary productivity and energy budget, 1186–91
  secondary productivity and 1191
  trophic efficiency, ecological pyramids, and, 1192–93
  trophic relationships as determinants of, 1185
Energy levels of electrons, **36**, 37f
Energy processing as property of life, 3f
Energy transformation, laws of, 143–44
Engelmann, Theodor, 187f, 188
Engelmann's experiment, 187f, 188
Enhancers, 364, **365f**
  function of, in control of eukaryotic gene expression, 365, 366f
Entamoebas, 564
Enteric division of autonomic nervous system, **1028**
Enterocoelous development, **632**
Enthalpy, change in (ΔH), 145
Entropy, **144**
  change in (ΔS), 145
Environment. *See also* Biosphere; Earth
  cancer caused by factors in, 370 (*see also* Cancer)
  carrying capacity of, 1145, 1155
  emerging viruses linked to changes in, 344
  genetics combined with influences of, resulting in animal behaviors, 1113–18
  impact of, on phenotype, 264
  influence of, on animal nitrogenous waste, 928
  interactions between organisms and, 27f, 1080–83 (*see also* Ecology)
  regulation of animal's internal, 831–33
  signal transduction pathways in plant responses to, 788–91
  water and fitness of, 47–57
Environmental education, 437, 1079
Environmentalism, 1083
Environmental problems
  acid precipitation, 55–56, 1201–2

atmospheric carbon dioxide and climate change, 1203–5
decline in amphibian species due to, 686
disrupted chemical cycling, 1200–1206
DNA technology applied to solving, 405–6
ecology and evaluation of, 1083
human population growth as, 1155–56
humans as cause of (see Human environmental impact)
ozone depletion, 1205–6
soil depletion, 1200
toxins in ecosystems, 1202–3
Environmental stress, plant responses to, 810–12
Enzymatic hydrolysis, **853**
Enzyme(s), **77**, **150**–55
activation energy barrier and, 150–51
active site of, 152, 153*f*
allosteric regulation of, 156–57
catalysis in active site of, 152, 153*f*, 154
as catalysts, 77, 150
catalytic cycle of, 78*f*, 153*f*
cofactors of, 155
conformation of lysozyme, 81*f*
digestive, 857, 858, 859*f*
DNA proofreading and damage repair by, 305–6
DNA replication and role of, 301–3, 304*t*
effects of temperature and pH on, 154
induced fit between substrate and, 152, 153*f*
inhibitors of, 155f
localization of, within cells, 157
lowering of activation energy barrier by, 152
membrane protein as, 128*f*
normal binding of, 155*f*
one gene-one enzyme hypothesis, 310, 311*f*
repressible, and inducible, 354–56
substrates of, 152
substrate specificity of, 81, 152
Enzyme-substrate complex, **152**
Eosinophils, 880*f*, 881, **900**
Ephedra, 594*f*
Ephrussi, Boris, 310
Epiblast, 994, 1000
Epicotyl, **778**
Epidermis
animal, 835*f*
plant, **717**, 724, 813
Epididymis, **971**
Epigenesis, 987
Epigenetic inheritance, **364**
Epiglottis, **856**
Epilepsy, 1032
Epinephrine, 944, **956**, **1025**
cell signaling and glycogen breakdown by, 203–4, 213*f*, 946
different responses to, 947*f*
effects of, on cardiovascular and respiratory systems, 874, 957
stress and secretion of, 874, 946, 947*f*
Epiphytes, 767, **768**
Episome, **349**
Epistasis, **262**, 263*f*
Epithalamus, **1030**
Epithelial tissue, **823**
transport, 926–27
Epitope, **903**
Epstein-Barr virus, 370, 910
Equatorial-polar gradients in biodiversity, 1176
Equilibrium
free energy and, 145, 146*f*
metabolism and, 147–48
organs of, in vertebrates, 1052, 1053f
work and, in closed and open systems, 147*f*
Equilibrium potential (E_{ion}), **1016**
Equisetum arvense field horsetail, 587*f*
Erectile dysfunction, 211, 948, 1025
Ericksson, Peter, 1038–39
Ergotism, 622
Erythrocytes, 825*f*, **880**–81. *See also* Red blood cells (erythrocytes)
Erythromycin, 537
Erythropoietin (EPO), **881**
Escherichia coli (E. coli)
chaperone protein in, 85
chemotaxis in, 537
chromosome and DNA of, 537*f*
chromosome length of, 226
DNA replication in, 300–305, 347*f*
genetic recombination in, 460, 346–51
genome and replication of, 346, 347*f*
genome map, 394
Hershey and Chase experiment on viral DNA infection of, 294, 295*f*, 296
in human colon, 542*f*, 862
as pathogen, 460, 546
restriction site, 386*f*
reproduction rates of, 346, 538
T2 bacteriophage infection of, 294, 295f
T4 bacteriophage infection of, 334*f*
transcription and translation in, 328*f*
transposition of genetic elements in, 351–52
E site, tRNA binding, **322***f*
Esophagus, **856**
peristalsis in, 856*f*
Essential amino acids, **849**–50
in vegetarian diet, 850*f*
Essential element, **757**
Essential fatty acids, **850**
Essential nutrients, **849**
in animals, 849–52
in plants, 757, 758*t*
Estivation, **841**
Estradiol, 63*f*
Estrogen, **958**
human pregnancy and childbirth and role of, 981
intracellular receptors for, 947
Estrous cycles, **973**
Estrus, **973**
Estuary(ies), **1095***f*
partial food web for Chesapeake Bay, 1167*f*
Etges, William, 1113
Ethane, molecular shape of, 60*f*
Ethanethiol, **65***f*
Ethanol, **64***f*
Ethene, molecular shape of, 60*f*
Ethics
DNA technology and, 26, 404, 407–8
human cloning and, 417
Ethology, **1107**–9
Ethylene, **799**–801
fruit ripening and, 203, 801
leaf abscission and, 800, 801*f*
programmed cell death and, 800
root apoptosis and, 811
triple response induced by, 799–80
Etiolation in plants, **789**
eto (ethylene-overproducing) mutants, 800
Euchromatin, **360**
Eudicots, **602**, 603*f*
embryo development, 777*f*
monocots compared with, 603*f*
stem tissue in, 724*f*
Euglena sp., 529, 554*f*
Euglenids, **554***f*
Euglenozoans, 553–54
Eukarya (domain), **14***f*, 507*f*, 521
compared with Archaea and Bacteria, 541*t*
Eukaryotes
animals as (*see* Animal(s))
cells of (*see* Eukaryotic cell)
cilia of, 15*f* (*see also* Cilia)
cloning and expression genes of, 390–91
cloning genes of, in bacterial plasmid, 386, 387*f*, 388, 389*f*
endosymbiosis in evolution of, 523–25, 550–51
gene expression in (*see* Gene expression)
genome (*see* Eukaryotic genome)
fungi as (*see* Fungi)
multicellularity and evolution of, 525–28
origins of, and relationship to prokaryotes, 523, 524*f*, 525
phylogeny of, 552*f*
plants as (see Plant(s))
protists as, 549–51 (*see also* Protists)
Eukaryotic cell, **8**, **98**–109
animal, 100*f* (*see also* Animal cell)
chromosomes of, 219*f*
cytoskeleton of, 112–18
DNA replication in, 301*f*, 306–7
endomembrane system of, 104–9
extracellular components of, 118–20, 121*f*
gene expression in, differential, 362–63
gene expression in prokaryotic cell compared with, 327, 328*f*
as genetic chimera, 525
infection of, by microsporidia, 615*f*
as living unit greater than sum of parts, 120
mitochondria and chloroplasts of, 109–11
modification of RNA after transcription in, 317–19
nucleus of, 102, 103*f*
panoramic view of, 99, 100*f*, 101*f*
plant, 101*f* (*see also* Plant cell)
prokaryotic cell compared with, 8*f*, 98–99, 312*f*
protein synthesis in, versus in prokaryotes, 327–28
ribosomes of, 102, 103*f*, 104, 322
RNA polymerases of, 316
RNA processing in, 312*f*
RNA types in, 327*t*
size of, 98
transcription in, 311, 312*f*, 315, 316*f*, 317, 328*f*, 331*f*
translation in, 311, 312*f*, 328*f*, 331*f*
Eukaryotic genome, 359–81. *See also* Genome
cancer resulting from genetic changes affecting cell cycle controls in, 370–74
complexity of, 374–75
chromatin structure based on DNA packing in, 359–61
control of gene expression in (*see* Gene expression)
DNA duplications, rearrangement, and mutations contributing to evolution of, 378–81
gene and multigene families in, 377–78
gene expression, regulation of, 362–70
human (*see* Human genome)
noncoding DNA sequences in, 374–78
overview, 359
as product of genetic annealing, 525
relationship between organismal complexity and complexity of, 374–75
transposable elements in, 375–76
Eumetazoans, **633**, 643
Europa, 514
Europe
acid precipitation in, 55*f*, 1201*f*
blackcap migration in, 1120, 1121*f*
European kestrel, parental caregiving by, 1142f
Euryarchaeota, 544
Euryhaline animals, **923**
Eurypterids, **658**
Eustachian tube, **1051**
Eutherian (placental) mammals, **697**–701
biogeography and evolution of, 528
convergent evolution of marsupials and, 450, 494*f*
reproduction in, 967–84
Eutrophication in freshwater ecosystems, **1190**
Eutrophic lakes, **1094**
experimental creation of, 1190
human impact and creation of, 1201
Evaporation, 50, **835**
Evaporative cooling, **50**
in animals, 837
transpiration in plants and, 749
Evapotranspiration, **1176**
actual, 1190
species richness and, 1176*f*
"Evo-devo," 431–32, 484–86
origin of flowers and, 601–2

Index

- Evolution, 15–19, **438**
  - adaptation and (*see* Evolutionary adaptation)
  - anagenesis and cladogenesis as patterns of, 472f
  - of angiosperms, 601–2
  - of animals, 628–30
  - of behavioral traits, 1118–21
  - of birds, 692–93
  - chordate, 674–75
  - coevolution, 1164–65
  - convergent (*see* Convergent evolution)
  - C. Darwin's field research and his ideas about, 441–43
  - C. Darwin's theory of, 438, 443–53
  - of development, leading to morphological diversity, 431–33
  - DNA and proteins as measure of, 89
  - effects of differential predation on guppy, 446, 447f
  - of eukaryotes, 523–28
  - evolutionary novelty, 482–83
  - fossil, homologies, and biogeography as evidence of, 448–51
  - of genes controlling development, 484–86
  - of genetic code, 314
  - gradualism theory, 440
  - of gymnosperms, 596
  - human culture and, 1132–33
  - influence of, on animal nitrogenous waste, 928
  - of jaw and ear bones in mammals, 695f
  - J. B. Lamarck's theory of, 440–41
  - E. Lander on genomics and, 237
  - of leaves, 585, 586f
  - macroevolution, 472, 482–88
  - of mammals, 694–95
  - microevolution (*see* Microevolution)
  - of mitosis, 227f, 228
  - modern synthesis of Mendelian and Darwinian ideas, 455
  - mosaic, 703
  - of multicellularity, 525–28
  - natural selection and, 16f, 17f (*see also* Natural selection)
  - obesity and, 848
  - of organic compounds, 59
  - origin of species (*see* Speciation; Species)
  - of plants, from charophyceans (green algae), 573–74
  - of plants, highlights of, 579f
  - of plants, seed plants, 591–93
  - of plants, vascular, 584–86
  - of populations, 454–70
  - of prokaryotes, 521–23
  - of reptiles, 689–90
  - resistance to idea of, 439–40
  - of roots in plants, 585–85
  - of tetrapods, 684, 685f
  - vertebrate digestive systems, 862–64
  - of viruses, 342–43
- Evolutionary adaptation, **16**, **438**. *See also* Natural selection
  - camouflage as example of, 446f
  - C. Darwin's theory of natural selection and, 16, 442–43, 444–46
  - in prokaryotes, 537–40
  - as property of life, 3f
- Evolutionary biology. *See also* Evolution
  - ecology and, 1081
  - K. Kaneshiro on, 436–37
  - modern synthesis of Darwinism and Mendelism in, 455
- Evolutionary fitness, 464–65
- Exaptation, 483
- Excitatory postsynaptic potential (EPSP), **1022**
- Excretion, **922**
- Excretory systems, 928–39
  - excretory processes and, 929
  - homeostasis and, 937f
  - key functions of, 929f
  - kidneys, 931–39
  - Malpighian tubules as, 930f–31
  - mammalian (human), 931–38
  - metanephridia as, 930f
  - nitrogenous waste and, 927–28
  - protonephridia (flame-bulb) as, 929f–30
  - water conservation and, 862
- Exercise
  - blood pressure during, 876
  - health benefits of, 883
- Exergonic reaction, **146**
  - cellular respiration as, 146, 163f
  - coupling of, to endergonic reactions by ATP, 148, 149f
  - energy profile of, 151f
  - DNA replication, 302f
  - free energy changes in, 146, 147f
- Exit tunnel, ribosome, 323
- Exocytosis, transport of large molecules by, **137**
- Exoenzymes, **609**
- Exons, **318**
  - correspondence between protein domains and, 319f
  - duplication of, 380
- Exon shuffling, 319, 380–81
  - evolution of new gene by, 380f
- Exoskeleton, **657**, **1064**
  - arthropod, 74f, 657f
  - molluscs, 1064
- Exotic (introduced) species, 1213–14
- Exotoxins, **546**
- Expansins, **795**
- Exponential population growth, **1144**–45
  - global human, 1152, 1153f
  - intrinsic rate of increase (r_{max}) and, 1144f
- Expression vector, **390**
- Express sequence tags (ESTs), 399
- External fertilization, **967**
- Exteroreceptors, **1046**
- Extinction of species, 1209–10
  - mass, 518–21
  - overexploitation and, 1214
- Extinction vortex, **1215**
  - greater prairie chicken case study, 1216
  - grizzly bear case study, 1217–18
  - process culminating in, 1215f
- Extracellular components, 118–20
- Extracellular digestion, **853**–55
- Extracellular fluid, ionic gradients across neuron plasma membrane and, 1016f
- Extracellular matrix (ECM), **119**–20
  - animal morphogenesis and role of, 1002–3
  - membrane proteins and, 128f
  - structure, 119f
- Extraembryonic membranes, **688**, **999**
  - in birds/reptiles, 688, 999f
  - as derived character of amniotes, 688f
  - in human embryo, 1000f
- Extranuclear genes, 289–90
- Extraterrestrial sources of life, 514
- Extreme halophiles, **541**, 544f
- Extreme thermophiles, 534f, **541**
- Extremophiles, **541**
- Eye
  - compound, 1058
  - of *Drosophila* fruit fly, 276f, 277f, 280, 281f, 411f
  - evolution of, 483
  - invertebrate, 1057f, 1058f
  - lens cells of, gene expression in, 367f
  - molluscs, 483f
  - resolving power of, 95
  - single-lens, 1058
  - vertebrate, 1058, 1059f, 1060f
- Eye cup, 1058 *See also* Ocellus
- Eyespot, 1058 *See also* Ocellus

- F_1 generation, **253**
  - incomplete dominance and, 261f
  - law of segregation and, 253f, 254t, 255f, 257f
- F_2 generation, **253**
  - incomplete dominance and, 261f
  - law of segregation and, 253f, 254t, 255f, 257f
- Facilitated diffusion, **133**, 134f
- Facilitators in communities, **1170**
- Facultative anaerobes, **176**, **539**
- $FADH_2$, as source of electrons for electron transport chain, 170, 171f
- Fairy ring, 618f
- Family (taxonomy), **496**
- Famine, 558, 1156
- Fasciated (f) allele, 430
- *Fass* mutants, plant growth and, 730, 731f
- Fast block to polyspermy, **989**
- Fast muscle fibers, 1071, 1072t
- Fat(s), **75**–76
  - catabolism of, 177f
  - digestion and absorption of, 859f, 861f
  - hydrocarbons and, 61f
  - saturated and unsaturated, 75f, 76
  - synthesis and structure of, 75f
- Fat, body
  - adipose tissue, 61f, 76
  - obesity and, 847–48
- Fate maps, **1004**
  - of frog and tunicate embryos, 1004f
- Fatty acids, **75**
  - digestion of, 861f
  - essential dietary, 850
  - saturated and unsaturated, 75f, 76
- Feathers, bird, 472, 692f
- Feather star (Crinoidea), 666, 667f
- Feces, **862**
- Feedback inhibition of enzyme activity, **157**f
  - in cellular respiration, 177, 178f
- Feedback mechanisms
  - in biological systems, 11–12
  - in cellular respiration, 177, 178f
  - in endocrine and nervous system pathways, 944
  - negative, 11f
  - positive, 12f
  - regulation of homeostasis by, 832–33
  - regulation of kidney function by, 937f
  - in thermoregulation, 838–39
- Feeding
  - bulk, 845f
  - energy costs and benefits of foraging, 1122, 1123f
  - fluid, 845f
  - mineralization of bone and mechanisms for, 679
  - pseudopodia for, 563, 564f
  - substrate, 845f
  - suspension, 845f
- Females, human
  - conception, embryonic development and childbirth in, 978–82
  - contraception and abortion in, 982–83
  - hormonal regulation of reproductive system of, 958–59, 973, 976–77
  - menopause in, 977
  - menstruation in, 973
  - oogenesis in, 973, 974f
  - pregnancy in, 915–16, 978–81
  - reproductive anatomy, 969, 970f, 971
  - reproductive cycle of, 973, 976–77
  - semen in reproductive tract of, 972
  - sex hormones of, 63f, 958–59
  - sexual dimorphism and, 468
  - sexual response in, 972–73
  - X chromosome inactivation in, 284f
- Females, mate choice by, 1125–26
- Fermentation, **161**, 174–76
  - alcohol, 175
  - cellular respiration compared with, 175–76
  - glycolysis and, 174
  - lactic acid, 175
  - types of, 175
- Ferns (Pterophyta), 2f, 588
  - life cycle of, 585f
  - sporophylls of 586
- Fertilization, 238, **241**–42, **967**, 988–91
  - acrosomal reaction in, 988, 989f
  - in animals, 967–69

contraception against human, 982–83
cortical reaction in, 989, 990*f*
double, in angiosperm plants, 599, 600*f*
egg activation in process of, 990
internal versus external, 967
in vitro, 984
in mammals, 990, 991*f*
human, 978, 979*f*, 991
in plants, and seed development, 592, 593*f*, 777*f*–78
prezygotic barriers to, 473, 474*f*
random, as source of genetic variation, 248
in sea urchins, 988–90
segregation of alleles and, as chance events, 259*f*, 275*f*
Fertilization envelope, **989**
Fertilizers, agricultural, 761–62
Fetoscopy, 269
Fetus, **979**
birth of, 981
circulation in placenta and, 980*f*
development of human, 980*f*
maternal antibodies, Rh factor and, 92, 915–16
passive immunity in, 914
testing of, for genetic disorders, 268–70, 983–84
Fever, 85, 874, 948
F factor, **349**
Fibers, sclerenchyma, **718***f*
Fibrin, **882**
Fibrinogen, 880, **882**
Fibroblast, **823**
platelet-derived growth factor and stimulation of human, 231*f*
Fibroblast growth factor (FGF), 1007
Fibronectins, **119**
cell migration in animal morphogenesis and role of, 1002*f*
in extracellular matrix of animal cell, 119*f*, 1002–3
Fibrous connective tissue, **825***f*
Fibrous root system, **713**
Fiji, intertidal reserve in, 1224*f*
Filaments, flower, **598**
Filter feeders. *See* Suspension feeders
Filtrate, production of urine from, **929**, 931–34
Filtration, **929**
Fimbriae, **536***f*
Finches
adaptive radiation of, on Galápagos Islands, 17, 18*f*
beak variations in, 443*f*
character displacement in communities of, 1161*f*
C. Darwin's studies of Galápagos Islands, 442, 443*f*, 450
disruptive selection in case of, black-bellied seed-cracker, 466
speciation in, 478
zebrafinches, mate selection in, 1125–26
Finland, biomanipulation of Lake Vesijärvi in, 1171
Fire
as community disturbance, 1172, 1173*f*
wildfire as component of terrestrial biomes, 1099
Firefly, 314*f*
First law of thermodynamics, **143**
Fish(es)
bony, 682–84
circulatory system of, 870*f*, 871
countercurrent exchange in, 885, 886*f*
effect of differential predation on evolution of guppies, 446, 447*f*
gills of, 885, 886*f*
hearing and equilibrium in, 1053–54
hermaphroditic, 966*f*
jawed, 679, 680f, 682–84
jawless, 678–79
kidney adaptations in, 939*f*
lateral line system of, 680, 1054*f*
lobe-finned and lungfishes, 683–84
mutualism between bacteria and, 545*f*
nitrogenous wastes of, 927*f*
osmoregulation in, 923*f*, 924–25
ray-finned, 682–83
rheotaxis in, 1110*f*
sequential hermaphroditism in species of, 966
speciation of cichlid, 479, 480*f*
territoriality of, 1107*f*, 1108*f*
thermoregulation in, 834*f*, 836, 837*f*
Fisher, R. A., 455
Fission
binary, in bacteria, 226, 227*f*, 346, 347*f*, 537, 558
invertebrate, **964**, 965*f*
Fitness, **464**–65
relative, 464
Fitzroy, Robert, 441
Five-kingdom taxonomic system, 529*f*, 530
Fixed action pattern (FAP), **1108**
Flaccid plant cell, **133**, **741**, 750*f*
Flagella, **114**–16
beating of, versus beating of cilia, 115*f*
dynein "walking" and movement of, 116*f*
fungi and evolutionary loss of, 613*f*
motion of, 115*f*
prokaryotic, 536*f*
of protist Euglenozoa, 553*f*
of stramenopile, 558*f*
ultrastructure of, 115*f*
Flagellates, protistan, 628*f*
Flagellin, 536*f*
Flame-bulb system, 929*f*, 930
Flashlight fish (*Photoblepharon palpebratus*), 545*f*
Flatworms (Platyhelminthes), 639*f*, 646–48
as acoelomate, 631*f*
blood fluke, 647*f*, 1163
classes of, 646*t*
hydrostatic skeleton in, 1063
monogeneans and trematodes, 647
nervous system, 1012*f*
planarian anatomy, 647*f*
reproductive anatomy of, 968*f*
tapeworm anatomy, 648*f*
turbellarians, 646–47
Flavonoids, 575, 766
Flavoprotein, 170
Fletcher, W. J., 1085, 1086*f*
Flight
in birds, 483, 692*f*
energy cost of, 1074*f*
in insects, 660–61
theropods and, 692–93
Flooding, plant response to, 811
Florigen, **808**
Flower, **598**
cell signaling in development of, 429–31
complete versus incomplete, 773*f*
development of, 429*f*
diversity in types and structures, 773*f*
environmental effect of phenotype on, 264*f*
evo-devo hypothesis on origins of, 601–2
monoecious versus dioecious, 773*f*
pattern formation in development of, 429*f*, 430*f*, 431*f*
"pin" and "thrum" types of, 775*f*
pollinators of, 604*f*
self-pollinating, versus self-incompatible, 775–76
structure of, 598*f*, 772*f*
Flower fly, 22*f*
Flowering
evidence for hormonal control of, 808*f*
genetic control of, 734–35
photoperiodic control of, 806–8
vernalization and, 807
Flowering plants. *See* Angiosperms
Fluid feeders, **845***f*
Fluid mosaic model of plasma membrane, **124**, 125*f*
Flukes, 639*f*
Fluorescence, 189*f*
Fluorescence microscopy, 94*f*, 96*f*
Fly agaric (*Amanita muscaria*), 618*f*
Flying squirrel, 450*f*
Flying squirrel, northern (*Glaucomys sabrinus*), 1031*f*
Follicle, **969**
Follicle-stimulating hormone (FSH), **952**, 959
female reproductive cycle and, 976, 977
Follicular phase, ovarian cycle, **976**
Food
angiosperms as, 605
caloric imbalance in, 846–48
catabolism of, in cellular respiration, 161–65
energy of, available at different tropic levels, 1193*f*
feeding (*see* Feeding)
as fuel and nutrition for animals, 828–31, 844–46
functions of, 177
fungi as, 623
fungi as spoilers of, 622
fungi in production of, 623
genetically modified organisms as, 407–8
photosynthesis, and conversion of light energy into chemical energy of, 182–85
plant biotechnology and increased supply of, 784
seaweed as, 561*f*
Food chain, **1166**
energy partitioning, 1192*f*
examples of terrestrial and marine, 1166*f*
limits on length of, 1167–68
Food plants. *See* Crop plants
Food poisoning, 460, 542*f*
Food processing in animals, 853–55
Food vacuoles, 107, **108**
as digestive compartment, 853
in *Paramecium*, 557*f*
Food web, **1166**–67
biological magnification of PCBs (polychlorinated biphenyls) in, 1202*f*
in marine ecosystems, 1166*f*
partial, in Chesapeake Bay estuary, 1167*f*
Foot, **650**
bird, 694*f*
moss sporophyte, **580**
mollusc, 650
Foraging behavior, **1119**, 1122–23
energy costs and benefits of, 1122
Foraminiferans (forams), **563**
Forebrain, 680
human, **1028**
Forelimb, homologous structures in mammalian, 449*f*
Forensic science, DNA technology and, 26*f*, 404–5
Forest
canopy of, 1098
coniferous, 1102*f*
deforestation of, 606, 1078–79, 1198, 1199*f*, 1209*f*, 1212, 1221*f*
determining species diversity in, 1165*f*
effects of acid precipitation on, 55*f*
effects of climate change on, 1092*f*, 1204
management of, based on Hubbard Brook Experimental Forest findings, 1078–79
temperate broadleaf, 1103*f*
tropical, 1100*f*
Forest Atmosphere Carbon Transfer and Storage (FACTS-1) experiment, effects of elevated CO_2 on, 1204
Form and function
bat anatomy and, 17*f*
fluid mosaic model of plasma membrane and, 130
Fossil(s), 15*f*, **439**. *See also* Fossil record
amniotes, 688
angiosperm, 601f
birds, 693*f*
from Burgess Shale, Canada, 629*f*
from China, 676, 677*f*
dating of, 517–18
of dragonfly, 491*f*
of early chordates, 676, 677*f*
of early vertebrates, 678, 679*f*
Ediacaran, 628*f*
examples of, 493*f*
fungi, 612*f*
gnathostome, 680
human, 706*f*

  index, 491f, 517f
  insect, 660
  mass extinctions and, 518–21
  primate, 697
  Proterozoic, 526f
  in sedimentary rocks, 439, 440f, 492f
  stromatolites, 521
  transitional, 451f
  trilobite, 657f
Fossil fuel, 61
  acid precipitation caused by burning, 55–56
  increased atmospheric carbon dioxide due to combustion of, 1196f
Fossil record, **492**. *See also* Fossil(s)
  mass extinctions revealed in, 518–21
  evidence for natural selection in, 451
  phylogenies inferred from, 492
  in sedimentary rock, 439, 440f, 492f
Foundation species, 1169–70
Founder cells, 1004
Founder effect, **462**
Fovea, **1060**
FOXP2 gene, language ability in humans linked to, 706
F plasmid, **349**
  bacterial conjugation and, 349, 350f
Fragmentation, **781**
  asexual reproduction beginning with, **965**
  of ecosystems, 1220–21
Frameshift mutation, **329**
Franklin, Rosalind, discovery of DNA structure and contribution of, 296, 297f, 298
Free energy, **145**–48
  change in (ΔG), 145
  electron transport and changes in, 171f
  in exergonic and endergonic reactions, 146, 147f
  metabolism and, 146–48
  relationship of, to stability, work capacity, and spontaneous change, 146f
  stability (equilibrium) and, 145
Free energy of activation, **151**
  lowering barrier of, by enzymes, 152f
Free ribosomes, 102, 103f, 105, 325
Freeze-fracture method of cell preparation, 126, 126f
Frequency-dependent selection, **467**–68
Freshwater animals, osmoregulation in, 924–25
Freshwater biomes
  estuaries as, 1095f
  lakes as, 1093f, 1094f
  streams and rivers as, 1095f
  wetlands as, 1094f
Freshwater ecosystems, primary productivity of, 1187f, 1188–91
Fringe wetlands, 1094
Fritillaria assyriaca, 399
Frogs (Anura), 685
  aposematic coloration in poison arrow, 1162
  body axes in, 1004
  cadherins and formation of blastula in, 1003f
  cleavage in embryo of, 992, 993f
  cryptic coloration in canyon, 1162f
  fate map of embryo of, 1004f
  gastrulation in, 995, 996f
  metamorphosis in, 686f
  organogenesis in, 997f
  poisonous, 685
  reproduction in, 967f
  skin color, 685, 946
Frohlich, Michael, 601
Fructose, hydrolysis of sucrose to glucose and, 150, 151f
Fruit, **598**–99, 774, **778**
  development of, 778, 779f
  ethylene and ripening of, 801
  gibberellins and growth of, 797f
  seed dispersal and adaptations of, 599f
  variations in structures of, 599f
Fruit fly (*Drosophila* sp.)
  associative learning in, 1116
  axis establishment in development of, 423–25
  average heterozygosity of, 463
  biological clocks in, 1030
  body color and wing size, 277–78, 279f, 280, 281f
  chromosome number in, 102
  courtship and mating in, 436–37
  defense system of, 902
  diet and mate choice in, 1113f, 1114
  effect of bicoid gene in, 424f
  evolution of, on Hawaiian Islands, 436–37
  eye color, 276f, 277f, 280, 281f
  eye development, 411f
  foraging behavior in, 1120
  genetic analysis of early development in, 423
  genome, 394, 398, 399t, 401
  hierarchy of gene activity in development of, 425
  homeotic genes in, 425, 431f
  identity of body parts and homeotic genes in development of, 425
  interactions among proteins in cell of, 10f
  life cycle of, 421, 422f, 423
  logistic model of population growth and life history of, 1147
  as model organism, 412f
  T. Morgan's studies of, 276, 277f, 277–78, 279f
  natural selection and evolution of, 506
  radiation and mutations in, 329–30
  reproductive isolation of populations of, 477f
  segmentation pattern in development of, 425
  sex differences in, due to RNA splicing, 319
  sex-linked inheritance in, 276, 277f
  songs of, 1111–12
  zygote, 994
Functional groups, **63**–66
  ATP and, 66
  in female and male sex hormones, 63f
  important, in organic compounds, 64–65f
  RNA and, 327
  significance of, to life, 63
Fungi, 608–25
  ascomycetes (Ascomycota), 616–17, 622
  basidiomycetes (Basidiomycota), 618–19
  bioluminescent, 141f
  chytrids (Chytridiomycota), 613
  club, 618–19
  as decomposers, 620, 1186f
  ecological impact of, 620–22
  evolution of, 612
  as food and in food production, 623
  fossils, 612f
  generalized life cycle, 611f
  glomeromycetes (Glomeromycota), 615
  land colonization by, 527, 612
  lichens, 621–22
  nutrition of, 608–10
  as pathogens, 622–23
  phyla of, 612–19
  phylogeny of, 612, 613f
  reproduction of, 610, 611f, 614f, 617f, 619f
  sac fungi (Ascomycota),
  sexual life cycle in, 242f
  structure of, 609f
  zygomycetes (Zygomycota), 613–15
Funnel web spider, feeding and aggressive behavior in, 1119f, 1120
Fusiform initials, **727**

G_0 phase, molecular control of cell cycle and, **229**
G_1 phase of cell cycle, **221**f
G_2 phase of cell cycle, **221**f, 222f, 228
  molecular control of cell cycle at, 230f
G3P (glyceraldehyde-3-phosphate), **194**
Gage, Fred, 1038
Gage, Phineas, 1035
β-Galactosidase, 355f, 387
Galápagos ground finch (*Geospiza dificilis*), 478
Galápagos Islands
  adaptive radiation and evolution of finches on, 17, 18f, 443f, 450
  C. Darwin's studies of species on, 442–43
Gallbladder, **855**
Gametangia (gametangium), **577**f
Gametes, **219**, **239**, 241, **964**. *See also* Egg; Sperm
  animal, 968–69
  chromosome number in, 219
  crossing over and recombinant chromosomes in, 248, 249f
  fertilization and formation of, 241–42 (*see also* Fertilization)
  genetic variation in populations of, 459–60
  independent assortment of chromosomes and genetic variation in, 247, 248f
  isolation of, as reproductive barrier, 475f
  law of segregation and alleles for genetic character in separate, 254, 255f
  meiosis as cell division of, 220
  plant, 771–76 (*see also* Gametophyte)
  random fertilization and genetic variation in, 248
Game theory, **1127**
  applied to reciprocal altruism, 1130–31
  applied to survival and mating strategies, **1127**–28
Gametogenesis, **973**
Gametophore, **580**, 581f
Gametophyte, **242**, **576**
  of bryophytes, 580, 581f
  as haploid stage, 242
  reduced, in seed plants, 591–93, 772, 774–75
  relationship of sporophyte to, 592f
Gamma aminobutyric acid (GABA), **1025**
Ganglia (ganglion) cells, **1012**, 1061f, **1062**
Gannets, territoriality in, 1149f
Gap genes, 425
Gap junctions, 120, **121**f, 627
  electrical synapses containing, 1021
Garrod, Archibald, 309–10
Garstang, William, 675
Garter snake (*Thamnophis elegans*), 474f
  behavioral variation in selection of prey by, 1118, 1119f
Gas exchange in animals, **884**–95
  in arthropods, 657, 658
  bioenergetics and, 884f
  body size and shape and, 821–22
  breathing and, 888–90
  in elite athletes and deep-diving mammals, 894–95
  gills and, 884, 885f, 886f
  lungs and, 886–87, 888f
  oxygen uptake and carbon dioxide disposal, 867, 892–94
  partial pressure gradients and, 891
  in protists, 884
  respiratory pigments and gas transport in blood, 892–94
  respiratory surfaces for, 884–87
  tracheal systems in insects and, 886, 887f
Gastric gland, 857f
Gastric juice, **857**
Gastric ulcers, 858f
Gastropods (Gastropoda), 651–52
  torsion in, 651f
Gastrovascular cavity, **643**, **854**
  cnidarian, 643, 868f
  internal transport and circulation through invertebrate, 868
Gastrula, **627**, **994**
Gastrula organizer, 1006
Gastrulation, **627**, **994**–97
  in bird embryo, 995–96, 997f
  embryonic tissue layers formed by, 994
  in frog embryo, 995, 996f
  in human embryo, 1000f
  of sea urchin embryo, 995f
Gated channels, **133**
Gated ion channels, action potentials in nervous system and, **1017**–21, 1046
Gause, G. F., competitive exclusion principle and, 1160
Gecko lizard, van der Waals interactions and climbing ability of, 42

Geese, imprinting of graylag, 1108, 1109f
Gel electrophoresis, **392**
of restriction fragments, 392, 393f
Gene(s), 7, **86**, **239**. *See also* Chromosome(s); DNA (deoxyribonucleic acid)
activation of, by growth-factor, 213f
allele as alternate form of, 253–54, 255f (*see also* Alleles)
avirulence (*avr*) gene, 813, 814f
BRCA1 and BRCA2, and breast cancer, 374
cloning of (*see* Gene cloning)
coordinately controlled, 367
definition of, 330
determining function of, 400
duplications of, 378f, 380, 459
egg-polarity, 424
evidence for association of specific chromsosome with specific, 276–77
evolution of, controlling development, 484–86
exon duplication and exon shuffling in, 380
expression of (*see* Gene expression)
extending Mendelian genetics for single, 260–62
extending Mendelian genetics for two or more, 262–63
extranuclear, 289–90
FOXP2, 706
gap, 425
genetic recombination of unlinked and linked, 278–79, 280f
hedgehog, 499f
homeotic, 485
homologous, 505f
horizontal transfer of, in prokaryotes, 538, 541, 546
Hox (*see Hox* genes)
in human genome (*see* Human genome)
immunoglobulin, gene rearrangement in, 906f
jumping, 351–52, 371, 375–76
E. Lander on identifying, 236–37
linked, 277–82
location of, on chromosomes (*see* Gene locus)
maternal effect, 424
molecular homologies in, 494, 495f
multigene families, 377–78
mutations in (*see* Mutations)
nod, 766
with novel functions, 380
number of, in select genomes, 399t
organelle, inheritance of, 289–90
organ-identity, 430
organization of typical eukaryotic, 364, 365f
orthologous, 505f
pair-rule, 425
paralogous, 505f
point mutations in, 328–30, 459
R, 813, 814f
rearrangement of, in lymphocytes, 906–7
relationship of proteins to (*see also* Protein synthesis)
RNA splicing and split, 318f, 319
S-, and self-incompatibility in plants, 776
segmentation, 425
segment polarity, 425
sex-linked, 282–84
transcription factors and controls on, 205, 364–66
transfer of, in bacteria, 348–51
Gene clones, **385**
identification of, 388, 389f
libraries of, 388, 389f
Gene cloning, **385**
bacterial plasmids and production of, 385, 386, 387f, 388, 389f
overview of, 385f–86
recombinant DNA and, 385f, 386f
Gene expression, 309–33
cellular differentiation in plants and, 732, 733f
cloning and, in eukaryotic genes, 390–91
different cell types resulting from differential, 415–20
expression systems, in bacteria, 390
genetic code and, 312–14
genomics and study of interacting groups of genes, 400, 401f
flow of genetic information and, 309
identifying protein-coding genes in DNA sequences, 399
point mutations and, 328–30
in prokaryotes (bacteria), and metabolic adjustments, 352–56
in prokaryotes compared with eukaryotes, 327, 328f
regulation of, during development, 418–20
relationship between genes and proteins and, 309–10
roles of RNA in, 327
summary, 331f
translation, 311–12, 320–26
transcription, 311–12, 315–17
transcription in eukaryotic cells, 317–19
transcription in eukaryotic cells, regulation of, 362–70
viruses and, 339
Gene flow, **462**
lack of, 458
Gene-for-gene recognition, **813**–14
Gene locus, **239**
mapping of, using recombination data, 279–81
Gene pool, **455**
allele frequencies in, 455–56
of nonevolving population, 457–58
mutations and sexual recombination as cause of change in, 459–60
natural selection, genetic drift, and gene flow as cause of change in, 460–62
reproductive barriers and isolation of, 473, 474–75f
Generalized transduction, 348f, 349
Generative cell, 599
Gene regulation
in eukaryotic cells, 362–70
negative, 354–56
positive, 356
in prokaryotic cells, 352–56
Gene segments (*J*), 906, 907
Gene therapy, **403**–4
Genetic annealing, **525**
Genetically modified (GM) organisms, **407**–8
plants as, 784–86
Genetic code, 8, 312–14, 459
deciphering, 313–14
dictionary of, 314f
evolution of, 314
reading frame for, 314
redundancy in, 314
triplet code of, 312, 313f
Genetic disorders. *See also names of specific disorders*
cancer as, 305 (*see also* Cancer)
carriers of, 266
chromosomal alterations as cause of, 285–88
color blindness, 283, 1061
cystinuria, 134
diagnosis of, 269, 270f, 983–84
dominantly inherited, 267–68
exchange of (*see* Genetic recombination)
extranuclear genes and, 289–90
gene therapy for, 403–4
genetic drift, founder effect and, 462
genomic imprinting and, 289
genomics and, 401–2
growth disorders, 952
karyotyping for determining, 287f
Mendelian patterns of inheritance and, 266–68
metabolic, 309–10
multifactorial, 268
phenotype, recessive traits, and, 261
population genetics and, 458
relay proteins in cell signaling, and, 215
sex-linked, 283–84
testing and counseling for, 268–70
Genetic diversity. *See* Genetic variation
Genetic drift, **460**–62
bottleneck effect and, 461f
example of, in wildflower population, 461f
founder effect and, 462f
lack of, 458
Genetic engineering, **384**. *See also* DNA technology
in animals, 406, 416
in plants, 783–86
Genetic map, **279**
cytogenetic map as, 281, 396f
partial, of *Drosophila* chromosome, 281f
Genetic markers, RFLPs as, 394, 396f, 403f
Genetic polymorphisms, **463**
Genetic prospecting, 541
Genetic recombination, **278**. *See also* Recombinant DNA
in bacteria, gene transfer and, 348–51
in bacteria, mutations and, 346–48
in *Drosophila* fruit fly, 278–81
mutations and, 459–60 (*see also* Mutations)
produced by crossing over, 278, 279f, 280f
produced by independent assortment of chromosomes, 278
Genetics, **238**. *See also* Gene(s); Gene expression; Genome
of bacteria (prokaryotes), 334, 346–52
of behavior, 1109–13
chromosomes and (*see* Chromosomal basis of inheritance)
of development (*see* Development)
environmental factors combined with, effects on behavior, 1113–18
of eukaryotes (*see* Eukaryotic genome)
genetic variation (*see* Genetic variation)
inheritance of genes and, 239
E. Lander on research in, 236-37
meiosis, sexual life cycle, and, 240–47
Mendelian (see Mendelian inheritance)
molecular (*see* Molecular basis of inheritance)
of populations (*see* Population genetics)
sexual and asexual reproduction and, 239
of viruses, 334–46
Genetics problems
laws of probability used for solving, 259–60
Punnett square for solving, 254, 255f
Genetic testing
carrier recognition, 269
fetal testing, 269, 270f
newborn screening, 269–70
Genetic variation, 238, 247–49, 462–64
benefits of, to human welfare, 1211–12
biodiversity crisis and loss of, 1210
independent assortment of chromosomes as source of, 247, 248f, 278
crossing over as source of, 248, 249f
evolutionary significance of, in populations, 248–49, 459–60
measuring, 463
mutation as source of, 459–60
natural selection and role of, 462–64
between populations, 463–64
preservation of, by diploidy and balanced polymorphism, 466–68
random fertilization as source of, 248
sexual recombination as source of, 460
Genome, **8**, **219**. See also Gene(s)
analysis of, 392–94
of *Arabidopsis*, 729f
comparing, of different species, 399t, 400–402
determining gene function in, 400
eukaryotic (*see* Eukaryotic genome)
evolutionary history of organisms found in, 504–6
evolution of, 378–81, 505–6
future studies in, 402
human (*see* Human genome)
identifying protein-coding genes in DNA sequences, 399
mapping of, at DNA level, 394–98
prokaryotic (*see* Prokaryotic genome)

relationship between composition of, and organismal complexity, 374–75
size of, and estimated number of genes in select, 399*t*
studying expression of interacting gene groups, 400, 401*f*
viral, 335
Genome mapping, **279**, 394–98
cytogenetic map as, 281, 396*f*
DNA sequencing as, 396–98
linkage map, 279–81, 396
partial, of *Drosophila* chromosome, 281*f*
physical, 396
whole-genome shotgun approach to, 398*f*
Genomic equivalence, 415–18
animal stem cells and, 418
nuclear transplantation in animals and, 415–17
totipotency in plants and, 415*f*
Genomic imprinting, **288**–89, **364**
of mouse *Igf2* gene, 289*f*
Genomic library, **388**, 389*f*
Genomics, **398**–402
animal phylogeny based on molecular data and, 635*f*
comparing genomes, 399*t*, 400–402
determining gene function, 400
future directions in, 402
genome mapping, 394–98
identifying protein-coding genes in DNA sequences, 399
E. Lander on evolution and, 237
studying expression of gene groups, 400
Genotype, **256**
Hardy-Weinberg theorem and calculating frequencies of, 457–58
norm of reaction, 264, 1109
phenotype versus, 256*f*
Genus (taxonomy), **496**
Geographic barriers, speciation with and without, 476–82
Geographic distribution of species, 1083
abiotic factors and, 1086–87
behavior and habitat selection and, 1085
biotic factors and, 1085–86
climate and, 1087–92
dispersal and distribution factors, 1084–85
factors limiting, 1084*f*
Geographic variation, **463**–64
in mice populations, 463*f*
in yarrow plants, 464*f*
Geologic record, **518**, 519*t*
Geometric isomers, **62***f*
Geospiza fortis, 1161*f*
Geospiza fuliginosa, 1161*f*
Geospiza magnirostris (large ground finch), 443*f*
Geospiza scandens (cactus ground finch), 443*f*
Germination, 779, 780*f*
dormancy and, 779–80, 798, 799*f*
gibberellins and, 798*f*
light-induced, 803*f*
phytochromes and, 803–4
stages of, 780
two types of, 780*f*
Germ layers, **631**, **994**
adult derivatives of embryonic, 999*t*
gastrulation of zygote and formation of, 994–97
Gestation, **978**–79. *See also* Pregnancy
Ghost crabs (Ocypode), 664*f*, 829*f*
Ghrelin hormone, 847*f*
Giant panda, 160
Giardia intestinalis, 552, 553*f*
Gibberellins, **797**–98
fruit growth and, 797*f*
seed germination and, 798*f*
stem elongation and, 309, 797
Gibbons, 701*f*
Gibbs, J. Willard, 145
GIFT (gamete intrafallopian transfer), 984
Gigantism, 952
Gill(s), **884**, 885*f*
gas exchange and, 884–85
fish, 886*f*
invertebrate, 650*f*, 652, 885*f*
Gill circulation, **871**
Ginkgo biloba, 593, 594*f*
Giraffe, blood pressure in, 876
Giurfa, Martin, 1117
GLABRA-2 homeotic gene, 732, 733*f*
Glaciers, 588
ecological succession following retreat of, 1174*f*, 1175*f*
Glands. *See* Endocrine glands
Glandular epithelia, **823**
Glans penis, **972**
Gleason, H. A., 1179
Glia cells, **1014**–15
Global warming, 1079, 1204–5
Globigerina, 563*f*
Globin genes and proteins
evolution of α-globin and β-globin gene families, 379*f*
multigene families encoding of, 377*f*, 378
similarity of amino acid sequence in human, 379*f*
Glomeromycetes (Glomeromycota), **615**
Glomerulus, **931**, 932*f*
Glomus mosseae, 615*f*
Glucagon, **955**
maintenance of glucose homeostasis by insulin and, 955*f*, 956
Glucocorticoids, **958**
Glucose, 70
α and β ring forms of, 72, 73*f*
ATP yield from, in cellular respiration, 173*f*, 174
blood levels of, homeostatic regulation of, 846
cellular respiration and degradation of, 161, 162, 164f, 165–74
control of blood levels of, 846*f*, 955*f*, 956
conversion of carbon dioxide to, in Calvin cycle, 193–95
disaccharide synthesis from, 71*f*
gene regulation in bacterial metabolism of, 356*f*
hydrolysis of sucrose to fructose and, 150, 151*f*
linear and ring forms of, 71*f*
oxidation of, during glycolysis, 165–67
synthesis of, from muscle proteins, 958
transport of, in blood, 130
Glutamate
light reception, vision, and, 1061
receptors for, and long-term potentiation in vertebrate brain, 1036, 1037*f*
receptors for, and schizophrenia, 1039–40
taste reception and, 1055
Glutamic acid, 149
Glutamine, synthesis of, 149
Glyceraldehyde-3-phosphate (G3P), **194**
Glycerol phosphate, **65***f*
Glycine, **65***f*
Glycogen, **72**
blood glucose homeostasis and role of, 846
breakdown of, by epinephrine, 204–5, 213*f*, 946
storage of, in animal, 72*f*
Glycolipids, **129**
Glycolysis, **164**, 165–67
connection of, to other metabolic pathways, 176–78
energy input and output of, 165*f*
evolutionary significance of, 176
fermentation and, 174, 175
junction between citric acid cycle and, 168*f*
steps of, 166–67*f*
Glycoproteins, **105**, 119, **129**. *See also* Collagen
in viral envelopes, 336, 340
Glycosidic linkage, **70**, 71*f*
Glyoxysomes, 111
Gnathostomes, **679**–84
chondrichthyans as, 680–82
derived characters of, 679–80
early, 680*f*
evolution of vertebrate jaws and, 680*f*
fossil, 680
osteichthyans (ray-finned and lobe-finned fish) as, 682–84
tetrapods as, 684–86 (*see also* Tetrapods)
Gnetophytes (Gnetophyta), 593, 594*f*, 600
Gnetum, 594*f*
Gnom mutant, and axial polarity of plants, 732*f*
Goiter, 33*f*, 954
Golden algae (Chrysophyta), **560**
Golgi apparatus, **105**, 106*f*, 107
cis and trans face of, 105, 106*f*
cisternae of, 105, 106*f*
origins of, 525
Gonadotropin-releasing hormone (GnRH), 976, 977
Gonadotropins, **952**. *See also* Follicle-stimulating hormone (FSH); Luteinizing hormone (LH)
Gonads, 241, **968**
female, 958–59
hormones produced by, 949*t*, 958–59
male, 958
Goodman, Corey, 1037–38
Gorillas, 701*f*
Gorter, E., 125
Gould, Stephen Jay, 481
G protein, **206**
acetylcholine, cardiac muscle and, 1024
human olfaction and, 1056
inhibitory, 211
ras gene and Ras, 371, 372*f*
receptor proteins linked to, 206*f*
G-protein-linked receptor, **206***f*
cAMP as second messenger in pathway of, 210, 211*f*
Grackle, great-tailed, distribution of, 1084, 1085*f*
Grade, animal phylogenetic trees and, **630**
Graded potentials, **1018**
action potential and, in neurons, 1018*f*
Gradualism, **440**
Gradualism model for speciation tempo, 482*f*
Grafting, vegetative reproduction of plants by, 782
Graft versus host reaction, **916**
Graham, Linda, on evolution of land plants, 510–11
Gram, Hans Christian, 535
Gram-negative bacteria, **535***f*, 541
Gram-positive bacteria, **535***f*, 543*f*
Gram stain, **535***f*
Grana (granum), **110**, 111*f*
Grand Canyon
allopatric speciation at, 477*f*
fossils from strata of sedimentary rock in, 440*f*
Grand Teton National Park, Wyoming, biotic and legal boundaries of, 1222, 1223*f*
Granzymes, 911*f*
Grape, effect of gibberellin on, 797*f*
Grasshopper
anatomy of, 660*f*
digestive tract in, 854*f*
muscle and skeleton interaction in, 1066*f*
open circulatory system in, 869*f*
tracheal system, 887*f*
Grasslands
temperate, 1102*f*
wildfire in, 1099
Grave's disease, 953*f*
Gravitropism, **809**
Gravity
blood pressure and, 876
free energy and motion of, 146*f*
invertebrate detection of, 1050
plant responses to, 809*f*
Gray crescent, 1005
Gray matter, nervous system, **1026**
Greater prairie chicken (*Tympanuchus cupido*), extinction vortex and case study of, 1216
Great Lakes food web, biological magnification of PCBs in, 1202*f*
Green algae, 550*f*, **567**–69
colonial and multicellular, 568*f*
evolution of green plants from, 573–74

life cycle of *Chlamydomonas*, 569*f*
lichens and, 621–22
Greenhouse effect and global warming, **1204**–5
Green warbler finch (*Certhidea olivacea*), 443*f*
Green world hypothesis, **1193**–94
Grendel, F., 125
Grizzly bear
extinction vortex and case study of, 1217–18
nature reserves and, 1222, 1223*f*
Gross primary productivity (GPP), **1187**
Groth, James, 1149
Ground tissue system, **717**
Growth
allometric, 484*f*
cancer and faulty cell, 232–33
disorders of, in humans, 952
in plants (*see* Plant growth)
as property of life, 3*f*
Growth cone, **1038**
Growth factor, **231**, **948**
activation of specific gene by, 213*f*
bone morphogenetic protein BMP-4, 1006
as local regulator, 202–3
regulation of cell cycle and cell division by, 231*f*
Sonic hedgehog, 1008
Growth hormones (GH), **952**
Growth regulators, 203
GTP (guanosine triphosphate), energy for polypeptide synthesis provided by, 323
Guanine (G), 88
DNA structure and, 296*f*, 297*f*
Guanosine triphosphate. *See* GTP (guanosine triphosphate)
Guard cells, **724**
control of transpiration by opening and closing of stomata by, 750*f*, 751
Guppy (*Poecilia reticulata*)
differential predation and natural selection in populations of, 446, 447*f*
mate choice copying, 1131*f*
Gurdon, John, 416
Gustation, **1055**
Guttation, **746***f*
Gymnamoebas, **564***f*
Gymnosperms, **579**
evolution of, 596
four phyla of extant, 593, 594–95*f*
pine life cycle, 596, 597*f*
seed, 593, 596, 597*f*

Habitat
behavior and selection of, 1085
biodiversity crisis and loss/fragmentation of, 1212–13
corridors between, 1221*f*–22
differentiation of, and speciation of cichlids, 479, 480*f*
edges between, 1220–21
isolation of, as prezygotic reproductive barrier, 474*f*
of protists, 550
of red-cockaded woodpecker, 1218, 1219*f*
Habitat selection, species dispersal and, 1085
Habituation as learning, **1115**
Hacker, Sally, 1170
Hagfishes (Myxini), 641*f*, 676–77
Haikouella and *Haikouichthys* fossils, 676, 677*f*
Hair, mammalian, 694
Hair cell, mechanoreception by, **1046**, 1047*f*, 1052, 1053*f*
Haldane, J. B. S., 455, 513, 1130
Hales, Stephen, 756
Half-life of radioactive isotopes, **517**
Halophiles, extreme, **541**, 544*f*
Hamilton, William, 1129, 1130
Hamilton's rule, kin selection and, 1129, **1130**
Hantavirus, 344
Haploid cell, **241**. *See also* Gametes
reduction of chromosome from diploid cell to, 242, 243–47
Haploid-diploid system of sex determination, 282*f*
Hardy-Weinberg equation, 457
Hardy-Weinberg equilibrium, **457***f*, 458
conditions for, 458
Hardy-Weinberg theorem, **456**–58
Hare, population cycles of lynx and snowshoe, 1151, 1152*f*
Harper, John, 1148
Harris, William, 1116
Haustoria, **610**
Hawaiian Islands
evolution and adaptive radiation on, 436, 480, 481*f*
evolution of fruit fly *Drosophila* on, 436–37
Hawkmoth larva, mimicry in, 1162*f*
Health, human. *See also* Diseases and disorders; Genetic disorders
benefits of exercise to, 883
risks of anabolic steroids to, 959*f*
Health of organisms, population density and, 1149
Health science. *See* Medicine
Hearing
human ear, 1051*f*
in mammals, 1050–52
pitch distinguished by cochlea, 1052*f*
sound detection in invertebrates and, 1050
Heart, **868**, 871–74
cardiac cycle of, 872, 873*f*
cardiac muscle of, 1024
craniate, 676
maintaining rhythm of, 873, 874*f*
structure of mammalian, 872*f*
Heart attack, **883**
Heart murmur, **873**
Heart rate, **873**
Heartwood, **727**
Heat, **49**, **142**
activation energy supplied by, 151
exchanges of, between organisms and environment, 834, 835*f*
plant response to excessive, 812
protein denaturation caused by, 85
specific, 49
thermoregulation and (*see* Thermoregulation)
Heat of vaporization, **50**
Heat-shock proteins, **812**, **840**
Heavy chains, B cell receptors, **903**, 904*f*
Hedgehog gene, phylogram comparing, 499*f*
Hedrick, Ann, 1119–20
Heinrich, Bernd, 1117
Helicase, **303**, 304*t*
Helicobacter pylori, 542*f*, 858*f*
Helium, models of, 34*f*
Helper T cells, **909**
effect of HIV virus on, 918, 919*f*
interaction of, with MHC molecules, 905*f*
role of, in humoral and cell-mediated immunity, 909, 910*f*
Heme, 83
Heme group, 170
Hemichordates (acorn worm), 641*f*
Hemizygous organisms, 283
Hemocyanin, **892**
Hemocytes, 902
Hemoglobin, 170, **880**–81
comparison of amino acid sequence of, in humans and select animals, 89, 449, 450*f*
evolution of α- and β-globin gene families, 379*f*
heterozygote advantage in, 466
loading and unloading of oxygen by, 892*f*
mRNA degradation in, and eukaryotic gene expression, 368, 369*f*
multigene familes in α- and β-globin, 377f
oxygen dissociation curves for, 892*f*
point mutations in, 329*f*
quaternary structure of protein in, 83*f*
as respiratory pigment, 892
sickle-cell disease and, 84*f* (*see also* Sickle-cell disease)
Hemolymph, 657, **868**, 869*f*
Hemophilia, **284**, 403, **882**
Hemophilus influenzae, genome mapping of, 398
Hemorrhagic fevers, 344
Henry, Charles, 1112
Henslow, John, 441
Hepatic portal vein, **861**, 869–70
Hepatophyta (liverworts), 580, 582*f*
Herbaceous plants, **720**
Herbicide, auxin as, 795–96
Herbivores, **844**
digestion and dentition in, 863*f*
green world hypothesis regarding checks on, and, 1194
plant defenses against, 813, 1163
Herbivory, **1163**
Hercules scarab beetle (*Dynastes hercules*), 1080*f*
Heredity, **238**–39. *See also* Genetics; Inheritance
blending hypothesis of, 251
C. Darwin's theory of evolution and, 438, 443–51
DNA and (*see* Molecular basis of inheritance)
genes, chromosomes, and, 238–39, 274–77 (*see also* Chromosomal basis of inheritance; Mendelian inheritance)
information conveyed by, 7*f*, 27*f*
J. B. Lamarck's theory of, 440–41
Hermaphrodites, **642**, **966**
bluehead wrasse as, 966*f*
sequential, 966
sponges as, 642
Herpesvirus, 340
Hershey, Alfred, experiments on genetic material by, 294, 295*f*, 296
Hesiolyra bergi, 655*f*
Heterochromatin, **360**
Heterochrony, **484**–85
Heterocytes, nitrogen fixation by, 539*f*
Heterokaryon, 611
Heteromorphic generations, **562**
Heterosporous plants, **586**
among seed plants, 592
Heterotrophs, **181**, 538, 539*t*
fungi as, 608–10
prokaryotes as, 538–39
protists as, 550
Heterozygote advantage, **466**
Heterozygotes, **255**–56
average, 463
Hexapods (insects), **658**, 660–64
Hexose sugars, 70*f*
Hfr cell, 349, 350*f*
H^+ gradients and pump, 171, 172*f*, 173, 521–22, 739–40, 794
Hibernation, **840**
High-density lipoproteins (HDLs), **883**
High-throughput technology, **11**
Hindbrain, **1028**
Hinge joint, 1065*f*
Hippocampus, 1034, 1035*f*
Hippocrates, 605
Hirudinea (leeches), 655*f*, 1012*f*
Histamine, immune response and release of, **901**, 916–17
Histone(s), **360**
association of DNA and, in nucleosomes, 360, 361*f*
modifications of, as regulation of eukaryotic gene expression, 363*f*, 364
Histone acetylation, **363***f*, 364
Histone code hypothesis, 364
HIV (human immunodeficiency virus), 337, **341**, 344, **918**–19
detection of, 403
drug resistance in, 447, 448*f*
infection by, 918–19
molecular clock used to date origin of, 507
mutation rates of, 459–60
reproductive cycle, 342*f*
transmission, 919
Hoatzin, 863

Index

Hoiland, Wade, 1149
Holdfast, **560**, 561*f*
Holoblastic cleavage, **994**
Holothuroidea (sea cucumbers), 666*t*, 667*f*
Holsinger, Kent, 1112
Homeobox, **431**–32
Homeodomains, 431–32
Homeostasis, **832**
  animal nutrition and mechanisms of, 844–48
  in body temperature (thermoregulation), 833–41
  calcium, 954*f*, 955
  endocrine system, hormones, and, 953
  excretory system and maintenance of, 927–42
  as feedback system, 832–33
  of glucose blood levels, 846*f*, 955*f*, 956
  of glucose synthesis from proteins, 958
  kidney function, 936, 937*f*, 938
  osmoregulation (water balance/waste disposal), 922–27
  peripheral nervous system and maintenance of, 1028
  regulating versus conforming and, 832
Homeotherm, 833–34
Homeotic genes, **425**, **485**. *See also Hox* genes
  conservation of, in animals, 431*f*–32
  identity of body parts and, 425
  in plants, 430, 431*f*, 732*f*, 733*f*, 734, 735*f*
  spatial patterns of body parts and, 485–86
Hominids, **702**–3
  timeline for species of, 702*f*
Hominoids, **700**, 701*f*
Homo erectus, 706
Homo floresiensis, 706
Homo ergaster, 704, 705*f*
Homo habilis, 704
Homo heidelbergensis, 705
Homologous chromosomes, **240**
  nondisjunction of, 285*f*
  separation of, in meiosis, 243*f*, 244*f*, 247, 274
Homologous genes, two types of, 505*f*
Homologous structures, **448**
Homology, **448**
  analogy versus, 493–94, 504*f*
  anatomical, 448, 449*f*
  embryological, 448, 449*f*
  evidence for natural selection in, 448–49
  molecular, 449, 492–95
  phylogenies based on evidence of, 492–95
  tree of life and, 449, 450*f*
Homo neanderthalensis, 705
Homoplasies, **494**
  molecular, 495*f*
Homo sapiens, 703, 705–7. *See also* Human(s)
Homosporous plants, **586**
Homozygotes, **255**, 456
Homunculus, 987*f*
Honey mushrooms (*Armillaria ostoyae*), 608
Hooke, Robert, 95
Horizons, soil, **759**, 760*f*
Horizontal cells, 1061*f*, **1062**
Horizontal transmission of viruses in plants, 345
Hormonal control pathways, 944, 945*f*
Hormones, **203**, **791**, **943**–63
  animal (human) reproduction and regulation by, 973–78
  appetite-regulating, 847*f*
  binding of, to receptors, 945–48
  cell signaling by, 203*f*
  cell-surface receptors for water-soluble, 946
  control pathways of, and feedback loops, 944, 945*f*
  digestion and role of, 860*f*
  endocrine system and, 943–44, 948–59
  endocrine glands and secretions of, 949*t*
  insect, 960*f*
  insulin (*see* Insulin)
  intracellular receptors for lipid-soluble, 947
  invertebrate, 959–61
  kidney function regulated by, 936, 937*f*, 938
  mechanisms of signaling by, 946*f*
  nervous system chemical signals and, 943, 944
  nonpituitary, 953–59
  paracrine signaling and, 947–48
  pituitary, 948–52
  plant (*see* Plant hormones)
  sex (*see* Sex hormones)
  steroid, 205*f*, 947, 959*f*
Hornworts (Anthocerophyta), **580**, 582*f*
Horse, 604
  evolution of, 486, 487*f*
Horseshoe crabs (*Limulus polyphemus*), 658*f*
Horsetails, 587*f*, 588
Horvitz, H. Robert, 429
Host, **545**
  of parasites, **1163**
  symbiotic, 545
Host range of viruses, **337**
Hox genes, 629
  animal development and, 432*f*, 485–86, 627
  evolution of tetrapod limbs and, 485*f*
  expression of, in crustaceans and insects, 432*f*
  vertebrate brain and, 675*f*
  vertebrate origins and, 486*f*
HTLV (human T-cell leukemia virus), 370
Hubbard Brook Experimental Forest, New Hampshire
  acid precipitation and, 1199
  G. Likens on research studies at, 1078–79
  vegetation and nutrient cycling in, 1198, 1199*f*
Hudson Bay, limiting nutrient in salt marsh of, 1191*f*
Hughes, Kimberly, 1120
Human(s)
  biodiversity linked to welfare of, 1211–12
  chromosome number in (*see* Chromosome number in humans)
  circadian rhythms and biological clocks in, 1030–31
  cloning of, 417
  consciousness in, 1036–37
  derived characters of, 701–2
  diseases of (*see* Diseases and disorders)
  embryo (*see* Human embryo)
  emotions of, 1034–35
  energy budget for, 830, 831*f*
  environmental impact of (*see* Human environmental impact)
  evolution of (*see* Human evolution)
  genetic disorders of (*see* Genetic disorders)
  genome of chimpanzees and, 702
  family tree of tulips, mushrooms, and, 491*f*, 501*f*
  language and speech in, 706, 1011*f*, 1034
  life cycle of, 241*f*
  memory and learning in, 1035–36
  Mendelian inheritance patterns in, 265–70
  sex determination in, 282*f*
  sexes (*see* Females, human; Males, human)
Human body
  automatic control of breathing by, 890*f*
  basal metabolic rate (BMR) in, 829
  blood (*see* Blood groups, human)
  brain (*see* Human brain)
  circulatory and cardiovascular system, 871–83
  digestive system, 855*f*–62
  ear, and hearing, 1051*f*
  elements in, 33*t*
  endocrine system, 949t, 950t, 951–59 (*see also* Endocrine system)
  excretory system and kidney of, 931–38
  eye, and vision, 1058, 1059*f*, 1060*f*
  gas exchange in, 886–95
  genome (*see* Human genome)
  growth disorders in, 952
  growth rates in, 484*f*
  hemoglobin, 83*f*, 84*f* (*see also* Hemoglobin)
  hormones and development of, 952, 953 (*see also* Hormones)
  immune system, 899
  joints of, 1065*f*
  lymphatic system of, 878–79, 901*f*
  metabolic rate, 834
  muscles, 1066*f*–72
  nutrition (*see* Human nutrition)
  reproductive system (*see* Human reproduction)
  respiratory system, 888f, 890
  sex characteristics in, 283
  skeleton, 1065*f*
  stabilizing selection and infant size, 466
  taste and smell in, 1056*f*, 1057*f*
  thermoregulation in, 50, 839*f*
  water balance and osmoregulation in, 925f
Human brain
  adult, 1028*f*
  arousal and sleep in, 1029–30
  auditory and visual systems of, 1029
  brainstem, 1029–30
  cerebellum, 1030
  cerebral cortex, 1029, 1031, 1032–37
  cerebrum, 1031–32
  circadian rhythms and, 1030–31
  consciousness, 1036–37
  diencephalons, 1030–31
  embryonic development of, 1028*f*, 1029
  emotions and, 1034–35
  information processing in, 1032–33
  language and speech processing in, 706, 1011*f*, 1034
  lateralization of function in, 1033–34
  memory and learning in, 1035–36
  mechanism of long-term potentiation in, 1037*f*
  primary motor and somatosensory cortices of, 1033*f*
  thalamus and hypothalamus of, 1030
Human chorionic gonadotropin (HCG), **978**, 982
Human culture, 1131
Human embryo, 987*f*
  development of, 978, 979*f*, 980*f*, 987*f*, 991, 999, 1000*f*, 1001
  homologous structures in chick and, 449*f*
Human environmental impact, 1200–1206
  acid precipitation as, 55–56, 1201–2
  on aquatic biomes, 1200–1201
  climate change as, 1204–5
  disruption of chemical cycling as, 1200–1201
  disturbances in communities as, 1204
  human population growth and, 1155–56
  rising levels of atmospheric carbon dioxide as, 1203–4
  on terrestrial biomes, 1200
  toxin in environment, 1202–3
Human evolution, 701–7
  Australopiths and, 703, 704*f*
  bipedalism and, 703–4
  derived characters of humans, 701–2
  earliest hominids and, 702*f*, 703
  early *Homo* genus, 704–5
  Homo sapiens and, 705–7
  Neanderthals and, 705
  tool use and, 704
Human gene, cloning in bacterial plasmid, 387*f*, 388
Human genome, 359, 505
  E. Lander on, 237
  mapping, 394, 398 (*see also* Human Genome Project)
  number of genes in, 374
  types of DNA sequences in, 375*f*
Human Genome Project, 9*f*, 11, 319, **394**, 398
  E. Lander on, 236–37
Human growth hormone, DNA technology and, 404
Human health
  disease and (*see* Diseases and disorders)
  genetically modified organisms and, 784–85
  human population density, pathogens, and, 1149
Human immunodeficiency virus (HIV), **918**. *See also* HIV (human immunodeficiency virus)
Human life cycle, behavior of chromosome sets in, 241–42
Human nutrition
  caloric imbalance, 846–48
  digestive system and, 855–62
  nutritional deficiencies in, 33*f*, 34, 849–52
  plant biotechnology and, 784–86

Human population, 1136, 1152–56
age structure of, 1154–55
carrying capacity of Earth and, 1155–56
demographic transition and regional patterns of change in, 1153–54
effects of, on environment (*see* Human environmental impact)
global exponential growth of, 1152, 1153*f*
infant mortality and life expectancy in, 1155
T. Malthus on, 445
Human reproduction, 969–84
conception, embryonic development, and birth, 978–82
contraception and abortion, 982–83
female reproductive anatomy, 969–71
female reproductive cycle, hormonal regulation of, 973–77
hormone regulation and gametogenesis, 973, 974f, 975*f*, 976–78
male reproductive anatomy, 971–72
male reproductive system, hormonal regulation of, 977–78
pregnancy and fetal development, 915–16
sexual response, 972–73
spermatogenesis and oogenesis in, 974*f*, 975*f*
technologies applied to problems of, 984
Human skull
allometric growth in, 484*f*
comparison to chimpanzee, 484*f*, 494
Humoral immune response, 899*f*, 908, **909**, 910–14
antibody classes in, 912*f*, 913
antibody-mediated disposal of antigens, 913*f*, 914
B cell response in, 910, 911*f*, 912–14
overview of, 909*f*, 911*f*
role of helper T lymphocytes in, 909, 910*f*
T-dependent and T-independent antigens in, 912
Humus, soil, **759**, 760
Hundred Heartbeat Club, 1211
Hunger, plant biotechnology and reduction of human, 784
Huntington's disease, **268**, 403, 623
Hutchinson Cancer Research Center, 558
Hutton, James, 440
Hybridization, **253**
nucleic acid, 388, 389*f*
plant, 252–53
Hybrids
breakdown of, 475*f*
reduced viability and fertility of as reproductive barrier, 475*f*
Hydra sp., 639*f*
asexual reproduction in, 239*f*
cnidocyte of, 643*f*
digestion in, 854*f*
environmental exchange and body structure of, 821*f*
hormonal regulation in, 959–60
nervous system of, 1012*f*
Hydration shell, **51**
Hydrocarbons, **61**–62
fats and, 61*f*
Hydrochloric acid (HCl), 53, 54
Hydroelectric system, closed and open, 147*f*
Hydrogen
covalent bonding in, 39*f*, 40*f*
H^+ gradients and pump, 171, 172*f*, 173, 521–22, 739–40, 794
valence for, 60*f*
Hydrogen bond, **42***f*
RNA and, 327
between water molecules, 47, 48*f*
Hydrogen ion, **53**
Hydrolysis, **69**
of polymers, 69*f*
Hydrophilic substances, **52**
Hydrophobic interactions, **83**
Hydrophobic substances, **52**
Hydroponic culture, **757**
identifying essential plant nutrients using, 757*f*
Hydrostatic skeleton, **1063**–64
Hydrothermal vents, deep-sea, 514*f*, 541, **1097***f*
Hydroxide ion, **53**
Hydroxyl functional group, **64***f*
Hydrozoans (Hydrozoa), 644*t*, 645*f*
Hymen, **970**
Hypercholesterolemia, 137
Hyperpolarization, membrane potential and, **1018***f*
Hypersensitivity response (HR) in plants, **814**, 815*f*
Hypertension, **883**
Hyperthyroidism, 953
Hypertonic solution, **132**, 133*f*
Hyphae, fungal, **609***f*
adaptions for trapping and killing prey, 610*f*
specialized haustoria, 610*f*
Hypoblast, 994, 1000
Hypocotyl, **778**
Hypodermis, animal, 835*f*
Hypothalamus, 948, 949*t*, **950**, **1030**
neurosecretory cells of, 944
relationship between pituitary gland and, 950–51
secretion of thyroid hormones and role of, 953
thermoregulation and role of, 839*f*
Hypothesis, **20**
scientific inquiry and, 20, 21
Hypothyroidism, 954
Hypotonic solution, **132**, 133*f*
Hyracotherium, 486, 487*f*

I band, muscle, **1066**, 1067*f*
Ice
insulation of bodies of water by floating, 50–51
structure of, 51*f*
Iguana
desert (*Dipsosaurus dorsalis*), 939*f*
Galápagos, 438*f*
Imbibition, **780**
Immigration, population density and influence of, **1137**
Immune system, 343, 898–921
acquired immunity, 898, 903–14
active and passive immunization, 914
diseases caused by abnormal function of, 916–19
humoral and cell-mediated immunity, 908–14
innate immunity, 898–903
in invertebrates, 902–3
lymphocytes and function of (*see* Lymphocytes)
pregnancy and, 915–16, 982
self-nonself recognition by, 915–16
suppression of, by glucocorticoids, 958
Immunization, **914**
vaccines and, 343, 914
Immunodeficiency disease, 917–19
acquired (secondary), 917, 918–19
inborn (primary), 917, 918
HIV virus and AIDS as, 918–19
stress and, 918
Immunoglobulins (Igs), 880, **904**
classes of, 906*f*
gene rearrangement in, 906*f*, 907
IgA, 912*f*, 914
IgD, 912*f*
IgE, 912*f*, 916, 917*f*
IgG, 912*f*, 914
IgM, 912*f*, 913
Immunological memory, 903, 908*f*
Imperfect fungi, **612**
Imprinting, **1108**–9
of graylag geese, 1109*f*
sexual selection influenced by, in zebra finch, 1126*f*
of whooping crane mating and migration, 1109*f*
Imprinting stimulus, 1108
Inclusive fitness, altruism and, **1129**–31
Incomplete dominance, **260**, 261*f*
Incomplete flowers, **773**
Incomplete metamorphosis, **661**
Incus, **1051**
Independent assortment of alleles to gametes, 256–58, 274, 275*f*, 278
Indeterminate cleavage, **632**
Indeterminate growth, **720**
Index fossils, 517*f*
India, birth rate and population growth in, 1154*f*
Indian pipe (*Monotropa uniflora*), 768*f*
Indirect synaptic transmission, 1023–24
Individualistic hypothesis of community structure, **1179**
Individual variation, natural selection and, **16**
Indoleacetic acid (IAA), 793. *See also* Auxin
Induced fit, enzyme-substrate, **152**, 153*f*
Inducer, **355**
Inducible enzymes, 354, 355
lac operon and regulated synthesis of, 355*f*
Induction, **420**, **1003**
cell signaling and, in cell differentiation and development, 420, 421*f*, 426, 427*f*
developmental fate of cells and role of, 1003, 1006–8
Inductive reasoning, **20**
Infant, passive immunity in, 914
Infant mortality, **1155**
in Costa Rica, 1229*f*
Infection. *See* Diseases and conditions; Pathogens
Infection thread, 765
Inflammatory response, **901**, 948
major events of, 902*f*
Inflorescences, **773**
Influenza virus, 344
structure of, 336*f*
Information flow in cells
DNA → RNA → protein, 86*f*, 87
heritable, in cells, 7–8
Information processing in nervous system, 1013*f*
brain and, 1032–33
Ingestion (eating), **853**
Ingroup, **498**
Inheritance. *See also* Genetics
chromosomal basis of, 274–92
extranuclear genes and exceptions to standard, 289–90
of genes, 238–39
genome and (*see* Genome)
Mendelian, 251–73
molecular basis of, 293–308
sex-linked, 282–84
Inherited disorders. *See* Genetic disorders
Inhibitory G protein, 211
Inhibitory postsynaptic potential (IPSP), **1022**
Initials, plant cell, **721**
Initiation factors, polypeptide synthesis, 323
Injuries, brain, 1035, 1037–41
Innate behavior, **1110**–13
animal signals as, 1111–12
directed movements as, 1110–11
mating, parental behavior and, 1112–13
Innate immunity, **898**–903
acquired immunity versus, 899*f*
external defenses as, 899
internal cellular and chemical defenses, 899–902
in invertebrates, 902–3
Inner cell mass, **1000**
Inner ear, **1051**
equilibrium and, 1052, 1053*f*
hearing and, 1051*f*, 1052, 1053*f*
Inositol triphosphate (IP_3), role in signaling pathways, **212***f*
Inquiry, **19**
abiotic synthesis of organic compounds, 59*f*
bacterial acquisition of genes from another bacterium, 347f
body color and wing size in *Drosophila* (parental and recombinant types), 279*f*
cadherins and formation of blastula, 1003*f*
case study of scientific, 21–24
cell cycle regulation by cytoplasmic signals, 228*f*
cell migration and role of fibronectin, 1002*f*
competition and niches of barnacles, 1160*f*
determining reasons for decline of Illinois greater prairie chicken, 1216*f*

Index

development of differentiated plant cell into whole plant, 415*f*
development of floral organs in plants, 430*f*
diet and mate choice in *Drosphila* fruit fly, 1113*f*, 1114
digger wasp nest-locating behavior, 1115*f*
DNA replication models, 300*f*
dorsal lip of blastopore and cell fate, 1006*f*
egg fertilization and effect of sperm binding on Ca^{2+} distribution, 990*f*
eye color in *Drosophila* (wild and mutant phenotype), 277*f*
flower color in crosses of true-breeding plants, 253*f*
formation of organic molecules in reducing atmosphere, 513*f*
gene specification of enzymes in arginine biosynthesis, 311*f*
genetic component to geographic variation in yarrow plants, 464*f*
genetic control of lacewing courtship songs, 1112*f*
genetic material in bacteriophages, 295*f*
gray crescent distribution and, cellular potency, 1005*f*
hydroponic culture and essential plant nutrients, 757*f*
hypothesis-based scientific, 20–21
kinetochore microtubules during mitotic anaphase, 225*f*
limiting nutrient in Hudson Bay salt marsh, 1191*f*
moose population size on Isle Royale, Michigan, 1150*f*
movement in membrane proteins, 127*f*
nuclear transplantion and development of organisms, 416*f*
nutrients limiting phytoplankton production off Long Island shore, 1189*f*
parental caregiving by European kestrels and offspring survival, 1142*f*
phloem sap flow from source to sink, 753*f*
photoperiodism in plants and flowering, 807*f*
photosynthesis and light wavelengths, 187*f*
phototropism in plants, 792*f*, 793*f*, 803*f*
platelet-derived growth factor and stimulation of human fibroblast cells, 231*f*
polar transport of auxin, 795*f*
predation and natural selection in guppy populations, 447*f*
predation and mimicry of warning coloration in snakes, 23*f*
relationship of species richness to area, 1178*f*
on reproductive isolation in divergent *Drosophila* populations, 477*f*
role of fur in camel water conservation, 926*f*
scientific, as biological theme, 27*f*
seaweed distribution and feeding by sea urchins and limpets, 1086*f*
seed color and shape and independent assortment of alleles, 257*f*
transfer of genetic material between bacteria, 295*f*
triple response to stress in plants, 799*f*
Insects, 641*f*
ants (*see* Ants)
bees (*see* Bees)
butterflies (*see* Butterflies)
camouflage in, 446*f*
chemoreceptors in, 1048, 1049*f*
compound eyes in, 1058*f*
development in, hormonal regulation of, 960*f*
digestive tract in, 854*f*
as disease vectors, 554*f*, 555, 664
digger wasp nest-locating behavior, 1115*f*
ears of, and mechanoreceptors in, 1050*f*
flight in, 660–61
as fluid feeders, 845*f*
fossils, 660
fruit flies (*see* Fruit flies (*Drosophila* sp.))
gas exchange in tracheal system of, 886, 887*f*
genetic control of lacewing courtship song, 1112*f*
grasshopper, 660*f*, 663*f*
hormonal regulation of development in, 960*f*
Hox gene expression in, 432*f*
kinesis in, 1110*f*
major orders of, 662–63*f*
Malpighian tubules (excretory system) of, 930*f*, 931
metamorphosis in, 661*f*, 943*f*
nervous system, 1012*f*
parasitoid wasp, 813*f*
pheromones in, 437, 1048, 1409*f*
reproduction in, 661, 968, 969*f*
substrate feeding by, 845*f*
taste detection in, 1055*f*
thermoregulation in, 836, 837*f*, 838*f*, 839*f*
Insel, Thomas R., 1112–13
Insertions, nucleotide, as point mutation, **329**, 330*f*
Insertion sequences as transposable elements, **351**, 352*f*
Institute of Ecosystem Studies (IES), New York, 1078
Insulation
animal body, and thermoregulation, 834, 835*f*
of bodies of water by floating ice, 50, 51*f*
Insulin, 8, 105, **955**
amino acid sequence in, 80–81
diabetes mellitus and deficiency of, 956
DNA technology and synthesis of, 404
homeostasis of blood glucose and role of, 846*f*, 955*f*, 956
target tissues for, 956
Insulin-like growth factors, 289, **952**
Integral proteins, **128**
Integrated hypothesis of community structure, **1178**–79
Integration of sensory information, 1047–48
Integrins, **119**
Integumentary system, **834**–35. *See also* Skin
Integuments, seed plant, **592**, 593*f*
Interaction networks, biodiversity crisis and disruption of, 1214
Intercalated disks, **1072**
Intercellular junctions, 120
in animal tissue, 120, 121*f*
membrane proteins and, 128*f*
in plant tissue, 120*f*
Interdisciplinary research teams, **11**
Interferons (α and β), **900**
Intermediate disturbance hypothesis, **1172**
Intermediate filaments, **113**, 118
structure and function of, 113*t*
Internal environment, regulation of. *See* Animal regulatory systems
Internal fertilization, **967**
International Union for Conservation of Nature and Natural Resources, 1212
Interneurons, **1013**
axon growth direction, 1038*f*
structure of, 1014*f*
Internodes, **715**
Interoreceptors, **1046**
Interphase
cell cycle, **221***f*, 222–23*f*
meiosis and 244*f*
Intersexual selection, **468**, 1125
Interspecific competition, **1160**. *See also* Competition
Interspecific interactions in communities, **1159**–65
coevolution and, 1164–65
commensalism as, 1164
competition as, 1160–61
disruption of, 1214
dominant and keystone species and, 1165–71
herbivory as, 1163
mutualism as, 1164
parasitism as, 1163
pathogens and disease as, 1163
predation as, 1161–62
Interstitial fluid, **831**
capillary exchange between blood and, 875, 879*f*
lymphatic system and, 901*f*
Intertidal zone, 1093*f*, **1096***f*
keystone predator in, 1168*f*, 1169*f*
Intracellular digestion, **853**
Intracellular receptors, 205
for lipid-soluble hormones, 947
steroid hormone interaction with, 205*f*, 947
Intracellular recording for measurement of membrane potential, 1016*f*
Intrasexual selection, **468**, 1125
male competition for mates and, 1126–27
Intrinsic rate of increase in populations (r_{max}), 1144*f*
Introduced species, **1213**
as threat to biodiversity, 1213–14
Introns, **318**. *See also* Noncoding DNA sequences
crossing over facilitated by, 319
functional and evolutionary importance of, 319
Invagination, **995**
Invasive species, 437, **1168**
Inversion, chromosomal, **286***f*, 287
Invertebrates, **638**–68
animal phyla including, 668*f*
animal phylogeny and, 638*f*
annelids (Annelida), 653–55
arthropods (Arthropoda), 656–65
bilateral symmetry in, 646
cephalochordates (Cephalochordata), 674, 675*f*
chordates (Chordata) as, 667
circulatory system in, 868, 869*f*, 870
cnidarians (Cnidaria), 643–46
diversity of, 639–41*f*
echinoderms (Echinodermata), 665–66
eyes of, 1057*f*, 1058*f*
flatworms (Platyhelminthes), 646–48
gas exchange in, 884, 885*f*
gastrovascular cavities of, 868
gills of, 650*f*, 652, 885*f*
gravity and sound sensors in, 1050
immune system of, 902–3
insects (*see* Insect)
lophophorates (ectoprocts, phoronids, brachiopods), 649
molluscs (Mollusca), 650–53
muscles of, 1072
nematodes (Nematoda), 655–56
nemerteans (Nemertea), 649–50
nervous system, 1012*f*, 1020
osmoregulation in, 924
parental care among, 968*f*
regulatory systems of, 959–61
rotifers (Rotifera), 648–49
sponges (Porifera), 642–43
thermoregulation in, 833, 836, 837*f*, 838*f*, 839*f*
vision in, 1057*f*, 1058
In vitro fertilization (IVF), **984**
In vitro mutagenesis, **400**
Involution, **995**
Iodine deficiency, 33*f*, 34, 954
Ion(s), **41**
calcium, 211*f*, 212*f*
electrical charge and, 42
hydrogen, 53
as second messengers, 210–12
Ion channels, **133**
gated, in nervous system, 1017
ligand-gated, 208*f*, 1017, 1022–25
membrane potential and, 134–36
stretch-gated, 1017
voltage-gated, 1017
voltage-gated, role in neural action potential, 1018, 1019*f*
Ionic bonds, **41**–42
electron transfer and, 41*f*
Ionic compounds (salts), **41**
Ion pumps
membrane potential maintained by, 134–36
IP_3 (inositol triphosphate), role in signaling pathways, **212***f*
Iris, of the eye, **1058**, 1059*f*
Irish famine of 19th century, and potato blight, 558
Iron as limiting nutrient in ecosystems, 1189
Irrigation, 762

Index

Island
adaptive radiation on chains of, 436, 480, 481*f*
biogeography and species richness on, 1177*f*, 1178*f*
Island equilibrium model, 1177–78
Islets of Langerhans, **955**
Isoetes gunnii quillwort, 587*f*
Isoleucine, feedback inhibition in synthesis of, 157*f*
Isomers, **62**–63
three types of, 62*f*
Isomorphic generations, **562**
Isopods, **664**
Isotonic solution, **132**, 133*f*
Isotopes, **35**
half-life of radioactive, 517
radioactive, 35–36
Italy, population age structure of, 1154*f*
Iteroparity, **1141**
Ivanowsky, Dimitri, 335

Janzen, Daniel, 1223
Japan
infant mortality in, 1155
restoration project in coastal, 1228*f*
Jarvis, Erich, on genetics of songbird behaviors, 818–19
Jasmonic acid, **813**
Javan rhinoceros, 1211*f*
Jawfish, paternal care by, 1125*f*
Jawless vertebrates, 678–79
Jaws
evolution of mammalian, 695*f*
vertebrate, 679, 680*f*, 682–84
Jellies (jellyfish), 639*f*, 644*f*
internal transport in, 868*f*
Jenner, Edward, 914
Joints, 1065*f*
Joule (J), **49**
J-shaped curve of exponential population growth, 1144
Jukes, Thomas, 506
Juvenile hormone, **960**
Juxtaglomerular apparatus (JGA), **937**
Juxtamedullary nephrons, **931**, 936

Kamil, Alan, 467*f*, 1116
Kandel, Eric, 1036
Kaneshiro, Kenneth, 480
on evolution and mating behavior in *Drosophila*, 436–37
Kaneshiro Hypothesis, 437
Kangaroo (marsupials), 695
distribution and abundance of red, in Australia, 1081*f*
locomotion in, 1073*f*
Kangaroo rats, 925*f*, 939*f*
Karyogamy, **611**
Karyotype, **240**
geographic variation in mice populations, 463*f*
preparation of, 240*f*
of trisomy 21, 287*f*
Kelp (*Macrocystis*), 561*f*
Keratins, 688
Ketones, **64***f*
Ketoses, 70*f*
Keystone species, **1168**–69
disruption of interaction networks involving, 1214
Kidneys
adaptations of, to diverse habitats, 938, 939*f*
blood filtration and urine production by, 931–32, 933*f*, 934
hormonal regulation of, 936, 937*f*, 938
nephron and associated structures, anatomy and function, 931–34
urine concentration by (two-solute model), 935*f*
water conservation by, 934–38
Kilocalorie (kcal), **49**
Kinesis, **1110**
Kinetic energy, **49**, 50, **142**
transformations between potential energy and, 142*f*
Kinetochore, **221**, 224, 225*f*
Kinetoplast, 553
Kinetoplastids, **553**, 554*f*
King, Thomas, 416, 506
Kingdom(s) (taxonomy), **496**
five-kingdom system, 529*f*, 530
three-domain system and, 14*f*, 530–31*f*, 532
Kinorhyncha, 639*f*
Kin selection, **1130**
altruism and, in Belding's ground squirrel, 1130*f*
Hamilton's rule and, 1129–30
Kipukas, in Hawaii, 436
Kissimmee River, Florida, restoration project in, 1227*f*
Klinefelter syndrome, 287
Knee-jerk reflex, 1013*f*
KNOTTED-1 homeotic gene, 732*f*
Koala, 695
digestive tract of, 863*f*
Kohler, Steven, 1149
Kolbe, Hermann, 59
Komodo dragon, 689*f*
Korarchaeota, 544
Korpi, Nichole, 1116, 11117
Krebs, Han, 168
Krebs cycle. *See* Citric acid cycle
Krill, 664*f*
Kruger National Park, South Africa, elephant populations in, 1144*f*
K-selected (density-dependent) populations, **1147**
Kudzu as introduced species, 1213*f*
Kurosawa, Ewiti, 797
Kwashiorkor, 849*f*, 850

Labia majora, **970**
Labia minora, **970**
Labor, **981**
model for induction of, 981*f*
stages of childbirth and, 981*f*
LacA gene, 355*f*
Lacewing (*Chrysoperla* sp.), genetic control of courtship song in, 1112*f*
LacI gene, 355*f*
Lac operon
positive control of, by catabolite activator protein, 356*f*
regulated synthesis of inducible enzymes and, 354, 355*f*
α-Lactalbumin, 380
Lactate, 175
muscle fatigue and accumulation of, 1070–71
Lactation, 380, **981**
in marsupials, 695, 696*f*
Lacteal, **860**
Lactic acid fermentation, **175**
Lactose, 70
regulation of metabolism of, in bacteria, 354, 355*f*
LacY gene, 355*f*
LacZ gene, 355*f*, 387*f*, 388
Lagging strand, DNA replication, **302**f, 303f
proteins and functions related to, 304*t*
Lakes, 1094*f*
biomanipulation and restoration of, 1171
eutrophication of, 1190, 1201
evaporative cooling and temperature stability in, 50
experimental eutrophication of, 1190*f*
human impact on, 1201
oligotrophic, 1094
seasonal temperature and turnover in, 1090, 1091*f*
zonation in, 1093*f*, 1094
Lamarck, Jean Baptiste, evolutionary views of, 440–41
Laminaria sp., 561
life cycle of, 562*f*
Lampreys (Cephalaspidomorphi), 678*f*
Lamp shells (brachiopods), 640*f*, 649*f*
Lancelets (Cephalochordata), 498, 641*f*, **674**, 675f
Land. *See also* Landscape(s); Terrestrial biomes
colonization of, by animals 527
colonization of, by fungi, 527, 612
colonization of, by plants, 527, 573, 574, 575–79
locomotion on, 1073–74
osmoregulation in animals living on, 925–26
role of skeleton in physical support of animal body on, 1064–65
Lander, Eric, on genetics research, 236–37
Landmarks, **1115**
Landscape(s)
biophilia concept and, 1230
corridors connecting habitat fragments in, 1221–22
fragmentation and edges of, 1220–21
nature reserves as protected areas of, 1222–24
restoring degraded, 1224–28
sustainable development and conservation of, 1229–30
Landscape ecology, **1082**, **1220**
Language, human brain and, 706, 1011*f*, 1034*f*
Large ground finch (*Geospiza magnirostris*), 443*f*
Large intestine, **861**–62
Larva, **627**
amphibian, 686
insect, 1152*f*
trochophore, 635
tunicate, 673, 674*f*
Larynx, **887**
Lascaux, France, Paleolithic art, 1230*f*
Lateral geniculate nuclei, **1063**
Lateral inhibition, **1062**
Lateralization, brain, **1033**–34
Lateral line system, **680**, **1054**
Lateral meristems, **720**
secondary growth and, 720*f*, 721, 725–28
Lateral roots, **713**
Latitude, global climate and Earth's, 1088*f*
Law of continuity, 875
Law of independent assortment, 256, 257, **258**, 274, 275*f*
Law of segregation, 253–56, 274, 275*f*
statement of, **254**
L-dopa, 62, 63*f*, 1040
Leading strand, DNA replication, **302**f
proteins and functions related to, 304*t*
Leaf (leaves), **585**, **715**
abscission of, 800, 801*f*
anatomy of, 725*f*
apoptosis in, 800
in C_4 plants, 196*f*
effects of transpiration on temperature of, 749
evolution of, 585, 586*f*
extranuclear genes and color of, 289, 290*f*
green color of, 186*f*
guttation in, 746
modified, 716*f*
overexpression of homeotic gene in formation of, 732*f*
peroxisomes in, 111*f*
simple versus compound, 716*f*
as site of photosynthesis, 183*f* (*see also* Chloroplasts)
tissue organization of, 724, 725*f*
transpirational pull in, 747*f*
xerophyte, 751*f*
Leaf primordia, **723***f*
Leaf traces, **724**
Learning, **1114**–18
associative, 1116–17
of bird songs, 1118
cellular mechanisms of, 1036, 1037*f*
cognition, problem solving and, 1117–18
cognitive maps and, 1116
genetic and environmental interactions in, 1118
habituation as, 1115
imprinting as, 1108, 1109*f*
memory and, 1035–36
sensitization as simple, 1036*f*
spatial, 1115–16
Leeches (Hirudinea), 655*f*
nervous system, 1012*f*
Leeuwenhoek, Anton van, 549
Leg, size of animal, and skeletal support, 1065

Legal boundary versus biotic boundary of reserves, 1222, 1223*f*
Legionnaire's disease, 542*f*
Legumes, symbiotic nitrogen fixation between bacteria and, 764, 765*f*, 766
Lemurs, 468, 697*f*
Lens, of the eye, **1059**
  regulation of gene expression in cells of, 367*f*, 419
Lenski, Richard, 538
Lenticels, **728**
Leopold, Aldo, 1083
Lepidosaurs, **689**, 690, 691*f*
Lepidoptera (butterflies, moths). *See* Butterflies
Leptin, obesity and defective gene for, 847–48
Lettuce seed, response of, to red light, 803*f*
Leukemia, 288, 370, 404, 882
Leukocytes, 825*f*, 880*f*, **881**. *See also* White blood cells (leukocytes)
Lewis, Edward, 423, 425
Leydig cells, **971**
Lichens, **621**–22
Life, introduction to the study of, 2–29
  biology and study of, 2–8
  biological diversity of (*see* Biodiversity)
  cells as basic units of, 6–8
  chemical context of (*see* Chemistry)
  classification of, (*see also* Systematics; Taxonomy)
  continuum of, on Earth, 512
  ecosystems and, 6
  elements required by, 33–34
  evolution of, 15–19 (*see also* Evolution)
  extraterrestrial sources of, 514
  grouping into species and domains, 12–14
  order as characteristic of, 144*f*
  properties and processes of, 3*f*
  protobionts and origins of, 515
  scientific inquiry into, 19–26
  systems concept applied to, 9–12
  themes connecting biological study of, 26, 27*t*
  tree of (*see* Tree of life)
  unity underlying all, 14, 15*f*
Life cycle, **240**. *See also* Sexual life cycle
  of angiosperm, 600*f*, 772*f*
  of apicomplexan *Plasmodium* sp.,556*f*
  of ascomycete *Neurospora crassa*, 617*f*
  of basidiomycete, 619*f*
  of cellular slime mold, 566*f*
  of chlorophyte *Chlamydomonas*, 569*f*
  of *Drosophila* fruit fly, 421, 422*f*, 423
  of fern, 585*f*
  of fungi, 611*f*, 614*f*
  of hydrozoan *Obelia*, 645*f*
  of moss *Polytrichum*, 581*f*
  of pine (gymnosperm), 596, 597*f*
  of plants, 576, 581*f*, 585*f*, 597*f*, 600*f*
  of plasmodial slime mold, 565*f*
  of protist *Laminaria*, 562*f*
  of trematode blood fluke, 647*f*
  of water mold, 559*f*
Life expectancy, **1155**
  in Costa Rica, 1229
Life history, **1141**–43
  diversity of, 1141–42
  logistic population growth model and, 1147
  reproduction versus survival trade-offs in, 1142–43
Life history traits, natural selection and production of, 1141–43
Life tables, population, **1139**
  of Belding's ground squirrel, 1139*t*
Ligaments, **825***f*
Ligand, **137**, **205**
Ligand-gated ion channels, **208**, **1017**
  chemical synapse and, 1022f, 1022–25
  as membrane receptors, 208*f*
Light. *See also* Sunlight
  as abiotic factor, 1087
  animal breeding and, 1107
  animal hibernation and estivation related to, 840, 841
  characteristics of, 186
  circadian rhythms and, 1030, 1031*f*
  conversion of energy in, into chemical energy, 182–85
  energy levels of electrons and, 36–37
  as limiting factor for ecosystem productivity, 1188
  sensory receptors for (*see* Photoreceptors)
  vision and (*see* Vision)
Light, plant responses to, 788*f*, 802–8
  biological clocks and circadian rhythms associated with, 805
  blue-light photoreceptors and, 750, 802
  greening of plants in response to, 789*f*, 790*f*, 791
  harvesting of, in thylakoid membrane, 189–92
  interaction of chloroplasts with, 110–11
  opening/closing of leaf stomata and, 750
  photoperiodism as, 806–8
  phytochromes and reception of, 789–91, 802–5
  phototropism as, 792–93, 803*f*
  red light and, 803*f*, 806, 807*f*
Light chain, B cell receptors, **903**, 904*f*
Light-harvesting complexes, **189***f*, 190
Light microscopes and light micrographs (LMs), **95**
  types of, 96*f*
Light reactions, photosynthesis, **184**–85, 186–93, 198*f*
  chemiosmosis in chloroplasts, compared to mitochondria, 192–93
  cooperation between Calvin cycle and, 185*f*
  cyclic electron flow during, 191*f*–92
  excitation of chlorophyll by light and, 188, 189*f*
  mechanical analogy for, 191*f*
  nature of sunlight and, 186, 187*f*
  noncyclic electron flow during, 190*f*–91
  organization of thylakoid membrane and, 193*f*
  photosynthetic pigments as light receptors and, 186–88
  photosystems of, 189–90
Lignin, **584**
Likens, Gene, 1081, 1083, 1199
  on ecological studies of Hubbard Brook Experimental Forest, 1078–79
Lilac (*Syringa*) leaf, 725*f*
Limb
  homologous structures, 448, 449*f*
  Hox gene and evolution of tetrapod, 485*f*
  pattern formation during development of vertebrate, 1006, 1007*f*, 1008*f*
Limb bud, 1007*f*, 1008*f*
Limbic system, emotions and, **1034**, 1035*f*
Limiting nutrients, **1188**–90
Limnetic zone, lake, 1093*f*, **1094**
Limpet, seaweed distribution and feeding by, 1086*f*
Lindstedt, Stan, 894
Linkage map, **279**–81, **396**
  construction of, 281*f*
  of *Drosophila* genes, 281*f*
Linked genes, **277**–82
  evidence for inheritance and, in *Drosophila*, 277–78
  genetic recombination (crossing over) of, 278–79, 280*f*
  linkage map, 279–80, 281*f*
Linker DNA, 360
Linnaeus, Carolus, as founder of taxonomy, 439, 444
Lion, African (*Panthera leo*), hierarchical classification of, 496*f*
Lipase, 861*f*
Lipid(s), **74**–77
  fats, 75–76
  hydrophobic behavior of, 75
  phospholipids, 76–77
  steroids, 77
Lipid-soluble hormones, intracellular receptors for, 947
Littoral zone, lake, 1093*f*, **1094**
Lively, Curt, 1127–28
Liver, **855**
  blood circulation in, 861, 869–70
  digestive function, 858
  regulation of gene expression in cells of, 367*f*, 419
Liverworts (Hepatophyta), **580**, 582*f*
  gametangia of, 577*f*
Lizards (Squamata), 690
  game theory applied to mating success of side-blotched, 1127–28
  parthenogenic reproduction in, 965, 966*f*
  resource partitioning among *Anolis*, 1161*f*
Loams, **759**
Lobe-fins (Sarcopterygii), **683**–84
Lobotomy, frontal, 1035
Lobsters, 664
  anatomy of, 657*f*
Local regulators, chemical signals as, **202**, 203*f*, 947–48
Locomotion, **1073**–74. *See also* Movement
  energy costs of, 1074*f*
  flying, 1074 (*see also* Flight)
  on land, 1073–74
  muscles and, 1066–72
  pseudopodia and, 563
  skeleton, role of, 1064–65
  swimming, 821*f*
Locus, gene, **239**
  mapping of, using recombination data, 279–81
Logistic population growth model, **1145**–46
  applied to real populations, 1146, 1147*f*
  carrying capacity and, 1145
  hypothetical example of, 1145*t*
  life histories and, 1147
  logistic growth equation and, 1145
  population growth predicted by, 1146*f*
Long-day plants, **806**
Long-distance cell signaling, 203*f*
Long Island, New York, nutrients limiting phytoplankton production off shore of, 1189*f*
Long-term memory, **1035**
Long-term potentiation (LTP), **1036**, 1037*f*
Looped domains, DNA and chromatin packing in, 360, 361*f*
Loop of Henle, **931**, 932*f*
  descending and ascending limbs, 934
Loose connective tissue, **825***f*
Lophophorates, 649*f*
Lophophore, **635**
Lophotrochozoans, **634**, 635
  characteristics of, 635*f*
Lorenz, Konrad, 1108–9
Loriciferans, 640*f*
Low-density lipoproteins (LDLs), 137, **883**
Luciferase, 805
"Lucy" (*Australopithecus afarensis*), 703, 704*f*
Lugo, Ariel, 1226
Lungs, **886**
  breathing and ventilation of, 888–90
  evolution of, 682
  gas exchange and, 886–87, 888*f*
  mammalian, 887, 888*f*
Lupine inflorescence, 773*f*
Luteal phase of, ovarian cycle, **977**
Luteinizing hormone (LH), **952**, 959
  female reproductive cycle and, 976, 977
Lycophytes, **578**, 586, 587*f*
Lyell, Charles, 440, 443
Lyme disease, 543*f*, 545*f*, 546
Lymph, **878**
Lymphatic system, **878**
  defense against infection and, 878–79, 901
  human, 878, 879*f*, 901*f*
Lymph nodes, **878**, 901*f*
Lymphocytes, 880*f*, 881, **898**
  antigen recognition by, 903–5
  B cells, 903–4 (*see also* B lymphocytes (B cells))
  clonal selection of, 907*f*, 908
  development of, 905–8
  gene rearrangement in, 906*f*, 907
  self and nonself recognition and, 915–16
  T cell, 904–5 (*see also* T lymphocytes (T cells))
  testing and removal of self-reactive, 907

Lynx, population cycles of snowshoe hare and, 1151, 1152*f*
Lyon, Mary, 284
Lysergic acid (LSD), 622
Lysine, 783
Lysogenic cycle, **338**, 339*f*
Lysosomes, 102, **107**–8
digestion of materials by, 107*f*
Lysozyme, 380, **899**
conformation of, 81*f*
Lytic cycle, **337**, 338*f*

M, N, and MN blood groups, 260
MAC (membrane attack complex), **914**
MacArthur, Robert, on island biogeography, 1177–78
Macaques, 700*f*

Macroclimate, **1087**
Macroevolution, **472**, 482–88
evolution of genes controlling development, 484–85
origin of evolutionary novelty and, 482–83
speciation and (*see* Speciation)
Macromolecules, **68**–91
carbohydrates as, 69–74
classes of, 68
digestion and enzymatic hydrolysis of, 853
Golgi apparatus and production of, 106
lipids as, 74–77
nucleic acids as, 86–89
as polymers, 68–69
proteins as, 77–86
Macronutrients, plant, **757**, 758*t*
Macrophage, **823**, **900**
coordinated activity of, 121*f*
ingestion of yeast cell by, 898*f*
phagocytosis by, 107*f*
Mad cow disease, 345
Mads-box genes, 433
Magnetic field, reception of Earth's, 1049*f*
Magnetic resonance imaging (MRI), 1011*f*
Magnetic reversals, **518**
Magnification, of microscopes, 95
Magnolia (*Magnolia grandiflora*), 602*f*
Magnoliids, **602***f*
Maiden veil fungus (*Dictyphora*), 618*f*
Maize (*Zea mays*)
artificial selection and, 783*f*
germination of, 780*f*
B. McClintock's research on genetics of kernel color in, 375*f*
mineral deficiencies in, 33*f*, 759*f*
as monoecious species, 773*f*
parasitoid wasp defense of, 813*f*
phototropism in, 803*f*
root response to oxygen deprivation, 811*f*
seed, 778*f*
Major depression, **1040**
Major histocompatibility complex (MHC), **904**
rejection of tissue grafts and organ transplants and, 916
T cell interactions with, 904, 905*f*
Makhubu, Lydia, 32
on chemistry and medicine, 30–31
Malaria
heterozygote advantage in hemoglobin and, 466*f*
mosquitos as vector for, 555, 664, 1085
Plasmodium life history and, 555, 556*f*
sickle-cell disease and, 267, 466*f*
Males, human
contraception for, 982*f*, 983
erectile dysfunction in, 211, 948
hormonal regulation of reproductive system in, 977, 978*f*
reproductive anatomy of human, 971*f*
sex hormones of, 63*f*, 958–59, 977–78
sexual dimorphism and, 468
sexual response in, 948
spermatogenesis in, 975*f*
Malheur National Forest, Oregon, 608
Malignant tumor, **232**
Mallards, 693*f*
Malleus, **1051**
Malnourishment, **849**, 1156
diseases caused by, 849*f*
reducing human, 784
Malpighian tubules, **930***f*, 931
Malthus, Thomas, 445
Maltose, dehydration synthesis of, 70, 71*f*
Mammals (Mammalia), **694**–701
biological clocks in, 1030, 1031*f*
breathing in, 888–89
circulatory and cardiovascular system of, 870*f*, 871–83
countercurrent heat exchange in, 836*f*
digestive system of, 855–62
ear of, hearing and equilibrium in, 1050–53
early embryonic development in, 999–1001
eutherian (placental), 697–701, 967
evolution of, 694–95
excretory system and kidney of, 931–38
eye of, 1058–60
fertilization of eggs in, 990, 991*f*
hearing and equilibrium in, 1050–53
integumenary system, 835*f*
jaw and earbones of, 695*f*
kidney adaptations by, 939*f*
marsupials, 695, 696*f*, 697, 967
monotremes, 695, 696*f*
nitrogenous wastes of, 927*f*, 928
orders of, 699*f*
phylogenic relationships of, 698*f*
primates, 697, 700*f*, 701*f*
reproduction in, 967, 969–84
reproductive cloning of, 416, 417*f*
respiratory adaptations of endurance runners and deep-diving, 894–95
respiratory system and gas exchange in, 887–95
resting potential in neurons of, 1016*f*
taste and smell in, 1054–56
thermoregulation in, 834, 835*f*, 836*f*, 837, 839*f*
Mammary glands, **970**
lactation and, 380, 695, 696*f*, **981**
Mandibles, myriapod, **659**
Mangold, Hilde, 1006
Mangrove, 811
Mantid, camouflage in, 446*f*
Mantle, mollusc, **650**, 1064
Mantle cavity, mollusc, **650**
Map butterflies (*Araschnia levana*), 462*f*
Map units, **280**
Marchantia liverwort
archegonia and antheridia of, 577*f*
embryo and placental transfer cell of, 577*f*
sporophyte of, 582*f*
Margulis, Lynn, 522*f*
Marine animals
coral reefs, 1097
at deep-sea vent communities, 514*f*, 541, **1097***f*
invertebrates (*see* Invertebrates)
kidney adaptations in bony fishes, 939*f*
osmoregulation in, 923f, 924–25
respiratory adaptations of diving mammals, 894–95
salt excretion in birds, 926*f*
thermoregulation in, 834*f*, 835, 836*f*, 837*f*
Marine benthic zone, **1097**
Marine biomes
animals of (*see* Marine animals)
benthic zone, 1097*f*
coral reefs, 1097*f*
estuaries, 1095*f*
intertidal zone, 1095*f*
nature reserves in, 1224
oceanic pelagic biome, 1095*f*
zonation in, 1093*f*
Marine ecosystems
food chains in, 1166*f*
food web in, 1166*f*
habitat loss in, 1213
primary productivity in, 1187f, 1188–90
Marine worms, 640*f*, 649, 655*f*
gills, 885*f*
Markow, Therese, 1114
Mark-recapture method, estimating wildlife populations using, **1137**
Marler, Catherine, 1114
Mars, 514
Marsupials, **695**, 696*f*, 697
biogeography and evolution of, 450*f*, 528
convergent evolution of eutherians and, 450, 494*f*, 495
reproduction in, 967
Martínez, Lucía, 1112
Mass and weight, terms, 32
Mass extinctions, 518–21
Mass number, **34**–35
Mast cells, **901**
Mate choice. *See also* Mating; Sexual selection
copying, 1131*f*
diet and, in *Drosophila*, 1113*f*, 1114
by females, 1125–26
male competition for mates and, 1126–27
Mate choice copying, **1131**
Maternal effect gene, **424**
Mating. *See also* Sexual selection
of close relatives, and genetic disorders, 267
courtship behaviors and, in *Drosophila*, 436–37
genetic influences on, 1112–13
human sexual response and, 972–73
mate choice, 1123–25
populations with no random, 458
random, as source of genetic variation, 248
Mating bridge, conjugation, 349
Matsuzawa, Tetsuro, 1117–1118
Matter, **32**
chemical bonding and molecule formation, 39–42
chemical reactions and rearrangement of, 44–46
elements and compounds making up, 32–34
laws of thermodynamics and, 143–44
transformation of, by metabolic processes, 141–44
Maximum likelihood, **501**
Maximum parsimony, **501**
analogy-versus-homology pitfall and, 504*f*
applied to molecular systematics problem, 502–3*f*
Mayer, Alfred, 334–35
Mayr, Ernst, 455, 473
McClintock, Barbara, genetics research on transposons by, 375*f*
Meadowlark, similarity between two species of, 473*f*
Measles virus, 337
Mechanical isolation as prezygotic reproductive barrier, 474f
Mechanical stimuli, plant response to, 809–10
Mechanical stress, plant response to, 799–80
Mechanical work, cellular, 148
driven by ATP, 149*f*
Mechanoreceptors, **1048**, 1050–54
for equilibrium in mammals, 1046, 1047f, 1052–53
for equilibrium in nonmammals, 1053–54
for gravity and sound in invertebrates, 1050*f*
for hearing in mammals, 1050–52
for hearing in nonmammals, 1053–54
lateral line system as, 1054*f*
for muscle stretching, 1047*f*
for touch, 1048
Mediator protein, 365
Medicines
chitin used in, 74*f*
derived from plants, 594*f*, 595*f*, 605*t*
DNA technology applications in, 402–4
Hirudinea (leeches) used in, 655*f*
L. Makhubu on, 30–31
radioactive isotopes used in diagnostic, 36
stem cells and, 418

Mediterranean climate, 1090
Medulla, **1029**
control center for breathing in, 890*f*
Medulla oblongata, **1029**
functions controlled by, 1029
Medusa (cnidarian), **643**
Megapascals, **740**
Megaphylls, **585**, 586*f*
Megasporangium, 592, 593*f*
Megaspores, **586**, **775**
in seed plants, 592, 593*f*
Meiosis, **220**, 238, 241, **242**
alternation of fertilization and, in sexual life cycle, 242*f*, 243
in human life cycle, 241*f*
mitosis compared to, 246*f*, 247
nondisjunction of chromosomes in, 285*f*
overview of, 243*f*
reduction of chromosome number during, 242, 243–47
stages of, 243, 244–45*f*
Meiosis I, **243**, 244*f*
Meiosis II, **243**, 245*f*
Melanocyte-stimulating hormone (MSH), 946, **952**
Melatonin, **959**, 1030
Mello, Claudio, 819
Membranes, biological. *See also* Plasma membrane
extraembryonic, 688, 999
fluidity of, 126*f*
models of, 125
nuclear envelope as, 102, 103*f*
structure of, 99
transport across, 126–27
Membrane antibodies (membrane immunoglobulins), 904
Membrane attack complex (MAC), **914**
Membrane potentials, **135**, **739**, **1015**–17
basis of, 134–35
cation uptake and, in transport in plants, 740*f*
intracellular recording for measurement of, 1016*f*
maintenance of, by ion pumps, 134–36
resting, 1016–17
Membrane proteins
functions of, 127, 128*f*, 129
movement in, 127f
Memory
human brain, learning, and, 1035–36
immunological, 903, 908*f*
short-term and long-term, 1035
Memory cells, **907**
Mendel, Gregor Johann, 249, 251*f*–53, 293. *See also* Mendelian inheritance
Mendelian inheritance, 251–73
basis of, in chromosomal behavior, 274, 275*f*, 276–77
genetic variation preserved by, 456*f*
genetic testing and counseling based on, 268–70
genotype and phenotype relationship, 256*f*
in humans, 265–70
inheritance patterns more complex than explained by, 260–64
integrating, 264
law of independent assortment and, 256–58, 274, 275*f*
law of segregation, 253–56, 274, 275*f*
laws of probability governing, 258–60
Mendel's research approach, 251–53
modern synthesis of Darwinism and, 455
Menopause, **977**
Menstrual cycle, **973**
Menstrual flow phase, of menstrual cycle, **977**
Menstruation, **973**
Mercury, biological magnification of, 1203
Meristems, **720**
apical, and plant primary growth, 576*f*, 720*f*, 721–25
flower development and, 429*f*
lateral, and secondary growth, 720*f*, 725–28
transition of, and flowering, 808
Meristem identity genes, **734**
Meroblastic cleavage, **994**
Meselson, Matthew
models of DNA replication by F. Stahl and, 299–300
research on parthenogenic reproduction in rotifers, 649
Mesencephalon, 1028
Mesenchyme cells, 994–95
Mesenteries, **827**
Mesoderm, **631**, **994**
gastrulation and formation of, 994
organs and tissues formed from, 999*t*
Mesohyl, **642**
Mesophyll, **182**, **724**
Mesophyll cells, **196***f*
loading sucrose into phloem from, 752*f*
Mesozoic era, 519*t*, 520, 521*f*, 527–28, 596, 629
Messenger RNA (mRNA), 86*f*, 87, 102, **311**, 327*t*
codons of, 313, 314*f*
degradation of, and regulation of gene expression, 368, 369*f*
DNA library, 390
transcription of, 311, 315–17, 790–91
translation of polypeptides under direction of, 320–26
Metabolic defects, evidence for gene-directed protein synthesis from, 309–10
Metabolic pathways, **141**–42
Metabolic rate, **828**
activity and, 829–30
adjustment of, for thermoregulation, 838
animal bioenergetic strategies and, 829
animal body size and, 829
basal, 829
calculating, 829*f*
cardiovascular system and, 870
human, 834
maximum, over select time spans, 830*f*
standard, 829
Metabolism, **141**–59
allosteric regulation of enzymes and controls on, 156*f*, 157
anabolic pathways in, 142 (*see also* Protein synthesis)
ATP in, 148–50, 828 (*see also* ATP (adenosine triphosphate))
catabolic pathways in, 142 (*see also* Cellular respiration)
energy transformations (laws of thermodynamics), 143–44
enzymes and, 150–55
enzymes, regulation of, and controls on, 156–57
forms of energy and, 142–43
free energy and, 145–48
gene expression and adjustments of, in bacteria, 352–56
metabolic pathways in, 141–42
organelles and structural order in, 157*f*
in prokaryotes, 352–56, 539–40
protobionts and simple, 515
regulation of, 352–56
thermoregulation and changes in, 838
transformation of matter and energy in, 141–44
Metamorphosis, **627**
in frogs, 686*f*
in insects, 661*f*, 943*f*
Metanephridia, **930***f*
Metaphase
meiosis, 244*f*, 245*f*
mitosis, **221**, 223*f*, 224*f*, 226*f*
Metaphase plate, **224***f*
tetrads on, 247
Metapopulations, **1151**
Metastasis, **233**
Metencephalon, 1028
Methane
combustion of, as redox reaction, 161, 162*f*
covalent bonding in, 40*f*
as greenhouse gas, 540
molecular shape of, 43*f*, 60*f*
production of, by methanogens, 544
Methanogens, **544**
Methionine, 313, 323
Methylation, DNA, 284
genomic imprinting and, 289, 364
regulation of eukaryotic gene expression and, 364
Methylation, histone, 363
Mexico, demographic transition in, 1153*f*
MHC. *See* Major histocompatibility complex (MHC)
Microclimate, **1087**, 1091
Microevolution, **454**, **472**
gene flow as cause of, 462
genetic drift as cause of, 460–62
mutations as cause of, 459–60
natural selection as cause of, 460, 462–70
sexual recombination and, 460
Microfibrils in plant cellulose, 72, 73*f*
Microfilaments, **113***t*, 116–18
actin and, 116, 226, 1068*f*
cell motility and, 117*f*
cytoplasmic streaming and, 117*f*, 118
structural role of, 117*f*
structure and function, 113*t*
Micronutrients, plant, **757**, 758*t*
Microphylls, **585**, 586*f*
Micropyle, **599**
MicroRNA (miRNA), 327*t*, **368**
regulation of gene expression by, 369*f*
Microscopy, 94–97
Microsporangia, 774*f*, 775
Microspores, **586**, **774***f*–75
Microsporidia, 615*f*
Microtubule-organizing center, 221, 222*f*, 223*f*
Microtubules, **113***t*, 114–16
cell motility and, 112–18
centrosomes and centrioles, 114*f*
cilia and flagella and, 114, 115*f*, 116
kinetochore, 221, 224, 225*f*
of mitotic spindle, and chromosome movement, 224
9 + 2 pattern of, 114, 115*f*, 525
plant growth and cytoplasmic, 730, 731*f*
polar, 224
structure and function, 113*t*
Microvilli, **860**
Midbrain, **1028**, 1029
Middle ear, **1051**
Middle lamella, **118**, 119*f*
Migration
bird, 1106, 1109*f*, 1110–11
morphogenesis and cell, 1001–2
Migratory restlessness, 1110–11, 1121
Milk, 694
α-lactalbumin and production of mammalian, 380
passive immunity provided by, 914
prolactin and production of, 952, 981
Miller, Carlos O., 796
Miller, Stanley, 59*f*, 513*f*, 514
Millipedes (Diplopoda), 659*f*
Mimicry
H. Bates hypothesis regarding, 22
case study of, in snake populations, 21–24
Mimosa pudica (sensitive plant), rapid turgor movements in, 810*f*
Mimulus sp., 474*f*, 481
Minerals, **851**
absorption of, by plant roots, 744–46
availability of soil, to plants, 761–62
deficiencies of, in plants, 758, 759*f*
as essential nutritional requirement, 851, 852*t*
as plant nutrients, 756–59
Mineral nutrients, **756**
Mineralocorticoids, **958**
Minichromosomes, 340
Minimum viable population size (MVP), **1216**
Mining industry, use of prokaryotes in, 405, 546–47
miRNA (microRNA), 327*t*, **368**, **369***f*
Mismatch repair of DNA, **305**
Missense mutations, **328**, 329*f*

- Mistletoe (*Phoradendron*), 768f
- Mitchell, Charles, 1149
- Mites, 659f
- Mitochondria, **109**–11
  - chemical energy conversion in, 109–10
  - chemiosmosis in, versus in chloroplasts, 109, 173, 192–93
  - endosymbiosis and origins of, 523–25
  - as site of cellular respiration, 109, 110f, 164–65, 168–74
  - structure, 110f
- Mitochondrial DNA (mtDNA), human genetic disorders and, 290
- Mitochondrial matrix, **110**
  - ATP synthase as molecular mill in, 171f
  - citric acid cycle enzymes in, 169f
- Mitochondrial myopathy, 290
- Mitosis, **220**
  - in animal cells, 222–23f
  - chromosome duplication and distribution during, 219–20
  - cytokinesis and, 224–26
  - evolution of, 227f, 228
  - meiosis compared to, 246f, 247
  - mitotic spindle in, 221, 224f
  - phases of, 221, 222–23f
  - in plant cells, 226f
- Mitotic (M) phase of cell cycle, **221**f
- Mitotic spindle, **221**, 222f, 223f
- *Mixotricha paradoxa*, symbiotic relationships of, 525
- Mixotrophs, **550**
- Model-building in science, 24, **25**f
- Model organisms, study of development using, **412**–13f
- Model system, microbial, 334
- Modern synthesis of evolutionary theory, **455**
- Molarity, **53**
- Molar mass, 52
- Molds, **611**f
  - black bread mold (*Rhizopus stolonifer*), 614f
  - bread (*Neurospora crassa*), 616f, 617f
  - as pathogens, 622–23
- Mole (mol), **52**–53
- Mole(s), convergent evolution of analogous burrowing characteristics in Australian and North American, 494f, 495
- Molecular basis of inheritance, 293–308
  - DNA as genetic material, evidence for, 293–98
  - DNA proofreading and repair, 305–6
  - DNA replication, base pairing to template strands in, 299f, 300f
  - DNA replication, models of, 299f, 300f
  - DNA replication, proteins assisting in, 303, 304t
  - DNA replication, steps in, 300–305
  - DNA replication, summary of bacterial, 304f
  - DNA structural model, scientific process of building, 296–98
  - replicating ends of DNA molecules, 306–7
- Molecular clocks, **506**–7
  - applying, to origin of HIV, 507
  - difficulties with, 506–7
  - of multicellular eukaryotes, 526
  - neutral theory and, 506
  - of plants, 578
- Molecular formula, **40**
- Molecular geneaology, 89
- Molecular homologies, 449
- Molecular mass, **52**
- Molecular systematics, **491**
  - animal phylogeny based on, 635f
  - applying parsimony to problem of, 502–3f
  - prokaryotic phylogeny and lessons of, 540–41
- Molecules, 5f, 39–43. *See also* Macromolecules
  - amphipathic, 124
  - chemical bonds and formation of, 39–42
  - defined, **40**
  - shape and function of, 42–43
  - small, as second messengers, 210–12
- Mole rats, naked (*Heterocephalus glaber*), altruistic behavior in, 1128, 1129f
- Molluscs (Mollusca), 639f, 650–53
  - basic body plan, 650f
  - Bivalvia (clams, oysters), 652
  - Cephalopoda (cephalopods), 652–53f
  - exoskeleton of, 650, 1064
  - Gastropoda (snails, slugs), 651f–52
  - hormonal regulation in, 960
  - major classes of, 651t
  - nervous system of, 1012f
  - Polyplacophora (chitons), 651f
- Molting, **657**
  - nutritional requirements during, 850f
- Monkey flower (*Mimulus*), 474f, 481
- Monkeys
  - Old World, and New World, 697, 700f
  - social learning of alarm calls in vervet, 1131–32
- Monoclonal antibodies, **913**
- Monocots (Monocotyledones), **602**, 603f
  - dicots compared to, 603f
  - germination and cotyledons of, 780f
  - stem tissue in, 724f
- Monocytes, 880f, 881, **900**
- Monoecious plants, **773**
- Monogamous mating, **1123**, 1124f
- Monogeneans, 646t, 647
- Monohybrid, **256**–57
- Monohybrid cross, 256
  - multiplication and addition rules applied to, 258, 259f
- Monomers, **68**
  - amino acid, 78–80
  - nucleotide, 87f, 88
  - polymers built from, 68, 69f
- Mononucleosis, 370
- Monophyletic grouping, **498**f
- Monosaccharides, **70**–71. *See also* Glucose
  - disaccharide synthesis from, 70, 71f
  - structure and classification of, 70f
- Monosomic cells, **285**
- Monotremes, **695**, 696f
- Montreal Protocol, 1206
- Moose, stability and fluctuation of population size on Isle Royale, Michigan, 1150f
- Moray eel, fine-spotted (*Gymnothorax dovii*), 683f
- *Morchella esculenta*, 616f
- Morgan, Thomas Hunt
  - experimental evidence for chromosome-gene association from, 276–77, 293
  - experimental evidence for linked genes from, 277–78, 279f
- *Morganucodon*, jaw and ear bones of, 695f
- Morphogen(s), **424**
- Morphogenesis, development and, **413**, **987**
  - in animals, 987, 1001–3
  - in plants, **728**–35
- Morphological species concept, **476**
- Morphology
  - animal phyogeny based on, 634f
  - phylogeny and evidence of, 492–93
  - plant, **712**
- Morton, Martin, 1138, 1139
- Morula, **992**
- Mosaic evolution, **703**
- Mosquitos
  - as disease vector, 555, 556f, 664, 1085, 1163
  - as fluid feeders, 845f
  - habitat and distribution of, 1085
  - mechanoreceptors in, 1050
- Mosses (Bryophyta), **580**, 582f, 586, 587f
  - life cycle of *Polytrichum*, 581f
  - sporophytes of, 577f
- Moths
  - bat predation of, 1045f
  - chemoreceptors in, 1049f
  - predator-prey relationship with blue jays, 467f
  - sphinx, 820f
  - thermoregulation in, 837f, 839f
- Motility. *See* Movement
- Motor mechanisms, 1063–74
- Motor neurons, **1013**f
  - axon growth direction, 1038f
  - muscle contraction stimulated by, 1069–71
  - recruitment of, 1070
  - structure of, 1014f
- Motor output, nervous system function and, 1013f. *See also* Locomotion; Movement
- Motor proteins, 78t, **112**f, 114
  - mechanical work in cells and, 149f
- Motor unit of vertebrate muscle, **1070**, 1071f
- Mountain, effects of, on climate, 1090f
- Mountain lion (*Puma concolor*), 1123
- Mouse (*Mus musculus*)
  - apoptosis and paw formation in, 428, 429f
  - effect of social environment on aggressive behavior in California, 1114t
  - energy budget for deer, 830, 831f
  - genome, 394, 401
  - genomic imprinting of *Igf2* gene, 289f
  - geographic variation of chromosomal mutations in, 463f, 464
  - homeotic genes in, 431f
  - lack of V_{1a} receptor in male wild-type, 1113
  - as model organism, 413f
  - obesity in, 848f
- Movement. *See also* Locomotion
  - cell crawling during morphogenesis, 1001–2
  - cell motility, 112–18
  - in plants, 810f
  - prokaryotic cell, 536f, 537
  - in protists, 117f, 118
- Movement corridor, **1221**
- MPF (maturation promoting factor), **230**f
- Mucin, 856
- Mucosa tissue layer, of stomach, **827**f
- Mucous membranes, **823**
  - as body defense mechanism, 899f
- Mucus, 899
- Mule, as hybrid between horse and donkey, 475f
- Mule deer (*Odocoileus hemionus*), foraging and risk of predation, 1123f
- Muller, Hermann, 329
- Müllerian mimicry, **1162**
- Multicellularity, origins of, 525–28
- Multifactorial characters, **264**
  - as genetic disorders, 268
- Multigene family, **377**–78
- Multiple alleles, 262
- Multiple fruit, **779**
- Multiple sclerosis, 917, 1015
- Multiplication rule and Mendelian inheritance, 258, 259f
- Muscle(s)
  - cardiac, **826**f, 1024, **1072**
  - contraction (*see* Muscle contraction)
  - effects of anabolic steroids on, 959f
  - fatigue in, and accumulation of lactate, 1070–71
  - interaction between skeleton and, 1066f
  - locomotion and contraction of, 1066f
  - neural control of tension in, 1069–71
  - skeletal, 1066f–72
  - smooth, 1072
  - stretch receptors for, 1047f
  - structure, 1066, 1067f
- Muscle cell, 130, 627
  - determination and cell differentiation of, 419f, 420
  - microfilaments and motililty in, 116, 117f
  - lactic acid fermentation in, 175
- Muscle contraction
  - myosin and actin interactions and, 116, 117f, 1068f
  - neural controls and, 1069–71
  - role of calcium and regulatory proteins in, 1068, 1069f
  - role of sarcoplasmic reticulum and T tubules in, 1069f
  - sliding-filament model of, 1067f, 1068f
- Muscle fiber
  - contraction of (*see* Muscle contraction)
  - fast, and slow, 1071, 1072t

neural control of tension in, 1069–71
rate of stimulation in, 1071
relaxed and contracted, 1067*f*
types of, 1071, 1072*t*
Muscle spindles, **1048**
Muscle tissue, **823**
Muscularis tissue layer, of stomach, **827***f*
Mushrooms, 608, 618–19
Mustard, vegetables artificially selected from wild, 445*f*
Mutagens, **329**–30
Mutant phenotype, 276
fruit fly eye color, 277*f*
Mutations, **328**, **459**
in bacteria, 344–48
base-pair substitutions as, 328, 329*f*
cancer and, 370
as cause of microevolution in populations, 459–60
chromosomal, 285–88, 463–64
deletions, 329, 330*f*
duplications, 459
as embryonic lethals, 423
frameshifts, 329
gene number and sequence alterations as, 459
genetic variation in populations caused by, 459–60
genomic evolution and role of, 378–81
insertions, 329, 330*f*
missense, 328, 329*f*
nonsense, 328, 329*f*
point, 328–30
populations lacking, 458
proto-oncogenes transformed into oncogenes by, 371*f*
rates of, 459–60
translocations, **286***f*, 287, 288*f*, 459
Mutualism, **545**, **1164**
roots and mycorrhizal fungi, 620
Myasthenia gravis, 1069
Mycelium, fungal, **609**
Mycetozoans (slime molds), 564–66
Mycoplasmas, 98, 543*f*
Mycorrhizae, **610**, **744**, **766**–67
agricultural importance of, 620
arbuscular, 615*f*
symbiotic relationship of root nodules to, 620
two types of, 766, 767*f*
water absorption in plants and role of, 744, 745*f*
Mycosis, **622**
Myelencephalon, 1028
Myelin sheath, **1014**
conduction of action potentials and, 1020–21
multiple sclerosis and destruction of, 917
Schwann cells and, 1015*f*
Myoblasts, 419*f*, 420
MyoD gene, 419*f*, 420
Myofibrils, **1066**, 1067*f*
Myofilaments, **1066**, 1067*f*
Myogenic heart, **873**
Myoglobin, **895**, **1071**
in oxidative muscle fibers, 1071, 1072*f*
Myosin, **117**
cell motility, actin filaments and, 117*f*
cytokinesis and, 226
muscle contraction and, 117*f*, 1068*f*, 1069*f*
Myotonia, **972**
Myriapods (Myriapoda), **658***t*, 659*f*–60*f*

NAD+ (nicotinamide adenine dinucleotide), **162**
fermentation and, 174–75
as oxidizing agent and electron shuttle during cellular respiration, 162, 163*f*, 170, 175
NADH
citric acid cycle and, 168*f*, 169*f*, 170
fermentation and, 174, 175*f*
glycolysis and, 165*f*, 166*f*
as source of electrons for electron transport chain, 170, 171*f*
$NADP^+$ (nicotinamide adenine dinucleotide phosphate), photosynthetic light reactions and role of, *184*, 185*f*, 193*f*
NADPH
Calvin cycle and conversion of carbon dioxide to sugar using, 193–95
photosynthetic light reactions, and conversion of solar energy to, 185, 190, 191, 193*f*
Naked mole rat (*Heterocephalus glaber*), altruism in, 1128, 1129*f*
Nash, John, 1127
National parks
in Costa Rica, 1223*f*, 1224
Grand Canyon, 440*f*, 477*f*
Grand Teton, 1222, 1223*f*
Yellowstone, 544, 1172, 1173*f*, 1217*f*, 1218, 1220*f*, 1222, 1223*f*
Native Americans, genome of, 505
Natural family planning, **982**
Natural killer (NK) cells, **902**, 910
Natural selection, 16, 17*f*, **438**, 444–46, 460, 462–70
artificial selection versus, 445
behavioral traits evolving from, 1118–28
C. Darwin's theory of, 444–46
directional, disruptive, and stabilizing modes of, 465*f*–66
effects of differential predation on, 446, 447*f*
evolutionary fitness and, 464–65
evolution in populations due to, 460, 462–70
evolution of drug-resistant HIV as example of, 447, 448*f*
evolution of sex and sexual reproduction, 469
in finch populations, 443*f*
frequency-dependent, 467–68
genetic variation and, 462–64
perfection not created by, 469–70
populations with no, 458
preservation of genetic variation despite effects of, 466–68
RNA, protobionts, and dawn of, 515–16
sexual selection and, 468
summary of, 445–46
survival/reproductive success increased by behaviors favored by, 1121–28
Nature reserves, 1222–24
philosophy of, 1222–23
zoned, 1223–24
Nealson, Kenneth, 522*f*
Neanderthals, 705
Negative feedback, **11***f*, **832**, **944**
kidney regulation by, 937*f*
as mechanism of homeostasis, 832*f*
Negative pressure breathing, **888**, 889*f*
Nematocysts, **644**
Nematodes (roundworms), 640f, 655–56. *See also* *Caenorhabditis elegans* (nematode)
hydrostatic skeleton in, 1063
as model organism, 412*f*
as pseudocoelomate, 631*f*
Trichinella spiralis, 656*f*
Nemerteans (proboscis worms/ribbon worms), 639*f*, 649*f*, 650
Neocortex, **1031**, 1035
Neon, electron orbitals of, 38*f*
Neoproterozoic era, 628
Nephron, **931**, 932*f*
blood vessels associated with, 933
production of urine from blood filtrate in, 933*f*, 934
regional functions of transport epithelium and, 933*f*
structure and function of, 931–33
Neritic zone, 1093*f*, **1097**
Nernst equation, 1016–17
Nerve(s), 627, **1012**. *See also* Neuron(s)
cranial, 1026*f*
development of, 1037–38
herpesvirus in, 340
spinal, 1026*f*
Nerve cord, dorsal and hollow, in chordates, 673
Nerve net, **1012**
Nervous system, 1011–44
action potentials in, production and conduction, 1017–21
adrenal gland and, 957
central nervous system and brain, 1012, 1026f, 1028–37
central nervous system injuries and diseases, 1037–41
as command and control center, 1011–12
endocrine system overlap with, 943–44
human brain (*see* Human brain)
information processing by, 1013*f*
invertebrate, 1012*f*
maintenance of resting potential in neurons by ion pumps and ion channels of, 1015–17
membrane potentials in, 1015
neuron communication at synapses in, 1021–25
neuron structure in, 1013, 1014*f*
nerve signals (impulses) in, 203
organization of, 1012*f*, 1013
peripheral nervous system, 1012, 1026*f*, 1027*f*, 1028
senses and (*see* Sensory reception)
supporting cells (glia) of, 1014–15
vertebrate, 1026–32
Nervous tissue in animals, **823**. *See also* Nervous system; Neuron(s)
Nest, digger wasps and behaviors for locating, 1115*f*
Net primary productivity (NPP), **1187**
idealized pyramid of, 1192*f*
Neural crest, **676**, **998**
as embryonic source of vertebrate characters, 676*f*
Neural stem cells, 1038–39
Neural tube, **998**
Neuroendocrine pathway, 945*f*
Neurofibrillary tangles, 1040*f*
Neurogenic heart, **873**
Neurohormone, 944
Neurohormone pathway, 945*f*
Neurohypophysis, **950**
Neuromasts, 1054
Neurons, **826***f*, 943–44, **1011**
action potentials of, 1017–21
development of, 1037–38
gated ion channels of, 1017
interneurons, 1013*f*, 1014*f*
ion pumps/ion channels and, 1015–17
membrane potential of, 1015–16
motor, 1013*f*, 1014*f* (*see also* Motor neurons)
production of new, in adults, 1038–39
resting potential of, 1016–17
sensory, 1013, 1014*f*
structure of, 1013, 1014*f*
supporting cells for, 1014–15
synapses and, 1021–25
Neuropeptides as neurotransmitters, **1025**
Neurosecretory cells, **944**
Neurospora crassa (bread mold)
G. Beadle and E. Tatum experiment on gene-enzyme relationships in, 310, 311*f*
life cycle of, 616, 617*f*
Neurotransmitters, **1014**, 1024–25
acetylcholine as, 1024*t*
amino acids and peptides as, 1024*t*, 1025
biogenic amines as, 1024*t*, 1025
as intracellular messenger at neural synapses, 1022, 1024–25
as local regulators, 947–48
major, 1024*t*
migration of vesicles containing, 112*f*
Neutral theory, **506**
Neutral variations in populations, **468**
Neutrons, **34**
mass number and, 34–35
Neutrophils, 880*f*, 881, **900**
Newborns, screening for genetic disorders, 269–70
New World monkeys, 697, 700*f*
New Zealand, 690
Nicolson, G., 125
Nicotine, 1024, 1163
Night, plant photoperiodism and length of, 806, 807*f*

Index

9 + 2 structural microtubule pattern in flagella and cilia, 114, 115*f*, 525
Nirenberg, Marshall, 313
Nitric oxide (NO), 205, **948**
effects on male sexual function, 948
as local regulator, 948
as neurotransmitter, 205, 1025
Nitrogen
improving protein yield in crops and, 764
as limiting nutrient in ecosystems, 1188–89
plant requirement for, and deficiencies of, 33*f*, 759*f*, 761
prokaryotic metabolism and, 539
soil bacteria and fixing of, for plant use, 763*f*, 764–66
valences of, 60*f*
Nitrogenase, **764**
Nitrogen cycle in ecosystems, 1197*f*
critical load and, 1200–1201
human disturbance of, 1200–1201
Nitrogen fixation, **539**, **764**
by bacteria, 539, 763f, 764–66
Nitrogen-fixing bacteria, **764**
Nitrogenous bases
base-pair insertions and deletions, 329, 330*f*
base-pair substitutions, 328, 329*f*
Chargaff's rules on pairing of, 296, 298, 313*f*
genetic code and triplets of, 312–14
nucleic acid structure and, 87*f*, 88
pairing of, and DNA replication, 87–88
pairing of, and DNA structure, 87*f*, 296, 297*f*, 298*f*
pyrimidine and purine families of, 88, 298
sequences (*see* DNA sequences)
structure of DNA strand and, 296*f*
wobble in pairing of, 322
Nitrogenous wastes, animal phylogeny correlated with form of, 927–28
Nitrosomonas, 542*f*
NMDA receptors, 1036, 1037*f*
Nociceptors, **1049**
Nodes, of stem, **715**
Nodes of Ranvier, 1015*f*, 1021*f*
Nod genes, 766
Nodules, root, **764**, 765*f*
molecular biology of formation of, 766
Noncoding DNA sequences, 374–78
restriction fragment length polymorphisms and, 394
Noncompetitive enzyme inhibitors, **155*f***
Noncyclic electron flow, photosynthesis, **190***f*, 191
Nondisjunction of chromosomes, **285**
genetic disorders associated with, 285*f*, 286
Nonequilibrium model of communities, **1172**
Nonhomologous chromosomes, independent assortment of, 258, 274, 275*f*
Nonpolar covalent bond, **40**
Nonsense mutations, **328**, 329*f*
Nonshivering thermogenesis (NST), **838**
Nontemplate DNA strand, 313
Nontropic hormones, 952
Norepinephrine, **956**, **1025**
stress and secretion of, 957*f*
Norm of reaction, **264**
behavior and, 1109
North America
acid precipitation in, 55–56, 1201*f*
climograph for terrestrial biomes of, 1098*f*
plate tectonics of, 527, 528*f*
species richness of birds in, 1177*f*
species richness of trees and vertebrates in, 1177*f*
Northern fur seal (*Callorhinus ursinus*), 1136*f*
Notochord, 498, **673**, **997**
chordate, 673
organogenesis and formation of, 997*f*, 998
Nottebohm, Fernando, 818
N-P-K ratio, 761
NTS (nonshivering thermogenesis), **838**
Nuclear DNA, 109
Nuclear envelope, **102**, 103*f*
Nuclear lamina, **102**, 103*f*
Nuclear magnetic resonance (NMR), 85
Nuclear matrix, 102
Nuclear pore, 102, 103*f*
Nuclear transplantation in animals, 415–17
Nucleases, **305**
Nucleic acid(s), **86**–89. *See also* DNA (deoxyribonucleic acid); RNA (ribonucleic acid)
components of, 87*f*
digestion of, 859*f*
hereditary information transmitted by, 293
as polymer of nucleotides, 87–88 (*see also* Nucleotide(s))
roles of, 86–87
structure of, 87f, 88
Nucleic acid hybridization, **388**, 389*f*
Nucleic acid probe, **388**, 389*f*
Nucleoid, **98*f***, **346**
Nucleoid region, **537**
Nucleolus, **102**, 103*f*
Nucleomorph, 551
Nucleoside, structure of, 87*f*
Nucleoside triphosphate, DNA strand elongation and, 301, 302*f*
Nucleosomes, **360**, 361*f*
Nucleotide(s), 7–8, **87**. *See also* DNA sequences
amino acids specified by triplets of, 312–14
base pairing of, in DNA strands, 89, 296 (*see also* Nitrogenous bases)
as component of nucleic acid, 87*f*, 88
excision repair of DNA damage by, 305*f*
incorporation of, into 3' end of DNA strand, 302*f*
mapping sequences of, 396–98
point mutations in, 328–29
repairing mismatched and damaged, 305*f*, 306
restriction fragment analysis of, 392–94, 395*f*
structure, 87*f*
triplet code of, 312, 313*f*
Nucleotide excision repair, **305*f***
Nucleus, eukaryotic cell, 98, **102**
chemical signaling and responses in, 213*f*, 214
division of (*see* Mitosis)
structure of, 102, 103*f*
transplantation of, 415–17
Nudibranch, 651*f*
Numbers, pyramid of, 1193*f*
Nüsslein-Volhard, Christiane, 423
Nutrient cycling in ecosystems. *See also* Chemical cycling in ecosystems
general model of, 1195*f*
vegetation and, 1198, 1199*f*
Nutrients
absorption of, in small intestine, 859–61
essential animal, 849–52
essential plant, 756–59
as limiting factor in ecosystems, 1188–90
plant root absorption of, 744–46
translocation of, in plant phloem tissue, 751–53
Nutrition. *See also* Diet; Food
absorptive, in fungi, 608–10
animal (*see* Animal nutrition)
fiber, 74
in plants (*see* Plant nutrition)
in prokaryotes, 538, 539*t*

Obelia, life cycle, 645*f*
Obesity, 847–48
evolution and, 848
Obligate aerobes, **539**
Obligate anaerobes, **539**
Occam's Razor, 501
Ocean
benthic zone, 1097*f*
as biome (*see* Marine biomes)
coral reefs, 1097*f*
El Niño Southern Oscillation in, 1170, 1171*f*
global warming and increased sea level, 1205
pelagic zone, 1096*f*
primary production and nutrients in, 1188, 1189*t*, 1190
upwelling, 1190
zones of, 1093*f*
Oceanic pelagic biome, **1096**
Oceanic zone, 1093*f*
Ocellus, 1057*f*–1058
Octopuses, 639*f*, 653*f*
Odling-Smee, Lucy, 1115–16
Odum, Eugene, 1160
Oil spills, bioremediation of, 406, 546*f*
Okazaki fragments of DNA lagging strands, **302*f***, 303*f*, 305
Old World monkeys, 697, 700*f*
Oleander (*Nerium oleander*), 751*f*
Olfaction, **1055**
in humans, 1034, 1035*f*, 1056, 1057*f*
Olfactory bulb, 1034, 1035*f*, 1057*f*
Olfactory receptors, genes for, 459, 505
Oligochaeta (segmented worms), 654*f*. *See also* Earthworms
Oligodendrocytes, **1015**
Oligosaccharins in plants, **814**
Oligotrophic lakes, **1094**
Ommatidia, **1058*f***
Omnivores, **844**
dentition in, 863*f*
Oncogenes, **371**
conversion of proto-oncogenes into, 371*f*
One gene–one enzyme hypothesis, 310, 311*f*
One gene–one polypeptide hypothesis, **310**, 330
One gene–one protein hypothesis, 310
On the Origin of Species (Darwin), 15, 438, 439, 443–46, 455
Onychophorans (velvet worms), 641*f*
Oogenesis, **973**, 974*f*
Oogonia, 974
Oomycetes, **558**, 559*f*
Opaque-2 mutant, 783
Oparin, A. I., 513
Open circulatory system, **657**, **868**, 869*f*
Open system, 143
equilibrium and work in, 147*f*
Operant conditioning, **1117*f***
Operator, **353**
Operculum, **682**
Operons, **353**–54
positive control of *lac*, 356*f*
regulated synthesis of inducible enzymes by *lac*, 355*f*
regulated synthesis of repressible enzymes by *trp*, 354*f*
repressible versus inducible, 354–56
repressor protein and regulatory gene control of, 353
Ophiuroidea (brittle star), 666*t*, 667*f*
Opiates as mimics of endorphins, 43, 81, 1025
Opisthokonts, **612**
Opossums, 695, 696*f*
Opposable thumb, **697**
Opsin, **1060**
Opsonization, 913
Optic chiasm, **1063**
Optimal foraging theory, **1122**
Oral cavity, digestion and, **856**
Orangutans, 701*f*
Orbitals, electron, **38*f***–39
hybridization of, 43*f*
Orchid
bilateral symmetry in flower of, 773*f*
unity and diversity in family of, 16*f*
Order
evolution of biological, 144
as properties of life, 3*f*
Orders (taxonomy), **496**
mammalian, 699*f*
Organ, 5*f*, **712**
Organ, animal, **827**. *See also* Organ systems, animal
digestive, 855–62
reproductive, 969–73

tissues of, **823**
transplants of, and immune response, 916
vestigial, 448
Organ, plant. *See* Leaf (leaves); Root system; Shoot system
Organelles, 5*f*, **95**
of endomembrane system, 104–9
inheritance of genes located in, and genetic disorders, 289–90
isolating, by cell fractionation, 97
microscopy and views of, 95–97
mitochondria and chloroplasts, 109–11, 192–93
peroxisomes, 109, 110–11
relationship among endomembrane system, 109*f*
structural order in metabolism and, 157*f*
Organic chemistry, **58**–59
Organic compounds
abiotic synthesis of, 59*f*
carbon atoms as basis of, 59–61
carbon skeleton variations and diversity of, 61–63
in Earth's early atmosphere, 59*f*, 513–14
extraterrestrial sources of, 514
functional groups of, 63–66
shapes of three simple, 60*f*
Organic phosphates, **65***f*
Organ identity gene, **430**, **734**
ABC hypothesis for flower formation and, 734, 735*f*
Organisms, 4*f*. *See also* Species
anatomy and physiology of, 820
binomial nomenclature for naming, 496
classification of (*see* Systematics; Taxonomy)
colonies of, and origin of multicellular, 526*f*
communities of (*see* Community(ies))
complexity of, related to genomic composition, 374–75
as conformers and as regulators, 832
development of individual (*see* Development)
evolutionary history of (*see* Phylogeny)
genetically modified, 407–8
homeostasis in (*see* Homeostasis)
interactions between environment and, 27*f*, 1080–83 (*see also* Ecology)
life cycles of (*see* Life cycle)
life history, 1141–43, 1147
model, for research studies, 412–13*f*
overproduction of, 445*f*
populations of (*see* Population(s); Population ecology)
thermoregulation in, 833–41
transgenic, 406, 782, 783, 784–86
Organismal ecology, **1082**
Organizer (gastrula organizer), 1006
Organogenesis, **979**, **997**
in bird embryo, 998*f*
in frog embryo, 997*f*
from three germ layers, 997–98, 999*t*
Organ systems, **827**
animal, 827*t*
animal reproductive, 969–73
Orgasm, **972**–73
Origins of replication (DNA), **226**, 227*f*
in eukaryotes, **301***f*
Orrorin tugenensis, 703*f*
Orthologous genes, **505**
Osculum, **642**
Osmoconformers, **923**
Osmolarity, **923**
Osmoregulation, **132**, **922**–27
kidneys and, 936, 937*f*, 938
osmosis and, 922–23
role of transport epithelia in, 926–27
two strategies for challenges of, 923–26
water balance and, 131, 132*f*, 133
Osmoregulators, **923**
Osmosis, **132**, **740**
water balance 131, 132*f*, 133
water conservation, kidneys, and role of, 936
Osmotic potential, **741**
Osteichthyans, **682**–84
Osteoblasts, **825***f*
Osteons, **825***f*
Ostia, 868
Ostracoderms, 679
Otoliths, 1052
Outer ear, **1051**
Outgroup, **498**
Ova. *See* Ovum
Oval window, **1051**
Ovarian cycle, **973**
hormonal regulation of, 976*f*, 977
Ovaries, animal (human), 241, **969**
Ovary, plant, **598**, **772**, 774*f*, 775
fruit development from, 778, 779*f*
variation in locations of, 773*f*
Overexploitation of species, 1214
Overnourishment, **846**–48
Overproduction, natural selection and, **16**
Oviduct, **969**
Oviparous species, **681**
Oviraptor, 690*f*
Ovoviviparous species, **681**
Ovulation, **969**, **977**
Ovule, **592**, **772**, 774
fertilization and development of seed from, 592, 593*f*, 777*f*–78
Ovum, **964**. *See also* Egg
development of human, 973, 974*f*
plant, 592, 593*f*, 772*f*
Oxidation, **161**–62
beta, 177
NAD$^+$ as agent of, 162–63
peroxisomes and, 110–11
Oxidative phosphorylation, **164**
ATP synthesis during, 164, 165, 170–74
photophosphorylation compared to, 192
Oxidizing agent, **161**
Oxygen
covalent bonding of, 40*f*
deep-diving mammals and storage of, 894–95
dissociation curves of, for hemoglobin, 892*f*
loading/unloading of, by hemoglobin, 892*f*
myoglobin and storage of, 895
origins of, in Earth's atmosphere, 522–23
photosynthesis and production of, 183, 184*f*, 522–23
prokaryotic metabolism and, 539
respiratory pigments and transport of, 892*f*, 893
valences of, 60*f*
Oxytocin, **951**
human pregnancy and childbirth and role of, 981
secretion of, by posterior pituitary, 950*f*, 951
Ozone, depletion of atmospheric, 1205–6

p53 gene, **373**
cancer development and faulty, 373
P 680 Photosystem II, 190
P 700 Photosystem I, 190
Pace, Norman, 541
Pacemaker, heart, **873**
Pacific yew (*Taxa brevifolia*), 595*f*
Paedomorphosis, **485**, 685
Pain
endorphins and reduction of, 43, 81, 952
prostaglandins and sensation of, 948, 1049
sensory receptors for, 1049
Paine, Robert, 1169
Pain receptors (nociceptors), **1049**
Pair-rule genes, 425
Paleoanthropology, **702**
Paleontological species concept, **476**
Paleontology, **440**
Paleozoic era, 519*t*, 520, 521*f*, 527, 596, 629
Palisade mesophyll, **724**
Panama, 1080*f*
Pancreas, **855**, **955**
digestive function, 858, 859*f*
glucose homeostasis and role of, 846*f*, 955–56
hormones secretions by, 949*t*, 955*f*, 956
Pangaea, **527**, 529*f*, 696
Papaya, genetically engineered, 784*f*
Papillae, 1055
Papilloma virus, 370
Parabasalids (trichomonads), **553**
Parabronchi, **889**
Paracerceis sculpta, mating behavior and male polymorphism in, 1126, 1127*f*
Paracrine signaling, 203*f*, 947–48
Paralogous genes, **505**
Paramecium sp., 556
cilia of, 15*f*
conjugation and reproduction in, 557*f*, 558
interspecific competition between species of, 1160
logistic population growth model applied to, 1147*f*
structure and function of, 557*f*
water balance and role of contractile vacuole in, 132, 133*f*
Paramyosin, 1072
Paraphyletic grouping, **498***f*
Parareptiles, **689**
Parasites, 545, **1163**
blood flukes as, 647*f*
brown-headed cowbird as, 1221
flowers as, 771
leeches as, 655
nematodes as, 656*f*
plants as, 767, 768*f*
tapeworms as, 648*f*
viruses as (*see* Viruses)
Parasitism, **545**, **1163**
Parasitoidism, **1163**
Parasitoid wasp, 813*f*
Parasympathetic division of autonomic nervous system, 1027*f*, **1028**
Parathion pesticide, 155
Parathyroid glands, **954**
hormones secreted by, 949*t*, 954–55
Parathyroid hormone (PTH), **954**
Parazoan organization, **633**
Parenchyma cells, **718**, 722*f*
Parental behavior
care of offspring, 690*f*, 967, 968*f*, 1124, 1125*f*, 1142*f*, 1143
genetic influences on, 1112–13
Parental types, **278**
Parietal cells, 857*f*
Parker, John, 93
Parkinson's disease, 623, **1041**
E. Jarvis on, 819
lack of dopamine and, 1025
L-dopa and treatment of, 62, 63*f*
Parsimony, principle of maximum, 501
analogy-versus-homology pitfall and, 504*f*
applied to problem in molecular systematics, 502–3*f*
Parthenogenesis, **648**, **965**
in rotifers, 648–49
in whiptail lizards, 965, 966*f*
Partial pressure, loading/unloading of respiratory gases and, **891***f*
Parturition, **981**
Passive immunity, **914**
Passive transport, 130–34, **738**
active transport compared to, 135*f*
defined, **131**
diffusion as, 130
facilitated diffusion as, 133–34
osmosis as, 131–33
Patchiness, environmental, **1082**, 1172, 1173*f*
Paternity
certainty of, 1124–25
DNA technology and establishment of, 405
Pathogens, **1163**
agglutination of, 913
bacteria as, 295*f*, 339, 460, 545–46
B and T lymphocytes response to, 910–14
drug-resistant, 447, 448*f*

fungi as, 622–23
immunity to (*see* Immune system)
insects as vectors for, 554*f*, 555, 556*f*, 664
parasites as (*see* Parasite(s))
plant (*see* Plant disease)
plant defenses against, 813–15
population density of hosts and, 1149
prokaryotes as, 545–46
protists as, 552, 553*f*, 554*f*, 555, 556*f*, 564
viruses as, 337, 339, 343–45
Pattern formation, **421**–31, **731**, **1007**
cell fate determination and, by inductive signals, 1006–8
cell signaling and, in *C. elegans*, 425–27
cell signaling and transcriptional regulation in plants, 429–30
gene-activation cascade and, in *Drosophila*, 421–25
homeobox sequences and, 431–32
homeotic genes and, 425
positional information and, 421, 731, 732–33
in plant cell development, 730–32
programmed cell death (apoptosis) in, 427–29
during vertebrate limb development, 1006–8
Pauling, Linus, 296
PCBs (polychlorinated biphenyls), biological magnification of, in food web, 1202*f*
Peacocks, sexual dimorphism in, 468*f*
Pea plants, 603*f*
fruit development in, 779*f*
gibberellin and stem elongation in, 797
Mendel's studies on inheritance in, 251–58, 309
Rhizobium bacteria and nitrogen fixation in, 765*f*
triple response to stress in seedlings of, 799*f*
Peat, **583**
Peat bogs, L. Graham on ecological function of, 511
Pectin, 106
Pediastrum, 526*f*
Pedigree, Mendelian inheritance patterns in humans revealed in analysis of, **265**–66
Pelagic zone, 1093*f*
Penguins, 693, 821*f*
energy budget for, 830, 831*f*
nutritional requirements during molting, 850*f*
population dispersion patterns of, 1138*f*
Penicillin, 155, 611*f*, 623*f*
Penicillium sp., 611*f*, 623*f*
Penis, **972**
erectile dysfunction in, 211, 948, 1025
Pentose sugars, 70*f*
PEP carboxylase, **196**
Pepsin, **857**
Pepsinogen, **857**
Peptides, as neurotransmitters, 1024t, 1025
Peptide bond, **80**
Peptidoglycan, **535**
Perception, sensory input and, **1046**. *See also* Sensory mechanisms
Pereira, Sofia, 1120
Perennials, plant, **720**
Pericarp, **598**
Pericycle, **723**
Periderm, **717**, 814
cork cambia and production of, 728
Periodic table of elements, 37
Peripheral nervous system (PNS), **1012**, 1026*f*
enteric division of, 1028
functional hierarchy of, 1027*f*
homestasis and, 1028
parasympathetic divisions of, 1027*f*, 1028
role of, in homeostasis, 1028
Schwann cells of, 1015*f*
sympathetic division of, 1027*f*
Peripheral proteins, **128**
Peripheral resistance, **876**
Peristalsis, **855**, **1064**
in digestive tract, 855
esophageal, 856*f*
movement by, in annelids, 1064
Peristome, **581**
Peritubular capillaries, **933**
Perivitelline space, 989
Permian period, 596
mass extinctions in, 518*f*, 520
Peroxisome, **109**, 110, 111*f*
Peroxisome enzymes, **574**
Pert, Candace, 1025
Pertussis (whooping cough), 206*f*
Petals, flower, 429*f*, 430, **598**, **772**
Petiole, **715**
Petrels, excessive weight in chicks of, 848*f*
PET scans, 36*f*
Pettit, Robert, 643
Pfenning, David and Karin, inquiry into mimicry of warning coloration in snakes, 22, 23*f*, 24
Pfiesteria shumwayae, 555*f*
P generation, **253**
incomplete dominance and, 261*f*
law of segregation and, 253*f*, 255*f*, 257*f*, 275*f*
pH, **54**
acids,bases,and, 53–54
buffers, 55
enzyme activity and effects of, 154*f*
pH scale, 53, 54*f*, 55
Phaeophyta (brown algae), 560–62
Phage(s), **294**, **336**. *See also* Bacteriophage (phage)
Phage λ, lytic and lysogenic cycles of, 338, 339*f*
Phagocytes (phagocytic cells), 899–900
Phagocytosis, **107**, 137, **138***f*, 881, **899**, 900*f*
antibody-mediated disposal of antigens and, 913*f*
lysosomes and, 107*f*
four leukocytes capable of, 900
Pharmaceutical products. *See also* Drug(s); Medicine
biodiversity and development of new, 1211–12
derived from plants, 30–31, 594*f*, 595*f*, 605*t*, 1211, 1212*f*
DNA technology and development of, 404
enantiomers and, 62, 63*f*
G-protein pathways and, 206
"Pharm" animals, 406*f*
Pharyngeal clefts, **673**, 676
Pharyngeal slits, **673**
Pharyngolepis, 679*f*
Pharynx, **856**
Phase changes, **733**
Phase-contrast microscopy, 96*f*
Phelloderm, 728
Phenotype, **256**
effects of natural selection on, 462–70
genotype versus, 256*f*
impact of environment on, 264, 462
multiple effects in (pleiotropy), 262
mutant, 276, 277*f*
norm of reaction in, 264
relationship of, to allele dominance, 261
wild type, 276*f*, 277*f*
Phenotypic polymorphism, **463**
Phenylalanine, 313, 320
Phenylketonuria (PKU), 269–70, 458
Pheromones, **610**, 945, **967**, **1111**
in insects, 437, 1048, 1409*f*
Philadelphia chromosome, 288
Philippine eagle, 1211*f*
Phloem, **584**, **717**, 751–53
loading and unloading of, 752*f*
pressure flow of sap in, 753*f*
in roots, 722*f*
sieve-tube members and companion cells of, 719
transport (translocation) of sap through, 584, 751–53
Phoronids (marine worms), 640*f*, **649**
Phoronis hippocrepia, 649*f*
Phosphate bonds, high-energy, 148–49
Phosphate functional group, **65***f*
Phosphofructokinase, stimulation of, 177*f*
Phospholipids, **76**–77
bilayer structure formed by self-assembly of, in aqueous environments, 77*f*
membrane models and, 125
steroids as, 77
structure of, 76*f*
Phospholipid bilayer, 125*f*
permeability of, 126, 130
Phosphorus as plant nutrient, 761
deficiency of, 759*f*, 761*f*
Phosphorylation, **149**
ATP hydrolysis and, 149–50
cascade, 209*f*
oxidative, 164
photophosphorylation, 185
of proteins, as mechanism of signal transduction, 209–10
substrate-level, 165*f*
Photic zone, 1093
Photoautotrophs, 181, 182*f*
prokaryotes as, 538, 539*f*
protists as, 550
Photoexcitation of chlorophyll, 188, 189*f*
Photoheterotrophs, **539***t*
Photomorphogenesis, **802**
Photons, light, **186**
photoexcitation of chlorophyll and absorption of, 188, 189*f*
Photoperiodism, **806**
flowering and, 806–8
meristem transition and, 808
Photophosphorylation, **185**, 189–92
oxidative phosphorylation compared to, 192
Photoprotection, 188
Photopsins, **1061**
Photoreceptors, **1048**
blue-light, in plants, 750, 802
brain processing of visual information and, 1061–63
for infrared light, 1049*f*
in invertebrates, 1057–58
phototropism in plants and, 792–93
phytochromes as plant, 789–91, 802–5
in vertebrate eye, 1057–63
Photorespiration, **195**–96
Photosynthesis, **181**–200
absorption and action spectra for, 187*f*, 188
alternative mechanisms of, for hot and arid climates, 195–98
Calvin cycle and conversion of CO_2 to sugar in, 184, 185*f*, 193–95
chemical equation for, 44, 183
chloroplasts as site of, 109, 111*f*, 182–83
conversion of light energy into chemical energy in, 182–85
importance of, 197, 198*f*
light reactions in, 184, 185*f*, 186–93
origins of, 522–23
overview of two stages of, 184, 185*f*
role of, in biosphere, 181, 182*f*
splitting of water and redox reactions of, 184
tracking atoms through, 183, 184*f*
Photosystem I (PSI), **190**, 191*f*
Photosystem II (PSII), **190**, 191*f*
Photosystems, photosynthetic, **189**–90
light harvesting by, 189*f*
Phototrophs, 538–39
Phototropin, 802
Phototropism, **792**
early studies of, and discovery of plant hormones, 792–93
light wavelengths and, 803*f*
Phragmoplasts, **574**
pH scale, 53, **54***f*, 55
soil pH, 762
Phyla (taxonomy), **496**
Phylogenetic species concept, **476**
Phylogenetic trees, **496**, 497–504
of amniotes, 687*f*
of angiosperms, 602*f*
of animals, 633, 634*f*, 635*f*, 638*f*
of biological diversity, 530–31f

of chordates, 672*f*
cladograms, cladistics, and, 497–99
of eukaryotes, 552*f*
of fungi, 613*f*
of Galápagos finches, 18*f*
as hypotheses, 501–4
mammalian, 698*f*
maximum parsimony, maximum likelihood, and, 501, 502–3*f*
of plants, 579*f*
of prokaryotes, 540*f*
timing, chronologies, and, 499–500
Phylogeny, **491**–95
of amniotes, 697*f*
animal, 633–36, 638*f*
chordate, 671, 672f, 673
circulatory system as reflection of, 867–71
connection between classification and, 496, 497*f*
continental drift and biogeographical basis of, 528
of eukaryotes, 552*f*
fossil record as evidence for, 492
of fungi, 612, 613*f*
genome and, 504–6
mammals, 698–99*f*
morphological and molecular homologies and, 492–95
phylogenetic trees (*see* Phylogenetic trees)
plants, 575–59
of prokaryotes, 540–44
systematics (*see* Systematics)
taxonomic classification and (*see* Taxonomy)
tree of life and, 491, 507
Phylogram, 499*f*, 500
Physarum polycephalum, 564*f*
Physical map, DNA fragments, **396**
Physiology, **820**. *See also* Animal anatomy and physiology
Phytoalexins, **814**
Phytochemicals, 188
Phytochromes, **802**
as molecular switching mechanism, 804*f*
as photoreceptors, in plants, 789, 790*f*, 802–5
structure of, 804*f*
Phytolacca dodecandra, medicinal properties of, 30
Phytophthora infestans, 558
Phytophthora ramorum (sudden oak death disease), 1163
Phytoplankton, 550
nutrients for, and ecosystem productivity, 1189*f*
turnover time for, 1192
Phytoremediation, **763**
Pigmentation, skin, 952
Pigments
carotenoids, 188, 189
chlorophyll (*see* Chlorophyll(s))
photosynthetic, 186–88
respiratory, 892–94
Pili (pilus), 349, **536**
Pilobolus, 614, 615*f*
Pine, life cycle of, 596, 597*f*
Pineal gland, hormone produced by, 949*t*, **959**
Pin flower type, 775*f*
Pinocytosis, 137, **138*f***
Pinyon jays (*Gymnorhinus cyanocephalus*), 1116
Pisonia, seed dispersal in, 480*f*
Pistil, **598**, **772**
Pitch (sound), detection of, **1052**
Pitcher plant, 768*f*
Pith, plant tissue, **717**, 722
Pittsburgh Compound-B (PIB), 1041
Pituitary gland, 948, **950**
hormones of anterior, 949t, 951*f*, 952, 976*f*
hormones of posterior, 949t, 950*f*, 951
production and release of hormones by, 950*f*
relationship of, to hypothalamus, 950–51
Pivot joint, 1065*f*
PKU (phenylketonuria), 269–70, 458
Placenta, **695**, 915, **979**
circulation in fetus and, 980*f*
Placental mammals. *See* Eutherian (placental) mammals
Placental transfer cells, **577**
Placoderms, **680**
Placozoa, 639*f*
Plagiochila deltoidea liverwort, 582*f*
Plague, 546
Planarians, 639*f*, **646**
anatomy of, 647*f*
ocelli and orientation behavior of, 1057*f*
nervous system, 1012*f*
protonephridium (flame-bulb) of, 929*f*, 930
Plankton, 550
Plants, 573–90
adaptations of, for life on land, 573, 574, 575–79
alternation of generations in, 242, 576*f*
angiosperms and evolution of fruits and flowers (*see* Angiosperms)
in arid and hot climates, 195–98
artificial selection and selective breeding of, 783–84
biotechnology applied to, 783–86
bryophytes, 580–83
carnivorous, 767, 768*f*
cells of (*see* Plant cell)
colonization of land by, 527
crop (*see* Crop plants)
defenses of, against predators, 813
derived traits of, 575, 576–77*f*
development of (*see* Development; Plant development)
diseases in (*see* Plant disease)
epiphytes, 768*f*
evolution of, from charophyceans (green algae), 573–74
evolution of, highlights, 579*f*
extranuclear genes and leaf color in, 289, 290*f*
genetic engineering in, 406–7, 784–86
L. Graham on evolution of terrestrial, 510–11
greening of (de-etiolation), 789*f*, *790*f, *791*
growth in (*see* Plant growth)
gymnosperms (*see* Gymnosperms)
hybridization of, 252–53
kudzu as introduced species of, 1213*f*
medicines derived from, 594*f*, 595*f*, 605*t*
movements in, 805f, 810*f*
nutrition in (*see* Plant nutrition)
origin and diversification of, 575–79
parasitic, 768*f*
photosynthesis in (*see* Photosynthesis)
phyla of extant, 578*t*
polyploidy in, 285
reproduction in (*see* Plant reproduction)
responses of, to signals (*see* Plant responses)
seedless vascular plants, 584–88
seed vascular plants (*see* Seed plants)
self-recognition and self-incompatibility in, 775–76
sexual life cycle in, 242*f*
structure (*see* Plant structure)
sympatric speciation and evolution of, 478*f*, 479*f*
temperature moderation in, 50
threats to, 606
totipotency in, 415
transgenic, 406–7, 782, 783, 784–86
transport (*see* Transport in vascular plants)
vascular (*see* Vascular plants)
vascular tissues (*see* Phloem; Xylem)
water transport in, 48*f*
Plantae (kingdom), 575*f*. *See also* Plant(s)
Plantain, seed production and population density of, 1149*f*
Plant anatomy. *See* Plant cell; Plant structure
Plant cells, 717, 718–19*f*
abscisic acid effects on, 798–99
auxin and elongation of, 794, 795*f*
brassinosteroid effects on, 798
cell plate of, 225*f*, **226**
cell walls of, 118, 119*f*, 743*f*
central vacuole of, 108*f*
chloroplasts in (*see* Chloroplasts)
collenchyma cells, 718*f*
companion cells, 719*f*
compartments of, and short-distance transport, 743*f*
cytokinesis and cell plate formation in, 225*f*, 226
cytokinin and division/differentiation of, 796
cytoplasmic streaming in, 117*f*, 118
ethylene effects on, 799–801
expansion of, orientation, 730, 731*f*
flaccid and turgid, 132–33, 750*f*
gibberellin effects on, 309, 797
location and fate of, 732–33
mitosis in, 226*f*
orientation of expansion of, 730, 731*f*
parenchyma, 718*f*
pattern formation and development of, 730–32
plane and symmetry of division and growth of, 729*f*, 730*f*
plasmodesmata of, 120*f*
positional information and development of, 731, 732–33
protoplast of, 782*f*
sclerenchyma cells, 718*f*
sieve-tube members, 719*f*
storage of starch in, 71, 72*f*
structure of generalized, 101*f*
tracheids, 719*f*
vacuoles of, 108*f*
vessel elements, 719*f*
water balance in, 132, 133*f*
water relations in, 742*f*
Plant control systems. *See* Plant responses
Plant development, 728–35. *See also* Plant growth
apical meristems and, 576*f*
animal development compared to, 433
cell division and cell expansion in, 729*f*, 730*f*, 796
cell location and developmental fate of cells in, 732–33
cell signaling, induction and, 420, 421*f*
cell signaling, transcriptional regulation, and, 429–30
embryo development, 777*f*, 778
gene expression and control of cell differentiation in, 732, 733*f*
mechanisms of, 429
model organism *Arabidopsis* for study of, 413*f*, 728, 729*f*
molecular biology and study of, 728–29
organ-identity genes in, 430, 431*f*
pattern formation and, 730–32
pattern formation in flowers and, 429*f*, 430*f*, 431*f*
phase changes and shifts in, 733–34
totipotency, 415
Plant disease
chestnut blight, 1168
fungi as cause of, 622, 797
plant defenses against pathogens, 813–15
protistan, 1163
viral, 334, 335*f*, 345
white rusts and downy mildew, 558
Plant growth, 720–32. *See also* Plant development
apical meristems and primary, 576*f*, 720*f*, 721–25
auxin and, 793*f*, 794, 795*f*
cell division, cell expansion, and, 729*f*, 730
determinate, 720
indeterminate, 720
inhibitors of, 793
lateral meristems and secondary, 720*f*, 721, 725–28
morphogenesis, differentiation, and, 728–35
pattern formation and, 730–32
Plant hormones, **791**–802
abscisic acid, 794*t*, 798–99
auxin, 794*t*–97
brassinosteroids, 791, 794*t*, 798
cytokinins, 794*t*, 796–97
discovery of, 792–93

ethylene, 794*t*, 799–801
gibberellins, 309, 794*t*, 797–98
survey of, 793, 794*t*
systems biology and interactions of, 801–2
Plant morphology. *See* Plant structure
Plant nutrition, 756–70
adaptations of, involving relationships with other organisms, 764–67, 768*f*
mycorrhizae and, 620, 766–67
nitrogen as essential in, 33*f*, 759*f*, 763–64
nutritional requirements, 756–59
soil quality/characteristics, and, 759–63
"smart plants" and, 762*f*
uptake of nutrients, 757*f* (*see also* Transport in vascular plants)
Plant reproduction
alternation of generations and, 242*f*, 576*f*
in angiosperms, 598–601, 771–80
asexual, 781–83
biotechnology and, 783–86
in bryophytes, 580–83
double fertilization, 599, 600*f*, 776–77
embryophyte and, 575*f*, **577*f***
flowers and, 598*f*, 772*f*, 773*f*, 775*f*
fruit development and, 598, 599*f*, 778, 779*f*
gametophyte, 576, 772, 774*f*–75
in gymnosperms, 596, 597*f*
pollen and sperm production in seed plants, 592, 593*f*, 774*f*, 775
pollination, 592, 771–76
prevention of self-fertilization in, 775–76
seed, evolutionary advantage of, 593
seed development, 593*f*, 777–78
seed germination, 779, 780*f*
in seedless vascular plants, 585f, 586
sexual, 576, 580, 581*f*, 585*f*, 586, 592–93, 596, 597*f*, 599, 600*f*, 601, 771–80
sporophyte, 576, 771–72
Plant responses, 788–817
as defense against pathogens and herbivores, 812–15
to environmental stimuli other than light, 808–12
to light, 788*f*, 802–8
plant hormones and, 791–802
signal-transduction pathways involved in, 788–91
Plant structure, 712–19
cell division and cell expansion affecting, 729–30
cell types of, 717–19 (*see also* Plant cell)
developmental fate of cells affecting, 732–33
gene expression and cell differentiation affecting, 732, 733*f*
growth and (*see* Plant growth)
monocot and dicot, 603*f*
overview, 713*f*
pattern formation and, 730–32
phase changes and, 633–34
plasticity of, 712
roots, stems, and leaves of, 713–16
tissue systems, 717
Plasma, blood, **879**, 880*t*
Plasma cells, **908**
Plasma membrane, 98, **99**, 124–40
active transport across, 134–36
cell-cell recognition and carbohydrates of, 129
cotransport across, 136
exocytosis, endocytosis,and bulk transport across, 137, 138*f*
fluidity of, 126*f*–27
ionic gradients across neuron, 1016*f*
membrane potentials of, 1015–21
models of, 125
osmosis and transport of water across, 131–33
passive transport across, 130–34
phospholipid bilayer of, 125*f*
protein facilitation of transport across (*see also* Membrane proteins)
receptor proteins in, 205, 206–8*f*
selective permeability of, 124, 130
structure of, 99*f*, 130
synthesis and sidedness of, 129*f*
voltage across (membrane potential), 134–35
water balance and, 131–33
Plasmids, 343, **349**, **537**
cloning genes using, 385*f*–89
as cloning vector, 386, 387*f*, 388, 389*f*
F, and bacterial conjugation, 349, 350*f*
genetic recombination in bacteria and, 349–51
in prokaryotic cells, 349–51, 537
R, and antibiotic resistance, 351
Ti, 406, 407*f*
in yeasts, 390
Plasmodesmata, **120*f***
Plasmodial slime molds, **564**
life cycle of, 565*f*
Plasmodium (apicomplexan), 555, 556*f*. *See also* Malaria
Plasmodium (slime mold), **564**
Plasmogamy, **610**, 614*f*
Plasmolysis, **133**
Plasmolyzed cell, **742**
Plasticity, **712**
Plastids, **110**. *See also* Chloroplasts; Statoliths
apicoplast, 555
origins of, 523–25, 550–51
secondary endosymbiosis and diversity of eukaryotes, 551*f*
starch storage in, 71, 72*f*
Platelet, in blood, **880**, 881
blood clotting and, 880*f*, 882f
effect of prostaglandins on, 948
Platelet-derived growth factor (PDGF), cell division and, 231*f*
Plate tectonics, 527–29
Platyhelminthes (flatworms), 631*f*, 639*f*, 646–48
Pleiotropy, **262**
Plomin, Robert, 1133
Pluripotent stem cells, **418**
replacement of cellular elements in blood and, 881*f*
Pneumatophores, 714*f*
Pneumonia, 295*f*, 348
Poikilotherms, 833–34
Point mutations, **328**–30
mutagens causing, 329–30
proto-oncogene transformed into oncogene by, 371
as source of genetic variation, 459–60
types and consequences of, 328–29
Poisons. *See also* Toxins
enzyme inhibitors as, 155
in food, 460
smooth endoplasmic reticulum and detoxification of, 104
snake, 690, 1162
Poison arrow frog, aposematic coloration in, 1162*f*
Polar bears (*Ursus maritimus*), agonistic behavior in, 1127*f*
Polar covalent bond, **40**, 41*f*
Polarity in plants, **731**
Polar molecule, water as, **47**–48
Polar transport of auxin, 794, 795*f*
Polio virus, 337, 343
Pollen grains, **592**, 774–75
development of, 592–93, 774*f*
Pollen tube, growth of, 776*f*
Pollination, **592**, 771–76
cross-, 599
floral variations and, 773*f*
flower structure and, 772*f*
gametophyte development and, 774*f*, 775
self-incompatibility mechanisms, 775–76
Pollinators, relationship of angiosperms to, 604*f*
Polyandry, **1124**, 1125*f*
Poly-A tail of mRNA, **317**
Polychaetes (marine worms), 655*f*
Polygamous mating, **1123**, 1124*f*
Polygenic inheritance, **263**
Polygyny, **1123**
Polymers, **68**
abiotic synthesis of, 514–15
amino acid, 80
condensation and hydrolysis of, 68, 69*f*
diversity of, 69
nucleotide, 88
synthesis and breakdown of, 68, 69*f*
Polymerase chain reaction (PCR), **391***f*, 392
medical applications, 403
Polymorphism, 463
balanced, 466–68
phenotypic and genetic, 463
Polynucleotides, **87**
assembly of, by RNA polymerases in transcription, 315*f*
as component of nucleic acids, 87*f*
Polyp (cnidarian), **643**
Polypeptides, **78–81**. *See also* Amino acid(s); Protein(s)
amino acid monomers and, 78, 79*f*, 80
amino acid polymers and, 80
determining amino acid sequence of, 80–81
forming chains of, 80*f*, 323–24
one gene-directed production of one, 310
post-translational modifications to, 324–25, 791
translation and synthesis of, 320–26
Polyphyletic grouping, **498*f***
Polyplacophora (chitons), 651*f*
Polyploidy, **285**, **478**
genomic evolution and, 378
sympatric speciation by, 478*f*, 479*f*
Polyribosomes (polysomes), **324**, 325*f*, 326
Polysaccharides, **71**–74
storage, 71, 72*f*
structural, 72–74 (*see also* Cellulose; Chitin)
Polyspermy, fast and slow blocks to, 989
Polytrichum commune moss, 582*f*
Ponds. *See also* Lakes
protists in water of, 549
turnover in, 1091*f*
Pons (brain), **1029**
breathing controlled in medulla oblongata and, 1029
Populations, 4*f*, 454–70, **1082**, **1136**
biotic and abiotic influences on, 1148–52
communities of (*see* Community(ies))
conservation of species using approach of small, 1215–18
conservation of species using approach of declining, 1218–19
cycles in, 1151–52
defined, **455**
demographic studies of changes in, over time, 1139–41
density, and dispersion patterns in, 1137–39
density-dependent and density-independent, 1148–50
dynamics of, 1150–51
evolution of, 454 (*see also* Population genetics)
human (*see* Human population)
life histories of organisms in, 1141–43
life tables and survivorship curves in, 1139–40
models of growth in, 1143–47
reproductive rates in, 1140–41
species in (*see* Organisms; Species)
studying natural selection in virtual, 467*f*
variation within and between, 444, 445*f*
Population cycles, 1151–52
Population density, 1137–38
density-dependent birth/death rates, 1148–50
density-independent birth/death rates, 1148
determining equilibrium for, 1148*f*
dispersion patterns and, 1137–39
population change and, 1148
Population dynamics, 1137*f*, **1150**–51
Population ecology, **1082**, **1136**–58
biotic and abiotic influences on populations, 1148–52
carry capacity of environment and, 1145
exponential model of growth in, 1143–45
human population growth and, 1136, 1152–56
life history traits as product of natural selection and, 1141–43

  - logistic growth model of, 1145–47
  - population density, dispersion and demography, 1136–41
- Population genetics, **455**, 1136
  - gene flow and, 462
  - gene pools and allele frequencies, 455–56
  - genetic drift and, 460–62
  - genetic variation due to mutation in, 459–60
  - genetic variation due to sexual recombination in, 247–49, 460
  - Hardy-Weinberg theorem and equilibrium in, 456–58
  - human health and, 458
  - microevolution and, 454
  - modern synthesis of Darwinian and Mendelian ideas and development of, 455
  - natural selection and, 446, 447f, 454–55, 460, 462–70
- Population growth
  - cycles of, 1151–52
  - density-dependent and density-independent factors in, 1148–50
  - exponential model of, 1143–45
  - human, 1136
  - logistic model of, and concept of carrying capacity, 1145–47
  - overproduction of offspring in, 445f
  - zero, 1144
- Population size
  - change in (ΔN), 1143
  - declining-population approach to conservation, 1218–19
  - density-dependent and density-independent regulation of, 1148–50
  - estimating effective, 1216–17
  - minimum viable, 1216
  - per capita rate of increase in, 1143–44, 1145
  - small-population approach to conservation, 1215–18
- Population viability analysis (PVA), **1216**
- Pore complex, 102, 103f
- Porifera. *See* Sponges (Porifera)
- *Porphyra*, 561
- Positional information, **421**, **731**, **1007**
  - plant cell development and, 731, 732–33
  - vertebrate limbs and, 1007
- Positive feedback, **11**, 12f, **833**
  - as mechanism of homeostasis, 833
- Positive pressure breathing, **888**
- Positron-emission tomography (PET), 36f
- Posterior pituitary gland, **950**
  - hormones of, 949t, 950f, 951
- Posterior side of body, **630**
- Postsynaptic cell, **1014**
  - effect of neurotransmitters on, 1022f, 1023
  - structure of vertebrate neuron and, 1014f
  - synaptic terminals on, 1021f
- Postsynaptic potentials
  - excitatory, and inhibitory, 1022
  - summation of, 1023f
- Postzygotic barriers, **473**, 475f
- Potassium (K^+)
  - deficiency of, in plants, 759f
  - ligand-gated channels and, 1022f
  - regulation of blood concentrations of, in kidney, 934
  - role of, in opening/closing of stomata, 750
  - selective channels allowing passage of, 739, 740f
  - selective permeability of neuron membrane to, 1017f
  - sodium-potassium pump, 134, 135f, 1016–17
  - voltage-gated channels for, 1018, 1019f, 1020f, 1021
- Potassium channels, 739, 740f
- Potato, diseases of, 558
- Potential energy, **36**, **142**
  - transformations between kinetic and, 142f
- Potential evapotranspiration, 1176
- Precautionary principle, **1083**
- Prairie voles (*Microtus ochrogaster*), mating behavior, 1112f, 1113
- Precapillary sphincters, 877, 878f
- Precipitation
  - effects of mountains on, 1090
  - global air and wind patterns and, 1089f
  - of pathogens, 913
- Predation, **1161**–62
  - effects of differential, on guppy evolution, 446, 447f
  - foraging behavior and risk of, 1122–23
  - by fungi, 610f
  - herbivory as, 1163
  - by parasites and pathogens, 1163
  - population cycles and, 1151, 1152f
  - as population limiting factor, 1150
- Predators
  - adaptations of, 1045f
  - behavioral variation in prey selection by, 1118, 1119f
  - defenses against, 23f, 32f, 813, 1162, 1163
  - fungi as, 610f
  - keystone species as, 1168f, 1169f
  - mimicry as defense against, 23f, 1162
  - seed, 1161
- Predator-prey relationship
  - blue jay and moth, 467f
  - lynx and snowshoe hare, 1151, 1152f
- Preformation, development and concept of, 987
- Pregnancy, **978**–81
  - contraception against, 982–83
  - in humans, 915–16, 978–81
  - immunology of, 915–16, 982
  - reproductive technology and, 983–84
- Pre-mRNA, 312
- Preprophase band, **730**
- Prepuce, **972**
- Pressure flow, translocation of sap in angiosperms by, 753f
- Pressure potential (ψ_P), **741**
- Presynaptic cell, **1014**
- Prey, defenses of, 1162. *See also* Predation; Predators
- Prezygotic barriers, **473**, 474f
- Priapulans, 640f
- Primary cell wall, of plants, **118**, 119f
- Primary consumers, **1185**
- Primary electron acceptor, **189**
- Primary growth, plants, **720**, 721–25
  - of roots, 721f, 722f, 723f
  - of shoots and stems, 723f, 724, 726f
- Primary immune response, **908**
- Primary motor cortex, brain, 1033f
- Primary oocytes, **974**
- Primary plant body, **721**
- Primary producers, **1185**, **1186**
- Primary production in ecosystems, 1187–88
  - in aquatic ecosystems, 1188–90
  - energy budget and, 1186–87
  - gross, 1187–88
  - net, 1187f, 1188
  - regional annual net, for Earth, 1188f
  - in terrestrial and wetland ecosystems, 1190–91
- Primary somatosensory cortex, brain, 1032, 1033f
- Primary structure of protein, 81, **82**f
  - sickle-cell disease related to, 84f
- Primary succession, ecological, **1173**
- Primary transcript, **312**, 327t
- Primary visual cortex, 1032, **1063**
- Primase, **303**, 304t
- Primates, 697–701
  - derived characters of, 697
  - humans (*see* Human(s); Human evolution)
  - living, 697–701
  - phylogenetic tree of, 700f
- Primer, DNA synthesis, **303**
- Primitive streak, **996**
- Prions, **345**
  - propagation of, 346f
- Probability, Mendelian inheritance and laws of, 258–60, 268–69
- Problem solving in animals, 1117–18
- Proboscis worms (Nemertea), 639f
- Producers, **6**
- Production, pyramid of, 1192
- Production efficiency in ecosystems, **1191**–92
- Products of chemical reactions, **44**
- Pregenitor cells, 1039
- Progesterone, intracellular receptors for, 947
- Progestins, **959**
- Programmed cell death (apoptosis), 108, 427–29
  - ethylene and plant leaf, 800
- Progymnosperms, **596**f
- Prokaryotes, 534–48. *See also* Archaea; Bacteria
  - aerobic, and photosynthetic, 537f
  - bacteria and archaea as as evolutionary branches of, 541–44
  - chemiosmosis in early, 521–22
  - ecological roles of, 544–45
  - evolution of, and changes affecting Earth, 521–23
  - H^+ gradients and ATP synthesis in, 173
  - internal and genomic organization, 537
  - major groups of, 542–43f
  - motility of, 536–37
  - nutritional and metabolic diversity among, 538–40
  - as pathogens, 545–46
  - phylogeny of, 540–44
  - prevalence of, 534
  - protein synthesis in eukaryotes versus, 327–28
  - reproduction and adaptation by, 537–38
  - in research and technology, 546–47
  - structure and shapes of, 534, 535f, 536f
- Prokaryotic cell, **8**, **98**. *See also* Archaea; Bacteria
  - cell wall, 132, 534–36
  - chromosome of, 537f
  - DNA replication in, 300–305
  - eukaryotic cell compared to, 8f, 98–99, 312f
  - flagella of, 536f
  - membranes of, 537f
  - motility in, 536–37
  - ribosomes of, 322
  - RNA polymerase of, 315
  - structure and shapes of, 98f, 534, 535f
  - transcription in, 311, 312f, 315, 316, 317
  - translation in, 311, 312f
- Prokaryotic genome, 346–56, 537
  - bacterial replication of, 346, 347f
  - complexity of, 374
- Prolactin (PRL), **952**, 981
- Proliferative phase, menstrual cycle, **977**
- Prometaphase, mitosis, **221**, 222f, 226f
- Promiscuous mating, **1123**
- Promoter DNA sequence, **315**f, 316f
- Pronghorn, respiratory system of, 894
- Propanal, **64**f
- Prophage, **339**, 349
- Prophase
  - meiosis, 244f, 245f
  - mitosis, **221**, 222f, 226f
- Prop roots, 714f
- Prostaglandins (PGs), **948**
  - as local regulators, 948
  - pain reception and, 948, 1049
  - platelets and, 948f
- Prostate gland, **972**
- Prosthetic group, 170
- Protease activation, 859f
- Proteasome, **369**
  - degradation of protein by, 369, 370f
- Protein(s), **77**–86
  - amino acids of, 78, 79f, 80 (*see also* Amino acid(s))
  - antimicrobial, 900–901
  - aquaporins (*see* Aquaporins)
  - carrier, 133, 134f
  - catabolism of, 177f
  - catalytic (*see* Enzyme(s))
  - cell determination/differentiation and production of specific, 419
  - channel, 133, 134f
  - chaperone, 85, 325

conformation, structure, and function of, 81–86
denaturation and renaturation of, 84, 85*f*
defined, **78**
digestion of, 859*f*
DNA replication and repair, and role of, 301–3, 304*t*
domains of (*see* Domains, protein)
enzymes as, 77, 78*f* (*see also* Enzyme(s))
folding of, 85, 325, 345–46
functions of, linked to structure, 77, 78*t*
G (*see* G protein)
heat-shock, 812, 840
improving yield of, in crop plants, 764
infectious (prions), 345, 346*f*
integral, 128
interactions among, in cell, 10*f*
membrane, 127–29
motor (*see* Motor proteins)
as nutritional requirement, 849–50
peripheral, 128
as polypeptide chain, **78** (*see also* Polypeptides)
posttranslational processing of, as regulation of eukaryotic gene expression, 369, 370*f*, 791
PR, 814
R, 813, 814*f*
receptor (*see* Receptor proteins)
relay, 215
repressor, 353
respiratory pigments, 892–94
scaffolding, 215*f*
secretory, 105
stress-induced, 840
structural levels of, 81, 82–83*f*
synthesis of (*see* Protein synthesis)
tissue-specific, 419
transmembrane, 128*f*
transport (*see* Transport proteins)
Protein folding, 85, 325, 345–46
Protein kinase(s), **209**
role in signal transduction pathway, 1024
Protein kinase A, 210, 211*f*
Protein phosphatases, **210**
Protein phosphorylation, 209–10
Protein synthesis, 309–33
as anabolic pathway, 142
DNA-directed, 86*f*
gene expression as specification of, 309–14
genetic code and, 312–14
overview of transcription and translation steps in, 311–12
point mutations affecting, 328–30
in prokaryotes versus in eukaryotes, 327–28
RNA, multiple roles in, 327
RNA modification following transcription as phase of, 317–19
summary of, 331*f*
transcription and RNA synthesis as phase of, 311–12, 315–17
translation and polypeptide synthesis as phase of, 311–12, 320–26
Proteobacteria, 542*f*
Proteoglycans, **119**
in extracellular matrix of animal cells, 119*f*
Proteomics, **402**
Proterozoic era, 527
fossils from, 526*f*
Protists, **549**–73
alternation of generations in some, 561–62
alveolates, 555–58
amoebozoans, 564–66
animal evolution from, 628*f*
cercozoans, 563
contractile vacuoles of, 108
diplomonads and parabasalids, 552, 553*f*
diversity of, 549–51
endosymbiosis and, 550–51
euglenozoans, 553–54
evolution of fungi from, 612
exchange with environment and body of, 821*f*
gas exchange in, 884
motility of, 117*f*, 118
multicellularity and,
as pathogens, 552, 553*f*, 554*f*, 555, 556*f*, 564
phylogeny of eukaryotes and place of, 552*f*
pseudopodia and movement in, 117*f*
radiolarians, 563
red and green algae, 567–69
sexual life cycle in some, 242*f*
stramenopiles, 558–62
Protista (kingdom), 549. *See also* Protists
Protobionts, 513, **515**
Protons, **34**
atomic number and, 34–35
Protonema, **580**, 581*f*
Protonephridium, **929**, 930*f*
Proton-motive force, **173**
Proton pump, **136**, **739**
plant cell growth in reponse to auxin and, 794–95
transport in plants and role of, 739, 740*f*
Proto-oncogenes, **371**
conversion of,into oncogenes, 371*f*
Protoplast, **717**, **782**
Protoplast fusion, **782**
Protostomes
cleavage in, 631–32
coelom formation in, 632
deuterostome development verus, 632*f*
fate of blastophore in, 632–33
Protostome development mode, **631**–33
Protozoan *Mixotricha paradoxa*, symbiotic relationships of, 525
Provirus, **341**
Proximal control elements, 364
Proximal tubule, **931**, 932*f*, 933
Proximate causes of behavior, 1106, 1107, 1108*f*, 1109*f*
Proximate questions, **1107**
Prozac (drug), 1025, 1040
PR (pathogenesis-related) proteins, **814**
Pseudocoelom, 631
Pseudocoelomates, **631***f*
Pseudogenes, **378**, **468**
in globin gene family, 377*f*, 378
Pseudopodia, **117***f*, **563**, 564*f*
Pseudostratified ciliated columnar epithelium, **824***f*
Psilotum nudum whisk fern, 587*f*
P site, tRNA binding, **322***f*
Pteraspis, 679*f*
Pterophytes, **578**, 587*f*, 588
Pterosaurs, **689**
Public sanitation, 546
Puffballs, 618*f*
Pulmocutaneous circuit, **871**
Pulmonary circuit, **871**
Pulse, heart and, **873**
Punctuated equilibrium model for speciation tempo, **481**, 482*f*
Pundamilia pundamilia and *Pundamilia nyererei*, speciation of, 479, 480*f*
Punnett square, **254**, 255*f*, 263*f*
Pupil, of eye, **1059**
Purines, **88**, 298
Pyloric sphincter, 855*f*, **858**
Pyramid of biomass, 1192, 1193*f*
Pyramid of numbers, 1193
Pyramid of net production, 1192*f*
Pyrimidines, **88**, 298
Pyrococcus furiosus, 541
Pyrolobus fumarii, 541
Pyruvate
conversion of, to acetyl CoA, 168*f*
as key juncture in catabolism, 176*f*
oxidation of glucose to, during glycolysis, 164, 166–67*f*
Python
bulk feeding by, 845*f*
energy budget in, 830, 831*f*
PYY hormone, 847*f*
Quantitative characters, **263**, 463
Quaternary structure of protein, 81, **83***f*
Quillworts, 586, 587*f*
Quinn, William, 1116

RAAS (renin-angiotensin-aldosterone system), **937**, 938
Rabies, 914
Radial cleavage, **632**
Radial glia, **1015**
Radial symmetry, **630**
in animal body plan, 630*f*
in flowers, 773*f*
Radiation
heat and, **835**
mutations caused by, 329–30
Radicle, **778**, 780
Radioactive isotopes, **35**–36
use in biological research, 35*f*
uses and dangers of, 36
Radiolarians, **563***f*
Radiometric dating of fossils, **517**
Radula, of mollusc, **650**
Rafflesia arnoldii, 771*f*
Raikhel, Natasha, 712, 729, 802
on systems biology and study of *Arabidopsis*, 710–11
Rainbow trout (*Oncorrhynchus mykiss*), 939*f*
Rain shadow, 1090
Random dispersion of populations, 1138*f*
Random fertilization, genetic variation from, 248
Range of species, 1081*f*, 1084, 1085*f*
climate change and, 1092*f*
ras gene, **371**
development of cancer and role of, 371, 372*f*, 373
Rat
kangaroo, 925*f*, 939*f*
naked mole, 1128, 1129*f*
tetraploidy in, 286*f*
Ratfish, spotted (*Hydrolagus colliei*), 681*f*
Ratites (flightless birds), **693**
Ravens, problem solving in, 1117
Ray-finned fishes (Actinopterygii), 682, **683***f*
Ray initials, **727**
Rays, 680–82
Reactants in chemical reactions, **44**
key, and chemical work in cells, 149*f*
Reaction center, photosynthesis, **189**
Reading frame, **314**
Receptacle, flower, **598**, **772**
Reception of cell signaling, 204–8
intracellular, 205
intracellular, for lipid-soluble hormones, 947
by plants, 429–30, 789
in plasma membrane, 205, 206–8*f*
Receptive field of ganglion cells, 1062
Receptors, 944. *See also* Receptor proteins
cell-surface, for water-soluble hormones, 946
of internal environment, 832
intracellular, 205
intracellular, for lipid-soluble hormones, 947
in plasma membrane, 205, 206–8*f*
Receptor-ligand model of gene-for-gene plant resistance to pathogens, 814*f*
Receptor-mediated endocytosis, 137, **138***f*
Receptor molecules. *See* Receptor proteins
Receptor potential, **1046**
Receptor proteins, 78*t*
G-protein-linked, 206*f*
ligand-gated ion channel as, 208*f*
in plasma membrane, 205, 206–8*f*
receptor tyrosine kinase, 207*f*, 231
Receptor tyrosine kinase, **207***f*, 231
Recessive allele, **254**
sex-linked, 283–84
Recessive trait, 253
attached earlobes as, 265*f*
genetic disorders as, 266–67
sex-linked, 283*f*–84
Reciprocal altruism, **1130**–31
Reclinomonas americana, 525

Index

Recombinant chromosomes, **248**, 249*f*, 278, 279*f*, 280*f*
Recombinant DNA, **384**. *See also* Genetic recombination
bacterial conjugation, plasmids, and, 349–51
production of, overview, 385
restriction enzymes and production of, 386*f*
Recombinant frequency, 279–81
Recombinants (recombinant types), **278**
Recombinase, 906
Recruitment of motor neurons, **1070**
Rectum, **862**
Red algae (rhodophytes), 550*f*, 561*f*, **567**, 575*f*
Red blood cells (erythrocytes), 825*f*, **880**–82
carbohydrates on surface of, 129
glucose transport in, 130
hemoglobin and (*see* Hemoglobin)
normal, and sickled, 84*f*
plasma membrane of, 99*f*, 128
sickle-cell anemia in, 84 (*see also* Sickle-cell disease)
Red-cockaded woodpecker, 1218, 1219*f*
Red light, plant responses to, 803*f*, 806, 807*f*
Redox reactions, **161**–64
energy harvesting via NAD^+ and electron transport chain, 162, 163*f*, 164
methane combustion as energy-yielding, 161, 162*f*
oxidation of organic fuel during cellular respiration, 162
photosynthesis as, 184
principle of, 161–62
Red tide, 555
Reducing agent, **161**
Reduction, **161**
Reductionism, **9**–10
Redundancy model of community, **1180**
Reflexes, **1013**
knee-jerk, 1013*f*
Refractory period following action potential, **1020**
Regeneration, asexual reproduction by, **965**
Regulation
as biological theme, 27*f*
of cell cycle, 228–33
by feedback, in biological systems, 11–12
as property of life, 3*f*
Regulators versus conformers, **832**
Regulatory genes, 420
homeobox sequences and, 431–32, 627
homeotic genes as, 425 (*see also* Homeotic genes)
lacI, 355*f*
myoD, 419*f*, 420
trpR, **353**–54
Regulatory proteins, muscle contraction and role of calcium and, 1068, 1069*f*
Relative abundance of species in communities, **1165**
Relative fitness, **464**
Relay proteins, 215
Release factor, 324
Renal artery, **931**, 932*f*
Renal cortex, of kidney, **931**, 932*f*
Renal medulla, of kidney, **931**, 932*f*
Renal pelvis, **931**, 932*f*
Renal vein, **931**, 932*f*
Renin-angiotensin-aldosterone system (RAAS), **937**, 938
Repeated reproduction, **1141**
Repetitive DNA, **375**, 376–77
Replication. *See* DNA replication
Replication bubble, 301*f*
Replication fork, **301***f*
Repressible enzymes
inducible enzymes compared with, 355
trp operon and regulated synthesis of, 353, 354*f*
Repressor, **353**, **365**
control of eukaryotic gene expression and role of, 365
trp, in bacteria, 353
Reproduction. *See also* Asexual reproduction; Sexual reproduction
in animals (*see* Animal reproduction)
in bacteria (prokaryotes), 346–51, 537–38
of cells (*see* Cell division)
cycles and patterns of, 965–66
in fungi, 610–12
in insects, 661, 968, 969*f*
parthenogenesis, 648, 965, 966*f*
in plants (*see* Plant reproduction)
as property of life, 3*f*
in protists, 242*f*
protobionts and simple, 515*f*
semelparity versus iteroparity in, 1141
of viruses, 336–42
Reproductive barriers, 474–75*f*
Reproductive isolation, **473**
in divergent *Drosophila* populations, 477*f*
reproductive barriers and, 474–75*f*
Reproductive leaves, 716*f*
Reproductive rates in populations, 1140–41
Reproductive table, **1140**
for Belding's ground squirrel, 1141*f*
Reptiles, **688**–94
alligators and crocodiles, 691
birds, 691–94
characteristics of, 688–89
circulatory system, 870*f*, 871
dinosaurs, archosaurs, pterosaurs, and theropods, 689, 690*f*
kidney adaptations in, 939*f*
lepidosaurs, 689, 690, 691*f*
nitrogenous wastes of, 927*f*, 928
origin and evolutionary radiations of, 689–90
snakes (*see* Snakes (Squamata))
thermoregulation in, 835*f*, 836, 838
turtles, 690, 691*f*
Research methods
absorption spectrum, 187*f*
cell fractionation, 97*f*
cloning human genes in bacterial plasmid, 387*f*
dideoxy chain-termination method for sequencing DNA, 397*f*, 398
DNA microarray assay, 401*f*
electron microscopy, 96*f*
freeze-fracture preparation of cells for, 126*f*
intracellular recording for measurement of membrane potential, 1016*f*
karyotype preparation, 240*f*
light microscopy, 96*f*
linkage map, 281*f*
Mendel's plant breeding, 252*f*
parsimony principle applied to molecular systematics, 502–3*f*
polymerase chain reaction (PCR), 391*f*
radioactive tracers, 35*f*
studying natural selection in virtual populations, 467*f*
testcross, 256*f*
X-ray crystallography, 86*f*
Residual volume, breathing, **889**
Resolution, of microscope, 95
Resource competition, population regulation through, 1148–49
Resource partitioning, **1161**
Respiration. *See* Cellular respiration
Respiratory medium, **884**
Respiratory pigments and gas transport, **892**–95
carbon dioxide transport, 893–94
oxygen transport, 892*f*, 893
role of partial pressure gradients in, 891*f*
Respiratory surface, **884**
Respiratory system
breathing and, 888–90
in elite animal athletes, 894–95
mammalian, 887, 888*f*
Response as property of life, 3*f*
Response to cell signaling, 945–46
by regulation, or by transcription, 204, 212–15
by plants, 429–30, 790–91
Resting potential, **1016**
Restoration ecology, **1209**, 1224–28
biological augmentation as, 1226
bioremediation as, 1226
community recovery from disasters and, 1226*f*
exploring global projects for, 1226, 1227–28*f*
Restriction endonucleases, 338
Restriction enzymes, **338**, **386**
recombinant DNA produced using, 386*f*
Restriction fragments, DNA, **386**
analysis of, 392–94
Restriction fragment length polymorphisms (RFLPs), **394**
forensic analysis using, 405
linkage mapping, 396*f*
as markers for disease-causing alleles, 403*f*
noncoding DNA and analysis of, 394
Restriction site, **386***f*
Reticular fibers, **823**
Reticular formation, **1029***f*
Retina, **1059**
cellular organization of vertebrate, 1062*f*
rod and cone cells of, 1059, 1060*f*, 1061*f*, 1062*f*
visual information processing and, 1061–63
Retinal, **1060***f*
Retinitis pigmentosa, 462
Retrotransposons, **375**–76
copy and movement of, 376*f*
Retrovirus, **340**–41, 370
retrotransposons and, 376
as vector for gene therapy, 403*f*
Reverse transcriptase, **341**
Reznick, David, research on guppy evolution by, 446, 447*f*
Rheotaxis in fish, 1110*f*
Rheumatoid arthritis, 917*f*
Rh factor, **915**–16
Rhine River, Europe, restoration project in, 1228
Rhizobium sp., 542*f*
nitrogen-fixation by, 764, 765*f*, 766
Rhizoids, **580**
Rhizomes, 715*f*
Rhizopus stolonifer (black bread mold), life cycle of, 614*f*
Rhodopsin, **1060***f*
Rhythm method, contraceptive, **982**
Ribbon worms (Nemertea), 639*f*, 649*f*
Ribonuclease, determining structure of, 86*f*
Ribonucleic acid. *See* RNA (ribonucleic acid)
Ribose, **88**
Ribosomal RNA (rRNA), 102, **322**, 327*t*
multigene family for, 377*f*
small-subunit (SSU-rRNA), 524, 540
Ribosome(s), 98, **102**–4, 309*f*, **311**
anatomy of functioning, 322*f*
computer model of, 322*f*
free versus bound, 102, 325
protein synthesis and, 86*f*, 87, 323–24
as site of translation, 311
structure of, 103*f*
Ribozymes, **318**, **515**
as biological catalyst, 318
RNA replication by, 515, 516*f*
Rice (*Oryza sativa*), 475*f*, 802
diseases of, 797
transgenic, 784, 785*f*
Riechert, Susan, 1119–20
Ringworm, 622
Rivera Gonzales, Carlos, 1230*f*
Riverine wetlands, 1094
Rivers. *See* Streams and rivers
Rivet model of communities, **1180**
RNA (ribonucleic acid), **86**. *See also* Messenger RNA (mRNA); Ribosomal RNA (rRNA); Transfer RNA (tRNA)
as earliest form of genetic material, 515–16
modification of, after transcription, 317–19
priming DNA synthesis with, 303
protein synthesis and role of, 86*f*
ribozymes and replication of, 515, 516*f*
small nuclear RNA (snRNA), 318, 327t
transcription and synthesis of, in antiparallel direction, 313, 315*f*, 316*f*, 317

translation of polypeptides directed by, 320–26
types of, in eukaryotic cell, 327*t*
as viral genetic material, 340–42
RNA interference (RNAi), **368**, 369, **400**
RNA polymerase, **315**
attachment of, and initiation of transcription, 315*f*, 316
control of eukaryotic gene expression and, 364, 365*f*
RNA processing, **312**, 317–19
alteration of mRNA ends (addition of 5' cap and poly-A tail), 317*f*
intron function and importance in, 319
regulation of eukaryotic gene expression and, 368
ribozyme function during, 318
split genes and RNA splicing during, 318f, 319
RNA splicing, **318***f*
alternative, 319, 368, 399
RNA strands, antiparallel, 321*f*
RNA viruses, 340–42, 344
reproductive cycle of enveloped, 341*f*
Roadrunner (*Geococcyx californianus*), 939*f*
Rocks
as abiotic factor, 1087
fossils in (*see* Fossil(s))
magnetic reversals affecting, 518
Rod cells, eye, **1059**
production of receptor potential in, 1061*f*
structure and light absorption by, 1060*f*
Rodents, biological clocks in, 1030, 1031*f*
Rod-shaped (bacilli) prokaryotes, 98*f*, 535*f*
Rolling DNA replication, 350*f*
Roots, 577*f*, **584**, **713**. *See also* Root system
adventitious, 713
auxin and formation of, 795
evolution of, 584–85
fibrous, 713
gravitropism in, 809*f*
lateral, 723*f*
lateral transport in, 745*f*
modified, 714*f*
nitrogen-fixing bacterial nodules on, 764, 765*f*, 766
primary growth of, 721*f*, 722*f*, 723*f*
secondary growth of, 725–28
taproot, 713
water and mineral absorption by, 744–46
Root cap, **721**
Root hairs, 713*f*, **714**
homeotic gene GLABRA-2 and, 732, 733*f*
water and nutrient absorption and role of, 744
Root pressure, **746**
ascent of xylem sap and, 746
Root system, **713**–14. *See also* Root
primary growth in, 721–23
secondary growth in, 725–28
water and mineral absorption by, 744–46
Root tip, 713*f*, 721
Rose-shaped complexes for cellulose synthesis, **574**
Rosette cellulose-synthesizing complexes, **574**
Rosy periwinkle (*Catharanthus roseus*), 1211, 1212*f*
Rotifers (Rotifera), 639*f*, 648–49
Rough ER (endoplasmic reticulum), **104**, 105*f*
functions of, 105
secretory protein synthesis and, 105
Round window, ear, **1052**
Rous, Peyton, 370
R plasmids, **351**
antibiotic resistance in bacteria and, 351
R proteins and R genes, 813, 814*f*
R-selection (density-independent) populations, **1147**
RT-PRC, 403
Rubisco (RuBP carboxylase), **194**
Rudner, Rivka, 818
Ruffled grouse (*Bonasa umbellatus*), 1221
Ruminants, digestion in, **864***f*
Running, energy cost of, 1074*f*
Rwanda, 1155
Saccharomyces cerevisiae. *See* Yeast (*Saccharomyces cerevisiae*)
Saccule, **1052**, 1053*f*
Sac fungi (Ascomycota), **616**–17
Sagittaria latifolia (common arrowhead), as dioecious species, 773*f*
Sahelanthropus tchadensis, *702*
Salamanders (Urodela), 475*f*, 685
genome, 377*f*
heterochrony and evolution of feet of, 484*f*
nervous system, 1012*f*
paedomorphosis in, 485*f*
Salicylic acid, **815**
Saliva, 856
Salivary amylase, **856**
Salivary glands, **855**
Salmon, Pacific, life history pattern of, 1141–42
Salmonella sp., 542*f*
Salts. *See also* Sodium chloride (NaCl)
dietary, 852
excessive, in soils, 762
excretion of, by birds, 926*f*
ionic compounds as, **41**
plant response to excessive, 711–12
Saltatory conduction, **1021**
Salt marsh
black rush as facilitator species in, 1170*f*
limiting nutrient in Hudson Bay, 1191*f*
Sand dollars, 641*f*
cleavage in embryo of, 992*f*
Sanderson, Michael, 578
Sandpipers (*Calidris mauri*), 1110*f*
Sanger, Frederick, 80, 396
Sapwood, **727**
Sarcomeres, **1066**, 1067*f*
Sarcoplasmic reticulum (SR), **1069**
Sarcopterygii (lobe-fins), 683–84
Sargasso Sea, nutrient enrichment experiment in, 1188, 1189*t*
Sarin nerve gas, 155
SARS (severe acute respiratory syndrome), 344
Satellite DNA, 377
Saturated fatty acid, **75**
Savanna, **1101***f*
Sawka, Nadia, 1125
Scaffolding proteins, **215***f*
Scala naturae (scale of nature), 439
Scales
fish, 682
reptile, 688
Scallops (Bivalvia), 652*f*, 885*f*
Scanning electron microscopes and micrographs (SEMs), **96***f*
Scarlet fever, 339
Scarlet king snake, case study on mimicry of warning coloration in, 21–24
Schemske, Douglas, 481
Schindler, David, 1190
Schistosomiasis, blood fluke and, 30, 647*f*, 1163
Schistosoma mansoni (blood fluke), 30, 1163
life cycle, 647*f*
Schizocoelous development, **632**
Schizophrenia, **1039**–40
Schwann cells, **1015**
myelin sheath and, 1015*f*
Science
controlled experiments in, 23–24
culture of, 25
discovery, 19–20
hypothesis-based, 20*f*–21
inquiry at heart of, 19
limitations of, 24
model-building in, 24, 25f
technology, society, and, 25–26
theory in, 24
Scientific method, 21. *See also* Biological methods
Scion, of plant, **782**
Sclera, **1058**, 1059*f*
Sclereids, **718**
Sclerenchyma cells, **718***f*
Scolex, 648
Scorpions, 641*f*, 659*f*
Scrapie, 345
Scrotum, **971**
Scutellum, **778**
Scyphozoans (jellies, sea nettles), 644*t*
Sea anemones (*Anthopleura elegantissima*), 644*f*, 645
asexual reproduction in, 965*f*
Sea cucumbers (Holothuroidea), 666, 667*f*
Sea hare (Aplysia californica), cellular basis of learning in, 1036*f*
Sea horse (*Hippocampus ramulosus*), 683*f*
Seal, 821*f*
blubber, 835
bottleneck effect in populations of northern elephant, 461*f*
circulatory and respiratory adaptions of Weddell, 894–95
northern fur, 1136*f*
Sea level, global warming and rising, 1205
Sea lilies (Crinoidea), 666
Sea mat (*Membranipora membranacea*), 649*f*
Sea palm (*Postelsia*), 561*f*
Seasons, climate and effects of, 1088f, 1090, 1091*f*
Sea otters (*Enhydra lutris*) as keystone predator, 1169*f*, 1214
Sea squirt (tunicate), 673–74
Sea stars (*Pisaster ochraceous*), 641*f*, 666, 667*f*
anatomy, 665*f*
gills, 885*f*
innate defenses of, 902
as keystone predator, 1168*f*, 1169
nervous system, 1012*f*
Sea urchins (Echinoidea), 475*f*, 641*f*, 666, 667*f*
fertilization of eggs in, 988, 989*f*, 990*f*
gastrulation, 995*f*
seaweed distribution affected by feeding of, 1085, 1086*f*
Sea wasp (*Chironex fleckeri*), 644*f*
Seaweeds
alternation of generations in life cycle of, 561, 562*f*
brown algae as, 560–62
distribution of, affected by sea urchin feeding, 1085, 1086*f*
green algae as, 567–69
as human food source, 561*f*
red algae as, 567
Sebaceous glands, 899
Secondary plant body, **725**
Secondary cell wall, of plant, **118**, 119*f*
Secondary compounds, 575
Secondary consumers, **1185**
Secondary endosymbiosis, **551**
plastids produced by, 551*f*
Secondary growth, plants, **720**–21, 725–28
auxin as stimulant to, 796
cork cambia and periderm production, 728
vascular cambium and, 725, 726*f*, 727
wood as, 605, 727
Secondary immune response, **908**
Secondary oocyte, **974**
Secondary plant body, **725**
Secondary production in ecosystems, **1191**
green world hypothesis and, 1193–93
production efficiency of, 1191–93
Secondary structure of protein, 81, **82***f*
Secondary succession, ecological, **1173**
Second law of thermodynamics, 143, **144**
Second messengers, **210**, **790**
calcium ions and inositol trisphosphate as, 211*f*, 212*f*
cyclic AMP as, 210*f*, 211*f*, 1056
cyclic GMP (cGMP) as, 790
small molecules and ions as, 210–12
Secretion, excretory system and, **929**
Secretory phase, menstrual cycle, **977**
Secretory proteins, rough endoplasmic reticulum and synthesis of, 105

Sedimentary rocks, fossils in, **439**, 440*f*
Seed, **579**, **593**
  asexual production of (apomixis), 781
  crop size of, 1142*f*
  development of, from ovule, 593*f*
  dispersion of, 480*f*, 599*f*, 1138*f*
  dormancy of, 779–80, 798, 799*f*
  germination of, 779, 780*f*, 798
  of gynosperms, 593, 596, 597*f*
  structure of mature, 778*f*
Seed coat, **778**
Seedless vascular plants, **578***t*, 584–88
  carboniferous, 588*f*, 596
  classification of, 586–88
  diversity of, 587*f*
  ferns, life cycle, 585*f*
  origins of vascular plants, 584–86
  significance of, 588
Seed plants, **578***t*, 591–606
  angiosperms, 579, 598–604
  gymnosperms, 579, 593–97
  importance of, to human welfare, 591, 605–6
  reduced gametophytes and seeds of, 591–93
Seed predators, 1161
Segmentation genes, 423, **425**
Segment polarity genes, 425
Segregation of alleles, 254
  fertilization and, as chance events, 259*f*
  law of, 254–55, 274, 275*f*
Selaginella apoda spike moss, 587*f*
Selective permeability of membranes, **124**, 130
Selective reabsorption/secretion of solutes, **929**
Self-nonself recognition
  in animal immune system, 915–16
  human pregnancy and, 915
  in plants, 775–76
Semelparity, **1141**
Semen, 948, **972**
Semicircular canals, **1051**
Semiconservative model of DNA replication, **299**, 300*f*
Semilunar valves, **873**
Seminal vesicles, **972**
Seminiferous tubules, **971**
Senile plaques, 1040*f*
Sensations, perception of, **1046**. *See also* Sensory receptor
Sensilla, 1055
Sensitive period for learning, **1108**–9
  bird song and, 1118
Sensitization as learning, 1036*f*
Sensory adaptation, **1047**
Sensory input, 1013*f*. *See also* Sensory mechanisms
Sensory mechanisms, 1045–63
  functions of sensory receptors, 1046–48
  for gravity and sound in invertebrates, 1050
  for hearing and equilibrium, 1046, 1047*f*, 1050–54
  for magnetism and light, 1048–49, 1057–63
  mechanoreceptors, 1046, 1047*f*, 1050–54
  overview of, 1045
  for pain, 1049
  for taste and smell, 1048, 1054–57
  for temperature, 1049
  types of sensory receptors, 1048–49
  vision, 1057–63
Sensory neurons, **1013***f*
  structure, 1014*f*
Sensory reception, **1046**
  amplification of stimulus energy in, 1046
  general function of, 1046–48
  integration of sensory information, 1047–48
  transduction of signals, 1046
  transmission of action potentials following, 1046–47
  two mechanisms of, 1047*f*
Sensory receptors, **1046**
  chemoreceptors, 1048
  electromagnetic receptors, 1048, 1049*f*
  exteroreceptors, 1046
  functions performed by, 1046–48
  interoreceptors, 1046
  mechanoreceptors, 1046, 1047*f*, 1048
  pain receptors, 1049
  photoreceptors, 1057–63
  stretch, 1046, 1047*f*
  thermoreceptors, 1048*f*, 1049
  transduction/transmission of stimulus by, 1046
Sensory transduction, **1046**
  in cochlea of ear, 1052*f*
  in eye, 1060, 1061*f*
  for taste, 1055, 1056*f*
Sepals, flower, 429*f*, 430, **598**, **772**
Septa, fungal hyphae, **609***f*
Septic shock, 902
Sequential hermaphroditism, **966**
Sequoia (*Sequoiadendron giganteum*), 595*f*
Sereno, Paul, 15*f*
Serial endosymbiosis, **524**
Serosa tissue layer, of stomach, **827***f*
Serotonin, **1025**
Seta, of moss, **580**
Severe combined immunodeficiency (SCID), 404, 918
Sex, reproductive handicap of, 469*f*. *See also* Sexual reproduction
Sex chromosomes, **240**
  aneuploidy of, and human genetic disorders, 287
  sex determination and, 282*f*
  sex-linked genes on, 283–84
Sex determination, chromosomal basis of, 282*f*–83
Sex hormones, 77
  comparison of functional groups of, 63*f*
  female, 63*f*, 958–59
  gonads and secretion of, 958–59
  male, 63*f*, 958–59, 977–78
Sex-linked genes, **283**
  chromosomal basis of sex and, 282–83
  inheritance of, 283*f*, 284
  X-chromosome inactivation and, 284
Sexual dimorphism, **468**
  human evolution and, 704–5
Sexual life cycle, **240**. *See also* Sexual reproduction
  alternation of fertilization and meiosis in, 240–43
  angiosperm, 600*f*, 772*f*
  of conifer (pine), 596, 597*f*
  genetic variation produced by, 247–49
  of humans, 241*f*
  meiosis and, 243–47
  of protists, 550
  varieties of, 242*f*
  of water mold, 559*f*
Sexual recombination, genetic variation and, 247–49, 460
Sexual reproduction, **239**, 627, **964**. *See also* Sexual life cycle
  in animals, 627, 964–86
  in arthropods (insects), 661, 968, 969*f*
  asexual reproduction compared to, 239, 964
  fertilization mechanisms in, 967–68
  in fungi, 610, 611*f*, 614*f*, 617*f*, 619*f*
  gamete production and delivery in, 968–69
  genetic variation produced by, 247–49
  hormonal control of gametogenesis in animal, 973–78
  in invertebrates (earthworms), 964*f*
  natural selection and advantages/disadvantages of, 469
  in plants, 576, 580, 581*f*,, 585*f*, 586, 592–93, 596, 597*f*, 599, 600*f*, 601
  organs of animal, 969–73
  reduction of chromosome numbers via meiosis in, 242, 243–47
  reproductive cycles and patterns, 965–66
Sexual response, human, 972–73
  neurotransmitters and, 1025
Sexual selection, **468**. *See also* Mating
  imprinting and, in zebrafinch, 1126*f*
  mate choice and, 1125–27
  speciation and, in cichlid fishes, 479, 480*f*
Seyfarth, Richard, 1131–32
S-genes in plants, 776
Shade avoidance in plants, 804–5
Shaffer, Mark, 1217
Shared derived characters, **498**
Shared primitive character, **498**
Sharks, 821*f*, 680–82
  osmoregulation in, 924
  reproduction in, 681–82
  senses, 681
  thermoregulation in, 837*f*
Sheep, cloning of, 416
Shelf fungi, 618*f*
Shell
  amniote, 688
  mollusc, 650
Sherman, Paul, 1138, 1139
Shigella, antibiotic resistance in, 351
Shoot, 577*f*
Shoot system, plant, **713**
  cell fate and position in, 733*f*
  effect of phase changes on, 733*f*
  leaves, 725*f*
  primary growth in, 720*f*, 723*f*, 724
  secondary growth in, 720*f*, 725–28
  stems, 715, 724, 725–28
Short-day plant, **806**
Short-term memory, **1035**
Shrimp, 664
Shuster, Stephen M., 1126–27
Sibling, coefficient of relatedness between, 1129*f*
Sickle-cell disease, **267**, 403, 404
  amino-acid substitute in hemoglobin and, 84*f*
  heterozygote advantage, malaria, and, 267, 466*f*
  identifying sickle-cell allele mutation, 392, 393*f*
  point mutation as cause of, 328, 329*f*
  as recessively inherited disorder, 267
Side-blotched lizard (*Uta stansburiana*), game theory applied to mating successes of, 1127f, 1128
Sieve plates, **719**
Sieve-tube members, **719**
Signal, animal communication by, **1111**–12
Signal peptide, **326**
Signal-recognition particle (SRP), **326**, 327*t*
Signal transduction, 204, 208–12
  in cochlea of ear, 1052*f*
  hormones and, 945, 946*f*, 947*f*
  membrane proteins and, 128*f*
  in plants, 788–91
  protein phosphorylation and dephosporylation in, 209f, 210
  sensory reception in animals and, 1046 (*see also* Sensory transduction)
  small molecules and ions as second messengers in, 210–12
Signal transduction pathway, **202**, 204, 208, **946**
  in egg fertilization (cortical reactions), 989*f*, 990*f*
  general model of, 789*f*
  hormonal control, 945*f*, 946*f*
  of long-term potentiation, 1037*f*
  neural synaptic transmissions and, 1024
  in plants, 788–91
  protein phosphorylation/dephosphorylation and, 209f, 210
  relay proteins in, 215
  second messengers in, 210–12
Sign stimulus, fixed action patterns and, **1108**
Silencing, control of eukaryotic gene expression and, 365
Silent Spring (Carson), 1083*f*
Silkworm moth (*Bombyx mori*), chemoreceptors in, 1049*f*
Silurian period, 629
Silversword alliance, adaptive radiation and speciation in, 481*f*, 495
Simple epithelium, **823**
Simple fruit, **779**
Simple leaf, 716*f*
Simple sequence DNA, 376–77
Simpson, George Gaylord, 455

- Sinervo, Barry, 1127–28
- Singer, S. J., 125
- Single bond, **40**
- Single-lens eye, **1058**
- Single nucleotide polymorphisms (SNPs), **402**
- Single-strand binding protein, **303**, 304*t*
- Sinoatrial (SA) node, **873**–74
- Sinuses, open circulatory system, **868**, 869*f*
- siRNA (small interfering RNA), 327t, **368**, 369*f*
- Sister chromatids, **219**
  - mitosis and, 219, 220*f*
  - separation, in meiosis, 245*f*
- Skeletal muscle, **826***f*, **1066**
- Skeleton
  - animal, 1063–65
  - endoskeleton, 1064
  - exoskeleton, 657, 1064
  - human, 1065*f*
  - hydrostatic, 1063, 1064*f*
  - locomotion and, 1073–74
  - muscles and, 1066–72
  - origins of bone in, 679, 680, 682
  - support of, and size of animal, 1064–65
- Skin
  - in aquatic osteichthyans, 682
  - as body defense mechanism, 899
  - color of, melanocytes and, 946, 952
  - color of, and polygenic inheritance, 263*f*
  - role of, in thermoregulation, 835
  - sensory receptors in human, 1048*f*
  - ultraviolet damage to, 305*f*, 306
- Skoog, Folke, 796
- Skull, comparisons of human and chimpanzee, 484*f*, 494
- Skunk (*Spilogale* sp.), 474*f*
- Sleep
  - biological clocks and human disorders in, 1030–31
  - human brain and state of, 1029–30
- Sleeping sickness, 401, 554*f*, 664
- Sleep movements in plants, 805*f*
- Sliding-filament model of muscle contraction, **1067***f*, 1068
- Slime molds, 564–66
  - cellular, 565–66
  - plasmodial, 564–65
- Slow block to polyspermy, **989**
- Slow muscle fibers, 1071, 1072*t*
- Slugs (Gastropoda), 651–52
- Small interfering RNA (siRNA), 327t, **368**, 369*f*
- Small intestine, **858**–61
  - absorption of nutrients in, 859–61
  - enzymatic digestion in, 858–59
  - structure of, 860*f*
  - villi in, 859–60
- Small nuclear ribonucleoproteins (snRNPs), role in RNA splicing, 318, 319*f*
- Small nuclear RNA (snRNA), 318, 327*t*
- Small nucleolar RNA (snoRNA), 327*t*
- Small-population approach to conservation biology, 1215–18
  - case study of greater prairie chicken, 1216
  - case study of grizzly bear, 1217–18
  - effective population size and, 1216–17
  - extinction vortex and, 1215–16
  - minimum viable population size and, 1216
- Smallpox, 914
- Small subunit ribosomal RNA (SSU-rRNA), 540
  - origins of mitochondria and plastids, and, 524
- Smell, human sense of, 1056, 1057*f*
  - olfactory bulb, 1034, 1035*f*, 1057*f*
- Smooth ER (endoplasmic reticulum), **104**, 105*f*
  - functions of, 104–5
- Smooth muscle, **826***f*, **1072**
- Snails, 639*f*, 651–52
  - Cuban tree, 454*f*
- Snakes (Squamata), 690, 691*f*
  - brown, as introduced species into Guam, 1213*f*
  - bulk-feeding by, 845*f*
  - energy budget in, 830, 831*f*
  - fangs, 862
  - garter (*Thamnophis*), 474*f*, 1118, 1119*f*
  - infrared receptors in, 1049*f*
  - poisonous, 690
  - predation by, 1161–62
  - python, 830, 831*f*, 845*f*
  - scientific inquiry into mimicry in populations of, 21–24
- snoRNA (small nucleolar RNA), 327*t*
- Snowball Earth hypothesis, **526**
- SnRNA (small nuclear RNA), 318, 327*t*
- SnRNPs (small nuclear ribonucleoproteins), 318, 319*f*
- Snyder, Solomon, 1025
- Social behavior, 1128–33
  - altruistic, and inclusive fitness, 1128–31
  - competitive, 1160–61
  - cooperative, 1164
  - evolution, human culture and, 1132–33
  - mating, 1123–28
  - social learning and, 1131–32
- Social environment, aggressive behavior and influence of, 1114
- Social learning, **1131**–32
  - of alarm calls, 1131–32
  - mate choice copying, 1131*f*
- Sociobiology, **1132**–33. *See also* Social behavior
- *Sociobiology: The New Synthesis* (Wilson), 1132
- Sodium (Na^+)
  - ligand-gated channels and, 1022*f*
  - selective permeability of neuron membrane to, 1017*f*
  - voltage-gated channels for, 1018, 1019*f*, 1020*f*, 1021
- Sodium chloride (NaCl), 33*f*. *See also* Salt(s)
  - dissolving, in water, 51*f*
  - ionic bonds in crystals of, 41*f*
- Sodium hydroxide (NaOH), 54
- Sodium-potassium pump, **134**
  - active transport by way of, **134**, 135*f*
  - neuron resting potential and, 1016–17
- Soil, 759–63
  - as abiotic factor, 1087
  - agricultural impact on, 1200*f*
  - ammonifying bacteria in, 763*f*, 764
  - availability of water and minerals in, 761*f*
  - erosion of, 762
  - fertilizers for, 761–62
  - horizons, 760*f*
  - irrigation of, 762
  - nitrogen fixation by bacteria in, 764–66
  - pH of, 762
  - reclamation of, 763
  - saline, 762, 811
  - sustainable agriculture and conservation of, 761–63
  - texture and composition of, 759–60
  - topsoil and humus in, 759–60
- Soil conservation, 761–63
- Sokolowski, Marla, 1120
- Solar radiation. *See* Sunlight
- Solute, **51**
  - active transport of, 134–36
  - concentration of, in aqueous solutions, 52–53
  - concentration of, and water conservation in mammalians, 935–36
  - cotransport of, 136, 740*f*
  - diffusion of, 131*f*
  - osmoregulation and balance of water and, 922–27
  - passive transport of, across membranes, 130–34
  - pressure and concentration of, effects on water potential, 740–41
  - proton pumps and transport of, 739*f*
  - transport of, in plant cells, 740*f*
- Solute potential (ψ_s), **741**
- Solution, **51**
  - hyperosmotic and hypoosmotic, 923
  - hypertonic, hypotonic, and isotonic, 132, 133*f*
  - isoosmotic, 923, 936
  - pH of select aqueous, 54*f*, 55
- Solvent, **51**
  - water as most significant, 51–53
- Somatic cells, chromosomal number in, **219**, **240**
  - division of (*see* Cell division)
  - totipotency in differentiated, 415*f*
- Somatic nervous system, **1026**
- Somites, **674**, **998**
- Sonar, 1045
- Songbirds. *See* Bird songs
- Song sparrow
  - immigration and populations of, on Mandarte Island, 1151*f*
  - logistic model of population growth applied to, 1146, 1147*f*
  - resource competition and population of, 1149*f*
- Sonic hedgehog growth factor, 1008
- Soredia, **621**
- Sori, **586**
- Sound, volume and pitch, 1052
- Sound detection, mammalian, 1050–52
- South Africa
  - elephant populations in Kruger National Park, 1144*f*
  - Succulent Karoo restoration project in, 1228*f*
- Southern blotting, restriction fragment analysis using, **394**, 395*f*, 405
- Soybeans, bacteroids and nitrogen fixation in, 765*f*
- Spatial heterogeneity, 1086
- Spatial learning, **1115**–16
- Spatial summation, **1023**
- Specialized transduction, 349
- Speciation, **472**–90
  - adaptive radiation and, 480–81
  - allopatric, 476*f*, 477, 480
  - biological diversity linked to, 472
  - biological species concept and, 473–76
  - continental drift and, 528
  - genetics and, 481
  - macroevolution and, 472, 482–88
  - patterns of evolutionary change and (anagenesis and cladogenesis), 472*f*
  - sympatric, 476*f*, 478–80
  - tempo of, two models, 481–82*f*
- Species, 12, **473**
  - benefits of diversity in, to human welfare, 1211–12
  - biodiversity crisis and loss of, 1165–66
  - biological species concept, 473–76
  - biome composition of, 1099
  - classifying (*see* Taxonomy)
  - in communities (*see* Community (communities))
  - comparing genomes of, 399t, 400–402
  - conservation of (*see* Conservation biology)
  - distribution of, 1083–92
  - diversity of (*see* Biodiversity)
  - dominant, 1168
  - endangered and threatened, 1210–11
  - endemic, 450
  - extinctions of (*see* Extinction of species)
  - introduced (exotic), 1213–14
  - invasive, 1168
  - keystone, 1168*f*, 1169*f*
  - morphological concept of, 476
  - origin of new (*see* Speciation)
  - paleontological concept of, 476
  - phylogenetic concept of, 476
  - populations of (*see* Populations)
  - range of, 1081*f*, 1084, 1085*f*
  - reproductive barriers separating, 474–75*f*
  - richness and relative abundance of, 1165, 1176–78
  - sibling, 476
  - as transplants, 1084–85
- Species-area curve, **1176**, 1177*f*
- Species diversity, **1165**–66, 1210–11. *See also* Biodiversity
- Species richness, **1165**. *See also* Biodiversity
  - area effects on, 1176–77
  - energy, water, and, 1176*f*
  - equatorial-polar gradients of, 1176
  - on islands, 1177–78
- Species selection, **486**

Specific epithet (taxonomy), **496**
Specific heat, **49**
Spectrophotometer, **186**
determining absorption spectra with, 187*f*
Spectrum of dominance of alleles, 260–62
Speech, human brain, language, and, 706, 1011*f*, 1034*f*
Spemann, Hans, 1005, 1006
Spemann's organizer, 1006
Sperm, **964**
development of human, 973, 975*f*
fertilization of ovum by, 988–91
flagellated plant and charophycean, **574**
flagellate motion in, 115*f*
plant, 592, 593*f*
Spermatheca, **968**
Spermatogenesis, **973**
Spermatogonia, **975**
Sphagnum moss, 511, 577*f*, 583*f*
S phase of cell cycle, **221***f*, 228
Sphincters
digestive tract, **855**
pyloric, 855*f*, 858
precapillary, 877, 878*f*
Sphinx moth (*Xanthopan morgani*), 820*f*
Sphygomanometer, 877*f*
Spider monkeys, 700*f*
Spiders, 658–59
anatomy, 659*f*
feeding and aggressive behavior in funnel web, 1119*f*, 1120
Spider silk, 82*f*, 658–59
Spiderwort (*Tradescantia*) leaf, 725*f*
Spike moss, 586, 587*f*
Spinal cord, 1026*f*
Spinal nerves, **1026**
Spindle poles, 221
Spines, 498, 716*f*
Spiral cleavage, **631**–32
Spiral-shaped prokaryotes, 535*f*
Spiral valve, **681**
Spirochetes (bacteria), 543*f*
Spirogyra, 550*f*, 569
Spliceosome, **318**, 319*f*
Sponges (Porifera), 633, 639*f*, 642–43, 903
anatomy, 642*f*
food digestion in, 853
origins of, 526–27
Spongocoel, **642**
Spongy mesophyll, **724**
Spontaneous change, free energy and, 146*f*
Spontaneous mutations, 329
Sporangium (sporangia), **577**, 774
in fungi, 614, 615*f*
walled spores produced in, 577*f*
Spores, **242**, **576**
fossilized plant, 578*f*
homosporous, and heterosporous, 586
of plants, 576, 577*f*
Sporocytes, **577**
Sporophylls, **586**
Sporophyte, **242**, **576**
in bryophytes, 580, 581*f*
as diploid stage in life cycle, 242
fossilized, 578*f*
relationship of gametophyte to, 592*f*, 772
in seedless vascular plants, 584, 585*f*
in seed plants (angiosperms), 592, 592*f*, 593*f*, 771, 772*f*
self-incompatibility in, 776
Sporopollenin, **574**, 557*f*
Sporozoites, **555**
Squamous epithelial cells, **823**, **824***f*
Squids, 639*f*, 652, 653*f*
nervous system, 1012*f*
Squirrels
allopatric speciation of antelope, 477*f*
hibernation of Belding's ground, 840*f*
immigration of Belding's ground, 1138, 1151
flying, 450
influence of light on biological clocks of, 1030, 1031*f*
life table of Belding's ground, 1139*t*
reproductive table for Belding's, 1141*t*
survivorship curve for Belding's, 1140*f*
SSU-rRNA sequences, 524, 540
Stability, relationship of free energy to, 145, 146*f*
Stabilizing selection, 465f, **466**
Staghorn fern (*Platycerium*), 768*f*
Stahl, Franklin, models of DNA replication by M. Meselson and, 299–300
Stalk-eyed flies, mate selection in, 1125, 1126*f*
Stamens, flower, 429*f*, 430, **598**, **772**
Standard metabolic rate (SMR), **829**
Standing crop biomass, 1187, 1194*f*
pyramid of, 1193*f*
Stanley, Steven, 486
Stanley, Wendell, 335
Stapes, **1051**
Staphylococcus sp., 543*f*
Star anise (*Illicium floridanum*), 602*f*
Starch, **71**
amylose and amylopectin forms of, 71
storage of, in plants, 71, 72*f*
structure of, compared to cellulose, 73*f*
Starlings as introduced species, 1214
Start (initiation) codon, 313, 314*f*
Statocysts, **1050***f*
Statoliths, **809**, **1050**
Stebbins, G. Ledyard, 455
Stele, plant tissue, **717**, 722
Stem, plant, **715**
gibberellins and elongation of, 309, 797
modified, 715*f*
primary growth in, 723–24, 726*f*
secondary growth in, 725–28
terminal, 715
tissue organization of, 724*f*
Stem cells, **418**, **881**–82
embryonic, 418*f*, *1039*
neural, 1038–39
pluripotent, 418
replacement of blood cellular elements and, 881*f*, 882
therapeutic cloning and, 418
Stenohaline animals, **923**
Stentor, 550*f*
Steroid hormones, 77
anabolic, 959*f*
intracellular receptors and, 205*f*, 947
Steroids, **77**
Stethoscope, 877*f*
Stickle-back fish
sign stimuli in fixed action pattern in, 1108*f*
spatial learning and use of landmarks by, 1115–16
territorial behavior in, 1108*f*
Sticky ends, DNA, **386***f*
Stigma, flower, **598**, **772**
Stingray, southern (*Dasyatis americana*), 681*f*
Stipe, **560**, 561*f*
Stock, plant, **782**
Stolons, 715*f*
Stomach, **857**–58
defenses of, 899
evolutionary adaptations of, 863*f*
interior surface of, 857*f*
secretion of gastric juice in, 857
tissue layers of, 827*f*
ulcers in, 858*f*
Stomata (stoma), **182**, 183*f*, **581**, **724**
evolution of, 581, 583
guard cells and opening/closing of, 750*f*, 751
rate of transpiration regulated by, 749–51
Stop codon, 313, 314*f*, 324
Storage leaves, 716*f*
Storage polysaccharides, 71, 72*f*
Storage roots, 714*f*
Stramenopiles, 558–62
brown algae, 560–62
diatoms, 559–60
flagella, 558*f*
golden algae, 560
oomycetes, 558, 559*f*
"Strangling" aerial roots, 714*f*
Stratified epithelium, **823**, 824*f*
Strawberry, guttation in leaf of, 746*f*
Streams and rivers, **1095***f*
restoration projects in, 1227*f*, 1228*f*
Streptococcus sp., 543*f*, 643
capsule of, 536*f*
pneumoniae, genetic transformation of, by DNA, 295*f*, 348
Streptomycin, 322
Streptophyta (kingdom), 575*f*
Stress
adrenal gland secretions as response to, 957*f*
effects of, on immune system, 918
plant response to environmental, 799–800, 811
Stress-induced proteins, **840**
Stretch-gated ion channels, **1017**
Stretch receptors, 1047*f*
Striated muscle, **826***f*
Strobili, **586**
Stroke, **883**
Stroke volume, heart, **873**
Stroma, **110**, 111*f*, **182**, 183*f*, 185*f*
Stromatolites, **521**, 522*f*
Structural formula, **40**
Structural isomers, **62***f*
Structural polysaccharides, 72–74. *See also* Cellulose; Chitin
Structure and function, biological organization and correlation of, 27*f*
Strychonos toxifera, 1163
Sturna magna (eastern meadowlark), 473*f*
Sturnella neglecta (western meadowlark), 473*f*
Sturtevant, Alfred H., 279–81
Style, flower, **598**, **772**
Subatomic particles, 34
Suberin, 728
Submucosa tissue layer, of stomach, **827***f*
Substance P, **1025**
Substitutions, point mutations and nucleotide base-pair, 328, 329*f*
Substrate, 81, **152**
catalysis in active site of enzyme and, 152–54
induced fit of enzyme and, 153*f*
Substrate feeders, **845***f*
Substrate-level phosphorylation, **165***f*
Sucrase, 152
hydrolysis of sucrose by, 150, 151*f*
Sucrose
dehydration synthesis of, 70, 71*f*
hydrolysis of, by enzyme sucrase, 150, 151*f*
loading/unloading of, in plant phloem, 752*f*
molecular mass of, 52
Sugars
as component of nucleic acids, 87f, 88
conduction of, in phloem tissue, 719*f*, 752
conversion of carbon dioxide to, in Calvin cycle, 193–95
disaccharides, 70, 71*f*
monosaccharides, 70*f* (*see also* Glucose)
polysaccharides, 71–74
Sugar glider, 450
Sugar sink, **752**
Sugar source, **752**
Sulfhydryl functional group, **65***f*
Sulfolbus archaea, 541
Sulfur dioxide emissions, 1202
Sulston, John E., 429
Summation of muscle cell contractions, 1071*f*
Summation of postsynaptic potential, temporal and spatial, 1023
Sundews (*Drosera*), 768*f*
Sunflower inflorescence, 773*f*

- Sunlight
  - as abiotic factor, 1087
  - characteristics of, 186
  - Earth latitudes and, 1088f
  - global energy budget and, 1186–87
  - as limiting factor in ecosystems, 1188
  - plant response to (*see* Light, plant responses to)
- Suprachiasmatic nuclei (SCN), 959, **1030**
- Surface tension, **49**
- Surface-to-volume ratio
  - animal body structure and, 821–22
  - of cells, 99f
- Survivorship curves, **1139**–40
  - idealized, 1140f
- Suspension feeders, **642**, 681, **845**f
- Sustainable agriculture, 761, **762**, 763
- Sustainable Biosphere Initiative, 1229
- Sustainable development, **1229**–30
- Sutherland, Earl W., 203
- Sutton, Walter S., 274
- Swallowing reflex, 856f
- Swaziland, traditional and modern medicine in, 30–31
- Sweat glands
  - as immune response, 899
  - thermoregulation and, 837
  - X-chromosome mutation and, 284
- Sweden, demographic transition in, 1153f
- Sweet taste, sensory transduction for reception of, 1056f
- Swim bladder, **682**
- Swimming, 1073
  - body shape and fast, 821f
  - energy cost of, 1074f
- *Symbion pandora* (Cycliophoran), 641f
- Symbiont, **545**
- Symbiosis, **545**
  - between algae and fungus (lichens), 621f
  - between animals and fungus, 620
  - commensalism as, 545
  - herbivores and bacteria, 863–64
  - mutualism as, 1164
  - parasitism as, 1163
  - between plant roots and fungi, 620
  - between protozoan and bacteria, 525f
- Sympathetic division of autonomic nervous system, **1027**f
  - adrenal gland regulation by, 957
- Sympatric speciation, **478**–80
- Symplast, **743**
- Synapses, **1014**
  - chemical, 1021, 1022f, 1023–25
  - electrical, 1021
  - between rod cells and bipolar cells, 1061f
- Synapsids, **694**
- Synapsis, **247**
- Synaptic cleft, **1022**
- Synaptic signaling, 203f
- Synaptic terminals, **1014**, 1021f
- Synaptic vesicles, **1021**
  - neurotransmitters in, 1021 (*see also* Neurotransmitters)
- Synaptonemal complex, 247
- α-Synuclein, 1041
- Syphilis (*Treponema pallidum*), 543f
- System(s), **9**–12
  - emergent properties of, 9
  - feedback regulation in biological systems, 11–12
  - power and limitations of reductionism in, 9–10
  - systems biology, 10–11, 400
- Systematics, **491**
  - animal phylogeny and, 633–36
  - classification and phylogenetic, 495–97
  - evolutionary history in genomes and, 504–6
  - molecular, 491, 502–3f
  - molecular clocks and, 506–7
  - phylogenetic trees and, 496, 497–504 (*see also* Phylogenetic trees)
  - taxonomy, classification, and, 529–32 (*see also* Taxonomy)
- Systemic acquired resistance (SAR), **814**, 815f
- Systemic circuit, **871**
- Systemic lupus erythematosus (lupus), 917
- Systems biology, **10**–11, 400
  - feedback regulation in, 11–12
  - plant hormones and, 801–2
  - plant morphogenesis and, **729**
- Systole, cardiac cycle, **872**
- Systolic pressure, **876**

- T2 phage, infection of *E. coli* by, 294, 295f
- T4 phage
  - infection of *E. coli* by, 334f
  - lytic cycle of, 338f
  - structure, 336f
- Table salt. *See* Sodium chloride (NaCL)
- Table sugar. *See* Sucrose
- Tadpole, frog, 686f
- Taiga, 1102
- Tail, muscular and postanal, in chordates, 673f
- Tapeworms, 639f, 646t
  - anatomy, 648f
- Taproot system, **713**
- Tardigrades (water bears), 641f
  - anhydrobiosis in, 925
- Tardive dyskinesia, 1040
- Taste, receptors for, 1054–56
  - in insects, 1055f
- Taste buds, **1055**
- TATA box, **316**, 364
- Tatum, Edward, one gene–one enzyme hypothesis of G. Beadle and, 310, 311f
- Taxis, **536**–37, **1110**
- Taxon (taxonomy), **496**
- Taxonomy, **439**, **495**. *See also* Systematics
  - of animals, 13f
  - binomial nomenclature of, 496
  - connection between phylogeny and classification of, 496–97
  - continuing development of, 531–32
  - domains, 13, 14f
  - hierarchical classification system of, 496f
  - Linnaeus on, 439, 495
  - older systems of, 529–30
  - of plants, 578t
  - of seedless vascular plants, 586, 587f, 588
  - three-domain system, 14f, 530–31f, 532
- Tay-Sachs disease, **261**
  - dominant allele, phenotype, and, 261
  - as recessive trait, 266
- T cell receptors, **904**
  - for antigens, and role of MHC, 904, 905f
- T cells. *See* T lymphocytes (T cells)
- Technology, **25**–26. *See also* DNA technology
  - ethics and, 26
  - high-throughput, 11
  - human carrying capacity and, 1156
- Teeth
  - biofilm on surface of, 539f, 540
  - mammalian, 694
  - origins of, 679
- Telencephalon, 1028
- Telomerase, **307**, 373
- Telomeres, **306**f, 307
- Telophase
  - meiosis, 245f
  - mitosis, **221**, 223f, 226f
- Temperate broadleaf forest, **1103**
- Temperate grasslands, **1102**f
- Temperate phage, **338**, 349
- Temperature, **49**
  - as abiotic factor, 1086
  - cold, and plant flowering, 807
  - enzyme activity, and effects of, 154f
  - maintenance of body (*see* Thermoregulation)
  - membrane fluidity responsive to, 126f
  - moderating effect of water on, 49–50
  - negative feedback and control of, 832f
  - plant responses to excessive, 195–97, 749
- Template strand, DNA, **313**
- Temporal heterogeneity, 1086
- Temporal isolation as prezygotic reproductive barrier, 474f
- Temporal summation, **1023**
- Tendons, **825**f
- Tendrils, 716f
- Terminal bud, **715**
- Termination
  - of transcription, 317
  - of translation, 324, 325f
- Terminator DNA sequence, **315**f
- Terrestrial biomes, 1098–1104
  - chaparral, 1101f
  - climate and, 1098
  - coniferous forest, 1102f
  - desert, 1100f
  - distribution of major, 1099f
  - general features of, 1098–99
  - savanna, 1101f
  - temperate broadleaf forest, 1103f
  - temperate grassland, 1102f
  - tropical forest, 1100f
  - tundra, 1103f
- Terrestrial ecosystems
  - food chain in, 1166f
  - primary productivity of, 1187f, 1190–91
- Territoriality, **1138**
  - chemical markers of territories, 1149
  - in fish, 1107f, 1108f
  - population regulation and, 1149
- Tertiary consumers, **1185**
- Tertiary structure of protein, 81, **83**f
- Tessier-Lavigne, Marc, 1037–38
- Testcross, **256**f
- Testes, 241, **971**
  - hormonal control of, 978f
  - hormone synthesis by, 958
- Testosterone, 63f, **958**
  - as chemical messenger, 205
- Tetanus, muscle, **1071**
- Tetany, 954
- Tetracycline, 322, 537
- Tetrad, chromatid, **247**
- *Tetrahymena*, 318
- Tetraploidy, 285, 286f
- Tetrapods, **684**–86
  - amniotic egg in, 687–94
  - amphibians, 685–86
  - derived characters of, 684
  - homologies among, 449
  - humans as, 701–7
  - limb evolution in, 485f
  - mammals, 694–701
  - origin of, 527, 684, 685f
  - origin of lungs in, 682
  - reptiles, 688–94
- Tetravalence, 60
- Thalamus, **1030**, 1032
- Thallus, **560**, 561f
- Theory, scientific, **24**
- Therapeutic cloning, 418
- Thermal energy, **142**. *See also* Heat
- Thermocline, 1091f, **1093**
- Thermodynamics, energy transformations and laws of, **143**–44
- Thermogenesis, 838
- Thermophiles, extreme, 534f, **541**
- Thermoreceptors, **1049**
  - in skin, 1048f
- Thermoregulation, **833**–41
  - acclimatization and, 839
  - adjustment to changing temperatures, 839–40
  - adjustment to metabolic heat production, 838
  - in amphibians and reptiles, 835f
  - balancing heat loss and gain, 834–38
  - behavior responses for, 837–38A
  - circulatory adaptations for, 835, 836f, 837f
  - ectothermic, 829, 833–34

endothermic, 829, 833–34
evaporative cooling and, 50, 837
feedback mechanisms in, 838–39
in fishes, 834*f*, 837*f*
in humans, 839*f*
insulation as adaptation for, 834
in invertebrates, 833–34, 836, 837*f*, 838*f*, 839*f*
in mammals and birds, 834*f*, 836*f*, 837*f*
modes of heat exchange and, 834, 835*f*
thermoreceptors and, 1048*f*, 1049
torpor and energy conservation, 840–41
Theropods, **689**
flight and, 692–93
Thick filaments, myofilaments, **1066**, 1067*f*
Thigmomorphogenesis, **809**
Thigmotropism, **810**
Thimann, Kenneth, 793
Thin filaments, myofilaments, **1066**, 1067*f*
Thiols, **65***f*
Thiomargarita namibiensis, 434
Third World Organization for Women in Science (TWOWS), 31
Thoracic cavity, **827**
Thorny-headed worms (acanthocephalans), 640*f*
Threatened species, **1210**–11
Three-domain system of taxonomy, 14*f*, 530*f*, **531**, 532
Threshold, **1018**
Thrombus, **882**, 883
Thrum flower type, 775*f*
Thylakoids, **110**, 111*f*, **182**, 183*f*
light reactions in, 184, 185*f*, 193*f*
photosystems as light-harvesting complexes in, 189–90
in prokaryotes, 537*f*
Thylakoid space, 182, 183*f*
Thymine (T), 88
DNA structure and, 296*f*, 297*f*, 298*f*
repair of DNA damage from dimers of, 306
Thyroid gland, **953**
hormones secreted by, 949*t*, 953–54
iodine deficiency and, 33*f*, 34, 954
Thyroid-stimulating hormone (TSH), **952**, 953*f*
Thyroxine (T_4), 947, **953**
Tidal volume, breathing, **889**
Tight junctions, 120, **121***f*, 627
Tilapia (*Tilapia mossambica*), 923*f*
Tilman, David, 1149
Time, using molecular clocks to measure absolute, 506–7
Tinbergen, Niko
questions on behavior formulated by, 1107
on spatial learning in digger wasps, 1115
Ti plasmid, **406**, 407*f*
Tissue, 5*f*, **712**, **823**
connective, 823, 825*f*
epithelial, 823, 824*f*
muscle, 823, 826*f*
nervous, 823, 826*f*
Tissue plasminogen activator (TPA), 380, 404
Tissues, animal
body plan and organization of, 630–31
connective, 823, 825*f*
embryonic, 994–97
epithelial, 823
grafts/transplants of, and immune response, 916
muscle, 1066–72 (*see also* Muscles)
nervous, 823 (*see also* Neurons)
organ systems and, 827
Tissue-specific proteins, 419
Tissue systems, plant, **717***f*
in stems, 724*f*
transport in, 717 (*see also* Vascular tissue system)
Tit-for-tat strategy, reciprocal altruism and, 1130–31
T lymphocytes (T cells), 881, **903**
development of, 905*f*
as receptors for antigens, and role of MHC, 904–5
Tobacco mosaic disease (TMV), 334, 335*f*
structure, 336*f*
Tobacco plant, firefly gene and light energy in, 314*f*
Tomato, *aurea* mutant, and greening of, 789
Tongue, taste receptors on, 1056
Tonicity, **132**
Tonoplast, **108**, **743**
Toolson, Eric, 1114
Tool use, human evolution and, 704
Top-down model of community structure, **1170**, 1171*f*
Topoisomerase, **303**, 304*t*
Topsoil, **759**
Torpor, **840**–41
Torsion in gastropods, **651**
Totipotency
in animal cells, 1005
in plant cells, **415**
Totipotent zygote, **1005**
Touch
detection of, by mechanoreceptors, 1048
turgor movements in plants in response to, 810*f*
Toxic substances
as human-caused environmental problem, 1202–3
as population-limiting factor, 1150
Toxins. *See also* Poison(s)
bacterial (prokaryotes), 339, 535
biological magnification of, 1202–3
botulism, 1069
as defense against predation, 1163
from dinoflagellates, 555
enzyme inhibitors as, 155
in food webs, 1202*f*
fungal, 622
G proteins modified by, 211
phytoremediation of, 763
Trace elements, **33**
Trachea (windpipe), **887**, 899*f*
Tracheal systems, insect, **886**, 887*f*
Tracheids, **584**, **719***f*, 727
water conduction by, 719, 746
Trait, genetic, **252**. *See also* Character, genetic
dominant, 253, 265*f*
pedigrees for tracing, 265–66
recessive, 253, 265*f*, 283–84
Transacetylase, 355f
Transcription, **311**, 315–17
activator of, 356
basic principles of, 311–12*f*
cell-type specific, 367*f*
initiation of, and regulation of eukaryotic gene expression, 364–67
initiation of, RNA polymerase binding and, 316*f*
modification of RNA after, in eukaryotic cells, 317–19
molecular components of, 315–16
of operon genes, 353, 354*f*
post-transcriptional gene regulation, 368–70
in prokaryotic versus eukaryotic cells, 312*f*, 537
regulation of eukaryotic gene expression at, 362–70
regulation of gene expression during development at, 418–20
RNA strand elongation, 315*f*, 316
synthesis of RNA transcript, 316–17
stages of, 315*f*
summary of, in eukaryotic cell, 331*f*
termination of, 317
Transcription factors, **316**, **364**
cell response to cell signaling and, 205, 213
Dlx gene family as, 678
general, 364
role of, in control of eukaryotic gene expression, 316*f*, 364–66
specific, and enhancers, 364, 365*f*
Transcription initiation complex, **316***f*
Transcription unit, **315**
Transduction (signal-transduction pathway), 204, 208–12. *See also* Signal transduction
in animals, sensory reception, 1046
in plants, 789–90
protein phosphorylation and dephosporylation in, 209f, 210
signal-transduction pathways in, 208
small molecules and ions as second messengers in, 210–12
Transduction in bacteria, **348***f*–49
Transfer cells, **752**
Transfer RNA (tRNA), **320**, 327*t*
binding of specific amino acid to, 321*f*
structure and function of, 320, 321*f*, 322
Transformation
cancerous, **232**
genetic, of bacteria, **294**, **348**
Transgenes, 406
plants, 784–86
Transgenic animal, **406**
Transgenic plant, 406–7*f*, **782**, 783, 784–86
Transition state, activation of chemical reactions and, 151
Transitional endoplasmic reticulum, 105
Transitional fossil, 451*f*
Translation, **311**, 320–26
basic principles of, 311–12*f*, 320*f*
initiation of, 323*f*
initiation of, regulation of eukaryotic gene expression at, 369
molecular components of, 320–23
polypeptide synthesis in, 323–24
polyribosomes formed during, 324, 325*f*
in prokaryotes versus eukaryotes, 312*f*, 537
ribosome function in, 322*f*, 323
role of aminoacyl-tRNA synthetase in joining tRNA to amino acid during, 321*f*
summary of, in eukaryotic cell, 331*f*
targeting polypeptides to specific locations, 325–26*f*
termination of, 324, 325*f*
transfer RNA (tRNA) structure and function in, 320, 321*f*, 322
transformation of polypeptides to functional proteins following, 324, 325
Translocation
chromosomal, **286***f*, 287, 288*f*
of phloem sap, **752**, 751–53
Transmembrane proteins, structure of, 128*f*
Transmission of sensory receptor potential, 1046–47
Transmission electron microscope (TEM), **96***f*
Transpiration, **746**
ascent of xylem sap dependent on, 747*f*
effects of, on wilting and leaf temperature, 749
stomata and regulation of, 749f, 750*f*–51
xerophyte adaptations for reducing, 751
Transpirational pull, 747*f*
Transpiration-cohesion-tension mechanism, 746–48
Transplanted species, 1084–85
Transplants, tissue and organ, 916
Transport across membranes, 126–27
Transport epithelium
kidney nephron and role of, 933*f*
osmoregulation and, **926**
salt excretion in marine birds and, 926f, 927
Transport in animals. *See* Circulatory system
Transport in vascular plants, 584, 738–58. *See also* Plant nutrition
absorption of water and nutrients through roots as, 744–46
aquaporins and water, 742
ascension of water and minerals through xylem, 746–49
bulk flow functions in long-distance, 743–44
overview of, 739*f*
physical forces driving, 48*f*, 738–44
proton pumps, role of, 739–40
plant cell compartments and short-distance, 743*f*
stomata and regulation of transpiration rates, 749–51
symplast and apoplast function in, 743
translocation of nutrients through phloem, 751–53
water potential differences and, 740–42
Transport proteins, 78*t*, **130**, **739**
carrier proteins as, 130, 134*f*
channel proteins as, 130, 134*f*

cotransporter as, 136*f*
as electrogenic pump and proton pump, 136*f*
facilitated diffusion and role of, 133, 134*f*
mechanical work in cells and, 149*f*
role of, in membranes, 128*f*, 130
role of, in plant transport, 739, 740*f*
Transport vesicles, **105**, 107
Transport work, cellular, 148
driven by ATP, 149*f*
Transposable elements, **351**–52, 371, 375–76
genome evolution and, 380–81
insertion sequences as, 351, 352*f*
movement of eukaryotic, 376*f*
repetitive DNA as, 375, 376–77
sequences related to, 376
transposons as, 343, **352, 375**–76)
Transposase, 351, 352*f*
Transposons, 343, **352, 375**–76
cut-and-paste method of copying, 376*f*
Transthyretin, 82, 83
Transverse (T) tubules, **1069**
Tree(s). See also Forest
American chestnut, 1168
anatomy of trunk of, 727*f*
aspen, 781*f*
bonsai, 441*f*
effect of climate change on American beech, 1092*f*
effect of winds on, 1087*f*
energy, water, and species richness of North American, 1176*f*
individualist hypothesis of communities applied to, 1179*f*
mutalism between ants and acacia, 1164*f*
wood and, 605, 727
Tree of life, 17–19, 491, 507, 512–33. *See also* Biodiversity
continuum of life on earth and, 512
early Earth conditions and origins of life, 513–16
evolution of prokaryotes and, 521–23
fossil record as chronicle of life, 517–21
homology and, 449, 450*f*
new information and understanding of, 529–32
origin of eukaryotes, 523–25
Trematodes (flukes), 646*t*, 647
life cycle of blood fluke (*Schistosoma*), 647*f*
Triacylglycerol, **75**
Tricarboxylic acid cycle, 168
Trichinella spiralis (trichinosis), 656*f*
Trichomonas vaginalis, 553*f*
Trichoplax adhaerens, 639*f*
Triglyceride, 75
Triiodothyronine (T_3), **953**
Trilobites (Trilobita), **656**, 657*f*
Trimesters of pregnancy, **978**–81
Trimethylamine oxide (TMAO), 924
Triose sugars, 70*f*
Triple response of plants to mechanical stress, **799**–80
Triplet code (DNA), **312**
Triploblastic animals, **631**
body plans of, 631*f*
Triploidy, 285
Trisomic cells, **285**
Trisomy 21, 287
Tristan da Cunha, 462
Trochophore, molluscs, **650**
Trochophore larva, **635**
Trophic cascade model, 1170
Trophic efficiency, ecological pyramids and, **1192**, 1193*f*
Trophic levels in ecosystems, 1185
decomposition and, 1185–86
energy transfer between, 1191–94
food energy available to humans at different, 1193*f*
Trophic structure in communities, **1166**–68
bottom-up and top-down controls and, 1170–71
food chains, 1166*f*
food chains, limits on length of, 1167–68
food webs and, 1166–67
Trophoblast, **979**, 982, **1000**
Tropical forests, 1100*f*
deforestation of, 1209*f*, 1212, 1221*f*
productivity of, 1187
Tropic hormones, **951**, 952
Tropisms, **792**
gravitropism, 809
phototropism, 792–93, 803*f*
thigmotropism, 810
Tropomyosin, **1068**
Troponin complex, **1068**
Trout
anatomy of, 682*f*
rainbow (*Oncorrhynchus mykiss*), 939*f*
Trp operon, regulated synthesis of repressible enzymes and, 353, 354*f*
Truckee River, Nevada, restoration project in, 1227*f*
True-breeding plants, **252**–53
Trumpeter swans, mating behaviors, 1124*f*
Trypanosoma (trypanosomes), 554*f*. *See also* Sleeping sickness
Tryptophan, 783
genetic recombination of bacteria capable of synthesizing, 347*f*
levels of, and gene expression in bacteria, 352–56
Tsetse fly (*Glossina palpalis*), 554
genome, 401
Tuataras (*Sphenodon punctatus*), 690, 691*f*
Tubal ligation, **983**
Tube cell, 599
Tube feet, echinoderms, **665**
Tuberculosis (*Mycobacterium tuberculosis*), 545, 1149, 1155
Tuber melanosporum, 616*f*
Tubers, 715*f*
Tulips, 491*f*, 501*f*
Tumor, 232–33. *See also* Cancer
Tumor-suppressor genes, **371**
faulty and development of cancer, 371, 373
Tumor viruses, 370
Tuna, 821*f*
northern bluefin (*Thunnus thynnus*), 939*f*, 1214*f*
thermoregulation in, 837*f*
yellowfin (*Thunnus albacares*), 683*f*
Tundra, **1103***f*
Tunicates (Urochordata), 641*f*, **673**, 674*f*
fate map of cells in, 1004*f*
larva, 674, 674*f*
Turbellarians (Turbellaria), 646–47
Turgid plant cell, **132**, **742**, 750*f*
Turgor movements in plants, 810*f*
Turgor pressure, **741**
opening/closing of stomata and, 750*f*
Turnover in lakes and ponds, **1091***f*
Turnover time, phytoplankton, **1192**
Turtles, 690, 691*f*
Tympanic membrane
human, 1051*f*
insect, 1050*f*
Type I diabetes mellitus, 956
Type II diabetes mellitus, 956

Ubiquinone, 170
Ubiquitin, 370*f*
Ultimate causes of behavior, 1106, 1107, 1108*f*, 1109*f*
Ultimate questions, **1107**
Ultracentrifuge, **97**
Ultrametric tree, **500***f*
Ultrasound imaging, fetal testing by, 269, 984*f*
Ultraviolet (UV) light
atmospheric ozone depletion and increased, 1206
as mutagen, 330
skin damage caused by, 305*f*, 306
Ulva (sea lettuce), 568*f*, 569
Umami taste, 1055
Undernourishment, **846**
Unequal reproductive success, natural selection and, **16**
Unger, Christian, 252
Uniformitarianism, **440**
Uniform pattern of population dispersion, 1138*f*
United States
acid precipitation in, 55–56, 1202*f*
age structure of population in, 1154*f*, 1155
aquatic biomes in, 1094*f*, 1095*f*, 1096*f*
populations of great-tailed grackles in, 1085*f*
range of American beech tree in, 1092*f*
terrestrial biomes in, 1100*f*, 1101*f*, 1102*f*, 1103*f*
Unity underlying life's diversity, 14, 15*f*
as biological theme, 27*f*
Unsaturated fatty acid, **75**–76
Uracil (U), 88
Urea, 59, 60–61, **928**. *See also* Urine
as nitrogenous waste, 927*f*, 928
Ureter, **931**, 932*f*
Urethra, **931**, **972**
Urey, Harold, 513*f*, 514
Uric acid as nitrogenous waste, 927*f*, **928**
Urinary bladder, **931**
Urination, 936–37
Urine
filtration of blood by kidneys and production of, 931–34
kidney concentration of (two-solute model), 935*f*
Urodeles. *See* Salamanders (Urodela)
Use and disuse idea, 441
Uterine cycle, **973**
hormonal regulation of, 976*f*, 977
Uterus, 959, **969**
Utricle, **1052**, 1053

V_{1a} receptor, 1113
Vaccination, **914**
Vaccines, **343**
DNA technology and production of, 404
Vacuolar membrane, **743**
Vacuoles, **108**
contractile, 108
food, 107, 108
in plant cell, 108*f*, 743
Vagina, **970**
Valence, **40**, 60*f*
Valence electrons, **38**
Valence shell, **38**
Vampire bat (*Desmodus rotundas*), 938*f*
Van der Waals interactions, **42**
in tertiary structure of proteins, 83f
Van Helmont, Jan Baptista, 756
Van Niel, C. B., 184
Van Overbeek, Johannes, 796
Vaporization, heat of, 50
Variable (V) regions, lymphocytes, 904
Variation, genetic, **238**. *See also* Genetic variation
Vasa recta, **933**, 936
Vascular bundles, **717**
Vascular cambium, **720**
production of secondary vascular tissue by, 725–27
Vascular cylinder, **717**
Vascular plants, **578***t*
origin and traits of, 584–86
seed (*see* Seed plants)
seedless, 586–88
structure of (*see* Plant structure)
transport in, 738–58
Vascular tissue system, **578**, **717**
phloem, 584, 717, 719, 722*f*, 751–53
vascular cambium and secondary, 725–27
xylem, 584, 717, 722f, 724f, 746–49
Vas deferens, **971**
Vasectomy, **983**
Vasocongestion, **972**
Vasoconstriction, **835**
Vasodilation, **835**
Vectors. *See* Cloning vectors; Disease vectors
Vegetal plate, 995
Vegetal pole, **992**
Vegetation, regulation of chemical cycling by, 1198, 1199*f*
Vegetative reproduction in plants, **781**–83

Veins, **715**, **869**
blood flow velocity, 875*f*
countercurrent heat exchanger in, 836*f*
structure and function of, 874, 875*f*
Velvet worms, 641*f*
Ventilation, **885**
Ventral side, body, **630**
Ventricles of brain, **1026***f*
Ventricles of heart, **869**
Venules, **869**, 874*f*
Venus flytrap (*Dionaea muscipula*), 768*f*, 810
Vernalization, **807**
Vertebrae, 671
Vertebrates, 629, **671**, 678–708
amniotes and development of land-adapted egg, 687–94
amphibians, 685–86
apoptosis in, 428, 429*f*
birds, 691–94
bones and teeth in, origins of, 679
brain (*see* Brain)
circulatory and cardiovascular system of, 869–71
chondrichthyans (sharks, rays), 680–82
chordates and origin of, 671–77
comparison of amino acid sequence in select, 449, 450*f*
clade Deuterostomia and, 633
craniates and development of head in, 675–77
derived characteristics of, 678
development in (*see* Animal development)
digestive system, evolutionary adaptations of, 862–64
endocrine system of, 943–59
evapotranspiration and species richness of North American, 1176*f*
fossils of early, 678, 679*f*
gas exchange in, 886–95
gnathostomes and development of jaws in, 679–84
hagfishes, 676, 677*f*
homologous structures in embryos of, 448, 449*f*
hormone action in (*see* Hormones)
Hox genes and evolution of, 486*f*
humans as, 701–7 (*see also* Human(s))
kidneys and excretory system of, 934–38
lampreys, 678
lancelets as, 674, 675*f*
long-term potentiation in, 1036, 1037*f*
mammals as, 694–701 (*see also* Mammals)
nervous system of, 1012–13, 1020–21, 1026–41
pattern formation during limb development in, 1006–8
ray-finned and lobe-finned fishes, 682–84
reproduction in (*see* Animal reproduction)
reptiles, 688–94
skeleton of, 1064, 1065*f*
tetrapods and development of limbs and feet in, 684–86
tunicates as, 673, 674f
vision in, 1058–63
Vertical transmission of viruses in plants, 345
Vervet monkeys (*Cercopithecus aethiops*), social learning of alarm calls in, 1131, 1132*f*
Vesicles, **104**
migration of, 112*f*
Vessel, **719**
Vessel elements, **719**, 727
water conduction by, 719*f*, 746
Vestibule, **970**, 1053
Vestigial organs, **448**
Viagra (drug), 201, 211, 948, 1025
Villi in small intestine, **859**, 860*f*
Viral envelopes, **336***f*
of animal viruses, 340, 341*f*
Viral neutralization, 913
Virchow, Rudolf, 218
Viridiplantae (Kingdom), 567, 575*f*
Viroids, **345**–46
Virulent pathogens, **813**
Virulent phage, **338**
Viruses, 334–43
in bacteria (*see* Bacteriophage (phage))
cancer-causing, 370, 374
capsids and envelopes of, 335–36
classes of animal, 340*t*
discovery of, 334–35
emerging, 344
evolution of, 342–43
genome of, 335
Hershey-Chase experiment on infection of *E. coli* by, 294, 295*f*, 296
infection of bacteria by, 295*f*
lytic and lysogenic cycles, 337, 338*f*, 339*f*
as pathogens, 337, 343–45
reproduction of, general features, 336, 337*f*
reproductive cycle, 337–39
reproductive cycle of animal viruses, 339–42
size of, 335*f*
structure of, 335–36*f*
viroids and prions, 345–46
Visceral mass, mollusc, **650**
Visible light, **186**
Vision, 1057–63
color, 1059–60, 1061
geometric isomers and biochemistry of, 62
in invertebrates, 1057–58
neural pathways for, 1062*f*
processing visual information, 1061–63
sensory transduction in, 1060–61
vertebrate eye and, 1058, 1059f, 1060
Vital capacity, **889**
Vitalism, 59
Vitamins, **850**
essential nutritional requirements, 850, 851*t*
vitamin D, **954**
Vitelline layer, 988
Vitreous humor, **1059**
Viviparous species, **682**, 689
Vocal chords, **887**
Vogt, W., 1004
Voltage, cell, 134–35
membrane potential and, 1015–17
Voltage-gated ion channels, **1017**
chemical synapse and, 1022*f*
production/conduction of action potentials and role of, 1018, 1019*f*, 1020*f*, 1021
Volvox sp., 568*f*
Volume (loudness) of sound, 1052
Von Frisch, Karl, 1107
Von Humboldt, Alexander, 1176
Vulva, **970**

Wade, Michael, 1127
Wagler's pit viper (*Tropidolaemus wagleri*), 691
Walker, Brian, 1180
Wallace, Alfred, evolutionary theory of, 443, 1175
Water
as abiotic factor, 1086
availability of soil, to plants, 761*f*
balance of (*see* Water balance)
climate and effect of bodies of, 1087, 1090*f*
cohesion and adhesion of, 48–49
contaminated drinking, 552
covalent bonds in, 40*f*
freshwater biomes (*see* Freshwater biomes)
human ecological footprint and overuse of, 1156
hydrogen bonding and polarity of, 47–48
hydrogen bonds of, 48*f*
insulation of bodies of, by floating ice, 50, 51*f*
as key to all life, 47
marine biomes (*see* Marine biomes)
molecular dissociation of, and acid/base conditions, 53–56
molecular shape of, 43*f*
as solvent of life, 51–53
species richness, energy, and, 1176*f*
specific heat of, 49–50
splitting of, during photosynthesis, 183, 184*f*
temperature moderation by, 49–50
transport of, in plants (*see* Transport in vascular plants)
Water balance, 131–33, 922–27
in animals living in temporary waters, 925
in cells with walls, 132, 133*f*
in cells without walls, 132, 133*f*
environment and adaptations for,
in freshwater animals, 924–25
in kangaroo rat versus in humans, 925*f*
in land animals, 925–26
in marine animals, 924, 926*f*
osmoconformers and osmoregulators as response to challenge of, 923–26
osmosis and, 131–33, 922–23
transport epithelia and, 926–27
Water bears (tardigrades), 641*f*, 925
Water buffalo, commensalisms between cattle egrets and, 1164*f*
Water channel proteins, 93
Water conduction in plant tissues, 719*f*
Water conservation, 862
in camels, 926*f*
mammalian kidney as terrestrial adaptation, 934–38
Water cycle in ecosystems, 1196*f*
Water flea (*Daphnia* sp.), logistic population growth model applied to, 1146, 1147*f*
Water lily (*Nymphaea* "Rene Gerard"), 602*f*
Watermelon snow (*Chlamydomonas nivalis*), 568*f*
Water molds (Oomycotes), 558, 559*f*
Water pollution, acid precipitation and, 55–56
Water potential (ψ), **740**–42
aquaporin proteins, water transport in plants, and, 742
effect of solutes and pressure on, 740–41
model of water movement and, 741*f*
quantitative analysis of, 741–42
Water potential equation, 740
Water soluble hormones, cell-surface receptors for, 946
Water strider, 49*f*
Water vascular system, echinoderms, **665**
Watkinson, Andrew, 1148
Watson, James, 9
on DNA replication, 299
modeling of DNA structure by F. Crick and, 88–89, 293*f*
Wavelength of light, **186***f*
photosynthetically important, 187*f*
phototropism and, 803*f*
Weak chemical bonds, 42
Weapons, 546
Weberian apparatus, 1054
Weddell seal, circulatory and respiratory adaptions of, 894–95
Weight and mass, terms, 32 *n*.
Welwitschia mirabilis, 594*f*
Went, Frits, discovery of auxin by, 792, 793f
Wernicke, Karl, 1034
Wernicke's area, brain, 1034, 1035
West Nile virus, 337, 344, 1163
Wetlands, **1094***f*
primary production in, 1190–91
Wexler, Nancy, 268*f*
Whales
blubber, 835
body plan of, 822
electromagnetic reception in beluga, 1049*f*
size of blue, 671
suspension-feeding in, 845*f*
Wheat (*Triticum aestivum*), 478
Whiptail lizards (*Cnemidophorus uniparens*), 965, 966*f*
Whisk fern, 587*f*, 588
White blood cells (leukocytes), 825*f*, **880**, 881
apoptosis and development of, 428*f*
body's defense system and, 878, 900
lymphocytes (*see* Lymphocytes)

lysosomes of, 107, 108
phagocytic, 900
White matter, nervous system, **1026**
White rot (*Phanerochaete chrysoporium*), 623
White rusts (oomycota), 558
Whittaker, Robert H., 529–30, 1179f
Whooping cranes, imprinting in, 1109*f*
Widow's peak as dominant trait, 265*f*
Wieschaus, Eric, 423
Wildlife
effects of acid precipitation on, 686
mark-recapture method of estimating population size of, 1137
Wild type phenotype, **276***f*
fruit fly eye color and, 276, 277*f*
Wilkins, Maurice, 296, 297
William of Occam, 501
Wilson, E. O.
on biophilia, 1230
on endangered species, 1211
on island biogeography, 1177–78
on sociobiology, 1132–33
Wilson's phalaropes, mating behavior in, 1124*f*
Wilt in plant leaves, **742***f*
effects of transpiration on, 749
Wind
effect of, on trees, 1087*f*
global patterns of, 1089*f*
Wings,
bird, 692*f*
insect, 661
Wisenden, Brian, 1116, 1117
Wiskott-Aldrich syndrome (WAS), 215
Witte, Klaudia, 1125
Wobble, base pair-rule relaxation, **322**
Woese, Carol, 525, 540–41
Wöhler,Friedrich, 59
Wolf, population dispersion patterns, 1138
Wollemia pine (*Wollemia nobilis*), 595*f*
Wood, 605
heartwood and sapwood, 727
production of, by vascular cambium, 727
Work
ATP as energy source for cellular, 148–50
equilibrium and, in closed and open systems, 147f
stability, spontaneous change, and capacity for, 146*f*
Worms
acorn, 641*f*
flatworms, 639*f*, 646–48, 968*f*
marine, 640*f*
proboscis/ribbon worms, 639*f*
roundworms, 641*f*
segmented (annelids), 640*f* (*see also* Earthworms)
tapeworm, 639f, 646*t*, 648*f*
thorny-headed, 640*f*
velvet, 641*f*
Wright, Sewall, 455

X chromosome, 240, 282*f*
fruit fly eye color and, 276, 277*f*
inactivation of, in females, 284*f*
sex-linked genes on, 282–84
Xerophytes, **751**
reduction of transpiration in, 751*f*
X-O system of sex determination, 282*f*
X-ray crystallography, **85**, 86*f*
X-rays as mutagens, 329
X-Y system of sex determination, 282*f*
Xylem, **584**, **717**
secondary (*see* Wood)
in roots, 722*f*
in stems, 724*f*
transport in, 584, 746–49
water conduction by vessel elements and tracheids of, 719*f*

Yarrow plants (*Achillea*), geographic variation and genetics in, 464*f*
Y chromosome, 240, 282*f*
fruit fly eye color and, 276, 277*f*
sex-linked genes on, 282–84
Yeast (*Saccharomyces cerevisiae*), 505–6, **611**
alcohol and bread production and, 175, 623
budding by, 611*f*
cell signaling mechanisms in, 201, 202*f*
gene cloning and, 390
genome, 394, 400
ingestion of, by macrophage, 898*f*
Yeast artificial chromosomes (YACs), **390**, 396
Yellow jacket wasp, 1162*f*
Yellowstone National Park, Wyoming
archaea in hot springs of, 544
fires in, 1172, 1173*f*
grizzly bear populations in, 1217*f*, 1218, 1223*f*
legal and biotic boundaries of, 1222, 1223*f*
natural edges in ecosystem of, 1220*f*
Yolk, egg, **992**
Yolk plug, **995**
Yolk sac, **1000**

Zach, Reto, 1122
Zeatin, 796
Zeaxanthin, 802
Zebrafinches
E. Jarvis on genetics of song in, 819
mate selection in, 1125, 1226*f*
Zebrafish (*Danio rerio*)
associative learning in, 1116, 1117*f*
as model organism, 413*f*
Zero population growth (ZPG), **1144**
age structure and, 1155
demographic transition and, 1153–54
ZIFT (zygote intrafallopian transfer), 984
Z lines, **1066**, 1067*f*
Zona pellucida, **991***f*
Zonation
in lakes, 1093*f*
in marine communities, 1093*f*
Zoned reserves, **1223***f*, **1224**
Zone of cell division, plant, **722**
Zone of elongation, **722**
Zone of maturation, **722**
Zone of polarizing activity (ZPA), **1007***f*
limb pattern formation and role of, 1008*f*
Zoospores, 562, **613***f*
Z-W system of sex determination, 282*f*
Zygomycetes (Zygomycota), 613–15
life cycle, 614*f*
Zygosporangium, **614**, 615
Zygote, **241**, **964**
animal, 627, 978–79, 987–1001
cleavage in, 627*f*, 992–94 (*see also* Cleavage)
human, 979*f*
plant, 732*f*, 777–78
totipotency of, 1005